VOLUME I
A–F

RISK MANAGEMENT FOR HAZARDOUS CHEMICALS

VOLUME I
A – F

RISK MANAGEMENT FOR HAZARDOUS CHEMICALS

Jeffrey W. Vincoli, CSP

LEWIS PUBLISHERS

Boca Raton New York London Tokyo

Acquiring Editor:	Ken McCombs
Project Editor:	Joan Moscrop
Marketing Manager:	Greg Daurelle
Direct Marketing Manager:	Arline Massey
Cover design:	Jonathan Pennell
Manufacturing:	Sheri Schwartz

Library of Congress Cataloging-in-Publication Data

Vincoli, Jeffrey W.
Risk management for hazardous chemicals / Jeffrey W. Vincoli.
p. cm.
Includes bibliographical references and index.
ISBN 1-56670-200-3
1. Hazardous substances--Safety measures--Handbooks, manuals, etc.
2. Hazardous substances--Risk assessment--Handbooks, manuals, etc.
I. Title.
T55.3.H3V55 1996
604.7--DC20 96-35282
CIP

International Standard Book Number 1-56670-200-3
Library of Congress Card Number 96-35282
Printed in the United States of America 1 2 3 4 5 6 7 8 9 0
Printed on acid-free paper

THE AUTHOR

Jeffrey W. Vincoli, CSP, CHCM has over 15 years of experience in the occupational safety, health, and environmental compliance professions. This experience has included the development, implementation, administration, and management of occupational safety and health programs. Currently, Mr. Vincoli is President and Principal Consultant of his own consulting firm, J.W. Vincoli and Associates, located in Titusville, Florida. The firm specializes in providing professional safety, health, and environmental training and consulting services to clients on the national and international levels. He has provided safety, health, and environmental training and management consulting services for literally thousands of professionals across the United States and in more than 15 countries. He also specializes in providing expert witness testimony on matters of fact pertaining to occupational safety, health, and the environment.

Prior to beginning his own consulting practice, Mr. Vincoli spent 14 years working in our nation's missile, space, and strategic defense programs. With more than 10 years with McDonnell Douglas Corporation, he worked first as a Safety Engineer and then as a Manager of occupational safety and health, system safety engineering, industrial hygiene, hazardous waste management, and environmental compliance programs. Mr. Vincoli has worked on programs such as Space Shuttle, Space Station, and unmanned launch vehicle operations, as well as the Tomahawk Cruise Missile and other specialized weapon systems for the United States and allied governments. Mr. Vincoli also worked for companies such as EG&G, Inc., and United Technologies Corporation, always with focus on ensuring a safe and healthy work environment for several thousand employees.

Mr. Vincoli received his undergraduate degrees from the Florida Institute of Technology and completed his Master of Science (Aeronautical Science) and Master of Business Administration from Embry-Riddle Aeronautical University. He is a Certified Safety Professional, a Certified Hazard Control Manager, a Registered Environmental Professional, and a World Safety Organization-Certified Safety Specialist.

Mr. Vincoli is a member of many recognized professional organizations, including the System Safety Society, the National Environmental Health Association, and the American Society of Safety Engineers. He has published more than two dozen articles in professional trade journals such as *Professional Safety, Occupational Health and Safety, Hazard Prevention, Green Cross* (Hong Kong), and *Notícias de Seguridad* (Mexico). *Risk Management for Hazardous Chemicals* is his sixth published text in the field of safety, health, and the environment. He has served on the Editorial Board for *Occupational Hazards* magazine (1995-1997) and on the Advisory Committee for the Bureau of Business Practice (1995-1997).

Mr. Vincoli is an active member of the American Society of Safety Engineers. He has held offices on the Chapter level, including President, and has served on numerous Regional and National Committees and Special Task Forces focusing on the Society's service to its members and to the safety profession. He has received numerous awards from professional societies, including the System Safety Society's Manager of the Year (1994) and the

American Society of Safety Engineers' Regional Safety Professional of the Year (1987 and 1993). His personal and professional biographies have appeared in several noted Who's Who publications, including *Who's Who in the Safety Profession, Who's Who in Science and Engineering, Who's Who in America,* and *International Men of Achievement* (Cambridge, England).

Mr. Vincoli is a noted speaker, lecturer, trainer, and published author on subjects that extend across the broad scope of the occupational safety, health, and environmental industries.

Table of Contents

To my wife Rosemary.
Without her love, patience, understanding, sacrifice, and commitment this work would simply not have been possible.

Acknowledgments

There are always numerous associates, friends, and colleagues who deserve a note of thanks following the completion of a project of this magnitude. Unfortunately, space and time could not possibly permit an individual acknowledgment of everyone involved. However, there are some specific acknowledgments that must be mentioned.

First, I would like to acknowledge and thank the many safety, health, and environmental professionals who responded to surveys and/or agreed to be interviewed at the beginning of this endeavor. They provided valuable insight and information concerning their individual and collective concerns and frustrations regarding the practice of risk management for hazardous chemicals.

Second, I would like to express my great appreciation and thanks to the anonymous reviewers who gave of their valuable time to first evaluate the proposal, and then the final manuscript. Their endorsements, advice, recommendations, and suggestions provided the encouragement needed to see this project through to its completion.

Third, to the highly competent and professional staff at CRC Press/Lewis Publishers for their devotion to excellence and their commitment to perfection. While all those who had involvement with this work are certainly appreciated, I would like to specifically thank Greg Daurelle, Tom York, Joan Moscrop, Victoria Held, and especially Susan Alfieri.

Finally, a very special thanks to Kenneth P. McCombs, my Acquisitions Editor, at CRC Press/Lewis Publishers. For his persistent and devoted attention to detail; for his firm but understanding style; for his advice, council, and encouragement; and for his frank and honest approach to the world of publishing, I am forever in his debt. He is truly a professional's professional. But he is much more than that, he is my friend.

PREFACE

Risk Management for Hazardous Chemicals

The use of chemicals across virtually every type of industry has always been an issue of importance for the practicing safety and health professional. Risks to employee safety and health and the preservation of the environment associated with the use of chemical substances have been and will continue to be a major concern, and rightfully so. The myriad of obvious and insidious hazards to life and health posed by the use of such chemicals cannot be overstated. It has always been the obligation, both morally and professionally, of the safety and health practitioner to ensure adequate protection of life, health, property, and the environment whenever and wherever hazardous chemicals are used.

The key to the successful accomplishment of this objective is *information*. Having the right information, readily available, in easy-to-read, non-technical language can literally save a life and/or prevent costly and devastating environmental contamination or property loss. However, anyone who has practiced in the field of occupational or environmental safety and health even for a short period of time has certainly suffered the frustration of having inadequate or unavailable information. This shortcoming has been a personal source of frustration for this author during the past fifteen years. There is nothing more aggravating than going to a "noted" source for chemical information, especially during an emergency situation, to find missing, incomplete, or even incorrect data. Over the years, I have researched a number of common and uncommon hazardous chemicals using a variety of well-known reference sources. More often than not, I have discovered conflicting information between sources that range from minor chemical property information errors to major discrepancies associated with the administration of health care protocol following an exposure. Even a chemical manufacturer's own material safety data sheet (MSDS) has been incomplete, inconsistent, and incorrect on too many occasions to be considered merely incidental. The manufacturer's MSDS is supposed to provide all the information about a particular chemical that would be needed to properly assess and manage the hazards posed by that chemical. Such errors, discrepancies, and inconsistencies leave the user dismayed and confused as to the proper course of action to be taken in any given circumstance. Be it occupational or environmental safety and health that is at issue, consistent and accurate information will always be a critical tool. *Understanding* of hazard risks is essential before management of those risks can be accomplished. Improper understanding based upon faulty or incomplete information can be more harmful to employee and environmental health than no information at all.

Risk Management for Hazardous Chemicals has been compiled to provide the quick and accurate reference information needed for safety and health professionals to accomplish their duties more effectively, efficiently, and with more confidence. Obviously, saving lives, preserving the environment, and protecting property and resources are no tasks to be taken lightly. The information these professionals require to meet this charter should not be taken lightly either. For this reason, *Risk Management for Hazardous Chemicals* has been written using over sixty of the most well-known and well-researched reference sources available today. In most cases, discrepancies and inconsistencies on important chemical data were discovered routinely between these many sources. Before the data were entered into this text, verifications were made on *all* such inconsistencies. By using "referee" sources such as the National Institute for Occupational Safety and Health

(NIOSH), the American Conference of Governmental Industrial Hygienists (ACGIH), the National Toxicology Program (NTP), the Integrated Risk Information System (IRIS), the Royal Society of Chemistry (United Kingdom), the U.S. Environmental Protection Agency (EPA), and many other equally renowned authorities, the data provided for each chemical are as accurate as possible. Also, when specific data points could not be found either because they have not been determined or because they are simply not applicable to a given chemical substance, the reader will see the words "Not Found," "Not Determined," "Not Applicable," None Reported," or "No Citation." If these terms have been used, it means that a great deal of effort went into researching this information and all available reference sources were checked and re-checked *before* the entry was finalized for print. Unlike other data sheets, even those prepared by manufacturers, that leave the data fields blank or simply omit these data points altogether, the information provided on an MSDS in this text will leave no reader guessing. Simply stated, if the information is not available, then the reader should know this immediately so he or she can move on to the next issue.

The user can now put this information into practice without having to hesitate over any concern regarding the accuracy of one source over another. The information on each chemical contained in this reference required from two to five hours to compile, verify, reverify, and document. This is mentioned only so users will know that they can now save at least that much time when researching the hazards of a given chemical. The CD-ROM version will provide even quicker access to this essential and even critical information.

Each chemical profiled in the following pages will provide the reader with a simple, accurate, and user-friendly MSDS. These data sheets have been developed based somewhat on the Occupational Safety and Health Administration (OSHA) format without regard to information that is normally considered mildly interesting but seldom useful in everyday practice. Only the most essential and necessary information has been included on these data sheets. One may ask, *what determines "essential and necessary" information?* The answer is contained in the experience of numerous practicing safety and environmental professionals who were contacted, surveyed, and/or interviewed about their frustrations associated with the use of manufacturer-provided data sheets. The primary complaint was in format inconsistencies. Each manufacturer's MSDS is always different from the next. Some are better than others but few make it very easy to find the desired information quickly. There are often missing data fields or simply incomplete entries on many of the data sheets. The second most cited complaint focused on the type of information presented on a MSDS. For example, while it may be interesting to the scientific community, very few safety and health professionals in the "real world" of everyday practice find data on the lethal concentration or dose required for killing rats to be very useful in their assessment of employee health risk. This is not to suggest or even imply that such information and data are not important. It is only an indication of what the user community really looks for on a day-to-day basis when using a MSDS. Therefore, only the most usable information has been provided on the sheets contained in this volume. While the data sheets provided herein may not be cited as the most complete, they are by far the most accurate and, subsequently, the most user-friendly available to industry today.

The information that follows each MSDS addresses three of the most serious issues facing the safety, health, and environmental practitioner. First, *risks to health* are addressed in terms of general, acute, and chronic health effects. By first obtaining a basic understanding of these risks, they can then be managed more effectively. Established and proven risk reduction measures are then provided to help the user achieve this goal. Second, *risks to the environment* are discussed with focus on the general, acute, and chronic ecological hazards. Once again, risk reduction methods are suggested to assist the reader in environmental risk management. Finally, *risks to business operations* resulting from the use of chemicals and their hazards are discussed in brief to ensure the user is aware of the legal and financial risks associated with hazardous chemical management and mismanagement.

It is my sincere hope that the effort involved in compiling this text will eliminate the need for similar effort from countless practicing safety, health, and environmental professionals. With this text, they can now perform their responsibilities more efficiently and more effectively. To ensure consistent accuracy and usability, this volume will grow annually. Existing chemical data will be updated and new chemicals will be added to the database.

Jeffrey W. Vincoli, CSP

AN INTRODUCTION

Risk Management and Risk Assessment

INTRODUCTION

Risks associated with the use of hazardous chemicals can be *assessed* and *managed* in terms of their effect on human health, environmental health, and business operations in general. When discussing a particular chemical substance, each of these categories should be examined to ensure a comprehensive understanding of total risk and to provide the basis for an acceptable risk management program.

The use or simple presence of a chemical commodity in the work place or the environment in general presents some level of risk. Assessing the nature or severity of this risk is dependent upon a number of factors, all of which focus on one common element: *exposure*. In assessing risk, the questions of exposure must consistently be asked. Are personnel being exposed? Is there an exposure to the environment? What is the risk to continued success in business operations if there is an exposure to personnel and/or the environment? Therefore, *risk assessment,* which is the basis for *risk management,* is partially contingent upon an understanding of the term *exposure.*

However, an exposure to a chemical does not always mean that the results will be detrimental. If such exposures occur (i.e., those with no detrimental or adverse effects to human health, the environment, or business operations), then additional questions must be asked: What is the nature of the risk associated with such exposures? Are these exposures hazardous or toxic? Hence, it is not always enough to have experienced an exposure to a chemical to accurately assess the risks posed by such. One must determine (or assess) if the exposure was also *hazardous* before the level or nature of risk can be properly identified. In other words, the specific hazards of an exposure that present risk to persons and/or the environment must be examined. It becomes clear that the risk assessor or risk manager must understand the principles of hazard, exposure, and risk.

THE RISK EQUATION

Hazard can be defined as an inherent property or ability of a substance to cause an adverse or detrimental effect or event. For ease of discussion with regard to chemical substances, hazards can be divided into two separate categories or classifications. There are chemicals that pose a *physical hazard* (such as explosives, compressed gases, flammable commodities, unstable and reactive materials), and there are those which will pose a health hazard (such as asphyxiants, irritants, carcinogens, mutagens, reproductive toxins). Health hazards are such because they are toxic (i.e., poisonous) in some degree to living organisms. The potential hazards to human and environmental health associated with chemicals can range from mild to severe, with each level along the way having equally adverse effects on business operations. For example, a chemical may cause mild skin irritation upon exposure which, under most circumstances, will have little effect on business operations in general. However, if exposure results in a miscarriage, decreased fertility, or death, the negative publicity generated in response to such emotionally charged issues can be devastating to business operations. Similarly, the uncontrolled discharge of a chemical into the environment may result in little measurable effects on aquatic or terrestrial life. However, if a major fish kill occurs or bird eggs stop hatching, the effect on business operations could conceivably be extremely adverse. It should be noted that these examples are not meant to diminish the value of human and/or environmental health over the importance of successful business operations. Most would agree there is nothing more important than the preservation of human and environmental health, whatever the cost. These examples are only used here to illustrate the fact that *everything* a business does and every decision it makes, either positive or negative, including the use of chemical substances, affects the bottom line of profitability. When it comes to assessing the real or potential risks resulting from the use of chemical substances, the risk to business operations must also be considered or the assessment will not be complete and the resultant management of those risks will be flawed.

Exposure describes the contact of a hazard with a specific set of conditions. With a physical hazard, exposure defines the conditions of contact (e.g., flash point) which will cause the hazard (e.g., flammability). For health hazards (toxicity), exposure defines contact of a biologic system (e.g., a person, an organ, a specific cell of an organ, a plant, an animal) with a hazardous substance or condition. Among the exposure components that will affect toxicity are:

a. The route of exposure
b. Frequency of exposure
c. Duration of each and all exposures
d. Magnitude or level of exposure
e. Genetic parameters (e.g., enzymes, hormones)
f. Dietary status and condition
g. Concurrent exposures
h. Differences between individuals being exposed.

Differences between individuals that can affect the toxic outcome of an exposure include age, sex, physical condition and health status, sensitivity, and ADME (absorption, distribution, metabolism, and excretion) rates.

With human exposure, chemical substances can enter the body by a number of routes. The most common is *inhalation,* since people have to breathe in order to live. Unfortunately, anything that happens to be in the air and is small enough to enter the lungs (10 microns in diameter or smaller) may enter the body as well through the respiration process. Another common route with many chemicals is *absorption.* Numerous substances have the ability to penetrate unbroken skin and enter the bloodstream to be distributed throughout the body through the circulatory system. A third route is *ingestion.* Swallowing a chemical substance can occur when eating, drinking, or, under highly contaminated conditions, when talking. A fourth, less common route, is *injection.* While more of a concern in the health care profession, hazardous chemicals can be present on the end of sharp objects in any industry. A chemical can literally be injected into the body when such objects pierce or puncture the skin. In the environment, routes of entry include *air* (distribution of chemical vapors, fumes, mists, fogs, aerosols, etc., directly into the atmosphere), *water* (discharge of chemical pollutants directly into the effluent), or *soil* (spillage or dumping of chemicals onto soil).

The frequency and duration of exposure to toxic chemicals play an important role in toxicity. For many chemicals, the adverse health effects resultant from a single (acute, or short-term) exposure are quite different than those associated with repeated (chronic, or long-term) exposures. Therefore, minimizing the duration of exposure and maximizing the time between exposures may reduce toxicity. In humans, such controlled conditions of exposure will allow the body's natural defense mechanisms, including the ability to detoxify and/or excrete foreign substances, to work more efficiently thereby diminishing the toxic effect of the exposure. In the environment, controlling the timing and level of chemical release into water, for example, will allow certain aquatic organisms the necessary time they may need to adequately detoxify the chemical substance before any significant harm results. However, with both human and environmental exposures, natural defense mechanisms can be overwhelmed when the frequency, magnitude and/or duration of exposure is/are too great.

The magnitude (i.e., dose or concentration) also plays a significant role in toxicity of an exposure. Toxic responses generally result in a wide range of effects on living organisms. This range if often described as the *dose-response relationship.* The magnitude of exposure controls this relationship because it determines the amount of toxin that will eventually end up inside the body or in the environment. Therefore, magnitude of exposure controls the response (will it happen?) and the degree of the response (how severe will it be?).

Risk is the probability that a hazard will occur (i.e., that an adverse effect or event will result from a given set of exposure conditions). Since risk is typically expressed as a mathematical probability, the range of risk can be stated as *zero* (having no possibility of adverse effect or event) to *one* (having a certainty that an adverse effect or event will result). Having established this, it is important to note that *risk* is the mathematical product of *hazard* and *exposure*. This relationship can be expressed in the following simple formula:

RISK = HAZARD x EXPOSURE

Simple math tells us that multiplying any number by zero forces a product of zero. Therefore, the above equation means that an extremely hazardous substance can be present with little risk of adverse effect if it is handled under safe and proper conditions (i.e., when the exposure component of the risk equation is driven towards zero). Similarly, risk can be reduced towards zero by driving the hazard component of the equation towards zero (e.g., changing the process design, substituting less hazardous commodities, using a lesser amount of the chemical, etc.), even if there is still a high probability of exposure. Of course, the ultimate risk management solution would be to drive both the exposure *and* the hazard components of the equation to as low a probability as possible. Such measures would virtually guarantee a low or no risk scenario. However, in the real world of everyday operations, it is not always practical, feasible, or possible to reduce the elements of risk to a zero level of probability. For this reason, risk assessment and risk management have become extremely vital elements to successful business operations in recent years. More importantly, the proper assessment and management of risks which may be posed by the use, transport, storage, or disposal of hazardous chemicals can literally save lives, prevent illness and injury, and preserve precious environmental resources.

RISK ASSESSMENT

It has been established that risk is a product of hazard and exposure to those hazards. When it comes to chemical substances, the risk assessor must also take into consideration any particular peculiarities of a given exposure which may influence the risks to humans and the environment posed by such an exposure. This is especially true when attempting to evaluate the long-term (chronic) effects of exposure to hazardous chemicals. In most instances, exposure data on chronic effects are determined under specific laboratory conditions using test animals. There are difficulties in trying to apply toxicity assessments performed under such conditions. More specifically, carcinogenic (i.e., cancer-causing) risk assessments are generally performed by exposing laboratory animals to the same dose every day for a lifetime. This type of exposure rarely if ever occurs in humans. Even in the workplace, exposure to chemical substances is interrupted by vacations, weekends, sick days, shift changes, job rotation, and work assignment changes. As uncertain as the consistency of work place exposures may be, environmental exposures are even more inconsistent. Even when the level of a contaminant in drinking water is known, for example, exposure to the hazard is still not a constant. Water consumption changes seasonally, it varies from person to person, and people change their source of drinking water throughout the day (work, home, when dinning out, and so on). In addition, it is rare to find a person who has lived in the same community, much less the same house, for an entire life time.

Risk assessment is, therefore, filled with difficulties and uncertainties, especially with respect to chemical risk assessment. The uncertainties in this process can lead to two questions of which the risk manager should be aware. The first is the scientific question of how to arrive at the best risk value, taking into account all of the uncertainty. The second is the social question of how best to deal with risks when it is not possible to determine with great certainty the magnitude of the risk and how it might affect very specific populations or the environment. Admittedly, these questions have no concrete answers. It is best to acknowledge them during the risk assessment process and attempt to establish, to the greatest extent possible, that the assessments performed are based upon the best available data.

This text has been prepared to provide data which can be used to facilitate the risk assessment and risk management process. As stated at the beginning of this chapter, understanding the hazards and the exposure potential posed by a specific chemical is key to understanding risk. Using these data, the reader should be able to better assess and manage the risks which are inherent with the use of hazardous chemical substances.

RISK MANAGEMENT

Once the level of risk has been assessed, exposure potential identified, and hazards determined, it becomes a matter of policy that these risks be managed. This means that the assessed risks must be either accepted or rejected by the responsible management function. Risk management is by no means an exact science since the criteria for risk acceptance and risk rejection can differ dramatically from one organization to another, from one type of business to another, and from one person to another. There is potential risk associated with virtually everything that is done. There are very few activities or tasks that can be said to be totally without risk. When hazardous chemicals are involved, the risks can be extremely serious to life and health. Even the simple act of drinking water, if done in excess, runs the risk of causing severe renal problems in most cases.

The question *"How safe is safe enough?"* has plagued mankind for ages and, unfortunately, has no simple answer. Risk managers must attempt to define a level of risk that, with all things considered, is acceptable to the success of their business enterprise. Examples of how truly difficult this concept of risk acceptance can be are around us every day. For example, safety razors or safety matches are not entirely without risk. They are only *safer* than their alternatives because modifications have been made to reduce the risk of harm from using these devices to a level acceptable to the user. In effect, the user becomes a risk manager by making the decision whether or not to accept the reduced level of risk and use the product. A more vivid example of risk management involves the sport of skydiving. Most sane skydivers would never jump out of an airplane without a parachute. The parachute provides a measure of control intended to reduce the level of risk. However, even with the parachute strapped in place, the jumper must still accept the risk of parachute failure, injury, or death. The point is that the jumper has established a level of risk that is acceptable and it is that acceptance which forms the basis of risk management.

By definition, the process of risk management can be summarized as risk assessment, risk reduction, and risk acceptance (or rejection). *Assessment* can be difficult at best. *Reduction* can also prove to be complex since the methods available may not always be practical for a given situation. However, since nothing could probably ever occur without a decision by someone to accept risk, *acceptance* is the key to the entire process.

EXPOSURE REDUCTION

As stated above, reducing risk is an important and complex component of risk management. Reducing the risk of exposure to hazardous chemicals actually facilitates the overall risk acceptance process. Where there is a potential for exposure to hazardous chemicals, there are some recommended methods to reduce the risk which have proven to be generally successful. First and foremost, hazards should be engineered out of a process. Engineering controls are always the best method of exposure reduction or elimination. These include redesigning a process to reduce the hazard exposure, substituting a hazardous chemical with a less hazardous one, and so on. Unfortunately, while engineering controls are the most desired method of risk reduction, they are not always the most practical. Therefore, administrative controls should also be considered in the risk reduction equation. Implementing specific procedural controls, for example, may help reduce the likelihood of exposure. When this is not enough, additional measures such as specific training of the applicable personnel and the provision of personal protective equipment may help reduce exposure risk even further. The goal is to reach a level of risk that can be determined to be acceptable by the responsible management function.

Additional steps for risk reduction are offered here for consideration:

- ☑ Where possible, enclose operations and use local exhaust ventilation at the site of chemical release. If local ventilation or enclosure is not used, respiratory protection should be provided.
- ☑ Protective clothing suitable to the chemical hazards should be provided and worn.
- ☑ Personnel should wash thoroughly immediately after exposure to chemical substances and at the end of each work shift. Good work practices can help reduce the risk of chemical exposure.
- ☑ Workers whose clothing has been contaminated should change into clean clothing promptly.
- ☑ Contaminated clothing should be laundered by persons informed of the hazards of chemical exposure.
- ☑ Eye wash fountains should be provided in the immediate work area for emergency use. If portable units are used, they should be capable of providing a minimum of 15 minutes of continuous water flow.
- ☑ If there is the possibility of skin or body exposure, emergency showers should be provided in the area.
- ☑ Personnel should never eat, drink, or smoke in areas where hazardous chemicals are used or stored. Hands should be washed carefully before eating or smoking.

These simple measures are by no means comprehensive but will form the beginning of a risk reduction program.

More specific measures can be included depending on the hazards at a particular location or facility.

CONCLUSIONS

When dealing with hazardous chemicals, there will always be some level of risk which must be considered. The proper assessment of this risk, in terms of a chemical's effect on human health, environmental health, and overall business health is crucial to the long-term success of any enterprise.

Understanding risk requires an understanding of the terms exposure and hazard. *Hazard* can be defined as an inherent property or ability of a substance to cause an adverse or detrimental effect or event. The data provided in this text are intended to assist in the determination of the hazards associated with a given chemical. *Exposure* is the contact that occurs with a hazard under a specific set of circumstances or conditions. While not all exposures are hazardous, there is a level of risk which must be assessed. If the outcome of such exposures can be deemed acceptable (i.e., the level of risk is low enough to warrant its acceptance), then risk management has occurred.

The information and data provided in this text have been derived from a variety of widely known and used chemical reference sources. The material safety data sheet (MSDS) provided for each chemical has been developed to assist in the risk assessment process. It should be noted that these data sheets have been purposely designed to simplify the presentation of information considered vital to the risk assessment task. No attempt has been made to include information which may be interesting to some but useless to most. For example, data usually presented as LD_{50} or LC_{50} (lethal dose/50 percent kill and lethal concentration/50 percent kill, respectively) has not been provided. Such information may be somewhat interesting to scientists and toxicologists but has little practical application in the real world of risk management. Hence, these data sheets are not in the fourteen-section format currently recommended by the American National Standards Institute (ANSI Z14). They are, however, useful tools to assist in the assessment, reduction, and management of the risks posed by the given chemical substance.

True risk assessment, which is key to proper risk management, cannot occur without understanding. The risk assessor must obtain an understanding of hazard as well as the real and potential consequences of exposure to that hazard. With this information, the reader should be capable of more accurately assessing the risk of exposure to a specific chemical with regard to human health, to the environment, and to business operations in general. Using this data together with the risk reduction techniques discussed in this chapter, the chances of successful risk management should be greatly improved.

REFERENCES

Ahlbom, A. 1993. *Biostatistics for Epidemiologists.* New York: CRC Press/Lewis Publishers.

Amdur, M.O., J. Doull, and C.D. Klaassen (eds.). 1991. *Cassarett and Doull's Toxicology. The Basic Science of Poisons,* 4th Edition. New York: Pergamon Press.

American Conference of Governmental Industrial Hygienists. 1993. *Documentation of the Threshold Limit Values and Biological Exposure Indices,* 6th Edition. Cincinnati: ACGIH.

Asante-Duah, D. K. 1993. *Hazardous Waste Risk Assessment.* New York: CRC Press/Lewis Publishers.

Bloom, A. D. and F. J. DeSerres. 1995. *Ecotoxicity and Human Health.* New York: CRC Press/Lewis Publishers.

Brooks, L. A risk assessment in support of a facility wind tunnel study. *Journal of Environmental Health.* June 1995, Vol. 57, No. 10. pp 14-18.

Brusick, D. J. 1994. *Methods for Genetic Risk Assessment.* New York: CRC Press/Lewis Publishers.

Calabrese, E. J., E. M. Kenyon. *Air Toxics and Risk Assessment.* New York: CRC Press/Lewis Publishers.

Calabrese, E. J. 1994. *Biological Effects of Low Level Exposures.* New York: CRC Press/Lewis Publishers.

Cockerman, L. G., B. S. Shane. 1994. *Basic Environmental Toxicology.* New York: CRC Press/Lewis Publishers.

Cothern, C. R., N. P. Ross. 1994. *Environmental Statistics, Assessment, and Forecasting.* New York: CRC Press/Lewis Publishers.

Cothern, R. R. 1993. *Comparative Environmental Risk Assessment.* New York: CRC Press/Lewis Publishers.

Cothern, C. R. 1995. *Handbook for Environmental Risk Decisionmaking.* New York: CRC Press/Lewis Publishers.

Covello, V.T. and M.W. Merkhofer. 1993. *Risk Assessment Methods. Approaches for Assessing Health and Environmental Risks.* New York: Plenum Press.

Danse, I.R. 1991. *Common Sense Toxics in the Workplace.* New York: Van Nostrand Reinhold.

Environmental Protection Agency. 1983. *Chemical Hazard Information Profiles.* Washington, D.C.: U.S. EPA.

Fodden, C. R., and J. L. Weddell. 1992. *Hazardous Materials Emergency Action Data.* New York: CRC Press/Lewis Publishers.

Forseberg, K., and L. H. Keith. 1993. *Chemical; Protective Clothing Permeation and Degradation Database.* Chelsea, MI: Lewis Publishers.

Gardner, R. W. 1994. *Chemical Intolerance.* New York: CRC Press/Lewis Publishers.

Gerarde, H. W. 1960. *Toxicology and Biochemistry of Aromatic Hydrocarbons.* New York: Elsevier.

Hallenbeck, W. H. 1993. *Quantitative Risk Assessment for Environmental and Occupational Health,* 2nd Edition. New York: CRC Press/Lewis Publishers.

Howard, P. H. 1995. *Handbook of Environmental Fate and Exposure Data for Organic Chemicals,* Vol. I-IV. New York: CRC Press/Lewis Publishers.

Howard, P. H., et. al. 1991. *Handbook of Environmental Degradation Rates.* New York: CRC Press/Lewis Publishers.

Hosty, J. W., and P. Foster. 1990. *A Practical Guide to Chemical Spill Response.* New York: Van Nostrand Reinhold.

International Agency for Research on Cancer (IARC). 1987. *IARC Monographs on the Evaluation of Carcinogenic Risk to Humans, Overall Evaluations of Carcinogenicity: An updating of IARC Monographs Volumes 1-42.* Lyon: IARC.

James, A., and M. Lord. 1992. *Index of Chemical and Physical Data.* New York: Van Nostrand Reinhold.

Jones, F. E. 1994. *Toxic Organic Vapors in the Workplace.* New York: CRC Press/Lewis Publishers.

Kamrin, M.A. 1988. *Toxicology, A primer on Principles and Applications.* Chelsea, MI: Lewis Publishers.

Keith, L. H. , M. T. Johnston, C. E. Hudak, and M. Conoley. 1992. *National Toxicology Program's Gloves+ Database.* Chelsea, MI: Lewis Publishers.

Keith, L. H. and M. M. Walker. 1995. *Handbook of Air Toxics.* New York: CRC Press/Lewis Publishers.

Keith, L. H. (Editor). 1992. *Integrated Risk Information System (IRIS) Chemical Information Database.* Prepared by the U.S. Environmental Protection Agency. Boca Raton, FL: CRC Press/Lewis Publishers.

Keith, L. H. and D. B. Walters. (Editors) 1992. *National Toxicology Program's Chemical Database, Volumes I-VII.* Chelsea, MI: Lewis Publishers.

Landis, W. G., M. H. Yu. 1995. *Introduction to Environmental Toxicology.* New York: CRC Press/Lewis Publishers.

Langerman, N. 1994 *Precautionary Labels for Chemical Containers.* New York: CRC Press/Lewis Publishers.

Lauwerys, R. R., P. Hoet. 1993. *Industrial Chemical Exposure.* New York: CRC Press/Lewis Publishers.

Lide, D. R. 1995. *Handbook of Chemistry and Physics,* 76th Edition. New York: CRC Press/Lewis Publishers.

Lisella, F. S. 1994. *The VNR Dictionary of Environmental Health and Safety.* New York: Van Nostrand Reinhold.

Mackay, D., W. Y. Shiu, and K. C. Ma. 1992. *Illustrated Handbook of Chemical-Physical Properties and Environmental Fate for Organic Chemicals.* New York: CRC Press/Lewis Publishers.

Manahan, S. E. 1994. *Environmental Chemistry,* 6th Edition. New York: CRC Press/Lewis Publishers.

Manahan, S. E. 1992. *Toxicological Chemistry,* 2nd Edition. New York: CRC Press/Lewis Publishers.

Mellström, G. A., J. E. Wahlberg, and H. I. Maibach. 1994. *Protective Gloves for Occupational Use.* New York: CRC Press/Lewis Publishers.

Meyer, E. 1990. *Chemistry of Hazardous Materials.* Englewood Cliffs, NJ: Prentice Hall.

Miller, B. F., M.D., and C. B. Keane, R.N. 1972. *Encyclopedia and Dictionary of Medicine and Nursing.* Philadelphia: W.B. Sanders Company.

Mycock, J. C., J. D. McKenna, and L. Theodore. 1995. *Handbook of Air Pollution Control Engineering and Technology.* New York: CRC Press/Lewis Publishers.

Ott, W. R. 1995. *Environmental Statistics and Data Analysis.* New York: CRC Press/Lewis Publishers.

Patnaik, P. 1992. *A Comprehensive Guide to the Hazardous Properties of Chemical Substances.* New York: Van Nostrand Reinhold.

Philp, R. B. 1995. *Environmental Hazards and Human Health.* New York: CRC Press/Lewis Publishers.

Proctor, N. H., and J. P. Hughes. 1991. *Chemical Hazards of the Workplace.* New York: Van Nostrand Reinhold.

Quigley, D. R. 1994. *Handbook of Emergency Chemical Management.* New York: CRC Press/Lewis Publishers.

Rea, W. J. 1995. *Chemical Sensitivity,* Vol. I-III. New York: CRC Press/Lewis Publishers.

Reeves, A. L. 1979. *Handbook on the Toxicology of Metals.* New York: Elsevier North-Holland.

Richardson, M. 1995. *Dictionary of Substance and Their Effects,* Vol. I-VI. New York: CRC Press/Lewis Publishers.

Royal Society of Chemistry. 1992. *Chemical Safety Data Sheets, Volumes 1-5.* Newcastle-upon-Tyne, UK: Athenaeum Press..

Rylander, R., R. R. Jacobs. 1994. *Organic Dusts.* New York: CRC Press/Lewis Publishers.

Sacarello, H. L. A. 1994. *The Comprehensive Handbook of Hazardous Materials.* New York: CRC Press/Lewis Publishers.

Sheftel, V. O. 1995. *Handbook of Toxic Properties of Monomers and Additives.* New York: CRC Press/Lewis Publishers.

Talbott, E., G. Craun. 1995. *An Introduction to Environmental Epidemiology.* New York: CRC Press/Lewis Publishers.

Torkelson, T. R. V. K. Rowe. 1981. *Patty's Industrial Hygiene and Toxicology,* Vol. 2B. New York: Wiley-Interscience.

U.S. Department of Health and Human Services, Public Health Service, Agency for Toxic Substances and Disease Registry. 1992. *Public Health Assessment Guidance Manual.* Chelsea, MI: Lewis Publishers.

Vincoli, J.W. 1993. *Basic Guide to Environmental Compliance.* New York: Van Nostrand Reinhold.

Vincoli, J.W. 1995. *Basic Guide to Industrial Hygiene.* New York: Van Nostrand Reinhold.

Vincoli, J. W. 1993. *Basic Guide to System Safety.* New York: Van Nostrand Reinhold.

MATERIAL SAFETY DATA SHEET

CHEMICAL NAME

ACETALDEHYDE

HAZARD WARNING INFORMATION

HEALTH	FIRE	REACTIVE	OTHER	DEGREE OF HAZARD	COLOR CODING	OTHER CODES
2	4	2		0 = Minimum Hazard 1 = Slight Hazard 2 = Moderate Hazard 3 = Serious Hazard 4 = Severe Hazard	HEALTH = BLUE FIRE = RED REACTIVITY = YELLOW OTHER = WHITE	OX = Oxidizer ACID = Acid ALK = Alkali COR = Corrosive W = Use No Water

SECTION I - GENERAL INFORMATION

Characterization: **Aldehyde**	RCRA Number: **U001** — EPA Class: **Toxic Waste**
DOT Proper Shipping Name: **Acetaldehyde**	Chemical Abstract Service (CAS) Number: **75-07-0**
DOT Hazard Class and Label Requirements: **Flammable Liquid**	DOT Emergency Guide Code: **26**
DOT Identification Number: **UN 1090**	Chemical Formula: **CH_3CHO**

Synonyms

Ethanal; acetic aldehyde; ethyladehyde; methyl formaldehyde.

SECTION II - HAZARDOUS INGREDIENTS AND IDENTITY INFORMATION

Hazardous Components (specific identity)	OSHA Exposure Criteria	NIOSH Exposure Criteria	Immediately Dangerous to Life and Health (IDLH)	ACGIH Exposure Criteria
Acetaldehyde (derivation: From oxidation of ethylene; vapor phase oxidation of ethanol, propane and butane; catalytic reaction of acetylene and water). **1 ppm = 1.83 mg/m^3**	**PEL:** **100 ppm** **180 mg/m^3** **STEL:** **150 ppm** **270 mg/m^3**	**REL:** **Not Established** **Possible Cancer Agent**	**10,000 ppm**	**TLV:** **100 ppm** **180 mg/m^3** **STEL:** **150 ppm** **270 mg/m^3**

SECTION III - PHYSICAL AND CHEMICAL CHARACTERISTICS

Boiling Point: **69°F (20°C)**	Specific Gravity (H_2O = 1): **0.79**
Vapor Pressure (mm Hg): **740 at 69°F (20°C)**	Molecular Weight: **44.06**
Vapor Density (Air = 1): **1.52**	Freezing Point: **-190°F (-123°C)**

Solubility

Miscible (also with ether, benzene, gasoline, solvent naphtha, toluene, xylene, turpentine, and acetone).

Appearance and Odor

Colorless liquid or fuming gas (above 69°F) with a pungent, fruity odor. Odor Threshold = 0.21 ppm.

SECTION IV - FIRE AND EXPLOSION HAZARD DATA

Flash Point (method used): **-36°F (-57°C) closed cup**	Explosive Limits in Air % by Volume: **LEL: 4.0% UEL: 60%**
NFPA Classification: **Class 1A Flammable Liquid**	Autoignition Temperature: **347°F (175°C)**

Extinguishing Media

Carbon dioxide, dry chemical, water mist, or fog.

Special Fire Fighting Procedures

Wear full protective clothing and self-contained breathing apparatus (SCBA). Heat will build pressure and may rupture closed storage containers. Keep fire-exposed containers cool with water spray. Vapors are heavier than air; they may travel a distance to cause fire.

Unusual Fire and Explosion Hazards

Prolonged contact with air may cause formation of peroxides which may explode and burst containers; easily undergoes polymerization. Poisonous gases are produced in fire.

SECTION V - REACTIVITY DATA

Stability		Conditions to Avoid **Dangerous fire hazard when exposed to heat and flame; can react vigorously with oxidizing materials. Forms heat-sensitive explosive peroxides on contact with air.**
Stable **X**	Unstable	Incompatibility *(materials to avoid)* **Strong oxidizers, acids, bases, alcohols, ammonia and amines, phenols, ketones, hydrogen cyanide, hydrogen sulfide, acid anhydrides, and halogens.**
Hazardous Polymerization		Conditions to Avoid **Contact with mineral acids (H_2SO_4, HCL, H_3PO_4) will form paraldehyde, which is toxic to humans and animals.**
May Occur	Will Not Occur **X**	Hazardous Decomposition or By-products **When heated to decomposition, acetaldehyde emits acrid, irritating smoke and fumes.**

SECTION VI - HEALTH HAZARD DATA

Primary Route(s) of Entry:	Inhalation? X	Absorption (skin)? X	Ingestion? X

Health Hazards

INHALATION: Eye, nose, and throat irritant. Produces narcotic effects. Headache, sore throat, and paralysis of the respiratory system. High exposures can cause pulmonary edema.

ABSORPTION: Skin contact can result in irritating rash or burning feeling. Can cause severe eye burns leading to permanent damage. Can result in sensitization following repeated contact.

INGESTION: Central nervous system (CNS) depression. Toxic to liver and kidney.

Carcinogenicity	NTP Listed?	IARC Cancer Review Group?	OSHA Regulated?	Target Organs?
Suspected Human Confirmed Animal	**5th Annual Report**	**Group 2B**	**29 CFR 1910.1000 (Table Z-1)**	**Respiratory system; skin; eyes; kidneys**

Medical Conditions Generally Aggravated by Exposure

Respiratory impairments (asthma); Renal function abnormalities; Skin conditions (dermatitis).

Emergency and First-aid Procedures

Eye contact: Flush immediately with water for 30 minutes (minimum); seek medical attention. Skin contact: Remove all contaminated clothing. Immediately wash area with large amounts of soap and water. Seek medical assistance. For inhalation: Remove the person from exposure. Provide respiratory assistance and CPR. Transfer to medical facility. Medical observation is recommended for 24 to 48 hours after inhalation exposure (pulmonary edema). If swallowed, seek medical attention immediately.

SECTION VII - PRECAUTIONS FOR SAFE HANDLING AND USE

Steps to be Taken in Case Material is Released or Spilled

Contain spills using absorbent material (vermiculite, or other). Ventilate area of spill. Have water source available in case of fire. Store materials in DOT-approved containers. Restrict those not involved in cleanup from entering area. Notify appropriate authorities, as required.

Preferred Waste Disposal Method

Incineration.

Precautions to be Taken in Handling and Storage

Acetaldehyde is highly volatile. Do not store with incompatible chemicals. Store in tightly closed containers in a cool, dark, well-ventilated area. Keep fire and flame away. Use non-sparking tools to open or close containers. Ground and bond all metal containers during transfer operations.

Other Precautions and Warnings

Bulk storage of acetaldehyde is not recommended. Nitrogen or other inert gas should be used as an "inert blanket" over liquid acetaldehyde in storage containers.

SECTION VIII - CONTROL MEASURES AND PERSONAL PROTECTIVE EQUIPMENT

Respiratory Protection *(specify type)*

Exposures below 100 ppm: MSHA/NIOSH approved full-facepiece respirator with organic vapor cartridge. Exposure above 100 ppm: Self-contained breathing apparatus (SCBA) with full facepiece and pressure demand or other positive pressure mode.

Ventilation

Local exhaust or general mechanical systems recommended.

Protective Gloves	Eye Protection	Other Protective Clothing
Butyl Rubber	**Chemical Goggles or Face Mask**	**Rubber Apron**

Work/Hygiene Practices

Always wash hands thoroughly after using chemical; never bring food, drink, or smoking materials into vicinity of chemicals.

ACETALDEHYDE

CH_3CHO CAS: 75-07-0

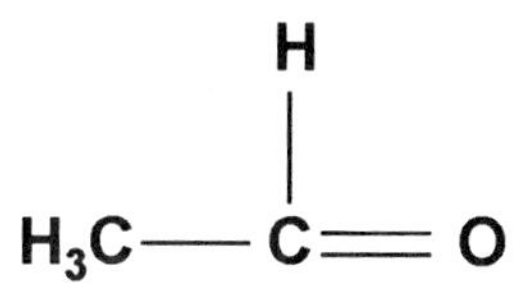

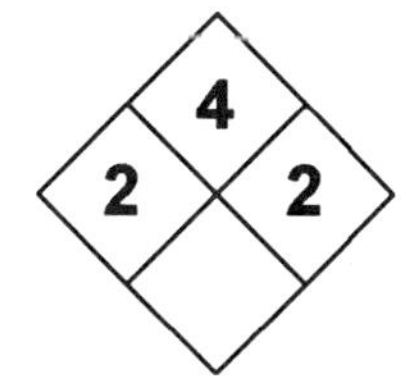

IDENTIFICATION AND TYPICAL USES

Acetaldehyde is a colorless liquid or gas with a strong fruity odor. It is used in the production of acetic acid, acetic anhydride, cellulose acetate, synthetic pyridine derivatives, pentaerythritol, terphtalic acid, and many other raw materials. It is also used in the production of perfumes, polyester resins, and dyes, and as a food preservative and flavoring agent. It is also used in silvering mirrors and in hardening gelatin fibers.

RISK ASSESSMENT: HEALTH

General Assessment

Acetaldehyde is carcinogenic in experimental animals and is considered to be a probable human carcinogen. It is a mutagen and a suspected teratogen. Mutagens may have a cancer risk. All contact with this chemical should be reduced to the lowest possible level. Human exposure to acetaldehyde can occur through three routes of entry into the body: *inhalation, absorption,* and *ingestion.* Each presents a moderate level of toxicity. Absorption through the skin and direct ingestion can result in the development of conjunctivitis (inflammation of the conjunctiva tissue of the eye), central nervous system depression, eye and skin burns, and dermatitis. Large doses can be fatal. Because its metabolic structure is closely linked to that of ethanol (alcohol), exposure can result in an intoxication effect similar to that of chronic ethanol intoxication. Inhalation can result in eye, nose, and throat irritation and will produce narcotic effects. At higher concentrations, individuals may experience headache, sore throat, and, if exposure continues, paralysis of the respiratory muscles. Prolonged exposures can cause a rise in blood pressure and a decrease in red and white blood cells.

Acetaldehyde can bind to liver protein and hemoglobin. This binding probably results in an alteration of the biological functions of protein and hemoglobin, thereby contributing to the toxicity of acetaldehyde. At ambient temperatures, acetaldehyde easily polymerizes on exposure to mineral acids such as hydrochloric acid (HCL) to form toxic paraldehyde, which can cause respiratory depression and cardiovascular collapse. At lower temperatures, acetaldehyde can polymerize to form metaldehyde, which is toxic to the liver, kidney, and intestines.

Acute Health Effects

The following acute (short-term) health effects may occur immediately or shortly after exposure to acetaldehyde:

Skin: Irritation and rash or burning sensation on contact.

Eye: Burns leading to permanent damage.

Lung: Irritation causing coughing and/or shortness of breath. Higher exposures can cause a build-up of fluid in the lungs (pulmonary edema), which is a medical emergency.

CNS: Inhalation of high concentrations of the vapor may cause sleepiness, dizziness, loss of consciousness, or death.

Chronic Health Effects

The following chronic (long-term) health effects may occur at some time after exposure to acetaldehyde and can last for months or even years:

Cancer Hazards: Acetaldehyde causes mutations (genetic changes). Such chemicals may have a cancer or reproductive risk. There is limited evidence that acetaldehyde causes cancer in animals. Acetaldehyde may be a teratogen in humans since it has been shown to cause birth defects in animals.

Sensitivity: Exposure to the skin may result in allergy (sensitization). If such an allergy develops, very low future exposure can result in skin irritation and rash. Repeated exposure can cause chronic irritation of the eyes and lungs. It is not known if lung damage will occur over prolonged periods of exposure.

Recommended Risk-Reduction Measures

Personnel should avoid direct contact with acetaldehyde. Unless a less toxic chemical can be substituted for a hazardous substance, *engineering controls* are the most effective method of reducing exposures. The best protection is to enclose operations and/or provide local exhaust ventilation at the site of chemical release. While not always operationally feasible, isolating operations can also reduce exposure.

Using respiratory protection is less effective than the controls mentioned above, but is still advisable whenever working with or around acetaldehyde. A self-contained breathing apparatus (SCBA) with full facepiece and pressure demand is the recommended respiratory protection of choice. If a full facepiece is not available, then chemical goggles should be worn to protect the eyes. Whenever a chemical splash hazard exists, a face shield and a rubber apron should be worn. To prevent hand and skin exposures, butyl rubber gloves should be worn.

Administrative controls should also be in place to minimize the potential for human exposures. These may include written procedures or policies which specify the methods and techniques that will be practiced whenever personnel are to work with acetaldehyde.

All personnel should receive *training* on the use, hazards, protective measures, emergency actions, and other precautions per 29 CFR 1910.1200 (Hazard Communication), prior to the first assignment in an area where acetaldehyde is used or stored.

Other methods to reduce exposure include:

- ☑ Where possible, enclose operations and use local exhaust ventilation at the site of chemical release. If local exhaust ventilation or enclosure is not used, respiratory protection should be mandatory.
- ☑ Always ensure that proper protective clothing is worn when using chemical substances.
- ☑ Wash thoroughly immediately after exposure to acetaldehyde and at the end of the work shift or before eating, drinking, or smoking.
- ☑ Hazard warning information should be posted in the work area. In addition, as part of an on-going education and training program, all information on the health and safety hazards of acetaldehyde should be communicated to all potentially exposed workers.
- ☑ Never eat, drink, or smoke in areas were acetaldehyde is used, handled or stored.

RISK ASSESSMENT: ENVIRONMENT

General Assessment

The environment is at risk of exposure during transportation, storage, disposal, or destruction of acetaldehyde. In almost every scenario, the threat of environmental exposure is contingent upon the proper handling of the chemical substance. Accidental spills, large or small, can result in fire, explosion, and possible contamination of the surrounding environmental mediums (water, soil, and air). Acetaldehyde is considered a highly flammable liquid. Because of its low flash point and low boiling point, it is classified as a Class 1A flammable liquid (per OSHA 29 CFR 1910.106). Its low auto-ignition temperature and low flash point, together with its high vapor pressure make it a dangerous fire and explosion hazard. It is also susceptible to forming peroxide, which is highly unstable and can catch fire or explode with little provocation. These characteristics require special consideration during any emergency situation involving a leak or spill of acetaldehyde. Should acetaldehyde ever come into contact with incompatible substances such as oxidizers, acids, bases, alcohols, ammonia, phenols, or ketones either during use, transportation, or storage, the formation of highly toxic and/or highly explosive commodities is extremely possible. The proper disposal/destruction method for acetaldehyde is to burn it in a chemical incinerator equipped with an afterburner and air scrubber. Acetaldehyde can enter the environment through manufacturing, unchecked discharge into effluents, and through spills.

Acute Ecological Effects

Acute (short-term) toxic effects may include the death of animals, birds, or fish, and death or low growth rate in plants. Acute effects are seen 2 to 4 days after animals or plants are exposed to acetaldehyde. This chemical has high acute toxicity to aquatic life. No data are available on the short-term effects of acetaldehyde to terrestrial animals.

Chronic Ecological Effects

Chronic toxic effects may include shortened life span, reproductive problems, lower fertility, and changes in appearance or behavior in exposed animals. These effects can be seen long after first exposure(s) to toxic chemicals. Acetaldehyde has high chronic toxicity in aquatic life. No data are available on the long-term effects of acetaldehyde to plants, birds, or land animals.

♦ *Water Solubility*

Acetaldehyde is highly soluble in water. Concentrations of 1000 milligrams and more will mix with a liter of water.

⌛ *Persistence in the Environment*

Acetaldehyde is moderately persistent in water, with a half-life of between 2 to 20 days. The half-life of a pollutant is the amount of time it takes for one-half of the chemical to be degraded. Approximately 73% of acetaldehyde will eventually end up in air; the rest will end up in water.

Bioaccumulation in Aquatic Organisms

Some substances increase in concentration, or *bioaccumulate,* in living organisms as they breathe contaminated air, drink contaminated water, or eat contaminated food. These chemicals can become concentrated in the tissues and internal organs of animals as well as humans.

The concentration of acetaldehyde found in fish tissues is expected to be about the same as the average concentration of acetaldehyde in water from which the fish was taken.

Recommended Risk-Reduction Measures

Proper training of all transporters will reduce the likelihood of a mishap or accident resulting in a leak or spill to the environment. The correct use of DOT labeling (per the MSDS) on all containers, trucks, and rail cars will enable emergency responders to react properly and quickly to any disaster thereby reducing the potential risk to the environment and to personnel.

Storage of acetaldehyde should be segregated from incompatible chemicals to minimize the risk of cross-contamination or contact. Buildings designated for storage should be equipped with appropriate fire protection systems (alarms, sprinklers, emergency lighting, portable extinguishers).

If a spill or leak to the environment has occurred, fire department, emergency response, and/or hazardous materials spill personnel should be notified immediately. Cleanup should be attempted only by those trained in proper spill containment procedures. Contaminated soils should be removed for incineration and replaced with clean soil.

If acetaldehyde should contact the water table, aquifer, or navigable waterway, time is of the essence. It is readily miscible in water and, therefore, total containment and remediation may not be entirely possible. When such spills occur, the local and/or state emergency response authorities must be notified. A comprehensive emergency response or disaster preparedness/recovery plan should be in place prior to any operations involving the use, transportation, storage, or disposal of acetaldehyde. If acetaldehyde is spilled or leaked, the following specific steps are recommended:

- ☑ Restrict persons not wearing protective clothing from area of spill or leak until cleanup is complete and area can be opened for normal work.
- ☑ Remove all ignition sources.
- ☑ Ventilate area of spill or leak.
- ☑ Absorb liquids in vermiculite, dry sand, earth, or a similar material and deposit in sealed containers.
- ☑ If applicable, stop flow of leaking liquid or gas. If leak source is a cylinder and the leak can not be stopped in place, remove leaking cylinder to a safe place in the open air and repair or allow cylinder to empty.
- ☑ Keep acetaldehyde out of a confined space, such as a sewer, because of the possibility of explosion (unless the sewer is designed to prevent the build-up of explosive concentrations).
- ☑ It may be necessary to dispose of acetaldehyde as a hazardous waste. The responsible state agency or the regional office of the federal Environmental Protection Agency (EPA) should be contacted for specific recommendations.

RISK ASSESSMENT: BUSINESS

General Assessment

Accidents or mishaps involving acetaldehyde can present a significant threat to business operations. The loss or damage of equipment or facilities can significantly affect fiscal viability. Lawsuits that may result from personnel injury/death, public exposures, and/or environmental contamination will also require a serious expenditure of resources.

Media attention surrounding an injury, death, or environmental damage can also result in a loss of profits and loss of current as well as future business.

Recommended Risk-Reduction Measures

Company attorneys, safety and health professionals, and environmental specialists should be involved in the

development of any procedures dealing with safety or emergency response. A company official should be pre-designated as a public relations officer with specific training in dealing with the press. Corporate plans and policies should be developed, approved, and implemented long before any need for such arises.

REFERENCES

Ahlbom, A. 1993. *Biostatistics for Epidemiologists.* New York: CRC Press/Lewis Publishers.

American Conference of Governmental Industrial Hygienists. 1988. *Documentation of the Threshold Limit Values and Biological Exposure Indices,* 5th Edition (with updates). Cincinnati: ACGIH

Anthony, C. P., and N. J. Kolthoff. 1971. *Textbook of Anatomy and Physiology.* St. Louis: C. V. Mosby Company.

Asante-Duah, D. K. 1993. *Hazardous Waste Risk Assessment.* New York: CRC Press/Lewis Publishers.

Bloom, A. D. and F. J. DeSerres. 1995. *Ecotoxicity and Human Health.* New York: CRC Press/Lewis Publishers.

Brusick, D. J. 1994. *Methods for Genetic Risk Assessment.* New York: CRC Press/Lewis Publishers.

Bueche, J. R. 1972. *Principles of Physics,* 2nd Edition. New York: McGraw-Hill.

Burrell, R., D. K. Flaherty, and L. J. Sauers. 1992. *Toxicology of the Immune System.* New York: Van Nostrand Reinhold.

Calabrese, E. J., E. M. Kenyon. *Air Toxics and Risk Assessment.* New York: CRC Press/Lewis Publishers.

Calabrese, E. J. 1994. *Biological Effects of Low Level Exposures.* New York: CRC Press/Lewis Publishers.

Cockerman, L. G., B. S. Shane. 1994. *Basic Environmental Toxicology.* New York: CRC Press/Lewis Publishers.

Cothern, C. R., N. P. Ross. 1994. *Environmental Statistics, Assessment, and Forecasting.* New York: CRC Press/Lewis Publishers.

Cothern, R. R. 1993. *Comparative Environmental Risk Assessment.* New York: CRC Press/Lewis Publishers.

Cothern, C. R. 1995. *Handbook for Environmental Risk Decisionmaking.* New York: CRC Press/Lewis Publishers.

Danse, I. R. 1991. *Common Sense Toxics in the Workplace.* New York: Van Nostrand Reinhold.

Environmental Protection Agency. 1983. *Chemical Hazard Information Profile: Acetaldehyde.* Washington, D.C.: U.S. EPA.

Gardner, R. W. 1994. *Chemical Intolerance.* New York: CRC Press/Lewis Publishers.

Hallenbeck, W. H. 1993. *Quantitative Risk Assessment for Environmental and Occupational Health,* 2nd Edition. New York: CRC Press/Lewis Publishers.

Hodgson, E., R. B. Mailman, and J. E. Chambers. 1988. *Dictionary of Toxicology.* London: The MacMillian Company.

Heck, H. 1985. Mechanisms of aldehyde toxicity: Structure activity Studies. *CIIT Activities,* 5(10):106.

Hosty, J. W., and P. Foster. 1990. *A Practical Guide to Chemical Spill Response.* New York: Van Nostrand Reinhold.

Howard, P. H. 1995. *Handbook of Environmental Fate and Exposures Data for Organic Chemicals,* Vol. I-IV. New York: CRC Press/Lewis Publishers.

International Agency for Research on Cancer (IARC). 1985. *IARC Monographs on the Evaluation of Carcinogenic Risk of Chemicals to Humans,* Vol. 36, *Allyl Compounds, Aldehydes, Epoxides, and Peroxides,* pp 101-136. Lyon: IARC.

International Agency for Research on Cancer (IARC). 1987. *IARC Monographs on the Evaluation of Carcinogenic Risk to Humans, Overall Evaluations of Carcinogenicity: An updating of IARC Monographs Volumes 1-42,* Supplement 7, pp 77-78. Lyon: IARC.

James, A., and M. Lord. 1992. *Index of Chemical and Physical Data.* New York: Van Nostrand Reinhold.

Jones, F. E. 1994. *Toxic Organic Vapors in the Workplace.* New York: CRC Press/Lewis Publishers.

Kamrin, M. 1 A. 1988. *Toxicology: A Primer on Principles and Applications.* Chelsea, Michigan: Lewis Publishers.

Keith, L. H. and M. M. Walker. 1995. *Handbook of Air Toxics.* New York: CRC Press/Lewis Publishers.

Landis, W. G., M. H. Yu. 1995. *Introduction to Environmental Toxicology.* New York: CRC Press/Lewis Publishers.

Lewis, .J., Sr. 1993. *Hawley's Condensed Chemical Dictionary,* 12th Edition. New York: Van Nostrand Reinhold.

Lewis, . J., Sr. 1992. *Sax's Dangerous Properties of Industrial Materials,* Volumes 1, 2, and 3, 8th Edition. New York: Van Nostrand Reinhold.

Lisella, F. S. 1994. *The VNR Dictionary of Environmental Health and Safety.* New York: Van Nostrand Reinhold.

Meyer, E. 1990. *Chemistry of Hazardous Materials.* Englewood Cliffs, NJ: Prentice Hall.

Patnaik, P. 1992. *A Comprehensive Guide to the Hazardous Properties of Chemical Substances.* New York: Van Nostrand Reinhold.

Proctor, N. H., and J. P. Hughes. 1991. *Chemical Hazards of the Workplace.* New York: Van Nostrand Reinhold.

Silverman, L., H. F. Schuttle, and M. W. First. 1946. Further studies on sensory response to certain industrial solvent vapors. *Journal of Industrial Hygiene Toxicology,* 28:262-266.

Sherman, J. D. 1988. *Chemical Exposure and Disease.* New York: Van Nostrand Reinhold.

U.S. Department of Health and Human Services, National Institute for Occupational Safety and Health. 1994. *NIOSH Pocket Guide to Chemical Hazards.* Washington, D.C.: U.S. Government Printing Office.

MATERIAL SAFETY DATA SHEET

CHEMICAL NAME	CRC PRESS/LEWIS PUBLISHERS DISCLAIMER
ACETONE	CRC PRESS/LEWIS PUBLISHERS DISCLAIMER: THE INFORMATION AND RECOMMENDATIONS PRESENTED HEREIN ARE BASED ON SOURCES BELIEVED TO BE RELIABLE. CRC MAKES NO REPRESENTATION ON ITS COMPLETENESS OR ACCURACY. IT IS THE RESPONSIBILITY OF THE USER TO DETERMINE THE CHEMICAL'S SUITABILITY FOR ITS INTENDED USE, THE CHEMICAL'S SAFE USE, AND THE CHEMICAL'S PROPER DISPOSAL. NO REPRESENTATIONS AND/OR WARRANTIES, EITHER EXPRESSED OR IMPLIED, OF THE MERCHANTABILITY OR FITNESS FOR A PARTICULAR PURPOSE, OR OF ANY OTHER NATURE, ARE MADE WITH RESPECT TO THE INFORMATION PROVIDED IN THIS MSDS OR TO THE CHEMICAL TO WHICH INFORMATION MAY REFER. CRC NEITHER ASSUMES NOR AUTHORIZES ANY OTHER PERSON TO ASSUME FOR IT, ANY OTHER ADDITIONAL RESPONSIBILITY OR LIABILITY FOR THE USE OF, OR RELIANCE UPON, THIS INFORMATION.

HAZARD WARNING INFORMATION

HEALTH	FIRE	REACTIVE	OTHER	DEGREE OF HAZARD	COLOR CODING	OTHER CODES
1	3	0		0 = Minimum Hazard 1 = Slight Hazard 2 = Moderate Hazard 3 = Serious Hazard 4 = Severe Hazard	HEALTH = BLUE FIRE = RED REACTIVITY = YELLOW OTHER = WHITE	OX = Oxidizer ACID = Acid ALK = Alkali COR = Corrosive W = Use No Water

SECTION I - GENERAL INFORMATION

Characterization	**Ketone**	RCRA Number	**U002**
		EPA Class	**Toxic Waste**
DOT Proper Shipping Name	**Acetone**	Chemical Abstract Service (CAS) Number	**67-64-1**
DOT Hazard Class and Label Requirements	**Flammable Liquid**	DOT Emergency Guide Code	**26**
DOT Identification Number	**UN 1090/UN 1091**	Chemical Formula	**C_3H_6CO**

Synonyms

Dimethylketone; 2-propanone; pyroacetic acid; pyroacetic ether; methyl ketone.

SECTION II - HAZARDOUS INGREDIENTS AND IDENTITY INFORMATION

Hazardous Components (specific identity)	OSHA Exposure Criteria	NIOSH Exposure Criteria	Immediately Dangerous to Life and Health (IDLH)	ACGIH Exposure Criteria
Acetone (derivation: From oxidation of cumene; dehydrogenation or oxidation of isopropyl alcohol with metallic catalyst; vapor phase oxidation of butane; by-product of synthetic glycerol production). **1 ppm = 2.42 mg/m³**	**PEL:** **1000 ppm** **2400 mg/m³** **STEL:** **1000 ppm** **2400 mg/m³**	**REL (10 hour):** **245 ppm** **590 mg/m³** **STEL:** **Not Established**	**20,000 ppm**	**TLV:** **750 ppm** **1815 mg/m³** **STEL:** **1000 ppm** **2400 mg/m³**

SECTION III - PHYSICAL AND CHEMICAL CHARACTERISTICS

Boiling Point	**133°F (56°C)**	Specific Gravity (H_2O = 1)	**0.79**
Vapor Pressure (mm Hg)	**180 at 69°F (20°C)**	Molecular Weight	**58.08**
Vapor Density (Air = 1)	**2.00**	Freezing Point	**-140°F**

Solubility

Miscible (also miscible with alcohol, ether, and chloroform).

Appearance and Odor

Colorless liquid with a fragrant, mint-like odor.

SECTION IV - FIRE AND EXPLOSION HAZARD DATA

Flash Point (method used)	**1.4°F (-17°C) closed cup**	Explosive Limits in Air % by Volume	**LEL: 2.5% UEL: 12.8%**
NFPA Classification	**Class 1B Flammable Liquid**	Autoignition Temperature	**869°F (465°C)**

Extinguishing Media

Carbon dioxide, dry chemical, alcohol foam.

Special Fire Fighting Procedures

Wear full protective clothing and self-contained breathing apparatus (SCBA). Heat will build pressure and may rupture closed storage containers. Fight fire from a distance if possible. Keep fire-exposed containers cool with water spray. Vapors are heavier than air; they may travel a distance to flashback and cause fire.

Unusual Fire and Explosion Hazards

Explodes when mixed with nitric-sulfuric acid mixture in narrow container. Violent explosion may occur when mixed with permonosulfuric acid. Poisonous gases are produced in fire.

SECTION V - REACTIVITY DATA

Stability		Conditions to Avoid
Stability		Do not allow acetone to come into contact with any of the incompatible chemicals listed below. Avoid contact with fire or high heat sources. Acetone is highly volatile. Do not use in confined spaces.
Stable	Unstable	Incompatibility *(materials to avoid)*
X		Nitric-sulfuric acid mixtures, oxidizers, acids, chromium trioxide, sulfuric acid-potassium dichromate; chloroform and a base (caustic soda); hydrogen peroxide; nitrosyl perchlorate; sodium hypobromate.
Hazardous Polymerization		Conditions to Avoid Working in confined spaces or closed, poorly ventilated areas with acetone; combining with acids or oxidizers.
May Occur	Will Not Occur	Hazardous Decomposition or By-products
	X	When heated to decomposition, acetone emits acrid, irritating smoke and fumes. Reacts to form explosive peroxide products with 2-methyl-1,3-butadiene, hydrogen peroxide.

SECTION VI - HEALTH HAZARD DATA

	Inhalation?	Absorption (skin)?	Ingestion?
Primary Route(s) of Entry:	X	X	X

Health Hazards

INHALATION: Eye, nose, and throat irritant. Produces narcotic effects. Headache, nausea, dizziness, muscle weakness, loss of coordinated speech, drowsiness, dryness in the mouth.

ABSORPTION: Skin contact can result in irritating rash or burning feeling. Can cause severe eye burns leading to permanent damage. Can result in sensitization following repeated contact.

INGESTION: Headache, dizziness, dermatitis, stupor, vomiting, and possibly coma.

Carcinogenicity	NTP Listed?	IARC Cancer Review Group?	OSHA Regulated?	Target Organs?
Unknown Human Unknown Animal	No	No	29 CFR 1910.1000 (Table Z-1)	Respiratory system; skin; eyes; CNS.

Medical Conditions Generally Aggravated by Exposure

Respiratory function impairments (asthma, etc.); Nervous disorders; Skin conditions (dermatitis).

Emergency and First-aid Procedures

Eye contact: Immediately flush large amounts of water for 15 minutes (minimum), occasionally lifting eyelids, seek medical attention. Skin contact: Remove all contaminated clothing. Immediately wash area with large amounts of soap and water. For inhalation: Remove the person from exposure. Provide respiratory assistance and CPR. Transfer promptly to medical facility. If swallowed, seek medical attention immediately. Never try to give an unconscious person anything by mouth.

SECTION VII - PRECAUTIONS FOR SAFE HANDLING AND USE

Steps to be Taken in Case Material is Released or Spilled

Remove all ignition sources. Restrict persons not wearing protective equipment from entering area of spill or leak until clean-up is complete. Ventilate area. Absorb liquid in dry sand or other material and deposit in sealed containers. Alert appropriate state or federal response agencies.

Preferred Waste Disposal Method

Incineration (with afterburner and scrubber).

Precautions to be Taken in Handling and Storage

Acetone is highly volatile. Do not store with incompatible chemicals since violent reactions can occur. Store in tightly closed containers in a cool, dark, well-ventilated area away from fire, sparks, and flame. Use non-sparking tools on containers. Ground and bond metal containers during transfer operations.

Other Precautions and Warnings

Bulk storage of acetone is not recommended. Containers may explode in fire or under conditions of extreme heat (do not store outdoors or in direct sunlight).

SECTION VIII - CONTROL MEASURES AND PERSONAL PROTECTIVE EQUIPMENT

Respiratory Protection *(specify type)*

Exposures above 250ppm: MSHA/NIOSH approved full-facepiece respirator with organic vapor cartridge. Greater protection is obtained from a self-contained breathing apparatus (SCBA) with full facepiece and pressure demand or other positive pressure mode.

Ventilation

Local exhaust or general mechanical systems recommended.

Protective Gloves	Eye Protection	Other Protective Clothing
Solvent Resistant Butyl Rubber	Chemical Goggles and/or Face Mask	Rubber Apron

Work/Hygiene Practices

Always wash hands thoroughly after using chemical; never bring food, drink, or smoking materials into vicinity of chemicals.

ACETONE

C_2H_6CO CAS: 67-64-1

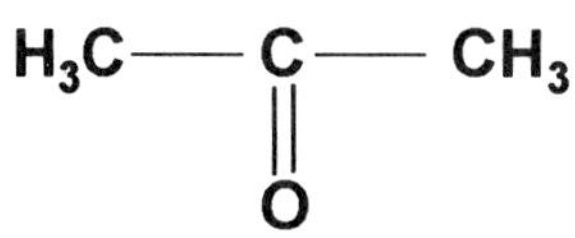

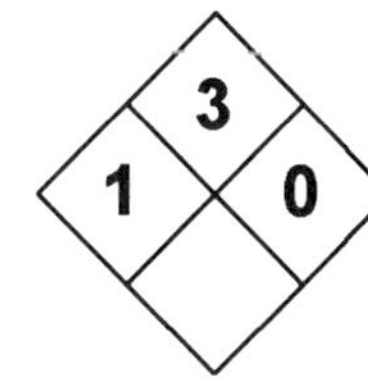

IDENTIFICATION AND TYPICAL USES

Acetone is a colorless liquid with a sweet, fragrant, mint-like odor. It is used in the manufacture of a large number of compounds, such as acetic acid, chloroform, meityl oxide, and methyl isobutyl ketone. It is also used in the manufacture of rayon, photographic films, and explosives. As a common solvent, it is found in paint and varnish removers, and for purifying paraffins. It is used as a spinning solvent to clean and dry parts of precision equipment. It is used in specification testing of vulcanized rubber products. Acetone is also used as a chemical intermediate in the production of pharmaceuticals and plastic or resin materials. Acetone is also used in the production of drugs of abuse.

RISK ASSESSMENT: HEALTH

General Assessment

While no carcinogenic or mutagenic effects in humans have been reported, human exposure to acetone can result in harmful effects. Such exposures can occur in three ways: *inhalation, absorption,* and *ingestion.* Each presents a low-to-moderate level of toxicity. The acute toxicity in test animals was found to be low. Reports of human exposures show a need for some concern.

Inhalation of acetone vapors in high concentrations (above 10,000 ppm) produces dryness of the mouth and throat, dizziness, nausea, uncoordinated movements, loss of coordinated speech, drowsiness, and in extreme cases, coma. Human systemic effects from inhalation include changes in brain wave signatures (EEG), changes in carbohydrate metabolism, nasal effects, conjunctiva irritation, respiratory system effects, weakness, and vomiting.

Ingestion can result in vomiting, stupor, shallow respiration, and possibly coma. Systemic effects include kidney damage and metabolic changes. Skin or eye contact produces severe irritation and possibly damage (to eyes).

Acute Health Effects

The following acute (short-term) health effects may occur immediately or shortly after exposure to acetone:

Skin: Irritation/rash or burning sensation on contact.

Eye: Burns which can lead to permanent damage.

Lung: Irritation causing coughing and/or shortness of breath. Higher exposures can cause drowsiness, headache, and, possibly, coma.

CNS: Inhalation of high concentrations of the vapor may cause dizziness, lightheadedness, and possible loss of consciousness.

Chronic Health Effects

The following chronic (long-term) health effects may occur at some time after exposure to acetone and can last for months or even years:

Cancer Hazards: According to information presented in the references, exposure to acetone does not currently lead to carcinogenic or mutagenic effects. There are no reports to support any claims of reproductive hazards.

Sensitivity: Exposure to the skin may result in allergy (sensitization). If allergy develops, very low future exposure can result in skin irritation and rash. Repeated exposure can cause chronic irritation of the eyes and lungs.

Other Chronic Effects: High exposures over time may damage the liver and kidneys. Long-term exposures can also cause chronic nose and throat irritation.

Recommended Risk-Reduction Measures

Personnel should avoid direct contact with acetone. Unless a less toxic chemical can be substituted for a hazardous substance, *engineering controls* are the most effective method of reducing exposures. The best protection is to enclose operations and/or provide local exhaust ventilation at the site of chemical release.

While not always operationally feasible, isolating operations can also reduce exposure. Using respiratory protection is less effective than the controls mentioned above, but is still advisable whenever working with acetone. A NIOSH-approved, full-facepiece organic vapor respirator is sufficient for low exposures. A self-contained breathing apparatus (SCBA) with full facepiece and pressure demand is the recommended respiratory protection of choice. If a full facepiece is not available, then chemical goggles should be worn to protect the eyes. Whenever a chemical splash hazard exists, a face shield and a rubber apron should be worn. To prevent hand and skin exposures, butyl rubber gloves should be worn.

Administrative controls should also be in place to minimize the potential for human exposures. These may include written procedures or policies which specify the methods and techniques that will be practiced whenever personnel are to work with acetone.

All personnel should receive *training* on the use, hazards, protective measures, emergency actions, and other precautions per 29 CFR 1910.1200 (Hazard Communication), prior to the first assignment in an area where acetone is used or stored.

Other methods to reduce exposure include:

☑ Where possible, enclose operations and use local exhaust ventilation at the site of chemical release. If local exhaust ventilation or enclosure is not used, respiratory protection should be mandatory.

☑ Always ensure that proper protective clothing is worn when using chemical substances.

☑ Wash thoroughly immediately after exposure to acetone and at the end of the work shift or before eating, drinking, or smoking. Never wear clothes contaminated with acetone at home. Family members can be exposed.

☑ Hazard warning information should be posted in the work area. As part of an on-going education and training program, all information on the health and safety hazards of acetone should be communicated to all potentially exposed workers.

RISK ASSESSMENT: ENVIRONMENT

General Assessment

The environment is at risk of exposure during transportation, storage, disposal, or destruction of acetone. In almost every scenario, the threat of environmental exposure is contingent upon the proper handling of the chemical substance. Accidental spills, large or small, can result in fire, explosion, and possible contamination of the surrounding environmental mediums (water, soil, and air).

Acetone is considered a highly flammable liquid. Because of its low flash point and relatively high boiling point, it is classified as a Class 1B flammable liquid (per OSHA 29 CFR 1910.106). Its low auto-ignition temperature and low flash point, together with its high vapor pressure make it a dangerous fire and explosion hazard. It is also incompatible with a number of commodities. Potentially explosive reactions can occur if acetone is mixed or comes into contact with oxidizers and/or acids. These characteristics require special consideration during any emergency involving a leak or spill of acetone.

The proper disposal/destruction method for acetone is to burn it in a chemical incinerator equipped with an afterburner and air scrubber.

Acetone can enter the environment through manufacturing, unchecked discharge into effluents from industrial or municipal waste treatment plants, and through spills.

Acute Ecological Effects

Acute (short-term) toxic effects may include the death of animals, birds, or fish, and death or low growth rate in plants. Acute effects are seen 2 to 4 days after animals or plants are exposed to acetone. This chemical has a slight acute toxicity to aquatic life. Acetone has caused membrane damage, size decrease, and germination decrease in various agricultural and ornamental crops. No data are available on the short-term effects of acetone on birds or land animals.

Chronic Ecological Effects

Chronic (long-term) toxic effects may include shortened life span, reproductive problems, lower fertility, and changes in appearance or behavior in exposed animals. These effects can be seen long after first exposure(s) to toxic chemicals. Acetone has slight chronic toxicity in aquatic life. No data are available to evaluate the long-term effects of acetone to plants, birds, or land animals.

Water Solubility

Acetone is highly soluble in water. Concentrations of up to a 1000 milligrams and possibly more will readily mix with a liter of water.

Persistence in the Environment

Acetone is slightly persistent in water, with a half-life of between 2 to 20 days. The half-life of a pollutant is the amount of time it takes for one half of the chemical to be degraded. Approximately 50% of acetone will eventually end up in air; the rest will end up in water.

Bioaccumulation in Aquatic Organisms

Some substances increase in concentration, or *bioaccumulate,* in living organisms as they breathe contaminated air, drink contaminated water, or eat contaminated food. These chemicals can become concentrated in the tissues and internal organs of animals as well as humans.

The concentration of acetone found in fish tissues is expected to be about the same as the average concentration of acetone in water from which the fish was taken.

Recommended Risk-Reduction Measures

Proper training of all transporters will reduce the likelihood of a mishap or accident resulting in a leak or spill to the environment. The correct use of DOT labeling (per the MSDS) on all containers, trucks, and rail cars will enable emergency responders to react properly and quickly to any disaster thereby reducing the potential risk to the environment and to personnel.

Storage of acetone should be segregated from incompatible chemicals to minimize the risk of cross-contamination or contact. Buildings designated for storage should be equipped with appropriate fire protection systems (alarms, sprinklers, emergency lighting, portable extinguishers).

If a spill or leak to the environment has occurred, fire department, emergency response, and/or hazardous materials spill personnel should be notified immediately. Cleanup should be attempted only by those trained in proper spill containment procedures. Contaminated soils should be removed for incineration and replaced with clean soil. If acetone should contact the water table, aquifer, or navigable waterway, time is of the essence. It is readily miscible in water and, therefore, total containment and remediation may not be entirely possible. When such spills occur, the local and/or state emergency response authorities must be notified. A comprehensive emergency response or disaster preparedness/recovery plan should be in place prior to any operations involving the use, transportation, storage, or disposal of acetone. If acetone is spilled or leaked, the following specific steps are recommended:

- ☑ Restrict persons not wearing protective clothing from area of spill or leak until cleanup is complete.
- ☑ Remove all ignition sources.
- ☑ Ventilate area of spill or leak.
- ☑ Absorb liquids in vermiculite, dry sand, earth, or a similar material and deposit in sealed containers.
- ☑ If applicable, stop flow of leaking liquid or gas. If leak source is a cylinder and the leak cannot be stopped in place, remove leaking cylinder to a safe place in the open air, and repair or allow cylinder to empty.
- ☑ Keep acetone out of a confined space, such as a sewer, because of the possibility of explosion (unless the sewer is designed to prevent the build-up of explosive concentrations).
- ☑ It may be necessary to dispose of acetone as a hazardous waste. The responsible state agency or the regional office of the federal Environmental Protection Agency (EPA) should be contacted for specific recommendations.

RISK ASSESSMENT: BUSINESS

General Assessment

Accidents or mishaps involving acetone can present a significant threat to business operations. The loss or damage of equipment or facilities can significantly effect fiscal viability. Lawsuits that may result from personnel injury, public exposures, and/or environmental contamination will also require a serious expenditure of resources.

Media attention surrounding an injury, illness, or environmental damage can also result in a loss of profits and loss of current as well as future business.

Recommended Risk-Reduction Measures

Company attorneys, safety and health professionals, and environmental specialists should be involved in the development of any procedures that deal with or address safety and health issues. A company official should be pre-designated as a public relations officer with specific training in dealing with the press. Corporate plans and policies should be developed, approved, and implemented long before any need for such arises.

REFERENCES

Ahlbom, A. 1993. *Biostatistics for Epidemiologists.* New York: CRC Press/Lewis Publishers.

American Conference of Governmental Industrial Hygienists. 1988. *Documentation of the Threshold Limit Values and Biological Exposure Indices,* 5th Edition (with updates). Cincinnati: ACGIH

Anthony, C. P., and N. J. Kolthoff. 1971. *Textbook of Anatomy and Physiology.* St. Louis: C. V. Mosby Company.

Asante-Duah, D. K. 1993. *Hazardous Waste Risk Assessment.* New York: CRC Press/Lewis Publishers.

Bloom, A. D. and F. J. DeSerres. 1995. *Ecotoxicity and Human Health.* New York: CRC Press/Lewis Publishers.

Brusick, D. J. 1994. *Methods for Generic Risk Assessment.* New York: CRC Press/Lewis Publishers.

Bueche, J. R. 1972. *Principles of Physics,* 2nd Edition. New York: McGraw-Hill.

Burrell, R., D. K. Flaherty, and L. J. Sauers. 1992. *Toxicology of the Immune System.* New York: Van Nostrand Reinhold.

Calabrese, E. J., E. M. Kenyon. *Air Toxics and Risk Assessment.* New York: CRC Press/Lewis Publishers.

Calabrese, E. J. 1994. *Biological Effects of Low Level Exposures.* New York: CRC Press/Lewis Publishers.

Cockerman, L. G., B. S. Shane. 1994. *Basic Environmental Toxicology.* New York: CRC Press/Lewis Publishers.

Cothern, R. R. 1993. *Comparative Environmental Risk Assessment.* New York: CRC Press/Lewis Publishers.

Cothern, C. R. 1995. *Handbook for Environmental Risk Decisionmaking.* New York: CRC Press/Lewis Publishers.

DiVincenzo, G. O., F. J. Yanno, and B. D. Astill. 1973. Exposure of man and dog to low concentrations of acetone vapor. *American Industrial Hygiene Journal.* 34:329-336.

Gardner, R. W. 1994. *Chemical Intolerance.* New York: CRC Press/Lewis Publishers.

Hallenbeck, W. H. 1993. *Quantitative Risk Assessment for Environmental and Occupational Health,* 2nd Edition. New York: CRC Press/Lewis Publishers.

Hodgson, E., R. B. Mailman, and J. E. Chambers. 1988. *Dictionary of Toxicology.* London: The MacMillian Company.

Hosty, J. W., and P. Foster. 1990. *A Practical Guide to Chemical Spill Response.* New York: Van Nostrand Reinhold.

Howard, P. H. 1995. *Handbook of Environmental Fate and Exposures Data for Organic Chemicals,* Vol. I-IV. New York: CRC Press/Lewis Publishers.

Howard, P. H., et. al. 1991. *Handbook of Environmental Degradation Rates.* New York: CRC Press/Lewis Publishers.

James, A., and M. Lord. 1992. *Index of Chemical and Physical Data.* New York: Van Nostrand Reinhold.

Jones, F. E. 1994. *Toxic Organic Vapors in the Workplace.* New York: CRC Press/Lewis Publishers.

Kamrin, M. A. 1988. *Toxicology: A Primer on Principles and Applications.* Chelsea, Michigan: Lewis Publishers.

Keith, L. H. and M. M. Walker. 1995. *Handbook of Air Toxics.* New York: CRC Press/Lewis Publishers.

Landis, W. G., M. H. Yu. 1995. *Introduction to Environmental Toxicology.* New York: CRC Press/Lewis Publishers.

Langerman, N. 1994 *Precautionary Labels for Chemical Containers.* New York: CRC Press/Lewis Publishers.

Lauwerys, R. R., P. Hoet. 1993. *Industrial Chemical Exposure.* New York: CRC Press/Lewis Publishers.

Lewis, R. J., Sr. 1993. *Hawley's Condensed Chemical Dictionary,* 12th Edition. New York: Van Nostrand Reinhold.

Lewis, R. J., Sr. 1992. *Sax's Dangerous Properties of Industrial Materials,* Volumes 1, 2, and 3, 8th Edition. New York: Van Nostrand Reinhold.

Lide, D. R. 1995. *Handbook of Chemistry and Physics,* 76th Edition. New York: CRC Press/Lewis Publishers.

Manahan, S. E. 1994. *Environmental Chemistry,* 6th Edition. New York: CRC Press/Lewis Publishers.

Manahan, S. E. 1992. *Toxicological Chemistry,* 2nd Edition. New York: CRC Press/Lewis Publishers.

Mellström, G. A., J. E. Wahlberg, and H. I. Maibach. 1994. *Protective Gloves for Occupational Use.* New York: CRC Press/Lewis Publishers.

Meyer, E. 1990. *Chemistry of Hazardous Materials.* Englewood Cliffs, NJ: Prentice Hall.

Patnaik, P. 1992. *A Comprehensive Guide to the Hazardous Properties of Chemical Substances.* New York: Van Nostrand Reinhold.

Philp, R. B. 1995. *Environmental Hazards and Human Health.* New York: CRC Press/Lewis Publishers.

Proctor, N. H., and J. P. Hughes. 1991. *Chemical Hazards of the Workplace.* New York: Van Nostrand Reinhold.

Quigley, D. R. 1994. *Handbook of Emergency Chemical Management.* New York: CRC Press/Lewis Publishers.

Raleigh, R. L. and W. A. McGee. 1972. Effects of short, high concentration exposures to acetone as determined by observation in the work area. *Journal of Occupational Medicine.* 14:607-610.

Rea, W. J. 1995. *Chemical Sensitivity,* Vol. I-III. New York: CRC Press/Lewis Publishers.

Richardson, M. 1995. *Dictionary of Substance and Their Effects,* Vol. I-VI. New York: CRC Press/Lewis Publishers.

U.S. Department of Health and Human Services, National Institute for Occupational Safety and Health. 1994. *NIOSH Pocket Guide to Chemical Hazards.* Washington, D.C.: U.S. Government Printing Office.

U.S. Department of Health and Human Services, National Institute for Occupational Safety and Health. 1978. *Criteria for a Recommended Standard - Occupational Exposure to Ketones.* DHEW Publication Number: 78-173. Washington, D.C.: U.S. Government Printing Office.

MATERIAL SAFETY DATA SHEET

CHEMICAL NAME

ACETONITRILE

HAZARD WARNING INFORMATION

HEALTH	FIRE	REACTIVE	OTHER	DEGREE OF HAZARD	COLOR CODING	OTHER CODES
2	3	0		0 = Minimum Hazard 1 = Slight Hazard 2 = Moderate Hazard 3 = Serious Hazard 4 = Severe Hazard	HEALTH = BLUE FIRE = RED REACTIVITY = YELLOW OTHER = WHITE	OX = Oxidizer ACID = Acid ALK = Alkali COR = Corrosive ~~W~~ = Use No Water

SECTION I - GENERAL INFORMATION

Characterization	RCRA Number	EPA Class
Aliphatic Nitrile	**U003**	**Toxic Waste**

DOT Proper Shipping Name **Acetonitrile**	Chemical Abstract Service (CAS) Number **75-05-8**
DOT Hazard Class and Label Requirements **Flammable Liquid**	DOT Emergency Guide Code **28**
DOT Identification Number **UN 1648**	Chemical Formula **CH_3CN**

Synonyms
Methyl cyanide; ethanenitrile; ethyl nitrile; methanecarbonitrile; cyanomethane.

SECTION II - HAZARDOUS INGREDIENTS AND IDENTITY INFORMATION

Hazardous Components (specific identity)	OSHA Exposure Criteria	NIOSH Exposure Criteria	Immediately Dangerous to Life and Health (IDLH)	ACGIH Exposure Criteria
Acetonitrile (derivation: By-product of the propylene-ammonia process during acrylonitrile manufacturing). **1 ppm = 1.71 mg/m³**	**PEL:** **40 ppm** **70 mg/m³** **STEL:** **60 ppm** **103 mg/m³**	**REL:** **20 ppm** **30 mg/m³** **STEL:** **Not Established**	**500 ppm**	**TLV:** **40 ppm** **70 mg/m³** **STEL:** **60 ppm** **103 mg/m³**

SECTION III - PHYSICAL AND CHEMICAL CHARACTERISTICS

Boiling Point **179°F (82°C)**	Specific Gravity (H_2O = 1) **0.78**
Vapor Pressure (mm Hg) **73 at 69°F (20°C)**	Molecular Weight **41.10**
Vapor Density (Air = 1) **1.42**	Freezing Point **- 49°F (- 45.7°C)**

Solubility
Miscible in water (also miscible with alcohol, ether, and most organic solvents). Immiscible with petroleum fractions and some saturated hydrocarbons.

Appearance and Odor
Colorless liquid with ether-like smell (NOTE: Forms cyanide once in the body).

SECTION IV - FIRE AND EXPLOSION HAZARD DATA

Flash Point (method used) **42°F (6°C) open cup**	Explosive Limits in Air % by Volume **LEL: 3.0% UEL: 16%**
NFPA Classification **Class 1B Flammable Liquid**	Autoignition Temperature **975°F (524°C)**

Extinguishing Media
Carbon dioxide, dry chemical, foam.

Special Fire Fighting Procedures
Wear full protective clothing and self-contained breathing apparatus (SCBA). Dangerous fire hazard. Use a water spray to keep fire-exposed containers cool. Poisonous gases are produced in fire. Vapors are heavier than air; flashback is likely.

Unusual Fire and Explosion Hazards
Potentially explosive reaction with lanthanide perchlorates and nitrogen-fluorine compounds. Will react with water, steam, acids to produce toxic flammable vapors.

SECTION V - REACTIVITY DATA

Stability		Conditions to Avoid
		Exothermic reaction with sulfuric acid at 53°C. Reacts with strong oxidizers, acids (contact in a closed container can cause rupture of the container). Acetonitrile vapors form explosive mixtures in air.
Stable	Unstable	Incompatibility *(materials to avoid)*
X		Strong oxidizers, acids, oleum, chlorosulfonic acid, perchlorates, nitrating agents, indium, dinitrogen tetroxide, N-fluoro compounds (i.e., perfluorourea + acetonitrile) HNO_3, SO_3.
Hazardous Polymerization		Conditions to Avoid
		Keep away from fire, heat, ignition sources. Poisonous gases are produced during fire, including hydrogen cyanide. Metals, such as lithium, react exothermically at ambient temperatures.
May Occur	Will Not Occur	Hazardous Decomposition or By-products
	X	When heated to decomposition, acetonitrile emits highly toxic fumes of CN^- and NO_x.

SECTION VI - HEALTH HAZARD DATA

Primary Route(s) of Entry:	Inhalation?	Absorption (skin)?	Ingestion?
	X	X	X

Health Hazards

INHALATION: At high concentrations (400-500 ppm) causes asphyxia, nausea, vomiting, and tightness of chest. At lower concentrations (100 ppm), only slight adverse effect may be noted.

ABSORPTION: Skin contact can result in irritating rash. Can cause severe eye burns and/or irritation. Can result in sensitization following repeated contact.

INGESTION: Convulsions, nausea or vomiting, metabolic acidosis, gastrointestinal pain, stupor.

Carcinogenicity	NTP Listed?	IARC Cancer Review Group?	OSHA Regulated?	Target Organs?
Human: Not tested Animal: Possible	5th Annual Report	No	29 CFR 1910.1000 (Table Z-1)	Kidney, liver, CNS, lungs, skin, eyes, CVS.

Medical Conditions Generally Aggravated by Exposure

Respiratory impairments, (asthma); Renal function abnormalities; Skin conditions (dermatitis).

Emergency and First-aid Procedures

Eye contact: Flush immediately with water for 15 minutes (minimum); seek medical attention. Skin contact: Remove all contaminated clothing. Immediately wash area with large amounts of soap and water. Seek medical assistance. For inhalation: Remove the person from exposure. Provide respiratory assistance and CPR. Transfer to medical facility. Acetonitrile can cause fatal cyanide poisoning. If swallowed, seek medical attention immediately.

SECTION VII - PRECAUTIONS FOR SAFE HANDLING AND USE

Steps to be Taken in Case Material is Released or Spilled

Contain spills using absorbent material (vermiculite, or other). Ventilate area of spill. Have water source available in case of fire. Store materials in DOT-approved containers. Restrict those not involved in cleanup from entering area. Notify appropriate authorities, as required by SARA III, if applicable.

Preferred Waste Disposal Method

Incineration (with after burner and scrubber).

Precautions to be Taken in Handling and Storage

Acetonitrile is highly volatile. Store to avoid contact with strong oxidizers (chlorine, bromine, and fluorine). Store in tightly closed containers in cool, well-ventilated area away from heat. Drums used for storage should be equipped with self-closing valves, pressure vacuum bungs, and flame arresters.

Other Precautions and Warnings

Bulk storage of acetonitrile is not recommended. Prior to working with acetonitrile, personnel should be trained on its proper handling and storage.

SECTION VIII - CONTROL MEASURES AND PERSONAL PROTECTIVE EQUIPMENT

Respiratory Protection *(specify type)*

For low exposures (below 100 ppm), NIOSH approved full-facepiece respirator with organic vapor cartridge. Exposures above 100 ppm: Self-contained breathing apparatus (SCBA) with full facepiece and pressure demand or other positive pressure mode.

Ventilation

Local exhaust or general mechanical systems recommended.

Protective Gloves	Eye Protection	Other Protective Clothing
Butyl Rubber	Chemical Goggles or Face Mask	Rubber Apron

Work/Hygiene Practices

Always wash hands thoroughly after using chemical; never bring food, drink, or smoking materials into vicinity of chemicals.

ACETONITRILE

CH_3CN CAS: 75-05-8

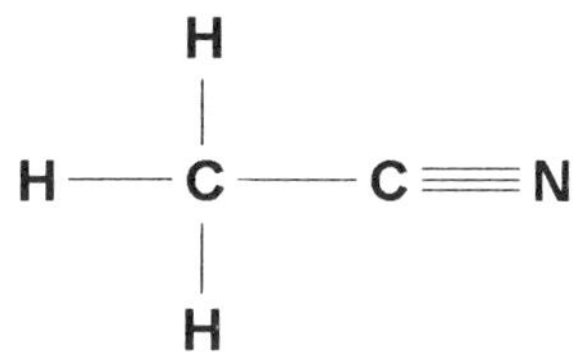

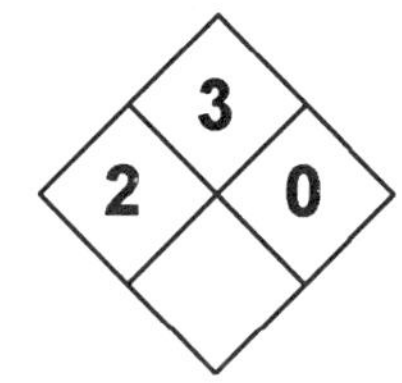

IDENTIFICATION AND TYPICAL USES

Acetonitrile is a colorless liquid with an ether-like odor. It is used as a solvent and in the making of pharmaceuticals. It is used for polymers, spinning fibers, casting and molding plastics, and HPLC analyses. It is used for extraction of butadiene and other olefins from hydrocarbon streams; in dyeing and coating textiles; in the photographic industry; in the extraction and refining of copper; and as a stabilizer for chlorinated solvents. It occurs in coal tar and forms as a by-product when acrylonitrile is manufactured. It is also used as an extractant for animal and vegetable oils and in the perfume industry.

RISK ASSESSMENT: HEALTH

General Assessment

Acetonitrile causes headache, dizziness, and nausea. At extremely high concentrations, it can cause convulsions, coma, and death. No data exist on the chemical's carcinogenicity in humans. At high concentrations (above 500 ppm) for extended periods, there appears to be some affect on the liver and kidneys, as well as some pulmonary distress. Extremely high concentrations (above 5000 ppm) produce some teratogenic effects in experimental animals.

The toxic effects of acetonitrile are attributed to the metabolic release of cyanide, which in turn acts by inhibiting cytochrome oxidase and thus impairs cellular respiration.

Human exposure to acetonitrile can occur through three routes of entry into the body: *inhalation, absorption,* and *ingestion.* The toxicity of acetonitrile to human and test animals is considerably lower than that of other nitriles. However, at high concentrations, acetonitrile can produce severe adverse effects. The target organs are the kidney, liver, central nervous system (CNS), lungs, cardiovascular system (CVS), skin, and eyes. In humans, inhalation of acetonitrile vapors can result in asphyxia, nausea, vomiting, and tightness of the chest. These effects will most likely manifest at several hours exposure to concentrations above 400 ppm. Lower concentrations result in only slight adverse effect. It is excreted in the urine as a cyanate. The blood cyanide concentration does not always show any significant increase in cyanide at low exposure concentrations. Ingestion of acetonitrile results in gastrointestinal pain, nausea, vomiting, stupor, convulsion, and weakness. These effects may become highly marked in humans from ingestion of 40-50 ml. Large doses can be fatal. Skin contact results in absorption and systemic effects. Severe eye irritation is likely.

Acute Health Effects

The following acute (short-term) health effects may occur immediately or shortly after exposure to acetonitrile. (Note: Exposure can cause fatal cyanide poisoning. Symptoms of flushing of face, chest tightness, nausea, and vomiting, weakness, and shortness of breath may lead to death. These reactions may begin hours after an overexposure).

Skin: Irritation/rash on contact (contact dermatitis).

Eye: Severe irritation, conjunctivitis, damage.

Lung: Irritation of the nose, throat, and lungs following exposure.

CNS: Inhalation of high concentrations of the vapor may cause sleepiness, dizziness, loss of consciousness, or death.

Chronic Health Effects

The following chronic (long-term) health effects may occur at some time after exposure to acetonitrile and can last for months or even years:

Cancer Hazards: According to information presented in the references, acetonitrile has not been adequately tested for its ability to cause cancer in test animals. Chronic effects of high exposures have not been documented in humans studies.

Reproductive Hazard: According to the references, acetonitrile has not been significantly tested for its ability to adversely affect reproduction. However, at high concentrations (5000-8000 ppm), some disorders were shown in axial skeletal development of hamsters.

Other Chronic Effects: Repeated exposures may affect the thyroid gland, causing enlargement. High exposures may damage the nervous system, liver, and kidneys. Very irritating substances may effect the lungs. It is not known whether acetonitrile causes permanent lung damage.

Recommended Risk-Reduction Measures

Personnel should avoid direct contact with acetonitrile. Unless a less toxic chemical can be substituted for a hazardous substance, *engineering controls* are the most effective method of reducing exposures. The best protection is to enclose operations and/or provide local exhaust ventilation at the site of chemical release. While not always operationally feasible, isolating operations can also reduce exposure. Using respiratory protection is less effective than the controls mentioned above, but is still advisable whenever working with or around acetonitrile. A self-contained breathing apparatus (SCBA) with full facepiece and pressure demand is the recommended respiratory protection of choice. If a full facepiece is not available, then chemical goggles should be worn to protect the eyes. Whenever a chemical splash hazard exists, a face shield and a rubber apron should be worn. To prevent hand and skin exposures, butyl rubber gloves should be worn.

Administrative controls should also be in place to minimize the potential for human exposures. These may include written procedures or policies which specify the methods and techniques that will be practiced whenever personnel are to work with acetonitrile.

All personnel should receive *training* on the use, hazards, protective measures, emergency actions, and other precautions per 29 CFR 1910.1200 (Hazard Communication), prior to the first assignment in an area where acetonitrile is used or stored.

Medical testing of personnel with frequent or potentially high exposure (half the PEL or greater), should be offered before beginning work and at regular intervals thereafter (e.g., annually). These tests should include:

☑ Lung function (pulmonary function) test.

If symptoms develop or overexposure is suspected, the following tests may be useful:

☑ Blood cyanide test and/or urine thiocyanate test (currently, a blood cyanide level over 0.1 mg/l or urine thiocyanate over 20 mg/l indicates overexposure).

☑ Maintain close medical surveillance. Slow release of cyanide from absorbed acetonitrile may cause delayed symptoms.

☑ Liver and kidney function tests.

Other methods to reduce exposure include:

☑ Where possible, enclose operations and use local exhaust ventilation at the site of chemical release. If local exhaust ventilation or enclosure is not used, respiratory protection should be mandatory.

☑ Always ensure that proper protective clothing is worn when using chemical substances.

☑ Wash thoroughly immediately after exposure to acetonitrile and at the end of the work shift or before eating, drinking, or smoking.

☑ Hazard warning information should be posted in the work area. In addition, as part of an on-going education and training program, all information on the health and safety hazards of acetonitrile should be communicated to all exposed or potentially exposed workers.

RISK ASSESSMENT: ENVIRONMENT

General Assessment

The environment is at risk of exposure during transportation, storage, disposal, or destruction of acetonitrile. In almost every scenario, the threat of environmental exposure is contingent upon the proper handling of the chemical substance. Accidental spills, large or small, can result in fire, explosion, and possible contamination of the surrounding environmental mediums (water, soil, and air).

Acetonitrile is considered a highly flammable liquid and a poison. Because of its low flash point and high boiling point, it is classified as a Class 1B flammable liquid (per OSHA 29 CFR 1910.106). Acetonitrile vapors form explosive mixtures in air. It reacts with strong oxidizers and acids, liberating heat along with pressure increase. Thus, contact with either under closed conditions can cause explosion. Metals, such as lithium, react exothermically with acetonitrile at ambient temperatures. These characteristics require special consideration during any emergency situation involving a leak or spill of acetonitrile. Should acetonitrile ever come into contact with incompatible substances such as oxidizers, acids, oleum, chlorosulfonic acid, perchlorates, nitrating agents, indium, dinitrogen tetroxide, or N-fluoro compounds either during use, transportation, or storage, the formation of highly toxic and/or highly explosive commodities is extremely possible.

The proper disposal/destruction method for acetonitrile is to burn it in a chemical incinerator equipped with an afterburner and air scrubber. Acetonitrile can enter the environment through manufacturing effluents, municipal waste treatment discharges, and spills.

Acute Ecological Effects

Acute (short-term) toxic effects may include the death of animals, birds, or fish and death or low growth rate in plants. Acute effects are seen 2 to 4 days after animals or plants are exposed to acetonitrile.

This chemical has slight acute toxicity to aquatic life. No data are available on the short-term effects of acetonitrile to plants or land animals.

Chronic Ecological Effects

Chronic toxic effects may include shortened life span, reproductive problems, lower fertility, and changes in appearance or behavior in exposed animals. These effects can be seen long after first exposure(s) to toxic chemicals.

Acetonitrile has slight chronic toxicity in aquatic life. No data are available on the long-term effects of acetonitrile to plants, birds, or land animals.

Water Solubility

Acetonitrile is highly soluble in water. Concentrations of 1000 milligrams and more will mix with a liter of water.

Persistence in the Environment

Acetonitrile is slightly persistent in water, with a half-life of between 2 to 20 days. The half-life of a pollutant is the amount of time it takes for one half of the chemical to be degraded. Approximately 21% of acetonitrile will eventually end up in air; 79% will end up in water.

Bioaccumulation in Aquatic Organisms

Some substances increase in concentration, or *bioaccumulate,* in living organisms as they breathe contaminated air, drink contaminated water, or eat contaminated food. These chemicals can become concentrated in the tissues and internal organs of animals as well as humans.

The concentration of acetonitrile found in fish tissues is expected to be about the same as the average concentration of acetonitrile in water from which the fish was taken.

Recommended Risk-Reduction Measures

Proper training of all transporters will reduce the likelihood of a mishap or accident resulting in a leak or spill to the environment. The correct use of DOT labeling (per the MSDS) on all containers, trucks, and rail cars will enable emergency responders to react properly and quickly to any disaster thereby reducing the potential risk to the environment and to personnel.

Storage of acetonitrile should be segregated from incompatible chemicals to minimize the risk of cross-contamination or contact. Buildings designated for storage should be equipped with appropriate fire protection systems (alarms, sprinklers, emergency lighting, portable extinguishers).

If a spill or leak to the environment has occurred, fire department, emergency response, and/or hazardous materials spill personnel should be notified immediately. Cleanup should be attempted only by those trained in proper spill containment procedures. Contaminated soils should be removed for incineration and replaced with clean soil. If acetonitrile should contact the water table, aquifer, or navigable waterway, time is of the essence. It is readily miscible in water and, therefore, total containment and remediation may not be entirely possible. When such spills occur, the local and/or state emergency response authorities must be notified. A comprehensive emergency response or disaster preparedness/recovery plan should be in place prior to any operations involving the use, transportation, storage, or disposal of acetonitrile. If acetonitrile is spilled or leaked, the following specific steps are recommended:

- ☑ Restrict persons not wearing protective clothing from area of spill or leak until cleanup is complete and area can be opened for normal work.
- ☑ Remove all ignition sources.
- ☑ Ventilate area of spill or leak.
- ☑ Absorb liquids in vermiculite, dry sand, earth, or a similar material and deposit in sealed containers.
- ☑ If applicable, stop flow of leaking liquid or gas. If leak source is a cylinder and the leak cannot be

stopped in place, remove leaking cylinder to a safe place in the open air, and repair or allow cylinder to empty.

☑ Keep acetonitrile out of a confined space, such as a sewer, because of the possibility of explosion (unless the sewer is designed to prevent the build-up of explosive concentrations).

☑ It may be necessary to dispose of acetonitrile as a hazardous waste. The responsible state agency or the regional office of the federal Environmental Protection Agency (EPA) should be contacted for specific recommendations.

RISK ASSESSMENT: BUSINESS

General Assessment

Accidents or mishaps involving acetonitrile can present a significant threat to business operations. The loss or damage of equipment or facilities can significantly effect fiscal viability. Lawsuits that may result from personnel injury/death, public exposures, and/or environmental contamination will also require a serious expenditure of resources. Media attention surrounding an injury, death, or environmental damage can also result in a loss of profits and loss of current as well as future business.

Recommended Risk-Reduction Measures

Company attorneys, safety and health professionals, and environmental specialists should be involved in the development of any procedures dealing with safety or emergency response. A company official should be pre-designated as a public relations officer with specific training in dealing with the press. Corporate plans and policies should be developed, approved, and implemented long before any need for such arises.

REFERENCES

Ahlbom, A. 1993. *Biostatistics for Epidemiologists.* New York: CRC Press/Lewis Publishers.

American Conference of Governmental Industrial Hygienists. 1988. *Documentation of the Threshold Limit Values and Biological Exposure Indices,* 5th Edition (with updates). Cincinnati: ACGIH

Amdur, M. L. 1959. Accidental group exposure to acetonitrile. *Journal of Occupational Medicine.* 1:627-633.

Asante-Duah, D. K. 1993. *Hazardous Waste Risk Assessment.* New York: CRC Press/Lewis Publishers.

Brusick, D. J. 1994. *Methods for Genetic Risk Assessment.* New York: CRC Press/Lewis Publishers.

Bueche, J. R. 1972. *Principles of Physics,* 2nd Edition. New York: McGraw-Hill.

Burrell, R., D. K. Flaherty, and L. J. Sauers. 1992. *Toxicology of the Immune System.* New York: Van Nostrand Reinhold.

Calabrese, E. J., E. M. Kenyon. *Air Toxics and Risk Assessment.* New York: CRC Press/Lewis Publishers.

Calabrese, E. J. 1994. *Biological Effects of Low Level Exposures.* New York: CRC Press/Lewis Publishers.

Cockerman, L. G., B. S. Shane. 1994. *Basic Environmental Toxicology.* New York: CRC Press/Lewis Publishers.

Cothern, R. R. 1993. *Comparative Environmental Risk Assessment.* New York: CRC Press/Lewis Publishers.

Cothern, C. R. 1995. *Handbook for Environmental Risk Decisionmaking.* New York: CRC Press/Lewis Publishers.

Gardner, R. W. 1994. *Chemical Intolerance.* New York: CRC Press/Lewis Publishers.

Hallenbeck, W. H. 1993. *Quantitative Risk Assessment for Environmental and Occupational Health,* 2nd Edition. New York: CRC Press/Lewis Publishers.

Hosty, J. W., and P. Foster. 1990. *A Practical Guide to Chemical Spill Response.* New York: Van Nostrand Reinhold.

Howard, P. H. 1995. *Handbook of Environmental Fate and Exposures Data for Organic Chemicals,* Vol. I-IV. New York: CRC Press/Lewis Publishers.

Jones, F. E. 1994. *Toxic Organic Vapors in the Workplace.* New York: CRC Press/Lewis Publishers.

Langerman, N. 1994 *Precautionary Labels for Chemical Containers.* New York: CRC Press/Lewis Publishers.

Lewis, R. J., Sr. 1993. *Hawley's Condensed Chemical Dictionary,* 12th Edition. New York: Van Nostrand Reinhold.

Lewis, R. J., Sr. 1992. *Sax's Dangerous Properties of Industrial Materials,* Volumes 1, 2, and 3, 8th Edition. New York: Van Nostrand Reinhold.

Meyer, E. 1990. *Chemistry of Hazardous Materials.* Englewood Cliffs, NJ: Prentice Hall.

Patnaik, P. 1992. *A Comprehensive Guide to the Hazardous Properties of Chemical Substances.* New York: Van Nostrand Reinhold.

Proctor, N. H., and James P. Hughes. 1991. *Chemical Hazards of the Workplace.* New York: Van Nostrand Reinhold.

Sherman, J. D. 1988. *Chemical Exposure and Disease.* New York: Van Nostrand Reinhold.

Union Carbide Corporation. 1965. *Toxicology Studies, Acetonitrile.* New York: Union Carbide Corporation.

U.S. Department of Health and Human Services, National Institute for Occupational Safety and Health. 1978. *Criteria for a Recommended Standard - Occupational Exposure to Nitriles.* DHEW (NIOSH) Publication Number: 78-212. Washington, D.C.: U.S. Government Printing Office.

U.S. Department of Health and Human Services, National Institute for Occupational Safety and Health. 1994. *NIOSH Pocket Guide to Chemical Hazards.* Washington, D.C.: U.S. Government Printing Office.

MATERIAL SAFETY DATA SHEET

CHEMICAL NAME

ACROLEIN

CRC PRESS/LEWIS PUBLISHERS DISCLAIMER: THE INFORMATION AND RECOMMENDATIONS PRESENTED HEREIN ARE BASED ON SOURCES BELIEVED TO BE RELIABLE. CRC MAKES NO REPRESENTATION ON ITS COMPLETENESS OR ACCURACY. IT IS THE RESPONSIBILITY OF THE USER TO DETERMINE THE CHEMICAL'S SUITABILITY FOR ITS INTENDED USE, THE CHEMICAL'S SAFE USE, AND THE CHEMICAL'S PROPER DISPOSAL. NO REPRESENTATIONS AND/OR WARRANTIES, EITHER EXPRESSED OR IMPLIED, OF THE MERCHANTABILITY OR FITNESS FOR A PARTICULAR PURPOSE, OR OF ANY OTHER NATURE, ARE MADE WITH RESPECT TO THE INFORMATION PROVIDED IN THIS MSDS OR TO THE CHEMICAL TO WHICH INFORMATION MAY REFER. CRC NEITHER ASSUMES NOR AUTHORIZES ANY OTHER PERSON TO ASSUME FOR IT, ANY OTHER ADDITIONAL RESPONSIBILITY OR LIABILITY FOR THE USE OF, OR RELIANCE UPON, THIS INFORMATION.

HAZARD WARNING INFORMATION

HEALTH	FIRE	REACTIVE	OTHER	DEGREE OF HAZARD	COLOR CODING	OTHER CODES
3	3	3		0 = Minimum Hazard 1 = Slight Hazard 2 = Moderate Hazard 3 = Serious Hazard 4 = Severe Hazard	HEALTH = BLUE FIRE = RED REACTIVITY = YELLOW OTHER = WHITE	OX = Oxidizer ACID = Acid ALK = Alkali COR = Corrosive W = Use No Water

SECTION I - GENERAL INFORMATION

Characterization: **Aldehyde**	RCRA Number: **P003** / EPA Class: **Acute Hazardous Waste**
DOT Proper Shipping Name: **Acrolein, inhibited**	Chemical Abstract Service (CAS) Number: **107-02-8**
DOT Hazard Class & Label Requirements: **Flammable Liquid; IMO: Poison B**	DOT Emergency Guide Code: **30**
DOT Identification Number: **UN 1092**	Chemical Formula: **C_2H_3CHO**

Synonyms

2-Propenal; 2-propen-1-one; allyl aldehyde; acrylaldehyde; acraldehyde; ethylene aldehyde; aqualin; biocide.

SECTION II - HAZARDOUS INGREDIENTS AND IDENTITY INFORMATION

Hazardous Components (specific identity)	OSHA Exposure Criteria	NIOSH Exposure Criteria	Immediately Dangerous to Life and Health (IDLH)	ACGIH Exposure Criteria
Acrolein (derivation: Oxidation of allyl alcohol or propylene; by heating glycerol with magnesium sulfate; from propylene with bismuth-phosphorous-molybdenum catalyct). **1 ppm = 2.33 mg/m^3**	**PEL:** **0.1 ppm** **0.25 mg/m^3** **STEL:** **0.3 ppm** **0.80 mg/m^3**	**REL:** **0.1 ppm** **0.25 mg/m^3** **STEL:** **0.3 ppm** **0.80 mg/m^3**	**2 ppm**	**TLV:** **0.1 ppm** **0.25 mg/m^3** **STEL:** **0.3 ppm** **0.80 mg/m^3**

SECTION III - PHYSICAL AND CHEMICAL CHARACTERISTICS

Boiling Point: **127°F (53°C)**	Specific Gravity (H_2O = 1): **0.84**
Vapor Pressure (mm Hg): **210 at 69°F (20°C)**	Molecular Weight: **56.07**
Vapor Density (Air = 1): **1.94**	Freezing Point: **-126°F**

Solubility

Moderately soluble in water (40%); readily mixes with alcohol and ether.

Appearance and Odor

Colorless or yellowish liquid with a piercing, disagreeable odor. Odor threshold = 0.16 ppm.

SECTION IV - FIRE AND EXPLOSION HAZARD DATA

Flash Point (method used): **-15°F (-18°C) open cup**	Explosive Limits in Air % by Volume: **LEL: 2.8% UEL: 31%**
NFPA Classification: **Class 1B Flammable Liquid**	Autoignition Temperature: **455°F (235°C)**

Extinguishing Media

Carbon dioxide, dry chemical, "alcohol" foam.

Special Fire Fighting Procedures

Wear full protective clothing and self-contained breathing apparatus (SCBA). Dangerous fire hazard. Use a water spray to keep fire-exposed containers cool. Poisonous gases are produced in fire. Vapors are heavier than air; flashback is likely.

Unusual Fire and Explosion Hazards

Can react vigorously with oxidizing materials. Dangerous fire hazard when exposed to heat, flame, or oxidizers. An explosion hazard as well. Containers may explode in fire.

SECTION V - REACTIVITY DATA

Stability		Conditions to Avoid: Do not use storage containers consisting of copper, zinc, or polyethylene materials as violent reaction may occur. It is unstable at elevated temperatures. Closed containers may rupture violently.
Stable	Unstable: X	Incompatibility *(materials to avoid)*: Strong oxidizers; alkaline substances such as caustic soda, amines, or ammonia. Also avoid contact with SO_2, metal salts, oxidants (light and heat), weak acid conditions, thiourea, or dimethylamine.
Hazardous Polymerization		Conditions to Avoid: Violent polymerization on contact with strong acid, strong base. Self-polymerizes at high temperatures. Polymerizes readily unless inhibited. May form shock-sensitive peroxides over time.
May Occur: X	Will Not Occur	Hazardous Decomposition or By-products: Heat is liberated during polymerization. Highly toxic and irritating fumes are emitted during polymerization.

SECTION VI - HEALTH HAZARD DATA

Primary Route(s) of Entry:	Inhalation?	Absorption (skin)?	Ingestion?
	X	X	

Health Hazards

INHALATION: Lacrimation, delayed hypersensitivity with multiple organ involvement, and respiratory system damage. Strong nasal irritant. Severe irritant to eyes.

ABSORPTION: Skin contact can result in irritating or burning rash. Can cause severe eye burns and/or irritation. Can also result in chronic respiratory disease and delayed pulmonary edema.

INGESTION: If ingested, may cause acute gastrointestinal pain with pulmonary congestion.

Carcinogenicity	NTP Listed?	IARC Cancer Review Group?	OSHA Regulated?	Target Organs?
Unknown Human Unknown Animal	NSC 8819	Group 3	29 CFR 1910.1000 (Table Z-1)	Eyes, respiratory system, skin, heart

Medical Conditions Generally Aggravated by Exposure

Respiratory impairments (bronchitis); heart/hypertension conditions; Skin conditions (dermatitis).

Emergency and First-aid Procedures

Eye contact: Flush immediately with water for 30 minutes (minimum); seek medical attention. Skin contact: Remove all contaminated clothing. Immediately wash area with large amounts of soap and water. Seek medical assistance. For inhalation: Remove the person from exposure. Provide respiratory assistance and CPR. Transfer to medical facility. Acrolein can cause pulmonary edema (fluid in the lungs), a medical emergency, hours after exposure. If swallowed, seek medical attention immediately.

SECTION VII - PRECAUTIONS FOR SAFE HANDLING AND USE

Steps to be Taken in Case Material is Released or Spilled

Contain spills using absorbent material (vermiculite, or other). Ventilate area of spill. Have water source available in case of fire. Store materials in DOT-approved containers. Restrict those not involved in cleanup from entering area. Notify appropriate authorities, as required.

Preferred Waste Disposal Method

Controlled Incineration (with after burner and scrubber). Biodegradation using dilute NaOH.

Precautions to be Taken in Handling and Storage

Acrolein is volatile. Store to avoid contact with strong oxidizers such as chlorine, bromine, and fluorine. Store in tightly closed containers in cool, well-ventilated area away from heat. Drums used for storage should be equipped with self-closing valves, pressure vacuum bungs, and flame arresters.

Other Precautions and Warnings

Bulk storage is not recommended. Personnel should be trained on proper handling and storage.

SECTION VIII - CONTROL MEASURES AND PERSONAL PROTECTIVE EQUIPMENT

Respiratory Protection *(specify type)*

For exposures over 0.1 ppm, use MSHA/NIOSH supplied-air, full-facepiece operated in positive pressure mode or with a full facepiece, hood, or helmet in the continuous flow mode. Or, use a self-contained breathing apparatus (SCBA) in pressure demand or other positive pressure mode.

Ventilation

Local exhaust or general mechanical systems to ensure vapor concentration below OSHA PEL.

Protective Gloves	Eye Protection	Other Protective Clothing
Neoprene or Butyl Rubber	Chemical Goggles and Face Mask	Rubber Apron

Work/Hygiene Practices

Always wash hands thoroughly after using chemical; never bring food, drink, or smoking materials into vicinity of chemicals.

ACROLEIN

C_2H_3CHO CAS: 107-02-8

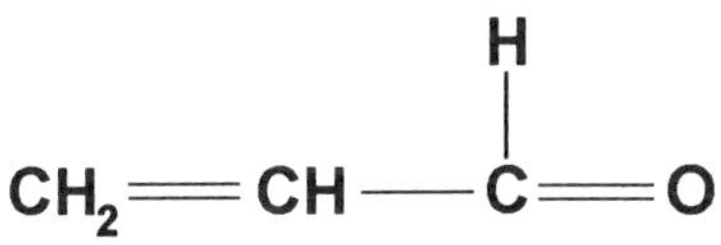

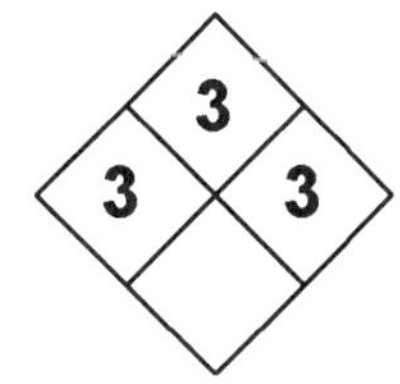

IDENTIFICATION AND TYPICAL USES

Acrolein is a colorless or yellow liquid with a disagreeable, choking odor. It is used in the manufacture of pharmaceuticals, perfumes, food supplements, and resins; as a warning agent in methyl chloride refrigerating systems; as an anti-microbial agent to prevent the growth of microbe against plugging and corrosion; to control the aquatic weed and algae; in slime control in paper manufacturing; as a tissue fixative; and in leather tanning.

RISK ASSESSMENT: HEALTH

General Assessment

Acrolein is a highly toxic poison compound that can severely damage the eyes and respiratory system and cause burns to the skin on contact. Ingestion can lead to acute gastrointestinal pain with pulmonary congestion.

At extremely high concentrations, *inhalation* of its vapors may cause tracheobronchitis and pulmonary edema. Exposure to concentrations as low as 0.25 ppm for as little as 5 minutes can irritate mucous membranes in humans. Fatalities can occur with exposures as low as 10 ppm. This violent effect usually prevents chronic toxicity (i.e., prolonged exposures usually prove fatal thereby eliminating nearly all potential for chronic effects through inhalation). Acrolein can be easily *absorbed* through the skin. Dermal contact may lead to chronic pulmonary edema. Prolonged or repeated skin contact usually produces irritation, burns, and sometimes sensitization. Eye splashes cause corneal damage, palpebral edema, blepharoconjunctivitis, and fibrous or purulent discharge. No data exist on the chemical's carcinogenicity in humans. However, one of its potential metabolites, glycidaldehyde, is considered to be carcinogenic. Acrolein has been found to be mutagenic to bacteria and some human mutagenic effects have been reported. It is a questionable carcinogen in both animals and humans.

Acute Health Effects

The following acute (short-term) health effects may occur immediately or shortly after exposure to Acrolein:

Skin: Irritation/rash upon contact with skin (contact dermatitis).

Eye: Severe irritation, burn, permanent damage.

Lung: Irritation of the nose, throat, and lungs following exposure. Can cause coughing or shortness of breath. Higher exposures may lead to pulmonary edema, which can be fatal.

CNS: Inhalation of high concentrations of the vapor may cause dizziness, nausea, headaches, and, at higher concentrations, loss of consciousness, or death.

Chronic Health Effects

The following chronic (long-term) health effects may occur at some time after exposure to acrolein and can last for months or even years:

Cancer Hazards: According to the references, insufficient data exist to determine the human carcinogenicity of acrolein. However, acrolein is a known mutagen and mutagens may have a cancer risk. Also, one of its known metabolites, glycidaldehyde, is considered to be carcinogenic. The chronic effects of high exposures to acrolein have not been documented in humans because long-term exposures generally prove to be fatal.

Reproductive Hazard: According to the references, there is insufficient evidence to evaluate acrolein for its ability to cause reproductive damage.

Other Chronic Effects: Acrolein may cause a skin allergy. If an allergy develops (sensitization), very low future exposures can cause itching and skin rash. Long-term exposure can cause drying and cracking of skin. Permanent lung damage can occur after a high exposure or repeated low exposures.

⓾ *Recommended Risk-Reduction Measures*

Personnel should avoid direct contact with acrolein. Unless a less toxic chemical can be substituted for a hazardous substance, *engineering controls* are the most effective method of reducing exposures. The best protection is to enclose operations and/or provide local exhaust ventilation at the site of chemical release. While not always operationally feasible, isolating operations can also reduce exposure. Using respiratory protection is less effective than the controls mentioned above, but is still advisable whenever working with or around acrolein. A self-contained breathing apparatus (SCBA) with full facepiece and pressure demand is the recommended respiratory protection of choice. If a full facepiece is not available, then chemical goggles should be worn to protect the eyes. Whenever a chemical splash hazard exists, a face shield and a rubber apron should be worn. To prevent hand and skin exposures, butyl rubber gloves should be worn.

Administrative controls should also be in place to minimize the potential for human exposures. These may include written procedures or policies which specify the methods and techniques that will be practiced whenever personnel are to work with acrolein.

All personnel should receive *training* on the use, hazards, protective measures, emergency actions, and other precautions per 29 CFR 1910.1200 (Hazard Communication), prior to the first assignment in an area where acrolein is used or stored.

Medical testing of personnel with frequent or potentially high exposure (half the PEL or greater), should be offered before beginning work and at regular intervals thereafter. These tests should include:

☑ Lung function test.

If symptoms develop or overexposure is suspected, the following tests may be useful:

☑ Evaluation by a qualified allergist, including careful exposure history and special testing.

☑ Maintain close medical surveillance. Pulmonary edema may result 24 to 48 hours following high exposures. A chest X-ray should be considered after acute overexposure (may be negative if taken immediately after exposure).

Other methods to reduce exposure include:

☑ Where possible, enclose operations and use local exhaust ventilation at the site of chemical release. If local exhaust ventilation or enclosure is not used, respiratory protection should be mandatory.

☑ Always ensure that proper protective clothing is worn when using chemical substances.

☑ Wash thoroughly immediately after exposure to acrolein and at the end of the work shift or before eating, drinking, or smoking.

☑ Hazard warning information should be posted in the work area. In addition, as part of an on-going education and training program, all information on the health and safety hazards of acrolein should be communicated to all potentially exposed workers.

It should be noted that *mixed exposures* can significantly increase the effect of the chemical's reaction in humans. Because smoking can cause heart disease, as well as lung cancer, emphysema, and other respiratory problems, and two of the target organs of acrolein are the heart and lungs, smokers may react more severely/quickly to acrolein exposures than the non-smoker exposed to the same levels.

RISK ASSESSMENT: ENVIRONMENT

General Assessment

The environment is at risk of exposure during transportation, storage, disposal, or destruction of acrolein. In almost every scenario, the threat of environmental exposure is contingent upon the proper handling of the chemical substance. Accidental spills, large or small, can result in fire, explosion, and possible contamination of the surrounding environmental mediums (water, soil, and air). Acrolein is considered a highly flammable liquid and a poison. Because of its low flash point and high boiling point, it is classified as a Class 1B flammable liquid (per OSHA 29 CFR 1910.106). Acrolein vapors form explosive mixtures in air. It reacts with strong oxidizers and acids, liberating heat along with pressure increase. Thus, contact with either under closed conditions can cause explosion. These characteristics require special consideration during any emergency situation involving a leak or spill of acrolein.

Should acrolein ever come into contact with incompatible substances such as strong oxidizers or acids, weak acid conditions (nitrous fumes, sulfur dioxide, carbon dioxide), thiourea, or dimethylamine either during use, transportation, or storage, an extremely violent reaction may result.

The proper disposal/destruction method for acrolein is by controlled burning in an incinerator equipped with an afterburner and air scrubber. Acrolein is used commonly as an herbicide for the control of weeds in irrigation canals. It can enter the environment through its main use as an aquatic herbicide or by spills. Because of its widespread use, it occurs in the environment in both air and water. After formaldehyde, it is the second most abundant aldehyde, constituting 5% of total aldehydes in air.

Acute Ecological Effects

Acute (short-term) toxic effects may include the death of animals, birds, or fish and death or low growth rate in plants. Acute effects are seen 2 to 4 days after animals or plants are exposed to acrolein. This chemical has high acute toxicity to water plants and may be expected to adversely effect land plants. It has high acute toxicity to birds and moderate acute toxicity to land animals. A concentration of 55 ppb may be toxic to aquatic life in fresh and salt water fish. Acrolein is reported to be more toxic to aquatic life than are phenol, chloro- and nitrophenols, aniline, and *o*-xylene.

Chronic Ecological Effects

Chronic toxic effects may include shortened life span, reproductive problems, lower fertility, and changes in appearance or behavior in exposed animals. These effects can be seen long after first exposure(s) to toxic chemicals. Acrolein has high chronic toxicity in aquatic life. A concentration as low as 21 ppb may produce chronic toxicity in fresh water aquatic life. No data are available on the long-term effects of acrolein to plants, birds, or land animals.

Water Solubility

Acrolein is moderately soluble in water. Concentrations of between 1 and 1000 milligrams and more will mix with a liter of water.

Persistence in the Environment

Acrolein is slightly persistent in the aquatic environment, with a half-life of between 2 to 20 days. The half-life of a pollutant is the amount of time it takes for one half of the chemical to be degraded. Approximately 77% of acrolein will eventually end up in air; the rest will end up in water.

Bioaccumulation in Aquatic Organisms

Some substances increase in concentration, or *bioaccumulate,* in living organisms as they breathe contaminated air, drink contaminated water, or eat contaminated food. These chemicals can become concentrated in the tissues and internal organs of animals as well as humans.

The concentration of acrolein found in fish tissues is expected to be somewhat higher than the average concentration of acrolein in the water from which the fish was taken.

Recommended Risk-Reduction Measures

Proper training of all transporters will reduce the likelihood of a mishap or accident resulting in a leak or spill to the environment. The correct use of DOT labeling (per the MSDS) on all containers, trucks, and rail cars will enable emergency responders to react properly and quickly to any disaster thereby reducing the potential risk to the environment and to personnel.

Storage of acrolein should be segregated from incompatible chemicals to minimize the risk of cross-contamination or contact. Buildings designated for storage should be equipped with appropriate fire protection systems (alarms, sprinklers, emergency lighting, portable extinguishers). Acrolein should be stored under an inert atmosphere below 100°F (38°C). Steel, ceramic, glass, Teflon, silicone rubber, and containers with baked phenolic coatings are suitable for storage.

If a spill or leak to the environment has occurred, fire department, emergency response, and/or hazardous materials spill personnel should be notified immediately. Clean-up should be attempted only by those trained in proper spill containment procedures. Contaminated soils should be removed for incineration and replaced with clean soil. If acrolein should contact the water table, aquifer, or navigable waterway, time is of the essence. It is moderately miscible in water and, therefore, total containment and remediation may not be entirely possible. When such spills occur, the local and/or state emergency response authorities must be notified. A comprehensive emergency response or disaster preparedness/recovery plan should be in place prior to any operations involving the use, transportation, storage, or disposal of acrolein.

If acrolein is spilled or leaked, the following specific steps are recommended:

- ☑ Restrict persons not wearing protective clothing from area of spill or leak until cleanup is complete and area can be opened for normal work.
- ☑ Remove all ignition sources from area. Use only non-sparking tools during cleanup operations
- ☑ Ventilate area of spill or leak.

- ☑ Absorb liquids in vermiculite, dry sand, earth, or a similar material and deposit in sealed containers.
- ☑ If applicable, stop flow of leaking liquid or gas. If leak source is a cylinder and the leak cannot be stopped in place, remove leaking cylinder to a safe place in the open air, and repair or allow cylinder to empty.
- ☑ Keep acrolein out of a confined space, such as a sewer, because of the possibility of explosion.
- ☑ It may be necessary to dispose of acrolein as a hazardous waste. The responsible state agency or the regional office of the federal Environmental Protection Agency (EPA) should be contacted for specific recommendations.

RISK ASSESSMENT: BUSINESS

General Assessment

Accidents or mishaps involving acrolein can present a significant threat to business operations. The loss or damage of equipment or facilities can significantly effect fiscal viability. Lawsuits that may result from personnel injury/death, public exposures, and/or environmental contamination will also require a serious expenditure of resources.

Media attention surrounding an injury, death, or environmental damage can also result in a loss of profits and loss of current as well as future business.

Recommended Risk-Reduction Measures

Company attorneys, safety and health professionals, and environmental specialists should be involved in the development of any procedures dealing with safety or emergency response. A company official should be pre-designated as a public relations officer with specific training in dealing with the press. Corporate plans and policies should be developed, approved, and implemented long before any need for such arises.

REFERENCES

Ahlbom, A. 1993. *Biostatistics for Epidemiologists.* New York: CRC Press/Lewis Publishers.

American Conference of Governmental Industrial Hygienists. 1988. *Documentation of the Threshold Limit Values and Biological Exposure Indices,* 5th Edition (with updates). Cincinnati: ACGIH

Bloom, A. D. and F. J. DeSerres. 1995. *Ecotoxicity and Human Health.* New York: CRC Press/Lewis Publishers.

Bueche, J R. 1972. *Principles of Physics,* 2nd Edition. New York: McGraw-Hill.

Burrell, R., D. K. Flaherty, and L. J. Sauers. 1992. *Toxicology of the Immune System.* New York: Van Nostrand Reinhold.

Calabrese, E. J., E. M. Kenyon. *Air Toxics and Risk Assessment.* New York: CRC Press/Lewis Publishers.

Calabrese, E. J. 1994. *Biological Effects of Low Level Exposures.* New York: CRC Press/Lewis Publishers.

Cockerman, L. G., B. S. Shane. 1994. *Basic Environmental Toxicology.* New York: CRC Press/Lewis Publishers.

Cothern, C. R., N. P. Ross. 1994. *Environmental Statistics, Assessment, and Forecasting.* New York: CRC Press/Lewis Publishers.

Cothern, R. R. 1993. *Comparative Environmental Risk Assessment.* New York: CRC Press/Lewis Publishers.

Cothern, C. R. 1995. *Handbook for Environmental Risk Decisionmaking.* New York: CRC Press/Lewis Publishers.

Gardner, R. W. 1994. *Chemical Intolerance.* New York: CRC Press/Lewis Publishers.

Grant, W. M. 1986. *Toxicology of the Eye,* 3rd Edition. Springfield, IL: Charles C. Thomas.

Hosty, J. W., and Patricia Foster. 1990. *A Practical Guide to Chemical Spill Response.* New York: Van Nostrand Reinhold.

International Agency for Research on Cancer (IARC). 1987. *IARC Monographs on the Evaluation of the Carcinogenic Risk to Humans,* Supplement 7, p 78. Lyon: IARC.

International Agency for Research on Cancer (IARC). 1979. *IARC Monographs on the Evaluation of the Carcinogenic Risk of Chemicals to Man,* Vol. 19, *Some Monomers, Plastics and Synthetic Elastomers, and Acrolein.* pp 579-594. Lyon: IARC.

James, A., and M. Lord. 1992. *Index of Chemical and Physical Data.* New York: Van Nostrand Reinhold.

Kutzman, R. S., et. al. 1985. Changes in rat lung structure and composition as a result of subchronic exposure to acrolein. *Toxicology.* 34:139-151.

Lewis, R. J., Sr. 1993. *Hawley's Condensed Chemical Dictionary,* 12th Edition. New York: Van Nostrand Reinhold.

Lewis, R. J., Sr. 1992. *Sax's Dangerous Properties of Industrial Materials,* Volumes 1, 2, and 3, 8th Edition. New York: Van Nostrand Reinhold.

Meyer, E. 1990. *Chemistry of Hazardous Materials.* Englewood Cliffs, NJ: Prentice Hall.

Patnaik, P. 1992. *A Comprehensive Guide to the Hazardous Properties of Chemical Substances.* New York: Van Nostrand Reinhold.

Proctor, N. H., and James P. Hughes. 1991. *Chemical Hazards of the Workplace.* New York: Van Nostrand Reinhold.

Quigley, D. R. 1994. *Handbook of Emergency Chemical Management.* New York: CRC Press/Lewis Publishers.

U.S. Department of Health and Human Services, National Institute for Occupational Safety and Health. 1994. *NIOSH Pocket Guide to Chemical Hazards.* Washington, D.C.: U.S. Government Printing Office.

MATERIAL SAFETY DATA SHEET

CHEMICAL NAME	
ACRYLAMIDE	**CRC PRESS/LEWIS PUBLISHERS DISCLAIMER:** THE INFORMATION AND RECOMMENDATIONS PRESENTED HEREIN ARE BASED ON SOURCES BELIEVED TO BE RELIABLE. CRC MAKES NO REPRESENTATION ON ITS COMPLETENESS OR ACCURACY. IT IS THE RESPONSIBILITY OF THE USER TO DETERMINE THE CHEMICAL'S SUITABILITY FOR ITS INTENDED USE, THE CHEMICAL'S SAFE USE, AND THE CHEMICAL'S PROPER DISPOSAL. NO REPRESENTATIONS AND/OR WARRANTIES, EITHER EXPRESSED OR IMPLIED, OF THE MERCHANTABILITY OR FITNESS FOR A PARTICULAR PURPOSE, OR OF ANY OTHER NATURE, ARE MADE WITH RESPECT TO THE INFORMATION PROVIDED IN THIS MSDS OR TO THE CHEMICAL TO WHICH INFORMATION MAY REFER. CRC NEITHER ASSUMES NOR AUTHORIZES ANY OTHER PERSON TO ASSUME FOR IT, ANY OTHER ADDITIONAL RESPONSIBILITY OR LIABILITY FOR THE USE OF, OR RELIANCE UPON, THIS INFORMATION.

HAZARD WARNING INFORMATION

HEALTH	FIRE	REACTIVE	OTHER	DEGREE OF HAZARD	COLOR CODING	OTHER CODES
3	2	2		0 = Minimum Hazard 1 = Slight Hazard 2 = Moderate Hazard 3 = Serious Hazard 4 = Severe Hazard	HEALTH = BLUE FIRE = RED REACTIVITY = YELLOW OTHER = WHITE	OX = Oxidizer ACID = Acid ALK = Alkali COR = Corrosive W = Use No Water

SECTION I - GENERAL INFORMATION

Characterization **Aldehyde**	RCRA Number **U007**	EPA Class **Toxic Waste**
DOT Proper Shipping Name **Acrylamide**	Chemical Abstract Service (CAS) Number **79-06-1**	
DOT Hazard Class and Label Requirements **Combustible Solid; IMO: Poison B**	DOT Emergency Guide Code **55**	
DOT Identification Number **UN 2074**	Chemical Formula C_3H_5NO	

Synonyms
Acrylic amide; ethylenecarboxamide; propenamide; 2-propenamide; acrylamide monomer.

SECTION II - HAZARDOUS INGREDIENTS AND IDENTITY INFORMATION

Hazardous Components (specific identity)	OSHA Exposure Criteria	NIOSH Exposure Criteria	Immediately Dangerous to Life and Health (IDLH)	ACGIH Exposure Criteria
Acrylamide (derivation: Reaction of acrylonitrile with sulfuric acid (84.5%) and neutralization).	**PEL:** 0.3 mg/m³ (skin) **STEL:** Not Established	**REL:** 0.3 mg/m³ (skin) Possible Cancer Agent	60 mg/m³	**TLV:** 0.3 mg/m³ (skin) **STEL:** Not Established

SECTION III - PHYSICAL AND CHEMICAL CHARACTERISTICS

Boiling Point **347-572°F (decomposes)**	Specific Gravity (H_2O = 1) **1.12**
Vapor Pressure (mm Hg) **0.007 at 69°F (20°C)**	Molecular Weight **71.10**
Vapor Density (Air = 1) **2.45**	Melting Point **184°F (85°C)**

Solubility
Very miscible (216% at 85°F). Also miscible with alcohol, ether, and acetone; insoluble in benzene and heptane.

Appearance and Odor
Colorless to white, odorless crystalline solid.

SECTION IV - FIRE AND EXPLOSION HAZARD DATA

Flash Point (method used) **280°F (138°C) closed cup**	Explosive Limits in Air % by Volume **LEL: N/A UEL: N/A**
NFPA Classification **Combustible Solid**	Autoignition Temperature **Not Established**

Extinguishing Media
Carbon dioxide, dry chemical, water spray or foam.

Special Fire Fighting Procedures
Wear full protective clothing and self-contained breathing apparatus (SCBA). A combustible solid (may also be dissolved in flammable liquids. Use a water spray to keep fire-exposed containers cool. Poisonous gases are produced in fire.

Unusual Fire and Explosion Hazards
Potentially explosive reaction with strong oxidizing compounds. Containers may explode in fire. Firefighters should avoid contact with vapors produced during fire.

SECTION V - REACTIVITY DATA

Stability		Conditions to Avoid: Normally stable, however it can polymerize when heated to its melting point, exposed to ultra-violet light, or in contact with a catalyst. Contact in a closed container can cause rupture of the container.
Stable	Unstable	Incompatibility *(materials to avoid)*: Strong oxidizers.
X		
Hazardous Polymerization		Conditions to Avoid: Keep away from fire, heat, ignition sources. Poisonous gases are produced during fire. Polymerizes violently at its melting point.
May Occur	Will Not Occur	Hazardous Decomposition or By-products: When heated to decomposition, acrylamide emits highly toxic fumes of NO_x.
X		

SECTION VI - HEALTH HAZARD DATA

Primary Route(s) of Entry:	Inhalation?	Absorption (skin)?	Ingestion?
	X	X	X

Health Hazards

INHALATION: Nervous system effects, including numbness and weakness in the hands and feet, loss of balance, slurred speech, and heavy sweating. These changes may be permanent.

ABSORPTION: Skin contact can result in irritating rash. Can cause severe eye burns and/or irritation. Readily absorbed through unbroken skin. A highly toxic poison through absorption.

INGESTION: Central nervous system disorders; thyroid function, muscular weakness.

Carcinogenicity	NTP Listed?	IARC Cancer Review Group?	OSHA Regulated?	Target Organs?
Suspected Human Confirmed Animal	No	Group 2B	29 CFR 1910.1000 (Table Z-1)	Lungs, eyes, skin, CNS, thyroid, reproductive sys.

Medical Conditions Generally Aggravated by Exposure

Respiratory impairments (bronchitis, asthma); Hypertension; Skin conditions; Reproductive disorders.

Emergency and First-aid Procedures

Eye contact: Flush immediately with water for 15 minutes (minimum), seek medical attention. Skin contact: Remove all contaminated clothing. Immediately wash area with large amounts of soap and water. Seek medical assistance. For inhalation: Remove the person from exposure. Provide respiratory assistance and CPR. Transfer to medical facility. If swallowed, seek medical attention immediately. Time onset of symptoms may be delayed 1-24 months, depending on exposure route and degree).

SECTION VII - PRECAUTIONS FOR SAFE HANDLING AND USE

Steps to be Taken in Case Material is Released or Spilled

Collect powdered material in most convenient way possible and deposit in sealed containers. Have water source available in case of fire. Restrict those not involved in clean-up from entering area. Notify appropriate authorities, as required by SARA III, if applicable.

Preferred Waste Disposal Method

Incineration (with after burner and scrubber).

Precautions to be Taken in Handling and Storage

Acrylamide is a combustible solid. Store to avoid contact with strong oxidizers such as chlorine, dioxide and bromine since violent reactions occur. Store in tightly closed containers in cool, well-ventilated area away from heat.

Other Precautions and Warnings

Sources of ignition are prohibited where acrylamide is used, handled, or stored. Prior to working with acrylamide, personnel should be trained on its proper handling and storage.

SECTION VIII - CONTROL MEASURES AND PERSONAL PROTECTIVE EQUIPMENT

Respiratory Protection *(specify type)*

When potential for exposures above 0.3 mg/m3 exist, use a MSHA/NIOSH approved supplied air respirator with full facepiece, hood, or helmet in continuous flow mode, or use a full face piece operated in pressure demand or other positive pressure mode.

Ventilation

Local exhaust or general mechanical systems recommended.

Protective Gloves	Eye Protection	Other Protective Clothing
Neoprene	Chemical/Dust Goggles or Face Mask	Rubber Apron

Work/Hygiene Practices

Always wash hands thoroughly after using chemical; never bring food, drink, or smoking materials into vicinity of chemicals.

ACRYLAMIDE

C_3H_5NO CAS: 79-06-1

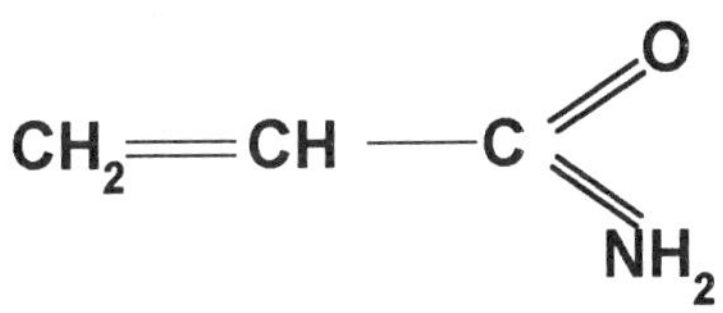

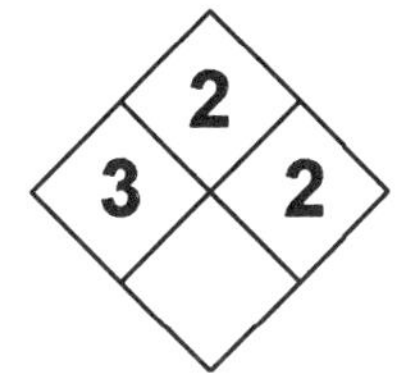

IDENTIFICATION AND TYPICAL USES

Acrylamide is a colorless-to-white, crystalline solid material. Its uses include waterproofing; as a soil stabilizer; in the production of polymers used in paper manufacturing and flocculating processes; in grout used to repair leaking sewer pipes; and as an additive to enhance oil recovery. It is also used in the synthesis of dyes; as a cross-linking agent; in the making of adhesives; in sewage and waste treatment; and in permanent-press fabrics.

RISK ASSESSMENT: HEALTH

General Assessment

Exposure to acrylamide results in neurological effects through systemic exposures and skin and mucous membrane irritation from local exposures. It is a confirmed carcinogen in animals and is suspected of having carcinogenic potential in humans as well. Exposure can occur by inhalation, skin absorption, and ingestion. In industry, most common route of entry is absorption, followed by inhalation, then ingestion. Known exposures in humans have resulted in symptoms which indicate involvement of the central, peripheral, and autonomic nervous systems.

Inhalation effects to the central nervous system include symptoms of intoxication such as confusion; disorientation; and hallucinations. Other symptoms include abnormal fatigue, memory difficulties, and dizziness. The effects on the peripheral nervous system include muscle weakness; numbness in feet, hands, lower legs, and lower arms; stupor; and difficulties in walking and standing. Autonomic effects include excessive sweating and difficulties in defecation. Acute exposures seem to affect the central nervous system while repeated, long-term exposures effect the peripheral nervous system.

Most exposures occur through skin *absorption.* Symptoms include local irritation and blistering. The skin on the palms of the hands and heels of the feet may begin to peel in conjunction with a blue coloring of the hands and feet.

Time of symptom onset may vary from one to 24 months to as much as 8 years. In the work place, acrylamide is an extremely dangerous poison because it can be absorbed through unbroken skin.

☠ *Acute Health Effects*

The following acute (short-term) health effects may occur immediately or shortly after exposure to acrylamide:

Skin: Irritation/rash or burning feeling upon contact with skin. Possible dermatitis.

Eye: Severe irritation; may cause burns.

Lung: Irritation of the nose, throat, and lungs following exposure.

CNS: Inhalation of acrylamide powders may cause sleepiness, dizziness, weakness, stupor, fatigue.

Chronic Health Effects

The following chronic (long-term) health effects may occur at some time after exposure to acrylamide and can last for months or even years:

Cancer Hazards: Acrylamide is a known carcinogen in experimental animals. Its carcinogenicity is considered suspect in humans. It may cause skin tumors and also tumors of the lung.

Reproductive Hazard: There is some evidence to show that exposure to acrylamide causes damage to the testes (male reproductive glands) in laboratory animals.

Other Chronic Effects: Acrylamide can cause damage to the central nervous system, peripheral nervous system, and autonomic nervous system. There may be permanent symptoms such as tingling of the hands and feet following long-term exposures.

Recommended Risk-Reduction Measures

Personnel should avoid direct contact with acrylamide. Unless a less toxic chemical can be substituted for a hazardous substance, *engineering controls* are the most effective method of reducing exposures. The best protection is to enclose operations and/or provide local exhaust ventilation at the site of chemical release. While not always operationally feasible, isolating operations can also reduce exposure. Using respiratory protection is less effective than the controls mentioned above, but is still advisable whenever working with or around acrylamide. A self-contained breathing apparatus (SCBA) with full facepiece and pressure demand is the recommended respiratory protection of choice. If a full facepiece is not available, then chemical/dust goggles should be worn to protect the eyes. Whenever a chemical splash hazard exists, a face shield and a rubber apron should be worn. To prevent hand and skin exposures, neoprene or butyl rubber gloves should be worn.

Administrative controls should also be in place to minimize the potential for human exposures. These may include written procedures or policies which specify the methods and techniques that will be practiced whenever personnel are to work with acrylamide.

All personnel should receive *training* on the use, hazards, protective measures, emergency actions, and other precautions per 29 CFR 1910.1200 (Hazard Communication), prior to the first assignment in an area where acrylamide is used or stored.

Medical testing of personnel with frequent or potentially high exposure (half the PEL or greater or significant skin contact), should be offered before beginning work and at regular intervals thereafter. These tests should include:

- ☑ Examination of the nervous systems. Nerve conduction tests should be considered.

Other methods to reduce exposure include:

- ☑ Where possible, enclose operations and use local exhaust ventilation at the site of chemical release. If local exhaust ventilation or enclosure is not used, respiratory protection should be mandatory.
- ☑ Always ensure that proper protective clothing is worn when using chemical substances and that personnel are trained on its use and care.
- ☑ Wash thoroughly immediately after exposure to acrylamide and at the end of the work shift or before eating, drinking, or smoking.
- ☑ Hazard warning information should be posted in the work area. In addition, as part of an on-going education and training program, all information on the health and safety hazards of acrylamide should be communicated to all exposed or potentially exposed workers.

RISK ASSESSMENT: ENVIRONMENT

General Assessment

The environment is at risk of exposure during transportation, storage, disposal, or destruction of acrylamide. In almost every scenario, the threat of environmental exposure is contingent upon the proper handling of the chemical substance. Accidental spills, large or small, can result in fire, explosion, and possible contamination of the surrounding environmental mediums (water, soil, and, to a lesser degree, air).

Acrylamide is considered a combustible solid and a poison. It is incompatible with strong oxidizing agents and contact can cause fire or explosion. These characteristics require special consideration during any emergency situation involving a leak or spill of acrylamide. Should acrylamide ever come into contact with incompatible substances such as chlorine dioxide or bromine either during use, transportation, or storage, the formation of highly toxic and/or highly explosive commodities is extremely possible.

The proper disposal/destruction method for acrylamide is to burn it in a chemical incinerator equipped with an afterburner and air scrubber. Acrylamide can enter the environment mainly through the effluents of industrial process, and spills.

Acute Ecological Effects

Acute (short-term) toxic effects may include the death of animals, birds, or fish and death or low growth rate in plants. Acute effects are seen 2 to 4 days after animals or plants are exposed to acrylamide. This chemical has moderate acute toxicity to aquatic life. No data are available on the short-term effects of acrylamide to plants, birds, or land animals.

Chronic Ecological Effects

Chronic (long-term) toxic effects may include shortened life span, reproductive problems, lower fertility, and changes in appearance or behavior in exposed animals. These effects can be seen long after first exposure(s) to toxic chemicals.

Acrylamide has moderate chronic toxicity in aquatic life. No data are available on the long-term effects of acrylamide to plants, birds, or land animals.

Water Solubility

Acrylamide is highly soluble in water. Concentrations of 1000 milligrams and more will mix with a liter of water.

Persistence in the Environment

Acrylamide is slightly persistent in the aquatic environment, with a half-life of between 2 to 20 days. The half-life of a pollutant is the amount of time it takes for one half of the chemical to be degraded. Approximately 95% of acrylamide will eventually end up in water; the remainder will end up in air.

Bioaccumulation in Aquatic Organisms

Some substances increase in concentration, or *bioaccumulate,* in living organisms as they breathe contaminated air, drink contaminated water, or eat contaminated food. These chemicals can become concentrated in the tissues and internal organs of animals as well as humans.

The concentration of acrylamide found in fish tissues is expected to be about the same as the average concentration of acrylamide in water from which the fish was taken.

Recommended Risk-Reduction Measures

Proper training of all transporters will reduce the likelihood of a mishap or accident resulting in a leak or spill to the environment. The correct use of DOT labeling (per the MSDS) on all containers, trucks, and rail cars will enable emergency responders to react properly and quickly to any disaster thereby reducing the potential risk to the environment and to personnel.

Storage of acrylamide should be segregated from incompatible chemicals to minimize the risk of cross-contamination or contact. Buildings designated for storage should be equipped with appropriate fire protection systems (alarms, sprinklers, emergency lighting, portable extinguishers).

If a spill or leak to the environment has occurred, fire department, emergency response and/or hazardous materials spill personnel should be notified immediately. Clean-up should be attempted only by those trained in proper spill containment procedures. Contaminated soils should be removed for incineration and replaced with clean soil. If acrylamide should contact the water table, aquifer, or navigable waterway, time is of the essence. It is highly miscible in water and, therefore, total containment and remediation may not be entirely possible. When such spills occur, the local or state emergency response authorities must be notified. A comprehensive emergency response or disaster preparedness/recovery plan should be in place prior to any operations involving the use, transportation, storage, or disposal of acrylamide.

If acrylamide is spilled or leaked, the following specific steps are recommended:

- ☑ Restrict persons not wearing protective clothing from area of spill or leak until cleanup is complete and area can be opened for normal work.
- ☑ Remove all ignition sources and ventilate area.
- ☑ Collect powdered material in the most convenient manner possible and deposit in sealed containers. Use non-sparking tools.
- ☑ It may be necessary to dispose of acrylamide as a hazardous waste. The responsible state agency or the regional office of the federal Environmental Protection Agency (EPA) should be contacted for specific recommendations.

RISK ASSESSMENT: BUSINESS

General Assessment

Accidents or mishaps involving acrylamide can present a significant threat to business operations. The loss or damage of equipment or facilities can significantly effect fiscal viability. Lawsuits that may result from personnel injury/death, public exposures, and/or environmental contamination will also require a serious expenditure of resources.

Media attention surrounding an injury, death, or environmental damage can also result in a loss of profits and loss of current as well as future business.

Recommended Controls for Risk-Reduction

Company attorneys, safety and health professionals, and environmental specialists should be involved in the development of any safety, health, or environmental procedures. A company official should be predesignated as a public relations officer with specific training in dealing with the press. Corporate plans and policies should be developed, approved, and implemented long before any need for such arises.

REFERENCES

Ahlbom, A. 1993. *Biostatistics for Epidemiologists.* New York: CRC Press/Lewis Publishers.

American Conference of Governmental Industrial Hygienists. 1988. *Documentation of the Threshold Limit Values and Biological Exposure Indices,* 5th Edition (with updates). Cincinnati: ACGIH

Asante-Duah, D. K. 1993. *Hazardous Waste Risk Assessment.* New York: CRC Press/Lewis Publishers.

Bloom, A. D. and F. J. DeSerres. 1995. *Ecotoxicity and Human Health.* New York: CRC Press/Lewis Publishers.

Brusick, D. J. 1994. *Methods for Genetic Risk Assessment.* New York: CRC Press/Lewis Publishers.

Bueche, J. R. 1972. *Principles of Physics,* 2nd Edition. New York: McGraw-Hill.

Burrell, R., D. K. Flaherty, and L. J. Sauers. 1992. *Toxicology of the Immune System.* New York: Van Nostrand Reinhold.

Calabrese, E. J., E. M. Kenyon. *Air Toxics and Risk Assessment.* New York: CRC Press/Lewis Publishers.

Calabrese, E. J. 1994. *Biological Effects of Low Level Exposures.* New York: CRC Press/Lewis Publishers.

Cockerman, L. G., B. S. Shane. 1994. *Basic Environmental Toxicology.* New York: CRC Press/Lewis Publishers.

Cothern, C. R., N. P. Ross. 1994. *Environmental Statistics, Assessment, and Forecasting.* New York: CRC Press/Lewis Publishers.

Cothern, R. R. 1993. *Comparative Environmental Risk Assessment.* New York: CRC Press/Lewis Publishers.

Cothern, C. R. 1995. *Handbook for Environmental Risk Decisionmaking.* New York: CRC Press/Lewis Publishers.

Danse, I. R. 1991. *Common Sense Toxics in the Workplace.* New York: Van Nostrand Reinhold.

Gardner, R. W. 1994. *Chemical Intolerance.* New York: CRC Press/Lewis Publishers.

Hallenbeck, W. H. 1993. *Quantitative Risk Assessment for Environmental and Occupational Health,* 2nd Edition. New York: CRC Press/Lewis Publishers.

Hogan, K. A. 1990. Mortality patterns and acrylamide exposures. *Journal of Occupational Medicine,* 32:947-949.

Hosty, J. W., and P. Foster. 1990. *A Practical Guide to Chemical Spill Response.* New York: Van Nostrand Reinhold.

Howard, P. H. 1995. *Handbook of Environmental Fate and Exposures Data for Organic Chemicals,* Vol. I-IV. New York: CRC Press/Lewis Publishers.

Howard, P. H., et. al. 1991. *Handbook of Environmental Degradation Rates.* New York: CRC Press/Lewis Publishers.

James, A., and M. Lord. 1992. *Index of Chemical and Physical Data.* New York: Van Nostrand Reinhold.

Johnson, K. A., et. al. 1984. *Acrylamide: A Two-Year Drinking Water Chronic Toxicity-Oncogenicity Study in Fischer 344 Rats.* Sponsored by American Cyanamid Company, Dow Chemical U.S.A., Nalco Chemical Company and Standard Oil Company, Ohio.

Jones, F E. 1994. *Toxic Organic Vapors in the Workplace.* New York: CRC Press/Lewis Publishers.

Keith, L. H. and M. M. Walker. 1995. *Handbook of Air Toxics.* New York: CRC Press/Lewis Publishers.

Landis, W. G., M. H. Yu. 1995. *Introduction to Environmental Toxicology.* New York: CRC Press/Lewis Publishers.

Langerman, N. 1994 *Precautionary Labels for Chemical Containers.* New York: CRC Press/Lewis Publishers.

Lauwerys, R. R., P. Hoet. 1993. *Industrial Chemical Exposure.* New York: CRC Press/Lewis Publishers.

Lewis, R. J., Sr. 1993. *Hawley's Condensed Chemical Dictionary,* 12th Edition. New York: Van Nostrand Reinhold.

Lewis, R. J., Sr. 1992. *Sax's Dangerous Properties of Industrial Materials,* Volumes 1, 2, and 3, 8th Edition. New York: Van Nostrand Reinhold.

Lide, D. R. 1995. *Handbook of Chemistry and Physics,* 76th Edition. New York: CRC Press/Lewis Publishers.

Manahan, S. E. 1994. *Environmental Chemistry,* 6th Edition. New York: CRC Press/Lewis Publishers.

Manahan, S. E. 1992. *Toxicological Chemistry,* 2nd Edition. New York: CRC Press/Lewis Publishers.

Mellström, G. A., J. E. Wahlberg, and H. I. Maibach. 1994. *Protective Gloves for Occupational Use.* New York: CRC Press/Lewis Publishers.

Meyer, E. 1990. *Chemistry of Hazardous Materials.* Englewood Cliffs, NJ: Prentice Hall.

Mycock, J. C., J. D. McKenna, and L. Theodore. 1995. *Handbook of Air Pollution Control Engineering and Technology.* New York: CRC Press/Lewis Publishers.

Nelson, G. O. 1992. *Gas Mixtures.* New York: CRC Press/Lewis Publishers.

Ott, W. R. 1995. *Environmental Statistics and Data Analysis.* New York: CRC Press/Lewis Publishers.

Patnaik, P. 1992. *A Comprehensive Guide to the Hazardous Properties of Chemical Substances.* New York: Van Nostrand Reinhold.

Philp, R. B. 1995. *Environmental Hazards and Human Health.* New York: CRC Press/Lewis Publishers.

Proctor, N. H., and J. P. Hughes. 1991. *Chemical Hazards of the Workplace.* New York: Van Nostrand Reinhold.

Quigley, D. R. 1994. *Handbook of Emergency Chemical Management.* New York: CRC Press/Lewis Publishers.

Sherman, J. D. 1988. *Chemical Exposure and Disease.* New York: Van Nostrand Reinhold.

U.S. Department of Health and Human Services, National Institute for Occupational Safety and Health. 1994. *NIOSH Pocket Guide to Chemical Hazards.* Washington, D.C.: U.S. Government Printing Office.

World Health Organization. 1985. *Environmental Health Criteria 49: Acrylamide,* pp 1-121. The International Programme on Chemical Safety. WHO: Geneva.

MATERIAL SAFETY DATA SHEET

CHEMICAL NAME	
ACRYLIC ACID	**CRC PRESS/LEWIS PUBLISHERS DISCLAIMER:** THE INFORMATION AND RECOMMENDATIONS PRESENTED HEREIN ARE BASED ON SOURCES BELIEVED TO BE RELIABLE. CRC MAKES NO REPRESENTATION ON ITS COMPLETENESS OR ACCURACY. IT IS THE RESPONSIBILITY OF THE USER TO DETERMINE THE CHEMICAL'S SUITABILITY FOR ITS INTENDED USE, THE CHEMICAL'S SAFE USE, AND THE CHEMICAL'S PROPER DISPOSAL. NO REPRESENTATIONS AND/OR WARRANTIES, EITHER EXPRESSED OR IMPLIED, OF THE MERCHANTABILITY OR FITNESS FOR A PARTICULAR PURPOSE, OR OF ANY OTHER NATURE, ARE MADE WITH RESPECT TO THE INFORMATION PROVIDED IN THIS MSDS OR TO THE CHEMICAL TO WHICH INFORMATION MAY REFER. CRC NEITHER ASSUMES NOR AUTHORIZES ANY OTHER PERSON TO ASSUME FOR IT, ANY OTHER ADDITIONAL RESPONSIBILITY OR LIABILITY FOR THE USE OF, OR RELIANCE UPON, THIS INFORMATION.

HAZARD WARNING INFORMATION

HEALTH	FIRE	REACTIVE	OTHER	DEGREE OF HAZARD	COLOR CODING	OTHER CODES
3	2	2	ACID	0 = Minimum Hazard 1 = Slight Hazard 2 = Moderate Hazard 3 = Serious Hazard 4 = Severe Hazard	HEALTH = BLUE FIRE = RED REACTIVITY = YELLOW OTHER = WHITE	OX = Oxidizer ACID = Acid ALK = Alkali COR = Corrosive W = Use No Water

SECTION I - GENERAL INFORMATION

Characterization	**Acid, Carboxylic**	RCRA Number	**U008**
		EPA Class	**Toxic Waste**
DOT Proper Shipping Name	**Acrylic Acid**	Chemical Abstract Service (CAS) Number	**79-10-7**
DOT Hazard Class and Label Requirements	**Corrosive; Corrosive, Flammable Liquid**	DOT Emergency Guide Code	**29**
DOT Identification Number	**UN 2218**	Chemical Formula	**$C_3H_4O_2$**

Synonyms

Acroleic acid; propenpoic acid; glacial acrylic acid; ethylenecarboxylic acid; vinylformic acid.

SECTION II - HAZARDOUS INGREDIENTS AND IDENTITY INFORMATION

Hazardous Components (specific identity)	OSHA Exposure Criteria	NIOSH Exposure Criteria	Immediately Dangerous to Life and Health (IDLH)	ACGIH Exposure Criteria
Acrylic acid (derivation: Through the condensation of ethylene oxide with hydrocyanic acid followed by reaction with sulfuric acid at 320°F; acetylene, CO, and water, with nickel catalyst). **1 ppm = 3.00 mg/m³**	**PEL (skin):** **10 ppm** **30 mg/m³** **STEL:** **Not Established**	**REL (skin):** **2 ppm** **6 mg/m³** **STEL:** **Not Established**	**Not Established**	**TLV (skin):** **2 ppm** **6 mg/m³** **STEL:** **Not Established**

SECTION III - PHYSICAL AND CHEMICAL CHARACTERISTICS

Boiling Point	**286°F (141°C)**	Specific Gravity (H_2O = 1)	**1.05**
Vapor Pressure (mm Hg)	**3 at 69°F (20°C)**	Molecular Weight	**72.10**
Vapor Density (Air = 1)	**2.45**	Melting Point	**55°F (13°C)**

Solubility

Very miscible. Also miscible with benzene, chloroform, alcohol, ether, acetone, and organic solvents.

Appearance and Odor

Colorless liquid with an acrid odor. Odor threshold = <1 ppm.

SECTION IV - FIRE AND EXPLOSION HAZARD DATA

Flash Point (method used)	**130°F (54°C) closed cup**	Explosive Limits in Air % by Volume	**LEL: 2.4% UEL: 8%**
NFPA Classification	**Class II Combustible Liquid**	Autoignition Temperature	**774°F (360°C)**

Extinguishing Media

Carbon dioxide, dry chemical, water spray or foam.

Special Fire Fighting Procedures

Wear full protective clothing and self-contained breathing apparatus (SCBA). Use water spray to flush and dilute a spill and to disperse the vapors. Also, use a water spray to keep fire-exposed containers cool. Poisonous gases are produced in fire.

Unusual Fire and Explosion Hazards

Potentially explosive reaction with strong oxidizing compounds, especially at elevated temperatures. Containers may explode in fire. Firefighters should avoid contact with vapors produced during fire.

SECTION V - REACTIVITY DATA

Stability		Conditions to Avoid
		Reacts with strong oxidizers. Contact in a closed container can cause rupture of the container. A serious fire hazard when exposed to heat or flame. It is corrosive to many materials.
Stable	Unstable	Incompatibility *(materials to avoid)*
	X	Strong oxidizers, amines, alkalis, ammonium hydroxide, chlorosulfonic acid, oleum, ethylene diamine, ethyleneimine, 2-aminoethanol. Reactions to amines, imines, and oleum are exothermic, not violent.
Hazardous Polymerization		Conditions to Avoid
		In the presence of a catalyst or at an elevated temperature, polymerization rate may accelerate, causing an explosion. Acrylic acid may readily polymerize at room temperatures.
May Occur	Will Not Occur	Hazardous Decomposition or By-products
X		When heated to decomposition, acrylic acid can emit highly toxic/poisonous gases.

SECTION VI - HEALTH HAZARD DATA

Primary Route(s) of Entry:	Inhalation?	Absorption (skin)?	Ingestion?
	X	X	X

Health Hazards

INHALATION: Extremely irritating to the eyes, nose, throat, and respiratory system. Inhaled vapors may be absorbed in the gastrointestinal tract.

ABSORPTION: Skin contact can result in burns. Can cause severe eye burns and/or irritation. Absorbed acrylic acid is rapidly metabolized as a short chain fatty acid. Possible skin sensitizer.

INGESTION: Poisonous by ingestion. An experimental teratogen. Questionable carcinogenic effects.

Carcinogenicity	NTP Listed?	IARC Cancer Review Group?	OSHA Regulated?	Target Organs?
Questioned Human Suspected Animal	No	Group 3	No Specific Listing	Respiratory system, eyes, skin, kidneys, liver.

Medical Conditions Generally Aggravated by Exposure

Respiratory impairments (bronchitis, asthma); Skin conditions (dermatitis); Renal or liver disorders.

Emergency and First-aid Procedures

Eye contact: Flush immediately with water for 15 minutes (minimum), seek medical attention. Skin contact: Remove all contaminated clothing. Immediately wash area with large amounts of soap and water. Seek medical assistance. For inhalation: Remove the person from exposure. Provide respiratory assistance and CPR. Transfer to medical facility. If swallowed, seek medical attention.

SECTION VII - PRECAUTIONS FOR SAFE HANDLING AND USE

Steps to be Taken in Case Material is Released or Spilled

Absorb liquids in vermiculite, dry sand, earth, or similar material and deposit in sealed containers. Ventilate area of spill or leak. Restrict those not involved in cleanup from entering area. Notify appropriate authorities, as required by SARA III, if applicable.

Preferred Waste Disposal Method

Incineration (with after burner and scrubber).

Precautions to be Taken in Handling and Storage

Acrylic acid is a combustible liquid. Avoid contact with strong oxidizers such as chlorine since violent reactions occur. Store in tightly closed containers in cool, well-ventilated area away from heat.

Other Precautions and Warnings

Sources of ignition are prohibited where acrylic acid is used, handled, or stored. Storing acrylic acid at temperatures at or below 57°F (13.9°C), the explosion hazard is significantly reduced.

SECTION VIII - CONTROL MEASURES AND PERSONAL PROTECTIVE EQUIPMENT

Respiratory Protection *(specify type)*

When potential for exposures above 10 ppm exist, use a MSHA/NIOSH approved supplied air respirator with full facepiece, hood, or helmet in continuous flow mode, or use a full facepiece operated in pressure demand or other positive pressure mode. Use half-mask respirator with acid cartridge for low or transient exposures (cartridge should be designed to protect against acid vapor or mist).

Ventilation

Local exhaust or general mechanical systems recommended.

Protective Gloves	Eye Protection	Other Protective Clothing
Acid resistant synthetic rubber	Chemical Goggles or Face Mask	Rubber Apron

Work/Hygiene Practices

Always wash hands thoroughly after using chemical; never bring food, drink, or smoking materials into vicinity of chemicals.

ACRYLIC ACID

$C_3H_4O_2$ CAS: 79-10-7

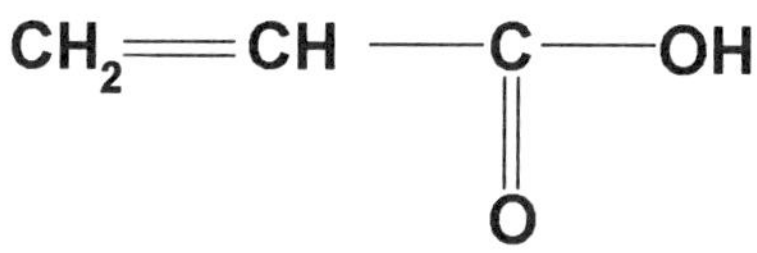

IDENTIFICATION AND TYPICAL USES

Acrylic acid is a colorless liquid with an acrid odor. It serves as a building block for the synthesis of other chemicals. It is produced by oxidation of acrolein or hydrolysis of acrylonitrile. It is used in the manufacture of plastics; in paints, polishes, and adhesives; and as coatings on leather.

RISK ASSESSMENT: HEALTH

General Assessment

Exposure to acrylic acid is possible through *inhalation* of its vapors and by *ingestion.* Also, the gastrointestinal tract appears to be the site of *absorption* of inhaled acrylic acid vapors. In addition, acrylic acid is very irritating to and readily penetrates the skin.

Inhalation of its vapors can irritate the nose and throat as well as the lungs. It is a serious eye hazard. Splashing into the eyes can cause damage and loss of vision. Experimental animals (rats) have shown the effects of exposure to include nasal discharge, weight loss, lethargy, and kidney congestion. High exposures (excess of 4000 ppm for 4 hours) has proven to be lethal in laboratory experiments with test animals.

Laboratory testing of oral exposures were inconclusive. Although no significant embryotoxicity or teratogenicity could be documented, teratogenic studies in animals are still suspected (animal testing is limited, difficult to quantify, and lack conclusive evidence of teratogenicity). It is still recommended that acrylic acid be treated as a questionable teratogen until further studies verify otherwise.

High, excessive, or long-term exposures may damage the kidneys and lungs. Repeated or high exposure to the skin may cause an allergic reaction in some individuals (sensitization). Once sensitized, even very small future exposures to acrylic acid liquid on the skin's surface can result in a recurrence of symptoms, which include rash and inflammation.

Acute Health Effects

The following acute (short-term) health effects may occur immediately or shortly after exposure to acrylic acid:

Skin: Irritation/rash or burning feeling on contact.

Eye: Severe irritation; may cause burns and damage.

Lung: Irritation of the nose, throat, and lungs following exposure.

CNS: Inconclusive (possible lethargy, weight loss).

Chronic Health Effects

The following chronic (long-term) health effects may occur at some time after exposure to acrylic acid and can last for months or even years:

Cancer Hazards: Acrylic acid is a questionable carcinogen in experimental animals. Its carcinogenicity in humans may be suspect, but the data are inconclusive.

Reproductive Hazard: There is limited evidence to show that exposure to acrylic acid causes teratogenic effects in laboratory animals.

Recommended Risk-Reduction Measures

Personnel should avoid direct contact with acrylic acid. If a less toxic chemical cannot be substituted for a hazardous substance, then *engineering controls* are the most effective method of reducing exposures.

The best protection is to enclose operations and/or provide local exhaust ventilation at the site of chemical release. While not always operationally feasible, isolating operations can also reduce exposure. Using respiratory protection is less effective than the controls mentioned above, but is still advisable whenever working with or around acrylic acid. A self-contained breathing apparatus (SCBA) with full facepiece and pressure demand is the recommended respiratory protection of choice. If a full facepiece is not available, then chemical goggles should be worn to protect the eyes. For low concentrations (below 10 ppm), an acid gas cartridge respirator (equipped with a mist/vapor filter) is acceptable protection. Whenever a chemical splash hazard exists, a face shield and a rubber apron should be worn. To prevent hand and skin exposures, acid resistant gloves should be worn.

Administrative controls should also be in place to minimize the potential for human exposures. These may include written procedures or policies which specify the methods and techniques that will be practiced whenever personnel are to work with acrylic acid.

All personnel should receive *training* on the use, hazards, protective measures, emergency actions, and other precautions per 29 CFR 1910.1200 (Hazard Communication), prior to the first assignment in an area where acrylic acid is used or stored. Other methods to reduce exposure include:

☑ Where possible, enclose operations and use local exhaust ventilation at the site of chemical release. If local exhaust ventilation or enclosure is not used, respiratory protection should be mandatory.

☑ Always ensure that proper protective clothing is worn when using chemical substances and that personnel are trained in its use, care, and maintenance.

☑ Wash thoroughly immediately after exposure to acrylic acid and at the end of the work shift or before eating, drinking, or smoking.

☑ Hazard warning information should be posted in the work area. In addition, as part of an on-going education and training program, all information on the health and safety hazards of acrylic acid should be communicated to all exposed and potentially exposed workers.

RISK ASSESSMENT: ENVIRONMENT

General Assessment

The environment is at risk of exposure during transportation, storage, disposal, or destruction of acrylic acid. In almost every scenario, the threat of environmental exposure is contingent upon the proper handling of the chemical substance. Accidental spills, large or small, can result in fire, explosion, and possible contamination of the surrounding environmental mediums (water, soil, and, to a lesser degree, air).

Acrylic acid is considered a combustible liquid. It is also corrosive to many materials. It is incompatible with oxidizing agents and contact can cause fire or explosion. These characteristics require special consideration during any emergency situation involving a leak or spill of acrylic acid.

Should acrylic acid ever come into contact with incompatible substances such as oxidizers, amines, alkalis, ammonium hydroxide, chlorosulfonic acid, oleum, ethylene diamine, or ethyleneimine either during use, transportation, or storage, the formation of highly toxic and/or highly explosive commodities is extremely possible.

The proper disposal/destruction method for acrylic acid waste is to burn it in a chemical incinerator equipped with an afterburner and air scrubber.

Acrylic acid can enter the environment mainly industrial discharges or spills.

Acute Ecological Effects

Acute (short-term) toxic effects may include the death of animals, birds, or fish and death or low growth rate in plants. Acute effects are seen 2 to 4 days after animals or plants are exposed to acrylic acid.

This chemical has slight acute toxicity to aquatic life and high toxicity to birds. Insufficient data are available on the short-term effects of acrylic acid exposure to plants or land animals.

Chronic Ecological Effects

Chronic (long-term) toxic effects may include shortened life span, reproductive problems, lower fertility, and changes in appearance or behavior in exposed animals. These effects can be seen long after first exposure(s) to toxic chemicals. Acrylic acid has slight chronic toxicity in aquatic life. Insufficient data are available on the long-term effects of acrylic acid to plants, birds, or land animals.

Water Solubility

Acrylic acid is highly soluble in water. Concentrations of 1000 milligrams and more will mix with a liter of water.

Persistence in the Environment

Acrylic acid is slightly persistent in the aquatic environment, with a half-life of between 2 to 20 days. The half-life of a pollutant is the amount of time it takes for one half of the chemical to be degraded.

Approximately 90% of acrylic acid will eventually end up in water; the remainder will end up in air.

Bioaccumulation in Aquatic Organisms

Some substances increase in concentration, or *bioaccumulate,* in living organisms as they breathe contaminated air, drink contaminated water, or eat contaminated food. These chemicals can become concentrated in the tissues and internal organs of animals as well as humans.

The concentration of acrylic acid found in fish tissues is expected to be about the same as the average concentration of acrylic acid in the water from which the fish was taken.

Recommended Risk-Reduction Measures

Proper training of all transporters will reduce the likelihood of a mishap or accident resulting in a leak or spill to the environment. The correct use of DOT labeling (per the MSDS) on all containers, trucks, and rail cars will enable emergency responders to react properly and quickly to any disaster thereby reducing the potential risk to the environment and to personnel.

Storage of acrylic acid should be segregated from incompatible chemicals to minimize the risk of cross-contamination or contact. Buildings designated for storage should be equipped with appropriate fire protection systems (alarms, sprinklers, emergency lighting, portable extinguishers).

If a spill or leak to the environment has occurred, fire department, emergency response and/or hazardous materials spill personnel should be notified immediately. Clean-up should be attempted only by those trained in proper spill containment procedures. Contaminated soils should be removed for incineration and replaced with clean soil. If acrylic acid should contact the water table, aquifer, or navigable waterway, time is of the essence. It is highly miscible in water and, therefore, total containment and remediation may not be entirely possible. When such spills occur, the local and/or state emergency response authorities must be notified. A comprehensive emergency response or disaster preparedness/recovery plan should be in place prior to any operations involving the use, transportation, storage, or disposal of acrylic acid. If acrylic acid is spilled or leaked, the following specific steps are recommended:

- ☑ Restrict persons not wearing protective clothing from area of spill or leak until clean-up is complete.
- ☑ Remove all ignition sources.
- ☑ Absorb liquids in vermiculite, dry sand, earth, or similar material and deposit in sealed containers.
- ☑ It may be necessary to dispose of acrylic acid as a hazardous waste. The responsible state agency or the regional office of the federal Environmental Protection Agency (EPA) should be contacted for specific recommendations.

RISK ASSESSMENT: BUSINESS

General Assessment

Accidents or mishaps involving acrylic acid can present a significant threat to business operations. The loss or damage of equipment or facilities can significantly effect fiscal viability. Lawsuits that may result from personnel injury/death, public exposures, and/or environmental contamination will also require a serious expenditure of resources. Media attention surrounding an injury, death, or environmental damage can also result in a loss of profits and loss of current as well as future business.

Recommended Risk-Reduction Measures

Company attorneys, safety and health professionals, and environmental specialists should be involved in the development of any procedures dealing with safety or emergency response. A company official should be pre-designated as a public relations officer with specific training in dealing with the press. Corporate plans and policies should be developed, approved, and implemented long before any need for such arises.

REFERENCES

Ahlbom, A. 1993. *Biostatistics for Epidemiologists.* New York: CRC Press/Lewis Publishers.

American Conference of Governmental Industrial Hygienists. 1988. *Documentation of the Threshold Limit Values and Biological Exposure Indices,* 5th Edition (with updates). Cincinnati: ACGIH

Asante-Duah, D. K. 1993. *Hazardous Waste Risk Assessment.* New York: CRC Press/Lewis Publishers.

Bloom, A. D. and F. J. DeSerres. 1995. *Ecotoxicity and Human Health.* New York: CRC Press/Lewis Publishers.

Brusick, D. J. 1994. *Methods for Genetic Risk Assessment.* New York: CRC Press/Lewis Publishers.

Bueche, J. R. 1972. *Principles of Physics,* 2nd Edition. New York: McGraw-Hill.

Burrell, R., D. K. Flaherty, and Leonard J. Sauers. 1992. *Toxicology of the Immune System.* New York: Van Nostrand Reinhold.

Calabrese, E. J., E. M. Kenyon. *Air Toxics and Risk Assessment.* New York: CRC Press/Lewis Publishers.

Calabrese, E. J. 1994. *Biological Effects of Low Level Exposures.* New York: CRC Press/Lewis Publishers.

Cockerman, L. G., B. S. Shane. 1994. *Basic Environmental Toxicology.* New York: CRC Press/Lewis Publishers.

Cothern, C. R., N. P. Ross. 1994. *Environmental Statistics, Assessment, and Forecasting.* New York: CRC Press/Lewis Publishers.

Cothern, R. R. 1993. *Comparative Environmental Risk Assessment.* New York: CRC Press/Lewis Publishers.

Cothern, C. R. 1995. *Handbook for Environmental Risk Decisionmaking.* New York: CRC Press/Lewis Publishers.

Danse, I. R. 1991. *Common Sense Toxics in the Workplace.* New York: Van Nostrand Reinhold.

Gardner, R. W. 1994. *Chemical Intolerance.* New York: CRC Press/Lewis Publishers.

Hallenbeck, W. H. 1993. *Quantitative Risk Assessment for Environmental and Occupational Health,* 2nd Edition. New York: CRC Press/Lewis Publishers.

Hosty, J. W., P. Foster. 1990. *A Practical Guide to Chemical Spill Response.* New York: Van Nostrand Reinhold.

Howard, P. H. 1995. *Handbook of Environmental Fate and Exposures Data for Organic Chemicals,* Vol. I-IV. New York: CRC Press/Lewis Publishers.

Howard, P. H., et. al. 1991. *Handbook of Environmental Degradation Rates.* New York: CRC Press/Lewis Publishers.

James, A., M. Lord. 1992. *Index of Chemical and Physical Data.* New York: Van Nostrand Reinhold.

Jones, F. E. 1994. *Toxic Organic Vapors in the Workplace.* New York: CRC Press/Lewis Publishers.

Keith, L. H. and M. M. Walker. 1995. *Handbook of Air Toxics.* New York: CRC Press/Lewis Publishers.

Landis, W. G., M. H. Yu. 1995. *Introduction to Environmental Toxicology.* New York: CRC Press/Lewis Publishers.

Langerman, N. 1994 *Precautionary Labels for Chemical Containers.* New York: CRC Press/Lewis Publishers.

Lauwerys, R. R., P. Hoet. 1993. *Industrial Chemical Exposure.* New York: CRC Press/Lewis Publishers.

Lewis, R. J., Sr. 1993. *Hawley's Condensed Chemical Dictionary,* 12th Edition. New York: Van Nostrand Reinhold.

Lewis, R. J., Sr. 1992. *Sax's Dangerous Properties of Industrial Materials,* Volumes 1, 2, and 3, 8th Edition. New York: Van Nostrand Reinhold.

Lide, D. R. 1995. *Handbook of Chemistry and Physics,* 76th Edition. New York: CRC Press/Lewis Publishers.

Manahan, S. E. 1994. *Environmental Chemistry,* 6th Edition. New York: CRC Press/Lewis Publishers.

Manahan, S. E. 1992. *Toxicological Chemistry,* 2nd Edition. New York: CRC Press/Lewis Publishers.

Mellström, G. A., J. E. Wahlberg, and H. I. Maibach. 1994. *Protective Gloves for Occupational Use.* New York: CRC Press/Lewis Publishers.

Meyer, E. 1990. *Chemistry of Hazardous Materials.* Englewood Cliffs, NJ: Prentice Hall.

Mycock, J. C., J. D. McKenna, and L. Theodore. 1995. *Handbook of Air Pollution Control Engineering and Technology.* New York: CRC Press/Lewis Publishers.

Nelson, G. O. 1992. *Gas Mixtures.* New York: CRC Press/Lewis Publishers.

Ott, W. R. 1995. *Environmental Statistics and Data Analysis.* New York: CRC Press/Lewis Publishers.

Patnaik, P. 1992. *A Comprehensive Guide to the Hazardous Properties of Chemical Substances.* New York: Van Nostrand Reinhold.

Philp, R. B. 1995. *Environmental Hazards and Human Health.* New York: CRC Press/Lewis Publishers.

Proctor, N. H., J. P. Hughes. 1991. *Chemical Hazards of the Workplace.* New York: Van Nostrand Reinhold.

Quigley, D. R. 1994. *Handbook of Emergency Chemical Management.* New York: CRC Press/Lewis Publishers.

Rea, W. J. 1995. *Chemical Sensitivity,* Vol. I-III. New York: CRC Press/Lewis Publishers.

Richardson, M. 1995. *Dictionary of Substance and Their Effects,* Vol. I-VI. New York: CRC Press/Lewis Publishers.

Rylander, R., R. R. Jacobs. 1994. *Organic Dusts.* New York: CRC Press/Lewis Publishers.

Sacarello, H. L. A. 1994. *The Comprehensive Handbook of Hazardous Materials.* New York: CRC Press/Lewis Publishers.

Sheftel, V. O. 1995. *Handbook of Toxic Properties of Monomers and Additives.* New York: CRC Press/Lewis Publishers.

Sherman, J. D. 1988. *Chemical Exposure and Disease.* New York: Van Nostrand Reinhold.

Talbott, E., G. Craun. 1995. *An Introduction to Environmental Epidemiology.* New York: CRC Press/Lewis Publishers.

The Merck Index, 10th edition. 1983. Rahway, NJ: Merck & Company.

U.S. Department of Health and Human Services, National Institute for Occupational Safety and Health. 1980. *Acrylic Acid. Registry of Toxic Effects of Chemical Substances,* Vol. 1. Washington, D.C.: U.S. Government Printing Office.

U.S. Department of Health and Human Services, National Institute for Occupational Safety and Health. 1994. *NIOSH Pocket Guide to Chemical Hazards.* Washington, D.C.: U.S. Government Printing Office.

Witorsch, P. and S. V. Spagnolo. 1994. *Air Pollution and Lung Disease in Adults.* New York: CRC Press/Lewis Publishers.

MATERIAL SAFETY DATA SHEET

CHEMICAL NAME	
ACRYLONITRILE	**CRC PRESS/LEWIS PUBLISHERS DISCLAIMER:** THE INFORMATION AND RECOMMENDATIONS PRESENTED HEREIN ARE BASED ON SOURCES BELIEVED TO BE RELIABLE. CRC MAKES NO REPRESENTATION ON ITS COMPLETENESS OR ACCURACY. IT IS THE RESPONSIBILITY OF THE USER TO DETERMINE THE CHEMICAL'S SUITABILITY FOR ITS INTENDED USE, THE CHEMICAL'S SAFE USE, AND THE CHEMICAL'S PROPER DISPOSAL. NO REPRESENTATIONS AND/OR WARRANTIES, EITHER EXPRESSED OR IMPLIED, OF THE MERCHANTABILITY OR FITNESS FOR A PARTICULAR PURPOSE, OR OF ANY OTHER NATURE, ARE MADE WITH RESPECT TO THE INFORMATION PROVIDED IN THIS MSDS OR TO THE CHEMICAL TO WHICH INFORMATION MAY REFER. CRC NEITHER ASSUMES NOR AUTHORIZES ANY OTHER PERSON TO ASSUME FOR IT, ANY OTHER ADDITIONAL RESPONSIBILITY OR LIABILITY FOR THE USE OF, OR RELIANCE UPON, THIS INFORMATION.

HAZARD WARNING INFORMATION

HEALTH	FIRE	REACTIVE	OTHER	DEGREE OF HAZARD	COLOR CODING	OTHER CODES
4	3	2		0 = Minimum Hazard 1 = Slight Hazard 2 = Moderate Hazard 3 = Serious Hazard 4 = Severe Hazard	HEALTH = BLUE FIRE = RED REACTIVITY = YELLOW OTHER = WHITE	OX = Oxidizer ACID = Acid ALK = Alkali COR = Corrosive W = Use No Water

SECTION I - GENERAL INFORMATION

Characterization	**Organic Cyanides (Nitriles)**	RCRA Number	**U009**
		EPA Class	**Toxic Waste**
DOT Proper Shipping Name	**Acrylonitrile**	Chemical Abstract Service (CAS) Number	**107-13-1**
DOT Hazard Class and Label Requirements	**Flammable Liquid and Poison**	DOT Emergency Guide Code	**30**
DOT Identification Number	**UN 1093**	Chemical Formula	**C_2H_3CN**

Synonyms
Vinyl cyanide; 2-propenenitrile; cyanoethylene; Acrylon; Acritet; Ventox; carbacryl.

SECTION II - HAZARDOUS INGREDIENTS AND IDENTITY INFORMATION

Hazardous Components (specific identity)	OSHA Exposure Criteria	NIOSH Exposure Criteria	Immediately Dangerous to Life and Health (IDLH)	ACGIH Exposure Criteria
Acrylonitrile (derivation: From propylene oxygen and ammonia with ether bismuth phosphomolybate or uranium-based compounds as catalysts; addition of hydrogen cyanide to acetelene with cuprous chloride catalyst). **1 ppm = 2.21 mg/m³**	**PEL:** **2 ppm** **4.42 mg/m³** **STEL (skin):** **10 ppm** **22 mg/m³**	**REL:** **1 ppm** **2.21 mg/m³** **Confirmed Human Carcinogen**	**85 ppm**	**TLV:** **2 ppm (skin)** **4.42 mg/m³** **STEL:** **Not Established**

SECTION III - PHYSICAL AND CHEMICAL CHARACTERISTICS

Boiling Point	**171°F (77°C)**	Specific Gravity (H_2O = 1)	**0.81**
Vapor Pressure (mm Hg)	**83 at 69°F (20°C)**	Molecular Weight	**53.10**
Vapor Density (Air = 1)	**1.83**	Freezing Point	**-116°F (-83°C)**

Solubility
Partially soluble (7%) in water (readily miscible with most organic solvents).

Appearance and Odor
Clear, colorless or slightly yellowish, mobile liquid with a faint, mild odor. Odor threshold = 17 ppm.

SECTION IV - FIRE AND EXPLOSION HAZARD DATA

Flash Point (method used)	**30°F (-1°C) closed cup**	Explosive Limits in Air % by Volume	**LEL: 3.0% UEL: 17%**
NFPA Classification	**Class 1B Flammable Liquid**	Autoignition Temperature	**898°F (481°C)**

Extinguishing Media
Carbon dioxide, dry chemical, alcohol foam.

Special Fire Fighting Procedures
Wear full protective clothing and self-contained breathing apparatus (SCBA). Heat will build pressure and may rupture closed storage containers. Keep fire-exposed containers cool with water spray. Vapors are heavier than air, they may travel a distance to cause fire or explosion due to flashback phenomenon.

Unusual Fire and Explosion Hazards
Dilute solutions are also flammable. Vapors form explosive mixtures in air. Violent rupture of glass containers can occur due to polymerization. Poisonous gases are produced in fire.

SECTION V - REACTIVITY DATA

Stability		Conditions to Avoid: Dangerous fire hazard when exposed to heat and flame. Avoid contact with incompatible materials. When mixed with bromine, acrylonitrile can explode.
Stable	Unstable: X	Incompatibility *(materials to avoid)*: Reactions can occur with strong oxidizers, strong bases, strong acids, ammonia, amines, and copper and its alloys can be violent. Lithium reacts exothermically.
Hazardous Polymerization		Conditions to Avoid: Contact with mineral acids (H_2SO_4, HCL, H_3PO_4) will form paraldehyde, which is toxic to humans and animals.
May Occur: X	Will Not Occur	Hazardous Decomposition or By-products: Polymerizes violently in presence of concentrated alkali or at elevated temperature. It undergoes photochemical polymerization, liberating heat. Violent rupture of glass containers can occur.

SECTION VI - HEALTH HAZARD DATA

Primary Route(s) of Entry:	Inhalation?	Absorption (skin)?	Ingestion?
	X	X	

Health Hazards

INHALATION: Asphyxia, headache, chest pains, shortness of breath, lightheadedness, coughing, and the peeling of skin from lips and hands. High concentrations can result in symptoms such as somnolence (sleepiness), diarrhea, nausea, and vomiting.

ABSORPTION: Headache, lightheadedness, sneezing, weakness, nausea, and vomiting. Severe intoxication can result in loss of consciousness, convulsions, respiratory arrest, and death. Acrylonitrile is a skin irritant. It may cause blistering after a few hours.

Carcinogenicity	NTP Listed?	IARC Cancer Review Group?	OSHA Regulated?	Target Organs?
Confirmed Human Confirmed Animal	5th Annual Report	Group 2A	29 CFR 1910.1045	Respiratory tract; liver; skin; eyes; CNS; kidneys.

Medical Conditions Generally Aggravated by Exposure

Respiratory impairments, (asthma); Renal function abnormalities; Skin conditions (dermatitis).

Emergency and First-aid Procedures

Eye contact: Flush immediately with water for 30 minutes (minimum), seek medical attention. Skin contact: Remove all contaminated clothing. Immediately wash area with large amounts of soap and water. Seek medical assistance. For inhalation: Remove the person from exposure. Provide respiratory assistance and CPR. Transfer to medical facility. Medical observation for 24 to 48 hours after inhalation exposure (pulmonary edema). If swallowed, seek medical attention immediately.

SECTION VII - PRECAUTIONS FOR SAFE HANDLING AND USE

Steps to be Taken in Case Material is Released or Spilled

Contain spills using absorbent material (vermiculite, or other). Ventilate area of spill. Have water source available in case of fire. Place materials in DOT-approved containers. Restrict those not involved in clean-up from entering area. Notify appropriate authorities, as required.

Preferred Waste Disposal Method

Chemical incinerator equipped with scrubber and afterburner.

Precautions to be Taken in Handling and Storage

Acrylonitrile is highly volatile. Do not store with incompatible chemicals. Store in tightly closed containers in a cool, dark, well-ventilated area. Keep fire and flame away. Use non-sparking tools to open or close containers. Ground and bond all metal containers during transfer operations.

Other Precautions and Warnings

If exposed to heat or light, uninhibited acrylonitrile can react violently by itself, leading to an explosion of closed containers.

SECTION VIII - CONTROL MEASURES AND PERSONAL PROTECTIVE EQUIPMENT

Respiratory Protection *(specify type)*

At any exposure level: Self-contained breathing apparatus (SCBA) with full facepiece and pressure demand or other positive pressure mode.

Ventilation

Local exhaust or general mechanical systems recommended.

Protective Gloves	Eye Protection	Other Protective Clothing
Neoprene Rubber	Chemical Goggles or Face Mask	Rubber Apron

Work/Hygiene Practices

Always wash hands thoroughly after using chemical; never bring food, drink, or smoking materials into vicinity of chemicals.

ACRYLONITRILE

C_2H_3CN CAS: 107-13-1

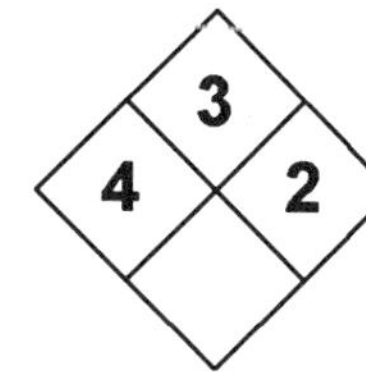

IDENTIFICATION AND TYPICAL USES

Acrylonitrile is a clear, colorless or slightly yellowish mobile liquid with a faint, mild odor. It is used in the production of acrylic fibers, resins, nitrile rubber, and surface coatings; as an intermediate in the production of pharmaceuticals and dyes; as a polymer modifier; and as a pesticide fumigant for stored grains.

RISK ASSESSMENT: HEALTH

General Assessment

Acrylonitrile is carcinogenic in experimental animals and is considered to be a probable human carcinogen. Acrylonitrile is a mutagen. All contact with this chemical should be reduced to the lowest possible level. Human exposure to acrylonitrile can occur through two primary routes of entry into the body: *inhalation,* and *absorption.* Each present a moderate level of toxicity. Acrylonitrile is readily absorbed through the respiratory tract, the gastrointestinal tract, and through unbroken skin. Absorption through the skin can result in toxic symptoms such as headache, lightheadedness, sneezing, weakness, nausea, and vomiting. In humans, many symptoms are non-specific but appear to be related to the central nervous system (CNS), the respiratory tract, gastrointestinal tract, and the skin.

Inhalation of this compound can cause asphyxia, headache, mild to severe chest pains, shortness of breath, lightheadedness, and coughing. Severe intoxication can cause loss of consciousness, respiratory arrest, convulsions, and death. There is also evidence to show that acrylonitrile can cause congestion in all types of organs and damage to the kidneys, CNS, lungs, and liver.

Acute Health Effects

The following acute (short-term) health effects may occur immediately or shortly after exposure to acrylonitrile:

Skin: Severe irritation and blistering a few hours after exposure.

Eye: Burns leading to permanent damage.

Lung: Irritation causing coughing and/or shortness of breath. Higher exposures can cause a build-up of fluid in the lungs (pulmonary edema), which is a medical emergency.

CNS: Inhalation of high concentrations of the vapor may cause sleepiness, dizziness, loss of consciousness, or death. Lower exposures can cause weakness, headache, confusion, nausea, vomiting, and can also lead to death.

Chronic Health Effects

The following chronic (long-term) health effects may occur at some time after exposure to acrylonitrile and can last for months or even years:

Cancer Hazards: Acrylonitrile causes mutations (genetic changes). There is strong evidence that acrylonitrile causes cancer (of the lung and large intestines) in humans. It has been shown to cause brain and stomach cancer in animals.

Reproductive Hazard: There is no evidence that acrylonitrile adversely affects reproduction.

Other Chronic Effects: Exposure may cause the thyroid gland to enlarge and interfere with normal thyroid function. Repeated exposure can irritate the nasal membranes causing discharge, nose bleeds, and sores inside the nose. Very irritating substances, such as acrylonitrile, may effect the lungs. It is not known whether chronic exposure to acrylonitrile will damage the lungs since such exposures usually prove fatal.

Recommended Risk-Reduction Measures

Personnel should avoid direct contact with acrylonitrile. Unless a less toxic chemical can be substituted for a hazardous substance, *engineering controls* are the most effective method of reducing exposures. The best protection is to enclose operations and/or provide local exhaust ventilation at the site of chemical release. While not always operationally feasible, isolating operations can also reduce exposure.

Using respiratory protection is less effective than the controls mentioned above, but is still advisable whenever working with or around acrylonitrile. A self-contained breathing apparatus (SCBA) with full facepiece and pressure demand is the recommended respiratory protection of choice. If a full facepiece is not available, then chemical goggles should be worn to protect the eyes. Whenever a chemical splash hazard exists, a face shield and a rubber apron should be worn. To prevent hand and skin exposures, neoprene rubber gloves should be worn.

Administrative controls should also be in place to minimize the potential for human exposures. These may include written procedures or policies which specify the methods and techniques that will be practiced whenever personnel are to work with acrylonitrile.

All personnel should receive *training* on the use, hazards, protective measures, emergency actions, and other precautions per 29 CFR 1910.1200 (Hazard Communication), prior to the first assignment in an area where acrylonitrile is used or stored.

Medical testing of personnel should be performed before the first exposure and at least annually thereafter. As per OSHA requirements, these tests must be provided by the employer and include:

- ☑ Complete physical examination.
- ☑ Chest X-ray.
- ☑ Stool test for blood (employees over 40).

In addition, for those with frequent or potentially high exposure (half the TLV or greater, or significant skin contact), the following are recommended before beginning work and at regular intervals thereafter:

- ☑ Urine thiocyanate levels.
- ☑ Lung function (pulmonary function) tests.

If symptoms develop or overexposure is suspected, the following additional tests may be useful:

- ☑ Blood cyanide level.
- ☑ Thyroid function test.
- ☑ Consider chest X-ray after acute overexposure.

Any evaluation should include a careful medical history of past and present symptoms with an examination. However, medical tests that evaluate existing damage are not a substitute for controlling exposure. Antidotes following exposure to cyanides include:

- ☑ Administration of 4-dimethylaminophenol plus sodium thiosulfate shows some protection only after oral intake (ineffective against inhalation).
- ☑ Cystein and *N*-acetyl cystein are somewhat effective in combating cyanide poisoning.

Other methods to reduce exposure include:

- ☑ Use amyl nitrate capsules as antidote if symptoms develop.
- ☑ Employees should be trained regularly in emergency measures for cyanide poisoning and in CPR.
- ☑ A cyanide antidote kit must be rapidly available and ingredients replaced every 1-2 years to ensure freshness.
- ☑ Medical personnel trained in treatment of cyanide poisoning should be quickly available
- ☑ Where possible, enclose operations and use local exhaust ventilation at the site of chemical release. If local exhaust ventilation or enclosure is not used, respiratory protection should be mandatory.
- ☑ Always ensure that proper protective clothing is worn when using chemical substances.
- ☑ Wash thoroughly immediately after exposure to acrylonitrile and at the end of the work shift or before eating, drinking, or smoking.
- ☑ Hazard warning information should be posted in the work area. In addition, as part of an on-going education and training program, all information on the health and safety hazards of acrylonitrile should be communicated to all potentially exposed workers.

RISK ASSESSMENT: ENVIRONMENT

General Assessment

The environment is at risk of exposure during transportation, storage, disposal, or destruction of acrylonitrile. In almost every scenario, the threat of environmental exposure is contingent upon the proper handling of the chemical substance. Accidental spills, large or small, can result in fire, explosion, and possible contamination of the surrounding environmental mediums (water, soil, and air).

Acrylonitrile is considered a highly flammable liquid. Because of its low flash point and high boiling point, it is classified as a Class 1B flammable liquid (per OSHA 29 CFR 1910.106). It is also a dangerous fire and explosion hazard, highly reactive, and it polymerizes easily. These characteristics require special consideration during any emergency situation involving a leak or spill of acrylonitrile. Should acrylonitrile ever come into contact with incompatible substances such as oxidizers, acids, bases, bromine, and amines either during use, transportation, or storage, the formation of highly toxic and/or highly explosive commodities is extremely possible. The proper disposal/destruction method for acrylonitrile is to burn it in a chemical incinerator equipped with an afterburner and air scrubber. Acrylonitrile can enter the environment through industrial effluents, and through spills.

Acute Ecological Effects

Acute (short-term) toxic effects may include the death of animals, birds, or fish and death or low growth rate in plants. Acute effects are seen 2 to 4 days after animals or plants are exposed to acrylonitrile.

This chemical has high acute toxicity to aquatic life. No data are available on the short-term effects of acrylonitrile to plants, animals, or birds.

Chronic Ecological Effects

Chronic toxic effects may include shortened life span, reproductive problems, lower fertility, and changes in appearance or behavior in exposed animals. These effects can be seen long after first exposure(s) to toxic chemicals.

Acrylonitrile has high chronic toxicity in aquatic life. No data are available on the long-term effects of acrylonitrile to plants, birds, or land animals.

Water Solubility

Acrylonitrile is highly soluble in water. Concentrations of 1000 milligrams and more will mix with a liter of water.

Persistence in the Environment

Acrylonitrile is slightly persistent in water, with a half-life of between 2 to 20 days. The half-life of a pollutant is the amount of time it takes for one half of the chemical to be degraded. Approximately 63% of acrylonitrile will eventually end up in air; the rest will end up in water.

Bioaccumulation in Aquatic Organisms

Some substances increase in concentration, or *bioaccumulate,* in living organisms as they breathe contaminated air, drink contaminated water, or eat contaminated food. These chemicals can become concentrated in the tissues and internal organs of animals as well as humans.

The concentration of acrylonitrile found in fish tissues is expected to be somewhat higher than the average concentration of acrylonitrile in the water from which the fish was taken.

Recommended Risk-Reduction Measures

Proper training of all transporters will reduce the likelihood of a mishap or accident resulting in a leak or spill to the environment. The correct use of DOT labeling (per the MSDS) on all containers, trucks, and rail cars will enable emergency responders to react properly and quickly to any disaster thereby reducing the potential risk to the environment and to personnel.

Storage of acrylonitrile should be segregated from incompatible chemicals to minimize the risk of cross-contamination or contact. Buildings designated for storage should be equipped with appropriate fire protection systems (alarms, sprinklers, emergency lighting, portable extinguishers).

If a spill or leak to the environment has occurred, fire department, emergency response and/or hazardous materials spill personnel should be notified immediately. Clean-up should be attempted only by those trained in proper spill containment procedures. Contaminated soils should be removed for incineration and replaced with clean soil. If acrylonitrile should contact the water table, aquifer, or navigable waterway, time is of the essence. It is readily miscible in water and, therefore, total containment and remediation may not be entirely possible. When such spills occur, the local and/or state emergency response authorities must be notified. A comprehensive emergency response or disaster preparedness/recovery plan should be in place prior to any operations involving the use, transportation, storage, or disposal of acrylonitrile. If acrylonitrile is spilled or leaked, the following specific steps are recommended:

- ☑ Restrict persons not wearing protective clothing from area of spill or leak until cleanup is complete and area can be opened for normal work.
- ☑ Remove all ignition sources.
- ☑ Ventilate area of spill or leak.

☑ Absorb liquids in vermiculite, dry sand, earth, or a similar material and deposit in sealed containers.

☑ If applicable, stop flow of leaking liquid or gas. If leak source is a cylinder and the leak can not be stopped in place, remove leaking cylinder to a safe place in the open air, and repair or allow cylinder to empty.

☑ Keep acrylonitrile out of a confined space, such as a sewer, because of the possibility of explosion (unless the sewer is designed to prevent the build-up of explosive concentrations).

☑ It may be necessary to dispose of acrylonitrile as a hazardous waste. The responsible state agency or the regional office of the federal Environmental Protection Agency (EPA) should be contacted for specific recommendations.

RISK ASSESSMENT: BUSINESS

General Assessment

Accidents or mishaps involving acrylonitrile can present a significant threat to business operations. The loss or damage of equipment or facilities can significantly effect fiscal viability. Lawsuits that may result from personnel injury/death, public exposures, and/or environmental contamination will also require a serious expenditure of resources. Media attention surrounding an injury, death, or environmental damage can also result in a loss of profits and loss of current as well as future business.

Recommended Risk-Reduction Measures

Company attorneys, safety and health professionals, and environmental specialists should be involved in the development of any procedures dealing with safety or emergency response. A company official should be pre-designated as a public relations officer with specific training in dealing with the press. Corporate plans and policies should be developed, approved, and implemented long before any need for such arises.

REFERENCES

American Conference of Governmental Industrial Hygienists. 1988. *Documentation of the Threshold Limit Values and Biological Exposure Indices,* 5th Edition (with updates). Cincinnati: ACGIH

Anthony, C. P., and N. J. Kolthoff. 1971. *Textbook of Anatomy and Physiology.* St. Louis: C. V. Mosby Company.

Bloom, A. D. and F. J. DeSerres. 1995. *Ecotoxicity and Human Health.* New York: CRC Press/Lewis Publishers.

Brusick, D. J. 1994. *Methods for Genetic Risk Assessment.* New York: CRC Press/Lewis Publishers.

Bueche, J. R. 1972. *Principles of Physics,* 2nd Edition. New York: McGraw-Hill.

Burrell, R., D. K. Flaherty, and L. J. Sauers. 1992. *Toxicology of the Immune System.* New York: Van Nostrand Reinhold.

Calabrese, E. J. 1994. *Biological Effects of Low Level Exposures.* New York: CRC Press/Lewis Publishers.

Cockerman, L. G., B. S. Shane. 1994. *Basic Environmental Toxicology.* New York: CRC Press/Lewis Publishers.

Cothern, C. R., N. P. Ross. 1994. *Environmental Statistics, Assessment, and Forecasting.* New York: CRC Press/Lewis Publishers.

Cothern, R. R. 1993. *Comparative Environmental Risk Assessment.* New York: CRC Press/Lewis Publishers.

Cothern, C. R. 1995. *Handbook for Environmental Risk Decisionmaking.* New York: CRC Press/Lewis Publishers.

Environmental Protection Agency. October 1983. *Health Assessment for Acrylonitrile.* Final Report: EPA 600/8-82-007F. NTIS PB84-149152. Springfield, VA: National Technical Information Service.

Gardner, R. W. 1994. *Chemical Intolerance.* New York: CRC Press/Lewis Publishers.

Hodgson, E., R. B. Mailman, and Janice E. Chambers. 1988. *Dictionary of Toxicology.* London: The MacMillian Company.

Jones, F. E. 1994. *Toxic Organic Vapors in the Workplace.* New York: CRC Press/Lewis Publishers.

Lewis, R. J., Sr. 1993. *Hawley's Condensed Chemical Dictionary,* 12th Edition. New York: Van Nostrand Reinhold.

Lewis, R. J., Sr. 1992. *Sax's Dangerous Properties of Industrial Materials,* Volumes 1, 2, and 3, 8th Edition. New York: Van Nostrand Reinhold.

Meyer, E. 1990. *Chemistry of Hazardous Materials.* Englewood Cliffs, NJ: Prentice Hall.

Patnaik, P. 1992. *A Comprehensive Guide to the Hazardous Properties of Chemical Substances.* New York: Van Nostrand Reinhold.

Proctor, N. H., and James P. Hughes. 1991. *Chemical Hazards of the Workplace.* New York: Van Nostrand Reinhold.

Quigley, D. R. 1994. *Handbook of Emergency Chemical Management.* New York: CRC Press/Lewis Publishers.

U.S. Department of Health and Human Services, National Institute for Occupational Safety and Health. 1978. *Criteria Recommended Standard - Occupational Exposure to Acrylonitrile.* Washington, D.C.: U.S. Government Printing Office.

U.S. Department of Health and Human Services, National Institute for Occupational Safety and Health. 1994. *NIOSH Pocket Guide to Chemical Hazards.* Washington, D.C.: U.S. Government Printing Office.

Willhite, C. C. Toxicological updates. Acrylonitrile. *Journal of Applied Toxicology.* 2:54-56, 1982.

MATERIAL SAFETY DATA SHEET

CHEMICAL NAME	
ALDRIN	**CRC PRESS/LEWIS PUBLISHERS DISCLAIMER:** THE INFORMATION AND RECOMMENDATIONS PRESENTED HEREIN ARE BASED ON SOURCES BELIEVED TO BE RELIABLE. CRC MAKES NO REPRESENTATION ON ITS COMPLETENESS OR ACCURACY. IT IS THE RESPONSIBILITY OF THE USER TO DETERMINE THE CHEMICAL'S SUITABILITY FOR ITS INTENDED USE, THE CHEMICAL'S SAFE USE, AND THE CHEMICAL'S PROPER DISPOSAL. NO REPRESENTATIONS AND/OR WARRANTIES, EITHER EXPRESSED OR IMPLIED, OF THE MERCHANTABILITY OR FITNESS FOR A PARTICULAR PURPOSE, OR OF ANY OTHER NATURE, ARE MADE WITH RESPECT TO THE INFORMATION PROVIDED IN THIS MSDS OR TO THE CHEMICAL TO WHICH INFORMATION MAY REFER. CRC NEITHER ASSUMES NOR AUTHORIZES ANY OTHER PERSON TO ASSUME FOR IT, ANY OTHER ADDITIONAL RESPONSIBILITY OR LIABILITY FOR THE USE OF, OR RELIANCE UPON, THIS INFORMATION.

HAZARD WARNING INFORMATION

HEALTH	FIRE	REACTIVE	OTHER	DEGREE OF HAZARD	COLOR CODING	OTHER CODES
N/A	N/A	N/A		0 = Minimum Hazard 1 = Slight Hazard 2 = Moderate Hazard 3 = Serious Hazard 4 = Severe Hazard	HEALTH = BLUE FIRE = RED REACTIVITY = YELLOW OTHER = WHITE	OX = Oxidizer ACID = Acid ALK = Alkali COR = Corrosive W = Use No Water

SECTION I - GENERAL INFORMATION

Characterization	**Pesticide; Insecticide**	RCRA Number	**P004**
		EPA Class	**Acute Hazardous Waste**
DOT Proper Shipping Name	**Aldrin**	Chemical Abstract Service (CAS) Number	**309-00-2**
DOT Hazard Class and Label Requirements	**Poison B; ORM-A**	DOT Emergency Guide Code	**55**
DOT Identification Number	**UN 2761; NA 2762**	Chemical Formula	$C_{12}H_8Cl_6$

Synonyms

1,4:5,8 Dimethanonaphthalene; 1,2,3,4,10,10-Hexachloro-1,4,4A,5,8,8A-Hexahydro; Octalene; Aldrex; Aldrite; exodimethanonaphthalene

SECTION II - HAZARDOUS INGREDIENTS AND IDENTITY INFORMATION

Hazardous Components (specific identity)	OSHA Exposure Criteria	NIOSH Exposure Criteria	Immediately Dangerous to Life and Health (IDLH)	ACGIH Exposure Criteria
Aldrin	**PEL:** 0.25 mg/m³ (skin)	**REL:** 0.15 mg/m³ (skin) Suspected Human Carcinogen	**25 mg/m³**	**TLV:** 0.25 mg/m³ (skin)

SECTION III - PHYSICAL AND CHEMICAL CHARACTERISTICS

Boiling Point	**Decomposes**	Specific Gravity (H_2O = 1)	**1.60**
Vapor Pressure (mm Hg)	**7.5×10^{-5} at 69°F (20°C)**	Molecular Weight	**364.9**
Vapor Density (Air = 1)	**N/A**	Freezing Point	**N/A (melts at 219°F)**

Solubility

Insoluble in water (0.003%); readily miscible with most organic solvents, aromatics, esters, ketones, paraffins, and halogenated solvents.

Appearance and Odor

Colorless to dark tan or brown solid crystals with a mild chemical odor.

SECTION IV - FIRE AND EXPLOSION HAZARD DATA

Flash Point (method used)	**Not Applicable**	Explosive Limits in Air % by Volume	**LEL: N/A UEL: N/A**
NFPA Classification	**Noncombustible Solid**	Autoignition Temperature	**Not Applicable**

Extinguishing Media

A noncombustible solid. Extinguish fire using an agent suitable for the type of surrounding fire.

Special Fire Fighting Procedures

Since poisonous gases are produced in fire, wear full protective clothing and self-contained breathing apparatus (SCBA).

Unusual Fire and Explosion Hazards

Containers may explode in fire due to build-up of excessive internal pressures. Exercise extreme caution when attempting to move or relocate containers that may have been exposed to high heat. Aldrin is an acute hazardous waste; by-products of fire fighting operations may have to be handled as acutely hazardous as well.

SECTION V - REACTIVITY DATA

Stability		Conditions to Avoid
		Avoid contact with incompatible materials. Although not a flammable material itself, commercial solutions of Aldrin may contain flammable or combustible liquids as additives.
Stable	Unstable	Incompatibility *(materials to avoid)*
X		Concentrated mineral acids (sulfuric, hydrochloric, nitric); active metals; acid catalysts; acid oxidizing agents (chlorine); phenols, and reactive metals (sodium, potassium, magnesium).
Hazardous Polymerization		Conditions to Avoid
		Contact with mineral acids (H_2SO_4, HCL, H_3PO_4) will form paraldehyde, which is toxic to humans and animals.
May Occur	Will Not Occur	Hazardous Decomposition or By-products
	X	Poisonous gases are produced when aldrin is involved in fire.

SECTION VI - HEALTH HAZARD DATA

Primary Route(s) of Entry:	Inhalation?	Absorption (skin)?	Ingestion?
	X	X	

Health Hazards

INHALATION: Entry route by inhalation possible when working in dusty or dust-generating conditions. Symptoms can be as those described below or absorption and ingestion.

ABSORPTION: Headache, dizziness, nausea, vomiting, tremor, ataxia, convulsions, CNS depression, respiratory failure, and may cause renal damage. May cause skin allergy.

INGESTION: May cause liver and lung damage, including cancer (data inconclusive in humans); tremors; nausea; vomiting. An experimental teratogen.

Carcinogenicity	NTP Listed?	IARC Cancer Review Group?	OSHA Regulated?	Target Organs?
Suspected Human Confirmed Animal	5th Annual Report	Group 3	29 CFR 1910.1000 Table Z-1 (skin)	CNS, liver, kidneys, skin, possible thyroid, adrenal.

Medical Conditions Generally Aggravated by Exposure

Renal and/or liver function problems; Skin conditions (dermatitis).

Emergency and First-aid Procedures

<u>Eye contact</u>: Flush immediately with water for 15 minutes (minimum), seek medical attention. <u>Skin contact</u>: Remove all contaminated clothing. Immediately wash area with large amounts of soap and water. Seek medical assistance. For <u>inhalation</u>: Remove the person from exposure. Provide respiratory assistance and CPR. Transfer to medical facility. Medical observation is recommended for 24 to 48 hours after inhalation exposure (pulmonary edema). If <u>swallowed</u>, seek medical attention immediately. Never attempt to give an unconscious or convulsing person anything by mouth.

SECTION VII - PRECAUTIONS FOR SAFE HANDLING AND USE

Steps to be Taken in Case Material is Released or Spilled

Collect powdered material in the most convenient (safe) way possible and deposit in DOT-approved drum. Restrict those not involved in cleanup from entering area. Notify appropriate authorities, as required by SARA III, if applicable.

Preferred Waste Disposal Method

Chemical incinerator equipped with scrubber and afterburner.

Precautions to be Taken in Handling and Storage

Do not store aldrin in the presence of incompatible chemicals or materials. Store in tightly closed containers in a cool, dark, well-ventilated area. Personnel working with aldrin should be trained on its proper handing and storage *prior* to being assigned to such responsibilities.

Other Precautions and Warnings

Bulk storage of aldrin is not recommended.

SECTION VIII - CONTROL MEASURES AND PERSONAL PROTECTIVE EQUIPMENT

Respiratory Protection *(specify type)*

At any exposure level: Use a MSHA/NIOSH-approved air-supplied respirator with full facepiece set in positive pressure mode or use a self-contained breathing apparatus (SCBA) with full facepiece and pressure demand or other positive pressure mode.

Ventilation

Local exhaust or general mechanical systems recommended.

Protective Gloves	Eye Protection	Other Protective Clothing
Rubber or Latex	Chemical/Dust Goggles or Face Mask	Apron

Work/Hygiene Practices

Always wash hands thoroughly after using chemical; never bring food, drink, or smoking materials into vicinity of chemicals.

ALDRIN

$C_{12}H_8Cl_6$ CAS: 309-00-2

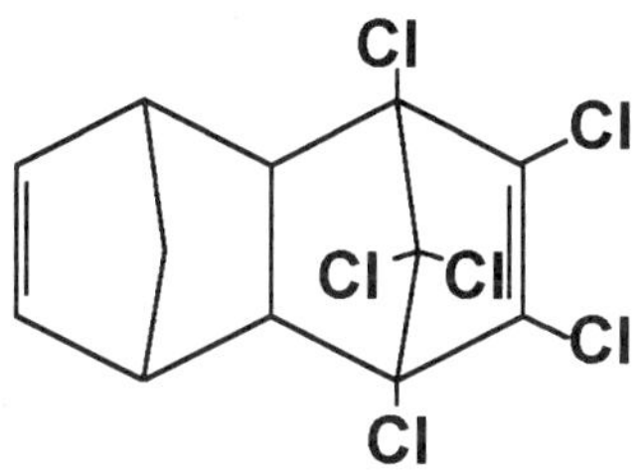

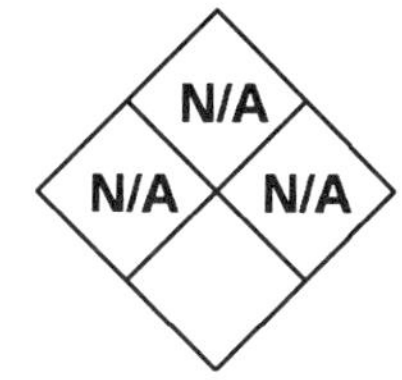

IDENTIFICATION AND TYPICAL USES

Aldrin is a colorless or sometimes tan to dark brown solid crystal. It has had extensive use as an insecticide. However, its use has been sharply curtailed partially on the basis of its persistence in the environment and bioconcentration in the food chain. It is no longer available for use as an insecticide in the United States and, therefore, much of the data on aldrin are dated.

RISK ASSESSMENT: HEALTH

General Assessment

Aldrin is highly toxic to humans and animals. It is carcinogenic in experimental animals and may cause cancer in humans as well; however, the data are inconclusive. Human mutagenic data has been reported. All contact with this chemical should be reduced to the lowest possible level. Human exposure to aldrin can occur through two primary routes of entry into the body: *ingestion* and *absorption* (although *inhalation* is always a possible, especially when working in extremely dusty conditions). Each of the two primary routes present a moderate to high level of toxicity. Aldrin appears to be absorbed through the skin in significant amounts, particular on a chronic basis Absorption through the skin can result in toxic symptoms such as headache, dizziness, nausea, vomiting, myoclonic jerks of the limbs (muscle twitching), clonic and tonic convulsions, malaise, and sometimes coma. Once aldrin is absorbed, either through the skin or through the digestive tract, it is rapidly metabolized to dieldrin.

Acute Health Effects

The following acute (short-term) health effects may occur immediately or shortly after exposure to aldrin:

Skin: Dermatitis associated with aldrin is unusual. Most reaction from skin contact is internal due to absorption. Symptoms include vomiting, nausea, malaise, and muscular twitching at high doses (above the PEL).

Eye: Possible irritation.

Lung: Irritation causing coughing and/or shortness of breath. Higher exposures can lead to respiratory problems.

CNS: Dizziness, headaches, nausea, vomiting, muscle jerks, severe seizures, and death.

Chronic Health Effects

The following chronic (long-term) health effects may occur at some time after exposure to aldrin and can last for months or even years:

Cancer Hazards: Aldrin may be a carcinogen in humans since it has been shown to cause liver cancer in some test animals.

Reproductive Hazard. Aldrin may be a teratogen in humans since it has been shown to be a teratogen in some test animals.

Other Chronic Effects: Aldrin may cause a skin allergy (sensitization). If an allergy develops, very low future skin-contact exposures to aldrin crystals or dust can cause itching and rash. Prolonged exposures may damage the liver.

Recommended Risk-Reduction Measures

Personnel should avoid direct contact with aldrin. Unless a less toxic chemical can be substituted for a hazardous substance, *engineering controls* are the most effective method of reducing exposures. The best protection is to enclose operations and/or provide local exhaust ventilation at the site of chemical release. While not always operationally feasible, isolating operations can also reduce exposure.

Using respiratory protection is less effective than the controls mentioned above, but is still advisable whenever working with or around aldrin. A self-contained breathing apparatus (SCBA) with full facepiece and pressure demand is the recommended respiratory protection of choice. If a full facepiece is not available, then chemical/dust goggles should be worn to protect the eyes. Whenever aldrin is being mixed with liquids, a chemical splash hazard exists and a face shield and rubber apron should be worn. To prevent hand and skin exposures, rubber gloves should be worn.

Administrative controls should also be in place to minimize the potential for human exposures. These may include written procedures or policies which specify the methods and techniques that will be practiced whenever personnel are to work with aldrin.

All personnel should receive *training* on the use, hazards, protective measures, emergency actions, and other precautions per 29 CFR 1910.1200 (Hazard Communication), prior to the first assignment in an area where aldrin is used or stored.

If symptoms develop or overexposure is suspected, the following medical tests may be useful:

- ☑ Electroencephalograph (EEG) to test brain activity/waves (identify seizure potential).
- ☑ Liver function tests.
- ☑ Evaluation by a qualified allergist, including careful exposure history and special testing (may help diagnose skin allergy).

Any evaluation should include a careful medical history of past and present symptoms with an examination. However, medical tests that evaluate existing damage are not a substitute for controlling exposure.

Other methods to reduce exposure include:

- ☑ Where possible, enclose operations and use local exhaust ventilation at the site of chemical release. If local exhaust ventilation or enclosure is not used, respiratory protection should be mandatory.
- ☑ Always ensure that proper protective clothing is worn when using chemical substances and that personnel are trained in its use and care.
- ☑ Wash thoroughly immediately after exposure to aldrin and at the end of the work shift or before eating, drinking, or smoking.
- ☑ Hazard warning information should be posted in the work area. In addition, as part of an on-going education and training program, all information on the health and safety hazards of aldrin should be communicated to all exposed and potentially exposed workers.
- ☑ Eye wash stations should be provided in work areas. If the potential for whole body exposure exists, then safety showers should also be provided.

RISK ASSESSMENT: ENVIRONMENT

General Assessment

The environment is at risk of exposure during transportation, storage, disposal, or destruction of aldrin. In almost every scenario, the threat of environmental exposure is contingent upon the proper handling of the chemical substance. Accidental spills, large or small, can result in fire, explosion, and possible contamination of the surrounding environmental mediums (water, soil, and air). Aldrin is considered a noncombustible solid. While it is not a fire hazard in and of itself, it is often mixed with liquids which are either flammable or combustible. These characteristics require special consideration during any emergency situation involving a leak or spill of aldrin or aldrin mixtures. Should aldrin ever come into contact with incompatible substances such as concentrated mineral acids (sulfuric, hydrochloric, or nitric acids, for example), acid catalysts, acid oxidizing agents (chlorine), phenols, or reactive metals (such as sodium, potassium, and magnesium) either during use, transportation, or storage, violent reaction can occur.

The proper disposal/destruction method for aldrin is to burn it in a chemical incinerator equipped with an afterburner and air scrubber.

Aldrin can enter the environment through industrial discharges, urban runoff, and through spills.

☠ *Acute Ecological Effects*

Acute (short-term) toxic effects may include the death of animals, birds, or fish and death or low growth rate in plants. Acute effects are seen 2 to 4 days after animals or plants are exposed to aldrin. This chemical has high acute toxicity to aquatic life and to birds. It has caused a variety injurious effects to both agricultural and ornamental crops. Insufficient data are available to evaluate or predict the short-term effects of aldrin to land animals.

Chronic Ecological Effects

Chronic toxic effects may include shortened life span, reproductive problems, lower fertility, and changes in appearance or behavior in exposed animals. These effects can be seen long after first exposure(s) to toxic chemicals. Aldrin has high chronic toxicity to aquatic life. Insufficient data are available to evaluate or predict the long-term effects of aldrin to plants, birds, land animals.

Water Solubility

Aldrin is slightly soluble in water. Concentrations of 1 milligram or less will mix with a liter of water.

Persistence in the Environment

Aldrin is highly persistent in water, with a half-life greater than 200 days. The half-life of a pollutant is the amount of time it takes for one-half of the chemical to be degraded. Under aerobic (oxygenated) conditions, aldrin is oxidized to the highly persistent insecticide, dieldrin. Approximately 49% of aldrin will eventually end in terrestrial soils; about 46% in aquatic sediments, and the rest will end up in water.

Bioaccumulation in Aquatic Organisms

Some substances increase in concentration, or *bioaccumulate,* in living organisms as they breathe contaminated air, drink contaminated water, or eat contaminated food. These chemicals can become concentrated in the tissues and internal organs of animals as well as humans.

The concentration of aldrin found in fish tissues is expected to be much higher than the average concentration of aldrin in the water from which the fish was taken.

Recommended Risk-Reduction Measures

Proper training of all transporters will reduce the likelihood of a mishap or accident resulting in a leak or spill to the environment. The correct use of DOT labeling (per the MSDS) on all containers, trucks, and rail cars will enable emergency responders to react properly and quickly to any disaster thereby reducing the potential risk to the environment and to personnel.

Storage of aldrin should be segregated from incompatible chemicals to minimize the risk of cross-contamination or contact. If a spill or leak to the environment has occurred, fire department, emergency response and/or hazardous materials spill personnel should be notified immediately. Cleanup should be attempted only by those trained in proper spill containment procedures. Contaminated soils should be removed for incineration and replaced with clean soil. If aldrin should contact the water table, aquifer, or navigable waterway, remediation activities should be prompt to ensure protection of aquatic sediments. When such spills occur, the local and/or state emergency response authorities must be notified. A comprehensive emergency response or disaster preparedness/recovery plan should be in place prior to any operations involving the use, transportation, storage, or disposal of aldrin.

If aldrin is spilled or leaked, the following specific steps are recommended:

- ☑ Restrict persons not wearing protective clothing from area of spill or leak until cleanup is complete.
- ☑ Collect powdered material in the most convenient (safe) manner possible and deposit in sealed containers.
- ☑ It may be necessary to dispose of aldrin as a hazardous waste. The responsible state agency or the regional office of the federal Environmental Protection Agency (EPA) should be contacted for specific recommendations.

RISK ASSESSMENT: BUSINESS

General Assessment

Accidents or mishaps involving aldrin can present a significant threat to business operations. The loss or damage of equipment or facilities can significantly effect fiscal viability. Lawsuits that may result from personnel injury/death, public exposures, and/or environmental contamination will also require a serious expenditure of resources.

Media attention surrounding an injury, death, or environmental damage can also result in a loss of profits and loss of current as well as future business.

Recommended Risk-Reduction Measures

Company attorneys, safety and health professionals, and environmental specialists should be involved in the development of any such procedures. A company official should be pre-designated as a public relations officer with specific training in dealing with the press. Corporate plans and policies should be developed, approved, and implemented long before any need for such arises.

REFERENCES

Ahlbom, A. 1993. *Biostatistics for Epidemiologists.* New York: CRC Press/Lewis Publishers.

Anthony, C. P., and N. J. Kolthoff. 1971. *Textbook of Anatomy and Physiology.* St. Louis: C. V. Mosby Company.

Asante-Duah, D. K. 1993. *Hazardous Waste Risk Assessment.* New York: CRC Press/Lewis Publishers.

Bloom, A. D. and F. J. DeSerres. 1995. *Ecotoxicity and Human Health.* New York: CRC Press/Lewis Publishers.

Brusick, D. J. 1994. *Methods for Genetic Risk Assessment.* New York: CRC Press/Lewis Publishers.

Bueche, J. R. 1972. *Principles of Physics,* 2nd Edition. New York: McGraw-Hill.

Burrell, R., D. K. Flaherty, and L. J. Sauers. 1992. *Toxicology of the Immune System.* New York: Van Nostrand Reinhold.

Calabrese, E. J. 1994. *Biological Effects of Low Level Exposures.* New York: CRC Press/Lewis Publishers.

Cockerman, L. G., B. S. Shane. 1994. *Basic Environmental Toxicology.* New York: CRC Press/Lewis Publishers.

Cothern, C. R., N. P. Ross. 1994. *Environmental Statistics, Assessment, and Forecasting.* New York: CRC Press/Lewis Publishers.

Cothern, R. R. 1993. *Comparative Environmental Risk Assessment.* New York: CRC Press/Lewis Publishers.

Cothern, C. R. 1995. *Handbook for Environmental Risk Decisionmaking.* New York: CRC Press/Lewis Publishers.

Environmental Protection Agency. October 1983. *Health Assessment for Aldrin.* Final Report: EPA 600/8-82-007F. NTIS PB84-149152. Springfield, VA: National Technical Information Service.

Gardner, R. W. 1994. *Chemical Intolerance.* New York: CRC Press/Lewis Publishers.

Hallenbeck, W. H. 1993. *Quantitative Risk Assessment for Environmental and Occupational Health,* 2nd Edition. New York: CRC Press/Lewis Publishers.

Hodgson, E., R. B. Mailman, and Janice E. Chambers. 1988. *Dictionary of Toxicology.* London: The MacMillian Company.

Howard, P. H. 1995. *Handbook of Environmental Fate and Exposures Data for Organic Chemicals,* Vol. I-IV. New York: CRC Press/Lewis Publishers.

Howard, P. H., et. al. 1991. *Handbook of Environmental Degradation Rates.* New York: CRC Press/Lewis Publishers.

James, A., and M. Lord. 1992. *Index of Chemical and Physical Data.* New York: Van Nostrand Reinhold.

Jones, F. E. 1994. *Toxic Organic Vapors in the Workplace.* New York: CRC Press/Lewis Publishers.

Kamrin, M. A. 1988. *Toxicology: A Primer on Principles and Applications.* Chelsea, Michigan: Lewis Publishers.

Landis, W. G., M. H. Yu. 1995. *Introduction to Environmental Toxicology.* New York: CRC Press/Lewis Publishers.

Langerman, N. 1994 *Precautionary Labels for Chemical Containers.* New York: CRC Press/Lewis Publishers.

Lauwerys, R. R., P. Hoet. 1993. *Industrial Chemical Exposure.* New York: CRC Press/Lewis Publishers.

Lide, D. R. 1995. *Handbook of Chemistry and Physics,* 76th Edition. New York: CRC Press/Lewis Publishers.

Lewis, R. J., Sr. 1993. *Hawley's Condensed Chemical Dictionary,* 12th Edition. New York: Van Nostrand Reinhold.

Lewis, R. J., Sr. 1992. *Sax's Dangerous Properties of Industrial Materials,* Volumes 1, 2, and 3, 8th Edition. New York: Van Nostrand Reinhold.

Manahan, S. E. 1994. *Environmental Chemistry,* 6th Edition. New York: CRC Press/Lewis Publishers.

Manahan, S. E. 1992. *Toxicological Chemistry,* 2nd Edition. New York: CRC Press/Lewis Publishers.

Mellström, G. A., J. E. Wahlberg, and H. I. Maibach. 1994. *Protective Gloves for Occupational Use.* New York: CRC Press/Lewis Publishers.

Meyer, E. 1990. *Chemistry of Hazardous Materials.* Englewood Cliffs, NJ: Prentice Hall.

Mycock, J. C., J. D. McKenna, and L. Theodore. 1995. *Handbook of Air Pollution Control Engineering and Technology.* New York: CRC Press/Lewis Publishers.

Nelson, G. O. 1992. *Gas Mixtures.* New York: CRC Press/Lewis Publishers.

Ott, W. R. 1995. *Environmental Statistics and Data Analysis.* New York: CRC Press/Lewis Publishers.

Patnaik, P. 1992. *A Comprehensive Guide to the Hazardous Properties of Chemical Substances.* New York: Van Nostrand Reinhold.

Philp, R. B. 1995. *Environmental Hazards and Human Health.* New York: CRC Press/Lewis Publishers.

Proctor, N. H., and James P. Hughes. 1991. *Chemical Hazards of the Workplace.* New York: Van Nostrand Reinhold.

Quigley, D. R. 1994. *Handbook of Emergency Chemical Management.* New York: CRC Press/Lewis Publishers.

Rea, W. J. 1995. *Chemical Sensitivity,* Vol. I-III. New York: CRC Press/Lewis Publishers.

Richardson, M. 1995. *Dictionary of Substance and Their Effects,* Vol. I-VI. New York: CRC Press/Lewis Publishers.

Rylander, R., R. R. Jacobs. 1994. *Organic Dusts.* New York: CRC Press/Lewis Publishers.

Sacarello, H. L. A. 1994. *The Comprehensive Handbook of Hazardous Materials.* New York: CRC Press/Lewis Publishers.

U.S. Department of Health and Human Services, National Institute for Occupational Safety and Health. 1978. *Criteria Recommended Standard - Occupational Exposure to Aldrin.* Washington, D.C.: U.S. Government Printing Office.

U.S. Department of Health and Human Services, National Institute for Occupational Safety and Health. 1994. *NIOSH Pocket Guide to Chemical Hazards.* Washington, D.C.: U.S. Government Printing Office.

Willhite, C. C. Toxicological updates. Aldrin. *Journal of Applied Toxicology.* 2:54-56, 1982.

MATERIAL SAFETY DATA SHEET

CHEMICAL NAME

ALLYL CHLORIDE

CRC PRESS/LEWIS PUBLISHERS DISCLAIMER: THE INFORMATION AND RECOMMENDATIONS PRESENTED HEREIN ARE BASED ON SOURCES BELIEVED TO BE RELIABLE. CRC MAKES NO REPRESENTATION ON ITS COMPLETENESS OR ACCURACY. IT IS THE RESPONSIBILITY OF THE USER TO DETERMINE THE CHEMICAL'S SUITABILITY FOR ITS INTENDED USE, THE CHEMICAL'S SAFE USE, AND THE CHEMICAL'S PROPER DISPOSAL. NO REPRESENTATIONS AND/OR WARRANTIES, EITHER EXPRESSED OR IMPLIED, OF THE MERCHANTABILITY OR FITNESS FOR A PARTICULAR PURPOSE, OR OF ANY OTHER NATURE, ARE MADE WITH RESPECT TO THE INFORMATION PROVIDED IN THIS MSDS OR TO THE CHEMICAL TO WHICH INFORMATION MAY REFER. CRC NEITHER ASSUMES NOR AUTHORIZES ANY OTHER PERSON TO ASSUME FOR IT, ANY OTHER ADDITIONAL RESPONSIBILITY OR LIABILITY FOR THE USE OF, OR RELIANCE UPON, THIS INFORMATION.

HAZARD WARNING INFORMATION

HEALTH	FIRE	REACTIVE	OTHER	DEGREE OF HAZARD	COLOR CODING	OTHER CODES
3	3	1		0 = Minimum Hazard 1 = Slight Hazard 2 = Moderate Hazard 3 = Serious Hazard 4 = Severe Hazard	HEALTH = BLUE FIRE = RED REACTIVITY = YELLOW OTHER = WHITE	OX = Oxidizer ACID = Acid ALK = Alkali COR = Corrosive W = Use No Water

SECTION I - GENERAL INFORMATION

Characterization **Halogenated Aliphatic Hydrocarbon**	RCRA Number **None**	EPA Class **Not Applicable**
DOT Proper Shipping Name **Allyl Chloride**	Chemical Abstract Service (CAS) Number **107-05-1**	
DOT Hazard Class and Label Requirements **Flammable Liquid**	DOT Emergency Guide Code **57**	
DOT Identification Number **UN 1100**	Chemical Formula **C_3H_5Cl**	

Synonyms
Chlorallylene; 3-chloroprene;1-chloro-2-propene; 3-chloropropylene; 2-propenyl chloride.

SECTION II - HAZARDOUS INGREDIENTS AND IDENTITY INFORMATION

Hazardous Components (specific identity)	OSHA Exposure Criteria	NIOSH Exposure Criteria	Immediately Dangerous to Life and Health (IDLH)	ACGIH Exposure Criteria
Allyl chloride **1 ppm = 3.18 mg/m³**	**PEL:** **1 ppm** **3 mg/m³** **STEL:** **2 ppm** **6.2 mg/m³**	**REL:** **1 ppm** **3 mg/m³** **STEL:** **3 ppm** **9.5 mg/m³**	**300 ppm**	**TLV:** **1 ppm** **3 mg/m³** **STEL:** **2 ppm** **6.2 mg/m³**

SECTION III - PHYSICAL AND CHEMICAL CHARACTERISTICS

Boiling Point **113°F (44.6°C)**	Specific Gravity (H_2O = 1) **0.94**
Vapor Pressure (mm Hg) **295 at 69°F (20°C)**	Molecular Weight **76.5**
Vapor Density (Air = 1) **2.64**	Melting Point **-210°F (-136.4°C)**

Solubility
Only slightly soluble in water (0.4%). Will mix (miscible) with alcohol, chloroform, ether, and petroleum ether.

Appearance and Odor
Colorless, yellow, or purple liquid with an unpleasant, pungent odor. Odor threshold = 1.2 ppm.

SECTION IV - FIRE AND EXPLOSION HAZARD DATA

Flash Point (method used) **-25°F (-32°C) closed cup**	Explosive Limits in Air % by Volume **LEL: 2.9% UEL: 11.1%**
NFPA Classification **Class IB Flammable Liquid**	Autoignition Temperature **737°F (392°C)**

Extinguishing Media
Carbon dioxide, dry chemical, water spray, or alcohol foam.

Special Fire Fighting Procedures
Allyl chloride is an extremely volatile liquid. Wear full protective clothing and self-contained breathing apparatus (SCBA). Use a water spray to keep fire-exposed containers cool. Poisonous gases are produced in fire.

Unusual Fire and Explosion Hazards
Dangerous fire and explosion hazard when exposed to heat, flame, or oxidizers. Containers may explode in fire. Firefighters should avoid contact with vapors produced during fire.

SECTION V - REACTIVITY DATA

Stability		Conditions to Avoid
		Reacts with strong oxidizers. Vigorous or explosive reaction above -70°C with alkyl aluminum chlorides + aromatic hydrocarbons. Decomposes on storage to release hydrogen chloride gas.
Stable	Unstable	Incompatibility *(materials to avoid)*
	X	Strong oxidizers, acids, amines, iron and aluminum chlorides, magnesium, zinc, ethylene imine, ethylenediamine, chlorosulfonic acid, oleum, NaOH, and HNO_3.
Hazardous Polymerization		Conditions to Avoid
		Violent exothermic polymerization with aluminum chloride, boron trifluoride, or sulfuric acid and with metals (aluminum, magnesium, zinc, or galvanized metals).
May Occur	Will Not Occur	Hazardous Decomposition or By-products
X		When heated to decomposition, allyl chloride can emit highly toxic/poisonous gases.

SECTION VI - HEALTH HAZARD DATA

Primary Route(s) of Entry:	Inhalation?	Absorption (skin)?	Ingestion?
	X	X	

Health Hazards

INHALATION: Extremely irritating to the lungs, throat, nose and eyes. Coughing and shortness of breath that may persist for hours after exposure. Can cause permanent damage to the lungs. Fluid can build up in the lungs (pulmonary edema), which is a medical emergency.

ABSORPTION: Contact with the liquid can cause severe skin burns, deep aching, and "bone pain." The liquid can cause severe eye burns leading to permanent damage.

Carcinogenicity	NTP Listed?	IARC Cancer Review Group?	OSHA Regulated?	Target Organs?
Suspected Human Suspected Animal	5th Annual Report	Group 3	29 CFR 1910.1000 Table Z-1	Respiratory system, eyes, skin, kidneys, liver.

Medical Conditions Generally Aggravated by Exposure

Respiratory impairments (bronchitis, asthma); Skin conditions (dermatitis); Renal or liver disorders.

Emergency and First-aid Procedures

Eye contact: Flush immediately with water for 15 minutes (minimum), seek medical attention. Skin contact: Remove all contaminated clothing. Immediately wash area with large amounts of soap and water. Seek medical assistance. For inhalation: Remove the person from exposure. Provide respiratory assistance and CPR. Transfer to medical facility. Observe for 24-48 hours since pulmonary edema may be a delayed reaction. If swallowed, seek medical attention immediately.

SECTION VII - PRECAUTIONS FOR SAFE HANDLING AND USE

Steps to be Taken in Case Material is Released or Spilled

Absorb liquids in vermiculite, dry sand, earth, or similar material and deposit in sealed containers. Ventilate area of spill or leak. Restrict those not involved in cleanup from entering area. Notify appropriate authorities, as required by SARA III, if applicable.

Preferred Waste Disposal Method

Incineration (with after burner and scrubber).

Precautions to be Taken in Handling and Storage

Allyl chloride is a combustible liquid. Store to avoid contact with strong oxidizers, acids, amines, and metals since violent reactions can occur. Store in tightly closed containers in cool, well-ventilated area away from heat. Ground and bond all containers prior to transfer.

Other Precautions and Warnings

Sources of ignition are prohibited where allyl chloride is used, handled, or stored. Use only non-sparking tools and equipment, especially when opening or closing containers of allyl chloride.

SECTION VIII - CONTROL MEASURES AND PERSONAL PROTECTIVE EQUIPMENT

Respiratory Protection *(specify type)*

For exposures below the PEL, use a half mask respirator with organic vapor cartridge. For all other exposures, use a MSHA/NIOSH approved supplied air respirator with full facepiece, hood, or helmet in continuous flow mode, or use a full facepiece operated in pressure demand or other positive pressure mode.

Ventilation

Local exhaust or general mechanical systems recommended.

Protective Gloves	Eye Protection	Other Protective Clothing
Solvent resistant rubber	Chemical Goggles or Face Mask	Rubber Apron

Work/Hygiene Practices

Always wash hands thoroughly after using chemical; never bring food, drink, or smoking materials into vicinity of chemicals.

ALLYL CHLORIDE

C_3H_5Cl CAS: 107-05-1

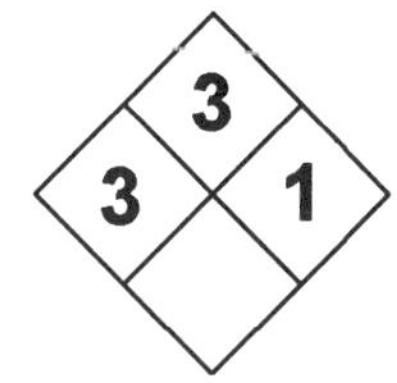

$H_2C{=}CHCH_2Cl$

IDENTIFICATION AND TYPICAL USES

Allyl chloride is a colorless or yellow, or sometimes purple liquid with a strong, disagreeable odor. It is used in the organic synthesis of allyl compounds. It is used to manufacture epichlorohydrin, epoxy resin, glycerin pesticides, and sodium allyl sulfonate.

RISK ASSESSMENT: HEALTH

General Assessment

Exposure to allyl chloride is possible through *inhalation* of its vapors and by *absorption* of the liquid through unbroken skin. Allyl chloride is an irritant of the eyes, mucous membranes, and skin. In animals, it causes renal (kidney), hepatic (liver), and pulmonary damage and, at high concentrations, central nervous system depression.

Inhalation can cause headache, dizziness, and in high concentrations, loss of consciousness. Its vapors can cause conjunctiva irritation and eye pain with photophobia. Eye irritation occurs between 50 and 100 ppm. Irritation of the nose occurs at levels below 25 ppm. Skin *absorption* results in surface burning and deep-seated ("bone") pain. It is rapidly absorbed and distributed throughout the body.

Allyl chloride is a suspected carcinogen with experimental tumorigenic data reported. While the primary routes are inhalation and absorption, it is also moderately toxic by ingestion. Human mutation data has also been reported.

Acute Health Effects

The following acute (short-term) health effects may occur immediately or shortly after exposure to allyl chloride:

Skin: Irritation or burning and deep-seated (bone) pain.

Eye: Severe irritation; may cause burns and permanent damage.

Lung: Irritation of the nose, throat, and lungs following exposure. Pulmonary edema is possible (24 to 48 hours following a high exposure).

CNS: Headache, dizziness, loss of consciousness. Possible mild neuropathy, weakness, paresthesia, cramping pain, and numbness in the extremities with sensory impairment and diminished ankle reflexes.

Chronic Health Effects

The following chronic (long-term) health effects may occur at some time after exposure to allyl chloride and can last for months or even years:

Cancer Hazards: Allyl chloride may cause mutations (genetic changes) in living cells. The data on its carcinogenicity are inconclusive.

Reproductive Hazard: There is limited evidence to show that exposure to allyl chloride may damage the developing fetus.

Other Chronic Effects: Allyl chloride may damage the liver and kidneys. Long-term exposures may cause drying and cracking of the skin. It is very irritating to the lungs (it is not known whether chronic exposure causes lung damage).

Recommended Risk-Reduction Measures

Personnel should avoid direct contact with allyl chloride. If a less toxic chemical can not be substituted for a hazardous substance, then *engineering controls* are the most effective method of reducing exposures. The best protection is to enclose operations and/or provide local exhaust ventilation at the site of chemical release. While not always operationally feasible, isolating operations can also reduce exposure. Using respiratory protection is less effective than the controls mentioned above, but is still advisable whenever working with or around allyl chloride. A self-contained breathing apparatus (SCBA) with full facepiece and pressure demand is the recommended respiratory protection of choice. If a full facepiece is not available, then chemical goggles should be worn to protect the eyes. For low concentrations (below the PEL), an organic vapor cartridge respirator (equipped with a mist/vapor filter) is acceptable protection. Whenever a chemical splash hazard exists, a face shield and a rubber apron should be worn. To prevent hand and skin exposures, acid resistant gloves should be worn.

Administrative controls should also be in place to minimize the potential for human exposures. These may include written procedures or policies which specify the methods and techniques that will be practiced whenever personnel are to work with allyl chloride.

All personnel should receive *training* on the use, hazards, protective measures, emergency actions, and other precautions per 29 CFR 1910.1200 (Hazard Communication), prior to the first assignment in an area where allyl chloride is used or stored.

For those personnel with frequent or potentially high exposure (half the PEL or greater), the following are recommended medical tests to be provided before initial assignment and at regular intervals thereafter (e.g., annually):

☑ Liver and kidney function tests.

☑ Lung function tests.

Any evaluation should include a careful history of past and present symptoms with an examination. Medical tests that simply look for existing damage are not a substitute for controlling exposures. Also, because smoking can cause heart disease, as well as lung cancer, emphysema, and other respiratory problems, exposure to allyl chloride may be more pronounced in smokers than in non-smokers under the same conditions of exposure. Prudent risk management requires careful consideration of *all* possible exposure factors.

Other methods to reduce exposure include:

☑ Where possible, enclose operations and use local exhaust ventilation at the site of chemical release. If local exhaust ventilation or enclosure is not used, respiratory protection should be mandatory.

☑ Always ensure that proper protective clothing is worn when using chemical substances.

☑ Wash thoroughly immediately after exposure to allyl chloride and at the end of the work shift or before eating, drinking, or smoking.

☑ Hazard warning information should be posted in the work area. In addition, as part of an on-going education and training program, all information on the health and safety hazards of allyl chloride should be communicated to all potentially exposed workers.

☑ Eye wash stations should be provided in the immediate work area for emergency use. If there is a possibility of skin exposure to allyl chloride, emergency shower facilities should also be provided.

☑ Personnel should never wear contaminated clothing home (family members can be exposed). Clothing should be laundered by personnel who have been trained on the hazards of allyl chloride.

RISK ASSESSMENT: ENVIRONMENT

General Assessment

The environment is at risk of exposure during transportation, storage, disposal, or destruction of allyl chloride. In almost every scenario, the threat of environmental exposure is contingent upon the proper handling of the chemical substance. Accidental spills, large or small, can result in fire, explosion, and possible contamination of the surrounding environmental mediums (water, soil, and air).

Allyl chloride is considered a class 1B flammable liquid. Because of its low flash point and high autoignition temperature, it requires extreme caution in handling, storage, transportation, and disposal. It is incompatible with oxidizing agents and contact can cause fire or explosion. These characteristics require special consideration during any emergency situation involving a leak or spill of allyl chloride. Should allyl chloride ever come into contact with incompatible substances such as oxidizers, acids, amines, iron, and aluminum chlorides, magnesium, zinc, and galvanized metals, chlorosulfonic acid, and oleum either during use, transportation, storage, or disposal, the formation of highly toxic and/or highly explosive commodities is extremely possible. The proper disposal/destruction method for allyl chloride waste is to burn it in a chemical incinerator equipped with an afterburner and air scrubber. Allyl chloride can enter the environment mainly industrial effluents or spills.

☠ *Acute Ecological Effects*

Acute (short-term) toxic effects may include the death of animals, birds, or fish and death or low growth rate in plants. Acute effects are seen 2 to 4 days after animals or plants are exposed to allyl chloride. This chemical has moderate acute toxicity to aquatic life. No data are available on the short-term effects of allyl chloride exposure to plants, birds, or land animals.

Chronic Ecological Effects

Chronic (long-term) toxic effects may include shortened life span, reproductive problems, lower fertility, and changes in appearance or behavior in exposed animals. These effects can be seen long after first exposure(s) to toxic chemicals. Allyl chloride has moderate chronic toxicity in aquatic life. No data are available on the long-term effects of allyl chloride to plants, birds, or land animals.

Water Solubility

Allyl chloride is slightly soluble in water. Concentrations of 1 milligram will mix with a liter of water.

Persistence in the Environment

Allyl chloride is non-persistent in water, with a half-life of less than 2 days. The half-life of a pollutant is the amount of time it takes for one half of the chemical to be degraded. Approximately 99.3% of allyl chloride will eventually end up in water; the remainder will end up in air.

Bioaccumulation in Aquatic Organisms

Some substances increase in concentration, or *bioaccumulate,* in living organisms as they breathe contaminated air, drink contaminated water, or eat contaminated food. These chemicals can become concentrated in the tissues and internal organs of animals as well as humans.

The concentration of allyl chloride found in fish tissues is expected to be about the same as the average concentration of allyl chloride in the water from which the fish was taken.

Recommended Risk-Reduction Measures

Proper training of all transporters will reduce the likelihood of a mishap or accident resulting in a leak or spill to the environment. The correct use of DOT labeling (per the MSDS) on all containers, trucks, and rail cars will enable emergency responders to react properly and quickly to any disaster thereby reducing the potential risk to the environment and to personnel.

Storage of allyl chloride should be segregated from incompatible chemicals to minimize the risk of cross-contamination or contact. Buildings designated for storage should be equipped with appropriate fire protection systems (alarms, sprinklers, emergency lighting, portable extinguishers).

If a spill or leak to the environment has occurred, fire department, emergency response and/or hazardous materials spill personnel should be notified immediately. Cleanup should be attempted only by those trained in proper spill containment procedures. Contaminated soils should be removed for incineration and replaced with clean soil. If allyl chloride should contact the water table, aquifer, or navigable waterway, the local and/or state emergency response authorities must be notified. A comprehensive emergency response or disaster preparedness/recovery plan should be in place prior to any operations involving the use, transportation, storage, or disposal of allyl chloride. If allyl chloride is spilled or leaked, the following specific steps are recommended:

- ☑ Restrict persons not wearing protective clothing from area of spill or leak until cleanup is complete and area can be opened for normal work.
- ☑ Remove all ignition sources.
- ☑ Absorb liquids in vermiculite, dry sand, earth, or similar material and deposit in sealed containers.
- ☑ It may be necessary to dispose of allyl chloride as a hazardous waste. The responsible state agency or the regional office of the federal Environmental Protection Agency (EPA) should be contacted for specific recommendations.

RISK ASSESSMENT: BUSINESS

General Assessment

Accidents or mishaps involving allyl chloride can present a significant threat to business operations. The loss or damage of equipment or facilities can significantly effect fiscal viability. Lawsuits that may result from personnel injury/death, public exposures, and/or environmental contamination will also require a serious expenditure of resources. Media attention surrounding an injury, death, or environmental damage can also result in a loss of profits and loss of current as well as future business.

Recommended Risk-Reduction Measures

Company attorneys, safety and health professionals, and environmental specialists should be involved in the

development of any procedures dealing with safety or emergency response. A company official should be pre-designated as a public relations officer with specific training in dealing with the press. Corporate plans and policies should be developed, approved, and implemented long before any need for such arises.

REFERENCES

Ahlbom, A. 1993. *Biostatistics for Epidemiologists.* New York: CRC Press/Lewis Publishers.

American Conference of Governmental Industrial Hygienists. 1988. *Documentation of the Threshold Limit Values and Biological Exposure Indices,* 5th Edition (with updates). Cincinnati: ACGIH

Asante-Duah, D. K. 1993. *Hazardous Waste Risk Assessment.* New York: CRC Press/Lewis Publishers.

Bloom, A. D. and F. J. DeSerres. 1995. *Ecotoxicity and Human Health.* New York: CRC Press/Lewis Publishers.

Brusick, D. J. 1994. *Methods for Genetic Risk Assessment.* New York: CRC Press/Lewis Publishers.

Bueche, J. R. 1972. *Principles of Physics,* 2nd Edition. New York: McGraw-Hill.

Calabrese, E. J. 1994. *Biological Effects of Low Level Exposures.* New York: CRC Press/Lewis Publishers.

Cockerman, L. G., B. S. Shane. 1994. *Basic Environmental Toxicology.* New York: CRC Press/Lewis Publishers.

Cothern, R. R. 1993. *Comparative Environmental Risk Assessment.* New York: CRC Press/Lewis Publishers.

Gardner, R. W. 1994. *Chemical Intolerance.* New York: CRC Press/Lewis Publishers.

Hallenbeck, W. H. 1993. *Quantitative Risk Assessment for Environmental and Occupational Health,* 2nd Edition. New York: CRC Press/Lewis Publishers.

Howard, P. H. 1995. *Handbook of Environmental Fate and Exposures Data for Organic Chemicals,* Vol. I-IV. New York: CRC Press/Lewis Publishers.

Jones, F. E. 1994. *Toxic Organic Vapors in the Workplace.* New York: CRC Press/Lewis Publishers.

Keith, L. H. and M. M. Walker. 1995. *Handbook of Air Toxics.* New York: CRC Press/Lewis Publishers.

Landis, W. G., M. H. Yu. 1995. *Introduction to Environmental Toxicology.* New York: CRC Press/Lewis Publishers.

Langerman, N. 1994 *Precautionary Labels for Chemical Containers.* New York: CRC Press/Lewis Publishers.

Lauwerys, R. R., P. Hoet. 1993. *Industrial Chemical Exposure.* New York: CRC Press/Lewis Publishers.

Lewis, R. J., Sr. 1993. *Hawley's Condensed Chemical Dictionary,* 12th Edition. New York: Van Nostrand Reinhold.

Lewis, R. J., Sr. 1992. *Sax's Dangerous Properties of Industrial Materials,* Volumes 1, 2, and 3, 8th Edition. New York: Van Nostrand Reinhold.

Lide, D. R. 1995. *Handbook of Chemistry and Physics,* 76th Edition. New York: CRC Press/Lewis Publishers.

Manahan, S. E. 1994. *Environmental Chemistry,* 6th Edition. New York: CRC Press/Lewis Publishers.

Manahan, S. E. 1992. *Toxicological Chemistry,* 2nd Edition. New York: CRC Press/Lewis Publishers.

Mellström, G. A., J. E. Wahlberg, and H. I. Maibach. 1994. *Protective Gloves for Occupational Use.* New York: CRC Press/Lewis Publishers.

Meyer, E. 1990. *Chemistry of Hazardous Materials.* Englewood Cliffs, NJ: Prentice Hall.

Mycock, J. C., J. D. McKenna, and L. Theodore. 1995. *Handbook of Air Pollution Control Engineering and Technology.* New York: CRC Press/Lewis Publishers.

National Cancer Institute. 1978. *Carcinogenesis Technical Report Series. Bioassay of Allyl Chloride for Possible Carcinogenicity.* DEHW (NIH) Publication Number: 78-1323, p 53. Washington, DC: U.S. Government Printing Office.

Ott, W. R. 1995. *Environmental Statistics and Data Analysis.* New York: CRC Press/Lewis Publishers.

Patnaik, P. 1992. *A Comprehensive Guide to the Hazardous Properties of Chemical Substances.* New York: Van Nostrand Reinhold.

Philp, R. B. 1995. *Environmental Hazards and Human Health.* New York: CRC Press/Lewis Publishers.

Proctor, N. H., J. P. Hughes. 1991. *Chemical Hazards of the Workplace.* New York: Van Nostrand Reinhold.

Quigley, D. R. 1994. *Handbook of Emergency Chemical Management.* New York: CRC Press/Lewis Publishers.

Rea, W. J. 1995. *Chemical Sensitivity,* Vol. I-III. New York: CRC Press/Lewis Publishers.

Richardson, M. 1995. *Dictionary of Substance and Their Effects,* Vol. I-VI. New York: CRC Press/Lewis Publishers.

Sacarello, H. L. A. 1994. *The Comprehensive Handbook of Hazardous Materials.* New York: CRC Press/Lewis Publishers.

Talbott, E., G. Craun. 1995. *An Introduction to Environmental Epidemiology.* New York: CRC Press/Lewis Publishers.

Torkelson, T. R. V. K. Rowe. 1981. Halogenated aliphatic hydrocarbons. *Patty's Industrial Hygiene and Toxicology,* Vol. 2B, *Toxicology,* pp 3568-3572. New York: Wiley-Interscience.

The Merck Index, 10th edition. 1983. Rahway, NJ: Merck & Company.

U.S. Department of Health, Education, and Welfare, National Institute for Occupational Safety and Health. 1976. *Criteria for a Recommended Standard: Occupational Exposure to Allyl Chloride.* DEHW (NIOSH) Publication Number: 76-204. Washington, D.C.: U.S. Government Printing Office.

U.S. Department of Health and Human Services, National Institute for Occupational Safety and Health. 1994. *NIOSH Pocket Guide to Chemical Hazards.* Washington, D.C.: U.S. Government Printing Office.

MATERIAL SAFETY DATA SHEET

CHEMICAL NAME	CRC PRESS/LEWIS PUBLISHERS DISCLAIMER
ALUMINUM	**CRC PRESS/LEWIS PUBLISHERS DISCLAIMER:** THE INFORMATION AND RECOMMENDATIONS PRESENTED HEREIN ARE BASED ON SOURCES BELIEVED TO BE RELIABLE. CRC MAKES NO REPRESENTATION ON ITS COMPLETENESS OR ACCURACY. IT IS THE RESPONSIBILITY OF THE USER TO DETERMINE THE CHEMICAL'S SUITABILITY FOR ITS INTENDED USE, THE CHEMICAL'S SAFE USE, AND THE CHEMICAL'S PROPER DISPOSAL. NO REPRESENTATIONS AND/OR WARRANTIES, EITHER EXPRESSED OR IMPLIED, OF THE MERCHANTABILITY OR FITNESS FOR A PARTICULAR PURPOSE, OR OF ANY OTHER NATURE, ARE MADE WITH RESPECT TO THE INFORMATION PROVIDED IN THIS MSDS OR TO THE CHEMICAL TO WHICH INFORMATION MAY REFER. CRC NEITHER ASSUMES NOR AUTHORIZES ANY OTHER PERSON TO ASSUME FOR IT, ANY OTHER ADDITIONAL RESPONSIBILITY OR LIABILITY FOR THE USE OF, OR RELIANCE UPON, THIS INFORMATION.

HAZARD WARNING INFORMATION

HEALTH	FIRE	REACTIVE	OTHER	DEGREE OF HAZARD	COLOR CODING	OTHER CODES
2	1	1		0 = Minimum Hazard 1 = Slight Hazard 2 = Moderate Hazard 3 = Serious Hazard 4 = Severe Hazard	HEALTH = BLUE FIRE = RED REACTIVITY = YELLOW OTHER = WHITE	OX = Oxidizer ACID = Acid ALK = Alkali COR = Corrosive W = Use No Water

SECTION I - GENERAL INFORMATION

Characterization	**Metal, Reactive**	RCRA Number	**None**
		EPA Class	**Not Applicable**
DOT Proper Shipping Name	**Aluminum**	Chemical Abstract Service (CAS) Number	**7429-90-5**
DOT Hazard Class and Label Requirements	**Flammable Solid**	DOT Emergency Guide Code	**40**
DOT Identification Number	**UN 1390/1396/1383**	Chemical Symbol	**Al**

Synonyms
Aluminum metal; aluminum powder; elemental aluminum.

SECTION II - HAZARDOUS INGREDIENTS AND IDENTITY INFORMATION

Hazardous Components (specific identity)	OSHA Exposure Criteria	NIOSH Exposure Criteria	Immediately Dangerous to Life and Health (IDLH)	ACGIH Exposure Criteria
Aluminum (derivation: From bauxite and subsequent electrolytic reduction).	**PEL:** 15 mg/m^3 (total dusts) 5 mg/m^3 (respirable) 2 mg/m^3 (soluble salts)	**REL:** 10 mg/m^3 (total dusts) 5 mg/m^3 (respirable)	**Not Established**	**TLV:** 10 mg/m^3 (total dusts) 5 mg/m^3 (powders/fumes) 2 mg/m^3 (soluble salts)

SECTION III - PHYSICAL AND CHEMICAL CHARACTERISTICS

Boiling Point	**4221°F (2327°C)**	Specific Gravity (H_2O = 1)	**2.70**
Vapor Pressure (mm Hg)	**1 at 2343°F (1284°C)**	Molecular Weight	**27**
Vapor Density (Air = 1)	**Not Applicable**	Melting Point	**1220°F (660°C)**

Solubility
Insoluble. Soluble in HCl, H_2SO_4, and alkalis.

Appearance and Odor
White, malleable metal with bluish tint. May also be present as a powder.

SECTION IV - FIRE AND EXPLOSION HAZARD DATA

Flash Point (method used)	**Not Applicable**	Explosive Limits in Air % by Volume	**LEL: N/A UEL: N/A**
NFPA Classification	**Combustible Powder**	Autoignition Temperature	**Not Applicable**

Extinguishing Media
Sand, soda ash, lime, or dry chemical extinguishers.

Special Fire Fighting Procedures
Powdered aluminum or aluminum dusts are potentially explosive and reactive. Great caution should be exercised when fighting fires involving aluminum powders/dusts.

Unusual Fire and Explosion Hazards
Powdered aluminum undergoes explosive reactions in the presence of numerous chemical substances. It also forms sensitive explosive mixtures with some oxidants. Dust is moderately flammable/explosive by heat, flame, or chemical reaction with powerful oxidizers.

SECTION V - REACTIVITY DATA

Stability		Conditions to Avoid
		Aluminum is a reactive metal. The greatest industrial hazards are with chemical reactions. As with other metals, the powder and dusts are the most dangerous forms. Avoid contact with incompatibles.
Stable	Unstable	Incompatibility *(materials to avoid)*
X		Strong oxidizers and acids; halogenated hydrocarbons; metal oxides; oxosalts; sulfides; barium peroxide; sodium acetylide; sodium peroxide; bromine pentafluoride; chlorine fluoride.
Hazardous Polymerization		Conditions to Avoid
		Violent reaction with sulfur when heated. Exothermic reaction with iron powder and water (releases explosive hydrogen gas).
May Occur	Will Not Occur	Hazardous Decomposition or By-products
X		Vigorous amalgamation reactions with mercury salts and moisture. Violent exothermic reaction above 1112°F (600°C) with sodium diuranate.

SECTION VI - HEALTH HAZARD DATA

Primary Route(s) of Entry:	Inhalation?	Absorption (skin)?	Ingestion?
	X		

Health Hazards

INHALATION: Inhaling finely divided powder may cause pulmonary fibrosis (scarring of lung tissue). Long-term exposure to dusts may cause dyspnea, cough, and weakness.

SKIN & EYES: Skin and eye contact with powdered fibers can result in scratches and minor irritation.

INGESTION: Ingested aluminum is poorly absorbed, there appears to be no retention in organs.

Carcinogenicity	NTP Listed?	IARC Cancer Review Group?	OSHA Regulated?	Target Organs?
No Evidence.	No	No	29 CFR 1910.1000 Table Z-1	Respiratory system, eyes, skin.

Medical Conditions Generally Aggravated by Exposure

Respiratory impairments (bronchitis, asthma); Skin conditions (dermatitis).

Emergency and First-aid Procedures

Eye contact: Flush immediately with water for 15 minutes (minimum), seek medical attention. Skin contact: Remove all contaminated clothing. Wash contaminated skin with soap and water. For inhalation: Remove the person from exposure. Provide respiratory assistance and CPR. Transfer to medical facility. Never attempt to give anything by mouth to an unconscious or convulsing person.

SECTION VII - PRECAUTIONS FOR SAFE HANDLING AND USE

Steps to be Taken in Case Material is Released or Spilled

Collect powdered material in the most convenient way possible and deposit in sealed containers. Remove all sources of ignition. Restrict those not involved in cleanup from entering area. Notify appropriate authorities, as required.

Preferred Waste Disposal Method

Aluminum is a naturally occurring metal and does not pose a significant threat to the environment.

Precautions to be Taken in Handling and Storage

Containers should be protected against physical damage. Store in tightly closed containers in a cool, dry, well ventilated area away from acids, caustics, chlorinated hydrocarbons, and combustible metals. Keep aluminum dry.

Other Precautions and Warnings

Whenever bulk or finely powdered aluminum is used, handled, manufactured, or stored, use explosion-proof electrical equipment and fittings.

SECTION VIII - CONTROL MEASURES AND PERSONAL PROTECTIVE EQUIPMENT

Respiratory Protection *(specify type)*

For exposures near or over 2 mg/m^3: NIOSH approved respirator with particulate (dust/fume/mist) filters. Otherwise, use a self-contained breathing apparatus in continuous flow mode.

Ventilation

Local exhaust or general mechanical systems recommended.

Protective Gloves	Eye Protection	Other Protective Clothing
Suitable to protect from skin contact	Chemical/Dust Goggles	N/A

Work/Hygiene Practices

Always wash hands thoroughly after working with aluminum powders and dusts; never bring food, drink, or smoking materials into vicinity of aluminum powders.

ALUMINUM

Al CAS: 7429-90-5

IDENTIFICATION AND TYPICAL USES

Aluminum is a silvery white metallic solid. It may have a bluish tint. It may be present as a fine or coarse powder or dust. It is mined from bauxite and has many uses in construction and in industry. It may be present in metal alloys and is used in the paper and printing industries, in medicine, electronics, and in packaging. Fine powder is used in explosives, fireworks, as flashlights in photography, and in aluminum paints. It is also used in combination with other chemicals to serve many purposes such as water purification, for medicinal purposes, catalysts, dyes, baking powder, fireproofing, and tanning.

RISK ASSESSMENT: HEALTH

General Assessment

Although aluminum is not generally considered to be an industrial poison, *inhalation* of finely divided powder has resulted in the scarring of lung tissues (pulmonary fibrosis). Skin contact with fibers of aluminum may result in scratching and scarring. The same holds true for eye contact. *Ingested* aluminum is poorly absorbed, and there appears to be no retention of aluminum from nutritional sources in individuals with normal kidneys. Dusts of metallic aluminum and aluminum oxide are not significantly absorbed systemically, although fumes from welding aluminum is absorbed through the lung, producing an increase in aluminum levels in plasma and in urine.

The biological half-life of absorbed aluminum, as measured by urinary excretion, is approximately 9 days in workers exposed less than 1 year, and increases to 6 months after 10 years of occupational exposure. The neurotoxicity of aluminum and its salts is not known.

Acute Health Effects

The following acute (short-term) health effects may occur immediately or shortly after exposure to aluminum:

Skin: Irritation/scratching on contact with particles.

Eye: Irritation/scratching on contact with particles.

Lung: Possible irritation of the nose, throat, and lungs following exposure to fibers and dusts.

CNS: Inconclusive.

Chronic Health Effects

The following chronic (long-term) health effects may occur at some time after exposure to aluminum and can last for months or even years:

Cancer Hazards: There is limited evidence of an increase in bladder, lung, and other cancers among aluminum smelter workers. However, the increase appears to be due to exposure to polycyclic aromatic hydrocarbons, not to aluminum compounds.

Reproductive Hazard: Aluminum salts such as aluminum chloride and aluminum nitrate are toxic to the animal fetus and cause fetal damage. No data are available on the teratogenicity of aluminum in humans.

Other Chronic Effects: Long-term exposure to the fine dusts of aluminum can cause scarring of the lungs (pulmonary fibrosis) with symptoms of cough and shortness of breath. Very high levels of aluminum may damage the brain, although the data are not conclusive in this regard.

Recommended Risk-Reduction Measures

Personnel should avoid direct contact with aluminum. *Engineering controls* are always the most effective method of reducing exposures. The best protection, especially when working with fine powders and dusts of aluminum, is to enclose operations and/or provide explosion-proof, local exhaust ventilation at the site of powder/dust release. While not always operationally feasible, isolating operations can also reduce exposure. Using respiratory protection is less effective than the controls mentioned above, but is still advisable whenever working with or around aluminum. A self-contained breathing apparatus (SCBA) with full facepiece and pressure demand is the best respiratory protection. However, for exposures near or over the 2 mg/m^3 PEL for dusts, a respirator with a dust/mist/fume filter would suffice. If a full facepiece respirator is not available, then dust goggles or safety glasses with side shields attached should be worn to protect the eyes. To prevent hand and skin exposures, suitable gloves should be worn.

Administrative controls should also be in place to minimize the potential for human exposures. These may include written procedures or policies which specify the methods and techniques that will be practiced whenever personnel are to work with aluminum.

All personnel should receive *training* on the use, hazards, protective measures, emergency actions, and other precautions per 29 CFR 1910.1200 (Hazard Communication), prior to the first assignment in an area where aluminum is used or stored. Before beginning work with aluminum, and at regular intervals thereafter (e.g., annually), the following medical tests are recommended:

☑ Lung function tests.

☑ Chest X-ray should be considered.

☑ Urinalysis (monitor aluminum content).

Other methods to reduce exposure include:

☑ Where possible, enclose operations and use local exhaust ventilation at the site where aluminum powder or dust is worked. If local exhaust ventilation or enclosure is not used, respiratory protection should be mandatory.

☑ Always ensure that proper protective clothing is worn when working with aluminum compounds.

☑ Wash thoroughly immediately after exposure to aluminum and at the end of the work shift or before eating, drinking, or smoking. Clothing contaminated with aluminum dusts or powders should not be taken home.

☑ Hazard warning information should be posted in the work area. In addition, as part of an on-going education and training program, all information on the health and safety hazards of aluminum should be communicated to all exposed workers.

RISK ASSESSMENT: ENVIRONMENT

General Assessment

The environment is at risk of exposure during transportation, storage, disposal, or destruction of aluminum. In almost every scenario, the threat of environmental exposure is contingent upon the proper handling of aluminum materials, especially the dusts and powders. Accidental contact with a variety of incompatible materials can result in fire, explosion, and possible contamination of the surrounding environmental mediums (water, soil, and, to a lesser degree, air).

Aluminum is considered a combustible solid. It is incompatible with oxidizing agents and contact can cause fire or explosion. These characteristics require special consideration during any emergency situation involving a leak or spill of aluminum. Should aluminum ever come into contact with incompatible substances such as oxidizers, acids, halogenated hydrocarbons, or other reactive metals either during use, transportation, or storage, the formation of highly explosive commodities is extremely possible.

Elemental aluminum is one of the most abundant metals in the earth's crust. It is used extensively across a broad spectrum of industries for a large variety of purposes. Because of its frequent use and common occurrence, aluminum enters the environment from point and non-point sources.

Acute Ecological Effects

Acute (short-term) toxic effects may include the death of animals, birds, or fish and death or low growth rate in plants. Acute effects are seen 2 to 4 days after animals or plants are exposed to aluminum.

This metal has moderate acute toxicity to aquatic life and high toxicity to birds. Insufficient data are available on the short-term effects of aluminum exposure to plants or land animals.

Chronic Ecological Effects

Chronic (long-term) toxic effects may include shortened life span, reproductive problems, lower fertility, and changes in appearance or behavior in exposed animals. These effects can be seen long after first exposure(s) to toxic chemicals.

Aluminum has high chronic toxicity in aquatic life. Insufficient data are available on the long-term effects of aluminum to plants, birds, or land animals.

Water Solubility

Aluminum solubility in water is dependent upon pH. As pH increases or decreases from near neutrality, solubility in water increases.

⌛ *Persistence in the Environment*

Aluminum is highly persistent in water, with a half-life greater than 200 days. The half-life of a pollutant is the amount of time it takes for one half of the chemical to be degraded.

Bioaccumulation in Aquatic Organisms

Some substances increase in concentration, or *bioaccumulate,* in living organisms as they breathe contaminated air, drink contaminated water, or eat contaminated food. These chemicals can become concentrated in the tissues and internal organs of animals as well as humans.

Aluminum will not accumulate in fish tissues.

Recommended Risk-Reduction Measures

Proper training of all transporters will reduce the likelihood of a mishap or accident resulting in the contamination of the environment. The correct use of DOT labeling (per the MSDS) on all containers, trucks, and rail cars will enable emergency responders to react properly and quickly to any disaster thereby reducing the potential risk to the environment and to personnel.

Storage of aluminum should be segregated from incompatible chemicals to minimize the risk of cross-contamination or contact. Buildings designated for storage should be equipped with appropriate fire protection systems (alarms, sprinklers, emergency lighting, portable extinguishers).

If a spill or leak to the environment has occurred, fire department, emergency response and/or hazardous materials spill personnel should be notified immediately. Cleanup should be attempted only by those trained in proper spill containment procedures. Contaminated soils should be removed and replaced with clean soil. When environmental contamination occurs, the local and/or state emergency response authorities may require notification. A comprehensive emergency response or disaster preparedness/recovery plan should be in place prior to any operations involving the use, transportation, storage, or disposal of aluminum. If aluminum is spilled or leaked, the following specific steps are recommended:

- ☑ Restrict persons not wearing protective clothing from area of spill or leak until cleanup is complete and area can be opened for normal work.
- ☑ Remove all ignition sources and ventilate area of aluminum powder spill or leak.
- ☑ Collect powdered materials in the safest manner possible and deposit in sealed containers.
- ☑ It may be necessary to dispose of aluminum as a hazardous waste. The responsible state agency or the regional office of the federal Environmental Protection Agency (EPA) should be contacted for specific recommendations.

RISK ASSESSMENT: BUSINESS

General Assessment

Accidents or mishaps involving aluminum can present a significant threat to business operations. The loss or damage of equipment or facilities due to explosion or fire can significantly effect fiscal viability. Lawsuits that may result from personnel injury/death, public exposures, and/or environmental contamination will also require a serious expenditure of resources.

Media attention surrounding an injury, death, or environmental damage can also result in a loss of profits and loss of current as well as future business.

Recommended Risk-Reduction Measures

Company attorneys, safety and health professionals, and environmental specialists should be involved in the development of any procedures dealing with safety or emergency response. A company official should be pre-designated as a public relations officer with specific training in dealing with the press. Corporate plans and policies should be developed, approved, and implemented long before any need for such arises.

REFERENCES

Ahlbom, A. 1993. *Biostatistics for Epidemiologists.* New York: CRC Press/Lewis Publishers.

Alfrey, A. C., G. R. LeGrandre, W. D. Koehny. 1976, The dialysis encephalopathy syndrome: Possible aluminum intoxication. *New England Journal of Medicine.* 294:184-188.

American Conference of Governmental Industrial Hygienists. 1988. *Documentation of the Threshold Limit Values and Biological Exposure Indices,* 5th Edition (with updates). Cincinnati: ACGIH

Asante-Duah, D. K. 1993. *Hazardous Waste Risk Assessment.* New York: CRC Press/Lewis Publishers.

Bloom, A. D. and F. J. DeSerres. 1995. *Ecotoxicity and Human Health.* New York: CRC Press/Lewis Publishers.

Brusick, D. J. 1994. *Methods for Genetic Risk Assessment.* New York: CRC Press/Lewis Publishers.

Bueche, J. R. 1972. *Principles of Physics,* 2nd Edition. New York: McGraw-Hill.

Burrell, R., D. K. Flaherty, and Leonard J. Sauers. 1992. *Toxicology of the Immune System.* New York: Van Nostrand Reinhold.

Calabrese, E. J., E. M. Kenyon. *Air Toxics and Risk Assessment.* New York: CRC Press/Lewis Publishers.

Calabrese, E. J. 1994. *Biological Effects of Low Level Exposures.* New York: CRC Press/Lewis Publishers.

Cockerman, L. G., B. S. Shane. 1994. *Basic Environmental Toxicology.* New York: CRC Press/Lewis Publishers.

Cothern, C. R., N. P. Ross. 1994. *Environmental Statistics, Assessment, and Forecasting.* New York: CRC Press/Lewis Publishers.

Cothern, R. R. 1993. *Comparative Environmental Risk Assessment.* New York: CRC Press/Lewis Publishers.

Cothern, C. R. 1995. *Handbook for Environmental Risk Decisionmaking.* New York: CRC Press/Lewis Publishers.

Danse, I. R. 1991. *Common Sense Toxics in the Workplace.* New York: Van Nostrand Reinhold.

Gardner, R. W. 1994. *Chemical Intolerance.* New York: CRC Press/Lewis Publishers.

Hallenbeck, W. H. 1993. *Quantitative Risk Assessment for Environmental and Occupational Health,* 2nd Edition. New York: CRC Press/Lewis Publishers.

Hosty, J. W., P. Foster. 1990. *A Practical Guide to Chemical Spill Response.* New York: Van Nostrand Reinhold.

Howard, P. H. 1995. *Handbook of Environmental Fate and Exposures Data for Organic Chemicals,* Vol. I-IV. New York: CRC Press/Lewis Publishers.

Howard, P. H., et. al. 1991. *Handbook of Environmental Degradation Rates.* New York: CRC Press/Lewis Publishers.

Hughes, T. 1989. Aluminum and the human brain. *Practitioner.* 233:920-923.

James, A., M. Lord. 1992. *Index of Chemical and Physical Data.* New York: Van Nostrand Reinhold.

Jones, F. E. 1994. *Toxic Organic Vapors in the Workplace.* New York: CRC Press/Lewis Publishers.

Keith, L. H. and M. M. Walker. 1995. *Handbook of Air Toxics.* New York: CRC Press/Lewis Publishers.

Landis, W. G., M. H. Yu. 1995. *Introduction to Environmental Toxicology.* New York: CRC Press/Lewis Publishers.

Langerman, N. 1994 *Precautionary Labels for Chemical Containers.* New York: CRC Press/Lewis Publishers.

Lauwerys, R. R., P. Hoet. 1993. *Industrial Chemical Exposure.* New York: CRC Press/Lewis Publishers.

Lewis, R. J., Sr. 1993. *Hawley's Condensed Chemical Dictionary,* 12th Edition. New York: Van Nostrand Reinhold.

Lewis, R. J., Sr. 1992. *Sax's Dangerous Properties of Industrial Materials,* Volumes 1, 2, and 3, 8th Edition. New York: Van Nostrand Reinhold.

Lide, D. R. 1995. *Handbook of Chemistry and Physics,* 76th Edition. New York: CRC Press/Lewis Publishers.

Manahan, S. E. 1994. *Environmental Chemistry,* 6th Edition. New York: CRC Press/Lewis Publishers.

Manahan, S. E. 1992. *Toxicological Chemistry,* 2nd Edition. New York: CRC Press/Lewis Publishers.

Mellström, G. A., J. E. Wahlberg, and H. I. Maibach. 1994. *Protective Gloves for Occupational Use.* New York: CRC Press/Lewis Publishers.

Meyer, E. 1990. *Chemistry of Hazardous Materials.* Englewood Cliffs, NJ: Prentice Hall.

Mycock, J. C., J. D. McKenna, and L. Theodore. 1995. *Handbook of Air Pollution Control Engineering and Technology.* New York: CRC Press/Lewis Publishers.

Nelson, G. O. 1992. *Gas Mixtures.* New York: CRC Press/Lewis Publishers.

Ott, W. R. 1995. *Environmental Statistics and Data Analysis.* New York: CRC Press/Lewis Publishers.

Patnaik, P. 1992. *A Comprehensive Guide to the Hazardous Properties of Chemical Substances.* New York: Van Nostrand Reinhold.

Philp, R. B. 1995. *Environmental Hazards and Human Health.* New York: CRC Press/Lewis Publishers.

Proctor, N. H., J. P. Hughes. 1991. *Chemical Hazards of the Workplace.* New York: Van Nostrand Reinhold.

Quigley, D. R. 1994. *Handbook of Emergency Chemical Management.* New York: CRC Press/Lewis Publishers.

Rea, W. J. 1995. *Chemical Sensitivity,* Vol. I-III. New York: CRC Press/Lewis Publishers.

Richardson, M. 1995. *Dictionary of Substance and Their Effects,* Vol. I-VI. New York: CRC Press/Lewis Publishers.

Rylander, R., R. R. Jacobs. 1994. *Organic Dusts.* New York: CRC Press/Lewis Publishers.

Sacarello, H. L. A. 1994. *The Comprehensive Handbook of Hazardous Materials.* New York: CRC Press/Lewis Publishers.

Sheftel, V. O. 1995. *Handbook of Toxic Properties of Monomers and Additives.* New York: CRC Press/Lewis Publishers.

Sherman, J. D. 1988. *Chemical Exposure and Disease.* New York: Van Nostrand Reinhold.

Talbott, E., G. Craun. 1995. *An Introduction to Environmental Epidemiology.* New York: CRC Press/Lewis Publishers.

The Merck Index, 10th edition. 1983. Rahway, NJ: Merck & Company.

U.S. Department of Health and Human Services, National Institute for Occupational Safety and Health. 1994. *NIOSH Pocket Guide to Chemical Hazards.* Washington, D.C.: U.S. Government Printing Office.

Witorsch, P. and S. V. Spagnolo. 1994. *Air Pollution and Lung Disease in Adults.* New York: CRC Press/Lewis Publishers.

MATERIAL SAFETY DATA SHEET

CHEMICAL NAME

ALUMINUM BROMIDE

HAZARD WARNING INFORMATION

HEALTH	FIRE	REACTIVE	OTHER	DEGREE OF HAZARD	COLOR CODING	OTHER CODES
2	0	1	~~W~~ COR	0 = Minimum Hazard 1 = Slight Hazard 2 = Moderate Hazard 3 = Serious Hazard 4 = Severe Hazard	HEALTH = BLUE FIRE = RED REACTIVITY = YELLOW OTHER = WHITE	OX = Oxidizer ACID = Acid ALK = Alkali COR = Corrosive ~~W~~ = Use No Water

SECTION I - GENERAL INFORMATION

Characterization	**Metal**	RCRA Number	**None**
		EPA Class	**Not Applicable**
DOT Proper Shipping Name	**Aluminum Bromide Solution**	Chemical Abstract Service (CAS) Number	**7727-15-3**
DOT Hazard Class and Label Requirements	**Corrosive**	DOT Emergency Guide Code	**No Citation**
DOT Identification Number	**UN 1725 (anhydrous) UN 2580 (solution)**	Chemical Symbol	**$AlBr_3$**
Synonyms	**Aluminum tribromide, tribromoaluminum.**		

SECTION II - HAZARDOUS INGREDIENTS AND IDENTITY INFORMATION

Hazardous Components (specific identity)	OSHA Exposure Criteria	NIOSH Exposure Criteria	Immediately Dangerous to Life and Health (IDLH)	ACGIH Exposure Criteria
Aluminum bromide (derivation: By passing bromine over heated aluminum; reaction of hydrogen bromide with aluminum hydroxide).	**PEL:** 2 mg/m^3 **STEL:** Not Established	**REL:** Not Established	**Not Established**	**TLV:** 2 mg/m^3 **STEL:** Not Established

SECTION III - PHYSICAL AND CHEMICAL CHARACTERISTICS

Boiling Point	**493°F (256°C)**	Specific Gravity (H_2O = 1)	**3.2**
Vapor Pressure (mm Hg)	**1 at 178°F (81°C)**	Molecular Weight	**266.71**
Vapor Density (Air = 1)	**Not Found**	Melting Point	**207°F (97°C)**

Solubility

Soluble in ether, alcohol, carbon disulfide, and many organic solvents, including benzene, toluene, and simple hydrocarbons.

Appearance and Odor

White to yellowish-red lumps or deliquescent crystals.

SECTION IV - FIRE AND EXPLOSION HAZARD DATA

Flash Point (method used)	**Not Applicable**	Explosive Limits in Air % by Volume	**LEL: N/A UEL: N/A**
NFPA Classification	**Non-Flammable Solid**	Autoignition Temperature	**Not Applicable**

Extinguishing Media

Carbon dioxide, dry chemical, alcohol, or foam. DO NOT use water.

Special Fire Fighting Procedures

Powdered aluminum or aluminum dusts are potentially explosive and reactive. Great caution should be exercised when fighting fires involving aluminum powders/dusts.

Unusual Fire and Explosion Hazards

Powdered aluminum undergoes explosive reactions in the presence of numerous chemical substances. It also forms sensitive explosive mixtures with some oxidants. Dust is moderately flammable/explosive by heat, flame, or chemical reaction with powerful oxidizers.

SECTION V - REACTIVITY DATA

Stability		Conditions to Avoid **Aluminum is a reactive metal. The greatest industrial hazards are with chemical reactions. As with other metals, the powder and dusts are the most dangerous forms. Avoid contact with incompatibles.**
Stable **X**	Unstable	Incompatibility *(materials to avoid)* **Strong oxidizers (such as chlorine, bromine, fluorine), and strong acids (such as hydrochloric, sulfuric, and nitric), alcohols, and mixtures of sodium or potassium. Water and moist air.**
Hazardous Polymerization		Conditions to Avoid **Under normal conditions of temperature and pressure, aluminum bromide is not expected to undergo hazardous polymerization.**
May Occur	Will Not Occur **X**	Hazardous Decomposition or By-products **Aluminum bromide react with water or moist air to evolve heat and hydrogen bromide, which appears as white fumes in air.**

SECTION VI - HEALTH HAZARD DATA

Primary Route(s) of Entry:	Inhalation?	Absorption (skin)?	Ingestion?
	X	X	

Health Hazards

INHALATION: The powders and vapors are highly irritating. The liquid causes severe burns. Symptoms include shortness of breath, wheezing, nausea, aching muscles, and slight fever. May be fatal by inhalation due to spasm, inflammation, and edema of the larynx and bronchi, chemical pneumonitis and pulmonary edema.

SKIN & EYES: Skin and eye contact with powder can result in severe irritation and burns.

INGESTION: Toxic and destructive to the mucus membranes. Can lead to bromide poisoning.

Carcinogenicity	NTP Listed?	IARC Cancer Review Group?	OSHA Regulated?	Target Organs?
Unknown Human Unknown Animal	**No**	**No**	**No Specific Listing**	**Respiratory system, eyes, skin, liver, kidney.**

Medical Conditions Generally Aggravated by Exposure

Respiratory impairments (bronchitis, asthma); Skin conditions (dermatitis).

Emergency and First-aid Procedures

Eye contact: Flush immediately with water for 15 minutes (minimum), seek medical attention. Skin contact: Remove all contaminated clothing. Wash contaminated skin with soap and water. For inhalation: Remove the person from exposure. Provide respiratory assistance and CPR. Transfer to medical facility. Observe 24-48 hours for lung effects. If swallowed: Do NOT induce vomiting. Give one pint of water immediately and repeat every 10 minutes. Seek medical attention immediately.

SECTION VII - PRECAUTIONS FOR SAFE HANDLING AND USE

Steps to be Taken in Case Material is Released or Spilled

Collect powdered material in the most convenient way possible and deposit in sealed containers. Remove all sources of ignition. Ventilate area of spill. Restrict those not involved in cleanup from entering area. Keep water away. Hydrolysis is very violent.

Preferred Waste Disposal Method

Incineration in a chemical incinerator equipped with an afterburner and scrubber.

Precautions to be Taken in Handling and Storage

Containers should be protected against physical damage. Store in tightly closed containers in a cool, dry, well ventilated area preferably under a nitrogen purge. Keep aluminum bromide dry. It is advisable to refrigerate the anhydrous salt.

Other Precautions and Warnings

Whenever bulk or finely powdered aluminum bromide is used, handled, manufactured, or stored, use explosion-proof electrical equipment and fittings.

SECTION VIII - CONTROL MEASURES AND PERSONAL PROTECTIVE EQUIPMENT

Respiratory Protection *(specify type)*

For exposures near or over 2 mg/m^3: NIOSH approved respirator with high efficiency particulate air (HEPA) filter and a dust/fume/mist pre-filter. Greater protection is provided with a SCBA.

Ventilation

Local exhaust or general mechanical systems recommended.

Protective Gloves	Eye Protection	Other Protective Clothing
Suitable to protect from skin contact	**Chemical/Dust Goggles**	**N/A**

Work/Hygiene Practices

Always wash hands thoroughly after working with aluminum powders and dusts; never bring food, drink, or smoking materials into vicinity of aluminum powders.

ALUMINUM BROMIDE

$AlBr_3$ CAS: 7727-15-3

IDENTIFICATION AND TYPICAL USES

Aluminum bromide is a white to yellowish-red deliquescent crystalline solid. The solid crystal is used as an acid catalyst in organic synthesis. It is similar to aluminum chloride, but is more reactive and more soluble in organic media.

RISK ASSESSMENT: HEALTH

General Assessment

Aluminum bromide is a toxic and corrosive material. It has a high affinity for water and moisture which can cause severe tissue burns. Due to the water content in the eyes and perspiration on the skin, aluminum bromide can be extremely destructive and damaging on contact. It is toxic by *inhalation* of finely divided powders and by *ingestion.*

Prolonged exposures can lead to inorganic bromide poisoning with symptoms of depression, emaciation, and in severe cases, psychoses and mental deterioration.

Inhalation of aluminum bromide powders or vapors is extremely irritating to the mucous-lined, moisture-rich respiratory tract. The liquid can cause severe burns to these areas. Symptoms may take several hours to develop and usually include shortness of breath, coughing, wheezing, nausea, aching muscles, and slight fever. Inhalation may prove fatal in some cases due to spasm, inflammation, and edema of the larynx and bronchi, chemical pneumonitis, and delayed pulmonary edema (a medical emergency that can be fatal. Skin contact causes corrosive damage, especially when moist. Prolonged contact may lead to bromide rashes.

On ingestion, aluminum bromide is extremely toxic and destructive to the mucous membranes. Prolonged exposure may lead to bromide poisoning. Mild poisoning is characterized by nausea, vomiting, and diarrhea. Severe poisoning can cause convulsions and liver and kidney damage.

Acute Health Effects

The following acute (short-term) health effects may occur immediately or shortly after exposure to aluminum:

Skin: Strong irritation and corrosive damage. Moist or wet skin will be affected more severely.

Eye: Highly irritating and destructive. Liquid will cause severe burns. Possible damage or loss of vision.

Lung: Extremely irritating to the respiratory tract. May result in coughing, wheezing, nausea, aching muscles, and slight fever. Can also cause pulmonary edema (fluid in the lungs).

CNS: Inconclusive. However symptoms such as depression, emaciation, and, in severe cases, psychoses and mental deterioration have been reported as a result of bromide poisoning.

Chronic Health Effects

The following chronic (long-term) health effects may occur at some time after exposure to aluminum and can last for months or even years:

Cancer Hazards: Insufficient data exists to properly evaluate or predict the carcinogenicity of aluminum bromide in humans or test animals. While some studies suggest no carcinogenicity, the results of numerous other studies were inconclusive.

Reproductive Hazard: Other aluminum salts such as aluminum chloride and aluminum nitrate are toxic to the animal fetus and cause fetal damage. No data are available on the teratogenicity of aluminum bromide in humans.

Other Chronic Effects: Prolonged exposure may lead to bromide rashes on the skin (an irritating and difficult to heal form of dermatitis). Severe poisoning by ingestion can lead to liver and kidney damage.

Recommended Risk-Reduction Measures

Personnel should avoid direct contact with aluminum bromide. *Engineering controls* are always the most effective method of reducing exposures. The best protection, especially when working with powders and dusts of aluminum bromide, is to enclose operations and provide explosion-proof, local exhaust ventilation at the site of powder/dust release. While not always operationally feasible, isolating operations can also reduce exposure. Using respiratory protection is less effective than the controls mentioned above, but is still advisable whenever working with or around aluminum bromide. A MSHA/NIOSH approved self-contained breathing apparatus (SCBA) with full facepiece and pressure demand is the best respiratory protection. However, for exposures near or over the 2 mg/m^3 PEL for dusts, a full facepiece respirator equipped with a high efficiency particulate air (HEPA) filter and a dust/mist/fume pre-filter would suffice. If a full facepiece respirator is not available, then dust goggles or safety glasses with side shields attached should be worn to protect the eyes. To prevent hand and skin exposures, suitable gloves should be worn.

Administrative controls should also be in place to minimize the potential for human exposures. These may include written procedures or policies which specify the methods and techniques that will be practiced whenever personnel are to work with aluminum bromide.

All personnel should receive *training* on the use, hazards, protective measures, emergency actions, and other precautions per 29 CFR 1910.1200 (Hazard Communication), prior to the first assignment in an area where aluminum bromide is used, handled, or stored. Before beginning work with aluminum, and at regular intervals thereafter (e.g., annually), the following medical tests are recommended:

☑ Lung function tests.

☑ Chest X-ray should be considered (may be negative if taken immediately after exposure due to delayed onset of pulmonary edema).

Other methods to reduce exposure include:

☑ Where possible, enclose operations and use local exhaust ventilation at the site where aluminum powder or dust is worked. If local exhaust ventilation or enclosure is not used, respiratory protection should be mandatory.

☑ Always ensure that proper protective clothing is worn when working with aluminum compounds.

☑ Wash thoroughly immediately after exposure to aluminum and at the end of the work shift or before eating, drinking, or smoking. Clothing contaminated with aluminum dusts or powders should not be taken home.

☑ Hazard warning information should be posted in the work area. In addition, as part of an on-going education and training program, all information on the health and safety hazards of aluminum bromide should be communicated to all exposed and potentially exposed workers.

☑ A safety shower and emergency eyewash station should be readily available for emergency use in or near areas where aluminum bromide is used, handled, or stored.

RISK ASSESSMENT: ENVIRONMENT

General Assessment

The environment is at risk of exposure during transportation, storage, disposal, or destruction of aluminum bromide. In almost every scenario, the threat of environmental exposure is contingent upon the proper handling of aluminum bromide, especially the dusts and powders. Accidental contact with a variety of incompatible materials, including water, can result in fire, explosion, generation of other hazardous commodities, and possible contamination of the surrounding environmental mediums (water, soil, and, to a lesser degree, air).

Aluminum bromide is considered a non-combustible solid. However, it is incompatible with oxidizing agents and contact can cause fire or explosion. Mixtures with sodium or potassium can cause extremely violent explosions. These characteristics require special consideration during any emergency situation involving a leak or spill of aluminum. Should aluminum bromide ever come into contact with incompatible substances either during use, transportation, or storage, the formation of highly explosive commodities is extremely possible. On contact with water or moist air, it evolves flammable and corrosive hydrogen bromide along with heat.

Aluminum bromide can enter the environment through industrial discharges and from spills.

Acute Ecological Effects

Acute (short-term) toxic effects may include the death of animals, birds, or fish and death or low growth rate

in plants. Acute effects are seen 2 to 4 days after animals or plants are exposed to aluminum bromide. This material has moderate acute toxicity to aquatic life and high toxicity to birds. Insufficient data are available on the short-term effects of exposure to plants or land animals.

Chronic Ecological Effects

Chronic (long-term) toxic effects may include shortened life span, reproductive problems, lower fertility, and changes in appearance or behavior in exposed animals. These effects can be seen long after first exposure(s) to toxic chemicals. Aluminum bromide has high chronic toxicity in aquatic life. Insufficient data are available on the long-term effects of aluminum bromide to plants, birds, or land animals.

Water Solubility

Aluminum bromide solubility in water is dependent upon pH. As pH increases or decreases from near neutrality, solubility in water increases.

Persistence in the Environment

Aluminum bromide is highly persistent in water, with a half-life greater than 200 days. The half-life of a pollutant is the amount of time it takes for one half of the chemical to be degraded.

Bioaccumulation in Aquatic Organisms

Some substances increase in concentration, or *bioaccumulate,* in living organisms as they breathe contaminated air, drink contaminated water, or eat contaminated food. These chemicals can become concentrated in the tissues and internal organs of animals as well as humans.

Aluminum bromide will not accumulate in fish tissues.

Recommended Risk-Reduction Measures

Proper training of all transporters will reduce the likelihood of a mishap or accident resulting in the contamination of the environment. The correct use of DOT labeling (per the MSDS) on all containers, trucks, and rail cars will enable emergency responders to react properly and quickly to any disaster thereby reducing the potential risk to the environment and to personnel.

Storage of aluminum bromide should be segregated from incompatible chemicals to minimize the risk of cross-contamination or contact. It should be kept in tightly closed containers, preferably under a nitrogen purge. It must be kept free of moisture and away from water. Buildings designated for storage should be equipped with appropriate fire protection systems (alarms, sprinklers, emergency lighting, portable extinguishers).

If a spill or leak to the environment has occurred, fire department, emergency response and/or hazardous materials spill personnel should be notified immediately. Cleanup should be attempted only by those trained in proper spill containment procedures. Contaminated soils should be removed and replaced with clean soil. When environmental contamination occurs, the local and/or state emergency response authorities may require notification. A comprehensive emergency response or disaster preparedness/recovery plan should be in place prior to any operations involving the use, transportation, storage, or disposal of aluminum bromide.

The proper destruction or disposal method for aluminum bromide is to burn it in a chemical incinerator equipped with and afterburner and scrubber.

If aluminum bromide is spilled or leaked, the following specific steps are recommended:

- ☑ Restrict persons not wearing protective clothing from area of spill or leak until cleanup is complete and area can be opened for normal work.
- ☑ Remove all ignition sources and ventilate area of spill or leak.
- ☑ Collect powdered materials in the safest manner possible and deposit in sealed containers. Use a vacuum equipped with a HEPA filter for cleanup. Dry sweeping is not recommended (generates airborne dusts). Keep water away. Hydrolysis can be extremely violent and destructive.
- ☑ It may be necessary to dispose of aluminum bromide as a hazardous waste. The responsible state agency or the regional office of the federal Environmental Protection Agency (EPA) should be contacted for specific recommendations.

RISK ASSESSMENT: BUSINESS

General Assessment

Accidents or mishaps involving aluminum bromide can present a significant threat to business operations. The loss or damage of equipment or facilities due to explosion or fire can significantly effect fiscal viability.

Lawsuits that may result from personnel injury/death, public exposures, and/or environmental contamination will also require a serious expenditure of resources. Media attention surrounding an injury, death, or environmental damage can also result in a loss of profits and loss of current as well as future business.

Recommended Risk-Reduction Measures

Company attorneys, safety and health professionals, and environmental specialists should be involved in the development of any procedures dealing with safety or emergency response. A company official should be pre-designated as a public relations officer with specific training in dealing with the press. Corporate plans and policies should be developed, approved, and implemented long before any need for such arises.

REFERENCES

Ahlbom, A. 1993. *Biostatistics for Epidemiologists.* New York: CRC Press/Lewis Publishers.

Alfrey, A. C., G. R. LeGrandre, W. D. Koehny. 1976, The dialysis encephalopathy syndrome: Possible aluminum intoxication. *New England Journal of Medicine.* 294:184-188.

American Conference of Governmental Industrial Hygienists. 1988. *Documentation of the Threshold Limit Values and Biological Exposure Indices,* 5th Edition (with updates). Cincinnati: ACGIH

Asante-Duah, D. K. 1993. *Hazardous Waste Risk Assessment.* New York: CRC Press/Lewis Publishers.

Bloom, A. D. and F. J. DeSerres. 1995. *Ecotoxicity and Human Health.* New York: CRC Press/Lewis Publishers.

Brusick, D. J. 1994. *Methods for Genetic Risk Assessment.* New York: CRC Press/Lewis Publishers.

Bueche, J. R. 1972. *Principles of Physics,* 2nd Edition. New York: McGraw-Hill.

Burrell, R., D. K. Flaherty, and Leonard J. Sauers. 1992. *Toxicology of the Immune System.* New York: Van Nostrand Reinhold.

Calabrese, E. J. 1994. *Biological Effects of Low Level Exposures.* New York: CRC Press/Lewis Publishers.

Cockerman, L. G., B. S. Shane. 1994. *Basic Environmental Toxicology.* New York: CRC Press/Lewis Publishers.

Cothern, R. R. 1993. *Comparative Environmental Risk Assessment.* New York: CRC Press/Lewis Publishers.

Danse, I. R. 1991. *Common Sense Toxics in the Workplace.* New York: Van Nostrand Reinhold.

Gardner, R. W. 1994. *Chemical Intolerance.* New York: CRC Press/Lewis Publishers.

Hallenbeck, W. H. 1993. *Quantitative Risk Assessment for Environmental and Occupational Health,* 2nd Edition. New York: CRC Press/Lewis Publishers.

Hosty, J. W., P. Foster. 1990. *A Practical Guide to Chemical Spill Response.* New York: Van Nostrand Reinhold.

Howard, P. H., et. al. 1991. *Handbook of Environmental Degradation Rates.* New York: CRC Press/Lewis Publishers.

James, A., M. Lord. 1992. *Index of Chemical and Physical Data.* New York: Van Nostrand Reinhold.

Landis, W. G., M. H. Yu. 1995. *Introduction to Environmental Toxicology.* New York: CRC Press/Lewis Publishers.

Langerman, N. 1994 *Precautionary Labels for Chemical Containers.* New York: CRC Press/Lewis Publishers.

Lauwerys, R. R., P. Hoet. 1993. *Industrial Chemical Exposure.* New York: CRC Press/Lewis Publishers.

Lewis, R. J., Sr. 1993. *Hawley's Condensed Chemical Dictionary,* 12th Edition. New York: Van Nostrand Reinhold.

Lewis, R. J., Sr. 1992. *Sax's Dangerous Properties of Industrial Materials,* Volumes 1, 2, and 3, 8th Edition. New York: Van Nostrand Reinhold.

Lide, D. R. 1995. *Handbook of Chemistry and Physics,* 76th Edition. New York: CRC Press/Lewis Publishers.

Manahan, S. E. 1992. *Toxicological Chemistry,* 2nd Edition. New York: CRC Press/Lewis Publishers.

Meyer, E. 1990. *Chemistry of Hazardous Materials.* Englewood Cliffs, NJ: Prentice Hall.

Mycock, J. C., J. D. McKenna, and L. Theodore. 1995. *Handbook of Air Pollution Control Engineering and Technology.* New York: CRC Press/Lewis Publishers.

Nelson, G. O. 1992. *Gas Mixtures.* New York: CRC Press/Lewis Publishers.

Ott, W. R. 1995. *Environmental Statistics and Data Analysis.* New York: CRC Press/Lewis Publishers.

Philp, R. B. 1995. *Environmental Hazards and Human Health.* New York: CRC Press/Lewis Publishers.

Quigley, D. R. 1994. *Handbook of Emergency Chemical Management.* New York: CRC Press/Lewis Publishers.

Rea, W. J. 1995. *Chemical Sensitivity,* Vol. I-III. New York: CRC Press/Lewis Publishers.

Richardson, M. 1995. *Dictionary of Substance and Their Effects,* Vol. I-VI. New York: CRC Press/Lewis Publishers.

Sacarello, H. L. A. 1994. *The Comprehensive Handbook of Hazardous Materials.* New York: CRC Press/Lewis Publishers.

Sheftel, V. O. 1995. *Handbook of Toxic Properties of Monomers and Additives.* New York: CRC Press/Lewis Publishers.

Sherman, J. D. 1988. *Chemical Exposure and Disease.* New York: Van Nostrand Reinhold.

Talbott, E., G. Craun. 1995. *An Introduction to Environmental Epidemiology.* New York: CRC Press/Lewis Publishers.

The Merck Index, 10th edition. 1983. Rahway, NJ: Merck & Company.

Witorsch, P. and S. V. Spagnolo. 1994. *Air Pollution and Lung Disease in Adults.* New York: CRC Press/Lewis Publishers.

MATERIAL SAFETY DATA SHEET

CHEMICAL NAME	
ALUMINUM CHLORIDE	**CRC PRESS/LEWIS PUBLISHERS DISCLAIMER:** THE INFORMATION AND RECOMMENDATIONS PRESENTED HEREIN ARE BASED ON SOURCES BELIEVED TO BE RELIABLE. CRC MAKES NO REPRESENTATION ON ITS COMPLETENESS OR ACCURACY. IT IS THE RESPONSIBILITY OF THE USER TO DETERMINE THE CHEMICAL'S SUITABILITY FOR ITS INTENDED USE, THE CHEMICAL'S SAFE USE, AND THE CHEMICAL'S PROPER DISPOSAL. NO REPRESENTATIONS AND/OR WARRANTIES, EITHER EXPRESSED OR IMPLIED, OF THE MERCHANTABILITY OR FITNESS FOR A PARTICULAR PURPOSE, OR OF ANY OTHER NATURE, ARE MADE WITH RESPECT TO THE INFORMATION PROVIDED IN THIS MSDS OR TO THE CHEMICAL TO WHICH INFORMATION MAY REFER. CRC NEITHER ASSUMES NOR AUTHORIZES ANY OTHER PERSON TO ASSUME FOR IT, ANY OTHER ADDITIONAL RESPONSIBILITY OR LIABILITY FOR THE USE OF, OR RELIANCE UPON, THIS INFORMATION.

HAZARD WARNING INFORMATION

HEALTH	FIRE	REACTIVE	OTHER	DEGREE OF HAZARD	COLOR CODING	OTHER CODES
3	0	2	~~W~~ COR	0 = Minimum Hazard 1 = Slight Hazard 2 = Moderate Hazard 3 = Serious Hazard 4 = Severe Hazard	HEALTH = BLUE FIRE = RED REACTIVITY = YELLOW OTHER = WHITE	OX = Oxidizer ACID = Acid ALK = Alkali COR = Corrosive ~~W~~ = Use No Water

SECTION I - GENERAL INFORMATION

Characterization	RCRA Number	EPA Class
Chloride	**Not Listed**	**Not Applicable**

DOT Proper Shipping Name **Aluminum Chloride Solution**	Chemical Abstract Service (CAS) Number **7446-70-0**
DOT Hazard Class and Label Requirements **Corrosive Material**	DOT Emergency Guide Code **No Citation**
DOT Identification Number **UN 1725 (anhydrous) UN 2581 (solution)**	Chemical Symbol **$AlCl_3$**

Synonyms
Aluminum trichloride, tribromoaluminum.

SECTION II - HAZARDOUS INGREDIENTS AND IDENTITY INFORMATION

Hazardous Components (specific identity)	OSHA Exposure Criteria	NIOSH Exposure Criteria	Immediately Dangerous to Life and Health (IDLH)	ACGIH Exposure Criteria
Aluminum chloride (derivation: By reaction of purified gaseous chlorine with molten aluminum; by reaction of bauxite with coke and chlorine at approximately 875°C).	**PEL:** 2 mg(Al)/m^3 **STEL:** Not Established	**REL:** Not Established	**Not Established**	**TLV:** 2 mg(Al)/m^3 **STEL:** Not Established

SECTION III - PHYSICAL AND CHEMICAL CHARACTERISTICS

Boiling Point **358°F (181°C) Decomposes**	Specific Gravity (H_2O = 1) **2.44**
Vapor Pressure (mm Hg) **1 at 212°F (100°C)**	Molecular Weight **133.33**
Vapor Density (Air = 1) **1.0**	Melting Point **381°F (194°C) sublimes**

Solubility
Violently soluble in water; soluble in absolute alcohol, ether, tetrachloromethane, benzene.

Appearance and Odor
White to yellowish deliquescent crystalline solid. May have an odor of hydrogen chloride.

SECTION IV - FIRE AND EXPLOSION HAZARD DATA

Flash Point (method used) **Not Applicable**	Explosive Limits in Air % by Volume **LEL: N/A UEL: N/A**
NFPA Classification **Non-Combustible Powder**	Autoignition Temperature **Not Applicable**

Extinguishing Media
Carbon dioxide, dry chemical, alcohol, or foam. DO NOT use water.

Special Fire Fighting Procedures
Wear full protective clothing with self-contained breathing apparatus (SCBA). Structural fire fighting clothing is permeable and will not provide adequate protection. Poisonous gases are produced in fire. Move containers from fire area if it can be done without risk.

Unusual Fire and Explosion Hazards
Keep water away from fires involving aluminum chloride. Reacts with water forming hydrochloric acid and explosive hydrogen gas. Aluminum chloride presents a fire hazard when stored in metal containers and exposed to moisture.

SECTION V - REACTIVITY DATA

Stability		Conditions to Avoid **Keep away from incompatible materials, especially water and moisture, since violent and even explosive reactions can occur.**
Stable **X**	Unstable	Incompatibility *(materials to avoid)* **Water, allyl chloride, ethylene, ethylene oxide, most metals, benzoyl chloride/naphthalene, carbon dioxide or ethylenimine, most peroxides, ethyl oxide, nitromethane, oxygen difluoride, perchlorylbenzene.**
Hazardous Polymerization		Conditions to Avoid **Keep away from heat and alkenes.**
May Occur	Will Not Occur **X**	Hazardous Decomposition or By-products **Highly exothermic reaction with water or steam produces toxic fumes of hydrogen chloride (a corrosive) and hydrogen gas (a flammable).**

SECTION VI - HEALTH HAZARD DATA

Primary Route(s) of Entry:	Inhalation?	Absorption (skin)?	Ingestion?
	X	X	X

Health Hazards

INHALATION: **Destructive to the mucous membranes and upper respiratory tract. Dusts and vapors are are extremely irritating. The liquid solution can cause severe burns. Symptoms include burning sensation in the lungs, coughing, wheezing, headache, nausea, vomiting, and inflammation of the larynx and bronchi. Pulmonary edema also possible.**

SKIN & EYES: Skin and eye contact with powder can result in severe irritation and burns.

INGESTION: Toxic and destructive to the mucus membranes. Can cause nausea, vomiting, burning in the throat, and diarrhea.

Carcinogenicity	NTP Listed?	IARC Cancer Review Group?	OSHA Regulated?	Target Organs?
No Evidence.	**No**	**No**	**No Specific Listing**	**Respiratory system, eyes, skin, liver, kidney.**

Medical Conditions Generally Aggravated by Exposure

Respiratory impairments (bronchitis, asthma); Skin conditions (dermatitis).

Emergency and First-aid Procedures

Eye contact: Flush immediately with water for 15 minutes (minimum), seek medical attention. Skin contact: Remove all contaminated clothing. Wash contaminated skin with soap and water. For inhalation: Remove the person from exposure. Provide respiratory assistance and CPR. Transfer to medical facility. Observe 24-48 hours for lung effects. If swallowed: Do NOT induce vomiting. Give one pint of water immediately and repeat every 10 minutes. Seek medical attention at immediately.

SECTION VII - PRECAUTIONS FOR SAFE HANDLING AND USE

Steps to be Taken in Case Material is Released or Spilled

Collect powdered material in the most convenient way possible and deposit in sealed containers. Remove all sources of ignition. Restrict those not involved in cleanup from entering area.

Preferred Waste Disposal Method

Incineration in a chemical incinerator equipped with an afterburner and scrubber.

Precautions to be Taken in Handling and Storage

Containers should be protected against physical damage. Store in tightly closed containers in a cool, dry, well ventilated area preferably under a nitrogen purge. Keep aluminum chloride dry. Old containers may explode on opening, due to spontaneous decomposition in the presence of moisture.

Other Precautions and Warnings

Whenever bulk or finely powdered aluminum chloride is used, handled, manufactured, or stored, use explosion-proof electrical equipment and fittings.

SECTION VIII - CONTROL MEASURES AND PERSONAL PROTECTIVE EQUIPMENT

Respiratory Protection *(specify type)*

For exposures near or over 2 mg/m^3: NIOSH approved respirator with high efficiency particulate air (HEPA) filter and a dust/fume/mist pre-filter. Greater protection is provided with a SCBA.

Ventilation

Local exhaust or general mechanical systems recommended.

Protective Gloves	Eye Protection	Other Protective Clothing
Neoprene, Viton, Butyl Rubber	**Chemical/Dust Goggles**	**Full Protective Suit**

Work/Hygiene Practices

Always wash hands thoroughly after working with aluminum powders and dusts; never bring food, drink, or smoking materials into vicinity of aluminum powders.

ALUMINUM CHLORIDE

$AlCl_3$ CAS: 7446-70-0

IDENTIFICATION AND TYPICAL USES

Aluminum chloride (anhydrous) is a white to yellow crystalline powder that may have no odor or may have an odor of hydrogen chloride due to the formation of that gas in the presence of moist air. Aluminum chloride (hydrate) appears as a white or yellowish deliquescent crystalline solid, nearly odorless, with a sweat astringent taste. The anhydrous form is often used as an ethylbenzene catalyst, as a dyestuff intermediate, a detergent alkylate, in making ethyl chloride, in pharmaceuticals and organics, in the manufacture of butyl rubber, petroleum refining, hydrocarbon resins, and as a nucleating agent for titanium dioxide pigments. The hydrate is used in antiperspirants, and as roofing granules.

RISK ASSESSMENT: HEALTH

General Assessment

Aluminum chloride is a toxic and corrosive material. It has a high affinity for water and moisture which can cause severe tissue burns. Due to the water content in the eyes and perspiration on the skin, aluminum chloride can be extremely destructive and damaging on contact. It is toxic by *inhalation* of finely divided powders and by *ingestion.* The acute biological hazards of aluminum chloride are mostly due to the extremely acid products generated when it contacts water which it will take from the tissues, and the heat generated in that reaction.

Inhalation of aluminum chloride powders or vapors is extremely irritating to the mucous-lined, moisture-rich respiratory tract. The liquid can cause severe burns to these areas. Symptoms may take several hours to develop and usually include a burning sensation in the chest, coughing, wheezing, headache, nausea, and vomiting. Inhalation may prove fatal in some cases due to spasm, inflammation, and edema of the larynx and bronchi, chemical pneumonitis, and pulmonary edema. Prolonged exposures may lead to lung damage.

On ingestion, aluminum chloride is extremely toxic and destructive to the mucous membranes. It is an oral irritant. Symptoms of mild poisoning include nausea, vomiting, burning in the throat, and diarrhea. Severe poisoning can cause convulsions and liver and kidney damage. Exposure to the hydrate form may cause stomach irritation, and relaxation and swelling of the duodenum and small intestine (animal studies confirm).

Acute Health Effects

The following acute (short-term) health effects may occur immediately or shortly after exposure to aluminum:

Skin: Dusts and vapors are strong irritants and the liquid may cause sever burns. Moist or wet skin will be affected more severely.

Eye: Highly irritating and destructive. Liquid causes severe burns. Possible damage or loss of vision.

Lung: Extremely irritating to the respiratory tract and destructive to the mucous membranes. May cause burning sensation, coughing, wheezing, nausea, headache, and vomiting. Can also cause pulmonary edema (fluid buildup in the lungs) and/or chemical pneumonitis.

Other: Ingestion may result in severe gastrointestinal irritation with symptoms of nausea, vomiting, burning in the throat, and diarrhea. Severe poisoning may cause convulsions and kidney and liver damage.

Chronic Health Effects

The following chronic (long-term) health effects may occur at some time after exposure to aluminum and can last for months or even years:

Cancer Hazards: Insufficient data exist to properly evaluate or predict the carcinogenicity of aluminum chloride in humans or test animals. While some studies suggest no carcinogenicity, the results of numerous other studies were inconclusive (some studies suggest action on the chromosomal level).

Reproductive Hazard: Some aluminum salts, including aluminum chloride and aluminum nitrate, are toxic to the animal fetus and cause fetal damage. No data are available on the teratogenicity of aluminum chloride in humans.

Other Chronic Effects: Prolonged exposure to aluminum chloride may result in lung damage. Animal studies using hydrated aluminum chloride dust have reported changes in alveoli and the development of granule growths in the lung. The hydrated form also has been found to inhibit sweating and cause clinical irritation after prolonged exposure. Aluminum chloride may cause dermatitis.

Recommended Risk-Reduction Measures

Personnel should avoid direct contact with aluminum chloride. *Engineering controls* are always the most effective method of reducing exposures. The best protection, especially when working with powders and dusts of aluminum chloride, is to enclose operations and provide explosion-proof, local exhaust ventilation at the site of powder/dust release. While not always operationally feasible, isolating operations can also reduce exposure. Using respiratory protection is less effective than the controls mentioned above, but is still advisable whenever working with or around aluminum chloride. A MSHA/NIOSH approved self-contained breathing apparatus (SCBA) with full facepiece and pressure demand is the best respiratory protection. However, for exposures near or over the 2 mg/m^3 PEL for dusts, a full facepiece respirator equipped with a high efficiency particulate air (HEPA) filter and a dust/mist/fume pre-filter would suffice. If a full facepiece respirator is not available, then dust goggles or safety glasses with side shields attached should be worn to protect the eyes. If splash hazards exist, then a face shield is required. Eye contact must be prevented since severe and perhaps permanent damage is likely on contact with aluminum chloride powders or solutions. To prevent hand and skin exposures, suitable gloves should be worn.

Administrative controls should also be in place to minimize the potential for human exposures. These may include written procedures or policies which specify the methods and techniques that will be practiced whenever personnel are to work with aluminum chloride.

All personnel should receive *training* on the use, hazards, protective measures, emergency actions, and other precautions per 29 CFR 1910.1200 (Hazard Communication), prior to the first assignment in an area where aluminum chloride is used, handled, or stored. Before beginning employment with aluminum, and at regular intervals thereafter (e.g., annually), the following medical tests are recommended:

☑ Lung function tests.

☑ Chest X-ray should be considered.

If symptoms occur or an overexposure is suspected, the following additional tests may be useful:

☑ Kidney and liver function tests.

Other methods to reduce exposure include:

☑ Where possible, enclose operations and use local exhaust ventilation at the site where aluminum powder or dust is worked. If local exhaust ventilation or enclosure is not used, respiratory protection should be mandatory.

☑ Always ensure that proper protective clothing is worn when working with aluminum compounds.

☑ Wash thoroughly immediately after exposure to aluminum and at the end of the work shift or before eating, drinking, or smoking. Clothing contaminated with aluminum dusts or powders should not be taken home.

☑ Hazard warning information should be posted in the work area. In addition, as part of an on-going education and training program, all information on the health and safety hazards of aluminum chloride should be communicated to all potentially exposed workers.

☑ A safety shower and emergency eyewash station should be readily available for emergency use in or near areas where aluminum chloride is used, handled, or stored.

RISK ASSESSMENT: ENVIRONMENT

General Assessment

The environment is at risk of exposure during transportation, storage, disposal, or destruction of aluminum chloride. In almost every scenario, the threat of environmental exposure is contingent upon the proper handling of aluminum chloride, especially the dusts and powders (anhydrous form). Accidental contact with a variety of incompatible materials, including water, can have adverse and potentially dangerous results. This includes fire, explosion, generation of other hazardous commodities, and possible contamination of the surrounding environmental mediums (water, soil, and, to a lesser degree, air).

Aluminum chloride is considered a non-combustible solid. However, it is incompatible with many other commodities including chemically active metals (potassium and sodium, for example), and oxidizing agents. Contact can cause fire or explosion. Contact with water or moisture will result in a violent evolution of heat and the generation of hydrochloric acid fumes. These characteristics require special consideration during any emergency situation involving a leak or spill of aluminum. Should aluminum chloride ever come into contact with such incompatible substances either during use, transportation, or storage, the formation of highly explosive commodities is extremely possible.

Aluminum chloride can enter the environment through industrial discharges and from spills.

Acute Ecological Effects

Acute (short-term) toxic effects may include the death of animals, birds, or fish and death or low growth rate in plants. Acute effects are seen 2 to 4 days after animals or plants are exposed to aluminum chloride. This material has moderate acute toxicity to aquatic life. Insufficient data are available on the short-term effects of exposure to plants, birds, or land animals.

Chronic Ecological Effects

Chronic (long-term) toxic effects may include shortened life span, reproductive problems, lower fertility, and changes in appearance or behavior in exposed animals. These effects can be seen long after first exposure(s) to toxic chemicals. Aluminum chloride has high chronic toxicity in aquatic life. Insufficient data are available on the long-term effects of aluminum chloride to plants, birds, or land animals.

Water Solubility

Aluminum chloride is violently soluble in water. It will react on contact to form toxic fumes of hydrochloric acid and evolve heat.

Hydrochloric acid is highly soluble in water. Concentrations of 1000 milliliters and more will mix with a liter of water.

Persistence in the Environment

Aluminum chloride is practically non-persistent in water, since it undergoes exothermic reaction to evolve heat and hydrochloric acid. Hydrochloric acid is moderately persistent in the aquatic environment with a half life between 20 and 200 days. The half-life of a pollutant is the amount of time it takes for one half of the chemical to be degraded.

Bioaccumulation in Aquatic Organisms

Some substances increase in concentration, or *bioaccumulate,* in living organisms as they breathe contaminated air, drink contaminated water, or eat contaminated food. These chemicals can become concentrated in the tissues and internal organs of animals as well as humans.

Aluminum chloride will not accumulate in fish tissues.

Recommended Risk-Reduction Measures

Proper training of all transporters will reduce the likelihood of a mishap or accident resulting in the contamination of the environment. The correct use of DOT labeling (per the MSDS) on all containers, trucks, and rail cars will enable emergency responders to react properly and quickly to any disaster thereby reducing the potential risk to the environment and to personnel.

Storage of aluminum chloride should be segregated from incompatible chemicals to minimize the risk of cross-contamination or contact. It should be kept in tightly closed containers, preferably under a nitrogen purge. It must be kept free of moisture and away from water. Buildings designated for storage should be equipped with appropriate fire protection systems (alarms, sprinklers, emergency lighting, portable extinguishers).

If a spill or leak to the environment has occurred, fire department, emergency response and/or hazardous materials spill personnel should be notified immediately. Cleanup should be attempted only by those trained in proper spill containment procedures. Contaminated soils should be removed and replaced with clean soil. When environmental contamination occurs, the local and/or state emergency response authorities may require notification. A comprehensive emergency response or disaster preparedness/recovery plan should be in place prior to any operations involving the use, transportation, storage, or disposal of aluminum chloride.

Spills of aluminum chloride can be neutralized by rinsing with sodium bicarbonate and lime solution. If crushed limestone, soda ash or lime is used, flush spill area with water.

The proper destruction or disposal method for aluminum chloride is to burn it in a chemical incinerator equipped with and afterburner and scrubber.

If aluminum chloride is spilled or leaked, the following specific steps are recommended:

☑ Restrict persons not wearing protective clothing from area of spill or leak until cleanup is complete.

☑ Remove all ignition sources and ventilate area of spill or leak.

☑ Collect powdered materials in the safest manner possible and deposit in sealed containers. Use a vacuum equipped with a HEPA filter for cleanup. Dry sweeping is not recommended (generates airborne dusts). Keep water away.

☑ It may be necessary to dispose of aluminum chloride as a hazardous waste. The responsible state agency or the regional office of the federal Environmental Protection Agency (EPA) should be contacted for specific recommendations.

RISK ASSESSMENT: BUSINESS

General Assessment

Accidents or mishaps involving aluminum chloride can present a significant threat to business operations. The loss or damage of equipment or facilities due to explosion or fire can significantly affect fiscal viability. Lawsuits that may result from personnel injury/death, public exposures, and/or environmental contamination will also require a serious expenditure of resources. Media attention surrounding an injury, death, or environmental damage can also result in a loss of profits and loss of current as well as future business.

Recommended Risk-Reduction Measures

Company attorneys, safety and health professionals, and environmental specialists should be involved in the development of any such procedures. A company official should be pre-designated as a public relations officer with specific training in dealing with the press. Corporate plans and policies should be developed, approved, and implemented long before any need for such arises.

REFERENCES

American Conference of Governmental Industrial Hygienists. 1988. *Documentation of the Threshold Limit Values and Biological Exposure Indices,* 5th Edition (with updates). Cincinnati: ACGIH

Bloom, A. D. and F. J. DeSerres. 1995. *Ecotoxicity and Human Health.* New York: CRC Press/Lewis Publishers.

Bueche, J. R. 1972. *Principles of Physics,* 2nd Edition. New York: McGraw-Hill.

Burrell, R., D. K. Flaherty, and Leonard J. Sauers. 1992. *Toxicology of the Immune System.* New York: Van Nostrand Reinhold.

Calabrese, E. J. 1994. *Biological Effects of Low Level Exposures.* New York: CRC Press/Lewis Publishers.

Cockerman, L. G., B. S. Shane. 1994. *Basic Environmental Toxicology.* New York: CRC Press/Lewis Publishers.

Gardner, R. W. 1994. *Chemical Intolerance.* New York: CRC Press/Lewis Publishers.

Hallenbeck, W. H. 1993. *Quantitative Risk Assessment for Environmental and Occupational Health,* 2nd Edition. New York: CRC Press/Lewis Publishers.

Hosty, J. W., P. Foster. 1990. *A Practical Guide to Chemical Spill Response.* New York: Van Nostrand Reinhold.

Lauwerys, R. R., P. Hoet. 1993. *Industrial Chemical Exposure.* New York: CRC Press/Lewis Publishers.

Lewis, R. J., Sr. 1992. *Sax's Dangerous Properties of Industrial Materials,* Volumes 1, 2, and 3, 8th Edition. New York: Van Nostrand Reinhold.

Manahan, S. E. 1992. *Toxicological Chemistry,* 2nd Edition. New York: CRC Press/Lewis Publishers.

Meyer, E. 1990. *Chemistry of Hazardous Materials.* Englewood Cliffs, NJ: Prentice Hall.

Quigley, D. R. 1994. *Handbook of Emergency Chemical Management.* New York: CRC Press/Lewis Publishers.

Rea, W. J. 1995. *Chemical Sensitivity,* Vol. I-III. New York: CRC Press/Lewis Publishers.

Richardson, M. 1995. *Dictionary of Substance and Their Effects,* Vol. I-VI. New York: CRC Press/Lewis Publishers.

Sacarello, H. L. A. 1994. *The Comprehensive Handbook of Hazardous Materials.* New York: CRC Press/Lewis Publishers.

Sherman, J. D. 1988. *Chemical Exposure and Disease.* New York: Van Nostrand Reinhold.

Talbott, E., G. Craun. 1995. *An Introduction to Environmental Epidemiology.* New York: CRC Press/Lewis Publishers.

MATERIAL SAFETY DATA SHEET

CHEMICAL NAME

ALUMINUM FLUORIDE

HAZARD WARNING INFORMATION

HEALTH	FIRE	REACTIVE	OTHER	DEGREE OF HAZARD	COLOR CODING	OTHER CODES
1	0	0		0 = Minimum Hazard 1 = Slight Hazard 2 = Moderate Hazard 3 = Serious Hazard 4 = Severe Hazard	HEALTH = BLUE FIRE = RED REACTIVITY = YELLOW OTHER = WHITE	OX = Oxidizer ACID = Acid ALK = Alkali COR = Corrosive W = Use No Water

SECTION I - GENERAL INFORMATION

Characterization: **Metal**	RCRA Number: **Not Listed**	EPA Class: **Not Applicable**
DOT Proper Shipping Name: **Aluminum Fluoride Solution**	Chemical Abstract Service (CAS) Number: **7784-18-1**	
DOT Hazard Class and Label Requirements: **No Citation**	DOT Emergency Guide Code: **No Citation**	
DOT Identification Number: **No Citation**	Chemical Symbol: **AlF_3**	

Synonyms

Aluminum trifluoride.

SECTION II - HAZARDOUS INGREDIENTS AND IDENTITY INFORMATION

Hazardous Components (specific identity)	OSHA Exposure Criteria	NIOSH Exposure Criteria	Immediately Dangerous to Life and Health (IDLH)	ACGIH Exposure Criteria
Aluminum fluoride (derivation: Action of hydrogen fluoride gas on alumina trihydrate; reaction of hydrogen fluoride on a suspension of aluminum trihydrate, followed by calcining the hydrate formed; reaction of fluosilic acid on aluminum hydrate).	**PEL:** 2 mg(Al)/m^3 **STEL:** Not Established	**REL:** Not Established	**Not Established**	**TLV:** 2 mg(Al)/m^3 **STEL:** Not Established

SECTION III - PHYSICAL AND CHEMICAL CHARACTERISTICS

Boiling Point: **2800°F (1537°C)**	Specific Gravity (H_2O = 1): **2.88**
Vapor Pressure (mm Hg): **1 at 2260°F (1238°C)**	Molecular Weight: **83.98**
Vapor Density (Air = 1): **Not Available**	Melting Point: **2300°F (1260°C) sublimes**

Solubility in Water

Sparingly soluble in water, acids, and alkalis. Insoluble in most organic solvents.

Appearance and Odor

White crystals or white crystalline powder.

SECTION IV - FIRE AND EXPLOSION HAZARD DATA

Flash Point (method used): **Not Applicable**	Explosive Limits in Air % by Volume: **LEL: N/A UEL: N/A**
NFPA Classification: **Non-Combustible Powder**	Autoignition Temperature: **Not Applicable**

Extinguishing Media

Carbon dioxide, dry chemical, water fog. Use extinguishing media appropriate to surrounding fires.

Special Fire Fighting Procedures

Wear full protective clothing with self-contained breathing apparatus (SCBA). Structural fire fighting clothing is permeable and will not provide adequate protection. Poisonous gases are produced in fire, including toxic hydrogen fluoride. Move containers from fire area if it can be done without risk. Dike area for control and containment to prevent water run-off into sewers. Evacuate non-essential personnel a minimum of 2500 feet from danger area (consider down-wind conditions).

Unusual Fire and Explosion Hazards

Reacts violently with chlorinated rubber.

SECTION V - REACTIVITY DATA

Stability		Conditions to Avoid **Aluminum fluoride is generally considered to be quite inert, but violent reactions are still possible in contact with incompatible materials (e.g., violently impact sensitive in contact with sodium or potassium).**
Stable **X**	Unstable	Incompatibility *(materials to avoid)* **Potassium and sodium, strong acids (such as sulfuric, nitric, and hydrochloric) and acid fumes. May react very slowly in contact with water to evolve heat and possible hydrofluoric acid fumes.**
Hazardous Polymerization		Conditions to Avoid **Under normal conditions of temperature and pressure, hazardous polymerization of aluminum fluoride is not expected to occur.**
May Occur	Will Not Occur **X**	Hazardous Decomposition or By-products **When heated to decomposition, toxic fumes of hydrogen fluoride and aluminum oxide are released.**

SECTION VI - HEALTH HAZARD DATA

Primary Route(s) of Entry:	Inhalation?	Absorption (skin)?	Ingestion?
	X	X	X

Health Hazards

INHALATION: **Symptoms of fluoride poisoning, including shortness of breath, congestion of the lungs, muscle spasm, asthma-like problems, and convulsions. Exposures can prove fatal. May cause perforation of the nasal septum. May affect the brain, liver, kidneys, and lungs.**

ABSORPTION: **Skin and eye contact with powder can result in severe irritation and burns. Can pass through unbroken skin to cause toxic systemic effects of fluoride poisoning.**

INGESTION: **Symptoms of toxic systemic fluoride poisoning. May also cause shock, blurred vision, shallow breathing, convulsions, kidney failure, and death.**

Carcinogenicity	NTP Listed?	IARC Cancer Review Group?	OSHA Regulated?	Target Organs?
No Evidence.	**No**	**No**	**No Specific Listing**	**Respiratory system, eyes, skin, liver, kidney, brain.**

Medical Conditions Generally Aggravated by Exposure

Respiratory impairments (bronchitis, asthma); Skin conditions (dermatitis).

Emergency and First-aid Procedures

Eye contact: Flush immediately with water for 15 minutes, followed by irrigation with normal saline for 30 minutes. Do NOT use calcium gluconate gel. Seek medical attention. Skin contact: Remove all contaminated clothing. If calcium gluconate gel is available, rub into effected areas until pain subsides. If not available, wash contaminated skin with soap and water for 30 minutes. For inhalation: Remove the person from exposure. Provide respiratory assistance and CPR. Transfer to medical facility. If swallowed: Do NOT induce vomiting. If conscious, give milk or calcium gluconate by mouth. Seek medical attention immediately. Never try to give an unconscious person anything by mouth.

SECTION VII - PRECAUTIONS FOR SAFE HANDLING AND USE

Steps to be Taken in Case Material is Released or Spilled

Collect powdered material in the most convenient way possible (use HEPA vacuum). Remove all sources of ignition and ventilate area. Restrict those not involved in cleanup from entering area.

Preferred Waste Disposal Method

Neutralize with water and sodium carbonate. Dispose in landfill regulated for hazardous wastes.

Precautions to be Taken in Handling and Storage

Aluminum fluoride should not be stored in glass containers. Use tightly closed containers stored in a cool, dry, well-ventilated area. Always wear full-protective clothing to avoid contact.

Other Precautions and Warnings

Do not breathe dusts and avoid contact with eyes, skin, and clothing.

SECTION VIII - CONTROL MEASURES AND PERSONAL PROTECTIVE EQUIPMENT

Respiratory Protection *(specify type)*

For exposures near or over 2 mg/m^3: NIOSH approved respirator with high efficiency particulate air (HEPA) filter and a dust/fume/mist pre-filter. Greater protection is provided with a SCBA.

Ventilation

Local exhaust or general mechanical systems recommended.

Protective Gloves	Eye Protection	Other Protective Clothing
Impervious Rubber (not chlorinated)	**Chemical/Dust Goggles**	**Full Protective Clothing**

Work/Hygiene Practices

Always wash hands thoroughly after working with aluminum powders and dusts; never bring food, drink, or smoking materials into vicinity of aluminum powders.

ALUMINUM FLUORIDE

AlF_3 CAS: 7784-18-1

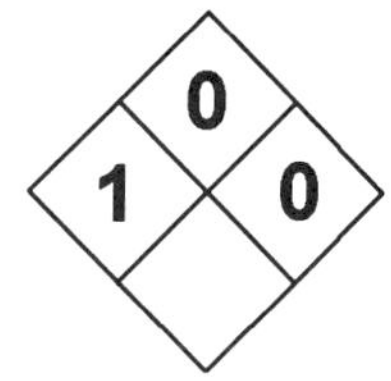

IDENTIFICATION AND TYPICAL USES

Aluminum fluoride is a white crystalline solid or a white powder. The anhydrous form is used in the production of aluminum to lower the melting point and increase the conductivity of the electrolyte. It is also used as a flux in ceramic glazes and enamels, in the manufacture of aluminum silicate, and as a catalyst. The hydrate is primarily used in the production of white enamels used in ceramics.

RISK ASSESSMENT: HEALTH

General Assessment

Aluminum fluoride is generally less toxic that most fluorides because of its slight water solubility. However, it is a human poison by *ingestion, inhalation,* and *absorption.* Exposures can be fatal in extreme cases. General symptoms of overexposure may include salivation, nausea, vomiting, abdominal pain, fever, and difficulty breathing. Exposure to high concentrations of fluoride results in hypocalcemia (low calcium content).

Inhalation causes symptoms of fluoride poisoning and may be fatal if large quantities are inhaled. Symptoms of severe poisoning include shortness of breath, congestion of the lungs, muscle spasm, and convulsions. Aluminum chloride has also been shown to be associated with asthma-like problems. Prolonged exposure to the fumes and dusts can cause perforations in the nasal septum.

Contact with the eye is severely irritating and may result in corneal burns. It is also extremely irritating to the skin and may result in deep and painful burns. The material may be absorbed through unbroken skin, resulting in symptoms of fluoride poisoning. There may or may not be any noticeable pain on initial contact. However, as the fluoride ions pass to deeper layers of skin, the pain will become quite severe and may be extremely difficult to remedy.

Ingestion of aluminum fluoride results in symptoms of fluoride poisoning. Severe exposures may cause shock, blurred vision, muscle spasms, shallow breathing, and convulsions. Kidney failure may also result.

Acute Health Effects

The following acute (short-term) health effects may occur immediately or shortly after exposure to aluminum:

Skin: Dusts and vapors are strong irritants and the liquid may cause severe burns. May pass through unbroken skin to cause toxic systemic effects. Fluoride will also attack bones because of its high affinity for calcium.

Eye: Highly irritating and destructive. Liquid causes severe burns. Possible damage to cornea and loss of vision.

Lung: Extremely irritating to the respiratory tract and destructive to the mucous membranes. May result in shortness of breath, coughing, lung congestion, muscle spasms, and convulsions.

CNS: Inconclusive. However, ingestion and inhalation can lead to symptoms of blurred vision, muscle spasms, shallow breathing, and convulsions.

Chronic Health Effects

The following chronic (long-term) health effects may occur at some time after exposure to aluminum and can last for months or even years:

Cancer Hazards: According to the information presented in the references, aluminum fluoride has not been shown to cause cancer in test animals.

Reproductive Hazard: Some aluminum salts have been shown to be toxic to the animal fetus and cause fetal damage. No data are available on the teratogenicity of aluminum fluoride in humans.

Other Chronic Effects: Prolonged exposures to powders and dusts can cause perforations to appear in the nasal septa. Degenerative changes in the brain, liver, and kidneys were reported in animal tests. Histological changes are also possible in the lungs following chronic exposures. Chronic exposures also cause excessive calcification of the bones, ligaments, and tendons. Exposures to high concentrations can result in hypocalcemia.

Recommended Risk-Reduction Measures

Personnel should avoid direct contact with aluminum fluoride. It is extremely irritating to tissues and, because of it ability to cause toxic fluoride poisoning, any form of contact is considered hazardous. *Engineering controls* are always the most effective method of reducing exposures. The best protection, especially when working with powders and dusts of aluminum fluoride, is to enclose operations and provide local exhaust ventilation at the site of powder/dust release. While not always operationally feasible, isolating operations can also reduce exposure. Using respiratory protection is less effective than the controls mentioned above, but is still advisable whenever working with or around aluminum fluoride. An MSHA/NIOSH approved self-contained breathing apparatus (SCBA) with full facepiece and pressure demand is the best respiratory protection. However, for exposures near or over the PEL (2 mg/m^3 for dusts), a full facepiece respirator equipped with a high efficiency particulate air (HEPA) filter and a dust/mist/fume pre-filter would suffice. If a full facepiece respirator is not available, then dust goggles or safety glasses with side shields attached should be worn to protect the eyes. If splash hazards exist, then a face shield is required. Eye contact must be prevented since severe and perhaps permanent damage is likely on contact with aluminum fluoride powders or solutions. To prevent hand and skin exposures, suitable gloves should be worn.

Administrative controls should also be in place to minimize the potential for human exposures. These may include written procedures or policies which specify the methods and techniques that will be practiced whenever personnel are to work with aluminum fluoride.

All personnel should receive *training* on the use, hazards, protective measures, emergency actions, and other precautions per 29 CFR 1910.1200 (Hazard Communication), prior to the first assignment in an area where aluminum fluoride is used or stored.

If symptoms occur or an overexposure is suspected, the following tests may be useful:

- ☑ Kidney and liver function tests.
- ☑ Lung function tests.
- ☑ Chest X-ray should be considered.

Other methods to reduce exposure include:

- ☑ Have calcium gluconate gel available for emergency use. On contact with skin, rub the gel into affected area until pain subsides. The fluoride ion will attack calcium. The gluconate will provide a calcium source to deter bone effects or damage. If swallowed, administer calcium gluconate by mouth (conscious victims only). Never apply gluconate gel to the eyes.
- ☑ Where possible, enclose operations and use local exhaust ventilation at the site where aluminum powder or dust is worked. If local exhaust ventilation or enclosure is not used, respiratory protection should be mandatory.
- ☑ Always ensure that proper protective clothing is worn when working with aluminum compounds.
- ☑ Wash thoroughly immediately after exposure to aluminum and at the end of the work shift or before eating, drinking, or smoking. Clothing contaminated with aluminum dusts or powders should not be taken home.
- ☑ Hazard warning information should be posted in the work area. In addition, as part of an on-going education and training program, all information on the health and safety hazards of aluminum fluoride should be communicated to all exposed or potentially exposed workers.
- ☑ A safety shower and emergency eyewash station should be readily available for emergency use in or near areas where aluminum fluoride is used, handled, or stored.

RISK ASSESSMENT: ENVIRONMENT

General Assessment

The environment is at risk of exposure during transportation, storage, disposal, or destruction of aluminum fluoride. In almost every scenario, the threat of environmental exposure is contingent upon the proper handling of aluminum fluoride, especially the dusts and powders (anhydrous form). Accidental contact with a variety of incompatible materials can have adverse and potentially dangerous results. This includes fire, explosion, generation of other hazardous commodities, and possible contamination of the surrounding environmental mediums (water, soil, and, to a lesser degree, air).

Aluminum fluoride is considered a non-combustible solid. However, it is incompatible with many other commodities including chemically active metals (potassium and sodium, for example), acids, and acid fumes. Contact can cause fire or explosion. When heated, it can emit toxic fumes of fluoride (including extremely toxic hydrogen fluoride). These characteristics require special consideration during any emergency situation involving a leak or spill of aluminum. Should aluminum fluoride ever come into contact with any incompatible substances either during use, transportation, or storage, extremely hazardous results can occur.

Aluminum fluoride can enter the environment through industrial discharges and from spills.

Acute Ecological Effects

Acute (short-term) toxic effects may include the death of animals, birds, or fish and death or low growth rate in plants. Acute effects are seen 2 to 4 days after animals or plants are exposed to aluminum fluoride. Insufficient data are available on the short-term effects of aluminum fluoride exposure aquatic life, plants, birds, or land animals.

Chronic Ecological Effects

Chronic (long-term) toxic effects may include shortened life span, reproductive problems, lower fertility, and changes in appearance or behavior in exposed animals. These effects can be seen long after first exposure(s) to toxic chemicals. Insufficient data are available on the long-term effects of aluminum fluoride to aquatic life, plants, birds, or land animals.

Water Solubility

Aluminum fluoride is very slightly soluble in water. Concentrations of less than 100 milliliters will mix with a liter of water.

Persistence in the Environment

Aluminum fluoride is slightly non-persistent in the aquatic environment with a half life between 2 and 20 days. The half-life of a pollutant is the amount of time it takes for one half of the chemical to be degraded.

Bioaccumulation in Aquatic Organisms

Some substances increase in concentration, or *bioaccumulate,* in living organisms as they breathe contaminated air, drink contaminated water, or eat contaminated food. These chemicals can become concentrated in the tissues and internal organs of animals as well as humans.

The concentration of aluminum fluoride found in fish tissue is expected to be about the same as that found in the water from which the fish was taken.

Recommended Risk-Reduction Measures

Proper training of all transporters will reduce the likelihood of a mishap or accident resulting in the contamination of the environment. The correct use of DOT labeling (per the MSDS) on all containers, trucks, and rail cars will enable emergency responders to react properly and quickly to any disaster thereby reducing the potential risk to the environment and to personnel.

Storage of aluminum fluoride should be segregated from incompatible chemicals to minimize the risk of cross-contamination or contact. It should be kept in tightly closed containers, never in glass. Buildings designated for storage should be equipped with appropriate fire protection systems (alarms, sprinklers, emergency lighting, portable extinguishers).

If a spill or leak to the environment has occurred, fire department, emergency response and/or hazardous materials spill personnel should be notified immediately. Cleanup should be attempted only by those trained in proper spill containment procedures. Contaminated soils should be removed and replaced with clean soil. When environmental contamination occurs, the local and/or state emergency response authorities may require notification. A comprehensive emergency response or disaster preparedness/recovery plan should be in place prior to any operations involving the use, transportation, storage, or disposal of aluminum fluoride.

Small spills of aluminum fluoride may be disposed of by cautiously adding water and immediately neutralizing with sodium carbonate. A little hydrochloric acid may be followed immediately by sodium carbonate if the material has not completely dissolved. Add calcium chloride in excess of the amount needed to precipitate the fluoride and/or carbonate. Separate the solids and dispose of in a landfill site approved for hazardous waste.

If aluminum fluoride is spilled or leaked, the following specific steps are recommended:

- ☑ Restrict persons not wearing protective clothing from area of spill or leak until cleanup is complete.
- ☑ Remove all ignition sources and ventilate area of spill or leak.
- ☑ Collect powdered materials in the safest manner possible and deposit in sealed containers. Use a vacuum equipped with a HEPA filter for cleanup. Dry sweeping is not recommended (generates airborne dusts). Keep water away.
- ☑ It may be necessary to dispose of aluminum fluoride as a hazardous waste. The responsible state agency or the regional office of the federal Environmental Protection Agency (EPA) should be contacted for specific recommendations.

RISK ASSESSMENT: BUSINESS

General Assessment

Accidents or mishaps involving aluminum fluoride can present a significant threat to business operations. The loss or damage of equipment or facilities due to explosion or fire can significantly affect fiscal viability. Lawsuits that may result from personnel injury/death, public exposures, and/or environmental contamination will also require a serious expenditure of resources. Media attention surrounding an injury, death, or environmental damage can also result in a loss of profits and loss of current as well as future business.

Recommended Risk-Reduction Measures

Company attorneys, safety and health professionals, and environmental specialists should be involved in the development of any procedures dealing with safety or emergency response. A company official should be pre-designated as a public relations officer with specific training in dealing with the press. Corporate plans and policies should be developed, approved, and implemented long before any need for such arises.

REFERENCES

American Conference of Governmental Industrial Hygienists. 1988. *Documentation of the Threshold Limit Values and Biological Exposure Indices*, 5th Edition (with updates). Cincinnati: ACGIH

Bloom, A. D. and F. J. DeSerres. 1995. *Ecotoxicity and Human Health.* New York: CRC Press/Lewis Publishers.

Bueche, J. R. 1972. *Principles of Physics*, 2nd Edition. New York: McGraw-Hill.

Burrell, R., D. K. Flaherty, and Leonard J. Sauers. 1992. *Toxicology of the Immune System.* New York: Van Nostrand Reinhold.

Calabrese, E. J. 1994. *Biological Effects of Low Level Exposures.* New York: CRC Press/Lewis Publishers.

Cockerman, L. G., B. S. Shane. 1994. *Basic Environmental Toxicology.* New York: CRC Press/Lewis Publishers.

Gardner, R. W. 1994. *Chemical Intolerance.* New York: CRC Press/Lewis Publishers.

Hallenbeck, W. H. 1993. *Quantitative Risk Assessment for Environmental and Occupational Health,* 2nd Edition. New York: CRC Press/Lewis Publishers.

Hosty, J. W., P. Foster. 1990. *A Practical Guide to Chemical Spill Response.* New York: Van Nostrand Reinhold.

Howard, P. H., et. al. 1991. *Handbook of Environmental Degradation Rates.* New York: CRC Press/Lewis Publishers.

Landis, W. G., M. H. Yu. 1995. *Introduction to Environmental Toxicology.* New York: CRC Press/Lewis Publishers.

Lauwerys, R. R., P. Hoet. 1993. *Industrial Chemical Exposure.* New York: CRC Press/Lewis Publishers.

Lewis, R. J., Sr. 1993. *Hawley's Condensed Chemical Dictionary,* 12th Edition. New York: Van Nostrand Reinhold.

Lewis, R. J., Sr. 1992. *Sax's Dangerous Properties of Industrial Materials,* Volumes 1, 2, and 3, 8th Edition. New York: Van Nostrand Reinhold.

Manahan, S. E. 1992. *Toxicological Chemistry,* 2nd Edition. New York: CRC Press/Lewis Publishers.

Meyer, E. 1990. *Chemistry of Hazardous Materials.* Englewood Cliffs, NJ: Prentice Hall.

Quigley, D. R. 1994. *Handbook of Emergency Chemical Management.* New York: CRC Press/Lewis Publishers.

Rea, W. J. 1995. *Chemical Sensitivity,* Vol. I-III. New York: CRC Press/Lewis Publishers.

Richardson, M. 1995. *Dictionary of Substance and Their Effects,* Vol. I-VI. New York: CRC Press/Lewis Publishers.

Sacarello, H. L. A. 1994. *The Comprehensive Handbook of Hazardous Materials.* New York: CRC Press/Lewis Publishers.

Sherman, J. D. 1988. *Chemical Exposure and Disease.* New York: Van Nostrand Reinhold.

Talbott, E., G. Craun. 1995. *An Introduction to Environmental Epidemiology.* New York: CRC Press/Lewis Publishers.

MATERIAL SAFETY DATA SHEET

CHEMICAL NAME	CRC PRESS/LEWIS PUBLISHERS DISCLAIMER
ALUMINUM HYDROXIDE	THE INFORMATION AND RECOMMENDATIONS PRESENTED HEREIN ARE BASED ON SOURCES BELIEVED TO BE RELIABLE. CRC MAKES NO REPRESENTATION ON ITS COMPLETENESS OR ACCURACY. IT IS THE RESPONSIBILITY OF THE USER TO DETERMINE THE CHEMICAL'S SUITABILITY FOR ITS INTENDED USE, THE CHEMICAL'S SAFE USE, AND THE CHEMICAL'S PROPER DISPOSAL. NO REPRESENTATIONS AND/OR WARRANTIES, EITHER EXPRESSED OR IMPLIED, OF THE MERCHANTABILITY OR FITNESS FOR A PARTICULAR PURPOSE, OR OF ANY OTHER NATURE, ARE MADE WITH RESPECT TO THE INFORMATION PROVIDED IN THIS MSDS OR TO THE CHEMICAL TO WHICH INFORMATION MAY REFER. CRC NEITHER ASSUMES NOR AUTHORIZES ANY OTHER PERSON TO ASSUME FOR IT, ANY OTHER ADDITIONAL RESPONSIBILITY OR LIABILITY FOR THE USE OF, OR RELIANCE UPON, THIS INFORMATION.

HAZARD WARNING INFORMATION

HEALTH	FIRE	REACTIVE	OTHER	DEGREE OF HAZARD	COLOR CODING	OTHER CODES
1	0	0		0 = Minimum Hazard 1 = Slight Hazard 2 = Moderate Hazard 3 = Serious Hazard 4 = Severe Hazard	HEALTH = BLUE FIRE = RED REACTIVITY = YELLOW OTHER = WHITE	OX = Oxidizer ACID = Acid ALK = Alkali COR = Corrosive W = Use No Water

SECTION I - GENERAL INFORMATION

Characterization	**Metal Hydroxide**	RCRA Number	**None**
		EPA Class	**Not Applicable**
DOT Proper Shipping Name	**No Citation**	Chemical Abstract Service (CAS) Number	**21645-51-2**
DOT Hazard Class and Label Requirements	**No Citation**	DOT Emergency Guide Code	**No Citation**
DOT Identification Number	**None Established**	Molecular Formula	**$Al(OH)_3$**

Synonyms

Alumgel, aluminum hydrate, aluminic acid, Alusal, hydrated alumina, Higilite, Alcoa, alumina trihydrate, aluminum oxide hydrate, aluminum trihydroxide, aluminum oxide trihydrate, martinal.

SECTION II - HAZARDOUS INGREDIENTS AND IDENTITY INFORMATION

Hazardous Components (specific identity)	OSHA Exposure Criteria	NIOSH Exposure Criteria	Immediately Dangerous to Life and Health (IDLH)	ACGIH Exposure Criteria
Aluminum hydroxide (derivation: From bauxite. The ore is dissolved in strong caustic and aluminum hydroxide precipitated from the sodium aluminate solution by neutralization (as with carbon dioxide) or by autoprecipitation).	**PEL:** Not Established **STEL:** Not Established	**REL:** Not Established **STEL:** Not Established	**Not Determined**	**TLV:** 2 mg(Al)/m³ **STEL:** Not Established

SECTION III - PHYSICAL AND CHEMICAL CHARACTERISTICS

Boiling Point	**Not Available**	Specific Gravity (H_2O = 1)	**2.42**
Vapor Pressure (mm Hg)	**Not Available**	Molecular Weight	**78.01**
Vapor Density (Air = 1)	**Not Available**	Melting Point	**Not Available**

Solubility

Insoluble in water. Soluble in mineral acid and caustic soda. Will release water when heated.

Appearance and Odor

White, odorless, crystalline powder or granules. Depending upon method of preparation, appearance can also be as white balls, or lumps of various mesh.

SECTION IV - FIRE AND EXPLOSION HAZARD DATA

Flash Point (method used)	**Not Applicable**	Explosive Limits in Air % by Volume	**LEL: N/A UEL: N/A**
NFPA Classification	**Non-Combustible Solid**	Autoignition Temperature	**Not Applicable**

Extinguishing Media

Use carbon dioxide, dry chemical, alcohol foam, high expansion or low expansion foam, water.

Special Fire Fighting Procedures

Move containers from fire if it can be done without risk. Evacuate non-essential personnel 2500 feet from fire area. Fire fighters should take care to avoid dusty situations or wear appropriate respiratory protection where aluminum hydroxide may be present.

Unusual Fire and Explosion Hazards

Aluminum hydroxide is incompatible with chlorinated rubber products. Poisonous gases are produced in fire. Keep away from smoke and runoff.

SECTION V - REACTIVITY DATA

Stability		Conditions to Avoid
		Aluminum hydroxide should be kept away from contact with hot chlorinated rubber. Also, when it is co-precipitated with bismuth hydroxide and reduced by water, it is violently flammable in air.
Stable	Unstable	Incompatibility *(materials to avoid)*
X		**Aluminum hydroxide is not compatible with hot chlorinated rubber and strong acids (such as sulfuric, nitric, and hydrochloric).**
Hazardous Polymerization		Conditions to Avoid **Under normal conditions of temperature and pressure, hazardous polymerization of aluminum hydroxide is not expected to occur.**
May Occur	Will Not Occur	Hazardous Decomposition or By-products
	X	**Thermal decomposition may yield acrid smoke and irritating fumes and vapors. Also, it decomposes to form aluminum oxide.**

SECTION VI - HEALTH HAZARD DATA

Primary Route(s) of Entry:	Inhalation?	Absorption (skin/eye)?	Ingestion?
	X		

Health Hazards

INHALATION: **Irritation of the mucous membranes. May cause fibrosis in lung tissue, emphysema, and pneumothorax.**

SKIN & EYES: **Contact with the skin or eyes may cause irritation. Skin allergy may develop.**

INGESTION: **Generally harmless, although some persons may experience vomiting, nausea, and constipation, upset stomach.**

Carcinogenicity	NTP Listed?	IARC Cancer Review Group?	OSHA Regulated?	Target Organs?
No Citation	**No**	**No**	**No**	**Eyes, skin, respiratory system**

Medical Conditions Generally Aggravated by Exposure

None reported.

Emergency and First-aid Procedures

Eye contact: Do not allow victim to keep eyes tightly shut. Flush immediately with water for 15 minutes (minimum), seek medical attention. Skin contact: Remove all contaminated clothing. Immediately wash area with large amounts of soap and water. For inhalation: Remove the person from exposure. Provide respiratory assistance and CPR. Transfer to medical facility. If swallowed, remove person to fresh air and seek medical attention immediately. Do not attempt to give an unconscious or convulsing person anything by mouth. Do NOT induce vomiting. If conscious, give one pint of water immediately.

SECTION VII - PRECAUTIONS FOR SAFE HANDLING AND USE

Steps to be Taken in Case Material is Released or Spilled

Ventilate area of spill or leak. Remove sources of ignition. Restrict those not involved in cleanup from entering area. Collect spilled powdered materials in safest manner possible and deposit in sealed drum for disposal. Notify appropriate authorities, as required by applicable environmental regulations.

Preferred Waste Disposal Method

Not listed. Check applicable federal, state, and/or local environmental regulations.

Precautions to be Taken in Handling and Storage

Aluminum hydroxide is a non-combustible solid. Store to avoid contact with strong acids. Store in tightly closed containers in cool, well-ventilated area.

Other Precautions and Warnings

Always wear appropriate protective equipment when working with aluminum hydroxide.

SECTION VIII - CONTROL MEASURES AND PERSONAL PROTECTIVE EQUIPMENT

Respiratory Protection *(specify type)*

If airborne concentrations exceed the TLV, a dust/mist respirator is recommended. For high exposures, or if concentrations exceed the capacity of the respirator, then a self-contained breathing apparatus (SCBA) with full facepiece operated in positive pressure mode is advised.

Ventilation

Local exhaust (preferred) at site of chemical work or general ventilation.

Protective Gloves	Eye Protection	Other Protective Clothing
Impervious Rubber (non-chlorinated)	**Safety Glasses with Sides Shields**	**Protective Clothing**

Work/Hygiene Practices

Always wash hands thoroughly after using chemical; never bring food, drink, or smoking materials into vicinity of chemicals.

ALUMINUM HYDROXIDE

$Al(OH)_3$ CAS: 21645-51-2

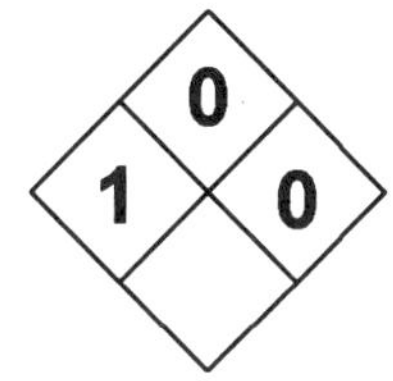

IDENTIFICATION AND TYPICAL USES

Aluminum hydroxide is a white crystalline sand-like powder with no odor. Depending upon its preparation, appearance can also be as white balls or lumps of various sizes. It is used in the production of glass and ceramics, in iron-free aluminum and aluminum salts, in the manufacture of activated alumina, and in flame retardant materials (mattress bedding, for example). Finely divided particles are used as a rubber reinforcing agent, in paper coatings, as fillers, and in cosmetics. May also be used in the manufacture of some antiperspirants.

RISK ASSESSMENT: HEALTH

General Assessment

The aluminas are generally considered nuisance dusts and under some conditions of human exposure may cause a minimal pulmonary nodular response. However, excessive *inhalation* of finely divided particles may lead to lung damage (Shaver's disease) characterized by scarred tissue (pulmonary fibrosis) with shortness of breath, tightness and pain in the chest, and coughing. The condition can be fatal.

It should be noted that population studies of potentially exposed workers have shown minimal evidence of pulmonary fibrosis or pneumoconiosis. Also, animal experiments with aluminum hydroxide have shown that the type of reaction in lung tissue varies with the form of alumina used and its particle size, the species of animal tested, and the route of administration. However, the data from both human and animal studies seem to indicate that alumina is efficiently eliminated from the lung and has a low degree of fibrogenicity.

This does not mean there is no health risk associated with aluminum hydroxide. Symptoms of exposure can include nausea, vomiting, and constipation. There is no way of estimating exact human response under all possible types of exposure conditions. Prudent risk management requires caution when working with or around aluminum hydroxide.

Acute Health Effects

The following acute (short-term) health effects may occur immediately or shortly after exposure to aluminum hydroxide:

Skin: Irritation with a potential for allergic reaction with itching and redness.

Eye: Severe eye irritation likely.

Lung: Can irritate the respiratory tract, in large doses, may cause pulmonary fibrosis, emphysema, and pneumothorax.

Chronic Health Effects

The following chronic (long-term) health effects may occur at some time after exposure to aluminum hydroxide and can last for months or even years:

Cancer Hazards: According to the information presented in the references, aluminum hydroxide has been tested and has not been shown to cause cancer in animals. No data were found on its human carcinogenic capabilities. However, some sources reported limited mutagenic data on test animals.

Reproductive Hazard: According to the information presented in the references, aluminum hydroxide has not been adequately tested for its ability to affect reproduction. Some aluminum compounds have been shown to adversely affect the developing fetus in test animals. Human data are inadequate to evaluate the effects of exposure to aluminum hydroxide on human reproduction.

Other Chronic Effects: Repeated exposure to large amounts of aluminum hydroxide dust may damage the lungs, leading to scarring of the tissue (pulmonary fibrosis), with shortness of breath and possible death.

☉ *Recommended Risk-Reduction Measures*

Personnel should avoid direct contact with aluminum hydroxide. The exact nature of its harmful capacities is not clearly understood in the references. Therefore, it is recommended that it be handled as a toxic material. If a less toxic material cannot be substituted for aluminum hydroxide, then *engineering controls* are the most effective method of reducing exposures. The best protection is to enclose operations and/or provide local exhaust ventilation at the site of aluminum hydroxide release. While not always operationally feasible, isolating operations can also reduce exposure.

Using respiratory protection and other personal protective equipment (PPE) is less effective than the engineering controls mentioned above, but is still advisable whenever working with or around aluminum hydroxide. Under normal operating conditions of low exposure potential, a dust/mist respirator will suffice. However, for maximum protection in large concentrations or when the capacity of the dust/mist respirator is exceeded, personnel should wear a self-contained breathing apparatus (SCBA) with full facepiece operated in pressure demand mode. If a full facepiece is not available, then chemical/dust goggles, safety glass with side shields, or a face shield should be worn to protect the eyes. To prevent hand and skin exposures, impervious rubber gloves should be worn.

Administrative controls should also be in place to minimize the potential for human exposures. These may include written procedures or policies which specify the methods and techniques that will be practiced whenever personnel are to work with aluminum hydroxide.

All personnel should receive *training* on the use, hazards, protective measures, emergency actions, and other precautions per 29 CFR 1910.1200 (Hazard Communication), prior to the first assignment in an area where aluminum hydroxide is used or stored.

If there has been a long history of high exposure, or if lung symptoms develop, the following medical tests are recommended:

- ☑ Consider chest X-ray following an acute overexposure to aluminum hydroxide.
- ☑ Pulmonary function tests.

Any evaluation should include a careful history of past and present symptoms with an examination. Medical tests that simply look for existing damage are not a substitute for controlling exposures. Also, because smoking can cause lung cancer, emphysema, and a host of other medical problems, working with or around aluminum hydroxide may aggravate these conditions further.

Other methods to reduce exposure to aluminum hydroxide include:

- ☑ Where possible, enclose operations and use local exhaust ventilation at the site of chemical release. If local exhaust ventilation or enclosure is not used, then respiratory protection should be required.
- ☑ Always ensure that proper protective clothing is worn when using chemical substances. Never allow contaminated clothing to be taken home from the work place.
- ☑ Hazard warning information should be posted in the work area. In addition, as part of an on-going education and training program, all information on the health and safety hazards of aluminum hydroxide should be communicated to all potentially exposed workers.
- ☑ Eye wash stations should be provided in the immediate work area for emergency use.

RISK ASSESSMENT: ENVIRONMENT

General Assessment

The environment is at risk of exposure during transportation, storage, disposal, or destruction of aluminum hydroxide. In almost every scenario, the threat of environmental exposure is contingent upon the proper handling of the chemical substance. Accidental spills, large or small, where aluminum hydroxide contacts incompatible commodities can result in violent reactions as well as the possible contamination of the surrounding environmental mediums (water, soil, and air).

Aluminum hydroxide is considered a noncombustible solid. However, because it is incompatible with strong oxidizers, extreme caution is required in handling, storage, transportation, and disposal of aluminum hydroxide. These characteristics also require special consideration during any emergency situation involving a leak or spill of aluminum hydroxide. Should aluminum hydroxide ever come into contact with incompatible substances either during use, transportation, storage, or disposal, dangerous reactions are possible.

Aluminum hydroxide occurs naturally in the environment in the minerals bauxite, baerites, boehmite, corundum, diaspore, and gibbsite. It therefore occurs naturally in the aquatic environment but can also enter it through industrial discharges and/or spills.

Acute Ecological Effects

Acute (short-term) toxic effects may include the death of animals, birds, or fish and death or low growth rate in plants. Acute effects are seen 2 to 4 days after animals or plants are exposed to aluminum hydroxide.

Aluminum hydroxide has slight acute toxicity to aquatic life, but its toxicity will be increased under alkaline conditions. Insufficient data are available to evaluate or predict the short-term toxic effects of aluminum hydroxide to plants, birds, and land animals.

Chronic Ecological Effects

Chronic (long-term) toxic effects may include shortened life span, reproductive problems, lower fertility, and changes in appearance or behavior in exposed animals. These effects can be seen long after first exposure to toxic chemicals.

Aluminum hydroxide has slight chronic toxicity to aquatic life, but this toxicity will increase under alkaline conditions. Insufficient data are available to evaluate or predict the long-term toxic effects of aluminum hydroxide to plants, birds, or land animals.

Water Solubility

Aluminum hydroxide is slightly soluble in water. Concentrations of 1 milligram or less will mix with a liter of water.

Persistence in the Environment

Aluminum hydroxide is highly persistent in water, with a half-life greater than 200 days. The half-life of a pollutant is the amount of time it takes for one-half of the chemical to be degraded.

Bioaccumulation in Aquatic Organisms

Some substances increase in concentration, or *bioaccumulate,* in living organisms as they breathe contaminated air, drink contaminated water, or eat contaminated food. These chemicals can become concentrated in the tissues and internal organs of animals as well as humans.

Aluminum hydroxide is not expected to accumulate in the edible tissues of aquatic life that are normally consumed by humans.

Recommended Risk-Reduction Measures

Proper training of all transporters will reduce the likelihood of a mishap or accident resulting in a leak or spill of aluminum hydroxide into the environment. The correct use of labeling on all containers, trucks, and rail cars will enable emergency responders to react properly and quickly to any disaster thereby reducing the potential risk to the environment and to personnel.

Storage of aluminum hydroxide should be segregated from incompatible chemicals to minimize the risk of cross-contamination or contact.

If a spill or leak to the environment has occurred, fire department, emergency response and/or hazardous materials spill personnel should be notified immediately. Cleanup should be attempted only by those trained in proper spill containment procedures. If aluminum hydroxide should contact the water table, aquifer, or navigable waterway, the local and/or state emergency response authorities must be notified. A comprehensive emergency response or disaster preparedness/recovery plan should be in place prior to any operations involving the use, transportation, storage, or disposal of aluminum hydroxide.

If aluminum hydroxide is spilled or leaked, the following specific steps are recommended:

- ☑ Restrict persons not wearing protective clothing from area of spill or leak until cleanup is complete.
- ☑ Do not allow exposure to incompatible materials.
- ☑ Close or remove all ignition sources and ventilate area of spill.
- ☑ Collect powdered materials in the safest and most efficient manner possible. Place materials in a sealed drum. Do not dry sweep (generates dusts). Use a vacuum equipped with a high efficiency particulate air (HEPA) filter instead.
- ☑ It may be necessary to dispose of aluminum hydroxide as a hazardous waste. Consult state, local, or federal environmental regulations.

RISK ASSESSMENT: BUSINESS

General Assessment

Accidents or mishaps involving aluminum hydroxide can present a moderate threat to business operations. Although not likely, equipment and facilities may be lost or their use disrupted as a result of an incident involving aluminum hydroxide which can significantly effect fiscal viability. Lawsuits that may result from personnel injury, illness, and/or death, public exposures, and/or environmental contamination will also require a serious expenditure of resources. Media attention surrounding an injury, death, or environmental damage can also result in a loss of profits and loss of current as well as future business.

Recommended Risk-Reduction Measures

Company attorneys, safety and health professionals, and environmental specialists should be involved in the development of any procedures that focus on emergency response, disaster preparedness, and disaster recovery.

A company official should be pre-designated as a public relations officer with specific training in dealing with the press. Corporate plans and policies should be developed, approved, and implemented long before any need for such arises.

REFERENCES

Ahlbom, A. 1993. *Biostatistics for Epidemiologists.* New York: CRC Press/Lewis Publishers.

American Conference of Governmental Industrial Hygienists. 1988. *Documentation of the Threshold Limit Values and Biological Exposure Indices,* 5th Edition (with updates). Cincinnati: ACGIH

Bloom, A. D. and F. J. DeSerres. 1995. *Ecotoxicity and Human Health.* New York: CRC Press/Lewis Publishers.

Brusick, D. J. 1994. *Methods for Genetic Risk Assessment.* New York: CRC Press/Lewis Publishers.

Bueche, J. R. 1972. *Principles of Physics,* 2nd Edition. New York: McGraw-Hill.

Calabrese, E. J., E. M. Kenyon. *Air Toxics and Risk Assessment.* New York: CRC Press/Lewis Publishers.

Calabrese, E. J. 1994. *Biological Effects of Low Level Exposures.* New York: CRC Press/Lewis Publishers.

Cockerman, L. G., B. S. Shane. 1994. *Basic Environmental Toxicology.* New York: CRC Press/Lewis Publishers.

Cothern, R. R. 1993. *Comparative Environmental Risk Assessment.* New York: CRC Press/Lewis Publishers.

Dinman, B. D. 1988. Alumina-related pulmonary disease. *Journal of Occupational Medicine.* 30:328-335.

Gardner, R. W. 1994. *Chemical Intolerance.* New York: CRC Press/Lewis Publishers.

Gibbs, G. W. 1985. Mortality of aluminum reduction plant workers, 1950 through 1977. *Journal of Occupational Medicine.* 27:761-770.

Hallenbeck, W. H. 1993. *Quantitative Risk Assessment for Environmental and Occupational Health,* 2nd Edition. New York: CRC Press/Lewis Publishers.

Howard, P. H. 1995. *Handbook of Environmental Fate and Exposures Data for Organic Chemicals,* Vol. I-IV. New York: CRC Press/Lewis Publishers.

Howard, P. H., et. al. 1991. *Handbook of Environmental Degradation Rates.* New York: CRC Press/Lewis Publishers.

Jones, F. E. 1994. *Toxic Organic Vapors in the Workplace.* New York: CRC Press/Lewis Publishers.

Keith, L. H. and M. M. Walker. 1995. *Handbook of Air Toxics.* New York: CRC Press/Lewis Publishers.

Landis, W. G., M. H. Yu. 1995. *Introduction to Environmental Toxicology.* New York: CRC Press/Lewis Publishers.

Langerman, N. 1994 *Precautionary Labels for Chemical Containers.* New York: CRC Press/Lewis Publishers.

Lauwerys, R. R., P. Hoet. 1993. *Industrial Chemical Exposure.* New York: CRC Press/Lewis Publishers.

Lewis, R. J., Sr. 1993. *Hawley's Condensed Chemical Dictionary,* 12th Edition. New York: Van Nostrand Reinhold.

Lewis, R. J., Sr. 1992. *Sax's Dangerous Properties of Industrial Materials,* Volumes 1, 2, and 3, 8th Edition. New York: Van Nostrand Reinhold.

Lide, D. R. 1995. *Handbook of Chemistry and Physics,* 76th Edition. New York: CRC Press/Lewis Publishers.

Manahan, S. E. 1994. *Environmental Chemistry,* 6th Edition. New York: CRC Press/Lewis Publishers.

Mellström, G. A., J. E. Wahlberg, and H. I. Maibach. 1994. *Protective Gloves for Occupational Use.* New York: CRC Press/Lewis Publishers.

Meyer, E. 1990. *Chemistry of Hazardous Materials.* Englewood Cliffs, NJ: Prentice Hall.

Mycock, J. C., J. D. McKenna, and L. Theodore. 1995. *Handbook of Air Pollution Control Engineering and Technology.* New York: CRC Press/Lewis Publishers.

Patnaik, P. 1992. *A Comprehensive Guide to the Hazardous Properties of Chemical Substances.* New York: Van Nostrand Reinhold.

Proctor, N. H., J. P. Hughes. 1991. *Chemical Hazards of the Workplace.* New York: Van Nostrand Reinhold.

Quigley, D. R. 1994. *Handbook of Emergency Chemical Management.* New York: CRC Press/Lewis Publishers.

Rea, W. J. 1995. *Chemical Sensitivity,* Vol. I-III. New York: CRC Press/Lewis Publishers.

Richardson, M. 1995. *Dictionary of Substance and Their Effects,* Vol. I-VI. New York: CRC Press/Lewis Publishers.

Sacarello, H. L. A. 1994. *The Comprehensive Handbook of Hazardous Materials.* New York: CRC Press/Lewis Publishers.

Sheftel, V. O. 1995. *Handbook of Toxic Properties of Monomers and Additives.* New York: CRC Press/Lewis Publishers.

Talbott, E., G. Craun. 1995. *An Introduction to Environmental Epidemiology.* New York: CRC Press/Lewis Publishers.

Torkelson, T. R. V. K. Rowe. 1981. *Patty's Industrial Hygiene and Toxicology,* Vol. 2B. New York: Wiley-Interscience.

MATERIAL SAFETY DATA SHEET

CHEMICAL NAME	CRC PRESS/LEWIS PUBLISHERS DISCLAIMER:
ALUMINUM NITRATE	THE INFORMATION AND RECOMMENDATIONS PRESENTED HEREIN ARE BASED ON SOURCES BELIEVED TO BE RELIABLE. CRC MAKES NO REPRESENTATION ON ITS COMPLETENESS OR ACCURACY. IT IS THE RESPONSIBILITY OF THE USER TO DETERMINE THE CHEMICAL'S SUITABILITY FOR ITS INTENDED USE, THE CHEMICAL'S SAFE USE, AND THE CHEMICAL'S PROPER DISPOSAL. NO REPRESENTATIONS AND/OR WARRANTIES, EITHER EXPRESSED OR IMPLIED, OF THE MERCHANTABILITY OR FITNESS FOR A PARTICULAR PURPOSE, OR OF ANY OTHER NATURE, ARE MADE WITH RESPECT TO THE INFORMATION PROVIDED IN THIS MSDS OR TO THE CHEMICAL TO WHICH INFORMATION MAY REFER. CRC NEITHER ASSUMES NOR AUTHORIZES ANY OTHER PERSON TO ASSUME FOR IT, ANY OTHER ADDITIONAL RESPONSIBILITY OR LIABILITY FOR THE USE OF, OR RELIANCE UPON, THIS INFORMATION.

HAZARD WARNING INFORMATION

HEALTH	FIRE	REACTIVE	OTHER	DEGREE OF HAZARD	COLOR CODING	OTHER CODES
1	0	0		0 = Minimum Hazard 1 = Slight Hazard 2 = Moderate Hazard 3 = Serious Hazard 4 = Severe Hazard	HEALTH = BLUE FIRE = RED REACTIVITY = YELLOW OTHER = WHITE	OX = Oxidizer ACID = Acid ALK = Alkali COR = Corrosive W = Use No Water

SECTION I - GENERAL INFORMATION

Characterization: **Metal Nitrate**	RCRA Number: **Not Listed**	EPA Class: **Not Applicable**
DOT Proper Shipping Name: **Aluminum Nitrate**	Chemical Abstract Service (CAS) Number: **13473-90-0**	
DOT Hazard Class and Label Requirements: **Oxidizer**	DOT Emergency Guide Code: **35**	
DOT Identification Number: **UN 1438**	Chemical Symbol: $Al(NO_3)_3$	

Synonyms
Aluminum trinitrate, nitric acid aluminum salt, aluminum nitrate nonhydrate.

SECTION II - HAZARDOUS INGREDIENTS AND IDENTITY INFORMATION

Hazardous Components (specific identity)	OSHA Exposure Criteria	NIOSH Exposure Criteria	Immediately Dangerous to Life and Health (IDLH)	ACGIH Exposure Criteria
Aluminum nitrate (derivation: Formed by the action of nitric acid on aluminum and crystallization of the product).	**PEL: Not Established** **STEL: Not Established**	**REL: Not Established** **STEL: Not Established**	**Not Established**	**TLV: 2 mg(Al)/m^3** **STEL: Not Established**

SECTION III - PHYSICAL AND CHEMICAL CHARACTERISTICS

Boiling Point: **275°F (135°C) Decomposes**	Specific Gravity (H_2O = 1): **<1 at 69°F (20°C)**
Vapor Pressure (mm Hg): **Not Available**	Molecular Weight: **133.33**
Vapor Density (Air = 1): **13.0**	Melting Point: **163°F (73°C)**

Solubility
Soluble in water and alcohol. Slightly soluble in acetone. Insoluble in ethyl acetate and pyradine.

Appearance and Odor
White crystalline solid, usually appearing in powdered form.

SECTION IV - FIRE AND EXPLOSION HAZARD DATA

Flash Point (method used): **Not Applicable**	Explosive Limits in Air % by Volume: **LEL: N/A UEL: N/A**
NFPA Classification: **Non-Combustible Powder**	Autoignition Temperature: **Not Applicable**

Extinguishing Media
Use water or water spray only.

Special Fire Fighting Procedures
Wear full protective clothing with self-contained breathing apparatus (SCBA). Structural fire fighting clothing is permeable and will not provide adequate protection. Poisonous gases are produced in fire. Move containers from fire area if it can be done without risk. For large fires, use flooding amounts of water. Water spray should be used to keep fire-exposed, closed containers cool.

Unusual Fire and Explosion Hazards
Fire intensity may increase when in contact with combustible materials (wood, paper, oil). A powerful oxidizing agent.

SECTION V - REACTIVITY DATA

Stability		Conditions to Avoid **Keep away from incompatible materials, including ordinary combustibles (paper, wood, oil), since violent and even explosive reactions can occur. Aqueous solutions are slightly acidic.**
Stable **X**	Unstable	Incompatibility *(materials to avoid)* **alkyl esters, other metal nitrates and phosphinates, phosphorus, tin(II) chloride, other reducing agents, mixtures containing other nitrates and organic materials, acids, heavy metals, cyanides, thiocyanates.**
Hazardous Polymerization		Conditions to Avoid **Under normal conditions of temperature and pressure, hazardous polymerization of aluminum nitrate is not expected to occur.**
May Occur	Will Not Occur **X**	Hazardous Decomposition or By-products **When heated to decomposition, aluminum nitrate emits toxic and poisonous fumes of nitrogen and aluminum oxides.**

SECTION VI - HEALTH HAZARD DATA

Primary Route(s) of Entry:	Inhalation?	Absorption (skin)?	Ingestion?
	X	X	X

Health Hazards

INHALATION: May severely irritate the upper respiratory tract with symptoms of coughing, tightness in in the chest, labored breathing. May also cause dizziness and headaches.

SKIN & EYES: Skin contact produces mild irritation. Aluminum nitrate may be absorbed through the skin. Eye contact with powder can result in severe irritation and burns.

INGESTION: Causes symptoms of dizziness and headaches. In severe cases, there may be abdominal cramps, vomiting, bloody diarrhea, flushing of the skin, shock, convulsions, and a loss of consciousness.

Carcinogenicity	NTP Listed?	IARC Cancer Review Group?	OSHA Regulated?	Target Organs?
Unconfirmed Human Questioned Animal	**No**	**No**	**No Specific Listing**	**Respiratory system, eyes, skin.**

Medical Conditions Generally Aggravated by Exposure

Respiratory impairments (bronchitis, asthma); Skin conditions (dermatitis).

Emergency and First-aid Procedures

Eye contact: Flush immediately with water for 15 minutes (minimum), seek medical attention. Skin contact: Remove all contaminated clothing. Wash contaminated skin with soap and water. For inhalation: Remove the person from exposure. Provide respiratory assistance and CPR. Transfer to medical facility. If swallowed: Do NOT induce vomiting. Give one pint of water immediately. Seek medical attention immediately.

SECTION VII - PRECAUTIONS FOR SAFE HANDLING AND USE

Steps to be Taken in Case Material is Released or Spilled

Collect powdered material in the most convenient way possible and deposit in sealed containers. Remove all sources of ignition. Restrict those not involved in cleanup from entering area.

Preferred Waste Disposal Method

No Citation.

Precautions to be Taken in Handling and Storage

Containers should be protected against physical damage. Store in tightly closed containers in a cool, dry, well ventilated area away from heat, sparks, or open flame. Personnel assigned to area must be properly trained on the hazards associated with aluminum nitrate.

Other Precautions and Warnings

Whenever bulk or finely powdered aluminum nitrate is used, handled, manufactured, or stored, use explosion-proof electrical equipment and fittings.

SECTION VIII - CONTROL MEASURES AND PERSONAL PROTECTIVE EQUIPMENT

Respiratory Protection *(specify type)*

For exposures near or over 2 mg/m^3: NIOSH approved respirator with high efficiency particulate air (HEPA) filter and a dust/fume/mist pre-filter. Greater protection is provided with a SCBA.

Ventilation

Local exhaust or general mechanical systems recommended.

Protective Gloves	Eye Protection	Other Protective Clothing
Neoprene, Viton, Butyl Rubber	**Chemical/Dust Goggles**	**Full Protective Suit**

Work/Hygiene Practices

Always wash hands thoroughly after working with aluminum powders and dusts; never bring food, drink, or smoking materials into vicinity of aluminum powders.

ALUMINUM NITRATE

$Al(NO_3)_3$ CAS: 7784-27-2

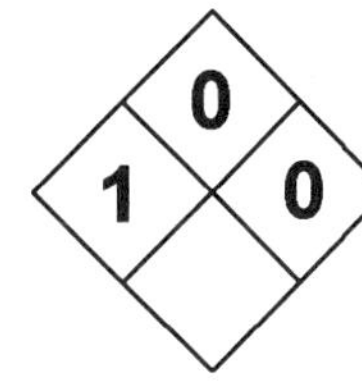

IDENTIFICATION AND TYPICAL USES

Aluminum nitrate is a white, crystalline powder that is usually present in a powder (sugar or sand-like) form. It is used as a mordant in textiles, leather tanning, manufacture of incandescent filaments, as a catalyst in petroleum refining, in nucleonics, as an anticorrosion agent, as a transformer core laminate, in heating elements, in the preparation of insulating papers, and in the manufacture of antiperspirant.

RISK ASSESSMENT: HEALTH

General Assessment

Aluminum nitrate is a human poison. The biological hazards of aluminum nitrate appear to be mainly due to the toxicity of the nitrate group. Exposures can occur through *inhalation* of the fine powders, *ingestion* of the material, or through skin *absorption.*

Inhaling the dusts of aluminum nitrate is a common route of exposure in the occupational setting. Risk managers should look for those work practices that may have a tendency to generate airborne dusts and powders. Aluminum nitrate will irritate the upper respiratory tract causing sore throat, coughing, and a possibility of congestion if exposure continues. It may cause symptoms such as dizziness and headaches.

Skin contact produces mild irritation, possible rash, and, in prolonged or repeated exposures, may cause dermatitis in sensitive people. Aluminum nitrate will also pass through unbroken skin to cause symptoms of systemic exposure, such as those occurring after inhalation or ingestion. Eye contact will cause severe irritation and must be avoided. Prolonged contact can cause damage to vision.

Ingestion of nitrates causes symptoms of dizziness and headaches. More severe poisoning may cause abdominal pain, vomiting, bloody diarrhea, flushing of the skin, shock, convulsions, and could cause a loss of consciousness, coma, and possibly death.

Acute Health Effects

The following acute (short-term) health effects may occur immediately or shortly after exposure to aluminum:

Skin: Dusts and vapors can be strong irritants and the liquid (hydrate) may cause sever burns. Can pass through unbroken skin to cause toxic systemic effects.

Eye: Highly irritating and destructive. Liquid causes severe burns. Possible damage or loss of vision.

Lung: Mildly irritating to the upper respiratory tract. May result in burning sensation, cough, headache, and dizziness.

Other: Ingestion may result in severe gastrointestinal irritation with symptoms of dizziness, headache, nausea, vomiting, abdominal cramps, bloody diarrhea, flushing of the skin, and shock. Severe poisoning may cause convulsions and loss of consciousness.

Chronic Health Effects

The following chronic (long-term) health effects may occur at some time after exposure to aluminum and can last for months or even years:

Cancer Hazards: Insufficient data exist to properly evaluate or predict the carcinogenicity of aluminum nitrate in humans or test animals. While some studies suggest no carcinogenicity, the results of numerous other studies were inconclusive (some studies suggest possible anti-tumor activity). Further research is required before any determinations on the carcinogenic properties of aluminum nitrate can be made.

Reproductive Hazard: Some aluminum salts, including aluminum nitrate, are toxic to the animal fetus and cause fetal damage. No data are available on the teratogenicity of aluminum nitrate in humans.

Other Chronic Effects: Repeated exposures to aluminum nitrate can lead to weakness, general depression, headache, and mental impairment. Aluminum nitrate may cause dermatitis.

Recommended Risk-Reduction Measures

Personnel should avoid direct contact with aluminum nitrate. *Engineering controls* are always the most effective method of reducing exposures. The best protection, especially when working with powders and dusts of aluminum nitrate, is to enclose operations and provide explosion-proof, local exhaust ventilation at the site of powder or dust release. While not always operationally feasible, isolating operations can also reduce exposure. Using respiratory protection is less effective than the controls mentioned above, but is still advisable whenever working with or around aluminum nitrate. A MSHA/NIOSH approved self-contained breathing apparatus (SCBA) with full facepiece and pressure demand is the best respiratory protection. However, for exposures near or over the 2 mg/m^3 TLV for dusts, a full facepiece respirator equipped with a high efficiency particulate air (HEPA) filter and a dust/mist/fume prefilter would suffice. If a full facepiece respirator is not available, then dust goggles or safety glasses with side shields attached should be worn to protect the eyes. If splash hazards exist, then a face shield is required. Eye contact must be prevented since severe and perhaps permanent damage is likely on contact with aluminum nitrate powders or solutions. To prevent hand and skin exposures, suitable gloves should be worn.

Administrative controls should also be in place to minimize the potential for human exposures. These may include written procedures or policies which specify the methods and techniques that will be practiced whenever personnel are to work with aluminum nitrate.

All personnel should receive *training* on the use, hazards, protective measures, emergency actions, and other precautions per 29 CFR 1910.1200 (Hazard Communication), prior to the first assignment in an area where aluminum nitrate is used, handled, or stored. Before beginning employment with aluminum, and at regular intervals thereafter (e.g., annually), the following medical tests are recommended:

- ☑ Lung function tests.
- ☑ Chest X-ray should be considered.
- ☑ Evaluation by a qualified allergist to determine sensitivity and potential for dermatitis.

Other methods to reduce exposure include:

- ☑ Where possible, enclose operations and use local exhaust ventilation at the site where aluminum powder or dust is worked. If local exhaust ventilation or enclosure is not used, respiratory protection should be mandatory.
- ☑ Always ensure that proper protective clothing is worn when working with aluminum compounds.
- ☑ Wash thoroughly immediately after exposure to aluminum and at the end of the work shift or before eating, drinking, or smoking. Clothing contaminated with aluminum dusts or powders should not be taken home.
- ☑ Hazard warning information should be posted in the work area. In addition, as part of an on-going education and training program, all information on the health and safety hazards of aluminum nitrate should be communicated to all exposed and potentially exposed workers.
- ☑ A safety shower and emergency eyewash station should be readily available for emergency use in or near areas where aluminum nitrate is used, handled, or stored.

RISK ASSESSMENT: ENVIRONMENT

General Assessment

The environment is at risk of exposure during transportation, storage, disposal, or destruction of aluminum nitrate. In almost every scenario, the threat of environmental exposure is contingent upon the proper handling of aluminum nitrate, especially the dusts and powders (anhydrous form). Accidental contact with a variety of incompatible materials, including ordinary combustibles (wood, paper, oil, fuel), can have adverse and potentially dangerous results. This includes fire, explosion, generation of other hazardous commodities, and possible contamination of the surrounding environmental mediums (water, soil, and, air).

Aluminum nitrate is considered a non-combustible solid. However, it is incompatible with many other commodities including other finely divided metal powders, strong reducing agents, strong acids, cyanides, thiocyanates, isothiocyanates, and hypophosphites. Contact can cause fire or explosion. The aqueous solutions are slightly corrosive. These characteristics require special consideration during any emergency situation involving a leak or spill of aluminum. Should aluminum nitrate ever come into contact with such incompatible substances either during use, transportation, or storage, the formation of highly explosive commodities is extremely possible.

Aluminum nitrate can enter the environment through industrial discharges and from spills.

Acute Ecological Effects

Acute (short-term) toxic effects may include the death of animals, birds, or fish and death or low growth rate in plants. Acute effects are seen 2 to 4 days after animals or plants are exposed to aluminum nitrate.

This material has moderate acute toxicity to aquatic life. Insufficient data are available on the short-term effects of exposure to plants, birds, or land animals.

Chronic Ecological Effects

Chronic (long-term) toxic effects may include shortened life span, reproductive problems, lower fertility, and changes in appearance or behavior in exposed animals. These effects can be seen long after first exposure(s) to toxic chemicals.

Aluminum nitrate has high chronic toxicity in aquatic life. Insufficient data are available on the long-term effects of aluminum nitrate to plants, birds, or land animals.

Water Solubility

Aluminum nitrate is slightly soluble in water. Concentrations of 1 - 100 milliliters will mix with a liter of water.

Persistence in the Environment

Aluminum nitrate is slightly-persistent in water, with a half life between 2 and 20 days. The half-life of a pollutant is the amount of time it takes for one half of the chemical to be degraded.

Bioaccumulation in Aquatic Organisms

Some substances increase in concentration, or *bioaccumulate,* in living organisms as they breathe contaminated air, drink contaminated water, or eat contaminated food. These chemicals can become concentrated in the tissues and internal organs of animals as well as humans.

The concentration of aluminum nitrate found in the tissue of fish is expected to be slightly lower than that found in the water from which the fish was taken.

Recommended Risk-Reduction Measures

Proper training of all transporters will reduce the likelihood of a mishap or accident resulting in the contamination of the environment. The correct use of DOT labeling (per the MSDS) on all containers, trucks, and rail cars will enable emergency responders to react properly and quickly to any disaster thereby reducing the potential risk to the environment and to personnel.

Storage of aluminum nitrate should be segregated from incompatible chemicals to minimize the risk of cross-contamination or contact. It should be kept in tightly closed containers, in a cool, well ventilated area, away from sources of heat or ignition. Buildings designated for storage should be equipped with appropriate fire protection systems (alarms, sprinklers, emergency lighting, portable extinguishers).

If a spill or leak to the environment has occurred, fire department, emergency response and/or hazardous materials spill personnel should be notified immediately. Cleanup should be attempted only by those trained in proper spill containment procedures. Contaminated soils should be removed and replaced with clean soil. When environmental contamination occurs, the local and/or state emergency response authorities may require notification. A comprehensive emergency response or disaster preparedness/recovery plan should be in place prior to any operations involving the use, transportation, storage, or disposal of aluminum nitrate.

Small spills of aluminum nitrate powder can be neutralized by cautiously mixing with large amounts of water while stirring. Any insoluble solids or liquids should be separated for disposal. Aqueous solutions can be flushed to sanitary sewer system with flooding amounts of water.

If aluminum nitrate is spilled or leaked, the following specific steps are recommended:

- ☑ Restrict persons not wearing protective clothing from area of spill or leak until cleanup is complete.
- ☑ Remove all ignition sources and ventilate area of spill or leak.
- ☑ Collect powdered materials in the safest manner possible and deposit in sealed containers. Use a vacuum equipped with a HEPA filter for cleanup. Dry sweeping is not recommended (generates airborne dusts). Keep water away.
- ☑ It may be necessary to dispose of aluminum nitrate as a hazardous waste. The responsible state agency or the regional office of the federal Envi-

ronmental Protection Agency (EPA) should be contacted for specific recommendations.

RISK ASSESSMENT: BUSINESS

General Assessment

Accidents or mishaps involving aluminum nitrate can present a significant threat to business operations. The loss or damage of equipment or facilities due to explosion or fire can significantly affect fiscal viability. Lawsuits that may result from personnel injury/death, public exposures, and/or environmental contamination will also require a serious expenditure of resources. Media attention surrounding an injury, death, or environmental damage can also result in a loss of profits and loss of current as well as future business.

ʋ *Recommended Risk-Reduction Measures*

Company attorneys, safety and health professionals, and environmental specialists should be involved in the development of any procedures dealing with safety or emergency response. A company official should be pre-designated as a public relations officer with specific training in dealing with the press. Corporate plans and policies should be developed, approved, and implemented long before any need for such arises.

REFERENCES

Ahlbom, A. 1993. *Biostatistics for Epidemiologists.* New York: CRC Press/Lewis Publishers.

American Conference of Governmental Industrial Hygienists. 1988. *Documentation of the Threshold Limit Values and Biological Exposure Indices,* 5th Edition (with updates). Cincinnati: ACGIH

Asante-Duah, D. K. 1993. *Hazardous Waste Risk Assessment.* New York: CRC Press/Lewis Publishers.

Bloom, A. D. and F. J. DeSerres. 1995. *Ecotoxicity and Human Health.* New York: CRC Press/Lewis Publishers.

Bueche, J. R. 1972. *Principles of Physics,* 2nd Edition. New York: McGraw-Hill.

Brusick, D. J. 1994. *Methods for Genetic Risk Assessment.* New York: CRC Press/Lewis Publishers.

Burrell, R., D. K. Flaherty, and Leonard J. Sauers. 1992. *Toxicology of the Immune System.* New York: Van Nostrand Reinhold.

Calabrese, E. J., E. M. Kenyon. *Air Toxics and Risk Assessment.* New York: CRC Press/Lewis Publishers.

Calabrese, E. J. 1994. *Biological Effects of Low Level Exposures.* New York: CRC Press/Lewis Publishers.

Cockerman, L. G., B. S. Shane. 1994. *Basic Environmental Toxicology.* New York: CRC Press/Lewis Publishers.

Cothern, C. R., N. P. Ross. 1994. *Environmental Statistics, Assessment, and Forecasting.* New York: CRC Press/Lewis Publishers.

Cothern, R. R. 1993. *Comparative Environmental Risk Assessment.* New York: CRC Press/Lewis Publishers.

Gardner, R. W. 1994. *Chemical Intolerance.* New York: CRC Press/Lewis Publishers.

Hallenbeck, W. H. 1993. *Quantitative Risk Assessment for Environmental and Occupational Health,* 2nd Edition. New York: CRC Press/Lewis Publishers.

Hosty, J. W., P. Foster. 1990. *A Practical Guide to Chemical Spill Response.* New York: Van Nostrand Reinhold.

Howard, P. H., et. al. 1991. *Handbook of Environmental Degradation Rates.* New York: CRC Press/Lewis Publishers.

Jones, F. E. 1994. *Toxic Organic Vapors in the Workplace.* New York: CRC Press/Lewis Publishers.

Keith, L. H. and M. M. Walker. 1995. *Handbook of Air Toxics.* New York: CRC Press/Lewis Publishers.

Landis, W. G., M. H. Yu. 1995. *Introduction to Environmental Toxicology.* New York: CRC Press/Lewis Publishers.

Langerman, N. 1994 *Precautionary Labels for Chemical Containers.* New York: CRC Press/Lewis Publishers.

Lauwerys, R. R., P. Hoet. 1993. *Industrial Chemical Exposure.* New York: CRC Press/Lewis Publishers.

Lewis, R. J., Sr. 1992. *Sax's Dangerous Properties of Industrial Materials,* Volumes 1, 2, and 3, 8th Edition. New York: Van Nostrand Reinhold.

Manahan, S. E. 1992. *Toxicological Chemistry,* 2nd Edition. New York: CRC Press/Lewis Publishers.

Meyer, E. 1990. *Chemistry of Hazardous Materials.* Englewood Cliffs, NJ: Prentice Hall.

Quigley, D. R. 1994. *Handbook of Emergency Chemical Management.* New York: CRC Press/Lewis Publishers.

Rea, W. J. 1995. *Chemical Sensitivity,* Vol. I-III. New York: CRC Press/Lewis Publishers.

Richardson, M. 1995. *Dictionary of Substance and Their Effects,* Vol. I-VI. New York: CRC Press/Lewis Publishers.

Sacarello, H. L. A. 1994. *The Comprehensive Handbook of Hazardous Materials.* New York: CRC Press/Lewis Publishers.

Sherman, J. D. 1988. *Chemical Exposure and Disease.* New York: Van Nostrand Reinhold.

Talbott, E., G. Craun. 1995. *An Introduction to Environmental Epidemiology.* New York: CRC Press/Lewis Publishers.

MATERIAL SAFETY DATA SHEET

CHEMICAL NAME	CRC PRESS/LEWIS PUBLISHERS DISCLAIMER
ALUMINUM OXIDE	**CRC PRESS/LEWIS PUBLISHERS DISCLAIMER:** THE INFORMATION AND RECOMMENDATIONS PRESENTED HEREIN ARE BASED ON SOURCES BELIEVED TO BE RELIABLE. CRC MAKES NO REPRESENTATION ON ITS COMPLETENESS OR ACCURACY. IT IS THE RESPONSIBILITY OF THE USER TO DETERMINE THE CHEMICAL'S SUITABILITY FOR ITS INTENDED USE, THE CHEMICAL'S SAFE USE, AND THE CHEMICAL'S PROPER DISPOSAL. NO REPRESENTATIONS AND/OR WARRANTIES, EITHER EXPRESSED OR IMPLIED, OF THE MERCHANTABILITY OR FITNESS FOR A PARTICULAR PURPOSE, OR OF ANY OTHER NATURE, ARE MADE WITH RESPECT TO THE INFORMATION PROVIDED IN THIS MSDS OR TO THE CHEMICAL TO WHICH INFORMATION MAY REFER. CRC NEITHER ASSUMES NOR AUTHORIZES ANY OTHER PERSON TO ASSUME FOR IT, ANY OTHER ADDITIONAL RESPONSIBILITY OR LIABILITY FOR THE USE OF, OR RELIANCE UPON, THIS INFORMATION.

HAZARD WARNING INFORMATION

HEALTH	FIRE	REACTIVE	OTHER	DEGREE OF HAZARD	COLOR CODING	OTHER CODES
1	0	0		0 = Minimum Hazard 1 = Slight Hazard 2 = Moderate Hazard 3 = Serious Hazard 4 = Severe Hazard	HEALTH = BLUE FIRE = RED REACTIVITY = YELLOW OTHER = WHITE	OX = Oxidizer ACID = Acid ALK = Alkali COR = Corrosive W̶ = Use No Water

SECTION I - GENERAL INFORMATION

Characterization	**Metal**	RCRA Number / EPA Class	**Not Listed** / **Not Listed**
DOT Proper Shipping Name	**Chemicals, N.O.S.**	Chemical Abstract Service (CAS) Number	**1344-28-1**
DOT Hazard Class and Label Requirements	**Non-Regulated**	DOT Emergency Guide Code	**No Citation**
DOT Identification Number	**None Established**	Molecular Formula	**Al_2O_3**

Synonyms

Alumina; α-alumina (OSHA); β-alumina; Γ-alumina; aluminum trioxide; activated aluminum oxide; almite; alon; alumite; α-aluminum oxide.

SECTION II - HAZARDOUS INGREDIENTS AND IDENTITY INFORMATION

Hazardous Components (specific identity)	OSHA Exposure Criteria	NIOSH Exposure Criteria	Immediately Dangerous to Life and Health (IDLH)	ACGIH Exposure Criteria
Aluminum oxide (derivation: Leaching of bauxite with caustic soda and precipitation of hydrated aluminum oxide by hydrolysis and seeding of the solution. The alumina hydrate is then washed, filtered, and calcined to remove water and obtain the anhydrous oxide. Also, coal mine waste waters are used to get aluminum sulfate, which is then reduced to alumina).	**PEL:** 15 mg/m^3 (total) 5 mg/m^3 (respiratory)	**REL:** 10 mg/m^3 (proposed)	**Not Determined**	**TLV:** 10 mg/m^3 **STEL:** Not Established

SECTION III - PHYSICAL AND CHEMICAL CHARACTERISTICS

Boiling Point	**5396°F (2977°C)**	Specific Gravity (H_2O = 1)	**4.0**
Vapor Pressure (mm Hg)	**1 (approx.) at 2158°C**	Molecular Weight	**101.96**
Vapor Density (Air = 1)	**Not Applicable**	Melting Point	**3686°F (2030°C)**

Solubility

Negligible solubility (less than 0.1%). Difficult, but soluble in mineral acids and strong alkali.

Appearance and Odor

White, odorless, crystalline powder. Depending upon method of preparation, appearance can also be as white balls, or lumps of various mesh.

SECTION IV - FIRE AND EXPLOSION HAZARD DATA

Flash Point (method used)	**Not Applicable**	Explosive Limits in Air % by Volume	**LEL: N/A UEL: N/A**
NFPA Classification	**Noncombustible Solid**	Autoignition Temperature	**Not Applicable**

Extinguishing Media

Use extinguishing media suitable to surrounding fires.

Special Fire Fighting Procedures

None noted. But, aluminum oxide dust is toxic by inhalation. Fire fighters should take care to avoid dusty situations or wear appropriate respiratory protection where aluminum oxide may be present.

Unusual Fire and Explosion Hazards

Aluminum oxide is a noncombustible solid. But, its dusts may form explosive mixtures in air. Caution is warranted when approaching situations that may involve aluminum oxide dusts.

SECTION V - REACTIVITY DATA

Stability		Conditions to Avoid None documented. However, aluminum oxide should be kept away from contact with hot chlorinated rubber. It is considered stable under normal conditions of temperature and pressure.
Stable X	Unstable	Incompatibility *(materials to avoid)* Aluminum oxide is not compatible with hot chlorinated rubber; acids; oxidizers; and chlorine trifluoride.
Hazardous Polymerization		Conditions to Avoid Explosive hydrogen gas may be formed when finely divided iron contacts moisture during crushing and milling operations. Hazardous polymerization cannot occur.
May Occur	Will Not Occur X	Hazardous Decomposition or By-products Exothermic reaction above 200°C with halocarbon vapors produces toxic hydrogen chloride and phosgene gases and fumes.

SECTION VI - HEALTH HAZARD DATA

Primary Route(s) of Entry:	Inhalation?	Absorption (skin/eye)?	Ingestion?
	X		

Health Hazards

INHALATION: Inhaling finely divided particles may cause lung damage (Shaver's disease). In general, dust inhalation may cause tightness and pain in chest, coughing, difficulty breathing.

SKIN & EYES: Contact with the skin or eyes may cause irritation.

Carcinogenicity	NTP Listed?	IARC Cancer Review Group?	OSHA Regulated?	Target Organs?
Unknown Human Unknown Animal	Not Listed	No	Not Listed	Eyes, Skin, Respiratory System

Medical Conditions Generally Aggravated by Exposure

Pre-existing lung disease or dysfunction may be aggravated by exposure to ammonium oxide dusts.

Emergency and First-aid Procedures

Eye contact: Do not allow victim to keep eyes tightly shut. Flush immediately with water for 15 minutes (minimum), seek medical attention. Skin contact: Remove all contaminated clothing. Immediately wash area with large amounts of soap and water. For inhalation: Remove the person from exposure. Provide respiratory assistance and CPR. Transfer to medical facility. If swallowed, remove person to fresh air and seek medical attention immediately. Begin rescue breathing if breathing has stopped and CPR if heart action has stopped. Do not attempt to give an unconscious or convulsing person anything by mouth.

SECTION VII - PRECAUTIONS FOR SAFE HANDLING AND USE

Steps to be Taken in Case Material is Released or Spilled

Ventilate area of spill or leak. Restrict those not involved in cleanup from entering area. Collect spilled powdered materials in safest manner possible and deposit in sealed drum for disposal. Notify appropriate authorities, as required by applicable environmental regulations.

Preferred Waste Disposal Method

Not listed. Check applicable federal, state, and/or local environmental regulations.

Precautions to be Taken in Handling and Storage

Aluminum oxide is a noncombustible solid. Store to avoid contact with strong acids and oxidizers. Store in tightly closed containers in cool, well-ventilated area.

Other Precautions and Warnings

Aluminum oxide is suitable for storage in general chemical storage areas.

SECTION VIII - CONTROL MEASURES AND PERSONAL PROTECTIVE EQUIPMENT

Respiratory Protection *(specify type)*

None required where adequate ventilation conditions exist. If airborne concentrations exceed the TLV, a dust/mist respirator is recommended. For high exposures, or if concentrations exceed the capacity of the respirator, then a self-contained breathing apparatus (SCBA) with full facepiece operated in positive pressure mode is advised.

Ventilation

Local exhaust (preferred) at site of chemical work or general ventilation.

Protective Gloves	Eye Protection	Other Protective Clothing
Impervious Rubber	Safety Glasses with Side Shields	Not Applicable

Work/Hygiene Practices

Always wash hands thoroughly after using chemical; never bring food, drink, or smoking materials into vicinity of chemicals.

ALUMINUM OXIDE

Al_2O_3 CAS: 1344-28-1

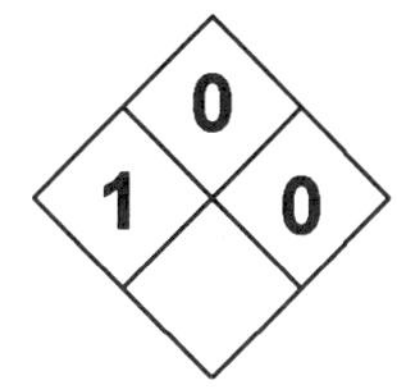

IDENTIFICATION AND TYPICAL USES

Aluminum oxide is a white crystalline sand-like powder with no odor. The mineral corundum is natural aluminum oxide. Emery, ruby, and sapphire are impure crystalline varieties. The mixed mineral bauxite is a hydrated aluminum oxide. It is used in the production of aluminum; in the manufacture of abrasives, refractories, ceramics, and electrical insulators; as catalysts and catalyst supports; in some paper products; spark plugs; crucible and laboratory wares; as adsorbent for gases and water vapors, in chromatographic analysis; in fluxes and light bulbs; as artificial gems; in heat-resistant fibers; and as a dispersing agent in food additives.

RISK ASSESSMENT: HEALTH

General Assessment

The aluminas are generally considered nuisance dusts and under some conditions of human exposure may cause a minimal pulmonary nodular response. However, excessive *inhalation* of finely divided particles may lead to lung damage (Shaver's disease) characterized by scarred tissue (pulmonary fibrosis) with shortness of breath, tightness and pain in the chest, and coughing. The condition can be fatal.

It should be noted that population studies of potentially exposed workers have shown minimal evidence of pulmonary fibrosis or pneumoconiosis. Also, animal experiments with aluminum oxide have shown that the type of reaction in lung tissue varies with the form of alumina used and its particle size, the species of animal tested, and the route of administration. However, the data from both human and animal studies seem to indicate that alumina is efficiently eliminated from the lung and has a low degree of fibrogenicity.

This does not mean there is no health risk associated with aluminum oxide. There is no way of estimating exact human response under all possible types of exposure conditions. Prudent risk management requires caution when working with or around aluminum oxide.

Acute Health Effects

The following acute (short-term) health effects may occur immediately or shortly after exposure to aluminum oxide:

Skin: Irritation.

Eye: Severe eye irritation likely.

Lung: Can irritate the respiratory tract, in large doses, may cause pulmonary fibrosis.

Chronic Health Effects

The following chronic (long-term) health effects may occur at some time after exposure to aluminum oxide and can last for months or even years:

Cancer Hazards: According to the information presented in the references, aluminum oxide has been tested and has not been shown to cause cancer in animals. No data were found on its human carcinogenic capabilities.

Reproductive Hazard: According to the information presented in the references, aluminum oxide has been tested and has not been shown to affect reproduction.

Other Chronic Effects: Repeated exposure to large amounts of aluminum oxide dust may damage the lungs, leading to scarring of the tissue (pulmonary fibrosis), with shortness of breath and possible death.

Recommended Risk-Reduction Measures

Personnel should avoid direct contact with aluminum oxide. The exact nature of its harmful capacities is not clearly understood in the references. Therefore, it is recommended that it be handled as a toxic material. If a less toxic material cannot be substituted for aluminum oxide, then *engineering controls* are the most effective method of reducing exposures. The best protection is to enclose operations and/or provide local exhaust ventilation at the site of aluminum oxide release. While not always operationally feasible, isolating operations can also reduce exposure.

Using respiratory protection and other personal protective equipment (PPE) is less effective than the engineering controls mentioned above, but is still advisable whenever working with or around aluminum oxide. Under normal operating conditions of low exposure potential, a dust/mist respirator will suffice. However, for maximum protection in large concentrations or when the capacity of the dust/mist respirator is exceeded, personnel should wear a self-contained breathing apparatus (SCBA) with full facepiece and pressure demand. If a full facepiece is not available, then chemical/dust goggles, safety glass with side shields, or a face shield should be worn to protect the eyes. To prevent hand and skin exposures, impervious rubber gloves should be worn.

Administrative controls should also be in place to minimize the potential for human exposures. These may include written procedures or policies which specify the methods and techniques that will be practiced whenever personnel are to work with aluminum oxide.

All personnel should receive *training* on the use, hazards, protective measures, emergency actions, and other precautions per 29 CFR 1910.1200 (Hazard Communication), prior to the first assignment in an area where aluminum oxide is used or stored.

If there has been a long history of high exposure, or if lung symptoms develop, the following medical tests are recommended:

- ☑ Chest X-ray.
- ☑ Pulmonary function tests.

Any evaluation should include a careful history of past and present symptoms with an examination. Medical tests that simply look for existing damage are not a substitute for controlling exposures. Also, because smoking can cause lung cancer, emphysema, and a host of other medical problems, working with or around aluminum oxide may aggravate these conditions further.

Other methods to reduce exposure to aluminum oxide include:

- ☑ Where possible, enclose operations and use local exhaust ventilation at the site of chemical release. If local exhaust ventilation or enclosure is not used, then respiratory protection should be required.
- ☑ Always ensure that proper protective clothing is worn when using chemical substances. Never allow contaminated clothing to be taken home from the work place.
- ☑ Hazard warning information should be posted in the work area. In addition, as part of an on-going education and training program, all information on the health and safety hazards of aluminum oxide should be communicated to all potentially exposed workers.
- ☑ Eye wash stations should be provided in the immediate work area for emergency use.

RISK ASSESSMENT: ENVIRONMENT

General Assessment

The environment is at risk of exposure during transportation, storage, disposal, or destruction of aluminum oxide. In almost every scenario, the threat of environmental exposure is contingent upon the proper handling of the chemical substance. Accidental spills, large or small, where aluminum oxide contacts incompatible commodities can result in violent reactions as well as the possible contamination of the surrounding environmental mediums (water, soil, and air).

Aluminum oxide is considered a noncombustible solid. However, its dusts may form explosive mixtures in air. Also, because it is incompatible with strong oxidizers and strong acids, extreme caution is required in handling, storage, transportation, and disposal of aluminum oxide. These characteristics also require special consideration during any emergency situation involving a leak or spill of aluminum oxide. Should aluminum oxide ever come into contact with incompatible substances either during use, transportation, storage, or disposal, dangerous reactions are possible.

Aluminum oxide occurs naturally in the environment in the minerals bauxite, baerites, boehmite, corundum, diaspore, and gibbsite. It therefore occurs naturally in the aquatic environment but can also enter it through industrial discharges and/or spills.

☠ *Acute Ecological Effects*

Acute (short-term) toxic effects may include the death of animals, birds, or fish and death or low growth rate in plants. Acute effects are seen 2 to 4 days after animals or plants are exposed to aluminum oxide.

Aluminum oxide has slight acute toxicity to aquatic life, but its toxicity will be increased under alkaline conditions. Insufficient data are available to evaluate or predict the short-term toxic effects of aluminum oxide to plants, birds, and land animals.

Chronic Ecological Effects

Chronic (long-term) toxic effects may include shortened life span, reproductive problems, lower fertility, and changes in appearance or behavior in exposed animals. These effects can be seen long after first exposure to toxic chemicals.

Aluminum oxide has slight chronic toxicity to aquatic life, but this toxicity will increase under alkaline conditions. Insufficient data are available to evaluate or predict the long-term toxic effects of aluminum oxide to plants, birds, or land animals.

Water Solubility

Aluminum oxide is slightly solubility in water. Concentrations of 1 milligram or less will mix with a liter of water.

Persistence in the Environment

Aluminum oxide is highly persistent in water, with a half-life greater than 200 days. The half-life of a pollutant is the amount of time it takes for one-half of the chemical to be degraded.

Bioaccumulation in Aquatic Organisms

Some substances increase in concentration, or *bioaccumulate,* in living organisms as they breathe contaminated air, drink contaminated water, or eat contaminated food. These chemicals can become concentrated in the tissues and internal organs of animals as well as humans.

Aluminum oxide will not accumulate in the edible tissues of aquatic life that are consumed by humans.

Recommended Risk-Reduction Measures

Proper training of all transporters will reduce the likelihood of a mishap or accident resulting in a leak or spill of aluminum oxide into the environment. The correct use of labeling on all containers, trucks, and rail cars will enable emergency responders to react properly and quickly to any disaster thereby reducing the potential risk to the environment and to personnel.

Storage of aluminum oxide should be segregated from incompatible chemicals to minimize the risk of cross-contamination or contact.

If a spill or leak to the environment has occurred, fire department, emergency response and/or hazardous materials spill personnel should be notified immediately. Cleanup should be attempted only by those trained in proper spill containment procedures. If aluminum oxide should contact the water table, aquifer, or navigable waterway, the local and/or state emergency response authorities must be notified. A comprehensive emergency response or disaster preparedness/recovery plan should be in place prior to any operations involving the use, transportation, storage, or disposal of aluminum oxide.

If aluminum oxide is spilled or leaked, the following specific steps are recommended:

- ☑ Restrict persons not wearing protective clothing from area of spill or leak until cleanup is complete and area can be opened for normal work.
- ☑ Do not allow exposure to incompatible materials.
- ☑ Collect powdered materials in the safest and most efficient manner possible. Place materials in a sealed drum.

RISK ASSESSMENT: BUSINESS

General Assessment

Accidents or mishaps involving aluminum oxide can present a moderate threat to business operations. Although not likely, equipment and facilities may be lost or their use disrupted as a result of an incident involving aluminum oxide which can significantly affect fiscal viability. Lawsuits that may result from personnel injury, illness, and/or death, public exposures, and/or environmental contamination will also require a serious expenditure of resources. Media attention surrounding an injury, death, or environmental damage can also result in a loss of profits and loss of current as well as future business.

Recommended Risk-Reduction Measures

Company attorneys, safety and health professionals, and environmental specialists should be involved in the development of any procedures that focus on emergency response, disaster preparedness, and disaster recovery. A company official should be pre-designated as a public relations officer with specific training in dealing with the press. Corporate plans and policies should be developed, approved, and implemented long before any need for such arises.

REFERENCES

Ahlbom, A. 1993. *Biostatistics for Epidemiologists.* New York: CRC Press/Lewis Publishers.

American Conference of Governmental Industrial Hygienists. 1988. *Documentation of the Threshold Limit Values and Biological Exposure Indices,* 5th Edition (with updates). Cincinnati: ACGIH

Bloom, A. D. and F. J. DeSerres. 1995. *Ecotoxicity and Human Health.* New York: CRC Press/Lewis Publishers.

Brusick, D. J. 1994. *Methods for Genetic Risk Assessment.* New York: CRC Press/Lewis Publishers.

Bueche, J. R. 1972. *Principles of Physics,* 2nd Edition. New York: McGraw-Hill.

Calabrese, E. J., E. M. Kenyon. *Air Toxics and Risk Assessment.* New York: CRC Press/Lewis Publishers.

Calabrese, E. J. 1994. *Biological Effects of Low Level Exposures.* New York: CRC Press/Lewis Publishers.

Cockerman, L. G., B. S. Shane. 1994. *Basic Environmental Toxicology.* New York: CRC Press/Lewis Publishers.

Cothern, R. R. 1993. *Comparative Environmental Risk Assessment.* New York: CRC Press/Lewis Publishers.

Dinman, B. D. 1988. Alumina-related pulmonary disease. *Journal of Occupational Medicine.* 30:328-335.

Gardner, R. W. 1994. *Chemical Intolerance.* New York: CRC Press/Lewis Publishers.

Gibbs, G. W. 1985. Mortality of aluminum reduction plant workers, 1950 through 1977. *Journal of Occupational Medicine.* 27:761-770.

Hallenbeck, W. H. 1993. *Quantitative Risk Assessment for Environmental and Occupational Health,* 2nd Edition. New York: CRC Press/Lewis Publishers.

Howard, P. H. 1995. *Handbook of Environmental Fate and Exposures Data for Organic Chemicals,* Vol. I-IV. New York: CRC Press/Lewis Publishers.

Howard, P. H., et. al. 1991. *Handbook of Environmental Degradation Rates.* New York: CRC Press/Lewis Publishers.

Jones, F. E. 1994. *Toxic Organic Vapors in the Workplace.* New York: CRC Press/Lewis Publishers.

Keith, L. H. and M. M. Walker. 1995. *Handbook of Air Toxics.* New York: CRC Press/Lewis Publishers.

Landis, W. G., M. H. Yu. 1995. *Introduction to Environmental Toxicology.* New York: CRC Press/Lewis Publishers.

Langerman, N. 1994 *Precautionary Labels for Chemical Containers.* New York: CRC Press/Lewis Publishers.

Lauwerys, R. R., P. Hoet. 1993. *Industrial Chemical Exposure.* New York: CRC Press/Lewis Publishers.

Lewis, R. J., Sr. 1993. *Hawley's Condensed Chemical Dictionary,* 12th Edition. New York: Van Nostrand Reinhold.

Lewis, R. J., Sr. 1992. *Sax's Dangerous Properties of Industrial Materials,* Volumes 1, 2, and 3, 8th Edition. New York: Van Nostrand Reinhold.

Lide, D. R. 1995. *Handbook of Chemistry and Physics,* 76th Edition. New York: CRC Press/Lewis Publishers.

Manahan, S. E. 1994. *Environmental Chemistry,* 6th Edition. New York: CRC Press/Lewis Publishers.

Mellström, G. A., J. E. Wahlberg, and H. I. Maibach. 1994. *Protective Gloves for Occupational Use.* New York: CRC Press/Lewis Publishers.

Meyer, E. 1990. *Chemistry of Hazardous Materials.* Englewood Cliffs, NJ: Prentice Hall.

Mycock, J. C., J. D. McKenna, and L. Theodore. 1995. *Handbook of Air Pollution Control Engineering and Technology.* New York: CRC Press/Lewis Publishers.

Patnaik, P. 1992. *A Comprehensive Guide to the Hazardous Properties of Chemical Substances.* New York: Van Nostrand Reinhold.

Proctor, N. H., J. P. Hughes. 1991. *Chemical Hazards of the Workplace.* New York: Van Nostrand Reinhold.

Quigley, D. R. 1994. *Handbook of Emergency Chemical Management.* New York: CRC Press/Lewis Publishers.

Rea, W. J. 1995. *Chemical Sensitivity,* Vol. I-III. New York: CRC Press/Lewis Publishers.

Richardson, M. 1995. *Dictionary of Substance and Their Effects,* Vol. I-VI. New York: CRC Press/Lewis Publishers.

Sacarello, H. L. A. 1994. *The Comprehensive Handbook of Hazardous Materials.* New York: CRC Press/Lewis Publishers.

Sheftel, V. O. 1995. *Handbook of Toxic Properties of Monomers and Additives.* New York: CRC Press/Lewis Publishers.

Talbott, E., G. Craun. 1995. *An Introduction to Environmental Epidemiology.* New York: CRC Press/Lewis Publishers.

Torkelson, T. R. V. K. Rowe. 1981. *Patty's Industrial Hygiene and Toxicology,* Vol. 2B. New York: Wiley-Interscience.

MATERIAL SAFETY DATA SHEET

CHEMICAL NAME	
4-AMINODIPHENYL	**CRC PRESS/LEWIS PUBLISHERS DISCLAIMER:** THE INFORMATION AND RECOMMENDATIONS PRESENTED HEREIN ARE BASED ON SOURCES BELIEVED TO BE RELIABLE. CRC MAKES NO REPRESENTATION ON ITS COMPLETENESS OR ACCURACY. IT IS THE RESPONSIBILITY OF THE USER TO DETERMINE THE CHEMICAL'S SUITABILITY FOR ITS INTENDED USE, THE CHEMICAL'S SAFE USE, AND THE CHEMICAL'S PROPER DISPOSAL. NO REPRESENTATIONS AND/OR WARRANTIES, EITHER EXPRESSED OR IMPLIED, OF THE MERCHANTABILITY OR FITNESS FOR A PARTICULAR PURPOSE, OR OF ANY OTHER NATURE, ARE MADE WITH RESPECT TO THE INFORMATION PROVIDED IN THIS MSDS OR TO THE CHEMICAL TO WHICH INFORMATION MAY REFER. CRC NEITHER ASSUMES NOR AUTHORIZES ANY OTHER PERSON TO ASSUME FOR IT, ANY OTHER ADDITIONAL RESPONSIBILITY OR LIABILITY FOR THE USE OF, OR RELIANCE UPON, THIS INFORMATION.

HAZARD WARNING INFORMATION

HEALTH	FIRE	REACTIVE	OTHER	DEGREE OF HAZARD	COLOR CODING	OTHER CODES
3	1	0		0 = Minimum Hazard 1 = Slight Hazard 2 = Moderate Hazard 3 = Serious Hazard 4 = Severe Hazard	HEALTH = BLUE FIRE = RED REACTIVITY = YELLOW OTHER = WHITE	OX = Oxidizer ACID = Acid ALK = Alkali COR = Corrosive W = Use No Water

SECTION I - GENERAL INFORMATION

Characterization **Aromatic Amine**	RCRA Number **Not Listed**	EPA Class **Not Listed**
DOT Proper Shipping Name **Chemicals, N.O.S.**	Chemical Abstract Service (CAS) Number **97-67-1**	
DOT Hazard Class and Label Requirements **Non-Regulated**	DOT Emergency Guide Code **No Citation**	
DOT Identification Number **None Established**	Molecular Formula $C_6H_5C_6H_4NH_2$	

Synonyms

4-aminobiphenyl; p-aminobiphenyl; p-aminodiphenyl; 4-phenylaniline; p-xenylamine; biphenylamine; (1,1'-biphenyl)-4-amine; 4-biphenylamine.

SECTION II - HAZARDOUS INGREDIENTS AND IDENTITY INFORMATION

Hazardous Components (specific identity)	OSHA Exposure Criteria	NIOSH Exposure Criteria	Immediately Dangerous to Life and Health (IDLH)	ACGIH Exposure Criteria
4-Aminodiphenyl	**PEL: Not Established** **Cancer Producing Agent**	**REL: Not Established** **Confirmed Human Carcinogen**	**Not Determined**	**TLV: Not Established** **Confirmed Human Carcinogen**

SECTION III - PHYSICAL AND CHEMICAL CHARACTERISTICS

Boiling Point **576°F (302°C)**	Specific Gravity (H_2O = 1) **1.16**
Vapor Pressure (mm Hg) **1 (approx.) at 227°F**	Molecular Weight **169.2**
Vapor Density (Air = 1) **Not Found**	Melting Point **127°F (53°C)**

Solubility

Moderately soluble in water.

Appearance and Odor

Colorless or yellowish-brown crystalline (sand-like) solid material with a floral-like odor (Note: Turns purple on contact with air due to oxidation process). It can also exist in liquid solutions.

SECTION IV - FIRE AND EXPLOSION HAZARD DATA

Flash Point (method used) **Not Found**	Explosive Limits in Air % by Volume **LEL: Not Found UEL: Not Found**
NFPA Classification **Combustible Solid**	Autoignition Temperature **842°F (450°C)**

Extinguishing Media

Use extinguishing media suitable to surrounding fires. Recommend water spray/mist or dry chemical.

Special Fire Fighting Procedures

Poisonous gases may be emitted during fire. Fire fighters should take precautions to avoid inhalation or skin contact by wearing full body protection and use a self-contained breathing apparatus (SCBA).

Unusual Fire and Explosion Hazards

4-Aminodiphenyl must be preheated before ignition is possible. It is a slight to moderate fire hazard when exposed to heat, flames, sparks, or powerful oxidizers.

SECTION V - REACTIVITY DATA

Stability		Conditions to Avoid: Normally stable but avoid contact with air. 4-Aminodiphenyl will oxidize on contact with air.
Stable	Unstable	Incompatibility *(materials to avoid)*: May react explosively on contact with powerful oxidizers.
X		
Hazardous Polymerization		Conditions to Avoid: Hazardous polymerization cannot occur. Ignition may occur when mixed with red fuming nitric acid.
May Occur	Will Not Occur	Hazardous Decomposition or By-products: When heated to decomposition, 4-aminodiphenyl emits toxic oxides of nitrogen.
	X	

SECTION VI - HEALTH HAZARD DATA

Primary Route(s) of Entry:	Inhalation?	Absorption (skin/eye)?	Ingestion?
	X	X	X

Health Hazards

INHALATION: Exposure may interfere with the ability of the blood to carry oxygen causing headache, fatigue, difficulty in breathing, and blue or violet discoloration of the lips and skin due to depleted oxygen content in blood. Also causes dizziness, lethargy, and dyspnea.

ABSORPTION: Ataxia, weakness, and interference in the oxygen-carrying capabilities of the blood (see symptoms noted above for inhalation.

INGESTION: May cause acute urinary burning due to bladder irritation. Blood may be present as well.

Carcinogenicity	NTP Listed?	IARC Cancer Review Group?	OSHA Regulated?	Target Organs?
Confirmed Human Confirmed Animal	5th Annual Report	Group 1	29 CFR 1910.1011	Bladder (causes bladder cancer), skin

Medical Conditions Generally Aggravated by Exposure

None reported. However, it may be assumed that any previous skin conditions might be aggravated.

Emergency and First-aid Procedures

Eye contact: Do not allow victim to keep eyes tightly shut. Flush immediately with water for 15 minutes (minimum), seek medical attention. Skin contact: Remove all contaminated clothing. Immediately wash area with large amounts of soap and water. For inhalation: Remove the person from exposure. Provide respiratory assistance and CPR. Transfer to medical facility. If swallowed, remove person to fresh air and seek medical attention immediately. Begin rescue breathing if breathing has stopped and CPR if heart action has stopped. Do not give an unconscious or convulsing person anything by mouth.

SECTION VII - PRECAUTIONS FOR SAFE HANDLING AND USE

Steps to be Taken in Case Material is Released or Spilled

Ventilate area of spill or leak. Restrict those not involved in cleanup from entering area. Collect spilled powdered materials in safest manner possible and deposit in sealed drum for disposal. Notify appropriate authorities, as required by applicable environmental regulations.

Preferred Waste Disposal Method

Dissolve in a combustible solvent, burn in chemical incinerator equipped with afterburner & scrubber.

Precautions to be Taken in Handling and Storage

Prior to working with 4-aminodiphenyl, personnel must, by law, be trained. A regulated, marked area must be established where it is handled, used, or stored that is maintained under a negative pressure.

Other Precautions and Warnings

Bulk storage is not recommended. Refer to OSHA 29 CFR 1910.1011 for regulatory requirements.

SECTION VIII - CONTROL MEASURES AND PERSONAL PROTECTIVE EQUIPMENT

Respiratory Protection *(specify type)*

At any exposure level, use a supplied air respirator with a full facepiece operated in positive pressure mode; or a full facepiece, hood, or helmet in the continuous flow mode; or a MSHA/NIOSH approved self-contained breathing apparatus (SCBA) with a full facepiece operated in pressure demand mode.

Ventilation

Local exhaust (preferred) at site of chemical work or general ventilation.

Protective Gloves	Eye Protection	Other Protective Clothing
Impervious Polyvinyl Chloride (PVC)	Chemical/dust Goggles or Face Shield	Full Body Protection Recommended

Work/Hygiene Practices

Always wash hands thoroughly after using chemical; never bring food, drink, or smoking materials into vicinity of chemicals.

4-AMINODIPHENYL

$C_6H_5C_6H_4NH_2$ CAS: 97-67-1

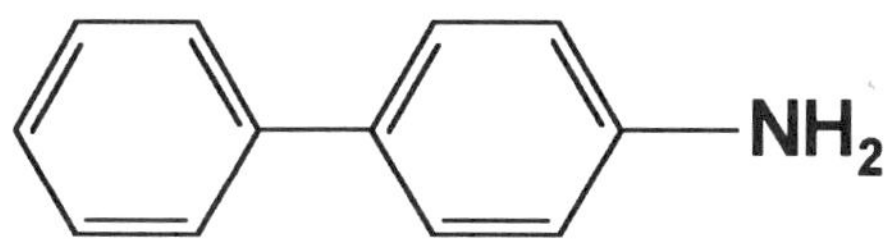

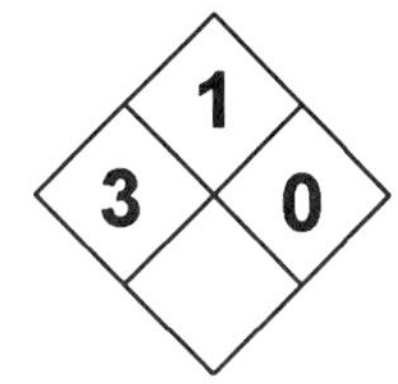

IDENTIFICATION AND TYPICAL USES

4-Aminodiphenyl is a colorless or yellowish-brown crystalline (sand-like) solid material with a sweet, floral odor. It may also exist in liquid solutions. It is used primarily in research laboratories. It was previously used as a rubber antioxidant. However, it is no longer produced on a commercial scale.

RISK ASSESSMENT: HEALTH

General Assessment

4-Aminodiphenyl is a confirmed carcinogen in humans. It is also a mutagen and there is evidence of animal teratogenic effects reported. Many scientists believe there may be no safe level of exposure to human carcinogens and, therefore, all contact (including skin) should be reduced to the lowest possible level. No exposure limits have been established. However, there is a full OSHA standard (29 CFR 1910.1011) that provides detailed information and requirements on how 4-aminodiphenyl must be handled.

Exposure is possible by all routes (*inhalation,* skin *absorption,* and *ingestion*), all having similar systemic effects as well as localized irritation at point of contact. Inhalation exposures and skin absorption may result in a condition known as methemoglobinemia (a decrease in the bloods oxygen-carrying capabilities). This condition is characterized by a blue or violet discoloration of the lips and skin. Other symptoms of exposure include headaches, fatigue, difficulty in breathing, lethargy, dizziness, ataxia, and weakness. 4-Aminodiphenyl is a known bladder carcinogen, especially on ingestion. Related symptoms may include burning urination with possible blood present in urine.

Acute Health Effects

The following acute (short-term) health effects may occur immediately or shortly after exposure to 4-aminodiphenyl:

Skin: Irritation.

Eye: Severe eye irritation likely.

Lung: Can irritate the respiratory tract and cause difficulty in breathing. Will inhibit the oxygen carrying capabilities of the blood and lead to methemoglobinemia.

CNS: Symptoms of dizziness, weakness, fatigue, ataxia, lethargy, and headaches are likely.

Chronic Health Effects

The following chronic (long-term) health effects may occur at some time after exposure to 4-aminodiphenyl and can last for months or even years:

Cancer Hazards: 4-Aminodiphenyl is a confirmed carcinogen in humans. It has been shown to cause bladder cancer in humans and animals. Human and animal mutation data also reported.

Reproductive Hazard: There is limited evidence that 4-aminodiphenyl may damage the developing fetus in test animals. Human teratogenic data are inconclusive.

Other Chronic Effects: This chemical has not been adequately evaluated to determine whether brain or other nerve damage could occur with repeated exposure. However, many solvents and other petroleum based chemicals have been shown to cause such damage. Effects may include reduced memory and concentration, personality changes (withdrawal, irritability), fatigue, sleep disturbances, reduced coordination, and/or effects on nerves supplying internal organs (autonomic nervous system), and/or nerves to the arms and legs characterized by weakness and a feeling of "pins and needles."

Recommended Risk-Reduction Measures

Personnel should avoid direct contact with 4-aminodiphenyl. This chemical is extremely dangerous and should therefore be handled with caution in all situations. If a less toxic material cannot be substituted for 4-aminodiphenyl, then specific *engineering controls* as required by OSHA are the most effective method of reducing exposures. For example, OSHA requires a *regulated area* be established wherever 4-aminodiphenyl is manufactured, used, or stored. Only specifically trained and authorized personnel can be allowed in this area. Other special engineering controls are established for this chemical by OSHA at 29 CFR 1910.1011. The best protection is to enclose operations and/or provide local exhaust ventilation at the site of 4-aminodiphenyl release. While not always operationally feasible, isolating operations within the regulated area can also reduce exposure potential further.

Using respiratory protection and other personal protective equipment (PPE) is less effective than the engineering controls mentioned above, but is still required whenever working with or around 4-aminodiphenyl. At any exposure level to a known carcinogen, personnel should wear a self-contained breathing apparatus (SCBA) with full facepiece and pressure demand. If a full facepiece is not available, then chemical goggles or a face shield should be worn to protect the eyes. To prevent hand and skin exposures, impervious polyvinyl chloride (PVC) gloves should be used. Glove manufacturers should be contacted and permeation studies obtained *before* selecting gloves

Administrative controls must also be in place to minimize the potential for human exposures. These may include written procedures or policies which specify the methods and techniques that will be practiced whenever personnel are to work with 4-aminodiphenyl.

All personnel should receive *training* on the use, hazards, protective measures, emergency actions, and other precautions per 29 CFR 1910.1200 (Hazard Communication) and 29 CFR 1910.1011 (4-Aminodiphenyl), prior to the first assignment in an area where 4-aminodiphenyl is used or stored.

At regular intervals (at least twice a year) after employment, the following medical tests are recommended for those who work with 4-aminodiphenyl:

- ☑ Urine cytology (a test for abnormal cells in urine).
- ☑ Interview for possible brain effects, noting any accounts of recent memory loss, mood shifts (such as irritability or withdrawal), concentration problems, headaches, malaise, and altered sleeping patterns. Consider performing cerebellar, autonomic, and peripheral nervous system evaluation. Positive and borderline individuals should be referred for neuropsychological testing.

If symptoms develop or overexposure is suspected, the following may be useful:

- ☑ Blood methemoglobin level test.

It is recommended that the most recent version of OSHA 29 CFR 1910.1011 be consulted for information on specific tests required by law for those exposed to 4-aminodiphenyl.

Any evaluation should include a careful history of past and present symptoms with an examination. Medical tests that simply look for existing damage are not a substitute for controlling exposures.

Other methods to reduce exposure to 4-aminodiphenyl include:

- ☑ Where possible, enclose operations and use local exhaust ventilation at the site of chemical release. Respiratory protection is required whenever exposure to a carcinogen is possible.
- ☑ Ensure that proper, full-body protective clothing is worn. Never allow contaminated clothing to be taken home from the work place. Clothing contaminated by 4-aminodiphenyl should only be laundered by those who have been informed of the hazards of exposure to 4-aminodiphenyl.
- ☑ Hazard warning information should be posted in the work area. As part of an on-going education and training program, all information on the health and safety hazards of 4-aminodiphenyl should be communicated to all potentially exposed workers.
- ☑ Eye wash stations should be provided in the immediate work area for emergency use. If there is a possibility of skin exposure, emergency showers must also be provided in the immediate work area.
- ☑ On skin contact, immediately wash or shower to remove the chemical. Wash any areas of the body that may have contacted 4-aminodiphenyl prior to each exit from the regulated area.
- ☑ Do not eat, drink, or smoke in areas where 4-aminodiphenyl is handled, processed, or stored since the chemical can be swallowed. Wash hands carefully before eating or smoking.

RISK ASSESSMENT: ENVIRONMENT

General Assessment

The environment is at risk of exposure during transportation, storage, disposal, or destruction of 4-aminodiphenyl. In almost every scenario, the threat of environmental exposure is contingent upon the proper handling of the chemical substance. Accidental spills, large or small, where 4-aminodiphenyl contacts incompatible commodities can result in violent reactions as well as the possible contamination of the surrounding environmental mediums (water, soil, and air).

4-Aminodiphenyl is considered a combustible solid. Also, because it is incompatible with strong oxidizers and because it will oxidize itself upon contact with air, extreme caution is required in handling, storage, transportation, and disposal of 4-aminodiphenyl. These characteristics also require special consideration during any emergency situation involving a leak or spill of 4-aminodiphenyl. Should 4-aminodiphenyl ever come into contact with incompatible substances either during use, transportation, storage, or disposal, dangerous reactions are possible.

4-Aminodiphenyl can enter it through industrial discharges and/or spills.

☠ *Acute Ecological Effects*

Acute (short-term) toxic effects may include the death of animals, birds, or fish and death or low growth rate in plants. Acute effects are seen 2 to 4 days after animals or plants are exposed to 4-aminodiphenyl.

4-Aminodiphenyl has moderate acute toxicity to aquatic life. Insufficient data are available to evaluate or predict the short-term toxic effects of 4-aminodiphenyl to plants, birds, and land animals.

Chronic Ecological Effects

Chronic (long-term) toxic effects may include shortened life span, reproductive problems, lower fertility, and changes in appearance or behavior in exposed animals. These effects can be seen long after first exposure to toxic chemicals.

4-Aminodiphenyl has moderate chronic toxicity to aquatic life. Insufficient data are available to evaluate or predict the long-term toxic effects of 4-aminodiphenyl to plants, birds, or land animals.

Water Solubility

4-Aminodiphenyl is moderately soluble in water. Concentrations of 1 to 1000 milligrams will mix with a liter of water.

⌛ *Persistence in the Environment*

4-Aminodiphenyl is moderately persistent in water, with a half-life between 2 and 200 days. The half-life of a pollutant is the amount of time it takes for one-half of the chemical to be degraded. About 90% of 4-aminodiphenyl will eventually end up in water; approximately 4.5% will end up in terrestrial soils; about 4.25% will end up in aquatic sediments; and the remainder (1.25%) will end up in the air.

Bioaccumulation in Aquatic Organisms

Some substances increase in concentration, or *bioaccumulate,* in living organisms as they breathe contaminated air, drink contaminated water, or eat contaminated food. These chemicals can become concentrated in the tissues and internal organs of animals as well as humans.

The concentration of 4-aminodiphenyl found in fish is expected to be somewhat higher than the average concentration of 4-aminodiphenyl in the water from which the fish was taken.

Recommended Risk-Reduction Measures

Proper training of all transporters will reduce the likelihood of a mishap or accident resulting in a leak or spill of 4-aminodiphenyl into the environment. The correct use of labeling (per the MSDS) on all containers, trucks, and rail cars will enable emergency responders to react properly and quickly to any disaster thereby reducing the potential risk to the environment and to personnel.

Storage of 4-aminodiphenyl should be segregated from incompatible chemicals to minimize the risk of cross-contamination or contact. Bulk storage is not recommended.

If a spill or leak to the environment has occurred, fire department, emergency response and/or hazardous materials spill personnel should be notified immediately. Cleanup should be attempted only by those trained in proper spill containment procedures. If 4-aminodiphenyl should contact the water table, aquifer, or navigable waterway, the local and/or state emergency response authorities must be notified. A comprehensive emergency response or disaster preparedness/recovery plan should be in place prior to any operations involving the use, transportation, storage, or disposal of 4-aminodiphenyl.

If 4-aminodiphenyl is spilled or leaked, the following specific steps are recommended:

☑ Restrict persons not wearing protective clothing from area of spill or leak until cleanup is complete and area is cleared for normal operations.

☑ Do not allow exposure to incompatible materials.

☑ Do not dry sweep for cleanup. Use a vacuum equipped with a high efficiency particulate air (HEPA) filter, not a standard shop vacuum.

RISK ASSESSMENT: BUSINESS

General Assessment

Accidents or mishaps involving 4-aminodiphenyl can present a serious threat to business operations. Although not likely, equipment and facilities may be lost or their use disrupted as a result of an incident involving 4-aminodiphenyl which can significantly affect fiscal viability. Lawsuits that may result from personnel injury, illness, and/or death, public exposures, and/or environmental contamination will also require a serious expenditure of resources. Media attention surrounding an injury, death, or environmental damage can also result in a loss of profits and loss of current as well as future business.

Recommended Risk-Reduction Measures

Company attorneys, safety and health professionals, and environmental specialists should be involved in the development of any procedures that focus on emergency response, disaster preparedness, and disaster recovery. A company official should be pre-designated as a public relations officer with specific training in dealing with the press. Corporate plans and policies should be developed, approved, and implemented long before any need for such arises.

REFERENCES

American Conference of Governmental Industrial Hygienists. 1988. *Documentation of the Threshold Limit Values and Biological Exposure Indices,* 5th Edition (with updates). Cincinnati: ACGIH

Bloom, A. D. and F. J. DeSerres. 1995. *Ecotoxicity and Human Health.* New York: CRC Press/Lewis Publishers.

Brusick, D. J. 1994. *Methods for Genetic Risk Assessment.* New York: CRC Press/Lewis Publishers.

Bueche, J. R. 1972. *Principles of Physics,* 2nd Edition. New York: McGraw-Hill.

Calabrese, E. J. 1994. *Biological Effects of Low Level Exposures.* New York: CRC Press/Lewis Publishers.

Cockerman, L. G., B. S. Shane. 1994. *Basic Environmental Toxicology.* New York: CRC Press/Lewis Publishers.

Cothern, R. R. 1993. *Comparative Environmental Risk Assessment.* New York: CRC Press/Lewis Publishers.

Gardner, R. W. 1994. *Chemical Intolerance.* New York: CRC Press/Lewis Publishers.

Hallenbeck, W. H. 1993. *Quantitative Risk Assessment for Environmental and Occupational Health,* 2nd Edition. New York: CRC Press/Lewis Publishers.

Howard, P. H., et. al. 1991. *Handbook of Environmental Degradation Rates.* New York: CRC Press/Lewis Publishers.

International Agency for Research on Cancer (IARC). 1987. *Monographs on the Evaluation of Carcinogenic Risks to Humans,* Supplement 7, pp 91-92. Lyon: IARC.

Keith, L. H. and M. M. Walker. 1995. *Handbook of Air Toxics.* New York: CRC Press/Lewis Publishers.

Landis, W. G., M. H. Yu. 1995. *Introduction to Environmental Toxicology.* New York: CRC Press/Lewis Publishers.

Langerman, N. 1994 *Precautionary Labels for Chemical Containers.* New York: CRC Press/Lewis Publishers.

Lauwerys, R. R., P. Hoet. 1993. *Industrial Chemical Exposure.* New York: CRC Press/Lewis Publishers.

Lewis, R. J., Sr. 1993. *Hawley's Condensed Chemical Dictionary,* 12th Edition. New York: Van Nostrand Reinhold.

Lewis, R. J., Sr. 1992. *Sax's Dangerous Properties of Industrial Materials,* Volumes 1, 2, and 3, 8th Edition. New York: Van Nostrand Reinhold.

Mellström, G. A., J. E. Wahlberg, and H. I. Maibach. 1994. *Protective Gloves for Occupational Use.* New York: CRC Press/Lewis Publishers.

Meyer, E. 1990. *Chemistry of Hazardous Materials.* Englewood Cliffs, NJ: Prentice Hall.

Mycock, J. C., J. D. McKenna, and L. Theodore. 1995. *Handbook of Air Pollution Control Engineering and Technology.* New York: CRC Press/Lewis Publishers.

Patnaik, P. 1992. *A Comprehensive Guide to the Hazardous Properties of Chemical Substances.* New York: Van Nostrand Reinhold.

Proctor, N. H., J. P. Hughes. 1991. *Chemical Hazards of the Workplace.* New York: Van Nostrand Reinhold.

Quigley, D. R. 1994. *Handbook of Emergency Chemical Management.* New York: CRC Press/Lewis Publishers.

Rea, W. J. 1995. *Chemical Sensitivity,* Vol. I-III. New York: CRC Press/Lewis Publishers.

Richardson, M. 1995. *Dictionary of Substance and Their Effects,* Vol. I-VI. New York: CRC Press/Lewis Publishers.

Sacarello, H. L. A. 1994. *The Comprehensive Handbook of Hazardous Materials.* New York: CRC Press/Lewis Publishers.

Torkelson, T. R. V. K. Rowe. 1981. *Patty's Industrial Hygiene and Toxicology,* Vol. 2B. New York: Wiley-Interscience.

U.S. Department of Labor. 1974. Occupational safety and health standards, carcinogens. *Federal Register.* 39:3756, 3781-3784.

Walpole, A. L., et. al. 1952. The carcinogenic action of 4-aminodiphenyl and 3,2'-dimethyl-4-aminodiphenyl. *British Journal of Industrial Medicine.* 9:255-261.

MATERIAL SAFETY DATA SHEET

CHEMICAL NAME

AMMONIA

CRC PRESS/LEWIS PUBLISHERS DISCLAIMER: THE INFORMATION AND RECOMMENDATIONS PRESENTED HEREIN ARE BASED ON SOURCES BELIEVED TO BE RELIABLE. CRC MAKES NO REPRESENTATION ON ITS COMPLETENESS OR ACCURACY. IT IS THE RESPONSIBILITY OF THE USER TO DETERMINE THE CHEMICAL'S SUITABILITY FOR ITS INTENDED USE, THE CHEMICAL'S SAFE USE, AND THE CHEMICAL'S PROPER DISPOSAL. NO REPRESENTATIONS AND/OR WARRANTIES, EITHER EXPRESSED OR IMPLIED, OF THE MERCHANTABILITY OR FITNESS FOR A PARTICULAR PURPOSE, OR OF ANY OTHER NATURE, ARE MADE WITH RESPECT TO THE INFORMATION PROVIDED IN THIS MSDS OR TO THE CHEMICAL TO WHICH INFORMATION MAY REFER. CRC NEITHER ASSUMES NOR AUTHORIZES ANY OTHER PERSON TO ASSUME FOR IT, ANY OTHER ADDITIONAL RESPONSIBILITY OR LIABILITY FOR THE USE OF, OR RELIANCE UPON, THIS INFORMATION.

HAZARD WARNING INFORMATION

HEALTH	FIRE	REACTIVE	OTHER	DEGREE OF HAZARD	COLOR CODING	OTHER CODES
2	1	0		0 = Minimum Hazard 1 = Slight Hazard 2 = Moderate Hazard 3 = Serious Hazard 4 = Severe Hazard	HEALTH = BLUE FIRE = RED REACTIVITY = YELLOW OTHER = WHITE	OX = Oxidizer ACID = Acid ALK = Alkali COR = Corrosive W = Use No Water

SECTION I - GENERAL INFORMATION

Characterization	**Gas**	RCRA Number	**None**
		EPA Class	**Not Applicable**
DOT Proper Shipping Name	**Ammonia, anhydrous**	Chemical Abstract Service (CAS) Number	**7664-41-7**
DOT Hazard Class and Label Requirements	**Non Flammable Gas**	DOT Emergency Guide Code	**15**
DOT Identification Number	**UN 1005/UN 2073/UN 2672**	Chemical Formula	**NH_3**

Synonyms

Anhydrous ammonia, aqua ammonia, aqueous ammonia, ammonia gas.

SECTION II - HAZARDOUS INGREDIENTS AND IDENTITY INFORMATION

Hazardous Components (specific identity)	OSHA Exposure Criteria	NIOSH Exposure Criteria	Immediately Dangerous to Life and Health (IDLH)	ACGIH Exposure Criteria
Ammonia (derivation from: numerous processes, including from synthesis gas, a mixture of CO, hydrogen, CO_2, and nitrogen (from air) obtained by steam re- forming or by partial combustion of natural gas; an end product of animal metabolism). **1 ppm = 0.71 mg/m^3**	**PEL:** **35 ppm** **27 mg/m^3** **STEL:** **50 ppm** **35 mg/m^3**	**REL:** **25 ppm** **18 mg/m^3** **STEL:** **35 ppm** **27 mg/m^3**	**300 ppm**	**TLV:** **25 ppm** **18 mg/m^3** **STEL:** **35 ppm** **27 mg/m^3**

SECTION III - PHYSICAL AND CHEMICAL CHARACTERISTICS

Boiling Point	**-28°F (-3.5°C)**	Specific Gravity (H_2O = 1)	**0.77 at 0°F**
Vapor Pressure (atmospheres)	**8.5 at 69°F (20°C)**	Molecular Weight	**17**
Vapor Density (Air = 1)	**0.6**	Freezing Point	**-108°F (-77.7°C)**

Solubility

Very miscible in water, alcohol, and ether; aqueous solution is highly alkaline.

Appearance and Odor

Colorless gas with extremely pungent, intense, irritating, and even suffocating odor.

SECTION IV - FIRE AND EXPLOSION HAZARD DATA

Flash Point (method used)	**Not Applicable (gas)**	Explosive Limits in Air % by Volume	**LEL: 15% UEL: 28%**
NFPA Classification	**Class 1B Flammable Liquid**	Autoignition Temperature	**1204°F (650°C)**

Extinguishing Media

Reactive Gas.

Special Fire Fighting Procedures

Ammonia gas is difficult to ignite but presents an explosion hazard when exposed to flame or in fire. Gas is extremely irritating: wear full protective clothing and self-contained breathing apparatus (SCBA). Heat will build pressure and may rupture closed storage containers. Keep fire-exposed containers cool with water spray.

Unusual Fire and Explosion Hazards

Emits toxic fumes of NH_3 and NO_x when exposed to heat. To stop fire, stop the flow of gas. Use extinguishing agent suitable for the type of surrounding fire.

SECTION V - REACTIVITY DATA

Stability		Conditions to Avoid
		Ammonia is corrosive to copper and galvanized surfaces. Forms sensitive explosive mixtures with air and hydrocarbons. Explosive reaction with silver chloride, silver nitrate, silver azide, and silver oxide.
Stable	Unstable	Incompatibility *(materials to avoid)*
X		Ammonia is incompatible with numerous commodities, primarily strong oxidizers, acids, halogens, salts of silver and zinc, gold, mercury, calcium, and hypochlorite bleaches.
Hazardous Polymerization		Conditions to Avoid
		Potentially violent or explosive reactions on contact with interhalogens. Do not mix with other materials, especially for cleaning purposes (never mix ammonia with bleach as toxic gas emissions can occur).
May Occur	Will Not Occur	Hazardous Decomposition or By-products
X		Ammonia catalyzes the polymerization of acrolein and other unsaturates, causing an increase in temperatures and pressure, which may explode containers. Emits toxic NH_3 and NO_x in heat conditions.

SECTION VI - HEALTH HAZARD DATA

Primary Route(s) of Entry:	Inhalation?	Absorption (skin)?	Ingestion?
	X		X

Health Hazards

INHALATION: Eye, nose, and throat irritant. Toxic effects include lachrymation, respiratory distress, chest pain, and pulmonary edema.

SKIN & EYES: Ammonia is corrosive to skin because it reacts with moisture to form caustic ammonium hydroxide. Long exposure may result in destruction of tissue.

INGESTION: Headache, dizziness, nausea, may cause vomiting.

Carcinogenicity	NTP Listed?	IARC Cancer Review Group?	OSHA Regulated?	Target Organs?
Insufficient data	No	No	29 CFR 1910.1000 (Table Z-1)	Respiratory system; skin; eyes.

Medical Conditions Generally Aggravated by Exposure

Respiratory function impairments (bronchitis, asthma, etc.); Skin conditions (dermatitis).

Emergency and First-aid Procedures

Eye contact: Immediately flush large amounts of water for 15 minutes (minimum), occasionally lifting eyelids, seek medical attention. Skin contact: Remove all contaminated clothing. Immediately wash area with large amounts of soap and water. For inhalation: Remove the person from exposure. Provide respiratory assistance and CPR. Transfer promptly to medical facility. Observation for 24 - 48 hours since pulmonary edema is possible. If swallowed, seek medical attention immediately.

SECTION VII - PRECAUTIONS FOR SAFE HANDLING AND USE

Steps to be Taken in Case Material is Released or Spilled

Remove all ignition sources. Ventilate area. Stop flow of gas. If the source of leak is a cylinder and cannot be stopped, remove to safe place in open air; repair leak or let cylinder empty. For liquid spills, neutralize with hydrochloric acid. Wipe with mop or use water aspirator.

Preferred Waste Disposal Method

None specified in the references (incineration of a gas may be the only alternative).

Precautions to be Taken in Handling and Storage

Do not store with incompatible chemicals since violent reactions can occur. Store in tightly closed containers in a cool, dark, well-ventilated area away from sunlight. Keep away from ignition sources such as fire, sparks, and flame.

Other Precautions and Warnings

Bulk storage of ammonia is not recommended. Containers may explode in fire or under conditions of extreme heat (do not store outdoors or in direct sunlight).

SECTION VIII - CONTROL MEASURES AND PERSONAL PROTECTIVE EQUIPMENT

Respiratory Protection *(specify type)*

Exposures above 50 ppm: MSHA/NIOSH approved full-facepiece respirator with an ammonia vapor cartridge. Greater protection is obtained from a self-contained breathing apparatus (SCBA) with full facepiece and pressure demand or other positive pressure mode.

Ventilation

Local exhaust or general mechanical systems recommended.

Protective Gloves	Eye Protection	Other Protective Clothing
Butyl Rubber	Chemical Goggles and/or Face Mask	Rubber Apron

Work/Hygiene Practices

Always wash hands thoroughly after using chemical; never bring food, drink, or smoking materials into vicinity of chemicals.

AMMONIA

NH_3 CAS: 7664-41-7

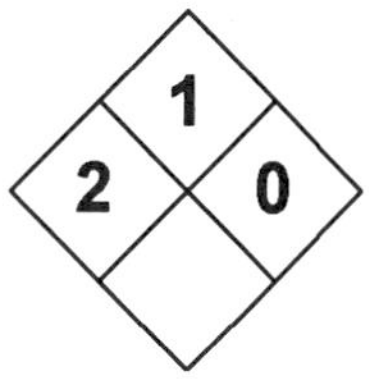

NH_3

IDENTIFICATION AND TYPICAL USES

Ammonia is a colorless gas with a strong, intense, suffocating odor. It is a common molecule given off by living organisms. It is used to make fertilizers, animal foods, synthetic fibers, glues and explosives. It is also used in the treatment and refining of metals; in the manufacture of nitric acid, hydrazine hydrate, and acrylonitrile. It is often used in refrigeration.

RISK ASSESSMENT: HEALTH

General Assessment

Ammonia is a severe irritant of the eyes, respiratory tract, and skin. Exposures to and *inhalation* of large concentrations of liquid anhydrous ammonia causes severe corneal irritation, dyspnea, bronchospasm, chest pain, and pulmonary edema, which may be fatal. Exposure of liquid ammonia to the eyes may cause serious injury to the cornea and deeper structures and sometimes loss of sight. Skin contact causes first and second degree burns that are severe and, if extensive, may be fatal. Extremely high concentrations of vapor (10,000 ppm) can still result in skin burns. With skin and mucous membrane contact, burns are of three types: *cryogenic* (from liquid ammonia), *thermal* (from exothermic dissociation of ammonium hydroxide), and *chemical* (alkaline).

Although no carcinogenic data are reported, there are inconclusive reports of mutagenic reactions.

☠ *Acute Health Effects*

The following acute (short-term) health effects may occur immediately or shortly after exposure to ammonia:

Skin: Irritation and burning on contact with liquid.

Eye: Burns which can lead to permanent damage.

Lung: Irritation of the mouth, nose, and throat causing coughing and/or shortness of breath. Higher exposures can lead to pulmonary edema (fluid in the lungs), a medical emergency which can be fatal.

CNS: Data on the acute effects of ammonia on the central nervous system are inconclusive. However, exposure can result in headaches and the loss of sense of smell (exact mechanism for reaction not completely understood).

💣 *Chronic Health Effects*

The following chronic (long-term) health effects may occur at some time after exposure to ammonia and can last for months or even years:

Cancer Hazards: According to information presented in the references, exposure to ammonia does not currently lead to carcinogenic effects (although some mutagenic data are reported).

Reproductive Hazards: There are no reports to support any claims of reproductive hazards.

Other Chronic Effects: Repeated exposures can cause chronic irritation of the eyes, nose, throat, and airways. Very irritating to the lungs. It is not known whether ammonia causes lung damage in the long-term.

Recommended Risk-Reduction Measures

Personnel should avoid direct contact with ammonia. Unless a less toxic chemical can be substituted for a hazardous substance, *engineering controls* are the most effective method of reducing exposures. The best protection is to enclose operations and/or provide local exhaust ventilation at the site of chemical release. While not always operationally feasible, isolating operations can also reduce exposure. Using respiratory protection is less effective than the controls mentioned above, but is still advisable whenever working with ammonia. A NIOSH-approved, full-facepiece respirator with ammonia vapor cartridge or canister is sufficient for low exposures. A self-contained breathing apparatus (SCBA) with full facepiece and pressure demand is the recommended respiratory protection of choice. If a full facepiece is not available, then chemical goggles should be worn to protect the eyes. Whenever a chemical splash hazard exists, a face shield and a rubber apron should be worn. To prevent hand and skin exposures, butyl rubber gloves should be worn.

Administrative controls should also be in place to minimize the potential for human exposures. These may include written procedures or policies which specify the methods and techniques that will be practiced whenever personnel are to work with ammonia.

All personnel should receive *training* on the use, hazards, protective measures, emergency actions, and other precautions per 29 CFR 1910.1200 (Hazard Communication), prior to the first assignment in an area where ammonia is used or stored. Before beginning employment and at regular intervals thereafter (e.g., annually), the following medical tests are recommended:

☑ Lung function tests.

If symptoms develop or overexposure is suspected, the following may be useful:

☑ Consider chest X-ray after acute overexposure (may be negative if taken immediately after exposure due to delayed onset of pulmonary edema).

Other methods to reduce exposure to chemicals include:

☑ Where possible, enclose operations and use local exhaust ventilation at the site of chemical release. If local exhaust ventilation or enclosure is not used, respiratory protection should be mandatory.

☑ Always ensure that proper protective clothing is worn when using chemical substances.

☑ Wash thoroughly immediately after exposure to ammonia and at the end of the work shift or before eating, drinking, or smoking. Work clothing contaminated with ammonia should never be taken home for laundering. Only personnel trained in the hazards of exposure should launder contaminated clothing.

☑ Safety shower and eyewash stations should be readily available in work areas where ammonia is used or stored.

☑ Hazard warning information should be posted in the work area. In addition, as part of an on-going education and training program, all information on the health and safety hazards of ammonia should be communicated to all exposed workers.

RISK ASSESSMENT: ENVIRONMENT

General Assessment

The environment is at risk of exposure during transportation, storage, disposal, or destruction of ammonia. In almost every scenario, the threat of environmental exposure is contingent upon the proper handling of the chemical substance. Accidental releases, large or small, can result in fire, explosion, and possible contamination of the surrounding environmental mediums (water, soil, and air). Ammonia is not considered to be a flammable gas. However, because it is incompatible with so very many other commodities, contact with which can result in violent explosion and fire, ammonia requires special consideration during any emergency involving a leak or release of ammonia gas or liquid.

Ammonia can enter the environment through natural organic matter decomposition, run-off from agricultural fields or feedlots, municipal waste treatment plant discharges, oil refinery and chemical manufacturing effluents, or from atmospheric fallout.

Acute Ecological Effects

Acute (short-term) toxic effects may include the death of animals, birds, or fish, and death or low growth rate in plants. Acute effects are seen 2 to 4 days after animals or plants are exposed to ammonia. "Total ammonia" refers to the sum of ammonia (NH_3) plus the ionized form (NH_4+).

The toxicity of liquid ammonia solutions is primarily due to the presence of NH_3, the percentage of which increases with pH and temperature. Therefore, the toxicity of "total ammonia" is greater in more alkaline waters at higher temperatures. It is also more toxic under conditions of decreased oxygen concentrations.

Under natural conditions of pH and temperature, ammonia has a moderate acute toxicity to aquatic life. No data are available on the short-term effects of ammonia on plants, birds, or land animals.

Chronic Ecological Effects

Chronic (long-term) toxic effects may include shortened life span, reproductive problems, lower fertility, and changes in appearance or behavior in exposed animals. These effects can be seen long after first exposure(s) to toxic chemicals. Under natural conditions of pH and temperature, total ammonia has moderate chronic toxicity in aquatic life. No data are available to evaluate the long-term effects of ammonia to plants, birds, or land animals.

Water Solubility

Ammonia is highly soluble in water. Concentrations of 1000 milligrams and more will mix with a liter of water.

Persistence in the Environment

Total ammonia is non-persistent in water, with a half-life of less than 2 days. The half-life of a pollutant is the amount of time it takes for one half of the chemical to be degraded.

Bioaccumulation in Aquatic Organisms

Some substances increase in concentration, or *bioaccumulate,* in living organisms as they breathe contaminated air, drink contaminated water, or eat contaminated food. These chemicals can become concentrated in the tissues and internal organs of animals as well as humans.

Ammonia is not expected to accumulate in the tissues of fish.

Recommended Risk-Reduction Measures

Proper training of all transporters will reduce the likelihood of a mishap or accident resulting in a leak or release to the environment. The correct use of DOT labeling (per the MSDS) on all containers, trucks, and rail cars will enable emergency responders to react properly and quickly to any disaster thereby reducing the potential risk to the environment and to personnel.

Storage of ammonia should be segregated from incompatible chemicals to minimize the risk of cross-contamination or contact. Buildings designated for storage should be equipped with appropriate fire protection systems (alarms, sprinklers, emergency lighting, portable extinguishers).

If a spill or release to the environment has occurred, fire department, emergency response, and/or hazardous materials spill personnel should be notified immediately. Cleanup should be attempted only by those trained in proper containment procedures. Contaminated soils should be removed for incineration and replaced with clean soil. If ammonia should contact the water table, aquifer, or navigable waterway, time is of the essence. It is readily miscible in water and total containment and remediation may not be entirely possible. When such spills occur, the local and/or state emergency response authorities must be notified. A comprehensive emergency response or disaster preparedness/recovery plan should be in place prior to any operations involving the use, transportation, storage, or disposal of ammonia. If ammonia is spilled or leaked, the following specific steps are recommended:

- ☑ Restrict persons not wearing protective clothing from area of spill or leak until cleanup is complete and area can be opened for normal work.
- ☑ Remove all ignition sources.
- ☑ Ventilate area of spill or leak.
- ☑ If applicable, stop flow of leaking liquid or gas. If leak source is a cylinder and the leak cannot be stopped in place, remove leaking cylinder to a safe place in the open air, and repair or allow cylinder to empty.
- ☑ Keep ammonia out of a confined space, such as a sewer, because of the possibility of explosion (unless the sewer is designed to prevent the build-up of explosive concentrations).
- ☑ It may be necessary to dispose of ammonia as a hazardous waste. The responsible state agency or the regional office of the federal Environmental Protection Agency (EPA) should be contacted for specific recommendations.

RISK ASSESSMENT: BUSINESS

General Assessment

Accidents or mishaps involving ammonia can present a significant threat to business operations. The loss or damage of equipment or facilities can significantly affect fiscal viability. Lawsuits that may result from personnel injury, public exposures, and environmental contamination will also require a serious expenditure of resources. Media attention surrounding an injury, illness, or environmental damage can also result in a loss of profits and loss of current as well as future business.

Recommended Risk-Reduction Measures

Company attorneys, safety and health professionals, and environmental specialists should be involved in the development of any procedures dealing with safety or emergency response. A company official should be pre-designated as a public relations officer with specific training in dealing with the press. Corporate plans and policies should be developed, approved, and implemented long before any need for such arises.

REFERENCES

Ahlbom, A. 1993. *Biostatistics for Epidemiologists.* New York: CRC Press/Lewis Publishers.

American Conference of Governmental Industrial Hygienists. 1988. *Documentation of the Threshold Limit Values and Biological Exposure Indices,* 5th Edition (with updates). Cincinnati: ACGIH

Anthony, C. P., and N. J. Kolthoff. 1971. *Textbook of Anatomy and Physiology.* St. Louis: C. V. Mosby Company.

Asante-Duah, D. K. 1993. *Hazardous Waste Risk Assessment.* New York: CRC Press/Lewis Publishers.

Bloom, A. D. and F. J. DeSerres. 1995. *Ecotoxicity and Human Health.* New York: CRC Press/Lewis Publishers.

Brusick, D. J. 1994. *Methods for Genetic Risk Assessment.* New York: CRC Press/Lewis Publishers.

Bueche, J. R. 1972. *Principles of Physics,* 2nd Edition. New York: McGraw-Hill.

Burrell, R., D. K. Flaherty, and L. J. Sauers. 1992. *Toxicology of the Immune System.* New York: Van Nostrand Reinhold.

Calabrese, E. J., E. M. Kenyon. *Air Toxics and Risk Assessment.* New York: CRC Press/Lewis Publishers.

Calabrese, E. J. 1994. *Biological Effects of Low Level Exposures.* New York: CRC Press/Lewis Publishers.

Cockerman, L. G., B. S. Shane. 1994. *Basic Environmental Toxicology.* New York: CRC Press/Lewis Publishers.

Cothern, R. R. 1993. *Comparative Environmental Risk Assessment.* New York: CRC Press/Lewis Publishers.

Cothern, C. R. 1995. *Handbook for Environmental Risk Decisionmaking.* New York: CRC Press/Lewis Publishers.

DiVincenzo, G. O., F. J. Yanno, and B. D. Astill. 1973. Exposure of man and dog to low concentrations of ammonia vapor. *American Industrial Hygiene Journal.* 34:329-336.

Gardner, R. W. 1994. *Chemical Intolerance.* New York: CRC Press/Lewis Publishers.

Hallenbeck, W. H. 1993. *Quantitative Risk Assessment for Environmental and Occupational Health,* 2nd Edition. New York: CRC Press/Lewis Publishers.

Hodgson, E., R. B. Mailman, and J. E. Chambers. 1988. *Dictionary of Toxicology.* London: The MacMillian Company.

Hosty, J. W., and P. Foster. 1990. *A Practical Guide to Chemical Spill Response.* New York: Van Nostrand Reinhold.

Howard, P. H. 1995. *Handbook of Environmental Fate and Exposures Data for Organic Chemicals,* Vol. I-IV. New York: CRC Press/Lewis Publishers.

Howard, P. H., et. al. 1991. *Handbook of Environmental Degradation Rates.* New York: CRC Press/Lewis Publishers.

James, A., and M. Lord. 1992. *Index of Chemical and Physical Data.* New York: Van Nostrand Reinhold.

Jones, F. E. 1994. *Toxic Organic Vapors in the Workplace.* New York: CRC Press/Lewis Publishers.

Kamrin, M. A. 1988. *Toxicology: A Primer on Principles and Applications.* Chelsea, Michigan: Lewis Publishers.

Keith, L. H. and M. M. Walker. 1995. *Handbook of Air Toxics.* New York: CRC Press/Lewis Publishers.

Landis, W. G., M. H. Yu. 1995. *Introduction to Environmental Toxicology.* New York: CRC Press/Lewis Publishers.

Langerman, N. 1994 *Precautionary Labels for Chemical Containers.* New York: CRC Press/Lewis Publishers.

Lauwerys, R. R., P. Hoet. 1993. *Industrial Chemical Exposure.* New York: CRC Press/Lewis Publishers.

Lewis, R. J., Sr. 1993. *Hawley's Condensed Chemical Dictionary,* 12th Edition. New York: Van Nostrand Reinhold.

Lewis, R. J., Sr. 1992. *Sax's Dangerous Properties of Industrial Materials,* Volumes 1, 2, and 3, 8th Edition. New York: Van Nostrand Reinhold.

Lide, D. R. 1995. *Handbook of Chemistry and Physics,* 76th Edition. New York: CRC Press/Lewis Publishers.

Manahan, S. E. 1992. *Toxicological Chemistry,* 2nd Edition. New York: CRC Press/Lewis Publishers.

Mellström, G. A., J. E. Wahlberg, and H. I. Maibach. 1994. *Protective Gloves for Occupational Use.* New York: CRC Press/Lewis Publishers.

Meyer, E. 1990. *Chemistry of Hazardous Materials.* Englewood Cliffs, NJ: Prentice Hall.

Patnaik, P. 1992. *A Comprehensive Guide to the Hazardous Properties of Chemical Substances.* New York: Van Nostrand Reinhold.

Philp, R. B. 1995. *Environmental Hazards and Human Health.* New York: CRC Press/Lewis Publishers.

Proctor, N. H., and J. P. Hughes. 1991. *Chemical Hazards of the Workplace.* New York: Van Nostrand Reinhold.

Quigley, D. R. 1994. *Handbook of Emergency Chemical Management.* New York: CRC Press/Lewis Publishers.

Raleigh, R. L. and W. A. McGee. 1972. Effects of short, high concentration exposures to ammonia as determined by observation in the work area. *Journal of Occupational Medicine.* 14:607-610.

Rea, W. J. 1995. *Chemical Sensitivity,* Vol. I-III. New York: CRC Press/Lewis Publishers.

Richardson, M. 1995. *Dictionary of Substance and Their Effects,* Vol. I-VI. New York: CRC Press/Lewis Publishers.

U.S. Department of Labor (OSHA). 1975. Exposure to ammonia, proposed standard. *Federal Register.* 40:54684-54693.

U.S. Department of Health and Human Services, National Institute for Occupational Safety and Health. 1994. *NIOSH Pocket Guide to Chemical Hazards.* Washington, D.C.: U.S. Government Printing Office.

U.S. Department of Health and Human Services, National Institute for Occupational Safety and Health. 1974. *Criteria for a Recommended Standard - Occupational Exposure to Ammonia.* DHEW Publication Number: 74-136. Washington, D.C.: U.S. Government Printing Office.

MATERIAL SAFETY DATA SHEET

CHEMICAL NAME	CRC PRESS/LEWIS PUBLISHERS DISCLAIMER
AMMONIUM ACETATE	THE INFORMATION AND RECOMMENDATIONS PRESENTED HEREIN ARE BASED ON SOURCES BELIEVED TO BE RELIABLE. CRC MAKES NO REPRESENTATION ON ITS COMPLETENESS OR ACCURACY. IT IS THE RESPONSIBILITY OF THE USER TO DETERMINE THE CHEMICAL'S SUITABILITY FOR ITS INTENDED USE, THE CHEMICAL'S SAFE USE, AND THE CHEMICAL'S PROPER DISPOSAL. NO REPRESENTATIONS AND/OR WARRANTIES, EITHER EXPRESSED OR IMPLIED, OF THE MERCHANTABILITY OR FITNESS FOR A PARTICULAR PURPOSE, OR OF ANY OTHER NATURE, ARE MADE WITH RESPECT TO THE INFORMATION PROVIDED IN THIS MSDS OR TO THE CHEMICAL TO WHICH INFORMATION MAY REFER. CRC NEITHER ASSUMES NOR AUTHORIZES ANY OTHER PERSON TO ASSUME FOR IT, ANY OTHER ADDITIONAL RESPONSIBILITY OR LIABILITY FOR THE USE OF, OR RELIANCE UPON, THIS INFORMATION.

HAZARD WARNING INFORMATION

HEALTH	FIRE	REACTIVE	OTHER	DEGREE OF HAZARD	COLOR CODING	OTHER CODES
1	1	0		0 = Minimum Hazard 1 = Slight Hazard 2 = Moderate Hazard 3 = Serious Hazard 4 = Severe Hazard	HEALTH = BLUE FIRE = RED REACTIVITY = YELLOW OTHER = WHITE	OX = Oxidizer ACID = Acid ALK = Alkali COR = Corrosive ~~W~~ = Use No Water

SECTION I - GENERAL INFORMATION

Characterization	**Ester**	RCRA Number	**None**
		EPA Class	**Not Applicable**
DOT Proper Shipping Name	**No Citation**	Chemical Abstract Service (CAS) Number	**631-61-8**
DOT Hazard Class and Label Requirements	**ORM-E; Label: None**	DOT Emergency Guide Code	**No Citation**
DOT Identification Number	**NA 9079**	Chemical Formula	**$NH_4(C_2H_3O_2)$**
Synonyms	**Acetic acid, ammonium salt.**		

SECTION II - HAZARDOUS INGREDIENTS AND IDENTITY INFORMATION

Hazardous Components (specific identity)	OSHA Exposure Criteria	NIOSH Exposure Criteria	Immediately Dangerous to Life and Health (IDLH)	ACGIH Exposure Criteria
Ammonium acetate (derivation from: The interaction of glacial acetic acid and ammonia).	**PEL:** Not Established **STEL:** Not Established	**REL:** Not Established **STEL:** Not Established	Not Established	**TLV:** Not Established **STEL:** Not Established

SECTION III - PHYSICAL AND CHEMICAL CHARACTERISTICS

Boiling Point	**Decomposes**	Specific Gravity (H_2O = 1)	**1.17**
Vapor Pressure (mm Hg)	**Not Applicable**	Molecular Weight	**77.10**
Vapor Density (Air = 1)	**2.6**	Freezing Point	**237°F (114°C)**
Solubility	**Soluble in water, alcohol, and methanol. Slightly soluble in acetone.**		
Appearance and Odor	**White, deliquescent crystals with a slightly acetic odor.**		

SECTION IV - FIRE AND EXPLOSION HAZARD DATA

Flash Point (method used)	**Not Applicable**	Explosive Limits in Air % by Volume	**LEL: N/A UEL: N/A**
NFPA Classification	**Combustible Solid**	Autoignition Temperature	**Not Applicable**

Extinguishing Media

Dry chemical, carbon dioxide, carbon dioxide, Halon, water spray, standard foam, water fog or spray.

Special Fire Fighting Procedures

Toxic and poisonous gases are produced in fire. Wear full protective clothing to protect skin and eyes and self-contained breathing apparatus (SCBA). Move containers from fire if it can be done without risk. Do NOT use a direct water stream since it will scatter material (and fire) to a larger area. Do not allow run-off to reach sewers or waterways.

Unusual Fire and Explosion Hazards

None reported.

SECTION V - REACTIVITY DATA

Stability		Conditions to Avoid: Normally stable under ordinary conditions of use and storage. Ammonia is released when ammonium acetate is exposed to air. Keep containers tightly closed. Avoid excessive dust generation.
Stable	Unstable	Incompatibility *(materials to avoid)*: Strong oxidizers (such as chlorine, bromine, fluorine), and oxidizing materials. Will decompose upon contact with sodium hypochlorite.
X		
Hazardous Polymerization		Conditions to Avoid: Under normal conditions of temperature and pressure, hazardous polymerization of ammonium acetate is not expected to occur.
May Occur	Will Not Occur	Hazardous Decomposition or By-products: When heated to decomposition, ammonium acetate emits toxic nitric oxide and ammonia fumes.
	X	

SECTION VI - HEALTH HAZARD DATA

Primary Route(s) of Entry:	Inhalation?	Absorption (skin)?	Ingestion?
	X		X

Health Hazards

INHALATION: Dusts are irritating to the nose, mouth, and respiratory tract. General symptoms of exposure may include loss of facial muscle firmness, tremor, general discomfort, and impairment of motor functions and recognition. Inhalation toxicity seems to be low.

SKIN & EYES: Skin contact may cause mild irritation. Absorption has NOT been reported.

INGESTION: Irritation of the digestive tract, diarrhea, diuresis, and systemic ammonia poisoning.

Carcinogenicity	NTP Listed?	IARC Cancer Review Group?	OSHA Regulated?	Target Organs?
Not Listed Human Not Listed Animal	No	No	Not Listed.	Respiratory system; skin; eyes, digestive tract.

Medical Conditions Generally Aggravated by Exposure

None identified, however pre-existing liver disease may be aggravated by exposure.

Emergency and First-aid Procedures

Eye contact: Flush immediately with water for 30 minutes (minimum); seek medical attention. Skin contact: Remove all contaminated clothing. Immediately wash area with large amounts of soap and water. Seek medical assistance. For inhalation: Remove the person from exposure. Provide respiratory assistance and CPR. Transfer to medical facility. If swallowed, Give large amounts of water and seek medical attention immediately. Never give an unconscious or convulsing person anything by mouth.

SECTION VII - PRECAUTIONS FOR SAFE HANDLING AND USE

Steps to be Taken in Case Material is Released or Spilled

Evacuate area of non-essential personnel. Remove all sources of ignition. Wear SCBA, heavy rubber boots and gloves. Minimize dust generation, use a vacuum equipped with HEPA filter and damp mop residue. Collect and contain in tightly closed drum. Ventilate area of spill site.

Preferred Waste Disposal Method

No citation.

Precautions to be Taken in Handling and Storage

Keep in tightly closed container, stored in a cool, dry, ventilated area. Protect against physical damage. Store separate from ignition sources and strong oxidizers. Keep airborne concentrations to the lowest possible level using adequate dilution ventilation.

Other Precautions and Warnings

All employees assigned to work with, use, handle, or store ammonium acetate must be trained on its hazards *prior* to their first assignment.

SECTION VIII - CONTROL MEASURES AND PERSONAL PROTECTIVE EQUIPMENT

Respiratory Protection *(specify type)*

Where exposure to dusts are anticipated, a full facepiece respirator equipped with a dust/mist pre-filter should be worn. For emergency situations or very high transient exposures, use a supplied-air respirator or a self-contained breathing apparatus (SCBA) may be necessary.

Ventilation

Local exhaust systems recommended (explosion-proof electrical equipment may be necessary).

Protective Gloves	Eye Protection	Other Protective Clothing
Chemical Resistant Rubber	Chemical/Dust Goggles or Face Mask	Rubber Apron

Work/Hygiene Practices

Always wash hands thoroughly after using chemical; never bring food, drink, or smoking materials into vicinity of chemicals.

AMMONIUM ACETATE

$NH_4(C_2H_3O_2)$ CAS: 631-61-8

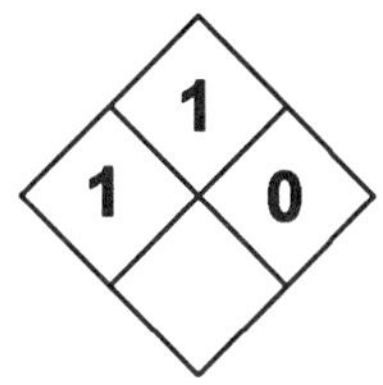

IDENTIFICATION AND TYPICAL USES

Ammonium acetate is a white, deliquescent crystalline solid with a slightly acetic odor. It is used as a reagent in analytical chemistry for determining lead and iron content, a chemical intermediate in manufacturing acetamide. It is also a diuretic in veterinary medicine, in other drugs, in textile dyeing, in meat preservative, in foam rubbers, in vinyl plastics, in stripping explosives, and in determining lead and iron content.

RISK ASSESSMENT: HEALTH

General Assessment

Ammonium acetate is not rated for carcinogenic, mutagenic, or teratogenic effects. No data are available to substantiate claims to the contrary. However, prudent risk management requires that all contact with this chemical be reduced to the lowest possible level. The references contain very little information on the toxic nature of this chemical; however, it is believed to be relatively low. Human exposure to ammonium acetate can occur primarily through two routes of entry into the body: *inhalation* and *ingestion*. Each can present a mild to moderate level of toxicity. Absorption through the skin has not been reported.

Inhalation can cause irritation of the nose, throat, and upper respiratory tract. Systemic toxicity by this route of entry has not been fully researched and is therefore, not clearly understood.

Ingestion can lead to irritation of the digestive tract, diarrhea, diuresis (urine secretion indicating increased kidney function), and systemic ammonia poisoning. Other possible symptoms include flaccid facial muscles (loss of firmness and tension), tremor, general discomfort, anxiety, and impairment of motor performance and recognition. In the long term, exposure may cause liver dysfunction and/or disease.

Acute Health Effects

The following acute (short-term) health effects may occur immediately or shortly after exposure to ammonium acetate:

Skin: Mild irritation.

Eye: Mild irritation and redness.

Lung: Irritation of the mucous membranes of the nose and throat as well as mild irritation of the upper respiratory tract.

Other: Ingestion can cause irritation of the digestive tract with diarrhea and general discomfort.

Chronic Health Effects

The following chronic (long-term) health effects may occur at some time after exposure to ammonium acetate and can last for months or even years:

Cancer Hazards: According to the references, ammonium acetate has not been tested for its ability to cause cancer.

Reproductive Hazard: According to the references, ammonium acetate has not been tested for its ability to adversely affect reproduction.

Recommended Risk-Reduction Measures

Personnel should avoid direct contact with ammonium acetate. Unless a another chemical can be substituted for a hazardous substance, *engineering controls* are the most effective method of reducing exposures. The best protection is to enclose operations and/or provide local exhaust ventilation at the site of chemical release. While not always operationally feasible, isolating operations can also reduce exposure.

Using respiratory protection is less effective than the controls mentioned above, but is still advisable whenever working with or around ammonium acetate.

For low exposures, a respirator fitted with a dust/fume/mist pre-filter is acceptable protection. At higher concentrations of exposure, a self-contained breathing apparatus (SCBA) with full facepiece and pressure demand is the recommended respiratory protection of choice. If a full facepiece is not available, then chemical/dust goggles should be worn to protect the eyes. Whenever a potential for airborne release of dusts exists, a face shield and a rubber apron should be worn. To prevent hand and skin exposures, heavy rubber gloves should be worn.

Administrative controls should also be in place to minimize the potential for human exposures. These may include written procedures or policies which specify the methods and techniques that will be practiced whenever personnel are to work with ammonium acetate.

All personnel should receive *training* on the use, hazards, protective measures, emergency actions, and other precautions per 29 CFR 1910.1200 (Hazard Communication), prior to the first assignment in an area where ammonium acetate is used or stored. If symptoms develop or overexposure is suspected, the following may be useful:

☑ Liver function tests.

Other methods to reduce exposure include:

☑ Where possible, enclose operations and use local exhaust ventilation at the site of chemical release. If local exhaust ventilation or enclosure is not used, respiratory protection should be mandatory.

☑ Always ensure that proper protective clothing is worn when using chemical substances. Clothing contaminated with ammonium acetate should never be laundered at home. Only personnel who have been trained on exposure hazards should launder contaminated clothing.

☑ Wash thoroughly immediately after exposure to ammonium acetate and at the end of the work shift or before eating, drinking, or smoking.

☑ Hazard warning information should be posted in the work area. In addition, as part of an on-going education and training program, all information on the health and safety hazards of ammonium acetate should be communicated to all exposed and potentially exposed workers.

☑ Never eat, drink, or smoke in areas were ammonium acetate is used, handled or stored.

RISK ASSESSMENT: ENVIRONMENT

General Assessment

The environment is at risk of exposure during transportation, storage, disposal, or destruction of ammonium acetate. In almost every scenario, the threat of environmental exposure is contingent upon the proper handling of the chemical substance. Accidental spills, large or small, can result in fire, explosion, and possible contamination of the surrounding environmental mediums (water, soil, and air).

Ammonium acetate is a combustible solid that can react in contact with incompatible materials to cause fire or explosion. Its dusts may form explosive mixtures in air. At high temperatures, its decomposition products of nitrous oxides are highly toxic. These characteristics require special consideration during any emergency situation involving a leak or spill of ammonium acetate. Should ammonium acetate ever come into contact with incompatible substances, such as oxidizing materials, either during use, transportation, or storage, explosive reactions may occur. All federal, state, and local environmental regulations must be observed during transportation and disposal of ammonium acetate.

Ammonium acetate can enter the environment through industrial or municipal waste treatment plant discharges, agricultural runoff, or spills.

☠ *Acute Ecological Effects*

Acute (short-term) toxic effects may include the death of animals, birds, or fish, and death or low growth rate in plants. Acute effects are seen 2 to 4 days after animals or plants are exposed to ammonium acetate.

This chemical has moderate acute toxicity to aquatic life. No data are available to evaluate the short-term effects of ammonium acetate to plants, animals, or birds.

Chronic Ecological Effects

Chronic toxic effects may include shortened life span, reproductive problems, lower fertility, and changes in appearance or behavior in exposed animals. These effects can be seen long after first exposure(s) to toxic chemicals.

Ammonium acetate has moderate chronic toxicity in aquatic life. Insufficient data are available to evaluate or predict the long-term effects of ammonium acetate to plants, birds, or land animals.

♦ *Water Solubility*

Crystalline ammonium acetate is highly soluble in water. Concentrations of 1000 milligrams and more will mix with a liter of water.

⌛ *Persistence in the Environment*

Ammonium acetate is moderately persistent in the aquatic environment with a half-life between 2 and 20 days. The half-life of a pollutant is the amount of time it takes for one-half of the chemical to be degraded.

Bioaccumulation in Aquatic Organisms

Some substances increase in concentration, or *bioaccumulate,* in living organisms as they breathe contaminated air, drink contaminated water, or eat contaminated food. These chemicals can become concentrated in the tissues and internal organs of animals as well as humans.

Most ammonium compounds will probably not accumulate in edible tissues of aquatic species that are eaten by humans.

Recommended Risk-Reduction Measures

Proper training of all transporters will reduce the likelihood of a mishap or accident resulting in a leak or spill to the environment. The correct use of DOT labeling (per the MSDS) on all containers, trucks, and rail cars will enable emergency responders to react properly and quickly to any disaster thereby reducing the potential risk to the environment and to personnel.

Storage of ammonium acetate should be segregated from incompatible chemicals to minimize the risk of cross-contamination or contact. Buildings designated for storage should be equipped with appropriate fire protection systems (alarms, sprinklers, emergency lighting, portable extinguishers).

If a spill or leak to the environment has occurred, fire department, emergency response, and/or hazardous materials spill personnel should be notified immediately. Cleanup should be attempted only by those trained in proper spill containment procedures. Contaminated soils should be removed and replaced with clean soil.

If ammonium acetate should contact the water table, aquifer, or navigable waterway, time is of the essence. It is readily miscible in water and, therefore, total containment and remediation may not be entirely possible. The ammonia constituent will probably evaporate quickly. When such spills occur, the local and/or state emergency response authorities must be notified. A comprehensive emergency response or disaster preparedness/recovery plan should be in place prior to any operations involving the use, transportation, storage, or disposal of ammonium acetate.

If ammonium acetate is spilled or leaked, the following specific steps are recommended:

- ☑ Restrict persons not wearing protective clothing from area of spill or leak until cleanup is complete.
- ☑ Remove all ignition sources.
- ☑ Ventilate area of spill or leak.
- ☑ Cover solutions with vermiculite, sand, earth, or other absorbent and collect/contain in sealed drum or container. Collect dry spills for reclamation or disposal. Avoid generating dusts. Wet mop area to collect residue after bulk materials have been collected.
- ☑ It may be necessary to dispose of ammonium acetate as a hazardous waste. The responsible state agency or the regional office of the federal Environmental Protection Agency (EPA) should be contacted for specific recommendations.

RISK ASSESSMENT: BUSINESS

General Assessment

Accidents or mishaps involving ammonium acetate can present a significant threat to business operations. The loss or damage of equipment or facilities due to fire or explosion can significantly affect fiscal viability. Lawsuits that may result from personnel injury/death, public exposures, and/or environmental contamination will also require a serious expenditure of resources.

Media attention surrounding an injury, death, or environmental damage can also result in a loss of profits and loss of current as well as future business.

Recommended Risk-Reduction Measures

Company attorneys, safety and health professionals, and environmental specialists should be involved in the development of safety, health, or environmental procedures. A company official should be pre-designated as a public relations officer with specific training in dealing with the press. Corporate plans and policies should be developed, approved, and implemented long before any need for such arises.

REFERENCES

Ahlbom, A. 1993. *Biostatistics for Epidemiologists.* New York: CRC Press/Lewis Publishers.

American Conference of Governmental Industrial Hygienists. 1988. *Documentation of the Threshold Limit Values and Biological Exposure Indices,* 5th Edition (with updates). Cincinnati: ACGIH

Anthony, C. P., and N. J. Kolthoff. 1971. *Textbook of Anatomy and Physiology.* St. Louis: C. V. Mosby Company.

Asante-Duah, D. K. 1993. *Hazardous Waste Risk Assessment.* New York: CRC Press/Lewis Publishers.

Bloom, A. D. and F. J. DeSerres. 1995. *Ecotoxicity and Human Health.* New York: CRC Press/Lewis Publishers.

Brusick, D. J. 1994. *Methods for Genetic Risk Assessment.* New York: CRC Press/Lewis Publishers.

Bueche, J. R. 1972. *Principles of Physics,* 2nd Edition. New York: McGraw-Hill.

Burrell, R., D. K. Flaherty, and L. J. Sauers. 1992. *Toxicology of the Immune System.* New York: Van Nostrand Reinhold.

Calabrese, E. J. 1994. *Biological Effects of Low Level Exposures.* New York: CRC Press/Lewis Publishers.

Cockerman, L. G., B. S. Shane. 1994. *Basic Environmental Toxicology.* New York: CRC Press/Lewis Publishers.

Cothern, R. R. 1993. *Comparative Environmental Risk Assessment.* New York: CRC Press/Lewis Publishers.

Cothern, C. R. 1995. *Handbook for Environmental Risk Decisionmaking.* New York: CRC Press/Lewis Publishers.

Danse, I. R. 1991. *Common Sense Toxics in the Workplace.* New York: Van Nostrand Reinhold.

Environmental Protection Agency. 1983. *Chemical Hazard Information Profiles.* Washington, D.C.: U.S. EPA.

Fodden, C. R., and J. L. Weddell. 1992. *Hazardous Materials Emergency Action Data.* New York: CRC Press/Lewis Publishers.

Forseberg, K., and L. H. Keith. 1993. *Chemical; Protective Clothing Permeation and Degradation Database.* Chelsea, MI: Lewis Publishers.

Gardner, R. W. 1994. *Chemical Intolerance.* New York: CRC Press/Lewis Publishers.

Hallenbeck, W. H. 1993. *Quantitative Risk Assessment for Environmental and Occupational Health,* 2nd Edition. New York: CRC Press/Lewis Publishers.

Hodgson, E., R. B. Mailman, and J. E. Chambers. 1988. *Dictionary of Toxicology.* London: The MacMillian Company.

Heck, H. 1985. Mechanisms of aldehyde toxicity: Structure activity Studies. *CIIT Activities,* 5(10):106.

Hosty, J. W., and P. Foster. 1990. *A Practical Guide to Chemical Spill Response.* New York: Van Nostrand Reinhold.

Howard, P. H. 1995. *Handbook of Environmental Fate and Exposures Data for Organic Chemicals,* Vol. I-IV. New York: CRC Press/Lewis Publishers.

International Agency for Research on Cancer (IARC). 1987. *IARC Monographs on the Evaluation of Carcinogenic Risk to Humans, Overall Evaluations of Carcinogenicity: An updating of IARC Monographs Volumes 1-42.* Lyon: IARC.

James, A., and M. Lord. 1992. *Index of Chemical and Physical Data.* New York: Van Nostrand Reinhold.

Jones, F. E. 1994. *Toxic Organic Vapors in the Workplace.* New York: CRC Press/Lewis Publishers.

Kamrin, M. 1 A. 1988. *Toxicology: A Primer on Principles and Applications.* Chelsea, Michigan: Lewis Publishers.

Keith, L. H. , M. T. Johnston, C. E. Hudak, and M. Conoley. 1992. *National Toxicology Program's Gloves+ Database.* Chelsea, MI: Lewis Publishers.

Keith, L. H. (Editor). 1992. *Integrated Risk Information System (IRIS) Chemical Information Database.* Prepared by the U.S. Environmental Protection Agency. Boca Raton, FL: CRC Press/Lewis Publishers.

Keith, L. H. and D. B. Walters. (Editors) 1992. *National Toxicology Program's Chemical Database, Volumes I-VII.* Chelsea, MI: Lewis Publishers.

Landis, W. G., M. H. Yu. 1995. *Introduction to Environmental Toxicology.* New York: CRC Press/Lewis Publishers.

Langerman, N. 1994 *Precautionary Labels for Chemical Containers.* New York: CRC Press/Lewis Publishers.

Lauwerys, R. R., P. Hoet. 1993. *Industrial Chemical Exposure.* New York: CRC Press/Lewis Publishers.

Lewis, R. J., Sr. 1993. *Hawley's Condensed Chemical Dictionary,* 12th Edition. New York: Van Nostrand Reinhold.

Lewis, R. J., Sr. 1992. *Sax's Dangerous Properties of Industrial Materials,* Volumes 1, 2, and 3, 8th Edition. New York: Van Nostrand Reinhold.

Lisella, F. S. 1994. *The VNR Dictionary of Environmental Health and Safety.* New York: Van Nostrand Reinhold.

Meyer, E. 1990. *Chemistry of Hazardous Materials.* Englewood Cliffs, NJ: Prentice Hall.

Patnaik, P. 1992. *A Comprehensive Guide to the Hazardous Properties of Chemical Substances.* New York: Van Nostrand Reinhold.

Proctor, N. H., and J. P. Hughes. 1991. *Chemical Hazards of the Workplace.* New York: Van Nostrand Reinhold.

Silverman, L., H. F. Schuttle, and M. W. First. 1946. Further studies on sensory response to certain industrial solvent vapors. *Journal of Industrial Hygiene Toxicology,* 28:262-266.

Sherman, J. D. 1988. *Chemical Exposure and Disease.* New York: Van Nostrand Reinhold.

U.S. Department of Health and Human Services, National Institute for Occupational Safety and Health. 1994. *NIOSH Pocket Guide to Chemical Hazards.* Washington, D.C.: U.S. Government Printing Office.

U.S. Department of Labor. 1995. *Code of Federal Regulations, Title 29, Part 1910. Regulations Relating to Labor: Safety for General Industry.* Washington, D.C.: U.S. Government Printing Office.

MATERIAL SAFETY DATA SHEET

CHEMICAL NAME

AMMONIUM BIFLUORIDE

HAZARD WARNING INFORMATION

HEALTH	FIRE	REACTIVE	OTHER	DEGREE OF HAZARD	COLOR CODING	OTHER CODES
3	0	0		0 = Minimum Hazard 1 = Slight Hazard 2 = Moderate Hazard 3 = Serious Hazard 4 = Severe Hazard	HEALTH = BLUE FIRE = RED REACTIVITY = YELLOW OTHER = WHITE	OX = Oxidizer ACID = Acid ALK = Alkali COR = Corrosive W = Use No Water

SECTION I - GENERAL INFORMATION

Characterization **Fluorides (inorganic)**	RCRA Number **None**	EPA Class **Not Applicable**
DOT Proper Shipping Name **Ammonium Hydrogen Fluoride solid, solution**	Chemical Abstract Service (CAS) Number **1341-49-7**	
DOT Hazard Class and Label Requirements **ORM-B Corrosive; Corrosive Poison**	DOT Emergency Guide Code **15 (solid) 59 (solution)**	
DOT Identification Number **UN 1727 (solid); UN 2817 (solution)**	Chemical Formula NH_4HF_2	

Synonyms
Ammonium acid fluoride; ammonium hydrogen fluoride; acid ammonium fluoride; ammonium hydrogen bifluoride.

SECTION II - HAZARDOUS INGREDIENTS AND IDENTITY INFORMATION

Hazardous Components (specific identity)	OSHA Exposure Criteria	NIOSH Exposure Criteria	Immediately Dangerous to Life and Health (IDLH)	ACGIH Exposure Criteria
Ammonium bifluoride (derivation: By action of ammonium hydroxide on hydrofluoric acid with subsequent crystallization).	**PEL:** 2.5 mg/m³ (as Fluorides) **STEL:** Not Established	**REL (10 hour):** 2.5 mg/m³ (as Fluorides) **STEL:** Not Established	**500 mg/m³**	**TLV:** 2.5 mg/m³ (as Fluorides) **STEL:** Not Established

SECTION III - PHYSICAL AND CHEMICAL CHARACTERISTICS

Boiling Point **463°F (239°C)**	Specific Gravity (H_2O = 1) **1.5**
Vapor Pressure (mm Hg) **Not Applicable**	Molecular Weight **57.06**
Vapor Density (Air = 1) **2.0**	Melting Point **256°F (125°C)**

Solubility
Soluble in water (forms hydrofluoric acid solution) and alcohol.

Appearance and Odor
White, odorless, orthorhombic crystals (hygroscopic in humidity >50%).

SECTION IV - FIRE AND EXPLOSION HAZARD DATA

Flash Point (method used) **Not Applicable**	Explosive Limits in Air % by Volume **LEL: Not Determined UEL: Not Determined**
NFPA Classification **Non Combustible Solid**	Autoignition Temperature **Not Applicable**

Extinguishing Media
Dry chemical or carbon dioxide. Do NOT use water.

Special Fire Fighting Procedures
Toxic and poisonous gases are produced in fire. Wear full protective clothing to protect skin and eyes, and a self-contained breathing apparatus (SCBA). Move containers from fire if it can be done without risk. Do not apply water to adjacent fires since it may liberate hydrogen fluoride gas capable of etching glass, cement, and many metals.

Unusual Fire and Explosion Hazards
None reported.

SECTION V - REACTIVITY DATA

Stability		Conditions to Avoid **Normally stable under ordinary conditions of use and storage. Keep away from moisture and water (it will etch glass, cement, and many metals. Flammable hydrogen gas may be liberated if exposed to water.**
Stable **X**	Unstable	Incompatibility *(materials to avoid)* **Strong oxidizers (such as chlorine, bromine, fluorine), and oxidizing materials, strong acids (sulfuric, nitric), and water.**
Hazardous Polymerization		Conditions to Avoid **Under normal conditions of temperature and pressure, hazardous polymerization of ammonium bifluoride is not expected to occur.**
May Occur	Will Not Occur **X**	Hazardous Decomposition or By-products **When heated to decomposition, ammonium bifluoride emits toxic and poisonous gases of fluorine, nitrogen oxide, and ammonia.**

SECTION VI - HEALTH HAZARD DATA

Primary Route(s) of Entry:	Inhalation?	Absorption (skin)?	Ingestion?
	X	X	X

Health Hazards

INHALATION: Severe irritation of the respiratory tract with initial stimulation of the respiratory system followed by depressed respiration and possible death due to respiratory paralysis.

SKIN & EYES: Irritation and rash (may be large with numerous pustules). Moist skin may form hydrochloric acid with deep, severe burns. Eye contact causes serious burns and vision loss.

INGESTION: Irritation or corrosion of the digestive tract, nausea, bloody vomiting, bloody diarrhea, abdominal pain, muscle weakness, spasms, dehydration, convulsions, progressive CNS depression, cardiac arrhythmia, and excessive potassium and calcium levels in blood.

Carcinogenicity	NTP Listed?	IARC Cancer Review Group?	OSHA Regulated?	Target Organs?
Not Listed Human **Not Listed Animal**	**No**	**No**	**29 CFR 1910.1000** **(As Fluorides, F-)**	**Skin, eyes, teeth, bones, kidneys, CNS.**

Medical Conditions Generally Aggravated by Exposure

Pre-existing kidney or bone disorders may be aggravated by exposure.

Emergency and First-aid Procedures

Eye contact: Flush immediately with water for 30 minutes (minimum); seek medical attention. Skin contact: Remove all contaminated clothing. Immediately wash area with large amounts of soap and water. Rub a 2.5% solution of calcium gluconate gel until pain subsides. Seek medical assistance. For inhalation: Remove the person from exposure. Provide respiratory assistance and CPR. Transfer to medical facility. If swallowed: Give large amounts of water to dilute and seek medical attention immediately. Never give an unconscious or convulsing person anything by mouth.

SECTION VII - PRECAUTIONS FOR SAFE HANDLING AND USE

Steps to be Taken in Case Material is Released or Spilled

Evacuate area of non-essential personnel. Wear SCBA, heavy rubber boots and gloves. Neutralize with sodium bicarbonate, soda ash, slacked lime, or crushed limestone. Do NOT dry sweep, use a vacuum equipped with HEPA filter and damp mop residue. Collect and contain in tightly closed drum.

Preferred Waste Disposal Method

No Citation.

Precautions to be Taken in Handling and Storage

Store in plastic, rubber, or paraffin containers. Do NOT use metal (will corrode and generate flammable hydrogen gas). Store in cool dry, well-ventilated area.

Other Precautions and Warnings

Do NOT use in moist or humid areas.

SECTION VIII - CONTROL MEASURES AND PERSONAL PROTECTIVE EQUIPMENT

Respiratory Protection *(specify type)*

Where low exposure to dusts are anticipated, use a full facepiece respirator equipped with a dust/mist pre-filter. For emergency situations or very high transient exposures, use a supplied-air respirator or a self-contained breathing apparatus (SCBA) may be necessary.

Ventilation

Local exhaust or general dilution systems recommended.

Protective Gloves	Eye Protection	Other Protective Clothing
Chemical Resistant Rubber	**Chemical/Dust Goggles or Face Mask**	**Rubber Apron**

Work/Hygiene Practices

Always wash hands thoroughly after using chemical; never bring food, drink, or smoking materials into vicinity of chemicals.

AMMONIUM BIFLUORIDE

NH_4HF_2 CAS: 1341-49-7

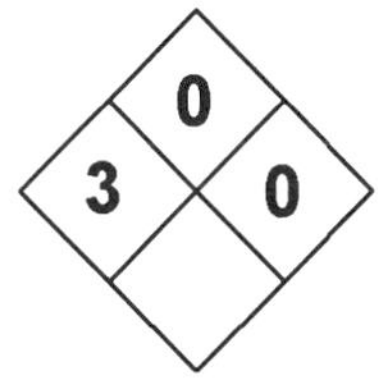

IDENTIFICATION AND TYPICAL USES

Ammonium bifluoride is a white, odorless, orthorhombic, crystalline solid. It is used as a chemical reagent, in the production of ceramics, as a sour in laundering cloth, in electroplating processing beryllium, and for etching glass (white acid). It is also used as a sterilizing agent for brewery, dairy, and other equipment.

RISK ASSESSMENT: HEALTH

General Assessment

Ammonium bifluoride is not rated for carcinogenic, mutagenic, or teratogenic effects. No data are available to substantiate claims to the contrary. However, prudent risk management requires that all contact with this chemical be reduced to the lowest possible level. It is a caustic poison and extremely corrosive to human tissue and highly toxic through all routes of exposure (*inhalation, ingestion,* and *absorption*). Each can present a severe level of toxicity. While pure "absorption" (in the classical sense) leading to systemic effects is not likely, the chemical will still pass through intact skin to cause serious burns and possible bone damage. The primary mechanism for its local corrosive action as well as its systemic toxicity is its conversion to hydrogen fluoride and fluoride ion absorption. Fluoride ion is a direct cellular poison which has the ability to interfere with calcium and magnesium balances.

Inhalation can cause moderate to severe irritation of the respiratory tract. Initially, it will stimulate and then depress the respiration process, making breathing difficult and labored. Death can occur from respiratory paralysis.

Ingestion can lead to irritation or corrosion of the digestive tract, nausea, vomiting (with blood), bloody diarrhea, abdominal pain, muscle weakness and spasms, dehydration, convulsions, cardiac arrhythmia, and excessive blood levels of potassium and calcium. It will also depress the central nervous system with symptoms progressing from fatigue, coma to respiratory arrest (even in the absence of circulatory failure).

Skin contact is irritating and may cause an unusual, large, red rash with numerous puss-filled sections appearing in a balloon-like fashion on the surface of the rash. If the skin surface is moist (either through natural perspiration or from some other means), hydrochloric acid can form and cause serious, deep burns. These burns may initially seem mild, but may penetrate through to the bone. Eye contact will result in serious (dangerous) ocular burns and possible loss of vision.

Acute Health Effects

The following acute (short-term) health effects may occur immediately or shortly after exposure to ammonium bifluoride:

Skin: Severe irritation, unusual red rash with pustules. Possible deep burning (to the bone).

Eye: Dangerous burns and potential loss or damage to vision.

Lung: Irritation of the mucous membranes of the nose and throat as well as severe irritation of the upper respiratory tract leading to respiratory paralysis.

CNS: Fatigue, coma, and respiratory arrest.

Other: Ingestion can cause irritation of the digestive tract with bloody diarrhea, vomiting, convulsions, and death due to action on the CNS.

Chronic Health Effects

The following chronic (long-term) health effects may occur at some time after exposure to ammonium bifluoride and can last for months or even years:

Cancer Hazards: According to the references, ammonium bifluoride has not been adequately tested for its ability to cause cancer.

Reproductive Hazard: According to the references, ammonium bifluoride has not been adequately tested for its ability to adversely affect reproduction.

Other Chronic Hazards: Repeated or long-term exposure to and subsequent absorption of the fluoride ion can result in kidney damage as well as fluorosis, which is a serious illness characterized by brittle bones, calcified ligaments, and chronic anemia.

⓿ *Recommended Risk-Reduction Measures*

Personnel should avoid direct contact with ammonium bifluoride. It is extremely corrosive locally and highly toxic systemically. Unless a another chemical can be substituted for a hazardous substance, *engineering controls* are the most effective method of reducing exposures. The best protection is to enclose operations and/or provide local exhaust ventilation at the site of chemical release. While not always operationally feasible, isolating operations can also reduce exposure.

Using respiratory protection is less effective than the controls mentioned above, but is still advisable whenever working with or around ammonium bifluoride. For low exposures, a respirator fitted with a dust/fume/mist pre-filter is acceptable protection. At higher concentrations of exposure, a supplied-air respirator with full facepiece operated in positive pressure or continuous flow mode, or a self-contained breathing apparatus (SCBA) with full facepiece and pressure demand are the recommended respiratory protection methods of choice. If a full facepiece is not available, then chemical/dust goggles should be worn to protect the eyes. Whenever a potential for airborne release of dusts exists, a face shield and a rubber apron should be worn. To prevent hand and skin exposures, heavy rubber gloves should be worn.

Administrative controls should also be in place to minimize the potential for human exposures. These may include written procedures or policies which specify the methods and techniques that will be practiced whenever personnel are to work with ammonium bifluoride.

All personnel should receive *training* on the use, hazards, protective measures, emergency actions, and other precautions per 29 CFR 1910.1200 (Hazard Communication), prior to the first assignment in an area where ammonium bifluoride is used or stored. If symptoms develop or overexposure is suspected, the following may be useful:

- ☑ Kidney function tests (with urinalysis).
- ☑ Nervous system examination and evaluation by a qualified neurologist concentrating on motor functions and muscle control.
- ☑ Examination of the skin to ensure absence of burns or rash (reaction can be delayed).

To prevent skin burns from becoming serious, a 2.5% calcium gluconate gel should be gently massaged into burn area until pain is relieved.

Other methods to reduce exposure include:

- ☑ Where possible, enclose operations and use local exhaust ventilation at the site of chemical release. If local exhaust ventilation or enclosure is not used, respiratory protection should be mandatory.
- ☑ Always ensure that proper protective clothing is worn when using chemical substances. Clothing contaminated with ammonium bifluoride should never be laundered at home. Only personnel who have been trained on exposure hazards should launder contaminated clothing.
- ☑ Wash thoroughly immediately after exposure to ammonium bifluoride and at the end of the work shift or before eating, drinking, or smoking.
- ☑ Hazard warning information should be posted in the work area. In addition, as part of an on-going education and training program, all information on the health and safety hazards of ammonium bifluoride should be communicated to all potentially exposed workers.
- ☑ Never eat, drink, or smoke in areas were ammonium bifluoride is used, handled or stored.
- ☑ A safety shower and emergency eyewash station should be located in any area where ammonium bifluoride is used, handled, or stored.

RISK ASSESSMENT: ENVIRONMENT

General Assessment

The environment is at risk of exposure during transportation, storage, disposal, or destruction of ammonium bifluoride. In almost every scenario, the threat of environmental exposure is contingent upon the proper handling of the chemical substance. Accidental spills, large or small, can result in fire, explosion, and possi-

ble contamination of the surrounding environmental mediums (water, soil, and air).

Ammonium bifluoride is a non-combustible solid, but can react in contact with incompatible materials to liberate heat and other more toxic commodities. Its dusts may form explosive mixtures in air. At high temperatures, its decomposition products of nitrogen oxide, fluorine, and ammonia gasses are highly toxic. In sufficient levels, ammonia gas can form explosive mixtures in air. These characteristics require special consideration during any emergency situation involving a leak or spill of ammonium bifluoride. Should ammonium bifluoride ever come into contact with incompatible substances, such as oxidizing materials, acids, or water, either during use, transportation, or storage, dangerous reactions may occur. All federal, state, and local environmental regulations must be observed during transportation and disposal of ammonium bifluoride.

Ammonium bifluoride can enter the environment through industrial or municipal waste treatment plant discharges, agricultural runoff, or spills.

Acute Ecological Effects

Acute (short-term) toxic effects may include the death of animals, birds, or fish, and death or low growth rate in plants. Acute effects are seen 2 to 4 days after animals or plants are exposed to ammonium bifluoride. This chemical has slight acute toxicity to aquatic life.

It will react with water to form hydrochloric acid which has moderate acute toxicity to aquatic life. No data are available to evaluate the short-term effects of ammonium bifluoride to plants, animals, or birds.

Chronic Ecological Effects

Chronic toxic effects may include shortened life span, reproductive problems, lower fertility, and changes in appearance or behavior in exposed animals. These effects can be seen long after first exposure(s) to toxic chemicals. This chemical has slight chronic toxicity in aquatic life. It will react with water to form hydrochloric acid which has moderate acute toxicity to aquatic life. No data are available on the long-term effects of ammonium bifluoride to plants, birds, or land animals.

Water Solubility

Crystalline ammonium bifluoride is highly soluble in water. Concentrations of 1000 milligrams and more will mix with a liter of water to form hydrochloric acid.

Persistence in the Environment

Ammonium bifluoride not persistent in the aquatic environment since it will form hydrochloric acid when mixed with water. Hydrochloric acid is moderately persistent in the aquatic environment with a half-life between 2 and 20 days. The half-life of a pollutant is the amount of time it takes for one-half of the chemical to be degraded.

Bioaccumulation in Aquatic Organisms

Some substances increase in concentration, or *bioaccumulate,* in living organisms as they breathe contaminated air, drink contaminated water, or eat contaminated food. These chemicals can become concentrated in the tissues and internal organs of animals as well as humans.

Ammonium bifluoride is biodegradable and is not expected to accumulate in the edible tissues of aquatic species that are eaten by humans.

Recommended Risk-Reduction Measures

Proper training of all transporters will reduce the likelihood of a mishap or accident resulting in a leak or spill to the environment. The correct use of DOT labeling (per the MSDS) on all containers, trucks, and rail cars will enable emergency responders to react properly and quickly to any disaster thereby reducing the potential risk to the environment and to personnel.

Storage of ammonium bifluoride should be segregated from incompatible chemicals and moisture to minimize the risk of cross-contamination or contact. Buildings designated for storage should be equipped with appropriate fire protection systems (alarms, sprinklers, emergency lighting, portable extinguishers).

If a spill or leak to the environment has occurred, fire department, emergency response, and/or hazardous materials spill personnel should be notified immediately. Cleanup should be attempted only by those trained in proper spill containment procedures. Contaminated soils should be removed and replaced with clean soil.

If ammonium bifluoride should contact the water table, aquifer, or navigable waterway, time is of the essence. It is readily miscible in water and, therefore, total containment and remediation may not be possible. The formation of hydrochloric acid will probably occur quickly. When such spills occur, the local and/or state emergency response authorities must be notified. A comprehensive emergency response or disaster preparedness/recovery plan should be in place prior to any operations involving the use, transportation, storage, or disposal of ammonium bifluoride.

If ammonium bifluoride is spilled or leaked, the following specific steps are recommended:

- ☑ Restrict persons not wearing protective clothing from area of spill or leak until cleanup is complete and area can be opened for normal work.
- ☑ Remove all ignition sources.
- ☑ Ventilate area of spill or leak.
- ☑ Cover solutions with vermiculite, sand, earth, or other absorbent and collect/contain in sealed drum or container. Collect dry spills for reclamation or disposal. Avoid generating dusts. Wet mop area to collect residue after bulk materials have been collected. Use a dilute solution of sodium bicarbonate to neutralize residue.
- ☑ It may be necessary to dispose of ammonium bifluoride as a hazardous waste. The responsible state agency or the regional office of the federal Environmental Protection Agency (EPA) should be contacted for specific recommendations.

RISK ASSESSMENT: BUSINESS

General Assessment

Accidents or mishaps involving ammonium bifluoride can present a significant threat to business operations. The loss or damage of equipment or facilities due to fire or explosion can significantly affect fiscal viability. Lawsuits that may result from personnel injury/death, public exposures, and/or environmental contamination will also require a serious expenditure of resources.

Media attention surrounding an injury, death, or environmental damage can also result in a loss of profits and loss of current as well as future business.

Recommended Risk-Reduction Measures

Company attorneys, safety and health professionals, and environmental specialists should be involved in the development of safety, health, or environmental policies or procedures. A company official should be predesignated as a public relations officer with specific training in dealing with the press. Corporate plans and policies should be developed, approved, and implemented long before any need for such arises.

REFERENCES

Ahlbom, A. 1993. *Biostatistics for Epidemiologists.* New York: CRC Press/Lewis Publishers.

American Conference of Governmental Industrial Hygienists. 1988. *Documentation of the Threshold Limit Values and Biological Exposure Indices,* 5th Edition (with updates). Cincinnati: ACGIH.

Anthony, C. P., and N. J. Kolthoff. 1971. *Textbook of Anatomy and Physiology.* St. Louis: C. V. Mosby Company.

Asante-Duah, D. K. 1993. *Hazardous Waste Risk Assessment.* New York: CRC Press/Lewis Publishers.

Bloom, A. D. and F. J. DeSerres. 1995. *Ecotoxicity and Human Health.* New York: CRC Press/Lewis Publishers.

Brusick, D. J. 1994. *Methods for Genetic Risk Assessment.* New York: CRC Press/Lewis Publishers.

Bueche, J. R. 1972. *Principles of Physics,* 2nd Edition. New York: McGraw-Hill.

Burrell, R., D. K. Flaherty, and L. J. Sauers. 1992. *Toxicology of the Immune System.* New York: Van Nostrand Reinhold.

Calabrese, E. J. 1994. *Biological Effects of Low Level Exposures.* New York: CRC Press/Lewis Publishers.

Cockerman, L. G., B. S. Shane. 1994. *Basic Environmental Toxicology.* New York: CRC Press/Lewis Publishers.

Cothern, R. R. 1993. *Comparative Environmental Risk Assessment.* New York: CRC Press/Lewis Publishers.

Danse, I. R. 1991. *Common Sense Toxics in the Workplace.* New York: Van Nostrand Reinhold.

Environmental Protection Agency. 1983. *Chemical Hazard Information Profiles.* Washington, D.C.: U.S. EPA.

Fodden, C. R., and J. L. Weddell. 1992. *Hazardous Materials Emergency Action Data.* New York: CRC Press/Lewis Publishers.

Gardner, R. W. 1994. *Chemical Intolerance.* New York: CRC Press/Lewis Publishers.

Hallenbeck, W. H. 1993. *Quantitative Risk Assessment for Environmental and Occupational Health,* 2nd Edition. New York: CRC Press/Lewis Publishers.

Hodgson, E., R. B. Mailman, and J. E. Chambers. 1988. *Dictionary of Toxicology.* London: The MacMillian Company.

Jones, F. E. 1994. *Toxic Organic Vapors in the Workplace.* New York: CRC Press/Lewis Publishers.

Lewis, R. J., Sr. 1993. *Hawley's Condensed Chemical Dictionary,* 12th Edition. New York: Van Nostrand Reinhold.

Lewis, R. J., Sr. 1992. *Sax's Dangerous Properties of Industrial Materials,* Volumes 1, 2, and 3, 8th Edition. New York: Van Nostrand Reinhold.

Meyer, E. 1990. *Chemistry of Hazardous Materials.* Englewood Cliffs, NJ: Prentice Hall.

Sherman, J. D. 1988. *Chemical Exposure and Disease.* New York: Van Nostrand Reinhold.

MATERIAL SAFETY DATA SHEET

CHEMICAL NAME	
AMMONIUM NITRATE	**CRC PRESS/LEWIS PUBLISHERS DISCLAIMER:** THE INFORMATION AND RECOMMENDATIONS PRESENTED HEREIN ARE BASED ON SOURCES BELIEVED TO BE RELIABLE. CRC MAKES NO REPRESENTATION ON ITS COMPLETENESS OR ACCURACY. IT IS THE RESPONSIBILITY OF THE USER TO DETERMINE THE CHEMICAL'S SUITABILITY FOR ITS INTENDED USE, THE CHEMICAL'S SAFE USE, AND THE CHEMICAL'S PROPER DISPOSAL. NO REPRESENTATIONS AND/OR WARRANTIES, EITHER EXPRESSED OR IMPLIED, OF THE MERCHANTABILITY OR FITNESS FOR A PARTICULAR PURPOSE, OR OF ANY OTHER NATURE, ARE MADE WITH RESPECT TO THE INFORMATION PROVIDED IN THIS MSDS OR TO THE CHEMICAL TO WHICH INFORMATION MAY REFER. CRC NEITHER ASSUMES NOR AUTHORIZES ANY OTHER PERSON TO ASSUME FOR IT, ANY OTHER ADDITIONAL RESPONSIBILITY OR LIABILITY FOR THE USE OF, OR RELIANCE UPON, THIS INFORMATION.

HAZARD WARNING INFORMATION

HEALTH	FIRE	REACTIVE	OTHER	DEGREE OF HAZARD	COLOR CODING	OTHER CODES
2	**0**	**3**	**OX**	0 = Minimum Hazard 1 = Slight Hazard 2 = Moderate Hazard 3 = Serious Hazard 4 = Severe Hazard	HEALTH = BLUE FIRE = RED REACTIVITY = YELLOW OTHER = WHITE	OX = Oxidizer ACID = Acid ALK = Alkali COR = Corrosive W = Use No Water

SECTION I - GENERAL INFORMATION

Chemical Family	RCRA Number	EPA Class
Nitrate	**None**	**Toxic Substance (SARA 313)**

DOT Proper Shipping Name **Ammonium Nitrate**	Chemical Abstract Service (CAS) Number **6484-52-2**
DOT Hazard Class and Label Requirements **Oxidizer; Class A Explosive**	DOT Emergency Guide Code **43**
DOT Identification Number **UN 1942/NA 1942/UN 2426/UN 0222**	Chemical Formula **NH_4NO_3**

Synonyms
Ammonia salt; ammonium salt; nitric acid; Norway saltpeter; Varioform I; Herco Prills.

SECTION II - HAZARDOUS INGREDIENTS AND IDENTITY INFORMATION

Hazardous Components (specific identity)	OSHA Exposure Criteria	NIOSH Exposure Criteria	Immediately Dangerous to Life and Health (IDLH)	ACGIH Exposure Criteria
Ammonium Nitrate (derivation from: Action of ammonia vapor on nitric acid).	**PEL:** Not Established **STEL:** Not Established	**REL:** Not Established **STEL:** Not Established	**Not Established**	**TLV:** Not Established **STEL:** Not Established

SECTION III - PHYSICAL AND CHEMICAL CHARACTERISTICS

Boiling Point **Decomposes at 410°F (210°C)**	Specific Gravity (H_2O = 1) **Not Applicable**
Vapor Pressure (mm Hg) **Not Applicable**	Molecular Weight **80.06**
Vapor Density (Air = 1) **1.725 at 25°C**	Freezing Point **Not Applicable**

Solubility in Water
Soluble in water, alcohol, and alkalies.

Appearance and Odor
Colorless crystals; no appreciable odor.

SECTION IV - FIRE AND EXPLOSION HAZARD DATA

Flash Point (method used) **Not Applicable**	Explosive Limits in Air % by Volume **LEL: N/A UEL: N/A**
NFPA Classification **Not Classified**	Autoignition Temperature **Not Applicable**

Extinguishing Media
Water spray (do not use salt water, carbon dioxide, or foam extinguishers).

Special Fire Fighting Procedures
Wear full protective clothing and self-contained breathing apparatus (SCBA) to protect skin and eyes. A strong oxidizer; can cause combustible materials (wood, paper, and oil) to self-ignite. Never attempt to smother fire. Do not add steam. Fight fire from protected location as steam eruptions are common.

Unusual Fire and Explosion Hazards
Nitrogen oxide gases are emitted on decomposition are toxic. Contamination of ammonium nitrate with organics or other oxidizable substances may result in explosion. May undergo detonation if heated under confinement or if subjected to strong shock. Compound decomposes by heat, evolving oxygen to support combustion.

SECTION V - REACTIVITY DATA

Stability		Conditions to Avoid
		Stable under ordinary conditions of use and storage. At temperatures ranging 80 - 90°C, decomposes endothermically to ammonia and nitric acid. Above 210°C, exothermic reaction produces nitrous oxide.
Stable	Unstable	Incompatibility *(materials to avoid)*
X		Aluminum, antimony, chromium, copper, iron, lead, magnesium, manganese, nickel, zinc, brass, oil, charcoal, organic material, acetic acid, ammonium chloride, bismuth, cadmium, chlorides, cobalt, phosphorus, potassium and ammonium sulfate, sodium, and sulfur.
Hazardous Polymerization		Conditions to Avoid
		High temperatures (300°C and above), there is rapid evolution of brown fumes of nitrogen oxides, which are highly toxic.
May Occur	Will Not Occur	Hazardous Decomposition or By-products
	X	When heated to decomposition, ammonium nitrate emits toxic fumes of nitric oxides. It liberates ammonia in reaction with strong alkalis.

SECTION VI - HEALTH HAZARD DATA

Primary Route(s) of Entry:	Inhalation?	Absorption (skin)?	Ingestion?
	X	X	X

Health Hazards

INHALATION: May irritate mucous membranes; cough, sore throat, and shortness of breath. At high temperatures, exposure to toxic nitrous oxides can quickly cause respiratory problems.

ABSORPTION: Skin contact may cause mild irritation. May pass through unbroken skin to be absorbed into bloodstream.

INGESTION: Large doses may cause dizziness, abdominal pain, vomiting, bloody diarrhea, weakness, convulsions, and collapse.

Carcinogenicity	NTP Listed?	IARC Cancer Review Group?	OSHA Regulated?	Target Organs?
No data found.	No data.	No data.	Not Listed.	Respiratory system; skin; eyes.

Medical Conditions Generally Aggravated by Exposure

None identified (although skin contact may irritate previously existing skin conditions).

Emergency and First-aid Procedures

Eye contact: Flush immediately with water for 30 minutes (minimum); seek medical attention. Skin contact: Remove all contaminated clothing. Immediately wash area with large amounts of soap and water. Seek medical assistance. For inhalation: Remove the person from exposure. Provide respiratory assistance and CPR. Transfer to medical facility. If swallowed, seek medical attention immediately.

SECTION VII - PRECAUTIONS FOR SAFE HANDLING AND USE

Steps to be Taken in Case Material is Released or Spilled

Evacuate area of non-essential personnel. Remove all sources of ignition. Wear SCBA, heavy rubber boots and gloves. Cover material with dry lime or soda ash. Collect and contain in tightly closed drum. Ventilate area and wash spill site after cleanup is complete.

Preferred Waste Disposal Method

Hygroscopic materials can be mixed with large amounts of water, pH adjusted, and flushed to drain.

Precautions to be Taken in Handling and Storage

Keep in tightly closed container, stored in a cool, dry, ventilated area. Protect against physical damage. Separate from combustibles, organic, or other readily oxidizable materials. Do not store above 130°F (54°C).

Other Precautions and Warnings

Wherever ammonium nitrate is used, handled, manufactured, or stored, explosion-proof electrical equipment and fittings should be used.

SECTION VIII - CONTROL MEASURES AND PERSONAL PROTECTIVE EQUIPMENT

Respiratory Protection *(specify type)*

Where exposure to dusts are anticipated, a full facepiece respirator equipped with a dust/mist pre-filter should be worn. For emergency situations, a supplied-air respirator or self-contained breathing apparatus (SCBA) operated in continuous flow mode may be necessary.

Ventilation

Local exhaust systems recommended (explosion-proof electrical equipment may be necessary).

Protective Gloves	Eye Protection	Other Protective Clothing
Chemical Resistant Rubber	Chemical/Dust Goggles or Face Mask	Rubber Apron

Work/Hygiene Practices

Always wash hands thoroughly after using chemical; never bring food, drink, or smoking materials into vicinity of chemicals.

AMMONIUM NITRATE

NH_4NO_3 CAS: 75-07-0

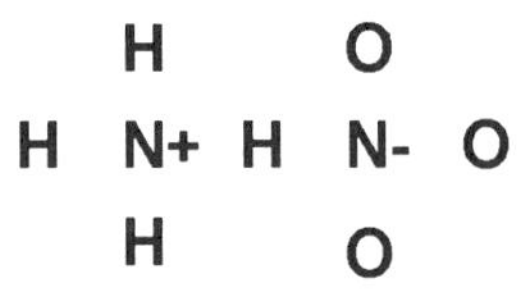

IDENTIFICATION AND TYPICAL USES

Ammonium nitrate is a colorless crystalline solid that occurs in two forms:

a). α-NH_4NO_3, tetragonal crystals that are stable between -16°C and 32°C (specific gravity = 1.66).

b). β-NH4NO3, rhombic or monoclinic crystals, stable between 32°C and 84°C (specific gravity = 1.725).

Ammonium nitrate is a synthetic inorganic compound which is produced in large volumes. Crystalline forms are a major constituent of nitrogen fertilizers, herbicides, and insecticides. Lesser amounts are used to make explosives (especially as prills/oil mixture, pyrotechnics) and nitrous oxide. It is also used as an anesthetic and as an ingredient for freezing mixtures. It may be the oxidizer in rocket propellants. It is also used a as nutrient for antibiotics and yeast.

RISK ASSESSMENT: HEALTH

General Assessment

Ammonium nitrate is not rated for carcinogenic, mutagenic, or teratogenic effects. No data are available to substantiate claims to the contrary. However, all contact with this chemical should be reduced to the lowest possible level. Human exposure to ammonium nitrate can occur through three routes of entry into the body: *inhalation, absorption,* and *ingestion.* Each presents a moderate level of toxicity. Absorption through the skin and direct ingestion may cause dizziness, abdominal pain, vomiting, bloody diarrhea, weakness, convulsions, and collapse. Large doses can be fatal. Skin contact can also cause irritation. Eye contact can result in irritation, redness, and pain. Inhalation may irritate the mucous membranes; symptoms may include coughing, sore throat, and shortness of breath. At high temperatures (above 300°C), exposure to toxic nitrogen oxides can quickly cause acute respiratory problems (pulmonary edema).

With severe overexposures, a condition known as methemoglobinemia (decrease in the oxygen-carrying ability of the blood) can result. This condition is characterized by a bluish color of the skin and lips (cyanosis), headaches, dizziness, and with higher exposures, collapse and death.

Acute Health Effects

The following acute (short-term) health effects may occur immediately or shortly after exposure to ammonium nitrate:

Skin: Irritation/rash, possible contact dermatitis.

Eye: Irritation, redness, and possible pain.

Lung: Irritation of the mucous membranes of the nose and throat causing coughing and shortness of breath. Higher concentrations can lead to sever respiratory problems.

CNS: Inhalation of high concentrations of the vapor may cause dizziness, weakness, loss of consciousness, or death.

Chronic Health Effects

The following chronic (long-term) health effects may occur at some time after exposure to ammonium nitrate and can last for months or even years:

Cancer Hazards: According to the references, ammonium nitrate has not been tested for its ability to cause cancer.

Reproductive Hazard: According to the references, ammonium nitrate has not been tested for its ability to adversely affect reproduction.

Other Chronic Effects: Small repeated oral doses of nitrates, including ammonium nitrate, may cause weakness, malaise, depression, headaches, and mental impairment. No other chronic, long-term effects are known at this time.

❶ *Recommended Risk-Reduction Measures*

Personnel should avoid direct contact with ammonium nitrate. Unless a another chemical can be substituted for a hazardous substance, *engineering controls* are the most effective method of reducing exposures. The best protection is to enclose operations and/or provide local exhaust ventilation at the site of chemical release. Electrical equipment should be rated as explosion-proof. While not always operationally feasible, isolating operations can also reduce exposure.

Using respiratory protection is less effective than the controls mentioned above, but is still advisable whenever working with or around ammonium nitrate. For low exposures, a respirator fitted with a dust/fume/mist pre-filter is acceptable protection. At higher concentrations of exposure, a self-contained breathing apparatus (SCBA) with full facepiece and pressure demand is the recommended respiratory protection of choice. If a full facepiece is not available, then chemical/dust goggles should be worn to protect the eyes. Whenever a potential for airborne release of dusts exists, a face shield and a rubber apron should be worn. To prevent hand and skin exposures, heavy rubber gloves should be worn.

Administrative controls should also be in place to minimize the potential for human exposures. These may include written procedures or policies which specify the methods and techniques that will be practiced whenever personnel are to work with ammonium nitrate.

All personnel should receive *training* on the use, hazards, protective measures, emergency actions, and other precautions per 29 CFR 1910.1200 (Hazard Communication), prior to the first assignment in an area where ammonium nitrate is used or stored. If symptoms develop or overexposure is suspected, the following may be useful:

☑ A blood test for methemoglobin levels.

Other methods to reduce exposure include:

☑ Where possible, enclose operations and use explosion-proof local exhaust ventilation at the site of chemical release. If local exhaust ventilation or enclosure is not used, respiratory protection should be mandatory.

☑ Always ensure that proper protective clothing is worn when using chemical substances. Clothing contaminated with ammonium nitrate should never be laundered at home. Only personnel who have been trained on exposure hazards should launder contaminated clothing.

☑ Wash thoroughly immediately after exposure to ammonium nitrate and at the end of the work shift or before eating, drinking, or smoking.

☑ Hazard warning information should be posted in the work area. In addition, as part of an on-going education and training program, all information on the health and safety hazards of ammonium nitrate should be communicated to all potentially exposed workers.

☑ Never eat, drink, or smoke in areas were ammonium nitrate is used, handled or stored.

RISK ASSESSMENT: ENVIRONMENT

General Assessment

The environment is at risk of exposure during transportation, storage, disposal, or destruction of ammonium nitrate. In almost every scenario, the threat of environmental exposure is contingent upon the proper handling of the chemical substance. Accidental spills, large or small, can result in fire, explosion, and possible contamination of the surrounding environmental mediums (water, soil, and air).

Ammonium nitrate is highly reactive and is therefore a dangerous fire and explosion hazard. At high temperatures, its decomposition products of nitrous oxides are highly toxic. These characteristics require special consideration during any emergency situation involving a leak or spill of ammonium nitrate. Should ammonium nitrate ever come into contact with incompatible substances such as other oxidizing materials, combustibles, or metal either during use, transportation, or storage, extremely explosive reactions may occur. Waste disposal of small quantities can be accomplished by cautiously adding ammonium nitrate to a large, stirred excess of water, adjusting for pH to neutral, and separating any insoluble solids. Aqueous solution can be flushed to drain with copious amounts of water. The hydrolysis and neutralization reactions may generate heat and fumes which can be controlled by rate of addition. All federal, state, and local environmental regulations must be observed.

Ammonium nitrate can enter the environment through industrial or municipal waste treatment plant discharges, agricultural runoff, or spills.

Acute Ecological Effects

Acute (short-term) toxic effects may include the death of animals, birds, or fish, and death or low growth rate in plants. Acute effects are seen 2 to 4 days after animals or plants are exposed to ammonium nitrate.

This chemical has moderate acute toxicity to aquatic life. No data are available to evaluate the short-term effects of ammonium nitrate to plants, animals, or birds.

Chronic Ecological Effects

Chronic toxic effects may include shortened life span, reproductive problems, lower fertility, and changes in appearance or behavior in exposed animals. These effects can be seen long after first exposure(s) to toxic chemicals.

Ammonium nitrate has moderate chronic toxicity in aquatic life. No data are available on the long-term effects of ammonium nitrate to plants, birds, or land animals.

Water Solubility

Crystalline ammonium nitrate is highly soluble in water and, because of its negative heat of solution in water, is used to prepare freezing mixtures, such as mixtures with salt and water. Concentrations of 1000 milligrams and more will mix with a liter of water.

Persistence in the Environment

Nitrate compounds are probably highly persistent in water. Due to volatilization, most ammonia will disappear from aquatic ecosystems in less than 2 days.

Bioaccumulation in Aquatic Organisms

Some substances increase in concentration, or *bioaccumulate,* in living organisms as they breathe contaminated air, drink contaminated water, or eat contaminated food. These chemicals can become concentrated in the tissues and internal organs of animals as well as humans.

Most ammonia and nitrate compounds will probably not accumulate in edible tissues of aquatic species that are eaten by humans.

Recommended Risk-Reduction Measures

Proper training of all transporters will reduce the likelihood of a mishap or accident resulting in a leak or spill to the environment. The correct use of DOT labeling (per the MSDS) on all containers, trucks, and rail cars will enable emergency responders to react properly and quickly to any disaster thereby reducing the potential risk to the environment and to personnel.

Storage of ammonium nitrate should be segregated from incompatible chemicals to minimize the risk of cross-contamination or contact. Buildings designated for storage should be equipped with appropriate fire protection systems (alarms, sprinklers, emergency lighting, portable extinguishers).

If a spill or leak to the environment has occurred, fire department, emergency response, and/or hazardous materials spill personnel should be notified immediately. Cleanup should be attempted only by those trained in proper spill containment procedures. Contaminated soils should be removed and replaced with clean soil.

If ammonium nitrate should contact the water table, aquifer, or navigable waterway, time is of the essence. It is readily miscible in water and, therefore, total containment and remediation may not be entirely possible. While ammonia will likely evaporate quickly, nitrate can remain. When such spills occur, the local and/or state emergency response authorities must be notified.

A comprehensive emergency response or disaster preparedness/recovery plan should be in place prior to any operations involving the use, transportation, storage, or disposal of ammonium nitrate.

If ammonium nitrate is spilled or leaked, the following specific steps are recommended:

- ☑ Restrict persons not wearing protective clothing from area of spill or leak until cleanup is complete.
- ☑ Remove all ignition sources.
- ☑ Ventilate area of spill or leak.
- ☑ Cover materials with dry lime or soda ash and collect/contain in sealed drum or container.
- ☑ Keep ammonium nitrate out of a confined space, such as a sewer, because of the possibility of explosion (unless the sewer is designed to prevent the buildup of explosive concentrations).
- ☑ It may be necessary to dispose of ammonium nitrate as a hazardous waste. The responsible state agency or the regional office of the federal Envi-

ronmental Protection Agency (EPA) should be contacted for specific recommendations.

RISK ASSESSMENT: BUSINESS

General Assessment

Accidents or mishaps involving ammonium nitrate can present a significant threat to business operations. The loss or damage of equipment or facilities due to fire or explosion can significantly affect fiscal viability. Lawsuits that may result from personnel injury/death, public exposures, and/or environmental contamination will also require a serious expenditure of resources.

Media attention surrounding an injury, death, or environmental damage can also result in a loss of profits and loss of current as well as future business.

Recommended Risk-Reduction Measures

Company attorneys, safety and health professionals, and environmental specialists should be involved in the development of any such procedures. A company official should be pre-designated as a public relations officer with specific training in dealing with the press. Corporate plans and policies should be developed, approved, and implemented long before any need for such arises.

REFERENCES

Ahlbom, A. 1993. *Biostatistics for Epidemiologists.* New York: CRC Press/Lewis Publishers.

American Conference of Governmental Industrial Hygienists. 1988. *Documentation of the Threshold Limit Values and Biological Exposure Indices,* 5th Edition (with updates). Cincinnati: ACGIH

Anthony, C. P., and N. J. Kolthoff. 1971. *Textbook of Anatomy and Physiology.* St. Louis: C. V. Mosby Company.

Asante-Duah, D. K. 1993. *Hazardous Waste Risk Assessment.* New York: CRC Press/Lewis Publishers.

Bloom, A. D. and F. J. DeSerres. 1995. *Ecotoxicity and Human Health.* New York: CRC Press/Lewis Publishers.

Brusick, D. J. 1994. *Methods for Genetic Risk Assessment.* New York: CRC Press/Lewis Publishers.

Bueche, J. R. 1972. *Principles of Physics,* 2nd Edition. New York: McGraw-Hill.

Burrell, R., D. K. Flaherty, and L. J. Sauers. 1992. *Toxicology of the Immune System.* New York: Van Nostrand Reinhold.

Calabrese, E. J. 1994. *Biological Effects of Low Level Exposures.* New York: CRC Press/Lewis Publishers.

Cockerman, L. G., B. S. Shane. 1994. *Basic Environmental Toxicology.* New York: CRC Press/Lewis Publishers.

Cothern, C. R. 1995. *Handbook for Environmental Risk Decisionmaking.* New York: CRC Press/Lewis Publishers.

Danse, I. R. 1991. *Common Sense Toxics in the Workplace.* New York: Van Nostrand Reinhold.

Gardner, R. W. 1994. *Chemical Intolerance.* New York: CRC Press/Lewis Publishers.

Hallenbeck, W. H. 1993. *Quantitative Risk Assessment for Environmental and Occupational Health,* 2nd Edition. New York: CRC Press/Lewis Publishers.

Hodgson, E., R. B. Mailman, and J. E. Chambers. 1988. *Dictionary of Toxicology.* London: The MacMillian Company.

Heck, H. 1985. Mechanisms of aldehyde toxicity: Structure activity Studies. *CIIT Activities,* 5(10):106.

Hosty, J. W., and P. Foster. 1990. *A Practical Guide to Chemical Spill Response.* New York: Van Nostrand Reinhold.

Howard, P. H. 1995. *Handbook of Environmental Fate and Exposures Data for Organic Chemicals,* Vol. I-IV. New York: CRC Press/Lewis Publishers.

International Agency for Research on Cancer (IARC). 1985. *IARC Monographs on the Evaluation of Carcinogenic Risk of Chemicals to Humans,* Vol. 36, *Allyl Compounds, Aldehydes, Epoxides, and Peroxides,* pp 101-136. Lyon: IARC.

Jones, F. E. 1994. *Toxic Organic Vapors in the Workplace.* New York: CRC Press/Lewis Publishers.

Kamrin, M. 1 A. 1988. *Toxicology: A Primer on Principles and Applications.* Chelsea, Michigan: Lewis Publishers.

Keith, L. H. and M. M. Walker. 1995. *Handbook of Air Toxics.* New York: CRC Press/Lewis Publishers.

Landis, W. G., M. H. Yu. 1995. *Introduction to Environmental Toxicology.* New York: CRC Press/Lewis Publishers.

Lewis, R. J., Sr. 1992. *Sax's Dangerous Properties of Industrial Materials,* Volumes 1, 2, and 3, 8th Edition. New York: Van Nostrand Reinhold.

Lisella, F. S. 1994. *The VNR Dictionary of Environmental Health and Safety.* New York: Van Nostrand Reinhold.

Meyer, E. 1990. *Chemistry of Hazardous Materials.* Englewood Cliffs, NJ: Prentice Hall.

Patnaik, P. 1992. *A Comprehensive Guide to the Hazardous Properties of Chemical Substances.* New York: Van Nostrand Reinhold.

Proctor, N. H., and J. P. Hughes. 1991. *Chemical Hazards of the Workplace.* New York: Van Nostrand Reinhold.

Silverman, L., H. F. Schuttle, and M. W. First. 1946. Further studies on sensory response to certain industrial solvent vapors. *Journal of Industrial Hygiene Toxicology,* 28:262-266.

Sherman, J. D. 1988. *Chemical Exposure and Disease.* New York: Van Nostrand Reinhold.

U.S. Department of Health and Human Services, National Institute for Occupational Safety and Health. 1994. *NIOSH Pocket Guide to Chemical Hazards.* Washington, D.C.: U.S. Government Printing Office.

MATERIAL SAFETY DATA SHEET

CHEMICAL NAME	CRC PRESS/LEWIS PUBLISHERS DISCLAIMER
ANILINE	

HAZARD WARNING INFORMATION

HEALTH	FIRE	REACTIVE	OTHER	DEGREE OF HAZARD	COLOR CODING	OTHER CODES
3	2	0		0 = Minimum Hazard 1 = Slight Hazard 2 = Moderate Hazard 3 = Serious Hazard 4 = Severe Hazard	HEALTH = BLUE FIRE = RED REACTIVITY = YELLOW OTHER = WHITE	OX = Oxidizer ACID = Acid ALK = Alkali COR = Corrosive ~~W~~ = Use No Water

SECTION I - GENERAL INFORMATION

Characterization	**Aromatic Amine**	RCRA Number	**U012**
		EPA Class	**Toxic Waste**
DOT Proper Shipping Name	**Aniline**	Chemical Abstract Service (CAS) Number	**62-53-3**
DOT Hazard Class and Label Requirements	**Poison B**	DOT Emergency Guide Code	**57**
DOT Identification Number	**UN 1547**	Chemical Formula	**$C_6H_5NH_2$**

Synonyms
Aminobenzene; aminophen; phenylamine; benzeamine; blue oil.

SECTION II - HAZARDOUS INGREDIENTS AND IDENTITY INFORMATION

Hazardous Components (specific identity)	OSHA Exposure Criteria	NIOSH Exposure Criteria	Immediately Dangerous to Life and Health (IDLH)	ACGIH Exposure Criteria
Aniline (derivation: By catalytic vapor-phase reduction of nitrobenzene with hydrogen; reduction of nitrobenzene with iron filings using hydrochloric acid as catalyst; catalytic reaction of chlorobenzene and aqueous ammonia). **1 ppm = 3.87 mg/m³**	**PEL (skin):** **5 ppm** **19 mg/m³** **STEL:** **Not Established**	**Suspected Human Carcinogen**	**100 ppm**	**TLV:** **2 ppm** **8 mg/m³** **STEL:** **Not Established**

SECTION III - PHYSICAL AND CHEMICAL CHARACTERISTICS

Boiling Point	**363°F (184.4°C)**	Specific Gravity (H_2O = 1)	**1.02**
Vapor Pressure (mm Hg)	**0.6 at 69°F (20°C)**	Molecular Weight	**93.1**
Vapor Density (Air = 1)	**3.22**	Freezing Point	**21°F (-20°C)**

Solubility
Moderate in water; miscible with most organic solvents, alcohol, benzene, and chloroform.

Appearance and Odor
Oily, colorless liquid which darkens on exposure to light and air; aromatic, amine-like odor and burning taste. Odor Threshold = 1 ppm.

SECTION IV - FIRE AND EXPLOSION HAZARD DATA

Flash Point (method used)	**158°F (70°C) closed cup**	Explosive Limits in Air % by Volume	**LEL: 1.3% UEL: 11%**
NFPA Classification	**Class III A Combustible Liquid**	Autoignition Temperature	**1418°F (770°C)**

Extinguishing Media
Carbon dioxide, dry chemical, water spray, or alcohol foam.

Special Fire Fighting Procedures
Wear full protective clothing and self-contained breathing apparatus (SCBA). Use a water spray to keep fire-exposed containers cool. Poisonous gases are produced in fire.

Unusual Fire and Explosion Hazards
Moderately flammable when exposed to heat. Containers may explode in fire. Firefighters should avoid contact with vapors produced during fire. Ignites on contact with sodium peroxide and water. Forms heat-sensitive or shock-sensitive explosive mixtures with a number of incompatible chemicals.

SECTION V - REACTIVITY DATA

Stability		Conditions to Avoid: Reacts with strong oxidizers, accompanied by flame and/or explosion. Vigorous or explosive reaction when mixed with hexachloromelanamine or trichloromelanamine.
Stable	Unstable	Incompatibility *(materials to avoid)*: Strong oxidizers (such as perchlorates, peroxides, permanganates, chlorates and nitrates); strong acids (such as hydrochloric, sulfuric, and nitric); toluene diisocyanate; alkali metal peroxides.
X		
Hazardous Polymerization		Conditions to Avoid: Many perchlorates form solvated adducts with organic solvents, including aniline, which explode on impact.
May Occur	Will Not Occur	Hazardous Decomposition or By-products: When heated to decomposition, aniline can emit highly toxic/poisonous gases (e.g., NO_X).
	X	

SECTION VI - HEALTH HAZARD DATA

Primary Route(s) of Entry:	Inhalation?	Absorption (skin)?	Ingestion?
	X	X	X

Health Hazards

INHALATION: Causes anoxia due to the formation of methemoglobin. Symptoms: headache, dizziness, ataxia, and cyanosis. Overexposure may lead to death due to respiratory paralysis.

ABSORPTION: Absorbs freely through intact skin. Symptoms include headache and cyanosis, overexposure can result in death. Skin contact with pure liquid can produce mild irritation.

INGESTION: Unspecified effects resulting from ingestion. Liver damage is suspect.

Carcinogenicity	NTP Listed?	IARC Cancer Review Group?	OSHA Regulated?	Target Organs?
Suspected Human Suspected Animal	NCI-C03736	Group 3	29 CFR 1910.1000 Table Z-1	Respiratory sys. eyes, skin, blood, kidney, liver.

Medical Conditions Generally Aggravated by Exposure

Respiratory impairments (bronchitis, asthma); Skin conditions (dermatitis); Renal or liver disorders.

Emergency and First-aid Procedures

Eye contact: Flush immediately with water for 15 minutes (minimum), seek medical attention. Skin contact: Remove all contaminated clothing. Immediately wash area with large amounts of soap and water. Seek medical assistance. For inhalation: Remove the person from exposure. Provide respiratory assistance and CPR. Transfer to medical facility. If swallowed, seek medical attention immediately.

SECTION VII - PRECAUTIONS FOR SAFE HANDLING AND USE

Steps to be Taken in Case Material is Released or Spilled

Absorb liquids in vermiculite, dry sand, earth, or similar material and deposit in sealed containers. Ventilate area of spill or leak. Restrict those not involved in cleanup from entering area. Notify appropriate authorities, as required by SARA III, if applicable.

Preferred Waste Disposal Method

Dissolve in combustible solvent and burn in chemical incinerator equipped with scrubber and afterburner.

Precautions to be Taken in Handling and Storage

Aniline is a combustible liquid. Store to avoid contact with strong oxidizers and acids, since violent reactions can occur. Store in tightly closed containers in cool, well-ventilated area away from heat. Ground and bond all containers prior to transferring liquids.

Other Precautions and Warnings

Sources of ignition are prohibited where aniline is used, handled, or stored. Bulk storage of aniline is not recommended.

SECTION VIII - CONTROL MEASURES AND PERSONAL PROTECTIVE EQUIPMENT

Respiratory Protection *(specify type)*

Use a supplied air respirator with full facepiece, hood, or helmet in continuous flow mode, or a self-contained breathing apparatus (SCBA) with full facepiece operated in positive pressure mode.

Ventilation

Local exhaust at site of chemical work.

Protective Gloves	Eye Protection	Other Protective Clothing
Natural Rubber	Chemical Goggles or Face Mask	Rubber Apron

Work/Hygiene Practices

Always wash hands thoroughly after using chemical; never bring food, drink, or smoking materials into vicinity of chemicals.

ANILINE

$C_6H_5NH_2$ CAS: 62-53-3

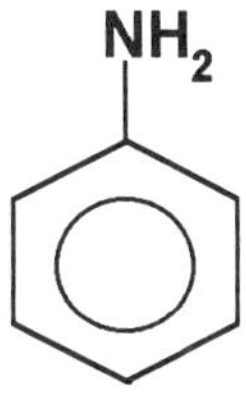

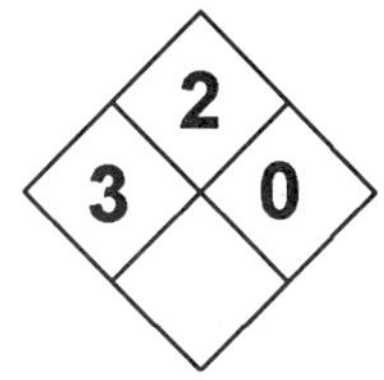

IDENTIFICATION AND TYPICAL USES

Aniline is an oily, colorless liquid with characteristic odor and burning taste; color darkens to brown or tan upon exposure to air or light. Aniline is used in the manufacture of dyes, pharmaceuticals, varnishes, resins, photographic chemicals, perfumes, shoe blacks, herbicides, and fungicides. It is also used in vulcanizing rubber and as a solvent. It occurs naturally in coal tar and is produced from the biodegradation of many pesticides. It is an intermediate in chemical synthesis and one of the most important of the organic bases. It is a metabolite of many toxic compounds, such as nitrobenzene, phenacetin, and phenylhydroxylamine.

RISK ASSESSMENT: HEALTH

General Assessment

Aniline is moderately toxic through *inhalation, ingestion,* and *absorption* through the skin. Toxic symptoms of aniline exposure include headache, weakness, dizziness, ataxia, and cyanosis (blue tinting of the lips, skin, and nail beds due to oxygen depravation). Cyanosis is due to the formation of methemoglobin which restricts the blood's oxygen carrying capabilities. Cyanosis occurs when the methemoglobin concentration is 15% or more. Blueness develops first in the lips, nose, and earlobes. At 40% concentration, there is weakness and dizziness. At up to 70% concentration, symptoms may include ataxia, dyspnea on mild exertion, and tachycardia. Coma may ensue at concentrations over 70%. Overexposure (between 85% and 90% concentrations) may lead to death from respiratory paralysis.

Contact with the pure liquid aniline on the skin can produce moderate irritation, while the effect on the eyes can be quite severe. Oral administration to laboratory animals resulted in tumors in the kidney and bladder. However, aniline carcinogenicity in both animals and humans is suspected but not confirmed due to inconclusive data. Aniline may cause mutations. Such substances may also have the potential to cause reproductive damage in humans.

Acute Health Effects

The following acute (short-term) health effects may occur immediately or shortly after exposure to aniline:

Skin: Irritation and possible color change around nails and fingers.

Eye: Severe irritation; may cause corneal damage.

Lung: Not known to be acutely irritating to lung tissue. However, overexposure can lead to death due to respiratory arrest.

Other: Aniline affects the blood's ability to transport oxygen and high exposure can cause death. Symptoms include headaches, weakness, irritability, drowsiness, and shortness of breath.

Chronic Health Effects

The following chronic (long-term) health effects may occur at some time after exposure to aniline and can last for months or even years:

Cancer Hazards: Aniline may cause mutations (genetic changes) in living cells. The data on its carcinogenicity are inconclusive.

Reproductive Hazard: There is limited evidence to show that exposure to aniline may damage the developing fetus.

Other Chronic Effects: Continuous exposure to small doses of aniline may produce anemia, loss of energy, digestive disturbance, and headache. Repeated exposures may cause anemia and brownish staining of the eyes can occur.

Recommended Risk-Reduction Measures

Personnel should avoid direct contact with aniline. If a less toxic chemical can not be substituted for a hazardous substance, then *engineering controls* are the most effective method of reducing exposures. The best protection is to enclose operations and/or provide local exhaust ventilation at the site of chemical release. While not always operationally feasible, isolating operations can also reduce exposure. Using respiratory protection is less effective than the controls mentioned above, but is still advisable whenever working with or around aniline. A self-contained breathing apparatus (SCBA) with full facepiece and pressure demand is the recommended respiratory protection of choice. If a full facepiece is not available, then chemical goggles should be worn to protect the eyes. Whenever a chemical splash hazard exists, a face shield and a rubber apron should be worn. To prevent hand and skin exposures, natural rubber gloves should be worn.

Administrative controls should also be in place to minimize the potential for human exposures. These may include written procedures or policies which specify the methods and techniques that will be practiced whenever personnel are to work with aniline.

All personnel should receive *training* on the use, hazards, protective measures, emergency actions, and other precautions per 29 CFR 1910.1200 (Hazard Communication), prior to the first assignment in an area where aniline is used or stored.

If symptoms develop or overexposure is suspected, the following medical tests may be useful:

☑ Methemoglobin tests.

☑ Urinalysis (for blood content).

Any evaluation should include a careful history of past and present symptoms with an examination. Medical tests that simply look for existing damage are not a substitute for controlling exposures.

Other methods to reduce exposure include:

☑ Where possible, enclose operations and use local exhaust ventilation at the site of chemical release. If local exhaust ventilation or enclosure is not used, respiratory protection should be mandatory.

☑ Always ensure that proper protective clothing is worn when using chemical substances.

☑ Safety showers should be provided in areas where work is performed around aniline.

☑ Wash thoroughly immediately after exposure to aniline and at the end of the work shift or before eating, drinking, or smoking.

☑ Hazard warning information should be posted in the work area. In addition, as part of an on-going education and training program, all information on the health and safety hazards of aniline should be communicated to all potentially exposed workers.

☑ Eye wash stations should be provided in the immediate work area for emergency use. If there is a possibility of skin exposure to aniline, emergency shower facilities should also be provided.

RISK ASSESSMENT: ENVIRONMENT

General Assessment

The environment is at risk of exposure during transportation, storage, disposal, or destruction of aniline. In almost every scenario, the threat of environmental exposure is contingent upon the proper handling of the chemical substance. Accidental spills, large or small, can result in fire, explosion, and possible contamination of the surrounding environmental mediums (water, soil, and air).

Aniline is considered a class IIIA combustible liquid. Alone, its high flash point and autoignition temperature do not present a serious fire or explosion hazard. However, because it is incompatible with so very many other commodities, extreme caution is required in handling, storage, transportation, and disposal of aniline. These characteristics also require special consideration during any emergency situation involving a leak or spill of aniline. Should aniline ever come into contact with incompatible substances such as oxidizers, acids, or alkalis either during use, transportation, storage, or disposal, the formation of highly toxic and/or highly explosive commodities is extremely possible.

The proper disposal/destruction method for aniline waste is to dissolve it in a combustible solvent and burn it in a chemical incinerator equipped with an afterburner and air scrubber. Aniline can enter the environment through direct disposal and indirect use, such as industrial discharges and non-point source from agricultural use.

☠ *Acute Ecological Effects*

Acute (short-term) toxic effects may include the death of animals, birds, or fish and death or low growth rate in plants. Acute effects are seen 2 to 4 days after animals or plants are exposed to aniline. This chemical has high acute toxicity to aquatic life and moderate acute toxicity to birds. Aniline causes germination decrease, stunting and size decreases on numerous agricultural crops. No data are available on the short-term effects of aniline to land animals.

Chronic Ecological Effects

Chronic (long-term) toxic effects may include shortened life span, reproductive problems, lower fertility, and changes in appearance or behavior in exposed animals. These effects can be seen long after first exposure(s) to toxic chemicals. Aniline has high chronic toxicity in aquatic life. No data are available on the long-term effects of aniline to plants, birds, or land animals.

Water Solubility

Aniline is highly soluble in water. Concentrations of 1000 milligrams and more will mix with a liter of water.

Persistence in the Environment

Aniline is slightly persistent in water, with a half-life between 2 and 20 days. The half-life of a pollutant is the amount of time it takes for one half of the chemical to be degraded. Approximately 93% of aniline will eventually end up in water; about 6.5% will end up in air; and less than 1% will end up in terrestrial soil and aquatic sediments.

Bioaccumulation in Aquatic Organisms

Some substances increase in concentration, or *bioaccumulate,* in living organisms as they breathe contaminated air, drink contaminated water, or eat contaminated food. These chemicals can become concentrated in the tissues and internal organs of animals as well as humans.

The concentration of aniline found in fish tissues is expected to be somewhat higher than the average concentration of aniline in the water from which the fish was taken.

Recommended Risk-Reduction Measures

Proper training of all transporters will reduce the likelihood of a mishap or accident resulting in a leak or spill to the environment. The correct use of DOT labeling (per the MSDS) on all containers, trucks, and rail cars will enable emergency responders to react properly and quickly to any disaster thereby reducing the potential risk to the environment and to personnel.

Storage of aniline should be segregated from incompatible chemicals to minimize the risk of cross-contamination or contact.

If a spill or leak to the environment has occurred, fire department, emergency response and/or hazardous materials spill personnel should be notified immediately. Cleanup should be attempted only by those trained in proper spill containment procedures. Contaminated soils should be removed for incineration and replaced with clean soil. If aniline should contact the water table, aquifer, or navigable waterway, the local and/or state emergency response authorities must be notified. A comprehensive emergency response or disaster preparedness/recovery plan should be in place prior to any operations involving the use, transportation, storage, or disposal of aniline. If aniline is spilled or leaked, the following specific steps are recommended:

- ☑ Restrict persons not wearing protective clothing from area of spill or leak until cleanup is complete and area can be opened for normal work.
- ☑ Remove all ignition sources.
- ☑ Absorb liquids in vermiculite, dry sand, earth, or similar material and deposit in sealed containers.
- ☑ It may be necessary to dispose of aniline as a hazardous waste. The responsible state agency or the regional office of the federal Environmental Protection Agency (EPA) should be contacted for specific recommendations.

RISK ASSESSMENT: BUSINESS

General Assessment

Accidents or mishaps involving aniline can present a significant threat to business operations. The loss or damage of equipment or facilities can significantly affect fiscal viability. Lawsuits that may result from personnel injury/death, public exposures, and/or environmental contamination will also require a serious expenditure of resources. Media attention surrounding an injury, death, or environmental damage can also result in a loss of profits and loss of current as well as future business.

Recommended Risk-Reduction Measures

Company attorneys, safety and health professionals, and environmental specialists should be involved in the development of any procedures dealing with safety or emergency response. A company official should be pre-designated as a public relations officer with specific training in dealing with the press. Corporate plans and policies should be developed, approved, and implemented long before any need for such arises.

REFERENCES

Ahlbom, A. 1993. *Biostatistics for Epidemiologists.* New York: CRC Press/Lewis Publishers.

American Conference of Governmental Industrial Hygienists. 1988. *Documentation of the Threshold Limit Values and Biological Exposure Indices,* 5th Edition (with updates). Cincinnati: ACGIH

Asante-Duah, D. K. 1993. *Hazardous Waste Risk Assessment.* New York: CRC Press/Lewis Publishers.

Bloom, A. D. and F. J. DeSerres. 1995. *Ecotoxicity and Human Health.* New York: CRC Press/Lewis Publishers.

Brusick, D. J. 1994. *Methods for Genetic Risk Assessment.* New York: CRC Press/Lewis Publishers.

Bueche, J. R. 1972. *Principles of Physics,* 2nd Edition. New York: McGraw-Hill.

Calabrese, E. J. 1994. *Biological Effects of Low Level Exposures.* New York: CRC Press/Lewis Publishers.

Cockerman, L. G., B. S. Shane. 1994. *Basic Environmental Toxicology.* New York: CRC Press/Lewis Publishers.

Cothern, R. R. 1993. *Comparative Environmental Risk Assessment.* New York: CRC Press/Lewis Publishers.

Gardner, R. W. 1994. *Chemical Intolerance.* New York: CRC Press/Lewis Publishers.

Hallenbeck, W. H. 1993. *Quantitative Risk Assessment for Environmental and Occupational Health,* 2nd Edition. New York: CRC Press/Lewis Publishers.

Howard, P. H. 1995. *Handbook of Environmental Fate and Exposures Data for Organic Chemicals,* Vol. I-IV. New York: CRC Press/Lewis Publishers.

Jones, F. E. 1994. *Toxic Organic Vapors in the Workplace.* New York: CRC Press/Lewis Publishers.

Keith, L. H. and M. M. Walker. 1995. *Handbook of Air Toxics.* New York: CRC Press/Lewis Publishers.

Landis, W. G., M. H. Yu. 1995. *Introduction to Environmental Toxicology.* New York: CRC Press/Lewis Publishers.

Langerman, N. 1994 *Precautionary Labels for Chemical Containers.* New York: CRC Press/Lewis Publishers.

Lauwerys, R. R., P. Hoet. 1993. *Industrial Chemical Exposure.* New York: CRC Press/Lewis Publishers.

Lewis, R. J., Sr. 1993. *Hawley's Condensed Chemical Dictionary,* 12th Edition. New York: Van Nostrand Reinhold.

Lewis, R. J., Sr. 1992. *Sax's Dangerous Properties of Industrial Materials,* Volumes 1, 2, and 3, 8th Edition. New York: Van Nostrand Reinhold.

Lide, D. R. 1995. *Handbook of Chemistry and Physics,* 76th Edition. New York: CRC Press/Lewis Publishers.

Manahan, S. E. 1994. *Environmental Chemistry,* 6th Edition. New York: CRC Press/Lewis Publishers.

Manahan, S. E. 1992. *Toxicological Chemistry,* 2nd Edition. New York: CRC Press/Lewis Publishers.

Mellström, G. A., J. E. Wahlberg, and H. I. Maibach. 1994. *Protective Gloves for Occupational Use.* New York: CRC Press/Lewis Publishers.

Meyer, E. 1990. *Chemistry of Hazardous Materials.* Englewood Cliffs, NJ: Prentice Hall.

Mycock, J. C., J. D. McKenna, and L. Theodore. 1995. *Handbook of Air Pollution Control Engineering and Technology.* New York: CRC Press/Lewis Publishers.

National Cancer Institute. 1978. *Carcinogenesis Technical Report Series. Bioassay of Aniline for Possible Carcinogenicity.* DEHW (NIH) Publication Number: 78-1323, p 53. Washington, DC: U.S. Government Printing Office.

Ott, W. R. 1995. *Environmental Statistics and Data Analysis.* New York: CRC Press/Lewis Publishers.

Patnaik, P. 1992. *A Comprehensive Guide to the Hazardous Properties of Chemical Substances.* New York: Van Nostrand Reinhold.

Philp, R. B. 1995. *Environmental Hazards and Human Health.* New York: CRC Press/Lewis Publishers.

Proctor, N. H., J. P. Hughes. 1991. *Chemical Hazards of the Workplace.* New York: Van Nostrand Reinhold.

Quigley, D. R. 1994. *Handbook of Emergency Chemical Management.* New York: CRC Press/Lewis Publishers.

Rea, W. J. 1995. *Chemical Sensitivity,* Vol. I-III. New York: CRC Press/Lewis Publishers.

Richardson, M. 1995. *Dictionary of Substance and Their Effects,* Vol. I-VI. New York: CRC Press/Lewis Publishers.

Sacarello, H. L. A. 1994. *The Comprehensive Handbook of Hazardous Materials.* New York: CRC Press/Lewis Publishers.

Sheftel, V. O. 1995. *Handbook of Toxic Properties of Monomers and Additives.* New York: CRC Press/Lewis Publishers.

Talbott, E., G. Craun. 1995. *An Introduction to Environmental Epidemiology.* New York: CRC Press/Lewis Publishers.

Torkelson, T. R. V. K. Rowe. 1981. *Patty's Industrial Hygiene and Toxicology,* Vol. 2B. New York: Wiley-Interscience.

The Merck Index, 10th edition. 1983. Rahway, NJ: Merck & Company.

U.S. Department of Health and Human Services, National Institute for Occupational Safety and Health. 1994. *NIOSH Pocket Guide to Chemical Hazards.* Washington, D.C.: U.S. Government Printing Office.

MATERIAL SAFETY DATA SHEET

CHEMICAL NAME

ANTHRACENE

CRC PRESS/LEWIS PUBLISHERS DISCLAIMER: THE INFORMATION AND RECOMMENDATIONS PRESENTED HEREIN ARE BASED ON SOURCES BELIEVED TO BE RELIABLE. CRC MAKES NO REPRESENTATION ON ITS COMPLETENESS OR ACCURACY. IT IS THE RESPONSIBILITY OF THE USER TO DETERMINE THE CHEMICAL'S SUITABILITY FOR ITS INTENDED USE, THE CHEMICAL'S SAFE USE, AND THE CHEMICAL'S PROPER DISPOSAL. NO REPRESENTATIONS AND/OR WARRANTIES, EITHER EXPRESSED OR IMPLIED, OF THE MERCHANTABILITY OR FITNESS FOR A PARTICULAR PURPOSE, OR OF ANY OTHER NATURE, ARE MADE WITH RESPECT TO THE INFORMATION PROVIDED IN THIS MSDS OR TO THE CHEMICAL TO WHICH INFORMATION MAY REFER. CRC NEITHER ASSUMES NOR AUTHORIZES ANY OTHER PERSON TO ASSUME FOR IT, ANY OTHER ADDITIONAL RESPONSIBILITY OR LIABILITY FOR THE USE OF, OR RELIANCE UPON, THIS INFORMATION.

NATIONAL FIRE PROTECTION ASSOCIATION (NFPA) LABELING INFORMATION

HEALTH	FIRE	REACTIVE	OTHER	DEGREE OF HAZARD	COLOR CODING	OTHER CODES
1	1	0		0 = Minimum Hazard 1 = Slight Hazard 2 = Moderate Hazard 3 = Serious Hazard 4 = Severe Hazard	HEALTH = BLUE FIRE = RED REACTIVITY = YELLOW OTHER = WHITE	OX = Oxidizer ACID = Acid ALK = Alkali COR = Corrosive W = Use No Water

SECTION I - GENERAL INFORMATION

Characterization: **Polyaromatic Hydrocarbon**

RCRA Number: **Not Listed**

EPA Class: **Not Listed**

DOT Proper Shipping Name: **Anthracene**

Chemical Abstract Service (CAS) Number: **120-12-7**

DOT Hazard Class and Label Requirements: **Hazardous Substance; Solid**

DOT Emergency Guide Code: **No Citation**

DOT Identification Number: **UN 3077**

Chemical Formula: **$C_{14}H_{10}$**

Synonyms: **Anthracin; green oil; paranaphthalene; Tetra Olive N2G.**

SECTION II - HAZARDOUS INGREDIENTS AND IDENTITY INFORMATION

Hazardous Components (specific identity)	OSHA Exposure Criteria	NIOSH Exposure Criteria	Immediately Dangerous to Life and Health (IDLH)	ACGIH Exposure Criteria
Anthracene (derivation: Distillation of crude anthracene oil with alkali carbonate in iron retorts; by salting-out from crude anthracene oil and draining; occurs naturally in smoke from gas, coal, cigarettes, etc., char-broiled foods, and coal tar pitch volatiles).	**PEL:** 0.2 mg/m^3 **STEL:** Not Established	**REL:** 0.1 mg/m^3 Suspected Human Carcinogen	700 ppm	**TLV:** 0.2 mg/m^3 **STEL:** Not Established

SECTION III - PHYSICAL AND CHEMICAL CHARACTERISTICS

Boiling Point: **644°F (340°C)**

Specific Gravity (H_2O = 1): **1.25**

Vapor Pressure (mm Hg): **1 at 293°F (145°C)**

Molecular Weight: **178.24**

Vapor Density (Air = 1): **6.15**

Melting Point: **423°F (217°C)**

Solubility: **Insoluble in water (1.29 mg/L at 77°F, distilled water); slightly soluble in absolute alcohol, benzene, chloroform, ether, carbon disulfide, carbon tetrachloride, toluene, and acetone.**

Appearance and Odor: **Colorless crystalline solid with a violet fluorescence (pure), or yellow crystals with a green fluorescence due to tetracene and napthacene.**

SECTION IV - FIRE AND EXPLOSION HAZARD DATA

Flash Point (method used): **250°F (121°C) closed cup**

Explosive Limits in Air % by Volume: **LEL: 0.6% UEL: Not Identified**

NFPA Classification: **Combustible Solid**

Autoignition Temperature: **1004°F (540°C)**

Extinguishing Media: **Carbon dioxide, dry chemical, water spray or mist, or alcohol foam.**

Special Fire Fighting Procedures: **Wear full protective clothing and self-contained breathing apparatus (SCBA). Poisonous gases are produced in fire.**

Unusual Fire and Explosion Hazards: **Anthracene may explode on contact with air. Moderately explosive when exposed to flame, $Ca(OCL)_2$, or chromic acid. Explodes on contact with fluorine.**

SECTION V - REACTIVITY DATA

Stability		Conditions to Avoid Reacts with strong oxidizers (such as chlorine, bromine, and fluorine), chromic acid, and calcium hypochlorite. Anthracene darkens upon exposure to sunlight and transforms to *para*-anthracene.
Stable X	Unstable	Incompatibility *(materials to avoid)* Avoid exposure to heat, ignition sources, sunlight, and incompatibilities.
Hazardous Polymerization		Conditions to Avoid Many perchlorates form solvated adducts with organic solvents, including anthracene, which explode on impact.
May Occur	Will Not Occur X	Hazardous Decomposition or By-products When heated to decomposition, anthracene can emit carbon dioxide and acrid, irritating smoke.

SECTION VI - HEALTH HAZARD DATA

Primary Route(s) of Entry:	Inhalation?	Absorption (skin/eye)?	Ingestion?
	X	X	

Health Hazards

INHALATION: Irritation of the respiratory tract, headache, nausea, vomiting, loss of appetite, slowed reactions, and adynamia (a lack of strength).

ABSORPTION: Coal tar derivatives may absorb through unbroken skin. Irritation with burning, itching, and edema (fluid buildup). Irritates the conjunctiva of the eye.

INGESTION: Gastrointestinal tract irritation.

Carcinogenicity	NTP Listed?	IARC Cancer Review Group?	OSHA Regulated?	Target Organs?
Suspected Human Suspected Animal	Not Cited	Group 3	29 CFR 1910.1000 Table Z-1	Respiratory sys., eyes, skin, digestive system

Medical Conditions Generally Aggravated by Exposure

Skin conditions (dermatitis).

Emergency and First-aid Procedures

Eye contact: Do not allow victim to keep eyes tightly shut. Flush immediately with water for 15 minutes (minimum), seek medical attention. Skin contact: Remove all contaminated clothing. Immediately wash area with large amounts of soap and water. For red or blistered skin, contact physician. For inhalation: Remove the person from exposure. Provide respiratory assistance and CPR. Transfer to medical facility. If swallowed, seek medical attention immediately. Never give anything by mouth to an unconscious or convulsing person.

SECTION VII - PRECAUTIONS FOR SAFE HANDLING AND USE

Steps to be Taken in Case Material is Released or Spilled

Collect powered materials in most convenient and safe manner and deposit in sealed containers. Ventilate area of spill or leak. Remove all ignition sources. Restrict those not involved in cleanup from entering area. Notify appropriate authorities, as required by applicable environmental regulations.

Preferred Waste Disposal Method

Incinerator equipped with scrubber and afterburner.

Precautions to be Taken in Handling and Storage

Anthracene is a combustible solid. Store to avoid contact with strong oxidizers, since violent reactions can occur. Store in tightly closed containers in cool, well-ventilated area away from heat.

Other Precautions and Warnings

Sources of ignition (heat, flame, cigarettes, incompatibles) are prohibited where anthracene is used, handled, or stored. Bulk storage of anthracene is not recommended.

SECTION VIII - CONTROL MEASURES AND PERSONAL PROTECTIVE EQUIPMENT

Respiratory Protection *(specify type)*

Use a MSHA/NIOSH approved supplied air respirator with full facepiece, hood, or helmet in continuous flow mode, or use a full facepiece operated in pressure demand or other positive pressure mode or use a self-contained breathing apparatus (SCBA) with full facepiece operated in positive pressure mode.

Ventilation

Local exhaust (preferred) at site of chemical work or general ventilation.

Protective Gloves	Eye Protection	Other Protective Clothing
Polyvinylchloride (PVC)	Chemical/Dust Goggles or Face Mask	Boots; Aprons; Gauntlets

Work/Hygiene Practices

Always wash hands thoroughly after using chemical; never bring food, drink, or smoking materials into vicinity of chemicals. Use photoprotective creams or pastes on bare skin areas.

ANTHRACENE

$C_{14}H_{10}$ CAS: 120-12-7

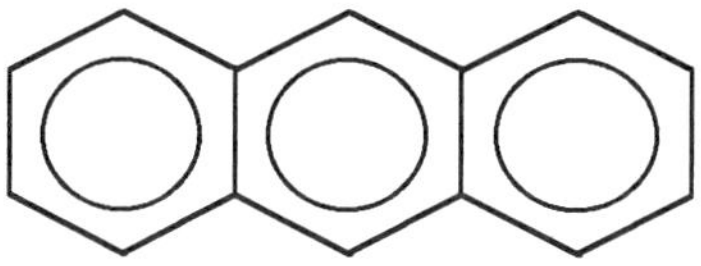

IDENTIFICATION AND TYPICAL USES

Pure anthracene is a colorless crystalline solid with a violet fluorescence. Also appears as yellow crystals with a green fluorescence due to the presence of tetracene and naphthacene. It is a naturally occurring compound found in coal-tar and is formed during the incomplete combustion of organic compounds. It occurs naturally in smoke (gasoline, coal, cigarette) and in char-broiled foods. It is used primarily in manufacturing of dyestuffs, and in chemical manufacturing (of phenanthrene, carbazole, and anthaquinone); in calico printing; as a component of dyes, scintillation fluids, and in smoke screens; and in organic semi-conductor research.

RISK ASSESSMENT: HEALTH

General Assessment

Anthracene is a skin irritant and allergen. It is a questionable (human) carcinogen with experimental neoplastigenic and tumorigenic data. Mutation data is also reported. It is moderately toxic by *inhalation* and *absorption* (skin/eye contact). Inhalation can result in symptoms such as headache, nausea and vomiting, loss of appetite, and slowed reactions coupled with weakness. Direct skin or eye contact can lead to itching and burning and possible allergic reactions at site of contact.

Anthracene often contains acridine as a contaminant. Acridine can also cause skin and eye allergies and irritates the skin, eyes, and air passages. Exposure to sunlight can further aggravate skin irritation and cause dermatitis.

Acute Health Effects

The following acute (short-term) health effects may occur immediately or shortly after (within hours or days) exposure to anthracene:

Skin: Irritation with burning, itching, and edema (buildup of fluid), redness and swelling.

Eye: Severe irritation of the conjunctiva with burning, itching, and watering. Direct contact with heated "fumes" can irritate the eyes.

Lung: Irritation of the bronchial tubes and other members of the respiratory tract. May cause headache, loss of appetite, nausea, vomiting, slowed reactions, and a loss of strength.

CNS: No conclusive evidence of effect on CNS.

Chronic Health Effects

The following chronic (long-term) health effects may occur at some time after exposure to anthracene and can last for months or even years:

Cancer Hazards: Anthracene may cause mutations (genetic changes) in living cells. The data on its carcinogenicity are inconclusive.

Reproductive Hazard: According to the references, anthracene has not been adequately tested for its ability to adversely affect reproduction.

Other Chronic Effects: Continuous exposures can cause allergies to the eyes and skin which are greatly aggravated upon exposure to sunlight during or shortly after exposure. Repeated skin exposures can cause patchy areas of increased brown pigment changes, loss of skin pigment, thinning or patchy thickening of the skin, skin warts, skin cancer, and pimples. Repeated breathing of "fumes" from heated anthracene may cause chronic bronchitis with cough and phlegm. Repeated exposure to male scrotum can cause skin thinning and increased pigment.

Recommended Risk-Reduction Measures

Personnel should avoid direct contact with anthracene. If a less toxic chemical can not be substituted for a hazardous substance, then *engineering controls* are the most effective method of reducing exposures. The best protection is to enclose operations and/or provide local exhaust ventilation at the site of chemical release. While not always operationally feasible, isolating operations can also reduce exposure. Using respiratory protection is less effective than the controls mentioned above, but is still advisable whenever working with or around anthracene. A self-contained breathing apparatus (SCBA) with full facepiece and pressure demand is the recommended respiratory protection of choice. If a full facepiece is not available, then chemical/dust goggles should be worn to protect the eyes. A face shield and a polyvinylchloride (PVC) apron should also be worn. To prevent hand and skin exposures, PVC gloves should be worn and photoprotective creams or pastes should be used on bare skin surfaces.

Administrative controls should also be in place to minimize the potential for human exposures. These may include written procedures or policies which specify the methods and techniques that will be practiced whenever personnel are to work with anthracene.

All personnel should receive *training* on the use, hazards, protective measures, emergency actions, and other precautions per 29 CFR 1910.1200 (Hazard Communication), prior to the first assignment in an area where anthracene is used or stored.

Before beginning employment, and at regular intervals thereafter (e.g., annually), the following medical tests are recommended for personnel working with anthracene:

- ☑ Skin examination for growths or changes in warts or moles (skin cancers are generally easily curable when removed early).
- ☑ Lung function test.

If symptoms develop or overexposure is suspected, the following may be useful:

- ☑ Evaluation by a qualified allergist, including careful exposure history and special testing (may help diagnose skin allergy).

Any evaluation should include a careful history of past and present symptoms with an examination. Medical tests that simply look for existing damage are not a substitute for controlling exposures.

Other methods to reduce exposure include:

- ☑ Where possible, enclose operations and use local exhaust ventilation at the site of chemical release. If local exhaust ventilation or enclosure is not used, respiratory protection should be mandatory.
- ☑ Always ensure that proper protective clothing is worn when using chemical substances.
- ☑ Wash thoroughly immediately after exposure to anthracene and at the end of the work shift or before eating, drinking, or smoking. Avoid contact with direct sunlight.
- ☑ Hazard warning information should be posted in the work area. In addition, as part of an on-going education and training program, all information on the health and safety hazards of anthracene should be communicated to all potentially exposed workers.
- ☑ Eye wash stations should be provided in the immediate work area for emergency use. If there is a possibility of skin exposure to anthracene, emergency shower facilities should also be provided.

RISK ASSESSMENT: ENVIRONMENT

General Assessment

The environment is at risk of exposure during transportation, storage, disposal, or destruction of anthracene. In almost every scenario, the threat of environmental exposure is contingent upon the proper handling of the chemical substance. Accidental spills, large or small, can result in fire, explosion (depending upon conditions of spill), and possible contamination of the surrounding environmental mediums (water, soil, and air).

Anthracene is considered a combustible solid. Alone, its high flash point and autoignition temperature do not present a serious fire or explosion hazard. However, because it is incompatible with many other commodities, extreme caution is required in handling, storage, transportation, and disposal of anthracene. These characteristics also require special consideration during any emergency situation involving a leak or spill of anthracene crystals. Should anthracene ever come into contact with incompatible substances (such as oxidizers) either during use, transportation, storage, or disposal, the formation of highly toxic and/or highly explosive commodities is extremely possible.

The proper disposal/destruction method for anthracene waste is to burn it in a chemical incinerator equipped with an afterburner and air scrubber.

Due to its widespread occurrence in nature and its use in the dye industries, anthracene can enter the environment from non-point sources, as well as industrial and municipal discharges.

Acute Ecological Effects

Acute (short-term) toxic effects may include the death of animals, birds, or fish and death or low growth rate in plants. Acute effects are seen 2 to 4 days after animals or plants are exposed to anthracene. This chemical has high acute toxicity to aquatic life and to birds. It showed no effect on corn root size or respiration in one study. No data are available on the short-term effects of anthracene to land animals.

Chronic Ecological Effects

Chronic (long-term) toxic effects may include shortened life span, reproductive problems, lower fertility, and changes in appearance or behavior in exposed animals. These effects can be seen long after first exposure(s) to toxic chemicals. Insufficient data are available to evaluate or predict the long-term effects of anthracene to plants, birds, or land animals.

Water Solubility

Anthracene is very slightly soluble in water. Concentrations of 1 milligram or less will mix with a liter of water.

Persistence in the Environment

Anthracene is moderately persistent in water, with a half-life between 20 and 200 days. The half-life of a pollutant is the amount of time it takes for one half of the chemical to be degraded. Approximately 77.5% of anthracene will eventually end up in air; about 9.5% will end up in terrestrial soil; about 9% in aquatic sediments; and the rest will end up in water.

Bioaccumulation in Aquatic Organisms

Some substances increase in concentration, or *bioaccumulate,* in living organisms as they breathe contaminated air, drink contaminated water, or eat contaminated food. These chemicals can become concentrated in the tissues and internal organs of animals as well as humans.

The concentration of anthracene found in fish tissues is expected to be much higher than the average concentration of anthracene in the water from which the fish was taken.

Recommended Risk-Reduction Measures

Proper training of all transporters will reduce the likelihood of a mishap or accident resulting in a leak or spill of anthracene into the environment. The correct use of DOT labeling (per the MSDS) on all containers, trucks, and rail cars will enable emergency responders to react properly and quickly to any disaster thereby reducing the potential risk to the environment and to personnel.

Storage of anthracene should be segregated from incompatible chemicals to minimize the risk of cross-contamination or contact.

If a spill or leak to the environment has occurred, fire department, emergency response and/or hazardous materials spill personnel should be notified immediately. Cleanup should be attempted only by those trained in proper spill containment procedures. Contaminated soils should be removed for incineration and replaced with clean soil. If anthracene should contact the water table, aquifer, or navigable waterway, the local and/or state emergency response authorities must be notified. A comprehensive emergency response or disaster preparedness/recovery plan should be in place prior to any operations involving the use, transportation, storage, or disposal of anthracene. If anthracene is spilled or leaked, the following specific steps are recommended:

- ☑ Restrict persons not wearing protective clothing from area of spill or leak until cleanup is complete and area can be opened for normal work.
- ☑ Remove all ignition sources.
- ☑ Collect powdered materials in the safest manner possible and deposit in sealed containers.
- ☑ It may be necessary to dispose of anthracene as a hazardous waste. The responsible state agency or the regional office of the federal Environmental Protection Agency (EPA) should be contacted for specific recommendations.

RISK ASSESSMENT: BUSINESS

General Assessment

Accidents or mishaps involving anthracene can present a significant threat to business operations. The loss or damage of equipment or facilities can significantly affect fiscal viability. Lawsuits that may result from personnel injury/death, public exposures, and/or environmental contamination will also require a serious expenditure of resources.

Media attention surrounding an injury, death, or environmental damage can also result in a loss of profits and loss of current as well as future business.

Ü *Recommended Risk-Reduction Measures*

Company attorneys, safety and health professionals, and environmental specialists should be involved in the development of any safety, health or environmental policies or procedures. A company official should be pre-designated as a public relations officer with specific training in dealing with the press. Corporate plans and policies should be developed, approved, and implemented long before any need for such arises.

REFERENCES

Ahlbom, A. 1993. *Biostatistics for Epidemiologists.* New York: CRC Press/Lewis Publishers.

American Conference of Governmental Industrial Hygienists. 1988. *Documentation of the Threshold Limit Values and Biological Exposure Indices,* 5th Edition (with updates). Cincinnati: ACGIH

Asante-Duah, D. K. 1993. *Hazardous Waste Risk Assessment.* New York: CRC Press/Lewis Publishers.

Bloom, A. D. and F. J. DeSerres. 1995. *Ecotoxicity and Human Health.* New York: CRC Press/Lewis Publishers.

Brusick, D. J. 1994. *Methods for Genetic Risk Assessment.* New York: CRC Press/Lewis Publishers.

Bueche, J. R. 1972. *Principles of Physics,* 2nd Edition. New York: McGraw-Hill.

Calabrese, E. J. 1994. *Biological Effects of Low Level Exposures.* New York: CRC Press/Lewis Publishers.

Cockerman, L. G., B. S. Shane. 1994. *Basic Environmental Toxicology.* New York: CRC Press/Lewis Publishers.

Cothern, R. R. 1993. *Comparative Environmental Risk Assessment.* New York: CRC Press/Lewis Publishers.

Gardner, R. W. 1994. *Chemical Intolerance.* New York: CRC Press/Lewis Publishers.

Hallenbeck, W. H. 1993. *Quantitative Risk Assessment for Environmental and Occupational Health,* 2nd Edition. New York: CRC Press/Lewis Publishers.

Howard, P. H. 1995. *Handbook of Environmental Fate and Exposures Data for Organic Chemicals,* Vol. I-IV. New York: CRC Press/Lewis Publishers.

Jones, F. E. 1994. *Toxic Organic Vapors in the Workplace.* New York: CRC Press/Lewis Publishers.

Keith, L. H. and M. M. Walker. 1995. *Handbook of Air Toxics.* New York: CRC Press/Lewis Publishers.

Landis, W. G., M. H. Yu. 1995. *Introduction to Environmental Toxicology.* New York: CRC Press/Lewis Publishers.

Langerman, N. 1994 *Precautionary Labels for Chemical Containers.* New York: CRC Press/Lewis Publishers.

Lauwerys, R. R., P. Hoet. 1993. *Industrial Chemical Exposure.* New York: CRC Press/Lewis Publishers.

Lewis, R. J., Sr. 1993. *Hawley's Condensed Chemical Dictionary,* 12th Edition. New York: Van Nostrand Reinhold.

Lewis, R. J., Sr. 1992. *Sax's Dangerous Properties of Industrial Materials,* Volumes 1, 2, and 3, 8th Edition. New York: Van Nostrand Reinhold.

Lide, D. R. 1995. *Handbook of Chemistry and Physics,* 76th Edition. New York: CRC Press/Lewis Publishers.

Manahan, S. E. 1994. *Environmental Chemistry,* 6th Edition. New York: CRC Press/Lewis Publishers.

Manahan, S. E. 1992. *Toxicological Chemistry,* 2nd Edition. New York: CRC Press/Lewis Publishers.

Mellström, G. A., J. E. Wahlberg, and H. I. Maibach. 1994. *Protective Gloves for Occupational Use.* New York: CRC Press/Lewis Publishers.

Meyer, E. 1990. *Chemistry of Hazardous Materials.* Englewood Cliffs, NJ: Prentice Hall.

Mycock, J. C., J. D. McKenna, and L. Theodore. 1995. *Handbook of Air Pollution Control Engineering and Technology.* New York: CRC Press/Lewis Publishers.

National Cancer Institute. 1978. *Carcinogenesis Technical Report Series. Bioassay of Anthracene for Possible Carcinogenicity.* DEHW (NIH) Publication Number: 78-1323, p 53. Washington, DC: U.S. Government Printing Office.

Ott, W. R. 1995. *Environmental Statistics and Data Analysis.* New York: CRC Press/Lewis Publishers.

Patnaik, P. 1992. *A Comprehensive Guide to the Hazardous Properties of Chemical Substances.* New York: Van Nostrand Reinhold.

Philp, R. B. 1995. *Environmental Hazards and Human Health.* New York: CRC Press/Lewis Publishers.

Proctor, N. H., J. P. Hughes. 1991. *Chemical Hazards of the Workplace.* New York: Van Nostrand Reinhold.

Quigley, D. R. 1994. *Handbook of Emergency Chemical Management.* New York: CRC Press/Lewis Publishers.

Rea, W. J. 1995. *Chemical Sensitivity,* Vol. I-III. New York: CRC Press/Lewis Publishers.

Richardson, M. 1995. *Dictionary of Substance and Their Effects,* Vol. I-VI. New York: CRC Press/Lewis Publishers.

Sacarello, H. L. A. 1994. *The Comprehensive Handbook of Hazardous Materials.* New York: CRC Press/Lewis Publishers.

Sheftel, V. O. 1995. *Handbook of Toxic Properties of Monomers and Additives.* New York: CRC Press/Lewis Publishers.

Talbott, E., G. Craun. 1995. *An Introduction to Environmental Epidemiology.* New York: CRC Press/Lewis Publishers.

Torkelson, T. R. V. K. Rowe. 1981. *Patty's Industrial Hygiene and Toxicology,* Vol. 2B. New York: Wiley-Interscience.

MATERIAL SAFETY DATA SHEET

CHEMICAL NAME	
ANTIMONY	**CRC PRESS/LEWIS PUBLISHERS DISCLAIMER:** THE INFORMATION AND RECOMMENDATIONS PRESENTED HEREIN ARE BASED ON SOURCES BELIEVED TO BE RELIABLE. CRC MAKES NO REPRESENTATION ON ITS COMPLETENESS OR ACCURACY. IT IS THE RESPONSIBILITY OF THE USER TO DETERMINE THE CHEMICAL'S SUITABILITY FOR ITS INTENDED USE, THE CHEMICAL'S SAFE USE, AND THE CHEMICAL'S PROPER DISPOSAL. NO REPRESENTATIONS AND/OR WARRANTIES, EITHER EXPRESSED OR IMPLIED, OF THE MERCHANTABILITY OR FITNESS FOR A PARTICULAR PURPOSE, OR OF ANY OTHER NATURE, ARE MADE WITH RESPECT TO THE INFORMATION PROVIDED IN THIS MSDS OR TO THE CHEMICAL TO WHICH INFORMATION MAY REFER. CRC NEITHER ASSUMES NOR AUTHORIZES ANY OTHER PERSON TO ASSUME FOR IT, ANY OTHER ADDITIONAL RESPONSIBILITY OR LIABILITY FOR THE USE OF, OR RELIANCE UPON, THIS INFORMATION.

HAZARD WARNING INFORMATION

HEALTH	FIRE	REACTIVE	OTHER	DEGREE OF HAZARD	COLOR CODING	OTHER CODES
3	1	1		0 = Minimum Hazard 1 = Slight Hazard 2 = Moderate Hazard 3 = Serious Hazard 4 = Severe Hazard	HEALTH = BLUE FIRE = RED REACTIVITY = YELLOW OTHER = WHITE	OX = Oxidizer ACID = Acid ALK = Alkali COR = Corrosive W = Use No Water

SECTION I - GENERAL INFORMATION

Characterization **Metal**	RCRA Number **Not Listed**	EPA Class **RCRA Priority Pollutant**
DOT Proper Shipping Name **Antimony**	Chemical Abstract Service (CAS) Number **7440-36-0**	
DOT Hazard Class and Label Requirements **Hazardous Substance; Solid**	DOT Emergency Guide Code **53**	
DOT Identification Number **UN 2871**	Atomic Symbol **Sb (from the Latin "Stibium")**	

Synonyms
Stibium; antimony black; antimony regulus.

SECTION II - HAZARDOUS INGREDIENTS AND IDENTITY INFORMATION

Hazardous Components (specific identity)	OSHA Exposure Criteria	NIOSH Exposure Criteria	Immediately Dangerous to Life and Health (IDLH)	ACGIH Exposure Criteria
Antimony (derivation: Reduction of stibnite from iron scrap; direct reduction of natural oxide ores; about 50% of the antimony used in the U.S. is recovered from lead base battery scrap metal).	**PEL:** 0.5 mg/m^3 **STEL:** Not Established	**REL (10-hr):** 0.5 mg/m^3 **Suspected Human Carcinogen**	**80 mg/m^3**	**TLV:** 0.5 mg/m^3 **STEL:** Not Established

SECTION III - PHYSICAL AND CHEMICAL CHARACTERISTICS

Boiling Point **2975°F (1635°C)**	Specific Gravity (H_2O = 1) **6.69**
Vapor Pressure (mm Hg) **0 (approx.) at 69°F (20°C)**	Molecular Weight (atomic weight) **121.8**
Vapor Density (Air = 1) **N/A (metal)**	Melting Point **1166°F (630°C)**

Solubility
Insoluble in water. Soluble in hot, concentrated H_2SO_4.

Appearance and Odor
Silvery white lustrous metal; hard and brittle solid, scale-like crystals, or dark gray lustrous powder; tarnishes slowly in moist air. No appreciable odor. Can also exist as a dust.

SECTION IV - FIRE AND EXPLOSION HAZARD DATA

Flash Point (method used) **Not Applicable**	Explosive Limits in Air % by Volume **LEL: N/A UEL: N/A**
NFPA Classification **Noncombustible Solid (bulk form)**	Autoignition Temperature **Not Applicable**

Extinguishing Media
Dry chemicals appropriate for extinguishing metal (Class D) fires. DO NOT USE WATER.

Special Fire Fighting Procedures
Wear full protective clothing and self-contained breathing apparatus (SCBA). Poisonous gases are produced in fire.

Unusual Fire and Explosion Hazards
Moderate fire and explosion hazard in the forms of dust and vapor, when exposed to heat or flame. Poisonous gases are produced in fire, including antimony oxide and stibine.

SECTION V - REACTIVITY DATA

Stability		Conditions to Avoid **Avoid heat and open flame, especially around dusts and powders of antimony. Do not allow contact with acids since toxic SbH_3 can be emitted due to reaction.**
Stable **X**	Unstable	Incompatibility *(materials to avoid)* **Oxidizers (perchlorates, peroxides, permanganates, chlorates, nitrates), halogens (chlorine, bromine); bromine trifluoride, chlorine trifluoride, chloric acid, chlorine monoxide, bromazide, and nitric acid.**
Hazardous Polymerization		Conditions to Avoid **Hazardous polymerization of antimony metal will not occur**
May Occur	Will Not Occur **X**	Hazardous Decomposition or By-products **When heated or mixed with strong acids, toxic stibine can be emitted.**

SECTION VI - HEALTH HAZARD DATA

Primary Route(s) of Entry:	Inhalation?	Absorption (skin/eye)?	Ingestion?
	X	X	

Health Hazards

INHALATION: Irritation of the respiratory tract. Can cause metallic taste, nausea, sore throat. Higher concentrations can cause lung congestion and may cause irregular heartbeat.

ABSORPTION: Skin contact causes irritation and itchy rash. Powders can be absorbed through the skin. May cause papules and pustules around sweat and sebaceous glands.

INGESTION: Possible gastrointestinal tract irritation.

Carcinogenicity	NTP Listed?	IARC Cancer Review Group?	OSHA Regulated?	Target Organs?
Suspected Human Suspected Animal	**5th Annual Report**	**No**	**29 CFR 1910.1000 Table Z-1**	**Eyes, respiratory system, skin, CVS.**

Medical Conditions Generally Aggravated by Exposure

Skin conditions (dermatitis); respiratory problems; heart or other cardiovascular conditions.

Emergency and First-aid Procedures

Eye contact: Do not allow victim to keep eyes tightly shut. Flush immediately with water for 15 minutes (minimum), seek medical attention. Skin contact: Remove all contaminated clothing. Immediately wash area with large amounts of soap and water. For inhalation: Remove the person from exposure. Provide respiratory assistance and CPR. Transfer to medical facility. Observe for 1 or 2 days after overexposure, as lung and heart conditions may be delayed. If swallowed, seek medical attention immediately.

SECTION VII - PRECAUTIONS FOR SAFE HANDLING AND USE

Steps to be Taken in Case Material is Released or Spilled

Collect powered materials in most convenient and safe manner and deposit in sealed containers. Ventilate area of spill or leak. Remove all ignition sources. Restrict those not involved in cleanup from entering area. Notify appropriate authorities, as required by applicable environmental regulations.

Preferred Waste Disposal Method

Incinerator equipped with scrubber and afterburner.

Precautions to be Taken in Handling and Storage

Antimony is a noncombustible solid, but its dusts is a moderate explosion hazard when exposed to heat. Store to avoid contact with strong oxidizers and acids, since violent reactions can occur. Store in tightly closed containers in cool, well-ventilated area away from heat. Do not use near heat or flame.

Other Precautions and Warnings

Sources of ignition (heat, flame, cigarettes, incompatibles) are prohibited where antimony is used, handled, or stored. Bulk storage of antimony powders or dusts is not recommended.

SECTION VIII - CONTROL MEASURES AND PERSONAL PROTECTIVE EQUIPMENT

Respiratory Protection *(specify type)*

A full facepiece respirator with a high efficiency particulate air (HEPA) filter. Greater protection is provided by a powered air purifying respirator. At high exposures, use a self-contained breathing apparatus (SCBA) with full facepiece operated in positive pressure mode.

Ventilation

Local exhaust (preferred) at site of chemical work or general ventilation.

Protective Gloves	Eye Protection	Other Protective Clothing
Rubber	**Chemical/Dust Goggles or Face Mask**	**Protective Apron, Boots**

Work/Hygiene Practices

Always wash hands thoroughly after using chemical; never bring food, drink, or smoking materials into vicinity of chemicals.

ANTIMONY

Sb CAS: 7440-36-0

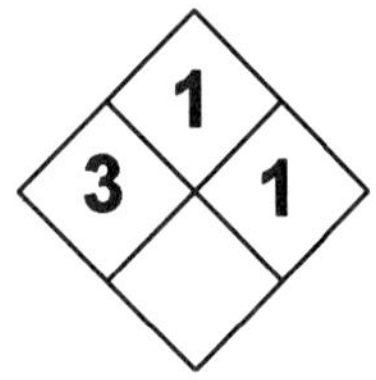

IDENTIFICATION AND TYPICAL USES

Antimony is a silvery white lustrous metal, appearing frequently as a hard and brittle solid. Can also exist as scale-like crystals or a dark gray lustrous powder. Antimony tarnishes slowly in moist air. There is no appreciable odor. Can also exist as a dust. It is used to make alloys such as Babbit metal, white metal, and hard lead, and for coating metals. It is a constituent of alloys of other metals such as copper and tin. Its sulfides are used in compounding of rubber and the manufacture of pyrotechniques, including bullets and fireworks. Its chlorides are used as coloring agents and as catalysts. It fluorides are used in organic synthesis and pottery manufacture.

RISK ASSESSMENT: HEALTH

General Assessment

Antimony is an experimental poison and a questionable human carcinogen with experimental carcinogenic data being reported. It is an irritant of the mucous membranes, eyes, and skin. Heavy exposure to antimony trioxide and pentoxide is associated with pulmonary injury while antimony trisulfide is considered cardiotoxic (harmful to the heart and cardiovascular system).

Inhalation or exposure to dusts or fumes can result in rhinitis, irritation of the eyes and sore throat. Higher levels (at the IDLH and above) could cause antimony poisoning with symptoms of headache, pain or tightness in the chest, shortness of breath, metallic taste, nausea, vomiting, diarrhea, and weight loss, and even death. Repeated exposure can cause an abnormal chest X-ray to develop and can also damage the heart and liver. Current data are inconclusive regarding any increase risk of lung cancer and reproductive disorders due to antimony exposures.

Skin contact with antimony or its compounds causes papules and pustules around sweat and sebaceous glands. Prolonged or repeated skin contact can cause skin, nose, and mouth ulcers and sores.

Acute Health Effects

The following acute (short-term) health effects may occur immediately or shortly after exposure to antimony:

Skin: Irritation with itchy rash and possible papules and pustules around sweat and sebaceous glands.

Eye: Severe irritation, possible itching or burning.

Lung: Can cause metallic taste, nausea, sore throat, and irritation of the air passages. Higher levels can cause congestion of the lungs.

CVS: Can cause damage to the heart. May cause irregular heartbeat and might even cause heart to stop.

Chronic Health Effects

The following chronic (long-term) health effects may occur at some time after exposure to antimony and can last for months or even years:

Cancer Hazards: Antimony is a questionable carcinogen in animals and humans. The data on its carcinogenicity are not totally conclusive.

Reproductive Hazard: There is limited evidence that antimony may decrease fertility in females.

Other Chronic Effects: Repeated exposure can cause headaches, poor appetite, dry throat, and loss of sleep. Damage to the liver and heart muscle with abnormal EKG may also occur, especially with higher or frequent exposures.

Recommended Risk-Reduction Measures

Personnel should avoid direct contact with antimony, especially its dusts or powders. If a less toxic material or compound can not be substituted for antimony, then *engineering controls* are the most effective method of reducing exposures. The best protection is to enclose operations and/or provide local exhaust ventilation at the site of antimony dust release. While not always operationally feasible, isolating operations can also reduce exposure. Using respiratory protection is less effective than the controls mentioned above, but is still advisable whenever working with or around antimony. A self-contained breathing apparatus (SCBA) with full facepiece and pressure demand is recommended for the greatest possible respiratory protection. However, for low or infrequent exposure, a MSHA/NIOSH approved full facepiece respirator with a high efficiency particulate air (HEPA) filter can be used. If a full facepiece is not available, then chemical/dust goggles should be worn to protect the eyes. A face shield and protective apron should also be worn. To prevent hand and skin exposures, gloves should be worn.

Administrative controls should also be in place to minimize the potential for human exposures. These may include written procedures or policies which specify the methods and techniques that will be practiced whenever personnel are to work with antimony.

All personnel should receive *training* on the use, hazards, protective measures, emergency actions, and other precautions per 29 CFR 1910.1200 (Hazard Communication), prior to the first assignment in an area where antimony is used or stored.

Before beginning employment working with antimony, and at regular intervals thereafter (e.g., annually), those personnel with frequent or potentially high exposures should be provided the following recommended medical tests:

- ☑ Urine test for antimony (and arsenic if arsenic contamination is suspect).

If symptoms develop or overexposure is suspected, the following additional tests may be useful:

- ☑ EKG (electrocardiogram).
- ☑ Liver function test.
- ☑ Consider chest X-ray following acute exposure.

Any evaluation should include a careful history of past and present symptoms with an examination. Medical tests that simply look for existing damage are not a substitute for controlling exposures. Also note that commercial grade antimony is often contaminated with arsenic, a toxic chemical and carcinogen. Use of antimony near acid or acid mist can cause the release of the deadly gas, stibine. Therefore, such potential for "mixed exposures" warrants additional caution when working with or around antimony.

Other methods to reduce exposure include:

- ☑ Where possible, enclose operations and use local exhaust ventilation at the site of chemical release. If local exhaust ventilation or enclosure is not used, respiratory protection should be mandatory.
- ☑ Always ensure that proper protective clothing is worn when using chemical substances.
- ☑ Wash thoroughly immediately after exposure to antimony and at the end of the work shift or before eating, drinking, or smoking.
- ☑ Hazard warning information should be posted in the work area. In addition, as part of an on-going education and training program, all information on the health and safety hazards of antimony should be communicated to all exposed or potentially exposed workers.
- ☑ Eye wash stations should be provided in the immediate work area for emergency use. If there is a possibility of skin exposure to antimony, emergency shower facilities should also be provided.

RISK ASSESSMENT: ENVIRONMENT

General Assessment

The environment is at risk of exposure during transportation, storage, disposal, or destruction of antimony. In almost every scenario, the threat of environmental exposure is contingent upon the proper handling of the chemical substance. Accidental spills, large or small, where antimony contacts incompatible commodities can result in fire, explosion (depending upon conditions of spill), and possible contamination of the surrounding environmental mediums (water, soil, and air).

Antimony is considered a noncombustible solid. However, in powder or dust form it can present a moderate fire and explosion hazard. In bulk form, it does not present a serious fire or explosion hazard. However, because it is incompatible with many other commodities, extreme caution is required in handling, storage, transportation, and disposal of antimony. These characteristics also require special consideration during any emergency situation involving a leak or spill of antimony powder or dust.

Should antimony ever come into contact with incompatible substances (such as oxidizers) either during use, transportation, storage, or disposal, the formation of highly toxic and/or reactive commodities is extremely possible.

Antimony may occur in the aquatic environment from natural weathering of mineral formations, from mining and manufacturing effluents, and from municipal wastes.

Acute Ecological Effects

Acute (short-term) toxic effects may include the death of animals, birds, or fish and death or low growth rate in plants. Acute effects are seen 2 to 4 days after animals or plants are exposed to antimony. Antimony has high acute toxicity to aquatic life. No data are available on the short-term effects of antimony to plants, birds, or land animals.

Chronic Ecological Effects

Chronic (long-term) toxic effects may include shortened life span, reproductive problems, lower fertility, and changes in appearance or behavior in exposed animals. These effects can be seen long after first exposure(s) to toxic chemicals. Insufficient data are available to evaluate or predict the long-term effects of antimony to plants, birds, or land animals.

Water Solubility

Antimony and its salts have varying degrees of solubility in water ranging from low to high.

Persistence in the Environment

Antimony is highly persistent in water, with a half-life of longer than 200 days. The half-life of a pollutant is the amount of time it takes for one half of the chemical to be degraded. No data are available on the estimated percentages of antimony persistence in the aquatic or terrestrial environments.

Bioaccumulation in Aquatic Organisms

Some substances increase in concentration, or *bioaccumulate,* in living organisms as they breathe contaminated air, drink contaminated water, or eat contaminated food. These chemicals can become concentrated in the tissues and internal organs of animals as well as humans.

The concentration of antimony found in fish tissues is expected to be about the same as the average concentration of antimony in the water from which the fish was taken.

Recommended Risk-Reduction Measures

Proper training of all transporters will reduce the likelihood of a mishap or accident resulting in a leak or spill of antimony dusts or powders into the environment. The correct use of DOT labeling (per the MSDS) on all containers, trucks, and rail cars will enable emergency responders to react properly and quickly to any disaster thereby reducing the potential risk to the environment and to personnel.

Storage of antimony should be segregated from incompatible chemicals to minimize the risk of cross-contamination or contact.

If a spill or leak to the environment has occurred, fire department, emergency response and/or hazardous materials spill personnel should be notified immediately. Cleanup should be attempted only by those trained in proper spill containment procedures. Contaminated soils should be removed for incineration and replaced with clean soil. If antimony should contact the water table, aquifer, or navigable waterway, the local and/or state emergency response authorities must be notified. A comprehensive emergency response or disaster preparedness/recovery plan should be in place prior to any operations involving the use, transportation, storage, or disposal of antimony. If antimony is spilled or leaked, the following specific steps are recommended:

- ☑ Restrict persons not wearing protective clothing from area of spill or leak until cleanup is complete.
- ☑ Remove all ignition sources.
- ☑ Collect powdered materials in the safest manner possible and deposit in sealed containers.
- ☑ It may be necessary to dispose of antimony as a hazardous waste. The responsible state agency or the regional office of the federal Environmental Protection Agency (EPA) should be contacted for specific recommendations.

RISK ASSESSMENT: BUSINESS

General Assessment

Accidents or mishaps involving antimony can present a significant threat to business operations. The loss or damage of equipment or facilities can significantly affect fiscal viability. Lawsuits that may result from personnel injury/death, public exposures, and/or environmental contamination will also require a serious expen-

diture of resources. Media attention surrounding an injury, death, or environmental damage can also result in a loss of profits and loss of current as well as future business.

Recommended Risk-Reduction Measures

Company attorneys, safety and health professionals, and environmental specialists should be involved in the development of any procedures dealing with safety or emergency response. A company official should be pre-designated as a public relations officer with specific training in dealing with the press. Corporate plans and policies should be developed, approved, and implemented long before any need for such arises.

REFERENCES

Ahlbom, A. 1993. *Biostatistics for Epidemiologists.* New York: CRC Press/Lewis Publishers.

American Conference of Governmental Industrial Hygienists. 1988. *Documentation of the Threshold Limit Values and Biological Exposure Indices,* 5th Edition (with updates). Cincinnati: ACGIH

Asante-Duah, D. K. 1993. *Hazardous Waste Risk Assessment.* New York: CRC Press/Lewis Publishers.

Bloom, A. D. and F. J. DeSerres. 1995. *Ecotoxicity and Human Health.* New York: CRC Press/Lewis Publishers.

Brusick, D. J. 1994. *Methods for Genetic Risk Assessment.* New York: CRC Press/Lewis Publishers.

Bueche, J. R. 1972. *Principles of Physics,* 2nd Edition. New York: McGraw-Hill.

Calabrese, E. J. 1994. *Biological Effects of Low Level Exposures.* New York: CRC Press/Lewis Publishers.

Cockerman, L. G., B. S. Shane. 1994. *Basic Environmental Toxicology.* New York: CRC Press/Lewis Publishers.

Cothern, R. R. 1993. *Comparative Environmental Risk Assessment.* New York: CRC Press/Lewis Publishers.

Gardner, R. W. 1994. *Chemical Intolerance.* New York: CRC Press/Lewis Publishers.

Hallenbeck, W. H. 1993. *Quantitative Risk Assessment for Environmental and Occupational Health,* 2nd Edition. New York: CRC Press/Lewis Publishers.

Howard, P. H. 1995. *Handbook of Environmental Fate and Exposures Data for Organic Chemicals,* Vol. I-IV. New York: CRC Press/Lewis Publishers.

Jones, F. E. 1994. *Toxic Organic Vapors in the Workplace.* New York: CRC Press/Lewis Publishers.

Keith, L. H. and M. M. Walker. 1995. *Handbook of Air Toxics.* New York: CRC Press/Lewis Publishers.

Landis, W. G., M. H. Yu. 1995. *Introduction to Environmental Toxicology.* New York: CRC Press/Lewis Publishers.

Langerman, N. 1994 *Precautionary Labels for Chemical Containers.* New York: CRC Press/Lewis Publishers.

Lauwerys, R. R., P. Hoet. 1993. *Industrial Chemical Exposure.* New York: CRC Press/Lewis Publishers.

Lewis, R. J., Sr. 1993. *Hawley's Condensed Chemical Dictionary,* 12th Edition. New York: Van Nostrand Reinhold.

Lewis, R. J., Sr. 1992. *Sax's Dangerous Properties of Industrial Materials,* Volumes 1, 2, and 3, 8th Edition. New York: Van Nostrand Reinhold.

Lide, D. R. 1995. *Handbook of Chemistry and Physics,* 76th Edition. New York: CRC Press/Lewis Publishers.

Manahan, S. E. 1994. *Environmental Chemistry,* 6th Edition. New York: CRC Press/Lewis Publishers.

Mellström, G. A., J. E. Wahlberg, and H. I. Maibach. 1994. *Protective Gloves for Occupational Use.* New York: CRC Press/Lewis Publishers.

Meyer, E. 1990. *Chemistry of Hazardous Materials.* Englewood Cliffs, NJ: Prentice Hall.

Mycock, J. C., J. D. McKenna, and L. Theodore. 1995. *Handbook of Air Pollution Control Engineering and Technology.* New York: CRC Press/Lewis Publishers.

Patnaik, P. 1992. *A Comprehensive Guide to the Hazardous Properties of Chemical Substances.* New York: Van Nostrand Reinhold.

Proctor, N. H., J. P. Hughes. 1991. *Chemical Hazards of the Workplace.* New York: Van Nostrand Reinhold.

Quigley, D. R. 1994. *Handbook of Emergency Chemical Management.* New York: CRC Press/Lewis Publishers.

Rea, W. J. 1995. *Chemical Sensitivity,* Vol. I-III. New York: CRC Press/Lewis Publishers.

Richardson, M. 1995. *Dictionary of Substance and Their Effects,* Vol. I-VI. New York: CRC Press/Lewis Publishers.

Sacarello, H. L. A. 1994. *The Comprehensive Handbook of Hazardous Materials.* New York: CRC Press/Lewis Publishers.

Sheftel, V. O. 1995. *Handbook of Toxic Properties of Monomers and Additives.* New York: CRC Press/Lewis Publishers.

Talbott, E., G. Craun. 1995. *An Introduction to Environmental Epidemiology.* New York: CRC Press/Lewis Publishers.

Taylor, P. J. 1966. Acute intoxication from antimony trichloride. *British Journal of Industrial Medicine.* 23:318-321.

Torkelson, T. R. V. K. Rowe. 1981. *Patty's Industrial Hygiene and Toxicology,* Vol. 2B. New York: Wiley-Interscience.

U.S. Department of Health, Education, and Welfare. National Institute for Occupational Safety and Health (NIOSH). 1978. *Criteria for a Recommended Standard. Occupational Exposure to Antimony.* DHEW (NIOSH) Publication Number: 78-216. Washington, DC: U.S. Government Printing Office.

U.S. Department of Labor. 1983. Antimony metal: antimony trioxide; and antimony sulfide response to the interagency testing committee. *Federal Register.* 48:717-724.

MATERIAL SAFETY DATA SHEET

CHEMICAL NAME	
ARSENIC	**CRC PRESS/LEWIS PUBLISHERS DISCLAIMER:** THE INFORMATION AND RECOMMENDATIONS PRESENTED HEREIN ARE BASED ON SOURCES BELIEVED TO BE RELIABLE. CRC MAKES NO REPRESENTATION ON ITS COMPLETENESS OR ACCURACY. IT IS THE RESPONSIBILITY OF THE USER TO DETERMINE THE CHEMICAL'S SUITABILITY FOR ITS INTENDED USE, THE CHEMICAL'S SAFE USE, AND THE CHEMICAL'S PROPER DISPOSAL. NO REPRESENTATIONS AND/OR WARRANTIES, EITHER EXPRESSED OR IMPLIED, OF THE MERCHANTABILITY OR FITNESS FOR A PARTICULAR PURPOSE, OR OF ANY OTHER NATURE, ARE MADE WITH RESPECT TO THE INFORMATION PROVIDED IN THIS MSDS OR TO THE CHEMICAL TO WHICH INFORMATION MAY REFER. CRC NEITHER ASSUMES NOR AUTHORIZES ANY OTHER PERSON TO ASSUME FOR IT, ANY OTHER ADDITIONAL RESPONSIBILITY OR LIABILITY FOR THE USE OF, OR RELIANCE UPON, THIS INFORMATION.

HAZARD WARNING INFORMATION

HEALTH	FIRE	REACTIVE	OTHER	DEGREE OF HAZARD	COLOR CODING	OTHER CODES
3	0	0		0 = Minimum Hazard 1 = Slight Hazard 2 = Moderate Hazard 3 = Serious Hazard 4 = Severe Hazard	HEALTH = BLUE FIRE = RED REACTIVITY = YELLOW OTHER = WHITE	OX = Oxidizer ACID = Acid ALK = Alkali COR = Corrosive ~~W~~ = Use No Water

SECTION I - GENERAL INFORMATION

Characterization: **Metal**	RCRA Number: **D004**	EPA Class: **Characteristic Toxic Waste**
DOT Proper Shipping Name: **Arsenic**	Chemical Abstract Service (CAS) Number: **7440-38-2**	
DOT Hazard Class and Label Requirements: **Hazardous Substance; Solid**	DOT Emergency Guide Code: **53**	
DOT Identification Number: **UN 1558**	Atomic Symbol: **As**	

Synonyms

Gray arsenic; metallic arsenic; arsenic trichloride; arsenic salts; collodial arsenic.

SECTION II - HAZARDOUS INGREDIENTS AND IDENTITY INFORMATION

Hazardous Components (specific identity)	OSHA Exposure Criteria	NIOSH Exposure Criteria	Immediately Dangerous to Life and Health (IDLH)	ACGIH Exposure Criteria
Arsenic (derivation from: Flue dust of copper and lead smelters from which it is obtained as white arsenic (arsenic trioxide) in varying degrees of purity. This is reduced with charcoal).	**PEL:** **0.01 mg/m³** **CEILING:** **0.002 mg/m³ (15 minutes)**	**REL:** **0.5 mg/m³** **Suspected Human Carcinogen**	**5 mg/m³**	**TLV:** **0.2 mg/m³** **STEL:** **Not Established**

SECTION III - PHYSICAL AND CHEMICAL CHARACTERISTICS

Boiling Point: **Sublimes at 1140°F (614°C)**	Specific Gravity (H_2O = 1): **5.73 (metal)**
Vapor Pressure (mm Hg): **0 at 69°F (20°C)**	Molecular Weight (atomic weight): **74.9**
Vapor Density (Air = 1): **Not Applicable (metal)**	Melting Point: **Sublimes at 1202°F (650°C)**

Solubility

Insoluble in water and in caustic and nonoxidizing acids. Soluble in HNO_3.

Appearance and Odor

Metallic arsenic is a steel-gray (silver-gray) brittle metal; arsenic trichloride is an oily liquid; arsenic trioxide is a crystalline solid that darkens in moist air.

SECTION IV - FIRE AND EXPLOSION HAZARD DATA

Flash Point (method used): **Not Applicable (Metal)**	Explosive Limits in Air % by Volume: **LEL: N/A UEL: N/A**
NFPA Classification: **Noncombustible Solid (bulk form)**	Autoignition Temperature: **Not Applicable**

Extinguishing Media

Use dry chemical, carbon dioxide, water spray, or foam extinguishers.

Special Fire Fighting Procedures

Wear full protective clothing and self-contained breathing apparatus (SCBA). Poisonous gases are produced in fire.

Unusual Fire and Explosion Hazards

In bulk form, a noncombustible solid. A slight explosion hazard in the forms of dust and vapor, when exposed to heat or flame. Poisonous gases are produced in fire.

SECTION V - REACTIVITY DATA

Stability		Conditions to Avoid
		Dusts are flammable when exposed to heat or flame or by chemical reaction with powerful oxidizers such as bromates, chlorates, iodates, peroxides, lithium, and nitric acid. Decomposes in moisture and sunlight.
Stable	Unstable	Incompatibility *(materials to avoid)*
X		Reacts vigorously on contact with oxidizers. Incompatible with bromine azide, dirubidium acetylide, halogens, palladium, zinc, platinum, Ncl3, AgNO3, CrO3, NaO2, hexafluro isopropylideneamino lithium.
Hazardous Polymerization		Conditions to Avoid Exposure to heat and contact with acid or acid fumes.
May Occur	Will Not Occur	Hazardous Decomposition or By-products
	X	Emits highly toxic fumes of arsine on contact with heat, flame, or acids.

SECTION VI - HEALTH HAZARD DATA

Primary Route(s) of Entry:	Inhalation?	Absorption (skin/eye)?	Ingestion?
	X	X	X

Health Hazards

INHALATION: Irritation of the upper respiratory tract. Conjunctivitis; visual disturbances, ulceration and perforation of nasal septum; pharyngitis. May cause cancer of the lung, larynx, lymphoid system, or viscera.

ABSORPTION: Epidermal carcinoma and arsenical dermatitis; toxic to liver, blood forming organs, the central and peripheral nervous systems, and the cardiovascular system.

INGESTION: Burning lips; throat constriction; dysphasia followed by severe stomach pain, nausea, and projectile vomiting. Convulsions, coma, death occurs within 24 hours (severe cases).

Carcinogenicity	NTP Listed?	IARC Cancer Review Group?	OSHA Regulated?	Target Organs?
Confirmed Human Confirmed Animal	5th Annual Report	Group 1	29 CFR 1910.1000 Table Z-1	Liver, kidneys, skin, lungs, lymphatic system.

Medical Conditions Generally Aggravated by Exposure

Skin conditions (dermatitis); respiratory problems; renal conditions and hepatic disorders.

Emergency and First-aid Procedures

Eye contact: Do not allow victim to keep eyes tightly shut. Flush immediately with water for 15 minutes (minimum), seek medical attention. Skin contact: Remove all contaminated clothing. Immediately wash area with large amounts of soap and water. For inhalation: Remove the person from exposure. Provide respiratory assistance and CPR. Transfer to medical facility. If swallowed, seek medical attention immediately. For severe cases, use BAL as antidote. For milder cases, use penicillamine (not penicillin).

SECTION VII - PRECAUTIONS FOR SAFE HANDLING AND USE

Steps to be Taken in Case Material is Released or Spilled

Collect powered materials in most convenient and safe manner and deposit in sealed containers. Ventilate area of spill or leak. Remove all ignition sources. Restrict those not involved in cleanup from entering area. Notify appropriate authorities, as required by applicable environmental regulations.

Preferred Waste Disposal Method

Incinerator equipped with scrubber and afterburner.

Precautions to be Taken in Handling and Storage

Arsenic is a noncombustible solid, but its dusts is a moderate explosion hazard when exposed to heat. Store to avoid contact with strong oxidizers and acid. Store in tightly closed containers in cool, well-ventilated area away from heat. Do not use near heat or flame.

Other Precautions and Warnings

Sources of ignition are prohibited where arsenic is used, handled, or stored.

SECTION VIII - CONTROL MEASURES AND PERSONAL PROTECTIVE EQUIPMENT

Respiratory Protection *(specify type)*

Arsenic is a human poison by inhalation. At any exposure level, use a MSHA/NIOSH approved self-contained breathing apparatus (SCBA) with full facepiece operated in positive pressure mode.

Ventilation

Local exhaust (preferred) at site of chemical work or general ventilation.

Protective Gloves	Eye Protection	Other Protective Clothing
Rubber	Chemical/Dust Goggles or Face Mask	Protective Apron, Boots

Work/Hygiene Practices

Always wash hands thoroughly after using chemical; never bring food, drink, or smoking materials into vicinity of chemicals.

ARSENIC

As CAS: 7440-38-2

IDENTIFICATION AND TYPICAL USES

Metallic arsenic is a silver-gray crystalline solid that darkens in moist air. It also exists in black and yellow amorphous forms. Arsenic trichloride is an oily liquid. Arsenic trioxide is a crystalline solid. It is used for hardening metals such as copper and lead, battery grids, cable sheaths, boiler tubes, and as a doping agent in solid-state products of silicon and germanium, special solders, and medicine. Its salts are used in making herbicides, rodenticides, in semiconductors, and in pyrotechniques.

RISK ASSESSMENT: HEALTH

General Assessment

All arsenic compounds are toxic. Arsenic compounds are irritants of the skin, mucous membranes, and eyes. Arsenic is absorbed into the body through a gastrointestinal route (*ingestion*), *absorption* through the skin, and through *inhalation*. Acute arsenic poisoning is rare in the occupational setting and results primarily from ingestion of contaminated foods. Symptoms include burning of the lips, constriction of the throat, and dysphasia followed by excruciating abdominal pain, sever nausea, projectile vomiting, and profuse diarrhea. Other toxic effects are on the liver, the blood-forming organs, the central and peripheral nervous systems, and the cardiovascular system. Convulsions, coma, and death follow with 24 hours of severe exposures.

Inhalation can irritate the upper respiratory tract leading to nasal perforations. Other symptoms of arsenic poisoning can include weakness, conjunctivitis, visual disturbances, pharyngitis, pulmonary irritation, peripheral neuropathy, hyperpigmentation of the skin, pulmar and plantar hyperkeratosis, dermatitis, and skin cancer.

Arsenic is a confirmed human carcinogen producing liver tumors. It may also cause cancer of the lungs, larynx, lymphoid system, or viscera. There are reported teratogenic effects, including exencephaly, skeletal defects, and genitourinary system defects.

Acute Health Effects

The following acute (short-term) health effects may occur immediately or shortly after exposure to arsenic:

Skin: Irritation, burning, itching, and rash. Contact dermatitis is possible.

Eye: Red, watery eyes and irritation. Conjunctivitis may occur in some people.

Lung: Nose and throat irritation and possible nasal perforation.

CVS: Can affect the rhythm of the heart and cause abnormal EKG at high exposures.

CNS: Can cause nausea, poor appetite, vomiting, and muscle cramps, cognitive impairments, and peripheral neuropathy.

Chronic Health Effects

The following chronic (long-term) health effects may occur at some time after exposure to arsenic and can last for months or even years:

Cancer Hazards: Arsenic is a carcinogen in humans. It has been shown to cause skin, lung, and liver cancer.

Reproductive Hazard: There is reported evidence that arsenic causes teratogenic effects (fetal damage) in laboratory test animals.

Other Chronic Effects: Repeated exposure can cause nerve damage with "pins and needles" effect, burning, numbness, and later weakness of arms and legs. Repeated skin contact can cause thickened skin and/or patchy areas of darkening and loss of pigment. Some persons develop white lines on the nails. Repeated exposures can also damage the liver, cause narrowing of the blood vessels, or interfere with the bone marrow's ability to make red blood cells.

ⓘ *Recommended Risk-Reduction Measures*

Personnel should avoid direct contact with arsenic, especially its dusts or powders. If a less toxic material or compound can not be substituted for arsenic, then *engineering controls* are the most effective method of reducing exposures. The best protection is to enclose operations and/or provide local exhaust ventilation at the site of arsenic dust release. While not always operationally feasible, isolating operations can also reduce exposure. Using respiratory protection is less effective than the controls mentioned above, but is still advisable whenever working with or around arsenic. At any exposure level, a self-contained breathing apparatus (SCBA) with full facepiece and pressure demand is recommended for the greatest possible respiratory protection. If a full facepiece is not available, then chemical/dust goggles should be worn to protect the eyes. A face shield and protective apron should also be worn. To prevent hand and skin exposures, gloves should be worn.

Administrative controls should also be in place to minimize the potential for human exposures. These may include written procedures or policies which specify the methods and techniques that will be practiced whenever personnel are to work with arsenic.

All personnel should receive *training* on the use, hazards, protective measures, emergency actions, and other precautions per 29 CFR 1910.1200 (Hazard Communication), prior to the first assignment in an area where arsenic is used or stored.

Before first exposure and every 6 to 12 months thereafter, a medical history and examination is recommended, including the following:

- ☑ Examination of the nose (nasal septum), skin, eyes, nails, and nervous system.
- ☑ Test for urine arsenic (may not be accurate within 2 days of eating shellfish or fish; most accurate at the end of a work shift) should not be greater than 100 micrograms per gram creatinine in the urine.

After suspected overexposure, repeat these tests and consider complete blood count (CBC) and liver function tests. Also examine skin periodically for abnormal growths (skin cancer from arsenic can be easily cured if detected early).

Any evaluation should include a careful history of past and present symptoms with an examination. Medical tests that simply look for existing damage are not a substitute for controlling exposures. Use of arsenic near acid or acid mist can cause the release of the deadly gas, arsine. Therefore, such potential for "mixed exposures" warrants additional caution when working with or around arsenic.

The antidote for severe poisoning is to administer BAL. For milder poisoning, use penicillamine (not penicillin) has been used. Side effects occur with such treatment and it is never a substitute for controlling exposure. It can only be done under strict medical care.

Other methods to reduce exposure include:

- ☑ Where possible, enclose operations and use local exhaust ventilation at the site of chemical release. If local exhaust ventilation or enclosure is not used, respiratory protection should be mandatory.
- ☑ Always ensure that proper protective clothing is worn when using chemical substances.
- ☑ Wash thoroughly immediately after exposure to arsenic and at the end of the work shift or before eating, drinking, or smoking.
- ☑ Hazard warning information should be posted in the work area. In addition, as part of an on-going education and training program, all information on the health and safety hazards of arsenic should be communicated to all potentially exposed workers.
- ☑ Eye wash stations should be provided in the immediate work area for emergency use. If there is a possibility of skin exposure to arsenic, emergency shower facilities should also be provided.

RISK ASSESSMENT: ENVIRONMENT

General Assessment

The environment is at risk of exposure during transportation, storage, disposal, or destruction of arsenic. In almost every scenario, the threat of environmental exposure is contingent upon the proper handling of the chemical substance. Accidental spills, large or small, where arsenic contacts incompatible commodities can result in fire, explosion (depending upon conditions of spill), and possible contamination of the surrounding environmental mediums (water, soil, and air).

Arsenic is considered a noncombustible solid. However, in powder or dust form it can present a moderate fire and explosion hazard. In bulk form, it does not present a serious fire or explosion hazard. However, because it is incompatible with many other commodities, extreme caution is required in handling, storage, transportation, and disposal of arsenic. These characteristics also require special consideration during any emergency situation involving a leak or spill of arsenic powder or dust. Should arsenic ever come into contact with incompatible substances (such as oxidizers) either during use, transportation, storage, or disposal, the formation of highly toxic and/or reactive commodities is extremely possible.

Arsenic enters the environment mainly from its use as a pesticide and from emissions from coal-fueled power plants.

Acute Ecological Effects

Acute (short-term) toxic effects may include the death of animals, birds, or fish and death or low growth rate in plants. Acute effects are seen 2 to 4 days after animals or plants are exposed to arsenic.

Arsenic metabolism and effects are significantly influenced by the animal/plant tested, the route of administration, the physical and chemical form of the arsenical, and the dose. Inorganic arsenic compounds are more toxic than organic arsenic compounds.

Arsenic has high acute toxicity to aquatic life, birds, and land animals. Except where soil arsenic content is high (e.g., around smelters and where arsenic-based pesticides are known to have been used), arsenic does not accumulate in plants to toxic levels. Where soil content is high, growth of crops and crop yields can be decreased.

Chronic Ecological Effects

Chronic (long-term) toxic effects may include shortened life span, reproductive problems, lower fertility, and changes in appearance or behavior in exposed animals. These effects can be seen long after first exposure(s) to toxic chemicals.

Arsenic has high chronic toxicity to aquatic life, and moderate chronic toxicity to birds and land animals.

Water Solubility

Arsenic and its salts have low solubility in water. Concentrations of less than 1 milligram will mix with a liter of water.

Persistence in the Environment

Arsenic is highly persistent in water, with a half-life of longer than 200 days. The half-life of a pollutant is the amount of time it takes for one half of the chemical to be degraded.

Bioaccumulation in Aquatic Organisms

Some substances increase in concentration, or *bioaccumulate,* in living organisms as they breathe contaminated air, drink contaminated water, or eat contaminated food. These chemicals can become concentrated in the tissues and internal organs of animals as well as humans.

The concentration of arsenic found in fish tissues is expected to be somewhat higher than the average concentration of arsenic in the water from which the fish was taken.

Recommended Risk-Reduction Measures

Proper training of all transporters will reduce the likelihood of a mishap or accident resulting in a leak or spill of arsenic dusts or powders into the environment. The correct use of DOT labeling (per the MSDS) on all containers, trucks, and rail cars will enable emergency responders to react properly and quickly to any disaster thereby reducing the potential risk to the environment and to personnel.

Storage of arsenic should be segregated from incompatible chemicals to minimize the risk of cross-contamination or contact.

If a spill or leak to the environment has occurred, fire department, emergency response and/or hazardous materials spill personnel should be notified immediately. Cleanup should be attempted only by those trained in proper spill containment procedures. Contaminated soils should be removed for incineration and replaced with clean soil. If arsenic should contact the water table, aquifer, or navigable waterway, time is of the essence since arsenic is so highly soluble. The local and/or state emergency response authorities must be notified. A comprehensive emergency response or disaster preparedness/recovery plan should be in place prior to any operations involving the use, transportation, storage, or disposal of arsenic. If arsenic is spilled or leaked, the following specific steps are recommended:

- ☑ Restrict persons not wearing protective clothing from area of spill or leak until cleanup is complete.
- ☑ Collect powdered materials in the safest manner possible and deposit in sealed containers.
- ☑ It may be necessary to dispose of arsenic as a hazardous waste. The responsible state agency or the regional office of the federal Environmental

Protection Agency (EPA) should be contacted for specific recommendations.

RISK ASSESSMENT: BUSINESS

General Assessment

Accidents or mishaps involving arsenic can present a significant threat to business operations. The loss or damage of equipment or facilities can significantly affect fiscal viability. Lawsuits that may result from personnel injury/death, public exposures, and/or environmental contamination will also require a serious expenditure of resources. Media attention surrounding an injury, death, or environmental damage can also result in a loss of profits and loss of current as well as future business. Always remember that anytime the terms "cancer," "carcinogen," or "reproductive hazard" are used, public hysteria, emotion, and ignorance can run equally high. This must be carefully considered whenever developing or implementing public relations policies.

Recommended Risk-Reduction Measures

Company attorneys, safety and health professionals, and environmental specialists should be involved in the development of any procedures dealing with safety or emergency response. A company official should be pre-designated as a public relations officer with specific training in dealing with the press. Corporate plans and policies should be developed, approved, and implemented long before any need for such arises.

REFERENCES

American Conference of Governmental Industrial Hygienists. 1988. *Documentation of the Threshold Limit Values and Biological Exposure Indices,* 5th Edition (with updates). Cincinnati: ACGIH

Bloom, A. D. and F. J. DeSerres. 1995. *Ecotoxicity and Human Health.* New York: CRC Press/Lewis Publishers.

Brusick, D. J. 1994. *Methods for Genetic Risk Assessment.* New York: CRC Press/Lewis Publishers.

Bueche, J. R. 1972. *Principles of Physics,* 2nd Edition. New York: McGraw-Hill.

Cockerman, L. G., B. S. Shane. 1994. *Basic Environmental Toxicology.* New York: CRC Press/Lewis Publishers.

Cothern, R. R. 1993. *Comparative Environmental Risk Assessment.* New York: CRC Press/Lewis Publishers.

Hallenbeck, W. H. 1993. *Quantitative Risk Assessment for Environmental and Occupational Health,* 2nd Edition. New York: CRC Press/Lewis Publishers.

Keith, L. H. and M. M. Walker. 1995. *Handbook of Air Toxics.* New York: CRC Press/Lewis Publishers.

Landis, W. G., M. H. Yu. 1995. *Introduction to Environmental Toxicology.* New York: CRC Press/Lewis Publishers.

Langerman, N. 1994 *Precautionary Labels for Chemical Containers.* New York: CRC Press/Lewis Publishers.

Lauwerys, R. R., P. Hoet. 1993. *Industrial Chemical Exposure.* New York: CRC Press/Lewis Publishers.

Lewis, R. J., Sr. 1993. *Hawley's Condensed Chemical Dictionary,* 12th Edition. New York: Van Nostrand Reinhold.

Lewis, R. J., Sr. 1992. *Sax's Dangerous Properties of Industrial Materials,* Volumes 1, 2, and 3, 8th Edition. New York: Van Nostrand Reinhold.

Lide, D. R. 1995. *Handbook of Chemistry and Physics,* 76th Edition. New York: CRC Press/Lewis Publishers.

Manahan, S. E. 1994. *Environmental Chemistry,* 6th Edition. New York: CRC Press/Lewis Publishers.

Mellström, G. A., J. E. Wahlberg, and H. I. Maibach. 1994. *Protective Gloves for Occupational Use.* New York: CRC Press/Lewis Publishers.

Meyer, E. 1990. *Chemistry of Hazardous Materials.* Englewood Cliffs, NJ: Prentice Hall.

Mycock, J. C., J. D. McKenna, and L. Theodore. 1995. *Handbook of Air Pollution Control Engineering and Technology.* New York: CRC Press/Lewis Publishers.

Patnaik, P. 1992. *A Comprehensive Guide to the Hazardous Properties of Chemical Substances.* New York: Van Nostrand Reinhold.

Proctor, N. H., J. P. Hughes. 1991. *Chemical Hazards of the Workplace.* New York: Van Nostrand Reinhold.

Quigley, D. R. 1994. *Handbook of Emergency Chemical Management.* New York: CRC Press/Lewis Publishers.

Rea, W. J. 1995. *Chemical Sensitivity,* Vol. I-III. New York: CRC Press/Lewis Publishers.

Richardson, M. 1995. *Dictionary of Substance and Their Effects,* Vol. I-VI. New York: CRC Press/Lewis Publishers.

Sacarello, H. L. A. 1994. *The Comprehensive Handbook of Hazardous Materials.* New York: CRC Press/Lewis Publishers.

Sheftel, V. O. 1995. *Handbook of Toxic Properties of Monomers and Additives.* New York: CRC Press/Lewis Publishers.

Talbott, E., G. Craun. 1995. *An Introduction to Environmental Epidemiology.* New York: CRC Press/Lewis Publishers.

Torkelson, T. R. V. K. Rowe. 1981. *Patty's Industrial Hygiene and Toxicology,* Vol. 2B. New York: Wiley-Interscience.

U.S. Department of Health, Education, and Welfare. National Institute for Occupational Safety and Health (NIOSH). 1975. *Criteria for a Recommended Standard. Occupational Exposure to Inorganic Arsenic-New Criteria.* DHEW (NIOSH) Publication Number: 75-149. Washington, DC: U.S. Government Printing Office.

U.S. Environmental Protection Agency (EPA). March 1984. *Health Assessment Document for Inorganic Arsenic: Final Report.* North Carolina: Research Triangle Park.

MATERIAL SAFETY DATA SHEET

CHEMICAL NAME	
ARSINE	**CRC PRESS/LEWIS PUBLISHERS DISCLAIMER:** THE INFORMATION AND RECOMMENDATIONS PRESENTED HEREIN ARE BASED ON SOURCES BELIEVED TO BE RELIABLE. CRC MAKES NO REPRESENTATION ON ITS COMPLETENESS OR ACCURACY. IT IS THE RESPONSIBILITY OF THE USER TO DETERMINE THE CHEMICAL'S SUITABILITY FOR ITS INTENDED USE, THE CHEMICAL'S SAFE USE, AND THE CHEMICAL'S PROPER DISPOSAL. NO REPRESENTATIONS AND/OR WARRANTIES, EITHER EXPRESSED OR IMPLIED, OF THE MERCHANTABILITY OR FITNESS FOR A PARTICULAR PURPOSE, OR OF ANY OTHER NATURE, ARE MADE WITH RESPECT TO THE INFORMATION PROVIDED IN THIS MSDS OR TO THE CHEMICAL TO WHICH INFORMATION MAY REFER. CRC NEITHER ASSUMES NOR AUTHORIZES ANY OTHER PERSON TO ASSUME FOR IT, ANY OTHER ADDITIONAL RESPONSIBILITY OR LIABILITY FOR THE USE OF, OR RELIANCE UPON, THIS INFORMATION.

HAZARD WARNING INFORMATION

HEALTH	FIRE	REACTIVE	OTHER	DEGREE OF HAZARD	COLOR CODING	OTHER CODES
4	4	2		0 = Minimum Hazard 1 = Slight Hazard 2 = Moderate Hazard 3 = Serious Hazard 4 = Severe Hazard	HEALTH = BLUE FIRE = RED REACTIVITY = YELLOW OTHER = WHITE	OX = Oxidizer ACID = Acid ALK = Alkali COR = Corrosive W = Use No Water

SECTION I - GENERAL INFORMATION

Characterization	Hydride	RCRA Number	Not Listed
		EPA Class	Not Listed
DOT Proper Shipping Name	Arsine	Chemical Abstract Service (CAS) Number	7784-42-1
DOT Hazard Class and Label Requirements	Poison A, Flammable Gas	DOT Emergency Guide Code	18
DOT Identification Number	UN 2188	Atomic Symbol	AsH_3

Synonyms

Arsenic hydride; arseniuretted hydrogen; arsenous hydride; hydrogen arsenide; arsenic trihydride.

SECTION II - HAZARDOUS INGREDIENTS AND IDENTITY INFORMATION

Hazardous Components (specific identity)	OSHA Exposure Criteria	NIOSH Exposure Criteria	Immediately Dangerous to Life and Health (IDLH)	ACGIH Exposure Criteria
Arsine (derivation from: Reaction of aluminum arsenide with water or hydrochloric acid; electrochemical reduction of arsenic compounds and arsenic solutions; produced when arsenic is exposed to heat or acid). **1 ppm = 3.24 mg/m³**	**PEL:** 0.05 ppm, 0.2 mg/m³ **STEL:** Not Established	**REL (ceiling):** 0.002 mg/m³ (15 minute) Confirmed Human Carcinogen	3 ppm	**TLV:** 0.05 ppm, 0.2 mg/m³ **STEL:** Not Established

SECTION III - PHYSICAL AND CHEMICAL CHARACTERISTICS

Boiling Point	-81°F (-62.5°C)	Specific Gravity (H_2O = 1)	2.69
Vapor Pressure (atmospheres)	14.9 at 69°F (20°C)	Molecular Weight (atomic weight)	78.0
Vapor Density (Air = 1)	2.66	Freezing Point	-179°F (-117°C)

Solubility

Moderately soluble in water (20%). 28 mg/100 at 20°C. Soluble in benzene and chloroform.

Appearance and Odor

Colorless gas, heavier than air, with a mild, disagreeable, garlic-like odor. Normally shipped as a liquefied compressed gas.

SECTION IV - FIRE AND EXPLOSION HAZARD DATA

Flash Point (method used)	-113°C, closed cup	Explosive Limits in Air % by Volume	LEL: 5.1% UEL: 78%
NFPA Classification	Flammable Gas	Autoignition Temperature	Not Found

Extinguishing Media

Use dry chemical, carbon dioxide, water spray, or foam extinguishers.

Special Fire Fighting Procedures

Warning properties of exposure to hazardous concentrations are inadequate. Wear full protective clothing and self-contained breathing apparatus (SCBA). Poisonous gases are produced in fire.

Unusual Fire and Explosion Hazards

Flammable gas. Moderately explosive when exposed to chlorine or nitric acid. Shock-sensitive (can ignite or explode). Can form explosive mixture with air. When heated, arsine can ignite at once or liberate explosive hydrogen.

SECTION V - REACTIVITY DATA

Stability		Conditions to Avoid Avoid contact with incompatible materials. Do not store near heat source. Avoid exposure to powerful shock since arsine is shock-sensitive and may explode under such conditions.
Stable	Unstable	Incompatibility *(materials to avoid)*
X		Reacts violently on contact with strong oxidizers, chlorine, and nitric acid.
Hazardous Polymerization		Conditions to Avoid When heated, arsine can ignite at once or liberate explosive hydrogen. Can react with moisture or acids to evolve heat and hydrogen. Often enough heat is liberated to cause ignition and/or explosion.
May Occur	Will Not Occur	Hazardous Decomposition or By-products
	X	When heated to decomposition, emits highly toxic fumes of arsenic.

SECTION VI - HEALTH HAZARD DATA

Primary Route(s) of Entry:	Inhalation?	Absorption (skin/eye)?	Ingestion?
	X		

Health Hazards

INHALATION: Headache, dizziness, weakness, vomiting, abdominal pain, and dyspnea. In severe cases of acute exposure, death can occur as a result of renal failure and pulmonary edema. Prolonged exposures can lead to jaundice, hemoglobinuria, hemolysis, and chronic renal damage. Arsine is a cancer-causing gas. There is sufficient evidence of its carcinogenicity in humans. Other symptoms of exposure can include malaise, nausea, and the production of dark red urine 4 to 6 hours following exposure. This often progresses to a brown urine, with jaundice appearing 24 to 48 hours after exposure. An unusual bronze skin color has been noted in some cases.

Carcinogenicity	NTP Listed?	IARC Cancer Review Group?	OSHA Regulated?	Target Organs?
Confirmed Human Confirmed Animal	5th Annual Report	Group 1	29 CFR 1910.1000 Table Z-1	Blood, kidney, liver (lung and lymphatic cancer)

Medical Conditions Generally Aggravated by Exposure

Respiratory problems; renal conditions and hepatic disorders; blood disorders (lupus, leukemia, etc.).

Emergency and First-aid Procedures

Eye contact: Frostbite can occur. Do not allow victim to keep eyes tightly shut. Flush immediately with water for 15 minutes (minimum), seek medical attention. Skin contact: Frostbite can occur. Remove all contaminated clothing. Immediately wash area with large amounts of soap and water. For inhalation: Remove the person from exposure. Provide respiratory assistance and CPR. Transfer to medical facility. Observation recommended for 24-48 hours (pulmonary edema possible)

SECTION VII - PRECAUTIONS FOR SAFE HANDLING AND USE

Steps to be Taken in Case Material is Released or Spilled

If liquefied compressed gas is leaked, take measures to stop leak. Remove ignition sources. Ventilate area of spill or leak. Restrict those not involved in cleanup from entering area. Notify appropriate authorities, as required by applicable environmental regulations.

Preferred Waste Disposal Method

Oxidation with aqueous copper sulfate under nitrogen atmosphere.

Precautions to be Taken in Handling and Storage

Arsine is a flammable gas. Store to avoid contact with strong oxidizers and acid. Store in tightly closed containers in cool, well-ventilated area away from heat. Do not use near heat or flame.

Other Precautions and Warnings

Sources of ignition are prohibited where arsine is used, handled, or stored.

SECTION VIII - CONTROL MEASURES AND PERSONAL PROTECTIVE EQUIPMENT

Respiratory Protection *(specify type)*

Arsine is a poison by inhalation. At any exposure level, use a powered air purifying respirator or a self-contained breathing apparatus (SCBA) with full facepiece operated in positive pressure mode.

Ventilation

Local exhaust (preferred) at site of chemical work or general ventilation.

Protective Gloves	Eye Protection	Other Protective Clothing
Neoprene or Butyl Rubber	Chemical Goggles or Face Mask	Not Applicable

Work/Hygiene Practices

Always wash hands thoroughly after using chemical; never bring food, drink, or smoking materials into vicinity of chemicals.

ARSINE

AsH_3 CAS: 7440-38-2

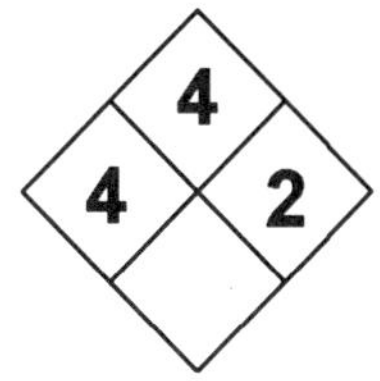

IDENTIFICATION AND TYPICAL USES

Arsine is a colorless gas with a mild but disagreeable garlic-like odor. It is used in organic synthesis, as a doping agent in the electronics industry to manufacture gallium arsenide and gallium arsenide phosphide for semiconductors and other such solid-state electronic equipment. It also has been used as a military poison gas.

RISK ASSESSMENT: HEALTH

General Assessment

Arsine is a highly toxic hydride. It is the most acutely toxic form of arsenic. It is particularly dangerous because of its volatility and ease of entry into the body. It is a confirmed human carcinogen and a poison by *inhalation*. Human blood cells, the gastrointestinal tract, the central nervous system, the liver, the kidneys, and the lungs are all affected by inhalation exposures. Symptoms of exposure include headache, weakness, malaise, dizziness, vomiting, abdominal pain, and dyspnea. Arsenic is excreted in the urine following exposure. Dark red urine may occur 4 to 6 hours after exposure. This often progresses to dark brown urine, with jaundice appearing 24 to 48 hours after exposure. In severe cases of acute exposure, death can occur due to renal failure or pulmonary edema (or a combination of both). Chronic poisoning can lead to jaundice, hemoglobinuria, hemolysis, and renal damage. The target organs are the blood, kidneys, and liver (although lung and lymphatic cancers can also occur). An unusual brown skin color has been noted in some patients. Symptoms of arsenic poisoning, in addition to those of arsine may be present. There may also be mild to severe anemia and peripheral neuropathy. The later resulting in numbness of the hands and feet, severe muscle weakness, and photophobia.

Inhaled arsine is oxidized to form trivalent arsenic (As^{+3}) and arsenous oxide (As_2O_3), which are both human carcinogens. Arsine is a non-irritating gas when inhaled. Although its garlic-like odor may be a warning of exposure, it is inadequate to warn of exposure to hazardous levels.

Acute Health Effects

The following acute (short-term) health effects may occur immediately or shortly after exposure to arsine:

Skin: Frostbite may occur upon exposure to compressed liquefied gas.

Eye: Frostbite and severe irritation on exposure to compressed liquefied gas.

Lung: May cause severe irritation of the respiratory tract resulting in pulmonary edema (a medical emergency) which can lead to death.

CNS: Can cause headache, malaise, weakness, vomiting, dizziness, and dyspnea.

Other: Abdominal pain, hematuria, jaundice, oliguria, anuria, and anemia.

Chronic Health Effects

The following chronic (long-term) health effects may occur at some time after exposure to arsine and can last for months or even years:

Cancer Hazards: Arsine is a carcinogen in humans. It has been shown to cause carcinogenic effects on the blood, liver, kidney, lung, and lymphatic system.

Reproductive Hazard: There is insufficient data in the references that arsine presents a reproductive hazard.

Other Chronic Effects: Chronic poisoning can lead to jaundice, hemoglobinuria, hemolysis, and renal (kidney) injury or damage.

Recommended Risk-Reduction Measures

Personnel should avoid direct contact with arsine. Although it is a gas, it is often shipped as a compressed liquefied gas. Therefore, frostbite can occur upon exposure. If a less toxic material cannot be substituted for arsine, then *engineering controls* are the most effective method of reducing exposures. The best protection is to enclose operations and/or provide local exhaust ventilation at the site of arsine gas release. While not always operationally feasible, isolating operations can also reduce exposure.

Using respiratory protection is less effective than the controls mentioned above, but is still advisable whenever working with or around arsine. At any exposure level, a self-contained breathing apparatus (SCBA) with full facepiece and pressure demand is recommended for the greatest possible respiratory protection. If a full facepiece is not available, then chemical goggles or face shield should be worn to protect the eyes. To prevent hand and skin exposures (frostbite), gloves should be worn.

Administrative controls should also be in place to minimize the potential for human exposures. These may include written procedures or policies which specify the methods and techniques that will be practiced whenever personnel are to work with arsine.

All personnel should receive *training* on the use, hazards, protective measures, emergency actions, and other precautions per 29 CFR 1910.1200 (Hazard Communication), prior to the first assignment in an area where arsine is used or stored.

Before first exposure and every 6 to 12 months thereafter, a medical history and examination is recommended, including the following:

- ☑ Pulmonary function test (baseline).
- ☑ Complete blood count (CBC) for baseline of plasma hemoglobin and white cell count.
- ☑ Urinalysis to test for urine arsine and/or arsenic (may not be accurate within 2 days of eating shellfish or fish; most accurate at the end of a work shift) should not be greater than 100 micrograms per gram creatinine in the urine. Also check urinary concentration of hemoglobin.

After suspected overexposure, repeat these tests and consider the following additional medical protocols:

- ☑ Liver function tests.
- ☑ Examine skin periodically for abnormal coloring (jaundice).

Any evaluation should include a careful history of past and present symptoms with an examination. Medical tests that simply look for existing damage are not a substitute for controlling exposures. Use of arsine near acid, oxidizers, or if heated can cause the release of arsenic fumes. Therefore, such potential for "mixed exposures" warrants additional caution when working with or around arsine.

The treatment of choice for acute and severe arsine poisoning is exchange transfusion and, if renal failure develops, hemodialysis. Dimercaprol (BAL), which is an antidote for arsenic poisoning is of no use in arsine poisoning because it affords no protection against hemolysis.

Other methods to reduce exposure to arsine include:

- ☑ Where possible, enclose operations and use local exhaust ventilation at the site of chemical release. If local exhaust ventilation or enclosure is not used, respiratory protection should be mandatory.
- ☑ Always ensure that proper protective clothing is worn when using chemical substances. Never allow contaminated clothing to be taken home from the work place. Clothing that has been contaminated should be laundered only by persons who have been trained on exposure hazards.
- ☑ Hazard warning information should be posted in the work area. In addition, as part of an on-going education and training program, all information on the health and safety hazards of arsine should be communicated to all potentially exposed workers.
- ☑ Eye wash stations should be provided in the immediate work area for emergency use. If there is a possibility of skin exposure to compressed liquefied arsine, emergency shower facilities should also be provided.

RISK ASSESSMENT: ENVIRONMENT

General Assessment

The environment is at risk of exposure during transportation, storage, disposal, or destruction of arsine. In almost every scenario, the threat of environmental exposure is contingent upon the proper handling of the chemical substance. Accidental spills, large or small, where arsine contacts incompatible commodities can result in fire, explosion (depending upon conditions of spill), and possible limited contamination of the surrounding environmental mediums (water and air).

Arsine is considered a flammable gas. It can present a moderate fire and explosion hazard. In compressed liquefied form as well as in gas form, it is shock-sensitive and can explode under powerful shock conditions that may occur during transport.

Also, because it is incompatible acids and oxidizers, extreme caution is required in handling, storage, transportation, and disposal of arsine. These characteristics also require special consideration during any emergency situation involving a leak or spill of arsine gas. Should arsine ever come into contact with incompatible substances (such as oxidizers) either during use, transportation, storage, or disposal, the formation of highly toxic commodities is extremely possible.

Arsine enters the environment mainly through accidents and industrial discharges.

Acute Ecological Effects

Acute (short-term) toxic effects may include the death of animals, birds, or fish and death or low growth rate in plants. Acute effects are seen 2 to 4 days after animals or plants are exposed to arsine.

As a gas, arsine presents little toxicity to terrestrial or aquatic plants. As a liquefied compressed gas, it can cause destruction on contact with plant life.

Arsine has minimal acute toxicity to aquatic life, birds, and land animals.

Chronic Ecological Effects

Chronic (long-term) toxic effects may include shortened life span, reproductive problems, lower fertility, and changes in appearance or behavior in exposed animals. These effects can be seen long after first exposure(s) to toxic chemicals.

There is insufficient data to establish whether arsine poses chronic toxicity to aquatic life, birds, or land animals.

Water Solubility

Arsine has slight solubility in water. Concentrations of 2 to 20 milligrams will mix with a liter of water.

Persistence in the Environment

As a gas, there is no evidence in the references to show that arsine has any measurable persistence in the environment.

Bioaccumulation in Aquatic Organisms

Some substances increase in concentration, or *bioaccumulate,* in living organisms as they breathe contaminated air, drink contaminated water, or eat contaminated food. These chemicals can become concentrated in the tissues and internal organs of animals as well as humans.

The concentration of arsine found in fish tissues is expected to be much lower than the average concentration of arsine in the water from which the fish was taken.

Recommended Risk-Reduction Measures

Proper training of all transporters will reduce the likelihood of a mishap or accident resulting in a leak or spill of arsine into the environment. The correct use of DOT labeling (per the MSDS) on all containers, trucks, and rail cars will enable emergency responders to react properly and quickly to any disaster thereby reducing the potential risk to the environment and to personnel.

Storage of arsine should be segregated from incompatible chemicals to minimize the risk of cross-contamination or contact.

If a spill or leak to the environment has occurred, fire department, emergency response and/or hazardous materials spill personnel should be notified immediately. Cleanup should be attempted only by those trained in proper spill containment procedures. If arsine should contact the water table, aquifer, or navigable waterway, the local and/or state emergency response authorities must be notified. A comprehensive emergency response or disaster preparedness/recovery plan should be in place prior to any operations involving the use, transportation, storage, or disposal of arsine.

Arsine may be destroyed by oxidation of aqueous copper sulfate under a nitrogen atmosphere. The reaction is carried out in a three-necked flask equipped with a stirrer, a gas inlet, and a nitrogen inlet.

If arsine gas is spilled or leaked, the following specific steps are recommended:

- ☑ Restrict persons not wearing protective clothing from area of spill or leak until cleanup is complete and area can be opened for normal work.
- ☑ If possible, remove leaking containers to a ventilated area and attempt to seal leak. Exercise extreme caution, arsine gas is shock-sensitive. If leak cannot be sealed, allow to vent until empty.
- ☑ Remove all sources of ignition and do not allow exposure to incompatible materials.
- ☑ Ventilate area of spill or leak.

RISK ASSESSMENT: BUSINESS

General Assessment

Accidents or mishaps involving arsine can present a significant threat to business operations. The loss or damage of equipment or facilities can significantly affect fiscal viability. Lawsuits that may result from personnel injury/death, public exposures, and/or environmental contamination will also require a serious expenditure of resources. Media attention surrounding an injury, death, or environmental damage can also result in a loss of profits and loss of current as well as future business.

Recommended Risk-Reduction Measures

Company attorneys, safety and health professionals, and environmental specialists should be involved in the development of any procedures dealing with safety, or emergency response. A company official should be pre-designated as a public relations officer with specific training in dealing with the press. Corporate plans and policies should be developed, approved, and implemented long before any need for such arises.

REFERENCES

American Conference of Governmental Industrial Hygienists. 1988. *Documentation of the Threshold Limit Values and Biological Exposure Indices,* 5th Edition (with updates). Cincinnati: ACGIH

Bloom, A. D. and F. J. DeSerres. 1995. *Ecotoxicity and Human Health.* New York: CRC Press/Lewis Publishers.

Brusick, D. J. 1994. *Methods for Genetic Risk Assessment.* New York: CRC Press/Lewis Publishers.

Bueche, J. R. 1972. *Principles of Physics,* 2nd Edition. New York: McGraw-Hill.

Calabrese, E. J., E. M. Kenyon. *Air Toxics and Risk Assessment.* New York: CRC Press/Lewis Publishers.

Cockerman, L. G., B. S. Shane. 1994. *Basic Environmental Toxicology.* New York: CRC Press/Lewis Publishers.

Cothern, R. R. 1993. *Comparative Environmental Risk Assessment.* New York: CRC Press/Lewis Publishers.

Gardner, R. W. 1994. *Chemical Intolerance.* New York: CRC Press/Lewis Publishers.

Hallenbeck, W. H. 1993. *Quantitative Risk Assessment for Environmental and Occupational Health,* 2nd Edition. New York: CRC Press/Lewis Publishers.

Hesdorffer, C. S., et. al. 1986. Arsine gas poisoning: The importance of exchange transfusion in severe cases. *British Journal of Industrial Medicine.* 43:353-355.

Howard, P. H., et. al. 1991. *Handbook of Environmental Degradation Rates.* New York: CRC Press/Lewis Publishers.

Jones, F. E. 1994. *Toxic Organic Vapors in the Workplace.* New York: CRC Press/Lewis Publishers.

Keith, L. H. and M. M. Walker. 1995. *Handbook of Air Toxics.* New York: CRC Press/Lewis Publishers.

Landis, W. G., M. H. Yu. 1995. *Introduction to Environmental Toxicology.* New York: CRC Press/Lewis Publishers.

Langerman, N. 1994 *Precautionary Labels for Chemical Containers.* New York: CRC Press/Lewis Publishers.

Lewis, R. J., Sr. 1993. *Hawley's Condensed Chemical Dictionary,* 12th Edition. New York: Van Nostrand Reinhold.

Lewis, R. J., Sr. 1992. *Sax's Dangerous Properties of Industrial Materials,* Volumes 1, 2, and 3, 8th Edition. New York: Van Nostrand Reinhold.

Lide, D. R. 1995. *Handbook of Chemistry and Physics,* 76th Edition. New York: CRC Press/Lewis Publishers.

Manahan, S. E. 1994. *Environmental Chemistry,* 6th Edition. New York: CRC Press/Lewis Publishers.

Mellström, G. A., J. E. Wahlberg, and H. I. Maibach. 1994. *Protective Gloves for Occupational Use.* New York: CRC Press/Lewis Publishers.

Meyer, E. 1990. *Chemistry of Hazardous Materials.* Englewood Cliffs, NJ: Prentice Hall.

Mycock, J. C., J. D. McKenna, and L. Theodore. 1995. *Handbook of Air Pollution Control Engineering and Technology.* New York: CRC Press/Lewis Publishers.

Patnaik, P. 1992. *A Comprehensive Guide to the Hazardous Properties of Chemical Substances.* New York: Van Nostrand Reinhold.

Pinto, S. S. 1976. Arsine poisoning: Evaluation of the acute phase. *Journal of Occupation Medicine.* 18:633-635.

Proctor, N. H., J. P. Hughes. 1991. *Chemical Hazards of the Workplace.* New York: Van Nostrand Reinhold.

Quigley, D. R. 1994. *Handbook of Emergency Chemical Management.* New York: CRC Press/Lewis Publishers.

Rea, W. J. 1995. *Chemical Sensitivity,* Vol. I-III. New York: CRC Press/Lewis Publishers.

Richardson, M. 1995. *Dictionary of Substance and Their Effects,* Vol. I-VI. New York: CRC Press/Lewis Publishers.

Sacarello, H. L. A. 1994. *The Comprehensive Handbook of Hazardous Materials.* New York: CRC Press/Lewis Publishers.

Sheftel, V. O. 1995. *Handbook of Toxic Properties of Monomers and Additives.* New York: CRC Press/Lewis Publishers.

Talbott, E., G. Craun. 1995. *An Introduction to Environmental Epidemiology.* New York: CRC Press/Lewis Publishers.

Torkelson, T. R. V. K. Rowe. 1981. *Patty's Industrial Hygiene and Toxicology,* Vol. 2B. New York: Wiley-Interscience.

U.S. Department of Health, Education, and Welfare. National Institute for Occupational Safety and Health (NIOSH). 1979. *Current Intelligence Bulletin 32: Arsine (arsenic Hydride) Poisoning in the Work Place.* DHEW (NIOSH) Publication Number: 79-142. Cincinnati, OH: NIOSH.

MATERIAL SAFETY DATA SHEET

CHEMICAL NAME	
ASBESTOS	**CRC PRESS/LEWIS PUBLISHERS DISCLAIMER;** THE INFORMATION AND RECOMMENDATIONS PRESENTED HEREIN ARE BASED ON SOURCES BELIEVED TO BE RELIABLE. CRC MAKES NO REPRESENTATION ON ITS COMPLETENESS OR ACCURACY. IT IS THE RESPONSIBILITY OF THE USER TO DETERMINE THE CHEMICAL'S SUITABILITY FOR ITS INTENDED USE, THE CHEMICAL'S SAFE USE, AND THE CHEMICAL'S PROPER DISPOSAL. NO REPRESENTATIONS AND/OR WARRANTIES, EITHER EXPRESSED OR IMPLIED, OF THE MERCHANTABILITY OR FITNESS FOR A PARTICULAR PURPOSE, OR OF ANY OTHER NATURE, ARE MADE WITH RESPECT TO THE INFORMATION PROVIDED IN THIS MSDS OR TO THE CHEMICAL TO WHICH INFORMATION MAY REFER. CRC NEITHER ASSUMES NOR AUTHORIZES ANY OTHER PERSON TO ASSUME FOR IT, ANY OTHER ADDITIONAL RESPONSIBILITY OR LIABILITY FOR THE USE OF, OR RELIANCE UPON, THIS INFORMATION.

HAZARD WARNING INFORMATION

HEALTH	FIRE	REACTIVE	OTHER	DEGREE OF HAZARD	COLOR CODING	OTHER CODES
3	0	0		0 = Minimum Hazard 1 = Slight Hazard 2 = Moderate Hazard 3 = Serious Hazard 4 = Severe Hazard	HEALTH = BLUE FIRE = RED REACTIVITY = YELLOW OTHER = WHITE	OX = Oxidizer ACID = Acid ALK = Alkali COR = Corrosive W = Use No Water

SECTION I - GENERAL INFORMATION

Characterization: **Mineral**	RCRA Number: **Not Listed**	EPA Class: **Nonhazardous Waste**
DOT Proper Shipping Name: **Asbestos**	Chemical Abstract Service (CAS) Number: **1332-21-4**	
DOT Hazard Class and Label Requirements: **ORM-C**	DOT Emergency Guide Code: **31**	
DOT Identification Number: **UN 2212 (blue) UN 2590 (white)**	Chemical Formula: **Various compositions**	

Synonyms

Asbestos is a generic term for a number of hydrated silicates, including amosite (CAS: 12172-73-5); actinolite (CAS: 13768-00-8); anthophyllite (CAS: 17068-78-9); chrysotile (CAS: 12001-29-5); crocidolite (CAS: 12001-28-4); and tremolite (CAS: 14567-73-8).

SECTION II - HAZARDOUS INGREDIENTS AND IDENTITY INFORMATION

Hazardous Components (specific identity)	OSHA Exposure Criteria	NIOSH Exposure Criteria	Immediately Dangerous to Life and Health (IDLH)	ACGIH Exposure Criteria
Asbestos (derivation: A group of impure magnesium silicate minerals which occur in fibrous form. Colors: white, gray, green, brown, blue. In the U.S., found in Vermont, Arizona, California, North Carolina. Also found in Africa, Italy, Yukon, Quebec, Mexico).	**PEL:** **2 million fb/m^3** **CEILING:** **10 million fb/m^3** **Cancer Hazard**	**REL:** **100,000 fb/m^3** **>5 μm in length** **Recognized Human Carcinogen**	**Cancer Hazard**	**TLV (cuurent):** **2 fb/cc** **TLV(proposed):** **0.2 fb/cc** **Confirmed Human Carcinogen**

SECTION III - PHYSICAL AND CHEMICAL CHARACTERISTICS

Boiling Point: **Decomposes**	Specific Gravity (H_2O = 1): **Varies depending on type of asbestos**
Vapor Pressure (atmospheres): **0 (approx) at 69°F (20°C)**	Molecular Weight (atomic weight): **Varies depending on type of asbestos**
Vapor Density (Air = 1): **Not Found**	Melting Point: **1112°F (600°C) (decomposes)**

Solubility

Insoluble in water.

Appearance and Odor

Mineral fibers of various sizes, colors, lengths, and textures. No odor.

SECTION IV - FIRE AND EXPLOSION HAZARD DATA

Flash Point (method used): **Not Applicable**	Explosive Limits in Air % by Volume: **LEL: N/A UEL: N/A**
NFPA Classification: **Noncombustible Solid**	Autoignition Temperature: **Not Applicable**

Extinguishing Media

Use agent suitable for type of surrounding fire, asbestos itself does not burn or support combustion.

Special Fire Fighting Procedures

Care should be taken to contain asbestos materials that may have been disturbed in fire. To avoid breathing fibers, wear full protective clothing and self-contained breathing apparatus (SCBA).

Unusual Fire and Explosion Hazards

No unusual fire or explosion hazards listed. Asbestos is a noncombustible solid. It can resist temperatures over 500°C. It can resist most acids, has a high tensile strength, and provides excellent thermal and acoustical insulation. It may therefore be encountered in many older structures during fire fighting activities. Caution is warranted when approaching any fire where the presence of asbestos is suspected.

SECTION V - REACTIVITY DATA

Stability		Conditions to Avoid **Avoid unprotected and/or unauthorized disturbance of asbestos materials. Do not allow asbestos or asbestos containing materials to become airborne. Asbestos is stable and inert.**
Stable X	Unstable	Incompatibility *(materials to avoid)* **None reported.**
Hazardous Polymerization		Conditions to Avoid **Hazardous polymerization will not occur.**
May Occur	Will Not Occur X	Hazardous Decomposition or By-products **None reported. However, it should be noted that asbestos fibers themselves present the primary hazard associated with these materials. In a friable, airborne state, asbestos is extremely hazardous.**

SECTION VI - HEALTH HAZARD DATA

Primary Route(s) of Entry:	Inhalation?	Absorption (skin/eye)?	Ingestion?
	X		

Health Hazards

INHALATION: Repeated, long-term exposures can cause asbestosis, a disease characterized by diffuse interstitial pulmonary fibrosis (a scarring of the lung tissues). Depending upon the nature of exposure (length of time in years, type of asbestos exposed to, size of fibers reaching the the lung tissue), more serious cancers, including mesothelioma, can occur.

Carcinogenicity	NTP Listed?	IARC Cancer Review Group?	OSHA Regulated?	Target Organs?
Confirmed Human Confirmed Animal	**5th Annual Report**	**Group 1**	**29 CFR 1910.1001**	**Eyes, respiratory system (lung cancer)**

Medical Conditions Generally Aggravated by Exposure

Pre-existing respiratory problems (bronchitis, asthma, etc.) may be aggravated by exposure to asbestos over the long-term (years); however, the data are inconclusive.

Emergency and First-aid Procedures

Eye contact: Do not allow victim to keep eyes tightly shut. Flush immediately with water for 15 minutes (minimum), seek medical attention. Skin contact: Remove all contaminated clothing. Immediately wash area with large amounts of water. For inhalation: Remove the person from exposure. Provide respiratory assistance and CPR. Transfer to medical facility. Respiratory problems associated with asbestos do not occur for many years following repeated exposures. All exposures should be documented.

SECTION VII - PRECAUTIONS FOR SAFE HANDLING AND USE

Steps to be Taken in Case Material is Released or Spilled

Restrict those not involved in cleanup from entering area. Proper procedures for the repair or removal of asbestos must be followed by trained personnel. Notify appropriate authorities, as required by applicable environmental regulations.

Preferred Waste Disposal Method

Asbestos wastes may be solidified prior to their landfill burial.

Precautions to be Taken in Handling and Storage

Prior to working with asbestos, persons must be trained (by law) on its proper handling and storage. A regulated, marked area should be established where asbestos is used or stored.

Other Precautions and Warnings

Airborne asbestos dust is difficult to remove. It is essential that asbestos work areas be closed or isolated to ensure controlled exposures.

SECTION VIII - CONTROL MEASURES AND PERSONAL PROTECTIVE EQUIPMENT

Respiratory Protection *(specify type)*

Asbestos is a human carcinogen. At any exposure level, use a MSHA/NIOSH combination respirator which includes a Type-C supplied air respirator with a full facepiece operated in pressure demand mode with either emergency HEPA back-up filter or auxiliary self-contained breathing apparatus (SCBA) with full facepiece operated in positive pressure mode.

Ventilation

Local exhaust (preferred) at site of asbestos work. Area should be isolated from other work areas.

Protective Gloves	Eye Protection	Other Protective Clothing
Impervious Rubber	**Dust Goggles and Face Mask**	**Full Body Protective Suit (disposable) with Boots and Head Protection**

Work/Hygiene Practices

Always wash thoroughly after working around asbestos; never bring food, drink, or smoking materials into vicinity of asbestos work or abatement projects.

ASBESTOS

CAS: 1332-21-4

ASBESTOS TYPE	CAS NUMBER	CHEMICAL COMPOSITION
Actinolite	13768-00-8	$2CaO \bullet 4MgO \bullet FeO \bullet 8SiO_2 \bullet H_2O$
Anthophyllite	17068-78-9	$7MgO \bullet 8SiO_2 \bullet H_2O$
Amosite	12172-73-5	$11FeO \bullet 3MgO \bullet 16SiO_2 \bullet 2H_2O$
Chrysotile	12007-29-5	$3MgO \bullet 2SiO_2 \bullet 2H_2O$
Crocidolite	12001-28-4	$Na_2O \bullet Fe_2O_3 \bullet 3FeO \bullet 8SiO_2 \bullet H_2O$
Tremolite	14567-73-8	$2CaO \bullet 5MgO \bullet 8SiO_2 \bullet H_2O$

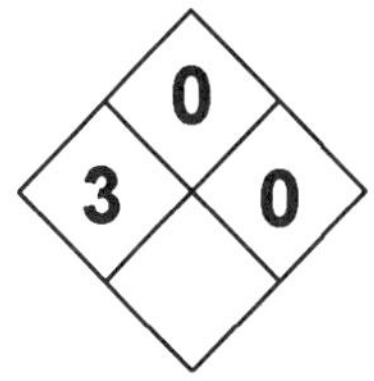

IDENTIFICATION AND TYPICAL USES

Asbestos is a general term used to describe a variety of naturally occurring hydrated silicates that produce mineral fibers upon mechanical processing. There are two varieties of asbestos: serpentine and amphibole minerals. Asbestos occurs in nature in fibrous stones, silicate minerals, rocks, open-pit quarries, and underground mines. It also occurs in burial pits and man-made sites where is was dumped in the era before regulations. Because of the unique physical and chemical properties of asbestos fibers (noncombustible and capable of withstanding temperatures over 500°C, high tensile strength, resistance to acids, and thermal and acoustic insulation), asbestos was used widely in many products. These include textiles, cement, paper, wicks, ropes, floor and roofing tiles, water pipe, wallboard, fire resistant clothing, gaskets, curtains, and many other products.

RISK ASSESSMENT: HEALTH

General Assessment

Health hazards associated with asbestos occur as a result of prolonged and repeated exposures over the course of many years. Asbestos is a confirmed human carcinogen producing lung tumors and other cancers. Exposure to airborne dusts can cause asbestosis, mesothelioma, lung cancer, and gastrointestinal cancer. These asbestos-related diseases usually show a long latency period, sometimes many years after the last known exposure episode. This latency time, however, will vary depending upon the species of asbestos, the degree of exposure, and the type of disease. *Inhalation* is the primary route of entry of asbestos particles into the body. Other than the pulmonary system, fibers may accumulate in the gastrointestinal tract from drinking water contaminated with asbestos or due to transportation from the nasal passage (i.e., when swallowing).

Asbestosis is a disease characterized by lung disorder and a diffuse interstitial fibrosis that is irreversible in nature. It can lead to respiratory disability and cause death from inability of the body to obtain needed oxygen or from cardiac failure.

Mesothelioma is a rare malignant tumor that can occur after a short intensive exposure to certain types of asbestos (as well as other minerals). Unlike asbestosis, mesothelioma develops in a shorter latent time and death may occur within a year of diagnosis. It is characterized by malignant cancer spread over the surfaces of the lung, heart, and abdominal organs (the area known as the mesothelium). Early symptoms include shortness of breath, chest pain, progressive weight loss, and abdominal swelling. Among the asbestos types, chrosidolite is most hazardous, followed by amosite and chrysotile, in causing mesotheliomas in humans.

Lung cancer may occur from moderate exposure to airborne asbestos fibers. The latency period may range from 15-35 years. Symptoms include irritating cough, blood-flecked sputum, chest pain, loss of weight and appetite.

Acute Health Effects

There are no known acute (short-term) health effects associated with exposure to asbestos. People who develop serious and even fatal diseases later in life may be asymptomatic (without symptoms) at the time of exposure and for several months or years thereafter.

Chronic Health Effects

The following chronic (long-term) health effects may occur at some time after exposure to asbestos and can last for months or even years before death occurs:

Cancer Hazards: Asbestos is a carcinogen in humans. It has been shown to cause lung cancers (including mesothelioma) as well as stomach, colon, rectal, vocal cord, and kidney cancers.

Reproductive Hazard: According to the references, asbestos has been tested and has not been shown to affect reproduction.

Other Chronic Effects: Repeated, long-term exposures to asbestos can cause the disease asbestosis, a scarring of the lungs. The higher the exposure, and the longer the time of the exposure, the greater the chance of developing serious illness. Asbestosis develops many years (from 7 to 30) after the period of exposure. It can progress to disability and death. Changes in chest X-ray can indicate the presence of fibrous tissues.

ʘ *Recommended Risk-Reduction Measures*

Personnel should avoid direct contact with asbestos. Its airborne dusts present a serious health hazard to unprotected, exposed individuals. By law, only trained and qualified persons are permitted to work with or remove asbestos or asbestos containing materials (ACM). *Engineering controls* are required to ensure maximum reduction of exposure potential. Specific procedures are required when working with asbestos. These include, but are not limited to: total containment of the affected work area; use of air showers and barriers; use of specific personnel protective equipment (PPE); use of specialized mechanical equipment to ensure total capture of all ACM; and use of specific waste disposal techniques and methods. During any work with asbestos (abatement, removal, or otherwise), the best protection is to enclose operations and/or provide local exhaust ventilation at the site of asbestos work. Isolating operations will also reduce exposure. Using respiratory protection is less effective than the controls mentioned above, but is still required whenever working with or around friable asbestos. At any exposure level, a self-contained breathing apparatus (SCBA) with full facepiece and pressure demand is recommended for the greatest possible respiratory protection. However, for non-abatement work, a powered respirator equipped with a high efficiency particulate air (HEPA) filter may suffice. If a full facepiece is not available, then dust goggles or face shield should be worn to protect the eyes. To prevent hand and skin exposures, impervious rubber gloves should be worn.

Administrative controls should also be in place to minimize the potential for human exposures. These are required by law and include written procedures or policies which specify the methods and techniques that will be practiced whenever personnel are to work with asbestos.

All personnel must receive *training* on the use, hazards, protective measures, emergency actions, and other precautions per 29 CFR 1910.1200 (Hazard Communication) and 29 CFR 1910.1001 (Asbestos), prior to the first assignment in an area where asbestos is used, removed, stored, or destroyed.

Before beginning employment with asbestos and at regular intervals thereafter (e.g., annually), a medical history and examination is recommended, including the following:

☑ Pulmonary function test (baseline and annually).

Any evaluation should include a careful history of past and present symptoms with an examination. Medical tests that simply look for existing damage are not a substitute for controlling exposures. Because smoking can cause heart disease, as well as lung cancer, emphysema, and other respiratory problems, respiratory conditions caused by asbestos exposure may be more severe. The risk of lung cancer may be as much as 92 times higher for people with asbestos exposure who also smoke than for those without both exposures.

Other methods to reduce exposure to asbestos include:

☑ Specific engineering controls are required by OSHA 29 CFR 1910.1001 and 29 CFR 1926.58. Also refer to NIOSH Criteria Document: *Occupational Exposure to Asbestos* (#77-169).

☑ Substitute the less toxic mineral wool and fiberglass for asbestos where possible. There are substitutes for almost every use of asbestos.

☑ Always ensure that proper protective clothing is worn when working with asbestos.

☑ Hazard warning information should be posted in the work area. In addition, as part of an on-going education and training program, all information on the health and safety hazards of asbestos must be communicated to all potentially exposed workers.

☑ Wash any areas of the body that may have contacted asbestos.

☑ Do not eat, drink, or smoke where asbestos is handled, processed, or stored since asbestos can be swallowed. Wash hands carefully before eating or smoking.

☑ On-going asbestos abatement projects in sealed areas become hot and humid. There is a risk of heat stress. Personnel should be trained to recognize the warning signs and the proper actions to take to avoid seriously dangerous working conditions.

☑ Do not sweep for cleanup. Use a vacuum equipped with a HEPA filter or a wet method to reduce dust generation.

☑ Eye wash stations should be provided in the immediate work area for emergency use. If there is a possibility of skin exposure to asbestos, emergency shower facilities should also be provided.

RISK ASSESSMENT: ENVIRONMENT

General Assessment

The environment is at risk of exposure during transportation, storage, disposal, or destruction of asbestos. In almost every scenario, the threat of environmental exposure is contingent upon the proper handling asbestos.

Accidental spills and damaged asbestos materials must be cleaned using wet methods. Asbestos spills must never be dry swept or shoveled. Such actions will cause the generation of dusts and airborne asbestos fibers. Since asbestos is a common air contaminant, its threat to the surrounding environment is minimal. However, this does not justify careless or unprofessional cleanup methods in the event of a spill.

Asbestos is considered a noncombustible solid. It was commonly used for a multitude of purposes for many years before becoming regulated by OSHA and before its carcinogenic properties were discovered. These characteristics mean that asbestos can unknowingly be present when other types of materials are being transported.

Asbestos is a naturally occurring mineral. In the United States, it is found primarily in Vermont, Arizona, California, and North Carolina. It also occurs in Africa, Italy, the Yukon, Quebec, and Mexico. It can enter the environment through run-off, or through accidental spill.

Acute Ecological Effects

Acute (short-term) ecological toxic effects may include the death of animals, birds, or fish and death or low growth rate in plants. Acute effects are seen 2 to 4 days after animals or plants are exposed to asbestos.

The data in the references are insufficient to determine if asbestos (in any form) poses any acute toxicity hazard to aquatic life, plants, birds, or terrestrial animals.

Chronic Ecological Effects

Chronic (long-term) toxic effects may include shortened life span, reproductive problems, lower fertility, and changes in appearance or behavior in exposed animals. These effects can be seen long after first exposure(s) to toxic chemicals.

Asbestos (fibers) can cause cancer if inhaled; birds or land animals which inhale asbestos fibers are therefore susceptible to cancer or other long-term effects. Asbestos fibers which are ingested (swallowed) are generally not toxic to birds or animals, but they are passed through their digestive systems.

Water Solubility

Asbestos minerals are slightly solubility in water. Concentrations of less than 1 milligram will mix with a liter of water.

Persistence in the Environment

Asbestos fibers are highly persistent in water, with a half-life greater than 200 days. The half-life of a pollutant is the amount of time it takes for one-half of the chemical to be degraded.

Bioaccumulation in Aquatic Organisms

Some substances increase in concentration, or *bioaccumulate,* in living organisms as they breathe contaminated air, drink contaminated water, or eat contaminated food. These chemicals can become concentrated in the tissues and internal organs of animals as well as humans.

The concentration of asbestos found in fish tissues is expected to be less than the average concentration of asbestos in the water from which the fish was taken.

Recommended Risk-Reduction Measures

Proper training of all transporters will reduce the likelihood of a mishap or accident resulting in a leak or spill of asbestos into the environment. The correct use of DOT labeling (per the MSDS) on all containers, trucks, and rail cars will enable emergency responders to react properly and quickly to any disaster thereby reducing the potential risk to the environment and to personnel.

If a spill or leak to the environment has occurred, hazardous materials spill response personnel should be notified immediately. Cleanup must be attempted only by those trained in proper asbestos spill containment procedures. If asbestos should contact the water table, aquifer, or navigable waterway, the local and/or state emergency response authorities must be notified. A comprehensive emergency response or disaster preparedness/recovery plan should be in place prior to any operations involving the use, transportation, storage, or disposal of asbestos.

Asbestos wastes may be solidified prior to their landfill burial. This may be achieved by a cementing process, such as that using pozzolanic concrete, which contains fly ash or kiln dust mixed with lime, water, and other additives. Other methods of solidification include thermoplastic and polymeric processes.

If asbestos is spilled or damaged, the following specific steps are recommended:

☑ Restrict persons not wearing protective clothing from area of spill until cleanup is complete.

☑ Applicable OSHA regulations must be strictly followed during any abatement procedures.

RISK ASSESSMENT: BUSINESS

General Assessment

Accidents or mishaps involving asbestos can present a significant threat to business operations. The use of the term asbestos can generate serious emotional response from employees as well as the public in surrounding communities. Lawsuits that may result from personnel injury/death, public exposures, and/or environmental contamination will also require a serious expenditure of resources. Media attention surrounding an injury, death, or environmental damage can also result in a loss of profits and loss of current as well as future business. Always remember that anytime the word "asbestos" or the terms "cancer," or "carcinogen" are used, public emotion, hysteria, and ignorance can run equally high. This should be carefully considered whenever developing or implementing public relations policies.

Recommended Risk-Reduction Measures

Company attorneys, safety and health professionals, and environmental specialists should be involved in the development of any procedures dealing with safety or emergency response. A company official should be pre-designated as a public relations officer with specific training in dealing with the press. Corporate plans and policies should be developed, approved, and implemented long before any need for such arises.

REFERENCES

American Conference of Governmental Industrial Hygienists (ACGIH). 1988. *Documentation of the Threshold Limit Values and Biological Exposure Indices,* 5th Edition (with updates). Cincinnati: ACGIH

Calabrese, E. J. 1994. *Biological Effects of Low Level Exposures.* New York: CRC Press/Lewis Publishers.

Cockerman, L. G., B. S. Shane. 1994. *Basic Environmental Toxicology.* New York: CRC Press/Lewis Publishers.

Cothern, R. R. 1993. *Comparative Environmental Risk Assessment.* New York: CRC Press/Lewis Publishers.

Hallenbeck, W. H. 1993. *Quantitative Risk Assessment for Environmental and Occupational Health,* 2nd Edition. New York: CRC Press/Lewis Publishers.

Health and Safety Commission, United Kingdom. 1978. *Measurement and Monitoring of Asbestos in Air,* Second report by the advisory committee on asbestos. London: Crown Publishers

Landis, W. G., M. H. Yu. 1995. *Introduction to Environmental Toxicology.* New York: CRC Press/Lewis Publishers.

Langerman, N. 1994 *Precautionary Labels for Chemical Containers.* New York: CRC Press/Lewis Publishers.

Levadie, B. editor. 1982. *Definitions for Asbestos and Other Health-Related Silicates.* American Society for Testing Materials (ASTM) Spec. Tech. Publication 834. Philadelphia: ASTM

Lewis, R. J., Sr. 1993. *Hawley's Condensed Chemical Dictionary,* 12th Edition. New York: Van Nostrand Reinhold.

Lewis, R. J., Sr. 1992. *Sax's Dangerous Properties of Industrial Materials,* Volume 1. 8th Edition. New York: Van Nostrand Reinhold.

Meyer, E. 1990. *Chemistry of Hazardous Materials.* Englewood Cliffs, NJ: Prentice Hall.

Patnaik, P. 1992. *A Comprehensive Guide to the Hazardous Properties of Chemical Substances.* New York: Van Nostrand Reinhold.

Proctor, N. H., J. P. Hughes. 1991. *Chemical Hazards of the Workplace.* New York: Van Nostrand Reinhold.

Richardson, M. 1995. *Dictionary of Substance and Their Effects,* Vol. I-VI. New York: CRC Press/Lewis Publishers.

Sacarello, H. L. A. 1994. *The Comprehensive Handbook of Hazardous Materials.* New York: CRC Press/Lewis Publishers.

Torkelson, T. R. V. K. Rowe. 1981. *Patty's Industrial Hygiene and Toxicology,* Vol. 2B. New York: Wiley-Interscience.

U.S. Department of Health and Human Services. National Institute for Occupational Safety and Health (NIOSH). 1984. *Manual of Analytical Methods,* 3rd. Edition. Cincinnati, OH: NIOSH.

U.S. Department of Health, Education, and Welfare (DHEW). National Institute for Occupational Safety and Health (NIOSH). 1972 *Criteria for a Recommended Standard...Occupational Exposure to Asbestos.* DHEW (HSM) Publication Number 72-10267. Washington, DC: U.S. Government Printing Office.

MATERIAL SAFETY DATA SHEET

CHEMICAL NAME	CRC PRESS/LEWIS PUBLISHERS DISCLAIMER
ATRAZINE	THE INFORMATION AND RECOMMENDATIONS PRESENTED HEREIN ARE BASED ON SOURCES BELIEVED TO BE RELIABLE. CRC MAKES NO REPRESENTATION ON ITS COMPLETENESS OR ACCURACY. IT IS THE RESPONSIBILITY OF THE USER TO DETERMINE THE CHEMICAL'S SUITABILITY FOR ITS INTENDED USE, THE CHEMICAL'S SAFE USE, AND THE CHEMICAL'S PROPER DISPOSAL. NO REPRESENTATIONS AND/OR WARRANTIES, EITHER EXPRESSED OR IMPLIED, OF THE MERCHANTABILITY OR FITNESS FOR A PARTICULAR PURPOSE, OR OF ANY OTHER NATURE, ARE MADE WITH RESPECT TO THE INFORMATION PROVIDED IN THIS MSDS OR TO THE CHEMICAL TO WHICH INFORMATION MAY REFER. CRC NEITHER ASSUMES NOR AUTHORIZES ANY OTHER PERSON TO ASSUME FOR IT, ANY OTHER ADDITIONAL RESPONSIBILITY OR LIABILITY FOR THE USE OF, OR RELIANCE UPON, THIS INFORMATION.

HAZARD WARNING INFORMATION

HEALTH	FIRE	REACTIVE	OTHER	DEGREE OF HAZARD	COLOR CODING	OTHER CODES
1	0	1		0 = Minimum Hazard 1 = Slight Hazard 2 = Moderate Hazard 3 = Serious Hazard 4 = Severe Hazard	HEALTH = BLUE FIRE = RED REACTIVITY = YELLOW OTHER = WHITE	OX = Oxidizer ACID = Acid ALK = Alkali COR = Corrosive W = Use No Water

SECTION I - GENERAL INFORMATION

Characteristic	RCRA Number	EPA Class
Herbicide	**Not Listed**	**Not Listed**

DOT Proper Shipping Name **Atrazine**	Chemical Abstract Service (CAS) Number **1912-24-9**
DOT Hazard Class and Label Requirements **Noncombustible Solid**	DOT Emergency Guide Code **55 (triazine pesticide)**
DOT Identification Number **UN 2763**	Molecular Formula $C_8H_{14}ClN_5$

Synonyms

2-chloro-4-ethylamino-6-isopropylamino-*s*-triazine; 2-chloro-4-ethylamino-6-isopropyl-1,3,5-triazine; 6-chloro-*N*-ethyl-*N'*-(1-methylethyl)1,3,5-triazine-2,4-diamine; Gesaprim; Primatol; Weedx A; Zeazine.

SECTION II - HAZARDOUS INGREDIENTS AND IDENTITY INFORMATION

Hazardous Components (specific identity)	OSHA Exposure Criteria	NIOSH Exposure Criteria	Immediately Dangerous to Life and Health (IDLH)	ACGIH Exposure Criteria
Atrazine (derivation: An s-triazine herbicide).	**PEL:** Not Established **STEL:** Not Established	**REL:** 5 mg/m³ Questionable Human Carcinogen	Not Determined	**TLV:** 5 mg/m³ **STEL:** Not Established

SECTION III - PHYSICAL AND CHEMICAL CHARACTERISTICS

Boiling Point **Decomposes**	Specific Gravity (H_2O = 1) **1.19**
Vapor Pressure (atmospheres) **3×10^{-7} at 69°F (20°C)**	Molecular Weight (atomic weight) **215.06**
Vapor Density (Air = 1) **Not Found**	Melting Point **340°F (173°C)**

Solubility

At 25°C: Insoluble in water (70 ppm); slightly soluble in ether (12,000 ppm) and methanol (18,000 ppm); soluble in chloroform (52,000 ppm).

Appearance and Odor

Colorless or white crystalline solid powder, no odor.

SECTION IV - FIRE AND EXPLOSION HAZARD DATA

Flash Point (method used) **Not Applicable**	Explosive Limits in Air % by Volume **LEL: N/A UEL: N/A**
NFPA Classification **Noncombustible Solid**	Autoignition Temperature **Not Applicable**

Extinguishing Media

Use extinguishing media suitable to surrounding fires.

Special Fire Fighting Procedures

Wear full protective clothing and self-contained breathing apparatus (SCBA) when fighting fires near or around atrazine since poisonous gases are produced in fire.

Unusual Fire and Explosion Hazards

Atrazine is a noncombustible solid. But, it common to find it mixed with flammable or combustible liquids. Fires involving atrazine mixtures present additional hazards caused by the release of toxic gases.

SECTION V - REACTIVITY DATA

<table>
<tr><td colspan="2">Stability</td><td>Conditions to Avoid
Atrazine is a very stable s-triazine herbicide. It should be kept away from heat, flame, and incompatible materials.</td></tr>
<tr><td>Stable
X</td><td>Unstable</td><td>Incompatibility (materials to avoid)
Can react violently and vigorously in the presence of strong acids and strong bases.</td></tr>
<tr><td colspan="2">Hazardous Polymerization</td><td>Conditions to Avoid
Hazardous polymerization of atrazine has not been reported.</td></tr>
<tr><td>May Occur</td><td>Will Not Occur
X</td><td>Hazardous Decomposition or By-products
When heated to decomposition, emits highly toxic Cl^- and NO_x fumes.</td></tr>
</table>

SECTION VI - HEALTH HAZARD DATA

Primary Route(s) of Entry:	Inhalation?	Absorption (skin/eye)?	Ingestion?
	X		X

Health Hazards

INHALATION: Human data are inconclusive. Symptoms may include ataxia, dypsnea, and convulsion, conjunctivitis, excessive salivation, sneezing, and muscle weakness.

INGESTION: Human data are inconclusive. Symptoms may include gradual reduction in hemoglobin content and erythrocyte and leukocyte count; an increase in glucose, cholesterol, total pro tiens, and the enzyme transaminases. There may be a loss of appetite, debility, progressive anemia, enteritis, and muscle weakness. There may also be visible effects on multiple organs. There is limited evidence to show that atrazine taken orally may affect the brain, the liver, and the kidney in experimental studies using test animals.

Carcinogenicity	NTP Listed?	IARC Cancer Review Group?	OSHA Regulated?	Target Organs?
Questioned Human Questioned Animal	Not Listed	No	Not Listed	Eyes, Skin, Respiratory System, CNS, Liver

Medical Conditions Generally Aggravated by Exposure

The data do not indicate any pre-existing medical conditions that might be aggravated by exposure.

Emergency and First-aid Procedures

Eye contact: Do not allow victim to keep eyes tightly shut. Flush immediately with water for 15 minutes (minimum), seek medical attention. Skin contact: Remove all contaminated clothing. Immediately wash area with large amounts of soap and water. For inhalation: Remove the person from exposure. Provide respiratory assistance and CPR. Transfer to medical facility. If swallowed, seek medical attention immediately. Do not attempt to give an unconscious or convulsing person anything by mouth.

SECTION VII - PRECAUTIONS FOR SAFE HANDLING AND USE

Steps to be Taken in Case Material is Released or Spilled

Remove ignition sources. Ventilate area of spill or leak. Restrict those not involved in cleanup from entering area. Collect spilled powdered materials in safest manner possible and deposit in sealed drum for disposal. Notify appropriate authorities, as required by applicable environmental regulations.

Preferred Waste Disposal Method

Incineration.

Precautions to be Taken in Handling and Storage

Atrazine is a nonflammable solid. Store to avoid contact with strong acids and bases. Store in tightly closed containers in cool, well-ventilated area away from heat. Do not use near heat or flame.

Other Precautions and Warnings

Sources of ignition are prohibited where atrazine is used, handled, or stored.

SECTION VIII - CONTROL MEASURES AND PERSONAL PROTECTIVE EQUIPMENT

Respiratory Protection *(specify type)*

Atrazine is mildly toxic by inhalation. An MSHA/NIOSH approved half-mask respirator with organic vapor cartridge and dust/mist pre-filter should be worn. For high exposures, a self-contained breathing apparatus (SCBA) with full facepiece operated in positive pressure mode is recommended.

Ventilation

Local exhaust (preferred) at site of chemical work or general ventilation.

Protective Gloves	Eye Protection	Other Protective Clothing
Impervious Rubber	Chemical/Dust Goggles or Face Mask	Rubber Apron

Work/Hygiene Practices

Always wash hands thoroughly after using chemical; never bring food, drink, or smoking materials into vicinity of chemicals.

ATRAZINE

$C_8H_{14}ClN_5$ CAS: 1912-24-9

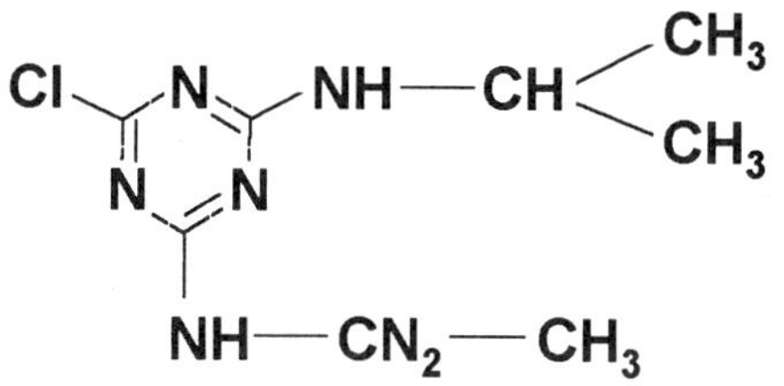

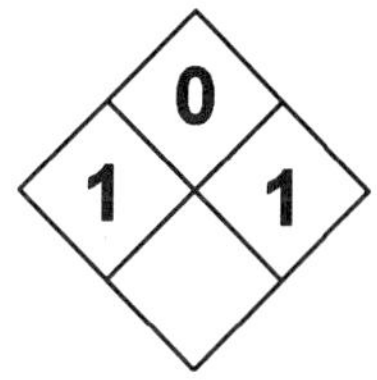

IDENTIFICATION AND TYPICAL USES

Atrazine is a colorless or white solid crystalline powder with no detectable odor. Of the *s*-triazine herbicides, atrazine is the best known example. It is used for the control of broad-leaf and grassy weeds, in corn, asparagus, tomatoes, potatoes, and ornamentals.

RISK ASSESSMENT: HEALTH

General Assessment

The acute and chronic toxicity of atrazine is considered to be low in both animal and human exposures. Although there have been no substantial reports of skin or eye irritation resulting from experimental or commercial applications of atrazine, its properties are conducive to causing such problems in animals and possibly in humans as well. No human poisoning cases have been reported from ingestion of atrazine.

However, from a *risk management* perspective, atrazine should be considered mildly toxic by *inhalation* and *skin contact* and moderately toxic by *ingestion*. This is recommended because animal studies have shown some adverse effects resulting from exposure by these routes. Specifically, animals studies have produced a variety of symptoms, including ataxia, dyspnea, and convulsions. There have also been documented reports of debility, loss of appetite, progressive anemia, enteritis, and muscle weakness following oral exposure to test animals. Although most organs were also affected in some way, the primary targets appear to be the eyes, liver, kidney, brain, and respiratory tract.

There is also limited evidence to show that exposure to atrazine effects the action of the central nervous system. The *s*-triazine herbicides are not retained very long in body tissues and fluids of animals. They are excreted in the urine and feces.

Acute Health Effects

The following acute (short-term) health effects may occur immediately or shortly after exposure to atrazine:

Skin: Irritation and possible rash or allergic reaction.

Eye: Severe eye irritation likely, conjunctivitis possible in some exposures.

Lung: Can irritate the respiratory tract, in large doses, may be absorbed through respiratory tissues.

CNS: May cause ataxia, headache, muscle weakness, dyspnea, and convulsions.

Chronic Health Effects

The following chronic (long-term) health effects may occur at some time after exposure to atrazine and can last for months or even years:

Cancer Hazards: Atrazine may be a questionable carcinogen humans, however, lengthy studies in animals have shown no substantial data to support any claims of carcinogenicity of atrazine. The data are inconclusive on the exact nature of any carcinogenic properties. However, some limited mutagenic data have been reported.

Reproductive Hazard: Atrazine is considered an experimental teratogen but the data are insufficient to confirm that atrazine presents a reproductive hazard.

Recommended Risk-Reduction Measures

Personnel should avoid direct contact with atrazine. The exact nature of its harmful capacities is not clearly understood in the references. Therefore, it should be handled as a toxic material. If a less toxic material cannot be substituted for atrazine, then *engineering controls* are the most effective method of reducing exposures. The best protection is to enclose operations and/or provide local exhaust ventilation at the site of atrazine release. While not always operationally feasible, isolating operations can also reduce exposure. These may be appropriate actions in manufacturing, production, and packing operations. However, since the primary application of atrazine is outdoors, isolation and local ventilation are not considered practical exposure control measures

Using respiratory protection and other personal protective equipment (PPE) is less effective than the engineering controls mentioned above, but is still advisable whenever working with or around atrazine. Under normal operating conditions of low exposure potential, A half mask respirator equipped with an organic vapor cartridge and dust/mist pre-filter will suffice. However, for maximum protection, a self-contained breathing apparatus (SCBA) with full facepiece and pressure demand is recommended. If a full facepiece is not available, then chemical/dust goggles or face shield should be worn to protect the eyes. To prevent hand and skin exposures, impervious rubber gloves should be worn.

Administrative controls should also be in place to minimize the potential for human exposures. These may include written procedures or policies which specify the methods and techniques that will be practiced whenever personnel are to work with atrazine.

All personnel should receive *training* on the use, hazards, protective measures, emergency actions, and other precautions per 29 CFR 1910.1200 (Hazard Communication), prior to the first assignment in an area where atrazine is used or stored.

After suspected overexposure, consider the following medical tests:

☑ Urinalysis to check for *s*-triazine herbicide metabolite excretion.

Any evaluation should include a careful history of past and present symptoms with an examination. Medical tests that simply look for existing damage are not a substitute for controlling exposures.

Other methods to reduce exposure to atrazine include:

☑ Where possible, enclose operations and use local exhaust ventilation at the site of chemical release, especially during manufacturing/production and packing or shipping operations. If local exhaust ventilation or enclosure is not used in these situations, respiratory protection should be required.

☑ Always ensure that proper protective clothing is worn when using chemical substances. Never allow contaminated clothing to be taken home from the work place.

☑ Hazard warning information should be posted in the work area. In addition, as part of an on-going education and training program, all information on the health and safety hazards of atrazine should be communicated to all potentially exposed workers.

☑ Eye wash stations should be provided in the immediate work area for emergency use. If there is a possibility of skin exposure to atrazine, emergency shower facilities should also be provided.

RISK ASSESSMENT: ENVIRONMENT

General Assessment

While the environment is at risk of exposure during transportation, storage, disposal, or destruction of atrazine, its ordinary use as an herbicide poses the greatest potential for environmental contamination. In almost every scenario, the threat of environmental exposure is contingent upon the proper handling of the chemical substance. Accidental spills, large or small, where atrazine contacts incompatible commodities can result in violent reactions as well as the possible contamination of the surrounding environmental mediums (water, soil, and air).

Atrazine is considered a noncombustible solid. However, it may often be mixed with other materials and commodities that are flammable. Such mixtures can present a moderate fire and explosion hazard. Also, because it is incompatible with strong bases and strong acids, extreme caution is required in handling, storage, transportation, and disposal of atrazine. These characteristics also require special consideration during any emergency situation involving a leak or spill of atrazine. Should atrazine ever come into contact with such incompatible substances either during use, transportation, storage, or disposal, dangerous reactions are possible.

Atrazine enters the environment mainly through its primary use as an herbicide but also as a result of accidents and industrial discharges.

☠ *Acute Ecological Effects*

Acute (short-term) toxic effects may include the death of animals, birds, or fish and death or low growth rate in plants. Acute effects are seen 2 to 4 days after animals or plants are exposed to atrazine.

Atrazine has minimal acute toxicity to aquatic life, birds, and land animals. As an herbicide, it poses high toxicity to some plant life.

Chronic Ecological Effects

Chronic (long-term) toxic effects may include shortened life span, reproductive problems, lower fertility, and changes in appearance or behavior in exposed animals. These effects can be seen long after first exposure(s) to toxic chemicals.

There are insufficient data to establish whether atrazine poses chronic toxicity to aquatic life, birds, or land animals. It poses high chronic toxicity to some forms of plant life.

Water Solubility

Atrazine has very slight solubility in water. Concentrations of 2 to 20 milligrams will mix with 1000 liters of water.

Persistence in the Environment

There is no evidence in the references to show that atrazine has any measurable persistence in the environment.

Bioaccumulation in Aquatic Organisms

Some substances increase in concentration, or *bioaccumulate,* in living organisms as they breathe contaminated air, drink contaminated water, or eat contaminated food. These chemicals can become concentrated in the tissues and internal organs of animals as well as humans.

The concentration of atrazine found in fish tissues is expected to be much lower than the average concentration of atrazine in the water from which the fish was taken.

Recommended Risk-Reduction Measures

Proper training of all transporters will reduce the likelihood of a mishap or accident resulting in a leak or spill of atrazine into the environment. The correct use of labeling on all containers, trucks, and rail cars will enable emergency responders to react properly and quickly to any disaster thereby reducing the potential risk to the environment and to personnel.

Storage of atrazine should be segregated from incompatible chemicals to minimize the risk of cross-contamination or contact.

If a spill or leak to the environment has occurred, fire department, emergency response and/or hazardous materials spill personnel should be notified immediately. Cleanup should be attempted only by those trained in proper spill containment procedures. If atrazine should contact the water table, aquifer, or navigable waterway, the local and/or state emergency response authorities must be notified. A comprehensive emergency response or disaster preparedness/recovery plan should be in place prior to any operations involving the use, transportation, storage, or disposal of atrazine.

Atrazine may be destroyed by incineration in a chemical incinerator equipped with an air scrubber.

If atrazine is spilled or leaked, the following specific steps are recommended:

- ☑ Restrict persons not wearing protective clothing from area of spill or leak until cleanup is complete.
- ☑ If possible, remove leaking containers to a ventilated area and attempt to seal leak.
- ☑ Do not allow exposure to incompatible materials.
- ☑ Collect powdered materials in the safest and most efficient manner possible. Place materials in a sealed drum.

RISK ASSESSMENT: BUSINESS

General Assessment

Accidents or mishaps involving atrazine can present a moderate threat to business operations. Although not likely, equipment and facilities may be lost or their use disrupted as a result of an incident involving atrazine which can significantly affect fiscal viability. Lawsuits that may result from personnel injury, public exposures, and/or environmental contamination will also require a serious expenditure of resources. Media attention surrounding an injury, illness, or environmental damage can also result in a loss of profits and loss of current as well as future business.

Recommended Risk-Reduction Measures

Company attorneys, safety and health professionals, and environmental specialists should be involved in the development of any procedures that focus on emergency response, disaster preparedness, and disaster recovery. A company official should be pre-designated as a public relations officer with specific training in dealing with the press. Corporate plans and policies should be developed, approved, and implemented long before any need for such arises.

REFERENCES

Ahlbom, A. 1993. *Biostatistics for Epidemiologists.* New York: CRC Press/Lewis Publishers.

American Conference of Governmental Industrial Hygienists. 1988. *Documentation of the Threshold Limit Values and Biological Exposure Indices,* 5th Edition (with updates). Cincinnati: ACGIH

Bloom, A. D. and F. J. DeSerres. 1995. *Ecotoxicity and Human Health.* New York: CRC Press/Lewis Publishers.

Brusick, D. J. 1994. *Methods for Genetic Risk Assessment.* New York: CRC Press/Lewis Publishers.

Bueche, J. R. 1972. *Principles of Physics,* 2nd Edition. New York: McGraw-Hill.

Calabrese, E. J., E. M. Kenyon. *Air Toxics and Risk Assessment.* New York: CRC Press/Lewis Publishers.

Calabrese, E. J. 1994. *Biological Effects of Low Level Exposures.* New York: CRC Press/Lewis Publishers.

Ciba-Geigy Corporation. 1 October 1971. *Atrazine.* Ciba-Geigy toxicology data. Ardsley, New York: Ciba-Geigy Corporation, Agricultural Division.

Cockerman, L. G., B. S. Shane. 1994. *Basic Environmental Toxicology.* New York: CRC Press/Lewis Publishers.

Cothern, R. R. 1993. *Comparative Environmental Risk Assessment.* New York: CRC Press/Lewis Publishers.

Gardner, R. W. 1994. *Chemical Intolerance.* New York: CRC Press/Lewis Publishers.

Hallenbeck, W. H. 1993. *Quantitative Risk Assessment for Environmental and Occupational Health,* 2nd Edition. New York: CRC Press/Lewis Publishers.

Howard, P. H. 1995. *Handbook of Environmental Fate and Exposures Data for Organic Chemicals,* Vol. I-IV. New York: CRC Press/Lewis Publishers.

Howard, P. H., et. al. 1991. *Handbook of Environmental Degradation Rates.* New York: CRC Press/Lewis Publishers.

Jones, F. E. 1994. *Toxic Organic Vapors in the Workplace.* New York: CRC Press/Lewis Publishers.

Keith, L. H. and M. M. Walker. 1995. *Handbook of Air Toxics.* New York: CRC Press/Lewis Publishers.

Landis, W. G., M. H. Yu. 1995. *Introduction to Environmental Toxicology.* New York: CRC Press/Lewis Publishers.

Langerman, N. 1994 *Precautionary Labels for Chemical Containers.* New York: CRC Press/Lewis Publishers.

Lauwerys, R. R., P. Hoet. 1993. *Industrial Chemical Exposure.* New York: CRC Press/Lewis Publishers.

Lewis, R. J., Sr. 1993. *Hawley's Condensed Chemical Dictionary,* 12th Edition. New York: Van Nostrand Reinhold.

Lewis, R. J., Sr. 1992. *Sax's Dangerous Properties of Industrial Materials,* Volumes 1, 2, and 3, 8th Edition. New York: Van Nostrand Reinhold.

Lide, D. R. 1995. *Handbook of Chemistry and Physics,* 76th Edition. New York: CRC Press/Lewis Publishers.

Manahan, S. E. 1994. *Environmental Chemistry,* 6th Edition. New York: CRC Press/Lewis Publishers.

Mellström, G. A., J. E. Wahlberg, and H. I. Maibach. 1994. *Protective Gloves for Occupational Use.* New York: CRC Press/Lewis Publishers.

Meyer, E. 1990. *Chemistry of Hazardous Materials.* Englewood Cliffs, NJ: Prentice Hall.

Mycock, J. C., J. D. McKenna, and L. Theodore. 1995. *Handbook of Air Pollution Control Engineering and Technology.* New York: CRC Press/Lewis Publishers.

Patnaik, P. 1992. *A Comprehensive Guide to the Hazardous Properties of Chemical Substances.* New York: Van Nostrand Reinhold.

Proctor, N. H., J. P. Hughes. 1991. *Chemical Hazards of the Workplace.* New York: Van Nostrand Reinhold.

Quigley, D. R. 1994. *Handbook of Emergency Chemical Management.* New York: CRC Press/Lewis Publishers.

Rea, W. J. 1995. *Chemical Sensitivity,* Vol. I-III. New York: CRC Press/Lewis Publishers.

Richardson, M. 1995. *Dictionary of Substance and Their Effects,* Vol. I-VI. New York: CRC Press/Lewis Publishers.

Sacarello, H. L. A. 1994. *The Comprehensive Handbook of Hazardous Materials.* New York: CRC Press/Lewis Publishers.

Sheftel, V. O. 1995. *Handbook of Toxic Properties of Monomers and Additives.* New York: CRC Press/Lewis Publishers.

Talbott, E., G. Craun. 1995. *An Introduction to Environmental Epidemiology.* New York: CRC Press/Lewis Publishers.

Torkelson, T. R. V. K. Rowe. 1981. *Patty's Industrial Hygiene and Toxicology,* Vol. 2B. New York: Wiley-Interscience.

MATERIAL SAFETY DATA SHEET

CHEMICAL NAME	
BARIUM	**CRC PRESS/LEWIS PUBLISHERS DISCLAIMER:** THE INFORMATION AND RECOMMENDATIONS PRESENTED HEREIN ARE BASED ON SOURCES BELIEVED TO BE RELIABLE. CRC MAKES NO REPRESENTATION ON ITS COMPLETENESS OR ACCURACY. IT IS THE RESPONSIBILITY OF THE USER TO DETERMINE THE CHEMICAL'S SUITABILITY FOR ITS INTENDED USE, THE CHEMICAL'S SAFE USE, AND THE CHEMICAL'S PROPER DISPOSAL. NO REPRESENTATIONS AND/OR WARRANTIES, EITHER EXPRESSED OR IMPLIED, OF THE MERCHANTABILITY OR FITNESS FOR A PARTICULAR PURPOSE, OR OF ANY OTHER NATURE, ARE MADE WITH RESPECT TO THE INFORMATION PROVIDED IN THIS MSDS OR TO THE CHEMICAL TO WHICH INFORMATION MAY REFER. CRC NEITHER ASSUMES NOR AUTHORIZES ANY OTHER PERSON TO ASSUME FOR IT, ANY OTHER ADDITIONAL RESPONSIBILITY OR LIABILITY FOR THE USE OF, OR RELIANCE UPON, THIS INFORMATION.

HAZARD WARNING INFORMATION

HEALTH	FIRE	REACTIVE	OTHER	DEGREE OF HAZARD	COLOR CODING	OTHER CODES
2	0	0	~~W~~	0 = Minimum Hazard 1 = Slight Hazard 2 = Moderate Hazard 3 = Serious Hazard 4 = Severe Hazard	HEALTH = BLUE FIRE = RED REACTIVITY = YELLOW OTHER = WHITE	OX = Oxidizer ACID = Acid ALK = Alkali COR = Corrosive ~~W~~ = Use No Water

SECTION I - GENERAL INFORMATION

Characterization	**Metal**	RCRA Number	**D005**
		EPA Class	**Characteristic Toxic Waste**
DOT Proper Shipping Name	**Barium**	Chemical Abstract Service (CAS) Number	**7440-39-3**
DOT Hazard Class and Label Requirements	**Flammable Solid; Dangerous When Wet**	DOT Emergency Guide Code	**40/37**
DOT Identification Number	**UN 2871/UN 1400/UN/ 1854**	Atomic Symbol	**Ba**

Synonyms
Not Applicable.

SECTION II - HAZARDOUS INGREDIENTS AND IDENTITY INFORMATION

Hazardous Components (specific identity)	OSHA Exposure Criteria	NIOSH Exposure Criteria	Immediately Dangerous to Life and Health (IDLH)	ACGIH Exposure Criteria
Barium (derivation: Alkaline-earth element of atomic number 56. Also derived from reduction of barium oxide with aluminum or silicon in a vacuum at high temperature).	**PEL:** 0.5 mg/m^3 **STEL:** Not Established	**REL:** 0.5 mg/m^3 **STEL:** Not Established	**250 mg/m^3 for Soluble Compounds**	**TLV:** 0.5 mg/m^3 **STEL:** Not Established

SECTION III - PHYSICAL AND CHEMICAL CHARACTERISTICS

Boiling Point	**2912°F (1600°C)**	Specific Gravity (H_2O = 1)	**3.6**
Vapor Pressure (mm Hg)	**10 at 1920°F (1049°C)**	Molecular Weight (atomic weight)	**137.34**
Vapor Density (Air = 1)	**N/A (metal)**	Melting Point	**1310°F (710°C)**

Solubility
Most of the barium salts are either highly or moderately soluble in water. Barium powder reacts readily in water. In most liquids, reactions may occur.

Appearance and Odor
Yellowish-white or silver-white, malleable metal or metal powder, no appreciable odor.

SECTION IV - FIRE AND EXPLOSION HAZARD DATA

Flash Point (method used)	**Not Found (Metal)**	Explosive Limits in Air % by Volume	**LEL: N/A UEL: N/A**
NFPA Classification	**Flammable Solid (in powder form)**	Autoignition Temperature	**Not Found**

Extinguishing Media
Dry chemicals appropriate for extinguishing metal (Class D) fires. DO NOT USE WATER.

Special Fire Fighting Procedures
Wear full protective clothing and self-contained breathing apparatus (SCBA). Poisonous gases are produced in fire. Do not use water mist to keep fire-exposed containers cool.

Unusual Fire and Explosion Hazards
Barium powder is a flammable solid. Containers may explode in fire. Fire may restart after it has been extinguished. Powder may ignite or explode spontaneously in air or other oxidizing gases. Dust is dangerous when exposed to heat, flame, or chemical reaction. Violent or explosive reaction on contact with water.

SECTION V - REACTIVITY DATA

Stability		Conditions to Avoid **Avoid heat and open flame, especially around dusts and powders of barium. Do not allow contact with water or other incompatible agents since violent and explosive reaction can occur.**
Stable X	Unstable	Incompatibility *(materials to avoid)* **Violent or explosive reactions in water, CCl_4, fluorotrichloromethane, tricloroethylene, and C_2Cl_4. Incompatible with acids, $C_2Cl_3F_3$, $C_2H_2FCl_3$, C_2HCl_3, 1,1,2-trichlorotrifluoro ethane, and fluorotrichloroethane.**
Hazardous Polymerization		Conditions to Avoid **Hazardous polymerization of barium metal will not occur**
May Occur	Will Not Occur X	Hazardous Decomposition or By-products **Poisonous gases are emitted during fire.**

SECTION VI - HEALTH HAZARD DATA

Primary Route(s) of Entry:	Inhalation?	Absorption (skin/eye)?	Ingestion?
	X		

Health Hazards

INHALATION: May produce benign pneumoconiosis (baritosis) which may cause X-ray changes. Other symptoms of exposure include eye, nose, and throat irritation, coughing.

EYES & SKIN: Exposure to the eyes may cause moderate to severe and perhaps permanent damage.

INGESTION: Water and stomach acids solubilize barium salts and can cause poisoning. Symptoms are vomiting, colic, diarrhea, slow irregular pulse, transient hypertension, convulsive tremors.

Carcinogenicity	NTP Listed?	IARC Cancer Review Group?	OSHA Regulated?	Target Organs?
Unknown Human Unknown Animal	No	No	29 CFR 1910.1000 Table Z-1	Respiratory system, eyes, CNS.

Medical Conditions Generally Aggravated by Exposure

None reported. However, persons with existing lung dysfunction may experience effects quicker.

Emergency and First-aid Procedures

Eye contact: Do not allow victim to keep eyes tightly shut. Flush immediately with water for 15 minutes (minimum), seek medical attention. Skin contact: Although no immediate reaction may be evident, barium powders may begin to burn skin when sweat begins (water reactive). Remove all contaminated clothing. Immediately wash area with flooding amounts of soap and water. For inhalation: Remove the person from exposure. Provide respiratory assistance and CPR. Transfer to medical facility. If swallowed, seek medical attention immediately. Never give anything by mouth to a convulsing person.

SECTION VII - PRECAUTIONS FOR SAFE HANDLING AND USE

Steps to be Taken in Case Material is Released or Spilled

Collect powdered materials in most convenient and safe manner and deposit in sealed containers. Dry sweep not recommended. Use HEPA vacuum. Remove all ignition sources. Restrict those not involved in cleanup from entering area. Notify appropriate authorities, as required by applicable regulations.

Preferred Waste Disposal Method

Reactive waste may have to be treated and rendered non-reactive prior to disposal.

Precautions to be Taken in Handling and Storage

Explosion risks vary with type of barium compound. Store to avoid contact with strong acids and halogenated hydrocarbons. Store powders in sealed containers under an inert gas or oxygen-free petroleum liquid away from water, oxidizers, acids, or ammonia.

Other Precautions and Warnings

Sources of ignition (heat, flame, cigarettes, incompatibles) are prohibited where barium is used, handled, or stored. Use only non-sparking tools or equipment, especially when opening or closing containers.

SECTION VIII - CONTROL MEASURES AND PERSONAL PROTECTIVE EQUIPMENT

Respiratory Protection *(specify type)*

For dust exposures below PEL, use a respirator with a high efficiency particulate air (HEPA) filter, otherwise use a self-contained breathing apparatus with full facepiece operated in positive pressure mode.

Ventilation

Local exhaust (preferred) at site of chemical work or general ventilation.

Protective Gloves	Eye Protection	Other Protective Clothing
Impervious Rubber	Chemical/Dust Goggles or Face Mask	Protective Apron, Boots

Work/Hygiene Practices

Always wash hands thoroughly after using chemical; never bring food, drink, or smoking materials into vicinity of chemicals.

BARIUM

Ba CAS: 7440-39-3

IDENTIFICATION AND TYPICAL USES

Barium is a silver white or yellowish white, slightly malleable metal. It can also exist as a powder or dust. It is used in the manufacture of electronic tubes and as a carrier for radium. It is also used as a catalyst for organic reactions; as a lubricating oil additive; as rat poison; in manufacture of paper electrodes; in fireworks; and in electroplating. Many barium salts are used for various purposes, such as in paints, ceramics, in tanning and finishing leathers, as a mordant for fabrics and dyes, in aluminum refining, in rubber manufacture, and in analytical work.

RISK ASSESSMENT: HEALTH

General Assessment

Certain compounds of barium are irritants of the eyes, mucous membranes, and skin. Primary exposure routes are *inhalation* and *ingestion*. Inhalation of barium dusts can cause irritation of the nose and upper respiratory tract and may produce a benign pneumoconiosis known as baritosis. There may be characteristic X-ray changes. All soluble salts of barium are acute poisons. The barium ion is toxic to muscle producing stimulation and then paralysis.

Ingestion can cause severe hypokalemia. Ingesting barium salts or powders produces initial symptoms related to gastrointestinal reaction. These include nausea, vomiting, colic, diarrhea, followed by myocardial and general muscular stimulation with tingling in the extremities. Severe cases continue to loss of tendon reflexes, general muscular paralysis, and death from respiratory arrest or ventricular fibrillation.

Acute Health Effects

The following acute (short-term) health effects may occur immediately or shortly after exposure to barium:

Skin: No initial symptoms. However, sweating may cause burning sensation as barium dusts react on contact with water.

Eye: Severe irritation and possible damage. Due to water content in the eye, reactions may be immediate and serious.

Lung: Can cause metallic taste, nausea, sore throat, and irritation of the air passages.

Other: Serious effects on musculatory system at high exposures. May cause tingling of the extremities followed by irregular heartbeat and might even cause heart to stop (ventricular fibrillation). Death is also possible as a result of respiratory arrest.

Chronic Health Effects

The following chronic (long-term) health effects may occur at some time after exposure to barium and can last for months or even years:

Cancer Hazards: According to the references, barium has not been tested for its ability to cause cancer in animals.

Reproductive Hazard: According to the references, barium has not been tested for its ability to affect reproduction.

Other Chronic Effects: Repeated exposure can cause abnormal chest X-ray. This usually takes years to develop. An abnormal chest X-ray from barium exposure does not necessarily mean the lung has been damaged but future exposures to irritating substances may affect the lungs. It is not currently known whether barium causes lung damage.

Recommended Risk-Reduction Measures

Personnel should avoid direct contact with barium, especially its dusts or powders. If a less toxic material or compound cannot be substituted for barium, then *engi-*

neering controls are the most effective method of reducing exposures. The best protection is to enclose operations and/or provide local exhaust ventilation at the site of barium dust release. While not always operationally feasible, isolating operations can also reduce exposure. Using respiratory protection is less effective than the controls mentioned above, but is still advisable whenever working with or around barium. A self-contained breathing apparatus (SCBA) with full facepiece and pressure demand is recommended for the greatest possible respiratory protection, especially when exposures near the IDLH level of 250 mg/m^3. However, for low or infrequent exposure, a MSHA/NIOSH approved full facepiece respirator with a high efficiency particulate air (HEPA) filter can be used. If a full facepiece is not available, then chemical/dust goggles should be worn to protect the eyes. A face shield and protective apron should also be worn. To prevent hand and skin exposures, gloves should be worn.

Administrative controls should also be in place to minimize the potential for human exposures. These may include written procedures or policies which specify the methods and techniques that will be practiced whenever personnel are to work with barium.

All personnel should receive *training* on the use, hazards, protective measures, emergency actions, and other precautions per 29 CFR 1910.1200 (Hazard Communication), prior to the first assignment in an area where barium is used or stored.

Before beginning employment working with barium, and at regular intervals thereafter (e.g., annually), those personnel with frequent or potentially high exposures should be provided the following recommended medical tests:

☑ Lung function test.

Any evaluation should include a careful history of past and present symptoms with an examination. Medical tests that simply look for existing damage are not a substitute for controlling exposures. Also, since smoking can cause heart disease as well as lung cancer, emphysema, and other known respiratory problems, exposure to barium may cause additional complications. Therefore, personnel who smoke should avoid any unprotected contact with barium powders.

Other methods to reduce exposure include:

☑ Where possible, enclose operations and use local exhaust ventilation at the site of chemical release. If local exhaust ventilation or enclosure is not used, respiratory protection should be mandatory.

☑ Always ensure that proper protective clothing is worn when using chemical substances.

☑ Wash thoroughly immediately after exposure to barium and at the end of the work shift or before eating, drinking, or smoking.

☑ Hazard warning information should be posted in the work area. In addition, as part of an on-going education and training program, all information on the health and safety hazards of barium should be communicated to all potentially exposed workers.

☑ Eye wash stations should be provided in the immediate work area for emergency use. If there is a possibility of skin exposure to barium, emergency shower facilities should also be provided.

☑ Workers whose clothing has been contaminated by barium should change into clean clothes before leaving work. Contaminated work clothing should be laundered only by individuals who have been informed of the hazards of exposure to barium.

RISK ASSESSMENT: ENVIRONMENT

General Assessment

The environment is at risk of exposure during transportation, storage, disposal, or destruction of barium. In almost every scenario, the threat of environmental exposure is contingent upon the proper handling of the chemical substance. Accidental spills, large or small, where barium contacts incompatible commodities can result in fire, explosion (depending upon conditions of spill), and possible contamination of the surrounding environmental mediums (water, soil, and air).

In powdered form, barium is considered a flammable solid. It can present a moderate fire and explosion hazard. In bulk form, it does not present a serious fire or explosion hazard. However, because it is incompatible with water and many other commodities, extreme caution is required in handling, storage, transportation, and disposal of barium. These characteristics also require special consideration during any emergency situation involving a leak or spill of barium powder or dust. Should barium ever come into contact with incompatible substances (such as water and strong acids) either during use, transportation, storage, or disposal, violent and even explosive reactions are possible.

Barium may enter the environment as a result of industrial and municipal waste treatment plant discharges and from spills.

☠ *Acute Ecological Effects*

Acute (short-term) toxic effects may include the death of animals, birds, or fish and death or low growth rate in plants. Acute effects are seen 2 to 4 days after animals or plants are exposed to barium. Barium and its salts have moderate acute toxicity to aquatic life. Insufficient data are available to evaluate the short-term effects of barium or its salts to plants, birds, or land animals.

Chronic Ecological Effects

Chronic (long-term) toxic effects may include shortened life span, reproductive problems, lower fertility, and changes in appearance or behavior in exposed animals. These effects can be seen long after first exposure(s) to toxic chemicals. Barium and its salts have moderate chronic toxicity to aquatic life. Insufficient data are available to evaluate or predict the long-term effects of barium to plants, birds, or land animals.

Water Solubility

Most of the barium salts are either highly or moderately soluble in water. Concentrations of 1 to 1000 milligrams and more will mix with a liter of water.

⌛ *Persistence in the Environment*

Barium is highly persistent in water, with a half-life greater than 200 days. The half-life of a pollutant is the amount of time it takes for one half of the chemical to be degraded. No data are available on the estimated percentages of barium persistence in the aquatic or terrestrial environments.

Bioaccumulation in Aquatic Organisms

Some substances increase in concentration, or *bioaccumulate,* in living organisms as they breathe contaminated air, drink contaminated water, or eat contaminated food. These chemicals can become concentrated in the tissues and internal organs of animals as well as humans.

The concentration of barium found in fish tissues is expected to be about the same as the average concentration of barium in the water from which the fish was taken.

Recommended Risk-Reduction Measures

Proper training of all transporters will reduce the likelihood of a mishap or accident resulting in a leak or spill of barium dusts or powders into the environment. The correct use of DOT labeling (per the MSDS) on all containers, trucks, and rail cars will enable emergency responders to react properly and quickly to any disaster thereby reducing the potential risk to the environment and to personnel.

Storage of barium should be segregated from incompatible chemicals to minimize the risk of cross-contamination or contact.

If a spill or leak to the environment has occurred, fire department, emergency response and/or hazardous materials spill personnel should be notified immediately. Cleanup should be attempted only by those trained in proper spill containment procedures. Contaminated soils should be removed for incineration and replaced with clean soil. If barium should contact the water table, aquifer, or navigable waterway, time is of the essence. It is highly soluble and total remediation may not be possible. The local and/or state emergency response authorities must be notified. A comprehensive emergency response or disaster preparedness/recovery plan should be in place prior to any operations involving the use, transportation, storage, or disposal of barium. If barium is spilled or leaked, the following specific steps are recommended:

- ☑ Restrict persons not wearing protective clothing from area of spill or leak until cleanup is complete.
- ☑ Remove all ignition sources.
- ☑ Collect powdered materials in the safest manner possible and deposit in sealed containers. Recommend using vacuum with HEPA filter. Do not dry sweep (generates airborne dusts).
- ☑ It may be necessary to dispose of barium as a hazardous waste. The responsible state agency or the regional office of the federal Environmental Protection Agency (EPA) should be contacted for specific recommendations.

RISK ASSESSMENT: BUSINESS

General Assessment

Accidents or mishaps involving barium can present a significant threat to business operations. The loss or damage of equipment or facilities can significantly affect fiscal viability. Lawsuits that may result from personnel injury/death, public exposures, and/or environmental contamination will also require a serious expenditure of resources. Media attention surrounding an

injury, death, or environmental damage can also result in a loss of profits and loss of current as well as future business.

Recommended Risk-Reduction Measures

Company attorneys, safety and health professionals, and environmental specialists should be involved in the development of any procedures for responding to chemical incidents. A company official should be predesignated as a public relations officer with specific training in dealing with the press. Corporate plans and policies should be developed, approved, and implemented long before any need for such arises.

REFERENCES

Ahlbom, A. 1993. *Biostatistics for Epidemiologists.* New York: CRC Press/Lewis Publishers.

American Conference of Governmental Industrial Hygienists. 1988. *Documentation of the Threshold Limit Values and Biological Exposure Indices,* 5th Edition (with updates). Cincinnati: ACGIH

Asante-Duah, D. K. 1993. *Hazardous Waste Risk Assessment.* New York: CRC Press/Lewis Publishers.

Bloom, A. D. and F. J. DeSerres. 1995. *Ecotoxicity and Human Health.* New York: CRC Press/Lewis Publishers.

Brusick, D. J. 1994. *Methods for Genetic Risk Assessment.* New York: CRC Press/Lewis Publishers.

Bueche, J. R. 1972. *Principles of Physics,* 2nd Edition. New York: McGraw-Hill.

Calabrese, E. J. 1994. *Biological Effects of Low Level Exposures.* New York: CRC Press/Lewis Publishers.

Cockerman, L. G., B. S. Shane. 1994. *Basic Environmental Toxicology.* New York: CRC Press/Lewis Publishers.

Cothern, R. R. 1993. *Comparative Environmental Risk Assessment.* New York: CRC Press/Lewis Publishers.

Gardner, R. W. 1994. *Chemical Intolerance.* New York: CRC Press/Lewis Publishers.

Hallenbeck, W. H. 1993. *Quantitative Risk Assessment for Environmental and Occupational Health,* 2nd Edition. New York: CRC Press/Lewis Publishers.

Howard, P. H. 1995. *Handbook of Environmental Fate and Exposures Data for Organic Chemicals,* Vol. I-IV. New York: CRC Press/Lewis Publishers.

Jones, F. E. 1994. *Toxic Organic Vapors in the Workplace.* New York: CRC Press/Lewis Publishers.

Keith, L. H. and M. M. Walker. 1995. *Handbook of Air Toxics.* New York: CRC Press/Lewis Publishers.

Landis, W. G., M. H. Yu. 1995. *Introduction to Environmental Toxicology.* New York: CRC Press/Lewis Publishers.

Langerman, N. 1994 *Precautionary Labels for Chemical Containers.* New York: CRC Press/Lewis Publishers.

Lauwerys, R. R., P. Hoet. 1993. *Industrial Chemical Exposure.* New York: CRC Press/Lewis Publishers.

Lewis, R. J., Sr. 1993. *Hawley's Condensed Chemical Dictionary,* 12th Edition. New York: Van Nostrand Reinhold.

Lewis, R. J., Sr. 1992. *Sax's Dangerous Properties of Industrial Materials,* Volumes 1, 2, and 3, 8th Edition. New York: Van Nostrand Reinhold.

Lide, D. R. 1995. *Handbook of Chemistry and Physics,* 76th Edition. New York: CRC Press/Lewis Publishers.

Manahan, S. E. 1994. *Environmental Chemistry,* 6th Edition. New York: CRC Press/Lewis Publishers.

Mellström, G. A., J. E. Wahlberg, and H. I. Maibach. 1994. *Protective Gloves for Occupational Use.* New York: CRC Press/Lewis Publishers.

Meyer, E. 1990. *Chemistry of Hazardous Materials.* Englewood Cliffs, NJ: Prentice Hall.

Mycock, J. C., J. D. McKenna, and L. Theodore. 1995. *Handbook of Air Pollution Control Engineering and Technology.* New York: CRC Press/Lewis Publishers.

Patnaik, P. 1992. *A Comprehensive Guide to the Hazardous Properties of Chemical Substances.* New York: Van Nostrand Reinhold.

Proctor, N. H., J. P. Hughes. 1991. *Chemical Hazards of the Workplace.* New York: Van Nostrand Reinhold.

Quigley, D. R. 1994. *Handbook of Emergency Chemical Management.* New York: CRC Press/Lewis Publishers.

Rea, W. J. 1995. *Chemical Sensitivity,* Vol. I-III. New York: CRC Press/Lewis Publishers.

Reeves, A. L. 1979. Barium. *Handbook on the Toxicology of Metals.* pp 321-328. New York: Elsevier North-Holland.

Richardson, M. 1995. *Dictionary of Substance and Their Effects,* Vol. I-VI. New York: CRC Press/Lewis Publishers.

Sacarello, H. L. A. 1994. *The Comprehensive Handbook of Hazardous Materials.* New York: CRC Press/Lewis Publishers.

Sheftel, V. O. 1995. *Handbook of Toxic Properties of Monomers and Additives.* New York: CRC Press/Lewis Publishers.

Talbott, E., G. Craun. 1995. *An Introduction to Environmental Epidemiology.* New York: CRC Press/Lewis Publishers.

Torkelson, T. R. V. K. Rowe. 1981. *Patty's Industrial Hygiene and Toxicology,* Vol. 2B. New York: Wiley-Interscience.

U.S. Department of Health, Education, and Welfare. National Institute for Occupational Safety and Health (NIOSH). 1978. *Criteria for a Recommended Standard. Occupational Exposure to Barium.* DHEW (NIOSH) Publication Number: 78-216. Washington, DC: U.S. Government Printing Office.

U.S. Department of Labor. 1983. Barium metal: barium trioxide; and barium sulfide response to the interagency testing committee. *Federal Register.* 48:717-724.

MATERIAL SAFETY DATA SHEET

CHEMICAL NAME	
BARIUM BROMATE	**CRC PRESS/LEWIS PUBLISHERS DISCLAIMER:** THE INFORMATION AND RECOMMENDATIONS PRESENTED HEREIN ARE BASED ON SOURCES BELIEVED TO BE RELIABLE. CRC MAKES NO REPRESENTATION ON ITS COMPLETENESS OR ACCURACY. IT IS THE RESPONSIBILITY OF THE USER TO DETERMINE THE CHEMICAL'S SUITABILITY FOR ITS INTENDED USE, THE CHEMICAL'S SAFE USE, AND THE CHEMICAL'S PROPER DISPOSAL. NO REPRESENTATIONS AND/OR WARRANTIES, EITHER EXPRESSED OR IMPLIED, OF THE MERCHANTABILITY OR FITNESS FOR A PARTICULAR PURPOSE, OR OF ANY OTHER NATURE, ARE MADE WITH RESPECT TO THE INFORMATION PROVIDED IN THIS MSDS OR TO THE CHEMICAL TO WHICH INFORMATION MAY REFER. CRC NEITHER ASSUMES NOR AUTHORIZES ANY OTHER PERSON TO ASSUME FOR IT, ANY OTHER ADDITIONAL RESPONSIBILITY OR LIABILITY FOR THE USE OF, OR RELIANCE UPON, THIS INFORMATION.

HAZARD WARNING INFORMATION

HEALTH	FIRE	REACTIVE	OTHER	DEGREE OF HAZARD	COLOR CODING	OTHER CODES
2	0	1	OX	0 = Minimum Hazard 1 = Slight Hazard 2 = Moderate Hazard 3 = Serious Hazard 4 = Severe Hazard	HEALTH = BLUE FIRE = RED REACTIVITY = YELLOW OTHER = WHITE	OX = Oxidizer ACID = Acid ALK = Alkali COR = Corrosive W = Use No Water

SECTION I - GENERAL INFORMATION

Characterization: **Metal**	RCRA Number: **Not Listed**	EPA Class: **Not Applicable**
DOT Proper Shipping Name: **Barium Bromate**	Chemical Abstract Service (CAS) Number: **13967-90-3**	
DOT Hazard Class and Label Requirements: **Oxidizer; Oxidizer and Poison**	DOT Emergency Guide Code: **42**	
DOT Identification Number: **UN 2719**	Atomic Symbol: **$Ba(BrO_3)_2 \cdot H_2O$**	

Synonyms
Not Applicable.

SECTION II - HAZARDOUS INGREDIENTS AND IDENTITY INFORMATION

Hazardous Components (specific identity)	OSHA Exposure Criteria	NIOSH Exposure Criteria	Immediately Dangerous to Life and Health (IDLH)	ACGIH Exposure Criteria
Barium bromate (derivation: Passing bromine into a solution of barium hydroxide will form barium bromide and barium bromate, which are then separated by crystallization).	**PEL:** 0.5 mg(Ba)/m^3 **STEL:** Not Established	**REL:** 0.5 mg(Ba)/m^3 **STEL:** Not Established	**1100 mg/m^3 for Soluble Compounds**	**TLV:** 0.5 mg(Ba)/m^3 **STEL:** Not Established

SECTION III - PHYSICAL AND CHEMICAL CHARACTERISTICS

Boiling Point: **Not Found**	Specific Gravity (H_2O = 1): **3.99**
Vapor Pressure (mm Hg): **1 at 1920°F (1049°C)**	Molecular Weight (atomic weight): **393.1**
Vapor Density (Air = 1): **Not Applicable**	Melting Point: **500°F (260°C) Decomposes**

Solubility
Most of the barium salts are either highly or moderately soluble in water.

Appearance and Odor
White crystals or crystalline powder, no appreciable odor.

SECTION IV - FIRE AND EXPLOSION HAZARD DATA

Flash Point (method used): **Not Found (Metal)**	Explosive Limits in Air % by Volume: **LEL: N/A UEL: N/A**
NFPA Classification: **Flammable Solid (in powder form)**	Autoignition Temperature: **Not Found**

Extinguishing Media
Use agent suitable to surrounding fire. Barium bromate by itself will not burn.

Special Fire Fighting Procedures
Wear full protective clothing and self-contained breathing apparatus (SCBA). Poisonous gases are produced in fire, including bromide. Use water mist from unmanned device to keep fire-exposed containers cool. Keep cooling containers after fire is extinguished.

Unusual Fire and Explosion Hazards
Containers may buildup pressure and explode in fire. Serious fire hazard by chemical reaction with easily oxidized materials. Barium bromate is non-combustible, but it will explode when temperatures reach 300°C (572°F).

SECTION V - REACTIVITY DATA

Stability		Conditions to Avoid **Avoid heat and open flame, especially around dusts and powders of barium. Do not allow contact with incompatible agents since violent and explosive reaction can occur.**
Stable	Unstable	Incompatibility *(materials to avoid)*
X		**Violent or explosive reactions in arsenic, ammonia, carbon, copper, aluminum, metallic sulfides, ordinary combustibles (wood, paper, oil, fuels), heat organics, phosphorus, and sulfur.**
Hazardous Polymerization		Conditions to Avoid **Hazardous polymerization of barium bromate will not occur.**
May Occur	Will Not Occur	Hazardous Decomposition or By-products
	X	**Poisonous gases, including bromide, are emitted during fire.**

SECTION VI - HEALTH HAZARD DATA

Primary Route(s) of Entry:	Inhalation?	Absorption (skin/eye)?	Ingestion?
	X		X

Health Hazards

INHALATION: Breathing the dusts of barium bromate may cause severe irritation of the nose, throat, and bronchial tubes, causing cough and phlegm. May also damage lung tissue.

EYES & SKIN: Exposure to the eyes may cause moderate to severe irritation and even burns.

INGESTION: Water and stomach acids solubilize some barium salts. Can cause poisoning. Symptoms are vomiting, diarrhea, slow irregular pulse, transient hypertension, convulsive tremors.

Carcinogenicity	NTP Listed?	IARC Cancer Review Group?	OSHA Regulated?	Target Organs?
Unknown Human Unknown Animal	**No**	**No**	**29 CFR 1910.1000 Table Z-1**	**Respiratory system, eyes, skin.**

Medical Conditions Generally Aggravated by Exposure

None reported. However, persons with existing lung dysfunction may experience effects quicker.

Emergency and First-aid Procedures

Eye contact: Do not allow victim to keep eyes tightly shut. Flush immediately with water for 15 minutes (minimum), seek medical attention. Skin contact: Remove all contaminated clothing. Immediately wash area with flooding amounts of soap and water. For inhalation: Remove the person from exposure. Provide respiratory assistance and CPR. Transfer to medical facility. If swallowed, seek medical attention immediately. Never give anything by mouth to an unconscious or convulsing person.

SECTION VII - PRECAUTIONS FOR SAFE HANDLING AND USE

Steps to be Taken in Case Material is Released or Spilled

Collect powdered materials in most convenient and safe manner and deposit in sealed containers. Dry sweep not recommended. Use HEPA vacuum. Remove all ignition sources. Restrict those not involved in cleanup from entering area. Notify appropriate authorities, as required by applicable regulations.

Preferred Waste Disposal Method

Reactive waste may have to be treated and rendered non-reactive prior to disposal.

Precautions to be Taken in Handling and Storage

Explosion risks vary with type of barium compound. Store to avoid contact with strong acids and halogenated hydrocarbons. Store powders in sealed containers under an inert gas or oxygen-free petroleum liquid away from oxidizers, acids, metals, or ammonia.

Other Precautions and Warnings

Sources of ignition (heat, flame, cigarettes, incompatibles) are prohibited where barium is used, handled, or stored. Use only non-sparking tools or equipment, especially when opening or closing containers.

SECTION VIII - CONTROL MEASURES AND PERSONAL PROTECTIVE EQUIPMENT

Respiratory Protection *(specify type)*

For dust exposures below PEL, use a respirator with a high efficiency particulate air (HEPA) filter, otherwise use a NIOSH/MSHA approved self-contained breathing apparatus with full facepiece operated in positive pressure mode.

Ventilation

Local exhaust (preferred) at site of chemical work or general ventilation.

Protective Gloves	Eye Protection	Other Protective Clothing
Impervious Rubber	**Chemical/Dust Goggles or Face Mask**	**Protective Apron, Boots**

Work/Hygiene Practices

Always wash hands thoroughly after using chemical; never bring food, drink, or smoking materials into vicinity of chemicals.

BARIUM BROMATE

$Ba(BrO_3)_2 \cdot H_2O$ CAS: 13967-90-3

IDENTIFICATION AND TYPICAL USES

Barium bromate is a white crystal or crystalline powder. Many barium compounds are used in the manufacture of electronic tubes and as carriers for radium. Barium is also used as a catalyst for organic reactions; as a lubricating oil additive; as rat poison; in manufacture of paper electrodes; in fireworks; and in electroplating. Many barium salts are used for various purposes, such as in paints, ceramics, in tanning and finishing leathers, as a mordant for fabrics and dyes, in aluminum refining, in rubber manufacture, and in analytical work.

RISK ASSESSMENT: HEALTH

General Assessment

Certain compounds of barium are irritants of the eyes, mucous membranes, and skin. Primary exposure routes are *inhalation* and *ingestion*. Inhalation of barium dusts can cause irritation of the nose and upper respiratory tract and may produce a benign pneumoconiosis known as baritosis. There may be characteristic X-ray changes. All soluble salts of barium are acute poisons. The barium ion is toxic to muscle producing stimulation and then paralysis.

Ingestion can cause severe hypokalemia. Ingesting barium salts or powders produces initial symptoms related to gastrointestinal reaction. These include nausea, vomiting, colic, diarrhea, followed by myocardial and general muscular stimulation with tingling in the extremities. Severe cases continue to loss of tendon reflexes, general muscular paralysis, and death from respiratory arrest or ventricular fibrillation.

Acute Health Effects

The following acute (short-term) health effects may occur immediately or shortly after exposure (within hours or within days) to barium bromate:

Skin: May cause severe irritation and even burns.

Eye: Severe irritation and possible damage.

Lung: Dusts will irritate the lungs causing cough and phlegm.

Other: Industrial grade barium compounds may be contaminated with silica. Exposure can lead to lung scarring and reduce lung function.

Chronic Health Effects

The following chronic (long-term) health effects may occur at some time after exposure to barium and can last for months or even years:

Cancer Hazards: According to the references, barium has not been tested for its ability to cause cancer in animals.

Reproductive Hazard: According to the references, barium has not been tested for its ability to affect reproduction.

Other Chronic Effects: Repeated exposure can cause abnormal chest X-ray. This usually takes years to develop. An abnormal chest X-ray from barium exposure does not necessarily mean the lung has been damaged but future exposures to irritating substances may affect the lungs. It is not currently known whether barium causes lung damage. Repeated skin contact can cause chronic dryness and cracking. Repeated inhalation can cause bromine to build up in the body.

Recommended Risk-Reduction Measures

Personnel should avoid direct contact with barium bromate dusts or powders. If a less toxic material or compound cannot be substituted for barium bromate, then *engineering controls* are the most effective method of reducing exposures. The best protection is to enclose operations and/or provide local exhaust ventilation at the site of barium dust release. While not always operationally feasible, isolating operations can also reduce exposure. Using respiratory protection is less effective than the controls mentioned above, but is still advisable whenever working with or around barium bromate. A self-contained breathing apparatus (SCBA) with full facepiece and pressure demand is recommended for the greatest possible respiratory protection, especially when exposures near the IDLH level of 1100 mg/m^3. However, for low or infrequent exposure, a MSHA/NIOSH approved full facepiece respirator with a high efficiency particulate air (HEPA) filter can be used. If a full facepiece is not available, then chemical/dust goggles should be worn to protect the eyes. A face shield and protective apron should also be worn. To prevent hand and skin exposures, gloves should be worn.

Administrative controls should also be in place to minimize the potential for human exposures. These may include written procedures or policies which specify the methods and techniques that will be practiced whenever personnel are to work with barium bromate.

All personnel should receive *training* on the use, hazards, protective measures, emergency actions, and other precautions per 29 CFR 1910.1200 (Hazard Communication), prior to the first assignment in an area where barium bromate is used or stored.

Before beginning employment working with barium bromate, and at regular intervals thereafter (e.g., annually), those personnel with frequent or potentially high exposures should be provided the following recommended medical tests:

- ☑ Lung function test.
- ☑ Serum bromide level.

Any evaluation should include a careful history of past and present symptoms with an examination. Medical tests that simply look for existing damage are not a substitute for controlling exposures. Also, since smoking can cause heart disease as well as lung cancer, emphysema, and other known respiratory problems, exposure to barium may cause additional complications. Therefore, personnel who smoke should avoid any unprotected contact with barium powders.

Other methods to reduce exposure include:

- ☑ Where possible, enclose operations and use local exhaust ventilation at the site of chemical release. If local exhaust ventilation or enclosure is not used, respiratory protection should be mandatory.
- ☑ Always ensure that proper protective clothing is worn when using chemical substances.
- ☑ Wash thoroughly immediately after exposure to barium bromate and at the end of the work shift or before eating, drinking, or smoking.
- ☑ Hazard warning information should be posted in the work area. In addition, as part of an on-going education and training program, all information on the health and safety hazards of barium bromate should be communicated to all exposed workers.
- ☑ Eye wash stations should be provided in the immediate work area for emergency use. If there is a possibility of skin exposure to barium bromate, emergency shower facilities should also be provided in the immediate area.
- ☑ Workers whose clothing has been contaminated by barium bromate should change into clean clothes before leaving work. Contaminated work clothing should be laundered only by individuals who have been informed of the hazards of exposure to barium bromate.

RISK ASSESSMENT: ENVIRONMENT

General Assessment

The environment is at risk of exposure during transportation, storage, disposal, or destruction of barium compounds. In almost every scenario, the threat of environmental exposure is contingent upon the proper handling of the chemical substance. Accidental spills, large or small, where barium bromate contacts incompatible commodities can result in fire, explosion (depending upon conditions of spill), and possible contamination of the surrounding environmental mediums (water, soil, and air).

In its powdered form, barium bromate is considered a non-combustible solid. However, it can react explosively and present a serious fire hazard. Because it is incompatible with many commodities, including other metals and most combustibles, extreme caution is required in handling, storage, transportation, and disposal of barium bromate. These characteristics also require special consideration during any emergency situation involving a leak or spill of barium bromate powder or dust. Should barium bromate ever come into contact with incompatible substances either during use, transportation, storage, or disposal, violent and even explosive reactions are possible.

Barium bromate can enter the environment from industrial and municipal waste treatment plant discharges and from spills.

Acute Ecological Effects

Acute (short-term) toxic effects may include the death of animals, birds, or fish and death or low growth rate in plants. Acute effects are seen 2 to 4 days after animals or plants are exposed to barium. Barium and its salts have moderate acute toxicity to aquatic life. Insufficient data are available to evaluate the short-term effects of barium or its salts to plants, birds, or land animals.

Chronic Ecological Effects

Chronic (long-term) toxic effects may include shortened life span, reproductive problems, lower fertility, and changes in appearance or behavior in exposed animals. These effects can be seen long after first exposure(s) to toxic chemicals. Barium and its salts have moderate chronic toxicity to aquatic life. Insufficient data are available to evaluate or predict the long-term effects of barium to plants, birds, or land animals.

Water Solubility

Most of the barium salts are either highly or moderately soluble in water. Concentrations of 1 to 1000 milligrams and more will mix with a liter of water.

Persistence in the Environment

Barium is highly persistent in water, with a half-life greater than 200 days. The half-life of a pollutant is the amount of time it takes for one half of the chemical to be degraded. No data are available on the estimated percentages of barium persistence in the aquatic or terrestrial environments.

Bioaccumulation in Aquatic Organisms

Some substances increase in concentration, or *bioaccumulate,* in living organisms as they breathe contaminated air, drink contaminated water, or eat contaminated food. These chemicals can become concentrated in the tissues and internal organs of animals as well as humans.

The concentration of barium found in fish tissues is expected to be about the same as the average concentration of barium in the water from which the fish was taken.

Recommended Risk-Reduction Measures

Proper training of all transporters will reduce the likelihood of a mishap or accident resulting in a leak or spill of barium bromate dusts or powders into the environment. The correct use of DOT labeling (per the MSDS) on all containers, trucks, and rail cars will enable emergency responders to react properly and quickly to any disaster thereby reducing the potential risk to the environment and to personnel.

Storage of barium bromate should be segregated from incompatible chemicals to minimize the risk of cross-contamination or contact.

If a spill or leak to the environment has occurred, fire department, emergency response and/or hazardous materials spill personnel should be notified immediately. Cleanup should be attempted only by those trained in proper spill containment procedures. Contaminated soils should be removed for incineration and replaced with clean soil. If barium bromate should contact the water table, aquifer, or navigable waterway, time is of the essence. It is highly soluble and total remediation may not be possible. The local and/or state emergency response authorities must be notified. A comprehensive emergency response or disaster preparedness/recovery plan should be in place prior to any operations involving the use, transportation, storage, or disposal of barium bromate. If barium bromate powder is spilled or leaked, the following specific steps are recommended:

- ☑ Restrict persons not wearing protective clothing from area of spill or leak until cleanup is complete.
- ☑ Remove all ignition sources.
- ☑ Collect powdered materials in the safest manner possible and deposit in sealed containers. Recommend using vacuum with HEPA filter. Do not dry sweep (generates airborne dusts).
- ☑ It may be necessary to dispose of barium bromate as a hazardous waste. The responsible state agency or the regional office of the federal Environmental Protection Agency (EPA) should be contacted for specific recommendations.

RISK ASSESSMENT: BUSINESS

General Assessment

Accidents or mishaps involving barium bromate can present a significant threat to business operations. The loss or damage of equipment or facilities can significantly effect fiscal viability. Lawsuits that may result from personnel injury/death, public exposures, and/or environmental contamination will also require a serious

expenditure of resources. Media attention surrounding an injury, death, or environmental damage can also result in a loss of profits and loss of current as well as future business.

❶ *Recommended Risk-Reduction Measures*

Company attorneys, safety and health professionals, and environmental specialists should be involved in the development of any procedures for responding to chemical incidents. A company official should be predesignated as a public relations officer with specific training in dealing with the press. Corporate plans and policies should be developed, approved, and implemented long before any need for such arises.

REFERENCES

Ahlbom, A. 1993. *Biostatistics for Epidemiologists.* New York: CRC Press/Lewis Publishers.

American Conference of Governmental Industrial Hygienists. 1988. *Documentation of the Threshold Limit Values and Biological Exposure Indices,* 5th Edition (with updates). Cincinnati: ACGIH

Asante-Duah, D. K. 1993. *Hazardous Waste Risk Assessment.* New York: CRC Press/Lewis Publishers.

Bloom, A. D. and F. J. DeSerres. 1995. *Ecotoxicity and Human Health.* New York: CRC Press/Lewis Publishers.

Brusick, D. J. 1994. *Methods for Genetic Risk Assessment.* New York: CRC Press/Lewis Publishers.

Bueche, J. R. 1972. *Principles of Physics,* 2nd Edition. New York: McGraw-Hill.

Calabrese, E. J. 1994. *Biological Effects of Low Level Exposures.* New York: CRC Press/Lewis Publishers.

Cockerman, L. G., B. S. Shane. 1994. *Basic Environmental Toxicology.* New York: CRC Press/Lewis Publishers.

Cothern, R. R. 1993. *Comparative Environmental Risk Assessment.* New York: CRC Press/Lewis Publishers.

Gardner, R. W. 1994. *Chemical Intolerance.* New York: CRC Press/Lewis Publishers.

Hallenbeck, W. H. 1993. *Quantitative Risk Assessment for Environmental and Occupational Health,* 2nd Edition. New York: CRC Press/Lewis Publishers.

Howard, P. H. 1995. *Handbook of Environmental Fate and Exposures Data for Organic Chemicals,* Vol. I-IV. New York: CRC Press/Lewis Publishers.

Jones, F. E. 1994. *Toxic Organic Vapors in the Workplace.* New York: CRC Press/Lewis Publishers.

Keith, L. H. and M. M. Walker. 1995. *Handbook of Air Toxics.* New York: CRC Press/Lewis Publishers.

Landis, W. G., M. H. Yu. 1995. *Introduction to Environmental Toxicology.* New York: CRC Press/Lewis Publishers.

Langerman, N. 1994 *Precautionary Labels for Chemical Containers.* New York: CRC Press/Lewis Publishers.

Lauwerys, R. R., P. Hoet. 1993. *Industrial Chemical Exposure.* New York: CRC Press/Lewis Publishers.

Lewis, R. J., Sr. 1993. *Hawley's Condensed Chemical Dictionary,* 12th Edition. New York: Van Nostrand Reinhold.

Lewis, R. J., Sr. 1992. *Sax's Dangerous Properties of Industrial Materials,* Volumes 1, 2, and 3, 8th Edition. New York: Van Nostrand Reinhold.

Lide, D. R. 1995. *Handbook of Chemistry and Physics,* 76th Edition. New York: CRC Press/Lewis Publishers.

Manahan, S. E. 1994. *Environmental Chemistry,* 6th Edition. New York: CRC Press/Lewis Publishers.

Mellström, G. A., J. E. Wahlberg, and H. I. Maibach. 1994. *Protective Gloves for Occupational Use.* New York: CRC Press/Lewis Publishers.

Meyer, E. 1990. *Chemistry of Hazardous Materials.* Englewood Cliffs, NJ: Prentice Hall.

Mycock, J. C., J. D. McKenna, and L. Theodore. 1995. *Handbook of Air Pollution Control Engineering and Technology.* New York: CRC Press/Lewis Publishers.

Patnaik, P. 1992. *A Comprehensive Guide to the Hazardous Properties of Chemical Substances.* New York: Van Nostrand Reinhold.

Proctor, N. H., J. P. Hughes. 1991. *Chemical Hazards of the Workplace.* New York: Van Nostrand Reinhold.

Quigley, D. R. 1994. *Handbook of Emergency Chemical Management.* New York: CRC Press/Lewis Publishers.

Rea, W. J. 1995. *Chemical Sensitivity,* Vol. I-III. New York: CRC Press/Lewis Publishers.

Reeves, A. L. 1979. Barium. *Handbook on the Toxicology of Metals.* pp 321-328. New York: Elsevier North-Holland.

Richardson, M. 1995. *Dictionary of Substance and Their Effects,* Vol. I-VI. New York: CRC Press/Lewis Publishers.

Sacarello, H. L. A. 1994. *The Comprehensive Handbook of Hazardous Materials.* New York: CRC Press/Lewis Publishers.

Sheftel, V. O. 1995. *Handbook of Toxic Properties of Monomers and Additives.* New York: CRC Press/Lewis Publishers.

Torkelson, T. R. V. K. Rowe. 1981. *Patty's Industrial Hygiene and Toxicology,* Vol. 2B. New York: Wiley-Interscience.

U.S. Department of Health, Education, and Welfare. National Institute for Occupational Safety and Health (NIOSH). 1978. *Criteria for a Recommended Standard. Occupational Exposure to Barium.* DHEW (NIOSH) Publication Number: 78-216. Washington, DC: U.S. Government Printing Office.

U.S. Department of Labor. 1983. Barium metal: barium trioxide; and barium sulfide response to the interagency testing committee. *Federal Register.* 48:717-724.

MATERIAL SAFETY DATA SHEET

CHEMICAL NAME	
BARIUM CARBONATE	**CRC PRESS/LEWIS PUBLISHERS DISCLAIMER:** THE INFORMATION AND RECOMMENDATIONS PRESENTED HEREIN ARE BASED ON SOURCES BELIEVED TO BE RELIABLE. CRC MAKES NO REPRESENTATION ON ITS COMPLETENESS OR ACCURACY. IT IS THE RESPONSIBILITY OF THE USER TO DETERMINE THE CHEMICAL'S SUITABILITY FOR ITS INTENDED USE, THE CHEMICAL'S SAFE USE, AND THE CHEMICAL'S PROPER DISPOSAL. NO REPRESENTATIONS AND/OR WARRANTIES, EITHER EXPRESSED OR IMPLIED, OF THE MERCHANTABILITY OR FITNESS FOR A PARTICULAR PURPOSE, OR OF ANY OTHER NATURE, ARE MADE WITH RESPECT TO THE INFORMATION PROVIDED IN THIS MSDS OR TO THE CHEMICAL TO WHICH INFORMATION MAY REFER. CRC NEITHER ASSUMES NOR AUTHORIZES ANY OTHER PERSON TO ASSUME FOR IT, ANY OTHER ADDITIONAL RESPONSIBILITY OR LIABILITY FOR THE USE OF, OR RELIANCE UPON, THIS INFORMATION.

HAZARD WARNING INFORMATION

HEALTH	FIRE	REACTIVE	OTHER	DEGREE OF HAZARD	COLOR CODING	OTHER CODES
2	0	0		0 = Minimum Hazard 1 = Slight Hazard 2 = Moderate Hazard 3 = Serious Hazard 4 = Severe Hazard	HEALTH = BLUE FIRE = RED REACTIVITY = YELLOW OTHER = WHITE	OX = Oxidizer ACID = Acid ALK = Alkali COR = Corrosive W = Use No Water

SECTION I - GENERAL INFORMATION

Characterization	**Metallic Carbonate**	RCRA Number	**None**
		EPA Class	**Not Applicable**
DOT Proper Shipping Name	**Barium Compound, N.O.S.**	Chemical Abstract Service (CAS) Number	**513-77-9**
DOT Hazard Class and Label Requirements	**Poison; St. Andrews Cross**	DOT Emergency Guide Code	**No Citation**
DOT Identification Number	**UN 1564**	Atomic Symbol	**$BaCO_3$**

Synonyms
Barium salt; carbonic acid; C.I. pigment white 10.

SECTION II - HAZARDOUS INGREDIENTS AND IDENTITY INFORMATION

Hazardous Components (specific identity)	OSHA Exposure Criteria	NIOSH Exposure Criteria	Immediately Dangerous to Life and Health (IDLH)	ACGIH Exposure Criteria
Barium carbonate (derivation: Precipitated barium carbonate is made by reaction of sodium carbonate or carbon dioxide with barium sulfide. Also occurs naturally as the mineral witherite).	**PEL:** 0.5 mg(Ba)/m^3 **STEL:** Not Established	**REL:** 0.5 mg(Ba)/m^3 **STEL:** Not Established	**Not Established for BaO_2 (250 mg/m^3 for Soluble Ba Compounds)**	**TLV:** 0.5 mg(Ba)/m^3 **STEL:** Not Established

SECTION III - PHYSICAL AND CHEMICAL CHARACTERISTICS

Boiling Point	**2372°F (1300°C) Decomposes**	Specific Gravity (H_2O = 1)	**4.3**
Vapor Pressure (mm Hg)	**1 at 68°F (20°C)**	Molecular Weight (atomic weight)	**197.35**
Vapor Density (Air = 1)	**4.43**	Melting Point	**1492°F (811°C)**

Solubility
Most of the barium salts are either highly or moderately soluble in water. Barium carbonate powder is slightly soluble (0.002g/100 cc at 20°C) in water.

Appearance and Odor
White to grayish-white fine granular powder, no appreciable odor.

SECTION IV - FIRE AND EXPLOSION HAZARD DATA

Flash Point (method used)	**Not Found**	Explosive Limits in Air % by Volume	**LEL: N/A UEL: N/A**
NFPA Classification	**Non-Combustible Solid**	Autoignition Temperature	**Not Found**

Extinguishing Media
Use extinguishing media suitable to surrounding fire. Use water spray to wet down material.

Special Fire Fighting Procedures
Wear full protective clothing and self-contained breathing apparatus (SCBA). Poisonous gases are produced in fire. Use water mist to keep fire-exposed containers cool. Fight fire from distance, if possible.

Unusual Fire and Explosion Hazards
None reported. However, when involved in a fire, barium carbonate produces barium oxide which is a strong alkaline material and exothermic upon contact with water. Hot barium oxide can react with oxygen to form barium peroxide, which can create a serious fire and explosion risk if in contact with organic or combustible materials. Remove containers from fire if it can be done without risk.

SECTION V - REACTIVITY DATA

Stability		Conditions to Avoid Barium carbonate is normally stable under routine conditions of handling and storage. It is not compatible with some acids and acid solutions.
Stable X	Unstable	Incompatibility *(materials to avoid)* Incompatible with bromine trifluoride and 2-furanpercarboxylic acid. It is soluble in most acids, producing carbon dioxide as a by-product gas from the mixture.
Hazardous Polymerization		Conditions to Avoid Hazardous polymerization of barium carbonate metal will not occur.
May Occur	Will Not Occur X	Hazardous Decomposition or By-products Poisonous gases are emitted during fire, including barium oxide and carbon dioxide.

SECTION VI - HEALTH HAZARD DATA

Primary Route(s) of Entry:	Inhalation?	Absorption (skin/eye)?	Ingestion?
	X		X

Health Hazards

INHALATION: May produce benign pneumoconiosis (baritosis) which may cause X-ray changes. Other symptoms of exposure include eye, nose, and throat irritation, coughing, and a potential for potassium deficiency that can effect the function of the heart and CNS.

SKIN & EYES: Exposure to the eyes and skin may cause moderate to severe irritation.

INGESTION: Water and stomach acids solubilize barium salts and can cause poisoning. Symptoms are vomiting, colic, diarrhea, slow irregular pulse, transient hypertension, convulsive tremors.

Carcinogenicity	NTP Listed?	IARC Cancer Review Group?	OSHA Regulated?	Target Organs?
Unknown Human Unknown Animal	No	No	No	Respiratory system, GI tract, CNS, muscles, heart.

Medical Conditions Generally Aggravated by Exposure

Persons with existing lung dysfunction, hypertension, CNS, or muscular disorders may be effected.

Emergency and First-aid Procedures

Eye contact: Do not allow victim to keep eyes tightly shut. Flush immediately with water for 15 minutes (minimum), seek medical attention. Skin contact: Remove all contaminated clothing. Immediately wash area with flooding amounts of water. For reddened or blistered skin, seek medical assistance. For inhalation: Remove the person from exposure. Provide respiratory assistance and CPR. Transfer to medical facility. If swallowed, seek medical attention. Never give anything by mouth to a convulsing person.

SECTION VII - PRECAUTIONS FOR SAFE HANDLING AND USE

Steps to be Taken in Case Material is Released or Spilled

Collect powdered materials in most convenient and safe manner and deposit in sealed containers. Dry sweep not recommended. Use HEPA vacuum. Ventilate area of spill. Restrict those not involved in cleanup from entering area. Keep combustible materials away.

Preferred Waste Disposal Method

No citation.

Precautions to be Taken in Handling and Storage

Explosion risks vary with type of barium compound. Store to avoid contact with incompatible materials, moisture, and combustible materials. Prevent physical damage to containers. Keep tightly closed.

Other Precautions and Warnings

Sources of ignition are prohibited where barium carbonate is used, handled, or stored. Store on floors constructed of impermeable and non-combustible materials.

SECTION VIII - CONTROL MEASURES AND PERSONAL PROTECTIVE EQUIPMENT

Respiratory Protection *(specify type)*

For dust exposures below PEL, use a respirator with a high efficiency particulate air (HEPA) filter, otherwise use a self-contained breathing apparatus with full facepiece operated in positive pressure mode.

Ventilation

Local exhaust (preferred) at site of chemical work or general ventilation.

Protective Gloves	Eye Protection	Other Protective Clothing
Impervious Rubber	Chemical/Dust Goggles or Face Mask	Protective Apron, Boots

Work/Hygiene Practices

Always wash hands thoroughly after using chemical; never bring food, drink, or smoking materials into vicinity of chemicals.

BARIUM CARBONATE

$BaCO_3$ CAS: 513-77-9

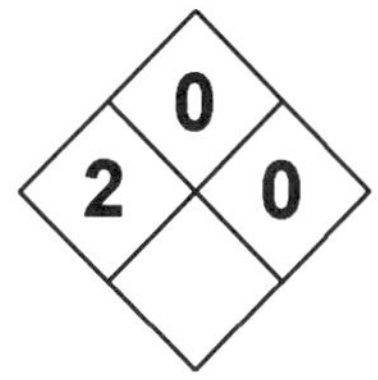

IDENTIFICATION AND TYPICAL USES

Barium carbonate is a white to grayish-white fine granular powder with no detectable odor. Used in rodenticides, ceramics, paints, enamels, marble substitutes, and rubber. Also used in the manufacturing of paper, electrodes, barium salts, optical glasses, as an analytical reagent, in the treatment of brines in chlorine-alkali cells to remove sulfates, in casehardening baths, and in radiation-resistant glass for color television tubes.

RISK ASSESSMENT: HEALTH

General Assessment

Certain compounds of barium are irritants of the eyes, mucous membranes, and skin. Primary exposure routes are *inhalation* and *ingestion.* It has not been established if exposure to barium poses any carcinogenic, teratogenic, or mutagenic effects.

Inhalation of barium dusts can cause hypokalemia (potassium deficiency in blood). It will cause irritation of the nose and upper respiratory tract and may produce a benign pneumoconiosis known as baritosis. There may be characteristic X-ray changes. All soluble salts of barium are acute poisons. The barium ion is toxic to muscle producing stimulation and then paralysis. Specifically, barium carbonate will irritate the nose, throat, and mucosa of the upper respiratory tract and bronchial tubes. There may be adverse effects on the heart and the function of the central nervous system (CNS).

Ingesting barium salts or powders produces initial symptoms related to gastrointestinal reaction. These include nausea, vomiting, colic, diarrhea, followed by myocardial and general muscular stimulation with tingling and paralysis in the extremities. Severe cases continue to loss of tendon reflexes, general muscular paralysis, and death from respiratory arrest or ventricular fibrillation. In addition to these effects, ingestion of barium carbonate specifically can cause stomach ulcers, muscle weakness, hypertension, bradycardia (slow pulse rate), paresthesis (tingling sensation of the skin), bloody diarrhea, dypsnea, salivation, nausea, vomiting, and paralysis. It can adversely effect the central nervous system (CNS) with giddiness, convulsions, and tremors.

Acute Health Effects

The following acute (short-term) health effects may occur immediately or shortly after exposure to barium carbonate:

Skin: May cause mild to moderate irritation, depending on the conditions of skin contact and length of exposure.

Eye: Severe irritation.

Lung: Can cause metallic taste, nausea, sore throat, and irritation of the air passages.

Other: Serious effects on musculatory system at high exposures. May cause tingling and paralysis of the extremities followed by irregular heartbeat and might even cause the heart to stop (ventricular fibrillation). Death is also possible as a result of respiratory arrest.

Chronic Health Effects

The following chronic (long-term) health effects may occur at some time after exposure to barium carbonate and can last for months or even years:

Cancer Hazards: According to the references, barium carbonate has not been tested for its ability to cause cancer in animals.

Reproductive Hazard: According to the references, there is limited evidence that exposure to barium carbonate causes adverse reproductive effects in test animals. In rats, it will impair the function of the female and male genitalia. It is not known if exposure to barium carbonate will have this effect on humans.

Other Chronic Effects: Repeated exposure can cause abnormal chest X-ray. This usually takes years to develop. An abnormal chest X-ray from barium carbonate exposure does not necessarily mean the lung has been damaged but future exposures to irritating substances may affect the lungs. It is not currently known whether barium carbonate causes lung damage.

Recommended Risk-Reduction Measures

Personnel should avoid direct contact with barium, especially its dusts or powders. If a less toxic material or compound cannot be substituted for barium carbonate, then *engineering controls* are the most effective method of reducing exposures. The best protection is to enclose operations and/or provide local exhaust ventilation at the site of barium carbonate dust release. While not always operationally feasible, isolating operations can also reduce exposure. Using respiratory protection is less effective than the controls mentioned above, but is still advisable whenever working with or around barium carbonate. A self-contained breathing apparatus (SCBA) with full facepiece and pressure demand is recommended for the greatest possible respiratory protection, especially when exposures near the IDLH level of 250 mg/m^3 (as barium). However, for low or infrequent exposure, a MSHA/NIOSH approved full facepiece respirator with a high efficiency particulate air (HEPA) filter can be used. If a full facepiece is not available, then chemical/dust goggles should be worn to protect the eyes. A face shield and protective apron should also be worn. To prevent hand and skin exposures, gloves should be worn.

Administrative controls should also be in place to minimize the potential for human exposures. These may include written procedures or policies which specify the methods and techniques that will be practiced whenever personnel are to work with barium carbonate.

All personnel should receive *training* on the use, hazards, protective measures, emergency actions, and other precautions per 29 CFR 1910.1200 (Hazard Communication), prior to the first assignment in an area where barium carbonate is used or stored.

Before beginning employment working with barium carbonate, and at regular intervals thereafter (e.g., annually), those personnel with frequent or potentially high exposures should be provided the following recommended medical tests:

☑ Lung function test (establish baseline).

If symptoms develop or overexposure is suspected, consider the following additional tests:

☑ Electrocardiogram (EKG).

☑ Neurological examination.

☑ Consider chest X-ray.

☑ Evaluation by a qualified allergist with careful consideration of exposure history and special testing (may help diagnose skin allergy).

Any evaluation should include a careful history of past and present symptoms with an examination. Medical tests that simply look for existing damage are not a substitute for controlling exposures. Also, since smoking can cause heart disease as well as lung cancer, emphysema, and other known respiratory problems, exposure to barium carbonate may cause additional complications. Therefore, personnel who smoke should avoid any unprotected contact with barium carbonate powders.

Other methods to reduce exposure include:

☑ Where possible, enclose operations and use local exhaust ventilation at the site of chemical release. If local exhaust ventilation or enclosure is not used, respiratory protection should be mandatory.

☑ Always ensure that proper protective clothing is worn when using chemical substances. Proper maintenance and care of protective equipment is essential to its proper function.

☑ Wash thoroughly immediately after exposure to barium carbonate and at the end of the work shift or before eating, drinking, or smoking.

☑ Hazard warning information should be posted in the work area. In addition, as part of an on-going education and training program, all information on the health and safety hazards of barium carbonate should be communicated to all potentially exposed workers.

☑ Eye wash stations should be provided in the immediate work area for emergency use. If there is a possibility of skin exposure to barium carbonate, emergency shower facilities should be provided.

☑ Workers whose clothing has been contaminated by barium carbonate should change into clean clothes before leaving work. Contaminated work clothing should be laundered only by individuals who have been informed of the hazards of exposure to barium carbonate.

RISK ASSESSMENT: ENVIRONMENT

General Assessment

The environment is at risk of exposure during transportation, storage, disposal, or destruction of barium carbonate. In almost every scenario, the threat of environmental exposure is contingent upon the proper handling of the chemical substance. Accidental spills, large or small, where barium carbonate contacts incompatible commodities can result in fire, explosion (depending upon conditions of spill), and possible contamination of the surrounding environmental mediums (water, soil, and air).

Barium carbonate is considered a non-combustible solid. However, finely divided dusts and powders may create an explosive mixture in air in closed spaces. It is incompatible with many acids and extreme caution is required in handling, storage, transportation, and disposal of barium carbonate. These characteristics also require special consideration during any emergency situation involving a leak or spill of barium carbonate powder or dust. Should barium carbonate ever come into contact with incompatible substances either during use, transportation, storage, or disposal, violent and even explosive reactions are possible.

Barium carbonate may enter the environment as a result of industrial and municipal waste treatment plant discharges and from spills.

Acute Ecological Effects

Acute (short-term) toxic effects may include the death of animals, birds, or fish and death or low growth rate in plants. Acute effects are seen 2 to 4 days after animals or plants are exposed to barium.

Barium and its salts have moderate acute toxicity to aquatic life. Insufficient data are available to evaluate the short-term effects of barium or its salts to plants, birds, or land animals.

Chronic Ecological Effects

Chronic (long-term) toxic effects may include shortened life span, reproductive problems, lower fertility, and changes in appearance or behavior in exposed animals. These effects can be seen long after first exposure(s) to toxic chemicals.

Barium and its salts have moderate chronic toxicity to aquatic life. Insufficient data are available to evaluate or predict the long-term effects of barium to plants, birds, or land animals.

Water Solubility

Barium carbonate is nearly insoluble (0.002 g/100 cc at 20°C) in water. Concentrations of less than 1 milligram may mix with a liter of water.

Persistence in the Environment

Barium is highly persistent in water, with a half-life greater than 200 days. The half-life of a pollutant is the amount of time it takes for one half of the chemical to be degraded. No data are available on the estimated percentages of barium persistence in the aquatic or terrestrial environments.

Bioaccumulation in Aquatic Organisms

Some substances increase in concentration, or *bioaccumulate,* in living organisms as they breathe contaminated air, drink contaminated water, or eat contaminated food. These chemicals can become concentrated in the tissues and internal organs of animals as well as humans.

The concentration of barium found in fish tissues is expected to be about the same as the average concentration of barium in the water from which the fish was taken.

Recommended Risk-Reduction Measures

Proper training of all transporters will reduce the likelihood of a mishap or accident resulting in a leak or spill of barium carbonate dusts or powders into the environment. The correct use of DOT labeling (per the MSDS) on all containers, trucks, and rail cars will enable emergency responders to react properly and quickly to any disaster thereby reducing the potential risk to the environment and to personnel.

Storage of barium carbonate should be segregated from incompatible chemicals to minimize the risk of cross-contamination or contact.

If a spill or leak to the environment has occurred, fire department, emergency response and/or hazardous materials spill personnel should be notified immediately. Cleanup should be attempted only by those trained in proper spill containment procedures. Contaminated soils should be removed for incineration and replaced with clean soil. If barium carbonate should contact the water table, aquifer, or navigable waterway, time is of the essence. It is slightly soluble and will decompose. Therefore, total remediation may not be possible. The local and/or state emergency response authorities must be notified. A comprehensive emergency response or disaster preparedness/recovery plan should be in place prior to any operations involving the use, transportation, storage, or disposal of barium carbonate. If barium carbonate is spilled or leaked, the following specific steps are recommended:

☑ Restrict persons not wearing protective clothing from area of spill or leak until cleanup is complete.

☑ Remove all ignition sources and ventilate area.

☑ Collect powdered materials in the safest manner possible and deposit in sealed containers. Recommend using vacuum with HEPA filter. Do not dry sweep (generates airborne dusts).

☑ It may be necessary to dispose of barium carbonate as a hazardous waste. The responsible state agency or the regional office of the federal Environmental Protection Agency (EPA) should be contacted for specific recommendations.

RISK ASSESSMENT: BUSINESS

General Assessment

Accidents or mishaps involving barium carbonate can present a significant threat to business operations. The loss or damage of equipment or facilities can significantly affect fiscal viability. Lawsuits that may result from personnel injury/death, public exposures, and/or environmental contamination will also require a serious expenditure of resources. Media attention surrounding an injury, death, or environmental damage can also result in a loss of profits and loss of current as well as future business.

Recommended Risk-Reduction Measures

Company attorneys, safety and health professionals, and environmental specialists should be involved in the development of any procedures for responding to chemical incidents. A company official should be predesignated as a public relations officer with specific training in dealing with the press. Corporate plans and policies should be developed, approved, and implemented long before any need for such arises.

REFERENCES

American Conference of Governmental Industrial Hygienists. 1988. *Documentation of the Threshold Limit Values and Biological Exposure Indices,* 5th Edition (with updates). Cincinnati: ACGIH

Bloom, A. D. and F. J. DeSerres. 1995. *Ecotoxicity and Human Health.* New York: CRC Press/Lewis Publishers.

Brusick, D. J. 1994. *Methods for Genetic Risk Assessment.* New York: CRC Press/Lewis Publishers.

Bueche, J. R. 1972. *Principles of Physics,* 2nd Edition. New York: McGraw-Hill.

Calabrese, E. J. 1994. *Biological Effects of Low Level Exposures.* New York: CRC Press/Lewis Publishers.

Cockerman, L. G., B. S. Shane. 1994. *Basic Environmental Toxicology.* New York: CRC Press/Lewis Publishers.

Gardner, R. W. 1994. *Chemical Intolerance.* New York: CRC Press/Lewis Publishers.

Howard, P. H. 1995. *Handbook of Environmental Fate and Exposures Data for Organic Chemicals,* Vol. I-IV. New York: CRC Press/Lewis Publishers.

Langerman, N. 1994 *Precautionary Labels for Chemical Containers.* New York: CRC Press/Lewis Publishers.

Lauwerys, R. R., P. Hoet. 1993. *Industrial Chemical Exposure.* New York: CRC Press/Lewis Publishers.

Lewis, R. J., Sr. 1993. *Hawley's Condensed Chemical Dictionary,* 12th Edition. New York: Van Nostrand Reinhold.

Lewis, R. J., Sr. 1992. *Sax's Dangerous Properties of Industrial Materials,* Volumes 1, 2, and 3, 8th Edition. New York: Van Nostrand Reinhold.

Meyer, E. 1990. *Chemistry of Hazardous Materials.* Englewood Cliffs, NJ: Prentice Hall.

Mycock, J. C., J. D. McKenna, and L. Theodore. 1995. *Handbook of Air Pollution Control Engineering and Technology.* New York: CRC Press/Lewis Publishers.

Quigley, D. R. 1994. *Handbook of Emergency Chemical Management.* New York: CRC Press/Lewis Publishers.

Reeves, A. L. 1979. Barium. *Handbook on the Toxicology of Metals.* pp 321-328. New York: Elsevier North-Holland.

Richardson, M. 1995. *Dictionary of Substance and Their Effects,* Vol. I-VI. New York: CRC Press/Lewis Publishers.

Sacarello, H. L. A. 1994. *The Comprehensive Handbook of Hazardous Materials.* New York: CRC Press/Lewis Publishers.

Talbott, E., G. Craun. 1995. *An Introduction to Environmental Epidemiology.* New York: CRC Press/Lewis Publishers.

Torkelson, T. R. V. K. Rowe. 1981. *Patty's Industrial Hygiene and Toxicology,* Vol. 2B. New York: Wiley-Interscience.

U.S. Department of Health, Education, and Welfare. National Institute for Occupational Safety and Health (NIOSH). 1978. *Criteria for a Recommended Standard. Occupational Exposure to Barium.* DHEW (NIOSH) Publication Number: 78-216. Washington, DC: U.S. Government Printing Office.

MATERIAL SAFETY DATA SHEET

CHEMICAL NAME	
BARIUM CYANIDE	**CRC PRESS/LEWIS PUBLISHERS DISCLAIMER:** THE INFORMATION AND RECOMMENDATIONS PRESENTED HEREIN ARE BASED ON SOURCES BELIEVED TO BE RELIABLE. CRC MAKES NO REPRESENTATION ON ITS COMPLETENESS OR ACCURACY. IT IS THE RESPONSIBILITY OF THE USER TO DETERMINE THE CHEMICAL'S SUITABILITY FOR ITS INTENDED USE, THE CHEMICAL'S SAFE USE, AND THE CHEMICAL'S PROPER DISPOSAL. NO REPRESENTATIONS AND/OR WARRANTIES, EITHER EXPRESSED OR IMPLIED, OF THE MERCHANTABILITY OR FITNESS FOR A PARTICULAR PURPOSE, OR OF ANY OTHER NATURE, ARE MADE WITH RESPECT TO THE INFORMATION PROVIDED IN THIS MSDS OR TO THE CHEMICAL TO WHICH INFORMATION MAY REFER. CRC NEITHER ASSUMES NOR AUTHORIZES ANY OTHER PERSON TO ASSUME FOR IT, ANY OTHER ADDITIONAL RESPONSIBILITY OR LIABILITY FOR THE USE OF, OR RELIANCE UPON, THIS INFORMATION.

HAZARD WARNING INFORMATION

HEALTH	FIRE	REACTIVE	OTHER	DEGREE OF HAZARD	COLOR CODING	OTHER CODES
Not Found	Not Found	Not Found		0 = Minimum Hazard 1 = Slight Hazard 2 = Moderate Hazard 3 = Serious Hazard 4 = Severe Hazard	HEALTH = BLUE FIRE = RED REACTIVITY = YELLOW OTHER = WHITE	OX = Oxidizer ACID = Acid ALK = Alkali COR = Corrosive W = Use No Water

SECTION I - GENERAL INFORMATION

Characterization: **Metal**	RCRA Number: **P013**	EPA Class: **Acute Hazardous Waste**
DOT Proper Shipping Name: **Barium Cyanide Solid**	Chemical Abstract Service (CAS) Number: **542-62-1**	
DOT Hazard Class and Label Requirements: **Poison B**	DOT Emergency Guide Code: **53**	
DOT Identification Number: **UN 1565**	Molecular Formula: **$Ba(CN)_2$**	

Synonyms
Barium cyanide solid; barium dicyanide.

SECTION II - HAZARDOUS INGREDIENTS AND IDENTITY INFORMATION

Hazardous Components (specific identity)	OSHA Exposure Criteria	NIOSH Exposure Criteria	Immediately Dangerous to Life and Health (IDLH)	ACGIH Exposure Criteria
Barium cyanide (derivation: By the action of hydrocyanic acid on barium hydroxide with subsequent crystallization).	**PEL:** 0.5 mg(Ba)/m^3 **STEL:** Not Established	**REL:** 0.5 mg(Ba)/m^3 **STEL:** Not Established	1100 mg(Ba)/m^3	**TLV:** 0.5 mg(Ba)/m^3 **STEL:** Not Established

SECTION III - PHYSICAL AND CHEMICAL CHARACTERISTICS

Boiling Point: **Not Applicable**	Specific Gravity (H_2O = 1): **Not Listed**
Vapor Pressure (mm Hg): **<1 at 69°F (20°C)**	Molecular Weight (atomic weight): **189.4**
Vapor Density (Air = 1): **6.5**	Melting Point: **Not Listed**

Solubility in Water
Soluble in water and alcohol.

Appearance and Odor
White crystalline (sugar or sand-like) solid powder, no appreciable odor.

SECTION IV - FIRE AND EXPLOSION HAZARD DATA

Flash Point (method used): **Not Found**	Explosive Limits in Air % by Volume: **LEL: N/A UEL: N/A**
NFPA Classification: **Not Found**	Autoignition Temperature: **Not Found**

Extinguishing Media
Barium cyanide itself does not burn. Extinguish fire with an agent suitable to surrounding fire. Avoid use of carbon dioxide extinguishers if possible.

Special Fire Fighting Procedures
Fire fighting clothing normally worn for structural fires is inadequate protection. Wear fully contained suit with self-contained breathing apparatus (SCBA) set on positive pressure or pressure demand mode. Poisonous gases are produced in fire.

Unusual Fire and Explosion Hazards
Even though barium cyanide by itself will not burn, if it should contact acids, acid salts, or carbon dioxide in air, it may produce highly flammable hydrogen cyanide gas. It will also emit toxic cyanides of barium oxides when it is involved in a fire.

SECTION V - REACTIVITY DATA

Stability		
Stability		Conditions to Avoid Avoid contact with incompatible materials. Keep away from heat and open flame.
Stable X	Unstable	Incompatibility *(materials to avoid)* Acids, acid salts (such as potassium bisulfate, calcium biphosphate, and calcium nitrate); carbon dioxide, and strong oxidizers (such as nitrates, chlorates, and chlorine).
Hazardous Polymerization		Conditions to Avoid Under normal conditions of temperature and pressure, hazardous polymerization of barium cyanide will not occur.
May Occur	Will Not Occur X	Hazardous Decomposition or By-products Contact with acids or acid salts forms deadly, flammable, hydrogen cyanide gas. Also, poisonous gases of cyanide are emitted during fire.

SECTION VI - HEALTH HAZARD DATA

Primary Route(s) of Entry:	Inhalation?	Absorption (skin/eye)?	Ingestion?
	X	X	

Health Hazards

INHALATION: Will cause weakness, headaches, nausea, vomiting, gasping for air, collapse, and even death from cyanide poisoning. After repeated exposure, it may show up as spots in the lungs on chest X-ray.

ABSORPTION: Irritation followed by desquamation and itching. High exposures may be fatal.

EYE CONTACT: Eye contact causes irritation.

Carcinogenicity	NTP Listed?	IARC Cancer Review Group?	OSHA Regulated?	Target Organs?
Unknown Human Unknown Animal	No	No	29 CFR 1910.1000 Table Z-1 (Barium)	Respiratory system, eyes, CNS, skin.

Medical Conditions Generally Aggravated by Exposure

None reported. However, persons with existing lung dysfunction may experience effects quicker.

Emergency and First-aid Procedures

Eye contact: Do not allow victim to keep eyes tightly shut. Flush immediately with water for 15 minutes (minimum), seek medical attention. Skin contact: Remove all contaminated clothing. Immediately wash area with flooding amounts of soap and water. For inhalation: Remove the person from exposure. Provide respiratory assistance and CPR. Transfer to medical facility. If swallowed and patient is conscious, give large amounts of water and induce vomiting. Seek medical attention immediately. Never give anything by mouth to a convulsing or unconscious person.

SECTION VII - PRECAUTIONS FOR SAFE HANDLING AND USE

Steps to be Taken in Case Material is Released or Spilled

Collect powdered materials in most convenient and safe manner and deposit in sealed containers. Dry sweep not recommended. Use HEPA vacuum. Remove all ignition sources. Restrict those not involved in cleanup from entering area. Notify appropriate authorities, as required by applicable regulations.

Preferred Waste Disposal Method

Incinerate in a chemical incinerator equipped with a scrubber and afterburner.

Precautions to be Taken in Handling and Storage

Explosion risks vary with type of barium compound. Store to avoid contact with strong acids and acid salts. Store powders in sealed containers away from heat, flame, or other sources of ignition. While barium cyanide itself will not burn, it will emit poisonous cyanide gas in a fire.

Other Precautions and Warnings

Use only non-sparking tools or equipment when working around barium cyanide, especially when opening or closing containers.

SECTION VIII - CONTROL MEASURES AND PERSONAL PROTECTIVE EQUIPMENT

Respiratory Protection *(specify type)*

Barium cyanide is a class B poison. Where potential exists for exposure above 0.5 mg/m^3, use a self-contained breathing apparatus with full facepiece operated in positive pressure mode.

Ventilation

Local exhaust (preferred) at site of chemical work or general ventilation.

Protective Gloves	Eye Protection	Other Protective Clothing
Impervious Rubber	Chemical/Dust Goggles or Face Mask	Full Protective Suit

Work/Hygiene Practices

Always wash hands thoroughly after using chemical; never bring food, drink, or smoking materials into vicinity of chemicals.

BARIUM CYANIDE

$Ba(CN)_2$ CAS: 542-62-1

IDENTIFICATION AND TYPICAL USES

Barium cyanide is a silver white or yellowish white crystalline powder. It is used primarily in metallurgy and electroplating. It may also be used as a catalyst for organic reactions; as a lubricating oil additive; as rat poison; in manufacture of paper electrodes and in fireworks. Many barium salts are used for various purposes, such as in paints, ceramics, in tanning and finishing leathers, as a mordant for fabrics and dyes, in aluminum refining, in rubber manufacture, and in analytical work.

RISK ASSESSMENT: HEALTH

General Assessment

Barium cyanide is a class B poison. On *inhalation* or *ingestion,* exposure to barium cyanide at low concentrations can cause headaches, vertigo, nausea, and vomiting as a result of cyanide poisoning. At higher concentrations, there can be difficulty in breathing, heart palpitation, paralysis, unconsciousness, respiratory arrest, cyanosis, and death. Similar symptoms may appear as a result of *absorption* through the skin. Industrial grade cyanide often contains silica. If so, overexposure can lead to gradual lung scarring which can resemble barium spots on chest X-ray and can also reduce lung function. Also, if barium cyanide is near acid or acid mist, deadly cyanide gas is released and rapid poisoning can occur.

Eye contact can result in severe irritation of the conjunctiva ion and possible damage to corneal tissues.

Acute Health Effects

The following acute (short-term) health effects may occur immediately or shortly after exposure to barium:

Skin: Irritation, rash, desquamation, and itching. Absorption is likely with resulting toxic systemic effects attributed to acute cyanide poisoning.

Eye: Severe irritation and possible damage.

Lung: Irritation of the respiratory tract with difficulty breathing and shortness of breath, cough, and possible death due to respiratory arrest as a result of acute cyanide poisoning.

CNS: Symptoms include headaches, vertigo, confusion, nausea, and vomiting at low concentrations. Higher acute concentrations cause heart palpitation, paralysis, unconsciousness, and cyanosis.

Chronic Health Effects

The following chronic (long-term) health effects may occur at some time after exposure to barium and can last for months or even years:

Cancer Hazards: According to the references, barium cyanide has not been tested for its ability to cause cancer in animals.

Reproductive Hazard: According to the references, barium has not been tested for its ability to affect reproduction.

Other Chronic Effects: Repeated exposure can cause abnormal chest X-ray (spots on lungs). This usually takes years to develop. An abnormal chest X-ray from barium cyanide exposure does not necessarily mean the lung has been damaged but future exposures to irritating substances may affect the lungs.

Repeated exposures can interfere with the thyroid gland's normal functioning, causing enlarged thyroid (goiter).

Long-term effects of repeated exposures to small amounts of barium cyanide also include loss of appetite, headaches, weakness, nausea, dizziness, and irritation of the upper respiratory tract and eyes.

ʊ *Recommended Risk-Reduction Measures*

Personnel should avoid direct contact with barium cyanide, especially its dusts or powders. If a less toxic material or compound cannot be substituted for barium cyanide, then *engineering controls* are the most effective method of reducing exposures. The best protection is to enclose operations and/or provide local exhaust ventilation at the site of barium cyanide dust release. While not always operationally feasible, isolating operations can also reduce exposure. Using respiratory protection is less effective than the controls mentioned above, but is still advisable whenever working with or around barium. A self-contained breathing apparatus (SCBA) with full facepiece and pressure demand is recommended for the greatest possible respiratory protection, especially when exposures near the IDLH level of 1100 mg/m^3. However, for low or infrequent exposure below the 0.5 mg/m^3 PEL, a MSHA/NIOSH approved full facepiece respirator with a high efficiency particulate air (HEPA) filter can be used. If a full facepiece is not available, then chemical/dust goggles should be worn to protect the eyes. A face shield and protective apron should also be worn. To prevent hand and skin exposures, gloves should be worn.

Administrative controls should also be in place to minimize the potential for human exposures. These may include written procedures or policies which specify the methods and techniques that will be practiced whenever personnel are to work with barium cyanide.

All personnel should receive *training* on the use, hazards, protective measures, emergency actions, and other precautions per 29 CFR 1910.1200 (Hazard Communication), prior to the first assignment in an area where barium is used or stored.

For those with frequent or potentially high exposure (half the PEL or greater), or significant skin contact, the following medical tests are recommended before initial assignment and at regular intervals (e.g., annually) thereafter:

- ☑ Urine thiocyanate test (most accurate if performed right after exposure).

If symptoms develop or overexposure is suspected, the following should also be considered:

- ☑ Blood cyanide level. (Note: Since cigarettes contain some cyanide, smokers may show an increase in cyanide level on testing).
- ☑ Evaluation of thyroid function
- ☑ Lung function test and chest X-ray (for baseline).

Any evaluation should include a careful history of past and present symptoms with an examination. Medical tests that simply look for existing damage are not a substitute for controlling exposures. Also, personnel who smoke should avoid any unprotected contact with barium cyanide powders.

Other methods to reduce exposure include:

- ☑ Where possible, enclose operations and use local exhaust ventilation at the site of chemical release. If local exhaust ventilation or enclosure is not used, respiratory protection should be mandatory.
- ☑ Always ensure that proper protective clothing is worn when using chemical substances.
- ☑ Wash thoroughly immediately after exposure to barium cyanide and at the end of the work shift or before eating, drinking, or smoking.
- ☑ Hazard warning information should be posted in the work area. In addition, as part of an on-going education and training program, all information on the health and safety hazards of barium cyanide should be communicated to all exposed and potentially exposed workers.
- ☑ Eye wash stations should be provided in the immediate work area for emergency use. If there is a possibility of skin exposure to barium cyanide, emergency shower facilities should also be provided in the immediate area.
- ☑ Workers whose clothing has been contaminated by barium cyanide should change into clean clothes before leaving work. Contaminated work clothing should be laundered only by individuals who have been informed of the hazards of exposure to barium cyanide.

RISK ASSESSMENT: ENVIRONMENT

General Assessment

The environment is at risk of exposure during transportation, storage, disposal, or destruction of barium cyanide. In almost every scenario, the threat of environmental exposure is contingent upon the proper handling of the chemical substance. Accidental spills, large or small, where barium cyanide contacts incompatible commodities can result in fire, explosion (depending upon conditions of spill), and possible contamination of the surrounding environmental mediums (water, soil, and air).

In its powdered form, barium cyanide is considered a noncombustible solid and by itself, it will not burn. However, if it should contact acid, acid mist, acid salts or carbon dioxide in air, it may produce highly flammable and extremely toxic hydrogen cyanide gas. Extreme caution is always required in handling, storage, transportation, and disposal of barium cyanide. These characteristics also require special consideration during any emergency situation involving a leak or spill of barium cyanide powder or dust. Should barium cyanide ever come into contact with incompatible substances (such as strong acids) either during use, transportation, storage, or disposal, poisonous cyanide gases may be formed.

Barium cyanide enters the environment from industrial and municipal waste treatment plant discharges and from spills.

Acute Ecological Effects

Acute (short-term) toxic effects may include the death of animals, birds, or fish and death or low growth rate in plants. Acute effects are seen 2 to 4 days after animals or plants are exposed to barium. Barium and its salts have moderate acute toxicity to aquatic life. Insufficient data are available to evaluate the short-term effects of barium or its salts to plants, birds, or land animals.

Chronic Ecological Effects

Chronic (long-term) toxic effects may include shortened life span, reproductive problems, lower fertility, and changes in appearance or behavior in exposed animals. These effects can be seen long after first exposure(s) to toxic chemicals. Barium and its salts have moderate chronic toxicity to aquatic life. Insufficient data are available to evaluate or predict the long-term effects of barium to plants, birds, or land animals.

Water Solubility

Most of the barium salts are either highly or moderately soluble in water. Concentrations of 1 to 1000 milligrams and more will mix with a liter of water.

Persistence in the Environment

Barium is highly persistent in water, with a half-life greater than 200 days. The half-life of a pollutant is the amount of time it takes for one half of the chemical to be degraded. No data are available on the estimated percentages of barium cyanide persistence in the aquatic or terrestrial environments.

Bioaccumulation in Aquatic Organisms

Some substances increase in concentration, or *bioaccumulate,* in living organisms as they breathe contaminated air, drink contaminated water, or eat contaminated food. These chemicals can become concentrated in the tissues and internal organs of animals as well as humans.

The concentration of barium found in fish tissues is expected to be about the same as the average concentration of barium in the water from which the fish was taken.

Recommended Risk-Reduction Measures

Proper training of all transporters will reduce the likelihood of a mishap or accident resulting in a leak or spill of barium cyanide dusts or powders into the environment. The correct use of DOT labeling (per the MSDS) on all containers, trucks, and rail cars will enable emergency responders to react properly and quickly to any disaster thereby reducing the potential risk to the environment and to personnel.

Storage of barium cyanide should be segregated from incompatible chemicals to minimize the risk of cross-contamination or contact.

If a spill or leak to the environment has occurred, fire department, emergency response and/or hazardous materials spill personnel should be notified immediately. Cleanup should be attempted only by those trained in proper spill containment procedures. Contaminated soils should be removed for incineration and replaced with clean soil. If barium cyanide should contact the water table, aquifer, or navigable waterway, time is of the essence. It is highly soluble and total remediation may not be possible. The local and/or state emergency response authorities must be notified. A comprehensive emergency response or disaster preparedness/recovery plan should be in place prior to any operations involving the use, transportation, storage, or disposal of barium. If barium cyanide is spilled or leaked, the following specific steps are recommended:

- ☑ Restrict persons not wearing protective clothing from area of spill or leak until cleanup is complete and area can be opened for normal work.
- ☑ Ventilate the area.
- ☑ Collect powdered materials in the safest manner possible and deposit in sealed containers. Rec-

ommend using vacuum with HEPA filter. Do not dry sweep (generates airborne dusts).

☑ It may be necessary to dispose of barium cyanide as a hazardous waste. The responsible state agency or the regional office of the federal Environmental Protection Agency (EPA) should be contacted for specific recommendations.

RISK ASSESSMENT: BUSINESS

General Assessment

Accidents or mishaps involving barium cyanide can present a significant threat to business operations. The loss or damage of equipment or facilities can significantly affect fiscal viability. Lawsuits that may result from personnel injury/death, public exposures, and/or environmental contamination will also require a serious expenditure of resources. Media attention surrounding an injury, death, or environmental damage can also result in a loss of profits and loss of current as well as future business.

Recommended Risk-Reduction Measures

Company attorneys, safety and health professionals, and environmental specialists should be involved in the development of any procedures for responding to chemical incidents. A company official should be predesignated as a public relations officer with specific training in dealing with the press. Corporate plans and policies should be developed, approved, and implemented long before any need for such arises.

REFERENCES

American Conference of Governmental Industrial Hygienists. 1988. *Documentation of the Threshold Limit Values and Biological Exposure Indices,* 5th Edition (with updates). Cincinnati: ACGIH

Bloom, A. D. and F. J. DeSerres. 1995. *Ecotoxicity and Human Health.* New York: CRC Press/Lewis Publishers.

Brusick, D. J. 1994. *Methods for Genetic Risk Assessment.* New York: CRC Press/Lewis Publishers.

Bueche, J. R. 1972. *Principles of Physics,* 2nd Edition. New York: McGraw-Hill.

Calabrese, E. J. 1994. *Biological Effects of Low Level Exposures.* New York: CRC Press/Lewis Publishers.

Cockerman, L. G., B. S. Shane. 1994. *Basic Environmental Toxicology.* New York: CRC Press/Lewis Publishers.

Cothern, R. R. 1993. *Comparative Environmental Risk Assessment.* New York: CRC Press/Lewis Publishers.

Gardner, R. W. 1994. *Chemical Intolerance.* New York: CRC Press/Lewis Publishers.

Hallenbeck, W. H. 1993. *Quantitative Risk Assessment for Environmental and Occupational Health,* 2nd Edition. New York: CRC Press/Lewis Publishers.

Howard, P. H. 1995. *Handbook of Environmental Fate and Exposures Data for Organic Chemicals,* Vol. I-IV. New York: CRC Press/Lewis Publishers.

Keith, L. H. and M. M. Walker. 1995. *Handbook of Air Toxics.* New York: CRC Press/Lewis Publishers.

Landis, W. G., M. H. Yu. 1995. *Introduction to Environmental Toxicology.* New York: CRC Press/Lewis Publishers.

Langerman, N. 1994 *Precautionary Labels for Chemical Containers.* New York: CRC Press/Lewis Publishers.

Lauwerys, R. R., P. Hoet. 1993. *Industrial Chemical Exposure.* New York: CRC Press/Lewis Publishers.

Lewis, R. J., Sr. 1993. *Hawley's Condensed Chemical Dictionary,* 12th Edition. New York: Van Nostrand Reinhold.

Lewis, R. J., Sr. 1992. *Sax's Dangerous Properties of Industrial Materials,* Volumes 1, 2, and 3, 8th Edition. New York: Van Nostrand Reinhold.

Lide, D. R. 1995. *Handbook of Chemistry and Physics,* 76th Edition. New York: CRC Press/Lewis Publishers.

Mellström, G. A., J. E. Wahlberg, and H. I. Maibach. 1994. *Protective Gloves for Occupational Use.* New York: CRC Press/Lewis Publishers.

Meyer, E. 1990. *Chemistry of Hazardous Materials.* Englewood Cliffs, NJ: Prentice Hall.

Mycock, J. C., J. D. McKenna, and L. Theodore. 1995. *Handbook of Air Pollution Control Engineering and Technology.* New York: CRC Press/Lewis Publishers.

Quigley, D. R. 1994. *Handbook of Emergency Chemical Management.* New York: CRC Press/Lewis Publishers.

Reeves, A. L. 1979. Barium. *Handbook on the Toxicology of Metals.* pp 321-328. New York: Elsevier North-Holland.

Richardson, M. 1995. *Dictionary of Substance and Their Effects,* Vol. I-VI. New York: CRC Press/Lewis Publishers.

Sacarello, H. L. A. 1994. *The Comprehensive Handbook of Hazardous Materials.* New York: CRC Press/Lewis Publishers.

Sheftel, V. O. 1995. *Handbook of Toxic Properties of Monomers and Additives.* New York: CRC Press/Lewis Publishers.

Torkelson, T. R. V. K. Rowe. 1981. *Patty's Industrial Hygiene and Toxicology,* Vol. 2B. New York: Wiley-Interscience.

U.S. Department of Health, Education, and Welfare. National Institute for Occupational Safety and Health (NIOSH). 1978. *Criteria for a Recommended Standard. Occupational Exposure to Barium.* DHEW (NIOSH) Publication Number: 78-216. Washington, DC: U.S. Government Printing Office.

U.S. Department of Labor. 1983. Barium metal: barium trioxide; and barium sulfide response to the interagency testing committee. *Federal Register.* 48:717-724.

MATERIAL SAFETY DATA SHEET

CHEMICAL NAME	
BARIUM FLUORIDE	**CRC PRESS/LEWIS PUBLISHERS DISCLAIMER:** THE INFORMATION AND RECOMMENDATIONS PRESENTED HEREIN ARE BASED ON SOURCES BELIEVED TO BE RELIABLE. CRC MAKES NO REPRESENTATION ON ITS COMPLETENESS OR ACCURACY. IT IS THE RESPONSIBILITY OF THE USER TO DETERMINE THE CHEMICAL'S SUITABILITY FOR ITS INTENDED USE, THE CHEMICAL'S SAFE USE, AND THE CHEMICAL'S PROPER DISPOSAL. NO REPRESENTATIONS AND/OR WARRANTIES, EITHER EXPRESSED OR IMPLIED, OF THE MERCHANTABILITY OR FITNESS FOR A PARTICULAR PURPOSE, OR OF ANY OTHER NATURE, ARE MADE WITH RESPECT TO THE INFORMATION PROVIDED IN THIS MSDS OR TO THE CHEMICAL TO WHICH INFORMATION MAY REFER. CRC NEITHER ASSUMES NOR AUTHORIZES ANY OTHER PERSON TO ASSUME FOR IT, ANY OTHER ADDITIONAL RESPONSIBILITY OR LIABILITY FOR THE USE OF, OR RELIANCE UPON, THIS INFORMATION.

HAZARD WARNING INFORMATION

HEALTH	FIRE	REACTIVE	OTHER	DEGREE OF HAZARD	COLOR CODING	OTHER CODES
3	1	0		0 = Minimum Hazard 1 = Slight Hazard 2 = Moderate Hazard 3 = Serious Hazard 4 = Severe Hazard	HEALTH = BLUE FIRE = RED REACTIVITY = YELLOW OTHER = WHITE	OX = Oxidizer ACID = Acid ALK = Alkali COR = Corrosive ~~W~~ = Use No Water

SECTION I - GENERAL INFORMATION

Characterization	**Soluble Metallic Fluoride**	RCRA Number	**D005 (as barium)**
		EPA Class	**Characteristic (T) Waste**
DOT Proper Shipping Name	**Barium Compounds, N.O.S.**	Chemical Abstract Service (CAS) Number	**7787-32-8**
DOT Hazard Class and Label Requirements	**Keep Away From Food**	DOT Emergency Guide Code	**53 (for compounds)**
DOT Identification Number	**UN 1564**	Atomic Symbol	**BaF_2**

Synonyms
Not Applicable.

SECTION II - HAZARDOUS INGREDIENTS AND IDENTITY INFORMATION

Hazardous Components (specific identity)	OSHA Exposure Criteria	NIOSH Exposure Criteria	Immediately Dangerous to Life and Health (IDLH)	ACGIH Exposure Criteria
Barium fluoride (derivation: By reaction of barium sulfide with hydrofluoric acid followed by crystallization or by dissolving barium carbonate in excess hydrofluoric acid, evaporating to dryness and then heating to red heat).	**PEL:** 0.5 mg(Ba)/m³ **STEL:** Not Established	**REL:** 0.5 mg(Ba)/m³ **STEL:** Not Established	**250 mg/m³ for Soluble Compounds**	**TLV:** 0.5 mg(Ba)/m³ **STEL:** Not Established

SECTION III - PHYSICAL AND CHEMICAL CHARACTERISTICS

Boiling Point	**3879°F (2137°C)**	Specific Gravity (H_2O = 1)	**4.89**
Vapor Pressure (mm Hg)	**Not Found**	Molecular Weight (atomic weight)	**175.4**
Vapor Density (Air = 1)	**Not Found**	Melting Point	**2457°F (1353°C)**

Solubility
Most of the barium salts are either highly or moderately soluble in water. Barium fluoride powder is only slightly soluble in water. Soluble in hydrochloric, nitric, and hydrofluoric acid and ammonia chloride.

Appearance and Odor
Either transparent cubic crystalline solid or a white powder with a strong, disagreeable odor.

SECTION IV - FIRE AND EXPLOSION HAZARD DATA

Flash Point (method used)	**Not Found**	Explosive Limits in Air % by Volume	**LEL: N/A UEL: N/A**
NFPA Classification	**Combustible Solid (in powder form)**	Autoignition Temperature	**Not Found**

Extinguishing Media
Use dry chemical, carbon monoxide, water spray, fog, or regular foam.

Special Fire Fighting Procedures
Wear full protective clothing and self-contained breathing apparatus (SCBA). Poisonous gases are produced in fire. Do not direct water stream on fire (will scatter material). Do not release water runoff to sewers or waterways. Contain for proper disposal as waste.

Unusual Fire and Explosion Hazards
Barium fluoride powder is a combustible solid but is difficult to ignite. Dust is dangerous when exposed to heat, flame, or chemical reaction. Violent or explosive reactions may occur.

SECTION V - REACTIVITY DATA

Stability		Conditions to Avoid **Avoid heat and open flame, especially around dusts and powders of barium fluoride. Avoid generation of excessive dusty conditions.**
Stable X	Unstable	Incompatibility *(materials to avoid)* **Incompatible with strong inorganic acids. Keep heat and flame away, especially in dusty conditions.**
Hazardous Polymerization		Conditions to Avoid **Hazardous polymerization of barium fluoride metal will not occur.**
May Occur	Will Not Occur X	Hazardous Decomposition or By-products **Poisonous gases are emitted during fire, including barium oxide gas, and toxic and corrosive hydrogen fluoride gas.**

SECTION VI - HEALTH HAZARD DATA

Primary Route(s) of Entry:	Inhalation?	Absorption (skin/eye)?	Ingestion?
	X		

Health Hazards

INHALATION: May produce benign pneumoconiosis (baritosis) which may cause X-ray changes. Other symptoms of exposure include eye, nose, and throat irritation, coughing. Exposure may also cause pulmonary edema (fluid in the lungs), which can be fatal.

EYES & SKIN: Exposure to the eyes may cause moderate to severe and perhaps permanent damage.

INGESTION: Water and stomach acids solubilize barium salts and can cause poisoning. Symptoms are vomiting, colic, diarrhea, slow irregular pulse, transient hypertension, convulsive tremors.

Carcinogenicity	NTP Listed?	IARC Cancer Review Group?	OSHA Regulated?	Target Organs?
Unknown Human Unknown Animal	No	No	29 CFR 1910.1000 Table Z-1	Respiratory system, eyes, CNS, skin, muscles.

Medical Conditions Generally Aggravated by Exposure

None reported. However, persons with existing lung dysfunction may experience effects quicker.

Emergency and First-aid Procedures

Eye contact: Do not allow victim to keep eyes tightly shut. Flush immediately with water for 15 minutes (minimum), seek medical attention. Skin contact: Although no immediate reaction may be evident, barium fluoride powders may begin to burn skin when sweat begins (water reactive). Remove all contaminated clothing. Immediately wash area with flooding amounts of soap and water. For inhalation: Remove the person from exposure. Provide respiratory assistance and CPR. Transfer to medical facility. Observe 48 hours for lung effects. If swallowed, seek medical attention immediately. Never give anything by mouth to a convulsing person.

SECTION VII - PRECAUTIONS FOR SAFE HANDLING AND USE

Steps to be Taken in Case Material is Released or Spilled

Collect powdered materials in most convenient and safe manner and deposit in sealed containers. Dry sweep not recommended. Use HEPA vacuum. Remove all ignition sources. Restrict those not involved in cleanup from entering area. Notify appropriate authorities, as required by applicable regulations.

Preferred Waste Disposal Method

No citation.

Precautions to be Taken in Handling and Storage

Store to avoid contact with strong inorganic acids. Store powders in sealed containers in cool, dry, well-ventilated areas away from heat and flame. Prevent physical damage to containers.

Other Precautions and Warnings

Sources of ignition are prohibited where barium fluoride is used, handled, or stored. Use only non-sparking tools or equipment, especially when opening or closing containers.

SECTION VIII - CONTROL MEASURES AND PERSONAL PROTECTIVE EQUIPMENT

Respiratory Protection *(specify type)*

For dust exposures below PEL, use a respirator with a high efficiency particulate air (HEPA) filter, otherwise use a self-contained breathing apparatus with full facepiece operated in positive pressure mode.

Ventilation

Local exhaust (preferred) at site of chemical work or general ventilation.

Protective Gloves	Eye Protection	Other Protective Clothing
Impervious Rubber	Chemical/Dust Goggles or Face Mask	Protective Apron, Boots

Work/Hygiene Practices

Always wash hands thoroughly after using chemical; never bring food, drink, or smoking materials into vicinity of chemicals.

BARIUM FLUORIDE

BaF_2 CAS: 7787-32-8

IDENTIFICATION AND TYPICAL USES

Barium fluoride can appear as a transparent, cube-shaped, crystalline solid or a white powder with a strong, disagreeable odor. It is used as a ceramic flux, in carbon brushes for electrical equipment, in glass making, in the manufacture of other fluorides, in crystals for spectroscopy, in electronics, and in dry-film lubricants.

RISK ASSESSMENT: HEALTH

General Assessment

Certain compounds of barium (including barium fluoride) are irritants of the eyes, mucous membranes, and skin. Primary exposure routes are *inhalation* and *ingestion.* Inhalation of barium fluoride dusts can cause irritation of the nose and upper respiratory tract and may produce a benign pneumoconiosis known as baritosis. Barium fluoride exposure can lead to a condition known as fluorosis with symptoms of joint pain, mottling of teeth, ossification on X-rays, and osteosclerosis. There may be characteristic X-ray changes. All soluble salts of barium are acute poisons. The barium fluoride ion is toxic to muscle producing stimulation and then paralysis.

Ingestion can cause severe hypokalemia. Ingesting barium fluoride salts or powders produces initial symptoms related to gastrointestinal reaction. These include nausea, vomiting, colic, diarrhea, excessive salivation, followed by myocardial and general muscular stimulation with tingling in the extremities, mouth, and face. There may be irregular heart beat, kidney dysfunction (decreased urine excretion) or injury, hypokalemia (potassium deficiency), convulsions, slowed pulse rate, and potentially fatal pulmonary edema (fluid in the lungs). Indication of pulmonary edema may be delayed up to 48 hours, creating a false sense of security with regard to health exposure risk. Severe cases continue to loss of tendon reflexes, general muscular paralysis, and death from respiratory arrest or ventricular fibrillation. There may also be vascular constriction, bladder contraction, and increased voluntary muscle tension leading to paralysis or rhabdomyolysis, a disease that can cause death as a result of the destruction or degeneration of skeletal muscle.

Acute Health Effects

The following acute (short-term) health effects may occur immediately or shortly after exposure to barium fluoride:

Skin: No initial symptoms. However, sweating may cause burning sensation as barium fluoride dusts react on contact with moist skin. Severe exposure may cause necrosis (localized tissue destruction).

Eye: Severe irritation with inflammation, burning, and watering.

Lung: Can cause metallic taste, nausea, sore throat, and irritation of the air passages. Pulmonary edema (fluid in the lungs) is possible.

Other: Serious effects on musculatory system at high exposures. May cause tingling of the extremities followed by irregular heartbeat and might even cause heart to stop (ventricular fibrillation). Death is also possible as a result of respiratory arrest.

Chronic Health Effects

The following chronic (long-term) health effects may occur at some time after exposure to barium fluoride and can last for months or even years:

Cancer Hazards: According to the information presented in the references, barium fluoride has not been adequately tested for its ability to cause cancer in laboratory test animals.

Reproductive Hazard: According to the references, there is limited evidence that exposure to barium fluoride may adversely affect the developing fetus.

Other Chronic Effects: Repeated exposure can cause abnormal chest X-ray. This usually takes years to develop. An abnormal chest X-ray from barium fluoride exposure does not necessarily mean the lung has been damaged but future exposures to irritating substances may affect the lungs. Fluorosis may develop in the long-term with symptoms of joint pain, mottling of teeth, ossification, and osteosclerosis.

Recommended Risk-Reduction Measures

Personnel should avoid direct contact with barium fluoride, especially its dusts or powders. If a less toxic material or compound cannot be substituted for barium fluoride, then *engineering controls* are the most effective method of reducing exposures. The best protection is to enclose operations and/or provide local exhaust ventilation at the site of barium fluoride dust release. While not always operationally feasible, isolating operations can also reduce exposure. Using respiratory protection is less effective than the controls mentioned above, but is still advisable whenever working with or around barium fluoride. A self-contained breathing apparatus (SCBA) with full facepiece and pressure demand is recommended for the greatest possible respiratory protection, especially when exposures near the IDLH level of 250 mg/m^3. However, for low or infrequent exposure, a MSHA/NIOSH approved full facepiece respirator with a high efficiency particulate air (HEPA) filter can be used. If a full facepiece is not available, then chemical/dust goggles should be worn to protect the eyes. A face shield and protective apron should also be worn. To prevent hand and skin exposures, gloves should be worn.

Administrative controls should also be in place to minimize the potential for human exposures. These may include written procedures or policies which specify the methods and techniques that will be practiced whenever personnel are to work with barium fluoride.

All personnel should receive *training* on the use, hazards, protective measures, emergency actions, and other precautions per 29 CFR 1910.1200 (Hazard Communication), prior to the first assignment in an area where barium fluoride is used or stored.

Before beginning employment working with barium fluoride, and at regular intervals thereafter (e.g., annually), those personnel with frequent or potentially high exposures should be provided the following recommended medical tests:

- ☑ Lung function test.

If symptoms develop or overexposure is suspected, consider the following additional tests:

- ☑ Electrocardiogram (EKG).
- ☑ Neurological examination.
- ☑ Consider chest X-ray.
- ☑ Evaluation by a qualified allergist with careful consideration of exposure history and special testing (may help diagnose skin allergy).

Any evaluation should include a careful history of past and present symptoms with an examination. Medical tests that simply look for existing damage are not a substitute for controlling exposures. Also, since smoking can cause heart disease as well as lung cancer, emphysema, and other known respiratory problems, exposure to barium fluoride may cause additional complications. Therefore, personnel who smoke should avoid any unprotected contact with barium fluoride powders.

Other methods to reduce exposure include:

- ☑ Where possible, enclose operations and use local exhaust ventilation at the site of chemical release. If local exhaust ventilation or enclosure is not used, respiratory protection should be mandatory.
- ☑ Always ensure that proper protective clothing is worn when using chemical substances. Adequate maintenance and care of protective equipment are essential to ensure proper function.
- ☑ Wash thoroughly immediately after exposure to barium fluoride and at the end of the work shift or before eating, drinking, or smoking.
- ☑ Hazard warning information should be posted in the work area. In addition, as part of an on-going education and training program, all information on the health and safety hazards of barium fluoride should be communicated to all exposed or potentially exposed workers.
- ☑ Eye wash stations should be provided in the immediate work area for emergency use. If there is a possibility of skin exposure to barium fluoride, emergency shower facilities should be provided.

☑ Workers whose clothing has been contaminated by barium fluoride should change into clean clothes before leaving work. Contaminated work clothing should be laundered only by individuals who have been informed of the hazards of exposure to barium fluoride.

RISK ASSESSMENT: ENVIRONMENT

General Assessment

The environment is at risk of exposure during transportation, storage, disposal, or destruction of barium fluoride. In almost every scenario, the threat of environmental exposure is contingent upon the proper handling of the chemical substance. Accidental spills, large or small, where barium fluoride contacts incompatible commodities can result in fire, explosion (depending upon conditions of spill), and possible contamination of the surrounding environmental mediums (water, soil, and air).

In powdered form, barium fluoride is considered a combustible solid. It can present a moderate fire and explosion hazard. It is incompatible with a number of materials, especially inorganic acids. Extreme caution is always required in handling, storage, transportation, and disposal of barium fluoride. These characteristics also require special consideration during any emergency situation involving a leak or spill of barium fluoride powder or dust. Should barium fluoride ever come into contact with incompatible substances either during use, transportation, storage, or disposal, violent and even explosive reactions are possible.

Barium fluoride may enter the environment as a result of industrial and municipal waste treatment plant discharges and from spills.

Acute Ecological Effects

Acute (short-term) toxic effects may include the death of animals, birds, or fish and death or low growth rate in plants. Acute effects are seen 2 to 4 days after animals or plants are exposed to barium. Barium and its salts have moderate acute toxicity to aquatic life. Insufficient data are available to evaluate the short-term effects of barium or its salts to plants, birds, or land animals.

Chronic Ecological Effects

Chronic (long-term) toxic effects may include shortened life span, reproductive problems, lower fertility, and changes in appearance or behavior in exposed animals. These effects can be seen long after first exposure(s) to toxic chemicals. Barium and its salts have moderate chronic toxicity to aquatic life. Insufficient data are available to evaluate or predict the long-term effects of barium to plants, birds, or land animals.

Water Solubility

Most of the barium fluoride is sparingly soluble in water. Concentrations of less than 1 milligram may mix with a liter of water.

Persistence in the Environment

Barium is highly persistent in water, with a half-life greater than 200 days. The half-life of a pollutant is the amount of time it takes for one half of the chemical to be degraded. No data are available on the estimated percentages of barium persistence in the aquatic or terrestrial environments.

Bioaccumulation in Aquatic Organisms

Some substances increase in concentration, or *bioaccumulate,* in living organisms as they breathe contaminated air, drink contaminated water, or eat contaminated food. These chemicals can become concentrated in the tissues and internal organs of animals as well as humans.

The concentration of barium found in fish tissues is expected to be about the same as the average concentration of barium in the water from which the fish was taken.

Recommended Risk-Reduction Measures

Proper training of all transporters will reduce the likelihood of a mishap or accident resulting in a leak or spill of barium dusts or powders into the environment. The correct use of DOT labeling (per the MSDS) on all containers, trucks, and rail cars will enable emergency responders to react properly and quickly to any disaster thereby reducing the potential risk to the environment and to personnel.

Storage of barium fluoride should be segregated from incompatible chemicals to minimize the risk of cross-contamination or contact.

If a spill or leak to the environment has occurred, fire department, emergency response and/or hazardous materials spill personnel should be notified immediately. Cleanup should be attempted only by those trained in proper spill containment procedures. Contaminated soils should be removed for incineration and replaced with clean soil. If barium fluoride should contact the water table, aquifer, or navigable waterway, time is of the essence. It is nearly insoluble and total remediation may be possible. The local and/or state emergency response authorities must be notified. A comprehensive emergency response or disaster preparedness/recovery plan should be in place prior to any operations involving the use, transportation, storage, or disposal of barium fluoride. If barium fluoride is spilled or leaked, the following specific steps are recommended:

- ☑ Restrict persons not wearing protective clothing from area of spill or leak until cleanup is complete.
- ☑ Remove all ignition sources, ventilate area.
- ☑ Collect powdered materials in the safest manner possible and deposit in sealed containers. Recommend using vacuum with HEPA filter. Do not dry sweep (generates airborne dusts).
- ☑ It may be necessary to dispose of barium fluoride as a hazardous waste. The responsible state agency or the regional office of the federal Environmental Protection Agency (EPA) should be contacted for specific recommendations.

RISK ASSESSMENT: BUSINESS

General Assessment

Accidents or mishaps involving barium fluoride can present a significant threat to business operations. The loss or damage of equipment or facilities can significantly affect fiscal viability. Lawsuits that may result from personnel injury/death, public exposures, and/or environmental contamination will also require a serious expenditure of resources. Media attention surrounding an injury, death, or environmental damage can also result in a loss of profits and loss of current as well as future business.

Recommended Risk-Reduction Measures

Company attorneys, safety and health professionals, and environmental specialists should be involved in the development of any procedures for responding to chemical incidents. A company official should be predesignated as a public relations officer with specific training in dealing with the press. Corporate plans and policies should be developed, approved, and implemented long before any need for such arises.

REFERENCES

Ahlbom, A. 1993. *Biostatistics for Epidemiologists.* New York: CRC Press/Lewis Publishers.

American Conference of Governmental Industrial Hygienists. 1988. *Documentation of the Threshold Limit Values and Biological Exposure Indices,* 5th Edition (with updates). Cincinnati: ACGIH

Asante-Duah, D. K. 1993. *Hazardous Waste Risk Assessment.* New York: CRC Press/Lewis Publishers.

Bloom, A. D. and F. J. DeSerres. 1995. *Ecotoxicity and Human Health.* New York: CRC Press/Lewis Publishers.

Brusick, D. J. 1994. *Methods for Genetic Risk Assessment.* New York: CRC Press/Lewis Publishers.

Bueche, J. R. 1972. *Principles of Physics,* 2nd Edition. New York: McGraw-Hill.

Calabrese, E. J. 1994. *Biological Effects of Low Level Exposures.* New York: CRC Press/Lewis Publishers.

Cockerman, L. G., B. S. Shane. 1994. *Basic Environmental Toxicology.* New York: CRC Press/Lewis Publishers.

Cothern, R. R. 1993. *Comparative Environmental Risk Assessment.* New York: CRC Press/Lewis Publishers.

Gardner, R. W. 1994. *Chemical Intolerance.* New York: CRC Press/Lewis Publishers.

Hallenbeck, W. H. 1993. *Quantitative Risk Assessment for Environmental and Occupational Health,* 2nd Edition. New York: CRC Press/Lewis Publishers.

Keith, L. H. and M. M. Walker. 1995. *Handbook of Air Toxics.* New York: CRC Press/Lewis Publishers.

Langerman, N. 1994 *Precautionary Labels for Chemical Containers.* New York: CRC Press/Lewis Publishers.

Lewis, R. J., Sr. 1993. *Hawley's Condensed Chemical Dictionary,* 12th Edition. New York: Van Nostrand Reinhold.

Lewis, R. J., Sr. 1992. *Sax's Dangerous Properties of Industrial Materials,* Volumes 1, 2, and 3, 8th Edition. New York: Van Nostrand Reinhold.

Reeves, A. L. 1979. Barium. *Handbook on the Toxicology of Metals.* pp 321-328. New York: Elsevier North-Holland.

Richardson, M. 1995. *Dictionary of Substance and Their Effects,* Vol. I-VI. New York: CRC Press/Lewis Publishers.

Torkelson, T. R. V. K. Rowe. 1981. *Patty's Industrial Hygiene and Toxicology,* Vol. 2B. New York: Wiley-Interscience.

U.S. Department of Health, Education, and Welfare. National Institute for Occupational Safety and Health (NIOSH). 1978. *Criteria for a Recommended Standard. Occupational Exposure to Barium.* DHEW (NIOSH) Publication Number: 78-216. Washington, DC: U.S. Government Printing Office.

MATERIAL SAFETY DATA SHEET

CHEMICAL NAME	
BARIUM HYDROXIDE (anhydrous)	**CRC PRESS/LEWIS PUBLISHERS DISCLAIMER:** THE INFORMATION AND RECOMMENDATIONS PRESENTED HEREIN ARE BASED ON SOURCES BELIEVED TO BE RELIABLE. CRC MAKES NO REPRESENTATION ON ITS COMPLETENESS OR ACCURACY. IT IS THE RESPONSIBILITY OF THE USER TO DETERMINE THE CHEMICAL'S SUITABILITY FOR ITS INTENDED USE, THE CHEMICAL'S SAFE USE, AND THE CHEMICAL'S PROPER DISPOSAL. NO REPRESENTATIONS AND/OR WARRANTIES, EITHER EXPRESSED OR IMPLIED, OF THE MERCHANTABILITY OR FITNESS FOR A PARTICULAR PURPOSE, OR OF ANY OTHER NATURE, ARE MADE WITH RESPECT TO THE INFORMATION PROVIDED IN THIS MSDS OR TO THE CHEMICAL TO WHICH INFORMATION MAY REFER. CRC NEITHER ASSUMES NOR AUTHORIZES ANY OTHER PERSON TO ASSUME FOR IT, ANY OTHER ADDITIONAL RESPONSIBILITY OR LIABILITY FOR THE USE OF, OR RELIANCE UPON, THIS INFORMATION.

HAZARD WARNING INFORMATION

HEALTH	FIRE	REACTIVE	OTHER	DEGREE OF HAZARD	COLOR CODING	OTHER CODES
4	0	0		0 = Minimum Hazard 1 = Slight Hazard 2 = Moderate Hazard 3 = Serious Hazard 4 = Severe Hazard	HEALTH = BLUE FIRE = RED REACTIVITY = YELLOW OTHER = WHITE	OX = Oxidizer ACID = Acid ALK = Alkali COR = Corrosive W = Use No Water

SECTION I - GENERAL INFORMATION

Characterization	RCRA Number	EPA Class
Inorganic (metallic) hydroxide	**D005**	**Characteristic (T) Waste**

DOT Proper Shipping Name **Barium Compounds, N.O.S.**	Chemical Abstract Service (CAS) Number **17194-00-2**
DOT Hazard Class and Label Requirements **Varies (depending on toxicity)**	DOT Emergency Guide Code **49**
DOT Identification Number **UN 1564**	Atomic Symbol **$Ba(OH)_2$**

Synonyms
Barium hydrate, anhydrous; barium dihydrate; caustic baryta.

SECTION II - HAZARDOUS INGREDIENTS AND IDENTITY INFORMATION

Hazardous Components (specific identity)	OSHA Exposure Criteria	NIOSH Exposure Criteria	Immediately Dangerous to Life and Health (IDLH)	ACGIH Exposure Criteria
Barium hydroxide (derivation: By dissolving barium oxide in water with subsequent crystallization or by precipitation from an aqueous solution of the sulfide by caustic soda).	**PEL:** 0.5 mg(Ba)/m^3 **STEL:** Not Established	**REL:** 0.5 mg(Ba)/m^3 **STEL:** Not Established	**Not Established for BaO_2 (250 mg/m^3 for Soluble Ba Compounds)**	**TLV:** 0.5 mg(Ba)/m^3 **STEL:** Not Established

SECTION III - PHYSICAL AND CHEMICAL CHARACTERISTICS

Boiling Point **Not Determined**	Specific Gravity (H_2O = 1) **2.2 at 61°F (16°C)**
Vapor Pressure (mm Hg) **227 at 172°F (80°C)**	Molecular Weight (atomic weight) **171.35**
Vapor Density (Air = 1) **Not Determined**	Melting Point **766°F (408°C)**

Solubility
Most of the barium salts are either highly or moderately soluble in water. Barium hydroxide powder is slightly soluble in water. Soluble in alcohol and ether, insoluble in acetone.

Appearance and Odor
Clear or white powder with no discernible odor.

SECTION IV - FIRE AND EXPLOSION HAZARD DATA

Flash Point (method used) **Not Found**	Explosive Limits in Air % by Volume **LEL: N/A UEL: N/A**
NFPA Classification **Not Listed**	Autoignition Temperature **Not Found**

Extinguishing Media
Use agents suitable to surrounding fire (alcohol foam, carbon dioxide, dry chemical, or water).

Special Fire Fighting Procedures
Poisonous gases are produced in fire. Structural fire-fighting clothing is permeable and may not provide adequate protection. Wear full protective clothing and self-contained breathing apparatus (SCBA). Use water mist to keep fire-exposed containers cool. Fight fire from distance, if possible.

Unusual Fire and Explosion Hazards
The material may become unstable at high temperatures. Do not touch spilled materials; barium hydroxide is an extremely dangerous neurotoxin. Move containers from fire if it can be done without risk. Material may explode in contact with acids or chlorinated rubber.

SECTION V - REACTIVITY DATA

Stability		Conditions to Avoid **Avoid heat and open flame, especially around dusts and powders of barium hydroxide. It is normally stable under normal conditions of handling and storage.**
Stable X	Unstable	Incompatibility *(materials to avoid)* **Incompatible with acids, oxidizers, and chlorinated rubber. It can be extremely corrosive to metals and generates extreme heat when combined with oxygen, nitrogen, hydrogen, ammonia, water, halogens, sulfides.**
Hazardous Polymerization		Conditions to Avoid **Hazardous polymerization of anhydrous barium hydroxide will not occur.**
May Occur	Will Not Occur X	Hazardous Decomposition or By-products **None reported for this barium compound.**

SECTION VI - HEALTH HAZARD DATA

Primary Route(s) of Entry:	Inhalation?	Absorption (skin/eye)?	Ingestion?
	X		X

Health Hazards

INHALATION: Dusts can irritate the nose, throat, and respiratory tract. High concentrations can also be swallowed along with inhalation, thereby increasing the overall toxicity of exposure.

EYES & SKIN: Exposure to the eyes and skin may cause moderate to severe irritation.

INGESTION: Can cause death. Symptoms include hypokalemia (low blood potassium) causing muscle stimulation and subsequent paralysis. There can be nausea, vomiting, diarrhea, tremors, leg cramps, facial twitching, speech impairment, bladder constriction, convulsions, paralysis in the extremities, cardiac arrhythmia and/or fibrillation, and respiratory failure.

Carcinogenicity	NTP Listed?	IARC Cancer Review Group?	OSHA Regulated?	Target Organs?
Unknown Human Unknown Animal	No	No	29 CFR 1910.1000 Table Z-1	Respiratory system, eyes, skin, CNS, heart.

Medical Conditions Generally Aggravated by Exposure

Persons with existing lung dysfunction, hypertension, CNS, or muscular disorders may be affected.

Emergency and First-aid Procedures

Eye contact: Do not allow victim to keep eyes tightly shut. Flush immediately with water for 15 minutes (minimum), seek medical attention. **Skin contact:** Remove all contaminated clothing. Immediately wash area with flooding amounts of water. For reddened or blistered skin, seek medical assistance. For **inhalation:** Remove the person from exposure. Provide respiratory assistance and CPR. Transfer to medical facility. If **swallowed:** Never give anything by mouth to a convulsing person. If conscious and alert, give 3-4 glasses of water. Seek medical attention. Inducing vomiting may or may not be advisable.

SECTION VII - PRECAUTIONS FOR SAFE HANDLING AND USE

Steps to be Taken in Case Material is Released or Spilled

Collect powdered materials in most convenient and safe manner and deposit in sealed containers. Dry sweep not recommended. Use HEPA vacuum. Remove all ignition sources. Restrict those not involved in cleanup from entering area. Keep combustible materials away.

Preferred Waste Disposal Method

No citation.

Precautions to be Taken in Handling and Storage

Explosion risks vary with type of barium compound. Store to avoid contact with incompatible materials, moisture, and combustible materials. Prevent physical damage to containers. Keep tightly closed.

Other Precautions and Warnings

Sources of ignition are prohibited where barium hydroxide is used, handled, or stored. Avoid any conditions which may cause the generation of airborne dusts.

SECTION VIII - CONTROL MEASURES AND PERSONAL PROTECTIVE EQUIPMENT

Respiratory Protection *(specify type)*

For dust exposures below PEL, use a respirator with a high efficiency particulate air (HEPA) filter, otherwise use a self-contained breathing apparatus with full facepiece operated in positive pressure mode.

Ventilation

Local exhaust (preferred) at site of chemical work or general ventilation.

Protective Gloves	Eye Protection	Other Protective Clothing
Impervious Rubber	Chemical/Dust Goggles or Face Mask	Protective Apron, Boots

Work/Hygiene Practices

Always wash hands thoroughly after using chemical; never bring food, drink, or smoking materials into vicinity of chemicals.

BARIUM HYDROXIDE

$Ba(OH)_2$ CAS: 17194-00-2

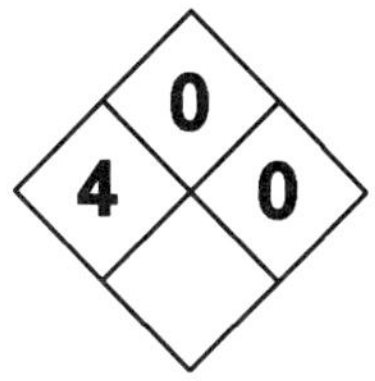

IDENTIFICATION AND TYPICAL USES

Barium hydroxide (anhydrous) is a clear to white powder with no detectable odor. It is used in the manufacture of alkalis, glass, oil and grease additives, barium soaps, and other barium compounds. It is also used in the vulcanization of synthetic rubber, as a corrosion inhibitor, in drilling fluids, as a lubricant, in pesticides, in the sugar industry, for softening water, in painting, as a agent for caustic soda, and as a steel carbonizing agent.

RISK ASSESSMENT: HEALTH

General Assessment

Certain compounds of barium are irritants of the eyes, mucous membranes, and skin. Primary exposure routes are *inhalation* and *ingestion*. It has not been established if exposure to barium hydroxide poses any carcinogenic, teratogenic, or mutagenic effects. Barium hydroxide is an extremely dangerous neurotoxin.

Inhalation of barium dusts can cause irritation of the nose and upper respiratory tract and may produce a benign pneumoconiosis known as baritosis. There may be characteristic X-ray changes. If exposed to large amounts, some barium hydroxide may be ingested as well as inhaled. This will increase the severity of the toxic exposure. All soluble salts of barium are acute poisons. The barium ion is toxic to muscle, especially the heart, producing stimulation and then paralysis. Specifically, inhalation barium hydroxide will irritate the nose, throat, and mucosa of the upper respiratory tract and bronchial tubes. There may be adverse effects on the heart and the function of the central nervous system (CNS).

Ingesting barium hydroxide can be fatal. As noted above for inhalation exposure, once inside the body, barium competes with blood potassium content to cause muscle stimulation followed by paralysis. Symptoms of exposure include nausea, vomiting, violent diarrhea, tremors, salivation, twitching of the neck and facial muscles (tightness or tingling also possible), abdominal pain, leg cramps, speech impairment, convulsions, weakness, bladder contraction, paralysis in the extremities, dypsnea (difficulty breathing), slowed pulse rate, cardiac arrhythmia and/or fibrillation, and respiratory failure.

Acute Health Effects

The following acute (short-term) health effects may occur immediately or shortly after exposure to barium hydroxide:

Skin: Extreme irritation at site of contact.

Eye: Severe irritation, damage may be serious.

Lung: Can cause metallic taste, nausea, sore throat, and irritation of the air passages.

Other: Serious effects on musculatory system at high exposures. May cause tingling and paralysis of the extremities followed by irregular heartbeat and might even cause the heart to stop (ventricular fibrillation). Death is also possible as a result of respiratory arrest. Also, kidney failure may occur as a result of circulatory dysfunction.

Chronic Health Effects

The following chronic (long-term) health effects may occur at some time after exposure to barium hydroxide and can last for months or even years:

Cancer Hazards: According to the information presented in the references, barium hydroxide has not been adequately tested for its ability to cause cancer in laboratory test animals.

Reproductive Hazard: According to the information presented in the references, barium hydroxide has not been adequately tested for its ability to adversely affect reproduction in animals.

Other Chronic Effects: Repeated exposure can cause abnormal chest X-ray. This usually takes years to develop. An abnormal chest X-ray from barium hydroxide exposure does not necessarily mean the lung has been damaged but future exposures to irritating substances may affect the lungs. It is not currently known whether barium hydroxide causes lung damage. Also, the barium ion can remain in bone tissue for up to 50 days.

Recommended Risk-Reduction Measures

Personnel should avoid direct contact with barium, especially its dusts or powders. If a less toxic material or compound cannot be substituted for barium hydroxide, then *engineering controls* are the most effective method of reducing exposures. The best protection is to enclose operations and/or provide local exhaust ventilation at the site of barium hydroxide dust release. While not always operationally feasible, isolating operations can also reduce exposure. Using respiratory protection is less effective than the controls mentioned above, but is still advisable whenever working with or around barium hydroxide. A self-contained breathing apparatus (SCBA) with full facepiece and pressure demand is recommended for the greatest possible respiratory protection, especially when exposures near the IDLH level of 250 mg/m^3 (as barium). However, for low or infrequent exposure, a MSHA/NIOSH approved full facepiece respirator with a high efficiency particulate air (HEPA) filter can be used. If a full facepiece is not available, then chemical/dust goggles should be worn to protect the eyes. A face shield and protective apron should also be worn. To prevent hand and skin exposures, gloves should be worn.

Administrative controls should also be in place to minimize the potential for human exposures. These may include written procedures or policies which specify the methods and techniques that will be practiced whenever personnel are to work with barium hydroxide.

All personnel should receive *training* on the use, hazards, protective measures, emergency actions, and other precautions per 29 CFR 1910.1200 (Hazard Communication), prior to the first assignment in an area where barium hydroxide is used or stored.

Before beginning employment working with barium hydroxide, and at regular intervals thereafter (e.g., annually), those personnel with frequent or potentially high exposures to barium hydroxide should be provided the following recommended medical tests:

- ☑ Lung function test (establish baseline).

If symptoms develop or overexposure is suspected, consider the following additional tests:

- ☑ Electrocardiogram (EKG).
- ☑ Neurological examination.
- ☑ Consider chest X-ray.
- ☑ Evaluation by a qualified allergist with careful consideration of exposure history and special testing (may help diagnose skin allergy).

Any evaluation should include a careful history of past and present symptoms with an examination. Medical tests that simply look for existing damage are not a substitute for controlling exposures. Also, since smoking can cause heart disease as well as lung cancer, emphysema, and other known respiratory problems, exposure to barium hydroxide may cause additional complications. Therefore, personnel who smoke should avoid any unprotected contact with barium hydroxide powders.

Other methods to reduce exposure include:

- ☑ Where possible, enclose operations and use local exhaust ventilation at the site of chemical release. If local exhaust ventilation or enclosure is not used, respiratory protection should be mandatory.
- ☑ Always ensure that proper protective clothing is worn when using chemical substances.
- ☑ Wash thoroughly immediately after exposure to barium hydroxide and at the end of the work shift or before eating, drinking, or smoking.
- ☑ Hazard warning information should be posted in the work area. In addition, as part of an on-going education and training program, all information on the health and safety hazards of barium hydroxide should be communicated to all exposed workers.
- ☑ Eye wash stations should be provided in the immediate work area for emergency use. If there is a possibility of skin exposure to barium hydroxide, emergency shower facilities should also be provided.
- ☑ Workers whose clothing has been contaminated by barium hydroxide should change into clean clothes before leaving work. Contaminated work clothing should be laundered only by individuals who have been informed of the hazards of exposure to barium hydroxide.

RISK ASSESSMENT: ENVIRONMENT

General Assessment

The environment is at risk of exposure during transportation, storage, disposal, or destruction of barium hydroxide. In almost every scenario, the threat of environmental exposure is contingent upon the proper handling of the chemical substance. Accidental spills, large or small, where barium hydroxide contacts incompatible commodities can result in fire, explosion (depending upon conditions of spill), and possible contamination of the surrounding environmental mediums (water, soil, and air).

Finely divided powders of barium hydroxide in air may present a moderate fire or explosion hazard. It is incompatible with acids, oxidizers, and chlorinated rubber. It is corrosive to metals can generate excessive heat when combined with oxygen, nitrogen, hydrogen, ammonia, water, and halogens, and sulfides. Extreme caution is required in handling, storage, transportation, and disposal of barium hydroxide. These characteristics also require special consideration during any emergency situation involving a leak or spill of barium hydroxide powder or dust. Should barium hydroxide ever come into contact with these incompatible substances either during use, transportation, storage, or disposal, violent and even explosive reactions are possible.

Barium hydroxide may enter the environment as a result of industrial and municipal waste treatment plant discharges and from spills.

Acute Ecological Effects

Acute (short-term) toxic effects may include the death of animals, birds, or fish and death or low growth rate in plants. Acute effects are seen 2 to 4 days after animals or plants are exposed to barium. Barium and its salts have moderate acute toxicity to aquatic life. Insufficient data are available to evaluate the short-term effects of barium or its salts to plants, birds, or land animals.

Chronic Ecological Effects

Chronic (long-term) toxic effects may include shortened life span, reproductive problems, lower fertility, and changes in appearance or behavior in exposed animals. These effects can be seen long after first exposure(s) to toxic chemicals. Barium and its salts have moderate chronic toxicity to aquatic life. Insufficient data are available to evaluate or predict the long-term effects of barium to plants, birds, or land animals.

Water Solubility

Barium hydroxide is slightly soluble in water. Concentrations of 1 to 100 milligrams will mix with a liter of water.

Persistence in the Environment

Barium is highly persistent in water, with a half-life greater than 200 days. The half-life of a pollutant is the amount of time it takes for one half of the chemical to be degraded. No data are available on the estimated percentages of barium persistence in the aquatic or terrestrial environments.

Bioaccumulation in Aquatic Organisms

Some substances increase in concentration, or *bioaccumulate,* in living organisms as they breathe contaminated air, drink contaminated water, or eat contaminated food. These chemicals can become concentrated in the tissues and internal organs of animals as well as humans.

The concentration of barium found in fish tissues is expected to be about the same as the average concentration of barium in the water from which the fish was taken.

Recommended Risk-Reduction Measures

Proper training of all transporters will reduce the likelihood of a mishap or accident resulting in a leak or spill of barium hydroxide dusts or powders into the environment. The correct use of DOT labeling (per the MSDS) on all containers, trucks, and rail cars will enable emergency responders to react properly and quickly to any disaster thereby reducing the potential risk to the environment and to personnel.

Storage of barium hydroxide should be segregated from incompatible chemicals to minimize the risk of cross-contamination or contact. Impermeable floors constructed of non-combustible materials are recommended.

If a spill or leak to the environment has occurred, fire department, emergency response and/or hazardous materials spill personnel should be notified immediately. Cleanup should be attempted only by those trained in proper spill containment procedures. Contaminated soils should be removed for incineration and replaced with clean soil. If barium hydroxide should contact the water table, aquifer, or navigable waterway, time is of the essence. It is only slightly soluble and total remediation may be possible. The local and/or state emergency response authorities must be notified. A comprehensive emergency response or disaster preparedness/recovery plan should be in place prior to any operations involving the use, transportation, storage, or disposal of barium hydroxide. If barium hydroxide is spilled or leaked, the following specific steps are recommended:

☑ Restrict persons not wearing protective clothing from area of spill or leak until cleanup is complete.

☑ Remove all ignition sources, ventilate area.

☑ Collect powdered materials in the safest manner possible and deposit in sealed containers. Recommend using vacuum with HEPA filter. Do not dry sweep (generates airborne dusts).

☑ It may be necessary to dispose of barium hydroxide as a hazardous waste. The responsible state agency or the regional office of the federal Environmental Protection Agency (EPA) should be contacted for specific recommendations.

RISK ASSESSMENT: BUSINESS

General Assessment

Accidents or mishaps involving barium hydroxide can present a significant threat to business operations. The loss or damage of equipment or facilities can significantly affect fiscal viability. Lawsuits that may result from personnel injury/death, public exposures, and/or environmental contamination will also require a serious expenditure of resources. Media attention surrounding an injury, death, or environmental damage can also result in a loss of profits and loss of current as well as future business.

Recommended Risk-Reduction Measures

Company attorneys, safety and health professionals, and environmental specialists should be involved in the development of any procedures for responding to chemical incidents. A company official should be predesignated as a public relations officer with specific training in dealing with the press. Corporate plans and policies should be developed, approved, and implemented long before any need for such arises.

REFERENCES

Ahlbom, A. 1993. *Biostatistics for Epidemiologists.* New York: CRC Press/Lewis Publishers.

American Conference of Governmental Industrial Hygienists. 1988. *Documentation of the Threshold Limit Values and Biological Exposure Indices,* 5th Edition (with updates). Cincinnati: ACGIH

Bloom, A. D. and F. J. DeSerres. 1995. *Ecotoxicity and Human Health.* New York: CRC Press/Lewis Publishers.

Brusick, D. J. 1994. *Methods for Genetic Risk Assessment.* New York: CRC Press/Lewis Publishers.

Bueche, J. R. 1972. *Principles of Physics,* 2nd Edition. New York: McGraw-Hill.

Cockerman, L. G., B. S. Shane. 1994. *Basic Environmental Toxicology.* New York: CRC Press/Lewis Publishers.

Gardner, R. W. 1994. *Chemical Intolerance.* New York: CRC Press/Lewis Publishers.

Howard, P. H. 1995. *Handbook of Environmental Fate and Exposures Data for Organic Chemicals,* Vol. I-IV. New York: CRC Press/Lewis Publishers.

Keith, L. H. and M. M. Walker. 1995. *Handbook of Air Toxics.* New York: CRC Press/Lewis Publishers.

Landis, W. G., M. H. Yu. 1995. *Introduction to Environmental Toxicology.* New York: CRC Press/Lewis Publishers.

Langerman, N. 1994 *Precautionary Labels for Chemical Containers.* New York: CRC Press/Lewis Publishers.

Lauwerys, R. R., P. Hoet. 1993. *Industrial Chemical Exposure.* New York: CRC Press/Lewis Publishers.

Lewis, R. J., Sr. 1993. *Hawley's Condensed Chemical Dictionary,* 12th Edition. New York: Van Nostrand Reinhold.

Lewis, R. J., Sr. 1992. *Sax's Dangerous Properties of Industrial Materials,* Volumes 1, 2, and 3, 8th Edition. New York: Van Nostrand Reinhold.

Meyer, E. 1990. *Chemistry of Hazardous Materials.* Englewood Cliffs, NJ: Prentice Hall.

Mycock, J. C., J. D. McKenna, and L. Theodore. 1995. *Handbook of Air Pollution Control Engineering and Technology.* New York: CRC Press/Lewis Publishers.

Patnaik, P. 1992. *A Comprehensive Guide to the Hazardous Properties of Chemical Substances.* New York: Van Nostrand Reinhold.

Reeves, A. L. 1979. Barium. *Handbook on the Toxicology of Metals.* pp 321-328. New York: Elsevier North-Holland.

Richardson, M. 1995. *Dictionary of Substance and Their Effects,* Vol. I-VI. New York: CRC Press/Lewis Publishers.

Sacarello, H. L. A. 1994. *The Comprehensive Handbook of Hazardous Materials.* New York: CRC Press/Lewis Publishers.

Talbott, E., G. Craun. 1995. *An Introduction to Environmental Epidemiology.* New York: CRC Press/Lewis Publishers.

Torkelson, T. R. V. K. Rowe. 1981. *Patty's Industrial Hygiene and Toxicology,* Vol. 2B. New York: Wiley-Interscience.

U.S. Department of Health, Education, and Welfare. National Institute for Occupational Safety and Health (NIOSH). 1978. *Criteria for a Recommended Standard. Occupational Exposure to Barium.* DHEW (NIOSH) Publication Number: 78-216. Washington, DC: U.S. Government Printing Office.

MATERIAL SAFETY DATA SHEET

CHEMICAL NAME	
BARIUM NITRATE	**CRC PRESS/LEWIS PUBLISHERS DISCLAIMER:** THE INFORMATION AND RECOMMENDATIONS PRESENTED HEREIN ARE BASED ON SOURCES BELIEVED TO BE RELIABLE. CRC MAKES NO REPRESENTATION ON ITS COMPLETENESS OR ACCURACY. IT IS THE RESPONSIBILITY OF THE USER TO DETERMINE THE CHEMICAL'S SUITABILITY FOR ITS INTENDED USE, THE CHEMICAL'S SAFE USE, AND THE CHEMICAL'S PROPER DISPOSAL. NO REPRESENTATIONS AND/OR WARRANTIES, EITHER EXPRESSED OR IMPLIED, OF THE MERCHANTABILITY OR FITNESS FOR A PARTICULAR PURPOSE, OR OF ANY OTHER NATURE, ARE MADE WITH RESPECT TO THE INFORMATION PROVIDED IN THIS MSDS OR TO THE CHEMICAL TO WHICH INFORMATION MAY REFER. CRC NEITHER ASSUMES NOR AUTHORIZES ANY OTHER PERSON TO ASSUME FOR IT, ANY OTHER ADDITIONAL RESPONSIBILITY OR LIABILITY FOR THE USE OF, OR RELIANCE UPON, THIS INFORMATION.

HAZARD WARNING INFORMATION

HEALTH	FIRE	REACTIVE	OTHER	DEGREE OF HAZARD	COLOR CODING	OTHER CODES
2	0	0	OX	0 = Minimum Hazard 1 = Slight Hazard 2 = Moderate Hazard 3 = Serious Hazard 4 = Severe Hazard	HEALTH = BLUE FIRE = RED REACTIVITY = YELLOW OTHER = WHITE	OX = Oxidizer ACID = Acid ALK = Alkali COR = Corrosive W = Use No Water

SECTION I - GENERAL INFORMATION

Characterization	RCRA Number	EPA Class
Inorganic Salt	**D001**	**Characteristic Ignitable Waste**

DOT Proper Shipping Name	Chemical Abstract Service (CAS) Number
Barium Nitrate	**10022-31-8**
DOT Hazard Class and Label Requirements	DOT Emergency Guide Code
Oxidizer; Oxidizer, Poison	**42**
DOT Identification Number	Molecular Formula
UN 1446	$N_2O_6 \cdot Ba$

Synonyms
Barium dinitrate; nitric acid; barium salt; nitroborate; nitrate of barium; barium (II) nitrate.

SECTION II - HAZARDOUS INGREDIENTS AND IDENTITY INFORMATION

Hazardous Components (specific identity)	OSHA Exposure Criteria	NIOSH Exposure Criteria	Immediately Dangerous to Life and Health (IDLH)	ACGIH Exposure Criteria
Barium nitrate (derivation: By the action of nitric acid on barium nitrate carbonate or sulfide).	**PEL:** 0.5 mg/m³ **STEL:** Not Established	**REL:** 0.5 mg/m³ **STEL:** Not Established	**250 mg/m³ for Soluble Compounds**	**TLV:** 0.5 mg/m³ **STEL:** Not Established

SECTION III - PHYSICAL AND CHEMICAL CHARACTERISTICS

Boiling Point	Specific Gravity (H_2O = 1)
Decomposes	**3.24**
Vapor Pressure (mm Hg)	Molecular Weight (atomic weight)
10 at 1920°F (1049°C)	**261.35**
Vapor Density (Air = 1)	Melting Point
N/A (metal)	**1098°F (592°C)**

Solubility
Soluble in water. Slightly soluble in acids; very soluble in acetone; insoluble in alcohol.

Appearance and Odor
Barium nitrate is an odorless, shiny, white crystalline (sugar or sand-like) solid. It may also appear as colorless, lustrous crystals.

SECTION IV - FIRE AND EXPLOSION HAZARD DATA

Flash Point (method used)	Explosive Limits in Air % by Volume
Not Found	**LEL: N/A UEL: N/A**
NFPA Classification	Autoignition Temperature
Oxidizer	**Not Found**

Extinguishing Media
Water only, do not use dry chemical, carbon dioxide, or halon. For larger fires, flood area with water from a distance. For massive fire in cargo areas, use unmanned hose holder or monitor nozzles.

Special Fire Fighting Procedures
Wear full protective clothing and self-contained breathing apparatus (SCBA). Poisonous gases are produced in fire. Move container from fire area if it can be done without risk. Apply cooling water to sides of containers exposed to flames until well after fire is out. Stay away from ends of tanks.

Unusual Fire and Explosion Hazards
Barium nitrate is an oxidizer. By itself, it will not burn. However, contact with organic or combustible materials such as wood, paper, or oil, can cause fires or explosions. If fires are out of control, consider evacuation of 1/2 mile radius.

SECTION V - REACTIVITY DATA

Stability		Conditions to Avoid: Avoid contact with combustible materials. Barium nitrate may increase the burning rate. Finely divided materials may ignite or explode on contact.
Stable	Unstable	Incompatibility (materials to avoid): Organic and combustible materials (such as wood, paper, oils, and fuels); and aluminum and magnesium alloys (violent reactions can occur). Possible explosion hazard on contact with citric acid and tin chloride.
X		
Hazardous Polymerization		Conditions to Avoid: Hazardous polymerization has not been reported to occur under normal conditions of temperature and pressure.
May Occur	Will Not Occur	Hazardous Decomposition or By-products: Poisonous gases, including oxides of nitrogen, are emitted during fire.
	X	

SECTION VI - HEALTH HAZARD DATA

Primary Route(s) of Entry:	Inhalation?	Absorption (skin/eye)?	Ingestion?
	X		X

Health Hazards

INHALATION: May cause irritation of the respiratory tract with symptoms of sore throat, coughing, and shortness of breath or labored breathing.

EYE & SKIN: Irritation with redness and pain. Repeated eye contact may cause conjunctivitis.

INGESTION: Severe gastrointestinal irritation with nausea, vomiting, diarrhea, colic, and excessive salivation. Dryness and constriction of mouth and throat with metallic taste.

Carcinogenicity	NTP Listed?	IARC Cancer Review Group?	OSHA Regulated?	Target Organs?
Unknown Human Unknown Animal	No	No	29 CFR 1910.1000 Table Z-1	Respiratory system, CNS, muscular system, kidneys.

Medical Conditions Generally Aggravated by Exposure

Persons with pre-existing chronic respiratory, cardiovascular, or skin diseases are at increased risk.

Emergency and First-aid Procedures

Eye contact: Do not allow victim to keep eyes tightly shut. Flush immediately with water for 15 minutes (minimum), seek medical attention. Skin contact: Remove all contaminated clothing, including shoes. Immediately wash area with flooding amounts of soap and water. Seek medical attention immediately. For inhalation: Remove the person from exposure. Provide respiratory assistance and CPR. Transfer to medical facility. If swallowed, Induce vomiting immediately, seek medical attention immediately.

SECTION VII - PRECAUTIONS FOR SAFE HANDLING AND USE

Steps to be Taken in Case Material is Released or Spilled

Keep combustibles away from spilled material. Do not touch spilled materials with bare skin. For small spills, use a clean shovel to place materials into clean, dry container and cover. For large spills, dike area far ahead of spill for later disposal. Restrict those not involved in cleanup from entering area.

Preferred Waste Disposal Method

Refer to local, state, and/or federal disposal regulations for characteristic waste disposal requirements.

Precautions to be Taken in Handling and Storage

Protect against physical damage. Separate from combustible, organic or other easily oxidized materials. Avoid storage on wood floors. Immediately remove and dispose of any spilled materials. Store in tightly closed containers in cool, well ventilated location.

Other Precautions and Warnings

Mixtures of barium nitrate with finely divided aluminum or magnesium alloys are easily ignitable and extremely sensitive to friction or impact.

SECTION VIII - CONTROL MEASURES AND PERSONAL PROTECTIVE EQUIPMENT

Respiratory Protection (specify type)

For dust exposures below PEL, use a respirator with a high efficiency particulate air (HEPA) filter. Greater protection is provided by a powered air purifying respirator or use a self-contained breathing apparatus (SCBA) with full facepiece operated in positive pressure mode.

Ventilation

Local exhaust (preferred) at site of chemical work or general ventilation.

Protective Gloves	Eye Protection	Other Protective Clothing
Impervious Rubber	Chemical/Dust Goggles or Face Mask	Protective Apron, Boots

Work/Hygiene Practices

Always wash hands thoroughly after using chemical; never bring food, drink, or smoking materials into vicinity of chemicals.

BARIUM NITRATE

$N_2O_6 \cdot Ba$ CAS: 7440-39-3

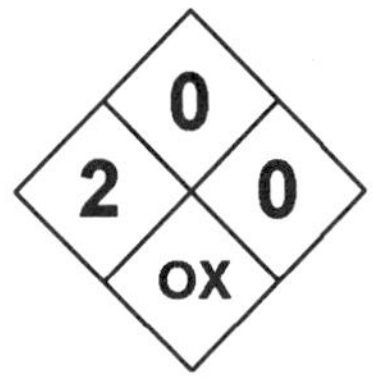

IDENTIFICATION AND TYPICAL USES

Barium nitrate is shiny yellowish-white or colorless crystalline (sugar or sand-like) solid with no odor. It can exist in a variety of salt forms. It is used in the manufacture of pyrotechniques and fireworks (it produces green light), incendiaries, chemicals (barium peroxide), ceramic glazes, rodenticides, and in the electronics industry.

RISK ASSESSMENT: HEALTH

General Assessment

Barium nitrate is a human poison via *ingestion* and *inhalation*. It is also a strong irritant to the eyes and skin. Symptoms of inhalation include nose and throat irritation, bronchial tube inflammation causing cough and phlegm.

The most severe cases of poisoning from barium nitrate exposure are those resulting from ingestion. The initial effects from soluble barium compounds are those of severe gastrointestinal irritation, including nausea, vomiting and diarrhea that may be accompanied by abdominal colic, and excessive salivation. There may also be dryness, a sense of constriction of the mouth and throat, and a metallic taste. Systemic effects follow and may include ringing in the ears, dizziness, elevated blood pressure, ocular changes causing blurred vision, and convulsive tremors. Hypokalemia may occur and result in abnormalities of myocardial function and impaired neuromuscular function, which may vary from minimal weakness to frank paralysis.

Other effects of exposure may include anxiety, weakness, and gradually increasing sleepiness with mental confusion. The central nervous system may be first stimulated and then depressed. Hemorrhages may occur in the stomach, intestines, and kidneys. Kidney damage and kidney failure have been attributed to barium nitrate exposure. Collapse and death from respiratory failure, severe hypokalemia, and cardiac arrest may occur. Rarely, inorganic nitrates may be converted to nitrates by nitrate-reducing bacteria in the digestive tract, resulting in methemoglobinemia.

Acute Health Effects

The following acute (short-term) health effects may occur immediately or shortly after exposure to barium nitrate:

Skin: Severe irritation and burns, redness, and pain.

Eye: Severe irritation, redness, and pain.

Lung: Respiratory tract irritation causing cough, labored breathing, and phlegm discharge.

CNS: First stimulated, then depressed. Causes a number of mild to serious systemic effects.

Chronic Health Effects

The following chronic (long-term) health effects may occur at some time after exposure to barium nitrate and can last for months or even years:

Cancer Hazards: According to the references, barium nitrate has not been tested for its ability to cause cancer in animals.

Reproductive Hazard: According to the references, barium nitrate has not been tested for its ability to affect reproduction.

Other Chronic Effects: Repeated barium exposure may cause spots to appear on lung X-rays. This usually takes 1 to 2 years to develop and can remain for years after exposure. It is not believed that barium itself affects lung function but exposure to other irritants may have more pronounced effects. Repeated skin contact may cause dryness and cracking. Prolonged eye contact may cause conjunctivitis. Very high exposures (such as swallowing or extremely high dust exposure) can cause barium poisoning with symptoms of vomiting, irregular heartbeat, paralysis, and death.

Recommended Risk-Reduction Measures

Personnel should avoid direct contact with barium nitrate dusts or powders. If a less toxic material or compound cannot be substituted for barium nitrate, then *engineering controls* are the most effective method of reducing exposures. The best protection is to enclose operations and/or provide local exhaust ventilation at the site of barium nitrate dust release. While not always operationally feasible, isolating operations can also reduce exposure. Using respiratory protection is less effective than the controls mentioned above, but is still advisable whenever working with or around barium nitrate. A self-contained breathing apparatus (SCBA) with full facepiece and pressure demand is recommended for the greatest possible respiratory protection, especially when exposures near the IDLH level of 250 mg/m^3. However, for low or infrequent exposure, a MSHA/NIOSH approved full facepiece respirator with a high efficiency particulate air (HEPA) filter can be used. If a full facepiece is not available, then chemical/dust goggles should be worn to protect the eyes. A face shield and protective apron should also be worn. To prevent hand and skin exposures, gloves should be worn.

Administrative controls should also be in place to minimize the potential for human exposures. These may include written procedures or policies which specify the methods and techniques that will be practiced whenever personnel are to work with barium nitrate.

All personnel should receive *training* on the use, hazards, protective measures, emergency actions, and other precautions per 29 CFR 1910.1200 (Hazard Communication), prior to the first assignment in an area where barium nitrate is used or stored.

Before beginning employment working with barium nitrate, and at regular intervals thereafter (e.g., annually), those personnel with frequent or potentially high exposures should be provided the following recommended medical tests:

☑ Lung function test.

Any evaluation should include a careful history of past and present symptoms with an examination. Medical tests that simply look for existing damage are not a substitute for controlling exposures.

When barium poisoning has occurred, the following antidote is recommended. However, the decision as to whether the severity of poisoning requires administration of any antidote and actual dose required should be made by qualified medical personnel:

☑ Give 30 grams of sodium sulfate in 250 milliliters of water orally and repeat in one hour. Give by gastric tube if symptoms have appeared. Administration of sulfate salts intravenously is extremely hazardous, since they induce the precipitation of barium sulfate in the kidney, with subsequent renal failure. Administration of potassium is critical. Any antidote must be administered by qualified medical personnel.

Other methods to reduce exposure include:

☑ Where possible, enclose operations and use local exhaust ventilation at the site of chemical release. If local exhaust ventilation or enclosure is not used, respiratory protection should be mandatory.

☑ Always ensure that proper protective clothing is worn when using chemical substances.

☑ Wash thoroughly immediately after exposure to barium nitrate and at the end of the work shift or before eating, drinking, or smoking.

☑ Hazard warning information should be posted in the work area. In addition, as part of an on-going education and training program, all information on the health and safety hazards of barium nitrate should be communicated to all exposed workers.

☑ Eye wash stations should be provided in the immediate work area for emergency use. If there is a possibility of skin exposure to barium nitrate, emergency shower facilities should also be provided.

☑ Workers whose clothing has been contaminated by barium nitrate should change into clean clothes before leaving work. Contaminated work clothing should be laundered only by individuals who have been informed of the hazards of exposure to barium nitrate.

RISK ASSESSMENT: ENVIRONMENT

General Assessment

The environment is at risk of exposure during transportation, storage, disposal, or destruction of barium nitrate. In almost every scenario, the threat of environmental exposure is contingent upon the proper handling of the chemical substance. Accidental spills, large or small, where barium nitrate contacts incompatible commodities can result in fire, explosion (depending upon conditions of spill), and possible contamination of the surrounding environmental mediums (water, soil, and air).

Barium nitrate is an oxidizer. By itself, it will not burn. However, it readily facilitates the combustion of other materials. It can therefore present a moderate fire and explosion hazard. Also, because it is incompatible so many commonly encountered materials such as wood and paper, as well as other commodities, extreme caution is required in handling, storage, transportation, and disposal of barium nitrate. These characteristics also require special consideration during any emergency situation involving a leak or spill of barium nitrate powder or dust. Should barium nitrate ever come into contact with incompatible substances (such as combustible materials and aluminum or magnesium alloys) either during use, transportation, storage, or disposal, violent and even explosive reactions are possible.

Barium nitrate enter the environment from industrial and municipal waste treatment plant discharges and from spills.

Acute Ecological Effects

Acute (short-term) toxic effects may include the death of animals, birds, or fish and death or low growth rate in plants. Acute effects are seen 2 to 4 days after animals or plants are exposed to barium nitrate.

Barium and its salts have moderate acute toxicity to aquatic life. Insufficient data are available to evaluate the short-term effects of barium or its salts to plants, birds, or land animals.

Chronic Ecological Effects

Chronic (long-term) toxic effects may include shortened life span, reproductive problems, lower fertility, and changes in appearance or behavior in exposed animals. These effects can be seen long after first exposure(s) to toxic chemicals.

Barium and its salts have moderate chronic toxicity to aquatic life. Insufficient data are available to evaluate or predict the long-term effects of barium or its salts to plants, birds, or terrestrial animals.

Water Solubility

Most of the barium salts are either highly or moderately soluble in water. Concentrations of 1 to 1000 milligrams and more will mix with a liter of water.

Persistence in the Environment

Barium nitrate is highly persistent in water, with a half-life greater than 200 days. The half-life of a pollutant is the amount of time it takes for one half of the chemical to be degraded. No data are available on the estimated percentages of barium nitrate persistence in the aquatic or terrestrial environments.

Bioaccumulation in Aquatic Organisms

Some substances increase in concentration, or *bioaccumulate,* in living organisms as they breathe contaminated air, drink contaminated water, or eat contaminated food. These chemicals can become concentrated in the tissues and internal organs of animals as well as humans.

The concentration of barium nitrate found in fish tissues is expected to be about the same as the average concentration of barium nitrate in the water from which the fish was taken.

Recommended Risk-Reduction Measures

Proper training of all transporters will reduce the likelihood of a mishap or accident resulting in a leak or spill of barium nitrate dusts or powders into the environment. The correct use of DOT labeling (per the MSDS) on all containers, trucks, and rail cars will enable emergency responders to react properly and quickly to any disaster thereby reducing the potential risk to the environment and to personnel.

Storage of barium nitrate should be segregated from incompatible chemicals and materials to minimize the risk of cross-contamination or contact.

If a spill or leak to the environment has occurred, fire department, emergency response and/or hazardous materials spill personnel should be notified immediately. Cleanup should be attempted only by those trained in proper spill containment procedures. Contaminated soils should be removed for incineration and replaced with clean soil. If barium nitrate should contact the water table, aquifer, or navigable waterway, time is of the essence. It is highly soluble and total remediation may not be possible. The local and/or state emergency response authorities must be notified. A comprehensive emergency response or disaster preparedness/recovery plan should be in place prior to any operations involving the use, transportation, storage, or disposal of barium nitrate.

If barium nitrate is spilled or leaked, the following actions are recommended:

☑ Restrict persons not wearing protective clothing from area of spill or leak until cleanup is complete and area can be opened for normal work.

☑ Collect powdered materials in the safest manner possible and deposit in sealed containers. Recommend using clean shovel for small spills, depositing materials in clean container. For larger spills, use a vacuum with HEPA filter. Do not dry sweep (generates airborne dusts).

☑ It may be necessary to dispose of barium nitrate as a hazardous waste. The responsible state agency or the regional office of the federal Environmental Protection Agency (EPA) should be contacted for specific recommendations.

RISK ASSESSMENT: BUSINESS

General Assessment

Accidents or mishaps involving barium nitrate can present a significant threat to business operations. The loss or damage of equipment or facilities can significantly affect fiscal viability. Lawsuits that may result from personnel injury/death, public exposures, and/or environmental contamination will also require a serious expenditure of resources. Media attention surrounding an injury, death, or environmental damage can also result in a loss of profits and loss of current as well as future business.

Recommended Risk-Reduction Measures

Company attorneys, safety and health professionals, and environmental specialists should be involved in the development of any procedures for responding to chemical incidents. A company official should be predesignated as a public relations officer with specific training in dealing with the press. Corporate plans and policies should be developed, approved, and implemented long before any need for such arises.

REFERENCES

American Conference of Governmental Industrial Hygienists. 1988. *Documentation of the Threshold Limit Values and Biological Exposure Indices,* 5th Edition (with updates). Cincinnati: ACGIH

Bloom, A. D. and F. J. DeSerres. 1995. *Ecotoxicity and Human Health.* New York: CRC Press/Lewis Publishers.

Brusick, D. J. 1994. *Methods for Genetic Risk Assessment.* New York: CRC Press/Lewis Publishers.

Bueche, J. R. 1972. *Principles of Physics,* 2nd Edition. New York: McGraw-Hill.

Calabrese, E. J. 1994. *Biological Effects of Low Level Exposures.* New York: CRC Press/Lewis Publishers.

Cockerman, L. G., B. S. Shane. 1994. *Basic Environmental Toxicology.* New York: CRC Press/Lewis Publishers.

Cothern, R. R. 1993. *Comparative Environmental Risk Assessment.* New York: CRC Press/Lewis Publishers.

Gardner, R. W. 1994. *Chemical Intolerance.* New York: CRC Press/Lewis Publishers.

Hallenbeck, W. H. 1993. *Quantitative Risk Assessment for Environmental and Occupational Health,* 2nd Edition. New York: CRC Press/Lewis Publishers.

Howard, P. H. 1995. *Handbook of Environmental Fate and Exposures Data for Organic Chemicals,* Vol. I-IV. New York: CRC Press/Lewis Publishers.

Landis, W. G., M. H. Yu. 1995. *Introduction to Environmental Toxicology.* New York: CRC Press/Lewis Publishers.

Langerman, N. 1994 *Precautionary Labels for Chemical Containers.* New York: CRC Press/Lewis Publishers.

Lauwerys, R. R., P. Hoet. 1993. *Industrial Chemical Exposure.* New York: CRC Press/Lewis Publishers.

Lewis, R. J., Sr. 1993. *Hawley's Condensed Chemical Dictionary,* 12th Edition. New York: Van Nostrand Reinhold.

Lewis, R. J., Sr. 1992. *Sax's Dangerous Properties of Industrial Materials,* Volumes 1, 2, and 3, 8th Edition. New York: Van Nostrand Reinhold.

Manahan, S. E. 1994. *Environmental Chemistry,* 6th Edition. New York: CRC Press/Lewis Publishers.

Meyer, E. 1990. *Chemistry of Hazardous Materials.* Englewood Cliffs, NJ: Prentice Hall.

Patnaik, P. 1992. *A Comprehensive Guide to the Hazardous Properties of Chemical Substances.* New York: Van Nostrand Reinhold.

Proctor, N. H., J. P. Hughes. 1991. *Chemical Hazards of the Workplace.* New York: Van Nostrand Reinhold.

Quigley, D. R. 1994. *Handbook of Emergency Chemical Management.* New York: CRC Press/Lewis Publishers.

Rea, W. J. 1995. *Chemical Sensitivity,* Vol. I-III. New York: CRC Press/Lewis Publishers.

Richardson, M. 1995. *Dictionary of Substance and Their Effects,* Vol. I-VI. New York: CRC Press/Lewis Publishers.

Sacarello, H. L. A. 1994. *The Comprehensive Handbook of Hazardous Materials.* New York: CRC Press/Lewis Publishers.

Talbott, E., G. Craun. 1995. *An Introduction to Environmental Epidemiology.* New York: CRC Press/Lewis Publishers.

Torkelson, T. R. V. K. Rowe. 1981. *Patty's Industrial Hygiene and Toxicology,* Vol. 2B. New York: Wiley-Interscience.

MATERIAL SAFETY DATA SHEET

CHEMICAL NAME	
BARIUM OCTAHYDRATE	**CRC PRESS/LEWIS PUBLISHERS DISCLAIMER:** THE INFORMATION AND RECOMMENDATIONS PRESENTED HEREIN ARE BASED ON SOURCES BELIEVED TO BE RELIABLE. CRC MAKES NO REPRESENTATION ON ITS COMPLETENESS OR ACCURACY. IT IS THE RESPONSIBILITY OF THE USER TO DETERMINE THE CHEMICAL'S SUITABILITY FOR ITS INTENDED USE, THE CHEMICAL'S SAFE USE, AND THE CHEMICAL'S PROPER DISPOSAL. NO REPRESENTATIONS AND/OR WARRANTIES, EITHER EXPRESSED OR IMPLIED, OF THE MERCHANTABILITY OR FITNESS FOR A PARTICULAR PURPOSE, OR OF ANY OTHER NATURE, ARE MADE WITH RESPECT TO THE INFORMATION PROVIDED IN THIS MSDS OR TO THE CHEMICAL TO WHICH INFORMATION MAY REFER. CRC NEITHER ASSUMES NOR AUTHORIZES ANY OTHER PERSON TO ASSUME FOR IT, ANY OTHER ADDITIONAL RESPONSIBILITY OR LIABILITY FOR THE USE OF, OR RELIANCE UPON, THIS INFORMATION.

HAZARD WARNING INFORMATION

HEALTH	FIRE	REACTIVE	OTHER	DEGREE OF HAZARD	COLOR CODING	OTHER CODES
4	0	0		0 = Minimum Hazard 1 = Slight Hazard 2 = Moderate Hazard 3 = Serious Hazard 4 = Severe Hazard	HEALTH = BLUE FIRE = RED REACTIVITY = YELLOW OTHER = WHITE	OX = Oxidizer ACID = Acid ALK = Alkali COR = Corrosive W = Use No Water

SECTION I - GENERAL INFORMATION

Characterization	RCRA Number	EPA Class
Inorganic (metallic) octahydrate	**D005**	**Characteristic (T) Waste**

DOT Proper Shipping Name **Barium Compounds, N.O.S.**	Chemical Abstract Service (CAS) Number **12230-71-6**
DOT Hazard Class and Label Requirements **Varies (depending on toxicity)**	DOT Emergency Guide Code **49**
DOT Identification Number **UN 1564**	Atomic Symbol **$Ba(OH)_2$**

Synonyms
Barium hydrate; barium dihydrate; caustic baryta.

SECTION II - HAZARDOUS INGREDIENTS AND IDENTITY INFORMATION

Hazardous Components (specific identity)	OSHA Exposure Criteria	NIOSH Exposure Criteria	Immediately Dangerous to Life and Health (IDLH)	ACGIH Exposure Criteria
Barium octahydrate (derivation: By dissolving barium oxide in water with subsequent crystallization, or by precipitation from an aqueous solution of the sulfide by caustic soda, or by heating barium sulfide in earthenware retorts into which a current of moist carbonic acid is passed).	**PEL:** 0.5 mg(Ba)/m^3 **STEL:** Not Established	**REL:** 0.5 mg(Ba)/m^3 **STEL:** Not Established	**Not Established for BaO_2 (250 mg/m^3 for Soluble Ba Compounds)**	**TLV:** 0.5 mg(Ba)/m^3 **STEL:** Not Established

SECTION III - PHYSICAL AND CHEMICAL CHARACTERISTICS

Boiling Point **Not Determined**	Specific Gravity (H_2O = 1) **2.18**
Vapor Pressure (mm Hg) **227 at 172°F (80°C)**	Molecular Weight (atomic weight) **315.48**
Vapor Density (Air = 1) **Not Determined**	Melting Point **172°F (78°C)**

Solubility
Most of the barium salts are either highly or moderately soluble in water. Barium octahydrate powder is slightly soluble in water. Soluble in alcohol and ether, insoluble in acetone.

Appearance and Odor
White crystalline solid with no discernible odor.

SECTION IV - FIRE AND EXPLOSION HAZARD DATA

Flash Point (method used) **Not Found**	Explosive Limits in Air % by Volume **LEL: N/A UEL: N/A**
NFPA Classification **Not Listed**	Autoignition Temperature **Not Found**

Extinguishing Media
Use agents suitable to surrounding fire (alcohol foam, carbon dioxide, dry chemical, or water).

Special Fire Fighting Procedures
Poisonous gases are produced in fire. Structural fire-fighting clothing is permeable and may not provide adequate protection. Wear full protective clothing and self-contained breathing apparatus (SCBA). Use water mist to keep fire-exposed containers cool. Fight fire from distance, if possible.

Unusual Fire and Explosion Hazards
The material may become unstable at high temperatures. Do not touch spilled materials; barium octahydrate is an extremely dangerous neurotoxin. Move containers from fire if it can be done without risk. Material may explode in contact with acids or chlorinated rubber.

SECTION V - REACTIVITY DATA

Stability		Conditions to Avoid **Avoid heat and open flame, especially around dusts and powders of barium octahydrate. It is normally stable under normal conditions of handling and storage.**
Stable X	Unstable	Incompatibility *(materials to avoid)* **Incompatible with acids, oxidizers, and chlorinated rubber. It can be extremely corrosive to metals and generates extreme heat when combined with oxygen, nitrogen, hydrogen, ammonia, water, halogens, sulfides.**
Hazardous Polymerization		Conditions to Avoid **Hazardous polymerization of barium octahydrate will not occur.**
May Occur	Will Not Occur X	Hazardous Decomposition or By-products **None reported for this barium compound.**

SECTION VI - HEALTH HAZARD DATA

Primary Route(s) of Entry:	Inhalation?	Absorption (skin/eye)?	Ingestion?
	X		X

Health Hazards

INHALATION: Dusts can irritate the nose, throat, and respiratory tract. High concentrations can also be swallowed along with inhalation, thereby increasing the overall toxicity of exposure.

EYES & SKIN: Exposure to the eyes and skin may cause moderate to severe irritation.

INGESTION: Can cause death. Symptoms include hypokalemia (low blood potassium) causing muscle stimulation and subsequent paralysis. There can be nausea, vomiting, diarrhea, tremors, leg cramps, facial twitching, speech impairment, bladder constriction, convulsions, paralysis in the extremities, cardiac arrhythmia and/or fibrillation, and respiratory failure.

Carcinogenicity	NTP Listed?	IARC Cancer Review Group?	OSHA Regulated?	Target Organs?
Unknown Human Unknown Animal	No	No	29 CFR 1910.1000 Table Z-1	Respiratory system, eyes, skin, CNS, heart.

Medical Conditions Generally Aggravated by Exposure

Persons with existing lung dysfunction, hypertension, CNS or muscular disorders may be effected.

Emergency and First-aid Procedures

Eye contact: Do not allow victim to keep eyes tightly shut. Flush immediately with water for 15 minutes (minimum), seek medical attention. **Skin contact:** Remove all contaminated clothing. Immediately wash area with flooding amounts of water. For reddened or blistered skin, seek medical assistance. For **inhalation:** Remove the person from exposure. Provide respiratory assistance and CPR. Transfer to medical facility. If **swallowed:** Never give anything by mouth to a convulsing person. If conscious and alert, give 3-4 glasses of water. Seek medical attention. Inducing vomiting may or may not be advisable.

SECTION VII - PRECAUTIONS FOR SAFE HANDLING AND USE

Steps to be Taken in Case Material is Released or Spilled

Collect powdered materials in most convenient and safe manner and deposit in sealed containers. Dry sweep not recommended. Use HEPA vacuum. Remove all ignition sources. Restrict those not involved in cleanup from entering area. Keep combustible materials away.

Preferred Waste Disposal Method

No citation.

Precautions to be Taken in Handling and Storage

Explosion risks vary with type of barium compound. Store to avoid contact with incompatible materials, moisture, and combustible materials. Prevent physical damage to containers. Keep tightly closed.

Other Precautions and Warnings

Sources of ignition are prohibited where barium octahydrate is used, handled, or stored. Avoid any conditions which may cause the generation of airborne dusts.

SECTION VIII - CONTROL MEASURES AND PERSONAL PROTECTIVE EQUIPMENT

Respiratory Protection *(specify type)*

For dust exposures below PEL, use a respirator with a high efficiency particulate air (HEPA) filter, otherwise use a self-contained breathing apparatus with full facepiece operated in positive pressure mode.

Ventilation

Local exhaust (preferred) at site of chemical work or general ventilation.

Protective Gloves	Eye Protection	Other Protective Clothing
Impervious Rubber	Chemical/Dust Goggles or Face Mask	Protective Apron, Boots

Work/Hygiene Practices

Always wash hands thoroughly after using chemical; never bring food, drink, or smoking materials into vicinity of chemicals.

BARIUM OCTAHYDRATE

$Ba(OH)_2$ CAS: 12230-71-6

IDENTIFICATION AND TYPICAL USES

Barium octahydrate is a white crystalline solid with no detectable odor. It is used in the manufacture of alkalis, glass, oil and grease additives, barium soaps, and other barium compounds. It is also used in organic preparations, in making barium salts, in analytical chemistry, in the vulcanization of synthetic rubber, as a corrosion inhibitor, in drilling fluids, as a lubricant, in pesticides, in the sugar industry, for softening water, in painting, as a agent for caustic soda, and as a steel carbonizing agent.

RISK ASSESSMENT: HEALTH

General Assessment

Certain compounds of barium are irritants of the eyes, mucous membranes, and skin. Primary exposure routes are *inhalation* and *ingestion*. It has not been established if exposure to barium octahydrate poses any carcinogenic, teratogenic, or mutagenic effects. Barium octahydrate is an extremely dangerous neurotoxin.

Inhalation of barium dusts can cause irritation of the nose and upper respiratory tract and may produce a benign pneumoconiosis known as baritosis. There may be characteristic X-ray changes. If exposed to large amounts, some barium octahydrate may be ingested as well as inhaled. This will increase the severity of the toxic exposure. All soluble salts of barium are acute poisons. The barium ion is toxic to muscle, especially the heart, producing stimulation and then paralysis. Specifically, inhalation of barium octahydrate will irritate the nose, throat, and mucosa of the upper respiratory tract and bronchial tubes. There may be adverse effects on the heart and the function of the central nervous system (CNS).

Ingesting barium octahydrate can be fatal. As noted above for inhalation exposure, once inside the body, barium competes with blood potassium content to cause muscle stimulation followed by paralysis. Symptoms of exposure include nausea, vomiting, violent diarrhea, tremors, salivation, twitching of the neck and facial muscles (tightness or tingling also possible), abdominal pain, leg cramps, speech impairment, convulsions, weakness, bladder contraction, paralysis in the extremities, dypsnea (difficulty breathing), slowed pulse rate, cardiac arrhythmia and/or fibrillation, and respiratory failure.

Acute Health Effects

The following acute (short-term) health effects may occur immediately or shortly after exposure to barium octahydrate:

Skin: Extreme irritation at site of contact.

Eye: Severe irritation, damage may be serious.

Lung: Can cause metallic taste, nausea, sore throat, and irritation of the air passages.

Other: Serious effects on musculatory system at high exposures. May cause tingling and paralysis of the extremities followed by irregular heartbeat and might even cause the heart to stop (ventricular fibrillation). Death is also possible as a result of respiratory arrest. Also, kidney failure may occur as a result of circulatory dysfunction.

Chronic Health Effects

The following chronic (long-term) health effects may occur at some time after exposure to barium octahydrate and can last for months or even years:

Cancer Hazards: According to the information presented in the references, barium octahydrate has not been adequately tested for its ability to cause cancer in laboratory test animals.

Reproductive Hazard: According to the references, barium octahydrate has not been tested for its ability to affect reproduction.

Other Chronic Effects: Repeated exposure can cause abnormal chest X-ray. This usually takes years to develop. An abnormal chest X-ray from barium octahydrate exposure does not necessarily mean the lung has been damaged but future exposures to irritating substances may affect the lungs. It is not currently known whether barium octahydrate causes lung damage. Also, the barium ion can remain in bone tissue for up to 50 days.

❶ *Recommended Risk-Reduction Measures*

Personnel should avoid direct contact with barium, especially its dusts or powders. If a less toxic material or compound cannot be substituted for barium octahydrate, then *engineering controls* are the most effective method of reducing exposures. The best protection is to enclose operations and/or provide local exhaust ventilation at the site of barium octahydrate dust release. While not always operationally feasible, isolating operations can also reduce exposure. Using respiratory protection is less effective than the controls mentioned above, but is still advisable whenever working with or around barium octahydrate. A self-contained breathing apparatus (SCBA) with full facepiece and pressure demand is recommended for the greatest possible respiratory protection, especially when exposures near the IDLH level of 250 mg/m^3 (as barium). However, for low or infrequent exposure, a MSHA/NIOSH approved full facepiece respirator with a high efficiency particulate air (HEPA) filter can be used. If a full facepiece is not available, then chemical/dust goggles should be worn to protect the eyes. A face shield and protective apron should also be worn. To prevent hand and skin exposures, gloves should be worn.

Administrative controls should also be in place to minimize the potential for human exposures. These may include written procedures or policies which specify the methods and techniques that will be practiced whenever personnel are to work with barium octahydrate.

All personnel should receive *training* on the use, hazards, protective measures, emergency actions, and other precautions per 29 CFR 1910.1200 (Hazard Communication), prior to the first assignment in an area where barium octahydrate is used or stored.

Before beginning employment working with barium octahydrate, and at regular intervals thereafter (e.g., annually), those personnel with frequent or potentially high exposures should be provided the following recommended medical tests:

☑ Lung function test (establish baseline).

If symptoms develop or overexposure is suspected, consider the following additional tests:

☑ Electrocardiogram (EKG).

☑ Neurological examination.

☑ Consider chest X-ray.

☑ Evaluation by a qualified allergist with careful consideration of exposure history and special testing (may help diagnose skin allergy).

Any evaluation should include a careful history of past and present symptoms with an examination. Medical tests that simply look for existing damage are not a substitute for controlling exposures. Also, since smoking can cause heart disease as well as lung cancer, emphysema, and other known respiratory problems, exposure to barium octahydrate may cause additional complications. Therefore, personnel who smoke should avoid any unprotected contact with barium octahydrate powders.

Other methods to reduce exposure include:

☑ Where possible, enclose operations and use local exhaust ventilation at the site of chemical release. If local exhaust ventilation or enclosure is not used, respiratory protection should be mandatory.

☑ Always ensure that proper protective clothing is worn when using chemical substances.

☑ Wash thoroughly immediately after exposure to barium octahydrate and at the end of the work shift or before eating, drinking, or smoking.

☑ Hazard warning information should be posted in the work area. In addition, as part of an on-going education and training program, all information on the health and safety hazards of barium octahydrate should be communicated to all potentially exposed workers.

☑ Eye wash stations should be provided in the immediate work area for emergency use. If there is a possibility of skin exposure to barium octahydrate, emergency shower facilities should also be provided.

☑ Workers whose clothing has been contaminated by barium octahydrate should change into clean clothes before leaving work. Contaminated work clothing should be laundered only by individuals who have been informed of the hazards of exposure to barium octahydrate.

RISK ASSESSMENT: ENVIRONMENT

General Assessment

The environment is at risk of exposure during transportation, storage, disposal, or destruction of barium octahydrate. In almost every scenario, the threat of environmental exposure is contingent upon the proper handling of the chemical substance. Accidental spills, large or small, where barium octahydrate contacts incompatible commodities can result in fire, explosion (depending upon conditions of spill), and possible contamination of the surrounding environmental mediums (water, soil, and air).

Finely divided powders of barium octahydrate in air may present a moderate fire or explosion hazard. It is incompatible with acids, oxidizers, and chlorinated rubber. It is corrosive to metals can generate excessive heat when combined with oxygen, nitrogen, hydrogen, ammonia, water, and halogens, and sulfides. Extreme caution is required in handling, storage, transportation, and disposal of barium octahydrate. These characteristics also require special consideration during any emergency situation involving a leak or spill of barium octahydrate powder or dust. Should barium octahydrate ever come into contact with these incompatible substances either during use, transportation, storage, or disposal, violent and even explosive reactions are possible.

Barium octahydrate may enter the environment as a result of industrial and municipal waste treatment plant discharges and from spills.

Acute Ecological Effects

Acute (short-term) toxic effects may include the death of animals, birds, or fish and death or low growth rate in plants. Acute effects are seen 2 to 4 days after animals or plants are exposed to barium.

Barium and its salts have moderate acute toxicity to aquatic life. Insufficient data are available to evaluate the short-term effects of barium or its salts to plants, birds, or land animals.

Chronic Ecological Effects

Chronic (long-term) toxic effects may include shortened life span, reproductive problems, lower fertility, and changes in appearance or behavior in exposed animals. These effects can be seen long after first exposure(s) to toxic chemicals.

Barium and its salts have moderate chronic toxicity to aquatic life. Insufficient data are available to evaluate or predict the long-term effects of barium to plants, birds, or land animals.

Water Solubility

Barium octahydrate is slightly soluble in water. Concentrations of 1 to 100 milligrams will mix with a liter of water.

Persistence in the Environment

Barium is highly persistent in water, with a half-life greater than 200 days. The half-life of a pollutant is the amount of time it takes for one half of the chemical to be degraded. No data are available on the estimated percentages of barium persistence in the aquatic or terrestrial environments.

Bioaccumulation in Aquatic Organisms

Some substances increase in concentration, or *bioaccumulate,* in living organisms as they breathe contaminated air, drink contaminated water, or eat contaminated food. These chemicals can become concentrated in the tissues and internal organs of animals as well as humans.

The concentration of barium found in fish tissues is expected to be about the same as the average concentration of barium in the water from which the fish was taken.

Recommended Risk-Reduction Measures

Proper training of all transporters will reduce the likelihood of a mishap or accident resulting in a leak or spill of barium octahydrate dusts or powders into the environment. The correct use of DOT labeling (per the MSDS) on all containers, trucks, and rail cars will enable emergency responders to react properly and quickly to any disaster thereby reducing the potential risk to the environment and to personnel.

Storage of barium octahydrate should be segregated from incompatible chemicals to minimize the risk of cross-contamination or contact. Impermeable floors constructed of non-combustible materials are recommended.

If a spill or leak to the environment has occurred, fire department, emergency response and/or hazardous materials spill personnel should be notified immediately. Cleanup should be attempted only by those trained in proper spill containment procedures. Contaminated soils should be removed for incineration and replaced with clean soil. If barium octahydrate should contact the water table, aquifer, or navigable waterway, time is of the essence. It is only slightly soluble and total remediation may be possible. The local and/or state emergency response authorities must be notified. A comprehensive emergency response or disaster preparedness/recovery plan should be in place prior to any operations involving the use, transportation, storage, or disposal of barium octahydrate. If barium octahydrate is spilled or leaked, the following specific steps are recommended:

☑ Restrict persons not wearing protective clothing from area of spill or leak until cleanup is complete and area can be opened for normal work.

☑ Remove all ignition sources, ventilate area.

☑ Collect powdered materials in the safest manner possible and deposit in sealed containers. Recommend using vacuum with HEPA filter. Do not dry sweep (generates airborne dusts).

☑ It may be necessary to dispose of barium octahydrate as a hazardous waste. The responsible state agency or the regional office of the federal Environmental Protection Agency (EPA) should be contacted for specific recommendations.

RISK ASSESSMENT: BUSINESS

General Assessment

Accidents or mishaps involving barium octahydrate can present a significant threat to business operations. The loss or damage of equipment or facilities can significantly affect fiscal viability. Lawsuits that may result from personnel injury/death, public exposures, and/or environmental contamination will also require a serious expenditure of resources. Media attention surrounding an injury, death, or environmental damage can also result in a loss of profits and loss of current as well as future business.

Recommended Risk-Reduction Measures

Company attorneys, safety and health professionals, and environmental specialists should be involved in the development of any procedures for responding to chemical incidents. A company official should be predesignated as a public relations officer with specific training in dealing with the press. Corporate plans and policies should be developed, approved, and implemented long before any need for such arises.

REFERENCES

Ahlbom, A. 1993. *Biostatistics for Epidemiologists.* New York: CRC Press/Lewis Publishers.

American Conference of Governmental Industrial Hygienists. 1988. *Documentation of the Threshold Limit Values and Biological Exposure Indices,* 5th Edition (with updates). Cincinnati: ACGIH

Bloom, A. D. and F. J. DeSerres. 1995. *Ecotoxicity and Human Health.* New York: CRC Press/Lewis Publishers.

Brusick, D. J. 1994. *Methods for Genetic Risk Assessment.* New York: CRC Press/Lewis Publishers.

Bueche, J. R. 1972. *Principles of Physics,* 2nd Edition. New York: McGraw-Hill.

Cockerman, L. G., B. S. Shane. 1994. *Basic Environmental Toxicology.* New York: CRC Press/Lewis Publishers.

Gardner, R. W. 1994. *Chemical Intolerance.* New York: CRC Press/Lewis Publishers.

Howard, P. H. 1995. *Handbook of Environmental Fate and Exposures Data for Organic Chemicals,* Vol. I-IV. New York: CRC Press/Lewis Publishers.

Keith, L. H. and M. M. Walker. 1995. *Handbook of Air Toxics.* New York: CRC Press/Lewis Publishers.

Landis, W. G., M. H. Yu. 1995. *Introduction to Environmental Toxicology.* New York: CRC Press/Lewis Publishers.

Langerman, N. 1994 *Precautionary Labels for Chemical Containers.* New York: CRC Press/Lewis Publishers.

Lauwerys, R. R., P. Hoet. 1993. *Industrial Chemical Exposure.* New York: CRC Press/Lewis Publishers.

Lewis, R. J., Sr. 1993. *Hawley's Condensed Chemical Dictionary,* 12th Edition. New York: Van Nostrand Reinhold.

Lewis, R. J., Sr. 1992. *Sax's Dangerous Properties of Industrial Materials,* Volumes 1, 2, and 3, 8th Edition. New York: Van Nostrand Reinhold.

Meyer, E. 1990. *Chemistry of Hazardous Materials.* Englewood Cliffs, NJ: Prentice Hall.

Mycock, J. C., J. D. McKenna, and L. Theodore. 1995. *Handbook of Air Pollution Control Engineering and Technology.* New York: CRC Press/Lewis Publishers.

Reeves, A. L. 1979. Barium. *Handbook on the Toxicology of Metals.* pp 321-328. New York: Elsevier North-Holland.

Richardson, M. 1995. *Dictionary of Substance and Their Effects,* Vol. I-VI. New York: CRC Press/Lewis Publishers.

Sacarello, H. L. A. 1994. *The Comprehensive Handbook of Hazardous Materials.* New York: CRC Press/Lewis Publishers.

Torkelson, T. R. V. K. Rowe. 1981. *Patty's Industrial Hygiene and Toxicology,* Vol. 2B. New York: Wiley-Interscience.

U.S. Department of Health, Education, and Welfare. National Institute for Occupational Safety and Health (NIOSH). 1978. *Criteria for a Recommended Standard. Occupational Exposure to Barium.* DHEW (NIOSH) Publication Number: 78-216. Washington, DC: U.S. Government Printing Office.

MATERIAL SAFETY DATA SHEET

CHEMICAL NAME	
BARIUM PEROXIDE	**CRC PRESS/LEWIS PUBLISHERS DISCLAIMER:** THE INFORMATION AND RECOMMENDATIONS PRESENTED HEREIN ARE BASED ON SOURCES BELIEVED TO BE RELIABLE. CRC MAKES NO REPRESENTATION ON ITS COMPLETENESS OR ACCURACY. IT IS THE RESPONSIBILITY OF THE USER TO DETERMINE THE CHEMICAL'S SUITABILITY FOR ITS INTENDED USE, THE CHEMICAL'S SAFE USE, AND THE CHEMICAL'S PROPER DISPOSAL. NO REPRESENTATIONS AND/OR WARRANTIES, EITHER EXPRESSED OR IMPLIED, OF THE MERCHANTABILITY OR FITNESS FOR A PARTICULAR PURPOSE, OR OF ANY OTHER NATURE, ARE MADE WITH RESPECT TO THE INFORMATION PROVIDED IN THIS MSDS OR TO THE CHEMICAL TO WHICH INFORMATION MAY REFER. CRC NEITHER ASSUMES NOR AUTHORIZES ANY OTHER PERSON TO ASSUME FOR IT, ANY OTHER ADDITIONAL RESPONSIBILITY OR LIABILITY FOR THE USE OF, OR RELIANCE UPON, THIS INFORMATION.

HAZARD WARNING INFORMATION

HEALTH	FIRE	REACTIVE	OTHER	DEGREE OF HAZARD	COLOR CODING	OTHER CODES
1	**0**	**0**	**OX**	0 = Minimum Hazard 1 = Slight Hazard 2 = Moderate Hazard 3 = Serious Hazard 4 = Severe Hazard	HEALTH = BLUE FIRE = RED REACTIVITY = YELLOW OTHER = WHITE	OX = Oxidizer ACID = Acid ALK = Alkali COR = Corrosive W = Use No Water

SECTION I - GENERAL INFORMATION

Characterization	RCRA Number	EPA Class
Inorganic (metallic) peroxide	**D005**	**Characteristic (T) Waste**

DOT Proper Shipping Name	Chemical Abstract Service (CAS) Number
Barium peroxide	**1304-29-6**
DOT Hazard Class and Label Requirements	DOT Emergency Guide Code
Oxidizer	**49**
DOT Identification Number	Atomic Symbol
UN 1449	**BaO_2**

Synonyms
Barium binoxide; barium dioxide; barium superoxide.

SECTION II - HAZARDOUS INGREDIENTS AND IDENTITY INFORMATION

Hazardous Components (specific identity)	OSHA Exposure Criteria	NIOSH Exposure Criteria	Immediately Dangerous to Life and Health (IDLH)	ACGIH Exposure Criteria
Barium peroxide (derivation: By heating barium oxide in oxygen or air at approximately 1000°F or 538°C; or from reaction of barium hydroxide with hydrogen peroxide or from a solution of barium salts, ammonium hydroxide, and hydrogen peroxide).	**PEL:** 0.5 mg(Ba)/m^3 **STEL:** Not Established	**REL:** 0.5 mg(Ba)/m^3 **STEL:** Not Established	**Not Established for BaO_2 (250 mg/m^3 for Soluble Ba Compounds)**	**TLV:** 0.5 mg(Ba)/m^3 **STEL:** Not Established

SECTION III - PHYSICAL AND CHEMICAL CHARACTERISTICS

Boiling Point	Specific Gravity (H_2O = 1)
1472°F (800°C) Decomposes	**4.96**
Vapor Pressure (mm Hg)	Molecular Weight (atomic weight)
Not Determined	**169.36**
Vapor Density (Air = 1)	Melting Point
Not Determined	**842°F (450°C)**

Solubility
Most of the barium salts are either highly or moderately soluble in water. Barium peroxide powder is slightly soluble in water. Combines in water to form octahydrate.

Appearance and Odor
Grayish-white heavy powder, no appreciable odor.

SECTION IV - FIRE AND EXPLOSION HAZARD DATA

Flash Point (method used)	Explosive Limits in Air % by Volume
Not Found	**LEL: N/A UEL: N/A**
NFPA Classification	Autoignition Temperature
Flammable Solid (in powder form)	**Not Found**

Extinguishing Media
Dry chemicals for (Class D) fires, G-1 powder, or flooding amounts of water.

Special Fire Fighting Procedures
Wear full protective clothing and self-contained breathing apparatus (SCBA). Poisonous gases are produced in fire. Use water mist to keep fire-exposed containers cool. Fight fire from distance, if possible.

Unusual Fire and Explosion Hazards
Barium peroxide powder is a flammable solid. Containers may explode in fire. Fire may restart after it has been extinguished. Powder may ignite or explode spontaneously in air or other oxidizing gases. Dust is dangerous when exposed to heat, flame, or chemical reaction. Will ignite other combustible materials (wood, paper, oil) on contact. Slowly decomposes in air. It may react violently with fuels.

SECTION V - REACTIVITY DATA

Stability		Conditions to Avoid **Avoid heat and open flame, especially around dusts and powders of barium peroxide. It is normally stable under normal conditions of handling and storage.**
Stable	Unstable	Incompatibility *(materials to avoid)* **Acetic anhydride; propane; calcium-silicone alloys, powdered aluminum, powdered magnesium, water with organic compounds, hydrogen sulfide, peroxyformic acid, hydroxylamine solution; organic matter.**
X		
Hazardous Polymerization		Conditions to Avoid **Hazardous polymerization of barium peroxide metal will not occur.**
May Occur	Will Not Occur	Hazardous Decomposition or By-products **Poisonous gases are emitted during fire, including barium oxide.**
	X	

SECTION VI - HEALTH HAZARD DATA

Primary Route(s) of Entry:	Inhalation?	Absorption (skin/eye)?	Ingestion?
	X		X

Health Hazards

INHALATION: May produce benign pneumoconiosis (baritosis) which may cause X-ray changes. Other symptoms of exposure include eye, nose, and throat irritation, coughing, and a potential for potassium deficiency that can effect the function of the heart and CNS.

EYES & SKIN: Exposure to the eyes and skin may cause moderate to severe irritation.

INGESTION: Water and stomach acids solubilize barium salts and can cause poisoning. Symptoms are vomiting, colic, diarrhea, slow irregular pulse, transient hypertension, convulsive tremors.

Carcinogenicity	NTP Listed?	IARC Cancer Review Group?	OSHA Regulated?	Target Organs?
Unknown Human Unknown Animal	No	No	29 CFR 1910.1000 Table Z-1	Respiratory system, eyes, CNS, heart.

Medical Conditions Generally Aggravated by Exposure

Persons with existing lung dysfunction, hypertension, CNS, or muscular disorders may be effected.

Emergency and First-aid Procedures

Eye contact: Do not allow victim to keep eyes tightly shut. Flush immediately with water for 15 minutes (minimum), seek medical attention. **Skin contact:** Remove all contaminated clothing. Immediately wash area with flooding amounts of water. For reddened or blistered skin, seek medical assistance. For **inhalation**: Remove the person from exposure. Provide respiratory assistance and CPR. Transfer to medical facility. If **swallowed**, seek medical attention immediately. Never give anything by mouth to a convulsing person. If conscious and alert, give 1-2 glasses of water and induce vomiting.

SECTION VII - PRECAUTIONS FOR SAFE HANDLING AND USE

Steps to be Taken in Case Material is Released or Spilled

Collect powdered materials in most convenient and safe manner and deposit in sealed containers. Dry sweep not recommended. Use HEPA vacuum. Remove all ignition sources. Restrict those not involved in cleanup from entering area. Keep combustible materials away.

Preferred Waste Disposal Method

No citation.

Precautions to be Taken in Handling and Storage

Explosion risks vary with type of barium compound. Store to avoid contact with incompatible materials, moisture, and combustible materials. Prevent physical damage to containers. Keep tightly closed.

Other Precautions and Warnings

Sources of ignition are prohibited where barium peroxide is used, handled, or stored. Store on floors constructed of impermeable and non-combustible materials.

SECTION VIII - CONTROL MEASURES AND PERSONAL PROTECTIVE EQUIPMENT

Respiratory Protection *(specify type)*

For dust exposures below PEL, use a respirator with a high efficiency particulate air (HEPA) filter, otherwise use a self-contained breathing apparatus with full facepiece operated in positive pressure mode.

Ventilation

Local exhaust (preferred) at site of chemical work or general ventilation.

Protective Gloves	Eye Protection	Other Protective Clothing
Impervious Rubber	Chemical/Dust Goggles or Face Mask	Protective Apron, Boots

Work/Hygiene Practices

Always wash hands thoroughly after using chemical; never bring food, drink, or smoking materials into vicinity of chemicals.

BARIUM PEROXIDE

BaO_2 CAS: 1304-29-6

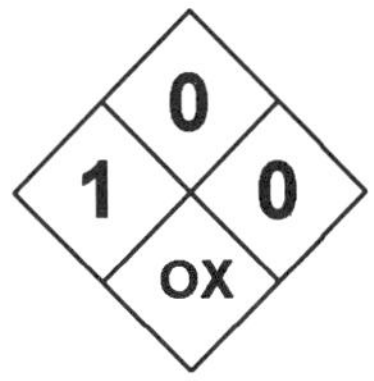

IDENTIFICATION AND TYPICAL USES

Barium peroxide is a grayish-white, heavy powder with no appreciable odor. It is used as a bleaching agent for animal substances, vegetable fibers, and straw. It is also used to strip the color from glass, in dyeing textiles, in the manufacture of pyrotechnics and tracer bullet formulations, in the manufacture of hydrogen peroxide, as a oxidizing agent, and as a polysulfide curing agent.

RISK ASSESSMENT: HEALTH

General Assessment

Certain compounds of barium are irritants of the eyes, mucous membranes, and skin. Primary exposure routes are *inhalation* and *ingestion*. It has not been established if exposure to barium poses any carcinogenic, teratogenic, or mutagenic effects.

Inhalation of barium dusts can cause irritation of the nose and upper respiratory tract and may produce a benign pneumoconiosis known as baritosis. There may be characteristic X-ray changes. All soluble salts of barium are acute poisons. The barium ion is toxic to muscle producing stimulation and then paralysis. Specifically, barium peroxide will irritate the nose, throat, and mucosa of the upper respiratory tract and bronchial tubes. There may be adverse effects on the heart and the function of the central nervous system (CNS).

Ingesting barium salts or powders produces initial symptoms related to gastrointestinal reaction. These include nausea, vomiting, colic, diarrhea, followed by myocardial and general muscular stimulation with tingling and paralysis in the extremities. Severe cases continue to loss of tendon reflexes, general muscular paralysis, and death from respiratory arrest or ventricular fibrillation. In addition to these effects, ingestion of barium peroxide specifically can cause excessive salivation, giddiness, convulsions, bloody and watery diarrhea, dilated pupils, vertigo, and dizziness.

Acute Health Effects

The following acute (short-term) health effects may occur immediately or shortly after exposure to barium peroxide:

Skin: No initial symptoms. However, sweating may cause burning sensation as barium dusts react on contact with water.

Eye: Severe irritation. Due to water content in the eye, reactions may be immediate and serious.

Lung: Can cause metallic taste, nausea, sore throat, and irritation of the air passages.

Other: Serious effects on musculatory system at high exposures. May cause tingling and paralysis of the extremities followed by irregular heartbeat and might even cause the heart to stop (ventricular fibrillation). Death is also possible as a result of respiratory arrest.

Chronic Health Effects

The following chronic (long-term) health effects may occur at some time after exposure to barium peroxide and can last for months or even years:

Cancer Hazards: According to the references, barium peroxide has not been tested for its ability to cause cancer in animals.

Reproductive Hazard: According to the references, barium peroxide has not been tested for its ability to affect reproduction.

Other Chronic Effects: Repeated exposure can cause abnormal chest X-ray. This usually takes years to develop. An abnormal chest X-ray from barium peroxide exposure does not necessarily mean the lung has been damaged but future exposures to irritating substances may affect the lungs. It is not currently known whether barium peroxide causes lung damage. Also, the barium ion can remain in bone tissue for up to 50 days.

↻ *Recommended Risk-Reduction Measures*

Personnel should avoid direct contact with barium, especially its dusts or powders. If a less toxic material or compound cannot be substituted for barium peroxide, then *engineering controls* are the most effective method of reducing exposures. The best protection is to enclose operations and/or provide local exhaust ventilation at the site of barium peroxide dust release. While not always operationally feasible, isolating operations can also reduce exposure. Using respiratory protection is less effective than the controls mentioned above, but is still advisable whenever working with or around barium peroxide. A self-contained breathing apparatus (SCBA) with full facepiece and pressure demand is recommended for the greatest possible respiratory protection, especially when exposures near the IDLH level of 250 mg/m^3. However, for low or infrequent exposure, a MSHA/NIOSH approved full facepiece respirator with a high efficiency particulate air (HEPA) filter can be used. If a full facepiece is not available, then chemical/dust goggles should be worn to protect the eyes. A face shield and protective apron should also be worn. To prevent hand and skin exposures, gloves should be worn.

Administrative controls should also be in place to minimize the potential for human exposures. These may include written procedures or policies which specify the methods and techniques that will be practiced whenever personnel are to work with barium peroxide.

All personnel should receive *training* on the use, hazards, protective measures, emergency actions, and other precautions per 29 CFR 1910.1200 (Hazard Communication), prior to the first assignment in an area where barium peroxide is used or stored.

Before beginning employment working with barium peroxide, and at regular intervals thereafter (e.g., annually), those personnel with frequent or potentially high exposures should be provided the following recommended medical tests:

☑ Lung function test (establish baseline).

If symptoms develop or overexposure is suspected, consider the following additional tests:

☑ Electrocardiogram (EKG).

☑ Neurological examination.

☑ Consider chest X-ray.

☑ Evaluation by a qualified allergist with careful consideration of exposure history and special testing (may help diagnose skin allergy).

Any evaluation should include a careful history of past and present symptoms with an examination. Medical tests that simply look for existing damage are not a substitute for controlling exposures. Also, since smoking can cause heart disease as well as lung cancer, emphysema, and other known respiratory problems, exposure to barium peroxide may cause additional complications. Therefore, personnel who smoke should avoid any unprotected contact with barium peroxide powders.

Other methods to reduce exposure include:

☑ Where possible, enclose operations and use local exhaust ventilation at the site of chemical release. If local exhaust ventilation or enclosure is not used, respiratory protection should be mandatory.

☑ Always ensure that proper protective clothing is worn when using chemical substances.

☑ Wash thoroughly immediately after exposure to barium peroxide and at the end of the work shift or before eating, drinking, or smoking.

☑ Hazard warning information should be posted in the work area. In addition, as part of an on-going education and training program, all information on the health and safety hazards of barium peroxide should be communicated to all potentially exposed workers.

☑ Eye wash stations should be provided in the immediate work area for emergency use. If there is a possibility of skin exposure to barium peroxide, emergency shower facilities should also be provided.

☑ Workers whose clothing has been contaminated by barium peroxide should change into clean clothes before leaving work. Contaminated work clothing should be laundered only by individuals who have been informed of the hazards of exposure to barium peroxide.

RISK ASSESSMENT: ENVIRONMENT

General Assessment

The environment is at risk of exposure during transportation, storage, disposal, or destruction of barium peroxide. In almost every scenario, the threat of environmental exposure is contingent upon the proper handling of the chemical substance. Accidental spills, large or small, where barium peroxide contacts incompatible commodities can result in fire, explosion (depending upon conditions of spill), and possible contamination of the surrounding environmental mediums (water, soil, and air).

In powdered form, barium is considered a flammable solid. It can present a moderate fire and explosion hazard. It is incompatible with many other commodities and extreme caution is required in handling, storage, transportation, and disposal of barium peroxide. These characteristics also require special consideration during any emergency situation involving a leak or spill of barium peroxide powder or dust. It is incompatible with numerous substances such as hydrogen sulfide, water, peroxyformic acid, hydroxylamine solution, organic matter, and mixtures of manganese, zinc, and barium nitrate. It is divalent and combines with oxygen, nitrogen, hydrogen, ammonia, water, halogens, and sulfides, generating large amounts of heat. Should barium peroxide ever come into contact with these incompatible substances either during use, transportation, storage, or disposal, violent and even explosive reactions are possible.

Barium peroxide may enter the environment as a result of industrial and municipal waste treatment plant discharges and from spills.

Acute Ecological Effects

Acute (short-term) toxic effects may include the death of animals, birds, or fish and death or low growth rate in plants. Acute effects are seen 2 to 4 days after animals or plants are exposed to barium. Barium and its salts have moderate acute toxicity to aquatic life. Insufficient data are available to evaluate the short-term effects of barium or its salts to plants, birds, or land animals.

Chronic Ecological Effects

Chronic (long-term) toxic effects may include shortened life span, reproductive problems, lower fertility, and changes in appearance or behavior in exposed animals. These effects can be seen long after first exposure(s) to toxic chemicals. Barium and its salts have moderate chronic toxicity to aquatic life. Insufficient data are available to evaluate or predict the long-term effects of barium to plants, birds, or land animals.

Water Solubility

Most of the barium salts are either highly or moderately soluble in water. Concentrations of 1 to 1000 milligrams and more will mix with a liter of water.

Persistence in the Environment

Barium is highly persistent in water, with a half-life greater than 200 days. The half-life of a pollutant is the amount of time it takes for one half of the chemical to be degraded. No data are available on the estimated percentages of barium persistence in the aquatic or terrestrial environments.

Bioaccumulation in Aquatic Organisms

Some substances increase in concentration, or *bioaccumulate,* in living organisms as they breathe contaminated air, drink contaminated water, or eat contaminated food. These chemicals can become concentrated in the tissues and internal organs of animals as well as humans.

The concentration of barium found in fish tissues is expected to be about the same as the average concentration of barium in the water from which the fish was taken.

Recommended Risk-Reduction Measures

Proper training of all transporters will reduce the likelihood of a mishap or accident resulting in a leak or spill of barium peroxide dusts or powders into the environment. The correct use of DOT labeling (per the MSDS) on all containers, trucks, and rail cars will enable emergency responders to react properly and quickly to any disaster thereby reducing the potential risk to the environment and to personnel.

Storage of barium peroxide should be segregated from incompatible chemicals to minimize the risk of cross-contamination or contact. Impermeable floors constructed of non-combustible materials are recommended.

If a spill or leak to the environment has occurred, fire department, emergency response and/or hazardous materials spill personnel should be notified immediately. Cleanup should be attempted only by those trained in proper spill containment procedures. Contaminated soils should be removed for incineration and replaced with clean soil. If barium peroxide should contact the water table, aquifer, or navigable waterway, time is of the essence. It is slightly soluble and will decompose. Therefore, total remediation may not be possible. The local and/or state emergency response authorities must be notified. A comprehensive emergency response or disaster preparedness/recovery plan should be in place prior to any operations involving the use, transportation, storage, or disposal of barium peroxide. If barium peroxide is spilled or leaked, the following specific steps are recommended:

- ☑ Restrict persons not wearing protective clothing from area of spill or leak until cleanup is complete.
- ☑ Remove all ignition sources.
- ☑ Collect powdered materials in the safest manner possible and deposit in sealed containers. Recommend using vacuum with HEPA filter. Do not dry sweep (generates airborne dusts).
- ☑ It may be necessary to dispose of barium peroxide as a hazardous waste. The responsible state agency or the regional office of the federal Environmental Protection Agency (EPA) should be contacted for specific recommendations.

RISK ASSESSMENT: BUSINESS

General Assessment

Accidents or mishaps involving barium peroxide can present a significant threat to business operations. The loss or damage of equipment or facilities can significantly affect fiscal viability. Lawsuits that may result from personnel injury/death, public exposures, and/or environmental contamination will also require a serious expenditure of resources. Media attention surrounding an injury, death, or environmental damage can also result in a loss of profits and loss of current as well as future business.

Recommended Risk-Reduction Measures

Company attorneys, safety and health professionals, and environmental specialists should be involved in the development of any procedures for responding to chemical incidents. A company official should be predesignated as a public relations officer with specific training in dealing with the press. Corporate plans and policies should be developed, approved, and implemented long before any need for such arises.

REFERENCES

Ahlbom, A. 1993. *Biostatistics for Epidemiologists.* New York: CRC Press/Lewis Publishers.

American Conference of Governmental Industrial Hygienists. 1988. *Documentation of the Threshold Limit Values and Biological Exposure Indices,* 5th Edition (with updates). Cincinnati: ACGIH

Bloom, A. D. and F. J. DeSerres. 1995. *Ecotoxicity and Human Health.* New York: CRC Press/Lewis Publishers.

Brusick, D. J. 1994. *Methods for Genetic Risk Assessment.* New York: CRC Press/Lewis Publishers.

Bueche, J. R. 1972. *Principles of Physics,* 2nd Edition. New York: McGraw-Hill.

Calabrese, E. J. 1994. *Biological Effects of Low Level Exposures.* New York: CRC Press/Lewis Publishers.

Cockerman, L. G., B. S. Shane. 1994. *Basic Environmental Toxicology.* New York: CRC Press/Lewis Publishers.

Gardner, R. W. 1994. *Chemical Intolerance.* New York: CRC Press/Lewis Publishers.

Howard, P. H. 1995. *Handbook of Environmental Fate and Exposures Data for Organic Chemicals,* Vol. I-IV. New York: CRC Press/Lewis Publishers.

Keith, L. H. and M. M. Walker. 1995. *Handbook of Air Toxics.* New York: CRC Press/Lewis Publishers.

Landis, W. G., M. H. Yu. 1995. *Introduction to Environmental Toxicology.* New York: CRC Press/Lewis Publishers.

Langerman, N. 1994 *Precautionary Labels for Chemical Containers.* New York: CRC Press/Lewis Publishers.

Lauwerys, R. R., P. Hoet. 1993. *Industrial Chemical Exposure.* New York: CRC Press/Lewis Publishers.

Lewis, R. J., Sr. 1993. *Hawley's Condensed Chemical Dictionary,* 12th Edition. New York: Van Nostrand Reinhold.

Lewis, R. J., Sr. 1992. *Sax's Dangerous Properties of Industrial Materials,* Volumes 1, 2, and 3, 8th Edition. New York: Van Nostrand Reinhold.

Meyer, E. 1990. *Chemistry of Hazardous Materials.* Englewood Cliffs, NJ: Prentice Hall.

Mycock, J. C., J. D. McKenna, and L. Theodore. 1995. *Handbook of Air Pollution Control Engineering and Technology.* New York: CRC Press/Lewis Publishers.

Patnaik, P. 1992. *A Comprehensive Guide to the Hazardous Properties of Chemical Substances.* New York: Van Nostrand Reinhold.

Quigley, D. R. 1994. *Handbook of Emergency Chemical Management.* New York: CRC Press/Lewis Publishers.

Reeves, A. L. 1979. Barium. *Handbook on the Toxicology of Metals.* pp 321-328. New York: Elsevier North-Holland.

Richardson, M. 1995. *Dictionary of Substance and Their Effects,* Vol. I-VI. New York: CRC Press/Lewis Publishers.

Sacarello, H. L. A. 1994. *The Comprehensive Handbook of Hazardous Materials.* New York: CRC Press/Lewis Publishers.

Talbott, E., G. Craun. 1995. *An Introduction to Environmental Epidemiology.* New York: CRC Press/Lewis Publishers.

Torkelson, T. R. V. K. Rowe. 1981. *Patty's Industrial Hygiene and Toxicology,* Vol. 2B. New York: Wiley-Interscience.

U.S. Department of Health, Education, and Welfare. National Institute for Occupational Safety and Health (NIOSH). 1978. *Criteria for a Recommended Standard. Occupational Exposure to Barium.* DHEW (NIOSH) Publication Number: 78-216. Washington, DC: U.S. Government Printing Office.

MATERIAL SAFETY DATA SHEET

CHEMICAL NAME

BARIUM SULFATE

CRC PRESS/LEWIS PUBLISHERS DISCLAIMER: THE INFORMATION AND RECOMMENDATIONS PRESENTED HEREIN ARE BASED ON SOURCES BELIEVED TO BE RELIABLE. CRC MAKES NO REPRESENTATION ON ITS COMPLETENESS OR ACCURACY. IT IS THE RESPONSIBILITY OF THE USER TO DETERMINE THE CHEMICAL'S SUITABILITY FOR ITS INTENDED USE, THE CHEMICAL'S SAFE USE, AND THE CHEMICAL'S PROPER DISPOSAL. NO REPRESENTATIONS AND/OR WARRANTIES, EITHER EXPRESSED OR IMPLIED, OF THE MERCHANTABILITY OR FITNESS FOR A PARTICULAR PURPOSE, OR OF ANY OTHER NATURE, ARE MADE WITH RESPECT TO THE INFORMATION PROVIDED IN THIS MSDS OR TO THE CHEMICAL TO WHICH INFORMATION MAY REFER. CRC NEITHER ASSUMES NOR AUTHORIZES ANY OTHER PERSON TO ASSUME FOR IT, ANY OTHER ADDITIONAL RESPONSIBILITY OR LIABILITY FOR THE USE OF, OR RELIANCE UPON, THIS INFORMATION.

HAZARD WARNING INFORMATION

HEALTH	FIRE	REACTIVE	OTHER	DEGREE OF HAZARD	COLOR CODING	OTHER CODES
1	0	0		0 = Minimum Hazard 1 = Slight Hazard 2 = Moderate Hazard 3 = Serious Hazard 4 = Severe Hazard	HEALTH = BLUE FIRE = RED REACTIVITY = YELLOW OTHER = WHITE	OX = Oxidizer ACID = Acid ALK = Alkali COR = Corrosive W = Use No Water

SECTION I - GENERAL INFORMATION

Characterization **Metal**	RCRA Number **None**	EPA Class **Not Applicable**
DOT Proper Shipping Name **Barium Compounds, N.O.S.**	Chemical Abstract Service (CAS) Number **7727-43-7**	
DOT Hazard Class and Label Requirements **No Citation**	DOT Emergency Guide Code **55**	
DOT Identification Number **UN 1564**	Atomic Symbol **$BaSO_4$**	

Synonyms
Barium sulphate; barytes (natural); blanc fixe (artificial, precipitated); basofor; artificial barite.

SECTION II - HAZARDOUS INGREDIENTS AND IDENTITY INFORMATION

Hazardous Components (specific identity)	OSHA Exposure Criteria	NIOSH Exposure Criteria	Immediately Dangerous to Life and Health (IDLH)	ACGIH Exposure Criteria
Barium sulfate (derivation: By treating a solution of barium salt with sodium sulfate; by-product in manufacture of hydrogen peroxide; occurs in nature as the mineral barite, which is found in Arkansas, Missouri, Georgia, Nevada, Canada, and Mexico).	**PEL (8-hour):** 15 mg/m³ (total dust) 5 mg/m³ (respirable) **STEL:** Not Established	**REL (10-hour):** 10 mg/m³ (total dust) 5 mg/m³ (respirable) **STEL:** Not Established	**Not Determined**	**TLV:** 10 mg/m³ **STEL:** Not Established

SECTION III - PHYSICAL AND CHEMICAL CHARACTERISTICS

Boiling Point **2912°F (1600°C) Decomposes**	Specific Gravity (H_2O = 1) **4.25-4.5**
Vapor Pressure (mm Hg) **0 (approx.) at 68°F (20°C)**	Molecular Weight (atomic weight) **233.4**
Vapor Density (Air = 1) **Not Found**	Melting Point **2876°F (1580°C)**

Solubility
Practically insoluble in water (0.0002% at 64°F). Soluble in hot, concentrated sulfuric acid.

Appearance and Odor
White or yellow-tinted powder that is odorless and tasteless.

SECTION IV - FIRE AND EXPLOSION HAZARD DATA

Flash Point (method used) **Not Found**	Explosive Limits in Air % by Volume **LEL: N/A UEL: N/A**
NFPA Classification **Non-Combustible Solid**	Autoignition Temperature **Not Found**

Extinguishing Media
Use agent suitable to surrounding fire. Barium sulfate by itself will not burn.

Special Fire Fighting Procedures
Wear full protective clothing and self-contained breathing apparatus (SCBA). Poisonous gases are produced in fire, including bromide. Remain clear of smoke and fumes. Use water mist from unmanned device to keep fire-exposed containers cool. Keep cooling containers after fire is extinguished.

Unusual Fire and Explosion Hazards
No specific explosion hazards noted. However, heating with aluminum can produce an explosion. Move fire-exposed containers from fire if it can be done without risk.

SECTION V - REACTIVITY DATA

Stability		Conditions to Avoid Very stable at room temperature and pressure under normal conditions of handling and storage. Avoid contact with incompatible materials and do not create dusty conditions.
Stable X	Unstable	Incompatibility *(materials to avoid)* Incompatible with aluminum and potassium. When heated with aluminum, it can produce an extremely violent explosion.
Hazardous Polymerization		Conditions to Avoid Hazardous polymerization of barium sulfate will not occur.
May Occur	Will Not Occur X	Hazardous Decomposition or By-products Poisonous gases, including oxides of sulfur and barium fumes, are emitted during fire.

SECTION VI - HEALTH HAZARD DATA

Primary Route(s) of Entry:	Inhalation?	Absorption (skin/eye)?	Ingestion?
	X		X

Health Hazards

INHALATION: The insolubility of barium sulfate in water makes for low-level toxicity. It does not exhibit the systemic toxicity noted for other barium salts. However, any soluble impurities can lead to toxic effects. Breathing fine dusts may irritate the nose, throat, and respiratory tract leading to a benign form of pneumoconiosis (baritosis).

EYES & SKIN: Exposure to the eyes may cause moderate to severe irritation and even burns.

INGESTION: No specific effects noted.

Carcinogenicity	NTP Listed?	IARC Cancer Review Group?	OSHA Regulated?	Target Organs?
Unknown Human Unknown Animal	No	No	29 CFR 1910.1000 Table Z-1	Respiratory system, eyes, skin.

Medical Conditions Generally Aggravated by Exposure

None reported.

Emergency and First-aid Procedures

Eye contact: Do not allow victim to keep eyes tightly shut. Flush immediately with water for 15 minutes (minimum), seek medical attention. Skin contact: Remove all contaminated clothing. Immediately wash area with flooding amounts of soap and water. For inhalation: Remove the person from exposure. Provide respiratory assistance and CPR. Transfer to medical facility. If swallowed, seek medical attention immediately. Never give anything by mouth to an unconscious or convulsing person.

SECTION VII - PRECAUTIONS FOR SAFE HANDLING AND USE

Steps to be Taken in Case Material is Released or Spilled

Collect powdered materials in most convenient and safe manner and deposit in sealed containers. Dry sweep not recommended. Use HEPA vacuum. Remove all ignition sources. Restrict those not involved in cleanup from entering area. Notify appropriate authorities, as required by applicable regulations.

Preferred Waste Disposal Method

No citation.

Precautions to be Taken in Handling and Storage

Store to avoid contact with aluminum or potassium. Store powders in sealed containers in cool, dry, well-ventilated area. Protect containers from physical damage.

Other Precautions and Warnings

Avoid actions or practices that may create dusty conditions, such as moving opened containers or sweeping up storage areas.

SECTION VIII - CONTROL MEASURES AND PERSONAL PROTECTIVE EQUIPMENT

Respiratory Protection *(specify type)*

For dust exposures below PEL, use a respirator with a high efficiency particulate air (HEPA) filter, otherwise use a NIOSH/MSHA approved self-contained breathing apparatus or a supplied-air respirator with full facepiece operated in positive pressure or some other continuous flow mode.

Ventilation

Local exhaust (preferred) at site of chemical work or general ventilation.

Protective Gloves	Eye Protection	Other Protective Clothing
Impervious Rubber	Chemical/Dust Goggles or Face Mask	Protective Apron, Boots

Work/Hygiene Practices

Always wash hands thoroughly after using chemical; never bring food, drink, or smoking materials into vicinity of chemicals.

BARIUM SULFATE

$BaSO_4$ CAS: 7727-43-7

IDENTIFICATION AND TYPICAL USES

Barium sulfate is a white or yellowish, odorless powder. It is used in the manufacture of photographic papers, artificial ivory, cellophane, paper coatings, and paints. It is also used as a filler and delustrant for textiles, rubber, plastic, linoleum, oil cloth, paper, lithograph inks, polymeric fibers, and resins. It is a base for lake colors; it is used in X-ray photography, it is an opaque medium for gastrointestinal radiography; and is used in battery plate expanders.

RISK ASSESSMENT: HEALTH

General Assessment

Barium sulfate is unlike most other barium salts in that its low water solubility makes for low-level toxicity under most conditions of exposure. However, commercial grades may contain insoluble impurities that can be quite toxic. Primary exposure routes are *inhalation* and, to a lesser degree, *ingestion.* While there is no evidence in the references of its carcinogenic potential, the data are inconclusive and conflicting in this regard.

Inhalation can cause irritation of the nose, throat, and respiratory system. High exposures may lead to a benign form of pneumoconiosis known as baritosis.

Ingestion of barium salts can cause severe hypokalemia (potassium deficiency). However, ingesting barium sulfate is not thought to cause this effect. More research is required based upon the amount of conflicting and inconclusive data presented in the references. Ingesting other barium salts or powders produces initial symptoms related to gastrointestinal reaction. These include nausea, vomiting, colic, diarrhea, followed by myocardial and general muscular stimulation with tingling in the extremities. Severe cases continue to loss of tendon reflexes, general muscular paralysis, and death from respiratory arrest or ventricular fibrillation. Before dismissing these potential health risks as unrelated to barium sulfate, additional research is required to ensure proper risk management of barium sulfate exposure.

Acute Health Effects

The following acute (short-term) health effects may occur immediately or shortly after exposure (within hours or within days) to barium sulfate:

Skin: May cause moderate irritation on contact.

Eye: Severe irritation and possible damage.

Lung: Dust exposures may cause no effect or irritation of the lungs with cough and phlegm.

Other: Industrial grade barium compounds may be contaminated with impurities (such as silica) which can produce toxic effects. Exposure can lead to lung scarring and reduce lung function.

Chronic Health Effects

The following chronic (long-term) health effects may occur at some time after exposure to barium and can last for months or even years:

Cancer Hazards: According to the references, barium has not been tested for its ability to cause cancer in animals. The data are conflicting and inconclusive with regard to its specific ability to cause cancer in test animals. Some tumorigenic data have been reported.

Reproductive Hazard: According to the references, barium has not been tested for its ability to affect reproduction.

Other Chronic Effects: Repeated exposure can cause abnormal chest X-ray with the appearance of small nodules despite otherwise normal respiratory function. This usually takes years to develop. An abnormal chest X-ray from barium exposure does not necessarily mean the lung has been damaged but future exposures to irritating substances may affect the lungs. It is not currently known whether barium causes lung damage.

ʘ *Recommended Risk-Reduction Measures*

Personnel should avoid direct contact with barium sulfate dusts or powders. The specific toxicity of barium sulfate is poorly understood as compared to other barium compounds. However, as a barium salt, caution is suggested whenever working with or around barium sulfate. If a less toxic material or compound can not be substituted for barium sulfate, then *engineering controls* are the most effective method of reducing exposures. The best protection is to enclose operations and/or provide local exhaust ventilation at the site of barium dust release. While not always operationally feasible, isolating operations can also reduce exposure. Using respiratory protection is less effective than the controls mentioned above, but is still advisable whenever working with or around barium sulfate. A supplied-air respirator or a self-contained breathing apparatus (SCBA) with full facepiece and operated in pressure demand or other positive pressure mode is recommended for the greatest possible respiratory protection. However, for low or infrequent exposure, a MSHA/NIOSH approved full facepiece respirator with a high efficiency particulate air (HEPA) filter can be used. If a full facepiece is not available, then chemical/dust goggles should be worn to protect the eyes. A face shield and protective apron should also be worn. To prevent hand and skin exposures, gloves should be worn.

Administrative controls should also be in place to minimize the potential for human exposures. These may include written procedures or policies which specify the methods and techniques that will be practiced whenever personnel are to work with barium sulfate.

All personnel should receive *training* on the use, hazards, protective measures, emergency actions, and other precautions per 29 CFR 1910.1200 (Hazard Communication), prior to the first assignment in an area where barium sulfate is used or stored.

Before beginning employment working with barium sulfate, and at regular intervals thereafter (e.g., annually), those personnel with frequent or potentially high exposures should be provided the following recommended medical tests:

- ☑ Lung function test.
- ☑ Consider chest X-ray, especially if chronic exposure is alleged or suspected.
- ☑ Serum bromide level.

Any evaluation should include a careful history of past and present symptoms with an examination. Medical tests that simply look for existing damage are not a substitute for controlling exposures. Also, since smoking can cause heart disease as well as lung cancer, emphysema, and other known respiratory problems, exposure to barium may cause additional complications. Therefore, personnel who smoke should avoid any unprotected contact with barium powders.

Other methods to reduce exposure include:

- ☑ Where possible, enclose operations and use local exhaust ventilation at the site of chemical release. If local exhaust ventilation or enclosure is not used, respiratory protection should be mandatory.
- ☑ Always ensure that proper protective clothing is worn when using chemical substances.
- ☑ Wash thoroughly immediately after exposure to barium sulfate and at the end of the work shift or before eating, drinking, or smoking.
- ☑ Hazard warning information should be posted in the work area. In addition, as part of an on-going education and training program, all information on the health and safety hazards of barium sulfate should be communicated to all potentially exposed workers.
- ☑ Eye wash stations should be provided in the immediate work area for emergency use. If there is a possibility of skin exposure to barium sulfate, emergency shower facilities should also be provided.
- ☑ Workers whose clothing has been contaminated by barium sulfate should change into clean clothes before leaving work. Contaminated work clothing should be laundered only by individuals who have been informed of the hazards of exposure to barium sulfate.

RISK ASSESSMENT: ENVIRONMENT

General Assessment

The environment is at risk of exposure during transportation, storage, disposal, or destruction of barium compounds. In almost every scenario, the threat of environmental exposure is contingent upon the proper handling of the chemical substance. Accidental spills, large or small, where barium sulfate contacts incompatible commodities can result in fire, explosion (depending upon conditions of spill), and possible contamination of the surrounding environmental mediums (water, soil, and air).

In its powdered form, barium sulfate is considered a non-combustible solid. However, it can react explosively and present a serious fire hazard. Because it is incompatible with many commodities, including other metals (such as aluminum and potassium), extreme caution is required in handling, storage, transportation, and disposal of barium sulfate. These characteristics also require special consideration during any emergency situation involving a leak or spill of barium sulfate powder or dust. Should barium sulfate ever come into contact with incompatible substances either during use, transportation, storage, or disposal, violent and even explosive reactions are possible.

Barium sulfate can enter the environment from industrial and municipal waste treatment plant discharges and from spills.

Acute Ecological Effects

Acute (short-term) toxic effects may include the death of animals, birds, or fish and death or low growth rate in plants. Acute effects are seen 2 to 4 days after animals or plants are exposed to barium. Barium and its salts have moderate acute toxicity to aquatic life. Insufficient data are available to evaluate the short-term effects of barium or its salts to plants, birds, or land animals.

Chronic Ecological Effects

Chronic (long-term) toxic effects may include shortened life span, reproductive problems, lower fertility, and changes in appearance or behavior in exposed animals. These effects can be seen long after first exposure(s) to toxic chemicals. Barium and its salts have moderate chronic toxicity to aquatic life. Insufficient data are available to evaluate or predict the long-term effects of barium to plants, birds, or land animals.

Water Solubility

Most of the barium salts are either highly or moderately soluble in water. Concentrations of 1 to 1000 milligrams and more will mix with a liter of water.

Persistence in the Environment

Barium is highly persistent in water, with a half-life greater than 200 days. The half-life of a pollutant is the amount of time it takes for one half of the chemical to be degraded. No data are available on the estimated percentages of barium persistence in the aquatic or terrestrial environments.

Bioaccumulation in Aquatic Organisms

Some substances increase in concentration, or *bioaccumulate,* in living organisms as they breathe contaminated air, drink contaminated water, or eat contaminated food. These chemicals can become concentrated in the tissues and internal organs of animals as well as humans.

The concentration of barium found in fish tissues is expected to be about the same as the average concentration in the water from which the fish was taken.

Recommended Risk-Reduction Measures

Proper training of all transporters will reduce the likelihood of a mishap or accident resulting in a leak or spill of barium sulfate dusts or powders into the environment. The correct use of DOT labeling (per the MSDS) on all containers, trucks, and rail cars will enable emergency responders to react properly and quickly to any disaster thereby reducing the potential risk to the environment and to personnel.

Storage of barium sulfate should be segregated from incompatible chemicals to minimize the risk of cross-contamination or contact. Containers should be protected from physical damage.

If a spill or leak to the environment has occurred, fire department, emergency response and/or hazardous materials spill personnel should be notified immediately. Cleanup should be attempted only by those trained in proper spill containment procedures. Contaminated soils should be removed for incineration and replaced with clean soil. If barium sulfate should contact the water table, aquifer, or navigable waterway, time is of the essence. It is nearly insoluble and total remediation may be possible. The local and/or state emergency response authorities must be notified. A comprehensive emergency response or disaster preparedness/recovery plan should be in place prior to any operations involving the use, transportation, storage, or disposal of barium sulfate.

If barium sulfate powder is spilled or leaked, the following specific steps are recommended:

- ☑ Restrict persons not wearing protective clothing from area of spill or leak until cleanup is complete and area can be opened for normal work.
- ☑ Remove all ignition sources, ventilate area.
- ☑ Collect powdered materials in the safest manner possible and deposit in sealed containers. Recommend using vacuum with HEPA filter. Do not dry sweep (generates airborne dusts).

☑ It may be necessary to dispose of barium sulfate as a hazardous waste. The responsible state agency or the regional office of the federal Environmental Protection Agency (EPA) should be contacted for specific recommendations.

RISK ASSESSMENT: BUSINESS

General Assessment

Accidents or mishaps involving barium sulfate can present a significant threat to business operations. The loss or damage of equipment or facilities can significantly affect fiscal viability. Lawsuits that may result from personnel injury/death, public exposures, and/or environmental contamination will also require a serious expenditure of resources. Media attention surrounding an injury, death, or environmental damage can also result in a loss of profits and loss of current as well as future business.

Recommended Risk-Reduction Measures

Company attorneys, safety and health professionals, and environmental specialists should be involved in the development of any procedures for responding to chemical incidents. A company official should be predesignated as a public relations officer with specific training in dealing with the press. Corporate plans and policies should be developed, approved, and implemented long before any need for such arises.

REFERENCES

Ahlbom, A. 1993. *Biostatistics for Epidemiologists.* New York: CRC Press/Lewis Publishers.

American Conference of Governmental Industrial Hygienists. 1988. *Documentation of the Threshold Limit Values and Biological Exposure Indices,* 5th Edition (with updates). Cincinnati: ACGIH

Asante-Duah, D. K. 1993. *Hazardous Waste Risk Assessment.* New York: CRC Press/Lewis Publishers.

Bloom, A. D. and F. J. DeSerres. 1995. *Ecotoxicity and Human Health.* New York: CRC Press/Lewis Publishers.

Brusick, D. J. 1994. *Methods for Genetic Risk Assessment.* New York: CRC Press/Lewis Publishers.

Bueche, J. R. 1972. *Principles of Physics,* 2nd Edition. New York: McGraw-Hill.

Calabrese, E. J. 1994. *Biological Effects of Low Level Exposures.* New York: CRC Press/Lewis Publishers.

Cothern, R. R. 1993. *Comparative Environmental Risk Assessment.* New York: CRC Press/Lewis Publishers.

Gardner, R. W. 1994. *Chemical Intolerance.* New York: CRC Press/Lewis Publishers.

Howard, P. H. 1995. *Handbook of Environmental Fate and Exposures Data for Organic Chemicals,* Vol. I-IV. New York: CRC Press/Lewis Publishers.

Keith, L. H. and M. M. Walker. 1995. *Handbook of Air Toxics.* New York: CRC Press/Lewis Publishers.

Landis, W. G., M. H. Yu. 1995. *Introduction to Environmental Toxicology.* New York: CRC Press/Lewis Publishers.

Lauwerys, R. R., P. Hoet. 1993. *Industrial Chemical Exposure.* New York: CRC Press/Lewis Publishers.

Lewis, R. J., Sr. 1993. *Hawley's Condensed Chemical Dictionary,* 12th Edition. New York: Van Nostrand Reinhold.

Lewis, R. J., Sr. 1992. *Sax's Dangerous Properties of Industrial Materials,* Volumes 1, 2, and 3, 8th Edition. New York: Van Nostrand Reinhold.

Manahan, S. E. 1994. *Environmental Chemistry,* 6th Edition. New York: CRC Press/Lewis Publishers.

Meyer, E. 1990. *Chemistry of Hazardous Materials.* Englewood Cliffs, NJ: Prentice Hall.

Mycock, J. C., J. D. McKenna, and L. Theodore. 1995. *Handbook of Air Pollution Control Engineering and Technology.* New York: CRC Press/Lewis Publishers.

Patnaik, P. 1992. *A Comprehensive Guide to the Hazardous Properties of Chemical Substances.* New York: Van Nostrand Reinhold.

Proctor, N. H., J. P. Hughes. 1991. *Chemical Hazards of the Workplace.* New York: Van Nostrand Reinhold.

Quigley, D. R. 1994. *Handbook of Emergency Chemical Management.* New York: CRC Press/Lewis Publishers.

Rea, W. J. 1995. *Chemical Sensitivity,* Vol. I-III. New York: CRC Press/Lewis Publishers.

Reeves, A. L. 1979. Barium. *Handbook on the Toxicology of Metals.* pp 321-328. New York: Elsevier North-Holland.

Richardson, M. 1995. *Dictionary of Substance and Their Effects,* Vol. I-VI. New York: CRC Press/Lewis Publishers.

Sacarello, H. L. A. 1994. *The Comprehensive Handbook of Hazardous Materials.* New York: CRC Press/Lewis Publishers.

Sheftel, V. O. 1995. *Handbook of Toxic Properties of Monomers and Additives.* New York: CRC Press/Lewis Publishers.

Torkelson, T. R. V. K. Rowe. 1981. *Patty's Industrial Hygiene and Toxicology,* Vol. 2B. New York: Wiley-Interscience.

U.S. Department of Health, Education, and Welfare. National Institute for Occupational Safety and Health (NIOSH). 1978. *Criteria for a Recommended Standard. Occupational Exposure to Barium.* DHEW (NIOSH) Publication Number: 78-216. Washington, DC: U.S. Government Printing Office.

U.S. Department of Labor. 1983. Barium metal: barium trioxide; and barium sulfide response to the interagency testing committee. *Federal Register.* 48:717-724.

MATERIAL SAFETY DATA SHEET

CHEMICAL NAME

BENZAL CHLORIDE

CRC PRESS/LEWIS PUBLISHERS DISCLAIMER: THE INFORMATION AND RECOMMENDATIONS PRESENTED HEREIN ARE BASED ON SOURCES BELIEVED TO BE RELIABLE. CRC MAKES NO REPRESENTATION ON ITS COMPLETENESS OR ACCURACY. IT IS THE RESPONSIBILITY OF THE USER TO DETERMINE THE CHEMICAL'S SUITABILITY FOR ITS INTENDED USE, THE CHEMICAL'S SAFE USE, AND THE CHEMICAL'S PROPER DISPOSAL. NO REPRESENTATIONS AND/OR WARRANTIES, EITHER EXPRESSED OR IMPLIED, OF THE MERCHANTABILITY OR FITNESS FOR A PARTICULAR PURPOSE, OR OF ANY OTHER NATURE, ARE MADE WITH RESPECT TO THE INFORMATION PROVIDED IN THIS MSDS OR TO THE CHEMICAL TO WHICH INFORMATION MAY REFER. CRC NEITHER ASSUMES NOR AUTHORIZES ANY OTHER PERSON TO ASSUME FOR IT, ANY OTHER ADDITIONAL RESPONSIBILITY OR LIABILITY FOR THE USE OF, OR RELIANCE UPON, THIS INFORMATION.

HAZARD WARNING INFORMATION

HEALTH	FIRE	REACTIVE	OTHER	DEGREE OF HAZARD	COLOR CODING	OTHER CODES
3	1	2		0 = Minimum Hazard 1 = Slight Hazard 2 = Moderate Hazard 3 = Serious Hazard 4 = Severe Hazard	HEALTH = BLUE FIRE = RED REACTIVITY = YELLOW OTHER = WHITE	OX = Oxidizer ACID = Acid ALK = Alkali COR = Corrosive W = Use No Water

SECTION I - GENERAL INFORMATION

Characterization	RCRA Number	EPA Class
Aromatic Chlorinated Hydrocarbon	**U017**	**Toxic Waste**

DOT Proper Shipping Name	Chemical Abstract Service (CAS) Number
Benzal Chloride	**98-87-3**
DOT Hazard Class and Label Requirements	DOT Emergency Guide Code
Poison B	**No Citation**
DOT Identification Number	Chemical Formula
UN 1886	**$C_6H_5CHCl_2$**

Synonyms

Benzyl dichloride; benzylidene chloride; chlorobenzal; α,α-dichlorotoluene.

SECTION II - HAZARDOUS INGREDIENTS AND IDENTITY INFORMATION

Hazardous Components (specific identity)	OSHA Exposure Criteria	NIOSH Exposure Criteria	Immediately Dangerous to Life and Health (IDLH)	ACGIH Exposure Criteria
Benzal chloride (derivation: Chlorination of toluene until two formula weights of chlorine are absorbed, in absence of catalysts but presence of light).	**PEL:** Not Established **STEL:** Not Established	**Suspected Human Carcinogen**	**Not Determined**	**TLV:** Not Established **STEL:** Not Established

SECTION III - PHYSICAL AND CHEMICAL CHARACTERISTICS

Boiling Point	Specific Gravity (H_2O = 1)
417°F (214°C)	**1.295 (at 16°C)**
Vapor Pressure (mm Hg)	Molecular Weight
1.55 at 69°F (20°C)	**161.03**
Vapor Density (Air = 1)	Freezing Point
Not Found	**-16.1°C**

Solubility

Moderately soluble in water. Soluble in alcohol, ether, and dilute alkali.

Appearance and Odor

Very refractive colorless, oily liquid with a faint aromatic odor.

SECTION IV - FIRE AND EXPLOSION HAZARD DATA

Flash Point (method used)	Explosive Limits in Air % by Volume
Not Found	**LEL: Not Found** **UEL: Not Found**
NFPA Classification	Autoignition Temperature
Combustible Liquid	**Not Found**

Extinguishing Media

Carbon dioxide, dry chemical, water spray, or alcohol foam.

Special Fire Fighting Procedures

Wear full protective clothing and self-contained breathing apparatus (SCBA). Use a water spray to keep fire-exposed containers cool. Poisonous gases are produced in fire, including hydrogen chloride and phosgene.

Unusual Fire and Explosion Hazards

Benzal chloride is a combustible liquid. Containers may explode in fire due to pressure from heat buildup. Contact with acid fumes can also cause the release of toxic gases. Firefighters should avoid contact with vapors produced during fire.

SECTION V - REACTIVITY DATA

Stability		Conditions to Avoid
		Avoid contact with incompatible materials since toxic gases of chlorine can be released. Benzal chloride should be stabilized with an additive such as propylene oxide to help reduce possible dangerous reactions.
Stable	Unstable	Incompatibility *(materials to avoid)*
	X	Strong acids and acid fumes.
Hazardous Polymerization		Conditions to Avoid
		There are no reports of hazardous polymerization of benzal chloride.
May Occur	Will Not Occur	Hazardous Decomposition or By-products
	X	When heated to decomposition, benzal chloride can emit highly toxic/poisonous gases including hydrogen chloride and phosgene.

SECTION VI - HEALTH HAZARD DATA

Primary Route(s) of Entry:	Inhalation?	Absorption (skin)?	Ingestion?
	X	X	X

Health Hazards

INHALATION: Irritation of the nose, throat, and lungs. Higher levels may cause a buildup of fluid in the lungs, a condition known as pulmonary edema, which can be fatal. Other symptoms include lightheadedness, dizziness, shortness of breath, cough, and loss of consciousness.

ABSORPTION: Absorbs freely through intact skin causing dizziness and other neurologic symptoms. Contact with eyes can cause severe irritation and excessive tearing.

INGESTION: Moderately toxic from ingestion. Gastrointestinal irritation is possible.

Carcinogenicity	NTP Listed?	IARC Cancer Review Group?	OSHA Regulated?	Target Organs?
Suspected Human Suspected Animal	No	Group 3	No	Respiratory system, eyes, skin, CNS.

Medical Conditions Generally Aggravated by Exposure

Respiratory impairments (bronchitis) and skin conditions (dermatitis) may be aggravated by exposure.

Emergency and First-aid Procedures

Eye contact: Flush immediately with water for 15 minutes (minimum), seek medical attention. Skin contact: Remove all contaminated clothing. Immediately wash area with large amounts of soap and water. Seek medical assistance. For inhalation: Remove the person from exposure. Provide respiratory assistance and CPR. Transfer to medical facility. Medical observation for 24 to 48 hours is recommended since pulmonary edema may be a delayed reaction. If swallowed, seek medical attention immediately.

SECTION VII - PRECAUTIONS FOR SAFE HANDLING AND USE

Steps to be Taken in Case Material is Released or Spilled

Absorb liquids in vermiculite, dry sand, earth, or similar material and deposit in sealed containers. Ventilate area of spill or leak. Restrict those not involved in cleanup from entering area. Remove all ignition sources. Use water spray to reduce vapors. Notify authorities, as required by SARA III, if applicable.

Preferred Waste Disposal Method

Burn in chemical incinerator equipped with scrubber and afterburner.

Precautions to be Taken in Handling and Storage

Benzal chloride is a combustible liquid. Store to avoid contact with strong acids or acid fumes since violent reactions can occur. Store in tightly closed containers in cool, well-ventilated area away from heat. Ground and bond all containers prior to transferring liquids.

Other Precautions and Warnings

Sources of ignition are prohibited where benzal chloride is used, handled, or stored.

SECTION VIII - CONTROL MEASURES AND PERSONAL PROTECTIVE EQUIPMENT

Respiratory Protection *(specify type)*

Use a MSHA/NIOSH approved supplied air respirator with full facepiece, hood, or helmet in continuous flow mode, or use a full facepiece operated in pressure demand or other positive pressure mode or use a self-contained breathing apparatus (SCBA) with full facepiece operated in positive pressure mode.

Ventilation

Local exhaust at site of chemical work.

Protective Gloves	Eye Protection	Other Protective Clothing
Impervious Rubber	Chemical Goggles or Face Mask	Rubber Apron

Work/Hygiene Practices

Always wash hands thoroughly after using chemical; never bring food, drink, or smoking materials into vicinity of chemicals.

BENZAL CHLORIDE

$C_6H_5CHCl_2$ CAS: 98-87-3

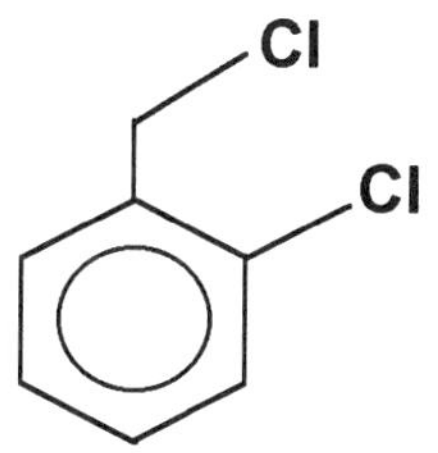

IDENTIFICATION AND TYPICAL USES

Benzal chloride is an oily, colorless liquid with a faint but characteristic aromatic odor. Benzal chloride is used in the manufacture of dyes, in the manufacture of benzaldehyde and cinnamic acid.

RISK ASSESSMENT: HEALTH

General Assessment

Benzal chloride is a poison through *inhalation* and *absorption* and is moderately toxic by *ingestion*. Inhalation causes irritation of the eyes, nose, throat, and lungs. Excessive exposures can cause fluid to buildup in the lungs in a condition known as pulmonary edema, which can be fatal. Eye contact causes severe irritation and excessive tearing. Toxic symptoms of benzal chloride exposure through inhalation and through absorption include headache, weakness, dizziness, lightheadedness, shortness of breath, cough, and possible loss of consciousness at high exposures.

Benzal chloride is a suspected human carcinogenic. It is a strong irritant and lachrymator. Exposure causes central nervous system depression with symptoms as noted above primarily occurring subsequent to inhalation. Mutation data have also been reported but not well substantiated.

Acute Health Effects

The following acute (short-term) health effects may occur immediately or shortly after exposure to benzal chloride:

Skin: Severe irritation, rash. Possible dermatitis.

Eye: Severe irritation with excessive tearing.

Lung: Severe irritation of the nose and throat. Can cause cough, congestion, difficulty in breathing, and, in high exposures, fluid buildup in the lungs (pulmonary edema), which is a medical emergency and can be fatal.

CNS: Benzal chloride is a CNS depressant. Can cause dizziness, lightheadedness, and loss of consciousness.

Chronic Health Effects

The following chronic (long-term) health effects may occur at some time after exposure to benzal chloride and can last for months or even years:

Cancer Hazards: There is limited evidence that benzal chloride causes cancer in animals. Human carcinogenicity is suspect but not confirmed. It may cause mutations (genetic changes) in living cells. The data, however, are inconclusive.

Reproductive Hazard: According to information available in the references, benzal chloride has not been tested for its ability to adversely affect reproduction.

Other Chronic Effects: Benzal chloride is very irritating to lung tissue. However, while pulmonary edema usually occurs subsequent to high acute exposures, it is not known whether long-term exposures can damage the lungs.

Recommended Risk-Reduction Measures

Personnel should avoid direct contact with benzal chloride. If a less toxic chemical cannot be substituted for a hazardous substance, then *engineering controls* are the most effective method of reducing exposures. The best protection is to enclose operations and/or provide local exhaust ventilation at the site of chemical release. While not always operationally feasible, isolating operations can also reduce exposure. Using respiratory protection is less effective than the controls mentioned above, but is still advisable whenever working with or around benzal chloride. Since no exposure limits have been established for this chemical and its human carcinogenic properties are suspect, a self-contained breathing apparatus (SCBA) with full facepiece and pressure demand is the recommended respiratory protection of choice. If a full facepiece is not available, then chemical goggles should be worn to protect the eyes. Whenever a chemical splash hazard exists, a face shield and a rubber apron should be worn. To prevent hand and skin exposures, natural, impervious rubber gloves should be worn.

Administrative controls should also be in place to minimize the potential for human exposures. These may include written procedures or policies which specify the methods and techniques that will be practiced whenever personnel are to work with benzal chloride.

All personnel should receive *training* on the use, hazards, protective measures, emergency actions, and other precautions per 29 CFR 1910.1200 (Hazard Communication), prior to the first assignment in an area where benzal chloride is used or stored.

Before beginning employment and at regular times thereafter, those personnel with frequent or potentially frequent exposures should be provided with the following medical tests:

☑ Pulmonary function tests.

If symptoms develop or overexposure is suspected, the following medical tests may be useful:

☑ Consider chest X-ray following acute overexposure.

Any evaluation should include a careful history of past and present symptoms with an examination. Medical tests that simply look for existing damage are not a substitute for controlling exposures. Also, because smoking can cause emphysema, lung cancer, and other respiratory problems, exposure to benzal chloride may affect smokers quicker and with more pronounced symptoms. Therefore, smokers should avoid unprotected contact with benzal chloride.

Other methods to reduce exposure include:

☑ Where possible, enclose operations and use local exhaust ventilation at the site of chemical release. If local exhaust ventilation or enclosure is not used, respiratory protection should be mandatory.

☑ Always ensure that proper protective clothing is worn when using chemical substances and that personnel are trained in its use and care.

☑ Safety showers and eye wash stations should be provided in areas where work is performed with or around benzal chloride.

☑ Wash thoroughly immediately after exposure to benzal chloride and at the end of the work shift or before eating, drinking, or smoking.

☑ Hazard warning information should be posted in the work area. In addition, as part of an on-going education and training program, all information on the health and safety hazards of benzal chloride should be communicated to all potentially exposed workers.

☑ Eye wash stations should be provided in the immediate work area for emergency use. If there is a possibility of skin exposure to benzal chloride, emergency shower facilities should also be provided.

RISK ASSESSMENT: ENVIRONMENT

General Assessment

The environment is at risk of exposure during transportation, storage, disposal, or destruction of benzal chloride. In almost every scenario, the threat of environmental exposure is contingent upon the proper handling of the chemical substance. Accidental spills, large or small, can result in fire, explosion, and possible contamination of the surrounding environmental mediums (water, soil, and air).

Benzal chloride is considered a combustible liquid. Alone, it is not considered a serious fire hazard. However, it is incompatible with acid and acid fumes and violent reactions can occur upon contact. Also, in fire situations, toxic gases of chlorine can be emitted, including hydrogen chloride and phosgene. Therefore, extreme caution is required in handling, storage, transportation, and disposal of benzal chloride. These characteristics also require special consideration during any emergency situation involving a leak or spill of benzal chloride. Once again, should benzal chloride ever come into contact with incompatible substances either during use, transportation, storage, or disposal, the formation of highly toxic commodities is extremely possible.

The proper disposal/destruction method for benzal chloride waste is to burn it in a chemical incinerator equipped with an afterburner and air scrubber. Benzal chloride can enter the environment through manufacturing effluents or from spills.

☠ *Acute Ecological Effects*

Acute (short-term) toxic effects may include the death of animals, birds, or fish and death or low growth rate

in plants. Acute effects are seen 2 to 4 days after animals or plants are exposed to benzal chloride. This chemical has moderate acute toxicity to aquatic life. No data are available on the short-term toxicity of benzal chloride to plants, birds, or land animals.

Chronic Ecological Effects

Chronic (long-term) toxic effects may include shortened life span, reproductive problems, lower fertility, and changes in appearance or behavior in exposed animals. These effects can be seen long after first exposure(s) to toxic chemicals. Benzal chloride has moderate chronic toxicity in aquatic life. No data are available on the long-term effects of benzal chloride to plants, birds, or land animals.

Water Solubility

Benzal chloride is moderately soluble in water. Concentrations of between 1 and 1000 milligrams and more will mix with a liter of water.

Persistence in the Environment

Benzal chloride is slightly persistent in water, with a half-life between 2 and 20 days. The half-life of a pollutant is the amount of time it takes for one half of the chemical to be degraded. Approximately 93% of benzal chloride will eventually end up in air; about 5% will end up in water; and less than 0.6% will end up in terrestrial soils and aquatic sediments, respectively.

Bioaccumulation in Aquatic Organisms

Some substances increase in concentration, or *bioaccumulate,* in living organisms as they breathe contaminated air, drink contaminated water, or eat contaminated food. These chemicals can become concentrated in the tissues and internal organs of animals as well as humans.

The concentration of benzal chloride found in fish tissues is expected to be somewhat higher than the average concentration of benzal chloride in the water from which the fish was taken.

Recommended Risk-Reduction Measures

Proper training of all transporters will reduce the likelihood of a mishap or accident resulting in a leak or spill to the environment. The correct use of DOT labeling (per the MSDS) on all containers, trucks, and rail cars will enable emergency responders to react properly and quickly to any disaster thereby reducing the potential risk to the environment and to personnel.

Storage of benzal chloride should be segregated from incompatible chemicals to minimize the risk of cross-contamination or contact. Bulk storage is not recommended.

If a spill or leak to the environment has occurred, fire department, emergency response and/or hazardous materials spill personnel should be notified immediately. Cleanup should be attempted only by those trained in proper spill containment procedures. Contaminated soils should be removed for incineration and replaced with clean soil. If benzal chloride should contact the water table, aquifer, or navigable waterway, the local and/or state emergency response authorities must be notified. A comprehensive emergency response or disaster preparedness/recovery plan should be in place prior to any operations involving the use, transportation, storage, or disposal of benzal chloride. If benzal chloride is spilled or leaked, the following specific steps are recommended:

- ☑ Restrict persons not wearing protective clothing from area of spill or leak until cleanup is complete and area can be opened for normal work.
- ☑ Remove all ignition sources.
- ☑ Ventilate area of spill.
- ☑ Absorb liquids in vermiculite, dry sand, earth, or similar material and deposit in sealed containers.
- ☑ It may be necessary to dispose of benzal chloride as a hazardous waste. The responsible state agency or the regional office of the federal Environmental Protection Agency (EPA) should be contacted for specific recommendations.

RISK ASSESSMENT: BUSINESS

General Assessment

Accidents or mishaps involving benzal chloride can present a significant threat to business operations. The loss or damage of equipment or facilities can significantly affect fiscal viability. Lawsuits that may result from personnel injury/death, public exposures, and/or environmental contamination will also require a serious expenditure of resources. Media attention surrounding an injury, death, or environmental damage can also result in a loss of profits and loss of current as well as future business.

Recommended Risk-Reduction Measures

Company attorneys, safety and health professionals, and environmental specialists should be involved in the development of any procedures or policies associated with chemical substances. A company official should be pre-designated as a public relations officer with specific training in dealing with the press. Corporate plans and policies should be developed, approved, and implemented long before any need for such arises.

REFERENCES

Ahlbom, A. 1993. *Biostatistics for Epidemiologists.* New York: CRC Press/Lewis Publishers.

American Conference of Governmental Industrial Hygienists. 1988. *Documentation of the Threshold Limit Values and Biological Exposure Indices,* 5th Edition (with updates). Cincinnati: ACGIH

Asante-Duah, D. K. 1993. *Hazardous Waste Risk Assessment.* New York: CRC Press/Lewis Publishers.

Bloom, A. D. and F. J. DeSerres. 1995. *Ecotoxicity and Human Health.* New York: CRC Press/Lewis Publishers.

Brusick, D. J. 1994. *Methods for Genetic Risk Assessment.* New York: CRC Press/Lewis Publishers.

Bueche, J. R. 1972. *Principles of Physics,* 2nd Edition. New York: McGraw-Hill.

Calabrese, E. J. 1994. *Biological Effects of Low Level Exposures.* New York: CRC Press/Lewis Publishers.

Cockerman, L. G., B. S. Shane. 1994. *Basic Environmental Toxicology.* New York: CRC Press/Lewis Publishers.

Cothern, R. R. 1993. *Comparative Environmental Risk Assessment.* New York: CRC Press/Lewis Publishers.

Gardner, R. W. 1994. *Chemical Intolerance.* New York: CRC Press/Lewis Publishers.

Hallenbeck, W. H. 1993. *Quantitative Risk Assessment for Environmental and Occupational Health,* 2nd Edition. New York: CRC Press/Lewis Publishers.

Howard, P. H. 1995. *Handbook of Environmental Fate and Exposures Data for Organic Chemicals,* Vol. I-IV. New York: CRC Press/Lewis Publishers.

Jones, F. E. 1994. *Toxic Organic Vapors in the Workplace.* New York: CRC Press/Lewis Publishers.

Keith, L. H. and M. M. Walker. 1995. *Handbook of Air Toxics.* New York: CRC Press/Lewis Publishers.

Landis, W. G., M. H. Yu. 1995. *Introduction to Environmental Toxicology.* New York: CRC Press/Lewis Publishers.

Langerman, N. 1994 *Precautionary Labels for Chemical Containers.* New York: CRC Press/Lewis Publishers.

Lauwerys, R. R., P. Hoet. 1993. *Industrial Chemical Exposure.* New York: CRC Press/Lewis Publishers.

Lewis, R. J., Sr. 1993. *Hawley's Condensed Chemical Dictionary,* 12th Edition. New York: Van Nostrand Reinhold.

Lewis, R. J., Sr. 1992. *Sax's Dangerous Properties of Industrial Materials,* Volumes 1, 2, and 3, 8th Edition. New York: Van Nostrand Reinhold.

Lide, D. R. 1995. *Handbook of Chemistry and Physics,* 76th Edition. New York: CRC Press/Lewis Publishers.

Manahan, S. E. 1994. *Environmental Chemistry,* 6th Edition. New York: CRC Press/Lewis Publishers.

Manahan, S. E. 1992. *Toxicological Chemistry,* 2nd Edition. New York: CRC Press/Lewis Publishers.

Mellström, G. A., J. E. Wahlberg, and H. I. Maibach. 1994. *Protective Gloves for Occupational Use.* New York: CRC Press/Lewis Publishers.

Meyer, E. 1990. *Chemistry of Hazardous Materials.* Englewood Cliffs, NJ: Prentice Hall.

Mycock, J. C., J. D. McKenna, and L. Theodore. 1995. *Handbook of Air Pollution Control Engineering and Technology.* New York: CRC Press/Lewis Publishers.

Ott, W. R. 1995. *Environmental Statistics and Data Analysis.* New York: CRC Press/Lewis Publishers.

Patnaik, P. 1992. *A Comprehensive Guide to the Hazardous Properties of Chemical Substances.* New York: Van Nostrand Reinhold.

Philp, R. B. 1995. *Environmental Hazards and Human Health.* New York: CRC Press/Lewis Publishers.

Proctor, N. H., J. P. Hughes. 1991. *Chemical Hazards of the Workplace.* New York: Van Nostrand Reinhold.

Quigley, D. R. 1994. *Handbook of Emergency Chemical Management.* New York: CRC Press/Lewis Publishers.

Rea, W. J. 1995. *Chemical Sensitivity,* Vol. I-III. New York: CRC Press/Lewis Publishers.

Richardson, M. 1995. *Dictionary of Substance and Their Effects,* Vol. I-VI. New York: CRC Press/Lewis Publishers.

Sacarello, H. L. A. 1994. *The Comprehensive Handbook of Hazardous Materials.* New York: CRC Press/Lewis Publishers.

Sheftel, V. O. 1995. *Handbook of Toxic Properties of Monomers and Additives.* New York: CRC Press/Lewis Publishers.

Talbott, E., G. Craun. 1995. *An Introduction to Environmental Epidemiology.* New York: CRC Press/Lewis Publishers.

Torkelson, T. R. V. K. Rowe. 1981. *Patty's Industrial Hygiene and Toxicology,* Vol. 2B. New York: Wiley-Interscience.

The Merck Index, 10th edition. 1983. Rahway, NJ: Merck & Company.

MATERIAL SAFETY DATA SHEET

CHEMICAL NAME	
BENZALDEHYDE	**CRC PRESS/LEWIS PUBLISHERS DISCLAIMER:** THE INFORMATION AND RECOMMENDATIONS PRESENTED HEREIN ARE BASED ON SOURCES BELIEVED TO BE RELIABLE. CRC MAKES NO REPRESENTATION ON ITS COMPLETENESS OR ACCURACY. IT IS THE RESPONSIBILITY OF THE USER TO DETERMINE THE CHEMICAL'S SUITABILITY FOR ITS INTENDED USE, THE CHEMICAL'S SAFE USE, AND THE CHEMICAL'S PROPER DISPOSAL. NO REPRESENTATIONS AND/OR WARRANTIES, EITHER EXPRESSED OR IMPLIED, OF THE MERCHANTABILITY OR FITNESS FOR A PARTICULAR PURPOSE, OR OF ANY OTHER NATURE, ARE MADE WITH RESPECT TO THE INFORMATION PROVIDED IN THIS MSDS OR TO THE CHEMICAL TO WHICH INFORMATION MAY REFER. CRC NEITHER ASSUMES NOR AUTHORIZES ANY OTHER PERSON TO ASSUME FOR IT, ANY OTHER ADDITIONAL RESPONSIBILITY OR LIABILITY FOR THE USE OF, OR RELIANCE UPON, THIS INFORMATION.

HAZARD WARNING INFORMATION

HEALTH	FIRE	REACTIVE	OTHER	DEGREE OF HAZARD	COLOR CODING	OTHER CODES
2	2	0		0 = Minimum Hazard 1 = Slight Hazard 2 = Moderate Hazard 3 = Serious Hazard 4 = Severe Hazard	HEALTH = BLUE FIRE = RED REACTIVITY = YELLOW OTHER = WHITE	OX = Oxidizer ACID = Acid ALK = Alkali COR = Corrosive W = Use No Water

SECTION I - GENERAL INFORMATION

Characterization	Aldehyde	RCRA Number	Not Listed
		EPA Class	Not Applicable
DOT Proper Shipping Name	Benzaldehyde	Chemical Abstract Service (CAS) Number	100-52-7
DOT Hazard Class and Label Requirements	Combustible Liquid	DOT Emergency Guide Code	No Citation
DOT Identification Number	UN 1989	Chemical Formula	C_6H_5CHO

Synonyms

Benzoic aldehyde; phenyl methanal; oil of bitter almond; benzene carbaldehyde; benzene carbonal; artificial almond oil.

SECTION II - HAZARDOUS INGREDIENTS AND IDENTITY INFORMATION

Hazardous Components (specific identity)	OSHA Exposure Criteria	NIOSH Exposure Criteria	Immediately Dangerous to Life and Health (IDLH)	ACGIH Exposure Criteria
Benzaldehyde (derivation from: air oxidation of toluene with uranium or molybdenum oxides as catalysts; reaction of benzyl dichloride with lime; extraction from oil of bitter almond).	PEL: Not Established STEL: Not Established	REL: Questionable Human Carcinogen	Not Determined	TLV: Not Established STEL: Not Established

SECTION III - PHYSICAL AND CHEMICAL CHARACTERISTICS

Boiling Point	179°F (82°C)	Specific Gravity (H_2O = 1)	1.1
Vapor Pressure (mm Hg)	1 at 69°F (20°C)	Molecular Weight	106.1
Vapor Density (Air = 1)	3.7	Melting Point	-14.8°F (-26°C)

Solubility

Slightly soluble in water; readily soluble in alcohol, ether; moderately soluble in acetone, benzene.

Appearance and Odor

Colorless liquid with a burning taste and bitter almond odor.

SECTION IV - FIRE AND EXPLOSION HAZARD DATA

Flash Point (method used)	148°F (64°C) closed cup	Explosive Limits in Air % by Volume	LEL: Not Determined UEL: Not Determined
NFPA Classification	Class IIIA Combustible Liquid	Autoignition Temperature	377°F (191°C)

Extinguishing Media

Carbon dioxide, alcohol resistant foam, dry chemical, water spray, or foam.

Special Fire Fighting Procedures

Wear full protective clothing and self-contained breathing apparatus (SCBA). Heat will build pressure and may rupture closed storage containers. Keep fire-exposed containers cool with water spray. Continue to cool containers after fire is extinguished.

Unusual Fire and Explosion Hazards

Immediately withdraw if rising sound from venting device is heard of if fire is causing discoloration to the tank. If possible, allow larger fires to burn out. Forms explosive mixtures in air.

SECTION V - REACTIVITY DATA

Stability		Conditions to Avoid **Avoid contact with incompatible materials. Although explosive limits have not been determined, benzaldehyde does form explosive mixtures in air. It can explode when treated with performic acid.**
Stable **X**	Unstable	Incompatibility *(materials to avoid)* **Strong oxidizers (such as chlorine, bromine, and fluorine), and acids (such as performic and peroxyformic acids).**
Hazardous Polymerization		Conditions to Avoid **Under normal conditions of temperature and pressure, hazardous polymerization of benzaldehyde will not occur.**
May Occur **X**	Will Not Occur	Hazardous Decomposition or By-products **When heated to decomposition, benzaldehyde emits acrid, irritating smoke and fumes.**

SECTION VI - HEALTH HAZARD DATA

Primary Route(s) of Entry:	Inhalation? X	Absorption (skin)?	Ingestion? X

Health Hazards

INHALATION: Possible irritation of the upper respiratory tract with of nausea and headache.

ABSORPTION: May irritate the skin. Benzaldehyde is an allergen. May act like a local anesthetic.

INGESTION: Causes central nervous system depression in low doses, convulsions in large doses.

Carcinogenicity	NTP Listed?	IARC Cancer Review Group?	OSHA Regulated?	Target Organs?
Questioned Human **Questioned Animal**	**5th Annual Report**	**No**	**No**	**Respiratory system, skin, eyes, CNS**

Medical Conditions Generally Aggravated by Exposure

None reported.

Emergency and First-aid Procedures

Eye contact: Flush immediately with water for 30 minutes (minimum); seek medical attention. Skin contact: Remove all contaminated clothing. Immediately wash area with large amounts of soap and water. Seek medical assistance. For inhalation: Remove the person from exposure. Provide respiratory assistance and CPR. Transfer to medical facility. If swallowed, seek medical attention immediately. Induce vomiting if person is conscious. Do not make an unconscious person vomit.

SECTION VII - PRECAUTIONS FOR SAFE HANDLING AND USE

Steps to be Taken in Case Material is Released or Spilled

Contain spills using absorbent material (vermiculite, or other). Ventilate area of spill. Have water source available in case of fire. Store materials in DOT-approved containers. Restrict those not involved in cleanup from entering area. Notify appropriate authorities, as required.

Preferred Waste Disposal Method

Incineration in chemical incinerator equipped with scrubber and afterburner.

Precautions to be Taken in Handling and Storage

Do not store with incompatible chemicals. Store in tightly closed stainless steel, glass, Teflon, or phenolic resin-lined containers in a cool, dark, well-ventilated area. Keep fire and flame away. Use non-sparking tools to open or close containers. Ground and bond all metal containers during transfer operations. If possible, automatically transfer liquids between containers.

Other Precautions and Warnings

Bulk storage of benzaldehyde is not recommended.

SECTION VIII - CONTROL MEASURES AND PERSONAL PROTECTIVE EQUIPMENT

Respiratory Protection *(specify type)*

No exposure limits have been established. Recommend a self-contained breathing apparatus (SCBA) with full facepiece and pressure demand or other positive pressure mode.

Ventilation

Local exhaust or general mechanical systems recommended.

Protective Gloves	Eye Protection	Other Protective Clothing
Butyl Rubber, Polyvinyl Alcohol, or Viton	**Chemical Goggles or Face Mask**	**Impermeable Apron**

Work/Hygiene Practices

Always wash hands thoroughly after using chemical; never bring food, drink, or smoking materials into vicinity of chemicals.

BENZALDEHYDE

C_6H_5CHO CAS: 100-52-7

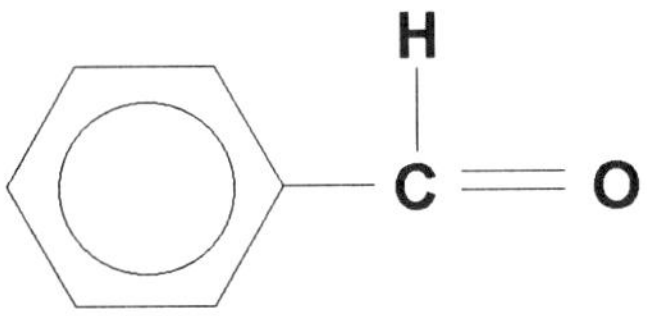

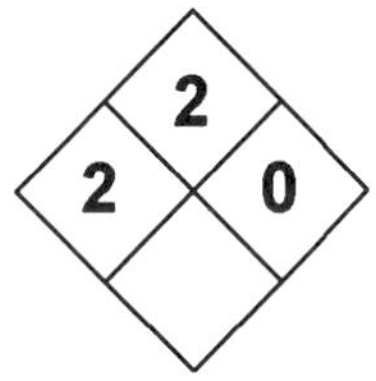

IDENTIFICATION AND TYPICAL USES

Benzaldehyde is a colorless liquid with a bitter almond taste and odor. It is used as an intermediate in the production of flavoring chemicals, such as cinnamaldehyde, cinnamalcohol, and amyl- and hexylcinnamaldehyde for perfume, soap, and food flavor; synthetic penicillin, ampicillin, and ephedrine; and as a raw material for the herbicide Avenge. Its also occurs in nature in the seeds of almonds, apricots, cherries, and peaches. It occurs in trace amounts in corn oil. Other uses include odor additive in aromatic alcohols; solvent for oils, resins, some cellulose ethers, cellulose acetate and nitrate; in pharmaceuticals; and in photographic chemicals.

RISK ASSESSMENT: HEALTH

General Assessment

Benzaldehyde is a poison by *ingestion* and moderately toxic by other routes of exposure. It acts as a mild anesthetic on contact with skin. It may cause contact dermatitis as well as irritation. It is an allergen. Persons allergic to benzaldehyde may therefore display more pronounced reactions than those not allergic to this chemical.

Inhalation of its vapors may cause irritation of the eyes and throat and can cause eye and lung injury on prolonged exposure. However, their vapors cannot normally be tolerated for long periods of time and chronic exposure is therefore not common. Direct contact with the liquid in the eyes causes immediate and potentially severe irritation. Benzaldehyde causes central nervous system depression in small doses and can cause convulsions when exposed to large amounts.

It is a questionable human carcinogen. Mutation data have also been reported, although the references are inconclusive.

Acute Health Effects

The following acute (short-term) health effects may occur immediately or shortly after exposure to benzaldehyde:

Skin: Irritation and possible contact dermatitis.

Eye: Irritation which can lead to injury if allowed to remain in contact with the eyes.

Lung: Irritation of the throat and eyes causing coughing, nausea, and possible vomiting.

CNS: Depression of the CNS at low doses and possible convulsions at higher doses. May act as a narcotic at high concentrations.

Chronic Health Effects

The following chronic (long-term) health effects may occur at some time after exposure to benzaldehyde and can last for months or even years:

Cancer Hazards: Benzaldehyde is a questionable carcinogen and mutagen, although the data are inconclusive.

Reproduction: According to the references, benzaldehyde has not been adequately tested for its ability to adversely affect reproduction in test animals.

Recommended Risk-Reduction Measures

Personnel should avoid direct contact with benzaldehyde. Even though benzaldehyde is considered to have low to moderate toxicity, a less toxic chemical should be substituted for a hazardous substance. If this is not possible or feasible, then *engineering controls* are the most effective method of reducing exposures. The best protection is to enclose operations and/or provide local exhaust ventilation at the site of chemical release. While not always operationally feasible, isolating operations can also reduce exposure.

Using respiratory protection is less effective than the controls mentioned above, but is still advisable whenever working with or around benzaldehyde. Since no exposure levels have been established and this chemical has questionable carcinogenic properties, a self-contained breathing apparatus (SCBA) with full facepiece and pressure demand is the recommended respiratory protection of choice. If a full facepiece is not available, then chemical goggles should be worn to protect the eyes. Whenever a chemical splash hazard exists, a face shield and a rubber apron should be worn. To prevent hand and skin exposures, butyl rubber gloves should be worn.

Administrative controls should also be in place to minimize the potential for human exposures. These may include written procedures or policies which specify the methods and techniques that will be practiced whenever personnel are to work with benzaldehyde.

All personnel should receive *training* on the use, hazards, protective measures, emergency actions, and other precautions per 29 CFR 1910.1200 (Hazard Communication), prior to the first assignment in an area where benzaldehyde is used or stored.

Other methods to reduce exposure include:

- ☑ Where possible, enclose operations and use local exhaust ventilation at the site of chemical release. If local exhaust ventilation or enclosure is not used, respiratory protection should be mandatory.
- ☑ Always ensure that proper protective clothing is worn when using chemical substances.
- ☑ Wash thoroughly immediately after exposure to benzaldehyde and at the end of the work shift or before eating, drinking, or smoking.
- ☑ Hazard warning information should be posted in the work area. In addition, as part of an on-going education and training program, all information on the health and safety hazards of benzaldehyde should be communicated to all potentially exposed workers.
- ☑ Never eat, drink, or smoke in areas were benzaldehyde is used, handled or stored.

RISK ASSESSMENT: ENVIRONMENT

General Assessment

The environment is at risk of exposure during transportation, storage, disposal, or destruction of benzaldehyde. In almost every scenario, the threat of environmental exposure is contingent upon the proper handling of the chemical substance.

Accidental spills, large or small, can result in fire, explosion, and possible contamination of the surrounding environmental mediums (water, soil, and air). Benzaldehyde is considered a moderate fire hazard. Because of its flash point and boiling point, it is classified as a Class IIIA combustible liquid (per OSHA 29 CFR 1910.106). These characteristics require special consideration during any emergency situation involving a leak or spill of benzaldehyde. Should benzaldehyde ever come into contact with incompatible substances such as oxidizers or acids, either during use, transportation, or storage, violent reactions are extremely possible. The proper disposal/destruction method for benzaldehyde is to burn it in a chemical incinerator equipped with an afterburner and air scrubber. Benzaldehyde can occur in the environment naturally as well as through manufacturing, unchecked discharge into effluents, and through spills.

Acute Ecological Effects

Acute (short-term) toxic effects may include the death of animals, birds, or fish, and death or low growth rate in plants. Acute effects are seen 2 to 4 days after animals or plants are exposed to benzaldehyde. This chemical has low acute toxicity to aquatic life. No data are available on the short-term effects of benzaldehyde to birds, plants or terrestrial animals.

Chronic Ecological Effects

Chronic toxic effects may include shortened life span, reproductive problems, lower fertility, and changes in appearance or behavior in exposed animals. These effects can be seen long after first exposure(s) to toxic chemicals. Benzaldehyde has low chronic toxicity in aquatic life. No data are available on the long-term effects of benzaldehyde to plants, birds, or land animals.

Water Solubility

Benzaldehyde is highly soluble in water. Concentrations of 1000 milligrams and more will mix with a liter of water.

Persistence in the Environment

Benzaldehyde is moderately persistent in water, with a half-life of between 2 to 20 days. The half-life of a pollutant is the amount of time it takes for one-half of the chemical to be degraded. No data are available on the percentages of benzaldehyde that will eventually end-up in water.

Bioaccumulation in Aquatic Organisms

Some substances increase in concentration, or *bioaccumulate,* in living organisms as they breathe contaminated air, drink contaminated water, or eat contaminated food. These chemicals can become concentrated in the tissues and internal organs of animals as well as humans.

The concentration of benzaldehyde found in fish tissues is expected to be about the same as the average concentration of benzaldehyde in water from which the fish was taken.

Recommended Risk-Reduction Measures

Proper training of all transporters will reduce the likelihood of a mishap or accident resulting in a leak or spill to the environment. The correct use of DOT labeling (per the MSDS) on all containers, trucks, and rail cars will enable emergency responders to react properly and quickly to any disaster thereby reducing the potential risk to the environment and to personnel.

Storage of benzaldehyde should be segregated from incompatible chemicals to minimize the risk of cross-contamination or contact. Benzaldehyde is stored in stainless steel, glass, Teflon, or phenolic resin-lined containers. Buildings designated for storage should be equipped with appropriate fire protection systems (alarms, sprinklers, emergency lighting, portable extinguishers).

If a spill or leak to the environment has occurred, fire department, emergency response, and/or hazardous materials spill personnel should be notified immediately. Cleanup should be attempted only by those trained in proper spill containment procedures. Contaminated soils should be removed for incineration and replaced with clean soil.

If benzaldehyde should contact the water table, aquifer, or navigable waterway, time is of the essence. It is moderately miscible in water and, therefore, total containment and remediation may not be entirely possible. When such spills occur, the local and/or state emergency response authorities must be notified. A comprehensive emergency response or disaster preparedness/recovery plan should be in place prior to any operations involving the use, transportation, storage, or disposal of benzaldehyde. If benzaldehyde is spilled or leaked, the following specific steps are recommended:

- ☑ Restrict persons not wearing protective clothing from area of spill or leak until cleanup is complete and area can be opened for normal work.
- ☑ Remove all ignition sources.
- ☑ Ventilate area of spill or leak.
- ☑ Absorb liquids in vermiculite, dry sand, earth, or a similar material and deposit in sealed containers.
- ☑ If applicable, stop flow of leaking liquid. If leak source is a cylinder and the leak cannot be stopped in place, remove leaking cylinder to a safe place in the open air and repair or allow cylinder to empty.
- ☑ Keep benzaldehyde out of a confined space, such as a sewer, because of the possibility of explosion (unless the sewer is designed to prevent the buildup of explosive concentrations).
- ☑ It may be necessary to dispose of benzaldehyde as a hazardous waste. The responsible state agency or the regional office of the federal Environmental Protection Agency (EPA) should be contacted for specific recommendations.

RISK ASSESSMENT: BUSINESS

General Assessment

Accidents or mishaps involving benzaldehyde can present a moderate threat to business operations. The loss or damage of equipment or facilities can significantly affect fiscal viability. Lawsuits that may result from personnel injury/death, public exposures, and/or environmental contamination will also require a serious expenditure of resources.

Media attention surrounding an injury, death, or environmental damage can also result in a loss of profits and loss of current as well as future business. Always remember that any time the terms "cancer" or "carcinogen are used, public emotion, ignorance, and hysteria can run equally high. This must be carefully considered whenever drafting or implementing public relations policies.

Recommended Risk-Reduction Measures

Company attorneys, safety and health professionals, and environmental specialists should be involved in the development of any procedures or policies intended to manage the use of chemicals in the workplace. A company official should be pre-designated as a public relations officer with specific training in dealing with the press. Corporate plans and policies should be developed, approved, and implemented long before any need for such arises.

REFERENCES

Ahlbom, A. 1993. *Biostatistics for Epidemiologists.* New York: CRC Press/Lewis Publishers.

American Conference of Governmental Industrial Hygienists. 1988. *Documentation of the Threshold Limit Values and Biological Exposure Indices,* 5th Edition (with updates). Cincinnati: ACGIH

Anthony, C. P., and N. J. Kolthoff. 1971. *Textbook of Anatomy and Physiology.* St. Louis: C. V. Mosby Company.

Asante-Duah, D. K. 1993. *Hazardous Waste Risk Assessment.* New York: CRC Press/Lewis Publishers.

Bloom, A. D. and F. J. DeSerres. 1995. *Ecotoxicity and Human Health.* New York: CRC Press/Lewis Publishers.

Brusick, D. J. 1994. *Methods for Genetic Risk Assessment.* New York: CRC Press/Lewis Publishers.

Bueche, J. R. 1972. *Principles of Physics,* 2nd Edition. New York: McGraw-Hill.

Burrell, R., D. K. Flaherty, and L. J. Sauers. 1992. *Toxicology of the Immune System.* New York: Van Nostrand Reinhold.

Calabrese, E. J. 1994. *Biological Effects of Low Level Exposures.* New York: CRC Press/Lewis Publishers.

Cockerman, L. G., B. S. Shane. 1994. *Basic Environmental Toxicology.* New York: CRC Press/Lewis Publishers.

Cothern, C. R., N. P. Ross. 1994. *Environmental Statistics, Assessment, and Forecasting.* New York: CRC Press/Lewis Publishers.

Cothern, C. R. 1995. *Handbook for Environmental Risk Decisionmaking.* New York: CRC Press/Lewis Publishers.

Danse, I. R. 1991. *Common Sense Toxics in the Workplace.* New York: Van Nostrand Reinhold.

Environmental Protection Agency. 1983. *Chemical Hazard Information Profile: Benzaldehyde.* Washington, D.C.: U.S. EPA.

Gardner, R. W. 1994. *Chemical Intolerance.* New York: CRC Press/Lewis Publishers.

Hallenbeck, W. H. 1993. *Quantitative Risk Assessment for Environmental and Occupational Health,* 2nd Edition. New York: CRC Press/Lewis Publishers.

Hodgson, E., R. B. Mailman, and J. E. Chambers. 1988. *Dictionary of Toxicology.* London: The MacMillian Company.

Heck, H. 1985. Mechanisms of benzaldehyde toxicity: Structure activity Studies. *CIIT Activities,* 5(10):106.

Hosty, J. W., and P. Foster. 1990. *A Practical Guide to Chemical Spill Response.* New York: Van Nostrand Reinhold.

Howard, P. H. 1995. *Handbook of Environmental Fate and Exposures Data for Organic Chemicals,* Vol. I-IV. New York: CRC Press/Lewis Publishers.

Howard, P. H., et. al. 1991. *Handbook of Environmental Degradation Rates.* New York: CRC Press/Lewis Publishers.

Jones, F. E. 1994. *Toxic Organic Vapors in the Workplace.* New York: CRC Press/Lewis Publishers.

Kamrin, M. 1 A. 1988. *Toxicology: A Primer on Principles and Applications.* Chelsea, Michigan: Lewis Publishers.

Keith, L. H. and M. M. Walker. 1995. *Handbook of Air Toxics.* New York: CRC Press/Lewis Publishers.

Landis, W. G., M. H. Yu. 1995. *Introduction to Environmental Toxicology.* New York: CRC Press/Lewis Publishers.

Langerman, N. 1994 *Precautionary Labels for Chemical Containers.* New York: CRC Press/Lewis Publishers.

Lauwerys, R. R., P. Hoet. 1993. *Industrial Chemical Exposure.* New York: CRC Press/Lewis Publishers.

Lide, D. R. 1995. *Handbook of Chemistry and Physics,* 76th Edition. New York: CRC Press/Lewis Publishers.

Lewis, Richard J., Sr. 1993. *Hawley's Condensed Chemical Dictionary,* 12th Edition. New York: Van Nostrand Reinhold.

Lewis, Richard J., Sr. 1992. *Sax's Dangerous Properties of Industrial Materials,* Volumes 1, 2, and 3, 8th Edition. New York: Van Nostrand Reinhold.

Lisella, Frank S. 1994. *The VNR Dictionary of Environmental Health and Safety.* New York: Van Nostrand Reinhold.

Manahan, S. E. 1994. *Environmental Chemistry,* 6th Edition. New York: CRC Press/Lewis Publishers.

Manahan, S. E. 1992. *Toxicological Chemistry,* 2nd Edition. New York: CRC Press/Lewis Publishers.

Mellström, G. A., J. E. Wahlberg, and H. I. Maibach. 1994. *Protective Gloves for Occupational Use.* New York: CRC Press/Lewis Publishers.

Meyer, Eugene. 1990. *Chemistry of Hazardous Materials.* Englewood Cliffs, NJ: Prentice Hall.

Mycock, J. C., J. D. McKenna, and L. Theodore. 1995. *Handbook of Air Pollution Control Engineering and Technology.* New York: CRC Press/Lewis Publishers.

Nelson, G. O. 1992. *Gas Mixtures.* New York: CRC Press/Lewis Publishers.

Ott, W. R. 1995. *Environmental Statistics and Data Analysis.* New York: CRC Press/Lewis Publishers.

Patnaik, Pradyot. 1992. *A Comprehensive Guide to the Hazardous Properties of Chemical Substances.* New York: Van Nostrand Reinhold.

Philp, R. B. 1995. *Environmental Hazards and Human Health.* New York: CRC Press/Lewis Publishers.

Proctor, N. H., and J. P. Hughes. 1991. *Chemical Hazards of the Workplace.* New York: Van Nostrand Reinhold.

Silverman, L., H. F. Schuttle, and M. W. First. 1946. Further studies on sensory response to certain industrial solvent vapors. *Journal of Industrial Hygiene Toxicology,* 28:262-266.

Sherman, J. D. 1988. *Chemical Exposure and Disease.* New York: Van Nostrand Reinhold.

MATERIAL SAFETY DATA SHEET

CHEMICAL NAME	
BENZAMIDE	**CRC PRESS/LEWIS PUBLISHERS DISCLAIMER:** THE INFORMATION AND RECOMMENDATIONS PRESENTED HEREIN ARE BASED ON SOURCES BELIEVED TO BE RELIABLE. CRC MAKES NO REPRESENTATION ON ITS COMPLETENESS OR ACCURACY. IT IS THE RESPONSIBILITY OF THE USER TO DETERMINE THE CHEMICAL'S SUITABILITY FOR ITS INTENDED USE, THE CHEMICAL'S SAFE USE, AND THE CHEMICAL'S PROPER DISPOSAL. NO REPRESENTATIONS AND/OR WARRANTIES, EITHER EXPRESSED OR IMPLIED, OF THE MERCHANTABILITY OR FITNESS FOR A PARTICULAR PURPOSE, OR OF ANY OTHER NATURE, ARE MADE WITH RESPECT TO THE INFORMATION PROVIDED IN THIS MSDS OR TO THE CHEMICAL TO WHICH INFORMATION MAY REFER. CRC NEITHER ASSUMES NOR AUTHORIZES ANY OTHER PERSON TO ASSUME FOR IT, ANY OTHER ADDITIONAL RESPONSIBILITY OR LIABILITY FOR THE USE OF, OR RELIANCE UPON, THIS INFORMATION.

HAZARD WARNING INFORMATION

HEALTH	FIRE	REACTIVE	OTHER	DEGREE OF HAZARD	COLOR CODING	OTHER CODES
1	1	1		0 = Minimum Hazard 1 = Slight Hazard 2 = Moderate Hazard 3 = Serious Hazard 4 = Severe Hazard	HEALTH = BLUE FIRE = RED REACTIVITY = YELLOW OTHER = WHITE	OX = Oxidizer ACID = Acid ALK = Alkali COR = Corrosive ~~W~~ = Use No Water

SECTION I - GENERAL INFORMATION

Characterization **Amide**	RCRA Number **None**	EPA Class **Not Applicable**
DOT Proper Shipping Name **Chemicals, N.O.S. (non-regulated)**	Chemical Abstract Service (CAS) Number **55-21-0**	
DOT Hazard Class and Label Requirements **Non-regulated**	DOT Emergency Guide Code **No Citation**	
DOT Identification Number **Not Listed**	Molecular Formula **$C_6H_5CONH_2$**	

Synonyms
Benzoylamide; benzoic acid amide; phenylcarboxyamide

SECTION II - HAZARDOUS INGREDIENTS AND IDENTITY INFORMATION

Hazardous Components (specific identity)	OSHA Exposure Criteria	NIOSH Exposure Criteria	Immediately Dangerous to Life and Health (IDLH)	ACGIH Exposure Criteria
Benzamide (derivation: From benzoyl chloride and ammonia or ammonium carbonate).	**PEL:** Not Established **STEL:** Not Established	**REL:** Not Established	**Not Determined**	**TLV:** Not Established **STEL:** Not Established

SECTION III - PHYSICAL AND CHEMICAL CHARACTERISTICS

Boiling Point **550°F (288°C)**	Specific Gravity (H_2O = 1) **1.34**
Vapor Pressure (mm Hg) **Not Found**	Molecular Weight (atomic weight) **121.14**
Vapor Density (Air = 1) **Not Found**	Melting Point **261°F (127°C)**

Solubility in Water
Moderately soluble in water (1 to 10%), especially hot water. Soluble in hot benzene, alcohol, and ether.

Appearance and Odor
Benzamide is a colorless crystalline (sand-like) material. Slight aromatic or benzene-like odor.

SECTION IV - FIRE AND EXPLOSION HAZARD DATA

Flash Point (method used) **Not Applicable**	Explosive Limits in Air % by Volume **LEL: N/A UEL: N/A**
NFPA Classification **Combustible solid**	Autoignition Temperature **Not Found**

Extinguishing Media
Use extinguishing media appropriate to surrounding fires.

Special Fire Fighting Procedures
Wear full protective clothing and self-contained breathing apparatus (SCBA). Poisonous gases are produced in fire. Move container from fire area if it can be done without risk. Apply cooling water to sides of containers exposed to flames until well after fire is out.

Unusual Fire and Explosion Hazards
Benzamide is a combustible solid. Finely divided airborne dusts may present an explosion hazard. Highly toxic and poisonous gases are released during fires, including nitrogen oxide, carbon dioxide, and carbon monoxide.

SECTION V - REACTIVITY DATA

Stability		Conditions to Avoid
		None documented.
Stable	Unstable	Incompatibility *(materials to avoid)*
X		None documented.
Hazardous Polymerization		Conditions to Avoid
		Hazardous polymerization has not been reported to occur under normal conditions of temperature and pressure.
May Occur	Will Not Occur	Hazardous Decomposition or By-products
	X	Poisonous gases, including oxides of nitrogen, carbon monoxide, and carbon dioxide are emitted during fire.

SECTION VI - HEALTH HAZARD DATA

Primary Route(s) of Entry:	Inhalation?	Absorption (skin/eye)?	Ingestion?
	X	X	X

Health Hazards

INHALATION: May cause irritation of the nose, throat, and respiratory tract with symptoms of sore throat, coughing, and shortness of breath. May also affect blood's ability to carry oxygen.

EYES & SKIN: Irritation with possible redness and pain. Also, absorbs through the skin and may result in a disruption of the blood's oxygen carrying capabilities (methemoglobinemia).

INGESTION: May cause gastrointestinal irritation with symptoms of nausea, vomiting and diarrhea.

Carcinogenicity	NTP Listed?	IARC Cancer Review Group?	OSHA Regulated?	Target Organs?
Unknown Human Unknown Animal	No	No	No	Blood, respiratory system, liver.

Medical Conditions Generally Aggravated by Exposure

None reported.

Emergency and First-aid Procedures

Eye contact: Do not allow victim to keep eyes tightly shut. Flush immediately with water for 15 minutes (minimum), seek medical attention. Skin contact: Remove all contaminated clothing, including shoes. Immediately wash area with flooding amounts of soap and water. Seek medical attention immediately. For inhalation: Remove the person from exposure. Provide respiratory assistance and CPR. Transfer to medical facility. If swallowed: Seek medical attention immediately. If person is conscious, immediately give large amounts of water. Never give anything by mouth to an unconscious or convulsing person.

SECTION VII - PRECAUTIONS FOR SAFE HANDLING AND USE

Steps to be Taken in Case Material is Released or Spilled

Remove all ignition sources. Collect powder materials in safest and most convenient manner possible. Avoid generating airborne dust concentrations. Restrict those not wearing protective equipment and who are not involved in cleanup from entering area. Avoid skin contact.

Preferred Waste Disposal Method

Not Listed. Refer to local, state, or federal requirements for the proper disposal methods of benzamide.

Precautions to be Taken in Handling and Storage

Protect containers against physical damage. Store in tightly closed containers in cool, well ventilated location. General chemical storage of benzamide in bulk is permissible.

Other Precautions and Warnings

Sources of ignition such as smoking, is prohibited where benzamide is used or stored. Where possible, automatically transfer liquid benzamide from drums or other containers to process containers.

SECTION VIII - CONTROL MEASURES AND PERSONAL PROTECTIVE EQUIPMENT

Respiratory Protection *(specify type)*

Use a MSHA/NIOSH approved respirator with a high efficiency particulate air (HEPA) filter. Greater protection is provided by a powered air purifying respirator or use a self-contained breathing apparatus (SCBA) with full facepiece operated in positive pressure mode.

Ventilation

Local exhaust (preferred) at site of chemical work or general ventilation.

Protective Gloves	Eye Protection	Other Protective Clothing
Impervious Rubber	Chemical/Dust Goggles or Face Mask	Not Applicable

Work/Hygiene Practices

Always wash hands thoroughly after using chemical; never bring food, drink, or smoking materials into vicinity of chemicals.

BENZAMIDE

$C_6H_5CONH_2$ CAS: 55-21-0

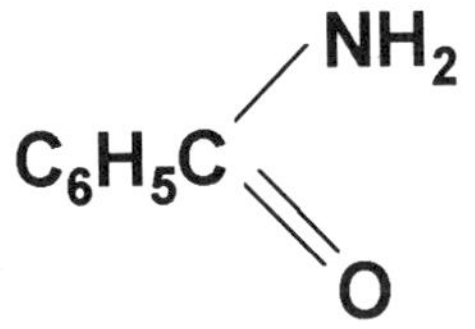

IDENTIFICATION AND TYPICAL USES

Benzamide is a colorless, crystalline (sand-like) solid with a slight aromatic or benzene-like odor. Its primary use is in organic synthesis in the manufacture of other chemicals.

RISK ASSESSMENT: HEALTH

General Assessment

Benzamide is moderately toxic by ingestion. It can also enter the body by inhalation and by absorption through unbroken skin. While its carcinogenic properties are not known, some human mutation data have been reported. Chemicals that are similar in structure can adversely affect the blood's capability to transport oxygen. This condition, known as methemoglobinemia, can lead to further complications, including death from respiratory arrest. It is not known whether exposure to benzamide causes this condition but caution is warranted. Also, similar chemicals can cause liver damage as well but, once again, there is insufficient data to validate the effect of benzamide exposure on the liver.

Other effects of exposure may include irritation of the eyes, nose, and throat. There may be also coughing and shortness of breath.

While no exposure limits have been established for benzamide, it does not mean that exposure to this chemical is without risk to human health. Sufficiently similar compounds can adversely affect health and, therefore, safe work practices should always be followed when working with or around benzamide.

Acute Health Effects

The following acute (short-term) health effects may occur immediately or shortly after exposure to benzamide:

Skin: Irritation with potential for redness and rash.

Eye: Severe irritation; redness and pain possible.

Lung: Nose, throat, and respiratory tract irritation with possible difficulty in breathing.

Other: Possible liver damage since other sufficiently similar compounds can target the liver. Also, acute overexposure may cause a disruption in the blood's ability to transport oxygen (methemoglobinemia) since other quite similar compounds have this effect.

Chronic Health Effects

The following chronic (long-term) health effects may occur at some time after exposure to benzamide and can last for months or even years:

Cancer Hazards: According to the references, benzamide has not been tested for its ability to cause cancer in animals.

Reproductive Hazard: According to the references, benzamide has not been tested for its ability to affect reproduction.

Other Chronic Effects: Similar chemicals (amides) cause liver damage. It is not known if benzamide has this effect.

Recommended Risk-Reduction Measures

Personnel should avoid direct contact with benzamide dusts or powders. If a less toxic material or compound cannot be substituted for benzamide, then *engineering controls* are the most effective method of reducing exposures. The best protection is to enclose operations and/or provide local exhaust ventilation at the site of benzamide dust release. While not always operationally feasible, isolating operations can also reduce exposure. Using respiratory protection is less effective than the controls mentioned above, but is still advisable whenever working with or around benzamide. A MSHA/NIOSH approved full facepiece respirator with a high efficiency particulate air (HEPA) filter is recommended. For extremely dusty conditions, a self-contained breathing apparatus (SCBA) with full facepiece and pressure demand is more appropriate. If a full facepiece is not available, then chemical/dust goggles should be worn to protect the eyes. A face shield should also be considered. To prevent hand and skin exposures, impervious gloves should be worn.

Administrative controls should also be in place to minimize the potential for human exposures. These may include written procedures or policies which specify the methods and techniques that will be practiced whenever personnel are to work with benzamide.

All personnel should receive *training* on the use, hazards, protective measures, emergency actions, and other precautions per 29 CFR 1910.1200 (Hazard Communication), prior to the first assignment in an area where benzamide is used or stored.

If symptoms develop or overexposure is suspected, the following recommended medical tests should be considered:

☑ Liver function test.

☑ Blood methemoglobin test.

Any evaluation should include a careful history of past and present symptoms with an examination. Medical tests that simply look for existing damage are not a substitute for controlling exposures.

Other methods to reduce exposure include:

☑ Where possible, enclose operations and use local exhaust ventilation at the site of chemical release. If local exhaust ventilation or enclosure is not used, respiratory protection should be mandatory.

☑ Always ensure that proper protective clothing is worn when using chemical substances.

☑ Wash thoroughly immediately after exposure to benzamide and at the end of the work shift or before eating, drinking, or smoking.

☑ Hazard warning information should be posted in the work area. In addition, as part of an on-going education and training program, all information on the health and safety hazards of benzamide should be communicated to all exposed and potentially exposed workers.

☑ Eye wash stations should be provided in the immediate work area for emergency use. If there is a possibility of skin exposure to benzamide, emergency shower facilities should also be provided.

☑ Workers whose clothing has been contaminated by benzamide should change into clean clothes before leaving work. Contaminated work clothing should be laundered only by individuals who have been informed of the hazards of exposure to benzamide.

☑ Where possible, automatically transfer liquids containing benzamide from drums or other containers to process containers.

RISK ASSESSMENT: ENVIRONMENT

General Assessment

The environment is at risk of exposure during transportation, storage, disposal, or destruction of benzamide. In almost every scenario, the threat of environmental exposure is contingent upon the proper handling of the chemical substance. Accidental spills, large or small, can result in contamination of the surrounding environmental mediums (water, soil, and air).

Benzamide is a combustible solid. By itself, it does not present a serious fire hazard. However, it readily supports the combustion of other materials. It can therefore present a moderate fire hazard. Also, finely divided airborne dusts can present an explosion hazard. Caution is always required in handling, storage, transportation, and disposal of benzamide. Emergency responders should be made aware of the presence of benzamide at any emergency response situation.

Benzamide can enter the environment from industrial discharges and from spills.

Acute Ecological Effects

Acute (short-term) toxic effects may include the death of animals, birds, or fish and death or low growth rate in plants. Acute effects are seen 2 to 4 days after animals or plants are exposed to benzamide.

Benzamide has slight acute toxicity to aquatic life. It has caused injury to agricultural crops. Insufficient data are available to evaluate the short-term effects of benzamide on birds or land animals.

Chronic Ecological Effects

Chronic (long-term) toxic effects may include shortened life span, reproductive problems, lower fertility, and changes in appearance or behavior in exposed animals. These effects can be seen long after first exposure(s) to toxic chemicals.

Benzamide has slight chronic toxicity to aquatic life. Insufficient data are available to evaluate or predict the long-term effects of benzamide to plants, birds, or terrestrial animals.

♦ *Water Solubility*

Benzamide is moderately soluble in water. Concentrations of 1000 milligrams and more will mix with a liter of water.

Persistence in the Environment

Benzamide is slightly persistent in water, with a half-life of between 2 to 20 days. The half-life of a pollutant is the amount of time it takes for one half of the chemical to be degraded. About 99.9% of benzamide will eventually end up in water; the rest will be divided about equally among air, terrestrial soils, and aquatic sediments.

Bioaccumulation in Aquatic Organisms

Some substances increase in concentration, or *bioaccumulate,* in living organisms as they breathe contaminated air, drink contaminated water, or eat contaminated food. These chemicals can become concentrated in the tissues and internal organs of animals as well as humans.

The concentration of benzamide found in fish tissues is expected to be about the same as the average concentration of benzamide in the water from which the fish was taken.

Recommended Risk-Reduction Measures

Proper training of all transporters will reduce the likelihood of a mishap or accident resulting in a leak or spill of benzamide dusts or powders into the environment. While benzamide is a non-regulated substance under DOT requirements, all containers, trucks, and rail cars should still carry nomenclature on their contents to enable emergency responders to react properly and quickly to any disaster thereby reducing the potential risk to the environment and to personnel.

Storage of benzamide should be segregated from other chemicals and materials to minimize the risk of cross-contamination.

If a spill or leak to the environment has occurred, fire department, emergency response and/or hazardous materials spill personnel should be notified immediately. Cleanup should be attempted only by those trained in proper spill containment procedures. Contaminated soils should be removed for incineration and replaced with clean soil. If benzamide should contact the water table, aquifer, or navigable waterway, time is of the essence. It is moderately soluble and total remediation may not be possible. The local and/or state emergency response authorities must be notified. A comprehensive emergency response or disaster preparedness/recovery plan should be in place prior to any operations involving the use, transportation, storage, or disposal of benzamide. If benzamide is spilled or leaked, the following specific steps are recommended:

- ☑ Restrict persons not wearing protective clothing from area of spill or leak until cleanup is complete and area can be opened for normal work.
- ☑ Collect powdered materials in the safest, most convenient manner possible and deposit in sealed containers. Recommend using clean shovel for small spills, depositing materials in clean container for disposal or reclamation. For larger spills, use a vacuum with high-efficiency particulate air (HEPA) filter. Do NOT dry sweep (generates airborne dusts).
- ☑ It may be necessary to dispose of benzamide as a hazardous waste. The responsible state agency or the regional office of the federal Environmental Protection Agency (EPA) should be contacted for specific recommendations.

RISK ASSESSMENT: BUSINESS

General Assessment

Accidents or mishaps involving benzamide can present a moderate threat to business operations. The loss or damage of equipment or facilities can significantly affect fiscal viability. While the specific effects of benzamide on human health or the environment may not be clearly understood, litigious situations can still arise. Lawsuits that may result from personnel injury/death, public exposures, and/or environmental contamination will require a serious expenditure of resources.

Media attention surrounding an injury, death, or environmental damage can also result in a loss of profits and loss of current as well as future business.

Recommended Risk-Reduction Measures

Company attorneys, safety and health professionals, and environmental specialists should be involved in the development of any procedures for responding to chemical incidents.

A company official should be pre-designated as a public relations officer with specific training in dealing with the press. Corporate plans and policies should be developed, approved, and implemented long before any need for such arises.

REFERENCES

Ahlbom, A. 1993. *Biostatistics for Epidemiologists.* New York: CRC Press/Lewis Publishers.

American Conference of Governmental Industrial Hygienists. 1988. *Documentation of the Threshold Limit Values and Biological Exposure Indices,* 5th Edition (with updates). Cincinnati: ACGIH

Asante-Duah, D. K. 1993. *Hazardous Waste Risk Assessment.* New York: CRC Press/Lewis Publishers.

Bloom, A. D. and F. J. DeSerres. 1995. *Ecotoxicity and Human Health.* New York: CRC Press/Lewis Publishers.

Brusick, D. J. 1994. *Methods for Genetic Risk Assessment.* New York: CRC Press/Lewis Publishers.

Bueche, J. R. 1972. *Principles of Physics,* 2nd Edition. New York: McGraw-Hill.

Calabrese, E. J., E. M. Kenyon. *Air Toxics and Risk Assessment.* New York: CRC Press/Lewis Publishers.

Calabrese, E. J. 1994. *Biological Effects of Low Level Exposures.* New York: CRC Press/Lewis Publishers.

Cockerman, L. G., B. S. Shane. 1994. *Basic Environmental Toxicology.* New York: CRC Press/Lewis Publishers.

Cothern, R. R. 1993. *Comparative Environmental Risk Assessment.* New York: CRC Press/Lewis Publishers.

Cothern, C. R. 1995. *Handbook for Environmental Risk Decisionmaking.* New York: CRC Press/Lewis Publishers.

Gardner, R. W. 1994. *Chemical Intolerance.* New York: CRC Press/Lewis Publishers.

Hallenbeck, W. H. 1993. *Quantitative Risk Assessment for Environmental and Occupational Health,* 2nd Edition. New York: CRC Press/Lewis Publishers.

Howard, P. H. 1995. *Handbook of Environmental Fate and Exposures Data for Organic Chemicals,* Vol. I-IV. New York: CRC Press/Lewis Publishers.

Howard, P. H., et. al. 1991. *Handbook of Environmental Degradation Rates.* New York: CRC Press/Lewis Publishers.

Jones, F. E. 1994. *Toxic Organic Vapors in the Workplace.* New York: CRC Press/Lewis Publishers.

Keith, L. H. and M. M. Walker. 1995. *Handbook of Air Toxics.* New York: CRC Press/Lewis Publishers.

Landis, W. G., M. H. Yu. 1995. *Introduction to Environmental Toxicology.* New York: CRC Press/Lewis Publishers.

Langerman, N. 1994 *Precautionary Labels for Chemical Containers.* New York: CRC Press/Lewis Publishers.

Lauwerys, R. R., P. Hoet. 1993. *Industrial Chemical Exposure.* New York: CRC Press/Lewis Publishers.

Lewis, R. J., Sr. 1993. *Hawley's Condensed Chemical Dictionary,* 12th Edition. New York: Van Nostrand Reinhold.

Lewis, R. J., Sr. 1992. *Sax's Dangerous Properties of Industrial Materials,* Volumes 1, 2, and 3, 8th Edition. New York: Van Nostrand Reinhold.

Lide, D. R. 1995. *Handbook of Chemistry and Physics,* 76th Edition. New York: CRC Press/Lewis Publishers.

Manahan, S. E. 1994. *Environmental Chemistry,* 6th Edition. New York: CRC Press/Lewis Publishers.

Mellström, G. A., J. E. Wahlberg, and H. I. Maibach. 1994. *Protective Gloves for Occupational Use.* New York: CRC Press/Lewis Publishers.

Meyer, E. 1990. *Chemistry of Hazardous Materials.* Englewood Cliffs, NJ: Prentice Hall.

Mycock, J. C., J. D. McKenna, and L. Theodore. 1995. *Handbook of Air Pollution Control Engineering and Technology.* New York: CRC Press/Lewis Publishers.

Ott, W. R. 1995. *Environmental Statistics and Data Analysis.* New York: CRC Press/Lewis Publishers.

Patnaik, P. 1992. *A Comprehensive Guide to the Hazardous Properties of Chemical Substances.* New York: Van Nostrand Reinhold.

Philp, R. B. 1995. *Environmental Hazards and Human Health.* New York: CRC Press/Lewis Publishers.

Proctor, N. H., J. P. Hughes. 1991. *Chemical Hazards of the Workplace.* New York: Van Nostrand Reinhold.

Quigley, D. R. 1994. *Handbook of Emergency Chemical Management.* New York: CRC Press/Lewis Publishers.

Rea, W. J. 1995. *Chemical Sensitivity,* Vol. I-III. New York: CRC Press/Lewis Publishers.

Richardson, M. 1995. *Dictionary of Substance and Their Effects,* Vol. I-VI. New York: CRC Press/Lewis Publishers.

Rylander, R., R. R. Jacobs. 1994. *Organic Dusts.* New York: CRC Press/Lewis Publishers.

Sacarello, H. L. A. 1994. *The Comprehensive Handbook of Hazardous Materials.* New York: CRC Press/Lewis Publishers.

Sheftel, V. O. 1995. *Handbook of Toxic Properties of Monomers and Additives.* New York: CRC Press/Lewis Publishers.

Talbott, E., G. Craun. 1995. *An Introduction to Environmental Epidemiology.* New York: CRC Press/Lewis Publishers.

Torkelson, T. R. V. K. Rowe. 1981. *Patty's Industrial Hygiene and Toxicology,* Vol. 2B. New York: Wiley-Interscience.

Witorsch, P. and S. V. Spagnolo. 1994. *Air Pollution and Lung Disease in Adults.* New York: CRC Press/Lewis Publishers.

MATERIAL SAFETY DATA SHEET

CHEMICAL NAME	CRC PRESS/LEWIS PUBLISHERS DISCLAIMER:
BENZENE	THE INFORMATION AND RECOMMENDATIONS PRESENTED HEREIN ARE BASED ON SOURCES BELIEVED TO BE RELIABLE. CRC MAKES NO REPRESENTATION ON ITS COMPLETENESS OR ACCURACY. IT IS THE RESPONSIBILITY OF THE USER TO DETERMINE THE CHEMICAL'S SUITABILITY FOR ITS INTENDED USE, THE CHEMICAL'S SAFE USE, AND THE CHEMICAL'S PROPER DISPOSAL. NO REPRESENTATIONS AND/OR WARRANTIES, EITHER EXPRESSED OR IMPLIED, OF THE MERCHANTABILITY OR FITNESS FOR A PARTICULAR PURPOSE, OR OF ANY OTHER NATURE, ARE MADE WITH RESPECT TO THE INFORMATION PROVIDED IN THIS MSDS OR TO THE CHEMICAL TO WHICH INFORMATION MAY REFER. CRC NEITHER ASSUMES NOR AUTHORIZES ANY OTHER PERSON TO ASSUME FOR IT, ANY OTHER ADDITIONAL RESPONSIBILITY OR LIABILITY FOR THE USE OF, OR RELIANCE UPON, THIS INFORMATION.

HAZARD WARNING INFORMATION

HEALTH	FIRE	REACTIVE	OTHER	DEGREE OF HAZARD	COLOR CODING	OTHER CODES
2	3	0		0 = Minimum Hazard 1 = Slight Hazard 2 = Moderate Hazard 3 = Serious Hazard 4 = Severe Hazard	HEALTH = BLUE FIRE = RED REACTIVITY = YELLOW OTHER = WHITE	OX = Oxidizer ACID = Acid ALK = Alkali COR = Corrosive ~~W~~ = Use No Water

SECTION I - GENERAL INFORMATION

Characterization	**Aromatic Hydrocarbon**	RCRA Number	**U019**
		EPA Class	**Toxic Waste**
DOT Proper Shipping Name	**Benzene (Benzol)**	Chemical Abstract Service (CAS) Number	**71-43-2**
DOT Hazard Class and Label Requirements	**Flammable Liquid**	DOT Emergency Guide Code	**27**
DOT Identification Number	**UN 1114**	Molecular Formula	**C_6H_6**

Synonyms

Benzol; benzole; phenyl hydride; bicarburet of hydrogen; carbon oil; coal naphtha; cyclohexatriene; pyrobenzole; benzolene.

SECTION II - HAZARDOUS INGREDIENTS AND IDENTITY INFORMATION

Hazardous Components (specific identity)	OSHA Exposure Criteria	NIOSH Exposure Criteria	Immediately Dangerous to Life and Health (IDLH)	ACGIH Exposure Criteria
Benzene (derivation: Hydrodealkylation of toluene or pyrolysis of gasoline; transalkylation of toluene by disproportionation reaction; catalytic reforming of petroleum; fractional distillation of coal tar). **1 ppm = 3.25 mg/m^3**	**PEL:** 1 ppm 3.25 mg/m^3 **STEL (10 min):** 5 ppm 16.25 mg/m^3	**REL:** 0.1 ppm .325 mg/m^3 **STEL (60 min):** 1 ppm 3.25 mg/m^3	**500 ppm Human Carcinogen**	**TLV (proposed):** 0.1 ppm .325 mg/m^3 **STEL:** 10 ppm 30 mg/m^3

SECTION III - PHYSICAL AND CHEMICAL CHARACTERISTICS

Boiling Point	**176°F (80°C)**	Specific Gravity (H_2O = 1)	**0.88**
Vapor Pressure (mm Hg)	**74.6 at 69°F (20°C)**	Molecular Weight (atomic weight)	**78.1**
Vapor Density (Air = 1)	**2.77**	Melting Point	**43°F (6°C)**

Solubility

Negligible solubility in water (less than 1%). Miscible with alcohol; ether; acetone; carbon tetrachloride, carbon disulfide; acetic acid.

Appearance and Odor

Colorless to light yellow, mobile, nonpolar liquid of highly refractive nature; characteristic aromatic odor. Odor threshold = 12 ppm.

SECTION IV - FIRE AND EXPLOSION HAZARD DATA

Flash Point (method used)	**12°F (-11°C) closed cup**	Explosive Limits in Air % by Volume	**LEL: 1.2% UEL: 7.8%**
NFPA Classification	**Class 1B Flammable Liquid**	Autoignition Temperature	**1044°F (562°C)**

Extinguishing Media

Use alcohol foam, dry chemical, or carbon dioxide (water may be ineffective).

Special Fire Fighting Procedures

Wear self-contained breathing apparatus (SCBA). Poisonous gases are produced in fire.

Unusual Fire and Explosion Hazards

Moderate explosion hazard when exposed to heat or flame. Containers may explode in fire. The vapor is heavier than air and may travel for some distance to cause fire in flashback reaction to source.

SECTION V - REACTIVITY DATA

Stability		Conditions to Avoid Heat, flame, other sources of ignition, and contact with incompatible materials and chemicals.
Stable	Unstable	Incompatibility *(materials to avoid)* Strong oxidizers, mainly fluorides and perchlorates, and strong acids such as sulfuric and nitric acid.
X		
Hazardous Polymerization		Conditions to Avoid Hazardous polymerization has not been reported to occur under normal conditions of temperature and pressure.
May Occur	Will Not Occur	Hazardous Decomposition or By-products Poisonous gases, including carbon monoxide and carbon dioxide are emitted during fire.
	X	

SECTION VI - HEALTH HAZARD DATA

Primary Route(s) of Entry:	Inhalation?	Absorption (skin/eye)?	Ingestion?
	X	X	X

Health Hazards

INHALATION: May cause nausea, headache, vomiting, dizziness, narcosis, suffocation, lower blood pressure, CNS depression. Vapors may cause severe irritation or burns of the respiratory system, pulmonary edema, or lung inflammation.

ABSORPTION: Irritation with possible redness and pain to the skin and eyes. Also, absorbs through the skin and may cause dermatitis. Eye contact may result in temporary corneal damage.

INGESTION: May cause nausea, vomiting, headache, dizziness, gastrointestinal irritation; blurred vision, lowering of blood pressure.

Carcinogenicity	NTP Listed?	IARC Cancer Review Group?	OSHA Regulated?	Target Organs?
Confirmed Human Confirmed Animal	5th Annual Report	Group 1	29 CFR 1910.1028	Eyes, skin, respiratory sys. blood, CNS, bone marrow (leukemia).

Medical Conditions Generally Aggravated by Exposure

None identified.

Emergency and First-aid Procedures

Eye contact: Do not allow victim to keep eyes tightly shut. Flush immediately with water for 15 minutes (minimum), seek medical attention. Skin contact: Remove all contaminated clothing, including shoes. Immediately wash area with flooding amounts of soap and water. Seek medical attention immediately. For inhalation: Remove the person from exposure. Provide respiratory assistance and CPR. Transfer to medical facility. If swallowed: Seek medical attention immediately. Do not induce vomiting.

SECTION VII - PRECAUTIONS FOR SAFE HANDLING AND USE

Steps to be Taken in Case Material is Released or Spilled

Remove all ignition sources. Wear SCBA respiratory protection. Restrict those not wearing protective equipment and who are not involved in cleanup from entering area. Avoid skin contact. Stop leak if it can be done without risk. Ventilate area of spill. Absorb with vermiculite or other similar material and deposit in sealed containers.

Preferred Waste Disposal Method

Burn in chemical incinerator equipped with an afterburner and scrubber.

Precautions to be Taken in Handling and Storage

Store in tightly closed containers in cool, well ventilated location. Keep away from sources of heat.

Other Precautions and Warnings

Sources of ignition such as open flame or smoking, is prohibited where benzene is used or stored. Metal containers should be bonded and grounded during transfer operations. Bulk storage not recommended.

SECTION VIII - CONTROL MEASURES AND PERSONAL PROTECTIVE EQUIPMENT

Respiratory Protection *(specify type)*

Use a MSHA/NIOSH approved powered air purifying respirator or use a self-contained breathing apparatus (SCBA) with full facepiece operated in positive pressure mode.

Ventilation

Local exhaust (preferred) at site of chemical work or general ventilation.

Protective Gloves	Eye Protection	Other Protective Clothing
Polyvinyl Alcohol Gloves	Chemical Goggles or Face Mask	Protective Apron if Splash is Likely

Work/Hygiene Practices

Always wash hands thoroughly after using chemical; never bring food, drink, or smoking materials into vicinity of chemicals.

BENZENE

C_6H_6 CAS: 71-43-2

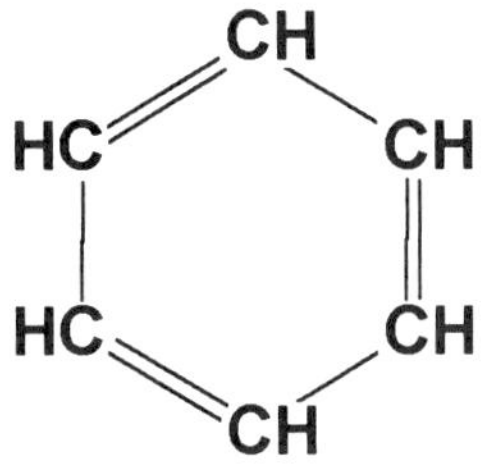

IDENTIFICATION AND TYPICAL USES

Benzene is a clear, colorless to light yellow, mobile, nonpolar liquid of highly refractive nature with a characteristic hydrocarbon odor. Benzene occurs in coal and coal-tar distillation products and in petroleum products such as gasoline.

It is used as a solvent for waxes, resins, and oils; in the manufacture of dyes, pharmaceuticals, varnishes, and linoleum; and as a raw material to produce a number of organic compounds. It is also used in the manufacture of other chemicals such as ethylbenzene (for styrene monomer), dodecylbenzene (for detergent), cyclohexane (for nylon), phenol, nitrobenzene (for aniline), maleic anhydride, chlorobenzene, diphenyl, benzene hexachloride, and benzene-sulfonic acid.

RISK ASSESSMENT: HEALTH

General Assessment

Benzene is an acute as well as chronic toxicant. It is a confirmed human carcinogen producing myeloid leukemia, Hodgkin's disease, and lymphomas by inhalation. It is also believed to be a human poison by skin contact. It is moderately toxic by ingestion. Benzene can severely irritate the eyes, nose, and respiratory system and a cause moderate irritation on skin contact. Chronic effects are much more severe than its acute toxicity. The target organs to acute and chronic poisoning are the blood, bone marrow, central nervous system, respiratory system, eyes, and skin.

Human systemic effects by *inhalation* and *ingestion* include blood changes and increased body temperature. Heavy occupational exposures to benzene can cause bone marrow depression and anemia. In rare cases, leukemia has been attributed to benzene exposure. Lung cancer has also been reported.

Other effects of inhalation include headaches, nausea, vomiting, dizziness, narcosis, suffocation, CNS depression, pulmonary edema or lung inflammation, blurring of vision, convulsions, weakness, loss of appetite, abdominal discomfort, and dyspnea on exertion.

Skin contact may cause dermatitis and absorption can occur with harmful effects. It is a comparatively strong irritant, producing erythema, and burning, and, in severe cases, edema and even blistering.

Ingestion may cause nausea, vomiting, headaches, dizziness, gastrointestinal irritation, blurred vision, and lowering of blood pressure.

Acute Health Effects

The following acute (short-term) health effects may occur immediately or shortly after exposure to benzene:

Skin: Irritation with potential for redness, blistering, and burning.

Eye: Severe irritation; pain and permanent damage possible.

Lung: Nose, throat, and respiratory tract irritation causing difficulty in breathing, possible pulmonary edema on high exposure (a medical emergency) which can be fatal.

CNS: Confusion, dizziness, tightening of the leg muscles, cranial pressure (headaches), possible coma.

Chronic Health Effects

The following chronic (long-term) health effects may occur at some time after exposure to benzene and can last for months or even years:

Cancer Hazards: Benzene is a known carcinogen in humans. It has been shown to cause leukemia, Hodgkin's disease, and lymphoma by inhalation. Human mutation data have also been reported.

Reproductive Hazard: There is limited evidence that benzene is a teratogen in animals. Until further tests are performed, it should be treated as a possible human teratogen.

Other Chronic Effects: Heavy exposures over a long period of time can cause bone marrow depression with symptoms of anemia with a tendency for hemorraghic symptoms such as easy bruising, epistaxis, bleeding from the gums, nosebleeds, menorrhagia, petachiae, purpora, and other symptoms of fatigue, headache, dizziness, nausea and loss of appetite, loss of weight and weakness, pallor. Repeated exposures can cause irreversible damage to the blood-forming organs resulting in aplastic anemia and possibly leukemia. There is great variation in individual response to chronic exposure to benzene.

Recommended Risk-Reduction Measures

Personnel should avoid direct contact with benzene. Occupational poisoning occurs most commonly by inhalation. If a less toxic material or compound cannot be substituted for benzene, then *engineering controls* are the most effective method of reducing exposures. The best protection is to enclose operations and/or provide local exhaust ventilation at the site of benzene release. While not always operationally feasible, isolating operations can also reduce exposure. Using respiratory protection is less effective than the controls mentioned above, but is still advisable whenever working with or around benzene. A MSHA/NIOSH approved self-contained breathing apparatus (SCBA) with full facepiece operated in pressure demand is the appropriate level of respiratory protection when exposure to benzene is anticipated. If a full facepiece is not available, then chemical goggles should be worn to protect the eyes. A face shield should also be considered. To prevent hand and skin exposures, polyvinyl alcohol gloves should be worn.

Administrative controls should also be in place to minimize the potential for human exposures. These may include written procedures or policies which specify the methods and techniques that will be practiced whenever personnel are to work with benzene.

All personnel should receive *training* on the use, hazards, protective measures, emergency actions, and other precautions per 29 CFR 1910.1200 (Hazard Communication), prior to the first assignment in an area where benzene is used or stored.

Before beginning employment and at regular times thereafter, the following recommended medical tests should be considered:

- ☑ Complete blood count (CBC).
- ☑ Urinary phenol (a test to see if benzene is in the body).

Any evaluation should include a careful history of past and present symptoms with an examination. Medical tests that simply look for existing damage are not a substitute for controlling exposures.

Other methods to reduce exposure include:

- ☑ Where possible, enclose operations and use local exhaust ventilation at the site of chemical release. If local exhaust ventilation or enclosure is not used, respiratory protection should be mandatory.
- ☑ Always ensure that proper protective clothing is worn when using chemical substances.
- ☑ Wash thoroughly immediately after exposure to benzene and at the end of the work shift or before eating, drinking, or smoking.
- ☑ Hazard warning information should be posted in the work area. In addition, as part of an on-going education and training program, all information on the health and safety hazards of benzene should be communicated to all potentially exposed workers.
- ☑ Eye wash stations should be provided in the immediate work area for emergency use. If there is a possibility of skin exposure to benzene, emergency shower facilities should also be provided.
- ☑ Workers whose clothing has been contaminated by benzene should change into clean clothes before leaving work. Contaminated work clothing should be laundered only by individuals who have been informed of the hazards of exposure to benzene.
- ☑ Where possible, automatically transfer liquids containing benzene from drums or other containers to process containers.

RISK ASSESSMENT: ENVIRONMENT

General Assessment

The environment is at risk of exposure during transportation, storage, disposal, or destruction of benzene. In almost every scenario, the threat of environmental exposure is contingent upon the proper handling of the chemical substance. Accidental spills, large or small, can result in contamination of the surrounding environmental mediums (water, soil, and air).

Benzene is a moderately flammable liquid. Because of its low flash point and high boiling point, it is classified as a Class 1B flammable liquid (per OSHA 29 CFR 1910.106). It is a dangerous fire hazard and moderate explosion hazard when exposed to heat or flame. Caution is always required in handling, storage, transportation, and disposal of benzene. Emergency responders should be made aware of the presence of benzene at any emergency response situation.

Benzene can enter the environment from industrial effluents and from spills.

Acute Ecological Effects

Acute (short-term) toxic effects may include the death of animals, birds, or fish and death or low growth rate in plants. Acute effects are seen 2 to 4 days after animals or plants are exposed to benzene.

Benzene has high acute toxicity to aquatic life. It can cause death in plants and roots and membrane damage in leaves of various agricultural crops. Insufficient data are available to evaluate the short-term effects of benzene on birds or land animals.

Chronic Ecological Effects

Chronic (long-term) toxic effects may include shortened life span, reproductive problems, lower fertility, and changes in appearance or behavior in exposed animals. These effects can be seen long after first exposure(s) to toxic chemicals.

Benzene has high chronic toxicity to aquatic life. Insufficient data are available to evaluate or predict the long-term effects of benzene to plants, birds, or terrestrial animals.

Water Solubility

Benzene is moderately soluble in water. Concentrations between 1 and 1000 milligrams and more will mix with a liter of water.

Persistence in the Environment

Benzene is slightly persistent in water, with a half-life of between 2 to 20 days. The half-life of a pollutant is the amount of time it takes for one half of the chemical to be degraded. About 99.5% of benzene will eventually end up in air; the rest will end up in water.

Bioaccumulation in Aquatic Organisms

Some substances increase in concentration, or *bioaccumulate,* in living organisms as they breathe contaminated air, drink contaminated water, or eat contaminated food. These chemicals can become concentrated in the tissues and internal organs of animals as well as humans.

The concentration of benzene found in fish tissues is expected to be somewhat higher than the average concentration of benzene in the water from which the fish was taken.

Recommended Risk-Reduction Measures

Proper training of all transporters will reduce the likelihood of a mishap or accident resulting in a leak or spill of benzene into the environment. Labels on all containers, trucks, and rail cars must meet DOT requirements and accurately reflect their contents to enable emergency responders to react properly and quickly to any disaster thereby reducing the potential risk to the environment and to personnel.

Storage of benzene should be segregated from other chemicals and materials to minimize the risk of cross-contamination.

If a spill or leak to the environment has occurred, fire department, emergency response and/or hazardous materials spill personnel should be notified immediately. Cleanup should be attempted only by those trained in proper spill containment procedures. Contaminated soils should be removed for incineration and replaced with clean soil. If benzene should contact the water table, aquifer, or navigable waterway, time is of the essence. It is moderately soluble and total remediation may not be possible. The local and/or state emergency response authorities must be notified. A comprehensive emergency response or disaster preparedness/recovery plan should be in place prior to any operations involving the use, transportation, storage, or disposal of benzene.

For disposal, benzene may be burned in a chemical incinerator equipped with an afterburner and scrubber.

If benzene is spilled or leaked, the following specific steps are recommended:

- ☑ Restrict persons not wearing protective clothing from area of spill or leak until cleanup is complete and area can be opened for normal work.
- ☑ Absorb liquids in vermiculite, sand, earth, or similar material and deposit in sealed drum.

- ☑ Ventilate area of spill or leak. Container should be removed to safer area if it can be done without increasing the risk.
- ☑ Use water spray to knock-down vapors.
- ☑ It may be necessary to dispose of benzene as a hazardous waste. The responsible state agency or the regional office of the federal Environmental Protection Agency (EPA) should be contacted for specific recommendations.

RISK ASSESSMENT: BUSINESS

General Assessment

Accidents or mishaps involving benzene can present a serious threat to business operations. The loss or damage of equipment or facilities can significantly affect fiscal viability. Lawsuits that may result from personnel illness (leukemia), injury/death, public exposures, and/or environmental contamination will require a serious expenditure of resources. Media attention surrounding an injury, death, or environmental damage can also result in a loss of profits and loss of current as well as future business.

Recommended Risk-Reduction Measures

Company attorneys, safety and health professionals, and environmental specialists should be involved in the development of any procedures for responding to chemical incidents. A company official should be predesignated as a public relations officer with specific training in dealing with the press. Corporate plans and policies should be developed, approved, and implemented long before any need for such arises.

REFERENCES

American Conference of Governmental Industrial Hygienists. 1988. *Documentation of the Threshold Limit Values and Biological Exposure Indices,* 5th Edition (with updates). Cincinnati: ACGIH

Bloom, A. D. and F. J. DeSerres. 1995. *Ecotoxicity and Human Health.* New York: CRC Press/Lewis Publishers.

Brusick, D. J. 1994. *Methods for Genetic Risk Assessment.* New York: CRC Press/Lewis Publishers.

Bueche, J. R. 1972. *Principles of Physics,* 2nd Edition. New York: McGraw-Hill.

Cockerman, L. G., B. S. Shane. 1994. *Basic Environmental Toxicology.* New York: CRC Press/Lewis Publishers.

Gerarde, H. W. 1960. *Toxicology and Biochemistry of Aromatic Hydrocarbons.* New York: Elsevier.

Hallenbeck, W. H. 1993. *Quantitative Risk Assessment for Environmental and Occupational Health,* 2nd Edition. New York: CRC Press/Lewis Publishers.

Howard, P. H., et. al. 1991. *Handbook of Environmental Degradation Rates.* New York: CRC Press/Lewis Publishers.

Jones, F. E. 1994. *Toxic Organic Vapors in the Workplace.* New York: CRC Press/Lewis Publishers.

Keith, L. H. and M. M. Walker. 1995. *Handbook of Air Toxics.* New York: CRC Press/Lewis Publishers.

Landis, W. G., M. H. Yu. 1995. *Introduction to Environmental Toxicology.* New York: CRC Press/Lewis Publishers.

Langerman, N. 1994 *Precautionary Labels for Chemical Containers.* New York: CRC Press/Lewis Publishers.

Lauwerys, R. R., P. Hoet. 1993. *Industrial Chemical Exposure.* New York: CRC Press/Lewis Publishers.

Lewis, R. J., Sr. 1993. *Hawley's Condensed Chemical Dictionary,* 12th Edition. New York: Van Nostrand Reinhold.

Lewis, R. J., Sr. 1992. *Sax's Dangerous Properties of Industrial Materials,* Volumes 1, 2, and 3, 8th Edition. New York: Van Nostrand Reinhold.

Manahan, S. E. 1994. *Environmental Chemistry,* 6th Edition. New York: CRC Press/Lewis Publishers.

Mellström, G. A., J. E. Wahlberg, and H. I. Maibach. 1994. *Protective Gloves for Occupational Use.* New York: CRC Press/Lewis Publishers.

Meyer, E. 1990. *Chemistry of Hazardous Materials.* Englewood Cliffs, NJ: Prentice Hall.

Ott, W. R. 1995. *Environmental Statistics and Data Analysis.* New York: CRC Press/Lewis Publishers.

Patnaik, P. 1992. *A Comprehensive Guide to the Hazardous Properties of Chemical Substances.* New York: Van Nostrand Reinhold.

Philp, R. B. 1995. *Environmental Hazards and Human Health.* New York: CRC Press/Lewis Publishers.

Proctor, N. H., J. P. Hughes. 1991. *Chemical Hazards of the Workplace.* New York: Van Nostrand Reinhold.

Quigley, D. R. 1994. *Handbook of Emergency Chemical Management.* New York: CRC Press/Lewis Publishers.

Sacarello, H. L. A. 1994. *The Comprehensive Handbook of Hazardous Materials.* New York: CRC Press/Lewis Publishers.

Torkelson, T. R. V. K. Rowe. 1981. *Patty's Industrial Hygiene and Toxicology,* Vol. 2B. New York: Wiley-Interscience.

U.S. Department of Health and Human Services, National Institute for Occupational Safety and Health. 1994. *NIOSH Pocket Guide to Chemical Hazards.* Washington, D.C.: U.S. Government Printing Office.

U.S. Department of Labor. 1977. Occupational exposure to benzene. *Federal Register.* 42:22516-22529.

U.S. Department of Labor. Occupational Safety and Health Administration (OSHA). 1987. Occupational exposure to benzene: final rule. *Federal Register.* 52(176):34460.

Yin, S. N., et. al. 1987. Leukemia in benzene workers: A retrospective cohort study. *British Journal of Industrial Medicine.* 44:124-128.

MATERIAL SAFETY DATA SHEET

CHEMICAL NAME	
BENZIDINE	**CRC PRESS/LEWIS PUBLISHERS DISCLAIMER:** THE INFORMATION AND RECOMMENDATIONS PRESENTED HEREIN ARE BASED ON SOURCES BELIEVED TO BE RELIABLE. CRC MAKES NO REPRESENTATION ON ITS COMPLETENESS OR ACCURACY. IT IS THE RESPONSIBILITY OF THE USER TO DETERMINE THE CHEMICAL'S SUITABILITY FOR ITS INTENDED USE, THE CHEMICAL'S SAFE USE, AND THE CHEMICAL'S PROPER DISPOSAL. NO REPRESENTATIONS AND/OR WARRANTIES, EITHER EXPRESSED OR IMPLIED, OF THE MERCHANTABILITY OR FITNESS FOR A PARTICULAR PURPOSE, OR OF ANY OTHER NATURE, ARE MADE WITH RESPECT TO THE INFORMATION PROVIDED IN THIS MSDS OR TO THE CHEMICAL TO WHICH INFORMATION MAY REFER. CRC NEITHER ASSUMES NOR AUTHORIZES ANY OTHER PERSON TO ASSUME FOR IT, ANY OTHER ADDITIONAL RESPONSIBILITY OR LIABILITY FOR THE USE OF, OR RELIANCE UPON, THIS INFORMATION.

HAZARD WARNING INFORMATION

HEALTH	FIRE	REACTIVE	OTHER	DEGREE OF HAZARD	COLOR CODING	OTHER CODES
4	0	0		0 = Minimum Hazard 1 = Slight Hazard 2 = Moderate Hazard 3 = Serious Hazard 4 = Severe Hazard	HEALTH = BLUE FIRE = RED REACTIVITY = YELLOW OTHER = WHITE	OX = Oxidizer ACID = Acid ALK = Alkali COR = Corrosive ~~W~~ = Use No Water

SECTION I - GENERAL INFORMATION

Characterization	RCRA Number	EPA Class
Aromatic Amine	**U021**	**Characteristic Toxic Waste**

DOT Proper Shipping Name **Benzidine**	Chemical Abstract Service (CAS) Number **92-87-5**
DOT Hazard Class and Label Requirements **Poison B; Poison**	DOT Emergency Guide Code **53**
DOT Identification Number **UN 1885**	Molecular Formula **$NH_2(C_6H_4)_2NH_6$**

Synonyms

4,4'-biphenyldiamine; 4,4'-diamino-1,1'-biphenyl; 4,4'-diphenylenediamine; *p,p'*-dianiline; *p,p'*-bianiline; 4,4'-bianiline.

SECTION II - HAZARDOUS INGREDIENTS AND IDENTITY INFORMATION

Hazardous Components (specific identity)	OSHA Exposure Criteria	NIOSH Exposure Criteria	Immediately Dangerous to Life and Health (IDLH)	ACGIH Exposure Criteria
Benzidine (derivation: By reducing nitrobenzidine with zinc dust in alkaline solution followed by distillation; by electrolysis of nitrobenzidine, followed by distillation; by nitration of diphenyl followed by reduction of the product with zinc dust in alkaline solution with subsequent distillation).	**PEL:** Cancer Suspect Agent	**REL:** Confirmed Human Carcinogen	**Not Determined (Carcinogen)**	**TLV:** Confirmed Human Carcinogen

SECTION III - PHYSICAL AND CHEMICAL CHARACTERISTICS

Boiling Point **752°F (400°C)**	Specific Gravity (H_2O = 1) **1.25**
Vapor Pressure (mm Hg) **Low**	Molecular Weight (atomic weight) **184.3**
Vapor Density (Air = 1) **Not Found**	Melting Point **239°F (120°C)**

Solubility

Slightly soluble in cold water (0.04% at 54°F); moderately soluble in hot water, alcohol, and ether.

Appearance and Odor

Grayish-yellow crystalline powder, or white to slightly reddish crystals (sand or sugar-like), powder, or leaf-like solid.

SECTION IV - FIRE AND EXPLOSION HAZARD DATA

Flash Point (method used) **Not Found**	Explosive Limits in Air % by Volume **LEL: Not Found UEL: Not Found**
NFPA Classification **Combustible Solid**	Autoignition Temperature **Not Found**

Extinguishing Media

Benzidine by itself is difficult to burn. Extinguish fire using an agent suitable for surrounding fires.

Special Fire Fighting Procedures

Wear full protective clothing and self-contained breathing apparatus (SCBA). Poisonous gases are produced in fire. Move container from fire area if it can be done without risk.

Unusual Fire and Explosion Hazards

Benzidine is a combustible solid. No unusual fire or explosion hazards have been reported. However, firefighters must be aware that exposure to any level can be extremely hazardous.

SECTION V - REACTIVITY DATA

Stability		Conditions to Avoid **None reported.**
Stable X	Unstable	Incompatibility *(materials to avoid)* **Reactions on contact with red fuming nitric acid may be violent.**
Hazardous Polymerization		Conditions to Avoid **Hazardous polymerization has not been reported to occur under normal conditions of temperature and pressure.**
May Occur	Will Not Occur X	Hazardous Decomposition or By-products **When heated to decomposition, benzidine emits highly toxic fumes of nitrous oxide.**

SECTION VI - HEALTH HAZARD DATA

Primary Route(s) of Entry:	Inhalation?	Absorption (skin/eye)?	Ingestion?
	X	X	X

Health Hazards

INHALATION: May cause irritation of the eyes, nose, and throat. Benzidine is a confirmed human carcinogen (bladder cancer).

ABSORPTION: Possible skin irritation, rash, and dermatitis. Can pass through skin and present cancer hazard in humans. May cause tumors on the skin as well.

INGESTION: May cause nausea, vomiting, kidney and liver damage, lung tumors, and may cause damage to the blood including hemolysis and bone marrow depression.

Carcinogenicity	NTP Listed?	IARC Cancer Review Group?	OSHA Regulated?	Target Organs?
Confirmed Human **Confirmed Animal**	**5th Annual Report**	**Group 1**	**29 CFR 1910.1010**	**Bladder, skin, kidney, liver, blood, bone marrow.**

Medical Conditions Generally Aggravated by Exposure

None identified.

Emergency and First-aid Procedures

Eye contact: Do not allow victim to keep eyes tightly shut. Flush immediately with water for 15 minutes (minimum), seek medical attention. Skin contact: Remove all contaminated clothing, including shoes. Immediately wash area with flooding amounts of soap and water. Seek medical attention immediately. For inhalation: Remove the person from exposure. Provide respiratory assistance and CPR. Transfer to medical facility. If swallowed: Seek medical attention immediately. Never give anything by mouth to an unconscious or convulsing person.

SECTION VII - PRECAUTIONS FOR SAFE HANDLING AND USE

Steps to be Taken in Case Material is Released or Spilled

Wear self-contained breathing apparatus (SCBA) for respiratory protection. Restrict those not wearing protective equipment and who are not involved in cleanup from entering area. Avoid skin contact. Ventilate area of spill. Do not dry sweep (generates airborne dusts); collect powdered materials in most efficient and safest manner possible. Recommend use of vacuum equipped with HEPA filter.

Preferred Waste Disposal Method

Dissolve in a combustible solvent, burn in chemical incinerator equipped with afterburner and scrubber.

Precautions to be Taken in Handling and Storage

Benzidine should be handled and stored only in a regulated area, as required by OSHA 29 CFR 1910.1010.

Other Precautions and Warnings

Prior to working with benzidine, the law requires personnel to be trained on its proper handling and storage requirements. Bulk storage not recommended.

SECTION VIII - CONTROL MEASURES AND PERSONAL PROTECTIVE EQUIPMENT

Respiratory Protection *(specify type)*

At any exposure to a carcinogen, use a MSHA/NIOSH approved powered air purifying respirator or use a self-contained breathing apparatus (SCBA) with full facepiece operated in positive pressure mode.

Ventilation

Local exhaust (preferred) at site of chemical work or general ventilation.

Protective Gloves	Eye Protection	Other Protective Clothing
Impervious Gloves	**Chemical/Dust Goggles or Face Mask**	**Skin Protection**

Work/Hygiene Practices

Always wash hands thoroughly after using chemical; never bring food, drink, or smoking materials into vicinity of chemicals.

BENZIDINE

$NH_2(C_6H_4)_2NH_2$ CAS: 92-87-5

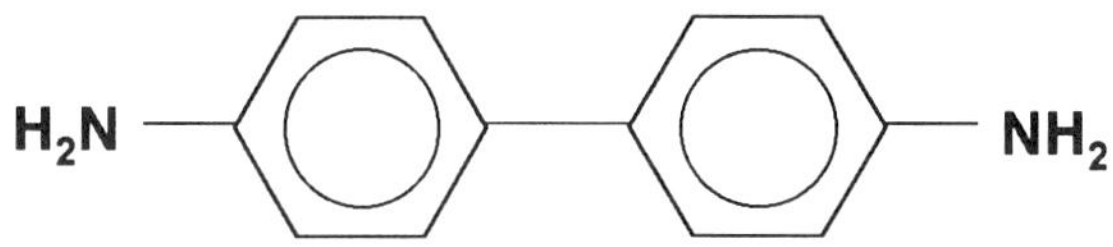

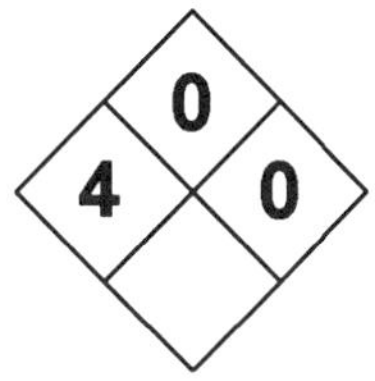

IDENTIFICATION AND TYPICAL USES

Benzidine is a grayish-yellow or white to reddish crystalline solid (powder or dust). It has been used extensively in the manufacture of dyes. Because of its carcinogenic effects on humans, its application has been extremely curtailed. Other uses of its compound are in chemical analysis, as a reagent for the determination of hydrogen peroxide in milk, and in the analysis of nicotine. Its hydrochloride is used as a reagent to analyze metals and sulfate. Benzidine has been used as a stiffening agent in rubber compounding. In microscopy, it is used as a stain or to aid in the detection of blood stains.

RISK ASSESSMENT: HEALTH

General Assessment

Benzidine is known to cause cancer of the bladder in humans. There are numerous reports in the references on its carcinogenicity in both humans and animals. Human mutation data have also been reported. The routes of entry into the body are primarily the *inhalation* of its dusts and *absorption* through the skin. However, its physical properties (dust, powder) also make *ingestion* an extremely possible exposure route. Any exposure to benzidine by any route should be considered extremely hazardous.

Inhalation causes irritation to the eyes, nose, throat, and respiratory tract. Once in the body, it can cause damage to the blood including hemolysis and bone marrow depression.

Skin contact may cause dermatitis and absorption can occur with harmful effects to the blood, blood-forming organs, the liver, the kidneys, and the bladder. Dermatitis may occur and there may also be the presence of skin tumors subsequent to prolonged or repeated contact.

Ingestion may cause nausea, vomiting, kidney, and liver damage. There can also be adverse effects on the blood, including bone marrow changes.

Acute Health Effects

The following acute (short-term) health effects may occur immediately or shortly after exposure to benzidine:

Skin: Irritation with potential for dermatitis and even skin tumors.

Eye: Severe irritation; pain and permanent damage possible.

Lung: Nose, throat, and respiratory tract irritation causing difficulty in breathing.

Other: Heavy exposures can cause bone marrow depression and anemia with a tendency for hemorraghic symptoms such as easy bruising, epistaxis, bleeding from the gums, or menorrhagia, and leukemia. The skin and mucous membranes may appear pale.

Chronic Health Effects

The following chronic (long-term) health effects may occur at some time after exposure to benzidine and can last for months or even years:

Cancer Hazards: Benzidine is a known carcinogen in humans. It has been shown to cause bladder cancer. Other tumorigenic effects suspect (liver and kidney).

Reproductive Hazard: According to the information available in the references, benzidine has not been tested for its ability to adversely affect reproduction.

Other Chronic Effects: Benzidine may cause skin allergy (sensitization). Once sensitized to benzidine skin contact, even the slightest future exposures can cause itching and skin rash.

ʊ *Recommended Risk-Reduction Measures*

Personnel should avoid direct contact with benzidine. Occupational poisoning can occur through all common exposure routes (inhalation, absorption, and ingestion). If a less toxic material or compound cannot be substituted for benzidine, then *engineering controls* (as required by OSHA 29 CFR 1910.1010) are the most effective method of reducing exposures. The best protection is to enclose operations and/or provide local exhaust ventilation at the site of benzidine dust release. While not always operationally feasible, isolating operations can also reduce exposure. Using respiratory protection is less effective than the controls mentioned above, but is still required whenever working with or around benzidine. A MSHA/NIOSH approved self-contained breathing apparatus (SCBA) with full facepiece and pressure demand is the appropriate level of respiratory protection, when any exposure to benzidine is anticipated. If a full facepiece is not available, then chemical/dust goggles should be worn to protect the eyes. A face shield should also be considered. To prevent hand and skin exposures, polyvinyl alcohol gloves should be worn.

Administrative controls should also be in place to minimize the potential for human exposures. These may include written procedures or policies which specify the methods and techniques that will be practiced whenever personnel are to work with benzidine.

All personnel should receive *training* on the use, hazards, protective measures, emergency actions, and other precautions per 29 CFR 1910.1200 (Hazard Communication) and OSHA 29 CFR 1910.1010 (Benzidine) prior to the first assignment in an area where benzidine is used or stored.

Before beginning employment and at regular times thereafter (e.g., annually), the following recommended medical tests should be considered:

- ☑ Urine cytology (a test for abnormal cells in the urine).

If symptoms develop or overexposure is suspected, the following may be useful:

- ☑ Evaluation by a qualified allergist, including careful evaluation of exposure history and special testing (may help diagnose skin allergy).

It should be noted that medical tests that simply look for existing damage are not a substitute for controlling exposures. Medical histories are extremely important when assessing exposure risk.

Other methods to reduce exposure include:

- ☑ According to OSHA, benzidine should be handled in an enclosed system. There should be local exhaust ventilation at the site of chemical release. Even with local exhaust ventilation or enclosure, respiratory protection must be used.
- ☑ Always ensure that proper protective clothing is worn when using chemical substances.
- ☑ Wash thoroughly immediately after exposure to benzidine and at the end of the work shift or before eating, drinking, or smoking.
- ☑ Hazard warning information should be posted in the work area. In addition, as part of an on-going education and training program, all information on the health and safety hazards of benzidine should be communicated to all exposed and potentially exposed workers.
- ☑ Eye wash stations should be provided in the immediate work area for emergency use. If there is a possibility of skin exposure to benzidine, emergency shower facilities should also be provided.
- ☑ Workers whose clothing has been contaminated by benzidine should change into clean clothes before leaving work. Contaminated work clothing should be laundered only by individuals who have been informed of the hazards of exposure to benzidine.
- ☑ A regulated work area must be established. No unauthorized personnel can enter the area where benzidine is used, handled, or stored.

RISK ASSESSMENT: ENVIRONMENT

General Assessment

The environment is at risk of exposure during transportation, storage, disposal, or destruction of benzidine. In almost every scenario, the threat of environmental exposure is contingent upon the proper handling of the chemical substance. Accidental spills, large or small, can result in contamination of the surrounding environmental mediums (water, soil, and air).

Benzidine is a combustible solid that does not easily burn. However, caution is always required in handling, storage, transportation, and disposal of benzidine. Emergency responders should be made aware of the presence of benzidine at any emergency response situation.

Benzidine can enter the environment from industrial effluents and from spills.

Acute Ecological Effects

Acute (short-term) toxic effects may include the death of animals, birds, or fish and death or low growth rate in plants. Acute effects are seen 2 to 4 days after animals or plants are exposed to benzidine. Benzidine and its salts have moderate acute toxicity to aquatic life. Insufficient data are available to evaluate the short-term effects of benzidine on birds or land animals.

Chronic Ecological Effects

Chronic (long-term) toxic effects may include shortened life span, reproductive problems, lower fertility, and changes in appearance or behavior in exposed animals. These effects can be seen long after first exposure(s) to toxic chemicals. Benzidine and its salts have moderate chronic toxicity to aquatic life. Insufficient data are available to evaluate or predict the long-term effects of benzidine to plants, birds, or terrestrial animals.

Water Solubility

Benzidine and its salts are moderately soluble in water. Concentrations between 1 and 1000 will mix with a liter of water.

Persistence in the Environment

Benzidine is highly persistent in water, with a half-life of greater than 200 days. The half-life of a pollutant is the amount of time it takes for one half of the chemical to be degraded. About 99.3% of benzidine will eventually end up in water; 0.3% will end up in terrestrial soil, aquatic sediments, respectively.

Bioaccumulation in Aquatic Organisms

Some substances increase in concentration, or *bioaccumulate,* in living organisms as they breathe contaminated air, drink contaminated water, or eat contaminated food. These chemicals can become concentrated in the tissues and internal organs of animals as well as humans.

The concentration of benzidine found in fish tissues is expected to be about the same as the average concentration of benzidine in the water from which the fish was taken.

Recommended Risk-Reduction Measures

Proper training of all transporters will reduce the likelihood of a mishap or accident resulting in a leak or spill of benzidine into the environment. Labels on all containers, trucks, and rail cars must meet DOT requirements and accurately reflect their contents to enable emergency responders to react properly and quickly to any disaster thereby reducing the potential risk to the environment and to personnel.

Storage of benzidine should be segregated from other chemicals and materials to minimize the risk of cross-contamination.

If a spill or leak to the environment has occurred, fire department, emergency response and/or hazardous materials spill personnel should be notified immediately. Cleanup should be attempted only by those trained in proper spill containment procedures. Contaminated soils should be removed for incineration and replaced with clean soil. If benzidine should contact the water table, aquifer, or navigable waterway, time is of the essence. It is moderately soluble and total remediation may not be possible. The local and/or state emergency response authorities must be notified. A comprehensive emergency response or disaster preparedness/recovery plan should be in place prior to any operations involving the use, transportation, storage, or disposal of benzidine.

For disposal, benzidine may be dissolved in a combustible solvent and burned in a chemical incinerator equipped with an afterburner and scrubber.

If benzidine is spilled or leaked, the following specific steps are recommended:

- ☑ Restrict persons not wearing protective clothing from area of spill or leak until cleanup is complete.
- ☑ Collect spilled powders in the safest and most convenient manner possible and place in a sealed drum for disposal. Do not dry sweep (generates airborne dusts). Recommend use of a vacuum equipped with a high efficiency particulate air (HEPA) filter.
- ☑ Ventilate area of spill or leak.
- ☑ It may be necessary to dispose of benzidine as a hazardous waste. The responsible state agency or the regional office of the federal Environmental Protection Agency (EPA) should be contacted for specific recommendations.

RISK ASSESSMENT: BUSINESS

General Assessment

Accidents or mishaps involving benzidine can present a serious threat to business operations. The loss or damage of equipment or facilities can significantly affect fiscal viability. Lawsuits that may result from personnel illness (cancer), injury/death, public exposures, and/or environmental contamination will require a serious expenditure of resources. Media attention surrounding an injury, death, or environmental damage can also result in a loss of profits and loss of current as well as future business.

✪ *Recommended Risk-Reduction Measures*

Company attorneys, safety and health professionals, and environmental specialists should be involved in the development of any procedures for responding to chemical incidents. A company official should be predesignated as a public relations officer with specific training in dealing with the press. Corporate plans and policies should be developed, approved, and implemented long before any need for such arises.

REFERENCES

Ahlbom, A. 1993. *Biostatistics for Epidemiologists.* New York: CRC Press/Lewis Publishers.

American Conference of Governmental Industrial Hygienists. 1988. *Documentation of the Threshold Limit Values and Biological Exposure Indices,* 5th Edition (with updates). Cincinnati: ACGIH

Anthony, C. P., and N. J. Kolthoff. 1971. *Textbook of Anatomy and Physiology.* St. Louis: C. V. Mosby Company.

Asante-Duah, D. K. 1993. *Hazardous Waste Risk Assessment.* New York: CRC Press/Lewis Publishers.

Bloom, A. D. and F. J. DeSerres. 1995. *Ecotoxicity and Human Health.* New York: CRC Press/Lewis Publishers.

Brusick, D. J. 1994. *Methods for Generic Risk Assessment.* New York: CRC Press/Lewis Publishers.

Bueche, J. R. 1972. *Principles of Physics,* 2nd Edition. New York: McGraw-Hill.

Burrell, R., D. K. Flaherty, and L. J. Sauers. 1992. *Toxicology of the Immune System.* New York: Van Nostrand Reinhold.

Calabrese, E. J. 1994. *Biological Effects of Low Level Exposures.* New York: CRC Press/Lewis Publishers.

Cockerman, L. G., B. S. Shane. 1994. *Basic Environmental Toxicology.* New York: CRC Press/Lewis Publishers.

Gardner, R. W. 1994. *Chemical Intolerance.* New York: CRC Press/Lewis Publishers.

Hallenbeck, W. H. 1993. *Quantitative Risk Assessment for Environmental and Occupational Health,* 2nd Edition. New York: CRC Press/Lewis Publishers.

Hodgson, E., R. B. Mailman, and J. E. Chambers. 1988. *Dictionary of Toxicology.* London: The MacMillian Company.

Howard, P. H., et. al. 1991. *Handbook of Environmental Degradation Rates.* New York: CRC Press/Lewis Publishers.

International Agency for the Research on Cancer (IARC). 1987. *Monographs on the Evaluation of Carcinogenic Risks to Humans.* Supplement 7, pp-123-125. Lyon: IARC.

Jones, F. E. 1994. *Toxic Organic Vapors in the Workplace.* New York: CRC Press/Lewis Publishers.

Keith, L. H. and M. M. Walker. 1995. *Handbook of Air Toxics.* New York: CRC Press/Lewis Publishers.

Landis, W. G., M. H. Yu. 1995. *Introduction to Environmental Toxicology.* New York: CRC Press/Lewis Publishers.

Langerman, N. 1994 *Precautionary Labels for Chemical Containers.* New York: CRC Press/Lewis Publishers.

Lauwerys, R. R., P. Hoet. 1993. *Industrial Chemical Exposure.* New York: CRC Press/Lewis Publishers.

Lewis, R. J., Sr. 1993. *Hawley's Condensed Chemical Dictionary,* 12th Edition. New York: Van Nostrand Reinhold.

Lewis, R. J., Sr. 1992. *Sax's Dangerous Properties of Industrial Materials,* Volumes 1, 2, and 3, 8th Edition. New York: Van Nostrand Reinhold.

Meigs, J. W., et. al. 1986. Bladder tumor incidence among workers exposed to benzidine: A thirty year follow-up. *Journal of the National Cancer Institute.* 76:1-8.

Mellström, G. A., J. E. Wahlberg, and H. I. Maibach. 1994. *Protective Gloves for Occupational Use.* New York: CRC Press/Lewis Publishers.

Meyer, E. 1990. *Chemistry of Hazardous Materials.* Englewood Cliffs, NJ: Prentice Hall.

Patnaik, P. 1992. *A Comprehensive Guide to the Hazardous Properties of Chemical Substances.* New York: Van Nostrand Reinhold.

Philp, R. B. 1995. *Environmental Hazards and Human Health.* New York: CRC Press/Lewis Publishers.

Proctor, N. H., J. P. Hughes. 1991. *Chemical Hazards of the Workplace.* New York: Van Nostrand Reinhold.

Quigley, D. R. 1994. *Handbook of Emergency Chemical Management.* New York: CRC Press/Lewis Publishers.

Richardson, M. 1995. *Dictionary of Substance and Their Effects,* Vol. I-VI. New York: CRC Press/Lewis Publishers.

Sacarello, H. L. A. 1994. *The Comprehensive Handbook of Hazardous Materials.* New York: CRC Press/Lewis Publishers.

Torkelson, T. R. V. K. Rowe. 1981. *Patty's Industrial Hygiene and Toxicology,* Vol. 2B. New York: Wiley-Interscience.

U.S. Department of Health and Human Services, National Institute for Occupational Safety and Health. 1994. *NIOSH Pocket Guide to Chemical Hazards.* Washington, D.C.: U.S. Government Printing Office.

U.S. Department of Labor. 1977. Occupational exposure to benzidine. *Federal Register.* 42:22516-22529.

U.S. Department of Labor. Occupational Safety and Health Administration (OSHA). 1987. Occupational exposure to benzidine: final rule. *Federal Register.* 52(176):34460.

MATERIAL SAFETY DATA SHEET

CHEMICAL NAME	CRC PRESS/LEWIS PUBLISHERS DISCLAIMER:
BENZOTRICHLORIDE	THE INFORMATION AND RECOMMENDATIONS PRESENTED HEREIN ARE BASED ON SOURCES BELIEVED TO BE RELIABLE. CRC MAKES NO REPRESENTATION ON ITS COMPLETENESS OR ACCURACY. IT IS THE RESPONSIBILITY OF THE USER TO DETERMINE THE CHEMICAL'S SUITABILITY FOR ITS INTENDED USE, THE CHEMICAL'S SAFE USE, AND THE CHEMICAL'S PROPER DISPOSAL. NO REPRESENTATIONS AND/OR WARRANTIES, EITHER EXPRESSED OR IMPLIED, OF THE MERCHANTABILITY OR FITNESS FOR A PARTICULAR PURPOSE, OR OF ANY OTHER NATURE, ARE MADE WITH RESPECT TO THE INFORMATION PROVIDED IN THIS MSDS OR TO THE CHEMICAL TO WHICH INFORMATION MAY REFER. CRC NEITHER ASSUMES NOR AUTHORIZES ANY OTHER PERSON TO ASSUME FOR IT, ANY OTHER ADDITIONAL RESPONSIBILITY OR LIABILITY FOR THE USE OF, OR RELIANCE UPON, THIS INFORMATION.

HAZARD WARNING INFORMATION

HEALTH	FIRE	REACTIVE	OTHER	DEGREE OF HAZARD	COLOR CODING	OTHER CODES
3	1	0	COR	0 = Minimum Hazard 1 = Slight Hazard 2 = Moderate Hazard 3 = Serious Hazard 4 = Severe Hazard	HEALTH = BLUE FIRE = RED REACTIVITY = YELLOW OTHER = WHITE	OX = Oxidizer ACID = Acid ALK = Alkali COR = Corrosive W = Use No Water

SECTION I - GENERAL INFORMATION

Characterization	RCRA Number	EPA Class
Aromatic Chlorinated Hydrocarbon	**U023**	**Toxic Waste**

DOT Proper Shipping Name	Chemical Abstract Service (CAS) Number
Benzotrichloride	**98-07-7**
DOT Hazard Class and Label Requirements	DOT Emergency Guide Code
IMO: Corrosive Material	**60**
DOT Identification Number	Molecular Formula
UN 2226	$C_7H_5Cl_3$

Synonyms

Benzenyl chloride; benzoic trichloride; benzylidyne chloride; benzyl trichloride; phenylchloroform; toluene trichloride; trichlorotoluene

SECTION II - HAZARDOUS INGREDIENTS AND IDENTITY INFORMATION

Hazardous Components (specific identity)	OSHA Exposure Criteria	NIOSH Exposure Criteria	Immediately Dangerous to Life and Health (IDLH)	ACGIH Exposure Criteria
Benzotrichloride: (derivation: By chlorination of boiling toluene)	**PEL:** Not Established **STEL:** Not Established	**REL:** Not Established Possible Cancer Agent	Not Established	**TLV:** Not Established **STEL:** Not Established

SECTION III - PHYSICAL AND CHEMICAL CHARACTERISTICS

Boiling Point	Specific Gravity (H_2O = 1)
428°F (220°C)	**1.38**
Vapor Pressure (mm Hg)	Molecular Weight (atomic weight)
1.55 at 69°F (20°C)	**195.47**
Vapor Density (Air = 1)	Melting Point
6.77	**23°F (-5°C)**

Solubility

Negligible in water (decomposes). Soluble in alcohol and ether.

Appearance and Odor

Colorless to slightly yellow liquid, fumes in air, with a pungent, penetrating aromatic odor.

SECTION IV - FIRE AND EXPLOSION HAZARD DATA

Flash Point (method used)	Explosive Limits in Air % by Volume
207°F (97°C) closed cup	**LEL: Not Found UEL: Not Found**
NFPA Classification	Autoignition Temperature
Combustible Liquid	**1085°F (585°C)**

Extinguishing Media

Use dry chemical, foam, or carbon dioxide, or water spray.

Special Fire Fighting Procedures

Poisonous gases are produced in fire. Wear full protective clothing, self-contained breathing apparatus (SCBA) operated in pressure-demand mode. Move container from fire area if it can be done without risk.

Unusual Fire and Explosion Hazards

Benzotrichloride is a combustible liquid. It may burn, but does not easily ignite. It may ignite combustible materials such as wood and paper.

SECTION V - REACTIVITY DATA

Stability		Conditions to Avoid Keep away from combustible materials such as wood, paper, oil, etc. Avoid contact with moisture.
Stable X	Unstable	Incompatibility *(materials to avoid)* None noted. However, in the presence of moisture benzotrichloride may release fumes of hydrogen chloride and benzoic acid.
Hazardous Polymerization		Conditions to Avoid Under normal conditions of temperature and pressure, benzotrichloride is not known to undergo hazardous polymerization.
May Occur	Will Not Occur X	Hazardous Decomposition or By-products When heated to decomposition, releases toxic fumes of hydrogen chloride, carbon monoxide, and carbon dioxide.

SECTION VI - HEALTH HAZARD DATA

Primary Route(s) of Entry:	Inhalation?	Absorption (skin/eye)?	Ingestion?
	X	X	X

Health Hazards

INHALATION: May cause severe irritation to the respiratory system. High doses can result in pulmonary edema (fluid buildup in lungs), which can be fatal.

SKIN & EYES: Contact with liquid or vapor can cause severe irritation or burns to the skin, eyes, and mucous membranes. Liquid may cause permanent eye damage.

INGESTION: May cause central nervous system depression, headaches, weakness, fatigue, irritability, and poor appetite.

Carcinogenicity	NTP Listed?	IARC Cancer Review Group?	OSHA Regulated?	Target Organs?
Suspected Human Confirmed Animal	5th Annual Report	Limited Human	No	Skin, eyes, respiratory system, CNS.

Medical Conditions Generally Aggravated by Exposure

None reported; however, skin conditions (dermatitis) and respiratory problems may be aggravated.

Emergency and First-aid Procedures

Eye contact: Do not allow victim to keep eyes tightly shut. Flush immediately with water for 15 minutes (minimum), seek medical attention. Skin contact: Remove all contaminated clothing, including shoes. Immediately wash area with flooding amounts of soap and water. Seek medical attention immediately. For inhalation: Remove the person from exposure. Provide respiratory assistance and CPR. Transfer to medical facility. Observe for 24- 48 hours (pulmonary edema possible). If swallowed: Seek medical attention immediately. Do not attempt to give an unconscious or convulsing person anything by mouth.

SECTION VII - PRECAUTIONS FOR SAFE HANDLING AND USE

Steps to be Taken in Case Material is Released or Spilled

Wear self-contained breathing apparatus (SCBA) for respiratory protection. Restrict those not wearing protective equipment and who are not involved in cleanup from entering area. Avoid skin contact. Ventilate area of spill. Do not use water. Take up with sand or vermiculite and deposit in sealed drum.

Preferred Waste Disposal Method

Burn in a chemical incinerator equipped with an afterburner and scrubber.

Precautions to be Taken in Handling and Storage

A regulated, controlled area should be established where benzotrichloride is used, handled, or stored.

Other Precautions and Warnings

Keep away from sources of ignition such as smoking and open flames. Also prevent contact with moisture.

SECTION VIII - CONTROL MEASURES AND PERSONAL PROTECTIVE EQUIPMENT

Respiratory Protection *(specify type)*

At any concentration, use a NIOSH/MSHA approved self-contained breathing apparatus (SCBA) with full facepiece operated in positive pressure mode.

Ventilation

Local exhaust (preferred) at site of chemical work or general ventilation.

Protective Gloves	Eye Protection	Other Protective Clothing
Rubber Gloves	Chemical Goggles and Face Mask	Rubber Apron

Work/Hygiene Practices

Always wash hands thoroughly after using chemical; never bring food, drink, or smoking materials into vicinity of chemicals.

BENZOTRICHLORIDE

$C_7H_5Cl_3$ CAS: 98-07-7

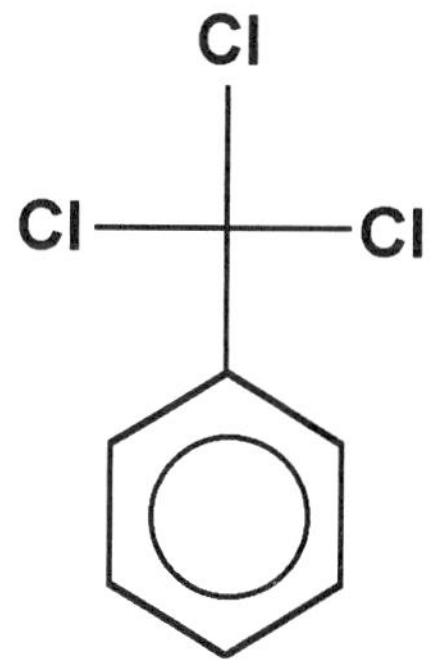

IDENTIFICATION AND TYPICAL USES

Benzotrichloride is a colorless to slightly yellow oily liquid with a pungent, irritating odor. It is used primarily as an intermediate in the production of other chemicals such as benzoyl chloride, benzotrifluoride, dyes, and ultraviolet light stabilizers.

RISK ASSESSMENT: HEALTH

General Assessment

Benzotrichloride is a corrosive liquid. It is an irritant and a suspected human carcinogen. Carcinogenic data have been reported following exposures. Mutation data have also been reported. It can enter the body through inhalation, absorption, and/or ingestion.

Toxic effects of exposure may include central nervous system excitation followed by depression, irritation of the eyes, nose, mucous membranes, and upper respiratory tract, and slowed respiration. Contact with the eyes can cause corneal injury. Upon exposure to 1000 mg/m^3, test animals developed motor automatism and twitching of the peripheral muscles. Other potential symptoms include leukopenia, mild anemia, and possible decreases in renal function in chronic cases. Although there are no data clearly linking exposure to benzotrichloride to cancer in humans, an excess in respiratory cancers (six cases) has been reported in benzoyl chloride manufacturing workers who were potentially exposed to benzotrichloride.

Animal tests have reported skin carcinomas (squamous cell), lung carcinomas, pulmonary adenomas, and lymphomas. Higher concentrations may cause pulmonary edema (fluid buildup in the lungs), which is a medical emergency and can lead to death.

Acute Health Effects

The following acute (short-term) health effects may occur immediately or shortly after exposure to benzotrichloride:

Skin: Severe irritation. Skin tumors may be possible.

Eye: Severe irritation, burns, pain, and possible permanent damage to vision.

Lung: Nose, throat, and respiratory irritation causing difficulty in breathing. High-level acute exposures can lead to pulmonary edema, which may be fatal.

CNS: Benzotrichloride may act to depress the central nervous system causing weakness, fatigue, irritability, headache, poor appetite, loss of consciousness, sleeplessness, and other symptoms of CNS depression.

Chronic Health Effects

The following chronic (long-term) health effects may occur at some time after exposure to benzotrichloride and can last for months or even years:

Cancer Hazards: Benzotrichloride may cause mutations (genetic changes) in living cells. It is considered a probable human carcinogen since there is limited evidence in the references that exposure may cause cancer in animals. There is insufficient data to evaluate the exact nature of its carcinogenic potential in humans.

Reproductive Hazard: According to the references, benzotrichloride has not been tested for its ability to adversely affect reproduction.

Other Chronic Effects: Over the long-term, exposure to benzotrichloride can affect the nervous system causing weakness, fatigue, headache, irritability, sleeplessness, and loss of appetite. Long-term effects on the brain and other organs are unclear. However, repeated exposures may damage the liver and kidneys.

Recommended Risk-Reduction Measures

Personnel should avoid direct contact with benzotrichloride. If a less toxic material or compound cannot be substituted for benzotrichloride, then *engineering controls* are the most effective method of reducing exposure risk. The best protection is to enclose operations and/or provide local exhaust ventilation at the site of benzotrichloride release. While not always operationally feasible, isolating operations can also reduce exposure. Using respiratory protection is less effective than the controls mentioned above, but is still required whenever working with or around benzotrichloride. For any exposure level, a MSHA/NIOSH approved self-contained breathing apparatus (SCBA) with full facepiece and pressure demand is the appropriate level of respiratory protection. If a full facepiece is not available, then chemical goggles should be worn to protect the eyes, especially when a splash hazard exists. A face shield should also be considered. To prevent hand and skin exposures, rubber gloves should be worn.

Administrative controls should also be in place to minimize the potential for human exposures. These may include written procedures or policies which specify the methods and techniques that will be practiced whenever personnel are to work with benzotrichloride.

All personnel should receive *training* on the use, hazards, protective measures, emergency actions, and other precautions per 29 CFR 1910.1200 (Hazard Communication) prior to the first assignment in an area where benzotrichloride is used or stored.

Before beginning employment and at regular times thereafter, the following recommended medical tests should be considered for those with frequent or potentially high exposures:

☑ Lung function tests.

If symptoms develop or overexposure is suspected, the following may be useful:

☑ Liver and kidney function tests.

☑ Consider chest X-ray following acute exposure.

☑ Examination of the nervous system.

It should be noted that medical tests that simply look for existing damage are not a substitute for controlling exposures. Medical histories are extremely important when assessing exposure risk.

Other methods to reduce exposure include:

☑ Benzotrichloride should be handled in an enclosed system. There should be local exhaust ventilation at the site of chemical release.

☑ Always ensure that proper protective clothing is worn when using chemical substances.

☑ Wash thoroughly immediately after exposure to benzotrichloride and at the end of the work shift or before eating, drinking, or smoking.

☑ Hazard warning information should be posted in the work area. In addition, as part of an on-going education and training program, all information on the health and safety hazards of benzotrichloride should be communicated to all exposed and potentially exposed workers.

☑ Eye wash stations should be provided in the immediate work area for emergency use. If there is a possibility of skin exposure to benzotrichloride, emergency shower facilities should also be provided.

☑ Workers whose clothing has been contaminated by benzotrichloride should change into clean clothes before leaving work. Contaminated work clothing should be laundered only by individuals who have been informed of the hazards of exposure to benzotrichloride.

RISK ASSESSMENT: ENVIRONMENT

General Assessment

The environment is at risk of exposure during transportation, storage, disposal, or destruction of benzotrichloride. In almost every scenario, the threat of environmental exposure is contingent upon the proper handling of the chemical substance. Accidental spills, large or small, can result in contamination of the surrounding environmental mediums (water, soil, and air).

Benzotrichloride is a combustible liquid. However, caution is always required in handling, storage, transportation, and disposal of benzotrichloride. It can react with moisture to produce irritating fumes of hydrogen chloride. It is a powerful corrosive and contact with combustible materials such as wood, paper, or oil may cause ignition. Emergency responders should be made aware of the presence of benzotrichloride at any emergency response situation.

Benzotrichloride can enter the environment from industrial discharges and from spills.

Acute Ecological Effects

Acute (short-term) toxic effects may include the death of animals, birds, or fish and death or low growth rate in plants. Acute effects are seen 2 to 4 days after animals or plants are exposed to benzotrichloride Benzotrichloride has moderate acute toxicity to aquatic life and high acute toxicity to birds. Insufficient data are available to evaluate or predict the acute (short-term) effects of benzotrichloride to plants or land animals.

Chronic Ecological Effects

Chronic (long-term) toxic effects may include shortened life span, reproductive problems, lower fertility, and changes in appearance or behavior in exposed animals. These effects can be seen long after first exposure(s) to toxic chemicals. Benzotrichloride has high chronic toxicity to aquatic life. Insufficient data are available to evaluate the chronic (long-term) effects of benzotrichloride on plants, birds, or land animals.

Water Solubility

Benzotrichloride is moderately soluble in water. Concentrations between 1 and 1000 milligrams will mix with a liter of water.

Persistence in the Environment

Benzotrichloride is non-persistent in water, with a half-life of less than 2 days. The half-life of a pollutant is the amount of time it takes for one-half of the chemical to be degraded. The chemical properties of benzotrichloride indicates that approximately 98% will enter the air and about 0.7% will end up in water, terrestrial soil, and aquatic sediments.

Bioaccumulation in Aquatic Organisms

Some substances increase in concentration, or *bioaccumulate,* in living organisms as they breathe contaminated air, drink contaminated water, or eat contaminated food. These chemicals can become concentrated in the tissues and internal organs of animals as well as humans.

The concentration of benzotrichloride found in edible fish tissue is expected to be somewhat higher than the average concentration of benzotrichloride found in the water from which the fish was taken.

Recommended Risk-Reduction Measures

Proper training of all transporters will reduce the likelihood of a mishap or accident resulting in a leak or spill of benzotrichloride into the environment. Labels on all containers, trucks, and rail cars must meet DOT requirements and accurately reflect their contents to enable emergency responders to react properly and quickly to any disaster thereby reducing the potential risk to the environment and to personnel.

Storage of benzotrichloride should be segregated from other chemicals and materials to minimize the risk of cross-contamination.

If a spill or leak to the environment has occurred, fire department, emergency response and/or hazardous materials spill personnel should be notified immediately. Cleanup should be attempted only by those trained in proper spill containment procedures. Contaminated soils should be removed for incineration and replaced with clean soil. If benzotrichloride should contact the water table, aquifer, or navigable waterway, time is of the essence. It is moderately soluble and total remediation may not be possible. The local and/or state emergency response authorities must be notified. A comprehensive emergency response or disaster preparedness/recovery plan should be in place prior to any operations involving the use, transportation, storage, or disposal of benzotrichloride.

For disposal, benzotrichloride should be burned in a chemical incinerator equipped with a scrubber and afterburner.

If benzotrichloride is spilled or leaked, the following specific steps are recommended:

- ☑ Restrict persons not wearing protective clothing from area of spill or leak until cleanup is complete and area can be opened for normal work. Avoid skin and eye contact.
- ☑ Absorb liquids in vermiculite or other similar materials and deposit in sealed drum for disposal. Use non-sparking tools
- ☑ A regulated, marked area should be established where benzotrichloride is handled, stored, or used. No unauthorized personnel should enter the area.
- ☑ Ventilate area of spill or leak.
- ☑ It may be necessary to dispose of benzotrichloride as a hazardous waste. The responsible state agency or the regional office of the federal Environmental Protection Agency (EPA) should be contacted for specific recommendations.

RISK ASSESSMENT: BUSINESS

General Assessment

Accidents or mishaps involving benzotrichloride can present a serious threat to business operations. The loss or damage of equipment or facilities can significantly affect fiscal viability. Lawsuits that may result from personnel illness, injury/death, public exposures, and/or environmental contamination will require a serious expenditure of resources.

Media attention surrounding an injury, death, or environmental damage can also result in a loss of profits and loss of current as well as future business. Always remember that anytime the words "cancer" or "carcinogen" are used, public emotion, hysteria, and ignorance can run equally high. This should be carefully considered whenever developing or implementing public relations policies.

➊ Recommended Risk-Reduction Measures

Company attorneys, safety and health professionals, and environmental specialists should be involved in the development of any procedures for responding to chemical incidents. A company official should be predesignated as a public relations officer with specific training in dealing with the press. Corporate plans and policies should be developed, approved, and implemented long before any need for such arises.

REFERENCES

American Conference of Governmental Industrial Hygienists. 1988. *Documentation of the Threshold Limit Values and Biological Exposure Indices,* 5th Edition (with updates). Cincinnati: ACGIH

Anthony, C. P., and N. J. Kolthoff. 1971. *Textbook of Anatomy and Physiology.* St. Louis: C. V. Mosby Company.

Asante-Duah, D. K. 1993. *Hazardous Waste Risk Assessment.* New York: CRC Press/Lewis Publishers.

Bloom, A. D. and F. J. DeSerres. 1995. *Ecotoxicity and Human Health.* New York: CRC Press/Lewis Publishers.

Brusick, D. J. 1994. *Methods for Genetic Risk Assessment.* New York: CRC Press/Lewis Publishers.

Bueche, J. R. 1972. *Principles of Physics,* 2nd Edition. New York: McGraw-Hill.

Calabrese, E. J. 1994. *Biological Effects of Low Level Exposures.* New York: CRC Press/Lewis Publishers.

Cockerman, L. G., B. S. Shane. 1994. *Basic Environmental Toxicology.* New York: CRC Press/Lewis Publishers.

Danse, I. R. 1991. *Common Sense Toxics in the Workplace.* New York: Van Nostrand Reinhold.

Gardner, R. W. 1994. *Chemical Intolerance.* New York: CRC Press/Lewis Publishers.

Hallenbeck, W. H. 1993. *Quantitative Risk Assessment for Environmental and Occupational Health,* 2nd Edition. New York: CRC Press/Lewis Publishers.

Hodgson, E., R. B. Mailman, and J. E. Chambers. 1988. *Dictionary of Toxicology.* London: The MacMillian Company.

Howard, P. H., et. al. 1991. *Handbook of Environmental Degradation Rates.* New York: CRC Press/Lewis Publishers.

International Agency for the Research on Cancer (IARC). 1987. *Monographs on the Evaluation of Carcinogenic Risks to Humans, Volume 36,* pp-267-283. Lyon: IARC.

Jones, F. E. 1994. *Toxic Organic Vapors in the Workplace.* New York: CRC Press/Lewis Publishers.

Keith, L. H. and M. M. Walker. 1995. *Handbook of Air Toxics.* New York: CRC Press/Lewis Publishers.

Landis, W. G., M. H. Yu. 1995. *Introduction to Environmental Toxicology.* New York: CRC Press/Lewis Publishers.

Langerman, N. 1994 *Precautionary Labels for Chemical Containers.* New York: CRC Press/Lewis Publishers.

Lauwerys, R. R., P. Hoet. 1993. *Industrial Chemical Exposure.* New York: CRC Press/Lewis Publishers.

Lewis, R. J., Sr. 1993. *Hawley's Condensed Chemical Dictionary,* 12th Edition. New York: Van Nostrand Reinhold.

Lewis, R. J., Sr. 1992. *Sax's Dangerous Properties of Industrial Materials,* Volumes 1, 2, and 3, 8th Edition. New York: Van Nostrand Reinhold.

Mellström, G. A., J. E. Wahlberg, and H. I. Maibach. 1994. *Protective Gloves for Occupational Use.* New York: CRC Press/Lewis Publishers.

Meyer, E. 1990. *Chemistry of Hazardous Materials.* Englewood Cliffs, NJ: Prentice Hall.

Patnaik, P. 1992. *A Comprehensive Guide to the Hazardous Properties of Chemical Substances.* New York: Van Nostrand Reinhold.

Philp, R. B. 1995. *Environmental Hazards and Human Health.* New York: CRC Press/Lewis Publishers.

Proctor, N. H., J. P. Hughes. 1991. *Chemical Hazards of the Workplace.* New York: Van Nostrand Reinhold.

Quigley, D. R. 1994. *Handbook of Emergency Chemical Management.* New York: CRC Press/Lewis Publishers.

Richardson, M. 1995. *Dictionary of Substance and Their Effects,* Vol. I-VI. New York: CRC Press/Lewis Publishers.

Sacarello, H. L. A. 1994. *The Comprehensive Handbook of Hazardous Materials.* New York: CRC Press/Lewis Publishers.

Torkelson, T. R. V. K. Rowe. 1981. *Patty's Industrial Hygiene and Toxicology,* Vol. 2B. New York: Wiley-Interscience.

U.S. Department of Health and Human Services, National Institute for Occupational Safety and Health. 1994. *NIOSH Pocket Guide to Chemical Hazards.* Washington, D.C.: U.S. Government Printing Office.

MATERIAL SAFETY DATA SHEET

CHEMICAL NAME	
BENZOYL CHLORIDE	**CRC PRESS/LEWIS PUBLISHERS DISCLAIMER:** THE INFORMATION AND RECOMMENDATIONS PRESENTED HEREIN ARE BASED ON SOURCES BELIEVED TO BE RELIABLE. CRC MAKES NO REPRESENTATION ON ITS COMPLETENESS OR ACCURACY. IT IS THE RESPONSIBILITY OF THE USER TO DETERMINE THE CHEMICAL'S SUITABILITY FOR ITS INTENDED USE, THE CHEMICAL'S SAFE USE, AND THE CHEMICAL'S PROPER DISPOSAL. NO REPRESENTATIONS AND/OR WARRANTIES, EITHER EXPRESSED OR IMPLIED, OF THE MERCHANTABILITY OR FITNESS FOR A PARTICULAR PURPOSE, OR OF ANY OTHER NATURE, ARE MADE WITH RESPECT TO THE INFORMATION PROVIDED IN THIS MSDS OR TO THE CHEMICAL TO WHICH INFORMATION MAY REFER. CRC NEITHER ASSUMES NOR AUTHORIZES ANY OTHER PERSON TO ASSUME FOR IT, ANY OTHER ADDITIONAL RESPONSIBILITY OR LIABILITY FOR THE USE OF, OR RELIANCE UPON, THIS INFORMATION.

HAZARD WARNING INFORMATION

HEALTH	FIRE	REACTIVE	OTHER	DEGREE OF HAZARD	COLOR CODING	OTHER CODES
3	2	2	COR	0 = Minimum Hazard 1 = Slight Hazard 2 = Moderate Hazard 3 = Serious Hazard 4 = Severe Hazard	HEALTH = BLUE FIRE = RED REACTIVITY = YELLOW OTHER = WHITE	OX = Oxidizer ACID = Acid ALK = Alkali COR = Corrosive W = Use No Water

SECTION I - GENERAL INFORMATION

Characterization: **Chlorinated Aldehyde**	RCRA Number: **D002**	EPA Class: **Corrosive Waste**
DOT Proper Shipping Name: **Benzoyl Chloride**	Chemical Abstract Service (CAS) Number: **98-88-4**	
DOT Hazard Class and Label Requirements: **Corrosive Material (Liquid)**	DOT Emergency Guide Code: **39**	
DOT Identification Number: **UN 1736**	Molecular Formula: **C_6H_5COCl**	

Synonyms

Benzoic acid, chloride; α-chlorobenzaldehyde; benzene carbonyl chloride.

SECTION II - HAZARDOUS INGREDIENTS AND IDENTITY INFORMATION

Hazardous Components (specific identity)	OSHA Exposure Criteria	NIOSH Exposure Criteria	Immediately Dangerous to Life and Health (IDLH)	ACGIH Exposure Criteria
Benzoyl chloride (derivation: Interaction of benzoic acid and sulfuryl chloride; benzotrichloride and water in the presence of zinc chloride; phosphorus tri- or pentachloride and benzoic acid). **1 ppm = 5 mg/m^3**	**PEL:** Not Established **STEL:** Not Established	**REL:** Not Established **STEL:** Not Established	**Not Determined** (Questionable Carcinogen)	**TLV:** 1 ppm 5 mg/m^3 **STEL:** Not Established

SECTION III - PHYSICAL AND CHEMICAL CHARACTERISTICS

Boiling Point: **387°F (197°C)**	Specific Gravity (H_2O = 1): **1.21**
Vapor Pressure (mm Hg): **1 at 69°F (20°C)**	Molecular Weight (atomic weight): **140.57**
Vapor Density (Air = 1): **4.88**	Melting Point: **30°F (-1°C)**

Solubility

Decomposes in water. Soluble in ether, carbon disulfide.

Appearance and Odor

Transparent, colorless, fuming liquid with a pungent, penetrating odor. Vapors cause tearing.

SECTION IV - FIRE AND EXPLOSION HAZARD DATA

Flash Point (method used): **162°F (72.2°C) closed cup**	Explosive Limits in Air % by Volume: **LEL: Not Found** **UEL: Not Found**
NFPA Classification: **Combustible Liquid**	Autoignition Temperature: **Not Found**

Extinguishing Media

Use dry chemical, foam, or carbon dioxide to extinguish fire. DO NOT USE WATER (poisonous gases are released on contact with water).

Special Fire Fighting Procedures

Severely corrosive to skin on contact. Poisonous gases are produced in fire. Wear full protective clothing and self-contained breathing apparatus (SCBA) operated in pressure-demand mode. Move container from fire area if it can be done without risk.

Unusual Fire and Explosion Hazards

Benzoyl chloride is a combustible liquid. It reacts violently with water or steam releasing heat, phosgene, and corrosive fumes.

SECTION V - REACTIVITY DATA

Stability		Conditions to Avoid **Avoid contact with water, moisture, heat, flame, or other sources of ignition.**
Stable X	Unstable	Incompatibility *(materials to avoid)* **Violent reactions on contact with water or steam, strong oxidizing agents (such as chlorine, bromine, and fluorine), and alcohol. Also incompatible with dimethyl sulfoxide.**
Hazardous Polymerization		Conditions to Avoid **Hazardous polymerization has not been reported to occur under normal conditions of temperature and pressure.**
May Occur	Will Not Occur X	Hazardous Decomposition or By-products **When heated to decomposition, benzoyl chloride releases water and toxic fumes of phosgene and hydrogen chloride.**

SECTION VI - HEALTH HAZARD DATA

Primary Route(s) of Entry:	Inhalation?	Absorption (skin/eye)?	Ingestion?
	X	X	

Health Hazards

INHALATION: Vapor can severely irritate the nose, throat, and bronchial tubes. High exposures can cause a dangerous buildup of fluid in the lungs (pulmonary edema), which can be fatal. Chronic exposures to low levels may cause reduction in sense of smell.

SKIN & EYES: Irritation on contact with skin and eyes. Strong corrosive action causing burns. Repeated exposure could cause chronic rash, dermatitis, and warts.

INGESTION: If swallowed, will irritate digestive tract possibly causing ulceration.

Carcinogenicity	NTP Listed?	IARC Cancer Review Group?	OSHA Regulated?	Target Organs?
Questioned Human Questioned Animal	No	Group 3	No	Respiratory system, skin, eyes.

Medical Conditions Generally Aggravated by Exposure

Respiratory function disorders (bronchitis); skin conditions (dermatitis, psoriasis).

Emergency and First-aid Procedures

Eye contact: Do not allow victim to keep eyes tightly shut. Flush immediately with water for 25 minutes (minimum), seek medical attention. **Skin contact:** Remove all contaminated clothing, including shoes. Immediately wash area with flooding amounts of soap and water. Seek medical attention immediately. For **inhalation**: Remove the person from exposure. Provide respiratory assistance and CPR. Transfer to medical facility. Observation recommended for 24 - 48 hours after exposure (pulmonary edema can be delayed reaction). If **swallowed**: Seek medical attention immediately.

SECTION VII - PRECAUTIONS FOR SAFE HANDLING AND USE

Steps to be Taken in Case Material is Released or Spilled

Wear self-contained breathing apparatus (SCBA) for respiratory protection. Restrict those not wearing protective equipment and who are not involved in cleanup from entering area. Avoid skin contact. Ventilate area of spill. Absorb liquids in vermiculite or other suitable material and place in sealed drums.

Preferred Waste Disposal Method

Burn in chemical incinerator equipped with an afterburner and scrubber. Verify disposal methods in accordance with local, state, and/or federal environmental regulations.

Precautions to be Taken in Handling and Storage

Store in closed containers in cool, dry, well ventilated area away from water, moisture, and combustibles.

Other Precautions and Warnings

Avoid storage with strong oxidizers, alcohols, and water since violent reactions can occur.

SECTION VIII - CONTROL MEASURES AND PERSONAL PROTECTIVE EQUIPMENT

Respiratory Protection *(specify type)*

None required where adequate ventilation exists. If airborne concentrations are high, use a MSHA/NIOSH approved chemical cartridge respirator with organic vapor cartridge. Greater protection is provided by a self-contained breathing apparatus (SCBA) with full facepiece operated in positive pressure mode.

Ventilation

Local exhaust (preferred) at site of chemical work or general ventilation.

Protective Gloves	Eye Protection	Other Protective Clothing
Rubber Gloves	**Chemical Goggles and Face Mask**	**Rubber Apron**

Work/Hygiene Practices

Always wash hands thoroughly after using chemical; never bring food, drink, or smoking materials into vicinity of chemicals.

BENZOYL CHLORIDE

C_6H_5COCl CAS: 98-88-4

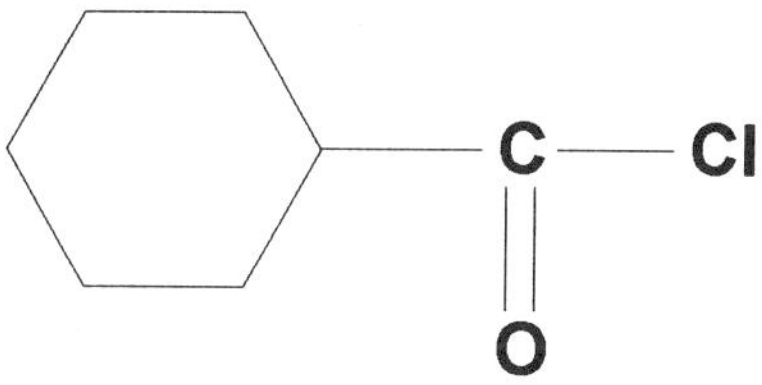

IDENTIFICATION AND TYPICAL USES

Benzoyl chloride is a transparent or colorless liquid with a strong, penetrating, pungent odor. Its vapors cause tearing. It is an analytical reagent and is used in the manufacture of benzoyl peroxide and other various products including dye intermediates, resins, perfumes, pharmaceuticals, and polymers.

RISK ASSESSMENT: HEALTH

General Assessment

Benzoyl chloride is a questionable carcinogen with experimental tumorigenic data by skin contact also reported. Exposure occurs primarily through inhalation. It is an extremely corrosive material. Immediate irritation and burns on contact with skin and eyes. Eye damage can be irreversible. Also irritating to mucous membranes on inhalation. High exposures are extremely dangerous since benzoyl chloride vapors can irritate the bronchial tubes and cause a buildup of fluid in the lungs (pulmonary edema), which is a medical emergency and can be fatal. Repeated low exposures can cause a reduction in the sense of smell which may be permanent. Effects of long-term exposures on respiratory tract not documented since pulmonary edema usually develops in the short term.

Repeated or prolonged contact with skin can lead to chronic rash, irritation, and possible dermatitis in sensitive persons.

Ingestion is not normally an occupational exposure route. If swallowed, benzoyl chloride could cause irritation of the digestive tract and possibly result in ulceration of the mucous membrane.

Acute Health Effects

The following acute (short-term) health effects may occur immediately or shortly after exposure to benzoyl chloride:

Skin: Contact could cause burns and severe irritation.

Eye: Severe irritation, burns, pain, and possible permanent damage to vision.

Lung: Nose, throat, and respiratory tract (bronchial tubes) irritation causing difficulty in breathing. May cause pulmonary edema on high exposures, which can be fatal.

Chronic Health Effects

The following chronic (long-term) health effects may occur at some time after exposure to benzoyl chloride and can last for months or even years:

Cancer Hazards: There is limited evidence that employees working in benzoyl chloride manufacturing have an increased risk of developing lung cancer. It is not known what factors cause this risk.

Reproductive Hazard: According to the information available in the references, benzoyl chloride has not been tested for its ability to adversely affect reproduction.

Other Chronic Effects: Repeated exposures could lead to chronic sore throat, chronic sinus symptoms, loss or reduced sense of smell (thereby affecting the taste of foods), and/or chronic rash. Very irritating substances may affect the lungs. It is not known whether benzoyl chloride causes lung damage.

Recommended Risk-Reduction Measures

Personnel should avoid direct contact with benzoyl chloride. Occupational exposure can occur primarily through inhalation. Skin and eye contact also present risks to heath. If a less toxic material or compound cannot be substituted for benzoyl chloride, then *engineering controls* are the most effective method of reducing exposure risk. The best protection is to enclose operations and/or provide local exhaust ventilation at the site of benzoyl chloride dust release. While not always operationally feasible, isolating operations can also reduce exposure. Using respiratory protection is less effective than the controls mentioned above, but is still recommended whenever working with or around benzoyl chloride. For low exposures, a chemical cartridge respirator with organic vapor cartridge is recommended. For higher exposures, a MSHA/NIOSH approved self-contained breathing apparatus (SCBA) with full facepiece and pressure demand is the appropriate level of respiratory protection. If a full facepiece is not available, then chemical goggles should be worn to protect the eyes. A face shield should also be considered. To prevent hand and skin exposures, rubber gloves should be worn.

Administrative controls should also be in place to minimize the potential for human exposures. These may include written procedures or policies which specify the methods and techniques that will be practiced whenever personnel are to work with benzoyl chloride.

All personnel should receive *training* on the use, hazards, protective measures, emergency actions, and other precautions per 29 CFR 1910.1200 (Hazard Communication) prior to the first assignment in an area where benzoyl chloride is used or stored.

Before beginning employment and at regular times thereafter, the following recommended medical tests should be considered for those with frequent or potentially high exposures:

☑ Lung function tests.

If symptoms develop or overexposure is suspected, the following may be useful:

☑ Consider chest X-ray after acute overexposure.

It should be noted that medical tests that simply look for existing damage are not a substitute for controlling exposures. Medical histories are extremely important when assessing exposure risk. Also, persons who smoke are already susceptible to lung cancer, emphysema, and other respiratory problems and may have a more rapid and pronounced adverse response on exposure to benzoyl chloride.

Other methods to reduce exposure include:

☑ Benzoyl chloride should be handled in an enclosed system. There should be local exhaust ventilation at the site of chemical release.

☑ Always ensure that proper protective clothing is worn when using chemical substances.

☑ Wash thoroughly immediately after exposure to benzoyl chloride and at the end of the work shift or before eating, drinking, or smoking.

☑ Hazard warning information should be posted in the work area. In addition, as part of an on-going education and training program, all information on the health and safety hazards of benzoyl chloride should be communicated to all exposed workers.

☑ Eye wash stations should be provided in the immediate work area for emergency use. If there is a possibility of skin exposure to benzoyl chloride, emergency shower facilities should be provided.

☑ Workers whose clothing has been contaminated by benzoyl chloride should change into clean clothes before leaving work. Contaminated work clothing should be laundered only by individuals who have been informed of the hazards of exposure to benzoyl chloride.

☑ Where possible, automatically pump liquid benzoyl chloride from drums or other storage containers to process containers.

RISK ASSESSMENT: ENVIRONMENT

General Assessment

The environment is at risk of exposure during transportation, storage, disposal, or destruction of benzoyl chloride. In almost every scenario, the threat of environmental exposure is contingent upon the proper handling of the chemical substance. Accidental spills, large or small, can result in contamination of the surrounding environmental mediums (water, soil, and air).

Benzoyl chloride is a combustible liquid. However, caution is always required in handling, storage, transportation, and disposal of benzoyl chloride. It will react violently on contact with water or steam to produce heat and toxic and corrosive fumes. It will also react violently on contact with strong oxidizers (such as bromine, chlorine, and fluorine). Emergency responders should be made aware of the presence of benzoyl chloride at any emergency response situation.

Benzoyl chloride can enter the environment from industrial discharges and from spills.

☠ *Acute Ecological Effects*

Acute (short-term) toxic effects may include the death of animals, birds, or fish and death or low growth rate in plants. Acute effects are seen 2 to 4 days after ani-

mals or plants are exposed to benzoyl chloride. Benzoyl chloride has moderate acute toxicity to aquatic life. Insufficient data are available to evaluate the short-term effects of benzoyl chloride on birds or land animals.

Chronic Ecological Effects

Chronic (long-term) toxic effects may include shortened life span, reproductive problems, lower fertility, and changes in appearance or behavior in exposed animals. These effects can be seen long after first exposure(s) to toxic chemicals. Benzoyl chloride has moderate chronic toxicity to aquatic life. Insufficient data are available to evaluate or predict the long-term effects of benzoyl chloride to plants, birds, or terrestrial animals.

Water Solubility

Benzoyl chloride is moderately soluble in water. Concentrations between 1 and 1000 will mix with a liter of water.

Persistence in the Environment

Benzoyl chloride will react rapidly with water and therefore is not persistent in the aquatic environment.

Bioaccumulation in Aquatic Organisms

Some substances increase in concentration, or *bioaccumulate,* in living organisms as they breathe contaminated air, drink contaminated water, or eat contaminated food. These chemicals can become concentrated in the tissues and internal organs of animals as well as humans.

Benzoyl chloride will react rapidly with water and therefore it will not accumulate in aquatic organisms or the environment.

Recommended Risk-Reduction Measures

Proper training of all transporters will reduce the likelihood of a mishap or accident resulting in a leak or spill of benzoyl chloride into the environment. Labels on all containers, trucks, and rail cars must meet DOT requirements and accurately reflect their contents to enable emergency responders to react properly and quickly to any disaster thereby reducing the potential risk to the environment and to personnel.

Storage of benzoyl chloride should be segregated from other chemicals and materials, especially water, to minimize the risk of cross-contamination and violent reactions.

If a spill or leak to the environment has occurred, fire department, emergency response and/or hazardous materials spill personnel should be notified immediately. Cleanup should be attempted only by those trained in proper spill containment procedures. Contaminated soils should be removed for incineration and replaced with clean soil. If benzoyl chloride should contact the water table, aquifer, or navigable waterway, time is of the essence. It is moderately soluble and total remediation may not be possible. The local and/or state emergency response authorities must be notified. A comprehensive emergency response or disaster preparedness/recovery plan should be in place prior to any operations involving the use, transportation, storage, or disposal of benzoyl chloride.

For disposal, benzoyl chloride may be burned in a chemical incinerator equipped with an afterburner and scrubber.

If benzoyl chloride is spilled or leaked, the following specific steps are recommended:

- ☑ Restrict persons not wearing protective clothing from area of spill or leak until cleanup is complete. Avoid skin contact. Benzoyl chloride is extremely corrosive.
- ☑ Absorb liquids in vermiculite or other similar materials and deposit in sealed drum for disposal. Do not allow contact with water, steam, or moisture.
- ☑ Ventilate area of spill or leak.
- ☑ It may be necessary to dispose of benzoyl chloride as a hazardous waste. The responsible state agency or the regional office of the federal Environmental Protection Agency (EPA) should be contacted for specific recommendations.

RISK ASSESSMENT: BUSINESS

General Assessment

Accidents or mishaps involving benzoyl chloride can present a serious threat to business operations. The loss or damage of equipment or facilities can significantly affect fiscal viability. Lawsuits that may result from personnel illness (cancer), injury/death, public exposures, and/or environmental contamination will require a serious expenditure of resources. Media attention surrounding an injury, death, or environmental damage can also result in a loss of profits and loss of current as well as future business.

Recommended Risk-Reduction Measures

Company attorneys, safety and health professionals, and environmental specialists should be involved in the development of any procedures for responding to chemical incidents. A company official should be pre-designated as a public relations officer with specific training in dealing with the press. Corporate plans and policies should be developed, approved, and implemented long before any need for such arises.

REFERENCES

Ahlbom, A. 1993. *Biostatistics for Epidemiologists.* New York: CRC Press/Lewis Publishers.

American Conference of Governmental Industrial Hygienists. 1988. *Documentation of the Threshold Limit Values and Biological Exposure Indices,* 5th Edition (with updates). Cincinnati: ACGIH

Anthony, C. P., and N. J. Kolthoff. 1971. *Textbook of Anatomy and Physiology.* St. Louis: C. V. Mosby Company.

Asante-Duah, D. K. 1993. *Hazardous Waste Risk Assessment.* New York: CRC Press/Lewis Publishers.

Bloom, A. D. and F. J. DeSerres. 1995. *Ecotoxicity and Human Health.* New York: CRC Press/Lewis Publishers.

Brusick, D. J. 1994. *Methods for Genetic Risk Assessment.* New York: CRC Press/Lewis Publishers.

Bueche, J. R. 1972. *Principles of Physics,* 2nd Edition. New York: McGraw-Hill.

Burrell, R., D. K. Flaherty, and L. J. Sauers. 1992. *Toxicology of the Immune System.* New York: Van Nostrand Reinhold.

Calabrese, E. J. 1994. *Biological Effects of Low Level Exposures.* New York: CRC Press/Lewis Publishers.

Cockerman, L. G., B. S. Shane. 1994. *Basic Environmental Toxicology.* New York: CRC Press/Lewis Publishers.

Gardner, R. W. 1994. *Chemical Intolerance.* New York: CRC Press/Lewis Publishers.

Hallenbeck, W. H. 1993. *Quantitative Risk Assessment for Environmental and Occupational Health,* 2nd Edition. New York: CRC Press/Lewis Publishers.

Hodgson, E., R. B. Mailman, and J. E. Chambers. 1988. *Dictionary of Toxicology.* London: The MacMillian Company.

Howard, P. H., et. al. 1991. *Handbook of Environmental Degradation Rates.* New York: CRC Press/Lewis Publishers.

International Agency for the Research on Cancer (IARC). 1987. *Monographs on the Evaluation of Carcinogenic Risks to Humans.* Supplement 7, pp-123-125. Lyon: IARC.

Jones, F. E. 1994. *Toxic Organic Vapors in the Workplace.* New York: CRC Press/Lewis Publishers.

Keith, L. H. and M. M. Walker. 1995. *Handbook of Air Toxics.* New York: CRC Press/Lewis Publishers.

Landis, W. G., M. H. Yu. 1995. *Introduction to Environmental Toxicology.* New York: CRC Press/Lewis Publishers.

Langerman, N. 1994 *Precautionary Labels for Chemical Containers.* New York: CRC Press/Lewis Publishers.

Lauwerys, R. R., P. Hoet. 1993. *Industrial Chemical Exposure.* New York: CRC Press/Lewis Publishers.

Lewis, R. J., Sr. 1993. *Hawley's Condensed Chemical Dictionary,* 12th Edition. New York: Van Nostrand Reinhold.

Lewis, R. J., Sr. 1992. *Sax's Dangerous Properties of Industrial Materials,* Volumes 1, 2, and 3, 8th Edition. New York: Van Nostrand Reinhold.

Manahan, S. E. 1994. *Environmental Chemistry,* 6th Edition. New York: CRC Press/Lewis Publishers.

Mellström, G. A., J. E. Wahlberg, and H. I. Maibach. 1994. *Protective Gloves for Occupational Use.* New York: CRC Press/Lewis Publishers.

Meyer, E. 1990. *Chemistry of Hazardous Materials.* Englewood Cliffs, NJ: Prentice Hall.

Ott, W. R. 1995. *Environmental Statistics and Data Analysis.* New York: CRC Press/Lewis Publishers.

Patnaik, P. 1992. *A Comprehensive Guide to the Hazardous Properties of Chemical Substances.* New York: Van Nostrand Reinhold.

Philp, R. B. 1995. *Environmental Hazards and Human Health.* New York: CRC Press/Lewis Publishers.

Proctor, N. H., J. P. Hughes. 1991. *Chemical Hazards of the Workplace.* New York: Van Nostrand Reinhold.

Quigley, D. R. 1994. *Handbook of Emergency Chemical Management.* New York: CRC Press/Lewis Publishers.

Richardson, M. 1995. *Dictionary of Substance and Their Effects,* Vol. I-VI. New York: CRC Press/Lewis Publishers.

Sacarello, H. L. A. 1994. *The Comprehensive Handbook of Hazardous Materials.* New York: CRC Press/Lewis Publishers.

Torkelson, T. R. V. K. Rowe. 1981. *Patty's Industrial Hygiene and Toxicology,* Vol. 2B. New York: Wiley-Interscience.

U.S. Department of Health and Human Services, National Institute for Occupational Safety and Health. 1994. *NIOSH Pocket Guide to Chemical Hazards.* Washington, D.C.: U.S. Government Printing Office.

MATERIAL SAFETY DATA SHEET

CHEMICAL NAME	
BENZOYL PEROXIDE	**CRC PRESS/LEWIS PUBLISHERS DISCLAIMER:** THE INFORMATION AND RECOMMENDATIONS PRESENTED HEREIN ARE BASED ON SOURCES BELIEVED TO BE RELIABLE. CRC MAKES NO REPRESENTATION ON ITS COMPLETENESS OR ACCURACY. IT IS THE RESPONSIBILITY OF THE USER TO DETERMINE THE CHEMICAL'S SUITABILITY FOR ITS INTENDED USE, THE CHEMICAL'S SAFE USE, AND THE CHEMICAL'S PROPER DISPOSAL. NO REPRESENTATIONS AND/OR WARRANTIES, EITHER EXPRESSED OR IMPLIED, OF THE MERCHANTABILITY OR FITNESS FOR A PARTICULAR PURPOSE, OR OF ANY OTHER NATURE, ARE MADE WITH RESPECT TO THE INFORMATION PROVIDED IN THIS MSDS OR TO THE CHEMICAL TO WHICH INFORMATION MAY REFER. CRC NEITHER ASSUMES NOR AUTHORIZES ANY OTHER PERSON TO ASSUME FOR IT, ANY OTHER ADDITIONAL RESPONSIBILITY OR LIABILITY FOR THE USE OF, OR RELIANCE UPON, THIS INFORMATION.

HAZARD WARNING INFORMATION

HEALTH	FIRE	REACTIVE	OTHER	DEGREE OF HAZARD	COLOR CODING	OTHER CODES
1	4	4	OX	0 = Minimum Hazard 1 = Slight Hazard 2 = Moderate Hazard 3 = Serious Hazard 4 = Severe Hazard	HEALTH = BLUE FIRE = RED REACTIVITY = YELLOW OTHER = WHITE	OX = Oxidizer ACID = Acid ALK = Alkali COR = Corrosive W = Use No Water

SECTION I - GENERAL INFORMATION

Characterization: **Organic Peroxide**	RCRA Number: **D001, D002, D003**	EPA Class: **Ignitable, Corrosive, Reactive Waste**
DOT Proper Shipping Name: **Benzoyl Peroxide**	Chemical Abstract Service (CAS) Number: **94-36-0**	
DOT Hazard Class and Label Requirements: **Organic Peroxide**	DOT Emergency Guide Code: **49**	
DOT Identification Number: **UN 2085/2087/2088/2089/2090**	Molecular Formula: **$(C_6H_5CO)_2O_2$**	

Synonyms
Benzoperoxide; dibenzoyl peroxide; benzoyl superoxide; benzoyl; acetoxyl; Benzac; Oxylite; Superox.

SECTION II - HAZARDOUS INGREDIENTS AND IDENTITY INFORMATION

Hazardous Components (specific identity)	OSHA Exposure Criteria	NIOSH Exposure Criteria	Immediately Dangerous to Life and Health (IDLH)	ACGIH Exposure Criteria
Benzoyl peroxide (derivation: By interaction of benzoyl chloride and a cooled solution of sodium peroxide.	PEL: 5 mg/m^3 STEL: Not Established	REL: 5 mg/m^3	1500 mg/m^3	TLV: 5 mg/m^3 STEL: Not Established

SECTION III - PHYSICAL AND CHEMICAL CHARACTERISTICS

Boiling Point: **Decomposes Explosively**	Specific Gravity (H_2O = 1): **1.33**
Vapor Pressure (mm Hg): **<1 at 69°F (20°C)**	Molecular Weight (atomic weight): **242.2**
Vapor Density (Air = 1): **Not Applicable**	Melting Point: **217°F (105°C)**

Solubility
Negligible in water (less than 1%). Soluble in benzene, acetone, chloroform. Slightly soluble in alcohol.

Appearance and Odor
White, granular, crystalline solid with a faint, benzaldehyde odor; tasteless; may also appear as liquid.

SECTION IV - FIRE AND EXPLOSION HAZARD DATA

Flash Point (method used): **105°F (41°C) closed cup**	Explosive Limits in Air % by Volume: **LEL: Not Found**	**UEL: Not Found**
NFPA Classification: **Combustible Solid**	Autoignition Temperature: **176°F (80°C)**	

Extinguishing Media
Use water spray or foam only. Do NOT use carbon dioxide or dry chemical.

Special Fire Fighting Procedures
Use water spray to keep containers cool. Poisonous gases are produced in fire. Wear full protective clothing and self-contained breathing apparatus (SCBA) operated in pressure-demand mode. Move container from fire area if it can be done without risk. Continue flushing area with water until cool to prevent re-ignition.

Unusual Fire and Explosion Hazards
Benzoyl peroxide is a combustible solid. At high temperatures it can explode. Containers may explode in fire. Dust can form explosive mixture with air. Can react violently with shock, friction, or heat. It may attack some forms of rubber, plastics, and coatings to produce fires or explosions.

SECTION V - REACTIVITY DATA

Stability		Conditions to Avoid Avoid contact with heat, flame, other sources of ignition, friction, and shock. Store to avoid contact with combustible materials and other incompatible materials.
Stable	Unstable X	Incompatibility *(materials to avoid)* Polymerization catalysts and accelerators; strong acids; strong oxidizing agents; strong reducing agents; most common metals, metal oxides, amines, combustible materials (wood, paper), alcohols, dimethyl aniline.
Hazardous Polymerization		Conditions to Avoid Hazardous polymerization has not been reported to occur under normal conditions of temperature and pressure.
May Occur	Will Not Occur X	Hazardous Decomposition or By-products When heated to decomposition, releases toxic fumes or carbon dioxide and carbon monoxide. Also produces dense white smoke of benzoic acid, phenyl benzoate, terphenyls, biphenyls, and benzene.

SECTION VI - HEALTH HAZARD DATA

Primary Route(s) of Entry:	Inhalation?	Absorption (skin/eye)?	Ingestion?
	X		

Health Hazards

INHALATION: Can severely irritate the nose, throat, and lungs. May cause asthmatic effects.

SKIN & EYES: Causes both primary irritation to skin and sensitized dermatitis. Skin tumors may also result. A serious eye irritant causing possible permanent damage.

INGESTION: A human poison by ingestion. May cause gene damage and DNA inhibition

Carcinogenicity	NTP Listed?	IARC Cancer Review Group?	OSHA Regulated?	Target Organs?
Questioned Human Questioned Animal	No	Group 3	29 CFR 1910.1000 Table Z-1	Skin, eyes, respiratory system.

Medical Conditions Generally Aggravated by Exposure

None reported; however, existing skin conditions (dermatitis, psoriasis) may be aggravated.

Emergency and First-aid Procedures

Eye contact: Do not allow victim to keep eyes tightly shut. Flush immediately with water for 25 minutes (minimum), seek medical attention. Skin contact: Remove all contaminated clothing, including shoes. Immediately wash area with flooding amounts of soap and water. Seek medical attention immediately. For inhalation: Remove the person from exposure. Provide respiratory assistance and CPR. Transfer to medical facility. If swallowed: Seek medical attention immediately. Do not attempt to give anything by mouth to an unconscious or convulsing person.

SECTION VII - PRECAUTIONS FOR SAFE HANDLING AND USE

Steps to be Taken in Case Material is Released or Spilled

Wear self-contained breathing apparatus (SCBA) for respiratory protection. Restrict those not wearing protective equipment and who are not involved in cleanup from entering area. Avoid skin contact. Ventilate area of spill. Soak with water or cover with wet vermiculite or sand or other suitable material and place in sealed drums.

Preferred Waste Disposal Method

Benzoyl peroxide is diluted in water and disposed of in a large flowing stream of water.

Precautions to be Taken in Handling and Storage

Protect containers from impact, blows, shock, or friction. Avoid contact with heat or flame.

Other Precautions and Warnings

Use only non-sparking tools and equipment. Use explosion-proof electrical fittings in storage areas.

SECTION VIII - CONTROL MEASURES AND PERSONAL PROTECTIVE EQUIPMENT

Respiratory Protection *(specify type)*

For low exposures, use an MSHA/NIOSH approved cartridge or canister respirator with a dust/mist filter containing non-oxidizable sorbents to prevent fires or explosion. For higher exposures, use self-contained breathing apparatus (SCBA) with full facepiece operated in positive pressure mode.

Ventilation

Local exhaust (preferred) at site of chemical work or general ventilation.

Protective Gloves	Eye Protection	Other Protective Clothing
Rubber Gloves	Chemical/Dust Goggles and Face Mask	Rubber Apron, Disposable Uniform

Work/Hygiene Practices

Always wash hands thoroughly after using chemical; never bring food, drink, or smoking materials into vicinity of chemicals.

BENZOYL PEROXIDE

$(C_6H_5CO)_2O_2$ CAS: 94-36-0

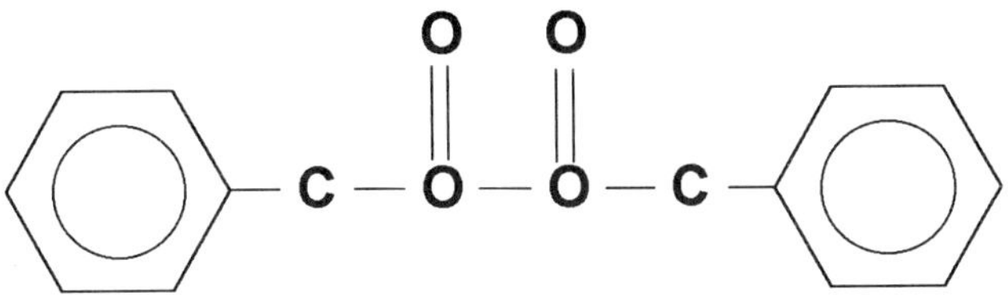

IDENTIFICATION AND TYPICAL USES

Benzoyl peroxide is a colorless to white crystalline solid or a powder with a faint, benzaldehyde odor and no taste. It may also appear as a liquid. It is used as a source of free radicals in many organic syntheses and to initiate polymerization of styrene, vinyl chloride, vinyl acetate, and acrylics. Also used to cure thermoset polyester resins and silicone rubbers; in medicine for treating acne; for bleaching vegetable oil, waxes, cheeses, flour, and fats. Also used as a drying agent for unsaturated oils.

RISK ASSESSMENT: HEALTH

General Assessment

Benzoyl peroxide is a questionable carcinogen with experimental tumorigenic data by skin contact also reported (animal studies). Exposure occurs primarily through *inhalation* and *ingestion*; however, skin and eye contact will also result in adverse health effects. Can cause dermatitis, asthmatic effects, testicular atrophy, and vasodilatation. It is a skin allergen, with possible tumors resulting from exposure. Benzoyl peroxide can severely irritate the eyes on contact with possibility for permanent damage. Some evidence of human mutation data also reported, but systemic effects of human exposures have not been accurately documented.

Dusts may be irritating to the eyes, skin, nose, throat, and lungs. Benzoyl peroxide may cause gene damage and DNA inhibition. While it has been reported to cause skin tumors, evidence of its carcinogenicity is inadequate.

Acute Health Effects

The following acute (short-term) health effects may occur immediately or shortly after exposure to benzoyl peroxide:

Skin: Irritation with possible allergenic effects (dermatitis). Skin tumors may be possible.

Eye: Severe irritation, burns, pain, and possible permanent damage to vision.

Lung: Nose, throat, and respiratory irritation causing difficulty in breathing.

Chronic Health Effects

The following chronic (long-term) health effects may occur at some time after exposure to benzoyl peroxide and can last for months or even years:

Cancer Hazards: According to the information available in the references, benzoyl peroxide has not been tested for its ability to cause cancer in animals.

Reproductive Hazard: According to the information available in the references, benzoyl peroxide has not been tested for its ability to adversely affect reproduction.

Other Chronic Effects: Over the long-term, exposure to benzoyl peroxide can lead to skin allergy. If an allergy develops, very low future exposures can cause itching and rash. Very irritating substances such as benzoyl peroxide can affect the lungs. It is not known whether benzoyl peroxide causes lung damage.

Recommended Risk-Reduction Measures

Personnel should avoid direct contact with benzoyl peroxide. Occupational exposure can occur primarily through inhalation and ingestion. Skin and eye contact also present risks to heath. If a less toxic material or compound cannot be substituted for benzoyl peroxide, then *engineering controls* are the most effective method of reducing exposure risk. The best protection is to enclose operations and/or provide local exhaust ventilation at the site of benzoyl peroxide dust release. While not always operationally feasible, isolating operations can also reduce exposure. Using respiratory protection is less effective than the controls mentioned above, but is still recommended whenever working with or around benzoyl peroxide. For low exposures, a chemical cartridge respirator with a dust/mist filter is recommended. For higher exposures, a MSHA/NIOSH approved self-contained breathing apparatus (SCBA) with full facepiece and pressure demand is the appropriate level of respiratory protection. If a full facepiece is not available, then chemical goggles should be worn to protect the eyes. A face shield should also be considered. To prevent hand and skin exposures, rubber gloves should be worn.

Administrative controls should also be in place to minimize the potential for human exposures. These may include written procedures or policies which specify the methods and techniques that will be practiced whenever personnel are to work with benzoyl peroxide.

All personnel should receive *training* on the use, hazards, protective measures, emergency actions, and other precautions per 29 CFR 1910.1200 (Hazard Communication) prior to the first assignment in an area where benzoyl peroxide is used or stored.

Before beginning employment and at regular times thereafter, the following recommended medical tests should be considered for those with frequent or potentially high exposures:

☑ Lung function tests.

If symptoms develop or overexposure is suspected, the following may be useful:

☑ Skin testing with dilute benzoyl peroxide may help diagnose allergy, if performed by a qualified allergists.

It should be noted that medical tests that simply look for existing damage are not a substitute for controlling exposures. Medical histories are extremely important when assessing exposure risk.

Other methods to reduce exposure include:

☑ Benzoyl peroxide should be handled in an enclosed system. There should be local exhaust ventilation at the site of chemical release.

☑ Always ensure that proper protective clothing is worn when using chemical substances.

☑ Wash thoroughly immediately after exposure to benzoyl peroxide and at the end of the work shift or before eating, drinking, or smoking.

☑ Hazard warning information should be posted in the work area. In addition, as part of an on-going education and training program, all information on the health and safety hazards of benzoyl peroxide should be communicated to all exposed workers.

☑ Eye wash stations should be provided in the immediate work area for emergency use. If there is a possibility of skin exposure to benzoyl peroxide, emergency shower facilities should also be provided in the immediate area.

☑ Workers whose clothing has been contaminated by benzoyl peroxide should change into clean clothes before leaving work. Contaminated work clothing should be laundered only by individuals who have been informed of the hazards of exposure to benzoyl peroxide.

☑ Specific engineering controls are recommended for this chemical by NIOSH. Refer to NIOSH Criteria Document Number 77-166 on benzoyl peroxide.

RISK ASSESSMENT: ENVIRONMENT

General Assessment

The environment is at risk of exposure during transportation, storage, disposal, or destruction of benzoyl peroxide. In almost every scenario, the threat of environmental exposure is contingent upon the proper handling of the chemical substance. Accidental spills, large or small, can result in contamination of the surrounding environmental mediums (water, soil, and air).

Benzoyl peroxide is a combustible solid. However, caution is always required in handling, storage, transportation, and disposal of benzoyl peroxide. It is a moderate fire hazard by spontaneous chemical reaction in contact with reducing agents. It ignites readily and burns rapidly. It is a powerful oxidizer. It presents a dangerous explosion hazard and may explode spontaneously when heated. It is moderately sensitive to heat, shock, friction, or contact with combustible materials. It will also react violently on contact with strong oxidizers, strong acids, strong reducing agents, most metals, and amines. Emergency responders should be made aware of the presence of benzoyl peroxide at any emergency response situation.

Benzoyl peroxide can enter the environment from industrial discharges and from spills.

☠ *Acute Ecological Effects*

Acute (short-term) toxic effects may include the death of animals, birds, or fish and death or low growth rate in plants. Acute effects are seen 2 to 4 days after ani-

mals or plants are exposed to benzoyl peroxide Insufficient data are available to evaluate the acute (short-term) effects of benzoyl peroxide on aquatic life, birds or land animals.

Chronic Ecological Effects

Chronic (long-term) toxic effects may include shortened life span, reproductive problems, lower fertility, and changes in appearance or behavior in exposed animals. These effects can be seen long after first exposure(s) to toxic chemicals. Insufficient data are available to evaluate or predict the chronic (long-term) effects of benzoyl peroxide to aquatic life, plants, birds, or terrestrial animals.

Water Solubility

Benzoyl peroxide is moderately soluble in water. Concentrations between 1 and 1000 milligrams will mix with a liter of water.

Persistence in the Environment

Benzoyl peroxide will react rapidly with water and therefore is not persistent in the aquatic environment.

Bioaccumulation in Aquatic Organisms

Some substances increase in concentration, or *bioaccumulate,* in living organisms as they breathe contaminated air, drink contaminated water, or eat contaminated food. These chemicals can become concentrated in the tissues and internal organs of animals as well as humans.

Benzoyl peroxide will react rapidly with water and therefore it will not accumulate in aquatic organisms or the environment.

Recommended Risk-Reduction Measures

Proper training of all transporters will reduce the likelihood of a mishap or accident resulting in a leak or spill of benzoyl peroxide into the environment. Labels on all containers, trucks, and rail cars must meet DOT requirements and accurately reflect their contents to enable emergency responders to react properly and quickly to any disaster thereby reducing the potential risk to the environment and to personnel.

Storage of benzoyl peroxide should be segregated from other chemicals and materials and combustible materials to minimize the risk of cross-contamination and violent reactions.

If a spill or leak to the environment has occurred, fire department, emergency response and/or hazardous materials spill personnel should be notified immediately. Cleanup should be attempted only by those trained in proper spill containment procedures. Contaminated soils should be removed for incineration and replaced with clean soil. If benzoyl peroxide should contact the water table, aquifer, or navigable waterway, time is of the essence. It is moderately soluble and total remediation may not be possible. The local and/or state emergency response authorities must be notified. A comprehensive emergency response or disaster preparedness/recovery plan should be in place prior to any operations involving the use, transportation, storage, or disposal of benzoyl peroxide.

For disposal, benzoyl peroxide is diluted in water and disposed of in a large flowing stream of water. Another common disposal method is to spread the material over a large area on the downwind side of a pit. The waste solvent is fed to the pit and ignited carefully from a remote position.

If benzoyl peroxide is spilled or leaked, the following specific steps are recommended:

- ☑ Restrict persons not wearing protective clothing from area of spill or leak until cleanup is complete. Avoid skin contact.
- ☑ Absorb liquids in vermiculite or other similar materials and deposit in sealed drum for disposal. Collect powdered materials in the most convenient manner using wet methods and deposit in sealed containers.
- ☑ Ventilate area of spill or leak.
- ☑ It may be necessary to dispose of benzoyl peroxide as a hazardous waste. The responsible state agency or the regional office of the federal Environmental Protection Agency (EPA) should be contacted for specific recommendations.

RISK ASSESSMENT: BUSINESS

General Assessment

Accidents or mishaps involving benzoyl peroxide can present a serious threat to business operations. The loss or damage of equipment or facilities can significantly affect fiscal viability. Lawsuits that may result from personnel illness, injury/death, public exposures, and/or environmental contamination will require a serious expenditure of resources. Media attention surrounding an injury, death, or environmental damage can also result in a loss of profits and loss of current as well as future business.

Recommended Risk-Reduction Measures

Company attorneys, safety and health professionals, and environmental specialists should be involved in the development of any procedures for responding to chemical incidents. A company official should be pre-designated as a public relations officer with specific training in dealing with the press. Corporate plans and policies should be developed, approved, and implemented long before any need for such arises.

REFERENCES

Ahlbom, A. 1993. *Biostatistics for Epidemiologists.* New York: CRC Press/Lewis Publishers.

American Conference of Governmental Industrial Hygienists. 1988. *Documentation of the Threshold Limit Values and Biological Exposure Indices,* 5th Edition (with updates). Cincinnati: ACGIH

Anthony, C. P., and N. J. Kolthoff. 1971. *Textbook of Anatomy and Physiology.* St. Louis: C. V. Mosby Company.

Asante-Duah, D. K. 1993. *Hazardous Waste Risk Assessment.* New York: CRC Press/Lewis Publishers.

Bloom, A. D. and F. J. DeSerres. 1995. *Ecotoxicity and Human Health.* New York: CRC Press/Lewis Publishers.

Brusick, D. J. 1994. *Methods for Genetic Risk Assessment.* New York: CRC Press/Lewis Publishers.

Bueche, J. R. 1972. *Principles of Physics,* 2nd Edition. New York: McGraw-Hill.

Burrell, R., D. K. Flaherty, and L. J. Sauers. 1992. *Toxicology of the Immune System.* New York: Van Nostrand Reinhold.

Calabrese, E. J. 1994. *Biological Effects of Low Level Exposures.* New York: CRC Press/Lewis Publishers.

Cockerman, L. G., B. S. Shane. 1994. *Basic Environmental Toxicology.* New York: CRC Press/Lewis Publishers.

Danse, I. R. 1991. *Common Sense Toxics in the Workplace.* New York: Van Nostrand Reinhold.

Gardner, R. W. 1994. *Chemical Intolerance.* New York: CRC Press/Lewis Publishers.

Hallenbeck, W. H. 1993. *Quantitative Risk Assessment for Environmental and Occupational Health,* 2nd Edition. New York: CRC Press/Lewis Publishers.

Hodgson, E., R. B. Mailman, and J. E. Chambers. 1988. *Dictionary of Toxicology.* London: The MacMillian Company.

Howard, P. H., et. al. 1991. *Handbook of Environmental Degradation Rates.* New York: CRC Press/Lewis Publishers.

International Agency for the Research on Cancer (IARC). 1987. *Monographs on the Evaluation of Carcinogenic Risks to Humans, Volume 36,* pp-267-283. Lyon: IARC.

Jones, F. E. 1994. *Toxic Organic Vapors in the Workplace.* New York: CRC Press/Lewis Publishers.

Keith, L. H. and M. M. Walker. 1995. *Handbook of Air Toxics.* New York: CRC Press/Lewis Publishers.

Landis, W. G., M. H. Yu. 1995. *Introduction to Environmental Toxicology.* New York: CRC Press/Lewis Publishers.

Langerman, N. 1994 *Precautionary Labels for Chemical Containers.* New York: CRC Press/Lewis Publishers.

Lauwerys, R. R., P. Hoet. 1993. *Industrial Chemical Exposure.* New York: CRC Press/Lewis Publishers.

Lewis, R. J., Sr. 1993. *Hawley's Condensed Chemical Dictionary,* 12th Edition. New York: Van Nostrand Reinhold.

Lewis, R. J., Sr. 1992. *Sax's Dangerous Properties of Industrial Materials,* Volumes 1, 2, and 3, 8th Edition. New York: Van Nostrand Reinhold.

Manahan, S. E. 1994. *Environmental Chemistry,* 6th Edition. New York: CRC Press/Lewis Publishers.

Mellström, G. A., J. E. Wahlberg, and H. I. Maibach. 1994. *Protective Gloves for Occupational Use.* New York: CRC Press/Lewis Publishers.

Meyer, E. 1990. *Chemistry of Hazardous Materials.* Englewood Cliffs, NJ: Prentice Hall.

Ott, W. R. 1995. *Environmental Statistics and Data Analysis.* New York: CRC Press/Lewis Publishers.

Patnaik, P. 1992. *A Comprehensive Guide to the Hazardous Properties of Chemical Substances.* New York: Van Nostrand Reinhold.

Philp, R. B. 1995. *Environmental Hazards and Human Health.* New York: CRC Press/Lewis Publishers.

Proctor, N. H., J. P. Hughes. 1991. *Chemical Hazards of the Workplace.* New York: Van Nostrand Reinhold.

Quigley, D. R. 1994. *Handbook of Emergency Chemical Management.* New York: CRC Press/Lewis Publishers.

Richardson, M. 1995. *Dictionary of Substance and Their Effects,* Vol. I-VI. New York: CRC Press/Lewis Publishers.

Sacarello, H. L. A. 1994. *The Comprehensive Handbook of Hazardous Materials.* New York: CRC Press/Lewis Publishers.

Torkelson, T. R. V. K. Rowe. 1981. *Patty's Industrial Hygiene and Toxicology,* Vol. 2B. New York: Wiley-Interscience.

U.S. Department of Health and Human Services, National Institute for Occupational Safety and Health. 1994. *NIOSH Pocket Guide to Chemical Hazards.* Washington, D.C.: U.S. Government Printing Office.

U.S. Department of Health ,Education, and Welfare (DHEW), National Institute for Occupational Safety and Health (NIOSH). 1977. *Criteria for a Recommended Standard: Occupational Exposure to Benzoyl Peroxide.* Publication Number 77-166. Washington, D.C.: U.S. Government Printing Office.

MATERIAL SAFETY DATA SHEET

CHEMICAL NAME	
BENZYL ACETATE	**CRC PRESS/LEWIS PUBLISHERS DISCLAIMER:** THE INFORMATION AND RECOMMENDATIONS PRESENTED HEREIN ARE BASED ON SOURCES BELIEVED TO BE RELIABLE. CRC MAKES NO REPRESENTATION ON ITS COMPLETENESS OR ACCURACY. IT IS THE RESPONSIBILITY OF THE USER TO DETERMINE THE CHEMICAL'S SUITABILITY FOR ITS INTENDED USE, THE CHEMICAL'S SAFE USE, AND THE CHEMICAL'S PROPER DISPOSAL. NO REPRESENTATIONS AND/OR WARRANTIES, EITHER EXPRESSED OR IMPLIED, OF THE MERCHANTABILITY OR FITNESS FOR A PARTICULAR PURPOSE, OR OF ANY OTHER NATURE, ARE MADE WITH RESPECT TO THE INFORMATION PROVIDED IN THIS MSDS OR TO THE CHEMICAL TO WHICH INFORMATION MAY REFER. CRC NEITHER ASSUMES NOR AUTHORIZES ANY OTHER PERSON TO ASSUME FOR IT, ANY OTHER ADDITIONAL RESPONSIBILITY OR LIABILITY FOR THE USE OF, OR RELIANCE UPON, THIS INFORMATION.

HAZARD WARNING INFORMATION

HEALTH	FIRE	REACTIVE	OTHER	DEGREE OF HAZARD	COLOR CODING	OTHER CODES
1	1	0		0 = Minimum Hazard 1 = Slight Hazard 2 = Moderate Hazard 3 = Serious Hazard 4 = Severe Hazard	HEALTH = BLUE FIRE = RED REACTIVITY = YELLOW OTHER = WHITE	OX = Oxidizer ACID = Acid ALK = Alkali COR = Corrosive W = Use No Water

SECTION I - GENERAL INFORMATION

Characterization **Aliphatic Ester**	RCRA Number **None**	EPA Class **No Applicable**
DOT Proper Shipping Name **Combustible Liquid, N.O.S.**	Chemical Abstract Service (CAS) Number **140-11-4**	
DOT Hazard Class and Label Requirements **Combustible Liquid**	DOT Emergency Guide Code **26**	
DOT Identification Number **NA 1883**	Chemical Formula **$CH_3COOCH_2C_6H_5$**	

Synonyms
Acetic ether; benzyl ester of acetic acid; benzyl ethanoate; acetic acid, phenyl methyl ester; α-acetoxy toluene; phenyl methyl acetate.

SECTION II - HAZARDOUS INGREDIENTS AND IDENTITY INFORMATION

Hazardous Components (specific identity)	OSHA Exposure Criteria	NIOSH Exposure Criteria	Immediately Dangerous to Life and Health (IDLH)	ACGIH Exposure Criteria
Benzyl acetate (derivation: By treating benzyl chloride with sodium acetate in various solvents; by esterification of benzyl alcohol with acetic anhydride or acetic acid).	**PEL:** Not Established **STEL:** Not Established	**REL:** Not Established **STEL:** Not Established	**Not Determined**	**TLV:** Not Established **STEL:** Not Established

SECTION III - PHYSICAL AND CHEMICAL CHARACTERISTICS

Boiling Point **417°F (214°C)**	Specific Gravity (H_2O = 1) **1.1**
Vapor Pressure (mm Hg) **1 at 113°F (45°C)**	Molecular Weight **150.19**
Vapor Density (Air = 1) **5.1**	Melting Point **-60°F (-51°C)**

Solubility
Slightly soluble in water. Soluble in ethyl alcohol, benzene, chloroform, ether, glycerin, most fixed oils.

Appearance and Odor
Colorless or water-white liquid with a floral, fruity, or pear-like odor and a pungent, bitter taste.

SECTION IV - FIRE AND EXPLOSION HAZARD DATA

Flash Point (method used) **216°F (102°C) closed cup**	Explosive Limits in Air % by Volume **LEL: Not Determined** **UEL: Not Determined**
NFPA Classification **Class IIIA Combustible Liquid**	Autoignition Temperature **862°F (461°C)**

Extinguishing Media
Dry chemical, carbon dioxide, water spray or foam. Use water to keep fire-exposed containers cool.

Special Fire Fighting Procedures
Poisonous gases are produced in fire. Wear full protective clothing and self-contained breathing apparatus (SCBA). Use blanketing effect with foam to smother large fires. Continue to cool containers with water after fire is extinguished. Be aware of fire control measure by-products. Do NOT release runoff to sewers or waterways. Dike and control for later disposal.

Unusual Fire and Explosion Hazards
None reported.

SECTION V - REACTIVITY DATA

Stability		Conditions to Avoid Benzyl acetate is stable under normal conditions of storage and operation. Avoid contact with incompatible materials. Keep away from heat, flame, and other sources of ignition.
Stable X	Unstable	Incompatibility *(materials to avoid)* Reacts violently in contact with strong oxidizers, such as chlorine, bromine, fluorine.
Hazardous Polymerization		Conditions to Avoid Under normal conditions of temperature and pressure, hazardous polymerization of benzyl acetate is not expected to occur.
May Occur	Will Not Occur X	Hazardous Decomposition or By-products When heated to decomposition, benzyl acetate emits acrid, poisonous, and irritating oxides of carbon.

SECTION VI - HEALTH HAZARD DATA

Primary Route(s) of Entry:	Inhalation?	Absorption (skin)?	Ingestion?
	X		X

Health Hazards

INHALATION: Chemical's low volatility makes vapor inhalation unlikely but possible. Antipsychotic effects are possible, unspecified respiratory effects, and possible urinary system effects.

SKIN & EYES: Irritation of the skin with sensitization in some individuals. Eye irritation is likely.

INGESTION: Gastrointestinal tract irritation, nausea, and vomiting.

Carcinogenicity	NTP Listed?	IARC Cancer Review Group?	OSHA Regulated?	Target Organs?
Questioned Human Suspected Animal	No	Group 3	No	Respiratory system; skin; eyes.

Medical Conditions Generally Aggravated by Exposure

None reported.

Emergency and First-aid Procedures

Eye contact: Flush immediately with water for 15 minutes (minimum); seek medical attention. Skin contact: Remove all contaminated clothing. Immediately wash area with large amounts of soap and water. Seek medical assistance. For inhalation: Remove the person from exposure. Provide respiratory assistance and CPR. Transfer to medical facility. If swallowed: Seek medical attention immediately. Call poison control center for advice. Give 1-2 glasses of water and induce vomiting unless advised otherwise. Never give an unconscious or convulsing person anything by mouth.

SECTION VII - PRECAUTIONS FOR SAFE HANDLING AND USE

Steps to be Taken in Case Material is Released or Spilled

Contain spills using absorbent material (vermiculite, or other). Ventilate area of spill. Have water source available in case of fire. Store materials in DOT-approved containers. Restrict those not involved in cleanup from entering area. Notify appropriate authorities, as required.

Preferred Waste Disposal Method

Incineration in chemical incinerator equipped with scrubber and afterburner.

Precautions to be Taken in Handling and Storage

Store in tightly closed containers in a cool, dark, well-ventilated area. Protect containers from physical damage. Keep heat, fire, and flame away. If possible, automatically transfer liquids between containers.

Other Precautions and Warnings

Use non-sparking tools to open and close containers. Ground and bond all metal containers.

SECTION VIII - CONTROL MEASURES AND PERSONAL PROTECTIVE EQUIPMENT

Respiratory Protection *(specify type)*

No exposure levels established and benzyl acetate may be a carcinogen. For best protection, recommend using a MSHA/NIOSH approved supplied-air respirator or a self-contained breathing apparatus (SCBA) with full facepiece operated in pressure demand or other positive pressure mode.

Ventilation

Local exhaust or general mechanical systems recommended.

Protective Gloves	Eye Protection	Other Protective Clothing
Impervious Gloves (avoid rubber)	Chemical Goggles or Face Mask	Impermeable Apron

Work/Hygiene Practices

Always wash hands thoroughly after using chemical; never bring food, drink, or smoking materials into vicinity of chemicals.

BENZYL ACETATE

$CH_3COOCH_2C_6H_5$ CAS: 140-11-4

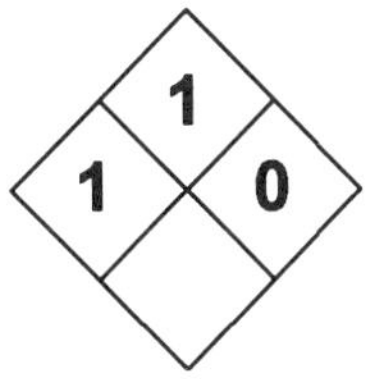

IDENTIFICATION AND TYPICAL USES

Benzyl acetate is a colorless or water-white liquid with a floral, fruity, or pear-like odor and pungent, bitter taste. It is used in the production of fragrances (artificial jasmine and other perfumes); in soap perfumes; as a flavoring agent; as a solvent and high boiler for cellulose acetate and nitrate, and natural and synthetic resins; in oils, lacquers, and polishes; in printing inks; and in varnish removers.

RISK ASSESSMENT: HEALTH

General Assessment

Benzyl acetate is a human poison by *inhalation* causing toxic systemic effects. It is mildly toxic by *ingestion*. Skin contact can produce localized effects but absorption has not been reported. It has been shown to cause cancer in some laboratory studies with test animals and its human carcinogenicity should therefore be considered suspect.

Inhalation of benzyl acetate vapors is not likely to occur under normal operating conditions because of its rather low volatility. However, inhalation is still possible and, therefore, presents a health exposure risk that must be managed. Should an exposure occur by inhalation, it will likely result in irritation of the eyes, nose, throat, and upper respiratory tract, including the mucosa (particularly the gums and respiratory passages). In high concentrations, it may cause unspecified antipsychotic effects with action on the central nervous system. Urinary system effects have also been reported.

Skin contact results in the removal of the skin's natural protective oily layer on its surface with subsequent drying, cracking, and possible sensitization. Eye contact can cause mild to moderate irritation.

Ingestion of benzyl acetate may cause gastrointestinal tract irritation with nausea and vomiting.

Acute Health Effects

The following acute (short-term) health effects may occur immediately or shortly after exposure to benzyl acetate:

Skin: Mild to moderate irritation. May cause sensitization in some individuals.

Eye: Mild to moderate irritation.

Lung: Irritation of the eyes, nose, throat and lungs (especially the mucus membranes).

Other: Possible antipsychotic effects and urinary system reactions.

Chronic Health Effects

The following chronic (long-term) health effects may occur at some time after exposure to benzyl acetate and can last for months or even years:

Cancer Hazards: According to the references, benzyl acetate has been shown to produce tumors of the pancreas, liver, and stomach in some test animals. Human data are inconclusive but carcinogenicity is considered questionable at this time.

Reproduction: According to the references, benzyl acetate has not been adequately tested for its ability to adversely affect reproduction in test animals.

Other Known Effects: Once inside the body, benzyl acetate will hydrolyze to benzyl alcohol, then oxidize to benzoic acid, and is excreted as hippuric acid.

Recommended Risk-Reduction Measures

Personnel should avoid direct contact with benzyl acetate. Its human carcinogenic properties are questioned but not completely understood at this time. The best risk reduction measure is to use a less toxic chemical as a substitute for benzyl acetate. If this is not possible or feasible, then *engineering controls* are the most effective method of reducing exposures. The best protection is to enclose operations and/or provide local exhaust ventilation at the site of chemical release. While not always operationally feasible, isolating operations can also reduce exposure risk.

Using respiratory protection is less effective than the controls mentioned above, but is still advisable whenever working with or around benzyl acetate. No exposure levels have been established for benzyl acetate. This does not mean that exposure is without risk. It is a questionable human carcinogen and no exposure level may be entirely safe. For the best protection, use a self-contained breathing apparatus (SCBA) with full facepiece and operated in pressure demand mode or a supplied-air respirator with full facepiece operated in pressure demand or other positive pressure mode. If a full facepiece is not available, then chemical goggles should be worn to protect the eyes. Whenever a chemical splash hazard exists, a face shield and a protective apron should be worn. To prevent hand and skin exposures, impervious gloves should be used. Glove manufacturers should be contacted to obtain permeation studies *before* gloves are selected.

Administrative controls should also be in place to minimize the potential for human exposures. These may include written procedures or policies which specify the methods and techniques that will be practiced whenever personnel are to work with benzyl acetate.

All personnel should receive *training* on the use, hazards, protective measures, emergency actions, and other precautions per 29 CFR 1910.1200 (Hazard Communication), prior to the first assignment in an area where benzyl acetate is used or stored.

If symptoms develop or overexposure is suspected, the following medical tests are recommended:

☑ Liver and kidney function tests.

☑ Urinalysis with specific attention on excreted hippuric acid.

☑ Lung function tests.

☑ Skin testing with dilute benzyl acetate to help diagnose allergy (performed by a qualified allergist).

Any medical evaluation should include a careful history of past and present symptoms with an examination. Medical tests that look for existing damage are not a substitute for controlling exposures. Also, since smoking can cause heart disease, emphysema, and other respiratory disorders, smokers exposed to benzyl acetate may experience symptoms more quickly than nonsmokers under the same conditions of exposure. Prudent risk management requires careful consideration of *all* possible factors that may be causing the appearance of exposure symptoms in the workplace.

Other methods to reduce exposure include:

☑ Where possible, enclose operations and use local exhaust ventilation at the site of chemical release. If local exhaust ventilation or enclosure is not used, respiratory protection should be mandatory.

☑ Always ensure that proper protective clothing is worn when using chemical substances.

☑ Wash thoroughly immediately after exposure to benzyl acetate and at the end of the work shift or before eating, drinking, or smoking.

☑ Hazard warning information should be posted in the work area. In addition, as part of an on-going education and training program, all information on the health and safety hazards of benzyl acetate should be communicated to all exposed and potentially exposed workers.

☑ Never eat, drink, or smoke in areas were benzyl acetate is used, handled or stored.

RISK ASSESSMENT: ENVIRONMENT

General Assessment

The environment is at risk of exposure during transportation, storage, disposal, or destruction of benzyl acetate. In almost every scenario, the threat of environmental exposure is contingent upon the proper handling of the chemical substance. Accidental spills, large or small, can result in fire, explosion, and possible contamination of the surrounding environmental mediums (water, soil, and air).

Benzyl acetate is considered a class IIIA combustible liquid (according to OSHA 29 CFR 1910.106). This means that it can be ignited as well as support the combustion of other materials. Also, because it is incompatible with a number of common oxidizers, contact can result in violent and explosive reactions. These characteristics require special consideration during any emergency situation involving a leak or spill of benzyl acetate.

The proper disposal/destruction method for benzyl acetate is to burn it in a chemical incinerator equipped with an afterburner and air scrubber. Benzyl acetate can enter the environment through unchecked industrial discharges into effluents and through spills.

Acute Ecological Effects

Acute (short-term) toxic effects may include the death of animals, birds, or fish, and death or low growth rate in plants. Acute effects are seen 2 to 4 days after animals or plants are exposed to benzyl acetate. This chemical has moderate acute toxicity to aquatic life. Insufficient data are available to evaluate or predict the short-term effects of benzyl acetate to plants, birds, or terrestrial animals.

Chronic Ecological Effects

Chronic toxic effects may include shortened life span, reproductive problems, lower fertility, and changes in appearance or behavior in exposed animals. These effects can be seen long after first exposure(s) to toxic chemicals. Benzyl acetate has moderate chronic toxicity to aquatic life. Insufficient data are available to evaluate or predict the long-term effects of benzyl acetate to plants, birds, or land animals.

Water Solubility

Benzyl acetate is slightly soluble in water. Concentrations of 1 to 100 milligrams will mix with a liter of water.

Persistence in the Environment

Benzyl acetate is slightly persistent in water, with a half-life of between 2 to 20 days. The half-life of a pollutant is the amount of time it takes for one-half of the chemical to be degraded.

Bioaccumulation in Aquatic Organisms

Some substances increase in concentration, or *bioaccumulate,* in living organisms as they breathe contaminated air, drink contaminated water, or eat contaminated food. These chemicals can become concentrated in the tissues and internal organs of animals as well as humans.

The concentration of benzyl acetate found in fish tissues is expected to be somewhat higher then the average concentration of benzyl acetate in water from which the fish was taken.

Recommended Risk-Reduction Measures

Proper training of all transporters will reduce the likelihood of a mishap or accident resulting in a leak or spill to the environment. The correct use of DOT labeling (per the MSDS) on all containers, trucks, and rail cars will enable emergency responders to react properly and quickly to any disaster thereby reducing the potential risk to the environment and to personnel.

Storage of benzyl acetate should be segregated from incompatible chemicals to minimize the risk of cross-contamination or contact. Buildings designated for storage should be equipped with appropriate fire protection systems (alarms, sprinklers, emergency lighting, portable extinguishers). If a spill or leak to the environment has occurred, fire department, emergency response, and/or hazardous materials spill personnel should be notified immediately. Cleanup should be attempted only by those trained in proper spill containment procedures. Contaminated soils should be removed for incineration and replaced with clean soil. If benzyl acetate should contact the water table, aquifer, or navigable waterway, time is of the essence. It is slightly soluble in water and, therefore, total containment and remediation may not be possible. When such spills occur, the local and/or state emergency response authorities must be notified. A comprehensive emergency response or disaster preparedness/recovery plan should be in place prior to any operations involving the use, transportation, storage, or disposal of benzyl acetate. If benzyl acetate is spilled or leaked, the following specific steps are recommended:

- ☑ Restrict persons not wearing protective clothing from area of spill or leak until cleanup is complete and area can be opened for normal work.
- ☑ Ventilate area of spill, remove ignition sources.
- ☑ Absorb liquids in vermiculite, dry sand, earth, or a similar material and deposit in sealed containers. Use non-sparking tools.
- ☑ It may be necessary to dispose of benzyl acetate as a hazardous waste. The responsible state agency or the regional office of the federal Environmental Protection Agency (EPA) should be contacted for specific recommendations.

RISK ASSESSMENT: BUSINESS

General Assessment

Accidents or mishaps involving benzyl acetate can present a moderate threat to business operations. The loss or damage of equipment or facilities can significantly affect fiscal viability. Lawsuits that may result from personnel injury/death, public exposures, and/or environmental contamination will also require a serious expenditure of resources.

Media attention surrounding an injury, death, or environmental damage can also result in a loss of profits and loss of current as well as future business. Always remember that anytime the words "cancer" or "carcinogen" are used, public hysteria, emotion, and ignorance can run equally high. This is a matter for careful consideration during the development and implementation of any public relations policies.

U *Recommended Risk-Reduction Measures*

Company attorneys, safety and health professionals, and environmental specialists should be involved in the development of any procedures or policies intended to manage the use of chemicals in the workplace. A company official should be pre-designated as a public relations officer with specific training in dealing with the press. Corporate plans and policies should be developed, approved, and implemented long before any need for such arises.

REFERENCES

American Conference of Governmental Industrial Hygienists. 1988. *Documentation of the Threshold Limit Values and Biological Exposure Indices,* 5th Edition (with updates). Cincinnati: ACGIH

Anthony, C. P., and N. J. Kolthoff. 1971. *Textbook of Anatomy and Physiology.* St. Louis: C. V. Mosby Company.

Bloom, A. D. and F. J. DeSerres. 1995. *Ecotoxicity and Human Health.* New York: CRC Press/Lewis Publishers.

Brusick, D. J. 1994. *Methods for Genetic Risk Assessment.* New York: CRC Press/Lewis Publishers.

Bueche, J. R. 1972. *Principles of Physics,* 2nd Edition. New York: McGraw-Hill.

Burrell, R., D. K. Flaherty, and L. J. Sauers. 1992. *Toxicology of the Immune System.* New York: Van Nostrand Reinhold.

Calabrese, E. J. 1994. *Biological Effects of Low Level Exposures.* New York: CRC Press/Lewis Publishers.

Cockerman, L. G., B. S. Shane. 1994. *Basic Environmental Toxicology.* New York: CRC Press/Lewis Publishers.

Cothern, C. R., N. P. Ross. 1994. *Environmental Statistics, Assessment, and Forecasting.* New York: CRC Press/Lewis Publishers.

Danse, I. R. 1991. *Common Sense Toxics in the Workplace.* New York: Van Nostrand Reinhold.

Environmental Protection Agency. 1983. *Chemical Hazard Information Profile: Benzyl acetate.* Washington, D.C.: U.S. EPA.

Gardner, R. W. 1994. *Chemical Intolerance.* New York: CRC Press/Lewis Publishers.

Hallenbeck, W. H. 1993. *Quantitative Risk Assessment for Environmental and Occupational Health,* 2nd Edition. New York: CRC Press/Lewis Publishers.

Hodgson, E., R. B. Mailman, and J. E. Chambers. 1988. *Dictionary of Toxicology.* London: The MacMillian Company.

Hosty, J. W., and P. Foster. 1990. *A Practical Guide to Chemical Spill Response.* New York: Van Nostrand Reinhold.

Howard, P. H., et. al. 1991. *Handbook of Environmental Degradation Rates.* New York: CRC Press/Lewis Publishers.

International Agency for Research on Cancer (IARC). 1987. *IARC Monographs on the Evaluation of Carcinogenic Risk to Humans, Overall Evaluations of Carcinogenicity: An updating of IARC Monographs Volumes 1-42.* Lyon: IARC.

James, A., and M. Lord. 1992. *Index of Chemical and Physical Data.* New York: Van Nostrand Reinhold.

Jones, F. E. 1994. *Toxic Organic Vapors in the Workplace.* New York: CRC Press/Lewis Publishers.

Kamrin, M. 1 A. 1988. *Toxicology: A Primer on Principles and Applications.* Chelsea, Michigan: Lewis Publishers.

Keith, L. H. and M. M. Walker. 1995. *Handbook of Air Toxics.* New York: CRC Press/Lewis Publishers.

Landis, W. G., M. H. Yu. 1995. *Introduction to Environmental Toxicology.* New York: CRC Press/Lewis Publishers.

Langerman, N. 1994 *Precautionary Labels for Chemical Containers.* New York: CRC Press/Lewis Publishers.

Lauwerys, R. R., P. Hoet. 1993. *Industrial Chemical Exposure.* New York: CRC Press/Lewis Publishers.

Lide, D. R. 1995. *Handbook of Chemistry and Physics,* 76th Edition. New York: CRC Press/Lewis Publishers.

Lewis, Richard J., Sr. 1992. *Sax's Dangerous Properties of Industrial Materials,* Volumes 1, 2, and 3, 8th Edition. New York: Van Nostrand Reinhold.

Lisella, Frank S. 1994. *The VNR Dictionary of Environmental Health and Safety.* New York: Van Nostrand Reinhold.

Manahan, S. E. 1994. *Environmental Chemistry,* 6th Edition. New York: CRC Press/Lewis Publishers.

Manahan, S. E. 1992. *Toxicological Chemistry,* 2nd Edition. New York: CRC Press/Lewis Publishers.

Meyer, Eugene. 1990. *Chemistry of Hazardous Materials.* Englewood Cliffs, NJ: Prentice Hall.

Patnaik, P. 1992. *A Comprehensive Guide to the Hazardous Properties of Chemical Substances.* New York: Van Nostrand Reinhold.

Proctor, N. H., and J. P. Hughes. 1991. *Chemical Hazards of the Workplace.* New York: Van Nostrand Reinhold.

Reeves, A. L. 1979. *Handbook on the Toxicology of Metals.* New York: Elsevier North-Holland.

Richardson, M. 1995. *Dictionary of Substance and Their Effects,* Vol. I-VI. New York: CRC Press/Lewis Publishers.

Royal Society of Chemistry. 1992. *Chemical Safety Data Sheets, Volumes 1-5.* Newcastle-upon-Tyne, UK: Athenaeum Press..

Rylander, R., R. R. Jacobs. 1994. *Organic Dusts.* New York: CRC Press/Lewis Publishers.

MATERIAL SAFETY DATA SHEET

CHEMICAL NAME	
BENZYL ALCOHOL	**CRC PRESS/LEWIS PUBLISHERS DISCLAIMER:** THE INFORMATION AND RECOMMENDATIONS PRESENTED HEREIN ARE BASED ON SOURCES BELIEVED TO BE RELIABLE. CRC MAKES NO REPRESENTATION ON ITS COMPLETENESS OR ACCURACY. IT IS THE RESPONSIBILITY OF THE USER TO DETERMINE THE CHEMICAL'S SUITABILITY FOR ITS INTENDED USE, THE CHEMICAL'S SAFE USE, AND THE CHEMICAL'S PROPER DISPOSAL. NO REPRESENTATIONS AND/OR WARRANTIES, EITHER EXPRESSED OR IMPLIED, OF THE MERCHANTABILITY OR FITNESS FOR A PARTICULAR PURPOSE, OR OF ANY OTHER NATURE, ARE MADE WITH RESPECT TO THE INFORMATION PROVIDED IN THIS MSDS OR TO THE CHEMICAL TO WHICH INFORMATION MAY REFER. CRC NEITHER ASSUMES NOR AUTHORIZES ANY OTHER PERSON TO ASSUME FOR IT, ANY OTHER ADDITIONAL RESPONSIBILITY OR LIABILITY FOR THE USE OF, OR RELIANCE UPON, THIS INFORMATION.

HAZARD WARNING INFORMATION

HEALTH	FIRE	REACTIVE	OTHER	DEGREE OF HAZARD	COLOR CODING	OTHER CODES
2	1	0		0 = Minimum Hazard 1 = Slight Hazard 2 = Moderate Hazard 3 = Serious Hazard 4 = Severe Hazard	HEALTH = BLUE FIRE = RED REACTIVITY = YELLOW OTHER = WHITE	OX = Oxidizer ACID = Acid ALK = Alkali COR = Corrosive W = Use No Water

SECTION I - GENERAL INFORMATION

Characterization: **Alcohol**	RCRA Number: **None**	EPA Class: **Not Applicable**
DOT Proper Shipping Name: **Combustible Liquid, N.O.S.**	Chemical Abstract Service (CAS) Number: **100-51-6**	
DOT Hazard Class and Label Requirements: **Combustible Liquid**	DOT Emergency Guide Code: **26**	
DOT Identification Number: **NA 1993**	Chemical Formula: **$C_6H_5CH_2OH$**	

Synonyms

Benzal alcohol; benzenecarbinol; α-hydroxytoluene; benzene methanol; phenolcarbinol; α-toluene.

SECTION II - HAZARDOUS INGREDIENTS AND IDENTITY INFORMATION

Hazardous Components (specific identity)	OSHA Exposure Criteria	NIOSH Exposure Criteria	Immediately Dangerous to Life and Health (IDLH)	ACGIH Exposure Criteria
Benzyl alcohol (derivation: By hydrolysis of benzyl chloride; from benzaldehyde by catalytic reduction or Cannizzaro reaction).	**PEL:** Not Established **STEL:** Not Established	**REL:** Not Established **STEL:** Not Established	**Not Determined**	**TLV:** Not Established **STEL:** Not Established

SECTION III - PHYSICAL AND CHEMICAL CHARACTERISTICS

Boiling Point: **403°F (206°C)**	Specific Gravity (H_2O = 1): **1.04**
Vapor Pressure (mm Hg): **.15 at 77°F (25°C)**	Molecular Weight: **108.15**
Vapor Density (Air = 1): **3.7**	Melting Point: **4.5°F (-15.3°C)**

Solubility

Somewhat soluble in water (3.5%). Soluble in acetone, benzene, chloroform, ether, aromatic hydrocarbons, and vegetable oils.

Appearance and Odor

Colorless or water-white liquid with a faint, aromatic odor and a sharp, pungent, burning taste.

SECTION IV - FIRE AND EXPLOSION HAZARD DATA

Flash Point (method used): **200°F (93°C) closed cup**	Explosive Limits in Air % by Volume: **LEL: Not Determined** **UEL: Not Determined**
NFPA Classification: **Class IIIA Combustible Liquid**	Autoignition Temperature: **817°F (436°C)**

Extinguishing Media

Dry chemical, carbon dioxide, or alcohol-resistant foam. Water may cause frothing.

Special Fire Fighting Procedures

Poisonous gases are produced in fire. Wear full protective clothing and self-contained breathing apparatus (SCBA). Continue to cool containers with water after fire is extinguished. Move containers from fire area if it can be done without risk.

Unusual Fire and Explosion Hazards

Vapors are heavier than air and may travel a distances to ignition source to flashback. Vapors may explode if ignited in an enclosed area.

SECTION V - REACTIVITY DATA

Stability		Conditions to Avoid Benzyl alcohol slowly oxidizes in air to benzaldehyde and benzoic acid. Keep stable in air-tight containers. Avoid contact with incompatible materials. Keep away from heat, flame, and other ignition sources.
Stable X	Unstable	Incompatibility *(materials to avoid)* Incompatible with acids (especially sulfuric), hydrogen bromides + iron at temperatures above 212°F (100°C), and oxidizers (such as chlorine, bromine, and fluorine). It will attack some plastic materials.
Hazardous Polymerization		Conditions to Avoid Under normal conditions of temperature and pressure, benzyl alcohol is not expected to undergo hazardous polymerization.
May Occur	Will Not Occur X	Hazardous Decomposition or By-products When heated to decomposition, benzyl alcohol emits acrid and irritating fumes, including oxides of carbon (monoxide and dioxide).

SECTION VI - HEALTH HAZARD DATA

Primary Route(s) of Entry:	Inhalation?	Absorption (skin)?	Ingestion?
	X	X	X

Health Hazards

INHALATION: There may be slight irritation of the eyes, nose, and mucosa of the respiratory tract. Exposure can also cause stupor, fatigue, and sleepiness. There can be headache, nausea, vomiting, diarrhea, respiratory paralysis, convulsion, unconsciousness, and death. It can also cause neurologic, liver, kidney, and blood disturbances.

ABSORPTION: As a solvent, causes skin dryness, cracking, possible dermatitis. Eye contact causes in burning, stinging, tearing, corneal injury, and hyperemia of the conjunctiva. May pass through intact skin to cause toxic systemic effects.

INGESTION: Gastrointestinal irritation with nausea, stomach cramps, diarrhea, and tissue ulceration.

Carcinogenicity	NTP Listed?	IARC Cancer Review Group?	OSHA Regulated?	Target Organs?
Unknown Human Unknown Animal	No	No	No	Skin, eyes, CNS, liver, blood, kidneys.

Medical Conditions Generally Aggravated by Exposure

Pre-existing skin, liver, kidney, nervous system, or blood disorders may be aggravated by exposure.

Emergency and First-aid Procedures

Eye contact: Flush immediately with water for 15 minutes (minimum); seek medical attention. Skin contact: Remove all contaminated clothing. Immediately wash area with large amounts of soap and water. Seek medical assistance. For inhalation: Remove the person from exposure. Provide respiratory assistance and CPR. Transfer to medical facility. If swallowed: Seek medical attention immediately. Give 1-2 glasses of water to dilute. Do NOT induce vomiting. Seek medical attention immediately.

SECTION VII - PRECAUTIONS FOR SAFE HANDLING AND USE

Steps to be Taken in Case Material is Released or Spilled

Contain spills using vermiculite (or other absorbent). Ventilate area of spill. Remove ignition sources. Have extinguishing agent available in case of fire. Store materials in DOT-approved containers.

Preferred Waste Disposal Method

Incineration in chemical incinerator equipped with scrubber and afterburner.

Precautions to be Taken in Handling and Storage

Store in air-tight closed containers in a cool, dark, well-ventilated area. Protect containers from physical damage. Keep heat, fire, and flame away. Outside or underground storage is recommended.

Other Precautions and Warnings

Ground and bond all metal containers.

SECTION VIII - CONTROL MEASURES AND PERSONAL PROTECTIVE EQUIPMENT

Respiratory Protection *(specify type)*

No exposure limits have been established. For low exposures, an organic vapor air-purifying respirator may suffice. Greater protection is obtained using a supplied-air respirator or a self-contained breathing apparatus (SCBA) with full facepiece operated in pressure demand or other positive pressure mode.

Ventilation

Local exhaust or general mechanical systems recommended.

Protective Gloves	Eye Protection	Other Protective Clothing
Impervious Butyl Rubber Gloves	Chemical Goggles or Face Mask	Impermeable Apron or Protective Suit

Work/Hygiene Practices

Always wash hands thoroughly after using chemical; never bring food, drink, or smoking materials into vicinity of chemicals.

BENZYL ALCOHOL

$C_6H_5CH_2OH$ CAS: 100-51-6

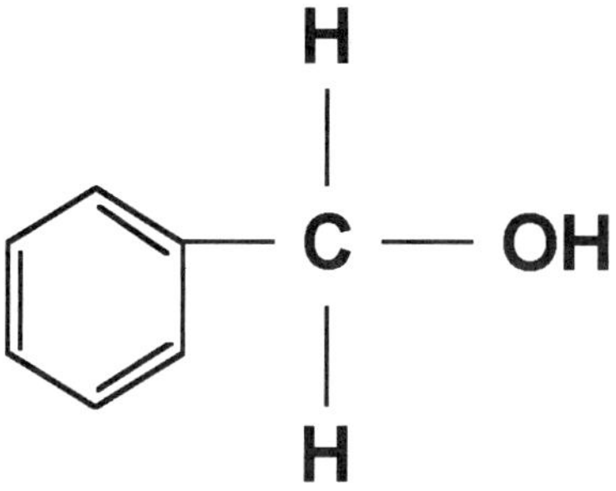

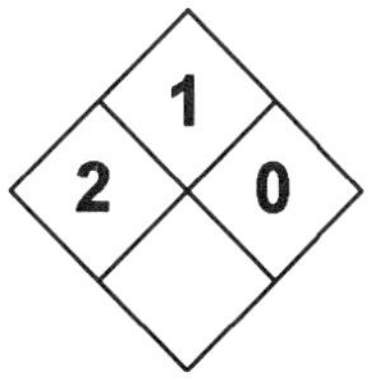

IDENTIFICATION AND TYPICAL USES

Benzyl alcohol is a colorless or water-white liquid with a faint, aromatic odor and a sharp, pungent, burning taste. It is used in the manufacture of perfumes and flavors; in photographic development for color movie films; for dyeing nylon filament, other textiles, and sheet plastics; as a solvent for dyestuffs, cellulose esters, casein, and waxes; in heat-sealing polyethylene films; as an intermediate for benzyl esters and ethers; as a bacteriostat; in the manufacture of some cosmetics, ointments, and emulsions; and in ball-point pen inks and stencil inks.

RISK ASSESSMENT: HEALTH

General Assessment

Benzyl alcohol is a human poison by *ingestion* and is considered moderately toxic by *inhalation* and skin *absorption*. There are no reports of any carcinogenic or teratogenic effects subsequent to exposure to benzyl alcohol. However, mutation data have been reported and some scientists believe that such chemicals may pose a cancer risk in the long-term. Additional research is required in this regard.

Inhalation can cause irritation of the eyes, nose, throat, and respiratory tract. There can be headache, nausea, vomiting, diarrhea, respiratory and muscular paralysis, convulsions, unconsciousness, and death. Exposure can cause neurological, liver, kidney, and blood disturbances. Human data are lacking for adequate evaluation of health exposure risk. The majority of data appear to follow from exposures of infants given medication containing benzyl alcohol as a preservative. When administered rectally, it caused gradual neurological deterioration, severe acidosis, gasping breath, skin disorders, decreased blood pressure, and liver and kidney failure followed by cardiovascular collapse and death. It is not known if occupational exposure by inhalation or any other route will have these effects.

Liquid contact with the eyes causes immediate burning and stinging with lachrymation and reflex closure of the lids. There may be injury to the corneal epithelium and possible hyperemia (excessive blood) of the conjunctiva. Skin contact results in drying and cracking which can lead to secondary infections and dermatitis.

Ingestion of benzyl alcohol is not likely to occur in the industrial environment. However, if it does, symptoms can include gastrointestinal tract irritation with nausea, vomiting, cramps, and diarrhea. There may be ulceration of stomach tissue or other tissues in the digestive system.

Acute Health Effects

The following acute (short-term) health effects may occur immediately or shortly after exposure to benzyl alcohol:

Skin: Causes dryness and cracking leading to dermatitis and possible infection.

Eye: Severe irritation with burning and stinging with possible damage to the cornea and conjunctiva.

Lung: Irritation of the eyes, nose, throat, and respiratory tract.

Other: High concentrations can cause neurological, liver, and kidney problems and possible blood disturbances.

Chronic Health Effects

The following chronic (long-term) health effects may occur at some time after exposure to benzyl alcohol and can last for months or even years:

Cancer Hazards: It is not known if exposure to benzyl alcohol will lead to cancer in humans. There are some mutation data reported but the references are limited in this area of study.

Reproduction: According to information presented in the references, it is not currently known if exposure to benzyl alcohol will have teratogenic or other reproductive effects in humans or test animals.

Other Chronic Effects: Very high or prolonged skin exposure may cause contact dermatitis.

Recommended Risk-Reduction Measures

The best risk-reduction measure is to use a less toxic chemical as a substitute for a benzyl alcohol. If substitution is not possible or operationally feasible, *engineering controls* are the most effective method of reducing exposures. The best protection is to enclose operations and/or provide local exhaust ventilation at the site of chemical release. While not always operationally feasible, isolating operations can also reduce exposure risk.

Using respiratory protection is less effective than the controls mentioned above, but is still advisable whenever working with or around benzyl alcohol. No exposure levels have been established for benzyl alcohol. This does not mean that exposure is without risk. For relatively low or infrequent exposure, an air-purifying respirator with an organic vapor cartridge may suffice. For higher exposures, a supplied-air respirator with full facepiece operated in positive pressure mode, or a self-contained breathing apparatus (SCBA) with full facepiece and operated in pressure demand mode are the recommended respiratory protection methods of choice. If a full facepiece is not available, then chemical goggles should be worn to protect the eyes. Whenever a chemical splash hazard exists, a face shield and a protective apron should be worn. To prevent hand and skin exposures, impervious gloves should be used. Glove manufacturers should be contacted and permeation data obtained *before* final glove selections are made.

Administrative controls should also be in place to minimize the potential for human exposures. These may include written procedures or policies which specify the methods and techniques that will be practiced whenever personnel are to work with benzyl alcohol.

All personnel should receive *training* on the use, hazards, protective measures, emergency actions, and other precautions per 29 CFR 1910.1200 (Hazard Communication), prior to the first assignment in an area where benzyl alcohol is used or stored.

If symptoms develop or overexposure is suspected, the following medical tests are recommended:

- ☑ Liver and kidney function tests.
- ☑ Skin testing with dilute benzyl alcohol to help diagnose allergy (performed by a qualified allergist).
- ☑ Evaluation of nervous system function by a qualified neurologist.
- ☑ Complete blood count (CBC) with blood gases.

Any medical evaluation should include a careful history of past and present symptoms with an examination. Medical tests that look for existing damage are not a substitute for controlling exposures. Also, since consuming large quantities of alcoholic beverages can lead to liver dysfunction and even cancer, persons with alcohol addiction who are exposed to benzyl alcohol on the job may develop symptoms much quicker and with greater intensity than those who do not drink under identical exposure conditions. Prudent risk management requires careful consideration of *all* possible factors which may be causing the appearance of exposure symptoms.

Other methods to reduce exposure include:

- ☑ Where possible, enclose operations and use local exhaust ventilation at the site of chemical release. If local exhaust ventilation or enclosure is not used, respiratory protection should be mandatory. Also, automatic transfer of benzyl alcohol from storage to process containers is recommended.
- ☑ Always ensure that proper protective clothing is worn when using chemical substances.
- ☑ Wash thoroughly immediately after exposure to benzyl alcohol and at the end of the work shift or before eating, drinking, or smoking.
- ☑ Hazard warning information should be posted in the work area. In addition, as part of an on-going education and training program, all information on the health and safety hazards of benzyl alcohol should be communicated to all exposed or potentially exposed workers.

RISK ASSESSMENT: ENVIRONMENT

General Assessment

The environment is at risk of exposure during transportation, storage, disposal, or destruction of benzyl alcohol. In almost every scenario, the threat of environmental exposure is contingent upon the proper handling of the chemical substance. Accidental spills, large or small, can result in fire, explosion, and possible contamination of the surrounding environmental mediums (water, soil, and air).

Benzyl alcohol is considered a class IIIA combustible liquid (according to OSHA 29 CFR 1910.106). This means it will ignite and burn as well as accelerate the burning rate of other combustible materials. Also, because it is incompatible with a number of common materials, especially strong oxidizers and many acids, contact can result in violent and explosive reactions. It is also known to attack some plastic materials. These characteristics require special consideration during any emergency situation involving a leak or spill of benzyl alcohol.

Benzyl alcohol can enter the environment through unchecked industrial discharges into effluents and through spills.

Acute Ecological Effects

Acute (short-term) toxic effects may include the death of animals, birds, or fish, and death or low growth rate in plants. Acute effects are seen 2 to 4 days after animals or plants are exposed to benzyl alcohol. This chemical has moderate acute toxicity to aquatic life. Insufficient data are available to evaluate or predict the short-term effects of benzyl alcohol to plants, birds, or terrestrial animals.

Chronic Ecological Effects

Chronic toxic effects may include shortened life span, reproductive problems, lower fertility, and changes in appearance or behavior in exposed animals. These effects can be seen long after first exposure(s) to toxic chemicals. Benzyl alcohol has moderate chronic toxicity to aquatic life. Insufficient data are available to evaluate or predict the long-term effects of benzyl alcohol to plants, birds, or land animals.

Water Solubility

Benzyl alcohol is slightly soluble in water. Concentrations ranging between 1 to 100 milligrams will mix with a liter of water.

Persistence in the Environment

Benzyl alcohol is slightly persistent in water, with a half-life of between 2 to 20 days. The half-life of a pollutant is the amount of time it takes for one-half of the chemical to be degraded. About 93% of benzyl alcohol will eventually end up in the air; the remainder will end up in water.

Bioaccumulation in Aquatic Organisms

Some substances increase in concentration, or *bioaccumulate,* in living organisms as they breathe contaminated air, drink contaminated water, or eat contaminated food. These chemicals can become concentrated in the tissues and internal organs of animals as well as humans.

The concentration of benzyl alcohol found in fish tissues is expected to be about the same or possibly lower than the average concentration of benzyl alcohol in water from which the fish was taken.

Recommended Risk-Reduction Measures

Proper training of all transporters will reduce the likelihood of a mishap or accident resulting in a leak or spill to the environment. The correct use of DOT labeling (per the MSDS) on all containers, trucks, and rail cars will enable emergency responders to react properly and quickly to any disaster thereby reducing the potential risk to the environment and to personnel.

Storage of benzyl alcohol should be segregated from incompatible chemicals to minimize the risk of cross-contamination or contact. Buildings designated for storage should be equipped with appropriate fire protection systems (alarms, sprinklers, emergency lighting, portable extinguishers).

If a spill or leak to the environment has occurred, fire department, emergency response, and/or hazardous materials spill personnel should be notified immediately. Cleanup should be attempted only by those trained in proper spill containment procedures. Contaminated soils should be removed for incineration and replaced with clean soil. If benzyl alcohol should contact the water table, aquifer, or navigable waterway, time is of the essence. It is slightly soluble in water and, therefore, total containment and remediation may not be entirely possible. When such spills occur, the local and/or state emergency response authorities must be notified. A comprehensive emergency response or disaster preparedness/recovery plan should be in place prior to any operations involving the use, transportation, storage, or disposal of benzyl alcohol. If benzyl alcohol is spilled or leaked, the following specific steps are recommended:

☑ Restrict persons not wearing protective clothing from area of spill or leak until cleanup is complete and area can be opened for normal work.

☑ Ventilate area and remove ignition sources.

☑ Absorb liquids in vermiculite, dry sand, earth, or a similar material and deposit in sealed containers. Use non-sparking tools. Avoid plastic shovels and other plastic tools since benzyl alcohol has been know to attack plastics. Use brass or other non-sparking metal tools instead.

☑ It may be necessary to dispose of benzyl alcohol as a hazardous waste. The responsible state agency or the regional office of the federal Environmental Protection Agency (EPA) should be contacted for specific recommendations.

RISK ASSESSMENT: BUSINESS

General Assessment

Accidents or mishaps involving benzyl alcohol can present a moderate threat to business operations. The loss or damage of equipment or facilities can significantly affect fiscal viability. Lawsuits that may result from personnel injury/death, public exposures, and/or environmental contamination will also require a serious expenditure of resources. Media attention surrounding an injury, death, or environmental damage can also result in a loss of profits and loss of current as well as future business.

Recommended Risk-Reduction Measures

Company attorneys, safety and health professionals, and environmental specialists should be involved in the development of any procedures or policies intended to manage the use of chemicals in the workplace. A company official should be pre-designated as a public relations officer with specific training in dealing with the press. Corporate plans and policies should be developed, approved, and implemented long before any need for such arises.

REFERENCES

American Conference of Governmental Industrial Hygienists. 1988. *Documentation of the Threshold Limit Values and Biological Exposure Indices*, 5th Edition (with updates). Cincinnati: ACGIH

Anthony, C. P., and N. J. Kolthoff. 1971. *Textbook of Anatomy and Physiology.* St. Louis: C. V. Mosby Company.

Bloom, A. D. and F. J. DeSerres. 1995. *Ecotoxicity and Human Health.* New York: CRC Press/Lewis Publishers.

Brusick, D. J. 1994. *Methods for Generic Risk Assessment.* New York: CRC Press/Lewis Publishers.

Bueche, J. R. 1972. *Principles of Physics,* 2nd Edition. New York: McGraw-Hill.

Burrell, R., D. K. Flaherty, and L. J. Sauers. 1992. *Toxicology of the Immune System.* New York: Van Nostrand Reinhold.

Calabrese, E. J. 1994. *Biological Effects of Low Level Exposures.* New York: CRC Press/Lewis Publishers.

Cothern, C. R., N. P. Ross. 1994. *Environmental Statistics, Assessment, and Forecasting.* New York: CRC Press/Lewis Publishers.

Danse, I. R. 1991. *Common Sense Toxics in the Workplace.* New York: Van Nostrand Reinhold.

Environmental Protection Agency. 1983. *Chemical Hazard Information Profiles.* Washington, D.C.: U.S. EPA.

Fodden, C. R., and J. L. Weddell. 1992. *Hazardous Materials Emergency Action Data.* New York: CRC Press/Lewis Publishers.

Forseberg, K., and L. H. Keith. 1993. *Chemical; Protective Clothing Permeation and Degradation Database.* Chelsea, MI: Lewis Publishers.

Gardner, R. W. 1994. *Chemical Intolerance.* New York: CRC Press/Lewis Publishers.

Hodgson, E., R. B. Mailman, and J. E. Chambers. 1988. *Dictionary of Toxicology.* London: The MacMillian Company.

Hosty, J. W., and P. Foster. 1990. *A Practical Guide to Chemical Spill Response.* New York: Van Nostrand Reinhold.

Howard, P. H., et. al. 1991. *Handbook of Environmental Degradation Rates.* New York: CRC Press/Lewis Publishers.

International Agency for Research on Cancer (IARC). 1987. *IARC Monographs on the Evaluation of Carcinogenic Risk to Humans, Overall Evaluations of Carcinogenicity: An updating of IARC Monographs Volumes 1-42.* Lyon: IARC.

Keith, L. H. and D. B. Walters. (Editors) 1992. *National Toxicology Program's Chemical Database, Volumes I-VII.* Chelsea, MI: Lewis Publishers.

Landis, W. G., M. H. Yu. 1995. *Introduction to Environmental Toxicology.* New York: CRC Press/Lewis Publishers.

Lauwerys, R. R., P. Hoet. 1993. *Industrial Chemical Exposure.* New York: CRC Press/Lewis Publishers.

Lewis, Richard J., Sr. 1992. *Sax's Dangerous Properties of Industrial Materials,* Volumes 1, 2, and 3, 8th Edition. New York: Van Nostrand Reinhold.

Manahan, S. E. 1992. *Toxicological Chemistry,* 2nd Edition. New York: CRC Press/Lewis Publishers.

Meyer, Eugene. 1990. *Chemistry of Hazardous Materials.* Englewood Cliffs, NJ: Prentice Hall.

Patnaik, P. 1992. *A Comprehensive Guide to the Hazardous Properties of Chemical Substances.* New York: Van Nostrand Reinhold.

MATERIAL SAFETY DATA SHEET

CHEMICAL NAME	
BENZYL BROMIDE	**CRC PRESS/LEWIS PUBLISHERS DISCLAIMER:** THE INFORMATION AND RECOMMENDATIONS PRESENTED HEREIN ARE BASED ON SOURCES BELIEVED TO BE RELIABLE. CRC MAKES NO REPRESENTATION ON ITS COMPLETENESS OR ACCURACY. IT IS THE RESPONSIBILITY OF THE USER TO DETERMINE THE CHEMICAL'S SUITABILITY FOR ITS INTENDED USE, THE CHEMICAL'S SAFE USE, AND THE CHEMICAL'S PROPER DISPOSAL. NO REPRESENTATIONS AND/OR WARRANTIES, EITHER EXPRESSED OR IMPLIED, OF THE MERCHANTABILITY OR FITNESS FOR A PARTICULAR PURPOSE, OR OF ANY OTHER NATURE, ARE MADE WITH RESPECT TO THE INFORMATION PROVIDED IN THIS MSDS OR TO THE CHEMICAL TO WHICH INFORMATION MAY REFER. CRC NEITHER ASSUMES NOR AUTHORIZES ANY OTHER PERSON TO ASSUME FOR IT, ANY OTHER ADDITIONAL RESPONSIBILITY OR LIABILITY FOR THE USE OF, OR RELIANCE UPON, THIS INFORMATION.

HAZARD WARNING INFORMATION

HEALTH	FIRE	REACTIVE	OTHER	DEGREE OF HAZARD	COLOR CODING	OTHER CODES
3	2	1	COR ~~W~~	0 = Minimum Hazard 1 = Slight Hazard 2 = Moderate Hazard 3 = Serious Hazard 4 = Severe Hazard	HEALTH = BLUE FIRE = RED REACTIVITY = YELLOW OTHER = WHITE	OX = Oxidizer ACID = Acid ALK = Alkali COR = Corrosive ~~W~~ = Use No Water

SECTION I - GENERAL INFORMATION

Characterization **Halogenated Hydrocarbon**	RCRA Number **None**	EPA Class **Not Applicable**
DOT Proper Shipping Name **Benzyl Bromide**	Chemical Abstract Service (CAS) Number **100-39-0**	
DOT Hazard Class and Label Requirements **Corrosive Material; Poison**	DOT Emergency Guide Code **59**	
DOT Identification Number **UN 1737**	Molecular Formula **C_7H_7Br**	

Synonyms
α-Bromotoluene; bromomethylbenzene; bromophenylmethane.

SECTION II - HAZARDOUS INGREDIENTS AND IDENTITY INFORMATION

Hazardous Components (specific identity)	OSHA Exposure Criteria	NIOSH Exposure Criteria	Immediately Dangerous to Life and Health (IDLH)	ACGIH Exposure Criteria
Benzyl bromide: (derivation: By bromination of toluene; interaction of benzyl alcohol and hydrobromic acid; by reacting bromine and toluene in ultra-violet light) **1 ppm = 3.33 mg/m^3**	**PEL:** 3 ppm 10 mg/m^3 **STEL:** Not Established	**REL (ceiling):** 3 ppm 10 mg/m^3 **STEL:** Not Established	**50 ppm** **166 mg/m^3**	**TLV (ceiling):** 3 ppm 9.9 mg/m^3 **STEL:** Not Established

SECTION III - PHYSICAL AND CHEMICAL CHARACTERISTICS

Boiling Point **390°F (199°C)**	Specific Gravity (H_2O = 1) **1.38**
Vapor Pressure (mm Hg) **1 at 90°F (32°C)**	Molecular Weight (atomic weight) **171.05**
Vapor Density (Air = 1) **5.8**	Melting Point **390°F (199°C)**

Solubility
Slowly decomposes in water to liberate hydrogen bromide gas. Soluble in alcohol, benzene, and ether.

Appearance and Odor
Colorless liquid with a pleasant, aromatic odor.

SECTION IV - FIRE AND EXPLOSION HAZARD DATA

Flash Point (method used) **174°F (79°C) closed cup**	Explosive Limits in Air % by Volume **LEL: Not Determined UEL: Not Determined**
NFPA Classification **Class IIIA Combustible Liquid**	Autoignition Temperature **Not Reported**

Extinguishing Media
Use dry chemical, foam, or carbon dioxide. Only use water in flooding quantities as fog.

Special Fire Fighting Procedures
Poisonous gases are produced in fire. Structural fire-fighting clothing is permeable and may not provide adequate protection. Wear full protective clothing and self-contained breathing apparatus (SCBA) operated in pressure-demand mode. Move container from fire area if it can be done without risk.

Unusual Fire and Explosion Hazards
Benzyl bromide is a combustible liquid. It is flammable and moderately explosive when exposed to heat or flame. When heated, it can release explosive hydrogen bromide gas. Will react with water or steam to produce toxic and corrosive fumes. Containers may explode in fire.

SECTION V - REACTIVITY DATA

Stability		Conditions to Avoid **Normally stable under routine conditions of handing and storage. Do not store near heat. The buildup of hydrogen bromide gas can cause an explosion.**
Stable X	Unstable	Incompatibility *(materials to avoid)* **Benzyl bromide is incompatible with water and most metals, except lead and nickel, to liberate dangerous and toxic hydrogen bromide gas.**
Hazardous Polymerization		Conditions to Avoid **Decomposition rate can reach explosive violence in presence of metals such as iron. Catalytic impurities (e.g., aluminum, iron, rust) may cause violent polymerization.**
May Occur X	Will Not Occur	Hazardous Decomposition or By-products **When heated to decomposition, benzyl bromide produces toxic and poisonous gases and fumes, including toxic hydrogen bromide gas.**

SECTION VI - HEALTH HAZARD DATA

Primary Route(s) of Entry:	Inhalation?	Absorption (skin/eye)?	Ingestion?
	X		X

Health Hazards

INHALATION: May cause severe irritation or burns of the respiratory system. Can result in pulmonary edema (fluid buildup in lungs), which can be fatal. Causes CNS depression.

SKIN & EYES: Contact with liquid or vapor can cause severe irritation or burns to the skin, eyes, and mucous membranes. Liquid may cause permanent eye damage.

INGESTION: May cause central nervous system depression, nausea, vomiting, headaches, dizziness, gastrointestinal irritation. Chronic effects may include kidney and/or liver damage.

Carcinogenicity	NTP Listed?	IARC Cancer Review Group?	OSHA Regulated?	Target Organs?
Questioned Human Questioned Animal	No	No	29 CFR 1910.1000 Table Z-1	Skin, eyes, respiratory system, CNS.

Medical Conditions Generally Aggravated by Exposure

None reported, however, skin conditions (dermatitis) and respiratory problems may be aggravated.

Emergency and First-aid Procedures

Eye contact: Do not allow victim to keep eyes tightly shut. Flush immediately with water for 15 minutes (minimum), seek medical attention. Skin contact: Remove all contaminated clothing, including shoes. Immediately wash area with flooding amounts of soap and water. Seek medical attention immediately. For inhalation: Remove the person from exposure. Provide respiratory assistance and CPR. Transfer to medical facility. Observe for 24- 48 hours (pulmonary edema possible). If swallowed: Seek medical attention immediately. If conscious, give large amounts of water to dilute. Do NOT induce vomiting.

SECTION VII - PRECAUTIONS FOR SAFE HANDLING AND USE

Steps to be Taken in Case Material is Released or Spilled

Wear self-contained breathing apparatus (SCBA) for respiratory protection. Restrict those not wearing protective equipment and who are not involved in cleanup from entering area. Avoid skin contact. Ventilate area of spill. Do not use water. Take up with sand or vermiculite and deposit in sealed drum.

Preferred Waste Disposal Method

Dissolve in a combustible solvent and burn in an incinerator with an afterburner and scrubber.

Precautions to be Taken in Handling and Storage

Avoid contact with heat, sparks, or flame. Keep away from water. Use with adequate ventilation.

Other Precautions and Warnings

Use only non-sparking tools and equipment. Keep away from sunlight and moisture (can react violently).

SECTION VIII - CONTROL MEASURES AND PERSONAL PROTECTIVE EQUIPMENT

Respiratory Protection *(specify type)*

At concentrations up to 10 ppm, use a chemical respirator with an acid/organic cartridge. For higher exposures, use a supplied-air respirator or a self-contained breathing apparatus (SCBA) with full facepiece operated in positive pressure or continuous flow mode.

Ventilation

Local exhaust (preferred) at site of chemical work or general ventilation.

Protective Gloves	Eye Protection	Other Protective Clothing
Butyl Rubber Gloves	Chemical Goggles and Face Mask	Rubber Apron

Work/Hygiene Practices

Always wash hands thoroughly after using chemical; never bring food, drink, or smoking materials into vicinity of chemicals.

BENZYL BROMIDE

C_7H_7Br CAS: 100-39-0

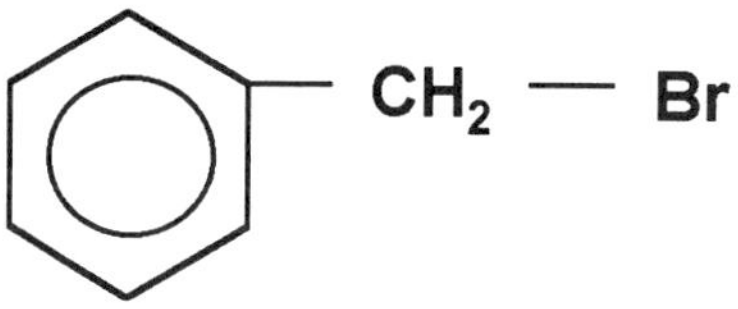

IDENTIFICATION AND TYPICAL USES

Benzyl bromide is a colorless or clear liquid with a pleasant odor. It is used to make foaming and frothing agents and in organic synthesis.

RISK ASSESSMENT: HEALTH

General Assessment

Benzyl bromide is a dangerously corrosive liquid. Contact with the eyes can cause corneal injury. It is a severe irritant to the eyes, mucous membranes, and the skin. Skin contact may produce dermatitis and possibly severe, deep burns. Its carcinogenicity and teratogenicity have not been determined. However, human mutation data have been reported. It is a human poison by *inhalation*. It is also moderately toxic by *ingestion*.

Exposure to its vapors can produce intense and painful irritation of the eyes, nose, and throat. Higher concentrations may cause depression of the central nervous system and pulmonary edema (fluid buildup in the lungs), which is a medical emergency and can lead to death. Symptoms of congestion, cough, dypsnea, and tightness in chest may be delayed up to 48 hours creating a false sense of security with regard to health exposure risk.

Ingestion will result in irritation and burns to the mouth, throat, and gastrointestinal tract. It can cause nausea, abdominal pain, and possibly lead to CNS depression.

Acute Health Effects

The following acute (short-term) health effects may occur immediately or shortly after exposure to benzyl bromide:

Skin: Irritation with possible allergenic effects (dermatitis). Skin burns are likely.

Eye: Severe irritation, burns, pain, and possible permanent damage to vision.

Lung: Nose, throat, and respiratory irritation causing difficulty in breathing. High-level acute exposures can lead to pulmonary edema, which may be fatal.

CNS: Benzyl bromide may act to depress the central nervous system causing weakness, irritability, headache, and other symptoms of CNS depression.

Other: Ingestion of benzyl bromide may cause nausea, vomiting, headaches, dizziness, and gastrointestinal irritation.

Chronic Health Effects

The following chronic (long-term) health effects may occur at some time after exposure to benzyl bromide and can last for months or even years:

Cancer Hazards: Benzyl bromide may cause mutations (genetic changes) in living cells. There is limited evidence in the references that exposure to mutagenic chemicals may cause cancer in animals. There is insufficient data to evaluate its carcinogenic potential in humans. No agency (NTP, IARC, NIOSH, OSHA) lists benzyl bromide as a carcinogen.

Reproductive Hazard: There are no reports in the references to indicate that benzyl bromide may affect the developing fetus.

Other Chronic Effects: Over the long-term, exposure to benzyl bromide can lead to skin allergy. If an allergy develops, very low future exposures can cause itching and rash. Very irritating substances such as benzyl bromide can affect the lungs. It is not known whether benzyl bromide causes lung damage. Chronic exposure may also affect the liver and cause damage.

Recommended Risk-Reduction Measures

Personnel should avoid direct contact with benzyl bromide. Occupational exposure can occur primarily through inhalation and ingestion. Skin and eye contact also present risks to health. If a less toxic material or compound cannot be substituted for benzyl bromide, then *engineering controls* are the most effective method of reducing exposure risk. The best protection is to enclose operations and/or provide local exhaust ventilation at the site of benzyl bromide release. While not always operationally feasible, isolating operations can also reduce exposure. Using respiratory protection is less effective than the controls mentioned above, but is still recommended whenever working with or around benzyl bromide. For low exposures, a chemical cartridge respirator with an acid/organic vapor cartridge is recommended. For higher exposures, a MSHA/NIOSH approved self-contained breathing apparatus (SCBA) or a supplied-air respirator with full facepiece and operated in pressure demand or other continuous flow mode are the appropriate levels of respiratory protection. If a full facepiece is not available, then chemical goggles should be worn to protect the eyes, especially when a splash hazard exists. A face shield should also be considered. To prevent hand and skin exposures, butyl rubber gloves should be worn. Glove manufacturers should be contracted and permeation studies obtained *before* glove selections are made.

Administrative controls should also be in place to minimize the potential for human exposures. These may include written procedures or policies which specify the methods and techniques that will be practiced whenever personnel are to work with benzyl bromide.

All personnel should receive *training* on the use, hazards, protective measures, emergency actions, and other precautions per 29 CFR 1910.1200 (Hazard Communication) prior to the first assignment in an area where benzyl bromide is used or stored.

Before beginning employment and at regular times thereafter, the following recommended medical tests should be considered for those with frequent or potentially high exposures:

☑ Lung function tests.

If symptoms develop or overexposure is suspected, the following may be useful:

☑ Skin testing with dilute benzyl bromide may help diagnose allergy, if performed by a qualified allergist.

☑ Consider chest X-ray following acute exposure (may be negative if taken immediately after exposure due to the delayed onset of pulmonary edema).

☑ Liver function tests.

It should be noted that medical tests that simply look for existing damage are not a substitute for controlling exposures. Medical histories are extremely important when assessing exposure risk. Also, since smoking can lead to heart disease, emphysema, and other respiratory disorders, smokers exposed to benzyl bromide may experience symptoms more quickly than non-smokers under the same conditions of exposure. Prudent risk management requires careful consideration of *all* possible exposure factors.

Other methods to reduce exposure include:

☑ Benzyl bromide should be handled in an enclosed system. There should be local exhaust ventilation at the site of chemical release.

☑ Always ensure that proper protective clothing is worn when using chemical substances.

☑ Wash thoroughly immediately after exposure to benzyl bromide and at the end of the work shift or before eating, drinking, or smoking.

☑ Hazard warning information should be posted in the work area. In addition, as part of an on-going education and training program, all information on the health and safety hazards of benzyl bromide should be communicated to all potentially exposed workers.

☑ Eye wash stations should be provided in the immediate work area for emergency use. If there is a possibility of skin exposure to benzyl bromide, emergency shower facilities should also be provided.

☑ Workers whose clothing has been contaminated by benzyl bromide should change into clean clothes before leaving work. Contaminated work clothing should be laundered only by individuals who have been informed of the hazards of exposure to benzyl bromide.

RISK ASSESSMENT: ENVIRONMENT

General Assessment

The environment is at risk of exposure during transportation, storage, disposal, or destruction of benzyl bromide. In almost every scenario, the threat of environmental exposure is contingent upon the proper handling of the chemical substance. Accidental spills, small, can result in contamination of the surrounding environmental mediums (water, soil, and air).

Benzyl bromide is a combustible liquid. However, caution is always required in handling, storage, transportation, and disposal of benzyl bromide. It is flammable and moderately explosive when exposed to heat or flame. It can react vigorously with many metals (copper, aluminum, magnesium, iron, zinc, and tin). It can decompose violently in the presence of some metals and, therefore, do not store it in metal containers. It is a powerful oxidizer and can react violently on contact with water or moisture. It presents a dangerous explosion hazard and may explode spontaneously when heated. Emergency responders should be made aware of the presence of benzyl bromide at any emergency response situation.

Benzyl bromide can enter the environment from industrial effluents and from spills.

Acute Ecological Effects

Acute (short-term) toxic effects may include the death of animals, birds, or fish and death or low growth rate in plants. Acute effects are seen 2 to 4 days after animals or plants are exposed to benzyl bromide Benzyl bromide has moderate acute toxicity to aquatic life. Insufficient data are available to evaluate the acute (short-term) effects of benzyl bromide on plants, birds, or land animals.

Chronic Ecological Effects

Chronic (long-term) toxic effects may include shortened life span, reproductive problems, lower fertility, and changes in appearance or behavior in exposed animals. These effects can be seen long after first exposure(s) to toxic chemicals. Benzyl bromide has moderate chronic toxicity to aquatic life. Insufficient data are available to evaluate the chronic (long-term) effects of benzyl bromide on plants, birds, or land animals.

Water Solubility

Benzyl bromide slowly decomposes and reacts in water to produce hydrogen bromide gas. Concentrations between 1 and 1000 milligrams will mix with a liter of water.

Persistence in the Environment

Benzyl bromide is slightly persistent in the environment, with a half-life of between 2 and 20 days. The half-life of a pollutant is the amount of time it takes for one-half of the chemical to be degraded. The chemical properties of benzyl bromide indicate that approximately 93% will enter the air with the remaining 7% ending up in water.

Bioaccumulation in Aquatic Organisms

Some substances increase in concentration, or *bioaccumulate,* in living organisms as they breathe contaminated air, drink contaminated water, or eat contaminated food. These chemicals can become concentrated in the tissues and internal organs of animals as well as humans.

The concentration of benzyl bromide found in edible fish tissue is expected to be somewhat higher than the average concentration of benzyl bromide found in the water from which the fish was taken.

Recommended Risk-Reduction Measures

Proper training of all transporters will reduce the likelihood of a mishap or accident resulting in a leak or spill of benzyl bromide into the environment. Labels on all containers, trucks, and rail cars must meet DOT requirements and accurately reflect their contents to enable emergency responders to react properly and quickly to any disaster thereby reducing the potential risk to the environment and to personnel.

Storage of benzyl bromide should be segregated from other chemicals and materials to minimize the risk of cross-contamination and violent reactions. Water should not be kept in the same area since contact can cause violent reaction.

If a spill or leak to the environment has occurred, fire department, emergency response and/or hazardous materials spill personnel should be notified immediately. Cleanup should be attempted only by those trained in proper spill containment procedures. Contaminated soils should be removed for incineration and replaced with clean soil. If benzyl bromide should contact the water table, aquifer, or navigable waterway, time is of the essence. It will decompose and may react violently in water releasing hydrogen bromide gas. Total remediation may not be possible. The local and/or state emergency response authorities must be notified. A comprehensive emergency response or disaster preparedness/recovery plan should be in place prior to any operations involving the use, transportation, storage, or disposal of benzyl bromide.

For disposal, halogenated hydrocarbons may be incinerated, or disposed of by hydrolysis with caustic alkalis, and/or by biodegradation.

If benzyl bromide is spilled or leaked, the following specific steps are recommended:

☑ Restrict persons not wearing protective clothing from area of spill or leak until cleanup is complete. Avoid skin and eye contact.

☑ Absorb liquids in vermiculite or other similar materials and deposit in sealed drum for disposal.

☑ Ventilate area of spill or leak and remove all ignition sources.

☑ It may be necessary to dispose of benzyl bromide as a hazardous waste. The responsible state agency or the regional office of the federal Environmental Protection Agency (EPA) should be contacted for specific recommendations.

RISK ASSESSMENT: BUSINESS

General Assessment

Accidents or mishaps involving benzyl bromide can present a serious threat to business operations. The loss or damage of equipment or facilities can significantly affect fiscal viability. Lawsuits that may result from personnel illness, injury/death, public exposures, and/or environmental contamination will require a serious expenditure of resources. Media attention surrounding an injury, death, or environmental damage can also result in a loss of profits and loss of current as well as future business.

Recommended Risk-Reduction Measures

Company attorneys, safety and health professionals, and environmental specialists should be involved in the development of any procedures for responding to chemical incidents. A company official should be predesignated as a public relations officer with specific training in dealing with the press. Corporate plans and policies should be developed, approved, and implemented long before any need for such arises.

REFERENCES

American Conference of Governmental Industrial Hygienists. 1988. *Documentation of the Threshold Limit Values and Biological Exposure Indices*, 5th Edition (with updates). Cincinnati: ACGIH

Anthony, C. P., and N. J. Kolthoff. 1971. *Textbook of Anatomy and Physiology.* St. Louis: C. V. Mosby Company.

Asante-Duah, D. K. 1993. *Hazardous Waste Risk Assessment.* New York: CRC Press/Lewis Publishers.

Bloom, A. D. and F. J. DeSerres. 1995. *Ecotoxicity and Human Health.* New York: CRC Press/Lewis Publishers.

Brusick, D. J. 1994. *Methods for Genetic Risk Assessment.* New York: CRC Press/Lewis Publishers.

Bueche, J. R. 1972. *Principles of Physics,* 2nd Edition. New York: McGraw-Hill.

Calabrese, E. J. 1994. *Biological Effects of Low Level Exposures.* New York: CRC Press/Lewis Publishers.

Cockerman, L. G., B. S. Shane. 1994. *Basic Environmental Toxicology.* New York: CRC Press/Lewis Publishers.

Danse, I. R. 1991. *Common Sense Toxics in the Workplace.* New York: Van Nostrand Reinhold.

Gardner, R. W. 1994. *Chemical Intolerance.* New York: CRC Press/Lewis Publishers.

Hallenbeck, W. H. 1993. *Quantitative Risk Assessment for Environmental and Occupational Health,* 2nd Edition. New York: CRC Press/Lewis Publishers.

Hodgson, E., R. B. Mailman, and J. E. Chambers. 1988. *Dictionary of Toxicology.* London: The MacMillian Company.

Howard, P. H., et. al. 1991. *Handbook of Environmental Degradation Rates.* New York: CRC Press/Lewis Publishers.

International Agency for the Research on Cancer (IARC). 1987. *Monographs on the Evaluation of Carcinogenic Risks to Humans, Volume 36*, pp-267-283. Lyon: IARC.

Landis, W. G., M. H. Yu. 1995. *Introduction to Environmental Toxicology.* New York: CRC Press/Lewis Publishers.

Langerman, N. 1994 *Precautionary Labels for Chemical Containers.* New York: CRC Press/Lewis Publishers.

Lauwerys, R. R., P. Hoet. 1993. *Industrial Chemical Exposure.* New York: CRC Press/Lewis Publishers.

Lewis, R. J., Sr. 1993. *Hawley's Condensed Chemical Dictionary,* 12th Edition. New York: Van Nostrand Reinhold.

Lewis, R. J., Sr. 1992. *Sax's Dangerous Properties of Industrial Materials,* Volumes 1, 2, and 3, 8th Edition. New York: Van Nostrand Reinhold.

Meyer, E. 1990. *Chemistry of Hazardous Materials.* Englewood Cliffs, NJ: Prentice Hall.

Mikhailova, T. V. 1971. Benzyl bromide in international labour office. *Encyclopedia of Occupational Health and Safety,* Volume 1. New York: McGraw-Hill.

Richardson, M. 1995. *Dictionary of Substance and Their Effects,* Vol. I-VI. New York: CRC Press/Lewis Publishers.

Sacarello, H. L. A. 1994. *The Comprehensive Handbook of Hazardous Materials.* New York: CRC Press/Lewis Publishers.

Torkelson, T. R. V. K. Rowe. 1981. *Patty's Industrial Hygiene and Toxicology,* Vol. 2B. New York: Wiley-Interscience.

U.S. Department of Health and Human Services, National Institute for Occupational Safety and Health. 1994. *NIOSH Pocket Guide to Chemical Hazards.* Washington, D.C.: U.S. Government Printing Office.

U.S. Department of Health ,Education, and Welfare (DHEW), National Institute for Occupational Safety and Health (NIOSH). 1978. *Criteria for a Recommended Standard: Occupational Exposure to Benzyl bromide.* Publication Number 78-182. Washington, D.C.: U.S. Government Printing Office.

MATERIAL SAFETY DATA SHEET

CHEMICAL NAME	
BENZYL CHLORIDE	**CRC PRESS/LEWIS PUBLISHERS DISCLAIMER:** THE INFORMATION AND RECOMMENDATIONS PRESENTED HEREIN ARE BASED ON SOURCES BELIEVED TO BE RELIABLE. CRC MAKES NO REPRESENTATION ON ITS COMPLETENESS OR ACCURACY. IT IS THE RESPONSIBILITY OF THE USER TO DETERMINE THE CHEMICAL'S SUITABILITY FOR ITS INTENDED USE, THE CHEMICAL'S SAFE USE, AND THE CHEMICAL'S PROPER DISPOSAL. NO REPRESENTATIONS AND/OR WARRANTIES, EITHER EXPRESSED OR IMPLIED, OF THE MERCHANTABILITY OR FITNESS FOR A PARTICULAR PURPOSE, OR OF ANY OTHER NATURE, ARE MADE WITH RESPECT TO THE INFORMATION PROVIDED IN THIS MSDS OR TO THE CHEMICAL TO WHICH INFORMATION MAY REFER. CRC NEITHER ASSUMES NOR AUTHORIZES ANY OTHER PERSON TO ASSUME FOR IT, ANY OTHER ADDITIONAL RESPONSIBILITY OR LIABILITY FOR THE USE OF, OR RELIANCE UPON, THIS INFORMATION.

HAZARD WARNING INFORMATION

HEALTH	FIRE	REACTIVE	OTHER	DEGREE OF HAZARD	COLOR CODING	OTHER CODES
3	2	1	~~W~~	0 = Minimum Hazard 1 = Slight Hazard 2 = Moderate Hazard 3 = Serious Hazard 4 = Severe Hazard	HEALTH = BLUE FIRE = RED REACTIVITY = YELLOW OTHER = WHITE	OX = Oxidizer ACID = Acid ALK = Alkali COR = Corrosive ~~W~~ = Use No Water

SECTION I - GENERAL INFORMATION

Characterization	Halogenated Hydrocarbon	RCRA Number	P028
		EPA Class	Acute Hazardous Waste
DOT Proper Shipping Name	Benzyl Chloride	Chemical Abstract Service (CAS) Number	100-44-7
DOT Hazard Class and Label Requirements	Corrosive Material (Liquid)	DOT Emergency Guide Code	59
DOT Identification Number	UN 1738	Molecular Formula	C_7H_7Cl

Synonyms

α-Chlorotoluene; ω-Chlorotoluene; (chloromethyl) benzene; chloromethylbenzene; chlorophenylmethane.

SECTION II - HAZARDOUS INGREDIENTS AND IDENTITY INFORMATION

Hazardous Components (specific identity)	OSHA Exposure Criteria	NIOSH Exposure Criteria	Immediately Dangerous to Life and Health (IDLH)	ACGIH Exposure Criteria
Benzyl Chloride: (derivation: By passing chlorine over boiling toluene until it has increased 38% in weight. The product is washed with water and separated by fractional distillation) 1 ppm = 5.26 mg/m^3	PEL: 1 ppm 5 mg/m^3 STEL: Not Established	REL (ceiling): 1 ppm 5 mg/m^3 Possible Cancer Agent	10 ppm 52 mg/m^3	TLV: 1 ppm 5 mg/m^3 STEL: Not Established

SECTION III - PHYSICAL AND CHEMICAL CHARACTERISTICS

Boiling Point	354°F (179°C)	Specific Gravity (H_2O = 1)	1.10
Vapor Pressure (mm Hg)	1 at 69°F (20°C)	Molecular Weight (atomic weight)	126.6
Vapor Density (Air = 1)	4.4	Melting Point	-49°F (-45°C)

Solubility

Negligible in water (less than 1%). Soluble in alcohol and ether.

Appearance and Odor

Colorless to slightly yellow liquid with a pungent, aromatic odor. Odor threshold = 0.044 ppm.

SECTION IV - FIRE AND EXPLOSION HAZARD DATA

Flash Point (method used)	153°F (67°C) closed cup	Explosive Limits in Air % by Volume	LEL: 1.1% UEL: Not Found
NFPA Classification	Class IIIA Combustible Liquid	Autoignition Temperature	1085°F (585°C)

Extinguishing Media

Use dry chemical, foam, or carbon dioxide. DO NOT use water.

Special Fire Fighting Procedures

Poisonous gases are produced in fire. Wear full protective clothing and self-contained breathing apparatus (SCBA) operated in pressure-demand mode. Move container from fire area if it can be done without risk. Stay clear of smoke. Do not release runoff to sewers or waterways.

Unusual Fire and Explosion Hazards

Benzyl chloride is a combustible liquid. Flammable and moderately explosive when exposed to heat or flame. Can react vigorously with oxidizing materials. Will react with water or steam to produce toxic and corrosive fumes. Containers may explode in fire.

SECTION V - REACTIVITY DATA

Stability		
Stability		Conditions to Avoid **Heat, flame, other sources of ignition, moisture, light. May explode during distillation. Can react vigorously with oxidizing materials.**
Stable	Unstable **X**	Incompatibility *(materials to avoid)* **Oxidizers (such as perchlorates, peroxides, chlorates, nitrates, and permanganates), acids, copper, aluminum, magnesium, iron, zinc, tin, water, dimethyl sulfoxide.**
Hazardous Polymerization		Conditions to Avoid **Decomposition rate can reach explosive violence in presence of metals such as iron. Catalytic impurities (e.g., aluminum, iron, rust) or sodium acetate + pyradine + iron (115°C) may cause violent polymerization.**
May Occur **X**	Will Not Occur	Hazardous Decomposition or By-products **When heated to decomposition, releases toxic fumes of hydrogen chloride, carbon monoxide, and carbon dioxide.**

SECTION VI - HEALTH HAZARD DATA

Primary Route(s) of Entry:	Inhalation?	Absorption (skin/eye)?	Ingestion?
	X		X

Health Hazards

INHALATION: **May cause severe irritation or burns of the respiratory system. Can result in pulmonary edema (fluid buildup in lungs), which can be fatal.**

SKIN & EYES: **Contact with liquid or vapor can cause severe irritation or burns to the skin, eyes, and mucous membranes. Liquid may cause permanent eye damage.**

INGESTION: **May cause central nervous system depression, nausea, vomiting, headaches, dizziness, gastrointestinal irritation. Chronic effects may include kidney and/or liver damage.**

Carcinogenicity	NTP Listed?	IARC Cancer Review Group?	OSHA Regulated?	Target Organs?
Questioned Human **Questioned Animal**	**No**	**Group 3**	**29 CFR 1910.1000** **Table Z-1**	**Skin, eyes, respiratory system, CNS.**

Medical Conditions Generally Aggravated by Exposure

None reported, however, skin conditions (dermatitis) and respiratory problems may be aggravated.

Emergency and First-aid Procedures

Eye contact: Do not allow victim to keep eyes tightly shut. Flush immediately with water for 15 minutes (minimum), seek medical attention. Skin contact: Remove all contaminated clothing, including shoes. Immediately wash area with flooding amounts of soap and water. Seek medical attention immediately. For inhalation: Remove the person from exposure. Provide respiratory assistance and CPR. Transfer to medical facility. Observe for 24- 48 hours (pulmonary edema possible). If swallowed: Seek medical attention immediately. If conscious, give large amounts of water and induce vomiting.

SECTION VII - PRECAUTIONS FOR SAFE HANDLING AND USE

Steps to be Taken in Case Material is Released or Spilled

Wear self-contained breathing apparatus (SCBA) for respiratory protection. Restrict those not wearing protective equipment and who are not involved in cleanup from entering area. Avoid skin contact. Ventilate area of spill. Do not use water. Take up with sand or vermiculite and deposit in sealed drum.

Preferred Waste Disposal Method

Destroy by incineration, hydrolysis with caustic alkalis, or by biodegradation.

Precautions to be Taken in Handling and Storage

Avoid contact with heat, sparks, or flame. Keep away from water. Use with adequate ventilation.

Other Precautions and Warnings

Use only non-sparking tools and equipment. Keep away from sunlight and moisture (can react violently).

SECTION VIII - CONTROL MEASURES AND PERSONAL PROTECTIVE EQUIPMENT

Respiratory Protection *(specify type)*

At concentrations up to 10 ppm, use a chemical respirator with an acid/organic cartridge. For higher exposures, use a self-contained breathing apparatus (SCBA) with full facepiece operated in positive pressure mode.

Ventilation

Local exhaust (preferred) at site of chemical work or general ventilation.

Protective Gloves	Eye Protection	Other Protective Clothing
Rubber Gloves	**Chemical Goggles and Face Mask**	**Rubber Apron**

Work/Hygiene Practices

Always wash hands thoroughly after using chemical; never bring food, drink, or smoking materials into vicinity of chemicals.

BENZYL CHLORIDE

C_7H_7Cl CAS: 100-44-7

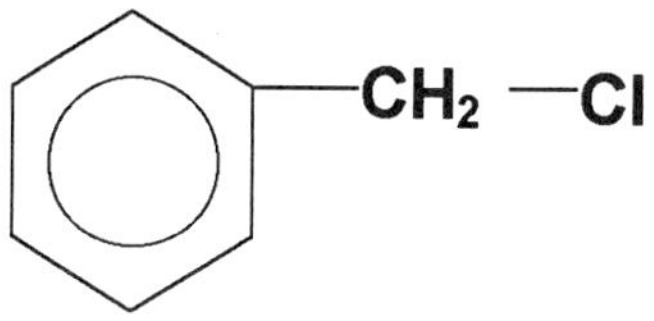

IDENTIFICATION AND TYPICAL USES

Benzyl chloride is a colorless to slightly yellow liquid with a pungent, aromatic odor. It is used in the manufacture of dyes, artificial resins, tanning agents, pharmaceuticals, plasticizers, perfumes, lubricants, and miscellaneous benyl compounds. Also used in the manufacture of photographic developer, gasoline gum inhibitors, penicillin precursors, and quaternary ammonium compounds.

RISK ASSESSMENT: HEALTH

General Assessment

Benzyl chloride is a corrosive liquid. Contact with the eyes can cause corneal injury. It is a severe irritant to the eyes, mucous membranes, and the skin. Skin contact may produce dermatitis. It is a suspected carcinogen with experimental carcinogenic and tumorigenic data reported. Human mutation data have also been reported. It is a human poison by *inhalation*. It is also moderately toxic by ingestion. There is limited evidence of reproductive effects in experimental animals.

Exposure to its vapors can produce intense and painful irritation of the eyes, nose, and throat. Higher concentrations may cause depression of the central nervous system and pulmonary edema (fluid buildup in the lungs), which is a medical emergency and can lead to death.

Acute Health Effects

The following acute (short-term) health effects may occur immediately or shortly after exposure to benzyl chloride:

Skin: Irritation with possible allergenic effects (dermatitis). Skin tumors may be possible.

Eye: Severe irritation, burns, pain, and possible permanent damage to vision.

Lung: Nose, throat, and respiratory irritation causing difficulty in breathing. High-level acute exposures can lead to pulmonary edema, which may be fatal.

CNS: Benzyl chloride may act to depress the central nervous system causing weakness, irritability, headache, and other symptoms of CNS depression.

Other: Ingestion of benzyl chloride may cause nausea, vomiting, headaches, dizziness, and gastrointestinal irritation.

Chronic Health Effects

The following chronic (long-term) health effects may occur at some time after exposure to benzyl chloride and can last for months or even years:

Cancer Hazards: Benzyl chloride may cause mutations (genetic changes) in living cells. There is limited evidence in the references that exposure may cause cancer in animals. There are insufficient data to evaluate its carcinogenic potential in humans.

Reproductive Hazard: There is limited evidence in the references to indicate that benzyl chloride may affect the developing fetus.

Other Chronic Effects: Over the long-term, exposure to benzyl chloride can lead to skin allergy. If an allergy develops, very low future exposures can cause itching and rash. Very irritating substances such as benzyl chloride can affect the lungs. It is not known whether benzyl chloride causes lung damage. Chronic exposure may also affect the liver and cause damage.

Recommended Risk-Reduction Measures

Personnel should avoid direct contact with benzyl chloride. Occupational exposure can occur primarily through *inhalation* and *ingestion*. Skin and eye contact also present risks to heath. If a less toxic material or compound cannot be substituted for benzyl chloride, then *engineering controls* are the most effective method of reducing exposure risk. The best protection is to enclose operations and/or provide local exhaust ventilation at the site of benzyl chloride release. While not always operationally feasible, isolating operations can also reduce exposure. Using respiratory protection is less effective than the controls mentioned above, but is still recommended whenever working with or around benzyl chloride. For low exposures, a chemical cartridge respirator with an acid/organic vapor cartridge is recommended. For higher exposures, a MSHA/NIOSH approved self-contained breathing apparatus (SCBA) with full facepiece and pressure demand is the appropriate level of respiratory protection. If a full facepiece is not available, then chemical goggles should be worn to protect the eyes, especially when a splash hazard exists. A face shield should also be considered. To prevent hand and skin exposures, rubber gloves should be worn.

Administrative controls should also be in place to minimize the potential for human exposures. These may include written procedures or policies which specify the methods and techniques that will be practiced whenever personnel are to work with benzyl chloride.

All personnel should receive *training* on the use, hazards, protective measures, emergency actions, and other precautions per 29 CFR 1910.1200 (Hazard Communication) prior to the first assignment in an area where benzyl chloride is used or stored.

Before beginning employment and at regular times thereafter, the following recommended medical tests should be considered for those with frequent or potentially high exposures:

☑ Lung function tests.

If symptoms develop or overexposure is suspected, the following may be useful:

☑ Skin testing with dilute benzyl chloride may help diagnose allergy, if performed by a qualified allergists.

☑ Consider chest X-ray following acute exposure.

☑ Liver function tests.

It should be noted that medical tests that simply look for existing damage are not a substitute for controlling exposures. Medical histories are extremely important when assessing exposure risk.

Other methods to reduce exposure include:

☑ Benzyl chloride should be handled in an enclosed system. There should be local exhaust ventilation at the site of chemical release.

☑ Always ensure that proper protective clothing is worn when using chemical substances.

☑ Wash thoroughly immediately after exposure to benzyl chloride and at the end of the work shift or before eating, drinking, or smoking.

☑ Hazard warning information should be posted in the work area. In addition, as part of an on-going education and training program, all information on the health and safety hazards of benzyl chloride should be communicated to all potentially exposed workers.

☑ Eye wash stations should be provided in the immediate work area for emergency use. If there is a possibility of skin exposure to benzyl chloride, emergency shower facilities should also be provided.

☑ Workers whose clothing has been contaminated by benzyl chloride should change into clean clothes before leaving work. Contaminated work clothing should be laundered only by individuals who have been informed of the hazards of exposure to benzyl chloride.

RISK ASSESSMENT: ENVIRONMENT

General Assessment

The environment is at risk of exposure during transportation, storage, disposal, or destruction of benzyl chloride. In almost every scenario, the threat of environmental exposure is contingent upon the proper handling of the chemical substance. Accidental spills, large or small, can result in contamination of the surrounding environmental mediums (water, soil, and air).

Benzyl chloride is a combustible liquid. However, caution is always required in handling, storage, transportation, and disposal of benzyl chloride. It is flammable and moderately explosive when exposed to heat or flame. It can react vigorously with oxidizing materials (such as perchlorates, peroxides, chlorates, nitrates, and permanganates) and active metals (copper, aluminum, magnesium, iron, zinc, and tin). It can decompose violently in the presence of some metals. It is a powerful oxidizer and can react violently on contact with water or moisture. It presents a dangerous explosion hazard and may explode spontaneously when heated. Emergency responders should be made aware of the presence of benzyl chloride at any emergency response situation.

Benzyl chloride can enter the environment from industrial effluents and from spills.

Acute Ecological Effects

Acute (short-term) toxic effects may include the death of animals, birds, or fish and death or low growth rate in plants. Acute effects are seen 2 to 4 days after animals or plants are exposed to benzyl chloride Benzyl chloride has moderate acute toxicity to aquatic life. Insufficient data are available to evaluate the acute (short-term) effects of benzyl chloride on plants, birds, or land animals.

Chronic Ecological Effects

Chronic (long-term) toxic effects may include shortened life span, reproductive problems, lower fertility, and changes in appearance or behavior in exposed animals. These effects can be seen long after first exposure(s) to toxic chemicals. Benzyl chloride has moderate chronic toxicity to aquatic life. Insufficient data are available to evaluate the chronic (long-term) effects of benzyl chloride on plants, birds, or land animals.

Water Solubility

Benzyl chloride is moderately soluble in water. Concentrations between 1 and 1000 milligrams will mix with a liter of water.

Persistence in the Environment

Benzyl chloride is slightly persistent in the environment, with a half-life of between 2 and 20 days. The half-life of a pollutant is the amount of time it takes for one-half of the chemical to be degraded. The chemical properties of benzyl chloride indicate that approximately 93% will enter the air with the remaining 7% ending up in water.

Bioaccumulation in Aquatic Organisms

Some substances increase in concentration, or *bioaccumulate,* in living organisms as they breathe contaminated air, drink contaminated water, or eat contaminated food. These chemicals can become concentrated in the tissues and internal organs of animals as well as those of humans.

The concentration of benzyl chloride found in edible fish tissue is expected to be somewhat higher than the average concentration of benzyl chloride found in the water from which the fish was taken.

Recommended Risk-Reduction Measures

Proper training of all transporters will reduce the likelihood of a mishap or accident resulting in a leak or spill of benzyl chloride into the environment. Labels on all containers, trucks, and rail cars must meet DOT requirements and accurately reflect their contents to enable emergency responders to react properly and quickly to any disaster thereby reducing the potential risk to the environment and to personnel.

Storage of benzyl chloride should be segregated from other chemicals and materials to minimize the risk of cross-contamination and violent reactions.

If a spill or leak to the environment has occurred, fire department, emergency response and/or hazardous materials spill personnel should be notified immediately. Cleanup should be attempted only by those trained in proper spill containment procedures. Contaminated soils should be removed for incineration and replaced with clean soil. If benzyl chloride should contact the water table, aquifer, or navigable waterway, time is of the essence. It is moderately soluble and total remediation may not be possible. The local and/or state emergency response authorities must be notified. A comprehensive emergency response or disaster preparedness/recovery plan should be in place prior to any operations involving the use, transportation, storage, or disposal of benzyl chloride.

For disposal, halogenated hydrocarbons may be incinerated, or disposed of by hydrolysis with caustic alkalis, and/or biodegradation.

If benzyl chloride is spilled or leaked, the following specific steps are recommended:

- ☑ Restrict persons not wearing protective clothing from area of spill or leak until cleanup is complete. Avoid skin and eye contact.
- ☑ Absorb liquids in vermiculite or other similar materials and deposit in sealed drum for disposal.
- ☑ Ventilate area of spill or leak.
- ☑ It may be necessary to dispose of benzyl chloride as a hazardous waste. The responsible state agency or the regional office of the federal Environmental Protection Agency (EPA) should be contacted for specific recommendations.

RISK ASSESSMENT: BUSINESS

General Assessment

Accidents or mishaps involving benzyl chloride can present a serious threat to business operations. The loss or damage of equipment or facilities can significantly affect fiscal viability. Lawsuits that may result from personnel illness, injury/death, public exposures, and/or environmental contamination will require a serious expenditure of resources. Media attention surrounding an injury, death, or environmental damage can also result in a loss of profits and loss of current as well as future business.

Recommended Risk-Reduction Measures

Company attorneys, safety and health professionals, and environmental specialists should be involved in the development of any procedures for responding to chemical incidents. A company official should be predesignated as a public relations officer with specific training in dealing with the press. Corporate plans and policies should be developed, approved, and implemented long before any need for such arises.

REFERENCES

American Conference of Governmental Industrial Hygienists. 1988. *Documentation of the Threshold Limit Values and Biological Exposure Indices,* 5th Edition (with updates). Cincinnati: ACGIH

Anthony, C. P., and N. J. Kolthoff. 1971. *Textbook of Anatomy and Physiology.* St. Louis: C. V. Mosby Company.

Asante-Duah, D. K. 1993. *Hazardous Waste Risk Assessment.* New York: CRC Press/Lewis Publishers.

Bloom, A. D. and F. J. DeSerres. 1995. *Ecotoxicity and Human Health.* New York: CRC Press/Lewis Publishers.

Brusick, D. J. 1994. *Methods for Genetic Risk Assessment.* New York: CRC Press/Lewis Publishers.

Bueche, J. R. 1972. *Principles of Physics,* 2nd Edition. New York: McGraw-Hill.

Calabrese, E. J. 1994. *Biological Effects of Low Level Exposures.* New York: CRC Press/Lewis Publishers.

Cockerman, L. G., B. S. Shane. 1994. *Basic Environmental Toxicology.* New York: CRC Press/Lewis Publishers.

Danse, I. R. 1991. *Common Sense Toxics in the Workplace.* New York: Van Nostrand Reinhold.

Gardner, R. W. 1994. *Chemical Intolerance.* New York: CRC Press/Lewis Publishers.

Hallenbeck, W. H. 1993. *Quantitative Risk Assessment for Environmental and Occupational Health,* 2nd Edition. New York: CRC Press/Lewis Publishers.

Hodgson, E., R. B. Mailman, and J. E. Chambers. 1988. *Dictionary of Toxicology.* London: The MacMillian Company.

Howard, P. H., et. al. 1991. *Handbook of Environmental Degradation Rates.* New York: CRC Press/Lewis Publishers.

International Agency for the Research on Cancer (IARC). 1987. *Monographs on the Evaluation of Carcinogenic Risks to Humans, Volume 36,* pp-267-283. Lyon: IARC.

Jones, F. E. 1994. *Toxic Organic Vapors in the Workplace.* New York: CRC Press/Lewis Publishers.

Keith, L. H. and M. M. Walker. 1995. *Handbook of Air Toxics.* New York: CRC Press/Lewis Publishers.

Landis, W. G., M. H. Yu. 1995. *Introduction to Environmental Toxicology.* New York: CRC Press/Lewis Publishers.

Langerman, N. 1994 *Precautionary Labels for Chemical Containers.* New York: CRC Press/Lewis Publishers.

Lauwerys, R. R., P. Hoet. 1993. *Industrial Chemical Exposure.* New York: CRC Press/Lewis Publishers.

Lewis, R. J., Sr. 1993. *Hawley's Condensed Chemical Dictionary,* 12th Edition. New York: Van Nostrand Reinhold.

Lewis, R. J., Sr. 1992. *Sax's Dangerous Properties of Industrial Materials,* Volumes 1, 2, and 3, 8th Edition. New York: Van Nostrand Reinhold.

Mellström, G. A., J. E. Wahlberg, and H. I. Maibach. 1994. *Protective Gloves for Occupational Use.* New York: CRC Press/Lewis Publishers.

Meyer, E. 1990. *Chemistry of Hazardous Materials.* Englewood Cliffs, NJ: Prentice Hall.

Mikhailova, T. V. 1971. Benzyl chloride in international labour office. *Encyclopedia of Occupational Health and Safety,* Volume 1. New York: McGraw-Hill.

Patnaik, P. 1992. *A Comprehensive Guide to the Hazardous Properties of Chemical Substances.* New York: Van Nostrand Reinhold.

Philp, R. B. 1995. *Environmental Hazards and Human Health.* New York: CRC Press/Lewis Publishers.

Proctor, N. H., J. P. Hughes. 1991. *Chemical Hazards of the Workplace.* New York: Van Nostrand Reinhold.

Quigley, D. R. 1994. *Handbook of Emergency Chemical Management.* New York: CRC Press/Lewis Publishers.

Richardson, M. 1995. *Dictionary of Substance and Their Effects,* Vol. I-VI. New York: CRC Press/Lewis Publishers.

Sacarello, H. L. A. 1994. *The Comprehensive Handbook of Hazardous Materials.* New York: CRC Press/Lewis Publishers.

Torkelson, T. R. V. K. Rowe. 1981. *Patty's Industrial Hygiene and Toxicology,* Vol. 2B. New York: Wiley-Interscience.

U.S. Department of Health and Human Services, National Institute for Occupational Safety and Health. 1994. *NIOSH Pocket Guide to Chemical Hazards.* Washington, D.C.: U.S. Government Printing Office.

U.S. Department of Health ,Education, and Welfare (DHEW), National Institute for Occupational Safety and Health (NIOSH). 1978. *Criteria for a Recommended Standard: Occupational Exposure to Benzyl chloride.* Publication Number 78-182. Washington, D.C.: U.S. Government Printing Office.

MATERIAL SAFETY DATA SHEET

CHEMICAL NAME	
BENZYL ETHER	**CRC PRESS/LEWIS PUBLISHERS DISCLAIMER:** THE INFORMATION AND RECOMMENDATIONS PRESENTED HEREIN ARE BASED ON SOURCES BELIEVED TO BE RELIABLE. CRC MAKES NO REPRESENTATION ON ITS COMPLETENESS OR ACCURACY. IT IS THE RESPONSIBILITY OF THE USER TO DETERMINE THE CHEMICAL'S SUITABILITY FOR ITS INTENDED USE, THE CHEMICAL'S SAFE USE, AND THE CHEMICAL'S PROPER DISPOSAL. NO REPRESENTATIONS AND/OR WARRANTIES, EITHER EXPRESSED OR IMPLIED, OF THE MERCHANTABILITY OR FITNESS FOR A PARTICULAR PURPOSE, OR OF ANY OTHER NATURE, ARE MADE WITH RESPECT TO THE INFORMATION PROVIDED IN THIS MSDS OR TO THE CHEMICAL TO WHICH INFORMATION MAY REFER. CRC NEITHER ASSUMES NOR AUTHORIZES ANY OTHER PERSON TO ASSUME FOR IT, ANY OTHER ADDITIONAL RESPONSIBILITY OR LIABILITY FOR THE USE OF, OR RELIANCE UPON, THIS INFORMATION.

HAZARD WARNING INFORMATION

HEALTH	FIRE	REACTIVE	OTHER	DEGREE OF HAZARD	COLOR CODING	OTHER CODES
0	1	0		0 = Minimum Hazard 1 = Slight Hazard 2 = Moderate Hazard 3 = Serious Hazard 4 = Severe Hazard	HEALTH = BLUE FIRE = RED REACTIVITY = YELLOW OTHER = WHITE	OX = Oxidizer ACID = Acid ALK = Alkali COR = Corrosive W = Use No Water

SECTION I - GENERAL INFORMATION

Characterization	RCRA Number	EPA Class
Ether	**None**	**Not Applicable**

DOT Proper Shipping Name **Not Listed**	Chemical Abstract Service (CAS) Number **103-50-4**
DOT Hazard Class and Label Requirements **No Citation**	DOT Emergency Guide Code **No Citation**
DOT Identification Number **No Citation**	Molecular Formula **$(C_6H_5CH_2)_2O$**

Synonyms
Dibenzyl ether; benzyl oxide; dibenyl ether; 1,1'-[oxybis(methylene)]bis[benzene].

SECTION II - HAZARDOUS INGREDIENTS AND IDENTITY INFORMATION

Hazardous Components (specific identity)	OSHA Exposure Criteria	NIOSH Exposure Criteria	Immediately Dangerous to Life and Health (IDLH)	ACGIH Exposure Criteria
Benzyl ether: (derivation: By reduction of benzaldehyde in the presence of dicobalt octacarbonyl; or self-condensation of benzyl alcohol in the presence of acid or benzyl chloride in the presence of) alkali; a by-product in preparation of benzyl alcohol by hydrolysis of benzyl chloride).	**PEL:** Not Established **STEL:** Not Established	**REL:** Not Established **STEL:** Not Established	**Not Determined**	**TLV:** Not Established **STEL:** Not Established

SECTION III - PHYSICAL AND CHEMICAL CHARACTERISTICS

Boiling Point **568°F (298°C)**	Specific Gravity (H_2O = 1) **1.04**
Vapor Pressure (mm Hg) **Not Found**	Molecular Weight (atomic weight) **198.28**
Vapor Density (Air = 1) **6.84**	Melting Point **41°F (5°C)**

Solubility
Practically insoluble in water. Soluble in ethanol, ether, chloroform, and acetone.

Appearance and Odor
A colorless or pale-yellow, unstable liquid with a faint almond-like odor and a mushroom-like taste.

SECTION IV - FIRE AND EXPLOSION HAZARD DATA

Flash Point (method used) **275°F (135°C) closed cup**	Explosive Limits in Air % by Volume **LEL: Not Determined UEL: Not Determined**
NFPA Classification **Class IIIB Combustible Liquid**	Autoignition Temperature **Not Determined**

Extinguishing Media
Dry chemical, foam, water spray, or carbon dioxide. Use water spray to keep fire-exposed containers cool. Keep cooling containers until long after fire is out.

Special Fire Fighting Procedures
Wear full protective clothing and self-contained breathing apparatus (SCBA). Move container from fire area if it can be done without risk. Use water spray to flush spills from ignition sources and to knock down vapors that have not yet ignited.

Unusual Fire and Explosion Hazards
None reported.

SECTION V - REACTIVITY DATA

Stability		Conditions to Avoid: Benzyl ether is NOT normally stable. It can form unstable, shock-sensitive peroxides if exposed to air, oxygen, or sunlight for prolonged periods of time. Peroxide can explode if distilled or heated.
Stable	Unstable: X	Incompatibility *(materials to avoid)*: Incompatible with strong oxidizers, peroxides, permanganates, air, liquid oxygen, and aluminum dichloride hydride.
Hazardous Polymerization		Conditions to Avoid: Under normal conditions of temperature and pressure, hazardous polymerization of benzyl ether is not expected to occur.
May Occur	Will Not Occur: X	Hazardous Decomposition or By-products: When heated to decomposition, benzyl ether emits acrid smoke and fumes, including toxic oxides of carbon.

SECTION VI - HEALTH HAZARD DATA

Primary Route(s) of Entry:	Inhalation?	Absorption (skin/eye)?	Ingestion?
	X	X	

Health Hazards

INHALATION: Causes headache, nausea, vomiting, loss of consciousness, and, at high concentrations, death. Depression of the central nervous system (CNS) causing anesthesia, exhaustion, loss of appetite, sleepiness, dizziness, excitation, anorexia, giddiness, narcotic effects.

SKIN & EYES: An eye and skin irritant. Eye irritation may be severe by direct contact with the liquid and mild to moderate with the vapor. Direct contact of the liquid on the skin is irritating.

INGESTION: Moderately toxic, causing nausea and vomiting.

Carcinogenicity	NTP Listed?	IARC Cancer Review Group?	OSHA Regulated?	Target Organs?
Unknown Human No evidence Animal	No	No	No	Skin, eyes, CNS.

Medical Conditions Generally Aggravated by Exposure

None reported.

Emergency and First-aid Procedures

Eye contact: Do not allow victim to keep eyes tightly shut. Flush immediately with water for 15 minutes (minimum), seek medical attention. Skin contact: Contaminated clothing should be removed immediately. Wash area with large amounts of soap and water. For reddened or blistered skin, seek medical attention as soon as possible. For inhalation: Remove the person from exposure. Provide respiratory assistance and CPR. Transfer to medical facility. If swallowed: Seek medical attention immediately. Contact poison control center. Unless advised otherwise, give 1 to 2 glasses of water and induce vomiting. Never attempt to give an unconscious or convulsing person anything by mouth.

SECTION VII - PRECAUTIONS FOR SAFE HANDLING AND USE

Steps to be Taken in Case Material is Released or Spilled

Wear self-contained breathing apparatus (SCBA). Restrict those not wearing protective equipment and who are not involved in cleanup from area. Avoid skin contact. Ventilate area of spill and remove ignition sources. Absorb liquids in vermiculite, collect with non-sparking tools, and deposit in sealed drum.

Preferred Waste Disposal Method

Mix with a higher-boiling solvent and burn in chemical incinerator with afterburner and scrubber.

Precautions to be Taken in Handling and Storage

Store in tightly closed (air-tight), metal containers in a cool, dry, well ventilated area away from heat and incompatible materials. Control inventories and avoid long-term storage of previously opened containers. Once opened, the formation of peroxides is possible. Protect containers from physical damage.

Other Precautions and Warnings

Before using bulk material, test small quantity to ascertain quality and peroxide content.

SECTION VIII - CONTROL MEASURES AND PERSONAL PROTECTIVE EQUIPMENT

Respiratory Protection *(specify type)*

Exposure level not established. Best protection is obtained using a supplied air respirator or a self-contained breathing apparatus (SCBA) with full facepiece operated in positive pressure mode.

Ventilation

Local exhaust (preferred) at site of chemical work or general ventilation.

Protective Gloves	Eye Protection	Other Protective Clothing
Impervious Polyvinyl	Chemical Goggles and Face Mask	Protective Uniform or Apron

Work/Hygiene Practices

Always wash hands thoroughly after using chemical; never bring food, drink, or smoking materials into vicinity of chemicals.

BENZYL ETHER

$(C_6H_5CH_2)_2O$ CAS: 103-50-4

IDENTIFICATION AND TYPICAL USES

Benzyl ether appears as a clear, colorless, or pale-yellow unstable liquid with a faint, almond-like odor and a mushroom-like taste. It is used as a solvent in the production of nitro-musk perfumes, as a plasticizer for nitrocellulose, in the manufacture of synthetic rubber, and various resins. Also used in flavorings and insect repellents.

RISK ASSESSMENT: HEALTH

General Assessment

Benzyl ether is a narcotic substance and is considered mildly to moderately toxic by *inhalation* and skin *absorption*. Although unlikely under normal operating conditions, *ingestion* may also be an exposure route. There are no reports to indicate that this chemical poses a carcinogenic, mutagenic, or teratogenic effect in humans. However, other similar ethers have been shown to cause mutations (genetic changes in living cells) in test animals. It is not known whether benzyl ether will cause this effect in humans.

Inhalation of high concentrations of benzyl ether vapors causes anesthetic effects because of the chemical's ability to quickly and effectively depress the central nervous system. Symptoms of exposure include exhaustion, loss of appetite, sleepiness, dizziness, headache, giddiness, excitation, nausea, vomiting, anorexia, loss of consciousness, possible seizures and convulsions, psychic disturbances, and death due to respiratory arrest. There may also be an occurrence of polycythemia, a condition characterized by an abnormal increase in circulatory red blood cells. Exposure can cause irritation of the eyes, nose, throat, and respiratory system with subsequent changes (numbing) of the sense of smell. At low concentrations, there can still be nausea due to the chemical's strong and disagreeable odor.

Skin contact with the liquid poses no serious threat because benzyl ether will evaporate rather quickly. However, if allowed to remain in contact (in wet clothing or in soaked shoes, for example) for extended periods, there can be drying, cracking, even some burning of the skin's surface. The vapors will cause irritation and slight smarting of the eyes. The liquid will cause more severe irritation and possible inflammation of the eyes and surrounding tissue areas.

Acute Health Effects

The following acute (short-term) health effects may occur immediately or shortly after exposure to benzyl ether:

Skin: Mild irritation and, on prolonged exposure, drying, cracking, secondary infection, and possible burns

Eye: Irritation and smarting of eye tissues and associated skin surfaces (lids).

Lung: Irritation to respiratory tract causing cough and sore throat. May also cause olfactory changes (numbing of the sense of smell). On high concentrations, inhalation causes anesthetic effects and narcosis (deep unconsciousness).

CNS: Rapid depression of the central nervous system causing a variety of toxic effects including loss of coordination and consciousness, and even death due to respiratory arrest.

Other: Ingestion will cause nausea, possible vomiting, and other related gastrointestinal effects.

Chronic Health Effects

The following chronic (long-term) health effects may occur at some time after exposure to benzyl ether and can last for months or even years:

Cancer Hazards: There is no evidence in the references that exposure to benzyl ether causes cancer in animals or humans. However, mutation data have been reported for similar ethers in test animals. Some scientists believe that mutagenic chemicals may pose a cancer risk in the long-term. Additional research is required in this area of study before any conclusions can be made regarding the specific carcinogenic or mutagenic properties of benzyl ether.

Reproductive Hazard: According to the references, benzyl ether has not been adequately tested for its ability to adversely affect reproduction.

Other Chronic Effects: Repeated low-dose exposures can cause anorexia, chronic fatigue or exhaustion, persistent headache, drowsiness, dizziness, excitation, and psychic disturbances due to its depression of the CNS.

Recommended Risk-Reduction Measures

Personnel should avoid direct contact with benzyl ether. If a less toxic material or compound cannot be substituted for benzyl ether, then *engineering controls* are the most effective method of reducing exposure risk. The best protection is to enclose operations and/or provide local exhaust ventilation at the site of benzyl ether release. While not always operationally feasible, isolating operations can also reduce exposure. Using respiratory protection is less effective than the controls mentioned above, but is still required whenever working with or around benzyl ether. No exposure levels have been established for benzyl ether. This does not mean that exposure is without health risk. While an air-purifying respirator may suffice for low concentrations, there is no way to determine if sufficient protection is being provided since no occupational exposure levels have been established. Better protection is provided using a MSHA/NIOSH approved supplied air respirator or a self-contained breathing apparatus (SCBA) with full facepiece and operated in pressure demand mode. If a full facepiece is not available, then chemical goggles should be worn to protect the eyes, especially when a splash hazard exists. A face shield should also be considered. To prevent hand and skin exposures, polyvinyl or other non-permeable gloves should be worn. Glove manufacturers should be contacted and permeation studies obtained *before* selecting gloves.

Administrative controls should also be in place to minimize the potential for human exposures. These may include written procedures or policies which specify the methods and techniques that will be practiced whenever personnel are to work with benzyl ether.

All personnel should receive *training* on the use, hazards, protective measures, emergency actions, and other precautions per 29 CFR 1910.1200 (Hazard Communication) prior to the first assignment in an area where benzyl ether is used or stored.

If symptoms develop or overexposure is suspected, the following may be useful:

- ☑ Liver, kidney, and lung function tests.
- ☑ Evaluation of the central nervous system.

It should be noted that medical tests that simply look for existing damage are not a substitute for controlling exposures. Medical histories are extremely important when assessing exposure risk.

Other methods to reduce exposure include:

- ☑ Always ensure that proper protective clothing is worn when using chemical substances.
- ☑ Wash thoroughly immediately after exposure to benzyl ether and at the end of the work shift or before eating, drinking, or smoking.
- ☑ Hazard warning information should be posted in the work area. In addition, as part of an on-going education and training program, all information on the health and safety hazards of benzyl ether should be communicated to all exposed or potentially exposed workers.
- ☑ Eye wash stations should be provided in the immediate work area for emergency use. If there is a possibility of skin exposure to benzyl ether, emergency shower facilities should also be provided.
- ☑ Workers whose clothing has been contaminated by benzyl ether should change into clean clothes before leaving work. Contaminated work clothing should be laundered only by individuals who have been informed of the hazards of exposure to benzyl ether.

RISK ASSESSMENT: ENVIRONMENT

General Assessment

The environment is at risk of exposure during transportation, storage, disposal, or destruction of benzyl ether. In almost every scenario, the threat of environmental exposure is contingent upon the proper handling of the chemical substance. Accidental spills, large or small, can result in contamination of the surrounding environmental mediums (primarily air, but water and soil contamination can also occur).

Benzyl ether is class IIIB combustible liquid (per OSHA 29 CFR 1910.106). This presents a moderate fire hazard. However, if involved in a fire where combustible organic materials are also involved, the presence of benzyl ether may greatly accelerate the burning rate of the combustible materials (wood, paper, oil, plastics). It can form explosive, shock-sensitive peroxides in contact with air or oxygen. It is incompatible with many oxidizers and caution is always required in handling, storage, transportation, and disposal of benzyl ether. When heated to decomposition it emits very acrid and irritating smoke and fumes. Emergency responders should be made aware of the presence of benzyl ether at any emergency response situation.

Benzyl ether may enter the environment through industrial effluents and spills.

Acute Ecological Effects

Acute (short-term) toxic effects may include the death of animals, birds, or fish and death or low growth rate in plants. Acute effects are seen 2 to 4 days after animals or plants are exposed to benzyl ether.

Insufficient data are available to evaluate or predict the acute (short-term) effects of benzyl ether to aquatic life, plants, birds, or land animals.

Chronic Ecological Effects

Chronic (long-term) toxic effects may include shortened life span, reproductive problems, lower fertility, and changes in appearance or behavior in exposed animals. These effects can be seen long after first exposure(s) to toxic chemicals.

Insufficient data are available to evaluate the chronic (long-term) effects of benzyl ether on aquatic life, plants, birds, or land animals.

Water Solubility

Benzyl ether is nearly insoluble in water. Concentrations of 1 milligram or less may mix with a liter of water.

Persistence in the Environment

Benzyl ether is non-persistent in water, with a half-life of less than 2 days. The half-life of a pollutant is the amount of time it takes for one half of the chemical to be degraded. Approximately 100% of benzyl ether will eventually end up in air.

Bioaccumulation in Aquatic Organisms

Some substances increase in concentration, or *bioaccumulate,* in living organisms as they breathe contaminated air, drink contaminated water, or eat contaminated food. These chemicals can become concentrated in the tissues and internal organs of animals as well as humans.

The references do not address the bioaccumulation capabilities of benzyl ether in aquatic organisms. However, considering the extremely volatile nature of this chemical, bioaccumulation may not be possible.

Recommended Risk-Reduction Measures

Proper training of all transporters will reduce the likelihood of a mishap or accident resulting in a leak or spill of benzyl ether into the environment. Labels on all containers, trucks, and rail cars must meet DOT requirements and accurately reflect their contents to enable emergency responders to react properly and quickly to any disaster, thereby reducing the potential risk to the environment and to personnel.

Storage of benzyl ether should be segregated from other chemicals to minimize the risk of cross-contamination. Benzyl ether must be stored to avoid contact with strong oxidizers, such as chlorine and bromine, since violent reactions can occur. Containers should be protected from physical damage and stored to avoid contact with heat, including sunlight. Storage containers should be kept air-tight. Once opened, shelf-life is limited and should be closely monitored for the formation of explosive peroxides. Any indication of peroxide formation, such as the appearance of crystals around the container opening, should be approached with great caution. The friction from opening the lid could cause an explosion.

If a spill or leak to the environment has occurred, fire department, emergency response and/or hazardous materials spill personnel should be notified immediately. Cleanup should be attempted only by those trained in proper spill containment procedures. Contaminated soils should be removed for incineration and replaced with clean soil. If benzyl ether should contact the water table, aquifer, or navigable waterway, time is of the essence. Its rapid evaporation may make total remediation impossible. The local and/or state emergency response authorities must be notified. A comprehensive emergency response or disaster preparedness/recovery plan should be in place prior to any task or operations involving the use, transportation, storage, or disposal of benzyl ether.

Benzyl ether may be disposed of by mixing with an excess of a higher-boiling solvent and then burning in a chemical incinerator equipped with an afterburner and scrubber.

If benzyl ether is spilled or leaked, the following specific steps are recommended:

- ☑ Restrict persons not wearing protective clothing from area of spill or leak until cleanup is complete. Avoid skin and eye contact.
- ☑ Absorb spilled liquids with vermiculite, dry sand, or a similar material and deposit in sealed drum for disposal. Use non-sparking tools!
- ☑ Ventilate area and remove all ignition sources.
- ☑ It may be necessary to dispose of benzyl ether as a hazardous waste. The responsible state agency or the regional office of the federal Environmental Protection Agency (EPA) should be contacted for specific recommendations.

RISK ASSESSMENT: BUSINESS

General Assessment

Accidents or mishaps involving benzyl ether can present a serious threat to business operations. The loss or damage of equipment or facilities can significantly affect fiscal viability. Lawsuits that may result from personnel illness, injury/death, public exposures, and/or environmental contamination will require a serious expenditure of resources. Media attention surrounding an injury, death, or environmental damage can also result in a loss of profits and loss of current as well as future business.

Recommended Risk-Reduction Measures

Company attorneys, safety and health professionals, and environmental specialists should be involved in the development of any procedures for responding to chemical incidents. A company official should be predesignated as a public relations officer with specific training in dealing with the press. Corporate plans and policies should be developed, approved, and implemented long before any need for such arises.

REFERENCES

American Conference of Governmental Industrial Hygienists. 1988. *Documentation of the Threshold Limit Values and Biological Exposure Indices,* 5th Edition (with updates). Cincinnati: ACGIH

Anthony, C. P., and N. J. Kolthoff. 1971. *Textbook of Anatomy and Physiology.* St. Louis: C. V. Mosby Company.

Bloom, A. D. and F. J. DeSerres. 1995. *Ecotoxicity and Human Health.* New York: CRC Press/Lewis Publishers.

Brusick, D. J. 1994. *Methods for Genetic Risk Assessment.* New York: CRC Press/Lewis Publishers.

Calabrese, E. J. 1994. *Biological Effects of Low Level Exposures.* New York: CRC Press/Lewis Publishers.

Cothern, C. R., N. P. Ross. 1994. *Environmental Statistics, Assessment, and Forecasting.* New York: CRC Press/Lewis Publishers.

Danse, I. R. 1991. *Common Sense Toxics in the Workplace.* New York: Van Nostrand Reinhold.

Fodden, C. R., and J. L. Weddell. 1992. *Hazardous Materials Emergency Action Data.* New York: CRC Press/Lewis Publishers.

Gardner, R. W. 1994. *Chemical Intolerance.* New York: CRC Press/Lewis Publishers.

Hallenbeck, W. H. 1993. *Quantitative Risk Assessment for Environmental and Occupational Health,* 2nd Edition. New York: CRC Press/Lewis Publishers.

Hodgson, E., R. B. Mailman, and J. E. Chambers. 1988. *Dictionary of Toxicology.* London: The MacMillian Company.

Howard, P. H., et. al. 1991. *Handbook of Environmental Degradation Rates.* New York: CRC Press/Lewis Publishers.

Keith, L. H. and M. M. Walker. 1995. *Handbook of Air Toxics.* New York: CRC Press/Lewis Publishers.

Landis, W. G., M. H. Yu. 1995. *Introduction to Environmental Toxicology.* New York: CRC Press/Lewis Publishers.

Langerman, N. 1994 *Precautionary Labels for Chemical Containers.* New York: CRC Press/Lewis Publishers.

Lauwerys, R. R., P. Hoet. 1993. *Industrial Chemical Exposure.* New York: CRC Press/Lewis Publishers.

Lewis, R. J., Sr. 1993. *Hawley's Condensed Chemical Dictionary,* 12th Edition. New York: Van Nostrand Reinhold.

Lewis, R. J., Sr. 1992. *Sax's Dangerous Properties of Industrial Materials,* Volumes 1, 2, and 3, 8th Edition. New York: Van Nostrand Reinhold.

Meyer, E. 1990. *Chemistry of Hazardous Materials.* Englewood Cliffs, NJ: Prentice Hall.

Patnaik, P. 1992. *A Comprehensive Guide to the Hazardous Properties of Chemical Substances.* New York: Van Nostrand Reinhold.

Philp, R. B. 1995. *Environmental Hazards and Human Health.* New York: CRC Press/Lewis Publishers.

Richardson, M. 1995. *Dictionary of Substance and Their Effects,* Vol. I-VI. New York: CRC Press/Lewis Publishers.

Torkelson, T. R. V. K. Rowe. 1981. *Patty's Industrial Hygiene and Toxicology,* Vol. 2B. New York: Wiley-Interscience.

MATERIAL SAFETY DATA SHEET

CHEMICAL NAME	CRC PRESS/LEWIS PUBLISHERS DISCLAIMER
BERYLLIUM	THE INFORMATION AND RECOMMENDATIONS PRESENTED HEREIN ARE BASED ON SOURCES BELIEVED TO BE RELIABLE. CRC MAKES NO REPRESENTATION ON ITS COMPLETENESS OR ACCURACY. IT IS THE RESPONSIBILITY OF THE USER TO DETERMINE THE CHEMICAL'S SUITABILITY FOR ITS INTENDED USE, THE CHEMICAL'S SAFE USE, AND THE CHEMICAL'S PROPER DISPOSAL. NO REPRESENTATIONS AND/OR WARRANTIES, EITHER EXPRESSED OR IMPLIED, OF THE MERCHANTABILITY OR FITNESS FOR A PARTICULAR PURPOSE, OR OF ANY OTHER NATURE, ARE MADE WITH RESPECT TO THE INFORMATION PROVIDED IN THIS MSDS OR TO THE CHEMICAL TO WHICH INFORMATION MAY REFER. CRC NEITHER ASSUMES NOR AUTHORIZES ANY OTHER PERSON TO ASSUME FOR IT, ANY OTHER ADDITIONAL RESPONSIBILITY OR LIABILITY FOR THE USE OF, OR RELIANCE UPON, THIS INFORMATION.

HAZARD WARNING INFORMATION

HEALTH	FIRE	REACTIVE	OTHER	DEGREE OF HAZARD	COLOR CODING	OTHER CODES
3	1	0		0 = Minimum Hazard 1 = Slight Hazard 2 = Moderate Hazard 3 = Serious Hazard 4 = Severe Hazard	HEALTH = BLUE FIRE = RED REACTIVITY = YELLOW OTHER = WHITE	OX = Oxidizer ACID = Acid ALK = Alkali COR = Corrosive W = Use No Water

SECTION I - GENERAL INFORMATION

Characterization **Metal**	RCRA Number: **P015** — EPA Class: **Acute Hazardous Waste**
DOT Proper Shipping Name **Beryllium, Metal Powder**	Chemical Abstract Service (CAS) Number **7440-41-7**
DOT Hazard Class and Label Requirements **Poison B, Flammable Solid Powder**	DOT Emergency Guide Code **53 (compounds) 32 (powder)**
DOT Identification Number **UN 1566 (compounds) UN 1567 (powder)**	Atomic Formula **Be**

Synonyms
Beryllium metal; glucinium; beryllium oxide; beryllium salts.

SECTION II - HAZARDOUS INGREDIENTS AND IDENTITY INFORMATION

Hazardous Components (specific identity)	OSHA Exposure Criteria	NIOSH Exposure Criteria	Immediately Dangerous to Life and Health (IDLH)	ACGIH Exposure Criteria
Beryllium: (derivation: The ore is converted to the oxide or hydroxide, then to the chloride or fluoride. The halide may be reduced in a furnace by magnesium metal, or reduced by electrolysis. Liquid-liquid extraction with an organophosphate chelating agent can be used as a method of purification, or as an alternative process on the ore itself).	PEL: .002 mg(Be)/m^3 STEL: .005 mg(Be)/m^3 .025 mg(Be)/m^3 Peak (30 min)	REL (ceiling): .0005 mg/m^3 (as Be) Possible Cancer Agent	4 mg(Be)/m^3 (as Be)	TLV: .002 mg(Be)/m^3 Suspected Human Carcinogen

SECTION III - PHYSICAL AND CHEMICAL CHARACTERISTICS

Boiling Point **4532°F (2500°C)**	Specific Gravity (H_2O = 1) **1.848**
Vapor Pressure (mm Hg) **0 (approx.) at 69°F (20°C)**	Molecular Weight (atomic weight) **9.0**
Vapor Density (Air = 1) **Not Found**	Melting Point **2349°F (1287°C)**

Solubility
Insoluble to slightly soluble in water. Soluble in acids, except nitric, and in alkalis.

Appearance and Odor
A hard, brittle, gray-white solid lightweight metal. Can exist as a dust or powder.

SECTION IV - FIRE AND EXPLOSION HAZARD DATA

Flash Point (method used) **Not Applicable**	Explosive Limits in Air % by Volume **LEL: Not Applicable** **UEL: Not Applicable**
NFPA Classification **Noncombustible Solid (bulk form)**	Autoignition Temperature **Not Applicable**

Extinguishing Media
In powder form, beryllium is a combustible solid. Use dry chemical, foam, or carbon dioxide. DO NOT USE WATER.

Special Fire Fighting Procedures
Poisonous gases are produced in fire. Wear full protective clothing, self-contained breathing apparatus (SCBA) operated in pressure-demand mode. Move container from fire area if it can be done without risk.

Unusual Fire and Explosion Hazards
In powdered form, beryllium is a moderate fire hazard. Will also burn vigorously on contact with flame or by chemical reaction. Powders and dusts form explosive mixtures in air.

SECTION V - REACTIVITY DATA

Stability		Conditions to Avoid
		Avoid contact with incompatible materials, water, and moisture. Do not mix powders with CCl_4 or trichloroethylene (mixture will flash or spark on impact). Reacts incandescently with fluorine or chlorine.
Stable	Unstable	Incompatibility *(materials to avoid)*
X		Incompatible with halocarbons, acids (hydrochloric, sulfuric), caustics, chlorinated hydrocarbons, oxidizers (such as perchlorates, permanganates, peroxides, chlorates, and nitrates), potassium, and molten lithium.
Hazardous Polymerization		Conditions to Avoid
		Under normal conditions of temperature and pressure, beryllium is not known to undergo hazardous polymerization.
May Occur	Will Not Occur	Hazardous Decomposition or By-products
	X	When heated to decomposition in air it emits highly toxic fumes of beryllium oxide.

SECTION VI - HEALTH HAZARD DATA

Primary Route(s) of Entry:	Inhalation?	Absorption (skin/eye)?	Ingestion?
	X		

Health Hazards

INHALATION: May cause severe irritation to the respiratory system. High chronic exposures can result in the onset of berylliosis with early symptoms of slight cough and fatigue followed by progressive pulmonary insufficiency, anorexia, weight loss, weakness, chest pain, and a constant hacking cough. Cyanosis and clubbing of fingers may also be seen.

SKIN & EYES: Eye contact will result in irritation. When particles get under cuts in the skin, ulcers or lumps may develop.

Carcinogenicity	NTP Listed?	IARC Cancer Review Group?	OSHA Regulated?	Target Organs?
Suspected Human Confirmed Animal	5th Annual Report	Group 2A	No	Skin, eyes, respiratory system (lung cancer).

Medical Conditions Generally Aggravated by Exposure

None reported. However, existing respiratory problems (bronchitis, asthma) may be aggravated.

Emergency and First-aid Procedures

Eye contact: Do not allow victim to keep eyes tightly shut. Flush immediately with water for 15 minutes (minimum), seek medical attention. Skin contact: Avoid skin contact in areas with cuts or scratches. If such contact occurs, wash with large amounts of water, scrubbing to remove chemical particles. For inhalation: Remove the person from exposure. Provide respiratory assistance and CPR. Transfer to medical facility. Observe for 24- 48 hours for lung effects. If swallowed: Seek medical attention immediately. Do not attempt to give an unconscious or convulsing person anything by mouth.

SECTION VII - PRECAUTIONS FOR SAFE HANDLING AND USE

Steps to be Taken in Case Material is Released or Spilled

Wear self-contained breathing apparatus (SCBA) for respiratory protection. Restrict those not wearing protective equipment and who are not involved in cleanup from entering area. Avoid skin contact. Ventilate area of spill. Do not use water. Collect powder materials and deposit in sealed drum.

Preferred Waste Disposal Method

Burn in a chemical incinerator equipped with an afterburner and scrubber.

Precautions to be Taken in Handling and Storage

Store in tightly closed containers in a cool, well ventilated area away from heat. Protect containers from physical damage. Avoid contact in storage with incompatible materials as violent reactions can occur.

Other Precautions and Warnings

Use only non-sparking equipment and tools, especially when opening or closing containers of beryllium.

SECTION VIII - CONTROL MEASURES AND PERSONAL PROTECTIVE EQUIPMENT

Respiratory Protection *(specify type)*

Beryllium is a suspected human carcinogen by the inhalation route. At any concentration, use a NIOSH/MSHA approved self-contained breathing apparatus (SCBA) with full facepiece operated in positive pressure mode.

Ventilation

Local exhaust (preferred) at site of chemical work or general ventilation.

Protective Gloves	Eye Protection	Other Protective Clothing
Impervious Gloves	Chemical Goggles and Face Mask	Protective Uniform or Apron

Work/Hygiene Practices

Always wash hands thoroughly after using chemical; never bring food, drink, or smoking materials into vicinity of chemicals.

BERYLLIUM

Be CAS: 7440-41-7

IDENTIFICATION AND TYPICAL USES

Beryllium is a metallic element of atomic number 4, Group IIA of the periodic system. It has no stable isotopes. It appears as a grayish-white, hard, light-weight solid metal and can be present in bulk or powder/dust forms. It is used as a structural material in space technology; as a moderator and reflector of neutrons in nuclear reactors; as a source of neutrons when bombarded by alpha particles; in special windows for X-ray tubes; in gyroscopes, computer parts, inertial guidance systems; and as an additive is solid propellant rocket fuels. It is also used in beryllium-copper alloys.

RISK ASSESSMENT: HEALTH

General Assessment

Beryllium is a highly toxic metal. Primary exposure route is through inhalation of its dusts. It is a confirmed carcinogen in animals and highly suspected as such in humans. Experimental carcinogenic, neoplastigenic, and tumorigenic data are reported.

Human systemic effects following *inhalation* exposure include lung fibrosis, dyspnea, weight loss, and fatigue. The primary acute effect from single exposure to beryllium dusts is pneumonitis. Severe bronchitis or pneumonia can occur within 48 hours after high exposure and may cause death in severe cases. Chronic exposures can cause pulmonary disease. High or repeated low-level exposures can cause fibrosis (scarring of the tissue) of the lung and other organs. In severe cases, grave disability and heart failure are possible. Beryllium disease is regarded as chronic if it persists for a year or more. The onset of "berylliosis" may be insidious with only slight cough and fatigue, which can occur as early as 1 year or as late as 25 years after exposure. Progressive pulmonary insufficiency, anorexia, weight loss, chest pain, and persistent hacking cough are symptoms of the advanced diseases.

Skin contact may produce irritation or sensitization dermatitis. If dusts get under cuts or abrasions, ulcers or lumps can develop.

Acute Health Effects

The following acute (short-term) health effects may occur immediately or shortly after exposure to beryllium:

Skin: Severe irritation. Skin ulcers may be possible.

Eye: Severe irritation, itching, and burning on contact. Sometimes an allergic eye problem develops, breaking out again with future exposures.

Lung: Overexposure can severely irritate the airways and lungs, causing nasal discharge, tightness in the chest, cough, shortness of breath, and/or fever. Future exposures can cause further attacks. Death can occur in severe cases. High level acute exposures can lead to pneumonitis.

Chronic Health Effects

The following chronic (long-term) health effects may occur at some time after exposure to beryllium and can last for months or even years:

Cancer Hazards: Beryllium is a probable cancer causing agent in humans. There is some evidence that it causes lung and bone cancer in humans and it has been shown to cause lung and bone cancers in animals.

Reproductive Hazard: There is no evidence in the published studies to show that beryllium has the ability to adversely affect reproduction.

Other Chronic Effects: A single high exposure or repeated lower exposures can cause permanent scars in the lungs or other body organs. Symptoms may include fatigue, shortness of breath, weight loss, and poor appetite. These effects may occur months or years after exposure. Severe cases develop disability and even heart failure. If chemical particles get under cuts in the skin, ulcers or lumps can develop. Surgery will be needed to remove these chemical particles. Allergic skin rashes can also occur, with itching and redness. These improve away from exposure and can come back with future exposures. Kidney stones can occur from high or repeated exposures.

Recommended Risk-Reduction Measures

Personnel should avoid direct contact with beryllium. If a less toxic material or compound cannot be substituted for beryllium, then *engineering controls* are the most effective method of reducing exposure risk. The best protection is to enclose operations and/or provide local exhaust ventilation at the site of beryllium release. While not always operationally feasible, isolating operations can also reduce exposure. Using respiratory protection is less effective than the controls mentioned above, but is still required whenever working with or around beryllium. For any exposure level, a MSHA/NIOSH approved self-contained breathing apparatus (SCBA) with full facepiece and pressure demand is the appropriate level of respiratory protection. If a full facepiece is not available, then chemical goggles should be worn to protect the eyes, especially when a splash hazard exists. A face shield should also be considered. To prevent hand and skin exposures, rubber gloves should be worn.

Administrative controls should also be in place to minimize the potential for human exposures. These may include written procedures or policies which specify the methods and techniques that will be practiced whenever personnel are to work with beryllium.

All personnel should receive *training* on the use, hazards, protective measures, emergency actions, and other precautions per 29 CFR 1910.1200 (Hazard Communication) prior to the first assignment in an area where beryllium is used or stored.

Before beginning employment and at regular times thereafter, the following recommended medical tests should be considered for those with frequent or potentially high exposures:

☑ Lung function tests.

☑ Consider chest X-ray.

It should be noted that medical tests that simply look for existing damage are not a substitute for controlling exposures. Medical histories are extremely important when assessing exposure risk.

Other methods to reduce exposure include:

☑ Specific engineering controls are recommended for this chemical by NIOSH. Refer to the NIOSH criteria document: Occupational Exposure to Beryllium (#72-10268).

☑ Always ensure that proper protective clothing is worn when using chemical substances.

☑ Wash thoroughly immediately after exposure to beryllium and at the end of the work shift or before eating, drinking, or smoking.

☑ Hazard warning information should be posted in the work area. In addition, as part of an on-going education and training program, all information on the health and safety hazards of beryllium should be communicated to all potentially exposed workers.

☑ Eye wash stations should be provided in the immediate work area for emergency use. If there is a possibility of skin exposure to beryllium, emergency shower facilities should also be provided.

☑ Workers whose clothing has been contaminated by beryllium should change into clean clothes before leaving work. Contaminated work clothing should be laundered only by individuals who have been informed of the hazards of exposure to beryllium.

RISK ASSESSMENT: ENVIRONMENT

General Assessment

The environment is at risk of exposure during transportation, storage, disposal, or destruction of beryllium. In almost every scenario, the threat of environmental exposure is contingent upon the proper handling of the chemical substance. Accidental spills, large or small, can result in contamination of the surrounding environmental mediums (water, soil, and air).

Beryllium is a combustible solid. However, caution is always required in handling, storage, transportation, and disposal of beryllium. It presents a slight to moderate explosion hazard in the form of powders and dusts. It is incompatible with halocarbons and reacts incandescently with fluorine or chlorine. Mixtures of the powder with $CCl4$ or trichloroethylene will flash or spark on impact. When heated to decomposition in air it emits very toxic fumes of BeO. Emergency responders should be made aware of the presence of beryllium at any emergency response situation.

The major source of beryllium in the environment is burning fossil fuels, some of which may enter the aquatic environment through atmospheric fallout. Beryllium may also enter the aquatic environment through the weathering of rocks and soils and through discharge from industrial and municipal operations.

Acute Ecological Effects

Acute (short-term) toxic effects may include the death of animals, birds, or fish and death or low growth rate in plants. Acute effects are seen 2 to 4 days after animals or plants are exposed to beryllium Beryllium is more toxic in soft water than in hard water. Beryllium and its compounds have high acute toxicity to aquatic life. Insufficient data are available to evaluate or predict the acute (short-term) effects of beryllium to birds, plants or land animals.

Chronic Ecological Effects

Chronic (long-term) toxic effects may include shortened life span, reproductive problems, lower fertility, and changes in appearance or behavior in exposed animals. These effects can be seen long after first exposure(s) to toxic chemicals. Beryllium and its compounds are considered to have high chronic toxicity to aquatic life. Insufficient data are available to evaluate the chronic (long-term) effects of beryllium on plants, birds, or land animals.

Water Solubility

Beryllium is highly soluble in water. Concentrations of 1000 milligrams and more will mix with a liter of water.

Persistence in the Environment

Beryllium is highly persistent in water, with a half-life of greater than 200 days. The half-life of a pollutant is the amount of time it takes for one-half of the chemical to be degraded.

Bioaccumulation in Aquatic Organisms

Some substances increase in concentration, or *bioaccumulate,* in living organisms as they breathe contaminated air, drink contaminated water, or eat contaminated food. These chemicals can become concentrated in the tissues and internal organs of animals as well as humans.

The concentration of beryllium found in edible fish tissue is expected to be somewhat higher than the average concentration of beryllium found in the water from which the fish was taken.

Recommended Risk-Reduction Measures

Proper training of all transporters will reduce the likelihood of a mishap or accident resulting in a leak or spill of beryllium into the environment. Labels on all containers, trucks, and rail cars must meet DOT requirements and accurately reflect their contents to enable emergency responders to react properly and quickly to any disaster thereby reducing the potential risk to the environment and to personnel.

Storage of beryllium should be segregated from other chemicals and materials to minimize the risk of cross-contamination. Beryllium must be stored to avoid contact with oxidizers (such as perchlorates, peroxides, permanganates, chlorates, and nitrates), and strong acids (such as hydrochloric, sulfuric, and nitric) since violent reactions can occur. Containers should be protected from physical damage and stored to avoid contact with heat.

If a spill or leak to the environment has occurred, fire department, emergency response and/or hazardous materials spill personnel should be notified immediately. Cleanup should be attempted only by those trained in proper spill containment procedures. Contaminated soils should be removed for incineration and replaced with clean soil. If beryllium should contact the water table, aquifer, or navigable waterway, time is of the essence. It is highly soluble and total remediation may not be possible. The local and/or state emergency response authorities must be notified. A comprehensive emergency response or disaster preparedness/recovery plan should be in place prior to any operations involving the use, transportation, storage, or disposal of beryllium.

If beryllium is spilled or leaked, the following specific steps are recommended:

- ☑ Restrict persons not wearing protective clothing from area of spill or leak until cleanup is complete. Avoid skin and eye contact.
- ☑ Collect powdered material in the most convenient and safe manner possible and deposit in sealed drum for disposal.
- ☑ Ventilate area of spill or leak.
- ☑ It may be necessary to dispose of beryllium as a hazardous waste. The responsible state agency or the regional office of the federal Environmental

Protection Agency (EPA) should be contacted for specific recommendations.

RISK ASSESSMENT: BUSINESS

General Assessment

Accidents or mishaps involving beryllium can present a serious threat to business operations. The loss or damage of equipment or facilities can significantly affect fiscal viability. Lawsuits that may result from personnel illness, injury/death, public exposures, and/or environmental contamination will require a serious expenditure of resources. Media attention surrounding an injury, death, or environmental damage can also result in a loss of profits and loss of current as well as future business.

Recommended Risk-Reduction Measures

Company attorneys, safety and health professionals, and environmental specialists should be involved in the development of any procedures for responding to chemical incidents. A company official should be predesignated as a public relations officer with specific training in dealing with the press. Corporate plans and policies should be developed, approved, and implemented long before any need for such arises.

REFERENCES

American Conference of Governmental Industrial Hygienists. 1988. *Documentation of the Threshold Limit Values and Biological Exposure Indices,* 5th Edition (with updates). Cincinnati: ACGIH

Anthony, C. P., and N. J. Kolthoff. 1971. *Textbook of Anatomy and Physiology.* St. Louis: C. V. Mosby Company.

Asante-Duah, D. K. 1993. *Hazardous Waste Risk Assessment.* New York: CRC Press/Lewis Publishers.

Bloom, A. D. and F. J. DeSerres. 1995. *Ecotoxicity and Human Health.* New York: CRC Press/Lewis Publishers.

Brusick, D. J. 1994. *Methods for Genetic Risk Assessment.* New York: CRC Press/Lewis Publishers.

Bueche, J. R. 1972. *Principles of Physics,* 2nd Edition. New York: McGraw-Hill.

Calabrese, E. J. 1994. *Biological Effects of Low Level Exposures.* New York: CRC Press/Lewis Publishers.

Cockerman, L. G., B. S. Shane. 1994. *Basic Environmental Toxicology.* New York: CRC Press/Lewis Publishers.

Danse, I. R. 1991. *Common Sense Toxics in the Workplace.* New York: Van Nostrand Reinhold.

Gardner, R. W. 1994. *Chemical Intolerance.* New York: CRC Press/Lewis Publishers.

Hallenbeck, W. H. 1993. *Quantitative Risk Assessment for Environmental and Occupational Health,* 2nd Edition. New York: CRC Press/Lewis Publishers.

Hodgson, E., R. B. Mailman, and J. E. Chambers. 1988. *Dictionary of Toxicology.* London: The MacMillian Company.

Howard, P. H., et. al. 1991. *Handbook of Environmental Degradation Rates.* New York: CRC Press/Lewis Publishers.

International Agency for the Research on Cancer (IARC). 1987. *Monographs on the Evaluation of Carcinogenic Risks to Humans, Supplement 4,* pp 60-62. Lyon: IARC.

Jones, F. E. 1994. *Toxic Organic Vapors in the Workplace.* New York: CRC Press/Lewis Publishers.

Keith, L. H. and M. M. Walker. 1995. *Handbook of Air Toxics.* New York: CRC Press/Lewis Publishers.

Landis, W. G., M. H. Yu. 1995. *Introduction to Environmental Toxicology.* New York: CRC Press/Lewis Publishers.

Langerman, N. 1994 *Precautionary Labels for Chemical Containers.* New York: CRC Press/Lewis Publishers.

Lauwerys, R. R., P. Hoet. 1993. *Industrial Chemical Exposure.* New York: CRC Press/Lewis Publishers.

Lewis, R. J., Sr. 1993. *Hawley's Condensed Chemical Dictionary,* 12th Edition. New York: Van Nostrand Reinhold.

Lewis, R. J., Sr. 1992. *Sax's Dangerous Properties of Industrial Materials,* Volumes 1, 2, and 3, 8th Edition. New York: Van Nostrand Reinhold.

Mellström, G. A., J. E. Wahlberg, and H. I. Maibach. 1994. *Protective Gloves for Occupational Use.* New York: CRC Press/Lewis Publishers.

Meyer, E. 1990. *Chemistry of Hazardous Materials.* Englewood Cliffs, NJ: Prentice Hall.

Patnaik, P. 1992. *A Comprehensive Guide to the Hazardous Properties of Chemical Substances.* New York: Van Nostrand Reinhold.

Philp, R. B. 1995. *Environmental Hazards and Human Health.* New York: CRC Press/Lewis Publishers.

Proctor, N. H., J. P. Hughes. 1991. *Chemical Hazards of the Workplace.* New York: Van Nostrand Reinhold.

Quigley, D. R. 1994. *Handbook of Emergency Chemical Management.* New York: CRC Press/Lewis Publishers.

Richardson, M. 1995. *Dictionary of Substance and Their Effects,* Vol. I-VI. New York: CRC Press/Lewis Publishers.

Sacarello, H. L. A. 1994. *The Comprehensive Handbook of Hazardous Materials.* New York: CRC Press/Lewis Publishers.

Torkelson, T. R. V. K. Rowe. 1981. *Patty's Industrial Hygiene and Toxicology,* Vol. 2B. New York: Wiley-Interscience.

U.S. Department of Health, Education, and Welfare (DHEW), National Institute for Occupational Safety and Health (NIOSH). 1972. *Criteria for a Recommended Standard... Occupational Exposure to Beryllium.* Publication 72-10268. Washington, D.C.: U.S. Government Printing Office.

U.S. Department of Health and Human Services, National Institute for Occupational Safety and Health. 1994. *NIOSH Pocket Guide to Chemical Hazards.* Washington, D.C.: U.S. Government Printing Office.

MATERIAL SAFETY DATA SHEET

CHEMICAL NAME	
BERYLLIUM CHLORIDE	**CRC PRESS/LEWIS PUBLISHERS DISCLAIMER:** THE INFORMATION AND RECOMMENDATIONS PRESENTED HEREIN ARE BASED ON SOURCES BELIEVED TO BE RELIABLE. CRC MAKES NO REPRESENTATION ON ITS COMPLETENESS OR ACCURACY. IT IS THE RESPONSIBILITY OF THE USER TO DETERMINE THE CHEMICAL'S SUITABILITY FOR ITS INTENDED USE, THE CHEMICAL'S SAFE USE, AND THE CHEMICAL'S PROPER DISPOSAL. NO REPRESENTATIONS AND/OR WARRANTIES, EITHER EXPRESSED OR IMPLIED, OF THE MERCHANTABILITY OR FITNESS FOR A PARTICULAR PURPOSE, OR OF ANY OTHER NATURE, ARE MADE WITH RESPECT TO THE INFORMATION PROVIDED IN THIS MSDS OR TO THE CHEMICAL TO WHICH INFORMATION MAY REFER. CRC NEITHER ASSUMES NOR AUTHORIZES ANY OTHER PERSON TO ASSUME FOR IT, ANY OTHER ADDITIONAL RESPONSIBILITY OR LIABILITY FOR THE USE OF, OR RELIANCE UPON, THIS INFORMATION.

HAZARD WARNING INFORMATION

HEALTH	FIRE	REACTIVE	OTHER	DEGREE OF HAZARD	COLOR CODING	OTHER CODES
Not Found	Not Found	Not Found		0 = Minimum Hazard 1 = Slight Hazard 2 = Moderate Hazard 3 = Serious Hazard 4 = Severe Hazard	HEALTH = BLUE FIRE = RED REACTIVITY = YELLOW OTHER = WHITE	OX = Oxidizer ACID = Acid ALK = Alkali COR = Corrosive ~~W~~ = Use No Water

SECTION I - GENERAL INFORMATION

Characterization: **Metal**	RCRA Number: **None**	EPA Class: **Not Listed**
DOT Proper Shipping Name: **Beryllium Compound**	Chemical Abstract Service (CAS) Number: **7787-47-5**	
DOT Hazard Class and Label Requirements: **Poison B**	DOT Emergency Guide Code: **53 (compounds)**	
DOT Identification Number: **UN 1566 (compounds)**	Atomic Formula: **$BeCl_2$**	

Synonyms
Beryllium dichloride.

SECTION II - HAZARDOUS INGREDIENTS AND IDENTITY INFORMATION

Hazardous Components (specific identity)	OSHA Exposure Criteria	NIOSH Exposure Criteria	Immediately Dangerous to Life and Health (IDLH)	ACGIH Exposure Criteria
Beryllium chloride: (derivation: By passing chlorine over a mixture of beryllium oxide and carbon).	PEL (as Be): .002 mg/m^3 STEL (as Be): .005 mg/m^3 .025 mg/m^3 Peak (30 min)	REL (ceiling): .0005 mg/m^3 (as Be) Possible Cancer Agent	Not Determined	TLV (as Be): .002 mg/m^3 Suspected Human Carcinogen

SECTION III - PHYSICAL AND CHEMICAL CHARACTERISTICS

Boiling Point: **968°F (520°C)**	Specific Gravity (H_2O = 1): **1.90**
Vapor Pressure (mm Hg): **1 (approx.) at 556°F (291°C)**	Molecular Weight (atomic weight): **79.91**
Vapor Density (Air = 1): **2.00**	Melting Point: **824°F (440°C)**

Solubility
Very soluble in water; soluble in alcohol, benzene, ether, carbon disulfide. Readily hydrolyzed.

Appearance and Odor
Can be colorless or white, slightly yellow to green deliquescent needle-like crystals; sweetish taste.

SECTION IV - FIRE AND EXPLOSION HAZARD DATA

Flash Point (method used): **Not Applicable**	Explosive Limits in Air % by Volume: **LEL: Not Applicable UEL: Not Applicable**
NFPA Classification: **Noncombustible Solid (bulk form)**	Autoignition Temperature: **Not Applicable**

Extinguishing Media
Use extinguishing agent suitable for the type of surrounding fire. Beryllium chloride itself does not burn. DO NOT USE WATER.

Special Fire Fighting Procedures
Poisonous gases and fumes are produced in fire. Wear full protective clothing, self-contained breathing apparatus (SCBA) operated in pressure-demand mode. Move container from fire area if it can be done without risk. Remain clear of smoke, water fall-out and water run-off. Evacuate non-essential personnel 1500 feet from fire area.

Unusual Fire and Explosion Hazards
With water, forms beryllium oxide and hydrochloric acid solution.

SECTION V - REACTIVITY DATA

Stability		Conditions to Avoid **Avoid contact with incompatible materials.**
Stable	Unstable	Incompatibility *(materials to avoid)*
X		**Incompatible with acids and magnesium.**
Hazardous Polymerization		Conditions to Avoid **Under normal conditions of temperature and pressure, beryllium fluoride is not known to undergo hazardous polymerization.**
May Occur	Will Not Occur	Hazardous Decomposition or By-products
	X	**When heated to decomposition in air it emits highly toxic beryllium oxide and fluoride fumes. Reaction with acid yields hydrofluoric acid vapors.**

SECTION VI - HEALTH HAZARD DATA

Primary Route(s) of Entry:	Inhalation?	Absorption (skin/eye)?	Ingestion?
	X		X

Health Hazards

INHALATION: **Causes severe irritation to the respiratory system with symptoms of pneumonitis, and pulmonary edema. Sever bronchitis can occur days after high exposure with fever, cough, shortness of breath, and death.**

SKIN & EYES: **Eye contact will result in conjunctival irritation. When particles get under cuts in the skin, persistent, non-healing ulcers or lumps may develop. Allergic skin rash may occur.**

INGESTION: **Poison by ingestion. Causes fatigue, weakness, and loss of appetite.**

Carcinogenicity	NTP Listed?	IARC Cancer Review Group?	OSHA Regulated?	Target Organs?
Suspected Human Confirmed Animal	**5th Annual Report**	**Group 2A**	**No**	**Skin, eyes, lungs, liver, spleen, bone, kidneys.**

Medical Conditions Generally Aggravated by Exposure

None reported. However, existing respiratory problems (bronchitis, asthma) may be aggravated.

Emergency and First-aid Procedures

Eye contact: Do not allow victim to keep eyes tightly shut. Flush immediately with water for 15 minutes (minimum), seek medical attention. Skin contact: Avoid skin contact in areas with cuts or scratches. If such contact occurs, wash with large amounts of water, scrubbing to remove chemical particles. For inhalation: Remove the person from exposure. Provide respiratory assistance and CPR. Transfer to medical facility. Observe for 24 - 48 hours for lung effects. If swallowed: Give large amounts of water and induce vomiting if person is conscious. Seek medical attention immediately. Do not induce vomiting or attempt to give an unconscious or convulsing person anything by mouth.

SECTION VII - PRECAUTIONS FOR SAFE HANDLING AND USE

Steps to be Taken in Case Material is Released or Spilled

Wear self-contained breathing apparatus (SCBA) for respiratory protection. Restrict those not wearing protective equipment and who are not involved in cleanup from entering area. Avoid skin contact. Ventilate area of spill. Do not use water. Collect powder materials and deposit in sealed drum.

Preferred Waste Disposal Method

Burn in a chemical incinerator equipped with an afterburner and scrubber.

Precautions to be Taken in Handling and Storage

Store in tightly closed containers in a cool, well ventilated area away from heat. Protect containers from physical damage.

Other Precautions and Warnings

A regulated, marked area should be established where beryllium fluoride is handled, used, or stored.

SECTION VIII - CONTROL MEASURES AND PERSONAL PROTECTIVE EQUIPMENT

Respiratory Protection *(specify type)*

Beryllium fluoride is a suspected human carcinogen. Use a NIOSH/MSHA approved self-contained breathing apparatus (SCBA) with full facepiece operated in positive pressure mode.

Ventilation

Local exhaust (preferred) at site of chemical work or general ventilation.

Protective Gloves	Eye Protection	Other Protective Clothing
Butyl Rubber (or other non-permeable)	**Chemical Dust Goggles and Face Mask**	**Protective Uniform or Apron**

Work/Hygiene Practices

Always wash hands thoroughly after using chemical; never bring food, drink, or smoking materials into vicinity of chemicals.

BERYLLIUM CHLORIDE

$BeCl_2$ CAS: 7787-47-5

IDENTIFICATION AND TYPICAL USES

Beryllium chloride appears as a colorless or white to yellow-green needle-like crystalline solid. It has chemical properties similar to aluminum. In dry form, it is used as a catalyst for organic reactions. It is used in refining beryllium ores and as a chemical reagent. Some beryllium compounds are used to make light alloys, in nuclear reactors, in radio tube parts, in aerospace structures, and in inertial guidance systems.

RISK ASSESSMENT: HEALTH

General Assessment

Beryllium chloride is a confirmed carcinogen with experimental tumorigenic effects. It is poisonous by *ingestion* and can causes toxic effects upon *inhalation.*

Inhalation causes pneumonitis, nasopharyngitis, tracheobronchitis, dyspnea, chronic cough, shortness of breath, and death. High or repeated exposures can lead to scars in the lungs and other organs. Fatigue, weight loss, and heart failure can occur years later. Beryllium chloride is also an experimental teratogen. Mutation data have also been reported. Beryllium chloride produces chronic systemic disease that primarily affects the lungs, but can also involve other organs such as the lymph nodes, liver, bones, and kidneys.

Ingestion causes irritation of the mouth and stomach. Eye contact with dusts causes conjunctival irritation and skin contact may cause severe irritation as well, leading to skin rash. Particles that get under cuts in skin may cause skin ulcers at site of contact.

Acute Health Effects

The following acute (short-term) health effects may occur immediately or shortly after exposure to beryllium chloride:

Skin: Severe irritation and possible burns. Localized skin ulcers may also be possible.

Eye: Severe irritation, itching, and burning on contact. Sometimes an allergic eye problem develops, breaking-out again with future exposures.

Lung: Overexposure can severely irritate the airways and lungs, causing nasal discharge, tightness in the chest, cough, shortness of breath, and/or fever. Future exposures can cause further attacks. Death can occur in severe cases. High acute exposures can lead to pneumonitis. Symptoms may be delayed for days after exposure. Some persons later develop lung scarring after such exposures.

Chronic Health Effects

The following chronic (long-term) health effects may occur at some time after exposure to beryllium chloride and can last for months or even years:

Cancer Hazards: Beryllium chloride is a probable cancer causing agent in humans. There is some evidence that it causes lung cancer in humans and it has been shown to cause lung cancers in animals.

Reproductive Hazard: There is no evidence in the published studies to show that beryllium chloride has the ability to adversely affect reproduction. Some experimental data on animals indicate inconclusive teratogenic effects.

Other Chronic Effects: A single high exposure or repeated lower exposures can cause permanent scars in the lungs or other body organs. Symptoms may include fatigue, shortness of breath, weight loss, and poor appetite. These effects may occur months or years after exposure. Severe cases result in disability and even heart failure. If chemical particles get under cuts in the skin, ulcers or lumps can develop. Surgery will be needed to remove these chemical particles. Allergic skin rashes can also occur, with itching and redness. These improve away from exposure and can come back with future exposures.

Ʊ Recommended Risk-Reduction Measures

Personnel should avoid direct contact with beryllium chloride. If a less toxic material or compound cannot be substituted for beryllium, then *engineering controls* are the most effective method of reducing exposure risk. The best protection is to enclose operations and/or provide local exhaust ventilation at the site of beryllium chloride release. While not always operationally feasible, isolating operations can also reduce exposure. Using respiratory protection is less effective than the controls mentioned above, but is still required whenever working with or around beryllium. For any exposure level, a MSHA/NIOSH approved self-contained breathing apparatus (SCBA) with full facepiece and pressure demand is the appropriate level of respiratory protection. If a full facepiece is not available, then chemical goggles should be worn to protect the eyes, especially when a splash hazard exists. A face shield should also be considered. To prevent hand and skin exposures, rubber gloves should be worn.

Administrative controls should also be in place to minimize the potential for human exposures. These may include written procedures or policies which specify the methods and techniques that will be practiced whenever personnel are to work with beryllium chloride.

All personnel should receive *training* on the use, hazards, protective measures, emergency actions, and other precautions per 29 CFR 1910.1200 (Hazard Communication) prior to the first assignment in an area where beryllium is used or stored.

Before beginning employment and at regular times thereafter, the following recommended medical tests should be considered for those with frequent or potentially high exposures:

☑ Lung function tests (spirometry and gas transfer testing). For high exposures, such as with refinery, alloy, or ceramic workers, more frequent tests should be performed, some experts recommend monthly.

☑ Chest X-ray (establish baseline).

☑ Careful physical examination of the skin and lungs.

If symptoms develop or overexposure is suspected, the following may be useful:

☑ Consider chest X-ray after acute exposure.

It should be noted that medical tests that simply look for existing damage are not a substitute for controlling exposures. Medical histories are extremely important when assessing exposure risk.

Other methods to reduce exposure include:

☑ Specific engineering controls are recommended for this chemical by NIOSH. Refer to the NIOSH criteria document: Occupational Exposure to Beryllium (#72-10268).

☑ Always ensure that proper protective clothing is worn when using chemical substances.

☑ Wash thoroughly immediately after exposure to beryllium chloride and at the end of the work shift or before eating, drinking, or smoking.

☑ Hazard warning information should be posted in the work area. In addition, as part of an on-going education and training program, all information on the health and safety hazards of beryllium should be communicated to all potentially exposed workers.

☑ Eye wash stations should be provided in the immediate work area for emergency use. If there is a possibility of skin exposure to beryllium chloride, emergency shower facilities should also be provided.

☑ Workers whose clothing has been contaminated by beryllium chloride should change into clean clothes before leaving work. Contaminated work clothing should be laundered only by individuals who have been informed of the hazards of exposure to beryllium chloride.

RISK ASSESSMENT: ENVIRONMENT

General Assessment

The environment is at risk of exposure during transportation, storage, disposal, or destruction of beryllium chloride. In almost every scenario, the threat of environmental exposure is contingent upon the proper handling of the chemical substance. Accidental spills, large or small, can result in contamination of the surrounding environmental mediums (water, soil, and air).

Beryllium chloride is a non-combustible solid. However, caution is always required in handling, storage, transportation, and disposal of beryllium chloride. When heated to decomposition in air it emits very toxic fumes of BeO. Emergency responders should be made aware of the presence of beryllium chloride at any emergency response situation.

The major source of beryllium chloride in the environment is burning fossil fuels, some of which may enter the aquatic environment through atmospheric fallout. Beryllium may also enter the aquatic environment through the weathering of rocks and soils and through discharge from industrial and municipal operations.

Acute Ecological Effects

Acute (short-term) toxic effects may include the death of animals, birds, or fish and death or low growth rate in plants. Acute effects are seen 2 to 4 days after animals or plants are exposed to beryllium chloride. Beryllium chloride is more toxic in soft water than in hard water. Beryllium and its compounds have high acute toxicity to aquatic life. Insufficient data are available to evaluate or predict the acute (short-term) effects of beryllium to birds, plants or land animals.

Chronic Ecological Effects

Chronic (long-term) toxic effects may include shortened life span, reproductive problems, lower fertility, and changes in appearance or behavior in exposed animals. These effects can be seen long after first exposure(s) to toxic chemicals. Beryllium and its compounds are considered to have high chronic toxicity to aquatic life. Insufficient data are available to evaluate the chronic (long-term) effects of beryllium on plants, birds, or land animals.

Water Solubility

Beryllium is highly soluble in water. Concentrations of 1000 milligrams and more will mix with a liter of water.

Persistence in the Environment

Beryllium is highly persistent in water, with a half-life of greater than 200 days. The half-life of a pollutant is the amount of time it takes for one-half of the chemical to be degraded.

Bioaccumulation in Aquatic Organisms

Some substances increase in concentration, or *bioaccumulate,* in living organisms as they breathe contaminated air, drink contaminated water, or eat contaminated food. These chemicals can become concentrated in the tissues and internal organs of animals as well as those of humans.

The concentration of beryllium found in edible fish tissue is expected to be somewhat higher than the average concentration of beryllium found in the water from which the fish was taken.

Recommended Risk-Reduction Measures

Proper training of all transporters will reduce the likelihood of a mishap or accident resulting in a leak or spill of beryllium chloride into the environment. Labels on all containers, trucks, and rail cars must meet DOT requirements and accurately reflect their contents to enable emergency responders to react properly and quickly to any disaster thereby reducing the potential risk to the environment and to personnel.

Storage of beryllium chloride should be segregated from other chemicals and materials to minimize the risk of cross-contamination. Beryllium chloride must be stored to avoid contact with water and finely divided metal dusts since violent reactions can occur. Containers should be protected from physical damage and stored to avoid contact with heat.

If a spill or leak to the environment has occurred, fire department, emergency response and/or hazardous materials spill personnel should be notified immediately. Cleanup should be attempted only by those trained in proper spill containment procedures. Neutralizing agents are agricultural lime (slaked lime), crushed limestone, and sodium bicarbonate. Contaminated soils should be removed for incineration and replaced with clean soil. If beryllium chloride should contact the water table, aquifer, or navigable waterway, time is of the essence. It is highly soluble and total remediation may not be possible. The local and/or state emergency response authorities must be notified. A comprehensive emergency response or disaster preparedness/recovery plan should be in place prior to any operations involving the use, transportation, storage, or disposal of beryllium chloride.

If beryllium chloride is spilled or leaked, the following specific steps are recommended:

- ☑ Restrict persons not wearing protective clothing from area of spill or leak until cleanup is complete. Avoid skin and eye contact.
- ☑ Collect powdered material in the most convenient and safe manner possible and deposit in sealed drum for disposal.
- ☑ Ventilate area of spill or leak.
- ☑ It may be necessary to dispose of beryllium chloride as a hazardous waste. The responsible state

agency or the regional office of the federal Environmental Protection Agency (EPA) should be contacted for specific recommendations.

☑ Use a high-efficiency particulate air (HEPA) filter vacuum system for cleanup.

RISK ASSESSMENT: BUSINESS

General Assessment

Accidents or mishaps involving beryllium chloride can present a serious threat to business operations. The loss or damage of equipment or facilities can significantly affect fiscal viability. Lawsuits that may result from personnel illness, injury/death, public exposures, and/or environmental contamination will require a serious expenditure of resources. Media attention surrounding an injury, death, or environmental damage can also result in a loss of profits and loss of current as well as future business.

Recommended Risk-Reduction Measures

Company attorneys, safety and health professionals, and environmental specialists should be involved in the development of any procedures for responding to chemical incidents. A company official should be predesignated as a public relations officer with specific training in dealing with the press. Corporate plans and policies should be developed, approved, and implemented long before any need for such arises.

REFERENCES

American Conference of Governmental Industrial Hygienists. 1988. *Documentation of the Threshold Limit Values and Biological Exposure Indices,* 5th Edition (with updates). Cincinnati: ACGIH

Anthony, C. P., and N. J. Kolthoff. 1971. *Textbook of Anatomy and Physiology.* St. Louis: C. V. Mosby Company.

Asante-Duah, D. K. 1993. *Hazardous Waste Risk Assessment.* New York: CRC Press/Lewis Publishers.

Brusick, D. J. 1994. *Methods for Genetic Risk Assessment.* New York: CRC Press/Lewis Publishers.

Bueche, J. R. 1972. *Principles of Physics,* 2nd Edition. New York: McGraw-Hill.

Calabrese, E. J. 1994. *Biological Effects of Low Level Exposures.* New York: CRC Press/Lewis Publishers.

Cockerman, L. G., B. S. Shane. 1994. *Basic Environmental Toxicology.* New York: CRC Press/Lewis Publishers.

Danse, I. R. 1991. *Common Sense Toxics in the Workplace.* New York: Van Nostrand Reinhold.

Gardner, R. W. 1994. *Chemical Intolerance.* New York: CRC Press/Lewis Publishers.

Hallenbeck, W. H. 1993. *Quantitative Risk Assessment for Environmental and Occupational Health,* 2nd Edition. New York: CRC Press/Lewis Publishers.

Hodgson, E., R. B. Mailman, and J. E. Chambers. 1988. *Dictionary of Toxicology.* London: The MacMillian Company.

Howard, P. H., et. al. 1991. *Handbook of Environmental Degradation Rates.* New York: CRC Press/Lewis Publishers.

International Agency for the Research on Cancer (IARC). 1987. *Monographs on the Evaluation of Carcinogenic Risks to Humans, Supplement 4,* pp 60-62. Lyon: IARC.

Jones, F. E. 1994. *Toxic Organic Vapors in the Workplace.* New York: CRC Press/Lewis Publishers.

Keith, L. H. and M. M. Walker. 1995. *Handbook of Air Toxics.* New York: CRC Press/Lewis Publishers.

Landis, W. G., M. H. Yu. 1995. *Introduction to Environmental Toxicology.* New York: CRC Press/Lewis Publishers.

Langerman, N. 1994 *Precautionary Labels for Chemical Containers.* New York: CRC Press/Lewis Publishers.

Lauwerys, R. R., P. Hoet. 1993. *Industrial Chemical Exposure.* New York: CRC Press/Lewis Publishers.

Lewis, R. J., Sr. 1993. *Hawley's Condensed Chemical Dictionary,* 12th Edition. New York: Van Nostrand Reinhold.

Lewis, R. J., Sr. 1992. *Sax's Dangerous Properties of Industrial Materials,* Volumes 1, 2, and 3, 8th Edition. New York: Van Nostrand Reinhold.

Mellström, G. A., J. E. Wahlberg, and H. I. Maibach. 1994. *Protective Gloves for Occupational Use.* New York: CRC Press/Lewis Publishers.

Meyer, E. 1990. *Chemistry of Hazardous Materials.* Englewood Cliffs, NJ: Prentice Hall.

Patnaik, P. 1992. *A Comprehensive Guide to the Hazardous Properties of Chemical Substances.* New York: Van Nostrand Reinhold.

Philp, R. B. 1995. *Environmental Hazards and Human Health.* New York: CRC Press/Lewis Publishers.

Proctor, N. H., J. P. Hughes. 1991. *Chemical Hazards of the Workplace.* New York: Van Nostrand Reinhold.

Quigley, D. R. 1994. *Handbook of Emergency Chemical Management.* New York: CRC Press/Lewis Publishers.

Richardson, M. 1995. *Dictionary of Substance and Their Effects,* Vol. I-VI. New York: CRC Press/Lewis Publishers.

Torkelson, T. R. V. K. Rowe. 1981. *Patty's Industrial Hygiene and Toxicology,* Vol. 2B. New York: Wiley-Interscience.

U.S. Department of Health, Education, and Welfare (DHEW), National Institute for Occupational Safety and Health (NIOSH). 1972. *Criteria for a Recommended Standard... Occupational Exposure to Beryllium.* Publication 72-10268. Washington, D.C.: U.S. Government Printing Office.

U.S. Department of Health and Human Services, National Institute for Occupational Safety and Health. 1994. *NIOSH Pocket Guide to Chemical Hazards.* Washington, D.C.: U.S. Government Printing Office.

MATERIAL SAFETY DATA SHEET

CHEMICAL NAME	
BERYLLIUM FLUORIDE	**CRC PRESS/LEWIS PUBLISHERS DISCLAIMER:** THE INFORMATION AND RECOMMENDATIONS PRESENTED HEREIN ARE BASED ON SOURCES BELIEVED TO BE RELIABLE. CRC MAKES NO REPRESENTATION ON ITS COMPLETENESS OR ACCURACY. IT IS THE RESPONSIBILITY OF THE USER TO DETERMINE THE CHEMICAL'S SUITABILITY FOR ITS INTENDED USE, THE CHEMICAL'S SAFE USE, AND THE CHEMICAL'S PROPER DISPOSAL. NO REPRESENTATIONS AND/OR WARRANTIES, EITHER EXPRESSED OR IMPLIED, OF THE MERCHANTABILITY OR FITNESS FOR A PARTICULAR PURPOSE, OR OF ANY OTHER NATURE, ARE MADE WITH RESPECT TO THE INFORMATION PROVIDED IN THIS MSDS OR TO THE CHEMICAL TO WHICH INFORMATION MAY REFER. CRC NEITHER ASSUMES NOR AUTHORIZES ANY OTHER PERSON TO ASSUME FOR IT, ANY OTHER ADDITIONAL RESPONSIBILITY OR LIABILITY FOR THE USE OF, OR RELIANCE UPON, THIS INFORMATION.

HAZARD WARNING INFORMATION

HEALTH	FIRE	REACTIVE	OTHER	DEGREE OF HAZARD	COLOR CODING	OTHER CODES
Not Found	Not Found	Not Found		0 = Minimum Hazard 1 = Slight Hazard 2 = Moderate Hazard 3 = Serious Hazard 4 = Severe Hazard	HEALTH = BLUE FIRE = RED REACTIVITY = YELLOW OTHER = WHITE	OX = Oxidizer ACID = Acid ALK = Alkali COR = Corrosive W = Use No Water

SECTION I - GENERAL INFORMATION

Characterizarion	**Metal**	RCRA Number	**None**
		EPA Class	**Not Listed**
DOT Proper Shipping Name	**Beryllium Compound**	Chemical Abstract Service (CAS) Number	**7787-49-7**
DOT Hazard Class and Label Requirements	**Poison B**	DOT Emergency Guide Code	**53 (compounds)**
DOT Identification Number	**UN 1566 (compounds)**	Atomic Formula	**BeF_2**

Synonyms
Beryllium difluoride.

SECTION II - HAZARDOUS INGREDIENTS AND IDENTITY INFORMATION

Hazardous Components (specific identity)	OSHA Exposure Criteria	NIOSH Exposure Criteria	Immediately Dangerous to Life and Health (IDLH)	ACGIH Exposure Criteria
Beryllium fluoride: (derivation: By the thermal decomposition [900-950°C] of ammonium beryllium fluoride).	**PEL (as Be):** .002 mg/m^3 **STEL (as Be):** .005 mg/m^3 .025 mg/m^3 Peak (30 min)	**REL (ceiling):** .0005 mg/m^3 (as Be) Possible Cancer Agent	**Not Determined**	**TLV (as Be):** .002 mg/m^3 Suspected Human Carcinogen

SECTION III - PHYSICAL AND CHEMICAL CHARACTERISTICS

Boiling Point	**Not Found**	Specific Gravity (H_2O = 1)	**1.986 at 25°C**
Vapor Pressure (mm Hg)	**1 (approx.)**	Molecular Weight (atomic weight)	**47.01**
Vapor Density (Air = 1)	**2.00**	Melting Point	**1472°F (800°C)**

Solubility
Readily soluble in water; sparingly soluble in alcohol and benzene.

Appearance and Odor
Hygroscopic solid. Can be colorless or white, slightly yellow to green deliquescent needle-like crystals; sweetish taste.

SECTION IV - FIRE AND EXPLOSION HAZARD DATA

Flash Point (method used)	**Not Applicable**	Explosive Limits in Air % by Volume	**LEL: Not Applicable UEL: Not Applicable**
NFPA Classification	**Noncombustible Solid (bulk form)**	Autoignition Temperature	**Not Applicable**

Extinguishing Media
Use extinguishing agent suitable for the type of surrounding fire. Beryllium fluoride itself does not burn.

Special Fire Fighting Procedures
Poisonous gases and fumes are produced in fire. Structural fire-fighting protective clothing will not provide adequate protection. Wear full protective clothing, self-contained breathing apparatus (SCBA) operated in pressure-demand mode. Move container from fire area if it can be done without risk. Remain clear of smoke, water fall-out and water run-off. Evacuate non-essential personnel 1500 feet from fire area. Isolate for ½ mile if rail or tank car is involved in fire.

Unusual Fire and Explosion Hazards
Toxic and irritating vapor of unburned material may form in fire.

SECTION V - REACTIVITY DATA

Stability		Conditions to Avoid: **Avoid contact with incompatible materials.**
Stable	Unstable	Incompatibility *(materials to avoid)*: **Incompatible with acids and magnesium.**
X		
Hazardous Polymerization		Conditions to Avoid: **Under normal conditions of temperature and pressure, beryllium fluoride is not known to undergo hazardous polymerization.**
May Occur	Will Not Occur	Hazardous Decomposition or By-products: **When heated to decomposition in air it emits highly toxic beryllium oxide and fluoride fumes. Reaction with acid yields hydrofluoric acid vapors.**
	X	

SECTION VI - HEALTH HAZARD DATA

Primary Route(s) of Entry:	Inhalation?	Absorption (skin/eye)?	Ingestion?
	X		X

Health Hazards

INHALATION: Causes severe irritation to the respiratory system with symptoms of pneumonitis, and pulmonary edema. Sever bronchitis can occur days after high exposure with fever, cough, shortness of breath, and death.

SKIN & EYES: Eye contact will result in conjunctival irritation. When particles get under cuts in the skin, persistent, non-healing ulcers or lumps may develop. Allergic skin rash may occur.

INGESTION: Poison by ingestion. Causes fatigue, weakness, and loss of appetite.

Carcinogenicity	NTP Listed?	IARC Cancer Review Group?	OSHA Regulated?	Target Organs?
Suspected Human Confirmed Animal	5th Annual Report	Group 2A	No	Skin, eyes, lungs, liver, spleen, bone, kidneys.

Medical Conditions Generally Aggravated by Exposure

None reported. However, existing respiratory problems (bronchitis, asthma) may be aggravated.

Emergency and First-aid Procedures

Eye contact: Do not allow victim to keep eyes tightly shut. Flush immediately with water for 15 minutes (minimum), seek medical attention. Skin contact: Avoid skin contact in areas with cuts or scratches. If such contact occurs, wash with large amounts of water, scrubbing to remove chemical particles. For inhalation: Remove the person from exposure. Provide respiratory assistance and CPR. Transfer to medical facility. Observe for 24- 48 hours for lung effects. If swallowed: Give large amounts of water and induce vomiting if person is conscious. Seek medical attention immediately. Do not induce vomiting or attempt to give an unconscious or convulsing person anything by mouth.

SECTION VII - PRECAUTIONS FOR SAFE HANDLING AND USE

Steps to be Taken in Case Material is Released or Spilled

Wear self-contained breathing apparatus (SCBA) for respiratory protection. Restrict those not wearing protective equipment and who are not involved in cleanup from entering area. Avoid skin contact. Ventilate area of spill. Do not use water. Collect powder materials and deposit in sealed drum.

Preferred Waste Disposal Method

Burn in a chemical incinerator equipped with an afterburner and scrubber.

Precautions to be Taken in Handling and Storage

Store in tightly closed containers in a cool, well ventilated area away from heat. Protect containers from physical damage.

Other Precautions and Warnings

A regulated, marked area should be established where beryllium fluoride is handled, used, or stored.

SECTION VIII - CONTROL MEASURES AND PERSONAL PROTECTIVE EQUIPMENT

Respiratory Protection *(specify type)*

Beryllium fluoride is a suspected human carcinogen. Use a NIOSH/MSHA approved self-contained breathing apparatus (SCBA) with full facepiece operated in positive pressure mode.

Ventilation

Local exhaust (preferred) at site of chemical work or general ventilation.

Protective Gloves	Eye Protection	Other Protective Clothing
Butyl Rubber (or other non-permeable)	Chemical Dust Goggles and Face Mask	Protective Uniform or Apron

Work/Hygiene Practices

Always wash hands thoroughly after using chemical; never bring food, drink, or smoking materials into vicinity of chemicals.

BERYLLIUM FLUORIDE

BeF_2 CAS: 7787-49-7

IDENTIFICATION AND TYPICAL USES

Beryllium fluoride appears as a colorless or white to yellow deliquescent crystalline solid. It has chemical, physical, and toxicological properties similar to beryllium chloride (CAS: 7787-47-5). It is used in the production of beryllium metal by reduction with magnesium metal. Also used in nuclear reactors, in glass manufacturing, and as a chemical reagent.

RISK ASSESSMENT: HEALTH

General Assessment

Beryllium fluoride is a confirmed carcinogen with experimental tumorigenic effects. It is poisonous by *ingestion* and can cause toxic effects upon *inhalation.*

Inhalation can cause severe bronchitis and/or pneumonia within 24 to 48 hours of high exposure. Symptoms include fever, cough, and shortness of breath. Death can also occur. Pulmonary edema (fluid buildup in the lungs) can occur following high acute exposures. High or repeated exposures can lead to scars in the lungs and other organs. Fatigue, weight loss, and heart failure can occur years later.

Beryllium fluoride is also an experimental teratogen, but the data are inconclusive in this area. Mutation data have also been reported. Beryllium fluoride produces chronic systemic disease that primarily affects the lungs as berylliosis, but it can also involve other organs such as the lymph nodes, liver, spleen, bones, and kidneys.

Ingestion causes irritation of the mouth and stomach with symptoms of fatigue, weakness, and loss of appetite.

Eye contact with dusts causes conjunctival irritation and skin contact may cause severe irritation as well, leading to skin rash. Particles that get under cuts in skin may cause persistent, non-healing skin ulcers at site of contact.

Acute Health Effects

The following acute (short-term) health effects may occur immediately or shortly after exposure to beryllium fluoride:

Skin: Severe irritation and possible burns. Localized skin ulcers may also be possible.

Eye: Severe irritation, itching, and burning on contact. Sometimes an allergic eye problem develops, breaking out again with future exposures.

Lung: Overexposure can severely irritate the airways and lungs, causing nasal discharge, tightness in the chest, cough, shortness of breath, and/or fever. Future exposures can cause further attacks. Death can occur in severe cases. High acute exposures can lead to pulmonary edema. Symptoms may be delayed for days after exposure. Some persons later develop lung scarring after such exposures.

Chronic Health Effects

The following chronic (long-term) health effects may occur at some time after exposure to beryllium fluoride and can last for months or even years:

Cancer Hazards: Beryllium fluoride is a probable cancer causing agent in humans. There is some evidence that it causes lung cancer in humans and it has been shown to cause lung cancers in animals.

Reproductive Hazard: There is no evidence in the published studies to show that beryllium fluoride has the ability to adversely affect reproduction. Some experimental data on animals indicate inconclusive teratogenic effects.

Other Chronic Effects: A single high exposure or repeated lower exposures can cause permanent scars in the lungs or other body organs. Symptoms may include fatigue, shortness of breath, weight loss, and poor appetite. These effects may occur months or years after exposure. Severe cases result in disability and even heart failure. If chemical particles get under cuts in the skin, ulcers or lumps can develop. Surgery will be needed to remove these chemical particles. Allergic skin rashes can also occur, with itching and redness. These improve away from exposure and can come back with future exposures.

Recommended Risk-Reduction Measures

Personnel should avoid direct contact with beryllium fluoride. If a less toxic material or compound cannot be substituted for beryllium, then *engineering controls* are the most effective method of reducing exposure risk. The best protection is to enclose operations and/or provide local exhaust ventilation at the site of beryllium fluoride release. While not always operationally feasible, isolating operations can also reduce exposure. Using respiratory protection is less effective than the controls mentioned above, but is still required whenever working with or around beryllium. For any exposure level, a MSHA/NIOSH approved self-contained breathing apparatus (SCBA) with full facepiece and pressure demand is the appropriate level of respiratory protection. If a full facepiece is not available, then chemical goggles should be worn to protect the eyes, especially when a splash hazard exists. A face shield should also be considered. To prevent hand and skin exposures, rubber or other non-permeable gloves should be worn.

Administrative controls should also be in place to minimize the potential for human exposures. These may include written procedures or policies which specify the methods and techniques that will be practiced whenever personnel are to work with beryllium fluoride.

All personnel should receive *training* on the use, hazards, protective measures, emergency actions, and other precautions per 29 CFR 1910.1200 (Hazard Communication) prior to the first assignment in an area where beryllium fluoride is used or stored.

Before beginning employment and at regular times thereafter, the following recommended medical tests should be considered for those with frequent or potentially high exposures:

- ☑ Lung function tests (spirometry and gas transfer testing). For high exposures, such as with refinery, alloy, or ceramic workers, more frequent tests should be performed, some experts recommend monthly.
- ☑ Consider chest X-ray after overexposure.
- ☑ Careful physical examination of the skin and lungs.

If symptoms develop or overexposure is suspected, the following may be useful:

- ☑ Consider chest X-ray after acute exposure.

It should be noted that medical tests that simply look for existing damage are not a substitute for controlling exposures. Medical histories are extremely important when assessing exposure risk.

Other methods to reduce exposure include:

- ☑ Specific engineering controls are recommended for this chemical by NIOSH. Refer to the NIOSH criteria documents: Occupational Exposure to Beryllium (#72-10268) and Inorganic Fluoride (#76-103).
- ☑ Always ensure that proper protective clothing is worn when using chemical substances.
- ☑ Wash thoroughly immediately after exposure to beryllium fluoride and at the end of the work shift or before eating, drinking, or smoking.
- ☑ Hazard warning information should be posted in the work area. In addition, as part of an on-going education and training program, all information on the health and safety hazards of beryllium should be communicated to all potentially exposed workers.
- ☑ Eye wash stations should be provided in the immediate work area for emergency use. If there is a possibility of skin exposure to beryllium fluoride, emergency shower facilities should also be provided.
- ☑ Workers whose clothing has been contaminated by beryllium fluoride should change into clean clothes before leaving work. Contaminated work clothing should be laundered only by individuals who have been informed of the hazards of exposure to beryllium fluoride.

RISK ASSESSMENT: ENVIRONMENT

General Assessment

The environment is at risk of exposure during transportation, storage, disposal, or destruction of beryllium fluoride. In almost every scenario, the threat of environmental exposure is contingent upon the proper handling of the chemical substance. Accidental spills, large or small, can result in contamination of the surrounding environmental mediums (water, soil, and air).

Beryllium fluoride is a non-combustible solid. However, caution is always required in handling, storage, transportation, and disposal of beryllium fluoride. When heated to decomposition in air it emits very toxic fumes of beryllium oxide and fluoride. Reaction with acid yields very toxic hydrofluoric acid vapors. Emergency responders should be made aware of the presence of beryllium fluoride at any emergency response situation.

The major source of beryllium fluoride in the environment is burning fossil fuels, some of which may enter the aquatic environment through atmospheric fallout. Beryllium compounds may also enter the aquatic environment through the weathering of rocks and soils and through discharge from industrial and municipal operations.

Acute Ecological Effects

Acute (short-term) toxic effects may include the death of animals, birds, or fish and death or low growth rate in plants. Acute effects are seen 2 to 4 days after animals or plants are exposed to beryllium fluoride. Beryllium fluoride is more toxic in soft water than in hard water. Beryllium and its compounds have high acute toxicity to aquatic life. Insufficient data are available to evaluate or predict the acute (short-term) effects of beryllium to birds, plants or land animals.

Chronic Ecological Effects

Chronic (long-term) toxic effects may include shortened life span, reproductive problems, lower fertility, and changes in appearance or behavior in exposed animals. These effects can be seen long after first exposure(s) to toxic chemicals. Beryllium and its compounds are considered to have high chronic toxicity to aquatic life. Insufficient data are available to evaluate the chronic (long-term) effects of beryllium on plants, birds, or land animals.

Water Solubility

Beryllium is highly soluble in water. Concentrations of 1000 milligrams and more will mix with a liter of water.

Persistence in the Environment

Beryllium is highly persistent in water, with a half-life of greater than 200 days. The half-life of a pollutant is the amount of time it takes for one-half of the chemical to be degraded.

Bioaccumulation in Aquatic Organisms

Some substances increase in concentration, or *bioaccumulate,* in living organisms as they breathe contaminated air, drink contaminated water, or eat contaminated food. These chemicals can become concentrated in the tissues and internal organs of animals as well as humans.

The concentration of beryllium found in edible fish tissue is expected to be somewhat higher than the average concentration of beryllium found in the water from which the fish was taken.

Recommended Risk-Reduction Measures

Proper training of all transporters will reduce the likelihood of a mishap or accident resulting in a leak or spill of beryllium fluoride into the environment. Labels on all containers, trucks, and rail cars must meet DOT requirements and accurately reflect their contents to enable emergency responders to react properly and quickly to any disaster thereby reducing the potential risk to the environment and to personnel.

Storage of beryllium fluoride should be segregated from other chemicals and materials to minimize the risk of cross-contamination. Beryllium fluoride must be stored to avoid contact with acids and magnesium since violent reactions can occur. Containers should be protected from physical damage and stored to avoid contact with heat.

If a spill or leak to the environment has occurred, fire department, emergency response and/or hazardous materials spill personnel should be notified immediately. Cleanup should be attempted only by those trained in proper spill containment procedures. Neutralizing agents are agricultural lime (slaked lime), crushed limestone, and sodium bicarbonate. Contaminated soils should be removed for incineration and replaced with clean soil.

If beryllium fluoride should contact the water table, aquifer, or navigable waterway, time is of the essence. It is highly soluble and total remediation may not be possible. The local and/or state emergency response authorities must be notified. A comprehensive emergency response or disaster preparedness/recovery plan should be in place prior to any operations involving the use, transportation, storage, or disposal of beryllium fluoride. If beryllium fluoride is spilled or leaked, the following specific steps are recommended:

☑ Restrict persons not wearing protective clothing from area of spill or leak until cleanup is complete. Avoid skin and eye contact.

- ☑ Collect powdered material in the most convenient and safe manner possible and deposit in sealed drum for disposal.
- ☑ Ventilate area of spill or leak.
- ☑ It may be necessary to dispose of beryllium fluoride as a hazardous waste. The responsible state agency or the regional office of the federal Environmental Protection Agency (EPA) should be contacted for specific recommendations.
- ☑ Use a high-efficiency particulate air (HEPA) filter vacuum system for cleanup.

RISK ASSESSMENT: BUSINESS

General Assessment

Accidents or mishaps involving beryllium fluoride can present a serious threat to business operations. The loss or damage of equipment or facilities can significantly affect fiscal viability. Lawsuits that may result from personnel illness, injury/death, public exposures, and/or environmental contamination will require a serious expenditure of resources. Media attention surrounding an injury, death, or environmental damage can also result in a loss of profits and loss of current as well as future business.

Recommended Risk-Reduction Measures

Company attorneys, safety and health professionals, and environmental specialists should be involved in the development of any procedures for responding to chemical incidents. A company official should be predesignated as a public relations officer with specific training in dealing with the press. Corporate plans and policies should be developed, approved, and implemented long before any need for such arises.

REFERENCES

American Conference of Governmental Industrial Hygienists. 1988. *Documentation of the Threshold Limit Values and Biological Exposure Indices,* 5th Edition (with updates). Cincinnati: ACGIH

Anthony, C. P., and N. J. Kolthoff. 1971. *Textbook of Anatomy and Physiology.* St. Louis: C. V. Mosby Company.

Brusick, D. J. 1994. *Methods for Genetic Risk Assessment.* New York: CRC Press/Lewis Publishers.

Calabrese, E. J. 1994. *Biological Effects of Low Level Exposures.* New York: CRC Press/Lewis Publishers.

Cockerman, L. G., B. S. Shane. 1994. *Basic Environmental Toxicology.* New York: CRC Press/Lewis Publishers.

Danse, I. R. 1991. *Common Sense Toxics in the Workplace.* New York: Van Nostrand Reinhold.

Hallenbeck, W. H. 1993. *Quantitative Risk Assessment for Environmental and Occupational Health,* 2nd Edition. New York: CRC Press/Lewis Publishers.

Hodgson, E., R. B. Mailman, and J. E. Chambers. 1988. *Dictionary of Toxicology.* London: The MacMillian Company.

Jones, F. E. 1994. *Toxic Organic Vapors in the Workplace.* New York: CRC Press/Lewis Publishers.

Keith, L. H. and M. M. Walker. 1995. *Handbook of Air Toxics.* New York: CRC Press/Lewis Publishers.

Landis, W. G., M. H. Yu. 1995. *Introduction to Environmental Toxicology.* New York: CRC Press/Lewis Publishers.

Langerman, N. 1994 *Precautionary Labels for Chemical Containers.* New York: CRC Press/Lewis Publishers.

Lauwerys, R. R., P. Hoet. 1993. *Industrial Chemical Exposure.* New York: CRC Press/Lewis Publishers.

Lewis, R. J., Sr. 1993. *Hawley's Condensed Chemical Dictionary,* 12th Edition. New York: Van Nostrand Reinhold.

Lewis, R. J., Sr. 1992. *Sax's Dangerous Properties of Industrial Materials,* Volumes 1, 2, and 3, 8th Edition. New York: Van Nostrand Reinhold.

Meyer, E. 1990. *Chemistry of Hazardous Materials.* Englewood Cliffs, NJ: Prentice Hall.

Patnaik, P. 1992. *A Comprehensive Guide to the Hazardous Properties of Chemical Substances.* New York: Van Nostrand Reinhold.

Philp, R. B. 1995. *Environmental Hazards and Human Health.* New York: CRC Press/Lewis Publishers.

Proctor, N. H., J. P. Hughes. 1991. *Chemical Hazards of the Workplace.* New York: Van Nostrand Reinhold.

Quigley, D. R. 1994. *Handbook of Emergency Chemical Management.* New York: CRC Press/Lewis Publishers.

Richardson, M. 1995. *Dictionary of Substance and Their Effects,* Vol. I-VI. New York: CRC Press/Lewis Publishers.

Torkelson, T. R. V. K. Rowe. 1981. *Patty's Industrial Hygiene and Toxicology,* Vol. 2B. New York: Wiley-Interscience.

U.S. Department of Health, Education, and Welfare (DHEW), National Institute for Occupational Safety and Health (NIOSH). 1972. *Criteria for a Recommended Standard... Occupational Exposure to Beryllium.* Publication 72-10268. Washington, D.C.: U.S. Government Printing Office.

U.S. Department of Health, Education, and Welfare (DHEW), National Institute for Occupational Safety and Health (NIOSH). 1972. *Inorganic Fluorides.* Publication 76-103. Washington, D.C.: U.S. Government Printing Office.

U.S. Department of Health and Human Services, National Institute for Occupational Safety and Health. 1994. *NIOSH Pocket Guide to Chemical Hazards.* Washington, D.C.: U.S. Government Printing Office.

MATERIAL SAFETY DATA SHEET

CHEMICAL NAME	CRC PRESS/LEWIS PUBLISHERS DISCLAIMER:
BERYLLIUM NITRATE	

HAZARD WARNING INFORMATION

HEALTH	FIRE	REACTIVE	OTHER	DEGREE OF HAZARD	COLOR CODING	OTHER CODES
Not Found	Not Found	Not Found		0 = Minimum Hazard 1 = Slight Hazard 2 = Moderate Hazard 3 = Serious Hazard 4 = Severe Hazard	HEALTH = BLUE FIRE = RED REACTIVITY = YELLOW OTHER = WHITE	OX = Oxidizer ACID = Acid ALK = Alkali COR = Corrosive W = Use No Water

SECTION I - GENERAL INFORMATION

Chemical Family	RCRA Number	EPA Class
Metal	**None**	**Not Applicable**

DOT Proper Shipping Name **Beryllium Compound**	Chemical Abstract Service (CAS) Number **13597-99-4**
DOT Hazard Class and Label Requirements **Oxidizer and Poison**	DOT Emergency Guide Code **42**
DOT Identification Number **UN 2464**	Molecular Formula **$Be(NO_3)_2$**

Synonyms
Beryllium dinitrate; nitric acid, beryllium salt; barium nitrate trihydrate.

SECTION II - HAZARDOUS INGREDIENTS AND IDENTITY INFORMATION

Hazardous Components (specific identity)	OSHA Exposure Criteria	NIOSH Exposure Criteria	Immediately Dangerous to Life and Health (IDLH)	ACGIH Exposure Criteria
Beryllium nitrate: (derivation: By action of nitric acid on beryllium oxide, with subsequent evaporation and crystallization; reaction of beryllium sulfate with barium nitrate).	**PEL (as Be):** .002 mg/m^3 **STEL (as Be):** .005 mg/m^3 .025 mg/m^3 Peak (30 min)	**REL (ceiling):** .0005 mg/m^3 (as Be) Possible Cancer Agent	Not Determined	**TLV (as Be):** .002 mg/m^3 Suspected Human Carcinogen

SECTION III - PHYSICAL AND CHEMICAL CHARACTERISTICS

Boiling Point **212°F (100°C) decomposes**	Specific Gravity (H_2O = 1) **1.56 at 20°C**
Vapor Pressure (mm Hg) **Not Applicable**	Molecular Weight (atomic weight) **133.03**
Vapor Density (Air = 1) **6.4**	Melting Point **140°F (60°C)**

Solubility in Water
Highly soluble in water; slightly soluble in alcohol.

Appearance and Odor
Can be colorless or white, slightly yellow to green deliquescent needle-like crystals; has no odor, but may have a bitter taste.

SECTION IV - FIRE AND EXPLOSION HAZARD DATA

Flash Point (method used) **Not Applicable**	Explosive Limits in Air % by Volume **LEL: Not Applicable** **UEL: Not Applicable**
NFPA Classification **Noncombustible Solid (bulk form)**	Autoignition Temperature **Not Applicable**

Extinguishing Media
Use extinguishing agent suitable for the type of surrounding fire. Beryllium nitrate itself does not burn.

Special Fire Fighting Procedures
Poisonous gases and fumes are produced in fire. Structural fire-fighting protective clothing will not provide adequate protection. Wear full protective clothing, self-contained breathing apparatus (SCBA) operated in pressure-demand mode. Move container from fire area if it can be done without risk. Remain clear of smoke, water fall-out and water run-off. Evacuate non-essential personnel 1500 feet from fire area. Isolate for ½ mile if rail or tank car is involved in fire.

Unusual Fire and Explosion Hazards
As an oxidizer, beryllium nitrate will increase the intensity of a fire.

SECTION V - REACTIVITY DATA

Stability		Conditions to Avoid: Avoid contact with incompatible materials.
Stable	Unstable	Incompatibility *(materials to avoid)*: Incompatible with fuels and flammable and combustible materials. Will also form weak solutions of nitric acid on contact with water.
X		
Hazardous Polymerization		Conditions to Avoid: Under normal conditions of temperature and pressure, beryllium nitrate is not known to undergo hazardous polymerization.
May Occur	Will Not Occur	Hazardous Decomposition or By-products: When heated to decomposition in air it emits highly toxic beryllium oxide and nitrogen oxides.
	X	

SECTION VI - HEALTH HAZARD DATA

Primary Route(s) of Entry:	Inhalation?	Absorption (skin/eye)?	Ingestion?
	X		X

Health Hazards

INHALATION: Causes severe irritation to the respiratory system with symptoms of pneumonitis and bronchitis. Sever pneumonia can occur days after high exposure with fever, cough, shortness of breath, and death.

SKIN & EYES: Eye contact will result in conjunctival irritation. When particles get under cuts in the skin, persistent, non-healing ulcers or lumps may develop. Allergic skin rash may occur.

INGESTION: Poison by ingestion. Causes fatigue, weakness, weight loss, and loss of appetite.

Carcinogenicity	NTP Listed?	IARC Cancer Review Group?	OSHA Regulated?	Target Organs?
Suspected Human Confirmed Animal	5th Annual Report	Group 2A	No	Skin, eyes, lungs, liver, spleen, bone, kidneys.

Medical Conditions Generally Aggravated by Exposure

None reported. However, existing respiratory problems (bronchitis, asthma) may be aggravated.

Emergency and First-aid Procedures

Eye contact: Do not allow victim to keep eyes tightly shut. Flush immediately with water for 15 minutes (minimum), seek medical attention. Skin contact: Avoid skin contact in areas with cuts or scratches. If such contact occurs, wash with large amounts of water, scrubbing to remove chemical particles. For inhalation: Remove the person from exposure. Provide respiratory assistance and CPR. Transfer to medical facility. Observe for 24- 48 hours for lung effects. If swallowed: Give large amounts of water and induce vomiting if person is conscious. Seek medical attention immediately. Do not induce vomiting or attempt to give an unconscious or convulsing person anything by mouth.

SECTION VII - PRECAUTIONS FOR SAFE HANDLING AND USE

Steps to be Taken in Case Material is Released or Spilled

Wear self-contained breathing apparatus (SCBA) for respiratory protection. Restrict those not wearing protective equipment and who are not involved in cleanup from entering area. Avoid skin contact. Ventilate area of spill. Do not generate dusts. Collect powder materials and deposit in sealed drum.

Preferred Waste Disposal Method

Burn in a chemical incinerator equipped with an afterburner and scrubber.

Precautions to be Taken in Handling and Storage

Store in tightly closed containers in a cool, well ventilated area away from heat. Protect containers from physical damage.

Other Precautions and Warnings

A regulated, marked area should be established where beryllium nitrate is handled, used, or stored.

SECTION VIII - CONTROL MEASURES AND PERSONAL PROTECTIVE EQUIPMENT

Respiratory Protection *(specify type)*

Beryllium nitrate is a suspected human carcinogen. Use a NIOSH/MSHA approved self-contained breathing apparatus (SCBA) with full facepiece operated in positive pressure mode.

Ventilation

Local exhaust (preferred) at site of chemical work or general ventilation.

Protective Gloves	Eye Protection	Other Protective Clothing
Butyl Rubber (or other non-permeable)	Chemical Dust Goggles and Face Mask	Protective Uniform or Apron

Work/Hygiene Practices

Always wash hands thoroughly after using chemical; never bring food, drink, or smoking materials into vicinity of chemicals.

BERYLLIUM NITRATE

$Be(NO_3)_2$ CAS: 13597-99-4

IDENTIFICATION AND TYPICAL USES

Beryllium nitrate appears as a colorless or white to yellow deliquescent crystalline solid. It is used as a chemical reagent and as a gas mantle hardener. It also may be used in the production of beryllium metal by reduction with magnesium metal. Also used in nuclear reactors and in glass manufacturing.

RISK ASSESSMENT: HEALTH

General Assessment

Beryllium nitrate is a confirmed carcinogen with experimental reproductive effects. It is poisonous by *ingestion* and can cause toxic effects upon *inhalation.*

Inhalation can cause severe bronchitis and/or pneumonia within 24 to 48 hours of high exposure. Symptoms include fever, cough, and shortness of breath. Death can also occur. Pulmonary edema (fluid buildup in the lungs) can occur following high acute exposures. High or repeated exposures can lead to scars in the lungs and other organs. Fatigue, weight loss, and heart failure can occur years later. Beryllium nitrate is also an experimental teratogen, but the data are inconclusive in this area.

Beryllium nitrate produces chronic systemic disease that primarily effects the lungs as berylliosis, a fatal lung disease, but it can also involve other organs such as the lymph nodes, liver, spleen, bones, and kidneys.

Ingestion causes irritation of the mouth and stomach with symptoms of fatigue, weakness, and loss of appetite.

Eye contact with dusts causes conjunctiva irritation with redness, itching, and swelling of the eyelids. Allergies may develop. Skin contact may cause severe irritation, leading to skin rash. Skin burns are also possible, depending on amount of exposure. Particles that get under cuts in skin may cause persistent, non-healing skin ulcers at site of contact.

Acute Health Effects

The following acute (short-term) health effects may occur immediately or shortly after exposure to beryllium nitrate:

Skin: Severe irritation and possible burns. Localized skin ulcers may also be possible.

Eye: Severe irritation, itching, and burning on contact. Sometimes an allergic eye problem develops, breaking out again with future exposures.

Lung: Overexposure can severely irritate the airways and lungs, causing nasal discharge, tightness in the chest, cough, shortness of breath, and/or fever. Future exposures can cause further attacks. Death can occur in severe cases. High acute exposures can lead to pulmonary edema. Symptoms may be delayed for days after exposure. Some persons later develop lung scarring after such exposures.

Chronic Health Effects

The following chronic (long-term) health effects may occur at some time after exposure to beryllium nitrate and can last for months or even years:

Cancer Hazards: Beryllium nitrate is a probable cancer causing agent in humans. There is some evidence that it causes lung cancer in humans and it has been shown to cause lung cancers in animals.

Reproductive Hazard: Beryllium nitrate may damage the testes (male reproductive glands).

Other Chronic Effects: A single high exposure or repeated lower exposures can cause permanent scars in the lungs or other body organs. Symptoms may include fatigue, shortness of breath, weight loss, and poor appetite. These effects may occur months or years after exposure. Severe cases result in disability and even heart failure. If chemical particles get under cuts in the skin, ulcers or lumps can develop. Surgery will be needed to remove these chemical particles. Allergic skin rashes can also occur, with itching and redness. These improve away from exposure and can come back with future exposures. Kidney stones can occur from high or repeated exposure.

Recommended Risk-Reduction Measures

Personnel should avoid direct contact with beryllium nitrate. If a less toxic material or compound cannot be substituted for beryllium, then *engineering controls* are the most effective method of reducing exposure risk. The best protection is to enclose operations and/or provide local exhaust ventilation at the site of beryllium nitrate release. While not always operationally feasible, isolating operations can also reduce exposure. Using respiratory protection is less effective than the controls mentioned above, but is still required whenever working with or around beryllium. For any exposure level, a MSHA/NIOSH approved self-contained breathing apparatus (SCBA) with full facepiece and pressure demand is the appropriate level of respiratory protection. If a full facepiece is not available, then chemical goggles should be worn to protect the eyes, especially when a splash hazard exists. A face shield should also be considered. To prevent hand and skin exposures, rubber or other non-permeable gloves should be worn.

Administrative controls should also be in place to minimize the potential for human exposures. These may include written procedures or policies which specify the methods and techniques that will be practiced whenever personnel are to work with beryllium nitrate.

All personnel should receive *training* on the use, hazards, protective measures, emergency actions, and other precautions per 29 CFR 1910.1200 (Hazard Communication) prior to the first assignment in an area where beryllium nitrate is used or stored.

Before first exposure and yearly, a medical history and exam are recommended, with:

- ☑ Lung function tests (spirometry and gas transfer testing). For high exposures more frequent tests should be performed, some experts recommend monthly.
- ☑ Chest X-ray.
- ☑ Careful physical examination of the skin and lungs.

If symptoms develop or overexposure is suspected, the following may be useful:

- ☑ Consider chest X-ray after acute exposure.

It should be noted that medical tests that simply look for existing damage are not a substitute for controlling exposures. Medical histories are extremely important when assessing exposure risk.

Other methods to reduce exposure include:

- ☑ Specific engineering controls are recommended for this chemical by NIOSH. Refer to the NIOSH criteria documents: Occupational Exposure to Beryllium (#72-10268).
- ☑ Always ensure that proper protective clothing is worn when using chemical substances.
- ☑ Wash thoroughly immediately after exposure to beryllium nitrate and at the end of the work shift or before eating, drinking, or smoking.
- ☑ Hazard warning information should be posted in the work area. In addition, as part of an on-going education and training program, all information on the health and safety hazards of beryllium should be communicated to all exposed and potentially exposed workers.
- ☑ Eye wash stations should be provided in the immediate work area for emergency use. If there is a possibility of skin exposure to beryllium nitrate, emergency shower facilities should also be provided.
- ☑ Workers whose clothing has been contaminated by beryllium nitrate should change into clean clothes before leaving work. Contaminated work clothing should be laundered only by individuals who have been informed of the hazards of exposure to beryllium nitrate.

RISK ASSESSMENT: ENVIRONMENT

General Assessment

The environment is at risk of exposure during transportation, storage, disposal, or destruction of beryllium nitrate. In almost every scenario, the threat of environmental exposure is contingent upon the proper handling of the chemical substance. Accidental spills, large or small, can result in contamination of the surrounding environmental mediums (water, soil, and air).

Beryllium nitrate is a non-combustible solid. However, it is an oxidizer and caution is always required in handling, storage, transportation, and disposal of beryllium nitrate. When heated to decomposition in air it emits very toxic fumes of beryllium oxide and nitrogen oxides. Emergency responders should be made aware of the presence of beryllium nitrate at any emergency responsc situation.

The major source of beryllium nitrate in the environment is burning fossil fuels, some of which may enter the aquatic environment through atmospheric fallout. Beryllium compounds may also enter the aquatic environment through the weathering of rocks and soils and through discharge from industrial and municipal operations.

Acute Ecological Effects

Acute (short-term) toxic effects may include the death of animals, birds, or fish and death or low growth rate in plants. Acute effects are seen 2 to 4 days after animals or plants are exposed to beryllium nitrate. Beryllium nitrate is more toxic in soft water than in hard water. Beryllium and its compounds have high acute toxicity to aquatic life. Insufficient data are available to evaluate or predict the acute (short-term) effects of beryllium to birds, plants or land animals.

Chronic Ecological Effects

Chronic (long-term) toxic effects may include shortened life span, reproductive problems, lower fertility, and changes in appearance or behavior in exposed animals. These effects can be seen long after first exposure(s) to toxic chemicals. Beryllium and its compounds are considered to have high chronic toxicity to aquatic life. Insufficient data are available to evaluate the chronic (long-term) effects of beryllium on plants, birds, or land animals.

Water Solubility

Beryllium is highly soluble in water. Concentrations of 1000 milligrams and more will mix with a liter of water.

Persistence in the Environment

Beryllium is highly persistent in water, with a half-life of greater than 200 days. The half-life of a pollutant is the amount of time it takes for one-half of the chemical to be degraded.

Bioaccumulation in Aquatic Organisms

Some substances increase in concentration, or *bioaccumulate,* in living organisms as they breathe contaminated air, drink contaminated water, or eat contaminated food. These chemicals can become concentrated in the tissues and internal organs of animals as well as humans.

The concentration of beryllium found in edible fish tissue is expected to be somewhat higher than the average concentration of beryllium found in the water from which the fish was taken.

Recommended Risk-Reduction Measures

Proper training of all transporters will reduce the likelihood of a mishap or accident resulting in a leak or spill of beryllium nitrate into the environment. Labels on all containers, trucks, and rail cars must meet DOT requirements and accurately reflect their contents to enable emergency responders to react properly and quickly to any disaster thereby reducing the potential risk to the environment and to personnel.

Storage of beryllium nitrate should be segregated from other chemicals and materials to minimize the risk of cross-contamination. Beryllium nitrate must be stored to avoid contact with fuels and flammable and combustible materials. Containers should be protected from physical damage and stored to avoid contact with heat.

If a spill or leak to the environment has occurred, fire department, emergency response and/or hazardous materials spill personnel should be notified immediately. Cleanup should be attempted only by those trained in proper spill containment procedures. Neutralizing agents are agricultural lime (slaked lime), crushed limestone, and sodium bicarbonate. Contaminated soils should be removed for incineration and replaced with clean soil.

If beryllium nitrate should contact the water table, aquifer, or navigable waterway, time is of the essence. It is highly soluble and total remediation may not be possible. The local and/or state emergency response authorities must be notified. A comprehensive emergency response or disaster preparedness/recovery plan should be in place prior to any operations involving the use, transportation, storage, or disposal of beryllium nitrate.

If beryllium nitrate is spilled or leaked, the following specific steps are recommended:

- ☑ Restrict persons not wearing protective clothing from area of spill or leak until cleanup is complete. Avoid skin and eye contact.
- ☑ Collect powdered material in the most convenient and safe manner possible and deposit in sealed drum for disposal. Use a high efficiency particulate air (HEPA) filter vacuum during cleanup.

- ☑ Ventilate area of spill or leak.
- ☑ It may be necessary to dispose of beryllium nitrate as a hazardous waste. The responsible state agency or the regional office of the federal Environmental Protection Agency (EPA) should be contacted for specific recommendations.
- ☑ Do not generate dusts.

RISK ASSESSMENT: BUSINESS

General Assessment

Accidents or mishaps involving beryllium nitrate can present a serious threat to business operations. The loss or damage of equipment or facilities can significantly affect fiscal viability. Lawsuits that may result from personnel illness, injury/death, public exposures, and/or environmental contamination will require a serious expenditure of resources. Media attention surrounding an injury, death, or environmental damage can also result in a loss of profits and loss of current as well as future business.

Recommended Risk-Reduction Measures

Company attorneys, safety and health professionals, and environmental specialists should be involved in the development of any procedures for responding to chemical incidents. A company official should be predesignated as a public relations officer with specific training in dealing with the press. Corporate plans and policies should be developed, approved, and implemented long before any need for such arises.

REFERENCES

American Conference of Governmental Industrial Hygienists. 1988. *Documentation of the Threshold Limit Values and Biological Exposure Indices,* 5th Edition (with updates). Cincinnati: ACGIH

Anthony, C. P., and N. J. Kolthoff. 1971. *Textbook of Anatomy and Physiology.* St. Louis: C. V. Mosby Company.

Brusick, D. J. 1994. *Methods for Genetic Risk Assessment.* New York: CRC Press/Lewis Publishers.

Calabrese, E. J. 1994. *Biological Effects of Low Level Exposures.* New York: CRC Press/Lewis Publishers.

Cockerman, L. G., B. S. Shane. 1994. *Basic Environmental Toxicology.* New York: CRC Press/Lewis Publishers.

Danse, I. R. 1991. *Common Sense Toxics in the Workplace.* New York: Van Nostrand Reinhold.

Hallenbeck, W. H. 1993. *Quantitative Risk Assessment for Environmental and Occupational Health,* 2nd Edition. New York: CRC Press/Lewis Publishers.

Hodgson, E., R. B. Mailman, and J. E. Chambers. 1988. *Dictionary of Toxicology.* London: The MacMillian Company.

Jones, F. E. 1994. *Toxic Organic Vapors in the Workplace.* New York: CRC Press/Lewis Publishers.

Keith, L. H. and M. M. Walker. 1995. *Handbook of Air Toxics.* New York: CRC Press/Lewis Publishers.

Landis, W. G., M. H. Yu. 1995. *Introduction to Environmental Toxicology.* New York: CRC Press/Lewis Publishers.

Langerman, N. 1994 *Precautionary Labels for Chemical Containers.* New York: CRC Press/Lewis Publishers.

Lauwerys, R. R., P. Hoet. 1993. *Industrial Chemical Exposure.* New York: CRC Press/Lewis Publishers.

Lewis, R. J., Sr. 1993. *Hawley's Condensed Chemical Dictionary,* 12th Edition. New York: Van Nostrand Reinhold.

Lewis, R. J., Sr. 1992. *Sax's Dangerous Properties of Industrial Materials,* Volumes 1, 2, and 3, 8th Edition. New York: Van Nostrand Reinhold.

Meyer, E. 1990. *Chemistry of Hazardous Materials.* Englewood Cliffs, NJ: Prentice Hall.

Patnaik, P. 1992. *A Comprehensive Guide to the Hazardous Properties of Chemical Substances.* New York: Van Nostrand Reinhold.

Philp, R. B. 1995. *Environmental Hazards and Human Health.* New York: CRC Press/Lewis Publishers.

Proctor, N. H., J. P. Hughes. 1991. *Chemical Hazards of the Workplace.* New York: Van Nostrand Reinhold.

Quigley, D. R. 1994. *Handbook of Emergency Chemical Management.* New York: CRC Press/Lewis Publishers.

Richardson, M. 1995. *Dictionary of Substance and Their Effects,* Vol. I-VI. New York: CRC Press/Lewis Publishers.

Torkelson, T. R. V. K. Rowe. 1981. *Patty's Industrial Hygiene and Toxicology,* Vol. 2B. New York: Wiley-Interscience.

U.S. Department of Health, Education, and Welfare (DHEW), National Institute for Occupational Safety and Health (NIOSH). 1972. *Criteria for a Recommended Standard... Occupational Exposure to Beryllium.* Publication 72-10268. Washington, D.C.: U.S. Government Printing Office.

U.S. Department of Health and Human Services, National Institute for Occupational Safety and Health. 1994. *NIOSH Pocket Guide to Chemical Hazards.* Washington, D.C.: U.S. Government Printing Office.

MATERIAL SAFETY DATA SHEET

CHEMICAL NAME	CRC PRESS/LEWIS PUBLISHERS DISCLAIMER
BIS(2-CHLORETHYL) ETHER	

HAZARD WARNING INFORMATION

HEALTH	FIRE	REACTIVE	OTHER	DEGREE OF HAZARD	COLOR CODING	OTHER CODES
3	2	0		0 = Minimum Hazard 1 = Slight Hazard 2 = Moderate Hazard 3 = Serious Hazard 4 = Severe Hazard	HEALTH = BLUE FIRE = RED REACTIVITY = YELLOW OTHER = WHITE	OX = Oxidizer ACID = Acid ALK = Alkali COR = Corrosive W = Use No Water

SECTION I - GENERAL INFORMATION

Characterization	**Haloether**	RCRA Number	**U025**
		EPA Class	**Toxic Waste**
DOT Proper Shipping Name	**Dichloroethyl Ether**	Chemical Abstract Service (CAS) Number	**111-44-4**
DOT Hazard Class and Label Requirements	**IMO: Poison B**	DOT Emergency Guide Code	**57**
DOT Identification Number	**UN 1916**	Molecular Formula	**$(C_2H_4Cl)_2O$**

Synonyms

Dichloroethyl ether; 2,2'-dichloroethyl ether; di(2-chloroethyl)ether; dichloroethyl oxide; 1,1'-oxybis(2-chloroethane); Chlorex; DCEE; dichlorodiethyl ether; sym-dichloroethyl ether.

SECTION II - HAZARDOUS INGREDIENTS AND IDENTITY INFORMATION

Hazardous Components (specific identity)	OSHA Exposure Criteria	NIOSH Exposure Criteria	Immediately Dangerous to Life and Health (IDLH)	ACGIH Exposure Criteria
Bis(2-chloroethyl)ether: (derivation: By chlorination of ethyl ether). 1 ppm = 6 mg/m^3	**PEL (skin):** 15 ppm 90 mg/m^3 5 ppm (TWA) **STEL (skin):** 10 ppm	Possible Cancer Agent	250 ppm	**TLV (skin):** 5 ppm 30 mg/m^3 5 ppm (TWA) **STEL (skin):** 10 ppm

SECTION III - PHYSICAL AND CHEMICAL CHARACTERISTICS

Boiling Point	**352°F (178°C)**	Specific Gravity (H_2O = 1)	**1.22 at 20°C**
Vapor Pressure (mm Hg)	**2.00**	Molecular Weight (atomic weight)	**143**
Vapor Density (Air = 1)	**4.93**	Melting Point	**-62°F (-52°C)**

Solubility

Insoluble in water. Miscible with most organic solvents.

Appearance and Odor

Colorless, clear liquid with a pungent odor like that of ethylene dichloride.

SECTION IV - FIRE AND EXPLOSION HAZARD DATA

Flash Point (method used)	**131°F (55°C) closed cup**	Explosive Limits in Air % by Volume	**LEL: 2.7% UEL: Not Determined**
NFPA Classification	**Class II Combustible**	Autoignition Temperature	**696°F (396°C)**

Extinguishing Media

Dry chemical, foam, water spray, or carbon dioxide. Use water spray to keep closed containers cool.

Special Fire Fighting Procedures

Poisonous gases of hydrogen chloride and phosgene are produced in fire. Structural fire-fighting protective clothing will not provide adequate protection. Wear full protective clothing, self-contained breathing apparatus (SCBA) operated in pressure-demand mode. Move container from fire area if it can be done without risk. Will attack some plastics, rubber, and coatings. Evacuate non-essential personnel 1500 feet from fire area. Isolate for ½ mile if rail or tank car is involved in fire.

Unusual Fire and Explosion Hazards

Immediately withdraw if rising sound from venting tank is heard or if fire causes discoloration to the tank.

SECTION V - REACTIVITY DATA

Stability		Conditions to Avoid
		Avoid contact with incompatible materials since violent reactions can occur. Also avoid contact with water or steam. May form unstable peroxides on exposure to air.
Stable	Unstable	Incompatibility *(materials to avoid)*
X		Incompatible with strong oxidizers, such as chlorine, bromine, and chlorine dioxide, and strong acids, such as sulfuric and chlorosulfonic. Reacts with water or steam to emit toxic and corrosive fumes.
Hazardous Polymerization		Conditions to Avoid
		Under normal conditions of temperature and pressure, bis (2-chloroethyl) ether is not known to undergo hazardous polymerization.
May Occur	Will Not Occur	Hazardous Decomposition or By-products
	X	When heated to decomposition in air it emits highly toxic chloride fumes.

SECTION VI - HEALTH HAZARD DATA

Primary Route(s) of Entry:	Inhalation?	Absorption (skin/eye)?	Ingestion?
	X	X	X

Health Hazards

INHALATION: Causes severe irritation to the respiratory system with symptoms of cough and shortness of breath. Can lead to pulmonary edema, which is a medical emergency and can be fatal.

SKIN & EYES: A severe eye and skin irritant. Can pass through unbroken skin causing toxic systemic effects.

INGESTION: Produces nausea and vomiting.

Carcinogenicity	NTP Listed?	IARC Cancer Review Group?	OSHA Regulated?	Target Organs?
Suspected Human Confirmed Animal	No	Group 3	No	Skin, eyes, respiratory system.

Medical Conditions Generally Aggravated by Exposure

None reported. However, existing respiratory problems (bronchitis, asthma) may be aggravated.

Emergency and First-aid Procedures

Eye contact: Do not allow victim to keep eyes tightly shut. Flush immediately with water for 15 minutes (minimum), seek medical attention. Skin contact: Wash with large amounts of soap and water for a minimum of 15 minutes. For inhalation: Remove the person from exposure. Provide respiratory assistance and CPR. Transfer to medical facility. Observe for 24- 48 hours for lung effects. If swallowed: Induce vomiting if person is conscious. Seek medical attention immediately. Do not induce vomiting or attempt to give an unconscious or convulsing person anything by mouth.

SECTION VII - PRECAUTIONS FOR SAFE HANDLING AND USE

Steps to be Taken in Case Material is Released or Spilled

Wear self-contained breathing apparatus (SCBA) for respiratory protection. Restrict those not wearing protective equipment and who are not involved in cleanup from entering area. Avoid skin contact. Ventilate area of spill. Do not generate dusts. Collect powder materials and deposit in sealed drum.

Preferred Waste Disposal Method

Burn in a chemical incinerator equipped with an afterburner and scrubber.

Precautions to be Taken in Handling and Storage

Store in tightly closed containers in a cool, well ventilated area away from heat. Protect containers from physical damage.

Other Precautions and Warnings

Bis (2-chloroethyl) ether must be stored to avoid contact with strong oxidizers since violent reactions can occur. Sources of ignition should be prohibited in areas where this chemical is used or stored.

SECTION VIII - CONTROL MEASURES AND PERSONAL PROTECTIVE EQUIPMENT

Respiratory Protection *(specify type)*

Bis (2-chloroethyl) ether is a suspected human carcinogen. Use a NIOSH/MSHA approved self-contained breathing apparatus (SCBA) with full facepiece operated in positive pressure mode.

Ventilation

Local exhaust (preferred) at site of chemical work or general ventilation.

Protective Gloves	Eye Protection	Other Protective Clothing
Butyl Rubber (or other non-permeable)	Chemical Goggles and Face Mask	Protective Uniform or Apron

Work/Hygiene Practices

Always wash hands thoroughly after using chemical; never bring food, drink, or smoking materials into vicinity of chemicals.

BIS (2-CHLOROETHYL) ETHER

$(C_2H_4Cl)_2O$ CAS: 111-44-4

$Cl — CH_2 — CH_2 — O — CH_2 — CH_2 — Cl$

IDENTIFICATION AND TYPICAL USES

Bis (2-chloroethyl) ether appears as a colorless, clear liquid with a pungent odor like that of ethylene dichloride. It is one of several chloroalkyl ethers. It is a general solvent as well as a selective solvent for producing high-grade lubricating oils, textile scouring and cleansing, fulling compounds, wetting and penetrating compounds, paints, varnishes, lacquers, finish removers, spotting and dry cleaning, soil fumigant, intermediate and cross-linker in organic synthesis. It is also used in the production of glycol products, rubber, and insecticides.

RISK ASSESSMENT: HEALTH

General Assessment

Bis (2-chloroethyl) ether is a questionable carcinogen (has cancer-causing properties) with experimental mutagenic and tumorigenic data reported. It is poisonous by *ingestion, inhalation,* and by *absorption* through the skin.

Exposure to its vapors can cause severe irritation to the eyes, nose, and respiratory tract. Exposure to 1000 ppm for 30 to 60 minutes may cause death within days as a result of lung damage or possibly pulmonary edema. At lower concentrations, death due to lung damage is still possible. Other organs, such as the kidney, liver, and brain, can also be affected to a lesser degree.

Eye contact with liquid bis (2 chloroethyl) ether can cause conjunctival irritation and injury to the cornea. On the skin, its irritation effect can range from mild irritation to first degree burns on short exposures. However, on prolonged contact, the liquid may be absorbed through the skin causing second degree burns and toxic systemic effects.

Ingestion of this compound in small amounts may produce nausea and vomiting. It can irritate the mouth and stomach and manifest symptoms of systemic poisoning. Large doses of 50 to 75 mL is expected to be fatal in humans.

☠ *Acute Health Effects*

The following acute (short-term) health effects may occur immediately or shortly after exposure to bis (2-chloroethyl) ether:

Skin: Severe irritation and possible burns. On prolonged exposure, can be absorbed though intact skin causing toxic systemic effects.

Eye: Severe conjunctival irritation and possible damage to the cornea.

Lung: Severe irritation to respiratory tract causing shortness of breath and cough. High acute exposures can be fatal due to a buildup of fluid in the lungs (pulmonary edema).

Chronic Health Effects

The following chronic (long-term) health effects may occur at some time after exposure to bis (2-chloroethyl) ether and can last for months or even years:

Cancer Hazards: Bis (2-chloroethyl) ether may be a cancer causing agent in humans since it has been shown to cause liver cancer in animals.

Reproductive Hazard: According to the references, bis (2-chloroethyl) ether has not been adequately tested for its ability to adversely affect reproduction.

Other Chronic Effects: Repeated low-dose exposures can irritate the lungs and result in bronchitis with symptoms of cough, phlegm, and shortness of breath. It may damage the liver and kidneys.

Recommended Risk-Reduction Measures

Personnel should avoid direct contact with bis (2-chloroethyl) ether. If a less toxic material or compound cannot be substituted for bis (2-chloroethyl) ether, then *engineering controls* are the most effective method of reducing exposure risk. The best protection is to enclose operations and/or provide local exhaust ventilation at the site of bis (2-chloroethyl) ether release. While not always operationally feasible, isolating operations can also reduce exposure. Using respiratory protection is less effective than the controls mentioned above, but is still required whenever working with or around bis (2-chloroethyl) ether. For any exposure level, a MSHA/NIOSH approved self-contained breathing apparatus (SCBA) with full facepiece and pressure demand is the appropriate level of respiratory protection. If a full facepiece is not available, then chemical goggles should be worn to protect the eyes, especially when a splash hazard exists. A face shield should also be considered. To prevent hand and skin exposures, rubber or other non-permeable gloves should be worn.

Administrative controls should also be in place to minimize the potential for human exposures. These may include written procedures or policies which specify the methods and techniques that will be practiced whenever personnel are to work with bis (2-chloroethyl) ether.

All personnel should receive *training* on the use, hazards, protective measures, emergency actions, and other precautions per 29 CFR 1910.1200 (Hazard Communication) prior to the first assignment in an area where bis (2-chloroethyl) ether is used or stored.

Before beginning employment and at regular intervals thereafter (e.g., annually), the following medical test is recommended:

☑ Lung function tests.

If symptoms develop or overexposure is suspected, the following may be useful:

☑ Liver, kidney, and lung function tests.

☑ Consider chest X-ray after acute exposure.

It should be noted that medical tests that simply look for existing damage are not a substitute for controlling exposures. Medical histories are extremely important when assessing exposure risk. Also, since smoking can cause heart disease as well as lung cancer, emphysema, and other respiratory ailments, any exposure to bis (2-chloroethy) ether can result in quicker and more devastating symptoms. Smokers should therefore avoid unprotected contact with this chemical.

Other methods to reduce exposure include:

☑ Always ensure that proper protective clothing is worn when using chemical substances.

☑ Wash thoroughly immediately after exposure to bis (2-chloroethyl) ether and at the end of the work shift or before eating, drinking, or smoking.

☑ Hazard warning information should be posted in the work area. In addition, as part of an on-going education and training program, all information on the health and safety hazards of bis (2-chloroethyl) ether should be communicated to all potentially exposed workers.

☑ Eye wash stations should be provided in the immediate work area for emergency use. If there is a possibility of skin exposure to bis (2-chloroethyl) ether, emergency shower facilities should also be provided.

☑ Workers whose clothing has been contaminated by bis (2-chloroethyl) ether should change into clean clothes before leaving work. Contaminated work clothing should be laundered only by individuals who have been informed of the hazards of exposure to bis (2-chloroethyl) ether.

RISK ASSESSMENT: ENVIRONMENT

General Assessment

The environment is at risk of exposure during transportation, storage, disposal, or destruction of bis (2-chloroethyl) ether. In almost every scenario, the threat of environmental exposure is contingent upon the proper handling of the chemical substance. Accidental spills, large or small, can result in contamination of the surrounding environmental mediums (water, soil, and air).

Bis (2-chloroethyl) ether is class II combustible. It will support combustion as well as intensify any fire. It is incompatible with oxidizers and can also react in the presence of water or steam to evolve toxic and corrosive fumes. Caution is always required in handling, storage, transportation, and disposal of bis (2-chloroethyl) ether. When heated to decomposition in air it emits very toxic chloride fumes. Emergency responders should be made aware of the presence of bis (2-chloroethyl) ether at any emergency response situation.

Bis (2-chloroethyl) ether may enter the environment through industrial effluents, spills, and possibly from the chlorination of water.

Acute Ecological Effects

Acute (short-term) toxic effects may include the death of animals, birds, or fish and death or low growth rate in plants. Acute effects are seen 2 to 4 days after animals or plants are exposed to bis (2-chloroethyl) ether. Bis (2-chloroethyl) ether has slight acute toxicity to aquatic life. It can kill or injure plants. Insufficient data are available to evaluate or predict the acute (short-term) effects of bis (2-chloroethyl) ether to birds or land animals.

Chronic Ecological Effects

Chronic (long-term) toxic effects may include shortened life span, reproductive problems, lower fertility, and changes in appearance or behavior in exposed animals. These effects can be seen long after first exposure(s) to toxic chemicals. Bis (2-chloroethyl) ether has slight chronic toxicity to aquatic life. Insufficient data are available to evaluate the chronic (long-term) effects of bis (2-chloroethyl) ether on plants, birds, or land animals.

Water Solubility

Bis (2-chloroethyl) ether is highly soluble in water. Concentrations of 1000 milligrams and more will mix with a liter of water.

Persistence in the Environment

Bis (2-chloroethyl) ether is moderately persistent in water, with a half-life of between 20 and 200 days. The half-life of a pollutant is the amount of time it takes for one-half of the chemical to be degraded. Approximately 91.5% of bis (2-chloroethyl) ether will eventually end up in water, the remainder will end up in the air.

Bioaccumulation in Aquatic Organisms

Some substances increase in concentration, or *bioaccumulate,* in living organisms as they breathe contaminated air, drink contaminated water, or eat contaminated food. These chemicals can become concentrated in the tissues and internal organs of animals as well as humans.

The concentration of bis (2-chloroethyl) ether found in edible fish tissue is expected to be somewhat higher than the average concentration of bis (2-chloroethyl) ether found in the water from which the fish was taken.

Recommended Risk-Reduction Measures

Proper training of all transporters will reduce the likelihood of a mishap or accident resulting in a leak or spill of bis (2-chloroethyl) ether into the environment. Labels on all containers, trucks, and rail cars must meet DOT requirements and accurately reflect their contents to enable emergency responders to react properly and quickly to any disaster thereby reducing the potential risk to the environment and to personnel.

Storage of bis (2-chloroethyl) ether should be segregated from other chemicals and to minimize the risk of cross-contamination. Bis (2-chloroethyl) ether must be stored to avoid contact with strong oxidizers, such as chlorine and bromine, and strong acids, such as sulfuric and chlorosulfonic, since violent reactions can occur. Containers should be protected from physical damage and stored to avoid contact with heat.

If a spill or leak to the environment has occurred, fire department, emergency response and/or hazardous materials spill personnel should be notified immediately. Cleanup should be attempted only by those trained in proper spill containment procedures. Contaminated soils should be removed for incineration and replaced with clean soil.

If bis (2-chloroethyl) ether should contact the water table, aquifer, or navigable waterway, time is of the essence. It is highly soluble and total remediation may not be possible. The local and/or state emergency response authorities must be notified. A comprehensive emergency response or disaster preparedness/recovery plan should be in place prior to any operations involving the use, transportation, storage, or disposal of bis (2-chloroethyl) ether.

If bis (2-chloroethyl) ether is spilled or leaked, the following specific steps are recommended:

- ☑ Restrict persons not wearing protective clothing from area of spill or leak until cleanup is complete. Avoid skin and eye contact.
- ☑ Absorb spilled liquids with vermiculite, dry sand, or a similar material and deposit in sealed drum for disposal.
- ☑ Ventilate area of spill or leak.
- ☑ It may be necessary to dispose of bis (2-chloroethyl) ether as a hazardous waste. The responsible state agency or the regional office of the federal Environmental Protection Agency (EPA) should be contacted for specific recommendations.
- ☑ Remove all ignition sources.

RISK ASSESSMENT: BUSINESS

General Assessment

Accidents or mishaps involving bis (2-chloroethyl) ether can present a serious threat to business operations. The loss or damage of equipment or facilities can significantly affect fiscal viability. Lawsuits that may result from personnel illness, injury/death, public exposures, and/or environmental contamination will require a serious expenditure of resources. Media attention surrounding an injury, death, or environmental damage can also result in a loss of profits and loss of current as well as future business.

Recommended Risk-Reduction Measures

Company attorneys, safety and health professionals, and environmental specialists should be involved in the development of any procedures for responding to chemical incidents. A company official should be predesignated as a public relations officer with specific training in dealing with the press. Corporate plans and policies should be developed, approved, and implemented long before any need for such arises.

REFERENCES

Ahlbom, A. 1993. *Biostatistics for Epidemiologists.* New York: CRC Press/Lewis Publishers.

American Conference of Governmental Industrial Hygienists. 1988. *Documentation of the Threshold Limit Values and Biological Exposure Indices,* 5th Edition (with updates). Cincinnati: ACGIH

Anthony, C. P., and N. J. Kolthoff. 1971. *Textbook of Anatomy and Physiology.* St. Louis: C. V. Mosby Company.

Asante-Duah, D. K. 1993. *Hazardous Waste Risk Assessment.* New York: CRC Press/Lewis Publishers.

Bloom, A. D. and F. J. DeSerres. 1995. *Ecotoxicity and Human Health.* New York: CRC Press/Lewis Publishers.

Brusick, D. J. 1994. *Methods for Genetic Risk Assessment.* New York: CRC Press/Lewis Publishers.

Calabrese, E. J., E. M. Kenyon. *Air Toxics and Risk Assessment.* New York: CRC Press/Lewis Publishers.

Calabrese, E. J. 1994. *Biological Effects of Low Level Exposures.* New York: CRC Press/Lewis Publishers.

Cockerman, L. G., B. S. Shane. 1994. *Basic Environmental Toxicology.* New York: CRC Press/Lewis Publishers.

Cothern, C. R., N. P. Ross. 1994. *Environmental Statistics, Assessment, and Forecasting.* New York: CRC Press/Lewis Publishers.

Cothern, R. R. 1993. *Comparative Environmental Risk Assessment.* New York: CRC Press/Lewis Publishers.

Cothern, C. R. 1995. *Handbook for Environmental Risk Decisionmaking.* New York: CRC Press/Lewis Publishers.

Danse, I. R. 1991. *Common Sense Toxics in the Workplace.* New York: Van Nostrand Reinhold.

Fodden, C. R., and J. L. Weddell. 1992. *Hazardous Materials Emergency Action Data.* New York: CRC Press/Lewis Publishers.

Gardner, R. W. 1994. *Chemical Intolerance.* New York: CRC Press/Lewis Publishers.

Hallenbeck, W. H. 1993. *Quantitative Risk Assessment for Environmental and Occupational Health,* 2nd Edition. New York: CRC Press/Lewis Publishers.

Hodgson, E., R. B. Mailman, and J. E. Chambers. 1988. *Dictionary of Toxicology.* London: The MacMillian Company.

Howard, P. H. 1995. *Handbook of Environmental Fate and Exposures Data for Organic Chemicals,* Vol. I-IV. New York: CRC Press/Lewis Publishers.

Howard, P. H., et. al. 1991. *Handbook of Environmental Degradation Rates.* New York: CRC Press/Lewis Publishers.

Jones, F. E. 1994. *Toxic Organic Vapors in the Workplace.* New York: CRC Press/Lewis Publishers.

Keith, L. H. and M. M. Walker. 1995. *Handbook of Air Toxics.* New York: CRC Press/Lewis Publishers.

Landis, W. G., M. H. Yu. 1995. *Introduction to Environmental Toxicology.* New York: CRC Press/Lewis Publishers.

Langerman, N. 1994 *Precautionary Labels for Chemical Containers.* New York: CRC Press/Lewis Publishers.

Lauwerys, R. R., P. Hoet. 1993. *Industrial Chemical Exposure.* New York: CRC Press/Lewis Publishers.

Lewis, R. J., Sr. 1993. *Hawley's Condensed Chemical Dictionary,* 12th Edition. New York: Van Nostrand Reinhold.

Lewis, R. J., Sr. 1992. *Sax's Dangerous Properties of Industrial Materials,* Volumes 1, 2, and 3, 8th Edition. New York: Van Nostrand Reinhold.

Lide, D. R. 1995. *Handbook of Chemistry and Physics,* 76th Edition. New York: CRC Press/Lewis Publishers.

Meyer, E. 1990. *Chemistry of Hazardous Materials.* Englewood Cliffs, NJ: Prentice Hall.

Patnaik, P. 1992. *A Comprehensive Guide to the Hazardous Properties of Chemical Substances.* New York: Van Nostrand Reinhold.

Philp, R. B. 1995. *Environmental Hazards and Human Health.* New York: CRC Press/Lewis Publishers.

Proctor, N. H., J. P. Hughes. 1991. *Chemical Hazards of the Workplace.* New York: Van Nostrand Reinhold.

Quigley, D. R. 1994. *Handbook of Emergency Chemical Management.* New York: CRC Press/Lewis Publishers.

Richardson, M. 1995. *Dictionary of Substance and Their Effects,* Vol. I-VI. New York: CRC Press/Lewis Publishers.

Torkelson, T. R. V. K. Rowe. 1981. *Patty's Industrial Hygiene and Toxicology,* Vol. 2B. New York: Wiley-Interscience.

MATERIAL SAFETY DATA SHEET

CHEMICAL NAME	
BIS(2-CHLOROMETHYL) ETHER	**CRC PRESS/LEWIS PUBLISHERS DISCLAIMER:** THE INFORMATION AND RECOMMENDATIONS PRESENTED HEREIN ARE BASED ON SOURCES BELIEVED TO BE RELIABLE. CRC MAKES NO REPRESENTATION ON ITS COMPLETENESS OR ACCURACY. IT IS THE RESPONSIBILITY OF THE USER TO DETERMINE THE CHEMICAL'S SUITABILITY FOR ITS INTENDED USE, THE CHEMICAL'S SAFE USE, AND THE CHEMICAL'S PROPER DISPOSAL. NO REPRESENTATIONS AND/OR WARRANTIES, EITHER EXPRESSED OR IMPLIED, OF THE MERCHANTABILITY OR FITNESS FOR A PARTICULAR PURPOSE, OR OF ANY OTHER NATURE, ARE MADE WITH RESPECT TO THE INFORMATION PROVIDED IN THIS MSDS OR TO THE CHEMICAL TO WHICH INFORMATION MAY REFER. CRC NEITHER ASSUMES NOR AUTHORIZES ANY OTHER PERSON TO ASSUME FOR IT, ANY OTHER ADDITIONAL RESPONSIBILITY OR LIABILITY FOR THE USE OF, OR RELIANCE UPON, THIS INFORMATION.

HAZARD WARNING INFORMATION

HEALTH	FIRE	REACTIVE	OTHER	DEGREE OF HAZARD	COLOR CODING	OTHER CODES
4	3	1	W	0 = Minimum Hazard 1 = Slight Hazard 2 = Moderate Hazard 3 = Serious Hazard 4 = Severe Hazard	HEALTH = BLUE FIRE = RED REACTIVITY = YELLOW OTHER = WHITE	OX = Oxidizer ACID = Acid ALK = Alkali COR = Corrosive W = Use No Water

SECTION I - GENERAL INFORMATION

Characterization **Haloether**	RCRA Number **P016**	EPA Class **Acute Hazardous Waste**
DOT Proper Shipping Name **Dichloromethyl Ether**	Chemical Abstract Service (CAS) Number **542-88-1**	
DOT Hazard Class and Label Requirements **Poison B; Flammable Liquid and Poison**	DOT Emergency Guide Code **55**	
DOT Identification Number **UN 2249**	Molecular Formula $(CH_2Cl)_2O$	

Synonyms
Dichloromethyl ether; chlorodimethyl ether; dimethyl-1:1-dichloroether; oxybis (chloromethane); chloro (chloromethoxy) methane; bis(chloromethyl)ether; bis BCME.

SECTION II - HAZARDOUS INGREDIENTS AND IDENTITY INFORMATION

Hazardous Components (specific identity)	OSHA Exposure Criteria	NIOSH Exposure Criteria	Immediately Dangerous to Life and Health (IDLH)	ACGIH Exposure Criteria
Bis(2-chloromethyl)ether: (derivation: By chlorination of methyl ether; also reported to form spontaneously from formaldehyde and chloride ions in moist air). 1 ppm = 4.7 mg/m^3	**PEL:** Suspected Cancer Agent	**REL:** Sufficient Evidence: Possible Human Cancer Agent	Not Determined	**TLV:** 0.001 ppm 0.0047 mg/m^3 Suspected Human Carcinogen

SECTION III - PHYSICAL AND CHEMICAL CHARACTERISTICS

Boiling Point **223°F (106°C)**	Specific Gravity (H_2O = 1) **1.315 at 20°C**
Vapor Pressure (mm Hg) **Not Found**	Molecular Weight (atomic weight) **114.96**
Vapor Density (Air = 1) **4.00**	Melting Point **-43°F (-41.5°C)**

Solubility
Decomposes in water to hydrochloric acid and formaldehyde; soluble in most organic solvents.

Appearance and Odor
Colorless, clear liquid with a suffocating odor.

SECTION IV - FIRE AND EXPLOSION HAZARD DATA

Flash Point (method used) **95°F (35°C) closed cup (calculated)**	Explosive Limits in Air % by Volume **LEL: Not Determined UEL: Not Determined**
NFPA Classification **Class IC Flammable Liquid**	Autoignition Temperature **Not Found**

Extinguishing Media
Dry chemical, foam, or carbon dioxide. DO NOT USE WATER OR WATER SPRAY/MIST.

Special Fire Fighting Procedures
A poison by inhalation. Wear full protective clothing, self-contained breathing apparatus (SCBA) operated in pressure-demand mode. Move container from fire area if it can be done without risk. Will react with water to form hydrochloric acid and formaldehyde. Evacuate non-essential personnel 1500 feet from fire area. Isolate for ½ mile if rail or tank car is involved in fire.

Unusual Fire and Explosion Hazards
Immediately withdraw if rising sound from venting tank is heard or if fire causes discoloration to the tank.

SECTION V - REACTIVITY DATA

Stability		Conditions to Avoid: Keep away from water or moist areas.
Stable	Unstable	Incompatibility (materials to avoid): Reacts with water and moist air to form hydrochloric acid and formaldehyde.
X		
Hazardous Polymerization		Conditions to Avoid: Under normal conditions of temperature and pressure, bis (2-chloromethyl) ether is not known to undergo hazardous polymerization.
May Occur	Will Not Occur	Hazardous Decomposition or By-products: When heated to decomposition in air it emits highly toxic chloride fumes.
	X	

SECTION VI - HEALTH HAZARD DATA

Primary Route(s) of Entry:	Inhalation?	Absorption (skin/eye)?	Ingestion?
	X	X	X

Health Hazards

INHALATION: Causes severe irritation to the eyes, nose, and throat in humans. Eye irritation can be moderate to severe with conjunctival inflammation likely and possible burns.

SKIN & EYES: A severe skin irritant. Can pass through unbroken skin causing toxic systemic effects. Symptoms may include loss of appetite, weakness, nausea, and fatigue.

INGESTION: Moderate poisoning effects; may cause nausea and vomiting.

Carcinogenicity	NTP Listed?	IARC Cancer Review Group?	OSHA Regulated?	Target Organs?
Confirmed Human Confirmed Animal	5th Annual Report	Group 1	29 CFR 1910.1008	Skin, eyes, respiratory system; CNS.

Medical Conditions Generally Aggravated by Exposure

None reported. However, existing respiratory problems (bronchitis, asthma) may be aggravated.

Emergency and First-aid Procedures

Eye contact: Do not allow victim to keep eyes tightly shut. Flush immediately with water for 15 minutes (minimum), seek medical attention. Skin contact: Wash with large amounts of soap and water for a minimum of 15 minutes. For inhalation: Remove the person from exposure. Provide respiratory assistance and CPR. Transfer to medical facility. Observe for 24-48 hours for lung effects (pulmonary edema). If swallowed: Induce vomiting if person is conscious. Seek medical attention immediately. If unconscious or convulsing, do not induce vomiting or attempt to give anything by mouth.

SECTION VII - PRECAUTIONS FOR SAFE HANDLING AND USE

Steps to be Taken in Case Material is Released or Spilled

Wear self-contained breathing apparatus (SCBA) for respiratory protection. Restrict those not wearing protective equipment and who are not involved in cleanup from entering area. Avoid skin contact. Ventilate area of spill. Absorb liquids in vermiculite (or other absorbent) and place in sealed drum.

Preferred Waste Disposal Method

Burn in a chemical incinerator equipped with an afterburner and scrubber.

Precautions to be Taken in Handling and Storage

Store in tightly closed containers in a cool, well ventilated area away from heat and moisture. Protect containers from physical damage. A regulated work area is required (per OSHA 29 CFR 1910.1008).

Other Precautions and Warnings

Use only non-sparking tools and equipment. Use explosion-proof electrical equipment. Drinking fountains are not permitted in regulated areas. Automatic transfer of liquids should be done, if possible. Sources of ignition should be prohibited in areas where this chemical is used or stored.

SECTION VIII - CONTROL MEASURES AND PERSONAL PROTECTIVE EQUIPMENT

Respiratory Protection (specify type)

Bis (2-chloromethyl) ether is a confirmed carcinogen. For any exposure, use a NIOSH/MSHA approved self-contained breathing apparatus (SCBA) with full facepiece operated in positive pressure mode.

Ventilation

Local exhaust (preferred) at site of chemical work or general ventilation.

Protective Gloves	Eye Protection	Other Protective Clothing
Butyl Rubber (or other non-permeable)	Chemical Goggles and Face Mask	Protective Uniform or Apron

Work/Hygiene Practices

Always wash hands thoroughly after using chemical; never bring food, drink, or smoking materials into vicinity of chemicals.

BIS (2-CHLOROMETHYL) ETHER

$(CH_2Cl)_2O$ CAS: 542-88-1

Cl — CH_2 — O — CH_2 — Cl

IDENTIFICATION AND TYPICAL USES

Bis (2-chloromethyl) ether appears as a colorless, clear liquid with a suffocating odor. It is used as an intermediate in anionic-exchange quaternary resins. Its use as a chloromethylation reagent in industry is being discontinued because of its high carcinogenic properties. Exposure risks to this compound can arise during the use or production of chloromethyl methyl ether when the latter comes into contact with traces of water in the presence of hydrogen or hydroxyl ions. It may occur in trace amounts in chloromethyl methyl ether.

RISK ASSESSMENT: HEALTH

General Assessment

Bis (2-chloromethyl) ether is highly toxic and a confirmed carcinogenic compound (has cancer-causing properties) with experimental testing also showing neoplastigenic and tumorigenic data. It is poisonous by *ingestion, inhalation,* and by *absorption* through the skin.

Exposure to its vapors can cause severe irritation to the eyes, nose, and respiratory tract. The primary target organ is the lungs. Exposure to 100 ppm of this compound in air for only a few minutes can cause death within days as a result of lung damage or possibly pulmonary edema. At lower concentrations, death due to lung damage is still possible.

Eye contact with liquid bis (2 chloromethyl) ether can cause conjunctival irritation and injury to the cornea. On the skin, its irritation effect can range from mild irritation to first degree burns on short exposures. However, on prolonged contact, the liquid may be absorbed through the skin causing second degree burns and toxic systemic effects.

Ingestion of this compound in small amounts may produce nausea and vomiting. It can irritate the mouth and stomach and manifest symptoms of systemic poisoning. Symptoms may include weakness, vomiting, loss of appetite, fatigue, irritability, and anxiety.

Acute Health Effects

The following acute (short-term) health effects may occur immediately or shortly after exposure to bis (2-chloromethyl) ether:

Skin: Severe irritation and possible burns. On prolonged exposure, can be absorbed though intact skin causing toxic systemic effects.

Eye: Severe conjunctival irritation and possible damage to the cornea.

Lung: Severe irritation to respiratory tract and lungs causing shortness of breath and cough. High level acute exposures can be fatal due to a buildup of fluid in the lungs (pulmonary edema).

CNS: Effects include weakness, fatigue, nausea, loss of appetite, irritability, and anxiety.

Chronic Health Effects

The following chronic (long-term) health effects may occur at some time after exposure to bis (2-chloromethyl) ether and can last for months or even years:

Cancer Hazards: Bis (2-chloromethyl) ether is a cancer causing agents in humans. It has been shown to cause lung cancer in humans and other cancers in experimental animals.

Reproductive Hazard: According to the references, bis (2-chloromethyl) ether has not been adequately tested for its ability to adversely affect reproduction.

Other Chronic Effects: Other than its known carcinogenic properties, bis (2-cholormethyl) ether has not been adequately tested to accurately determine its long-term health effects in humans.

❶ *Recommended Risk-Reduction Measures*

Personnel should avoid direct contact with bis (2-chloromethyl) ether. If a less toxic material or compound cannot be substituted for bis (2-chloromethyl) ether, then *engineering controls* are the most effective method of reducing exposure risk. An established, regulated control area must be enforced wherever this chemical is used, handled, or stored (per OSHA 29 CFR 1910.1008). The best protection is to enclose operations and/or provide local exhaust ventilation at the site of bis (2-chloromethyl) ether release. While not always operationally feasible, isolating operations within the regulated control area can also reduce exposure risk. Using respiratory protection is less effective than the controls mentioned above, but is still required whenever working with or around bis (2-chloromethyl) ether. For any exposure level, a MSHA/NIOSH approved self-contained breathing apparatus (SCBA) with full facepiece and pressure demand is the appropriate level of respiratory protection. If a full facepiece is not available, then chemical goggles should be worn to protect the eyes, especially when a splash hazard exists. A face shield should also be considered. To prevent hand and skin exposures, rubber or other non-permeable gloves should be worn.

Administrative controls should also be in place to minimize the potential for human exposures. These may include written procedures or policies which specify the methods and techniques that will be practiced whenever personnel are to work with or around bis (2-chloromethyl) ether.

All personnel should receive *training* on the use, hazards, protective measures, emergency actions, and other precautions as per OSHA 29 CFR 1910.1200 (Hazard Communication) and 29 CFR 1910.1008 (bis-Chloromethyl Ether) prior to the first assignment in an area where bis (2-chloromethyl) ether is used or stored.

Before beginning employment and at regular intervals thereafter (e.g., annually), the following medical test is recommended:

☑ Lung function tests.

If symptoms develop or overexposure is suspected, the following may be useful:

☑ Consider chest X-ray after acute exposure.

It should be noted that medical tests that simply look for existing damage are not a substitute for controlling exposures. Medical histories are extremely important when assessing exposure risk. Also, since smoking can cause heart disease as well as lung cancer, emphysema, and other respiratory ailments, any exposure to bis (2-chloroethy) ether can result in quicker and more devastating symptoms. Smokers should therefore avoid unprotected contact with this chemical.

Other methods to reduce exposure include:

☑ Always ensure that proper protective clothing is worn when using chemical substances.

☑ Wash thoroughly immediately after exposure to bis (2-chloromethyl) ether and at the end of the work shift or before eating, drinking, or smoking.

☑ Hazard warning information should be posted in the work area. In addition, as part of an on-going education and training program, all information on the health and safety hazards of bis (2-chloromethyl) ether should be communicated to all potentially exposed workers.

☑ Eye wash stations should be provided in the immediate work area for emergency use. If there is a possibility of skin exposure to bis (2-chloromethyl) ether, emergency shower facilities should also be provided.

☑ Workers whose clothing has been contaminated by bis (2-chloromethyl) ether should change into clean clothes before leaving work. Contaminated work clothing should be laundered only by individuals who have been informed of the hazards of exposure to bis (2-chloromethyl) ether.

☑ An established and regulated work area must be enforced wherever this chemical is used, handled, or stored. Drinking fountains should not be located in regulated areas. Additional precautions and requirements are established at OSHA 29 CFR 1910.1008.

RISK ASSESSMENT: ENVIRONMENT

General Assessment

The environment is at risk of exposure during transportation, storage, disposal, or destruction of bis (2-chloromethyl) ether. In almost every scenario, the threat of environmental exposure is contingent upon the proper handling of the chemical substance. Accidental spills, large or small, can result in contamination of the surrounding environmental mediums (water, soil, and air).

Bis (2-chloromethyl) ether is class IC flammable liquid presenting a moderate fire hazard. Also, on contact with water and moist air, toxic vapors of formaldehyde and corrosive hydrochloric acid are produced. Caution is always required in handling, storage, transportation, and disposal of bis (2-chloromethyl) ether. When heated to decomposition in air it emits very toxic chloride fumes. Emergency responders should be made aware of the presence of bis (2-chloromethyl) ether at any emergency response situation.

Bis (2-chloromethyl) ether may enter the environment through industrial effluents and spills.

Acute Ecological Effects

Acute (short-term) toxic effects may include the death of animals, birds, or fish and death or low growth rate in plants. Acute effects are seen 2 to 4 days after animals or plants are exposed to bis (2-chloromethyl) ether. Insufficient data are available to evaluate or predict the acute (short-term) effects of bis (2-chloromethyl) ether to aquatic life, plants, birds, or land animals.

Chronic Ecological Effects

Chronic (long-term) toxic effects may include shortened life span, reproductive problems, lower fertility, and changes in appearance or behavior in exposed animals. These effects can be seen long after first exposure(s) to toxic chemicals. Insufficient data are available to evaluate the chronic (long-term) effects of bis (2-chloromethyl) ether on aquatic life, plants, birds, or land animals.

Water Solubility

Bis (2-chloromethyl) ether is highly soluble in water. Concentrations of 1000 milligrams and more will mix with a liter of water.

Persistence in the Environment

Bis (2-chloromethyl) ether is slightly persistent in water, with a half-life of between 2 and 20 days. The half-life of a pollutant is the amount of time it takes for one-half of the chemical to be degraded. Approximately 84% of bis (2-chloromethyl) ether will eventually end up in air, the remainder will end up in water.

Bioaccumulation in Aquatic Organisms

Some substances increase in concentration, or *bioaccumulate,* in living organisms as they breathe contaminated air, drink contaminated water, or eat contaminated food. These chemicals can become concentrated in the tissues and internal organs of animals as well as humans.

The concentration of bis (2-chloromethyl) ether found in edible fish tissue is expected to be approximately the same as the average concentration of bis (2-chloromethyl) ether found in the water from which the fish was taken.

Recommended Risk-Reduction Measures

Proper training of all transporters will reduce the likelihood of a mishap or accident resulting in a leak or spill of bis (2-chloromethyl) ether into the environment. Labels on all containers, trucks, and rail cars must meet DOT requirements and accurately reflect their contents to enable emergency responders to react properly and quickly to any disaster thereby reducing the potential risk to the environment and to personnel.

Storage of bis (2-chloromethyl) ether should be segregated from other chemicals to minimize the risk of cross-contamination. Bis (2-chloromethyl) ether must be stored to avoid contact with water and moisture. Containers should be protected from physical damage and stored to avoid contact with heat. Transfer of liquids between containers should be done automatically, if possible. Non-sparking tools and explosion-proof equipment should be used around bis (2-chloromethyl) ether.

If a spill or leak to the environment has occurred, fire department, emergency response and/or hazardous materials spill personnel should be notified immediately. Cleanup should be attempted only by those trained in proper spill containment procedures. Contaminated soils should be removed for incineration and replaced with clean soil.

If bis (2-chloromethyl) ether should contact the water table, aquifer, or navigable waterway, time is of the essence. It is highly soluble and total remediation may not be possible. The local and/or state emergency response authorities must be notified. A comprehensive emergency response or disaster preparedness/recovery plan should be in place prior to any operations involving the use, transportation, storage, or disposal of bis (2-chloromethyl) ether.

If bis (2-chloromethyl) ether is spilled or leaked, the following specific steps are recommended:

☑ Restrict persons not wearing protective clothing from area of spill or leak until cleanup is complete. Avoid skin and eye contact.

- ☑ Absorb spilled liquids with vermiculite, dry sand, or a similar material and deposit in sealed drum for disposal.
- ☑ Ventilate area of spill or leak.
- ☑ It may be necessary to dispose of bis (2-chloromethyl) ether as a hazardous waste. The responsible state agency or the regional office of the federal Environmental Protection Agency (EPA) should be contacted for specific recommendations.
- ☑ Remove all ignition sources.

RISK ASSESSMENT: BUSINESS

General Assessment

Accidents or mishaps involving bis (2-chloromethyl) ether can present a serious threat to business operations. The loss or damage of equipment or facilities can significantly affect fiscal viability. Lawsuits that may result from personnel illness, injury/death, public exposures, and/or environmental contamination will require a serious expenditure of resources. Media attention surrounding an injury, death, or environmental damage can also result in a loss of profits and loss of current as well as future business.

Recommended Risk-Reduction Measures

Company attorneys, safety and health professionals, and environmental specialists should be involved in the development of any procedures for responding to chemical incidents. A company official should be predesignated as a public relations officer with specific training in dealing with the press. Corporate plans and policies should be developed, approved, and implemented long before any need for such arises.

REFERENCES

American Conference of Governmental Industrial Hygienists. 1988. *Documentation of the Threshold Limit Values and Biological Exposure Indices,* 5th Edition (with updates). Cincinnati: ACGIH

Anthony, C. P., and N. J. Kolthoff. 1971. *Textbook of Anatomy and Physiology.* St. Louis: C. V. Mosby Company.

Bloom, A. D. and F. J. DeSerres. 1995. *Ecotoxicity and Human Health.* New York: CRC Press/Lewis Publishers.

Brusick, D. J. 1994. *Methods for Genetic Risk Assessment.* New York: CRC Press/Lewis Publishers.

Calabrese, E. J., E. M. Kenyon. *Air Toxics and Risk Assessment.* New York: CRC Press/Lewis Publishers.

Cockerman, L. G., B. S. Shane. 1994. *Basic Environmental Toxicology.* New York: CRC Press/Lewis Publishers.

Cothern, C. R., N. P. Ross. 1994. *Environmental Statistics, Assessment, and Forecasting.* New York: CRC Press/Lewis Publishers.

Danse, I. R. 1991. *Common Sense Toxics in the Workplace.* New York: Van Nostrand Reinhold.

Fodden, C. R., and J. L. Weddell. 1992. *Hazardous Materials Emergency Action Data.* New York: CRC Press/Lewis Publishers.

Gardner, R. W. 1994. *Chemical Intolerance.* New York: CRC Press/Lewis Publishers.

Hallenbeck, W. H. 1993. *Quantitative Risk Assessment for Environmental and Occupational Health,* 2nd Edition. New York: CRC Press/Lewis Publishers.

Hodgson, E., R. B. Mailman, and J. E. Chambers. 1988. *Dictionary of Toxicology.* London: The MacMillian Company.

Howard, P. H. 1995. *Handbook of Environmental Fate and Exposures Data for Organic Chemicals,* Vol. I-IV. New York: CRC Press/Lewis Publishers.

Howard, P. H., et. al. 1991. *Handbook of Environmental Degradation Rates.* New York: CRC Press/Lewis Publishers.

Jones, F. E. 1994. *Toxic Organic Vapors in the Workplace.* New York: CRC Press/Lewis Publishers.

Keith, L. H. and M. M. Walker. 1995. *Handbook of Air Toxics.* New York: CRC Press/Lewis Publishers.

Landis, W. G., M. H. Yu. 1995. *Introduction to Environmental Toxicology.* New York: CRC Press/Lewis Publishers.

Langerman, N. 1994 *Precautionary Labels for Chemical Containers.* New York: CRC Press/Lewis Publishers.

Lauwerys, R. R., P. Hoet. 1993. *Industrial Chemical Exposure.* New York: CRC Press/Lewis Publishers.

Lewis, R. J., Sr. 1993. *Hawley's Condensed Chemical Dictionary,* 12th Edition. New York: Van Nostrand Reinhold.

Lewis, R. J., Sr. 1992. *Sax's Dangerous Properties of Industrial Materials,* Volumes 1, 2, and 3, 8th Edition. New York: Van Nostrand Reinhold.

Lide, D. R. 1995. *Handbook of Chemistry and Physics,* 76th Edition. New York: CRC Press/Lewis Publishers.

Meyer, E. 1990. *Chemistry of Hazardous Materials.* Englewood Cliffs, NJ: Prentice Hall.

Patnaik, P. 1992. *A Comprehensive Guide to the Hazardous Properties of Chemical Substances.* New York: Van Nostrand Reinhold.

Proctor, N. H., J. P. Hughes. 1991. *Chemical Hazards of the Workplace.* New York: Van Nostrand Reinhold.

Quigley, D. R. 1994. *Handbook of Emergency Chemical Management.* New York: CRC Press/Lewis Publishers.

Richardson, M. 1995. *Dictionary of Substance and Their Effects,* Vol. I-VI. New York: CRC Press/Lewis Publishers.

Torkelson, T. R. V. K. Rowe. 1981. *Patty's Industrial Hygiene and Toxicology,* Vol. 2B. New York: Wiley-Interscience.

MATERIAL SAFETY DATA SHEET

CHEMICAL NAME	CRC PRESS/LEWIS PUBLISHERS DISCLAIMER:
BROMINE	THE INFORMATION AND RECOMMENDATIONS PRESENTED HEREIN ARE BASED ON SOURCES BELIEVED TO BE RELIABLE. CRC MAKES NO REPRESENTATION ON ITS COMPLETENESS OR ACCURACY. IT IS THE RESPONSIBILITY OF THE USER TO DETERMINE THE CHEMICAL'S SUITABILITY FOR ITS INTENDED USE, THE CHEMICAL'S SAFE USE, AND THE CHEMICAL'S PROPER DISPOSAL. NO REPRESENTATIONS AND/OR WARRANTIES, EITHER EXPRESSED OR IMPLIED, OF THE MERCHANTABILITY OR FITNESS FOR A PARTICULAR PURPOSE, OR OF ANY OTHER NATURE, ARE MADE WITH RESPECT TO THE INFORMATION PROVIDED IN THIS MSDS OR TO THE CHEMICAL TO WHICH INFORMATION MAY REFER. CRC NEITHER ASSUMES NOR AUTHORIZES ANY OTHER PERSON TO ASSUME FOR IT, ANY OTHER ADDITIONAL RESPONSIBILITY OR LIABILITY FOR THE USE OF, OR RELIANCE UPON, THIS INFORMATION.

HAZARD WARNING INFORMATION

HEALTH	FIRE	REACTIVE	OTHER	DEGREE OF HAZARD	COLOR CODING	OTHER CODES
4	0	0	OX	0 = Minimum Hazard 1 = Slight Hazard 2 = Moderate Hazard 3 = Serious Hazard 4 = Severe Hazard	HEALTH = BLUE FIRE = RED REACTIVITY = YELLOW OTHER = WHITE	OX = Oxidizer ACID = Acid ALK = Alkali COR = Corrosive W = Use No Water

SECTION I - GENERAL INFORMATION

Characterization	**Halogen**	RCRA Number	**None**
		EPA Class	**Not Applicable**
DOT Proper Shipping Name	**Bromine**	Chemical Abstract Service (CAS) Number	**7726-95-6**
DOT Hazard Class and Label Requirements	**Corrosive Material**	DOT Emergency Guide Code	**59**
DOT Identification Number	**UN 1744**	Chemical Formula	Br_2

Synonyms
Molecular bromine; bromine solution.

SECTION II - HAZARDOUS INGREDIENTS AND IDENTITY INFORMATION

Hazardous Components (specific identity)	OSHA Exposure Criteria	NIOSH Exposure Criteria	Immediately Dangerous to Life and Health (IDLH)	ACGIH Exposure Criteria
Bromine (derivation: From sea water and natural brines by oxidation of bromine salts with chlorine; solar evaporation (Great Salt Lake); from salt beds at Stassfurt, and the Dead Sea). **1 ppm = 6.64 mg/m^3**	**PEL:** 0.1 ppm, 0.7 mg/m^3 **STEL:** 0.3 ppm, 2.0 mg/m^3	**REL:** 0.1 ppm, 0.7 mg/m^3 **STEL:** 0.3 ppm, 2.0 mg/m^3	**3 ppm**	**TLV:** 0.1 ppm, 0.7 mg/m^3 **STEL:** 0.3 ppm, 2.0 mg/m^3

SECTION III - PHYSICAL AND CHEMICAL CHARACTERISTICS

Boiling Point	**138°F (59°C)**	Specific Gravity (H_2O = 1)	**3.12**
Vapor Pressure (mm Hg)	**175 at 70°F (21°C)**	Molecular Weight	**159.8**
Vapor Density (Air = 1)	**5.5**	Melting Point	**-78°F (-61°C)**

Solubility
Slightly soluble in water (4%-16%). Soluble in common organic solvents.

Appearance and Odor
Dark reddish-brown, fuming liquid with suffocating, irritating fumes.

SECTION IV - FIRE AND EXPLOSION HAZARD DATA

Flash Point (method used)	**Not Applicable**	Explosive Limits in Air % by Volume	**LEL: Not Determined UEL: Not Determined**
NFPA Classification	**Non-combustible Liquid**	Autoignition Temperature	**Not Applicable**

Extinguishing Media
Use agent suitable to surrounding fire. Bromine itself will not burn, but it can greatly accelerate the burning of combustible materials.

Special Fire Fighting Procedures
Poisonous and corrosive fumes are emitted during fire. Wear full protective clothing and self-contained breathing apparatus. Heat will build pressure and may rupture closed storage containers. Keep fire-exposed containers cool with water spray. Continue to cool containers after fire is extinguished.

Unusual Fire and Explosion Hazards
Immediately withdraw if rising sound from venting device is heard of if fire is causing discoloration to the tank. Incompatible with a number of other chemicals. Violent and explosive reactions possible.

SECTION V - REACTIVITY DATA

Stability		Conditions to Avoid
Stable	Unstable	Avoid contact with incompatible materials. Mixtures with some chemicals can produce violent explosive reactions and fires. Avoid contact with heat or flame.
X		**Incompatibility** *(materials to avoid)* Organics, reducing agents, alkali metals, finely divided metals, hydrogen metal azides, organo metallics, ammonia, ordinary combustible (wood, paper, cotton, etc.), acetylene, acetaldehyde, acrolein, methanol.
Hazardous Polymerization		**Conditions to Avoid** Under normal conditions of temperature and pressure, hazardous polymerization of bromine will not occur.
May Occur	Will Not Occur	**Hazardous Decomposition or By-products** When heated to decomposition, bromine emits acrid, irritating smoke and fumes.
	X	

SECTION VI - HEALTH HAZARD DATA

Primary Route(s) of Entry:	Inhalation?	Absorption (skin)?	Ingestion?
	X	X	X

Health Hazards

INHALATION: Irritation of the respiratory passages and damaging to the lungs (pulmonary edema).

ABSORPTION: Corrosive. Will irritate and even burn the skin and eyes. May cause acne and slow healing skin ulcers.

INGESTION: Nausea, abdominal pain, diarrhea, headache. May be fatal.

Carcinogenicity	NTP Listed?	IARC Cancer Review Group?	OSHA Regulated?	Target Organs?
Unknown Human Unknown Animal	No	No	29 CFR 1910.1000 Table Z-1	Respiratory system; skin; eyes; CNS

Medical Conditions Generally Aggravated by Exposure

None reported. However, existing respiratory problems and skin disorders may be aggravated.

Emergency and First-aid Procedures

Eye contact: Flush immediately with water for 30 minutes (minimum); seek medical attention. Skin contact: Remove all contaminated clothing. Immediately wash area with large amounts of soap and water. Seek medical assistance. For inhalation: Remove the person from exposure. Provide respiratory assistance and CPR. Transfer to medical facility. Observe for 24-48 hours for lung effects (pulmonary edema). If swallowed: Seek medical attention immediately. Give the person all the water and milk they can drink. Do NOT induce vomiting. Keep person warm and at rest.

SECTION VII - PRECAUTIONS FOR SAFE HANDLING AND USE

Steps to be Taken in Case Material is Released or Spilled

Contain spills using absorbent material (vermiculite, or other). Ventilate area of spill. Have water source available in case of fire. Store materials in DOT-approved containers. Restrict those not involved in cleanup from entering area. Notify appropriate authorities, as required.

Preferred Waste Disposal Method

Add slowly to sodium hydroxide with stirring to form non-toxic, water soluble sodium bromide and bromate. Flush down drain with excess water.

Precautions to be Taken in Handling and Storage

Do not store with incompatible chemicals. Store in tightly closed containers in a cool, dark, well-ventilated area. Keep fire and flame away. Keep separate from organic, combustible, and readily oxidizable substances. Keep out of direct sunlight.

Other Precautions and Warnings

Bulk storage of bromine is not recommended. If spilled, use anhydrous ammonia vapors from a safe distance to neutralize large quantities of bromine vapors.

SECTION VIII - CONTROL MEASURES AND PERSONAL PROTECTIVE EQUIPMENT

Respiratory Protection *(specify type)*

Extremely corrosive and damaging to lung tissue. For any exposure, use a self-contained breathing apparatus (SCBA) with full facepiece and pressure demand or other positive pressure mode.

Ventilation

Local exhaust or general mechanical systems recommended.

Protective Gloves	Eye Protection	Other Protective Clothing
Butyl Rubber	Chemical Goggles or Face Mask	Impermeable Apron

Work/Hygiene Practices

Always wash hands thoroughly after using chemical; never bring food, drink, or smoking materials into vicinity of chemicals.

BROMINE

Br_2 CAS: 7226-95-6

IDENTIFICATION AND TYPICAL USES

Bromine is a dark, reddish brown liquid with fuming vapors and a suffocating, irritating odor. It is used for bleaching fibers and silk; as a disinfectant for purifying water; in the manufacture of bromo compounds for dyes and pharmaceutical uses; as an analytical reagent; and in organic synthesis. It occurs in igneous rock and sea water. It is also used in the manufacture of ethylene dibromide (anti-knock gasoline); as a solvent, as an intermediate for fumigants (methyl bromide); as a fire retardant for plastics; in photography; and in shrink-proofing of wool.

RISK ASSESSMENT: HEALTH

General Assessment

Bromine is a corrosive liquid and a moderately toxic substance. The target organs are the respiratory system (primarily the lungs), the eyes, and the central nervous system. The primary routes of exposure are *inhalation* of its vapors, *ingestion* of the liquid, and skin *contact.*

On inhalation, bromine is an irritant to respiratory passages and can cause injury to the lungs. Inhalation can be fatal by causing glottis to spasm and suffocation. Exposure may also result in chemical pneumonitis and pulmonary edema. A brief exposure (15 minutes) of 1000 ppm may be fatal in humans. However, the irritant qualities of bromine usually force the worker to leave the exposure area before serious poisoning can occur. Other symptoms are dizziness, headache, coughing, epistaxis, and lachrymation.

Ingestion of the liquid can cause nausea, abdominal pain, and diarrhea. It is corrosive to the skin and eyes, causing burns. Skin contact with the liquid or vapor produces a mild cooling sensation on first contact followed by a burning. If prolonged contact, it can produce deep burns. It may also cause acne or measles-like eruptions on the face, trunk, and extremities that may turn to slow-healing, ulcer-like wounds.

Acute Health Effects

The following acute (short-term) health effects may occur immediately or shortly after exposure to bromine:

Skin: Irritation, burns, possible ulcer-like wounds.

Eye: Irritation which can lead to permanent damage if allowed to remain in contact with the eyes.

Lung: Severe irritation of respiratory passages and lung tissue with coughing, nausea, and possible pulmonary edema (fluid on the lungs), which is a medical emergency and can be fatal.

CNS: Causes CNS depression with dizziness, nausea, headache, and related symptoms.

Chronic Health Effects

The following chronic (long-term) health effects may occur at some time after exposure to bromine and can last for months or even years:

Cancer Hazards: According to the references, bromine is not known to cause cancer in humans or test animals. Long-term effects are not well documented since exposures to even low levels can not be tolerated for any appreciable length of time. The irritating properties of bromine make chronic effects improbable.

Reproduction: According to the references, bromine has not been adequately tested for its ability to adversely affect reproduction in test animals.

Recommended Risk-Reduction Measures

Personnel should avoid direct contact with bromine under any circumstances. A less toxic chemical should be substituted for a hazardous substance. If this is not possible or feasible, then *engineering controls* are the most effective method of reducing exposures. The best protection is to enclose operations and/or provide local exhaust ventilation at the site of chemical release. While not always operationally feasible, isolating operations can also reduce exposure.

Using respiratory protection is less effective than the controls mentioned above, but is still advisable whenever working with or around bromine. At virtually any exposure level, a self-contained breathing apparatus (SCBA) with full facepiece and pressure demand is the recommended respiratory protection of choice. If a full facepiece is not available, then chemical goggles should be worn to protect the eyes. Whenever a chemical splash hazard exists, a face shield and a rubber apron should be worn. To prevent hand and skin exposures, butyl rubber gloves should be worn.

Administrative controls should also be in place to minimize the potential for human exposures. These may include written procedures or policies which specify the methods and techniques that will be practiced whenever personnel are to work with bromine.

All personnel should receive *training* on the use, hazards, protective measures, emergency actions, and other precautions per 29 CFR 1910.1200 (Hazard Communication), prior to the first assignment in an area where bromine is used or stored. If an overexposure is suspected, the following is recommended:

- ☑ Lung function test.
- ☑ Chest X-ray.

Other methods to reduce exposure include:

- ☑ Where possible, enclose operations and use local exhaust ventilation at the site of chemical release. If local exhaust ventilation or enclosure is not used, respiratory protection should be mandatory.
- ☑ Always ensure that proper protective clothing is worn when using chemical substances.
- ☑ Wash thoroughly immediately after exposure to bromine and at the end of the work shift or before eating, drinking, or smoking.
- ☑ Hazard warning information should be posted in the work area. In addition, as part of an on-going education and training program, all information on the health and safety hazards of bromine should be communicated to all potentially exposed workers.
- ☑ Never eat, drink, or smoke in areas were bromine is used, handled or stored.

RISK ASSESSMENT: ENVIRONMENT

General Assessment

The environment is at risk of exposure during transportation, storage, disposal, or destruction of bromine. In almost every scenario, the threat of environmental exposure is contingent upon the proper handling of the chemical substance.

Accidental spills, large or small, can result in fire, explosion, and possible contamination of the surrounding environmental mediums (water, soil, and air). Bromine is considered a non-combustible liquid. However, as a powerful oxidizing agent, bromine will accelerate the burning of combustibles. These characteristics require special consideration during any emergency situation involving a leak or spill of bromine.

Should bromine ever come into contact with incompatible substances such as acids, fuels, aluminum, ammonia, hydrogen, acetylene, phosphorus, potassium, sodium, acetaldehyde, acrolein, acrylonitrile, or dimethyl formamide either during use, transportation, or storage, violent and explosive reactions are extremely possible. When heated with ammonia, the solution produces shock-sensitive nitrogen tribromide.

Destruction of bromine can be accomplished by:
(1). Cautiously acidifying to pH 2 with sulfuric acid. Add a 50% excess of sodium bisulfite solution gradually with stirring at room temperature. Reaction should occur with an increase of temperature. If no increase occurs, add 10% excess of bisulfite followed by cautious addition of more sulfuric acid. Neutralize to pH 7 and flush the solution down the drain.
(2). Add bromine slowly to sodium hydroxide solution with stirring. This results in the formation of mixtures of sodium bromide and bromate, which are water soluble and nontoxic and can be flushed down the drain with excess water.

Bromine can occur in the environment naturally as well as through manufacturing, unchecked discharge into effluents, and through spills.

☠ *Acute Ecological Effects*

Acute (short-term) toxic effects may include the death of animals, birds, or fish, and death or low growth rate in plants. Acute effects are seen 2 to 4 days after animals or plants are exposed to bromine. There is insufficient data available to evaluate the short-term effects of bromine on aquatic life, birds, plants or terrestrial animals.

Chronic Ecological Effects

Chronic toxic effects may include shortened life span, reproductive problems, lower fertility, and changes in appearance or behavior in exposed animals. These effects can be seen long after first exposure(s) to toxic chemicals. There is insufficient data available to evaluate the long-term effects of bromine on aquatic life, birds, plants or terrestrial animals.

Water Solubility

Bromine is moderately soluble in water. Concentrations of up to 1000 milligrams will mix with a liter of water.

Persistence in the Environment

Bromine is highly persistent in water, with a half-life of between 20 to 200 days. The half-life of a pollutant is the amount of time it takes for one-half of the chemical to be degraded. No data are available on the percentages of bromine that will eventually end up in water.

Bioaccumulation in Aquatic Organisms

Some substances increase in concentration, or *bioaccumulate,* in living organisms as they breathe contaminated air, drink contaminated water, or eat contaminated food. These chemicals can become concentrated in the tissues and internal organs of animals as well as humans.

The concentration of bromine found in fish tissues is expected to be about the same as the average concentration of bromine in water from which the fish was taken.

Recommended Risk-Reduction Measures

Proper training of all transporters will reduce the likelihood of a mishap or accident resulting in a leak or spill to the environment. The correct use of DOT labeling (per the MSDS) on all containers, trucks, and rail cars will enable emergency responders to react properly and quickly to any disaster thereby reducing the potential risk to the environment and to personnel.

Storage of bromine should be segregated from incompatible chemicals to minimize the risk of cross-contamination or contact. Bromine should be shipped in quart glass bottles, nickel or Monel drums, or lead-lined tank cars. Buildings designated for storage should be equipped with appropriate fire protection systems (alarms, sprinklers, emergency lighting, portable extinguishers).

If a spill or leak to the environment has occurred, fire department, emergency response, and/or hazardous materials spill personnel should be notified immediately. Cleanup should be attempted only by those trained in proper spill containment procedures. Contaminated soils should be removed for incineration and replaced with clean soil.

If bromine should contact the water table, aquifer, or navigable waterway, time is of the essence. It is moderately soluble in water and, therefore, total containment and remediation may not be entirely possible. When such spills occur, the local and/or state emergency response authorities must be notified. A comprehensive emergency response or disaster preparedness/recovery plan should be in place prior to any operations involving the use, transportation, storage, or disposal of bromine. If bromine is spilled or leaked, the following specific steps are recommended:

- ☑ Restrict persons not wearing protective clothing from area of spill or leak until cleanup is complete and area can be opened for normal work.
- ☑ Remove all ignition sources.
- ☑ Ventilate area of spill or leak.
- ☑ Absorb liquids in vermiculite, dry sand, earth, or a similar material and deposit in sealed containers. Use non-sparking tools.
- ☑ If applicable, stop flow of leaking liquid. If leak source is a cylinder and the leak cannot be stopped in place, remove leaking cylinder to a safe place in the open air and repair or allow cylinder to empty.
- ☑ Keep bromine out of a confined space, such as a sewer, because of the possibility of explosion (unless the sewer is designed to prevent the buildup of explosive concentrations).
- ☑ It may be necessary to dispose of bromine as a hazardous waste. The responsible state agency or the regional office of the federal Environmental Protection Agency (EPA) should be contacted for specific recommendations.

RISK ASSESSMENT: BUSINESS

General Assessment

Accidents or mishaps involving bromine can present a moderate threat to business operations. The loss or damage of equipment or facilities can significantly affect fiscal viability. Lawsuits that may result from personnel injury/death, public exposures, and/or environmental contamination will also require a serious expenditure of resources.

Media attention surrounding an injury, death, or environmental damage can also result in a loss of profits and loss of current as well as future business.

Recommended Risk-Reduction Measures

Company attorneys, safety and health professionals, and environmental specialists should be involved in the development of any procedures or policies intended to manage the use of chemicals in the workplace. A company official should be pre-designated as a public relations officer with specific training in dealing with the press. Corporate plans and policies should be developed, approved, and implemented long before any need for such arises.

REFERENCES

Ahlbom, A. 1993. *Biostatistics for Epidemiologists.* New York: CRC Press/Lewis Publishers.

Anthony, C. P., and N. J. Kolthoff. 1971. *Textbook of Anatomy and Physiology.* St. Louis: C. V. Mosby Company.

Asante-Duah, D. K. 1993. *Hazardous Waste Risk Assessment.* New York: CRC Press/Lewis Publishers.

Bloom, A. D. and F. J. DeSerres. 1995. *Ecotoxicity and Human Health.* New York: CRC Press/Lewis Publishers.

Brusick, D. J. 1994. *Methods for Genetic Risk Assessment.* New York: CRC Press/Lewis Publishers.

Bueche, J. R. 1972. *Principles of Physics,* 2nd Edition. New York: McGraw-Hill.

Burrell, R., D. K. Flaherty, and L. J. Sauers. 1992. *Toxicology of the Immune System.* New York: Van Nostrand Reinhold.

Calabrese, E. J. 1994. *Biological Effects of Low Level Exposures.* New York: CRC Press/Lewis Publishers.

Cockerman, L. G., B. S. Shane. 1994. *Basic Environmental Toxicology.* New York: CRC Press/Lewis Publishers.

Cothern, R. R. 1993. *Comparative Environmental Risk Assessment.* New York: CRC Press/Lewis Publishers.

Danse, I. R. 1991. *Common Sense Toxics in the Workplace.* New York: Van Nostrand Reinhold.

Gardner, R. W. 1994. *Chemical Intolerance.* New York: CRC Press/Lewis Publishers.

Hallenbeck, W. H. 1993. *Quantitative Risk Assessment for Environmental and Occupational Health,* 2nd Edition. New York: CRC Press/Lewis Publishers.

Hodgson, E., R. B. Mailman, and J. E. Chambers. 1988. *Dictionary of Toxicology.* London: The MacMillian Company.

Hosty, J. W., and P. Foster. 1990. *A Practical Guide to Chemical Spill Response.* New York: Van Nostrand Reinhold.

Howard, P. H., et. al. 1991. *Handbook of Environmental Degradation Rates.* New York: CRC Press/Lewis Publishers.

James, A., and M. Lord. 1992. *Index of Chemical and Physical Data.* New York: Van Nostrand Reinhold.

Jones, F. E. 1994. *Toxic Organic Vapors in the Workplace.* New York: CRC Press/Lewis Publishers.

Kamrin, M. 1 A. 1988. *Toxicology: A Primer on Principles and Applications.* Chelsea, Michigan: Lewis Publishers.

Keith, L. H. and M. M. Walker. 1995. *Handbook of Air Toxics.* New York: CRC Press/Lewis Publishers.

Landis, W. G., M. H. Yu. 1995. *Introduction to Environmental Toxicology.* New York: CRC Press/Lewis Publishers.

Langerman, N. 1994 *Precautionary Labels for Chemical Containers.* New York: CRC Press/Lewis Publishers.

Lauwerys, R. R., P. Hoet. 1993. *Industrial Chemical Exposure.* New York: CRC Press/Lewis Publishers.

Lide, D. R. 1995. *Handbook of Chemistry and Physics,* 76th Edition. New York: CRC Press/Lewis Publishers.

Lewis, Richard J., Sr. 1993. *Hawley's Condensed Chemical Dictionary,* 12th Edition. New York: Van Nostrand Reinhold.

Lewis, Richard J., Sr. 1992. *Sax's Dangerous Properties of Industrial Materials,* Volumes 1, 2, and 3, 8th Edition. New York: Van Nostrand Reinhold.

Lisella, Frank S. 1994. *The VNR Dictionary of Environmental Health and Safety.* New York: Van Nostrand Reinhold.

Manahan, S. E. 1992. *Toxicological Chemistry,* 2nd Edition. New York: CRC Press/Lewis Publishers.

Mellström, G. A., J. E. Wahlberg, and H. I. Maibach. 1994. *Protective Gloves for Occupational Use.* New York: CRC Press/Lewis Publishers.

Meyer, Eugene. 1990. *Chemistry of Hazardous Materials.* Englewood Cliffs, NJ: Prentice Hall.

Mycock, J. C., J. D. McKenna, and L. Theodore. 1995. *Handbook of Air Pollution Control Engineering and Technology.* New York: CRC Press/Lewis Publishers.

Nelson, G. O. 1992. *Gas Mixtures.* New York: CRC Press/Lewis Publishers.

Ott, W. R. 1995. *Environmental Statistics and Data Analysis.* New York: CRC Press/Lewis Publishers.

Patnaik, P. 1992. *A Comprehensive Guide to the Hazardous Properties of Chemical Substances.* New York: Van Nostrand Reinhold.

Philp, R. B. 1995. *Environmental Hazards and Human Health.* New York: CRC Press/Lewis Publishers.

Proctor, N. H., and J. P. Hughes. 1991. *Chemical Hazards of the Workplace.* New York: Van Nostrand Reinhold.

Sherman, J. D. 1988. *Chemical Exposure and Disease.* New York: Van Nostrand Reinhold.

MATERIAL SAFETY DATA SHEET

CHEMICAL NAME

BROMINE PENTAFLUORIDE

HAZARD WARNING INFORMATION

HEALTH	FIRE	REACTIVE	OTHER	DEGREE OF HAZARD	COLOR CODING	OTHER CODES
4	0	3	OX ~~W~~	0 = Minimum Hazard 1 = Slight Hazard 2 = Moderate Hazard 3 = Serious Hazard 4 = Severe Hazard	HEALTH = BLUE FIRE = RED REACTIVITY = YELLOW OTHER = WHITE	OX = Oxidizer ACID = Acid ALK = Alkali COR = Corrosive ~~W~~ = Use No Water

SECTION I - GENERAL INFORMATION

Characterization: **Interhalogen Compound**	RCRA Number: **D001 and D003** — EPA Class: **Characteristic (I) and (R) Waste**
DOT Proper Shipping Name: **Bromine pentafluoride**	Chemical Abstract Service (CAS) Number: **7789-30-2**
DOT Hazard Class and Label Requirements: **Oxidizer**	DOT Emergency Guide Code: **44**
DOT Identification Number: **UN 1745**	Chemical Formula: **BrF_5**

Synonyms

Bromine fluoride.

SECTION II - HAZARDOUS INGREDIENTS AND IDENTITY INFORMATION

Hazardous Components (specific identity)	OSHA Exposure Criteria	NIOSH Exposure Criteria	Immediately Dangerous to Life and Health (IDLH)	ACGIH Exposure Criteria
Bromine pentafluoride (derivation: By reacting bromine, diluted with nitrogen and fluorine, in an iron or copper vessel at 392°F/200°C). **1 ppm = 7.27 mg/m^3**	**PEL(proposed):** **0.1 ppm** **0.7 mg/m^3** **STEL:** **0.3 ppm** **2.0 mg/m^3**	**REL (10-hour):** **0.1 ppm** **0.7 mg/m^3** **STEL:** **0.3 ppm** **2.0 mg/m^3**	**Not Determined**	**TLV:** **0.1 ppm** **0.7 mg/m^3** **STEL:** **0.3 ppm** **2.0 mg/m^3**

SECTION III - PHYSICAL AND CHEMICAL CHARACTERISTICS

Boiling Point: **105°F (40°C)**	Specific Gravity (H_2O = 1): **2.48**
Vapor Pressure (mm Hg): **14.2 at 70°F (21°C)**	Molecular Weight: **174.9**
Vapor Density (Air = 1): **6.05**	Melting Point: **-78°F (-61°C)**

Solubility

Violently decomposes in cold water. Reacts violently with many acids, alkalis, and solvents.

Appearance and Odor

Colorless to pale yellow fuming liquid with a pungent odor. A colorless gas above 105°F (40°C).

SECTION IV - FIRE AND EXPLOSION HAZARD DATA

Flash Point (method used): **None, but will decompose at 860°F (400°C)**	Explosive Limits in Air % by Volume: **LEL: Not Determined** **UEL: Not Determined**
NFPA Classification: **Non-combustible Liquid**	**Autoignition Temperature**: **Not Applicable**

Extinguishing Media

Use agent suitable to surrounding fire. Bromine pentafluoride itself will not burn, but it can greatly accelerate the burning of combustible materials. It is a powerful oxidizer. Use water only on large fires.

Special Fire Fighting Procedures

Poisonous and corrosive fumes are emitted during fire. Wear full protective clothing and self-contained breathing apparatus. Heat will build pressure and may rupture closed storage containers. Keep fire-exposed containers cool with water spray. Continue to cool containers after fire is extinguished.

Unusual Fire and Explosion Hazards

Immediately withdraw if rising sound from venting device is heard of if fire is causing discoloration to the tank. Incompatible with a number of other chemicals. Violent and explosive reactions possible.

SECTION V - REACTIVITY DATA

Stability		Conditions to Avoid **Avoid contact with incompatible materials. Mixtures with some chemicals can produce violent explosive reactions and fires. Avoid contact with heat or flame.**
Stable **X**	Unstable	Incompatibility *(materials to avoid)* **Acids, halogens, arsenic, selenium, sulfur, glass, organic materials, water. Reacts with all elements except inert gases, nitrogen, and oxygen.**
Hazardous Polymerization		Conditions to Avoid **Under normal conditions of temperature and pressure, hazardous polymerization of bromine pentafluoride will not occur.**
May Occur **X**	Will Not Occur	Hazardous Decomposition or By-products **When heated to decomposition, bromine pentafluoride emits acrid, irritating smoke and fumes, including hydrogen fluoride and hydrogen bromide gases.**

SECTION VI - HEALTH HAZARD DATA

Primary Route(s) of Entry:	Inhalation?	Absorption (skin)?	Ingestion?
	X		X

Health Hazards

INHALATION: Irritation of the respiratory passages and damaging to the lungs. There can be gasping, salivation, and respiratory distress subsequent to lung injury (pulmonary edema).

EYES & SKIN: Corrosive. Will irritate and even burn the skin and eyes. May cause swelling of the eyelids, clouding of the cornea, lachrymation, and corneal necrosis (tissue destruction).

INGESTION: Burning of the mouth and throat with nausea, abdominal pain, diarrhea, headache.

Carcinogenicity	NTP Listed?	IARC Cancer Review Group?	OSHA Regulated?	Target Organs?
Unknown Human Unknown Animal	No	No	29 CFR 1910.1000 Table Z-1	Respiratory system; skin; eyes; liver, kidney.

Medical Conditions Generally Aggravated by Exposure

None reported. However, existing respiratory problems and skin disorders may be aggravated.

Emergency and First-aid Procedures

Eye contact: Flush immediately with water for 30 minutes (minimum); seek medical attention. Skin contact: Remove all contaminated clothing. Immediately wash area with large amounts of soap and water. Seek medical assistance. For inhalation: Remove the person from exposure. Provide respiratory assistance and CPR. Transfer to medical facility. Observe for 24-48 hours for lung effects (pulmonary edema). If swallowed: Seek medical attention immediately. Give the person copious amounts of water or milk to drink. Do NOT induce vomiting. Keep person warm and at rest.

SECTION VII - PRECAUTIONS FOR SAFE HANDLING AND USE

Steps to be Taken in Case Material is Released or Spilled

Contain spills using absorbent material (vermiculite or other). Ventilate area of spill. Have water source available in case of fire. Store materials in DOT-approved containers. Restrict those not involved in cleanup from entering area. Notify appropriate authorities, as required.

Preferred Waste Disposal Method

No citation.

Precautions to be Taken in Handling and Storage

Do not store with incompatible chemicals. Store in tightly closed containers in a cool, dark, well-ventilated area. Keep fire and flame away. Keep separate from organic, combustible, and readily oxidizable substances. Keep out of direct sunlight. Do NOT store near water.

Other Precautions and Warnings

Bulk storage of bromine pentafluoride is not recommended. Keep cylinders restrained properly and protect from physical damage.

SECTION VIII - CONTROL MEASURES AND PERSONAL PROTECTIVE EQUIPMENT

Respiratory Protection *(specify type)*

Extremely corrosive and damaging to lung tissue. For any exposure, use a self-contained breathing apparatus (SCBA) with full facepiece and pressure demand or other positive pressure mode.

Ventilation

Local exhaust or general mechanical systems recommended.

Protective Gloves	Eye Protection	Other Protective Clothing
Butyl Rubber	Chemical Goggles or Face Mask	Impermeable Apron

Work/Hygiene Practices

Always wash hands thoroughly after using chemical; never bring food, drink, or smoking materials into vicinity of chemicals.

BROMINE PENTAFLUORIDE

BrF_5 CAS: 7789-30-2

IDENTIFICATION AND TYPICAL USES

Bromine pentafluoride is a colorless to pale-yellow, fuming liquid with a pungent odor. It will appear as a gas at temperatures above 105°F (40°C). It is usually shipped as a compressed gas. There is very little commercial usage of bromine pentafluoride. It is sometimes used as a fluorinating agent and an oxidizer.

RISK ASSESSMENT: HEALTH

General Assessment

Bromine pentafluoride is a corrosive liquid and a moderately toxic substance. The target organs are the respiratory system (primarily the lungs), the eyes, skin, liver, and the kidneys. The primary routes of exposure are *inhalation* of its vapors, *ingestion* of the liquid, and skin contact (although absorption as an entry route has not been documented in the references).

On inhalation, bromine pentafluoride is an irritant to respiratory passages and can cause injury to the lungs. Inhalation can be fatal by causing the glottis to spasm and suffocation. Exposure may also result in chemical pneumonitis and pulmonary edema (fluid in the lungs) which is a medical emergency and can be fatal. Symptoms may be delayed up to 48 hours, thereby creating a false sense of security with regard to health exposure risk. However, the irritant qualities of bromine pentafluoride usually force the worker to leave the exposure area before serious poisoning can occur. Other symptoms of exposure may include dizziness, headache, coughing, dypsnea, and lachrymation. The liver and/or kidneys may also be damaged as a result of exposure.

Ingestion of the liquid can cause serious mouth and esophageal burns. If swallowed, there may be nausea, abdominal pain, and diarrhea. It is corrosive to the skin and eyes, causing burns. Skin contact with the liquid or vapor produces a mild cooling sensation on first contact followed by a burning. With prolonged contact, it can produce deep burns. Eye contact with the vapors or liquid may cause clouding of the cornea, swelling of the eyelids, and corneal necrosis (tissue death). The damage may be permanent.

Acute Health Effects

The following acute (short-term) health effects may occur immediately or shortly after exposure to bromine pentafluoride:

Skin: Irritation, burns, possible ulcer-like wounds. Contact with the compressed gas may cause frostbite burns.

Eye: Irritation which can lead to permanent damage if allowed to remain in contact with the eyes.

Lung: Severe irritation of respiratory passages and lung tissue with coughing, nausea, and possible pulmonary edema (fluid on the lungs).

Other: Ingestion causes burning of the mouth and throat.

Chronic Health Effects

The following chronic (long-term) health effects may occur at some time after exposure to bromine pentafluoride and can last for months or even years:

Cancer Hazards: According to information presented in the references, bromine pentafluoride is not known to cause cancer in humans or test animals. Long-term effects are not well documented since exposures to even low levels cannot be tolerated for any appreciable length of time. The irritating properties of bromine pentafluoride make chronic exposure unlikely.

Reproduction: According to information presented in the references, bromine pentafluoride has not been adequately tested for its ability to adversely affect reproduction in test animals.

Other: Although unlikely, chromic exposure to low levels of bromine pentafluoride may lead to nephrosis and hepatosis. Exposure to levels above 3 ppm has caused liver, kidney, and lung damage in test animals. It is not known if these effects will occur in humans.

Recommended Risk-Reduction Measures

Personnel should avoid direct contact with bromine pentafluoride under any circumstances. A less toxic chemical should be substituted for a hazardous substance. If this is not possible or feasible, then *engineering controls* are the most effective method of reducing exposures. The best protection is to enclose operations and/or provide local exhaust ventilation at the site of chemical release. While not always operationally feasible, isolating operations can also reduce exposure.

Using respiratory protection is less effective than the controls mentioned above, but is still advisable whenever working with or around bromine pentafluoride. At virtually any exposure level, a self-contained breathing apparatus (SCBA) or a supplied-air respirator with full facepiece operated in pressure demand or other continuous flow mode are the recommended respiratory protection methods of choice. If a full facepiece is not available, then chemical goggles should be worn to protect the eyes. Whenever a chemical splash hazard exists, a face shield and a rubber apron should be worn. To prevent hand and skin exposures, chemically protective gloves should be worn. Glove manufacturers should be contacted and permeation studies obtained *before* making final glove selections.

Administrative controls should also be in place to minimize the potential for human exposures. These may include written procedures or policies which specify the methods and techniques that will be practiced whenever personnel are to work with bromine pentafluoride.

All personnel should receive *training* on the use, hazards, protective measures, emergency actions, and other precautions per 29 CFR 1910.1200 (Hazard Communication), prior to the first assignment in an area where bromine pentafluoride is used or stored.

If an overexposure is suspected, the following is recommended:

☑ Lung function test.

☑ Consider chest X-ray after acute overexposure (may be negative if taken immediately following exposure due to delayed pulmonary edema).

Also, since smoking can cause heart disease, emphysema, and other respiratory disorders, smokers exposed to bromine pentafluoride may experiences symptoms more quickly than non-smokers under the same conditions of exposure. Prudent risk management requires careful consideration of *all* possible exposure factors.

Other methods to reduce exposure include:

☑ Where possible, enclose operations and use local exhaust ventilation at the site of chemical release. If local exhaust ventilation or enclosure is not used, respiratory protection should be mandatory.

☑ Always ensure that proper protective clothing is worn when using chemical substances.

☑ Wash thoroughly immediately after exposure to bromine pentafluoride and at the end of the work shift or before eating, drinking, or smoking.

☑ Hazard warning information should be posted in the work area. In addition, as part of an on-going education and training program, all information on the health and safety hazards of bromine pentafluoride should be communicated to all potentially exposed workers.

☑ Never eat, drink, or smoke in areas were bromine pentafluoride is used, handled or stored.

RISK ASSESSMENT: ENVIRONMENT

General Assessment

The environment is at risk of exposure during transportation, storage, disposal, or destruction of bromine pentafluoride. In almost every scenario, the threat of environmental exposure is contingent upon the proper handling of the chemical substance.

Accidental spills, large or small, can result in fire, explosion, and possible contamination of the surrounding environmental mediums (water, soil, and air). Bromine pentafluoride is considered a non-combustible liquid. However, as a powerful oxidizing agent, bromine pentafluoride will accelerate the burning of combustibles. These characteristics require special consideration during any emergency situation involving a leak or spill of bromine pentafluoride.

Should bromine pentafluoride ever come into contact with incompatible substances such as acids, halogens, arsenic, selenium, sulfur, glass, organic materials, or water either during use, transportation, or storage, violent and explosive reactions are extremely possible.

Bromine pentafluoride can occur in the environment naturally as well as through manufacturing, unchecked discharge into effluents, and through spills.

Acute Ecological Effects

Acute (short-term) toxic effects may include the death of animals, birds, or fish, and death or low growth rate in plants. Acute effects are seen 2 to 4 days after animals or plants are exposed to bromine. There is insufficient data available to evaluate the short-term effects of bromine pentafluoride on aquatic life, birds, plants or terrestrial animals.

Chronic Ecological Effects

Chronic toxic effects may include shortened life span, reproductive problems, lower fertility, and changes in appearance or behavior in exposed animals. These effects can be seen long after first exposure(s) to toxic chemicals. There is insufficient data available to evaluate the long-term effects of bromine pentafluoride on aquatic life, birds, plants or terrestrial animals.

Water Solubility

Bromine pentafluoride is highly reactive in water.

Persistence in the Environment

Since it is so readily reactive in contact with water, bromine pentafluoride is virtually non-persistent in the aquatic environment. No specific data are available on the percentages of bromine pentafluoride that will eventually end up in the environment.

Bioaccumulation in Aquatic Organisms

Some substances increase in concentration, or *bioaccumulate,* in living organisms as they breathe contaminated air, drink contaminated water, or eat contaminated food. These chemicals can become concentrated in the tissues and internal organs of animals as well as humans.

The concentration of bromine found in fish tissues is expected to be about the same as the average concentration of bromine in water from which the fish was taken.

Recommended Risk-Reduction Measures

Proper training of all transporters will reduce the likelihood of a mishap or accident resulting in a leak or spill to the environment. The correct use of DOT labeling (per the MSDS) on all containers, trucks, and rail cars will enable emergency responders to react properly and quickly to any disaster thereby reducing the potential risk to the environment and to personnel.

Storage of bromine pentafluoride should be segregated from incompatible chemicals to minimize the risk of cross-contamination or contact. Bromine pentafluoride is normally shipped as a compressed gas in cylinders. These should be properly restrained when in storage and protected from physical damage. Buildings designated for storage should be equipped with appropriate fire protection systems (alarms, sprinklers, emergency lighting, portable extinguishers).

If a spill or leak to the environment has occurred, fire department, emergency response, and/or hazardous materials spill personnel should be notified immediately. Cleanup should be attempted only by those trained in proper spill containment procedures. Contaminated soils should be removed for incineration and replaced with clean soil.

If bromine pentafluoride should contact the water table, aquifer, or navigable waterway, time is of the essence. It is highly reactive in water and, therefore, total containment and remediation may not be entirely possible. When such spills occur, the local and/or state emergency response authorities must be notified. A comprehensive emergency response or disaster preparedness/recovery plan should be in place prior to any operations involving the use, transportation, storage, or disposal of bromine pentafluoride. If bromine pentafluoride is spilled or leaked, the following specific steps are recommended:

- ☑ Restrict persons not wearing protective clothing from area of spill or leak until cleanup is complete.
- ☑ Remove all ignition sources.
- ☑ Ventilate area of spill or leak.
- ☑ Absorb liquids in vermiculite, dry sand, earth, or a similar material and deposit in sealed containers.
- ☑ If applicable, stop flow of leaking liquid. If leak source is a cylinder and the leak cannot be stopped in place, remove leaking cylinder to a safe place in the open air and repair or allow cylinder to empty.
- ☑ Keep bromine pentafluoride out of a confined space, such as a sewer, because of the possibility of explosion (unless the sewer is designed to prevent the buildup of explosive concentrations).

☑ It may be necessary to dispose of bromine pentafluoride as a hazardous waste. The responsible state agency or the regional office of the federal Environmental Protection Agency (EPA) should be contacted for specific recommendations.

RISK ASSESSMENT: BUSINESS

General Assessment

Accidents or mishaps involving bromine pentafluoride can present a moderate threat to business operations. The loss or damage of equipment or facilities can significantly affect fiscal viability. Lawsuits that may result from personnel injury/death, public exposures, and/or environmental contamination will also require a serious expenditure of resources. Media attention surrounding an injury, death, or environmental damage can also result in a loss of profits and loss of current as well as future business.

Recommended Risk-Reduction Measures

Company attorneys, safety and health professionals, and environmental specialists should be involved in the development of any procedures or policies intended to manage the use of chemicals in the workplace. A company official should be pre-designated as a public relations officer with specific training in dealing with the press. Corporate plans and policies should be developed, approved, and implemented long before any need for such arises.

REFERENCES

Ahlbom, A. 1993. *Biostatistics for Epidemiologists.* New York: CRC Press/Lewis Publishers.

Anthony, C. P., and N. J. Kolthoff. 1971. *Textbook of Anatomy and Physiology.* St. Louis: C. V. Mosby Company.

Asante-Duah, D. K. 1993. *Hazardous Waste Risk Assessment.* New York: CRC Press/Lewis Publishers.

Bloom, A. D. and F. J. DeSerres. 1995. *Ecotoxicity and Human Health.* New York: CRC Press/Lewis Publishers.

Brusick, D. J. 1994. *Methods for Genetic Risk Assessment.* New York: CRC Press/Lewis Publishers.

Bueche, J. R. 1972. *Principles of Physics,* 2nd Edition. New York: McGraw-Hill.

Burrell, R., D. K. Flaherty, and L. J. Sauers. 1992. *Toxicology of the Immune System.* New York: Van Nostrand Reinhold.

Calabrese, E. J. 1994. *Biological Effects of Low Level Exposures.* New York: CRC Press/Lewis Publishers.

Cockerman, L. G., B. S. Shane. 1994. *Basic Environmental Toxicology.* New York: CRC Press/Lewis Publishers.

Cothern, R. R. 1993. *Comparative Environmental Risk Assessment.* New York: CRC Press/Lewis Publishers.

Gardner, R. W. 1994. *Chemical Intolerance.* New York: CRC Press/Lewis Publishers.

Hallenbeck, W. H. 1993. *Quantitative Risk Assessment for Environmental and Occupational Health,* 2nd Edition. New York: CRC Press/Lewis Publishers.

Hodgson, E., R. B. Mailman, and J. E. Chambers. 1988. *Dictionary of Toxicology.* London: The MacMillian Company.

Hosty, J. W., and P. Foster. 1990. *A Practical Guide to Chemical Spill Response.* New York: Van Nostrand Reinhold.

Howard, P. H., et. al. 1991. *Handbook of Environmental Degradation Rates.* New York: CRC Press/Lewis Publishers.

Jones, F. E. 1994. *Toxic Organic Vapors in the Workplace.* New York: CRC Press/Lewis Publishers.

Kamrin, M. 1 A. 1988. *Toxicology: A Primer on Principles and Applications.* Chelsea, Michigan: Lewis Publishers.

Keith, L. H. and M. M. Walker. 1995. *Handbook of Air Toxics.* New York: CRC Press/Lewis Publishers.

Landis, W. G., M. H. Yu. 1995. *Introduction to Environmental Toxicology.* New York: CRC Press/Lewis Publishers.

Langerman, N. 1994 *Precautionary Labels for Chemical Containers.* New York: CRC Press/Lewis Publishers.

Lauwerys, R. R., P. Hoet. 1993. *Industrial Chemical Exposure.* New York: CRC Press/Lewis Publishers.

Lide, D. R. 1995. *Handbook of Chemistry and Physics,* 76th Edition. New York: CRC Press/Lewis Publishers.

Lewis, Richard J., Sr. 1993. *Hawley's Condensed Chemical Dictionary,* 12th Edition. New York: Van Nostrand Reinhold.

Lewis, Richard J., Sr. 1992. *Sax's Dangerous Properties of Industrial Materials,* Volumes 1, 2, and 3, 8th Edition. New York: Van Nostrand Reinhold.

Meyer, Eugene. 1990. *Chemistry of Hazardous Materials.* Englewood Cliffs, NJ: Prentice Hall.

Mycock, J. C., J. D. McKenna, and L. Theodore. 1995. *Handbook of Air Pollution Control Engineering and Technology.* New York: CRC Press/Lewis Publishers.

Nelson, G. O. 1992. *Gas Mixtures.* New York: CRC Press/Lewis Publishers.

Ott, W. R. 1995. *Environmental Statistics and Data Analysis.* New York: CRC Press/Lewis Publishers.

Patnaik, P. 1992. *A Comprehensive Guide to the Hazardous Properties of Chemical Substances.* New York: Van Nostrand Reinhold.

Philp, R. B. 1995. *Environmental Hazards and Human Health.* New York: CRC Press/Lewis Publishers.

Sherman, J. D. 1988. *Chemical Exposure and Disease.* New York: Van Nostrand Reinhold.

MATERIAL SAFETY DATA SHEET

CHEMICAL NAME	
BROMINE TRIFLUORIDE	**CRC PRESS/LEWIS PUBLISHERS DISCLAIMER:** THE INFORMATION AND RECOMMENDATIONS PRESENTED HEREIN ARE BASED ON SOURCES BELIEVED TO BE RELIABLE. CRC MAKES NO REPRESENTATION ON ITS COMPLETENESS OR ACCURACY. IT IS THE RESPONSIBILITY OF THE USER TO DETERMINE THE CHEMICAL'S SUITABILITY FOR ITS INTENDED USE, THE CHEMICAL'S SAFE USE, AND THE CHEMICAL'S PROPER DISPOSAL. NO REPRESENTATIONS AND/OR WARRANTIES, EITHER EXPRESSED OR IMPLIED, OF THE MERCHANTABILITY OR FITNESS FOR A PARTICULAR PURPOSE, OR OF ANY OTHER NATURE, ARE MADE WITH RESPECT TO THE INFORMATION PROVIDED IN THIS MSDS OR TO THE CHEMICAL TO WHICH INFORMATION MAY REFER. CRC NEITHER ASSUMES NOR AUTHORIZES ANY OTHER PERSON TO ASSUME FOR IT, ANY OTHER ADDITIONAL RESPONSIBILITY OR LIABILITY FOR THE USE OF, OR RELIANCE UPON, THIS INFORMATION.

HAZARD WARNING INFORMATION

HEALTH	FIRE	REACTIVE	OTHER	DEGREE OF HAZARD	COLOR CODING	OTHER CODES
4	0	3	OX ~~W~~	0 = Minimum Hazard 1 = Slight Hazard 2 = Moderate Hazard 3 = Serious Hazard 4 = Severe Hazard	HEALTH = BLUE FIRE = RED REACTIVITY = YELLOW OTHER = WHITE	OX = Oxidizer ACID = Acid ALK = Alkali COR = Corrosive ~~W~~ = Use No Water

SECTION I - GENERAL INFORMATION

Characterization	RCRA Number	EPA Class
Interhalogen Compound	**D003**	**Characteristic (R) Waste**

DOT Proper Shipping Name: **Bromine trifluoride**	Chemical Abstract Service (CAS) Number: **7787-71-5**
DOT Hazard Class and Label Requirements: **Oxidizer, Corrosive, Poison**	DOT Emergency Guide Code: **44**
DOT Identification Number: **UN 1746**	Chemical Formula: **BrF_3**

Synonyms
Not applicable.

SECTION II - HAZARDOUS INGREDIENTS AND IDENTITY INFORMATION

Hazardous Components (specific identity)	OSHA Exposure Criteria	NIOSH Exposure Criteria	Immediately Dangerous to Life and Health (IDLH)	ACGIH Exposure Criteria
Bromine trifluoride (derivation: By fluorination of bromine at 176°F/80°C). Bromine trifluoride may contain some free bromine	**PEL(fluorides):** **2.5 mg/m^3** **STEL:** **Not Established**	**REL (fluorides):** **2.5 mg/m^3** **STEL:** **Not Established**	**Not Determined**	**TLV (fluorides):** **2.5 mg/m^3** **STEL:** **Not Established**

SECTION III - PHYSICAL AND CHEMICAL CHARACTERISTICS

Boiling Point: **275°F (135°C)**	Specific Gravity (H_2O = 1): **2.8**
Vapor Pressure (mm Hg): **18 at 102°F (39°C)**	Molecular Weight: **136.9**
Vapor Density (Air = 1): **Not Found**	Melting Point: **48°F (9°C)**

Solubility
Violently decomposes in cold water (releases hydrogen fluoride gas).

Appearance and Odor
Colorless to pale yellow fuming liquid with a pungent odor. Long prisms may be solid.

SECTION IV - FIRE AND EXPLOSION HAZARD DATA

Flash Point (method used): **Not Applicable**	Explosive Limits in Air % by Volume: **LEL: Not Determined UEL: Not Determined**
NFPA Classification: **Non-combustible Liquid**	Autoignition Temperature: **Not Applicable**

Extinguishing Media
Use agent suitable to surrounding fire. Bromine trifluoride itself will not burn, but it can greatly accelerate the burning of combustible materials. It is a powerful oxidizer. Use water only on large fires.

Special Fire Fighting Procedures
Poisonous and corrosive fumes are emitted during fire. Wear full protective clothing and self-contained breathing apparatus. Heat will build pressure and may rupture closed storage containers. Keep fire-exposed containers cool with water spray. Continue to cool containers after fire is extinguished.

Unusual Fire and Explosion Hazards
Immediately withdraw if rising sound from venting device is heard of if fire is causing discoloration to the tank. Incompatible with a number of other chemicals. Violent and explosive reactions possible.

SECTION V - REACTIVITY DATA

Stability		
Stability		Conditions to Avoid: **Avoid contact with incompatible materials. Mixtures with some chemicals can produce violent explosive reactions and fires. Avoid contact with heat or flame.**
Stable	Unstable	Incompatibility *(materials to avoid)*: **Water, ammonium halides, antimony trichloride oxide, most metallic chlorides, bismuth pentoxide, manganous iodate, rhodium tetrabromide, vanadium, sulfur, carbon monoxide, carbon tetrachloride.**
X		
Hazardous Polymerization		Conditions to Avoid: **Under normal conditions of temperature and pressure, hazardous polymerization of bromine trifluoride will not occur.**
May Occur	Will Not Occur	Hazardous Decomposition or By-products: **When heated to decomposition, bromine trifluoride emits acrid, irritating smoke and fumes, including hydrogen fluoride and hydrogen bromide gases.**
X		

SECTION VI - HEALTH HAZARD DATA

Primary Route(s) of Entry:	Inhalation?	Absorption (skin)?	Ingestion?
	X		X

Health Hazards

INHALATION: **Irritation of the respiratory passages and damaging to the lungs. There can be gasping, salivation, and respiratory distress subsequent to lung injury (pulmonary edema).**

EYES & SKIN: **Corrosive. Will irritate and even burn the skin and eyes. May cause swelling of the eyelids, clouding of the cornea, lachrymation, and corneal necrosis (tissue destruction).**

INGESTION: **Burning of the mouth and throat with nausea, abdominal pain, diarrhea, headache.**

Carcinogenicity	NTP Listed?	IARC Cancer Review Group?	OSHA Regulated?	Target Organs?
Unknown Human Unknown Animal	**No**	**No**	**29 CFR 1910.1000 Table Z-1**	**Respiratory system; skin; eyes; liver, kidney.**

Medical Conditions Generally Aggravated by Exposure

None reported. However, existing respiratory problems and skin disorders may be aggravated.

Emergency and First-aid Procedures

Eye contact: Flush immediately with water for 30 minutes (minimum); seek medical attention. Skin contact: Remove all contaminated clothing. Immediately wash area with large amounts of soap and water. Seek medical assistance. For inhalation: Remove the person from exposure. Provide respiratory assistance and CPR. Transfer to medical facility. Observe for 24-48 hours for lung effects (pulmonary edema). If swallowed: seek medical attention immediately. Give the person all the water and milk they can drink. Do NOT induce vomiting. Keep person warm and at rest.

SECTION VII - PRECAUTIONS FOR SAFE HANDLING AND USE

Steps to be Taken in Case Material is Released or Spilled

Contain spills using absorbent material (vermiculite, or other). Ventilate area of spill. Have water source available in case of fire. Store materials in DOT-approved containers. Restrict those not involved in cleanup from entering area. Notify appropriate authorities, as required.

Preferred Waste Disposal Method

No citation.

Precautions to be Taken in Handling and Storage

Do not store with incompatible chemicals. Store in tightly closed containers in a cool, dark, well-ventilated area. Keep fire and flame away. Keep separate from organic, combustible, and readily oxidizable substances. Keep out of direct sunlight. Do NOT store near water.

Other Precautions and Warnings

Bulk storage of bromine trifluoride is not recommended. Keep cylinders restrained properly and protect from physical damage.

SECTION VIII - CONTROL MEASURES AND PERSONAL PROTECTIVE EQUIPMENT

Respiratory Protection *(specify type)*

Extremely corrosive and damaging to lung tissue. For any exposure, use a self-contained breathing apparatus (SCBA) with full facepiece and pressure demand or other positive pressure mode.

Ventilation

Local exhaust or general mechanical systems recommended.

Protective Gloves	Eye Protection	Other Protective Clothing
Butyl Rubber	**Chemical Goggles or Face Mask**	**Impermeable Apron**

Work/Hygiene Practices

Always wash hands thoroughly after using chemical; never bring food, drink, or smoking materials into vicinity of chemicals.

BROMINE TRIFLUORIDE

BrF_3 CAS: 7787-71-5

IDENTIFICATION AND TYPICAL USES

Bromine trifluoride is a colorless to pale-yellow, fuming liquid with a pungent odor. The commercial uses of bromine trifluoride are very limited. It is sometimes used as a fluorinating agent, as a solvent, and as an oxidizer.

RISK ASSESSMENT: HEALTH

General Assessment

Bromine trifluoride is a corrosive liquid and a moderately toxic substance. The target organs are the respiratory system (primarily the lungs), the eyes, skin, liver, and the kidneys. The primary routes of exposure are *inhalation* of its vapors, *ingestion* of the liquid, and skin contact (although absorption as an entry route has not been documented in the references).

On inhalation, the action of bromine trifluoride is similar to that of bromine pentafluoride, but not quite as toxic. It is an irritant to respiratory passages and can cause injury to the lungs. Inhalation can be fatal by causing the glottis to spasm resulting in suffocation. Exposure may also result in chemical pneumonitis and pulmonary edema (fluid in the lungs) which is a medical emergency and can be fatal. Symptoms may be delayed up to 48 hours, thereby creating a false sense of security with regard to health exposure risk. However, the irritant qualities of bromine trifluoride usually force the worker to leave the exposure area before serious poisoning can occur. Other symptoms of exposure may include dizziness, headache, coughing, dypsnea, and lachrymation. The liver and/or kidneys may also be damaged as a result of exposure.

Ingestion of the liquid can cause serious mouth and esophageal burns. If swallowed, there may be nausea, abdominal pain, and diarrhea. It is corrosive to the skin and eyes, causing burns. Skin contact with the liquid or vapor produces a mild cooling sensation on first contact followed by a burning. With prolonged contact, it can produce deep burns. Eye contact with the vapors or liquid may cause clouding of the cornea, swelling of the eyelids, and corneal necrosis (tissue death). The damage may be permanent.

Acute Health Effects

The following acute (short-term) health effects may occur immediately or shortly after exposure to bromine trifluoride:

Skin: Irritation, burns, possible ulcer-like wounds. Contact with the compressed gas may cause frostbite burns.

Eye: Irritation which can lead to permanent damage if allowed to remain in contact with the eyes.

Lung: Severe irritation of respiratory passages and lung tissue with coughing, nausea, and possible pulmonary edema (fluid on the lungs).

Other: Ingestion causes burning of the mouth and throat.

Chronic Health Effects

The following chronic (long-term) health effects may occur at some time after exposure to bromine trifluoride and can last for months or even years:

Cancer Hazards: According to information presented in the references, bromine trifluoride is not known to cause cancer in humans or test animals. Long-term effects are not well documented since exposures to even low levels cannot be tolerated for any appreciable length of time. The irritating properties of bromine trifluoride make chronic exposure unlikely.

Reproduction: According to information presented in the references, bromine trifluoride has not been adequately tested for its ability to adversely affect reproduction in test animals.

Other: Although unlikely, chromic exposure to low levels of bromine trifluoride may lead to nephrosis and hepatosis. Exposure to levels above 3 ppm has caused liver, kidney, and lung damage in test animals. It is not known if these effects will occur in humans.

ʊ *Recommended Risk-Reduction Measures*

Personnel should avoid direct contact with bromine trifluoride under any circumstances. A less toxic chemical should be substituted for a hazardous substance. If this is not possible or feasible, then *engineering controls* are the most effective method of reducing exposures. The best protection is to enclose operations and/or provide local exhaust ventilation at the site of chemical release. While not always operationally feasible, isolating operations can also reduce exposure.

Using respiratory protection is less effective than the controls mentioned above, but is still advisable whenever working with or around bromine trifluoride. At virtually any exposure level, a self-contained breathing apparatus (SCBA) or a supplied-air respirator with full facepiece operated in pressure demand or other continuos flow mode are the recommended respiratory protection methods of choice. If a full facepiece is not available, then chemical goggles should be worn to protect the eyes. Whenever a chemical splash hazard exists, a face shield and a rubber apron should be worn. To prevent hand and skin exposures, chemically protective gloves should be worn. Glove manufacturers should be contacted and permeation studies obtained *before* making final glove selections.

Administrative controls should also be in place to minimize the potential for human exposures. These may include written procedures or policies which specify the methods and techniques that will be practiced whenever personnel are to work with bromine trifluoride.

All personnel should receive *training* on the use, hazards, protective measures, emergency actions, and other precautions per 29 CFR 1910.1200 (Hazard Communication), prior to the first assignment in an area where bromine trifluoride is used or stored.

If an overexposure is suspected, the following is recommended:

- ☑ Lung function test.
- ☑ Consider chest X-ray after acute overexposure (may be negative if taken immediately following exposure due to delayed pulmonary edema).

Also, since smoking can cause heart disease, emphysema, and other respiratory disorders, smokers exposed to bromine trifluoride may experiences symptoms more quickly than non-smokers under the same conditions of exposure. Prudent risk management requires careful consideration of *all* possible exposure factors.

Other methods to reduce exposure include:

- ☑ Where possible, enclose operations and use local exhaust ventilation at the site of chemical release. If local exhaust ventilation or enclosure is not used, respiratory protection should be mandatory.
- ☑ Always ensure that proper protective clothing is worn when using chemical substances.
- ☑ Wash thoroughly immediately after exposure to bromine trifluoride and at the end of the work shift or before eating, drinking, or smoking.
- ☑ Hazard warning information should be posted in the work area. In addition, as part of an on-going education and training program, all information on the health and safety hazards of bromine trifluoride should be communicated to all potentially exposed workers.
- ☑ Never eat, drink, or smoke in areas were bromine trifluoride is used, handled or stored.

RISK ASSESSMENT: ENVIRONMENT

General Assessment

The environment is at risk of exposure during transportation, storage, disposal, or destruction of bromine trifluoride. In almost every scenario, the threat of environmental exposure is contingent upon the proper handling of the chemical substance.

Accidental spills, large or small, can result in fire, explosion, and possible contamination of the surrounding environmental mediums (water, soil, and air). Bromine trifluoride is considered a non-combustible liquid. However, as a powerful oxidizing agent, bromine trifluoride will accelerate the burning of combustibles. These characteristics require special consideration during any emergency situation involving a leak or spill of bromine trifluoride.

Should bromine trifluoride ever come into contact with incompatible substances such as acids, halogens, arsenic, selenium, sulfur, glass, organic materials, or water either during use, transportation, or storage, violent and explosive reactions are extremely possible.

Bromine trifluoride can occur in the environment through manufacturing, unchecked discharge into effluents, and through spills.

Acute Ecological Effects

Acute (short-term) toxic effects may include the death of animals, birds, or fish, and death or low growth rate in plants. Acute effects are seen 2 to 4 days after animals or plants are exposed to bromine. There is insufficient data available to evaluate the short-term effects of bromine trifluoride on aquatic life, birds, plants or terrestrial animals.

Chronic Ecological Effects

Chronic toxic effects may include shortened life span, reproductive problems, lower fertility, and changes in appearance or behavior in exposed animals. These effects can be seen long after first exposure(s) to toxic chemicals. There are insufficient data available to evaluate the long-term effects of bromine trifluoride on aquatic life, birds, plants or terrestrial animals.

Water Solubility

Bromine trifluoride is highly reactive in water.

Persistence in the Environment

Since it is so readily reactive in contact with water, bromine trifluoride is virtually non-persistent in the aquatic environment. No specific data are available on the percentages of bromine trifluoride that will eventually end up in the environment.

Bioaccumulation in Aquatic Organisms

Some substances increase in concentration, or *bioaccumulate,* in living organisms as they breathe contaminated air, drink contaminated water, or eat contaminated food. These chemicals can become concentrated in the tissues and internal organs of animals as well as humans.

The concentration of bromine found in fish tissues is expected to be about the same as the average concentration of bromine in water from which the fish was taken.

Recommended Risk-Reduction Measures

Proper training of all transporters will reduce the likelihood of a mishap or accident resulting in a leak or spill to the environment. The correct use of DOT labeling (per the MSDS) on all containers, trucks, and rail cars will enable emergency responders to react properly and quickly to any disaster thereby reducing the potential risk to the environment and to personnel.

Storage of bromine trifluoride should be segregated from incompatible chemicals to minimize the risk of cross-contamination or contact. Bromine trifluoride is normally shipped as a compressed gas in cylinders. These should be properly restrained when in storage and protected from physical damage. Buildings designated for storage should be equipped with appropriate fire protection systems (alarms, sprinklers, emergency lighting, portable extinguishers).

If a spill or leak to the environment has occurred, fire department, emergency response, and/or hazardous materials spill personnel should be notified immediately. Cleanup should be attempted only by those trained in proper spill containment procedures. Contaminated soils should be removed for incineration and replaced with clean soil.

If bromine trifluoride should contact the water table, aquifer, or navigable waterway, time is of the essence. It is highly reactive in water and, therefore, total containment and remediation may not be entirely possible. When such spills occur, the local and/or state emergency response authorities must be notified. A comprehensive emergency response or disaster preparedness/recovery plan should be in place prior to any operations involving the use, transportation, storage, or disposal of bromine trifluoride. If bromine trifluoride is spilled or leaked, the following specific steps are recommended:

- ☑ Restrict persons not wearing protective clothing from area of spill or leak until cleanup is complete and area can be opened for normal work.
- ☑ Remove all ignition sources.
- ☑ Ventilate area of spill or leak.
- ☑ Absorb liquids in vermiculite, dry sand, earth, or a similar material and deposit in sealed containers.
- ☑ If applicable, stop flow of leaking liquid. If leak source is a cylinder and the leak cannot be stopped in place, remove leaking cylinder to a safe place in the open air and repair or allow cylinder to empty.
- ☑ Keep bromine trifluoride out of a confined space, such as a sewer, because of the possibility of explosion (unless the sewer is designed to prevent the buildup of explosive concentrations).
- ☑ It may be necessary to dispose of bromine trifluoride as a hazardous waste. The responsible state

agency or the regional office of the federal Environmental Protection Agency (EPA) should be contacted for specific recommendations.

RISK ASSESSMENT: BUSINESS

General Assessment

Accidents or mishaps involving bromine trifluoride can present a moderate threat to business operations. The loss or damage of equipment or facilities can significantly affect fiscal viability. Lawsuits that may result from personnel injury/death, public exposures, and/or environmental contamination will also require a serious expenditure of resources. Media attention surrounding an injury, death, or environmental damage can also result in a loss of profits and loss of current as well as future business.

Recommended Risk-Reduction Measures

Company attorneys, safety and health professionals, and environmental specialists should be involved in the development of any procedures or policies intended to manage the use of chemicals in the workplace. A company official should be pre-designated as a public relations officer with specific training in dealing with the press. Corporate plans and policies should be developed, approved, and implemented long before any need for such arises.

REFERENCES

Ahlbom, A. 1993. *Biostatistics for Epidemiologists.* New York: CRC Press/Lewis Publishers.

Anthony, C. P., and N. J. Kolthoff. 1971. *Textbook of Anatomy and Physiology.* St. Louis: C. V. Mosby Company.

Asante-Duah, D. K. 1993. *Hazardous Waste Risk Assessment.* New York: CRC Press/Lewis Publishers.

Bloom, A. D. and F. J. DeSerres. 1995. *Ecotoxicity and Human Health.* New York: CRC Press/Lewis Publishers.

Brusick, D. J. 1994. *Methods for Genetic Risk Assessment.* New York: CRC Press/Lewis Publishers.

Bueche, J. R. 1972. *Principles of Physics,* 2nd Edition. New York: McGraw-Hill.

Burrell, R., D. K. Flaherty, and L. J. Sauers. 1992. *Toxicology of the Immune System.* New York: Van Nostrand Reinhold.

Calabrese, E. J. 1994. *Biological Effects of Low Level Exposures.* New York: CRC Press/Lewis Publishers.

Cockerman, L. G., B. S. Shane. 1994. *Basic Environmental Toxicology.* New York: CRC Press/Lewis Publishers.

Cothern, R. R. 1993. *Comparative Environmental Risk Assessment.* New York: CRC Press/Lewis Publishers.

Gardner, R. W. 1994. *Chemical Intolerance.* New York: CRC Press/Lewis Publishers.

Hallenbeck, W. H. 1993. *Quantitative Risk Assessment for Environmental and Occupational Health,* 2nd Edition. New York: CRC Press/Lewis Publishers.

Hodgson, E., R. B. Mailman, and J. E. Chambers. 1988. *Dictionary of Toxicology.* London: The MacMillian Company.

Hosty, J. W., and P. Foster. 1990. *A Practical Guide to Chemical Spill Response.* New York: Van Nostrand Reinhold.

Howard, P. H., et. al. 1991. *Handbook of Environmental Degradation Rates.* New York: CRC Press/Lewis Publishers.

Jones, F. E. 1994. *Toxic Organic Vapors in the Workplace.* New York: CRC Press/Lewis Publishers.

Kamrin, M. I A. 1988. *Toxicology: A Primer on Principles and Applications.* Chelsea, Michigan: Lewis Publishers.

Keith, L. H. and M. M. Walker. 1995. *Handbook of Air Toxics.* New York: CRC Press/Lewis Publishers.

Landis, W. G., M. H. Yu. 1995. *Introduction to Environmental Toxicology.* New York: CRC Press/Lewis Publishers.

Langerman, N. 1994 *Precautionary Labels for Chemical Containers.* New York: CRC Press/Lewis Publishers.

Lauwerys, R. R., P. Hoet. 1993. *Industrial Chemical Exposure.* New York: CRC Press/Lewis Publishers.

Lide, D. R. 1995. *Handbook of Chemistry and Physics,* 76th Edition. New York: CRC Press/Lewis Publishers.

Lewis, Richard J., Sr. 1993. *Hawley's Condensed Chemical Dictionary,* 12th Edition. New York: Van Nostrand Reinhold.

Lewis, Richard J., Sr. 1992. *Sax's Dangerous Properties of Industrial Materials,* Volumes 1, 2, and 3, 8th Edition. New York: Van Nostrand Reinhold.

Meyer, Eugene. 1990. *Chemistry of Hazardous Materials.* Englewood Cliffs, NJ: Prentice Hall.

Mycock, J. C., J. D. McKenna, and L. Theodore. 1995. *Handbook of Air Pollution Control Engineering and Technology.* New York: CRC Press/Lewis Publishers.

Nelson, G. O. 1992. *Gas Mixtures.* New York: CRC Press/Lewis Publishers.

Ott, W. R. 1995. *Environmental Statistics and Data Analysis.* New York: CRC Press/Lewis Publishers.

Patnaik, P. 1992. *A Comprehensive Guide to the Hazardous Properties of Chemical Substances.* New York: Van Nostrand Reinhold.

Philp, R. B. 1995. *Environmental Hazards and Human Health.* New York: CRC Press/Lewis Publishers.

Sherman, J. D. 1988. *Chemical Exposure and Disease.* New York: Van Nostrand Reinhold.

MATERIAL SAFETY DATA SHEET

CHEMICAL NAME	
BROMOFORM	**CRC PRESS/LEWIS PUBLISHERS DISCLAIMER:** THE INFORMATION AND RECOMMENDATIONS PRESENTED HEREIN ARE BASED ON SOURCES BELIEVED TO BE RELIABLE. CRC MAKES NO REPRESENTATION ON ITS COMPLETENESS OR ACCURACY. IT IS THE RESPONSIBILITY OF THE USER TO DETERMINE THE CHEMICAL'S SUITABILITY FOR ITS INTENDED USE, THE CHEMICAL'S SAFE USE, AND THE CHEMICAL'S PROPER DISPOSAL. NO REPRESENTATIONS AND/OR WARRANTIES, EITHER EXPRESSED OR IMPLIED, OF THE MERCHANTABILITY OR FITNESS FOR A PARTICULAR PURPOSE, OR OF ANY OTHER NATURE, ARE MADE WITH RESPECT TO THE INFORMATION PROVIDED IN THIS MSDS OR TO THE CHEMICAL TO WHICH INFORMATION MAY REFER. CRC NEITHER ASSUMES NOR AUTHORIZES ANY OTHER PERSON TO ASSUME FOR IT, ANY OTHER ADDITIONAL RESPONSIBILITY OR LIABILITY FOR THE USE OF, OR RELIANCE UPON, THIS INFORMATION.

HAZARD WARNING INFORMATION

HEALTH	FIRE	REACTIVE	OTHER	DEGREE OF HAZARD	COLOR CODING	OTHER CODES
3	0	0		0 = Minimum Hazard 1 = Slight Hazard 2 = Moderate Hazard 3 = Serious Hazard 4 = Severe Hazard	HEALTH = BLUE FIRE = RED REACTIVITY = YELLOW OTHER = WHITE	OX = Oxidizer ACID = Acid ALK = Alkali COR = Corrosive W = Use No Water

SECTION I - GENERAL INFORMATION

Characterization	**Halogenated Hydrocarbon**	RCRA Number	**U225**
		EPA Class	**Toxic Waste**
DOT Proper Shipping Name	**Bromoform**	Chemical Abstract Service (CAS) Number	**75-25-2**
DOT Hazard Class and Label Requirements	**Poison B**	DOT Emergency Guide Code	**58**
DOT Identification Number	**UN 2515**	Chemical Formula	**$CHBr_3$**

Synonyms
Methyl tribromide; tribromomethane; methenyl tribromide.

SECTION II - HAZARDOUS INGREDIENTS AND IDENTITY INFORMATION

Hazardous Components (specific identity)	OSHA Exposure Criteria	NIOSH Exposure Criteria	Immediately Dangerous to Life and Health (IDLH)	ACGIH Exposure Criteria
Bromoform (derivation from: By heating acetone or ethanol with bromine and alkali hydroxide and recovery by distillation, similar to acetone process of chloroform). **1 ppm = 10.51 mg/m^3**	**PEL (skin):** **0.5 ppm** **5.2 mg/m^3** **STEL:** **Not Determined**	**REL (skin):** **0.5 ppm** **5.2 mg/m^3** **STEL:** **Not Determined**	**850 ppm**	**TLV (skin):** **0.5 ppm** **5.2 mg/m^3** **STEL:** **Not Determined**

SECTION III - PHYSICAL AND CHEMICAL CHARACTERISTICS

Boiling Point	**291°F (144°C)**	Specific Gravity (H_2O = 1)	**2.89**
Vapor Pressure (mm Hg)	**4.4 at 69°F (20°C)**	Molecular Weight	**252.8**
Vapor Density (Air = 1)	**Not Found**	Melting Point	**46°F (8°C)**

Solubility in Water
Slightly soluble in water; soluble in alcohol, ether, chloroform, benzene, solvent naptha, fixed and volatile oils.

Appearance and Odor
Colorless, heavy liquid with an odor and taste similar to that of chloroform.

SECTION IV - FIRE AND EXPLOSION HAZARD DATA

Flash Point (method used)	**Not Applicable**	Explosive Limits in Air % by Volume	**LEL: Not Determined** **UEL: Not Determined**
NFPA Classification	**Non combustible Liquid**	Autoignition Temperature	**Not Applicable**

Extinguishing Media
Extinguish fire using agent suitable for surrounding fire. Bromoform itself will not burn.

Special Fire Fighting Procedures
Poisonous gases are produced in fire. Wear full protective clothing and self-contained breathing apparatus (SCBA). Heat will build pressure and may rupture closed storage containers. Keep fire-exposed containers cool with water spray. Continue to cool containers after fire is extinguished.

Unusual Fire and Explosion Hazards
Immediately withdraw if rising sound from venting device is heard of if fire is causing discoloration to the tank. Mixing with sodium/potassium alloy or lithium can cause explosion.

SECTION V - REACTIVITY DATA

Stability		
Stability		Conditions to Avoid **Avoid contact with incompatible materials. Keep away from heat. Explosive reaction in contact with crown ethers or potassium hydroxide. Violent reaction with acetone or bases.**
Stable	Unstable	Incompatibility *(materials to avoid)*
X		**Chemically active metals (such as lithium, sodium, potassium, calcium, powdered aluminum, zinc, magnesium), strong caustics, acetone, crown ethers, strong oxidizers and strong acids.**
Hazardous Polymerization		Conditions to Avoid **Under normal conditions of temperature and pressure, hazardous polymerization of bromoform will not occur. However, gradually decomposes, acquiring yellow color; air and light accelerate decomposition.**
May Occur	Will Not Occur	Hazardous Decomposition or By-products
	X	**When heated to decomposition, bromoform emits highly toxic bromide fumes.**

SECTION VI - HEALTH HAZARD DATA

Primary Route(s) of Entry:	Inhalation?	Absorption (skin)?	Ingestion?
	X	X	X

Health Hazards

INHALATION: Low exposures cause tearing, salivation, and the reddening of the face. Higher exposures cause dizziness, lightheadedness, and loss of consciousness.

ABSORPTION: May irritate the skin causing rash or burning on contact. Will also burn the eyes.

INGESTION: Poison by ingestion. Can be fatal. May adversely affect the liver on high exposures.

Carcinogenicity	NTP Listed?	IARC Cancer Review Group?	OSHA Regulated?	Target Organs?
Unknown Human Questioned Animal	5th Annual Report	No	29 CFR 1910.1000 Table Z-1	Respiratory system; skin; eyes; CNS; liver, kidney

Medical Conditions Generally Aggravated by Exposure

None reported. However, pre-existing respiratory ailments or skin disorders may be aggravated.

Emergency and First-aid Procedures

Eye contact: Flush immediately with water for 15 minutes (minimum); seek medical attention. Skin contact: Remove all contaminated clothing. Immediately wash area with large amounts of soap and water. Seek medical assistance. For inhalation: Remove the person from exposure. Provide respiratory assistance and CPR. Transfer to medical facility. If swallowed: Seek medical attention immediately. If person is conscious, give 2-3 glasses of milk or water. DO NOT induce vomiting unless directed to do so by a physician. Keep victim warm and at rest.

SECTION VII - PRECAUTIONS FOR SAFE HANDLING AND USE

Steps to be Taken in Case Material is Released or Spilled

Contain spills using absorbent material (vermiculite, or other). Ventilate area of spill. Have water source available in case of fire. Store materials in DOT-approved containers. Restrict those not involved in cleanup from entering area. Notify appropriate authorities, as required.

Preferred Waste Disposal Method

Incineration in chemical incinerator equipped with scrubber and afterburner.

Precautions to be Taken in Handling and Storage

Do not store with incompatible chemicals. Store in tightly closed containers in a cool, dark, well-ventilated area. Do not store in direct sunlight. Keep heat, fire, and flame away. If possible, automatically transfer liquids between containers.

Other Precautions and Warnings

Bulk storage of bromoform is not recommended. Explosive reactions can occur when bromoform comes into contact with bases, acids, or oxidizers.

SECTION VIII - CONTROL MEASURES AND PERSONAL PROTECTIVE EQUIPMENT

Respiratory Protection *(specify type)*

Where potential exists for exposure over 0.5 ppm, use a MSHA/NIOSH approved full facepiece respirator with an organic vapor cartridge. For high exposures, use a self-contained breathing apparatus (SCBA) with full facepiece and pressure demand or other positive pressure mode.

Ventilation

Local exhaust or general mechanical systems recommended.

Protective Gloves	Eye Protection	Other Protective Clothing
Butyl Rubber	**Chemical Goggles or Face Mask**	**Impermeable Apron**

Work/Hygiene Practices

Always wash hands thoroughly after using chemical; never bring food, drink, or smoking materials into vicinity of chemicals.

BROMOFORM

$CHBr_3$ CAS: 75-25-2

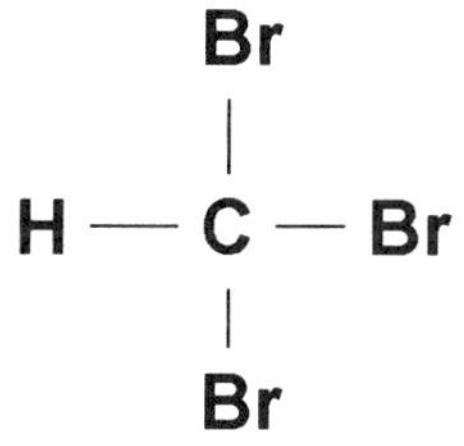

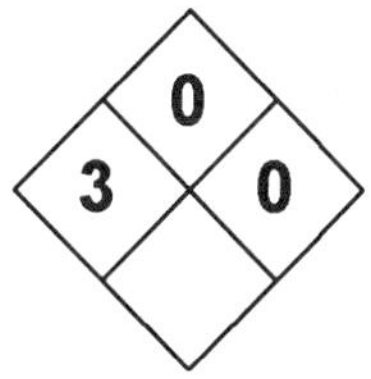

IDENTIFICATION AND TYPICAL USES

Bromoform is a colorless, heavy liquid with a chloroform-like odor and taste. Used as an intermediate in organic synthesis of pharmaceuticals (sedative), in geological assaying, as a solvent for waxes, greases, and oils. It is used as an ingredient in fire-resistant chemicals and gauge fluid, and in separating mixtures of minerals. It is also used in the shipbuilding and aircraft industries.

RISK ASSESSMENT: HEALTH

General Assessment

Bromoform is a poison by *ingestion* and moderately toxic by other routes of exposure. It is a questionable carcinogen with human mutation data also reported. It can damage the liver to a serious degree and even cause death. It has anesthetic properties similar to those of chloroform, but is not sufficiently volatile for inhalation purposes and is far too toxic for human use. As a sedative and an antitussive, its medicinal uses have caused numerous poisonings. It is a respiratory irritant and causes central nervous system depression.

Inhalation of its vapors in small amounts may cause irritation of the eyes, provoking lachrymation (tearing), salivation, and reddening of the face. Abuse can lead to addiction and related serious consequences.

Accidental *ingestion* has caused central nervous system depression with coma and loss of reflexes. Smaller doses have led to listlessness, lightheadedness, headache, and vertigo.

Bromoform can pass through unbroken skin (absorption) and enter the blood stream to cause toxic systemic effects including narcosis, lachrymation, loss of consciousness, dizziness, and salivation. Long-term exposure by any route may damage the liver and adversely affect the kidneys.

Acute Health Effects

The following acute (short-term) health effects may occur immediately or shortly after exposure to bromoform:

Skin: Irritation and burning on contact. Will pass through unbroken skin.

Eye: Irritation which can lead to injury if allowed to remain in contact with the eyes.

Lung: Irritation of the eyes, nose, and throat causing tears and drooling. Can also irritate the air passages causing cough and or phlegm.

CNS: Depression of the CNS causing symptoms of dizziness, listlessness, headache, nausea, loss of consciousness, and possible death.

Chronic Health Effects

The following chronic (long-term) health effects may occur at some time after exposure to bromoform and can last for months or even years:

Cancer Hazards: There is inconclusive evidence that bromoform causes cancer in test animals and no evidence that it causes cancer in humans, based upon studies to date.

Reproduction: According to the references, bromoform has not been adequately tested for its ability to adversely affect reproduction in test animals.

Other Chronic Effects: Long-term exposure may lead to liver damage. There may also be adverse kidney effects resulting from overexposure. Repeated exposure can cause skin rash. Very irritating substances such as bromoform may adversely affect the lungs. However, it is not known whether bromoform causes lung damage on long-term exposure.

ⓘ *Recommended Risk-Reduction Measures*

Personnel should avoid direct contact with bromoform. The best risk reduction measure is to use a less toxic chemical as a substitute for a bromoform. If this is not possible or feasible, then *engineering controls* are the most effective method of reducing exposures. The best protection is to enclose operations and/or provide local exhaust ventilation at the site of chemical release. While not always operationally feasible, isolating operations can also reduce exposure risk.

Using respiratory protection is less effective than the controls mentioned above, but is still advisable whenever working with or around bromoform. For exposures over 0.5 ppm (8 hours), an MSHA/NIOSH approved chemical respirator with an organic vapor cartridge should be used. For higher exposures, a self-contained breathing apparatus (SCBA) with full facepiece and pressure demand is the recommended respiratory protection of choice. If a full facepiece is not available, then chemical goggles should be worn to protect the eyes. Whenever a chemical splash hazard exists, a face shield and a rubber apron should be worn. To prevent hand and skin exposures, butyl rubber or neoprene gloves should be worn.

Administrative controls should also be in place to minimize the potential for human exposures. These may include written procedures or policies which specify the methods and techniques that will be practiced whenever personnel are to work with bromoform.

All personnel should receive *training* on the use, hazards, protective measures, emergency actions, and other precautions per 29 CFR 1910.1200 (Hazard Communication), prior to the first assignment in an area where bromoform is used or stored.

For those with frequent or potentially high exposure, (half the PEL or greater, or significant skin contact), the following are recommended before beginning work and at regular times thereafter:

☑ Liver function tests.

☑ Lung function tests.

Any medical evaluation should include a careful history of past and present symptoms with an examination. Medical tests that look for existing damage are not a substitute for controlling exposures.

Other methods to reduce exposure include:

☑ Where possible, enclose operations and use local exhaust ventilation at the site of chemical release. If local exhaust ventilation or enclosure is not used, respiratory protection should be mandatory.

☑ Always ensure that proper protective clothing is worn when using chemical substances.

☑ Wash thoroughly immediately after exposure to bromoform and at the end of the work shift or before eating, drinking, or smoking.

☑ Hazard warning information should be posted in the work area. In addition, as part of an on-going education and training program, all information on the health and safety hazards of bromoform should be communicated to all exposed and potentially exposed workers.

☑ Never eat, drink, or smoke in areas were bromoform is used, handled or stored.

RISK ASSESSMENT: ENVIRONMENT
General Assessment

The environment is at risk of exposure during transportation, storage, disposal, or destruction of bromoform. In almost every scenario, the threat of environmental exposure is contingent upon the proper handling of the chemical substance.

Accidental spills, large or small, can result in fire, explosion, and possible contamination of the surrounding environmental mediums (water, soil, and air). Bromoform is considered a non-flammable liquid. However, it is incompatible with a number of common materials (acids, bases, oxidizers, powdered metals, crown ethers, alkali earth metals) and contact can result in violent and explosive reactions. These characteristics require special consideration during any emergency situation involving a leak or spill of bromoform.

The proper disposal/destruction method for bromoform is to burn it in a chemical incinerator equipped with an afterburner and air scrubber. Bromoform can enter the environment through unchecked industrial discharges into effluents, municipal waste treatment plant discharges, and through spills.

☠ *Acute Ecological Effects*

Acute (short-term) toxic effects may include the death of animals, birds, or fish, and death or low growth rate in plants. Acute effects are seen 2 to 4 days after animals or plants are exposed to bromoform. This chemical has moderate acute toxicity to aquatic life. No data are available on the short-term effects of bromoform to birds, plants or terrestrial animals.

Chronic Ecological Effects

Chronic toxic effects may include shortened life span, reproductive problems, lower fertility, and changes in appearance or behavior in exposed animals. These effects can be seen long after first exposure(s) to toxic chemicals. Bromoform has moderate chronic toxicity to aquatic life. No data are available on the long-term effects of bromoform to plants, birds, or land animals.

Water Solubility

Bromoform is highly soluble in water. Concentrations of 1000 milligrams and more will mix with a liter of water.

Persistence in the Environment

Bromoform is moderately persistent in water, with a half-life of between 20 to 200 days. The half-life of a pollutant is the amount of time it takes for one-half of the chemical to be degraded. Approximately 90% of bromoform will eventually end up in air with the remainder ending up in water.

Bioaccumulation in Aquatic Organisms

Some substances increase in concentration, or *bioaccumulate,* in living organisms as they breathe contaminated air, drink contaminated water, or eat contaminated food. These chemicals can become concentrated in the tissues and internal organs of animals as well as humans.

The concentration of bromoform found in fish tissues is expected to be somewhat higher then the average concentration of bromoform in water from which the fish was taken.

Recommended Risk-Reduction Measures

Proper training of all transporters will reduce the likelihood of a mishap or accident resulting in a leak or spill to the environment. The correct use of DOT labeling (per the MSDS) on all containers, trucks, and rail cars will enable emergency responders to react properly and quickly to any disaster thereby reducing the potential risk to the environment and to personnel.

Storage of bromoform should be segregated from incompatible chemicals to minimize the risk of cross-contamination or contact. Buildings designated for storage should be equipped with appropriate fire protection systems (alarms, sprinklers, emergency lighting, portable extinguishers).

If a spill or leak to the environment has occurred, fire department, emergency response, and/or hazardous materials spill personnel should be notified immediately. Cleanup should be attempted only by those trained in proper spill containment procedures. Contaminated soils should be removed for incineration and replaced with clean soil.

If bromoform should contact the water table, aquifer, or navigable waterway, time is of the essence. It is highly miscible in water and, therefore, total containment and remediation may not be entirely possible. When such spills occur, the local and/or state emergency response authorities must be notified. A comprehensive emergency response or disaster preparedness/recovery plan should be in place prior to any operations involving the use, transportation, storage, or disposal of bromoform. If bromoform is spilled or leaked, the following specific steps are recommended:

- ☑ Restrict persons not wearing protective clothing from area of spill or leak until cleanup is complete.
- ☑ Ventilate area of spill or leak.
- ☑ Absorb liquids in vermiculite, dry sand, earth, or a similar material and deposit in sealed containers.
- ☑ If applicable, stop flow of leaking liquid. If leak source is a cylinder and the leak cannot be stopped in place, remove leaking cylinder to a safe place in the open air and repair or allow cylinder to empty.
- ☑ It may be necessary to dispose of bromoform as a hazardous waste. The responsible state agency or the regional office of the federal Environmental Protection Agency (EPA) should be contacted for specific recommendations.

RISK ASSESSMENT: BUSINESS

General Assessment

Accidents or mishaps involving bromoform can present a moderate threat to business operations. The loss or damage of equipment or facilities can significantly affect fiscal viability. Lawsuits that may result from personnel injury/death, public exposures, and/or environmental contamination will also require a serious expenditure of resources.

Media attention surrounding an injury, death, or environmental damage can also result in a loss of profits and loss of current as well as future business.

Recommended Risk-Reduction Measures

Company attorneys, safety and health professionals, and environmental specialists should be involved in the development of any procedures or policies intended to manage the use of chemicals in the workplace. A company official should be pre-designated as a public relations officer with specific training in dealing with the press. Corporate plans and policies should be developed, approved, and implemented long before any need for such arises.

REFERENCES

American Conference of Governmental Industrial Hygienists. 1988. *Documentation of the Threshold Limit Values and Biological Exposure Indices,* 5th Edition (with updates). Cincinnati: ACGIH

Anthony, C. P., and N. J. Kolthoff. 1971. *Textbook of Anatomy and Physiology.* St. Louis: C. V. Mosby Company.

Asante-Duah, D. K. 1993. *Hazardous Waste Risk Assessment.* New York: CRC Press/Lewis Publishers.

Bloom, A. D. and F. J. DeSerres. 1995. *Ecotoxicity and Human Health.* New York: CRC Press/Lewis Publishers.

Brusick, D. J. 1994. *Methods for Genetic Risk Assessment.* New York: CRC Press/Lewis Publishers.

Bueche, J. R. 1972. *Principles of Physics,* 2nd Edition. New York: McGraw-Hill.

Burrell, R., D. K. Flaherty, and L. J. Sauers. 1992. *Toxicology of the Immune System.* New York: Van Nostrand Reinhold.

Calabrese, E. J. 1994. *Biological Effects of Low Level Exposures.* New York: CRC Press/Lewis Publishers.

Cockerman, L. G., B. S. Shane. 1994. *Basic Environmental Toxicology.* New York: CRC Press/Lewis Publishers.

Cothern, C. R., N. P. Ross. 1994. *Environmental Statistics, Assessment, and Forecasting.* New York: CRC Press/Lewis Publishers.

Danse, I. R. 1991. *Common Sense Toxics in the Workplace.* New York: Van Nostrand Reinhold.

Environmental Protection Agency. 1983. *Chemical Hazard Information Profile: Bromoform.* Washington, D.C.: U.S. EPA.

Gardner, R. W. 1994. *Chemical Intolerance.* New York: CRC Press/Lewis Publishers.

Hallenbeck, W. H. 1993. *Quantitative Risk Assessment for Environmental and Occupational Health,* 2nd Edition. New York: CRC Press/Lewis Publishers.

Hodgson, E., R. B. Mailman, and J. E. Chambers. 1988. *Dictionary of Toxicology.* London: The MacMillian Company.

Heck, H. 1985. Mechanisms of bromoform toxicity: Structure activity Studies. *CIIT Activities,* 5(10):106.

Hosty, J. W., and P. Foster. 1990. *A Practical Guide to Chemical Spill Response.* New York: Van Nostrand Reinhold.

Howard, P. H. 1995. *Handbook of Environmental Fate and Exposures Data for Organic Chemicals,* Vol. I-IV. New York: CRC Press/Lewis Publishers.

Howard, P. H., et. al. 1991. *Handbook of Environmental Degradation Rates.* New York: CRC Press/Lewis Publishers.

James, A., and M. Lord. 1992. *Index of Chemical and Physical Data.* New York: Van Nostrand Reinhold.

Jones, F. E. 1994. *Toxic Organic Vapors in the Workplace.* New York: CRC Press/Lewis Publishers.

Kamrin, M. 1 A. 1988. *Toxicology: A Primer on Principles and Applications.* Chelsea, Michigan: Lewis Publishers.

Keith, L. H. and M. M. Walker. 1995. *Handbook of Air Toxics.* New York: CRC Press/Lewis Publishers.

Landis, W. G., M. H. Yu. 1995. *Introduction to Environmental Toxicology.* New York: CRC Press/Lewis Publishers.

Langerman, N. 1994 *Precautionary Labels for Chemical Containers.* New York: CRC Press/Lewis Publishers.

Lauwerys, R. R., P. Hoet. 1993. *Industrial Chemical Exposure.* New York: CRC Press/Lewis Publishers.

Lide, D. R. 1995. *Handbook of Chemistry and Physics,* 76th Edition. New York: CRC Press/Lewis Publishers.

Lewis, Richard J., Sr. 1993. *Hawley's Condensed Chemical Dictionary,* 12th Edition. New York: Van Nostrand Reinhold.

Lewis, Richard J., Sr. 1992. *Sax's Dangerous Properties of Industrial Materials,* Volumes 1, 2, and 3, 8th Edition. New York: Van Nostrand Reinhold.

Lisella, Frank S. 1994. *The VNR Dictionary of Environmental Health and Safety.* New York: Van Nostrand Reinhold.

Manahan, S. E. 1994. *Environmental Chemistry,* 6th Edition. New York: CRC Press/Lewis Publishers.

Manahan, S. E. 1992. *Toxicological Chemistry,* 2nd Edition. New York: CRC Press/Lewis Publishers.

Mellström, G. A., J. E. Wahlberg, and H. I. Maibach. 1994. *Protective Gloves for Occupational Use.* New York: CRC Press/Lewis Publishers.

Meyer, Eugene. 1990. *Chemistry of Hazardous Materials.* Englewood Cliffs, NJ: Prentice Hall.

Patnaik, P. 1992. *A Comprehensive Guide to the Hazardous Properties of Chemical Substances.* New York: Van Nostrand Reinhold.

Philp, R. B. 1995. *Environmental Hazards and Human Health.* New York: CRC Press/Lewis Publishers.

Proctor, N. H., and J. P. Hughes. 1991. *Chemical Hazards of the Workplace.* New York: Van Nostrand Reinhold.

MATERIAL SAFETY DATA SHEET

CHEMICAL NAME	
BUTYL ACRYLATE	**CRC PRESS/LEWIS PUBLISHERS DISCLAIMER:** THE INFORMATION AND RECOMMENDATIONS PRESENTED HEREIN ARE BASED ON SOURCES BELIEVED TO BE RELIABLE. CRC MAKES NO REPRESENTATION ON ITS COMPLETENESS OR ACCURACY. IT IS THE RESPONSIBILITY OF THE USER TO DETERMINE THE CHEMICAL'S SUITABILITY FOR ITS INTENDED USE, THE CHEMICAL'S SAFE USE, AND THE CHEMICAL'S PROPER DISPOSAL. NO REPRESENTATIONS AND/OR WARRANTIES, EITHER EXPRESSED OR IMPLIED, OF THE MERCHANTABILITY OR FITNESS FOR A PARTICULAR PURPOSE, OR OF ANY OTHER NATURE, ARE MADE WITH RESPECT TO THE INFORMATION PROVIDED IN THIS MSDS OR TO THE CHEMICAL TO WHICH INFORMATION MAY REFER. CRC NEITHER ASSUMES NOR AUTHORIZES ANY OTHER PERSON TO ASSUME FOR IT, ANY OTHER ADDITIONAL RESPONSIBILITY OR LIABILITY FOR THE USE OF, OR RELIANCE UPON, THIS INFORMATION.

HAZARD WARNING INFORMATION

HEALTH	FIRE	REACTIVE	OTHER	DEGREE OF HAZARD	COLOR CODING	OTHER CODES
2	2	2		0 = Minimum Hazard 1 = Slight Hazard 2 = Moderate Hazard 3 = Serious Hazard 4 = Severe Hazard	HEALTH = BLUE FIRE = RED REACTIVITY = YELLOW OTHER = WHITE	OX = Oxidizer ACID = Acid ALK = Alkali COR = Corrosive ~~W~~ = Use No Water

SECTION I - GENERAL INFORMATION

Characterization **Ester**	RCRA Number **None**	EPA Class **Not Applicable**
DOT Proper Shipping Name **Butylacrelate (inhibited)**	Chemical Abstract Service (CAS) Number **141-32-2**	
DOT Hazard Class and Label Requirements **Flammable Liquid**	DOT Emergency Guide Code **26**	
DOT Identification Number **UN 2348**	Chemical Formula $C_7H_{12}O_2$	

Synonyms
Butyl-2-propenoate, 2-propenoic acid butyl ester, *n*-butyl acrylate, acrylic acid *n*-butyl ester.

SECTION II - HAZARDOUS INGREDIENTS AND IDENTITY INFORMATION

Hazardous Components (specific identity)	OSHA Exposure Criteria	NIOSH Exposure Criteria	Immediately Dangerous to Life and Health (IDLH)	ACGIH Exposure Criteria
Butyl acrylate (derivation: Reaction of acrylic acid or methyl acrylate with butanol). **1 ppm = 5.33 mg/m^3**	**PEL:** **10 ppm** **55 mg/m^3** **STEL:** **Not Determined**	**REL:** **10 ppm** **55 mg/m^3** **STEL:** **Not Determined**	**Not Determined**	**TLV:** **10 ppm** **55 mg/m^3** **STEL:** **Not Determined**

SECTION III - PHYSICAL AND CHEMICAL CHARACTERISTICS

Boiling Point **300°F (149°C)**	Specific Gravity (H_2O = 1) **0.89**
Vapor Pressure (mm Hg) **10 at 96°F (35.5°C)**	Molecular Weight **128.2**
Vapor Density (Air = 1) **4.4**	Melting Point **-85°F (-65°C)**

Solubility
Slightly soluble in water (0.14% at 20°C), soluble in alcohol and ether.

Appearance and Odor
Clear colorless liquid with a strong fruity odor.

SECTION IV - FIRE AND EXPLOSION HAZARD DATA

Flash Point (method used) **103°F (39°C) closed cup**	Explosive Limits in Air % by Volume **LEL: 1.5%** **UEL: 9.9%**
NFPA Classification **Class II Combustible Liquid**	Autoignition Temperature **534°F (279°C)**

Extinguishing Media
Dry chemical, foam, carbon dioxide. Use water to keep fire-exposed containers cool.

Special Fire Fighting Procedures
Poisonous gases are produced in fire. Wear full protective clothing and self-contained breathing apparatus (SCBA). Heat will build pressure and may rupture closed storage containers. Continue to cool containers with water after fire is extinguished. Water on fire itself may be ineffective.

Unusual Fire and Explosion Hazards
Immediately withdraw if rising sound from venting device is heard of if fire is causing discoloration to the tank. Evacuate 1500 feet radius if fire becomes uncontrollable or container is exposed to direct flame. Uncontrolled bulk polymerization can be explosive.

SECTION V - REACTIVITY DATA

Stability		Conditions to Avoid Butyl acrylate is a reactive chemical. Avoid contact with incompatible materials. Keep away from heat, flame, and other sources of ignition.
Stable	Unstable X	Incompatibility *(materials to avoid)* Reacts violently in contact with strong acids and alkalis, amines, halogens, hydrogen compounds, oxidizers, heat, flame, and sunlight.
Hazardous Polymerization		Conditions to Avoid Butyl acrylate will polymerize readily on heating. The reaction is exothermic and may cause pressure buildup and violent rupture of closed containers. May form unstable peroxides on exposure to air.
May Occur X	Will Not Occur	Hazardous Decomposition or By-products When heated to decomposition, butyl acrylate emits acrid and irritating fumes.

SECTION VI - HEALTH HAZARD DATA

Primary Route(s) of Entry:	Inhalation?	Absorption (skin)?	Ingestion?
	X	X	

Health Hazards

INHALATION: Irritation to the mucous membranes. Can also cause a buildup of fluid in the lungs (pulmonary edema), which is a medical emergency and can be fatal. May also cause dizziness and lightheadedness.

SKIN & EYE: Will irritate the skin. May cause contact dermatitis. Eye contact may cause corneal necrosis. Possible skin allergies can also develop.

INGESTION: May cause vomiting, nausea, headache, and adverse systemic effects.

Carcinogenicity	NTP Listed?	IARC Cancer Review Group?	OSHA Regulated?	Target Organs?
Unknown Human Unknown Animal	No	Group 3	No	Respiratory system; skin; eyes.

Medical Conditions Generally Aggravated by Exposure

None reported.

Emergency and First-aid Procedures

Eye contact: Flush immediately with water for 15 minutes (minimum); seek medical attention. Skin contact: Remove all contaminated clothing. Immediately wash area with large amounts of soap and water. Seek medical assistance. For inhalation: Remove the person from exposure. Provide respiratory assistance and CPR. Transfer to medical facility. Observe for 24-48 hours for lung effects (pulmonary edema). If swallowed: Seek medical attention immediately. Call poison control center for advice.

SECTION VII - PRECAUTIONS FOR SAFE HANDLING AND USE

Steps to be Taken in Case Material is Released or Spilled

Contain spills using absorbent material (vermiculite, or other). Ventilate area of spill. Have water source available in case of fire. Store materials in DOT-approved containers. Restrict those not involved in cleanup from entering area. Notify appropriate authorities, as required.

Preferred Waste Disposal Method

Incineration in chemical incinerator equipped with scrubber and afterburner.

Precautions to be Taken in Handling and Storage

Do not store with incompatible chemicals. Store in tightly closed containers in a cool, dark, well-ventilated area. Do not store in direct sunlight. Keep heat, fire, and flame away. If possible, automatically transfer liquids between containers.

Other Precautions and Warnings

Bulk storage of butyl acrylate is not recommended. Polymerization can be explosive.

SECTION VIII - CONTROL MEASURES AND PERSONAL PROTECTIVE EQUIPMENT

Respiratory Protection *(specify type)*

Where potential exists for exposure over 10 ppm, use a MSHA/NIOSH approved full facepiece respirator with an organic vapor cartridge. For high exposures, use a self-contained breathing apparatus (SCBA) with full facepiece and pressure demand or other positive pressure mode.

Ventilation

Local exhaust or general mechanical systems recommended.

Protective Gloves	Eye Protection	Other Protective Clothing
Butyl Rubber	Chemical Goggles or Face Mask	Impermeable Apron

Work/Hygiene Practices

Always wash hands thoroughly after using chemical; never bring food, drink, or smoking materials into vicinity of chemicals.

BUTYL ACRYLATE

$C_7H_{12}O_2$ CAS: 141-32-2

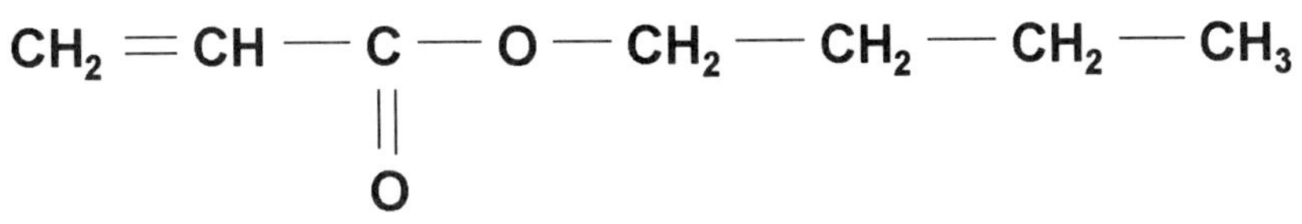

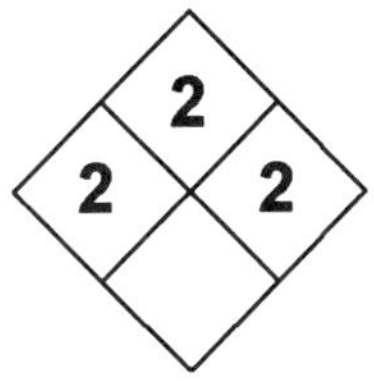

IDENTIFICATION AND TYPICAL USES

Butyl acrylate is a colorless liquid with a rancid, plastic-like odor. It is used primarily as an intermediate in organic synthesis in the manufacture of certain polymers and copolymers which are, in turn, used as resins for textile and leather finishes, and in solvents, paints, binders, and emulsifiers.

RISK ASSESSMENT: HEALTH

General Assessment

Butyl acrylate is moderately toxic by *ingestion, inhalation,* and by skin contact (*absorption*). The toxic and irritant properties of the compound are similar to those of methyl acrylate. The adverse health effects, however, may not be as severe.

Inhalation of its vapor will irritate the mucous membranes, nose, and throat. Very high exposures may damage the lungs, causing fluid buildup in a condition known as pulmonary edema. This can be fatal. Liver damage is also possible following high exposures.

Ingestion may cause upset stomach, nausea, vomiting, diarrhea, headache, and other adverse gastrointestinal reactions. Its toxicity on ingestion appears to be of low order, as compared to the other routes of exposure. Large doses, however, can cause serious and potentially damaging effects in the gastrointestinal tract and the liver.

Butyl acrylate is a skin and eye irritant. The liquid can cause corneal necrosis and other damage to the eye. It may also cause skin allergy. Once such an allergy develops, very low future exposures can cause itching and a skin rash. It can also pass through unbroken skin and enter the bloodstream to cause systemic toxic effects. Symptoms of exposure may include fatigue, sleep disturbances, malaise, general weakness, headache, nausea and vomiting.

Acute Health Effects

The following acute (short-term) health effects may occur immediately or shortly after exposure to butyl acrylate:

Skin: Irritation and possible allergic reaction (rash and itching. Will pass through unbroken skin.

Eye: Irritation which can lead to injury and permanent damage if allowed to remain in contact with the eyes.

Lung: Irritation of the eyes, nose, throat and lungs. Can cause pulmonary edema (fluid buildup in the lungs) which is a medical emergency and can be fatal.

CNS: Insufficient evidence of nerve action on acute exposures. However, exposure to other chemicals having similar properties has resulted in adverse effects on the CNS.

Chronic Health Effects

The following chronic (long-term) health effects may occur at some time after exposure to butyl acrylate and can last for months or even years:

Cancer Hazards: According to the references, butyl acrylate has not been adequately tested for its ability to cause cancer in test animals.

Reproduction: According to the references, butyl acrylate has not been adequately tested for its ability to adversely affect reproduction in test animals.

Other Chronic Effects: Very high or prolonged exposure may lead to lung and liver damage. This chemical has not been adequately evaluated to determine whether brain or other nerve damage could occur as a result of repeated exposure. However, exposure to many solvents and other petroleum-based chemicals causes a reduction in memory and concentration, personality changes including withdrawal, irritability, fatigue, sleep disturbances, reduced coordination, and/or effects on nerves supplying internal organs (autonomic nerves), and/or nerves to the arms and legs (causing weakness and "pins and needles").

ʋ *Recommended Risk-Reduction Measures*

Personnel should avoid direct contact with butyl acrylate. The best risk reduction measure is to use a less toxic chemical as a substitute for a butyl acrylate. If this is not possible or feasible, then *engineering controls* are the most effective method of reducing exposures. The best protection is to enclose operations and/or provide local exhaust ventilation at the site of chemical release. While not always operationally feasible, isolating operations can also reduce exposure risk.

Using respiratory protection is less effective than the controls mentioned above, but is still advisable whenever working with or around butyl acrylate. For exposures over 10 ppm (8 hours), an MSHA/NIOSH approved chemical respirator with an organic vapor cartridge should be used. For higher exposures, a self-contained breathing apparatus (SCBA) with full facepiece and pressure demand is the recommended respiratory protection of choice. If a full facepiece is not available, then chemical goggles should be worn to protect the eyes. Whenever a chemical splash hazard exists, a face shield and a rubber apron should be worn. To prevent hand and skin exposures, butyl rubber or neoprene gloves should be worn.

Administrative controls should also be in place to minimize the potential for human exposures. These may include written procedures or policies which specify the methods and techniques that will be practiced whenever personnel are to work with butyl acrylate.

All personnel should receive *training* on the use, hazards, protective measures, emergency actions, and other precautions per 29 CFR 1910.1200 (Hazard Communication), prior to the first assignment in an area where butyl acrylate is used or stored.

If symptoms develop or overexposure is suspected, the following medical tests are recommended:

- ☑ Liver function tests.
- ☑ Lung function tests.
- ☑ Skin testing with dilute butyl acrylate to help diagnose allergy (performed by a qualified allergist).
- ☑ Interview for brain effects, including recent memory, mood (irritability, withdrawal), concentration, headaches, malaise, and altered sleep patterns. Consider cerebellar, autonomic, and peripheral nervous system evaluation. Positive and borderline individuals should be referred for neuropsychological testing.

Any medical evaluation should include a careful history of past and present symptoms with an examination. Medical tests that look for existing damage are not a substitute for controlling exposures.

Other methods to reduce exposure include:

- ☑ Where possible, enclose operations and use local exhaust ventilation at the site of chemical release. If local exhaust ventilation or enclosure is not used, respiratory protection should be mandatory.
- ☑ Always ensure that proper protective clothing is worn when using chemical substances.
- ☑ Wash thoroughly immediately after exposure to butyl acrylate and at the end of the work shift or before eating, drinking, or smoking.
- ☑ Hazard warning information should be posted in the work area. In addition, as part of an on-going education and training program, all information on the health and safety hazards of butyl acrylate should be communicated to all exposed workers.
- ☑ Never eat, drink, or smoke in areas were butyl acrylate is used, handled or stored.

RISK ASSESSMENT: ENVIRONMENT

General Assessment

The environment is at risk of exposure during transportation, storage, disposal, or destruction of butyl acrylate. In almost every scenario, the threat of environmental exposure is contingent upon the proper handling of the chemical substance.

Accidental spills, large or small, can result in fire, explosion, and possible contamination of the surrounding environmental mediums (water, soil, and air). Butyl acrylate is considered a class II combustible liquid (according to OSHA 29 CFR 1910.106) which presents a low to moderate fire hazard. However, because it is incompatible with a number of common materials, especially strong oxidizers and acids, contact can result in violent and explosive reactions. It can also form explosive peroxides in air and can react on exposure to sunlight. These characteristics require special consideration during any emergency situation involving a leak or spill of butyl acrylate.

The proper disposal/destruction method for butyl acrylate is to burn it in a chemical incinerator equipped with an afterburner and air scrubber. Butyl acrylate can enter the environment through unchecked industrial discharges into effluents and through spills.

Acute Ecological Effects

Acute (short-term) toxic effects may include the death of animals, birds, or fish, and death or low growth rate in plants. Acute effects are seen 2 to 4 days after animals or plants are exposed to butyl acrylate. This chemical has moderate acute toxicity to aquatic life and high acute toxicity to birds. Insufficient data are available to evaluate or predict the short-term effects of butyl acrylate to plants or terrestrial animals.

Chronic Ecological Effects

Chronic toxic effects may include shortened life span, reproductive problems, lower fertility, and changes in appearance or behavior in exposed animals. These effects can be seen long after first exposure(s) to toxic chemicals. Butyl acrylate has moderate chronic toxicity to aquatic life. Insufficient data are available to evaluate or predict the long-term effects of butyl acrylate to plants, birds, or land animals.

Water Solubility

Butyl acrylate is moderately soluble in water. Concentrations of 1 to 1000 milligrams will mix with a liter of water.

Persistence in the Environment

Butyl acrylate is slightly persistent in water, with a half-life of between 2 to 20 days. The half-life of a pollutant is the amount of time it takes for one-half of the chemical to be degraded. Approximately 98% of butyl acrylate will eventually end up in air with the remainder ending up in water.

Bioaccumulation in Aquatic Organisms

Some substances increase in concentration, or *bioaccumulate,* in living organisms as they breathe contaminated air, drink contaminated water, or eat contaminated food. These chemicals can become concentrated in the tissues and internal organs of animals as well as in those of humans.

The concentration of butyl acrylate found in fish tissues is expected to be somewhat higher then the average concentration of butyl acrylate in water from which the fish was taken.

Recommended Risk-Reduction Measures

Proper training of all transporters will reduce the likelihood of a mishap or accident resulting in a leak or spill to the environment. The correct use of DOT labeling (per the MSDS) on all containers, trucks, and rail cars will enable emergency responders to react properly and quickly to any disaster thereby reducing the potential risk to the environment and to personnel.

Storage of butyl acrylate should be segregated from incompatible chemicals to minimize the risk of cross-contamination or contact. Buildings designated for storage should be equipped with appropriate fire protection systems (alarms, sprinklers, emergency lighting, portable extinguishers).

If a spill or leak to the environment has occurred, fire department, emergency response, and/or hazardous materials spill personnel should be notified immediately. Cleanup should be attempted only by those trained in proper spill containment procedures. Contaminated soils should be removed for incineration and replaced with clean soil.

If butyl acrylate should contact the water table, aquifer, or navigable waterway, time is of the essence. It is moderately soluble in water and, therefore, total containment and remediation may not be entirely possible. When such spills occur, the local and/or state emergency response authorities must be notified. A comprehensive emergency response or disaster preparedness/recovery plan should be in place prior to any operations involving the use, transportation, storage, or disposal of butyl acrylate. If butyl acrylate is spilled or leaked, the following specific steps are recommended:

- ☑ Restrict persons not wearing protective clothing from area of spill or leak until cleanup is complete and area can be opened for normal work.
- ☑ Ventilate area of spill or leak.
- ☑ Absorb liquids in vermiculite, dry sand, earth, or a similar material and deposit in sealed containers.
- ☑ If applicable, stop flow of leaking liquid. If leak source is a cylinder and the leak can not be stopped in place, remove leaking cylinder to a safe place in the open air and repair or allow cylinder to empty. Send back to manufacturer for repair.
- ☑ It may be necessary to dispose of butyl acrylate as a hazardous waste. The responsible state agency or the regional office of the federal Environmental

Protection Agency (EPA) should be contacted for specific recommendations.

RISK ASSESSMENT: BUSINESS

General Assessment

Accidents or mishaps involving butyl acrylate can present a moderate threat to business operations. The loss or damage of equipment or facilities can significantly affect fiscal viability. Lawsuits that may result from personnel injury/death, public exposures, and/or environmental contamination will also require a serious expenditure of resources.

Media attention surrounding an injury, death, or environmental damage can also result in a loss of profits and loss of current as well as future business.

Recommended Risk-Reduction Measures

Company attorneys, safety and health professionals, and environmental specialists should be involved in the development of any procedures or policies intended to manage the use of chemicals in the workplace. A company official should be pre-designated as a public relations officer with specific training in dealing with the press. Corporate plans and policies should be developed, approved, and implemented long before any need for such arises.

REFERENCES

American Conference of Governmental Industrial Hygienists. 1988. *Documentation of the Threshold Limit Values and Biological Exposure Indices,* 5th Edition (with updates). Cincinnati: ACGIH

Anthony, C. P., and N. J. Kolthoff. 1971. *Textbook of Anatomy and Physiology.* St. Louis: C. V. Mosby Company.

Bloom, A. D. and F. J. DeSerres. 1995. *Ecotoxicity and Human Health.* New York: CRC Press/Lewis Publishers.

Brusick, D. J. 1994. *Methods for Genetic Risk Assessment.* New York: CRC Press/Lewis Publishers.

Bueche, J. R. 1972. *Principles of Physics,* 2nd Edition. New York: McGraw-Hill.

Burrell, R., D. K. Flaherty, and L. J. Sauers. 1992. *Toxicology of the Immune System.* New York: Van Nostrand Reinhold.

Calabrese, E. J. 1994. *Biological Effects of Low Level Exposures.* New York: CRC Press/Lewis Publishers.

Cockerman, L. G., B. S. Shane. 1994. *Basic Environmental Toxicology.* New York: CRC Press/Lewis Publishers.

Cothern, C. R., N. P. Ross. 1994. *Environmental Statistics, Assessment, and Forecasting.* New York: CRC Press/Lewis Publishers.

Danse, I. R. 1991. *Common Sense Toxics in the Workplace.* New York: Van Nostrand Reinhold.

Environmental Protection Agency. 1983. *Chemical Hazard Information Profile: Butyl acrylate.* Washington, D.C.: U.S. EPA.

Gardner, R. W. 1994. *Chemical Intolerance.* New York: CRC Press/Lewis Publishers.

Hallenbeck, W. H. 1993. *Quantitative Risk Assessment for Environmental and Occupational Health,* 2nd Edition. New York: CRC Press/Lewis Publishers.

Hodgson, E., R. B. Mailman, and J. E. Chambers. 1988. *Dictionary of Toxicology.* London: The MacMillian Company.

Heck, H. 1985. Mechanisms of butyl acrylate toxicity: Structure activity Studies. *CIIT Activities,* 5(10):106.

Hosty, J. W., and P. Foster. 1990. *A Practical Guide to Chemical Spill Response.* New York: Van Nostrand Reinhold.

Howard, P. H., et. al. 1991. *Handbook of Environmental Degradation Rates.* New York: CRC Press/Lewis Publishers.

James, A., and M. Lord. 1992. *Index of Chemical and Physical Data.* New York: Van Nostrand Reinhold.

Jones, F. E. 1994. *Toxic Organic Vapors in the Workplace.* New York: CRC Press/Lewis Publishers.

Kamrin, M. 1 A. 1988. *Toxicology: A Primer on Principles and Applications.* Chelsea, Michigan: Lewis Publishers.

Keith, L. H. and M. M. Walker. 1995. *Handbook of Air Toxics.* New York: CRC Press/Lewis Publishers.

Landis, W. G., M. H. Yu. 1995. *Introduction to Environmental Toxicology.* New York: CRC Press/Lewis Publishers.

Langerman, N. 1994 *Precautionary Labels for Chemical Containers.* New York: CRC Press/Lewis Publishers.

Lauwerys, R. R., P. Hoet. 1993. *Industrial Chemical Exposure.* New York: CRC Press/Lewis Publishers.

Lide, D. R. 1995. *Handbook of Chemistry and Physics,* 76th Edition. New York: CRC Press/Lewis Publishers.

Lewis, Richard J., Sr. 1992. *Sax's Dangerous Properties of Industrial Materials,* Volumes 1, 2, and 3, 8th Edition. New York: Van Nostrand Reinhold.

Lisella, Frank S. 1994. *The VNR Dictionary of Environmental Health and Safety.* New York: Van Nostrand Reinhold.

Manahan, S. E. 1994. *Environmental Chemistry,* 6th Edition. New York: CRC Press/Lewis Publishers.

Manahan, S. E. 1992. *Toxicological Chemistry,* 2nd Edition. New York: CRC Press/Lewis Publishers.

Meyer, Eugene. 1990. *Chemistry of Hazardous Materials.* Englewood Cliffs, NJ: Prentice Hall.

Patnaik, P. 1992. *A Comprehensive Guide to the Hazardous Properties of Chemical Substances.* New York: Van Nostrand Reinhold.

Proctor, N. H., and J. P. Hughes. 1991. *Chemical Hazards of the Workplace.* New York: Van Nostrand Reinhold.

MATERIAL SAFETY DATA SHEET

CHEMICAL NAME	
***n*-BUTYRALDEHYDE**	**CRC PRESS/LEWIS PUBLISHERS DISCLAIMER:** THE INFORMATION AND RECOMMENDATIONS PRESENTED HEREIN ARE BASED ON SOURCES BELIEVED TO BE RELIABLE. CRC MAKES NO REPRESENTATION ON ITS COMPLETENESS OR ACCURACY. IT IS THE RESPONSIBILITY OF THE USER TO DETERMINE THE CHEMICAL'S SUITABILITY FOR ITS INTENDED USE, THE CHEMICAL'S SAFE USE, AND THE CHEMICAL'S PROPER DISPOSAL. NO REPRESENTATIONS AND/OR WARRANTIES, EITHER EXPRESSED OR IMPLIED, OF THE MERCHANTABILITY OR FITNESS FOR A PARTICULAR PURPOSE, OR OF ANY OTHER NATURE, ARE MADE WITH RESPECT TO THE INFORMATION PROVIDED IN THIS MSDS OR TO THE CHEMICAL TO WHICH INFORMATION MAY REFER. CRC NEITHER ASSUMES NOR AUTHORIZES ANY OTHER PERSON TO ASSUME FOR IT, ANY OTHER ADDITIONAL RESPONSIBILITY OR LIABILITY FOR THE USE OF, OR RELIANCE UPON, THIS INFORMATION.

HAZARD WARNING INFORMATION

HEALTH	FIRE	REACTIVE	OTHER	DEGREE OF HAZARD	COLOR CODING	OTHER CODES
2	3	2		0 = Minimum Hazard 1 = Slight Hazard 2 = Moderate Hazard 3 = Serious Hazard 4 = Severe Hazard	HEALTH = BLUE FIRE = RED REACTIVITY = YELLOW OTHER = WHITE	OX = Oxidizer ACID = Acid ALK = Alkali COR = Corrosive W = Use No Water

SECTION I - GENERAL INFORMATION

Characterization: **Aldehyde**	RCRA Number: **None**	EPA Class: **Not Applicable**
DOT Proper Shipping Name: **Butyraldehyde**	Chemical Abstract Service (CAS) Number: **123-72-8**	
DOT Hazard Class and Label Requirements: **Flammable Liquid**	DOT Emergency Guide Code: **26**	
DOT Identification Number: **UN 1129**	Chemical Formula: **C_3H_7CHO**	

Synonyms
***n*-Butyl aldehyde; butanal; *n*-butanal; butyric aldehyde; butaldehyde.**

SECTION II - HAZARDOUS INGREDIENTS AND IDENTITY INFORMATION

Hazardous Components (specific identity)	OSHA Exposure Criteria	NIOSH Exposure Criteria	Immediately Dangerous to Life and Health (IDLH)	ACGIH Exposure Criteria
***n*-Butyraldehyde** (derivation: Dehydrogentaing butanol vapors over a catalyst, the butyraldehyde being separated by distillation; partial reduction of crotonaldehyde).	**PEL:** Not Established **STEL:** Not Established	**REL:** Not Established **STEL:** Not Established	**Not Determined**	**TLV:** Not Established **STEL:** Not Established

SECTION III - PHYSICAL AND CHEMICAL CHARACTERISTICS

Boiling Point: **167°F (75°C)**	Specific Gravity (H_2O = 1): **0.9**
Vapor Pressure (mm Hg): **6 at 69°F (20°C)**	Molecular Weight: **72.1**
Vapor Density (Air = 1): **2.5**	Melting Point: **-148°F (-100°C)**

Solubility
Slightly soluble in water, soluble in alcohol, acetone, and ether.

Appearance and Odor
Clear colorless mobile liquid with a pungent, nutty odor.

SECTION IV - FIRE AND EXPLOSION HAZARD DATA

Flash Point (method used): **20°F (-6°C) closed cup**	Explosive Limits in Air % by Volume: **LEL: 1.9% UEL: 12.5%**
NFPA Classification: **Class IB Flammable Liquid**	Autoignition Temperature: **446°F (230°C)**

Extinguishing Media
Dry chemical, foam, carbon dioxide. Use a water spray to flush the spill and disperse the vapors.

Special Fire Fighting Procedures
Poisonous gases are produced in fire. Wear full protective clothing and self-contained breathing apparatus (SCBA). Heat will build pressure and may rupture closed storage containers. Continue to cool containers with water after fire is extinguished. Water on fire itself may be ineffective.

Unusual Fire and Explosion Hazards
Immediately withdraw if rising sound from venting device is heard of if fire is causing discoloration to the tank. Fight fire from a distance or protected location, if possible. For large fires, withdraw and allow to burn, if possible. Vapors may travel to a source of ignition and flash back.

SECTION V - REACTIVITY DATA

Stability		Conditions to Avoid *n*-butyraldehyde forms explosive mixtures in air. It generates heat when mixed with concentrated acids. It is extremely flammable, avoid contact with heat, flame, or other sources of ignition.
Stable	Unstable X	Incompatibility *(materials to avoid)* Reacts violently in contact with strong oxidizers (perchlorates, peroxides, permanganates, chlorates, and nitrates), strong bases, strong reducing agents, and strong acids (hydrochloric, sulfuric, nitric).
Hazardous Polymerization		Conditions to Avoid Under normal conditions of temperature and pressure, *n*-butyraldehyde is not known to undergo hazardous polymerization.
May Occur X	Will Not Occur	Hazardous Decomposition or By-products When heated to decomposition, *n*-butyraldehyde emits acrid and irritating smoke and fumes.

SECTION VI - HEALTH HAZARD DATA

Primary Route(s) of Entry:	Inhalation?	Absorption (skin)?	Ingestion?
	X	X	X

Health Hazards

INHALATION: Vapors can irritate the nose, throat, and lungs. High exposures may cause a buildup of fluid in the lungs (pulmonary edema), a medical emergency. May also cause symptoms of dizziness and lightheadedness.

SKIN & EYE: Will cause severe skin and eye irritation and even burns. Can be absorbed through unbroken skin and enter the bloodstream and cause toxic systemic effects.

INGESTION: May cause vomiting and nausea. Large doses can cause adverse systemic effects.

Carcinogenicity	NTP Listed?	IARC Cancer Review Group?	OSHA Regulated?	Target Organs?
Unknown Human Unknown Animal	5th Annual Report	No	No	Respiratory system; skin; eyes.

Medical Conditions Generally Aggravated by Exposure

None reported. However, pre-existing lung ailments or skin disorders may be aggravated by exposure.

Emergency and First-aid Procedures

Eye contact: Flush immediately with water for 15 minutes (minimum); seek medical attention. Skin contact: Remove all contaminated clothing. Immediately wash area with large amounts of soap and water. Seek medical assistance. For inhalation: Remove the person from exposure. Provide respiratory assistance and CPR. Transfer to medical facility. Observe for 24-48 hours for lung effects (pulmonary edema). If swallowed: Seek medical attention immediately. Call poison control center for advice.

SECTION VII - PRECAUTIONS FOR SAFE HANDLING AND USE

Steps to be Taken in Case Material is Released or Spilled

Contain spills using absorbent material (vermiculite, or other). Ventilate area of spill. Remove all potential ignition sources. Store materials in DOT-approved containers. Restrict those not involved in cleanup from entering area. Notify appropriate authorities, as required.

Preferred Waste Disposal Method

Not cited.

Precautions to be Taken in Handling and Storage

Do not store with incompatible chemicals. Store in tightly closed containers in a cool, dark, well-ventilated area. Use explosion-proof equipment in areas where *n*-butyraldehyde is used or stored. Keep heat, fire, and flame away. If possible, automatically transfer liquids between containers.

Other Precautions and Warnings

Bulk storage of *n*-butyraldehyde is not recommended. Use non-sparking tools and equipment.

SECTION VIII - CONTROL MEASURES AND PERSONAL PROTECTIVE EQUIPMENT

Respiratory Protection *(specify type)*

No exposure limits have been established. This does not mean that unprotected exposure to *n*-butyraldehyde is permissible. For high exposures, use a self-contained breathing apparatus (SCBA) with full facepiece and pressure demand or other positive pressure mode.

Ventilation

Local exhaust or general mechanical systems recommended.

Protective Gloves	Eye Protection	Other Protective Clothing
Butyl Rubber	Chemical Goggles or Face Mask	Impermeable Apron

Work/Hygiene Practices

Always wash hands thoroughly after using chemical; never bring food, drink, or smoking materials into vicinity of chemicals.

n-BUTYRALDEHYDE

C_3H_7CHO CAS: 123-72-8

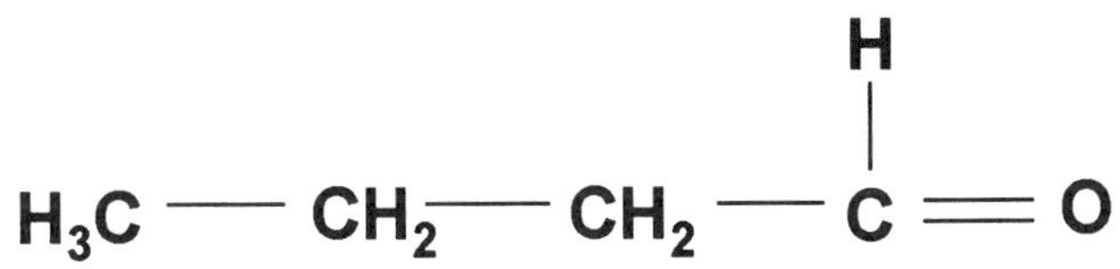

IDENTIFICATION AND TYPICAL USES

n-Butyraldehyde is a clear, colorless mobile liquid with a pungent, nutty odor. It is used to make rubber accelerators, synthetic resins, high polymers, and plasticizers. It is also used as a solvent.

RISK ASSESSMENT: HEALTH

General Assessment

n-Butyraldehyde is moderately toxic *by ingestion, inhalation,* and by skin contact (*absorption*). The toxicity of *n*-butyraldehyde is very low. The effect is primarily narcotic. Exposures to high levels (above 175 g/m^3) for periods of 30 minutes or more may produce a general anesthetic effect, with possible action on the liver and bladder.

Inhalation may cause irritation of the nose and throat with possible nausea, vomiting, headaches, and loss of consciousness. High or prolonged exposures cause dizziness, delayed hypersensitivity, and lightheadedness. Such exposures may also cause fluid to buildup in the lungs (a condition known as pulmonary edema), which is a medical emergency and can be fatal. Inhalation may also cause immunological effects in humans.

n-Butyraldehyde is a mild skin and eye irritant. However, prolonged contact may result in burns to the skin and eyes. Eye contact can also cause severe irritation and possible damage.

Acute Health Effects

The following acute (short-term) health effects may occur immediately or shortly after exposure to *n*-butyraldehyde:

Skin: Irritation and possible burns.

Eye: Irritation and possible burns which can lead to injury and permanent damage if allowed to remain in contact with the eyes.

Lung: Irritation of the nose, throat, and lungs. Can cause pulmonary edema (fluid buildup in the lungs) which can be fatal.

CNS: Insufficient evidence of nerve action on acute exposures. However, it is reported that exposures to high levels results in an anesthetic effect on test animals.

Chronic Health Effects

The following chronic (long-term) health effects may occur at some time after exposure to *n*-butyraldehyde and can last for months or even years:

Cancer Hazards: According to the references, *n*-butyraldehyde has not been adequately tested for its ability to cause cancer in test animals.

Reproduction: According to the references, *n*-butyraldehyde has not been adequately tested for its ability to adversely affect reproduction in test animals.

Other Chronic Effects: *n*-Butyraldehyde has not been tested for other chronic (long-term) health effects.

Recommended Risk-Reduction Measures

Personnel should avoid direct contact with *n*-butyraldehyde, especially skin and eye contact. The best risk reduction measure is to use a less toxic chemical as a substitute for a *n*-butyraldehyde. If this is not possible or feasible, then *engineering controls* are the most effective method of reducing exposures. The best protection is to enclose operations and/or provide local exhaust ventilation at the site of chemical release. While not always operationally feasible, isolating operations can also reduce exposure risk.

Using respiratory protection is less effective than the controls mentioned above, but is still advisable whenever working with or around *n*-butyraldehyde. No exposure levels have been established for this chemical. However, this does not mean that unprotected exposure is acceptable. For low level, infrequent exposures, an MSHA/NIOSH approved chemical respirator with an organic vapor cartridge should be used. For higher exposures or when exposure levels are not known, a self-contained breathing apparatus (SCBA) with full facepiece and pressure demand is the recommended respiratory protection of choice. If a full facepiece is not available, then chemical goggles should be worn to protect the eyes. Whenever a chemical splash hazard exists, a face shield and a rubber apron should be worn. To prevent hand and skin exposures, butyl rubber or neoprene gloves should be worn.

Administrative controls should also be in place to minimize the potential for human exposures. These may include written procedures or policies which specify the methods and techniques that will be practiced whenever personnel are to work with *n*-butyraldehyde.

All personnel should receive *training* on the use, hazards, protective measures, emergency actions, and other precautions per 29 CFR 1910.1200 (Hazard Communication), prior to the first assignment in an area where *n*-butyraldehyde is used or stored.

If symptoms develop or overexposure is suspected, the following medical tests are recommended:

☑ Consider chest X-ray after acute exposure.

Any medical evaluation should include a careful history of past and present symptoms with an examination. Medical tests that look for existing damage are not a substitute for controlling exposures.

Other methods to reduce exposure include:

☑ Where possible, enclose operations and use local exhaust ventilation at the site of chemical release. If local exhaust ventilation or enclosure is not used, respiratory protection should be mandatory.

☑ Always ensure that proper protective clothing is worn when using chemical substances and that personnel are trained in its use and care.

☑ Wash thoroughly immediately after exposure to *n*-butyraldehyde and at the end of the work shift or before eating, drinking, or smoking.

☑ Hazard warning information should be posted in the work area. In addition, as part of an on-going education and training program, all information on the health and safety hazards of *n*-butyraldehyde should be communicated to all exposed and potentially exposed workers.

☑ Never eat, drink, or smoke in areas were *n*-butyraldehyde is used, handled or stored.

☑ Where possible, automatically pump liquid *n*-butyraldehyde from drums or other containers to process containers.

☑ Before entering a confined space where *n*-butyraldehyde may be present, verify that an explosive concentration does not exist.

RISK ASSESSMENT: ENVIRONMENT

General Assessment

The environment is at risk of exposure during transportation, storage, disposal, or destruction of *n*-butyraldehyde. In almost every scenario, the threat of environmental exposure is contingent upon the proper handling of the chemical substance.

Accidental spills, large or small, can result in fire, explosion, and possible contamination of the surrounding environmental mediums (water, soil, and air). *n*-Butyraldehyde is considered a class IB flammable liquid (according to OSHA 29 CFR 1910.106) which presents a serious fire hazard concern. Also, because it is incompatible with a number of common materials, especially strong oxidizers (such as perchlorates, peroxides, permanganates, chlorates, and nitrates) and acids (such as hydrochloric, chlorosulfonic, sulfuric, and nitric), contact can result in violent and explosive reactions. It can also form explosive mixtures in air. These characteristics require special consideration during any emergency situation involving a leak or spill of *n*-butyraldehyde.

n-Butyraldehyde can enter the environment through unchecked industrial discharges into effluents and through spills.

☠ *Acute Ecological Effects*

Acute (short-term) toxic effects may include the death of animals, birds, or fish, and death or low growth rate in plants. Acute effects are seen 2 to 4 days after animals or plants are exposed to *n*-butyraldehyde. This chemical has moderate acute toxicity to aquatic life. Insufficient data are available to evaluate or predict the short-term effects of *n*-butyraldehyde to plants, birds, or terrestrial animals.

Chronic Ecological Effects

Chronic toxic effects may include shortened life span, reproductive problems, lower fertility, and changes in appearance or behavior in exposed animals. These effects can be seen long after first exposure(s) to toxic chemicals.

n-Butyraldehyde has moderate chronic toxicity to aquatic life. Insufficient data are available to evaluate or predict the long-term effects of *n*-butyraldehyde to plants, birds, or land animals.

Water Solubility

n-Butyraldehyde is highly soluble in water. Concentrations of 1000 milligrams and more will mix with a liter of water.

Persistence in the Environment

n-Butyraldehyde is slightly persistent in water, with a half-life of between 2 to 20 days. The half-life of a pollutant is the amount of time it takes for one-half of the chemical to be degraded. Approximately 88% of *n*-butyraldehyde will eventually end up in air with the remainder ending up in water.

Bioaccumulation in Aquatic Organisms

Some substances increase in concentration, or *bioaccumulate,* in living organisms as they breathe contaminated air, drink contaminated water, or eat contaminated food. These chemicals can become concentrated in the tissues and internal organs of animals as well as humans.

The concentration of *n*-butyraldehyde found in fish tissues is expected to be about the same the average concentration of *n*-butyraldehyde in water from which the fish was taken.

Recommended Risk-Reduction Measures

Proper training of all transporters will reduce the likelihood of a mishap or accident resulting in a leak or spill to the environment. The correct use of DOT labeling (per the MSDS) on all containers, trucks, and rail cars will enable emergency responders to react properly and quickly to any disaster thereby reducing the potential risk to the environment and to personnel.

Storage of *n*-butyraldehyde should be segregated from incompatible chemicals to minimize the risk of cross-contamination or contact. Buildings designated for storage should be equipped with appropriate fire protection systems (alarms, sprinklers, emergency lighting, portable extinguishers). Explosion-proof equipment is recommended in areas where *n*-butyraldehyde is used, handled, or stored.

If a spill or leak to the environment has occurred, fire department, emergency response, and/or hazardous materials spill personnel should be notified immediately. Cleanup should be attempted only by those trained in proper spill containment procedures. Contaminated soils should be removed for incineration and replaced with clean soil.

If *n*-butyraldehyde should contact the water table, aquifer, or navigable waterway, time is of the essence. It is highly soluble in water and, therefore, total containment and remediation may not be entirely possible. When such spills occur, the local and/or state emergency response authorities must be notified. A comprehensive emergency response or disaster preparedness/recovery plan should be in place prior to any operations involving the use, transportation, storage, or disposal of *n*-butyraldehyde. If *n*-butyraldehyde is spilled or leaked, the following specific steps are recommended:

- ☑ Restrict persons not wearing protective clothing from area of spill or leak until cleanup is complete and area can be opened for normal work.
- ☑ Remove all ignition sources.
- ☑ Ventilate area of spill or leak.
- ☑ Absorb liquids in vermiculite, dry sand, earth, or a similar material and deposit in sealed containers.
- ☑ If applicable, stop flow of leaking liquid. If leak source is a cylinder and the leak cannot be stopped in place, remove leaking cylinder to a safe place in the open air and repair or allow cylinder to empty.
- ☑ It may be necessary to dispose of *n*-butyraldehyde as a hazardous waste. The responsible state agency or the regional office of the federal Environmental Protection Agency (EPA) should be contacted for specific recommendations.

RISK ASSESSMENT: BUSINESS

General Assessment

Accidents or mishaps involving *n*-butyraldehyde can present a serious threat to business operations. The loss or damage of equipment or facilities can significantly affect fiscal viability. Lawsuits that may result from personnel injury/death, public exposures, and/or environmental contamination will also require a serious expenditure of resources.

Media attention surrounding an injury, death, or environmental damage can also result in a loss of profits and loss of current as well as future business.

Recommended Risk-Reduction Measures

Company attorneys, safety and health professionals, and environmental specialists should be involved in the development of any procedures or policies intended to manage the use of chemicals in the workplace. A company official should be pre-designated as a public relations officer with specific training in dealing with the press. Corporate plans and policies should be developed, approved, and implemented long before any need for such arises.

REFERENCES

American Conference of Governmental Industrial Hygienists. 1988. *Documentation of the Threshold Limit Values and Biological Exposure Indices,* 5th Edition (with updates). Cincinnati: ACGIH

Ahlbom, A. 1993. *Biostatistics for Epidemiologists.* New York: CRC Press/Lewis Publishers.

Anthony, C. P., and N. J. Kolthoff. 1971. *Textbook of Anatomy and Physiology.* St. Louis: C. V. Mosby Company.

Asante-Duah, D. K. 1993. *Hazardous Waste Risk Assessment.* New York: CRC Press/Lewis Publishers.

Bloom, A. D. and F. J. DeSerres. 1995. *Ecotoxicity and Human Health.* New York: CRC Press/Lewis Publishers.

Brusick, D. J. 1994. *Methods for Genetic Risk Assessment.* New York: CRC Press/Lewis Publishers.

Bueche, J. R. 1972. *Principles of Physics,* 2nd Edition. New York: McGraw-Hill.

Burrell, R., D. K. Flaherty, and L. J. Sauers. 1992. *Toxicology of the Immune System.* New York: Van Nostrand Reinhold.

Calabrese, E. J., E. M. Kenyon. *Air Toxics and Risk Assessment.* New York: CRC Press/Lewis Publishers.

Calabrese, E. J. 1994. *Biological Effects of Low Level Exposures.* New York: CRC Press/Lewis Publishers.

Cockerman, L. G., B. S. Shane. 1994. *Basic Environmental Toxicology.* New York: CRC Press/Lewis Publishers.

Cothern, C. R., N. P. Ross. 1994. *Environmental Statistics, Assessment, and Forecasting.* New York: CRC Press/Lewis Publishers.

Cothern, R. R. 1993. *Comparative Environmental Risk Assessment.* New York: CRC Press/Lewis Publishers.

Cothern, C. R. 1995. *Handbook for Environmental Risk Decisionmaking.* New York: CRC Press/Lewis Publishers.

Danse, I. R. 1991. *Common Sense Toxics in the Workplace.* New York: Van Nostrand Reinhold.

Gardner, R. W. 1994. *Chemical Intolerance.* New York: CRC Press/Lewis Publishers.

Hallenbeck, W. H. 1993. *Quantitative Risk Assessment for Environmental and Occupational Health,* 2nd Edition. New York: CRC Press/Lewis Publishers.

Hodgson, E., R. B. Mailman, and J. E. Chambers. 1988. *Dictionary of Toxicology.* London: The MacMillian Company.

Hosty, J. W., and P. Foster. 1990. *A Practical Guide to Chemical Spill Response.* New York: Van Nostrand Reinhold.

Howard, P. H., et. al. 1991. *Handbook of Environmental Degradation Rates.* New York: CRC Press/Lewis Publishers.

Howard, P. H. 1995. *Handbook of Environmental Fate and Exposures Data for Organic Chemicals,* Vol. I-IV. New York: CRC Press/Lewis Publishers.

James, A., and M. Lord. 1992. *Index of Chemical and Physical Data.* New York: Van Nostrand Reinhold.

Jones, F. E. 1994. *Toxic Organic Vapors in the Workplace.* New York: CRC Press/Lewis Publishers.

Kamrin, M. 1 A. 1988. *Toxicology: A Primer on Principles and Applications.* Chelsea, Michigan: Lewis Publishers.

Keith, L. H. and M. M. Walker. 1995. *Handbook of Air Toxics.* New York: CRC Press/Lewis Publishers.

Landis, W. G., M. H. Yu. 1995. *Introduction to Environmental Toxicology.* New York: CRC Press/Lewis Publishers.

Langerman, N. 1994 *Precautionary Labels for Chemical Containers.* New York: CRC Press/Lewis Publishers.

Lauwerys, R. R., P. Hoet. 1993. *Industrial Chemical Exposure.* New York: CRC Press/Lewis Publishers.

Lide, D. R. 1995. *Handbook of Chemistry and Physics,* 76th Edition. New York: CRC Press/Lewis Publishers.

Lewis, Richard J., Sr. 1992. *Sax's Dangerous Properties of Industrial Materials,* Volumes 1, 2, and 3, 8th Edition. New York: Van Nostrand Reinhold.

Lisella, Frank S. 1994. *The VNR Dictionary of Environmental Health and Safety.* New York: Van Nostrand Reinhold.

Manahan, S. E. 1994. *Environmental Chemistry,* 6th Edition. New York: CRC Press/Lewis Publishers.

Manahan, S. E. 1992. *Toxicological Chemistry,* 2nd Edition. New York: CRC Press/Lewis Publishers.

Mellström, G. A., J. E. Wahlberg, and H. I. Maibach. 1994. *Protective Gloves for Occupational Use.* New York: CRC Press/Lewis Publishers.

Meyer, Eugene. 1990. *Chemistry of Hazardous Materials.* Englewood Cliffs, NJ: Prentice Hall.

Patnaik, P. 1992. *A Comprehensive Guide to the Hazardous Properties of Chemical Substances.* New York: Van Nostrand Reinhold.

Proctor, N. H., and J. P. Hughes. 1991. *Chemical Hazards of the Workplace.* New York: Van Nostrand Reinhold.

MATERIAL SAFETY DATA SHEET

CHEMICAL NAME	
CADMIUM	**CRC PRESS/LEWIS PUBLISHERS DISCLAIMER:** THE INFORMATION AND RECOMMENDATIONS PRESENTED HEREIN ARE BASED ON SOURCES BELIEVED TO BE RELIABLE. CRC MAKES NO REPRESENTATION ON ITS COMPLETENESS OR ACCURACY. IT IS THE RESPONSIBILITY OF THE USER TO DETERMINE THE CHEMICAL'S SUITABILITY FOR ITS INTENDED USE, THE CHEMICAL'S SAFE USE, AND THE CHEMICAL'S PROPER DISPOSAL. NO REPRESENTATIONS AND/OR WARRANTIES, EITHER EXPRESSED OR IMPLIED, OF THE MERCHANTABILITY OR FITNESS FOR A PARTICULAR PURPOSE, OR OF ANY OTHER NATURE, ARE MADE WITH RESPECT TO THE INFORMATION PROVIDED IN THIS MSDS OR TO THE CHEMICAL TO WHICH INFORMATION MAY REFER. CRC NEITHER ASSUMES NOR AUTHORIZES ANY OTHER PERSON TO ASSUME FOR IT, ANY OTHER ADDITIONAL RESPONSIBILITY OR LIABILITY FOR THE USE OF, OR RELIANCE UPON, THIS INFORMATION.

HAZARD WARNING INFORMATION

HEALTH	FIRE	REACTIVE	OTHER	DEGREE OF HAZARD	COLOR CODING	OTHER CODES
3	0	0		0 = Minimum Hazard 1 = Slight Hazard 2 = Moderate Hazard 3 = Serious Hazard 4 = Severe Hazard	HEALTH = BLUE FIRE = RED REACTIVITY = YELLOW OTHER = WHITE	OX = Oxidizer ACID = Acid ALK = Alkali COR = Corrosive W = Use No Water

SECTION I - GENERAL INFORMATION

Characterization	**Metal**	RCRA Number	**None**
		EPA Class	**Not Applicable**
DOT Proper Shipping Name	**Cadmium**	Chemical Abstract Service (CAS) Number	**7440-43-9**
DOT Hazard Class and Label Requirements	**Poisonous Solid, N.O.S.**	DOT Emergency Guide Code	**53**
DOT Identification Number	**UN 2570**	Atomic Symbol	**Cd**

Synonyms
Not Applicable.

SECTION II - HAZARDOUS INGREDIENTS AND IDENTITY INFORMATION

Hazardous Components (specific identity)	OSHA Exposure Criteria	NIOSH Exposure Criteria	Immediately Dangerous to Life and Health (IDLH)	ACGIH Exposure Criteria
Cadmium (derivation: Dust or fume from roasting zinc ores is collected, mixed with coal or coke and sodium or zinc chloride, and sintered. The cadmium fume is collected in an electrostatic precipitator, leached, fractionally precipitated, and distilled; by direct distillation of cadmium-bearing zinc; recovery from electrolytic zinc process).	**PEL (Cd):** **fume:** **0.005 mg/m³** **dust:** **0.002 mg/m³** **STEL:** **0.6 mg/m³**	**REL:** **Cancer Causing Agent** **Reduce to Lowest Possible Level**	**9 mg/m³**	**TLV (Cd):** **dusts & salts:** **0.05 mg/m³** **STEL:** **Not Established**

SECTION III - PHYSICAL AND CHEMICAL CHARACTERISTICS

Boiling Point	**1409°F (765°C)**	Specific Gravity (H_2O = 1)	**8.65**
Vapor Pressure (mm Hg)	**0.095 at 610°F (321°C)**	Molecular Weight (atomic weight)	**112.4**
Vapor Density (Air = 1)	**Not Applicable**	Melting Point	**610°F (321°C)**

Solubility
Insoluble to slightly soluble in water.

Appearance and Odor
Bluish-white or silver-white, malleable metal. May also appear as a gray-black powder. Odorless.

SECTION IV - FIRE AND EXPLOSION HAZARD DATA

Flash Point (method used)	**Not Found (Metal)**	Explosive Limits in Air % by Volume	**LEL: N/A UEL: N/A**
NFPA Classification	**Flammable Solid (in powder form)**	Autoignition Temperature	**Not Found**

Extinguishing Media
For dust, use carbon dioxide or dry chemical. For metal, use agent suitable to surrounding fire. Do NOT use water. A class "D" extinguisher may be required.

Special Fire Fighting Procedures
Wear full protective clothing and self-contained breathing apparatus (SCBA). Poisonous gases are produced in fire.

Unusual Fire and Explosion Hazards
The dust ignites spontaneously in air and is flammable and explosive when exposed to heat, flame, or by chemical reaction with oxidizing agents and other metals.

SECTION V - REACTIVITY DATA

Stability		Conditions to Avoid Avoid contact with incompatible materials. Keep cadmium powders away from open flame, heat, or other sources of ignition.
Stable	Unstable	Incompatibility *(materials to avoid)*
X		Violent or explosive reaction in contact with oxidizers (such as perchlorates, peroxides, permanganates, chlorates, and nitrates), acids (sulfuric, hydroazoic, nitric), sulfur, selenium, tellurium, ammonia nitrate, zinc.
Hazardous Polymerization		Conditions to Avoid Hazardous polymerization of cadmium will not occur.
May Occur	Will Not Occur	Hazardous Decomposition or By-products
	X	Toxic cadmium fumes are produced in fires.

SECTION VI - HEALTH HAZARD DATA

Primary Route(s) of Entry:	Inhalation?	Absorption (skin/eye)?	Ingestion?
	X		X

Health Hazards

INHALATION: Breathing the dusts of cadmium may cause cough, tightness of chest, respiratory distress, congestion of the lungs, and broncho-pneumonia. Large doses can be fatal. May cause fluid buildup in the lungs (pulmonary edema).

EYE & SKIN: Possible irritation to the skin and eyes. An allergen.

INGESTION: A poison by ingestion. May cause toxic systemic effects.

Carcinogenicity	NTP Listed?	IARC Cancer Review Group?	OSHA Regulated?	Target Organs?
Confirmed Human Confirmed Animal	5th Annual Report	Group 2A	29 CFR 1910.1027	Lungs, liver, kidney, bone, skin, eye, prostate.

Medical Conditions Generally Aggravated by Exposure

None reported. However, persons with existing lung dysfunction may experience effects quicker.

Emergency and First-aid Procedures

Eye contact: Do not allow victim to keep eyes tightly shut. Flush immediately with water for 15 minutes (minimum), seek medical attention. Skin contact: Remove all contaminated clothing. Immediately wash area with flooding amounts of soap and water. For inhalation: Remove the person from exposure. Provide respiratory assistance and CPR. Transfer to medical facility. Observe 24 - 48 hours for lung effects (pulmonary edema likely). If swallowed: Seek medical attention immediately. Call poison control center for advice. Never give anything by mouth to an unconscious or convulsing person.

SECTION VII - PRECAUTIONS FOR SAFE HANDLING AND USE

Steps to be Taken in Case Material is Released or Spilled

Collect powdered materials in most convenient and safe manner and deposit in sealed containers. Dry sweep not recommended. Use HEPA vacuum. Remove all ignition sources. Restrict those not involved in cleanup from entering area. Notify appropriate authorities, as required by applicable regulations.

Preferred Waste Disposal Method

No citation.

Precautions to be Taken in Handling and Storage

Do not store near heat or other sources of ignition. A regulated, marked area must be established where cadmium is used, handled, or stored.

Other Precautions and Warnings

Store in tightly sealed drums in a cool, dry, well ventilated place.

SECTION VIII - CONTROL MEASURES AND PERSONAL PROTECTIVE EQUIPMENT

Respiratory Protection *(specify type)*

For any exposure level use a NIOSH/MSHA approved self-contained breathing apparatus with full face-piece operated in positive pressure mode. Cadmium is a carcinogen. Inhalation of its vapors must be avoided.

Ventilation

Local exhaust (preferred) at site of chemical work or general ventilation.

Protective Gloves	Eye Protection	Other Protective Clothing
Impervious Material	Chemical/Dust Goggles or Face Mask	Protective Apron, Boots (avoid skin contact)

Work/Hygiene Practices

Always wash hands thoroughly after using chemical; never bring food, drink, or smoking materials into vicinity of chemicals.

CADMIUM

Cd CAS: 13967-90-3

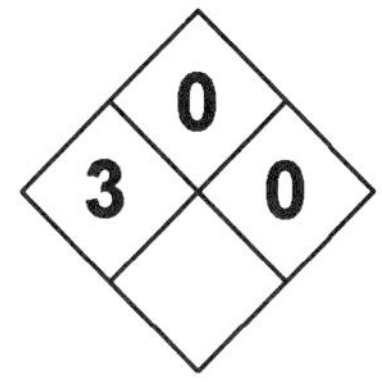

IDENTIFICATION AND TYPICAL USES

Cadmium is a silver-white, blue-tinged lustrous, malleable metal with no odor. Also appears as a gray-black powder. It is used in electroplating, in nickel-cadmium storage batteries, as a coating agent for other metals, in bearing and low-melting point alloys, and as control rods in nuclear reactors. Cadmium compounds have numerous applications, including dyeing and printing textiles, as TV phosphors, as pigments and enamels, in fire protection systems, in power transmission wires, in photography, as a fungicide additive, in electrodes for cadmium vapor lamps and photoelectric cells, and in semiconductors and solar cells.

RISK ASSESSMENT: HEALTH

General Assessment

There are several reports in the references of cadmium poisoning and human death. Cadmium is a confirmed human carcinogen with experimental carcinogenic, tumorigenic, mutagenic, and neoplastigenic data also being reported. Cadmium can enter the body by *inhalation* of its dusts or fumes, or by *ingestion.* In humans, the toxic symptoms of acute exposure may include nausea, vomiting, diarrhea, abdominal pain, muscular ache, salivation, and shock. In addition, inhalation of its fumes and dusts can cause coughing, tightness in the chest, respiratory distress, congestion of the lungs, and broncho-pneumonia. Inhalation causes an excess of protein to be present in the urine. Exposure to its fumes or dusts of 30 minutes to amounts of as little as 50 mg/m^3 can be fatal to humans.

Cadmium is a poison that is accumulated in the liver and kidneys. Chronic poisoning can therefore lead to liver and kidney damage. Its biological half-life in humans is estimated at approximately 20-30 years. Cadmium is also known to cause the so-called "itai, itai" disease, (a bone disease with kidney malfunction).

During welding or grinding operations, cadmium fumes are generated. Inhalation can cause a flu-like illness with chills, headache, aching and/or fever. This can lead to more serious illness.

Acute Health Effects

The following acute (short-term) health effects may occur immediately or shortly after exposure to cadmium:

Skin: Possible irritation may occur.

Eye: Irritation and possible damage on contact.

Lung: High exposures can cause rapid and severe lung damage, with shortness of breath, chest pain, cough, and a possible buildup of fluids in the lungs (pulmonary edema), which is a medical emergency. In serious cases, death or permanent lung damage can occur. Illness can be delayed for 48 hours, allowing overexposures to occur without warning signs.

Other: High exposure to cadmium may cause nausea, cramps, salivation, vomiting, and diarrhea.

Chronic Health Effects

The following chronic (long-term) health effects may occur at some time after exposure to cadmium and can last for months or even years:

Cancer Hazards: Cadmium, and especially cadmium oxide, is a confirmed human carcinogen. There is sufficient evidence that it causes prostate and kidney cancer in humans. It has also been shown to cause lung and testes cancer in animals.

Reproductive Hazard: Cadmium is a known teratogen in animals and a probable teratogen in humans. It may damage the testes (male reproductive glands) and may affect the female reproductive cycle.

Other Chronic Effects: Repeated low level exposures can cause permanent kidney damage which can go unnoticed for some time without proper testing. The kidney damage can lead to kidney stones and other more serious health problems. Emphysema or lung scarring can occur from a single high exposure or repeated low exposures. Long-term exposure can cause anemia, loss of sense of smell, fatigue, and/or yellow staining of the teeth.

Recommended Risk-Reduction Measures

Personnel should avoid direct contact with cadmium dusts or powders. If a less toxic material or compound cannot be substituted for cadmium, then *engineering controls* are the most effective method of reducing exposures. The best protection is to enclose operations and/or provide local exhaust ventilation at the site of cadmium dust release. While not always operationally feasible, isolating operations can also reduce exposure. Using respiratory protection is less effective than the controls mentioned above, but is still advisable whenever working with or around cadmium. At any exposure level, a self-contained breathing apparatus (SCBA) with full facepiece and pressure demand is recommended for the greatest possible respiratory protection. If a full facepiece is not available, then chemical/dust goggles should be worn to protect the eyes. A face shield and protective apron should also be worn. To prevent hand and skin exposures, gloves should be worn.

Administrative controls should also be in place to minimize the potential for human exposures. These may include written procedures or policies which specify the methods and techniques that will be practiced whenever personnel are to work with cadmium. As per OSHA 29 CFR 1910.1027 (Cadmium), a regulated and controlled work area must be established wherever cadmium is used, handled, or stored. Unauthorized personnel are not permitted in this area.

All personnel should receive *training* on the use, hazards, protective measures, emergency actions, and other precautions per 29 CFR 1910.1200 (Hazard Communication), prior to the first assignment in an area where cadmium is used or stored.

Before beginning employment working with cadmium, and at regular intervals thereafter (e.g., annually), those personnel with frequent or potentially high exposures should be provided the following recommended medical tests:

☑ Lung function test.

☑ Urine test for cadmium (levels should be less than 10 micrograms per liter of urine).

☑ Urine test for "low molecular weight proteins" to detect kidney damage.

☑ Complete urinalysis (UA).

If overexposure is suspected or symptoms develop, the following additional test is recommended:

☑ Complete blood count (CBC).

Any evaluation should include a careful history of past and present symptoms with an examination. Medical tests that simply look for existing damage are not a substitute for controlling exposures.

Cigarette smoke contains some cadmium. Because it is hard for the body to eliminate cadmium, it tends to build up in the body. Any workplace exposure will therefore add to any existing levels. Smoking or carrying cigarettes near cadmium increases release of toxic fumes. Also, since smoking can cause heart disease as well as lung cancer, emphysema, and other known respiratory problems, exposure to cadmium may cause additional complications. Therefore, personnel who smoke should avoid any unprotected contact with cadmium powders.

Other methods to reduce exposure include:

☑ Where possible, enclose operations and use local exhaust ventilation at the site of chemical release. If local exhaust ventilation or enclosure is not used, respiratory protection should be mandatory.

☑ Always ensure that proper protective clothing is worn when using chemical substances.

☑ Wash thoroughly immediately after exposure to cadmium and at the end of the work shift or before eating, drinking, or smoking.

☑ Hazard warning information should be posted in the work area. In addition, as part of an on-going education and training program, all information on the health and safety hazards of cadmium should be communicated to all exposed and potentially exposed workers.

☑ Eye wash stations should be provided in the immediate work area for emergency use. If there is a possibility of skin exposure to cadmium, emergency shower facilities should also be provided.

☑ Workers whose clothing has been contaminated by cadmium should change into clean clothes before leaving work. Contaminated work clothing should be laundered only by individuals who have been informed of the hazards of exposure to cadmium.

RISK ASSESSMENT: ENVIRONMENT

General Assessment

Cadmium is a naturally occurring element. However, industrial uses place the environment at risk of exposure during transportation, storage, disposal, or destruction of cadmium compounds. In almost every scenario, the threat of environmental exposure is contingent upon the proper handling of the chemical substance. Accidental spills, large or small, where cadmium contacts incompatible commodities can result in fire, explosion (depending upon conditions of spill), and possible contamination of the surrounding environmental mediums (water, soil, and air).

In its bulk form, cadmium is considered a noncombustible solid. However, its dusts can react explosively and present a serious fire hazard. Because it is incompatible with many commodities, including some other common metals, extreme caution is required in handling, storage, transportation, and disposal of cadmium. These characteristics also require special consideration during any emergency situation involving a leak or spill of cadmium powder or dust. Should cadmium ever come into contact with any incompatible substances either during use, transportation, storage, or disposal, violent and even explosive reactions are possible. The dust ignites spontaneously in air and is flammable and explosive when exposed to heat, flame, or by chemical reaction with oxidizing agents, metals (zinc, selenium, tellurium, elements of sulfur), strong oxidizers, and acids (explodes on contact with hydrazoic acid).

Cadmium can enter the environment from industrial and municipal waste treatment plant discharges and from spills.

Acute Ecological Effects

Acute (short-term) toxic effects may include the death of animals, birds, or fish and death or low growth rate in plants. Acute effects are seen 2 to 4 days after animals or plants are exposed to cadmium. In fresh water, cadmium toxicity is influenced by water hardness. The harder the water, the lower the toxicity. Cadmium has high acute toxicity to aquatic life. Insufficient data are available to evaluate the short-term effects of cadmium to plants, birds, or land animals.

Chronic Ecological Effects

Chronic (long-term) toxic effects may include shortened life span, reproductive problems, lower fertility, and changes in appearance or behavior in exposed animals. These effects can be seen long after first exposure(s) to toxic chemicals. Cadmium has high chronic toxicity to aquatic life. Insufficient data are available to evaluate or predict the long-term effects of cadmium to plants, birds, or land animals.

Water Solubility

Most of the cadmium compounds are slightly soluble in water. Concentrations of less than 1 milligram will mix with a liter of water.

Persistence in the Environment

Cadmium is highly persistent in water, with a half-life greater than 200 days. The half-life of a pollutant is the amount of time it takes for one half of the chemical to be degraded. No data are available on the estimated percentages of cadmium persistence in the aquatic or terrestrial environments.

Bioaccumulation in Aquatic Organisms

Some substances increase in concentration, or *bioaccumulate,* in living organisms as they breathe contaminated air, drink contaminated water, or eat contaminated food. These chemicals can become concentrated in the tissues and internal organs of animals as well as humans.

The concentration of cadmium found in fish tissues is expected to be much higher than the average concentration of cadmium in the water from which the fish was taken.

Recommended Risk-Reduction Measures

Proper training of all transporters will reduce the likelihood of a mishap or accident resulting in a leak or spill of cadmium dusts or powders into the environment. The correct use of DOT labeling (per the MSDS) on all containers, trucks, and rail cars will enable emergency responders to react properly and quickly to any disaster thereby reducing the potential risk to the environment and to personnel.

Storage of cadmium should be segregated from incompatible chemicals to minimize the risk of cross-contamination or contact.

If a spill or leak to the environment has occurred, fire department, emergency response and/or hazardous materials spill personnel should be notified immediately. Cleanup should be attempted only by those trained in proper spill containment procedures. If cadmium should contact the water table, aquifer, or navigable waterway, time is of the essence. It is highly soluble and total remediation may not be possible. The local and/or state emergency response authorities must be notified. A comprehensive emergency response or disaster preparedness/recovery plan should be in place prior to any operations involving the use, transportation, storage, or disposal of cadmium. If cadmium powder is spilled or leaked, the following specific steps are recommended:

- ☑ Restrict persons not wearing protective clothing from area of spill or leak until cleanup is complete.
- ☑ Remove all ignition sources.
- ☑ Collect powdered materials in the safest manner possible and deposit in sealed containers. Recommend using vacuum with HEPA filter. Do not dry sweep (generates airborne dusts).
- ☑ It may be necessary to dispose of cadmium as a hazardous waste. The responsible state agency or the regional office of the federal Environmental Protection Agency (EPA) should be contacted for specific recommendations.

RISK ASSESSMENT: BUSINESS

General Assessment

Accidents or mishaps involving cadmium can present a significant threat to business operations. The loss or damage of equipment or facilities can significantly affect fiscal viability. Lawsuits that may result from personnel injury/death, public exposures, and/or environmental contamination will also require a serious expenditure of resources. Media attention surrounding an injury, death, or environmental damage can also result in a loss of profits and loss of current as well as future business.

Recommended Risk-Reduction Measures

Company attorneys, safety and health professionals, and environmental specialists should be involved in the development of any procedures for responding to chemical incidents. A company official should be designated as a public relations officer with specific training in dealing with the press. Corporate plans and policies should be developed, approved, and implemented long before any need for such arises.

REFERENCES

American Conference of Governmental Industrial Hygienists. 1988. *Documentation of the Threshold Limit Values and Biological Exposure Indices,* 5th Edition (with updates). Cincinnati: ACGIH

Brusick, D. J. 1994. *Methods for Genetic Risk Assessment.* New York: CRC Press/Lewis Publishers.

Bueche, J. R. 1972. *Principles of Physics,* 2nd Edition. New York: McGraw-Hill.

Calabrese, E. J. 1994. *Biological Effects of Low Level Exposures.* New York: CRC Press/Lewis Publishers.

Cothern, R. R. 1993. *Comparative Environmental Risk Assessment.* New York: CRC Press/Lewis Publishers.

Gardner, R. W. 1994. *Chemical Intolerance.* New York: CRC Press/Lewis Publishers.

Hallenbeck, W. H. 1993. *Quantitative Risk Assessment for Environmental and Occupational Health,* 2nd Edition. New York: CRC Press/Lewis Publishers.

Howard, P. H. 1995. *Handbook of Environmental Fate and Exposures Data for Organic Chemicals,* Vol. I-IV. New York: CRC Press/Lewis Publishers.

Lewis, R. J., Sr. 1992. *Sax's Dangerous Properties of Industrial Materials,* Volumes 1, 2, and 3, 8th Edition. New York: Van Nostrand Reinhold.

Lide, D. R. 1995. *Handbook of Chemistry and Physics,* 76th Edition. New York: CRC Press/Lewis Publishers.

Manahan, S. E. 1989. *Toxicological Chemistry.* New York: CRC Press/Lewis Publishers.

Meyer, E. 1990. *Chemistry of Hazardous Materials.* Englewood Cliffs, NJ: Prentice Hall.

Patnaik, P. 1992. *A Comprehensive Guide to the Hazardous Properties of Chemical Substances.* New York: Van Nostrand Reinhold.

Proctor, N. H., J. P. Hughes. 1991. *Chemical Hazards of the Workplace.* New York: Van Nostrand Reinhold.

Quigley, D. R. 1994. *Handbook of Emergency Chemical Management.* New York: CRC Press/Lewis Publishers.

Reeves, A. L. 1979. Cadmium. *Handbook on the Toxicology of Metals.* pp 321-328. New York: Elsevier North-Holland.

Richardson, M. 1995. *Dictionary of Substance and Their Effects,* Vol. I-VI. New York: CRC Press/Lewis Publishers.

Sacarello, H. L. A. 1994. *The Comprehensive Handbook of Hazardous Materials.* New York: CRC Press/Lewis Publishers.

Torkelson, T. R. V. K. Rowe. 1981. *Patty's Industrial Hygiene and Toxicology,* Vol. 2B. New York: Wiley-Interscience.

MATERIAL SAFETY DATA SHEET

CHEMICAL NAME

CADMIUM ACETATE

CRC PRESS/LEWIS PUBLISHERS DISCLAIMER: THE INFORMATION AND RECOMMENDATIONS PRESENTED HEREIN ARE BASED ON SOURCES BELIEVED TO BE RELIABLE. CRC MAKES NO REPRESENTATION ON ITS COMPLETENESS OR ACCURACY. IT IS THE RESPONSIBILITY OF THE USER TO DETERMINE THE CHEMICAL'S SUITABILITY FOR ITS INTENDED USE, THE CHEMICAL'S SAFE USE, AND THE CHEMICAL'S PROPER DISPOSAL. NO REPRESENTATIONS AND/OR WARRANTIES, EITHER EXPRESSED OR IMPLIED, OF THE MERCHANTABILITY OR FITNESS FOR A PARTICULAR PURPOSE, OR OF ANY OTHER NATURE, ARE MADE WITH RESPECT TO THE INFORMATION PROVIDED IN THIS MSDS OR TO THE CHEMICAL TO WHICH INFORMATION MAY REFER. CRC NEITHER ASSUMES NOR AUTHORIZES ANY OTHER PERSON TO ASSUME FOR IT, ANY OTHER ADDITIONAL RESPONSIBILITY OR LIABILITY FOR THE USE OF, OR RELIANCE UPON, THIS INFORMATION.

HAZARD WARNING INFORMATION

HEALTH	FIRE	REACTIVE	OTHER	DEGREE OF HAZARD	COLOR CODING	OTHER CODES
3	0	0		0 = Minimum Hazard 1 = Slight Hazard 2 = Moderate Hazard 3 = Serious Hazard 4 = Severe Hazard	HEALTH = BLUE FIRE = RED REACTIVITY = YELLOW OTHER = WHITE	OX = Oxidizer ACID = Acid ALK = Alkali COR = Corrosive ~~W~~ = Use No Water

SECTION I - GENERAL INFORMATION

Chemical Family	**Metal**	RCRA Number	**Not Listed**
		EPA Class	**Not Applicable**
DOT Proper Shipping Name	**Cadmium Acetate**	Chemical Abstract Service (CAS) Number	**543-90-8**
DOT Hazard Class and Label Requirements	**Not Determined**	DOT Emergency Guide Code	**53**
DOT Identification Number	**NA 2570**	Atomic Symbol	**$C_2H_4O_2$•1/2Cd**

Synonyms
Cadmium acetate dihydrate, bis(acetoxy)cadmium, cadmium diacetate.

SECTION II - HAZARDOUS INGREDIENTS AND IDENTITY INFORMATION

Hazardous Components (specific identity)	OSHA Exposure Criteria	NIOSH Exposure Criteria	Immediately Dangerous to Life and Health (IDLH)	ACGIH Exposure Criteria
Cadmium (derivation: Interaction of acetic acid and cadmium oxide).	**PEL (Cd):** **fume:** **0.005 mg/m³** **dust:** **0.002 mg/m³** **STEL:** **0.6 mg/m³**	**REL:** **Cancer Causing Agent** **Reduce to Lowest Possible Level**	**9 mg(Cd)/m³**	**TLV (Cd):** **dusts & salts:** **0.05 mg/m³** **STEL:** **Not Established**

SECTION III - PHYSICAL AND CHEMICAL CHARACTERISTICS

Boiling Point	**Decomposes**	Specific Gravity (H_2O = 1)	**2.341**
Vapor Pressure (mm Hg)	**Not Applicable**	Molecular Weight (atomic weight)	**116.25**
Vapor Density (Air = 1)	**8.0**	Melting Point	**493°F (256°C)**

Solubility in Water
Soluble in water and alcohols.

Appearance and Odor
Monoclinic colorless crystalline solid with odor of acetic acid.

SECTION IV - FIRE AND EXPLOSION HAZARD DATA

Flash Point (method used)	**Not Applicable (Metal)**	Explosive Limits in Air % by Volume	**LEL: N/A UEL: N/A**
NFPA Classification	**Flammable Solid (in powder form)**	Autoignition Temperature	**Not Found**

Extinguishing Media
For dust, use carbon dioxide or dry chemical. For metal, use agent suitable to surrounding fire. Do NOT use water. A class "D" extinguisher may be required.

Special Fire Fighting Procedures
Wear full protective clothing and self-contained breathing apparatus (SCBA). Poisonous gases are produced in fire.

Unusual Fire and Explosion Hazards
The dust ignites spontaneously in air and is flammable and explosive when exposed to heat, flame, or by chemical reaction with oxidizing agents and other metals.

SECTION V - REACTIVITY DATA

Stability		Conditions to Avoid
		Avoid contact with incompatible materials. Keep cadmium powders away from open flame, heat, or other sources of ignition.
Stable	Unstable	Incompatibility *(materials to avoid)*
X		Violent or explosive reaction in contact with oxidizers (such as perchlorates, peroxides, permanganates, chlorates, and nitrates), acids (sulfuric, hydroazoic, nitric), sulfur, selenium, tellurium, ammonia nitrate, zinc.
Hazardous Polymerization		Conditions to Avoid Hazardous polymerization of cadmium will not occur.
May Occur	Will Not Occur	Hazardous Decomposition or By-products
	X	Toxic cadmium fumes are produced in fires.

SECTION VI - HEALTH HAZARD DATA

Primary Route(s) of Entry:	Inhalation?	Absorption (skin/eye)?	Ingestion?
	X		X

Health Hazards

INHALATION: Breathing the dusts of cadmium may cause cough, tightness of chest, respiratory distress, congestion of the lungs, and broncho-pneumonia. Large doses can be fatal. May cause fluid buildup in the lungs (pulmonary edema). Repeated low exposures can cause permanent kidney damage, emphysema, low blood sugar, and/or loss of smell.

EYES & SKIN: Possible irritation to the skin and eyes. An allergen.

INGESTION: A poison by ingestion. May cause toxic systemic effects.

Carcinogenicity	NTP Listed?	IARC Cancer Review Group?	OSHA Regulated?	Target Organs?
Confirmed Human Confirmed Animal	5th Annual Report	Group 2A	Refer to: 29 CFR 1910.1027	Lungs, liver, kidney, bone, skin, eye, prostate.

Medical Conditions Generally Aggravated by Exposure

None reported. However, persons with existing lung dysfunction may experience effects quicker.

Emergency and First-aid Procedures

Eye contact: Do not allow victim to keep eyes tightly shut. Flush immediately with water for 15 minutes (minimum), seek medical attention. Skin contact: Remove all contaminated clothing. Immediately wash area with flooding amounts of soap and water. For inhalation: Remove the person from exposure. Provide respiratory assistance and CPR. Transfer to medical facility. Observe 24-48 hours for lung effects (pulmonary edema likely). If swallowed: Seek medical attention immediately. Call poison control center for advice. Never give anything by mouth to an unconscious or convulsing person.

SECTION VII - PRECAUTIONS FOR SAFE HANDLING AND USE

Steps to be Taken in Case Material is Released or Spilled

Collect powdered materials in most convenient and safe manner and deposit in sealed containers. Dry sweep not recommended. Use HEPA vacuum. Remove all ignition sources. Restrict those not involved in cleanup from entering area. Notify appropriate authorities, as required by applicable regulations.

Preferred Waste Disposal Method

No citation.

Precautions to be Taken in Handling and Storage

Do not store near heat or other sources of ignition. A regulated, marked area must be established where cadmium is used, handled, or stored.

Other Precautions and Warnings

Store in tightly sealed drums in a cool, dry, well ventilated place.

SECTION VIII - CONTROL MEASURES AND PERSONAL PROTECTIVE EQUIPMENT

Respiratory Protection *(specify type)*

For any exposure level use a NIOSH/MSHA approved self-contained breathing apparatus with full facepiece operated in positive pressure mode. Cadmium is a carcinogen. Inhalation of its vapors must be avoided.

Ventilation

Local exhaust (preferred) at site of chemical work or general ventilation.

Protective Gloves	Eye Protection	Other Protective Clothing
Impervious Material	Chemical/Dust Goggles or Face Mask	Protective Apron, Boots (avoid skin contact)

Work/Hygiene Practices

Always wash hands thoroughly after using chemical; never bring food, drink, or smoking materials into vicinity of chemicals.

CADMIUM ACETATE

$C_2H_4O_2 \cdot 1/2Cd$ CAS: 543-90-8

IDENTIFICATION AND TYPICAL USES

Cadmium acetate appears as a monoclinic, sugar or sand-like colorless, crystalline solid with an odor of acetic acid. It is used in electroplating, dyeing and printing textiles, in ceramics, and to make other acetate compounds.

RISK ASSESSMENT: HEALTH

General Assessment

There are several reports in the references of cadmium poisoning and human death. Cadmium acetate is a confirmed human carcinogen with experimental carcinogenic, tumorigenic, mutagenic, and neoplastigenic data also being reported. Cadmium acetate can enter the body by *inhalation* of its dusts or fumes, or by *ingestion.* In humans, the toxic symptoms of acute exposure may include nausea, vomiting, diarrhea, abdominal pain, muscular ache, salivation, and shock. In addition, inhalation of its fumes and dusts can cause cough, tightness in the chest, respiratory distress, congestion of the lungs, and broncho-pneumonia. Inhalation causes an excess of protein to be present in the urine. Exposure to its fumes or dusts of 30 minutes to amounts of as little as 50 mg/m^3 can be fatal to humans.

Cadmium acetate is a poison that accumulates in the liver and kidneys. Chronic poisoning can therefore lead to liver and kidney damage, emphysema, low blood sugar, and/or loss of smell. Serious damage can occur at levels below the OSHA PEL. Its biological half-life in humans is estimated at approximately 20-30 years. Cadmium is also know to cause the so-called "itai, itai" disease, which is a bone disease together with kidney malfunction.

During welding or grinding operations, cadmium fumes are generated. Inhalation can cause a flu-like illness with chills, headache, aching and/or fever. This can lead to more serious illness.

Acute Health Effects

The following acute (short-term) health effects may occur immediately or shortly after exposure to cadmium:

Skin: Possible irritation may occur.

Eye: Irritation and possible damage on contact.

Lung: High exposures can cause rapid and severe lung damage, with shortness of breath, chest pain, cough, and a possible buildup of fluids in the lungs (pulmonary edema), which is a medical emergency. In serious cases, death or permanent lung damage can occur. Illness can be delayed for 4 to 8 hours, allowing overexposures to occur without warning signs.

Other: High exposure to cadmium may cause nausea, cramps, salivation, vomiting, and diarrhea.

Chronic Health Effects

The following chronic (long-term) health effects may occur at some time after exposure to cadmium and can last for months or even years:

Cancer Hazards: Cadmium acetate is a confirmed human carcinogen. There is sufficient evidence that it causes prostate and kidney cancer in humans. It has also been shown to cause lung and testes cancer in test animals.

Reproductive Hazard: Cadmium acetate should be handled as a potential teratogenic agent and reproductive hazard since several related cadmium compounds are known teratogens and decreases fertility in males and females. It may damage the testes (male reproductive glands) and may affect the female reproductive cycle in humans.

Other Chronic Effects: Repeated low level exposures can cause permanent kidney damage which can go unnoticed for some time without proper testing. The kidney damage can lead to kidney stones and other more serious health problems. Emphysema or lung scarring can occur from a single high exposure or repeated low exposures. Long-term exposure can cause anemia, loss of sense of smell, fatigue, and/or yellow staining of the teeth.

↻ *Recommended Risk-Reduction Measures*

Personnel should avoid direct contact with cadmium acetate dusts or powders. If a less toxic material or compound cannot be substituted for cadmium acetate, then *engineering controls* are the most effective method of reducing exposures. The best protection is to enclose operations and/or provide local exhaust ventilation at the site of cadmium dust release. While not always operationally feasible, isolating operations can also reduce exposure. Using respiratory protection is less effective than the controls mentioned above, but is still advisable whenever working with or around cadmium. At any exposure level, a self-contained breathing apparatus (SCBA) with full facepiece and pressure demand is recommended for the greatest possible respiratory protection. If a full facepiece is not available, then chemical/dust goggles should be worn to protect the eyes. A face shield and protective apron should also be worn. To prevent hand and skin exposures, gloves should be worn.

Administrative controls should also be in place to minimize the potential for human exposures. These may include written procedures or policies which specify the methods and techniques that will be practiced whenever personnel are to work with cadmium acetate. As per OSHA 29 CFR 1910.1027 (Cadmium), a regulated and controlled work area must be established wherever cadmium is used, handled, or stored. Unauthorized personnel are not permitted in this area.

All personnel should receive *training* on the use, hazards, protective measures, emergency actions, and other precautions per 29 CFR 1910.1200 (Hazard Communication), prior to the first assignment in an area where cadmium acetate is used or stored.

Before beginning employment working with cadmium acetate, and at regular intervals thereafter (e.g., annually), those personnel with frequent or potentially high exposures should be provided the following recommended medical tests:

- ☑ Lung function test.
- ☑ Urine test for cadmium (levels should be less than 10 micrograms per liter of urine).
- ☑ Urine test for "low molecular weight proteins" to detect kidney damage.
- ☑ Complete urinalysis (UA).

If overexposure is suspected or symptoms develop, the following additional test is recommended:

- ☑ Complete blood count (CBC).

Any evaluation should include a careful history of past and present symptoms with an examination. Medical tests that simply look for existing damage are not a substitute for controlling exposures.

Cigarette smoke contains some cadmium. Because it is hard for the body to eliminate cadmium, it tends to buildup in the body. Any workplace exposure will therefore add to any existing levels. Smoking or carrying cigarettes near cadmium acetate increases release of toxic fumes. Also, since smoking can cause heart disease as well as lung cancer, emphysema, and other known respiratory problems, exposure to cadmium acetate may cause additional complications. Therefore, personnel who smoke should avoid any unprotected contact with cadmium acetate powders.

Other methods to reduce exposure include:

- ☑ Where possible, enclose operations and use local exhaust ventilation at the site of chemical release. If local exhaust ventilation or enclosure is not used, respiratory protection should be mandatory.
- ☑ Always ensure that proper protective clothing is worn when using chemical substances.
- ☑ Wash thoroughly immediately after exposure to cadmium acetate and at the end of the work shift or before eating, drinking, or smoking.
- ☑ Hazard warning information should be posted in the work area. In addition, as part of an on-going education and training program, all information on the health and safety hazards of cadmium acetate should be communicated to all potentially exposed workers.
- ☑ Eye wash stations should be provided in the immediate work area for emergency use. If there is a possibility of skin exposure to cadmium acetate, emergency shower facilities should also be provided in the immediate area.
- ☑ Workers whose clothing has been contaminated by cadmium acetate should change into clean clothes before leaving work. Contaminated work clothing should be laundered only by individuals who have

been informed of the hazards of exposure to cadmium acetate.

RISK ASSESSMENT: ENVIRONMENT

General Assessment

Cadmium are naturally occurring element. However, industrial uses place the environment at risk of exposure during transportation, storage, disposal, or destruction of cadmium compounds, including cadmium acetate. In almost every scenario, the threat of environmental exposure is contingent upon the proper handling of the chemical substance. Accidental spills, large or small, where cadmium acetate contacts incompatible commodities can result in fire, explosion (depending upon conditions of spill), and possible contamination of the surrounding environmental mediums (water, soil, and air).

In its bulk form, cadmium acetate is considered a non-combustible solid. However, its dusts can react explosively and present a serious fire hazard. Because it is incompatible with many commodities, including some other common metals, extreme caution is required in handling, storage, transportation, and disposal of cadmium. These characteristics also require special consideration during any emergency situation involving a leak or spill of cadmium acetate powder or dust. Should cadmium acetate ever come into contact with any incompatible substances either during use, transportation, storage, or disposal, violent and even explosive reactions are possible. Cadmium acetate dust is flammable and explosive when exposed to heat, flame, or by chemical reaction with oxidizing agents, metals (zinc, selenium, tellurium, elements of sulfur), strong oxidizers, and acids.

Cadmium can enter the environment from industrial and municipal waste treatment plant discharges and from spills.

Acute Ecological Effects

Acute (short-term) toxic effects may include the death of animals, birds, or fish and death or low growth rate in plants. Acute effects are seen 2 to 4 days after animals or plants are exposed to cadmium. In fresh water, cadmium toxicity is influenced by water hardness. The harder the water, the lower the toxicity. Cadmium has high acute toxicity to aquatic life. Insufficient data are available to evaluate the short-term effects of cadmium to plants, birds, or land animals.

Chronic Ecological Effects

Chronic (long-term) toxic effects may include shortened life span, reproductive problems, lower fertility, and changes in appearance or behavior in exposed animals. These effects can be seen long after first exposure(s) to toxic chemicals. Cadmium has high chronic toxicity to aquatic life. Insufficient data are available to evaluate or predict the long-term effects of cadmium to plants, birds, or land animals.

Water Solubility

Most of the cadmium compounds are slightly soluble in water. Concentrations of less than 1 milligram will mix with a liter of water.

Persistence in the Environment

Cadmium is highly persistent in water, with a half-life greater than 200 days. The half-life of a pollutant is the amount of time it takes for one half of the chemical to be degraded. No data are available on the estimated percentages of cadmium persistence in the aquatic or terrestrial environments.

Bioaccumulation in Aquatic Organisms

Some substances increase in concentration, or *bioaccumulate,* in living organisms as they breathe contaminated air, drink contaminated water, or eat contaminated food. These chemicals can become concentrated in the tissues and internal organs of animals as well as humans.

The concentration of cadmium found in fish tissues is expected to be much higher than the average concentration of cadmium in the water from which the fish was taken.

Recommended Risk-Reduction Measures

Proper training of all transporters will reduce the likelihood of a mishap or accident resulting in a leak or spill of cadmium acetate dusts or powders into the environment. The correct use of DOT labeling (per the MSDS) on all containers, trucks, and rail cars will enable emergency responders to react properly and quickly to any disaster thereby reducing the potential risk to the environment and to personnel.

Storage of cadmium acetate should be segregated from incompatible chemicals to minimize the risk of cross-contamination or contact.

If a spill or leak to the environment has occurred, fire department, emergency response and/or hazardous materials spill personnel should be notified immediately. Cleanup should be attempted only by those trained in proper spill containment procedures. If cadmium acetate should contact the water table, aquifer, or navigable waterway, time is of the essence. It is highly soluble and total remediation may not be possible. The local and/or state emergency response authorities must be notified. A comprehensive emergency response or disaster preparedness/recovery plan should be in place prior to any operations involving the use, transportation, storage, or disposal of cadmium acetate. If cadmium acetate powder is spilled or leaked, the following specific steps are recommended:

☑ Restrict persons not wearing protective clothing from area of spill or leak until cleanup is complete and area can be opened for normal work.

☑ Remove all ignition sources.

☑ Collect powdered materials in the safest manner possible and deposit in sealed containers. Recommend using vacuum with HEPA filter. Do not dry sweep (generates airborne dusts).

☑ It may be necessary to dispose of cadmium acetate as a hazardous waste. The responsible state agency or the regional office of the federal Environmental Protection Agency (EPA) should be contacted for specific recommendations.

RISK ASSESSMENT: BUSINESS

General Assessment

Accidents or mishaps involving cadmium acetate can present a significant threat to business operations. The loss or damage of equipment or facilities can significantly affect fiscal viability. Lawsuits that may result from personnel injury/death, public exposures, and/or environmental contamination will also require a serious expenditure of resources. Media attention surrounding an injury, death, or environmental damage can also result in a loss of profits and loss of current as well as future business.

Recommended Risk-Reduction Measures

Company attorneys, safety and health professionals, and environmental specialists should be involved in the development of any procedures for responding to chemical incidents. A company official should be predesignated as a public relations officer with specific training in dealing with the press. Corporate plans and policies should be developed, approved, and implemented long before any need for such arises.

REFERENCES

American Conference of Governmental Industrial Hygienists. 1988. *Documentation of the Threshold Limit Values and Biological Exposure Indices,* 5th Edition (with updates). Cincinnati: ACGIH

Brusick, D. J. 1994. *Methods for Genetic Risk Assessment.* New York: CRC Press/Lewis Publishers.

Bueche, J. R. 1972. *Principles of Physics,* 2nd Edition. New York: McGraw-Hill.

Calabrese, E. J. 1994. *Biological Effects of Low Level Exposures.* New York: CRC Press/Lewis Publishers.

Cothern, R. R. 1993. *Comparative Environmental Risk Assessment.* New York: CRC Press/Lewis Publishers.

Gardner, R. W. 1994. *Chemical Intolerance.* New York: CRC Press/Lewis Publishers.

Hallenbeck, W. H. 1993. *Quantitative Risk Assessment for Environmental and Occupational Health,* 2nd Edition. New York: CRC Press/Lewis Publishers.

Howard, P. H. 1995. *Handbook of Environmental Fate and Exposures Data for Organic Chemicals,* Vol. I-IV. New York: CRC Press/Lewis Publishers.

Lewis, R. J., Sr. 1992. *Sax's Dangerous Properties of Industrial Materials,* Volumes 1, 2, and 3, 8th Edition. New York: Van Nostrand Reinhold.

Lide, D. R. 1995. *Handbook of Chemistry and Physics,* 76th Edition. New York: CRC Press/Lewis Publishers.

Manahan, S. E. 1989. *Toxicological Chemistry.* New York: CRC Press/Lewis Publishers.

Meyer, E. 1990. *Chemistry of Hazardous Materials.* Englewood Cliffs, NJ: Prentice Hall.

Patnaik, P. 1992. *A Comprehensive Guide to the Hazardous Properties of Chemical Substances.* New York: Van Nostrand Reinhold.

Proctor, N. H., J. P. Hughes. 1991. *Chemical Hazards of the Workplace.* New York: Van Nostrand Reinhold.

Quigley, D. R. 1994. *Handbook of Emergency Chemical Management.* New York: CRC Press/Lewis Publishers.

Reeves, A. L. 1979. Cadmium. *Handbook on the Toxicology of Metals.* pp 321-328. New York: Elsevier North-Holland.

Richardson, M. 1995. *Dictionary of Substance and Their Effects,* Vol. I-VI. New York: CRC Press/Lewis Publishers.

Sacarello, H. L. A. 1994. *The Comprehensive Handbook of Hazardous Materials.* New York: CRC Press/Lewis Publishers.

Torkelson, T. R. V. K. Rowe. 1981. *Patty's Industrial Hygiene and Toxicology,* Vol. 2B. New York: Wiley-Interscience.

MATERIAL SAFETY DATA SHEET

CHEMICAL NAME	CRC PRESS/LEWIS PUBLISHERS DISCLAIMER:
CADMIUM BROMIDE	THE INFORMATION AND RECOMMENDATIONS PRESENTED HEREIN ARE BASED ON SOURCES BELIEVED TO BE RELIABLE. CRC MAKES NO REPRESENTATION ON ITS COMPLETENESS OR ACCURACY. IT IS THE RESPONSIBILITY OF THE USER TO DETERMINE THE CHEMICAL'S SUITABILITY FOR ITS INTENDED USE, THE CHEMICAL'S SAFE USE, AND THE CHEMICAL'S PROPER DISPOSAL. NO REPRESENTATIONS AND/OR WARRANTIES, EITHER EXPRESSED OR IMPLIED, OF THE MERCHANTABILITY OR FITNESS FOR A PARTICULAR PURPOSE, OR OF ANY OTHER NATURE, ARE MADE WITH RESPECT TO THE INFORMATION PROVIDED IN THIS MSDS OR TO THE CHEMICAL TO WHICH INFORMATION MAY REFER. CRC NEITHER ASSUMES NOR AUTHORIZES ANY OTHER PERSON TO ASSUME FOR IT, ANY OTHER ADDITIONAL RESPONSIBILITY OR LIABILITY FOR THE USE OF, OR RELIANCE UPON, THIS INFORMATION.

HAZARD WARNING INFORMATION

HEALTH	FIRE	REACTIVE	OTHER	DEGREE OF HAZARD	COLOR CODING	OTHER CODES
Not Found	Not Found	Not Found		0 = Minimum Hazard 1 = Slight Hazard 2 = Moderate Hazard 3 = Serious Hazard 4 = Severe Hazard	HEALTH = BLUE FIRE = RED REACTIVITY = YELLOW OTHER = WHITE	OX = Oxidizer ACID = Acid ALK = Alkali COR = Corrosive W̶ = Use No Water

SECTION I - GENERAL INFORMATION

Characterizaion **Metal**	RCRA Number **None**	EPA Class **Not Applicable**
DOT Proper Shipping Name **Cadmium Bromide**	Chemical Abstract Service (CAS) Number **7789-42-6**	
DOT Hazard Class and Label Requirements **Not Determined**	DOT Emergency Guide Code **53**	
DOT Identification Number **NA 2570**	Atomic Symbol **Br_2Cd**	

Synonyms
Cadmium dibromide; cadmium bromide tetrahydrate.

SECTION II - HAZARDOUS INGREDIENTS AND IDENTITY INFORMATION

Hazardous Components (specific identity)	OSHA Exposure Criteria	NIOSH Exposure Criteria	Immediately Dangerous to Life and Health (IDLH)	ACGIH Exposure Criteria
Cadmium (derivation: By heating cadmium in bromide vapor).	**PEL (Cd):** **fume:** 0.005 mg/m³ **dust:** 0.002 mg/m³ **STEL:** 0.6 mg/m³	**REL:** **Cancer Causing Agent** **Reduce to Lowest Possible Level**	**9 mg(Cd)/m³**	**TLV (Cd):** **dusts & salts:** 0.05 mg/m³ **STEL:** Not Established

SECTION III - PHYSICAL AND CHEMICAL CHARACTERISTICS

Boiling Point **1765°F (963°C)**	Specific Gravity (H_2O = 1) **1.1**
Vapor Pressure (mm Hg) **Not Applicable**	Molecular Weight (atomic weight) **272.22**
Vapor Density (Air = 1) **8.0**	Melting Point **1054°F (568°C)**

Solubility
Soluble in water, acid, and alcohol; moderately soluble in acetone; slightly soluble in ethanol.

Appearance and Odor
White to yellow efflorescent, pearly hexagonal crystalline solid, appearing as a powder. Hygroscopic.

SECTION IV - FIRE AND EXPLOSION HAZARD DATA

Flash Point (method used) **Not Applicable (Metal)**	Explosive Limits in Air % by Volume **LEL: N/A UEL: N/A**
NFPA Classification **Flammable Solid (in powder form)**	Autoignition Temperature **Not Found**

Extinguishing Media
For dust, use carbon dioxide or dry chemical. For metal, use agent suitable to surrounding fire. Do NOT use water. A class "D" extinguisher may be required.

Special Fire Fighting Procedures
Wear full protective clothing and self-contained breathing apparatus (SCBA). Poisonous gases are produced in fire.

Unusual Fire and Explosion Hazards
The dust ignites spontaneously in air and is flammable and explosive when exposed to heat, flame, or by chemical reaction with oxidizing agents and other metals.

SECTION V - REACTIVITY DATA

Stability		Conditions to Avoid **Avoid contact with incompatible materials. Keep cadmium powders away from open flame, heat, or other sources of ignition.**
Stable **X**	Unstable	Incompatibility *(materials to avoid)* **Violent or explosive reaction in contact with oxidizers (such as perchlorates, peroxides, permanganates, chlorates, and nitrates), acids (sulfuric, hydroazoic, nitric), sulfur, selenium, tellurium, ammonia nitrate, zinc.**
Hazardous Polymerization		Conditions to Avoid **Hazardous polymerization of cadmium will not occur.**
May Occur	Will Not Occur **X**	Hazardous Decomposition or By-products **Toxic cadmium fumes are produced in fires.**

SECTION VI - HEALTH HAZARD DATA

Primary Route(s) of Entry:	Inhalation?	Absorption (skin/eye)?	Ingestion?
	X		X

Health Hazards

INHALATION: Breathing the dusts of cadmium may cause cough, tightness of chest, respiratory distress, congestion of the lungs, and broncho-pneumonia. Large doses can be fatal. May cause fluid buildup in the lungs (pulmonary edema). Repeated low exposures can cause permanent kidney damage, emphysema, low blood sugar, and/or loss of smell.

EYE & SKIN: Possible irritation to the skin and eyes. An allergen.

INGESTION: A poison by ingestion. May cause toxic systemic effects.

Carcinogenicity	NTP Listed?	IARC Cancer Review Group?	OSHA Regulated?	Target Organs?
Confirmed Human Confirmed Animal	5th Annual Report	Group 2A	Refer to: 29 CFR 1910.1027	Lungs, liver, kidney, bone, skin, eye, prostate.

Medical Conditions Generally Aggravated by Exposure

None reported. However, persons with existing lung dysfunction may experience effects quicker.

Emergency and First-aid Procedures

Eye contact: Do not allow victim to keep eyes tightly shut. Flush immediately with water for 15 minutes (minimum), seek medical attention. Skin contact: Remove all contaminated clothing. Immediately wash area with flooding amounts of soap and water. For inhalation: Remove the person from exposure. Provide respiratory assistance and CPR. Transfer to medical facility. Observe 24-48 hours for lung effects (pulmonary edema likely). If swallowed: Seek medical attention immediately. Call poison control center for advice. Never give anything by mouth to an unconscious or convulsing person.

SECTION VII - PRECAUTIONS FOR SAFE HANDLING AND USE

Steps to be Taken in Case Material is Released or Spilled

Collect powdered materials in most convenient and safe manner and deposit in sealed containers. Dry sweep not recommended. Use HEPA vacuum. Remove all ignition sources. Restrict those not involved in cleanup from entering area. Notify appropriate authorities, as required by applicable regulations.

Preferred Waste Disposal Method

No citation.

Precautions to be Taken in Handling and Storage

Do not store near heat or other sources of ignition. A regulated, marked area must be established where cadmium is used, handled, or stored.

Other Precautions and Warnings

Store in tightly sealed drums in a cool, dry, well ventilated place.

SECTION VIII - CONTROL MEASURES AND PERSONAL PROTECTIVE EQUIPMENT

Respiratory Protection *(specify type)*

For any exposure level use a NIOSH/MSHA approved self-contained breathing apparatus with full face-piece operated in positive pressure mode. Cadmium is a carcinogen. Inhalation of its vapors must be avoided.

Ventilation

Local exhaust (preferred) at site of chemical work or general ventilation.

Protective Gloves	Eye Protection	Other Protective Clothing
Impervious Material	**Chemical/Dust Goggles or Face Mask**	**Protective Apron, Boots (avoid skin contact)**

Work/Hygiene Practices

Always wash hands thoroughly after using chemical; never bring food, drink, or smoking materials into vicinity of chemicals.

CADMIUM BROMIDE

Br_2Cd CAS: 7789-42-6

IDENTIFICATION AND TYPICAL USES

Cadmium bromide appears as a white to yellow efflorescent, hygrocopic, pearly hexagonal, crystalline solid in powder form. It is used in photography, process engraving, and lithography.

RISK ASSESSMENT: HEALTH

General Assessment

There are several reports in the references of cadmium compounds causing poisoning and human death. Cadmium bromide is a confirmed human carcinogen with experimental carcinogenic, tumorigenic, mutagenic, and neoplastigenic data also being reported. Cadmium bromide can enter the body by *inhalation* of its dusts or fumes, or by *ingestion.* In humans, the toxic symptoms of acute exposure may include nausea, vomiting, diarrhea, abdominal pain, muscular ache, salivation, and shock. In addition, inhalation of its fumes and dusts can cause cough, tightness in the chest, respiratory distress, congestion of the lungs, and broncho-pneumonia. Inhalation causes an excess of protein to be present in the urine. Exposure to its fumes or dusts of 30 minutes to amounts of as little as 50 mg/m^3 can be fatal to humans.

Cadmium bromide is a poison that accumulates in the liver and kidneys. Chronic poisoning can therefore lead to liver and kidney damage, emphysema, low blood sugar, and/or loss of smell. Serious damage can occur at levels below the OSHA PEL. Its biological half-life in humans is estimated at approximately 20-30 years. Cadmium is also know to cause the so-called "itai, itai" disease, which is a bone disease together with kidney malfunction.

During welding or grinding operations, cadmium fumes are generated. Inhalation can cause a flu-like illness with chills, headache, aching and/or fever. This can lead to more serious illness.

Acute Health Effects

The following acute (short-term) health effects may occur immediately or shortly after exposure to cadmium:

Skin: Possible irritation may occur.

Eye: Irritation and possible damage on contact.

Lung: High exposures can cause rapid and severe lung damage, with shortness of breath, chest pain, cough, and a possible buildup of fluids in the lungs (pulmonary edema), which is a medical emergency. In serious cases, death or permanent lung damage can occur. Illness can be delayed for 4 to 8 hours, allowing overexposures to occur without warning signs.

Other: High exposure to cadmium may cause nausea, cramps, salivation, vomiting, and diarrhea.

Chronic Health Effects

The following chronic (long-term) health effects may occur at some time after exposure to cadmium and can last for months or even years:

Cancer Hazards: While cadmium bromide has not specifically been tested for its ability to cause cancer, it should be handled as a carcinogen since several related cadmium compounds are known cancer agents.

Reproductive Hazard: Cadmium bromide should be handled as a potential teratogenic agent and reproductive hazard since several related cadmium compounds are known teratogens and decreases fertility in males and females. It may damage the testes (male reproductive glands) and may affect the female reproductive cycle.

Other Chronic Effects: Repeated low level exposures can cause permanent kidney damage which can go unnoticed for some time without proper testing. The kidney damage can lead to kidney stones and other more serious health problems. Emphysema or lung scarring can occur from a single high exposure or repeated low exposures. Long-term exposure can cause anemia, loss of sense of smell, fatigue, and/or yellow staining of the teeth.

ʊ Recommended Risk-Reduction Measures

Personnel should avoid direct contact with cadmium bromide dusts or powders. If a less toxic material or compound cannot be substituted for cadmium bromide, then *engineering controls* are the most effective method of reducing exposures. The best protection is to enclose operations and/or provide local exhaust ventilation at the site of cadmium dust release. While not always operationally feasible, isolating operations can also reduce exposure. Using respiratory protection is less effective than the controls mentioned above, but is still advisable whenever working with or around cadmium. At any exposure level, a self-contained breathing apparatus (SCBA) with full facepiece and pressure demand is recommended for the greatest possible respiratory protection. If a full facepiece is not available, then chemical/dust goggles should be worn to protect the eyes. A face shield and protective apron should also be worn. To prevent hand and skin exposures, gloves should be used. Glove manufacturers should be contacted and permeation studies obtained *before* final glove selection are made.

Administrative controls should also be in place to minimize the potential for human exposures. These may include written procedures or policies which specify the methods and techniques that will be practiced whenever personnel are to work with cadmium bromide. As per OSHA 29 CFR 1910.1027 (Cadmium), a regulated and controlled work area must be established wherever cadmium is used, handled, or stored. Unauthorized personnel are not permitted in this area.

All personnel should receive *training* on the use, hazards, protective measures, emergency actions, and other precautions per 29 CFR 1910.1200 (Hazard Communication), prior to the first assignment in an area where cadmium bromide is used or stored.

Before beginning employment working with cadmium bromide, and at regular intervals thereafter (e.g., annually), those personnel with frequent or potentially high exposures should be provided the following recommended medical tests:

- ☑ Lung function test.
- ☑ Urine test for cadmium (levels should be less than 10 micrograms per liter of urine).
- ☑ Urine test for "low molecular weight proteins" to detect kidney damage.
- ☑ Complete urinalysis (UA).

If overexposure is suspected or symptoms develop, the following additional test is recommended:

- ☑ Complete blood count (CBC).

Any evaluation should include a careful history of past and present symptoms with an examination. Medical tests that simply look for existing damage are not a substitute for controlling exposures.

Cigarette smoke contains some cadmium. Because it is hard for the body to eliminate cadmium, it tends to buildup in the body. Any workplace exposure will therefore add to any existing levels. Smoking or carrying cigarettes near cadmium bromide increases release of toxic fumes. Also, since smoking can cause heart disease as well as lung cancer, emphysema, and other known respiratory problems, exposure to cadmium bromide may cause additional complications. Therefore, personnel who smoke should avoid any unprotected contact with cadmium bromide powders.

Other methods to reduce exposure include:

- ☑ Where possible, enclose operations and use local exhaust ventilation at the site of chemical release. If local exhaust ventilation or enclosure is not used, respiratory protection should be mandatory.
- ☑ Always ensure that proper protective clothing is worn when using chemical substances.
- ☑ Wash thoroughly immediately after exposure to cadmium bromide and at the end of the work shift or before eating, drinking, or smoking.
- ☑ Hazard warning information should be posted in the work area. In addition, as part of an on-going education and training program, all information on the health and safety hazards of cadmium bromide should be communicated to all exposed and potentially exposed workers.
- ☑ Eye wash stations should be provided in the immediate work area for emergency use. If there is a possibility of skin exposure to cadmium bromide, emergency shower facilities should be provided.
- ☑ Workers whose clothing has been contaminated by cadmium bromide should change into clean clothes before leaving work. Contaminated work clothing should be laundered only by individuals who have been informed of the hazards of exposure to cadmium bromide.

RISK ASSESSMENT: ENVIRONMENT

General Assessment

Cadmium are naturally occurring element. However, industrial uses place the environment at risk of exposure during transportation, storage, disposal, or destruction of cadmium compounds, including cadmium bromide. In almost every scenario, the threat of environmental exposure is contingent upon the proper handling of the chemical substance. Accidental spills, large or small, where cadmium bromide contacts incompatible commodities can result in fire, explosion (depending upon conditions of spill), and possible contamination of the surrounding environmental mediums (water, soil, and air).

In its bulk form, cadmium bromide is considered a non-combustible solid. However, its dusts can react explosively and present a serious fire hazard. Because it is incompatible with many commodities, including some other common metals, extreme caution is required in handling, storage, transportation, and disposal of cadmium. These characteristics also require special consideration during any emergency situation involving a leak or spill of cadmium bromide powder or dust. Should cadmium bromide ever come into contact with any incompatible substances either during use, transportation, storage, or disposal, violent and even explosive reactions are possible. Cadmium bromide dust is flammable and explosive when exposed to heat, flame, or by chemical reaction with oxidizing agents, metals (zinc, selenium, tellurium, elements of sulfur), strong oxidizers, and acids.

Cadmium can enter the environment from industrial and municipal waste treatment plant discharges and from spills.

Acute Ecological Effects

Acute (short-term) toxic effects may include the death of animals, birds, or fish and death or low growth rate in plants. Acute effects are seen 2 to 4 days after animals or plants are exposed to cadmium. In fresh water, cadmium toxicity is influenced by water hardness. The harder the water, the lower the toxicity. Cadmium has high acute toxicity to aquatic life. Insufficient data are available to evaluate the short-term effects of cadmium to plants, birds, or land animals.

Chronic Ecological Effects

Chronic (long-term) toxic effects may include shortened life span, reproductive problems, lower fertility, and changes in appearance or behavior in exposed animals. These effects can be seen long after first exposure(s) to toxic chemicals. Cadmium has high chronic toxicity to aquatic life. Insufficient data are available to evaluate or predict the long-term effects of cadmium to plants, birds, or land animals.

Water Solubility

Most of the cadmium compounds are slightly soluble in water. Concentrations of less than 1 milligram will mix with a liter of water.

Persistence in the Environment

Cadmium is highly persistent in water, with a half-life greater than 200 days. The half-life of a pollutant is the amount of time it takes for one half of the chemical to be degraded. No data are available on the estimated percentages of cadmium persistence in the aquatic or terrestrial environments.

Bioaccumulation in Aquatic Organisms

Some substances increase in concentration, or *bioaccumulate,* in living organisms as they breathe contaminated air, drink contaminated water, or eat contaminated food. These chemicals can become concentrated in the tissues and internal organs of animals as well as humans.

The concentration of cadmium found in fish tissues is expected to be much higher than the average concentration of cadmium in the water from which the fish was taken.

Recommended Risk-Reduction Measures

Proper training of all transporters will reduce the likelihood of a mishap or accident resulting in a leak or spill of cadmium bromide dusts or powders into the environment. The correct use of DOT labeling (per the MSDS) on all containers, trucks, and rail cars will enable emergency responders to react properly and quickly to any disaster thereby reducing the potential risk to the environment and to personnel.

Storage of cadmium bromide should be segregated from incompatible chemicals to minimize the risk of cross-contamination or contact.

If a spill or leak to the environment has occurred, fire department, emergency response and/or hazardous materials spill personnel should be notified immediately. Cleanup should be attempted only by those trained in proper spill containment procedures. If cadmium bromide should contact the water table, aquifer, or navigable waterway, time is of the essence. It is highly soluble and total remediation may not be possible. The local and/or state emergency response authorities must be notified. A comprehensive emergency response or disaster preparedness/recovery plan should be in place prior to any operations involving the use, transportation, storage, or disposal of cadmium bromide. If cadmium bromide powder is spilled or leaked, the following specific steps are recommended:

☑ Restrict persons not wearing protective clothing from area of spill or leak until cleanup is complete.

☑ Remove all ignition sources.

☑ Collect powdered materials in the safest manner possible and deposit in sealed containers. Recommend using vacuum with HEPA filter. Do not dry sweep (generates airborne dusts).

☑ It may be necessary to dispose of cadmium bromide as a hazardous waste. The responsible state agency or the regional office of the federal Environmental Protection Agency (EPA) should be contacted for specific recommendations.

RISK ASSESSMENT: BUSINESS

General Assessment

Accidents or mishaps involving cadmium bromide can present a significant threat to business operations. The loss or damage of equipment or facilities can significantly affect fiscal viability. Lawsuits that may result from personnel injury/death, public exposures, and/or environmental contamination will also require a serious expenditure of resources. Media attention surrounding an injury, death, or environmental damage can also result in a loss of profits and loss of current as well as future business.

Recommended Risk-Reduction Measures

Company attorneys, safety and health professionals, and environmental specialists should be involved in the development of any procedures for responding to chemical incidents. A company official should be predesignated as a public relations officer with specific training in dealing with the press. Corporate plans and policies should be developed, approved, and implemented long before any need for such arises.

REFERENCES

American Conference of Governmental Industrial Hygienists. 1988. *Documentation of the Threshold Limit Values and Biological Exposure Indices,* 5th Edition (with updates). Cincinnati: ACGIH

Brusick, D. J. 1994. *Methods for Genetic Risk Assessment.* New York: CRC Press/Lewis Publishers.

Bueche, J. R. 1972. *Principles of Physics,* 2nd Edition. New York: McGraw-Hill.

Calabrese, E. J. 1994. *Biological Effects of Low Level Exposures.* New York: CRC Press/Lewis Publishers.

Cothern, R. R. 1993. *Comparative Environmental Risk Assessment.* New York: CRC Press/Lewis Publishers.

Gardner, R. W. 1994. *Chemical Intolerance.* New York: CRC Press/Lewis Publishers.

Hallenbeck, W. H. 1993. *Quantitative Risk Assessment for Environmental and Occupational Health,* 2nd Edition. New York: CRC Press/Lewis Publishers.

Howard, P. H. 1995. *Handbook of Environmental Fate and Exposures Data for Organic Chemicals,* Vol. I-IV. New York: CRC Press/Lewis Publishers.

Lewis, R. J., Sr. 1992. *Sax's Dangerous Properties of Industrial Materials,* Volumes 1, 2, and 3, 8th Edition. New York: Van Nostrand Reinhold.

Lide, D. R. 1995. *Handbook of Chemistry and Physics,* 76th Edition. New York: CRC Press/Lewis Publishers.

Manahan, S. E. 1989. *Toxicological Chemistry.* New York: CRC Press/Lewis Publishers.

Meyer, E. 1990. *Chemistry of Hazardous Materials.* Englewood Cliffs, NJ: Prentice Hall.

Patnaik, P. 1992. *A Comprehensive Guide to the Hazardous Properties of Chemical Substances.* New York: Van Nostrand Reinhold.

Proctor, N. H., J. P. Hughes. 1991. *Chemical Hazards of the Workplace.* New York: Van Nostrand Reinhold.

Quigley, D. R. 1994. *Handbook of Emergency Chemical Management.* New York: CRC Press/Lewis Publishers.

Reeves, A. L. 1979. Cadmium. *Handbook on the Toxicology of Metals.* pp 321-328. New York: Elsevier North-Holland.

Richardson, M. 1995. *Dictionary of Substance and Their Effects,* Vol. I-VI. New York: CRC Press/Lewis Publishers.

Sacarello, H. L. A. 1994. *The Comprehensive Handbook of Hazardous Materials.* New York: CRC Press/Lewis Publishers.

Torkelson, T. R. V. K. Rowe. 1981. *Patty's Industrial Hygiene and Toxicology,* Vol. 2B. New York: Wiley-Interscience.

MATERIAL SAFETY DATA SHEET

CHEMICAL NAME	CRC PRESS/LEWIS PUBLISHERS DISCLAIMER
CADMIUM NITRATE	THE INFORMATION AND RECOMMENDATIONS PRESENTED HEREIN ARE BASED ON SOURCES BELIEVED TO BE RELIABLE. CRC MAKES NO REPRESENTATION ON ITS COMPLETENESS OR ACCURACY. IT IS THE RESPONSIBILITY OF THE USER TO DETERMINE THE CHEMICAL'S SUITABILITY FOR ITS INTENDED USE, THE CHEMICAL'S SAFE USE, AND THE CHEMICAL'S PROPER DISPOSAL. NO REPRESENTATIONS AND/OR WARRANTIES, EITHER EXPRESSED OR IMPLIED, OF THE MERCHANTABILITY OR FITNESS FOR A PARTICULAR PURPOSE, OR OF ANY OTHER NATURE, ARE MADE WITH RESPECT TO THE INFORMATION PROVIDED IN THIS MSDS OR TO THE CHEMICAL TO WHICH INFORMATION MAY REFER. CRC NEITHER ASSUMES NOR AUTHORIZES ANY OTHER PERSON TO ASSUME FOR IT, ANY OTHER ADDITIONAL RESPONSIBILITY OR LIABILITY FOR THE USE OF, OR RELIANCE UPON, THIS INFORMATION.

HAZARD WARNING INFORMATION

HEALTH	FIRE	REACTIVE	OTHER	DEGREE OF HAZARD	COLOR CODING	OTHER CODES
2	0	0		0 = Minimum Hazard 1 = Slight Hazard 2 = Moderate Hazard 3 = Serious Hazard 4 = Severe Hazard	HEALTH = BLUE FIRE = RED REACTIVITY = YELLOW OTHER = WHITE	OX = Oxidizer ACID = Acid ALK = Alkali COR = Corrosive W = Use No Water

SECTION I - GENERAL INFORMATION

Characterization	**Metallic Nitrate**
RCRA Number	**D006**
EPA Class	**Characteristic (T) Waste**
DOT Proper Shipping Name	**Cadmium Compound**
Chemical Abstract Service (CAS) Number	**10325-94-7** (anhydrous) **10022-68-1** (tetrahydrate)
DOT Hazard Class and Label Requirements	**Keep Away From Food**
DOT Emergency Guide Code	**No Citation.**
DOT Identification Number	**UN 2570**
Atomic Symbol	$Cd(NO_3)_2$ (anhydrous) $Cd(NO_3)_2{\bullet}4HOH$ (tetrahydrate)

Synonyms
Cadmium dinitrate; cadmium nitrate tetrahydrate; nitric acid, cadmium salt.

SECTION II - HAZARDOUS INGREDIENTS AND IDENTITY INFORMATION

Hazardous Components (specific identity)	OSHA Exposure Criteria	NIOSH Exposure Criteria	Immediately Dangerous to Life and Health (IDLH)	ACGIH Exposure Criteria
Cadmium nitrate (derivation: By reacting nitric acid with cadmium or cadmium oxide followed by subsequent crystallization).	**PEL (Cd):** **fume:** **0.005 mg/m³** **dust:** **0.002 mg/m³** **STEL:** **0.6 mg/m³**	**REL:** **Cancer Causing Agent** **Reduce to Lowest Possible Level**	**9 mg/m³** **(for Cd)**	**TLV (Cd):** **dusts & salts:** **0.05 mg/m³** **STEL:** **Not Established**

SECTION III - PHYSICAL AND CHEMICAL CHARACTERISTICS

Boiling Point	**132°F (270°C)** (tetrahydrate)
Specific Gravity (H_2O = 1)	**2.455** (tetrahydrate)
Vapor Pressure (mm Hg)	**Not Found**
Molecular Weight (atomic weight)	**236.42** (anhydrous) **308.5** (tetrahydrate)
Vapor Density (Air = 1)	**Not Found**
Melting Point	**662°F (350°C)** (anhydrous) **138°F (59°C)** (tetrahydrate)

Solubility
Soluble in water, alcohol, and ammonia.

Appearance and Odor
The tetrahydrate appears as odorless, white, amorphous pieces. The anhydrous is hygroscopic needles.

SECTION IV - FIRE AND EXPLOSION HAZARD DATA

Flash Point (method used)	**Not Applicable**
Explosive Limits in Air % by Volume	**LEL: Not Determined** **UEL: Not Determined**
NFPA Classification	**Non-Combustible Solid**
Autoignition Temperature	**Not Found**

Extinguishing Media
For dust, use carbon dioxide or dry chemical. For metal, use agent suitable to surrounding fire. Avoid use of water. A class "D" extinguisher may be required.

Special Fire Fighting Procedures
Wear full protective clothing and self-contained breathing apparatus (SCBA). Poisonous gases are produced in fire. Do not release runoff to sewers or waterways.

Unusual Fire and Explosion Hazards
The dust may form flammable mixtures air and can be flammable and explosive when exposed to heat, flame. Avoid contact with other metals.

SECTION V - REACTIVITY DATA

Stability		
Stability		Conditions to Avoid **Avoid contact with incompatible materials. Keep cadmium nitrate powders away from open flame, heat, or other sources of ignition.**
Stable **X**	Unstable	Incompatibility *(materials to avoid)* **No specific incompatible materials cited. However, keep away from other metals and sources of ignition, heat, and flame.**
Hazardous Polymerization		Conditions to Avoid **Hazardous polymerization of cadmium nitrate will not occur.**
May Occur	Will Not Occur **X**	Hazardous Decomposition or By-products **When heated to decomposition, cadmium nitrate can produce toxic and irritating fumes of cadmium and nitrogen oxides.**

SECTION VI - HEALTH HAZARD DATA

Primary Route(s) of Entry:	Inhalation?	Absorption (skin/eye)?	Ingestion?
	X		X

Health Hazards

INHALATION: Breathing the dusts of cadmium may cause cough, tightness in chest, respiratory distress, congestion of the lungs, vomiting, nausea, and headache. Large doses can be fatal. May cause fluid buildup in the lungs (pulmonary edema), liver, and kidney damage.

EYES & SKIN: Possible irritation to the skin and eyes. An potential allergen.

INGESTION: A poison by ingestion. May cause toxic systemic effects with nausea and vomiting.

Carcinogenicity	NTP Listed?	IARC Cancer Review Group?	OSHA Regulated?	Target Organs?
Confirmed Human Confirmed Animal	5th Annual Report	Group 2A	29 CFR 1910.1027	Lungs, liver, kidney, skin, eyes, prostate.

Medical Conditions Generally Aggravated by Exposure

None reported. However, persons with existing lung dysfunction may experience effects quicker.

Emergency and First-aid Procedures

Eye contact: Do not allow victim to keep eyes tightly shut. Flush immediately with water for 15 minutes (minimum), seek medical attention. Skin contact: Remove all contaminated clothing. Immediately wash area with flooding amounts of soap and water. For inhalation: Remove the person from exposure. Provide respiratory assistance and CPR. Transfer to medical facility. Observe 24-48 hours for lung effects (pulmonary edema likely). If swallowed: Seek medical attention immediately. Call poison control center for advice. Never give anything by mouth to an unconscious or convulsing person.

SECTION VII - PRECAUTIONS FOR SAFE HANDLING AND USE

Steps to be Taken in Case Material is Released or Spilled

Collect powdered materials in most convenient and safe manner and deposit in sealed containers. Dry sweep not recommended. Use HEPA vacuum. Remove all ignition sources. Restrict those not involved in cleanup from entering area. Notify appropriate authorities, as required by applicable regulations.

Preferred Waste Disposal Method

No citation.

Precautions to be Taken in Handling and Storage

Do not store near heat or other sources of ignition. A regulated, marked area must be established where cadmium nitrate is used, handled, or stored. Keep away from moisture.

Other Precautions and Warnings

Store in tightly sealed drums in a cool, dry, well ventilated place.

SECTION VIII - CONTROL MEASURES AND PERSONAL PROTECTIVE EQUIPMENT

Respiratory Protection *(specify type)*

For any exposure use self-contained breathing apparatus or a supplied-air respirator with full facepiece operated in positive pressure mode. Cadmium is a carcinogen. Inhalation of its vapors must be avoided.

Ventilation

Local exhaust (preferred) at site of chemical work or general ventilation.

Protective Gloves	Eye Protection	Other Protective Clothing
Impervious Material	Chemical/Dust Goggles or Face Mask	Protective Apron, Boots (avoid skin contact)

Work/Hygiene Practices

Always wash hands thoroughly after using chemical; never bring food, drink, or smoking materials into vicinity of chemicals.

CADMIUM NITRATE

$Cd(NO_3)_2$ (anhydrous) CAS: 13967-90-3
$Cd(NO_3)_2 \bullet 4HOH$ (tetrahydrate) CAS: 10022-68-1

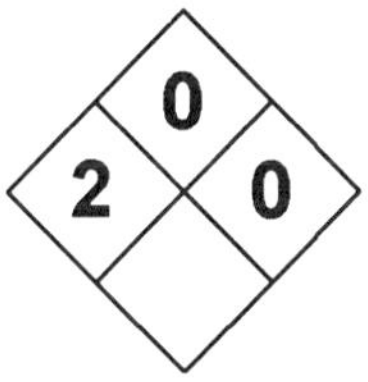

IDENTIFICATION AND TYPICAL USES

Cadmium nitrate (anhydrous) appears as odorless, white amorphous pieces and the tertrahydrate appears as hygroscopic needles. It is used in the manufacture of cadmium salts and photographic emulsions, for the cooling of glass and porcelain, and as a laboratory reagent.

RISK ASSESSMENT: HEALTH

General Assessment

There are several reports in the references of cadmium poisoning and human death. Cadmium is a confirmed human carcinogen with experimental carcinogenic, tumorigenic, mutagenic, and neoplastigenic data also being reported. Cadmium nitrate can enter the body by *inhalation* of its dusts or fumes, or by *ingestion*. In humans, the toxic symptoms of acute exposure may include nausea, vomiting, diarrhea, abdominal pain, muscular ache, salivation, and shock. In addition, inhalation of its fumes and dusts can cause cough, tightness in the chest, respiratory distress, congestion of the lungs, broncho-pneumonia, and/or pulmonary edema (fluid in the lungs), which can be fatal. Symptoms may be delayed up to 48 hours thereby creating a false sense of security with regard to health exposure risk. Inhalation of cadmium causes an excess of protein to be present in the urine (proteinuria). Exposure to its fumes or dusts of 30 minutes to amounts of as little as 50 mg/m^3 can be fatal to humans.

Cadmium is a poison that is accumulated in the liver and kidneys. Chronic poisoning can therefore lead to liver and kidney damage. Its biological half-life in humans is estimated at approximately 20-30 years. Cadmium is also known to cause the so-called "itai, itai" disease, which is a bone disease together with kidney malfunction.

During welding or grinding operations, cadmium fumes are generated. Inhalation can cause a flu-like illness with chills, headache, aching and/or fever. This can lead to more serious illness.

Acute Health Effects

The following acute (short-term) health effects may occur immediately or shortly after exposure to cadmium nitrate:

<u>Skin</u>: Possible irritation may occur.

<u>Eye</u>: Irritation and possible damage on contact.

<u>Lung</u>: High exposures can cause rapid and severe lung damage, with shortness of breath, chest pain, cough, and a possible buildup of fluids in the lungs (pulmonary edema), which is a medical emergency. In serious cases, death or permanent lung damage may occur.

<u>Other</u>: High exposure to cadmium nitrate may cause nausea, cramps, salivation, vomiting, and diarrhea.

Chronic Health Effects

The following chronic (long-term) health effects may occur at some time after exposure to cadmium nitrate and can last for months or even years:

<u>Cancer Hazards</u>: Cadmium, and especially cadmium oxide, is a confirmed human carcinogen. There is sufficient evidence that it causes prostate and kidney cancer in humans. It has also been shown to cause lung and testes cancer in animals.

<u>Reproductive Hazard</u>: Cadmium is a probable teratogen in humans. It may damage the testes (male reproductive glands) and may affect the female reproductive cycle. Cadmium nitrate teratogenicity should be considered suspect based upon its cadmium base.

Other Chronic Effects: Repeated low level exposures can cause permanent kidney damage which can go unnoticed for some time without proper testing. The kidney damage can lead to kidney stones and other more serious health problems. Emphysema or lung scarring can occur from a single high exposure or repeated low exposures. Long-term exposure can cause anemia, loss of sense of smell, fatigue, and/or yellow staining of the teeth.

Recommended Risk-Reduction Measures

Cadmium and its compounds should be considered human carcinogens. Personnel should avoid direct contact with cadmium nitrate dusts or powders. If a less toxic material or compound cannot be substituted for cadmium nitrate, then *engineering controls* are the most effective method of reducing exposures. The best protection is to enclose operations and/or provide local exhaust ventilation at the site of cadmium nitrate dust release. While not always operationally feasible, isolating operations can also reduce exposure. Using respiratory protection is less effective than the controls mentioned above, but is still advisable whenever working with or around cadmium nitrate. At any exposure to a known carcinogen, a supplied-air respirator or a self-contained breathing apparatus (SCBA) with full facepiece operated in pressure demand or other continuous flow mode are recommended for the greatest possible respiratory protection. If a full facepiece is not available, then chemical/dust goggles should be worn to protect the eyes. A face shield and protective apron should also be worn. To prevent hand and skin exposures, gloves should be worn.

Administrative controls should also be in place to minimize the potential for human exposures. These may include written procedures or policies which specify the methods and techniques that will be practiced whenever personnel are to work with cadmium nitrate. As per OSHA 29 CFR 1910.1027 (Cadmium), a regulated and controlled work area must be established wherever cadmium is used, handled, or stored. Unauthorized personnel are not permitted in this area.

All personnel should receive *training* on the use, hazards, protective measures, emergency actions, and other precautions per 29 CFR 1910.1200 (Hazard Communication) and the OSHA Standard for cadmium (29 CFR 1910.1027), prior to the first assignment in an area where cadmium nitrate is used or stored.

Before beginning employment working with cadmium nitrate, and at regular intervals thereafter (e.g., annually), those personnel with frequent or potentially high exposures should be provided the following recommended medical tests:

- ☑ Lung function test.
- ☑ Urine test for cadmium nitrate (levels should be less than 10 micrograms per liter of urine).
- ☑ Urine test for "low molecular weight proteins" to detect kidney damage.
- ☑ Complete urinalysis (UA).

If overexposure is suspected or symptoms develop, the following additional test is recommended:

- ☑ Complete blood count (CBC).
- ☑ Consider chest X-ray following acute overexposure (may be negative if taken immediately after exposure due to delayed onset of pulmonary edema).

Any evaluation should include a careful history of past and present symptoms with an examination. Medical tests that simply look for existing damage are not a substitute for controlling exposures.

Cigarette smoke contains some cadmium. Because it is hard for the body to eliminate cadmium, it tends to buildup in the body. Any workplace exposure will therefore add to any existing levels. Smoking or carrying cigarettes near cadmium increases release of toxic fumes. Also, since smoking can cause heart disease as well as lung cancer, emphysema, and other known respiratory problems, exposure to cadmium nitrate may cause additional complications. Therefore, personnel who smoke should avoid any unprotected contact with cadmium nitrate powders.

Other methods to reduce exposure include:

- ☑ Where possible, enclose operations and use local exhaust ventilation at the site of chemical release. If local exhaust ventilation or enclosure is not used, respiratory protection should be mandatory.
- ☑ Always ensure that proper protective clothing is worn when using chemical substances and that personnel are adequately trained on its use, care, and maintenance.
- ☑ Wash thoroughly immediately after exposure to cadmium nitrate and at the end of the work shift or before eating, drinking, or smoking.
- ☑ Hazard warning information should be posted in the work area. In addition, as part of an on-going education and training program, all information on the health and safety hazards of cadmium nitrate

should be communicated to all exposed and potentially exposed workers.

☑ Eye wash stations should be provided in the immediate work area for emergency use. If there is a possibility of skin exposure to cadmium nitrate, emergency shower facilities should also be provided in t he immediate area.

☑ Workers whose clothing has been contaminated by cadmium nitrate should change into clean clothes before leaving work. Contaminated work clothing should be laundered only by individuals who have been informed of the hazards of exposure to cadmium nitrate.

RISK ASSESSMENT: ENVIRONMENT

General Assessment

Cadmium is a naturally occurring element. However, industrial uses place the environment at risk of exposure during transportation, storage, disposal, or destruction of cadmium compounds, including cadmium nitrate. In almost every scenario, the threat of environmental exposure is contingent upon the proper handling of the chemical substance. Accidental spills, large or small, where cadmium nitrate contacts incompatible commodities can result in fire, explosion (depending upon conditions of spill), and possible contamination of the surrounding environmental mediums (water, soil, and air).

In its bulk form, cadmium nitrate is considered a non-combustible solid. However, its dusts may form explosive mixtures in air and can react explosively and present a serious fire hazard. Because it is incompatible with many other common metals, extreme caution is required in handling, storage, transportation, and disposal of cadmium nitrate. These characteristics also require special consideration during any emergency situation involving a leak or spill of cadmium nitrate powder or dust.

Should cadmium nitrate ever come into contact with any incompatible substances either during use, transportation, storage, or disposal, violent and even explosive reactions are possible. The dust may be flammable and explosive when exposed to heat, flame, or by chemical reaction with other metals (zinc, selenium, tellurium, elements of sulfur).

Cadmium nitrate can enter the environment from industrial and municipal waste treatment plant discharges and from spills.

Acute Ecological Effects

Acute (short-term) toxic effects may include the death of animals, birds, or fish and death or low growth rate in plants. Acute effects are seen 2 to 4 days after animals or plants are exposed to cadmium. In fresh water, cadmium toxicity is influenced by water hardness. The harder the water, the lower the toxicity. Cadmium has high acute toxicity to aquatic life. Insufficient data are available to evaluate the short-term effects of cadmium to plants, birds, or land animals.

Chronic Ecological Effects

Chronic (long-term) toxic effects may include shortened life span, reproductive problems, lower fertility, and changes in appearance or behavior in exposed animals. These effects can be seen long after first exposure(s) to toxic chemicals. Cadmium has high chronic toxicity to aquatic life. Insufficient data are available to evaluate or predict the long-term effects of cadmium to plants, birds, or land animals.

Water Solubility

Most of the cadmium compounds are slightly soluble in water. Concentrations of less than 1 milligram will mix with a liter of water.

Persistence in the Environment

Cadmium is highly persistent in water, with a half-life greater than 200 days. The half-life of a pollutant is the amount of time it takes for one half of the chemical to be degraded. No data are available on the estimated percentages of cadmium persistence in the aquatic or terrestrial environments.

Bioaccumulation in Aquatic Organisms

Some substances increase in concentration, or *bioaccumulate,* in living organisms as they breathe contaminated air, drink contaminated water, or eat contaminated food. These chemicals can become concentrated in the tissues and internal organs of animals as well as humans. The concentration of cadmium found in fish tissues is expected to be much higher than the average concentration of cadmium in the water from which the fish was taken.

Recommended Risk-Reduction Measures

Proper training of all transporters will reduce the likelihood of a mishap or accident resulting in a leak or spill of cadmium nitrate dusts or powders into the environment. The correct use of DOT labeling (per the MSDS) on all containers, trucks, and rail cars will enable emergency responders to react properly and quickly to any disaster thereby reducing the potential risk to the environment and to personnel.

Storage of cadmium nitrate should be segregated from incompatible chemicals to minimize the risk of cross-contamination or contact. Water and moisture should be kept away.

If a spill or leak to the environment has occurred, fire department, emergency response and/or hazardous materials spill personnel should be notified immediately. Cleanup should be attempted only by those trained in proper spill containment procedures. If cadmium nitrate should contact the water table, aquifer, or navigable waterway, time is of the essence. It is slightly soluble and total remediation may be possible. The local and/or state emergency response authorities must be notified. A comprehensive emergency response or disaster preparedness/recovery plan should be in place prior to any operations involving the use, transportation, storage, or disposal of cadmium nitrate. If cadmium nitrate powder is spilled or leaked, the following specific steps are recommended:

☑ Restrict persons not wearing protective clothing from area of spill or leak until cleanup is complete.

☑ Remove all ignition sources and ventilate area of spill.

☑ Collect powdered materials in the safest manner possible and deposit in sealed containers. Recommend using vacuum with HEPA filter. Do not dry sweep (generates airborne dusts). Use non-sparking tools.

☑ It may be necessary to dispose of cadmium nitrate as a hazardous waste. The responsible state agency or the regional office of the federal Environmental Protection Agency (EPA) should be contacted for specific recommendations.

RISK ASSESSMENT: BUSINESS

General Assessment

Accidents or mishaps involving cadmium nitrate can present a significant threat to business operations. The loss or damage of equipment or facilities can significantly affect fiscal viability. Lawsuits that may result from personnel injury/death, public exposures, and/or environmental contamination will also require a serious expenditure of resources. Media attention surrounding an injury, death, or environmental damage can also result in a loss of profits and loss of current as well as future business. Always remember that anytime the terms "cancer," or "carcinogen" are used, public hysteria, emotion, and ignorance can run equally high. This should be carefully considered whenever developing or implementing public relations policies.

Recommended Risk-Reduction Measures

Company attorneys, safety and health professionals, and environmental specialists should be involved in the development of any procedures for responding to chemical incidents. A company official should be predesignated as a public relations officer with specific training in dealing with the press. Corporate plans and policies should be developed, approved, and implemented long before any need for such arises.

REFERENCES

American Conference of Governmental Industrial Hygienists. 1988. *Documentation of the Threshold Limit Values and Biological Exposure Indices,* 5th Edition (with updates). Cincinnati: ACGIH

Brusick, D. J. 1994. *Methods for Genetic Risk Assessment.* New York: CRC Press/Lewis Publishers.

Bueche, J. R. 1972. *Principles of Physics,* 2nd Edition. New York: McGraw-Hill.

Cothern, R. R. 1993. *Comparative Environmental Risk Assessment.* New York: CRC Press/Lewis Publishers.

Gardner, R. W. 1994. *Chemical Intolerance.* New York: CRC Press/Lewis Publishers.

Lewis, R. J., Sr. 1992. *Sax's Dangerous Properties of Industrial Materials,* Volumes 1, 2, and 3, 8th Edition. New York: Van Nostrand Reinhold.

Meyer, E. 1990. *Chemistry of Hazardous Materials.* Englewood Cliffs, NJ: Prentice Hall.

Quigley, D. R. 1994. *Handbook of Emergency Chemical Management.* New York: CRC Press/Lewis Publishers.

Reeves, A. L. 1979. Cadmium nitrate. *Handbook on the Toxicology of Metals.* pp 321-328. New York: Elsevier North-Holland.

Richardson, M. 1995. *Dictionary of Substance and Their Effects,* Vol. I-VI. New York: CRC Press/Lewis Publishers.

Sacarello, H. L. A. 1994. *The Comprehensive Handbook of Hazardous Materials.* New York: CRC Press/Lewis Publishers.

Torkelson, T. R. V. K. Rowe. 1981. *Patty's Industrial Hygiene and Toxicology,* Vol. 2B. New York: Wiley-Interscience.

MATERIAL SAFETY DATA SHEET

CHEMICAL NAME	CRC PRESS/LEWIS PUBLISHERS DISCLAIMER
CADMIUM OXIDE	**CRC PRESS/LEWIS PUBLISHERS DISCLAIMER:** THE INFORMATION AND RECOMMENDATIONS PRESENTED HEREIN ARE BASED ON SOURCES BELIEVED TO BE RELIABLE. CRC MAKES NO REPRESENTATION ON ITS COMPLETENESS OR ACCURACY. IT IS THE RESPONSIBILITY OF THE USER TO DETERMINE THE CHEMICAL'S SUITABILITY FOR ITS INTENDED USE, THE CHEMICAL'S SAFE USE, AND THE CHEMICAL'S PROPER DISPOSAL. NO REPRESENTATIONS AND/OR WARRANTIES, EITHER EXPRESSED OR IMPLIED, OF THE MERCHANTABILITY OR FITNESS FOR A PARTICULAR PURPOSE, OR OF ANY OTHER NATURE, ARE MADE WITH RESPECT TO THE INFORMATION PROVIDED IN THIS MSDS OR TO THE CHEMICAL TO WHICH INFORMATION MAY REFER. CRC NEITHER ASSUMES NOR AUTHORIZES ANY OTHER PERSON TO ASSUME FOR IT, ANY OTHER ADDITIONAL RESPONSIBILITY OR LIABILITY FOR THE USE OF, OR RELIANCE UPON, THIS INFORMATION.

HAZARD WARNING INFORMATION

HEALTH	FIRE	REACTIVE	OTHER	DEGREE OF HAZARD	COLOR CODING	OTHER CODES
3	0	0		0 = Minimum Hazard 1 = Slight Hazard 2 = Moderate Hazard 3 = Serious Hazard 4 = Severe Hazard	HEALTH = BLUE FIRE = RED REACTIVITY = YELLOW OTHER = WHITE	OX = Oxidizer ACID = Acid ALK = Alkali COR = Corrosive ~~W~~ = Use No Water

SECTION I - GENERAL INFORMATION

Characterization	**Metallic Oxide**	RCRA Number	**D006**
		EPA Class	**Characteristic (T) Waste**
DOT Proper Shipping Name	**Cadmium Compounds**	Chemical Abstract Service (CAS) Number	**1306-19-0**
DOT Hazard Class and Label Requirements	**Poison; Keep Away From Food**	DOT Emergency Guide Code	**53**
DOT Identification Number	**UN 2570**	Atomic Symbol	**CdO**

Synonyms
Cadmium monoxide; cadmium oxide fume.

SECTION II - HAZARDOUS INGREDIENTS AND IDENTITY INFORMATION

Hazardous Components (specific identity)	OSHA Exposure Criteria	NIOSH Exposure Criteria	Immediately Dangerous to Life and Health (IDLH)	ACGIH Exposure Criteria
Cadmium oxide (derivation: Cadmium metal is distilled in a retort, the vapor reacted with air, and the oxide collected in a baghouse). Impurities may include chloride, nitrate, sulfate, copper, iron, and lead.	**PEL (Cd):** **fume:** **0.005 mg/m³** **dust:** **0.002 mg/m³** **CEILING:** **0.6 mg/m³**	**REL:** **Cancer Causing Agent** **Reduce to Lowest Possible Level**	**9 mg/m³** **(for cadmium)**	**TLV (Cd):** **dusts & salts:** **0.05 mg/m³** **CEILING:** **fume:** **0.05 mg/m³**

SECTION III - PHYSICAL AND CHEMICAL CHARACTERISTICS

Boiling Point	**2832°F (1559°C), sublimes**	Specific Gravity (H_2O = 1)	**6.95**
Vapor Pressure (mm Hg)	**1 at 1832°F (1000°C)**	Molecular Weight (atomic weight)	**128.4**
Vapor Density (Air = 1)	**Not Found**	Melting Point	**<2598°F (<1426°C)**

Solubility
Insoluble in water. Soluble in acids and ammonia salts, slightly soluble in ammonia hydroxide.

Appearance and Odor
Colorless to white amorphous powder or a brownish-red crystalline solid. The fume appears as odorless, tiny brownish yellow particulates in air.

SECTION IV - FIRE AND EXPLOSION HAZARD DATA

Flash Point (method used)	**Not Applicable**	Explosive Limits in Air % by Volume	**LEL: Not Determined UEL: Not Determined**
NFPA Classification	**Non-Combustible Solid**	Autoignition Temperature	**Non-Combustible**

Extinguishing Media
For dust, use carbon dioxide or dry chemical. For metal, use agent suitable to surrounding fire. Avoid use of water. A class "D" extinguisher may be required.

Special Fire Fighting Procedures
Wear full protective clothing and self-contained breathing apparatus (SCBA). Poisonous gases are produced in fire.

Unusual Fire and Explosion Hazards
Cadmium oxide dusts present a moderate explosion hazard. Keep clear of ignition sources.

SECTION V - REACTIVITY DATA

Stability		Conditions to Avoid Avoid contact with incompatible materials. Keep cadmium oxide powders away from open flame, heat, or other sources of ignition. Normally stable under routine conditions of storage and handling.
Stable X	Unstable	Incompatibility *(materials to avoid)* Violent or explosive reaction in contact with oxidizers (such as perchlorates, peroxides, permanganates, chlorates, and nitrates), acids (sulfuric, hydroazoic, nitric), sulfur, selenium, tellurium, ammonia nitrate, zinc.
Hazardous Polymerization		Conditions to Avoid Under normal conditions of temperature and pressure, hazardous polymerization of cadmium oxide is not expected to occur.
May Occur	Will Not Occur X	Hazardous Decomposition or By-products When heated to decomposition, cadmium oxide produces toxic and irritating cadmium fumes.

SECTION VI - HEALTH HAZARD DATA

Primary Route(s) of Entry:	Inhalation?	Absorption (skin/eye)?	Ingestion?
	X		X

Health Hazards

INHALATION: Breathing the dusts of cadmium may cause cough, tightness in chest, respiratory distress, congestion of the lungs, and broncho-pneumonia. Large doses can be fatal. May cause flue-like symptoms, chills, muscle pain, and fluid buildup in the lungs (pulmonary edema).

EYE & SKIN: Possible irritation to the skin and eyes. An allergen.

INGESTION: A poison by ingestion. May cause toxic systemic effects.

Carcinogenicity	NTP Listed?	IARC Cancer Review Group?	OSHA Regulated?	Target Organs?
Confirmed Human Confirmed Animal	5th Annual Report	Group 2A	29 CFR 1910.1027	Lungs, liver, kidney, bone, skin, eye, prostate.

Medical Conditions Generally Aggravated by Exposure

None reported. However, persons with existing lung dysfunction may experience effects quicker.

Emergency and First-aid Procedures

Eye contact: Do not allow victim to keep eyes tightly shut. Flush immediately with water for 15 minutes (minimum), seek medical attention. Skin contact: Remove all contaminated clothing. Immediately wash area with flooding amounts of soap and water. For inhalation: Remove the person from exposure. Provide respiratory assistance and CPR. Transfer to medical facility. Observe 24-48 hours for lung effects (pulmonary edema likely). If swallowed: Call poison control center for advice. Unless told otherwise, give 1-2 glasses of water to dilute. Do NOT induce vomiting. Seek medical attention immediately.

SECTION VII - PRECAUTIONS FOR SAFE HANDLING AND USE

Steps to be Taken in Case Material is Released or Spilled

Collect powdered materials in most convenient and safe manner and deposit in sealed containers. Dry sweep not recommended. Use HEPA vacuum. Remove all ignition sources. Restrict those not involved in cleanup from entering area. Notify appropriate authorities, as required by applicable regulations.

Preferred Waste Disposal Method

No citation.

Precautions to be Taken in Handling and Storage

Do not store near heat or other sources of ignition. A regulated, marked area must be established where cadmium oxide is used, handled, or stored.

Other Precautions and Warnings

Store in tightly sealed drums in a cool, dry, well ventilated place.

SECTION VIII - CONTROL MEASURES AND PERSONAL PROTECTIVE EQUIPMENT

Respiratory Protection *(specify type)*

For any exposure level use a NIOSH/MSHA approved self-contained breathing apparatus with full face-piece operated in positive pressure mode. Cadmium oxide is a carcinogen. Inhalation of its vapors must be avoided.

Ventilation

Local exhaust (preferred) at site of chemical work or general ventilation.

Protective Gloves	Eye Protection	Other Protective Clothing
Impervious Material	Chemical/Dust Goggles or Face Mask	Protective Apron, Boots (avoid skin contact)

Work/Hygiene Practices

Always wash hands thoroughly after using chemical; never bring food, drink, or smoking materials into vicinity of chemicals.

CADMIUM OXIDE

CdO CAS: 1306-19-0

IDENTIFICATION AND TYPICAL USES

Cadmium oxide can appear as a colorless or white amorphous (without definite form or shape) powder or a brownish-red crystalline solid. Its fumes appear as odorless, tiny, brownish-yellow particulates in air. It is used in cadmium plating baths, electrodes for storage batteries, cadmium salts, as a catalyst, in ceramic glazes, in phosphors, and as a nematocide.

RISK ASSESSMENT: HEALTH

General Assessment

There are several reports in the references of cadmium poisoning and human death. Cadmium is a confirmed human carcinogen with experimental carcinogenic, tumorigenic, mutagenic, and neoplastigenic data also being reported. Cadmium oxide can enter the body by *inhalation* of its dusts or fumes, or by *ingestion*. In humans, the toxic symptoms of acute exposure may include nausea, vomiting, diarrhea, abdominal pain, muscular ache, salivation, and shock. In addition, inhalation of its fumes and dusts can cause cough, tightness in chest, respiratory distress, congestion of the lungs, broncho-pneumonia, and pulmonary edema (fluid in the lungs), which is a medical emergency and can be fatal. Symptoms may be delayed up to 48 hours creating a false sense of security with regard to health exposure risk. Inhalation causes an excess of protein to be present in the urine (proteinuria). Exposure to its fumes or dusts of 30 minutes to amounts of as little as 50 mg/m^3 can be fatal to humans. Bone demineralization similar to that of osteoporosis may occur.

Cadmium is a poison that is accumulated in the liver and kidneys. Chronic poisoning can therefore lead to liver and kidney damage. Its biological half-life in humans is estimated at approximately 20-30 years. Cadmium is also known to cause the so-called "itai, itai" disease, which is a bone disease together with kidney malfunction.

During welding or grinding operations, cadmium fumes are generated. Inhalation can cause a flu-like illness with chills, headache, aching and/or fever. This can lead to more serious illness.

Acute Health Effects

The following acute (short-term) health effects may occur immediately or shortly after exposure to cadmium oxide:

Skin: Possible irritation may occur.

Eye: Irritation and possible damage on contact.

Lung: High exposures can cause rapid and severe lung damage, with shortness of breath, chest pain, cough, and a possible buildup of fluids in the lungs (pulmonary edema), which is a medical emergency. In serious cases, death or permanent lung damage may occur.

Other: High exposure to cadmium oxide may cause nausea, cramps, salivation, vomiting, and diarrhea.

Chronic Health Effects

The following chronic (long-term) health effects may occur at some time after exposure to cadmium oxide and can last for months or even years:

Cancer Hazards: Cadmium, and especially cadmium oxide, is a confirmed human carcinogen. There is sufficient evidence that it causes prostate and kidney cancer in humans. It has also been shown to cause lung and testes cancer in animals.

Reproductive Hazard: Cadmium is a probable teratogen in humans. It may damage the testes (male reproductive glands) and may affect the female reproductive cycle. Cadmium oxide may cause these effects as well.

Other Chronic Effects: Repeated low level exposures can cause permanent kidney damage which can go unnoticed for some time without proper testing. The kidney damage can lead to kidney stones and other more serious health problems. Emphysema or lung scarring (fibrosis) can occur from a single high exposure or repeated low exposures. Long-term exposure can cause anemia, loss of sense of smell, fatigue, and/or yellow staining of the teeth.

Recommended Risk-Reduction Measures

Cadmium is a known carcinogen. Personnel should avoid direct contact with cadmium and its compounds, especially cadmium oxide dusts or powders. If a less toxic material or compound cannot be substituted for cadmium oxide, then *engineering controls* are the most effective method of reducing exposures. The best protection is to enclose operations and/or provide local exhaust ventilation at the site of cadmium dust release. While not always operationally feasible, isolating operations can also reduce exposure. Using respiratory protection is less effective than the controls mentioned above, but is still advisable whenever working with or around cadmium oxide. At any exposure level to a known carcinogen, a supplied-air respirator or a self-contained breathing apparatus (SCBA) with full facepiece operated in pressure demand or other positive pressure mode are recommended for the greatest possible respiratory protection. If a full facepiece is not available, then chemical/dust goggles should be worn to protect the eyes. A face shield and protective apron should also be worn. To prevent hand and skin exposures, gloves should be worn.

Administrative controls should also be in place to minimize the potential for human exposures. These may include written procedures or policies which specify the methods and techniques that will be practiced whenever personnel are to work with cadmium oxide. As per OSHA 29 CFR 1910.1027 (Cadmium oxide), a regulated and controlled work area must be established wherever cadmium oxide is used, handled, or stored. Unauthorized personnel are not permitted in this area.

All personnel should receive *training* on the use, hazards, protective measures, emergency actions, and other precautions per 29 CFR 1910.1200 (Hazard Communication) and the OSHA Standard for cadmium (29 CFR 1910.1027), prior to the first assignment in an area where cadmium oxide is used or stored.

Before beginning employment working with cadmium oxide, and at regular intervals thereafter (e.g., annually), those personnel with frequent or potentially high exposures should be provided the following recommended medical tests:

- ☑ Lung function test.
- ☑ Urine test for cadmium oxide (levels should be less than 10 micrograms per liter of urine).
- ☑ Urine test for "low molecular weight proteins" to detect kidney damage.
- ☑ Complete urinalysis (UA).

If overexposure is suspected or symptoms develop, the following additional test is recommended:

- ☑ Complete blood count (CBC).
- ☑ Consider chest X-ray after acute overexposure (may be negative if taken immediately after exposure due to the delayed development of pulmonary edema).

Any evaluation should include a careful history of past and present symptoms with an examination. Medical tests that simply look for existing damage are not a substitute for controlling exposures.

Cigarette smoke contains some cadmium oxide. Because it is hard for the body to eliminate cadmium oxide, it tends to buildup in the body. Any workplace exposure will therefore add to any existing levels. Smoking or carrying cigarettes near cadmium oxide increases release of toxic fumes. Also, since smoking can cause heart disease as well as lung cancer, emphysema, and other known respiratory problems, exposure to cadmium oxide may cause additional complications. Therefore, personnel who smoke should avoid any unprotected contact with cadmium oxide powders.

Other methods to reduce exposure include:

- ☑ Where possible, enclose operations and use local exhaust ventilation at the site of chemical release. If local exhaust ventilation or enclosure is not used, respiratory protection should be mandatory.
- ☑ Always ensure that proper protective clothing is worn when using chemical substances.
- ☑ Wash thoroughly immediately after exposure to cadmium oxide and at the end of the work shift or before eating, drinking, or smoking.
- ☑ Hazard warning information should be posted in the work area. In addition, as part of an on-going education and training program, all information on the health and safety hazards of cadmium oxide should be communicated to all exposed workers.

☑ Eye wash stations should be provided in the immediate work area for emergency use. If there is a possibility of skin exposure to cadmium oxide, emergency shower facilities should also be provided in the immediate area.

☑ Workers whose clothing has been contaminated by cadmium oxide should change into clean clothes before leaving work. Contaminated work clothing should be laundered only by individuals who have been informed of the hazards of exposure to cadmium oxide.

RISK ASSESSMENT: ENVIRONMENT

General Assessment

Cadmium is a naturally occurring element. However, industrial uses place the environment at risk of exposure during transportation, storage, disposal, or destruction of cadmium oxide compounds. In almost every scenario, the threat of environmental exposure is contingent upon the proper handling of the chemical substance. Accidental spills, large or small, where cadmium oxide contacts incompatible commodities can result in fire, explosion (depending upon conditions of spill), and possible contamination of the surrounding environmental mediums (water, soil, and air).

In its bulk form, cadmium is considered a noncombustible solid. However, its dusts may create explosive mixtures in air and can present a moderate fire hazard. Because it is incompatible with many commodities, including some other common metals, extreme caution is required in handling, storage, transportation, and disposal of cadmium oxide. These characteristics also require special consideration during any emergency situation involving a leak or spill of cadmium oxide powder or dust. Should cadmium oxide ever come into contact with any incompatible substances either during use, transportation, storage, or disposal, violent and even explosive reactions are possible. Airborne dust concentrations may be flammable and explosive when exposed to heat, flame, or by chemical reaction with metals (zinc, selenium, tellurium, elements of sulfur), and acids. Cadmium oxide can enter the environment from industrial and municipal waste treatment plant discharges and from spills.

☠ *Acute Ecological Effects*

Acute (short-term) toxic effects may include the death of animals, birds, or fish and death or low growth rate in plants. Acute effects are seen 2 to 4 days after animals or plants are exposed to cadmium oxide. In fresh water, cadmium oxide toxicity is influenced by water hardness. The harder the water, the lower the toxicity. Cadmium has high acute toxicity to aquatic life. Insufficient data are available to evaluate the short-term effects of cadmium to plants, birds, or land animals.

Chronic Ecological Effects

Chronic (long-term) toxic effects may include shortened life span, reproductive problems, lower fertility, and changes in appearance or behavior in exposed animals. These effects can be seen long after first exposure(s) to toxic chemicals. Cadmium has high chronic toxicity to aquatic life. Insufficient data are available to evaluate or predict the long-term effects of cadmium to plants, birds, or land animals.

Water Solubility

Most of the cadmium compounds are slightly soluble in water. Cadmium oxide is insoluble in water. Concentrations of less than 1 milligram will mix with a liter of water.

⌛ *Persistence in the Environment*

Cadmium is highly persistent in water, with a half-life greater than 200 days. The half-life of a pollutant is the amount of time it takes for one half of the chemical to be degraded. No data are available on the estimated percentages of cadmium oxide persistence in the aquatic or terrestrial environments.

Bioaccumulation in Aquatic Organisms

Some substances increase in concentration, or *bioaccumulate,* in living organisms as they breathe contaminated air, drink contaminated water, or eat contaminated food. These chemicals can become concentrated in the tissues and internal organs of animals as well as humans.

The concentration of cadmium found in fish tissues is expected to be much higher than the average concentration of cadmium in the water from which the fish was taken.

Recommended Risk-Reduction Measures

Proper training of all transporters will reduce the likelihood of a mishap or accident resulting in a leak or spill of cadmium oxide dusts or powders into the environment. The correct use of DOT labeling (per the MSDS) on all containers, trucks, and rail cars will enable emergency responders to react properly and quickly to any disaster thereby reducing the potential risk to the environment and to personnel.

Storage of cadmium oxide should be segregated from incompatible chemicals to minimize the risk of cross-contamination or contact.

If a spill or leak to the environment has occurred, fire department, emergency response and/or hazardous materials spill personnel should be notified immediately. Cleanup should be attempted only by those trained in proper spill containment procedures. If cadmium oxide should contact the water table, aquifer, or navigable waterway, time is of the essence. It is highly soluble and total remediation may not be possible. The local and/or state emergency response authorities must be notified. A comprehensive emergency response or disaster preparedness/recovery plan should be in place prior to any operations involving the use, transportation, storage, or disposal of cadmium oxide. If cadmium oxide powder is spilled or leaked, the following specific steps are recommended:

- ☑ Restrict persons not wearing protective clothing from area of spill or leak until cleanup is complete and area can be opened for normal work.
- ☑ Remove all ignition sources and ventilate area.
- ☑ Collect powdered materials in the safest manner possible and deposit in sealed containers. Recommend using vacuum with HEPA filter. Do not dry sweep (generates airborne dusts). Use non-sparking tools only.
- ☑ It may be necessary to dispose of cadmium oxide as a hazardous waste. The responsible state agency or the regional office of the federal Environmental Protection Agency (EPA) should be contacted for specific recommendations.

RISK ASSESSMENT: BUSINESS

General Assessment

Accidents or mishaps involving cadmium oxide can present a significant threat to business operations. The loss or damage of equipment or facilities can significantly affect fiscal viability. Lawsuits that may result from personnel injury/death, public exposures, and/or environmental contamination will also require a serious expenditure of resources. Media attention surrounding an injury, death, or environmental damage can also result in a loss of profits and loss of current as well as future business. Always remember that anytime the terms "cancer," or "carcinogen" are used, public hysteria, emotion, and ignorance can run equally high. This should be carefully considered whenever developing or implementing public relations policies.

Recommended Risk-Reduction Measures

Company attorneys, safety and health professionals, and environmental specialists should be involved in the development of any procedures for responding to chemical incidents. A company official should be predesignated as a public relations officer with specific training in dealing with the press. Corporate plans and policies should be developed, approved, and implemented long before any need for such arises.

REFERENCES

American Conference of Governmental Industrial Hygienists. 1988. *Documentation of the Threshold Limit Values and Biological Exposure Indices,* 5th Edition (with updates). Cincinnati: ACGIH

Brusick, D. J. 1994. *Methods for Genetic Risk Assessment.* New York: CRC Press/Lewis Publishers.

Bueche, J. R. 1972. *Principles of Physics,* 2nd Edition. New York: McGraw-Hill.

Calabrese, E. J. 1994. *Biological Effects of Low Level Exposures.* New York: CRC Press/Lewis Publishers.

Cothern, R. R. 1993. *Comparative Environmental Risk Assessment.* New York: CRC Press/Lewis Publishers.

Gardner, R. W. 1994. *Chemical Intolerance.* New York: CRC Press/Lewis Publishers.

Howard, P. H. 1995. *Handbook of Environmental Fate and Exposures Data for Organic Chemicals,* Vol. I-IV. New York: CRC Press/Lewis Publishers.

Lewis, R. J., Sr. 1992. *Sax's Dangerous Properties of Industrial Materials,* Volumes 1, 2, and 3, 8th Edition. New York: Van Nostrand Reinhold.

Meyer, E. 1990. *Chemistry of Hazardous Materials.* Englewood Cliffs, NJ: Prentice Hall.

Quigley, D. R. 1994. *Handbook of Emergency Chemical Management.* New York: CRC Press/Lewis Publishers.

Reeves, A. L. 1979. Cadmium oxide. *Handbook on the Toxicology of Metals.* pp 321-328. New York: Elsevier North-Holland.

Richardson, M. 1995. *Dictionary of Substance and Their Effects,* Vol. I-VI. New York: CRC Press/Lewis Publishers.

Sacarello, H. L. A. 1994. *The Comprehensive Handbook of Hazardous Materials.* New York: CRC Press/Lewis Publishers.

Torkelson, T. R. V. K. Rowe. 1981. *Patty's Industrial Hygiene and Toxicology,* Vol. 2B. New York: Wiley-Interscience.

MATERIAL SAFETY DATA SHEET

CHEMICAL NAME	CRC PRESS/LEWIS PUBLISHERS DISCLAIMER
CADMIUM SULFATE	

HAZARD WARNING INFORMATION

HEALTH	FIRE	REACTIVE	OTHER	DEGREE OF HAZARD	COLOR CODING	OTHER CODES
2	0	0		0 = Minimum Hazard 1 = Slight Hazard 2 = Moderate Hazard 3 = Serious Hazard 4 = Severe Hazard	HEALTH = BLUE FIRE = RED REACTIVITY = YELLOW OTHER = WHITE	OX = Oxidizer ACID = Acid ALK = Alkali COR = Corrosive W = Use No Water

SECTION I - GENERAL INFORMATION

Characterization	**Metal**	RCRA Number	**D006**
		EPA Class	**Characteristic (T) Waste**
DOT Proper Shipping Name	**Cadmium Compounds**	Chemical Abstract Service (CAS) Number	**10124-36-4**
DOT Hazard Class and Label Requirements	**Not Determined**	DOT Emergency Guide Code	**53 (as compound)**
DOT Identification Number	**UN 2570**	Atomic Symbol	**$CdSO_4$**

Synonyms
Cadmium monosulfate; cadmium sulfate (1:1); cadmium sulphate; sulfuric acid, cadmium salt (1:1).

SECTION II - HAZARDOUS INGREDIENTS AND IDENTITY INFORMATION

Hazardous Components (specific identity)	OSHA Exposure Criteria	NIOSH Exposure Criteria	Immediately Dangerous to Life and Health (IDLH)	ACGIH Exposure Criteria
Cadmium sulfate (derivation: By action of dilute sulfuric acid on cadmium, cadmium carbonate, or cadmium oxide; produced as an intermediate in recovery of cadmium from zinc ore. Anhydrous cadmium sulfate is prepared by the oxidation of the sulfide or sulfite at elevated temperatures).	**PEL (as Cd):** **fume:** **0.005 mg/m³** **dust:** **0.002 mg/m³** **STEL:** **0.6 mg/m³**	**REL:** **Cancer Causing Agent** **Reduce to Lowest Possible Level**	**9 mg/m³ (as Cd)**	**TLV (Cd):** **dusts & salts:** **0.01 mg/m³** **respirable** **0.002 mg/m³** **STEL:** **Not Established**

SECTION III - PHYSICAL AND CHEMICAL CHARACTERISTICS

Boiling Point	**Not Found**	Specific Gravity (H_2O = 1)	**4.691**
Vapor Pressure (mm Hg)	**Not Found**	Molecular Weight (atomic weight)	**208.46**
Vapor Density (Air = 1)	**Not Applicable**	Melting Point	**1832°F (1000°C)**

Solubility
Soluble in water. Insoluble in alcohol, acids, acetone, and ammonia.

Appearance and Odor
Colorless to white odorless crystals.

SECTION IV - FIRE AND EXPLOSION HAZARD DATA

Flash Point (method used)	**Not Found (Metal)**	Explosive Limits in Air % by Volume	**LEL: N/A UEL: N/A**
NFPA Classification	**Flammable Solid (in powder form)**	Autoignition Temperature	**Not Found**

Extinguishing Media
For dust, use carbon dioxide or dry chemical. For metal, use agent suitable to surrounding fire. Do NOT use water. A class "D" extinguisher may be required.

Special Fire Fighting Procedures
Cadmium sulfate is a moderate fire hazard in the form of dust. Wear full protective clothing and self-contained breathing apparatus (SCBA). Poisonous gases are produced in fire.

Unusual Fire and Explosion Hazards
The dust may ignite spontaneously in air and is flammable and explosive when exposed to heat, flame, or by chemical reaction with oxidizing agents and other metals.

SECTION V - REACTIVITY DATA

Stability		Conditions to Avoid
		Avoid contact with incompatible materials. Keep cadmium sulfate powders away from open flame, heat, or other sources of ignition. It is normally stable at room temperature under routine conditions.
Stable	Unstable	Incompatibility *(materials to avoid)*
X		Violent or explosive reaction in contact with oxidizers (such as perchlorates, peroxides, permanganates, chlorates, and nitrates), metals, hydrogen azide, selenium, tellurium, zinc.
Hazardous Polymerization		Conditions to Avoid
		Hazardous polymerization of cadmium sulfate will not occur.
May Occur	Will Not Occur	Hazardous Decomposition or By-products
	X	If cadmium sulfate is involved in fire, hazardous and toxic fumes and gases are produced, including oxides of sulfur and cadmium.

SECTION VI - HEALTH HAZARD DATA

Primary Route(s) of Entry:	Inhalation?	Absorption (skin/eye)?	Ingestion?
	X		X

Health Hazards

INHALATION: Breathing dusts may cause metallic taste, nausea, vomiting, chills, weakness, leg pains, diarrhea, throat dryness, cough, dypsnea, chest pains, kidney injury, extreme restlessness, irritability, pneumonitis, bronchitis, and potentially fatal pulmonary edema (fluid in lungs).

EYES & SKIN: Possible irritation to the skin and eyes. An allergen.

INGESTION: A poison causing nausea, salivation, choking, vomiting, diarrhea, abdominal pain, headache, muscular cramps, vertigo, possible convulsions, exhaustion, shock, and death.

Carcinogenicity	NTP Listed?	IARC Cancer Review Group?	OSHA Regulated?	Target Organs?
Confirmed Human Confirmed Animal	5th Annual Report	Group 2A	29 CFR 1910.1027	Respiratory and G.I. tract, kidneys, skin, eyes.

Medical Conditions Generally Aggravated by Exposure

None reported. However, persons with existing lung dysfunction may experience effects quicker.

Emergency and First-aid Procedures

Eye contact: Do not allow victim to keep eyes tightly shut. Flush immediately with water for 15 minutes (minimum), seek medical attention. Skin contact: Remove all contaminated clothing. Immediately wash area with flooding amounts of soap and water. For inhalation: Remove the person from exposure. Provide respiratory assistance and CPR. Transfer to medical facility. Observe 24-48 hours for lung effects (pulmonary edema likely). If swallowed: Seek medical attention immediately. Call poison control center for advice. Never give anything by mouth to an unconscious or convulsing person.

SECTION VII - PRECAUTIONS FOR SAFE HANDLING AND USE

Steps to be Taken in Case Material is Released or Spilled

Collect powdered materials in most convenient and safe manner and deposit in sealed containers. Dry sweep not recommended. Use HEPA vacuum. Remove all ignition sources. Restrict those not involved in cleanup from entering area. Notify appropriate authorities, as required by applicable regulations.

Preferred Waste Disposal Method

No citation.

Precautions to be Taken in Handling and Storage

Do not store near heat or other sources of ignition. A regulated, marked area must be established where cadmium sulfate is used, handled, or stored.

Other Precautions and Warnings

Store in tightly sealed drums in a cool, dry, well ventilated place.

SECTION VIII - CONTROL MEASURES AND PERSONAL PROTECTIVE EQUIPMENT

Respiratory Protection *(specify type)*

For any exposure level use a NIOSH/MSHA approved self-contained breathing apparatus with full facepiece operated in positive pressure mode. Cadmium sulfate is a carcinogen. Inhalation of its vapors must be avoided.

Ventilation

Local exhaust (preferred) at site of chemical work or general ventilation.

Protective Gloves	Eye Protection	Other Protective Clothing
Impervious Material	Chemical/Dust Goggles or Face Mask	Protective Apron, Boots (avoid skin contact)

Work/Hygiene Practices

Always wash hands thoroughly after using chemical; never bring food, drink, or smoking materials into vicinity of chemicals.

CADMIUM SULFATE

$CdSO_4$ CAS: 10124-36-4

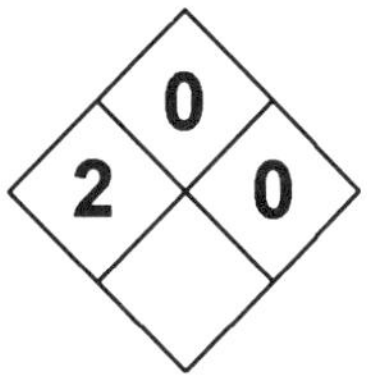

IDENTIFICATION AND TYPICAL USES

Cadmium sulfate is a colorless to white, rhombic, crystalline solid with no odor. It is used in the making of pigments and inks, in ceramic glazes, in the manufacture of pyrotechnics, phosphorus, in fluorescent screens, in scintillation counters, in rectifiers, as an accelerator in cement formation, as a photoconductor in xerography, in transistors, in photovoltain cells, in solar cells, and as a catalyst in photo-decomposition of hydrogen sulfide.

RISK ASSESSMENT: HEALTH

General Assessment

There are several reports in the references of cadmium poisoning and human death. Cadmium compounds (including the sulfate) are considered confirmed human carcinogens with experimental carcinogenic, tumorigenic, mutagenic, and neoplastigenic data also being reported. Cadmium sulfate can enter the body by *inhalation* of its dusts or fumes, or by *ingestion.*

In humans, the toxic symptoms of acute exposure may include headache, chills, leg cramps, nausea, vomiting, diarrhea, dry throat, dypsnea, chest and abdominal pain, muscular ache, salivation, kidney damage, extreme irritability and restlessness, and shock. In addition, inhalation of its fumes and dusts can cause metallic taste, cough, tightness in chest, respiratory distress, congestion of the lungs, pneumonitis, and bronchitis. Severe exposures may cause pulmonary edema (fluid in the lungs) which is a medical emergency and can be fatal. Symptoms may be delayed up to 48 hours, thereby creating a false sense of security with regard to health exposure risk. Inhalation of cadmium dust causes an excess of protein to be present in the urine. Exposure to its fumes or dusts of 30 minutes to amounts of as little as 50 mg/m^3 can be fatal to humans. Ingestion causes severe nausea, salivation, choking, vomiting, diarrhea, abdominal pain, headache, muscular cramps, vertigo (dizziness), possible convulsions, exhaustion, collapse, shock, and death.

Cadmium is a poison that is accumulated in the liver and kidneys. Chronic poisoning can therefore lead to liver and kidney damage. Its biological half-life in humans is estimated at approximately 20-30 years. Cadmium is also known to cause the so-called "itai, itai" disease, which is a bone disease together with kidney malfunction.

During welding or grinding operations, cadmium fumes are generated. Inhalation can cause a flu-like illness with chills, headache, aching and/or fever. This can lead to more serious illness.

Acute Health Effects

The following acute (short-term) health effects may occur immediately or shortly after exposure to cadmium sulfate:

Skin: Possible irritation may occur.

Eye: Irritation and possible damage on contact.

Lung: High exposures can cause rapid and severe lung damage, with shortness of breath, chest pain, cough, and a possible buildup of fluids in the lungs (pulmonary edema), which is a medical emergency. In serious cases, death or permanent lung damage is possible. Symptoms of illness can be delayed for up to 48 hours or more, allowing overexposures to occur without warning signs.

Other: High exposure to cadmium sulfate may cause nausea, cramps, salivation, vomiting, and diarrhea.

Chronic Health Effects

The following chronic (long-term) health effects may occur at some time after exposure to cadmium sulfate and can last for months or even years:

Cancer Hazards: Cadmium, and especially cadmium oxide, is a confirmed human carcinogen. There is sufficient evidence that it causes prostate and kidney cancer in humans. However, there are also conflicting studies to contradict this assertion. It has also been shown to cause lung and testes cancer in animals.

Reproductive Hazard: Cadmium sulfate is a probable teratogen in humans. It may damage the testes (male reproductive glands) and may affect the female reproductive cycle.

Other Chronic Effects: Repeated low level exposures can cause permanent kidney damage which can go unnoticed for some time without proper testing. The kidney damage can lead to kidney stones and other more serious health problems. Emphysema or lung scarring (pulmonary fibrosis) can occur from a single high exposure or repeated low exposures. Long-term exposure can cause anemia, chronic bronchitis, rhinitis with loss of sense of smell, gastrointestinal symptoms, fatigue, bone changes, and/or yellow staining of the teeth.

Recommended Risk-Reduction Measures

Personnel should avoid direct contact with cadmium sulfate dusts or powders. If a less toxic material or compound cannot be substituted for cadmium sulfate, then *engineering controls* are the most effective method of reducing exposures. The best protection is to enclose operations and/or provide local exhaust ventilation at the site of cadmium sulfate dust release. While not always operationally feasible, isolating operations can also reduce exposure. Using respiratory protection is less effective than the controls mentioned above, but is still advisable whenever working with or around cadmium sulfate. At any exposure level, a self-contained breathing apparatus (SCBA) or a supplied air respirator with full facepiece and operated in pressure demand or continuous flow mode are recommended for the greatest possible respiratory protection. If a full facepiece is not available, then chemical/dust goggles should be worn to protect the eyes. A face shield and protective apron should also be worn. To prevent hand and skin exposures, gloves should be worn.

Administrative controls should also be in place to minimize the potential for human exposures. These may include written procedures or policies which specify the methods and techniques that will be practiced whenever personnel are to work with cadmium sulfate. As per OSHA 29 CFR 1910.1027 (Cadmium), a regulated and controlled work area must be established wherever cadmium sulfate is used, handled, or stored. Unauthorized personnel are not permitted in this area.

All personnel should receive *training* on the use, hazards, protective measures, emergency actions, and other precautions per 29 CFR 1910.1200 (Hazard Communication), prior to the first assignment in an area where cadmium sulfate is used or stored.

Before beginning employment working with cadmium sulfate, and at regular intervals thereafter (e.g., annually), those personnel with frequent or potentially high exposures should be provided the following recommended medical tests:

- ☑ Lung function test.
- ☑ Urine test for cadmium sulfate (levels should be less than 10 micrograms per liter of urine).
- ☑ Urine test for "low molecular weight proteins" to detect kidney damage.
- ☑ Complete urinalysis (UA).

If overexposure is suspected or symptoms develop, the following additional test is recommended:

- ☑ Complete blood count (CBC).
- ☑ Consider chest X-ray following acute overexposure (may be negative if taken immediately after exposure due to delayed development of pulmonary edema).

Any evaluation should include a careful history of past and present symptoms with an examination. Medical tests that simply look for existing damage are not a substitute for controlling exposures.

Cigarette smoke contains some cadmium. Because it is hard for the body to eliminate cadmium, it tends to build up in the body. Any workplace exposure will therefore add to any existing levels. Smoking or carrying cigarettes near cadmium compounds increases release of toxic fumes. Also, since smoking can cause heart disease as well as lung cancer, emphysema, and other known respiratory problems, exposure to cadmium sulfate may cause additional complications. Therefore, personnel who smoke should avoid any unprotected contact with cadmium sulfate powders.

Other methods to reduce exposure include:

- ☑ Where possible, enclose operations and use local exhaust ventilation at the site of chemical release. If local exhaust ventilation or enclosure is not used, respiratory protection should be mandatory.
- ☑ Always ensure that proper protective clothing is worn when using chemical substances.

- ☑ Wash thoroughly immediately after exposure to cadmium sulfate and at the end of the work shift or before eating, drinking, or smoking.
- ☑ Hazard warning information should be posted in the work area. In addition, as part of an on-going education and training program, all information on the health and safety hazards of cadmium sulfate should be communicated to all potentially exposed workers.
- ☑ Eye wash stations should be provided in the immediate work area for emergency use. If there is a possibility of skin exposure to cadmium sulfate, emergency shower facilities should also be provided.
- ☑ Workers whose clothing has been contaminated by cadmium sulfate should change into clean clothes before leaving work. Contaminated work clothing should be laundered only by individuals who have been informed of the hazards of exposure to cadmium sulfate.

RISK ASSESSMENT: ENVIRONMENT

General Assessment

Cadmium is a naturally occurring element. However, industrial uses of its compounds, including cadmium sulfate, place the environment at risk of exposure during transportation, storage, disposal, or destruction of cadmium sulfate compounds. In almost every scenario, the threat of environmental exposure is contingent upon the proper handling of the chemical substance. Accidental spills, large or small, where cadmium sulfate contacts incompatible commodities can result in fire, explosion (depending upon conditions of spill), and possible contamination of the surrounding environmental mediums (water, soil, and air).

In its bulk form, cadmium sulfate is considered a non-combustible solid. However, its dusts can react explosively with incompatible materials as well as create explosive mixtures with air in closed spaces. This presents a serious fire hazard. Because it is incompatible with many commodities, including some other common metals and common oxidizing agents, extreme caution is required in handling, storage, transportation, and disposal of cadmium sulfate. These characteristics also require special consideration during any emergency situation involving a leak or spill of cadmium sulfate powder or dust. Should cadmium sulfate ever come into contact with any incompatible substances either during use, transportation, storage, or disposal, violent and even explosive reactions are possible. The dust ignites spontaneously in air and is flammable and explosive when exposed to heat, flame, or by chemical reaction with oxidizing agents, metals (zinc, selenium, tellurium, elements of sulfur), strong oxidizers, and acids (explodes on contact with hydrazoic acid).

Cadmium sulfate can enter the environment from industrial and municipal waste treatment plant discharges and from spills.

Acute Ecological Effects

Acute (short-term) toxic effects may include the death of animals, birds, or fish and death or low growth rate in plants. Acute effects are seen 2 to 4 days after animals or plants are exposed to cadmium sulfate. In fresh water, cadmium toxicity is influenced by water hardness. The harder the water, the lower the toxicity. Cadmium has high acute toxicity to aquatic life. Insufficient data are available to evaluate the short-term effects of cadmium to plants, birds, or land animals.

Chronic Ecological Effects

Chronic (long-term) toxic effects may include shortened life span, reproductive problems, lower fertility, and changes in appearance or behavior in exposed animals. These effects can be seen long after first exposure(s) to toxic chemicals. Cadmium has high chronic toxicity to aquatic life. Insufficient data are available to evaluate or predict the long-term effects of cadmium to plants, birds, or land animals.

Water Solubility

Most of the cadmium sulfate compounds are slightly soluble in water. Concentrations of less than 1 milligram will mix with a liter of water.

Persistence in the Environment

Cadmium is highly persistent in water, with a half-life greater than 200 days. The half-life of a pollutant is the amount of time it takes for one half of the chemical to be degraded. No data are available on the estimated percentages of cadmium persistence in the aquatic or terrestrial environments.

Bioaccumulation in Aquatic Organisms

Some substances increase in concentration, or *bioaccumulate,* in living organisms as they breathe contami-

nated air, drink contaminated water, or eat contaminated food. These chemicals can become concentrated in the tissues and internal organs of animals as well as humans.

The concentration of cadmium found in fish tissues is expected to be much higher than the average concentration of cadmium in the water from which the fish was taken.

ʘ *Recommended Risk-Reduction Measures*

Proper training of all transporters will reduce the likelihood of a mishap or accident resulting in a leak or spill of cadmium sulfate dusts or powders into the environment. The correct use of DOT labeling (per the MSDS) on all containers, trucks, and rail cars will enable emergency responders to react properly and quickly to any disaster thereby reducing the potential risk to the environment and to personnel.

Storage of cadmium sulfate should be segregated from incompatible chemicals to minimize the risk of cross-contamination or contact.

If a spill or leak to the environment has occurred, fire department, emergency response and/or hazardous materials spill personnel should be notified immediately. Cleanup should be attempted only by those trained in proper spill containment procedures. If cadmium sulfate should contact the water table, aquifer, or navigable waterway, time is of the essence. It is highly soluble and total remediation may not be possible. The local and/or state emergency response authorities must be notified. A comprehensive emergency response or disaster preparedness/recovery plan should be in place prior to any operations involving the use, transportation, storage, or disposal of cadmium sulfate. If cadmium sulfate powder is spilled or leaked, the following specific steps are recommended:

- ☑ Restrict persons not wearing protective clothing from area of spill or leak until cleanup is complete.
- ☑ Remove all ignition sources and ventilate area.
- ☑ Collect powdered materials in the safest manner possible and deposit in sealed containers. Recommend using vacuum with HEPA filter. Do not dry sweep (generates airborne dusts).
- ☑ It may be necessary to dispose of cadmium sulfate as a hazardous waste. The responsible state agency or the regional office of the federal Environmental Protection Agency (EPA) should be contacted for specific recommendations.

RISK ASSESSMENT: BUSINESS

General Assessment

Accidents or mishaps involving cadmium sulfate can present a significant threat to business operations. The loss or damage of equipment or facilities can significantly affect fiscal viability. Lawsuits that may result from personnel injury/death, public exposures, and/or environmental contamination will also require a serious expenditure of resources. Media attention surrounding an injury, death, or environmental damage can also result in a loss of profits and loss of current as well as future business. Always remember that anytime the term "carcinogen" or "cancer" are used, public emotion, hysteria, and ignorance can run equally high. This should be carefully considered whenever developing or implementing public relations policies.

ʘ *Recommended Risk-Reduction Measures*

Company attorneys, safety and health professionals, and environmental specialists should be involved in the development of any procedures for responding to chemical incidents. A company official should be predesignated as a public relations officer with specific training in dealing with the press. Corporate plans and policies should be developed, approved, and implemented long before any need for such arises.

REFERENCES

American Conference of Governmental Industrial Hygienists. 1988. *Documentation of the Threshold Limit Values and Biological Exposure Indices,* 5th Edition (with updates). Cincinnati: ACGIH

Howard, P. H. 1995. *Handbook of Environmental Fate and Exposures Data for Organic Chemicals,* Vol. I-IV. New York: CRC Press/Lewis Publishers.

International Agency for Research on Cancer (IARC). 1987. *IARC Monographs on the Evaluation of Carcinogenic Risk to Humans, Overall Evaluations of Carcinogenicity: An updating of IARC Monographs Volumes 1-42.* Lyon: IARC.

Lewis, R. J., Sr. 1992. *Sax's Dangerous Properties of Industrial Materials,* Volumes 1, 2, and 3, 8th Edition. New York: Van Nostrand Reinhold.

Meyer, E. 1990. *Chemistry of Hazardous Materials.* Englewood Cliffs, NJ: Prentice Hall.

Reeves, A. L. 1979. Cadmium. *Handbook on the Toxicology of Metals.* pp 321-328. New York: Elsevier North-Holland.

Sacarello, H. L. A. 1994. *The Comprehensive Handbook of Hazardous Materials.* New York: CRC Press/Lewis Publishers.

Torkelson, T. R. V. K. Rowe. 1981. *Patty's Industrial Hygiene and Toxicology,* Vol. 2B. New York: Wiley-Interscience.

MATERIAL SAFETY DATA SHEET

CHEMICAL NAME	
CALCIUM	**CRC PRESS/LEWIS PUBLISHERS DISCLAIMER:** THE INFORMATION AND RECOMMENDATIONS PRESENTED HEREIN ARE BASED ON SOURCES BELIEVED TO BE RELIABLE. CRC MAKES NO REPRESENTATION ON ITS COMPLETENESS OR ACCURACY. IT IS THE RESPONSIBILITY OF THE USER TO DETERMINE THE CHEMICAL'S SUITABILITY FOR ITS INTENDED USE, THE CHEMICAL'S SAFE USE, AND THE CHEMICAL'S PROPER DISPOSAL. NO REPRESENTATIONS AND/OR WARRANTIES, EITHER EXPRESSED OR IMPLIED, OF THE MERCHANTABILITY OR FITNESS FOR A PARTICULAR PURPOSE, OR OF ANY OTHER NATURE, ARE MADE WITH RESPECT TO THE INFORMATION PROVIDED IN THIS MSDS OR TO THE CHEMICAL TO WHICH INFORMATION MAY REFER. CRC NEITHER ASSUMES NOR AUTHORIZES ANY OTHER PERSON TO ASSUME FOR IT, ANY OTHER ADDITIONAL RESPONSIBILITY OR LIABILITY FOR THE USE OF, OR RELIANCE UPON, THIS INFORMATION.

HAZARD WARNING INFORMATION

HEALTH	FIRE	REACTIVE	OTHER	DEGREE OF HAZARD	COLOR CODING	OTHER CODES
1	1	2	~~W~~	0 = Minimum Hazard 1 = Slight Hazard 2 = Moderate Hazard 3 = Serious Hazard 4 = Severe Hazard	HEALTH = BLUE FIRE = RED REACTIVITY = YELLOW OTHER = WHITE	OX = Oxidizer ACID = Acid ALK = Alkali COR = Corrosive ~~W~~ = Use No Water

SECTION I - GENERAL INFORMATION

Characterization **Metal**	RCRA Number **D003**	EPA Class **Characteristic (R) Waste**
DOT Proper Shipping Name **Calcium Metal**	Chemical Abstract Service (CAS) Number **7440-70-2**	
DOT Hazard Class and Label Requirements **Flammable Solid; Dangerous When Wet**	DOT Emergency Guide Code **60**	
DOT Identification Number **UN 1401**	Atomic Formula **Ca**	

Synonyms
Calcicat; calcium metal; calcium (non-pyrophoric); calcium (pyrophoric).

SECTION II - HAZARDOUS INGREDIENTS AND IDENTITY INFORMATION

Hazardous Components (specific identity)	OSHA Exposure Criteria	NIOSH Exposure Criteria	Immediately Dangerous to Life and Health (IDLH)	ACGIH Exposure Criteria
Calcium (derivation: From electrolysis of fused calcium chloride by thermal process under high vacuum from lime reduced with aluminum. Does NOT occur free in nature).	**PEL:** Not Established **STEL:** Not Established	**REL:** Not Established **STEL:** Not Established	**Not Determined**	**TLV:** Not Established **STEL:** Not Established

SECTION III - PHYSICAL AND CHEMICAL CHARACTERISTICS

Boiling Point **2703°F (1484°C)**	Specific Gravity (H_2O = 1) **1.5**
Vapor Pressure (mm Hg) **10 at 1801°F (983°C)**	Molecular Weight **40.08**
Vapor Density (Air = 1) **Not Found**	Melting Point **1542°F (839°C)**

Solubility
Soluble (decomposes) in water to liberate the hydroxide and hydrogen. Soluble in acids (forms salts).

Appearance and Odor
Moderately soft, silver-colored, crystalline, lustrous metal. Oxidizes or tarnishes on exposure to air to form a blue-gray-white adherent protective film.

SECTION IV - FIRE AND EXPLOSION HAZARD DATA

Flash Point (method used) **Not Determined**	Explosive Limits in Air % by Volume **LEL: Not Applicable UEL: Not Applicable**
NFPA Classification **Flammable Solid**	Autoignition Temperature **Not Determined**

Extinguishing Media
Do NOT use water (reacts!), foam or halogenated hydrocarbons. Use class "D" extinguishing agent.

Special Fire Fighting Procedures
Poisonous gases are produced in fire. Wear full protective clothing and self-contained breathing apparatus (SCBA). Only use a water spray to keep fire-exposed containers cool. Do NOT get water inside containers. Burning calcium metal may reach temperatures in excess of 5000°F.

Unusual Fire and Explosion Hazards
Can cause violent explosions when wet or upon contact with strong oxidizers. In contact with alkali hydroxides causes detonation. Potentially explosive in contact with carbonate compounds.

SECTION V - REACTIVITY DATA

Stability		Conditions to Avoid
		Calcium metal is NOT stable under normal conditions of temperature and pressure. Keep away from incompatible materials since violent and or explosive reactions can occur.
Stable	Unstable	Incompatibility *(materials to avoid)*
	X	Incompatible with strong oxidizers (such as perchlorates, permanganates, peroxides, chlorates, and nitrates), air, moist air, water, alcohols, alkali hydroxides, and halogens.
Hazardous Polymerization		Conditions to Avoid
		Under normal conditions of temperature and pressure, hazardous polymerization of calcium is not known to occur.
May Occur	Will Not Occur	Hazardous Decomposition or By-products
	X	When heated to decomposition, calcium can emit toxic/poisonous fumes and gases. These include calcium oxide fumes, which can react with water to produce heat, calcium hydroxide, and hydrogen gas.

SECTION VI - HEALTH HAZARD DATA

Primary Route(s) of Entry:	Inhalation?	Absorption (skin)?	Ingestion?
	X	X	X

Health Hazards

INHALATION: Irritation of the respiratory system. Symptoms of cough, dypsnea, and chemical pneumonitis. Possible pulmonary edema (fluid in the lungs) may be a delayed reaction.

ABSORPTION: Eye contact can cause severe irritation and can damage vision or cause blindness. Skin absorption of the dust or fume is possible, causing toxic systemic effects.

INGESTION: A potential fatal poison with gastrointestinal effects. Will cause nausea and vomiting.

Carcinogenicity	NTP Listed?	IARC Cancer Review Group?	OSHA Regulated?	Target Organs?
Unknown Human Unknown Animal	No	No	29 CFR 1910.1000 Table Z-1	Respiratory system, skin, eyes.

Medical Conditions Generally Aggravated by Exposure

None reported.

Emergency and First-aid Procedures

Eye contact: Flush immediately with water for 15 minutes (minimum), seek medical attention. Skin contact: Remove all contaminated clothing. Immediately wash area with large amounts of soap and water. For inhalation: Remove the person from exposure. Provide respiratory assistance and CPR. Transfer to medical facility. If swallowed: Provide breathing assistance if necessary. Never attempt to give a convulsing or unconscious person anything by mouth. If victim is conscious, give 3 to 4 glasses of water to dilute, do NOT induce vomiting. Seek medical attention immediately.

SECTION VII - PRECAUTIONS FOR SAFE HANDLING AND USE

Steps to be Taken in Case Material is Released or Spilled

Collect powdered material in most convenient manner possible and deposit in sealed drum. Keep those not involved in cleanup from entering area. Do NOT dry sweep (creates airborne dusts). Use a vacuum equipped with a high efficiency particulate air (HEPA) filter. Do NOT use water.

Preferred Waste Disposal Method

No Citation.

Precautions to be Taken in Handling and Storage

Store in tightly closed containers in cool, well-ventilated area and protect from physical damage. Sources of ignition are prohibited where calcium is used, handled, or stored. Keep away from incompatible materials and especially water. An inert blanket of nitrogen inside containers is suggested.

Other Precautions and Warnings

Outside or detached storage facility is recommended.

SECTION VIII - CONTROL MEASURES AND PERSONAL PROTECTIVE EQUIPMENT

Respiratory Protection *(specify type)*

No exposure levels established. A full facepiece respirator with high efficiency particulate air (HEPA) filter may suffice. For best protection, use a supplied air respirator with a full facepiece, hood, or helmet or a self-contained breathing apparatus (SCBA) operated in positive pressure mode.

Ventilation

Local exhaust or general mechanical systems recommended.

Protective Gloves	Eye Protection	Other Protective Clothing
Impervious Material	Chemical/Dust Goggles or Face Mask	Protective Suit or Clothing

Work/Hygiene Practices

Always wash hands thoroughly after using chemical; never bring food, drink, or smoking materials into vicinity of chemicals.

CALCIUM

Ca CAS: 7440-70-2

IDENTIFICATION AND TYPICAL USES

Calcium is an alkaline-earth element of atomic number 20, group II of the periodic system It does not occur free in nature. It is used as an alloying agent for aluminum, copper, and lead. It is a reducing agent for beryllium, and a deoxidizer for alloys. It is also a dehydrating oil, used to in the processing of iron and its alloys, and in the manufacture of vacuum tubes. It assists in the separation of nitrogen from argon, a reducing agent in preparation of chromium metal powder, thorium, zirconium, and uranium. It is also an ingredient in fertilizer preparations.

RISK ASSESSMENT: HEALTH

General Assessment

Calcium is a common air contaminant. It is generally considered toxic when combined with other toxic elements or compounds (such as arsenic), as an oxide or hydroxide. However, it can be harmful or even fatal if *ingested* or if *inhaled* as a mist, dust, or fume. It can also be *absorbed* through the skin in toxic amounts. When combined with any other route, absorption can present a significant health threat.

Inhalation of the dusts, mists, and especially the heated fumes of calcium can cause serious irritation of the nose, throat, and mucosa of the respiratory tract. It can cause coughing, dypsnea (difficulty breathing), and chemical pneumonitis. In severe cases, pulmonary edema (fluid in the lungs) can develop which is a medical emergency and can be fatal. Symptoms may be delayed up to 48 hours creating a false sense of security with regard to health exposure risk.

It is corrosive and irritating to body tissues, especially the skin and eyes. It may cause skin rash, irritation, and even burns. Eye contact may result in corneal damage on contact. If not removed immediately, the damage may be permanent.

Acute Health Effects

The following acute (short-term) health effects may occur immediately or shortly after exposure to calcium:

Skin: Irritation, rash, and possible burns.

Eye: Calcium particles and dusts can be extremely irritating to the eyes. Water content in eyes causes the formation of corrosive calcium hydroxide. Exposure can damage vision and lead corneal damage.

Lung: Inhalation of calcium dusts may cause irritation of the mucous membranes in the respiratory tract. High exposures may cause significant irritation leading to a dangerous buildup of fluids in the lungs (pulmonary edema), which can be fatal.

Chronic Health Effects

The following chronic (long-term) health effects may occur at some time after exposure to calcium and can last for months or even years:

Cancer Hazards: According to the information presented in the references, calcium metal has been tested and does not appear to present a cancer hazard. However, many of its compounds are suspected or known to cause cancer in test animals and, in some cases, humans. Carcinogenicity is thought to be due to the presence of the other elements and not specific to the calcium base.

Reproductive Hazard: According to information available in the references, calcium has not been adequately tested for its ability to adversely affect reproduction in laboratory animals.

Other Chronic Effects: Repeated exposures can cause severe irritation of the mucous membranes in the nose, throat, and respiratory system leading to chronic chemical pneumonitis.

℧ *Recommended Risk-Reduction Measures*

Personnel should avoid direct contact with calcium. If a less toxic chemical cannot be substituted for a hazardous material, then *engineering controls* are the most effective method of reducing exposures.

The best protection is to enclose operations and/or provide local exhaust ventilation at the site of chemical release. While not always operationally feasible, isolating operations can also reduce exposure. Using respiratory protection is less effective than the controls mentioned above, but is still advisable whenever working with or around calcium. No exposure levels have been established for calcium metal. This does not mean that exposure is without health risk. A full facepiece respirator equipped with a high efficiency particulate air (HEPA) filter and a dust/mist/fume pre-filter may suffice. For the best protection, use a supplied-air respirator with a full facepiece operated in positive pressure mode or a MSHA/NIOSH approved self-contained breathing apparatus (SCBA) with full facepiece operated in pressure demand or other continuous flow mode. If a full facepiece is not available, then chemical/dust goggles should be worn to protect the eyes. Whenever a chemical dust hazard exists, a face shield and a rubber apron should be worn. To prevent hand and skin exposures, impervious rubber gloves should be worn.

Administrative controls should also be in place to minimize the potential for human exposures. These may include written procedures or policies which specify the methods and techniques that will be practiced whenever personnel are to work with calcium.

All personnel should receive *training* on the use, hazards, protective measures, emergency actions, and other precautions per 29 CFR 1910.1200 (Hazard Communication), prior to the first assignment in an area where calcium is used or stored.

If symptoms develop or overexposure is suspected, the following may also be useful:

- ☑ Urine and serum calcium levels.
- ☑ Lung function tests.
- ☑ Consider chest X-ray following acute overexposure (may be negative if taken immediately after exposure due to delayed development of pulmonary edema).

Any evaluation should include a careful history of past and present symptoms with a medical examination. Medical tests that look for damage already done are not a substitute for controlling exposure. Also, since smoking can cause heart disease, emphysema, and a number of other respiratory disorders, smokers exposed to lung irritants such as calcium metal dusts and fumes may experience symptoms much more quickly than non-smokers under the same conditions of exposure. Prudent risk management requires proper consideration of *all* factors which may be causing the appearance of exposures symptoms in the work place.

Other methods to reduce exposure include:

- ☑ Where possible, enclose operations and use local exhaust ventilation at the site of chemical release. If local exhaust ventilation or enclosure is not used, respiratory protection should be mandatory.
- ☑ Always ensure that proper protective clothing is worn when using chemical substances.
- ☑ Wash thoroughly immediately after exposure to calcium and at the end of the work shift or before eating, drinking, or smoking.
- ☑ Hazard warning information should be posted in the work area. In addition, as part of an on-going education and training program, all information on the health and safety hazards of calcium should be communicated to all potentially exposed workers.
- ☑ Eye wash stations and emergency showers should be located in the immediate work area.
- ☑ Work clothing that has been contaminated with calcium should never be worn home where family members can be exposed as well. Contaminated clothing should be laundered at work or by an authorized service by personnel who have been trained in the hazards associated with exposure to calcium metal and calcium compounds.

RISK ASSESSMENT: ENVIRONMENT

General Assessment

The environment is at risk of exposure during transportation, storage, disposal, or destruction of calcium metal. In almost every scenario, the threat of environmental exposure is contingent upon the proper handling of the chemical substance. Accidental spills, large or small, can result in fire, explosion, and possible contamination of the surrounding environmental mediums (water, soil, and air).

Calcium can ignite spontaneously if exposed to air or if kept in the presence of water. It can detonate violently in contact with alkali hydroxides and may react violently in contact with oxidizers, alcohols, and halogens. Reactions with water and heat produce toxic and flammable calcium hydroxide and flammable hydrogen gas. In a fire, calcium will emit toxic and poisonous gases. These characteristics require special consideration during any emergency situation involving a leak or spill of calcium.

Although calcium is an alkaline-earth element (atomic number 20), it does not occur free in nature. It can enter the environment through industrial discharge, spills, or leaks.

Acute Ecological Effects

Acute (short-term) toxic effects may include the death of animals, birds, or fish and death or low growth rate in plants. Acute effects are seen 2 to 4 days after animals or plants are exposed to calcium.

Calcium metal has low acute toxicity to aquatic life. Insufficient data are available to evaluate the short-term effects of calcium exposure to plants, birds or land animals.

Chronic Ecological Effects

Chronic (long-term) toxic effects may include shortened life span, reproductive problems, lower fertility, and changes in appearance or behavior in exposed animals. These effects can be seen long after first exposure(s) to toxic chemicals.

Calcium has low chronic toxicity to aquatic life. Insufficient data are available to evaluate the long-term effects of calcium exposure to plants, birds or land animals.

Water Solubility

Calcium metal is insoluble in water. It reacts in water to produce heat, calcium hydroxide (a corrosive), and flammable hydrogen gas. Concentrations of less than 1 milligram will not mix with a liter of water.

Persistence in the Environment

Calcium is highly persistent in the aquatic environment, with a half-life of greater than 200 days. The half-life of a pollutant is the amount of time it takes for one half of the chemical to be degraded.

Bioaccumulation in Aquatic Organisms

Some substances increase in concentration, or *bioaccumulate,* in living organisms as they breathe contaminated air, drink contaminated water, or eat contaminated food. These chemicals can become concentrated in the tissues and internal organs of animals as well as humans.

The concentration of calcium found in fish is expected to be considerably higher than the average concentration of calcium in the water from which the fish was taken.

Recommended Risk-Reduction Measures

Proper training of all transporters will reduce the likelihood of a mishap or accidents resulting in a leak or spill to the environment. The correct use of DOT labeling (per the MSDS) on all containers, trucks, and rail cars will enable emergency responders to react properly and quickly to any disaster thereby reducing the potential risk to the environment and to personnel.

If a spill or leak to the environment has occurred, fire department, emergency response and/or hazardous materials spill personnel should be notified immediately. Do not dry sweep. When vacuuming, a high efficiency particulate air (HEPA) filter should be used, not a standard shop vacuum. Do NOT use water to rinse residue. Deposit collected materials in sealed drum for disposal. Cleanup should be attempted only by those trained in proper spill containment procedures. Contaminated soils may need to be removed and replaced with clean soil. If calcium should contact the water table, aquifer, or navigable waterway, time is of the essence. Calcium is insoluble but reactive in water, therefore, total containment and remediation may not be entirely possible. When such spills occur, the local and/or state emergency response authorities must be notified. A comprehensive emergency response or disaster preparedness/recovery plan should be in place prior to any operations involving the use, transportation, storage, or disposal of calcium. If calcium is spilled or leaked, the following specific steps are recommended:

- ☑ Restrict persons not wearing protective clothing from area of spill or leak until cleanup is complete and area can be opened for normal work.
- ☑ Collect powdered materials in most convenient and safe manner possible and place in sealed drums for disposal. Do not dry sweep, use HEPA vacuum instead.
- ☑ It may be necessary to dispose of calcium as a hazardous waste. The responsible state agency or the regional office of the federal Environmental Protection Agency (EPA) should be contacted for specific recommendations.

RISK ASSESSMENT: BUSINESS

General Assessment

Accidents or mishaps involving calcium can present a significant threat to business operations. The loss or damage of equipment or facilities can significantly affect fiscal viability. Lawsuits that may result from personnel injury/death, public exposures, and/or environmental contamination will also require a serious expenditure of resources. Media attention surrounding an injury, death, or environmental damage can also result in a loss of profits and loss of current as well as future business.

Recommended Risk-Reduction Measures

Company attorneys, safety and health professionals, and environmental specialists should be involved in the development of any such procedures. A company official should be pre-designated as a public relations officer with specific training in dealing with the press. Corporate plans and policies should be developed, approved, and implemented long before any need for such arises.

REFERENCES

Ahlbom, A. 1993. *Biostatistics for Epidemiologists.* New York: CRC Press/Lewis Publishers.

American Conference of Governmental Industrial Hygienists. 1988. *Documentation of the Threshold Limit Values and Biological Exposure Indices,* 5th Edition (with updates). Cincinnati: ACGIH

Asante-Duah, D. K. 1993. *Hazardous Waste Risk Assessment.* New York: CRC Press/Lewis Publishers.

Bloom, A. D. and F. J. DeSerres. 1995. *Ecotoxicity and Human Health.* New York: CRC Press/Lewis Publishers.

Brusick, D. J. 1994. *Methods for Genetic Risk Assessment.* New York: CRC Press/Lewis Publishers.

Calabrese, E. J., E. M. Kenyon. *Air Toxics and Risk Assessment.* New York: CRC Press/Lewis Publishers.

Calabrese, E. J. 1994. *Biological Effects of Low Level Exposures.* New York: CRC Press/Lewis Publishers.

Cockerman, L. G., B. S. Shane. 1994. *Basic Environmental Toxicology.* New York: CRC Press/Lewis Publishers.

Cothern, C. R., N. P. Ross. 1994. *Environmental Statistics, Assessment, and Forecasting.* New York: CRC Press/Lewis Publishers.

Cothern, R. R. 1993. *Comparative Environmental Risk Assessment.* New York: CRC Press/Lewis Publishers.

Cothern, C. R. 1995. *Handbook for Environmental Risk Decisionmaking.* New York: CRC Press/Lewis Publishers.

Gardner, R. W. 1994. *Chemical Intolerance.* New York: CRC Press/Lewis Publishers.

Hallenbeck, W. H. 1993. *Quantitative Risk Assessment for Environmental and Occupational Health,* 2nd Edition. New York: CRC Press/Lewis Publishers.

Howard, P. H., et. al. 1991. *Handbook of Environmental Degradation Rates.* New York: CRC Press/Lewis Publishers.

Howard, P. H. 1995. *Handbook of Environmental Fate and Exposures Data for Organic Chemicals,* Vol. I-IV. New York: CRC Press/Lewis Publishers.

Jones, F. E. 1994. *Toxic Organic Vapors in the Workplace.* New York: CRC Press/Lewis Publishers.

Keith, L. H. and M. M. Walker. 1995. *Handbook of Air Toxics.* New York: CRC Press/Lewis Publishers.

Landis, W. G., M. H. Yu. 1995. *Introduction to Environmental Toxicology.* New York: CRC Press/Lewis Publishers.

Langerman, N. 1994 *Precautionary Labels for Chemical Containers.* New York: CRC Press/Lewis Publishers.

Lauwerys, R. R., P. Hoet. 1993. *Industrial Chemical Exposure.* New York: CRC Press/Lewis Publishers.

Lewis, R. J., Sr. 1992. *Sax's Dangerous Properties of Industrial Materials,* Volumes 1, 2, and 3, 8th Edition. New York: Van Nostrand Reinhold.

Manahan, S. E. 1994. *Environmental Chemistry,* 6th Edition. New York: CRC Press/Lewis Publishers.

Manahan, S. E. 1992. *Toxicological Chemistry,* 2nd Edition. New York: CRC Press/Lewis Publishers.

Meyer, E. 1990. *Chemistry of Hazardous Materials.* Englewood Cliffs, NJ: Prentice Hall.

Patnaik, P. 1992. *A Comprehensive Guide to the Hazardous Properties of Chemical Substances.* New York: Van Nostrand Reinhold.

Quigley, D. R. 1994. *Handbook of Emergency Chemical Management.* New York: CRC Press/Lewis Publishers.

Rea, W. J. 1995. *Chemical Sensitivity,* Vol. I-III. New York: CRC Press/Lewis Publishers.

Sacarello, H. L. A. 1994. *The Comprehensive Handbook of Hazardous Materials.* New York: CRC Press/Lewis Publishers.

Sherman, J. D. 1988. *Chemical Exposure and Disease.* New York: Van Nostrand Reinhold.

Talbott, E., G. Craun. 1995. *An Introduction to Environmental Epidemiology.* New York: CRC Press/Lewis Publishers.

U.S. Department of Health and Human Services (DHHS), National Institute for Occupational Health (NIOHS). 1994. *NIOSH Pocket Guide to Hazardous Chemicals.* Publication Number 94-116. Washington, D.C.; U.S. Government Printing Office

MATERIAL SAFETY DATA SHEET

CHEMICAL NAME	
CALCIUM CARBIDE	**CRC PRESS/LEWIS PUBLISHERS DISCLAIMER:** THE INFORMATION AND RECOMMENDATIONS PRESENTED HEREIN ARE BASED ON SOURCES BELIEVED TO BE RELIABLE. CRC MAKES NO REPRESENTATION ON ITS COMPLETENESS OR ACCURACY. IT IS THE RESPONSIBILITY OF THE USER TO DETERMINE THE CHEMICAL'S SUITABILITY FOR ITS INTENDED USE, THE CHEMICAL'S SAFE USE, AND THE CHEMICAL'S PROPER DISPOSAL. NO REPRESENTATIONS AND/OR WARRANTIES, EITHER EXPRESSED OR IMPLIED, OF THE MERCHANTABILITY OR FITNESS FOR A PARTICULAR PURPOSE, OR OF ANY OTHER NATURE, ARE MADE WITH RESPECT TO THE INFORMATION PROVIDED IN THIS MSDS OR TO THE CHEMICAL TO WHICH INFORMATION MAY REFER. CRC NEITHER ASSUMES NOR AUTHORIZES ANY OTHER PERSON TO ASSUME FOR IT, ANY OTHER ADDITIONAL RESPONSIBILITY OR LIABILITY FOR THE USE OF, OR RELIANCE UPON, THIS INFORMATION.

HAZARD WARNING INFORMATION

HEALTH	FIRE	REACTIVE	OTHER	DEGREE OF HAZARD	COLOR CODING	OTHER CODES
1	4	2	~~W~~	0 = Minimum Hazard 1 = Slight Hazard 2 = Moderate Hazard 3 = Serious Hazard 4 = Severe Hazard	HEALTH = BLUE FIRE = RED REACTIVITY = YELLOW OTHER = WHITE	OX = Oxidizer ACID = Acid ALK = Alkali COR = Corrosive ~~W~~ = Use No Water

SECTION I - GENERAL INFORMATION

Characterization **Calcium Compound (Metal)**	RCRA Number **None**	EPA Class **Not Applicable**
DOT Proper Shipping Name **Calcium Carbide**	Chemical Abstract Service (CAS) Number **75-20-7**	
DOT Hazard Class and Label Requirements **Flammable Solid; Dangerous When Wet**	DOT Emergency Guide Code **No Citation**	
DOT Identification Number **UN 1402**	Atomic Formula **C_2Ca**	

Synonyms
Acetylenogen; calcium acetylide; calcium dicarbide.

SECTION II - HAZARDOUS INGREDIENTS AND IDENTITY INFORMATION

Hazardous Components (specific identity)	OSHA Exposure Criteria	NIOSH Exposure Criteria	Immediately Dangerous to Life and Health (IDLH)	ACGIH Exposure Criteria
Calcium carbide (derivation: By interaction of pulverized limestone or quicklime with crushed coke or anthracite in an electric furnace).	**PEL:** Not Established **STEL:** Not Established	**REL:** Not Established **STEL:** Not Established	Not Determined	**TLV:** Not Established **STEL:** Not Established

SECTION III - PHYSICAL AND CHEMICAL CHARACTERISTICS

Boiling Point **Not Found**	Specific Gravity (H_2O = 1) **2.22**
Vapor Pressure (mm Hg) **Not Found**	Molecular Weight **64.10**
Vapor Density (Air = 1) **2.2**	Melting Point **4172°F (2300°C)**

Solubility
Decomposes in hot and cold water to form acetylene, calcium hydroxide, and heat.

Appearance and Odor
A grayish-black, hard solid with irregular shaped lumps. Moisture creates a garlic-like odor.

SECTION IV - FIRE AND EXPLOSION HAZARD DATA

Flash Point (method used) **Not Determined**	Explosive Limits in Air % by Volume **LEL: Not Applicable UEL: Not Applicable**
NFPA Classification **Non-Flammable Solid (when dry)**	Autoignition Temperature **Not Determined**

Extinguishing Media
Do NOT use water (reacts!), foam or halogenated hydrocarbons. Use class "D" extinguishing agent, dry powders to smother, dry sand, or G-1 graphite powder. Carbon dioxide may be ineffective.

Special Fire Fighting Procedures
Poisonous gases are produced in fire. Wear full protective clothing and self-contained breathing apparatus (SCBA). If possible and without risk, move containers from fire. If necessary, use a water spray to keep fire-exposed containers cool with extreme caution. Do NOT get water inside containers! Do NOT release runoff to sewers or waterways.

Unusual Fire and Explosion Hazards
When dry, a non-flammable solid. When wet, liberates flammable acetylene gas. Reaction with water may also liberate enough heat to ignite the acetylene gas. Isolate fire area and deny entry.

SECTION V - REACTIVITY DATA

Stability		Conditions to Avoid **Calcium carbide is stable under normal conditions of temperature and pressure. Keep away from incompatible materials, especially water and moisture, since violent and or explosive reactions can occur.**
Stable X	Unstable	Incompatibility *(materials to avoid)* **Incompatible with strong oxidizers water, selenium, sliver nitrate, stannous chloride, sulfur, sodium peroxide, potassium hydroxide, and chlorine. Reacts with calcium and brass to form explosive compounds.**
Hazardous Polymerization		Conditions to Avoid **Under normal conditions of temperature and pressure, hazardous polymerization of calcium carbide is not known to occur.**
May Occur	Will Not Occur X	Hazardous Decomposition or By-products **When heated to decomposition, calcium carbide can emit toxic/poisonous fumes and gases. These include acetylene, hydrogen chloride, and hydrogen gas.**

SECTION VI - HEALTH HAZARD DATA

Primary Route(s) of Entry:	Inhalation?	Absorption (skin)?	Ingestion?
	X		

Health Hazards

INHALATION: Severe irritation of the eyes, lips, mouth, throat, and the mucosa of the respiratory tract. Acetylene gas is formed in contact with moisture and can cause heart injury and CNS depression with impaired judgment, dizziness, headache, and other systemic effects.

EYES & SKIN: Burns and blindness likely. Conjunctivitis, swelling of the lids, corneal ulceration, and possible degeneration to corneal opacities. Deep skin burns on contact.

INGESTION: Not likely. If occurs, may cause serious mouth, esophagus, and stomach burns.

Carcinogenicity	NTP Listed?	IARC Cancer Review Group?	OSHA Regulated?	Target Organs?
Unknown Human Unknown Animal	No	No	No	Respiratory system, skin, eyes.

Medical Conditions Generally Aggravated by Exposure

None reported.

Emergency and First-aid Procedures

Eye contact: Blindness is likely! Flush immediately with water for 30 minutes (minimum), seek medical attention. Skin contact: Remove all contaminated clothing. Immediately wash area with large amounts of water. For inhalation: Remove the person from exposure. Provide respiratory assistance and CPR. Transfer to medical facility. If swallowed: Provide breathing assistance if necessary. Never attempt to give a convulsing or unconscious person anything by mouth. If victim is conscious, give 1 to 2 glasses of water to dilute, do NOT induce vomiting. Treat as caustic ingestion. Get medical assistance.

SECTION VII - PRECAUTIONS FOR SAFE HANDLING AND USE

Steps to be Taken in Case Material is Released or Spilled

Collect powdered material in most convenient manner possible and deposit in sealed drum. Keep those not involved in cleanup from entering area. Do NOT dry sweep (creates airborne dusts). Use a vacuum equipped with a high efficiency particulate air (HEPA) filter. Do NOT use water.

Preferred Waste Disposal Method

No Citation.

Precautions to be Taken in Handling and Storage

Store in tightly closed containers in cool, well-ventilated area and protect from physical damage. Sources of ignition are prohibited where calcium carbide is used, handled, or stored. Keep away from incompatible materials and especially water. An inert blanket of nitrogen in containers is suggested.

Other Precautions and Warnings

Outside or detached storage facility is recommended.

SECTION VIII - CONTROL MEASURES AND PERSONAL PROTECTIVE EQUIPMENT

Respiratory Protection *(specify type)*

No exposure levels established. A full facepiece respirator with high efficiency particulate air (HEPA) filter may suffice. For best protection, use a supplied air respirator with a full facepiece, hood, or helmet or a self-contained breathing apparatus (SCBA) operated in positive pressure mode.

Ventilation

Local exhaust or general mechanical systems recommended.

Protective Gloves	Eye Protection	Other Protective Clothing
Water-proof Material	Chemical/Dust Goggles or Face Mask	Protective Suit or Clothing

Work/Hygiene Practices

Always wash hands thoroughly after using chemical; never bring food, drink, or smoking materials into vicinity of chemicals. Use Vaseline barrier to protect bare or exposed skin.

CALCIUM CARBIDE

C_2Ca CAS: 75-20-7

IDENTIFICATION AND TYPICAL USES

Calcium carbide appears as a grayish-black solid material with irregular-shaped lumps. When dry, there is no detectable odor. Trace amounts of moisture will cause a garlic-like odor. It is used to generate acetylene gas and oxyacetylene for welding and cutting, in acetylene generators for acetylene lamps, in the manufacture of calcium cyanamide, and chloroethylenes. It is a vinyl acetate monomer. It is used in the manufacture of acetylene chemicals, calcium, iron, lampblack, and alloys. It is a reddening agent and is used in fire signals for marine service.

RISK ASSESSMENT: HEALTH

General Assessment

Calcium carbide is dangerous and corrosive on contact with moisture. This characteristic creates a serious health exposure risk with regard to eye, skin, and mucous membrane contact. Human exposures occur primarily through *inhalation* and skin contact.

Because it forms corrosive calcium hydroxide and acetylene in the presence of moisture, it can cause severe and thermal caustic burns. It is especially damaging to the eyes due to the water content and exposure can cause conjunctivitis with corneal ulceration and possible degeneration to corneal opacities. Blindness is extremely possible. Skin contact will cause severe burns, especially if skin is moist (sweat). Burns can be deep and painful (up to 5 millimeters in depth).

In the presence of moisture, calcium carbide will liberate acetylene gas. If inhaled in high concentrations, acetylene can cause depression of the central nervous system (CNS) with lightheadedness, headache, dizziness, impaired judgment, and other unspecified effects. There can be inflammation of respiratory organs and mucosa leading to ulceration and scarring, and dryness, cracking, and ulceration of the lips.

Acute Health Effects

The following acute (short-term) health effects may occur immediately or shortly after exposure to calcium carbide:

Skin: Irritation, inflammation, and deep, painful burns.

Eye: Corrosive destruction to eyes and eye tissue. Moisture causes formation of caustic calcium hydroxide and can cause blindness.

Lung: Inhalation of calcium carbide causes severe irritation of the mucous membranes in the respiratory tract. In contact with moisture, it releases acetylene gas which will effectively depress the CNS.

Chronic Health Effects

The following chronic (long-term) health effects may occur at some time after exposure to calcium carbide and can last for months or even years:

Cancer Hazards: According to the information presented in the references, calcium carbide has not been adequately tested for its ability to cause cancer in test animals.

Reproductive Hazard: According to information available in the references, calcium carbide has not been adequately tested for its ability to adversely affect reproduction in laboratory animals.

Other Chronic Effects: Long-term exposures mail cause nail and eye lesions, with mucopurulent (mucous and pus) secretions and hyperaemia (excessive blood) of the lids and conjunctiva of the eyes.

Recommended Risk-Reduction Measures

Personnel should avoid direct contact with calcium carbide. If a less toxic chemical cannot be substituted for a hazardous material, then *engineering controls* are the most effective method of reducing exposures.

The best protection is to enclose operations and/or provide local exhaust ventilation at the site of chemical release. While not always operationally feasible, isolating operations can also reduce exposure. Using respiratory protection is less effective than the controls mentioned above, but is still advisable whenever working with or around calcium carbide. No exposure levels have been established for calcium carbide metal. This does not mean that exposure is without health risk. A full facepiece respirator equipped with a high efficiency particulate air (HEPA) filter equipped with a dust/mist/fume pre-filter may suffice. For the best protection, use a supplied-air respirator with a full facepiece operated in positive pressure mode or a MSHA/NIOSH approved self-contained breathing apparatus (SCBA) with full facepiece operated in pressure demand or other continuous flow mode. If a full facepiece is not available, then chemical/dust goggles should be worn to protect the eyes. Whenever a chemical dust hazard exists, a face shield and a rubber apron should be worn. To prevent hand and skin exposures, impervious, water-proof rubber gloves should be worn. A thin layer of Vaseline will afford additional protection. Moist skin must not contact calcium carbide.

Administrative controls should also be in place to minimize the potential for human exposures. These may include written procedures or policies which specify the methods and techniques that will be practiced whenever personnel are to work with calcium carbide.

All personnel should receive *training* on the use, hazards, protective measures, emergency actions, and other precautions per 29 CFR 1910.1200 (Hazard Communication), prior to the first assignment in an area where calcium carbide is used or stored.

If symptoms develop or overexposure is suspected, the following may also be useful:

- ☑ Urine and serum calcium levels.
- ☑ Lung function tests.
- ☑ Consider chest X-ray following acute overexposure.

Any evaluation should include a careful history of past and present symptoms with a medical examination. Medical tests that look for damage already done are not a substitute for controlling exposure. Also, since smoking can cause heart disease, emphysema, and a number of other respiratory disorders, smokers exposed to lung irritants such as calcium carbide dusts and fumes may experience symptoms much more quickly than non-smokers under the same conditions of exposure. Prudent risk management requires proper consideration of *all* factors which may be causing the appearance of exposures symptoms in the work place.

Other methods to reduce exposure include:

- ☑ Where possible, enclose operations and use local exhaust ventilation at the site of chemical release. If local exhaust ventilation or enclosure is not used, respiratory protection should be mandatory.
- ☑ Always ensure that proper protective clothing is worn when using chemical substances.
- ☑ Wash thoroughly immediately after exposure to calcium carbide and at the end of the work shift or before eating, drinking, or smoking.
- ☑ Hazard warning information should be posted in the work area. In addition, as part of an on-going education and training program, all information on the health and safety hazards of calcium carbide should be communicated to all potentially exposed workers.
- ☑ Eye wash stations and emergency showers should be located in the immediate work area.
- ☑ Work clothing that has been contaminated with calcium carbide should never be worn home where family members can be exposed as well. Contaminated clothing should be laundered at work or by an authorized service by personnel who have been trained in the hazards associated with exposure to calcium carbide metal and calcium carbide compounds.

RISK ASSESSMENT: ENVIRONMENT

General Assessment

The environment is at risk of exposure during transportation, storage, disposal, or destruction of calcium carbide. In almost every scenario, the threat of environmental exposure is contingent upon the proper handling of the chemical substance. Accidental spills, large or small, can result in fire, explosion, and possible contamination of the surrounding environmental mediums (water, soil, and air).

Calcium carbide is incompatible with water. When hydrolyzed with water, impurities such as calcium phosphate or calcium arsenate may produce toxic phosgene or arsine. At high temperatures, an incandescent reaction occurs with chlorine, bromine, iodine, lead fluoride, hydrogen chloride gas, magnesium, and heat Addition to silver nitrate solutions precipitates explosive silver acetylide. Mixtures with iron or tin chlorides easily ignite. Reactions with water and heat produce toxic and flammable acetylene gas. In a fire, calcium carbide will emit toxic and poisonous gases. These characteristics require special consideration during any emergency situation involving a leak or spill of calcium carbide.

Calcium carbide can enter the environment through industrial discharge, spills, or leaks.

Acute Ecological Effects

Acute (short-term) toxic effects may include the death of animals, birds, or fish and death or low growth rate in plants. Acute effects are seen 2 to 4 days after animals or plants are exposed to calcium carbide.

Insufficient data are available to evaluate the short-term effects of calcium carbide exposure to aquatic life, plants, birds or land animals.

Chronic Ecological Effects

Chronic (long-term) toxic effects may include shortened life span, reproductive problems, lower fertility, and changes in appearance or behavior in exposed animals. These effects can be seen long after first exposure(s) to toxic chemicals.

Insufficient data are available to evaluate the long-term effects of calcium carbide exposure to aquatic life, plants, birds or land animals.

Water Solubility

Calcium carbide decomposes in water to produce calcium hydroxide and flammable acetylene gas.

Persistence in the Environment

Calcium carbide is non-persistent in the aquatic environment, with a half-life of less than 2 days. The half-life of a pollutant is the amount of time it takes for one half of the chemical to be degraded.

Bioaccumulation in Aquatic Organisms

Some substances increase in concentration, or *bioaccumulate,* in living organisms as they breathe contaminated air, drink contaminated water, or eat contaminated food. These chemicals can become concentrated in the tissues and internal organs of animals as well as humans who have been exposed.

The concentration of calcium found in fish is expected to be considerably higher than the average concentration of calcium in the water from which the fish was taken.

Recommended Risk-Reduction Measures

Proper training of all transporters will reduce the likelihood of a mishap or accidents resulting in a leak or spill to the environment. The correct use of DOT labeling (per the MSDS) on all containers, trucks, and rail cars will enable emergency responders to react properly and quickly to any disaster thereby reducing the potential risk to the environment and to personnel.

If a spill or leak to the environment has occurred, fire department, emergency response and/or hazardous materials spill personnel should be notified immediately. Do not dry sweep. When vacuuming, a high efficiency particulate air (HEPA) filter should be used, not a standard shop vacuum. Do NOT use water to rinse residue. Deposit collected materials in sealed drum for disposal. Cleanup should be attempted only by those trained in proper spill containment procedures. Contaminated soils may need to be removed and replaced with clean soil. If calcium carbide should contact the water table, aquifer, or navigable waterway, time is of the essence. Calcium carbide is insoluble but reactive in water, therefore, total containment and remediation may not be entirely possible. When such spills occur, the local and/or state emergency response authorities must be notified. A comprehensive emergency response or disaster preparedness/recovery plan should be in place prior to any operations involving the use, transportation, storage, or disposal of calcium carbide. If calcium carbide is spilled or leaked, the following specific steps are recommended:

- ☑ Restrict persons not wearing protective clothing from area of spill or leak until cleanup is complete and area can be opened for normal work.
- ☑ Remove all ignition sources and ventilate area.
- ☑ Collect powdered materials in most convenient and safe manner possible and place in sealed drums for disposal. Do not dry sweep, use HEPA vacuum instead.
- ☑ It may be necessary to dispose of calcium carbide as a hazardous waste. The responsible state agency or the regional office of the federal Environmental Protection Agency (EPA) should be contacted for specific recommendations.

RISK ASSESSMENT: BUSINESS

General Assessment

Accidents or mishaps involving calcium carbide can present a significant threat to business operations. The loss or damage of equipment or facilities can significantly affect fiscal viability. Lawsuits that may result from personnel injury/death, public exposures, and/or environmental contamination will also require a serious expenditure of resources. Media attention surrounding an injury, death, or environmental damage can also result in a loss of profits and loss of current as well as future business.

Recommended Risk-Reduction Measures

Company attorneys, safety and health professionals, and environmental specialists should be involved in the development of any such procedures. A company official should be pre-designated as a public relations officer with specific training in dealing with the press. Corporate plans and policies should be developed, approved, and implemented long before any need for such arises.

REFERENCES

Ahlbom, A. 1993. *Biostatistics for Epidemiologists.* New York: CRC Press/Lewis Publishers.

American Conference of Governmental Industrial Hygienists. 1988. *Documentation of the Threshold Limit Values and Biological Exposure Indices,* 5th Edition (with updates). Cincinnati: ACGIH

Asante-Duah, D. K. 1993. *Hazardous Waste Risk Assessment.* New York: CRC Press/Lewis Publishers.

Bloom, A. D. and F. J. DeSerres. 1995. *Ecotoxicity and Human Health.* New York: CRC Press/Lewis Publishers.

Brusick, D. J. 1994. *Methods for Genetic Risk Assessment.* New York: CRC Press/Lewis Publishers.

Calabrese, E. J., E. M. Kenyon. *Air Toxics and Risk Assessment.* New York: CRC Press/Lewis Publishers.

Calabrese, E. J. 1994. *Biological Effects of Low Level Exposures.* New York: CRC Press/Lewis Publishers.

Cockerman, L. G., B. S. Shane. 1994. *Basic Environmental Toxicology.* New York: CRC Press/Lewis Publishers.

Cothern, C. R., N. P. Ross. 1994. *Environmental Statistics, Assessment, and Forecasting.* New York: CRC Press/Lewis Publishers.

Cothern, R. R. 1993. *Comparative Environmental Risk Assessment.* New York: CRC Press/Lewis Publishers.

Cothern, C. R. 1995. *Handbook for Environmental Risk Decisionmaking.* New York: CRC Press/Lewis Publishers.

Gardner, R. W. 1994. *Chemical Intolerance.* New York: CRC Press/Lewis Publishers.

Hallenbeck, W. H. 1993. *Quantitative Risk Assessment for Environmental and Occupational Health,* 2nd Edition. New York: CRC Press/Lewis Publishers.

Howard, P. H., et. al. 1991. *Handbook of Environmental Degradation Rates.* New York: CRC Press/Lewis Publishers.

Howard, P. H. 1995. *Handbook of Environmental Fate and Exposures Data for Organic Chemicals,* Vol. I-IV. New York: CRC Press/Lewis Publishers.

Jones, F. E. 1994. *Toxic Organic Vapors in the Workplace.* New York: CRC Press/Lewis Publishers.

Keith, L. H. and M. M. Walker. 1995. *Handbook of Air Toxics.* New York: CRC Press/Lewis Publishers.

Landis, W. G., M. H. Yu. 1995. *Introduction to Environmental Toxicology.* New York: CRC Press/Lewis Publishers.

Langerman, N. 1994 *Precautionary Labels for Chemical Containers.* New York: CRC Press/Lewis Publishers.

Lauwerys, R. R., P. Hoet. 1993. *Industrial Chemical Exposure.* New York: CRC Press/Lewis Publishers.

Lewis, R. J., Sr. 1992. *Sax's Dangerous Properties of Industrial Materials,* Volumes 1, 2, and 3, 8th Edition. New York: Van Nostrand Reinhold.

Manahan, S. E. 1994. *Environmental Chemistry,* 6th Edition. New York: CRC Press/Lewis Publishers.

Manahan, S. E. 1992. *Toxicological Chemistry,* 2nd Edition. New York: CRC Press/Lewis Publishers.

Meyer, E. 1990. *Chemistry of Hazardous Materials.* Englewood Cliffs, NJ: Prentice Hall.

Patnaik, P. 1992. *A Comprehensive Guide to the Hazardous Properties of Chemical Substances.* New York: Van Nostrand Reinhold.

Quigley, D. R. 1994. *Handbook of Emergency Chemical Management.* New York: CRC Press/Lewis Publishers.

Rea, W. J. 1995. *Chemical Sensitivity,* Vol. I-III. New York: CRC Press/Lewis Publishers.

Sacarello, H. L. A. 1994. *The Comprehensive Handbook of Hazardous Materials.* New York: CRC Press/Lewis Publishers.

Sherman, J. D. 1988. *Chemical Exposure and Disease.* New York: Van Nostrand Reinhold.

Talbott, E., G. Craun. 1995. *An Introduction to Environmental Epidemiology.* New York: CRC Press/Lewis Publishers.

U.S. Department of Health and Human Services (DHHS), National Institute for Occupational Health (NIOHS). 1994. *NIOSH Pocket Guide to Hazardous Chemicals.* Publication Number 94-116. Washington, D.C.; U.S. Government Printing Office

MATERIAL SAFETY DATA SHEET

CHEMICAL NAME	
CALCIUM CHLORIDE	**CRC PRESS/LEWIS PUBLISHERS DISCLAIMER:** THE INFORMATION AND RECOMMENDATIONS PRESENTED HEREIN ARE BASED ON SOURCES BELIEVED TO BE RELIABLE. CRC MAKES NO REPRESENTATION ON ITS COMPLETENESS OR ACCURACY. IT IS THE RESPONSIBILITY OF THE USER TO DETERMINE THE CHEMICAL'S SUITABILITY FOR ITS INTENDED USE, THE CHEMICAL'S SAFE USE, AND THE CHEMICAL'S PROPER DISPOSAL. NO REPRESENTATIONS AND/OR WARRANTIES, EITHER EXPRESSED OR IMPLIED, OF THE MERCHANTABILITY OR FITNESS FOR A PARTICULAR PURPOSE, OR OF ANY OTHER NATURE, ARE MADE WITH RESPECT TO THE INFORMATION PROVIDED IN THIS MSDS OR TO THE CHEMICAL TO WHICH INFORMATION MAY REFER. CRC NEITHER ASSUMES NOR AUTHORIZES ANY OTHER PERSON TO ASSUME FOR IT, ANY OTHER ADDITIONAL RESPONSIBILITY OR LIABILITY FOR THE USE OF, OR RELIANCE UPON, THIS INFORMATION.

HAZARD WARNING INFORMATION

HEALTH	FIRE	REACTIVE	OTHER	DEGREE OF HAZARD	COLOR CODING	OTHER CODES
2	0	1	~~W~~	0 = Minimum Hazard 1 = Slight Hazard 2 = Moderate Hazard 3 = Serious Hazard 4 = Severe Hazard	HEALTH = BLUE FIRE = RED REACTIVITY = YELLOW OTHER = WHITE	OX = Oxidizer ACID = Acid ALK = Alkali COR = Corrosive ~~W~~ = Use No Water

SECTION I - GENERAL INFORMATION

Characterization	RCRA Number	EPA Class
Calcium Compound	**None**	**Not Applicable**

DOT Proper Shipping Name **Not Listed**	Chemical Abstract Service (CAS) Number **10043-52-4**
DOT Hazard Class and Label Requirements **No Citation**	DOT Emergency Guide Code **No Citation**
DOT Identification Number **No Citation**	Atomic Formula **$CaCl_2$**

Synonyms
Calpus; Caltac; Dowflake; Liquidow; Peladow; Snowmelt; Superflake (anhydrous).

SECTION II - HAZARDOUS INGREDIENTS AND IDENTITY INFORMATION

Hazardous Components (specific identity)	OSHA Exposure Criteria	NIOSH Exposure Criteria	Immediately Dangerous to Life and Health (IDLH)	ACGIH Exposure Criteria
Calcium chloride (derivation: By action of hydrochloric acid on calcium carbonate and subsequent crystallization; commercially obtained as a by-product in the Solvay soda process and other processes).	**PEL:** Not Established **STEL:** Not Established	**REL:** Not Established **STEL:** Not Established	**Not Determined**	**TLV:** Not Established **STEL:** Not Established

SECTION III - PHYSICAL AND CHEMICAL CHARACTERISTICS

Boiling Point **>2912°F (1484°C)**	Specific Gravity (H_2O = 1) **2.15**
Vapor Pressure (mm Hg) **Not Found**	Molecular Weight **110.98**
Vapor Density (Air = 1) **Not Found**	Melting Point **1422°F (772°C)**

Solubility
Soluble in water, acetone, alcohol, and acetic acid. Mixing with water creates exothermic reaction.

Appearance and Odor
White, deliquescent (i.e., moisture-absorbing) crystals, flakes, granules, or lumps with no odor.

SECTION IV - FIRE AND EXPLOSION HAZARD DATA

Flash Point (method used) **Not Determined**	Explosive Limits in Air % by Volume **LEL: Not Applicable UEL: Not Applicable**
NFPA Classification **Non-Flammable Solid**	Autoignition Temperature **Not Determined**

Extinguishing Media
Use agent suitable to surrounding fire. Do NOT use water (reaction is exothermic). Dry chemical or foam may suffice for small fires. For large fires, try fog or regular foam.

Special Fire Fighting Procedures
Poisonous gases are produced in fire. Wear full protective clothing and self-contained breathing apparatus (SCBA). Only use a water spray to keep fire-exposed containers cool. Do NOT get water inside containers. Do NOT release runoff from fire-fighting measures to sewers or waterways.

Unusual Fire and Explosion Hazards
Can release extreme heat when wet or upon contact with incompatible materials. When combined with some incompatible materials, explosion and fire can be severe.

SECTION V - REACTIVITY DATA

Stability		
Stability		Conditions to Avoid Calcium chloride is stable under normal conditions of temperature and pressure. Because it will readily absorb moisture, containers must be kept tightly closed. Keep away from incompatible materials.
Stable X	Unstable	Incompatibility *(materials to avoid)* Incompatible with boric acid + calcium oxide; bromine trifluoride; zinc (produces hydrogen gas); causes exothermic polymerization of vinyl ether; explodes when added to furan-2-peroxycarbolic acid; water.
Hazardous Polymerization		Conditions to Avoid Under normal conditions of temperature and pressure, hazardous polymerization of calcium chloride is not known to occur.
May Occur	Will Not Occur X	Hazardous Decomposition or By-products When heated to decomposition, calcium chloride can emit toxic/poisonous fumes and gases. These include toxic chloride fumes.

SECTION VI - HEALTH HAZARD DATA

Primary Route(s) of Entry:	Inhalation?	Absorption (skin)?	Ingestion?
	X		X

Health Hazards

INHALATION: Irritation and slight corrosion of the mucosa of the respiratory tract. Rapid respiration with sneezing, coughing, and irritated throat. There may be burning and pain in the nasal cavity with nosebleeds and possible perforation of the nasal septum.

EYES & SKIN: Skin contact causes irritation and inflammation with burns and small ulcerations. Eye contact with solutions causes burns, lachrymation, and inflammation. Possible eye discharge, skin erythema (redness), and peeling of facial skin.

INGESTION: Foul taste in mouth, nausea, vomiting, G.I. tract irritation, fever, muscle twitching, and possible seizures, rapid breathing, slowed heartbeat, and death (severe cases).

Carcinogenicity	NTP Listed?	IARC Cancer Review Group?	OSHA Regulated?	Target Organs?
Unknown Human Unknown Animal	No	No	No	Skin, eyes, and respiratory system (mucosa).

Medical Conditions Generally Aggravated by Exposure

None reported.

Emergency and First-aid Procedures

Eye contact: Flush immediately with water for 15 minutes (minimum), seek medical attention. Skin contact: Remove all contaminated clothing. Immediately wash area with large amounts of soap and water. For inhalation: Remove the person from exposure. Provide respiratory assistance and CPR. Transfer to medical facility. If swallowed: Provide breathing assistance if necessary. Never attempt to give a convulsing or unconscious person anything by mouth. If victim is conscious, give 1-2 glasses of water to dilute and induce vomiting with Syrup of Ipecac. Seek medical attention immediately.

SECTION VII - PRECAUTIONS FOR SAFE HANDLING AND USE

Steps to be Taken in Case Material is Released or Spilled

Collect powdered material in most convenient manner possible and deposit in sealed drum. Keep those not involved in cleanup from entering area. Do NOT dry sweep (creates airborne dusts). Use a vacuum equipped with a high efficiency particulate air (HEPA) filter. Do NOT use water.

Preferred Waste Disposal Method

No Citation.

Precautions to be Taken in Handling and Storage

Store in tightly closed containers in cool, well-ventilated area. Sources of ignition are prohibited where calcium chloride is stored. Keep away from water. Storage with desiccant is recommended.

Other Precautions and Warnings

Outside or detached storage facility is recommended. Protect from physical damage.

SECTION VIII - CONTROL MEASURES AND PERSONAL PROTECTIVE EQUIPMENT

Respiratory Protection *(specify type)*

No exposure levels established. For best protection, use a supplied air respirator with a full facepiece, hood, or helmet or a self-contained breathing apparatus (SCBA) operated in positive pressure mode.

Ventilation

Local exhaust or general mechanical systems recommended.

Protective Gloves	Eye Protection	Other Protective Clothing
Impervious Material	Chemical/Dust Goggles or Face Mask	Protective Suit or Clothing

Work/Hygiene Practices

Always wash hands thoroughly after using chemical; never bring food, drink, or smoking materials into vicinity of chemicals.

CALCIUM CHLORIDE

$CaCl_2$ CAS: 10043-52-4

IDENTIFICATION AND TYPICAL USES

Calcium chloride appears as white, deliquescent, crystals, flakes, granules, or lumps. There is no detectable odor. It has many uses, including de-icing and dust control on roads, in drilling muds, and in dust-proofing, freeze-proofing and thawing coal, coke, stone, sand, and ore. It is used in concrete conditioning, in the paper and pulp industry, in fungicides, in refrigeration brines, as a drying and desiccating agent, as a sequestrant in foods, a firming agent in tomato canning, in tire weighting, in the manufacture of some pharmaceuticals, and in the production of electrolytic cells.

RISK ASSESSMENT: HEALTH

General Assessment

Calcium chloride is moderately toxic by *ingestion* and *inhalation*. Skin contact with a solution of calcium chloride is also dangerous due to the corrosive and destructive nature of this compound (absorption has not been reported as an entry route). Its carcinogenic potential is not clearly understood in the references. Some mutation data have been reported.

Inhalation causes sneezing, cough, and irritation of the nose and throat. There may be lachrymation (tearing), and a burning sensation in the nasal cavity often accompanied by nosebleeds. In some cases, there may be deterioration and/or perforation of the nasal septum.

Skin contact with a concentrated solution causes irritation, inflammation, and the appearance of small ulcerations on the skin's surface. There may be erythema (redness) and possible peeling of the facial skin. Eye contact will cause tearing, irritation, and inflammation of the eyelids.

Ingestion can cause nausea, foul taste in mouth, vomiting, gastrointestinal tract irritation, fever, muscular twitching, possible seizures, rapid breathing, slowed or diminished heartbeat, and death in cases of severe exposure.

Acute Health Effects

The following acute (short-term) health effects may occur immediately or shortly after exposure to calcium chloride:

Skin: Irritation, rash, and possible burns.

Eye: Irritation, inflammation, watering, and swelling.

Lung: Inhalation of calcium chloride dusts may cause irritation of the mucous membranes in the respiratory tract. There may be burning in the nasal cavity with nosebleeds and possible perforation of the nasal septum.

Other: Ingestion can cause gastrointestinal irritation with systemic reaction up to and including death.

Chronic Health Effects

The following chronic (long-term) health effects may occur at some time after exposure to calcium chloride and can last for months or even years:

Cancer Hazards: According to the information presented in the references, calcium chloride has not been adequately tested for its ability to cause cancer in test animals. The data are conflicting and inconclusive in this regard. Mutation (causes genetic changes) data have been documented and many scientists believe that such chemicals may pose a cancer risk in the long-term. Addition study and research is required.

Reproductive Hazard: According to information available in the references, calcium chloride has not been adequately tested for its ability to adversely affect reproduction in laboratory animals.

Recommended Risk-Reduction Measures

Personnel should avoid direct contact with calcium chloride. If a less toxic chemical cannot be substituted for a hazardous material, then *engineering controls* are the most effective method of reducing exposures.

The best protection is to enclose operations and/or provide local exhaust ventilation at the site of chemical release. While not always operationally feasible, isolating operations can also reduce exposure. Using respiratory protection is less effective than the controls mentioned above, but is still advisable whenever working with or around calcium chloride. No exposure levels have been established for calcium chloride. This does not mean that exposure is without health risk. A full facepiece respirator equipped with a high efficiency particulate air (HEPA) filter equipped with a dust/mist/fume pre-filter may suffice. However, for the best protection use a supplied-air respirator with a full facepiece operated in positive pressure mode or a MSHA/NIOSH approved self-contained breathing apparatus (SCBA) with full facepiece operated in pressure demand or other continuous flow mode. If a full facepiece is not available, then chemical/dust goggles should be worn to protect the eyes. Whenever a chemical dust hazard exists, a face shield and a rubber apron should be worn. To prevent hand and skin exposures, impervious rubber gloves should be worn.

Administrative controls should also be in place to minimize the potential for human exposures. These may include written procedures or policies which specify the methods and techniques that will be practiced whenever personnel are to work with calcium chloride.

All personnel should receive *training* on the use, hazards, protective measures, emergency actions, and other precautions per 29 CFR 1910.1200 (Hazard Communication), prior to the first assignment in an area where calcium chloride is used or stored.

If symptoms develop or overexposure is suspected, the following may also be useful:

☑ Urine and serum calcium levels.

☑ Lung function tests.

☑ Consider chest X-ray following acute overexposure.

Any evaluation should include a careful history of past and present symptoms with a medical examination. Medical tests that look for damage already done are not a substitute for controlling exposure. Also, since smoking can cause heart disease, emphysema, and a number of other respiratory disorders, smokers exposed to lung irritants such as calcium chloride dusts and fumes may experience symptoms much more quickly than non-smokers under the same conditions of exposure. Prudent risk management requires proper consideration of *all* factors which may be causing the appearance of exposures symptoms in the work place.

Other methods to reduce exposure include:

☑ Where possible, enclose operations and use local exhaust ventilation at the site of chemical release. If local exhaust ventilation or enclosure is not used, respiratory protection should be mandatory.

☑ Always ensure that proper protective clothing is worn when using chemical substances.

☑ Wash thoroughly immediately after exposure to calcium chloride and at the end of the work shift or before eating, drinking, or smoking.

☑ Hazard warning information should be posted in the work area. In addition, as part of an on-going education and training program, all information on the health and safety hazards of calcium chloride should be communicated to all potentially exposed workers.

☑ Eye wash stations and emergency showers should be located in the immediate work area.

☑ Work clothing that has been contaminated with calcium chloride should never be worn home where family members can be exposed as well. Contaminated clothing should be laundered at work or by an authorized service by personnel who have been trained in the hazards associated with exposure to calcium chloride and calcium compounds.

RISK ASSESSMENT: ENVIRONMENT

General Assessment

The environment is at risk of exposure during transportation, storage, disposal, or destruction of calcium chloride. In almost every scenario, the threat of environmental exposure is contingent upon the proper handling of the chemical substance. Accidental spills, large or small, can result in fire, explosion, and possible contamination of the surrounding environmental mediums (water, soil, and air).

Calcium chloride may react violently in contact with boric oxide, calcium oxide, and bromine trifluoride. Reactions with water evolves extreme heat. In contact with zinc it will produce explosive hydrogen gas. In a fire, calcium chloride will emit toxic and poisonous gases. These characteristics require special consideration during any emergency situation involving a leak or spill of calcium chloride.

Calcium chloride can enter the environment through industrial discharge, spills, or leaks.

Acute Ecological Effects

Acute (short-term) toxic effects may include the death of animals, birds, or fish and death or low growth rate in plants. Acute effects are seen 2 to 4 days after animals or plants are exposed to calcium chloride. Insufficient data are available to evaluate the short-term effects of calcium chloride exposure to aquatic life, plants, birds or land animals.

Chronic Ecological Effects

Chronic (long-term) toxic effects may include shortened life span, reproductive problems, lower fertility, and changes in appearance or behavior in exposed animals. These effects can be seen long after first exposure(s) to toxic chemicals. Insufficient data are available to evaluate the long-term effects of calcium chloride exposure to aquatic life, plants, birds or land animals.

Water Solubility

Calcium chloride is soluble in water, decomposing and producing heat. Concentrations of 1000 milligrams or more will mix with a liter of water.

Persistence in the Environment

Calcium chloride is only slightly persistent in the aquatic environment, with a half-life of less than 2 days. The half-life of a pollutant is the amount of time it takes for one half of the chemical to be degraded.

Bioaccumulation in Aquatic Organisms

Some substances increase in concentration, or *bioaccumulate,* in living organisms as they breathe contaminated air, drink contaminated water, or eat contaminated food. These chemicals can become concentrated in the tissues and internal organs of animals as well as humans.

The concentration of calcium chloride found in fish is expected to be much lower than the average concentration of calcium chloride in the water from which the fish was taken.

Recommended Risk-Reduction Measures

Proper training of all transporters will reduce the likelihood of a mishap or accidents resulting in a leak or spill to the environment. Proper labeling on all containers, trucks, and rail cars will enable emergency responders to react properly and quickly to any disaster thereby reducing the potential risk to the environment and to personnel.

If a spill or leak to the environment has occurred, fire department, emergency response and/or hazardous materials spill personnel should be notified immediately. Do not dry sweep. When vacuuming, a high efficiency particulate air (HEPA) filter should be used, not a standard shop vacuum. Do NOT use water to rinse residue. Deposit collected materials in sealed drum for disposal. Cleanup should be attempted only by those trained in proper spill containment procedures. Contaminated soils may need to be removed and replaced with clean soil. If calcium chloride should contact the water table, aquifer, or navigable waterway, time is of the essence. Calcium chloride is soluble in water and will decompose. Therefore, total containment and remediation may not be entirely possible. When such spills occur, the local and/or state emergency response authorities must be notified. A comprehensive emergency response or disaster preparedness/recovery plan should be in place prior to any operations involving the use, transportation, storage, or disposal of calcium chloride. If calcium chloride is spilled or leaked, the following specific steps are recommended:

- ☑ Restrict persons not wearing protective clothing from area of spill or leak until cleanup is complete and area can be opened for normal work.
- ☑ Ventilate area of spill or leak and remove all ignition sources.
- ☑ Collect powdered materials in most convenient and safe manner possible and place in sealed drums for disposal. Do NOT dry sweep (generates dangerous airborne dusts), use vacuum equipped with a high-efficiency particulate air (HEPA) filter instead. Do NOT use water to clean residue or damp mop.
- ☑ It may be necessary to dispose of calcium chloride as a hazardous waste. The responsible state agency or the regional office of the federal Environmental Protection Agency (EPA) should be contacted for specific recommendations.

RISK ASSESSMENT: BUSINESS

General Assessment

Accidents or mishaps involving calcium chloride can present a significant threat to business operations. The loss or damage of equipment or facilities can significantly affect fiscal viability. Lawsuits that may result from personnel injury/death, public exposures, and/or environmental contamination will also require a serious expenditure of resources. Media attention surrounding an injury, death, or environmental damage can also result in a loss of profits and loss of current as well as future business.

Recommended Risk-Reduction Measures

Company attorneys, safety and health professionals, and environmental specialists should be involved in the development of any such procedures. A company official should be pre-designated as a public relations officer with specific training in dealing with the press. Corporate plans and policies should be developed, approved, and implemented long before any need for such arises.

REFERENCES

Ahlbom, A. 1993. *Biostatistics for Epidemiologists.* New York: CRC Press/Lewis Publishers.

American Conference of Governmental Industrial Hygienists. 1988. *Documentation of the Threshold Limit Values and Biological Exposure Indices,* 5th Edition (with updates). Cincinnati: ACGIH

Asante-Duah, D. K. 1993. *Hazardous Waste Risk Assessment.* New York: CRC Press/Lewis Publishers.

Bloom, A. D. and F. J. DeSerres. 1995. *Ecotoxicity and Human Health.* New York: CRC Press/Lewis Publishers.

Brusick, D. J. 1994. *Methods for Genetic Risk Assessment.* New York: CRC Press/Lewis Publishers.

Calabrese, E. J., E. M. Kenyon. *Air Toxics and Risk Assessment.* New York: CRC Press/Lewis Publishers.

Calabrese, E. J. 1994. *Biological Effects of Low Level Exposures.* New York: CRC Press/Lewis Publishers.

Cockerman, L. G., B. S. Shane. 1994. *Basic Environmental Toxicology.* New York: CRC Press/Lewis Publishers.

Cothern, C. R., N. P. Ross. 1994. *Environmental Statistics, Assessment, and Forecasting.* New York: CRC Press/Lewis Publishers.

Cothern, R. R. 1993. *Comparative Environmental Risk Assessment.* New York: CRC Press/Lewis Publishers.

Cothern, C. R. 1995. *Handbook for Environmental Risk Decisionmaking.* New York: CRC Press/Lewis Publishers.

Gardner, R. W. 1994. *Chemical Intolerance.* New York: CRC Press/Lewis Publishers.

Hallenbeck, W. H. 1993. *Quantitative Risk Assessment for Environmental and Occupational Health,* 2nd Edition. New York: CRC Press/Lewis Publishers.

Howard, P. H., et. al. 1991. *Handbook of Environmental Degradation Rates.* New York: CRC Press/Lewis Publishers.

Howard, P. H. 1995. *Handbook of Environmental Fate and Exposures Data for Organic Chemicals,* Vol. I-IV. New York: CRC Press/Lewis Publishers.

Jones, F. E. 1994. *Toxic Organic Vapors in the Workplace.* New York: CRC Press/Lewis Publishers.

Keith, L. H. and M. M. Walker. 1995. *Handbook of Air Toxics.* New York: CRC Press/Lewis Publishers.

Landis, W. G., M. H. Yu. 1995. *Introduction to Environmental Toxicology.* New York: CRC Press/Lewis Publishers.

Langerman, N. 1994 *Precautionary Labels for Chemical Containers.* New York: CRC Press/Lewis Publishers.

Lauwerys, R. R., P. Hoet. 1993. *Industrial Chemical Exposure.* New York: CRC Press/Lewis Publishers.

Lewis, R. J., Sr. 1992. *Sax's Dangerous Properties of Industrial Materials,* Volumes 1, 2, and 3, 8th Edition. New York: Van Nostrand Reinhold.

Manahan, S. E. 1994. *Environmental Chemistry,* 6th Edition. New York: CRC Press/Lewis Publishers.

Manahan, S. E. 1992. *Toxicological Chemistry,* 2nd Edition. New York: CRC Press/Lewis Publishers.

Meyer, E. 1990. *Chemistry of Hazardous Materials.* Englewood Cliffs, NJ: Prentice Hall.

Patnaik, P. 1992. *A Comprehensive Guide to the Hazardous Properties of Chemical Substances.* New York: Van Nostrand Reinhold.

Quigley, D. R. 1994. *Handbook of Emergency Chemical Management.* New York: CRC Press/Lewis Publishers.

Rea, W. J. 1995. *Chemical Sensitivity,* Vol. I-III. New York: CRC Press/Lewis Publishers.

Sacarello, H. L. A. 1994. *The Comprehensive Handbook of Hazardous Materials.* New York: CRC Press/Lewis Publishers.

Sherman, J. D. 1988. *Chemical Exposure and Disease.* New York: Van Nostrand Reinhold.

Talbott, E., G. Craun. 1995. *An Introduction to Environmental Epidemiology.* New York: CRC Press/Lewis Publishers.

U.S. Department of Health and Human Services (DHHS), National Institute for Occupational Health (NIOHS). 1994. *NIOSH Pocket Guide to Hazardous Chemicals.* Publication Number 94-116. Washington, D.C.; US. Government Printing Office

MATERIAL SAFETY DATA SHEET

CHEMICAL NAME	CRC PRESS/LEWIS PUBLISHERS DISCLAIMER
CALCIUM CHROMATE	**CRC PRESS/LEWIS PUBLISHERS DISCLAIMER:** THE INFORMATION AND RECOMMENDATIONS PRESENTED HEREIN ARE BASED ON SOURCES BELIEVED TO BE RELIABLE. CRC MAKES NO REPRESENTATION ON ITS COMPLETENESS OR ACCURACY. IT IS THE RESPONSIBILITY OF THE USER TO DETERMINE THE CHEMICAL'S SUITABILITY FOR ITS INTENDED USE, THE CHEMICAL'S SAFE USE, AND THE CHEMICAL'S PROPER DISPOSAL. NO REPRESENTATIONS AND/OR WARRANTIES, EITHER EXPRESSED OR IMPLIED, OF THE MERCHANTABILITY OR FITNESS FOR A PARTICULAR PURPOSE, OR OF ANY OTHER NATURE, ARE MADE WITH RESPECT TO THE INFORMATION PROVIDED IN THIS MSDS OR TO THE CHEMICAL TO WHICH INFORMATION MAY REFER. CRC NEITHER ASSUMES NOR AUTHORIZES ANY OTHER PERSON TO ASSUME FOR IT, ANY OTHER ADDITIONAL RESPONSIBILITY OR LIABILITY FOR THE USE OF, OR RELIANCE UPON, THIS INFORMATION.

HAZARD WARNING INFORMATION

HEALTH	FIRE	REACTIVE	OTHER	DEGREE OF HAZARD	COLOR CODING	OTHER CODES
3	0	0	OX	0 = Minimum Hazard 1 = Slight Hazard 2 = Moderate Hazard 3 = Serious Hazard 4 = Severe Hazard	HEALTH = BLUE FIRE = RED REACTIVITY = YELLOW OTHER = WHITE	OX = Oxidizer ACID = Acid ALK = Alkali COR = Corrosive W = Use No Water

SECTION I - GENERAL INFORMATION

Characterization	**Calcium Compound**	RCRA Number / EPA Class	**U032** / **Toxic Waste**
DOT Proper Shipping Name	**Hazardous Substance (liquid or solid), n.o.s.**	Chemical Abstract Service (CAS) Number	**13765-19-0**
DOT Hazard Class and Label Requirements	**ORM-E**	DOT Emergency Guide Code	**No Citation**
DOT Identification Number	**NA 9188**	Atomic Formula	**$CaCrO_4$**

Synonyms
Calcium chromate (VI); calcium chromate dihydrate; chromic acid, calcium salt (1:1); Gelbin Yellow Ultramarine; Steinbuhl Yellow.

SECTION II - HAZARDOUS INGREDIENTS AND IDENTITY INFORMATION

Hazardous Components (specific identity)	OSHA Exposure Criteria	NIOSH Exposure Criteria	Immediately Dangerous to Life and Health (IDLH)	ACGIH Exposure Criteria
Calcium chromate (derivation: By action of hydrochloric acid on calcium carbonate and subsequent crystallization; commercially obtained as a by-product in the Solvay soda process and other processes).	**PEL (ceiling):** 0.1 mg/m^3 (as chromium) **STEL:** Not Established	**REL (10-hour):** 0.001 mg/m^3 (as chromium) **STEL:** Not Established	**Not Determined**	**TLV:** 0.001 mg/m^3 (as chromium) **Suspected Human Carcinogen**

SECTION III - PHYSICAL AND CHEMICAL CHARACTERISTICS

Boiling Point	**Not Found**	Specific Gravity (H_2O = 1)	**2.89**
Vapor Pressure (mm Hg)	**Not Found**	Molecular Weight	**156.08**
Vapor Density (Air = 1)	**5.4**	Melting Point	**Not Found**

Solubility
Slightly soluble in water. Soluble in dilute acids, insoluble in alcohol.

Appearance and Odor
Bright yellow powder, or monoclinic or rhombic crystals. The hydrate will lose water at 392°F (200°C).

SECTION IV - FIRE AND EXPLOSION HAZARD DATA

Flash Point (method used)	**Not Determined**	Explosive Limits in Air % by Volume	**LEL: Not Applicable UEL: Not Applicable**
NFPA Classification	**Non-Combustible Solid**	Autoignition Temperature	**Not Determined**

Extinguishing Media
Use agent suitable to surrounding fire. Calcium chloride itself does not burn.

Special Fire Fighting Procedures
Poisonous gases are produced in fire. Wear full protective clothing and self-contained breathing apparatus (SCBA). Do NOT release runoff from fire-fighting measures to sewers or waterways. Use water spray to cool fire exposed containers. Move containers from fire if it can be done without risk.

Unusual Fire and Explosion Hazards
Although a non-combustible solid, calcium chromate is a powerful oxidizer that can greatly accelerate the combustion of other materials such as wood, paper, and plastics. When combined with some incompatible materials (boron), resulting fire can be severe.

SECTION V - REACTIVITY DATA

Stability		Conditions to Avoid Calcium chromate is stable under normal conditions of temperature and pressure. Avoid conditions which generate excessive airborne dusts and keep calcium chromate away from incompatible materials.
Stable X	Unstable	Incompatibility (materials to avoid) Will react with most organic matter (paper, wood, plastics, oil), reducing agents (aluminum, sulfur) to form chromium (III). Reacts slowly with acids and ethanol. It will explosively decompose hydrazine.
Hazardous Polymerization		Conditions to Avoid Under normal conditions of temperature and pressure, hazardous polymerization of calcium chromate is not known to occur.
May Occur	Will Not Occur X	Hazardous Decomposition or By-products When heated to decomposition, calcium chromate can emit toxic/poisonous fumes and gases. These may include toxic chromium fumes.

SECTION VI - HEALTH HAZARD DATA

Primary Route(s) of Entry:	Inhalation?	Absorption (skin)?	Ingestion?
	X		

Health Hazards

INHALATION: Severe irritation of the eyes, nose, throat, bronchial tubes, and lungs. Inhaling the dusts or fumes can lead to progressive deterioration and eventual perforation of the nasal septum, pulmonary edema, bronchospasm, and blood changes. Cancer is also possible.

EYES & SKIN: Skin contact causes contact dermatitis and "chrome holes" (usually painless but deep, slow healing holes in the skin). Eye contact causes irritation and conjunctivitis.

INGESTION: Ingestion is not likely, but can occur. Causes violent gastroenteritis, severe circulatory collapse, chronic nephritis, and a possibility for liver damage.

Carcinogenicity	NTP Listed?	IARC Cancer Review Group?	OSHA Regulated?	Target Organs?
Confirmed Human Confirmed Animal	5th Annual Report	Group 1	19 CFR 1910.1000 Table Z-1-A	Skin, eye, resp. sys., liver, blood, circ. sys., kidneys.

Medical Conditions Generally Aggravated by Exposure

Skin disorders may be aggravated by exposure.

Emergency and First-aid Procedures

Eye contact: Flush immediately with water for 15 minutes (minimum), seek medical attention. Skin contact: Remove all contaminated clothing. Immediately wash area with large amounts of soap and water. For inhalation: Remove the person from exposure. Provide respiratory assistance and CPR. Transfer to medical facility. If swallowed: Provide breathing assistance if necessary. Never attempt to give a convulsing or unconscious person anything by mouth. If victim is conscious, give 1-2 glasses of milk. Do NOT induce vomiting. Seek medical attention immediately.

SECTION VII - PRECAUTIONS FOR SAFE HANDLING AND USE

Steps to be Taken in Case Material is Released or Spilled

Collect powdered material in most convenient manner possible and deposit in sealed drum. Keep those not involved in cleanup from entering area. Do NOT dry sweep (creates airborne dusts). Use a vacuum equipped with a high efficiency particulate air (HEPA) filter. Damp mop residue.

Preferred Waste Disposal Method

No Citation.

Precautions to be Taken in Handling and Storage

Store in tightly closed containers in cool, well-ventilated area. Sources of ignition are prohibited where calcium chromate is stored. Storage near use areas is recommended to minimize transport distance.

Other Precautions and Warnings

Protect from physical damage. Work areas should be posted as containing a suspected carcinogen.

SECTION VIII - CONTROL MEASURES AND PERSONAL PROTECTIVE EQUIPMENT

Respiratory Protection (specify type)

For low exposures, use an MSHA/NIOSH approved full facepiece respirator with HEPA filters. For best protection, use a supplied air respirator with a full facepiece, hood, or helmet or a self-contained breathing apparatus (SCBA) operated in positive pressure mode.

Ventilation

Local exhaust or general mechanical systems recommended.

Protective Gloves	Eye Protection	Other Protective Clothing
Impervious Material	Chemical/Dust Goggles or Face Mask	Protective Suit or Clothing

Work/Hygiene Practices

Always wash hands thoroughly after using chemical; never bring food, drink, or smoking materials into vicinity of chemicals.

CALCIUM CHROMATE

$CaCrO_4$ CAS: 13765-19-0

IDENTIFICATION AND TYPICAL USES

Calcium chromate can appear as a bright yellow powder, or monoclinic or rhombic crystals. It is used as a pigment in paints, cement, paper, and rubber. It is also used as a corrosion inhibitor, an oxidizing agent, a depolarlizer for batteries, a coating for light metal alloys, an electroplating cleaning agent, in photochemical processing, and in the treatment of industrial wastes.

RISK ASSESSMENT: HEALTH

General Assessment

Calcium chromate is highly toxic primarily by *inhalation,* although *ingestion* is also possible but not likely. Skin contact poses a moderate health risk (absorption has not been reported as an entry route). It is a known human carcinogen and mutagen.

Inhalation of calcium chromate dusts will result in severe and corrosive irritation to the eyes, nose, throat, bronchial tubes, and lungs. A primary target of inhaled dusts is the nasal cavity and, specifically, the nasal septum (thin tissue between the nostrils). Deterioration followed by perforation of the nasal septum is possible, especially on frequent, high, or prolonged exposures. Injury to the lungs is also possible. Specifically, pulmonary edema (fluid in the lungs) can occur which can be fatal. Symptoms of coughing, congestion, tightness in chest, and difficulty breathing may be delayed up to 48 hours creating a false sense of security with regard to health exposure risk. There may also be blood changes such as leukocytosis (increased leukocyte level in blood) and leukopenia (reduced leukocyte circulation in blood).

Skin contact will cause irritation, redness, defatting, and contact dermatitis. Tiny lesions, referred to as "chrome holes," can also occur. These are deep but typically painless, slow healing holes in the skin that can penetrate to the joints. Eye contact will cause irritation and conjunctivitis with tearing, irritation, and inflammation of the eyelids.

Ingestion is not likely but can occur. Symptoms may include violent gastroenteritis and a possibility of circulatory collapse, chronic nephritis (inflammation of the kidney), and liver injury.

Acute Health Effects

The following acute (short-term) health effects may occur immediately or shortly after exposure to calcium chromate:

Skin: Irritation, rash, and possible contact dermatitis.

Eye: Irritation, conjunctivitis, inflammation, watering, and swelling.

Lung: Inhalation of calcium chromate dusts may cause irritation of the mucous membranes in the respiratory tract. There may be burning in the nasal cavity with nosebleeds and possible perforation of the nasal septum. Pulmonary edema is possible on high exposures.

Other: Ingestion can cause gastrointestinal irritation with a potential for fatal circulatory system collapse.

Chronic Health Effects

The following chronic (long-term) health effects may occur at some time after exposure to calcium chromate and can last for months or even years:

Cancer Hazards: According to the information presented in the references, calcium chromate has been shown to cause cancer and mutations in humans and animals. Nasal cancers and lung cancers have been reported.

Reproductive Hazard: According to information available in the references, calcium chromate has not been adequately tested for its ability to adversely affect reproduction in laboratory animals.

Other Chronic Hazards: Repeated, long-term skin exposures can cause the development of deep but relatively painless lesions in the skin that can progress to the joints and take long to heal. Prolonged contact can also cause lung cancer.

Recommended Risk-Reduction Measures

Personnel should avoid direct contact with calcium chromate. It is a known human carcinogen. If a less toxic chemical cannot be substituted for a hazardous material, then *engineering controls* are the most effective method of reducing exposures.

The best protection is to enclose operations and/or provide local exhaust ventilation at the site of chemical release. While not always operationally feasible, isolating operations can also reduce exposure. Using respiratory protection is less effective than the controls mentioned above, but is required whenever working with or around calcium chromate. For low or infrequent exposures, a full facepiece respirator equipped with a high efficiency particulate air (HEPA) filter equipped with a dust/mist/fume pre-filter may suffice. However, because calcium chromate is a carcinogen, the best protection is provided using a supplied-air respirator with a full facepiece operated in positive pressure mode or a MSHA/NIOSH approved self-contained breathing apparatus (SCBA) with full facepiece operated in pressure demand or other continuous flow mode. If a full facepiece is not available, then chemical/dust goggles should be worn to protect the eyes. Whenever a chemical dust hazard exists, a face shield and a rubber apron should be worn. To prevent hand and skin exposures, impervious rubber gloves should be used. Glove manufacturers should be contacted and permeation studies obtained *before* selecting gloves.

Administrative controls should also be in place to minimize the potential for human exposures. These may include written procedures or policies which specify the methods and techniques that will be practiced whenever personnel are to work with calcium chromate.

All personnel should receive *training* on the use, hazards, protective measures, emergency actions, and other precautions per 29 CFR 1910.1200 (Hazard Communication), prior to the first assignment in an area where calcium chromate is used or stored.

Before beginning employment with calcium chromate and at regular intervals thereafter (e.g., annually), the following medical testing is recommended:

- ☑ Lung function tests (establish baseline).
- ☑ Examination of the nasal septum (establish a baseline of nasal cavity condition).

If symptoms develop or overexposure is suspected, the following may also be useful:

- ☑ Urine and serum calcium levels.
- ☑ Lung function tests (compare to baseline).
- ☑ Examination of the nasal septum for deterioration or perforation.
- ☑ Consider chest X-ray following acute overexposure (may be negative if taken immediately after exposures due to the delayed development of pulmonary edema).

Any evaluation should include a careful history of past and present symptoms with a medical examination. Medical tests that look for damage already done are not a substitute for controlling exposure. Also, since smoking can cause heart disease, emphysema, and a number of other respiratory disorders, smokers exposed to lung irritants such as calcium chromate dusts and fumes may experience symptoms much more quickly than non-smokers under the same conditions of exposure. Prudent risk management requires proper consideration of *all* factors which may be causing the appearance of exposures symptoms in the work place.

Other methods to reduce exposure include:

- ☑ Where possible, enclose operations and use local exhaust ventilation at the site of chemical release. If local exhaust ventilation or enclosure is not used, respiratory protection should be mandatory.
- ☑ Always ensure that proper protective clothing is worn when using chemical substances.
- ☑ Work areas should be posted with signs indicated a suspected human carcinogen is present. Access to the area must be controlled.
- ☑ Wash thoroughly immediately after exposure to calcium chromate and at the end of the work shift or before eating, drinking, or smoking.
- ☑ Hazard warning information should be posted in the work area. In addition, as part of an on-going education and training program, all information on the health and safety hazards of calcium chromate should be communicated to all potentially exposed workers.
- ☑ Eye wash stations and emergency showers should be located in the immediate work area.

☑ Work clothing that has been contaminated with calcium chromate should never be worn home where family members can be exposed as well. Contaminated clothing should be laundered at work or by an authorized service by personnel who have been trained in the hazards associated with exposure to calcium chromate and calcium compounds.

RISK ASSESSMENT: ENVIRONMENT

General Assessment

The environment is at risk of exposure during transportation, storage, disposal, or destruction of calcium chromate. In almost every scenario, the threat of environmental exposure is contingent upon the proper handling of the chemical substance. Accidental spills, large or small, can result in fire, explosion, and possible contamination of the surrounding environmental mediums (water, soil, and air).

Calcium chromate can react with organic matter (such as paper, wood, plastics, and oil) and reducing agents (such as aluminum and sulfur) to form chromium (III). It will explosively decompose hydrazine and burns violently when mixed with boron and then ignited. In a fire, calcium chromate will emit toxic and poisonous gases. These characteristics require special consideration during any emergency situation involving a leak or spill of calcium chromate.

Calcium chromate can enter the environment through industrial discharge, spills, or leaks.

Acute Ecological Effects

Acute (short-term) toxic effects may include the death of animals, birds, or fish and death or low growth rate in plants. Acute effects are seen 2 to 4 days after animals or plants are exposed to calcium chromate. Insufficient data are available to evaluate the short-term effects of calcium chromate exposure to aquatic life, plants, birds or land animals.

Chronic Ecological Effects

Chronic (long-term) toxic effects may include shortened life span, reproductive problems, lower fertility, and changes in appearance or behavior in exposed animals. These effects can be seen long after first exposure(s) to toxic chemicals. Insufficient data are available to evaluate the long-term effects of calcium chromate exposure to aquatic life, plants, birds or land animals.

Water Solubility

Calcium chromate is slightly soluble in water. Concentrations of 1 to 100 milligrams will mix with a liter of water.

Persistence in the Environment

Calcium chromate is highly persistent in the aquatic environment, with a half-life of greater than 200 days. The half-life of a pollutant is the amount of time it takes for one half of the chemical to be degraded.

Bioaccumulation in Aquatic Organisms

Some substances increase in concentration, or *bioaccumulate,* in living organisms as they breathe contaminated air, drink contaminated water, or eat contaminated food. These chemicals can become concentrated in the tissues and internal organs of animals as well as humans.

The concentration of calcium chromate found in fish is expected to about the same as the average concentration of calcium chromate in the water from which the fish was taken.

Recommended Risk-Reduction Measures

Proper training of all transporters will reduce the likelihood of a mishap or accidents resulting in a leak or spill to the environment. Proper labeling on all containers, trucks, and rail cars will enable emergency responders to react properly and quickly to any disaster thereby reducing the potential risk to the environment and to personnel.

If a spill or leak to the environment has occurred, fire department, emergency response and/or hazardous materials spill personnel should be notified immediately. Do not dry sweep. When vacuuming, a high efficiency particulate air (HEPA) filter should be used, not a standard shop vacuum. Damp mop residue. Deposit collected materials in sealed drum for disposal. Cleanup should be attempted only by those trained in proper spill containment procedures. Contaminated soils may need to be removed and replaced with clean soil. If calcium chromate should contact the water table, aquifer, or navigable waterway, time is of the essence. Calcium chromate is slightly soluble in water and total containment and remediation may not be entirely possible. When such spills occur, the local and/or state emergency response authorities must be notified. A comprehensive emergency response or disaster preparedness/recovery plan should be in place prior to any operations involving the use, transportation, storage, or disposal of calcium chromate. If calcium chromate is spilled or leaked, the following specific steps are recommended:

☑ Restrict persons not wearing protective clothing from area of spill or leak until clean-up is complete and area can be opened for normal work.

☑ Ventilate area of spill and remove ignition sources.

☑ Collect powdered materials in most convenient and safe manner possible and place in sealed drums for disposal. Do NOT dry sweep, use HEPA vacuum instead. Use water to flush residue, then damp mop.

☑ It may be necessary to dispose of calcium chromate as a hazardous waste. The responsible state agency or the regional office of the federal Environmental Protection Agency (EPA) should be contacted for specific recommendations.

RISK ASSESSMENT: BUSINESS

General Assessment

Accidents or mishaps involving calcium chromate can present a significant threat to business operations. The loss or damage of equipment or facilities can significantly affect fiscal viability. Lawsuits that may result from personnel injury/death, public exposures, and/or environmental contamination will also require a serious expenditure of resources. Media attention surrounding an injury, death, or environmental damage can also result in a loss of profits and loss of current as well as future business. Always remember that anytime the terms "cancer," or "carcinogen" are used, public emotion, hysteria, and ignorance can run equally high. This must be carefully considered whenever developing or implementing public relations policies.

Recommended Risk-Reduction Measures

Company attorneys, safety and health professionals, and environmental specialists should be involved in the development of any procedures dealing with safety or emergency response. A company official should be pre-designated as a public relations officer with specific training in dealing with the press. Corporate plans and policies should be developed, approved, and implemented long before any need for such arises.

REFERENCES

Ahlbom, A. 1993. *Biostatistics for Epidemiologists.* New York: CRC Press/Lewis Publishers.

American Conference of Governmental Industrial Hygienists. 1988. *Documentation of the Threshold Limit Values and Biological Exposure Indices,* 5th Edition (with updates). Cincinnati: ACGIH

Brusick, D. J. 1994. *Methods for Genetic Risk Assessment.* New York: CRC Press/Lewis Publishers.

Calabrese, E. J., E. M. Kenyon. *Air Toxics and Risk Assessment.* New York: CRC Press/Lewis Publishers.

Cockerman, L. G., B. S. Shane. 1994. *Basic Environmental Toxicology.* New York: CRC Press/Lewis Publishers.

Cothern, C. R. 1995. *Handbook for Environmental Risk Decisionmaking.* New York: CRC Press/Lewis Publishers.

Gardner, R. W. 1994. *Chemical Intolerance.* New York: CRC Press/Lewis Publishers.

Howard, P. H., et. al. 1991. *Handbook of Environmental Degradation Rates.* New York: CRC Press/Lewis Publishers.

Jones, F. E. 1994. *Toxic Organic Vapors in the Workplace.* New York: CRC Press/Lewis Publishers.

Keith, L. H. and M. M. Walker. 1995. *Handbook of Air Toxics.* New York: CRC Press/Lewis Publishers.

Langerman, N. 1994 *Precautionary Labels for Chemical Containers.* New York: CRC Press/Lewis Publishers.

Lauwerys, R. R., P. Hoet. 1993. *Industrial Chemical Exposure.* New York: CRC Press/Lewis Publishers.

Lewis, R. J., Sr. 1992. *Sax's Dangerous Properties of Industrial Materials,* Volumes 1, 2, and 3, 8th Edition. New York: Van Nostrand Reinhold.

Manahan, S. E. 1992. *Toxicological Chemistry,* 2nd Edition. New York: CRC Press/Lewis Publishers.

Meyer, E. 1990. *Chemistry of Hazardous Materials.* Englewood Cliffs, NJ: Prentice Hall.

Rea, W. J. 1995. *Chemical Sensitivity,* Vol. I-III. New York: CRC Press/Lewis Publishers.

Sacarello, H. L. A. 1994. *The Comprehensive Handbook of Hazardous Materials.* New York: CRC Press/Lewis Publishers.

Sherman, J. D. 1988. *Chemical Exposure and Disease.* New York: Van Nostrand Reinhold.

Talbott, E., G. Craun. 1995. *An Introduction to Environmental Epidemiology.* New York: CRC Press/Lewis Publishers.

U.S. Department of Health and Human Services (DHHS), National Institute for Occupational Health (NIOHS). 1994. *NIOSH Pocket Guide to Hazardous Chemicals.* Publication Number 94-116. Washington, D.C.; U.S. Government Printing Office

MATERIAL SAFETY DATA SHEET

CHEMICAL NAME	CRC PRESS/LEWIS PUBLISHERS DISCLAIMER:
CALCIUM CYANAMIDE	

HAZARD WARNING INFORMATION

HEALTH	FIRE	REACTIVE	OTHER	DEGREE OF HAZARD	COLOR CODING	OTHER CODES
3	1	1	W	0 = Minimum Hazard 1 = Slight Hazard 2 = Moderate Hazard 3 = Serious Hazard 4 = Severe Hazard	HEALTH = BLUE FIRE = RED REACTIVITY = YELLOW OTHER = WHITE	OX = Oxidizer ACID = Acid ALK = Alkali COR = Corrosive W = Use No Water

SECTION I - GENERAL INFORMATION

Chemical Family	RCRA Number	EPA Class
Metal (powder)	**Not Listed**	**Not Applicable**

DOT Proper Shipping Name	Chemical Abstract Service (CAS) Number
Lime Nitrogen	**156-62-7**
DOT Hazard Class and Label Requirements	**DOT Emergency Guide Code**
ORM-C; Dangerous When Wet	**40**
DOT Identification Number	**Chemical Formula**
UN 1403	**$CaCN_2$**

Synonyms
Lime nitrogen, calcium carbimide, aero-cyanamide, cyanimid granular, nitrogen lime, nitrolime.

SECTION II - HAZARDOUS INGREDIENTS AND IDENTITY INFORMATION

Hazardous Components (specific identity)	OSHA Exposure Criteria	NIOSH Exposure Criteria	Immediately Dangerous to Life and Health (IDLH)	ACGIH Exposure Criteria
Calcium cyanamide (derivation: Calcium carbide powder is heated in an electric oven into which nitrogen is passed (24 to 26 hours). Any uncombined calcium carbide is leached-out after removal).	**PEL:** 0.5 mg/m^3 **STEL:** Not Established	**REL:** 0.5 mg/m^3 **STEL:** Not Established	**Not Determined**	**TLV:** 0.5 mg/m^3 **STEL:** Not Established

SECTION III - PHYSICAL AND CHEMICAL CHARACTERISTICS

Boiling Point	Specific Gravity (H_2O = 1)
Sublimes	**2.29**
Vapor Pressure (mm Hg)	Molecular Weight (atomic weight)
0 (approx.) at 69°F (20°C)	**80.11**
Vapor Density (Air = 1)	Melting Point
Not Applicable	**2444°F (1540°C)**

Solubility in Water
Insoluble in water (will decompose).

Appearance and Odor
Colorless or gray to grayish-black crystalline solid (sugar or sand-like powder).

SECTION IV - FIRE AND EXPLOSION HAZARD DATA

Flash Point (method used)	Explosive Limits in Air % by Volume
Not Applicable	**LEL: N/A UEL: N/A**
NFPA Classification	Autoignition Temperature
Non-combustible Solid	**Not Found**

Extinguishing Media
Do NOT use foam or water. Use dry chemical, soda ash, or lime.

Special Fire Fighting Procedures
Wear full protective clothing and self-contained breathing apparatus (SCBA). Poisonous cyanide and nitric oxide fumes are produced in fire. Structural protective clothing is permeable and may not provide adequate protection. Avoid contact with skin. Decomposes in water to form acetylene and ammonia which are extremely flammable. Fight fire from distance or protected location, if possible.

Unusual Fire and Explosion Hazards
Finely divided dusts may ignite in air and may be flammable and explosive when exposed to heat, flame, or by chemical reaction with water, moisture, or steam (forms explosive acetylene gas and ammonia).

SECTION V - REACTIVITY DATA

Stability		Conditions to Avoid
		Avoid contact with incompatible materials, primarily water.
Stable	Unstable	Incompatibility *(materials to avoid)*
X		Reacts on contact with water. If combined with calcium carbide, creates serious fire hazard.
Hazardous Polymerization		Conditions to Avoid
		Do not allow calcium cyanamide to become wet or moist. Decomposes to form acetylene and ammonia.
May Occur	Will Not Occur	Hazardous Decomposition or By-products
X		When heated to decomposition, calcium cyanamide releases toxic fumes of cyanide and nitric oxide. May polymerize in water or alkaline solutions to dicyanamide.

SECTION VI - HEALTH HAZARD DATA

Primary Route(s) of Entry:	Inhalation?	Absorption (skin/eye)?	Ingestion?
	X	X	X

Health Hazards

INHALATION: Breathing the dusts can cause feelings of dizziness, flushed skin, and lightheadedness. A nose and throat irritant. Drinking alcohol before or after exposure can make symptoms seem worse and may cause nausea, vomiting, shock, and possibly death.

ABSORPTION: Will pass through unbroken skin and enter the bloodstream. Contact may severely irritate the skin and eyes. Prolonged contact can cause skin ulcers. May cause allergy.

INGESTION: A poison by ingestion. Moderately toxic to humans resulting in toxic systemic effects.

Carcinogenicity	NTP Listed?	IARC Cancer Review Group?	OSHA Regulated?	Target Organs?
Unknown Human Questioned Animal	5th Annual Report	No	No	Respiratory tract, skin, eyes, vasomotor system.

Medical Conditions Generally Aggravated by Exposure

None reported. However, persons with existing lung dysfunction may experience effects quicker.

Emergency and First-aid Procedures

Eye contact: Do not allow victim to keep eyes tightly shut. Flush immediately with water for 15 minutes (minimum), seek medical attention. Skin contact: Remove all contaminated clothing. Immediately wash area with flooding amounts of soap and water. For inhalation: Remove the person from exposure. Provide respiratory assistance and CPR. Transfer to medical facility. If swallowed: Provide CPR if necessary. Seek medical attention immediately. Call poison control center for advice. Never give anything by mouth to an unconscious or convulsing person.

SECTION VII - PRECAUTIONS FOR SAFE HANDLING AND USE

Steps to be Taken in Case Material is Released or Spilled

Collect powdered materials in safest manner possible and deposit in sealed containers. Remove all sources of ignition. Ventilate area of spill. Use HEPA vacuum. Restrict those not involved in cleanup from entering area. Notify appropriate authorities. Absorb any liquids containing calcium cyanamide in vermiculite, dry sand, earth, or similar material and deposit in sealed drum for disposal or reclamation.

Preferred Waste Disposal Method

No Citation.

Precautions to be Taken in Handling and Storage

Store in tightly closed containers free from moisture. Keep away from calcium carbide.

Other Precautions and Warnings

Reaction with water forms explosive acetylene gas.

SECTION VIII - CONTROL MEASURES AND PERSONAL PROTECTIVE EQUIPMENT

Respiratory Protection *(specify type)*

For high exposures, use a NIOSH/MSHA approved self-contained breathing apparatus with full facepiece operated in positive pressure mode. For low exposure, use respirator equipped with particulate filter.

Ventilation

Local exhaust (preferred) at site of chemical work or general ventilation.

Protective Gloves	Eye Protection	Other Protective Clothing
Impervious Material	Chemical/Dust Goggles or Face Mask	Protective Apron, Boots (avoid skin contact)

Work/Hygiene Practices

Always wash hands thoroughly after using chemical; never bring food, drink, or smoking materials into vicinity of chemicals.

CALCIUM CYANAMIDE

$CaCN_2$ CAS: 156-62-7

IDENTIFICATION AND TYPICAL USES

Calcium cyanamide is a white or colorless powder in pure form and a gray or grayish-black solid (sugar or sand-like powder) in the commercial grade. It is used, primarily as a fertilizer, defoliant, and a pesticide. Also used to manufacture some calcium cyanide, melamine, dicyandiamide, and some nitrogen products. Also used in the manufacture and refining (as a hardening agent) of iron and steel.

RISK ASSESSMENT: HEALTH

General Assessment

Calcium cyanamide is a poison by *ingestion, inhalation,* and skin contact (*absorption*). It is moderately toxic to humans through ingestion. It is a questionable carcinogen in animals with some experimental tumorigenic data reported. Mutation data also reported. Carcinogenic properties in humans are not known.

Breathing its fine dusts can irritate the respiratory system, especially the nose and throat. Exposed persons may feel dizzy, flushed, and lightheaded. Drinking alcohol before or within 1-2 days after of an exposure may make these symptoms seem more pronounced, and can cause nausea, vomiting, shock, and possibly death.

Contact with skin and eyes can be severely irritating. Prolonged contact can cause skin ulcers. Skin allergy can also occur. Symptoms of toxic systemic exposure (dizziness, vomiting, nausea) can also occur as a result of calcium cyanamide passing through unbroken skin and directly entering the bloodstream.

Acute Health Effects

The following acute (short-term) health effects may occur immediately or shortly after exposure to calcium cyanamide:

Skin: Possible irritation and skin ulcers may occur. Absorption is likely.

Eye: Severe irritation and possible damage on contact.

Lung: Exposures can cause severe irritation of the nose and throat with a potential for lung damage. Symptoms include cough and there may be congestion and shortness of breath.

CNS: Affects the vasomotor system (may cause dilation of the blood vessels) resulting in flushed skin. Other effects include dizziness and lightheadedness. There may also be vomiting, nausea, shock, and possibly death when combined with exposures to alcohol (drinking).

Chronic Health Effects

The following chronic (long-term) health effects may occur at some time after exposure to calcium cyanamide and can last for months or even years:

Cancer Hazards: According to the information in the references, the data on the cancer-causing properties of calcium cyanamide in animals are inconclusive. However, its carcinogenicity is still questioned in some studies.

Reproductive Hazard: It is not known if exposure to calcium cyanamide will adversely affect the reproductive system of males or females.

Other Chronic Effects: Calcium cyanamide may damage the nervous system, causing numbness, "pins and needles," and/or weakness in the hands and feet. Prolonged contact with skin may cause allergic reaction.

Recommended Risk-Reduction Measures

Personnel should avoid direct contact with calcium cyanamide dusts or powders. If a less toxic material or compound cannot be substituted for calcium cyanamide, then *engineering controls* are the most effective method of reducing exposures. The best protection is to enclose operations and/or provide local exhaust ventilation at the site of calcium cyanamide dust release. While not always operationally feasible, isolating operations involving calcium cyanamide manufacturing can also reduce exposure. Using respiratory protection is less effective than the controls mentioned above, but is still advisable whenever working with or around calcium cyanamide. At exposures above the OSHA PEL of 0.5 mg/m^3 (TWA), use a MSHA/NIOSH approved respirator equipped with a particulate (dust/fume/mist) pre-filter. For high or frequent exposures, use a self-contained breathing apparatus (SCBA) with full face-piece and pressure demand is recommended for the greatest possible respiratory protection. If a full face-piece is not available, then chemical/dust goggles should be worn to protect the eyes. A face shield and protective apron should also be worn. To prevent hand and skin exposures, gloves should be worn.

Administrative controls should also be in place to minimize the potential for human exposures. These may include written procedures or policies which specify the methods and techniques that will be practiced whenever personnel are to work with calcium cyanamide.

All personnel should receive *training* on the use, hazards, protective measures, emergency actions, and other precautions per 29 CFR 1910.1200 (Hazard Communication), prior to the first assignment in an area where calcium cyanamide is used or stored.

If symptoms develop or overexposure is suspected, the following medical testing may be useful:

- ☑ Examination of the nervous system.
- ☑ Evaluation by a qualified allergist, including careful consideration of exposure history and special testing (may help diagnose skin allergy).

Any evaluation should include a careful history of past and present symptoms with an examination. Medical tests that simply look for existing damage are not a substitute for controlling exposures. Also, drinking alcohol (beer, wine, etc.) before or within 1-2 days after exposure to calcium cyanamide can cause severe, possibly fatal reaction.

Other methods to reduce exposure include:

- ☑ Where possible, enclose operations and use local exhaust ventilation at the site of chemical release. If local exhaust ventilation or enclosure is not used, respiratory protection should be mandatory.
- ☑ Always ensure that proper protective clothing is worn when using chemical substances.
- ☑ Wash thoroughly immediately after exposure to calcium cyanamide and at the end of the work shift or before eating, drinking, or smoking.
- ☑ Hazard warning information should be posted in the work area. In addition, as part of an on-going education and training program, all information on the health and safety hazards of calcium cyanamide should be communicated to all potentially exposed workers.
- ☑ Eye wash stations should be provided in the immediate work area for emergency use. If there is a possibility of skin exposure to calcium cyanamide, emergency shower facilities should also be provided in the immediate area
- ☑ Workers whose clothing has been contaminated by calcium cyanamide should change into clean clothes before leaving work. Contaminated work clothing should be laundered only by individuals who have been informed of the hazards of exposure to calcium cyanamide.

RISK ASSESSMENT: ENVIRONMENT

General Assessment

Because of its use as a fertilizer, pesticide, and defoliant, calcium cyanamide can be present in the environment at any given time and at a variety of levels. The environment is also at risk of exposure during transportation, storage, disposal, or destruction of calcium cyanamide. In almost every scenario, the threat of environmental exposure is contingent upon the proper handling of the chemical substance. Accidental spills, large or small, where calcium cyanamide contacts incompatible commodities can result in the release of hazardous fumes and gases (depending upon conditions of spill), and possible contamination of the surrounding environmental mediums (water, soil, and air). In contact with water, it will form explosive acetylene gas and ammonia.

In its bulk form, calcium cyanamide is considered a non-combustible solid. However, it is a fire risk if it contains calcium carbide (most commercial grades do). Because it is incompatible with water, steam, and moisture, extreme caution is required in handling, storage, transportation, and disposal of calcium cyanamide. These characteristics also require special consideration during any emergency situation involving a leak or spill of calcium cyanamide powder or dust.

Calcium cyanamide can enter the environment from industrial and municipal discharges, through agricultural run-off, and from spills. Also, as a fertilizer, defoliant, and pesticide, it will also be present in the environment as a result of its intended use.

Acute Ecological Effects

Acute (short-term) toxic effects may include the death of animals, birds, or fish and death or low growth rate in plants. Acute effects are seen 2 to 4 days after animals or plants are exposed to calcium cyanamide.

Calcium cyanamide caused germination decrease in agricultural crops. Insufficient data are available to evaluate the short-term effects of calcium cyanamide to aquatic organisms, birds, or land animals.

Chronic Ecological Effects

Chronic (long-term) toxic effects may include shortened life span, reproductive problems, lower fertility, and changes in appearance or behavior in exposed animals. These effects can be seen long after first exposure(s) to toxic chemicals.

Insufficient data are available to evaluate or predict the long-term effects of calcium cyanamide to aquatic organisms, plants, birds, or land animals.

Water Solubility

In environmental aquifers, calcium cyanamide is considered to be insoluble in water because it will decompose before mixing.

Persistence in the Environment

Calcium cyanamide is non-volatile, so very little will occur in air. In acid solutions and in most soils, it hydrolyzes to urea. It also produces hydrogen cyanamide in moist soils. In pure water or alkaline solutions, it may polymerize to dicyandiamide. Thus, it would appear calcium cyanamide has little persistence in the environment.

Bioaccumulation in Aquatic Organisms

Some substances increase in concentration, or *bioaccumulate,* in living organisms as they breathe contaminated air, drink contaminated water, or eat contaminated food. These chemicals can become concentrated in the tissues and internal organs of animals as well as those of humans.

Insufficient data are available to evaluate or predict the bioaccumulation of calcium cyanamide in the aquatic environment.

Recommended Risk-Reduction Measures

Proper training of all transporters will reduce the likelihood of a mishap or accident resulting in a leak or spill of calcium cyanamide dusts or powders into the environment. The correct use of DOT labeling (per the MSDS) on all containers, trucks, and rail cars will enable emergency responders to react properly and quickly to any disaster thereby reducing the potential risk to the environment and to personnel.

Storage of calcium cyanamide should be segregated from incompatible materials to minimize the risk of cross-contamination or contact. Proper ventilation and protective equipment should be used while handing the powders or while preparing aqueous solutions.

If a spill or leak to the environment has occurred, fire department, emergency response and/or hazardous materials spill personnel should be notified immediately. Cleanup should be attempted only by those trained in proper spill containment procedures. If calcium cyanamide should contact the water table, aquifer, or navigable waterway, time is of the essence. It decomposes rapidly and total remediation will not be possible. The local and/or state emergency response authorities must be notified. A comprehensive emergency response or disaster preparedness/recovery plan should be in place prior to any operations involving the use, transportation, storage, or disposal of calcium cyanamide. If calcium cyanamide powder is spilled or leaked, the following specific steps are recommended:

- ☑ Restrict persons not wearing protective clothing from area of spill or leak until cleanup is complete and area can be opened for normal work.
- ☑ Remove all ignition sources.
- ☑ Collect powdered materials in the safest manner possible and deposit in sealed containers. Recommend using vacuum with HEPA filter. Do not dry sweep (generates airborne dusts). If liquid solutions are spilled, absorb in vermiculite or other material and place in sealed drum for disposal or reclamation.
- ☑ It may be necessary to dispose of calcium cyanamide as a hazardous waste. The responsible state agency or the regional office of the federal Environmental Protection Agency (EPA) should be contacted for specific recommendations.

RISK ASSESSMENT: BUSINESS

General Assessment

Accidents or mishaps involving calcium cyanamide can present a significant threat to business operations. The loss or damage of equipment or facilities can significantly affect fiscal viability. Lawsuits that may result from personnel injury/death, public exposures, and/or environmental contamination will also require a serious expenditure of resources. Media attention surrounding an injury, death, or environmental damage can also result in a loss of profits and loss of current as well as future business.

Recommended Risk-Reduction Measures

Company attorneys, safety and health professionals, and environmental specialists should be involved in the development of any procedures for responding to chemical incidents. A company official should be predesignated as a public relations officer with specific training in dealing with the press. Corporate plans and policies should be developed, approved, and implemented long before any need for such arises.

REFERENCES

Ahlbom, A. 1993. *Biostatistics for Epidemiologists.* New York: CRC Press/Lewis Publishers.

American Conference of Governmental Industrial Hygienists. 1988. *Documentation of the Threshold Limit Values and Biological Exposure Indices,* 5th Edition (with updates). Cincinnati: ACGIH

Asante-Duah, D. K. 1993. *Hazardous Waste Risk Assessment.* New York: CRC Press/Lewis Publishers.

Brusick, D. J. 1994. *Methods for Genetic Risk Assessment.* New York: CRC Press/Lewis Publishers.

Bloom, A. D. and F. J. DeSerres. 1995. *Ecotoxicity and Human Health.* New York: CRC Press/Lewis Publishers.

Bueche, J. R. 1972. *Principles of Physics,* 2nd Edition. New York: McGraw-Hill.

Calabrese, E. J. 1994. *Biological Effects of Low Level Exposures.* New York: CRC Press/Lewis Publishers.

Cothern, C. R., N. P. Ross. 1994. *Environmental Statistics, Assessment, and Forecasting.* New York: CRC Press/Lewis Publishers.

Gardner, R. W. 1994. *Chemical Intolerance.* New York: CRC Press/Lewis Publishers.

Hallenbeck, W. H. 1993. *Quantitative Risk Assessment for Environmental and Occupational Health,* 2nd Edition. New York: CRC Press/Lewis Publishers.

Howard, P. H., et. al. 1991. *Handbook of Environmental Degradation Rates.* New York: CRC Press/Lewis Publishers.

Jones, F. E. 1994. *Toxic Organic Vapors in the Workplace.* New York: CRC Press/Lewis Publishers.

Keith, L. H. and M. M. Walker. 1995. *Handbook of Air Toxics.* New York: CRC Press/Lewis Publishers.

Landis, W. G., M. H. Yu. 1995. *Introduction to Environmental Toxicology.* New York: CRC Press/Lewis Publishers.

Langerman, N. 1994 *Precautionary Labels for Chemical Containers.* New York: CRC Press/Lewis Publishers.

Lauwerys, R. R., P. Hoet. 1993. *Industrial Chemical Exposure.* New York: CRC Press/Lewis Publishers.

Lewis, R. J., Sr. 1992. *Sax's Dangerous Properties of Industrial Materials,* Volumes 1, 2, and 3, 8th Edition. New York: Van Nostrand Reinhold.

Lide, D. R. 1995. *Handbook of Chemistry and Physics,* 76th Edition. New York: CRC Press/Lewis Publishers.

Manahan, S. E. 1992. *Toxicological Chemistry,* 2nd Edition. New York: CRC Press/Lewis Publishers.

Mellström, G. A., J. E. Wahlberg, and H. I. Maibach. 1994. *Protective Gloves for Occupational Use.* New York: CRC Press/Lewis Publishers.

Meyer, E. 1990. *Chemistry of Hazardous Materials.* Englewood Cliffs, NJ: Prentice Hall.

Mycock, J. C., J. D. McKenna, and L. Theodore. 1995. *Handbook of Air Pollution Control Engineering and Technology.* New York: CRC Press/Lewis Publishers.

Nelson, G. O. 1992. *Gas Mixtures.* New York: CRC Press/Lewis Publishers.

Ott, W. R. 1995. *Environmental Statistics and Data Analysis.* New York: CRC Press/Lewis Publishers.

Patnaik, P. 1992. *A Comprehensive Guide to the Hazardous Properties of Chemical Substances.* New York: Van Nostrand Reinhold.

Proctor, N. H., J. P. Hughes. 1991. *Chemical Hazards of the Workplace.* New York: Van Nostrand Reinhold.

Quigley, D. R. 1994. *Handbook of Emergency Chemical Management.* New York: CRC Press/Lewis Publishers.

Reeves, A. L. 1979. Calcium cyanamide. *Handbook on the Toxicology of Metals.* pp 321-328. New York: Elsevier North-Holland.

Richardson, M. 1995. *Dictionary of Substance and Their Effects,* Vol. I-VI. New York: CRC Press/Lewis Publishers.

Sacarello, H. L. A. 1994. *The Comprehensive Handbook of Hazardous Materials.* New York: CRC Press/Lewis Publishers.

Torkelson, T. R. V. K. Rowe. 1981. *Patty's Industrial Hygiene and Toxicology,* Vol. 2B. New York: Wiley-Interscience.

MATERIAL SAFETY DATA SHEET

CHEMICAL NAME

CALCIUM CYANIDE

HAZARD WARNING INFORMATION

HEALTH	FIRE	REACTIVE	OTHER	DEGREE OF HAZARD	COLOR CODING	OTHER CODES
3	0	1	~~W~~	0 = Minimum Hazard 1 = Slight Hazard 2 = Moderate Hazard 3 = Serious Hazard 4 = Severe Hazard	HEALTH = BLUE FIRE = RED REACTIVITY = YELLOW OTHER = WHITE	OX = Oxidizer ACID = Acid ALK = Alkali COR = Corrosive ~~W~~ = Use No Water

SECTION I - GENERAL INFORMATION

Characterization **Cyanides, Inorganic**	RCRA Number **P021**	EPA Class **Acute Hazardous Waste**
DOT Proper Shipping Name **Calcium Cyanide Mixture, Solid**	Chemical Abstract Service (CAS) Number **592-01-8**	
DOT Hazard Class and Label Requirements **Poison B**	DOT Emergency Guide Code **55**	
DOT Identification Number **UN 1575**	Chemical Formula **$Ca(CN)_2$**	

Synonyms

Calcyanid; cyanogas; black cyanide; cyanide of calcium.

SECTION II - HAZARDOUS INGREDIENTS AND IDENTITY INFORMATION

Hazardous Components (specific identity)	OSHA Exposure Criteria	NIOSH Exposure Criteria	Immediately Dangerous to Life and Health (IDLH)	ACGIH Exposure Criteria
Calcium cyanide (derivation: Inorganic cyanides are the metal salts of hydrocyanic acid, HCN).	**PEL (CN):** **skin:** **5 mg/m³** **STEL:** **Not Established**	**REL (CN):** **ceiling:** **5 mg/m³** **(10 minutes)**	**50 mg(CN)m³**	**TLV (CN):** **skin:** **5 mg/m³** **STEL:** **Not Established**

SECTION III - PHYSICAL AND CHEMICAL CHARACTERISTICS

Boiling Point **Not established**	Specific Gravity (H_2O = 1) **1.85**
Vapor Pressure (mm Hg) **Not Applicable**	Molecular Weight (atomic weight) **92.1**
Vapor Density (Air = 1) **3.2**	Melting Point **662°F (350°C) Decomposes**

Solubility

Soluble in water, formamide, weak acids, and hydroxylamine; slightly soluble in ammonia.

Appearance and Odor

White crystalline solid or white powder. Technical grade may appear as a gray-black powder.

SECTION IV - FIRE AND EXPLOSION HAZARD DATA

Flash Point (method used) **Not Applicable**	Explosive Limits in Air % by Volume **LEL: N/A UEL: N/A**
NFPA Classification **Not Applicable**	Autoignition Temperature **Not Found**

Extinguishing Media

Use agent suitable to surrounding fire. Do NOT use carbon dioxide or water.

Special Fire Fighting Procedures

Wear full protective clothing and self-contained breathing apparatus (SCBA). Poisonous cyanide and nitric oxide fumes are produced in fire. Structural protective clothing is permeable and may not provide adequate protection. Avoid contact with skin. Toxic hydrogen cyanide gas may be released when calcium cyanide contacts moist or humid air. Fight fire from distance or protected location, if possible.

Unusual Fire and Explosion Hazards

Finely divided dusts may ignite in air and may be flammable and explosive when exposed to heat, flame, or by chemical reaction with water, acid, acid fumes, or steam.

SECTION V - REACTIVITY DATA

Stability		Conditions to Avoid **Avoid contact with incompatible materials. Keep calcium cyanide powders away from open flame, heat, or other sources of ignition.**
Stable **X**	Unstable	Incompatibility *(materials to avoid)* **Reacts on contact with water, acids (sulfuric, hydroazoic, nitric), acid fumes, acidic salts, carbon dioxide, and steam.**
Hazardous Polymerization		Conditions to Avoid **Under normal conditions of temperature and pressure, hazardous polymerization of calcium cyanide will not occur.**
May Occur	Will Not Occur **X**	Hazardous Decomposition or By-products **When heated to decomposition, calcium cyanide releases toxic fumes of cyanide and nitric oxide. On contact with water or steam, toxic hydrogen cyanide will be released.**

SECTION VI - HEALTH HAZARD DATA

Primary Route(s) of Entry:	Inhalation?	Absorption (skin/eye)?	Ingestion?
	X	X	X

Health Hazards

INHALATION: Breathing the dusts can react with the water content in the respiratory tract to release hydrogen cyanide in the lungs. This can cause death through systemic poisoning. Symptoms include cough, headache, nausea, weakness, malaise, vomiting, loss of consciousness, and death.

ABSORPTION: Will pass through unbroken skin and enter the bloodstream. Large exposures can be fatal.

INGESTION: A poison by ingestion. May cause toxic systemic effects and death.

Carcinogenicity	NTP Listed?	IARC Cancer Review Group?	OSHA Regulated?	Target Organs?
Unknown Human Unknown Animal	No	No	No	Respiratory tract, skin, eyes.

Medical Conditions Generally Aggravated by Exposure

None reported. However, persons with existing lung dysfunction may experience effects quicker.

Emergency and First-aid Procedures

Eye contact: Do not allow victim to keep eyes tightly shut. Flush immediately with water for 15 minutes (minimum), seek medical attention. Skin contact: Remove all contaminated clothing. Immediately wash area with flooding amounts of soap and water. Do not allow water to enter nose or mouth. For inhalation: Remove the person from exposure. Break amyl nitrate pearl in cloth and hold lightly under nose for 15 seconds; repeat 5 times at 15 second intervals. Give oxygen. Provide respiratory assistance and CPR. Transfer to medical facility. If swallowed: Break amyl nitrate pearl in cloth and hold lightly under nose for 15 seconds. If conscious, induce vomiting. Repeat until vomit is clear. Repeat inhalation of amyl nitrate 5 times at 15 second intervals. Provide CPR if necessary. Seek medical attention immediately. Call poison control center for advice. Never give anything by mouth to an unconscious or convulsing person.

SECTION VII - PRECAUTIONS FOR SAFE HANDLING AND USE

Steps to be Taken in Case Material is Released or Spilled

Collect powdered materials in safest manner possible and deposit in sealed containers. Use HEPA vacuum. Restrict those not involved in cleanup from entering area. Notify appropriate authorities.

Preferred Waste Disposal Method

Destroy by oxidizing it to calcium cyanate using sodium hypochlorite or other oxidizing compounds.

Precautions to be Taken in Handling and Storage

Store in tightly closed containers free from moisture. It is shipped in mild-steel or fiber drums.

Other Precautions and Warnings

Protective equipment must be worn when handling or preparing aqueous solutions.

SECTION VIII - CONTROL MEASURES AND PERSONAL PROTECTIVE EQUIPMENT

Respiratory Protection *(specify type)*

For any exposure level use a NIOSH/MSHA approved self-contained breathing apparatus with full facepiece operated in positive pressure mode. Inhalation of dusts must be avoided.

Ventilation

Local exhaust (preferred) at site of chemical work or general ventilation.

Protective Gloves	Eye Protection	Other Protective Clothing
Impervious Material	**Chemical/Dust Goggles or Face Mask**	**Protective Apron, Boots (avoid skin contact)**

Work/Hygiene Practices

Always wash hands thoroughly after using chemical; never bring food, drink, or smoking materials into vicinity of chemicals.

CALCIUM CYANIDE

$Ca(CN)_2$ CAS: 592-01-8

IDENTIFICATION AND TYPICAL USES

Calcium cyanide is a white crystalline solid or white powder. Its technical grade may appear as a black powder. It is used for the extraction of gold and silver from their ores, in the froth flotation of minerals, as a fumigant, and as a rodenticide. It is also used as a fumigant for greenhouses, flour mills, grain, seed, citrus trees under tents for control of scale insects, and in the manufacture of other cyanides.

RISK ASSESSMENT: HEALTH

General Assessment

Calcium cyanide is a highly poisonous compound. The toxic routes of entry are *inhalation* and *ingestion* of its dusts, and skin contact (*absorption*). Exposure by either route in large enough doses can lead to death. Calcium cyanide reacts on exposure to moisture to release highly toxic hydrogen cyanide. Since skin and the linings of the respiratory tract and lungs, as well as the digestive system contain moisture, exposure can be lethal.

Poisonous reactions on exposure may cause early symptoms of headache, nausea, vomiting, and weakness. Loss of consciousness and death can follow. High concentrations can be rapidly fatal.

Acute Health Effects

The following acute (short-term) health effects may occur immediately or shortly after exposure to calcium cyanide:

Skin: Possible irritation may occur. Absorption is likely.

Eye: Irritation and possible damage on contact.

Lung: High exposures can cause rapid and severe lung damage, with shortness of breath, chest pain, cough, loss of consciousness, and death.

Other: High exposure to calcium cyanide may be rapidly fatal.

Chronic Health Effects

The following chronic (long-term) health effects may occur at some time after exposure to calcium cyanide and can last for months or even years:

Cancer Hazards: According to the information in the references, the data on the cancer-causing properties of calcium cyanide are inconclusive. However, this is probably due to the fact that it is highly poisonous and long-term exposures seldom occur (death is more likely).

Reproductive Hazard: It is not known if exposure to calcium cyanide will adversely affect the reproductive system of males or females.

Recommended Risk-Reduction Measures

Personnel should avoid direct contact with calcium cyanide dusts or powders. If a less toxic material or compound cannot be substituted for calcium cyanide, then *engineering controls* are the most effective method of reducing exposures. The best protection is to enclose operations and/or provide local exhaust ventilation at the site of calcium cyanide dust release. While not always operationally feasible, isolating operations can also reduce exposure. Using respiratory protection is less effective than the controls mentioned above, but is still advisable whenever working with or around calcium cyanide. At anv exposure level, a self-contained breathing apparatus (SCBA) with full facepiece and pressure demand is recommended for the greatest possible respiratory protection. If a full facepiece is not available, then chemical/dust goggles should be worn to protect the eyes. A face shield and protective apron should also be worn. To prevent hand and skin exposures, gloves should be worn.

Administrative controls should also be in place to minimize the potential for human exposures. These may include written procedures or policies which specify the methods and techniques that will be practiced whenever personnel are to work with calcium cyanide.

All personnel should receive *training* on the use, hazards, protective measures, emergency actions, and other precautions per 29 CFR 1910.1200 (Hazard Communication), prior to the first assignment in an area where calcium cyanide is used or stored.

If exposure occurs, break an amyl nitrate pearl into a cloth and hold it lightly under the victim's nose for 15 seconds. This should be repeated five times, at fifteen second intervals.

For those with frequent or potentially high exposure (half the PEL or greater), or significant skin contact, the following medical tests are recommended before initial assignment and at regular time thereafter:

- ☑ Urine thiocyanate test (most accurate if performed right after exposure).

If symptoms develop or overexposure is suspected, the following should also be considered:

- ☑ Blood cyanide level. (Note: Since cigarettes contain some cyanide, smokers may show an increase in cyanide level on testing).
- ☑ Evaluation of thyroid function
- ☑ Lung function test and chest X-ray (for baseline).

Any evaluation should include a careful history of past and present symptoms with an examination. Medical tests that simply look for existing damage are not a substitute for controlling exposures. Also, personnel who smoke should avoid any unprotected contact with calcium cyanide powders.

Other methods to reduce exposure include:

- ☑ Where possible, enclose operations and use local exhaust ventilation at the site of chemical release. If local exhaust ventilation or enclosure is not used, respiratory protection should be mandatory.
- ☑ Always ensure that proper protective clothing is worn when using chemical substances.
- ☑ Wash thoroughly immediately after exposure to calcium cyanide and at the end of the work shift or before eating, drinking, or smoking.
- ☑ Hazard warning information should be posted in the work area. In addition, as part of an on-going education and training program, all information on the health and safety hazards of calcium cyanide should be communicated to all potentially exposed workers.
- ☑ Eye wash stations should be provided in the immediate work area for emergency use. If there is a possibility of skin exposure to calcium cyanide, emergency shower facilities should also be provided.
- ☑ Workers whose clothing has been contaminated by calcium cyanide should change into clean clothes before leaving work. Contaminated work clothing should be laundered only by individuals who have been informed of the hazards of exposure to calcium cyanide.

RISK ASSESSMENT: ENVIRONMENT

General Assessment

Calcium cyanide is a naturally occurring element. However, industrial uses place the environment at risk of exposure during transportation, storage, disposal, or destruction of calcium cyanide compounds. In almost every scenario, the threat of environmental exposure is contingent upon the proper handling of the chemical substance. Accidental spills, large or small, where calcium cyanide contacts incompatible commodities can result in the release of hazardous fumes and gases (depending upon conditions of spill), and possible contamination of the surrounding environmental mediums (water, soil, and air).

In its bulk form, calcium cyanide is considered a non-combustible solid. However, its dusts can react violently and may present a serious fire hazard. Because it is incompatible with many commodities, including water, steam, acids, and carbon dioxide, extreme caution is required in handling, storage, transportation, and disposal of calcium cyanide. These characteristics also require special consideration during any emergency situation involving a leak or spill of calcium cyanide powder or dust.

Should calcium cyanide ever come into contact with any incompatible substances either during use, transportation, storage, or disposal, release of poisonous fumes are released and possible violent reactions are possible.

Calcium cyanide can enter the environment from industrial and municipal waste treatment plant discharges and from spills. Also, as a fumigant and insecticide, it will also enter the environment as a result of its intended use.

Acute Ecological Effects

Acute (short-term) toxic effects may include the death of animals, birds, or fish and death or low growth rate in plants. Acute effects are seen 2 to 4 days after animals or plants are exposed to calcium cyanide. Calcium cyanide has high acute toxicity to aquatic life and terrestrial animals. Insufficient data are available to evaluate the short-term effects of calcium cyanide to plants and birds.

Chronic Ecological Effects

Chronic (long-term) toxic effects may include shortened life span, reproductive problems, lower fertility, and changes in appearance or behavior in exposed animals. These effects can be seen long after first exposure(s) to toxic chemicals. Calcium cyanide has high chronic toxicity to aquatic life and terrestrial animals. Insufficient data are available to evaluate or predict the long-term effects of calcium cyanide to plants and birds.

Water Solubility

Calcium cyanide is moderately soluble in water. Concentrations of 1 to 1000 milligrams will mix with a liter of water.

Persistence in the Environment

Calcium cyanide is highly persistent in water, with a half-life greater than 200 days. The half-life of a pollutant is the amount of time it takes for one half of the chemical to be degraded. No data are available on the estimated percentages of calcium cyanide persistence in the aquatic or terrestrial environments.

Bioaccumulation in Aquatic Organisms

Some substances increase in concentration, or *bioaccumulate,* in living organisms as they breathe contaminated air, drink contaminated water, or eat contaminated food. These chemicals can become concentrated in the tissues and internal organs of animals as well as humans.

The concentration of calcium cyanide found in fish tissues is expected to be much higher than the average concentration of calcium cyanide in the water from which the fish was taken.

Recommended Risk-Reduction Measures

Proper training of all transporters will reduce the likelihood of a mishap or accident resulting in a leak or spill of calcium cyanide dusts or powders into the environment. The correct use of DOT labeling (per the MSDS) on all containers, trucks, and rail cars will enable emergency responders to react properly and quickly to any disaster thereby reducing the potential risk to the environment and to personnel.

Storage of calcium cyanide should be segregated from incompatible chemicals to minimize the risk of cross-contamination or contact. It should be shipped in mild-steel or fiber drums. Proper ventilation and protective equipment should be used while handing the solid or while preparing aqueous solutions.

Calcium cyanide is destroyed by oxidizing it to calcium cyanate using sodium hypochlorite or other oxidizing compounds.

If a spill or leak to the environment has occurred, fire department, emergency response and/or hazardous materials spill personnel should be notified immediately. Cleanup should be attempted only by those trained in proper spill containment procedures. If calcium cyanide should contact the water table, aquifer, or navigable waterway, time is of the essence. It is moderately soluble and total remediation may not be possible. The local and/or state emergency response authorities must be notified. A comprehensive emergency response or disaster preparedness/recovery plan should be in place prior to any operations involving the use, transportation, storage, or disposal of calcium cyanide. If calcium cyanide powder is spilled or leaked, the following specific steps are recommended:

- ☑ Restrict persons not wearing protective clothing from area of spill or leak until cleanup is complete and area can be opened for normal work.
- ☑ Remove all ignition sources.
- ☑ Collect powdered materials in the safest manner possible and deposit in sealed containers. Recommend using vacuum with HEPA filter. Do not dry sweep (generates airborne dusts).
- ☑ It may be necessary to dispose of calcium cyanide as a hazardous waste. The responsible state agency or the regional office of the federal Environmental Protection Agency (EPA) should be contacted for specific recommendations.

RISK ASSESSMENT: BUSINESS

General Assessment

Accidents or mishaps involving calcium cyanide can present a significant threat to business operations. The loss or damage of equipment or facilities can significantly affect fiscal viability. Any incident involving a chemical containing the word "cyanide" in its description can evoke an emotional reaction from the public, even when no exposures have occurred. Lawsuits that may result from personnel injury/death, public exposures, and/or environmental contamination will also require a serious expenditure of resources. Media attention surrounding an injury, death, or environmental damage can also result in a loss of profits and loss of current as well as future business.

Recommended Risk-Reduction Measures

Company attorneys, safety and health professionals, and environmental specialists should be involved in the development of any procedures for responding to chemical incidents. A company official should be predesignated as a public relations officer with specific training in dealing with the press. Corporate plans and policies should be developed, approved, and implemented long before any need for such arises.

REFERENCES

Ahlbom, A. 1993. *Biostatistics for Epidemiologists.* New York: CRC Press/Lewis Publishers.

American Conference of Governmental Industrial Hygienists. 1988. *Documentation of the Threshold Limit Values and Biological Exposure Indices,* 5th Edition (with updates). Cincinnati: ACGIH

Asante-Duah, D. K. 1993. *Hazardous Waste Risk Assessment.* New York: CRC Press/Lewis Publishers.

Brusick, D. J. 1994. *Methods for Genetic Risk Assessment.* New York: CRC Press/Lewis Publishers.

Bloom, A. D. and F. J. DeSerres. 1995. *Ecotoxicity and Human Health.* New York: CRC Press/Lewis Publishers.

Bueche, J. R. 1972. *Principles of Physics,* 2nd Edition. New York: McGraw-Hill.

Calabrese, E. J., E. M. Kenyon. *Air Toxics and Risk Assessment.* New York: CRC Press/Lewis Publishers.

Calabrese, E. J. 1994. *Biological Effects of Low Level Exposures.* New York: CRC Press/Lewis Publishers.

Cothern, C. R., N. P. Ross. 1994. *Environmental Statistics, Assessment, and Forecasting.* New York: CRC Press/Lewis Publishers.

Cothern, R. R. 1993. *Comparative Environmental Risk Assessment.* New York: CRC Press/Lewis Publishers.

Gardner, R. W. 1994. *Chemical Intolerance.* New York: CRC Press/Lewis Publishers.

Hallenbeck, W. H. 1993. *Quantitative Risk Assessment for Environmental and Occupational Health,* 2nd Edition. New York: CRC Press/Lewis Publishers.

Howard, P. H. 1995. *Handbook of Environmental Fate and Exposures Data for Organic Chemicals,* Vol. I-IV. New York: CRC Press/Lewis Publishers.

Howard, P. H., et. al. 1991. *Handbook of Environmental Degradation Rates.* New York: CRC Press/Lewis Publishers.

Jones, F. E. 1994. *Toxic Organic Vapors in the Workplace.* New York: CRC Press/Lewis Publishers.

Keith, L. H. and M. M. Walker. 1995. *Handbook of Air Toxics.* New York: CRC Press/Lewis Publishers.

Landis, W. G., M. H. Yu. 1995. *Introduction to Environmental Toxicology.* New York: CRC Press/Lewis Publishers.

Langerman, N. 1994 *Precautionary Labels for Chemical Containers.* New York: CRC Press/Lewis Publishers.

Lauwerys, R. R., P. Hoet. 1993. *Industrial Chemical Exposure.* New York: CRC Press/Lewis Publishers.

Lewis, R. J., Sr. 1992. *Sax's Dangerous Properties of Industrial Materials,* Volumes 1, 2, and 3, 8th Edition. New York: Van Nostrand Reinhold.

Lide, D. R. 1995. *Handbook of Chemistry and Physics,* 76th Edition. New York: CRC Press/Lewis Publishers.

Manahan, S. E. 1992. *Toxicological Chemistry,* 2nd Edition. New York: CRC Press/Lewis Publishers.

Mellström, G. A., J. E. Wahlberg, and H. I. Maibach. 1994. *Protective Gloves for Occupational Use.* New York: CRC Press/Lewis Publishers.

Meyer, E. 1990. *Chemistry of Hazardous Materials.* Englewood Cliffs, NJ: Prentice Hall.

Mycock, J. C., J. D. McKenna, and L. Theodore. 1995. *Handbook of Air Pollution Control Engineering and Technology.* New York: CRC Press/Lewis Publishers.

Nelson, G. O. 1992. *Gas Mixtures.* New York: CRC Press/Lewis Publishers.

Ott, W. R. 1995. *Environmental Statistics and Data Analysis.* New York: CRC Press/Lewis Publishers.

Patnaik, P. 1992. *A Comprehensive Guide to the Hazardous Properties of Chemical Substances.* New York: Van Nostrand Reinhold.

Proctor, N. H., J. P. Hughes. 1991. *Chemical Hazards of the Workplace.* New York: Van Nostrand Reinhold.

Quigley, D. R. 1994. *Handbook of Emergency Chemical Management.* New York: CRC Press/Lewis Publishers.

Reeves, A. L. 1979. Calcium cyanide. *Handbook on the Toxicology of Metals.* pp 321-328. New York: Elsevier North-Holland.

Richardson, M. 1995. *Dictionary of Substance and Their Effects,* Vol. I-VI. New York: CRC Press/Lewis Publishers.

Sacarello, H. L. A. 1994. *The Comprehensive Handbook of Hazardous Materials.* New York: CRC Press/Lewis Publishers.

Torkelson, T. R. V. K. Rowe. 1981. *Patty's Industrial Hygiene and Toxicology,* Vol. 2B. New York: Wiley-Interscience.

MATERIAL SAFETY DATA SHEET

CHEMICAL NAME	CRC PRESS/LEWIS PUBLISHERS DISCLAIMER
CALCIUM FLUORIDE	**CRC PRESS/LEWIS PUBLISHERS DISCLAIMER:** THE INFORMATION AND RECOMMENDATIONS PRESENTED HEREIN ARE BASED ON SOURCES BELIEVED TO BE RELIABLE. CRC MAKES NO REPRESENTATION ON ITS COMPLETENESS OR ACCURACY. IT IS THE RESPONSIBILITY OF THE USER TO DETERMINE THE CHEMICAL'S SUITABILITY FOR ITS INTENDED USE, THE CHEMICAL'S SAFE USE, AND THE CHEMICAL'S PROPER DISPOSAL. NO REPRESENTATIONS AND/OR WARRANTIES, EITHER EXPRESSED OR IMPLIED, OF THE MERCHANTABILITY OR FITNESS FOR A PARTICULAR PURPOSE, OR OF ANY OTHER NATURE, ARE MADE WITH RESPECT TO THE INFORMATION PROVIDED IN THIS MSDS OR TO THE CHEMICAL TO WHICH INFORMATION MAY REFER. CRC NEITHER ASSUMES NOR AUTHORIZES ANY OTHER PERSON TO ASSUME FOR IT, ANY OTHER ADDITIONAL RESPONSIBILITY OR LIABILITY FOR THE USE OF, OR RELIANCE UPON, THIS INFORMATION.

HAZARD WARNING INFORMATION

HEALTH	FIRE	REACTIVE	OTHER	DEGREE OF HAZARD	COLOR CODING	OTHER CODES
2	0	0		0 = Minimum Hazard 1 = Slight Hazard 2 = Moderate Hazard 3 = Serious Hazard 4 = Severe Hazard	HEALTH = BLUE FIRE = RED REACTIVITY = YELLOW OTHER = WHITE	OX = Oxidizer ACID = Acid ALK = Alkali COR = Corrosive W = Use No Water

SECTION I - GENERAL INFORMATION

Characterization	RCRA Number	EPA Class
Calcium Compound	**None**	**Not Applicable**

DOT Proper Shipping Name	**Not Listed**	Chemical Abstract Service (CAS) Number	**7789-75-5**
DOT Hazard Class and Label Requirements	**No Citation**	DOT Emergency Guide Code	**No Citation**
DOT Identification Number	**No Citation**	Atomic Formula	**CaF_2**

Synonyms

Acid-Spar; calcium difluoride; fluorite; fluorspar; Itran 3; Liparite; Met-Spar.

SECTION II - HAZARDOUS INGREDIENTS AND IDENTITY INFORMATION

Hazardous Components (specific identity)	OSHA Exposure Criteria	NIOSH Exposure Criteria	Immediately Dangerous to Life and Health (IDLH)	ACGIH Exposure Criteria
Calcium fluoride (derivation: By powdering pure fluorite or fluorspar; by the interaction of soluble calcium salt and sodium fluoride. Occurs in nature as the mineral fluorite or fluorspar).	**PEL (8-hour):** 2.5 mg/m^3 (as fluoride) **STEL:** Not Established	**REL (10-hour):** 2.5 mg/m^3 (as fluoride) **STEL:** Not Established	**500 mg/m^3**	**TLV:** 2.5 mg/m^3 (as fluoride) **STEL:** Not Established

SECTION III - PHYSICAL AND CHEMICAL CHARACTERISTICS

Boiling Point	**4532°F (2500°C)**	Specific Gravity (H_2O = 1)	**3.18**
Vapor Pressure (mm Hg)	**Not Found**	Molecular Weight	**78.08**
Vapor Density (Air = 1)	**Not Found**	Melting Point	**2556°F (1402°C)**

Solubility

Nearly insoluble in water. Soluble in ammonia salts, slightly soluble in dilute mineral acids.

Appearance and Odor

White powder or cubic, colorless crystals that become luminous with heat.

SECTION IV - FIRE AND EXPLOSION HAZARD DATA

Flash Point (method used)	**Not Determined**	Explosive Limits in Air % by Volume	**LEL: Not Determined UEL: Not Determined**
NFPA Classification	**Non-Combustible Solid**	Autoignition Temperature	**Not Determined**

Extinguishing Media

Use agent suitable to surrounding fire. Dry chemical, carbon dioxide, water spray, or regular foam should be adequate for fire suppression.

Special Fire Fighting Procedures

Poisonous gases are produced in fire. Wear full protective clothing and self-contained breathing apparatus (SCBA). Cool fire-exposed containers using water spray. Move containers from fire if it can be done without risk. Remain clear of smoke, water fall-out and water runoff. Do NOT release runoff from fire-fighting measures to sewers or waterways.

Unusual Fire and Explosion Hazards

None reported.

SECTION V - REACTIVITY DATA

Stability		Conditions to Avoid: Calcium fluoride is stable under normal conditions of temperature and pressure. Keep away from incompatible materials. Avoid contact with heat.
Stable	Unstable	Incompatibility *(materials to avoid)*
X		Incompatible with hot, concentrated sulfuric acid (liberates hydrogen fluoride).
Hazardous Polymerization		Conditions to Avoid: Under normal conditions of temperature and pressure, hazardous polymerization of calcium fluoride is not known to occur.
May Occur	Will Not Occur	Hazardous Decomposition or By-products
	X	When heated to decomposition, calcium fluoride can emit toxic/poisonous fumes and gases. These include toxic fumes of fluorides.

SECTION VI - HEALTH HAZARD DATA

Primary Route(s) of Entry:	Inhalation?	Absorption (skin)?	Ingestion?
	X		X

Health Hazards

INHALATION: Initial increased respiration with coughing, choking, shortness of breath, followed by a decrease in respiration rate. Respiratory arrest, cardiac arrhythmia, hyperkalemia, and hypocalcemia (may cause hyperactive reflexes and painful muscular spasms). Prolonged exposure may cause fluorosis. Silica content may cause other lung disorders.

EYES & SKIN: Possible irritation at sight of contact. Eye irritation and inflammation also possible.

INGESTION: Gastrointestinal irritation. Abdominal pain, difficulty swallowing, nausea, vomiting, and increased salivation. There may be accumulation or buildup in bones and teeth due to the fluoride content.

Carcinogenicity	NTP Listed?	IARC Cancer Review Group?	OSHA Regulated?	Target Organs?
Unknown Human Unknown Animal	No	No	29 CFR 1910.1000 Table Z-1	Bones, teeth, G.I. tract, resp. sys., kidney, heart.

Medical Conditions Generally Aggravated by Exposure

Due to silica content, respiratory disorders may be aggravated by exposure.

Emergency and First-aid Procedures

Eye contact: Flush immediately with water for 15 minutes (minimum), seek medical attention. Skin contact: Remove all contaminated clothing. Immediately wash area with large amounts of soap and water. For inhalation: Remove the person from exposure. Provide respiratory assistance and CPR. Transfer to medical facility. If swallowed: Provide breathing assistance if necessary. Never attempt to give a convulsing or unconscious person anything by mouth. If victim is conscious, give 1-2 glasses of water to dilute and induce vomiting with Syrup of Ipecac. Seek medical attention immediately.

SECTION VII - PRECAUTIONS FOR SAFE HANDLING AND USE

Steps to be Taken in Case Material is Released or Spilled

Collect powdered material in most convenient manner possible and deposit in sealed drum. Keep those not involved in cleanup from entering area. Do NOT dry sweep (creates airborne dusts). Use a vacuum equipped with a high efficiency particulate air (HEPA) filter. Use wet mop to clean residues.

Preferred Waste Disposal Method

No Citation.

Precautions to be Taken in Handling and Storage

Store in tightly closed containers in cool, well-ventilated area away from sulfuric acid. Sources of ignition are prohibited where calcium fluoride is stored.

Other Precautions and Warnings

Protect from physical damage.

SECTION VIII - CONTROL MEASURES AND PERSONAL PROTECTIVE EQUIPMENT

Respiratory Protection *(specify type)*

For low exposures, a full facepiece respirator with HEPA filter may suffice. For best protection, use a supplied air respirator with a full facepiece, or an SCBA operated in positive pressure mode.

Ventilation

Local exhaust or general mechanical systems recommended.

Protective Gloves	Eye Protection	Other Protective Clothing
Impervious Material	Chemical/Dust Goggles or Face Mask	Protective Suit or Clothing

Work/Hygiene Practices

Always wash hands thoroughly after using chemical; never bring food, drink, or smoking materials into vicinity of chemicals.

CALCIUM FLUORIDE

CaF_2 CAS: 7789-75-5

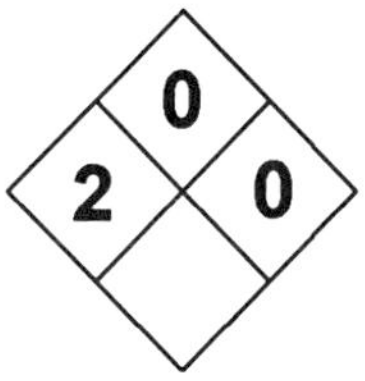

IDENTIFICATION AND TYPICAL USES

Calcium fluoride appears as a white powder or as cubic, colorless crystals which may become luminous with the introduction of heat. It is used as a raw material in the production of hydrofluoric acid, as a catalyst in dehydration and dehydrogenation, as a flux in the iron and steel industry, in etching glass, and to fluoridate drinking water. Single pure crystals (99.93%) are also produced for use in spectroscopy, electronics, lasers, and high-temperature dry-film lubricants.

RISK ASSESSMENT: HEALTH

General Assessment

Calcium fluoride is mildly toxic by *ingestion* and *inhalation*. It is considered "mildly" toxic due to its low solubility and low ionization. Its toxicity appears to result primarily from its fluorine and silica content. Its carcinogenic potential has not been adequately investigated. However, it is known to cause mutations (genetic changes) in living cells. Experimental teratogenic effects have also been reported.

Inhalation may cause an initial increase in respiration followed by coughing, choking, dypsnea, and then a decrease in the rate of respiration. Respiratory arrest and subsequent death are possible. There may be cardiac arrhythmia, hyperkalemia, and hypocalcemia which, in turn, can lead to hyperactive reflexes, and painful muscular spasms. Pulmonary lesions may occur on repeated inhalation of calcium fluoride dusts which are thought to be due to the silica content in calcium fluoride, usually 3.5%.

Ingestion causes gastrointestinal irritation with abdominal pain, difficulty swallowing, nausea, vomiting, and increased salivation which may be delayed in onset. Ingestion of large amounts can cause accumulation of fluorine deposits in bones and teeth.

Acute Health Effects

The following acute (short-term) health effects may occur immediately or shortly after exposure to calcium fluoride:

Skin: Irritation and possible rash.

Eye: Irritation.

Lung: Causes respiratory changes that can lead to respiratory arrest and death. There may also be cardiac arrhythmia and other effects on bones, teeth, the kidney, and electrolytes.

Other: Ingestion can cause gastrointestinal irritation with abdominal pain, nausea, and vomiting.

Chronic Health Effects

The following chronic (long-term) health effects may occur at some time after exposure to calcium fluoride and can last for months or even years:

Cancer Hazards: According to the information presented in the references, calcium fluoride has not been adequately tested for its ability to cause cancer in test animals. Mutation (causes genetic changes) data have been documented and many scientists believe that such chemicals may pose a cancer risk in the long-term. Addition study and research is required.

Reproductive Hazard: According to information available in the references, calcium fluoride has not been adequately tested for its ability to adversely affect reproduction in laboratory animals.

Other Chronic Effects: Repeated exposures may cause the development of pulmonary lesions as a result of calcium fluoride's silica content. Long-term exposures can cause bronchitis and silicosis characterized by extreme fibrosis in the lungs with shortness of breath. Fluorosis is also possible. Symptoms include brittle bones, calcification of ligaments, and other changes in the bones and teeth. Long-term exposure may lead to liver damage and may cause blood clotting.

Recommended Risk-Reduction Measures

Personnel should avoid direct contact with calcium fluoride. If a less toxic chemical cannot be substituted for a hazardous material, then *engineering controls* are the most effective method of reducing exposures.

The best protection is to enclose operations and/or provide local exhaust ventilation at the site of chemical release. While not always operationally feasible, isolating operations can also reduce exposure. Using respiratory protection is less effective than the controls mentioned above, but is still advisable whenever working with or around calcium fluoride. For low exposures, a full facepiece respirator equipped with a high efficiency particulate air (HEPA) filter equipped with a dust/mist/fume pre-filter may suffice. However, for the best protection use a supplied-air respirator with a full facepiece operated in positive pressure mode or a MSHA/NIOSH approved self-contained breathing apparatus (SCBA) with full facepiece operated in pressure demand or other continuous flow mode. If a full facepiece is not available, then chemical/dust goggles should be worn to protect the eyes. Whenever a chemical dust hazard exists, a face shield and a rubber apron should be worn. To prevent hand and skin exposures, impervious rubber gloves should be worn.

Administrative controls should also be in place to minimize the potential for human exposures. These may include written procedures or policies which specify the methods and techniques that will be practiced whenever personnel are to work with calcium fluoride.

All personnel should receive *training* on the use, hazards, protective measures, emergency actions, and other precautions per 29 CFR 1910.1200 (Hazard Communication), prior to the first assignment in an area where calcium fluoride is used or stored.

If symptoms develop or overexposure is suspected, the following may also be useful:

- ☑ Urine and serum fluorine levels.
- ☑ Lung function tests.
- ☑ Consider chest X-ray following acute overexposure (may be negative if taken immediately after exposure due to delayed onset of pulmonary edema).
- ☑ Examination of the teeth for fluoride accumulation.

Any evaluation should include a careful history of past and present symptoms with a medical examination. Medical tests that look for damage already done are not a substitute for controlling exposure. Also, since smoking can cause heart disease, emphysema, and a number of other respiratory disorders, smokers exposed to lung irritants such as calcium fluoride dusts and fumes may experience symptoms much more quickly than non-smokers under the same conditions of exposure. Prudent risk management requires proper consideration of *all* factors which may be causing the appearance of exposures symptoms in the work place.

Other methods to reduce exposure include:

- ☑ Where possible, enclose operations and use local exhaust ventilation at the site of chemical release. If local exhaust ventilation or enclosure is not used, respiratory protection should be mandatory.
- ☑ Always ensure that proper protective clothing is worn when using chemical substances.
- ☑ Wash thoroughly immediately after exposure to calcium fluoride and at the end of the work shift or before eating, drinking, or smoking.
- ☑ Hazard warning information should be posted in the work area. In addition, as part of an on-going education and training program, all information on the health and safety hazards of calcium fluoride should be communicated to all potentially exposed workers.
- ☑ Eye wash stations and emergency showers should be located in the immediate work area.
- ☑ Work clothing that has been contaminated with calcium fluoride should never be worn home where family members can be exposed as well. Contaminated clothing should be laundered at work or by an authorized service by personnel who have been trained in the hazards associated with exposure to calcium fluoride and calcium compounds.

RISK ASSESSMENT: ENVIRONMENT

General Assessment

The environment is at risk of exposure during transportation, storage, disposal, or destruction of calcium fluoride. In almost every scenario, the threat of environmental exposure is contingent upon the proper handling of the chemical substance. Accidental spills, large or small, can result in fire, explosion, and possible contamination of the surrounding environmental mediums (water, soil, and air).

Calcium fluoride may react violently in contact with hot sulfuric acid to liberate toxic and corrosive hydrogen fluoride. In a fire, calcium fluoride will emit toxic and poisonous gases. These characteristics require special consideration during any emergency situation involving a leak or spill of calcium fluoride.

Calcium fluoride can enter the environment through industrial discharge, spills, or leaks.

Acute Ecological Effects

Acute (short-term) toxic effects may include the death of animals, birds, or fish and death or low growth rate in plants. Acute effects are seen 2 to 4 days after animals or plants are exposed to calcium fluoride. Insufficient data are available to evaluate the short-term effects of calcium fluoride exposure to aquatic life, plants, birds or land animals.

Chronic Ecological Effects

Chronic (long-term) toxic effects may include shortened life span, reproductive problems, lower fertility, and changes in appearance or behavior in exposed animals. These effects can be seen long after first exposure(s) to toxic chemicals. Insufficient data are available to evaluate the long-term effects of calcium fluoride exposure to aquatic life, plants, birds or land animals.

Water Solubility

Calcium fluoride is only very slightly soluble in water. Concentrations of less than 1 milligram may mix with a liter of water.

Persistence in the Environment

Calcium fluoride is only moderately persistent in the aquatic environment, with a half-life between 20 and 200 days. The half-life of a pollutant is the amount of time it takes for one half of the chemical to be degraded.

Bioaccumulation in Aquatic Organisms

Some substances increase in concentration, or *bioaccumulate,* in living organisms as they breathe contaminated air, drink contaminated water, or eat contaminated food. These chemicals can become concentrated in the tissues and internal organs of animals as well as humans.

The concentration of calcium fluoride found in fish is expected to be about the same as the average concentration of calcium fluoride in the water from which the fish was taken.

Recommended Risk-Reduction Measures

Proper training of all transporters will reduce the likelihood of a mishap or accidents resulting in a leak or spill to the environment. Proper labeling on all containers, trucks, and rail cars will enable emergency responders to react properly and quickly to any disaster thereby reducing the potential risk to the environment and to personnel.

If a spill or leak to the environment has occurred, fire department, emergency response and/or hazardous materials spill personnel should be notified immediately. Do not dry sweep. When vacuuming, a high efficiency particulate air (HEPA) filter should be used, not a standard shop vacuum. Use damp mop to clean up residue. Deposit collected materials in sealed drum for disposal. Cleanup should be attempted only by those trained in proper spill containment procedures. Contaminated soils may need to be removed and replaced with clean soil. If calcium fluoride should contact the water table, aquifer, or navigable waterway, time is of the essence. Calcium fluoride is slightly soluble in water. Therefore, total containment and remediation may not be entirely possible. When such spills occur, the local and/or state emergency response authorities must be notified. A comprehensive emergency response or disaster preparedness/recovery plan should be in place prior to any operations involving the use, transportation, storage, or disposal of calcium fluoride. If calcium fluoride is spilled or leaked, the following specific steps are recommended:

- ☑ Restrict persons not wearing protective clothing from area of spill or leak until cleanup is complete and area can be opened for normal work.
- ☑ Ventilate area of spill or leak and remove all ignition sources.
- ☑ Collect powdered materials in most convenient and safe manner possible and place in sealed drums for disposal. Do NOT dry sweep, use vacuum equipped with a high-efficiency particulate air (HEPA) filter instead.
- ☑ It may be necessary to dispose of calcium fluoride as a hazardous waste. The responsible state agency or the regional office of the federal Environmental Protection Agency (EPA) should be contacted for specific recommendations.

RISK ASSESSMENT: BUSINESS

General Assessment

Accidents or mishaps involving calcium fluoride can present a significant threat to business operations. The loss or damage of equipment or facilities can significantly affect fiscal viability. Lawsuits that may result from personnel injury/death, public exposures, and/or environmental contamination will also require a serious expenditure of resources. Media attention surrounding an injury, death, or environmental damage can also result in a loss of profits and loss of current as well as future business.

Recommended Risk-Reduction Measures

Company attorneys, safety and health professionals, and environmental specialists should be involved in the development of any procedures dealing with safety or emergency response. A company official should be pre-designated as a public relations officer with specific training in dealing with the press. Corporate plans and policies should be developed, approved, and implemented long before any need for such arises.

REFERENCES

Ahlbom, A. 1993. *Biostatistics for Epidemiologists.* New York: CRC Press/Lewis Publishers.

American Conference of Governmental Industrial Hygienists. 1988. *Documentation of the Threshold Limit Values and Biological Exposure Indices,* 5th Edition (with updates). Cincinnati: ACGIH

Asante-Duah, D. K. 1993. *Hazardous Waste Risk Assessment.* New York: CRC Press/Lewis Publishers.

Bloom, A. D. and F. J. DeSerres. 1995. *Ecotoxicity and Human Health.* New York: CRC Press/Lewis Publishers.

Brusick, D. J. 1994. *Methods for Genetic Risk Assessment.* New York: CRC Press/Lewis Publishers.

Calabrese, E. J., E. M. Kenyon. *Air Toxics and Risk Assessment.* New York: CRC Press/Lewis Publishers.

Calabrese, E. J. 1994. *Biological Effects of Low Level Exposures.* New York: CRC Press/Lewis Publishers.

Cockerman, L. G., B. S. Shane. 1994. *Basic Environmental Toxicology.* New York: CRC Press/Lewis Publishers.

Cothern, C. R., N. P. Ross. 1994. *Environmental Statistics, Assessment, and Forecasting.* New York: CRC Press/Lewis Publishers.

Cothern, R. R. 1993. *Comparative Environmental Risk Assessment.* New York: CRC Press/Lewis Publishers.

Cothern, C. R. 1995. *Handbook for Environmental Risk Decisionmaking.* New York: CRC Press/Lewis Publishers.

Gardner, R. W. 1994. *Chemical Intolerance.* New York: CRC Press/Lewis Publishers.

Hallenbeck, W. H. 1993. *Quantitative Risk Assessment for Environmental and Occupational Health,* 2nd Edition. New York: CRC Press/Lewis Publishers.

Howard, P. H., et. al. 1991. *Handbook of Environmental Degradation Rates.* New York: CRC Press/Lewis Publishers.

Howard, P. H. 1995. *Handbook of Environmental Fate and Exposures Data for Organic Chemicals,* Vol. I-IV. New York: CRC Press/Lewis Publishers.

Jones, F. E. 1994. *Toxic Organic Vapors in the Workplace.* New York: CRC Press/Lewis Publishers.

Keith, L. H. and M. M. Walker. 1995. *Handbook of Air Toxics.* New York: CRC Press/Lewis Publishers.

Landis, W. G., M. H. Yu. 1995. *Introduction to Environmental Toxicology.* New York: CRC Press/Lewis Publishers.

Langerman, N. 1994 *Precautionary Labels for Chemical Containers.* New York: CRC Press/Lewis Publishers.

Lauwerys, R. R., P. Hoet. 1993. *Industrial Chemical Exposure.* New York: CRC Press/Lewis Publishers.

Lewis, R. J., Sr. 1992. *Sax's Dangerous Properties of Industrial Materials,* Volumes 1, 2, and 3, 8th Edition. New York: Van Nostrand Reinhold.

Manahan, S. E. 1994. *Environmental Chemistry,* 6th Edition. New York: CRC Press/Lewis Publishers.

Manahan, S. E. 1992. *Toxicological Chemistry,* 2nd Edition. New York: CRC Press/Lewis Publishers.

Meyer, E. 1990. *Chemistry of Hazardous Materials.* Englewood Cliffs, NJ: Prentice Hall.

Patnaik, P. 1992. *A Comprehensive Guide to the Hazardous Properties of Chemical Substances.* New York: Van Nostrand Reinhold.

Quigley, D. R. 1994. *Handbook of Emergency Chemical Management.* New York: CRC Press/Lewis Publishers.

Rea, W. J. 1995. *Chemical Sensitivity,* Vol. I-III. New York: CRC Press/Lewis Publishers.

Sacarello, H. L. A. 1994. *The Comprehensive Handbook of Hazardous Materials.* New York: CRC Press/Lewis Publishers.

Sherman, J. D. 1988. *Chemical Exposure and Disease.* New York: Van Nostrand Reinhold.

Talbott, E., G. Craun. 1995. *An Introduction to Environmental Epidemiology.* New York: CRC Press/Lewis Publishers.

U.S. Department of Health and Human Services (DHHS), National Institute for Occupational Health (NIOHS). 1994. *NIOSH Pocket Guide to Hazardous Chemicals.* Publication Number 94-116. Washington, D.C.; U.S. Government Printing Office

MATERIAL SAFETY DATA SHEET

CHEMICAL NAME	CRC PRESS/LEWIS PUBLISHERS DISCLAIMER
CALCIUM HYDROXIDE	**CRC PRESS/LEWIS PUBLISHERS DISCLAIMER:** THE INFORMATION AND RECOMMENDATIONS PRESENTED HEREIN ARE BASED ON SOURCES BELIEVED TO BE RELIABLE. CRC MAKES NO REPRESENTATION ON ITS COMPLETENESS OR ACCURACY. IT IS THE RESPONSIBILITY OF THE USER TO DETERMINE THE CHEMICAL'S SUITABILITY FOR ITS INTENDED USE, THE CHEMICAL'S SAFE USE, AND THE CHEMICAL'S PROPER DISPOSAL. NO REPRESENTATIONS AND/OR WARRANTIES, EITHER EXPRESSED OR IMPLIED, OF THE MERCHANTABILITY OR FITNESS FOR A PARTICULAR PURPOSE, OR OF ANY OTHER NATURE, ARE MADE WITH RESPECT TO THE INFORMATION PROVIDED IN THIS MSDS OR TO THE CHEMICAL TO WHICH INFORMATION MAY REFER. CRC NEITHER ASSUMES NOR AUTHORIZES ANY OTHER PERSON TO ASSUME FOR IT, ANY OTHER ADDITIONAL RESPONSIBILITY OR LIABILITY FOR THE USE OF, OR RELIANCE UPON, THIS INFORMATION.

HAZARD WARNING INFORMATION

HEALTH	FIRE	REACTIVE	OTHER	DEGREE OF HAZARD	COLOR CODING	OTHER CODES
2	0	0		0 = Minimum Hazard 1 = Slight Hazard 2 = Moderate Hazard 3 = Serious Hazard 4 = Severe Hazard	HEALTH = BLUE FIRE = RED REACTIVITY = YELLOW OTHER = WHITE	OX = Oxidizer ACID = Acid ALK = Alkali COR = Corrosive W = Use No Water

SECTION I - GENERAL INFORMATION

Characterization	**Calcium Compound**	RCRA Number	**None**
		EPA Class	**Not Applicable**
DOT Proper Shipping Name	**Not Listed**	Chemical Abstract Service (CAS) Number	**1305-62-0**
DOT Hazard Class and Label Requirements	**No Citation**	DOT Emergency Guide Code	**No Citation**
DOT Identification Number	**No Citation**	Atomic Formula	**CaH_2O_2**

Synonyms

Slaked lime; hydrated lime; calcium hydrate; caustic lime; bell lime; Kemikal; lime water.

SECTION II - HAZARDOUS INGREDIENTS AND IDENTITY INFORMATION

Hazardous Components (specific identity)	OSHA Exposure Criteria	NIOSH Exposure Criteria	Immediately Dangerous to Life and Health (IDLH)	ACGIH Exposure Criteria
Calcium hydroxide (derivation: From calcination of limestone followed by the addition of water to the resulting calcium oxide, or treating an aqueous calcium salt solution with an alkali).	**PEL:** 5 mg/m^3 (respirable dusts) **STEL:** 15 mg/m^3 (total dusts)	**REL:** 5 mg/m^3 **STEL:** Not Established	**Not Determined**	**TLV:** 5 mg/m^3 **STEL:** Not Established

SECTION III - PHYSICAL AND CHEMICAL CHARACTERISTICS

Boiling Point	**Decomposes**	Specific Gravity (H_2O = 1)	**2.24**
Vapor Pressure (mm Hg)	**Not Found**	Molecular Weight	**74.10**
Vapor Density (Air = 1)	**2.5**	Melting Point	**Loses water at 1076°F (580°C)**

Solubility

Slightly soluble in water. Soluble in glycerol, syrup, acids, ammonium salts. Insoluble in alcohol.

Appearance and Odor

Soft white granules or powder, or colorless, hexagonal crystals, no odor, bitter alkaline taste.

SECTION IV - FIRE AND EXPLOSION HAZARD DATA

Flash Point (method used)	**Not Determined**	Explosive Limits in Air % by Volume	**LEL: Not Determined UEL: Not Determined**
NFPA Classification	**Non-Flammable Solid**	Autoignition Temperature	**Not Determined**

Extinguishing Media

Use agent suitable to surrounding fire. Dry chemical, carbon dioxide, water spray, or regular foam should be adequate for fire suppression.

Special Fire Fighting Procedures

Poisonous gases are produced in fire. Wear full protective clothing and self-contained breathing apparatus (SCBA). Cool fire-exposed containers using water spray. Move containers from fire if it can be done without risk. Remain clear of smoke, water fall-out and water runoff. Do NOT release runoff from fire-fighting measures to sewers or waterways. Stay clear of ends of tanks.

Unusual Fire and Explosion Hazards

When heated, calcium hydroxide releases water to form calcium oxide, which is non-flammable.

SECTION V - REACTIVITY DATA

Stability		Conditions to Avoid Calcium hydroxide is stable under normal conditions of temperature and pressure. Keep away from incompatible materials. Avoid contact with heat. Keep lids tightly closed. Store above 77°F (25°C).
Stable X	Unstable	Incompatibility *(materials to avoid)* Causes violent decomposition of maleic anhydride; will form explosive products with nitroethane and water, nitromethane, nitroparaffins, and nitropropane; ammonia salts, phosphorus, chlorinated benzodioxin.
Hazardous Polymerization		Conditions to Avoid Under normal conditions of temperature and pressure, hazardous polymerization of calcium hydroxide is not known to occur.
May Occur	Will Not Occur X	Hazardous Decomposition or By-products When heated to decomposition (above 1075°F/580°C), calcium hydroxide will decompose, lose water, and form calcium oxide.

SECTION VI - HEALTH HAZARD DATA

Primary Route(s) of Entry:	Inhalation?	Absorption (skin)?	Ingestion?
	X		X

Health Hazards

INHALATION: Irritation of the eyes, nose, and throat with sneezing, dypsnea, coughing, and chemical bronchitis (in severe cases).

EYES & SKIN: Prolonged contact required for serious effects. Skin penetration is slow and may cause itching, burning, redness, and swelling. Eye contact may result in burns, pain, tearing, clouding of the eye, and loss of feeling (anesthesia) in the eye.

INGESTION: Gastrointestinal irritation with severe pain, diarrhea, vomiting with blood and mucous, collapse, and possible death within 24 hours if not treated. Delayed reactions include gastric and esophageal burns causing severe pain and hypotension.

Carcinogenicity	NTP Listed?	IARC Cancer Review Group?	OSHA Regulated?	Target Organs?
Unknown Human Unknown Animal	No	No	29 CFR 1910.1000 Table Z-1	Eyes, skin, mucous membranes, G.I. system.

Medical Conditions Generally Aggravated by Exposure

None reported.

Emergency and First-aid Procedures

Eye contact: Flush immediately with water for 15 minutes (minimum), seek medical attention. Skin contact: Remove all contaminated clothing. Immediately wash area with large amounts of water. Do NOT use soap. For inhalation: Remove the person from exposure. Provide respiratory assistance and CPR. Transfer to medical facility. If swallowed: Provide breathing assistance if necessary. Never attempt to give a convulsing or unconscious person anything by mouth. If victim is conscious, give 1-2 glasses of water or milk. Do NOT induce vomiting. Seek medical attention immediately.

SECTION VII - PRECAUTIONS FOR SAFE HANDLING AND USE

Steps to be Taken in Case Material is Released or Spilled

Collect powdered material in most convenient manner possible and deposit in sealed drum. Keep those not involved in cleanup from entering area. Do NOT dry sweep (creates airborne dusts). Use a vacuum equipped with a high efficiency particulate air (HEPA) filter. Use wet mop to clean residues.

Preferred Waste Disposal Method

No Citation.

Precautions to be Taken in Handling and Storage

Store in tightly closed containers in cool, well-ventilated area away from excessive heat and incompatible materials. Sources of ignition are prohibited where calcium hydroxide is stored.

Other Precautions and Warnings

Protect from physical damage.

SECTION VIII - CONTROL MEASURES AND PERSONAL PROTECTIVE EQUIPMENT

Respiratory Protection *(specify type)*

For low exposures, a full facepiece respirator with HEPA filter may suffice. For best protection, use a supplied air respirator with a full facepiece, or an SCBA operated in positive pressure mode.

Ventilation

Local exhaust or general mechanical systems recommended.

Protective Gloves	Eye Protection	Other Protective Clothing
Impervious Material	Chemical/Dust Goggles or Face Mask	Protective Suit or Clothing

Work/Hygiene Practices

Always wash hands thoroughly after using chemical; never bring food, drink, or smoking materials into vicinity of chemicals.

CALCIUM HYDROXIDE

CaH_2O_2 CAS: 1305-62-0

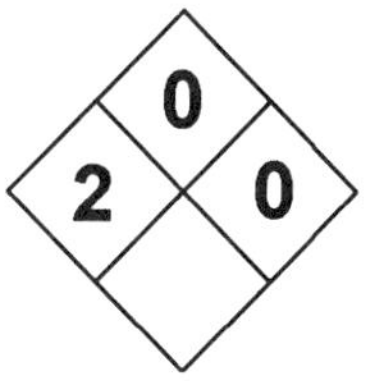

IDENTIFICATION AND TYPICAL USES

Calcium hydroxide appears as odorless, colorless, hexagonal crystals or as soft white granules or powder with a slightly bitter alkaline taste. It is used in mortar, plasters, cements, calcium salts, making caustic soda, depilatory (hair removing) for hides, as a whitewash, a soil conditioner, in ammonia recovery in gas manufacture, as a disinfectant, in water softening, in the purification of sugar juices, as a accelerator for low-grade rubber compounds, in petrochemicals, as a food additive buffer and neutralizing agent, a buffering agent for celluloid, metals, and pearls, as a shell-forming agent in the poultry industry, and as an antidote in veterinary medicine for tannin intoxication.

RISK ASSESSMENT: HEALTH

General Assessment

Calcium hydroxide is mildly toxic by *ingestion* and *inhalation*. It is a strong base and in solid form is moderately caustic to human tissues on contact. Its carcinogenic and teratogenic properties have not been adequately investigated. Mutation data have been reported.

Inhalation of calcium hydroxide dust will cause irritation to the nose, throat, and respiratory system with sneezing, coughing, and dypsnea (difficulty breathing). Severe cases may develop chemical bronchitis causing congestion, cough with phlegm, and possible progression into more serious respiratory dysfunction.

Prolonged contact with skin may produce itching, burning, redness (erythema), swelling, and possible dermatitis. Eye contact causes painful irritation with lachrymation (watering or tearing). Mild burns can produce clouding of the cornea and may cause anesthesia (loss of feeling) in the eye that can last for several hours or even days following contact exposures. Any solid materials in the eye may form difficult to remove "clumps" due to the eye's moisture content.

Ingestion causes serious gastrointestinal irritation with abdominal pain, diarrhea that may be bloody or even contain pieces of mucous membrane lining, and collapse. Death may occur within 24 hours. Survivors may experience delayed perforations to the gastrointestinal tract and/or esophagus which can cause additional and severe pain and possible hypotension (decreased blood pressure). Constriction of the esophagus may continue for several hours or days after exposure, making swallowing difficult.

Acute Health Effects

The following acute (short-term) health effects may occur immediately or shortly after exposure to calcium hydroxide:

<u>Skin</u>: Irritation and possible rash. Dermatitis is possible, but usually not in the short-term.

<u>Eye</u>: Irritation and possible clouding of the cornea and anesthesia.

<u>Lung</u>: Irritation of the eyes, nose, throat, and mucosa of the respiratory tract. There may be coughing, sneezing, and dypsnea.

<u>Other</u>: Ingestion is serious and can lead to death. Gastrointestinal irritation with abdominal pain and bloody diarrhea likely to occur.

Chronic Health Effects

The following chronic (long-term) health effects may occur at some time after exposure to calcium hydroxide and can last for months or even years:

<u>Cancer Hazards</u>: According to the information presented in the references, calcium hydroxide has not been adequately tested for its ability to cause cancer in test animals. Mutation (causes genetic changes) data

have been documented and many scientists believe that such chemicals may pose a cancer risk in the long-term. Additional study and research are required.

Reproductive Hazard: According to information available in the references, calcium hydroxide has not been adequately tested for its ability to adversely affect reproduction in laboratory animals.

Other Chronic Effects: Repeated skin exposures may cause chronic dermatitis in some individuals. Chemical bronchitis may develop in persons exposed to low doses over a prolonged period of time.

Recommended Risk-Reduction Measures

Personnel should avoid direct contact with calcium hydroxide. If a less toxic chemical cannot be substituted for a hazardous material, then *engineering controls* are the most effective method of reducing exposures.

The best protection is to enclose operations and/or provide local exhaust ventilation at the site of chemical release. While not always operationally feasible, isolating operations can also reduce exposure. Using respiratory protection is less effective than the controls mentioned above, but is still advisable whenever working with or around calcium hydroxide. For low exposures, a full facepiece respirator equipped with a high efficiency particulate air (HEPA) filter equipped with a dust/mist/fume pre-filter may suffice. However, for the best protection use a supplied-air respirator with a full facepiece operated in positive pressure mode or a MSHA/NIOSH approved self-contained breathing apparatus (SCBA) with full facepiece operated in pressure demand or other continuous flow mode. If a full facepiece is not available, then chemical/dust goggles should be worn to protect the eyes. Whenever a chemical dust hazard exists, a face shield and a rubber apron should be worn. To prevent hand and skin exposures, impervious rubber gloves should be worn.

Administrative controls should also be in place to minimize the potential for human exposures. These may include written procedures or policies which specify the methods and techniques that will be practiced whenever personnel are to work with calcium hydroxide.

All personnel should receive *training* on the use, hazards, protective measures, emergency actions, and other precautions per 29 CFR 1910.1200 (Hazard Communication), prior to the first assignment in an area where calcium hydroxide is used or stored.

Before beginning employment and at regular intervals thereafter (e.g., annually), the following medical tests are recommended for personnel with frequent or anticipated exposure to calcium hydroxide:

☑ Lung function tests (establish baseline).

If symptoms develop or overexposure is suspected, the following may also be useful:

☑ Lung function tests.

☑ Consider chest X-ray following acute overexposure.

☑ Examination by a qualified allergist with careful consideration of exposure history and special testing (may help diagnose skin allergy).

Any evaluation should include a careful history of past and present symptoms with a medical examination. Medical tests that look for damage already done are not a substitute for controlling exposure. Also, since smoking can cause heart disease, emphysema, and a number of other respiratory disorders, smokers exposed to lung irritants such as calcium hydroxide dusts and fumes may experience symptoms much more quickly than non-smokers under the same conditions of exposure. Prudent risk management requires proper consideration of *all* factors which may be causing the appearance of exposures symptoms in the work place.

Other methods to reduce exposure include:

☑ Where possible, enclose operations and use local exhaust ventilation at the site of chemical release. If local exhaust ventilation or enclosure is not used, respiratory protection should be mandatory.

☑ Always ensure that proper protective clothing is worn when using chemical substances.

☑ Wash thoroughly immediately after exposure to calcium hydroxide and at the end of the work shift or before eating, drinking, or smoking.

☑ Hazard warning information should be posted in the work area. In addition, as part of an on-going education and training program, all information on the health and safety hazards of calcium hydroxide should be communicated to all potentially exposed workers.

☑ Eye wash stations and emergency showers should be located in the immediate work area.

☑ Work clothing that has been contaminated with calcium hydroxide should never be worn home where family members can be exposed as well. Contaminated clothing should be laundered at work or by an authorized service by personnel who have been trained in the hazards associated with exposure to calcium hydroxide and calcium compounds.

RISK ASSESSMENT: ENVIRONMENT

General Assessment

The environment is at risk of exposure during transportation, storage, disposal, or destruction of calcium hydroxide. In almost every scenario, the threat of environmental exposure is contingent upon the proper handling of the chemical substance. Accidental spills, large or small, can result in fire, explosion, and possible contamination of the surrounding environmental mediums (water, soil, and air).

Calcium hydroxide causes violent decomposition of maleic anhydride and will form explosive products with nitroethane and water, nitromethane, nitroparaffins, and nitropropane. It will release ammonia when mixed with ammonia salts, and, when boiled with phosphorus, it releases mixed phosphines which can ignite spontaneously in air (pyrophoric). When heated to temperature above 1076°F (580°C), it will decompose to form calcium oxide. While it is not flammable, it can evolve toxic compounds if involved in a fire. These characteristics require special consideration during any emergency situation involving a leak or spill of calcium hydroxide.

Calcium hydroxide can enter the environment through industrial discharge, spills, or leaks.

Acute Ecological Effects

Acute (short-term) toxic effects may include the death of animals, birds, or fish and death or low growth rate in plants. Acute effects are seen 2 to 4 days after animals or plants are exposed to calcium hydroxide.

Insufficient data are available to evaluate the short-term effects of calcium hydroxide exposure to aquatic life, plants, birds or land animals.

Chronic Ecological Effects

Chronic (long-term) toxic effects may include shortened life span, reproductive problems, lower fertility, and changes in appearance or behavior in exposed animals. These effects can be seen long after first exposure(s) to toxic chemicals.

Insufficient data are available to evaluate the long-term effects of calcium hydroxide exposure to aquatic life, plants, birds or land animals.

Water Solubility

Calcium hydroxide is only very slightly soluble in water. Concentrations of less than 1 milligram may mix with a liter of water.

Persistence in the Environment

Calcium hydroxide is only moderately persistent in the aquatic environment, with a half-life between 20 and 200 days. The half-life of a pollutant is the amount of time it takes for one half of the chemical to be degraded.

Bioaccumulation in Aquatic Organisms

Some substances increase in concentration, or *bioaccumulate,* in living organisms as they breathe contaminated air, drink contaminated water, or eat contaminated food. These chemicals can become concentrated in the tissues and internal organs of animals as well as humans.

The concentration of calcium hydroxide found in fish is expected to be about the same as the average concentration of calcium hydroxide in the water from which the fish was taken.

Recommended Risk-Reduction Measures

Proper training of all transporters will reduce the likelihood of a mishap or accidents resulting in a leak or spill to the environment. Proper labeling on all containers, trucks, and rail cars will enable emergency responders to react properly and quickly to any disaster thereby reducing the potential risk to the environment and to personnel.

If a spill or leak to the environment has occurred, fire department, emergency response and/or hazardous materials spill personnel should be notified immediately. Do not dry sweep. When vacuuming, a high efficiency particulate air (HEPA) filter should be used, not a standard shop vacuum. Use damp mop to clean up residue. Deposit collected materials in sealed drum for disposal. Cleanup should be attempted only by those trained in proper spill containment procedures. Contaminated soils may need to be removed and replaced with clean soil. If calcium hydroxide should contact the water table, aquifer, or navigable waterway, time is of the essence. Calcium hydroxide is slightly soluble in water. Therefore, total containment and remediation may not be entirely possible. When such spills occur, the local and/or state emergency response authorities must be notified. A comprehensive emergency response or disaster preparedness/recovery plan should be in place prior to any operations involving the use, transportation, storage, or disposal of calcium hydroxide. If calcium hydroxide is spilled or leaked, the following specific steps are recommended:

☑ Restrict persons not wearing protective clothing from area of spill or leak until cleanup is complete and area can be opened for normal work.

☑ Ventilate area of spill or leak and remove all ignition sources.

☑ Collect powdered materials in most convenient and safe manner possible and place in sealed drums for disposal. Do NOT dry sweep, use HEPA vacuum instead. Damp mop residue.

☑ It may be necessary to dispose of calcium hydroxide as a hazardous waste. The responsible state agency or the regional office of the federal Environmental Protection Agency (EPA) should be contacted for specific recommendations.

RISK ASSESSMENT: BUSINESS

General Assessment

Accidents or mishaps involving calcium hydroxide can present a significant threat to business operations. The loss or damage of equipment or facilities can significantly affect fiscal viability. Lawsuits that may result from personnel injury/death, public exposures, and/or environmental contamination will also require a serious expenditure of resources. Media attention surrounding an injury, death, or environmental damage can also result in a loss of profits and loss of current as well as future business.

Recommended Risk-Reduction Measures

Company attorneys, safety and health professionals, and environmental specialists should be involved in the development of any procedures dealing with safety or emergency response. A company official should be pre-designated as a public relations officer with specific training in dealing with the press. Corporate plans and policies should be developed, approved, and implemented long before any need for such arises.

REFERENCES

Ahlbom, A. 1993. *Biostatistics for Epidemiologists.* New York: CRC Press/Lewis Publishers.

American Conference of Governmental Industrial Hygienists. 1988. *Documentation of the Threshold Limit Values and Biological Exposure Indices,* 5th Edition (with updates). Cincinnati: ACGIH

Asante-Duah, D. K. 1993. *Hazardous Waste Risk Assessment.* New York: CRC Press/Lewis Publishers.

Bloom, A. D. and F. J. DeSerres. 1995. *Ecotoxicity and Human Health.* New York: CRC Press/Lewis Publishers.

Brusick, D. J. 1994. *Methods for Genetic Risk Assessment.* New York: CRC Press/Lewis Publishers.

Calabrese, E. J. 1994. *Biological Effects of Low Level Exposures.* New York: CRC Press/Lewis Publishers.

Cockerman, L. G., B. S. Shane. 1994. *Basic Environmental Toxicology.* New York: CRC Press/Lewis Publishers.

Cothern, R. R. 1993. *Comparative Environmental Risk Assessment.* New York: CRC Press/Lewis Publishers.

Gardner, R. W. 1994. *Chemical Intolerance.* New York: CRC Press/Lewis Publishers.

Hallenbeck, W. H. 1993. *Quantitative Risk Assessment for Environmental and Occupational Health,* 2nd Edition. New York: CRC Press/Lewis Publishers.

Howard, P. H., et. al. 1991. *Handbook of Environmental Degradation Rates.* New York: CRC Press/Lewis Publishers.

Jones, F. E. 1994. *Toxic Organic Vapors in the Workplace.* New York: CRC Press/Lewis Publishers.

Keith, L. H. and M. M. Walker. 1995. *Handbook of Air Toxics.* New York: CRC Press/Lewis Publishers.

Landis, W. G., M. H. Yu. 1995. *Introduction to Environmental Toxicology.* New York: CRC Press/Lewis Publishers.

Langerman, N. 1994 *Precautionary Labels for Chemical Containers.* New York: CRC Press/Lewis Publishers.

Lauwerys, R. R., P. Hoet. 1993. *Industrial Chemical Exposure.* New York: CRC Press/Lewis Publishers.

Lewis, R. J., Sr. 1992. *Sax's Dangerous Properties of Industrial Materials,* Volumes 1, 2, and 3, 8th Edition. New York: Van Nostrand Reinhold.

Manahan, S. E. 1994. *Environmental Chemistry,* 6th Edition. New York: CRC Press/Lewis Publishers.

Meyer, E. 1990. *Chemistry of Hazardous Materials.* Englewood Cliffs, NJ: Prentice Hall.

Rea, W. J. 1995. *Chemical Sensitivity,* Vol. I-III. New York: CRC Press/Lewis Publishers.

Sacarello, H. L. A. 1994. *The Comprehensive Handbook of Hazardous Materials.* New York: CRC Press/Lewis Publishers.

Sherman, J. D. 1988. *Chemical Exposure and Disease.* New York: Van Nostrand Reinhold.

Talbott, E., G. Craun. 1995. *An Introduction to Environmental Epidemiology.* New York: CRC Press/Lewis Publishers.

U.S. Department of Health and Human Services (DHHS), National Institute for Occupational Health (NIOHS). 1994. *NIOSH Pocket Guide to Hazardous Chemicals.* Publication Number 94-116. Washington, D.C.; U.S. Government Printing Office.

MATERIAL SAFETY DATA SHEET

CHEMICAL NAME

CALCIUM PHOSPHATE, DIBASIC

HAZARD WARNING INFORMATION

HEALTH	FIRE	REACTIVE	OTHER	DEGREE OF HAZARD	COLOR CODING	OTHER CODES
1	0	0		0 = Minimum Hazard 1 = Slight Hazard 2 = Moderate Hazard 3 = Serious Hazard 4 = Severe Hazard	HEALTH = BLUE FIRE = RED REACTIVITY = YELLOW OTHER = WHITE	OX = Oxidizer ACID = Acid ALK = Alkali COR = Corrosive W = Use No Water

SECTION I - GENERAL INFORMATION

Characterization	**Calcium Compound**	RCRA Number	**None**
		EPA Class	**Not Applicable**
DOT Proper Shipping Name	**Not Listed**	Chemical Abstract Service (CAS) Number	**7757-93-9**
DOT Hazard Class and Label Requirements	**No Citation**	DOT Emergency Guide Code	**No Citation**
DOT Identification Number	**No Citation**	Atomic Formula	**$CaHPO_4 \bullet 2H_2O$**

Synonyms

Dicalcium orthophosphate; bicalcium phosphate; secondary calcium phosphate.

SECTION II - HAZARDOUS INGREDIENTS AND IDENTITY INFORMATION

Hazardous Components (specific identity)	OSHA Exposure Criteria	NIOSH Exposure Criteria	Immediately Dangerous to Life and Health (IDLH)	ACGIH Exposure Criteria
Calcium phosphate, dibasic (derivation: By interaction of fluorine-free phosphoric acid with milk and lime).	**PEL:** 5 mg/m^3 (respirable dusts) **STEL:** 15 mg/m^3 (total dusts)	**REL:** Not Established **STEL:** Not Established	**Not Determined**	**TLV (8-hour):** 10 mg/m^3 **STEL:** Not Established

SECTION III - PHYSICAL AND CHEMICAL CHARACTERISTICS

Boiling Point	**Decomposes**	Specific Gravity (H_2O = 1)	**2.306**
Vapor Pressure (mm Hg)	**Not Found**	Molecular Weight	**172.09**
Vapor Density (Air = 1)	**Not Found**	Melting Point	**Loses water at 228°F (109°C)**

Solubility

Slightly soluble in water. Soluble in dilute hydrochloric, nitric, and acetic acids. Insoluble in alcohol.

Appearance and Odor

White, odorless, tasteless, crystalline powder.

SECTION IV - FIRE AND EXPLOSION HAZARD DATA

Flash Point (method used)	**Not Determined**	Explosive Limits in Air % by Volume	**LEL: Not Determined UEL: Not Determined**
NFPA Classification	**Non-Flammable Solid**	Autoignition Temperature	**Not Determined**

Extinguishing Media

Use agent suitable to surrounding fire. Dry chemical, carbon dioxide, water spray, or regular foam should be adequate for fire suppression.

Special Fire Fighting Procedures

Structural protective clothing is permeable. Be sure to stay clear of smoke, water fall-out, and water run-off. Move containers from fire if it can be done without risk. Cool exposed containers with water spray. Dike area for control and containment to prevent entry into sewers, drains, and waterways. Evacuate non-essential personnel 1500 feet from the danger area. Consider down wind conditions.

Unusual Fire and Explosion Hazards

None reported. Avoid creating dusts during response procedures.

SECTION V - REACTIVITY DATA

Stability		Conditions to Avoid Calcium phosphate is stable under normal conditions of temperature and pressure. Avoid contact with heat. Keep lids tightly closed. Avoid generating dusty conditions.
Stable X	Unstable	Incompatibility *(materials to avoid)* May react in contact with strong acids (such as nitric, hydrochloric, and sulfuric).
Hazardous Polymerization		Conditions to Avoid Under normal conditions of temperature and pressure, hazardous polymerization of calcium phosphate is not known to occur.
May Occur	Will Not Occur X	Hazardous Decomposition or By-products When heated to decomposition calcium phosphate, dibasic will emit toxic and acrid smoke and fumes.

SECTION VI - HEALTH HAZARD DATA

Primary Route(s) of Entry:	Inhalation?	Absorption (skin)?	Ingestion?
	X		X

Health Hazards

INHALATION: Irritation of the eyes, nose, throat, and respiratory tract. Causes tearing. May cause pulmonary lesions on prolonged exposure. A nuisance dust.

EYES & SKIN: Skin contact causes irritation. Direct eye contact causes corneal burns.

INGESTION: Gastrointestinal irritation with severe pain, diarrhea, vomiting, and nausea.

Carcinogenicity	NTP Listed?	IARC Cancer Review Group?	OSHA Regulated?	Target Organs?
Unknown Human Unknown Animal	No	No	29 CFR 1910.1000 Table Z-1	Eyes, skin, mucous membranes, G.I. system.

Medical Conditions Generally Aggravated by Exposure

None reported.

Emergency and First-aid Procedures

Eye contact: Flush immediately with water for 15 minutes (minimum), gently lifting eyelids. Do not allow victim to keep eyes tightly shut. Seek medical attention. Skin contact: Immediately remove all contaminated clothing. Rinse in flooding amounts of water for 15 minutes (minimum). Wash area with large amounts of soap and water. For red or blistered skin, seek medical assistance. For inhalation: Remove the person from exposure. Provide respiratory assistance and CPR. Transfer to medical facility. If swallowed: Provide breathing assistance if necessary. Contact poison control center. Unless advised otherwise, give 1-2 glasses of water and induce vomiting. Never attempt to give a convulsing or unconscious person anything by mouth. Seek medical attention immediately. DOCTOR: Treatment should be symptomatic and supportive. Monitor calcium, phosphate, and magnesium levels.

SECTION VII - PRECAUTIONS FOR SAFE HANDLING AND USE

Steps to be Taken in Case Material is Released or Spilled

Collect powdered material in most convenient manner possible and deposit in sealed drum. Keep those not involved in cleanup from entering area. Do NOT dry sweep (creates airborne dusts). Use a vacuum equipped with a high efficiency particulate air (HEPA) filter. Use wet mop to clean residues.

Preferred Waste Disposal Method

No Citation.

Precautions to be Taken in Handling and Storage

Store in tightly closed containers in cool, well-ventilated area away from excessive heat and incompatible materials. Sources of ignition are prohibited where calcium phosphate is stored.

Other Precautions and Warnings

Protect from physical damage.

SECTION VIII - CONTROL MEASURES AND PERSONAL PROTECTIVE EQUIPMENT

Respiratory Protection *(specify type)*

For low exposures, a full facepiece respirator with HEPA filter may suffice. For best protection, use a supplied air respirator with a full facepiece, or an SCBA operated in positive pressure mode.

Ventilation

Local exhaust or general mechanical systems recommended.

Protective Gloves	Eye Protection	Other Protective Clothing
Impervious Material	Chemical/Dust Goggles or Face Mask	Protective Suit or Clothing

Work/Hygiene Practices

Always wash hands thoroughly after using chemical; never bring food, drink, or smoking materials into vicinity of chemicals.

CALCIUM PHOSPHATE, DIBASIC

$CaHPO_4 \bullet 2H_2O$ CAS: 7757-93-9

IDENTIFICATION AND TYPICAL USES

Calcium phosphate, dibasic, appears as a white, tasteless, crystalline powder with no odor. It is used as a animal feed supplement, a food supplement, as a dentifrice, in pharmaceuticals, in the manufacture of glass, as a fertilizer ingredient, as a stabilizer for plastics, as a dough conditioner, and in yeast foods.

RISK ASSESSMENT: HEALTH

General Assessment

Calcium phosphate is mildly toxic by *ingestion* and *inhalation*. There are no reports of carcinogenic, mutagenic, or teratogenic effects subsequent to exposure. While it is primarily characterized as a nuisance dust, there are some noted effects of exposure that must be assessed and managed if the dibasic form of calcium phosphate is used or handled in the work place.

Inhalation of calcium phosphates may cause irritation of the eyes (with tearing and redness), nose, throat, and respiratory tract. There is a possibility that inhalation may lead to the development of pulmonary lesions. However, the data are conflicting and inconclusive in this regard.

Skin contact can cause irritation with erythema (redness) and possible rash. Eye contact presents a more serious hazard since the phosphate may cause corneal burns, especially if the chemical is not removed immediately.

Ingestion can cause gastrointestinal irritation with nausea, vomiting, and diarrhea.

Acute Health Effects

The following acute (short-term) health effects may occur immediately or shortly after exposure to calcium phosphate, dibasic:

Skin: Irritation and possible rash.

Eye: Irritation and possible burning of the cornea.

Lung: Irritation of the eyes, nose, throat, and mucosa of the respiratory tract. There may be coughing, sneezing, and other unspecified pulmonary effects.

Other: Gastrointestinal irritation with abdominal cramps, vomiting, nausea, and diarrhea.

Chronic Health Effects

The following chronic (long-term) health effects may occur at some time after exposure to calcium phosphate and can last for months or even years:

Cancer Hazards: According to the information presented in the references, calcium phosphate has not been adequately tested for its ability to cause cancer in test animals.

Reproductive Hazard: According to information available in the references, calcium phosphate has not been adequately tested for its ability to adversely affect reproduction in laboratory animals.

Other Chronic Effects: Exposure may cause the development of pulmonary granulomatous lesions.

Recommended Risk-Reduction Measures

Personnel should avoid direct contact with calcium phosphate. While it is classified (by NIOSH and the ACGIH) as a nuisance dust, it can pose a toxic health hazard risk on unprotected exposure. If a less toxic chemical cannot be substituted for a hazardous material, then *engineering controls* are the most effective method of reducing exposures.

The best protection is to enclose operations and/or provide local exhaust ventilation at the site of chemical release. While not always operationally feasible, isolating operations can also reduce exposure. Using respiratory protection is less effective than the controls

mentioned above, but is still advisable whenever working with or around calcium phosphate. For low exposures, a full facepiece respirator equipped with a high efficiency particulate air (HEPA) filter equipped with a dust/mist/fume pre-filter may suffice. However, for the best protection use a supplied-air respirator with a full facepiece operated in positive pressure mode or a MSHA/NIOSH approved self-contained breathing apparatus (SCBA) with full facepiece operated in pressure demand or other continuous flow mode. If a full facepiece is not available, then chemical/dust goggles should be worn to protect the eyes. Whenever a chemical dust hazard exists, a face shield and a rubber apron should be worn. To prevent hand and skin exposures, impervious rubber gloves should be worn.

Administrative controls should also be in place to minimize the potential for human exposures. These may include written procedures or policies which specify the methods and techniques that will be practiced whenever personnel are to work with calcium phosphate.

All personnel should receive *training* on the use, hazards, protective measures, emergency actions, and other precautions per 29 CFR 1910.1200 (Hazard Communication), prior to the first assignment in an area where calcium phosphate is used or stored.

Before beginning employment and at regular intervals thereafter (e.g., annually), the following medical tests are recommended for personnel with frequent or anticipated exposure to calcium phosphate:

- ☑ Lung function tests (establish baseline).

If symptoms develop or overexposure is suspected, the following may also be useful:

- ☑ Lung function tests.
- ☑ Consider chest X-ray following acute overexposure.

Any evaluation should include a careful history of past and present symptoms with a medical examination. Medical tests that look for damage already done are not a substitute for controlling exposure. Also, since smoking can cause heart disease, emphysema, and a number of other respiratory disorders, smokers exposed to lung irritants such as calcium phosphate dusts and fumes may experience symptoms much more quickly than non-smokers under the same conditions of exposure. Prudent risk management requires proper consideration of *all* factors which may be causing the appearance of exposures symptoms in the work place.

Other methods to reduce exposure include:

- ☑ Where possible, enclose operations and use local exhaust ventilation at the site of chemical release. If local exhaust ventilation or enclosure is not used, respiratory protection should be mandatory.
- ☑ Always ensure that proper protective clothing is worn when using chemical substances.
- ☑ Wash thoroughly immediately after exposure to calcium phosphate and at the end of the work shift or before eating, drinking, or smoking.
- ☑ Hazard warning information should be posted in the work area. In addition, as part of an on-going education and training program, all information on the health and safety hazards of calcium phosphate should be communicated to all potentially exposed workers.
- ☑ Eye wash stations and emergency showers should be located in the immediate work area.
- ☑ Work clothing that has been contaminated with calcium phosphate should never be worn home where family members can be exposed as well. Contaminated clothing should be laundered at work or by an authorized service by personnel who have been trained in the hazards associated with exposure to calcium phosphate and calcium compounds.

RISK ASSESSMENT: ENVIRONMENT

General Assessment

The environment is at risk of exposure during transportation, storage, disposal, or destruction of calcium phosphate. In almost every scenario, the threat of environmental exposure is contingent upon the proper handling of the chemical substance. Accidental spills, large or small, can result in fire, explosion, and possible contamination of the surrounding environmental mediums (water, soil, and air).

Calcium phosphate is a non-flammable solid that can produce irritating and acrid smoke if it is involved in a fire. It may react violently in contact with strong acids, such as nitric, sulfuric, and hydrofluoric. These characteristics require special consideration during any emergency situation involving a leak or spill of calcium phosphate.

Calcium phosphate can enter the environment through industrial discharge, spills, or leaks.

Acute Ecological Effects

Acute (short-term) toxic effects may include the death of animals, birds, or fish and death or low growth rate in plants. Acute effects are seen 2 to 4 days after animals or plants are exposed to calcium phosphate.

Insufficient data are available to evaluate the short-term effects of calcium phosphate exposure to aquatic life, plants, birds or land animals.

Chronic Ecological Effects

Chronic (long-term) toxic effects may include shortened life span, reproductive problems, lower fertility, and changes in appearance or behavior in exposed animals. These effects can be seen long after first exposure(s) to toxic chemicals.

Insufficient data are available to evaluate the long-term effects of calcium phosphate exposure to aquatic life, plants, birds or land animals.

Water Solubility

Calcium phosphate is only very slightly soluble in water. Concentrations of less than 1 milligram may mix with a liter of water.

Persistence in the Environment

Calcium phosphate is only moderately persistent in the aquatic environment, with a half-life between 20 and 200 days. The half-life of a pollutant is the amount of time it takes for one half of the chemical to be degraded.

Bioaccumulation in Aquatic Organisms

Some substances increase in concentration, or *bioaccumulate,* in living organisms as they breathe contaminated air, drink contaminated water, or eat contaminated food. These chemicals can become concentrated in the tissues and internal organs of animals as well as humans.

The concentration of calcium phosphate found in fish is expected to be about the same as the average concentration of calcium phosphate in the water from which the fish was taken.

Recommended Risk-Reduction Measures

Proper training of all transporters will reduce the likelihood of a mishap or accidents resulting in a leak or spill to the environment. Proper labeling on all containers, trucks, and rail cars will enable emergency responders to react properly and quickly to any disaster thereby reducing the potential risk to the environment and to personnel.

If a spill or leak to the environment has occurred, fire department, emergency response and/or hazardous materials spill personnel should be notified immediately. Do not dry sweep. When vacuuming, a high efficiency particulate air (HEPA) filter should be used, not a standard shop vacuum. Use damp mop to clean up residue. Deposit collected materials in sealed drum for disposal. Cleanup should be attempted only by those trained in proper spill containment procedures. Contaminated soils may need to be removed and replaced with clean soil. If calcium phosphate should contact the water table, aquifer, or navigable waterway, time is of the essence. Calcium phosphate is slightly soluble in water. Therefore, total containment and remediation may not be entirely possible. When such spills occur, the local and/or state emergency response authorities must be notified. A comprehensive emergency response or disaster preparedness/recovery plan should be in place prior to any operations involving the use, transportation, storage, or disposal of calcium phosphate. If calcium phosphate is spilled or leaked, the following specific steps are recommended:

- ☑ Restrict persons not wearing protective clothing from area of spill or leak until cleanup is complete.
- ☑ Ventilate area of spill or leak and remove all ignition sources.
- ☑ Collect powdered materials in most convenient and safe manner possible and place in sealed drums for disposal. Do NOT dry sweep, use HEPA vacuum instead. Damp mop residue.
- ☑ It may be necessary to dispose of calcium phosphate as a hazardous waste. The responsible state agency or the regional office of the federal Environmental Protection Agency (EPA) should be contacted for specific recommendations.

RISK ASSESSMENT: BUSINESS

General Assessment

Accidents or mishaps involving calcium phosphate can present a moderate threat to business operations. The loss or damage of equipment or facilities can significantly affect fiscal viability. Lawsuits that may result from personnel injury/death, public exposures, and/or environmental contamination will also require a serious

expenditure of resources. Media attention surrounding an injury, death, or environmental damage can also result in a loss of profits and loss of current as well as future business.

❶ *Recommended Risk-Reduction Measures*

Company attorneys, safety and health professionals, and environmental specialists should be involved in the development of any procedures dealing with safety or emergency response. A company official should be pre-designated as a public relations officer with specific training in dealing with the press. Corporate plans and policies should be developed, approved, and implemented long before any need for such arises.

REFERENCES

Ahlbom, A. 1993. *Biostatistics for Epidemiologists.* New York: CRC Press/Lewis Publishers.

American Conference of Governmental Industrial Hygienists. 1988. *Documentation of the Threshold Limit Values and Biological Exposure Indices,* 5th Edition (with updates). Cincinnati: ACGIH

Asante-Duah, D. K. 1993. *Hazardous Waste Risk Assessment.* New York: CRC Press/Lewis Publishers.

Bloom, A. D. and F. J. DeSerres. 1995. *Ecotoxicity and Human Health.* New York: CRC Press/Lewis Publishers.

Brusick, D. J. 1994. *Methods for Genetic Risk Assessment.* New York: CRC Press/Lewis Publishers.

Calabrese, E. J. 1994. *Biological Effects of Low Level Exposures.* New York: CRC Press/Lewis Publishers.

Cockerman, L. G., B. S. Shane. 1994. *Basic Environmental Toxicology.* New York: CRC Press/Lewis Publishers.

Cothern, R. R. 1993. *Comparative Environmental Risk Assessment.* New York: CRC Press/Lewis Publishers.

Danse, I. R. 1991. *Common Sense Toxics in the Workplace.* New York: Van Nostrand Reinhold.

Environmental Protection Agency. 1983. *Chemical Hazard Information Profiles.* Washington, D.C.: U.S. EPA.

Fodden, C. R., and J. L. Weddell. 1992. *Hazardous Materials Emergency Action Data.* New York: CRC Press/Lewis Publishers.

Forseberg, K., and L. H. Keith. 1993. *Chemical; Protective Clothing Permeation and Degradation Database.* Chelsea, MI: Lewis Publishers.

Gerarde, H. W. 1960. *Toxicology and Biochemistry of Aromatic Hydrocarbons.* New York: Elsevier.

Hallenbeck, W. H. 1993. *Quantitative Risk Assessment for Environmental and Occupational Health,* 2nd Edition. New York: CRC Press/Lewis Publishers.

Hodgson, E., R. B. Mailman, and Janice E. Chambers. 1988. *Dictionary of Toxicology.* London: The MacMillian Company.

Howard, P. H. 1995. *Handbook of Environmental Fate and Exposures Data for Organic Chemicals,* Vol. I-IV. New York: CRC Press/Lewis Publishers.

Howard, P. H., et. al. 1991. *Handbook of Environmental Degradation Rates.* New York: CRC Press/Lewis Publishers.

Hosty, J. W., and P. Foster. 1990. *A Practical Guide to Chemical Spill Response.* New York: Van Nostrand Reinhold.

Jones, F. E. 1994. *Toxic Organic Vapors in the Workplace.* New York: CRC Press/Lewis Publishers.

Kamrin, M. A. 1988. *Toxicology: A Primer on Principles and Applications.* Chelsea, Michigan: Lewis Publishers.

Keith, L. H. , M. T. Johnston, C. E. Hudak, and M. Conoley. 1992. *National Toxicology Program's Gloves+ Database.* Chelsea, MI: Lewis Publishers.

Landis, W. G., M. H. Yu. 1995. *Introduction to Environmental Toxicology.* New York: CRC Press/Lewis Publishers.

Langerman, N. 1994 *Precautionary Labels for Chemical Containers.* New York: CRC Press/Lewis Publishers.

Lauwerys, R. R., P. Hoet. 1993. *Industrial Chemical Exposure.* New York: CRC Press/Lewis Publishers.

Lewis, R. J., Sr. 1992. *Sax's Dangerous Properties of Industrial Materials,* Volumes 1, 2, and 3, 8th Edition. New York: Van Nostrand Reinhold.

Mackay, D., W. Y. Shiu, and K. C. Ma. 1992. *Illustrated Handbook of Chemical-Physical Properties and Environmental Fate for Organic Chemicals.* New York: CRC Press/Lewis Publishers.

Manahan, S. E. 1994. *Environmental Chemistry,* 6th Edition. New York: CRC Press/Lewis Publishers.

Mellström, G. A., J. E. Wahlberg, and H. I. Maibach. 1994. *Protective Gloves for Occupational Use.* New York: CRC Press/Lewis Publishers.

Meyer, E. 1990. *Chemistry of Hazardous Materials.* Englewood Cliffs, NJ: Prentice Hall.

Miller, B. F., M.D., and C. B. Keane, R.N. 1972. *Encyclopedia and Dictionary of Medicine and Nursing.* Philadelphia: W.B. Sanders Company.

Mycock, J. C., J. D. McKenna, and L. Theodore. 1995. *Handbook of Air Pollution Control Engineering and Technology.* New York: CRC Press/Lewis Publishers.

Nelson, G. O. 1992. *Gas Mixtures.* New York: CRC Press/Lewis Publishers.

Ott, W. R. 1995. *Environmental Statistics and Data Analysis.* New York: CRC Press/Lewis Publishers.

Rea, W. J. 1995. *Chemical Sensitivity,* Vol. I-III. New York: CRC Press/Lewis Publishers.

Sacarello, H. L. A. 1994. *The Comprehensive Handbook of Hazardous Materials.* New York: CRC Press/Lewis Publishers.

Sherman, J. D. 1988. *Chemical Exposure and Disease.* New York: Van Nostrand Reinhold.

Talbott, E., G. Craun. 1995. *An Introduction to Environmental Epidemiology.* New York: CRC Press/Lewis Publishers.

MATERIAL SAFETY DATA SHEET

CHEMICAL NAME	
CALCIUM PHOSPHATE, MONOBASIC	**CRC PRESS/LEWIS PUBLISHERS DISCLAIMER:** THE INFORMATION AND RECOMMENDATIONS PRESENTED HEREIN ARE BASED ON SOURCES BELIEVED TO BE RELIABLE. CRC MAKES NO REPRESENTATION ON ITS COMPLETENESS OR ACCURACY. IT IS THE RESPONSIBILITY OF THE USER TO DETERMINE THE CHEMICAL'S SUITABILITY FOR ITS INTENDED USE, THE CHEMICAL'S SAFE USE, AND THE CHEMICAL'S PROPER DISPOSAL. NO REPRESENTATIONS AND/OR WARRANTIES, EITHER EXPRESSED OR IMPLIED, OF THE MERCHANTABILITY OR FITNESS FOR A PARTICULAR PURPOSE, OR OF ANY OTHER NATURE, ARE MADE WITH RESPECT TO THE INFORMATION PROVIDED IN THIS MSDS OR TO THE CHEMICAL TO WHICH INFORMATION MAY REFER. CRC NEITHER ASSUMES NOR AUTHORIZES ANY OTHER PERSON TO ASSUME FOR IT, ANY OTHER ADDITIONAL RESPONSIBILITY OR LIABILITY FOR THE USE OF, OR RELIANCE UPON, THIS INFORMATION.

HAZARD WARNING INFORMATION

HEALTH	FIRE	REACTIVE	OTHER	DEGREE OF HAZARD	COLOR CODING	OTHER CODES
1	0	0		0 = Minimum Hazard 1 = Slight Hazard 2 = Moderate Hazard 3 = Serious Hazard 4 = Severe Hazard	HEALTH = BLUE FIRE = RED REACTIVITY = YELLOW OTHER = WHITE	OX = Oxidizer ACID = Acid ALK = Alkali COR = Corrosive W = Use No Water

SECTION I - GENERAL INFORMATION

Characterization	**Calcium Compound**	RCRA Number	**None**
		EPA Class	**Not Applicable**
DOT Proper Shipping Name	**Not Listed**	Chemical Abstract Service (CAS) Number	**7758-23-8**
DOT Hazard Class and Label Requirements	**No Citation**	DOT Emergency Guide Code	**No Citation**
DOT Identification Number	**No Citation**	Atomic Formula	**$Ca(H_2PO_4)_2$**

Synonyms
Calcium biphosphate; acid calcium phosphate; calcium phosphate, primary; monocalcium phosphate.

SECTION II - HAZARDOUS INGREDIENTS AND IDENTITY INFORMATION

Hazardous Components (specific identity)	OSHA Exposure Criteria	NIOSH Exposure Criteria	Immediately Dangerous to Life and Health (IDLH)	ACGIH Exposure Criteria
Calcium phosphate, monobasic (derivation: By dissolving either dicalcium or tricalcium phosphate in phosphoric acid and allowing the solution to evaporate spontaneously).	**PEL:** 5 mg/m^3 (respirable dusts) **STEL:** 15 mg/m^3 (total dusts)	**REL:** Not Established **STEL:** Not Established	**Not Determined**	**TLV (8-hour):** 10 mg/m^3 **STEL:** Not Established

SECTION III - PHYSICAL AND CHEMICAL CHARACTERISTICS

Boiling Point	**392°F (200°C) Decomposes**	Specific Gravity (H_2O = 1)	**2.20**
Vapor Pressure (mm Hg)	**Not Found**	Molecular Weight	**234.05**
Vapor Density (Air = 1)	**Not Found**	Melting Point	**Loses water at 212°F (100°C)**

Solubility
Slightly soluble in water. Soluble in dilute hydrochloric, nitric, and acetic acids. Insoluble in alcohol.

Appearance and Odor
Colorless, pearly scales or powder, deliquescent in air.

SECTION IV - FIRE AND EXPLOSION HAZARD DATA

Flash Point (method used)	**Not Determined**	Explosive Limits in Air % by Volume	**LEL: Not Determined UEL: Not Determined**
NFPA Classification	**Non-Flammable Solid**	Autoignition Temperature	**Not Determined**

Extinguishing Media
Use agent suitable to surrounding fire. Dry chemical, carbon dioxide, water spray, or regular foam should be adequate for fire suppression.

Special Fire Fighting Procedures
Structural protective clothing is permeable. Be sure to stay clear of smoke, water fall-out, and water run-off. Move containers from fire if it can be done without risk. Cool exposed containers with water spray. Dike area for control and containment to prevent entry into sewers, drains, and waterways. Evacuate non-essential personnel 1500 feet from the danger area. Consider down wind conditions.

Unusual Fire and Explosion Hazards
None reported. Avoid creating dusts during response procedures.

SECTION V - REACTIVITY DATA

Stability		Conditions to Avoid
		Calcium phosphate is stable under normal conditions of temperature and pressure. Avoid contact with heat. Keep lids tightly closed. Avoid generating dusty conditions.
Stable	Unstable	Incompatibility *(materials to avoid)*
X		May react in contact with strong acids (such as nitric, hydrochloric, and sulfuric).
Hazardous Polymerization		Conditions to Avoid
		Under normal conditions of temperature and pressure, hazardous polymerization of calcium phosphate is not known to occur.
May Occur	Will Not Occur	Hazardous Decomposition or By-products
	X	When heated to decomposition calcium phosphate, monobasic will emit toxic and acrid smoke and fumes.

SECTION VI - HEALTH HAZARD DATA

Primary Route(s) of Entry:	Inhalation?	Absorption (skin)?	Ingestion?
	X		X

Health Hazards

INHALATION: Irritation of the eyes, nose, throat, and respiratory tract. Causes tearing. May cause pulmonary lesions on prolonged exposure. A nuisance dust.

EYES & SKIN: Skin contact causes irritation. Direct eye contact causes corneal burns.

INGESTION: Gastrointestinal irritation with severe pain, diarrhea, vomiting, and nausea.

Carcinogenicity	NTP Listed?	IARC Cancer Review Group?	OSHA Regulated?	Target Organs?
Unknown Human Unknown Animal	No	No	29 CFR 1910.1000 Table Z-1	Eyes, skin, mucous membranes, G.I. system.

Medical Conditions Generally Aggravated by Exposure

None reported.

Emergency and First-aid Procedures

Eye contact: Flush immediately with water for 15 minutes (minimum), gently lifting eyelids. Do not allow victim to keep eyes tightly shut. Seek medical attention. Skin contact: Immediately remove all contaminated clothing. Rinse in flooding amounts of water for 15 minutes (minimum). Wash area with large amounts of soap and water. For red or blistered skin, seek medical assistance. For inhalation: Remove the person from exposure. Provide respiratory assistance and CPR. Transfer to medical facility. If swallowed: Provide breathing assistance if necessary. Contact poison control center. Unless advised otherwise, give 1-2 glasses of water and induce vomiting. Never attempt to give a convulsing or unconscious person anything by mouth. Seek medical attention immediately. DOCTOR: Treatment should be symptomatic and supportive. Monitor calcium, phosphate, and magnesium levels.

SECTION VII - PRECAUTIONS FOR SAFE HANDLING AND USE

Steps to be Taken in Case Material is Released or Spilled

Collect powdered material in most convenient manner possible and deposit in sealed drum. Keep those not involved in cleanup from entering area. Do NOT dry sweep (creates airborne dusts). Use a vacuum equipped with a high efficiency particulate air (HEPA) filter. Use wet mop to clean residues.

Preferred Waste Disposal Method

No Citation.

Precautions to be Taken in Handling and Storage

Store in tightly closed containers in cool, well-ventilated area away from excessive heat and incompatible materials. Sources of ignition are prohibited where calcium phosphate is stored.

Other Precautions and Warnings

Protect from physical damage.

SECTION VIII - CONTROL MEASURES AND PERSONAL PROTECTIVE EQUIPMENT

Respiratory Protection *(specify type)*

For low exposures, a full facepiece respirator with HEPA filter may suffice. For best protection, use a supplied air respirator with a full facepiece, or an SCBA operated in positive pressure mode.

Ventilation

Local exhaust or general mechanical systems recommended.

Protective Gloves	Eye Protection	Other Protective Clothing
Impervious Material	Chemical/Dust Goggles or Face Mask	Protective Suit or Clothing

Work/Hygiene Practices

Always wash hands thoroughly after using chemical; never bring food, drink, or smoking materials into vicinity of chemicals.

CALCIUM PHOSPHATE, MONOBASIC

$Ca(H_2PO_4)_2$ CAS: 7758-23-8

IDENTIFICATION AND TYPICAL USES

Calcium phosphate, monobasic, appears as a colorless pearly scale or powder that is deliquescent (absorbs moisture and melts away or turns to a liquid) in air. It is used in baking powders, in fertilizer compositions, as a mineral supplement, as a stabilizer for plastics, to control the pH in malt, in the manufacture of glass, as a buffer in some foods, and as a firming agent.

RISK ASSESSMENT: HEALTH

General Assessment

Calcium phosphate is mildly toxic by *ingestion* and *inhalation*. There are no reports of carcinogenic, mutagenic, or teratogenic effects subsequent to exposure. While it is primarily characterized as a nuisance dust, there are some noted effects of exposure that must be assessed and managed if the monobasic form of calcium phosphate is used or handled in the work place.

Inhalation of calcium phosphates may cause irritation of the eyes (with tearing and redness), nose, throat, and respiratory tract. There is a possibility that inhalation may lead to the development of pulmonary lesions. However, the data are conflicting and inconclusive in this regard.

Skin contact can cause irritation with erythema (redness) and possible rash. Eye contact presents a more serious hazard since the phosphate may cause corneal burns, especially if the chemical is not removed immediately. Ingestion can cause gastrointestinal irritation with nausea, vomiting, and diarrhea.

Acute Health Effects

The following acute (short-term) health effects may occur immediately or shortly after exposure to calcium phosphate, monobasic:

Skin: Irritation and possible rash.

Eye: Irritation and possible burning of the cornea.

Lung: Irritation of the eyes, nose, throat, and mucosa of the respiratory tract. There may be coughing, sneezing, and other unspecified pulmonary effects.

Other: Gastrointestinal irritation with abdominal cramps, vomiting, nausea, and diarrhea.

Chronic Health Effects

The following chronic (long-term) health effects may occur at some time after exposure to calcium phosphate and can last for months or even years:

Cancer Hazards: According to the information presented in the references, calcium phosphate has not been adequately tested for its ability to cause cancer in test animals.

Reproductive Hazard: According to information available in the references, calcium phosphate has not been adequately tested for its ability to adversely affect reproduction in laboratory animals.

Other Chronic Effects: Exposure may cause the development of pulmonary granulomatous lesions.

Recommended Risk-Reduction Measures

Personnel should avoid direct contact with calcium phosphate. While it is classified (by NIOSH and the ACGIH) as a nuisance dust, it can pose a toxic health hazard risk on unprotected exposure. If a less toxic chemical cannot be substituted for a hazardous material, then *engineering controls* are the most effective method of reducing exposures.

The best protection is to enclose operations and/or provide local exhaust ventilation at the site of chemical release. While not always operationally feasible, isolating operations can also reduce exposure. Using respiratory protection is less effective than the controls

mentioned above, but is still advisable whenever working with or around calcium phosphate. For low exposures, a full facepiece respirator equipped with a high efficiency particulate air (HEPA) filter equipped with a dust/mist/fume pre-filter may suffice. However, for the best protection use a supplied-air respirator with a full facepiece operated in positive pressure mode or a MSHA/NIOSH approved self-contained breathing apparatus (SCBA) with full facepiece operated in pressure demand or other continuous flow mode. If a full facepiece is not available, then chemical/dust goggles should be worn to protect the eyes. Whenever a chemical dust hazard exists, a face shield and a rubber apron should be worn. To prevent hand and skin exposures, impervious rubber gloves should be worn.

Administrative controls should also be in place to minimize the potential for human exposures. These may include written procedures or policies which specify the methods and techniques that will be practiced whenever personnel are to work with calcium phosphate.

All personnel should receive *training* on the use, hazards, protective measures, emergency actions, and other precautions per 29 CFR 1910.1200 (Hazard Communication), prior to the first assignment in an area where calcium phosphate is used or stored.

Before beginning employment and at regular intervals thereafter (e.g., annually), the following medical tests are recommended for personnel with frequent or anticipated exposure to calcium phosphate:

- ☑ Lung function tests (establish baseline).

If symptoms develop or overexposure is suspected, the following may also be useful:

- ☑ Lung function tests.
- ☑ Consider chest X-ray following acute overexposure.

Any evaluation should include a careful history of past and present symptoms with a medical examination. Medical tests that look for damage already done are not a substitute for controlling exposure. Also, since smoking can cause heart disease, emphysema, and a number of other respiratory disorders, smokers exposed to lung irritants such as calcium phosphate dusts and fumes may experience symptoms much more quickly than non-smokers under the same conditions of exposure. Prudent risk management requires proper consideration of *all* factors which may be causing the appearance of exposures symptoms in the work place.

Other methods to reduce exposure include:

- ☑ Where possible, enclose operations and use local exhaust ventilation at the site of chemical release. If local exhaust ventilation or enclosure is not used, respiratory protection should be mandatory.
- ☑ Always ensure that proper protective clothing is worn when using chemical substances.
- ☑ Wash thoroughly immediately after exposure to calcium phosphate and at the end of the work shift or before eating, drinking, or smoking.
- ☑ Hazard warning information should be posted in the work area. In addition, as part of an on-going education and training program, all information on the health and safety hazards of calcium phosphate should be communicated to all potentially exposed workers.
- ☑ Eye wash stations and emergency showers should be located in the immediate work area.
- ☑ Work clothing that has been contaminated with calcium phosphate should never be worn home where family members can be exposed as well. Contaminated clothing should be laundered at work or by an authorized service by personnel who have been trained in the hazards associated with exposure to calcium phosphate and calcium compounds.

RISK ASSESSMENT: ENVIRONMENT

General Assessment

The environment is at risk of exposure during transportation, storage, disposal, or destruction of calcium phosphate. In almost every scenario, the threat of environmental exposure is contingent upon the proper handling of the chemical substance. Accidental spills, large or small, can result in fire, explosion, and possible contamination of the surrounding environmental mediums (water, soil, and air).

Calcium phosphate is a non-flammable solid that can produce irritating and acrid smoke if it is involved in a fire. It may react violently in contact with strong acids, such as nitric, sulfuric, and hydrofluoric. These characteristics require special consideration during any emergency situation involving a leak or spill of calcium phosphate.

Calcium phosphate can enter the environment through industrial discharge, spills, or leaks.

☠ *Acute Ecological Effects*

Acute (short-term) toxic effects may include the death of animals, birds, or fish and death or low growth rate in plants. Acute effects are seen 2 to 4 days after animals or plants are exposed to calcium phosphate.

Insufficient data are available to evaluate the short-term effects of calcium phosphate exposure to aquatic life, plants, birds or land animals.

Chronic Ecological Effects

Chronic (long-term) toxic effects may include shortened life span, reproductive problems, lower fertility, and changes in appearance or behavior in exposed animals. These effects can be seen long after first exposure(s) to toxic chemicals.

Insufficient data are available to evaluate the long-term effects of calcium phosphate exposure to aquatic life, plants, birds or land animals.

Water Solubility

Calcium phosphate is only very slightly soluble in water. Concentrations of less than 1 milligram may mix with a liter of water.

⌛ *Persistence in the Environment*

Calcium phosphate is only moderately persistent in the aquatic environment, with a half-life between 20 and 200 days. The half-life of a pollutant is the amount of time it takes for one half of the chemical to be degraded.

Bioaccumulation in Aquatic Organisms

Some substances increase in concentration, or *bioaccumulate,* in living organisms as they breathe contaminated air, drink contaminated water, or eat contaminated food. These chemicals can become concentrated in the tissues and internal organs of animals as well as humans.

The concentration of calcium phosphate found in fish is expected to be about the same as the average concentration of calcium phosphate in the water from which the fish was taken.

Recommended Risk-Reduction Measures

Proper training of all transporters will reduce the likelihood of a mishap or accidents resulting in a leak or spill to the environment. Proper labeling on all containers, trucks, and rail cars will enable emergency responders to react properly and quickly to any disaster thereby reducing the potential risk to the environment and to personnel.

If a spill or leak to the environment has occurred, fire department, emergency response and/or hazardous materials spill personnel should be notified immediately. Do not dry sweep. When vacuuming, a high efficiency particulate air (HEPA) filter should be used, not a standard shop vacuum. Use damp mop to clean up residue. Deposit collected materials in sealed drum for disposal. Cleanup should be attempted only by those trained in proper spill containment procedures. Contaminated soils may need to be removed and replaced with clean soil. If calcium phosphate should contact the water table, aquifer, or navigable waterway, time is of the essence. Calcium phosphate is slightly soluble in water. Therefore, total containment and remediation may not be entirely possible. When such spills occur, the local and/or state emergency response authorities must be notified. A comprehensive emergency response or disaster preparedness/recovery plan should be in place prior to any operations involving the use, transportation, storage, or disposal of calcium phosphate. If calcium phosphate is spilled or leaked, the following specific steps are recommended:

- ☑ Restrict persons not wearing protective clothing from area of spill or leak until clean-up is complete and area can be opened for normal work.
- ☑ Ventilate area of spill or leak and remove all ignition sources.
- ☑ Collect powdered materials in most convenient and safe manner possible and place in sealed drums for disposal. Do NOT dry sweep, use HEPA vacuum instead. Damp mop residue.
- ☑ It may be necessary to dispose of calcium phosphate as a hazardous waste. The responsible state agency or the regional office of the federal Environmental Protection Agency (EPA) should be contacted for specific recommendations.

RISK ASSESSMENT: BUSINESS

General Assessment

Accidents or mishaps involving calcium phosphate can present a moderate threat to business operations. The loss or damage of equipment or facilities can significantly affect fiscal viability. Lawsuits that may result from personnel injury/death, public exposures, and/or environmental contamination will also require a serious

expenditure of resources. Media attention surrounding an injury, death, or environmental damage can also result in a loss of profits and loss of current as well as future business.

Recommended Risk-Reduction Measures

Company attorneys, safety and health professionals, and environmental specialists should be involved in the development of any procedures dealing with safety or emergency response. A company official should be pre-designated as a public relations officer with specific training in dealing with the press. Corporate plans and policies should be developed, approved, and implemented long before any need for such arises.

REFERENCES

Ahlbom, A. 1993. *Biostatistics for Epidemiologists.* New York: CRC Press/Lewis Publishers.

American Conference of Governmental Industrial Hygienists. 1988. *Documentation of the Threshold Limit Values and Biological Exposure Indices,* 5th Edition (with updates). Cincinnati: ACGIH

Asante-Duah, D. K. 1993. *Hazardous Waste Risk Assessment.* New York: CRC Press/Lewis Publishers.

Bloom, A. D. and F. J. DeSerres. 1995. *Ecotoxicity and Human Health.* New York: CRC Press/Lewis Publishers.

Brusick, D. J. 1994. *Methods for Genetic Risk Assessment.* New York: CRC Press/Lewis Publishers.

Calabrese, E. J. 1994. *Biological Effects of Low Level Exposures.* New York: CRC Press/Lewis Publishers.

Cockerman, L. G., B. S. Shane. 1994. *Basic Environmental Toxicology.* New York: CRC Press/Lewis Publishers.

Cothern, R. R. 1993. *Comparative Environmental Risk Assessment.* New York: CRC Press/Lewis Publishers.

Danse, I. R. 1991. *Common Sense Toxics in the Workplace.* New York: Van Nostrand Reinhold.

Environmental Protection Agency. 1983. *Chemical Hazard Information Profiles.* Washington, D.C.: U.S. EPA.

Fodden, C. R., and J. L. Weddell. 1992. *Hazardous Materials Emergency Action Data.* New York: CRC Press/Lewis Publishers.

Forseberg, K., and L. H. Keith. 1993. *Chemical; Protective Clothing Permeation and Degradation Database.* Chelsea, MI: Lewis Publishers.

Gerarde, H. W. 1960. *Toxicology and Biochemistry of Aromatic Hydrocarbons.* New York: Elsevier.

Hallenbeck, W. H. 1993. *Quantitative Risk Assessment for Environmental and Occupational Health,* 2nd Edition. New York: CRC Press/Lewis Publishers.

Hodgson, E., R. B. Mailman, and Janice E. Chambers. 1988. *Dictionary of Toxicology.* London: The MacMillian Company.

Howard, P. H. 1995. *Handbook of Environmental Fate and Exposures Data for Organic Chemicals,* Vol. I-IV. New York: CRC Press/Lewis Publishers.

Howard, P. H., et. al. 1991. *Handbook of Environmental Degradation Rates.* New York: CRC Press/Lewis Publishers.

Hosty, J. W., and P. Foster. 1990. *A Practical Guide to Chemical Spill Response.* New York: Van Nostrand Reinhold.

Jones, F. E. 1994. *Toxic Organic Vapors in the Workplace.* New York: CRC Press/Lewis Publishers.

Kamrin, M. A. 1988. *Toxicology: A Primer on Principles and Applications.* Chelsea, Michigan: Lewis Publishers.

Keith, L. H. , M. T. Johnston, C. E. Hudak, and M. Conoley. 1992. *National Toxicology Program's Gloves+ Database.* Chelsea, MI: Lewis Publishers.

Landis, W. G., M. H. Yu. 1995. *Introduction to Environmental Toxicology.* New York: CRC Press/Lewis Publishers.

Langerman, N. 1994 *Precautionary Labels for Chemical Containers.* New York: CRC Press/Lewis Publishers.

Lauwerys, R. R., P. Hoet. 1993. *Industrial Chemical Exposure.* New York: CRC Press/Lewis Publishers.

Lewis, R. J., Sr. 1992. *Sax's Dangerous Properties of Industrial Materials,* Volumes 1, 2, and 3, 8th Edition. New York: Van Nostrand Reinhold.

Mackay, D., W. Y. Shiu, and K. C. Ma. 1992. *Illustrated Handbook of Chemical-Physical Properties and Environmental Fate for Organic Chemicals.* New York: CRC Press/Lewis Publishers.

Manahan, S. E. 1994. *Environmental Chemistry,* 6th Edition. New York: CRC Press/Lewis Publishers.

Mellström, G. A., J. E. Wahlberg, and H. I. Maibach. 1994. *Protective Gloves for Occupational Use.* New York: CRC Press/Lewis Publishers.

Meyer, E. 1990. *Chemistry of Hazardous Materials.* Englewood Cliffs, NJ: Prentice Hall.

Miller, B. F., M.D., and C. B. Keane, R.N. 1972. *Encyclopedia and Dictionary of Medicine and Nursing.* Philadelphia: W.B. Sanders Company.

Mycock, J. C., J. D. McKenna, and L. Theodore. 1995. *Handbook of Air Pollution Control Engineering and Technology.* New York: CRC Press/Lewis Publishers.

Nelson, G. O. 1992. *Gas Mixtures.* New York: CRC Press/Lewis Publishers.

Ott, W. R. 1995. *Environmental Statistics and Data Analysis.* New York: CRC Press/Lewis Publishers.

Rea, W. J. 1995. *Chemical Sensitivity,* Vol. I-III. New York: CRC Press/Lewis Publishers.

Sacarello, H. L. A. 1994. *The Comprehensive Handbook of Hazardous Materials.* New York: CRC Press/Lewis Publishers.

Sherman, J. D. 1988. *Chemical Exposure and Disease.* New York: Van Nostrand Reinhold.

Talbott, E., G. Craun. 1995. *An Introduction to Environmental Epidemiology.* New York: CRC Press/Lewis Publishers.

MATERIAL SAFETY DATA SHEET

CHEMICAL NAME	
CALCIUM PHOSPHATE, TRIBASIC	**CRC PRESS/LEWIS PUBLISHERS DISCLAIMER:** THE INFORMATION AND RECOMMENDATIONS PRESENTED HEREIN ARE BASED ON SOURCES BELIEVED TO BE RELIABLE. CRC MAKES NO REPRESENTATION ON ITS COMPLETENESS OR ACCURACY. IT IS THE RESPONSIBILITY OF THE USER TO DETERMINE THE CHEMICAL'S SUITABILITY FOR ITS INTENDED USE, THE CHEMICAL'S SAFE USE, AND THE CHEMICAL'S PROPER DISPOSAL. NO REPRESENTATIONS AND/OR WARRANTIES, EITHER EXPRESSED OR IMPLIED, OF THE MERCHANTABILITY OR FITNESS FOR A PARTICULAR PURPOSE, OR OF ANY OTHER NATURE, ARE MADE WITH RESPECT TO THE INFORMATION PROVIDED IN THIS MSDS OR TO THE CHEMICAL TO WHICH INFORMATION MAY REFER. CRC NEITHER ASSUMES NOR AUTHORIZES ANY OTHER PERSON TO ASSUME FOR IT, ANY OTHER ADDITIONAL RESPONSIBILITY OR LIABILITY FOR THE USE OF, OR RELIANCE UPON, THIS INFORMATION.

HAZARD WARNING INFORMATION

HEALTH	FIRE	REACTIVE	OTHER	DEGREE OF HAZARD	COLOR CODING	OTHER CODES
1	0	0		0 = Minimum Hazard 1 = Slight Hazard 2 = Moderate Hazard 3 = Serious Hazard 4 = Severe Hazard	HEALTH = BLUE FIRE = RED REACTIVITY = YELLOW OTHER = WHITE	OX = Oxidizer ACID = Acid ALK = Alkali COR = Corrosive ~~W~~ = Use No Water

SECTION I - GENERAL INFORMATION

Characterization: **Calcium Compound**	RCRA Number: **None** — EPA Class: **Not Applicable**
DOT Proper Shipping Name: **Not Listed**	Chemical Abstract Service (CAS) Number: **12167-74-7**
DOT Hazard Class and Label Requirements: **No Citation**	DOT Emergency Guide Code: **No Citation**
DOT Identification Number: **No Citation**	Atomic Formula: **$10CaO \cdot 3P_2O_5 \cdot H_2O$**

Synonyms
Calcium orthophosphate; tricalcium phosphate; precipitated calcium phosphate; tricalcium orthophosphate.

SECTION II - HAZARDOUS INGREDIENTS AND IDENTITY INFORMATION

Hazardous Components (specific identity)	OSHA Exposure Criteria	NIOSH Exposure Criteria	Immediately Dangerous to Life and Health (IDLH)	ACGIH Exposure Criteria
Calcium phosphate, tribasic (derivation: From phosphate rock, apatite, and phosphorite; by the interaction of soluble calcium chloride and sodium triphosphate with excess ammonia; by interaction of hydrated lime and phosphoric acid).	**PEL:** 5 mg/m^3 (respirable dusts) **STEL:** 15 mg/m^3 (total dusts)	**REL:** Not Established **STEL:** Not Established	**Not Determined**	**TLV (8-hour):** 10 mg/m^3 **STEL:** Not Established

SECTION III - PHYSICAL AND CHEMICAL CHARACTERISTICS

Boiling Point: **Not Found**	Specific Gravity (H_2O = 1): **3.18**
Vapor Pressure (mm Hg): **Not Found**	Molecular Weight: **1004.64**
Vapor Density (Air = 1): **Not Found**	Melting Point: **3038°F (1670°C)**

Solubility
Insoluble in water, alcohol, and acetic acid.

Appearance and Odor
A white, odorless, tasteless, crystalline powder.

SECTION IV - FIRE AND EXPLOSION HAZARD DATA

Flash Point (method used): **Not Determined**	Explosive Limits in Air % by Volume: **LEL: Not Determined — UEL: Not Determined**
NFPA Classification: **Non-Flammable Solid**	Autoignition Temperature: **Not Determined**

Extinguishing Media
Use agent suitable to surrounding fire. Dry chemical, carbon dioxide, water spray, or regular foam should be adequate for fire suppression.

Special Fire Fighting Procedures
Structural protective clothing is permeable. Be sure to stay clear of smoke, water fall-out, and water run-off. Move containers from fire if it can be done without risk. Cool exposed containers with water spray. Dike area for control and containment to prevent entry into sewers, drains, and waterways. Evacuate non-essential personnel 1500 feet from the danger area. Consider down wind conditions.

Unusual Fire and Explosion Hazards
None reported. Avoid creating dusts during response procedures.

SECTION V - REACTIVITY DATA

Stability		Conditions to Avoid
		Calcium phosphate is stable under normal conditions of temperature and pressure. Avoid contact with heat. Keep lids tightly closed. Avoid generating dusty conditions.
Stable	Unstable	Incompatibility *(materials to avoid)*
X		May react in contact with strong acids (such as nitric, hydrochloric, and sulfuric).
Hazardous Polymerization		Conditions to Avoid
		Under normal conditions of temperature and pressure, hazardous polymerization of calcium phosphate is not known to occur.
May Occur	Will Not Occur	Hazardous Decomposition or By-products
	X	When heated to decomposition calcium phosphate, tribasic will emit toxic and acrid smoke and fumes.

SECTION VI - HEALTH HAZARD DATA

Primary Route(s) of Entry:	Inhalation?	Absorption (skin)?	Ingestion?
	X		X

Health Hazards

INHALATION: Irritation of the eyes, nose, throat, and respiratory tract. Causes tearing. May cause pulmonary lesions on prolonged exposure. A nuisance dust.

EYES & SKIN: Skin contact causes irritation. Direct eye contact causes corneal burns.

INGESTION: Gastrointestinal irritation with severe pain, diarrhea, vomiting, and nausea.

Carcinogenicity	NTP Listed?	IARC Cancer Review Group?	OSHA Regulated?	Target Organs?
Unknown Human Unknown Animal	No	No	29 CFR 1910.1000 Table Z-1	Eyes, skin, mucous membranes, G.I. system.

Medical Conditions Generally Aggravated by Exposure

None reported.

Emergency and First-aid Procedures

Eye contact: Flush immediately with water for 15 minutes (minimum), gently lifting eyelids. Do not allow victim to keep eyes tightly shut. Seek medical attention. Skin contact: Immediately remove all contaminated clothing. Rinse in flooding amounts of water for 15 minutes (minimum). Wash area with large amounts of soap and water. For red or blistered skin, seek medical assistance. For inhalation: Remove the person from exposure. Provide respiratory assistance and CPR. Transfer to medical facility. If swallowed: Provide breathing assistance if necessary. Contact poison control center. Unless advised otherwise, give 1-2 glasses of water and induce vomiting. Never attempt to give a convulsing or unconscious person anything by mouth. Seek medical attention immediately. DOCTOR: Treatment should be symptomatic and supportive. Monitor calcium, phosphate, and magnesium levels.

SECTION VII - PRECAUTIONS FOR SAFE HANDLING AND USE

Steps to be Taken in Case Material is Released or Spilled

Collect powdered material in most convenient manner possible and deposit in sealed drum. Keep those not involved in cleanup from entering area. Do NOT dry sweep (creates airborne dusts). Use a vacuum equipped with a high efficiency particulate air (HEPA) filter. Use wet mop to clean residues.

Preferred Waste Disposal Method

No Citation.

Precautions to be Taken in Handling and Storage

Store in tightly closed containers in cool, well-ventilated area away from excessive heat and incompatible materials. Sources of ignition are prohibited where calcium phosphate is stored.

Other Precautions and Warnings

Protect from physical damage.

SECTION VIII - CONTROL MEASURES AND PERSONAL PROTECTIVE EQUIPMENT

Respiratory Protection *(specify type)*

For low exposures, a full face piece respirator with HEPA filter may suffice. For best protection, use a supplied air respirator with a full face piece, or an SCBA operated in positive pressure mode.

Ventilation

Local exhaust or general mechanical systems recommended.

Protective Gloves	Eye Protection	Other Protective Clothing
Impervious Material	Chemical/Dust Goggles or Face Mask	Protective Suit or Clothing

Work/Hygiene Practices

Always wash hands thoroughly after using chemical; never bring food, drink, or smoking materials into vicinity of chemicals.

CALCIUM PHOSPHATE, TRIBASIC

$10CaO \bullet 3P_2O_5 \bullet H_2O$ CAS: 12167-74-7

IDENTIFICATION AND TYPICAL USES

Calcium phosphate, tribasic, appears as a white, odorless, tasteless, crystalline powder. It is used in ceramics; in producing calcium acid phosphate, phosphorous, and phosphoric acid; as a polishing powder; in cattle foods; clarifying sugar syrups; medicines; mordant for dyeing textiles ("Turkey Red"); fertilizers; dentifrice; stabilizer for plastics; in meat tenderizers; as an anti-caking agent in foods; as a buffer; as a nutrient supplement; and in the removal of Sr^{90} from milk products.

RISK ASSESSMENT: HEALTH

General Assessment

Calcium phosphate is mildly toxic by *ingestion* and *inhalation*. There are no reports of carcinogenic, mutagenic, or teratogenic effects subsequent to exposure. While it is primarily characterized as a nuisance dust, there are some noted effects of exposure that must be assessed and managed if the tribasic form of calcium phosphate is used or handled in the work place.

Inhalation of calcium phosphates may cause irritation of the eyes (with tearing and redness), nose, throat, and respiratory tract. There is a possibility that inhalation may lead to the development of pulmonary lesions. However, the data are inconclusive in this regard.

Skin contact can cause irritation with erythema (redness) and possible rash. Eye contact presents a more serious hazard since the phosphate may cause corneal burns, especially if the chemical is not removed immediately. Ingestion can cause gastrointestinal irritation with nausea, vomiting, and diarrhea.

Acute Health Effects

The following acute (short-term) health effects may occur immediately or shortly after exposure to calcium phosphate, tribasic:

Skin: Irritation and possible rash.

Eye: Irritation and possible burning of the cornea.

Lung: Irritation of the eyes, nose, throat, and mucosa of the respiratory tract. There may be coughing, sneezing, and other unspecified pulmonary effects.

Other: Gastrointestinal irritation with abdominal cramps, vomiting, nausea, and diarrhea.

Chronic Health Effects

The following chronic (long-term) health effects may occur at some time after exposure to calcium phosphate and can last for months or even years:

Cancer Hazards: According to the information presented in the references, calcium phosphate has not been adequately tested for its ability to cause cancer in test animals.

Reproductive Hazard: According to information available in the references, calcium phosphate has not been adequately tested for its ability to adversely affect reproduction in laboratory animals.

Other Chronic Effects: Exposure may cause the development of pulmonary granulomatous lesions.

Recommended Risk-Reduction Measures

Personnel should avoid direct contact with calcium phosphate. While it is classified (by NIOSH and the ACGIH) as a nuisance dust, it can pose a toxic health hazard risk on unprotected exposure. If a less toxic chemical cannot be substituted for a hazardous material, then *engineering controls* are the most effective method of reducing exposures.

The best protection is to enclose operations and/or provide local exhaust ventilation at the site of chemical release. While not always operationally feasible, isolating operations can also reduce exposure. Using respiratory protection is less effective than the controls mentioned above, but is still advisable whenever working with or around calcium phosphate. For low exposures, a full face piece respirator equipped with a high efficiency particulate air (HEPA) filter equipped with a dust/mist/fume pre-filter may suffice. However, for the best protection use a supplied-air respirator with a full face piece operated in positive pressure mode or a MSHA/NIOSH approved self-contained breathing apparatus (SCBA) with full facepiece operated in pressure demand or other continuous flow mode. If a full facepiece is not available, then chemical/dust goggles should be worn to protect the eyes. Whenever a chemical dust hazard exists, a face shield and a rubber apron should be worn. To prevent hand and skin exposures, impervious rubber gloves should be worn.

Administrative controls should also be in place to minimize the potential for human exposures. These may include written procedures or policies which specify the methods and techniques that will be practiced whenever personnel are to work with calcium phosphate.

All personnel should receive *training* on the use, hazards, protective measures, emergency actions, and other precautions per 29 CFR 1910.1200 (Hazard Communication), prior to the first assignment in an area where calcium phosphate is used or stored.

Before beginning employment and at regular intervals thereafter (e.g., annually), the following medical tests are recommended for personnel with frequent or anticipated exposure to calcium phosphate:

☑ Lung function tests (establish baseline).

If symptoms develop or overexposure is suspected, the following may also be useful:

☑ Lung function tests.

☑ Consider chest X-ray following acute overexposure.

Any evaluation should include a careful history of past and present symptoms with a medical examination. Medical tests that look for damage already done are not a substitute for controlling exposure. Also, since smoking can cause heart disease, emphysema, and a number of other respiratory disorders, smokers exposed to lung irritants such as calcium phosphate dusts and fumes may experience symptoms much more quickly than non-smokers under the same conditions of exposure. Prudent risk management requires proper consideration of *all* factors which may be causing the appearance of exposures symptoms in the work place.

Other methods to reduce exposure include:

☑ Where possible, enclose operations and use local exhaust ventilation at the site of chemical release. If local exhaust ventilation or enclosure is not used, respiratory protection should be mandatory.

☑ Always ensure that proper protective clothing is worn when using chemical substances.

☑ Wash thoroughly immediately after exposure to calcium phosphate and at the end of the work shift or before eating, drinking, or smoking.

☑ Hazard warning information should be posted in the work area. In addition, as part of an on-going education and training program, all information on the health and safety hazards of calcium phosphate should be communicated to all potentially exposed workers.

☑ Eye wash stations and emergency showers should be located in the immediate work area.

☑ Work clothing that has been contaminated with calcium phosphate should never be worn home where family members can be exposed as well. Contaminated clothing should be laundered at work or by an authorized service by personnel who have been trained in the hazards associated with exposure to calcium phosphate and calcium compounds.

RISK ASSESSMENT: ENVIRONMENT

General Assessment

The environment is at risk of exposure during transportation, storage, disposal, or destruction of calcium phosphate. In almost every scenario, the threat of environmental exposure is contingent upon the proper handling of the chemical substance. Accidental spills, large or small, can result in fire, explosion, and possible contamination of the surrounding environmental mediums (water, soil, and air).

Calcium phosphate is a non-flammable solid that can produce irritating and acrid smoke if it is involved in a fire. It may react violently in contact with strong acids, such as nitric, sulfuric, and hydrofluoric. These characteristics require special consideration during any emergency situation involving a leak or spill of calcium phosphate.

Calcium phosphate can enter the environment through industrial discharge, spills, or leaks.

Acute Ecological Effects

Acute (short-term) toxic effects may include the death of animals, birds, or fish and death or low growth rate in plants. Acute effects are seen 2 to 4 days after animals or plants are exposed to calcium phosphate.

Insufficient data are available to evaluate the short-term effects of calcium phosphate exposure to aquatic life, plants, birds or land animals.

Chronic Ecological Effects

Chronic (long-term) toxic effects may include shortened life span, reproductive problems, lower fertility, and changes in appearance or behavior in exposed animals. These effects can be seen long after first exposure(s) to toxic chemicals.

Insufficient data are available to evaluate the long-term effects of calcium phosphate exposure to aquatic life, plants, birds or land animals.

Water Solubility

Calcium phosphate is only very slightly soluble in water. Concentrations of less than 1 milligram may mix with a liter of water.

Persistence in the Environment

Calcium phosphate is only moderately persistent in the aquatic environment, with a half-life between 20 and 200 days. The half-life of a pollutant is the amount of time it takes for one half of the chemical to be degraded.

Bioaccumulation in Aquatic Organisms

Some substances increase in concentration, or *bioaccumulate,* in living organisms as they breathe contaminated air, drink contaminated water, or eat contaminated food. These chemicals can become concentrated in the tissues and internal organs of animals as well as humans.

The concentration of calcium phosphate found in fish is expected to be about the same as the average concentration of calcium phosphate in the water from which the fish was taken.

Recommended Risk-Reduction Measures

Proper training of all transporters will reduce the likelihood of a mishap or accidents resulting in a leak or spill to the environment. Proper labeling on all containers, trucks, and rail cars will enable emergency responders to react properly and quickly to any disaster thereby reducing the potential risk to the environment and to personnel.

If a spill or leak to the environment has occurred, fire department, emergency response and/or hazardous materials spill personnel should be notified immediately. Do not dry sweep. When vacuuming, a high efficiency particulate air (HEPA) filter should be used, not a standard shop vacuum. Use damp mop to clean up residue. Deposit collected materials in sealed drum for disposal. Cleanup should be attempted only by those trained in proper spill containment procedures. Contaminated soils may need to be removed and replaced with clean soil. If calcium phosphate should contact the water table, aquifer, or navigable waterway, time is of the essence. Calcium phosphate is slightly soluble in water. Therefore, total containment and remediation may not be entirely possible. When such spills occur, the local and/or state emergency response authorities must be notified. A comprehensive emergency response or disaster preparedness/recovery plan should be in place prior to any operations involving the use, transportation, storage, or disposal of calcium phosphate. If calcium phosphate is spilled or leaked, the following specific steps are recommended:

- ☑ Restrict persons not wearing protective clothing from area of spill or leak until clean-up is complete.
- ☑ Ventilate area of spill or leak and remove all ignition sources.
- ☑ Collect powdered materials in most convenient and safe manner possible and place in sealed drums for disposal. Do NOT dry sweep, use HEPA vacuum instead. Damp mop residue.
- ☑ It may be necessary to dispose of calcium phosphate as a hazardous waste. The responsible state agency or the regional office of the federal Environmental Protection Agency (EPA) should be contacted for specific recommendations.

RISK ASSESSMENT: BUSINESS

General Assessment

Accidents or mishaps involving calcium phosphate can present a moderate threat to business operations. The loss or damage of equipment or facilities can significantly affect fiscal viability. Lawsuits that may result from personnel injury/death, public exposures, and/or environmental contamination will also require a serious

expenditure of resources. Media attention surrounding an injury, death, or environmental damage can also result in a loss of profits and loss of current as well as future business.

ʘ *Recommended Risk-Reduction Measures*

Company attorneys, safety and health professionals, and environmental specialists should be involved in the development of any such procedures. A company official should be pre-designated as a public relations officer with specific training in dealing with the press. Corporate plans and policies should be developed, approved, and implemented long before any need for such arises.

REFERENCES

Ahlbom, A. 1993. *Biostatistics for Epidemiologists.* New York: CRC Press/Lewis Publishers.

American Conference of Governmental Industrial Hygienists. 1988. *Documentation of the Threshold Limit Values and Biological Exposure Indices,* 5th Edition (with updates). Cincinnati: ACGIH

Asante-Duah, D. K. 1993. *Hazardous Waste Risk Assessment.* New York: CRC Press/Lewis Publishers.

Bloom, A. D. and F. J. DeSerres. 1995. *Ecotoxicity and Human Health.* New York: CRC Press/Lewis Publishers.

Brusick, D. J. 1994. *Methods for Genetic Risk Assessment.* New York: CRC Press/Lewis Publishers.

Calabrese, E. J. 1994. *Biological Effects of Low Level Exposures.* New York: CRC Press/Lewis Publishers.

Cockerman, L. G., B. S. Shane. 1994. *Basic Environmental Toxicology.* New York: CRC Press/Lewis Publishers.

Cothern, R. R. 1993. *Comparative Environmental Risk Assessment.* New York: CRC Press/Lewis Publishers.

Danse, I. R. 1991. *Common Sense Toxics in the Workplace.* New York: Van Nostrand Reinhold.

Environmental Protection Agency. 1983. *Chemical Hazard Information Profiles.* Washington, D.C.: U.S. EPA.

Fodden, C. R., and J. L. Weddell. 1992. *Hazardous Materials Emergency Action Data.* New York: CRC Press/Lewis Publishers.

Forseberg, K., and L. H. Keith. 1993. *Chemical; Protective Clothing Permeation and Degradation Database.* Chelsea, MI: Lewis Publishers.

Gerarde, H. W. 1960. *Toxicology and Biochemistry of Aromatic Hydrocarbons.* New York: Elsevier.

Hallenbeck, W. H. 1993. *Quantitative Risk Assessment for Environmental and Occupational Health,* 2nd Edition. New York: CRC Press/Lewis Publishers.

Hodgson, E., R. B. Mailman, and Janice E. Chambers. 1988. *Dictionary of Toxicology.* London: The MacMillian Company.

Howard, P. H. 1995. *Handbook of Environmental Fate and Exposures Data for Organic Chemicals,* Vol. I-IV. New York: CRC Press/Lewis Publishers.

Howard, P. H., et. al. 1991. *Handbook of Environmental Degradation Rates.* New York: CRC Press/Lewis Publishers.

Hosty, J. W., and P. Foster. 1990. *A Practical Guide to Chemical Spill Response.* New York: Van Nostrand Reinhold.

Jones, F. E. 1994. *Toxic Organic Vapors in the Workplace.* New York: CRC Press/Lewis Publishers.

Kamrin, M. A. 1988. *Toxicology: A Primer on Principles and Applications.* Chelsea, Michigan: Lewis Publishers.

Keith, L. H. , M. T. Johnston, C. E. Hudak, and M. Conoley. 1992. *National Toxicology Program's Gloves+ Database.* Chelsea, MI: Lewis Publishers.

Landis, W. G., M. H. Yu. 1995. *Introduction to Environmental Toxicology.* New York: CRC Press/Lewis Publishers.

Langerman, N. 1994 *Precautionary Labels for Chemical Containers.* New York: CRC Press/Lewis Publishers.

Lauwerys, R. R., P. Hoet. 1993. *Industrial Chemical Exposure.* New York: CRC Press/Lewis Publishers.

Lewis, R. J., Sr. 1992. *Sax's Dangerous Properties of Industrial Materials,* Volumes 1, 2, and 3, 8th Edition. New York: Van Nostrand Reinhold.

Mackay, D., W. Y. Shiu, and K. C. Ma. 1992. *Illustrated Handbook of Chemical-Physical Properties and Environmental Fate for Organic Chemicals.* New York: CRC Press/Lewis Publishers.

Manahan, S. E. 1994. *Environmental Chemistry,* 6th Edition. New York: CRC Press/Lewis Publishers.

Mellström, G. A., J. E. Wahlberg, and H. I. Maibach. 1994. *Protective Gloves for Occupational Use.* New York: CRC Press/Lewis Publishers.

Meyer, E. 1990. *Chemistry of Hazardous Materials.* Englewood Cliffs, NJ: Prentice Hall.

Miller, B. F., M.D., and C. B. Keane, R.N. 1972. *Encyclopedia and Dictionary of Medicine and Nursing.* Philadelphia: W.B. Sanders Company.

Mycock, J. C., J. D. McKenna, and L. Theodore. 1995. *Handbook of Air Pollution Control Engineering and Technology.* New York: CRC Press/Lewis Publishers.

Nelson, G. O. 1992. *Gas Mixtures.* New York: CRC Press/Lewis Publishers.

Ott, W. R. 1995. *Environmental Statistics and Data Analysis.* New York: CRC Press/Lewis Publishers.

Rea, W. J. 1995. *Chemical Sensitivity,* Vol. I-III. New York: CRC Press/Lewis Publishers.

Sacarello, H. L. A. 1994. *The Comprehensive Handbook of Hazardous Materials.* New York: CRC Press/Lewis Publishers.

Sherman, J. D. 1988. *Chemical Exposure and Disease.* New York: Van Nostrand Reinhold.

Talbott, E., G. Craun. 1995. *An Introduction to Environmental Epidemiology.* New York: CRC Press/Lewis Publishers.

MATERIAL SAFETY DATA SHEET

CHEMICAL NAME	CRC PRESS/LEWIS PUBLISHERS DISCLAIMER:
CALCIUM SILICATE	

HAZARD WARNING INFORMATION

HEALTH	FIRE	REACTIVE	OTHER	DEGREE OF HAZARD	COLOR CODING	OTHER CODES
0	0	0		0 = Minimum Hazard 1 = Slight Hazard 2 = Moderate Hazard 3 = Serious Hazard 4 = Severe Hazard	HEALTH = BLUE FIRE = RED REACTIVITY = YELLOW OTHER = WHITE	OX = Oxidizer ACID = Acid ALK = Alkali COR = Corrosive W = Use No Water

SECTION I - GENERAL INFORMATION

Characterization	**Calcium Compound**	RCRA Number	**None**
		EPA Class	**Not Applicable**
DOT Proper Shipping Name	**Not Listed**	Chemical Abstract Service (CAS) Number	**1344-95-2**
DOT Hazard Class and Label Requirements	**No Citation**	DOT Emergency Guide Code	**No Citation**
DOT Identification Number	**No Citation**	Atomic Formula	**$CaSiO_3$**

Synonyms

Calcium metasilicate; calcium hydrosilicate; calcium monosilicate; calcium salt of silicic acid; wollastonite (mineral).

SECTION II - HAZARDOUS INGREDIENTS AND IDENTITY INFORMATION

Hazardous Components (specific identity)	OSHA Exposure Criteria	NIOSH Exposure Criteria	Immediately Dangerous to Life and Health (IDLH)	ACGIH Exposure Criteria
Calcium silicate (derivation: The commercial product is prepared from diatomaceous earth and lime).	**PEL:** 5 mg/m^3 (respirable dusts) 15 mg/m^3 (for total dusts containing no asbestos)	**REL:** 5 mg/m^3 (respirable dusts) 10 mg/m^3 (for total dusts containing no asbestos)	**Not Determined**	**TLV (8-hour):** 10 mg/m^3 (for total dusts containing no asbestos)

SECTION III - PHYSICAL AND CHEMICAL CHARACTERISTICS

Boiling Point	**Decomposes**	Specific Gravity (H_2O = 1)	**2.9**
Vapor Pressure (mm Hg)	**0 (approx.) at 68°F (20°C)**	Molecular Weight	**116.2**
Vapor Density (Air = 1)	**Not Found**	Melting Point	**2804°F (1540°C) Decomposes**

Solubility

Very slightly soluble in water (0.01%).

Appearance and Odor

A white or cream-colored, free-flowing powder.

SECTION IV - FIRE AND EXPLOSION HAZARD DATA

Flash Point (method used)	**Not Determined**	Explosive Limits in Air % by Volume	**LEL: Not Determined UEL: Not Determined**
NFPA Classification	**Non-Combustible Solid**	Autoignition Temperature	**Not Determined**

Extinguishing Media

Use agent suitable to surrounding fire. Dry chemical, carbon dioxide, water spray, or regular foam should be adequate for fire suppression.

Special Fire Fighting Procedures

Structural protective clothing is permeable. Be sure to stay clear of smoke, water fall-out, and water run-off. Move containers from fire if it can be done without risk. Cool exposed containers with water spray. Dike area for control and containment to prevent entry into sewers, drains, and waterways. Evacuate non-essential personnel 1500 feet from the danger area. Consider down wind conditions.

Unusual Fire and Explosion Hazards

None reported. Avoid creating dusts during response procedures.

SECTION V - REACTIVITY DATA

Stability		Conditions to Avoid **Calcium silicate is stable under normal conditions of temperature and pressure. Avoid contact with heat. Keep lids tightly closed. Avoid generating dusty conditions.**
Stable X	Unstable	Incompatibility *(materials to avoid)* **None reported. After prolonged contact with water, the solution reverts to soluble calcium salts and amorphous silica.**
Hazardous Polymerization		Conditions to Avoid **Under normal conditions of temperature and pressure, hazardous polymerization of calcium silicate is not known to occur.**
May Occur	Will Not Occur X	Hazardous Decomposition or By-products **None reported. Calcium silicate is not expected to create any unusual reactivity hazards under normal conditions of handling or storage.**

SECTION VI - HEALTH HAZARD DATA

Primary Route(s) of Entry:	Inhalation?	Absorption (skin)?	Ingestion?
	X		

Health Hazards

INHALATION: **A nuisance dust causing minor irritation of the eyes, nose, throat, and upper respiratory tract. Long-term exposures to high concentrations may cause chronic cough.**

EYES & SKIN: **Possible eye and skin irritation that is minimal and temporary.**

INGESTION: **Not likely under normal conditions. Possible irritation of the mouth and esophagus.**

Carcinogenicity	NTP Listed?	IARC Cancer Review Group?	OSHA Regulated?	Target Organs?
Unknown Human Unknown Animal	**No**	**No**	**No**	**Upper respiratory tract, eyes, skin.**

Medical Conditions Generally Aggravated by Exposure

None reported.

Emergency and First-aid Procedures

Eye contact: Flush immediately with water for 15 minutes (minimum), gently lifting eyelids. Do not allow victim to keep eyes tightly shut. Seek medical attention. Skin contact: Immediately remove all contaminated clothing. Wash area with large amounts of soap and water. For inhalation: Remove the person from exposure. Provide respiratory assistance and CPR. Transfer to medical facility. If swallowed: Give 1-2 glasses of water and induce vomiting by giving an emetic. Never attempt to give a convulsing or unconscious person anything by mouth. Seek medical attention immediately.

SECTION VII - PRECAUTIONS FOR SAFE HANDLING AND USE

Steps to be Taken in Case Material is Released or Spilled

Collect powdered material in most convenient manner possible and deposit in sealed drum. Keep those not involved in cleanup from entering area. Do NOT dry sweep (creates airborne dusts). Use a vacuum equipped with a high efficiency particulate air (HEPA) filter. Use wet mop to clean residues.

Preferred Waste Disposal Method

No Citation.

Precautions to be Taken in Handling and Storage

Store in tightly closed containers in cool, well-ventilated area. Do not allow or cause the generation of dusty conditions during use, handling, storage, or disposal of calcium silicate. Always practice good housekeeping procedures.

Other Precautions and Warnings

Protect containers from physical damage.

SECTION VIII - CONTROL MEASURES AND PERSONAL PROTECTIVE EQUIPMENT

Respiratory Protection *(specify type)*

A nuisance dust. For low exposures, a full facepiece respirator with HEPA filter may suffice. For particularly dusty conditions, best protection is obtained using a supplied air respirator with a full facepiece, hood, or helmet, or a self-contained breathing apparatus (SCBA) with a full facepiece operated in positive pressure or other continuos flow mode.

Ventilation

Local exhaust or general mechanical systems recommended.

Protective Gloves	Eye Protection	Other Protective Clothing
Impervious Material	**Chemical/Dust Goggles or Face Mask**	**Protective Suit or Clothing**

Work/Hygiene Practices

Always wash hands thoroughly after using chemical; never bring food, drink, or smoking materials into vicinity of chemicals.

CALCIUM SILICATE

$CaSiO_3$ CAS: 1344-95-2

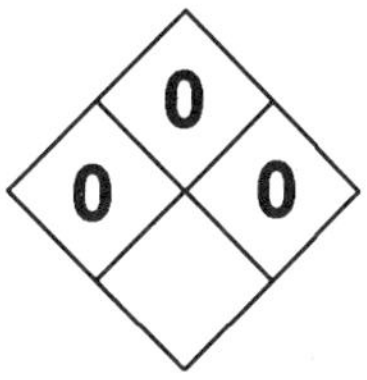

IDENTIFICATION AND TYPICAL USES

Calcium silicate is a finely powdered, white, or cream-colored free-flowing solid. It is used as a binder for refractory materials, as an anti-caking agent food additive, and a suspension agent. It is used in paint pigments and in road construction. It is an absorbent, an antacid, a filler for paper and paper coatings, in cosmetics, in the manufacture of glass and Portland cement.

RISK ASSESSMENT: HEALTH

General Assessment

Calcium silicate is considered a nuisance dust. Exposure occurs through *inhalation* of its dusts and powders. Its toxicity is considered relatively low. There are no reports of carcinogenic, mutagenic, or teratogenic effects subsequent to exposure. While it is primarily characterized as a nuisance dust, there are some noted effects of exposure that must be assessed and managed if calcium silicate is used or handled in the work place.

Inhalation of calcium silicates may cause minor or mild irritation of the eyes, nose, throat, and upper respiratory tract. Specific reaction of lung tissues is dependent upon conditions of exposure (level, duration) and individual response.

Skin contact can cause irritation and possible rash, depending upon individual sensitivity to the dusts. Eye contact can also cause irritation. Ingestion is not likely. It may cause irritation of the mouth and esophagus.

Acute Health Effects

The following acute (short-term) health effects may occur immediately or shortly after exposure to calcium silicate:

Skin: Irritation and possible rash.

Eye: Irritation.

Lung: Irritation of the eyes, nose, throat, and mucosa of the upper respiratory tract. There may be coughing, sneezing, and other unspecified pulmonary effects.

Other: If swallowed, may cause irritation of the mouth or esophagus.

Chronic Health Effects

The following chronic (long-term) health effects may occur at some time after exposure to calcium silicate and can last for months or even years:

Cancer Hazards: According to the information presented in the references, calcium silicate has not been adequately tested for its ability to cause cancer in test animals. As a nuisance dust, cancer is not considered likely to occur. More research is required.

Reproductive Hazard: According to information available in the references, calcium silicate has not been adequately tested for its ability to adversely affect reproduction in laboratory animals.

Other Chronic Effects: Damaging effects to the upper respiratory tract may occur. But the extent of this damage is contingent upon the duration and severity of exposure.

Recommended Risk-Reduction Measures

Personnel should avoid direct contact with calcium silicate. While it is classified as a nuisance dust, it may pose a health hazard risk on unprotected exposure. If a less hazardous chemical cannot be substituted for a hazardous material, then *engineering controls* are the most effective method of reducing exposures.

The best protection is to enclose operations and/or provide local exhaust ventilation at the site of chemical release. While not always operationally feasible, isolating operations can also reduce exposure. Using respiratory protection is less effective than the controls mentioned above, but is still advisable whenever working with or around calcium silicate. For low exposures, a full facepiece respirator equipped with a high efficiency particulate air (HEPA) filter equipped with a dust/mist/fume pre-filter may suffice. However, for the best protection use a supplied-air respirator with a full facepiece operated in positive pressure mode or a MSHA/NIOSH approved self-contained breathing apparatus (SCBA) with full facepiece operated in pressure demand or other continuous flow mode. If a full facepiece is not available, then chemical/dust goggles should be worn to protect the eyes. Whenever a chemical dust hazard exists, a face shield and a rubber apron should be worn. To prevent hand and skin exposures, impervious rubber gloves should be worn.

Administrative controls should also be in place to minimize the potential for human exposures. These may include written procedures or policies which specify the methods and techniques that will be practiced whenever personnel are to work with calcium silicate.

All personnel should receive *training* on the use, hazards, protective measures, emergency actions, and other precautions per 29 CFR 1910.1200 (Hazard Communication), prior to the first assignment in an area where calcium silicate is used or stored.

Before beginning employment and at regular intervals thereafter (e.g., annually), the following medical tests are recommended for personnel with frequent or anticipated exposure to calcium silicate:

☑ Lung function tests (establish baseline).

If symptoms develop or overexposure is suspected, the following may also be useful:

☑ Lung function tests.

☑ Consider chest X-ray following acute overexposure.

Any evaluation should include a careful history of past and present symptoms with a medical examination. Medical tests that look for damage already done are not a substitute for controlling exposure. Also, since smoking can cause heart disease, emphysema, and a number of other respiratory disorders, smokers exposed to lung irritants such as calcium silicate dusts may experience symptoms much more quickly than nonsmokers under the same conditions of exposure. Prudent risk management requires proper consideration of *all* factors which may be causing the appearance of exposures symptoms in the work place.

Other methods to reduce exposure include:

☑ Where possible, enclose operations and use local exhaust ventilation at the site of chemical release. If local exhaust ventilation or enclosure is not used, respiratory protection should be mandatory.

☑ Always ensure that proper protective clothing is worn when using chemical substances.

☑ Wash thoroughly immediately after exposure to calcium silicate and at the end of the work shift or before eating, drinking, or smoking.

☑ Hazard warning information should be posted in the work area. In addition, as part of an on-going education and training program, all information on the health and safety hazards of calcium silicate should be communicated to all potentially exposed workers.

☑ Eye wash stations and emergency showers should be located in the immediate work area.

☑ Work clothing that has been contaminated with calcium silicate should never be worn home where family members can be exposed as well. Contaminated clothing should be laundered at work or by an authorized service by personnel who have been trained in the hazards associated with exposure to calcium silicate and calcium compounds.

RISK ASSESSMENT: ENVIRONMENT

General Assessment

The environment is at risk of exposure during transportation, storage, disposal, or destruction of calcium silicate. In almost every scenario, the threat of environmental exposure is contingent upon the proper handling of the chemical substance. Accidental spills, large or small, can result in fire, explosion, and possible contamination of the surrounding environmental mediums (water, soil, and air).

Calcium silicate is a non-flammable solid. However, dusty conditions created during emergency response situations can pose a hazard to response personnel. These characteristics require special consideration during any emergency situation involving a leak or spill of calcium silicate.

If calcium silicate is released into the environment it poses no serious threat to ecosystems. However, this does not mean that uncontrolled emissions are permissible. Calcium silicate can enter the environment through industrial discharge, spills, or leaks.

☠ *Acute Ecological Effects*

Acute (short-term) toxic effects may include the death of animals, birds, or fish and death or low growth rate in plants. Acute effects are seen 2 to 4 days after animals or plants are exposed to calcium silicate.

Insufficient data are available to evaluate the short-term effects of calcium silicate exposure to aquatic life, plants, birds or land animals.

Chronic Ecological Effects

Chronic (long-term) toxic effects may include shortened life span, reproductive problems, lower fertility, and changes in appearance or behavior in exposed animals. These effects can be seen long after first exposure(s) to toxic chemicals.

Insufficient data are available to evaluate the long-term effects of calcium silicate exposure to aquatic life, plants, birds or land animals.

Water Solubility

Calcium silicate is only very slightly soluble in water. Concentrations of less than 1 milligram may mix with a liter of water. However, after prolonged contact with water, the solution reverts to soluble calcium salts and amorphous silica.

⌛ *Persistence in the Environment*

Calcium silicate is moderately persistent in the aquatic environment, with a half-life between 20 and 200 days. The half-life of a pollutant is the amount of time it takes for one half of the chemical to be degraded.

Bioaccumulation in Aquatic Organisms

Some substances increase in concentration, or *bioaccumulate,* in living organisms as they breathe contaminated air, drink contaminated water, or eat contaminated food. These chemicals can become concentrated in the tissues and internal organs of animals as well as humans.

The concentration of calcium found in fish is expected to be about the same as the average concentration of calcium in the water from which the fish was taken.

Recommended Risk-Reduction Measures

Proper training of all transporters will reduce the likelihood of a mishap or accidents resulting in a leak or spill to the environment. Proper labeling on all containers, trucks, and rail cars will enable emergency responders to react properly and quickly to any disaster thereby reducing the potential risk to the environment and to personnel.

If a spill or leak to the environment has occurred, fire department, emergency response and/or hazardous materials spill personnel should be notified immediately. Do not dry sweep. When vacuuming, a high efficiency particulate air (HEPA) filter should be used, not a standard shop vacuum. Use damp mop to clean up residue. Deposit collected materials in sealed drum for disposal. Cleanup should be attempted only by those trained in proper spill containment procedures. Contaminated soils may need to be removed and replaced with clean soil. If calcium silicate should contact the water table, aquifer, or navigable waterway, time is of the essence. Calcium silicate is nearly insoluble in water. But it can revert to calcium salts and amorphous silica on prolonged contact with water. Therefore, total containment and remediation may not be entirely possible. When such spills occur, the local and/or state emergency response authorities must be notified. A comprehensive emergency response or disaster preparedness/recovery plan should be in place prior to any operations involving the use, transportation, storage, or disposal of calcium silicate. If calcium silicate is spilled or leaked, the following specific steps are recommended:

- ☑ Restrict persons not wearing protective clothing from area of spill or leak until cleanup is complete.
- ☑ Ventilate area of spill or leak and remove all ignition sources.
- ☑ Collect powdered materials in most convenient and safe manner possible and place in sealed drums for disposal. Do NOT dry sweep, use HEPA vacuum instead. Damp mop residue.
- ☑ It may be necessary to dispose of calcium silicate as a hazardous waste. The responsible state agency or the regional office of the federal Environmental Protection Agency (EPA) should be contacted for specific recommendations.

RISK ASSESSMENT: BUSINESS

General Assessment

Accidents or mishaps involving calcium silicate can present a minor threat to business operations. The loss or damage of equipment or facilities is not likely, but possible and can significantly affect fiscal viability. Lawsuits that may result from personnel injury, public exposures, and/or environmental contamination will also require a serious expenditure of resources. Media attention surrounding an injury or environmental damage can also result in a loss of profits and loss of current as well as future business.

Recommended Risk-Reduction Measures

Company attorneys, safety and health professionals, and environmental specialists should be involved in the development of any procedures dealing with safety or emergency response. A company official should be pre-designated as a public relations officer with specific training in dealing with the press. Corporate plans and policies should be developed, approved, and implemented long before any need for such arises.

REFERENCES

Ahlbom, A. 1993. *Biostatistics for Epidemiologists.* New York: CRC Press/Lewis Publishers.

American Conference of Governmental Industrial Hygienists. 1988. *Documentation of the Threshold Limit Values and Biological Exposure Indices,* 5th Edition (with updates). Cincinnati: ACGIH

Asante-Duah, D. K. 1993. *Hazardous Waste Risk Assessment.* New York: CRC Press/Lewis Publishers.

Bloom, A. D. and F. J. DeSerres. 1995. *Ecotoxicity and Human Health.* New York: CRC Press/Lewis Publishers.

Brusick, D. J. 1994. *Methods for Genetic Risk Assessment.* New York: CRC Press/Lewis Publishers.

Calabrese, E. J. 1994. *Biological Effects of Low Level Exposures.* New York: CRC Press/Lewis Publishers.

Cockerman, L. G., B. S. Shane. 1994. *Basic Environmental Toxicology.* New York: CRC Press/Lewis Publishers.

Cothern, R. R. 1993. *Comparative Environmental Risk Assessment.* New York: CRC Press/Lewis Publishers.

Danse, I. R. 1991. *Common Sense Toxics in the Workplace.* New York: Van Nostrand Reinhold.

Environmental Protection Agency. 1983. *Chemical Hazard Information Profiles.* Washington, D.C.: U.S. EPA.

Fodden, C. R., and J. L. Weddell. 1992. *Hazardous Materials Emergency Action Data.* New York: CRC Press/Lewis Publishers.

Forseberg, K., and L. H. Keith. 1993. *Chemical; Protective Clothing Permeation and Degradation Database.* Chelsea, MI: Lewis Publishers.

Gerarde, H. W. 1960. *Toxicology and Biochemistry of Aromatic Hydrocarbons.* New York: Elsevier.

Hallenbeck, W. H. 1993. *Quantitative Risk Assessment for Environmental and Occupational Health,* 2nd Edition. New York: CRC Press/Lewis Publishers.

Hodgson, E., R. B. Mailman, and Janice E. Chambers. 1988. *Dictionary of Toxicology.* London: The MacMillian Company.

Howard, P. H. 1995. *Handbook of Environmental Fate and Exposures Data for Organic Chemicals,* Vol. I-IV. New York: CRC Press/Lewis Publishers.

Hosty, J. W., and P. Foster. 1990. *A Practical Guide to Chemical Spill Response.* New York: Van Nostrand Reinhold.

Jones, F. E. 1994. *Toxic Organic Vapors in the Workplace.* New York: CRC Press/Lewis Publishers.

Kamrin, M. A. 1988. *Toxicology: A Primer on Principles and Applications.* Chelsea, Michigan: Lewis Publishers.

Keith, L. H. , M. T. Johnston, C. E. Hudak, and M. Conoley. 1992. *National Toxicology Program's Gloves+ Database.* Chelsea, MI: Lewis Publishers.

Landis, W. G., M. H. Yu. 1995. *Introduction to Environmental Toxicology.* New York: CRC Press/Lewis Publishers.

Langerman, N. 1994 *Precautionary Labels for Chemical Containers.* New York: CRC Press/Lewis Publishers.

Lauwerys, R. R., P. Hoet. 1993. *Industrial Chemical Exposure.* New York: CRC Press/Lewis Publishers.

Lewis, R. J., Sr. 1992. *Sax's Dangerous Properties of Industrial Materials,* Volumes 1, 2, and 3, 8th Edition. New York: Van Nostrand Reinhold.

Mackay, D., W. Y. Shiu, and K. C. Ma. 1992. *Illustrated Handbook of Chemical-Physical Properties and Environmental Fate for Organic Chemicals.* New York: CRC Press/Lewis Publishers.

Manahan, S. E. 1994. *Environmental Chemistry,* 6th Edition. New York: CRC Press/Lewis Publishers.

Mellström, G. A., J. E. Wahlberg, and H. I. Maibach. 1994. *Protective Gloves for Occupational Use.* New York: CRC Press/Lewis Publishers.

Meyer, E. 1990. *Chemistry of Hazardous Materials.* Englewood Cliffs, NJ: Prentice Hall.

Miller, B. F., M.D., and C. B. Keane, R.N. 1972. *Encyclopedia and Dictionary of Medicine and Nursing.* Philadelphia: W.B. Sanders Company.

Mycock, J. C., J. D. McKenna, and L. Theodore. 1995. *Handbook of Air Pollution Control Engineering and Technology.* New York: CRC Press/Lewis Publishers.

Nelson, G. O. 1992. *Gas Mixtures.* New York: CRC Press/Lewis Publishers.

Ott, W. R. 1995. *Environmental Statistics and Data Analysis.* New York: CRC Press/Lewis Publishers.

Rea, W. J. 1995. *Chemical Sensitivity,* Vol. I-III. New York: CRC Press/Lewis Publishers.

Sacarello, H. L. A. 1994. *The Comprehensive Handbook of Hazardous Materials.* New York: CRC Press/Lewis Publishers.

Sherman, J. D. 1988. *Chemical Exposure and Disease.* New York: Van Nostrand Reinhold.

Talbott, E., G. Craun. 1995. *An Introduction to Environmental Epidemiology.* New York: CRC Press/Lewis Publishers.

MATERIAL SAFETY DATA SHEET

CHEMICAL NAME	
CALCIUM SULFATE	**CRC PRESS/LEWIS PUBLISHERS DISCLAIMER:** THE INFORMATION AND RECOMMENDATIONS PRESENTED HEREIN ARE BASED ON SOURCES BELIEVED TO BE RELIABLE. CRC MAKES NO REPRESENTATION ON ITS COMPLETENESS OR ACCURACY. IT IS THE RESPONSIBILITY OF THE USER TO DETERMINE THE CHEMICAL'S SUITABILITY FOR ITS INTENDED USE, THE CHEMICAL'S SAFE USE, AND THE CHEMICAL'S PROPER DISPOSAL. NO REPRESENTATIONS AND/OR WARRANTIES, EITHER EXPRESSED OR IMPLIED, OF THE MERCHANTABILITY OR FITNESS FOR A PARTICULAR PURPOSE, OR OF ANY OTHER NATURE, ARE MADE WITH RESPECT TO THE INFORMATION PROVIDED IN THIS MSDS OR TO THE CHEMICAL TO WHICH INFORMATION MAY REFER. CRC NEITHER ASSUMES NOR AUTHORIZES ANY OTHER PERSON TO ASSUME FOR IT, ANY OTHER ADDITIONAL RESPONSIBILITY OR LIABILITY FOR THE USE OF, OR RELIANCE UPON, THIS INFORMATION.

HAZARD WARNING INFORMATION

HEALTH	FIRE	REACTIVE	OTHER	DEGREE OF HAZARD	COLOR CODING	OTHER CODES
2	0	0		0 = Minimum Hazard 1 = Slight Hazard 2 = Moderate Hazard 3 = Serious Hazard 4 = Severe Hazard	HEALTH = BLUE FIRE = RED REACTIVITY = YELLOW OTHER = WHITE	OX = Oxidizer ACID = Acid ALK = Alkali COR = Corrosive W = Use No Water

SECTION I - GENERAL INFORMATION

Characterization	RCRA Number	EPA Class
Calcium Compound	**None**	**Not Applicable**

DOT Proper Shipping Name	Chemical Abstract Service (CAS) Number
Not Listed	**7778-18-9**
DOT Hazard Class and Label Requirements	DOT Emergency Guide Code
No Citation	**No Citation**
DOT Identification Number	Atomic Formula
No Citation	**$CaSO_4$**

Synonyms

Gypsum; mineral white; alabaster; Plaster of Paris; satin spar; terra alba; selenite; anhydrous calcium sulfate; anhydrous gypsum; anhydrous sulfate of lime; calcium salt of sulfuric acid.

SECTION II - HAZARDOUS INGREDIENTS AND IDENTITY INFORMATION

Hazardous Components (specific identity)	OSHA Exposure Criteria	NIOSH Exposure Criteria	Immediately Dangerous to Life and Health (IDLH)	ACGIH Exposure Criteria
Calcium sulfate (derivation: From natural sources, such as the mineral anhydrite, and as a by-product of many chemical operations).	**PEL:** 5 mg/m^3 (respirable fraction) 15 mg/m^3 (for total dusts containing no asbestos)	**REL:** 5 mg/m^3 (respirable fraction) 10 mg/m^3 (for total dusts containing no asbestos)	**Not Determined**	**TLV (8-hour):** 10 mg/m^3 (for total dusts containing no asbestos)

SECTION III - PHYSICAL AND CHEMICAL CHARACTERISTICS

Boiling Point	Specific Gravity (H_2O = 1)
2179°F (1193°C) Decomposes	**2.960**
Vapor Pressure (mm Hg)	Molecular Weight
0 (approx.) at 68°F (20°C)	**136.14**
Vapor Density (Air = 1)	Melting Point
Not Found	**2642°F (1450°C) Decomposes**

Solubility

The pure anhydrous is slightly soluble in water (0.31%).

Appearance and Odor

White, odorless powder or crystals. It may appear to have a blue, gray, or reddish tinge.

SECTION IV - FIRE AND EXPLOSION HAZARD DATA

Flash Point (method used)	Explosive Limits in Air % by Volume
Not Determined	**LEL: Not Determined UEL: Not Determined**
NFPA Classification	Autoignition Temperature
Non-Combustible Solid	**Not Determined**

Extinguishing Media

Use agent suitable to surrounding fire. Dry chemical, carbon dioxide, water spray, or regular foam should be adequate for fire suppression.

Special Fire Fighting Procedures

Structural protective clothing is permeable. Be sure to stay clear of smoke, water fall-out, and water run-off. Move containers from fire if it can be done without risk. Cool exposed containers with water spray. Dike area for control and containment to prevent entry into sewers, drains, and waterways. Evacuate non-essential personnel 1500 feet from the danger area. Consider down wind conditions.

Unusual Fire and Explosion Hazards

None reported. Avoid creating dusts during response procedures.

SECTION V - REACTIVITY DATA

Stability		Conditions to Avoid **Calcium sulfate is stable under normal conditions of temperature and pressure. Avoid contact with heat. Keep lids tightly closed. Avoid generating dusty conditions.**
Stable X	Unstable	Incompatibility *(materials to avoid)* **Calcium sulfate reacts violently with aluminum when heated. At high temperatures, mixtures of calcium sulfate and phosphorus will ignite with diazomethane, reacts exothermically and may explode.**
Hazardous Polymerization		Conditions to Avoid **Under normal conditions of temperature and pressure, hazardous polymerization of calcium sulfate is not known to occur.**
May Occur	Will Not Occur X	Hazardous Decomposition or By-products **When heated to decomposition, calcium sulfate can produce toxic and irritating smoke and fumes, including toxic fumes of sulfur oxides.**

SECTION VI - HEALTH HAZARD DATA

Primary Route(s) of Entry:	Inhalation?	Absorption (skin)?	Ingestion?
	X		

Health Hazards

INHALATION: A nuisance dust causing minor irritation of the eyes, nose, throat, and upper respiratory tract with inflammation of the mucosa, nosebleeds, laryngitis, rhinitis, and pharyngitis.

EYES & SKIN: Possible eye and skin irritation that is minimal and temporary. Conjunctivitis may occur.

INGESTION: In large quantities, there could be potential fatal obstruction of the intestinal tract.

Carcinogenicity	NTP Listed?	IARC Cancer Review Group?	OSHA Regulated?	Target Organs?
Unknown Human Unknown Animal	No	No	No	Upper respiratory tract, eyes, skin.

Medical Conditions Generally Aggravated by Exposure

Pulmonary diseases may be aggravated.

Emergency and First-aid Procedures

Eye contact: Flush immediately with water for 15 minutes (minimum), gently lifting eyelids. Do not allow victim to keep eyes tightly shut. Seek medical attention. Skin contact: Immediately remove all contaminated clothing. Wash area with large amounts of soap and water. For inhalation: Remove the person from exposure. Provide respiratory assistance and CPR. Transfer to medical facility. If swallowed: Give 1-2 glasses of water and induce vomiting by giving an emetic. Never attempt to give a convulsing or unconscious person anything by mouth. Seek medical attention immediately.

SECTION VII - PRECAUTIONS FOR SAFE HANDLING AND USE

Steps to be Taken in Case Material is Released or Spilled

Collect powdered material in most convenient manner possible and deposit in sealed drum. Keep those not involved in cleanup from entering area. Do NOT dry sweep (creates airborne dusts). Use a vacuum equipped with a high efficiency particulate air (HEPA) filter. Use wet mop to clean residues.

Preferred Waste Disposal Method

No Citation.

Precautions to be Taken in Handling and Storage

Store in tightly closed containers in cool, well-ventilated area. Do not allow or cause the generation of dusty conditions during use, handling, storage, or disposal of calcium sulfate. Always practice good housekeeping procedures. Keep away from aluminum, diazomethane, and phosphorus.

Other Precautions and Warnings

Protect containers from physical damage.

SECTION VIII - CONTROL MEASURES AND PERSONAL PROTECTIVE EQUIPMENT

Respiratory Protection *(specify type)*

A nuisance dust. For low exposures, a full facepiece respirator with HEPA filter may suffice. For particularly dusty conditions, best protection is obtained using a supplied air respirator with a full facepiece, hood, or helmet, or a self-contained breathing apparatus (SCBA) with a full facepiece operated in positive pressure or other continuos flow mode.

Ventilation

Local exhaust or general mechanical systems recommended.

Protective Gloves	Eye Protection	Other Protective Clothing
Impervious Material	Chemical/Dust Goggles or Face Mask	Protective Suit or Clothing

Work/Hygiene Practices

Always wash hands thoroughly after using chemical; never bring food, drink, or smoking materials into vicinity of chemicals.

CALCIUM SULFATE

$CaSO_4$ CAS: 7778-18-9

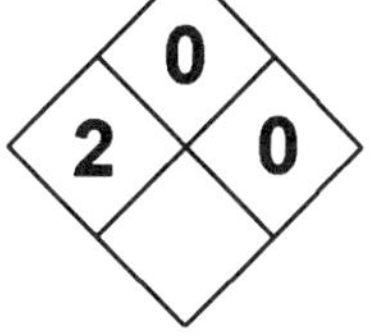

IDENTIFICATION AND TYPICAL USES

Calcium sulfate is an odorless white powder, or a colorless crystalline solid. It may have a blue, gray or reddish tinge. It is used as a retarding agent in Portland cement, in tiles and plaster, as a source of sulfur and sulfuric acid, in polishing compounds and powders, as a white pigment in paints (also as a filler or drier), in paper as a size filler or surface coating, in dyeing and calico printing, in metallurgy as a reducing agent for zinc minerals, for drying industrial gases, solids and many organic liquids, in granulated form as a soil conditioner, in quick-set cements, in molds and surgical casts, in wallboard (gypsum drywall), as a food additive, and as a desiccant.

RISK ASSESSMENT: HEALTH

General Assessment

Calcium sulfate is considered a nuisance dust. Exposure occurs through *inhalation* of its dusts and powders. Its toxicity is considered relatively low, however exposure still poses some degree of health risk that must be managed. There are no reports of carcinogenic, mutagenic, or teratogenic effects subsequent to exposure.

Inhalation of calcium sulfate may cause minor or mild irritation of the eyes, nose, throat, and upper respiratory tract, especially the larynx and pharynx. It can cause inflammation of the mucosa in the nose, which is referred to as rhinitis, accompanied by nosebleeds. Specific reaction of lung tissues is dependent upon conditions of exposure (level, duration) and individual response.

Skin contact can cause irritation and possible rash, depending upon individual sensitivity to the dusts. Eye contact can also cause irritation and possible conjunctivitis. Ingestion is not likely but can occur. It may cause irritation of the mouth and esophagus. If large quantities are ingested, there may be potentially fatal blockage in the gastrointestinal tract.

Acute Health Effects

The following acute (short-term) health effects may occur immediately or shortly after exposure to calcium sulfate:

Skin: Irritation and possible rash.

Eye: Irritation and conjunctivitis.

Lung: Irritation of the eyes, nose, throat, and mucosa of the upper respiratory tract. There may be coughing, sneezing, nosebleeds, and rhinitis.

Other: If swallowed, may cause irritation of the mouth or esophagus. In sufficient quantities, it can block the gastrointestinal tract and cause death.

Chronic Health Effects

The following chronic (long-term) health effects may occur at some time after exposure to calcium sulfate and can last for months or even years:

Cancer Hazards: According to the information presented in the references, calcium sulfate has not been adequately tested for its ability to cause cancer in test animals. As a nuisance dust, cancer is not considered likely to occur. More research is required.

Reproductive Hazard: According to information available in the references, calcium sulfate has not been adequately tested for its ability to adversely affect reproduction in laboratory animals.

Other Chronic Effects: Damaging effects to the upper respiratory tract may occur. Prolonged contact may cause laryngitis (inflammation of the larynx) and/or pharyngitis (inflammation of the pharynx). There may also be an impairment in the senses of smell and taste subsequent to repeated exposures.

ⓘ *Recommended Risk-Reduction Measures*

Personnel should avoid direct contact with calcium sulfate. While it is classified as a nuisance dust, it may pose a health hazard risk on unprotected exposure. If a less hazardous chemical cannot be substituted for a hazardous material, then *engineering controls* are the most effective method of reducing exposures.

The best protection is to enclose operations and/or provide local exhaust ventilation at the site of chemical release. While not always operationally feasible, isolating operations can also reduce exposure. Using respiratory protection is less effective than the controls mentioned above, but is still advisable whenever working with or around calcium sulfate. For low exposures, a full facepiece respirator equipped with a high efficiency particulate air (HEPA) filter equipped with a dust/mist/fume pre-filter may suffice. However, for the best protection use a supplied-air respirator with a full facepiece operated in positive pressure mode or a MSHA/NIOSH approved self-contained breathing apparatus (SCBA) with full facepiece operated in pressure demand or other continuous flow mode. If a full facepiece is not available, then chemical/dust goggles should be worn to protect the eyes. Whenever a chemical dust hazard exists, a face shield and a rubber apron should be worn. To prevent hand and skin exposures, impervious rubber gloves should be worn.

Administrative controls should also be in place to minimize the potential for human exposures. These may include written procedures or policies which specify the methods and techniques that will be practiced whenever personnel are to work with calcium sulfate.

All personnel should receive *training* on the use, hazards, protective measures, emergency actions, and other precautions per 29 CFR 1910.1200 (Hazard Communication), prior to the first assignment in an area where calcium sulfate is used or stored.

Before beginning employment and at regular intervals thereafter (e.g., annually), the following medical tests are recommended for personnel with frequent or anticipated exposure to calcium sulfate:

☑ Lung function tests (establish baseline).

If symptoms develop or overexposure is suspected, the following may also be useful:

☑ Lung function tests.

☑ Consider chest X-ray after acute overexposure.

☑ Examination of the nasal cavity, larynx, and pharynx for inflammation or irritation.

Any evaluation should include a careful history of past and present symptoms with a medical examination. Medical tests that look for damage already done are not a substitute for controlling exposure. Also, since smoking can cause heart disease, emphysema, and a number of other respiratory disorders, smokers exposed to lung irritants such as calcium sulfate dusts may experience symptoms much more quickly than nonsmokers under the same conditions of exposure. Prudent risk management requires proper consideration of *all* factors which may be causing the appearance of exposures symptoms in the work place.

Other methods to reduce exposure include:

☑ Where possible, enclose operations and use local exhaust ventilation at the site of chemical release. If local exhaust ventilation or enclosure is not used, respiratory protection should be mandatory.

☑ Always ensure that proper protective clothing is worn when using chemical substances.

☑ Wash thoroughly immediately after exposure to calcium sulfate and at the end of the work shift or before eating, drinking, or smoking.

☑ Hazard warning information should be posted in the work area. In addition, as part of an on-going education and training program, all information on the health and safety hazards of calcium sulfate should be communicated to all potentially exposed workers.

☑ Eye wash stations and emergency showers should be located in the immediate work area.

☑ Work clothing that has been contaminated with calcium sulfate should never be worn home where family members can be exposed as well. Contaminated clothing should be laundered at work or by an authorized service by personnel who have been trained in the hazards associated with exposure to calcium sulfate and calcium compounds.

RISK ASSESSMENT: ENVIRONMENT

General Assessment

The environment is at risk of exposure during transportation, storage, disposal, or destruction of calcium sulfate. In almost every scenario, the threat of environmental exposure is contingent upon the proper handling of the chemical substance. Accidental spills, large or small, can result in fire, explosion, and possi-

ble contamination of the surrounding environmental mediums (water, soil, and air).

Calcium sulfate is a non-flammable solid. However, dusty conditions created during emergency response situations can pose a hazard to response personnel. It is also incompatible and reacts violently in contact with aluminum under heat (as in a fire). Also, at high temperature mixtures of calcium sulfate and phosphorus will ignite and burn fiercely. These characteristics require special consideration during any emergency situation involving a leak or spill of calcium sulfate.

If calcium sulfate is released into the environment it poses no known threat to ecosystems. However, this does not mean that uncontrolled emissions are permissible. Calcium sulfate can enter the environment through industrial discharge, spills, or leaks.

Acute Ecological Effects

Acute (short-term) toxic effects may include the death of animals, birds, or fish and death or low growth rate in plants. Acute effects are seen 2 to 4 days after animals or plants are exposed to calcium sulfate.

Insufficient data are available to evaluate the short-term effects of calcium sulfate exposure to aquatic life, plants, birds or land animals.

Chronic Ecological Effects

Chronic (long-term) toxic effects may include shortened life span, reproductive problems, lower fertility, and changes in appearance or behavior in exposed animals. These effects can be seen long after first exposure(s) to toxic chemicals.

Insufficient data are available to evaluate the long-term effects of calcium sulfate exposure to aquatic life, plants, birds or land animals.

Water Solubility

Calcium sulfate is only very slightly soluble in water. Concentrations of less than 1 milligram may mix with a liter of water. However, after prolonged contact with water, it will react to form gypsum or Plaster of Paris, neither of which are considered hazardous or toxic to the aquatic environment.

Persistence in the Environment

Calcium sulfate is moderately persistent in the aquatic environment, with a half-life between 20 and 200 days. The half-life of a pollutant is the amount of time it takes for one half of the chemical to be degraded.

Bioaccumulation in Aquatic Organisms

Some substances increase in concentration, or *bioaccumulate,* in living organisms as they breathe contaminated air, drink contaminated water, or eat contaminated food. These chemicals can become concentrated in the tissues and internal organs of animals as well as humans.

The concentration of calcium found in fish is expected to be about the same as the average concentration of calcium in the water from which the fish was taken.

Recommended Risk-Reduction Measures

Proper training of all transporters will reduce the likelihood of a mishap or accidents resulting in a leak or spill to the environment. Proper labeling on all containers, trucks, and rail cars will enable emergency responders to react properly and quickly to any disaster thereby reducing the potential risk to the environment and to personnel.

If a spill or leak to the environment has occurred, fire department, emergency response and/or hazardous materials spill personnel should be notified immediately. Do not dry sweep. When vacuuming, a high efficiency particulate air (HEPA) filter should be used, not a standard shop vacuum. Use damp mop to clean up residue. Deposit collected materials in sealed drum for disposal. Cleanup should be attempted only by those trained in proper spill containment procedures. Contaminated soils may need to be removed and replaced with clean soil. If calcium sulfate should contact the water table, aquifer, or navigable waterway, time is of the essence. Calcium sulfate is nearly insoluble in water and total containment and remediation may be possible. When such spills occur, the local and/or state emergency response authorities must be notified. A comprehensive emergency response or disaster preparedness/recovery plan should be in place prior to any operations involving the use, transportation, storage, or disposal of calcium sulfate. If calcium sulfate is spilled or leaked, the following specific steps are recommended:

☑ Restrict persons not wearing protective clothing from area of spill or leak until cleanup is complete and area can be opened for normal work.

☑ Ventilate area of spill or leak and remove all ignition sources.

☑ Collect powdered materials in most convenient and safe manner possible and place in sealed drums for disposal. Do NOT dry sweep, use HEPA vacuum instead. Damp mop residue.

☑ It may be necessary to dispose of calcium sulfate as a hazardous waste. The responsible state agency or the regional office of the federal Environmental Protection Agency (EPA) should be contacted for specific recommendations.

RISK ASSESSMENT: BUSINESS

General Assessment

Accidents or mishaps involving calcium sulfate can present a minor threat to business operations. The loss or damage of equipment or facilities is not likely, but possible and can significantly affect fiscal viability. Lawsuits that may result from personnel injury, public exposures, and/or environmental contamination will also require a serious expenditure of resources. Media attention surrounding an injury or environmental damage can also result in a loss of profits and loss of current as well as future business.

Recommended Risk-Reduction Measures

Company attorneys, safety and health professionals, and environmental specialists should be involved in the development of any safety, health, or environmental procedures. An official company representative should be pre-designated as a public relations officer with specific training in dealing with the press. Corporate plans and policies should be developed, approved, and implemented long before any need for such arises.

REFERENCES

Ahlbom, A. 1993. *Biostatistics for Epidemiologists.* New York: CRC Press/Lewis Publishers.

American Conference of Governmental Industrial Hygienists. 1988. *Documentation of the Threshold Limit Values and Biological Exposure Indices,* 5th Edition (with updates). Cincinnati: ACGIH

Asante-Duah, D. K. 1993. *Hazardous Waste Risk Assessment.* New York: CRC Press/Lewis Publishers.

Bloom, A. D. and F. J. DeSerres. 1995. *Ecotoxicity and Human Health.* New York: CRC Press/Lewis Publishers.

Brusick, D. J. 1994. *Methods for Genetic Risk Assessment.* New York: CRC Press/Lewis Publishers.

Calabrese, E. J. 1994. *Biological Effects of Low Level Exposures.* New York: CRC Press/Lewis Publishers.

Cockerman, L. G., B. S. Shane. 1994. *Basic Environmental Toxicology.* New York: CRC Press/Lewis Publishers.

Cothern, R. R. 1993. *Comparative Environmental Risk Assessment.* New York: CRC Press/Lewis Publishers.

Danse, I. R. 1991. *Common Sense Toxics in the Workplace.* New York: Van Nostrand Reinhold.

Environmental Protection Agency. 1983. *Chemical Hazard Information Profiles.* Washington, D.C.: U.S. EPA.

Fodden, C. R., and J. L. Weddell. 1992. *Hazardous Materials Emergency Action Data.* New York: CRC Press/Lewis Publishers.

Forseberg, K., and L. H. Keith. 1993. *Chemical; Protective Clothing Permeation and Degradation Database.* Chelsea, MI: Lewis Publishers.

Gerarde, H. W. 1960. *Toxicology and Biochemistry of Aromatic Hydrocarbons.* New York: Elsevier.

Hodgson, E., R. B. Mailman, and Janice E. Chambers. 1988. *Dictionary of Toxicology.* London: The MacMillian Company.

Hosty, J. W., and P. Foster. 1990. *A Practical Guide to Chemical Spill Response.* New York: Van Nostrand Reinhold.

Jones, F. E. 1994. *Toxic Organic Vapors in the Workplace.* New York: CRC Press/Lewis Publishers.

Keith, L. H. , M. T. Johnston, C. E. Hudak, and M. Conoley. 1992. *National Toxicology Program's Gloves+ Database.* Chelsea, MI: Lewis Publishers.

Landis, W. G., M. H. Yu. 1995. *Introduction to Environmental Toxicology.* New York: CRC Press/Lewis Publishers.

Lauwerys, R. R., P. Hoet. 1993. *Industrial Chemical Exposure.* New York: CRC Press/Lewis Publishers.

Lewis, R. J., Sr. 1992. *Sax's Dangerous Properties of Industrial Materials,* Volumes 1, 2, and 3, 8th Edition. New York: Van Nostrand Reinhold.

Mackay, D., W. Y. Shiu, and K. C. Ma. 1992. *Illustrated Handbook of Chemical-Physical Properties and Environmental Fate for Organic Chemicals.* New York: CRC Press/Lewis Publishers.

Manahan, S. E. 1994. *Environmental Chemistry,* 6th Edition. New York: CRC Press/Lewis Publishers.

Meyer, E. 1990. *Chemistry of Hazardous Materials.* Englewood Cliffs, NJ: Prentice Hall.

Miller, B. F., M.D., and C. B. Keane, R.N. 1972. *Encyclopedia and Dictionary of Medicine and Nursing.* Philadelphia: W.B. Sanders Company.

Nelson, G. O. 1992. *Gas Mixtures.* New York: CRC Press/Lewis Publishers.

Rea, W. J. 1995. *Chemical Sensitivity,* Vol. I-III. New York: CRC Press/Lewis Publishers.

Sacarello, H. L. A. 1994. *The Comprehensive Handbook of Hazardous Materials.* New York: CRC Press/Lewis Publishers.

Sherman, J. D. 1988. *Chemical Exposure and Disease.* New York: Van Nostrand Reinhold.

MATERIAL SAFETY DATA SHEET

CHEMICAL NAME

CAMPHOR

CRC PRESS/LEWIS PUBLISHERS DISCLAIMER. THE INFORMATION AND RECOMMENDATIONS PRESENTED HEREIN ARE BASED ON SOURCES BELIEVED TO BE RELIABLE. CRC MAKES NO REPRESENTATION ON ITS COMPLETENESS OR ACCURACY. IT IS THE RESPONSIBILITY OF THE USER TO DETERMINE THE CHEMICAL'S SUITABILITY FOR ITS INTENDED USE, THE CHEMICAL'S SAFE USE, AND THE CHEMICAL'S PROPER DISPOSAL. NO REPRESENTATIONS AND/OR WARRANTIES, EITHER EXPRESSED OR IMPLIED, OF THE MERCHANTABILITY OR FITNESS FOR A PARTICULAR PURPOSE, OR OF ANY OTHER NATURE, ARE MADE WITH RESPECT TO THE INFORMATION PROVIDED IN THIS MSDS OR TO THE CHEMICAL TO WHICH INFORMATION MAY REFER. CRC NEITHER ASSUMES NOR AUTHORIZES ANY OTHER PERSON TO ASSUME FOR IT, ANY OTHER ADDITIONAL RESPONSIBILITY OR LIABILITY FOR THE USE OF, OR RELIANCE UPON, THIS INFORMATION.

HAZARD WARNING INFORMATION

HEALTH	FIRE	REACTIVE	OTHER	DEGREE OF HAZARD	COLOR CODING	OTHER CODES
0	2	0		0 = Minimum Hazard 1 = Slight Hazard 2 = Moderate Hazard 3 = Serious Hazard 4 = Severe Hazard	HEALTH = BLUE FIRE = RED REACTIVITY = YELLOW OTHER = WHITE	OX = Oxidizer ACID = Acid ALK = Alkali COR = Corrosive ~~W~~ = Use No Water

SECTION I - GENERAL INFORMATION

Characterization	**Ketone**	RCRA Number	**None**
		EPA Class	**Not Applicable**
DOT Proper Shipping Name	**Camphor (synthetic)**	Chemical Abstract Service (CAS) Number	**76-22-2**
DOT Hazard Class and Label Requirements	**Flammable Solid**	DOT Emergency Guide Code	**32**
DOT Identification Number	**UN 2717**	Chemical Formula	**$C_{10}H_{16}O$**

Synonyms

2-Camphonone; gum camphor; laurel camphor; synthetic camphor; 2-bornanone; formosa camphor.

SECTION II - HAZARDOUS INGREDIENTS AND IDENTITY INFORMATION

Hazardous Components (specific identity)	OSHA Exposure Criteria	NIOSH Exposure Criteria	Immediately Dangerous to Life and Health (IDLH)	ACGIH Exposure Criteria
Camphor (derivation: "Natural" camphor derived by steam distillation of camphor-tree wood and crystallization. "Synthetic" camphor, most of which is optically inactive, derived from pinene which is converted into camphene, treated with acetic acid and nitrobenzene to become camphor). **1 ppm = 6.66 mg/m³**	**PEL (8-hour):** **0.3 ppm** **2 mg/m³** **STEL:** **Not Established**	**REL (10-hour):** **0.3 ppm** **2 mg/m³** **STEL:** **Not Established**	**200 mg/m³**	**TLV:** **2 ppm** **12 mg/m³** **STEL:** **3 ppm** **19 mg/m³**

SECTION III - PHYSICAL AND CHEMICAL CHARACTERISTICS

Boiling Point	**399°F (204°C)**	Specific Gravity (H_2O = 1)	**0.99**
Vapor Pressure (mm Hg)	**0.2 at 69°F (20°C)**	Molecular Weight	**152.3**
Vapor Density (Air = 1)	**5.24**	Freezing Point	**345°F (174°C)**

Solubility

Slightly soluble (practically insoluble) in water. Soluble in aniline, benzene, alcohol, ether, chloroform, carbon disulfide, solvent naphtha, and fixed and volatile oils.

Appearance and Odor

Colorless or white crystals, granules, or easily broken masses with a penetrating, aromatic odor.

SECTION IV - FIRE AND EXPLOSION HAZARD DATA

Flash Point (method used)	**150°F (66°C) closed cup**	Explosive Limits in Air % by Volume	**LEL: 0.6% UEL: 3.5%**
NFPA Classification	**Combustible Solid**	Autoignition Temperature	**871°F (466°C)**

Extinguishing Media

Carbon dioxide, dry chemical, regular foam, water spray, or fog.

Special Fire Fighting Procedures

Wear full protective clothing and self-contained breathing apparatus (SCBA). Heat will build pressure and may rupture closed storage containers. Fight fire from a distance if possible. Keep fire-exposed containers cool with water spray. Vapor is explosive when exposed to heat or flame.

Unusual Fire and Explosion Hazards

Vapor burns rapidly with a flare-like effect. Vapors are heavier than air; they may travel some distance to flashback and cause fire or explosion.

SECTION V - REACTIVITY DATA

Stability		Conditions to Avoid
		Sublimes at room temperature and pressure (goes from liquid directly to gaseous state). It can be kept stable if held in airtight containers. Do not add salts to camphor water.
Stable	Unstable	Incompatibility *(materials to avoid)*
	X	Incompatible with potassium permanganate, chromic anhydride + naphthalene, oxidizers (such as chlorine, bromine, and fluoride) and steam (volatile reaction).
Hazardous Polymerization		Conditions to Avoid
		Under normal conditions of temperature and pressure, hazardous polymerization of camphor will not occur. Avoid exposures to heat, ignition sources, and incompatibles.
May Occur	Will Not Occur	Hazardous Decomposition or By-products
	X	When heated to decomposition, camphor emits acrid, irritating smoke and fumes. These include toxic carbon monoxide gas.

SECTION VI - HEALTH HAZARD DATA

Primary Route(s) of Entry:	Inhalation?	Absorption (skin)?	Ingestion?
	X	X	X

Health Hazards

INHALATION: Causes nausea, anxiety, headache, dizziness, dilation of the pupils and vision changes, corneal inflammation, hearing effects, confusion, facial twitching, spastic movements, convulsions, and coma. Possible liver effects and/or injury.

ABSORPTION: Skin or eye contact can result in localized irritation. Skin absorption is likely to cause systemic effects on the central nervous system. When combined with other exposure routes, effects can be serious.

INGESTION: Irritation of the mouth, throat, and stomach with fainting, cyanosis, hypotension, and possible cardiac arrhythmia, convulsions, and grand-mal seizures.

Carcinogenicity	NTP Listed?	IARC Cancer Review Group?	OSHA Regulated?	Target Organs?
Unknown Human Unknown Animal	No	No	29 CFR 1910.1000 (Table Z-1)	Respiratory system; skin; eyes; CNS, liver, G.I. tract.

Medical Conditions Generally Aggravated by Exposure

Respiratory function impairments (asthma, etc.); Nervous disorders; Skin conditions (dermatitis).

Emergency and First-aid Procedures

Eye contact: Immediately flush large amounts of water for 15 minutes (minimum), occasionally lifting eyelids, seek medical attention. Skin contact: Remove all contaminated clothing. Immediately wash area with large amounts of soap and water. For inhalation: Remove the person from exposure. Provide respiratory assistance and CPR. Transfer promptly to medical facility. If swallowed: Seek medical attention immediately. Do NOT induce vomiting. Do NOT give an unconscious person anything by mouth.

SECTION VII - PRECAUTIONS FOR SAFE HANDLING AND USE

Steps to be Taken in Case Material is Released or Spilled

Remove all ignition sources. Restrict persons not wearing protective equipment from entering area of spill or leak until cleanup is complete. Ventilate area. Collect solid materials in safest manner possible and deposit in sealed drum. Do NOT dry sweep (generates dusts), use HEPA vacuum instead.

Preferred Waste Disposal Method

Dissolve in combustible solvent and burn in a chemical incinerator with afterburner and scrubber.

Precautions to be Taken in Handling and Storage

Do not store with incompatible chemicals since violent reactions can occur. Store in tightly closed containers in a cool, dark, well-ventilated area away from fire, sparks, and flame. Use non-sparking tools on containers. Ground and bond metal containers during transfer operations.

Other Precautions and Warnings

Bulk storage of camphor is not recommended.

SECTION VIII - CONTROL MEASURES AND PERSONAL PROTECTIVE EQUIPMENT

Respiratory Protection *(specify type)*

For low exposures use a full facepiece respirator with organic vapor cartridge and HEPA filter. Greater protection is obtained from a supplied air respirator or a self-contained breathing apparatus (SCBA) with full facepiece operated in pressure demand or other positive pressure mode.

Ventilation

Local exhaust or general mechanical systems recommended.

Protective Gloves	Eye Protection	Other Protective Clothing
Butyl Rubber or Polycarbonate	Chemical Goggles and/or Face Mask	Protective Rubber Apron

Work/Hygiene Practices

Always wash hands thoroughly after using chemical; never bring food, drink, or smoking materials into vicinity of chemicals.

CAMPHOR

$C_{10}H_{16}O$ CAS: 76-22-2

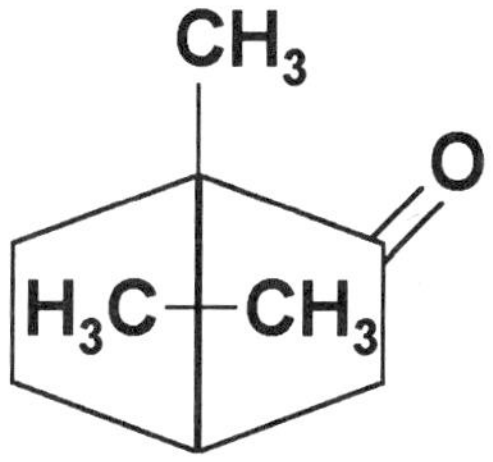

IDENTIFICATION AND TYPICAL USES

Camphor is a colorless or white crystalline or granular solid with a penetrating, aromatic odor. It is used in medicines (internal and external), as a plasticizer for cellulose nitrate, in explosives, lacquers, in insecticides, in moth and mildew proofing, in tooth powders, as a flavoring, in embalming, in the manufacture of pyrotechnics, and as an intermediate. It also is used as a topical anti-infective and anti-itching agent. A liquid variant (camphor oil) is produced almost exclusively in Taiwan and was formerly used in the manufacture of sassafras oil. It is now used chiefly as a fragrance of flavoring material and to some extent as a pharmaceutical product.

RISK ASSESSMENT: HEALTH

General Assessment

Camphor is a human poison by *ingestion* and possibly by *inhalation,* and skin contact (*absorption*). The central nervous system (CNS) seems to be the primary target of camphor exposure. There is no evidence of any carcinogenic or specific teratogenic effects resulting from exposure to camphor. However, mutation data have been reported and some scientists believe that mutagenic chemicals may pose a cancer risk in the long-term. The data are inconclusive in this regard and more research is required.

Inhalation of camphor may cause nausea, anxiety, headache, dizziness, excitement, confusion, facial twitching and spastic muscle movements, convulsions, and coma. There may also be visual disturbances (pupil dilation, corneal inflammation, and darkening of vision with flickering images) and hearing disturbances (ringing or noises in the ears). These effects appear to be primarily associated with camphor's specific ability to effect the CNS.

Ingestion causes irritation to the mouth, throat, and stomach. There may also be fainting, cyanosis due to lack of oxygenated blood in the body, hypotension, diarrhea, cardiac arrhythmia, convulsions, and, in some cases, grand-mal seizures following ingestion. The chemical is metabolized in the liver and may cause toxic action there as well.

Acute Health Effects

The following acute (short-term) health effects may occur immediately or shortly after exposure to camphor:

Skin: Irritation and possible rash.

Eye: Irritation and possible inflammation.

Lung: Irritation causing coughing and/or shortness of breath. Camphor is readily absorbed through the respiration process to cause toxic systemic effects.

CNS: Primary CNS toxicity. There can be excitement, dizziness, nausea, headaches, convulsions, and even epileptic-like seizures in some cases.

Chronic Health Effects

The following chronic (long-term) health effects may occur at some time after exposure to camphor and can last for months or even years:

Cancer Hazards: According to information presented in the references, camphor has not been adequately tested for its ability to cause cancer. Some mutagenic data have been recorded. Additional research is required.

Reproductive Hazards: There are no specific references to any teratogenic effects associated with exposure to camphor. However, it has been reported that camphor has the ability to cross the placenta barrier and it has been suggested that camphor caused at least one fetal death while other newborns were apparently not effected. Additional research is required to determine the specific reproductive and fetal toxicity of camphor.

Other Chronic Effects: High exposures over time may damage the liver to the point of developing hepatitis. There may also be symptoms similar to Reye's syndrome which is a potentially fatal condition with swelling of the brain and kidneys, fatty liver, fever, and vomiting.

Recommended Risk-Reduction Measures

Personnel should avoid direct contact with camphor. Its ability to rapidly affect the function of the central nervous system presents an extremely dangerous health hazard risk to exposed employees. Unless a less toxic chemical can be substituted for a hazardous substance, *engineering controls* are the most effective method of reducing exposures. The best protection is to enclose operations and/or provide local exhaust ventilation at the site of chemical release. While not always operationally feasible, isolating operations can also reduce exposure. Using respiratory protection is less effective than the controls mentioned above, but is still advisable whenever working with camphor. A NIOSH-approved, full facepiece organic vapor respirator is sufficient for low exposures. A supplied-air respirator or a self-contained breathing apparatus (SCBA) with full facepiece operated in pressure demand or other continuous flow mode are the recommended respiratory protection methods of choice. If a full facepiece is not available, then chemical goggles should be worn to protect the eyes. Whenever a chemical splash hazard exists, a face shield and a rubber apron should be worn. To prevent hand and skin exposures, butyl rubber or polycarbonate gloves should be worn when working with camphor.

Administrative controls should also be in place to minimize the potential for human exposures. These may include written procedures or policies which specify the methods and techniques that will be practiced whenever personnel are to work with camphor.

All personnel should receive *training* on the use, hazards, protective measures, emergency actions, and other precautions per 29 CFR 1910.1200 (Hazard Communication), prior to the first assignment in an area where camphor is used or stored.

If symptoms develop or overexposure is suspected, the following medical tests are suggested:

- ☑ Liver function tests.
- ☑ Evaluation of nervous system function by a qualified neurologist.

Other methods to reduce exposure include:

- ☑ Where possible, enclose operations and use local exhaust ventilation at the site of chemical release. If local exhaust ventilation or enclosure is not used, respiratory protection should be mandatory.
- ☑ Always ensure that proper protective clothing is worn when using chemical substances.
- ☑ Wash thoroughly immediately after exposure to camphor and at the end of the work shift or before eating, drinking, or smoking. Never wear clothes contaminated with camphor home. Family members can be exposed.
- ☑ Hazard warning information should be posted in the work area. As part of an on-going education and training program, all information on the health and safety hazards of camphor should be communicated to all potentially exposed workers.

RISK ASSESSMENT: ENVIRONMENT

General Assessment

The environment is at risk of exposure during transportation, storage, disposal, or destruction of camphor. In almost every scenario, the threat of environmental exposure is contingent upon the proper handling of the chemical substance. Accidental spills, large or small, can result in fire, explosion, and possible contamination of the surrounding environmental mediums (water, soil, and air).

Camphor is considered a combustible liquid. However, if ignited it may burn rapidly and fiercely in a manner similar to that of a road flare. Its vapor is explosive when exposed to heat or flame. It is also incompatible with a number of commodities. Potentially explosive reactions can occur if camphor is mixed or comes into contact with oxidizers. These characteristics require special consideration during any emergency involving a leak or spill of camphor. It will absorb into soils, especially those with high organic content, and may reach the water table.

The proper disposal/destruction method for camphor is to dissolve it in a combustible solvent and then burn it in a chemical incinerator equipped with an afterburner and air scrubber.

Camphor can enter the environment through manufacturing, unchecked discharge into effluents from industrial or municipal waste treatment plants, and through spills.

Acute Ecological Effects

Acute (short-term) toxic effects may include the death of animals, birds, or fish, and death or low growth rate in plants. Acute effects are seen 2 to 4 days after animals or plants are exposed to camphor.

Insufficient data are available to evaluate or predict the acute (short-term) effects of camphor to aquatic life, plants, birds, or land animals.

Chronic Ecological Effects

Chronic (long-term) toxic effects may include shortened life span, reproductive problems, lower fertility, and changes in appearance or behavior in exposed animals. These effects can be seen long after first exposure(s) to toxic chemicals.

Insufficient data are available to evaluate or predict the chronic (long-term) effects of camphor to aquatic life, plants, birds, or land animals.

Water Solubility

Camphor is practically insoluble in water. Concentrations of less than 1 milligram may mix with a liter of water.

Persistence in the Environment

Camphor quickly sublimes (goes from solid to gaseous state) at normal room temperature and pressure. It is slightly persistent in water, with a half-life of between 2 to 20 days. The half-life of a pollutant is the amount of time it takes for one half of the chemical to be degraded.

Bioaccumulation in Aquatic Organisms

Some substances increase in concentration, or *bioaccumulate,* in living organisms as they breathe contaminated air, drink contaminated water, or eat contaminated food. These chemicals can become concentrated in the tissues and internal organs of animals as well as humans.

The concentration of camphor found in fish tissues is expected to be about the same as the average concentration in water from which the fish was taken.

Recommended Risk-Reduction Measures

Proper training of all transporters will reduce the likelihood of a mishap or accident resulting in a leak or spill to the environment. The correct use of DOT labeling (per the MSDS) on all containers, trucks, and rail cars will enable emergency responders to react properly and quickly to any disaster thereby reducing the potential risk to the environment and to personnel.

Storage of camphor should be segregated from incompatible chemicals to minimize the risk of cross-contamination or contact. Buildings designated for storage should be equipped with appropriate fire protection systems (alarms, sprinklers, emergency lighting, portable extinguishers).

If a spill or leak to the environment has occurred, fire department, emergency response, and/or hazardous materials spill personnel should be notified immediately. Cleanup should be attempted only by those trained in proper spill containment procedures. Contaminated soils should be removed for incineration and replaced with clean soil. If camphor should contact the water table, aquifer, or navigable waterway, time is of the essence. It is readily miscible in water and, therefore, total containment and remediation may not be entirely possible. When such spills occur, the local and/or state emergency response authorities must be notified. A comprehensive emergency response or disaster preparedness/recovery plan should be in place prior to any operations involving the use, transportation, storage, or disposal of camphor. If camphor is spilled or leaked, the following specific steps are recommended:

- ☑ Restrict persons not wearing protective clothing from area of spill or leak until cleanup is complete.
- ☑ Remove all ignition sources.
- ☑ Ventilate area of spill or leak.
- ☑ Collect dry spills in safest manner possible. Do NOT dry sweep (generates dusts). Use a vacuum equipped with a high-efficiency particulate air (HEPA) filter instead. Damp mop residue. Absorb liquid spills in vermiculite, dry sand, earth, or a similar material and deposit in sealed containers.
- ☑ Keep camphor out of a confined space, such as a sewer, because of the possibility of explosion (unless the sewer is designed to prevent the buildup of explosive concentrations).

☑ It may be necessary to dispose of camphor as a hazardous waste. The responsible state agency or the regional office of the federal Environmental Protection Agency (EPA) should be contacted for specific recommendations.

RISK ASSESSMENT: BUSINESS

General Assessment

Accidents or mishaps involving camphor can present a significant threat to business operations. The loss or damage of equipment or facilities can significantly affect fiscal viability. Lawsuits that may result from personnel injury, public exposures, and/or environmental contamination will also require a serious expenditure of resources. Media attention surrounding an injury, illness, or environmental damage can also result in a loss of profits and loss of current as well as future business.

Recommended Risk-Reduction Measures

Company attorneys, safety and health professionals, and environmental specialists should be involved in the development of any procedures that deal with or address safety and health issues. A company official should be pre-designated as a public relations officer with specific training in dealing with the press. Corporate plans and policies should be developed, approved, and implemented long before any need for such arises.

REFERENCES

American Conference of Governmental Industrial Hygienists. 1988. *Documentation of the Threshold Limit Values and Biological Exposure Indices,* 5th Edition (with updates). Cincinnati: ACGIH

Anthony, C. P., and N. J. Kolthoff. 1971. *Textbook of Anatomy and Physiology.* St. Louis: C. V. Mosby Company.

Asante-Duah, D. K. 1993. *Hazardous Waste Risk Assessment.* New York: CRC Press/Lewis Publishers.

Bloom, A. D. and F. J. DeSerres. 1995. *Ecotoxicity and Human Health.* New York: CRC Press/Lewis Publishers.

Brusick, D. J. 1994. *Methods for Genetic Risk Assessment.* New York: CRC Press/Lewis Publishers.

Bueche, J. R. 1972. *Principles of Physics,* 2nd Edition. New York: McGraw-Hill.

Burrell, R., D. K. Flaherty, and L. J. Sauers. 1992. *Toxicology of the Immune System.* New York: Van Nostrand Reinhold.

Calabrese, E. J. 1994. *Biological Effects of Low Level Exposures.* New York: CRC Press/Lewis Publishers.

Cockerman, L. G., B. S. Shane. 1994. *Basic Environmental Toxicology.* New York: CRC Press/Lewis Publishers.

Cothern, C. R. 1995. *Handbook for Environmental Risk Decisionmaking.* New York: CRC Press/Lewis Publishers.

Gardner, R. W. 1994. *Chemical Intolerance.* New York: CRC Press/Lewis Publishers.

Hodgson, E., R. B. Mailman, and J. E. Chambers. 1988. *Dictionary of Toxicology.* London: The MacMillian Company.

Hosty, J. W., and P. Foster. 1990. *A Practical Guide to Chemical Spill Response.* New York: Van Nostrand Reinhold.

Howard, P. H., et. al. 1991. *Handbook of Environmental Degradation Rates.* New York: CRC Press/Lewis Publishers.

Jones, F. E. 1994. *Toxic Organic Vapors in the Workplace.* New York: CRC Press/Lewis Publishers.

Kamrin, M. A. 1988. *Toxicology: A Primer on Principles and Applications.* Chelsea, Michigan: Lewis Publishers.

Keith, L. H. and M. M. Walker. 1995. *Handbook of Air Toxics.* New York: CRC Press/Lewis Publishers.

Landis, W. G., M. H. Yu. 1995. *Introduction to Environmental Toxicology.* New York: CRC Press/Lewis Publishers.

Lauwerys, R. R., P. Hoet. 1993. *Industrial Chemical Exposure.* New York: CRC Press/Lewis Publishers.

Lewis, R. J., Sr. 1993. *Hawley's Condensed Chemical Dictionary,* 12th Edition. New York: Van Nostrand Reinhold.

Lewis, R. J., Sr. 1992. *Sax's Dangerous Properties of Industrial Materials,* Volumes 1, 2, and 3, 8th Edition. New York: Van Nostrand Reinhold.

Manahan, S. E. 1992. *Toxicological Chemistry,* 2nd Edition. New York: CRC Press/Lewis Publishers.

Mellström, G. A., J. E. Wahlberg, and H. I. Maibach. 1994. *Protective Gloves for Occupational Use.* New York: CRC Press/Lewis Publishers.

Meyer, E. 1990. *Chemistry of Hazardous Materials.* Englewood Cliffs, NJ: Prentice Hall.

Patnaik, P. 1992. *A Comprehensive Guide to the Hazardous Properties of Chemical Substances.* New York: Van Nostrand Reinhold.

Proctor, N. H., and J. P. Hughes. 1991. *Chemical Hazards of the Workplace.* New York: Van Nostrand Reinhold.

Quigley, D. R. 1994. *Handbook of Emergency Chemical Management.* New York: CRC Press/Lewis Publishers.

Rea, W. J. 1995. *Chemical Sensitivity,* Vol. I-III. New York: CRC Press/Lewis Publishers.

Richardson, M. 1995. *Dictionary of Substance and Their Effects,* Vol. I-VI. New York: CRC Press/Lewis Publishers.

U.S. Department of Health and Human Services, National Institute for Occupational Safety and Health. 1994. *NIOSH Pocket Guide to Chemical Hazards.* Washington, D.C.: U.S. Government Printing Office.

U.S. Department of Health and Human Services, National Institute for Occupational Safety and Health. 1978. *Criteria for a Recommended Standard - Occupational Exposure to Ketones.* DHEW Publication Number: 78-173. Washington, D.C.: U.S. Government Printing Office.

MATERIAL SAFETY DATA SHEET

CHEMICAL NAME	
CAPROLACTAM	**CRC PRESS/LEWIS PUBLISHERS DISCLAIMER:** THE INFORMATION AND RECOMMENDATIONS PRESENTED HEREIN ARE BASED ON SOURCES BELIEVED TO BE RELIABLE. CRC MAKES NO REPRESENTATION ON ITS COMPLETENESS OR ACCURACY. IT IS THE RESPONSIBILITY OF THE USER TO DETERMINE THE CHEMICAL'S SUITABILITY FOR ITS INTENDED USE, THE CHEMICAL'S SAFE USE, AND THE CHEMICAL'S PROPER DISPOSAL. NO REPRESENTATIONS AND/OR WARRANTIES, EITHER EXPRESSED OR IMPLIED, OF THE MERCHANTABILITY OR FITNESS FOR A PARTICULAR PURPOSE, OR OF ANY OTHER NATURE, ARE MADE WITH RESPECT TO THE INFORMATION PROVIDED IN THIS MSDS OR TO THE CHEMICAL TO WHICH INFORMATION MAY REFER. CRC NEITHER ASSUMES NOR AUTHORIZES ANY OTHER PERSON TO ASSUME FOR IT, ANY OTHER ADDITIONAL RESPONSIBILITY OR LIABILITY FOR THE USE OF, OR RELIANCE UPON, THIS INFORMATION.

HAZARD WARNING INFORMATION

HEALTH	FIRE	REACTIVE	OTHER	DEGREE OF HAZARD	COLOR CODING	OTHER CODES
2	1	0		0 = Minimum Hazard 1 = Slight Hazard 2 = Moderate Hazard 3 = Serious Hazard 4 = Severe Hazard	HEALTH = BLUE FIRE = RED REACTIVITY = YELLOW OTHER = WHITE	OX = Oxidizer ACID = Acid ALK = Alkali COR = Corrosive W = Use No Water

SECTION I - GENERAL INFORMATION

Characterization	RCRA Number	EPA Class
Aliphatic Amine	**None**	**Not Applicable**

DOT Proper Shipping Name **Not Listed**	Chemical Abstract Service (CAS) Number **105-60-2**
DOT Hazard Class and Label Requirements **Not Listed**	DOT Emergency Guide Code **No Citation**
DOT Identification Number **Not Listed**	Chemical Formula **$C_6H_{11}NO$**

Synonyms

Aminocaproic lactam; 2-oxohexamethyleneimine; alkamid; 6-aminohexanoic acid cyclic lactam; 2-ketohexamethylenimine; Bonamid; caprolactam monomer; Capron; Chemlon; Danamid.

SECTION II - HAZARDOUS INGREDIENTS AND IDENTITY INFORMATION

Hazardous Components (specific identity)	OSHA Exposure Criteria	NIOSH Exposure Criteria	Immediately Dangerous to Life and Health (IDLH)	ACGIH Exposure Criteria
Caprolactam (derivation: By catalytic oxidation of cyclohexane to cyclohexanol, reacting with peracetic acid to form caprolactone, and further reaction with ammonia; catalytic hydrogenation of phenol to cyclohexanone, reaction with ammonia to cyclohexanone oxime with Beckmann rearrangement with sulfuric acid catalyst). **1 ppm = 4.70 mg/m³**	**PEL (vapor):** **5 ppm** **20 mg/m³** **STEL (vapor):** **10 ppm** **40 mg/m³**	**REL (dust):** **1 ppm** **STEL (dust):** **3 mg/m³** **REL (vapor):** **0.22 ppm** **STEL (vapor):** **3 mg/m³**	**Not Determined**	**TLV (dust):** **1 ppm** **STEL (dust):** **3 mg/m³** **TLV (vapor):** **5 ppm** **STEL (vapor):** **46 mg/m³**

SECTION III - PHYSICAL AND CHEMICAL CHARACTERISTICS

Boiling Point **515°F (268°C)**	Specific Gravity (H_2O = 1) **1.01**
Vapor Pressure (mm Hg) **0.1 at 68°F (20°C)**	Molecular Weight **113.2**
Vapor Density (Air = 1) **Not Found**	Melting Point **156°F (69°C)**

Solubility

Soluble in water, benzene, chloroform, dimethyl formamide, cyclohexene, ether, ethanol, methanol, petroleum ether, tetrahydrofurfuryl alcohol, and most chlorinated hydrocarbons.

Appearance and Odor

White crystalline solid or flakes with an unpleasant odor. Odor Threshold = <1 ppm.

SECTION IV - FIRE AND EXPLOSION HAZARD DATA

Flash Point (method used) **257°F (125°C) closed cup**	Explosive Limits in Air % by Volume **LEL: 1.4% UEL: 8.0%**
NFPA Classification **Class IIIB Combustible Solid**	Autoignition Temperature **None Reported**

Extinguishing Media

Use dry chemical, carbon dioxide, foam, or water spray.

Special Fire Fighting Procedures

Use full protective clothing and self-contained breathing apparatus (SCBA). Fight fire from distance if possible. Move containers from fire area if it can be done without risk.

Unusual Fire and Explosion Hazards

Containers may explode in fire due to buildup of excessive internal pressures. Aqueous solutions are flammable unless diluted extensively. Cool exposed container with water.

SECTION V - REACTIVITY DATA

Stability		Conditions to Avoid Caprolactam is stable at room temperature under normal conditions of temperature and pressure. Avoid contact with incompatible materials. Keep away from heat.
Stable	Unstable	Incompatibility *(materials to avoid)* Reacts explosively with acetic acid + dinitrogen trioxide. It is hydrolyzed by strong mineral acids to form amino acids.
X		
Hazardous Polymerization		Conditions to Avoid Under normal conditions of temperature and pressure, hazardous polymerization of caprolactam is not expected to occur.
May Occur	Will Not Occur	Hazardous Decomposition or By-products Toxic and irritating fumes and vapors are produced when caprolactam is heated to decomposition. These include oxides of carbon and nitrogen.
	X	

SECTION VI - HEALTH HAZARD DATA

Primary Route(s) of Entry:	Inhalation?	Absorption (skin)?	Ingestion?
	X		X

Health Hazards

INHALATION: Symptoms of respiratory irritation with cough and delayed pulmonary edema (fluid in the lungs). Sensitization may occur. CNS effects include hypotension, fever, nausea, epigastric pain, and leukocytosis.

EYES & SKIN: Skin irritation with possible burns. Eye irritation and conjunctivitis may occur.

INGESTION: Gastrointestinal irritation, burns, and possible seizures.

Carcinogenicity	NTP Listed?	IARC Cancer Review Group?	OSHA Regulated?	Target Organs?
Unknown Human Unknown Animal	No	Group 4	29 CFR 1910.1000 Table Z-1	Eyes, skin, resp. sys., kidneys, liver, CNS, CVS.

Medical Conditions Generally Aggravated by Exposure

Liver, kidney, skin, and pulmonary diseases may be aggravated by exposure.

Emergency and First-aid Procedures

Eye contact: Flush immediately with water for 15 minutes (minimum), seek medical attention. Skin contact: Remove all contaminated clothing. Immediately wash area with large amounts of soap and water. Seek medical assistance. For inhalation: Remove the person from exposure. Provide respiratory assistance and CPR. Transfer to medical facility. Observe for 24-48 hours for lung effects. If swallowed: Call the poison control center and seek medical attention immediately. Unless advised otherwise, give 1-2 glasses of water to dilute. Do NOT induce vomiting. Never attempt to give an unconscious or convulsing person anything by mouth.

SECTION VII - PRECAUTIONS FOR SAFE HANDLING AND USE

Steps to be Taken in Case Material is Released or Spilled

Collect materials using vacuum equipped with HEPA filter and deposit in DOT-approved drum. Do NOT dry sweep (generates dusts). Ventilate area and remove ignition sources. Restrict those not involved in cleanup from entering area. Notify appropriate authorities, if applicable.

Preferred Waste Disposal Method

Burn in a chemical incinerator equipped with an afterburner and scrubber.

Precautions to be Taken in Handling and Storage

Do not store caprolactam in the presence of incompatible chemicals or materials. Store in tightly closed containers in a cool, well-ventilated area. Caprolactam is hygroscopic (absorbs moisture). Protect containers from physical damage and provide shade from radiant heat (including direct sunlight).

Other Precautions and Warnings

Electrically ground and bond equipment, especially during transfer operations. Detached facility storage is recommended.

SECTION VIII - CONTROL MEASURES AND PERSONAL PROTECTIVE EQUIPMENT

Respiratory Protection *(specify type)*

For low exposures, use an organic vapor respirator. Otherwise, use an MSHA/NIOSH approved supplied-air respirator with full facepiece operated in positive pressure mode or a self-contained breathing apparatus (SCBA) with full facepiece operated in pressure demand or other positive pressure mode.

Ventilation

Local exhaust or general mechanical systems recommended.

Protective Gloves	Eye Protection	Other Protective Clothing
Impervious Rubber Gloves	Chemical Goggles or Face Mask	Protective Apron

Work/Hygiene Practices

Always wash hands thoroughly after using chemical; never bring food, drink, or smoking materials into vicinity of chemicals.

CAPROLACTAM

$C_6H_{11}NO$ CAS: 105-60-2

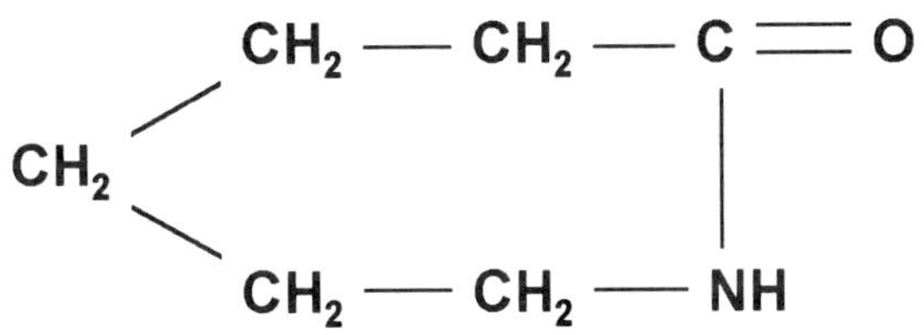

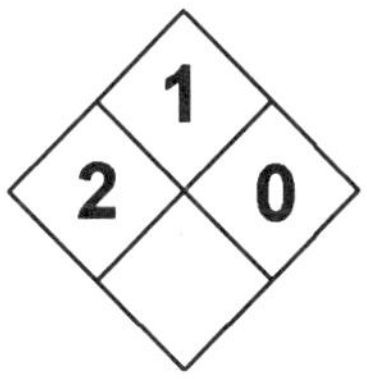

IDENTIFICATION AND TYPICAL USES

Caprolactam is a white, crystalline or flaked solid with an unpleasant odor. It is used to manufacture synthetic fibers (especially nylon 6), plastics, bristles, film, coatings, synthetic leather, plasticizers and paint vehicles, as a cross-linking agent for polyurethane, and in the synthesis of amino acid lysine.

RISK ASSESSMENT: HEALTH

General Assessment

Caprolactam is moderately toxic to humans by *ingestion* and *inhalation*. Skin contact also presents a mild to moderate health hazard risk. Its toxic effects appear to focus primarily on the respiratory system and the central nervous system (CNS). There are no reports to indicate that exposure to caprolactam poses a carcinogenic threat. However, mutation data have been reported and additional research is warranted before any definite determinations can be made with regard to cancer risk. Experimental teratogenic data have also been reported.

Inhalation of caprolactam dusts or vapors (from the solution) can cause immediate irritation to the eyes, nose, throat, and respiratory system with a pronounced and irritating cough and asthma. Respiratory sensitization may occur in some individuals. The development of pulmonary edema (fluid in the lungs) may be delayed up to 48 hours, thereby creating a false sense of security with regard to health exposure risk. Inhalation causes CNS depression with symptoms of headache, hypotension, fever, and nausea. Epigastric pain and possible leukocytosis may also develop.

Skin contact results in localized irritation and possible burns if allowed to remain in contact for prolonged periods. Eye contact causes intense irritation with possible progression to conjunctivitis. If ingested, caprolactam can cause irritation and burns to the mouth, throat (esophageal burns) and stomach. Seizures may occur in some individuals depending upon susceptibility, sensitivity, and the conditions of the ingested exposure (quantity and duration).

Acute Health Effects

The following acute (short-term) health effects may occur immediately or shortly after exposure to caprolactam:

Skin: Irritation, inflammation, redness, possible blistering, burns, and tissue damage.

Eye: Severe irritation, inflammation, with possible development of conjunctivitis.

Lung: Irritation causing coughing and/or shortness of breath. Higher exposures can lead to respiratory problems, including pulmonary edema (fluid buildup in lungs) which is a medical emergency and can be fatal.

CNS: Serious depression of the central nervous system causing fever, nausea, headache, hypotension, and other unspecified effects.

Chronic Health Effects

The following chronic (long-term) health effects may occur at some time after exposure to caprolactam and can last for months or even years:

Cancer Hazards: According to the information presented in the references, caprolactam has not been adequately tested for its ability to cause cancer in test animals. Human mutation (genetic changes) data have also been reported.

Reproductive Hazard: According to the information presented in the references, caprolactam may cause reproductive toxicity in test animals. Human data have not been reported in this regard.

Other Chronic Effects: Repeated or long-term exposure may cause sleeping, loss of appetite, chronic headache, chest pain or discomfort, mild hypotension, chronic fatigue, lymphocytosis (excessive white blood cells), seizures, nose bleeds, brittle nails, loss of sense of touch in fingerprints, flatulence, and a foul taste in the mouth. Repeated skin contact may cause chronic dermatitis and eczema. There may be a dermal reaction similar to sunburn resulting from hypersensitivity in some people. Animal studies have shown liver and kidney injury.

Ʊ *Recommended Risk-Reduction Measures*

Caprolactam is a moderately toxic chemical. Unless a less toxic chemical can be substituted for a hazardous substance, *engineering controls* are the most effective method of reducing exposures. The best protection is to enclose operations and/or provide local exhaust ventilation at the site of chemical release. To prevent static sparks, all containers and equipment should be electrically grounded and bonded. While not always operationally feasible, isolating operations can also reduce exposure.

Using respiratory protection is less effective than the controls mentioned above, but is still advisable whenever working with or around caprolactam. For relatively low, infrequent, or transient exposures, an organic vapor respirator may suffice. For higher or prolonged exposure, a supplied-air respirator or a self-contained breathing apparatus (SCBA) with full facepiece and operated in pressure demand or other continuous flow mode are the recommended respiratory protection methods of choice. If a full facepiece is not available, then chemical goggles should be worn to protect the eyes. Whenever a chemical splash or dust hazard exists, a face shield and apron should be worn. To prevent hand and skin exposures, impervious chemical gloves should be worn.

Administrative controls should also be in place to minimize the potential for human exposures. These may include written procedures or policies which specify the methods and techniques that will be practiced whenever personnel are to work with caprolactam.

All personnel should receive *training* on the use, hazards, protective measures, emergency actions, and other precautions per 29 CFR 1910.1200 (Hazard Communication), prior to the first assignment in an area where caprolactam is used or stored. Before beginning employment and at regular intervals thereafter (e.g. annually), the following tests are recommended:

- ☑ Lung function tests (establish a baseline).
- ☑ Liver and kidney function tests.

If symptoms develop or overexposure is suspected, the following medical tests may be useful:

- ☑ Lung function tests.
- ☑ Consider chest X-ray following acute overexposure (may be negative if taken immediately after exposure due to delayed onset of pulmonary edema).
- ☑ Complete blood count and blood gases (check BUN levels).
- ☑ For oral exposures, evaluate for esophageal burns.
- ☑ Evaluation by a qualified allergist, including careful consideration of exposure history and special testing (may help diagnose skin allergy).

Any evaluation should include a careful medical history of past and present symptoms with an examination. However, medical tests that evaluate existing damage are not a substitute for controlling exposure. Also, since smoking can cause heart diseases, emphysema, and numerous other respiratory disorders, smokers may be affected more rapidly than non-smokers to the same exposure conditions. Prudent risk assessment and management requires careful consideration of *all* possible risk factors that may be causing the appearance of symptoms in exposed workers.

Other methods to reduce exposure include:

- ☑ Where possible, enclose operations and use local exhaust ventilation at the site of chemical release. If local exhaust ventilation or enclosure is not used, respiratory protection should be mandatory. Also, if possible, automatically transfer liquids from storage containers to process containers using non-sparking tools.
- ☑ Always ensure that proper protective clothing is worn when using chemical substances and that personnel are adequately trained in its use, care, and maintenance.
- ☑ Wash thoroughly immediately after exposure to caprolactam and at the end of the work shift or before eating, drinking, or smoking.
- ☑ Hazard warning information should be posted in the work area. In addition, as part of an on-going education and training program, all information on the health and safety hazards of caprolactam should be communicated to all exposed and potentially exposed workers.

☑ Eye wash stations should be provided in work areas. If the potential for whole body exposure exists, then safety showers should also be provided.

RISK ASSESSMENT: ENVIRONMENT

General Assessment

The environment is at risk of exposure during intended use, transportation, storage, disposal, or destruction of caprolactam. In almost every scenario, the threat of environmental exposure is contingent upon the proper handling of the chemical substance. Accidental spills, large or small, can result in fire, explosion, and possible contamination of the surrounding environmental mediums (water, soil, and air).

Caprolactam is considered a class IIIB combustible solid (per OSHA 29 CFR 1910.106). It can react with some oxidizers (dinitrogen trioxide) with acetic acid to cause a fire or explosion. These characteristics require special consideration during any emergency situation involving a leak or spill of caprolactam. Should caprolactam ever come into contact with incompatible substances or heat either during use, transportation, or storage, violent reactions can occur.

Caprolactam can enter the environment through industrial discharges, municipal waste treatment discharges, and spills.

Acute Ecological Effects

Acute (short-term) toxic effects may include the death of animals, birds, or fish and death or low growth rate in plants. Acute effects are seen 2 to 4 days after animals or plants are exposed to caprolactam.

This chemical has moderate acute toxicity to aquatic life. Insufficient data are available to evaluate or predict the short-term effects of caprolactam to plant, birds, or land animals.

Chronic Ecological Effects

Chronic toxic effects may include shortened life span, reproductive problems, lower fertility, and changes in appearance or behavior in exposed animals. These effects can be seen long after first exposure(s) to toxic chemicals.

Caprolactam has moderate chronic toxicity to aquatic life. Insufficient data are available to evaluate or predict the long-term effects of caprolactam to plants, birds, or land animals.

Water Solubility

Caprolactam is highly soluble in water. Concentrations of 1000 milligrams and more will mix with a liter of water.

Persistence in the Environment

Caprolactam is slightly persistent in water, with a half-life of between 2 to 20 days. The half-life of a pollutant is the amount of time it takes for one half of the chemical to be degraded. It is not expected to accumulate in aquatic sediments or terrestrial soils.

Bioaccumulation in Aquatic Organisms

Some substances increase in concentration, or *bioaccumulate,* in living organisms as they breathe contaminated air, drink contaminated water, or eat contaminated food. These chemicals can become concentrated in the tissues and internal organs of animals as well as humans.

Bioaccumulation of caprolactam in aquatic tissues is not expected to occur.

Recommended Risk-Reduction Measures

Proper training of all transporters will reduce the likelihood of a mishap or accident resulting in a leak or spill to the environment. The correct use of DOT labeling (per the MSDS) on all containers, trucks, and rail cars will enable emergency responders to react properly and quickly to any disaster thereby reducing the potential risk to the environment and to personnel.

Storage of caprolactam should be segregated from incompatible chemicals to minimize the risk of cross-contamination or contact. Storage buildings should be equipped with the proper fire protection equipment (alarms, sprinklers, extinguishers, etc.). Static sparks should be prevented by electrically grounding and bonding all containers, especially during transfer operations.

If a spill or leak to the environment has occurred, fire department, emergency response and/or hazardous materials spill personnel should be notified immediately. Cleanup should be attempted only by those trained in proper spill containment procedures. Contaminated soils should be removed for incineration and replaced with clean soil. If caprolactam should contact the water table, aquifer, or navigable waterway, remediation activities should be prompt. When such spills occur, the local and/or state emergency response authorities must be notified. A comprehensive emergency response or disaster preparedness/recovery plan should be in place prior to any operations involving the use, transportation, storage, or disposal of caprolactam. If caprolactam is spilled or leaked, the following specific steps are recommended:

☑ Restrict persons not wearing protective clothing from area of spill or leak until cleanup is complete and area can be opened for normal work.

☑ Ventilate area and remove all ignition sources.

☑ Use water spray to disperse vapors and dilute spills to non-flammable mixtures. Flush spills away from exposures. Dispose of accordingly.

☑ It may be necessary to dispose of caprolactam as a hazardous waste. The responsible state agency or the regional office of the federal Environmental Protection Agency (EPA) should be contacted for specific recommendations.

RISK ASSESSMENT: BUSINESS

General Assessment

Accidents or mishaps involving caprolactam can present a significant threat to business operations. The loss or damage of equipment or facilities can significantly affect fiscal viability. Lawsuits that may result from personnel injury/death, public exposures, and/or environmental contamination will also require a serious expenditure of resources. Media attention surrounding an injury, death, or environmental damage can also result in a loss of profits and loss of current as well as future business.

Recommended Risk-Reduction Measures

Company attorneys, safety and health professionals, and environmental specialists should be involved in the development of any procedures dealing with safety or emergency response. A company official should be pre-designated as a public relations officer with specific training in dealing with the press. Corporate plans and policies should be developed, approved, and implemented long before any need for such arises.

REFERENCES

Anthony, C. P., and N. J. Kolthoff. 1971. *Textbook of Anatomy and Physiology.* St. Louis: C. V. Mosby Company.

Bueche, J. R. 1972. *Principles of Physics,* 2nd Edition. New York: McGraw-Hill.

Calabrese, E. J., E. M. Kenyon. *Air Toxics and Risk Assessment.* New York: CRC Press/Lewis Publishers.

Calabrese, E. J. 1994. *Biological Effects of Low Level Exposures.* New York: CRC Press/Lewis Publishers.

Cockerman, L. G., B. S. Shane. 1994. *Basic Environmental Toxicology.* New York: CRC Press/Lewis Publishers.

Cothern, C. R., N. P. Ross. 1994. *Environmental Statistics, Assessment, and Forecasting.* New York: CRC Press/Lewis Publishers.

Cothern, C. R. 1995. *Handbook for Environmental Risk Decisionmaking.* New York: CRC Press/Lewis Publishers.

Danse, I. R. 1991. *Common Sense Toxics in the Workplace.* New York: Van Nostrand Reinhold.

Environmental Protection Agency. 1983. *Chemical Hazard Information Profiles.* Washington, D.C.: U.S. EPA.

Fodden, C. R., and J. L. Weddell. 1992. *Hazardous Materials Emergency Action Data.* New York: CRC Press/Lewis Publishers.

Forseberg, K., and L. H. Keith. 1993. *Chemical; Protective Clothing Permeation and Degradation Database.* Chelsea, MI: Lewis Publishers.

Gardner, R. W. 1994. *Chemical Intolerance.* New York: CRC Press/Lewis Publishers.

Gerarde, H. W. 1960. *Toxicology and Biochemistry of Aromatic Hydrocarbons.* New York: Elsevier.

Hallenbeck, W. H. 1993. *Quantitative Risk Assessment for Environmental and Occupational Health,* 2nd Edition. New York: CRC Press/Lewis Publishers.

Hodgson, E., R. B. Mailman, and Janice E. Chambers. 1988. *Dictionary of Toxicology.* London: The MacMillian Company.

Howard, P. H., et. al. 1991. *Handbook of Environmental Degradation Rates.* New York: CRC Press/Lewis Publishers.

Hosty, J. W., and P. Foster. 1990. *A Practical Guide to Chemical Spill Response.* New York: Van Nostrand Reinhold.

International Agency for Research on Cancer (IARC). 1987. *IARC Monographs on the Evaluation of Carcinogenic Risk to Humans, Overall Evaluations of Carcinogenicity: An updating of IARC Monographs Volumes 1-42.* Lyon: IARC.

Lewis, R. J., Sr. 1993. *Hawley's Condensed Chemical Dictionary,* 12th Edition. New York: Van Nostrand Reinhold.

Lewis, R. J., Sr. 1992. *Sax's Dangerous Properties of Industrial Materials,* Volumes 1, 2, and 3, 8th Edition. New York: Van Nostrand Reinhold.

Meyer, E. 1990. *Chemistry of Hazardous Materials.* Englewood Cliffs, NJ: Prentice Hall.

Richardson, M. 1995. *Dictionary of Substance and Their Effects,* Vol. I-VI. New York: CRC Press/Lewis Publishers.

Sacarello, H. L. A. 1994. *The Comprehensive Handbook of Hazardous Materials.* New York: CRC Press/Lewis Publishers.

U.S. Department of Health and Human Services, National Institute for Occupational Safety and Health. 1994. *NIOSH Pocket Guide to Chemical Hazards.* Washington, D.C.: U.S. Government Printing Office.

MATERIAL SAFETY DATA SHEET

CHEMICAL NAME	
CAPTAN	**CRC PRESS/LEWIS PUBLISHERS DISCLAIMER:** THE INFORMATION AND RECOMMENDATIONS PRESENTED HEREIN ARE BASED ON SOURCES BELIEVED TO BE RELIABLE. CRC MAKES NO REPRESENTATION ON ITS COMPLETENESS OR ACCURACY. IT IS THE RESPONSIBILITY OF THE USER TO DETERMINE THE CHEMICAL'S SUITABILITY FOR ITS INTENDED USE, THE CHEMICAL'S SAFE USE, AND THE CHEMICAL'S PROPER DISPOSAL. NO REPRESENTATIONS AND/OR WARRANTIES, EITHER EXPRESSED OR IMPLIED, OF THE MERCHANTABILITY OR FITNESS FOR A PARTICULAR PURPOSE, OR OF ANY OTHER NATURE, ARE MADE WITH RESPECT TO THE INFORMATION PROVIDED IN THIS MSDS OR TO THE CHEMICAL TO WHICH INFORMATION MAY REFER. CRC NEITHER ASSUMES NOR AUTHORIZES ANY OTHER PERSON TO ASSUME FOR IT, ANY OTHER ADDITIONAL RESPONSIBILITY OR LIABILITY FOR THE USE OF, OR RELIANCE UPON, THIS INFORMATION.

HAZARD WARNING INFORMATION

HEALTH	FIRE	REACTIVE	OTHER	DEGREE OF HAZARD	COLOR CODING	OTHER CODES
3	0	1	~~W~~	0 = Minimum Hazard 1 = Slight Hazard 2 = Moderate Hazard 3 = Serious Hazard 4 = Severe Hazard	HEALTH = BLUE FIRE = RED REACTIVITY = YELLOW OTHER = WHITE	OX = Oxidizer ACID = Acid ALK = Alkali COR = Corrosive ~~W~~ = Use No Water

SECTION I - GENERAL INFORMATION

Characterization	RCRA Number	EPA Class
Pesticide	**Not Listed**	**Not Applicable**

DOT Proper Shipping Name **Captan**	Chemical Abstract Service (CAS) Number **133-06-2**
DOT Hazard Class and Label Requirements **Poison B**	DOT Emergency Guide Code **31**
DOT Identification Number **UN 9099**	Chemical Formula **$C_9H_8Cl_2NO_2S$**

Synonyms
N-trichloromethylmercapto-tetrahydrophthalimide; 1-2-dicarboximide; orthocide; vanicide.

SECTION II - HAZARDOUS INGREDIENTS AND IDENTITY INFORMATION

Hazardous Components (specific identity)	OSHA Exposure Criteria	NIOSH Exposure Criteria	Immediately Dangerous to Life and Health (IDLH)	ACGIH Exposure Criteria
Captan (derivation: Reaction product of tetrahydrophthaladimide and trichloromethyl-mercaptan).	**PEL:** Not Established **STEL:** Not Established	**REL:** 5 mg/m3 Possible Cancer Agent	Not Determined	**TLV:** 5 mg/m^3 **STEL:** Not Established

SECTION III - PHYSICAL AND CHEMICAL CHARACTERISTICS

Boiling Point **Decomposes**	Specific Gravity (H_2O = 1) **1.74**
Vapor Pressure (mm Hg) **0 (approx.) at 69°F (20°C)**	Molecular Weight (atomic weight) **300.6**
Vapor Density (Air = 1) **1.5**	Melting Point **352°F (178°C)**

Solubility
Practically insoluble in water (0.0003% at 77°F). Partially soluble in acetone, benzene, and toluene. Slightly soluble in ethylene dichloride and chloroform.

Appearance and Odor
Pure captan is a white to cream colored crystalline powder. It usually appears an odorless, yellow colored powder.

SECTION IV - FIRE AND EXPLOSION HAZARD DATA

Flash Point (method used) **Not Determined**	Explosive Limits in Air % by Volume **LEL: N/A UEL: N/A**
NFPA Classification **Combustible Solid**	Autoignition Temperature **Not Found**

Extinguishing Media
Do NOT use water (may cause formation of poisonous gases). Use carbon dioxide or foam.

Special Fire Fighting Procedures
Wear full protective clothing and self-contained breathing apparatus (SCBA). Structural fire-fighting equipment is permeable and may not provide sufficient protection. Evacuate personnel 2500 feet from fire area. Move containers from fire area if it can be done without risk. Cool exposed containers.

Unusual Fire and Explosion Hazards
Captan is a combustible solid. It may be dissolved in some flammable liquids, such as benzene.

SECTION V - REACTIVITY DATA

Stability		Conditions to Avoid
		Avoid contact with incompatible materials, primarily water and strong alkaline materials (e.g., hydrated lime).
Stable	Unstable	Incompatibility *(materials to avoid)*
X		Reacts on contact with water to give off poisonous gases. Corrosive to many metals. Keep away from heat, oil, strong alkalis, and hydrogen sulfide.
Hazardous Polymerization		Conditions to Avoid
		Do not allow captan to become wet or moist.
May Occur	Will Not Occur	Hazardous Decomposition or By-products
	X	At high temperatures, captan decomposes and may produce poisonous gases, including oxides of sulfur and nitrogen, hydrogen chloride, and deadly phosgene.

SECTION VI - HEALTH HAZARD DATA

Primary Route(s) of Entry:	Inhalation?	Absorption (skin/eye)?	Ingestion?
	X	X	X

Health Hazards

INHALATION: Primary irritant to the eyes and upper respiratory tract. May cause blurred vision, dyspnea, diarrhea, vomiting.

ABSORPTION: Will pass through unbroken skin and enter the bloodstream. Contact may severely irritate the skin and eyes. Prolonged contact can cause allergies (sensitization).

INGESTION: Moderately toxic to humans resulting in toxic systemic effects. May also cause severe gastroenteritis.

Carcinogenicity	NTP Listed?	IARC Cancer Review Group?	OSHA Regulated?	Target Organs?
Suspected Human Suspected Animal	5th Annual Report	Group 3	No	Resp. system, skin, eyes, GI tract, liver, kidneys.

Medical Conditions Generally Aggravated by Exposure

None reported. However, persons with existing lung dysfunction may experience effects quicker.

Emergency and First-aid Procedures

Eye contact: Do not allow victim to keep eyes tightly shut. Flush immediately with water for 15 minutes (minimum), seek medical attention. Skin contact: Remove all contaminated clothing. Immediately wash area with flooding amounts of soap and water. For inhalation: Remove the person from exposure. Maintain an open airway. Provide respiratory assistance and CPR. Transfer to medical facility. If swallowed: If victim is conscious, induce vomiting. Save some liquid and vomitus for laboratory examination. Seek medical attention immediately. Call poison control center for advice. Never give anything by mouth to an unconscious or convulsing person. Do not make an unconscious person vomit.

SECTION VII - PRECAUTIONS FOR SAFE HANDLING AND USE

Steps to be Taken in Case Material is Released or Spilled

Collect powdered materials in safest manner possible and deposit in sealed containers. Remove all sources of ignition. Ventilate area of spill. Use HEPA vacuum. Restrict those not involved in cleanup from entering area. Notify appropriate authorities. Absorb any liquids containing captan in vermiculite, dry sand, earth, or similar material and deposit in sealed drum for disposal or reclamation.

Preferred Waste Disposal Method

No Citation.

Precautions to be Taken in Handling and Storage

Store in tightly closed containers free from moisture. Keep in cool, well ventilated area.

Other Precautions and Warnings

Do not store near water, heat, oil, strong alkalis, and hydrogen sulfide.

SECTION VIII - CONTROL MEASURES AND PERSONAL PROTECTIVE EQUIPMENT

Respiratory Protection *(specify type)*

For exposures above 5.0 mg/m^3, use a respirator with a pesticide cartridge. For high exposures, use a self-contained breathing apparatus with full facepiece operated in positive pressure mode.

Ventilation

Local exhaust (preferred) at site of chemical work or general ventilation.

Protective Gloves	Eye Protection	Other Protective Clothing
Polycarbonate or Butyl Material	Chemical/Dust Goggles or Face Mask	Protective Apron, Boots (avoid skin contact)

Work/Hygiene Practices

Always wash hands thoroughly after using chemical; never bring food, drink, or smoking materials into vicinity of chemicals.

CAPTAN

$C_9H_8Cl_3NO_2S$ CAS: 133-06-2

IDENTIFICATION AND TYPICAL USES

Captan is a white to cream colored crystalline solid in its pure form. However, it normally appears as a yellow, odorless powder. It is used in seed treatments and on food crops, as a fungicide (kills molds and mildew) in paints, plastics, leather, and some fabrics. Also used in preserving fruits and as a bacterial retardant in soaps.

RISK ASSESSMENT: HEALTH

General Assessment

Captan is a poison by *ingestion, inhalation,* and skin contact (*absorption*). It is moderately toxic to humans through ingestion and inhalation. It is a questionable carcinogen in animals with some experimental teratogenic and reproductive effects reported. Human mutation, tumorigenic, and neoplastigenic data also reported.

Breathing its fine dusts can irritate the respiratory system, especially the nose and throat. It can cause blurred vision with other exposure symptoms such as diarrhea, vomiting, and dyspnea. The liver and kidneys may also be affected.

Contact with skin and eyes can be severely irritating. Prolonged contact can cause an allergic reaction (sensitization). Any exposures after that will result in a return of allergic symptoms (rash, irritation). Symptoms of toxic systemic exposure (dizziness, vomiting, nausea) can also occur as a result of captan passing through unbroken skin and directly entering the bloodstream.

Acute Health Effects

The following acute (short-term) health effects may occur immediately or shortly after exposure to captan:

Skin: Possible irritation allergic reaction. Absorption is likely to occur.

Eye: Severe irritation and possible damage on contact.

Lung: Irritation of the nose and throat.

Chronic Health Effects

The following chronic (long-term) health effects may occur at some time after exposure to captan and can last for months or even years:

Cancer Hazards: Captan is a mutagen (causes genetic changes). Such chemicals may have a cancer risk and there is limited evidence that captan does cause cancer in animals. Human carcinogenicity is suspected.

Reproductive Hazard: There is some evidence that exposure to captan cause damage to the developing fetus in test animals. Human teratogenicity is suspected.

Other Chronic Effects: Captan causes skin allergies. If allergy develops, very low future exposures can cause itching and skin rash.

Recommended Risk-Reduction Measures

Personnel should avoid direct contact with captan dusts or powders. If a less toxic material or compound cannot be substituted for captan, then *engineering controls* are the most effective method of reducing exposures. The best protection is to enclose operations and/or provide local exhaust ventilation at the site of captan dust release. This may be practical in manufacturing operations, but is not feasible in application since it is used outdoors. While not always operationally possible, isolating operations involving captan manufacturing can also reduce exposure. Using respiratory protection is less effective than the controls mentioned above, but is still advisable whenever working with or around captan. At exposures above the ACGIH TLV of 5 mg/m^3 (TWA), use a MSHA/NIOSH approved respirator equipped with a pesticide cartridge and dust pre-filter. For high or frequent exposures, use a self-contained breathing apparatus (SCBA) with full facepiece and pressure demand is recommended for the greatest possible respiratory protection. If a full face-piece is not available, then chemical/dust goggles should be worn to protect the eyes. A face shield and protective apron should also be worn. To prevent hand and skin exposures, gloves should be worn.

Administrative controls should also be in place to minimize the potential for human exposures. These may include written procedures or policies which specify the methods and techniques that will be practiced whenever personnel are to work with captan.

All personnel should receive *training* on the use, hazards, protective measures, emergency actions, and other precautions per 29 CFR 1910.1200 (Hazard Communication), prior to the first assignment in an area where captan is used or stored.

If symptoms develop or overexposure is suspected, the following medical testing may be useful:

☑ Evaluation by a qualified allergist, including careful consideration of exposure history and special testing. Skin testing with dilute captan may help diagnose skin allergy, if done by a qualified allergist.

Any evaluation should include a careful history of past and present symptoms with an examination. Medical tests that simply look for existing damage are not a substitute for controlling exposures.

Other methods to reduce exposure include:

☑ Where possible, enclose operations and use local exhaust ventilation at the site of chemical release. If local exhaust ventilation or enclosure is not used, respiratory protection should be mandatory.

☑ Always ensure that proper protective clothing is worn when using chemical substances and that personnel are trained in its use and care.

☑ Wash thoroughly immediately after exposure to captan and at the end of the work shift or before eating, drinking, or smoking.

☑ Hazard warning information should be posted in the work area. In addition, as part of an on-going education and training program, all information on the health and safety hazards of captan should be communicated to all potentially exposed workers.

☑ Eye wash stations should be provided in the immediate work area for emergency use. If there is a possibility of skin exposure to captan, emergency shower facilities should also be provided.

☑ Workers whose clothing has been contaminated by captan should change into clean clothes before leaving work. Contaminated work clothing should be laundered only by individuals who have been informed of the hazards of exposure to captan.

RISK ASSESSMENT: ENVIRONMENT

General Assessment

Because of its use as a fungicide and pesticide, captan can be present in the environment at any given time and at a variety of levels. The environment is also at risk of exposure during transportation, storage, disposal, or destruction of captan. In almost every scenario, the threat of environmental exposure is contingent upon the proper handling of the chemical substance. Accidental spills, large or small, where captan contacts incompatible commodities can result in the release of hazardous fumes and gases (depending upon conditions of spill), and possible contamination of the surrounding environmental mediums (water, soil, and air). In contact with water, it will form poisonous gases including oxides of nitrogen and sulfur, hydrogen chloride, and extremely hazardous phosgene.

In its powder form, captan is considered a combustible solid. However, it may be present (dissolved) in some flammable liquids such as benzene. Also, because it is incompatible with water and moisture, extreme caution is required in handling, storage, transportation, and disposal of captan. It is also incompatible with other common materials such as oil, and can react on exposure to heat or in the presence of hydrogen sulfide and strong alkalis. These characteristics also require special consideration during any emergency situation involving a leak or spill of captan powder or dust.

Captan can enter the environment from industrial and municipal discharges, through agricultural run-off, and from spills. Also, as a fungicide and pesticide, it will be present in the environment as a result of its intended use.

☠ *Acute Ecological Effects*

Acute (short-term) toxic effects may include the death of animals, birds, or fish and death or low growth rate in plants. Acute effects are seen 2 to 4 days after animals or plants are exposed to captan.

Captan has high acute toxicity to aquatic life and birds, and can cause injury to plants. No data are available on the short-term effects of captan to land animals.

Chronic Ecological Effects

Chronic (long-term) toxic effects may include shortened life span, reproductive problems, lower fertility, and changes in appearance or behavior in exposed animals. These effects can be seen long after first exposure(s) to toxic chemicals.

Captan has high chronic toxicity to aquatic life. No data are available to evaluate or predict the long-term effects of captan to plants, birds, or land animals.

Water Solubility

Captan is very slightly soluble (practically insoluble) in water. Concentrations of less than 1 milligram will mix with a liter of water.

Persistence in the Environment

Captan is highly persistent in the water, with a half-life of greater than 200 days. The half-life of a pollutant is the amount of time it takes for one-half of the chemical to be degraded. About 97% of captan will eventually end up in water; approximately 1.5% will end up in terrestrial soil and aquatic sediments.

Bioaccumulation in Aquatic Organisms

Some substances increase in concentration, or *bioaccumulate,* in living organisms as they breathe contaminated air, drink contaminated water, or eat contaminated food. These chemicals can become concentrated in the tissues and internal organs of animals as well as humans.

The concentration of captan found in fish tissues is expected to be somewhat higher that the average concentration of captan in the water from which the fish was taken.

Recommended Risk-Reduction Measures

Proper training of all transporters will reduce the likelihood of a mishap or accident resulting in a leak or spill of captan dusts or powders into the environment. The correct use of DOT labeling (per the MSDS) on all containers, trucks, and rail cars will enable emergency responders to react properly and quickly to any disaster thereby reducing the potential risk to the environment and to personnel.

Storage of captan should be segregated from incompatible materials to minimize the risk of cross-contamination or contact. Proper ventilation and protective equipment should be used while handing the powders or while preparing aqueous solutions.

If a spill or leak to the environment has occurred, fire department, emergency response and/or hazardous materials spill personnel should be notified immediately. Cleanup should be attempted only by those trained in proper spill containment procedures. If captan should contact the water table, aquifer, or navigable waterway, reclamation procedures should be initiated as soon as practical. It is only slightly soluble in water and total remediation may be possible. The local and/or state emergency response authorities must be notified. A comprehensive emergency response or disaster preparedness/recovery plan should be in place prior to any operations involving the use, transportation, storage, or disposal of captan. If captan powder is spilled or leaked, the following specific steps are recommended:

- ☑ Restrict persons not wearing protective clothing from area of spill or leak until cleanup is complete.
- ☑ Remove all ignition sources.
- ☑ Collect powdered materials in the safest manner possible and deposit in sealed containers. Recommend using vacuum with HEPA filter. Do not dry sweep (generates airborne dusts). If liquid solutions are spilled, absorb in vermiculite or other material and place in sealed drum for disposal or reclamation.
- ☑ It may be necessary to dispose of captan as a hazardous waste. The responsible state agency or the regional office of the federal Environmental Protection Agency (EPA) should be contacted for specific recommendations.

RISK ASSESSMENT: BUSINESS

General Assessment

Accidents or mishaps involving captan can present a significant threat to business operations. The loss or damage of equipment or facilities can significantly affect fiscal viability. Lawsuits that may result from personnel injury/death, public exposures, and/or environmental contamination will also require a serious expenditure of resources.

Media attention surrounding an injury, death, or environmental damage can also result in a loss of profits and loss of current as well as future business.

Recommended Risk-Reduction Measures

Company attorneys, safety and health professionals, and environmental specialists should be involved in the development of any procedures for responding to chemical incidents. A company official should be predesignated as a public relations officer with specific training in dealing with the press. Corporate plans and policies should be developed, approved, and implemented long before any need for such arises.

REFERENCES

Ahlbom, A. 1993. *Biostatistics for Epidemiologists.* New York: CRC Press/Lewis Publishers.

American Conference of Governmental Industrial Hygienists. 1988. *Documentation of the Threshold Limit Values and Biological Exposure Indices,* 5th Edition (with updates). Cincinnati: ACGIH

Asante-Duah, D. K. 1993. *Hazardous Waste Risk Assessment.* New York: CRC Press/Lewis Publishers.

Bloom, A. D. and F. J. DeSerres. 1995. *Ecotoxicity and Human Health.* New York: CRC Press/Lewis Publishers.

Brusick, D. J. 1994. *Methods for Genetic Risk Assessment.* New York: CRC Press/Lewis Publishers.

Bueche, J. R. 1972. *Principles of Physics,* 2nd Edition. New York: McGraw-Hill.

Calabrese, E. J., E. M. Kenyon. *Air Toxics and Risk Assessment.* New York: CRC Press/Lewis Publishers.

Calabrese, E. J. 1994. *Biological Effects of Low Level Exposures.* New York: CRC Press/Lewis Publishers.

Cockerman, L. G., B. S. Shane. 1994. *Basic Environmental Toxicology.* New York: CRC Press/Lewis Publishers.

Cothern, C. R., N. P. Ross. 1994. *Environmental Statistics, Assessment, and Forecasting.* New York: CRC Press/Lewis Publishers.

Cothern, C. R. 1995. *Handbook for Environmental Risk Decisionmaking.* New York: CRC Press/Lewis Publishers.

Gardner, R. W. 1994. *Chemical Intolerance.* New York: CRC Press/Lewis Publishers.

Hallenbeck, W. H. 1993. *Quantitative Risk Assessment for Environmental and Occupational Health,* 2nd Edition. New York: CRC Press/Lewis Publishers.

Howard, P. H. 1995. *Handbook of Environmental Fate and Exposures Data for Organic Chemicals,* Vol. I-IV. New York: CRC Press/Lewis Publishers.

Howard, P. H., et. al. 1991. *Handbook of Environmental Degradation Rates.* New York: CRC Press/Lewis Publishers.

Jones, F. E. 1994. *Toxic Organic Vapors in the Workplace.* New York: CRC Press/Lewis Publishers.

Keith, L. H. and M. M. Walker. 1995. *Handbook of Air Toxics.* New York: CRC Press/Lewis Publishers.

Landis, W. G., M. H. Yu. 1995. *Introduction to Environmental Toxicology.* New York: CRC Press/Lewis Publishers.

Langerman, N. 1994 *Precautionary Labels for Chemical Containers.* New York: CRC Press/Lewis Publishers.

Lauwerys, R. R., P. Hoet. 1993. *Industrial Chemical Exposure.* New York: CRC Press/Lewis Publishers.

Lewis, R. J., Sr. 1992. *Sax's Dangerous Properties of Industrial Materials,* Volumes 1, 2, and 3, 8th Edition. New York: Van Nostrand Reinhold.

Lide, D. R. 1995. *Handbook of Chemistry and Physics,* 76th Edition. New York: CRC Press/Lewis Publishers.

Manahan, S. E. 1994. *Environmental Chemistry,* 6th Edition. New York: CRC Press/Lewis Publishers.

Manahan, S. E. 1992. *Toxicological Chemistry,* 2nd Edition. New York: CRC Press/Lewis Publishers.

Mellström, G. A., J. E. Wahlberg, and H. I. Maibach. 1994. *Protective Gloves for Occupational Use.* New York: CRC Press/Lewis Publishers.

Meyer, E. 1990. *Chemistry of Hazardous Materials.* Englewood Cliffs, NJ: Prentice Hall.

Mycock, J. C., J. D. McKenna, and L. Theodore. 1995. *Handbook of Air Pollution Control Engineering and Technology.* New York: CRC Press/Lewis Publishers.

Nelson, G. O. 1992. *Gas Mixtures.* New York: CRC Press/Lewis Publishers.

Ott, W. R. 1995. *Environmental Statistics and Data Analysis.* New York: CRC Press/Lewis Publishers.

Patnaik, P. 1992. *A Comprehensive Guide to the Hazardous Properties of Chemical Substances.* New York: Van Nostrand Reinhold.

Proctor, N. H., J. P. Hughes. 1991. *Chemical Hazards of the Workplace.* New York: Van Nostrand Reinhold.

Quigley, D. R. 1994. *Handbook of Emergency Chemical Management.* New York: CRC Press/Lewis Publishers.

Rea, W. J. 1995. *Chemical Sensitivity,* Vol. I-III. New York: CRC Press/Lewis Publishers.

Reeves, A. L. 1979. Captan. *Handbook on the Toxicology of Metals.* pp 321-328. New York: Elsevier North-Holland.

Richardson, M. 1995. *Dictionary of Substance and Their Effects,* Vol. I-VI. New York: CRC Press/Lewis Publishers.

Sacarello, H. L. A. 1994. *The Comprehensive Handbook of Hazardous Materials.* New York: CRC Press/Lewis Publishers.

Talbott, E., G. Craun. 1995. *An Introduction to Environmental Epidemiology.* New York: CRC Press/Lewis Publishers.

Torkelson, T. R. V. K. Rowe. 1981. *Patty's Industrial Hygiene and Toxicology,* Vol. 2B. New York: Wiley-Interscience.

MATERIAL SAFETY DATA SHEET

CHEMICAL NAME	
CARBARYL	**CRC PRESS/LEWIS PUBLISHERS DISCLAIMER:** THE INFORMATION AND RECOMMENDATIONS PRESENTED HEREIN ARE BASED ON SOURCES BELIEVED TO BE RELIABLE. CRC MAKES NO REPRESENTATION ON ITS COMPLETENESS OR ACCURACY. IT IS THE RESPONSIBILITY OF THE USER TO DETERMINE THE CHEMICAL'S SUITABILITY FOR ITS INTENDED USE, THE CHEMICAL'S SAFE USE, AND THE CHEMICAL'S PROPER DISPOSAL. NO REPRESENTATIONS AND/OR WARRANTIES, EITHER EXPRESSED OR IMPLIED, OF THE MERCHANTABILITY OR FITNESS FOR A PARTICULAR PURPOSE, OR OF ANY OTHER NATURE, ARE MADE WITH RESPECT TO THE INFORMATION PROVIDED IN THIS MSDS OR TO THE CHEMICAL TO WHICH INFORMATION MAY REFER. CRC NEITHER ASSUMES NOR AUTHORIZES ANY OTHER PERSON TO ASSUME FOR IT, ANY OTHER ADDITIONAL RESPONSIBILITY OR LIABILITY FOR THE USE OF, OR RELIANCE UPON, THIS INFORMATION.

HAZARD WARNING INFORMATION

HEALTH	FIRE	REACTIVE	OTHER	DEGREE OF HAZARD	COLOR CODING	OTHER CODES
3	1	0		0 = Minimum Hazard 1 = Slight Hazard 2 = Moderate Hazard 3 = Serious Hazard 4 = Severe Hazard	HEALTH = BLUE FIRE = RED REACTIVITY = YELLOW OTHER = WHITE	OX = Oxidizer ACID = Acid ALK = Alkali COR = Corrosive W = Use No Water

SECTION I - GENERAL INFORMATION

Characterization	RCRA Number	EPA Class
Carbamate (Pesticide)	**Not Listed**	**Not Applicable**

DOT Proper Shipping Name **Carbaryl**	Chemical Abstract Service (CAS) Number **63-25-2**
DOT Hazard Class and Label Requirements **ORM-A**	DOT Emergency Guide Code **55**
DOT Identification Number **UN 2757**	Chemical Formula $C_{12}H_{11}NO_2$

Synonyms
Sevin®; 1-naphthyl-N-methylcarbamate; methylcarbamic acid 1-naphthyl carbamate.

SECTION II - HAZARDOUS INGREDIENTS AND IDENTITY INFORMATION

Hazardous Components (specific identity)	OSHA Exposure Criteria	NIOSH Exposure Criteria	Immediately Dangerous to Life and Health (IDLH)	ACGIH Exposure Criteria
Carbaryl (derivation: Synthesized directly from 1-naphthol and methyl isocyanate or from 1-naphthyl chloroformate (1-naphthol and phosgene) plus methylamine).	**PEL:** 5.0 mg/m3 **STEL:** Not Established	**REL:** 5.0 mg/m3 (10 hours) **STEL:** Not Established	**100 mg/m3**	**TLV:** 5.0 mg/m^3 **STEL:** Not Established

SECTION III - PHYSICAL AND CHEMICAL CHARACTERISTICS

Boiling Point **Decomposes**	Specific Gravity (H_2O = 1) **1.23**
Vapor Pressure (mm Hg) **0 (approx.) at 69°F (20°C)**	Molecular Weight (atomic weight) **200.1**
Vapor Density (Air = 1) **6.0**	Melting Point **293°F (145°C)**

Solubility
Slightly soluble in water (0.012% at 30°C), moderately soluble in acetone, cyclohexanone, and dimethylformamide. Hydrolyzed by alkalies.

Appearance and Odor
Carbaryl is a white or grayish-white odorless crystalline solid (sugar or sand-like).

SECTION IV - FIRE AND EXPLOSION HAZARD DATA

Flash Point (method used) **Not Determined**	Explosive Limits in Air % by Volume **LEL: N/A UEL: N/A**
NFPA Classification **Non-combustible solid**	Autoignition Temperature **Not Found**

Extinguishing Media
Use water, carbon dioxide, dry chemical, water spray, or foam.

Special Fire Fighting Procedures
Wear full protective clothing and self-contained breathing apparatus (SCBA). Structural fire-fighting equipment is permeable and may not provide sufficient protection (chemical can be absorbed through the skin). Remain clear of smoke, water fall-out and water run-off. Move containers from fire area if it can be done without risk. Cool exposed containers using water spray.

Unusual Fire and Explosion Hazards
Carbaryl is a non-combustible solid. However, in liquid form, it may be combustible. Also, it may be dissolved in some flammable liquids, such as benzene. Containers may explode in fire.

SECTION V - REACTIVITY DATA

Stability		Conditions to Avoid: Avoid contact with incompatible materials. If mixed with flammable liquids, carbaryl becomes combustible.
Stable	Unstable	Incompatibility *(materials to avoid)*: Violent reactions can occur on contact with strong oxidizers (such as chlorine, bromine, and fluorine) and strongly alkaline pesticides.
X		
Hazardous Polymerization		Conditions to Avoid: Under normal conditions of temperature and pressure, hazardous polymerization of carbaryl is not known to occur.
May Occur	Will Not Occur	Hazardous Decomposition or By-products: When heated to decomposition, carbaryl will produce poisonous fumes, including oxides of nitrogen.
	X	

SECTION VI - HEALTH HAZARD DATA

Primary Route(s) of Entry:	Inhalation?	Absorption (skin/eye)?	Ingestion?
	X	X	X

Health Hazards

INHALATION: Exposure can cause blurred vision, sweating, nausea and vomiting, and abdominal pain. Higher exposures can cause fluid to build up in the lungs (pulmonary edema) and death.

ABSORPTION: Will pass through unbroken skin and enter the bloodstream. Contact may severely irritate the skin and eyes. Skin absorption is slow. May cause rash or burning feeling.

INGESTION: Moderately toxic to humans resulting in toxic systemic effects. Symptoms include nausea, vomiting, diarrhea, abdominal cramps, miosis, lachrymation, and excessive salivation.

Carcinogenicity	NTP Listed?	IARC Cancer Review Group?	OSHA Regulated?	Target Organs?
Questioned Human Questioned Animal	No	Group 3	29 CFR 1910.1000 Table Z-1	Lungs, skin, eyes, CNS, CVS, blood, repro. sys.

Medical Conditions Generally Aggravated by Exposure

None reported. However, persons with existing lung dysfunction may experience effects quicker.

Emergency and First-aid Procedures

Eye contact: Do not allow victim to keep eyes tightly shut. Flush immediately with water for 15 minutes (minimum), seek medical attention. Skin contact: Remove all contaminated clothing. Immediately wash area with flooding amounts of soap and water. For inhalation: Remove the person from exposure. Maintain an open airway. Provide respiratory assistance and CPR. Transfer to medical facility and observe for 48 hours for lung effects. If swallowed: If victim is conscious, induce vomiting. Seek medical attention immediately. Call poison control center for advice. Never give anything by mouth to an unconscious or convulsing person. Do not make an unconscious person vomit.

SECTION VII - PRECAUTIONS FOR SAFE HANDLING AND USE

Steps to be Taken in Case Material is Released or Spilled

Collect powdered materials in safest manner possible and deposit in sealed containers. Remove all sources of ignition. Ventilate area of spill. Use HEPA vacuum. Restrict those not involved in cleanup from entering area. Notify appropriate authorities. Absorb any liquids containing carbaryl in vermiculite, dry sand, earth, or similar material and deposit in sealed drum for disposal or reclamation.

Preferred Waste Disposal Method

No Citation.

Precautions to be Taken in Handling and Storage

Store in tightly closed containers free from moisture. Keep in cool, well ventilated area.

Other Precautions and Warnings

Do not store near sources of heat or ignition. Smoking should be prohibited in carbaryl storage areas.

SECTION VIII - CONTROL MEASURES AND PERSONAL PROTECTIVE EQUIPMENT

Respiratory Protection *(specify type)*

For exposures above 5.0 mg/m^3, use a supplied air respirator or an MSHA/NIOSH approved self-contained breathing apparatus with full facepiece operated in positive pressure mode.

Ventilation

Local exhaust (preferred) at site of chemical work or general ventilation.

Protective Gloves	Eye Protection	Other Protective Clothing
Impervious Material	Chemical/Dust Goggles or Face Mask	Protective Apron, Boots (avoid skin contact)

Work/Hygiene Practices

Always wash hands thoroughly after using chemical; never bring food, drink, or smoking materials into vicinity of chemicals.

CARBARYL

$C_{12}H_{11}NO_2$ CAS: 63-25-2

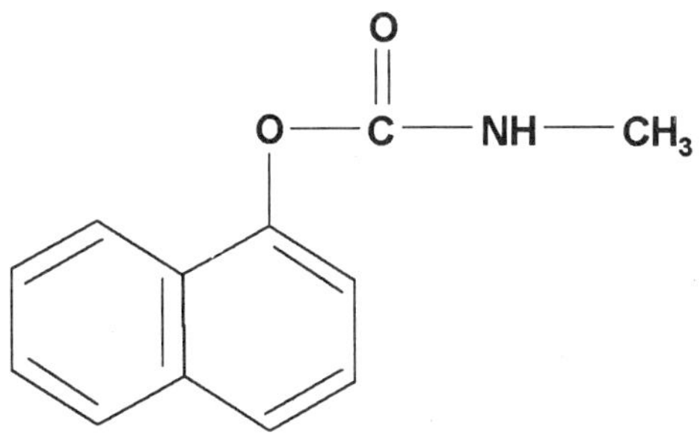

IDENTIFICATION AND TYPICAL USES

Carbaryl is a white or gray odorless crystalline solid powder (sugar or sand-like). It is a carbamate insecticide used in powder, liquid, or paste form.

RISK ASSESSMENT: HEALTH

General Assessment

Carbaryl is a poison by *ingestion, inhalation,* and skin contact (*absorption*). Human systemic effects by ingestion have been reported, including sensory changes involving peripheral nerves and muscle weakness. Skin absorption is reported to be very slow. Rash may develop along with burning sensation. Carbaryl is a mutagen in humans with questionable carcinogenic, tumorigenic, and teratogenic properties.

Symptoms of exposure by all routes can include nausea and vomiting, abdominal pain and cramps, blurred vision, lachrymation (tearing), nasal discharge and rhinitis, miosis, excessive salivation, sweating, cyanosis, muscle twitching, diarrhea, tremor and convulsions, and coma. Carbaryl is not reported to accumulate in body tissues.

Breathing dusts of carbaryl and, to a lesser extent, vapors of liquids containing carbaryl, can cause a fluid to build up in the lungs (pulmonary edema) 2-4 days following exposure. This is a medical emergency and can result in death.

Acute Health Effects

The following acute (short-term) health effects may occur immediately or shortly after exposure to carbaryl:

Skin: Possible irritation with burning and rash. Absorption is likely (although extremely slow).

Eye: Severe irritation and possible damage on contact.

Lung: Irritation of the respiratory system. Can result in pulmonary edema (fluid in lungs), which can be fatal.

CNS: Possible effects on peripheral nerves. Effects of exposure may include nausea, vomiting, diarrhea, abdominal cramps, cyanosis, convulsions, tremor, lachrymation, blurred vision, and coma.

Chronic Health Effects

The following chronic (long-term) health effects may occur at some time after exposure to carbaryl and can last for months or even years:

Cancer Hazards: Carbaryl is a mutagen (causes genetic changes) in humans. Such chemicals may have a cancer risk and there is limited evidence that carbaryl does cause cancer in animals. Human carcinogenicity is questioned but requires further study.

Reproductive Hazard: Carbaryl is a possible teratogen in humans and has been shown to be a teratogen in animals. There is limited evidence that it reduces fertility in both males and females.

Other Chronic Effects: Carbaryl may damage the kidneys and nervous systems.

Recommended Risk-Reduction Measures

Personnel should avoid direct contact with carbaryl dusts, powders, liquids, or pastes. If a less toxic material or compound cannot be substituted for carbaryl, then *engineering controls* are the most effective method of reducing exposures. The best protection is to enclose operations and/or provide local exhaust ventilation at the site of carbaryl release. This may be practical in manufacturing operations, but is not always feasible in application since it is used outdoors. While not always operationally possible, isolating operations involving carbaryl manufacturing can also reduce exposure. Using respiratory protection is less effective than the controls mentioned above, but is still advisable whenever working with or around carbaryl. At exposures above the OSHA PEL of 5.0 mg/m^3 (TWA), use a MSHA/NIOSH approved supplied-air respirator operated in positive pressure mode or use a self-contained breathing apparatus (SCBA) with full facepiece and pressure demand is recommended for the greatest possible respiratory protection. If a full facepiece is not available, then chemical/dust goggles should be worn to protect the eyes. A face shield and protective apron should also be worn. To prevent hand and skin exposures, gloves should be worn.

Administrative controls should also be in place to minimize the potential for human exposures. These may include written procedures or policies which specify the methods and techniques that will be practiced whenever personnel are to work with carbaryl.

All personnel should receive *training* on the use, hazards, protective measures, emergency actions, and other precautions per 29 CFR 1910.1200 (Hazard Communication), prior to the first assignment in an area where carbaryl is used or stored.

If symptoms develop or overexposure is suspected, the following medical testing may be useful:

☑ Kidney function test.

☑ Examination of the nervous system.

☑ If done within 2-3 hours of exposure, serum and red blood count cholinesterase levels may be helpful. Carbaryl is a cholinesterase inhibitor. Levels can return to normal before the person feels well again.

Any evaluation should include a careful history of past and present symptoms with an examination. Medical tests that simply look for existing damage are not a substitute for controlling exposures. Also, persons exposed to other chemicals which affect body cholinesterase, such as organophosphates and other carbamates, may be at increased risk.

Other methods to reduce exposure include:

☑ Where possible, enclose operations and use local exhaust ventilation at the site of chemical release. If local exhaust ventilation or enclosure is not used, respiratory protection should be mandatory.

☑ Always ensure that proper protective clothing is worn when using chemical substances.

☑ Wash thoroughly immediately after exposure to carbaryl and at the end of the work shift or before eating, drinking, or smoking.

☑ Hazard warning information should be posted in the work area. In addition, as part of an on-going education and training program, all information on the health and safety hazards of carbaryl should be communicated to all potentially exposed workers.

☑ Eye wash stations should be provided in the immediate work area for emergency use. If there is a possibility of skin exposure to carbaryl, emergency shower facilities should also be provided.

☑ Workers whose clothing has been contaminated by carbaryl should change into clean clothes before leaving work. Contaminated work clothing should be laundered only by individuals who have been informed of the hazards of exposure to carbaryl.

RISK ASSESSMENT: ENVIRONMENT

General Assessment

Because of its use as a pesticide, carbaryl can be present in the environment at any given time and at a variety of levels. The environment is also at risk of exposure during transportation, storage, disposal, or destruction of carbaryl. In almost every scenario, the threat of environmental exposure is contingent upon the proper handling of the chemical substance. Accidental spills, large or small, where carbaryl contacts incompatible commodities can result in violent reaction, including explosion and fire. Contamination of the surrounding environmental mediums (water, soil, and air) is also possible.

In its powder form, carbaryl is considered a noncombustible solid. However, it may be present (dissolved) in some flammable liquids. Also, because it is incompatible with strong oxidizers and strongly alkaline pesticides, extreme caution is required in handling, storage, transportation, and disposal of carbaryl. These characteristics require special consideration during any emergency situation involving a leak or spill of carbaryl powder, dust, or liquid solutions containing carbaryl.

Carbaryl is a widely used synthetic insecticide. It is a broad-spectrum contact insecticide registered on more than 100 crops. It can therefore be present in the environment as a result of its intended use. Also, it may enter the environment from industrial and municipal discharges, through agricultural run-off, and from spills.

Acute Ecological Effects

Acute (short-term) toxic effects may include the death of animals, birds, or fish and death or low growth rate in plants. Acute effects are seen 2 to 4 days after animals or plants are exposed to carbaryl.

Carbaryl has high acute toxicity to aquatic life and birds. It is highly-to-moderately toxic, acutely, to land animals. It has caused death and injuries of various kinds to agricultural and ornamental crops.

Chronic Ecological Effects

Chronic (long-term) toxic effects may include shortened life span, reproductive problems, lower fertility, and changes in appearance or behavior in exposed animals. These effects can be seen long after first exposure(s) to toxic chemicals.

Carbaryl has high chronic toxicity to aquatic life. Insufficient data are available to evaluate or predict the long-term effects of carbaryl to plants, birds, or land animals.

Water Solubility

Carbaryl is moderately soluble in water. Concentrations of 1 to 1000 milligrams will mix with a liter of water.

Persistence in the Environment

Carbaryl is moderately persistent in the water, with a half-life between 20 and 200 days. The half-life of a pollutant is the amount of time it takes for one-half of the chemical to be degraded. About 95% of carbaryl will eventually end up in water; approximately 1.5% will end up in air; and slightly less than 2% will end up in terrestrial soil and aquatic sediments.

Bioaccumulation in Aquatic Organisms

Some substances increase in concentration, or *bioaccumulate,* in living organisms as they breathe contaminated air, drink contaminated water, or eat contaminated food. These chemicals can become concentrated in the tissues and internal organs of animals as well as those of humans.

The concentration of carbaryl found in fish tissues is expected to be somewhat higher that the average concentration of carbaryl in the water from which the fish was taken.

Recommended Risk-Reduction Measures

Proper training of all transporters will reduce the likelihood of a mishap or accident resulting in a leak or spill of carbaryl dusts or powders into the environment. The correct use of DOT labeling (per the MSDS) on all containers, trucks, and rail cars will enable emergency responders to react properly and quickly to any disaster thereby reducing the potential risk to the environment and to personnel.

Storage of carbaryl should be segregated from incompatible materials to minimize the risk of cross-contamination or contact. Proper ventilation and protective equipment should be used while handing the powders or while preparing aqueous solutions.

If a spill or leak to the environment has occurred, fire department, emergency response and/or hazardous materials spill personnel should be notified immediately. Cleanup should be attempted only by those trained in proper spill containment procedures. If carbaryl should contact the water table, aquifer, or navigable waterway, time is of the essence. It is moderately soluble in water and total remediation may not be possible. The local and/or state emergency response authorities must be notified. A comprehensive emergency response or disaster preparedness/recovery plan should be in place prior to any operations involving the use, transportation, storage, or disposal of carbaryl. If carbaryl powder is spilled or leaked, the following specific steps are recommended:

- ☑ Restrict persons not wearing protective clothing from area of spill or leak until cleanup is complete.
- ☑ Remove all ignition sources.
- ☑ Collect powdered materials in the safest manner possible and deposit in sealed containers. Recommend using vacuum with HEPA filter. Do not dry sweep (generates airborne dusts). If liquid solutions are spilled, absorb in vermiculite or other material and place in sealed drum for disposal or reclamation.
- ☑ It may be necessary to dispose of carbaryl as a hazardous waste. The responsible state agency or the regional office of the federal Environmental

Protection Agency (EPA) should be contacted for specific recommendations.

RISK ASSESSMENT: BUSINESS

General Assessment

Accidents or mishaps involving carbaryl can present a significant threat to business operations. The loss or damage of equipment or facilities can significantly affect fiscal viability. Lawsuits that may result from personnel injury/death, public exposures, and/or environmental contamination will also require a serious expenditure of resources. Media attention surrounding an injury, death, or environmental damage can also result in a loss of profits and loss of current as well as future business.

Recommended Risk-Reduction Measures

Company attorneys, safety and health professionals, and environmental specialists should be involved in the development of any procedures for responding to chemical incidents. A company official should be predesignated as a public relations officer with specific training in dealing with the press. Corporate plans and policies should be developed, approved, and implemented long before any need for such arises.

REFERENCES

American Conference of Governmental Industrial Hygienists. 1988. *Documentation of the Threshold Limit Values and Biological Exposure Indices,* 5th Edition (with updates). Cincinnati: ACGIH

Asante-Duah, D. K. 1993. *Hazardous Waste Risk Assessment.* New York: CRC Press/Lewis Publishers.

Brusick, D. J. 1994. *Methods for Genetic Risk Assessment.* New York: CRC Press/Lewis Publishers.

Bloom, A. D. and F. J. DeSerres. 1995. *Ecotoxicity and Human Health.* New York: CRC Press/Lewis Publishers.

Bueche, J. R. 1972. *Principles of Physics,* 2nd Edition. New York: McGraw-Hill.

Calabrese, E. J. 1994. *Biological Effects of Low Level Exposures.* New York: CRC Press/Lewis Publishers.

Cockerman, L. G., B. S. Shane. 1994. *Basic Environmental Toxicology.* New York: CRC Press/Lewis Publishers.

Cothern, R. R. 1993. *Comparative Environmental Risk Assessment.* New York: CRC Press/Lewis Publishers.

Gardner, R. W. 1994. *Chemical Intolerance.* New York: CRC Press/Lewis Publishers.

Hallenbeck, W. H. 1993. *Quantitative Risk Assessment for Environmental and Occupational Health,* 2nd Edition. New York: CRC Press/Lewis Publishers.

Howard, P. H., et. al. 1991. *Handbook of Environmental Degradation Rates.* New York: CRC Press/Lewis Publishers.

Jones, F. E. 1994. *Toxic Organic Vapors in the Workplace.* New York: CRC Press/Lewis Publishers.

Keith, L. H. and M. M. Walker. 1995. *Handbook of Air Toxics.* New York: CRC Press/Lewis Publishers.

Landis, W. G., M. H. Yu. 1995. *Introduction to Environmental Toxicology.* New York: CRC Press/Lewis Publishers.

Langerman, N. 1994 *Precautionary Labels for Chemical Containers.* New York: CRC Press/Lewis Publishers.

Lauwerys, R. R., P. Hoet. 1993. *Industrial Chemical Exposure.* New York: CRC Press/Lewis Publishers.

Lewis, R. J., Sr. 1992. *Sax's Dangerous Properties of Industrial Materials,* Volumes 1, 2, and 3, 8th Edition. New York: Van Nostrand Reinhold.

Lide, D. R. 1995. *Handbook of Chemistry and Physics,* 76th Edition. New York: CRC Press/Lewis Publishers.

Manahan, S. E. 1992. *Toxicological Chemistry,* 2nd Edition. New York: CRC Press/Lewis Publishers.

Mellström, G. A., J. E. Wahlberg, and H. I. Maibach. 1994. *Protective Gloves for Occupational Use.* New York: CRC Press/Lewis Publishers.

Meyer, E. 1990. *Chemistry of Hazardous Materials.* Englewood Cliffs, NJ: Prentice Hall.

Mycock, J. C., J. D. McKenna, and L. Theodore. 1995. *Handbook of Air Pollution Control Engineering and Technology.* New York: CRC Press/Lewis Publishers.

Nelson, G. O. 1992. *Gas Mixtures.* New York: CRC Press/Lewis Publishers.

Ott, W. R. 1995. *Environmental Statistics and Data Analysis.* New York: CRC Press/Lewis Publishers.

Patnaik, P. 1992. *A Comprehensive Guide to the Hazardous Properties of Chemical Substances.* New York: Van Nostrand Reinhold.

Proctor, N. H., J. P. Hughes. 1991. *Chemical Hazards of the Workplace.* New York: Van Nostrand Reinhold.

Quigley, D. R. 1994. *Handbook of Emergency Chemical Management.* New York: CRC Press/Lewis Publishers.

Rea, W. J. 1995. *Chemical Sensitivity,* Vol. I-III. New York: CRC Press/Lewis Publishers.

Richardson, M. 1995. *Dictionary of Substance and Their Effects,* Vol. I-VI. New York: CRC Press/Lewis Publishers.

Sacarello, H. L. A. 1994. *The Comprehensive Handbook of Hazardous Materials.* New York: CRC Press/Lewis Publishers.

Talbott, E., G. Craun. 1995. *An Introduction to Environmental Epidemiology.* New York: CRC Press/Lewis Publishers.

MATERIAL SAFETY DATA SHEET

CHEMICAL NAME	
CARBON DISULFIDE	**CRC PRESS/LEWIS PUBLISHERS DISCLAIMER:** THE INFORMATION AND RECOMMENDATIONS PRESENTED HEREIN ARE BASED ON SOURCES BELIEVED TO BE RELIABLE. CRC MAKES NO REPRESENTATION ON ITS COMPLETENESS OR ACCURACY. IT IS THE RESPONSIBILITY OF THE USER TO DETERMINE THE CHEMICAL'S SUITABILITY FOR ITS INTENDED USE, THE CHEMICAL'S SAFE USE, AND THE CHEMICAL'S PROPER DISPOSAL. NO REPRESENTATIONS AND/OR WARRANTIES, EITHER EXPRESSED OR IMPLIED, OF THE MERCHANTABILITY OR FITNESS FOR A PARTICULAR PURPOSE, OR OF ANY OTHER NATURE, ARE MADE WITH RESPECT TO THE INFORMATION PROVIDED IN THIS MSDS OR TO THE CHEMICAL TO WHICH INFORMATION MAY REFER. CRC NEITHER ASSUMES NOR AUTHORIZES ANY OTHER PERSON TO ASSUME FOR IT, ANY OTHER ADDITIONAL RESPONSIBILITY OR LIABILITY FOR THE USE OF, OR RELIANCE UPON, THIS INFORMATION.

HAZARD WARNING INFORMATION

HEALTH	FIRE	REACTIVE	OTHER	DEGREE OF HAZARD	COLOR CODING	OTHER CODES
2	3	0		0 = Minimum Hazard 1 = Slight Hazard 2 = Moderate Hazard 3 = Serious Hazard 4 = Severe Hazard	HEALTH = BLUE FIRE = RED REACTIVITY = YELLOW OTHER = WHITE	OX = Oxidizer ACID = Acid ALK = Alkali COR = Corrosive ~~W~~ = Use No Water

SECTION I - GENERAL INFORMATION

Characterization **Organic, Sulfur-Containing**	RCRA Number **P022** — EPA Class **Acute Hazardous Waste**
DOT Proper Shipping Name **Carbon Bisulfide**	Chemical Abstract Service (CAS) Number **75-15-0**
DOT Hazard Class and Label Requirements **Flammable Liquid**	DOT Emergency Guide Code **28**
DOT Identification Number **UN 1131**	Chemical Formula **CS_2**

Synonyms
Carbon bisulfide; Dithiocarbonic anhydride; carbon sulfide.

SECTION II - HAZARDOUS INGREDIENTS AND IDENTITY INFORMATION

Hazardous Components (specific identity)	OSHA Exposure Criteria	NIOSH Exposure Criteria	Immediately Dangerous to Life and Health (IDLH)	ACGIH Exposure Criteria
Carbon disulfide (derivation: Reaction of natural gas or petroleum fraction with sulfur; from natural gas and hydrogen sulfide at very high temperature (plasma process); by heating sulfur and charcoal and condensing the carbon disulfide vapors). **1 ppm = 3.16 mg/m³**	**PEL:** **20 ppm** **63 mg/m³** **CEILING:** **30 ppm** **PEAK (30 min)** **100 ppm**	**REL (10 hour):** **1 ppm** **3 mg/m³** **STEL (skin):** **10 ppm** **30 mg/m³**	**500 ppm**	**TLV:** **10 ppm** **30 mg/m³** **STEL:** **Not Established**

SECTION III - PHYSICAL AND CHEMICAL CHARACTERISTICS

Boiling Point **116°F (241°C)**	Specific Gravity (H_2O = 1) **1.26**
Vapor Pressure (mm Hg) **297 at 69°F (20°C)**	Molecular Weight **76.1**
Vapor Density (Air = 1) **2.64**	Freezing Point **-169°F**

Solubility
Slightly soluble in water (0.22% at 20°C), miscible in alcohol, ether, benzene, and chloroform.

Appearance and Odor
Clear to faintly yellow liquid with a strong, foul, disagreeable odor. Odor threshold = 0.11 ppm.

SECTION IV - FIRE AND EXPLOSION HAZARD DATA

Flash Point (method used) **-22°F (-30°C) closed cup**	Explosive Limits in Air % by Volume **LEL: 1.3% UEL: 50.0%**
NFPA Classification **Class 1B Flammable Liquid**	**Autoignition Temperature** **194°F (90°C)**

Extinguishing Media
Carbon dioxide or dry chemical. Water and foam may be ineffective.

Special Fire Fighting Procedures
Wear full protective clothing and self-contained breathing apparatus (SCBA). Heat will build pressure and may rupture closed storage containers. Fight fire from a distance if possible. Keep fire-exposed containers cool with water spray. Vapors are heavier than air; they may travel a distance to flashback and cause fire.

Unusual Fire and Explosion Hazards
Severe explosion hazard when exposed to heat, flame, or other source of ignition. Ignition and potentially explosive reaction when heated in contact with rust or iron.

SECTION V - REACTIVITY DATA

Stability		Conditions to Avoid Do not allow carbon disulfide to come into contact with any of the incompatible chemicals listed below. Avoid contact with fire or high heat sources. Do not use in confined spaces.
Stable X	Unstable	Incompatibility *(materials to avoid)* Strong oxidizers, chemically-active metals (potassium, zinc), azides, rust, halogens, amines. Also incompatible with air. Vapors may ignite on contact with an ordinary lit light bulb.
Hazardous Polymerization		Conditions to Avoid Working in confined spaces or closed, poorly ventilated areas with carbon disulfide. Do not mix with sodium or potassium-sodium alloys. Mixtures are powerful, shock-sensitive explosives.
May Occur	Will Not Occur X	Hazardous Decomposition or By-products When heated to decomposition, carbon disulfide emits highly toxic oxides of sulfur.

SECTION VI. - HEALTH HAZARD DATA

Primary Route(s) of Entry:	Inhalation?	Absorption (skin)?	Ingestion?
	X	X	X

Health Hazards

INHALATION: Symptoms include narcosis, headache, nausea, pulmonary irritation, and convulsions. Other symptoms may be acute hemolytic anemia, methemoglobinemia, and cyanosis.

ABSORPTION: Skin contact can result in irritating rash or burning feeling. Can cause severe eye burns leading to permanent damage. Can also cause toxic systemic effects (CNS involvement).

INGESTION: Gastroenteritis, possible kidney and liver effects.

Carcinogenicity	NTP Listed?	IARC Cancer Review Group?	OSHA Regulated?	Target Organs?
Unknown Human Unknown Animal	No	No	No	Lung; skin; eyes; CNS, PNS, CVS, liver, kidneys.

Medical Conditions Generally Aggravated by Exposure

Respiratory function impairments (asthma, etc.); Nervous disorders; Skin conditions (dermatitis).

Emergency and First-aid Procedures

Eye contact: Immediately flush large amounts of water for 15 minutes (minimum), occasionally lifting eyelids, seek medical attention. Skin contact: Remove all contaminated clothing. Immediately wash area with large amounts of soap and water. For inhalation: Remove the person from exposure. Provide respiratory assistance and CPR. Transfer promptly to medical facility. If swallowed: Seek medical attention immediately. Never try to give an unconscious person anything by mouth.

SECTION VII - PRECAUTIONS FOR SAFE HANDLING AND USE

Steps to be Taken in Case Material is Released or Spilled

Remove all ignition sources. Restrict persons not wearing protective equipment from entering area of spill or leak until cleanup is complete. Ventilate area. Absorb liquid in dry sand or other material and deposit in sealed containers. Alert appropriate state or federal response agencies.

Preferred Waste Disposal Method

For small amounts (10mL) - incineration (with afterburner and scrubber). Dilute larger amounts first.

Precautions to be Taken in Handling and Storage

Do not store with incompatible chemicals since violent reactions can occur. Store in tightly closed containers in a cool, dark, well-ventilated area away from fire, sparks, and flame. Use non-sparking tools on containers. Ground and bond metal containers during transfer operations.

Other Precautions and Warnings

Bulk storage of carbon disulfide is not recommended. Containers may explode in fire or under conditions of extreme heat (do not store outdoors or in direct sunlight).

SECTION VIII - CONTROL MEASURES AND PERSONAL PROTECTIVE EQUIPMENT

Respiratory Protection *(specify type)*

Exposures above 1 ppm: MSHA/NIOSH approved full facepiece respirator with organic vapor cartridge. Greater protection is obtained from a self-contained breathing apparatus (SCBA) with full facepiece and pressure demand or other positive pressure mode.

Ventilation

Local exhaust or general mechanical systems recommended.

Protective Gloves	Eye Protection	Other Protective Clothing
Solvent Resistant Butyl Rubber	Chemical Goggles and/or Face Mask	Rubber Apron

Work/Hygiene Practices

Always wash hands thoroughly after using chemical; never bring food, drink, or smoking materials into vicinity of chemicals.

CARBON DISULFIDE

CS_2 CAS: 75-15-0

IDENTIFICATION AND TYPICAL USES

Carbon disulfide is a colorless to faintly yellow liquid with a strong, foul, disagreeable odor. It is used in the manufacture of viscose rayon, soil disinfectants, cellophane, electronic vacuum tubes, carbon tetrachloride, and flotation agents. It is also used as a solvent and as an eluant for organics adsorbed on charcoal in air analysis.

RISK ASSESSMENT: HEALTH

General Assessment

While carbon disulfide exhibits low toxicity in most experimental animals, its toxicity is relatively greater in humans. The primary route of exposure is *inhalation* of its vapors. But it may also enter the body through skin *absorption* and, in some cases, *ingestion*.

In both acute and chronic exposures, the toxic effect of exposure is primarily due to its action on the central and peripheral nervous systems. A single exposure may cause narcosis. It acts as a narcotic and anesthetic in acute poisoning with death following from respiratory failure. In chronic poisoning, damage to the nervous systems may be permanent and severe. Repeated exposures cause headache, dizziness, fatigue, nervousness, insomnia, psychosis, irritation, tremors, loss of appetite, indigestion, and gastric disturbances. Human reproductive and mutation effects by inhalation have also been reported. Overexposure to carbon disulfide has been associated with an increase in coronary heart disease. In extreme cases of intoxication, a Parkinson-like syndrome may result, characterized by speech disturbances, muscle spasms, tremor, memory loss, mental depression, and marked psychic symptoms with possible personality changes that may be permanent.

Liquid contact with the eyes causes immediate and severe irritation. May cause ocular changes (blind spot enlargement, contraction of peripheral field, corneal anesthesia, diminished pupil reflexes, nystagmus, and microscopic aneurysms in the retina). Skin contact will result in rash, burning, dermatitis, and vesiculation.

Although ingestion is not common, it may cause coma and convulsions.

Acute Health Effects

The following acute (short-term) health effects may occur immediately or shortly after exposure to carbon disulfide:

<u>Skin</u>: Irritation/rash or burning sensation on contact. There may be blistering and redness. Dermatitis may also appear.

<u>Eye</u>: Burns which can lead to permanent damage.

<u>Lung</u>: Irritation of the respiratory tract. Effects include headache, garlicky breath, vomiting, diarrhea (even after vapor exposure), and, occasionally, abdominal pain.

<u>CNS</u>: Inhalation of high concentrations of the vapor may cause weak pulse, heart palpitations, fatigue, weakness in legs, unsteady gait, vertigo, mania, hallucinations of sight, hearing, taste, and smell, dizziness, lightheadedness, central nervous system depression with respiratory paralysis, convulsion, loss of consciousness, and possible death.

Chronic Health Effects

The following chronic (long-term) health effects may occur at some time after exposure to carbon disulfide and can last for months or even years:

<u>Cancer Hazards</u>: According to information presented in the references, carbon disulfide has not been adequately tested for its ability to cause cancer in test animals.

<u>Reproductive Hazards</u>: According to information presented in the references, carbon disulfide has been

shown to have the ability to damage the developing fetus. It may decrease fertility in men and women, causing sperm abnormalities and spontaneous abortions.

Other Chronic Effects: Repeated exposures can cause severe changes in the brain and nervous system, causing tingling, pain, "pins and needles" feeling and weakness in the legs. There may also be stomach abnormalities and very severe mood, personality and thought changes, including total insanity. There may be nightmares, trouble concentrating and trouble with coordination and balance. Carbon disulfide may cause increased cholesterol and atherosclerosis (hardening of the arteries), high blood pressure, heart disease, and damage to the eyes and other organs from the effects on arteries. Carbon disulfide may also cause a skin allergy. If an allergy develops, very low future exposures can cause itching and skin rash.

Recommended Risk-Reduction Measures

Personnel should avoid direct contact with carbon disulfide. Unless a less toxic chemical can be substituted for a hazardous substance, *engineering controls* are the most effective method of reducing exposures. The best protection is to enclose operations and/or provide local exhaust ventilation at the site of chemical release. While not always operationally feasible, isolating operations can also reduce exposure. Using respiratory protection is less effective than the controls mentioned above, but is still advisable whenever working with carbon disulfide. A NIOSH-approved, full facepiece organic vapor respirator is sufficient for low exposures. A self-contained breathing apparatus (SCBA) with full facepiece and pressure demand is the recommended respiratory protection of choice. If a full facepiece is not available, then chemical goggles should be worn to protect the eyes. Whenever a chemical splash hazard exists, a face shield and a rubber apron should be worn. To prevent hand and skin exposures, butyl rubber gloves should be worn.

Administrative controls should also be in place to minimize the potential for human exposures. These may include written procedures or policies which specify the methods and techniques that will be practiced whenever personnel are to work with carbon disulfide.

All personnel should receive *training* on the use, hazards, protective measures, emergency actions, and other precautions per 29 CFR 1910.1200 (Hazard Communication), prior to the first assignment in an area where carbon disulfide is used or stored.

Before beginning employment and at regular times thereafter, the following medical tests are recommended:

- ☑ Examination of the nervous system, including mental status exam.
- ☑ Measurement of neurological impairment.
- ☑ Examination of the cardiovascular system for signs of arteriosclerosis.
- ☑ Iodine azide test to determine the intensity of an exposure to carbon disulfide (detects carbon disulfide metabolites present in urine).

Other methods to reduce exposure include:

- ☑ Where possible, enclose operations and use local exhaust ventilation at the site of chemical release. If local exhaust ventilation or enclosure is not used, respiratory protection should be mandatory.
- ☑ Always ensure that proper protective clothing is worn when using chemical substances.
- ☑ Wash thoroughly immediately after exposure to carbon disulfide and at the end of the work shift or before eating, drinking, or smoking. Never wear clothes contaminated with carbon disulfide home. Family members can be exposed.
- ☑ Hazard warning information should be posted in the work area. As part of an on-going education and training program, all information on the health and safety hazards of carbon disulfide should be communicated to all potentially exposed workers.

RISK ASSESSMENT: ENVIRONMENT

General Assessment

The environment is at risk of exposure during transportation, storage, disposal, or destruction of carbon disulfide. In almost every scenario, the threat of environmental exposure is contingent upon the proper handling of the chemical substance. Accidental spills, large or small, can result in fire, explosion, and possible contamination of the surrounding environmental mediums (water, soil, and air).

Carbon disulfide is considered a highly flammable liquid. Because of its low flash point and relatively high boiling point, it is classified as a Class 1B flammable liquid (per OSHA 29 CFR 1910.106). Its low auto-ignition temperature and low flash point, together with its high vapor pressure make it a dangerous fire and explosion hazard. It is also incompatible with a number of commodities. Potentially explosive reactions can occur if carbon disulfide is mixed or comes into contact with rust or iron, as well as any strong oxidizers. These characteristics require special consideration during any emergency involving a leak or spill of carbon disulfide.

The proper disposal/destruction method for small amounts of carbon disulfide (up to 10mL) is to burn it in a chemical incinerator equipped with an afterburner and air scrubber. Larger amounts should be mixed with a higher-boiling solvent and then burned.

Carbon disulfide can enter the environment through manufacturing, unchecked discharge into effluents from industrial or municipal waste treatment plants, and through spills.

Acute Ecological Effects

Acute (short-term) toxic effects may include the death of animals, birds, or fish, and death or low growth rate in plants. Acute effects are seen 2 to 4 days after animals or plants are exposed to carbon disulfide.

This chemical has moderate acute toxicity to aquatic life. No data are available on the short-term effects of carbon disulfide to plants, birds or land animals.

Chronic Ecological Effects

Chronic (long-term) toxic effects may include shortened life span, reproductive problems, lower fertility, and changes in appearance or behavior in exposed animals. These effects can be seen long after first exposure(s) to toxic chemicals.

Carbon disulfide has high chronic toxicity in aquatic life. No data are available to evaluate the long-term effects of carbon disulfide to plants, birds, or land animals.

Water Solubility

Carbon disulfide is moderately soluble in water. Concentrations between 1 and 1000 milligrams will mix with a liter of water.

Persistence in the Environment

Carbon disulfide is virtually non-persistent in water, with an estimated half-life of less than 2 days. The half-life of a pollutant is the amount of time it takes for one half of the chemical to be degraded. Approximately 99.8% of carbon disulfide will eventually end up in air; the rest will end up in water.

Bioaccumulation in Aquatic Organisms

Some substances increase in concentration, or *bioaccumulate,* in living organisms as they breathe contaminated air, drink contaminated water, or eat contaminated food. These chemicals can become concentrated in the tissues and internal organs of animals as well as humans.

The concentration of carbon disulfide found in fish tissues is expected to be somewhat higher than the average concentration of carbon disulfide in water from which the fish was taken.

Recommended Risk-Reduction Measures

Proper training of all transporters will reduce the likelihood of a mishap or accident resulting in a leak or spill to the environment. The correct use of DOT labeling (per the MSDS) on all containers, trucks, and rail cars will enable emergency responders to react properly and quickly to any disaster thereby reducing the potential risk to the environment and to personnel.

Storage of carbon disulfide should be segregated from incompatible chemicals to minimize the risk of cross-contamination or contact. Buildings designated for storage should be equipped with appropriate fire protection systems (alarms, sprinklers, emergency lighting, portable extinguishers).

If a spill or leak to the environment has occurred, fire department, emergency response, and/or hazardous materials spill personnel should be notified immediately. Cleanup should be attempted only by those trained in proper spill containment procedures. Contaminated soils should be removed for incineration and replaced with clean soil. If carbon disulfide should contact the water table, aquifer, or navigable waterway, time is of the essence. It is readily miscible in water and, therefore, total containment and remediation may not be entirely possible. When such spills occur, the local and/or state emergency response authorities must be notified. A comprehensive emergency response or disaster preparedness/recovery plan should be in place prior to any operations involving the use, transportation, storage, or disposal of carbon disulfide. If carbon disulfide is spilled or leaked, the following specific steps are recommended:

- ☑ Restrict persons not wearing protective clothing from area of spill or leak until cleanup is complete.
- ☑ Remove all ignition sources.
- ☑ Ventilate area of spill or leak.

- ☑ Absorb liquids in vermiculite, dry sand, earth, or a similar material and deposit in sealed containers.
- ☑ If applicable, stop flow of leaking liquid or gas. If leak source is a cylinder and the leak cannot be stopped in place, remove leaking cylinder to a safe place in the open air, and repair or allow cylinder to empty.
- ☑ Keep carbon disulfide out of a confined space, such as a sewer, because of the possibility of explosion (unless the sewer is designed to prevent the buildup of explosive concentrations).
- ☑ It may be necessary to dispose of carbon disulfide as a hazardous waste. The responsible state agency or the regional office of the federal Environmental Protection Agency (EPA) should be contacted for specific recommendations.

RISK ASSESSMENT: BUSINESS

General Assessment

Accidents or mishaps involving carbon disulfide can present a significant threat to business operations. The loss or damage of equipment or facilities can significantly affect fiscal viability. Lawsuits that may result from personnel injury, public exposures, and/or environmental contamination will also require a serious expenditure of resources.

Media attention surrounding an injury, illness, or environmental damage can also result in a loss of profits and loss of current as well as future business.

Recommended Risk-Reduction Measures

Company attorneys, safety and health professionals, and environmental specialists should be involved in the development of any procedures dealing with safety or emergency response. A company official should be pre-designated as a public relations officer with specific training in dealing with the press. Corporate plans and policies should be developed, approved, and implemented long before any need for such arises.

REFERENCES

American Conference of Governmental Industrial Hygienists. 1988. *Documentation of the Threshold Limit Values and Biological Exposure Indices,* 5th Edition (with updates). Cincinnati: ACGIH

Anthony, C. P., and N. J. Kolthoff. 1971. *Textbook of Anatomy and Physiology.* St. Louis: C. V. Mosby Company.

Bueche, J. R. 1972. *Principles of Physics,* 2nd Edition. New York: McGraw-Hill.

Burrell, R., D. K. Flaherty, and L. J. Sauers. 1992. *Toxicology of the Immune System.* New York: Van Nostrand Reinhold.

Calabrese, E. J. 1994. *Biological Effects of Low Level Exposures.* New York: CRC Press/Lewis Publishers.

Cothern, R. R. 1993. *Comparative Environmental Risk Assessment.* New York: CRC Press/Lewis Publishers.

Gardner, R. W. 1994. *Chemical Intolerance.* New York: CRC Press/Lewis Publishers.

Hallenbeck, W. H. 1993. *Quantitative Risk Assessment for Environmental and Occupational Health,* 2nd Edition. New York: CRC Press/Lewis Publishers.

Hodgson, E., R. B. Mailman, and J. E. Chambers. 1988. *Dictionary of Toxicology.* London: The MacMillian Company.

Howard, P. H., et. al. 1991. *Handbook of Environmental Degradation Rates.* New York: CRC Press/Lewis Publishers.

Jones, F. E. 1994. *Toxic Organic Vapors in the Workplace.* New York: CRC Press/Lewis Publishers.

Kamrin, M. A. 1988. *Toxicology: A Primer on Principles and Applications.* Chelsea, Michigan: Lewis Publishers.

Langerman, N. 1994 *Precautionary Labels for Chemical Containers.* New York: CRC Press/Lewis Publishers.

Lauwerys, R. R., P. Hoet. 1993. *Industrial Chemical Exposure.* New York: CRC Press/Lewis Publishers.

Lewis, R. J., Sr. 1992. *Sax's Dangerous Properties of Industrial Materials,* Volumes 1, 2, and 3, 8th Edition. New York: Van Nostrand Reinhold.

Lide, D. R. 1995. *Handbook of Chemistry and Physics,* 76th Edition. New York: CRC Press/Lewis Publishers.

Manahan, S. E. 1992. *Toxicological Chemistry,* 2nd Edition. New York: CRC Press/Lewis Publishers.

Mellström, G. A., J. E. Wahlberg, and H. I. Maibach. 1994. *Protective Gloves for Occupational Use.* New York: CRC Press/Lewis Publishers.

Meyer, E. 1990. *Chemistry of Hazardous Materials.* Englewood Cliffs, NJ: Prentice Hall.

Patnaik, P. 1992. *A Comprehensive Guide to the Hazardous Properties of Chemical Substances.* New York: Van Nostrand Reinhold.

Proctor, N. H., and J. P. Hughes. 1991. *Chemical Hazards of the Workplace.* New York: Van Nostrand Reinhold.

Quigley, D. R. 1994. *Handbook of Emergency Chemical Management.* New York: CRC Press/Lewis Publishers.

Richardson, M. 1995. *Dictionary of Substance and Their Effects,* Vol. I-VI. New York: CRC Press/Lewis Publishers.

Teisinger, J. 1974. *Encyclopedia of Environmental Health and Safety: Carbon Disulfide.* Vol. 1. New York: McGraw Hill.

U.S. Department of Health and Human Services, National Institute for Occupational Safety and Health. 1994. *NIOSH Pocket Guide to Chemical Hazards.* Washington, D.C.: U.S. Government Printing Office.

MATERIAL SAFETY DATA SHEET

CHEMICAL NAME	CRC PRESS/LEWIS PUBLISHERS DISCLAIMER
CARBON MONOXIDE	

HAZARD WARNING INFORMATION

HEALTH	FIRE	REACTIVE	OTHER	DEGREE OF HAZARD	COLOR CODING	OTHER CODES
3	4	0		0 = Minimum Hazard 1 = Slight Hazard 2 = Moderate Hazard 3 = Serious Hazard 4 = Severe Hazard	HEALTH = BLUE FIRE = RED REACTIVITY = YELLOW OTHER = WHITE	OX = Oxidizer ACID = Acid ALK = Alkali COR = Corrosive W = Use No Water

SECTION I - GENERAL INFORMATION

Characterization: **Gas**	RCRA Number: **None** — EPA Class: **Not Applicable**
DOT Proper Shipping Name: **Carbon monoxide**	Chemical Abstract Service (CAS) Number: **630-08-0**
DOT Hazard Class and Label Requirements: **Flammable Gas**	DOT Emergency Guide Code: **18** (gas) **67** (cryogenic liquid)
DOT Identification Number: **UN 1016** (gas) **NA9202** (cryogenic liquid)	Chemical Formula: **CO**

Synonyms
Carbon oxide; flue gas; monoxide; exhaust gas; carbonic oxide.

SECTION II - HAZARDOUS INGREDIENTS AND IDENTITY INFORMATION

Hazardous Components (specific identity)	OSHA Exposure Criteria	NIOSH Exposure Criteria	Immediately Dangerous to Life and Health (IDLH)	ACGIH Exposure Criteria
Carbon monoxide (derivation: When organic materials such as wood, coal, paper, oil, or gasoline are burned in limited air or oxygen; when flame contacts a surface colder than ignition temperature of its gaseous part; exhaust gas from internal combustion engines). **1 ppm = 1.16 mg/m³**	**PEL:** 50 ppm 55 mg/m³ **CEILING:** 200 ppm 229 mg/m³	**REL (10-hour):** 35 ppm 41 mg/m³ **CEILING:** 200 ppm 229 mg/m³	1200 ppm	**TLV:** 50 ppm 57 mg/m³ **STEL:** 400 ppm 458 mg/m³

SECTION III - PHYSICAL AND CHEMICAL CHARACTERISTICS

Boiling Point: **-313°F (-192°C)**	Specific Gravity (H_2O = 1): **0.97**
Vapor Pressure (atmospheres): **>760 at 69°F (20°C)**	Molecular Weight: **28.0**
Vapor Density (Air = 1): **0.97**	Freezing Point: **-337°F (-205°C)**

Solubility
Slightly soluble (2%) in water. Soluble in alcohol and benzene.

Appearance and Odor
Colorless gas, odorless gas. Shipped as a non-liquefied *or* liquefied compressed gas.

SECTION IV - FIRE AND EXPLOSION HAZARD DATA

Flash Point (method used): **Not Applicable (gas)**	Explosive Limits in Air % by Volume: **LEL: 12.5%** **UEL: 74%**
NFPA Classification: **Flammable Gas**	Autoignition Temperature: **1292°F (700°C)**

Extinguishing Media
Stop flow of gas before attempting to fight fire. Use water spray, fog, or regular foam.

Special Fire Fighting Procedures
Carbon monoxide is an extremely flammable gas and presents a serious explosion hazard when exposed to flame or in fire. Fires can produce toxic fumes: wear full protective clothing and self-contained breathing apparatus (SCBA). Heat will build pressure and may rupture closed storage containers. Keep fire-exposed containers cool with water spray.

Unusual Fire and Explosion Hazards
For large fires, use monitor nozzles or unmanned equipment, if possible. Boiling liquid expanding vapor explosion (BLEVE) is possible: immediately evacuate if rising sound is heard from venting device or fire is causing a discoloration in tanks. Isolate area 1500 -2000 feet in all directions.

SECTION V - REACTIVITY DATA

Stability		
		Conditions to Avoid **Carbon monoxide is stable under normal operating conditions. Avoid contact with incompatible materials, heat, and sources of ignition.**
Stable	Unstable	Incompatibility *(materials to avoid)* **Carbon monoxide is incompatible with strong oxidizers, bromine trifluoride, chlorine trifluoride, and most other halogen compounds, and lithium.**
X		
Hazardous Polymerization		Conditions to Avoid **Under normal conditions of temperature and pressure, hazardous polymerization of carbon monoxide cannot occur.**
May Occur	Will Not Occur	Hazardous Decomposition or By-products **Thermal oxidative decomposition of carbon monoxide can produce toxic and irritating fumes and gases, including carbon and carbon dioxide.**
	X	

SECTION VI - HEALTH HAZARD DATA

Primary Route(s) of Entry:	Inhalation?	Absorption (skin)?	Ingestion?
	X		

Health Hazards

INHALATION: **Extremely toxic gas. Causes chemical asphyxia with symptoms of headache, nausea, dizziness, loss of consciousness, convulsions, coma, infarction (heart attack), and death. In the bloodstream, it combines with hemoglobin 200 times more readily than oxygen creating carboxyhemoglobin and prevent oxygen distribution in the body.**

SKIN & EYES: **Contact with the liquid or compressed gas may cause burns (frostbite) or irritation.**

INGESTION: **Not likely.**

Carcinogenicity	NTP Listed?	IARC Cancer Review Group?	OSHA Regulated?	Target Organs?
No Evidence Human **No Evidence Animal**	**No**	**No**	**29 CFR 1910.1000 (Table Z-1)**	**Cardiovascular system, blood, CNS, lungs.**

Medical Conditions Generally Aggravated by Exposure

Blood disorders (anemia), heart diseases, chronic respiratory ailments.

Emergency and First-aid Procedures

<u>Eye contact</u>: Immediately flush large amounts of water for 15 minutes (minimum), occasionally lifting eyelids, seek medical attention. <u>Skin contact</u>: Remove all contaminated clothing. For frostbite, immediately wash area with large amounts of tepid water. Do NOT rub! For <u>inhalation</u>: Remove the person from exposure. Provide respiratory assistance and CPR. Transfer promptly to medical facility. DOCTOR: Treat aggressively with 100% oxygen and non-breather mask. If <u>swallowed</u>: Not likely.

SECTION VII - PRECAUTIONS FOR SAFE HANDLING AND USE

Steps to be Taken in Case Material is Released or Spilled

Wear SCBA. Remove all ignition sources. Ventilate area. Stop flow of gas. If the source of leak is a cylinder and cannot be stopped, remove to safe place in open air; repair leak or let cylinder empty.

Preferred Waste Disposal Method

None specified in the references (incineration of a gas may be the only alternative).

Precautions to be Taken in Handling and Storage

Do not store with incompatible chemicals since violent reactions can occur. Store in tightly closed containers in a cool, dark, well-ventilated area away heat. Keep away from ignition sources such as fire, sparks, and flame. Prevent damage to cylinders. Install carbon monoxide monitors with alarms.

Other Precautions and Warnings

Check containers for leaks using a soapy water solution upon arrival, whenever filling, and at least every three months if not used. Install Class I, Group C electrical equipment.

SECTION VIII - CONTROL MEASURES AND PERSONAL PROTECTIVE EQUIPMENT

Respiratory Protection *(specify type)*

Gas is odorless and tasteless. Exposures can occur without warning. Always use an MSHA/NIOSH approved full facepiece gas mask or respirator with an carbon monoxide gas cartridge. Greater protection is obtained from a supplied air respirator or a self-contained breathing apparatus (SCBA) with full facepiece and operated in pressure demand or other positive pressure or continuos flow mode.

Ventilation

Local exhaust or general mechanical systems recommended.

Protective Gloves	Eye Protection	Other Protective Clothing
Butyl Rubber	**Chemical Goggles and/or Face Mask**	**Rubber Apron**

Work/Hygiene Practices

Always wash hands thoroughly after using chemical; never bring food, drink, or smoking materials into vicinity of chemicals.

CARBON MONOXIDE

CO CAS: 630-08-0

IDENTIFICATION AND TYPICAL USES

Carbon monoxide is a colorless, odorless gas. It is usually shipped as a non-liquefied *or* liquefied compressed gas. It is used in the manufacture of metal carbonyls, zinc white pigments, as a reducing agent in metallurgy, and as an ingredient in many industrial gases used in heating boilers and furnaces. The primary source of industrial exposure is the use of internal combustion engines indoors or in enclosed areas.

RISK ASSESSMENT: HEALTH

General Assessment

Carbon monoxide is a deadly human poison primarily through *inhalation.* Skin or eye contact with the liquid also presents some exposure risk (frostbite and burns) but absorption is not likely since carbon monoxide is normally a gas. There are no reports of carcinogenic episodes in humans or animals. It is an experimental teratogen and has the ability to cross the placenta barrier.

Inhalation of carbon monoxide gas will cause progressive chemical asphyxia with symptoms of rapid, irregular breathing (tachypnea), feeling of air starvation or "air hunger," headache, fatigue, mental confusion, nausea, vomiting, giddiness, poor judgment, weakness, hallucinations, cyanosis, angina, syncope (sudden loss of strength and possible fainting), exhaustion, collapse, loss of consciousness, coma, convulsions, myocardial infarction, and death. In non-fatal exposures, there may be serious damage or injury to the central nervous system (CNS), and possible cerebral edema (fluid on the brain). Carbon monoxide has an affinity for blood hemoglobin some 200 times greater than that of oxygen. This means that it will bond much more quickly than oxygen and will form carboxyhemoglobin, thereby preventing the normal distribution of oxygen throughout the body. The gas may also combine with myoglobin to form carboxymyoglobin which may affect muscle metabolism and cause muscular disturbances, especially in the heart.

Acute Health Effects

The following acute (short-term) health effects may occur immediately or shortly after exposure to carbon monoxide:

Skin: Irritation and burning (frostbite) on contact with liquid.

Eye: Irritation and burns (frostbite) on contact with the liquid which can cause permanent damage.

Lung: Mild irritation of the respiratory tract. However, there is usually no indication of exposure and no noticeable effects until asphyxiation begins.

CVS: May cause angina, heart changes (depressed S-T segment on electrocardiogram), and myocardial infarction (heart attack).

CNS: Depression of the nervous system resulting from oxygen starvation.

Chronic Health Effects

The following chronic (long-term) health effects may occur at some time after exposure to carbon monoxide and can last for months or even years:

Cancer Hazards: According to information presented in the references, exposure to carbon monoxide does not currently lead to carcinogenic effects.

Reproductive Hazards: Since carbon monoxide has the ability to cross the placenta barrier, it may be able to harm a developing fetus. Experimental teratogenic effects have been reported but human data in this regard are inconclusive.

Other Chronic Effects: Repeated exposure may cause auditory and visual disturbances, glycosuria (high sugar content in urine), and cardiac irregularities.

There can also be cerebral congestion and edema with long-lasting mental or nervous system damage.

Recommended Risk-Reduction Measures

Personnel should avoid direct contact with carbon monoxide. It is extremely dangerous and provides no warning of its presence until after dangerous (life-threatening) exposures have occurred. Unless a less toxic chemical can be substituted for a hazardous substance, *engineering controls* are the most effective method of reducing exposures. The best protection is to enclose operations and/or provide local exhaust ventilation at the site of chemical release. While not always operationally feasible, isolating operations can also reduce exposure. Using respiratory protection is less effective than the controls mentioned above, but is still required whenever working with carbon monoxide. A NIOSH-approved, full facepiece respirator or gas mask with carbon monoxide gas cartridge or canister is sufficient for low exposures. A supplied-air respirator or a self-contained breathing apparatus (SCBA) with full facepiece operated in pressure demand or other positive pressure mode are the recommended respiratory protection methods of choice. If a full facepiece is not available, then chemical goggles should be worn to protect the eyes. Whenever a chemical splash hazard exists, a face shield and a rubber apron should be worn. To prevent hand and skin exposures, butyl rubber gloves should be worn.

Administrative controls should also be in place to minimize the potential for human exposures. These may include written procedures or policies which specify the methods and techniques that will be practiced whenever personnel are to work with carbon monoxide. The most common cause of occupational carbon monoxide exposures occurs as a result of the operation of liquid fuel engines inside a plant of facility. Administrative policies and procedures regarding the use of forklift trucks or other vehicles must be properly enforced to ensure adequate exposure control.

All personnel should receive *training* on the use, hazards, protective measures, emergency actions, and other precautions per 29 CFR 1910.1200 (Hazard Communication), prior to the first assignment in an area where carbon monoxide is used or stored.

Before beginning employment and at regular intervals thereafter (e.g., annually), the following medical tests are recommended:

☑ Lung function tests.

If symptoms develop or overexposure is suspected, the following may be useful:

☑ Lung function tests.

☑ Complete blood count (CBC) with blood gas measurement. Carboxyhemoglobin levels should be checked.

☑ Electrocardiogram (EKG) with specific attention on the S-T segment (may be unusually depressed).

☑ Evaluation of the nervous system by a neurologist.

Any medical evaluation should include a careful history of past and present symptoms with an examination. Medical tests that look for damage already present are not a substitute for controlling exposures. Also, the normal carboxyhemoglobin level in a non-smoker averages approximately 1%. It is common for smokers to carry a 5-10% level. Since exposure to carbon monoxide greatly increases the carboxyhemoglobin level, it is obvious that smokers will be affected more quickly by exposure than non-smokers under the same conditions of exposure. When levels reach 10-30%, there will be headache, nausea, and drowsiness (narcosis). At 40% and above, there can be confusion, weakness, and collapse. Levels of 50-60% can cause coma, convulsions, and death. Prudent risk management requires careful consideration of *all* possible exposure factors.

Other methods to reduce exposure to chemicals include:

☑ Where possible, enclose operations and use local exhaust ventilation at the site of chemical release. If local exhaust ventilation or enclosure is not used, respiratory protection should be mandatory.

☑ Always ensure that proper protective clothing is worn when using chemical substances and that personnel are adequately trained in its use, care, and maintenance.

☑ Hazard warning information should be posted in the work area. In addition, as part of an on-going education and training program, all information on the health and safety hazards of carbon monoxide should be communicated to all exposed and potentially exposed workers.

☑ Carbon monoxide gas monitors should be installed in work areas to provide adequate audible and visible alarms in the event of a leak or spill of carbon monoxide.

RISK ASSESSMENT: ENVIRONMENT

General Assessment

As an extremely volatile gas that is slightly lighter than air, the risk to the environment is primarily concerned with atmospheric pollution than with any other medium (soil or water). However, the environment is still at risk of exposure during transportation, storage, disposal, or destruction of carbon monoxide. In almost every scenario, the threat of environmental exposure is contingent upon the proper handling of the chemical substance. Accidental releases, large or small, can result in fire, explosion, and possible contamination of the air and atmosphere.

Carbon monoxide is an extremely dangerous flammable gas. Its vapors can explode violently in the presence of heat or flame. It can react with many metals and common oxidizers to cause fire or explosion. These characteristics require special consideration during any emergency involving a leak or spill of carbon monoxide.

Carbon monoxide can enter the environment through the use of any internal combustion, liquid-fueled engine as an exhaust product. It can also enter the environment through municipal waste treatment plant discharges, oil refinery and chemical manufacturing effluents, or from atmospheric fallout.

Acute Ecological Effects

Acute (short-term) toxic effects may include the death of animals, birds, or fish, and death or low growth rate in plants. Acute effects are seen 2 to 4 days after animals or plants are exposed to carbon monoxide.

The toxicity of liquid carbon monoxide is primarily due to its super-cold temperatures. The gas presents a similar hazard to animals as it does to humans. It is also more toxic under conditions of decreased oxygen concentrations.

Carbon monoxide has a moderate acute toxicity to birds and land animals. Insufficient data are available to evaluate or predict the short-term effects of carbon monoxide on plants or aquatic life.

Chronic Ecological Effects

Chronic (long-term) toxic effects may include shortened life span, reproductive problems, lower fertility, and changes in appearance or behavior in exposed animals. These effects can be seen long after first exposure(s) to toxic chemicals. Carbon monoxide has a moderate chronic toxicity to birds and land animals. Insufficient data are available to evaluate or predict the long-term effects of carbon monoxide on plants or aquatic life.

Water Solubility

Carbon monoxide is very slightly soluble in water. Concentrations of 1 milligram may mix with a liter of water.

Persistence in the Environment

Carbon monoxide is non-persistent in water, with a half-life of less than 2 days. The half-life of a pollutant is the amount of time it takes for one half of the chemical to be degraded. Virtually 100% of carbon monoxide will eventually end up in the air.

Bioaccumulation in Aquatic Organisms

Some substances increase in concentration, or *bioaccumulate,* in living organisms as they breathe contaminated air, drink contaminated water, or eat contaminated food. These chemicals can become concentrated in the tissues and internal organs of animals as well as humans.

Carbon monoxide is not expected to accumulate in the edible tissues of fish.

Recommended Risk-Reduction Measures

Proper training of all transporters will reduce the likelihood of a mishap or accident resulting in a leak or release to the environment. The correct use of DOT labeling (per the MSDS) on all containers, trucks, and rail cars will enable emergency responders to react properly and quickly to any disaster thereby reducing the potential risk to the environment and to personnel.

Storage of carbon monoxide should be segregated from incompatible chemicals to minimize the risk of cross-contamination or contact. Buildings designated for storage should be equipped with appropriate fire protection systems (alarms, sprinklers, emergency lighting, portable extinguishers). Explosion proof design (to Class 1, Group C standards) is recommended. All containers and equipment should be electrically grounded and bonded to prevent static sparks.

If a spill or release to the environment has occurred, fire department, emergency response, and/or hazardous materials spill personnel should be notified immediately. Cleanup should be attempted only by those trained in proper containment procedures. When such spills occur, it may be necessary to notify the local and/or state emergency response authorities. A comprehensive emergency response or disaster preparedness/recovery plan should be in place prior to any operations involving the use, transportation, storage, or disposal of carbon monoxide. If carbon monoxide is spilled or leaked, the following specific steps are recommended:

- ☑ Restrict persons not wearing protective clothing from area of spill or leak until cleanup is complete and area can be opened for normal work.
- ☑ Remove all ignition sources.
- ☑ Ventilate area of spill or leak.
- ☑ If applicable, stop flow of leaking liquid or gas. If leak source is a cylinder and the leak cannot be stopped in place, remove leaking cylinder to a safe place in the open air, and repair or allow cylinder to empty.
- ☑ Keep carbon monoxide out of a confined space, such as a sewer, because of the possibility of explosion (unless the sewer is designed to prevent the buildup of explosive concentrations).
- ☑ It may be necessary to dispose of carbon monoxide as a hazardous waste. The responsible state agency or the regional office of the federal Environmental Protection Agency (EPA) should be contacted for specific recommendations.

RISK ASSESSMENT: BUSINESS

General Assessment

Accidents or mishaps involving carbon monoxide can present a significant threat to business operations. The loss or damage of equipment or facilities can significantly affect fiscal viability. Lawsuits that may result from personnel injury, public exposures, and environmental contamination will also require a serious expenditure of resources. Media attention surrounding an injury, illness, or environmental damage can also result in a loss of profits and loss of current as well as future business.

Recommended Risk-Reduction Measures

Company attorneys, safety and health professionals, and environmental specialists should be involved in the development of any procedures dealing with safety or emergency response. A company official should be pre-designated as a public relations officer with specific training in dealing with the press. Corporate plans and policies should be developed, approved, and implemented long before any need for such arises.

REFERENCES

American Conference of Governmental Industrial Hygienists. 1988. *Documentation of the Threshold Limit Values and Biological Exposure Indices,* 5th Edition (with updates). Cincinnati: ACGIH

Anthony, C. P., and N. J. Kolthoff. 1971. *Textbook of Anatomy and Physiology.* St. Louis: C. V. Mosby Company.

Bueche, J. R. 1972. *Principles of Physics,* 2nd Edition. New York: McGraw-Hill.

Burrell, R., D. K. Flaherty, and L. J. Sauers. 1992. *Toxicology of the Immune System.* New York: Van Nostrand Reinhold.

Calabrese, E. J., E. M. Kenyon. *Air Toxics and Risk Assessment.* New York: CRC Press/Lewis Publishers.

Cothern, C. R. 1995. *Handbook for Environmental Risk Decisionmaking.* New York: CRC Press/Lewis Publishers.

Gardner, R. W. 1994. *Chemical Intolerance.* New York: CRC Press/Lewis Publishers.

Hodgson, E., R. B. Mailman, and J. E. Chambers. 1988. *Dictionary of Toxicology.* London: The MacMillian Company.

Hosty, J. W., and P. Foster. 1990. *A Practical Guide to Chemical Spill Response.* New York: Van Nostrand Reinhold.

Jones, F. E. 1994. *Toxic Organic Vapors in the Workplace.* New York: CRC Press/Lewis Publishers.

Keith, L. H. and M. M. Walker. 1995. *Handbook of Air Toxics.* New York: CRC Press/Lewis Publishers.

Landis, W. G., M. H. Yu. 1995. *Introduction to Environmental Toxicology.* New York: CRC Press/Lewis Publishers.

Langerman, N. 1994 *Precautionary Labels for Chemical Containers.* New York: CRC Press/Lewis Publishers.

Lauwerys, R. R., P. Hoet. 1993. *Industrial Chemical Exposure.* New York: CRC Press/Lewis Publishers.

Lewis, R. J., Sr. 1993. *Hawley's Condensed Chemical Dictionary,* 12th Edition. New York: Van Nostrand Reinhold.

Lewis, R. J., Sr. 1992. *Sax's Dangerous Properties of Industrial Materials,* Volumes 1, 2, and 3, 8th Edition. New York: Van Nostrand Reinhold.

Meyer, E. 1990. *Chemistry of Hazardous Materials.* Englewood Cliffs, NJ: Prentice Hall.

Patnaik, P. 1992. *A Comprehensive Guide to the Hazardous Properties of Chemical Substances.* New York: Van Nostrand Reinhold.

Rea, W. J. 1995. *Chemical Sensitivity,* Vol. I-III. New York: CRC Press/Lewis Publishers.

Richardson, M. 1995. *Dictionary of Substance and Their Effects,* Vol. I-VI. New York: CRC Press/Lewis Publishers.

U.S. Department of Health and Human Services, National Institute for Occupational Safety and Health. 1994. *NIOSH Pocket Guide to Chemical Hazards.* Washington, D.C.: U.S. Government Printing Office.

MATERIAL SAFETY DATA SHEET

CHEMICAL NAME	
CARBON TETRABROMIDE	**CRC PRESS/LEWIS PUBLISHERS DISCLAIMER:** THE INFORMATION AND RECOMMENDATIONS PRESENTED HEREIN ARE BASED ON SOURCES BELIEVED TO BE RELIABLE. CRC MAKES NO REPRESENTATION ON ITS COMPLETENESS OR ACCURACY. IT IS THE RESPONSIBILITY OF THE USER TO DETERMINE THE CHEMICAL'S SUITABILITY FOR ITS INTENDED USE, THE CHEMICAL'S SAFE USE, AND THE CHEMICAL'S PROPER DISPOSAL. NO REPRESENTATIONS AND/OR WARRANTIES, EITHER EXPRESSED OR IMPLIED, OF THE MERCHANTABILITY OR FITNESS FOR A PARTICULAR PURPOSE, OR OF ANY OTHER NATURE, ARE MADE WITH RESPECT TO THE INFORMATION PROVIDED IN THIS MSDS OR TO THE CHEMICAL TO WHICH INFORMATION MAY REFER. CRC NEITHER ASSUMES NOR AUTHORIZES ANY OTHER PERSON TO ASSUME FOR IT, ANY OTHER ADDITIONAL RESPONSIBILITY OR LIABILITY FOR THE USE OF, OR RELIANCE UPON, THIS INFORMATION.

HAZARD WARNING INFORMATION

HEALTH	FIRE	REACTIVE	OTHER	DEGREE OF HAZARD	COLOR CODING	OTHER CODES
4	0	0		0 = Minimum Hazard 1 = Slight Hazard 2 = Moderate Hazard 3 = Serious Hazard 4 = Severe Hazard	HEALTH = BLUE FIRE = RED REACTIVITY = YELLOW OTHER = WHITE	OX = Oxidizer ACID = Acid ALK = Alkali COR = Corrosive W = Use No Water

SECTION I - GENERAL INFORMATION

Characterization **Aliphatic Chlorinated Hydrocarbon**	RCRA Number **None**	EPA Class **Not Applicable**
DOT Proper Shipping Name **Carbon Tetrabromide**	Chemical Abstract Service (CAS) Number **558-13-4**	
DOT Hazard Class and Label Requirements **Poison B; St. Andrews Cross**	DOT Emergency Guide Code **53**	
DOT Identification Number **UN 2516**	Chemical Formula **CBr_4**	

Synonyms
Carbon bromide; tetrabromide methane; tetrabromomethane; methane tetrabromide.

SECTION II - HAZARDOUS INGREDIENTS AND IDENTITY INFORMATION

Hazardous Components (specific identity)	OSHA Exposure Criteria	NIOSH Exposure Criteria	Immediately Dangerous to Life and Health (IDLH)	ACGIH Exposure Criteria
Carbon tetrabromide (derivation: A brominated hydrocarbon). **1 ppm = 13.78 mg/m³**	**PEL:** **0.1 ppm** **1.4 mg/m³** **STEL:** **0.3 ppm** **4 mg/m³**	**REL:** **0.1 ppm** **1.4 mg/m³** **STEL:** **0.3 ppm** **4 mg/m³**	**Not Determined**	**TLV:** **0.1 ppm** **1.4 mg/m³** **STEL:** **0.3 ppm** **4 mg/m³**

SECTION III - PHYSICAL AND CHEMICAL CHARACTERISTICS

Boiling Point **373°F (189°C)**	Specific Gravity (H_2O = 1) **3.42**
Vapor Pressure (mm Hg) **40 at 205°F (96°C)**	Molecular Weight **331.7**
Vapor Density (Air = 1) **Not Found**	Freezing Point **194°F (90°C)**

Solubility
Practically insoluble (0.02%) in water. Soluble in alcohol, ether, and chloroform.

Appearance and Odor
Colorless to yellow-brown crystals with a slight odor.

SECTION IV - FIRE AND EXPLOSION HAZARD DATA

Flash Point (method used) **Not Applicable**	Explosive Limits in Air % by Volume **LEL: N/A UEL: N/A%**
NFPA Classification **Non-Combustible Solid**	Autoignition Temperature **Not Applicable**

Extinguishing Media
Use extinguishing media suitable to the surrounding fire. Carbon tetrabromide by itself is non-combustible and does not burn.

Special Fire Fighting Procedures
Wear full protective clothing and self-contained breathing apparatus (SCBA). Move containers from fire area if it can be done without risk. Fight fire from a distance if possible. Keep fire-exposed containers cool with water spray. Dike area for control and containment to prevent entry into sewers, drain, and waterways. Evacuate non-essential personnel 2500 feet from fire area. Consider down-wind conditions

Unusual Fire and Explosion Hazards
Containers may explode in fire.

SECTION V - REACTIVITY DATA

Stability		Conditions to Avoid: Normally stable under routine conditions of handling and storage. Avoid contact with fire or high heat sources. Do not use in confined spaces. Keep away from incompatible materials.
Stable	Unstable	Incompatibility *(materials to avoid)*: Strong oxidizers (chlorine, fluorine, bromine, bromine trifluoride, chlorine trifluoride) lithium, and hexacyclohexyldilead.
X		
Hazardous Polymerization		Conditions to Avoid: Under normal conditions of temperature and pressure, hazardous polymerization of carbon tetrabromide is not known to occur.
May Occur	Will Not Occur	Hazardous Decomposition or By-products: When heated to decomposition, carbon tetrabromide emits highly toxic fumes of carbon dioxide and vapors of bromine.
	X	

SECTION VI - HEALTH HAZARD DATA

Primary Route(s) of Entry:	Inhalation?	Absorption (skin)?	Ingestion?
	X		X

Health Hazards

INHALATION: Irritation of the eyes, nose, throat, and respiratory tract, with injury to the lungs, kidneys, and liver. There can be narcosis. Eye tearing (lachrymation) may also develop.

EYES & SKIN: Skin and eye irritation. Corneal damage may occur in some cases. Prolonged contact with skin (from contaminated clothing or shoes) may cause hyperemia (concentration of blood in a body part), and moderate edema.

INGESTION: Poisonous on ingestion.

Carcinogenicity	NTP Listed?	IARC Cancer Review Group?	OSHA Regulated?	Target Organs?
Unknown Human Unknown Animal	No	No	No	Eyes,; skin, liver, CNS, kidneys.

Medical Conditions Generally Aggravated by Exposure

None reported.

Emergency and First-aid Procedures

Eye contact: Immediately flush large amounts of water for 15 minutes (minimum), occasionally lifting eyelids, seek medical attention. Skin contact: Remove all contaminated clothing. Immediately wash area with large amounts of soap and water. For inhalation: Remove the person from exposure. Provide respiratory assistance and CPR. Transfer promptly to medical facility. If swallowed: Give 1 - 2 glasses of milk or water. Seek medical attention immediately. Unless advised otherwise, do NOT induce vomiting. Never try to give an unconscious are convulsing person anything by mouth.

SECTION VII - PRECAUTIONS FOR SAFE HANDLING AND USE

Steps to be Taken in Case Material is Released or Spilled

Restrict persons not wearing protective equipment from entering area of spill or leak until cleanup is complete. Ventilate area. Collect solid materials using a HEPA vacuum and deposit in sealed containers. Alert appropriate state or federal response agencies. Do NOT dry sweep (generates dusts).

Preferred Waste Disposal Method

No citation.

Precautions to be Taken in Handling and Storage

Do not store with incompatible chemicals since violent reactions can occur. Store in tightly closed containers in a cool, dark, well-ventilated area away heat. Provide explosion proof ventilation in storage areas. Smoking is prohibited in this area.

Other Precautions and Warnings

Bulk storage of carbon tetrabromide is not recommended. Empty containers can still contain sufficient residues to present a hazard exposure risk.

SECTION VIII - CONTROL MEASURES AND PERSONAL PROTECTIVE EQUIPMENT

Respiratory Protection *(specify type)*

At low exposures, use an organic vapor respirator with a dust/mist pre-filter. For higher exposures, use a supplied air respirator with full facepiece operated in positive pressure mode or a self-contained breathing apparatus (SCBA) with full facepiece in pressure demand or other positive pressure mode.

Ventilation

Local exhaust or general mechanical systems recommended.

Protective Gloves	Eye Protection	Other Protective Clothing
Solvent Resistant Butyl Rubber	Chemical Goggles and/or Face Mask	Rubber Apron

Work/Hygiene Practices

Always wash hands thoroughly after using chemical; never bring food, drink, or smoking materials into vicinity of chemicals.

CARBON TETRABROMIDE

CBr_4 CAS: 558-13-4

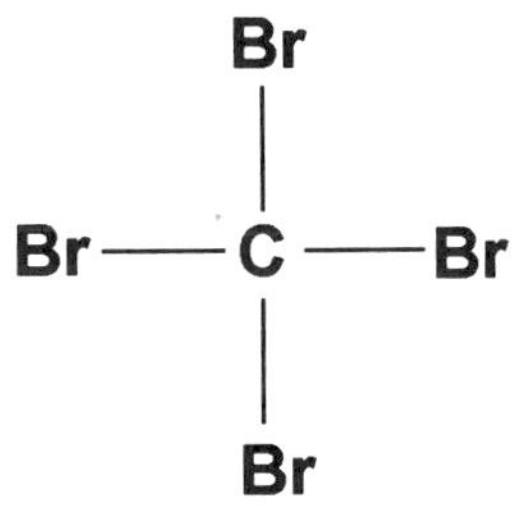

IDENTIFICATION AND TYPICAL USES

Carbon tetrabromide is a colorless to yellow-brown crystalline solid with a slight odor. It is used almost exclusively in organic synthesis.

RISK ASSESSMENT: HEALTH

General Assessment

Carbon tetrabromide is a moderately to highly toxic compound, primarily through *ingestion* and *inhalation.* Skin contact can also produce exposure risks and, in some instances *absorption* through intact skin has been reported (although occurrence is considered rare). It can have narcotic effects at high concentrations. There are no reports of any carcinogenic, teratogenic, or mutagenic hazards resulting from exposure to carbon tetrabromide.

Inhalation of high concentrations of carbon tetrabromide dusts or vapors (from the solution) can cause irritation of the eyes, nose, and upper respiratory tract producing coughing, congestion, labored breathing, lachrymation (tearing), and other symptoms of respiratory irritation.

Skin contact with the solid material or the liquid solution can produce severe localized irritation. Eye contact is more dangerous, resulting in severe irritation with burns and possible permanent damage to the cornea. Prolonged skin contact (from contaminated clothing or soaked/contaminated shoes) may cause hyperemia, a condition characterized by a congestion of blood in one area or part of the body.

Acute Health Effects

The following acute (short-term) health effects may occur immediately or shortly after exposure to carbon tetrabromide:

Skin: Irritation/rash on contact. Possible burns and the development of hyperemia if chemical is not quickly removed.

Eye: Irritation and possible damage to the cornea which may be permanent.

Lung: Irritation of the respiratory tract, possible lung damage.

CNS: At high concentrations, narcotic effects may occur. These include headache, dizziness, fatigue, nervousness, vomiting, diarrhea, renal damage, and liver injury.

Chronic Health Effects

The following chronic (long-term) health effects may occur at some time after exposure to carbon tetrabromide and can last for months or even years:

Cancer Hazards: According to information in the references, carbon tetrabromide has not been shown to cause cancer or mutations in test animals.

Reproductive Hazards: According to information presented in the references, carbon tetrabromide has not been adequately tested for its ability to adversely affect reproduction in test animals.

Other Chronic Effects: Repeated exposures can cause severe chronic liver, kidney, and lung damage.

Recommended Risk-Reduction Measures

Personnel should avoid direct contact with carbon tetrabromide. Unless a less toxic chemical can be substituted for a hazardous substance, *engineering controls* are the most effective method of reducing exposures. The best protection is to enclose operations and/or provide local exhaust ventilation at the site of chemical release. While not always operationally feasible, isolating operations can also reduce exposure.

Using respiratory protection is less effective than the controls mentioned above, but is still advisable whenever working with carbon tetrabromide. For relatively low or infrequent exposures, a organic vapor respirator equipped with a dust/mist pre-filter may provide suitable protection. Better respiratory protection is provided by a supplied-air respirator or a self-contained breathing apparatus (SCBA) with full facepiece operated in pressure demand or other continuous flow mode are the recommended respiratory protection methods of choice. If a full facepiece is not available, then chemical goggles should be worn to protect the eyes. Whenever a chemical splash hazard exists, a face shield and a rubber apron should be worn. To prevent hand and skin exposures, impervious rubber gloves should be worn. Glove manufacturers should be contacted and permeation studies obtained *before* final glove selections are made.

Administrative controls should also be in place to minimize the potential for human exposures. These may include written procedures or policies which specify the methods and techniques that will be practiced whenever personnel are to work with carbon tetrabromide.

All personnel should receive *training* on the use, hazards, protective measures, emergency actions, and other precautions per 29 CFR 1910.1200 (Hazard Communication), prior to the first assignment in an area where carbon tetrabromide is used or stored.

Before beginning employment and at regular times thereafter, the following medical tests are recommended:

☑ Liver function test.

☑ Lung function tests.

☑ Kidney function test, to include urinalysis.

Any medical evaluation should include a careful history of past and present symptoms with an examination. Medical tests that look for damage already present are not a substitute for controlling exposures. Also, because more than light or infrequent alcohol consumption can cause liver damage, it can increase the liver damage caused by exposure to carbon tetrabromide.

Other methods to reduce exposure include:

☑ Where possible, enclose operations and use local exhaust ventilation at the site of chemical release. If local exhaust ventilation or enclosure is not used, respiratory protection should be mandatory.

☑ Always ensure that proper protective clothing is worn when using chemical substances.

☑ Emergency eyewash and shower stations should be located in areas where exposure to carbon tetrabromide is possible.

☑ Wash thoroughly immediately after exposure to carbon tetrabromide and at the end of the work shift or before eating, drinking, or smoking. Never wear clothes contaminated with carbon tetrabromide home. Family members can be exposed.

☑ Hazard warning information should be posted in the work area. As part of an on-going education and training program, all information on the health and safety hazards of carbon tetrabromide should be communicated to all exposed or potentially exposed workers.

RISK ASSESSMENT: ENVIRONMENT

General Assessment

The environment is at risk of exposure during transportation, storage, disposal, or destruction of carbon tetrabromide. In almost every scenario, the threat of environmental exposure is contingent upon the proper handling of the chemical substance. Accidental spills, large or small, can result in fire, explosion, and possible contamination of the surrounding environmental mediums (water, soil, and air).

Carbon tetrabromide is considered a non-combustible solid. However, it is incompatible with a number of commodities. Potentially explosive reactions can occur if carbon tetrabromide is mixed or comes into contact with oxidizers or some metallic compounds. These characteristics require special consideration during any emergency involving a leak or spill of carbon tetrabromide.

Carbon tetrabromide can enter the environment through manufacturing, unchecked discharge into effluents from industrial or municipal waste treatment plants, and through spills.

☠ *Acute Ecological Effects*

Acute (short-term) toxic effects may include the death of animals, birds, or fish, and death or low growth rate in plants. Acute effects are seen 2 to 4 days after animals or plants are exposed to carbon tetrabromide.

This chemical has high acute toxicity to aquatic life. No data are available on the short-term effects of carbon tetrabromide to plants, birds or land animals.

Chronic Ecological Effects

Chronic (long-term) toxic effects may include shortened life span, reproductive problems, lower fertility, and changes in appearance or behavior in exposed animals. These effects can be seen long after first exposure(s) to toxic chemicals.

Carbon tetrabromide has high chronic toxicity in aquatic life. No data are available to evaluate the long-term effects of carbon tetrabromide to plants, birds, or land animals.

Water Solubility

Carbon tetrabromide is nearly insoluble in water. Concentrations between 1 and 10 milligrams will mix with a liter of water.

Persistence in the Environment

Carbon tetrabromide is moderately persistent in water, with a half-life between 20 and 200 days. The half-life of a pollutant is the amount of time it takes for one half of the chemical to be degraded.

Bioaccumulation in Aquatic Organisms

Some substances increase in concentration, or *bioaccumulate,* in living organisms as they breathe contaminated air, drink contaminated water, or eat contaminated food. These chemicals can become concentrated in the tissues and internal organs of animals as well as humans.

The concentration of carbon tetrabromide found in fish tissues is expected to be somewhat higher than the average concentration of carbon tetrabromide in water from which the fish was taken.

Recommended Risk-Reduction Measures

Proper training of all transporters will reduce the likelihood of a mishap or accident resulting in a leak or spill to the environment. The correct use of DOT labeling (per the MSDS) on all containers, trucks, and rail cars will enable emergency responders to react properly and quickly to any disaster thereby reducing the potential risk to the environment and to personnel.

Storage of carbon tetrabromide should be segregated from incompatible chemicals to minimize the risk of cross-contamination or contact. Storage facilities should be designed for maximum explosion-proof capability. Static sparks should be prevented, especially during transfer operations, with electrical grounds and bonds between all equipment involved.

If a spill or leak to the environment has occurred, fire department, emergency response, and/or hazardous materials spill personnel should be notified immediately. Cleanup should be attempted only by those trained in proper spill containment procedures using non-sparking tools and equipment. Contaminated soils should be removed for incineration and replaced with clean soil. If carbon tetrabromide should contact the water table, aquifer, or navigable waterway, time is of the essence. It is only slightly soluble in water and, therefore, total containment and remediation may be entirely possible. When such spills occur, the local and/or state emergency response authorities must be notified. A comprehensive emergency response or disaster preparedness/recovery plan should be in place prior to any operations involving the use, transportation, storage, or disposal of carbon tetrabromide.

If carbon tetrabromide is spilled or leaked, the following specific steps are recommended:

- ☑ Restrict persons not wearing protective clothing from area of spill or leak until cleanup is complete and area can be opened for normal work.
- ☑ Ventilate area of spill or leak. If in confined area, recommend using explosion-proof ventilation.
- ☑ Remove and restrict all ignition sources from area.
- ☑ Collect dry materials using safest and most convenient manner possible. Do NOT dry sweep (generates dangerous airborne dusts). Use a vacuum equipped with a high-efficiency particulate air (HEPA) filter instead. Damp mop residue.
- ☑ Absorb liquid solution spills in vermiculite, dry sand, earth, or a similar material and deposit in sealed containers. Use non-sparking tools!
- ☑ If applicable, stop flow of leaking liquid or gas. If leak source is a cylinder and the leak cannot be stopped in place, remove leaking cylinder to a safe place in the open air, and repair or allow cylinder to empty.
- ☑ It may be necessary to dispose of carbon tetrabromide as a hazardous waste. The responsible state agency or the regional office of the federal Environmental Protection Agency (EPA) should be contacted for specific recommendations.

RISK ASSESSMENT: BUSINESS

General Assessment

Accidents or mishaps involving carbon tetrabromide can present a significant threat to business operations. The loss or damage of equipment or facilities can significantly affect fiscal viability. Lawsuits that may result from personnel injury, public exposures, and/or environmental contamination will also require a serious expenditure of resources.

Media attention surrounding an injury, illness, or environmental damage can also result in a loss of profits and loss of current as well as future business.

℧ Recommended Risk-Reduction Measures

Company attorneys, safety and health professionals, and environmental specialists should be involved in the development of any procedures dealing with safety or emergency response. A company official should be pre-designated as a public relations officer with specific training in dealing with the press. Corporate plans and policies should be developed, approved, and implemented long before any need for such arises.

REFERENCES

Ahlbom, A. 1993. *Biostatistics for Epidemiologists.* New York: CRC Press/Lewis Publishers.

American Conference of Governmental Industrial Hygienists. 1988. *Documentation of the Threshold Limit Values and Biological Exposure Indices,* 5th Edition (with updates). Cincinnati: ACGIH

Anthony, C. P., and N. J. Kolthoff. 1971. *Textbook of Anatomy and Physiology.* St. Louis: C. V. Mosby Company.

Asante-Duah, D. K. 1993. *Hazardous Waste Risk Assessment.* New York: CRC Press/Lewis Publishers.

Bloom, A. D. and F. J. DeSerres. 1995. *Ecotoxicity and Human Health.* New York: CRC Press/Lewis Publishers.

Brusick, D. J. 1994. *Methods for Genetic Risk Assessment.* New York: CRC Press/Lewis Publishers.

Bueche, J. R. 1972. *Principles of Physics,* 2nd Edition. New York: McGraw-Hill.

Burrell, R., D. K. Flaherty, and L. J. Sauers. 1992. *Toxicology of the Immune System.* New York: Van Nostrand Reinhold.

Calabrese, E. J., E. M. Kenyon. *Air Toxics and Risk Assessment.* New York: CRC Press/Lewis Publishers.

Calabrese, E. J. 1994. *Biological Effects of Low Level Exposures.* New York: CRC Press/Lewis Publishers.

Cothern, R. R. 1993. *Comparative Environmental Risk Assessment.* New York: CRC Press/Lewis Publishers.

Danse, I. R. 1991. *Common Sense Toxics in the Workplace.* New York: Van Nostrand Reinhold.

Environmental Protection Agency. 1983. *Chemical Hazard Information Profiles.* Washington, D.C.: U.S. EPA.

Fodden, C. R., and J. L. Weddell. 1992. *Hazardous Materials Emergency Action Data.* New York: CRC Press/Lewis Publishers.

Forseberg, K., and L. H. Keith. 1993. *Chemical; Protective Clothing Permeation and Degradation Database.* Chelsea, MI: Lewis Publishers.

Gardner, R. W. 1994. *Chemical Intolerance.* New York: CRC Press/Lewis Publishers.

Hallenbeck, W. H. 1993. *Quantitative Risk Assessment for Environmental and Occupational Health,* 2nd Edition. New York: CRC Press/Lewis Publishers.

Hodgson, E., R. B. Mailman, and J. E. Chambers. 1988. *Dictionary of Toxicology.* London: The MacMillian Company.

Howard, P. H., et. al. 1991. *Handbook of Environmental Degradation Rates.* New York: CRC Press/Lewis Publishers.

Howard, P. H. 1995. *Handbook of Environmental Fate and Exposures Data for Organic Chemicals,* Vol. I-IV. New York: CRC Press/Lewis Publishers.

Jones, F. E. 1994. *Toxic Organic Vapors in the Workplace.* New York: CRC Press/Lewis Publishers.

Kamrin, M. A. 1988. *Toxicology: A Primer on Principles and Applications.* Chelsea, Michigan: Lewis Publishers.

Langerman, N. 1994 *Precautionary Labels for Chemical Containers.* New York: CRC Press/Lewis Publishers.

Lauwerys, R. R., P. Hoet. 1993. *Industrial Chemical Exposure.* New York: CRC Press/Lewis Publishers.

Lewis, R. J., Sr. 1992. *Sax's Dangerous Properties of Industrial Materials,* Volumes 1, 2, and 3, 8th Edition. New York: Van Nostrand Reinhold.

Lide, D. R. 1995. *Handbook of Chemistry and Physics,* 76th Edition. New York: CRC Press/Lewis Publishers.

Manahan, S. E. 1992. *Toxicological Chemistry,* 2nd Edition. New York: CRC Press/Lewis Publishers.

Mellström, G. A., J. E. Wahlberg, and H. I. Maibach. 1994. *Protective Gloves for Occupational Use.* New York: CRC Press/Lewis Publishers.

Meyer, E. 1990. *Chemistry of Hazardous Materials.* Englewood Cliffs, NJ: Prentice Hall.

Patnaik, P. 1992. *A Comprehensive Guide to the Hazardous Properties of Chemical Substances.* New York: Van Nostrand Reinhold.

Proctor, N. H., and J. P. Hughes. 1991. *Chemical Hazards of the Workplace.* New York: Van Nostrand Reinhold.

Quigley, D. R. 1994. *Handbook of Emergency Chemical Management.* New York: CRC Press/Lewis Publishers.

Rea, W. J. 1995. *Chemical Sensitivity,* Vol. I-III. New York: CRC Press/Lewis Publishers.

Richardson, M. 1995. *Dictionary of Substance and Their Effects,* Vol. I-VI. New York: CRC Press/Lewis Publishers.

U.S. Department of Health and Human Services, National Institute for Occupational Safety and Health. 1994. *NIOSH Pocket Guide to Chemical Hazards.* Washington, D.C.: U.S. Government Printing Office.

MATERIAL SAFETY DATA SHEET

CHEMICAL NAME

CARBON TETRACHLORIDE

CRC PRESS/LEWIS PUBLISHERS DISCLAIMER: THE INFORMATION AND RECOMMENDATIONS PRESENTED HEREIN ARE BASED ON SOURCES BELIEVED TO BE RELIABLE. CRC MAKES NO REPRESENTATION ON ITS COMPLETENESS OR ACCURACY. IT IS THE RESPONSIBILITY OF THE USER TO DETERMINE THE CHEMICAL'S SUITABILITY FOR ITS INTENDED USE, THE CHEMICAL'S SAFE USE, AND THE CHEMICAL'S PROPER DISPOSAL. NO REPRESENTATIONS AND/OR WARRANTIES, EITHER EXPRESSED OR IMPLIED, OF THE MERCHANTABILITY OR FITNESS FOR A PARTICULAR PURPOSE, OR OF ANY OTHER NATURE, ARE MADE WITH RESPECT TO THE INFORMATION PROVIDED IN THIS MSDS OR TO THE CHEMICAL TO WHICH INFORMATION MAY REFER. CRC NEITHER ASSUMES NOR AUTHORIZES ANY OTHER PERSON TO ASSUME FOR IT, ANY OTHER ADDITIONAL RESPONSIBILITY OR LIABILITY FOR THE USE OF, OR RELIANCE UPON, THIS INFORMATION.

HAZARD WARNING INFORMATION

HEALTH	FIRE	REACTIVE	OTHER	DEGREE OF HAZARD	COLOR CODING	OTHER CODES
3	0	0		0 = Minimum Hazard 1 = Slight Hazard 2 = Moderate Hazard 3 = Serious Hazard 4 = Severe Hazard	HEALTH = BLUE FIRE = RED REACTIVITY = YELLOW OTHER = WHITE	OX = Oxidizer ACID = Acid ALK = Alkali COR = Corrosive ~~W~~ = Use No Water

SECTION I - GENERAL INFORMATION

Characterization: **Halogenated Hydrocarbon**	RCRA Number: **U211** / EPA Class: **Toxic Waste**
DOT Proper Shipping Name: **Carbon Tetrachloride**	Chemical Abstract Service (CAS) Number: **56-23-5**
DOT Hazard Class and Label Requirements: **ORM-A; Poison**	DOT Emergency Guide Code: **53**
DOT Identification Number: **UN 1846**	Chemical Formula: **CCl_4**

Synonyms

Carbon chloride; tetrachloromethane; tetrachlorocarbon; tetraform; carbon tet; Freon® 10; Halon® 104.

SECTION II - HAZARDOUS INGREDIENTS AND IDENTITY INFORMATION

Hazardous Components (specific identity)	OSHA Exposure Criteria	NIOSH Exposure Criteria	Immediately Dangerous to Life and Health (IDLH)	ACGIH Exposure Criteria
Carbon tetrachloride (derivation: Interaction of carbon disulfide and chlorine in the presence of iron; chlorination of methane or higher hydrocarbons at 250-400°C). **1 ppm = 6.39 mg/m³**	**PEL:** **10 ppm** **64 mg/m³** **CEILING:** **25 ppm** **PEAK (5 min)** **200 ppm**	**REL:** **Possible Cancer Agent** **STEL (60 min):** **2 ppm** **13 mg/m³**	**200 ppm**	**TLV:** **Suspected Human Carcinogen** **5 ppm** **32 mg/m³** **30 ppm (skin)**

SECTION III - PHYSICAL AND CHEMICAL CHARACTERISTICS

Boiling Point: **171°F (77°C)**	Specific Gravity (H_2O = 1): **1.59**
Vapor Pressure (mm Hg): **91 at 69°F (20°C)**	Molecular Weight: **153.8**
Vapor Density (Air = 1): **5.3**	Freezing Point: **-9°F (-23°C)**

Solubility

Practically insoluble in water (0.05%). Miscible with alcohol, ether, chloroform, benzene, solvent naphtha, and most of the fixed and volatile oils.

Appearance and Odor

Colorless liquid with a characteristic ether-like odor. Odor threshold = 96 ppm.

SECTION IV - FIRE AND EXPLOSION HAZARD DATA

Flash Point (method used): **Not Applicable**	Explosive Limits in Air % by Volume: **LEL: N/A UEL: N/A%**
NFPA Classification: **Non-combustible Liquid**	Autoignition Temperature: **Not Applicable**

Extinguishing Media

Use extinguishing media suitable to the surrounding fire. Carbon tetrachloride by itself is non-combustible and does not burn.

Special Fire Fighting Procedures

Wear full protective clothing and self-contained breathing apparatus (SCBA). Move containers from fire area if it can be done without risk. Containers may explode in fire. Fight fire from a distance if possible. Keep fire-exposed containers cool with water spray.

Unusual Fire and Explosion Hazards

Explosions may occur when carbon tetrachloride is mixed with particles of many alkali metals, such as sodium, potassium, lithium, or their alloys, or finely divided aluminum, magnesium, or other metals.

SECTION V - REACTIVITY DATA

Stability		Conditions to Avoid **Do not allow carbon tetrachloride to come into contact with any of the incompatible chemicals listed below. Avoid contact with fire or high heat sources. Do not use in confined spaces.**
Stable	Unstable	Incompatibility *(materials to avoid)*
X		**Chemically active metals, such as sodium, potassium, and magnesium; barium, zinc, strong oxidizers (fluorine), halogens, peroxides, aluminum and aluminum trichloride, allyl alcohol.**
Hazardous Polymerization		Conditions to Avoid **Under normal conditions of temperature and pressure, hazardous polymerization of carbon tetrachloride is not known to occur.**
May Occur	Will Not Occur	Hazardous Decomposition or By-products
	X	**When heated to decomposition, carbon tetrachloride emits highly toxic fumes of chloride and phosgene.**

SECTION VI - HEALTH HAZARD DATA

Primary Route(s) of Entry:	Inhalation?	Absorption (skin)?	Ingestion?
	X	X	X

Health Hazards

INHALATION: **Symptoms include nausea, vomiting, pupillary constriction, coma, antipsychotic effects, tremors, somnolence, anorexia, unspecified respiratory system effects.**

ABSORPTION: **Skin contact can result in irritating rash or burning feeling. Can cause severe eye burns leading to permanent damage. Contact dermatitis is likely.**

INGESTION: **Same symptoms as those of inhalation with gastroenteritis also possible.**

Carcinogenicity	NTP Listed?	IARC Cancer Review Group?	OSHA Regulated?	Target Organs?
Suspected Human Confirmed Animal	**5Th Annual Report**	**Group 2B**	**No**	**Lung; skin; eyes; CNS, liver, kidneys.**

Medical Conditions Generally Aggravated by Exposure

Respiratory function impairments (asthma, etc.); Nervous disorders; Skin conditions (dermatitis).

Emergency and First-aid Procedures

Eye contact: Immediately flush large amounts of water for 15 minutes (minimum), occasionally lifting eyelids, seek medical attention. Skin contact: Remove all contaminated clothing. Immediately wash area with large amounts of soap and water. For inhalation: Remove the person from exposure. Provide respiratory assistance and CPR. Transfer promptly to medical facility. If swallowed: Induce vomiting if victim is conscious. Seek medical attention immediately. Never try to give an unconscious person anything by mouth or make an unconscious or convulsing person vomit.

SECTION VII - PRECAUTIONS FOR SAFE HANDLING AND USE

Steps to be Taken in Case Material is Released or Spilled

Restrict persons not wearing protective equipment from entering area of spill or leak until cleanup is complete. Ventilate area. Absorb liquid in dry sand or other material and deposit in sealed containers. Alert appropriate state or federal response agencies.

Preferred Waste Disposal Method

Mix with vermiculite, caustic soda, or slacked lime followed by burning in a chemical incinerator.

Precautions to be Taken in Handling and Storage

Do not store with incompatible chemicals since violent reactions can occur. Store in tightly closed containers in a cool, dark, well-ventilated area away heat. A regulated, marked area should be established where carbon tetrachloride is handled, used, or stored. Smoking is prohibited in this area.

Other Precautions and Warnings

Bulk storage of carbon tetrachloride is not recommended. Containers may explode in fire or under conditions of extreme heat (do not store outdoors or in direct sunlight).

SECTION VIII - CONTROL MEASURES AND PERSONAL PROTECTIVE EQUIPMENT

Respiratory Protection *(specify type)*

At any exposure level to a suspected human carcinogen, use a MSHA/NIOSH approved supplied air respirator with full facepiece operated in positive pressure mode or a self-contained breathing apparatus (SCBA) with full facepiece and pressure demand or other positive pressure mode.

Ventilation

Local exhaust or general mechanical systems recommended.

Protective Gloves	Eye Protection	Other Protective Clothing
Solvent Resistant Butyl Rubber	**Chemical Goggles and/or Face Mask**	**Rubber Apron**

Work/Hygiene Practices

Always wash hands thoroughly after using chemical; never bring food, drink, or smoking materials into vicinity of chemicals.

CARBON TETRACHLORIDE

CCl_4 CAS: 56-23-5

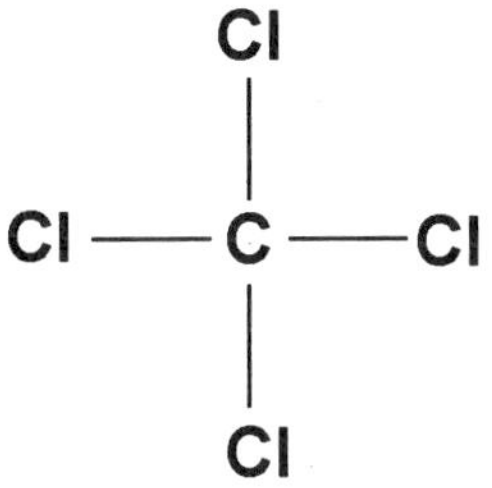

IDENTIFICATION AND TYPICAL USES

Carbon tetrachloride is a colorless, heavy liquid with a characteristic ether-like odor. It is used as a solvent, in fire extinguishers, in dry cleaning, and in the manufacture of fluorocarbon propellants. Also used in refrigerants, metal degreasing, as an agricultural fumigant, a chlorinating organic compound, and in the production of semiconductors. It is no longer permitted in products intended for home use.

RISK ASSESSMENT: HEALTH

General Assessment

Carbon tetrachloride is a confirmed carcinogen in animals and suspected in humans. It is a human poison by *ingestion* and possible other routes. It can also enter the body through skin *absorption* and *inhalation*. Human systemic effects by inhalation and ingestion include nausea or vomiting, pupillary constriction, coma, antipsychotic effects, tremors, somnolence, anorexia, unspecified respiratory system and gastrointestinal system effects. Experimental teratogenic and reproductive effects have also been reported. It is an eye irritant and skin irritant with dermatitis reaction possible. Mutation data have also been reported. It is known to damage the liver, kidney, and lungs.

Carbon tetrachloride has a narcotic action resembling that of chloroform, though not as strong. High exposures can cause loss of consciousness and death from respiratory failure.

Ingestion can be fatal in humans, death resulting from acute liver or kidney necrosis. Prolonged exposure to even small amounts has been reported to cause cirrhosis of the liver. There may be symptoms of dark urine, sometimes jaundice, and liver enlargement, followed by scanty urine, albumenuria and renal casts. Uremia may develop and cause death.

Acute Health Effects

The following acute (short-term) health effects may occur immediately or shortly after exposure to carbon tetrachloride:

<u>Skin</u>: Irritation/rash on contact. Dermatitis may also appear.

<u>Eye</u>: Irritation and possible damage which may be permanent.

<u>Lung</u>: Irritation of the respiratory tract, possible lung damage and death due to respiratory arrest. May also cause irregular heartbeat.

<u>CNS</u>: Effects include headache, dizziness, fatigue, nervousness, stupor, nausea, pupillary constriction, vomiting, diarrhea, renal damage, and liver injury. There may be coma, antipsychotic effects, tremors, somnolence, and anorexia.

Chronic Health Effects

The following chronic (long-term) health effects may occur at some time after exposure to carbon tetrachloride and can last for months or even years:

<u>Cancer Hazards</u>: Carbon tetrachloride is a probable human carcinogen. There is some evidence that it causes liver cancer in humans and it has been shown to cause liver cancer in animals.

<u>Reproductive Hazards</u>: According to information presented in the references, there is limited evidence that carbon tetrachloride has the ability to damage the developing fetus.

<u>Other Chronic Effects</u>: Repeated exposures can cause severe chronic liver, kidney, and lung damage. Repeated or prolonged contact with the skin may cause skin thickening and cracking. Prolonged exposure to small amounts of carbon tetrachloride may cause cirrhosis of the liver.

℧ *Recommended Risk-Reduction Measures*

Personnel should avoid direct contact with carbon tetrachloride. Unless a less toxic chemical can be substituted for a hazardous substance, *engineering controls* are the most effective method of reducing exposures. The best protection is to enclose operations and/or provide local exhaust ventilation at the site of chemical release. While not always operationally feasible, isolating operations can also reduce exposure. Using respiratory protection is less effective than the controls mentioned above, but is still advisable whenever working with carbon tetrachloride. For any exposure level to a suspected or probable human carcinogen, a self-contained breathing apparatus (SCBA) with full facepiece and pressure demand is the recommended respiratory protection of choice. If a full facepiece is not available, then chemical goggles should be worn to protect the eyes. Whenever a chemical splash hazard exists, a face shield and a rubber apron should be worn. To prevent hand and skin exposures, butyl rubber gloves should be worn.

Administrative controls should also be in place to minimize the potential for human exposures. These may include written procedures or policies which specify the methods and techniques that will be practiced whenever personnel are to work with carbon tetrachloride.

All personnel should receive *training* on the use, hazards, protective measures, emergency actions, and other precautions per 29 CFR 1910.1200 (Hazard Communication), prior to the first assignment in an area where carbon tetrachloride is used or stored.

Before beginning employment and at regular times thereafter, the following medical tests are recommended:

☑ Liver function test.

☑ Kidney function test, to include urinalysis.

Any medical evaluation should include a careful history of past and present symptoms with an examination. Medical tests that look for damage already present are not a substitute for controlling exposures. Also, because more than light or infrequent alcohol consumption can cause liver damage, it can increase the liver damage caused by exposure to carbon tetrachloride.

Other methods to reduce exposure include:

☑ Where possible, enclose operations and use local exhaust ventilation at the site of chemical release. If local exhaust ventilation or enclosure is not used, respiratory protection should be mandatory.

☑ Always ensure that proper protective clothing is worn when using chemical substances.

☑ Emergency eyewash and shower stations should be located in areas where exposure to carbon tetrachloride is possible.

☑ Wash thoroughly immediately after exposure to carbon tetrachloride and at the end of the work shift or before eating, drinking, or smoking. Never wear clothes contaminated with carbon tetrachloride home. Family members can be exposed.

☑ Hazard warning information should be posted in the work area. As part of an on-going education and training program, all information on the health and safety hazards of carbon tetrachloride should be communicated to all exposed workers.

RISK ASSESSMENT: ENVIRONMENT

General Assessment

The environment is at risk of exposure during transportation, storage, disposal, or destruction of carbon tetrachloride. In almost every scenario, the threat of environmental exposure is contingent upon the proper handling of the chemical substance. Accidental spills, large or small, can result in fire, explosion, and possible contamination of the surrounding environmental mediums (water, soil, and air).

Carbon tetrachloride is considered a non-combustible liquid (per OSHA 29 CFR 1910.106). However, it is incompatible with a number of commodities. Potentially explosive reactions can occur if carbon tetrachloride is mixed or comes into contact with fluorine, boranes, methanol, and most chemically-active metals. It forms impact-sensitive explosive mixtures with particulates of metals such as aluminum. These characteristics require special consideration during any emergency involving a leak or spill of carbon tetrachloride.

To dispose of carbon tetrachloride waste, it should be mixed with vermiculite, caustic soda, and slacked lime, followed by burning in a chemical incinerator equipped with an afterburner and scrubber. Incomplete combustion may produce chloroform. It may be mixed with a non-halogenated solvent (combustible) and incinerated. Carbon tetrachloride can enter the environment through manufacturing, unchecked discharge into effluents from industrial or municipal waste treatment plants, and through spills.

Acute Ecological Effects

Acute (short-term) toxic effects may include the death of animals, birds, or fish, and death or low growth rate in plants. Acute effects are seen 2 to 4 days after animals or plants are exposed to carbon tetrachloride.

This chemical has high acute toxicity to aquatic life. No data are available on the short-term effects of carbon tetrachloride to plants, birds or land animals.

Chronic Ecological Effects

Chronic (long-term) toxic effects may include shortened life span, reproductive problems, lower fertility, and changes in appearance or behavior in exposed animals. These effects can be seen long after first exposure(s) to toxic chemicals.

Carbon tetrachloride has high chronic toxicity in aquatic life. No data are available to evaluate the long-term effects of carbon tetrachloride to plants, birds, or land animals.

Water Solubility

Carbon tetrachloride is moderately soluble in water. Concentrations between 1 and 1000 milligrams will mix with a liter of water.

Persistence in the Environment

Carbon tetrachloride is non-persistent in water, with a half-life of less than 2 days. The half-life of a pollutant is the amount of time it takes for one half of the chemical to be degraded. Approximately 99.9% of carbon tetrachloride will eventually end up in air; the rest will end up in water.

Bioaccumulation in Aquatic Organisms

Some substances increase in concentration, or *bioaccumulate*, in living organisms as they breathe contaminated air, drink contaminated water, or eat contaminated food. These chemicals can become concentrated in the tissues and internal organs of animals as well as humans.

The concentration of carbon tetrachloride found in fish tissues is expected to be somewhat higher than the average concentration of carbon tetrachloride in water from which the fish was taken.

Recommended Risk-Reduction Measures

Proper training of all transporters will reduce the likelihood of a mishap or accident resulting in a leak or spill to the environment. The correct use of DOT labeling (per the MSDS) on all containers, trucks, and rail cars will enable emergency responders to react properly and quickly to any disaster thereby reducing the potential risk to the environment and to personnel.

Storage of carbon tetrachloride should be segregated from incompatible chemicals to minimize the risk of cross-contamination or contact.

If a spill or leak to the environment has occurred, fire department, emergency response, and/or hazardous materials spill personnel should be notified immediately. Cleanup should be attempted only by those trained in proper spill containment procedures. Contaminated soils should be removed for incineration and replaced with clean soil. If carbon tetrachloride should contact the water table, aquifer, or navigable waterway, time is of the essence. It is moderately soluble in water and, therefore, total containment and remediation may not be entirely possible. When such spills occur, the local and/or state emergency response authorities must be notified. A comprehensive emergency response or disaster preparedness/recovery plan should be in place prior to any operations involving the use, transportation, storage, or disposal of carbon tetrachloride. If carbon tetrachloride is spilled or leaked, the following specific steps are recommended:

- ☑ Restrict persons not wearing protective clothing from area of spill or leak until cleanup is complete and area can be opened for normal work.
- ☑ Ventilate area of spill or leak. If in confined area, recommend using maximum explosion-proof ventilation, if possible.
- ☑ Absorb liquids in vermiculite, dry sand, earth, or a similar material and deposit in sealed containers. Use non-sparking tools and equipment during cleanup operations.
- ☑ If applicable, stop flow of leaking liquid or gas. If leak source is a cylinder and the leak cannot be stopped in place, remove leaking cylinder to a safe place in the open air, and repair or allow cylinder to empty.
- ☑ It may be necessary to dispose of carbon tetrachloride as a hazardous waste. The responsible state agency or the regional office of the federal Environmental Protection Agency (EPA) should be contacted for specific recommendations.
- ☑ Carbon tetrachloride laboratory wastes may be placed in a lab pack and disposed of in an approved landfill.

RISK ASSESSMENT: BUSINESS

General Assessment

Accidents or mishaps involving carbon tetrachloride can present a significant threat to business operations. The loss or damage of equipment or facilities can significantly affect fiscal viability. Lawsuits that may result from personnel injury, public exposures, and/or environmental contamination will also require a serious expenditure of resources.

Media attention surrounding an injury, illness, or environmental damage can also result in a loss of profits and loss of current as well as future business.

Recommended Risk-Reduction Measures

Company attorneys, safety and health professionals, and environmental specialists should be involved in the development of any procedures dealing with safety or emergency response. A company official should be pre-designated as a public relations officer with specific training in dealing with the press. Corporate plans and policies should be developed, approved, and implemented long before any need for such arises.

REFERENCES

Ahlbom, A. 1993. *Biostatistics for Epidemiologists.* New York: CRC Press/Lewis Publishers.

American Conference of Governmental Industrial Hygienists. 1988. *Documentation of the Threshold Limit Values and Biological Exposure Indices,* 5th Edition (with updates). Cincinnati: ACGIH

Anthony, C. P., and N. J. Kolthoff. 1971. *Textbook of Anatomy and Physiology.* St. Louis: C. V. Mosby Company.

Asante-Duah, D. K. 1993. *Hazardous Waste Risk Assessment.* New York: CRC Press/Lewis Publishers.

Bloom, A. D. and F. J. DeSerres. 1995. *Ecotoxicity and Human Health.* New York: CRC Press/Lewis Publishers.

Brusick, D. J. 1994. *Methods for Genetic Risk Assessment.* New York: CRC Press/Lewis Publishers.

Bueche, J. R. 1972. *Principles of Physics,* 2nd Edition. New York: McGraw-Hill.

Burrell, R., D. K. Flaherty, and L. J. Sauers. 1992. *Toxicology of the Immune System.* New York: Van Nostrand Reinhold.

Calabrese, E. J., E. M. Kenyon. *Air Toxics and Risk Assessment.* New York: CRC Press/Lewis Publishers.

Calabrese, E. J. 1994. *Biological Effects of Low Level Exposures.* New York: CRC Press/Lewis Publishers.

Cothern, R. R. 1993. *Comparative Environmental Risk Assessment.* New York: CRC Press/Lewis Publishers.

Gardner, R. W. 1994. *Chemical Intolerance.* New York: CRC Press/Lewis Publishers.

Hallenbeck, W. H. 1993. *Quantitative Risk Assessment for Environmental and Occupational Health,* 2nd Edition. New York: CRC Press/Lewis Publishers.

Hodgson, E., R. B. Mailman, and J. E. Chambers. 1988. *Dictionary of Toxicology.* London: The MacMillian Company.

Howard, P. H., et. al. 1991. *Handbook of Environmental Degradation Rates.* New York: CRC Press/Lewis Publishers.

Howard, P. H. 1995. *Handbook of Environmental Fate and Exposures Data for Organic Chemicals,* Vol. I-IV. New York: CRC Press/Lewis Publishers.

Jones, F. E. 1994. *Toxic Organic Vapors in the Workplace.* New York: CRC Press/Lewis Publishers.

Kamrin, M. A. 1988. *Toxicology: A Primer on Principles and Applications.* Chelsea, Michigan: Lewis Publishers.

Langerman, N. 1994 *Precautionary Labels for Chemical Containers.* New York: CRC Press/Lewis Publishers.

Lauwerys, R. R., P. Hoet. 1993. *Industrial Chemical Exposure.* New York: CRC Press/Lewis Publishers.

Lewis, R. J., Sr. 1992. *Sax's Dangerous Properties of Industrial Materials,* Volumes 1, 2, and 3, 8th Edition. New York: Van Nostrand Reinhold.

Lide, D. R. 1995. *Handbook of Chemistry and Physics,* 76th Edition. New York: CRC Press/Lewis Publishers.

Manahan, S. E. 1992. *Toxicological Chemistry,* 2nd Edition. New York: CRC Press/Lewis Publishers.

Mellström, G. A., J. E. Wahlberg, and H. I. Maibach. 1994. *Protective Gloves for Occupational Use.* New York: CRC Press/Lewis Publishers.

Meyer, E. 1990. *Chemistry of Hazardous Materials.* Englewood Cliffs, NJ: Prentice Hall.

Patnaik, P. 1992. *A Comprehensive Guide to the Hazardous Properties of Chemical Substances.* New York: Van Nostrand Reinhold.

Proctor, N. H., and J. P. Hughes. 1991. *Chemical Hazards of the Workplace.* New York: Van Nostrand Reinhold.

Quigley, D. R. 1994. *Handbook of Emergency Chemical Management.* New York: CRC Press/Lewis Publishers.

Rea, W. J. 1995. *Chemical Sensitivity,* Vol. I-III. New York: CRC Press/Lewis Publishers.

Richardson, M. 1995. *Dictionary of Substance and Their Effects,* Vol. I-VI. New York: CRC Press/Lewis Publishers.

U.S. Department of Health and Human Services, National Institute for Occupational Safety and Health. 1976. *NIOSH Criteria Document: Carbon Tetrachloride,* Publication Number 76-122. Washington, D.C.: U.S. Government Printing Office.

U.S. Department of Health and Human Services, National Institute for Occupational Safety and Health. 1994. *NIOSH Pocket Guide to Chemical Hazards.* Washington, D.C.: U.S. Government Printing Office.

MATERIAL SAFETY DATA SHEET

CHEMICAL NAME	
CARBONYL SULFIDE	CRC PRESS/LEWIS PUBLISHERS DISCLAIMER: THE INFORMATION AND RECOMMENDATIONS PRESENTED HEREIN ARE BASED ON SOURCES BELIEVED TO BE RELIABLE. CRC MAKES NO REPRESENTATION ON ITS COMPLETENESS OR ACCURACY. IT IS THE RESPONSIBILITY OF THE USER TO DETERMINE THE CHEMICAL'S SUITABILITY FOR ITS INTENDED USE, THE CHEMICAL'S SAFE USE, AND THE CHEMICAL'S PROPER DISPOSAL. NO REPRESENTATIONS AND/OR WARRANTIES, EITHER EXPRESSED OR IMPLIED, OF THE MERCHANTABILITY OR FITNESS FOR A PARTICULAR PURPOSE, OR OF ANY OTHER NATURE, ARE MADE WITH RESPECT TO THE INFORMATION PROVIDED IN THIS MSDS OR TO THE CHEMICAL TO WHICH INFORMATION MAY REFER. CRC NEITHER ASSUMES NOR AUTHORIZES ANY OTHER PERSON TO ASSUME FOR IT, ANY OTHER ADDITIONAL RESPONSIBILITY OR LIABILITY FOR THE USE OF, OR RELIANCE UPON, THIS INFORMATION.

HAZARD WARNING INFORMATION

HEALTH	FIRE	REACTIVE	OTHER	DEGREE OF HAZARD	COLOR CODING	OTHER CODES
3	4	1		0 = Minimum Hazard 1 = Slight Hazard 2 = Moderate Hazard 3 = Serious Hazard 4 = Severe Hazard	HEALTH = BLUE FIRE = RED REACTIVITY = YELLOW OTHER = WHITE	OX = Oxidizer ACID = Acid ALK = Alkali COR = Corrosive W = Use No Water

SECTION I - GENERAL INFORMATION

Chemical Family **Carbonyl**	RCRA Number **None**	EPA Class **Not Applicable**
DOT Proper Shipping Name **Carbon Oxysulfide**	Chemical Abstract Service (CAS) Number **463-58-1**	
DOT Hazard Class and Label Requirements **Poison A; Poison Gas and Flammable Gas**	DOT Emergency Guide Code **18**	
DOT Identification Number **UN 2204**	Chemical Formula **COS**	

Synonyms
Carbon oxysulfide; carbon monoxide monosulfide; SCO.

SECTION II - HAZARDOUS INGREDIENTS AND IDENTITY INFORMATION

Hazardous Components (specific identity)	OSHA Exposure Criteria	NIOSH Exposure Criteria	Immediately Dangerous to Life and Health (IDLH)	ACGIH Exposure Criteria
Carbonyl sulfide (derivation: By hydrolysis of ammonium or potassium thiocyanate).	**PEL:** Not Established **STEL:** Not Established	**REL:** Not Established **STEL:** Not Established	**Not Determined**	**TLV:** Not Established **STEL:** Not Established

SECTION III - PHYSICAL AND CHEMICAL CHARACTERISTICS

Boiling Point **-58°F (-50°C)**	Specific Gravity (H_2O = 1) **1.2**
Vapor Pressure **160 psig at 69°F (20°C)**	Molecular Weight **60.1**
Vapor Density (Air = 1) **2.1**	Melting Point **-216°F (-138°C)**

Solubility in Water
Soluble in water and alcohol.

Appearance and Odor
Colorless gas or a cold liquid with typical sulfide odor (except in its pure state).

SECTION IV - FIRE AND EXPLOSION HAZARD DATA

Flash Point (method used) **Not Applicable (Gas)**	Explosive Limits in Air % by Volume **LEL: 12%** **UEL: 28.5%**
NFPA Classification **Not Applicable**	Autoignition Temperature **Not Applicable**

Extinguishing Media
Use dry chemical, carbon dioxide, water spray, or foam extinguishers.

Special Fire Fighting Procedures
Wear full protective clothing and self-contained breathing apparatus (SCBA). Move containers from fire area if it can be done without risk. Containers may explode in fire. Fight fire from a distance if possible. Keep fire-exposed containers cool with water spray.

Unusual Fire and Explosion Hazards
Carbonyl sulfide is a flammable gas or liquid. Fires may restart after they have been extinguished. Approach fire from upwind. Immediately withdraw if rising sound from venting device is heard or if fire is causing discoloration of a storage tank.

SECTION V - REACTIVITY DATA

Stability		Conditions to Avoid **Do not allow carbonyl sulfide to come into contact with any of the incompatible chemicals listed below. Avoid contact with fire or high heat sources. Do not use in confined spaces.**
Stable **X**	Unstable	Incompatibility *(materials to avoid)* **Do not store near or with oxidizers, (such as perchlorates, peroxides, permanganates, chlorates, and nitrates) and strong oxidizers (such as chlorine, bromine, and fluorine).**
Hazardous Polymerization		Conditions to Avoid **Under normal conditions of temperature and pressure, hazardous polymerization of carbonyl sulfide is not known to occur.**
May Occur	Will Not Occur **X**	Hazardous Decomposition or By-products **When heated to decomposition, carbonyl sulfide emits highly toxic fumes of carbon monoxide and hydrogen sulfide.**

SECTION VI - HEALTH HAZARD DATA

	Inhalation?	Absorption (skin)?	Ingestion?
Primary Route(s) of Entry:	X	X	

Health Hazards

INHALATION: Irritation of the nose, throat, and lungs causing cough and sneezing. High exposures may cause nausea, vomiting, weakness, and muscle cramps. It may also cause the heart to beat faster (tachyardia) or irregular (arrhythmia).

ABSORPTION: Skin contact can result in irritating rash or burning feeling. Can cause severe eye burns leading to permanent damage. Toxic systemic effects with CNS involvement are likely.

INGESTION: Not likely since carbonyl sulfide is a gas at normal temperature and pressure.

Carcinogenicity	NTP Listed?	IARC Cancer Review Group?	OSHA Regulated?	Target Organs?
Unknown Human **Unknown Animal**	**No**	**No**	**No**	**Respiratory system; skin; eyes; CNS, liver.**

Medical Conditions Generally Aggravated by Exposure

None reported.

Emergency and First-aid Procedures

<u>Eye contact</u>: Immediately flush large amounts of water for 15 minutes (minimum), occasionally lifting eyelids, seek medical attention. Contact lenses should not be worn when working with this material. <u>Skin contact</u>: Remove all contaminated clothing. Immediately wash area with large amounts of soap and water. For <u>inhalation</u>: Remove the person from exposure. Provide respiratory assistance and CPR. Transfer promptly to medical facility. If <u>swallowed</u>: Ingestion is not likely since this product is a gas at normal temperature and pressure.

SECTION VII - PRECAUTIONS FOR SAFE HANDLING AND USE

Steps to be Taken in Case Material is Released or Spilled

Restrict persons not wearing protective equipment from entering area of gas or liquid leak until cleanup is complete. Ventilate area. Stop flow of gas. If source is a cylinder and the leak cannot be stopped, remove to open air and allow to empty.

Preferred Waste Disposal Method

No Citation.

Precautions to be Taken in Handling and Storage

Do not store with incompatible chemicals (violent reactions can occur). Store in tightly closed containers in a cool, dark, well-ventilated area away from heat. Smoking is prohibited around this material.

Other Precautions and Warnings

Where possible, automatically transfer carbonyl sulfide from cylinders or other storage containers to process containers.

SECTION VIII - CONTROL MEASURES AND PERSONAL PROTECTIVE EQUIPMENT

Respiratory Protection *(specify type)*

At any exposure level, use a MSHA/NIOSH approved supplied air respirator with full facepiece operated in positive pressure mode or a self-contained breathing apparatus (SCBA) with full facepiece and pressure demand or other positive pressure mode.

Ventilation

Local exhaust or general mechanical systems recommended.

Protective Gloves	Eye Protection	Other Protective Clothing
Insulated, Impervious	**Chemical Goggles and/or Face Mask**	**Leather Apron**

Work/Hygiene Practices

Always wash hands thoroughly after using chemical; never bring food, drink, or smoking materials into vicinity of chemicals.

CARBONYL SULFIDE

COS CAS: 463-58-1

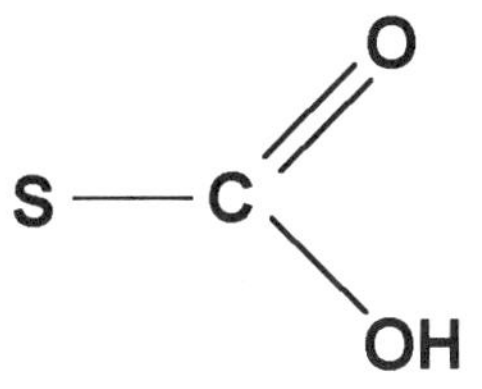

IDENTIFICATION AND TYPICAL USES

Carbonyl sulfide appears most commonly as a colorless gas with a typical sulfur-like odor. It may also be present as a cold liquid (boils at -58°F). It is used to make carbonate chemicals which are, in turn, used in many organic syntheses such as the manufacture of pharmaceuticals, solvents, agricultural chemicals, dyestuffs, and optical safety eye wear.

RISK ASSESSMENT: HEALTH

General Assessment

Carbonyl sulfide can enter the body either through *inhalation* or by passing through unbroken skin (*absorption*). Ingestion is not likely since carbonyl sulfide normally exists as a gas.

Breathing carbonyl sulfide gas can irritate the nose, throat, and lungs causing coughing and sneezing. High exposures cause nausea, vomiting, weakness, and muscle cramps. It may also cause the heart to beat faster (tachycardia) or irregularly (arrhythmia). Prolonged exposures to high concentrations may result in adverse effects to the nervous system, up to and including death. It may also damage the liver over prolonged periods of time and exposure.

On contact with the skin and eyes, it can cause burns. Even as a compressed gas, it is extremely cold and the resultant damage to tissues can therefore be quite severe.

Acute Health Effects

The following acute (short-term) health effects may occur immediately or shortly after exposure to carbonyl sulfide:

Skin: Irritation and burns (frostbite) on contact.

Eye: Irritation and possible damage which may be permanent. Tissue burning as a result from contact with cold gas or liquid can be quite severe.

Lung: Irritation of the respiratory tract and possible lung damage. May also cause irregular or fast heartbeat.

CNS: Effects include headache, dizziness, fatigue, confusion, salivation, nausea, vomiting, diarrhea, weakness, muscle cramps, and memory loss. Higher concentrations can cause loss of consciousness, convulsions, coma, sudden collapse, and death.

Chronic Health Effects

The following chronic (long-term) health effects may occur at some time after exposure to carbonyl sulfide and can last for months or even years:

Cancer Hazards: According to information presented in the references, carbonyl sulfide has not been adequately tested for its ability to cause cancer in animals.

Reproductive Hazards: According to information presented in the references, carbonyl sulfide has not been adequately tested for its ability to adversely affect reproduction.

Other Chronic Effects: Repeated exposures may cause brain damage. Effects could include reduced memory, inability to concentrate and/or personality changes (e.g., mood swings, irritability). Carbonyl sulfide may damage the liver. Repeated exposure may increase blood cholesterol and increase risk of cholesterol and fatty deposits in the blood vessels ("hardening of the arteries").

Recommended Risk-Reduction Measures

Personnel should avoid direct contact with carbonyl sulfide. Unless a less toxic chemical can be substituted for a hazardous substance, *engineering controls* are the most effective method of reducing exposures. The best protection is to enclose operations and/or provide local exhaust ventilation at the site of chemical release. While not always operationally feasible, isolating operations can also reduce exposure. Using respiratory protection is less effective than the controls mentioned above, but is still advisable whenever working with carbonyl sulfide. For any exposure level to a suspected or probable human carcinogen, a self-contained breathing apparatus (SCBA) with full facepiece and pressure demand is the recommended respiratory protection of choice. If a full facepiece is not available, then chemical goggles should be worn to protect the eyes. Whenever a chemical splash hazard exists, a face shield and a rubber apron should be worn. To prevent hand and skin exposures, insulated rubber gloves should be worn.

Administrative controls should also be in place to minimize the potential for human exposures. These may include written procedures or policies which specify the methods and techniques that will be practiced whenever personnel are to work with carbonyl sulfide.

All personnel should receive *training* on the use, hazards, protective measures, emergency actions, and other precautions per 29 CFR 1910.1200 (Hazard Communication), prior to the first assignment in an area where carbonyl sulfide is used or stored.

Before beginning employment and at regular times thereafter, for those with frequent or potentially high exposures, the following medical tests are recommended:

☑ Interview for brain effects, including recent memory, mood (irritability, withdrawal), concentration, headaches, malaise, and altered sleep patterns.

☑ Consider cerebellar, autonomic, and peripheral nervous system evaluation. Positive and borderline individuals should be referred for neuropsychological testing.

☑ Test for blood cholesterol levels.

If symptoms develop or overexposure is suspected, the following additional tests may be useful:

☑ Test for liver function.

☑ Complete nervous system evaluation by a qualified neurologist.

Any medical evaluation should include a careful history of past and present symptoms with an examination. Medical tests that look for damage already present are not a substitute for controlling exposures. Because more than light consumption of alcohol can cause liver damage, it can increase the liver damage potentially caused by exposure to carbonyl sulfide.

Other methods to reduce exposure include:

☑ Where possible, enclose operations and use local exhaust ventilation at the site of chemical release. If local exhaust ventilation or enclosure is not used, respiratory protection should be mandatory. Transfer of carbonyl sulfide from storage containers to process containers should be done automatically wherever possible.

☑ Always ensure that proper protective clothing is worn when using chemical substances.

☑ Emergency eyewash and shower stations should be located in areas where exposure to carbonyl sulfide is possible.

☑ Wash thoroughly immediately after exposure to carbonyl sulfide and at the end of the work shift or before eating, drinking, or smoking. Never wear clothes contaminated with carbonyl sulfide home. Family members can be exposed.

☑ Hazard warning information should be posted in the work area. As part of an on-going education and training program, all information on the health and safety hazards of carbonyl sulfide should be communicated to all potentially exposed workers.

RISK ASSESSMENT: ENVIRONMENT

General Assessment

The environment is at risk of exposure during transportation, storage, disposal, or destruction of carbonyl sulfide. In almost every scenario, the threat of environmental exposure is contingent upon the proper handling of the chemical substance. Accidental releases, large or small, can result in fire, explosion, and possible contamination of the surrounding environmental mediums (water and air).

Carbonyl sulfide is a very dangerous fire hazard and a moderate explosion hazard when exposed to heat or flame. Potentially explosive reactions can occur if carbonyl sulfide is mixed or comes into contact with oxidizing materials such as perchlorates, peroxides, permanganates, chlorates, and nitrates. These characteristics require special consideration during any emergency involving a liquid leak or gas release of carbonyl sulfide.

Carbonyl sulfide can enter the environment through acute industrial discharges and through spills.

Acute Ecological Effects

Acute (short-term) toxic effects may include the death of animals, birds, or fish, and death or low growth rate in plants. Acute effects are seen 2 to 4 days after animals or plants are exposed to carbonyl sulfide.

Insufficient data are available on the short-term effects of carbonyl sulfide to aquatic life, plants, birds or land animals.

Chronic Ecological Effects

Chronic (long-term) toxic effects may include shortened life span, reproductive problems, lower fertility, and changes in appearance or behavior in exposed animals. These effects can be seen long after first exposure(s) to toxic chemicals.

Insufficient data are available to evaluate the long-term effects of carbonyl sulfide to aquatic life, plants, birds, or land animals.

Water Solubility

Carbonyl sulfide is highly soluble in water. Concentrations of 1000 milligrams and more will mix with a liter of water.

Persistence in the Environment

Carbonyl sulfide is slightly in water, with a half-life between 2 and 20 days. The half-life of a pollutant is the amount of time it takes for one half of the chemical to be degraded. Carbonyl sulfide is very volatile and reacts slowly with water to form carbon dioxide and hydrogen sulfide. Due to its volatility, most carbonyl sulfide will escape to air.

Bioaccumulation in Aquatic Organisms

Some substances increase in concentration, or *bioaccumulate,* in living organisms as they breathe contaminated air, drink contaminated water, or eat contaminated food. These chemicals can become concentrated in the tissues and internal organs of animals as well as humans.

The concentration of carbonyl sulfide found in fish tissues is expected to be about the same as the average concentration of carbonyl sulfide in water from which the fish was taken.

Recommended Risk-Reduction Measures

Proper training of all transporters will reduce the likelihood of a mishap or accident resulting in a leak or spill to the environment. The correct use of DOT labeling (per the MSDS) on all containers, trucks, and rail cars will enable emergency responders to react properly and quickly to any disaster thereby reducing the potential risk to the environment and to personnel.

Storage of carbonyl sulfide should be segregated from incompatible chemicals to minimize the risk of cross-contamination or contact.

If a liquid spill or gas leak to the environment has occurred, fire department, emergency response, and/or hazardous materials spill personnel should be notified immediately. Cleanup should be attempted only by those trained in proper spill containment procedures. Although this chemical's volatility makes soil contamination highly unlikely, it can occur with massive liquid spills. Contaminated soils should be removed for incineration and replaced with clean soil. When such spills occur, the local and/or state emergency response authorities must be notified. A comprehensive emergency response or disaster preparedness/recovery plan should be in place prior to any operations involving the use, transportation, storage, or disposal of carbonyl sulfide. If carbonyl sulfide is spilled or leaked, the following specific steps are recommended:

- ☑ Restrict persons not wearing protective clothing from area of spill or leak until cleanup is complete and area can be opened for normal work.
- ☑ Ventilate area of spill or leak.
- ☑ If applicable, stop flow of leaking liquid or gas. If leak source is a cylinder and the leak cannot be stopped in place, remove leaking cylinder to a safe place in the open air, and repair or allow cylinder to empty.
- ☑ It may be necessary to dispose of carbonyl sulfide as a hazardous waste. The responsible state agency or the regional office of the federal Environmental Protection Agency (EPA) should be contacted for specific recommendations.

RISK ASSESSMENT: BUSINESS

General Assessment

Accidents or mishaps involving carbonyl sulfide can present a significant threat to business operations. The loss or damage of equipment or facilities can significantly affect fiscal viability. Lawsuits that may result from personnel injury, public exposures, and/or environmental contamination will also require a serious expenditure of resources.

Media attention surrounding an injury, illness, or environmental damage can also result in a loss of profits and loss of current as well as future business.

❶ *Recommended Risk-Reduction Measures*

Company attorneys, safety and health professionals, and environmental specialists should be involved in the development of any procedures dealing with safety or emergency response. A company official should be pre-designated as a public relations officer with specific training in dealing with the press. Corporate plans and policies should be developed, approved, and implemented long before any need for such arises.

REFERENCES

Ahlbom, A. 1993. *Biostatistics for Epidemiologists.* New York: CRC Press/Lewis Publishers.

American Conference of Governmental Industrial Hygienists. 1988. *Documentation of the Threshold Limit Values and Biological Exposure Indices,* 5th Edition (with updates). Cincinnati: ACGIH

Anthony, C. P., and N. J. Kolthoff. 1971. *Textbook of Anatomy and Physiology.* St. Louis: C. V. Mosby Company.

Asante-Duah, D. K. 1993. *Hazardous Waste Risk Assessment.* New York: CRC Press/Lewis Publishers.

Bloom, A. D. and F. J. DeSerres. 1995. *Ecotoxicity and Human Health.* New York: CRC Press/Lewis Publishers.

Brusick, D. J. 1994. *Methods for Genetic Risk Assessment.* New York: CRC Press/Lewis Publishers.

Bueche, J. R. 1972. *Principles of Physics,* 2nd Edition. New York: McGraw-Hill.

Burrell, R., D. K. Flaherty, and L. J. Sauers. 1992. *Toxicology of the Immune System.* New York: Van Nostrand Reinhold.

Calabrese, E. J., E. M. Kenyon. *Air Toxics and Risk Assessment.* New York: CRC Press/Lewis Publishers.

Calabrese, E. J. 1994. *Biological Effects of Low Level Exposures.* New York: CRC Press/Lewis Publishers.

Cothern, R. R. 1993. *Comparative Environmental Risk Assessment.* New York: CRC Press/Lewis Publishers.

Gardner, R. W. 1994. *Chemical Intolerance.* New York: CRC Press/Lewis Publishers.

Hallenbeck, W. H. 1993. *Quantitative Risk Assessment for Environmental and Occupational Health,* 2nd Edition. New York: CRC Press/Lewis Publishers.

Hodgson, E., R. B. Mailman, and J. E. Chambers. 1988. *Dictionary of Toxicology.* London: The MacMillian Company.

Howard, P. H., et. al. 1991. *Handbook of Environmental Degradation Rates.* New York: CRC Press/Lewis Publishers.

Howard, P. H. 1995. *Handbook of Environmental Fate and Exposures Data for Organic Chemicals,* Vol. I-IV. New York: CRC Press/Lewis Publishers.

Jones, F. E. 1994. *Toxic Organic Vapors in the Workplace.* New York: CRC Press/Lewis Publishers.

Kamrin, M. A. 1988. *Toxicology: A Primer on Principles and Applications.* Chelsea, Michigan: Lewis Publishers.

Langerman, N. 1994 *Precautionary Labels for Chemical Containers.* New York: CRC Press/Lewis Publishers.

Lauwerys, R. R., P. Hoet. 1993. *Industrial Chemical Exposure.* New York: CRC Press/Lewis Publishers.

Lewis, R. J., Sr. 1992. *Sax's Dangerous Properties of Industrial Materials,* Volumes 1, 2, and 3, 8th Edition. New York: Van Nostrand Reinhold.

Lide, D. R. 1995. *Handbook of Chemistry and Physics,* 76th Edition. New York: CRC Press/Lewis Publishers.

Manahan, S. E. 1992. *Toxicological Chemistry,* 2nd Edition. New York: CRC Press/Lewis Publishers.

Mellström, G. A., J. E. Wahlberg, and H. I. Maibach. 1994. *Protective Gloves for Occupational Use.* New York: CRC Press/Lewis Publishers.

Meyer, E. 1990. *Chemistry of Hazardous Materials.* Englewood Cliffs, NJ: Prentice Hall.

Patnaik, P. 1992. *A Comprehensive Guide to the Hazardous Properties of Chemical Substances.* New York: Van Nostrand Reinhold.

Proctor, N. H., and J. P. Hughes. 1991. *Chemical Hazards of the Workplace.* New York: Van Nostrand Reinhold.

Quigley, D. R. 1994. *Handbook of Emergency Chemical Management.* New York: CRC Press/Lewis Publishers.

Rea, W. J. 1995. *Chemical Sensitivity,* Vol. I-III. New York: CRC Press/Lewis Publishers.

Richardson, M. 1995. *Dictionary of Substance and Their Effects,* Vol. I-VI. New York: CRC Press/Lewis Publishers.

U.S. Department of Health and Human Services, National Institute for Occupational Safety and Health. 1994. *NIOSH Pocket Guide to Chemical Hazards.* Washington, D.C.: U.S. Government Printing Office.

MATERIAL SAFETY DATA SHEET

CHEMICAL NAME

CESIUM HYDROXIDE

HAZARD WARNING INFORMATION

HEALTH	FIRE	REACTIVE	OTHER	DEGREE OF HAZARD	COLOR CODING	OTHER CODES
3	0	1		0 = Minimum Hazard 1 = Slight Hazard 2 = Moderate Hazard 3 = Serious Hazard 4 = Severe Hazard	HEALTH = BLUE FIRE = RED REACTIVITY = YELLOW OTHER = WHITE	OX = Oxidizer ACID = Acid ALK = Alkali COR = Corrosive W = Use No Water

SECTION I - GENERAL INFORMATION

Characterization	**Metallic Hydroxide**	RCRA Number	**D002**
		EPA Class	**Characteristic Waste**
DOT Proper Shipping Name	**Cesium Hydroxide** (solid)	Chemical Abstract Service (CAS) Number	**21351-79-1**
DOT Hazard Class and Label Requirements	**Corrosive Material**	DOT Emergency Guide Code	**60**
DOT Identification Number	**UN 2682** (solid) **UN 2681** (solution)	Molecular Formula	**CsOH**

Synonyms

Cesium hydrate; cesium hydroxide dimer.

SECTION II - HAZARDOUS INGREDIENTS AND IDENTITY INFORMATION

Hazardous Components (specific identity)	OSHA Exposure Criteria	NIOSH Exposure Criteria	Immediately Dangerous to Life and Health (IDLH)	ACGIH Exposure Criteria
Cesium hydroxide (derivation: From adding barium hydroxide to an aqueous solution of cesium sulfate).	**PEL:** Not Established **STEL:** Not Established	**REL:** 2 mg/m$_3$ **STEL:** Not Established	Not Determined	**TLV:** 2 mg/m^3 **STEL:** Not Established

SECTION III - PHYSICAL AND CHEMICAL CHARACTERISTICS

Boiling Point	**Not Found**	Specific Gravity (H_2O = 1)	**3.68**
Vapor Pressure (mm Hg)	**0 at 68°F (20°C)**	Molecular Weight	**149.9**
Vapor Density (Air = 1)	**Not Found**	Melting Point	**522°F (272°C)**

Solubility

Highly soluble in water. Cesium hydroxide is the strongest known base compound.

Appearance and Odor

Colorless or yellowish, fused, crystalline mass. Strong alkaline reaction. Deliquescent and hygroscopic (can become liquid by absorbing moisture from the air). No detectable odor.

SECTION IV - FIRE AND EXPLOSION HAZARD DATA

Flash Point (method used)	**Not Applicable**	Explosive Limits in Air % by Volume	**LEL: N/A UEL: N/A**
NFPA Classification	**Non-Combustible Solid**	Autoignition Temperature	**Not Applicable**

Extinguishing Media

Use extinguishing agent suitable to surrounding fire (dry chemical, foam, or water).

Special Fire Fighting Procedures

Move containers from fire if it can be done without risk. Evacuate non-essential personnel 2500 feet from fire area. Fire fighters should take care to avoid dusty situations or wear appropriate respiratory protection where cesium hydroxide may be present. Avoid using carbon dioxide (may react!).

Unusual Fire and Explosion Hazards

Cesium hydroxide will melt at 522°F (272°C). The molten material will react violently with water and may ignite nearby combustible materials. Poisonous gases are produced in fire.

SECTION V - REACTIVITY DATA

Stability		Conditions to Avoid **Normally stable under routine conditions of handling and storage. Avoid contact with heat, metals, and other incompatible materials as reactions can be violent and explosive.**
Stable	Unstable	Incompatibility *(materials to avoid)*
X		**Cesium hydroxide may react in contact with water to liberate heat, carbon dioxide, oxygen, and many metals (such as aluminum, zinc, magnesium, potassium, and others) to produce explosive hydrogen gas.**
Hazardous Polymerization		Conditions to Avoid **Under normal conditions of temperature and pressure, hazardous polymerization of cesium hydroxide is not expected to occur.**
May Occur	Will Not Occur	Hazardous Decomposition or By-products
	X	**None reported.**

SECTION VI - HEALTH HAZARD DATA

Primary Route(s) of Entry:	Inhalation?	Absorption (skin/eye)?	Ingestion?
	X		X

Health Hazards

INHALATION: Irritating and corrosive to the eyes, nose, throat, and mucosa of the respiratory system. May cause cough, congestion, and sore throat.

SKIN & EYES: Contact with the skin or eyes can cause irritation and serious burns. Damage to eyes can be permanent up to and including loss of vision.

INGESTION: Burns to the mouth, esophagus, and stomach. Ingestion may be fatal.

Carcinogenicity	NTP Listed?	IARC Cancer Review Group?	OSHA Regulated?	Target Organs?
No Evidence Human No Evidence Animal	No	No	No	Eyes, skin, respiratory system.

Medical Conditions Generally Aggravated by Exposure

None reported.

Emergency and First-aid Procedures

Eye contact: Do not allow victim to keep eyes tightly shut. Flush immediately with water for 15 minutes (minimum), seek medical attention. Skin contact: Remove all contaminated clothing. Immediately wash area with large amounts of soap and water. For inhalation: Remove the person from exposure. Provide respiratory assistance and CPR. Transfer to medical facility. If swallowed: Remove person to fresh air and seek medical attention immediately. If conscious, immediately give large amounts of water or milk to dilute. Do NOT induce vomiting (corrosive liquid may burn the throat or be aspirated into the lungs). Do NOT attempt to give an unconscious or convulsing person anything by mouth.

SECTION VII - PRECAUTIONS FOR SAFE HANDLING AND USE

Steps to be Taken in Case Material is Released or Spilled

Ventilate area of spill or leak. Remove sources of ignition. Restrict those not involved in cleanup from entering area. Collect spilled powdered materials in safest manner possible and deposit in sealed drum for disposal. Do NOT dry sweep (generates dusts), use a HEPA vacuum instead.

Preferred Waste Disposal Method

Not listed. Check applicable federal, state, and/or local environmental regulations.

Precautions to be Taken in Handling and Storage

Cesium hydroxide is a non-combustible solid but a very powerful caustic. Store to avoid contact with acids. Store in tightly closed containers in cool, well-ventilated area away from heat or flame.

Other Precautions and Warnings

Always wear appropriate protective equipment when working with cesium hydroxide.

SECTION VIII - CONTROL MEASURES AND PERSONAL PROTECTIVE EQUIPMENT

Respiratory Protection *(specify type)*

If airborne concentrations exceed the TLV, a dust/mist respirator is recommended. For high exposures, or if concentrations exceed the capacity of the respirator, then a self-contained breathing apparatus (SCBA) with full facepiece operated in positive pressure mode is advised.

Ventilation

Local exhaust (preferred) at site of chemical work or general ventilation.

Protective Gloves	Eye Protection	Other Protective Clothing
Impervious Gloves	**Chemical/Dust Goggles or Safety Glasses with Sides Shields Installed**	**Protective Clothing**

Work/Hygiene Practices

Always wash hands thoroughly after using chemical; never bring food, drink, or smoking materials into vicinity of chemicals.

CESIUM HYDROXIDE

CsOH CAS: 21351-79-1

IDENTIFICATION AND TYPICAL USES

Cesium hydroxide is a colorless or yellowish, fused, crystalline mass with no odor. It makes a strong alkaline solution. It is hygroscopic and deliquescent (absorbs moisture from air an can become liquid). It is the strongest base compound known. It is used as an electrolyte in alkaline storage batteries at subzero temperatures. It is also a polymerization catalyst for siloxanes.

RISK ASSESSMENT: HEALTH

General Assessment

Cesium hydroxide is moderately toxic by *ingestion* and *inhalation.* Because of its extremely caustic nature, it can be dangerously corrosive and destructive to tissue on contact. There are no reports of any carcinogenic, mutagenic, or teratogenic effects subsequent to acute or chronic exposure to cesium hydroxide.

Inhalation of this material in dust or mist (from the solution) form can be immediately irritating to the eyes, nose, throat, and mucous membranes of the upper respiratory tract. There can be coughing, tearing, congestion, and sore throat. The data are limited with regard to lung injury or damage following high or concentrated exposures. However, it is an extremely corrosive material and caution is always warranted when using cesium hydroxide. Other less-caustic compounds have been known to cause serious lung injury, including potential fatal pulmonary edema (fluid in the lungs). It is not known whether exposure to cesium hydroxide will have this effect on humans.

Skin contact will cause localized irritation and possible burns. If allowed to remain in contact for long periods, the corrosive action may be quite severe. Eye contact will cause irritation and immediate burns. The damage to vision can be serious and permanent. Ingestion will cause burns to the mouth, esophagus, and stomach and can be fatal.

Acute Health Effects

The following acute (short-term) health effects may occur immediately or shortly after exposure to cesium hydroxide:

Skin: Irritation with a potential for serious burns and tissue destruction.

Eye: Severe eye irritation, burns, and permanent damage is likely.

Lung: Can irritate the nose, throat, and mucous membranes of the respiratory tract.

Chronic Health Effects

The following chronic (long-term) health effects may occur at some time after exposure to cesium hydroxide and can last for months or even years:

Cancer Hazards: According to the information presented in the references, cesium hydroxide has not been adequately tested for its ability to cause cancer in animals.

Reproductive Hazard: According to the information presented in the references, cesium hydroxide has not been adequately tested for its ability to affect reproduction. Human data are inadequate to evaluate the effects of exposure to cesium hydroxide on human reproduction.

Other Chronic Effects: Repeated exposure to large amounts of caustic materials such as cesium hydroxide may damage the lungs. It is not known if cesium hydroxide specifically will have this effect.

Recommended Risk-Reduction Measures

Personnel should avoid direct contact with cesium hydroxide. If a less toxic material cannot be substituted for cesium hydroxide, then *engineering controls* are the most effective method of reducing exposures. The best protection is to enclose operations and/or provide local exhaust ventilation at the site of cesium hydroxide release. While not always operationally feasible, isolating operations can also reduce exposure.

Using respiratory protection and other personal protective equipment (PPE) is less effective than the engineering controls mentioned above, but is still advisable whenever working with or around cesium hydroxide. Under normal operating conditions of low exposure potential, a dust/mist respirator should suffice. However, for maximum protection in large concentrations or when the capacity of the dust/mist respirator is exceeded, personnel should wear a self-contained breathing apparatus (SCBA) or a supplied-air respirator with full facepiece operated in pressure demand mode. If a full facepiece is not available, then chemical/dust goggles, safety glass with side shields, or a face shield should be worn to protect the eyes. To prevent hand and skin exposures, impervious rubber gloves should be worn. Glove manufacturers should be contacted and permeation studies obtained *before* final selection is made.

Administrative controls should also be in place to minimize the potential for human exposures. These may include written procedures or policies which specify the methods and techniques that will be practiced whenever personnel are to work with cesium hydroxide.

All personnel should receive *training* on the use, hazards, protective measures, emergency actions, and other precautions per 29 CFR 1910.1200 (Hazard Communication), prior to the first assignment in an area where cesium hydroxide is used or stored.

If there has been a long history of high exposure, or if lung symptoms develop, the following medical tests are recommended:

- ☑ Chest X-ray.
- ☑ Pulmonary function tests.

Any evaluation should include a careful history of past and present symptoms with an examination. Medical tests that simply look for existing damage are not a substitute for controlling exposures. Also, because smoking can cause lung cancer, emphysema, and a host of other medical problems, working with or around cesium hydroxide may aggravate these conditions further.

Other methods to reduce exposure to cesium hydroxide include:

- ☑ Where possible, enclose operations and use local exhaust ventilation at the site of chemical release. If local exhaust ventilation or enclosure is not used, then respiratory protection should be required.
- ☑ Always ensure that proper protective clothing is worn when using chemical substances. Never allow contaminated clothing to be taken home from the work place.
- ☑ Hazard warning information should be posted in the work area. In addition, as part of an on-going education and training program, all information on the health and safety hazards of cesium hydroxide should be communicated to all potentially exposed workers.
- ☑ Eye wash stations should be provided in the immediate work area for emergency use.

RISK ASSESSMENT: ENVIRONMENT

General Assessment

The environment is at risk of exposure during transportation, storage, disposal, or destruction of cesium hydroxide. In almost every scenario, the threat of environmental exposure is contingent upon the proper handling of the chemical substance. Accidental spills, large or small, where cesium hydroxide contacts incompatible commodities can result in violent reactions as well as the possible contamination of the surrounding environmental mediums (water, soil, and air).

Cesium hydroxide is considered a noncombustible solid. However, because of its extremely basic nature it is highly incompatible with strong acids and extreme caution is required in handling, storage, transportation, and disposal of cesium hydroxide. In contact with water it can liberate enough heat to ignite combustible materials. If involved in fire, it will melt at 522°F (272°C). If the hot molten liquid should contact water (highly probable in a fire-fighting activity), a violent reaction may occur. These characteristics also require special consideration during any emergency situation involving a leak or spill of cesium hydroxide. Should cesium hydroxide ever come into contact with incompatible substances either during use, transportation, storage, or disposal, dangerous reactions are possible.

Cesium hydroxide can enter the environment through industrial discharges and/or spills.

☠ *Acute Ecological Effects*

Acute (short-term) toxic effects may include the death of animals, birds, or fish and death or low growth rate

in plants. Acute effects are seen 2 to 4 days after animals or plants are exposed to cesium hydroxide.

Cesium hydroxide dissolves in water to create an strong alkaline solution. This may present a moderate acute toxicity to aquatic life. Insufficient data are available to evaluate or predict the short-term toxic effects of cesium hydroxide to plants, birds, and land animals.

Chronic Ecological Effects

Chronic (long-term) toxic effects may include shortened life span, reproductive problems, lower fertility, and changes in appearance or behavior in exposed animals. These effects can be seen long after first exposure to toxic chemicals.

Insufficient data are available to evaluate or predict the long-term toxic effects of cesium hydroxide to aquatic life, plants, birds, or land animals.

Water Solubility

Cesium hydroxide is highly soluble in water. Concentrations of 1000 milligrams or more will mix with a liter of water.

Persistence in the Environment

Cesium hydroxide is moderately persistent in water, with a half-life between 2 and 20 days. The half-life of a pollutant is the amount of time it takes for one-half of the chemical to be degraded.

Bioaccumulation in Aquatic Organisms

Some substances increase in concentration, or *bioaccumulate,* in living organisms as they breathe contaminated air, drink contaminated water, or eat contaminated food. These chemicals can become concentrated in the tissues and internal organs of animals as well as humans.

It is not known if cesium hydroxide will accumulate in the edible tissues of aquatic life that are consumed by humans.

Recommended Risk-Reduction Measures

Proper training of all transporters will reduce the likelihood of a mishap or accident resulting in a leak or spill of cesium hydroxide into the environment. The correct use of labeling on all containers, trucks, and rail cars will enable emergency responders to react properly and quickly to any disaster thereby reducing the potential risk to the environment and to personnel.

Storage of cesium hydroxide should be segregated from incompatible chemicals to minimize the risk of cross-contamination or contact.

If a spill or leak to the environment has occurred, fire department, emergency response and/or hazardous materials spill personnel should be notified immediately. Cleanup should be attempted only by those trained in proper spill containment procedures. If cesium hydroxide should contact the water table, aquifer, or navigable waterway, the local and/or state emergency response authorities must be notified. A comprehensive emergency response or disaster preparedness/recovery plan should be in place prior to any operations involving the use, transportation, storage, or disposal of cesium hydroxide.

If cesium hydroxide is spilled or leaked, the following specific steps are recommended:

- ☑ Restrict persons not wearing protective clothing from area of spill or leak until cleanup is complete.
- ☑ Do not allow exposure to incompatible materials.
- ☑ Close all ignition sources and ventilate area of spill.
- ☑ Collect powdered materials in the safest and most efficient manner possible. Place materials in a sealed drum. Do not dry sweep (generates dusts). Use a vacuum equipped with a high efficiency particulate air (HEPA) filter instead.
- ☑ It may be necessary to dispose of cesium hydroxide as a hazardous waste. Consult state, local, or federal environmental regulations.

RISK ASSESSMENT: BUSINESS

General Assessment

Accidents or mishaps involving cesium hydroxide can present a moderate threat to business operations. Although not likely, equipment and facilities may be lost or their use disrupted as a result of an incident involving cesium hydroxide which can significantly affect fiscal viability. Lawsuits that may result from personnel injury, illness, and/or death, public exposures, and/or environmental contamination will also require a serious expenditure of resources. Media attention surrounding an injury, death, or environmental damage can also result in a loss of profits and loss of current as well as future business.

Recommended Risk-Reduction Measures

Company attorneys, safety and health professionals, and environmental specialists should be involved in the development of any procedures that focus on emergency response, disaster preparedness, and disaster recovery. A company official should be pre-designated as a public relations officer with specific training in dealing with the press. Corporate plans and policies should be developed, approved, and implemented long before any need for such arises.

REFERENCES

Ahlbom, A. 1993. *Biostatistics for Epidemiologists.* New York: CRC Press/Lewis Publishers.

American Conference of Governmental Industrial Hygienists. 1988. *Documentation of the Threshold Limit Values and Biological Exposure Indices,* 5th Edition (with updates). Cincinnati: ACGIH

Bloom, A. D. and F. J. DeSerres. 1995. *Ecotoxicity and Human Health.* New York: CRC Press/Lewis Publishers.

Brusick, D. J. 1994. *Methods for Genetic Risk Assessment.* New York: CRC Press/Lewis Publishers.

Bueche, J. R. 1972. *Principles of Physics,* 2nd Edition. New York: McGraw-Hill.

Calabrese, E. J., E. M. Kenyon. *Air Toxics and Risk Assessment.* New York: CRC Press/Lewis Publishers.

Calabrese, E. J. 1994. *Biological Effects of Low Level Exposures.* New York: CRC Press/Lewis Publishers.

Cockerman, L. G., B. S. Shane. 1994. *Basic Environmental Toxicology.* New York: CRC Press/Lewis Publishers.

Cothern, R. R. 1993. *Comparative Environmental Risk Assessment.* New York: CRC Press/Lewis Publishers.

Danse, I. R. 1991. *Common Sense Toxics in the Workplace.* New York: Van Nostrand Reinhold.

Environmental Protection Agency. 1983. *Chemical Hazard Information Profiles.* Washington, D.C.: U.S. EPA.

Fodden, C. R., and J. L. Weddell. 1992. *Hazardous Materials Emergency Action Data.* New York: CRC Press/Lewis Publishers.

Forseberg, K., and L. H. Keith. 1993. *Chemical; Protective Clothing Permeation and Degradation Database.* Chelsea, MI: Lewis Publishers.

Gardner, R. W. 1994. *Chemical Intolerance.* New York: CRC Press/Lewis Publishers.

Hallenbeck, W. H. 1993. *Quantitative Risk Assessment for Environmental and Occupational Health,* 2nd Edition. New York: CRC Press/Lewis Publishers.

Howard, P. H. 1995. *Handbook of Environmental Fate and Exposures Data for Organic Chemicals,* Vol. I-IV. New York: CRC Press/Lewis Publishers.

Howard, P. H., et. al. 1991. *Handbook of Environmental Degradation Rates.* New York: CRC Press/Lewis Publishers.

Jones, F. E. 1994. *Toxic Organic Vapors in the Workplace.* New York: CRC Press/Lewis Publishers.

Keith, L. H. and M. M. Walker. 1995. *Handbook of Air Toxics.* New York: CRC Press/Lewis Publishers.

Landis, W. G., M. H. Yu. 1995. *Introduction to Environmental Toxicology.* New York: CRC Press/Lewis Publishers.

Langerman, N. 1994 *Precautionary Labels for Chemical Containers.* New York: CRC Press/Lewis Publishers.

Lauwerys, R. R., P. Hoet. 1993. *Industrial Chemical Exposure.* New York: CRC Press/Lewis Publishers.

Lewis, R. J., Sr. 1993. *Hawley's Condensed Chemical Dictionary,* 12th Edition. New York: Van Nostrand Reinhold.

Lewis, R. J., Sr. 1992. *Sax's Dangerous Properties of Industrial Materials,* Volumes 1, 2, and 3, 8th Edition. New York: Van Nostrand Reinhold.

Lide, D. R. 1995. *Handbook of Chemistry and Physics,* 76th Edition. New York: CRC Press/Lewis Publishers.

Manahan, S. E. 1994. *Environmental Chemistry,* 6th Edition. New York: CRC Press/Lewis Publishers.

Mellström, G. A., J. E. Wahlberg, and H. I. Maibach. 1994. *Protective Gloves for Occupational Use.* New York: CRC Press/Lewis Publishers.

Meyer, E. 1990. *Chemistry of Hazardous Materials.* Englewood Cliffs, NJ: Prentice Hall.

Mycock, J. C., J. D. McKenna, and L. Theodore. 1995. *Handbook of Air Pollution Control Engineering and Technology.* New York: CRC Press/Lewis Publishers.

Patnaik, P. 1992. *A Comprehensive Guide to the Hazardous Properties of Chemical Substances.* New York: Van Nostrand Reinhold.

Proctor, N. H., J. P. Hughes. 1991. *Chemical Hazards of the Workplace.* New York: Van Nostrand Reinhold.

Quigley, D. R. 1994. *Handbook of Emergency Chemical Management.* New York: CRC Press/Lewis Publishers.

Rea, W. J. 1995. *Chemical Sensitivity,* Vol. I-III. New York: CRC Press/Lewis Publishers.

Richardson, M. 1995. *Dictionary of Substance and Their Effects,* Vol. I-VI. New York: CRC Press/Lewis Publishers.

Sacarello, H. L. A. 1994. *The Comprehensive Handbook of Hazardous Materials.* New York: CRC Press/Lewis Publishers.

Sheftel, V. O. 1995. *Handbook of Toxic Properties of Monomers and Additives.* New York: CRC Press/Lewis Publishers.

Talbott, E., G. Craun. 1995. *An Introduction to Environmental Epidemiology.* New York: CRC Press/Lewis Publishers.

Torkelson, T. R. V. K. Rowe. 1981. *Patty's Industrial Hygiene and Toxicology,* Vol. 2B. New York: Wiley-Interscience.

MATERIAL SAFETY DATA SHEET

CHEMICAL NAME	CRC PRESS/LEWIS PUBLISHERS DISCLAIMER
CHLORDANE	THE INFORMATION AND RECOMMENDATIONS PRESENTED HEREIN ARE BASED ON SOURCES BELIEVED TO BE RELIABLE. CRC MAKES NO REPRESENTATION ON ITS COMPLETENESS OR ACCURACY. IT IS THE RESPONSIBILITY OF THE USER TO DETERMINE THE CHEMICAL'S SUITABILITY FOR ITS INTENDED USE, THE CHEMICAL'S SAFE USE, AND THE CHEMICAL'S PROPER DISPOSAL. NO REPRESENTATIONS AND/OR WARRANTIES, EITHER EXPRESSED OR IMPLIED, OF THE MERCHANTABILITY OR FITNESS FOR A PARTICULAR PURPOSE, OR OF ANY OTHER NATURE, ARE MADE WITH RESPECT TO THE INFORMATION PROVIDED IN THIS MSDS OR TO THE CHEMICAL TO WHICH INFORMATION MAY REFER. CRC NEITHER ASSUMES NOR AUTHORIZES ANY OTHER PERSON TO ASSUME FOR IT, ANY OTHER ADDITIONAL RESPONSIBILITY OR LIABILITY FOR THE USE OF, OR RELIANCE UPON, THIS INFORMATION.

HAZARD WARNING INFORMATION

HEALTH	FIRE	REACTIVE	OTHER	DEGREE OF HAZARD	COLOR CODING	OTHER CODES
2	1	0		0 = Minimum Hazard 1 = Slight Hazard 2 = Moderate Hazard 3 = Serious Hazard 4 = Severe Hazard	HEALTH = BLUE FIRE = RED REACTIVITY = YELLOW OTHER = WHITE	OX = Oxidizer ACID = Acid ALK = Alkali COR = Corrosive W = Use No Water

SECTION I - GENERAL INFORMATION

Characterization	RCRA Number	EPA Class
Pesticide (polycyclic, chlorinated)	**U036**	**Toxic Waste**

DOT Proper Shipping Name	Chemical Abstract Service (CAS) Number
Chlordane (liquid)	**57-74-9**
DOT Hazard Class and Label Requirements	**DOT Emergency Guide Code**
Flammable Liquid	**28**
DOT Identification Number	**Chemical Formula**
UN 2762	**$C_{10}H_6Cl_8$**

Synonyms

Chlordan; Toxichlor; Octaklor; Velsicol 1068; CD 68; Topichlor.

SECTION II - HAZARDOUS INGREDIENTS AND IDENTITY INFORMATION

Hazardous Components (specific identity)	OSHA Exposure Criteria	NIOSH Exposure Criteria	Immediately Dangerous to Life and Health (IDLH)	ACGIH Exposure Criteria
Chlordane (derivation: By Diels-Alder addition of hexachlorocyclopentadiene to cyclopentadiene, followed by reaction with chlorine). 1 ppm = 17.04 mg/m^3	PEL: 0.5 mg/m^3 (skin) STEL: Not Established	REL: 0.5 mg/m3 (skin) Possible Cancer Agent	100 mg/m^3	TLV: 0.5 mg/m^3 STEL: Not Established

SECTION III - PHYSICAL AND CHEMICAL CHARACTERISTICS

Boiling Point	Specific Gravity (H_2O = 1)
347°F (175°C)	**1.6 at 77°F (25°C)**
Vapor Pressure (mm Hg)	Molecular Weight (atomic weight)
0 (approx.) at 69°F (20°C)	**409.8**
Vapor Density (Air = 1)	Freezing Point
1.6	**217-228°F (103-109°C)**

Solubility

Insoluble in water (0.0001%). Soluble in most organic solvents. Miscible in deodorized kerosene. It releases chlorine and decomposes in the presence of alkalis.

Appearance and Odor

Pure chlordane is a colorless to amber, odorless, thick, viscous liquid. The commercial grade typically has a chlorine-like odor.

SECTION IV - FIRE AND EXPLOSION HAZARD DATA

Flash Point (method used)	Explosive Limits in Air % by Volume
132°F (55°C) closed cup	**LEL: 0.7% UEL: 5%**
NFPA Classification	Autoignition Temperature
Combustible Liquid	**410°F (210°C)**

Extinguishing Media

Alcohol resistant foam, dry chemical, carbon dioxide, water. (Water may be ineffective on solution fire).

Special Fire Fighting Procedures

Wear full protective clothing and self-contained breathing apparatus (SCBA). Structural fire-fighting equipment is permeable and may not provide sufficient protection. Remain clear of smoke, water fall-out and water run-off. Move containers from fire area if it can be done without risk. Cool exposed containers. Toxic hydrogen chloride and phosgene gases are formed in fire.

Unusual Fire and Explosion Hazards

Chlordane is a combustible liquid. It may be dissolved in some flammable liquids, such as benzene.

SECTION V - REACTIVITY DATA

Stability		Conditions to Avoid: Avoid contact with incompatible materials and do not store near heat or heat source.
Stable	Unstable	Incompatibility *(materials to avoid)*: Reacts on contact with strong oxidizers, such as perchlorates, peroxides, permanganates, chlorates, and nitrates. Also reacts in the presence of alkaline reagents.
X		
Hazardous Polymerization		Conditions to Avoid: Under normal conditions of temperature and pressure, hazardous polymerization of chlordane is not known to occur.
May Occur	Will Not Occur	Hazardous Decomposition or By-products: At high temperatures, chlordane decomposes and may produce poisonous gases, including toxic chloride fumes.
	X	

SECTION VI - HEALTH HAZARD DATA

Primary Route(s) of Entry:	Inhalation?	Absorption (skin/eye)?	Ingestion?
	X	X	X

Health Hazards

INHALATION: Breathing vapors can cause convulsions, unconsciousness and death. Can irritate the nose, eyes, mouth, and throat. Long-term exposure can damage the liver.

ABSORPTION: Will pass through unbroken skin and enter the bloodstream. Contact may severely irritate the skin and eyes. Prolonged contact can cause acne-like rash on skin.

INGESTION: Poisonous to humans by ingestion. Systemic effects due to central nervous system action include tremors, convulsion, excitement, ataxia, and gastritis.

Carcinogenicity	NTP Listed?	IARC Cancer Review Group?	OSHA Regulated?	Target Organs?
Suspected Human Confirmed Animal	5th Annual Report	Group 3	29 CFR 1910.1000 Table Z-1	Lungs, skin, eyes, CNS, liver, kidneys.

Medical Conditions Generally Aggravated by Exposure

None reported. However, persons with existing lung dysfunction may experience effects quicker.

Emergency and First-aid Procedures

Eye contact: Do not allow victim to keep eyes tightly shut. Flush immediately with water for 15 minutes, seek medical attention. Skin contact: Remove all contaminated clothing. Immediately wash area with flooding amounts of water. DO NOT scrub. Inhalation: Remove the person from exposure. Provide respiratory assistance and CPR. Transfer to medical facility. If swallowed: If victim is conscious, induce vomiting. Seek medical attention immediately. Never give anything by mouth to an unconscious or convulsing person. Do not make an unconscious person vomit. DOCTOR: Suggest gastric lavage and saline cathartics; ether and barbiturates may control convulsions. Oxygen and fluid therapy also recommended. Do not give epinephrine since it may induce ventricular fibrillation. Enforce complete rest.

SECTION VII - PRECAUTIONS FOR SAFE HANDLING AND USE

Steps to be Taken in Case Material is Released or Spilled

Remove all sources of ignition. Ventilate area of spill. Restrict those not involved in cleanup from entering area. Notify appropriate authorities. Absorb any liquids containing chlordane in vermiculite, dry sand, earth, or similar material and deposit in sealed drum for disposal or reclamation.

Preferred Waste Disposal Method

No Citation.

Precautions to be Taken in Handling and Storage

Store in tightly closed containers in cool, well ventilated area away from heat.

Other Precautions and Warnings

Avoid contact with oxidizers since violent reactions can occur.

SECTION VIII - CONTROL MEASURES AND PERSONAL PROTECTIVE EQUIPMENT

Respiratory Protection *(specify type)*

For exposures above 0.5 mg/m^3, use a NISOH approved supplied air respirator or a self-contained breathing apparatus (SCBA) with full facepiece operated in positive pressure mode.

Ventilation

Local exhaust (preferred) at site of chemical work or general ventilation.

Protective Gloves	Eye Protection	Other Protective Clothing
Rubber	Chemical Goggles or Face Mask	Protective Apron, Boots (avoid skin contact)

Work/Hygiene Practices

Always wash hands thoroughly after using chemical; never bring food, drink, or smoking materials into vicinity of chemicals.

CHLORDANE

$C_{10}H_6Cl_8$ CAS: 57-74-9

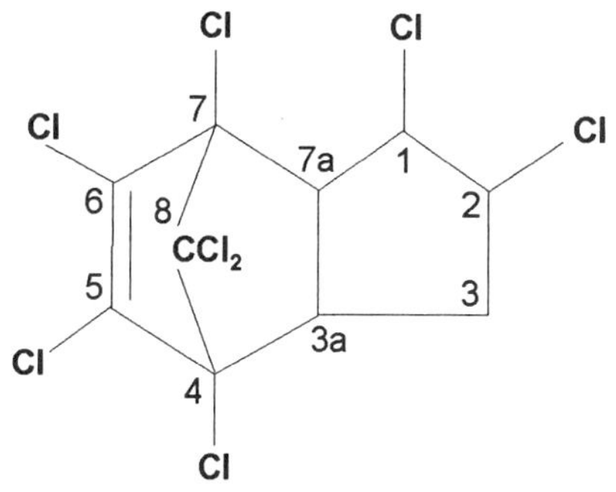

IDENTIFICATION AND TYPICAL USES

Chlordane is a colorless or amber thick, viscous liquid with no odor. However, its commercial grade may have a characteristic chlorine-like odor. It is used primarily as an insecticide for the control of termites and as a fumigant.

RISK ASSESSMENT: HEALTH

General Assessment

Chlordane is highly toxic to humans by *ingestion.* It is moderately toxic by *inhalation* and skin *absorption.* It is a suspected human carcinogen with experimental carcinogenic data reported. Chlordane has also been shown to have teratogenic properties in test animals. Human mutation data have been reported. This chemical is no longer permitted for use as a termiticide in residential properties.

Chlordane is a central nervous system stimulant. Symptoms of exposure by both ingestion and skin contact include nausea, vomiting, abdominal pain, irritation, confusion, ataxia (loss of muscle coordination), and gastritis. There may also be tremors, convulsions, blurred vision, and diarrhea.

Skin contact will result in moderate to severe irritation with rash or burning sensation. Eye contact may also result in irritation. Caution is warranted when inhalation and skin contact exposures are possible. Even though air levels may be within limits, personnel may be overexposed if chlordane should contact the skin. The dangerous chronic dose in humans is unknown. However, prolonged or long-term contact may lead to liver damage (fatty degeneration), a characteristic similar to that of chronic alcoholism.

Acute Health Effects

The following acute (short-term) health effects may occur immediately or shortly after exposure to chlordane:

Skin: Liquid can pass through unbroken skin. Contact may cause irritation and rash. Symptoms of exposure include headache, nausea, abdominal pain, and vomiting.

Eye: Severe irritation and possible damage on contact.

Lung: Irritation of the nose and throat. Breathing vapors can cause cough, convulsions, unconsciousness, and death.

CNS: A central nervous system stimulant. Symptoms include ataxia, excitement, tremors, convulsions, blurred vision, confusion, delirium, and irritability.

Chronic Health Effects

The following chronic (long-term) health effects may occur at some time after exposure to chlordane and can last for months or even years:

Cancer Hazards: Chlordane may be a carcinogen in humans since it has been shown to cause liver cancer in animals.

Reproductive Hazard: There is some evidence that exposure to chlordane causes damage to the developing fetus in test animals. Chlordane may decrease fertility in males and females.

Other Chronic Effects: Chlordane may damage the liver and kidneys. It may cause an acne-like rash following skin contact.

Recommended Risk-Reduction Measures

Personnel should avoid direct contact with chlordane. If a less toxic material or compound cannot be substituted for chlordane, then *engineering controls* are the most effective method of reducing exposures. The best protection is to enclose operations and/or provide local exhaust ventilation at the site of chlordane dust release. This may be practical in manufacturing operations, but is not feasible in application since it is used outdoors. While not always operationally possible, isolating operations involving chlordane manufacturing can also reduce exposure. Using respiratory protection is less effective than the controls mentioned above, but is still advisable whenever working with or around chlordane. At exposures above the ACGIH TLV of 5 mg/m^3 (TWA), use a MSHA/NIOSH approved self-contained breathing apparatus (SCBA) with full facepiece and pressure demand is recommended for the greatest possible respiratory protection. If a full facepiece is not available, then chemical goggles should be worn to protect the eyes. A face shield and protective apron should also be worn. To prevent hand and skin exposures, impervious gloves should be worn.

Administrative controls should also be in place to minimize the potential for human exposures. These may include written procedures or policies which specify the methods and techniques that will be practiced whenever personnel are to work with chlordane.

All personnel should receive *training* on the use, hazards, protective measures, emergency actions, and other precautions per 29 CFR 1910.1200 (Hazard Communication), prior to the first assignment in an area where chlordane is used or stored.

If symptoms develop or overexposure is suspected, the following medical testing may be useful:

☑ Liver and kidney function tests.

Any evaluation should include a careful history of past and present symptoms with an examination. Medical tests that simply look for existing damage are not a substitute for controlling exposures.

Other methods to reduce exposure include:

☑ Where possible, enclose operations and use local exhaust ventilation at the site of chemical release. If local exhaust ventilation or enclosure is not used, respiratory protection should be mandatory.

☑ Always ensure that proper protective clothing is worn when using chemical substances.

☑ Wash thoroughly immediately after exposure to chlordane and at the end of the work shift or before eating, drinking, or smoking.

☑ Hazard warning information should be posted in the work area. In addition, as part of an on-going education and training program, all information on the health and safety hazards of chlordane should be communicated to all potentially exposed workers.

☑ Eye wash stations should be provided in the immediate work area for emergency use. If there is a possibility of skin exposure to chlordane, emergency shower facilities should also be provided.

☑ Workers whose clothing has been contaminated by chlordane should change into clean clothes before leaving work. Contaminated work clothing should be laundered only by individuals who have been informed of the hazards of exposure to chlordane.

☑ Where possible, automatically pump liquid chlordane from drums or other storage containers to process containers.

☑ Specific engineering controls are recommended by NIOSH. Refer to NIOSH Criteria Document: Occupational Exposure During the Manufacture and Formulation of Pesticides (Publication Number 78-174).

RISK ASSESSMENT: ENVIRONMENT

General Assessment

Because of its use as a fumigant and pesticide, chlordane may be present in the environment at any given time and at a variety of levels. The environment is also at risk of exposure during transportation, storage, disposal, or destruction of chlordane. In almost every scenario, the threat of environmental exposure is contingent upon the proper handling of the chemical substance. Accidental spills, large or small, where chlordane contacts incompatible commodities can result in violent and even explosive reactions (depending upon conditions of spill), and possible contamination of the surrounding environmental mediums (water, soil, and air).

Chlordane is considered a flammable liquid. It may also be present (dissolved) in some other flammable liquids such as benzene. Also, because it is incompatible with many common oxidizers, such as chlorine, extreme caution is required in handling, storage, transportation, and disposal of chlordane. These characteristics also require special consideration during any emergency situation involving a leak or spill of chlordane powder or dust.

Chlordane is an insecticide of the polycyclic chlorinated hydrocarbon class of pesticides. It is a broad spectrum insecticide and was used for many years to control insects around the home, in gardens, in agricultural crops such as corn, and for termite control. In the U.S., chlordane is no longer permitted for residential use. However, because of its once widespread, general usage in the past in areas where termites are treated, chlordane enters the environment in agricultural and residential run-off.

Acute Ecological Effects

Acute (short-term) toxic effects may include the death of animals, birds, or fish and death or low growth rate in plants. Acute effects are seen 2 to 4 days after animals or plants are exposed to chlordane.

Chlordane has high acute toxicity to aquatic life. Chlordane has caused injury and germination decrease in various agricultural and ornamental crops. Insufficient data are available on the short-term effects of chlordane to birds and land animals.

Chronic Ecological Effects

Chronic (long-term) toxic effects may include shortened life span, reproductive problems, lower fertility, and changes in appearance or behavior in exposed animals. These effects can be seen long after first exposure(s) to toxic chemicals.

Chlordane has high chronic toxicity to aquatic life. Insufficient data are available to evaluate or predict the long-term effects of chlordane to plants, birds, or land animals.

Water Solubility

Chlordane is very slightly soluble (practically insoluble) in water. Concentrations of less than 1 milligram will mix with a liter of water.

Persistence in the Environment

Chlordane is highly persistent in the water, with a half-life of greater than 200 days. The half-life of a pollutant is the amount of time it takes for one-half of the chemical to be degraded. About 50.7% of chlordane will eventually end up in terrestrial soil; approximately 47.3% will end up and aquatic sediments; the remainder will end up in water.

Bioaccumulation in Aquatic Organisms

Some substances increase in concentration, or *bioaccumulate,* in living organisms as they breathe contaminated air, drink contaminated water, or eat contaminated food. These chemicals can become concentrated in the tissues and internal organs of animals as well as humans.

The concentration of chlordane found in fish tissues is expected to be considerably higher that the average concentration of chlordane in the water from which the fish was taken.

Recommended Risk-Reduction Measures

Proper training of all transporters will reduce the likelihood of a mishap or accident resulting in a leak or spill of chlordane into the environment. The correct use of DOT labeling (per the MSDS) on all containers, trucks, and rail cars will enable emergency responders to react properly and quickly to any disaster thereby reducing the potential risk to the environment and to personnel.

Storage of chlordane should be segregated from incompatible materials to minimize the risk of cross-contamination or contact. Buildings or storage areas should be equipped with appropriate fire protection systems (alarms, sprinklers). Proper ventilation and protective equipment should be used while handing the powders or while preparing aqueous solutions.

If a spill or leak to the environment has occurred, fire department, emergency response and/or hazardous materials spill personnel should be notified immediately. Cleanup should be attempted only by those trained in proper spill containment procedures. If chlordane should contact the water table, aquifer, or navigable waterway, reclamation procedures should be initiated as soon as practical. It is only slightly soluble in water and total remediation may be possible. The local and/or state emergency response authorities must be notified. A comprehensive emergency response or disaster preparedness/recovery plan should be in place prior to any operations involving the use, transportation, storage, or disposal of chlordane. If chlordane is spilled or leaked, the following specific steps are recommended:

- ☑ Restrict persons not wearing protective clothing from area of spill or leak until cleanup is complete and area can be opened for normal work.

- ☑ Remove all ignition sources.
- ☑ Absorb in vermiculite or other material and place in sealed drum for disposal or reclamation.
- ☑ It may be necessary to dispose of chlordane as a hazardous waste. The responsible state agency or the regional office of the federal Environmental Protection Agency (EPA) should be contacted for specific recommendations.

RISK ASSESSMENT: BUSINESS

General Assessment

Accidents or mishaps involving chlordane can present a significant threat to business operations. The loss or damage of equipment or facilities can significantly affect fiscal viability. Lawsuits that may result from personnel injury/death, public exposures, and/or environmental contamination will also require a serious expenditure of resources. Media attention surrounding an injury, death, or environmental damage can also result in a loss of profits and loss of current as well as future business.

Recommended Risk-Reduction Measures

Company attorneys, safety and health professionals, and environmental specialists should be involved in the development of any procedures for responding to chemical incidents. A company official should be predesignated as a public relations officer with specific training in dealing with the press. Corporate plans and policies should be developed, approved, and implemented long before any need for such arises.

REFERENCES

American Conference of Governmental Industrial Hygienists. 1988. *Documentation of the Threshold Limit Values and Biological Exposure Indices,* 5th Edition (with updates). Cincinnati: ACGIH

Brusick, D. J. 1994. *Methods for Genetic Risk Assessment.* New York: CRC Press/Lewis Publishers.

Bloom, A. D. and F. J. DeSerres. 1995. *Ecotoxicity and Human Health.* New York: CRC Press/Lewis Publishers.

Calabrese, E. J. 1994. *Biological Effects of Low Level Exposures.* New York: CRC Press/Lewis Publishers.

Cockerman, L. G., B. S. Shane. 1994. *Basic Environmental Toxicology.* New York: CRC Press/Lewis Publishers.

Cothern, C. R. 1995. *Handbook for Environmental Risk Decisionmaking.* New York: CRC Press/Lewis Publishers.

Gardner, R. W. 1994. *Chemical Intolerance.* New York: CRC Press/Lewis Publishers.

Hallenbeck, W. H. 1993. *Quantitative Risk Assessment for Environmental and Occupational Health,* 2nd Edition. New York: CRC Press/Lewis Publishers.

Howard, P. H., et. al. 1991. *Handbook of Environmental Degradation Rates.* New York: CRC Press/Lewis Publishers.

Keith, L. H. and M. M. Walker. 1995. *Handbook of Air Toxics.* New York: CRC Press/Lewis Publishers.

Landis, W. G., M. H. Yu. 1995. *Introduction to Environmental Toxicology.* New York: CRC Press/Lewis Publishers.

Langerman, N. 1994 *Precautionary Labels for Chemical Containers.* New York: CRC Press/Lewis Publishers.

Lauwerys, R. R., P. Hoet. 1993. *Industrial Chemical Exposure.* New York: CRC Press/Lewis Publishers.

Lewis, R. J., Sr. 1992. *Sax's Dangerous Properties of Industrial Materials,* Volumes 1, 2, and 3, 8th Edition. New York: Van Nostrand Reinhold.

Lide, D. R. 1995. *Handbook of Chemistry and Physics,* 76th Edition. New York: CRC Press/Lewis Publishers.

Manahan, S. E. 1992. *Toxicological Chemistry,* 2nd Edition. New York: CRC Press/Lewis Publishers.

Mellström, G. A., J. E. Wahlberg, and H. I. Maibach. 1994. *Protective Gloves for Occupational Use.* New York: CRC Press/Lewis Publishers.

Meyer, E. 1990. *Chemistry of Hazardous Materials.* Englewood Cliffs, NJ: Prentice Hall.

Mycock, J. C., J. D. McKenna, and L. Theodore. 1995. *Handbook of Air Pollution Control Engineering and Technology.* New York: CRC Press/Lewis Publishers.

Nelson, G. O. 1992. *Gas Mixtures.* New York: CRC Press/Lewis Publishers.

Ott, W. R. 1995. *Environmental Statistics and Data Analysis.* New York: CRC Press/Lewis Publishers.

Patnaik, P. 1992. *A Comprehensive Guide to the Hazardous Properties of Chemical Substances.* New York: Van Nostrand Reinhold.

Quigley, D. R. 1994. *Handbook of Emergency Chemical Management.* New York: CRC Press/Lewis Publishers.

Rea, W. J. 1995. *Chemical Sensitivity,* Vol. I-III. New York: CRC Press/Lewis Publishers.

Richardson, M. 1995. *Dictionary of Substance and Their Effects,* Vol. I-VI. New York: CRC Press/Lewis Publishers.

Sacarello, H. L. A. 1994. *The Comprehensive Handbook of Hazardous Materials.* New York: CRC Press/Lewis Publishers.

Torkelson, T. R. V. K. Rowe. 1981. *Patty's Industrial Hygiene and Toxicology,* Vol. 2B. New York: Wiley-Interscience.

U.S. Department of Health and Human Services, National Institute for Occupational Health (NIOSH). 1978. *NIOS Criteria Document: Occupational Exposure During the Manufacture and Formulation of Pesticides.* Publication Number 78-174. Washington, DC: U.S. Government Printing Office.

MATERIAL SAFETY DATA SHEET

CHEMICAL NAME	CRC PRESS/LEWIS PUBLISHERS DISCLAIMER:
CHLORINE	THE INFORMATION AND RECOMMENDATIONS PRESENTED HEREIN ARE BASED ON SOURCES BELIEVED TO BE RELIABLE. CRC MAKES NO REPRESENTATION ON ITS COMPLETENESS OR ACCURACY. IT IS THE RESPONSIBILITY OF THE USER TO DETERMINE THE CHEMICAL'S SUITABILITY FOR ITS INTENDED USE, THE CHEMICAL'S SAFE USE, AND THE CHEMICAL'S PROPER DISPOSAL. NO REPRESENTATIONS AND/OR WARRANTIES, EITHER EXPRESSED OR IMPLIED, OF THE MERCHANTABILITY OR FITNESS FOR A PARTICULAR PURPOSE, OR OF ANY OTHER NATURE, ARE MADE WITH RESPECT TO THE INFORMATION PROVIDED IN THIS MSDS OR TO THE CHEMICAL TO WHICH INFORMATION MAY REFER. CRC NEITHER ASSUMES NOR AUTHORIZES ANY OTHER PERSON TO ASSUME FOR IT, ANY OTHER ADDITIONAL RESPONSIBILITY OR LIABILITY FOR THE USE OF, OR RELIANCE UPON, THIS INFORMATION.

HAZARD WARNING INFORMATION

HEALTH	FIRE	REACTIVE	OTHER	DEGREE OF HAZARD	COLOR CODING	OTHER CODES
3	0	0	OX	0 = Minimum Hazard 1 = Slight Hazard 2 = Moderate Hazard 3 = Serious Hazard 4 = Severe Hazard	HEALTH = BLUE FIRE = RED REACTIVITY = YELLOW OTHER = WHITE	OX = Oxidizer ACID = Acid ALK = Alkali COR = Corrosive W = Use No Water

SECTION I - GENERAL INFORMATION

Characterization	**Halogen**	RCRA Number	**None**
		EPA Class	**Not Applicable**
DOT Proper Shipping Name	**Chlorine**	Chemical Abstract Service (CAS) Number	**7782-50-5**
DOT Hazard Class and Label Requirements	**Non-flammable Gas; Poison A**	DOT Emergency Guide Code	**20**
DOT Identification Number	**UN 1017**	Chemical Formula	**Cl_2**

Synonyms

Molecular chlorine; Bertholite.

SECTION II - HAZARDOUS INGREDIENTS AND IDENTITY INFORMATION

Hazardous Components (specific identity)	OSHA Exposure Criteria	NIOSH Exposure Criteria	Immediately Dangerous to Life and Health (IDLH)	ACGIH Exposure Criteria
Chlorine (derivation: Electrolysis of sodium chloride brine in either diaphragm or mercury cathode cells, chlorine is released at the anode; fused-salt electrolysis of sodium or magnesium chloride; electrolysis of hydrochloric acid; oxidation of hydrogen chloride with nitrogen oxide as catalyst and absorption of steam with sulfuric acid). **1 ppm = 2.95 mg/m³**	**PEL:** 0.5 ppm 1.45 mg/m³ **CEILING:** 1 ppm 3 mg/m³ **STEL:** 1 ppm	**REL (ceiling):** 0.5 ppm 1.45 mg/m³ **STEL:** Not Established	**10 ppm**	**TLV:** 1 ppm 3 mg/m³ **STEL:** 3 ppm 8.85 mg/m³

SECTION III - PHYSICAL AND CHEMICAL CHARACTERISTICS

Boiling Point	**-31°F (-35°C)**	Specific Gravity (H_2O = 1)	**2.47**
Vapor Pressure (gas)	**4800 at 69°F (20°C)**	Molecular Weight	**70.9**
Vapor Density (Air = 1)	**2.5**	Melting Point	**-150°F (-101°C)**

Solubility

Only slightly soluble in water (0.7%).

Appearance and Odor

Greenish-yellow gas, liquid, or rhombic crystals (commercial grades may appear white) with a strong, irritating, suffocating odor. Odor threshold = 0.31 ppm.

SECTION IV - FIRE AND EXPLOSION HAZARD DATA

Flash Point (method used)	**Not Applicable**	Explosive Limits in Air % by Volume	**LEL: N/A UEL: N/A**
NFPA Classification	**Non-flammable Gas**	Autoignition Temperature	**Not Applicable**

Extinguishing Media

Use extinguishing agent suitable to surrounding media. Chlorine itself does not burn.

Special Fire Fighting Procedures

Chlorine is a powerful oxidizer. Wear full protective clothing and self-contained breathing apparatus (SCBA). Use a water spray to keep fire-exposed containers cool. Poisonous gases are produced in fire.

Unusual Fire and Explosion Hazards

Dangerous fire and explosion hazard when exposed to heat or incompatible materials. Containers may explode in fire. Firefighters should avoid contact with vapors produced during fire.

SECTION V - REACTIVITY DATA

Stability		
Stability		Conditions to Avoid Keep away from incompatible materials. Violent and explosive reactions and fires extremely likely on contact with petroleum products. Keep away from heat (including sunlight).
Stable X	Unstable	Incompatibility *(materials to avoid)* Gasoline and other petroleum products, turpentine, alcohols, acetylene, hydrogen, ammonia, sulfur, finely divided metals. Explodes with ethylene, methane, propane, butane, propene, and benzene vapors.
Hazardous Polymerization		Conditions to Avoid Under normal conditions of temperature and pressure, hazardous polymerization of chlorine is not known to occur.
May Occur	Will Not Occur X	Hazardous Decomposition or By-products When heated to decomposition, chlorine can emit highly toxic/poisonous gases.

SECTION VI - HEALTH HAZARD DATA

Primary Route(s) of Entry:	Inhalation?	Absorption (skin)?	Ingestion?
	X		

Health Hazards

INHALATION: Extremely irritating to the lungs, throat, nose, and eyes. Coughing and shortness of breathe that may persist for hours after exposure. Can cause permanent damage to the lungs. Fluid can build up in the lungs (pulmonary edema), which is a medical emergency.

SKIN & EYES: Contact with the liquid can cause severe skin burns and persistent rash. The liquid can cause severe eye burns leading to permanent damage.

Carcinogenicity	NTP Listed?	IARC Cancer Review Group?	OSHA Regulated?	Target Organs?
Unknown Human Unknown Animal	No	No	29 CFR 1910.1000 Table Z-1	Respiratory system, eyes, skin.

Medical Conditions Generally Aggravated by Exposure

Tuberculosis, asthma, chronic obstructive pulmonary disease, or heart disease my be aggravated.

Emergency and First-aid Procedures

Eye contact: Flush immediately with water for 15 minutes (minimum), seek medical attention. Skin contact: Remove all contaminated clothing. Immediately wash area with large amounts of soap and water. Seek medical assistance. For inhalation: Remove the person from exposure. Provide respiratory assistance and CPR. Transfer to medical facility. Observe for 24-48 hours since pulmonary edema may be a delayed reaction. If swallowed: Seek medical attention immediately.

SECTION VII - PRECAUTIONS FOR SAFE HANDLING AND USE

Steps to be Taken in Case Material is Released or Spilled

Absorb liquids in vermiculite, dry sand, earth, or similar material and deposit in sealed containers. Ventilate area of spill or leak. Restrict those not involved in cleanup from entering area. If gas is leaked, stop source if possible or remove leaking cylinders to safe area outdoors and allow to empty.

Preferred Waste Disposal Method

No Citation.

Precautions to be Taken in Handling and Storage

Chlorine is a non-flammable gas but an extremely powerful oxidizer. Store to avoid contact with incompatible materials since violent reactions can occur. Store in tightly closed containers in cool, well-ventilated area away from heat and sunlight. Ship in steel pressure cylinders and tank cars.

Other Precautions and Warnings

Sources of ignition are prohibited where chlorine is used, handled, or stored. Heat may cause containers to buildup pressure and explode.

SECTION VIII - CONTROL MEASURES AND PERSONAL PROTECTIVE EQUIPMENT

Respiratory Protection *(specify type)*

For exposures below the PEL, use a full facepiece respirator with chlorine gas cartridge. For higher exposures, use a supplied air respirator with full facepiece, hood, or helmet in continuous flow mode, or use a SCBA operated in pressure demand or other positive pressure mode.

Ventilation

Local exhaust or general mechanical systems recommended.

Protective Gloves	Eye Protection	Other Protective Clothing
Neoprene rubber	Chemical Goggles or Face Mask	Neoprene Rubber Apron

Work/Hygiene Practices

Always wash hands thoroughly after using chemical; never bring food, drink, or smoking materials into vicinity of chemicals.

CHLORINE

Cl_2 CAS: 7782-50-5

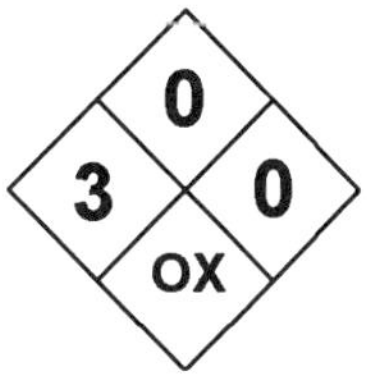

IDENTIFICATION AND TYPICAL USES

Chlorine is a greenish-yellow gas, liquid, or rhombic crystalline solid. Its commercial grade may sometimes appear as a white powder or solid. It is used as a disinfectant, for purifying and sterilizing water supplies and swimming pools, in metal fluxing, in the manufacture of bleaching powder, chlorinated hydrocarbons, synthetic rubber, pulp and paper, and plastics. It is also used as a chlorinating and oxidizing agent. It was used by the military as a poison under the name Bertholite.

RISK ASSESSMENT: HEALTH

General Assessment

Chlorine is a severely irritating gas, causing irritation of the eyes, nose, throat, and respiratory tract. It is considered moderately toxic to humans by *inhalation*. Human mutation data also reported. Other symptoms of exposure include burning of the mouth, coughing, choking, nausea, vomiting, headache, dizziness, pneumonia, muscle weakness, respiratory distress, changes in the trachea or bronchi, emphysema, chronic pulmonary edema, and congestion. A 30-minute exposure to 500-800 ppm can be fatal to humans.

Chlorine is extremely irrigating to the eyes and mucous membranes in the respiratory tract at concentrations as low as 3 ppm. Because of chlorine's intense and irritating properties, severe industrial exposures seldom occur (the worker is forced to leave the area before serious physical effects can occur). In cases where such exposures do occur, the worker will feel a burning sensation in the eyes, nose and throat, followed by tightness and pain in chest. Pulmonary edema (fluid buildup in the lungs) can follow which can be fatal.

Skin contact can cause irritating rash to develop. Chronic exposures to low levels may result in corrosion of the teeth, inflammation of the mucous membranes, respiratory disorders, and an increased susceptibility to tuberculosis (an infectious lung disease).

Acute Health Effects

The following acute (short-term) health effects may occur immediately or shortly after exposure to chlorine:

Skin: Immediate irritation and burn on contact. May cause irritating rash and possible permanent damage to skin tissue.

Eye: Severe irritation; may cause burns and permanent damage.

Lung: Irritation of the nose, throat, and lungs following exposure. Pulmonary edema is possible (24 to 48 hours following a high exposure). It can cause tearing, choking, coughing, sputum, bloody nose, and chest pain.

Other: Headache, dizziness, loss of consciousness, muscle weakness, burning of the mouth, nausea, vomiting, and pneumonia.

Chronic Health Effects

The following chronic (long-term) health effects may occur at some time after exposure to chlorine and can last for months or even years:

Cancer Hazards: According to information presented in the references, chlorine has not been adequately tested for its ability to cause cancer in test animals. Chlorine may cause mutations (genetic changes) in living cells and many scientists believe such chemicals may pose a cancer risk in the long-term.

Reproductive Hazard: According to information presented in the references, chlorine has not been adequately tested for its ability to adversely affect reproduction.

Other Chronic Effects: Chlorine can irritate the lungs. Repeated exposure may cause bronchitis to develop with cough, phlegm, and/or shortness of breath. Long-term exposure can cause corrosion of the teeth, inflammation of the mucous membranes, respiratory ailments, and an increased susceptibility to tuberculosis.

⊕ *Recommended Risk-Reduction Measures*

Personnel should avoid direct contact with chlorine. If a less toxic chemical cannot be substituted for a hazardous substance, then *engineering controls* are the most effective method of reducing exposures. The best protection is to enclose operations and/or provide local exhaust ventilation at the site of chemical release. While not always operationally feasible, isolating operations can also reduce exposure. Using respiratory protection is less effective than the controls mentioned above, but is still advisable whenever working with or around chlorine. A self-contained breathing apparatus (SCBA) with full facepiece and pressure demand is the recommended respiratory protection of choice. If a full facepiece is not available, then chemical goggles should be worn to protect the eyes. For low concentrations (below the PEL), an organic vapor cartridge respirator (equipped with a chlorine gas cartridge) is acceptable protection. Whenever a chemical splash hazard exists, a face shield and a rubber apron should be worn. To prevent hand and skin exposures, chemical resistant gloves should be worn.

Administrative controls should also be in place to minimize the potential for human exposures. These may include written procedures or policies which specify the methods and techniques that will be practiced whenever personnel are to work with chlorine.

All personnel should receive *training* on the use, hazards, protective measures, emergency actions, and other precautions per 29 CFR 1910.1200 (Hazard Communication), prior to the first assignment in an area where chlorine is used or stored.

For those personnel with frequent or potentially high exposure (half the PEL or greater), the following are recommended medical tests to be provided before initial assignment and at regular intervals thereafter (e.g., annually):

☑ Dental examination to check teeth for indications of erosion.

☑ Lung function tests.

If symptoms develop or overexposure is suspected, the following tests are recommended:

☑ Lung function tests.

☑ Consider chest X-ray after acute overexposures.

Any evaluation should include a careful history of past and present symptoms with an examination. Medical tests that simply look for existing damage are not a substitute for controlling exposures. Also, because smoking can cause heart disease, as well as lung cancer, emphysema, and other respiratory problems, exposure to chlorine may worsen pre-existing respiratory conditions.

Other methods to reduce exposure include:

☑ Where possible, enclose operations and use local exhaust ventilation at the site of chemical release. If local exhaust ventilation or enclosure is not used, respiratory protection should be mandatory.

☑ Always ensure that proper protective clothing is worn when using chemical substances.

☑ Wash thoroughly immediately after exposure to chlorine and at the end of the work shift or before eating, drinking, or smoking.

☑ Hazard warning information should be posted in the work area. In addition, as part of an on-going education and training program, all information on the health and safety hazards of chlorine should be communicated to all potentially exposed workers.

☑ Eye wash stations should be provided in the immediate work area for emergency use. If there is a possibility of skin exposure to chlorine, emergency shower facilities should also be provided.

☑ Personnel should never wear contaminated clothing home (family members can be exposed). Clothing should be laundered by personnel who have been trained on the hazards of chlorine.

☑ Specific engineering controls are recommended for chlorine by NIOSH. Refer to NIOSH Criteria Document, publication number 76-176.

RISK ASSESSMENT: ENVIRONMENT

General Assessment

The environment is at risk of exposure during transportation, storage, disposal, or destruction of chlorine. In almost every scenario, the threat of environmental exposure is contingent upon the proper handling of the chemical substance. Accidental spills, large or small, can result in fire, explosion, and possible contamination of the surrounding environmental mediums (water, soil, and air).

Chlorine is considered a non-flammable gas. However, it is a powerful oxidizing agent and reactions with incompatible commodities can result in fire and serious explosions. These characteristics require extreme caution in handling, storage, transportation, and disposal. It is incompatible with gasoline, and most petroleum products, turpentine, alcohols, acetylene, hydrogen, ammonia, and sulfur and contact can cause fire or explosion. These characteristics require special consideration during any emergency situation involving a leak or spill of chlorine. Should chlorine ever come into contact with incompatible substances during use, transportation, storage, or disposal, the formation of highly toxic and/or highly explosive commodities is extremely possible.

Chlorine is a natural element of common occurrence. It is also used extensively in many industrial and commercial processes. Discharges of chlorine to water are common because it is used to disinfect effluents and in various industrial processes (particularly in the food and paper industries). It can also enter the environment through spills.

Acute Ecological Effects

Acute (short-term) toxic effects may include the death of animals, birds, or fish and death or low growth rate in plants. Acute effects are seen 2 to 4 days after animals or plants are exposed to chlorine.

This chemical has high acute toxicity to aquatic life. No data are available on the short-term effects of chlorine exposure to plants, birds, or land animals.

Chronic Ecological Effects

Chronic (long-term) toxic effects may include shortened life span, reproductive problems, lower fertility, and changes in appearance or behavior in exposed animals. These effects can be seen long after first exposure(s) to toxic chemicals.

Chlorine has high chronic toxicity in aquatic life. No data are available on the long-term effects of chlorine to plants, birds, or land animals.

Water Solubility

Chlorine is highly soluble in water. Concentrations of 1000 milligrams and more will readily mix with a liter of water.

Persistence in the Environment

Chlorine is non-persistent in water, with a half-life of less than 2 days. The half-life of a pollutant is the amount of time it takes for one half of the chemical to be degraded.

No data are available on the amount of chlorine that will remain in air or water.

Bioaccumulation in Aquatic Organisms

Some substances increase in concentration, or *bioaccumulate,* in living organisms as they breathe contaminated air, drink contaminated water, or eat contaminated food. These chemicals can become concentrated in the tissues and internal organs of animals as well as humans.

Chlorine is not expected to accumulate in the tissues of fish.

Recommended Risk-Reduction Measures

Proper training of all transporters will reduce the likelihood of a mishap or accident resulting in a leak or spill to the environment. The correct use of DOT labeling (per the MSDS) on all containers, trucks, and rail cars will enable emergency responders to react properly and quickly to any disaster thereby reducing the potential risk to the environment and to personnel.

Storage of chlorine should be segregated from incompatible chemicals to minimize the risk of cross-contamination or contact. Buildings designated for storage should be equipped with appropriate fire protection systems (alarms, sprinklers, emergency lighting, portable extinguishers).

If a spill or leak to the environment has occurred, fire department, emergency response and/or hazardous materials spill personnel should be notified immediately. Cleanup should be attempted only by those trained in proper spill containment procedures. Contaminated soils should be removed for incineration and replaced with clean soil. If chlorine should contact the water table, aquifer, or navigable waterway, time is of the essence. It is highly soluble in water and total remediation and containment may not be possible. The local and/or state emergency response authorities must be notified. A comprehensive emergency response or disaster preparedness/recovery plan should be in place prior to any operations involving the use, transportation, storage, or disposal of chlorine. If chlorine is spilled or leaked, the following specific steps are recommended:

- ☑ Restrict persons not wearing protective clothing from area of spill or leak until cleanup is complete and area can be opened for normal work.
- ☑ Remove all ignition sources.
- ☑ Ventilate area of leak.
- ☑ Absorb liquids in vermiculite, dry sand, earth, or similar material and deposit in sealed containers.

- ☑ If gas is leaking, stop the flow of gas. If the source is a cylinder and leak can not be stopped in place, remove cylinder to a safe place in the open air and repair or allow to empty.
- ☑ It may be necessary to dispose of chlorine as a hazardous waste. The responsible state agency or the regional office of the federal Environmental Protection Agency (EPA) should be contacted for specific recommendations.

RISK ASSESSMENT: BUSINESS

General Assessment

Accidents or mishaps involving chlorine can present a significant threat to business operations. The loss or damage of equipment or facilities can significantly affect fiscal viability. Lawsuits that may result from personnel injury/death, public exposures, and/or environmental contamination will also require a serious expenditure of resources. Media attention surrounding an injury, death, or environmental damage can also result in a loss of profits and loss of current as well as future business.

Recommended Risk-Reduction Measures

Company attorneys, safety and health professionals, and environmental specialists should be involved in the development of any such procedures. A company official should be pre-designated as a public relations officer with specific training in dealing with the press. Corporate plans and policies should be developed, approved, and implemented long before any need for such arises.

REFERENCES

American Conference of Governmental Industrial Hygienists. 1988. *Documentation of the Threshold Limit Values and Biological Exposure Indices*, 5th Edition (with updates). Cincinnati: ACGIH

Bueche, J. R. 1972. *Principles of Physics*, 2nd Edition. New York: McGraw-Hill.

Calabrese, E. J. 1994. *Biological Effects of Low Level Exposures.* New York: CRC Press/Lewis Publishers.

Cockerman, L. G., B. S. Shane. 1994. *Basic Environmental Toxicology.* New York: CRC Press/Lewis Publishers.

Gardner, R. W. 1994. *Chemical Intolerance.* New York: CRC Press/Lewis Publishers.

Hallenbeck, W. H. 1993. *Quantitative Risk Assessment for Environmental and Occupational Health*, 2nd Edition. New York: CRC Press/Lewis Publishers.

Jones, F. E. 1994. *Toxic Organic Vapors in the Workplace.* New York: CRC Press/Lewis Publishers.

Keith, L. H. and M. M. Walker. 1995. *Handbook of Air Toxics.* New York: CRC Press/Lewis Publishers.

Landis, W. G., M. H. Yu. 1995. *Introduction to Environmental Toxicology.* New York: CRC Press/Lewis Publishers.

Langerman, N. 1994 *Precautionary Labels for Chemical Containers.* New York: CRC Press/Lewis Publishers.

Lauwerys, R. R., P. Hoet. 1993. *Industrial Chemical Exposure.* New York: CRC Press/Lewis Publishers.

Lewis, R. J., Sr. 1992. *Sax's Dangerous Properties of Industrial Materials*, Volumes 1, 2, and 3, 8th Edition. New York: Van Nostrand Reinhold.

Lide, D. R. 1995. *Handbook of Chemistry and Physics*, 76th Edition. New York: CRC Press/Lewis Publishers.

Manahan, S. E. 1994. *Environmental Chemistry*, 6th Edition. New York: CRC Press/Lewis Publishers.

Mellström, G. A., J. E. Wahlberg, and H. I. Maibach. 1994. *Protective Gloves for Occupational Use.* New York: CRC Press/Lewis Publishers.

Meyer, E. 1990. *Chemistry of Hazardous Materials.* Englewood Cliffs, NJ: Prentice Hall.

National Cancer Institute. 1978. *Carcinogenesis Technical Report Series. Bioassay of Chlorine for Possible Carcinogenicity.* DEHW (NIH) Publication Number: 78-1323, p 53. Washington, DC: U.S. Government Printing Office.

Patnaik, P. 1992. *A Comprehensive Guide to the Hazardous Properties of Chemical Substances.* New York: Van Nostrand Reinhold.

Proctor, N. H., J. P. Hughes. 1991. *Chemical Hazards of the Workplace.* New York: Van Nostrand Reinhold.

Quigley, D. R. 1994. *Handbook of Emergency Chemical Management.* New York: CRC Press/Lewis Publishers.

Rea, W. J. 1995. *Chemical Sensitivity*, Vol. I-III. New York: CRC Press/Lewis Publishers.

Richardson, M. 1995. *Dictionary of Substance and Their Effects*, Vol. I-VI. New York: CRC Press/Lewis Publishers.

Sacarello, H. L. A. 1994. *The Comprehensive Handbook of Hazardous Materials.* New York: CRC Press/Lewis Publishers.

Sheftel, V. O. 1995. *Handbook of Toxic Properties of Monomers and Additives.* New York: CRC Press/Lewis Publishers.

Torkelson, T. R. V. K. Rowe. 1981. Halogenated aliphatic hydrocarbons. *Patty's Industrial Hygiene and Toxicology*, Vol. 2B, *Toxicology*, pp 3568-3572. New York: Wiley-Interscience.

U.S. Department of Health, Education, and Welfare, National Institute for Occupational Safety and Health. 1976. *Criteria for a Recommended Standard: Occupational Exposure to Chlorine.* DEHW (NIOSH) Publication Number: 76-176. Washington, D.C.: U.S. Government Printing Office.

U.S. Department of Health and Human Services, National Institute for Occupational Safety and Health. 1994. *NIOSH Pocket Guide to Chemical Hazards.* Washington, D.C.: U.S. Government Printing Office.

MATERIAL SAFETY DATA SHEET

CHEMICAL NAME	
CHLORINE DIOXIDE	**CRC PRESS/LEWIS PUBLISHERS DISCLAIMER:** THE INFORMATION AND RECOMMENDATIONS PRESENTED HEREIN ARE BASED ON SOURCES BELIEVED TO BE RELIABLE. CRC MAKES NO REPRESENTATION ON ITS COMPLETENESS OR ACCURACY. IT IS THE RESPONSIBILITY OF THE USER TO DETERMINE THE CHEMICAL'S SUITABILITY FOR ITS INTENDED USE, THE CHEMICAL'S SAFE USE, AND THE CHEMICAL'S PROPER DISPOSAL. NO REPRESENTATIONS AND/OR WARRANTIES, EITHER EXPRESSED OR IMPLIED, OF THE MERCHANTABILITY OR FITNESS FOR A PARTICULAR PURPOSE, OR OF ANY OTHER NATURE, ARE MADE WITH RESPECT TO THE INFORMATION PROVIDED IN THIS MSDS OR TO THE CHEMICAL TO WHICH INFORMATION MAY REFER. CRC NEITHER ASSUMES NOR AUTHORIZES ANY OTHER PERSON TO ASSUME FOR IT, ANY OTHER ADDITIONAL RESPONSIBILITY OR LIABILITY FOR THE USE OF, OR RELIANCE UPON, THIS INFORMATION.

HAZARD WARNING INFORMATION

HEALTH	FIRE	REACTIVE	OTHER	DEGREE OF HAZARD	COLOR CODING	OTHER CODES
3	**0**	**2**	**OX**	0 = Minimum Hazard 1 = Slight Hazard 2 = Moderate Hazard 3 = Serious Hazard 4 = Severe Hazard	HEALTH = BLUE FIRE = RED REACTIVITY = YELLOW OTHER = WHITE	OX = Oxidizer ACID = Acid ALK = Alkali COR = Corrosive W = Use No Water

SECTION I - GENERAL INFORMATION

Characterization	RCRA Number	EPA Class
Halogen Oxide	**None**	**Not Applicable**

DOT Proper Shipping Name **Chlorine dioxide (not hydrated)**	Chemical Abstract Service (CAS) Number **10049-04-4**
DOT Hazard Class and Label Requirements **Forbidden (hydrated); Oxidizer & Poison**	DOT Emergency Guide Code **47**
DOT Identification Number **UN 9191 (frozen solution only)**	Chemical Formula **ClO_2**

Synonyms
Chlorine peroxide; chloroperoxyl; Alcide; anthium dioxide; chlorine (IV) oxide; Doxide 50.

SECTION II - HAZARDOUS INGREDIENTS AND IDENTITY INFORMATION

Hazardous Components (specific identity)	OSHA Exposure Criteria	NIOSH Exposure Criteria	Immediately Dangerous to Life and Health (IDLH)	ACGIH Exposure Criteria
Chlorine dioxide (derivation: Usually made at point of consumption from sodium chlorate, sulfuric acid, and methanol, or from sodium chlorate and sulfur dioxide. Concentration of gas is limited to 10% to reduce explosion hazard). **1 ppm = 2.81 mg/m^3**	**PEL:** **0.1 ppm** **0.3 mg/m^3** **STEL:** **0.3 ppm** **0.9 mg/m^3**	**REL:** **0.1 ppm** **0.3 mg/m^3** **STEL:** **0.3 ppm** **0.9 mg/m^3**	**5 ppm**	**TLV:** **0.1 ppm** **0.3 mg/m^3** **STEL:** **0.3 ppm** **0.9 mg/m^3**

SECTION III - PHYSICAL AND CHEMICAL CHARACTERISTICS

Boiling Point **50°F (10°C)**	Specific Gravity (H_2O = 1) **2.33**
Vapor Pressure (mmHg **>760 at 69°F (20°C)**	Molecular Weight **67.5**
Vapor Density (Air = 1) **3.09**	Melting Point **-74°F (-59°C)**

Solubility
Decomposes in water (produces hydrochloric acid); dissolves in alkalis.

Appearance and Odor
Yellow to orange gas at ordinary temperatures; pungent, irritating odor similar to that of chlorine. May appear as a reddish-brown liquid or yellowish-red crystals (frozen state). Odor threshold = 0.1 ppm.

SECTION IV - FIRE AND EXPLOSION HAZARD DATA

Flash Point (method used) **Not Applicable**	Explosive Limits in Air % by Volume **LEL: N/A UEL: N/A**
NFPA Classification **Flammable Gas, Combustible Liquid**	Autoignition Temperature **Not Applicable**

Extinguishing Media
Use water only. Do not use chemicals or carbon dioxide extinguishers.

Special Fire Fighting Procedures
Chlorine dioxide is a powerful oxidizer. Wear full protective clothing and self-contained breathing apparatus (SCBA). Use a water spray to keep fire-exposed containers cool. Poisonous gases are produced in fire. Isolate for ½ mile if rail or tank truck is involved. Containers may explode in fire or heat.

Unusual Fire and Explosion Hazards
Dangerous fire and explosion hazard when exposed to any form of energy, including sunlight, heat, or sparks. Explodes violently in air at concentrations over 10%.

SECTION V - REACTIVITY DATA

Stability		Conditions to Avoid
		Extremely reactive and unstable. Do not attempt shipment in liquid or gas form (frozen shipment only). Keep away from sunlight and other forms of energy. Do not expose to air.
Stable	Unstable	Incompatibility *(materials to avoid)*
	X	Organic materials, heat, phosphorus, potassium hydroxide, sulfur, mercury, carbon monoxide, dust, caustic potash, caustic soda, metal hydrides, fluorine, and difluoramine.
Hazardous Polymerization		Conditions to Avoid Under normal conditions of temperature and pressure, hazardous polymerization of chlorine dioxide is not known to occur.
May Occur	Will Not Occur	Hazardous Decomposition or By-products
	X	When heated to decomposition, chlorine dioxide can emit highly toxic/poisonous gases. Mixing with water or steam will produce toxic and corrosive fumes of hydrochloric acid.

SECTION VI - HEALTH HAZARD DATA

Primary Route(s) of Entry:	Inhalation?	Absorption (skin)?	Ingestion?
	X		

Health Hazards

INHALATION: Extremely irritating to the lungs, throat, nose, and eyes. Coughing and shortness of breath that may persist for hours after exposure. Can cause permanent damage to the lungs. Fluid can build up in the lungs (pulmonary edema), which is a medical emergency.

SKIN & EYES: Contact with the liquid can cause severe skin burns (due to its cold state). The liquid can cause severe eye burns leading to permanent damage.

Carcinogenicity	NTP Listed?	IARC Cancer Review Group?	OSHA Regulated?	Target Organs?
Unknown Human Unknown Animal	No	No	29 CFR 1910.1000 Table Z-1	Respiratory system, eyes.

Medical Conditions Generally Aggravated by Exposure

None reported.

Emergency and First-aid Procedures

Eye contact: Flush immediately with water for 30 minutes (minimum), seek medical attention. Skin contact: Remove all contaminated clothing. Immediately wash area with large amounts of soap and water. Seek medical assistance. For inhalation: Remove the person from exposure. Provide respiratory assistance and CPR. Transfer to medical facility. Observe for 24 - 48 hours since pulmonary edema may be a delayed reaction. If swallowed: Seek medical attention immediately.

SECTION VII - PRECAUTIONS FOR SAFE HANDLING AND USE

Steps to be Taken in Case Material is Released or Spilled

Absorb liquids in vermiculite, dry sand, earth, or similar material and deposit in sealed containers. Ventilate area of spill or leak. Restrict those not involved in cleanup from entering area. If gas is leaked, stop source if possible or remove leaking cylinders to safe area outdoors and allow to empty.

Preferred Waste Disposal Method

No Citation.

Precautions to be Taken in Handling and Storage

Chlorine dioxide is a flammable gas and an extremely powerful oxidizer. Store to avoid contact with incompatible materials since violent reactions can occur. Store in tightly closed containers in cool, well-ventilated area away from heat and sunlight. Keep temperature below 266°F (130°C).

Other Precautions and Warnings

Sources of ignition are prohibited where chlorine dioxide is used, handled, or stored. Heat may cause containers to build up pressure and explode.

SECTION VIII - CONTROL MEASURES AND PERSONAL PROTECTIVE EQUIPMENT

Respiratory Protection *(specify type)*

For exposures below the PEL, use a full facepiece respirator with chlorine dioxide gas cartridge. For higher exposures, use a supplied air respirator with full facepiece, hood, or helmet in continuous flow mode, or use a SCBA piece operated in pressure demand or other positive pressure mode.

Ventilation

Local exhaust or general mechanical systems recommended.

Protective Gloves	Eye Protection	Other Protective Clothing
Neoprene rubber	Chemical Goggles or Face Mask	Neoprene Rubber Apron

Work/Hygiene Practices

Always wash hands thoroughly after using chemical; never bring food, drink, or smoking materials into vicinity of chemicals.

CHLORINE DIOXIDE

ClO_2 CAS: 10049-04-4

IDENTIFICATION AND TYPICAL USES

Chlorine dioxide is a yellow to orange gas at ordinary temperatures with a pungent, irritating odor somewhat like that of chlorine. It liquefies at 52°F (11°C) to a reddish-brown liquid. It will solidify at -74°F (-59°C) to yellowish-red crystals. It is normally sold as a hydrate in its frozen form. Among its most common uses, chlorine dioxide is a bleaching agent to bleach fats, oils, textiles, cellulose, paper pulp, flour, and leather. It is also used for purifying water, as an oxidizing agent, as an antiseptic, and in the manufacture of many chlorite salts.

RISK ASSESSMENT: HEALTH

General Assessment

Chlorine dioxide is highly irritating to the eyes, nose, and throat. Moderately toxic by *inhalation.* There is also some evidence that it has the ability to adversely affect reproduction in test animals. Human mutation data have also been reported.

Breathing chlorine dioxide vapors can cause coughing, nasal discharge, wheezing, respiratory distress, bronchitis, and congestion of the lungs. Its irritant effects can be intense at a concentration in air as low as 5 ppm. High exposures can damage the lungs, causing a buildup of fluid (pulmonary edema) which is a medical emergency and can be fatal. The chronic toxicity signs are primarily dyspnea and asthmatic bronchitis and, in some cases, irritation of the gastrointestinal tract. Ingestion of the liquid may cause somnolence and respiratory stimulation.

Acute Health Effects

The following acute (short-term) health effects may occur immediately or shortly after exposure to chlorine dioxide:

<u>Skin</u>: Liquid may irritate the skin (cold temperature may cause burning sensation).

<u>Eye</u>: Severe irritation; may cause watery eyes.

<u>Lung</u>: Irritation of the nose, throat, and lungs following exposure. Pulmonary edema is possible (24 to 48 hours following a high exposure). It can cause chest pain, cough, bloody nose, and sputum.

<u>Other</u>: Headache, dizziness, loss of consciousness, muscle weakness, burning of the mouth, nausea, vomiting, and pneumonia.

Chronic Health Effects

The following chronic (long-term) health effects may occur at some time after exposure to chlorine dioxide and can last for months or even years:

<u>Cancer Hazards</u>: Chlorine dioxide may cause mutations (genetic changes) in living cells. It has not been adequately tested for its ability to cause cancer in test animals.

<u>Reproductive Hazard</u>: According to information presented in the references, there is some experimental evidence that chlorine dioxide can adversely affect reproduction in test animals.

<u>Other Chronic Effects</u>: Repeated exposures can cause bronchitis to develop with cough, phlegm, and/or shortness of breath. Permanent lung damage can occur, especially with repeated exposure to vapors.

Recommended Risk-Reduction Measures

Personnel should avoid direct contact with chlorine dioxide. If a less toxic chemical cannot be substituted for a hazardous substance, then *engineering controls* are the most effective method of reducing exposures. The best protection is to enclose operations and/or provide local exhaust ventilation at the site of chemical release. While not always operationally feasible, isolating operations can also reduce exposure. Using respiratory protection is less effective than the controls mentioned above, but is still advisable whenever working with or around chlorine dioxide. A self-contained breathing apparatus (SCBA) with full facepiece and pressure demand is the recommended respiratory protection of choice. If a full facepiece is not available, then chemical goggles should be worn to protect the eyes. For low concentrations (below the PEL), a full facepiece respirator equipped with a cartridge containing noncombustible sorbents and providing protection against chlorine dioxide gas is acceptable protection. Whenever working with the liquid and a chemical splash hazard exists, a face shield and a rubber apron should be worn. To prevent hand and skin exposures, acid resistant gloves should be worn.

Administrative controls should also be in place to minimize the potential for human exposures. These may include written procedures or policies which specify the methods and techniques that will be practiced whenever personnel are to work with chlorine dioxide.

All personnel should receive *training* on the use, hazards, protective measures, emergency actions, and other precautions per 29 CFR 1910.1200 (Hazard Communication), prior to the first assignment in an area where chlorine dioxide is used or stored.

For those personnel with frequent or potentially high exposure (half the PEL or greater), the following are recommended medical tests to be provided before initial assignment and at regular intervals thereafter (e.g., annually):

☑ Lung function tests.

If symptoms develop or overexposure is suspected, the following tests are recommended:

☑ Lung function tests.

☑ Consider chest X-ray after acute overexposures.

Any evaluation should include a careful history of past and present symptoms with an examination. Medical tests that simply look for existing damage are not a substitute for controlling exposures. Also, because smoking can cause heart disease, as well as lung cancer, emphysema, and other respiratory problems, exposure to chlorine dioxide may worsen pre-existing respiratory conditions.

Other methods to reduce exposure include:

☑ Where possible, enclose operations and use local exhaust ventilation at the site of chemical release. If local exhaust ventilation or enclosure is not used, respiratory protection should be mandatory.

☑ Always ensure that proper protective clothing is worn when using chemical substances.

☑ Wash thoroughly immediately after exposure to chlorine dioxide and at the end of the work shift or before eating, drinking, or smoking.

☑ Hazard warning information should be posted in the work area. In addition, as part of an on-going education and training program, all information on the health and safety hazards of chlorine dioxide should be communicated to all potentially exposed workers.

☑ Eye wash stations should be provided in the immediate work area for emergency use. If there is a possibility of skin exposure to chlorine dioxide, emergency shower facilities should also be provided.

☑ Personnel should never wear contaminated clothing home (family members can be exposed). Clothing should be laundered by personnel who have been trained on the hazards of chlorine dioxide.

RISK ASSESSMENT: ENVIRONMENT

General Assessment

The environment is at risk of exposure during transportation, storage, disposal, or destruction of chlorine dioxide. In almost every scenario, the threat of environmental exposure is contingent upon the proper handling of the chemical substance. Accidental spills, large or small, can result in fire, explosion, and possible contamination of the surrounding environmental mediums (water, soil, and air).

Chlorine dioxide is considered a flammable gas. It is a powerful oxidizing agent and extremely reactive with many commonly encountered commodities such as dust and air. Contact with incompatible materials can result in fire and serious explosions. These characteristics require extreme caution in handling, storage, transportation, and disposal. It is incompatible with organic matter, combustible materials, powdered metals, and ammonia compounds. It will explode violently upon exposure to sunlight or when subjected to a spark. Detonation occurs at concentrations above 10% in air in the presence of an energy source or a catalyst. It undergoes violent reactions with organic matter; explosion occurs when the mixture is subjected to shock or spark. It is extremely unstable and highly reactive. These characteristics require special consideration during any emergency situation involving a leak or spill of chlorine dioxide. Should chlorine dioxide ever come into contact with incompatible substances during use, transportation, storage, or disposal, the formation of highly toxic and/or highly explosive commodities is extremely possible.

Chlorine dioxide is used extensively in many industrial and commercial processes. Discharges of chlorine dioxide to water are common because it is used to disinfect effluents and in various industrial processes (particularly in the food and paper industries). It can also enter the environment through spills.

Acute Ecological Effects

Acute (short-term) toxic effects may include the death of animals, birds, or fish and death or low growth rate in plants. Acute effects are seen 2 to 4 days after animals or plants are exposed to chlorine dioxide.

This chemical has high acute toxicity to aquatic life. No data are available on the short-term effects of chlorine dioxide exposure to plants, birds, or land animals.

Chronic Ecological Effects

Chronic (long-term) toxic effects may include shortened life span, reproductive problems, lower fertility, and changes in appearance or behavior in exposed animals. These effects can be seen long after first exposure(s) to toxic chemicals.

Chlorine dioxide has high chronic toxicity in aquatic life. No data are available on the long-term effects of chlorine dioxide to plants, birds, or land animals.

Water Solubility

Chlorine dioxide decomposes in water to form a mixture of chlorite and chlorate.

Persistence in the Environment

Chlorine dioxide is non-persistent in water, with a half-life of less than 2 days. The half-life of a pollutant is the amount of time it takes for one half of the chemical to be degraded. No data are available on the amount of chlorine dioxide that will remain in air or water.

Bioaccumulation in Aquatic Organisms

Some substances increase in concentration, or *bioaccumulate,* in living organisms as they breathe contaminated air, drink contaminated water, or eat contaminated food. These chemicals can become concentrated in the tissues and internal organs of animals as well as humans.

Chlorine dioxide is not expected to accumulate in the tissues of fish.

Recommended Risk-Reduction Measures

Proper training of all transporters will reduce the likelihood of a mishap or accident resulting in a leak or spill to the environment. Transportation of pure chlorine dioxide is forbidden by the Department of Transportation (DOT). It must be kept frozen when shipped. The correct use of DOT labeling (per the MSDS) on all containers, trucks, and rail cars will enable emergency responders to react properly and quickly to any disaster thereby reducing the potential risk to the environment and to personnel.

When not is use, chlorine dioxide must be kept frozen. Storage of chlorine dioxide should be segregated from incompatible chemicals to minimize the risk of cross-contamination or contact. Containers should be kept tightly closed in a cool, dark, well-ventilated area at temperatures well below 266°F (130°C). Gas explosions occur above 266°F. Buildings designated for storage should be equipped with appropriate fire protection systems (alarms, sprinklers, emergency lighting, portable extinguishers).

If a spill or leak to the environment has occurred, fire department, emergency response and/or hazardous materials spill personnel should be notified immediately. Cleanup should be attempted only by those trained in proper spill containment procedures. Contaminated soils should be removed for incineration and replaced with clean soil. If chlorine dioxide should contact the water table, aquifer, or navigable waterway, time is of the essence. It is decomposes in water and total remediation and containment may not be possible. The local and/or state emergency response authorities must be notified. A comprehensive emergency response or disaster preparedness/recovery plan should be in place prior to any operations involving the use, transportation, storage, or disposal of chlorine dioxide. If chlorine dioxide is spilled or leaked, the following specific steps are recommended:

- ☑ Restrict persons not wearing protective clothing from area of spill or leak until cleanup is complete and area can be opened for normal work.
- ☑ Remove all ignition sources and ventilate area.
- ☑ Absorb liquids in vermiculite, dry sand, earth, or similar material and deposit in sealed containers.

☑ If gas is leaking, stop the flow of gas. If the source is a cylinder and leak cannot be stopped in place, remove cylinder to a safe place in the open air and repair or allow to empty.

☑ It may be necessary to dispose of chlorine dioxide as a hazardous waste. The responsible state agency or the regional office of the federal Environmental Protection Agency (EPA) should be contacted for specific recommendations.

RISK ASSESSMENT: BUSINESS

General Assessment

Accidents or mishaps involving chlorine dioxide can present a significant threat to business operations. The loss or damage of equipment or facilities can significantly affect fiscal viability. Lawsuits that may result from personnel injury/death, public exposures, and/or environmental contamination will also require a serious expenditure of resources. Media attention surrounding an injury, death, or environmental damage can also result in a loss of profits and loss of current as well as future business.

Recommended Risk-Reduction Measures

Company attorneys, safety and health professionals, and environmental specialists should be involved in the development of such procedures dealing with safety and emergency response. A company official should be pre-designated as a public relations officer with specific training in dealing with the press. Corporate plans and policies should be developed, approved, and implemented long before any need for such arises.

REFERENCES

American Conference of Governmental Industrial Hygienists. 1988. *Documentation of the Threshold Limit Values and Biological Exposure Indices,* 5th Edition (with updates). Cincinnati: ACGIH

Bueche, J. R. 1972. *Principles of Physics,* 2nd Edition. New York: McGraw-Hill.

Calabrese, E. J. 1994. *Biological Effects of Low Level Exposures.* New York: CRC Press/Lewis Publishers.

Cockerman, L. G., B. S. Shane. 1994. *Basic Environmental Toxicology.* New York: CRC Press/Lewis Publishers.

Gardner, R. W. 1994. *Chemical Intolerance.* New York: CRC Press/Lewis Publishers.

Hallenbeck, W. H. 1993. *Quantitative Risk Assessment for Environmental and Occupational Health,* 2nd Edition. New York: CRC Press/Lewis Publishers.

Jones, F. E. 1994. *Toxic Organic Vapors in the Workplace.* New York: CRC Press/Lewis Publishers.

Keith, L. H. and M. M. Walker. 1995. *Handbook of Air Toxics.* New York: CRC Press/Lewis Publishers.

Landis, W. G., M. H. Yu. 1995. *Introduction to Environmental Toxicology.* New York: CRC Press/Lewis Publishers.

Langerman, N. 1994 *Precautionary Labels for Chemical Containers.* New York: CRC Press/Lewis Publishers.

Lauwerys, R. R., P. Hoet. 1993. *Industrial Chemical Exposure.* New York: CRC Press/Lewis Publishers.

Lewis, R. J., Sr. 1992. *Sax's Dangerous Properties of Industrial Materials,* Volumes 1, 2, and 3, 8th Edition. New York: Van Nostrand Reinhold.

Lide, D. R. 1995. *Handbook of Chemistry and Physics,* 76th Edition. New York: CRC Press/Lewis Publishers.

Manahan, S. E. 1994. *Environmental Chemistry,* 6th Edition. New York: CRC Press/Lewis Publishers.

Mellström, G. A., J. E. Wahlberg, and H. I. Maibach. 1994. *Protective Gloves for Occupational Use.* New York: CRC Press/Lewis Publishers.

Meyer, E. 1990. *Chemistry of Hazardous Materials.* Englewood Cliffs, NJ: Prentice Hall.

National Cancer Institute. 1978. *Carcinogenesis Technical Report Series. Bioassay of Chlorine dioxide for Possible Carcinogenicity.* DEHW (NIH) Publication Number: 78-1323, p 53. Washington, DC: U.S. Government Printing Office.

Patnaik, P. 1992. *A Comprehensive Guide to the Hazardous Properties of Chemical Substances.* New York: Van Nostrand Reinhold.

Proctor, N. H., J. P. Hughes. 1991. *Chemical Hazards of the Workplace.* New York: Van Nostrand Reinhold.

Quigley, D. R. 1994. *Handbook of Emergency Chemical Management.* New York: CRC Press/Lewis Publishers.

Rea, W. J. 1995. *Chemical Sensitivity,* Vol. I-III. New York: CRC Press/Lewis Publishers.

Richardson, M. 1995. *Dictionary of Substance and Their Effects,* Vol. I-VI. New York: CRC Press/Lewis Publishers.

Sacarello, H. L. A. 1994. *The Comprehensive Handbook of Hazardous Materials.* New York: CRC Press/Lewis Publishers.

Sheftel, V. O. 1995. *Handbook of Toxic Properties of Monomers and Additives.* New York: CRC Press/Lewis Publishers.

Torkelson, T. R. V. K. Rowe. 1981. Halogenated aliphatic hydrocarbons. *Patty's Industrial Hygiene and Toxicology,* Vol. 2B, *Toxicology,* pp 3568-3572. New York: Wiley-Interscience.

U.S. Department of Health and Human Services, National Institute for Occupational Safety and Health. 1994. *NIOSH Pocket Guide to Chemical Hazards.* Washington, D.C.: U.S. Government Printing Office.

MATERIAL SAFETY DATA SHEET

CHEMICAL NAME	
CHLORINE PENTAFLUORIDE	**CRC PRESS/LEWIS PUBLISHERS DISCLAIMER:** THE INFORMATION AND RECOMMENDATIONS PRESENTED HEREIN ARE BASED ON SOURCES BELIEVED TO BE RELIABLE. CRC MAKES NO REPRESENTATION ON ITS COMPLETENESS OR ACCURACY. IT IS THE RESPONSIBILITY OF THE USER TO DETERMINE THE CHEMICAL'S SUITABILITY FOR ITS INTENDED USE, THE CHEMICAL'S SAFE USE, AND THE CHEMICAL'S PROPER DISPOSAL. NO REPRESENTATIONS AND/OR WARRANTIES, EITHER EXPRESSED OR IMPLIED, OF THE MERCHANTABILITY OR FITNESS FOR A PARTICULAR PURPOSE, OR OF ANY OTHER NATURE, ARE MADE WITH RESPECT TO THE INFORMATION PROVIDED IN THIS MSDS OR TO THE CHEMICAL TO WHICH INFORMATION MAY REFER. CRC NEITHER ASSUMES NOR AUTHORIZES ANY OTHER PERSON TO ASSUME FOR IT, ANY OTHER ADDITIONAL RESPONSIBILITY OR LIABILITY FOR THE USE OF, OR RELIANCE UPON, THIS INFORMATION.

HAZARD WARNING INFORMATION

HEALTH	FIRE	REACTIVE	OTHER	DEGREE OF HAZARD	COLOR CODING	OTHER CODES
4	0	3	OX ~~W~~	0 = Minimum Hazard 1 = Slight Hazard 2 = Moderate Hazard 3 = Serious Hazard 4 = Severe Hazard	HEALTH = BLUE FIRE = RED REACTIVITY = YELLOW OTHER = WHITE	OX = Oxidizer ACID = Acid ALK = Alkali COR = Corrosive ~~W~~ = Use No Water

SECTION I - GENERAL INFORMATION

Characterization	**Interhalogen Compound**	RCRA Number	**D003**
		EPA Class	**Characteristic (R) Waste**
DOT Proper Shipping Name	**Chlorine Pentafluoride**	Chemical Abstract Service (CAS) Number	**13637-63-3**
DOT Hazard Class and Label Requirements	**Oxidizer & Poison Gas**	DOT Emergency Guide Code	**44**
DOT Identification Number	**UN 2548**	Chemical Formula	**ClF_5**
Synonyms	**None**		

SECTION II - HAZARDOUS INGREDIENTS AND IDENTITY INFORMATION

Hazardous Components (specific identity)	OSHA Exposure Criteria	NIOSH Exposure Criteria	Immediately Dangerous to Life and Health (IDLH)	ACGIH Exposure Criteria
Chlorine pentafluoride (An inorganic fluoride).	**PEL:** 2.5 mg/m^3 (as fluorine) **STEL:** Not Established	**REL:** 2.5 mg/m^3 (as fluorine) **STEL:** Not Established	**Not Determined**	**TLV:** 2.5 mg/m^3 (as fluorine) **STEL:** Not Established

SECTION III - PHYSICAL AND CHEMICAL CHARACTERISTICS

Boiling Point	**Not Found**	Specific Gravity (H_2O = 1)	**1.77**
Vapor Pressure (atmospheres)	**Not Found**	Molecular Weight	**130.45**
Vapor Density (Air = 1)	**>1.0**	Melting Point	**Not Found**

Solubility
Reacts violently with water.

Appearance and Odor
Colorless gas with a suffocating odor.

SECTION IV - FIRE AND EXPLOSION HAZARD DATA

Flash Point (method used)	**Not Determined**	Explosive Limits in Air % by Volume	**LEL: Not Determined UEL: Not Determined**
NFPA Classification	**Non-Flammable Gas, Non-Combustible Liquid**	Autoignition Temperature	**Not Determined**

Extinguishing Media
Stop flow of gas or liquid. Use bicarbonate-based dry chemical. Do NOT use water (reacts explosively!)

Special Fire Fighting Procedures
Chlorine pentafluoride is a powerful oxidizer. Wear full protective clothing and self-contained breathing apparatus (SCBA). Use a water spray to keep fire-exposed containers cool. Poisonous gases are produced in fire. Isolate for ½ mile if rail or tank truck is involved. Containers may explode in fire or heat.

Unusual Fire and Explosion Hazards
Chlorine pentafluoride is hypergolic. It will ignite explosively in contact with a fuel or a spark. Dangerous fire and explosion hazard when exposed to any form of energy, including sunlight, heat, or sparks.

SECTION V - REACTIVITY DATA

Stability		Conditions to Avoid
		Extremely reactive but can be kept stable in special cylinders. Avoid contact with combustible materials such as wood, paper, oil, and plastics since spontaneous combustion can occur.
Stable	Unstable	Incompatibility *(materials to avoid)*
	X	Violent reacts with water, oxidizers, acids, combustible materials, sand, glass, and is corrosive to most metals. Reacts with water to form chlorine and extremely toxic and corrosive hydrofluoric acid.
Hazardous Polymerization		Conditions to Avoid
		Under normal conditions of temperature and pressure, hazardous polymerization of chlorine pentafluoride is not known to occur.
May Occur	Will Not Occur	Hazardous Decomposition or By-products
	X	When heated to decomposition, chlorine pentafluoride can emit highly toxic/poisonous gases. Mixing with water or steam will produce toxic and corrosive fumes of chlorine and fluorine.

SECTION VI - HEALTH HAZARD DATA

Primary Route(s) of Entry:	Inhalation?	Absorption (skin)?	Ingestion?
	X	X	

Health Hazards

INHALATION: Extremely irritating to the lungs, throat, nose, and eyes. Coughing and shortness of breath that may persist for hours after exposure. Can cause permanent damage to the lungs. Fluid can build up in the lungs (pulmonary edema), which is a medical emergency.

SKIN & EYES: Contact with the liquid can cause severe skin burns (due to its cold state). The liquid can cause severe eye burns leading to permanent damage.

Carcinogenicity	NTP Listed?	IARC Cancer Review Group?	OSHA Regulated?	Target Organs?
Unknown Human Unknown Animal	No	No	No	Respiratory system, eyes, and skin.

Medical Conditions Generally Aggravated by Exposure

Persons with existing pulmonary disease may be more susceptible to exposure symptoms.

Emergency and First-aid Procedures

Eye contact: Flush immediately with water for 30 minutes (minimum), seek medical attention. Skin contact: Quickly remove all contaminated clothing. Immediately wash area with large amounts water for 15 minutes (minimum). Apply topical hydrogen fluoride neutralizing agent (calcium gluconate gel). Seek medical assistance. For inhalation: Remove the person from exposure. Provide respiratory assistance and CPR. Transfer to medical facility. Observe for 24-48 hours for lung effects. If swallowed: Seek medical attention immediately. Drink large quantities of milk. Do NOT induce vomiting.

SECTION VII - PRECAUTIONS FOR SAFE HANDLING AND USE

Steps to be Taken in Case Material is Released or Spilled

Absorb liquids in vermiculite, or similar material and deposit in sealed containers. Ventilate area of spill or leak. Restrict those not involved in cleanup from entering area. If gas is leaked, stop source if possible or remove leaking cylinders to safe area outdoors and allow to empty.

Preferred Waste Disposal Method

No Citation.

Precautions to be Taken in Handling and Storage

Chlorine pentafluoride is a non-flammable gas but an extremely powerful oxidizer. Store to avoid contact with incompatible materials since violent reactions can occur. Store in tightly closed containers in cool, well-ventilated area away from heat and moisture (reacts violently with water).

Other Precautions and Warnings

Sources of ignition are prohibited around chlorine pentafluoride. Heat may cause containers to build up pressure and explode. An inert blanket of nitrogen inside containers is recommended.

SECTION VIII - CONTROL MEASURES AND PERSONAL PROTECTIVE EQUIPMENT

Respiratory Protection *(specify type)*

For exposures below the PEL, use a full facepiece respirator with organic vapor cartridge. For higher exposures, use a supplied air respirator with full facepiece, hood, or helmet in continuous flow mode, or use a SCBA operated in pressure demand or other positive pressure mode.

Ventilation

Local exhaust or general mechanical systems recommended.

Protective Gloves	Eye Protection	Other Protective Clothing
Impervious, Solvent-Resistant	Chemical Goggles or Face Mask	Protective Apron or Full-Body Suit

Work/Hygiene Practices

Always wash hands thoroughly after using chemical; never bring food, drink, or smoking materials into vicinity of chemicals.

CHLORINE PENTAFLUORIDE

ClF_5

CAS: 13637-63-3

IDENTIFICATION AND TYPICAL USES

Chlorine pentafluoride is a colorless gas with a suffocating odor. It is normally shipped as a liquefied compressed gas. Commercial applications are somewhat limited. It is sometimes used as a fluorinating and oxidizing agent.

RISK ASSESSMENT: HEALTH

General Assessment

Chlorine pentafluoride is human poison by *inhalation.* In liquid form, it can pose a moderate threat on contact with the skin and eyes (frostbite burns), although absorption has not been reported as an entry route. There are no reports of carcinogenic, mutagenic, or teratogenic cases subsequent to exposure to chlorine pentafluoride. It is a powerful oxidizer and is considered highly corrosive to human tissues.

Inhalation results in severe irritation and possible corrosive destruction to the mucous membranes in the nose, throat, and respiratory tract. Even relatively moderate concentrations can lead to a dangerous buildup of fluids in the lungs in a condition known as pulmonary edema. If this occurs, it is a medical emergency and can be fatal. Symptoms of cough with phlegm, congestion, tightness in the chest, headache, and difficulty in breathing (dypsnea) may be delayed up to 48 hours. This may create a false sense of security with regard to health exposure risk. If allowed to go untreated, death is very possible.

Eye contact can cause severe pain and irritation with watering (lachrymation), burns, corneal ulceration, and permanent blindness. Skin contact is seriously destructive to tissue and may cause injury equivalent to 2nd and 3rd degree burns. As a result of the fluoride ion and its affinity for calcium, deep corrosive destruction of tissues may reach the bone if it is not stopped in time (using calcium gluconate injections).

Acute Health Effects

The following acute (short-term) health effects may occur immediately or shortly after exposure to chlorine pentafluoride:

Skin: Liquid is destructive to skin tissue (cold temperature may cause burning frostbite). Deep and painful burns may result.

Eye: Severe irritation with may cause watery eyes. Corneal ulceration and blindness are extremely possible on contact with the liquid.

Lung: Irritation of the nose, throat, and lungs following exposure. Pulmonary edema is possible (24 to 48 hours following a high exposure). It can cause chest pain, cough, bloody nose, and sputum.

Chronic Health Effects

The following chronic (long-term) health effects may occur at some time after exposure to chlorine pentafluoride and can last for months or even years:

Cancer Hazards: According to information presented in the references, chlorine pentafluoride has not been adequately tested for its ability to cause cancer in test animals.

Reproductive Hazard: According to information presented in the references, chlorine pentafluoride has not been tested for its ability to adversely affect reproduction in test animals.

Other Chronic Effects: Repeated exposures can cause chronic coughing, sneezing, runny nose, frequent drooling, pneumonia, and death.

Recommended Risk-Reduction Measures

Personnel should avoid direct contact with chlorine pentafluoride. It is extremely corrosive and moderately toxic, even at low concentrations. If a less toxic chemical cannot be substituted for a hazardous substance, then *engineering controls* are the most effective method of reducing exposures. The best protection is to enclose operations and/or provide local exhaust ventilation at the site of chemical release. While not always operationally feasible, isolating operations can also reduce exposure. Using respiratory protection is less effective than the controls mentioned above, but is still advisable whenever working with or around chlorine pentafluoride. The PEL is relatively low (2.5 mg/m^3) and difficult to monitor. While a full facepiece respirator equipped with a cartridge containing non-combustible sorbents and providing protection against chlorine pentafluoride gas may suffice for low or infrequent exposures, it is not the most optimum choice for respiratory protection. A supplied-air respirator or a self-contained breathing apparatus (SCBA) with full facepiece operated in pressure demand mode are the recommended respiratory protection methods. If a full facepiece is not available, then chemical goggles should be worn to protect the eyes. is acceptable protection. Whenever working with the liquid and a chemical splash hazard exists, a face shield and a rubber apron should be worn. To prevent hand and skin exposures, acid resistant gloves should be worn. Glove manufactures should be contacted and permeation studies obtained *before* final glove selections are made.

Administrative controls should also be in place to minimize the potential for human exposures. These may include written procedures or policies which specify the methods and techniques that will be practiced whenever personnel are to work with chlorine pentafluoride.

All personnel should receive *training* on the use, hazards, protective measures, emergency actions, and other precautions per 29 CFR 1910.1200 (Hazard Communication), prior to the first assignment in an area where chlorine pentafluoride is used or stored.

For those personnel with frequent or potentially high exposure, the following are recommended medical tests to be provided before initial assignment and at regular intervals thereafter (e.g., annually):

☑ Lung function tests.

If symptoms develop or overexposure is suspected, the following tests are recommended:

☑ Lung function tests.

☑ Consider chest X-ray after acute overexposures (may be negative if taken immediately following exposure due to delayed development of pulmonary edema).

Any evaluation should include a careful history of past and present symptoms with an examination. Medical tests that simply look for existing damage are not a substitute for controlling exposures. Also, because smoking can cause heart disease, as well as lung cancer, emphysema, and other respiratory problems, smokers may experience symptoms much more quickly than non-smokers under the same conditions of exposure. Prudent risk management requires careful consideration of *all* exposure factors.

Other methods to reduce exposure include:

☑ Where possible, enclose operations and use local exhaust ventilation at the site of chemical release. If local exhaust ventilation or enclosure is not used, respiratory protection should be mandatory.

☑ Always ensure that proper protective clothing is worn when using chemical substances.

☑ Wash thoroughly immediately after exposure to chlorine pentafluoride and at the end of the work shift or before eating, drinking, or smoking.

☑ Hazard warning information should be posted in the work area. In addition, as part of an on-going education and training program, all information on the health and safety hazards of chlorine pentafluoride should be communicated to all potentially exposed workers.

☑ Eye wash stations should be provided in the immediate work area for emergency use. If there is a possibility of skin exposure to chlorine pentafluoride, emergency shower facilities should also be provided.

☑ Personnel should never wear contaminated clothing home (family members can be exposed). Clothing should be laundered by personnel who have been trained on the hazards of chlorine pentafluoride.

RISK ASSESSMENT: ENVIRONMENT

General Assessment

The environment is at risk of exposure during transportation, storage, disposal, or destruction of chlorine pentafluoride. In almost every scenario, the threat of environmental exposure is contingent upon the proper handling of the chemical substance. Accidental spills,

large or small, can result in fire, explosion, and possible contamination of the surrounding environmental mediums (water, soil, and air).

Chlorine pentafluoride is considered a non-flammable gas. However, it is an extremely powerful oxidizing agent and highly reactive with many commonly encountered commodities such as acids, oxidizers, many metal and metallic compounds, even sand and water. Contact with incompatible materials can result in fire and serious explosions. These characteristics require extreme caution in handling, storage, transportation, and disposal. It is incompatible with organic matter, combustible materials, powdered metals, and many other compounds and materials. It may explode violently upon exposure to air or when subjected to a spark. It is extremely unstable and highly reactive. These characteristics require special consideration during any emergency situation involving a leak or spill of chlorine pentafluoride. Should chlorine pentafluoride ever come into contact with incompatible substances during use, transportation, storage, or disposal, the formation of highly toxic and/or highly explosive commodities is extremely possible.

Chlorine pentafluoride can enter the environment through industrial discharges and through spills.

Acute Ecological Effects

Acute (short-term) toxic effects may include the death of animals, birds, or fish and death or low growth rate in plants. Acute effects are seen 2 to 4 days after animals or plants are exposed to chlorine pentafluoride.

Because it reacts violently in water to form hydrogen fluoride and chlorine, insufficient data are available on the short-term effects of chlorine pentafluoride exposure to aquatic life, plants, birds, or land animals. However, both hydrogen fluoride and chlorine are have moderate acute toxicity to aquatic life.

Chronic Ecological Effects

Chronic (long-term) toxic effects may include shortened life span, reproductive problems, lower fertility, and changes in appearance or behavior in exposed animals. These effects can be seen long after first exposure(s) to toxic chemicals.

Because it reacts violently in water to form hydrogen fluoride and chlorine, insufficient data are available on the long-term effects of chlorine pentafluoride to aquatic life, plants, birds, or land animals. However, both hydrogen fluoride and chlorine have slight chronic toxicity to aquatic life.

Water Solubility

Chlorine pentafluoride reacts violently in water to form a hydrogen fluoride and chlorine.

Persistence in the Environment

Chlorine pentafluoride is non-persistent in water, since it reacts violently on contact. It will form hydrogen fluoride and chlorine which are both slightly persistent in the aquatic environment with a half-life between than 2 and 20 days. The half-life of a pollutant is the amount of time it takes for one half of the chemical to be degraded.

Bioaccumulation in Aquatic Organisms

Some substances increase in concentration, or *bioaccumulate,* in living organisms as they breathe contaminated air, drink contaminated water, or eat contaminated food. These chemicals can become concentrated in the tissues and internal organs of animals as well as humans.

Chlorine pentafluoride is not expected to accumulate in the tissues of fish. However, the amount of hydrogen fluoride and chlorine (both by-products of the chlorine pentafluoride and water reaction) is expected to be about the same as the average concentration found in the water from which the fish was taken.

Recommended Risk-Reduction Measures

Proper training of all transporters will reduce the likelihood of a mishap or accident resulting in a leak or spill to the environment. Transportation of pure chlorine pentafluoride is forbidden by the Department of Transportation (DOT). It must be kept frozen when shipped. The correct use of DOT labeling (per the MSDS) on all containers, trucks, and rail cars will enable emergency responders to react properly and quickly to any disaster thereby reducing the potential risk to the environment and to personnel.

When not is use, chlorine pentafluoride must be kept in tightly sealed and controlled containers. Storage of chlorine pentafluoride should be segregated from incompatible chemicals to minimize the risk of cross-contamination or contact. Buildings designated for storage should be equipped with appropriate fire protection systems (alarms, sprinklers, emergency lighting, portable extinguishers).

If a spill or leak to the environment has occurred, fire department, emergency response and/or hazardous materials spill personnel should be notified immedi-

ately. Clean-up should be attempted only by those trained in proper spill containment procedures. Contaminated soils should be removed for incineration and replaced with clean soil. If chlorine pentafluoride should contact the water table, aquifer, or navigable waterway, time is of the essence. It is reacts violently in water and total remediation and containment may not be possible. The local and/or state emergency response authorities must be notified. A comprehensive emergency response or disaster preparedness/recovery plan should be in place prior to any operations involving the use, transportation, storage, or disposal of chlorine pentafluoride. If chlorine pentafluoride is spilled or leaked, the following specific steps are recommended:

- ☑ Restrict persons not wearing protective clothing from area of spill or leak until cleanup is complete. Use non-sparking tools in cleanup.
- ☑ Remove all ignition sources.
- ☑ Ventilate area of leak.
- ☑ Absorb liquids in vermiculite, earth, or similar material and deposit in sealed containers. Do NOT use sand (reacts!).
- ☑ If gas is leaking, stop the flow of gas. If the source is a cylinder and leak cannot be stopped in place, remove cylinder to a safe place in the open air and repair or allow to empty.
- ☑ It may be necessary to dispose of chlorine pentafluoride as a hazardous waste. The responsible state agency or the regional office of the federal Environmental Protection Agency (EPA) should be contacted for specific recommendations.

RISK ASSESSMENT: BUSINESS

General Assessment

Accidents or mishaps involving chlorine pentafluoride can present a significant threat to business operations. The loss or damage of equipment or facilities can significantly affect fiscal viability. Lawsuits that may result from personnel injury/death, public exposures, and/or environmental contamination will also require a serious expenditure of resources. Media attention surrounding an injury, death, or environmental damage can also result in a loss of profits and loss of current as well as future business.

Recommended Risk-Reduction Measures

Company attorneys, safety and health professionals, and environmental specialists should be involved in the development of any procedures dealing with safety and emergency response. A company official should be pre-designated as a public relations officer with specific training in dealing with the press. Corporate plans and policies should be developed, approved, and implemented long before any need for such arises.

REFERENCES

American Conference of Governmental Industrial Hygienists. 1988. *Documentation of the Threshold Limit Values and Biological Exposure Indices,* 5th Edition (with updates). Cincinnati: ACGIH

Bueche, J. R. 1972. *Principles of Physics,* 2nd Edition. New York: McGraw-Hill.

Calabrese, E. J. 1994. *Biological Effects of Low Level Exposures.* New York: CRC Press/Lewis Publishers.

Cockerman, L. G., B. S. Shane. 1994. *Basic Environmental Toxicology.* New York: CRC Press/Lewis Publishers.

Gardner, R. W. 1994. *Chemical Intolerance.* New York: CRC Press/Lewis Publishers.

Hallenbeck, W. H. 1993. *Quantitative Risk Assessment for Environmental and Occupational Health,* 2nd Edition. New York: CRC Press/Lewis Publishers.

Jones, F. E. 1994. *Toxic Organic Vapors in the Workplace.* New York: CRC Press/Lewis Publishers.

Landis, W. G., M. H. Yu. 1995. *Introduction to Environmental Toxicology.* New York: CRC Press/Lewis Publishers.

Langerman, N. 1994 *Precautionary Labels for Chemical Containers.* New York: CRC Press/Lewis Publishers.

Lauwerys, R. R., P. Hoet. 1993. *Industrial Chemical Exposure.* New York: CRC Press/Lewis Publishers.

Lewis, R. J., Sr. 1992. *Sax's Dangerous Properties of Industrial Materials,* Volumes 1, 2, and 3, 8th Edition. New York: Van Nostrand Reinhold.

Manahan, S. E. 1994. *Environmental Chemistry,* 6th Edition. New York: CRC Press/Lewis Publishers.

Meyer, E. 1990. *Chemistry of Hazardous Materials.* Englewood Cliffs, NJ: Prentice Hall.

Patnaik, P. 1992. *A Comprehensive Guide to the Hazardous Properties of Chemical Substances.* New York: Van Nostrand Reinhold.

Torkelson, T. R. V. K. Rowe. 1981. Halogenated aliphatic hydrocarbons. *Patty's Industrial Hygiene and Toxicology,* Vol. 2B, *Toxicology,* pp 3568-3572. New York: Wiley-Interscience.

U.S. Department of Health and Human Services, National Institute for Occupational Safety and Health. 1994. *NIOSH Pocket Guide to Chemical Hazards.* Washington, D.C.: U.S. Government Printing Office.

MATERIAL SAFETY DATA SHEET

CHEMICAL NAME	CRC PRESS/LEWIS PUBLISHERS DISCLAIMER
CHLORINE TRIFLUORIDE	**CRC PRESS/LEWIS PUBLISHERS DISCLAIMER:** THE INFORMATION AND RECOMMENDATIONS PRESENTED HEREIN ARE BASED ON SOURCES BELIEVED TO BE RELIABLE. CRC MAKES NO REPRESENTATION ON ITS COMPLETENESS OR ACCURACY. IT IS THE RESPONSIBILITY OF THE USER TO DETERMINE THE CHEMICAL'S SUITABILITY FOR ITS INTENDED USE, THE CHEMICAL'S SAFE USE, AND THE CHEMICAL'S PROPER DISPOSAL. NO REPRESENTATIONS AND/OR WARRANTIES, EITHER EXPRESSED OR IMPLIED, OF THE MERCHANTABILITY OR FITNESS FOR A PARTICULAR PURPOSE, OR OF ANY OTHER NATURE, ARE MADE WITH RESPECT TO THE INFORMATION PROVIDED IN THIS MSDS OR TO THE CHEMICAL TO WHICH INFORMATION MAY REFER. CRC NEITHER ASSUMES NOR AUTHORIZES ANY OTHER PERSON TO ASSUME FOR IT, ANY OTHER ADDITIONAL RESPONSIBILITY OR LIABILITY FOR THE USE OF, OR RELIANCE UPON, THIS INFORMATION.

HAZARD WARNING INFORMATION

HEALTH	FIRE	REACTIVE	OTHER	DEGREE OF HAZARD	COLOR CODING	OTHER CODES
4	0	3	OX ~~W~~	0 = Minimum Hazard 1 = Slight Hazard 2 = Moderate Hazard 3 = Serious Hazard 4 = Severe Hazard	HEALTH = BLUE FIRE = RED REACTIVITY = YELLOW OTHER = WHITE	OX = Oxidizer ACID = Acid ALK = Alkali COR = Corrosive ~~W~~ = Use No Water

SECTION I - GENERAL INFORMATION

Characterization	Interhalogen Compound	RCRA Number	D003
		EPA Class	Characteristic (R) Waste
DOT Proper Shipping Name	Chlorine Trifluoride	Chemical Abstract Service (CAS) Number	7790-91-2
DOT Hazard Class and Label Requirements	Oxidizer & Poison	DOT Emergency Guide Code	44
DOT Identification Number	UN 1749	Chemical Formula	ClF_3
Synonyms	Chlorine fluoride; chlorotrifluoride.		

SECTION II - HAZARDOUS INGREDIENTS AND IDENTITY INFORMATION

Hazardous Components (specific identity)	OSHA Exposure Criteria	NIOSH Exposure Criteria	Immediately Dangerous to Life and Health (IDLH)	ACGIH Exposure Criteria
Chlorine trifluoride (derivation: By reaction of chlorine and fluorine at 536°F/280°C and subsequent condensation of the product). **1 ppm = 3.85 mg/m³**	**PEL (ceiling):** 0.1 ppm 0.4 mg/m³ **STEL:** Not Applicable	**REL:** 0.1 ppm 0.4 mg/m³ **STEL:** Not Applicable	20 ppm	**TLV:** 0.1 ppm 0.38 mg/m³ **STEL:** Not Applicable

SECTION III - PHYSICAL AND CHEMICAL CHARACTERISTICS

Boiling Point	53°F (12°C)	Specific Gravity (H_2O = 1)	1.77 (liquid at 53°F)
Vapor Pressure (atmospheres)	1.4 at 69°F (20°C)	Molecular Weight	92.5
Vapor Density (Air = 1)	3.14	Melting Point	-105°F (-76°C)

Solubility
Reacts violently with water.

Appearance and Odor
Colorless gas or a greenish-yellow, volatile liquid (below 53°F) with a somewhat sweet, but irritating and suffocating odor. It will solidify to a white solid at -105°F (-76.3°C).

SECTION IV - FIRE AND EXPLOSION HAZARD DATA

Flash Point (method used)	Not Determined	Explosive Limits in Air % by Volume	LEL: Not Determined UEL: Not Determined
NFPA Classification	Non-Flammable Gas, Non-Combustible Liquid	Autoignition Temperature	Not Determined

Extinguishing Media
Stop flow of gas or liquid. Use bicarbonate-based dry chemical. Do NOT use water (reacts explosively!)

Special Fire Fighting Procedures
Chlorine trifluoride is a powerful oxidizer. Wear full protective clothing and self-contained breathing apparatus (SCBA). Use a water spray to keep fire-exposed containers cool. Poisonous gases are produced in fire. Isolate for ½ mile if rail or tank truck is involved. Containers may explode in fire or heat.

Unusual Fire and Explosion Hazards
Chlorine trifluoride is hypergolic. It will ignite explosively in contact with a fuel or a spark. Dangerous fire and explosion hazard when exposed to any form of energy, including sunlight, heat, or sparks.

SECTION V - REACTIVITY DATA

Stability		Conditions to Avoid **Extremely reactive but can be kept stable in special cylinders. It will thermally decompose at temperatures above 428°F (220°C).**
Stable	Unstable **X**	Incompatibility *(materials to avoid)* **Violent reacts with water, oxidizers, acids, combustible materials, sand, glass, and is corrosive to most metals. Reacts with water to form chlorine and extremely toxic and corrosive hydrofluoric acid.**
Hazardous Polymerization		Conditions to Avoid **Under normal conditions of temperature and pressure, hazardous polymerization of chlorine trifluoride is not known to occur.**
May Occur	Will Not Occur **X**	Hazardous Decomposition or By-products **When heated to decomposition, chlorine trifluoride can emit highly toxic/poisonous gases. Mixing with water or steam will produce toxic and corrosive fumes of chlorine and fluorine.**

SECTION VI - HEALTH HAZARD DATA

Primary Route(s) of Entry:	Inhalation?	Absorption (skin)?	Ingestion?
	X	X	

Health Hazards

INHALATION: Extremely irritating to the lungs, throat, nose, and eyes. Coughing and shortness of breath that may persist for hours after exposure. Can cause permanent damage to the lungs. Fluid can build up in the lungs (pulmonary edema), which is a medical emergency.

SKIN & EYES: Contact with the liquid can cause severe skin burns (due to its cold state). The liquid can cause severe eye burns leading to permanent damage.

Carcinogenicity	NTP Listed?	IARC Cancer Review Group?	OSHA Regulated?	Target Organs?
Unknown Human Unknown Animal	No	No	29 CFR 1910.1000 Table Z-1	Respiratory system, eyes, and skin.

Medical Conditions Generally Aggravated by Exposure

Persons with existing pulmonary disease may be more susceptible to exposure symptoms.

Emergency and First-aid Procedures

Eye contact: Flush immediately with water for 30 minutes (minimum), seek medical attention. Skin contact: Quickly remove all contaminated clothing. Immediately wash area with large amounts water for 15 minutes (minimum). Apply topical hydrogen fluoride neutralizing agent (calcium gluconate gel). Seek medical assistance. For inhalation: Remove the person from exposure. Provide respiratory assistance and CPR. Transfer to medical facility. Observe for 24-48 hours for lung effects. If swallowed: Seek medical attention immediately. Give large quantities of milk. Do NOT induce vomiting.

SECTION VII - PRECAUTIONS FOR SAFE HANDLING AND USE

Steps to be Taken in Case Material is Released or Spilled

Absorb liquids in vermiculite, or similar material and deposit in sealed containers. Ventilate area of spill or leak. Restrict those not involved in cleanup from entering area. If gas is leaked, stop source if possible or remove leaking cylinders to safe area outdoors and allow to empty.

Preferred Waste Disposal Method

No Citation.

Precautions to be Taken in Handling and Storage

Chlorine trifluoride is a non-flammable gas but an extremely powerful oxidizer. Store to avoid contact with incompatible materials since violent reactions can occur. Store in tightly closed containers in cool, well-ventilated area away from heat and moisture (reacts violently with water).

Other Precautions and Warnings

Sources of ignition are prohibited around chlorine trifluoride. Heat may cause containers to build up pressure and explode. An inert blanket of nitrogen inside containers is recommended.

SECTION VIII - CONTROL MEASURES AND PERSONAL PROTECTIVE EQUIPMENT

Respiratory Protection *(specify type)*

For exposures below the PEL, use a full facepiece respirator with organic vapor cartridge. For higher exposures, use a supplied air respirator with full facepiece, hood, or helmet in continuous flow mode, or use a SCBA operated in pressure demand or other positive pressure mode.

Ventilation

Local exhaust or general mechanical systems recommended.

Protective Gloves	Eye Protection	Other Protective Clothing
Impervious, Solvent-Resistant	Chemical Goggles or Face Mask	Protective Apron or Full-Body Suit

Work/Hygiene Practices

Always wash hands thoroughly after using chemical; never bring food, drink, or smoking materials into vicinity of chemicals.

CHLORINE TRIFLUORIDE

ClF_3 CAS: 7790-91-2

IDENTIFICATION AND TYPICAL USES

Chlorine trifluoride is a colorless gas or a greenish-yellow liquid (below 53°F/12°C) with a sweet but irritating and suffocating odor. It is normally shipped as a liquefied compressed gas. It is used as a fluorinating agent, as an igniting agent and propellant for liquid fueled rocket engines, a pyrolisis inhibitor for fluorocarbon polymers, and an incendiary. It is also used in nuclear reactor fuel processing.

RISK ASSESSMENT: HEALTH

General Assessment

Chlorine trifluoride is human poison by *inhalation*. In liquid form, it can pose a moderate threat on contact with the skin and eyes (frostbite burns), although absorption has not been reported as an entry route. There are no reports of carcinogenic, mutagenic, or teratogenic cases subsequent to exposure to chlorine trifluoride. It is a powerful oxidizer and is considered highly corrosive to human tissues.

Inhalation results in severe irritation and possible corrosive destruction to the mucous membranes in the nose, throat, and respiratory tract. Even relatively moderate concentrations can lead to the buildup of fluids in the lungs in a condition known as pulmonary edema. If this occurs, it is a medical emergency and can be fatal. Symptoms of cough with phlegm, congestion, tightness in the chest, headache, and difficulty in breathing (dypsnea) may be delayed up to 48 hours. This may create a false sense of security with regard to health exposure risk.

Eye contact can cause severe pain and irritation with watering (lachrymation), burns, corneal ulceration, and permanent blindness. Skin contact is destructive to tissue and may cause 2nd and 3rd degree burns. As a result of the fluoride ion and its affinity for calcium, corrosive destruction may reach the bone if it is not stopped in time.

Acute Health Effects

The following acute (short-term) health effects may occur immediately or shortly after exposure to chlorine trifluoride:

Skin: Liquid is destructive to skin tissue (cold temperature may cause burning frostbite). Deep and painful burns may result.

Eye: Severe irritation with may cause watery eyes. Corneal ulceration and blindness are extremely possible on contact with the liquid.

Lung: Irritation of the nose, throat, and lungs following exposure. Pulmonary edema is possible (24 to 48 hours following a high exposure). It can cause chest pain, cough, bloody nose, and sputum.

Chronic Health Effects

The following chronic (long-term) health effects may occur at some time after exposure to chlorine trifluoride and can last for months or even years:

Cancer Hazards: According to information presented in the references, chlorine trifluoride has not been adequately tested for its ability to cause cancer in test animals.

Reproductive Hazard: According to information presented in the references, chlorine trifluoride has not been tested for its ability to adversely affect reproduction in test animals.

Other Chronic Effects: Repeated exposures can cause chronic coughing, sneezing, runny nose, frequent drooling, pneumonia, and death.

Recommended Risk-Reduction Measures

Personnel should avoid direct contact with chlorine trifluoride. It is extremely corrosive and moderately toxic, even at low concentrations. If a less toxic chemical cannot be substituted for a hazardous substance, then *engineering controls* are the most effective method of reducing exposures. The best protection is to enclose operations and/or provide local exhaust ventilation at the site of chemical release. While not always operationally feasible, isolating operations can also reduce exposure. Using respiratory protection is less effective than the controls mentioned above, but is still advisable whenever working with or around chlorine trifluoride. The PEL (ceiling) is relatively low (0.1 ppm) and difficult to monitor. While a full facepiece respirator equipped with a cartridge containing noncombustible sorbents and providing protection against chlorine trifluoride gas may suffice for low or infrequent exposures, it is not the most optimum choice for respiratory protection. A supplied-air respirator or a self-contained breathing apparatus (SCBA) with full facepiece operated in pressure demand mode are the recommended respiratory protection methods. If a full facepiece is not available, then chemical goggles should be worn to protect the eyes. is acceptable protection. Whenever working with the liquid and a chemical splash hazard exists, a face shield and a rubber apron should be worn. To prevent hand and skin exposures, acid resistant gloves should be worn. Glove manufactures should be contacted and permeation studies obtained *before* final glove selections are made.

Administrative controls should also be in place to minimize the potential for human exposures. These may include written procedures or policies which specify the methods and techniques that will be practiced whenever personnel are to work with chlorine trifluoride.

All personnel should receive *training* on the use, hazards, protective measures, emergency actions, and other precautions per 29 CFR 1910.1200 (Hazard Communication), prior to the first assignment in an area where chlorine trifluoride is used or stored.

For those personnel with frequent or potentially high exposure, the following are recommended medical tests to be provided before initial assignment and at regular intervals thereafter (e.g., annually):

☑ Lung function tests.

If symptoms develop or overexposure is suspected, the following tests are recommended:

☑ Lung function tests.

☑ Consider chest X-ray after acute overexposures (may be negative if taken immediately following exposure due to delayed development of pulmonary edema).

Any evaluation should include a careful history of past and present symptoms with an examination. Medical tests that simply look for existing damage are not a substitute for controlling exposures. Also, because smoking can cause heart disease, as well as lung cancer, emphysema, and other respiratory problems, smokers may experience symptoms much more quickly than non-smokers under the same conditions of exposure. Prudent risk management requires careful consideration of *all* exposure factors.

Other methods to reduce exposure include:

☑ Where possible, enclose operations and use local exhaust ventilation at the site of chemical release. If local exhaust ventilation or enclosure is not used, respiratory protection should be mandatory.

☑ Always ensure that proper protective clothing is worn when using chemical substances.

☑ Wash thoroughly immediately after exposure to chlorine trifluoride and at the end of the work shift or before eating, drinking, or smoking.

☑ Hazard warning information should be posted in the work area. In addition, as part of an on-going education and training program, all information on the health and safety hazards of chlorine trifluoride should be communicated to all potentially exposed workers.

☑ Eye wash stations should be provided in the immediate work area for emergency use. If there is a possibility of skin exposure to chlorine trifluoride, emergency shower facilities should also be provided.

☑ Personnel should never wear contaminated clothing home (family members can be exposed). Clothing should be laundered by personnel who have been trained on the hazards of chlorine trifluoride.

RISK ASSESSMENT: ENVIRONMENT

General Assessment

The environment is at risk of exposure during transportation, storage, disposal, or destruction of chlorine trifluoride. In almost every scenario, the threat of environmental exposure is contingent upon the proper handling of the chemical substance. Accidental spills,

large or small, can result in fire, explosion, and possible contamination of the surrounding environmental mediums (water, soil, and air).

Chlorine trifluoride is considered a non-flammable gas. However, it is an extremely powerful oxidizing agent and highly reactive with many commonly encountered commodities such as acids, oxidizers, many metal and metallic compounds, even sand and water. Contact with incompatible materials can result in fire and serious explosions. These characteristics require extreme caution in handling, storage, transportation, and disposal. It is incompatible with organic matter, combustible materials, powdered metals, and many other compounds and materials. It may explode violently upon exposure to air or when subjected to a spark. It is extremely unstable and highly reactive. These characteristics require special consideration during any emergency situation involving a leak or spill of chlorine trifluoride. Should chlorine trifluoride ever come into contact with incompatible substances during use, transportation, storage, or disposal, the formation of highly toxic and/or highly explosive commodities is extremely possible.

Chlorine trifluoride can enter the environment through industrial discharges and through spills.

Acute Ecological Effects

Acute (short-term) toxic effects may include the death of animals, birds, or fish and death or low growth rate in plants. Acute effects are seen 2 to 4 days after animals or plants are exposed to chlorine trifluoride.

Because it reacts violently in water to form hydrogen fluoride and chlorine, insufficient data are available on the short-term effects of chlorine trifluoride exposure to aquatic life, plants, birds, or land animals. However, both hydrogen fluoride and chlorine are have moderate acute toxicity to aquatic life.

Chronic Ecological Effects

Chronic (long-term) toxic effects may include shortened life span, reproductive problems, lower fertility, and changes in appearance or behavior in exposed animals. These effects can be seen long after first exposure(s) to toxic chemicals.

Because it reacts violently in water to form hydrogen fluoride and chlorine, insufficient data are available on the long-term effects of chlorine trifluoride to aquatic life, plants, birds, or land animals. However, both hydrogen fluoride and chlorine have slight chronic toxicity to aquatic life.

Water Solubility

Chlorine trifluoride reacts violently in water to form a hydrogen fluoride and chlorine.

Persistence in the Environment

Chlorine trifluoride is non-persistent in water, since it reacts violently on contact. It will form hydrogen fluoride and chlorine which are both slightly persistent in the aquatic environment with a half-life between than 2 and 20 days. The half-life of a pollutant is the amount of time it takes for one half of the chemical to be degraded.

Bioaccumulation in Aquatic Organisms

Some substances increase in concentration, or *bioaccumulate,* in living organisms as they breathe contaminated air, drink contaminated water, or eat contaminated food. These chemicals can become concentrated in the tissues and internal organs of animals as well as humans.

Chlorine trifluoride is not expected to accumulate in the tissues of fish. However, the amount of hydrogen fluoride and chlorine (both by-products of the chlorine trifluoride and water reaction) is expected to be about the same as the average concentration found in the water from which the fish was taken.

Recommended Risk-Reduction Measures

Proper training of all transporters will reduce the likelihood of a mishap or accident resulting in a leak or spill to the environment. Transportation of pure chlorine trifluoride is forbidden by the Department of Transportation (DOT). It must be kept frozen when shipped. The correct use of DOT labeling (per the MSDS) on all containers, trucks, and rail cars will enable emergency responders to react properly and quickly to any disaster thereby reducing the potential risk to the environment and to personnel.

When not is use, chlorine trifluoride must be kept in tightly sealed and controlled containers. Storage of chlorine trifluoride should be segregated from incompatible chemicals to minimize the risk of cross-contamination or contact. Buildings designated for storage should be equipped with appropriate fire protection systems (alarms, sprinklers, emergency lighting, portable extinguishers).

If a spill or leak to the environment has occurred, fire department, emergency response and/or hazardous materials spill personnel should be notified immedi-

ately. Cleanup should be attempted only by those trained in proper spill containment procedures. Contaminated soils should be removed for incineration and replaced with clean soil. If chlorine trifluoride should contact the water table, aquifer, or navigable waterway, time is of the essence. It is reacts violently in water and total remediation and containment may not be possible. The local and/or state emergency response authorities must be notified. A comprehensive emergency response or disaster preparedness/recovery plan should be in place prior to any operations involving the use, transportation, storage, or disposal of chlorine trifluoride. If chlorine trifluoride is spilled or leaked, the following specific steps are recommended:

- ☑ Restrict persons not wearing protective clothing from area of spill or leak until cleanup is complete. Use non-sparking tools in cleanup.
- ☑ Remove all ignition sources.
- ☑ Ventilate area of leak.
- ☑ Absorb liquids in vermiculite, earth, or similar material and deposit in sealed containers. Do NOT use sand (reacts!).
- ☑ If gas is leaking, stop the flow of gas. If the source is a cylinder and leak cannot be stopped in place, remove cylinder to a safe place in the open air and repair or allow to empty.
- ☑ It may be necessary to dispose of chlorine trifluoride as a hazardous waste. The responsible state agency or the regional office of the federal Environmental Protection Agency (EPA) should be contacted for specific recommendations.

RISK ASSESSMENT: BUSINESS

General Assessment

Accidents or mishaps involving chlorine trifluoride can present a significant threat to business operations. The loss or damage of equipment or facilities can significantly affect fiscal viability. Lawsuits that may result from personnel injury/death, public exposures, and/or environmental contamination will also require a serious expenditure of resources. Media attention surrounding an injury, death, or environmental damage can also result in a loss of profits and loss of current as well as future business.

Recommended Risk-Reduction Measures

Company attorneys, safety and health professionals, and environmental specialists should be involved in the development of any procedures dealing with safety or emergency response. A company official should be pre-designated as a public relations officer with specific training in dealing with the press. Corporate plans and policies should be developed, approved, and implemented long before any need for such arises.

REFERENCES

American Conference of Governmental Industrial Hygienists. 1988. *Documentation of the Threshold Limit Values and Biological Exposure Indices,* 5th Edition (with updates). Cincinnati: ACGIH

Bueche, J. R. 1972. *Principles of Physics,* 2nd Edition. New York: McGraw-Hill.

Calabrese, E. J. 1994. *Biological Effects of Low Level Exposures.* New York: CRC Press/Lewis Publishers.

Cockerman, L. G., B. S. Shane. 1994. *Basic Environmental Toxicology.* New York: CRC Press/Lewis Publishers.

Gardner, R. W. 1994. *Chemical Intolerance.* New York: CRC Press/Lewis Publishers.

Hallenbeck, W. H. 1993. *Quantitative Risk Assessment for Environmental and Occupational Health,* 2nd Edition. New York: CRC Press/Lewis Publishers.

Jones, F. E. 1994. *Toxic Organic Vapors in the Workplace.* New York: CRC Press/Lewis Publishers.

Landis, W. G., M. H. Yu. 1995. *Introduction to Environmental Toxicology.* New York: CRC Press/Lewis Publishers.

Langerman, N. 1994 *Precautionary Labels for Chemical Containers.* New York: CRC Press/Lewis Publishers.

Lauwerys, R. R., P. Hoet. 1993. *Industrial Chemical Exposure.* New York: CRC Press/Lewis Publishers.

Lewis, R. J., Sr. 1992. *Sax's Dangerous Properties of Industrial Materials,* Volumes 1, 2, and 3, 8th Edition. New York: Van Nostrand Reinhold.

Manahan, S. E. 1994. *Environmental Chemistry,* 6th Edition. New York: CRC Press/Lewis Publishers.

Meyer, E. 1990. *Chemistry of Hazardous Materials.* Englewood Cliffs, NJ: Prentice Hall.

Patnaik, P. 1992. *A Comprehensive Guide to the Hazardous Properties of Chemical Substances.* New York: Van Nostrand Reinhold.

Torkelson, T. R. V. K. Rowe. 1981. Halogenated aliphatic hydrocarbons. *Patty's Industrial Hygiene and Toxicology,* Vol. 2B, *Toxicology,* pp 3568-3572. New York: Wiley-Interscience.

U.S. Department of Health and Human Services, National Institute for Occupational Safety and Health. 1994. *NIOSH Pocket Guide to Chemical Hazards.* Washington, D.C.: U.S. Government Printing Office.

MATERIAL SAFETY DATA SHEET

CHEMICAL NAME	CRC PRESS/LEWIS PUBLISHERS DISCLAIMER
CHLOROACETALDEHYDE	THE INFORMATION AND RECOMMENDATIONS PRESENTED HEREIN ARE BASED ON SOURCES BELIEVED TO BE RELIABLE. CRC MAKES NO REPRESENTATION ON ITS COMPLETENESS OR ACCURACY. IT IS THE RESPONSIBILITY OF THE USER TO DETERMINE THE CHEMICAL'S SUITABILITY FOR ITS INTENDED USE, THE CHEMICAL'S SAFE USE, AND THE CHEMICAL'S PROPER DISPOSAL. NO REPRESENTATIONS AND/OR WARRANTIES, EITHER EXPRESSED OR IMPLIED, OF THE MERCHANTABILITY OR FITNESS FOR A PARTICULAR PURPOSE, OR OF ANY OTHER NATURE, ARE MADE WITH RESPECT TO THE INFORMATION PROVIDED IN THIS MSDS OR TO THE CHEMICAL TO WHICH INFORMATION MAY REFER. CRC NEITHER ASSUMES NOR AUTHORIZES ANY OTHER PERSON TO ASSUME FOR IT, ANY OTHER ADDITIONAL RESPONSIBILITY OR LIABILITY FOR THE USE OF, OR RELIANCE UPON, THIS INFORMATION.

HAZARD WARNING INFORMATION

HEALTH	FIRE	REACTIVE	OTHER	DEGREE OF HAZARD	COLOR CODING	OTHER CODES
3	2	1		0 = Minimum Hazard 1 = Slight Hazard 2 = Moderate Hazard 3 = Serious Hazard 4 = Severe Hazard	HEALTH = BLUE FIRE = RED REACTIVITY = YELLOW OTHER = WHITE	OX = Oxidizer ACID = Acid ALK = Alkali COR = Corrosive W = Use No Water

SECTION I - GENERAL INFORMATION

Characterization	**Aldehyde**	RCRA Number	**P023**
		EPA Class	**Acute Toxic Waste**
DOT Proper Shipping Name	**2-Chloroacetaldehyde**	Chemical Abstract Service (CAS) Number	**107-20-0**
DOT Hazard Class and Label Requirements	**Poison**	DOT Emergency Guide Code	**55**
DOT Identification Number	**UN 2232**	Chemical Formula	**$ClCH_2CHO$**

Synonyms
2-Chloroacetaldehyde; 2-chloroethanal; monchloroacetaldehyde; 2-chloro-1-ethanal.

SECTION II - HAZARDOUS INGREDIENTS AND IDENTITY INFORMATION

Hazardous Components (specific identity)	OSHA Exposure Criteria	NIOSH Exposure Criteria	Immediately Dangerous to Life and Health (IDLH)	ACGIH Exposure Criteria
Chloroacetaldehyde (derivation: By carefully controlled chlorination of acetaldehyde; by reaction of vinyl chloride with triplet oxygen atoms which give high yields of carbon monoxide and chloroacetaldehyde; or by reaction of ozone with *cis*-1,3-dichloroproene and *trans*-1,3-dichloropropene during gas phase reactions of chemicals with ozone and hydroxyl radicals in Teflon chambers). **1 ppm = 3.26 mg/m³**	**PEL (ceiling):** 1 ppm 3 mg/m³ **STEL:** Not Applicable	**REL (ceiling):** 1 ppm 3 mg/m³ **STEL:** Not Applicable	**45 ppm**	**TLV (ceiling):** 1 ppm 3.2 mg/m³ **STEL:** Not Applicable

SECTION III - PHYSICAL AND CHEMICAL CHARACTERISTICS

Boiling Point	**186°F (85°C)**	Specific Gravity (H_2O = 1)	**1.19 (40% solution)**
Vapor Pressure (mm Hg)	**100 at 69°F (20°C)**	Molecular Weight	**78.5**
Vapor Density (Air = 1)	**Not Found**	Freezing Point	**2.7°F (-16°C) (40% solution)**

Solubility
Miscible (40% solution) in water. Soluble in acetone, methanol, ether. At greater than 50% solution in water, it forms an insoluble hemihydrate.

Appearance and Odor
Clear, colorless liquid with an acrid, penetrating odor, typically found as 40% aqueous solution.

SECTION IV - FIRE AND EXPLOSION HAZARD DATA

Flash Point (method used)	**190°F (88°C) closed cup (40% solution)**	Explosive Limits in Air % by Volume	**LEL: Not Determined UEL: Not Determined**
NFPA Classification	**Class 111A Combustible Liquid**	Autoignition Temperature	**None Reported**

Extinguishing Media
Carbon dioxide, dry chemical, water spray, regular foam, or fog.

Special Fire Fighting Procedures
Poisonous gases are produced in fire. Wear full protective clothing and self-contained breathing apparatus (SCBA). Heat will build pressure and may rupture closed storage containers. Keep fire-exposed containers cool with water spray.

Unusual Fire and Explosion Hazards
Move containers from fire if it can be done without risk. Do not release runoff into sewers or waterways.

SECTION V - REACTIVITY DATA

Stability		Conditions to Avoid: Normally stable in closed containers at room temperature under routine conditions of handling and storage. Avoid contact with heat, ignition sources, and incompatible materials.
Stable: X	Unstable:	Incompatibility (*materials to avoid*): Strong oxidizers (such as chlorine, bromine, fluorine), and strong acids (sulfuric, nitric, hydrochloric).
Hazardous Polymerization		Conditions to Avoid: The anhydrous substances polymerizes on standing but will revert to the monomer upon distillation.
May Occur: X	Will Not Occur:	Hazardous Decomposition or By-products: When heated to decomposition, chloroacetaldehyde emits acrid, irritating smoke and fumes, including carbon monoxide, carbon dioxide, and chloride.

SECTION VI - HEALTH HAZARD DATA

Primary Route(s) of Entry:	Inhalation?	Absorption (skin)?	Ingestion?
	X	X	X

Health Hazards

INHALATION: Irritation of the eyes, nose, throat, mucous membranes with coughing, choking, bronchial constriction, dypsnea, tachypnea, and pulmonary edema. Possibly toxic to the central nervous system (CNS) and may cause kidney and blood dysfunction.

ABSORPTION: Liquid contact causes skin and eye burns. Corneal damage and blindness are likely. The vapor will cause eye irritation, discomfort, and may cause swollen and sore eyelids.

INGESTION: Potentially serious burns of the mouth, throat, esophagus, and gastrointestinal tract.

Carcinogenicity	NTP Listed?	IARC Cancer Review Group?	OSHA Regulated?	Target Organs?
Unknown Human Unknown Animal	No	No	29 CFR 1910.1000 (Table Z-1)	Respiratory system; skin; eyes; kidneys, blood, CNS.

Medical Conditions Generally Aggravated by Exposure

Respiratory impairments (asthma); Renal function abnormalities; Skin conditions (dermatitis).

Emergency and First-aid Procedures

Eye contact: Flush immediately with water for 30 minutes (minimum); seek medical attention. Skin contact: Remove all contaminated clothing. Immediately wash area with large amounts of soap and water. Seek medical assistance. For inhalation: Remove the person from exposure. Provide respiratory assistance and CPR. Transfer to medical facility. Observe for 48 hours for lung effects (pulmonary edema). If swallowed: Seek medical attention immediately. Give 1-2 glasses of water. Do NOT induce vomiting. Never give anything by mouth to an unconscious or convulsing person.

SECTION VII - PRECAUTIONS FOR SAFE HANDLING AND USE

Steps to be Taken in Case Material is Released or Spilled

Contain spills using absorbent material (vermiculite, or other). Ventilate area of spill. Use water spray to knock-down vapors and dilute spills to nonflammable mixtures. Store materials in DOT-approved containers. Restrict those not involved in cleanup from entering area. Notify appropriate authorities.

Preferred Waste Disposal Method

Burn in a chemical incinerator equipped with an afterburner and scrubber.

Precautions to be Taken in Handling and Storage

Chloroacetaldehyde is highly volatile. Do not store with incompatible chemicals. Store in tightly closed containers in a cool, dark, well-ventilated area. Keep fire and flame away. Use non-sparking tools to open or close containers. Ground and bond all metal containers during transfer operations.

Other Precautions and Warnings

Bulk storage of chloroacetaldehyde is not recommended. Protect containers from physical damage.

SECTION VIII - CONTROL MEASURES AND PERSONAL PROTECTIVE EQUIPMENT

Respiratory Protection (*specify type*)

Exposures ceiling (1 ppm) is low and difficult to monitor: Use a MSHA/NIOSH approved full facepiece respirator with organic vapor cartridge. Best protection is obtained using a supplied air respirator or a self-contained breathing apparatus with full facepiece in pressure demand or positive pressure mode.

Ventilation

Local exhaust or general mechanical systems recommended.

Protective Gloves	Eye Protection	Other Protective Clothing
Butyl Rubber	Chemical Goggles or Face Mask	Rubber Apron

Work/Hygiene Practices

Always wash hands thoroughly after using chemical; never bring food, drink, or smoking materials into vicinity of chemicals.

CHLOROACETALDEHYDE

C_2H_3CHO CAS: 107-20-0

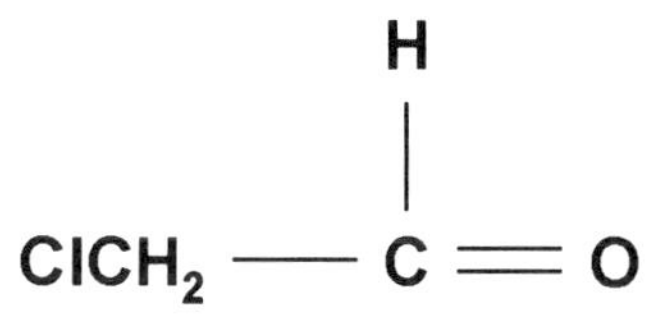

IDENTIFICATION AND TYPICAL USES

Chloroacetaldehyde is a colorless liquid with an acrid, penetrating odor that is typically found in 40% aqueous solution. It is used in the manufacture of other chemicals, such as 2-aminothiazole. It is also used to facilitate bark removal from tree trunks, as a fungicide, in dentistry, and in the spinning solution of poly B-alanine.

RISK ASSESSMENT: HEALTH

General Assessment

Chloroacetaldehyde is a human poison by *ingestion* and *inhalation,* and by skin *absorption.* While there are no reports on its carcinogenicity, it has been shown to cause mutations (genetic changes) in living cells. There are no data to indicate it poses any teratogenic effects in humans or animals.

Inhalation causes irritation of the eyes, nose, throat and mucous membranes of the respiratory tract with coughing, choking, constriction of bronchial tubes, increased respiratory rate (tachypnea). It can also cause respiratory tract sensitization. Pulmonary edema (fluid buildup in the lungs) may develop up to 48 hours after exposure. This is a dangerous medical emergency and can be fatal. Because of the delayed development of symptoms, it may create a false sense of security with regard to health exposure risk. As with other aldehydes, it may be toxic and injurious to the central nervous system (CNS) and possibly the kidneys. It may also cause blood changes.

Skin contact with the liquid causes burns, irritation and rash. There may be skin allergies following long-term exposures. The liquid can injure the eyes and cause permanent blindness. The vapor is also irritating and may cause pain and swollen eyelids.

Ingestion burns the mouth, throat, and gastrointestinal tract. Systemic effects may occur depending on the conditions of exposure (quantity ingested, potency of solution).

☠ *Acute Health Effects*

The following acute (short-term) health effects may occur immediately or shortly after exposure to chloroacetaldehyde:

Skin: Irritation and rash or burning sensation on contact.

Eye: Irritation, swollen lids, and burns leading to permanent damage, even blindness.

Lung: Irritation causing coughing and/or shortness of breath. Higher exposures can cause a buildup of fluid in the lungs (pulmonary edema).

CNS: Inhalation of high concentrations of the vapor may cause sleepiness, dizziness, loss of consciousness, and other unspecified effects.

Chronic Health Effects

The following chronic (long-term) health effects may occur at some time after exposure to chloroacetaldehyde and can last for months or even years:

Cancer Hazards: Chloroacetaldehyde causes mutations (genetic changes). Such chemicals may have a cancer or reproductive risk. However, there is no evidence that chloroacetaldehyde causes cancer in animals. Chloroacetaldehyde may be a teratogen in humans since it has been shown to cause birth defects in animals.

Reproductive Hazards: According to the information presented in the references, chloroacetaldehyde has not been adequately tested for its ability to cause reproductive effects in animals.

Sensitivity: Exposure to the skin may result in allergy (sensitization). If allergy develops, very low future exposure can result in skin irritation and rash. Repeated exposure can cause chronic irritation of the eyes and lungs. Sensitization of the respiratory system may occur. There is also a possibility that aldehydes cause kidney damage and blood changes in the long-term.

Recommended Risk-Reduction Measures

Personnel should avoid direct contact with chloroacetaldehyde. Unless a less toxic chemical can be substituted for a hazardous substance, *engineering controls* are the most effective method of reducing exposures. The best protection is to enclose operations and/or provide local exhaust ventilation at the site of chemical release. While not always operationally feasible, isolating operations can also reduce exposure.

Using respiratory protection is less effective than the controls mentioned above, but is still advisable whenever working with or around chloroacetaldehyde. The exposure ceiling of 1 ppm is extremely low and difficult to monitor. For low exposures, a full facepiece respirator with an organic vapor cartridge may suffice. However, for high exposures or when the exposure level is unknown (as in an emergency situation), a self-contained breathing apparatus (SCBA) with full facepiece operated pressure demand is the recommended respiratory protection method of choice. If a full facepiece is not available, then chemical goggles should be worn to protect the eyes. Whenever a chemical splash hazard exists, a face shield and a rubber apron should be worn. To prevent hand and skin exposures, butyl rubber gloves should be worn. Glove manufacturers should be contacted and permeation studies obtained *before* final glove selections are made.

Administrative controls should also be in place to minimize the potential for human exposures. These may include written procedures or policies which specify the methods and techniques that will be practiced whenever personnel are to work with chloroacetaldehyde.

All personnel should receive *training* on the use, hazards, protective measures, emergency actions, and other precautions per 29 CFR 1910.1200 (Hazard Communication), prior to the first assignment in an area where chloroacetaldehyde is used or stored.

Before beginning employment with chloroacetaldehyde and at regular intervals thereafter (e.g., annually), the following medical examinations are suggested:

- ☑ Lung function tests (establish baseline).

If symptoms develop or overexposure is suspected, the following additional tests are recommended:

- ☑ Lung function tests (compare to baseline).
- ☑ Consider chest X-ray after acute overexposure (may be negative if taken immediately after exposure due to the delayed onset of pulmonary edema).
- ☑ Evaluation by a qualified allergist with careful consideration of exposure history and special tests (may help diagnose skin allergy).
- ☑ Monitor arterial blood gases.
- ☑ Consider renal function tests and urinalysis for significant exposures.

Other methods to reduce exposure include:

- ☑ Where possible, enclose operations and use local exhaust ventilation at the site of chemical release. If local exhaust ventilation or enclosure is not used, respiratory protection should be mandatory.
- ☑ Always ensure that proper protective clothing is worn when using chemical substances.
- ☑ Wash thoroughly immediately after exposure to chloroacetaldehyde and at the end of the work shift or before eating, drinking, or smoking.
- ☑ Hazard warning information should be posted in the work area. In addition, as part of an on-going education and training program, all information on the health and safety hazards of chloroacetaldehyde should be communicated to all potentially exposed workers.
- ☑ Never eat, drink, or smoke in areas were chloroacetaldehyde is used, handled or stored.

RISK ASSESSMENT: ENVIRONMENT

General Assessment

The environment is at risk of exposure during transportation, storage, disposal, or destruction of chloroacetaldehyde. In almost every scenario, the threat of environmental exposure is contingent upon the proper handling of the chemical substance. Accidental spills, large or small, can result in fire, explosion, and possible contamination of the surrounding environmental mediums (water, soil, and air).

Chloroacetaldehyde is considered a combustible liquid. Because of its moderate flash point and boiling point, it is classified as a Class 111A combustible liquid (per OSHA 29 CFR 1910.106). It is also incompatible with commonly encountered oxidizers and acids. These characteristics require special consideration during any emergency situation involving a leak or spill of chloroacetaldehyde. Should chloroacetaldehyde ever come into contact with any incompatible substances either during use, transportation, or storage, the formation of highly toxic and/or highly explosive commodities is extremely possible.

The proper disposal/destruction method for chloroacetaldehyde is to burn it in a chemical incinerator equipped with an afterburner and air scrubber.

Chloroacetaldehyde can enter the environment through manufacturing, unchecked discharge into effluents, and through spills.

Acute Ecological Effects

Acute (short-term) toxic effects may include the death of animals, birds, or fish, and death or low growth rate in plants. Acute effects are seen 2 to 4 days after animals or plants are exposed to chloroacetaldehyde. This chemical has high acute toxicity to aquatic life. Insufficient data are available to evaluate or predict the short-term (acute) effects of chloroacetaldehyde to plants, birds, or terrestrial animals.

Chronic Ecological Effects

Chronic toxic effects may include shortened life span, reproductive problems, lower fertility, and changes in appearance or behavior in exposed animals. These effects can be seen long after first exposure(s) to toxic chemicals. Chloroacetaldehyde has high chronic toxicity in aquatic life. No data are available on the long-term effects of chloroacetaldehyde to plants, birds, or land animals.

Water Solubility

Chloroacetaldehyde is highly soluble in water. Concentrations of 1000 milligrams and more will mix with a liter of water.

Persistence in the Environment

Chloroacetaldehyde is moderately persistent in water, with a half-life of between 2 to 20 days. The half-life of a pollutant is the amount of time it takes for one-half of the chemical to be degraded. Approximately 73% of chloroacetaldehyde will eventually end up in air; the rest will end up in water or aquatic sediments.

Bioaccumulation in Aquatic Organisms

Some substances increase in concentration, or *bioaccumulate,* in living organisms as they breathe contaminated air, drink contaminated water, or eat contaminated food. These chemicals can become concentrated in the tissues and internal organs of animals as well as humans.

The concentration of chloroacetaldehyde found in fish tissues is expected to be about the same as the average concentration of chloroacetaldehyde in water from which the fish was taken.

Recommended Risk-Reduction Measures

Proper training of all transporters will reduce the likelihood of a mishap or accident resulting in a leak or spill to the environment. The correct use of DOT labeling (per the MSDS) on all containers, trucks, and rail cars will enable emergency responders to react properly and quickly to any disaster thereby reducing the potential risk to the environment and to personnel.

Storage of chloroacetaldehyde should be segregated from incompatible chemicals to minimize the risk of cross-contamination or contact. Buildings designated for storage should be equipped with appropriate fire protection systems (alarms, sprinklers, emergency lighting, portable extinguishers).

If a spill or leak to the environment has occurred, fire department, emergency response, and/or hazardous materials spill personnel should be notified immediately. Cleanup should be attempted only by those trained in proper spill containment procedures. Contaminated soils should be removed for incineration and replaced with clean soil. If chloroacetaldehyde should contact the water table, aquifer, or navigable waterway, time is of the essence. It is readily miscible in water and, therefore, total containment and remediation may not be entirely possible. When such spills occur, the local and/or state emergency response authorities must be notified. A comprehensive emergency response or disaster preparedness/recovery plan should be in place prior to any operations involving the use, transportation, storage, or disposal of chloroacetaldehyde. If chloroacetaldehyde is spilled or leaked, the following specific steps are recommended:

- ☑ Restrict persons not wearing protective clothing from area of spill or leak until cleanup is complete and area can be opened for normal work.
- ☑ Remove all ignition sources.
- ☑ Ventilate area of spill or leak.
- ☑ Absorb liquids in vermiculite, dry sand, earth, or a similar material and deposit in sealed containers.

☑ Keep chloroacetaldehyde out of a confined space, such as a sewer, because of the possibility of explosion (unless the sewer is designed to prevent the buildup of explosive concentrations).

☑ It may be necessary to dispose of chloroacetaldehyde as a hazardous waste. The responsible state agency or the regional office of the federal Environmental Protection Agency (EPA) should be contacted for specific recommendations.

RISK ASSESSMENT: BUSINESS

General Assessment

Accidents or mishaps involving chloroacetaldehyde can present a significant threat to business operations. The loss or damage of equipment or facilities can significantly affect fiscal viability. Lawsuits that may result from personnel injury/death, public exposures, and/or environmental contamination will also require a serious expenditure of resources. Media attention surrounding an injury, death, or environmental damage can also result in a loss of profits and loss of current as well as future business.

Recommended Risk-Reduction Measures

Company attorneys, safety and health professionals, and environmental specialists should be involved in the development of any procedures dealing with safety or emergency response. A company official should be pre-designated as a public relations officer with specific training in dealing with the press. Corporate plans and policies should be developed, approved, and implemented long before any need for such arises.

REFERENCES

American Conference of Governmental Industrial Hygienists. 1988. *Documentation of the Threshold Limit Values and Biological Exposure Indices,* 5th Edition (with updates). Cincinnati: ACGIH

Anthony, C. P., and N. J. Kolthoff. 1971. *Textbook of Anatomy and Physiology.* St. Louis: C. V. Mosby Company.

Asante-Duah, D. K. 1993. *Hazardous Waste Risk Assessment.* New York: CRC Press/Lewis Publishers.

Brusick, D. J. 1994. *Methods for Genetic Risk Assessment.* New York: CRC Press/Lewis Publishers.

Bueche, J. R. 1972. *Principles of Physics,* 2nd Edition. New York: McGraw-Hill.

Burrell, R., D. K. Flaherty, and L. J. Sauers. 1992. *Toxicology of the Immune System.* New York: Van Nostrand Reinhold.

Calabrese, E. J. 1994. *Biological Effects of Low Level Exposures.* New York: CRC Press/Lewis Publishers.

Cockerman, L. G., B. S. Shane. 1994. *Basic Environmental Toxicology.* New York: CRC Press/Lewis Publishers.

Cothern, C. R. 1995. *Handbook for Environmental Risk Decisionmaking.* New York: CRC Press/Lewis Publishers.

Danse, I. R. 1991. *Common Sense Toxics in the Workplace.* New York: Van Nostrand Reinhold.

Environmental Protection Agency. 1983. *Chemical Hazard Information Profile: Chloroacetaldehyde.* Washington, D.C.: U.S. EPA.

Gardner, R. W. 1994. *Chemical Intolerance.* New York: CRC Press/Lewis Publishers.

Hallenbeck, W. H. 1993. *Quantitative Risk Assessment for Environmental and Occupational Health,* 2nd Edition. New York: CRC Press/Lewis Publishers.

Hodgson, E., R. B. Mailman, and J. E. Chambers. 1988. *Dictionary of Toxicology.* London: The MacMillian Company.

Heck, H. 1985. Mechanisms of aldehyde toxicity: Structure activity Studies. *CIIT Activities,* 5(10):106.

Hosty, J. W., and P. Foster. 1990. *A Practical Guide to Chemical Spill Response.* New York: Van Nostrand Reinhold.

Howard, P. H. 1995. *Handbook of Environmental Fate and Exposures Data for Organic Chemicals,* Vol. I-IV. New York: CRC Press/Lewis Publishers.

Jones, F. E. 1994. *Toxic Organic Vapors in the Workplace.* New York: CRC Press/Lewis Publishers.

Kamrin, M. 1 A. 1988. *Toxicology: A Primer on Principles and Applications.* Chelsea, Michigan: Lewis Publishers.

Keith, L. H. and M. M. Walker. 1995. *Handbook of Air Toxics.* New York: CRC Press/Lewis Publishers.

Landis, W. G., M. H. Yu. 1995. *Introduction to Environmental Toxicology.* New York: CRC Press/Lewis Publishers.

Lewis, .J., Sr. 1993. *Hawley's Condensed Chemical Dictionary,* 12th Edition. New York: Van Nostrand Reinhold.

Lewis, . J., Sr. 1992. *Sax's Dangerous Properties of Industrial Materials,* Volumes 1, 2, and 3, 8th Edition. New York: Van Nostrand Reinhold.

Meyer, E. 1990. *Chemistry of Hazardous Materials.* Englewood Cliffs, NJ: Prentice Hall.

Patnaik, P. 1992. *A Comprehensive Guide to the Hazardous Properties of Chemical Substances.* New York: Van Nostrand Reinhold.

Silverman, L., H. F. Schuttle, and M. W. First. 1946. Further studies on sensory response to certain industrial solvent vapors. *Journal of Industrial Hygiene Toxicology,* 28:262-266.

Sherman, J. D. 1988. *Chemical Exposure and Disease.* New York: Van Nostrand Reinhold.

U.S. Department of Health and Human Services, National Institute for Occupational Safety and Health. 1994. *NIOSH Pocket Guide to Chemical Hazards.* Washington, D.C.: U.S. Government Printing Office.

MATERIAL SAFETY DATA SHEET

CHEMICAL NAME	CRC PRESS/LEWIS PUBLISHERS DISCLAIMER
CHLOROACETIC ACID	**CRC PRESS/LEWIS PUBLISHERS DISCLAIMER:** THE INFORMATION AND RECOMMENDATIONS PRESENTED HEREIN ARE BASED ON SOURCES BELIEVED TO BE RELIABLE. CRC MAKES NO REPRESENTATION ON ITS COMPLETENESS OR ACCURACY. IT IS THE RESPONSIBILITY OF THE USER TO DETERMINE THE CHEMICAL'S SUITABILITY FOR ITS INTENDED USE, THE CHEMICAL'S SAFE USE, AND THE CHEMICAL'S PROPER DISPOSAL. NO REPRESENTATIONS AND/OR WARRANTIES, EITHER EXPRESSED OR IMPLIED, OF THE MERCHANTABILITY OR FITNESS FOR A PARTICULAR PURPOSE, OR OF ANY OTHER NATURE, ARE MADE WITH RESPECT TO THE INFORMATION PROVIDED IN THIS MSDS OR TO THE CHEMICAL TO WHICH INFORMATION MAY REFER. CRC NEITHER ASSUMES NOR AUTHORIZES ANY OTHER PERSON TO ASSUME FOR IT, ANY OTHER ADDITIONAL RESPONSIBILITY OR LIABILITY FOR THE USE OF, OR RELIANCE UPON, THIS INFORMATION.

HAZARD WARNING INFORMATION

HEALTH	FIRE	REACTIVE	OTHER	DEGREE OF HAZARD	COLOR CODING	OTHER CODES
3	1	0	ACID	0 = Minimum Hazard 1 = Slight Hazard 2 = Moderate Hazard 3 = Serious Hazard 4 = Severe Hazard	HEALTH = BLUE FIRE = RED REACTIVITY = YELLOW OTHER = WHITE	OX = Oxidizer ACID = Acid ALK = Alkali COR = Corrosive W = Use No Water

SECTION I - GENERAL INFORMATION

Characterization	**Acid**	RCRA Number / EPA Class	**N/A** / **Not Applicable**
DOT Proper Shipping Name	**Chloroacetic acid (liquid)**	Chemical Abstract Service (CAS) Number	**79-11-8**
DOT Hazard Class and Label Requirements	**Corrosive Material**	DOT Emergency Guide Code	**60**
DOT Identification Number*	**UN 1750 (liquid) UN 1751 (solid)**	Chemical Formula	**$ClCH_2COOH$**

Synonyms
Monochloroacetic acid; α-chloroacetic acid.

SECTION II - HAZARDOUS INGREDIENTS AND IDENTITY INFORMATION

Hazardous Components (specific identity)	OSHA Exposure Criteria	NIOSH Exposure Criteria	Immediately Dangerous to Life and Health (IDLH)	ACGIH Exposure Criteria
Chloroacetic acid (derivation: Action of chlorine on acetic acid in the presence of acetic anhydride, phosphorus, or sulfur).	**PEL:** Not Established **STEL:** Not Established	**REL:** Not Established **STEL:** Not Established	Not Established	**TLV:** Not Established **STEL:** Not Established

SECTION III - PHYSICAL AND CHEMICAL CHARACTERISTICS

Boiling Point	**372°F (189°C)**	Specific Gravity (H_2O = 1)	**1.6**
Vapor Pressure (mm Hg)	**<1 at 69°F (20°C)**	Molecular Weight	**94.5**
Vapor Density (Air = 1)	**3.3**	Melting Point	**145°F (63°C)**

Solubility
Soluble in water, alcohol, ether, chloroform, carbon disulfide.

Appearance and Odor
White to light brown crystalline (sugar or sand-like) solid material. May also appear as a liquid.

SECTION IV - FIRE AND EXPLOSION HAZARD DATA

Flash Point (method used)	**259°F (126°C) closed cup**	Explosive Limits in Air % by Volume	**LEL: 8% UEL: Not Determined**
NFPA Classification	**Class IIIB Combustible Liquid**	Autoignition Temperature	**Not Determined**

Extinguishing Media
Carbon dioxide, dry chemical, water, or foam.

Special Fire Fighting Procedures
Wear full protective clothing and self-contained breathing apparatus (SCBA). Approach fire from upwind position. Use a water spray to keep fire-exposed containers cool. Poisonous gases are produced in fire.

Unusual Fire and Explosion Hazards
Potentially explosive reaction with strong oxidizing compounds, especially at elevated temperatures. Containers may explode in fire. Firefighters should avoid contact with vapors produced during fire.

SECTION V - REACTIVITY DATA

Stability		Conditions to Avoid
		Chloroacetic acid is an extremely corrosive compound. Avoid contact with metals and combustible materials. Keep away from other incompatible materials.
Stable	Unstable	Incompatibility *(materials to avoid)*
X		Strong oxidizers (such as chlorine, bromine , and fluorine), strong bases and strong reducing agents, most metals, and ordinary combustible materials (oil, paper, wood, etc.).
Hazardous Polymerization		Conditions to Avoid
		Under normal conditions of temperature and pressure, hazardous polymerization of chloroacetic acid is not known to occur.
May Occur	Will Not Occur	Hazardous Decomposition or By-products
	X	When heated to decomposition, chloroacetic acid can emit highly toxic/poisonous fumes including chlorine and phosgene.

SECTION VI - HEALTH HAZARD DATA

Primary Route(s) of Entry:	Inhalation?	Absorption (skin)?	Ingestion?
	X		X

Health Hazards

INHALATION: Extremely irritating to the eyes, nose, throat, and respiratory system. Inhaled vapors may cause feeling of anxiety, blurred vision, "pins and needles," muscle twitching, hallucinations and convulsions, and death.

EYES & SKIN: Skin contact can result in burns. Can cause severe eye burns and/or irritation.

INGESTION: Poisonous by ingestion. Questionable carcinogenic. May damage the kidneys.

Carcinogenicity	NTP Listed?	IARC Cancer Review Group?	OSHA Regulated?	Target Organs?
Questioned Human Suspected Animal	5th Annual Report	No	No Specific Listing	Respiratory system, eyes, skin, kidneys.

Medical Conditions Generally Aggravated by Exposure

Respiratory impairments (bronchitis, asthma); Skin conditions (dermatitis); Renal or liver disorders.

Emergency and First-aid Procedures

Eye contact: Flush immediately with water for 15 minutes (minimum), seek medical attention. Skin contact: Remove all contaminated clothing. Immediately wash area with large amounts of soap and water. Seek medical assistance. For inhalation: Remove the person from exposure. Provide respiratory assistance and CPR. Transfer to medical facility. If swallowed: Seek medical attention immediately.

SECTION VII - PRECAUTIONS FOR SAFE HANDLING AND USE

Steps to be Taken in Case Material is Released or Spilled

Collect powdered materials in most convenient and safe manner possible and deposit in sealed containers. Ventilate area of spill or leak. Restrict those not involved in cleanup from entering area. Notify appropriate authorities, as required by SARA III, if applicable.

Preferred Waste Disposal Method

No Citation.

Precautions to be Taken in Handling and Storage

Chloroacetic acid is a combustible solid. Avoid contact with strong oxidizers such as chlorine since violent reactions occur. Store in tightly closed containers in cool, well-ventilated area away from heat, metals, or combustibles.

Other Precautions and Warnings

Sources of ignition are prohibited where chloroacetic acid is used, handled, or stored.

SECTION VIII - CONTROL MEASURES AND PERSONAL PROTECTIVE EQUIPMENT

Respiratory Protection *(specify type)*

When potential for high exposures exist, use a MSHA/NIOSH approved supplied air respirator with full facepiece, hood, or helmet in continuous flow mode, or use a full facepiece operated in pressure demand or other positive pressure mode. Use half-mask respirator with acid cartridge and a high efficiency particulate air (HEPA) filter for low or transient exposures (cartridge should be designed to protect against acid vapor or mist).

Ventilation

Local exhaust or general mechanical systems recommended.

Protective Gloves	Eye Protection	Other Protective Clothing
Acid resistant synthetic rubber	Chemical/Dust Goggles or Face Mask	Rubber Apron

Work/Hygiene Practices

Always wash hands thoroughly after using chemical; never bring food, drink, or smoking materials into vicinity of chemicals.

CHLOROACETIC ACID

$ClCH_2COOH$ CAS: 79-11-8

IDENTIFICATION AND TYPICAL USES

Chloroacetic acid is a colorless to light brown crystalline (sugar or sand-like) solid. May also appear as a liquid. It is used primarily as an herbicide but also serves as a disinfectant, a preservative, a bacteriostat, as an intermediate in the production of carboxymethylcellulose, ethyl chloroacetate, glycine, synthetic caffeine, sacrosine, and thioglycolic acid. Also used in the manufacture of dyes and other chemicals.

RISK ASSESSMENT: HEALTH

General Assessment

Chloroacetic acid is a poison by *ingestion* and *inhalation.* It is also an extremely corrosive skin, eye, and mucous membrane irritant. It is a questionable carcinogen with human mutation data reported.

Inhalation of chloroacetic acid dusts or sprays can irritate the nose, throat, and bronchial tubes. This may cause nose bleeds, sore throat, hoarseness, and/or cough with phlegm. Exposure can cause severe and quite serious effects that may be delayed for several hours after exposure has ceased. Symptoms may begin with a feeling of anxiety, restlessness, "pins and needles," blurred vision, and/or hallucinations (hearing things). This can continue on to convulsions (seizures or "fits") and even death. These symptoms can occur even if the victim feels no effects during the actual exposure, and even after a delay of several hours. Death, if it occurs, is usually from irregular heartbeat (ventricular fibrillation). Other causes of death may be glottis spasm or suffocation. High or repeated exposure can damage the kidneys.

Acute Health Effects

The following acute (short-term) health effects may occur immediately or shortly after exposure to chloroacetic acid:

Skin: Possible irritation/rash or burning feeling.

Eye: Severe irritation; may cause burns and damage.

Lung: Irritation of the nose, throat, and lungs (bronchial tubes) following exposure. Symptoms may be delayed for several hours.

Other: Possible CNS effects by unknown mode. Symptoms include anxiety, blurred vision," pins and needles," muscle twitching, hallucinations, convulsions, and death.

Chronic Health Effects

The following chronic (long-term) health effects may occur at some time after exposure to chloroacetic acid and can last for months or even years:

Cancer Hazards: Chloroacetic acid may cause mutations in living cells. There is limited evidence that chloroacetic acid causes cancer in test animals. Its carcinogenicity in humans may be suspect, but the data are inconclusive.

Reproductive Hazard: According to the data presented in the references, chloroacetic acid has not been adequately tested for its ability to adversely affect reproduction.

Other Chronic Effects: Repeated or long-term skin contact may cause rash from irritation. Repeated exposure can also damage the kidneys and may affect liver and/or thyroid function. Very irritating substance can damage the lungs. It is not known whether long-term exposure to chloroacetic acid will cause permanent damage to the lungs.

Recommended Risk-Reduction Measures

Personnel should avoid direct contact with chloroacetic acid. If a less toxic chemical cannot be substituted for a hazardous substance, then *engineering controls* are the most effective method of reducing exposures.

The best protection is to enclose operations and/or provide local exhaust ventilation at the site of chemical release. While not always operationally feasible, isolating operations can also reduce exposure. Using respiratory protection is less effective than the controls mentioned above, but is still advisable whenever working with or around chloroacetic acid. A supplied-air respirator with a full facepiece operated in positive pressure mode or a MSHA/NIOSH approved self-contained breathing apparatus (SCBA) with full facepiece and pressure demand are the recommended respiratory protection methods of choice. For lower exposures, an MSHA/NIOSH approved full facepiece respirator with high efficiency particulate air (HEPA) filter may suffice. In either case, if a full facepiece is not available, then chemical goggles should be worn to protect the eyes. Whenever a chemical/dust splash hazard exists, a face shield and a rubber apron should be worn. To prevent hand and skin exposures, acid resistant gloves should be worn.

Administrative controls should also be in place to minimize the potential for human exposures. These may include written procedures or policies which specify the methods and techniques that will be practiced whenever personnel are to work with chloroacetic acid.

All personnel should receive *training* on the use, hazards, protective measures, emergency actions, and other precautions per 29 CFR 1910.1200 (Hazard Communication), prior to the first assignment in an area where chloroacetic acid is used or stored.

Before beginning employment and at regular intervals thereafter (e.g., annually), for those with frequent or potentially high exposures, the following medical tests are recommended:

☑ Lung function tests.

☑ Kidney function tests.

If symptoms develop or overexposure is suspected, the following may also be useful:

☑ Liver function tests.

☑ Consider chest X-ray after acute overexposure.

Other methods to reduce exposure include:

☑ Where possible, enclose operations and use local exhaust ventilation at the site of chemical release. If local exhaust ventilation or enclosure is not used, respiratory protection should be mandatory.

☑ Where possible, automatically transfer chloroacetic acid from drums or other storage containers to process containers.

☑ Always ensure that proper protective clothing is worn when using chemical substances.

☑ Wash thoroughly immediately after exposure to chloroacetic acid and at the end of the work shift or before eating, drinking, or smoking.

☑ Hazard warning information should be posted in the work area. In addition, as part of an on-going education and training program, all information on the health and safety hazards of chloroacetic acid should be communicated to all exposed workers.

RISK ASSESSMENT: ENVIRONMENT

General Assessment

The environment is at risk of exposure during transportation, storage, disposal, or destruction of chloroacetic acid. In almost every scenario, the threat of environmental exposure is contingent upon the proper handling of the chemical substance. Accidental spills, large or small, can result in fire, explosion, and possible contamination of the surrounding environmental mediums (water, soil, and air).

Chloroacetic acid is considered a combustible liquid. It is also corrosive to many materials. It is incompatible with oxidizing agents and contact can cause fire or explosion. These characteristics require special consideration during any emergency situation involving a leak or spill of chloroacetic acid. Should chloroacetic acid ever come into contact with incompatible substances such as strong oxidizers, strong bases, and strong reducing agents, violent and explosive reactions are possible. Chloroacetic acid can enter the environment mainly industrial discharges or spills.

☠ *Acute Ecological Effects*

Acute (short-term) toxic effects may include the death of animals, birds, or fish and death or low growth rate in plants. Acute effects are seen 2 to 4 days after animals or plants are exposed to chloroacetic acid.

Chloroacetic acid has slight acute toxicity to aquatic life. It has caused death and injury to several agricultural and ornamental crops. Insufficient data are available to evaluate the acute (short-term) effects of chloroacetic acid exposure to birds or land animals.

Chronic Ecological Effects

Chronic (long-term) toxic effects may include shortened life span, reproductive problems, lower fertility, and changes in appearance or behavior in exposed animals. These effects can be seen long after first exposure(s) to toxic chemicals.

Chloroacetic acid has slight chronic toxicity in aquatic life. Insufficient data are available on the long-term effects of chloroacetic acid to plants, birds, or land animals.

Water Solubility

Chloroacetic acid is highly soluble in water. Concentrations of 1000 milligrams and more will mix with a liter of water.

Persistence in the Environment

Chloroacetic acid is moderately persistent in the aquatic environment, with a half-life of between 20 to 200 days. The half-life of a pollutant is the amount of time it takes for one half of the chemical to be degraded.

Approximately 97% of chloroacetic acid will eventually end up in water; the remainder will end up in air.

Bioaccumulation in Aquatic Organisms

Some substances increase in concentration, or *bioaccumulate,* in living organisms as they breathe contaminated air, drink contaminated water, or eat contaminated food. These chemicals can become concentrated in the tissues and internal organs of animals as well as humans.

The concentration of chloroacetic acid found in fish tissues is expected to be about the same as the average concentration of chloroacetic acid in the water from which the fish was taken.

Recommended Risk-Reduction Measures

Proper training of all transporters will reduce the likelihood of a mishap or accident resulting in a leak or spill to the environment. The correct use of DOT labeling (per the MSDS) on all containers, trucks, and rail cars will enable emergency responders to react properly and quickly to any disaster thereby reducing the potential risk to the environment and to personnel.

Storage of chloroacetic acid should be segregated from incompatible chemicals to minimize the risk of cross-contamination or contact.

If a spill or leak to the environment has occurred, fire department, emergency response and/or hazardous materials spill personnel should be notified immediately. Cleanup should be attempted only by those trained in proper spill containment procedures. Contaminated soils should be removed for incineration and replaced with clean soil. If chloroacetic acid should contact the water table, aquifer, or navigable waterway, time is of the essence. It is highly miscible in water and, therefore, total containment and remediation may not be entirely possible. When such spills occur, the local and/or state emergency response authorities must be notified. A comprehensive emergency response or disaster preparedness/recovery plan should be in place prior to any operations involving the use, transportation, storage, or disposal of chloroacetic acid. If chloroacetic acid is spilled or leaked, the following specific steps are recommended:

- ☑ Restrict persons not wearing protective clothing from area of spill or leak until cleanup is complete and area can be opened for normal work.
- ☑ Remove all ignition sources.
- ☑ If liquid is spilled, absorb in vermiculite, dry sand, earth, or similar material and deposit in sealed containers. If powders are spilled, collect in most convenient manner possible and dispose of in sealed drum. Recommend use of vacuum equipped with a high efficiency particulate air (HEPA) filter. Do not dry sweep.
- ☑ It may be necessary to dispose of chloroacetic acid as a hazardous waste. The responsible state agency or the regional office of the federal Environmental Protection Agency (EPA) should be contacted for specific recommendations.

RISK ASSESSMENT: BUSINESS

General Assessment

Accidents or mishaps involving chloroacetic acid can present a significant threat to business operations. The loss or damage of equipment or facilities can significantly affect fiscal viability. Lawsuits that may result from personnel injury/death, public exposures, and/or environmental contamination will also require a serious expenditure of resources. Media attention surrounding an injury, death, or environmental damage can also result in a loss of profits and loss of current as well as future business.

Recommended Risk-Reduction Measures

Company attorneys, safety and health professionals, and environmental specialists should be involved in the

development of any procedures dealing with safety or emergency response. A company official should be pre-designated as a public relations officer with specific training in dealing with the press. Corporate plans and policies should be developed, approved, and implemented long before any need for such arises.

REFERENCES

Ahlbom, A. 1993. *Biostatistics for Epidemiologists.* New York: CRC Press/Lewis Publishers.

American Conference of Governmental Industrial Hygienists. 1988. *Documentation of the Threshold Limit Values and Biological Exposure Indices,* 5th Edition (with updates). Cincinnati: ACGIH

Asante-Duah, D. K. 1993. *Hazardous Waste Risk Assessment.* New York: CRC Press/Lewis Publishers.

Bloom, A. D. and F. J. DeSerres. 1995. *Ecotoxicity and Human Health.* New York: CRC Press/Lewis Publishers.

Brusick, D. J. 1994. *Methods for Genetic Risk Assessment.* New York: CRC Press/Lewis Publishers.

Bueche, J. R. 1972. *Principles of Physics,* 2nd Edition. New York: McGraw-Hill.

Burrell, R., D. K. Flaherty, and Leonard J. Sauers. 1992. *Toxicology of the Immune System.* New York: Van Nostrand Reinhold.

Calabrese, E. J. 1994. *Biological Effects of Low Level Exposures.* New York: CRC Press/Lewis Publishers.

Cockerman, L. G., B. S. Shane. 1994. *Basic Environmental Toxicology.* New York: CRC Press/Lewis Publishers.

Cothern, R. R. 1993. *Comparative Environmental Risk Assessment.* New York: CRC Press/Lewis Publishers.

Danse, I. R. 1991. *Common Sense Toxics in the Workplace.* New York: Van Nostrand Reinhold.

Gardner, R. W. 1994. *Chemical Intolerance.* New York: CRC Press/Lewis Publishers.

Hallenbeck, W. H. 1993. *Quantitative Risk Assessment for Environmental and Occupational Health,* 2nd Edition. New York: CRC Press/Lewis Publishers.

Hosty, J. W., P. Foster. 1990. *A Practical Guide to Chemical Spill Response.* New York: Van Nostrand Reinhold.

Howard, P. H., et. al. 1991. *Handbook of Environmental Degradation Rates.* New York: CRC Press/Lewis Publishers.

James, A., M. Lord. 1992. *Index of Chemical and Physical Data.* New York: Van Nostrand Reinhold.

Jones, F. E. 1994. *Toxic Organic Vapors in the Workplace.* New York: CRC Press/Lewis Publishers.

Keith, L. H. and M. M. Walker. 1995. *Handbook of Air Toxics.* New York: CRC Press/Lewis Publishers.

Landis, W. G., M. H. Yu. 1995. *Introduction to Environmental Toxicology.* New York: CRC Press/Lewis Publishers.

Langerman, N. 1994 *Precautionary Labels for Chemical Containers.* New York: CRC Press/Lewis Publishers.

Lauwerys, R. R., P. Hoet. 1993. *Industrial Chemical Exposure.* New York: CRC Press/Lewis Publishers.

Lewis, R. J., Sr. 1993. *Hawley's Condensed Chemical Dictionary,* 12th Edition. New York: Van Nostrand Reinhold.

Lewis, R. J., Sr. 1992. *Sax's Dangerous Properties of Industrial Materials,* Volumes 1, 2, and 3, 8th Edition. New York: Van Nostrand Reinhold.

Lide, D. R. 1995. *Handbook of Chemistry and Physics,* 76th Edition. New York: CRC Press/Lewis Publishers.

Manahan, S. E. 1994. *Environmental Chemistry,* 6th Edition. New York: CRC Press/Lewis Publishers.

Manahan, S. E. 1992. *Toxicological Chemistry,* 2nd Edition. New York: CRC Press/Lewis Publishers.

Mellström, G. A., J. E. Wahlberg, and H. I. Maibach. 1994. *Protective Gloves for Occupational Use.* New York: CRC Press/Lewis Publishers.

Meyer, E. 1990. *Chemistry of Hazardous Materials.* Englewood Cliffs, NJ: Prentice Hall.

Mycock, J. C., J. D. McKenna, and L. Theodore. 1995. *Handbook of Air Pollution Control Engineering and Technology.* New York: CRC Press/Lewis Publishers.

Nelson, G. O. 1992. *Gas Mixtures.* New York: CRC Press/Lewis Publishers.

Ott, W. R. 1995. *Environmental Statistics and Data Analysis.* New York: CRC Press/Lewis Publishers.

Patnaik, P. 1992. *A Comprehensive Guide to the Hazardous Properties of Chemical Substances.* New York: Van Nostrand Reinhold.

Philp, R. B. 1995. *Environmental Hazards and Human Health.* New York: CRC Press/Lewis Publishers.

Proctor, N. H., J. P. Hughes. 1991. *Chemical Hazards of the Workplace.* New York: Van Nostrand Reinhold.

Quigley, D. R. 1994. *Handbook of Emergency Chemical Management.* New York: CRC Press/Lewis Publishers.

Rea, W. J. 1995. *Chemical Sensitivity,* Vol. I-III. New York: CRC Press/Lewis Publishers.

Richardson, M. 1995. *Dictionary of Substance and Their Effects,* Vol. I-VI. New York: CRC Press/Lewis Publishers.

Rylander, R., R. R. Jacobs. 1994. *Organic Dusts.* New York: CRC Press/Lewis Publishers.

Sacarello, H. L. A. 1994. *The Comprehensive Handbook of Hazardous Materials.* New York: CRC Press/Lewis Publishers.

Sherman, J. D. 1988. *Chemical Exposure and Disease.* New York: Van Nostrand Reinhold.

Witorsch, P. and S. V. Spagnolo. 1994. *Air Pollution and Lung Disease in Adults.* New York: CRC Press/Lewis Publishers.

MATERIAL SAFETY DATA SHEET

CHEMICAL NAME	CRC PRESS/LEWIS PUBLISHERS DISCLAIMER
CHLOROBENZENE	THE INFORMATION AND RECOMMENDATIONS PRESENTED HEREIN ARE BASED ON SOURCES BELIEVED TO BE RELIABLE. CRC MAKES NO REPRESENTATION ON ITS COMPLETENESS OR ACCURACY. IT IS THE RESPONSIBILITY OF THE USER TO DETERMINE THE CHEMICAL'S SUITABILITY FOR ITS INTENDED USE, THE CHEMICAL'S SAFE USE, AND THE CHEMICAL'S PROPER DISPOSAL. NO REPRESENTATIONS AND/OR WARRANTIES, EITHER EXPRESSED OR IMPLIED, OF THE MERCHANTABILITY OR FITNESS FOR A PARTICULAR PURPOSE, OR OF ANY OTHER NATURE, ARE MADE WITH RESPECT TO THE INFORMATION PROVIDED IN THIS MSDS OR TO THE CHEMICAL TO WHICH INFORMATION MAY REFER. CRC NEITHER ASSUMES NOR AUTHORIZES ANY OTHER PERSON TO ASSUME FOR IT, ANY OTHER ADDITIONAL RESPONSIBILITY OR LIABILITY FOR THE USE OF, OR RELIANCE UPON, THIS INFORMATION.

HAZARD WARNING INFORMATION

HEALTH	FIRE	REACTIVE	OTHER	DEGREE OF HAZARD	COLOR CODING	OTHER CODES
2	3	0		0 = Minimum Hazard 1 = Slight Hazard 2 = Moderate Hazard 3 = Serious Hazard 4 = Severe Hazard	HEALTH = BLUE FIRE = RED REACTIVITY = YELLOW OTHER = WHITE	OX = Oxidizer ACID = Acid ALK = Alkali COR = Corrosive W = Use No Water

SECTION I - GENERAL INFORMATION

Characterization	**Halogen-Substituted Aromatic**	RCRA Number	**U037**
		EPA Class	**Toxic Waste**
DOT Proper Shipping Name	**Chlorobenzene**	Chemical Abstract Service (CAS) Number	**108-90-7**
DOT Hazard Class and Label Requirements	**Flammable Liquid**	DOT Emergency Guide Code	**27**
DOT Identification Number	**UN 1134**	Chemical Formula	**C_6H_5Cl**

Synonyms
Benzene chloride, chlorobenzol, MCB, monochlorobenzene, phenyl chloride.

SECTION II - HAZARDOUS INGREDIENTS AND IDENTITY INFORMATION

Hazardous Components (specific identity)	OSHA Exposure Criteria	NIOSH Exposure Criteria	Immediately Dangerous to Life and Health (IDLH)	ACGIH Exposure Criteria
Chlorobenzene (derivation: By passing dry chlorine into benzene with a catalyst). **1 ppm = 4.68 mg/m^3**	**PEL:** 75 ppm 350 mg/m^3 **STEL:** Not Established	**REL:** Not Established **STEL:** Not Established	**1000 ppm**	**TLV:** 10 ppm 49 mg/m^3 **STEL:** Not Established

SECTION III - PHYSICAL AND CHEMICAL CHARACTERISTICS

Boiling Point	**270°F (132°C)**	Specific Gravity (H_2O = 1)	**1.11**
Vapor Pressure (mm Hg)	**9 at 69°F (20°C)**	Molecular Weight	**112.6**
Vapor Density (Air = 1)	**3.88**	Melting Point	**-50°F (-45°C)**

Solubility
Insoluble in water. Miscible in most organic solvents.

Appearance and Odor
Colorless liquid with a faint almond-like odor. Odor Threshold = 0.68 ppm.

SECTION IV - FIRE AND EXPLOSION HAZARD DATA

Flash Point (method used)	**82°F (28°C) closed cup**	Explosive Limits in Air % by Volume	**LEL: 1.3% UEL: 9.6%**
NFPA Classification	**Class IC Flammable Liquid**	Autoignition Temperature	**1180°F (637°C)**

Extinguishing Media
Carbon dioxide, dry chemical, water, or foam.

Special Fire Fighting Procedures
Wear full protective clothing and self-contained breathing apparatus (SCBA). Approach fire from up-wind position. Use a water spray to keep fire-exposed containers cool. Poisonous gases are produced in fire, including hydrogen chloride gas.

Unusual Fire and Explosion Hazards
Potentially explosive reaction with strong oxidizing compounds, especially at elevated temperatures. Containers may explode in fire. Vapors may travel to a source of ignition and flashback.

SECTION V - REACTIVITY DATA

Stability		Conditions to Avoid: Chlorobenzene presents a dangerous fire hazard and a moderate explosion hazard when exposed to heat and flame. Vapors form explosive mixtures with air in the range 1.3% - 9.6%.
Stable	Unstable	Incompatibility *(materials to avoid)*
X		Strong oxidizers (chlorine, bromine, and fluorine). Also, in contact with many metal perchlorates, such as those of silver and mercury, may form shock-sensitive solvated perchlorates that may explode on impact.
Hazardous Polymerization		Conditions to Avoid: Under normal conditions of temperature and pressure, hazardous polymerization of chlorobenzene is not known to occur.
May Occur	Will Not Occur	Hazardous Decomposition or By-products
	X	When heated to decomposition, chlorobenzene can emit highly toxic/poisonous fumes including hydrogen chloride gas. Dimethyl sulfoxide decomposes violently in contact with chlorobenzene.

SECTION VI - HEALTH HAZARD DATA

Primary Route(s) of Entry:	Inhalation?	Absorption (skin)?	Ingestion?
	X	X	X

Health Hazards

INHALATION: Toxic symptoms include drowsiness, lack of coordination, and liver injuries. May also cause changes in the lungs, liver, and kidneys. Narcosis possible at high exposures.

ABSORPTION: Skin contact can result in rash and burns. Can cause severe eye burns and/or irritation.

INGESTION: Toxic by ingestion. Symptoms of ataxia, respiratory distress, and liver damage.

Carcinogenicity	NTP Listed?	IARC Cancer Review Group?	OSHA Regulated?	Target Organs?
Unknown Human Questioned Animal	5th Annual Report	No	29 CFR 1910.1000 Table Z-1	Respiratory system, eyes, skin, CNS, liver.

Medical Conditions Generally Aggravated by Exposure

Respiratory impairments (bronchitis, asthma); Skin conditions (dermatitis); Liver disorders.

Emergency and First-aid Procedures

Eye contact: Flush immediately with water for 15 minutes (minimum), seek medical attention. Skin contact: Remove all contaminated clothing. Immediately wash area with large amounts of soap and water. Seek medical assistance. For inhalation: Remove the person from exposure. Provide respiratory assistance and CPR. Transfer to medical facility. If swallowed: Seek medical attention immediately. Do not attempt to give a convulsing or unconscious person anything by mouth.

SECTION VII - PRECAUTIONS FOR SAFE HANDLING AND USE

Steps to be Taken in Case Material is Released or Spilled

Absorb liquids in vermiculite, dry sand, earth, or similar material and deposit in sealed containers. Ventilate area of spill or leak. Remove all ignition sources. Restrict those not involved in cleanup from entering area. Notify appropriate authorities, as required by SARA III, if applicable.

Preferred Waste Disposal Method

Chlorobenzene is burned in a chemical incinerator equipped with an afterburner and scrubber.

Precautions to be Taken in Handling and Storage

Chlorobenzene is a flammable liquid. Avoid contact with strong oxidizers such as chlorine since violent reactions occur. Store in tightly closed containers in cool, well-ventilated area away from heat, sparks, or flames. Sources of ignition are prohibited where chlorobenzene is used, handled, or stored.

Other Precautions and Warnings

Metal containers involving the transfer of 5 gallons or more should be grounded and bonded. Drums should be equipped with self-closing valves, pressure vacuum bungs, and flame arresters.

SECTION VIII - CONTROL MEASURES AND PERSONAL PROTECTIVE EQUIPMENT

Respiratory Protection *(specify type)*

When potential for exposures over 75 ppm, use a full facepiece respirator with an organic vapor cartridge or an MSHA/NIOSH approved supplied-air respirator with full facepiece, hood, or helmet in continuous flow mode.

Ventilation

Local exhaust or general mechanical systems recommended.

Protective Gloves	Eye Protection	Other Protective Clothing
Butyl Rubber	Chemical Goggles or Face Mask	Rubber Apron

Work/Hygiene Practices

Always wash hands thoroughly after using chemical; never bring food, drink, or smoking materials into vicinity of chemicals.

CHLOROBENZENE

C_6H_5Cl

CAS: 108-90-7

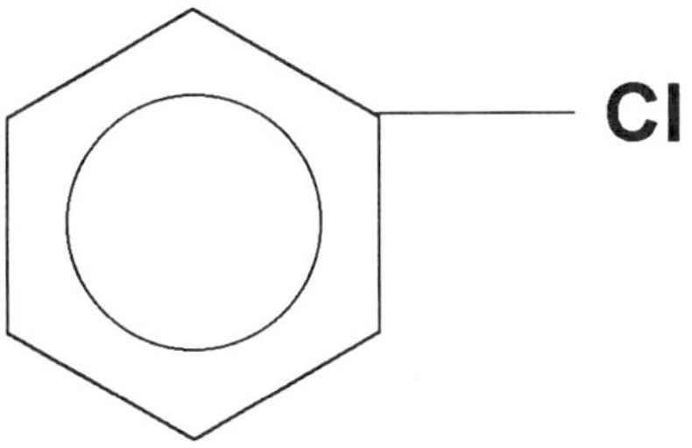

IDENTIFICATION AND TYPICAL USES

Chlorobenzene is a colorless liquid with a faint, almond-like odor. It is used primarily as a solvent for paints, as a heat transfer medium, and in the manufacture of phenol and aniline. It is also used as a feedstock to produce ortho- and para-nitrochlorobenzenes, and as a pesticide intermediate (DDT).

RISK ASSESSMENT: HEALTH

General Assessment

Chlorobenzene is a moderately toxic by *ingestion* and *inhalation*. It may also pass through unbroken skin (*absorption*) and enter the bloodstream directly. The liquid can irritate and even burn the skin if allowed to remain in contact (soaked clothing or shoes). Human mutation data have been reported. This compound may be carcinogenic to animals. There is some evidence in the references to show that a potential for carcinogenicity in some test animals, although this is not fully confirmed.

The acute inhalation and oral toxicity of chlorobenzene is considered to be relatively low. The target organs are the respiratory system, the skin, eyes, central nervous system, and the liver. The vapors will irritate the eyes and nose. Chlorobenzene is a CNS depressant. The toxic symptoms from inhalation of chlorobenzene vapors include drowsiness, loss of coordination (ataxia), and possible liver injury. Repeated exposures at 1000 ppm can cause changes in the lungs, liver, and kidneys. Exposures above 1000 ppm may cause narcosis. Ingestion of chlorobenzene can produce ataxia, respiratory distress, and liver damage. Little is known of the effects or repeated exposures at lower concentrations, but there is some evidence to show that it may cause liver and kidney damage.

Acute Health Effects

The following acute (short-term) health effects may occur immediately or shortly after exposure to chlorobenzene:

Skin: Possible irritation/rash or burning feeling.

Eye: Severe irritation; may cause burns and damage.

Lung: Irritation of the nose, throat, and lungs following exposure.

CNS: A CNS depressant causing systemic effects including drowsiness, ataxia, weakness, lethargy, malaise, fatigue, lightheadedness, and loss of consciousness.

Chronic Health Effects

The following chronic (long-term) health effects may occur at some time after exposure to chlorobenzene and can last for months or even years:

Cancer Hazards: Chlorobenzene may cause mutations in living cells. There is limited and inconclusive evidence that chlorobenzene may cause cancer in some test animals.

Reproductive Hazard: According to the data presented in the references, chlorobenzene has not been adequately tested for its ability to adversely affect reproduction.

Other Chronic Effects: Long-term exposure to chlorobenzene may cause liver damage. Repeated exposure to the liquid may cause skin burns and eye damage. Repeated exposures may also cause lung and kidney damage. This chemical has not been adequately evaluated to determine whether brain or other nerve damage could occur with repeated exposures. However, many organic solvents and other petroleum-based distillates have been shown to cause such damage. Effects may include reduced memory and concentration, personality changes (withdrawal, irritability), fatigue, sleep disturbances, reduced coordination, and/or effects on nerves supplying internal organs (autonomic nerves) and/or nerves to the arms and legs (weakness, "pins and needles").

Recommended Risk-Reduction Measures

Personnel should avoid direct contact with chlorobenzene. If a less toxic chemical cannot be substituted for a hazardous substance, then *engineering controls* are the most effective method of reducing exposures.

The best protection is to enclose operations and/or provide local exhaust ventilation at the site of chemical release. While not always operationally feasible, isolating operations can also reduce exposure. Using respiratory protection is less effective than the controls mentioned above, but is still advisable whenever working with or around chlorobenzene. A supplied-air respirator with a full facepiece operated in positive pressure mode or an MSHA/NIOSH approved self-contained breathing apparatus (SCBA) with full facepiece and pressure demand are the recommended respiratory protection methods of choice. For lower exposures, an MSHA/NIOSH approved full facepiece respirator with an organic vapor cartridge may suffice. In either case, if a full facepiece is not available, then chemical splash goggles should be worn to protect the eyes. Whenever a chemical splash hazard exists, a face shield and a rubber apron should be worn. To prevent hand and skin exposures, acid resistant gloves should be worn.

Administrative controls should also be in place to minimize the potential for human exposures. These may include written procedures or policies which specify the methods and techniques that will be practiced whenever personnel are to work with chlorobenzene.

All personnel should receive *training* on the use, hazards, protective measures, emergency actions, and other precautions per 29 CFR 1910.1200 (Hazard Communication), prior to the first assignment in an area where chlorobenzene is used or stored.

Before beginning employment and at regular intervals thereafter (e.g., annually), for those with frequent or potentially high exposures, the following medical tests are recommended:

☑ Lung function tests.

☑ Interview for brain effects, including recent memory, mood (irritability, withdrawal) changes, concentration problems, headaches, malaise, and altered sleep patterns.

☑ Consider cerebellar, autonomic and peripheral nervous system evaluation. Positive and borderline individuals should be referred for neuropsychological testing.

If symptoms develop or overexposure is suspected, the following may also be useful:

☑ Liver function tests.

☑ Kidney function tests and urinalysis.

Also, because smoking can cause heart disease as well as lung cancer, emphysema, and other respiratory problems, chemical exposures can seriously compound the effects of smoking on the human body. Therefore, smokers should avoid unprotected exposures to chlorobenzene. Heavy alcohol consumption can cause liver damage and may increase the liver damage potential caused by exposure to chlorobenzene.

Other methods to reduce exposure include:

☑ Where possible, enclose operations and use local exhaust ventilation at the site of chemical release. If local exhaust ventilation or enclosure is not used, respiratory protection should be mandatory.

☑ Where possible, automatically transfer chlorobenzene from drums or other storage containers to process containers. Metal containers involving transfers of 5 or more gallons of chlorobenzene should be grounded and bonded. Drums must be equipped with self-closing valves, pressure vacuum bungs, and flame arresters. Use only non-sparking tools and equipment, especially when opening and closing containers.

☑ Always ensure that proper protective clothing is worn when using chemical substances and that personnel are trained on its use, care, and maintenance.

☑ Wash thoroughly immediately after exposure to chlorobenzene and at the end of the work shift or before eating, drinking, or smoking.

☑ Hazard warning information should be posted in the work area. In addition, as part of an on-going education and training program, all information on the health and safety hazards of chlorobenzene should be communicated to all exposed and potentially exposed workers.

RISK ASSESSMENT: ENVIRONMENT

General Assessment

The environment is at risk of exposure during transportation, storage, disposal, or destruction of chlorobenzene. In almost every scenario, the threat of environmental exposure is contingent upon the proper handling of the chemical substance. Accidental spills, large or small, can result in fire, explosion, and possible contamination of the surrounding environmental mediums (water, soil, and air).

Chlorobenzene is considered a class 1C flammable liquid (per OSHA 29 CFR 1910.106). It will also form explosive mixtures with air in the range 1.3% to 9.6%. It is incompatible with oxidizing agents and contact can cause fire or explosion. These characteristics require special consideration during any emergency situation involving a leak or spill of chlorobenzene.

Should chlorobenzene ever come into contact with incompatible substances such as strong oxidizers, many metal perchlorates (such as mercury or silver), shock-sensitive compounds may form that may violently explode on impact. It is a dangerous fire hazard and a moderate explosive hazard when exposed to heat or flames. Potentially explosive reaction when mixed with powdered sodium or phosphorus trichloride plus sodium.

Chlorobenzene can enter the environment mainly through industrial and municipal discharges or spills.

Acute Ecological Effects

Acute (short-term) toxic effects may include the death of animals, birds, or fish and death or low growth rate in plants. Acute effects are seen 2 to 4 days after animals or plants are exposed to chlorobenzene.

Chlorobenzene has moderate acute toxicity to aquatic life. Insufficient data are available to evaluate the short-term effects of chlorobenzene exposure to plants, birds, or terrestrial animals.

Chronic Ecological Effects

Chronic (long-term) toxic effects may include shortened life span, reproductive problems, lower fertility, and changes in appearance or behavior in exposed animals. These effects can be seen long after first exposure(s) to toxic chemicals.

Chlorobenzene has moderate chronic toxicity in aquatic life. Insufficient data are available on the long-term effects of chlorobenzene to plants, birds, or terrestrial animals.

Water Solubility

Chlorobenzene is moderately soluble in water. Concentrations of 1 to 1000 milligrams will mix with a liter of water.

Persistence in the Environment

Chlorobenzene is slightly persistent in the aquatic environment, with a half-life of between 2 to 20 days. The half-life of a pollutant is the amount of time it takes for one half of the chemical to be degraded.

Approximately 99.25% of chlorobenzene will eventually end up in air; the remainder will end up in water.

Bioaccumulation in Aquatic Organisms

Some substances increase in concentration, or *bioaccumulate,* in living organisms as they breathe contaminated air, drink contaminated water, or eat contaminated food. These chemicals can become concentrated in the tissues and internal organs of animals as well as humans.

The concentration of chlorobenzene found in fish tissues is expected to be somewhat higher than the average concentration of chlorobenzene in the water from which the fish was taken.

Recommended Risk-Reduction Measures

Proper training of all transporters will reduce the likelihood of a mishap or accident resulting in a leak or spill to the environment. The correct use of DOT labeling (per the MSDS) on all containers, trucks, and rail cars will enable emergency responders to react properly and quickly to any disaster thereby reducing the potential risk to the environment and to personnel.

Storage of chlorobenzene should be segregated from incompatible chemicals to minimize the risk of cross-contamination or contact. Storage buildings or facilities should be equipped with proper fire protection equipment (sprinklers/extinguishers, alarms).

If a spill or leak to the environment has occurred, fire department, emergency response and/or hazardous materials spill personnel should be notified immediately. Cleanup should be attempted only by those trained in proper spill containment procedures. Contaminated soils should be removed for incineration and replaced with clean soil. If chlorobenzene should contact the water table, aquifer, or navigable waterway, time is of the essence. It is moderately soluble in water and, therefore, total containment and remediation may not be entirely possible. When such spills occur, the local and/or state emergency response authorities must be notified. A comprehensive emergency response or disaster preparedness/recovery plan should be in place prior to any operations involving the use, transportation, storage, or disposal of chlorobenzene. If chlorobenzene is spilled or leaked, the following specific steps are recommended:

- ☑ Restrict persons not wearing protective clothing from area of spill or leak until cleanup is complete.
- ☑ Remove all ignition sources.
- ☑ If liquid is spilled, absorb in vermiculite, dry sand, earth, or similar material and deposit in sealed containers.
- ☑ It may be necessary to dispose of chlorobenzene as a hazardous waste. The responsible state agency or the regional office of the federal Environmental Protection Agency (EPA) should be contacted for specific recommendations.

RISK ASSESSMENT: BUSINESS

General Assessment

Accidents or mishaps involving chlorobenzene can present a significant threat to business operations. The loss or damage of equipment or facilities can significantly affect fiscal viability. Lawsuits that may result from personnel injury/death, public exposures, and/or environmental contamination will also require a serious expenditure of resources. Media attention surrounding an injury, death, or environmental damage can also result in a loss of profits and loss of current as well as future business.

Recommended Risk-Reduction Measures

Company attorneys, safety and health professionals, and environmental specialists should be involved in the development of any procedures for responding to hazardous chemical incidents . A company official should be pre-designated as a public relations officer with specific training in dealing with the press. Corporate plans and policies should be developed, approved, and implemented long before any need for such arises.

REFERENCES

American Conference of Governmental Industrial Hygienists. 1988. *Documentation of the Threshold Limit Values and Biological Exposure Indices,* 5th Edition (with updates). Cincinnati: ACGIH

Bloom, A. D. and F. J. DeSerres. 1995. *Ecotoxicity and Human Health.* New York: CRC Press/Lewis Publishers.

Bueche, J. R. 1972. *Principles of Physics,* 2nd Edition. New York: McGraw-Hill.

Calabrese, E. J. 1994. *Biological Effects of Low Level Exposures.* New York: CRC Press/Lewis Publishers.

Cockerman, L. G., B. S. Shane. 1994. *Basic Environmental Toxicology.* New York: CRC Press/Lewis Publishers.

Cothern, R. R. 1993. *Comparative Environmental Risk Assessment.* New York: CRC Press/Lewis Publishers.

Gardner, R. W. 1994. *Chemical Intolerance.* New York: CRC Press/Lewis Publishers.

Hallenbeck, W. H. 1993. *Quantitative Risk Assessment for Environmental and Occupational Health,* 2nd Edition. New York: CRC Press/Lewis Publishers.

Howard, P. H., et. al. 1991. *Handbook of Environmental Degradation Rates.* New York: CRC Press/Lewis Publishers.

Jones, F. E. 1994. *Toxic Organic Vapors in the Workplace.* New York: CRC Press/Lewis Publishers.

Keith, L. H. and M. M. Walker. 1995. *Handbook of Air Toxics.* New York: CRC Press/Lewis Publishers.

Langerman, N. 1994 *Precautionary Labels for Chemical Containers.* New York: CRC Press/Lewis Publishers.

Lauwerys, R. R., P. Hoet. 1993. *Industrial Chemical Exposure.* New York: CRC Press/Lewis Publishers.

Lewis, R. J., Sr. 1992. *Sax's Dangerous Properties of Industrial Materials,* Volumes 1, 2, and 3, 8th Edition. New York: Van Nostrand Reinhold.

Manahan, S. E. 1992. *Toxicological Chemistry,* 2nd Edition. New York: CRC Press/Lewis Publishers.

Mellström, G. A., J. E. Wahlberg, and H. I. Maibach. 1994. *Protective Gloves for Occupational Use.* New York: CRC Press/Lewis Publishers.

Meyer, E. 1990. *Chemistry of Hazardous Materials.* Englewood Cliffs, NJ: Prentice Hall.

Patnaik, P. 1992. *A Comprehensive Guide to the Hazardous Properties of Chemical Substances.* New York: Van Nostrand Reinhold.

Proctor, N. H., J. P. Hughes. 1991. *Chemical Hazards of the Workplace.* New York: Van Nostrand Reinhold.

Rea, W. J. 1995. *Chemical Sensitivity,* Vol. I-III. New York: CRC Press/Lewis Publishers.

Sacarello, H. L. A. 1994. *The Comprehensive Handbook of Hazardous Materials.* New York: CRC Press/Lewis Publishers.

Sherman, J. D. 1988. *Chemical Exposure and Disease.* New York: Van Nostrand Reinhold.

MATERIAL SAFETY DATA SHEET

CHEMICAL NAME

CHLOROFORM

CRC PRESS/LEWIS PUBLISHERS DISCLAIMER: THE INFORMATION AND RECOMMENDATIONS PRESENTED HEREIN ARE BASED ON SOURCES BELIEVED TO BE RELIABLE. CRC MAKES NO REPRESENTATION ON ITS COMPLETENESS OR ACCURACY. IT IS THE RESPONSIBILITY OF THE USER TO DETERMINE THE CHEMICAL'S SUITABILITY FOR ITS INTENDED USE, THE CHEMICAL'S SAFE USE, AND THE CHEMICAL'S PROPER DISPOSAL. NO REPRESENTATIONS AND/OR WARRANTIES, EITHER EXPRESSED OR IMPLIED, OF THE MERCHANTABILITY OR FITNESS FOR A PARTICULAR PURPOSE, OR OF ANY OTHER NATURE, ARE MADE WITH RESPECT TO THE INFORMATION PROVIDED IN THIS MSDS OR TO THE CHEMICAL TO WHICH INFORMATION MAY REFER. CRC NEITHER ASSUMES NOR AUTHORIZES ANY OTHER PERSON TO ASSUME FOR IT, ANY OTHER ADDITIONAL RESPONSIBILITY OR LIABILITY FOR THE USE OF, OR RELIANCE UPON, THIS INFORMATION.

HAZARD WARNING INFORMATION

HEALTH	FIRE	REACTIVE	OTHER	DEGREE OF HAZARD	COLOR CODING	OTHER CODES
2	0	0		0 = Minimum Hazard 1 = Slight Hazard 2 = Moderate Hazard 3 = Serious Hazard 4 = Severe Hazard	HEALTH = BLUE FIRE = RED REACTIVITY = YELLOW OTHER = WHITE	OX = Oxidizer ACID = Acid ALK = Alkali COR = Corrosive W = Use No Water

SECTION I - GENERAL INFORMATION

Characterization	**Halogenated Hydrocarbon**	RCRA Number / EPA Class	**U044** / **Toxic Waste**
DOT Proper Shipping Name	**Chloroform**	Chemical Abstract Service (CAS) Number	**67-66-3**
DOT Hazard Class and Label Requirements	**ORM-A**	DOT Emergency Guide Code	**55**
DOT Identification Number	**UN 1888**	Chemical Formula	**$CHCl_3$**

Synonyms

Trichloromethane; methane trichloride.

SECTION II - HAZARDOUS INGREDIENTS AND IDENTITY INFORMATION

Hazardous Components (specific identity)	OSHA Exposure Criteria	NIOSH Exposure Criteria	Immediately Dangerous to Life and Health (IDLH)	ACGIH Exposure Criteria
Chloroform (derivation: Reaction of chlorinated lime with acetone, acetaldehyde, or ethanol; by-product from the chlorination of methane). **1 ppm = 4.96 mg/m^3**	**PEL:** **2 ppm** **10 mg/m^3** **STEL:** **(ceiling)** **50 ppm** **240 mg/m^3**	**REL:** **Possible Cancer Agent** **STEL (60 min):** **2 ppm** **10 mg/m^3**	**500 ppm**	**TLV:** **Suspected Cancer Agent** **STEL:** **10 ppm** **49 mg/m^3**

SECTION III - PHYSICAL AND CHEMICAL CHARACTERISTICS

Boiling Point	**142°F (61°C)**	Specific Gravity (H_2O = 1)	**1.5**
Vapor Pressure (mm Hg)	**7 at 69°F (20°C)**	Molecular Weight	**119.4**
Vapor Density (Air = 1)	**4.1**	Melting Point	**-81°F (-63°C)**

Solubility

Slightly soluble in water (0.5% at 77°F). Miscible with alcohol, ether, benzene, carbon disulfide, carbon tetrachloride, fixed and volatile oils.

Appearance and Odor

Colorless, highly, refractive, heavy, volatile liquid with a characteristic sweet odor and taste. Odor Threshold = 85 ppm.

SECTION IV - FIRE AND EXPLOSION HAZARD DATA

Flash Point (method used)	**Not Applicable**	Explosive Limits in Air % by Volume	**LEL: N/A UEL: N/A**
NFPA Classification	**Non-combustible Liquid**	Autoignition Temperature	**Not Applicable**

Extinguishing Media

Use extinguishing agent suitable to surrounding fire. Chloroform itself will not burn.

Special Fire Fighting Procedures

Wear full protective clothing and self-contained breathing apparatus (SCBA). Use a water spray to keep fire-exposed containers cool. Poisonous gases are produced in fire, including chlorine gas.

Unusual Fire and Explosion Hazards

Potentially explosive reaction with strong oxidizing compounds. Containers may explode in fire. Move containers from fire area if it can be done without risk.

SECTION V - REACTIVITY DATA

Stability		Conditions to Avoid
		Keep away from incompatible materials since explosions can occur.
Stable	Unstable	Incompatibility *(materials to avoid)*
X		Alkali metals, (sodium, potassium, lithium); finely divided magnesium or aluminum; sodium-potassium alloy; nitrogen tetroxide; and perchloric acid-phosphorus pentoxcide. Strong caustics and oxidizers.
Hazardous Polymerization		Conditions to Avoid Under normal conditions of temperature and pressure, hazardous polymerization of chloroform is not known to occur.
May Occur	Will Not Occur	Hazardous Decomposition or By-products
	X	When heated to decomposition, chloroform can emit highly toxic/poisonous fumes including chlorine and phosgene gas.

SECTION VI - HEALTH HAZARD DATA

Primary Route(s) of Entry:	Inhalation?	Absorption (skin)?	Ingestion?
	X	X	X

Health Hazards

INHALATION: Toxic symptoms include dizziness, lightheadedness, dullness, hallucination, nausea, headache, distorted perceptions, fatigue, and anesthesia. May cause irregular heartbeat.

ABSORPTION: Skin contact can result in irritation and rash. Can cause eye irritation with inflammation of the conjunctiva. Will pass through unbroken skin and cause toxic systemic effects.

INGESTION: Toxic by ingestion. May cause vomiting and other unspecified gastrointestinal effects.

Carcinogenicity	NTP Listed?	IARC Cancer Review Group?	OSHA Regulated?	Target Organs?
Suspected Human Confirmed Animal	5th Annual Report	Group 2B	29 CFR 1910.1000 Table Z-1	Liver, CNS, kidney, heart, skin, eyes.

Medical Conditions Generally Aggravated by Exposure

None Reported. However, existing skin conditions, heart and/or liver disorders may be affected.

Emergency and First-aid Procedures

Eye contact: Flush immediately with water for 15 minutes (minimum), seek medical attention. Skin contact: Remove all contaminated clothing. Immediately wash area with large amounts of soap and water. For inhalation: Remove the person from exposure. Provide respiratory assistance and CPR. Transfer to medical facility. If swallowed: If victim is conscious, induce vomiting. Seek medical attention immediately. Do not attempt to give a convulsing or unconscious person anything by mouth.

SECTION VII - PRECAUTIONS FOR SAFE HANDLING AND USE

Steps to be Taken in Case Material is Released or Spilled

Absorb liquids in vermiculite, dry sand, earth, or similar material and deposit in sealed containers. Ventilate area of spill or leak. Restrict those not involved in cleanup from entering area.

Preferred Waste Disposal Method

Mix with vermiculite, wrap in paper, burn in a chemical incinerator (with an afterburner and scrubber).

Precautions to be Taken in Handling and Storage

Chloroform is a non-combustible liquid. But contact with strong oxidizers such as chlorine will cause violent reactions. Store in tightly closed containers in cool, well-ventilated area. A regulated, marked work area should be established where chloroform is used, handled, or stored.

Other Precautions and Warnings

Prior to working with chloroform, personnel should be trained on its proper handling and storage.

SECTION VIII - CONTROL MEASURES AND PERSONAL PROTECTIVE EQUIPMENT

Respiratory Protection *(specify type)*

At any exposure level, use an MSHA/NIOSH approved supplied-air respirator with full facepiece, hood, or helmet in continuous flow mode or a self-contained breathing apparatus (SCBA) operated in positive pressure mode.

Ventilation

Local exhaust or general mechanical systems recommended.

Protective Gloves	Eye Protection	Other Protective Clothing
Polyvinyl Chloride, Viton, or Neoprene	Chemical Goggles or Face Mask	Protective Suit or Clothing

Work/Hygiene Practices

Always wash hands thoroughly after using chemical; never bring food, drink, or smoking materials into vicinity of chemicals.

CHLOROFORM

$CHCl_3$ CAS: 67-66-3

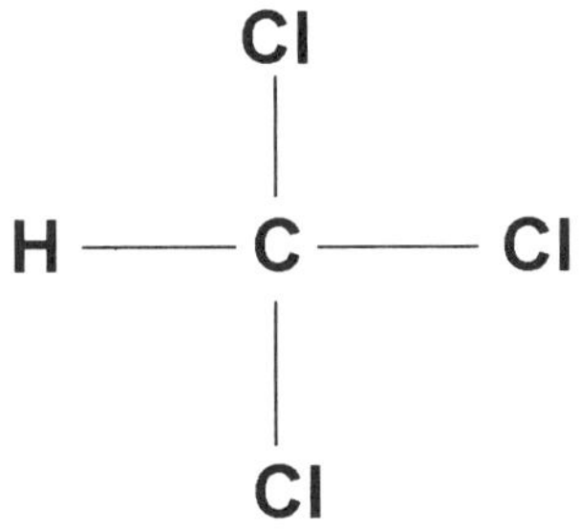

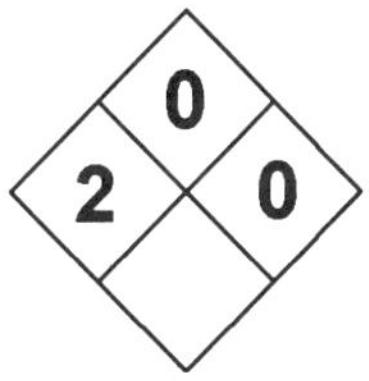

IDENTIFICATION AND TYPICAL USES

Chloroform is a colorless, highly refractive, heavy, volatile liquid with a characteristic sweet odor and taste. It is used in industry as a solvent, as a cleansing agent, in the manufacture of refrigerant, and in fire extinguishers. Also used in the manufacture of fluorocarbon plastics, in analytical chemistry, as a fumigant, and an insecticide. In the past, it was used extensively as an anesthetic. However, due to its toxic effects, this use has been abandoned.

RISK ASSESSMENT: HEALTH

General Assessment

Chloroform is a confirmed carcinogen with human mutation data reported. It is a human poison by *ingestion* and *inhalation*. It may also pass through unbroken skin (*absorption*) and enter the bloodstream directly. The liquid can irritate the skin if allowed to remain in contact. There is some evidence in the references to show that chloroform may cause damage to the developing fetus in some test animals, although this is not fully confirmed.

The acute inhalation toxicity of chloroform is considered to be relatively low. The target organs are the liver, kidney, central nervous system, skin, and heart. The vapors will irritate the eyes and nose. Chloroform is a CNS stimulant.

The toxic symptoms from inhalation of chloroform vapors include dizziness, lightheadedness, dullness, hallucination, nausea, headache, fatigue, and anesthesia. Inhalation of concentrated vapors causes dilation of the pupils with reduced reaction to light. In initial stages, there is a feeling of warmth on the face and body, then an irritation of the mucous membranes, conjunctiva, and skin; followed by excitation, loss of reflexes, sensation, and consciousness. Prolonged inhalation will bring on paralysis accompanied by cardiac respiratory failure and finally death.

Ingestion of chloroform can produce digestive disturbances, vomiting, necrosis of the liver, and kidney damage. Little is known of the effects of repeated exposures at lower concentrations, but there is some evidence to show that it may cause liver and kidney damage.

Acute Health Effects

The following acute (short-term) health effects may occur immediately or shortly after exposure to chloroform:

Skin: Irritation and possible rash or burning feeling on contact.

Eye: Severe irritation; may cause burns and damage.

Lung: Irritation of the nose and throat following exposure.

CNS: A CNS stimulant causing systemic effects including dizziness, lightheadedness, dullness, hallucination, nausea, headache, fatigue, and anesthesia.

Other: Exposure can cause the heart to beat irregularly or even stop which can be fatal.

Chronic Health Effects

The following chronic (long-term) health effects may occur at some time after exposure to chloroform and can last for months or even years:

Cancer Hazards: Chloroform is a probable carcinogen in humans. It has been shown to cause liver, kidney, and thyroid cancer in animals. It may cause mutations in living cells.

Reproductive Hazard: There is limited evidence that chloroform is a teratogen in animals. It has produced embryo toxicity causing a high incidence of fetal resorption and retarded fetal growth. Until further testing has been performed, it should be treated as a possible teratogen in humans.

Other Chronic Effects: Long-term exposure to chloroform may cause liver damage. Repeated exposure to the liquid may produce skin drying and cracking. Repeated exposures may affect the kidneys and nervous system.

Recommended Risk-Reduction Measures

Personnel should avoid direct contact with chloroform. If a less toxic chemical cannot be substituted for a hazardous substance, then *engineering controls* are the most effective method of reducing exposures.

The best protection is to enclose operations and/or provide local exhaust ventilation at the site of chemical release. While not always operationally feasible, isolating operations can also reduce exposure. Using respiratory protection is less effective than the controls mentioned above, but is still advisable whenever working with or around chloroform. For any exposure level, a supplied-air respirator with a full facepiece operated in positive pressure mode or an MSHA/NIOSH approved self-contained breathing apparatus (SCBA) with full facepiece and pressure demand are the recommended respiratory protection methods of choice. If a full facepiece is not available, then chemical splash goggles should be worn to protect the eyes. Whenever a chemical splash hazard exists, a face shield and a rubber apron should be worn. To prevent hand and skin exposures, solvent-resistant gloves should be worn.

Administrative controls should also be in place to minimize the potential for human exposures. These may include written procedures or policies which specify the methods and techniques that will be practiced whenever personnel are to work with chloroform. A regulated controlled work area should be established where chloroform is used, handled, or stored.

All personnel should receive *training* on the use, hazards, protective measures, emergency actions, and other precautions per 29 CFR 1910.1200 (Hazard Communication), prior to the first assignment in an area where chloroform is used or stored. Before beginning employment and at regular intervals thereafter (e.g., annually), for those with frequent or potentially high exposures (half the TLV or greater, or significant skin contact), the following medical tests are recommended:

☑ Lung function tests.

☑ Kidney function tests.

If symptoms develop or overexposure is suspected, the following may also be useful:

☑ Special 24 hour EKG (holster monitor) to examine for irregular heart beat.

Also, heavy alcohol consumption can cause liver damage and may increase the liver damage potential caused by exposure to chloroform.

Other methods to reduce exposure include:

☑ Where possible, enclose operations and use local exhaust ventilation at the site of chemical release. If local exhaust ventilation or enclosure is not used, respiratory protection should be mandatory.

☑ Always ensure that proper protective clothing is worn when using chemical substances.

☑ Wash thoroughly immediately after exposure to chloroform and at the end of the work shift or before eating, drinking, or smoking.

☑ Hazard warning information should be posted in the work area. In addition, as part of an on-going education and training program, all information on the health and safety hazards of chloroform should be communicated to all potentially exposed workers.

RISK ASSESSMENT: ENVIRONMENT

General Assessment

The environment is at risk of exposure during transportation, storage, disposal, or destruction of chloroform. In almost every scenario, the threat of environmental exposure is contingent upon the proper handling of the chemical substance. Accidental spills, large or small, can result in fire, explosion, and possible contamination of the surrounding environmental mediums (water, soil, and air).

Chloroform is considered a non-combustible liquid. However, mixing chloroform with incompatible substances can produce explosions. These materials include strong caustics, alkaline metals such as potassium or lithium, finely divided magnesium or aluminum, and other chemically-active metals. It will also react with strong oxidizers. These characteristics require special consideration during any emergency situation involving a leak or spill of chloroform.

Chloroform can enter the environment through industrial effluents, municipal waste treatment plant discharges, or spills.

Acute Ecological Effects

Acute (short-term) toxic effects may include the death of animals, birds, or fish and death or low growth rate in plants. Acute effects are seen 2 to 4 days after animals or plants are exposed to chloroform.

Chloroform has moderate acute toxicity to aquatic life. Chloroform has caused damage to various plants, including brittle roots and chromosomal damage. Insufficient data are available to evaluate the short-term effects of chloroform exposure to birds or land animals.

Chronic Ecological Effects

Chronic (long-term) toxic effects may include shortened life span, reproductive problems, lower fertility, and changes in appearance or behavior in exposed animals. These effects can be seen long after first exposure(s) to toxic chemicals.

Chloroform has moderate chronic toxicity in aquatic life. Insufficient data are available on the long-term effects of chloroform to plants, birds, or land animals.

Water Solubility

Chloroform is highly soluble in water. Concentrations of 1000 milligrams and more will mix with a liter of water.

Persistence in the Environment

Chloroform is non-persistent in the aquatic environment, with a half-life of less than 2 days. The half-life of a pollutant is the amount of time it takes for one half of the chemical to be degraded.

Approximately 99.67% of chloroform will eventually end up in air; the remainder will end up in water.

Bioaccumulation in Aquatic Organisms

Some substances increase in concentration, or *bioaccumulate,* in living organisms as they breathe contaminated air, drink contaminated water, or eat contaminated food. These chemicals can become concentrated in the tissues and internal organs of animals as well as humans.

The concentration of chloroform found in fish tissues is expected to be somewhat higher than the average concentration of chloroform in the water from which the fish was taken.

Recommended Risk-Reduction Measures

Proper training of all transporters will reduce the likelihood of a mishap or accident resulting in a leak or spill to the environment. The correct use of DOT labeling (per the MSDS) on all containers, trucks, and rail cars will enable emergency responders to react properly and quickly to any disaster thereby reducing the potential risk to the environment and to personnel.

Storage of chloroform should be segregated from incompatible chemicals to minimize the risk of cross-contamination or contact. To dispose of chloroform, it can be mixed with vermiculite, sodium carbonate, and slated lime, wrapped in paper, and burned in a chemical incinerator equipped with an afterburner and scrubber.

If a spill or leak to the environment has occurred, fire department, emergency response and/or hazardous materials spill personnel should be notified immediately. Cleanup should be attempted only by those trained in proper spill containment procedures. Contaminated soils should be removed for incineration and replaced with clean soil. If chloroform should contact the water table, aquifer, or navigable waterway, time is of the essence. It is highly soluble in water and, therefore, total containment and remediation may not be entirely possible. When such spills occur, the local and/or state emergency response authorities must be notified. A comprehensive emergency response or disaster preparedness/recovery plan should be in place prior to any operations involving the use, transportation, storage, or disposal of chloroform. If chloroform is spilled or leaked, the following specific steps are recommended:

- ☑ Restrict persons not wearing protective clothing from area of spill or leak until cleanup is complete and area can be opened for normal work.
- ☑ Collect liquid for reclamation or absorb in vermiculite, dry sand, earth, or similar material and deposit in sealed containers.
- ☑ It may be necessary to dispose of chloroform as a hazardous waste. The responsible state agency or the regional office of the federal Environmental

Protection Agency (EPA) should be contacted for specific recommendations.

RISK ASSESSMENT: BUSINESS

General Assessment

Accidents or mishaps involving chloroform can present a significant threat to business operations. The loss or damage of equipment or facilities can significantly affect fiscal viability. Lawsuits that may result from personnel injury/death, public exposures, and/or environmental contamination will also require a serious expenditure of resources. Media attention surrounding an injury, death, or environmental damage can also result in a loss of profits and loss of current as well as future business.

Recommended Risk-Reduction Measures

Company attorneys, safety and health professionals, and environmental specialists should be involved in the development of any procedures dealing with safety or emergency response. A company official should be pre-designated as a public relations officer with specific training in dealing with the press. Corporate plans and policies should be developed, approved, and implemented long before any need for such arises.

REFERENCES

Ahlbom, A. 1993. *Biostatistics for Epidemiologists.* New York: CRC Press/Lewis Publishers.

American Conference of Governmental Industrial Hygienists. 1988. *Documentation of the Threshold Limit Values and Biological Exposure Indices,* 5th Edition (with updates). Cincinnati: ACGIH

Asante-Duah, D. K. 1993. *Hazardous Waste Risk Assessment.* New York: CRC Press/Lewis Publishers.

Bloom, A. D. and F. J. DeSerres. 1995. *Ecotoxicity and Human Health.* New York: CRC Press/Lewis Publishers.

Brusick, D. J. 1994. *Methods for Genetic Risk Assessment.* New York: CRC Press/Lewis Publishers.

Bueche, J. R. 1972. *Principles of Physics,* 2nd Edition. New York: McGraw-Hill.

Calabrese, E. J., E. M. Kenyon. *Air Toxics and Risk Assessment.* New York: CRC Press/Lewis Publishers.

Calabrese, E. J. 1994. *Biological Effects of Low Level Exposures.* New York: CRC Press/Lewis Publishers.

Cockerman, L. G., B. S. Shane. 1994. *Basic Environmental Toxicology.* New York: CRC Press/Lewis Publishers.

Cothern, R. R. 1993. *Comparative Environmental Risk Assessment.* New York: CRC Press/Lewis Publishers.

Gardner, R. W. 1994. *Chemical Intolerance.* New York: CRC Press/Lewis Publishers.

Hallenbeck, W. H. 1993. *Quantitative Risk Assessment for Environmental and Occupational Health,* 2nd Edition. New York: CRC Press/Lewis Publishers.

Howard, P. H., et. al. 1991. *Handbook of Environmental Degradation Rates.* New York: CRC Press/Lewis Publishers.

Howard, P. H. 1995. *Handbook of Environmental Fate and Exposures Data for Organic Chemicals,* Vol. I-IV. New York: CRC Press/Lewis Publishers.

Jones, F. E. 1994. *Toxic Organic Vapors in the Workplace.* New York: CRC Press/Lewis Publishers.

Keith, L. H. and M. M. Walker. 1995. *Handbook of Air Toxics.* New York: CRC Press/Lewis Publishers.

Langerman, N. 1994 *Precautionary Labels for Chemical Containers.* New York: CRC Press/Lewis Publishers.

Lauwerys, R. R., P. Hoet. 1993. *Industrial Chemical Exposure.* New York: CRC Press/Lewis Publishers.

Lewis, R. J., Sr. 1992. *Sax's Dangerous Properties of Industrial Materials,* Volumes 1, 2, and 3, 8th Edition. New York: Van Nostrand Reinhold.

Manahan, S. E. 1992. *Toxicological Chemistry,* 2nd Edition. New York: CRC Press/Lewis Publishers.

Mellström, G. A., J. E. Wahlberg, and H. I. Maibach. 1994. *Protective Gloves for Occupational Use.* New York: CRC Press/Lewis Publishers.

Meyer, E. 1990. *Chemistry of Hazardous Materials.* Englewood Cliffs, NJ: Prentice Hall.

Patnaik, P. 1992. *A Comprehensive Guide to the Hazardous Properties of Chemical Substances.* New York: Van Nostrand Reinhold.

Proctor, N. H., J. P. Hughes. 1991. *Chemical Hazards of the Workplace.* New York: Van Nostrand Reinhold.

Quigley, D. R. 1994. *Handbook of Emergency Chemical Management.* New York: CRC Press/Lewis Publishers.

Rea, W. J. 1995. *Chemical Sensitivity,* Vol. I-III. New York: CRC Press/Lewis Publishers.

Sacarello, H. L. A. 1994. *The Comprehensive Handbook of Hazardous Materials.* New York: CRC Press/Lewis Publishers.

Sherman, J. D. 1988. *Chemical Exposure and Disease.* New York: Van Nostrand Reinhold.

Talbott, E., G. Craun. 1995. *An Introduction to Environmental Epidemiology.* New York: CRC Press/Lewis Publishers.

U.S. Department of Health and Human Services (DHHS), National Institute for Occupational Health (NIOHS). 1994. *NIOSH Pocket Guide to Hazardous Chemicals.* Publication Number 94-116. Washington, D.C.; U.S. Government Printing Office.

Witorsch, P. and S. V. Spagnolo. 1994. *Air Pollution and Lung Disease in Adults.* New York: CRC Press/Lewis Publishers.

MATERIAL SAFETY DATA SHEET

CHEMICAL NAME

CHLOROMETHYL METHYL ETHER

CRC PRESS/LEWIS PUBLISHERS DISCLAIMER: THE INFORMATION AND RECOMMENDATIONS PRESENTED HEREIN ARE BASED ON SOURCES BELIEVED TO BE RELIABLE. CRC MAKES NO REPRESENTATION ON ITS COMPLETENESS OR ACCURACY. IT IS THE RESPONSIBILITY OF THE USER TO DETERMINE THE CHEMICAL'S SUITABILITY FOR ITS INTENDED USE, THE CHEMICAL'S SAFE USE, AND THE CHEMICAL'S PROPER DISPOSAL. NO REPRESENTATIONS AND/OR WARRANTIES, EITHER EXPRESSED OR IMPLIED, OF THE MERCHANTABILITY OR FITNESS FOR A PARTICULAR PURPOSE, OR OF ANY OTHER NATURE, ARE MADE WITH RESPECT TO THE INFORMATION PROVIDED IN THIS MSDS OR TO THE CHEMICAL TO WHICH INFORMATION MAY REFER. CRC NEITHER ASSUMES NOR AUTHORIZES ANY OTHER PERSON TO ASSUME FOR IT, ANY OTHER ADDITIONAL RESPONSIBILITY OR LIABILITY FOR THE USE OF, OR RELIANCE UPON, THIS INFORMATION.

HAZARD WARNING INFORMATION

HEALTH	FIRE	REACTIVE	OTHER	DEGREE OF HAZARD	COLOR CODING	OTHER CODES
3	3	3	~~W~~	0 = Minimum Hazard 1 = Slight Hazard 2 = Moderate Hazard 3 = Serious Hazard 4 = Severe Hazard	HEALTH = BLUE FIRE = RED REACTIVITY = YELLOW OTHER = WHITE	OX = Oxidizer ACID = Acid ALK = Alkali COR = Corrosive ~~W~~ = Use No Water

SECTION I - GENERAL INFORMATION

Characterization	**Haloether**	RCRA Number	**U046**
		EPA Class	**Toxic Waste**
DOT Proper Shipping Name	**Methyl Chloromethyl Ether**	Chemical Abstract Service (CAS) Number	**107-30-2**
DOT Hazard Class and Label Requirements	**Flammable Liquid**	DOT Emergency Guide Code	**57**
DOT Identification Number	**UN 1239**	Chemical Formula	**CH_3OCH_2Cl**

Synonyms

Chlordimethyl ether, chloromethoxymethane, CMME, dimethylchloroether, methylchloromethyl ether.

SECTION II - HAZARDOUS INGREDIENTS AND IDENTITY INFORMATION

Hazardous Components (specific identity)	OSHA Exposure Criteria	NIOSH Exposure Criteria	Immediately Dangerous to Life and Health (IDLH)	ACGIH Exposure Criteria
Chloromethyl methyl ether **1 ppm = 3.35 mg/m^3**	**PEL:** Suspected Cancer Agent **STEL:** Not Applicable	**REL:** Possible Cancer Agent **STEL:** Not Applicable	**Carcinogen** **Not Determined**	**TLV:** Suspected Human Carcinogen **STEL:** Not Applicable

SECTION III - PHYSICAL AND CHEMICAL CHARACTERISTICS

Boiling Point	**135°F (57°C)**	Specific Gravity (H_2O = 1)	**1.06**
Vapor Pressure (mm Hg)	**192 at 69°F (20°C)**	Molecular Weight	**80.5**
Vapor Density (Air = 1)	**2.77**	Freezing Point	**-154°F**

Solubility

Reacts on contact with water to form hydrochloric acid and formaldehyde. Soluble in alcohol and ether.

Appearance and Odor

Colorless corrosive liquid with an ether-like odor.

SECTION IV - FIRE AND EXPLOSION HAZARD DATA

Flash Point (method used)	**60°F (15°C) closed cup**	Explosive Limits in Air % by Volume	**LEL: Not Determined** **UEL: Not Determined**
NFPA Classification	**Class 1B Flammable Liquid**	Autoignition Temperature	**Not Determined**

Extinguishing Media

Use carbon dioxide, dry chemical, foam, water-spray, or fog.

Special Fire Fighting Procedures

Wear full protective clothing and self-contained breathing apparatus (SCBA). Use a water spray to keep fire-exposed containers cool. Poisonous gases are produced in fire, including hydrogen chloride and phosgene gas. Water alone may be ineffective on fire.

Unusual Fire and Explosion Hazards

Containers may explode in fire. Move containers from fire area if it can be done without risk. Vapors are heavier than air and may travel for some distance to an ignition source to flash-back and cause fire.

SECTION V - REACTIVITY DATA

Stability		Conditions to Avoid **Keep away from incompatible materials, heat, flame, and water since violent or explosive reactions can occur or hazardous/toxic by-products may be produced.**
Stable **X**	Unstable	Incompatibility *(materials to avoid)* **Reacts with water to form hydrochloric acid and formaldehyde. Reaction with divalent metals forms a very reactive product.**
Hazardous Polymerization		Conditions to Avoid **Under normal conditions of temperature and pressure, hazardous polymerization of chloroform is not known to occur.**
May Occur	Will Not Occur **X**	Hazardous Decomposition or By-products **When heated to decomposition, chloromethyl methyl ether can emit highly toxic/poisonous fumes including hydrogen chloride (hydrochloric acid and phosgene gas.**

SECTION VI - HEALTH HAZARD DATA

Primary Route(s) of Entry:	Inhalation?	Absorption (skin)?	Ingestion?
	X	X	X

Health Hazards

INHALATION: **Irritation of the respiratory system, especially the lungs. Can cause buildup of fluid in the lungs (pulmonary edema) which can be fatal. Chronic bronchitis may develop.**

ABSORPTION: **Skin contact can result in irritation and rash. Can cause eye irritation with inflammation of the conjunctiva. Will pass through unbroken skin and cause toxic systemic effects.**

INGESTION: **Moderately toxic by ingestion. A human poison. May cause vomiting, nausea and other unspecified gastrointestinal effects.**

Carcinogenicity	NTP Listed?	IARC Cancer Review Group?	OSHA Regulated?	Target Organs?
Confirmed Human Confirmed Animal	**5th Annual Report**	**Group 1**	**29 CFR 1910.1006**	**Eyes, skin, respiratory system**

Medical Conditions Generally Aggravated by Exposure

None Reported. However, existing skin conditions or respiratory problems may be aggravated.

Emergency and First-aid Procedures

Eye contact: Flush immediately with water for 15 minutes (minimum), seek medical attention. Skin contact: Remove all contaminated clothing. Immediately wash area with large amounts of soap and water. For inhalation: Remove the person from exposure. Provide respiratory assistance and CPR. Transfer to medical facility. Observe for 24 - 48 hours for lung effects (pulmonary edema). If swallowed: Do NOT induce vomiting. If victim is conscious, give large amounts of water. Seek medical attention immediately. Do not attempt to give a convulsing or unconscious person anything by mouth.

SECTION VII - PRECAUTIONS FOR SAFE HANDLING AND USE

Steps to be Taken in Case Material is Released or Spilled

Absorb liquids in vermiculite, or similar material and deposit in sealed containers. Ventilate area of spill or leak. Remove ignition sources. Restrict those not involved in cleanup from entering area.

Preferred Waste Disposal Method

May be burned in a chemical incinerator equipped with an afterburner and scrubber.

Precautions to be Taken in Handling and Storage

Chloromethyl methyl ether is a flammable liquid. It presents a serious fire hazard when exposed to heat. Store in tightly closed containers in cool, well-ventilated area. A regulated, marked work area should be established where this chemical is used, handled, or stored (per OSHA 29 CFR 1910.1006).

Other Precautions and Warnings

Prior to working with chemicals, personnel should be trained on its proper handling and storage.

SECTION VIII - CONTROL MEASURES AND PERSONAL PROTECTIVE EQUIPMENT

Respiratory Protection *(specify type)*

At any exposure level, use an MSHA/NIOSH approved supplied-air respirator with full facepiece, hood, or helmet in continuous flow mode or a self-contained breathing apparatus (SCBA) operated in positive pressure mode.

Ventilation

Local exhaust or general mechanical systems recommended.

Protective Gloves	Eye Protection	Other Protective Clothing
Impervious Rubber	**Chemical Goggles or Face Mask**	**Protective Suit or Clothing**

Work/Hygiene Practices

Always wash hands thoroughly after using chemical; never bring food, drink, or smoking materials into vicinity of chemicals.

CHLOROMETHYL METHYL ETHER

CH_3OCH_3Cl CAS: 107-30-2

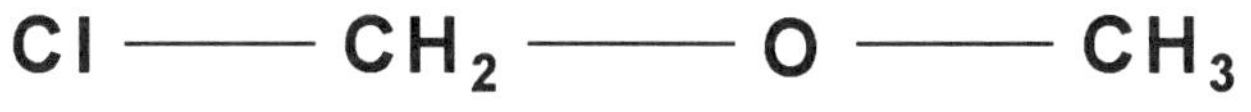

IDENTIFICATION AND TYPICAL USES

Chloromethyl methyl ether (CMME) is a colorless, corrosive liquid with an ether-like odor. It is one of several chloroalkyl ethers. It is used primarily as a methylating agent in the synthesis of chloromethylated compounds. It is also used in the manufacture of plastics and other chemicals (as an intermediate), in the treatment of textiles, to make polymers and insecticides, and in the preparation of ion-exchange resins.

RISK ASSESSMENT: HEALTH

General Assessment

CMME is a confirmed human carcinogen with human mutation data also reported. In humans, exposure can cause lung cancer. It is a human poison by *ingestion* and *inhalation*. It may also pass through unbroken skin (*absorption*) and enter the bloodstream. The liquid can irritate and possibly burn the skin if allowed to remain in contact. CMME is corrosive and forms hydrochloric acid in contact with water. Effects of skin exposure may therefore be more severe if there is excessive perspiration on the skin surface at the time of CMME contact.

The acute inhalation toxicity of CMME, like that of other lower aliphatic haloethers, is greater than the acute oral or dermal toxicity. The target organs are the eyes, skin, and respiratory system. The vapors will irritate the eyes and nose. High exposures can damage the lungs, causing fluid to build up (pulmonary edema), which is a medical emergency and can be fatal. Such reactions may be delayed up to 48 hours following exposure.

The toxic symptoms from inhalation of CMME vapors include irritation of the eyes, skin, mucous membranes, pulmonary edema, pulmonary congestion, pneumonia-like symptoms (chemical pneumonia), cough, wheeze, blood-stained sputum, and bronchial secretions.

Ingestion of CMME can produce digestive disturbances, vomiting, and other unspecified gastrointestinal disturbances. Little is known of the effects of repeated oral exposures at lower concentrations.

☠ *Acute Health Effects*

The following acute (short-term) health effects may occur immediately or shortly after exposure to CMME:

Skin: Irritation and possible rash or burning feeling on contact.

Eye: Severe irritation; may cause burns and permanent damage.

Lung: Irritation of the nose, throat, and lungs causing coughing and shortness of breath. Higher acute exposures can cause a buildup of fluid in the lungs (pulmonary edema) leading to death.

Chronic Health Effects

The following chronic (long-term) health effects may occur at some time after exposure to CMME and can last for months or even years:

Cancer Hazards: CMME is a confirmed carcinogen in humans. There is sufficient evidence to show that it causes lung cancer in humans. It has been shown to cause cancer of the skin and lungs in test animals. It may cause mutations in living cells.

Reproductive Hazard: According to information available in the references, CMME has not been adequately tested for its ability to adversely affect reproduction.

Other Chronic Effects: CMME can irritate the lungs. Repeated, low-level exposures may cause bronchitis to develop with symptoms of cough, phlegm, and/or shortness of breath.

Recommended Risk-Reduction Measures

Personnel should avoid direct contact with CMME. It is a confirmed human carcinogen and, therefore, there may be no safe exposure level. If a less toxic chemical cannot be substituted for a hazardous substance, then *engineering controls* are the most effective method of reducing exposures.

The best protection is to enclose operations and/or provide local exhaust ventilation at the site of chemical release. While not always operationally feasible, isolating operations can also reduce exposure. Using respiratory protection is less effective than the controls mentioned above, but is still advisable whenever working with or around CMME. For any exposure level, a supplied-air respirator with a full facepiece operated in positive pressure mode or an MSHA/NIOSH approved self-contained breathing apparatus (SCBA) with full facepiece and pressure demand are the recommended respiratory protection methods of choice. If a full facepiece is not available, then chemical splash goggles should be worn to protect the eyes. Whenever a chemical splash hazard exists, a face shield and a rubber apron should be worn. Because CMME reacts on contact with water to form corrosive hydrochloric acid, eye contact can result in serious and perhaps permanent damage. To prevent hand and skin exposures, acid resistant gloves should be worn.

Administrative controls should also be in place to minimize the potential for human exposures. These may include written procedures or policies which specify the methods and techniques that will be practiced whenever personnel are to work with CMME. A regulated controlled work area should be established where CMME is used, handled, or stored (per OSHA 29 CFR 1910.1006).

All personnel should receive *training* on the use, hazards, protective measures, emergency actions, and other precautions per 29 CFR 1910.1200 (Hazard Communication), prior to the first assignment in an area where CMME is used or stored.

Before beginning employment and at regular intervals thereafter (e.g., annually), for those with frequent or potentially high exposures (half the TLV or greater, or significant skin contact), the following medical tests are recommended:

☑ Lung function tests.

If symptoms develop or overexposure is suspected, the following may also be useful:

☑ Consider chest X-ray after acute overexposure.

Any evaluation should include a careful history of past and present symptoms with a medical examination. Medical tests that look for damage already done are not a substitute for controlling exposure. Also, since smoking can cause heart disease, as well as lung cancer, emphysema, and other related respiratory problems, the effects of exposure to CMME may be more pronounced in smokers than in non-smokers.

Other methods to reduce exposure include:

☑ Where possible, enclose operations and use local exhaust ventilation at the site of chemical release. If local exhaust ventilation or enclosure is not used, respiratory protection should be mandatory.

☑ Always ensure that proper protective clothing is worn when using chemical substances.

☑ Wash thoroughly immediately after exposure to CMME and at the end of the work shift or before eating, drinking, or smoking.

☑ Hazard warning information should be posted in the work area. In addition, as part of an on-going education and training program, all information on the health and safety hazards of CMME should be communicated to all potentially exposed workers.

☑ Additional, more detailed administrative and engineering control requirements are specified in OSHA 29 CFR 1910.1006 (Methyl Chloromethyl Ether).

RISK ASSESSMENT: ENVIRONMENT

General Assessment

The environment is at risk of exposure during transportation, storage, disposal, or destruction of CMME. In almost every scenario, the threat of environmental exposure is contingent upon the proper handling of the chemical substance. Accidental spills, large or small, can result in fire, explosion, and possible contamination of the surrounding environmental mediums (water, soil, and air).

CMME is considered a class 1B flammable liquid (per OSHA 29 CFR 1910.106). Its low flash point and relatively low boiling point make it a serious fire hazard. It will also react on contact with water to form hydrochloric acid and formaldehyde. In a fire, CMME will decompose and emit toxic and poisonous gases, including hydrogen chloride (hydrochloric acid) and phosgene. These characteristics require special consideration during any emergency situation involving a leak or spill of CMME. CMME can enter the environment through industrial effluents or spills.

Acute Ecological Effects

Acute (short-term) toxic effects may include the death of animals, birds, or fish and death or low growth rate in plants. Acute effects are seen 2 to 4 days after animals or plants are exposed to CMME.

Insufficient data are available to evaluate the short-term effects of CMME exposure to aquatic life, plants, birds or land animals.

Chronic Ecological Effects

Chronic (long-term) toxic effects may include shortened life span, reproductive problems, lower fertility, and changes in appearance or behavior in exposed animals. These effects can be seen long after first exposure(s) to toxic chemicals.

Insufficient data are available to evaluate the long-term effects of CMME exposure to aquatic life, plants, birds or land animals.

Water Solubility

CMME is highly soluble in water. Concentrations of 1000 milligrams and more will mix with a liter of water.

Persistence in the Environment

CMME is slightly persistent in the aquatic environment, with a half-life between 2 and 20 days. The half-life of a pollutant is the amount of time it takes for one half of the chemical to be degraded.

Approximately 91% of CMME will eventually end up in air; the remainder will end up in water.

Bioaccumulation in Aquatic Organisms

Some substances increase in concentration, or *bioaccumulate,* in living organisms as they breathe contaminated air, drink contaminated water, or eat contaminated food. These chemicals can become concentrated in the tissues and internal organs of animals as well as humans.

The concentration of CMME found in fish tissues is expected to be somewhat higher than the average concentration of CMME in the water from which the fish was taken.

Recommended Risk-Reduction Measures

Proper training of all transporters will reduce the likelihood of a mishap or accident resulting in a leak or spill to the environment. The correct use of DOT labeling (per the MSDS) on all containers, trucks, and rail cars will enable emergency responders to react properly and quickly to any disaster thereby reducing the potential risk to the environment and to personnel.

Storage of CMME should be segregated from incompatible chemicals to minimize the risk of cross-contamination or contact. To dispose of CMME, it can be burned in a chemical incinerator equipped with an afterburner and scrubber.

If a spill or leak to the environment has occurred, fire department, emergency response and/or hazardous materials spill personnel should be notified immediately. To neutralize a spill, flood area with copious amounts of water and rinse with sodium bicarbonate or lime solution. Cleanup should be attempted only by those trained in proper spill containment procedures. Contaminated soils should be removed for incineration and replaced with clean soil. If CMME should contact the water table, aquifer, or navigable waterway, time is of the essence. It is highly soluble in water and, therefore, total containment and remediation may not be entirely possible. When such spills occur, the local and/or state emergency response authorities must be notified. A comprehensive emergency response or disaster preparedness/recovery plan should be in place prior to any operations involving the use, transportation, storage, or disposal of CMME. If CMME is spilled or leaked, the following specific steps are recommended:

- ☑ Restrict persons not wearing protective clothing from area of spill or leak until cleanup is complete. Evacuate non-essential personnel a minimum of 1500 feet from the spill/hazard area. Wind direction must be considered for all evacuations (down-wind is preferable).
- ☑ Collect liquid for reclamation or absorb in vermiculite, dry sand, earth, or similar material and deposit in sealed containers.
- ☑ It may be necessary to dispose of CMME as a hazardous waste. The responsible state agency or the regional office of the federal Environmental Protection Agency (EPA) should be contacted for specific recommendations.

RISK ASSESSMENT: BUSINESS

General Assessment

Accidents or mishaps involving CMME can present a significant threat to business operations. The loss or damage of equipment or facilities can significantly affect fiscal viability. Lawsuits that may result from personnel injury/death, public exposures, and/or environmental contamination will also require a serious expenditure of resources. Media attention surrounding an injury, death, or environmental damage can also result in a loss of profits and loss of current as well as future business. Always remember that anytime the words "carcinogen" or "cancer" are used, public awareness and emotions can run high leading to irrational decisions and over-reaction to the situation.

Recommended Risk-Reduction Measures

Company attorneys, safety and health professionals, and environmental specialists should be involved in the development of any procedures dealing with safety or emergency response. A company official should be pre-designated as a public relations officer with specific training in dealing with the press. Corporate plans and policies should be developed, approved, and implemented long before any need for such arises.

REFERENCES

Ahlbom, A. 1993. *Biostatistics for Epidemiologists.* New York: CRC Press/Lewis Publishers.

American Conference of Governmental Industrial Hygienists. 1988. *Documentation of the Threshold Limit Values and Biological Exposure Indices,* 5th Edition (with updates). Cincinnati: ACGIH

Asante-Duah, D. K. 1993. *Hazardous Waste Risk Assessment.* New York: CRC Press/Lewis Publishers.

Bloom, A. D. and F. J. DeSerres. 1995. *Ecotoxicity and Human Health.* New York: CRC Press/Lewis Publishers.

Brusick, D. J. 1994. *Methods for Genetic Risk Assessment.* New York: CRC Press/Lewis Publishers.

Bueche, J. R. 1972. *Principles of Physics,* 2nd Edition. New York: McGraw-Hill.

Calabrese, E. J., E. M. Kenyon. *Air Toxics and Risk Assessment.* New York: CRC Press/Lewis Publishers.

Calabrese, E. J. 1994. *Biological Effects of Low Level Exposures.* New York: CRC Press/Lewis Publishers.

Cockerman, L. G., B. S. Shane. 1994. *Basic Environmental Toxicology.* New York: CRC Press/Lewis Publishers.

Calabrese, E. J., E. M. Kenyon. *Air Toxics and Risk Assessment.* New York: CRC Press/Lewis Publishers.

Cothern, R. R. 1993. *Comparative Environmental Risk Assessment.* New York: CRC Press/Lewis Publishers.

Gardner, R. W. 1994. *Chemical Intolerance.* New York: CRC Press/Lewis Publishers.

Hallenbeck, W. H. 1993. *Quantitative Risk Assessment for Environmental and Occupational Health,* 2nd Edition. New York: CRC Press/Lewis Publishers.

Howard, P. H., et. al. 1991. *Handbook of Environmental Degradation Rates.* New York: CRC Press/Lewis Publishers.

Howard, P. H. 1995. *Handbook of Environmental Fate and Exposures Data for Organic Chemicals,* Vol. I-IV. New York: CRC Press/Lewis Publishers.

Jones, F. E. 1994. *Toxic Organic Vapors in the Workplace.* New York: CRC Press/Lewis Publishers.

Keith, L. H. and M. M. Walker. 1995. *Handbook of Air Toxics.* New York: CRC Press/Lewis Publishers.

Langerman, N. 1994 *Precautionary Labels for Chemical Containers.* New York: CRC Press/Lewis Publishers.

Lauwerys, R. R., P. Hoet. 1993. *Industrial Chemical Exposure.* New York: CRC Press/Lewis Publishers.

Lewis, R. J., Sr. 1992. *Sax's Dangerous Properties of Industrial Materials,* Volumes 1, 2, and 3, 8th Edition. New York: Van Nostrand Reinhold.

Manahan, S. E. 1992. *Toxicological Chemistry,* 2nd Edition. New York: CRC Press/Lewis Publishers.

Mellström, G. A., J. E. Wahlberg, and H. I. Maibach. 1994. *Protective Gloves for Occupational Use.* New York: CRC Press/Lewis Publishers.

Meyer, E. 1990. *Chemistry of Hazardous Materials.* Englewood Cliffs, NJ: Prentice Hall.

Patnaik, P. 1992. *A Comprehensive Guide to the Hazardous Properties of Chemical Substances.* New York: Van Nostrand Reinhold.

Proctor, N. H., J. P. Hughes. 1991. *Chemical Hazards of the Workplace.* New York: Van Nostrand Reinhold.

Quigley, D. R. 1994. *Handbook of Emergency Chemical Management.* New York: CRC Press/Lewis Publishers.

Rea, W. J. 1995. *Chemical Sensitivity,* Vol. I-III. New York: CRC Press/Lewis Publishers.

Sacarello, H. L. A. 1994. *The Comprehensive Handbook of Hazardous Materials.* New York: CRC Press/Lewis Publishers.

Sherman, J. D. 1988. *Chemical Exposure and Disease.* New York: Van Nostrand Reinhold.

Talbott, E., G. Craun. 1995. *An Introduction to Environmental Epidemiology.* New York: CRC Press/Lewis Publishers.

U.S. Department of Health and Human Services (DHHS), National Institute for Occupational Health (NIOHS). 1994. *NIOSH Pocket Guide to Hazardous Chemicals.* Publication Number 94-116. Washington, D.C.; U.S. Government Printing Office.

Witorsch, P. and S. V. Spagnolo. 1994. *Air Pollution and Lung Disease in Adults.* New York: CRC Press/Lewis Publishers.

MATERIAL SAFETY DATA SHEET

CHEMICAL NAME	
CHLOROPRENE	**CRC PRESS/LEWIS PUBLISHERS DISCLAIMER:** THE INFORMATION AND RECOMMENDATIONS PRESENTED HEREIN ARE BASED ON SOURCES BELIEVED TO BE RELIABLE. CRC MAKES NO REPRESENTATION ON ITS COMPLETENESS OR ACCURACY. IT IS THE RESPONSIBILITY OF THE USER TO DETERMINE THE CHEMICAL'S SUITABILITY FOR ITS INTENDED USE, THE CHEMICAL'S SAFE USE, AND THE CHEMICAL'S PROPER DISPOSAL. NO REPRESENTATIONS AND/OR WARRANTIES, EITHER EXPRESSED OR IMPLIED, OF THE MERCHANTABILITY OR FITNESS FOR A PARTICULAR PURPOSE, OR OF ANY OTHER NATURE, ARE MADE WITH RESPECT TO THE INFORMATION PROVIDED IN THIS MSDS OR TO THE CHEMICAL TO WHICH INFORMATION MAY REFER. CRC NEITHER ASSUMES NOR AUTHORIZES ANY OTHER PERSON TO ASSUME FOR IT, ANY OTHER ADDITIONAL RESPONSIBILITY OR LIABILITY FOR THE USE OF, OR RELIANCE UPON, THIS INFORMATION.

HAZARD WARNING INFORMATION

HEALTH	FIRE	REACTIVE	OTHER	DEGREE OF HAZARD	COLOR CODING	OTHER CODES
3	3	0		0 = Minimum Hazard 1 = Slight Hazard 2 = Moderate Hazard 3 = Serious Hazard 4 = Severe Hazard	HEALTH = BLUE FIRE = RED REACTIVITY = YELLOW OTHER = WHITE	OX = Oxidizer ACID = Acid ALK = Alkali COR = Corrosive W = Use No Water

SECTION I - GENERAL INFORMATION

Characterization	RCRA Number	EPA Class
Chlorinated hydrocarbon	**None**	**Not Applicable**

DOT Proper Shipping Name	Chemical Abstract Service (CAS) Number
Chloroprene	**126-99-8**
DOT Hazard Class and Label Requirements	DOT Emergency Guide Code
Flammable Liquid	**30**
DOT Identification Number	Chemical Formula
UN 1991	**$H_2C{:}CHCCl{:}CH_2$**

Synonyms

2-chlorobutadiene-1,3; β-chloroprene; 2-chloro-1-3-butadiene; chlorobutadiene.

SECTION II - HAZARDOUS INGREDIENTS AND IDENTITY INFORMATION

Hazardous Components (specific identity)	OSHA Exposure Criteria	NIOSH Exposure Criteria	Immediately Dangerous to Life and Health (IDLH)	ACGIH Exposure Criteria
Chloroprene (derivation: Addition of cold hydrochloric acid to vinylacetylene and the chlorination of butadiene). **1 ppm = 3.68 mg/m³**	**PEL (skin):** **25 ppm** **92 mg/m³** **Suspected Human Carcinogen**	**REL (ceiling):** **1 ppm** **3.68 mg/m³** **(15 minutes)** **Possible Cancer Agent**	**300 ppm**	**TLV:** **10 ppm** **37 mg/m³** **Suspected Cancer Agent**

SECTION III - PHYSICAL AND CHEMICAL CHARACTERISTICS

Boiling Point	Specific Gravity (H_2O = 1)
138°F (59°C)	**.96**
Vapor Pressure (gas)	Molecular Weight
188 at 69°F (20°C)	**88.5**
Vapor Density (Air = 1)	Melting Point
3.0	**-153°F (-103°C)**

Solubility

Only slightly soluble in water, soluble in alcohol and ether.

Appearance and Odor

Colorless liquid with a pungent, ether-like odor. Odor threshold = 1 ppm (approximate).

SECTION IV - FIRE AND EXPLOSION HAZARD DATA

Flash Point (method used)	Explosive Limits in Air % by Volume
-4°F (-20°C) closed cup	**LEL: 4% UEL: 20%**
NFPA Classification	Autoignition Temperature
Class 1B Flammable Liquid	**Not Available**

Extinguishing Media

Use dry chemical, carbon dioxide, or alcohol foam. Water may be used to cool fire-exposed containers.

Special Fire Fighting Procedures

Chloroprene is a flammable liquid. Wear full protective clothing and self-contained breathing apparatus (SCBA). Poisonous gases are produced in fire. Water may be ineffective on fires.

Unusual Fire and Explosion Hazards

Dangerous fire and explosion hazard when exposed to heat or flame. Containers may explode in fire. It auto-oxidizes rapidly in air, even at 0°C, producing an unstable peroxide which catalyzes exothermic polymerization of the monomer.

SECTION V - REACTIVITY DATA

Stability		Conditions to Avoid Keep away from heat or flame. Exposure to air causes rapid auto-oxidation to produce unstable peroxides. Keep away from incompatible materials.
Stable	Unstable X	Incompatibility *(materials to avoid)* Peroxides and other oxidizers, such as permanganates, nitrates, chlorates, and perchlorates.
Hazardous Polymerization		Conditions to Avoid Polymerization is likely under normal conditions of temperature and pressure. Even at temperatures as low as 0°C, it will form "popcorn" polymers at a greater rate than butadiene.
May Occur X	Will Not Occur	Hazardous Decomposition or By-products When heated to decomposition, chloroprene can emit highly toxic/poisonous gases.

SECTION VI - HEALTH HAZARD DATA

Primary Route(s) of Entry:	Inhalation?	Absorption (skin)?	Ingestion?
	X	X	

Health Hazards

INHALATION: Fatigue, psychic changes, irritability, oppression in the chest, substernal pain, tachycardia on exertion. At high concentrations, may cause temporary unconsciousness and can lead to death. May cause nausea, giddiness, and possible temporary hair loss. CNS depression is likely and high exposures may result in lung, liver, and kidney damage.

SKIN & EYES: Eye contact with the liquid can cause conjunctivitis and focal cornea necrosis. Skin contact can cause chemical burns, irritation, and possible dermatitis.

Carcinogenicity	NTP Listed?	IARC Cancer Review Group?	OSHA Regulated?	Target Organs?
Suspected Human Confirmed Animal	5th Annual Report	No	29 CFR 1910.1000 Table Z-1	Respiratory system, eyes, skin, CNS, liver, kidney.

Medical Conditions Generally Aggravated by Exposure

None reported.

Emergency and First-aid Procedures

Eye contact: Flush immediately with water for 15 minutes (minimum), seek medical attention. Skin contact: Remove all contaminated clothing. Immediately wash area with large amounts of soap and water. Seek medical assistance. For inhalation: Remove the person from exposure. Provide respiratory assistance and CPR. Transfer to medical facility. Keep victim warm and at rest. If swallowed: If unconscious, do not give anything to drink. Seek medical attention immediately.

SECTION VII - PRECAUTIONS FOR SAFE HANDLING AND USE

Steps to be Taken in Case Material is Released or Spilled

Absorb liquids in vermiculite, dry sand, earth, or similar material and deposit in sealed containers. Ventilate area of spill or leak. Restrict those not involved in cleanup from entering area. Remove all sources of ignition.

Preferred Waste Disposal Method

No Citation.

Precautions to be Taken in Handling and Storage

Chloroprene is a flammable liquid. Store to avoid contact with incompatible materials since violent reactions can occur. Store in tightly closed containers in cool, well-ventilated area away from heat and flame. Ideal storage conditions in temperatures below 50°F (10°C).

Other Precautions and Warnings

Sources of ignition are prohibited where chloroprene is used, handled, or stored. Heat may cause containers to build up pressure and explode.

SECTION VIII - CONTROL MEASURES AND PERSONAL PROTECTIVE EQUIPMENT

Respiratory Protection *(specify type)*

For any exposures to a suspected carcinogen, use a supplied-air respirator with full facepiece, hood, or helmet in continuous flow mode, or use a self-contained breathing apparatus (SCBA) operated in pressure demand or other positive pressure mode.

Ventilation

Local exhaust or general mechanical systems recommended.

Protective Gloves	Eye Protection	Other Protective Clothing
Solvent-resistant Rubber	Chemical Goggles or Face Mask	Rubber Apron

Work/Hygiene Practices

Always wash hands thoroughly after using chemical; never bring food, drink, or smoking materials into vicinity of chemicals.

CHLOROPRENE

$H_2C{:}CHCCl{:}CH_2$ CAS: 126-99-8

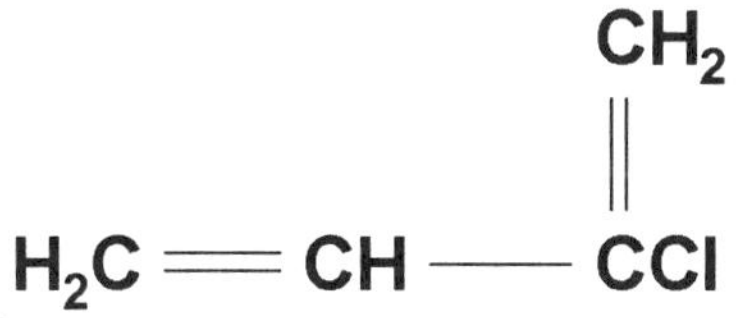

IDENTIFICATION AND TYPICAL USES

Chloroprene is a clear, colorless liquid with a pungent, ether-like odor. It is used in the manufacture of neoprene and other polymers.

RISK ASSESSMENT: HEALTH

General Assessment

Chloroprene is a highly toxic chemical by *inhalation,* and is considered moderately toxic by *ingestion* and skin *absorption.* Its carcinogenicity is suspect in humans, although studies have proven inconclusive in this regard. It has been confirmed that exposure to chloroprene can cause mutations in living cells and there have also been some studies to indicate that it may damage the developing fetus in test animals. A conservative approach to exposure risk management is recommended for chloroprene since the exact nature of its cancer risk is unresolved.

Inhalation of chloroprene vapors can act as a central nervous system (CNS) depressant. There will be symptoms of fatigue, changes in personality, nervousness, oppression of the chest, occasionally substernal pain, tachycardia on exertion, and reversible hair loss (after approximately one month after last exposure, hair begins to return). The primary effects of acute high-level exposures are CNS depression, lung, liver, and kidney damage. It is important to note that chloroprene vapors are very irritating such that personnel will not usually tolerate moderate or high exposures.

Liquid contact with the eyes may result in conjunctivitis, corneal necrosis, and edema of the eyelids. Skin contact may cause chemical burns, dermatitis, and temporary loss of hair. Chloroprene is readily absorbed through the skin to cause toxic systemic effects similar to those of inhalation.

Ingestion can produce gastrointestinal disorders along with headache, insomnia, and dizziness.

Acute Health Effects

The following acute (short-term) health effects may occur immediately or shortly after exposure to chloroprene:

Skin: Immediate irritation and burn on contact. May cause irritating rash, possible dermatitis, and reversible hair loss.

Eye: Severe irritation, conjunctivitis, and corneal necrosis. May cause burns and permanent damage.

Lung: Irritation of the respiratory tract. High-level exposures may cause CNS effects, lung, liver, and kidney damage. There may be oppression of the chest, chest pain, tachycardia upon exertion, nausea, headache, fatigue, giddiness, loss of consciousness, and death on high exposures. Exposure to vapors may also cause reversible hair loss.

CNS: A primary CNS depressant. May cause headache, dizziness, loss of consciousness, nausea, irritability, changes in personality, and nervousness.

Chronic Health Effects

The following chronic (long-term) health effects may occur at some time after exposure to chloroprene and can last for months or even years:

Cancer Hazards: Chloroprene may cause mutations (genetic changes) in living cells. Although there is insufficient and conflicting information available in the references to adequately evaluate chloroprene for its ability to cause cancer, there are some studies that suggest an increase in lung and skin cancer may occur in workers exposed to chloroprene.

Reproductive Hazard: According to information presented in the references, chloroprene may damage the developing fetus, cause spontaneous abortions, and interfere with sperm production.

Other Chronic Effects: Chronic exposure may cause headache, respiratory irritation, gastrointestinal disorders, dizziness, and insomnia. Repeated exposures caused CNS effects, lung edema, liver, and kidney damage in experimental animals. Chloroprene can also cause primary respiratory tract irritation, followed by respiratory depression and asphyxia.

Recommended Risk-Reduction Measures

Personnel should avoid direct contact with chloroprene or any other chemical with suspected or unconfirmed human carcinogenic characteristics. If a less toxic chemical cannot be substituted for a hazardous substance, then *engineering controls* are the most effective method of reducing exposures. The best protection is to enclose operations and/or provide local exhaust ventilation at the site of chemical release. While not always operationally feasible, isolating operations can also reduce exposure. Using respiratory protection is less effective than the controls mentioned above, but is still advisable whenever working with or around chloroprene. For any concentration, a self-contained breathing apparatus (SCBA) with full facepiece and pressure demand is the recommended respiratory protection of choice. If a full facepiece is not available, then chemical goggles should be worn to protect the eyes. Whenever a chemical splash hazard exists, a face shield and a rubber apron should be worn. To prevent hand and skin exposures, acid resistant gloves should be worn.

Administrative controls should also be in place to minimize the potential for human exposures. These may include written procedures or policies which specify the methods and techniques that will be practiced whenever personnel are to work with chloroprene.

All personnel should receive *training* on the use, hazards, protective measures, emergency actions, and other precautions per 29 CFR 1910.1200 (Hazard Communication), prior to the first assignment in an area where chloroprene is used or stored.

If symptoms develop or overexposure is suspected, the following tests are recommended:

- ☑ Lung function tests.
- ☑ Kidney and liver function tests.

Any evaluation should include a careful history of past and present symptoms with an examination. Medical tests that simply look for existing damage are not a substitute for controlling exposures.

Other methods to reduce exposure include:

- ☑ Where possible, enclose operations and use local exhaust ventilation at the site of chemical release. If local exhaust ventilation or enclosure is not used, respiratory protection should be mandatory.
- ☑ Always ensure that proper protective clothing is worn when using chemical substances.
- ☑ Wash thoroughly immediately after exposure to chloroprene and at the end of the work shift or before eating, drinking, or smoking.
- ☑ Hazard warning information should be posted in the work area. In addition, as part of an on-going education and training program, all information on the health and safety hazards of chloroprene should be communicated to all potentially exposed workers.
- ☑ Eye wash stations should be provided in the immediate work area for emergency use. If there is a possibility of skin exposure to chloroprene, emergency shower facilities should also be provided.
- ☑ Personnel should never wear contaminated clothing home (family members can be exposed). Clothing should be laundered by personnel who have been trained on the hazards of chloroprene.
- ☑ Specific engineering controls are recommended for chloroprene by NIOSH. Refer to NIOSH Criteria Document, Occupational Exposure to Chloroprene (Publication Number 77-210).

RISK ASSESSMENT: ENVIRONMENT

General Assessment

The environment is at risk of exposure during transportation, storage, disposal, or destruction of chloroprene. In almost every scenario, the threat of environmental exposure is contingent upon the proper handling of the chemical substance. Accidental spills, large or small, can result in fire, explosion, and possible contamination of the surrounding environmental mediums (water, soil, and air).

Chloroprene is considered a class 1B flammable liquid (per OSHA 29 CFR 1910.106). Its extremely low flash point and relatively low boiling point make this chemical a serious fire and explosion hazard concern. These characteristics require extreme caution in handling, storage, transportation, and disposal. It is incompatible with a number of peroxides and other oxidizers and contact can cause fire or explosion. Therefore, special consideration is required during any emergency situation involving a leak or spill of chloroprene. Should chloroprene ever come into contact with incompatible substances during use, transportation, storage, or disposal, the formation of highly toxic and/or highly explosive commodities (including hydrogen chloride) is extremely possible.

Chloroprene may enter the environment through industrial discharges or spills.

Acute Ecological Effects

Acute (short-term) toxic effects may include the death of animals, birds, or fish and death or low growth rate in plants. Acute effects are seen 2 to 4 days after animals or plants are exposed to chloroprene.

This chemical has high acute toxicity to aquatic life. No data are available on the short-term effects of chloroprene exposure to plants, birds, or land animals.

Chronic Ecological Effects

Chronic (long-term) toxic effects may include shortened life span, reproductive problems, lower fertility, and changes in appearance or behavior in exposed animals. These effects can be seen long after first exposure(s) to toxic chemicals.

Chloroprene has moderate chronic toxicity in aquatic life. No data are available on the long-term effects of chloroprene to plants, birds, or land animals.

Water Solubility

Chloroprene is highly soluble in water. Concentrations of 1000 milligrams and more will mix with a liter of water.

Persistence in the Environment

Chloroprene is non-persistent in water, with a half-life of less than 2 days. The half-life of a pollutant is the amount of time it takes for one half of the chemical to be degraded. About 99.25% of chloroprene will eventually end up in air; the remainder will end up in water.

Bioaccumulation in Aquatic Organisms

Some substances increase in concentration, or *bioaccumulate,* in living organisms as they breathe contaminated air, drink contaminated water, or eat contaminated food. These chemicals can become concentrated in the tissues and internal organs of animals as well as humans.

The concentration of chloroprene found in fish tissue is expected to be somewhat higher than the average concentration of chloroprene in the water from which the fish was taken.

Recommended Risk-Reduction Measures

Proper training of all transporters will reduce the likelihood of a mishap or accident resulting in a leak or spill to the environment. The correct use of DOT labeling (per the MSDS) on all containers, trucks, and rail cars will enable emergency responders to react properly and quickly to any disaster thereby reducing the potential risk to the environment and to personnel.

Storage of chloroprene should be segregated from incompatible chemicals to minimize the risk of cross-contamination or contact. Buildings designated for storage should be equipped with appropriate fire protection systems (alarms, sprinklers, emergency lighting, portable extinguishers).

If a spill or leak to the environment has occurred, fire department, emergency response and/or hazardous materials spill personnel should be notified immediately. Cleanup should be attempted only by those trained in proper spill containment procedures. Contaminated soils should be removed for incineration and replaced with clean soil. If chloroprene should contact the water table, aquifer, or navigable waterway, time is of the essence. It is highly soluble in water and total remediation and containment may not be possible. The local and/or state emergency response authorities must be notified. A comprehensive emergency response or disaster preparedness/recovery plan should be in place prior to any operations involving the use, transportation, storage, or disposal of chloroprene. If chloroprene is spilled or leaked, the following specific steps are recommended:

- ☑ Restrict persons not wearing protective clothing from area of spill or leak until cleanup is complete and area can be opened for normal work.
- ☑ Remove all ignition sources.
- ☑ Ventilate area of leak.
- ☑ Absorb liquids in vermiculite, dry sand, earth, or similar material and deposit in sealed containers.
- ☑ If gas is leaking, stop the flow of gas. If the source is a cylinder and leak cannot be stopped in

place, remove cylinder to a safe place in the open air and repair or allow to empty.

☑ It may be necessary to dispose of chloroprene as a hazardous waste. The responsible state agency or the regional office of the federal Environmental Protection Agency (EPA) should be contacted for specific recommendations.

RISK ASSESSMENT: BUSINESS

General Assessment

Accidents or mishaps involving chloroprene can present a significant threat to business operations. The loss or damage of equipment or facilities can significantly affect fiscal viability. Lawsuits that may result from personnel injury/death, public exposures, and/or environmental contamination will also require a serious expenditure of resources. Media attention surrounding an injury, death, or environmental damage can also result in a loss of profits and loss of current as well as future business.

Recommended Risk-Reduction Measures

Company attorneys, safety and health professionals, and environmental specialists should be involved in the development of any procedures dealing with safety or emergency response. A company official should be pre-designated as a public relations officer with specific training in dealing with the press. Corporate plans and policies should be developed, approved, and implemented long before any need for such arises.

REFERENCES

American Conference of Governmental Industrial Hygienists. 1988. *Documentation of the Threshold Limit Values and Biological Exposure Indices,* 5th Edition (with updates). Cincinnati: ACGIH

Bueche, J. R. 1972. *Principles of Physics,* 2nd Edition. New York: McGraw-Hill.

Calabrese, E. J. 1994. *Biological Effects of Low Level Exposures.* New York: CRC Press/Lewis Publishers.

Cockerman, L. G., B. S. Shane. 1994. *Basic Environmental Toxicology.* New York: CRC Press/Lewis Publishers.

Gardner, R. W. 1994. *Chemical Intolerance.* New York: CRC Press/Lewis Publishers.

Hallenbeck, W. H. 1993. *Quantitative Risk Assessment for Environmental and Occupational Health,* 2nd Edition. New York: CRC Press/Lewis Publishers.

Jones, F. E. 1994. *Toxic Organic Vapors in the Workplace.* New York: CRC Press/Lewis Publishers.

Keith, L. H. and M. M. Walker. 1995. *Handbook of Air Toxics.* New York: CRC Press/Lewis Publishers.

Landis, W. G., M. H. Yu. 1995. *Introduction to Environmental Toxicology.* New York: CRC Press/Lewis Publishers.

Langerman, N. 1994 *Precautionary Labels for Chemical Containers.* New York: CRC Press/Lewis Publishers.

Lauwerys, R. R., P. Hoet. 1993. *Industrial Chemical Exposure.* New York: CRC Press/Lewis Publishers.

Lewis, R. J., Sr. 1992. *Sax's Dangerous Properties of Industrial Materials,* Volumes 1, 2, and 3, 8th Edition. New York: Van Nostrand Reinhold.

Lide, D. R. 1995. *Handbook of Chemistry and Physics,* 76th Edition. New York: CRC Press/Lewis Publishers.

Manahan, S. E. 1994. *Environmental Chemistry,* 6th Edition. New York: CRC Press/Lewis Publishers.

Mellström, G. A., J. E. Wahlberg, and H. I. Maibach. 1994. *Protective Gloves for Occupational Use.* New York: CRC Press/Lewis Publishers.

Meyer, E. 1990. *Chemistry of Hazardous Materials.* Englewood Cliffs, NJ: Prentice Hall.

Patnaik, P. 1992. *A Comprehensive Guide to the Hazardous Properties of Chemical Substances.* New York: Van Nostrand Reinhold.

Proctor, N. H., J. P. Hughes. 1991. *Chemical Hazards of the Workplace.* New York: Van Nostrand Reinhold.

Quigley, D. R. 1994. *Handbook of Emergency Chemical Management.* New York: CRC Press/Lewis Publishers.

Rea, W. J. 1995. *Chemical Sensitivity,* Vol. I-III. New York: CRC Press/Lewis Publishers.

Richardson, M. 1995. *Dictionary of Substance and Their Effects,* Vol. I-VI. New York: CRC Press/Lewis Publishers.

Sacarello, H. L. A. 1994. *The Comprehensive Handbook of Hazardous Materials.* New York: CRC Press/Lewis Publishers.

Sheftel, V. O. 1995. *Handbook of Toxic Properties of Monomers and Additives.* New York: CRC Press/Lewis Publishers.

Torkelson, T. R. V. K. Rowe. 1981. Halogenated aliphatic hydrocarbons. *Patty's Industrial Hygiene and Toxicology,* Vol. 2B, *Toxicology,* pp 3568-3572. New York: Wiley-Interscience.

U.S. Department of Health, Education, and Welfare, National Institute for Occupational Safety and Health. 1976. *Criteria for a Recommended Standard: Occupational Exposure to Chloroprene.* DEHW (NIOSH) Publication Number: 77-210. Washington, D.C.: U.S. Government Printing Office.

U.S. Department of Health and Human Services, National Institute for Occupational Safety and Health. 1994. *NIOSH Pocket Guide to Chemical Hazards.* Washington, D.C.: U.S. Government Printing Office.

MATERIAL SAFETY DATA SHEET

CHEMICAL NAME

CHLOROSULFONIC ACID

CRC PRESS/LEWIS PUBLISHERS DISCLAIMER: THE INFORMATION AND RECOMMENDATIONS PRESENTED HEREIN ARE BASED ON SOURCES BELIEVED TO BE RELIABLE. CRC MAKES NO REPRESENTATION ON ITS COMPLETENESS OR ACCURACY. IT IS THE RESPONSIBILITY OF THE USER TO DETERMINE THE CHEMICAL'S SUITABILITY FOR ITS INTENDED USE, THE CHEMICAL'S SAFE USE, AND THE CHEMICAL'S PROPER DISPOSAL. NO REPRESENTATIONS AND/OR WARRANTIES, EITHER EXPRESSED OR IMPLIED, OF THE MERCHANTABILITY OR FITNESS FOR A PARTICULAR PURPOSE, OR OF ANY OTHER NATURE, ARE MADE WITH RESPECT TO THE INFORMATION PROVIDED IN THIS MSDS OR TO THE CHEMICAL TO WHICH INFORMATION MAY REFER. CRC NEITHER ASSUMES NOR AUTHORIZES ANY OTHER PERSON TO ASSUME FOR IT, ANY OTHER ADDITIONAL RESPONSIBILITY OR LIABILITY FOR THE USE OF, OR RELIANCE UPON, THIS INFORMATION.

HAZARD WARNING INFORMATION

HEALTH	FIRE	REACTIVE	OTHER	DEGREE OF HAZARD	COLOR CODING	OTHER CODES
3	0	2	~~W~~ COR	0 = Minimum Hazard 1 = Slight Hazard 2 = Moderate Hazard 3 = Serious Hazard 4 = Severe Hazard	HEALTH = BLUE FIRE = RED REACTIVITY = YELLOW OTHER = WHITE	OX = Oxidizer ACID = Acid ALK = Alkali COR = Corrosive ~~W~~ = Use No Water

SECTION I - GENERAL INFORMATION

Characterization **Acid**	RCRA Number: **None** — EPA Class: **Not Applicable**
DOT Proper Shipping Name **Chlorosulfonic Acid**	Chemical Abstract Service (CAS) Number **7790-94-5**
DOT Hazard Class and Label Requirements **Corrosive Material**	DOT Emergency Guide Code **39**
DOT Identification Number **UN 1754**	Chemical Formula $ClSO_3H$

Synonyms
Sulfuric chlorohydrin, monochlorosulfuric acid, monochloride sulfonic acid.

SECTION II - HAZARDOUS INGREDIENTS AND IDENTITY INFORMATION

Hazardous Components (specific identity)	OSHA Exposure Criteria	NIOSH Exposure Criteria	Immediately Dangerous to Life and Health (IDLH)	ACGIH Exposure Criteria
Chlorosulfonic acid (derivation: By treating sulfur trioxide or fuming sulfuric acid with hydrochloric acid).	**PEL:** Not Established **STEL:** Not Established	**REL:** Not Established **STEL:** Not Established	**Not Determined**	**TLV:** Not Established **STEL:** Not Established

SECTION III - PHYSICAL AND CHEMICAL CHARACTERISTICS

Boiling Point **304°F (151°C)**	Specific Gravity (H_2O = 1) **1.8**
Vapor Pressure (mm Hg) **1 at 69°F (20°C)**	Molecular Weight **116.5**
Vapor Density (Air = 1) **4.0**	Melting Point **-112°F (-80°C)**

Solubility
Decomposes in water to form sulfuric and hydrochloric acids. Also decomposed by alcohol and other acids.

Appearance and Odor
Colorless to light yellow, fuming, slightly cloudy liquid with a pungent odor.

SECTION IV - FIRE AND EXPLOSION HAZARD DATA

Flash Point (method used) **Not Applicable**	Explosive Limits in Air % by Volume **LEL: Not Applicable UEL: Not Applicable**
NFPA Classification **No Citation**	Autoignition Temperature **Not Determined**

Extinguishing Media
Avoid water. Use dry chemicals or carbon dioxide to fight fire.

Special Fire Fighting Procedures
Wear full protective clothing and self-contained breathing apparatus (SCBA). Use a water spray to keep fire-exposed containers cool. DO NOT get water inside containers. Poisonous gases are produced in fire, including chloride fumes.

Unusual Fire and Explosion Hazards
Containers may explode in fire. Move containers from fire area if it can be done without risk. Vapors are heavier than air and may travel for some distance to an ignition source to flash-back and cause fire.

SECTION V - REACTIVITY DATA

Stability		Conditions to Avoid Keep away from incompatible materials since violent and or explosive reactions can occur.
Stable	Unstable	Incompatibility *(materials to avoid)* Water, finely divided metals, alcohols, acids, organics, ordinary combustible materials, nitrates, amines, peroxides.
X		
Hazardous Polymerization		Conditions to Avoid Under normal conditions of temperature and pressure, hazardous polymerization of chlorosulfonic acid is not known to occur.
May Occur	Will Not Occur	Hazardous Decomposition or By-products When heated to decomposition, chlorosulfonic acid can emit highly toxic/poisonous gases of chloride and sulfur oxide.
	X	

SECTION VI - HEALTH HAZARD DATA

Primary Route(s) of Entry:	Inhalation?	Absorption (skin)?	Ingestion?
	X		X

Health Hazards

INHALATION: Irritation of the respiratory system, especially the lungs. Can cause buildup of fluid in the lungs (pulmonary edema) which can be fatal. May cause loss of consciousness.

SKIN & EYES: A severe irritant. Causes burns and possibly permanent damage to the skin and eyes. It may cause conjunctivitis, even in vapor exposures.

INGESTION: Moderately toxic by ingestion. A human poison. Serious irritation of the mouth, esophagus, and stomach.

Carcinogenicity	NTP Listed?	IARC Cancer Review Group?	OSHA Regulated?	Target Organs?
Unknown Human Unknown Animal	No	No	No	Eyes, skin, respiratory system

Medical Conditions Generally Aggravated by Exposure

None Reported. However, existing skin conditions or respiratory problems may be aggravated.

Emergency and First-aid Procedures

Eye contact: Flush immediately with water for 15 minutes (minimum), seek medical attention. Skin contact: Remove all contaminated clothing. Immediately wash area with large amounts of soap and water. For inhalation: Remove the person from exposure. Provide respiratory assistance and CPR. Transfer to medical facility. Observe for 24 - 48 hours for lung effects (pulmonary edema). If swallowed: If victim is conscious, DO NOT give any liquids unless under doctor orders. Seek medical attention immediately. Support breathing if necessary.

SECTION VII - PRECAUTIONS FOR SAFE HANDLING AND USE

Steps to be Taken in Case Material is Released or Spilled

Ventilate area of spill or leak. Neutralize using agricultural lime, soda ash, crushed limestone, or sodium bicarbonate and deposit in sealed containers. Keep those not involved in cleanup from area.

Preferred Waste Disposal Method

No Citation.

Precautions to be Taken in Handling and Storage

Presents a serious explosion hazard when exposed to water, alcohol, or acids. Store in tightly closed containers in cool, well-ventilated area. Vent stored drums twice per month to control pressure of hydrogen produced by action of acid on metal of drum. Bulk storage is not recommended.

Other Precautions and Warnings

Prior to working with chlorosulfonic acid, personnel should be trained on proper handling and storage.

SECTION VIII - CONTROL MEASURES AND PERSONAL PROTECTIVE EQUIPMENT

Respiratory Protection *(specify type)*

At any exposure level, use an MSHA/NIOSH approved supplied-air respirator with full facepiece, hood, or helmet in continuous flow mode or a self-contained breathing apparatus (SCBA) operated in positive pressure mode.

Ventilation

Local exhaust or general mechanical systems recommended.

Protective Gloves	Eye Protection	Other Protective Clothing
Impervious Rubber	Chemical Goggles or Face Mask	Protective Suit or Clothing

Work/Hygiene Practices

Always wash hands thoroughly after using chemical; never bring food, drink, or smoking materials into vicinity of chemicals.

CHLOROSULFONIC ACID

$ClSO_3H$ CAS: 7790-94-5

IDENTIFICATION AND TYPICAL USES

Chlorosulfonic acid is a colorless to light yellow fuming, slightly cloudy liquid with a pungent odor. It is used in the manufacture of synthetic detergents, in pharmaceuticals, as a sulfonating agent for dyes, in pesticides and intermediates, in ion-exchange resins, in the manufacture of hydrogen chloride and smoke-producing chemicals.

RISK ASSESSMENT: HEALTH

General Assessment

Chlorosulfonic acid is a poisonous irritant. Primary exposures occur through *inhalation* and *ingestion*. It may also absorb through the skin if allowed to remain in contact, although its corrosive properties make prolonged contact extremely painful and absorption unlikely. As a corrosive chemical, it can cause severe burns to human tissue. Its vapors are extremely irritating to the eyes, lungs, and mucous membranes. It can cause acute toxic effects either in the liquid or vapor state.

Inhalation of its vapors can result in loss of consciousness with serious, life-threatening damage to lung tissue. Its irritating and corrosive properties can cause fluid to buildup in the lungs in a condition known as pulmonary edema, which is a medical emergency and can be fatal. Exposure may also result in chemical pneumonitis with symptoms of cough, congestion and even pain in chest. Other symptoms include headache, nausea, laryngitis, and lachrymation.

Because chlorosulfonic acid reacts with water, contact of the liquid with the eyes can cause severe burns and, if not immediately and completely removed, permanent damage to vision. It may cause conjunctivitis, even in its vapor form. It also causes severe skin burns due to its highly corrosive action and may cause contact dermatitis as well. Perspiration levels on skin surface may affect the rapidity of corrosive action of chlorosulfonic acid on skin tissue.

Ingestion of chlorosulfonic acid will cause severe irritation of the mouth, esophagus, and stomach.

Acute Health Effects

The following acute (short-term) health effects may occur immediately or shortly after exposure to chlorosulfonic acid:

Skin: Irritation and/or burning feeling on contact. May produce contact dermatitis.

Eye: Severe irritation; may cause burns and permanent damage.

Lung: Irritation of the nose, throat, and lungs causing coughing, congestion, and shortness of breath. Higher acute exposures can cause a buildup of fluid in the lungs (pulmonary edema) leading to death.

Chronic Health Effects

The following chronic (long-term) health effects may occur at some time after exposure to chlorosulfonic acid and can last for months or even years:

Cancer Hazards: According to information available in the references, chlorosulfonic acid has not been adequately tested for its ability to cause cancer in test animals.

Reproductive Hazard: According to information available in the references, chlorosulfonic acid has not been adequately tested for its ability to adversely affect reproduction.

Other Chronic Effects: Chlorosulfonic acid can irritate the lungs. Repeated, low-level exposures may cause bronchitis to develop with symptoms of cough, phlegm, and/or shortness of breath.

ʊ *Recommended Risk-Reduction Measures*

Personnel should avoid direct contact with chlorosulfonic acid. If a less toxic chemical cannot be substituted for a hazardous substance, then *engineering controls* are the most effective method of reducing exposures.

The best protection is to enclose operations and/or provide local exhaust ventilation at the site of chemical release. While not always operationally feasible, isolating operations can also reduce exposure. Using respiratory protection is less effective than the controls mentioned above, but is still advisable whenever working with or around chlorosulfonic acid. For any exposure level, a supplied-air respirator with a full facepiece operated in continuous flow or other positive pressure mode or an MSHA/NIOSH approved self-contained breathing apparatus (SCBA) with full facepiece and pressure demand are the recommended respiratory protection methods of choice. If a full facepiece is not available, then chemical splash goggles should be worn to protect the eyes. Whenever a chemical splash hazard exists, a face shield and a rubber apron should be worn. Because chlorosulfonic acid reacts on contact with water, eye contact can result in serious and perhaps permanent damage. To prevent hand and skin exposures, acid resistant gloves should be worn.

Administrative controls should also be in place to minimize the potential for human exposures. These may include written procedures or policies which specify the methods and techniques that will be practiced whenever personnel are to work with chlorosulfonic acid.

All personnel should receive *training* on the use, hazards, protective measures, emergency actions, and other precautions per 29 CFR 1910.1200 (Hazard Communication), prior to the first assignment in an area where chlorosulfonic acid is used or stored.

Before beginning employment and at regular intervals thereafter (e.g., annually), for those with frequent or potentially high exposures (half the TLV or greater, or significant skin contact), the following medical tests are recommended:

☑ Lung function tests.

If symptoms develop or overexposure is suspected, the following may also be useful:

☑ Consider chest X-ray after acute overexposure (may be negative if taken immediately after exposure due to delayed onset of pulmonary edema).

Any evaluation should include a careful history of past and present symptoms with a medical examination. Medical tests that look for damage already done are not a substitute for controlling exposure. Also, since smoking can cause heart disease, as well as lung cancer, emphysema, and other related respiratory problems, the effects of exposure to chlorosulfonic acid may be more pronounced in smokers than in non-smokers.

Other methods to reduce exposure include:

☑ Where possible, enclose operations and use local exhaust ventilation at the site of chemical release. If local exhaust ventilation or enclosure is not used, respiratory protection should be mandatory.

☑ Always ensure that proper protective clothing is worn when using chemical substances.

☑ Wash thoroughly immediately after exposure to chlorosulfonic acid and at the end of the work shift or before eating, drinking, or smoking.

☑ Hazard warning information should be posted in the work area. In addition, as part of an on-going education and training program, all information on the health and safety hazards of chlorosulfonic acid should be communicated to all potentially exposed workers.

RISK ASSESSMENT: ENVIRONMENT

General Assessment

The environment is at risk of exposure during transportation, storage, disposal, or destruction of chlorosulfonic acid. In almost every scenario, the threat of environmental exposure is contingent upon the proper handling of the chemical substance. Accidental spills, large or small, can result in fire, explosion, and possible contamination of the surrounding environmental mediums (water, soil, and air).

Chlorosulfonic acid is considered a non-flammable liquid. However, because it is incompatible with so many other commodities, including water, it presents serious fire and explosion concerns. It decomposes explosively on contact with water, acids, or alcohols. It will also react explosively with phosphorus and violently with silver nitrate. In a fire, chlorosulfonic acid will decompose and emit toxic and poisonous gases, including toxic gases of chloride and sulfur oxide. It is also incompatible with many common combustible materials (such as paper, wood, and oil). These characteristics require special consideration during any emergency situation involving a leak or spill of chlorosulfonic acid.

Chlorosulfonic acid can enter the environment through industrial effluents or spills.

Acute Ecological Effects

Acute (short-term) toxic effects may include the death of animals, birds, or fish and death or low growth rate in plants. Acute effects are seen 2 to 4 days after animals or plants are exposed to chlorosulfonic acid. Insufficient data are available to evaluate the short-term effects of chlorosulfonic acid exposure to aquatic life, plants, birds or land animals.

Chronic Ecological Effects

Chronic (long-term) toxic effects may include shortened life span, reproductive problems, lower fertility, and changes in appearance or behavior in exposed animals. These effects can be seen long after first exposure(s) to toxic chemicals. Insufficient data are available to evaluate the long-term effects of chlorosulfonic acid exposure to aquatic life, plants, birds or land animals.

Water Solubility

Chlorosulfonic acid decomposes in water and, therefore, will not remain in the aquifer as chlorosulfonic acid. However, it will form sulfuric and hydrochloric acids on contact with water which are both highly soluble. Concentrations of 1000 milligrams and more of hydrochloric or sulfuric acid will mix with a liter of water.

Persistence in the Environment

Chlorosulfonic acid is non-persistent (decomposes) in the aquatic environment, with a half-life of less than 2 days). The half-life of a pollutant is the amount of time it takes for one half of the chemical to be degraded.

Bioaccumulation in Aquatic Organisms

Some substances increase in concentration, or *bioaccumulate,* in living organisms as they breathe contaminated air, drink contaminated water, or eat contaminated food. These chemicals can become concentrated in the tissues and internal organs of animals as well as humans. Chlorosulfonic acid is not expected to accumulate in fish tissues.

Recommended Risk-Reduction Measures

Proper training of all transporters will reduce the likelihood of a mishap or accident resulting in a leak or spill to the environment. The correct use of DOT labeling (per the MSDS) on all containers, trucks, and rail cars will enable emergency responders to react properly and quickly to any disaster thereby reducing the potential risk to the environment and to personnel.

Storage of chlorosulfonic acid should be segregated from incompatible chemicals to minimize the risk of cross-contamination or contact. Storage drums should be vented twice monthly to control the pressure of hydrogen gas which is produced by action of chlorosulfonic acid on most metal storage containers.

If a spill or leak to the environment has occurred, fire department, emergency response and/or hazardous materials spill personnel should be notified immediately. To neutralize a spill, mix with agricultural lime (slaked lime), sodium bicarbonate, crushed limestone, soda ash, or lime. Cleanup should be attempted only by those trained in proper spill containment procedures. Contaminated soils should be removed for incineration and replaced with clean soil. If chlorosulfonic acid should contact the water table, aquifer, or navigable waterway, time is of the essence. It decomposes in water to form hydrochloric acid and sulfuric acid and, therefore, total containment and remediation may not be entirely possible. When such spills occur, the local and/or state emergency response authorities must be notified. A comprehensive emergency response or disaster preparedness/recovery plan should be in place prior to any operations involving the use, transportation, storage, or disposal of chlorosulfonic acid. If chlorosulfonic acid is spilled or leaked, the following specific steps are recommended:

- ☑ Restrict persons not wearing protective clothing from area of spill or leak until cleanup is complete and area can be opened for normal work.
- ☑ Neutralize liquids and deposit in sealed containers.
- ☑ It may be necessary to dispose of chlorosulfonic acid as a hazardous waste. The responsible state agency or the regional office of the federal Environmental Protection Agency (EPA) should be contacted for specific recommendations.

RISK ASSESSMENT: BUSINESS

General Assessment

Accidents or mishaps involving chlorosulfonic acid can present a significant threat to business operations. The loss or damage of equipment or facilities can significantly affect fiscal viability. Lawsuits that may result

from personnel injury/death, public exposures, and/or environmental contamination will also require a serious expenditure of resources. Media attention surrounding an injury, death, or environmental damage can also result in a loss of profits and loss of current as well as future business.

Recommended Risk-Reduction Measures

Company attorneys, safety and health professionals, and environmental specialists should be involved in the development of any procedures dealing with safety or emergency response. A company official should be pre-designated as a public relations officer with specific training in dealing with the press. Corporate plans and policies should be developed, approved, and implemented long before any need for such arises.

REFERENCES

Ahlbom, A. 1993. *Biostatistics for Epidemiologists.* New York: CRC Press/Lewis Publishers.

American Conference of Governmental Industrial Hygienists. 1988. *Documentation of the Threshold Limit Values and Biological Exposure Indices,* 5th Edition (with updates). Cincinnati: ACGIH

Asante-Duah, D. K. 1993. *Hazardous Waste Risk Assessment.* New York: CRC Press/Lewis Publishers.

Bloom, A. D. and F. J. DeSerres. 1995. *Ecotoxicity and Human Health.* New York: CRC Press/Lewis Publishers.

Brusick, D. J. 1994. *Methods for Genetic Risk Assessment.* New York: CRC Press/Lewis Publishers.

Bueche, J. R. 1972. *Principles of Physics,* 2nd Edition. New York: McGraw-Hill.

Calabrese, E. J., E. M. Kenyon. *Air Toxics and Risk Assessment.* New York: CRC Press/Lewis Publishers.

Calabrese, E. J. 1994. *Biological Effects of Low Level Exposures.* New York: CRC Press/Lewis Publishers.

Cockerman, L. G., B. S. Shane. 1994. *Basic Environmental Toxicology.* New York: CRC Press/Lewis Publishers.

Cothern, C. R., N. P. Ross. 1994. *Environmental Statistics, Assessment, and Forecasting.* New York: CRC Press/Lewis Publishers.

Cothern, R. R. 1993. *Comparative Environmental Risk Assessment.* New York: CRC Press/Lewis Publishers.

Gardner, R. W. 1994. *Chemical Intolerance.* New York: CRC Press/Lewis Publishers.

Hallenbeck, W. H. 1993. *Quantitative Risk Assessment for Environmental and Occupational Health,* 2nd Edition. New York: CRC Press/Lewis Publishers.

Howard, P. H., et. al. 1991. *Handbook of Environmental Degradation Rates.* New York: CRC Press/Lewis Publishers.

Howard, P. H. 1995. *Handbook of Environmental Fate and Exposures Data for Organic Chemicals,* Vol. I-IV. New York: CRC Press/Lewis Publishers.

Jones, F. E. 1994. *Toxic Organic Vapors in the Workplace.* New York: CRC Press/Lewis Publishers.

Keith, L. H. and M. M. Walker. 1995. *Handbook of Air Toxics.* New York: CRC Press/Lewis Publishers.

Langerman, N. 1994 *Precautionary Labels for Chemical Containers.* New York: CRC Press/Lewis Publishers.

Lauwerys, R. R., P. Hoet. 1993. *Industrial Chemical Exposure.* New York: CRC Press/Lewis Publishers.

Lewis, R. J., Sr. 1992. *Sax's Dangerous Properties of Industrial Materials,* Volumes 1, 2, and 3, 8th Edition. New York: Van Nostrand Reinhold.

Manahan, S. E. 1994. *Environmental Chemistry,* 6th Edition. New York: CRC Press/Lewis Publishers.

Manahan, S. E. 1992. *Toxicological Chemistry,* 2nd Edition. New York: CRC Press/Lewis Publishers.

Mellström, G. A., J. E. Wahlberg, and H. I. Maibach. 1994. *Protective Gloves for Occupational Use.* New York: CRC Press/Lewis Publishers.

Meyer, E. 1990. *Chemistry of Hazardous Materials.* Englewood Cliffs, NJ: Prentice Hall.

Patnaik, P. 1992. *A Comprehensive Guide to the Hazardous Properties of Chemical Substances.* New York: Van Nostrand Reinhold.

Proctor, N. H., J. P. Hughes. 1991. *Chemical Hazards of the Workplace.* New York: Van Nostrand Reinhold.

Quigley, D. R. 1994. *Handbook of Emergency Chemical Management.* New York: CRC Press/Lewis Publishers.

Rea, W. J. 1995. *Chemical Sensitivity,* Vol. I-III. New York: CRC Press/Lewis Publishers.

Sacarello, H. L. A. 1994. *The Comprehensive Handbook of Hazardous Materials.* New York: CRC Press/Lewis Publishers.

Sherman, J. D. 1988. *Chemical Exposure and Disease.* New York: Van Nostrand Reinhold.

Talbott, E., G. Craun. 1995. *An Introduction to Environmental Epidemiology.* New York: CRC Press/Lewis Publishers.

U.S. Department of Health and Human Services (DHHS), National Institute for Occupational Health (NIOHS). 1994. *NIOSH Pocket Guide to Hazardous Chemicals.* Publication Number 94-116. Washington, D.C.; U.S. Government Printing Office.

Witorsch, P. and S. V. Spagnolo. 1994. *Air Pollution and Lung Disease in Adults.* New York: CRC Press/Lewis Publishers.

MATERIAL SAFETY DATA SHEET

CHEMICAL NAME

CHROMIC ACETATE

CRC PRESS/LEWIS PUBLISHERS DISCLAIMER: THE INFORMATION AND RECOMMENDATIONS PRESENTED HEREIN ARE BASED ON SOURCES BELIEVED TO BE RELIABLE. CRC MAKES NO REPRESENTATION ON ITS COMPLETENESS OR ACCURACY. IT IS THE RESPONSIBILITY OF THE USER TO DETERMINE THE CHEMICAL'S SUITABILITY FOR ITS INTENDED USE, THE CHEMICAL'S SAFE USE, AND THE CHEMICAL'S PROPER DISPOSAL. NO REPRESENTATIONS AND/OR WARRANTIES, EITHER EXPRESSED OR IMPLIED, OF THE MERCHANTABILITY OR FITNESS FOR A PARTICULAR PURPOSE, OR OF ANY OTHER NATURE, ARE MADE WITH RESPECT TO THE INFORMATION PROVIDED IN THIS MSDS OR TO THE CHEMICAL TO WHICH INFORMATION MAY REFER. CRC NEITHER ASSUMES NOR AUTHORIZES ANY OTHER PERSON TO ASSUME FOR IT, ANY OTHER ADDITIONAL RESPONSIBILITY OR LIABILITY FOR THE USE OF, OR RELIANCE UPON, THIS INFORMATION.

HAZARD WARNING INFORMATION

HEALTH	FIRE	REACTIVE	OTHER	DEGREE OF HAZARD	COLOR CODING	OTHER CODES
3	0	0		0 = Minimum Hazard 1 = Slight Hazard 2 = Moderate Hazard 3 = Serious Hazard 4 = Severe Hazard	HEALTH = BLUE FIRE = RED REACTIVITY = YELLOW OTHER = WHITE	OX = Oxidizer ACID = Acid ALK = Alkali COR = Corrosive W = Use No Water

SECTION I - GENERAL INFORMATION

Characterization: **Metal Compound**

RCRA Number: **None**

EPA Class: **Not Applicable**

DOT Proper Shipping Name: **Chromic acetate Mixture (dry)**

Chemical Abstract Service (CAS) Number: **1066-30-4**

DOT Hazard Class and Label Requirements: **ORM-E: No Label**

DOT Emergency Guide Code: **31**

DOT Identification Number: **NA 9101**

Chemical Formula: **$Cr(C_2H_3O_2)_3 \bullet H_2O$**

Synonyms: **Chromium acetate; chromium triacetate; chromic acetate (III).**

SECTION II - HAZARDOUS INGREDIENTS AND IDENTITY INFORMATION

Hazardous Components (specific identity)	OSHA Exposure Criteria	NIOSH Exposure Criteria	Immediately Dangerous to Life and Health (IDLH)	ACGIH Exposure Criteria
Chromic acetate (derivation: Action of acetic acid on chromium hydroxide. The solution is evaporated and crystallized).	**PEL: 0.5 mg(Cr)/m^3 (maximum)** **Suspected Cancer Agent**	**REL (10 hour): .001 mg(Cr)/m^3 (carcinogen)** **STEL (15 min): .05 mg(Cr)/m^3**	**15 mg(Cr)/m^3 as Cr(VI)**	**TLV: (8 hour) .5 mg(Cr)/m^3** **Confirmed Human Carcinogen**

SECTION III - PHYSICAL AND CHEMICAL CHARACTERISTICS

Boiling Point: **Not Available**

Specific Gravity (H_2O = 1): **1.3**

Vapor Pressure (mm Hg): **<1 at 69°F (20°C)**

Molecular Weight: **229.15**

Vapor Density (Air = 1): **1.0**

Melting Point: **Not Available**

Solubility in Water: **Soluble in water. Insoluble in alcohol.**

Appearance and Odor: **Grayish-green powder, or a bluish-green pasty mass.**

SECTION IV - FIRE AND EXPLOSION HAZARD DATA

Flash Point (method used): **Not Applicable**

Explosive Limits in Air % by Volume: **LEL: Not Applicable UEL: Not Applicable**

NFPA Classification: **No Citation**

Autoignition Temperature: **Not Determined**

Extinguishing Media: **Use dry chemical, carbon dioxide, water spray, or foam to fight fire.**

Special Fire Fighting Procedures: **A carcinogen. Chromic acetate may burn, but does not readily ignite. Wear full protective clothing (structural fire-fighting clothing is permeable and may not provide adequate protection). Use a self-contained breathing apparatus (SCBA). Use a water spray to keep fire-exposed containers cool. Remain clear of smoke, water fall-out and water run-off. Poisonous gases are produced in fire.**

Unusual Fire and Explosion Hazards: **Containers may explode in fire. Cool exposed containers.**

SECTION V - REACTIVITY DATA

Stability		Conditions to Avoid Not known to be reactive under normal conditions of temperature and pressure.
Stable	Unstable	Incompatibility *(materials to avoid)*
X		No incompatibilities cited.
Hazardous Polymerization		Conditions to Avoid Under normal conditions of temperature and pressure, hazardous polymerization of chromic acetate is not known to occur.
May Occur	Will Not Occur	Hazardous Decomposition or By-products
	X	When heated to decomposition, chromic acetate can emit highly toxic/poisonous fumes and gases.

SECTION VI - HEALTH HAZARD DATA

Primary Route(s) of Entry:	Inhalation?	Absorption (skin)?	Ingestion?
	X		X

Health Hazards

INHALATION: Irritation of the respiratory system. Can cause sores or "chrome holes" to appear in the nasal cavity. Lung allergy, with wheezing and shortness of breath, sometimes occurs. Exposure may result in chemical pneumonitis and/or bronchial carcinoma.

SKIN & EYES: A severe irritant. Exposure may cause burns. May cause dermatitis.

INGESTION: Can produce ulceration in the mouth, throat, and stomach.

Carcinogenicity	NTP Listed?	IARC Cancer Review Group?	OSHA Regulated?	Target Organs?
Confirmed Human Confirmed Animal	5th Annual Report	Group 3	29 CFR 1910.1000 Table Z-1	Respiratory system, skin, eyes.

Medical Conditions Generally Aggravated by Exposure

None Reported. However, existing skin conditions or respiratory problems may be aggravated.

Emergency and First-aid Procedures

Eye contact: Flush immediately with water for 15 minutes (minimum), seek medical attention. Skin contact: Remove all contaminated clothing. Immediately wash area with large amounts of soap and water. For inhalation: Remove the person from exposure. Provide respiratory assistance and CPR. Transfer to medical facility. If swallowed: If victim is conscious, DO NOT induce vomiting. Seek medical attention immediately. Support breathing if necessary. Keep victim warm and at rest. If breathing stops, give artificial respiration. Never give anything by mouth to an unconscious person.

SECTION VII - PRECAUTIONS FOR SAFE HANDLING AND USE

Steps to be Taken in Case Material is Released or Spilled

Ventilate area of spill or leak. Neutralize with agricultural lime (slaked lime), sodium bicarbonate solution, or crushed limestone and flood with water. Collect powdered material in most convenient manner possible and deposit in sealed drum. Keep those not involved in cleanup from area.

Preferred Waste Disposal Method

No Citation.

Precautions to be Taken in Handling and Storage

Store in tightly closed containers in cool, well-ventilated area and protect from physical damage. A regulated, marked area should be established where chromic acetate is used, handled, or stored. Bulk storage of chromic acetate is not recommended.

Other Precautions and Warnings

Prior to working with chromic acetate, personnel should be trained on its proper handling and storage.

SECTION VIII - CONTROL MEASURES AND PERSONAL PROTECTIVE EQUIPMENT

Respiratory Protection *(specify type)*

Confirmed carcinogen. At any exposure level, use a supplied-air respirator with full facepiece operated in positive pressure or continuous flow mode or, use an MSHA/NIOSH approved self-contained breathing apparatus (SCBA) operated in positive pressure mode.

Ventilation

Local exhaust or general mechanical systems recommended.

Protective Gloves	Eye Protection	Other Protective Clothing
Butyl Rubber	Chemical/Dust Goggles or Face Mask	Protective Suit or Clothing

Work/Hygiene Practices

Always wash hands thoroughly after using chemical; never bring food, drink, or smoking materials into vicinity of chemicals.

CHROMIC ACETATE

$Cr(C_2H_3O_2) \bullet H_2O$ CAS: 1066-30-4

IDENTIFICATION AND TYPICAL USES

Chromic acetate is a grayish-green powder or a bluish-green pasty mass. It is used to fix certain textile dyes, to harden photographic emulsions, in tanning, and as a polymerization and oxidation catalyst.

RISK ASSESSMENT: HEALTH

General Assessment

Chromic acetate is a confirmed human carcinogen with experimental tumorigenic data reported. Human mutation data also reported. Certain chromium compounds, known as hexavalent chromium (or chromium VI) compounds and trivalent (or chromium III) compounds, have been determined to be human carcinogens. Chromic acetate is a trivalent compound. Therefore, in terms of exposure risk assessment and management, it is recommended that chromic acetate be handled with extreme caution.

Primary exposures occur through *inhalation*. However, *ingestion* of chromic acetate powders or dusts is also a considered possibility and can result in toxic effects. Skin contact may cause irritation, especially if repeated or prolonged. Also, skin allergy (dermatitis) sometimes occurs with itching, redness, and/or eczema-like rash. If this happens, then any future exposures will prompt a return of symptoms.

Inhalation of chromic acetate can result in irritation of the nose, throat, and bronchial tubes with symptoms of coughing and wheezing. Prolonged inhalation of chromium compounds, including chromic acetate, may result in sores or even a perforation in the bones of the nasal septum (a condition known as "chrome holes"), which can cause bleeding, discharge, or crusting. May also cause pneumonitis and/or bronchial carcinoma.

Ingestion can produce ulceration in the mouth, throat, and stomach. The can be abdominal pain and unspecified gastrointestinal effects.

Acute Health Effects

The following acute (short-term) health effects may occur immediately or shortly after exposure to chromic acetate:

Skin: Irritation and/or burning feeling on contact. May produce contact dermatitis and/or an eczema-like rash with redness and itching.

Eye: Severe irritation; possible conjunctivitis, and may cause burns and permanent damage.

Lung: Large (high-level) acute exposures may cause irritation of the nose with a potential to cause sores or holes in the nasal septum. Also may irritate the throat and lungs causing coughing and shortness of breath.

Chronic Health Effects

The following chronic (long-term) health effects may occur at some time after exposure to chromic acetate and can last for months or even years:

Cancer Hazards: Chromic acetate has been shown to cause lung and throat cancers in humans.

Reproductive Hazard: According to information available in the references, chromic acetate has not been adequately tested for its ability to adversely affect reproduction.

Other Chronic Effects: Chromic acetate may irritate the nasal septum and cause holes or perforations to appear in the bones separating the nasal cavity ("chrome holes"). Irritation, discharge, bleeding and/or formation of a crust in the inner nose may occur. Exposed persons may develop a skin allergy with a rash similar to eczema. Once this occurs, the person is "sensitized" to chromic acetate. Any future skin contact, even slight, can trigger a severe rash. Repeated exposures may cause irritation of the bronchial tubes (bronchitis) with symptoms of cough with phlegm and may lead to bronchial carcinoma and/or pneumonitis. Lung allergy, with wheezing and shortness of breath, sometimes develops. In such cases, even small future exposures can cause symptoms to reappear.

Recommended Risk-Reduction Measures

Personnel should avoid direct contact with chromic acetate and all other known human carcinogens. If a less toxic chemical cannot be substituted for a hazardous substance, then *engineering controls* are the most effective method of reducing exposures.

The best protection is to enclose operations and/or provide local exhaust ventilation at the site of chemical release. While not always operationally feasible, isolating operations can also reduce exposure. Using respiratory protection is less effective than the controls mentioned above, but is still advisable whenever working with or around chromic acetate. For any exposures to a confirmed carcinogen, use a supplied-air respirator with a full facepiece operated in positive pressure mode or an MSHA/NIOSH approved self-contained breathing apparatus (SCBA) with full facepiece and operated in pressure demand as the recommended respiratory protection methods of choice. If a full facepiece is not available, then chemical splash goggles should be worn to protect the eyes. Whenever a chemical dust hazard exists, a face shield and a rubber apron should be worn. To prevent hand and skin exposures, acid resistant gloves should be worn.

Administrative controls should also be in place to minimize the potential for human exposures. These may include written procedures or policies which specify the methods and techniques that will be practiced whenever personnel are to work with chromic acetate.

All personnel should receive *training* on the use, hazards, protective measures, emergency actions, and other precautions per 29 CFR 1910.1200 (Hazard Communication), prior to the first assignment in an area where chromic acetate is used or stored. Before beginning employment and at regular intervals thereafter (e.g., annually), for those with frequent or potentially high exposures (half the TLV or greater, or significant skin contact), the following tests are suggested:

- ☑ Lung function tests.
- ☑ Examination of the mouth and larynx.
- ☑ Examination of the nasal septum.

If symptoms develop or overexposure is suspected, the following may also be useful:

- ☑ Examination of the nasal septum.
- ☑ Evaluation by a qualified allergist, including careful exposure history and special testing (may help diagnose skin allergy).

Any evaluation should include a careful history of past and present symptoms with a medical examination. Medical tests that look for damage already done are not a substitute for controlling exposure. Also, since smoking can cause heart disease, as well as lung cancer, emphysema, and other related respiratory problems, the effects of exposure to chromic acetate may be more pronounced in smokers than in non-smokers.

Other methods to reduce exposure include:

- ☑ Where possible, enclose operations and use local exhaust ventilation at the site of chemical release. If local exhaust ventilation or enclosure is not used, respiratory protection should be mandatory.
- ☑ Always ensure that proper protective clothing is worn when using chemical substances.
- ☑ Wash thoroughly immediately after exposure to chromic acetate and at the end of the work shift or before eating, drinking, or smoking.
- ☑ Hazard warning information should be posted in the work area. In addition, as part of an on-going education and training program, all information on the health and safety hazards of chromic acetate should be communicated to all potentially exposed workers.
- ☑ Where possible, automatically transfer chromic acetate from drums or other storage containers to process containers.

RISK ASSESSMENT: ENVIRONMENT

General Assessment

The environment is at risk of exposure during transportation, storage, disposal, or destruction of chromic acetate. In almost every scenario, the threat of environmental exposure is contingent upon the proper handling of the chemical substance. Accidental spills, large or small, can result in fire, explosion, and possible contamination of the surrounding environmental mediums (water, soil, and air).

Chromic acetate may burn, but will not easily ignite. It has no listed incompatibilities with other commodities. However, in a fire, chromic acetate will decompose and emit toxic and poisonous gases. It may also intensify burning of many common combustible materials (such as paper, wood, plastics, and oil). These characteristics require some consideration during any emergency situation involving a leak or spill of chromic acetate.

Chromic acetate can enter the environment through effluents from the metal plating industries, from municipal waste treatment plant discharges, or from spills.

Acute Ecological Effects

Acute (short-term) toxic effects may include the death of animals, birds, or fish and death or low growth rate in plants. Acute effects are seen 2 to 4 days after animals or plants are exposed to chromic acetate.

Chromium is more toxic in soft water than in hard water. Chromium (III) compounds, such as chromic acetate, have moderate acute toxicity to aquatic life. Insufficient data are available to evaluate the short-term effects of chromic acetate exposure to plants, birds or land animals.

Chronic Ecological Effects

Chronic (long-term) toxic effects may include shortened life span, reproductive problems, lower fertility, and changes in appearance or behavior in exposed animals. These effects can be seen long after first exposure(s) to toxic chemicals.

Chromium (III) compounds have high chronic toxicity to aquatic life. Insufficient data are available to evaluate the long-term effects of chromic acetate exposure to plants, birds or land animals.

Water Solubility

Water solubility of chromium and its salts ranges from low to high. Chromic acetate is moderately soluble in water. Concentrations of 1 to 1000 milligrams will mix with a liter of water.

Persistence in the Environment

Chromic acetate is highly persistent in the aquatic environment, with a half-life of greater than 200 days. The half-life of a pollutant is the amount of time it takes for one half of the chemical to be degraded.

Bioaccumulation in Aquatic Organisms

Some substances increase in concentration, or *bioaccumulate,* in living organisms as they breathe contaminated air, drink contaminated water, or eat contaminated food. These chemicals can become concentrated in the tissues and internal organs of animals as well as humans.

The concentration of chromium found in fish is expected to be somewhat higher than the average concentration of chromium in the water from which the fish was taken.

Recommended Risk-Reduction Measures

Proper training of all transporters will reduce the likelihood of a mishap or accidents resulting in a leak or spill to the environment. The correct use of DOT labeling (per the MSDS) on all containers, trucks, and rail cars will enable emergency responders to react properly and quickly to any disaster thereby reducing the potential risk to the environment and to personnel.

Chromic acetate should be stored in cool, dry, well-ventilated areas. Bulk storage of chromic acetate, or any other carcinogenic materials, should be avoided.

If a spill or leak to the environment has occurred, fire department, emergency response and/or hazardous materials spill personnel should be notified immediately. To neutralize a spill, rinse with sodium bicarbonate solution, flood with water, agricultural lime (slaked lime) crushed limestone, soda ash, or lime and collect material in safest manner possible. Deposit in sealed drum for disposal. Cleanup should be attempted only by those trained in proper spill containment procedures. Do not dry sweep (creates airborne dusts). Use vacuum equipped with HEPA filter instead. Contaminated soils should be removed for incineration and replaced with clean soil. If chromic acetate should contact the water table, aquifer, or navigable waterway, time is of the essence. It is moderately soluble in water and, therefore, total containment and remediation may not be entirely possible. When such spills occur, the local and/or state emergency response authorities must be notified. A comprehensive emergency response or disaster preparedness/recovery plan should be in place prior to any operations involving the use, transportation, storage, or disposal of chromic acetate. If chromic acetate is spilled or leaked, the following specific steps are recommended:

- ☑ Restrict persons not wearing protective clothing from area of spill or leak until cleanup is complete and area can be opened for normal work.
- ☑ Ventilate area of spill.
- ☑ If possible, neutralize and deposit in sealed containers for proper disposal.

☑ It may be necessary to dispose of chromic acetate as a hazardous waste. The responsible state agency or the regional office of the federal Environmental Protection Agency (EPA) should be contacted for specific recommendations.

RISK ASSESSMENT: BUSINESS

General Assessment

Accidents or mishaps involving chromic acetate can present a significant threat to business operations. The loss or damage of equipment or facilities can significantly affect fiscal viability. Lawsuits that may result from personnel injury/death, public exposures, and/or environmental contamination will also require a serious expenditure of resources. Media attention surrounding an injury, death, or environmental damage can also result in a loss of profits and loss of current as well as future business.

Always remember that anytime the terms "carcinogen" or "cancer" are used, public emotion and hysteria can run equally high. Risk managers must take this into consideration when preparing any release of information to the community.

Recommended Risk-Reduction Measures

Company attorneys, safety and health professionals, and environmental specialists should be involved in the development of any procedures dealing with safety or emergency response. A company official should be pre-designated as a public relations officer with specific training in dealing with the press. Corporate plans and policies should be developed, approved, and implemented long before any need for such arises.

REFERENCES

Ahlbom, A. 1993. *Biostatistics for Epidemiologists.* New York: CRC Press/Lewis Publishers.

American Conference of Governmental Industrial Hygienists. 1988. *Documentation of the Threshold Limit Values and Biological Exposure Indices,* 5th Edition (with updates). Cincinnati: ACGIH

Asante-Duah, D. K. 1993. *Hazardous Waste Risk Assessment.* New York: CRC Press/Lewis Publishers.

Bloom, A. D. and F. J. DeSerres. 1995. *Ecotoxicity and Human Health.* New York: CRC Press/Lewis Publishers.

Brusick, D. J. 1994. *Methods for Genetic Risk Assessment.* New York: CRC Press/Lewis Publishers.

Calabrese, E. J., E. M. Kenyon. *Air Toxics and Risk Assessment.* New York: CRC Press/Lewis Publishers.

Calabrese, E. J. 1994. *Biological Effects of Low Level Exposures.* New York: CRC Press/Lewis Publishers.

Cockerman, L. G., B. S. Shane. 1994. *Basic Environmental Toxicology.* New York: CRC Press/Lewis Publishers.

Cothern, C. R., N. P. Ross. 1994. *Environmental Statistics, Assessment, and Forecasting.* New York: CRC Press/Lewis Publishers.

Cothern, R. R. 1993. *Comparative Environmental Risk Assessment.* New York: CRC Press/Lewis Publishers.

Cothern, C. R. 1995. *Handbook for Environmental Risk Decisionmaking.* New York: CRC Press/Lewis Publishers.

Gardner, R. W. 1994. *Chemical Intolerance.* New York: CRC Press/Lewis Publishers.

Hallenbeck, W. H. 1993. *Quantitative Risk Assessment for Environmental and Occupational Health,* 2nd Edition. New York: CRC Press/Lewis Publishers.

Howard, P. H. 1995. *Handbook of Environmental Fate and Exposures Data for Organic Chemicals,* Vol. I-IV. New York: CRC Press/Lewis Publishers.

Howard, P. H., et. al. 1991. *Handbook of Environmental Degradation Rates.* New York: CRC Press/Lewis Publishers.

Lauwerys, R. R., P. Hoet. 1993. *Industrial Chemical Exposure.* New York: CRC Press/Lewis Publishers.

Lewis, R. J., Sr. 1992. *Sax's Dangerous Properties of Industrial Materials,* Volumes 1, 2, and 3, 8th Edition. New York: Van Nostrand Reinhold.

Manahan, S. E. 1992. *Toxicological Chemistry,* 2nd Edition. New York: CRC Press/Lewis Publishers.

Meyer, E. 1990. *Chemistry of Hazardous Materials.* Englewood Cliffs, NJ: Prentice Hall.

Patnaik, P. 1992. *A Comprehensive Guide to the Hazardous Properties of Chemical Substances.* New York: Van Nostrand Reinhold.

Quigley, D. R. 1994. *Handbook of Emergency Chemical Management.* New York: CRC Press/Lewis Publishers.

Rea, W. J. 1995. *Chemical Sensitivity,* Vol. I-III. New York: CRC Press/Lewis Publishers.

Sacarello, H. L. A. 1994. *The Comprehensive Handbook of Hazardous Materials.* New York: CRC Press/Lewis Publishers.

Sherman, J. D. 1988. *Chemical Exposure and Disease.* New York: Van Nostrand Reinhold.

Talbott, E., G. Craun. 1995. *An Introduction to Environmental Epidemiology.* New York: CRC Press/Lewis Publishers.

U.S. Department of Health and Human Services (DHHS), National Institute for Occupational Health (NIOHS). 1994. *NIOSH Pocket Guide to Hazardous Chemicals.* Publication Number 94-116. Washington, D.C.; U.S. Government Printing Office.

U.S. Department of Health, Education, and Welfare (DHEW), National Institute for Occupational Health (NIOHS). 1976. *NIOSH Criteria Document: Occupational Exposure to Chromium (VI).* Publication Number 76-129. Washington, D.C.; U.S. Government Printing Office.

MATERIAL SAFETY DATA SHEET

CHEMICAL NAME	
CHROMIC ACID	**CRC PRESS/LEWIS PUBLISHERS DISCLAIMER:** THE INFORMATION AND RECOMMENDATIONS PRESENTED HEREIN ARE BASED ON SOURCES BELIEVED TO BE RELIABLE. CRC MAKES NO REPRESENTATION ON ITS COMPLETENESS OR ACCURACY. IT IS THE RESPONSIBILITY OF THE USER TO DETERMINE THE CHEMICAL'S SUITABILITY FOR ITS INTENDED USE, THE CHEMICAL'S SAFE USE, AND THE CHEMICAL'S PROPER DISPOSAL. NO REPRESENTATIONS AND/OR WARRANTIES, EITHER EXPRESSED OR IMPLIED, OF THE MERCHANTABILITY OR FITNESS FOR A PARTICULAR PURPOSE, OR OF ANY OTHER NATURE, ARE MADE WITH RESPECT TO THE INFORMATION PROVIDED IN THIS MSDS OR TO THE CHEMICAL TO WHICH INFORMATION MAY REFER. CRC NEITHER ASSUMES NOR AUTHORIZES ANY OTHER PERSON TO ASSUME FOR IT, ANY OTHER ADDITIONAL RESPONSIBILITY OR LIABILITY FOR THE USE OF, OR RELIANCE UPON, THIS INFORMATION.

HAZARD WARNING INFORMATION

HEALTH	FIRE	REACTIVE	OTHER	DEGREE OF HAZARD	COLOR CODING	OTHER CODES
3	0	1	OX	0 = Minimum Hazard 1 = Slight Hazard 2 = Moderate Hazard 3 = Serious Hazard 4 = Severe Hazard	HEALTH = BLUE FIRE = RED REACTIVITY = YELLOW OTHER = WHITE	OX = Oxidizer ACID = Acid ALK = Alkali COR = Corrosive W = Use No Water

SECTION I - GENERAL INFORMATION

Chemical Family **Acid**	RCRA Number **None**	EPA Class **Not Applicable**
DOT Proper Shipping Name **Chromic Acid Mixture (dry)**	Chemical Abstract Service (CAS) Number **1333-82-0**	
DOT Hazard Class and Label Requirements **Oxidizer**	DOT Emergency Guide Code **42 (solid) 60 (solution)**	
DOT Identification Number **UN 1463 (solid) UN 1755 (solution)**	Chemical Formula CrO_3	

Synonyms
Chromium trioxide; chromic anhydride; chromium (VI) oxide (1:3); chromic oxide.

SECTION II - HAZARDOUS INGREDIENTS AND IDENTITY INFORMATION

Hazardous Components (specific identity)	OSHA Exposure Criteria	NIOSH Exposure Criteria	Immediately Dangerous to Life and Health (IDLH)	ACGIH Exposure Criteria
Chromic acid (derivation: Sulfuric acid is added to a solution of sodium dichromate and the product is crystallized out; chromite is fused with soda ash and limestone and then treated with sulfuric acid; electrolysis).	**PEL:** 0.1 mg(Cr)/m^3 (maximum) Suspected Cancer Agent	**REL (10 hour):** .025 mg(Cr)/m^3 (carcinogen) **STEL (15 min):** .05 mg(Cr)/m^3	15 mg(Cr)/m^3 as Cr(VI)	**TLV: (8 hour)** .05 mg(Cr)/m^3 Confirmed Human Carcinogen

SECTION III - PHYSICAL AND CHEMICAL CHARACTERISTICS

Boiling Point **482°F (250°C)**	Specific Gravity (H_2O = 1) **2.7**
Vapor Pressure (mm Hg) **<1 at 69°F (20°C)**	Molecular Weight **100.6**
Vapor Density (Air = 1) **3.4**	Melting Point **385°F (196°C)**

Solubility in Water
Soluble in water, alcohol, and mineral acids.

Appearance and Odor
Dark purple-to-red crystalline solid. Odorless.

SECTION IV - FIRE AND EXPLOSION HAZARD DATA

Flash Point (method used) **Not Applicable**	Explosive Limits in Air % by Volume **LEL: Not Applicable UEL: Not Applicable**
NFPA Classification **No Citation**	Autoignition Temperature **Not Determined**

Extinguishing Media
Use dry chemical, carbon dioxide, water spray, or foam to fight fire.

Special Fire Fighting Procedures
Wear full protective clothing and self-contained breathing apparatus (SCBA). Use a water spray to keep fire-exposed containers cool. Remain clear of smoke, water fall-out and water run-off. Poisonous gases are produced in fire.

Unusual Fire and Explosion Hazards
Containers may explode in fire. Cool exposed containers. Can cause violent explosions upon contact with powerful reducing agents. May react with organic materials to cause ignition.

SECTION V - REACTIVITY DATA

Stability		Conditions to Avoid Dangerously reactive. Keep away from incompatible materials since violent and or explosive reactions can occur.
Stable	Unstable X	Incompatibility *(materials to avoid)* Ordinary combustible materials (such as wood, paper, cloth, oil), halogens, reducing agents, organic compounds, hydrogen sulfide, phosphorus, heat.
Hazardous Polymerization		Conditions to Avoid Under normal conditions of temperature and pressure, hazardous polymerization of chromic acid is not known to occur.
May Occur	Will Not Occur X	Hazardous Decomposition or By-products When heated to decomposition, chromic acid can emit highly toxic/poisonous fumes and gases.

SECTION VI - HEALTH HAZARD DATA

Primary Route(s) of Entry:	Inhalation?	Absorption (skin)?	Ingestion?
	X	X	

Health Hazards

INHALATION: Irritation of the respiratory system. Can cause sores or "chrome holes" to appear in the nasal cavity. Lung allergy, with wheezing and shortness of breath, sometimes occurs. Exposure may result in chemical pneumonitis or pulmonary edema (fluid in lungs).

ABSORPTION: A severe irritant. Exposure may cause burns. Prolonged exposures will allow chromic acid to enter the body through the affected skin areas. Eye contact can damage vision.

Carcinogenicity	NTP Listed?	IARC Cancer Review Group?	OSHA Regulated?	Target Organs?
Confirmed Human Confirmed Animal	5th Annual Report	Group 3	29 CFR 1910.1000 Table Z-1	Blood, respiratory system, skin, eyes, liver, kidneys

Medical Conditions Generally Aggravated by Exposure

None Reported. However, existing skin conditions or respiratory problems may be aggravated.

Emergency and First-aid Procedures

Eye contact: Flush immediately with water for 15 minutes (minimum), seek medical attention. Skin contact: Remove all contaminated clothing. Immediately wash area with large amounts of soap and water. For inhalation: Remove the person from exposure. Provide respiratory assistance and CPR. Transfer to medical facility. Observe for 24 - 48 hours for lung effects (pulmonary edema). If swallowed: If victim is conscious, DO NOT induce vomiting. Seek medical attention immediately. Support breathing if necessary. Never attempt to give anything by mouth to an unconscious or convulsing person.

SECTION VII - PRECAUTIONS FOR SAFE HANDLING AND USE

Steps to be Taken in Case Material is Released or Spilled

Ventilate area of spill or leak. Rinse with sodium bicarbonate solution, flood with water, agricultural lime, crushed limestone, soda ash, or lime. Collect powdered material in most convenient manner possible and deposit in sealed drum. Keep those not involved in cleanup from area.

Preferred Waste Disposal Method

No Citation.

Precautions to be Taken in Handling and Storage

Store in tightly closed containers in cool, well-ventilated area and protect from physical damage. Sources of ignition, such as smoking and open flames, are prohibited where chromic acid is used, handled, or stored. Bulk storage of chromic acid is not recommended.

Other Precautions and Warnings

Prior to working with chromic acid, personnel should be trained on its proper handling and storage.

SECTION VIII - CONTROL MEASURES AND PERSONAL PROTECTIVE EQUIPMENT

Respiratory Protection *(specify type)*

For exposures above the PEL, use a full facepiece respirator with a high efficiency particulate air (HEPA) filter. Greater protection is provided by an MSHA/NIOSH approved self-contained breathing apparatus (SCBA) operated in positive pressure mode.

Ventilation

Local exhaust or general mechanical systems recommended.

Protective Gloves	Eye Protection	Other Protective Clothing
Butyl Rubber	Chemical/Dust Goggles or Face Mask	Protective Suit or Clothing

Work/Hygiene Practices

Always wash hands thoroughly after using chemical; never bring food, drink, or smoking materials into vicinity of chemicals.

CHROMIC ACID

CrO_3 CAS: 1333-82-0

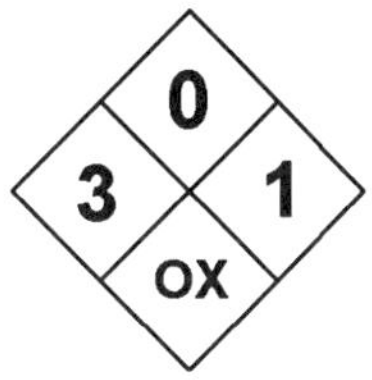

IDENTIFICATION AND TYPICAL USES

Chromic acid is a dark purplish-red, odorless, crystalline solid. It is used in the manufacture of chromates, oxidizing agents, and catalysts. It is also used as an intermediate in chromium plating, in medicine (caustic), process engraving, anodizing, ceramic glazes, colored glass, metal cleaning, inks, tanning, paints, textile mordant, and as an etchant for plastics.

RISK ASSESSMENT: HEALTH

General Assessment

Chromic acid is a poison and a powerful irritant of the eyes, skin, and mucous membranes. Some sources list chromic acid as a confirmed human carcinogen while others infer its carcinogenicity. Such inference appears to be based upon the fact the certain chromium compounds, known as hexavalent chromium (or chromium VI) compounds, have been determined to be human carcinogens. Chromic acid is such a compound. In terms of risk assessment and management, it is recommended that chromic acid be handled with extreme caution. Primary exposures occur through *inhalation* and *absorption,* although *ingestion* of chromic acid powders or dusts is also a considered possibility. It is a corrosive chemical and can pass through inner layers of skin and cause severe burns to human tissue. Skin contact may also cause deep, slow-healing ulcers or an allergic skin rash (dermatitis).

Inhalation of chromic acid can result in irritation of the nose, throat, and bronchial tubes with symptoms of coughing and wheezing. Prolonged inhalation of chromium compounds, including chromic acid, can result in sores or even a perforation in the bones of the nasal septum (a condition known as "chrome holes"), which can cause bleeding, discharge, or crusting. Prolonged or repeated exposures may damage the liver, kidneys, blood, and can cause lung cancers.

Acute Health Effects

The following acute (short-term) health effects may occur immediately or shortly after exposure to chromic acid:

Skin: Irritation and/or burning feeling on contact. May produce contact dermatitis and/or a deep skin ulceration.

Eye: Severe irritation; conjunctivitis, and may cause burns and permanent damage (loss of vision possible).

Lung: Large (high-level) acute exposures will cause irritation of the nose with a potential to cause sores or holes in the nasal septum. Also may irritate the throat and lungs causing coughing and shortness of breath.

Chronic Health Effects

The following chronic (long-term) health effects may occur at some time after exposure to chromic acid and can last for months or even years:

Cancer Hazards: Chromic acid is a hexavalent chromium compound. These compounds have been shown to cause cancer in humans and test animals.

Reproductive Hazard: According to information available in the references, chromic acid has not been adequately tested for its ability to adversely affect reproduction.

Other Chronic Effects: Chromic acid can irritate the nasal septum and cause holes or perforations to appear in the bones separating the nasal cavity ("chrome holes"). Irritation, discharge, bleeding and/or formation of a crust in the inner nose may occur. Exposed persons may develop a skin allergy with a rash similar to eczema. Once this occurs, the person is "sensitized" to chromic acid. Any future skin contact, even slight, can trigger a severe rash. When chromic acid enters skin cuts or sores, slow healing ulcers can result. Exposure can also erode or discolor the teeth and cause growths (polyps) on the voice box. Repeated exposures may cause irritation of the bronchial tubes (bronchitis) with symptoms of cough with phlegm. Lung allergy, with wheezing and shortness of breath, sometimes develops. In such cases, even small future exposures can cause symptoms to reappear. Repeated or severe exposures may cause kidney or liver damage.

Recommended Risk-Reduction Measures

Personnel should avoid direct contact with chromic acid. If a less toxic chemical cannot be substituted for a hazardous substance, then *engineering controls* are the most effective method of reducing exposures.

The best protection is to enclose operations and/or provide local exhaust ventilation at the site of chemical release. While not always operationally feasible, isolating operations can also reduce exposure. Using respiratory protection is less effective than the controls mentioned above, but is still advisable whenever working with or around chromic acid. For low exposures (up to the PEL), use a full facepiece respirator equipped with a high efficiency particulate air (HEPA) filter. Greater protection is provided with a supplied-air respirator with a full facepiece operated in positive pressure mode or an MSHA/NIOSH approved self-contained breathing apparatus (SCBA) with full facepiece and pressure demand are the recommended respiratory protection methods of choice. If a full facepiece is not available, then chemical splash goggles should be worn to protect the eyes. Whenever a chemical dust hazard exists, a face shield and a rubber apron should be worn. To prevent hand and skin exposures, acid resistant gloves should be worn.

Administrative controls should also be in place to minimize the potential for human exposures. These may include written procedures or policies which specify the methods and techniques that will be practiced whenever personnel are to work with chromic acid.

All personnel should receive *training* on the use, hazards, protective measures, emergency actions, and other precautions per 29 CFR 1910.1200 (Hazard Communication), prior to the first assignment in an area where chromic acid is used or stored.

Before beginning employment and at regular intervals thereafter (e.g., annually), for those with frequent or potentially high exposures (half the TLV or greater, or significant skin contact), the following medical tests are recommended:

- ☑ Lung function tests.
- ☑ Examination of the mouth and larynx.
- ☑ Examination of the nasal septum.

If symptoms develop or overexposure is suspected, the following may also be useful:

- ☑ Kidney and liver function tests.
- ☑ Examination of the nasal septum.
- ☑ Evaluation by a qualified allergist, including careful exposure history and special testing (may help diagnose skin allergy).

Workers should also check their skin daily for tiny bumps or blisters which are the first indications of "chrome ulcers." If not treated early, these can persist for years following exposure. Any evaluation should include a careful history of past and present symptoms with a medical examination. Medical tests that look for damage already done are not a substitute for controlling exposure. Also, since smoking can cause heart disease, as well as lung cancer, emphysema, and other related respiratory problems, the effects of exposure to chromic acid may be more pronounced in smokers than in non-smokers.

Other methods to reduce exposure include:

- ☑ Where possible, enclose operations and use local exhaust ventilation at the site of chemical release. If local exhaust ventilation or enclosure is not used, respiratory protection should be mandatory.
- ☑ Always ensure that proper protective clothing is worn when using chemical substances.
- ☑ Wash thoroughly immediately after exposure to chromic acid and at the end of the work shift or before eating, drinking, or smoking.
- ☑ Hazard warning information should be posted in the work area. In addition, as part of an on-going education and training program, all information on the health and safety hazards of chromic acid should be communicated to all exposed workers.
- ☑ Where possible, automatically transfer chromic acid from drums or other storage containers to process containers.
- ☑ Specific engineering controls are recommended by NIOSH. Refer to NIOSH Criteria Documents: Occupational Exposure to Chromic Acid, (Publication Number 73-11021) and Chromium VI Publication Number 76-129).

RISK ASSESSMENT: ENVIRONMENT

General Assessment

The environment is at risk of exposure during transportation, storage, disposal, or destruction of chromic acid. In almost every scenario, the threat of environmental exposure is contingent upon the proper handling of the chemical substance. Accidental spills, large or small, can result in fire, explosion, and possible contamination of the surrounding environmental mediums (water, soil, and air).

Chromic acid is considered a non-flammable liquid. However, it is an oxidizer and, because it is incompatible with so many other commodities, it presents serious fire and explosion concerns. In a fire, chromic acid will decompose and emit toxic and poisonous gases. It is also incompatible with many common combustible materials (such as paper, wood, plastics, and oil). These characteristics require special consideration during any emergency situation involving a leak or spill of chromic acid. Chromic acid can enter the environment through effluents from metal plating industry, from municipal waste treatment plants, or from spills.

Acute Ecological Effects

Acute (short-term) toxic effects may include the death of animals, birds, or fish and death or low growth rate in plants. Acute effects are seen 2 to 4 days after animals or plants are exposed to chromic acid.

Chromium is more toxic in soft water than in hard water. Chromium (VI) has high acute toxicity to aquatic life, and chromium (III) has moderate acute toxicity to aquatic life. Insufficient data are available to evaluate the short-term effects of chromic acid exposure to plants, birds or land animals.

Chronic Ecological Effects

Chronic (long-term) toxic effects may include shortened life span, reproductive problems, lower fertility, and changes in appearance or behavior in exposed animals. These effects can be seen long after first exposure(s) to toxic chemicals.

Chromium (III) and chromium (IV) both have high chronic toxicity to aquatic life. Insufficient data are available to evaluate the long-term effects of chromic acid exposure to plants, birds or land animals.

Water Solubility

Water solubility of chromium and its salts ranges from low to high. Chromic acid is moderately soluble in water. Concentrations of 1 to 1000 milligrams will mix with a liter of water.

Persistence in the Environment

Chromic acid is highly persistent in the aquatic environment, with a half-life of greater than 200 days. The half-life of a pollutant is the amount of time it takes for one half of the chemical to be degraded.

Bioaccumulation in Aquatic Organisms

Some substances increase in concentration, or *bioaccumulate,* in living organisms as they breathe contaminated air, drink contaminated water, or eat contaminated food. These chemicals can become concentrated in the tissues and internal organs of animals as well as humans.

The concentration of chromium found in fish is expected to be somewhat higher than the average concentration of chromium in the water from which the fish was taken.

Recommended Risk-Reduction Measures

Proper training of all transporters will reduce the likelihood of a mishap or accidents resulting in a leak or spill to the environment. The correct use of DOT labeling (per the MSDS) on all containers, trucks, and rail cars will enable emergency responders to react properly and quickly to any disaster thereby reducing the potential risk to the environment and to personnel.

Storage of chromic acid should be segregated from incompatible chemicals to minimize the risk of cross-contamination or contact. Bulk storage of chromic acid should be avoided.

If a spill or leak to the environment has occurred, fire department, emergency response and/or hazardous materials spill personnel should be notified immediately. To neutralize a spill, rinse with sodium bicarbonate solution, flood with water, agricultural lime (slaked lime) crushed limestone, soda ash, or lime and collect material in safest manner possible. Deposit in sealed drum for disposal. Cleanup should be attempted only by those trained in proper spill containment procedures. Do not dry sweep (creates airborne dusts). Use vacuum equipped with HEPA filter instead. Contaminated soils should be removed for incineration and replaced with clean soil. If chromic acid should contact the water table, aquifer, or navigable waterway, time is of the essence. It is moderately soluble in water and, therefore, total containment and remediation may not be entirely possible. When such spills occur, the local and/or state emergency response authorities must be notified. A comprehensive emergency response or disaster preparedness/recovery plan should be in place prior to any operations involving the use, transportation, storage, or disposal of chromic acid. If chromic acid is spilled or leaked, the following specific steps are recommended:

- ☑ Restrict persons not wearing protective clothing from area of spill or leak until cleanup is complete and area can be opened for normal work.
- ☑ Ventilate area of spill.
- ☑ Neutralize and deposit in sealed containers.
- ☑ It may be necessary to dispose of chromic acid as a hazardous waste. The responsible state agency or the regional office of the federal Environmental Protection Agency (EPA) should be contacted for specific recommendations.

RISK ASSESSMENT: BUSINESS

General Assessment

Accidents or mishaps involving chromic acid can present a significant threat to business operations. The loss or damage of equipment or facilities can significantly affect fiscal viability. Lawsuits that may result from personnel injury/death, public exposures, and/or environmental contamination will also require a serious expenditure of resources. Media attention surrounding an injury, death, or environmental damage can also result in a loss of profits and loss of current as well as future business.

Recommended Risk-Reduction Measures

Company attorneys, safety and health professionals, and environmental specialists should be involved in the development of any procedures dealing with safety or emergency response. A company official should be pre-designated as a public relations officer with specific training in dealing with the press. Corporate plans and policies should be developed, approved, and implemented long before any need for such arises.

REFERENCES

American Conference of Governmental Industrial Hygienists. 1988. *Documentation of the Threshold Limit Values and Biological Exposure Indices,* 5th Edition (with updates). Cincinnati: ACGIH.

Calabrese, E. J. 1994. *Biological Effects of Low Level Exposures.* New York: CRC Press/Lewis Publishers.

Cockerman, L. G., B. S. Shane. 1994. *Basic Environmental Toxicology* New York: CRC Press/Lewis Publishers.

Hallenbeck, W. H. 1993. *Quantitative Risk Assessment for Environmental and Occupational Health,* 2nd Edition. New York: CRC Press/Lewis Publishers.

Howard, P. H., et. al. 1991. *Handbook of Environmental Degradation Rates.* New York: CRC Press/Lewis Publishers.

Lauwerys, R. R., P. Hoet. 1993. *Industrial Chemical Exposure.* New York: CRC Press/Lewis Publishers.

Lewis, R. J., Sr. 1992. *Sax's Dangerous Properties of Industrial Materials,* Volumes 1, 2, and 3, 8th Edition. New York: Van Nostrand Reinhold.

Manahan, S. E. 1992. *Toxicological Chemistry,* 2nd Edition. New York: CRC Press/Lewis Publishers.

Meyer, E. 1990. *Chemistry of Hazardous Materials.* Englewood Cliffs, NJ: Prentice Hall.

Patnaik, P. 1992. *A Comprehensive Guide to the Hazardous Properties of Chemical Substances.* New York: Van Nostrand Reinhold.

Quigley, D. R. 1994. *Handbook of Emergency Chemical Management.* New York: CRC Press/Lewis Publishers.

Rea, W. J. 1995. *Chemical Sensitivity,* Vol. I-III. New York: CRC Press/Lewis Publishers.

Sacarello, H. L. A. 1994. *The Comprehensive Handbook of Hazardous Materials.* New York: CRC Press/Lewis Publishers.

Sherman, J. D. 1988. *Chemical Exposure and Disease.* New York: Van Nostrand Reinhold.

Talbott, E., G. Craun. 1995. *An Introduction to Environmental Epidemiology.* New York: CRC Press/Lewis Publishers.

U.S. Department of Health and Human Services (DHHS), National Institute for Occupational Health (NIOHS). 1994. *NIOSH Pocket Guide to Hazardous Chemicals.* Publication Number 94-116. Washington, D.C.; U.S. Government Printing Office.

U.S. Department of Health, Education, and Welfare (DHEW), National Institute for Occupational Health (NIOHS). 1976. *NIOSH Criteria Document: Occupational Exposure to Chromium (VI).* Publication Number 76-129. Washington, D.C.; U.S. Government Printing Office.

U.S. Department of Health, Education, and Welfare (DHEW), National Institute for Occupational Health (NIOHS). 1973. *NIOSH Criteria Document: Occupational Exposure to Chromic Acid.* Publication Number 73-11021. Washington, D.C.; U.S. Government Printing Office.

MATERIAL SAFETY DATA SHEET

CHEMICAL NAME	
CHROMIC NITRATE	**CRC PRESS/LEWIS PUBLISHERS DISCLAIMER:** THE INFORMATION AND RECOMMENDATIONS PRESENTED HEREIN ARE BASED ON SOURCES BELIEVED TO BE RELIABLE. CRC MAKES NO REPRESENTATION ON ITS COMPLETENESS OR ACCURACY. IT IS THE RESPONSIBILITY OF THE USER TO DETERMINE THE CHEMICAL'S SUITABILITY FOR ITS INTENDED USE, THE CHEMICAL'S SAFE USE, AND THE CHEMICAL'S PROPER DISPOSAL. NO REPRESENTATIONS AND/OR WARRANTIES, EITHER EXPRESSED OR IMPLIED, OF THE MERCHANTABILITY OR FITNESS FOR A PARTICULAR PURPOSE, OR OF ANY OTHER NATURE, ARE MADE WITH RESPECT TO THE INFORMATION PROVIDED IN THIS MSDS OR TO THE CHEMICAL TO WHICH INFORMATION MAY REFER. CRC NEITHER ASSUMES NOR AUTHORIZES ANY OTHER PERSON TO ASSUME FOR IT, ANY OTHER ADDITIONAL RESPONSIBILITY OR LIABILITY FOR THE USE OF, OR RELIANCE UPON, THIS INFORMATION.

HAZARD WARNING INFORMATION

HEALTH	FIRE	REACTIVE	OTHER	DEGREE OF HAZARD	COLOR CODING	OTHER CODES
3	**0**	**1**	**OX**	0 = Minimum Hazard 1 = Slight Hazard 2 = Moderate Hazard 3 = Serious Hazard 4 = Severe Hazard	HEALTH = BLUE FIRE = RED REACTIVITY = YELLOW OTHER = WHITE	OX = Oxidizer ACID = Acid ALK = Alkali COR = Corrosive W = Use No Water

SECTION I - GENERAL INFORMATION

Characterization: **Metallic Nitrate**	RCRA Number: **None**	EPA Class: **Not Applicable**
DOT Proper Shipping Name: **Chromium Nitrate**	Chemical Abstract Service (CAS) Number: **13548-38-4**	
DOT Hazard Class and Label Requirements: **Oxidizer**	DOT Emergency Guide Code: **35**	
DOT Identification Number: **UN 2720**	Chemical Formula: **$Cr(NO_3)_3$**	

Synonyms
Chromium (III) nitrate; chromium nitrate; chromium trinitrate; nitric acid; chromium (3+) salt.

SECTION II - HAZARDOUS INGREDIENTS AND IDENTITY INFORMATION

Hazardous Components (specific identity)	OSHA Exposure Criteria	NIOSH Exposure Criteria	Immediately Dangerous to Life and Health (IDLH)	ACGIH Exposure Criteria
Chromic nitrate (derivation: By action of nitric acid on chromium hydroxide).	**PEL: 0.5 mg(Cr)/m^3 (maximum)** **Suspected Cancer Agent**	**REL (10 hour): .025 mg(Cr)/m^3 (carcinogen)** **STEL (15 min): .05 mg(Cr)/m^3**	**15 mg(Cr)/m^3 as Cr(VI)**	**TLV: (8 hour) .5 mg(Cr)/m^3** **Confirmed Human Carcinogen**

SECTION III - PHYSICAL AND CHEMICAL CHARACTERISTICS

Boiling Point: **Not Applicable**	Specific Gravity (H_2O = 1): **Not Determined**
Vapor Pressure (mm Hg): **Not Applicable**	Molecular Weight: **238**
Vapor Density (Air = 1): **Not Applicable**	Melting Point: **140°F (60°C)**

Solubility
Soluble in water and alcohol.

Appearance and Odor
Dark blue to purple crystalline solid. Odorless.

SECTION IV - FIRE AND EXPLOSION HAZARD DATA

Flash Point (method used): **Not Applicable**	Explosive Limits in Air % by Volume: **LEL: Not Applicable UEL: Not Applicable**
NFPA Classification: **No Citation**	Autoignition Temperature: **Not Determined**

Extinguishing Media
Use extinguishing media suitable to surrounding fires. Chromium nitrate itself does not burn.

Special Fire Fighting Procedures
Wear full protective clothing and self-contained breathing apparatus (SCBA). Use a water spray to keep fire-exposed containers cool. Remain clear of smoke, water fall-out and water run-off. Poisonous gases are produced in fire, including nitrogen oxides and chromium oxide fumes.

Unusual Fire and Explosion Hazards
Containers may explode in fire. Cool exposed containers. Can cause violent explosions upon contact with powerful reducing agents. May react with organic materials to cause ignition.

SECTION V - REACTIVITY DATA

Stability		Conditions to Avoid **Keep away from incompatible materials since violent and or explosive reactions can occur.**
Stable	Unstable X	Incompatibility *(materials to avoid)* **Ordinary combustible materials (such as wood, paper, cloth, oil), halogens and ethers, reducing agents, organic compounds.**
Hazardous Polymerization		Conditions to Avoid **Under normal conditions of temperature and pressure, hazardous polymerization of chromic nitrate is not known to occur.**
May Occur	Will Not Occur X	Hazardous Decomposition or By-products **When heated to decomposition, chromic nitrate can emit highly toxic/poisonous fumes and gases, including nitrogen oxides and chromium oxides.**

SECTION VI - HEALTH HAZARD DATA

Primary Route(s) of Entry:	Inhalation?	Absorption (skin)?	Ingestion?
	X		X

Health Hazards

INHALATION: **Irritation of the upper respiratory tract. May cause headache and nausea. Lung allergy with wheezing and shortness of breath sometimes occurs.**

SKIN & EYES: **An irritant, especially on prolonged exposures. Skin allergy sometimes occurs, with itching, redness, and/or eczema-like rash. Eye contact can damage vision.**

INGESTION: **Gastrointestinal disturbances. A human poison by ingestion.**

Carcinogenicity	NTP Listed?	IARC Cancer Review Group?	OSHA Regulated?	Target Organs?
Questioned Human Suspected Animal	**5th Annual Report**	**Group 3**	**29 CFR 1910.1000 Table Z-1**	**Respiratory system, skin, eyes**

Medical Conditions Generally Aggravated by Exposure

None Reported. However, existing skin conditions or respiratory problems may be aggravated.

Emergency and First-aid Procedures

Eye contact: Flush immediately with water for 15 minutes (minimum), holding eyelids open (person may need assistance), seek medical attention. Skin contact: Remove all contaminated clothing. Immediately wash area with large amounts of soap and water. For inhalation: Remove the person from exposure. Provide respiratory assistance and CPR. Transfer to medical facility. If swallowed: Contact poison control center for information. Seek medical attention immediately. Support breathing and provide CPR if necessary. Do NOT give anything by mouth to an unconscious person.

SECTION VII - PRECAUTIONS FOR SAFE HANDLING AND USE

Steps to be Taken in Case Material is Released or Spilled

Ventilate area of spill or leak. Collect powdered material in most convenient manner possible and deposit in sealed drum. Do not dry sweep (creates airborne dusts). Use a vacuum equipped with a high efficiency particulate air (HEPA) filter. Keep those not involved in cleanup from area.

Preferred Waste Disposal Method

No Citation.

Precautions to be Taken in Handling and Storage

Store in tightly closed containers in cool, well-ventilated area and protect from physical damage. Sources of ignition, such as smoking and open flames, are prohibited where chromic nitrate is used, handled, or stored. Keep away from flammable or combustible materials.

Other Precautions and Warnings

Prior to working with chromic nitrate, personnel should be trained on its proper handling and storage.

SECTION VIII - CONTROL MEASURES AND PERSONAL PROTECTIVE EQUIPMENT

Respiratory Protection *(specify type)*

For exposures above the PEL, use a full facepiece respirator with a high efficiency particulate air (HEPA) filter. Greater protection is provided by an MSHA/NIOSH approved self-contained breathing apparatus (SCBA) operated in positive pressure mode.

Ventilation

Local exhaust or general mechanical systems recommended.

Protective Gloves	Eye Protection	Other Protective Clothing
Butyl Rubber	**Chemical/Dust Goggles or Face Mask**	**Protective Suit or Clothing**

Work/Hygiene Practices

Always wash hands thoroughly after using chemical; never bring food, drink, or smoking materials into vicinity of chemicals.

CHROMIC NITRATE

$Cr(NO_3)_3$ CAS: 13548-38-4

IDENTIFICATION AND TYPICAL USES

Chromic nitrate is a dark blue to purplish, odorless, crystalline solid. It is used in the preparation of chrome catalysts, in textile printing operations, and as a corrosion inhibitor.

RISK ASSESSMENT: HEALTH

General Assessment

Chromic nitrate is a poison and an irritant of the eyes, skin, and mucous membranes. It has been shown to cause mutations in living cells. While its specific carcinogenic properties require further study, several related chromium compounds have caused cancer in humans and animals. In terms of risk assessment and management, it is recommended that chromic nitrate be handled with extreme caution.

Primary exposures occur through *inhalation,* although *ingestion* of chromic nitrate powders or dusts is also a considered possibility. It is a skin and eye irritant as well. Prolonged contact can cause skin allergy to occur resulting in rash, itching, and an eczema-like reaction. Once this occurs, any future exposures will cause a return of symptoms. Chromic nitrate will also cause irritation to the eyes and may result in permanent damage to vision.

Inhalation of chromic nitrate can result in irritation of the nose, throat, and bronchial tubes with symptoms of coughing and wheezing.

Acute Health Effects

The following acute (short-term) health effects may occur immediately or shortly after exposure to chromic nitrate:

Skin: Irritation and/or burning feeling on contact. May produce contact dermatitis and/or a redness and rash.

Eye: Severe irritation; conjunctivitis, and may cause burns and permanent damage (loss of vision possible).

Lung: Irritation of the upper respiratory tract and mucous membranes. Also may irritate the throat and lungs causing coughing and shortness of breath.

Chronic Health Effects

The following chronic (long-term) health effects may occur at some time after exposure to chromic nitrate and can last for months or even years:

Cancer Hazards: Chromic nitrate may cause mutations in living cells. Whether or not it poses a cancer hazard risk requires further study. While chromium nitrate has not yet been specifically identified as a carcinogen, it should be handled with extreme caution since several related chromium compounds are known to cause cancers in humans and test animals.

Reproductive Hazard: According to information available in the references, chromic nitrate has not been adequately tested for its ability to adversely affect reproduction.

Other Chronic Effects: Exposed persons may develop a skin allergy with a rash similar to eczema. Once this occurs, the person is "sensitized" to chromic nitrate. Any future skin contact, even slight, can trigger a severe rash. Lung allergy, with wheezing and shortness of breath, sometimes develops. In such cases, even small future exposures can cause symptoms to reappear. Repeated or severe exposures may cause kidney or liver damage.

Recommended Risk-Reduction Measures

Personnel should avoid direct contact with chromic nitrate. If a less toxic chemical cannot be substituted for a hazardous substance, then *engineering controls* are the most effective method of reducing exposures.

The best protection is to enclose operations and/or provide local exhaust ventilation at the site of chemical release. While not always operationally feasible, isolating operations can also reduce exposure. Using respiratory protection is less effective than the controls mentioned above, but is still advisable whenever working with or around chromic nitrate. For low exposures (up to the PEL), use a full facepiece respirator equipped with a high efficiency particulate air (HEPA) filter and a dust/mist/fume pre-filter. Greater protection is provided with a supplied-air respirator with a full facepiece operated in positive pressure mode or an MSHA/NIOSH approved self-contained breathing apparatus (SCBA) with full facepiece and pressure demand are the recommended respiratory protection methods of choice. If a full facepiece is not available, then chemical splash/dust goggles should be worn to protect the eyes. Whenever a chemical dust hazard exists, a face shield and a rubber apron should be worn. To prevent hand and skin exposures, impervious rubber gloves should be worn.

Administrative controls should also be in place to minimize the potential for human exposures. These may include written procedures or policies which specify the methods and techniques that will be practiced whenever personnel are to work with chromic nitrate.

All personnel should receive *training* on the use, hazards, protective measures, emergency actions, and other precautions per 29 CFR 1910.1200 (Hazard Communication), prior to the first assignment in an area where chromic nitrate is used or stored.

If symptoms develop or overexposure is suspected, the following medical tests may be useful:

☑ Evaluation by a qualified allergist, including careful exposure history and special testing (may help diagnose skin allergy).

Any evaluation should include a careful history of past and present symptoms with a medical examination. Medical tests that look for damage already done are not a substitute for controlling exposure.

Other methods to reduce exposure include:

☑ Where possible, enclose operations and use local exhaust ventilation at the site of chemical release. If local exhaust ventilation or enclosure is not used, respiratory protection should be mandatory.

☑ Always ensure that proper protective clothing is worn when using chemical substances.

☑ Wash thoroughly immediately after exposure to chromic nitrate and at the end of the work shift or before eating, drinking, or smoking.

☑ Hazard warning information should be posted in the work area. In addition, as part of an on-going education and training program, all information on the health and safety hazards of chromic nitrate should be communicated to all potentially exposed workers.

☑ Where possible, automatically transfer chromic nitrate from drums or other storage containers to process containers.

RISK ASSESSMENT: ENVIRONMENT

General Assessment

The environment is at risk of exposure during transportation, storage, disposal, or destruction of chromic nitrate. In almost every scenario, the threat of environmental exposure is contingent upon the proper handling of the chemical substance. Accidental spills, large or small, can result in fire, explosion, and possible contamination of the surrounding environmental mediums (water, soil, and air).

Chromic nitrate is considered a non-flammable solid. However, it is an oxidizer and incompatible with many other commodities and presents serious fire and explosion concerns. In a fire, chromic nitrate will decompose and emit toxic and poisonous gases, including oxides of nitrogen and chromium. It is also incompatible with many common combustible materials (such as paper, wood, plastics, and oil). These characteristics require special consideration during any emergency situation involving a leak or spill of chromic nitrate.

Chromic nitrate can enter the environment through effluents from the metal plating industries, from municipal waste treatment plant discharges, or from spills.

☠ *Acute Ecological Effects*

Acute (short-term) toxic effects may include the death of animals, birds, or fish and death or low growth rate in plants. Acute effects are seen 2 to 4 days after animals or plants are exposed to chromic nitrate.

Chromium is more toxic in soft water than in hard water. Chromium (VI) has high acute toxicity to aquatic life, and chromium (III) has moderate acute toxicity to aquatic life. Insufficient data are available to evaluate the short-term effects of chromic nitrate exposure to plants, birds or land animals.

Chronic Ecological Effects

Chronic (long-term) toxic effects may include shortened life span, reproductive problems, lower fertility, and changes in appearance or behavior in exposed animals. These effects can be seen long after first exposure(s) to toxic chemicals.

Chromium (III) and chromium (IV) both have high chronic toxicity to aquatic life. Insufficient data are available to evaluate the long-term effects of chromic nitrate exposure to plants, birds or land animals.

Water Solubility

Water solubility of chromium and its salts ranges from low to high. Chromic nitrate is moderately soluble in water. Concentrations of 1 to 1000 milligrams will mix with a liter of water.

Persistence in the Environment

Chromic nitrate is highly persistent in the aquatic environment, with a half-life of greater than 200 days. The half-life of a pollutant is the amount of time it takes for one half of the chemical to be degraded.

Bioaccumulation in Aquatic Organisms

Some substances increase in concentration, or *bioaccumulate,* in living organisms as they breathe contaminated air, drink contaminated water, or eat contaminated food. These chemicals can become concentrated in the tissues and internal organs of animals as well as humans.

The concentration of chromium found in fish is expected to be somewhat higher than the average concentration of chromium in the water from which the fish was taken.

Recommended Risk-Reduction Measures

Proper training of all transporters will reduce the likelihood of a mishap or accidents resulting in a leak or spill to the environment. The correct use of DOT labeling (per the MSDS) on all containers, trucks, and rail cars will enable emergency responders to react properly and quickly to any disaster thereby reducing the potential risk to the environment and to personnel.

Storage of chromic nitrate should be segregated from incompatible chemicals to minimize the risk of cross-contamination or contact. Bulk storage of chromic nitrate should be avoided.

If a spill or leak to the environment has occurred, fire department, emergency response and/or hazardous materials spill personnel should be notified immediately. Cleanup should be attempted only by those trained in proper spill containment procedures. Do not dry sweep (creates airborne dusts). Use a vacuum equipped with HEPA filter instead. Contaminated soils should be removed for incineration and replaced with clean soil. If chromic nitrate should contact the water table, aquifer, or navigable waterway, time is of the essence. It is moderately soluble in water and, therefore, total containment and remediation may not be entirely possible. When such spills occur, the local and/or state emergency response authorities must be notified. A comprehensive emergency response or disaster preparedness/recovery plan should be in place prior to any operations involving the use, transportation, storage, or disposal of chromic nitrate. If chromic nitrate is spilled or leaked, the following specific steps are recommended:

- ☑ Restrict persons not wearing protective clothing from area of spill or leak until cleanup is complete and area can be opened for normal work.
- ☑ It may be necessary to dispose of chromic nitrate as a hazardous waste. The responsible state agency or the regional office of the federal Environmental Protection Agency (EPA) should be contacted for specific recommendations.

RISK ASSESSMENT: BUSINESS

General Assessment

Accidents or mishaps involving chromic nitrate can present a significant threat to business operations. The loss or damage of equipment or facilities can significantly affect fiscal viability. Lawsuits that may result from personnel injury/death, public exposures, and/or environmental contamination will also require a serious expenditure of resources. Media attention surrounding an injury, death, or environmental damage can also result in a loss of profits and loss of current as well as future business.

Recommended Risk-Reduction Measures

Company attorneys, safety and health professionals, and environmental specialists should be involved in the development of any such procedures. A company official should be pre-designated as a public relations officer with specific training in dealing with the press.

Corporate plans and policies should be developed, approved, and implemented long before any need for such arises.

REFERENCES

Ahlbom, A. 1993. *Biostatistics for Epidemiologists.* New York: CRC Press/Lewis Publishers.

American Conference of Governmental Industrial Hygienists. 1988. *Documentation of the Threshold Limit Values and Biological Exposure Indices,* 5th Edition (with updates). Cincinnati: ACGIH

Asante-Duah, D. K. 1993. *Hazardous Waste Risk Assessment.* New York: CRC Press/Lewis Publishers.

Bloom, A. D. and F. J. DeSerres. 1995. *Ecotoxicity and Human Health.* New York: CRC Press/Lewis Publishers.

Brusick, D. J. 1994. *Methods for Genetic Risk Assessment.* New York: CRC Press/Lewis Publishers.

Calabrese, E. J., E. M. Kenyon. *Air Toxics and Risk Assessment.* New York: CRC Press/Lewis Publishers.

Calabrese, E. J. 1994. *Biological Effects of Low Level Exposures.* New York: CRC Press/Lewis Publishers.

Cockerman, L. G., B. S. Shane. 1994. *Basic Environmental Toxicology.* New York: CRC Press/Lewis Publishers.

Cothern, C. R., N. P. Ross. 1994. *Environmental Statistics, Assessment, and Forecasting.* New York: CRC Press/Lewis Publishers.

Cothern, R. R. 1993. *Comparative Environmental Risk Assessment.* New York: CRC Press/Lewis Publishers.

Cothern, C. R. 1995. *Handbook for Environmental Risk Decisionmaking.* New York: CRC Press/Lewis Publishers.

Danse, I. R. 1991. *Common Sense Toxics in the Workplace.* New York: Van Nostrand Reinhold.

Environmental Protection Agency. 1983. *Chemical Hazard Information Profiles.* Washington, D.C.: U.S. EPA.

Fodden, C. R., and J. L. Weddell. 1992. *Hazardous Materials Emergency Action Data.* New York: CRC Press/Lewis Publishers.

Forseberg, K., and L. H. Keith. 1993. *Chemical; Protective Clothing Permeation and Degradation Database.* Chelsea, MI: Lewis Publishers.

Gardner, R. W. 1994. *Chemical Intolerance.* New York: CRC Press/Lewis Publishers.

Hallenbeck, W. H. 1993. *Quantitative Risk Assessment for Environmental and Occupational Health,* 2nd Edition. New York: CRC Press/Lewis Publishers.

Hodgson, E., R. B. Mailman, and Janice E. Chambers. 1988. *Dictionary of Toxicology.* London: The MacMillian Company.

Hosty, J. W., and P. Foster. 1990. *A Practical Guide to Chemical Spill Response.* New York: Van Nostrand Reinhold.

Howard, P. H. 1995. *Handbook of Environmental Fate and Exposures Data for Organic Chemicals,* Vol. I-IV. New York: CRC Press/Lewis Publishers.

Howard, P. H., et. al. 1991. *Handbook of Environmental Degradation Rates.* New York: CRC Press/Lewis Publishers.

International Agency for Research on Cancer (IARC). 1987. *IARC Monographs on the Evaluation of Carcinogenic Risk to Humans, Overall Evaluations of Carcinogenicity: An updating of IARC Monographs Volumes 1-42.* Lyon: IARC.

Jones, F. E. 1994. *Toxic Organic Vapors in the Workplace.* New York: CRC Press/Lewis Publishers.

Keith, L. H. and M. M. Walker. 1995. *Handbook of Air Toxics.* New York: CRC Press/Lewis Publishers.

Landis, W. G., M. H. Yu. 1995. *Introduction to Environmental Toxicology.* New York: CRC Press/Lewis Publishers.

Langerman, N. 1994 *Precautionary Labels for Chemical Containers.* New York: CRC Press/Lewis Publishers.

Lauwerys, R. R., P. Hoet. 1993. *Industrial Chemical Exposure.* New York: CRC Press/Lewis Publishers.

Lewis, .J., Sr. 1993. *Hawley's Condensed Chemical Dictionary,* 12th Edition. New York: Van Nostrand Reinhold.

Lewis, . J., Sr. 1992. *Sax's Dangerous Properties of Industrial Materials,* Volumes 1, 2, and 3, 8th Edition. New York: Van Nostrand Reinhold.

Lide, D. R. 1995. *Handbook of Chemistry and Physics,* 76th Edition. New York: CRC Press/Lewis Publishers.

Mackay, D., W. Y. Shiu, and K. C. Ma. 1992. *Illustrated Handbook of Chemical-Physical Properties and Environmental Fate for Organic Chemicals.* New York: CRC Press/Lewis Publishers.

Manahan, S. E. 1994. *Environmental Chemistry,* 6th Edition. New York: CRC Press/Lewis Publishers.

Manahan, S. E. 1992. *Toxicological Chemistry,* 2nd Edition. New York: CRC Press/Lewis Publishers.

Mellström, G. A., J. E. Wahlberg, and H. I. Maibach. 1994. *Protective Gloves for Occupational Use.* New York: CRC Press/Lewis Publishers.

Meyer, E. 1990. *Chemistry of Hazardous Materials.* Englewood Cliffs, NJ: Prentice Hall.

Miller, B. F., M.D., and C. B. Keane, R.N. 1972. *Encyclopedia and Dictionary of Medicine and Nursing.* Philadelphia: W.B. Sanders Company.

Mycock, J. C., J. D. McKenna, and L. Theodore. 1995. *Handbook of Air Pollution Control Engineering and Technology.* New York: CRC Press/Lewis Publishers.

Ott, W. R. 1995. *Environmental Statistics and Data Analysis.* New York: CRC Press/Lewis Publishers.

U.S. Department of Health, Education, and Welfare (DHEW), National Institute for Occupational Health (NIOHS). 1976. *NIOSH Criteria Document: Occupational Exposure to Chromium (VI).* Publication Number 76-129. Washington, D.C.; U.S. Government Printing Office.

U.S. Department of Health, Education, and Welfare (DHEW), National Institute for Occupational Health (NIOHS). 1973. *NIOSH Criteria Document: Occupational Exposure to Chromium.* Publication Number 73-11021. Washington, D.C.; U.S. Government Printing Office.

MATERIAL SAFETY DATA SHEET

CHEMICAL NAME	CRC PRESS/LEWIS PUBLISHERS DISCLAIMER
CHROMIUM	

HAZARD WARNING INFORMATION

HEALTH	FIRE	REACTIVE	OTHER	DEGREE OF HAZARD	COLOR CODING	OTHER CODES
3	0	0		0 = Minimum Hazard 1 = Slight Hazard 2 = Moderate Hazard 3 = Serious Hazard 4 = Severe Hazard	HEALTH = BLUE FIRE = RED REACTIVITY = YELLOW OTHER = WHITE	OX = Oxidizer ACID = Acid ALK = Alkali COR = Corrosive ~~W~~ = Use No Water

SECTION I - GENERAL INFORMATION

Characterization	**Metal**	RCRA Number	**None**
		EPA Class	**Not Applicable**
DOT Proper Shipping Name	**Chromium**	Chemical Abstract Service (CAS) Number	**7440-47-3**
DOT Hazard Class and Label Requirements	**No Citation**	DOT Emergency Guide Code	**No Citation**
DOT Identification Number	**No Citation**	Chemical Formula	**Cr**

Synonyms
Chromium metal; chrome.

SECTION II - HAZARDOUS INGREDIENTS AND IDENTITY INFORMATION

Hazardous Components (specific identity)	OSHA Exposure Criteria	NIOSH Exposure Criteria	Immediately Dangerous to Life and Health (IDLH)	ACGIH Exposure Criteria
Chromium (derivation: From chromite by direct reduction (ferrochrome), by reducing the oxide with finely divided aluminum or carbon, and by electrolysis of chromium solutions).	**PEL:** 1 mg(Cr)/m^3 **STEL:** Not Established	**REL:** 0.5 mg(Cr)/m^3 **STEL:** Not Established	250 mg/m^3 as Cr	**TLV:** 0.5 mg(Cr)/m^3 **STEL:** Not Established

SECTION III - PHYSICAL AND CHEMICAL CHARACTERISTICS

Boiling Point	**4788°F (2642°C)**	Specific Gravity (H_2O = 1)	**7.14**
Vapor Pressure (mm Hg)	**0 at 69°F (20°C)**	Molecular Weight	**52**
Vapor Density (Air = 1)	**Not Found**	Melting Point	**3452°F (1900°C)**

Solubility
Insoluble in water. Soluble in acids (except nitric) and strong alkalis.

Appearance and Odor
Hard, brittle, semi-gray, lustrous metal often found as a powder.

SECTION IV - FIRE AND EXPLOSION HAZARD DATA

Flash Point (method used)	**Not Applicable**	Explosive Limits in Air % by Volume	**LEL: Not Applicable UEL: Not Applicable**
NFPA Classification	**No Citation**	Autoignition Temperature	**Not Determined**

Extinguishing Media
Use class "D" extinguishing agent. Do NOT use carbon dioxide.

Special Fire Fighting Procedures
Wear full protective clothing and self-contained breathing apparatus (SCBA). Use a water spray to keep fire-exposed containers cool. Water may be ineffective on fire. Burning chromium metal may reach temperatures in excess of 5000°F. Remain clear of smoke, water fall-out and water run-off. Poisonous gases are produced in fire.

Unusual Fire and Explosion Hazards
Containers may explode in fire. Can cause violent explosions upon contact with strong oxidizers. Powders may explode in air. Potentially explosive in atmospheres of carbon dioxide.

SECTION V - REACTIVITY DATA

Stability		Conditions to Avoid Keep away from incompatible materials since violent and or explosive reactions can occur.
Stable X	Unstable	Incompatibility *(materials to avoid)* Incompatible with strong oxidizers (such as chlorine, bromine, hydrogen peroxide, and fluorine), and most alkali materials. Reacts explosive with carbon dioxide or when heated with aluminum nitrate.
Hazardous Polymerization		Conditions to Avoid Under normal conditions of temperature and pressure, hazardous polymerization of chromium is not known to occur.
May Occur	Will Not Occur X	Hazardous Decomposition or By-products When heated to decomposition, chromium can emit highly toxic/poisonous fumes and gases.

SECTION VI - HEALTH HAZARD DATA

Primary Route(s) of Entry:	Inhalation?	Absorption (skin)?	Ingestion?
	X		X

Health Hazards

INHALATION: Irritation of the respiratory system. Exposure to some chromium compounds can cause sores or "chrome holes" to appear in the nasal cavity. Exposure may cause "metal fume fever," a flu-like illness lasting about 24 hours with chills, aches, and cough.

SKIN & EYES: An irritant to the eyes and skin.

INGESTION: A poison with gastrointestinal effects.

Carcinogenicity	NTP Listed?	IARC Cancer Review Group?	OSHA Regulated?	Target Organs?
Confirmed Human Confirmed Animal	5th Annual Report	Group 3	29 CFR 1910.1000 Table Z-1	Respiratory system, skin, eyes

Medical Conditions Generally Aggravated by Exposure

None Reported. However, existing skin conditions or respiratory problems may be aggravated.

Emergency and First-aid Procedures

Eye contact: Flush immediately with water for 15 minutes (minimum), seek medical attention. Skin contact: Remove all contaminated clothing. Immediately wash area with large amounts of soap and water. For inhalation: Remove the person from exposure. Provide respiratory assistance and CPR. Transfer to medical facility. If swallowed: Provide breathing assistance if necessary. Never attempt to give a convulsing or unconscious person anything by mouth. Seek medical attention immediately. Contact poison control center for assistance.

SECTION VII - PRECAUTIONS FOR SAFE HANDLING AND USE

Steps to be Taken in Case Material is Released or Spilled

Collect powdered material in most convenient manner possible and deposit in sealed drum. Keep those not involved in cleanup from entering area. Do NOT dry sweep (creates airborne dusts). Use a vacuum equipped with a high efficiency particulate air (HEPA) filter.

Preferred Waste Disposal Method

No Citation.

Precautions to be Taken in Handling and Storage

Store in tightly closed containers in cool, well-ventilated area and protect from physical damage. Sources of ignition, such as smoking and open flames, are prohibited where chromium is used, handled, or stored.

Other Precautions and Warnings

A regulated, marked control area should be established where chromium is handled, used, or stored.

SECTION VIII - CONTROL MEASURES AND PERSONAL PROTECTIVE EQUIPMENT

Respiratory Protection *(specify type)*

At any exposure level, use an MSHA/NIOSH approved supplied-air respirator with a full facepiece, hood, or helmet operated in the continuous flow mode or a self-contained breathing apparatus (SCBA) operated in positive pressure mode.

Ventilation

Local exhaust or general mechanical systems recommended.

Protective Gloves	Eye Protection	Other Protective Clothing
Impervious Material	Chemical/Dust Goggles or Face Mask	Protective Suit or Clothing

Work/Hygiene Practices

Always wash hands thoroughly after using chemical; never bring food, drink, or smoking materials into vicinity of chemicals.

CHROMIUM

Cr CAS: 7440-47-3

IDENTIFICATION AND TYPICAL USES

Chromium is a hard, brittle, semi-gray, often lustrous metal. It commonly appears in industrial applications in powdered form. It is used as an alloying and plating element on metal and plastic substrates for corrosion resistance, in chromium-containing and stainless steels, as a protective coating for automotive and equipment accessories, in nuclear and high-temperature research, as a constituent of inorganic pigments, and in the tanning of leathers.

RISK ASSESSMENT: HEALTH

General Assessment

Chromium is a confirmed human carcinogen. It is a human poison by *ingestion* with gastrointestinal effects. It can also be toxic by *inhalation*. The toxicity of chromium alloys and compounds varies significantly. Chromium metal has not been shown to be toxic. However, divalent and trivalent compounds of chromium have a low order of toxicity in humans. Exposure to the dusts of chromite and ferrochrome alloys may cause lung diseases, including pulmonary fibrosis and pneumoconiosis.

Among the chromium compounds, only the hexavalent salts are considered to be extremely hazardous to human health. Occupational exposure can produce skin ulceration, dermatitis, perforation of the nasal septa, and kidney damage. Examples of hexavalent salts are the chromates and dichromates of sodium, potassium, and other metals. The water-soluble hexavalent chromium salts are absorbed into the bloodstream through inhalation. Many chromium (VI) compounds are confirmed carcinogens in humans and/or animals (causes lung cancers).

Inhalation of chromium fumes can cause "metal fume fever," which is a flu-like illness lasting up to 24 hours with characteristic symptoms of chills, aches, cough, and fever.

Acute Health Effects

The following acute (short-term) health effects may occur immediately or shortly after exposure to chromium:

Skin: Irritation, rash, and possible contact dermatitis.

Eye: Chromium particles can irritate the eyes.

Lung: Inhalation of chromium may cause a condition known as "metal fume fever" (flu-like symptoms).

Chronic Health Effects

The following chronic (long-term) health effects may occur at some time after exposure to chromium and can last for months or even years:

Cancer Hazards: Many chromium compounds are confirmed human carcinogens. They have been shown to cause cancer of the lung and throat in humans and test animals.

Reproductive Hazard: According to information available in the references, chromium has been tested for its ability to adversely affect reproduction in laboratory animals. There is no evidence to indicate that chromium adversely affects reproduction.

Other Chronic Effects: Chromium ore has been reported to cause lung allergy. Once allergy develops, even small future exposures may cause coughing, wheezing, or dypsnea. Repeated or severe exposures may cause lung cancer and/or thyroid cancers.

Recommended Risk-Reduction Measures

Personnel should avoid direct contact with chromium. If the specific chromium compound in question is unknown, adequate risk management dictates that it be handled and treated as though it were a carcinogen. If a less toxic chemical cannot be substituted for a hazardous substance, then *engineering controls* are the most effective method of reducing exposures.

The best protection is to enclose operations and/or provide local exhaust ventilation at the site of chemical release. While not always operationally feasible, isolating operations can also reduce exposure. Using respiratory protection is less effective than the controls mentioned above, but is still advisable whenever working with or around chromium. For any exposure level, use a supplied-air respirator with a full facepiece operated in positive pressure mode or an MSHA/NIOSH approved self-contained breathing apparatus (SCBA) with full facepiece and pressure demand as the recommended respiratory protection methods of choice. If a full facepiece is not available, then chemical splash goggles should be worn to protect the eyes. Whenever a chemical dust hazard exists, a face shield and a rubber apron should be worn. To prevent hand and skin exposures, impervious rubber gloves should be worn.

Administrative controls should also be in place to minimize the potential for human exposures. These may include written procedures or policies which specify the methods and techniques that will be practiced whenever personnel are to work with chromium.

All personnel should receive *training* on the use, hazards, protective measures, emergency actions, and other precautions per 29 CFR 1910.1200 (Hazard Communication), prior to the first assignment in an area where chromium is used or stored.

Before beginning employment and at regular intervals thereafter (e.g., annually), for those with frequent or potentially high exposures (half the TLV or greater, or significant skin contact), the following medical tests are recommended:

☑ Lung function tests (these may be normal if person is not displaying symptoms at the time of the test).

If symptoms develop or overexposure is suspected, the following may also be useful:

☑ Consider urine chromium levels. Persons without workplace exposure usually have levels below 15 micrograms per liter of urine.

☑ Lung function tests (these may be normal if person is not displaying symptoms at the time of the test).

Any evaluation should include a careful history of past and present symptoms with a medical examination. Medical tests that look for damage already done are not a substitute for controlling exposure. Also, since smoking can cause heart disease, as well as lung cancer, emphysema, and other related respiratory problems, the effects of exposure to chromium may be more pronounced in smokers than in non-smokers.

Other methods to reduce exposure include:

☑ Where possible, enclose operations and use local exhaust ventilation at the site of chemical release. If local exhaust ventilation or enclosure is not used, respiratory protection should be mandatory.

☑ Always ensure that proper protective clothing is worn when using chemical substances.

☑ Wash thoroughly immediately after exposure to chromium and at the end of the work shift or before eating, drinking, or smoking.

☑ Hazard warning information should be posted in the work area. In addition, as part of an on-going education and training program, all information on the health and safety hazards of chromium should be communicated to all exposed and potentially exposed workers.

☑ Where possible, automatically transfer chromium from drums or other storage containers to process containers.

☑ Specific engineering controls are recommended by NIOSH. Refer to NIOSH Criteria Documents: Occupational Exposure to Chromium VI. Publication Number 76-129.

RISK ASSESSMENT: ENVIRONMENT

General Assessment

The environment is at risk of exposure during transportation, storage, disposal, or destruction of chromium. In almost every scenario, the threat of environmental exposure is contingent upon the proper handling of the chemical substance. Accidental spills, large or small, can result in fire, explosion, and possible contamination of the surrounding environmental mediums (water, soil, and air).

Chromium powder will explode spontaneously in air. It will ignite and is potentially explosive in atmospheres of carbon dioxide. When heated in the presence of ammonium nitrate, violent or explosive reactions can result. In a fire, chromium will emit toxic and poisonous gases. It is also incompatible with strong oxidizers (such as chlorine, fluorine, bromine, and hydrogen peroxide), and many alkalis. These characteristics require special consideration during any emergency situation involving a leak or spill of chromium.

Chromium can enter the environment through effluents from the metal plating industries, from municipal waste treatment plant discharges, or from spills.

Acute Ecological Effects

Acute (short-term) toxic effects may include the death of animals, birds, or fish and death or low growth rate in plants. Acute effects are seen 2 to 4 days after animals or plants are exposed to chromium.

Chromium is more toxic in soft water than in hard water. Chromium (VI) has high acute toxicity to aquatic life, and chromium (III) has moderate acute toxicity to aquatic life. Insufficient data are available to evaluate the short-term effects of chromium exposure to plants, birds or land animals.

Chronic Ecological Effects

Chronic (long-term) toxic effects may include shortened life span, reproductive problems, lower fertility, and changes in appearance or behavior in exposed animals. These effects can be seen long after first exposure(s) to toxic chemicals.

Chromium (III) and chromium (IV) both have high chronic toxicity to aquatic life. Insufficient data are available to evaluate the long-term effects of chromium exposure to plants, birds or land animals.

Water Solubility

Water solubility of chromium and its salts ranges from low to high.

Persistence in the Environment

Chromium is highly persistent in the aquatic environment, with a half-life of greater than 200 days. The half-life of a pollutant is the amount of time it takes for one half of the chemical to be degraded.

Bioaccumulation in Aquatic Organisms

Some substances increase in concentration, or *bioaccumulate,* in living organisms as they breathe contaminated air, drink contaminated water, or eat contaminated food. These chemicals can become concentrated in the tissues and internal organs of animals as well as humans.

The concentration of chromium found in fish is expected to be somewhat higher than the average concentration of chromium in the water from which the fish was taken.

Recommended Risk-Reduction Measures

Proper training of all transporters will reduce the likelihood of a mishap or accidents resulting in a leak or spill to the environment. The correct use of DOT labeling (per the MSDS) on all containers, trucks, and rail cars will enable emergency responders to react properly and quickly to any disaster thereby reducing the potential risk to the environment and to personnel.

Storage of chromium should be segregated from incompatible chemicals to minimize the risk of cross-contamination or contact. Bulk storage of chromium should be avoided.

If a spill or leak to the environment has occurred, fire department, emergency response and/or hazardous materials spill personnel should be notified immediately. Use a vacuum or wet method to reduce dust during cleanup. Do not dry sweep. When vacuuming, a high efficiency particulate air (HEPA) filter should be used, not a standard shop vacuum. Deposit collected materials in sealed drum for disposal. Cleanup should be attempted only by those trained in proper spill containment procedures. Contaminated soils should be removed and replaced with clean soil. If chromium should contact the water table, aquifer, or navigable waterway, time is of the essence. Chromium compounds may be moderately soluble in water and, therefore, total containment and remediation may not be entirely possible. When such spills occur, the local and/or state emergency response authorities must be notified. A comprehensive emergency response or disaster preparedness/recovery plan should be in place prior to any operations involving the use, transportation, storage, or disposal of chromium.

If chromium is spilled or leaked, the following specific steps are recommended:

- ☑ Restrict persons not wearing protective clothing from area of spill or leak until cleanup is complete and area can be opened for normal work.
- ☑ Collect powdered materials in most convenient and safe manner possible and place in sealed drums for disposal. Do not dry sweep, use HEPA vacuum instead.
- ☑ It may be necessary to dispose of chromium as a hazardous waste. The responsible state agency or the regional office of the federal Environmental

Protection Agency (EPA) should be contacted for specific recommendations.

RISK ASSESSMENT: BUSINESS

General Assessment

Accidents or mishaps involving chromium can present a significant threat to business operations. The loss or damage of equipment or facilities can significantly affect fiscal viability. Lawsuits that may result from personnel injury/death, public exposures, and/or environmental contamination will also require a serious expenditure of resources. Media attention surrounding an injury, death, or environmental damage can also result in a loss of profits and loss of current as well as future business.

Recommended Risk-Reduction Measures

Company attorneys, safety and health professionals, and environmental specialists should be involved in the development of any procedures dealing with safety and emergency response. A company official should be pre-designated as a public relations officer with specific training in dealing with the press. Corporate plans and policies should be developed, approved, and implemented long before any need for such arises.

REFERENCES

Ahlbom, A. 1993. *Biostatistics for Epidemiologists.* New York: CRC Press/Lewis Publishers.

American Conference of Governmental Industrial Hygienists. 1988. *Documentation of the Threshold Limit Values and Biological Exposure Indices,* 5th Edition (with updates). Cincinnati: ACGIH

Asante-Duah, D. K. 1993. *Hazardous Waste Risk Assessment.* New York: CRC Press/Lewis Publishers.

Bloom, A. D. and F. J. DeSerres. 1995. *Ecotoxicity and Human Health.* New York: CRC Press/Lewis Publishers.

Brusick, D. J. 1994. *Methods for Genetic Risk Assessment.* New York: CRC Press/Lewis Publishers.

Calabrese, E. J., E. M. Kenyon. *Air Toxics and Risk Assessment.* New York: CRC Press/Lewis Publishers.

Calabrese, E. J. 1994. *Biological Effects of Low Level Exposures.* New York: CRC Press/Lewis Publishers.

Cockerman, L. G., B. S. Shane. 1994. *Basic Environmental Toxicology.* New York: CRC Press/Lewis Publishers.

Cothern, C. R., N. P. Ross. 1994. *Environmental Statistics, Assessment, and Forecasting.* New York: CRC Press/Lewis Publishers.

Cothern, R. R. 1993. *Comparative Environmental Risk Assessment.* New York: CRC Press/Lewis Publishers.

Cothern, C. R. 1995. *Handbook for Environmental Risk Decisionmaking.* New York: CRC Press/Lewis Publishers.

Gardner, R. W. 1994. *Chemical Intolerance.* New York: CRC Press/Lewis Publishers.

Hallenbeck, W. H. 1993. *Quantitative Risk Assessment for Environmental and Occupational Health,* 2nd Edition. New York: CRC Press/Lewis Publishers.

Howard, P. H., et. al. 1991. *Handbook of Environmental Degradation Rates.* New York: CRC Press/Lewis Publishers.

Howard, P. H. 1995. *Handbook of Environmental Fate and Exposures Data for Organic Chemicals,* Vol. I-IV. New York: CRC Press/Lewis Publishers.

Jones, F. E. 1994. *Toxic Organic Vapors in the Workplace.* New York: CRC Press/Lewis Publishers.

Keith, L. H. and M. M. Walker. 1995. *Handbook of Air Toxics.* New York: CRC Press/Lewis Publishers.

Landis, W. G., M. H. Yu. 1995. *Introduction to Environmental Toxicology.* New York: CRC Press/Lewis Publishers.

Langerman, N. 1994 *Precautionary Labels for Chemical Containers.* New York: CRC Press/Lewis Publishers.

Lauwerys, R. R., P. Hoet. 1993. *Industrial Chemical Exposure.* New York: CRC Press/Lewis Publishers.

Lewis, R. J., Sr. 1992. *Sax's Dangerous Properties of Industrial Materials,* Volumes 1, 2, and 3, 8th Edition. New York: Van Nostrand Reinhold.

Manahan, S. E. 1994. *Environmental Chemistry,* 6th Edition. New York: CRC Press/Lewis Publishers.

Manahan, S. E. 1992. *Toxicological Chemistry,* 2nd Edition. New York: CRC Press/Lewis Publishers.

Meyer, E. 1990. *Chemistry of Hazardous Materials.* Englewood Cliffs, NJ: Prentice Hall.

Patnaik, P. 1992. *A Comprehensive Guide to the Hazardous Properties of Chemical Substances.* New York: Van Nostrand Reinhold.

Quigley, D. R. 1994. *Handbook of Emergency Chemical Management.* New York: CRC Press/Lewis Publishers.

Rea, W. J. 1995. *Chemical Sensitivity,* Vol. I-III. New York: CRC Press/Lewis Publishers.

Sacarello, H. L. A. 1994. *The Comprehensive Handbook of Hazardous Materials.* New York: CRC Press/Lewis Publishers.

Sherman, J. D. 1988. *Chemical Exposure and Disease.* New York: Van Nostrand Reinhold.

Talbott, E., G. Craun. 1995. *An Introduction to Environmental Epidemiology.* New York: CRC Press/Lewis Publishers.

U.S. Department of Health and Human Services (DHHS), National Institute for Occupational Health (NIOHS). 1994. *NIOSH Pocket Guide to Hazardous Chemicals.* Publication Number 94-116. Washington, D.C.; U.S. Government Printing Office.

U.S. Department of Health, Education, and Welfare (DHEW), National Institute for Occupational Health (NIOHS). 1976. *NIOSH Criteria Document: Occupational Exposure to Chromium (VI).* Publication Number 76-129. Washington, D.C.; U.S. Government Printing Office.

MATERIAL SAFETY DATA SHEET

CHEMICAL NAME

CHROMIUM (III) OXIDE

CRC PRESS/LEWIS PUBLISHERS DISCLAIMER: THE INFORMATION AND RECOMMENDATIONS PRESENTED HEREIN ARE BASED ON SOURCES BELIEVED TO BE RELIABLE. CRC MAKES NO REPRESENTATION ON ITS COMPLETENESS OR ACCURACY. IT IS THE RESPONSIBILITY OF THE USER TO DETERMINE THE CHEMICAL'S SUITABILITY FOR ITS INTENDED USE, THE CHEMICAL'S SAFE USE, AND THE CHEMICAL'S PROPER DISPOSAL. NO REPRESENTATIONS AND/OR WARRANTIES, EITHER EXPRESSED OR IMPLIED, OF THE MERCHANTABILITY OR FITNESS FOR A PARTICULAR PURPOSE, OR OF ANY OTHER NATURE, ARE MADE WITH RESPECT TO THE INFORMATION PROVIDED IN THIS MSDS OR TO THE CHEMICAL TO WHICH INFORMATION MAY REFER. CRC NEITHER ASSUMES NOR AUTHORIZES ANY OTHER PERSON TO ASSUME FOR IT, ANY OTHER ADDITIONAL RESPONSIBILITY OR LIABILITY FOR THE USE OF, OR RELIANCE UPON, THIS INFORMATION.

HAZARD WARNING INFORMATION

HEALTH	FIRE	REACTIVE	OTHER	DEGREE OF HAZARD	COLOR CODING	OTHER CODES
3	0	0	OX	0 = Minimum Hazard 1 = Slight Hazard 2 = Moderate Hazard 3 = Serious Hazard 4 = Severe Hazard	HEALTH = BLUE FIRE = RED REACTIVITY = YELLOW OTHER = WHITE	OX = Oxidizer ACID = Acid ALK = Alkali COR = Corrosive ~~W~~ = Use No Water

SECTION I - GENERAL INFORMATION

Characterization: **Metal**	RCRA Number: **None** — EPA Class: **Not Applicable**
DOT Proper Shipping Name: **No Citation**	Chemical Abstract Service (CAS) Number: **1308-38-9**
DOT Hazard Class and Label Requirements: **No Citation**	DOT Emergency Guide Code: **No Citation**
DOT Identification Number: **No Citation**	Chemical Formula: **Cr_2O_3**

Synonyms

Chrome green; green cinnabar; chrome (III) oxide; anadomis green; chromic acid green; chromic oxide; ultramarine green, dichromium trioxide.

SECTION II - HAZARDOUS INGREDIENTS AND IDENTITY INFORMATION

Hazardous Components (specific identity)	OSHA Exposure Criteria	NIOSH Exposure Criteria	Immediately Dangerous to Life and Health (IDLH)	ACGIH Exposure Criteria
Chromic oxide (derivation: By heating chromium hydroxide; by heating dry ammonium dichromate; by heating sodium dichromate with sulfur and washing out the sodium sulfate).	**PEL:** 1 mg(Cr)/m^3 **STEL:** Not Established	**REL:** 0.5 mg(Cr)/m^3 **STEL:** Not Established	250mg/m^3 as Cr	**TLV:** 0.5 mg(Cr)/m^3 **STEL:** Not Established

SECTION III - PHYSICAL AND CHEMICAL CHARACTERISTICS

Boiling Point: **5432°F (3000°C)**	Specific Gravity (H_2O = 1): **5.2**
Vapor Pressure (mm Hg): **Not Applicable**	Molecular Weight: **152**
Vapor Density (Air = 1): **Not Applicable**	Melting Point: **4415°F (2435°C)**

Solubility

Insoluble in water, acids, and alkalis.

Appearance and Odor

A bright green, extremely hard, crystalline, odorless powder.

SECTION IV - FIRE AND EXPLOSION HAZARD DATA

Flash Point (method used): **Not Applicable**	Explosive Limits in Air % by Volume: **LEL: Not Applicable** **UEL: Not Applicable**
NFPA Classification: **No Citation**	Autoignition Temperature: **Not Determined**

Extinguishing Media

Use extinguishing agent suitable to surrounding fire. Chromium (III) oxide itself does not burn.

Special Fire Fighting Procedures

Wear full protective clothing and self-contained breathing apparatus (SCBA). Use a water spray to keep fire-exposed containers cool. A powerful oxidizing agent. Remain clear of smoke, water fall-out and water run-off. Poisonous gases are produced in fire. Treat all materials used or generated and equipment involved as contaminated by hazardous waste.

Unusual Fire and Explosion Hazards

Can cause violent explosions upon contact with strong oxidizers. Powders may explode in air.

SECTION V - REACTIVITY DATA

Stability		Conditions to Avoid **Keep away from incompatible materials since violent and or explosive reactions can occur.**
Stable	Unstable	Incompatibility *(materials to avoid)* **Incompatible with strong oxidizers (such as chlorine, bromine, hydrogen peroxide, and fluorine), glycerol, and lithium. Reacts violently with chlorine trifluoride.**
X		
Hazardous Polymerization		Conditions to Avoid **Under normal conditions of temperature and pressure, hazardous polymerization of chromium is not known to occur.**
May Occur	Will Not Occur	Hazardous Decomposition or By-products **When heated to decomposition, chromium can emit highly toxic/poisonous fumes and gases.**
	X	

SECTION VI - HEALTH HAZARD DATA

Primary Route(s) of Entry:	Inhalation?	Absorption (skin)?	Ingestion?
	X		X

Health Hazards

INHALATION: Irritation of the respiratory system. Exposure to some chromium compounds can cause sores or "chrome holes" to appear in the nasal cavity. It is not known if chromium (III) oxide will cause such damage.

SKIN & EYES: An irritant to the eyes and skin. Skin allergy may develop with itching, redness and/or an eczema-like rash.

INGESTION: A poison with gastrointestinal effects.

Carcinogenicity	NTP Listed?	IARC Cancer Review Group?	OSHA Regulated?	Target Organs?
Confirmed Human Confirmed Animal	5th Annual Report	Group 3	29 CFR 1910.1000 Table Z-1	Respiratory system, skin, eyes.

Medical Conditions Generally Aggravated by Exposure

None Reported. However, existing skin conditions or respiratory problems may be aggravated.

Emergency and First-aid Procedures

Eye contact: Flush immediately with water for 15 minutes (minimum), seek medical attention. Skin contact: Remove all contaminated clothing. Immediately wash area with large amounts of soap and water. For inhalation: Remove the person from exposure. Provide respiratory assistance and CPR. Transfer to medical facility. If swallowed: Provide breathing assistance if necessary. Never attempt to give a convulsing or unconscious person anything by mouth. Seek medical attention immediately.

SECTION VII - PRECAUTIONS FOR SAFE HANDLING AND USE

Steps to be Taken in Case Material is Released or Spilled

Collect powdered material in most convenient manner possible and deposit in sealed drum. Keep those not involved in cleanup from entering area. Do NOT dry sweep (creates airborne dusts). Use a vacuum equipped with a high efficiency particulate air (HEPA) filter.

Preferred Waste Disposal Method

No Citation.

Precautions to be Taken in Handling and Storage

Store in tightly closed containers in cool, well-ventilated area. Sources of ignition, such as smoking and open flames, are prohibited where chromium (III) oxide is used, handled, or stored.

Other Precautions and Warnings

A regulated, marked control area should be established where chromium is handled, used, or stored.

SECTION VIII - CONTROL MEASURES AND PERSONAL PROTECTIVE EQUIPMENT

Respiratory Protection *(specify type)*

At any exposure level, use an MSHA/NIOSH approved supplied-air respirator with a full facepiece, hood, or helmet operated in the continuous flow mode or a self-contained breathing apparatus (SCBA) operated in positive pressure mode.

Ventilation

Local exhaust or general mechanical systems recommended.

Protective Gloves	Eye Protection	Other Protective Clothing
Impervious Material	Chemical/Dust Goggles or Face Mask	Protective Suit or Clothing

Work/Hygiene Practices

Always wash hands thoroughly after using chemical; never bring food, drink, or smoking materials into vicinity of chemicals.

CHROMIUM (III) OXIDE

Cr_2O_3 CAS. 1308-38-9

IDENTIFICATION AND TYPICAL USES

Chromium (III) oxide is a bright green, extremely hard, crystalline, odorless powder. It is used primarily as a paint pigment, in metallurgy, as a fixative for certain textile dyes, in ceramics, as green granules in asphalt roofing, as a component in refractory brick, as a abrasive, and as a catalyst in organic synthesis.

RISK ASSESSMENT: HEALTH

General Assessment

Chromium (III) oxide is a confirmed human carcinogen with experimental teratogenic and mutagenic data also reported. It is a human poison by *ingestion* with gastrointestinal effects. It can also be toxic by *inhalation.*

Among the chromium compounds, the trivalent salts are normally considered to be of low toxicity. However, chromium (III) oxide has been shown to be extremely hazardous to human health. Occupational exposure can produce skin allergy (dermatitis), severe irritation to the eyes, nose, and mucous membranes of the respiratory tract. Chronic exposures may lead to lung and/or throat cancers.

Acute Health Effects

The following acute (short-term) health effects may occur immediately or shortly after exposure to chromium:

Skin: Irritation, rash, and possible contact dermatitis.

Eye: Chromium particles can irritate the eyes.

Lung: Acute affects on lung tissues appear to manifest as irritation of the mucous membranes of the respiratory tract. May cause congestion, cough with phlegm, tightness in chest, and other symptoms of respiratory irritation and inflammation. Possible nausea and headache.

Chronic Health Effects

The following chronic (long-term) health effects may occur at some time after exposure to chromium and can last for months or even years:

Cancer Hazards: Chromium (III) oxide is a confirmed human carcinogen. It has been shown to cause lung and throat cancers. Repeated or long-term exposures to this compound may also cause mutations in living cells.

Reproductive Hazard: Chromium (III) oxide may be a teratogen in humans since it has been shown to be a teratogen in test animals.

Other Chronic Effects: Skin allergy can develop with red, itching rash resembling that of eczema. When this occurs, even small future exposure can trigger symptoms. Repeated or severe exposures may cause lung cancer and/or throat cancers.

Recommended Risk-Reduction Measures

Personnel should avoid direct contact with chromium. It is a confirmed human carcinogen and many experts believe there is no safe exposure level to such compounds. Adequate risk management dictates that it be handled and treated with extreme caution. If a less toxic chemical cannot be substituted for a hazardous substance, then *engineering controls* are the most effective method of reducing exposures.

The best protection is to enclose operations and/or provide local exhaust ventilation at the site of chemical release. While not always operationally feasible, isolating operations can also reduce exposure. Using respiratory protection is less effective than the controls mentioned above, but is still advisable whenever working with or around chromium. For any exposure to a carcinogen, use a supplied-air respirator with a full facepiece operated in positive pressure mode or an MSHA/NIOSH approved self-contained breathing apparatus (SCBA) with full facepiece and pressure demand as the recommended respiratory protection methods of choice. If a full facepiece is not available, then chemical splash goggles should be worn to protect the eyes. Whenever a chemical dust hazard exists, a face shield and a rubber apron should be worn. To prevent hand and skin exposures, impervious rubber gloves should be worn.

Administrative controls should also be in place to minimize the potential for human exposures. These may include written procedures or policies which specify the methods and techniques that will be practiced whenever personnel are to work with chromium (III) oxide. A regulated, marked work area must be established wherever chromium (III) oxide is used, handled, or stored.

All personnel should receive *training* on the use, hazards, protective measures, emergency actions, and other precautions per 29 CFR 1910.1200 (Hazard Communication), prior to the first assignment in an area where chromium (III) oxide is used or stored.

Before beginning employment and at regular intervals thereafter (e.g., annually), for those with frequent or potentially high exposures (half the PEL or greater, or significant skin contact), the following medical tests are recommended:

- ☑ Evaluation by a qualified allergist, including careful exposure history and special testing (may help diagnose skin allergy).

If symptoms develop or overexposure is suspected, the following may also be useful:

- ☑ Consider urine chromium levels. Persons without workplace exposure usually have levels below 15 micrograms per liter of urine.
- ☑ Lung function tests (these may be normal if person is not displaying symptoms at the time of the test).

Any evaluation should include a careful history of past and present symptoms with a medical examination. Medical tests that look for damage already done are not a substitute for controlling exposure.

Other methods to reduce exposure include:

- ☑ Where possible, enclose operations and use local exhaust ventilation at the site of chemical release. If local exhaust ventilation or enclosure is not used, respiratory protection should be mandatory.
- ☑ Always ensure that proper protective clothing is worn when using chemical substances.
- ☑ Wash thoroughly immediately after exposure to chromium and at the end of the work shift or before eating, drinking, or smoking.
- ☑ Hazard warning information should be posted in the work area. In addition, as part of an on-going education and training program, all information on the health and safety hazards of chromium should be communicated to all potentially exposed workers.
- ☑ Where possible, automatically transfer chromium from drums or other storage containers to process containers.

RISK ASSESSMENT: ENVIRONMENT

General Assessment

The environment is at risk of exposure during transportation, storage, disposal, or destruction of chromium compounds, including chromium (III) oxide. In almost every scenario, the threat of environmental exposure is contingent upon the proper handling of the chemical substance. Accidental spills, large or small, can result in fire, explosion, and possible contamination of the surrounding environmental mediums (water, soil, and air).

Some chromium powders may explode spontaneously in air. In a fire, chromium will emit toxic and poisonous gases. It is also incompatible with strong oxidizers (such as chlorine, fluorine, bromine, and hydrogen peroxide), and will react violently in the presence of chlorine trifluoride. These characteristics require special consideration during any emergency situation involving a leak or spill of chromium.

Chromium can enter the environment through effluents from the metal plating industries, from municipal waste treatment plant discharges, or from spills.

☠ *Acute Ecological Effects*

Acute (short-term) toxic effects may include the death of animals, birds, or fish and death or low growth rate in plants. Acute effects are seen 2 to 4 days after animals or plants are exposed to chromium.

Chromium is more toxic in soft water than in hard water. Chromium (III) oxide has moderate acute toxicity to aquatic life. Insufficient data are available to evaluate the short-term effects of chromium exposure to plants, birds or land animals.

Chronic Ecological Effects

Chronic (long-term) toxic effects may include shortened life span, reproductive problems, lower fertility, and changes in appearance or behavior in exposed animals. These effects can be seen long after first exposure(s) to toxic chemicals.

Chromium (III) oxide has high chronic toxicity to aquatic life. Insufficient data are available to evaluate the long-term effects of chromium exposure to plants, birds or land animals.

Water Solubility

Water solubility of chromium and its salts ranges from low to high.

Persistence in the Environment

Chromium is highly persistent in the aquatic environment, with a half-life of greater than 200 days. The half-life of a pollutant is the amount of time it takes for one half of the chemical to be degraded.

Bioaccumulation in Aquatic Organisms

Some substances increase in concentration, or *bioaccumulate,* in living organisms as they breathe contaminated air, drink contaminated water, or eat contaminated food. These chemicals can become concentrated in the tissues and internal organs of animals as well as humans.

The concentration of chromium found in fish is expected to be somewhat higher than the average concentration of chromium in the water from which the fish was taken.

Recommended Risk-Reduction Measures

Proper training of all transporters will reduce the likelihood of a mishap or accidents resulting in a leak or spill to the environment. The correct use of DOT labeling (per the MSDS) on all containers, trucks, and rail cars will enable emergency responders to react properly and quickly to any disaster thereby reducing the potential risk to the environment and to personnel.

Storage of chromium (III) oxide should be segregated from incompatible chemicals to minimize the risk of cross-contamination or contact. Bulk storage should be avoided.

If a spill or leak to the environment has occurred, fire department, emergency response and/or hazardous materials spill personnel should be notified immediately. Use a vacuum or wet method to reduce dust during cleanup. Do not dry sweep. When vacuuming, a high efficiency particulate air (HEPA) filter should be used, not a standard shop vacuum. Deposit collected materials in sealed drum for disposal. Cleanup should be attempted only by those trained in proper spill containment procedures. Contaminated soils should be removed and replaced with clean soil. If chromium compounds should contact the water table, aquifer, or navigable waterway, time is of the essence. Chromium compounds may be moderately soluble in water and, therefore, total containment and remediation may not be entirely possible. When such spills occur, the local and/or state emergency response authorities must be notified. A comprehensive emergency response or disaster preparedness/recovery plan should be in place prior to any operations involving the use, transportation, storage, or disposal of chromium. If chromium is spilled or leaked, the following specific steps are recommended:

- ☑ Restrict persons not wearing protective clothing from area of spill or leak until cleanup is complete and area can be opened for normal work.
- ☑ Collect powdered materials in most convenient and safe manner possible and place in sealed drums for disposal. Do not dry sweep, use vacuum equipped with a high efficiency particulate air (HEPA) filter instead.
- ☑ It may be necessary to dispose of chromium as a hazardous waste. The responsible state agency or the regional office of the federal Environmental Protection Agency (EPA) should be contacted for specific recommendations.

RISK ASSESSMENT: BUSINESS

General Assessment

Accidents or mishaps involving chromium can present a significant threat to business operations. The loss or damage of equipment or facilities can significantly affect fiscal viability. Lawsuits that may result from personnel injury/death, public exposures, and/or environmental contamination will also require a serious expenditure of resources. Media attention surrounding an injury, death, or environmental damage can also result in a loss of profits and loss of current as well as future business. Always remember that anytime the words "carcinogen" or "cancer" are used, public reaction and emotion run equally high.

ⓤ *Recommended Risk-Reduction Measures*

Company attorneys, safety and health professionals, and environmental specialists should be involved in the development of any procedures dealing with safety and emergency response. A company official should be pre-designated as a public relations officer with specific training in dealing with the press. Corporate plans and policies should be developed, approved, and implemented long before any need for such arises.

REFERENCES

Ahlbom, A. 1993. *Biostatistics for Epidemiologists.* New York: CRC Press/Lewis Publishers.

American Conference of Governmental Industrial Hygienists. 1988. *Documentation of the Threshold Limit Values and Biological Exposure Indices,* 5th Edition (with updates). Cincinnati: ACGIH

Asante-Duah, D. K. 1993. *Hazardous Waste Risk Assessment.* New York: CRC Press/Lewis Publishers.

Bloom, A. D. and F. J. DeSerres. 1995. *Ecotoxicity and Human Health.* New York: CRC Press/Lewis Publishers.

Brusick, D. J. 1994. *Methods for Genetic Risk Assessment.* New York: CRC Press/Lewis Publishers.

Calabrese, E. J., E. M. Kenyon. *Air Toxics and Risk Assessment.* New York: CRC Press/Lewis Publishers.

Calabrese, E. J. 1994. *Biological Effects of Low Level Exposures.* New York: CRC Press/Lewis Publishers.

Cockerman, L. G., B. S. Shane. 1994. *Basic Environmental Toxicology.* New York: CRC Press/Lewis Publishers.

Cothern, C. R., N. P. Ross. 1994. *Environmental Statistics, Assessment, and Forecasting.* New York: CRC Press/Lewis Publishers.

Cothern, R. R. 1993. *Comparative Environmental Risk Assessment.* New York: CRC Press/Lewis Publishers.

Cothern, C. R. 1995. *Handbook for Environmental Risk Decisionmaking.* New York: CRC Press/Lewis Publishers.

Gardner, R. W. 1994. *Chemical Intolerance.* New York: CRC Press/Lewis Publishers.

Hallenbeck, W. H. 1993. *Quantitative Risk Assessment for Environmental and Occupational Health,* 2nd Edition. New York: CRC Press/Lewis Publishers.

Howard, P. H., et. al. 1991. *Handbook of Environmental Degradation Rates.* New York: CRC Press/Lewis Publishers.

Howard, P. H. 1995. *Handbook of Environmental Fate and Exposures Data for Organic Chemicals,* Vol. I-IV. New York: CRC Press/Lewis Publishers.

Jones, F. E. 1994. *Toxic Organic Vapors in the Workplace.* New York: CRC Press/Lewis Publishers.

Keith, L. H. and M. M. Walker. 1995. *Handbook of Air Toxics.* New York: CRC Press/Lewis Publishers.

Landis, W. G., M. H. Yu. 1995. *Introduction to Environmental Toxicology.* New York: CRC Press/Lewis Publishers.

Langerman, N. 1994 *Precautionary Labels for Chemical Containers.* New York: CRC Press/Lewis Publishers.

Lauwerys, R. R., P. Hoet. 1993. *Industrial Chemical Exposure.* New York: CRC Press/Lewis Publishers.

Lewis, R. J., Sr. 1992. *Sax's Dangerous Properties of Industrial Materials,* Volumes 1, 2, and 3, 8th Edition. New York: Van Nostrand Reinhold.

Lide, D. R. 1995. *Handbook of Chemistry and Physics,* 76th Edition. New York: CRC Press/Lewis Publishers.

Manahan, S. E. 1994. *Environmental Chemistry,* 6th Edition. New York: CRC Press/Lewis Publishers.

Manahan, S. E. 1992. *Toxicological Chemistry,* 2nd Edition. New York: CRC Press/Lewis Publishers.

Meyer, E. 1990. *Chemistry of Hazardous Materials.* Englewood Cliffs, NJ: Prentice Hall.

Mycock, J. C., J. D. McKenna, and L. Theodore. 1995. *Handbook of Air Pollution Control Engineering and Technology.* New York: CRC Press/Lewis Publishers.

Nelson, G. O. 1992. *Gas Mixtures.* New York: CRC Press/Lewis Publishers.

Ott, W. R. 1995. *Environmental Statistics and Data Analysis.* New York: CRC Press/Lewis Publishers.

Patnaik, P. 1992. *A Comprehensive Guide to the Hazardous Properties of Chemical Substances.* New York: Van Nostrand Reinhold.

Philp, R. B. 1995. *Environmental Hazards and Human Health.* New York: CRC Press/Lewis Publishers.

Quigley, D. R. 1994. *Handbook of Emergency Chemical Management.* New York: CRC Press/Lewis Publishers.

Rea, W. J. 1995. *Chemical Sensitivity,* Vol. I-III. New York: CRC Press/Lewis Publishers.

Sacarello, H. L. A. 1994. *The Comprehensive Handbook of Hazardous Materials.* New York: CRC Press/Lewis Publishers.

Sherman, J. D. 1988. *Chemical Exposure and Disease.* New York: Van Nostrand Reinhold.

Talbott, E., G. Craun. 1995. *An Introduction to Environmental Epidemiology.* New York: CRC Press/Lewis Publishers.

U.S. Department of Health and Human Services (DHHS), National Institute for Occupational Health (NIOHS). 1994. *NIOSH Pocket Guide to Hazardous Chemicals.* Publication Number 94-116. Washington, D.C.; U.S. Government Printing Office.

Witorsch, P. and S. V. Spagnolo. 1994. *Air Pollution and Lung Disease in Adults.* New York: CRC Press/Lewis Publishers.

MATERIAL SAFETY DATA SHEET

CHEMICAL NAME	CRC PRESS/LEWIS PUBLISHERS DISCLAIMER
COBALT	

HAZARD WARNING INFORMATION

HEALTH	FIRE	REACTIVE	OTHER	DEGREE OF HAZARD	COLOR CODING	OTHER CODES
3	0	0		0 = Minimum Hazard 1 = Slight Hazard 2 = Moderate Hazard 3 = Serious Hazard 4 = Severe Hazard	HEALTH = BLUE FIRE = RED REACTIVITY = YELLOW OTHER = WHITE	OX = Oxidizer ACID = Acid ALK = Alkali COR = Corrosive W = Use No Water

SECTION I - GENERAL INFORMATION

Characterization	**Metal**	RCRA Number	**None**
		EPA Class	**Not Applicable**
DOT Proper Shipping Name	**No Citation**	Chemical Abstract Service (CAS) Number	**7440-48-4**
DOT Hazard Class and Label Requirements	**No Citation**	DOT Emergency Guide Code	**No Citation**
DOT Identification Number	**No Citation**	Chemical Formula	**Co**

Synonyms

Super cobalt.

SECTION II - HAZARDOUS INGREDIENTS AND IDENTITY INFORMATION

Hazardous Components (specific identity)	OSHA Exposure Criteria	NIOSH Exposure Criteria	Immediately Dangerous to Life and Health (IDLH)	ACGIH Exposure Criteria
Cobalt (derivation: From ore concentration by roasting followed by thermal reduction by aluminum; by electrolytic reduction of solutions of metal, or; by leaching with either ammonia or acid in an autoclave under elevated temperatures and pressures and subsequent reduction by hydrogen).	**PEL:** 0.05 mg/m^3 **STEL:** Suspected Cancer Agent	**REL:** 0.05 mg/m^3 **STEL:** Possible Human Carcinogen	20 mg/m^3 as Co	**TLV:** 0.05 mg/m^3 **STEL:** Suspected Human Carcinogen

SECTION III - PHYSICAL AND CHEMICAL CHARACTERISTICS

Boiling Point	**5612°F (3100°C)**	Specific Gravity (H_2O = 1)	**8.92**
Vapor Pressure (mm Hg)	**0 (approx.) at 69°F (20°C)**	Molecular Weight	**58.9**
Vapor Density (Air = 1)	**Not Found**	Melting Point	**2719°F (1493°C)**

Solubility

Insoluble in water. Soluble in nitric acid, attacked by dilute hydrochloric and sulfuric acids.

Appearance and Odor

Steel-gray, shinning, hard, ductile, somewhat malleable metal; ferromagnetic and semi-permeable.

SECTION IV - FIRE AND EXPLOSION HAZARD DATA

Flash Point (method used)	**Not Applicable**	Explosive Limits in Air % by Volume	**LEL: Not Applicable** **UEL: Not Applicable**
NFPA Classification	**No Citation**	Autoignition Temperature	**Not Determined**

Extinguishing Media

Use class "D" extinguishing agent. Do NOT use water.

Special Fire Fighting Procedures

Wear full protective clothing and self-contained breathing apparatus (SCBA). Use a water spray to keep fire-exposed containers cool. Water may be ineffective on fire. Burning cobalt metal may reach temperatures in excess of 5000°F. Remain clear of smoke, water fall-out and water run-off. Poisonous gases are produced in fire.

Unusual Fire and Explosion Hazards

Powdered cobalt ignites spontaneously in air. Can cause violent explosions upon contact with strong oxidizers. Powders may explode in air. Flammable when exposed to heat or flame.

SECTION V - REACTIVITY DATA

Stability		Conditions to Avoid Keep away from incompatible materials since violent and or explosive reactions can occur.
Stable X	Unstable	Incompatibility *(materials to avoid)* Incompatible with strong oxidizers (such as perchlorates, peroxides, permanganates, chlorates, and nitrates), and acids (such as hydrochloric, sulfuric, and nitric), and ammonium nitrate.
Hazardous Polymerization		Conditions to Avoid Under normal conditions of temperature and pressure, hazardous polymerization of cobalt is not known to occur.
May Occur	Will Not Occur X	Hazardous Decomposition or By-products None noted.

SECTION VI - HEALTH HAZARD DATA

Primary Route(s) of Entry:	Inhalation?	Absorption (skin)?	Ingestion?
	X		X

Health Hazards

INHALATION: Irritation of the respiratory system. Exposure can cause allergic lung reaction with chest pain, coughing, wheezing, and shortness of breath. Repeated exposure may cause lung scarring (fibrosis), even if no symptoms are noticed. May also cause pulmonary edema.

SKIN & EYES: An irritant to the eyes and skin. The powder may cause contact dermatitis.

INGESTION: A poison with gastrointestinal effects with nausea and vomiting.

Carcinogenicity	NTP Listed?	IARC Cancer Review Group?	OSHA Regulated?	Target Organs?
Confirmed Human Confirmed Animal	5th Annual Report	No	No	Respiratory system, skin, eyes

Medical Conditions Generally Aggravated by Exposure

None Reported. However, existing skin conditions or respiratory problems may be aggravated.

Emergency and First-aid Procedures

Eye contact: Flush immediately with water for 15 minutes (minimum), seek medical attention. Skin contact: Remove all contaminated clothing. Immediately wash area with large amounts of soap and water. For inhalation: Remove the person from exposure. Provide respiratory assistance and CPR. Transfer to medical facility. Observe for 24 - 48 hours for lung effects (pulmonary edema). If swallowed: Provide breathing assistance if necessary. Never attempt to give a convulsing or unconscious person anything by mouth. Seek medical attention immediately. Contact poison control center for assistance.

SECTION VII - PRECAUTIONS FOR SAFE HANDLING AND USE

Steps to be Taken in Case Material is Released or Spilled

Collect powdered material in most convenient manner possible and deposit in sealed drum. Keep those not involved in cleanup from entering area. Do NOT dry sweep (creates airborne dusts). Use a vacuum equipped with a high efficiency particulate air (HEPA) filter. If spill involves radioactive cobalt, evacuate area and delay cleanup until instructed by qualified radiation authorities.

Preferred Waste Disposal Method

No Citation.

Precautions to be Taken in Handling and Storage

Store in tightly closed containers in cool, well-ventilated area and protect from physical damage. Sources of ignition are prohibited where cobalt is used, handled, or stored.

Other Precautions and Warnings

A regulated, marked area should be established where radioactive cobalt is handled, used, or stored.

SECTION VIII - CONTROL MEASURES AND PERSONAL PROTECTIVE EQUIPMENT

Respiratory Protection *(specify type)*

For exposure to non-radioactive cobalt, use a full facepiece respirator with a high efficiency particulate air (HEPA) filter. For exposure to radioactive cobalt, use a full facepiece supplied-air respirator or a self-contained breathing apparatus (SCBA) operated in positive pressure mode.

Ventilation

Local exhaust or general mechanical systems recommended.

Protective Gloves	Eye Protection	Other Protective Clothing
Impervious Material	Chemical/Dust Goggles or Face Mask	Protective Suit or Clothing

Work/Hygiene Practices

Always wash hands thoroughly after using chemical; never bring food, drink, or smoking materials into vicinity of chemicals.

COBALT

Co CAS: 7440-48-4

IDENTIFICATION AND TYPICAL USES

Cobalt is a steel-gray, shining, hard, ductile, somewhat malleable metal. It is ferromagnetic (highly magnetic, like iron), with permeability two-thirds that of iron. It is a natural element present in certain ores of the earth's crust, and is essential to life in trace amounts. It exists in the form of various salts. Cobalt and its salts are used in nuclear medicine, enamels and semiconductors, grinding wheels, painting on glass and porcelain, hygrometers and electroplating, as a foam stabilizer in beer, in vitamin B12 manufacture, as a drier for lacquers, varnishes, and paints, in steel alloys in jet engines, and as a catalyst for organic chemical reactions. Cobalt has several artificial radioactive isotopes, the most important being cobalt 60. Cobalt 60 is a beta and gamma emitter and is used in radiation therapy, in level gauges, and research.

RISK ASSESSMENT: HEALTH

General Assessment

Cobalt is a confirmed carcinogen. Human mutation data have also been reported. It is moderately toxic by *ingestion* with gastrointestinal irritation, nausea and vomiting. It can also be toxic by *inhalation,* with severe pulmonary damage (pulmonary edema) possible. Exposure to the dusts of cobalt may cause other lung effects, including pulmonary fibrosis and pneumoconiosis.

Occupational exposure can produce skin rash, itching, possible burning feeling on contact, and allergy (dermatitis). Eye contact will result in irritation, possible inflammation, and damage to vision which may be temporary or permanent, depending on the nature of the exposure.

Many synthetic versions of cobalt exist that are radioactive. Overexposure to such compounds can pose a significant hazard risk to human health.

Acute Health Effects

The following acute (short-term) health effects may occur immediately or shortly after exposure to cobalt:

Skin: Irritation, rash, and possible contact dermatitis.

Eye: Cobalt particles can irritate the eyes and may cause damage to vision.

Lung: Inhalation of the dust and fumes can irritate the nose, mouth, and throat. High levels can irritate the lungs, causing a buildup of fluid (pulmonary edema), which is a medical emergency and can be fatal.

Chronic Health Effects

The following chronic (long-term) health effects may occur at some time after exposure to cobalt and can last for months or even years:

Cancer Hazards: Cobalt compounds may cause mutations (genetic changes) in living cells. It is a confirmed carcinogen in animals and is suspected to cause cancer in humans, although its specific carcinogenicity requires further study. Some isotopes of cobalt do emit ionizing radiation. Exposure to ionizing radiation is associated with an increased risk of developing cancer.

Reproductive Hazard: According to information available in the references, cobalt has been tested and has not been shown to adversely affect reproduction in humans.

Other Chronic Effects: Cobalt can cause skin allergy. If an allergy develops, very low future exposures can cause itching and a skin rash. Cobalt may cause an asthma-like allergy. Future exposures can cause cough, shortness of breath, wheezing, and/or chest tightness. Cobalt can also damage the heart, leading to heart failure. Long-term exposure may damage the thyroid and liver. Repeated exposure can cause scarring of the lungs (pulmonary fibrosis), even if no symptoms are noticed. This condition can be disabling or may be fatal.

Recommended Risk-Reduction Measures

Personnel should avoid direct contact with cobalt. If the specific cobalt compound in question is unknown, adequate risk management dictates that it be handled and treated as though it were a carcinogen. If a less toxic chemical cannot be substituted for a hazardous substance, then *engineering controls* are the most effective method of reducing exposures.

The best protection is to enclose operations and/or provide local exhaust ventilation at the site of chemical release. While not always operationally feasible, isolating operations can also reduce exposure. Using respiratory protection is less effective than the controls mentioned above, but is still advisable whenever working with or around cobalt. For exposure to non-radioactive cobalt at levels above the PEL (0.05 mg/m^3), an MSHA/NIOSH approved respirator with a high efficiency particulate air (HEPA) filter is recommended. For higher exposures, or exposures to any levels of radioactive cobalt, use a supplied-air respirator with a full facepiece operated in positive pressure mode or an MSHA/NIOSH approved self-contained breathing apparatus (SCBA) with full facepiece and pressure demand. If a full facepiece is not available, then chemical dust goggles should be worn to protect the eyes. Whenever a chemical dust hazard exists, a face shield and a rubber apron should be worn. To prevent hand and skin exposures, impervious rubber gloves should be worn.

Administrative controls should also be in place to minimize the potential for human exposures. These may include written procedures or policies which specify the methods and techniques that will be practiced whenever personnel are to work with cobalt. A regulated, controlled work area should be established where cobalt is used, handled, or stored.

All personnel should receive *training* on the use, hazards, protective measures, emergency actions, and other precautions per 29 CFR 1910.1200 (Hazard Communication), prior to the first assignment in an area where cobalt is used or stored.

Before beginning employment and at regular intervals thereafter (e.g., annually), for those with frequent or potentially high exposures (half the PEL or greater, or significant skin contact), the following medical tests are recommended:

- ☑ Lung function tests.
- ☑ Chest X-ray for scarring should be performed every two to three years after five or more years of exposure.
- ☑ Evaluation for heart failure.

If symptoms develop or overexposure is suspected, the following may also be useful:

- ☑ Liver, thyroid, and lung function tests.
- ☑ Evaluation by a qualified allergist, including careful exposure history and special testing (may help diagnose skin allergy).

Any evaluation should include a careful history of past and present symptoms with a medical examination. Medical tests that look for damage already done are not a substitute for controlling exposure. Also, since smoking can cause heart disease, as well as lung cancer, emphysema, and other related respiratory problems, the effects of exposure to cobalt may be more pronounced in smokers than in non-smokers.

Other methods to reduce exposure include:

- ☑ Where possible, enclose operations and use local exhaust ventilation at the site of chemical release. If local exhaust ventilation or enclosure is not used, respiratory protection should be mandatory.
- ☑ Always ensure that proper protective clothing is worn when using chemical substances.
- ☑ Wash thoroughly immediately after exposure to cobalt and at the end of the work shift or before eating, drinking, or smoking.
- ☑ Hazard warning information should be posted in the work area. In addition, as part of an on-going education and training program, all information on the health and safety hazards of cobalt should be communicated to all potentially exposed workers.
- ☑ Where possible, automatically transfer cobalt from drums or other storage containers to process containers.
- ☑ Specific engineering controls are recommended by NIOSH. Refer to NIOSH Occupational Hazard Assessment: Criteria for Controlling Occupational Exposure to Cobalt (Publication Number 82-107).
- ☑ Guidelines and regulations for radioactive cobalt have been established by the Nuclear Regulatory Commission at 10 CFR 20.

RISK ASSESSMENT: ENVIRONMENT

General Assessment

The environment is at risk of exposure during transportation, storage, disposal, or destruction of cobalt. In almost every scenario, the threat of environmental exposure is contingent upon the proper handling of the chemical substance. Accidental spills, large or small, can result in fire, explosion, and possible contamination of the surrounding environmental mediums (water, soil, and air).

Cobalt powder will explode spontaneously in air. It will ignite and is potentially explosive in contact with bromine pentafluroide. When heated in the presence of ammonium nitrate or hydrazinium nitrate, violent or explosive reactions can result. It is also incompatible with strong oxidizers (such as perchlorates, peroxides, permanganates, chlorates, and nitrates), and strong acids (hydrochloric, sulfuric, and nitric). These characteristics require special consideration during any emergency situation involving a leak or spill of cobalt.

Cobalt can occur in the environment as a natural element in certain ores in the earth's crust. It can also enter the environment through industrial and municipal discharges, or from spills.

Acute Ecological Effects

Acute (short-term) toxic effects may include the death of animals, birds, or fish and death or low growth rate in plants. Acute effects are seen 2 to 4 days after animals or plants are exposed to cobalt.

Cobalt and its salts have high acute toxicity to aquatic life. Insufficient data are available to evaluate the short-term effects of cobalt exposure to plants, birds, or land animals.

Chronic Ecological Effects

Chronic (long-term) toxic effects may include shortened life span, reproductive problems, lower fertility, and changes in appearance or behavior in exposed animals. These effects can be seen long after first exposure(s) to toxic chemicals.

Cobalt and its salts have high chronic toxicity to aquatic life. Insufficient data are available to evaluate the long-term effects of cobalt exposure to plants, birds or land animals.

Water Solubility

Water solubility of cobalt and its salts ranges from highly soluble to practically insoluble.

Persistence in the Environment

Cobalt and its salts are highly persistent in the aquatic environment, with a half-life of greater than 200 days. The half-life of a pollutant is the amount of time it takes for one half of the chemical to be degraded.

Bioaccumulation in Aquatic Organisms

Some substances increase in concentration, or *bioaccumulate,* in living organisms as they breathe contaminated air, drink contaminated water, or eat contaminated food. These chemicals can become concentrated in the tissues and internal organs of animals as well as humans.

The concentration of cobalt found in fish is expected to be somewhat higher than the average concentration of cobalt in the water from which the fish was taken.

Recommended Risk-Reduction Measures

Proper training of all transporters will reduce the likelihood of a mishap or accidents resulting in a leak or spill to the environment. The correct use of DOT labeling (per the MSDS) on all containers, trucks, and rail cars will enable emergency responders to react properly and quickly to any disaster thereby reducing the potential risk to the environment and to personnel.

Storage of cobalt should be segregated from incompatible chemicals to minimize the risk of cross-contamination or contact. Bulk storage of radioactive cobalt should be avoided.

If a spill or leak to the environment has occurred, fire department, emergency response and/or hazardous materials spill personnel should be notified immediately. Use a vacuum method to reduce dust during cleanup. Do not dry sweep. When vacuuming, a high efficiency particulate air (HEPA) filter should be used, not a standard shop vacuum. Deposit collected materials in sealed drum for disposal. Cleanup should be attempted only by those trained in proper spill containment procedures. If radioactive cobalt is spilled, evacuate area and delay until instructions from qualified radiation personnel can be obtained. Contaminated soils should be removed and replaced with clean soil. If cobalt should contact the water table, aquifer, or navigable waterway, time is of the essence. Some cobalt compounds may be moderately to highly soluble in water and, therefore, total containment and remediation may not be entirely possible. When such spills occur, the local and/or state emergency response authorities must be notified. A comprehensive emergency response or disaster preparedness/recovery plan should be in place prior to any operations involving the use, transportation, storage, or disposal of cobalt. If cobalt is spilled or leaked, the following specific steps are recommended:

- ☑ Restrict persons not wearing protective clothing from area of spill or leak until cleanup is complete and area can be opened for normal work.
- ☑ Collect powdered materials in most convenient and safe manner possible and place in sealed drums for disposal. Do not dry sweep, use HEPA vacuum instead.
- ☑ It may be necessary to dispose of cobalt as a hazardous waste. The responsible state agency or the regional office of the federal Environmental Protection Agency (EPA) should be contacted for specific recommendations.

RISK ASSESSMENT: BUSINESS

General Assessment

Accidents or mishaps involving cobalt can present a significant threat to business operations. The loss or damage of equipment or facilities can significantly affect fiscal viability. Lawsuits that may result from personnel injury/death, public exposures, and/or environmental contamination will also require a serious expenditure of resources. Media attention surrounding an injury, death, or environmental damage can also result in a loss of profits and loss of current as well as future business.

Always remember that anytime the term "radiation" or "radioactive" are used, public emotion and anxiety can run equally high. This must be considered when dealing with the public following an incident involving cobalt.

Recommended Risk-Reduction Measures

Company attorneys, safety and health professionals, and environmental specialists should be involved in the development of any procedures dealing with safety or emergency response. A company official should be pre-designated as a public relations officer with specific training in dealing with the press. Corporate plans and policies should be developed, approved, and implemented long before any need for such arises.

REFERENCES

Ahlbom, A. 1993. *Biostatistics for Epidemiologists.* New York: CRC Press/Lewis Publishers.

American Conference of Governmental Industrial Hygienists. 1988. *Documentation of the Threshold Limit Values and Biological Exposure Indices,* 5th Edition (with updates). Cincinnati: ACGIH

Asante-Duah, D. K. 1993. *Hazardous Waste Risk Assessment.* New York: CRC Press/Lewis Publishers.

Bloom, A. D. and F. J. DeSerres. 1995. *Ecotoxicity and Human Health.* New York: CRC Press/Lewis Publishers.

Brusick, D. J. 1994. *Methods for Genetic Risk Assessment.* New York: CRC Press/Lewis Publishers.

Calabrese, E. J. 1994. *Biological Effects of Low Level Exposures.* New York: CRC Press/Lewis Publishers.

Cockerman, L. G., B. S. Shane. 1994. *Basic Environmental Toxicology.* New York: CRC Press/Lewis Publishers.

Cothern, C. R., N. P. Ross. 1994. *Environmental Statistics, Assessment, and Forecasting.* New York: CRC Press/Lewis Publishers.

Gardner, R. W. 1994. *Chemical Intolerance.* New York: CRC Press/Lewis Publishers.

Hallenbeck, W. H. 1993. *Quantitative Risk Assessment for Environmental and Occupational Health,* 2nd Edition. New York: CRC Press/Lewis Publishers.

Howard, P. H., et. al. 1991. *Handbook of Environmental Degradation Rates.* New York: CRC Press/Lewis Publishers.

Keith, L. H. and M. M. Walker. 1995. *Handbook of Air Toxics.* New York: CRC Press/Lewis Publishers.

Landis, W. G., M. H. Yu. 1995. *Introduction to Environmental Toxicology.* New York: CRC Press/Lewis Publishers.

Langerman, N. 1994 *Precautionary Labels for Chemical Containers.* New York: CRC Press/Lewis Publishers.

Lauwerys, R. R., P. Hoet. 1993. *Industrial Chemical Exposure.* New York: CRC Press/Lewis Publishers.

Lewis, R. J., Sr. 1992. *Sax's Dangerous Properties of Industrial Materials,* Volumes 1, 2, and 3, 8th Edition. New York: Van Nostrand Reinhold.

Manahan, S. E. 1994. *Environmental Chemistry,* 6th Edition. New York: CRC Press/Lewis Publishers.

Manahan, S. E. 1992. *Toxicological Chemistry,* 2nd Edition. New York: CRC Press/Lewis Publishers.

Meyer, E. 1990. *Chemistry of Hazardous Materials.* Englewood Cliffs, NJ: Prentice Hall.

U.S. Department of Health and Human Services (DHHS), National Institute for Occupational Health (NIOHS). 1994. *NIOSH Pocket Guide to Hazardous Chemicals.* Publication Number 94-116. Washington, D.C.; U.S. Government Printing Office.

U.S. Department of Health and Human Services (DHHS), National Institute for Occupational Health (NIOHS). 1982. *Criteria for Controlling Occupational Exposure to Cobalt.* Publication Number 82-107. Washington, D.C.; U.S. Government Printing Office.

MATERIAL SAFETY DATA SHEET

CHEMICAL NAME	
COBALT ACETATE	**CRC PRESS/LEWIS PUBLISHERS DISCLAIMER:** THE INFORMATION AND RECOMMENDATIONS PRESENTED HEREIN ARE BASED ON SOURCES BELIEVED TO BE RELIABLE. CRC MAKES NO REPRESENTATION ON ITS COMPLETENESS OR ACCURACY. IT IS THE RESPONSIBILITY OF THE USER TO DETERMINE THE CHEMICAL'S SUITABILITY FOR ITS INTENDED USE, THE CHEMICAL'S SAFE USE, AND THE CHEMICAL'S PROPER DISPOSAL. NO REPRESENTATIONS AND/OR WARRANTIES, EITHER EXPRESSED OR IMPLIED, OF THE MERCHANTABILITY OR FITNESS FOR A PARTICULAR PURPOSE, OR OF ANY OTHER NATURE, ARE MADE WITH RESPECT TO THE INFORMATION PROVIDED IN THIS MSDS OR TO THE CHEMICAL TO WHICH INFORMATION MAY REFER. CRC NEITHER ASSUMES NOR AUTHORIZES ANY OTHER PERSON TO ASSUME FOR IT, ANY OTHER ADDITIONAL RESPONSIBILITY OR LIABILITY FOR THE USE OF, OR RELIANCE UPON, THIS INFORMATION.

HAZARD WARNING INFORMATION

HEALTH	FIRE	REACTIVE	OTHER	DEGREE OF HAZARD	COLOR CODING	OTHER CODES
2	0	0		0 = Minimum Hazard 1 = Slight Hazard 2 = Moderate Hazard 3 = Serious Hazard 4 = Severe Hazard	HEALTH = BLUE FIRE = RED REACTIVITY = YELLOW OTHER = WHITE	OX = Oxidizer ACID = Acid ALK = Alkali COR = Corrosive W = Use No Water

SECTION I - GENERAL INFORMATION

Characterization	Cobalt Compound	RCRA Number	None
		EPA Class	Not Applicable
DOT Proper Shipping Name	No Citation	Chemical Abstract Service (CAS) Number	71-48-7
DOT Hazard Class and Label Requirements	No Citation	DOT Emergency Guide Code	No Citation
DOT Identification Number	No Citation	Chemical Formula	$(CH_3COO)_2Co$

Synonyms

Acetic acid, cobalt (2+) salt; bis(acetato)cobalt; cobalt (2+) acetate; cobalt (II) acetate); cobaltous acetate.

SECTION II - HAZARDOUS INGREDIENTS AND IDENTITY INFORMATION

Hazardous Components (specific identity)	OSHA Exposure Criteria	NIOSH Exposure Criteria	Immediately Dangerous to Life and Health (IDLH)	ACGIH Exposure Criteria
Cobalt acetate (derivation: From cobaltous hydroxide or carbonate and an excess of dilute acetic acid).	**PEL (8-hour):** 0.1 mg/m^3 (as cobalt) **STEL:** Not Established	**PEL (10-hour):** 0.05 mg/m^3 (as cobalt) **STEL:** Not Established	20 mg/m^3 as Co	**TLV:** 0.1 mg/m^3 (as cobalt) **STEL:** Not Established

SECTION III - PHYSICAL AND CHEMICAL CHARACTERISTICS

Boiling Point	Not Found	Specific Gravity (H_2O = 1)	1.7
Vapor Pressure (mm Hg)	0 (approx.) at 69°F (20°C)	Molecular Weight	177.03
Vapor Density (Air = 1)	Not Found	Melting Point	Not Found

Solubility

Readily soluble in water. Soluble in acids and alcohol.

Appearance and Odor

Reddish-violet, deliquescent crystals with a vinegar-like odor.

SECTION IV - FIRE AND EXPLOSION HAZARD DATA

Flash Point (method used)	Not Applicable	Explosive Limits in Air % by Volume	LEL: Not Applicable UEL: Not Applicable
NFPA Classification	Non-Combustible Solid	Autoignition Temperature	Not Determined

Extinguishing Media

Use extinguishing media suitable to surrounding fire. Cobalt acetate will not burn.

Special Fire Fighting Procedures

Poisonous gases are produced in fire. Wear full protective clothing and self-contained breathing apparatus (SCBA). Use a water spray to keep fire-exposed containers cool. Water may not be totally effective on fire by itself. Remain clear of smoke, water fall-out and water run-off.

Unusual Fire and Explosion Hazards

Can cause violent explosions upon contact with strong oxidizers. Finely divided powders in air may ignite and burn fiercely when exposed to heat or flame.

SECTION V - REACTIVITY DATA

Stability		Conditions to Avoid **Normally stable at room temperature and pressure in closed containers under routine conditions of handling and storage. Keep away from incompatible materials since violent or explosive reactions can occur.**
Stable **X**	Unstable	Incompatibility *(materials to avoid)* **Incompatible with strong oxidizers, such as perchlorates, peroxides, permanganates, chlorates, and nitrates.**
Hazardous Polymerization		Conditions to Avoid **Under normal conditions of temperature and pressure, hazardous polymerization of cobalt acetate is not known to occur.**
May Occur	Will Not Occur **X**	Hazardous Decomposition or By-products **When heated to decomposition, toxic and acrid fumes and gases are produced, including cobalt oxide.**

SECTION VI - HEALTH HAZARD DATA

Primary Route(s) of Entry:	Inhalation?	Absorption (skin)?	Ingestion?
	X		X

Health Hazards

INHALATION: **Irritation of the respiratory system. Exposure can cause allergic lung reaction with chest pain, coughing, wheezing, and shortness of breath. Repeated exposure may cause lung scarring (fibrosis), even if no symptoms are noticed. May also cause pulmonary edema.**

SKIN & EYES: **An irritant to the eyes and skin. The powder may cause contact dermatitis.**

INGESTION: **A poison with gastrointestinal effects with nausea and vomiting, flushing of the face and ears, hypertension (mild), rash, tinnitus, and nerve deafness.**

Carcinogenicity	NTP Listed?	IARC Cancer Review Group?	OSHA Regulated?	Target Organs?
Suspected Human Confirmed Animal	**5th Annual Report**	**No**	**No**	**Respiratory system, skin, eyes**

Medical Conditions Generally Aggravated by Exposure

None Reported. However, existing skin conditions or respiratory problems may be aggravated.

Emergency and First-aid Procedures

Eye contact: Flush immediately with water for 15 minutes (minimum), seek medical attention. Skin contact: Remove all contaminated clothing. Immediately wash area with large amounts of soap and water. For inhalation: Remove the person from exposure. Provide respiratory assistance and CPR. Transfer to medical facility. Observe for 24 - 48 hours for lung effects (pulmonary edema). If swallowed: Provide breathing assistance if necessary. Never attempt to give a convulsing or unconscious person anything by mouth. Seek medical attention immediately. Contact poison control center for assistance.

SECTION VII - PRECAUTIONS FOR SAFE HANDLING AND USE

Steps to be Taken in Case Material is Released or Spilled

Collect powdered material in most convenient manner possible and deposit in sealed drum. Keep those not involved in cleanup from entering area. Do NOT dry sweep (creates airborne dusts). Use a vacuum equipped with a high efficiency particulate air (HEPA) filter.

Preferred Waste Disposal Method

No Citation.

Precautions to be Taken in Handling and Storage

Store in tightly closed containers in cool, well-ventilated area and protect from physical damage. Sources of ignition are prohibited where cobalt acetate is used, handled, or stored.

Other Precautions and Warnings

A regulated, marked area should be established where radioactive cobalt compounds are handled, used, or stored.

SECTION VIII - CONTROL MEASURES AND PERSONAL PROTECTIVE EQUIPMENT

Respiratory Protection *(specify type)*

For relatively low exposures, use a full facepiece respirator with a high efficiency particulate air (HEPA) filter. For higher exposures, use a full facepiece supplied-air respirator or a self-contained breathing apparatus (SCBA) operated in positive pressure or other continuous flow mode.

Ventilation

Local exhaust or general mechanical systems recommended.

Protective Gloves	Eye Protection	Other Protective Clothing
Impervious Material	**Chemical/Dust Goggles or Face Mask**	**Protective Suit or Clothing**

Work/Hygiene Practices

Always wash hands thoroughly after using chemical; never bring food, drink, or smoking materials into vicinity of chemicals.

COBALT ACETATE

$(CH_3COO)_2Co$ CAS: 71-48-7

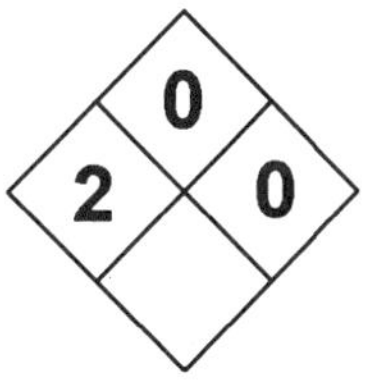

IDENTIFICATION AND TYPICAL USES

Cobalt acetate appears as light pink or reddish-violet deliquescent crystals with a mild, vinegar-like odor. It is used as a bleaching agent, as a drier for lacquers and varnishes, as an anodizing agent, a mineral supplement in feed additives, as a catalyst, and in the manufacture of sympathetic inks. It was also used in the beer production industry as a foam stabilizer for malts.

RISK ASSESSMENT: HEALTH

General Assessment

Cobalt acetate is a confirmed carcinogen in animal studies and cobalt itself is a known human carcinogen. Human mutation data have also been reported. Cobalt acetate is moderately toxic by *ingestion* and *inhalation.* Skin contact also produces a health risk but absorption is not likely to occur.

Inhalation causes severe, life-threatening pulmonary damage (pulmonary edema). The development of pulmonary edema, characterized by a fluid buildup in the lungs with cough, congestion, and tightness in chest, may be delayed up to 48 hours thereby creating a false sense of security with regard to health exposure risk. Exposure to the dusts of cobalt compounds, including cobalt acetate may cause other lung effects, including pulmonary fibrosis and pneumoconiosis. It can cause coughing and shortness of breath (dypsnea).

Occupational exposure can produce skin rash, itching, possible burning feeling on contact, and allergy (dermatitis). Eye contact will result in irritation, possible inflammation, and damage to vision which may be temporary or permanent, depending on the nature of the exposure.

Ingestion causes intense abdominal pain, vomiting, digestive tract irritation, and numerous systemic effects such as flushing of the face and ears, decreased blood pressure, rash, tinnitus (ringing in the ears), and even nerve deafness. It may also have the ability to produce myocardial dysfunction.

Acute Health Effects

The following acute (short-term) health effects may occur immediately or shortly after exposure to cobalt acetate:

Skin: Irritation, rash, and possible contact dermatitis.

Eye: Cobalt acetate particles can irritate the eyes and may cause damage to vision.

Lung: Inhalation of the dust and fumes can irritate the nose, mouth, and throat. High levels can irritate the lungs, causing a buildup of fluid (pulmonary edema), which is a medical emergency and can be fatal.

Other: Ingestion can cause a number of toxic systemic effects associated initially with gastrointestinal irritation and progressing to central nervous system (CNS) involvement.

Chronic Health Effects

The following chronic (long-term) health effects may occur at some time after exposure to cobalt acetate and can last for months or even years:

Cancer Hazards: Cobalt compounds may cause mutations (genetic changes) in living cells. Cobalt acetate is a confirmed carcinogen in animals and is suspected to cause cancer in humans, although its specific carcinogenicity requires further study. Some isotopes of cobalt do emit ionizing radiation. Exposure to ionizing radiation is associated with an increased risk of cancer.

Reproductive Hazard: According to information available in the references, cobalt has been tested and has not been shown to adversely affect reproduction in humans. The specific teratogenicity of cobalt acetate has not been determined and requires further study.

Other Chronic Effects: Cobalt acetate can cause skin allergy. If an allergy develops, very low future exposures can cause itching and a skin rash. Cobalt acetate may cause an asthma-like allergy. Future exposures can cause cough, shortness of breath, wheezing, and/or chest tightness. Cobalt acetate can also damage the heart, leading to heart failure. Long-term exposure may damage the thyroid and liver. Repeated exposure can cause scarring of the lungs (pulmonary fibrosis), even if no symptoms are noticed. This condition can be disabling or fatal.

Recommended Risk-Reduction Measures

Personnel should avoid direct contact with cobalt acetate. If a less toxic chemical cannot be substituted for a hazardous substance, then *engineering controls* are the most effective method of reducing exposures. The best protection is to enclose operations and/or provide local exhaust ventilation at the site of chemical release. While not always operationally feasible, isolating operations can also reduce exposure. Using respiratory protection is less effective than the controls mentioned above, but is still advisable whenever working with or around cobalt acetate. For exposure to cobalt acetate at levels above the established PEL (0.1 mg/m^3), an MSHA/NIOSH approved respirator with a high efficiency particulate air (HEPA) filter is recommended. For higher exposures use a supplied-air respirator with a full facepiece operated in positive pressure mode or an MSHA/NIOSH approved self-contained breathing apparatus (SCBA) with full facepiece and pressure demand. If a full facepiece is not available, then chemical splash goggles should be worn to protect the eyes. Whenever a chemical dust hazard exists, a face shield and a rubber apron should be worn. To prevent hand and skin exposures, impervious rubber gloves should be worn.

Administrative controls should also be in place to minimize the potential for human exposures. These may include written procedures or policies which specify the methods and techniques that will be practiced whenever personnel are to work with cobalt acetate. A regulated, controlled work area should be established where cobalt acetate is used, handled, or stored.

All personnel should receive *training* on the use, hazards, protective measures, emergency actions, and other precautions per 29 CFR 1910.1200 (Hazard Communication), prior to the first assignment in an area where cobalt acetate is used or stored. Before beginning employment and at regular intervals thereafter (e.g., annually), for those with frequent or potentially high exposures (half the PEL or greater, or significant skin contact), the following medical tests are recommended:

- ☑ Lung function tests.
- ☑ Chest X-ray for scarring should be performed every two to three years after five or more years of exposure.
- ☑ Evaluation for heart failure.

If symptoms develop or overexposure is suspected, the following may also be useful:

- ☑ Liver, thyroid, and lung function tests.
- ☑ Evaluation by a qualified allergist, including careful exposure history and special testing (may help diagnose skin allergy).

Any evaluation should include a careful history of past and present symptoms with a medical examination. Medical tests that look for damage already done are not a substitute for controlling exposure. Also, since smoking can cause heart disease, as well as lung cancer, emphysema, and other related respiratory problems, the effects of exposure to cobalt acetate may be more pronounced in smokers than in non-smokers.

Other methods to reduce exposure include:

- ☑ Where possible, enclose operations and use local exhaust ventilation at the site of chemical release. If local exhaust ventilation or enclosure is not used, respiratory protection should be mandatory.
- ☑ Always ensure that proper protective clothing is worn when using chemical substances.
- ☑ Wash thoroughly immediately after exposure to cobalt acetate and at the end of the work shift or before eating, drinking, or smoking.
- ☑ Hazard warning information should be posted in the work area. In addition, as part of an on-going education and training program, all information on the health and safety hazards of cobalt acetate should be communicated to all potentially exposed workers.
- ☑ Where possible, automatically transfer cobalt acetate from drums or other storage containers to process containers.
- ☑ Specific engineering controls are recommended by NIOSH. Refer to NIOSH Occupational Hazard Assessment: Criteria for Controlling Occupa-

tional Exposure to Cobalt (Publication Number 82-107).

RISK ASSESSMENT: ENVIRONMENT

General Assessment

The environment is at risk of exposure during transportation, storage, disposal, or destruction of cobalt acetate. In almost every scenario, the threat of environmental exposure is contingent upon the proper handling of the chemical substance. Accidental spills, large or small, can result in fire, explosion, and possible contamination of the surrounding environmental mediums (water, soil, and air).

Cobalt acetate is a non-combustible solid. However, it is incompatible with strong oxidizers (such as perchlorates, peroxides, permanganates, chlorates, and nitrates), and may ignite or explode on contact. Finely divided airborne dust particles may ignite explosively in confined areas. These characteristics require special consideration during any emergency situation involving a leak or spill of cobalt acetate. Cobalt can occur in the environment as a natural element in certain ores in the earth's crust. It can enter the environment through industrial and municipal discharges, or from spills.

Acute Ecological Effects

Acute (short-term) toxic effects may include the death of animals, birds, or fish and death or low growth rate in plants. Acute effects are seen 2 to 4 days after animals or plants are exposed to cobalt acetate. Cobalt and its salts have high acute toxicity to aquatic life. Insufficient data are available to evaluate the short-term effects of cobalt exposure to plants, birds, or land animals.

Chronic Ecological Effects

Chronic (long-term) toxic effects may include shortened life span, reproductive problems, lower fertility, and changes in appearance or behavior in exposed animals. These effects can be seen long after first exposure(s) to toxic chemicals. Cobalt and its salts have high chronic toxicity to aquatic life. Insufficient data are available to evaluate the long-term effects of cobalt exposure to plants, birds or land animals.

Water Solubility

Water solubility of cobalt and its salts ranges from highly soluble to practically insoluble. Cobalt acetate is considered highly soluble in the aquatic environment. Concentration of 1000 milligrams and more will mix with a liter of water.

Persistence in the Environment

Cobalt and its salts are highly persistent in the aquatic environment, with a half-life of greater than 200 days. The half-life of a pollutant is the amount of time it takes for one half of the chemical to be degraded.

Bioaccumulation in Aquatic Organisms

Some substances increase in concentration, or *bioaccumulate,* in living organisms as they breathe contaminated air, drink contaminated water, or eat contaminated food. These chemicals can become concentrated in the tissues and internal organs of animals as well as humans.

The concentration of cobalt found in fish is expected to be somewhat lower than the average concentration of cobalt in the water from which the fish was taken.

Recommended Risk-Reduction Measures

Proper training of all transporters will reduce the likelihood of a mishap or accidents resulting in a leak or spill to the environment. The correct use of DOT labeling (per the MSDS) on all containers, trucks, and rail cars will enable emergency responders to react properly and quickly to any disaster thereby reducing the potential risk to the environment and to personnel.

Storage of cobalt acetate should be segregated from incompatible chemicals to minimize the risk of cross-contamination or contact. Bulk storage of radioactive cobalt acetate should be avoided.

If a spill or leak to the environment has occurred, fire department, emergency response and/or hazardous materials spill personnel should be notified immediately. Use a vacuum method to reduce dust during cleanup. Do not dry sweep. When vacuuming, a high efficiency particulate air (HEPA) filter should be used, not a standard shop vacuum. Deposit collected materials in sealed drum for disposal. Cleanup should be attempted only by those trained in proper spill containment procedures. Contaminated soils should be removed and replaced with clean soil. If cobalt acetate should contact the water table, aquifer, or navigable waterway, time is of the essence. It is highly soluble in water and, therefore, total containment and remediation may not be possible. When such spills occur, the local and/or state emergency response authorities must be notified. A comprehensive emergency response or disaster preparedness/recovery plan should be in place prior to any operations involving the use, transportation, storage, or disposal of cobalt acetate. If cobalt acetate is spilled or leaked, the following specific steps are recommended:

☑ Restrict persons not wearing protective clothing from area of spill or leak until cleanup is complete and area can be opened for normal work.

☑ Collect powdered materials in most convenient and safe manner possible and place in sealed drums for disposal. Do not dry sweep, use HEPA vacuum instead.

☑ It may be necessary to dispose of cobalt acetate as a hazardous waste. The responsible state agency or the regional office of the federal Environmental Protection Agency (EPA) should be contacted for specific recommendations.

RISK ASSESSMENT: BUSINESS

General Assessment

Accidents or mishaps involving cobalt acetate can present a significant threat to business operations. The loss or damage of equipment or facilities can significantly affect fiscal viability. Lawsuits that may result from personnel injury/death, public exposures, and/or environmental contamination will also require a serious expenditure of resources. Media attention surrounding an injury, death, or environmental damage can also result in a loss of profits and loss of current as well as future business. Always remember that anytime the term "radiation" or "radioactive" are used, public emotion and anxiety can run equally high. This must be considered when dealing with the public following an incident involving cobalt acetate.

Recommended Risk-Reduction Measures

Company attorneys, safety and health professionals, and environmental specialists should be involved in the development of any procedures dealing with safety or emergency response. A company official should be pre-designated as a public relations officer with specific training in dealing with the press. Corporate plans and policies should be developed, approved, and implemented long before any need for such arises.

REFERENCES

Ahlbom, A. 1993. *Biostatistics for Epidemiologists.* New York: CRC Press/Lewis Publishers.

American Conference of Governmental Industrial Hygienists. 1988. *Documentation of the Threshold Limit Values and Biological Exposure Indices,* 5th Edition (with updates). Cincinnati: ACGIH

Asante-Duah, D. K. 1993. *Hazardous Waste Risk Assessment.* New York: CRC Press/Lewis Publishers.

Bloom, A. D. and F. J. DeSerres. 1995. *Ecotoxicity and Human Health.* New York: CRC Press/Lewis Publishers.

Brusick, D. J. 1994. *Methods for Genetic Risk Assessment.* New York: CRC Press/Lewis Publishers.

Calabrese, E. J. 1994. *Biological Effects of Low Level Exposures.* New York: CRC Press/Lewis Publishers.

Cockerman, L. G., B. S. Shane. 1994. *Basic Environmental Toxicology.* New York: CRC Press/Lewis Publishers.

Cothern, C. R., N. P. Ross. 1994. *Environmental Statistics, Assessment, and Forecasting.* New York: CRC Press/Lewis Publishers.

Gardner, R. W. 1994. *Chemical Intolerance.* New York: CRC Press/Lewis Publishers.

Hallenbeck, W. H. 1993. *Quantitative Risk Assessment for Environmental and Occupational Health,* 2nd Edition. New York: CRC Press/Lewis Publishers.

Howard, P. H., et. al. 1991. *Handbook of Environmental Degradation Rates.* New York: CRC Press/Lewis Publishers.

Keith, L. H. and M. M. Walker. 1995. *Handbook of Air Toxics.* New York: CRC Press/Lewis Publishers.

Landis, W. G., M. H. Yu. 1995. *Introduction to Environmental Toxicology.* New York: CRC Press/Lewis Publishers.

Langerman, N. 1994 *Precautionary Labels for Chemical Containers.* New York: CRC Press/Lewis Publishers.

Lauwerys, R. R., P. Hoet. 1993. *Industrial Chemical Exposure.* New York: CRC Press/Lewis Publishers.

Lewis, R. J., Sr. 1992. *Sax's Dangerous Properties of Industrial Materials,* Volumes 1, 2, and 3, 8th Edition. New York: Van Nostrand Reinhold.

Manahan, S. E. 1994. *Environmental Chemistry,* 6th Edition. New York: CRC Press/Lewis Publishers.

Manahan, S. E. 1992. *Toxicological Chemistry,* 2nd Edition. New York: CRC Press/Lewis Publishers.

Meyer, E. 1990. *Chemistry of Hazardous Materials.* Englewood Cliffs, NJ: Prentice Hall.

U.S. Department of Health and Human Services (DHHS), National Institute for Occupational Health (NIOHS). 1982. *Criteria for Controlling Occupational Exposure to Cobalt.* Publication Number 82-107. Washington, D.C.; U.S. Government Printing Office.

MATERIAL SAFETY DATA SHEET

CHEMICAL NAME	CRC PRESS/LEWIS PUBLISHERS DISCLAIMER
COBALT CARBONYL	THE INFORMATION AND RECOMMENDATIONS PRESENTED HEREIN ARE BASED ON SOURCES BELIEVED TO BE RELIABLE. CRC MAKES NO REPRESENTATION ON ITS COMPLETENESS OR ACCURACY. IT IS THE RESPONSIBILITY OF THE USER TO DETERMINE THE CHEMICAL'S SUITABILITY FOR ITS INTENDED USE, THE CHEMICAL'S SAFE USE, AND THE CHEMICAL'S PROPER DISPOSAL. NO REPRESENTATIONS AND/OR WARRANTIES, EITHER EXPRESSED OR IMPLIED, OF THE MERCHANTABILITY OR FITNESS FOR A PARTICULAR PURPOSE, OR OF ANY OTHER NATURE, ARE MADE WITH RESPECT TO THE INFORMATION PROVIDED IN THIS MSDS OR TO THE CHEMICAL TO WHICH INFORMATION MAY REFER. CRC NEITHER ASSUMES NOR AUTHORIZES ANY OTHER PERSON TO ASSUME FOR IT, ANY OTHER ADDITIONAL RESPONSIBILITY OR LIABILITY FOR THE USE OF, OR RELIANCE UPON, THIS INFORMATION.

HAZARD WARNING INFORMATION

HEALTH	FIRE	REACTIVE	OTHER	DEGREE OF HAZARD	COLOR CODING	OTHER CODES
2	0	0		0 = Minimum Hazard 1 = Slight Hazard 2 = Moderate Hazard 3 = Serious Hazard 4 = Severe Hazard	HEALTH = BLUE FIRE = RED REACTIVITY = YELLOW OTHER = WHITE	OX = Oxidizer ACID = Acid ALK = Alkali COR = Corrosive W = Use No Water

SECTION I - GENERAL INFORMATION

Characterization	RCRA Number	EPA Class
Cobalt Compound	**None**	**Not Applicable**

DOT Proper Shipping Name: **No Citation**	Chemical Abstract Service (CAS) Number: **10210-68-1**
DOT Hazard Class and Label Requirements: **No Citation**	DOT Emergency Guide Code: **No Citation**
DOT Identification Number: **No Citation**	Chemical Formula: **$C_8Co_2O_8$**

Synonyms

Cobalt octacarbonyl; di-mu-carbonylhexacarbonyldicobalt; cobalt tetracarbonyl dimer; dicobalt carbonyl; dicobalt octacarbonyl; octacarbobyldicobalt.

SECTION II - HAZARDOUS INGREDIENTS AND IDENTITY INFORMATION

Hazardous Components (specific identity)	OSHA Exposure Criteria	NIOSH Exposure Criteria	Immediately Dangerous to Life and Health (IDLH)	ACGIH Exposure Criteria
Cobalt carbonyl (derivation: By combining finely divided cobalt with carbon monoxide under pressure).	**PEL (8-hour): 0.1 mg/m^3 (as cobalt)** **STEL: Not Established**	**REL (10-hour): 0.1 mg/m^3 (as cobalt)** **STEL: Not Established**	**20 mg/m^3 as Co**	**TLV: 0.1 mg/m^3 (as cobalt)** **STEL: Not Established**

SECTION III - PHYSICAL AND CHEMICAL CHARACTERISTICS

Boiling Point: **126°F (52°C)**	Specific Gravity (H_2O = 1): **1.87**
Vapor Pressure (mm Hg): **0.7 at 59°F (15°C)**	Molecular Weight: **341.9**
Vapor Density (Air = 1): **Not Found**	Melting Point: **124°F (51°C)**

Solubility

Insoluble in water. Soluble in ether, alcohol, carbon disulfide, and naphtha.

Appearance and Odor

Orange to dark-brown, crystalline (platelets) solid. White when pure.

SECTION IV - FIRE AND EXPLOSION HAZARD DATA

Flash Point (method used): **Not Applicable**	Explosive Limits in Air % by Volume: **LEL: Not Applicable** **UEL: Not Applicable**
NFPA Classification: **Non-Combustible Solid**	Autoignition Temperature: **Not Determined**

Extinguishing Media

Use extinguishing media suitable to surrounding fire.

Special Fire Fighting Procedures

Poisonous gases are produced in fire. Wear full protective clothing and self-contained breathing apparatus (SCBA). Isolate area and deny entry. Use a water spray to keep fire-exposed containers cool. Water by itself may be ineffective on fire. Remain clear of smoke, water fall-out and water run-off.

Unusual Fire and Explosion Hazards

Cobalt carbonyl is a non-combustible solid. But in fire, it produces dangerously flammable carbon monoxide gas. Move containers from fire if it can be done without risk.

SECTION V - REACTIVITY DATA

Stability		Conditions to Avoid: Cobalt carbonyl can be kept stable in an atmosphere of hydrogen and carbon monoxide. It will decompose in air to form a product which ignites spontaneously in air. Avoid contact with heat.
Stable: X	Unstable	Incompatibility (*materials to avoid*): Incompatible with hydrochloric acid, sulfuric acid, nitric acid, and bromine. It will decompose on exposure to air or heat.
Hazardous Polymerization		Conditions to Avoid: Under normal conditions of temperature and pressure, hazardous polymerization of cobalt carbonyl is not known to occur.
May Occur	Will Not Occur: X	Hazardous Decomposition or By-products: When heated to decomposition, cobalt carbonyl will emit toxic and acrid fumes and gases, including carbon monoxide.

SECTION VI - HEALTH HAZARD DATA

Primary Route(s) of Entry:	Inhalation?	Absorption (skin)?	Ingestion?
	X	X	X

Health Hazards

INHALATION: Toxic on decomposition (produces carbon monoxide). Exposure causes symptoms of chemical asphyxia with headache, nausea, dizziness, loss of consciousness, convulsions, coma, infarction (heart attack), and death. Cobalt dust inhalation can cause an asthma-like disease with cough and dypsnea.

SKIN & EYES: An irritant to the eyes and skin. The powder may cause contact dermatitis. Absorption is possible. Eye contact causes conjunctivitis.

INGESTION: A poison with CNS and gastrointestinal effects with nausea and vomiting.

Carcinogenicity	NTP Listed?	IARC Cancer Review Group?	OSHA Regulated?	Target Organs?
Unknown Human Unknown Animal	No	No	No	Respiratory system, skin, eyes, blood, CNS.

Medical Conditions Generally Aggravated by Exposure

None Reported. However, existing skin conditions or respiratory problems may be aggravated.

Emergency and First-aid Procedures

Eye contact: Flush immediately with water for 15 minutes (minimum), seek medical attention. Skin contact: Remove all contaminated clothing. Immediately wash area with large amounts of soap and water. For inhalation: Remove the person from exposure. Provide respiratory assistance and CPR. Transfer to medical facility. If swallowed: Never attempt to give a convulsing or unconscious person anything by mouth. Seek medical attention immediately. Contact poison control center for assistance.

SECTION VII - PRECAUTIONS FOR SAFE HANDLING AND USE

Steps to be Taken in Case Material is Released or Spilled

Collect powdered material in most convenient manner possible and deposit in sealed drum. Keep those not involved in cleanup from entering area. Do NOT dry sweep (creates airborne dusts). Use a vacuum equipped with a high efficiency particulate air (HEPA) filter.

Preferred Waste Disposal Method

No Citation.

Precautions to be Taken in Handling and Storage

Store in tightly closed containers in cool, well-ventilated area and protect from physical damage. Store under a hydrogen or carbon monoxide blanket. Keep fire, heat, and ignitions sources away (hydrogen and carbon monoxide are extremely flammable). Protect containers from physical damage.

Other Precautions and Warnings

Avoid creating dusty conditions in storage and use areas.

SECTION VIII - CONTROL MEASURES AND PERSONAL PROTECTIVE EQUIPMENT

Respiratory Protection (*specify type*)

For low exposure cobalt carbonyl, use a full facepiece respirator with a high efficiency particulate air (HEPA) filter. For higher exposures, use a full facepiece supplied-air respirator or a self-contained breathing apparatus (SCBA) operated in positive pressure or other continuous flow mode.

Ventilation

Local exhaust or general mechanical systems recommended.

Protective Gloves	Eye Protection	Other Protective Clothing
Impervious Material	Chemical/Dust Goggles or Face Mask	Protective Suit or Clothing

Work/Hygiene Practices

Always wash hands thoroughly after using chemical; never bring food, drink, or smoking materials into vicinity of chemicals.

COBALT CARBONYL

$C_8Co_2O_8$ CAS: 10210-68-1

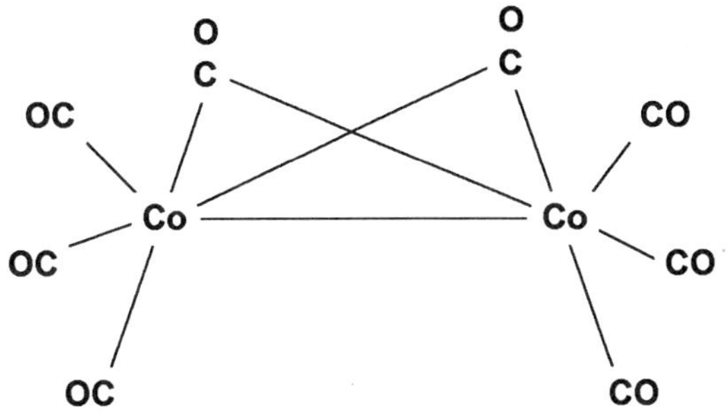

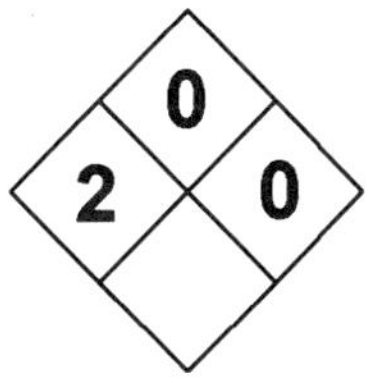

IDENTIFICATION AND TYPICAL USES

Cobalt carbonyl is a an orange to dark-brown crystalline solid. There is no odor. The pure substance is white. It is used as a catalyst for hydroformulation, hydrogenation, hydrosilation, isomerization, carboxylation, carbonylation, and polymerization. It is also a catalyst for amino acids, in anti-knock gasoline, and in high-purity cobalt salts.

RISK ASSESSMENT: HEALTH

General Assessment

Cobalt carbonyl is particularly toxic upon decomposition because it produces carbon monoxide, a deadly gas. Cobalt dust is also toxic by *inhalation* and can produce asthma-like symptoms with cough and dypsnea. There may be broncospasm and respiratory disturbances. Specific carcinogenicity for cobalt carbonyl has not been determined. However, cobalt itself is a confirmed carcinogen. Human mutation data have also been reported. It is a moderately toxic by *ingestion* with gastrointestinal irritation, nausea and vomiting. It can also be toxic by *inhalation,* with severe pulmonary damage (pulmonary edema) possible. Exposure to the dusts of cobalt and possibly cobalt carbonyl may cause other lung effects, including pulmonary fibrosis and pneumoconiosis.

Occupational exposure can produce skin rash, itching, possible burning feeling on contact, and allergy (dermatitis) with the appearance of urticaria (elevated and itching white patches) and erythema (red papules). Eye contact will result in irritation, possible inflammation, and damage to vision which may be temporary or permanent, depending on the nature of the exposure.

Inhalation of carbon monoxide gas will cause progressive chemical asphyxia with symptoms of rapid, irregular breathing (tachypnea), feeling of air starvation or "air hunger," headache, fatigue, mental confusion, nausea, vomiting, giddiness, poor judgment, weakness, hallucinations, cyanosis, angina, syncope (sudden loss of strength and possible fainting), exhaustion, collapse, loss of consciousness, coma, convulsions, myocardial infarction, and death. In non-fatal exposures, there may be serious damage or injury to the central nervous system (CNS), and possible cerebral edema (fluid on the brain). Carbon monoxide has an affinity for blood hemoglobin some 200 times greater than that of oxygen. This means that it will bond much more quickly than oxygen and will form carboxyhemoglobin, thereby preventing the normal distribution of oxygen throughout the body. The gas may also combine with myoglobin to form carboxymyoglobin which may affect muscle metabolism and cause muscular disturbances, especially in the heart.

Acute Health Effects

The following acute (short-term) health effects may occur immediately or shortly after exposure to cobalt carbonyl:

Skin: Irritation, rash, and possible contact dermatitis.

Eye: Cobalt carbonyl particles can irritate the eyes and may cause damage to vision.

Lung: Irritation of the dust and fumes can irritate the nose, mouth, and throat. High levels can irritate the lungs, causing a buildup of fluid (pulmonary edema), which is a medical emergency and can be fatal.

Other: On decomposition, carbon monoxide is produced. Inhalation can cause air-starvation, chemical asphyxia, and death.

Chronic Health Effects

The following chronic (long-term) health effects may occur at some time after exposure to cobalt carbonyl and can last for months or even years:

Cancer Hazards: Cobalt carbonyl is a questionable carcinogen in test animals. However, cobalt compounds may cause mutations (genetic changes) in living cells. Cobalt is a confirmed carcinogen in animals and is suspected to cause cancer in humans, although its specific carcinogenicity requires further study. Some isotopes of cobalt carbonyl do emit ionizing radiation. Exposure to ionizing radiation is associated with an increased risk of developing cancer.

Reproductive Hazard: According to information available in the references, cobalt carbonyl has been shown to cause teratogenic effects in limited studies.

Other Chronic Effects: Cobalt carbonyl can cause skin allergy. If an allergy develops, very low future exposures can cause itching and a skin rash. Cobalt carbonyl may cause an asthma-like allergy. Future exposures can cause cough, shortness of breath, wheezing, and/or chest tightness. Cobalt carbonyl can also damage the heart, leading to heart failure. Long-term exposure may damage the thyroid and liver. Repeated exposure can cause scarring of the lungs (pulmonary fibrosis), even if no symptoms are noticed. This condition can be disabling or fatal.

Recommended Risk-Reduction Measures

Personnel should avoid direct contact with cobalt carbonyl. If the specific cobalt carbonyl compound in question is unknown, adequate risk management dictates that it be handled and treated as though it were a carcinogen. If a less toxic chemical cannot be substituted for a hazardous substance, then *engineering controls* are the most effective method of reducing exposures.

The best protection is to enclose operations and/or provide local exhaust ventilation at the site of chemical release. While not always operationally feasible, isolating operations can also reduce exposure. Using respiratory protection is less effective than the controls mentioned above, but is still advisable whenever working with or around cobalt carbonyl. For exposure to cobalt carbonyl at levels above the PEL (0.1 mg/m^3), an MSHA/NIOSH approved respirator with a high efficiency particulate air (HEPA) filter is recommended. For higher exposures, use a supplied-air respirator with a full facepiece operated in positive pressure mode or an MSHA/NIOSH approved self-contained breathing apparatus (SCBA) with full facepiece operated in pressure demand mode. If a full facepiece is not available, then chemical splash goggles should be worn to protect the eyes. Whenever a chemical dust hazard exists, a face shield and a rubber apron should be worn. To prevent hand and skin exposures, impervious rubber gloves should be worn.

Administrative controls should also be in place to minimize the potential for human exposures. These may include written procedures or policies which specify the methods and techniques that will be practiced whenever personnel are to work with cobalt carbonyl. A regulated, controlled work area should be established where cobalt compounds are used, handled, or stored.

All personnel should receive *training* on the use, hazards, protective measures, emergency actions, and other precautions per 29 CFR 1910.1200 (Hazard Communication), prior to the first assignment in an area where cobalt carbonyl is used or stored.

Before beginning employment and at regular intervals thereafter (e.g., annually), for those with frequent or potentially high exposures (half the PEL or greater, or significant skin contact), the following medical tests are recommended:

☑ Lung function tests.

☑ Chest X-ray for scarring should be performed every two to three years after five or more years of exposure.

☑ Evaluation for heart failure.

If symptoms develop or overexposure is suspected, the following may also be useful:

☑ Liver, thyroid, and lung function tests.

☑ Evaluation by a qualified allergist, including careful exposure history and special testing (may help diagnose skin allergy).

Any evaluation should include a careful history of past and present symptoms with a medical examination. Medical tests that look for damage already done are not a substitute for controlling exposure. Also, since smoking can cause heart disease, as well as lung cancer, emphysema, and other related respiratory problems, the effects of exposure to cobalt carbonyl may be more pronounced in smokers than in non-smokers.

Other methods to reduce exposure include:

☑ Where possible, enclose operations and use local exhaust ventilation at the site of chemical release.

If local exhaust ventilation or enclosure is not used, respiratory protection should be mandatory.

- ☑ Always ensure that proper protective clothing is worn when using chemical substances.
- ☑ Wash thoroughly immediately after exposure to cobalt carbonyl and at the end of the work shift or before eating, drinking, or smoking.
- ☑ Hazard warning information should be posted in the work area. In addition, as part of an on-going education and training program, all information on the health and safety hazards of cobalt carbonyl should be communicated to all exposed workers.
- ☑ Where possible, automatically transfer cobalt carbonyl from drums or other storage containers to process containers.
- ☑ Specific engineering controls are recommended for cobalt by NIOSH. Refer to NIOSH Occupational Hazard Assessment: Criteria for Controlling Occupational Exposure to Cobalt (Publication Number 82-107).

RISK ASSESSMENT: ENVIRONMENT

General Assessment

The environment is at risk of exposure during transportation, storage, disposal, or destruction of cobalt carbonyl. In almost every scenario, the threat of environmental exposure is contingent upon the proper handling of the chemical substance. Accidental spills, large or small, can result in fire, explosion, and possible contamination of the surrounding environmental mediums (water, soil, and air).

Cobalt carbonyl is considered a non-combustible solid. However, when exposed to air or heat, or if involved in fire, it will decompose and produce highly flammable and explosive carbon monoxide gas. It is also incompatible with acids (hydrochloric, sulfuric, and nitric). These characteristics require special consideration during any emergency situation involving a leak or spill of cobalt carbonyl.

Cobalt can occur in the environment as a natural element in certain ores in the earth's crust. It can also enter the environment through industrial and municipal discharges, or from spills.

Acute Ecological Effects

Acute (short-term) toxic effects may include the death of animals, birds, or fish and death or low growth rate in plants. Acute effects are seen 2 to 4 days after animals or plants are exposed to cobalt.

Cobalt and its salts have high acute toxicity to aquatic life. Insufficient data are available to evaluate the short-term effects of cobalt exposure to plants, birds, or land animals.

Chronic Ecological Effects

Chronic (long-term) toxic effects may include shortened life span, reproductive problems, lower fertility, and changes in appearance or behavior in exposed animals. These effects can be seen long after first exposure(s) to toxic chemicals.

Cobalt and its salts have high chronic toxicity to aquatic life. Insufficient data are available to evaluate the long-term effects of cobalt exposure to plants, birds or land animals.

Water Solubility

Water solubility of cobalt and its salts ranges from highly soluble to practically insoluble. Cobalt carbonyl is considered insoluble. Concentrations of less than 1 milligram may not mix in a liter of water.

Persistence in the Environment

Cobalt and its salts are highly persistent in the aquatic environment, with a half-life of greater than 200 days. The half-life of a pollutant is the amount of time it takes for one half of the chemical to be degraded.

Bioaccumulation in Aquatic Organisms

Some substances increase in concentration, or *bioaccumulate,* in living organisms as they breathe contaminated air, drink contaminated water, or eat contaminated food. These chemicals can become concentrated in the tissues and internal organs of animals as well as humans.

The concentration of cobalt found in fish is expected to be somewhat higher than the average concentration of cobalt in the water from which the fish was taken.

Recommended Risk-Reduction Measures

Proper training of all transporters will reduce the likelihood of a mishap or accidents resulting in a leak or spill to the environment. The correct use of DOT labeling (per the MSDS) on all containers, trucks, and rail cars will enable emergency responders to react properly and quickly to any disaster thereby reducing the potential risk to the environment and to personnel.

Storage of cobalt compounds should be segregated from incompatible chemicals to minimize the risk of cross-contamination or contact.

If a spill or leak to the environment has occurred, fire department, emergency response and/or hazardous materials spill personnel should be notified immediately. Use a vacuum method to reduce dust during cleanup. Do not dry sweep. When vacuuming, a high efficiency particulate air (HEPA) filter should be used, not a standard shop vacuum. Deposit collected materials in sealed drum for disposal. Cleanup should be attempted only by those trained in proper spill containment procedures. Contaminated soils should be removed and replaced with clean soil. If cobalt carbonyl should contact the water table, aquifer, or navigable waterway, time is of the essence. It is insoluble in water and, therefore, total containment and remediation may be possible. When such spills occur, the local and/or state emergency response authorities must be notified. A comprehensive emergency response or disaster preparedness/recovery plan should be in place prior to any operations involving the use, transportation, storage, or disposal of cobalt carbonyl. If cobalt carbonyl is spilled or leaked, the following specific steps are recommended:

- ☑ Restrict persons not wearing protective clothing from area of spill or leak until cleanup is complete and area can be opened for normal work.
- ☑ Collect powdered materials in most convenient and safe manner possible and place in sealed drums for disposal. Do not dry sweep, use HEPA vacuum instead.
- ☑ It may be necessary to dispose of cobalt carbonyl as a hazardous waste. The responsible state agency or the regional office of the federal Environmental Protection Agency (EPA) should be contacted for specific recommendations.

RISK ASSESSMENT: BUSINESS

General Assessment

Accidents or mishaps involving cobalt carbonyl can present a significant threat to business operations. The loss or damage of equipment or facilities can significantly affect fiscal viability. Lawsuits that may result from personnel injury/death, public exposures, and/or environmental contamination will also require a serious expenditure of resources. Media attention surrounding an injury, death, or environmental damage can also result in a loss of profits and loss of current as well as future business. Always remember that anytime the term "radiation" or "radioactive" are used, public emotion and anxiety can run equally high. This must be considered when dealing with the public following an incident involving cobalt carbonyl.

Recommended Risk-Reduction Measures

Company attorneys, safety and health professionals, and environmental specialists should be involved in the development of any procedures dealing with safety or emergency response. A company official should be pre-designated as a public relations officer with specific training in dealing with the press. Corporate plans and policies should be developed, approved, and implemented long before any need for such arises.

REFERENCES

American Conference of Governmental Industrial Hygienists. 1988. *Documentation of the Threshold Limit Values and Biological Exposure Indices,* 5th Edition (with updates). Cincinnati: ACGIH

Brusick, D. J. 1994. *Methods for Genetic Risk Assessment.* New York: CRC Press/Lewis Publishers.

Cockerman, L. G., B. S. Shane. 1994. *Basic Environmental Toxicology.* New York: CRC Press/Lewis Publishers.

Gardner, R. W. 1994. *Chemical Intolerance.* New York: CRC Press/Lewis Publishers.

Howard, P. H., et. al. 1991. *Handbook of Environmental Degradation Rates.* New York: CRC Press/Lewis Publishers.

Keith, L. H. and M. M. Walker. 1995. *Handbook of Air Toxics.* New York: CRC Press/Lewis Publishers.

Lewis, R. J., Sr. 1992. *Sax's Dangerous Properties of Industrial Materials,* Volumes 1, 2, and 3, 8th Edition. New York: Van Nostrand Reinhold.

Meyer, E. 1990. *Chemistry of Hazardous Materials.* Englewood Cliffs, NJ: Prentice Hall.

U.S. Department of Health and Human Services (DHHS), National Institute for Occupational Health (NIOHS). 1994. *NIOSH Pocket Guide to Hazardous Chemicals.* Publication Number 94-116. Washington, D.C.; U.S. Government Printing Office.

U.S. Department of Health and Human Services (DHHS), National Institute for Occupational Health (NIOHS). 1982. *Criteria for Controlling Occupational Exposure to Cobalt.* Publication Number 82-107. Washington, D.C.; U.S. Government Printing Office.

MATERIAL SAFETY DATA SHEET

CHEMICAL NAME	CRC PRESS/LEWIS PUBLISHERS DISCLAIMER
COPPER	THE INFORMATION AND RECOMMENDATIONS PRESENTED HEREIN ARE BASED ON SOURCES BELIEVED TO BE RELIABLE. CRC MAKES NO REPRESENTATION ON ITS COMPLETENESS OR ACCURACY. IT IS THE RESPONSIBILITY OF THE USER TO DETERMINE THE CHEMICAL'S SUITABILITY FOR ITS INTENDED USE, THE CHEMICAL'S SAFE USE, AND THE CHEMICAL'S PROPER DISPOSAL. NO REPRESENTATIONS AND/OR WARRANTIES, EITHER EXPRESSED OR IMPLIED, OF THE MERCHANTABILITY OR FITNESS FOR A PARTICULAR PURPOSE, OR OF ANY OTHER NATURE, ARE MADE WITH RESPECT TO THE INFORMATION PROVIDED IN THIS MSDS OR TO THE CHEMICAL TO WHICH INFORMATION MAY REFER. CRC NEITHER ASSUMES NOR AUTHORIZES ANY OTHER PERSON TO ASSUME FOR IT, ANY OTHER ADDITIONAL RESPONSIBILITY OR LIABILITY FOR THE USE OF, OR RELIANCE UPON, THIS INFORMATION.

HAZARD WARNING INFORMATION

HEALTH	FIRE	REACTIVE	OTHER	DEGREE OF HAZARD	COLOR CODING	OTHER CODES
2	0	0		0 = Minimum Hazard 1 = Slight Hazard 2 = Moderate Hazard 3 = Serious Hazard 4 = Severe Hazard	HEALTH = BLUE FIRE = RED REACTIVITY = YELLOW OTHER = WHITE	OX = Oxidizer ACID = Acid ALK = Alkali COR = Corrosive ~~W~~ = Use No Water

SECTION I - GENERAL INFORMATION

Characterization	**Metal**	RCRA Number	**None**
		EPA Class	**Not Applicable**
DOT Proper Shipping Name	**Copper**	Chemical Abstract Service (CAS) Number	**7440-50-8**
DOT Hazard Class and Label Requirements	**No Citation**	DOT Emergency Guide Code	**No Citation**
DOT Identification Number	**No Citation**	Chemical Formula	**Cu**

Synonyms

Allbri natural copper; copper bronze; raney copper; copper-milled; c.i. pigment metal 2.

SECTION II - HAZARDOUS INGREDIENTS AND IDENTITY INFORMATION

Hazardous Components (specific identity)	OSHA Exposure Criteria	NIOSH Exposure Criteria	Immediately Dangerous to Life and Health (IDLH)	ACGIH Exposure Criteria
Copper (derivation: Varies with the type of ore. With sulfide ores, the steps may be concentration (of low grade ores) by floatation and leaching; roasting; formation of copper "matte" (40-50% Cu); reduction of matte to "blister" copper (96-98%); electrolytic refining to 99.9 + % copper).	**PEL:** **dust/mist:** **1.0 mg(Cu)/m^3** **fume** **0.1 mg(Cu)/m^3** **STEL:** **Not Established**	**REL:** **dust/mist:** **1.0 mg(Cu)/m^3** **fume** **0.1 mg(Cu)/m^3** **STEL:** **Not Established**	**100 mg/m^3 as Cu**	**TLV:** **dust/mist:** **1.0 mg(Cu)/m^3** **fume** **0.2 mg(Cu)/m^3** **STEL:** **Not Established**

SECTION III - PHYSICAL AND CHEMICAL CHARACTERISTICS

Boiling Point	**4703°F (2595°C)**	Specific Gravity (H_2O = 1)	**8.94**
Vapor Pressure (mm Hg)	**1 at 2962°F (1628°C)**	Molecular Weight	**63.5**
Vapor Density (Air = 1)	**Not Found**	Melting Point	**1981°F (1083°C)**

Solubility

Fumes are insoluble. Dusts are moderately to highly soluble. Dissolves readily in some acids.

Appearance and Odor

Reddish-brown, lustrous, malleable, odorless solid. May appear as a powder.

SECTION IV - FIRE AND EXPLOSION HAZARD DATA

Flash Point (method used)	**Not Applicable**	Explosive Limits in Air % by Volume	**LEL: Not Applicable UEL: Not Applicable**
NFPA Classification	**No Citation**	Autoignition Temperature	**Not Determined**

Extinguishing Media

Use class "D" extinguishing agent. Do NOT use water.

Special Fire Fighting Procedures

Wear full protective clothing and self-contained breathing apparatus (SCBA). Use a water spray to keep fire-exposed containers cool. Water may be ineffective on fire. Burning copper metal may reach temperatures in excess of 5000°F. Poisonous gases are produced in fire.

Unusual Fire and Explosion Hazards

Can cause violent explosions upon contact with strong oxidizers. Liquid copper explodes on contact with water. Potentially explosive in contact with acetylene compounds.

SECTION V - REACTIVITY DATA

Stability		Conditions to Avoid Keep away from incompatible materials since violent and or explosive reactions can occur.
Stable X	Unstable	Incompatibility *(materials to avoid)* Incompatible with strong oxidizers (such as perchlorates, permanganates, peroxides, chlorates, and nitrates), and chemically active metals (such as potassium, sodium, magnesium, and zinc).
Hazardous Polymerization		Conditions to Avoid Under normal conditions of temperature and pressure, hazardous polymerization of copper is not known to occur.
May Occur	Will Not Occur X	Hazardous Decomposition or By-products When heated to decomposition, copper can emit highly toxic/poisonous fumes and gases.

SECTION VI - HEALTH HAZARD DATA

Primary Route(s) of Entry:	Inhalation?	Absorption (skin)?	Ingestion?
	X		X

Health Hazards

INHALATION: Irritation of the respiratory system. Exposure to some copper compounds can cause severe throat and nose irritation and may cause nosebleeds. Exposure can cause "metal fume fever," a flu-like illness lasting about 24 hours with chills, aches, and cough.

SKIN & EYES: Eye contact can cause severe irritation and can damage vision or cause blindness.

INGESTION: A poison with gastrointestinal effects. Will cause nausea and vomiting.

Carcinogenicity	NTP Listed?	IARC Cancer Review Group?	OSHA Regulated?	Target Organs?
Confirmed Human Confirmed Animal	No	No	29 CFR 1910.1000 Table Z-1	Respiratory system, skin, eyes, liver, kidneys

Medical Conditions Generally Aggravated by Exposure

Wilson's disease.

Emergency and First-aid Procedures

Eye contact: Flush immediately with water for 15 minutes (minimum), seek medical attention. Skin contact: Remove all contaminated clothing. Immediately wash area with large amounts of soap and water. For inhalation: Remove the person from exposure. Provide respiratory assistance and CPR. Transfer to medical facility. If swallowed: Provide breathing assistance if necessary. Never attempt to give a convulsing or unconscious person anything by mouth. If victim is conscious, give 3 to 4 glasses of water and induce vomiting. Seek medical attention immediately.

SECTION VII - PRECAUTIONS FOR SAFE HANDLING AND USE

Steps to be Taken in Case Material is Released or Spilled

Collect powdered material in most convenient manner possible and deposit in sealed drum. Keep those not involved in cleanup from entering area. Do NOT dry sweep (creates airborne dusts). Use a vacuum equipped with a high efficiency particulate air (HEPA) filter.

Preferred Waste Disposal Method

No Citation.

Precautions to be Taken in Handling and Storage

Store in tightly closed containers in cool, well-ventilated area and protect from physical damage. Sources of ignition, such as smoking and open flames, are prohibited where copper is used, handled, or stored.

Other Precautions and Warnings

Keep away from acetylene gas since contact will produce flammable hydrogen gas.

SECTION VIII - CONTROL MEASURES AND PERSONAL PROTECTIVE EQUIPMENT

Respiratory Protection *(specify type)*

At exposures above PEL, use full facepiece respirator with high efficiency particulate air (HEPA) filter. For higher exposures, use a supplied-air respirator with a full facepiece, hood, or helmet or a self-contained breathing apparatus (SCBA) operated in positive pressure mode.

Ventilation

Local exhaust or general mechanical systems recommended.

Protective Gloves	Eye Protection	Other Protective Clothing
Impervious Material	Chemical/Dust Goggles or Face Mask	Protective Suit or Clothing

Work/Hygiene Practices

Always wash hands thoroughly after using chemical; never bring food, drink, or smoking materials into vicinity of chemicals.

COPPER

Cu CAS: 7440-50-8

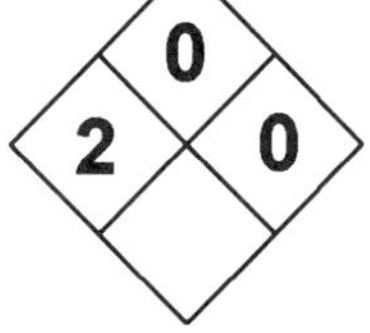

IDENTIFICATION AND TYPICAL USES

Copper is a commonly occurring element (atomic number 29). It appears as a reddish-brown, lustrous, malleable, odorless, solid. It also exists commonly in a powdered form. It is used extensively in electric wiring and switches, (an excellent conductor), plumbing, heating, roofing and building construction, chemical and pharmaceutical machinery, alloys (brass, bronze, Monel metal, beryllium-copper), electroplated protective coatings and undercoats for nickel, chromium, zinc, and others. It is used in cooking utensils, corrosion-resistant piping, insecticides, as a catalyst, and in antifouling paints. Copper flakes are sometimes used as insulation for liquid fuels. Copper "whiskers" are used in thermal and electrical composites.

RISK ASSESSMENT: HEALTH

General Assessment

Copper is a questionable carcinogen with tumorigenic data reported in test experiments. There is also some evidence in the references of teratogenic and reproductive effects in laboratory tests, but these results appear inconclusive especially regarding human exposures. It is a human poison by *ingestion* with systemic effects including vomiting and nausea. It can also be toxic by *inhalation.*

Occupational skin exposure can produce localized irritation and possible rash which can lead to dermatitis in sensitive individuals. Copper is also extremely irritating to the eyes. It can lead to temporary and/or permanent damage to vision up to and including blindness.

Inhalation of copper fumes can cause "metal fume fever," which is a flu-like illness lasting up to 24 hours with characteristic symptoms of chills, aches, cough, and fever. The dusts are extremely irritating to the nose and throat and may, on prolonged exposure, cause nasal ulcers with symptoms of extreme irritation and chronic nosebleeds.

Acute Health Effects

The following acute (short-term) health effects may occur immediately or shortly after exposure to copper:

Skin: Irritation, rash, and possible contact dermatitis.

Eye: Copper particles and dusts can be extremely irritating to the eyes. Exposure can damage vision and lead to total loss of sight.

Lung: Inhalation of copper may cause a condition known as "metal fume fever" (flu-like symptoms). Copper dusts and fumes can irritate the nose and throat. It may cause the appearance of ulcers in the nasal cavity accompanied by nosebleeds.

Chronic Health Effects

The following chronic (long-term) health effects may occur at some time after exposure to copper and can last for months or even years:

Cancer Hazards: There is evidence that workers in copper smelters have an increased risk of lung cancer, but this is thought to be due to arsenic trioxide exposure and not copper. There is also some evidence that exposure to copper dusts and fumes may cause cancer in laboratory test animals. Human data, however, are inconclusive in this regard.

Reproductive Hazard: According to information available in the references, there is some evidence to indicate that copper may adversely affect reproduction in laboratory animals.

Other Chronic Effects: Repeated exposures can cause chronic irritation of the nose and may cause ulcers. Copper may cause a skin allergy. If allergy develops, very low future exposures can cause itching and a skin rash. Repeated exposures can cause thickening of the skin and may cause a greenish color to the skin and hair. Repeated or very high exposures to copper can damage the liver and/or kidneys.

↻ *Recommended Risk-Reduction Measures*

Personnel should avoid direct contact with copper. If a less toxic chemical cannot be substituted for a hazardous substance, then *engineering controls* are the most effective method of reducing exposures.

The best protection is to enclose operations and/or provide local exhaust ventilation at the site of chemical release. While not always operationally feasible, isolating operations can also reduce exposure. Using respiratory protection is less effective than the controls mentioned above, but is still advisable whenever working with or around copper. For exposures above the PEL, use a full facepiece respirator equipped with a high efficiency particulate air (HEPA) filter and a dust/mist/fume pre-filter. For higher exposures, use a supplied-air respirator with a full facepiece operated in positive pressure mode or an MSHA/NIOSH approved self-contained breathing apparatus (SCBA) with full facepiece and pressure demand. If a full facepiece is not available, then chemical/dust goggles should be worn to protect the eyes. Whenever a chemical dust hazard exists, a face shield and a rubber apron should be worn. To prevent hand and skin exposures, impervious rubber gloves should be worn.

Administrative controls should also be in place to minimize the potential for human exposures. These may include written procedures or policies which specify the methods and techniques that will be practiced whenever personnel are to work with copper.

All personnel should receive *training* on the use, hazards, protective measures, emergency actions, and other precautions per 29 CFR 1910.1200 (Hazard Communication), prior to the first assignment in an area where copper is used or stored.

If symptoms develop or overexposure is suspected, the following may also be useful:

☑ Urine and serum copper levels.

Any evaluation should include a careful history of past and present symptoms with a medical examination. Medical tests that look for damage already done are not a substitute for controlling exposure. Copper metal often contains arsenic as an impurity. Therefore, exposure to copper dusts or fumes may also be mixed with varying levels of arsenic exposure as well. Also, "Wilson's Disease" is a rare condition that interferes with the body's ability to metabolize copper. Persons with this disease may be seriously affected by copper exposures.

Other methods to reduce exposure include:

☑ Where possible, enclose operations and use local exhaust ventilation at the site of chemical release. If local exhaust ventilation or enclosure is not used, respiratory protection should be mandatory.

☑ Always ensure that proper protective clothing is worn when using chemical substances.

☑ Wash thoroughly immediately after exposure to copper and at the end of the work shift or before eating, drinking, or smoking.

☑ Hazard warning information should be posted in the work area. In addition, as part of an on-going education and training program, all information on the health and safety hazards of copper should be communicated to all potentially exposed workers.

☑ Where possible, automatically transfer copper from drums or other storage containers to process containers.

RISK ASSESSMENT: ENVIRONMENT

General Assessment

Copper is a naturally occurring element and is therefore present in the environment. However, the environment is still at risk of exposure during transportation, storage, disposal, or destruction of copper. In almost every scenario, the threat of environmental exposure is contingent upon the proper handling of the chemical substance. Accidental spills, large or small, can result in fire, explosion, and possible contamination of the surrounding environmental mediums (water, soil, and air).

Liquid copper will explode on contact with water. It will ignite and is potentially explosive in contact with acetylene compounds; 3-bromopropyne; ethylene oxide; lead azide; and ammonium nitrate. It also ignites on contact with chlorine, chlorine trifluoride, fluorine (above 121°C), and hydrazinium nitrate (above 70°C). When heated in the presence of ammonium nitrate, violent or explosive reactions can result. In a fire, copper will emit toxic and poisonous gases. It is also incompatible with strong oxidizers (such as chlorine, fluorine, bromine, and hydrogen peroxide), and many chemically active metals (potassium, sodium, magnesium, and zinc). These characteristics require special consideration during any emergency situation involving a leak or spill of copper.

Copper is a naturally occurring element in natural water aquifers. At low concentrations, it is an essential element for both plants and animals. At slightly higher concentrations, it is toxic to aquatic life. The toxicity of copper and its compounds to aquatic life varies with the physical and chemical conditions of the water. Factors such as water hardness, alkalinity, and pH influence copper toxicity.

Acute Ecological Effects

Acute (short-term) toxic effects may include the death of animals, birds, or fish and death or low growth rate in plants. Acute effects are seen 2 to 4 days after animals or plants are exposed to copper.

Copper and its compounds have high acute toxicity to aquatic life. Insufficient data are available to evaluate the short-term effects of copper exposure to plants, birds or land animals.

Chronic Ecological Effects

Chronic (long-term) toxic effects may include shortened life span, reproductive problems, lower fertility, and changes in appearance or behavior in exposed animals. These effects can be seen long after first exposure(s) to toxic chemicals.

Copper and its compounds have high chronic toxicity to aquatic life. Insufficient data are available to evaluate the long-term effects of copper exposure to plants, birds or land animals.

Water Solubility

Copper and its salts are highly soluble in water. Concentrations of 1000 milligrams and more will mix with a liter of water.

Persistence in the Environment

Copper is highly persistent in the aquatic environment, with a half-life of greater than 200 days. The half-life of a pollutant is the amount of time it takes for one half of the chemical to be degraded.

Bioaccumulation in Aquatic Organisms

Some substances increase in concentration, or *bioaccumulate,* in living organisms as they breathe contaminated air, drink contaminated water, or eat contaminated food. These chemicals can become concentrated in the tissues and internal organs of animals as well as humans.

The concentration of copper found in fish is expected to be considerably higher than the average concentration of copper in the water from which the fish was taken.

Recommended Risk-Reduction Measures

Proper training of all transporters will reduce the likelihood of a mishap or accidents resulting in a leak or spill to the environment. The correct use of DOT labeling (per the MSDS) on all containers, trucks, and rail cars will enable emergency responders to react properly and quickly to any disaster thereby reducing the potential risk to the environment and to personnel.

If a spill or leak to the environment has occurred, fire department, emergency response and/or hazardous materials spill personnel should be notified immediately. Use a vacuum or wet method to reduce dust during cleanup. Do not dry sweep. When vacuuming, a high efficiency particulate air (HEPA) filter should be used, not a standard shop vacuum. Deposit collected materials in sealed drum for disposal. Cleanup should be attempted only by those trained in proper spill containment procedures. Contaminated soils may need to be removed and replaced with clean soil. If copper should contact the water table, aquifer, or navigable waterway, time is of the essence. Copper compounds may be moderately to highly soluble in water and, therefore, total containment and remediation may not be entirely possible. When such spills occur, the local and/or state emergency response authorities must be notified. A comprehensive emergency response or disaster preparedness/recovery plan should be in place prior to any operations involving the use, transportation, storage, or disposal of copper. If copper is spilled or leaked, the following specific steps are recommended:

- ☑ Restrict persons not wearing protective clothing from area of spill or leak until cleanup is complete and area can be opened for normal work.
- ☑ Collect powdered materials in most convenient and safe manner possible and place in sealed drums for disposal. Do not dry sweep, use HEPA vacuum instead.
- ☑ It may be necessary to dispose of copper as a hazardous waste. The responsible state agency or the regional office of the federal Environmental Protection Agency (EPA) should be contacted for specific recommendations.

RISK ASSESSMENT: BUSINESS

General Assessment

Accidents or mishaps involving copper can present a significant threat to business operations. The loss or

damage of equipment or facilities can significantly affect fiscal viability. Lawsuits that may result from personnel injury/death, public exposures, and/or environmental contamination will also require a serious expenditure of resources. Media attention surrounding an injury, death, or environmental damage can also result in a loss of profits and loss of current as well as future business.

❶ *Recommended Risk-Reduction Measures*

Company attorneys, safety and health professionals, and environmental specialists should be involved in the development of any procedures dealing with safety and emergency response. A company official should be pre-designated as a public relations officer with specific training in dealing with the press. Corporate plans and policies should be developed, approved, and implemented long before any need for such arises.

REFERENCES

Ahlbom, A. 1993. *Biostatistics for Epidemiologists.* New York: CRC Press/Lewis Publishers.

American Conference of Governmental Industrial Hygienists. 1988. *Documentation of the Threshold Limit Values and Biological Exposure Indices,* 5th Edition (with updates). Cincinnati: ACGIH

Asante-Duah, D. K. 1993. *Hazardous Waste Risk Assessment.* New York: CRC Press/Lewis Publishers.

Bloom, A. D. and F. J. DeSerres. 1995. *Ecotoxicity and Human Health.* New York: CRC Press/Lewis Publishers.

Brusick, D. J. 1994. *Methods for Genetic Risk Assessment.* New York: CRC Press/Lewis Publishers.

Calabrese, E. J., E. M. Kenyon. *Air Toxics and Risk Assessment.* New York: CRC Press/Lewis Publishers.

Calabrese, E. J. 1994. *Biological Effects of Low Level Exposures.* New York: CRC Press/Lewis Publishers.

Cockerman, L. G., B. S. Shane. 1994. *Basic Environmental Toxicology.* New York: CRC Press/Lewis Publishers.

Cothern, C. R., N. P. Ross. 1994. *Environmental Statistics, Assessment, and Forecasting.* New York: CRC Press/Lewis Publishers.

Cothern, R. R. 1993. *Comparative Environmental Risk Assessment.* New York: CRC Press/Lewis Publishers.

Cothern, C. R. 1995. *Handbook for Environmental Risk Decisionmaking.* New York: CRC Press/Lewis Publishers.

Gardner, R. W. 1994. *Chemical Intolerance.* New York: CRC Press/Lewis Publishers.

Hallenbeck, W. H. 1993. *Quantitative Risk Assessment for Environmental and Occupational Health,* 2nd Edition. New York: CRC Press/Lewis Publishers.

Howard, P. H., et. al. 1991. *Handbook of Environmental Degradation Rates.* New York: CRC Press/Lewis Publishers.

Howard, P. H. 1995. *Handbook of Environmental Fate and Exposures Data for Organic Chemicals,* Vol. I-IV. New York: CRC Press/Lewis Publishers.

Jones, F. E. 1994. *Toxic Organic Vapors in the Workplace.* New York: CRC Press/Lewis Publishers.

Keith, L. H. and M. M. Walker. 1995. *Handbook of Air Toxics.* New York: CRC Press/Lewis Publishers.

Landis, W. G., M. H. Yu. 1995. *Introduction to Environmental Toxicology.* New York: CRC Press/Lewis Publishers.

Langerman, N. 1994 *Precautionary Labels for Chemical Containers.* New York: CRC Press/Lewis Publishers.

Lauwerys, R. R., P. Hoet. 1993. *Industrial Chemical Exposure.* New York: CRC Press/Lewis Publishers.

Lewis, R. J., Sr. 1992. *Sax's Dangerous Properties of Industrial Materials,* Volumes 1, 2, and 3, 8th Edition. New York: Van Nostrand Reinhold.

Manahan, S. E. 1994. *Environmental Chemistry,* 6th Edition. New York: CRC Press/Lewis Publishers.

Manahan, S. E. 1992. *Toxicological Chemistry,* 2nd Edition. New York: CRC Press/Lewis Publishers.

Meyer, E. 1990. *Chemistry of Hazardous Materials.* Englewood Cliffs, NJ: Prentice Hall.

Patnaik, P. 1992. *A Comprehensive Guide to the Hazardous Properties of Chemical Substances.* New York: Van Nostrand Reinhold.

Quigley, D. R. 1994. *Handbook of Emergency Chemical Management.* New York: CRC Press/Lewis Publishers.

Rea, W. J. 1995. *Chemical Sensitivity,* Vol. I-III. New York: CRC Press/Lewis Publishers.

Sacarello, H. L. A. 1994. *The Comprehensive Handbook of Hazardous Materials.* New York: CRC Press/Lewis Publishers.

Sherman, J. D. 1988. *Chemical Exposure and Disease.* New York: Van Nostrand Reinhold.

Talbott, E., G. Craun. 1995. *An Introduction to Environmental Epidemiology.* New York: CRC Press/Lewis Publishers.

U.S. Department of Health and Human Services (DHHS), National Institute for Occupational Health (NIOHS). 1994. *NIOSH Pocket Guide to Hazardous Chemicals.* Publication Number 94-116. Washington, D.C.; U.S. Government Printing Office.

MATERIAL SAFETY DATA SHEET

CHEMICAL NAME	
COPPER CYANIDE	**CRC PRESS/LEWIS PUBLISHERS DISCLAIMER:** THE INFORMATION AND RECOMMENDATIONS PRESENTED HEREIN ARE BASED ON SOURCES BELIEVED TO BE RELIABLE. CRC MAKES NO REPRESENTATION ON ITS COMPLETENESS OR ACCURACY. IT IS THE RESPONSIBILITY OF THE USER TO DETERMINE THE CHEMICAL'S SUITABILITY FOR ITS INTENDED USE, THE CHEMICAL'S SAFE USE, AND THE CHEMICAL'S PROPER DISPOSAL. NO REPRESENTATIONS AND/OR WARRANTIES, EITHER EXPRESSED OR IMPLIED, OF THE MERCHANTABILITY OR FITNESS FOR A PARTICULAR PURPOSE, OR OF ANY OTHER NATURE, ARE MADE WITH RESPECT TO THE INFORMATION PROVIDED IN THIS MSDS OR TO THE CHEMICAL TO WHICH INFORMATION MAY REFER. CRC NEITHER ASSUMES NOR AUTHORIZES ANY OTHER PERSON TO ASSUME FOR IT, ANY OTHER ADDITIONAL RESPONSIBILITY OR LIABILITY FOR THE USE OF, OR RELIANCE UPON, THIS INFORMATION.

HAZARD WARNING INFORMATION

HEALTH	FIRE	REACTIVE	OTHER	DEGREE OF HAZARD	COLOR CODING	OTHER CODES
2	0	0		0 = Minimum Hazard 1 = Slight Hazard 2 = Moderate Hazard 3 = Serious Hazard 4 = Severe Hazard	HEALTH = BLUE FIRE = RED REACTIVITY = YELLOW OTHER = WHITE	OX = Oxidizer ACID = Acid ALK = Alkali COR = Corrosive ~~W~~ = Use No Water

SECTION I - GENERAL INFORMATION

Chemical Family	**Inorganic cyanide**	RCRA Number	**P029**
		EPA Class	**Acute Hazardous Waste**
DOT Proper Shipping Name	**Copper or cupric cyanide**	Chemical Abstract Service (CAS) Number	**14763-77-0**
DOT Hazard Class and Label Requirements	**Poison B**	DOT Emergency Guide Code	**53**
DOT Identification Number	**UN 1587**	Chemical Formula	**$Cu(CN)_2$**

Synonyms

Cupric cyanide; copper cyanamide.

SECTION II - HAZARDOUS INGREDIENTS AND IDENTITY INFORMATION

Hazardous Components (specific identity)	OSHA Exposure Criteria	NIOSH Exposure Criteria	Immediately Dangerous to Life and Health (IDLH)	ACGIH Exposure Criteria
Copper cyanide (derivation: Addition of potassium cyanide to a solution of copper sulfate; cupric cyanide is precipitated. This can be dried but is not stable).	**PEL:** dust/mist: 1.0 mg(Cu)/m^3 **STEL:** Not Established	**REL:** Not Established **STEL:** Not Established	**50 mg/m^3 as Cu**	**TLV:** dust/mist: 1.0 mg(Cu)/m^3 **STEL:** Not Established

SECTION III - PHYSICAL AND CHEMICAL CHARACTERISTICS

Boiling Point	**Not Applicable**	Specific Gravity (H_2O = 1)	**Not Determined**
Vapor Pressure (mm Hg)	**Not Applicable**	Molecular Weight	**115.58**
Vapor Density (Air = 1)	**Not Applicable**	Melting Point	**Decomposes**

Solubility in Water

Insoluble in water. Soluble in acids and alkalis.

Appearance and Odor

Green-colored powder, no distinguishable odor.

SECTION IV - FIRE AND EXPLOSION HAZARD DATA

Flash Point (method used)	**Not Applicable**	Explosive Limits in Air % by Volume	**LEL: Not Applicable UEL: Not Applicable**
NFPA Classification	**No Citation**	Autoignition Temperature	**Not Determined**

Extinguishing Media

Use dry chemical, carbon dioxide, water spray, or foam.

Special Fire Fighting Procedures

Wear full protective clothing and self-contained breathing apparatus (SCBA). Use a water spray to keep fire-exposed containers cool. Poisonous gases are produced in fire, including cyanide gas, nitrous oxide. In the presence of moisture, hydrogen cyanide may be emitted which is extremely flammable and toxic.

Unusual Fire and Explosion Hazards

Can cause violent explosions upon contact with chemically active metals. Potentially explosive in contact with acetylene compounds.

SECTION V - REACTIVITY DATA

Stability		Conditions to Avoid **Keep away from incompatible materials since violent and or explosive reactions can occur. Keep containers free of moisture since hydrogen cyanide gas will collect inside moist containers of copper cyanide.**
Stable **X**	Unstable	Incompatibility *(materials to avoid)* **Incompatible with strong oxidizers (such as perchlorates, permanganates, peroxides, chlorates, and nitrates), and chemically active metals (such as potassium, sodium, magnesium, and zinc).**
Hazardous Polymerization		Conditions to Avoid **Under normal conditions of temperature and pressure, hazardous polymerization of copper cyanide is not known to occur.**
May Occur	Will Not Occur **X**	Hazardous Decomposition or By-products **When heated to decomposition, copper cyanide can emit highly toxic/poisonous fumes and gases, including cyanide and nitrous oxide.**

SECTION VI - HEALTH HAZARD DATA

Primary Route(s) of Entry:	Inhalation?	Absorption (skin)?	Ingestion?
	X		X

Health Hazards

INHALATION: Irritation of the respiratory system. Exposure to some copper compounds can cause severe throat and nose irritation and may cause nosebleeds. Exposure can cause "metal fume fever," a flu-like illness lasting about 24 hours with chills, aches, and cough.

SKIN & EYES: Eye contact can cause severe irritation and can damage vision or cause blindness.

INGESTION: A poison with gastrointestinal effects. Will cause nausea and vomiting.

Carcinogenicity	NTP Listed?	IARC Cancer Review Group?	OSHA Regulated? As Copper	Target Organs?
Unknown Human Suspected Animal	**No**	**No**	**29 CFR 1910.1000 Table Z-1**	**Respiratory system, skin, eyes, liver, kidneys**

Medical Conditions Generally Aggravated by Exposure

Wilson's disease.

Emergency and First-aid Procedures

Eye contact: Flush immediately with water for 15 minutes (minimum), seek medical attention. Skin contact: Remove all contaminated clothing. Immediately wash area with large amounts of soap and water. For inhalation: Remove the person from exposure. Provide respiratory assistance and CPR. Transfer to medical facility. If swallowed: Provide breathing assistance if necessary. Never attempt to give a convulsing or unconscious person anything by mouth. If victim is conscious, administer inhalation of amyl nitrate pearls 15-30 seconds of every minute until medical attention is received.

SECTION VII - PRECAUTIONS FOR SAFE HANDLING AND USE

Steps to be Taken in Case Material is Released or Spilled

Collect powdered material in most convenient manner possible and deposit in sealed drum. Keep those not involved in cleanup from entering area. Do NOT dry sweep (creates airborne dusts). Use a vacuum equipped with a high efficiency particulate air (HEPA) filter.

Preferred Waste Disposal Method

No Citation.

Precautions to be Taken in Handling and Storage

Store in tightly closed containers in cool, well-ventilated area and protect from physical damage. Sources of ignition, such as smoking and open flames, are prohibited where copper cyanide is used, handled, or stored. Keep environment moisture-free to avoid generation of hydrogen cyanide.

Other Precautions and Warnings

Keep away from acetylene gas since contact will produce flammable hydrogen gas.

SECTION VIII - CONTROL MEASURES AND PERSONAL PROTECTIVE EQUIPMENT

Respiratory Protection *(specify type)*

At exposures above PEL, use an MSHA/NIOSH approved supplied-air respirator with a full facepiece, hood, or helmet operated in the continuos flow mode, or a self-contained breathing apparatus (SCBA) operated in positive pressure mode.

Ventilation

Local exhaust or general mechanical systems recommended.

Protective Gloves	Eye Protection	Other Protective Clothing
Impervious Material	**Chemical/Dust Goggles or Face Mask**	**Protective Suit or Clothing**

Work/Hygiene Practices

Always wash hands thoroughly after using chemical; never bring food, drink, or smoking materials into vicinity of chemicals.

COPPER CYANIDE

$Cu(CN)_2$ CAS: 14765-77-0

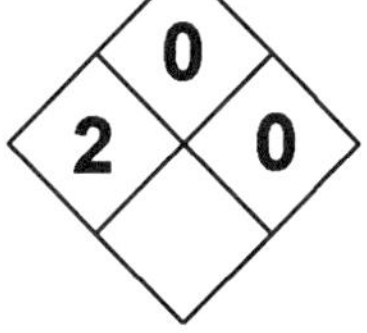

IDENTIFICATION AND TYPICAL USES

Copper cyanide is a greenish powder with no distinguishable odor. It is used in the electroplating of copper onto iron, as an intermediate in some aromatic organic compounds (introduction of the cyanide group in place of the amino radical). It may also be used as an insecticide and a catalyst.

RISK ASSESSMENT: HEALTH

General Assessment

Copper cyanide is not known to cause cancer in humans. However, many other copper compounds are either confirmed or suspected human and/or animal carcinogens. Proper risk management dictates treatment of copper compounds as extremely hazardous commodities unless firm data exist to the contrary. It is a human poison by *ingestion* with systemic effects including vomiting and nausea. It can also be toxic by *inhalation*.

Occupational skin exposure can produce localized irritation and possible rash which can lead to dermatitis in sensitive individuals. Copper cyanide is also extremely irritating to the eyes. It can lead to temporary and/or permanent damage to vision up to and including blindness.

Inhalation of copper fumes can cause "metal fume fever," which is a flu-like illness lasting up to 24 hours with characteristic symptoms of chills, aches, cough, and fever. The dusts are extremely irritating to the nose and throat and may, on prolonged exposure, cause nasal ulcers with symptoms of extreme irritation and chronic nosebleeds. Following severe exposure to the dusts of copper cyanide, symptoms of cyanide poisoning may develop.

Ingestion causes anxiety, confusion, dizziness, sudden loss of consciousness, odor of bitter almonds on breath or in vomitus, rapid or weak pulse, convulsions, and paralysis.

Acute Health Effects

The following acute (short-term) health effects may occur immediately or shortly after exposure to copper cyanide:

Skin: Irritation, rash, and possible contact dermatitis.

Eye: Copper cyanide particles and dusts can be extremely irritating to the eyes. Exposure can damage vision and lead to total loss of sight.

Lung: Inhalation of copper cyanide may cause a condition known as "metal fume fever" (flu-like symptoms). Copper cyanide dusts and fumes can irritate the nose and throat. It may cause the appearance of ulcers in the nasal cavity accompanied by nosebleeds.

Other: Possible CNS effects occurring upon ingestion may include anxiety, confusion, dizziness, loss of consciousness, convulsions, and paralysis.

Chronic Health Effects

The following chronic (long-term) health effects may occur at some time after exposure to copper cyanide and can last for months or even years:

Cancer Hazards: There is evidence that workers in copper smelters have an increased risk of lung cancer, but this is thought to be due to arsenic trioxide exposure and not copper. There is also some evidence that exposure to copper dusts and fumes may cause cancer in laboratory test animals. Human data, however, are inconclusive in this regard. According to information presented in the references, copper cyanide specifically has not been adequately tested for its ability to cause cancer in laboratory test animals.

Reproductive Hazard: According to information available in the references, copper cyanide has not been tested for its ability to adversely affect reproduction in laboratory animals. There is some evidence to indicate that other copper compounds may adversely affect reproduction.

Other Chronic Effects: Repeated exposures can cause copper to deposit in the liver and in other body organs, causing damage. Copper deposits in the skin and hair also occur, leaving a green color. Repeated exposure can also cause shrinking (atrophy) of the inner lining of the nose, with a watery discharge. There may also be a metallic taste. Exposure may cause skin allergy, with rash and itching. If allergy develops, even low future exposures may trigger a rash. Very irritating substances can irritate the lungs. It is not known whether prolonged exposure to copper cyanide will result in lung damage.

Recommended Risk-Reduction Measures

Personnel should avoid direct contact with copper cyanide. If a less toxic chemical cannot be substituted for a hazardous substance, then *engineering controls* are the most effective method of reducing exposures.

The best protection is to enclose operations and/or provide local exhaust ventilation at the site of chemical release. While not always operationally feasible, isolating operations can also reduce exposure. Using respiratory protection is less effective than the controls mentioned above, but is still advisable whenever working with or around copper cyanide. For exposures above the PEL, use a supplied-air respirator with a full facepiece operated in positive pressure or continuous flow mode or, use an MSHA/NIOSH approved self-contained breathing apparatus (SCBA) with full facepiece and pressure demand. If a full facepiece is not available, then chemical/dust goggles should be worn to protect the eyes. Whenever a chemical dust hazard exists, a face shield and a rubber apron should be worn. To prevent hand and skin exposures, impervious rubber gloves should be worn.

Administrative controls should also be in place to minimize the potential for human exposures. These may include written procedures or policies which specify the methods and techniques that will be practiced whenever personnel are to work with copper cyanide.

All personnel should receive *training* on the use, hazards, protective measures, emergency actions, and other precautions per 29 CFR 1910.1200 (Hazard Communication), prior to the first assignment in an area where copper cyanide is used or stored.

For those personnel with frequent or potentially high exposure (half the PEL or greater), the following medical tests are recommended at the time of initial assignment and at regular intervals thereafter (e.g., annually):

☑ Lung function tests (pulmonary functions, including Forced Vital Capacity or FVC test).

If symptoms develop or overexposure is suspected, the following may also be useful:

☑ Urine copper cyanide level tests.

Any evaluation should include a careful history of past and present symptoms with a medical examination. Medical tests that look for damage already done are not a substitute for controlling exposure. Because smoking can cause heart disease as well as lung cancer, emphysema, and other respiratory problems, exposure to copper cyanide may be more hazardous for smokers than for non-smokers. Proper risk assessment requires consideration of all exposure factors. Also, "Wilson's Disease" is a rare condition that interferes with the body's ability to metabolize copper cyanide. Persons with this disease may be seriously affected by copper cyanide exposures.

Other methods to reduce exposure risk include:

☑ Where possible, enclose operations and use local exhaust ventilation at the site of chemical release. If local exhaust ventilation or enclosure is not used, respiratory protection should be mandatory.

☑ Always ensure that proper protective clothing is worn when using chemical substances. Chemical dusts goggles or full facepiece should be worn unless a full facepiece respirator is used.

☑ Wash thoroughly immediately after exposure to copper cyanide and at the end of the work shift or before eating, drinking, or smoking.

☑ Hazard warning information should be posted in the work area. In addition, as part of an on-going education and training program, all information on the health and safety hazards of copper cyanide should be communicated to all potentially exposed workers.

☑ Eyewash stations and safety showers should be provided in the immediate work area for emergency use.

RISK ASSESSMENT: ENVIRONMENT

General Assessment

Copper is a naturally occurring element and is therefore present in the environment. However, copper cyanide is not particularly a "naturally occurring" compound. The environment is at risk of exposure during transportation, storage, disposal, or destruction of copper cyanide. In almost every scenario, the threat of environmental exposure is contingent upon the proper handling of the chemical substance. Accidental spills, large or small, can result in fire, explosion, and possible contamination of the surrounding environmental mediums (water, soil, and air).

Copper will ignite and is potentially explosive in contact with acetylene compounds; 3-bromopropyne; ethylene oxide; lead azide; and ammonium nitrate. It also ignites on contact with chlorine, chlorine trifluoride, fluorine (above 121°C), and hydrazinium nitrate (above 70°C), and other strong oxidizers. In a fire, copper cyanide will emit toxic and poisonous gases, including cyanide and nitrous oxide. In a moist atmosphere, hydrogen cyanide gas will collect inside sealed containers of copper cyanide. Hydrogen cyanide is extremely flammable and highly toxic. It is also incompatible with strong oxidizers (such as chlorine, fluorine, bromine, and hydrogen peroxide), and many chemically active metals such as potassium, sodium, magnesium, and zinc. These characteristics require special consideration during any emergency situation involving a leak or spill of copper cyanide.

Copper is a naturally occurring element in natural water aquifers. At low concentrations, it is an essential element for both plants and animals. At slightly higher concentrations, it is toxic to aquatic life. The toxicity of copper and its compounds to aquatic life varies with the physical and chemical conditions of the water. Factors such as water hardness, alkalinity, and pH influence copper toxicity.

Acute Ecological Effects

Acute (short-term) toxic effects may include the death of animals, birds, or fish and death or low growth rate in plants. Acute effects are seen 2 to 4 days after animals or plants are exposed to copper cyanide.

Copper and its compounds have high acute toxicity to aquatic life. Insufficient data are available to evaluate the short-term effects of copper exposure to plants, birds or land animals.

Chronic Ecological Effects

Chronic (long-term) toxic effects may include shortened life span, reproductive problems, lower fertility, and changes in appearance or behavior in exposed animals. These effects can be seen long after first exposure(s) to toxic chemicals.

Copper and its compounds have high chronic toxicity to aquatic life. Insufficient data are available to evaluate the long-term effects of copper exposure to plants, birds or land animals.

Water Solubility

Copper and its salts are highly soluble in water. Concentrations of 1000 milligrams and more will mix with a liter of water.

Persistence in the Environment

Copper is highly persistent in the aquatic environment, with a half-life of greater than 200 days. The half-life of a pollutant is the amount of time it takes for one half of the chemical to be degraded.

Bioaccumulation in Aquatic Organisms

Some substances increase in concentration, or *bioaccumulate,* in living organisms as they breathe contaminated air, drink contaminated water, or eat contaminated food. These chemicals can become concentrated in the tissues and internal organs of animals as well as humans.

The concentration of copper found in fish is expected to be considerably higher than the average concentration of copper cyanide in the water from which the fish was taken.

Recommended Risk-Reduction Measures

Proper training of all transporters will reduce the likelihood of a mishap or accidents resulting in a leak or spill to the environment. The correct use of DOT labeling (per the MSDS) on all containers, trucks, and rail cars will enable emergency responders to react properly and quickly to any disaster thereby reducing the potential risk to the environment and to personnel.

If a spill or leak to the environment has occurred, fire department, emergency response and/or hazardous materials spill personnel should be notified immediately. Use a vacuum or wet method to reduce dust during cleanup. Do not dry sweep. When vacuuming, a high efficiency particulate air (HEPA) filter should be used, not a standard shop vacuum. Deposit collected materials in sealed drum for disposal. Cleanup should be attempted only by those trained in proper spill containment procedures. Contaminated soils may need to be removed and replaced with clean soil. If copper should contact the water table, aquifer, or navigable waterway, time is of the essence. Copper compounds may be moderately to highly soluble in water and, therefore, total containment and remediation may not be entirely possible. When such spills occur, the local and/or state emergency response authorities must be notified. A comprehensive emergency response or disaster preparedness/recovery plan should be in place prior to any operations involving the use, transportation, storage, or disposal of copper cyanide. If copper cyanide is spilled or leaked, the following specific steps are recommended:

- ☑ Restrict persons not wearing protective clothing from area of spill or leak until cleanup is complete and area can be opened for normal work.
- ☑ Collect powdered materials in most convenient and safe manner possible and place in sealed drums for disposal. Do not dry sweep, use HEPA vacuum instead.
- ☑ It may be necessary to dispose of copper cyanide as a hazardous waste. The responsible state agency or the regional office of the federal Environmental Protection Agency (EPA) should be contacted for specific recommendations.

RISK ASSESSMENT: BUSINESS

General Assessment

Accidents or mishaps involving copper cyanide can present a significant threat to business operations. The loss or damage of equipment or facilities can significantly affect fiscal viability. Lawsuits that may result from personnel injury/death, public exposures, and/or environmental contamination will also require a serious expenditure of resources. Media attention surrounding an injury, death, or environmental damage can also result in a loss of profits and loss of current as well as future business.

Always remember that anytime the term "cyanide" is used, public anxiety and emotion will run equally high and the media will focus on this issue more than any other. This should be seriously considered in any risk management program or policy.

Recommended Risk-Reduction Measures

Company attorneys, safety and health professionals, and environmental specialists should be involved in the development of any procedures dealing with safety and emergency response. A company official should be pre-designated as a public relations officer with specific training in dealing with the press. Corporate plans and policies should be developed, approved, and implemented long before any need for such arises.

REFERENCES

Asante-Duah, D. K. 1993. *Hazardous Waste Risk Assessment.* New York: CRC Press/Lewis Publishers.

Brusick, D. J. 1994. *Methods for Genetic Risk Assessment.* New York: CRC Press/Lewis Publishers.

Calabrese, E. J., E. M. Kenyon. *Air Toxics and Risk Assessment.* New York: CRC Press/Lewis Publishers.

Cothern, C. R. 1995. *Handbook for Environmental Risk Decisionmaking.* New York: CRC Press/Lewis Publishers.

Gardner, R. W. 1994. *Chemical Intolerance.* New York: CRC Press/Lewis Publishers.

Hallenbeck, W. H. 1993. *Quantitative Risk Assessment for Environmental and Occupational Health,* 2nd Edition. New York: CRC Press/Lewis Publishers.

Howard, P. H., et. al. 1991. *Handbook of Environmental Degradation Rates.* New York: CRC Press/Lewis Publishers.

Keith, L. H. and M. M. Walker. 1995. *Handbook of Air Toxics.* New York: CRC Press/Lewis Publishers.

Landis, W. G., M. H. Yu. 1995. *Introduction to Environmental Toxicology.* New York: CRC Press/Lewis Publishers.

Lewis, R. J., Sr. 1992. *Sax's Dangerous Properties of Industrial Materials,* Volumes 1, 2, and 3, 8th Edition. New York: Van Nostrand Reinhold.

Manahan, S. E. 1992. *Toxicological Chemistry,* 2nd Edition. New York: CRC Press/Lewis Publishers.

Meyer, E. 1990. *Chemistry of Hazardous Materials.* Englewood Cliffs, NJ: Prentice Hall.

Quigley, D. R. 1994. *Handbook of Emergency Chemical Management.* New York: CRC Press/Lewis Publishers.

Rea, W. J. 1995. *Chemical Sensitivity,* Vol. I-III. New York: CRC Press/Lewis Publishers.

Sacarello, H. L. A. 1994. *The Comprehensive Handbook of Hazardous Materials.* New York: CRC Press/Lewis Publishers.

Sherman, J. D. 1988. *Chemical Exposure and Disease.* New York: Van Nostrand Reinhold.

U.S. Department of Health and Human Services (DHHS), National Institute for Occupational Health (NIOHS). 1994. *NIOSH Pocket Guide to Hazardous Chemicals.* Publication Number 94-116. Washington, D.C.; U.S. Government Printing Office.

MATERIAL SAFETY DATA SHEET

CHEMICAL NAME	
COPPER NITRATE	**CRC PRESS/LEWIS PUBLISHERS DISCLAIMER:** THE INFORMATION AND RECOMMENDATIONS PRESENTED HEREIN ARE BASED ON SOURCES BELIEVED TO BE RELIABLE. CRC MAKES NO REPRESENTATION ON ITS COMPLETENESS OR ACCURACY. IT IS THE RESPONSIBILITY OF THE USER TO DETERMINE THE CHEMICAL'S SUITABILITY FOR ITS INTENDED USE, THE CHEMICAL'S SAFE USE, AND THE CHEMICAL'S PROPER DISPOSAL. NO REPRESENTATIONS AND/OR WARRANTIES, EITHER EXPRESSED OR IMPLIED, OF THE MERCHANTABILITY OR FITNESS FOR A PARTICULAR PURPOSE, OR OF ANY OTHER NATURE, ARE MADE WITH RESPECT TO THE INFORMATION PROVIDED IN THIS MSDS OR TO THE CHEMICAL TO WHICH INFORMATION MAY REFER. CRC NEITHER ASSUMES NOR AUTHORIZES ANY OTHER PERSON TO ASSUME FOR IT, ANY OTHER ADDITIONAL RESPONSIBILITY OR LIABILITY FOR THE USE OF, OR RELIANCE UPON, THIS INFORMATION.

HAZARD WARNING INFORMATION

HEALTH	FIRE	REACTIVE	OTHER	DEGREE OF HAZARD	COLOR CODING	OTHER CODES
1	0	0	OX	0 = Minimum Hazard 1 = Slight Hazard 2 = Moderate Hazard 3 = Serious Hazard 4 = Severe Hazard	HEALTH = BLUE FIRE = RED REACTIVITY = YELLOW OTHER = WHITE	OX = Oxidizer ACID = Acid ALK = Alkali COR = Corrosive W = Use No Water

SECTION I - GENERAL INFORMATION

Chemical Family	**Metal**	RCRA Number	**None**
		EPA Class	**Not Applicable**
DOT Proper Shipping Name	**Cupric nitrate**	Chemical Abstract Service (CAS) Number	**3251-23-8**
DOT Hazard Class and Label Requirements	**Oxidizer**	DOT Emergency Guide Code	**35**
DOT Identification Number	**UN 1479**	Chemical Formula	**$Cu(NO_3)_2$**

Synonyms
Copper dinitrate; copper (2+) nitrate; cupric dinitrate.

SECTION II - HAZARDOUS INGREDIENTS AND IDENTITY INFORMATION

Hazardous Components (specific identity)	OSHA Exposure Criteria	NIOSH Exposure Criteria	Immediately Dangerous to Life and Health (IDLH)	ACGIH Exposure Criteria
Copper nitrate (derivation: By treating copper or copper oxide with nitric acid. The solution is evaporated and product recovered by crystallization).	**PEL:** dust/mist 1.0 mg(Cu)/m^3 **STEL:** Not Established	**REL:** dust/mist: 1.0 mg(Cu)/m^3 **STEL:** Not Established	100 mg/m^3 as Cu	**TLV:** dust/mist: 1.0 mg(Cu)/m^3 **STEL:** Not Established

SECTION III - PHYSICAL AND CHEMICAL CHARACTERISTICS

Boiling Point	**302-437°F (150-225°C) sublimes**	Specific Gravity (H_2O = 1)	**2.32**
Vapor Pressure (mm Hg)	**Not Available**	Molecular Weight	**187.55**
Vapor Density (Air = 1)	**8.3**	Melting Point	**493°F (256°C)**

Solubility in Water
Soluble in water and alcohol. Dissolves in and reacts vigorously with ether.

Appearance and Odor
Large blue-green, deliquescent, orthorhombic crystalline solid.

SECTION IV - FIRE AND EXPLOSION HAZARD DATA

Flash Point (method used)	**Not Applicable**	Explosive Limits in Air % by Volume	**LEL: Not Applicable UEL: Not Applicable**
NFPA Classification	**No Citation**	Autoignition Temperature	**Not Determined**

Extinguishing Media
Use flooding amounts of water in early stages of fire. Use dry chemical or carbon dioxide.

Special Fire Fighting Procedures
Wear full protective clothing and self-contained breathing apparatus (SCBA). Use a water spray to keep fire-exposed containers cool. Water may not be totally effective on fire. Poisonous oxides of nitrogen are produced in fire.

Unusual Fire and Explosion Hazards
Can cause violent explosions upon contact with strong oxidizers. Can increase intensity of fire in contact with combustible materials.

SECTION V - REACTIVITY DATA

Stability		Conditions to Avoid Keep away from incompatible materials since violent and or explosive reactions can occur.
Stable X	Unstable	Incompatibility *(materials to avoid)* Incompatible with combustible, organic, or other readily oxidizable materials and chemically active metals (such as potassium, sodium, magnesium, and zinc), nitro-methane, acetylene, and hydrazine.
Hazardous Polymerization		Conditions to Avoid Under normal conditions of temperature and pressure, hazardous polymerization of copper nitrate is not known to occur.
May Occur	Will Not Occur X	Hazardous Decomposition or By-products When heated to decomposition, copper nitrate can emit highly toxic and poisonous fumes and gases, including oxides of nitrogen.

SECTION VI - HEALTH HAZARD DATA

Primary Route(s) of Entry:	Inhalation?	Absorption (skin)?	Ingestion?
	X		X

Health Hazards

INHALATION: Irritation of the upper respiratory system, nose and throat. May cause nosebleeds and can cause "metal fume fever," a flu-like illness with chills, aches, and cough.

SKIN & EYES: Eye and skin contact can cause severe irritation. May damage vision.

INGESTION: A poison with gastrointestinal effects. Will cause nausea, violent vomiting and purging. Intense pain, collapse, coma, convulsions, and paralysis.

Carcinogenicity	NTP Listed?	IARC Cancer Review Group?	OSHA Regulated?	Target Organs?
Unknown Human Unknown Animal	No	No	29 CFR 1910.1000 Table Z-1	Respiratory system, skin, eyes, liver

Medical Conditions Generally Aggravated by Exposure

Wilson's disease.

Emergency and First-aid Procedures

Eye contact: Flush immediately with water for 15 minutes (minimum), seek medical attention. Skin contact: Remove all contaminated clothing. Immediately wash area with large amounts of soap and water. For inhalation: Remove the person from exposure. Provide respiratory assistance and CPR. Transfer to medical facility. If swallowed: If victim is conscious, give large amounts of water and induce vomiting. Never attempt to give a convulsing or unconscious person anything by mouth. Provide breathing assistance if necessary. Keep person at rest and quiet. Seek medical attention immediately.

SECTION VII - PRECAUTIONS FOR SAFE HANDLING AND USE

Steps to be Taken in Case Material is Released or Spilled

Collect powdered material in most convenient manner possible and deposit in sealed drum. Keep those not involved in cleanup from entering area. Do NOT dry sweep (creates airborne dusts). Use a vacuum equipped with a high efficiency particulate air (HEPA) filter.

Preferred Waste Disposal Method

No Citation.

Precautions to be Taken in Handling and Storage

Store in tightly closed containers in cool, well-ventilated area and protect from physical damage. Sources of ignition, such as smoking and open flames, are prohibited where copper nitrate is used, handled, or stored.

Other Precautions and Warnings

Keep away from acetylene gas since contact will produce flammable hydrogen gas.

SECTION VIII - CONTROL MEASURES AND PERSONAL PROTECTIVE EQUIPMENT

Respiratory Protection *(specify type)*

At exposures above PEL, use full facepiece respirator with high efficiency particulate air (HEPA) filter. For higher exposures, use a supplied-air respirator with a full facepiece, hood, or helmet or a self-contained breathing apparatus (SCBA) operated in positive pressure mode.

Ventilation

Local exhaust or general mechanical systems recommended.

Protective Gloves	Eye Protection	Other Protective Clothing
Impervious Material	Chemical/Dust Goggles or Face Mask	Protective Suit or Clothing

Work/Hygiene Practices

Always wash hands thoroughly after using chemical; never bring food, drink, or smoking materials into vicinity of chemicals.

COPPER NITRATE

$Cu(NO_3)_2$ CAS: 3251-23-8

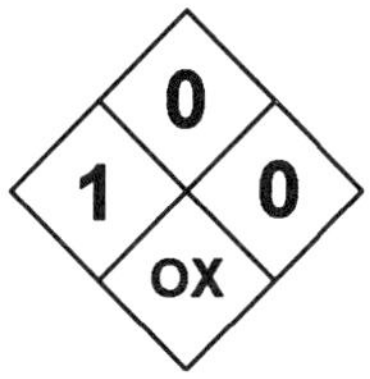

IDENTIFICATION AND TYPICAL USES

Copper nitrate appears as large, greenish-blue, deliquescent, orthorhombic crystals. It may also be in powdered (sugar or sand-like) form. It is used in the manufacture of light-sensitive papers, as an analytical reagent, as a mordant in textile dyeing, as a nitrating agent, an insecticide, herbicide, and fungicide for vines, for coloring copper metal black, in electroplating, and to produce a burnished effect on iron. It is also used in paints, varnishes, and enamels, in pharmaceutical preparations, and as a catalyst.

RISK ASSESSMENT: HEALTH

General Assessment

Copper nitrate is moderately toxic by *ingestion* producing systemic effects which appear to be consistent with that seen following exposure to other nitrates. This may indicate that the biological effects of this compound are linked directly to the toxicity of the nitrate group. It can also be toxic by *inhalation.*

Inhalation of copper nitrate fumes can cause "metal fume fever," which is a flu-like illness lasting up to 24 hours with characteristic symptoms of chills, aches, cough, and fever. The dusts are extremely irritating to the nose and throat and may, on prolonged exposure, cause nasal ulcers with symptoms of extreme irritation and chronic nosebleeds.

Ingestion of copper nitrate can cause nausea and headaches. Swallowing large amounts can cause violent vomiting and purging, intense abdominal pain, collapse, coma, convulsions, and paralysis.

Skin exposure can produce localized irritation and possible rash which can lead to dermatitis in sensitive individuals. Copper nitrate is also extremely irritating to the eyes. It can lead to temporary and/or permanent damage to vision.

Acute Health Effects

The following acute (short-term) health effects may occur immediately or shortly after exposure to copper nitrate:

Skin: Irritation, rash, and possible contact dermatitis.

Eye: Copper nitrate particles and dusts can be extremely irritating to the eyes. Exposure can damage vision and cause burns.

Lung: Inhalation of copper nitrate may cause a condition known as "metal fume fever" (flu-like symptoms). Copper nitrate dusts and fumes can irritate the nose and throat. It may cause the appearance of ulcers in the nasal cavity accompanied by nosebleeds.

Chronic Health Effects

The following chronic (long-term) health effects may occur at some time after exposure to copper nitrate and can last for months or even years:

Cancer Hazards: According to information presented in the references, copper nitrate has not been adequately tested for its ability to cause cancer in test animals. Other copper compounds are suspected to cause human cancers and still others are confirmed human carcinogens.

Reproductive Hazard: According to information available in the references, copper nitrate has not been tested for its ability to adversely affect reproduction in laboratory animals. There is some evidence to indicate that other copper compounds and other nitrates may adversely affect reproduction.

Other Chronic Effects: Repeated exposures can cause chronic irritation of the nose, a shrinking of the inner lining of the nasal cavity, and may cause sores to appear prompting watery discharge and/or nosebleeds.

Copper nitrate may cause a skin allergy. If allergy develops, very low future exposures can cause itching and a skin rash. Repeated exposures can cause thickening of the skin (not caused by the allergy). Repeated contact can also cause copper to deposit in various parts of the body. Large deposits may cause a greenish color to the skin and hair. Large deposits in the liver can cause liver damage. A metallic taste may also occur. Very irritating substances can damage lung tissues. It is not known whether copper nitrate dust and heated fumes cause lung damage.

Recommended Risk-Reduction Measures

Copper compounds are often used in "hot" operations (welding, brazing, soldering, plating, cutting, for example). At the high temperatures reached in these operations, metals often form metal fumes which have different health effects and exposure standards than the original compounds and, therefore, require specialized controls. Proper risk management requires evaluation for both dusts *and* fumes to ensure adequate controls in the workplace.

Personnel should avoid direct contact with copper nitrate. If a less toxic chemical cannot be substituted for a hazardous substance, then *engineering controls* are the most effective method of reducing exposures. The best protection is to enclose operations and/or provide local exhaust ventilation at the site of chemical release. While not always operationally feasible, isolating operations can also reduce exposure. Using respiratory protection is less effective than the controls mentioned above, but is still advisable whenever working with or around copper nitrate. For exposures above the PEL, use a full facepiece respirator equipped with a high efficiency particulate air (HEPA) filter and a dust/mist/fume pre-filter. For higher exposures, use a supplied-air respirator with a full facepiece operated in positive pressure mode or an MSHA/NIOSH approved self-contained breathing apparatus (SCBA) with full facepiece and pressure demand. If a full facepiece is not available, then chemical/dust goggles should be worn to protect the eyes. Whenever a chemical dust hazard exists, a face shield and a rubber apron should be worn. To prevent hand and skin exposures, impervious rubber gloves should be worn.

Administrative controls should also be in place to minimize the potential for human exposures. These may include written procedures or policies which specify the methods and techniques that will be practiced whenever personnel are to work with copper nitrate.

All personnel should receive *training* on the use, hazards, protective measures, emergency actions, and other precautions per 29 CFR 1910.1200 (Hazard Communication), prior to the first assignment in an area where copper nitrate is used or stored.

For those with frequent or potentially high exposures (half the PEL or greater), the following medical test is recommended before beginning employment and at regular intervals after that:

☑ Lung function tests, including Forced Vital Capacity (FVC).

If symptoms develop or overexposure is suspected, the following may also be useful:

☑ Urine test for copper (can measure recent exposure).

☑ Serum copper levels.

☑ Evaluation by a qualified allergist, including careful exposure history and special testing (may help diagnose skin allergy).

☑ Liver function test.

Any evaluation should include a careful history of past and present symptoms with a medical examination. Medical tests that look for damage already done are not a substitute for controlling exposure. Also, "Wilson's Disease" is a rare condition that interferes with the body's ability to metabolize copper nitrate. Persons with this disease may be seriously affected by copper nitrate exposures. Smokers may also be more seriously affected since smoking causes heart disease, emphysema, lung cancer, and other respiratory problems which may be further aggravated by exposure to copper nitrate.

Other methods to reduce exposure include:

☑ Where possible, enclose operations and use local exhaust ventilation at the site of chemical release. If local exhaust ventilation or enclosure is not used, respiratory protection should be mandatory.

☑ Always ensure that proper protective clothing is worn when using chemical substances.

☑ Wash thoroughly immediately after exposure to copper nitrate and at the end of the work shift or before eating, drinking, or smoking. Eyewash stations and safety showers should be readily available for emergency use.

☑ Hazard warning information should be posted in the work area. In addition, as part of an on-going

education and training program, all information on the health and safety hazards of copper nitrate should be communicated to all potentially exposed workers.

RISK ASSESSMENT: ENVIRONMENT

General Assessment

The environment is at risk of exposure during transportation, storage, disposal, or destruction of copper nitrate. In almost every scenario, the threat of environmental exposure is contingent upon the proper handling of the chemical substance. Accidental spills, large or small, can result in fire, explosion, and possible contamination of the surrounding environmental mediums (water, soil, and air).

Copper nitrate will ignite and is potentially explosive in contact with acetylene compounds. It will react violently in contact with acetic anhydride, nitromethane, and hydrazine. It can even ignite if allowed to remain in contact with paper for a prolonged period of time. Concentrated solutions can ignite in contact with common tin foil. In a fire, copper nitrate will emit toxic and poisonous gases, including oxides of nitrogen. It is also incompatible with strong reducing agents, ordinary combustible materials, other easily oxidized materials, and many chemically active metals (potassium, sodium, magnesium, and zinc). These characteristics require special consideration during any emergency situation involving a leak or spill of copper nitrate.

Copper is a naturally occurring element in natural water aquifers. At low concentrations, it is an essential element for both plants and animals. At slightly higher concentrations, it is toxic to aquatic life. The toxicity of copper and its compounds to aquatic life varies with the physical and chemical conditions of the water. Factors such as water hardness, alkalinity, and pH influence copper toxicity.

Acute Ecological Effects

Acute (short-term) toxic effects may include the death of animals, birds, or fish and death or low growth rate in plants. Acute effects are seen 2 to 4 days after animals or plants are exposed to copper nitrate.

Copper and its compounds have high acute toxicity to aquatic life. Insufficient data are available to evaluate the short-term effects of copper exposure to plants, birds or land animals.

Chronic Ecological Effects

Chronic (long-term) toxic effects may include shortened life span, reproductive problems, lower fertility, and changes in appearance or behavior in exposed animals. These effects can be seen long after first exposure(s) to toxic chemicals.

Copper and its compounds have high chronic toxicity to aquatic life. Insufficient data are available to evaluate the long-term effects of copper exposure to plants, birds or land animals.

Water Solubility

Copper and its salts are highly soluble in water. Concentrations of 1000 milligrams and more will mix with a liter of water.

Persistence in the Environment

Copper is highly persistent in the aquatic environment, with a half-life of greater than 200 days. The half-life of a pollutant is the amount of time it takes for one half of the chemical to be degraded.

Bioaccumulation in Aquatic Organisms

Some substances increase in concentration, or *bioaccumulate,* in living organisms as they breathe contaminated air, drink contaminated water, or eat contaminated food. These chemicals can become concentrated in the tissues and internal organs of animals as well as humans.

The concentration of copper found in fish is expected to be considerably higher than the average concentration of copper nitrate in the water from which the fish was taken.

Recommended Risk-Reduction Measures

Proper training of all transporters will reduce the likelihood of a mishap or accidents resulting in a leak or spill to the environment. The correct use of DOT labeling (per the MSDS) on all containers, trucks, and rail cars will enable emergency responders to react properly and quickly to any disaster thereby reducing the potential risk to the environment and to personnel.

If a spill or leak to the environment has occurred, fire department, emergency response and/or hazardous materials spill personnel should be notified immediately. Use a vacuum or wet method to reduce dust during cleanup. Do not dry sweep. When vacuuming, a high efficiency particulate air (HEPA) filter should be used, not a standard shop vacuum. Deposit collected materials in sealed drum for disposal. Cleanup should be attempted only by those trained in proper spill containment procedures. Contaminated soils may need to be removed and replaced with clean soil. If copper nitrate should contact the water table, aquifer, or navigable waterway, time is of the essence. Copper compounds, including copper nitrate, may be moderately to highly soluble in water and, therefore, total containment and remediation may not be entirely possible. When such spills occur, the local and/or state emergency response authorities must be notified. A comprehensive emergency response or disaster preparedness/recovery plan should be in place prior to any operations involving the use, transportation, storage, or disposal of copper nitrate. If copper nitrate is spilled or leaked, the following specific steps are recommended:

- ☑ Restrict persons not wearing protective clothing from area of spill or leak until cleanup is complete and area can be opened for normal work.
- ☑ Collect powdered materials in most convenient and safe manner possible and place in sealed drums for disposal.
- ☑ It may be necessary to dispose of copper nitrate as a hazardous waste. The responsible state agency or the regional office of the federal Environmental Protection Agency (EPA) should be contacted for specific recommendations.

RISK ASSESSMENT: BUSINESS

General Assessment

Accidents or mishaps involving copper nitrate can present a significant threat to business operations. The loss or damage of equipment or facilities can significantly affect fiscal viability. Lawsuits that may result from personnel injury/death, public exposures, and/or environmental contamination will also require a serious expenditure of resources. Media attention surrounding an injury, death, or environmental damage can also result in a loss of profits and loss of current as well as future business.

Recommended Risk-Reduction Measures

Company attorneys, safety and health professionals, and environmental specialists should be involved in the development of any procedures dealing with safety or emergency response. A company official should be pre-designated as a public relations officer with specific training in dealing with the press. Corporate plans and policies should be developed, approved, and implemented long before any need for such arises.

REFERENCES

Asante-Duah, D. K. 1993. *Hazardous Waste Risk Assessment.* New York: CRC Press/Lewis Publishers.

Brusick, D. J. 1994. *Methods for Genetic Risk Assessment.* New York: CRC Press/Lewis Publishers.

Calabrese, E. J. 1994. *Biological Effects of Low Level Exposures.* New York: CRC Press/Lewis Publishers.

Cothern, C. R., N. P. Ross. 1994. *Environmental Statistics, Assessment, and Forecasting.* New York: CRC Press/Lewis Publishers.

Gardner, R. W. 1994. *Chemical Intolerance.* New York: CRC Press/Lewis Publishers.

Hallenbeck, W. H. 1993. *Quantitative Risk Assessment for Environmental and Occupational Health,* 2nd Edition. New York: CRC Press/Lewis Publishers.

Howard, P. H., et. al. 1991. *Handbook of Environmental Degradation Rates.* New York: CRC Press/Lewis Publishers.

Jones, F. E. 1994. *Toxic Organic Vapors in the Workplace.* New York: CRC Press/Lewis Publishers.

Landis, W. G., M. H. Yu. 1995. *Introduction to Environmental Toxicology.* New York: CRC Press/Lewis Publishers.

Langerman, N. 1994 *Precautionary Labels for Chemical Containers.* New York: CRC Press/Lewis Publishers.

Lauwerys, R. R., P. Hoet. 1993. *Industrial Chemical Exposure.* New York: CRC Press/Lewis Publishers.

Lewis, R. J., Sr. 1992. *Sax's Dangerous Properties of Industrial Materials,* Volumes 1, 2, and 3, 8th Edition. New York: Van Nostrand Reinhold.

Manahan, S. E. 1992. *Toxicological Chemistry,* 2nd Edition. New York: CRC Press/Lewis Publishers.

Meyer, E. 1990. *Chemistry of Hazardous Materials.* Englewood Cliffs, NJ: Prentice Hall.

Rea, W. J. 1995. *Chemical Sensitivity,* Vol. I-III. New York: CRC Press/Lewis Publishers.

Sacarello, H. L. A. 1994. *The Comprehensive Handbook of Hazardous Materials.* New York: CRC Press/Lewis Publishers.

Sherman, J. D. 1988. *Chemical Exposure and Disease.* New York: Van Nostrand Reinhold.

Talbott, E., G. Craun. 1995. *An Introduction to Environmental Epidemiology.* New York: CRC Press/Lewis Publishers.

U.S. Department of Health and Human Services (DHHS), National Institute for Occupational Health (NIOHS). 1994. *NIOSH Pocket Guide to Hazardous Chemicals.* Publication Number 94-116. Washington, D.C.; U.S. Government Printing Office.

MATERIAL SAFETY DATA SHEET

CHEMICAL NAME	CRC PRESS/LEWIS PUBLISHERS DISCLAIMER
COPPER SULFATE	**CRC PRESS/LEWIS PUBLISHERS DISCLAIMER:** THE INFORMATION AND RECOMMENDATIONS PRESENTED HEREIN ARE BASED ON SOURCES BELIEVED TO BE RELIABLE. CRC MAKES NO REPRESENTATION ON ITS COMPLETENESS OR ACCURACY. IT IS THE RESPONSIBILITY OF THE USER TO DETERMINE THE CHEMICAL'S SUITABILITY FOR ITS INTENDED USE, THE CHEMICAL'S SAFE USE, AND THE CHEMICAL'S PROPER DISPOSAL. NO REPRESENTATIONS AND/OR WARRANTIES, EITHER EXPRESSED OR IMPLIED, OF THE MERCHANTABILITY OR FITNESS FOR A PARTICULAR PURPOSE, OR OF ANY OTHER NATURE, ARE MADE WITH RESPECT TO THE INFORMATION PROVIDED IN THIS MSDS OR TO THE CHEMICAL TO WHICH INFORMATION MAY REFER. CRC NEITHER ASSUMES NOR AUTHORIZES ANY OTHER PERSON TO ASSUME FOR IT, ANY OTHER ADDITIONAL RESPONSIBILITY OR LIABILITY FOR THE USE OF, OR RELIANCE UPON, THIS INFORMATION.

HAZARD WARNING INFORMATION

HEALTH	FIRE	REACTIVE	OTHER	DEGREE OF HAZARD	COLOR CODING	OTHER CODES
2	0	1		0 = Minimum Hazard 1 = Slight Hazard 2 = Moderate Hazard 3 = Serious Hazard 4 = Severe Hazard	HEALTH = BLUE FIRE = RED REACTIVITY = YELLOW OTHER = WHITE	OX = Oxidizer ACID = Acid ALK = Alkali COR = Corrosive W = Use No Water

SECTION I - GENERAL INFORMATION

Characterization	**Metal**	RCRA Number	**None**
		EPA Class	**Not Applicable**
DOT Proper Shipping Name	**Cupric sulfate**	Chemical Abstract Service (CAS) Number	**7758-98-7**
DOT Hazard Class and Label Requirements	**ORM-E; No Label**	DOT Emergency Guide Code	**31**
DOT Identification Number	**NA 9109**	Chemical Formula	**$CuSO_4$**

Synonyms

Cupric sulfate; blue vitriol; blue stone; blue copperas; copper sulfate pentahydrate; sulfate of copper.

SECTION II - HAZARDOUS INGREDIENTS AND IDENTITY INFORMATION

Hazardous Components (specific identity)	OSHA Exposure Criteria	NIOSH Exposure Criteria	Immediately Dangerous to Life and Health (IDLH)	ACGIH Exposure Criteria
Copper sulfate (derivation: Action of dilute sulfuric acid on copper or copper oxide (often as oxide ores) in large quantities with evaporation and crystallization).	**PEL: dust/mist 1.0 mg(Cu)/m^3** **STEL: Not Established**	**REL: dust/mist: 1.0 mg(Cu)/m^3** **STEL: Not Established**	**100 mg/m^3 as Cu**	**TLV: dust/mist: 1.0 mg(Cu)/m^3** **STEL: Not Established**

SECTION III - PHYSICAL AND CHEMICAL CHARACTERISTICS

Boiling Point	**1207°F (653°C) Decomposes**	Specific Gravity (H_2O = 1)	**2.284**
Vapor Pressure (mm Hg)	**Not Available**	Molecular Weight	**249.68**
Vapor Density (Air = 1)	**Not Available**	Melting Point	**297°F (147°C)**

Solubility

Soluble in water, acetone, ethanol, and methanol. Slightly soluble in alcohol and glycerol.

Appearance and Odor

Blue crystals or blue crystalline granules or powder. Slowly efflorescing in air, white when dehydrated. Odorless but may produce nauseous metallic taste.

SECTION IV - FIRE AND EXPLOSION HAZARD DATA

Flash Point (method used)	**Not Applicable**	Explosive Limits in Air % by Volume	**LEL: Not Applicable UEL: Not Applicable**
NFPA Classification	**No Citation**	Autoignition Temperature	**Not Determined**

Extinguishing Media

Copper sulfate may burn, but does not readily ignite. Use water, dry chemical, or carbon dioxide.

Special Fire Fighting Procedures

Wear full protective clothing and self-contained breathing apparatus (SCBA). Use a water spray to keep fire-exposed containers cool. Poisonous and toxic fumes are produced in fire, including oxides of sulfur. Move containers from fire area if it can be done without risk.

Unusual Fire and Explosion Hazards

Sulfur mixed with chlorates and copper are spontaneously explosive. Containers may build up pressure and explode in fire.

SECTION V - REACTIVITY DATA

Stability		Conditions to Avoid This material is sensitive to heat and moisture. It is stable indefinitely when kept dry. It will slowly effloresce in air. Aqueous solutions are stable for 96 hours when stored in the dark.
Stable X	Unstable	Incompatibility *(materials to avoid)* Hydroxylamine, finely powdered metals, chemically active metals (potassium, sodium, magnesium, and zinc), acetylene gas, alkalis, phosphates, hydrazine, nitromethane, sulfuric acid, and propylene glycol.
Hazardous Polymerization		Conditions to Avoid Under normal conditions of temperature and pressure, hazardous polymerization of copper sulfate is not known to occur.
May Occur	Will Not Occur X	Hazardous Decomposition or By-products When heated to decomposition, copper sulfate can emit highly toxic and poisonous fumes and gases, including oxides of sulfur.

SECTION VI - HEALTH HAZARD DATA

Primary Route(s) of Entry:	Inhalation?	Absorption (skin)?	Ingestion?
	X	X	X

Health Hazards

INHALATION: Irritation of the upper respiratory system, nose and throat. May cause nosebleeds and can cause "metal fume fever," a flu-like illness with chills, aches, and cough.

ABSORPTION: Skin contact may cause smarting and first degree burns. Second degree burns on prolonged contact. Can pass through unbroken skin and cause toxic systemic effects.

INGESTION: A poison with gastrointestinal effects. Will cause intense pain and local corrosion and hemorrhages, prostration, anuria, hematuria, anemia, coma, and death.

Carcinogenicity	NTP Listed?	IARC Cancer Review Group?	OSHA Regulated?	Target Organs?
Unknown Human Unknown Animal	No	No	29 CFR 1910.1000 Table Z-1	Respiratory system, skin, eyes, liver, kidney.

Medical Conditions Generally Aggravated by Exposure

Wilson's disease.

Emergency and First-aid Procedures

Eye contact: Flush immediately with water for 15 minutes (minimum), seek medical attention. Skin contact: Remove all contaminated clothing. Immediately wash area with large amounts of soap and water. For inhalation: Remove the person from exposure. Provide respiratory assistance and CPR. Transfer to medical facility. If swallowed: If victim is conscious, give large amounts of water and induce vomiting. Never attempt to give a convulsing or unconscious person anything by mouth. Provide breathing assistance if necessary. Keep person at rest and quiet. Seek medical attention immediately.

SECTION VII - PRECAUTIONS FOR SAFE HANDLING AND USE

Steps to be Taken in Case Material is Released or Spilled

Collect powdered material in most convenient manner possible and deposit in sealed drum. Keep those not involved in cleanup from entering area. Do NOT dry sweep (creates airborne dusts). Use a vacuum equipped with a high efficiency particulate air (HEPA) filter.

Preferred Waste Disposal Method

No Citation.

Precautions to be Taken in Handling and Storage

Store in tightly closed containers in cool, well-ventilated area. Keep dry and free of moisture. Recommend a nitrogen purge inside storage containers.

Other Precautions and Warnings

Keep away from acetylene gas and chlorates. Contact with either may cause explosion.

SECTION VIII - CONTROL MEASURES AND PERSONAL PROTECTIVE EQUIPMENT

Respiratory Protection *(specify type)*

At exposures above PEL, use full facepiece respirator with high efficiency particulate air (HEPA) filter. For higher exposures, use a supplied-air respirator with a full facepiece, hood, or helmet or a self-contained breathing apparatus (SCBA) operated in positive pressure mode.

Ventilation

Local exhaust or general mechanical systems recommended.

Protective Gloves	Eye Protection	Other Protective Clothing
Impervious Material	Chemical/Dust Goggles or Face Mask	Protective Suit or Clothing

Work/Hygiene Practices

Always wash hands thoroughly after using chemical; never bring food, drink, or smoking materials into vicinity of chemicals.

COPPER SULFATE

$CuSO_3$ CAS: 7758-98-7

IDENTIFICATION AND TYPICAL USES

Copper sulfate appears as blue crystals or blue, crystalline granules or powder. It is slowly efflorescent in air and turns white when dehydrated. It is generally odorless but may produce a nauseating metallic taste. It is used heavily in the agricultural industry as a soil additive, a pesticide, in Bordeaux mixture, as a feed additive, and a germicide. It is also used in the leather industry, in pigments, in electric batteries, in electroplating coatings, in copper salts, as a reagent in analytical chemistry, in medicines, as a wood preservative (preservation of pulp wood and ground pulp), in process engraving and lithography, for ore flotation, in the petroleum industry, in the manufacture of synthetic rubber, in steel manufacturing, and in the treatment of natural asphalt. The anhydrous salt is used as a dehydrating agent.

RISK ASSESSMENT: HEALTH

General Assessment

Copper sulfate is extremely toxic by *ingestion* producing systemic effects. It can also be toxic by *inhalation* and *absorption* through unbroken skin. Mutagenic and teratogenic data have been reported for this compound.

Inhalation of copper sulfate fumes can cause "metal fume fever," which is a flu-like illness lasting up to 24 hours with characteristic symptoms of chills, aches, cough, and fever. The dusts are extremely irritating to the nose and throat and may, on prolonged exposure, cause nasal ulcers with symptoms of extreme irritation and chronic nosebleeds.

Ingestion of copper sulfate can cause severe gastrointestinal irritation and pain. Symptoms include vomiting, localized corrosion, and hemorrhages. There may be prostration, anuria, hematuria, anemia, increased white blood cells, icterus, coma, respiratory difficulty, and circulatory failure resulting in death. It can also cause nausea, metallic taste, sweating, headache, burning sensation in the throat and stomach, hemorrhagic gastritis, weak pulse, hypotension, and turgidity. Other symptoms may include watery and bloody stools, burning in the mouth and throat, liver damage with jaundice, hemolysis, tenesmus, retching, collapse, and convulsions. Uremia, shock, and renal failure have also been reported.

Skin exposure can produce localized irritation, smarting, first degree burns, and, if prolonged exposure, second degree burns. Copper sulfate is also extremely irritating to the eyes. It can lead to temporary and/or permanent damage to vision.

Acute Health Effects

The following acute (short-term) health effects may occur immediately or shortly after exposure to copper sulfate:

Skin: Irritation, redness, and deep, painful burns.

Eye: Copper sulfate particles and dusts can be extremely irritating to the eyes. Exposure can damage vision and cause burns.

Lung: Inhalation of copper sulfate may cause a condition known as "metal fume fever" (flu-like symptoms). Copper sulfate dusts and fumes can irritate the nose and throat. It may cause the appearance of ulcers in the nasal cavity accompanied by nosebleeds.

Other: Ingestion of copper sulfate is extremely poisonous and can be fatal. Severe gastrointestinal irritation with toxic systemic effects.

Chronic Health Effects

The following chronic (long-term) health effects may occur at some time after exposure to copper sulfate and can last for months or even years:

Cancer Hazards: According to information presented in the references, copper sulfate has not been adequately tested for its ability to cause cancer in test animals. Other copper compounds are suspected to cause human cancers and still others are confirmed human carcinogens. Copper sulfate has been reported to cause mutations in some test animals. Human data are inconclusive in this regard.

Reproductive Hazard: According to information available in the references, there is some evidence to indicate that cupric sulfate may adversely affect the developing fetus and may decrease fertility in female animals.

Other Chronic Effects: Repeated exposures can cause chronic irritation of the nose, a shrinking of the inner lining of the nasal cavity, and may cause sores to appear prompting watery discharge and/or nosebleeds. Copper sulfate may cause a skin allergy. If allergy develops, very low future exposures can cause itching and a skin rash. Repeated exposures can cause thickening of the skin (not caused by the allergy). Repeated contact can also cause copper to deposit in various parts of the body. Large deposits may cause a greenish color to the skin and hair. Large deposits in the liver can cause liver damage. A metallic taste may also occur. Very irritating substances can damage lung tissues. It is not known whether copper sulfate dust and heated fumes cause lung damage.

Recommended Risk-Reduction Measures

Copper compounds are often used in "hot" operations (welding, brazing, soldering, plating, cutting, for example). At the high temperatures reached in these operations, metals often form metal fumes which have different health effects and exposure standards than the original compounds and, therefore, require specialized controls. Proper risk management requires evaluation for both dusts *and* fumes to ensure adequate controls in the workplace.

Personnel should avoid direct contact with copper sulfate. If a less toxic chemical cannot be substituted for a hazardous substance, then *engineering controls* are the most effective method of reducing exposures. The best protection is to enclose operations and/or provide local exhaust ventilation at the site of chemical release. While not always operationally feasible, isolating operations can also reduce exposure. Using respiratory protection is less effective than the controls mentioned above, but is still advisable whenever working with or around copper sulfate. For exposures above the PEL, use a full facepiece respirator equipped with a high efficiency particulate air (HEPA) filter and a dust/mist/fume pre-filter. For higher exposures, use a supplied-air respirator with a full facepiece operated in positive pressure mode or an MSHA/NIOSH approved self-contained breathing apparatus (SCBA) with full facepiece and pressure demand. If a full facepiece is not available, then chemical/dust goggles should be worn to protect the eyes. Whenever a chemical dust hazard exists, a face shield and a rubber apron should be worn. To prevent hand and skin exposures, impervious rubber gloves should be worn.

Administrative controls should also be in place to minimize the potential for human exposures. These may include written procedures or policies which specify the methods and techniques that will be practiced whenever personnel are to work with copper sulfate.

All personnel should receive *training* on the use, hazards, protective measures, emergency actions, and other precautions per 29 CFR 1910.1200 (Hazard Communication), prior to the first assignment in an area where copper sulfate is used or stored.

For those with frequent or potentially high exposures (half the PEL or greater), the following medical test is recommended before beginning employment and at regular intervals after that:

- ☑ Liver function test.

If symptoms develop or overexposure is suspected, the following may also be useful:

- ☑ Urine test for copper (can measure recent exposure).
- ☑ Serum copper levels.
- ☑ Evaluation by a qualified allergist, including careful exposure history and special testing (may help diagnose skin allergy).
- ☑ Lung function tests, including Forced Vital Capacity (FVC).

Any evaluation should include a careful history of past and present symptoms with a medical examination. Medical tests that look for damage already done are not a substitute for controlling exposure. Also, "Wilson's Disease" is a rare condition that interferes with the body's ability to metabolize copper sulfate. Persons with this disease may be affected by copper sulfate exposures. Smokers may also be more seriously affected since smoking causes heart disease, emphysema, lung cancer, and other respiratory problems which may be further aggravated by exposure to copper sulfate.

Other methods to reduce exposure include:

- ☑ Where possible, enclose operations and use local exhaust ventilation at the site of chemical release. If local exhaust ventilation or enclosure is not used, respiratory protection should be mandatory. Also, automatic transfer of copper sulfate from storage containers to process containers is recommended where possible.
- ☑ Always ensure that proper protective clothing is worn when using chemical substances.
- ☑ Wash thoroughly immediately after exposure to copper sulfate and at the end of the work shift or before eating, drinking, or smoking. Eyewash stations and safety showers should be readily available for emergency use.
- ☑ Hazard warning information should be posted in the work area. In addition, as part of an on-going education and training program, all information on the health and safety hazards of copper sulfate should be communicated to all exposed and potentially exposed workers.

RISK ASSESSMENT: ENVIRONMENT

General Assessment

The environment is at risk of exposure during transportation, storage, disposal, or destruction of copper sulfate. In almost every scenario, the threat of environmental exposure is contingent upon the proper handling of the chemical substance. Accidental spills, large or small, can result in fire, explosion, and possible contamination of the surrounding environmental mediums (water, soil, and air).

Copper sulfate will ignite and is potentially explosive in contact with acetylene compounds. It is also incompatible with finely divided metals and can corrode iron and steel. It can react in contact with alkalis, phosphates, hydrazine, or nitro-methane. It is sensitive to heat and moisture. In a fire, copper sulfate will emit toxic and poisonous gases, including oxides of sulfur. These characteristics require special consideration during any emergency situation involving a leak or spill of copper sulfate.

Copper is a naturally occurring element in natural water aquifers. At low concentrations, it is an essential element for both plants and animals. At slightly higher concentrations, it is toxic to aquatic life. The toxicity of copper and its compounds to aquatic life varies with the physical and chemical conditions of the water. Factors such as water hardness, alkalinity, and pH influence copper toxicity.

Acute Ecological Effects

Acute (short-term) toxic effects may include the death of animals, birds, or fish and death or low growth rate in plants. Acute effects are seen 2 to 4 days after animals or plants are exposed to copper sulfate.

Copper and its compounds have high acute toxicity to aquatic life. Insufficient data are available to evaluate the short-term effects of copper exposure to plants, birds or land animals.

Chronic Ecological Effects

Chronic (long-term) toxic effects may include shortened life span, reproductive problems, lower fertility, and changes in appearance or behavior in exposed animals. These effects can be seen long after first exposure(s) to toxic chemicals.

Copper and its compounds have high chronic toxicity to aquatic life. Insufficient data are available to evaluate the long-term effects of copper exposure to plants, birds or land animals.

Water Solubility

Copper and its salts are highly soluble in water. Concentrations of 1000 milligrams and more will mix with a liter of water.

Persistence in the Environment

Copper is highly persistent in the aquatic environment, with a half-life of greater than 200 days. The half-life of a pollutant is the amount of time it takes for one half of the chemical to be degraded.

Bioaccumulation in Aquatic Organisms

Some substances increase in concentration, or *bioaccumulate,* in living organisms as they breathe contaminated air, drink contaminated water, or eat contaminated food. These chemicals can become concentrated in the tissues and internal organs of animals as well as humans.

The concentration of copper found in fish is expected to be considerably higher than the average concentration of copper sulfate in the water from which the fish was taken.

ⓤ *Recommended Risk-Reduction Measures*

Proper training of all transporters will reduce the likelihood of a mishap or accidents resulting in a leak or spill to the environment. The correct use of DOT labeling (per the MSDS) on all containers, trucks, and rail cars will enable emergency responders to react properly and quickly to any disaster thereby reducing the potential risk to the environment and to personnel.

If a spill or leak to the environment has occurred, fire department, emergency response and/or hazardous materials spill personnel should be notified immediately. Use a vacuum or wet method to reduce dust during cleanup. Do not dry sweep. When vacuuming, a high efficiency particulate air (HEPA) filter should be used, not a standard shop vacuum. Deposit collected materials in sealed drum for disposal. Cleanup should be attempted only by those trained in proper spill containment procedures. Contaminated soils may need to be removed and replaced with clean soil. If copper sulfate should contact the water table, aquifer, or navigable waterway, time is of the essence. Copper compounds, including copper sulfate, may be moderately to highly soluble in water and, therefore, total containment and remediation may not be entirely possible. When such spills occur, the local and/or state emergency response authorities must be notified. A comprehensive emergency response or disaster preparedness/recovery plan should be in place prior to any operations involving the use, transportation, storage, or disposal of copper sulfate. If copper sulfate is spilled or leaked, the following specific steps are recommended:

- ☑ Restrict persons not wearing protective clothing from area of spill or leak until cleanup is complete and area can be opened for normal work.
- ☑ Collect powdered materials in most convenient and safe manner possible and place in sealed drums for disposal.
- ☑ It may be necessary to dispose of copper sulfate as a hazardous waste. The responsible state agency or the regional office of the federal Environmental Protection Agency (EPA) should be contacted for specific recommendations.

RISK ASSESSMENT: BUSINESS

General Assessment

Accidents or mishaps involving copper sulfate can present a significant threat to business operations. The loss or damage of equipment or facilities can significantly affect fiscal viability. Lawsuits that may result from personnel injury/death, public exposures, and/or environmental contamination will also require a serious expenditure of resources. Media attention surrounding an injury, death, or environmental damage can also result in a loss of profits and loss of current as well as future business.

ⓤ *Recommended Risk-Reduction Measures*

Company attorneys, safety and health professionals, and environmental specialists should be involved in the development of any procedures dealing with safety or emergency response. A company official should be pre-designated as a public relations officer with specific training in dealing with the press. Corporate plans and policies should be developed, approved, and implemented long before any need for such arises.

REFERENCES

Asante-Duah, D. K. 1993. *Hazardous Waste Risk Assessment.* New York: CRC Press/Lewis Publishers.

Calabrese, E. J. 1994. *Biological Effects of Low Level Exposures.* New York: CRC Press/Lewis Publishers.

Cothern, C. R., N. P. Ross. 1994. *Environmental Statistics, Assessment, and Forecasting.* New York: CRC Press/Lewis Publishers.

Gardner, R. W. 1994. *Chemical Intolerance.* New York: CRC Press/Lewis Publishers.

Hallenbeck, W. H. 1993. *Quantitative Risk Assessment for Environmental and Occupational Health,* 2nd Edition. New York: CRC Press/Lewis Publishers.

Howard, P. H., et. al. 1991. *Handbook of Environmental Degradation Rates.* New York: CRC Press/Lewis Publishers.

Landis, W. G., M. H. Yu. 1995. *Introduction to Environmental Toxicology.* New York: CRC Press/Lewis Publishers.

Lauwerys, R. R., P. Hoet. 1993. *Industrial Chemical Exposure.* New York: CRC Press/Lewis Publishers.

Lewis, R. J., Sr. 1992. *Sax's Dangerous Properties of Industrial Materials,* Volumes 1, 2, and 3, 8th Edition. New York: Van Nostrand Reinhold.

Manahan, S. E. 1992. *Toxicological Chemistry,* 2nd Edition. New York: CRC Press/Lewis Publishers.

Meyer, E. 1990. *Chemistry of Hazardous Materials.* Englewood Cliffs, NJ: Prentice Hall.

Quigley, D. R. 1994. *Handbook of Emergency Chemical Management.* New York: CRC Press/Lewis Publishers.

Rea, W. J. 1995. *Chemical Sensitivity,* Vol. I-III. New York: CRC Press/Lewis Publishers.

Sacarello, H. L. A. 1994. *The Comprehensive Handbook of Hazardous Materials.* New York: CRC Press/Lewis Publishers.

MATERIAL SAFETY DATA SHEET

CHEMICAL NAME	
CRESOL	

HAZARD WARNING INFORMATION

HEALTH	FIRE	REACTIVE	OTHER	DEGREE OF HAZARD	COLOR CODING	OTHER CODES
3	2	0		0 = Minimum Hazard 1 = Slight Hazard 2 = Moderate Hazard 3 = Serious Hazard 4 = Severe Hazard	HEALTH = BLUE FIRE = RED REACTIVITY = YELLOW OTHER = WHITE	OX = Oxidizer ACID = Acid ALK = Alkali COR = Corrosive W = Use No Water

SECTION I - GENERAL INFORMATION

Characterization	**Phenol**	RCRA Number / EPA Class	**U052** / **Toxic Waste**
DOT Proper Shipping Name	**Cresol**	Chemical Abstract Service (CAS) Number	**1319-77-3**
DOT Hazard Class and Label Requirements	**Corrosive; Poison B**	DOT Emergency Guide Code	**55**
DOT Identification Number	**UN2076**	Chemical Formula	**$CH_3C_6H_4OH$**

Synonyms
Cresylic acid, methylphenol, oxytoluene, hydroxytoluene, tar acid.

SECTION II - HAZARDOUS INGREDIENTS AND IDENTITY INFORMATION

Hazardous Components (specific identity)	OSHA Exposure Criteria	NIOSH Exposure Criteria	Immediately Dangerous to Life and Health (IDLH)	ACGIH Exposure Criteria
Cresol (derivation: From coal tar (from coke and gas works), also from toluene by sulfonation or oxidation). **1 ppm = 4.50 mg/m³**	**PEL (skin):** **5 ppm** **22 mg/m³** **STEL:** **Not Established**	**REL:** **2.3 ppm** **10 mg/m³** **STEL:** **Not Established**	**250 ppm**	**TLV:** **5 ppm** **22 mg/m³** **STEL:** **Not Established**

SECTION III - PHYSICAL AND CHEMICAL CHARACTERISTICS

Boiling Point	**376-396°F (191-203°C)**	Specific Gravity (H_2O = 1)	**1.03**
Vapor Pressure (gas)	**1 at 77°F (25°C)**	Molecular Weight	**108.15**
Vapor Density (Air = 1)	**3.72**	Melting Point	**52-95°F (10.9-35°C)**

Solubility
Soluble in alcohol, glycol, dilute alkalis, and water.

Appearance and Odor
Colorless to brown-yellow or pink liquid with a phenolic odor.

SECTION IV - FIRE AND EXPLOSION HAZARD DATA

Flash Point (method used)	**178°F (81°C) closed cup**	Explosive Limits in Air % by Volume	**LEL: 1.1%** **UEL: Not Found**
NFPA Classification	**Class IIIA Combustible Liquid**	Autoignition Temperature	**1138°F (558°C)**

Extinguishing Media
Use water to blanket fire, use dry chemical, foam, fog, or mist.

Special Fire Fighting Procedures
Wear full protective clothing and self-contained breathing apparatus (SCBA). Poisonous gases are produced in fire. Use water spray to keep fire-exposed containers cool.

Unusual Fire and Explosion Hazards
Dangerous fire and explosion hazard when exposed to heat or flame. Containers may explode in fire. Chemical will react on contact with oxidizable materials. Moderate explosion hazard when in the form of vapor. Vapors form explosive mixtures in air. The flammability range has not been established.

SECTION V - REACTIVITY DATA

Stability		Conditions to Avoid **Keep away from heat or flame and incompatible materials.**
Stable	Unstable	Incompatibility *(materials to avoid)* **Strong oxidizers, (such as chlorine, bromine, and fluorine), and many acids (such as chlorosulfonic and nitric), and oleum.**
X		
Hazardous Polymerization		Conditions to Avoid **Under normal conditions of temperature and pressure, hazardous polymerization of cresol is not expected to occur.**
May Occur	Will Not Occur	Hazardous Decomposition or By-products **When heated to decomposition, cresol can emit highly toxic/poisonous gases.**
	X	

SECTION VI - HEALTH HAZARD DATA

Primary Route(s) of Entry:	Inhalation?	Absorption (skin)?	Ingestion?
	X	X	X

Health Hazards

INHALATION: Symptoms include weakness, confusion, depression of the central nervous system, dyspnea, anemia, and respiratory failure. Possible liver and kidney damage.

ABSORPTION: Severe skin irritant causing burns and dermatitis. There may be skin eruptions. Eye contact can cause damage to and/or loss of vision. Will pass through skin to cause toxic systemic effects, including liver damage.

INGESTION: Gastrointestinal disorders, nervous disorders, tremor, convulsions, oliguria, jaundice.

Carcinogenicity	NTP Listed?	IARC Cancer Review Group?	OSHA Regulated?	Target Organs?
Questioned Human Questioned Animal	No	No	No	Respiratory system, eyes, skin, CNS, liver, kidney.

Medical Conditions Generally Aggravated by Exposure

None reported.

Emergency and First-aid Procedures

Eye contact: Flush immediately with water for 15 minutes (minimum), seek medical attention. Skin contact: Remove all contaminated clothing. Immediately wash area with large amounts of soap and water. Seek medical assistance. For inhalation: Remove the person from exposure. Provide respiratory assistance and CPR, if required. Transfer to medical facility. If swallowed: If conscious, give large amounts of water or milk, do NOT induce vomiting. Seek medical attention immediately.

SECTION VII - PRECAUTIONS FOR SAFE HANDLING AND USE

Steps to be Taken in Case Material is Released or Spilled

Absorb liquids in vermiculite, dry sand, earth, or other non-reactive material and deposit is sealed drum. Ventilate area of spill or leak. Restrict those not involved in cleanup from entering area. Remove all sources of ignition.

Preferred Waste Disposal Method

No Citation.

Precautions to be Taken in Handling and Storage

Store to avoid contact with incompatible materials in tightly closed containers in cool, well-ventilated area away from heat and flame.

Other Precautions and Warnings

Sources of ignition are prohibited where cresol is used, handled, or stored. Heat may cause solids to liquefy and containers to build up pressure and explode.

SECTION VIII - CONTROL MEASURES AND PERSONAL PROTECTIVE EQUIPMENT

Respiratory Protection *(specify type)*

For exposures above the PEL, use an MSHA/NIOSH-approved supplied-air respirator with full facepiece, hood, or helmet in continuous flow mode, or use a self-contained breathing apparatus (SCBA) operated in pressure demand or other positive pressure mode.

Ventilation

Local exhaust or general mechanical systems recommended.

Protective Gloves	Eye Protection	Other Protective Clothing
Impervious Rubber	Chemical/Dust Goggles or Face Mask	Rubber Apron

Work/Hygiene Practices

Always wash hands thoroughly after using chemical; never bring food, drink, or smoking materials into vicinity of chemicals.

CRESOL

$CH_3C_6H_4OH$ CAS: 1319-77-3

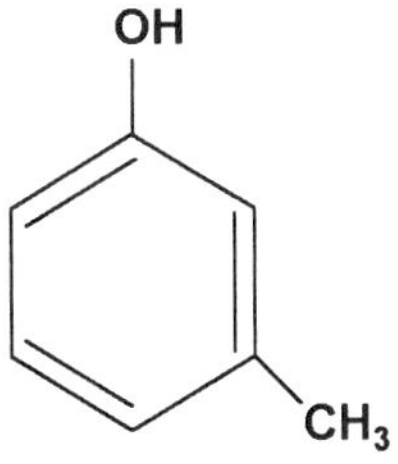

m-cresol

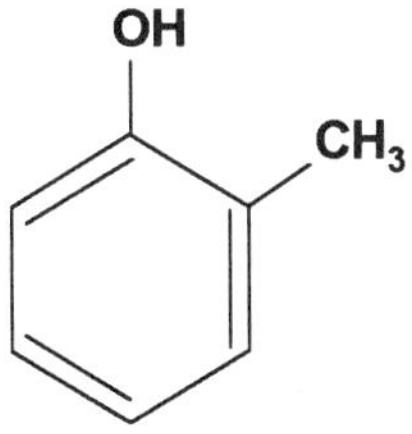

o-cresol

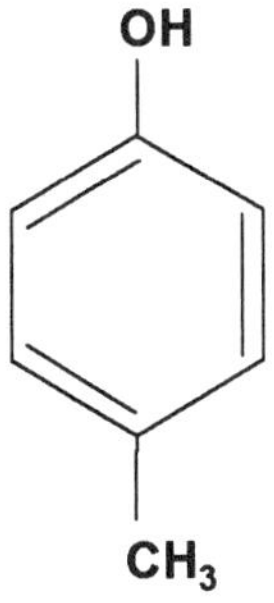

p-cresol

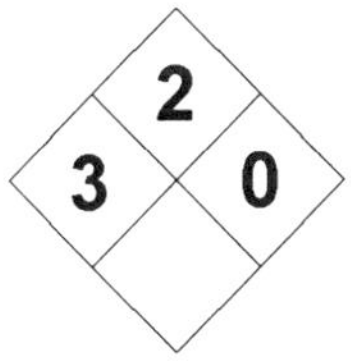

IDENTIFICATION AND TYPICAL USES

Cresol is a mixture of the three isomeric cresols, *m-(meta), o (ortho),* and *p (para).* It is obtained from coal tar and appears as a colorless or yellowish to brown-yellow or pink liquid with a phenolic-like odor. It is a white powder or crystalline solid with a sweet, tarry odor. It is used in making synthetic resins, photographic developers, explosives, and as a disinfectant and a fumigant.

RISK ASSESSMENT: HEALTH

General Assessment

Cresol is a human poison by *inhalation, ingestion,* and skin *absorption.* It is a severe irritant to the eyes, skin and respiratory system. Human mutation data have been reported for cresol. It is a questionable carcinogen in humans and animals, although the data are inconclusive in this area of study. The toxic actions of the cresol isomers are similar to that of phenol (the *"p"* or *para*-isomer is somewhat more toxic than the other two).

Inhalation can cause irritation to the respiratory system and associated mucosa. It can cause cough, congestion, with narcotic effects associated with the depression of the central nervous system (CNS). These include weakness, dyspnea, nausea, personality changes (irritability, mood changes), collapse, coma, and death (from respiratory failure).

Skin contact can produce burns, possible skin eruptions, and dermatitis. The liquid will pass through unbroken skin to cause toxic systemic effects similar to the effects on the CNS caused by inhalation. If cresol should contact the eyes, there may be serious damage up to and including loss of vision.

Ingestion of cresol can lead to gastrointestinal disorders, nervous disorders (CNS action), tremor, confusion, skin eruptions, oliguria (a diminishing of urinary secretion over a 24-hour period), jaundice, and liver damage. High doses will depress the CNS and may cause a reduction in body weight.

☠ *Acute Health Effects*

The following acute (short-term) health effects may occur immediately or shortly after exposure to cresol:

Skin: Severe irritation and possible burns. Will pass through unbroken skin to cause toxic systemic effects, including weakness, nausea, collapse, and death.

Eye: Severe irritation with damage and possible loss of vision likely.

Lung: Severe irritation to the respiratory tract and associated mucosa. Can cause weakness, nausea, confusion, collapse, coma, and death (respiratory failure).

CNS: A CNS depressant. Weakness, confusion, mood changes and other personality changes, nausea, collapse.

Other: Ingestion will cause irritation to the gastrointestinal system with nausea, vomiting, convulsions, dypsnea, oliguria, jaundice, tremor, confusion, and nervousness. It has also been found to be hepatotoxic (toxic to the liver) and nephrototoxic (toxic to the kidneys).

Chronic Health Effects

The following chronic (long-term) health effects may occur at some time after exposure to cresol and can last for months or even years:

Cancer Hazards: Some studies have suggested that exposure to cresol causes mutations (genetic changes) in humans. While there is no evidence of carcinogenicity of this chemical in humans or animals, it is still considered a *questioned* carcinogen primarily because of its mutagenic properties. It is suggested that mutagenic chemicals be regarded as carcinogens during use and handling.

Reproductive Hazard: According to information presented in the references, cresol has not been adequately tested for its ability to adversely affect reproduction in animals.

Other Chronic Effects: High or repeated lower skin or inhalation exposures can cause liver or kidney disorders. Repeated skin contact may cause sensitization, dermatitis, chronic rash, and associated irritation. Once an allergy develops, the slightest future exposures will cause a recurrence of the symptoms.

Recommended Risk-Reduction Measures

Personnel should avoid direct contact with cresol. If a less toxic chemical cannot be substituted for a hazardous substance, then *engineering controls* are the most effective method of reducing exposures. The best protection is to enclose operations and/or provide local exhaust ventilation at the site of chemical release. While not always operationally feasible, isolating operations can also reduce exposure. Using respiratory protection is less effective than the controls mentioned above, but is still advisable whenever working with or around cresol. For exposures above the PEL (5 ppm), use an MSHA/NIOSH approved full facepiece respirator equipped with an organic vapor cartridge. Increased protection is obtained from a supplied-air respirator with full facepiece operated in continuous flow mode, or a self-contained breathing apparatus (SCBA) with full facepiece and pressure demand. If a full facepiece is not available, then chemical goggles should be worn to protect the eyes. Whenever a chemical splash hazard exists, a face shield and a rubber apron should be worn. To prevent hand and skin exposures, impervious rubber gloves should be worn.

Administrative controls should also be in place to minimize the potential for human exposures. These may include written procedures or policies which specify the methods and techniques that will be practiced whenever personnel are to work with cresol.

All personnel should receive *training* on the use, hazards, protective measures, emergency actions, and other precautions per 29 CFR 1910.1200 (Hazard Communication), prior to the first assignment in an area where cresol is used or stored.

Before beginning employment with cresol and at regular intervals thereafter (e.g., annually), those persons with high or frequent exposures should be provided the following recommended medical tests:

☑ Kidney and liver function tests.

If symptoms develop or overexposure is suspected, the following tests are recommended:

☑ Consider chest X-ray.

☑ Evaluation by a qualified allergist, with careful evaluation of exposure history and special testing (may help diagnose allergy).

Any evaluation should include a careful history of past and present symptoms with an examination. Medical tests that simply look for existing damage are not a substitute for controlling exposures.

Other methods to reduce exposure include:

☑ Where possible, enclose operations and use local exhaust ventilation at the site of chemical release. If local exhaust ventilation or enclosure is not used, respiratory protection should be mandatory. Also, if possible, liquids should be transferred automatically from storage containers to process containers.

☑ Always ensure that proper protective clothing is worn when using chemical substances.

☑ Wash thoroughly immediately after exposure to cresol and at the end of the work shift or before eating, drinking, or smoking.

☑ Hazard warning information should be posted in the work area. In addition, as part of an on-going education and training program, all information on the health and safety hazards of cresol should be communicated to all potentially exposed workers.

☑ Eye wash stations should be provided in the immediate work area for emergency use. If there is a possibility of skin exposure to cresol, emergency shower facilities should also be provided.

☑ Personnel should never wear contaminated clothing home (family members can be exposed).

Clothing should be laundered by personnel who have been trained on the hazards of cresol.

☑ Specific engineering controls are recommended for this chemical by NIOSH. Refer to NIOSH Criteria Document: Occupational Exposure to Cresol (Publication Number 78-133).

RISK ASSESSMENT: ENVIRONMENT

General Assessment

The environment is at risk of exposure during transportation, storage, disposal, or destruction of cresol. In almost every scenario, the threat of environmental exposure is contingent upon the proper handling of the chemical substance. Accidental spills, large or small, can result in fire, explosion, and possible contamination of the surrounding environmental mediums (water, soil, and air).

Cresol is considered a class IIIA combustible liquid (per OSHA 29 CFR 1910.106). However, its vapors can form explosive mixtures in air. These characteristics require caution in handling, storage, transportation, and disposal. It is incompatible with strong oxidizers, oleum, and acids. Contact with either can cause fire or explosion. It is also a corrosive and, therefore, special consideration is required during any emergency situation involving a leak or spill of cresol. Should cresol ever come into contact with these incompatible substances during use, transportation, storage, or disposal, the formation of highly toxic and/or highly explosive commodities is extremely possible.

Cresol may enter the environment through industrial effluents or spills.

Acute Ecological Effects

Acute (short-term) toxic effects may include the death of animals, birds, or fish and death or low growth rate in plants. Acute effects are seen 2 to 4 days after animals or plants are exposed to cresol.

Cresol has moderate acute toxicity to aquatic life. Insufficient data are available on the short-term effects of cresol exposure to plants, birds or land animals.

Chronic Ecological Effects

Chronic (long-term) toxic effects may include shortened life span, reproductive problems, lower fertility, and changes in appearance or behavior in exposed animals. These effects can be seen long after first exposure(s) to toxic chemicals.

Cresol has moderate chronic toxicity to aquatic life. Insufficient data are available on the long-term effects of cresol to plants, birds, or land animals.

Water Solubility

Cresol is highly soluble in water. Concentrations of 1000 milligrams and more will mix with a liter of water.

Persistence in the Environment

Cresol is slightly persistent in water, with a half-life between 2 and 20 days. The half-life of a pollutant is the amount of time it takes for one half of the chemical to be degraded. About 56% of cresol will eventually end up in water; approximately 42.7% in air, and about 0.6% in terrestrial soil and aquatic sediments.

Bioaccumulation in Aquatic Organisms

Some substances increase in concentration, or *bioaccumulate,* in living organisms as they breathe contaminated air, drink contaminated water, or eat contaminated food. These chemicals can become concentrated in the tissues and internal organs of animals as well as humans.

The concentration of cresol found in fish tissue is expected to be somewhat higher than the average concentration of cresol in the water from which the fish was taken.

Recommended Risk-Reduction Measures

Proper training of all transporters will reduce the likelihood of a mishap or accident resulting in a leak or spill to the environment. The correct use of DOT labeling (per the MSDS) on all containers, trucks, and rail cars will enable emergency responders to react properly and quickly to any disaster thereby reducing the potential risk to the environment and to personnel.

Storage of cresol should be segregated from incompatible chemicals to minimize the risk of cross-contamination or contact. Buildings designated for storage should be equipped with appropriate fire protection systems (alarms, sprinklers, emergency lighting, portable extinguishers).

If a spill or leak to the environment has occurred, fire department, emergency response and/or hazardous materials spill personnel should be notified immediately. Cleanup should be attempted only by those trained in proper spill containment procedures. Contaminated soils should be removed for incineration and

replaced with clean soil. If cresol should contact the water table, aquifer, or navigable waterway, time is of the essence. It is highly soluble in the aquatic environment and total remediation may not be possible. The local and/or state emergency response authorities must be notified. A comprehensive emergency response or disaster preparedness/recovery plan should be in place prior to any operations involving the use, transportation, storage, or disposal of cresol. If cresol is spilled or leaked, the following specific steps are recommended:

☑ Restrict persons not wearing protective clothing from area of spill or leak until cleanup is complete and area can be opened for normal work.

☑ Remove all ignition sources, ventilate area of leak.

☑ Absorb liquid spills in vermiculite, dry sand, earth, or similar material and deposit in sealed containers.

☑ It may be necessary to dispose of cresol as a hazardous waste. The responsible state agency or the regional office of the federal Environmental Protection Agency (EPA) should be contacted for specific recommendations.

RISK ASSESSMENT: BUSINESS

General Assessment

Accidents or mishaps involving cresol can present a significant threat to business operations. The loss or damage of equipment or facilities can significantly affect fiscal viability. Lawsuits that may result from personnel injury/death, public exposures, and/or environmental contamination will also require a serious expenditure of resources. Media attention surrounding an injury, death, or environmental damage can also result in a loss of profits and loss of current as well as future business.

Recommended Risk-Reduction Measures

Company attorneys, safety and health professionals, and environmental specialists should be involved in the development of any procedures dealing with safety or emergency response. A company official should be pre-designated as a public relations officer with specific training in dealing with the press. Corporate plans and policies should be developed, approved, and implemented long before any need for such arises.

REFERENCES

Ahlbom, A. 1993. *Biostatistics for Epidemiologists.* New York: CRC Press/Lewis Publishers.

Asante-Duah, D. K. 1993. *Hazardous Waste Risk Assessment.* New York: CRC Press/Lewis Publishers.

Bloom, A. D. and F. J. DeSerres. 1995. *Ecotoxicity and Human Health.* New York: CRC Press/Lewis Publishers.

Brusick, D. J. 1994. *Methods for Genetic Risk Assessment.* New York: CRC Press/Lewis Publishers.

Bueche, J. R. 1972. *Principles of Physics,* 2nd Edition. New York: McGraw-Hill.

Calabrese, E. J. 1994. *Biological Effects of Low Level Exposures.* New York: CRC Press/Lewis Publishers.

Cockerman, L. G., B. S. Shane. 1994. *Basic Environmental Toxicology.* New York: CRC Press/Lewis Publishers.

Cothern, C. R., N. P. Ross. 1994. *Environmental Statistics, Assessment, and Forecasting.* New York: CRC Press/Lewis Publishers.

Cothern, C. R. 1995. *Handbook for Environmental Risk Decisionmaking.* New York: CRC Press/Lewis Publishers.

Hallenbeck, W. H. 1993. *Quantitative Risk Assessment for Environmental and Occupational Health,* 2nd Edition. New York: CRC Press/Lewis Publishers.

Howard, P. H. 1995. *Handbook of Environmental Fate and Exposures Data for Organic Chemicals,* Vol. I-IV. New York: CRC Press/Lewis Publishers.

Lauwerys, R. R., P. Hoet. 1993. *Industrial Chemical Exposure.* New York: CRC Press/Lewis Publishers.

Lewis, R. J., Sr. 1992. *Sax's Dangerous Properties of Industrial Materials,* Volumes 1, 2, and 3, 8th Edition. New York: Van Nostrand Reinhold.

Meyer, E. 1990. *Chemistry of Hazardous Materials.* Englewood Cliffs, NJ: Prentice Hall.

Patnaik, P. 1992. *A Comprehensive Guide to the Hazardous Properties of Chemical Substances.* New York: Van Nostrand Reinhold.

Quigley, D. R. 1994. *Handbook of Emergency Chemical Management.* New York: CRC Press/Lewis Publishers.

Torkelson, T. R. V. K. Rowe. 1981. Halogenated aliphatic hydrocarbons. *Patty's Industrial Hygiene and Toxicology,* Vol. 2B, *Toxicology,* pp 3568-3572. New York: Wiley-Interscience.

U.S. Department of Health and Human Services, National Institute for Occupational Safety and Health. 1978. *NIOSH Criteria Document: Occupational Exposure to Cresol.* (Publication Number: 78-133). Washington, D.C.: U.S. Government Printing Office.

U.S. Department of Health and Human Services, National Institute for Occupational Safety and Health. 1994. *NIOSH Pocket Guide to Chemical Hazards.* Washington, D.C.: U.S. Government Printing Office.

MATERIAL SAFETY DATA SHEET

CHEMICAL NAME	
***m*-CRESOL**	**CRC PRESS/LEWIS PUBLISHERS DISCLAIMER:** THE INFORMATION AND RECOMMENDATIONS PRESENTED HEREIN ARE BASED ON SOURCES BELIEVED TO BE RELIABLE. CRC MAKES NO REPRESENTATION ON ITS COMPLETENESS OR ACCURACY. IT IS THE RESPONSIBILITY OF THE USER TO DETERMINE THE CHEMICAL'S SUITABILITY FOR ITS INTENDED USE, THE CHEMICAL'S SAFE USE, AND THE CHEMICAL'S PROPER DISPOSAL. NO REPRESENTATIONS AND/OR WARRANTIES, EITHER EXPRESSED OR IMPLIED, OF THE MERCHANTABILITY OR FITNESS FOR A PARTICULAR PURPOSE, OR OF ANY OTHER NATURE, ARE MADE WITH RESPECT TO THE INFORMATION PROVIDED IN THIS MSDS OR TO THE CHEMICAL TO WHICH INFORMATION MAY REFER. CRC NEITHER ASSUMES NOR AUTHORIZES ANY OTHER PERSON TO ASSUME FOR IT, ANY OTHER ADDITIONAL RESPONSIBILITY OR LIABILITY FOR THE USE OF, OR RELIANCE UPON, THIS INFORMATION.

HAZARD WARNING INFORMATION

HEALTH	FIRE	REACTIVE	OTHER	DEGREE OF HAZARD	COLOR CODING	OTHER CODES
3	2	0		0 = Minimum Hazard 1 = Slight Hazard 2 = Moderate Hazard 3 = Serious Hazard 4 = Severe Hazard	HEALTH = BLUE FIRE = RED REACTIVITY = YELLOW OTHER = WHITE	OX = Oxidizer ACID = Acid ALK = Alkali COR = Corrosive W = Use No Water

SECTION I - GENERAL INFORMATION

Characterization	**Phenol**	RCRA Number	**U052**
		EPA Class	**Toxic Waste**
DOT Proper Shipping Name	**m-Cresol**	Chemical Abstract Service (CAS) Number	**108-39-4**
DOT Hazard Class and Label Requirements	**Poison B**	DOT Emergency Guide Code	**55**
DOT Identification Number	**UN2076**	Chemical Formula	**$CH_3C_6H_4OH$**

Synonyms

***m*-cresylic acid; 3-cresol; *m*-cresylic acid; meta-cresol; 1-hydroxy-3-methylbenzene; 3-methyl phenol.**

SECTION II - HAZARDOUS INGREDIENTS AND IDENTITY INFORMATION

Hazardous Components (specific identity)	OSHA Exposure Criteria	NIOSH Exposure Criteria	Immediately Dangerous to Life and Health (IDLH)	ACGIH Exposure Criteria
***m*-Cresol** (derivation: By fractional distillation of crude cresol from coal tar; also derived synthetically). **1 ppm = 4.50 mg/m³**	**PEL (skin):** **5 ppm** **22 mg/m³** **STEL:** **Not Established**	**REL:** **2.3 ppm** **10 mg/m³** **STEL:** **Not Established**	**250 ppm**	**TLV:** **5 ppm** **22 mg/m³** **STEL:** **Not Established**

SECTION III - PHYSICAL AND CHEMICAL CHARACTERISTICS

Boiling Point	**397°F (203°C)**	Specific Gravity (H_2O = 1)	**1.03**
Vapor Pressure (gas)	**.14 at 77°F (25°C)**	Molecular Weight	**108.2**
Vapor Density (Air = 1)	**3.72**	Melting Point	**52°F (11°C)**

Solubility

Soluble in alcohol, ether, chloroform, and water.

Appearance and Odor

Colorless to yellowish liquid with a sweet, tarry odor.

SECTION IV - FIRE AND EXPLOSION HAZARD DATA

Flash Point (method used)	**187°F (86°C) closed cup**	Explosive Limits in Air % by Volume	**LEL: 1.1% UEL: Not Found**
NFPA Classification	**Class IIIA Combustible Solid**	Autoignition Temperature	**1148°F (620°C)**

Extinguishing Media

Use water to blanket fire, use dry chemical, foam, fog, or mist.

Special Fire Fighting Procedures

Wear full protective clothing and self-contained breathing apparatus (SCBA). Poisonous gases are produced in fire. Use water spray to keep fire-exposed containers cool.

Unusual Fire and Explosion Hazards

Dangerous fire and explosion hazard when exposed to heat or flame. Containers may explode in fire. Chemical will react on contact with oxidizable materials. Moderate explosion hazard when in the form of vapor. Vapors form explosive mixtures in air. The flammability range has not been established.

SECTION V - REACTIVITY DATA

Stability		Conditions to Avoid Keep away from heat or flame and incompatible materials.
Stable	Unstable	Incompatibility *(materials to avoid)* Strong oxidizers, (such as chlorine, bromine, and fluorine), and many acids (such as chlorosulfonic and nitric), and oleum.
X		
Hazardous Polymerization		Conditions to Avoid Under normal conditions of temperature and pressure, hazardous polymerization of *m*-cresol is not expected to occur.
May Occur	Will Not Occur	Hazardous Decomposition or By-products When heated to decomposition, *m*-cresol can emit highly toxic/poisonous gases.
	X	

SECTION VI - HEALTH HAZARD DATA

Primary Route(s) of Entry:	Inhalation?	Absorption (skin)?	Ingestion?
	X	X	X

Health Hazards

INHALATION: Symptoms include weakness, confusion, depression of the central nervous system, dyspnea, anemia, and respiratory failure. Possible liver and kidney damage.

ABSORPTION: Severe skin irritant causing burns and dermatitis. There may be skin eruptions. Eye contact can cause damage to and/or loss of vision. Will pass through skin to cause toxic systemic effects, including liver damage.

INGESTION: Gastrointestinal disorders, nervous disorders, tremor, convulsions, oliguria, jaundice.

Carcinogenicity	NTP Listed?	IARC Cancer Review Group?	OSHA Regulated?	Target Organs?
Questioned Human Questioned Animal	No	No	No	Respiratory system, eyes, skin, CNS, liver, kidney.

Medical Conditions Generally Aggravated by Exposure

None reported.

Emergency and First-aid Procedures

Eye contact: Flush immediately with water for 15 minutes (minimum), seek medical attention. Skin contact: Remove all contaminated clothing. Immediately wash area with large amounts of soap and water. Seek medical assistance. For inhalation: Remove the person from exposure. Provide respiratory assistance and CPR, if required. Transfer to medical facility. If swallowed: If conscious, give large amounts of water or milk, do NOT induce vomiting. Seek medical attention immediately.

SECTION VII - PRECAUTIONS FOR SAFE HANDLING AND USE

Steps to be Taken in Case Material is Released or Spilled

Absorb liquids in vermiculite, dry sand, earth, or other non-reactive material and deposit in sealed drum. Ventilate area of spill or leak. Restrict those not involved in cleanup from entering area. Remove all sources of ignition.

Preferred Waste Disposal Method

No Citation.

Precautions to be Taken in Handling and Storage

Store to avoid contact with incompatible materials in tightly closed containers in cool, well-ventilated area away from heat and flame.

Other Precautions and Warnings

Sources of ignition are prohibited where *m*-cresol is used, handled, or stored. Heat may cause solids to liquefy and containers to build up pressure and explode.

SECTION VIII - CONTROL MEASURES AND PERSONAL PROTECTIVE EQUIPMENT

Respiratory Protection *(specify type)*

For exposures above the PEL, use an MSHA/NIOSH-approved supplied-air respirator with full facepiece, hood, or helmet in continuous flow mode, or use a self-contained breathing apparatus (SCBA) operated in pressure demand or other positive pressure mode.

Ventilation

Local exhaust or general mechanical systems recommended.

Protective Gloves	Eye Protection	Other Protective Clothing
Impervious Rubber	Chemical/Dust Goggles or Face Mask	Rubber Apron

Work/Hygiene Practices

Always wash hands thoroughly after using chemical; never bring food, drink, or smoking materials into vicinity of chemicals.

m-CRESOL

$CH_3C_6H_4OH$ CAS: 108-39-4

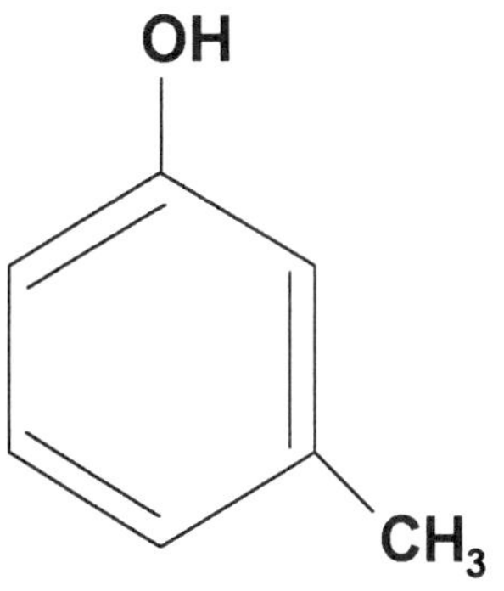

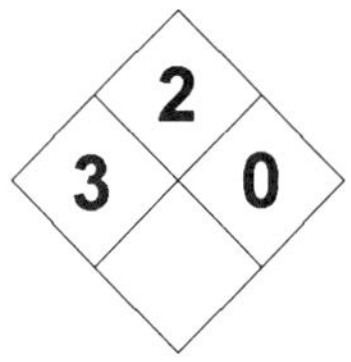

IDENTIFICATION AND TYPICAL USES

m-Cresol is a colorless to yellow liquid with a sweet, tarry odor. It is one of three isomers of the chemical cresol (cresylic acid). The other two, *o*-cresol and *p*-cresol, are similar in structure and hazard level. The *m*-cresol isomer is used in making synthetic resins, photographic developers, explosives, and as a disinfectant and a fumigant.

RISK ASSESSMENT: HEALTH

General Assessment

m-Cresol is a human poison by *inhalation, ingestion,* and skin *absorption.* It is a severe irritant to the eyes, skin and respiratory system. Human mutation data have been reported for *m*-cresol. It is a questionable carcinogen in humans and animals, although the data are inconclusive in this areas of study. The toxic actions of the cresol isomers are similar to that of phenol (the *"p"* or *para*-isomer is somewhat more toxic than the other two).

Inhalation can cause irritation to the respiratory system and associated mucosa. It can cause cough, congestion, with narcotic effects associated with the depression of the central nervous system (CNS). These include weakness, dyspnea, nausea, personality changes (irritability, mood changes), collapse, coma, and death (from respiratory failure).

Skin contact can produce burns, possible skin eruptions, and dermatitis. The liquid will pass through unbroken skin to cause toxic systemic effects similar to the effects on the CNS caused by inhalation. If *m*-cresol should contact the eyes, there may be serious damage up to and including loss of vision.

Ingestion of *m*-cresol can lead to gastrointestinal disorders, nervous disorders (CNS action), tremor, confusion, skin eruptions, oliguria (a diminishing of urinary secretion over a 24-hour period), jaundice, and liver damage. High doses will depress the CNS and may cause a reduction in body weight.

Acute Health Effects

The following acute (short-term) health effects may occur immediately or shortly after exposure to *m*-cresol:

Skin: Severe irritation and possible burns. Will pass through unbroken skin to cause toxic systemic effects (weakness, nausea, collapse, and death).

Eye: Severe irritation with damage and possible loss of vision likely.

Lung: Severe irritation to the respiratory tract and associated mucosa. Can cause weakness, nausea, confusion, collapse, coma, and death (respiratory failure).

CNS: A CNS depressant. Weakness, confusion, mood changes and other personality changes, nausea, collapse.

Other: Ingestion will cause irritation to the gastrointestinal system with nausea, vomiting, convulsions, dypsnea, oliguria, jaundice, tremor, confusion, and nervousness. It has also been found to be hepatotoxic (toxic to the liver) and nephrototoxic (toxic to the kidneys).

Chronic Health Effects

The following chronic (long-term) health effects may occur at some time after exposure to *m*-cresol and can last for months or even years:

Cancer Hazards: Some studies have suggested that exposure to *m*-cresol causes mutations (genetic changes) in humans. While there is no evidence of carcinogenicity of this chemical in humans or animals, it is still considered a *questioned* carcinogen primarily because of its mutagenic properties. It is suggested that mutagenic chemicals be regarded as carcinogens during use and handling.

Reproductive Hazard: According to information presented in the references, *m*-cresol has not been adequately tested for its ability to adversely affect reproduction in animals.

Other Chronic Effects: High or repeated lower skin or inhalation exposures can cause liver or kidney disorders. Repeated skin contact may cause sensitization, dermatitis, chronic rash, and associated irritation. Once an allergy develops, the slightest future exposures will cause a recurrence of the symptoms.

Recommended Risk-Reduction Measures

Personnel should avoid direct contact with *m*-cresol. If a less toxic chemical cannot be substituted for a hazardous substance, then *engineering controls* are the most effective method of reducing exposures. The best protection is to enclose operations and/or provide local exhaust ventilation at the site of chemical release. While not always operationally feasible, isolating operations can also reduce exposure. Using respiratory protection is less effective than the controls mentioned above, but is still advisable whenever working with or around *m*-cresol. For exposures above the PEL (5 ppm), use an MSHA/NIOSH approved full facepiece respirator equipped with an organic vapor cartridge. Increased protection is obtained from a supplied-air respirator with full facepiece operated in continuous flow mode, or a self-contained breathing apparatus (SCBA) with full facepiece and pressure demand. If a full facepiece is not available, then chemical goggles should be worn to protect the eyes. Whenever a chemical splash hazard exists, a face shield and a rubber apron should be worn. To prevent hand and skin exposures, impervious rubber gloves should be worn.

Administrative controls should also be in place to minimize the potential for human exposures. These may include written procedures or policies which specify the methods and techniques that will be practiced whenever personnel are to work with *m*-cresol.

All personnel should receive *training* on the use, hazards, protective measures, emergency actions, and other precautions per 29 CFR 1910.1200 (Hazard Communication), prior to the first assignment in an area where *m*-cresol is used or stored.

Before beginning employment with *m*-cresol and at regular intervals thereafter (e.g., annually), those persons with high or frequent exposures should be provided the following recommended medical tests:

☑ Kidney and liver function tests.

If symptoms develop or overexposure is suspected, the following tests are recommended:

☑ Consider chest X-ray.

☑ Evaluation by a qualified allergist, with careful evaluation of exposure history and special testing (may help diagnose allergy).

Any evaluation should include a careful history of past and present symptoms with an examination. Medical tests that simply look for existing damage are not a substitute for controlling exposures.

Other methods to reduce exposure include:

☑ Where possible, enclose operations and use local exhaust ventilation at the site of chemical release. If local exhaust ventilation or enclosure is not used, respiratory protection should be mandatory. Also, if possible, liquids should be transferred automatically from storage containers to process containers.

☑ Always ensure that proper protective clothing is worn when using chemical substances.

☑ Wash thoroughly immediately after exposure to *m*-cresol and at the end of the work shift or before eating, drinking, or smoking.

☑ Hazard warning information should be posted in the work area. In addition, as part of an on-going education and training program, all information on the health and safety hazards of *m*-cresol should be communicated to all exposed workers.

☑ Eye wash stations should be provided in the immediate work area for emergency use. If there is a possibility of skin exposure to *m*-cresol, emergency shower facilities should also be provided.

☑ Personnel should never wear contaminated clothing home (family members can be exposed).

Clothing should be laundered by personnel who have been trained on the hazards of *m*-cresol.

☑ Specific engineering controls are recommended for this chemical by NIOSH. Refer to NIOSH Criteria Document: Occupational Exposure to Cresol (Publication Number 78-133).

RISK ASSESSMENT: ENVIRONMENT

General Assessment

The environment is at risk of exposure during transportation, storage, disposal, or destruction of *m*-cresol. In almost every scenario, the threat of environmental exposure is contingent upon the proper handling of the chemical substance. Accidental spills, large or small, can result in fire, explosion, and possible contamination of the surrounding environmental mediums (water, soil, and air).

m-Cresol is considered a class IIIA combustible liquid (per OSHA 29 CFR 1910.106). Its vapors can form explosive mixtures in air. These characteristics require caution in handling, storage, transportation, and disposal. It is incompatible with strong oxidizers, oleum, and acids. Contact with either can cause fire or explosion. It is also a corrosive and, therefore, special consideration is required during any emergency situation involving a leak or spill of *m*-cresol. Should *m*-cresol ever come into contact with these incompatible substances during use, transportation, storage, or disposal, the formation of highly toxic and/or highly explosive commodities is extremely possible.

m-Cresol may enter the environment through industrial effluents or spills.

Acute Ecological Effects

Acute (short-term) toxic effects may include the death of animals, birds, or fish and death or low growth rate in plants. Acute effects are seen 2 to 4 days after animals or plants are exposed to *m*-cresol.

Cresol has moderate acute toxicity to aquatic life. Insufficient data are available on the short-term effects of *m*-cresol exposure to plants, birds or land animals.

Chronic Ecological Effects

Chronic (long-term) toxic effects may include shortened life span, reproductive problems, lower fertility, and changes in appearance or behavior in exposed animals. These effects can be seen long after first exposure(s) to toxic chemicals.

Cresol has moderate chronic toxicity to aquatic life. Insufficient data are available on the long-term effects of *m*-cresol to plants, birds, or land animals.

Water Solubility

Cresol is highly soluble in water. Concentrations of 1000 milligrams and more will mix with a liter of water.

Persistence in the Environment

Cresol is slightly persistent in water, with a half-life between 2 and 20 days. The half-life of a pollutant is the amount of time it takes for one half of the chemical to be degraded. About 56% of cresol will eventually end up in water; approximately 42.7% in air, and about 0.6% in terrestrial soil and aquatic sediments.

Bioaccumulation in Aquatic Organisms

Some substances increase in concentration, or *bioaccumulate,* in living organisms as they breathe contaminated air, drink contaminated water, or eat contaminated food. These chemicals can become concentrated in the tissues and internal organs of animals as well as humans.

The concentration of cresol found in fish tissue is expected to be somewhat higher than the average concentration of cresol in the water from which the fish was taken.

Recommended Risk-Reduction Measures

Proper training of all transporters will reduce the likelihood of a mishap or accident resulting in a leak or spill to the environment. The correct use of DOT labeling (per the MSDS) on all containers, trucks, and rail cars will enable emergency responders to react properly and quickly to any disaster thereby reducing the potential risk to the environment and to personnel.

Storage of *m*-cresol should be segregated from incompatible chemicals to minimize the risk of cross-contamination or contact. Buildings designated for storage should be equipped with appropriate fire protection systems (alarms, sprinklers, emergency lighting, portable extinguishers).

If a spill or leak to the environment has occurred, fire department, emergency response and/or hazardous materials spill personnel should be notified immediately. Cleanup should be attempted only by those trained in proper spill containment procedures. Contaminated soils should be removed for incineration and

replaced with clean soil. If *m*-cresol should contact the water table, aquifer, or navigable waterway, time is of the essence. It is highly soluble in the aquatic environment and total remediation may not be possible. The local and/or state emergency response authorities must be notified. A comprehensive emergency response or disaster preparedness/recovery plan should be in place prior to any operations involving the use, transportation, storage, or disposal of m-cresol. If *m*-cresol is spilled or leaked, the following specific steps are recommended:

☑ Restrict persons not wearing protective clothing from area of spill or leak until cleanup is complete and area can be opened for normal work.

☑ Remove all ignition sources, ventilate area of leak.

☑ Absorb liquid spills in vermiculite, dry sand, earth, or similar material and deposit in sealed containers.

☑ It may be necessary to dispose of *m*-cresol as a hazardous waste. The responsible state agency or the regional office of the federal Environmental Protection Agency (EPA) should be contacted for specific recommendations.

RISK ASSESSMENT: BUSINESS

General Assessment

Accidents or mishaps involving *m*-cresol can present a significant threat to business operations. The loss or damage of equipment or facilities can significantly affect fiscal viability. Lawsuits that may result from personnel injury/death, public exposures, and/or environmental contamination will also require a serious expenditure of resources. Media attention surrounding an injury, death, or environmental damage can also result in a loss of profits and loss of current as well as future business.

Recommended Risk-Reduction Measures

Company attorneys, safety and health professionals, and environmental specialists should be involved in the development of any procedures dealing with safety or emergency response. A company official should be pre-designated as a public relations officer with specific training in dealing with the press. Corporate plans and policies should be developed, approved, and implemented long before any need for such arises.

REFERENCES

Asante-Duah, D. K. 1993. *Hazardous Waste Risk Assessment.* New York: CRC Press/Lewis Publishers.

Bueche, J. R. 1972. *Principles of Physics,* 2nd Edition. New York: McGraw-Hill.

Cockerman, L. G., B. S. Shane. 1994. *Basic Environmental Toxicology.* New York: CRC Press/Lewis Publishers.

Cothern, C. R. 1995. *Handbook for Environmental Risk Decisionmaking.* New York: CRC Press/Lewis Publishers.

Hallenbeck, W. H. 1993. *Quantitative Risk Assessment for Environmental and Occupational Health,* 2nd Edition. New York: CRC Press/Lewis Publishers.

Howard, P. H. 1995. *Handbook of Environmental Fate and Exposures Data for Organic Chemicals,* Vol. I-IV. New York: CRC Press/Lewis Publishers.

Lauwerys, R. R., P. Hoet. 1993. *Industrial Chemical Exposure.* New York: CRC Press/Lewis Publishers.

Lewis, R. J., Sr. 1992. *Sax's Dangerous Properties of Industrial Materials,* Volumes 1, 2, and 3, 8th Edition. New York: Van Nostrand Reinhold.

Meyer, E. 1990. *Chemistry of Hazardous Materials.* Englewood Cliffs, NJ: Prentice Hall.

Patnaik, P. 1992. *A Comprehensive Guide to the Hazardous Properties of Chemical Substances.* New York: Van Nostrand Reinhold.

Quigley, D. R. 1994. *Handbook of Emergency Chemical Management.* New York: CRC Press/Lewis Publishers.

Rea, W. J. 1995. *Chemical Sensitivity,* Vol. I-III. New York: CRC Press/Lewis Publishers.

Sacarello, H. L. A. 1994. *The Comprehensive Handbook of Hazardous Materials.* New York: CRC Press/Lewis Publishers.

Torkelson, T. R. V. K. Rowe. 1981. Halogenated aliphatic hydrocarbons. *Patty's Industrial Hygiene and Toxicology,* Vol. 2B, *Toxicology,* pp 3568-3572. New York: Wiley-Interscience.

U.S. Department of Health and Human Services, National Institute for Occupational Safety and Health. 1978. *NIOSH Criteria Document: Occupational Exposure to Cresol.* (Publication Number: 78-133). Washington, D.C.: U.S. Government Printing Office.

U.S. Department of Health and Human Services, National Institute for Occupational Safety and Health. 1994. *NIOSH Pocket Guide to Chemical Hazards.* Washington, D.C.: U.S. Government Printing Office.

MATERIAL SAFETY DATA SHEET

CHEMICAL NAME	CRC PRESS/LEWIS PUBLISHERS DISCLAIMER:
o-CRESOL	THE INFORMATION AND RECOMMENDATIONS PRESENTED HEREIN ARE BASED ON SOURCES BELIEVED TO BE RELIABLE. CRC MAKES NO REPRESENTATION ON ITS COMPLETENESS OR ACCURACY. IT IS THE RESPONSIBILITY OF THE USER TO DETERMINE THE CHEMICAL'S SUITABILITY FOR ITS INTENDED USE, THE CHEMICAL'S SAFE USE, AND THE CHEMICAL'S PROPER DISPOSAL. NO REPRESENTATIONS AND/OR WARRANTIES, EITHER EXPRESSED OR IMPLIED, OF THE MERCHANTABILITY OR FITNESS FOR A PARTICULAR PURPOSE, OR OF ANY OTHER NATURE, ARE MADE WITH RESPECT TO THE INFORMATION PROVIDED IN THIS MSDS OR TO THE CHEMICAL TO WHICH INFORMATION MAY REFER. CRC NEITHER ASSUMES NOR AUTHORIZES ANY OTHER PERSON TO ASSUME FOR IT, ANY OTHER ADDITIONAL RESPONSIBILITY OR LIABILITY FOR THE USE OF, OR RELIANCE UPON, THIS INFORMATION.

HAZARD WARNING INFORMATION

HEALTH	FIRE	REACTIVE	OTHER	DEGREE OF HAZARD	COLOR CODING	OTHER CODES
3	2	0		0 = Minimum Hazard 1 = Slight Hazard 2 = Moderate Hazard 3 = Serious Hazard 4 = Severe Hazard	HEALTH = BLUE FIRE = RED REACTIVITY = YELLOW OTHER = WHITE	OX = Oxidizer ACID = Acid ALK = Alkali COR = Corrosive W = Use No Water

SECTION I - GENERAL INFORMATION

Characterization	RCRA Number	EPA Class
Phenol	**U052**	**Toxic Waste**

DOT Proper Shipping Name ***o*-Cresol**	Chemical Abstract Service (CAS) Number **95-48-7**
DOT Hazard Class and Label Requirements **Poison B**	DOT Emergency Guide Code **55**
DOT Identification Number **UN2076**	Chemical Formula $CH_3C_6H_4OH$

Synonyms
o-Cresylic acid; 3-methylphenol; 2-cresol; 1-hydroxy-2-methylbenzene; o-hydroxytoluene; ortho-cresol.

SECTION II - HAZARDOUS INGREDIENTS AND IDENTITY INFORMATION

Hazardous Components (specific identity)	OSHA Exposure Criteria	NIOSH Exposure Criteria	Immediately Dangerous to Life and Health (IDLH)	ACGIH Exposure Criteria
***o*-Cresol** (derivation: By fractional distillation of crude cresol from coal tar; interaction of methanol and phenol). **1 ppm = 4.50 mg/m³**	**PEL (skin):** **5 ppm** **22 mg/m³** **STEL:** **Not Established**	**REL:** **2.3 ppm** **10 mg/m³** **STEL:** **Not Established**	**250 ppm**	**TLV:** **5 ppm** **22 mg/m³** **STEL:** **Not Established**

SECTION III - PHYSICAL AND CHEMICAL CHARACTERISTICS

Boiling Point **376°F (191°C)**	Specific Gravity (H_2O = 1) **1.05**
Vapor Pressure (gas) **.29 at 77°F (25°C)**	Molecular Weight **108.2**
Vapor Density (Air = 1) **3.72**	Melting Point **88°F (31°C)**

Solubility
Soluble in alcohol, ether, chloroform, and hot water.

Appearance and Odor
White powder or crystalline solid with a sweet, tarry odor.

SECTION IV - FIRE AND EXPLOSION HAZARD DATA

Flash Point (method used) **178°F (81°C) closed cup**	Explosive Limits in Air % by Volume **LEL: 1.4% UEL: Not Found**
NFPA Classification **Combustible Solid; Class IIIA Combustible Liq.**	Autoignition Temperature **1110°F (598°C)**

Extinguishing Media
Use water to blanket fire, use dry chemical, foam, fog, or mist.

Special Fire Fighting Procedures
Wear full protective clothing and self-contained breathing apparatus (SCBA). Poisonous gases are produced in fire. Use water spray to keep fire-exposed containers cool.

Unusual Fire and Explosion Hazards
Dangerous fire and explosion hazard when exposed to heat or flame. Containers may explode in fire. Chemical will react on contact with oxidizable materials. Moderate explosion hazard when in the form of vapor. Vapors form explosive mixtures in air. The flammability range has not been established.

SECTION V - REACTIVITY DATA

Stability		Conditions to Avoid Keep away from heat or flame and incompatible materials.
Stable X	Unstable	Incompatibility *(materials to avoid)* Strong oxidizers, (such as chlorine, bromine, and fluorine), and many acids (such as chlorosulfonic and nitric), and oleum.
Hazardous Polymerization		Conditions to Avoid Under normal conditions of temperature and pressure, hazardous polymerization of *o*-cresol is not expected to occur.
May Occur	Will Not Occur X	Hazardous Decomposition or By-products When heated to decomposition, *o*-cresol can emit highly toxic/poisonous gases.

SECTION VI - HEALTH HAZARD DATA

Primary Route(s) of Entry:	Inhalation?	Absorption (skin)?	Ingestion?
	X	X	X

Health Hazards

INHALATION: Symptoms include weakness, confusion, depression of the central nervous system, dyspnea, anemia, and respiratory failure. Possible liver and kidney damage.

ABSORPTION: Severe skin irritant causing burns and dermatitis. There may be skin eruptions. Eye contact can cause damage to and/or loss of vision. Will pass through skin to cause toxic systemic effects, including liver damage.

INGESTION: Gastrointestinal disorders, nervous disorders, tremor, convulsions, oliguria, jaundice.

Carcinogenicity	NTP Listed?	IARC Cancer Review Group?	OSHA Regulated?	Target Organs?
Questioned Human Questioned Animal	No	No	No	Respiratory system, eyes, skin, CNS, liver, kidney.

Medical Conditions Generally Aggravated by Exposure

None reported.

Emergency and First-aid Procedures

Eye contact: Flush immediately with water for 15 minutes (minimum), seek medical attention. Skin contact: Remove all contaminated clothing. Immediately wash area with large amounts of soap and water. Seek medical assistance. For inhalation: Remove the person from exposure. Provide respiratory assistance and CPR, if required. Transfer to medical facility. If swallowed: If conscious, give large amounts of water or milk, do NOT induce vomiting. Seek medical attention immediately.

SECTION VII - PRECAUTIONS FOR SAFE HANDLING AND USE

Steps to be Taken in Case Material is Released or Spilled

Collect powdered materials in most convenient and safest manner possible and deposit in sealed drums. Do NOT dry sweep. Ventilate area of spill or leak. Restrict those not involved in cleanup from entering area. Remove all sources of ignition.

Preferred Waste Disposal Method

No Citation.

Precautions to be Taken in Handling and Storage

Store to avoid contact with incompatible materials in tightly closed containers in cool, well-ventilated area away from heat and flame. Will liquefy above 88°F (31°C).

Other Precautions and Warnings

Sources of ignition are prohibited where *o*-cresol is used, handled, or stored. Heat may cause solids to liquefy and containers to build up pressure and explode.

SECTION VIII - CONTROL MEASURES AND PERSONAL PROTECTIVE EQUIPMENT

Respiratory Protection *(specify type)*

For exposures above the PEL, use an MSHA/NIOSH-approved supplied-air respirator with full facepiece, hood, or helmet in continuous flow mode, or use a self-contained breathing apparatus (SCBA) operated in pressure demand or other positive pressure mode.

Ventilation

Local exhaust or general mechanical systems recommended.

Protective Gloves	Eye Protection	Other Protective Clothing
Impervious Rubber	Chemical/Dust Goggles or Face Mask	Rubber Apron

Work/Hygiene Practices

Always wash hands thoroughly after using chemical; never bring food, drink, or smoking materials into vicinity of chemicals.

o-CRESOL

$CH_3C_6H_4OH$ CAS: 95-48-7

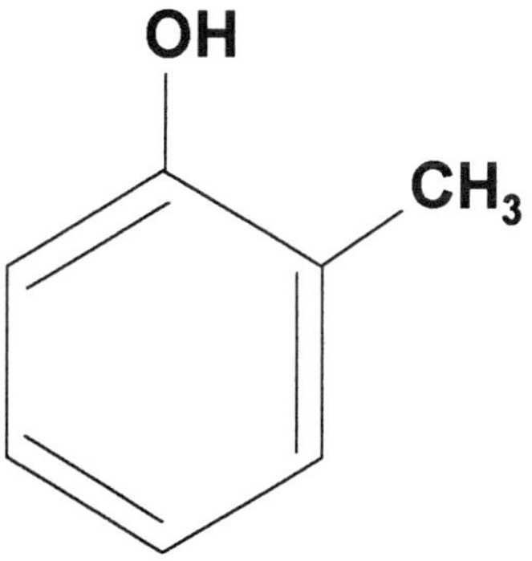

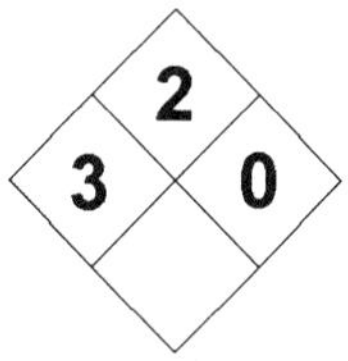

IDENTIFICATION AND TYPICAL USES

o-Cresol is a white powder or crystalline solid with a sweet, tarry odor. It will appear as a liquid at temperatures above 88°F (31°C). It is one of three isomers of the chemical cresol (cresylic acid). The other two, *m*-cresol and *p*-cresol, are similar in structure and hazard level. The *o*-cresol isomer is used in making synthetic resins, photographic developers, explosives, and as a disinfectant and a fumigant.

RISK ASSESSMENT: HEALTH

General Assessment

o-Cresol is a human poison by *inhalation, ingestion,* and skin *absorption.* It is a severe irritant to the eyes, skin and respiratory system. Human mutation data have been reported for *o*-cresol. It is a questionable carcinogen in humans and animals, although the data are inconclusive in this areas of study. The toxic actions of the cresol isomers are similar to that of phenol (the *"p"* or *para*-isomer is somewhat more toxic than the other two).

Inhalation can cause irritation to the respiratory system and associated mucosa. It can cause cough, congestion, with narcotic effects associated with the depression of the central nervous system (CNS). These include weakness, dyspnea, nausea, personality changes (irritability, mood changes), collapse, coma, and death (from respiratory failure).

Skin contact can produce burns, possible skin eruptions, and dermatitis. The liquid will pass through unbroken skin to cause toxic systemic effects similar to the effects on the CNS caused by inhalation. If *o*-cresol should contact the eyes, there may be serious damage up to and including loss of vision.

Ingestion of *o*-cresol can lead to gastrointestinal disorders, nervous disorders (CNS action), tremor, confusion, skin eruptions, oliguria (a diminishing of urinary secretion over a 24-hour period), jaundice, and liver damage. High doses will depress the CNS and may cause a reduction in body weight.

Acute Health Effects

The following acute (short-term) health effects may occur immediately or shortly after exposure to *o*-cresol:

Skin: Severe irritation and possible burns. Will pass through unbroken skin to cause toxic systemic effects, including weakness, nausea, collapse, and death.

Eye: Severe irritation with damage and possible loss of vision likely.

Lung: Severe irritation to the respiratory tract and associated mucosa. Can cause weakness, nausea, confusion, collapse, coma, and death (respiratory failure).

CNS: A CNS depressant. Weakness, confusion, mood changes and other personality changes, nausea, collapse.

Other: Ingestion will cause irritation to the gastrointestinal system with nausea, vomiting, convulsions, dypsnea, oliguria, jaundice, tremor, confusion, and nervousness. It has also been found to be hepatotoxic (toxic to the liver) and nephrototoxic (toxic to the kidneys).

Chronic Health Effects

The following chronic (long-term) health effects may occur at some time after exposure to *o*-cresol and can last for months or even years:

Cancer Hazards: Some studies have suggested that exposure to *o*-cresol causes mutations (genetic changes) in humans. While there is no evidence of carcinogenicity of this chemical in humans or animals, it is still considered a *questioned* carcinogen primarily because of its mutagenic properties. It is suggested that mutagenic chemicals be regarded as carcinogens during use and handling.

Reproductive Hazard: According to information presented in the references, *o*-cresol has not been adequately tested for its ability to adversely affect reproduction in animals.

Other Chronic Effects: High or repeated lower skin or inhalation exposures can cause liver or kidney disorders. Repeated skin contact may cause sensitization, dermatitis, chronic rash, and associated irritation. Once an allergy develops, the slightest future exposures will cause a recurrence of the symptoms.

Recommended Risk-Reduction Measures

Personnel should avoid direct contact with *o*-cresol. If a less toxic chemical cannot be substituted for a hazardous substance, then *engineering controls* are the most effective method of reducing exposures. The best protection is to enclose operations and/or provide local exhaust ventilation at the site of chemical release. While not always operationally feasible, isolating operations can also reduce exposure. Using respiratory protection is less effective than the controls mentioned above, but is still advisable whenever working with or around *o*-cresol. For exposures above the PEL (5 ppm), use an MSHA/NIOSH approved full facepiece respirator equipped with an organic vapor cartridge. Increased protection is obtained from a supplied-air respirator with full facepiece operated in continuous flow mode, or a self-contained breathing apparatus (SCBA) with full facepiece and pressure demand. If a full facepiece is not available, then chemical goggles should be worn to protect the eyes. Whenever a chemical splash hazard exists, a face shield and a rubber apron should be worn. To prevent hand and skin exposures, impervious rubber gloves should be worn.

Administrative controls should also be in place to minimize the potential for human exposures. These may include written procedures or policies which specify the methods and techniques that will be practiced whenever personnel are to work with *o*-cresol.

All personnel should receive *training* on the use, hazards, protective measures, emergency actions, and other precautions per 29 CFR 1910.1200 (Hazard Communication), prior to the first assignment in an area where *o*-cresol is used or stored.

Before beginning employment with *o*-cresol and at regular intervals thereafter (e.g., annually), those persons with high or frequent exposures should be provided the following recommended medical tests:

☑ Kidney and liver function tests.

If symptoms develop or overexposure is suspected, the following tests are recommended:

☑ Consider chest X-ray.

☑ Evaluation by a qualified allergist, with careful evaluation of exposure history and special testing (may help diagnose allergy).

Any evaluation should include a careful history of past and present symptoms with an examination. Medical tests that simply look for existing damage are not a substitute for controlling exposures.

Other methods to reduce exposure include:

☑ Where possible, enclose operations and use local exhaust ventilation at the site of chemical release. If local exhaust ventilation or enclosure is not used, respiratory protection should be mandatory. Also, if possible, liquids should be transferred automatically from storage containers to process containers.

☑ Always ensure that proper protective clothing is worn when using chemical substances.

☑ Wash thoroughly immediately after exposure to *o*-cresol and at the end of the work shift or before eating, drinking, or smoking.

☑ Hazard warning information should be posted in the work area. In addition, as part of an on-going education and training program, all information on the health and safety hazards of *o*-cresol should be communicated to all potentially exposed workers.

☑ Eye wash stations should be provided in the immediate work area for emergency use. If there is a possibility of skin exposure to *o*-cresol, emergency shower facilities should also be provided.

☑ Personnel should never wear contaminated clothing home (family members can be exposed).

Clothing should be laundered by personnel who have been trained on the hazards of *o*-cresol.

☑ Specific engineering controls are recommended for this chemical by NIOSH. Refer to NIOSH Criteria Document: Occupational Exposure to Cresol (Publication Number 78-133).

RISK ASSESSMENT: ENVIRONMENT

General Assessment

The environment is at risk of exposure during transportation, storage, disposal, or destruction of *o*-cresol. In almost every scenario, the threat of environmental exposure is contingent upon the proper handling of the chemical substance. Accidental spills, large or small, can result in fire, explosion, and possible contamination of the surrounding environmental mediums (water, soil, and air).

o-Cresol is considered a combustible solid. However, its vapors can form explosive mixtures in air and it will liquefy at temperatures above 88°F (31°C). In liquid form, volatility is higher and more explosive vapors may be present. These characteristics require caution in handling, storage, transportation, and disposal. It is incompatible with strong oxidizers, oleum, and acids. Contact with either can cause fire or explosion. It is also a corrosive and, therefore, special consideration is required during any emergency situation involving a leak or spill of *o*-cresol. Should *o*-cresol ever come into contact with these incompatible substances during use, transportation, storage, or disposal, the formation of highly toxic and/or highly explosive commodities is extremely possible.

o-Cresol may enter the environment through industrial effluents or spills.

Acute Ecological Effects

Acute (short-term) toxic effects may include the death of animals, birds, or fish and death or low growth rate in plants. Acute effects are seen 2 to 4 days after animals or plants are exposed to *o*-cresol.

Cresol has moderate acute toxicity to aquatic life. Insufficient data are available on the short-term effects of *o*-cresol exposure to plants, birds or land animals.

Chronic Ecological Effects

Chronic (long-term) toxic effects may include shortened life span, reproductive problems, lower fertility, and changes in appearance or behavior in exposed animals. These effects can be seen long after first exposure(s) to toxic chemicals.

Cresol has moderate chronic toxicity to aquatic life. Insufficient data are available on the long-term effects of *o*-cresol to plants, birds, or land animals.

Water Solubility

Cresol is highly soluble in water. Concentrations of 1000 milligrams and more will mix with a liter of water.

Persistence in the Environment

Cresol is slightly persistent in water, with a half-life between 2 and 20 days. The half-life of a pollutant is the amount of time it takes for one half of the chemical to be degraded. About 56% of cresol will eventually end up in water; approximately 42.7% in air, and about 0.6% in terrestrial soil and aquatic sediments.

Bioaccumulation in Aquatic Organisms

Some substances increase in concentration, or *bioaccumulate,* in living organisms as they breathe contaminated air, drink contaminated water, or eat contaminated food. These chemicals can become concentrated in the tissues and internal organs of animals as well as humans.

The concentration of cresol found in fish tissue is expected to be somewhat higher than the average concentration of cresol in the water from which the fish was taken.

Recommended Risk-Reduction Measures

Proper training of all transporters will reduce the likelihood of a mishap or accident resulting in a leak or spill to the environment. The correct use of DOT labeling (per the MSDS) on all containers, trucks, and rail cars will enable emergency responders to react properly and quickly to any disaster thereby reducing the potential risk to the environment and to personnel.

Storage of *o*-cresol should be segregated from incompatible chemicals to minimize the risk of cross-contamination or contact. Buildings designated for storage should be equipped with appropriate fire protection systems (alarms, sprinklers, emergency lighting, portable extinguishers).

If a spill or leak to the environment has occurred, fire department, emergency response and/or hazardous materials spill personnel should be notified immediately. Cleanup should be attempted only by those

trained in proper spill containment procedures. Contaminated soils should be removed for incineration and replaced with clean soil. If *o*-cresol should contact the water table, aquifer, or navigable waterway, time is of the essence. It is highly soluble in the aquatic environment and total remediation may not be possible. The local and/or state emergency response authorities must be notified. A comprehensive emergency response or disaster preparedness/recovery plan should be in place prior to any operations involving the use, transportation, storage, or disposal of *o*-cresol. If *o*-cresol is spilled or leaked, the following specific steps are recommended:

- ☑ Restrict persons not wearing protective clothing from area of spill or leak until cleanup is complete and area can be opened for normal work.
- ☑ Remove all ignition sources, ventilate area of leak.
- ☑ If powders are spilled, do not dry sweep. Use a vacuum equipped with a HEPA filter instead. Absorb liquid spills in vermiculite, dry sand, earth, or similar material and deposit in sealed containers.
- ☑ It may be necessary to dispose of *o*-cresol as a hazardous waste. The responsible state agency or the regional office of the federal Environmental Protection Agency (EPA) should be contacted for specific recommendations.

RISK ASSESSMENT: BUSINESS

General Assessment

Accidents or mishaps involving *m*-cresol can present a significant threat to business operations. The loss or damage of equipment or facilities can significantly affect fiscal viability. Lawsuits that may result from personnel injury/death, public exposures, and/or environmental contamination will also require a serious expenditure of resources. Media attention surrounding an injury, death, or environmental damage can also result in a loss of profits and loss of current as well as future business.

Recommended Risk-Reduction Measures

Company attorneys, safety and health professionals, and environmental specialists should be involved in the development of any procedures dealing with safety and emergency response. A company official should be pre-designated as a public relations officer with specific training in dealing with the press. Corporate plans and policies should be developed, approved, and implemented long before any need for such arises.

REFERENCES

Asante-Duah, D. K. 1993. *Hazardous Waste Risk Assessment.* New York: CRC Press/Lewis Publishers.

Bueche, J. R. 1972. *Principles of Physics,* 2nd Edition. New York: McGraw-Hill.

Cockerman, L. G., B. S. Shane. 1994. *Basic Environmental Toxicology.* New York: CRC Press/Lewis Publishers.

Cothern, C. R. 1995. *Handbook for Environmental Risk Decisionmaking.* New York: CRC Press/Lewis Publishers.

Hallenbeck, W. H. 1993. *Quantitative Risk Assessment for Environmental and Occupational Health,* 2nd Edition. New York: CRC Press/Lewis Publishers.

Howard, P. H. 1995. *Handbook of Environmental Fate and Exposures Data for Organic Chemicals,* Vol. I-IV. New York: CRC Press/Lewis Publishers.

Lauwerys, R. R., P. Hoet. 1993. *Industrial Chemical Exposure.* New York: CRC Press/Lewis Publishers.

Lewis, R. J., Sr. 1992. *Sax's Dangerous Properties of Industrial Materials,* Volumes 1, 2, and 3, 8th Edition. New York: Van Nostrand Reinhold.

Meyer, E. 1990. *Chemistry of Hazardous Materials.* Englewood Cliffs, NJ: Prentice Hall.

Patnaik, P. 1992. *A Comprehensive Guide to the Hazardous Properties of Chemical Substances.* New York: Van Nostrand Reinhold.

Quigley, D. R. 1994. *Handbook of Emergency Chemical Management.* New York: CRC Press/Lewis Publishers.

Torkelson, T. R. V. K. Rowe. 1981. Halogenated aliphatic hydrocarbons. *Patty's Industrial Hygiene and Toxicology,* Vol. 2B, *Toxicology,* pp 3568-3572. New York: Wiley-Interscience.

U.S. Department of Health and Human Services, National Institute for Occupational Safety and Health. 1978. *NIOSH Criteria Document: Occupational Exposure to Cresol.* (Publication Number: 78-133). Washington, D.C.: U.S. Government Printing Office.

U.S. Department of Health and Human Services, National Institute for Occupational Safety and Health. 1994. *NIOSH Pocket Guide to Chemical Hazards.* Washington, D.C.: U.S. Government Printing Office.

MATERIAL SAFETY DATA SHEET

CHEMICAL NAME	CRC PRESS/LEWIS PUBLISHERS DISCLAIMER
p-CRESOL	**CRC PRESS/LEWIS PUBLISHERS DISCLAIMER:** THE INFORMATION AND RECOMMENDATIONS PRESENTED HEREIN ARE BASED ON SOURCES BELIEVED TO BE RELIABLE. CRC MAKES NO REPRESENTATION ON ITS COMPLETENESS OR ACCURACY. IT IS THE RESPONSIBILITY OF THE USER TO DETERMINE THE CHEMICAL'S SUITABILITY FOR ITS INTENDED USE, THE CHEMICAL'S SAFE USE, AND THE CHEMICAL'S PROPER DISPOSAL. NO REPRESENTATIONS AND/OR WARRANTIES, EITHER EXPRESSED OR IMPLIED, OF THE MERCHANTABILITY OR FITNESS FOR A PARTICULAR PURPOSE, OR OF ANY OTHER NATURE, ARE MADE WITH RESPECT TO THE INFORMATION PROVIDED IN THIS MSDS OR TO THE CHEMICAL TO WHICH INFORMATION MAY REFER. CRC NEITHER ASSUMES NOR AUTHORIZES ANY OTHER PERSON TO ASSUME FOR IT, ANY OTHER ADDITIONAL RESPONSIBILITY OR LIABILITY FOR THE USE OF, OR RELIANCE UPON, THIS INFORMATION.

HAZARD WARNING INFORMATION

HEALTH	FIRE	REACTIVE	OTHER	DEGREE OF HAZARD	COLOR CODING	OTHER CODES
3	2	0		0 = Minimum Hazard 1 = Slight Hazard 2 = Moderate Hazard 3 = Serious Hazard 4 = Severe Hazard	HEALTH = BLUE FIRE = RED REACTIVITY = YELLOW OTHER = WHITE	OX = Oxidizer ACID = Acid ALK = Alkali COR = Corrosive W = Use No Water

SECTION I - GENERAL INFORMATION

Characterization: **Phenol**	RCRA Number: **U052**	EPA Class: **Toxic Waste**
DOT Proper Shipping Name: ***p*-Cresol**	Chemical Abstract Service (CAS) Number: **106-44-5**	
DOT Hazard Class and Label Requirements: **Poison B**	DOT Emergency Guide Code: **55**	
DOT Identification Number: **UN2076**	Chemical Formula: **$CH_3C_6H_4OH$**	

Synonyms
***p*-cresylic acid; 4-methylphenol; 4-cresol, 1-hydroxy-4-methylbenzene; *p*-hydroxytoluene; para-cresol.**

SECTION II - HAZARDOUS INGREDIENTS AND IDENTITY INFORMATION

Hazardous Components (specific identity)	OSHA Exposure Criteria	NIOSH Exposure Criteria	Immediately Dangerous to Life and Health (IDLH)	ACGIH Exposure Criteria
***p*-Cresol** (derivation: By fractional distillation of crude cresol from coal tar; interaction of methanol and phenol). **1 ppm = 4.60 mg/m³**	**PEL (skin):** 5 ppm, 22 mg/m³ **STEL:** Not Established	**REL:** 2.3 ppm, 10 mg/m³ **STEL:** Not Established	**250 ppm**	**TLV:** 5 ppm, 22 mg/m³ **STEL:** Not Established

SECTION III - PHYSICAL AND CHEMICAL CHARACTERISTICS

Boiling Point: **396°F (202°C)**	Specific Gravity (H_2O = 1): **1.04**
Vapor Pressure (gas): **.11 at 77°F (25°C)**	Molecular Weight: **108.2**
Vapor Density (Air = 1): **3.72**	Melting Point: **95°F (35°C)**

Solubility
Soluble in alcohol, ether, chloroform, and hot water.

Appearance and Odor
Crystalline solid with a sweet, tarry odor.

SECTION IV - FIRE AND EXPLOSION HAZARD DATA

Flash Point (method used): **187°F (86°C) closed cup**	Explosive Limits in Air % by Volume: **LEL: 1.1% UEL: Not Found**
NFPA Classification: **Combustible Solid; Class IIIA Combustible Liq.**	Autoignition Temperature: **1138°F (558°C)**

Extinguishing Media
Use water to blanket fire, use dry chemical, foam, fog, or mist.

Special Fire Fighting Procedures
Wear full protective clothing and self-contained breathing apparatus (SCBA). Poisonous gases are produced in fire. Use water spray to keep fire-exposed containers cool.

Unusual Fire and Explosion Hazards
Dangerous fire and explosion hazard when exposed to heat or flame. Containers may explode in fire. Chemical will react on contact with oxidizable materials. Moderate explosion hazard when in the form of vapor. Vapors form explosive mixtures in air. The flammability range has not been established.

SECTION V - REACTIVITY DATA

Stability		Conditions to Avoid: Keep away from heat or flame and incompatible materials.
Stable	Unstable	Incompatibility *(materials to avoid)*: Strong oxidizers, (such as chlorine, bromine, and fluorine), and many acids (such as chlorosulfonic and nitric), and oleum.
X		
Hazardous Polymerization		Conditions to Avoid: Under normal conditions of temperature and pressure, hazardous polymerization of *p*-cresol is not expected to occur.
May Occur	Will Not Occur	Hazardous Decomposition or By-products: When heated to decomposition, *p*-cresol can emit highly toxic/poisonous gases.
	X	

SECTION VI - HEALTH HAZARD DATA

Primary Route(s) of Entry:	Inhalation?	Absorption (skin)?	Ingestion?
	X	X	X

Health Hazards

INHALATION: Symptoms include weakness, confusion, depression of the central nervous system, dyspnea, anemia, and respiratory failure. Possible liver and kidney damage.

ABSORPTION: Severe skin irritant causing burns and dermatitis. There may be skin eruptions. Eye contact can cause damage to and/or loss of vision. Will pass through skin to cause toxic systemic effects, including liver damage.

INGESTION: Gastrointestinal disorders, nervous disorders, tremor, convulsions, oliguria, jaundice.

Carcinogenicity	NTP Listed?	IARC Cancer Review Group?	OSHA Regulated?	Target Organs?
Questioned Human Questioned Animal	No	No	No	Respiratory system, eyes, skin, CNS, liver, kidney.

Medical Conditions Generally Aggravated by Exposure

None reported.

Emergency and First-aid Procedures

Eye contact: Flush immediately with water for 15 minutes (minimum), seek medical attention. Skin contact: Remove all contaminated clothing. Immediately wash area with large amounts of soap and water. Seek medical assistance. For inhalation: Remove the person from exposure. Provide respiratory assistance and CPR, if required. Transfer to medical facility. If swallowed: If conscious, give large amounts of water or milk, do NOT induce vomiting. Seek medical attention immediately.

SECTION VII - PRECAUTIONS FOR SAFE HANDLING AND USE

Steps to be Taken in Case Material is Released or Spilled

Collect powdered materials in most convenient and safest manner possible and deposit in sealed drums. Do NOT dry sweep. Ventilate area of spill or leak. Restrict those not involved in cleanup from entering area. Remove all sources of ignition.

Preferred Waste Disposal Method

No Citation.

Precautions to be Taken in Handling and Storage

Store to avoid contact with incompatible materials in tightly closed containers in cool, well-ventilated area away from heat and flame. Will liquefy above 95°F (35°C).

Other Precautions and Warnings

Sources of ignition are prohibited where *p*-cresol is used, handled, or stored. Heat may cause solids to liquefy and containers to build up pressure and explode.

SECTION VIII - CONTROL MEASURES AND PERSONAL PROTECTIVE EQUIPMENT

Respiratory Protection *(specify type)*

For exposures above the PEL, use an MSHA/NIOSH-approved supplied-air respirator with full facepiece, hood, or helmet in continuous flow mode, or use a self-contained breathing apparatus (SCBA) operated in pressure demand or other positive pressure mode.

Ventilation

Local exhaust or general mechanical systems recommended.

Protective Gloves	Eye Protection	Other Protective Clothing
Impervious Rubber	Chemical/Dust Goggles or Face Mask	Rubber Apron

Work/Hygiene Practices

Always wash hands thoroughly after using chemical; never bring food, drink, or smoking materials into vicinity of chemicals.

p-CRESOL

$CH_3C_6H_4OH$ CAS: 106-44-5

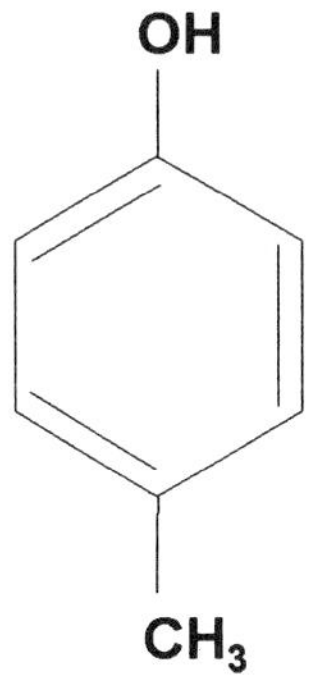

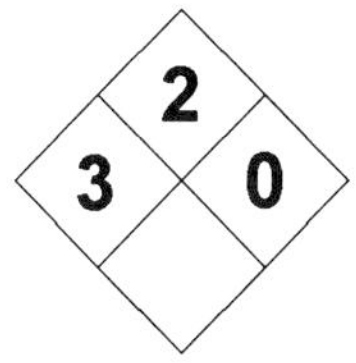

IDENTIFICATION AND TYPICAL USES

p-Cresol is a white powder or crystalline solid with a sweet, tarry odor. It will appear as a liquid at temperatures above 95°F (35°C). It is one of three isomers of the chemical cresol (cresylic acid). The other two, *m*-cresol and *o*-cresol, are similar in structure and hazard level. The *p*-cresol isomer is used in making synthetic resins, photographic developers, explosives, and as a disinfectant and a fumigant.

RISK ASSESSMENT: HEALTH

General Assessment

p-Cresol is a human poison by *inhalation, ingestion,* and skin *absorption.* It is a severe irritant to the eyes, skin and respiratory system. Human mutation data have been reported for *p*-cresol. It is a questionable carcinogen in humans and animals, although the data are inconclusive in this areas of study. The toxic actions of the cresol isomers are similar to that of phenol (the *"p"* or *para*-isomer is somewhat more toxic than the other two).

Inhalation can cause irritation to the respiratory system and associated mucosa. It can cause cough, congestion, with narcotic effects associated with the depression of the central nervous system (CNS). These include weakness, dyspnea, nausea, personality changes (irritability, mood changes), collapse, coma, and death (from respiratory failure).

Skin contact can produce burns, possible skin eruptions, and dermatitis. The liquid will pass through unbroken skin to cause toxic systemic effects similar to the effects on the CNS caused by inhalation. If *p*-cresol should contact the eyes, there may be serious damage up to and including loss of vision.

Ingestion of *p*-cresol can lead to gastrointestinal disorders, nervous disorders (CNS action), tremor, confusion, skin eruptions, oliguria (a diminishing of urinary secretion over a 24-hour period), jaundice, and liver damage. High doses will depress the CNS and may cause a reduction in body weight.

Acute Health Effects

The following acute (short-term) health effects may occur immediately or shortly after exposure to *p*-cresol:

Skin: Severe irritation and possible burns. Will pass through unbroken skin to cause toxic systemic effects, including weakness, nausea, collapse, and death.

Eye: Severe irritation with damage and possible loss of vision likely.

Lung: Severe irritation to the respiratory tract and associated mucosa. Can cause weakness, nausea, confusion, collapse, coma, and death (respiratory failure).

CNS: A CNS depressant. Weakness, confusion, mood changes and other personality changes, nausea, collapse.

Other: Ingestion will cause irritation to the gastrointestinal system with nausea, vomiting, convulsions, dypsnea, oliguria, jaundice, tremor, confusion, and nervousness. It has also been found to be hepatotoxic (toxic to the liver) and nephrototoxic (toxic to the kidneys).

Chronic Health Effects

The following chronic (long-term) health effects may occur at some time after exposure to *p*-cresol and can last for months or even years:

Cancer Hazards: Some studies have suggested that exposure to *p*-cresol causes mutations (genetic changes) in humans. While there is no evidence of carcinogenicity of this chemical in humans or animals, it is still considered a *questioned* carcinogen primarily because of its mutagenic properties. It is suggested that mutagenic chemicals be regarded as carcinogens during use and handling.

Reproductive Hazard: According to information presented in the references, *p*-cresol has not been adequately tested for its ability to adversely affect reproduction in animals.

Other Chronic Effects: High or repeated lower skin or inhalation exposures can cause liver or kidney disorders. Repeated skin contact may cause sensitization, dermatitis, chronic rash, and associated irritation. Once an allergy develops, the slightest future exposures will cause a recurrence of the symptoms.

Recommended Risk-Reduction Measures

Personnel should avoid direct contact with *p*-cresol. If a less toxic chemical cannot be substituted for a hazardous substance, then *engineering controls* are the most effective method of reducing exposures. The best protection is to enclose operations and/or provide local exhaust ventilation at the site of chemical release. While not always operationally feasible, isolating operations can also reduce exposure. Using respiratory protection is less effective than the controls mentioned above, but is still advisable whenever working with or around *p*-cresol. For exposures above the PEL (5 ppm), use an MSHA/NIOSH approved full facepiece respirator equipped with an organic vapor cartridge. Increased protection is obtained from a supplied-air respirator with full facepiece operated in continuous flow mode, or a self-contained breathing apparatus (SCBA) with full facepiece and pressure demand. If a full facepiece is not available, then chemical goggles should be worn to protect the eyes. Whenever a chemical splash hazard exists, a face shield and a rubber apron should be worn. To prevent hand and skin exposures, impervious rubber gloves should be worn.

Administrative controls should also be in place to minimize the potential for human exposures. These may include written procedures or policies which specify the methods and techniques that will be practiced whenever personnel are to work with *p*-cresol.

All personnel should receive *training* on the use, hazards, protective measures, emergency actions, and other precautions per 29 CFR 1910.1200 (Hazard Communication), prior to the first assignment in an area where *p*-cresol is used or stored.

Before beginning employment with *p*-cresol and at regular intervals thereafter (e.g., annually), those persons with high or frequent exposures should be provided the following recommended medical tests:

- ☑ Kidney and liver function tests.

If symptoms develop or overexposure is suspected, the following tests are recommended:

- ☑ Consider chest X-ray.
- ☑ Evaluation by a qualified allergist, with careful evaluation of exposure history and special testing (may help diagnose allergy).

Any evaluation should include a careful history of past and present symptoms with an examination. Medical tests that simply look for existing damage are not a substitute for controlling exposures.

Other methods to reduce exposure include:

- ☑ Where possible, enclose operations and use local exhaust ventilation at the site of chemical release. If local exhaust ventilation or enclosure is not used, respiratory protection should be mandatory. Also, if possible, liquids should be transferred automatically from storage containers to process containers.
- ☑ Always ensure that proper protective clothing is worn when using chemical substances.
- ☑ Wash thoroughly immediately after exposure to *p*-cresol and at the end of the work shift or before eating, drinking, or smoking.
- ☑ Hazard warning information should be posted in the work area. In addition, as part of an on-going education and training program, all information on the health and safety hazards of *p*-cresol should be communicated to all potentially exposed workers.
- ☑ Eye wash stations should be provided in the immediate work area for emergency use. If there is a possibility of skin exposure to *p*-cresol, emergency shower facilities should also be provided.
- ☑ Personnel should never wear contaminated clothing home (family members can be exposed).

Clothing should be laundered by personnel who have been trained on the hazards of *p*-cresol.

☑ Specific engineering controls are recommended for this chemical by NIOSH. Refer to NIOSH Criteria Document: Occupational Exposure to Cresol (Publication Number 78-133).

RISK ASSESSMENT: ENVIRONMENT

General Assessment

The environment is at risk of exposure during transportation, storage, disposal, or destruction of *p*-cresol. In almost every scenario, the threat of environmental exposure is contingent upon the proper handling of the chemical substance. Accidental spills, large or small, can result in fire, explosion, and possible contamination of the surrounding environmental mediums (water, soil, and air).

p-Cresol is considered a combustible solid. However, its vapors can form explosive mixtures in air and it will liquefy at temperatures above 95°F (35°C). In liquid form, volatility is higher and more explosive vapors may be present. These characteristics require caution in handling, storage, transportation, and disposal. It is incompatible with strong oxidizers, oleum, and acids. Contact with either can cause fire or explosion. It is also a corrosive and, therefore, special consideration is required during any emergency situation involving a leak or spill of *p*-cresol. Should *p*-cresol ever come into contact with these incompatible substances during use, transportation, storage, or disposal, the formation of highly toxic and/or highly explosive commodities is extremely possible.

p-Cresol may enter the environment through industrial effluents or spills.

Acute Ecological Effects

Acute (short-term) toxic effects may include the death of animals, birds, or fish and death or low growth rate in plants. Acute effects are seen 2 to 4 days after animals or plants are exposed to *p*-cresol.

Cresol has moderate acute toxicity to aquatic life. Insufficient data are available on the short-term effects of *p*-cresol exposure to plants, birds or land animals.

Chronic Ecological Effects

Chronic (long-term) toxic effects may include shortened life span, reproductive problems, lower fertility, and changes in appearance or behavior in exposed animals. These effects can be seen long after first exposure(s) to toxic chemicals.

Cresol has moderate chronic toxicity to aquatic life. Insufficient data are available on the long-term effects of *p*-cresol to plants, birds, or land animals.

Water Solubility

Cresol is highly soluble in water. Concentrations of 1000 milligrams and more will mix with a liter of water.

Persistence in the Environment

Cresol is slightly persistent in water, with a half-life between 2 and 20 days. The half-life of a pollutant is the amount of time it takes for one half of the chemical to be degraded. About 56% of cresol will eventually end up in water; approximately 42.7% in air, and about 0.6% in terrestrial soil and aquatic sediments.

Bioaccumulation in Aquatic Organisms

Some substances increase in concentration, or *bioaccumulate,* in living organisms as they breathe contaminated air, drink contaminated water, or eat contaminated food. These chemicals can become concentrated in the tissues and internal organs of animals as well as humans.

The concentration of cresol found in fish tissue is expected to be somewhat higher than the average concentration of cresol in the water from which the fish was taken.

Recommended Risk-Reduction Measures

Proper training of all transporters will reduce the likelihood of a mishap or accident resulting in a leak or spill to the environment. The correct use of DOT labeling (per the MSDS) on all containers, trucks, and rail cars will enable emergency responders to react properly and quickly to any disaster thereby reducing the potential risk to the environment and to personnel.

Storage of *p*-cresol should be segregated from incompatible chemicals to minimize the risk of cross-contamination or contact. Buildings designated for storage should be equipped with appropriate fire protection systems (alarms, sprinklers, emergency lighting, portable extinguishers).

If a spill or leak to the environment has occurred, fire department, emergency response and/or hazardous materials spill personnel should be notified immediately. Cleanup should be attempted only by those

trained in proper spill containment procedures. Contaminated soils should be removed for incineration and replaced with clean soil. If *p*-cresol should contact the water table, aquifer, or navigable waterway, time is of the essence. It is highly soluble in the aquatic environment and total remediation may not be possible. The local and/or state emergency response authorities must be notified. A comprehensive emergency response or disaster preparedness/recovery plan should be in place prior to any operations involving the use, transportation, storage, or disposal of *p*-cresol. If *p*-cresol is spilled or leaked, the following specific steps are recommended:

- ☑ Restrict persons not wearing protective clothing from area of spill or leak until cleanup is complete and area can be opened for normal work.
- ☑ Remove all ignition sources, ventilate area of leak.
- ☑ If powders are spilled, do not dry sweep. Use a vacuum equipped with a HEPA filter instead. Absorb liquid spills in vermiculite, dry sand, earth, or similar material and deposit in sealed containers.
- ☑ It may be necessary to dispose of *p*-cresol as a hazardous waste. The responsible state agency or the regional office of the federal Environmental Protection Agency (EPA) should be contacted for specific recommendations.

RISK ASSESSMENT: BUSINESS

General Assessment

Accidents or mishaps involving *p*-cresol can present a significant threat to business operations. The loss or damage of equipment or facilities can significantly affect fiscal viability. Lawsuits that may result from personnel injury/death, public exposures, and/or environmental contamination will also require a serious expenditure of resources. Media attention surrounding an injury, death, or environmental damage can also result in a loss of profits and loss of current as well as future business.

Recommended Risk-Reduction Measures

Company attorneys, safety and health professionals, and environmental specialists should be involved in the development of any such procedures. A company official should be pre-designated as a public relations officer with specific training in dealing with the press. Corporate plans and policies should be developed, approved, and implemented long before any need for such arises.

REFERENCES

Asante-Duah, D. K. 1993. *Hazardous Waste Risk Assessment.* New York: CRC Press/Lewis Publishers.

Bueche, J. R. 1972. *Principles of Physics,* 2nd Edition. New York: McGraw-Hill.

Cockerman, L. G., B. S. Shane. 1994. *Basic Environmental Toxicology.* New York: CRC Press/Lewis Publishers.

Cothern, C. R. 1995. *Handbook for Environmental Risk Decisionmaking.* New York: CRC Press/Lewis Publishers.

Hallenbeck, W. H. 1993. *Quantitative Risk Assessment for Environmental and Occupational Health,* 2nd Edition. New York: CRC Press/Lewis Publishers.

Howard, P. H. 1995. *Handbook of Environmental Fate and Exposures Data for Organic Chemicals,* Vol. I-IV. New York: CRC Press/Lewis Publishers.

Lauwerys, R. R., P. Hoet. 1993. *Industrial Chemical Exposure.* New York: CRC Press/Lewis Publishers.

Lewis, R. J., Sr. 1992. *Sax's Dangerous Properties of Industrial Materials,* Volumes 1, 2, and 3, 8th Edition. New York: Van Nostrand Reinhold.

Meyer, E. 1990. *Chemistry of Hazardous Materials.* Englewood Cliffs, NJ: Prentice Hall.

Patnaik, P. 1992. *A Comprehensive Guide to the Hazardous Properties of Chemical Substances.* New York: Van Nostrand Reinhold.

Quigley, D. R. 1994. *Handbook of Emergency Chemical Management.* New York: CRC Press/Lewis Publishers.

Torkelson, T. R. V. K. Rowe. 1981. Halogenated aliphatic hydrocarbons. *Patty's Industrial Hygiene and Toxicology,* Vol. 2B, *Toxicology,* pp 3568-3572. New York: Wiley-Interscience.

U.S. Department of Health and Human Services, National Institute for Occupational Safety and Health. 1978. *NIOSH Criteria Document: Occupational Exposure to Cresol.* (Publication Number: 78-133). Washington, D.C.: U.S. Government Printing Office.

U.S. Department of Health and Human Services, National Institute for Occupational Safety and Health. 1994. *NIOSH Pocket Guide to Chemical Hazards.* Washington, D.C.: U.S. Government Printing Office.

MATERIAL SAFETY DATA SHEET

CHEMICAL NAME	CRC PRESS/LEWIS PUBLISHERS DISCLAIMER:
CUMENE	THE INFORMATION AND RECOMMENDATIONS PRESENTED HEREIN ARE BASED ON SOURCES BELIEVED TO BE RELIABLE. CRC MAKES NO REPRESENTATION ON ITS COMPLETENESS OR ACCURACY. IT IS THE RESPONSIBILITY OF THE USER TO DETERMINE THE CHEMICAL'S SUITABILITY FOR ITS INTENDED USE, THE CHEMICAL'S SAFE USE, AND THE CHEMICAL'S PROPER DISPOSAL. NO REPRESENTATIONS AND/OR WARRANTIES, EITHER EXPRESSED OR IMPLIED, OF THE MERCHANTABILITY OR FITNESS FOR A PARTICULAR PURPOSE, OR OF ANY OTHER NATURE, ARE MADE WITH RESPECT TO THE INFORMATION PROVIDED IN THIS MSDS OR TO THE CHEMICAL TO WHICH INFORMATION MAY REFER. CRC NEITHER ASSUMES NOR AUTHORIZES ANY OTHER PERSON TO ASSUME FOR IT, ANY OTHER ADDITIONAL RESPONSIBILITY OR LIABILITY FOR THE USE OF, OR RELIANCE UPON, THIS INFORMATION.

HAZARD WARNING INFORMATION

HEALTH	FIRE	REACTIVE	OTHER	DEGREE OF HAZARD	COLOR CODING	OTHER CODES
2	3	1		0 = Minimum Hazard 1 = Slight Hazard 2 = Moderate Hazard 3 = Serious Hazard 4 = Severe Hazard	HEALTH = BLUE FIRE = RED REACTIVITY = YELLOW OTHER = WHITE	OX = Oxidizer ACID = Acid ALK = Alkali COR = Corrosive W = Use No Water

SECTION I - GENERAL INFORMATION

Characterization	RCRA Number	EPA Class
Solvent, Aromatic Hydrocarbon	**U055**	**Toxic Waste**

DOT Proper Shipping Name	Chemical Abstract Service (CAS) Number
Cumene	**98-82-8**
DOT Hazard Class and Label Requirements	**DOT Emergency Guide Code**
Flammable Liquid	**28**
DOT Identification Number	**Chemical Formula**
UN 1918	$C_2H_5CH(CH_3)_2$

Synonyms
Isopropylbenzene; 2-phenylpropane; (1-methylethyl)benzene; Cumol.

SECTION II - HAZARDOUS INGREDIENTS AND IDENTITY INFORMATION

Hazardous Components (specific identity)	OSHA Exposure Criteria	NIOSH Exposure Criteria	Immediately Dangerous to Life and Health (IDLH)	ACGIH Exposure Criteria
Cumene (derivation: Alkylation of benzene with propylene (phosphoric acid catalyst), distillation from coal tar naphtha fractions or from petroleum). **1 ppm = 5.00 mg/m³**	**PEL (skin):** **50 ppm** **250 mg/m³** **STEL:** **Not Established**	**REL (skin):** **50 ppm** **245 mg/m³** **STEL:** **Not Established**	**9000 ppm**	**TLV:** **50 ppm** **250 mg/m³** **STEL:** **Not Established**

SECTION III - PHYSICAL AND CHEMICAL CHARACTERISTICS

Boiling Point	Specific Gravity (H_2O = 1)
306°F (152°C)	**.09**
Vapor Pressure (gas)	Molecular Weight
8 at 69°F (20°C)	**120.2**
Vapor Density (Air = 1)	Melting Point
4.1	**-141°F (-96°C)**

Solubility
Insoluble in water. Soluble in alcohol, benzene, carbon tetrachloride, chlorinated solvents, and ether.

Appearance and Odor
Colorless liquid with a sharp, penetrating, aromatic odor. Odor threshold = 0.088 ppm.

SECTION IV - FIRE AND EXPLOSION HAZARD DATA

Flash Point (method used)	Explosive Limits in Air % by Volume
111°F (44°C)	**LEL: 0.9% UEL: 6.5%**
NFPA Classification	Autoignition Temperature
Class 1C Flammable Liquid	**795°F (424°C)**

Extinguishing Media
Use dry chemical, carbon dioxide, water spray, or alcohol foam.

Special Fire Fighting Procedures
Cumene is a flammable liquid. Wear full protective clothing and self-contained breathing apparatus (SCBA). Poisonous gases are produced in fire. Move containers from fire if it can be done without risk.

Unusual Fire and Explosion Hazards
Dangerous fire and explosion hazard when exposed to heat or flame. Containers may explode in fire. Immediately withdraw if rising sound from venting device is heard or if fire is causing a discoloration of tank. Vapors are heavier than air and may flash-back to ignition source to cause fire or explosion.

SECTION V - REACTIVITY DATA

Stability		Conditions to Avoid: Cumene is normally stable but can become unstable over a long period in aqueous solutions above 140°F). It may form explosive peroxides on prolonged contact with air. Keep away from incompatible materials.
Stable: X	Unstable:	Incompatibility *(materials to avoid)*: Strong oxidizers (such as permanganates, nitrates, peroxides, chlorates, and perchlorates), heat, strong acids (such as nitric, hydrochloric, sulfuric, chlorosulfonic), and oleum.
Hazardous Polymerization		Conditions to Avoid: Under normal conditions of temperature and pressure, hazardous polymerization of cumene is not expected to occur. However, it can react with oxidizers to form cumene hydroperoxide, which is explosive.
May Occur:	Will Not Occur: X	Hazardous Decomposition or By-products: When heated to decomposition, cumene can emit highly toxic/poisonous gases.

SECTION VI - HEALTH HAZARD DATA

Primary Route(s) of Entry:	Inhalation?	Absorption (skin)?	Ingestion?
	X	X	

Health Hazards

INHALATION: At high concentrations, causes headaches, dizziness, loss of coordination, and loss of consciousness. May cause lung damage. Central nervous system depression is also likely to occur. May also lead to liver and kidney damage. A potent narcotic.

ABSORPTION: Cumene is a skin irritant and may be absorbed slowly through unbroken skin to cause toxic systemic effects. Eye contact can cause irritation and conjunctivitis.

Carcinogenicity	NTP Listed?	IARC Cancer Review Group?	OSHA Regulated?	Target Organs?
Unknown Human Unknown Animal	No	No	No	Respiratory system, eyes, skin, CNS, liver, kidney.

Medical Conditions Generally Aggravated by Exposure

None reported.

Emergency and First-aid Procedures

Eye contact: Flush immediately with water for 15 minutes (minimum), lifting and holding eyelids open. Seek medical attention. Skin contact: Remove all contaminated clothing. Immediately wash area with large amounts of soap and water for 15 minutes (minimum). Seek medical assistance. For inhalation: Remove the person from exposure. Provide respiratory assistance and CPR. Transfer to medical facility. Keep victim warm and at rest. If swallowed: Seek medical attention immediately.

SECTION VII - PRECAUTIONS FOR SAFE HANDLING AND USE

Steps to be Taken in Case Material is Released or Spilled

Absorb liquids in vermiculite, dry sand, earth, or similar material and deposit in sealed containers. Ventilate area of spill or leak. Restrict those not involved in cleanup from entering area. Remove all sources of ignition.

Preferred Waste Disposal Method

No Citation.

Precautions to be Taken in Handling and Storage

Cumene is a flammable liquid. Store to avoid contact with incompatible materials since violent reactions can occur. Store in tightly closed containers in cool, well-ventilated area away from heat and flame. A nitrogen purge inside containers is recommended since cumene can react with air to form explosive cumene hydroperoxide.

Other Precautions and Warnings

Sources of ignition are prohibited where cumene is used, handled, or stored. Heat may cause containers to build up pressure and explode.

SECTION VIII - CONTROL MEASURES AND PERSONAL PROTECTIVE EQUIPMENT

Respiratory Protection *(specify type)*

For exposures over the PEL, use an organic vapor respirator. At higher exposures, use a supplied-air respirator with full facepiece, hood, or helmet in continuous flow mode, or use a self-contained breathing apparatus (SCBA) operated in pressure demand or other positive pressure mode.

Ventilation

Local exhaust or general mechanical systems recommended.

Protective Gloves	Eye Protection	Other Protective Clothing
Solvent-resistant Rubber	Chemical Goggles or Face Mask	Rubber Apron

Work/Hygiene Practices

Always wash hands thoroughly after using chemical; never bring food, drink, or smoking materials into vicinity of chemicals.

CUMENE

$C_6H_5CH(CH_3)_2$ CAS: 98-82-2

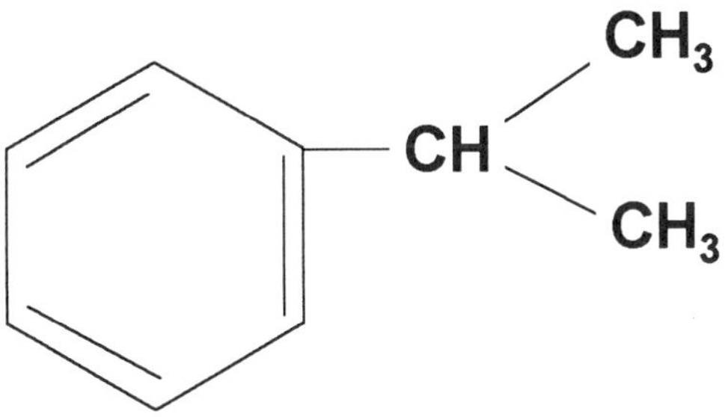

IDENTIFICATION AND TYPICAL USES

Cumene is a clear, colorless liquid with a sharp, penetrating, aromatic odor. It is used in the manufacture of phenol, acetone, styrene, and acetophenone. It is a high-octane blending component in some aviation fuels and is used as a thinner for cellulose paints. It is a constituent of many commercial petroleum solvents and a good solvent to cut fats and resins. It is also used in the perfume industry.

RISK ASSESSMENT: HEALTH

General Assessment

Cumene is a an irritant to the respiratory tract, eyes, and skin. On exposure, it has a tendency to accumulate in fatty tissues, the endocrine glands, spleen, and liver. However, it is metabolized without producing phenol metabolites and has not been shown to cause damage to bone marrow. Cumene can enter the body primarily through *inhalation* of its vapors and/or by slow *absorption* through unbroken skin.

Inhalation at high concentrations causes headaches, dizziness, loss of coordination, and loss of consciousness. It has caused lung damage in test animals. Vapor inhalation will also cause central nervous system (CNS) depression and there may be kidney and liver damage potential. Cumene is a potent narcotic and antipsychotic, characterized by slow induction into the system and long duration (relative to benzene and toluene). There may be unspecified changes in the sense of smell and in the respiratory system itself.

On skin contact, cumene will cause irritation and possibly rash. If allowed to remain in contact, slow absorption will occur to produce toxic systemic effects. Eye contact will also be irritating and conjunctivitis may occur.

The toxicity of ingestion is reported to be relatively low, with symptoms of nausea, mild narcosis, and gastritis. A major portion of ingested (and absorbed) cumene is metabolized by the liver and excreted. The urinary metabolites appear as conjugated alcohols and acids.

☠ *Acute Health Effects*

The following acute (short-term) health effects may occur immediately or shortly after exposure to cumene:

<u>Skin</u>: Immediate irritation, rash, and burning feeling on contact.

<u>Eye</u>: Irritation, conjunctivitis, and possible damage if allowed to remain in contact. May cause burns.

<u>Lung</u>: Irritation of the respiratory tract, including the nose and mouth. High level exposures may cause CNS effects, lung, liver, and kidney damage. There may be depression of the CNS, dizziness, lightheadedness, and loss of consciousness.

<u>CNS</u>: A primary CNS depressant. May cause headache, dizziness, loss of consciousness, nausea, irritability, changes in personality, and nervousness.

Chronic Health Effects

The following chronic (long-term) health effects may occur at some time after exposure to cumene and can last for months or even years:

Cancer Hazards: According to information presented in the references, cumene has not been adequately tested for its ability to cause cancer in test animals.

Reproductive Hazard: According to information presented in the references, cumene has not been adequately tested for its ability to adversely affect reproduction.

Other Chronic Effects: Long-term exposure can cause drying and cracking of the skin. Exposure may damage the lungs, liver, and kidneys. This chemical has not been adequately evaluated to determine whether brain or other nerve damage could occur with repeated exposure. However, many solvents and other petroleum-based chemicals have been shown to cause such damage. Effects may include reduced memory and concentration, personality changes (withdrawal, irritability), fatigue, sleep disturbances, reduced concentration and/or effects on the nerves supplying internal organs (autonomic nervous system) and/or nerves in the arms and legs (weakness, "pins and needles").

Recommended Risk-Reduction Measures

Personnel should avoid direct contact with cumene. If a less toxic chemical cannot be substituted for a hazardous substance, then *engineering controls* are the most effective method of reducing exposures. The best protection is to enclose operations and/or provide local exhaust ventilation at the site of chemical release. While not always operationally feasible, isolating operations can also reduce exposure. Using respiratory protection is less effective than the controls mentioned above, but is still advisable whenever working with or around cumene. For concentrations at or slightly above the PEL, use an organic vapor respirator. For higher or prolonged concentrations, a self-contained breathing apparatus (SCBA) with full facepiece and pressure demand is the recommended respiratory protection of choice. If a full facepiece is not available, then chemical goggles should be worn to protect the eyes. Whenever a chemical splash hazard exists, a face shield and a rubber apron should be worn. To prevent hand and skin exposures, acid resistant gloves should be worn.

Administrative controls should also be in place to minimize the potential for human exposures. These may include written procedures or policies which specify the methods and techniques that will be practiced whenever personnel are to work with cumene.

All personnel should receive *training* on the use, hazards, protective measures, emergency actions, and other precautions per 29 CFR 1910.1200 (Hazard Communication), prior to the first assignment in an area where cumene is used or stored.

If symptoms develop or overexposure is suspected, the following tests are recommended:

- ☑ Lung function tests.
- ☑ Interview for brain effects, including recent memory, mood, or concentration changes or problems. There may be headaches, malaise, and altered sleeping patterns as well. Consider cerebellar, autonomic, and peripheral nervous system evaluation. Positive or borderline individuals should be referred for neuropsychological testing.
- ☑ Urinalysis using gas chromatography to check for increased metabolites of cumene, primarily glucuronides of phenylpropanediol.
- ☑ Kidney and liver function tests.

Any evaluation should include a careful history of past and present symptoms with an examination. Medical tests that simply look for existing damage are not a substitute for controlling exposures.

Other methods to reduce exposure include:

- ☑ Where possible, enclose operations and use local exhaust ventilation at the site of chemical release. If local exhaust ventilation or enclosure is not used, respiratory protection should be mandatory. Also, automatic transfer of cumene from storage containers to process containers is highly desirable, if possible.
- ☑ Always ensure that proper protective clothing is worn when using chemical substances.
- ☑ Wash thoroughly immediately after exposure to cumene and at the end of the work shift or before eating, drinking, or smoking.
- ☑ Hazard warning information should be posted in the work area. In addition, as part of an on-going education and training program, all information on the health and safety hazards of cumene should be communicated to all potentially exposed workers.
- ☑ Eye wash stations should be provided in the immediate work area for emergency use. If there is a possibility of skin exposure to cumene, emergency shower facilities should also be provided.
- ☑ Personnel should never wear contaminated clothing home (family members can be exposed). Clothing should be laundered by personnel who have been trained on the hazards of cumene.

RISK ASSESSMENT: ENVIRONMENT

General Assessment

The environment is at risk of exposure during transportation, storage, disposal, or destruction of cumene. In almost every scenario, the threat of environmental exposure is contingent upon the proper handling of the chemical substance. Accidental spills, large or small, can result in fire, explosion, and possible contamination of the surrounding environmental mediums (water, soil, and air).

Cumene is considered a class 1C flammable liquid (per OSHA 29 CFR 1910.106). Its relatively low flash point and moderate boiling point make this chemical a moderate fire and explosion hazard concern. These characteristics require extreme caution in handling, storage, transportation, and disposal. It is incompatible with strong oxidizers, oleum, and acids and contact can cause violent fire or explosion. It reacts with oxidizers to form cumene hydroperoxide, which is extremely explosive. Vapors of cumene form explosive atmospheres above 88°F (31°C). Therefore, special consideration is required during any emergency situation involving a leak or spill of cumene. Should cumene ever come into contact with these incompatible substances during use, transportation, storage, or disposal, the formation of highly toxic and/or highly explosive commodities is extremely possible.

Cumene may enter the environment through industrial discharges or spills.

Acute Ecological Effects

Acute (short-term) toxic effects may include the death of animals, birds, or fish and death or low growth rate in plants. Acute effects are seen 2 to 4 days after animals or plants are exposed to cumene.

This chemical has moderate acute toxicity to aquatic life and high acute toxicity to birds. Insufficient data are available on the short-term effects of cumene exposure to plants or land animals.

Chronic Ecological Effects

Chronic (long-term) toxic effects may include shortened life span, reproductive problems, lower fertility, and changes in appearance or behavior in exposed animals. These effects can be seen long after first exposure(s) to toxic chemicals.

Cumene has moderate chronic toxicity in aquatic life. Insufficient data are available on the long-term effects of cumene to plants, birds, or land animals.

Water Solubility

Cumene is either very slightly soluble or totally insoluble in water (depending on concentration, water pH, and other variables). Concentrations of less than 100 milligrams may mix with a liter of water.

Persistence in the Environment

Cumene is non-persistent in water, with a half-life of less than 2 days. The half-life of a pollutant is the amount of time it takes for one half of the chemical to be degraded. About 99.6% of cumene will eventually end up in air; the remainder will end up in water.

Bioaccumulation in Aquatic Organisms

Some substances increase in concentration, or *bioaccumulate,* in living organisms as they breathe contaminated air, drink contaminated water, or eat contaminated food. These chemicals can become concentrated in the tissues and internal organs of animals as well as humans.

The concentration of cumene found in fish tissue is expected to be somewhat higher than the average concentration of cumene in the water from which the fish was taken.

Recommended Risk-Reduction Measures

Proper training of all transporters will reduce the likelihood of a mishap or accidents resulting in a leak or spill to the environment. The correct use of DOT labeling (per the MSDS) on all containers, trucks, and rail cars will enable emergency responders to react properly and quickly to any disaster thereby reducing the potential risk to the environment and to personnel.

Storage of cumene should be segregated from incompatible chemicals to minimize the risk of cross-contamination or contact. Buildings designated for storage should be equipped with appropriate fire protection systems (alarms, sprinklers, emergency lighting, portable extinguishers).

If a spill or leak to the environment has occurred, fire department, emergency response and/or hazardous materials spill personnel should be notified immediately. Cleanup should be attempted only by those trained in proper spill containment procedures. Contaminated soils should be removed for incineration and replaced with clean soil. If cumene should contact the water table, aquifer, or navigable waterway, the local and/or state emergency response authorities must be notified. A comprehensive emergency response or disaster preparedness/recovery plan should be in place prior to any operations involving the use, transportation, storage, or disposal of cumene. If cumene is spilled or leaked, the following specific steps are recommended:

- ☑ Restrict persons not wearing protective clothing from area of spill or leak until cleanup is complete and area can be opened for normal work.
- ☑ Remove all ignition sources and ventilate area of leak.
- ☑ Absorb liquids in vermiculite, dry sand, earth, or similar material and deposit in sealed containers.
- ☑ If gas is leaking, stop the flow of gas. If the source is a cylinder and leak cannot be stopped in place, remove cylinder to a safe place in the open air and repair or allow to empty.
- ☑ It may be necessary to dispose of cumene as a hazardous waste. The responsible state agency or the regional office of the federal Environmental Protection Agency (EPA) should be contacted for specific recommendations.

RISK ASSESSMENT: BUSINESS

General Assessment

Accidents or mishaps involving cumene can present a significant threat to business operations. The loss or damage of equipment or facilities can significantly affect fiscal viability. Lawsuits that may result from personnel injury/death, public exposures, and/or environmental contamination will also require a serious expenditure of resources. Media attention surrounding an injury, death, or environmental damage can also result in a loss of profits and loss of current as well as future business.

Recommended Risk-Reduction Measures

Company attorneys, safety and health professionals, and environmental specialists should be involved in the development of any procedures dealing with safety or emergency response. A company official should be pre-designated as a public relations officer with specific training in dealing with the press. Corporate plans and policies should be developed, approved, and implemented long before any need for such arises.

REFERENCES

American Conference of Governmental Industrial Hygienists. 1988. *Documentation of the Threshold Limit Values and Biological Exposure Indices,* 5th Edition (with updates). Cincinnati: ACGIH

Bueche, J. R. 1972. *Principles of Physics,* 2nd Edition. New York: McGraw-Hill.

Calabrese, E. J. 1994. *Biological Effects of Low Level Exposures.* New York: CRC Press/Lewis Publishers.

Cockerman, L. G., B. S. Shane. 1994. *Basic Environmental Toxicology.* New York: CRC Press/Lewis Publishers.

Gardner, R. W. 1994. *Chemical Intolerance.* New York: CRC Press/Lewis Publishers.

Hallenbeck, W. H. 1993. *Quantitative Risk Assessment for Environmental and Occupational Health,* 2nd Edition. New York: CRC Press/Lewis Publishers.

Jones, F. E. 1994. *Toxic Organic Vapors in the Workplace.* New York: CRC Press/Lewis Publishers.

Keith, L. H. and M. M. Walker. 1995. *Handbook of Air Toxics.* New York: CRC Press/Lewis Publishers.

Lauwerys, R. R., P. Hoet. 1993. *Industrial Chemical Exposure.* New York: CRC Press/Lewis Publishers.

Lewis, R. J., Sr. 1992. *Sax's Dangerous Properties of Industrial Materials,* Volumes 1, 2, and 3, 8th Edition. New York: Van Nostrand Reinhold.

Lide, D. R. 1995. *Handbook of Chemistry and Physics,* 76th Edition. New York: CRC Press/Lewis Publishers.

Manahan, S. E. 1994. *Environmental Chemistry,* 6th Edition. New York: CRC Press/Lewis Publishers.

Mellström, G. A., J. E. Wahlberg, and H. I. Maibach. 1994. *Protective Gloves for Occupational Use.* New York: CRC Press/Lewis Publishers.

Meyer, E. 1990. *Chemistry of Hazardous Materials.* Englewood Cliffs, NJ: Prentice Hall.

Patnaik, P. 1992. *A Comprehensive Guide to the Hazardous Properties of Chemical Substances.* New York: Van Nostrand Reinhold.

Quigley, D. R. 1994. *Handbook of Emergency Chemical Management.* New York: CRC Press/Lewis Publishers.

Rea, W. J. 1995. *Chemical Sensitivity,* Vol. I-III. New York: CRC Press/Lewis Publishers.

Richardson, M. 1995. *Dictionary of Substance and Their Effects,* Vol. I-VI. New York: CRC Press/Lewis Publishers.

Sacarello, H. L. A. 1994. *The Comprehensive Handbook of Hazardous Materials.* New York: CRC Press/Lewis Publishers.

Sheftel, V. O. 1995. *Handbook of Toxic Properties of Monomers and Additives.* New York: CRC Press/Lewis Publishers.

Torkelson, T. R. V. K. Rowe. 1981. Halogenated aliphatic hydrocarbons. *Patty's Industrial Hygiene and Toxicology,* Vol. 2B, *Toxicology,* pp 3568-3572. New York: Wiley-Interscience.

U.S. Department of Health and Human Services, National Institute for Occupational Safety and Health. 1994. *NIOSH Pocket Guide to Chemical Hazards.* Washington, D.C.: U.S. Government Printing Office.

MATERIAL SAFETY DATA SHEET

CHEMICAL NAME	CRC PRESS/LEWIS PUBLISHERS DISCLAIMER
CUMENE HYDROPEROXIDE	**CRC PRESS/LEWIS PUBLISHERS DISCLAIMER:** THE INFORMATION AND RECOMMENDATIONS PRESENTED HEREIN ARE BASED ON SOURCES BELIEVED TO BE RELIABLE. CRC MAKES NO REPRESENTATION ON ITS COMPLETENESS OR ACCURACY. IT IS THE RESPONSIBILITY OF THE USER TO DETERMINE THE CHEMICAL'S SUITABILITY FOR ITS INTENDED USE, THE CHEMICAL'S SAFE USE, AND THE CHEMICAL'S PROPER DISPOSAL. NO REPRESENTATIONS AND/OR WARRANTIES, EITHER EXPRESSED OR IMPLIED, OF THE MERCHANTABILITY OR FITNESS FOR A PARTICULAR PURPOSE, OR OF ANY OTHER NATURE, ARE MADE WITH RESPECT TO THE INFORMATION PROVIDED IN THIS MSDS OR TO THE CHEMICAL TO WHICH INFORMATION MAY REFER. CRC NEITHER ASSUMES NOR AUTHORIZES ANY OTHER PERSON TO ASSUME FOR IT, ANY OTHER ADDITIONAL RESPONSIBILITY OR LIABILITY FOR THE USE OF, OR RELIANCE UPON, THIS INFORMATION.

HAZARD WARNING INFORMATION

HEALTH	FIRE	REACTIVE	OTHER	DEGREE OF HAZARD	COLOR CODING	OTHER CODES
1	2	4	OX	0 = Minimum Hazard 1 = Slight Hazard 2 = Moderate Hazard 3 = Serious Hazard 4 = Severe Hazard	HEALTH = BLUE FIRE = RED REACTIVITY = YELLOW OTHER = WHITE	OX = Oxidizer ACID = Acid ALK = Alkali COR = Corrosive ~~W~~ = Use No Water

SECTION I - GENERAL INFORMATION

Characterization	RCRA Number	EPA Class
Peroxide	**None**	**Not Applicable**

DOT Proper Shipping Name **Cumene hydroperoxide**	Chemical Abstract Service (CAS) Number **80-15-9**
DOT Hazard Class and Label Requirements **Organic Peroxide**	DOT Emergency Guide Code **51**
DOT Identification Number **UN 2168**	Chemical Formula $C_6H_5C(CH_3)_2OOH$

Synonyms

α, α-dimethylbenzyl hydroperoxide; dimethylbenzene hydroperoxide; isopropylbenzene hydroperoxide; cumyl hydroperoxide; CHP.

SECTION II - HAZARDOUS INGREDIENTS AND IDENTITY INFORMATION

Hazardous Components (specific identity)	OSHA Exposure Criteria	NIOSH Exposure Criteria	Immediately Dangerous to Life and Health (IDLH)	ACGIH Exposure Criteria
Cumene hydroperoxide (derivation: A solution or emulsion of cumene is oxidized with air at approximately 130°C).	**PEL:** Not Established **STEL:** Not Established	**REL:** Not Established **STEL:** Not Established	**Not Determined**	**TLV:** Not Established **STEL:** Not Established

SECTION III - PHYSICAL AND CHEMICAL CHARACTERISTICS

Boiling Point **307°F (153°C)**	Specific Gravity (H_2O = 1) **1.1**
Vapor Pressure (gas) **<1 at 69°F (20°C)**	Molecular Weight **152.2**
Vapor Density (Air = 1) **5.2**	Melting Point **Not Determined**

Solubility

Slightly soluble in water. Soluble in alcohol, acetone, esters, hydrocarbons, chlorinated hydrocarbons.

Appearance and Odor

Colorless to pale yellow liquid.

SECTION IV - FIRE AND EXPLOSION HAZARD DATA

Flash Point (method used) **174°F (79°C)**	Explosive Limits in Air % by Volume **LEL: 0.9% UEL: 6.5%**
NFPA Classification **Class IIIB Combustible Liquid**	Autoignition Temperature **300°F (149°C)**

Extinguishing Media

Use dry chemical, carbon dioxide, water spray, or alcohol foam.

Special Fire Fighting Procedures

Wear full protective clothing and self-contained breathing apparatus (SCBA). Poisonous gases are produced in fire. Move containers from fire if it can be done without risk. Remain clear of smoke. Fight fires from a distance or protected location, if possible. Water by itself may be ineffective on fire.

Unusual Fire and Explosion Hazards

Dangerous fire and explosion hazard when exposed to heat or flame. Containers may explode in fire. May explosively decompose at elevated temperatures.

SECTION V - REACTIVITY DATA

Stability		Conditions to Avoid
Stable	Unstable	Cumene hydroperoxide is normally stable but is highly reactive with a large number of commodities and, therefore, it should be kept away from incompatible materials, heat and open flames.
X		**Incompatibility** *(materials to avoid)*: Strong oxidizers (chlorine, fluorine, and bromine), heat, strong acids (nitric, hydrochloric, sulfuric, chlorosulfonic), organic materials, reducing agents, copper, copper or lead alloys, cobalt, and mineral acids.

Hazardous Polymerization		Conditions to Avoid
May Occur	Will Not Occur	Under normal conditions of temperature and pressure, hazardous polymerization of cumene hydroperoxide is not expected to occur.
	X	**Hazardous Decomposition or By-products**: When heated to decomposition, explosion is possible. In fire, cumene hydroperoxide can emit highly toxic/poisonous gases, including phenol vapors.

SECTION VI - HEALTH HAZARD DATA

Primary Route(s) of Entry:	Inhalation?	Absorption (skin)?	Ingestion?
	X	X	

Health Hazards

INHALATION: Irritation of the nose, mouth, and lungs. Symptoms include nosebleeds, sore throat, hoarseness, cough with phlegm, increased saliva, and shortness of breath. Higher exposures can cause a buildup of fluid in the lungs (pulmonary edema) and death.

ABSORPTION: Cumene hydroperoxide is a skin irritant and may be absorbed through unbroken skin to cause toxic systemic effects. Eye contact can cause irritation and conjunctivitis.

Carcinogenicity	NTP Listed?	IARC Cancer Review Group?	OSHA Regulated?	Target Organs?
Unknown Human Questioned Animal	No	No	No	Respiratory system, eyes, skin, blood, liver, kidney.

Medical Conditions Generally Aggravated by Exposure

None reported.

Emergency and First-aid Procedures

Eye contact: Flush immediately with water for 15 minutes (minimum), lifting and holding eyelids open. Seek medical attention. **Skin contact:** Remove all contaminated clothing. Immediately wash area with large amounts of soap and water for 15 minutes (minimum). Seek medical assistance. For **inhalation:** Remove the person from exposure. Provide respiratory assistance and CPR. Transfer to medical facility. Observe for 24-48 hours for lung effects. If **swallowed:** Seek medical attention immediately.

SECTION VII - PRECAUTIONS FOR SAFE HANDLING AND USE

Steps to be Taken in Case Material is Released or Spilled

Absorb liquids in vermiculite, dry sand, earth, or similar material and deposit in sealed containers. Ventilate area of spill or leak. Restrict those not involved in cleanup from entering area. Remove all sources of ignition.

Preferred Waste Disposal Method

No Citation.

Precautions to be Taken in Handling and Storage

Store to avoid contact with incompatible materials since violent reactions can occur. Store in tightly closed containers in cool, well-ventilated area away from heat and flame. If possible, automatically transfer from storage to process containers. Use only non-sparking tools around this chemical.

Other Precautions and Warnings

Sources of ignition are prohibited where cumene hydroperoxide is used, handled, or stored. Heat may cause containers to build up pressure and explode.

SECTION VIII - CONTROL MEASURES AND PERSONAL PROTECTIVE EQUIPMENT

Respiratory Protection *(specify type)*

No exposure levels have been established. Recommend use of a supplied-air respirator with full facepiece, hood, or helmet in continuous flow mode, or use a self-contained breathing apparatus (SCBA) operated in pressure demand or other positive pressure mode.

Ventilation

Local exhaust or general mechanical systems recommended.

Protective Gloves	Eye Protection	Other Protective Clothing
Solvent-resistant Rubber	Chemical Goggles or Face Mask	Rubber Apron

Work/Hygiene Practices

Always wash hands thoroughly after using chemical; never bring food, drink, or smoking materials into vicinity of chemicals.

CUMENE HYDROPEROXIDE

$C_6H_5C(CH_3)_2OOH$ CAS: 80-15-9

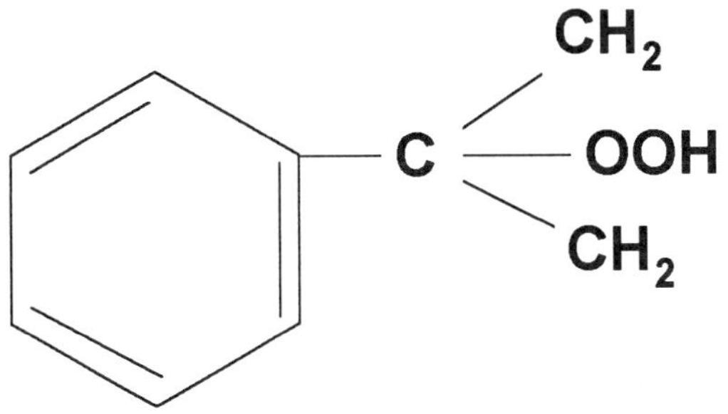

IDENTIFICATION AND TYPICAL USES

Cumene hydroperoxide is a colorless to pale yellow liquid. It is used in the production of acetone and phenol and as a polymerization catalyst, particularly in redox systems, to initiate rapid polymerization.

RISK ASSESSMENT: HEALTH

General Assessment

Cumene hydroperoxide is an irritant to the respiratory tract, eyes, and skin. Exposure may cause mutations in living cells and it should, therefore, be handled with extreme caution. Cumene hydroperoxide can enter the body primarily through *inhalation* of its vapors and/or by slow *absorption* through unbroken skin.

Inhalation can cause irritation of the nose, throat, and lungs. There may be headaches, dizziness, loss of coordination, and loss of consciousness. Other symptoms of exposure to vapors include nosebleeds, sore throat, hoarseness, cough and phlegm, increased saliva, and shortness of breath. Higher levels can cause fluid to buildup in the lungs (pulmonary edema), which is considered a medical emergency and can lead to death.

On skin contact, cumene hydroperoxide will cause irritation and possibly rash. If allowed to remain in contact, slow absorption will occur to produce toxic systemic effects. Eye contact will also be irritating and conjunctivitis may occur. Contact may also damage vision and can lead to permanent loss of some visual acuity.

Ingestion is not likely due to the sharp and irritating aromatic odor of cumene. The toxicity of ingestion is reported to be relatively low, with symptoms of nausea, mild narcosis, and gastritis. A major portion of ingested (and absorbed) cumene hydroperoxide is metabolized by the liver and excreted. The urinary metabolites appear as conjugated alcohols and acids.

☠ *Acute Health Effects*

The following acute (short-term) health effects may occur immediately or shortly after exposure to cumene hydroperoxide:

Skin: Immediate irritation, rash, and burning feeling on contact.

Eye: Irritation, conjunctivitis, and possible damage if allowed to remain in contact. May cause burns.

Lung: Irritation of the respiratory tract, including the nose and mouth. There may be nosebleeds, sore throat, hoarseness, cough with phlegm, salivation, and shortness of breath. Pulmonary edema possible at higher levels, which can be fatal. There may also be dizziness, lightheadedness, and loss of coordination and consciousness.

Other: High exposures *may* damage the bone marrow, causing reduced white blood cells (lower resistance to infection), blood platelets (bleeding tendency), and red blood cells (anemia). The studies on this effect, however, are conflicting and highly inconclusive.

Chronic Health Effects

The following chronic (long-term) health effects may occur at some time after exposure to cumene hydroperoxide and can last for months or even years:

Cancer Hazards: According to information presented in the references, cumene hydroperoxide may cause mutations (genetic changes) in living cells. Its carcinogenic properties are questioned in animals and unknown in humans and requires more testing.

Reproductive Hazard: According to information presented in the references, cumene hydroperoxide has not been adequately tested for its ability to adversely affect reproduction.

Other Chronic Effects: Repeated exposures may lead to lung damage (emphysema) and may damage the liver and kidneys. It can cause skin allergies with rash, itching, and redness. Once such an allergy develops, a person is "sensitized" to cumene hydroperoxide and any future small exposures can cause a recurrence of the symptoms.

Recommended Risk-Reduction Measures

Personnel should avoid direct contact with cumene hydroperoxide. If a less toxic chemical cannot be substituted for a hazardous substance, then *engineering controls* are the most effective method of reducing exposures. The best protection is to enclose operations and/or provide local exhaust ventilation at the site of chemical release. While not always operationally feasible, isolating operations can also reduce exposure. Using respiratory protection is less effective than the controls mentioned above, but is still advisable whenever working with or around cumene hydroperoxide. No exposure limits have been established for this chemical. However, this does not mean that it poses no risks to human health. For any concentration, a self-contained breathing apparatus (SCBA) with full facepiece and pressure demand is the recommended respiratory protection of choice. If a full facepiece is not available, then chemical goggles should be worn to protect the eyes. Whenever a chemical splash hazard exists, a face shield and a rubber apron should be worn. To prevent hand and skin exposures, protective gloves should be used. Glove manufacturers should be contacted and permeation studies obtained *before* final glove selections are made.

Administrative controls should also be in place to minimize the potential for human exposures. These may include written procedures or policies which specify the methods and techniques that will be practiced whenever personnel are to work with cumene hydroperoxide.

All personnel should receive *training* on the use, hazards, protective measures, emergency actions, and other precautions per 29 CFR 1910.1200 (Hazard Communication), prior to the first assignment in an area where cumene hydroperoxide is used or stored. Before beginning employment and at regular intervals thereafter (e.g., annually), or for those having frequent or potentially high exposures, the following medical tests are recommended:

☑ Lung function tests.

If symptoms develop or overexposure is suspected, the following tests are recommended:

☑ Lung function tests.

☑ Consider chest X-ray after acute overexposure (may be negative if taken immediately after exposure due to delayed onset of pulmonary edema).

☑ Evaluation by a qualified allergist, including careful exposure history and special testing (may help diagnose skin allergy).

☑ Kidney and liver function tests.

☑ Complete blood count (CBC) test.

Any evaluation should include a careful history of past and present symptoms with an examination. Medical tests that simply look for existing damage are not a substitute for controlling exposures. Also, because smoking can cause lung cancers, emphysema, heart disease, and a number of other respiratory problems, the effects of exposure to cumene hydroperoxide in smokers may appear to be significantly higher than that which is seen in non-smokers. This must be taken into consideration for proper risk assessment and management.

Other methods to reduce exposure include:

☑ Where possible, enclose operations and use local exhaust ventilation at the site of chemical release. If local exhaust ventilation or enclosure is not used, respiratory protection should be mandatory. Also, automatic transfer of cumene hydroperoxide from storage containers to process containers is highly desirable, if possible.

☑ Always ensure that proper protective clothing is worn when using chemical substances and that personnel have been trained on its use and care.

☑ Wash thoroughly immediately after exposure to cumene hydroperoxide and at the end of the work shift or before eating, drinking, or smoking.

☑ Hazard warning information should be posted in the work area. In addition, as part of an on-going education and training program, all information on

the health and safety hazards of cumene hydroperoxide should be communicated to all potentially exposed workers.

☑ Eye wash stations should be provided in the immediate work area for emergency use. If there is a possibility of skin exposure to cumene hydroperoxide, emergency shower facilities should also be provided.

☑ Personnel should never wear contaminated clothing home (family members can be exposed). Clothing should be laundered by personnel who have been trained on the hazards of cumene hydroperoxide.

RISK ASSESSMENT: ENVIRONMENT

General Assessment

The environment is at risk of exposure during transportation, storage, disposal, or destruction of cumene hydroperoxide. In almost every scenario, the threat of environmental exposure is contingent upon the proper handling of the chemical substance. Accidental spills, large or small, can result in fire, explosion, and possible contamination of the surrounding environmental mediums (water, soil, and air).

Cumene hydroperoxide is considered a class IIIB combustible liquid (per OSHA 29 CFR 1910.106). Its moderately low flash point presents a fire and explosion hazard concern. It is, however, highly reactive and can, on contact with incompatible materials, present a significant fire and explosion risk. These characteristics warrant extreme caution in handling, storage, transportation, and disposal of cumene hydroperoxide. It is incompatible with strong oxidizers, strong acids, organic materials, combustibles, reducing agents, copper, copper and lead alloys, cobalt, and mineral acids. Contact can cause violent fire or explosion. Therefore, special consideration is required during any emergency situation involving a leak or spill of cumene hydroperoxide. Should cumene hydroperoxide ever come into contact with these incompatible substances during use, transportation, storage, or disposal, the formation of highly toxic and/or highly explosive commodities is extremely possible. Cumene hydroperoxide may enter the environment through industrial discharges or spills.

Acute Ecological Effects

Acute (short-term) toxic effects may include the death of animals, birds, or fish and death or low growth rate in plants. Acute effects are seen 2 to 4 days after animals or plants are exposed to cumene hydroperoxide.

Cumene hydroperoxide has moderate acute toxicity to aquatic life. Insufficient data are available on the short-term effects of cumene hydroperoxide exposure to plants, birds, or land animals.

Chronic Ecological Effects

Chronic (long-term) toxic effects may include shortened life span, reproductive problems, lower fertility, and changes in appearance or behavior in exposed animals. These effects can be seen long after first exposure(s) to toxic chemicals.

Cumene hydroperoxide has moderate chronic toxicity in aquatic life. Insufficient data are available on the long-term effects of cumene hydroperoxide to plants, birds, or land animals.

Water Solubility

Insufficient data are available on the properties of cumene hydroperoxide to estimate or predict its water solubility.

Persistence in the Environment

Insufficient data are available on the properties of cumene hydroperoxide to estimate or predict is distribution in the environment. Limited data suggest that cumene hydroperoxide will be moderately persistent in water, with a half-life of between 2 and 20 days. The half-life of a pollutant is the amount of time it takes for one half of the chemical to be degraded.

Bioaccumulation in Aquatic Organisms

Some substances increase in concentration, or *bioaccumulate,* in living organisms as they breathe contaminated air, drink contaminated water, or eat contaminated food. These chemicals can become concentrated in the tissues and internal organs of animals as well as humans.

Insufficient data are available on the properties of cumene hydroperoxide to estimate or predict its bioaccumulation in aquatic organisms.

Recommended Risk-Reduction Measures

Proper training of all transporters will reduce the likelihood of a mishap or accidents resulting in a leak or spill to the environment. The correct use of DOT labeling (per the MSDS) on all containers, trucks, and rail cars will enable emergency responders to react properly and quickly to any disaster thereby reducing the potential risk to the environment and to personnel.

Storage of cumene hydroperoxide should be segregated from incompatible chemicals to minimize the risk of cross-contamination or contact. It is highly reactive, and such contact should be avoided. Buildings designated for storage should be equipped with appropriate fire protection systems (alarms, sprinklers, emergency lighting, portable extinguishers).

If a spill or leak to the environment has occurred, fire department, emergency response and/or hazardous materials spill personnel should be notified immediately. Cleanup should be attempted only by those trained in proper spill containment procedures. Contaminated soils should be removed for incineration and replaced with clean soil. If cumene hydroperoxide should contact the water table, aquifer, or navigable waterway, the local and/or state emergency response authorities must be notified. A comprehensive emergency response or disaster preparedness/recovery plan should be in place prior to any operations involving the use, transportation, storage, or disposal of cumene hydroperoxide. If cumene hydroperoxide is spilled or leaked, the following specific steps are recommended:

- ☑ Restrict persons not wearing protective clothing from area of spill or leak until cleanup is complete and area can be opened for normal work.
- ☑ Remove all ignition sources and ventilate area of leak.
- ☑ Absorb liquids in vermiculite, dry sand, earth, or similar material and deposit in sealed containers.
- ☑ It may be necessary to dispose of cumene hydroperoxide as a hazardous waste. The responsible state agency or the regional office of the federal Environmental Protection Agency (EPA) should be contacted for specific recommendations.

RISK ASSESSMENT: BUSINESS

General Assessment

Accidents or mishaps involving cumene hydroperoxide can present a significant threat to business operations. The loss or damage of equipment or facilities can significantly affect fiscal viability. Lawsuits that may result from personnel injury/death, public exposures, and/or environmental contamination will also require a serious expenditure of resources. Media attention surrounding an injury, death, or environmental damage can also result in a loss of profits and loss of current as well as future business.

Recommended Risk-Reduction Measures

Company attorneys, safety and health professionals, and environmental specialists should be involved in the development of any procedures dealing with safety or emergency response. A company official should be pre-designated as a public relations officer with specific training in dealing with the press. Corporate plans and policies should be developed, approved, and implemented long before any need for such arises.

REFERENCES

American Conference of Governmental Industrial Hygienists. 1988. *Documentation of the Threshold Limit Values and Biological Exposure Indices,* 5th Edition (with updates). Cincinnati: ACGIH

Bueche, J. R. 1972. *Principles of Physics,* 2nd Edition. New York: McGraw-Hill.

Calabrese, E. J. 1994. *Biological Effects of Low Level Exposures.* New York: CRC Press/Lewis Publishers.

Cockerman, L. G., B. S. Shane. 1994. *Basic Environmental Toxicology.* New York: CRC Press/Lewis Publishers.

Gardner, R. W. 1994. *Chemical Intolerance.* New York: CRC Press/Lewis Publishers.

Hallenbeck, W. H. 1993. *Quantitative Risk Assessment for Environmental and Occupational Health,* 2nd Edition. New York: CRC Press/Lewis Publishers.

Jones, F. E. 1994. *Toxic Organic Vapors in the Workplace.* New York: CRC Press/Lewis Publishers.

Keith, L. H. and M. M. Walker. 1995. *Handbook of Air Toxics.* New York: CRC Press/Lewis Publishers.

Lauwerys, R. R., P. Hoet. 1993. *Industrial Chemical Exposure.* New York: CRC Press/Lewis Publishers.

Meyer, E. 1990. *Chemistry of Hazardous Materials.* Englewood Cliffs, NJ: Prentice Hall.

Patnaik, P. 1992. *A Comprehensive Guide to the Hazardous Properties of Chemical Substances.* New York: Van Nostrand Reinhold.

Quigley, D. R. 1994. *Handbook of Emergency Chemical Management.* New York: CRC Press/Lewis Publishers.

Rea, W. J. 1995. *Chemical Sensitivity,* Vol. I-III. New York: CRC Press/Lewis Publishers.

Richardson, M. 1995. *Dictionary of Substance and Their Effects,* Vol. I-VI. New York: CRC Press/Lewis Publishers.

Sacarello, H. L. A. 1994. *The Comprehensive Handbook of Hazardous Materials.* New York: CRC Press/Lewis Publishers.

Torkelson, T. R. V. K. Rowe. 1981. Halogenated aliphatic hydrocarbons. *Patty's Industrial Hygiene and Toxicology,* Vol. 2B, *Toxicology,* pp 3568-3572. New York: Wiley-Interscience.

MATERIAL SAFETY DATA SHEET

CHEMICAL NAME	
CYANOGEN	

HAZARD WARNING INFORMATION

HEALTH	FIRE	REACTIVE	OTHER	DEGREE OF HAZARD	COLOR CODING	OTHER CODES
4	4	2		0 = Minimum Hazard 1 = Slight Hazard 2 = Moderate Hazard 3 = Serious Hazard 4 = Severe Hazard	HEALTH = BLUE FIRE = RED REACTIVITY = YELLOW OTHER = WHITE	OX = Oxidizer ACID = Acid ALK = Alkali COR = Corrosive W = Use No Water

SECTION I - GENERAL INFORMATION

Characterization	**Inorganic Cyanide**	RCRA Number	**P031**
		EPA Class	**Acute Hazardous Waste**
DOT Proper Shipping Name	**Cyanogen Gas**	Chemical Abstract Service (CAS) Number	**460-19-5**
DOT Hazard Class and Label Requirements	**Flammable Gas; Poisonous Gas**	DOT Emergency Guide Code	**18**
DOT Identification Number	**UN 1026**	Chemical Formula	**C_2N_2**

Synonyms
Carbon nitride; dicyan; dicyanogen; ethanedinitrile; oxalonitrile.

SECTION II - HAZARDOUS INGREDIENTS AND IDENTITY INFORMATION

Hazardous Components (specific identity)	OSHA Exposure Criteria	NIOSH Exposure Criteria	Immediately Dangerous to Life and Health (IDLH)	ACGIH Exposure Criteria
Cyanogen (derivation: Potassium cyanide solution is slowly dropped into copper sulfate solution; mercury cyanide is heated). **1 ppm = 2.16 mg/m³**	**PEL:** 10 ppm, 20 mg/m³ **STEL:** Not Established	**REL (ceiling):** 10 ppm, 20 mg/m³ **STEL:** Not Established	**Not Determined**	**TLV:** 10 ppm, 20 mg/m³ **STEL:** 3 ppm, 8.85 mg/m³

SECTION III - PHYSICAL AND CHEMICAL CHARACTERISTICS

Boiling Point	**-6°F (-21°C)**	Specific Gravity (H_2O = 1)	**1.82**
Vapor Pressure (gas)	**5.1 at 69°F (20°C)**	Molecular Weight	**52**
Vapor Density (Air = 1)	**1.8**	Melting Point	**-18°F (-30°C)**

Solubility
Only slightly soluble in water (1%). Soluble in alcohol and ether.

Appearance and Odor
Colorless gas with a pungent, almond-like odor. Odor Threshold = 250 ppm.

SECTION IV - FIRE AND EXPLOSION HAZARD DATA

Flash Point (method used)	**-16°F (-28°C) closed cup**	Explosive Limits in Air % by Volume	**LEL: 6.6% UEL: 32%**
NFPA Classification	**Flammable Gas**	Autoignition Temperature	**Not Applicable**

Extinguishing Media
Use a water spray to fight fire.

Special Fire Fighting Procedures
Cyanogen is an extremely dangerous fire hazard. Unburned vapors are highly toxic. Wear full protective clothing and self-contained breathing apparatus (SCBA). Use a water spray to keep fire-exposed containers cool. Stop flow of gas to stop fire.

Unusual Fire and Explosion Hazards
Dangerous fire and explosion hazard when exposed to heat or incompatible materials. Containers may explode in fire. Water provides heat to vaporize liquid cyanogen. Burns in a purple-tinged flame.

SECTION V - REACTIVITY DATA

Stability		Conditions to Avoid
		Reactive gas. Keep away from incompatible materials. Violent and explosive reactions and fires extremely likely on contact with oxidizers and acids. Keep away from heat (including sunlight).
Stable	Unstable	Incompatibility *(materials to avoid)*
	X	Acids, acid fumes, water, strong oxidizers (such as dichlorine oxide, fluorine, oxygen, ozone). Slowly reacts with water to form hydrogen cyanide, oxalic acid, or ammonia.
Hazardous Polymerization		Conditions to Avoid
		Under normal conditions of temperature and pressure, hazardous polymerization of cyanogen is not known to occur.
May Occur	Will Not Occur	Hazardous Decomposition or By-products
	X	When heated to decomposition, or in contact with acid, acid fumes, water, or steam, cyanogen can emit highly toxic/poisonous gases (including oxides of nitrogen and cyanide gas).

SECTION VI - HEALTH HAZARD DATA

Primary Route(s) of Entry:	Inhalation?	Absorption (skin)?	Ingestion?
	X	X	

Health Hazards

INHALATION: Symptoms of respiratory obstruction. Will cause irritation of the nose, throat, and upper respiratory system. Other symptoms include lachrymation, somnolence, red lips, headache, tachypnea, vertigo, convulsions, dizziness, confusion, loss of weight.

SKIN & EYES: Contact with the liquid can cause frostbite and eye damage. Gas is extremely irritating to the eyes and skin.

Carcinogenicity	NTP Listed?	IARC Cancer Review Group?	OSHA Regulated?	Target Organs?
Unknown Human Unknown Animal	No	No	No	Respiratory system, eyes, skin, CNS, CVS.

Medical Conditions Generally Aggravated by Exposure

Tuberculosis, asthma, chronic obstructive pulmonary disease, or heart disease may be aggravated.

Emergency and First-aid Procedures

Eye contact: Flush immediately with water for 15 minutes (minimum), seek medical attention. Skin contact: Remove all contaminated clothing. Immediately wash area with large amounts of soap and water. Seek medical assistance. For inhalation: Remove the person from exposure. Provide respiratory assistance and CPR. Transfer to medical facility. Antidote for hydrogen cyanide may be effective. Break an amyl nitrate pearl in a cloth and hold lightly under victim's nose for 15 seconds. Repeat five times at 15 second intervals. If swallowed: Seek medical attention immediately.

SECTION VII - PRECAUTIONS FOR SAFE HANDLING AND USE

Steps to be Taken in Case Material is Released or Spilled

Evacuate area and ventilate. Restrict those not involved in cleanup from entering area. Remove all sources of ignition. If gas leak cannot be stopped, remove leaking cylinders to safe area outdoors and allow to empty.

Preferred Waste Disposal Method

Dissolve in a combustible liquid; burn in a chemical incinerator with a scrubber and afterburner.

Precautions to be Taken in Handling and Storage

Cyanogen is a highly flammable gas. Store to avoid contact with incompatible materials since violent reactions can occur. Store in tightly closed containers in cool, well-ventilated area away from heat and sunlight. Ship in steel pressure cylinders and tank cars.

Other Precautions and Warnings

Sources of ignition are prohibited where cyanogen is used, handled, or stored. Heat may cause containers to build up pressure and explode.

SECTION VIII - CONTROL MEASURES AND PERSONAL PROTECTIVE EQUIPMENT

Respiratory Protection *(specify type)*

For any exposure level, use a supplied-air respirator with full facepiece, hood, or helmet set in continuous flow mode, or use a NISH/MSHA approved self-contained breathing apparatus (SCBA) operated in pressure demand or other positive pressure mode.

Ventilation

Local exhaust or general mechanical systems recommended.

Protective Gloves	Eye Protection	Other Protective Clothing
Insulated and Impervious	Chemical Goggles or Face Mask	Neoprene Rubber Apron

Work/Hygiene Practices

Always wash hands thoroughly after using chemical; never bring food, drink, or smoking materials into vicinity of chemicals.

CYANOGEN

C_2N_2 CAS: 460-19-5

IDENTIFICATION AND TYPICAL USES

Cyanogen is a colorless gas with an almond-like odor that becomes acrid and pungent at high concentrations. When it burns, it produces a purple-tinged flame. It is used as a fumigant, as a fuel gas for welding and cutting metals, as a propellant, and in organic synthesis. It occurs in blast furnaces.

RISK ASSESSMENT: HEALTH

General Assessment

Cyanogen is an extremely toxic and highly poisonous gas with symptoms of exposure similar to that of hydrogen cyanide. As a gas, the primary route of entry is through *inhalation.* Exposure can result in death by asphyxiation. It may also affect the body through skin contact with the liquid (frostbite is likely to occur). General symptoms of exposure may include vomiting, nausea, headache, confusion, and weakness. There is an apparent action to depress the central nervous system (CNS) and on the function of the cardiovascular system (CVS).

Inhalation and general exposure to the gas will produce irritation to the eyes, nose, throat, and upper respiratory system. Symptoms may include lachrymation, the appearance of cherry-red lips, tachypnea (rapid breathing), bradycardia (slow heart rate), headache, vertigo, convulsions, dizziness, loss of appetite, and low body weight.

Skin contact with the liquid will result in the destruction of tissue do to the extremely cold temperatures (frostbite). In gaseous form, it can still irritate the skin surface. Eye contact with the gas or liquid can be irritating and potentially destructive.

Acute Health Effects

The following acute (short-term) health effects may occur immediately or shortly after exposure to cyanogen:

Skin: Immediate irritation and possible burn on contact. The liquid will cause frostbite and possible permanent damage to skin tissue.

Eye: Severe irritation; conjunctivitis, and may cause burns and damage.

Lung: Irritation of the nose, throat, and lungs following exposure. May cause damage to the olfactory nerves. It can cause tearing, headache, nausea, and symptoms related to central nervous system (CNS) depression.

CNS: A depressant. May cause headache, dizziness, loss of appetite, weight loss, convulsions, vertigo, loss of consciousness.

CVS: Can affect the rate of heartbeat, slowing it to less than 60 beats per minute.

Chronic Health Effects

The following chronic (long-term) health effects may occur at some time after exposure to cyanogen and can last for months or even years:

Cancer Hazards: According to information presented in the references, cyanogen has not been adequately tested for its ability to cause cancer in test animals.

Reproductive Hazard: According to information presented in the references, cyanogen has not been adequately tested for its ability to adversely affect reproduction.

Other Chronic Effects: Cyanogen can irritate the lungs. Repeated exposure may cause bronchitis to develop with cough, phlegm, and/or shortness of breath.

Recommended Risk-Reduction Measures

Personnel should avoid direct contact with cyanogen. If a less toxic chemical cannot be substituted for a hazardous substance, then *engineering controls* are the most effective method of reducing exposures. The best protection is to enclose operations and/or provide local exhaust ventilation at the site of chemical release. While not always operationally feasible, isolating operations can also reduce exposure. Using respiratory protection is less effective than the controls mentioned above, but is still advisable whenever working with or around cyanogen. Because the odor threshold of cyanogen if 250 ppm, personnel may have no indications of lethal exposure. The PEL (8-hour time weighted average) is only 10 ppm. Therefore, the only recommended respiratory protection must be a self-contained breathing apparatus (SCBA) with full facepiece and pressure demand. If a full facepiece is not available, then chemical goggles should be worn to protect the eyes. Whenever a chemical splash hazard exists, a face shield and a rubber apron should be worn. To prevent hand and skin exposures, acid resistant gloves should be worn.

Administrative controls should also be in place to minimize the potential for human exposures. These may include written procedures or policies which specify the methods and techniques that will be practiced whenever personnel are to work with cyanogen.

All personnel should receive *training* on the use, hazards, protective measures, emergency actions, and other precautions per 29 CFR 1910.1200 (Hazard Communication), prior to the first assignment in an area where cyanogen is used or stored.

For those personnel with frequent or potentially high exposure (half the PEL or greater), the following are recommended medical tests to be provided before initial assignment and at regular intervals thereafter (e.g., annually):

☑ Lung function tests.

☑ Electrocardiogram (establish baseline).

If symptoms develop or overexposure is suspected, the following tests are recommended:

☑ Lung function tests.

☑ Electrocardiogram (compare to baseline).

Any evaluation should include a careful history of past and present symptoms with an examination. Medical tests that simply look for existing damage are not a substitute for controlling exposures. Also, because smoking can cause heart disease, as well as lung cancer, emphysema, and other respiratory problems, exposure to cyanogen may worsen pre-existing respiratory conditions.

Other methods to reduce exposure include:

☑ Where possible, enclose operations and use local exhaust ventilation at the site of chemical release. If local exhaust ventilation or enclosure is not used, respiratory protection should be mandatory.

☑ Always ensure that proper protective clothing is worn when using chemical substances.

☑ Wash thoroughly immediately after exposure to cyanogen and at the end of the work shift or before eating, drinking, or smoking.

☑ Hazard warning information should be posted in the work area. In addition, as part of an on-going education and training program, all information on the health and safety hazards of cyanogen should be communicated to all exposed and potentially exposed workers.

☑ Eye wash stations should be provided in the immediate work area for emergency use. If there is a possibility of skin exposure to cyanogen, emergency shower facilities should also be provided.

☑ Personnel should never wear contaminated clothing home (family members can be exposed). Clothing should be laundered by personnel who have been trained on the hazards of cyanogen.

RISK ASSESSMENT: ENVIRONMENT

General Assessment

The environment is at risk of exposure during transportation, storage, disposal, or destruction of cyanogen. In almost every scenario, the threat of environmental exposure is contingent upon the proper handling of the chemical substance. Accidental spills, large or small, can result in fire, explosion, and possible contamination of the surrounding environmental mediums (air, and to a lesser degree, water and soil).

Cyanogen is considered an extremely flammable gas. Its vapors are heavier than air and will travel, low to the ground, for a considerable distance to an ignition source and flashback. As a gas, it can easily form explosive mixtures in air. As a liquid, it can explode on contact when mixed with liquid oxygen. These characteristics require extreme caution in handling, storage, transportation, and disposal. It is incompatible with most acids, water, strong oxidizers (fluorine, for example) and contact can cause fire or explosion. It will slowly hydrolyze with water to form toxic hydrogen cyanide, oxalic acid, or ammonia. This can also occur if heated or mixed with acid. These characteristics require special consideration during any emergency situation involving a leak or spill of cyanogen. Should cyanogen ever come into contact with incompatible substances during use, transportation, storage, or disposal, the formation of highly toxic and/or highly explosive commodities is extremely possible.

Cyanogen can enter the environment through industrial discharges and through spills.

Acute Ecological Effects

Acute (short-term) toxic effects may include the death of animals, birds, or fish and death or low growth rate in plants. Acute effects are seen 2 to 4 days after animals or plants are exposed to cyanogen.

This chemical has high acute toxicity to aquatic life. No data are available on the short-term effects of cyanogen exposure to plants, birds, or land animals.

Chronic Ecological Effects

Chronic (long-term) toxic effects may include shortened life span, reproductive problems, lower fertility, and changes in appearance or behavior in exposed animals. These effects can be seen long after first exposure(s) to toxic chemicals.

Cyanogen has high chronic toxicity in aquatic life. No data are available on the long-term effects of cyanogen to plants, birds, or land animals.

Water Solubility

Cyanogen is highly soluble in the aquatic environment. Concentrations of 1000 milligrams and more will mix with a liter of water.

Persistence in the Environment

Cyanogen is non-persistent in water, with a half-life of less than 2 days. The half-life of a pollutant is the amount of time it takes for one half of the chemical to be degraded. No data are available on the amount of cyanogen that will remain in air or water. However, as a gas, nearly 100% of cyanogen might be expected to reach the air and virtually nothing will remain behind in water.

Bioaccumulation in Aquatic Organisms

Some substances increase in concentration, or *bioaccumulate,* in living organisms as they breathe contaminated air, drink contaminated water, or eat contaminated food. These chemicals can become concentrated in the tissues and internal organs of animals as well as humans.

Cyanogen is not expected to accumulate in the tissues of fish.

Recommended Risk-Reduction Measures

Proper training of all transporters will reduce the likelihood of a mishap or accident resulting in a leak or spill to the environment. The correct use of DOT labeling (per the MSDS) on all containers, trucks, and rail cars will enable emergency responders to react properly and quickly to any disaster thereby reducing the potential risk to the environment and to personnel.

Storage of cyanogen should be segregated from incompatible chemicals to minimize the risk of cross-contamination or contact. Buildings designated for storage should be equipped with appropriate fire protection systems (alarms, sprinklers, emergency lighting, portable extinguishers).

If a spill or leak to the environment has occurred, fire department, emergency response and/or hazardous materials spill personnel should be notified immediately. Cleanup should be attempted only by those trained in proper spill containment procedures. Contaminated soils should be removed for incineration and replaced with clean soil. If cyanogen should contact the water table, aquifer, or navigable waterway, time is of the essence. It is highly soluble in water and total remediation and containment may not be possible. The local and/or state emergency response authorities must be notified. A comprehensive emergency response or disaster preparedness/recovery plan should be in place prior to any operations involving the use, transportation, storage, or disposal of cyanogen. If cyanogen is spilled or leaked, the following specific steps are recommended:

- ☑ Restrict persons not wearing protective clothing from area of spill or leak until cleanup is complete and area can be opened for normal work.
- ☑ Remove all ignition sources.
- ☑ Ventilate area of leak.
- ☑ Absorb liquids in vermiculite, dry sand, earth, or similar material and deposit in sealed containers. Use non-sparking tools!
- ☑ If gas is leaking, stop the flow of gas. If the source is a cylinder and leak cannot be stopped in place, remove cylinder to a safe place in the open air and repair or allow to empty.

☑ It may be necessary to dispose of cyanogen as a hazardous waste. The responsible state agency or the regional office of the federal Environmental Protection Agency (EPA) should be contacted for specific recommendations.

RISK ASSESSMENT: BUSINESS

General Assessment

Accidents or mishaps involving cyanogen can present a significant threat to business operations. The loss or damage of equipment or facilities can significantly affect fiscal viability. Lawsuits that may result from personnel injury/death, public exposures, and/or environmental contamination will also require a serious expenditure of resources. Media attention surrounding an injury, death, or environmental damage can also result in a loss of profits and loss of current as well as future business.

Recommended Risk-Reduction Measures

Company attorneys, safety and health professionals, and environmental specialists should be involved in the development of any such procedures dealing with safety or emergency response. A company official should be pre-designated as a public relations officer with specific training in dealing with the press. Corporate plans and policies should be developed, approved, and implemented long before any need for such arises.

REFERENCES

American Conference of Governmental Industrial Hygienists. 1988. *Documentation of the Threshold Limit Values and Biological Exposure Indices,* 5th Edition (with updates). Cincinnati: ACGIH

Bueche, J. R. 1972. *Principles of Physics,* 2nd Edition. New York: McGraw-Hill.

Calabrese, E. J. 1994. *Biological Effects of Low Level Exposures.* New York: CRC Press/Lewis Publishers.

Cockerman, L. G., B. S. Shane. 1994. *Basic Environmental Toxicology.* New York: CRC Press/Lewis Publishers.

Gardner, R. W. 1994. *Chemical Intolerance.* New York: CRC Press/Lewis Publishers.

Hallenbeck, W. H. 1993. *Quantitative Risk Assessment for Environmental and Occupational Health,* 2nd Edition. New York: CRC Press/Lewis Publishers.

Jones, F. E. 1994. *Toxic Organic Vapors in the Workplace.* New York: CRC Press/Lewis Publishers.

Keith, L. H. and M. M. Walker. 1995. *Handbook of Air Toxics.* New York: CRC Press/Lewis Publishers.

Landis, W. G., M. H. Yu. 1995. *Introduction to Environmental Toxicology.* New York: CRC Press/Lewis Publishers.

Langerman, N. 1994 *Precautionary Labels for Chemical Containers.* New York: CRC Press/Lewis Publishers.

Lauwerys, R. R., P. Hoet. 1993. *Industrial Chemical Exposure.* New York: CRC Press/Lewis Publishers.

Lewis, R. J., Sr. 1992. *Sax's Dangerous Properties of Industrial Materials,* Volumes 1, 2, and 3, 8th Edition. New York: Van Nostrand Reinhold.

Lide, D. R. 1995. *Handbook of Chemistry and Physics,* 76th Edition. New York: CRC Press/Lewis Publishers.

Manahan, S. E. 1994. *Environmental Chemistry,* 6th Edition. New York: CRC Press/Lewis Publishers.

Mellström, G. A., J. E. Wahlberg, and H. I. Maibach. 1994. *Protective Gloves for Occupational Use.* New York: CRC Press/Lewis Publishers.

Meyer, E. 1990. *Chemistry of Hazardous Materials.* Englewood Cliffs, NJ: Prentice Hall.

National Cancer Institute. 1978. *Carcinogenesis Technical Report Series. Bioassay of* Cyanogen *for Possible Carcinogenicity.* DEHW (NIH) Publication Number: 78-1323, p 53. Washington, DC: U.S. Government Printing Office.

Patnaik, P. 1992. *A Comprehensive Guide to the Hazardous Properties of Chemical Substances.* New York: Van Nostrand Reinhold.

Proctor, N. H., J. P. Hughes. 1991. *Chemical Hazards of the Workplace.* New York: Van Nostrand Reinhold.

Quigley, D. R. 1994. *Handbook of Emergency Chemical Management.* New York: CRC Press/Lewis Publishers.

Rea, W. J. 1995. *Chemical Sensitivity,* Vol. I-III. New York: CRC Press/Lewis Publishers.

Richardson, M. 1995. *Dictionary of Substance and Their Effects,* Vol. I-VI. New York: CRC Press/Lewis Publishers.

Sacarello, H. L. A. 1994. *The Comprehensive Handbook of Hazardous Materials.* New York: CRC Press/Lewis Publishers.

Sheftel, V. O. 1995. *Handbook of Toxic Properties of Monomers and Additives.* New York: CRC Press/Lewis Publishers.

Torkelson, T. R. V. K. Rowe. 1981. Halogenated aliphatic hydrocarbons. *Patty's Industrial Hygiene and Toxicology,* Vol. 2B, *Toxicology,* pp 3568-3572. New York: Wiley-Interscience.

U.S. Department of Health, Education, and Welfare, National Institute for Occupational Safety and Health. 1976. *Criteria for a Recommended Standard: Occupational Exposure to* Cyanogen. DEHW (NIOSH) Publication Number: 76-176. Washington, D.C.: U.S. Government Printing Office.

U.S. Department of Health and Human Services, National Institute for Occupational Safety and Health. 1994. *NIOSH Pocket Guide to Chemical Hazards.* Washington, D.C.: U.S. Government Printing Office.

MATERIAL SAFETY DATA SHEET

CHEMICAL NAME	
CYANOGEN BROMIDE	**CRC PRESS/LEWIS PUBLISHERS DISCLAIMER:** THE INFORMATION AND RECOMMENDATIONS PRESENTED HEREIN ARE BASED ON SOURCES BELIEVED TO BE RELIABLE. CRC MAKES NO REPRESENTATION ON ITS COMPLETENESS OR ACCURACY. IT IS THE RESPONSIBILITY OF THE USER TO DETERMINE THE CHEMICAL'S SUITABILITY FOR ITS INTENDED USE, THE CHEMICAL'S SAFE USE, AND THE CHEMICAL'S PROPER DISPOSAL. NO REPRESENTATIONS AND/OR WARRANTIES, EITHER EXPRESSED OR IMPLIED, OF THE MERCHANTABILITY OR FITNESS FOR A PARTICULAR PURPOSE, OR OF ANY OTHER NATURE, ARE MADE WITH RESPECT TO THE INFORMATION PROVIDED IN THIS MSDS OR TO THE CHEMICAL TO WHICH INFORMATION MAY REFER. CRC NEITHER ASSUMES NOR AUTHORIZES ANY OTHER PERSON TO ASSUME FOR IT, ANY OTHER ADDITIONAL RESPONSIBILITY OR LIABILITY FOR THE USE OF, OR RELIANCE UPON, THIS INFORMATION.

HAZARD WARNING INFORMATION

HEALTH	FIRE	REACTIVE	OTHER	DEGREE OF HAZARD	COLOR CODING	OTHER CODES
3	0	2		0 = Minimum Hazard 1 = Slight Hazard 2 = Moderate Hazard 3 = Serious Hazard 4 = Severe Hazard	HEALTH = BLUE FIRE = RED REACTIVITY = YELLOW OTHER = WHITE	OX = Oxidizer ACID = Acid ALK = Alkali COR = Corrosive W = Use No Water

SECTION I - GENERAL INFORMATION

Characterization	**Inorganic Cyanide**	RCRA Number	**U246**
		EPA Class	**Toxic Waste**
DOT Proper Shipping Name	**Cyanogen bromide**	Chemical Abstract Service (CAS) Number	**506-68-3**
DOT Hazard Class and Label Requirements	**Poison B; Poison and Corrosive**	DOT Emergency Guide Code	**18**
DOT Identification Number	**UN 1889**	Chemical Formula	**CBrN**

Synonyms

Bromine cyanide; bromocyanide; bromocyan; bromocyanogen; cyanobromide; Campilit.

SECTION II - HAZARDOUS INGREDIENTS AND IDENTITY INFORMATION

Hazardous Components (specific identity)	OSHA Exposure Criteria	NIOSH Exposure Criteria	Immediately Dangerous to Life and Health (IDLH)	ACGIH Exposure Criteria
Cyanogen bromide (derivation: Action of bromine on potasssium cyanide; interaction of sodium bromide, sodium cyanide, sodium chlorate, and sulfuric acid).	**PEL:** Not Established **STEL:** Not Established	**REL:** Not Established **STEL:** Not Established	**Not Determined**	**TLV:** Not Established **STEL:** 3 ppm 8.85 mg/m^3

SECTION III - PHYSICAL AND CHEMICAL CHARACTERISTICS

Boiling Point	**143°F (61.4°C)**	Specific Gravity (H_2O = 1)	**2.015**
Vapor Pressure (gas)	**100 at 73°F (23°C)**	Molecular Weight	**105.93**
Vapor Density (Air = 1)	**3.6**	Melting Point	**126°F (52°C)**

Solubility

Soluble in water, alcohol, benzene, and ether (slowly decomposed by cold water).

Appearance and Odor

Needle-like crystalline solid with a penetrating odor.

SECTION IV - FIRE AND EXPLOSION HAZARD DATA

Flash Point (method used)	**Not Applicable**	Explosive Limits in Air % by Volume	**LEL: N/A UEL: N/A**
NFPA Classification	**Not Applicable**	Autoignition Temperature	**Not Applicable**

Extinguishing Media

Use water, dry chemical, or carbon dioxide.

Special Fire Fighting Procedures

In a fire or when heated it may emit extremely toxic cyanide fumes and vapors. Wear full protective clothing and self-contained breathing apparatus (SCBA). Use water spray to keep fire-exposed containers cool. Keep cooling until long after the fire has been extinguished. Evacuate non-essential personnel 1500 feet from fire area (ensure consideration of wind direction).

Unusual Fire and Explosion Hazards

None noted.

SECTION V - REACTIVITY DATA

Stability		Conditions to Avoid **May be unstable under normal conditions of temperature and pressure. Keep away from incompatible materials, to avoid violent reactions and possible fires/explosions.**
Stable	Unstable **X**	Incompatibility *(materials to avoid)* **Acids, acid fumes, water, strong oxidizers (such as dichlorine oxide, fluorine, oxygen, ozone). Slowly reacts with water to form hydrogen cyanide and oxides of bromine.**
Hazardous Polymerization		Conditions to Avoid **Under normal conditions of temperature and pressure, hazardous polymerization of cyanogen bromide is not known to occur. However, impure compounds decompose rapidly.**
May Occur	Will Not Occur **X**	Hazardous Decomposition or By-products **When heated to decomposition, or in contact with acid, acid fumes, water, or steam, cyanogen bromide can emit highly toxic/poisonous gases (including oxides of bromine and cyanide gas).**

SECTION VI - HEALTH HAZARD DATA

Primary Route(s) of Entry:	Inhalation?	Absorption (skin)?	Ingestion?
	X		

Health Hazards

INHALATION: **Irritation of the nose, throat, and upper respiratory system. Other symptoms include lachrymation, headache, tachypnea, vertigo, convulsions, dizziness, confusion, loss of weight. Pulmonary edema (fluid buildup in the lungs) is possible.**

SKIN & EYES: **Irritating to the skin and eyes on contact. May cause second and third degree burns.**

Carcinogenicity	NTP Listed?	IARC Cancer Review Group?	OSHA Regulated?	Target Organs?
Unknown Human Unknown Animal	**No**	**No**	**No**	**Respiratory system, eyes, skin, CNS, CVS.**

Medical Conditions Generally Aggravated by Exposure

None noted.

Emergency and First-aid Procedures

Eye contact: Flush immediately with water for 15 minutes (minimum), seek medical attention. Skin contact: Remove all contaminated clothing. Immediately wash area with large amounts of soap and water. For inhalation: Remove the person from exposure. Provide respiratory assistance and CPR. Antidote: Break amyl nitrate pearl in a cloth and hold under victim's nose for 15 seconds. Repeat five times at 15 second intervals. Transfer to medical facility. Observe for 24 - 48 hours for lung effects (pulmonary edema). If swallowed: If conscious, give large amounts of water or milk. Do NOT induce vomiting. Seek medical attention immediately. (Never try to give anything by mouth to an unconscious person).

SECTION VII - PRECAUTIONS FOR SAFE HANDLING AND USE

Steps to be Taken in Case Material is Released or Spilled

Evacuate area and ventilate. Restrict those not involved in cleanup from entering area. Remove all sources of ignition. Collect powdered materials in most convenient way possible and deposit in sealed drum. Do NOT dry sweep. Use a vacuum equipped with a HEPA filter.

Preferred Waste Disposal Method

Add slowly to an aqueous solution of caustic soda with an addition of excess laundry bleach. Nontoxic sodium thiocyanate and unreacted bleach are then washed down the drain with large amounts of water.

Precautions to be Taken in Handling and Storage

Store to avoid contact with incompatible materials since violent reactions can occur. Store in tightly closed containers in cool, well-ventilated area away from heat and moisture.

Other Precautions and Warnings

Sources of ignition are prohibited where cyanogen bromide is used, handled, or stored. Heat may cause emission of toxic cyanide and bromide fumes.

SECTION VIII - CONTROL MEASURES AND PERSONAL PROTECTIVE EQUIPMENT

Respiratory Protection *(specify type)*

For any exposure level, use a supplied-air respirator with full facepiece, hood, or helmet set in continuous flow mode, or use a NISH/MSHA approved self-contained breathing apparatus (SCBA) operated in pressure demand or other positive pressure mode.

Ventilation

Local exhaust or general mechanical systems recommended.

Protective Gloves	Eye Protection	Other Protective Clothing
Butyl or Neoprene Rubber	**Chemical Goggles or Face Mask**	**Neoprene Rubber Apron**

Work/Hygiene Practices

Always wash hands thoroughly after using chemical; never bring food, drink, or smoking materials into vicinity of chemicals.

CYANOGEN BROMIDE

$CBrN$ CAS: 506-68-3

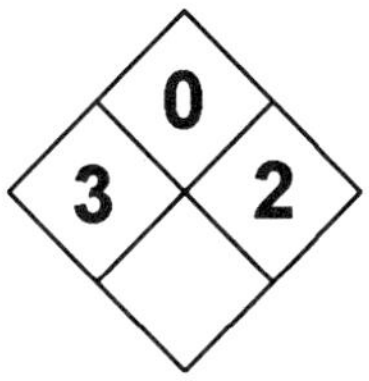

IDENTIFICATION AND TYPICAL USES

Cyanogen bromide appears as colorless, needle-like, crsytalline solid or powdered material with a pungent odor. It is used in organic synthesis, as a paraciticide, in fumigating compositions, in rodenticides, and as a cyaniding reagent in gold extraction processing.

RISK ASSESSMENT: HEALTH

General Assessment

Cyanogen bromide is an extremely toxic and highly poisonous gas with symptoms of exposure similar to that of hydrogen cyanide, although not as severe. Since it is readily volatile at ambient temperatures, the primary route of entry is through *inhalation.* It may also affect the body through skin contact, causing burns and tissue damage. Absorption, however, is not likely. General symptoms of exposure may include vomiting, nausea, headache, confusion, and weakness.

Inhalation will produce irritation to the eyes, nose, throat, and upper respiratory system. Symptoms may include irritation of the throat, palpitation, difficulty breathing, reddening of the eyes, salivation, nausea, headache, weakness in the arms and legs, giddiness, followed by collapse and convulsions.

Skin contact with the liquid will result in dermatitis at low concentrations and/or the destruction of tissue do to the extremely corrosive properties of cyanogen bromide. It may even cause second or third degree burns over a relatively short period of time. Eye contact can be irritating and potentially destructive, with permanent loss of vision likely.

Acute Health Effects

The following acute (short-term) health effects may occur immediately or shortly after exposure to cyanogen bromide:

Skin: Immediate irritation and possible serious burns on contact.

Eye: Severe irritation; conjunctivitis, and may cause burns and damage (blindness).

Lung: Irritation of the nose, throat, and lungs following exposure. Can instigate and lead to chronic lung disorders. May cause fluid to build up on the lungs (pulmonary edema) and death.

Other: A possible CNS depressant. May cause headache, dizziness, loss of appetite, weight loss, convulsions, vertigo, loss of consciousness. It may also affect the rate of heartbeat and breathing.

Chronic Health Effects

The following chronic (long-term) health effects may occur at some time after exposure to cyanogen bromide and can last for months or even years:

Cancer Hazards: According to information presented in the references, cyanogen bromide has not been adequately tested for its ability to cause cancer in test animals.

Reproductive Hazard: According to information presented in the references, cyanogen bromide has not been adequately tested for its ability to adversely affect reproduction.

Other Chronic Effects: Cyanogen bromide can irritate the lungs. Repeated exposure may cause bronchitis to develop with cough, phlegm, and/or shortness of breath. In weak solution, it may lead to dermatitis and rash.

Recommended Risk-Reduction Measures

Personnel should avoid direct contact with cyanogen bromide. If a less toxic chemical cannot be substituted for a hazardous substance, then *engineering controls* are the most effective method of reducing exposures. The best protection is to enclose operations and/or provide local exhaust ventilation at the site of chemical release. While not always operationally feasible, isolating operations can also reduce exposure. Using respiratory protection is less effective than the controls mentioned above, but is still advisable whenever working with or around cyanogen bromide. Because no exposure limits have been established for this chemical and, because exposure can be extremely hazardous to human health, prudent risk management requires a conservative approach to respiratory protection. Hence, the recommended respiratory protection is a self-contained breathing apparatus (SCBA) with full facepiece and pressure demand. If a full facepiece is not available, then chemical goggles should be worn to protect the eyes. Whenever a chemical splash hazard exists, a face shield and a rubber apron should be worn. To prevent hand and skin exposures, acid resistant gloves should be worn.

Administrative controls should also be in place to minimize the potential for human exposures. These may include written procedures or policies which specify the methods and techniques that will be practiced whenever personnel are to work with cyanogen bromide.

All personnel should receive *training* on the use, hazards, protective measures, emergency actions, and other precautions per 29 CFR 1910.1200 (Hazard Communication), prior to the first assignment in an area where cyanogen bromide is used or stored.

If symptoms develop or overexposure is suspected, the following tests are recommended:

☑ Lung function tests.

Any evaluation should include a careful history of past and present symptoms with an examination. Medical tests that simply look for existing damage are not a substitute for controlling exposures. Also, because smoking can cause heart disease, as well as lung cancer, emphysema, and other respiratory problems, exposure to cyanogen bromide may worsen pre-existing respiratory conditions.

Other methods to reduce exposure include:

☑ Where possible, enclose operations and use local exhaust ventilation at the site of chemical release. If local exhaust ventilation or enclosure is not used, respiratory protection should be mandatory.

☑ Always ensure that proper protective clothing is worn when using chemical substances.

☑ Wash thoroughly immediately after exposure to cyanogen bromide and at the end of the work shift or before eating, drinking, or smoking.

☑ Hazard warning information should be posted in the work area. In addition, as part of an on-going education and training program, all information on the health and safety hazards of cyanogen bromide should be communicated to all potentially exposed workers.

☑ Eye wash stations should be provided in the immediate work area for emergency use. If there is a possibility of skin exposure to cyanogen bromide, emergency shower facilities should also be provided.

☑ Personnel should never wear contaminated clothing home (family members can be exposed). Clothing should be laundered by personnel who have been trained on the hazards of cyanogen bromide.

RISK ASSESSMENT: ENVIRONMENT

General Assessment

The environment is at risk of exposure during transportation, storage, disposal, or destruction of cyanogen bromide. In almost every scenario, the threat of environmental exposure is contingent upon the proper handling of the chemical substance. Accidental spills, large or small, can result in fire, explosion, and possible contamination of the surrounding environmental mediums (air, and to a lesser degree, water and soil).

Cyanogen bromide is not a flammable material. However, it is known to be violently reactive with most acids and will decompose in water to release toxic gases. These characteristics require extreme caution in handling, storage, transportation, and disposal. It is also incompatible with strong oxidizers (fluorine, for example) and contact can cause fire or explosion. These characteristics require special consideration during any emergency situation involving a leak or spill of cyanogen bromide.

Cyanogen bromide can enter the environment through industrial discharges and through spills.

☠ *Acute Ecological Effects*

Acute (short-term) toxic effects may include the death of animals, birds, or fish and death or low growth rate in plants. Acute effects are seen 2 to 4 days after animals or plants are exposed to cyanogen bromide.

This chemical has high acute toxicity to aquatic life. No data are available on the short-term effects of cyanogen bromide exposure to plants, birds, or land animals.

Chronic Ecological Effects

Chronic (long-term) toxic effects may include shortened life span, reproductive problems, lower fertility, and changes in appearance or behavior in exposed animals. These effects can be seen long after first exposure(s) to toxic chemicals.

Cyanogen bromide has high chronic toxicity in aquatic life. No data are available on the long-term effects of cyanogen bromide to plants, birds, or land animals.

Water Solubility

Cyanogen bromide is highly soluble in the aquatic environment. Concentrations of 1000 milligrams and more will mix with a liter of water. It will decompose in water to release toxic gases.

Persistence in the Environment

Cyanogen bromide is non-persistent in water, with a half-life of less than 2 days. The half-life of a pollutant is the amount of time it takes for one half of the chemical to be degraded. No data are available on the amount of cyanogen bromide that will remain in air or water. However, as a gas, nearly 100% of cyanogen bromide might be expected to reach the air and virtually nothing will remain behind in water.

Bioaccumulation in Aquatic Organisms

Some substances increase in concentration, or *bioaccumulate,* in living organisms as they breathe contaminated air, drink contaminated water, or eat contaminated food. These chemicals can become concentrated in the tissues and internal organs of animals as well as humans.

Cyanogen bromide is not expected to accumulate in the tissues of fish.

Recommended Risk-Reduction Measures

Proper training of all transporters will reduce the likelihood of a mishap or accident resulting in a leak or spill to the environment. The correct use of DOT labeling (per the MSDS) on all containers, trucks, and rail cars will enable emergency responders to react properly and quickly to any disaster thereby reducing the potential risk to the environment and to personnel.

Storage of cyanogen bromide should be segregated from incompatible chemicals to minimize the risk of cross-contamination or contact. Buildings designated for storage should be equipped with appropriate fire protection systems (alarms, sprinklers, emergency lighting, portable extinguishers).

If a spill or leak to the environment has occurred, fire department, emergency response and/or hazardous materials spill personnel should be notified immediately. Cleanup should be attempted only by those trained in proper spill containment procedures. Contaminated soils should be removed for incineration and replaced with clean soil. If cyanogen bromide should contact the water table, aquifer, or navigable waterway, time is of the essence. It is highly soluble in water and will decompose. Total remediation and containment may not be possible. The local and/or state emergency response authorities must be notified. A comprehensive emergency response or disaster preparedness/recovery plan should be in place prior to any operations involving the use, transportation, storage, or disposal of cyanogen bromide. If cyanogen bromide is spilled or leaked, the following specific steps are recommended:

- ☑ Restrict persons not wearing protective clothing from area of spill or leak until cleanup is complete and area can be opened for normal work.
- ☑ Ventilate area of leak.
- ☑ Collect powdered materials in most convenient way possible and deposit in sealed drums. Do NOT dry sweep, use a HEPA vacuum instead. Do NOT use wet method.
- ☑ It may be necessary to dispose of cyanogen bromide as a hazardous waste. The responsible state agency or the regional office of the federal Environmental Protection Agency (EPA) should be contacted for specific recommendations.

RISK ASSESSMENT: BUSINESS

General Assessment

Accidents or mishaps involving cyanogen bromide can present a significant threat to business operations. The loss or damage of equipment or facilities can significantly affect fiscal viability. Lawsuits that may result from personnel injury/death, public exposures, and/or environmental contamination will also require a serious expenditure of resources. Media attention surrounding an injury, death, or environmental damage can also result in a loss of profits and loss of current as well as future business.

Recommended Risk-Reduction Measures

Company attorneys, safety and health professionals, and environmental specialists should be involved in the development of any procedures dealing with safety or emergency response. A company official should be pre-designated as a public relations officer with specific training in dealing with the press. Corporate plans and policies should be developed, approved, and implemented long before any need for such arises.

REFERENCES

Ahlbom, A. 1993. *Biostatistics for Epidemiologists.* New York: CRC Press/Lewis Publishers.

American Conference of Governmental Industrial Hygienists. 1988. *Documentation of the Threshold Limit Values and Biological Exposure Indices,* 5th Edition (with updates). Cincinnati: ACGIH

Asante-Duah, D. K. 1993. *Hazardous Waste Risk Assessment.* New York: CRC Press/Lewis Publishers.

Bloom, A. D. and F. J. DeSerres. 1995. *Ecotoxicity and Human Health.* New York: CRC Press/Lewis Publishers.

Brusick, D. J. 1994. *Methods for Genetic Risk Assessment.* New York: CRC Press/Lewis Publishers.

Bueche, J. R. 1972. *Principles of Physics,* 2nd Edition. New York: McGraw-Hill.

Calabrese, E. J., E. M. Kenyon. *Air Toxics and Risk Assessment.* New York: CRC Press/Lewis Publishers.

Calabrese, E. J. 1994. *Biological Effects of Low Level Exposures.* New York: CRC Press/Lewis Publishers.

Cockerman, L. G., B. S. Shane. 1994. *Basic Environmental Toxicology.* New York: CRC Press/Lewis Publishers.

Cothern, C. R., N. P. Ross. 1994. *Environmental Statistics, Assessment, and Forecasting.* New York: CRC Press/Lewis Publishers.

Cothern, R. R. 1993. *Comparative Environmental Risk Assessment.* New York: CRC Press/Lewis Publishers.

Cothern, C. R. 1995. *Handbook for Environmental Risk Decisionmaking.* New York: CRC Press/Lewis Publishers.

Cockerman, L. G., B. S. Shane. 1994. *Basic Environmental Toxicology.* New York: CRC Press/Lewis Publishers.

Gardner, R. W. 1994. *Chemical Intolerance.* New York: CRC Press/Lewis Publishers.

Hallenbeck, W. H. 1993. *Quantitative Risk Assessment for Environmental and Occupational Health,* 2nd Edition. New York: CRC Press/Lewis Publishers.

Howard, P. H. 1995. *Handbook of Environmental Fate and Exposures Data for Organic Chemicals,* Vol. I-IV. New York: CRC Press/Lewis Publishers.

Howard, P. H., et. al. 1991. *Handbook of Environmental Degradation Rates.* New York: CRC Press/Lewis Publishers.

Jones, F. E. 1994. *Toxic Organic Vapors in the Workplace.* New York: CRC Press/Lewis Publishers.

Keith, L. H. and M. M. Walker. 1995. *Handbook of Air Toxics.* New York: CRC Press/Lewis Publishers.

Landis, W. G., M. H. Yu. 1995. *Introduction to Environmental Toxicology.* New York: CRC Press/Lewis Publishers.

Langerman, N. 1994 *Precautionary Labels for Chemical Containers.* New York: CRC Press/Lewis Publishers.

Lauwerys, R. R., P. Hoet. 1993. *Industrial Chemical Exposure.* New York: CRC Press/Lewis Publishers.

Lewis, R. J., Sr. 1992. *Sax's Dangerous Properties of Industrial Materials,* Volumes 1, 2, and 3, 8th Edition. New York: Van Nostrand Reinhold.

Lide, D. R. 1995. *Handbook of Chemistry and Physics,* 76th Edition. New York: CRC Press/Lewis Publishers.

Manahan, S. E. 1994. *Environmental Chemistry,* 6th Edition. New York: CRC Press/Lewis Publishers.

Mellström, G. A., J. E. Wahlberg, and H. I. Maibach. 1994. *Protective Gloves for Occupational Use.* New York: CRC Press/Lewis Publishers.

Meyer, E. 1990. *Chemistry of Hazardous Materials.* Englewood Cliffs, NJ: Prentice Hall.

Patnaik, P. 1992. *A Comprehensive Guide to the Hazardous Properties of Chemical Substances.* New York: Van Nostrand Reinhold.

Proctor, N. H., J. P. Hughes. 1991. *Chemical Hazards of the Workplace.* New York: Van Nostrand Reinhold.

Quigley, D. R. 1994. *Handbook of Emergency Chemical Management.* New York: CRC Press/Lewis Publishers.

Rea, W. J. 1995. *Chemical Sensitivity,* Vol. I-III. New York: CRC Press/Lewis Publishers.

Richardson, M. 1995. *Dictionary of Substance and Their Effects,* Vol. I-VI. New York: CRC Press/Lewis Publishers.

Sacarello, H. L. A. 1994. *The Comprehensive Handbook of Hazardous Materials.* New York: CRC Press/Lewis Publishers.

Sheftel, V. O. 1995. *Handbook of Toxic Properties of Monomers and Additives.* New York: CRC Press/Lewis Publishers.

Torkelson, T. R. V. K. Rowe. 1981. Halogenated aliphatic hydrocarbons. *Patty's Industrial Hygiene and Toxicology,* Vol. 2B, *Toxicology,* pp 3568-3572. New York: Wiley-Interscience.

U.S. Department of Health and Human Services, National Institute for Occupational Safety and Health. 1994. *NIOSH Pocket Guide to Chemical Hazards.* Washington, D.C.: U.S. Government Printing Office.

MATERIAL SAFETY DATA SHEET

CHEMICAL NAME	
CYANOGEN CHLORIDE	**CRC PRESS/LEWIS PUBLISHERS DISCLAIMER:** THE INFORMATION AND RECOMMENDATIONS PRESENTED HEREIN ARE BASED ON SOURCES BELIEVED TO BE RELIABLE. CRC MAKES NO REPRESENTATION ON ITS COMPLETENESS OR ACCURACY. IT IS THE RESPONSIBILITY OF THE USER TO DETERMINE THE CHEMICAL'S SUITABILITY FOR ITS INTENDED USE, THE CHEMICAL'S SAFE USE, AND THE CHEMICAL'S PROPER DISPOSAL. NO REPRESENTATIONS AND/OR WARRANTIES, EITHER EXPRESSED OR IMPLIED, OF THE MERCHANTABILITY OR FITNESS FOR A PARTICULAR PURPOSE, OR OF ANY OTHER NATURE, ARE MADE WITH RESPECT TO THE INFORMATION PROVIDED IN THIS MSDS OR TO THE CHEMICAL TO WHICH INFORMATION MAY REFER. CRC NEITHER ASSUMES NOR AUTHORIZES ANY OTHER PERSON TO ASSUME FOR IT, ANY OTHER ADDITIONAL RESPONSIBILITY OR LIABILITY FOR THE USE OF, OR RELIANCE UPON, THIS INFORMATION.

HAZARD WARNING INFORMATION

HEALTH	FIRE	REACTIVE	OTHER	DEGREE OF HAZARD	COLOR CODING	OTHER CODES
3	0	1		0 = Minimum Hazard 1 = Slight Hazard 2 = Moderate Hazard 3 = Serious Hazard 4 = Severe Hazard	HEALTH = BLUE FIRE = RED REACTIVITY = YELLOW OTHER = WHITE	OX = Oxidizer ACID = Acid ALK = Alkali COR = Corrosive W = Use No Water

SECTION I - GENERAL INFORMATION

Characterization	RCRA Number	EPA Class
Inorganic Cyanide	**P033**	**Acute Hazardous Waste**

DOT Proper Shipping Name	Chemical Abstract Service (CAS) Number
Cyanogen chloride	**506-77-4**
DOT Hazard Class and Label Requirements	DOT Emergency Guide Code
Poison A; Flammable Gas & Poison Gas	**15**
DOT Identification Number	Chemical Formula
UN 1589	**ClCN**

Synonyms
Chlorocyan; chlorine cyanide; chlorocyanide; chlorocyanogen.

SECTION II - HAZARDOUS INGREDIENTS AND IDENTITY INFORMATION

Hazardous Components (specific identity)	OSHA Exposure Criteria	NIOSH Exposure Criteria	Immediately Dangerous to Life and Health (IDLH)	ACGIH Exposure Criteria
Cyanogen chloride (derivation: Action of chlorine on moist sodium cyanide suspended in carbon tetrachloride, and kept cooled to -3°C, followed by distillation). **1 ppm = 2.55 mg/m^3**	**PEL (ceiling):** **0.3 ppm** **0.6 mg/m^3** **STEL:** **Not Established**	**REL (ceiling):** **0.3 ppm** **0.6 mg/m^3** **STEL:** **Not Established**	**Not Determined**	**TLV:** **Not Established** **STEL:** **Not Established**

SECTION III - PHYSICAL AND CHEMICAL CHARACTERISTICS

Boiling Point	Specific Gravity (H_2O = 1)
55°F (13°C)	**2.16**
Vapor Pressure (gas)	Molecular Weight
1010 at 73°F (23°C)	**61.5**
Vapor Density (Air = 1)	Melting Point
1.98	**20°F (-6.5°C)**

Solubility
Soluble in water, alcohol, and ether, and other organic solvents.

Appearance and Odor
Colorless gas (or liquid below 55°F) with an irritating odor.

SECTION IV - FIRE AND EXPLOSION HAZARD DATA

Flash Point (method used)	Explosive Limits in Air % by Volume
Not Applicable	**LEL: N/A UEL: N/A**
NFPA Classification	Autoignition Temperature
Non-flammable Gas	**Not Applicable**

Extinguishing Media
Use water, dry chemical, or carbon dioxide.

Special Fire Fighting Procedures
In a fire or when heated it may emit extremely toxic fumes and vapors of cyanide and oxides of chlorine and nitrogen. Wear full protective clothing and self-contained breathing apparatus (SCBA). Use water spray to keep fire-exposed containers cool. Evacuate non-essential personnel 1500 feet from fire area (ensure consideration of wind direction).

Unusual Fire and Explosion Hazards
Vapors may become flammable at high temperatures or when exposed to heat or flame.

SECTION V - REACTIVITY DATA

Stability		Conditions to Avoid **Generally stable under normal conditions of temperature and pressure. Keep away from incompatible materials, to avoid violent reactions and possible fires/explosions.**
Stable **X**	Unstable	Incompatibility *(materials to avoid)* **Water, acids, alkalis, ammonia, and alcohols. May react very slowly with water to form toxic and explosive hydrogen cyanide.**
Hazardous Polymerization		Conditions to Avoid **Under normal conditions of temperature and pressure, hazardous polymerization of cyanogen chloride can be expected. May be stabilized to prevent polymerization.**
May Occur **X**	Will Not Occur	Hazardous Decomposition or By-products **When heated to decomposition, or in contact with acid, acid fumes, water, or steam, cyanogen chloride can emit highly toxic/poisonous gases (including oxides of chlorine, nitrogen, and cyanide gas).**

SECTION VI - HEALTH HAZARD DATA

Primary Route(s) of Entry:	Inhalation?	Absorption (skin)?	Ingestion?
	X		

Health Hazards

INHALATION: Irritation of the respiratory tract with hemorrhage of the bronchi and trachea, followed by pulmonary edema (fluid buildup in the lungs). Death is extremely possible at this point. Other symptoms: loss of appetite, headache, eye, irritation of the nose, and throat.

SKIN & EYES: Irritating to the skin and eyes on contact. May cause second and third degree burns. May Cause conjunctivitis and edema of the eyelids.

Carcinogenicity	NTP Listed?	IARC Cancer Review Group?	OSHA Regulated?	Target Organs?
Unknown Human Unknown Animal	**No**	**No**	**No**	**Respiratory system, eyes, skin, CNS, CVS.**

Medical Conditions Generally Aggravated by Exposure

None noted.

Emergency and First-aid Procedures

<u>Eye contact</u>: Flush immediately with water for 15 minutes (minimum), seek medical attention. <u>Skin contact</u>: Remove all contaminated clothing. Immediately wash area with large amounts of soap and water. For <u>inhalation</u>: Remove the person from exposure. Provide respiratory assistance and CPR. Antidote: Break amyl nitrate pearl in a cloth and hold under victim's nose for 15 seconds. Repeat five times at 15 second intervals. Transfer to medical facility. Observe for 24 - 48 hours for lung effects (pulmonary edema). If <u>swallowed</u>: If conscious, give large amounts of water or milk. Do NOT induce vomiting. Seek medical attention immediately. (Never try to give anything by mouth to an unconscious person).

SECTION VII - PRECAUTIONS FOR SAFE HANDLING AND USE

Steps to be Taken in Case Material is Released or Spilled

Evacuate area and ventilate. Restrict those not involved in cleanup from entering area. Remove all sources of ignition. If gas leak cannot be stopped, remove leaking cylinders to safe area outdoors and allow to empty.

Preferred Waste Disposal Method

Dissolve in combustible solvent and burn in chemical incinerator with an afterburner and scrubber.

Precautions to be Taken in Handling and Storage

Store to avoid contact with incompatible materials since violent reactions can occur. Store in tightly closed containers in cool, well-ventilated area away from heat and moisture.

Other Precautions and Warnings

Sources of ignition are prohibited where cyanogen chloride is used, handled, or stored. Heat may cause emission of toxic cyanide and bromide fumes.

SECTION VIII - CONTROL MEASURES AND PERSONAL PROTECTIVE EQUIPMENT

Respiratory Protection *(specify type)*

For any exposure level, use a supplied-air respirator with full facepiece, hood, or helmet set in continuous flow mode, or use a NISH/MSHA approved self-contained breathing apparatus (SCBA) operated in pressure demand or other positive pressure mode.

Ventilation

Local exhaust or general mechanical systems recommended.

Protective Gloves	Eye Protection	Other Protective Clothing
Butyl or Neoprene Rubber	**Chemical Goggles or Face Mask**	**Neoprene Rubber Apron**

Work/Hygiene Practices

Always wash hands thoroughly after using chemical; never bring food, drink, or smoking materials into vicinity of chemicals.

CYANOGEN CHLORIDE

ClCN CAS: 506-77-4

IDENTIFICATION AND TYPICAL USES

Cyanogen chloride is a colorless gas (or liquid, below 55°F) with an irritating odor. It is used in the organic synthesis of other chemicals, as a warning agent in fumigant gases, and as a military poisonous gas (tear gas).

RISK ASSESSMENT: HEALTH

General Assessment

Cyanogen chloride is an extremely toxic and highly poisonous gas with symptoms of exposure similar to that of hydrogen cyanide, although not as severe. As a gas, the primary route of entry is through *inhalation.* It may also affect the body through skin contact, causing burns and tissue damage, in either its gas or liquid state, although adverse effects are more likely with the latter. Absorption may also occur; however, the liquid is more prone to tissue destruction than simple absorption. The toxicity of cyanogen chloride is attributed to its relatively easy decomposition to cyanide ion in an aqueous environment (such as that which exists inside the human body). The cyanide then attacks the cells and interferes with cellular metabolism. General symptoms of exposure may include vomiting, nausea, headache, confusion, and weakness.

Inhalation will produce irritation to the eyes, nose, throat, and the respiratory tract followed by hemorrhage of the bronchi and trachea, and pulmonary edema (fluid buildup in the lungs), which can lead to death. Indications of illness may not appear for 24 - 48 hours following exposure, creating a false sense of security and well-being. Symptoms of exposure may include lachrymation, headache, congestion, convulsions, somnolence, dizziness, loss of appetite, and low body weight. It is improbable that anyone will remain in areas with high vapor concentrations for very long since the irritating properties of cyanogen chloride make it virtually impossible to withstand for more than short periods of time.

Skin contact with the liquid will result in the destruction of tissue do to the extremely corrosive properties of cyanogen chloride. It may even cause second or third degree burns over a relatively short period of time. Dermatitis is also a possibility over long periods of exposure to low concentrations. Eye contact can be irritating and potentially destructive, with conjunctivitis, edema of the eyelid and possible permanent loss of vision.

☠ *Acute Health Effects*

The following acute (short-term) health effects may occur immediately or shortly after exposure to cyanogen chloride:

Skin: Immediate irritation and possible serious burns on contact.

Eye: Severe irritation, conjunctivitis, and may cause burns and damage (blindness).

Lung: Irritation of the nose, throat, and lungs following exposure. Can instigate and lead to chronic lung disorders. May cause fluid to buildup on the lungs (pulmonary edema) and death.

Other: A possible CNS depressant and can affect the CVS. May cause headache, dizziness, loss of appetite, weight loss, convulsions, vertigo, loss of consciousness. It may also affect the rate of heartbeat and breathing.

💣 *Chronic Health Effects*

The following chronic (long-term) health effects may occur at some time after exposure to cyanogen chloride and can last for months or even years:

Cancer Hazards: According to information presented in the references, cyanogen chloride has not been adequately tested for its ability to cause cancer in test animals.

Reproductive Hazard: According to information presented in the references, cyanogen chloride has not been adequately tested for its ability to adversely affect reproduction.

Other Chronic Effects: Cyanogen chloride can irritate the lungs. Repeated exposure may cause bronchitis to develop with cough, phlegm, and/or shortness of breath. In weak solution, it may lead to dermatitis and rash.

Recommended Risk-Reduction Measures

Personnel should avoid direct contact with cyanogen chloride. It is extremely toxic and can cause death in short periods of time at high concentrations. If a less toxic chemical cannot be substituted for a hazardous substance, then *engineering controls* are the most effective method of reducing exposures. The best protection is to enclose operations and/or provide local exhaust ventilation at the site of chemical release. While not always operationally feasible, isolating operations can also reduce exposure. Using respiratory protection is less effective than the controls mentioned above, but is still advisable whenever working with or around cyanogen chloride. Because such low exposure limits have been established for this chemical and, because even the slightest exposure can be extremely hazardous to human health, prudent risk management requires a conservative approach to respiratory protection. Hence, the recommended respiratory protection is a self-contained breathing apparatus (SCBA) with full facepiece and pressure demand. If a full facepiece is not available, then chemical goggles should be worn to protect the eyes. This chemical is extremely corrosive and destructive to human tissues. Whenever a chemical splash hazard exists, a face shield and a rubber apron should be worn. To prevent hand and skin exposures, acid resistant gloves should be worn.

Administrative controls should also be in place to minimize the potential for human exposures. These may include written procedures or policies which specify the methods and techniques that will be practiced whenever personnel are to work with cyanogen chloride.

All personnel should receive *training* on the use, hazards, protective measures, emergency actions, and other precautions per 29 CFR 1910.1200 (Hazard Communication), prior to the first assignment in an area where cyanogen chloride is used or stored.

If symptoms develop or overexposure is suspected, the following tests are recommended:

- ☑ Lung function tests.
- ☑ Consider chest X-ray following acute over exposure.
- ☑ Evaluation by a qualified allergist with careful consideration of exposure history and special testing (may help diagnose skin allergy).

Any evaluation should include a careful history of past and present symptoms with an examination. Medical tests that simply look for existing damage are not a substitute for controlling exposures.

Also, because smoking can cause heart disease, as well as lung cancer, emphysema, and other respiratory problems, exposure to cyanogen chloride may worsen pre-existing respiratory conditions. Proper risk assessment requires consideration of *all* possible reasons for the appearance of exposure symptoms in workers.

Other methods to reduce exposure include:

- ☑ Where possible, enclose operations and use local exhaust ventilation at the site of chemical release. If local exhaust ventilation or enclosure is not used, respiratory protection should be mandatory. Also, automatic transfer of liquids from storage containers to process containers is recommended whenever possible.
- ☑ Always ensure that proper protective clothing is worn when using chemical substances.
- ☑ Wash thoroughly immediately after exposure to cyanogen chloride and at the end of the work shift or before eating, drinking, or smoking.
- ☑ Hazard warning information should be posted in the work area. In addition, as part of an on-going education and training program, all information on the health and safety hazards of cyanogen chloride should be communicated to all potentially exposed workers.
- ☑ Eye wash stations should be provided in the immediate work area for emergency use. If there is a possibility of skin exposure to cyanogen chloride, emergency shower facilities should also be provided.
- ☑ Personnel should never wear contaminated clothing home (family members can be exposed). Clothing should be laundered by personnel who have been trained on the specific hazards of working with cyanogen chloride.

RISK ASSESSMENT: ENVIRONMENT

General Assessment

The environment is at risk of exposure during transportation, storage, disposal, or destruction of cyanogen chloride. In almost every scenario, the threat of environmental exposure is contingent upon the proper handling of the chemical substance. Accidental spills, large or small, can result in fire, explosion, and possible contamination of the surrounding environmental mediums (air, and to a lesser degree, water and soil).

Cyanogen chloride is not a flammable material. However, it is known to be violently reactive with most acids and will decompose in water to release toxic cyanide. These characteristics require extreme caution in handling, storage, transportation, and disposal. It is also incompatible with strong oxidizers (fluorine, for example) and contact can cause fire or explosion. These characteristics require special consideration during any emergency situation involving a leak or spill of cyanogen chloride. Also, should cyanogen chloride ever come into contact with any incompatible substances during use, transportation, storage, or disposal, the formation of highly toxic and/or highly explosive commodities is extremely possible.

Cyanogen chloride can enter the environment through industrial discharges and through spills.

Acute Ecological Effects

Acute (short-term) toxic effects may include the death of animals, birds, or fish and death or low growth rate in plants. Acute effects are seen 2 to 4 days after animals or plants are exposed to cyanogen chloride. This chemical has high acute toxicity to aquatic life. No data are available on the short-term effects of cyanogen chloride exposure to plants, birds, or land animals.

Chronic Ecological Effects

Chronic (long-term) toxic effects may include shortened life span, reproductive problems, lower fertility, and changes in appearance or behavior in exposed animals. These effects can be seen long after first exposure(s) to toxic chemicals. Cyanogen chloride has high chronic toxicity in aquatic life. No data are available on the long-term effects of cyanogen chloride to plants, birds, or land animals.

Water Solubility

Cyanogen chloride is highly soluble in the aquatic environment. Concentrations of 1000 milligrams and more will mix with a liter of water. It will decompose in water to form cyanide.

Persistence in the Environment

Cyanogen chloride is very slightly persistent in water, with a half-life between 2 and 20 days. The half-life of a pollutant is the amount of time it takes for one half of the chemical to be degraded. No data are available on the amount of cyanogen chloride that will remain in air or water. However, as a gas, nearly 100% of cyanogen chloride might be expected to reach the air and virtually nothing will remain behind in water.

Bioaccumulation in Aquatic Organisms

Some substances increase in concentration, or *bioaccumulate,* in living organisms as they breathe contaminated air, drink contaminated water, or eat contaminated food. These chemicals can become concentrated in the tissues and internal organs of animals as well as humans.

Cyanogen chloride is not expected to accumulate in the tissues of fish. However, it may cause an increase in cyanide concentrations, but this is only a theory and requires additional testing.

Recommended Risk-Reduction Measures

Proper training of all transporters will reduce the likelihood of a mishap or accident resulting in a leak or spill to the environment. The correct use of DOT labeling (per the MSDS) on all containers, trucks, and rail cars will enable emergency responders to react properly and quickly to any disaster thereby reducing the potential risk to the environment and to personnel.

Storage of cyanogen chloride should be segregated from incompatible chemicals to minimize the risk of cross-contamination or contact. Buildings designated for storage should be equipped with appropriate fire protection systems (alarms, sprinklers, emergency lighting, portable extinguishers).

If a spill or leak to the environment has occurred, fire department, emergency response and/or hazardous materials spill personnel should be notified immediately. Cleanup should be attempted only by those trained in proper spill containment procedures. Contaminated soils should be removed for incineration and replaced with clean soil. If cyanogen chloride should contact the water table, aquifer, or navigable waterway, time is of the essence. It is slightly soluble in water but will decompose rapidly to form cyanide. Total remediation and containment may not be possible. The local and/or state emergency response authorities must be notified. A comprehensive emergency response or disaster preparedness/recovery plan should be in place prior to any operations involving the use, transportation, storage, or disposal of cyanogen chloride. If cyanogen chloride is spilled or leaked, the following specific steps are recommended:

- ☑ Restrict persons not wearing protective clothing from area of spill or leak until cleanup is complete and area can be opened for normal work.
- ☑ Remove ignition sources, ventilate area of leak.
- ☑ If possible, remove leaking cylinders to a safe location, preferably outdoors, and repair or allow to vent until empty.
- ☑ It may be necessary to dispose of cyanogen chloride as a hazardous waste. The responsible state agency or the regional office of the federal Environmental Protection Agency (EPA) should be contacted for specific recommendations.

RISK ASSESSMENT: BUSINESS

General Assessment

Accidents or mishaps involving cyanogen chloride can present a significant threat to business operations. The loss or damage of equipment or facilities can significantly affect fiscal viability. Lawsuits that may result from personnel injury/death, public exposures, and/or environmental contamination will also require a serious expenditure of resources. Media attention surrounding an injury, death, or environmental damage can also result in a loss of profits and loss of current as well as future business.

Recommended Risk-Reduction Measures

Company attorneys, safety and health professionals, and environmental specialists should be involved in the development of any procedures dealing with safety or emergency response. A company official should be pre-designated as a public relations officer with specific training in dealing with the press. Corporate plans and policies should be developed, approved, and implemented long before any need for such arises.

REFERENCES

American Conference of Governmental Industrial Hygienists. 1988. *Documentation of the Threshold Limit Values and Biological Exposure Indices,* 5th Edition (with updates). Cincinnati: ACGIH

Asante-Duah, D. K. 1993. *Hazardous Waste Risk Assessment.* New York: CRC Press/Lewis Publishers.

Brusick, D. J. 1994. *Methods for Genetic Risk Assessment.* New York: CRC Press/Lewis Publishers.

Bueche, J. R. 1972. *Principles of Physics,* 2nd Edition. New York: McGraw-Hill.

Calabrese, E. J. 1994. *Biological Effects of Low Level Exposures.* New York: CRC Press/Lewis Publishers.

Cockerman, L. G., B. S. Shane. 1994. *Basic Environmental Toxicology.* New York: CRC Press/Lewis Publishers.

Cothern, C. R., N. P. Ross. 1994. *Environmental Statistics, Assessment, and Forecasting.* New York: CRC Press/Lewis Publishers.

Cockerman, L. G., B. S. Shane. 1994. *Basic Environmental Toxicology.* New York: CRC Press/Lewis Publishers.

Gardner, R. W. 1994. *Chemical Intolerance.* New York: CRC Press/Lewis Publishers.

Hallenbeck, W. H. 1993. *Quantitative Risk Assessment for Environmental and Occupational Health,* 2nd Edition. New York: CRC Press/Lewis Publishers.

Howard, P. H. 1995. *Handbook of Environmental Fate and Exposures Data for Organic Chemicals,* Vol. I-IV. New York: CRC Press/Lewis Publishers.

Landis, W. G., M. H. Yu. 1995. *Introduction to Environmental Toxicology.* New York: CRC Press/Lewis Publishers.

Lewis, R. J., Sr. 1992. *Sax's Dangerous Properties of Industrial Materials,* Volumes 1, 2, and 3, 8th Edition. New York: Van Nostrand Reinhold.

Lide, D. R. 1995. *Handbook of Chemistry and Physics,* 76th Edition. New York: CRC Press/Lewis Publishers.

Manahan, S. E. 1994. *Environmental Chemistry,* 6th Edition. New York: CRC Press/Lewis Publishers.

Meyer, E. 1990. *Chemistry of Hazardous Materials.* Englewood Cliffs, NJ: Prentice Hall.

Patnaik, P. 1992. *A Comprehensive Guide to the Hazardous Properties of Chemical Substances.* New York: Van Nostrand Reinhold.

Rea, W. J. 1995. *Chemical Sensitivity,* Vol. I-III. New York: CRC Press/Lewis Publishers.

Richardson, M. 1995. *Dictionary of Substance and Their Effects,* Vol. I-VI. New York: CRC Press/Lewis Publishers.

Sacarello, H. L. A. 1994. *The Comprehensive Handbook of Hazardous Materials.* New York: CRC Press/Lewis Publishers.

Torkelson, T. R. V. K. Rowe. 1981. Halogenated aliphatic hydrocarbons. *Patty's Industrial Hygiene and Toxicology,* Vol. 2B, *Toxicology,* pp 3568-3572. New York: Wiley-Interscience.

U.S. Department of Health and Human Services, National Institute for Occupational Safety and Health. 1994. *NIOSH Pocket Guide to Chemical Hazards.* Washington, D.C.: U.S. Government Printing Office.

MATERIAL SAFETY DATA SHEET

CHEMICAL NAME	
CYCLOHEXANE	**CRC PRESS/LEWIS PUBLISHERS DISCLAIMER:** THE INFORMATION AND RECOMMENDATIONS PRESENTED HEREIN ARE BASED ON SOURCES BELIEVED TO BE RELIABLE. CRC MAKES NO REPRESENTATION ON ITS COMPLETENESS OR ACCURACY. IT IS THE RESPONSIBILITY OF THE USER TO DETERMINE THE CHEMICAL'S SUITABILITY FOR ITS INTENDED USE, THE CHEMICAL'S SAFE USE, AND THE CHEMICAL'S PROPER DISPOSAL. NO REPRESENTATIONS AND/OR WARRANTIES, EITHER EXPRESSED OR IMPLIED, OF THE MERCHANTABILITY OR FITNESS FOR A PARTICULAR PURPOSE, OR OF ANY OTHER NATURE, ARE MADE WITH RESPECT TO THE INFORMATION PROVIDED IN THIS MSDS OR TO THE CHEMICAL TO WHICH INFORMATION MAY REFER. CRC NEITHER ASSUMES NOR AUTHORIZES ANY OTHER PERSON TO ASSUME FOR IT, ANY OTHER ADDITIONAL RESPONSIBILITY OR LIABILITY FOR THE USE OF, OR RELIANCE UPON, THIS INFORMATION.

HAZARD WARNING INFORMATION

HEALTH	FIRE	REACTIVE	OTHER	DEGREE OF HAZARD	COLOR CODING	OTHER CODES
1	3	0		0 = Minimum Hazard 1 = Slight Hazard 2 = Moderate Hazard 3 = Serious Hazard 4 = Severe Hazard	HEALTH = BLUE FIRE = RED REACTIVITY = YELLOW OTHER = WHITE	OX = Oxidizer ACID = Acid ALK = Alkali COR = Corrosive W = Use No Water

SECTION I - GENERAL INFORMATION

Characterization	RCRA Number	EPA Class
Hydrocarbon, Aliphatic	**U056**	**Toxic Waste**

DOT Proper Shipping Name	Chemical Abstract Service (CAS) Number
Cyclohexane	**110-82-7**
DOT Hazard Class and Label Requirements	DOT Emergency Guide Code
Flammable Liquid	**26**
DOT Identification Number	Chemical Formula
UN 1145	C_6H_{12}

Synonyms

Hexamethylene; hexanaphthene; hexalhydrobenzene; benzene hexahydride.

SECTION II - HAZARDOUS INGREDIENTS AND IDENTITY INFORMATION

Hazardous Components (specific identity)	OSHA Exposure Criteria	NIOSH Exposure Criteria	Immediately Dangerous to Life and Health (IDLH)	ACGIH Exposure Criteria
Cyclohexane (derivation: Catalytic hydrogenation of benzene; a constituent of crude petroleum). **1 ppm = 3.50 mg/m³**	**PEL:** **300 ppm** **1050 mg/m³** **STEL:** **Not Established**	**REL:** **300 ppm** **1050 mg/m³** **STEL:** **Not Established**	**1300 ppm**	**TLV:** **300 ppm** **1050 mg/m³** **STEL:** **Not Established**

SECTION III - PHYSICAL AND CHEMICAL CHARACTERISTICS

Boiling Point	Specific Gravity (H_2O = 1)
178°F (81°C)	**0.8**
Vapor Pressure (gas)	Molecular Weight
77 at 69°F (20°C)	**84.2**
Vapor Density (Air = 1)	Melting Point
2.9	**45°F (7°C)**

Solubility

Insoluble in water, soluble in alcohol, benzene, acetone, and most organic solvents.

Appearance and Odor

Colorless, mobile liquid with sweet, chloroform-like odor. Odor threshold = 25 ppm.

SECTION IV - FIRE AND EXPLOSION HAZARD DATA

Flash Point (method used)	Explosive Limits in Air % by Volume
2°F (-17°C) closed cup	**LEL: 1.3% UEL: 8.4%**
NFPA Classification	Autoignition Temperature
Class 1B Flammable Liquid	**473°F (245°C)**

Extinguishing Media

Use dry chemical, carbon dioxide, water, or alcohol foam.

Special Fire Fighting Procedures

Cyclohexane is a flammable liquid. Wear full protective clothing and self-contained breathing apparatus (SCBA). Poisonous gases are produced in fire. Use water spray to keep fire-exposed containers cool.

Unusual Fire and Explosion Hazards

Dangerous fire and explosion hazard when exposed to heat or flame. Containers may explode in fire. Chemical will react on contact with oxidizable materials. Moderate explosion hazard when in the form of vapor. It is susceptible to forming explosive peroxide on long exposures to air.

SECTION V - REACTIVITY DATA

Stability		Conditions to Avoid
		Keep away from heat or flame. Normally stable, but do not allow to remain in prolonged contact with air (explosive and unstable peroxides may form).
Stable	Unstable	Incompatibility *(materials to avoid)*
X		Strong oxidizers, such as permanganates, peroxides, nitrates, chlorates, and perchlorates.
Hazardous Polymerization		Conditions to Avoid
		Under normal conditions of temperature and pressure, hazardous polymerization of cyclohexane is not expected to occur.
May Occur	Will Not Occur	Hazardous Decomposition or By-products
	X	When heated to decomposition, cyclohexane can emit highly toxic/poisonous gases.

SECTION VI - HEALTH HAZARD DATA

Primary Route(s) of Entry:	Inhalation?	Absorption (skin)?	Ingestion?
	X		X

Health Hazards

INHALATION: Low-order toxicity by inhalation. However, it does act as an irritant and a central nervous system depressant. Can cause nausea, dizziness, vomiting, and depression. Higher concentrations can lead to coma and death from respiratory failure.

SKIN & EYES: Irritating to the skin, may cause drying, cracking, and chapping. Eye irritation likely.

INGESTION: Irritation and narcotic effects, including drowsiness. Low-order toxicity by ingestion.

Carcinogenicity	NTP Listed?	IARC Cancer Review Group?	OSHA Regulated?	Target Organs?
Unknown Human Unknown Animal	No	No	29 CFR 1910.1000 Table Z-1	Respiratory system, eyes, skin, CNS.

Medical Conditions Generally Aggravated by Exposure

None reported.

Emergency and First-aid Procedures

Eye contact: Flush immediately with water for 15 minutes (minimum), seek medical attention. Skin contact: Remove all contaminated clothing. Immediately wash area with large amounts of soap and water. Seek medical assistance. For inhalation: Remove the person from exposure. Provide respiratory assistance and CPR, if required. Transfer to medical facility. Keep victim warm and at rest. If swallowed: Seek medical attention immediately.

SECTION VII - PRECAUTIONS FOR SAFE HANDLING AND USE

Steps to be Taken in Case Material is Released or Spilled

For small spills, absorb in paper towels and evaporate in a safe place (fume hood). Afterwards, burn paper away from combustible materials. Absorb large spills in vermiculite, dry sand, earth, or similar material and deposit in sealed containers. Ventilate area of spill or leak. Restrict those not involved in cleanup from entering area. Remove all sources of ignition.

Preferred Waste Disposal Method

No Citation.

Precautions to be Taken in Handling and Storage

Store to avoid contact with incompatible materials in tightly closed containers in cool, well-ventilated area away from heat and flame. Use only non-sparking tools to open and close containers.

Other Precautions and Warnings

Sources of ignition are prohibited where cyclohexane is used, handled, or stored. Heat may cause containers to build up pressure and explode. Containers must be grounded and bonded during transfer.

SECTION VIII - CONTROL MEASURES AND PERSONAL PROTECTIVE EQUIPMENT

Respiratory Protection *(specify type)*

For exposures above the PEL, use an MSHA/NIOSH-approved supplied-air respirator with full facepiece, hood, or helmet in continuous flow mode, or use a self-contained breathing apparatus (SCBA) operated in pressure demand or other positive pressure mode.

Ventilation

Local exhaust or general mechanical systems recommended.

Protective Gloves	Eye Protection	Other Protective Clothing
Nitrile Rubber, Viton, or Teflon	Chemical Goggles or Face Mask	Rubber Apron

Work/Hygiene Practices

Always wash hands thoroughly after using chemical; never bring food, drink, or smoking materials into vicinity of chemicals.

CYCLOHEXANE

C_6H_{12} CAS: 110-87-7

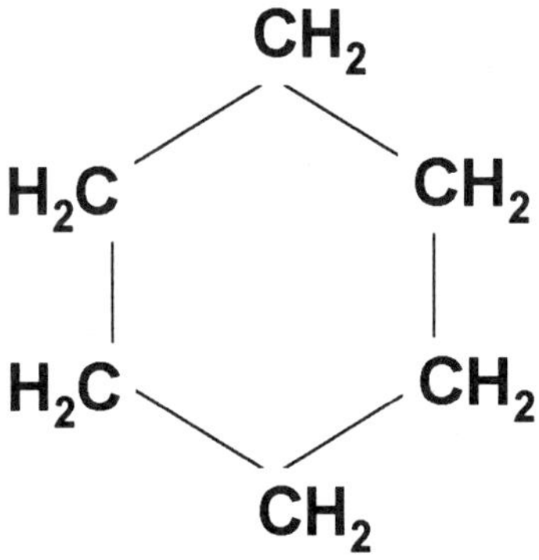

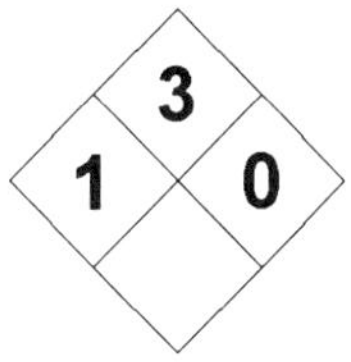

IDENTIFICATION AND TYPICAL USES

Cyclohexane is a colorless, mobile liquid with a pungent, sweet, chloroform-like odor. It is used in the manufacture of nylon, as a solvent for cellulose ethers, fats, oils, waxes, bitumen, resins, crude rubber, and for extracting essential oils from other products. It is also used in organic synthesis and as a re-crystallizing medium in the chemical industry. It is found in paint and varnish removers, in glass substitutes, in solid fuels (for camp stoves and catering applications), in fungicides, and is also used in analytical chemistry. It is used in the manufacture if benzene, cyclohexanol, caprolactam, and other chemicals.

RISK ASSESSMENT: HEALTH

General Assessment

Cyclohexane is an acute irritant of low-order toxicity by *inhalation* and *ingestion*. Skin contact can also produce some adverse topical effects. It is generally considered to be a safe solvent for industrial use when done so according to manufacturer's instructions. However, in sufficient or high concentrations, there can be toxic systemic effects through both routes of exposure.

Significant inhalation may cause irritation of the respiratory tract and depress the central nervous system. Symptoms include nausea, dizziness, vomiting, and depression. At high concentrations, it is a narcotic and can lead to coma and death due to respiratory failure. It is of low chronic toxicity due to its efficient metabolism and excretion. It does *not* produce toxic changes to nerve cells and peripheral neuropathy normally associated with exposure to *n*-hexane.

Skin and eye contact may produce mild to moderate irritation on contact. Ingestion can produce narcotic effects as those noted for inhalation above.

Acute Health Effects

The following acute (short-term) health effects may occur immediately or shortly after exposure to cyclohexane:

Skin: Mild irritation.

Eye: Mild to moderate irritation.

Lung: May cause dizziness, vomiting, and nausea. Primary result of inhalation of cyclohexane is its action on the CNS.

CNS: A CNS depressant causing headache, dizziness, depression, excitement, and changes in mood.

Chronic Health Effects

The following chronic (long-term) health effects may occur at some time after exposure to cyclohexane and can last for months or even years:

Cancer Hazards: According to information presented in the references, cyclohexane has not been adequately tested for its ability to cause cancer in animals. Limited studies indicate an increased incidence of Hodgkin's disease in workers exposed to organic solvents. It is not known, however, whether cyclohexane will have this effect.

Reproductive Hazard: According to information presented in the references, cyclohexane has not been tested for its ability to adversely affect reproduction.

Other Chronic Effects: Prolonged skin contact may cause drying, cracking, and chapping of the affected area. Repeated or long-term exposures to cyclohexane may affect the liver and kidneys.

⓿ *Recommended Risk-Reduction Measures*

Personnel should avoid direct contact with cyclohexane. Even though the toxicity of this chemical is considered to be of relatively low order, it can still cause significant health effects if improperly used or if overexposures occur. If a less toxic chemical cannot be substituted for a hazardous substance, then *engineering controls* are the most effective method of reducing exposures. The best protection is to enclose operations and/or provide local exhaust ventilation at the site of chemical release. While not always operationally feasible, isolating operations can also reduce exposure. Using respiratory protection is less effective than the controls mentioned above, but is still advisable whenever working with or around cyclohexane. For exposures above the PEL (300 ppm), a supplied-air respirator with full facepiece operated in continuous flow mode, or a self-contained breathing apparatus (SCBA) with full facepiece and pressure demand are the recommended respiratory protection methods of choice. If a full facepiece is not available, then chemical goggles should be worn to protect the eyes. Whenever a chemical splash hazard exists, a face shield and a rubber apron should be worn. To prevent hand and skin exposures, impervious gloves should be worn.

Administrative controls should also be in place to minimize the potential for human exposures. These may include written procedures or policies which specify the methods and techniques that will be practiced whenever personnel are to work with cyclohexane.

All personnel should receive *training* on the use, hazards, protective measures, emergency actions, and other precautions per 29 CFR 1910.1200 (Hazard Communication), prior to the first assignment in an area where cyclohexane is used or stored.

If symptoms develop or overexposure is suspected, the following tests are recommended:

☑ Kidney and liver function tests.

Any evaluation should include a careful history of past and present symptoms with an examination. Medical tests that simply look for existing damage are not a substitute for controlling exposures.

Other methods to reduce exposure include:

☑ Where possible, enclose operations and use local exhaust ventilation at the site of chemical release. If local exhaust ventilation or enclosure is not used, respiratory protection should be mandatory.

☑ Always ensure that proper protective clothing is worn when using chemical substances.

☑ Wash thoroughly immediately after exposure to cyclohexane and at the end of the work shift or before eating, drinking, or smoking.

☑ Hazard warning information should be posted in the work area. In addition, as part of an on-going education and training program, all information on the health and safety hazards of cyclohexane should be communicated to all exposed workers.

☑ Eye wash stations should be provided in the immediate work area for emergency use. If there is a possibility of skin exposure to cyclohexane, emergency shower facilities should also be provided.

☑ Personnel should never wear contaminated clothing home (family members can be exposed). Clothing should be laundered by personnel who have been trained on the hazards of cyclohexane.

RISK ASSESSMENT: ENVIRONMENT

General Assessment

The environment is at risk of exposure during transportation, storage, disposal, or destruction of cyclohexane. In almost every scenario, the threat of environmental exposure is contingent upon the proper handling of the chemical substance. Accidental spills, large or small, can result in fire, explosion, and possible contamination of the surrounding environmental mediums (water, soil, and air).

Cyclohexane is considered a class 1B flammable liquid (per OSHA 29 CFR 1910.106). Its extremely low flash point and relatively low boiling point make this chemical a serious fire and explosion hazard concern. These characteristics require extreme caution in handling, storage, transportation, and disposal. It is incompatible with common peroxides and other oxidizers and contact can cause fire or explosion. Therefore, special consideration is required during any emergency situation involving a leak or spill of cyclohexane.

Should cyclohexane ever come into contact with these incompatible substances during use, transportation, storage, or disposal, the formation of highly toxic and/or highly explosive commodities (including hydrogen chloride) is extremely possible.

Cyclohexane may enter the environment through industrial discharges or spills.

Acute Ecological Effects

Acute (short-term) toxic effects may include the death of animals, birds, or fish and death or low growth rate in plants. Acute effects are seen 2 to 4 days after animals or plants are exposed to cyclohexane.

This chemical has high acute toxicity to aquatic life. It causes membrane damage in ornamental crop species. Insufficient data are available on the short-term effects of cyclohexane exposure to birds or land animals.

Chronic Ecological Effects

Chronic (long-term) toxic effects may include shortened life span, reproductive problems, lower fertility, and changes in appearance or behavior in exposed animals. These effects can be seen long after first exposure(s) to toxic chemicals.

Cyclohexane has moderate chronic toxicity in aquatic life. Insufficient data are available on the long-term effects of cyclohexane to plants, birds, or land animals.

Water Solubility

Cyclohexane is nearly insoluble in water. Concentrations of less than 100 milligrams may mix with a liter of water, depending on water pH and chemical concentration of cyclohexane.

Persistence in the Environment

Cyclohexane is non-persistent in water, with a half-life of less than 2 days. The half-life of a pollutant is the amount of time it takes for one half of the chemical to be degraded. Virtually 100% of cyclohexane will eventually end up in air.

Bioaccumulation in Aquatic Organisms

Some substances increase in concentration, or *bioaccumulate,* in living organisms as they breathe contaminated air, drink contaminated water, or eat contaminated food. These chemicals can become concentrated in the tissues and internal organs of animals as well as humans.

The concentration of cyclohexane found in fish tissue is expected to be somewhat higher than the average concentration of cyclohexane in the water from which the fish was taken.

Recommended Risk-Reduction Measures

Proper training of all transporters will reduce the likelihood of a mishap or accident resulting in a leak or spill to the environment. The correct use of DOT labeling (per the MSDS) on all containers, trucks, and rail cars will enable emergency responders to react properly and quickly to any disaster thereby reducing the potential risk to the environment and to personnel.

Storage of cyclohexane should be segregated from incompatible chemicals to minimize the risk of cross-contamination or contact. Buildings designated for storage should be equipped with appropriate fire protection systems (alarms, sprinklers, emergency lighting, portable extinguishers).

If a spill or leak to the environment has occurred, fire department, emergency response and/or hazardous materials spill personnel should be notified immediately. Cleanup should be attempted only by those trained in proper spill containment procedures. Contaminated soils should be removed for incineration and replaced with clean soil. If cyclohexane should contact the water table, aquifer, or navigable waterway, the local and/or state emergency response authorities must be notified. A comprehensive emergency response or disaster preparedness/recovery plan should be in place prior to any operations involving the use, transportation, storage, or disposal of cyclohexane. If cyclohexane is spilled or leaked, the following specific steps are recommended:

- ☑ Restrict persons not wearing protective clothing from area of spill or leak until cleanup is complete and area can be opened for normal work.
- ☑ Remove all ignition sources, ventilate area of leak.
- ☑ For small spills (laboratory spills), absorb liquid in paper towels. Allow evaporation in a safe place (a fume hood or open air). After sufficient time has passed for evaporation, burn paper towels away from other combustibles. For larger spills, absorb liquids in vermiculite, dry sand, earth, or similar material and deposit in sealed containers.
- ☑ If gas is leaking, stop the flow of gas. If the source is a cylinder and leak cannot be stopped in

place, remove cylinder to a safe place in the open air and repair or allow to empty.

☑ It may be necessary to dispose of cyclohexane as a hazardous waste. The responsible state agency or the regional office of the federal Environmental Protection Agency (EPA) should be contacted for specific recommendations.

RISK ASSESSMENT: BUSINESS

General Assessment

Accidents or mishaps involving cyclohexane can present a significant threat to business operations. The loss or damage of equipment or facilities can significantly affect fiscal viability. Lawsuits that may result from personnel injury/death, public exposures, and/or environmental contamination will also require a serious expenditure of resources. Media attention surrounding an injury, death, or environmental damage can also result in a loss of profits and loss of current as well as future business.

Recommended Risk-Reduction Measures

Company attorneys, safety and health professionals, and environmental specialists should be involved in the development of any procedures dealing with safety and emergency response. A company official should be pre-designated as a public relations officer with specific training in dealing with the press. Corporate plans and policies should be developed, approved, and implemented long before any need for such arises.

REFERENCES

American Conference of Governmental Industrial Hygienists. 1988. *Documentation of the Threshold Limit Values and Biological Exposure Indices,* 5th Edition (with updates). Cincinnati: ACGIH

Bueche, J. R. 1972. *Principles of Physics,* 2nd Edition. New York: McGraw-Hill.

Calabrese, E. J. 1994. *Biological Effects of Low Level Exposures.* New York: CRC Press/Lewis Publishers.

Cockerman, L. G., B. S. Shane. 1994. *Basic Environmental Toxicology.* New York: CRC Press/Lewis Publishers.

Gardner, R. W. 1994. *Chemical Intolerance.* New York: CRC Press/Lewis Publishers.

Hallenbeck, W. H. 1993. *Quantitative Risk Assessment for Environmental and Occupational Health,* 2nd Edition. New York: CRC Press/Lewis Publishers.

Jones, F. E. 1994. *Toxic Organic Vapors in the Workplace.* New York: CRC Press/Lewis Publishers.

Keith, L. H. and M. M. Walker. 1995. *Handbook of Air Toxics.* New York: CRC Press/Lewis Publishers.

Landis, W. G., M. H. Yu. 1995. *Introduction to Environmental Toxicology.* New York: CRC Press/Lewis Publishers.

Langerman, N. 1994 *Precautionary Labels for Chemical Containers.* New York: CRC Press/Lewis Publishers.

Lauwerys, R. R., P. Hoet. 1993. *Industrial Chemical Exposure.* New York: CRC Press/Lewis Publishers.

Lewis, R. J., Sr. 1992. *Sax's Dangerous Properties of Industrial Materials,* Volumes 1, 2, and 3, 8th Edition. New York: Van Nostrand Reinhold.

Lide, D. R. 1995. *Handbook of Chemistry and Physics,* 76th Edition. New York: CRC Press/Lewis Publishers.

Manahan, S. E. 1994. *Environmental Chemistry,* 6th Edition. New York: CRC Press/Lewis Publishers.

Mellström, G. A., J. E. Wahlberg, and H. I. Maibach. 1994. *Protective Gloves for Occupational Use.* New York: CRC Press/Lewis Publishers.

Meyer, E. 1990. *Chemistry of Hazardous Materials.* Englewood Cliffs, NJ: Prentice Hall.

Patnaik, P. 1992. *A Comprehensive Guide to the Hazardous Properties of Chemical Substances.* New York: Van Nostrand Reinhold.

Proctor, N. H., J. P. Hughes. 1991. *Chemical Hazards of the Workplace.* New York: Van Nostrand Reinhold.

Quigley, D. R. 1994. *Handbook of Emergency Chemical Management.* New York: CRC Press/Lewis Publishers.

Rea, W. J. 1995. *Chemical Sensitivity,* Vol. I-III. New York: CRC Press/Lewis Publishers.

Richardson, M. 1995. *Dictionary of Substance and Their Effects,* Vol. I-VI. New York: CRC Press/Lewis Publishers.

Sacarello, H. L. A. 1994. *The Comprehensive Handbook of Hazardous Materials.* New York: CRC Press/Lewis Publishers.

Sheftel, V. O. 1995. *Handbook of Toxic Properties of Monomers and Additives.* New York: CRC Press/Lewis Publishers.

Torkelson, T. R. V. K. Rowe. 1981. Halogenated aliphatic hydrocarbons. *Patty's Industrial Hygiene and Toxicology,* Vol. 2B, *Toxicology,* pp 3568-3572. New York: Wiley-Interscience.

U.S. Department of Health, Education, and Welfare, National Institute for Occupational Safety and Health. 1976. *Criteria for a Recommended Standard: Occupational Exposure to Cyclohexane.* DEHW (NIOSH) Publication Number: 77-210. Washington, D.C.: U.S. Government Printing Office.

U.S. Department of Health and Human Services, National Institute for Occupational Safety and Health. 1994. *NIOSH Pocket Guide to Chemical Hazards.* Washington, D.C.: U.S. Government Printing Office.

MATERIAL SAFETY DATA SHEET

CHEMICAL NAME

CYCLOHEXANOL

HAZARD WARNING INFORMATION

HEALTH	FIRE	REACTIVE	OTHER	DEGREE OF HAZARD	COLOR CODING	OTHER CODES
1	2	0		0 = Minimum Hazard 1 = Slight Hazard 2 = Moderate Hazard 3 = Serious Hazard 4 = Severe Hazard	HEALTH = BLUE FIRE = RED REACTIVITY = YELLOW OTHER = WHITE	OX = Oxidizer ACID = Acid ALK = Alkali COR = Corrosive W = Use No Water

SECTION I - GENERAL INFORMATION

Characterization: **Alcohol**	RCRA Number: **None**	EPA Class: **Not Applicable**
DOT Proper Shipping Name: **Cyclohexanol**	Chemical Abstract Service (CAS) Number: **108-93-0**	
DOT Hazard Class and Label Requirements: **Combustible Liquid, N.O.S.**	DOT Emergency Guide Code: **27**	
DOT Identification Number: **UN 1993**	Chemical Formula: **$C_6H_{11}OH$**	

Synonyms

Anol; cyclohexyl alcohol; hexahydrophenol; hexalin; hydrain; hydroxycyclohexanol.

SECTION II - HAZARDOUS INGREDIENTS AND IDENTITY INFORMATION

Hazardous Components (specific identity)	OSHA Exposure Criteria	NIOSH Exposure Criteria	Immediately Dangerous to Life and Health (IDLH)	ACGIH Exposure Criteria
Cyclohexanol (derivation: Phenol is reduce with hydrogen over active nickel at 160 - 170°F (71 - 76.6°C). The cyclohexanone is removed by condensing with benzaldehyde in the presence of alkali). **1 ppm = 4.17 mg/m³**	**PEL (skin):** **50 ppm** **200 mg/m³** **STEL:** **Not Established**	**REL (skin):** **50 ppm** **200 mg/m³** **STEL:** **Not Established**	**400 ppm**	**TLV (skin):** **50 ppm** **200 mg/m³** **STEL:** **Not Established**

SECTION III - PHYSICAL AND CHEMICAL CHARACTERISTICS

Boiling Point: **324°F (162°C)**	Specific Gravity (H_2O = 1): **0.95**
Vapor Pressure (gas): **1 at 69°F (20°C)**	Molecular Weight: **100.2**
Vapor Density (Air = 1): **3.5**	Melting Point: **72°F (22°C)**

Solubility

Very slightly soluble in water (3-4%). Miscible with aromatic hydrocarbons, ethanol, ethyl acetate, and petroleum solvents.

Appearance and Odor

Colorless, hygroscopic, viscous liquid with a camphor- or menthol-like odor.

SECTION IV - FIRE AND EXPLOSION HAZARD DATA

Flash Point (method used): **154°F (68°C) closed cup**	Explosive Limits in Air % by Volume: **LEL: 2.4% UEL: Not Found**
NFPA Classification: **Class IIIA Combustible Liquid**	Autoignition Temperature: **572°F (300°C)**

Extinguishing Media

Use dry chemical, carbon dioxide, water, or alcohol foam.

Special Fire Fighting Procedures

Wear full protective clothing and self-contained breathing apparatus (SCBA). Poisonous gases are produced in fire. Use water spray to keep fire-exposed containers cool. Containers may explode in fire.

Unusual Fire and Explosion Hazards

Dangerous fire hazard when exposed to heat. Chemical will react on contact with oxidizable materials. Moderate explosion hazard when in the form of vapor.

SECTION V - REACTIVITY DATA

Stability		Conditions to Avoid Keep away from sources of heat (including direct sunlight) or flame. Large quantities should be stored in metal tanks or drums. Keep away from incompatible materials.
Stable X	Unstable	Incompatibility *(materials to avoid)* Strong oxidizers, such as permanganates, peroxides, nitrates, chlorates, and perchlorates. Can react with nitric acid to form violently explosive materials and ignites on contact with divided chromium trioxide.
Hazardous Polymerization		Conditions to Avoid Under normal conditions of temperature and pressure, hazardous polymerization of cyclohexanol is not expected to occur.
May Occur	Will Not Occur X	Hazardous Decomposition or By-products When heated to decomposition, cyclohexanol can emit highly toxic/poisonous and acrid fumes and gases.

SECTION VI - HEALTH HAZARD DATA

Primary Route(s) of Entry:	Inhalation?	Absorption (skin)?	Ingestion?
	X	X	X

Health Hazards

INHALATION: Can produce human systemic effects, including conjunctivitis and changes in the olfactory and respiratory systems. Can cause damage to the kidneys, liver, and blood vessels. Has narcotic-like action on high or prolonged exposures (headaches, dizziness). Also lachrymation, loss of coordination, convulsions, and possible effects on target organs.

ABSORPTION: Irritating to the skin, absorbed in toxic amounts. Eye contact may injure cornea.

INGESTION: Irritation and narcosis, including drowsiness. Toxic effects on liver and spleen.

Carcinogenicity	NTP Listed?	IARC Cancer Review Group?	OSHA Regulated?	Target Organs?
Unknown Human Questioned Animal	No	No	29 CFR 1910.1000 Table Z-1	Respiratory system, eyes, skin, liver, kidney, spleen.

Medical Conditions Generally Aggravated by Exposure

None reported. Skin conditions (dermatitis) or respiratory problems (asthma) may be aggravated.

Emergency and First-aid Procedures

Eye contact: Flush immediately with water for 15 minutes (minimum), seek medical attention. Skin contact: Remove all contaminated clothing. Immediately wash area with large amounts of soap and water. Seek medical assistance. For inhalation: Remove the person from exposure. Provide respiratory assistance and CPR, if required. Transfer to medical facility. Keep victim warm and at rest. If swallowed: Seek medical attention immediately.

SECTION VII - PRECAUTIONS FOR SAFE HANDLING AND USE

Steps to be Taken in Case Material is Released or Spilled

Absorb large spills in vermiculite, dry sand, earth, or similar material and deposit in sealed containers. Ventilate area of spill or leak. Restrict those not involved in cleanup from entering area. Remove all sources of ignition.

Preferred Waste Disposal Method

Burn in chemical incinerator equipped with an afterburner and scrubber.

Precautions to be Taken in Handling and Storage

Store to avoid contact with incompatible materials in tightly closed containers in cool, well-ventilated area away from heat and flame. Use only non-sparking tools to open and close containers.

Other Precautions and Warnings

Sources of ignition are prohibited where cyclohexanol is used, handled, or stored. Heat may cause containers to build up pressure and explode. Containers must be grounded and bonded during transfer.

SECTION VIII - CONTROL MEASURES AND PERSONAL PROTECTIVE EQUIPMENT

Respiratory Protection *(specify type)*

For exposures above the PEL, use an MSHA/NIOSH-approved supplied-air respirator with full facepiece, hood, or helmet in continuous flow mode, or use a self-contained breathing apparatus (SCBA) operated in pressure demand or other positive pressure mode.

Ventilation

Local exhaust or general mechanical systems recommended.

Protective Gloves	Eye Protection	Other Protective Clothing
Nitrile Rubber, Viton, or Teflon	Chemical Goggles or Face Mask	Rubber Apron

Work/Hygiene Practices

Always wash hands thoroughly after using chemical; never bring food, drink, or smoking materials into vicinity of chemicals.

CYCLOHEXANOL

$C_6H_{11}OH$ CAS: 110-87-7

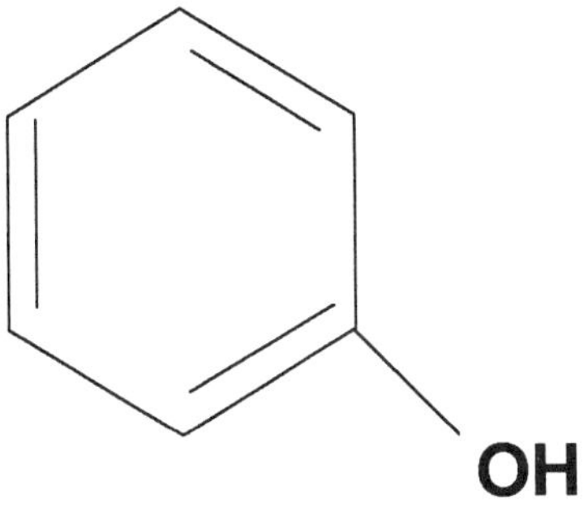

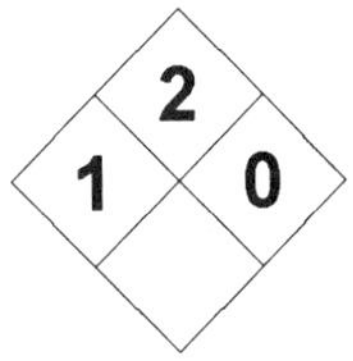

IDENTIFICATION AND TYPICAL USES

Cyclohexanol is a colorless, hygroscopic, viscous liquid with a camphor- or methanol-like odor. It is used in the manufacture of soap (to incorporate solvents and phenolic insecticides). It is a source of adipic acid for nylon 66. It is also used in textile finishing, as a solvent for alkyd and phenolic resins, as a blending agent in lacquers, paints, and varnishes, in finish removers and in emulsified products. Other applications are found in leather degreasing, in polishes, plasticizers, in plastics, and in germicidal agents.

RISK ASSESSMENT: HEALTH

General Assessment

Cyclohexanol is a human poison and is moderately toxic by *ingestion* and *inhalation.* It will also pass through unbroken skin (*absorption*) to cause toxic systemic effects. Its carcinogenicity in humans is not known but is questioned in animal studies. It has been shown to cause mutations in humans. Experimental teratogenic data have also been reported in animal tests.

Inhalation results in lachrymation, intoxication, loss of coordination, narcosis, and convulsions. It will irritate the nose, throat, and eyes. Narcotic effects include pronounced headache and may cause depression of the central nervous system, although this effect is not completely confirmed in the references.

Skin and eye contact may produce mild to moderate irritation on contact. It will pass through unbroken skin to cause toxic systemic effects including tremors and hypothermia. Absorption through the skin or inhalation of cyclohexanol vapors may cause toxic reactions in the brain, liver, kidney, and heart (in animal studies, it produced degenerative changes in the brain, heart, liver, and kidneys).

On ingestion, cyclohexanol is expected to produce symptoms of irritation and narcosis with toxic effects on target organs such as the liver and kidneys.

Acute Health Effects

The following acute (short-term) health effects may occur immediately or shortly after exposure to cyclohexanol:

Skin: Moderate irritation with redness and possibly painful itching. Will pass through unbroken skin.

Eye: Severe irritation with a potential for corneal injury which may be irreversible depending upon length of contact and concentration of solution.

Lung: Nose and throat irritation. Produces intoxicating effects and may cause lung effects as well.

CNS: Unconfirmed effects by all routes of exposure. There may be tremor, hypothermia, and other unspecified effects.

Chronic Health Effects

The following chronic (long-term) health effects may occur at some time after exposure to cyclohexanol and can last for months or even years:

Cancer Hazards: According to information presented in the references, cyclohexanol has not been adequately tested for its ability to cause cancer in animals. Some studies indicate that it causes mutations in humans, while others are inconclusive in this regard. Some studies have listed its cancer causing properties as questioned in animal studies.

Reproductive Hazard: According to information presented in the references, cyclohexanol has produced an increase in the mortality of the offspring of test animals (mice). The teratogenicity of cyclohexanol in humans has not been adequately studied.

Other Chronic Effects: Prolonged skin contact may cause drying, cracking, and chapping of the affected area. Repeated or long-term exposures to cyclohexanol may affect the liver and kidneys.

Recommended Risk-Reduction Measures

Personnel should avoid direct contact with cyclohexanol. The exact nature of the toxicity of this chemical is not entirely understood in the references. Therefore, prudent risk management requires it be treated as though it posed a significant health risk in the event of overexposure. If a less toxic chemical cannot be substituted for a hazardous substance, then *engineering controls* are the most effective method of reducing exposures. The best protection is to enclose operations and/or provide local exhaust ventilation at the site of chemical release. While not always operationally feasible, isolating operations can also reduce exposure. Using respiratory protection is less effective than the controls mentioned above, but is still advisable whenever working with or around cyclohexanol. For exposures above the PEL, a supplied-air respirator with full facepiece operated in continuous flow mode, or a self-contained breathing apparatus (SCBA) with full facepiece and pressure demand are the recommended respiratory protection methods of choice. If a full facepiece is not available, then chemical goggles should be worn to protect the eyes. Whenever a chemical splash hazard exists, a face shield and a rubber apron should be worn. To prevent hand and skin exposures, nitrile rubber, viton, or Teflon gloves should be worn.

Administrative controls should also be in place to minimize the potential for human exposures. These may include written procedures or policies which specify the methods and techniques that will be practiced whenever personnel are to work with cyclohexanol.

All personnel should receive *training* on the use, hazards, protective measures, emergency actions, and other precautions per 29 CFR 1910.1200 (Hazard Communication), prior to the first assignment in an area where cyclohexanol is used or stored.

If symptoms develop or overexposure is suspected, the following tests are recommended:

- ☑ Kidney and liver function tests.

Any evaluation should include a careful history of past and present symptoms with an examination. Medical tests that simply look for existing damage are not a substitute for controlling exposures.

Other methods to reduce exposure include:

- ☑ Where possible, enclose operations and use local exhaust ventilation at the site of chemical release. If local exhaust ventilation or enclosure is not used, respiratory protection should be mandatory.
- ☑ Always ensure that proper protective clothing is worn when using chemical substances.
- ☑ Wash thoroughly immediately after exposure to cyclohexanol and at the end of the work shift or before eating, drinking, or smoking.
- ☑ Hazard warning information should be posted in the work area. In addition, as part of an on-going education and training program, all information on the health and safety hazards of cyclohexanol should be communicated to all potentially exposed workers.
- ☑ Eye wash stations should be provided in the immediate work area for emergency use. If there is a possibility of skin exposure to cyclohexanol, emergency shower facilities should also be provided.
- ☑ Personnel should never wear contaminated clothing home (family members can be exposed). Clothing should be laundered by personnel who have been trained on the hazards of cyclohexanol.

RISK ASSESSMENT: ENVIRONMENT

General Assessment

The environment is at risk of exposure during transportation, storage, disposal, or destruction of cyclohexanol. In almost every scenario, the threat of environmental exposure is contingent upon the proper handling of the chemical substance. Accidental spills, large or small, can result in fire, explosion, and possible contamination of the surrounding environmental mediums (water, soil, and air).

Cyclohexanol is considered a class IIIA combustible liquid (per OSHA 29 CFR 1910.106). It presents a moderate fire hazard in contact with incompatible materials or in vapor form. These characteristics require extreme caution in handling, storage, transportation, and disposal. It is incompatible with common peroxides and other oxidizers and contact can cause fire or explosion. It will ignite on contact with chromium trioxide. Therefore, special consideration is required during any emergency situation involving a leak or spill of cyclohexanol. Should cyclohexanol ever come into contact with these incompatible substances during use, transportation, storage, or disposal, the formation of highly toxic and/or highly explosive commodities (including hydrogen chloride) is extremely possible.

The proper disposal method for cyclohexanol is to burn it in a chemical incinerator equipped with an afterburner and scrubber.

Cyclohexanol may enter the environment through industrial discharges or spills.

Acute Ecological Effects

Acute (short-term) toxic effects may include the death of animals, birds, or fish and death or low growth rate in plants. Acute effects are seen 2 to 4 days after animals or plants are exposed to cyclohexanol.

Insufficient data are available on the short-term effects of cyclohexanol exposure to aquatic life, birds, plants, or land animals.

Chronic Ecological Effects

Chronic (long-term) toxic effects may include shortened life span, reproductive problems, lower fertility, and changes in appearance or behavior in exposed animals. These effects can be seen long after first exposure(s) to toxic chemicals.

Insufficient data are available on the long-term effects of cyclohexanol exposure to aquatic life, birds, plants, or land animals.

Water Solubility

Cyclohexanol is nearly insoluble in water. Concentrations of less than 100 milligrams may mix with a liter of water, depending on water pH and chemical concentration of cyclohexanol.

Persistence in the Environment

Cyclohexanol is non-persistent in water, with a half-life of less than 2 days. The half-life of a pollutant is the amount of time it takes for one half of the chemical to be degraded. Virtually 100% of cyclohexanol will eventually end up in air.

Bioaccumulation in Aquatic Organisms

Some substances increase in concentration, or *bioaccumulate,* in living organisms as they breathe contaminated air, drink contaminated water, or eat contaminated food. These chemicals can become concentrated in the tissues and internal organs of animals as well as humans.

The concentration of cyclohexanol found in fish tissue is expected to be somewhat higher than the average concentration of cyclohexanol in the water from which the fish was taken.

Recommended Risk-Reduction Measures

Proper training of all transporters will reduce the likelihood of a mishap or accident resulting in a leak or spill to the environment. The correct use of DOT labeling (per the MSDS) on all containers, trucks, and rail cars will enable emergency responders to react properly and quickly to any disaster thereby reducing the potential risk to the environment and to personnel.

Storage of cyclohexanol should be segregated from incompatible chemicals to minimize the risk of cross-contamination or contact. Buildings designated for storage should be equipped with appropriate fire protection systems (alarms, sprinklers, emergency lighting, portable extinguishers).

If a spill or leak to the environment has occurred, fire department, emergency response and/or hazardous materials spill personnel should be notified immediately. Cleanup should be attempted only by those trained in proper spill containment procedures. Contaminated soils should be removed for incineration and replaced with clean soil. If cyclohexanol should contact the water table, aquifer, or navigable waterway, cleanup should be started immediately. It is only slightly soluble in water and total remediation may be possible. The local and/or state emergency response authorities must be notified. A comprehensive emergency response or disaster preparedness/recovery plan should be in place prior to any operations involving the use, transportation, storage, or disposal of cyclohexanol. If cyclohexanol is spilled or leaked, the following specific steps are recommended:

- ☑ Restrict persons not wearing protective clothing from area of spill or leak until cleanup is complete and area can be opened for normal work.
- ☑ Remove all ignition sources, ventilate area of leak.
- ☑ Absorb liquids in vermiculite, dry sand, earth, or similar material and deposit in sealed containers.

☑ If gas is leaking, stop the flow of gas. If the source is a cylinder and the leak cannot be stopped in place, remove cylinder to a safe place in the open air and repair or allow to empty.

☑ It may be necessary to dispose of cyclohexanol as a hazardous waste. The responsible state agency or the regional office of the federal Environmental Protection Agency (EPA) should be contacted for specific recommendations.

RISK ASSESSMENT: BUSINESS

General Assessment

Accidents or mishaps involving cyclohexanol can present a significant threat to business operations. The loss or damage of equipment or facilities can significantly affect fiscal viability. Lawsuits that may result from personnel injury/death, public exposures, and/or environmental contamination will also require a serious expenditure of resources. Media attention surrounding an injury, death, or environmental damage can also result in a loss of profits and loss of current as well as future business.

Recommended Risk-Reduction Measures

Company attorneys, safety and health professionals, and environmental specialists should be involved in the development of any procedures dealing with safety or emergency response. A company official should be pre-designated as a public relations officer with specific training in dealing with the press. Corporate plans and policies should be developed, approved, and implemented long before any need for such arises.

REFERENCES

American Conference of Governmental Industrial Hygienists. 1988. *Documentation of the Threshold Limit Values and Biological Exposure Indices*, 5th Edition (with updates). Cincinnati: ACGIH

Bueche, J. R. 1972. *Principles of Physics*, 2nd Edition. New York: McGraw-Hill.

Calabrese, E. J. 1994. *Biological Effects of Low Level Exposures.* New York: CRC Press/Lewis Publishers.

Cockerman, L. G., B. S. Shane. 1994. *Basic Environmental Toxicology.* New York: CRC Press/Lewis Publishers.

Gardner, R. W. 1994. *Chemical Intolerance.* New York: CRC Press/Lewis Publishers.

Hallenbeck, W. H. 1993. *Quantitative Risk Assessment for Environmental and Occupational Health,* 2nd Edition. New York: CRC Press/Lewis Publishers.

Jones, F. E. 1994. *Toxic Organic Vapors in the Workplace.* New York: CRC Press/Lewis Publishers.

Keith, L. H. and M. M. Walker. 1995. *Handbook of Air Toxics.* New York: CRC Press/Lewis Publishers.

Landis, W. G., M. H. Yu. 1995. *Introduction to Environmental Toxicology.* New York: CRC Press/Lewis Publishers.

Langerman, N. 1994 *Precautionary Labels for Chemical Containers.* New York: CRC Press/Lewis Publishers.

Lauwerys, R. R., P. Hoet. 1993. *Industrial Chemical Exposure.* New York: CRC Press/Lewis Publishers.

Lewis, R. J., Sr. 1992. *Sax's Dangerous Properties of Industrial Materials,* Volumes 1, 2, and 3, 8th Edition. New York: Van Nostrand Reinhold.

Lide, D. R. 1995. *Handbook of Chemistry and Physics,* 76th Edition. New York: CRC Press/Lewis Publishers.

Manahan, S. E. 1994. *Environmental Chemistry,* 6th Edition. New York: CRC Press/Lewis Publishers.

Mellström, G. A., J. E. Wahlberg, and H. I. Maibach. 1994. *Protective Gloves for Occupational Use.* New York: CRC Press/Lewis Publishers.

Meyer, E. 1990. *Chemistry of Hazardous Materials.* Englewood Cliffs, NJ: Prentice Hall.

Patnaik, P. 1992. *A Comprehensive Guide to the Hazardous Properties of Chemical Substances.* New York: Van Nostrand Reinhold.

Proctor, N. H., J. P. Hughes. 1991. *Chemical Hazards of the Workplace.* New York: Van Nostrand Reinhold.

Quigley, D. R. 1994. *Handbook of Emergency Chemical Management.* New York: CRC Press/Lewis Publishers.

Rea, W. J. 1995. *Chemical Sensitivity,* Vol. I-III. New York: CRC Press/Lewis Publishers.

Richardson, M. 1995. *Dictionary of Substance and Their Effects,* Vol. I-VI. New York: CRC Press/Lewis Publishers.

Sacarello, H. L. A. 1994. *The Comprehensive Handbook of Hazardous Materials.* New York: CRC Press/Lewis Publishers.

Sheftel, V. O. 1995. *Handbook of Toxic Properties of Monomers and Additives.* New York: CRC Press/Lewis Publishers.

Torkelson, T. R. V. K. Rowe. 1981. Halogenated aliphatic hydrocarbons. *Patty's Industrial Hygiene and Toxicology,* Vol. 2B, *Toxicology,* pp 3568-3572. New York: Wiley-Interscience.

U.S. Department of Health and Human Services, National Institute for Occupational Safety and Health. 1994. *NIOSH Pocket Guide to Chemical Hazards.* Washington, D.C.: U.S. Government Printing Office.

MATERIAL SAFETY DATA SHEET

CHEMICAL NAME	
CYCLOHEXANONE	**CRC PRESS/LEWIS PUBLISHERS DISCLAIMER:** THE INFORMATION AND RECOMMENDATIONS PRESENTED HEREIN ARE BASED ON SOURCES BELIEVED TO BE RELIABLE. CRC MAKES NO REPRESENTATION ON ITS COMPLETENESS OR ACCURACY. IT IS THE RESPONSIBILITY OF THE USER TO DETERMINE THE CHEMICAL'S SUITABILITY FOR ITS INTENDED USE, THE CHEMICAL'S SAFE USE, AND THE CHEMICAL'S PROPER DISPOSAL. NO REPRESENTATIONS AND/OR WARRANTIES, EITHER EXPRESSED OR IMPLIED, OF THE MERCHANTABILITY OR FITNESS FOR A PARTICULAR PURPOSE, OR OF ANY OTHER NATURE, ARE MADE WITH RESPECT TO THE INFORMATION PROVIDED IN THIS MSDS OR TO THE CHEMICAL TO WHICH INFORMATION MAY REFER. CRC NEITHER ASSUMES NOR AUTHORIZES ANY OTHER PERSON TO ASSUME FOR IT, ANY OTHER ADDITIONAL RESPONSIBILITY OR LIABILITY FOR THE USE OF, OR RELIANCE UPON, THIS INFORMATION.

HAZARD WARNING INFORMATION

HEALTH	FIRE	REACTIVE	OTHER	DEGREE OF HAZARD	COLOR CODING	OTHER CODES
1	2	0		0 = Minimum Hazard 1 = Slight Hazard 2 = Moderate Hazard 3 = Serious Hazard 4 = Severe Hazard	HEALTH = BLUE FIRE = RED REACTIVITY = YELLOW OTHER = WHITE	OX = Oxidizer ACID = Acid ALK = Alkali COR = Corrosive W = Use No Water

SECTION I - GENERAL INFORMATION

Characterization	**Ketone**	RCRA Number	**U057**
		EPA Class	**Toxic Waste**
DOT Proper Shipping Name	**Cyclohexanone**	Chemical Abstract Service (CAS) Number	**108-94-1**
DOT Hazard Class and Label Requirements	**Flammable or Combustible Liquid**	DOT Emergency Guide Code	**26**
DOT Identification Number	**UN 1915**	Chemical Formula	**$C_6H_{10}O$**

Synonyms
Cyclohexyl ketone; pimelic ketone; ketohexamethylene; Nadone; Anone; Hytrol O; Sextone.

SECTION II - HAZARDOUS INGREDIENTS AND IDENTITY INFORMATION

Hazardous Components (specific identity)	OSHA Exposure Criteria	NIOSH Exposure Criteria	Immediately Dangerous to Life and Health (IDLH)	ACGIH Exposure Criteria
Cyclohexanone (derivation: By passing cyclohexanol over copper with air at 280°F (138°C), also by oxidation of cyclohexanol with chromic acid or oxide). **1 ppm = 4.00 mg/m³**	**PEL (skin):** **50 ppm** **200 mg/m³** **STEL:** **Not Established**	**REL (skin):** **25 ppm** **100 mg/m³** **STEL:** **Not Established**	**700 ppm**	**TLV (skin):** **25 ppm** **100 mg/m³** **STEL:** **Not Established**

SECTION III - PHYSICAL AND CHEMICAL CHARACTERISTICS

Boiling Point	**312°F (155°C)**	Specific Gravity (H_2O = 1)	**0.9**
Vapor Pressure (gas)	**5 at 69°F (20°C)**	Molecular Weight	**98.2**
Vapor Density (Air = 1)	**3.4**	Melting Point	**-49°F (-45°C)**

Solubility
Sparingly soluble in water (15%), soluble in alcohol, ethanol, ether, acetone, benzene, and chloroform.

Appearance and Odor
Colorless or pale yellow oily liquid with an odor of acetone and peppermint.

SECTION IV - FIRE AND EXPLOSION HAZARD DATA

Flash Point (method used)	**146°F (63°C) closed cup**	Explosive Limits in Air % by Volume	**LEL: 1.1% UEL: 9.4%**
NFPA Classification	**Class IIIA Combustible Liquid**	Autoignition Temperature	**788°F (420°C)**

Extinguishing Media
Use dry chemical, carbon dioxide, water, or alcohol foam (best for large fires).

Special Fire Fighting Procedures
Wear full protective clothing and self-contained breathing apparatus (SCBA). Poisonous gases are produced in fire. Use water spray to keep fire-exposed containers cool. Containers may explode in fire.

Unusual Fire and Explosion Hazards
Combustible liquid but a dangerous fire hazard when exposed to heat. At high temperatures, forms explosive mixtures with air. Violent reactions can occur with nitric acid. Chemical will react on contact with oxidizable materials. Moderate explosion hazard when in the form of vapor.

SECTION V - REACTIVITY DATA

Stability		Conditions to Avoid **Keep away from sources of heat (including direct sunlight) or flame. Large quantities should be stored in metal tanks or drums. Keep away from incompatible materials.**
Stable	Unstable	Incompatibility *(materials to avoid)*
X		**Strong oxidizers (permanganates, peroxides, nitrates, chlorates, and perchlorates). Can react with nitric acid to form violently explosive materials. Mixed with hydrogen peroxide, produces explosive peroxides.**
Hazardous Polymerization		Conditions to Avoid **Under normal conditions of temperature and pressure, hazardous polymerization of cyclohexanone is not expected to occur.**
May Occur	Will Not Occur	Hazardous Decomposition or By-products
	X	**When heated to decomposition, cyclohexanone can emit highly toxic/poisonous and acrid fumes and gases.**

SECTION VI - HEALTH HAZARD DATA

Primary Route(s) of Entry:	Inhalation?	Absorption (skin)?	Ingestion?
	X		X

Health Hazards

INHALATION: Symptoms include salivation, hypothermia, decreased breathing rate, and anesthesia. May also cause central nervous system depression with lung congestion and cell proliferation in the spleen. There can be slight kidney and liver damage with mild narcosis. Can also lead to pulmonary edema (fluid buildup in the lungs).

SKIN & EYES: Irritating to the skin, may cause irritation and dermatitis. An eye irritant with possibility of corneal damage.

INGESTION: Irritation and narcosis, anesthesia, and respiratory difficulties.

Carcinogenicity	NTP Listed?	IARC Cancer Review Group?	OSHA Regulated?	Target Organs?
Unknown Human Questioned Animal	No	No	29 CFR 1910.1000 Table Z-1	Respiratory system, eyes, skin, liver, kidney, spleen.

Medical Conditions Generally Aggravated by Exposure

None reported. Skin conditions (dermatitis) or respiratory problems (asthma) may be aggravated.

Emergency and First-aid Procedures

Eye contact: Flush immediately with water for 15 minutes (minimum), seek medical attention. Skin contact: Remove all contaminated clothing. Immediately wash area with large amounts of soap and water. Seek medical assistance. For inhalation: Remove the person from exposure. Provide respiratory assistance and CPR, if required. Transfer to medical facility. Keep victim warm and at rest. Observe for 24-48 hours for lungs effects (pulmonary edema). If swallowed: Seek medical attention immediately.

SECTION VII - PRECAUTIONS FOR SAFE HANDLING AND USE

Steps to be Taken in Case Material is Released or Spilled

Absorb in vermiculite, earth, or similar material and deposit in sealed containers. Ventilate area of spill or leak. Restrict those not involved in cleanup from entering area. Remove all sources of ignition.

Preferred Waste Disposal Method

Burn in chemical incinerator equipped with an afterburner and scrubber.

Precautions to be Taken in Handling and Storage

Store to avoid contact with incompatible materials in tightly closed containers in cool, well-ventilated area away from heat and flame. Use only non-sparking tools to open and close containers.

Other Precautions and Warnings

Sources of ignition are prohibited where cyclohexanone is used, handled, or stored. Heat may cause containers to build up pressure and explode. Containers must be grounded and bonded during transfer.

SECTION VIII - CONTROL MEASURES AND PERSONAL PROTECTIVE EQUIPMENT

Respiratory Protection *(specify type)*

For exposures above the PEL, use an MSHA/NIOSH-approved supplied-air respirator with full facepiece, hood, or helmet in continuous flow mode, or use a self-contained breathing apparatus (SCBA) operated in pressure demand or other positive pressure mode.

Ventilation

Local exhaust or general mechanical systems recommended.

Protective Gloves	Eye Protection	Other Protective Clothing
Nitrile Rubber, Viton, or Teflon	Chemical Goggles or Face Mask	Rubber Apron

Work/Hygiene Practices

Always wash hands thoroughly after using chemical; never bring food, drink, or smoking materials into vicinity of chemicals.

CYCLOHEXANONE

$C_6H_{10}O$ CAS: 108-94-1

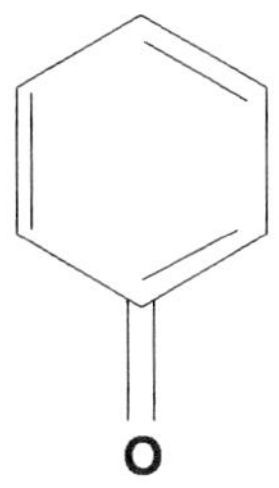

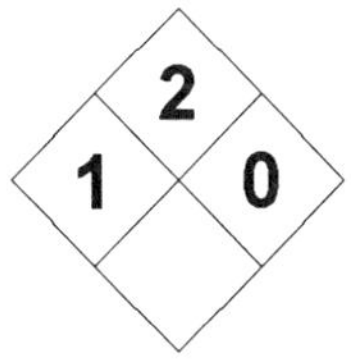

IDENTIFICATION AND TYPICAL USES

Cyclohexanone is a colorless to pale yellow, oily liquid with an acetone and peppermint-like odor. It is used in the production of adipid acid for making nylon. It is also used in the preparation of cyclohexane resins, fats, waxes, shellac and wood stains, rubber, paint and varnish removers, spot removers, for the degreasing of metals, in polishes, as a leveling agent, in dyeing and delustering silk, as a lubricating oil additive, and in the manufacture of some pesticides (e.g., DDT).

RISK ASSESSMENT: HEALTH

General Assessment

Cyclohexanone is moderately toxic by *ingestion* and *inhalation*. It will also pass through unbroken skin (*absorption*) to cause toxic systemic effects. Its carcinogenicity in humans is not known but is questioned in some animal studies. It has been shown to cause mutations (genetic changes) in human cells. Experimental teratogenic data have also been reported in animal tests.

Inhalation can causes changes in the sense of smell, conjunctiva irritation, and unspecified respiratory changes. It will irritate the nose, throat, and eyes. Mild narcotic properties have also been recorded, including pronounced headache and may cause depression of the central nervous system, although this effect is not completely confirmed in the references. Other symptoms on inhalation include salivation, hypothermia, decreased breathing rate, and anesthesia. There may also be lung congestion and cell proliferation in the spleen. Kidney and liver damage with mild narcosis (but no blood effects) may also be possible.

Skin and eye contact may produce mild to moderate irritation. There are also reports of skin absorption occurring on exposure to cyclohexanone. Pure liquid in contact with the eye may result in corneal damage.

On ingestion, cyclohexanone is expected to produce symptoms of irritation and narcosis, anesthesia, and some respiratory difficulties.

Acute Health Effects

The following acute (short-term) health effects may occur immediately or shortly after exposure to cyclohexanone:

Skin: Will defat the skin and cause irritation and dermatitis. Absorption is likely. Pure liquid will cause severe irritation.

Eye: Severe irritation with a potential for corneal injury or damage.

Lung: Nose and throat irritation. Produces intoxicating effects and may cause lung effects as well.

CNS: Unconfirmed effects by all routes of exposure. There may be depression, anesthesia, hypothermia, and other unspecified effects.

Chronic Health Effects

The following chronic (long-term) health effects may occur at some time after exposure to cyclohexanone and can last for months or even years:

Cancer Hazards: According to information presented in the references, cyclohexanone has not been adequately tested for its ability to cause cancer in animals. Some studies indicate that it causes mutations in humans, while others are inconclusive in this regard. Some studies have listed its cancer causing properties as questioned animal studies.

Reproductive Hazard: According to information presented in the references, cyclohexanone may have the ability to adversely affect reproduction in animals. The teratogenicity of cyclohexanone in humans has not been adequately studied.

Other Chronic Effects: Prolonged skin contact may cause drying, cracking, and chapping of the affected area. There may be skin sensitization and/or dermatitis in some individuals. Repeated or long-term exposures to cyclohexanone may affect the liver and kidneys.

Recommended Risk-Reduction Measures

Personnel should avoid direct contact with cyclohexanone. The exact nature of the toxicity of this chemical is not entirely understood in the references. Therefore, prudent risk management requires it be treated as though it posed a significant health risk in the event of overexposure. If a less toxic chemical cannot be substituted for a hazardous substance, then *engineering controls* are the most effective method of reducing exposures. The best protection is to enclose operations and/or provide local exhaust ventilation at the site of chemical release. While not always operationally feasible, isolating operations can also reduce exposure. Using respiratory protection is less effective than the controls mentioned above, but is still advisable whenever working with or around cyclohexanone. For exposures above the PEL, a supplied-air respirator with full facepiece operated in continuous flow mode, or a self-contained breathing apparatus (SCBA) with full facepiece and pressure demand are the recommended respiratory protection methods of choice. If a full facepiece is not available, then chemical goggles should be worn to protect the eyes. Whenever a chemical splash hazard exists, a face shield and a rubber apron should be worn. To prevent hand and skin exposures, impervious gloves should be worn. Glove manufacturers should be contacted to obtain permeation studies *before* glove selection has been made.

Administrative controls should also be in place to minimize the potential for human exposures. These may include written procedures or policies which specify the methods and techniques that will be practiced whenever personnel are to work with cyclohexanone.

All personnel should receive *training* on the use, hazards, protective measures, emergency actions, and other precautions per 29 CFR 1910.1200 (Hazard Communication), prior to the first assignment in an area where cyclohexanone is used or stored.

If symptoms develop or overexposure is suspected, the following tests are recommended:

☑ Kidney and liver function tests.

☑ Evaluation by a qualified allergist, with careful consideration of exposure history and special skin tests (may help diagnose allergy).

☑ Lung function tests, including Forced Vital Capacity (FVC).

Any evaluation should include a careful history of past and present symptoms with an examination. Medical tests that simply look for existing damage are not a substitute for controlling exposures. Also, since smoking can cause heart disease, emphysema, and other respiratory disorders, symptoms of exposure to cyclohexanone may be more pronounced (inflated) than those displayed by non-smokers. Proper risk assessment requires careful consideration of *all* possible factors that may be causing the appearance of symptoms in exposed workers.

Other methods to reduce exposure include:

☑ Where possible, enclose operations and use local exhaust ventilation at the site of chemical release. If local exhaust ventilation or enclosure is not used, respiratory protection should be mandatory. Also, if possible, automatic transfer of cyclohexanone from storage containers to process or work containers is recommended.

☑ Always ensure that proper protective clothing is worn when using chemical substances.

☑ Wash thoroughly immediately after exposure to cyclohexanone and at the end of the work shift or before eating, drinking, or smoking.

☑ Hazard warning information should be posted in the work area. In addition, as part of an on-going education and training program, all information on the health and safety hazards of cyclohexanone should be communicated to all potentially exposed workers.

☑ Eye wash stations should be provided in the immediate work area for emergency use. If there is a possibility of skin exposure to cyclohexanone, emergency shower facilities should also be provided.

☑ Personnel should never wear contaminated clothing home (family members can be exposed). Clothing should be laundered by personnel who have been trained on the hazards of cyclohexanone.

RISK ASSESSMENT: ENVIRONMENT

General Assessment

The environment is at risk of exposure during transportation, storage, disposal, or destruction of cyclohexanone. In almost every scenario, the threat of environmental exposure is contingent upon the proper handling of the chemical substance. Accidental spills, large or small, can result in fire, explosion, and possible contamination of the surrounding environmental mediums (water, soil, and air).

Cyclohexanone is considered a class IIIA combustible liquid (per OSHA 29 CFR 1910.106). It presents a moderate fire and explosion hazard risk, especially in contact with incompatible materials. These characteristics require extreme caution in handling, storage, transportation, and disposal. It is incompatible with common peroxides and other oxidizers and contact can cause fire or explosion. It explodes violently when added to nitric acid and, on mixing with hydrogen peroxide, it produces oily, explosives peroxides. Therefore, special consideration is required during any emergency situation involving a leak or spill of cyclohexanone. Should cyclohexanone ever come into contact with these incompatible substances during use, transportation, storage, or disposal, the formation of highly toxic and/or highly explosive commodities (including hydrogen chloride) is extremely possible.

The proper disposal method for cyclohexanone is to burn it in a chemical incinerator equipped with an afterburner and scrubber.

Cyclohexanone may enter the environment through industrial discharges or spills.

Acute Ecological Effects

Acute (short-term) toxic effects may include the death of animals, birds, or fish and death or low growth rate in plants. Acute effects are seen 2 to 4 days after animals or plants are exposed to cyclohexanone.

Insufficient data are available on the short-term effects of cyclohexanone exposure to aquatic life, birds, plants, or land animals.

Chronic Ecological Effects

Chronic (long-term) toxic effects may include shortened life span, reproductive problems, lower fertility, and changes in appearance or behavior in exposed animals. These effects can be seen long after first exposure(s) to toxic chemicals.

Insufficient data are available on the long-term effects of cyclohexanone exposure to aquatic life, birds, plants, or land animals.

Water Solubility

Cyclohexanone is nearly insoluble in water. Concentrations of less than 100 milligrams may mix with a liter of water, depending on water pH and chemical concentration of cyclohexanone.

Persistence in the Environment

Cyclohexanone is non-persistent in water, with a half-life of less than 2 days. The half-life of a pollutant is the amount of time it takes for one half of the chemical to be degraded. Virtually 100% of cyclohexanone will eventually end up in air.

Bioaccumulation in Aquatic Organisms

Some substances increase in concentration, or *bioaccumulate,* in living organisms as they breathe contaminated air, drink contaminated water, or eat contaminated food. These chemicals can become concentrated in the tissues and internal organs of animals as well as humans.

The concentration of cyclohexanone found in fish tissue is expected to be somewhat higher than the average concentration of cyclohexanone in the water from which the fish was taken.

Recommended Risk-Reduction Measures

Proper training of all transporters will reduce the likelihood of a mishap or accident resulting in a leak or spill to the environment. The correct use of DOT labeling (per the MSDS) on all containers, trucks, and rail cars will enable emergency responders to react properly and quickly to any disaster thereby reducing the potential risk to the environment and to personnel.

Storage of cyclohexanone should be segregated from incompatible chemicals to minimize the risk of cross-contamination or contact. Buildings designated for storage should be equipped with appropriate fire protection systems (alarms, sprinklers, emergency lighting, portable extinguishers).

If a spill or leak to the environment has occurred, fire department, emergency response and/or hazardous materials spill personnel should be notified immedi-

ately. Cleanup should be attempted only by those trained in proper spill containment procedures. Contaminated soils should be removed for incineration and replaced with clean soil. If cyclohexanone should contact the water table, aquifer, or navigable waterway, cleanup should be started immediately. It is only slightly soluble in water and total remediation may be possible. The local and/or state emergency response authorities must be notified. A comprehensive emergency response or disaster preparedness/recovery plan should be in place prior to any operations involving the use, transportation, storage, or disposal of cyclohexanone. If cyclohexanone is spilled or leaked, the following specific steps are recommended:

- ☑ Restrict persons not wearing protective clothing from area of spill or leak until cleanup is complete and area can be opened for normal work.
- ☑ Remove all ignition sources, ventilate area of leak.
- ☑ Absorb liquids in vermiculite, dry sand, earth, or similar material and deposit in sealed containers.
- ☑ It may be necessary to dispose of cyclohexanone as a hazardous waste. The responsible state agency or the regional office of the federal Environmental Protection Agency (EPA) should be contacted for specific recommendations.

RISK ASSESSMENT: BUSINESS

General Assessment

Accidents or mishaps involving cyclohexanone can present a significant threat to business operations. The loss or damage of equipment or facilities can significantly affect fiscal viability. Lawsuits that may result from personnel injury/death, public exposures, and/or environmental contamination will also require a serious expenditure of resources. Media attention surrounding an injury, death, or environmental damage can also result in a loss of profits and loss of current as well as future business.

Recommended Risk-Reduction Measures

Company attorneys, safety and health professionals, and environmental specialists should be involved in the development of any procedures dealing with safety or emergency response. A company official should be pre-designated as a public relations officer with specific training in dealing with the press. Corporate plans and policies should be developed, approved, and implemented long before any need for such arises.

REFERENCES

American Conference of Governmental Industrial Hygienists. 1988. *Documentation of the Threshold Limit Values and Biological Exposure Indices,* 5th Edition (with updates). Cincinnati: ACGIH

Bueche, J. R. 1972. *Principles of Physics,* 2nd Edition. New York: McGraw-Hill.

Calabrese, E. J. 1994. *Biological Effects of Low Level Exposures.* New York: CRC Press/Lewis Publishers.

Cockerman, L. G., B. S. Shane. 1994. *Basic Environmental Toxicology.* New York: CRC Press/Lewis Publishers.

Gardner, R. W. 1994. *Chemical Intolerance.* New York: CRC Press/Lewis Publishers.

Hallenbeck, W. H. 1993. *Quantitative Risk Assessment for Environmental and Occupational Health,* 2nd Edition. New York: CRC Press/Lewis Publishers.

Keith, L. H. and M. M. Walker. 1995. *Handbook of Air Toxics.* New York: CRC Press/Lewis Publishers.

Langerman, N. 1994 *Precautionary Labels for Chemical Containers.* New York: CRC Press/Lewis Publishers.

Lauwerys, R. R., P. Hoet. 1993. *Industrial Chemical Exposure.* New York: CRC Press/Lewis Publishers.

Lewis, R. J., Sr. 1992. *Sax's Dangerous Properties of Industrial Materials,* Volumes 1, 2, and 3, 8th Edition. New York: Van Nostrand Reinhold.

Meyer, E. 1990. *Chemistry of Hazardous Materials.* Englewood Cliffs, NJ: Prentice Hall.

Patnaik, P. 1992. *A Comprehensive Guide to the Hazardous Properties of Chemical Substances.* New York: Van Nostrand Reinhold.

Proctor, N. H., J. P. Hughes. 1991. *Chemical Hazards of the Workplace.* New York: Van Nostrand Reinhold.

Quigley, D. R. 1994. *Handbook of Emergency Chemical Management.* New York: CRC Press/Lewis Publishers.

Rea, W. J. 1995. *Chemical Sensitivity,* Vol. I-III. New York: CRC Press/Lewis Publishers.

Richardson, M. 1995. *Dictionary of Substance and Their Effects,* Vol. I-VI. New York: CRC Press/Lewis Publishers.

Sacarello, H. L. A. 1994. *The Comprehensive Handbook of Hazardous Materials.* New York: CRC Press/Lewis Publishers.

Torkelson, T. R. V. K. Rowe. 1981. Halogenated aliphatic hydrocarbons. *Patty's Industrial Hygiene and Toxicology,* Vol. 2B, *Toxicology,* pp 3568-3572. New York: Wiley-Interscience.

U.S. Department of Health and Human Services, National Institute for Occupational Safety and Health. 1994. *NIOSH Pocket Guide to Chemical Hazards.* Washington, D.C.: U.S. Government Printing Office.

MATERIAL SAFETY DATA SHEET

CHEMICAL NAME	
CYCLOHEXENE	**CRC PRESS/LEWIS PUBLISHERS DISCLAIMER:** THE INFORMATION AND RECOMMENDATIONS PRESENTED HEREIN ARE BASED ON SOURCES BELIEVED TO BE RELIABLE. CRC MAKES NO REPRESENTATION ON ITS COMPLETENESS OR ACCURACY. IT IS THE RESPONSIBILITY OF THE USER TO DETERMINE THE CHEMICAL'S SUITABILITY FOR ITS INTENDED USE, THE CHEMICAL'S SAFE USE, AND THE CHEMICAL'S PROPER DISPOSAL. NO REPRESENTATIONS AND/OR WARRANTIES, EITHER EXPRESSED OR IMPLIED, OF THE MERCHANTABILITY OR FITNESS FOR A PARTICULAR PURPOSE, OR OF ANY OTHER NATURE, ARE MADE WITH RESPECT TO THE INFORMATION PROVIDED IN THIS MSDS OR TO THE CHEMICAL TO WHICH INFORMATION MAY REFER. CRC NEITHER ASSUMES NOR AUTHORIZES ANY OTHER PERSON TO ASSUME FOR IT, ANY OTHER ADDITIONAL RESPONSIBILITY OR LIABILITY FOR THE USE OF, OR RELIANCE UPON, THIS INFORMATION.

HAZARD WARNING INFORMATION

HEALTH	FIRE	REACTIVE	OTHER	DEGREE OF HAZARD	COLOR CODING	OTHER CODES
1	3	0		0 = Minimum Hazard 1 = Slight Hazard 2 = Moderate Hazard 3 = Serious Hazard 4 = Severe Hazard	HEALTH = BLUE FIRE = RED REACTIVITY = YELLOW OTHER = WHITE	OX = Oxidizer ACID = Acid ALK = Alkali COR = Corrosive W = Use No Water

SECTION I - GENERAL INFORMATION

Characterization	RCRA Number	EPA Class
Hydrocarbon, Aliphatic	**None**	**Not Applicable**

DOT Proper Shipping Name	Chemical Abstract Service (CAS) Number
Cyclohexene	**110-83-8**
DOT Hazard Class and Label Requirements	DOT Emergency Guide Code
Flammable Liquid	**29**
DOT Identification Number	Chemical Formula
UN 2256	C_6H_{10}

Synonyms
Benzene tetrahydride; terahydrobenzene; hexanaphthylene.

SECTION II - HAZARDOUS INGREDIENTS AND IDENTITY INFORMATION

Hazardous Components (specific identity)	OSHA Exposure Criteria	NIOSH Exposure Criteria	Immediately Dangerous to Life and Health (IDLH)	ACGIH Exposure Criteria
Cyclohexene (derivation: Catalytic dehydration of cyclohexanol and dehydrogenation of cyclohexane). **1 ppm = 3.42 mg/m³**	**PEL:** **300 ppm** **1050 mg/m³** **STEL:** **Not Established**	**REL:** **300 ppm** **1050 mg/m³** **STEL:** **Not Established**	**2000 ppm**	**TLV:** **300 ppm** **1050 mg/m³** **STEL:** **Not Established**

SECTION III - PHYSICAL AND CHEMICAL CHARACTERISTICS

Boiling Point	Specific Gravity (H_2O = 1)
181°F (83°C)	**0.81**
Vapor Pressure (gas)	Molecular Weight
67 at 69°F (20°C)	**84.2**
Vapor Density (Air = 1)	Melting Point
2.8	**-154°F (-103°C)**

Solubility
Insoluble in water, soluble in alcohol, benzene, acetone, and most organic solvents.

Appearance and Odor
Colorless liquid with sweet odor.

SECTION IV - FIRE AND EXPLOSION HAZARD DATA

Flash Point (method used)	Explosive Limits in Air % by Volume
11°F (-11.6°C) closed cup	**LEL: 1% UEL: 5%**
NFPA Classification	Autoignition Temperature
Class 1B Flammable Liquid	**590°F (310°C)**

Extinguishing Media
Use dry chemical, carbon dioxide, water, or alcohol foam.

Special Fire Fighting Procedures
Cyclohexene is a flammable liquid. Wear full protective clothing and self-contained breathing apparatus (SCBA). Poisonous gases are produced in fire. Use water spray to keep fire-exposed containers cool.

Unusual Fire and Explosion Hazards
Dangerous fire and explosion hazard when exposed to heat or flame. Containers may explode in fire. Chemical will react on contact with oxidizable materials. Moderate explosion hazard when in the form of vapor. It is susceptible to forming explosive peroxide on long exposures to air.

SECTION V - REACTIVITY DATA

Stability		Conditions to Avoid **Keep away from heat or flame. Normally stable, but do not allow to remain in prolonged contact with air (explosive and unstable peroxides may form).**
Stable	Unstable	Incompatibility *(materials to avoid)*
X		**Strong oxidizers, such as permanganates, peroxides, nitrates, chlorates, and perchlorates.**
Hazardous Polymerization		Conditions to Avoid **Under normal conditions of temperature and pressure, hazardous polymerization of cyclohexene is not expected to occur.**
May Occur	Will Not Occur	Hazardous Decomposition or By-products
	X	**When heated to decomposition, cyclohexene can emit highly toxic/poisonous gases.**

SECTION VI - HEALTH HAZARD DATA

Primary Route(s) of Entry:	Inhalation?	Absorption (skin)?	Ingestion?
	X		X

Health Hazards

INHALATION: Low-order toxicity by inhalation. However, it does act as an irritant and a central nervous system depressant. Can cause nausea, dizziness, vomiting, and depression. Higher concentrations can lead to coma and death from respiratory failure.

SKIN & EYES: Irritating to the skin, may cause drying, cracking, and chapping. Eye irritation likely.

INGESTION: Irritation and narcotic effects, including drowsiness. Low-order toxicity by ingestion.

Carcinogenicity	NTP Listed?	IARC Cancer Review Group?	OSHA Regulated?	Target Organs?
Unknown Human Unknown Animal	No	No	29 CFR 1910.1000 Table Z-1	Respiratory system, eyes, skin, CNS.

Medical Conditions Generally Aggravated by Exposure

None reported.

Emergency and First-aid Procedures

Eye contact: Flush immediately with water for 15 minutes (minimum), seek medical attention. Skin contact: Remove all contaminated clothing. Immediately wash area with large amounts of soap and water. Seek medical assistance. For inhalation: Remove the person from exposure. Provide respiratory assistance and CPR, if required. Transfer to medical facility. Keep victim warm and at rest. If swallowed: Seek medical attention immediately.

SECTION VII - PRECAUTIONS FOR SAFE HANDLING AND USE

Steps to be Taken in Case Material is Released or Spilled

Absorb spills in vermiculite, dry sand, earth, or similar material and deposit in sealed containers. Ventilate area of spill or leak. Restrict those not involved in cleanup from entering area. Remove all sources of ignition.

Preferred Waste Disposal Method

No Citation.

Precautions to be Taken in Handling and Storage

Store to avoid contact with incompatible materials in tightly closed containers in cool, well-ventilated area away from heat and flame. Use only non-sparking tools to open and close containers.

Other Precautions and Warnings

Sources of ignition are prohibited where cyclohexene is used, handled, or stored. Heat may cause containers to build up pressure and explode. Containers must be grounded and bonded during transfer of liquids from storage containers to process containers.

SECTION VIII - CONTROL MEASURES AND PERSONAL PROTECTIVE EQUIPMENT

Respiratory Protection *(specify type)*

For exposures above the PEL, use an MSHA/NIOSH-approved supplied-air respirator with full facepiece, hood, or helmet in continuous flow mode, or use a self-contained breathing apparatus (SCBA) operated in pressure demand or other positive pressure mode.

Ventilation

Local exhaust or general mechanical systems recommended.

Protective Gloves	Eye Protection	Other Protective Clothing
Nitrile Rubber, Viton, or Teflon	**Chemical Goggles or Face Mask**	**Rubber Apron**

Work/Hygiene Practices

Always wash hands thoroughly after using chemical; never bring food, drink, or smoking materials into vicinity of chemicals.

CYCLOHEXENE

C_6H_{10} CAS: 110-83-8

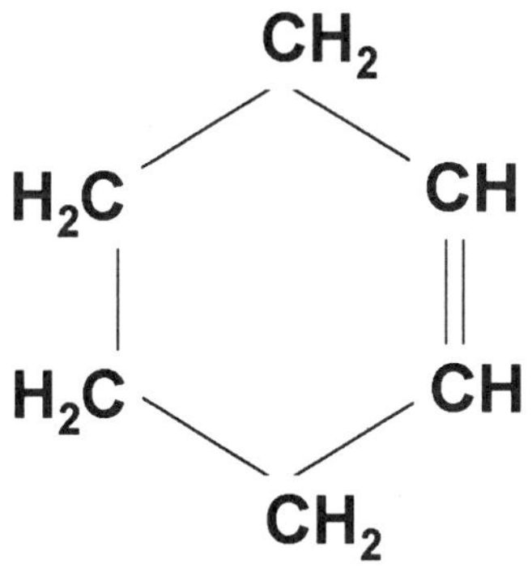

IDENTIFICATION AND TYPICAL USES

Cyclohexene is a colorless liquid with a sweet odor. It is used in the making of adipid acid, maleic acid, and butadiene. It is also used in oil extraction, as a stabilizer for high-octane gasoline, and in organic synthesis.

RISK ASSESSMENT: HEALTH

General Assessment

Cyclohexene is an acute irritant of moderate toxicity by *inhalation* and *ingestion*. The references contain relatively little information on the exact toxicological nature of this chemical. However, in sufficient or high concentrations, it has been shown to cause some toxic systemic effects through both routes of exposure. Skin contact can also produce some adverse topical effects and *absorption* has been reported but not confirmed.

Significant inhalation may cause irritation of the respiratory tract, eyes, and possible depression of the central nervous system. Symptoms may include nausea, dizziness, vomiting, and depression. At high concentrations, it is a narcotic and can lead to coma and death due to respiratory failure.

Skin and eye contact may produce mild to moderate irritation on contact. Ingestion can produce narcotic effects as those noted for inhalation above.

Acute Health Effects

The following acute (short-term) health effects may occur immediately or shortly after exposure to cyclohexene:

Skin: Mild to severe irritation possible.

Eye: Mild to moderate irritation. Damage is possible depending upon concentration of liquid and duration of contact.

Lung: May cause dizziness, vomiting, and nausea. Primary result of inhalation of cyclohexene is its action on the CNS.

CNS: A CNS depressant. Headache, dizziness, depression, and changes in mood.

Chronic Health Effects

The following chronic (long-term) health effects may occur at some time after exposure to cyclohexene and can last for months or even years:

Cancer Hazards: According to information presented in the references, cyclohexene has not been adequately tested for its ability to cause cancer in animals. Limited studies indicate an increased incidence of Hodgkin's disease in workers exposed to organic solvents. It is not known, however, whether cyclohexene will have this effect.

Reproductive Hazard: According to information presented in the references, cyclohexene has not been tested for its ability to adversely affect reproduction.

Other Chronic Effects: Prolonged skin contact may cause drying, cracking, and chapping of the affected area. Repeated or long-term exposures to cyclohexene may affect the liver and kidneys.

Recommended Risk-Reduction Measures

Personnel should avoid direct contact with cyclohexene. Even though the toxicity of this chemical is considered to be of relatively low order, it can still cause significant health effects if improperly used or if overexposures occur. If a less toxic chemical cannot be substituted for a hazardous substance, then *engineering controls* are the most effective method of reducing exposures. The best protection is to enclose operations and/or provide local exhaust ventilation at the site of chemical release. While not always operationally feasible, isolating operations can also reduce exposure. Using respiratory protection is less effective than the controls mentioned above, but is still advisable whenever working with or around cyclohexene. For exposures above the PEL (300 ppm), a supplied-air respirator with full facepiece operated in continuous flow mode, or a self-contained breathing apparatus (SCBA) with full facepiece and pressure demand are the recommended respiratory protection methods of choice. If a full facepiece is not available, then chemical goggles should be worn to protect the eyes. Whenever a chemical splash hazard exists, a face shield and a rubber apron should be worn. To prevent hand and skin exposures, acid resistant gloves should be worn.

Administrative controls should also be in place to minimize the potential for human exposures. These may include written procedures or policies which specify the methods and techniques that will be practiced whenever personnel are to work with cyclohexene.

All personnel should receive *training* on the use, hazards, protective measures, emergency actions, and other precautions per 29 CFR 1910.1200 (Hazard Communication), prior to the first assignment in an area where cyclohexene is used or stored.

If symptoms develop or overexposure is suspected, the following tests are recommended:

- ☑ Kidney and liver function tests.
- ☑ Lung function tests (including Forced Vital Capacity (FVC).

Any evaluation should include a careful history of past and present symptoms with an examination. Medical tests that simply look for existing damage are not a substitute for controlling exposures.

Other methods to reduce exposure include:

- ☑ Where possible, enclose operations and use local exhaust ventilation at the site of chemical release. If local exhaust ventilation or enclosure is not used, respiratory protection should be mandatory. Also, if possible, liquids should be transferred automatically from storage to process containers.
- ☑ Always ensure that proper protective clothing is worn when using chemical substances.
- ☑ Wash thoroughly immediately after exposure to cyclohexene and at the end of the work shift or before eating, drinking, or smoking.
- ☑ Hazard warning information should be posted in the work area. In addition, as part of an on-going education and training program, all information on the health and safety hazards of cyclohexene should be communicated to all potentially exposed workers.
- ☑ Eye wash stations should be provided in the immediate work area for emergency use. If there is a possibility of skin exposure to cyclohexene, emergency shower facilities should also be provided.
- ☑ Personnel should never wear contaminated clothing home (family members can be exposed). Clothing should be laundered by personnel who have been trained on the hazards of cyclohexene.

RISK ASSESSMENT: ENVIRONMENT

General Assessment

The environment is at risk of exposure during transportation, storage, disposal, or destruction of cyclohexene. In almost every scenario, the threat of environmental exposure is contingent upon the proper handling of the chemical substance. Accidental spills, large or small, can result in fire, explosion, and possible contamination of the surrounding environmental mediums (water, soil, and air).

Cyclohexene is considered a class 1B flammable liquid (per OSHA 29 CFR 1910.106). Its extremely low flash point and relatively low boiling point make this chemical a serious fire and explosion hazard concern. These characteristics require extreme caution in handling, storage, transportation, and disposal. It is incompatible with a common peroxides and other oxidizers and contact can cause fire or explosion. Therefore, special consideration is required during any emergency situation involving a leak or spill of cyclohexene. Should cyclohexene ever come into contact with these incompatible substances during use, transportation, storage, or disposal, the formation of highly toxic and/or highly explosive commodities is extremely possible.

Cyclohexene may enter the environment through industrial discharges or spills.

Acute Ecological Effects

Acute (short-term) toxic effects may include the death of animals, birds, or fish and death or low growth rate in plants. Acute effects are seen 2 to 4 days after animals or plants are exposed to cyclohexene.

This chemical has high acute toxicity to aquatic life. Insufficient data are available on the short-term effects of cyclohexene exposure to plants, birds or land animals.

Chronic Ecological Effects

Chronic (long-term) toxic effects may include shortened life span, reproductive problems, lower fertility, and changes in appearance or behavior in exposed animals. These effects can be seen long after first exposure(s) to toxic chemicals.

Cyclohexene has moderate chronic toxicity in aquatic life. Insufficient data are available on the long-term effects of cyclohexene to plants, birds, or land animals.

Water Solubility

Cyclohexene is nearly insoluble in water. Concentrations of less than 100 milligrams may mix with a liter of water.

Persistence in the Environment

Cyclohexene is non-persistent in water, with a half-life of less than 2 days. The half-life of a pollutant is the amount of time it takes for one half of the chemical to be degraded. Virtually 100% of cyclohexene will eventually end up in air.

Bioaccumulation in Aquatic Organisms

Some substances increase in concentration, or *bioaccumulate,* in living organisms as they breathe contaminated air, drink contaminated water, or eat contaminated food. These chemicals can become concentrated in the tissues and internal organs of animals as well as humans.

The concentration of cyclohexene found in fish tissue is expected to be somewhat higher than the average concentration of cyclohexene in the water from which the fish was taken.

Recommended Risk-Reduction Measures

Proper training of all transporters will reduce the likelihood of a mishap or accident resulting in a leak or spill to the environment. The correct use of DOT labeling (per the MSDS) on all containers, trucks, and rail cars will enable emergency responders to react properly and quickly to any disaster thereby reducing the potential risk to the environment and to personnel.

Storage of cyclohexene should be segregated from incompatible chemicals to minimize the risk of cross-contamination or contact. Buildings designated for storage should be equipped with appropriate fire protection systems (alarms, sprinklers, emergency lighting, portable extinguishers).

If a spill or leak to the environment has occurred, fire department, emergency response and/or hazardous materials spill personnel should be notified immediately. Cleanup should be attempted only by those trained in proper spill containment procedures. Contaminated soils should be removed for incineration and replaced with clean soil. If cyclohexene should contact the water table, aquifer, or navigable waterway, the local and/or state emergency response authorities must be notified. A comprehensive emergency response or disaster preparedness/recovery plan should be in place prior to any operations involving the use, transportation, storage, or disposal of cyclohexene. If cyclohexene is spilled or leaked, the following specific steps are recommended:

- ☑ Restrict persons not wearing protective clothing from area of spill or leak until cleanup is complete and area can be opened for normal work.
- ☑ Remove all ignition sources, ventilate area of leak.
- ☑ Absorb liquids in vermiculite, dry sand, earth, or similar material and deposit in sealed containers.
- ☑ If gas is leaking, stop the flow of gas. If the source is a cylinder and leak cannot be stopped in place, remove cylinder to a safe place in the open air and repair or allow to empty.
- ☑ It may be necessary to dispose of cyclohexene as a hazardous waste. The responsible state agency or the regional office of the federal Environmental Protection Agency (EPA) should be contacted for specific recommendations.

RISK ASSESSMENT: BUSINESS

General Assessment

Accidents or mishaps involving cyclohexene can present a significant threat to business operations. The loss or damage of equipment or facilities can significantly

affect fiscal viability. Lawsuits that may result from personnel injury/death, public exposures, and/or environmental contamination will also require a serious expenditure of resources. Media attention surrounding an injury, death, or environmental damage can also result in a loss of profits and loss of current as well as future business.

❶ *Recommended Risk-Reduction Measures*

Company attorneys, safety and health professionals, and environmental specialists should be involved in the development of any procedures dealing with safety or emergency response. A company official should be pre-designated as a public relations officer with specific training in dealing with the press. Corporate plans and policies should be developed, approved, and implemented long before any need for such arises.

REFERENCES

Ahlbom, A. 1993. *Biostatistics for Epidemiologists.* New York: CRC Press/Lewis Publishers.

American Conference of Governmental Industrial Hygienists. 1988. *Documentation of the Threshold Limit Values and Biological Exposure Indices,* 5th Edition (with updates). Cincinnati: ACGIH

Asante-Duah, D. K. 1993. *Hazardous Waste Risk Assessment.* New York: CRC Press/Lewis Publishers.

Bloom, A. D. and F. J. DeSerres. 1995. *Ecotoxicity and Human Health.* New York: CRC Press/Lewis Publishers.

Brusick, D. J. 1994. *Methods for Genetic Risk Assessment.* New York: CRC Press/Lewis Publishers.

Bueche, J. R. 1972. *Principles of Physics,* 2nd Edition. New York: McGraw-Hill.

Calabrese, E. J., E. M. Kenyon. *Air Toxics and Risk Assessment.* New York: CRC Press/Lewis Publishers.

Calabrese, E. J. 1994. *Biological Effects of Low Level Exposures.* New York: CRC Press/Lewis Publishers.

Cockerman, L. G., B. S. Shane. 1994. *Basic Environmental Toxicology.* New York: CRC Press/Lewis Publishers.

Cothern, C. R., N. P. Ross. 1994. *Environmental Statistics, Assessment, and Forecasting.* New York: CRC Press/Lewis Publishers.

Cothern, R. R. 1993. *Comparative Environmental Risk Assessment.* New York: CRC Press/Lewis Publishers.

Cothern, C. R. 1995. *Handbook for Environmental Risk Decisionmaking.* New York: CRC Press/Lewis Publishers.

Gardner, R. W. 1994. *Chemical Intolerance.* New York: CRC Press/Lewis Publishers.

Hallenbeck, W. H. 1993. *Quantitative Risk Assessment for Environmental and Occupational Health,* 2nd Edition. New York: CRC Press/Lewis Publishers.

Howard, P. H. 1995. *Handbook of Environmental Fate and Exposures Data for Organic Chemicals,* Vol. I-IV. New York: CRC Press/Lewis Publishers.

Howard, P. H., et. al. 1991. *Handbook of Environmental Degradation Rates.* New York: CRC Press/Lewis Publishers.

Jones, F. E. 1994. *Toxic Organic Vapors in the Workplace.* New York: CRC Press/Lewis Publishers.

Keith, L. H. and M. M. Walker. 1995. *Handbook of Air Toxics.* New York: CRC Press/Lewis Publishers.

Landis, W. G., M. H. Yu. 1995. *Introduction to Environmental Toxicology.* New York: CRC Press/Lewis Publishers.

Langerman, N. 1994 *Precautionary Labels for Chemical Containers.* New York: CRC Press/Lewis Publishers.

Lauwerys, R. R., P. Hoet. 1993. *Industrial Chemical Exposure.* New York: CRC Press/Lewis Publishers.

Lewis, R. J., Sr. 1992. *Sax's Dangerous Properties of Industrial Materials,* Volumes 1, 2, and 3, 8th Edition. New York: Van Nostrand Reinhold.

Lide, D. R. 1995. *Handbook of Chemistry and Physics,* 76th Edition. New York: CRC Press/Lewis Publishers.

Manahan, S. E. 1994. *Environmental Chemistry,* 6th Edition. New York: CRC Press/Lewis Publishers.

Manahan, S. E. 1992. *Toxicological Chemistry,* 2nd Edition. New York: CRC Press/Lewis Publishers.

Mellström, G. A., J. E. Wahlberg, and H. I. Maibach. 1994. *Protective Gloves for Occupational Use.* New York: CRC Press/Lewis Publishers.

Meyer, E. 1990. *Chemistry of Hazardous Materials.* Englewood Cliffs, NJ: Prentice Hall.

Patnaik, P. 1992. *A Comprehensive Guide to the Hazardous Properties of Chemical Substances.* New York: Van Nostrand Reinhold.

Proctor, N. H., J. P. Hughes. 1991. *Chemical Hazards of the Workplace.* New York: Van Nostrand Reinhold.

Quigley, D. R. 1994. *Handbook of Emergency Chemical Management.* New York: CRC Press/Lewis Publishers.

Rea, W. J. 1995. *Chemical Sensitivity,* Vol. I-III. New York: CRC Press/Lewis Publishers.

Richardson, M. 1995. *Dictionary of Substance and Their Effects,* Vol. I-VI. New York: CRC Press/Lewis Publishers.

Sacarello, H. L. A. 1994. *The Comprehensive Handbook of Hazardous Materials.* New York: CRC Press/Lewis Publishers.

Sheftel, V. O. 1995. *Handbook of Toxic Properties of Monomers and Additives.* New York: CRC Press/Lewis Publishers.

Torkelson, T. R. V. K. Rowe. 1981. Halogenated aliphatic hydrocarbons. *Patty's Industrial Hygiene and Toxicology,* Vol. 2B, *Toxicology,* pp 3568-3572. New York: Wiley-Interscience.

U.S. Department of Health and Human Services, National Institute for Occupational Safety and Health. 1994. *NIOSH Pocket Guide to Chemical Hazards.* Washington, D.C.: U.S. Government Printing Office.

MATERIAL SAFETY DATA SHEET

CHEMICAL NAME	
CYCLOHEXYLAMINE	**CRC PRESS/LEWIS PUBLISHERS DISCLAIMER:** THE INFORMATION AND RECOMMENDATIONS PRESENTED HEREIN ARE BASED ON SOURCES BELIEVED TO BE RELIABLE. CRC MAKES NO REPRESENTATION ON ITS COMPLETENESS OR ACCURACY. IT IS THE RESPONSIBILITY OF THE USER TO DETERMINE THE CHEMICAL'S SUITABILITY FOR ITS INTENDED USE, THE CHEMICAL'S SAFE USE, AND THE CHEMICAL'S PROPER DISPOSAL. NO REPRESENTATIONS AND/OR WARRANTIES, EITHER EXPRESSED OR IMPLIED, OF THE MERCHANTABILITY OR FITNESS FOR A PARTICULAR PURPOSE, OR OF ANY OTHER NATURE, ARE MADE WITH RESPECT TO THE INFORMATION PROVIDED IN THIS MSDS OR TO THE CHEMICAL TO WHICH INFORMATION MAY REFER. CRC NEITHER ASSUMES NOR AUTHORIZES ANY OTHER PERSON TO ASSUME FOR IT, ANY OTHER ADDITIONAL RESPONSIBILITY OR LIABILITY FOR THE USE OF, OR RELIANCE UPON, THIS INFORMATION.

HAZARD WARNING INFORMATION

HEALTH	FIRE	REACTIVE	OTHER	DEGREE OF HAZARD	COLOR CODING	OTHER CODES
2	3	0		0 = Minimum Hazard 1 = Slight Hazard 2 = Moderate Hazard 3 = Serious Hazard 4 = Severe Hazard	HEALTH = BLUE FIRE = RED REACTIVITY = YELLOW OTHER = WHITE	OX = Oxidizer ACID = Acid ALK = Alkali COR = Corrosive ~~W~~ = Use No Water

SECTION I - GENERAL INFORMATION

Characterization	**Aliphatic Amine**	RCRA Number	**None**
		EPA Class	**Not Applicable**
DOT Proper Shipping Name	**Cyclohexylamine**	Chemical Abstract Service (CAS) Number	**108-91-8**
DOT Hazard Class and Label Requirements	**Flammable Liquid; Flammable, Corrosive**	DOT Emergency Guide Code	**68**
DOT Identification Number	**UN 2357**	Chemical Formula	**$C_6H_{11}NH_2$**
Synonyms	**Cyclohexanamine; hexahydrobenzenamine; aminocyclohexane; hexahydroaniline.**		

SECTION II - HAZARDOUS INGREDIENTS AND IDENTITY INFORMATION

Hazardous Components (specific identity)	OSHA Exposure Criteria	NIOSH Exposure Criteria	Immediately Dangerous to Life and Health (IDLH)	ACGIH Exposure Criteria
Cyclohexylamine (derivation: By catalytic hydrogenation of aniline; ammonolysis of cyclohexyl chloride or cyclohexanol; reduction of nitrocyclohexane). **1 ppm = 4.12 mg/m³**	**PEL:** **10 ppm** **40 mg/m³** **STEL:** **Not Established**	**REL:** **10 ppm** **40 mg/m³** **STEL:** **Not Established**	**Not Determined**	**TLV:** **10 ppm** **40 mg/m³** **STEL:** **Not Established**

SECTION III - PHYSICAL AND CHEMICAL CHARACTERISTICS

Boiling Point	**275°F (135°C)**	Specific Gravity (H_2O = 1)	**0.9**
Vapor Pressure (gas)	**10 at 69°F (20°C)**	Molecular Weight	**99.2**
Vapor Density (Air = 1)	**3.4**	Melting Point	**0°F (-18°C)**
Solubility	**Soluble and water and miscible in most solvents (form azeotropic mixture in water).**		
Appearance and Odor	**Colorless or yellowish liquid with a strong fishy, amine odor.**		

SECTION IV - FIRE AND EXPLOSION HAZARD DATA

Flash Point (method used)	**83°F (28°C) closed cup**	Explosive Limits in Air % by Volume	**LEL: 1.5% UEL: 9.4%**
NFPA Classification	**Class 1C Flammable Liquid**	Autoignition Temperature	**560°F (293°C)**

Extinguishing Media
Use dry chemical, carbon dioxide, water, or alcohol foam.

Special Fire Fighting Procedures
Wear full protective clothing and self-contained breathing apparatus (SCBA). Poisonous gases are produced in fire. Use water spray to keep fire-exposed containers cool.

Unusual Fire and Explosion Hazards
Dangerous fire and explosion hazard when exposed to heat or flame. Containers may explode in fire. Chemical will react on contact with oxidizable materials. Moderate explosion hazard when in the form of vapor. Corrosive to many common metals.

SECTION V - REACTIVITY DATA

Stability		
Stability		Conditions to Avoid **Keep away from heat or flame and incompatible materials. Store to avoid contact with ordinary organic materials since reactions can occur. Sensitive to air and light.**
Stable **X**	Unstable	Incompatibility *(materials to avoid)* **Strong oxidizers, (such as permanganates, peroxides, nitrates, chlorates, and perchlorates), organic compounds, acid anhydrates, acid chlorides, acids, lead, copper, aluminum, zinc, and galvanized steel.**
Hazardous Polymerization		Conditions to Avoid **Under normal conditions of temperature and pressure, hazardous polymerization of cyclohexylamine is not expected to occur.**
May Occur	Will Not Occur **X**	Hazardous Decomposition or By-products **When heated to decomposition, cyclohexylamine can emit highly toxic/poisonous gases, including oxides of nitrogen.**

SECTION VI - HEALTH HAZARD DATA

Primary Route(s) of Entry:	Inhalation?	Absorption (skin)?	Ingestion?
	X	X	X

Health Hazards

INHALATION: **Irritation to the respiratory system and mucous membranes. Symptoms include cough, drowsiness, lightheadedness, dizziness, diarrhea, nausea, vomiting, and pulmonary edema (fluid buildup in the lungs) which can lead to death. Narcotic effects possible.**

ABSORPTION: **Severe skin irritant causing burns and sensitization. Eye contact with pure liquid can cause damage to and/or loss of vision. Will pass through skin to cause systemic effects.**

INGESTION: **Gastrointestinal irritation, diarrhea, vomiting, nausea, dizziness.**

Carcinogenicity	NTP Listed?	IARC Cancer Review Group?	OSHA Regulated?	Target Organs?
Questioned Human **Questioned Animal**	**No**	**Group 3**	**No**	**Respiratory system, eyes, skin, CNS.**

Medical Conditions Generally Aggravated by Exposure

None reported.

Emergency and First-aid Procedures

Eye contact: Flush immediately with water for 15 minutes (minimum), seek medical attention. Skin contact: Remove all contaminated clothing. Immediately wash area with large amounts of soap and water. Seek medical assistance. For inhalation: Remove the person from exposure. Provide respiratory assistance and CPR, if required. Transfer to medical facility. Observe for 24 - 48 hours for lung effects (pulmonary edema). If swallowed: Do NOT induce vomiting. Seek medical attention immediately.

SECTION VII - PRECAUTIONS FOR SAFE HANDLING AND USE

Steps to be Taken in Case Material is Released or Spilled

Absorb spills in vermiculite, dry sand, earth, or similar material and deposit in sealed containers using non-sparking tools. Ventilate area of spill or leak. Restrict those not involved in cleanup from entering area. Remove all sources of ignition. Decontaminate spill area using water (collect as waste).

Preferred Waste Disposal Method

No Citation.

Precautions to be Taken in Handling and Storage

Store to avoid contact with incompatible materials in tightly closed containers in cool, well-ventilated area away from heat, flame, air, and light. Use only non-sparking tools to open and close containers.

Other Precautions and Warnings

Sources of ignition are prohibited where cyclohexylamine is used, handled, or stored. Heat may cause containers to build up pressure and explode.

SECTION VIII - CONTROL MEASURES AND PERSONAL PROTECTIVE EQUIPMENT

Respiratory Protection *(specify type)*

For exposures above the PEL, use an MSHA/NIOSH-approved supplied-air respirator with full facepiece, hood, or helmet in continuous flow mode, or use a self-contained breathing apparatus (SCBA) operated in pressure demand or other positive pressure mode.

Ventilation

Local exhaust or general mechanical systems recommended.

Protective Gloves	Eye Protection	Other Protective Clothing
Nitrile Rubber, Viton, or Teflon	**Chemical Goggles or Face Mask**	**Rubber Apron**

Work/Hygiene Practices

Always wash hands thoroughly after using chemical; never bring food, drink, or smoking materials into vicinity of chemicals.

CYCLOHEXYLAMINE

$C_6H_{11}NH_2$ CAS: 108-91-8

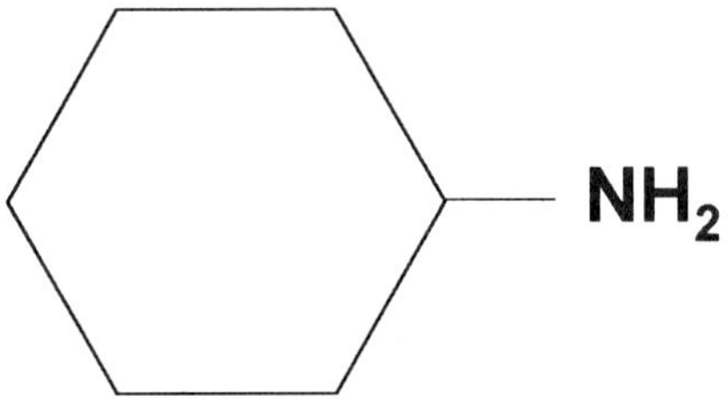

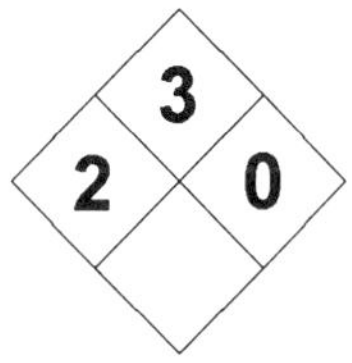

IDENTIFICATION AND TYPICAL USES

Cyclohexylamine is a colorless to yellow liquid with a strong, fishy, amine-like odor. It is used in the manufacture of numerous products, including plasticizers, dry-cleaning soaps, insecticides, and emulsifying agents. It is also used as a corrosion inhibitor and in organic synthesis.

RISK ASSESSMENT: HEALTH

General Assessment

Cyclohexylamine is a human poison by *inhalation, ingestion,* and skin *absorption.* It is a severe irritant to the eyes, skin and respiratory system. Human mutation data have been reported for cyclohexylamine. It is a questionable human and animal carcinogen as well as a suspected teratogen.

Inhalation can cause irritation to the respiratory system and associated mucous membranes. It can cause cough, congestion, with narcotic effects associated with the depression of the central nervous system (CNS). These include dizziness, drowsiness, lightheadedness, and possible convulsions. The most serious effect of inhalation is the potential for pulmonary edema (fluid buildup on the lungs), which is a medical emergency and may be fatal. Symptoms of congestion, cough with phlegm, nausea, vomiting, and loss of consciousness may not appear for 24 - 48 hours after exposure (giving a misleading or false sense of security).

Skin contact can produce burns and sensitization. The liquid will pass through unbroken skin to cause toxic systemic effects similar to the effects on the CNS caused by inhalation. Contact of the pure liquid or its concentrated solutions with the eyes may cause serious damage up to and including loss of vision.

The acute oral toxicity of cyclohexylamine (in test animals) appears to be moderate. Effects include nausea, vomiting, and degenerative changes in the brain, liver, and kidney.

Acute Health Effects

The following acute (short-term) health effects may occur immediately or shortly after exposure to cyclohexylamine:

Skin: Severe irritation and possible burns. Will pass through unbroken skin to cause toxic systemic effects.

Eye: Severe irritation with damage and possible loss of vision likely (liquid contact).

Lung: Severe irritation to the respiratory tract and associated mucosa. Can cause dizziness, vomiting, and nausea. Pulmonary edema (fluid in the lungs) is likely following acute exposure.

CNS: A CNS depressant. Headache, dizziness, drowsiness, possible convulsions.

Other: Ingestion will cause irritation to the gastrointestinal system with nausea, vomiting, convulsions, other toxic systemic effects. Liver and/or kidney damage possible.

Chronic Health Effects

The following chronic (long-term) health effects may occur at some time after exposure to cyclohexylamine and can last for months or even years:

Cancer Hazards: Some studies have suggested that cyclohexylamine causes mutations (genetic changes) in humans. While there is no evidence of carcinogenicity of this chemical in humans or animals, it is still considered a *questioned* carcinogen primarily because of its mutagenic properties. It has been suggested that mutagenic chemicals should be regarded as carcinogens during use and handling.

Reproductive Hazard: According to information presented in the references, cyclohexylamine has been shown to cause adverse effects on the reproductive systems of test animals. Testicular atrophy and changes in organ weights have also been reported. These studies, however, are inconclusive.

Other Chronic Effects: Repeated skin contact may cause sensitization, dermatitis, chronic rash, and associated irritation. Once an allergy develops, the slightest future exposures will cause a recurrence of the symptoms. Repeated or long-term exposures to cyclohexylamine may affect the liver and kidneys.

Recommended Risk-Reduction Measures

Personnel should avoid direct contact with cyclohexylamine. The long-term (chronic) toxicity of this chemical is not completely understood. However, it is known to cause serious health effects in the short-term and is also a know human mutagenic. Prudent risk management requires extreme caution and a conservative approach when assessing the hazard of such chemicals.

If a less toxic chemical cannot be substituted for a hazardous substance, then *engineering controls* are the most effective method of reducing exposures. The best protection is to enclose operations and/or provide local exhaust ventilation at the site of chemical release. While not always operationally feasible, isolating operations can also reduce exposure. Using respiratory protection is less effective than the controls mentioned above, but is still advisable whenever working with or around cyclohexylamine. For exposures above the PEL (10 ppm), a supplied-air respirator with full facepiece operated in continuous flow mode, or a self-contained breathing apparatus (SCBA) with full facepiece and pressure demand are the recommended respiratory protection methods of choice. If a full facepiece is not available, then chemical goggles should be worn to protect the eyes. Whenever a chemical splash hazard exists, a face shield and a rubber apron should be worn. To prevent hand and skin exposures, acid resistant gloves should be worn.

Administrative controls should also be in place to minimize the potential for human exposures. These may include written procedures or policies which specify the methods and techniques that will be practiced whenever personnel are to work with cyclohexylamine.

All personnel should receive *training* on the use, hazards, protective measures, emergency actions, and other precautions per 29 CFR 1910.1200 (Hazard Communication), prior to the first assignment in an area where cyclohexylamine is used or stored.

Before beginning employment with cyclohexylamine and at regular intervals thereafter (e.g., annually), those persons with high or frequent exposures should be provided the following recommended medical tests:

☑ Lung function tests (including Forced Vital Capacity, or "FVC").

If symptoms develop or overexposure is suspected, the following tests are recommended:

☑ Kidney and liver function tests.

☑ Lung function tests (including Forced Vital Capacity, or "FVC").

☑ Consider chest X-ray.

☑ Evaluation by a qualified allergist, with careful evaluation of exposure history and special testing (may help diagnose allergy).

Any evaluation should include a careful history of past and present symptoms with an examination. Medical tests that simply look for existing damage are not a substitute for controlling exposures.

Other methods to reduce exposure include:

☑ Where possible, enclose operations and use local exhaust ventilation at the site of chemical release. If local exhaust ventilation or enclosure is not used, respiratory protection should be mandatory. Also, if possible, liquids should be transferred automatically from storage containers to process containers.

☑ Always ensure that proper protective clothing is worn when using chemical substances.

☑ Wash thoroughly immediately after exposure to cyclohexylamine and at the end of the work shift or before eating, drinking, or smoking.

☑ Hazard warning information should be posted in the work area. In addition, as part of an on-going education and training program, all information on the health and safety hazards of cyclohexylamine should be communicated to all potentially exposed workers.

☑ Eye wash stations should be provided in the immediate work area for emergency use. If there is a possibility of skin exposure to cyclohexylamine, emergency shower facilities should be provided.

☑ Personnel should never wear contaminated clothing home (family members can be exposed). Clothing should be laundered by personnel who have been trained on the hazards of cyclohexylamine.

RISK ASSESSMENT: ENVIRONMENT

General Assessment

The environment is at risk of exposure during transportation, storage, disposal, or destruction of cyclohexylamine. In almost every scenario, the threat of environmental exposure is contingent upon the proper handling of the chemical substance. Accidental spills, large or small, can result in fire, explosion, and possible contamination of the surrounding environmental mediums (water, soil, and air).

Cyclohexylamine is considered a class 1C flammable liquid (per OSHA 29 CFR 1910.106). Its low flash point and boiling point make this chemical a moderate fire and explosion hazard concern. These characteristics require extreme caution in handling, storage, transportation, and disposal. It is incompatible with oxidizers, organic compounds, acids, and contact can cause fire or explosion. It is corrosive to many common metals (copper, aluminum, lead). Therefore, special consideration is required during any emergency situation involving a leak or spill of cyclohexylamine. Should cyclohexylamine ever come into contact with these incompatible substances during use, transportation, storage, or disposal, the formation of highly toxic and/or highly explosive commodities is extremely possible.

Cyclohexylamine may enter the environment through industrial discharges or spills.

Acute Ecological Effects

Acute (short-term) toxic effects may include the death of animals, birds, or fish and death or low growth rate in plants. Acute effects are seen 2 to 4 days after animals or plants are exposed to cyclohexylamine.

Insufficient data are available on the short-term effects of cyclohexylamine exposure to aquatic life, plants, birds or land animals.

Chronic Ecological Effects

Chronic (long-term) toxic effects may include shortened life span, reproductive problems, lower fertility, and changes in appearance or behavior in exposed animals. These effects can be seen long after first exposure(s) to toxic chemicals.

Insufficient data are available on the long-term effects of cyclohexylamine to aquatic life, plants, birds, or land animals.

Water Solubility

Cyclohexylamine is moderately soluble in water. Concentrations between 1 and 1000 milligrams will mix with a liter of water (forms an azeotropic solution in water).

Persistence in the Environment

Cyclohexylamine is slightly persistent in water, with a half-life between 2 and 20 days. The half-life of a pollutant is the amount of time it takes for one half of the chemical to be degraded.

Bioaccumulation in Aquatic Organisms

Some substances increase in concentration, or *bioaccumulate,* in living organisms as they breathe contaminated air, drink contaminated water, or eat contaminated food. These chemicals can become concentrated in the tissues and internal organs of animals as well as humans.

The concentration of cyclohexylamine found in fish tissue is expected to be about the same as the average concentration of cyclohexylamine in the water from which the fish was taken.

Recommended Risk-Reduction Measures

Proper training of all transporters will reduce the likelihood of a mishap or accident resulting in a leak or spill to the environment. The correct use of DOT labeling (per the MSDS) on all containers, trucks, and rail cars will enable emergency responders to react properly and quickly to any disaster thereby reducing the potential risk to the environment and to personnel.

Storage of cyclohexylamine should be segregated from incompatible chemicals to minimize the risk of cross-contamination or contact. Buildings designated for storage should be equipped with appropriate fire protection systems (alarms, sprinklers, emergency lighting, portable extinguishers).

If a spill or leak to the environment has occurred, fire department, emergency response and/or hazardous materials spill personnel should be notified immediately. Cleanup should be attempted only by those

trained in proper spill containment procedures. Contaminated soils should be removed for incineration and replaced with clean soil. If cyclohexylamine should contact the water table, aquifer, or navigable waterway, the local and/or state emergency response authorities must be notified. A comprehensive emergency response or disaster preparedness/recovery plan should be in place prior to any operations involving the use, transportation, storage, or disposal of cyclohexylamine. If cyclohexylamine is spilled or leaked, the following specific steps are recommended:

☑ Restrict persons not wearing protective clothing from area of spill or leak until cleanup is complete and area can be opened for normal work.

☑ Remove all ignition sources, ventilate area of leak.

☑ Absorb liquids in vermiculite, dry sand, earth, or similar material and deposit in sealed containers.

☑ It may be necessary to dispose of cyclohexylamine as a hazardous waste. The responsible state agency or the regional office of the federal Environmental Protection Agency (EPA) should be contacted for specific recommendations.

RISK ASSESSMENT: BUSINESS

General Assessment

Accidents or mishaps involving cyclohexylamine can present a significant threat to business operations. The loss or damage of equipment or facilities can significantly affect fiscal viability. Lawsuits that may result from personnel injury/death, public exposures, and/or environmental contamination will also require a serious expenditure of resources. Media attention surrounding an injury, death, or environmental damage can also result in a loss of profits and loss of current as well as future business. Remember that anytime the term "reproductive hazard" is used, public emotion and hysteria can run equally high. This requires careful consideration in the development of company policies.

❶ *Recommended Risk-Reduction Measures*

Company attorneys, safety and health professionals, and environmental specialists should be involved in the development of any procedures dealing with safety or emergency response. A company official should be pre-designated as a public relations officer with specific training in dealing with the press. Corporate plans and policies should be developed, approved, and implemented long before any need for such arises.

REFERENCES

Asante-Duah, D. K. 1993. *Hazardous Waste Risk Assessment.* New York: CRC Press/Lewis Publishers.

Bloom, A. D. and F. J. DeSerres. 1995. *Ecotoxicity and Human Health.* New York: CRC Press/Lewis Publishers.

Bueche, J. R. 1972. *Principles of Physics,* 2nd Edition. New York: McGraw-Hill.

Calabrese, E. J. 1994. *Biological Effects of Low Level Exposures.* New York: CRC Press/Lewis Publishers.

Cockerman, L. G., B. S. Shane. 1994. *Basic Environmental Toxicology.* New York: CRC Press/Lewis Publishers.

Cothern, C. R. 1995. *Handbook for Environmental Risk Decisionmaking.* New York: CRC Press/Lewis Publishers.

Gardner, R. W. 1994. *Chemical Intolerance.* New York: CRC Press/Lewis Publishers.

Hallenbeck, W. H. 1993. *Quantitative Risk Assessment for Environmental and Occupational Health,* 2nd Edition. New York: CRC Press/Lewis Publishers.

Howard, P. H. 1995. *Handbook of Environmental Fate and Exposures Data for Organic Chemicals,* Vol. I-IV. New York: CRC Press/Lewis Publishers.

Jones, F. E. 1994. *Toxic Organic Vapors in the Workplace.* New York: CRC Press/Lewis Publishers.

Keith, L. H. and M. M. Walker. 1995. *Handbook of Air Toxics.* New York: CRC Press/Lewis Publishers.

Lauwerys, R. R., P. Hoet. 1993. *Industrial Chemical Exposure.* New York: CRC Press/Lewis Publishers.

Lewis, R. J., Sr. 1992. *Sax's Dangerous Properties of Industrial Materials,* Volumes 1, 2, and 3, 8th Edition. New York: Van Nostrand Reinhold.

Manahan, S. E. 1992. *Toxicological Chemistry,* 2nd Edition. New York: CRC Press/Lewis Publishers.

Meyer, E. 1990. *Chemistry of Hazardous Materials.* Englewood Cliffs, NJ: Prentice Hall.

Patnaik, P. 1992. *A Comprehensive Guide to the Hazardous Properties of Chemical Substances.* New York: Van Nostrand Reinhold.

Rea, W. J. 1995. *Chemical Sensitivity,* Vol. I-III. New York: CRC Press/Lewis Publishers.

Sacarello, H. L. A. 1994. *The Comprehensive Handbook of Hazardous Materials.* New York: CRC Press/Lewis Publishers.

Torkelson, T. R. V. K. Rowe. 1981. Halogenated aliphatic hydrocarbons. *Patty's Industrial Hygiene and Toxicology,* Vol. 2B, *Toxicology,* pp 3568-3572. New York: Wiley-Interscience.

U.S. Department of Health and Human Services, National Institute for Occupational Safety and Health. 1994. *NIOSH Pocket Guide to Chemical Hazards.* Washington, D.C.: U.S. Government Printing Office.

MATERIAL SAFETY DATA SHEET

CHEMICAL NAME	
CYCLOPENTADIENE	**CRC PRESS/LEWIS PUBLISHERS DISCLAIMER:** THE INFORMATION AND RECOMMENDATIONS PRESENTED HEREIN ARE BASED ON SOURCES BELIEVED TO BE RELIABLE. CRC MAKES NO REPRESENTATION ON ITS COMPLETENESS OR ACCURACY. IT IS THE RESPONSIBILITY OF THE USER TO DETERMINE THE CHEMICAL'S SUITABILITY FOR ITS INTENDED USE, THE CHEMICAL'S SAFE USE, AND THE CHEMICAL'S PROPER DISPOSAL. NO REPRESENTATIONS AND/OR WARRANTIES, EITHER EXPRESSED OR IMPLIED, OF THE MERCHANTABILITY OR FITNESS FOR A PARTICULAR PURPOSE, OR OF ANY OTHER NATURE, ARE MADE WITH RESPECT TO THE INFORMATION PROVIDED IN THIS MSDS OR TO THE CHEMICAL TO WHICH INFORMATION MAY REFER. CRC NEITHER ASSUMES NOR AUTHORIZES ANY OTHER PERSON TO ASSUME FOR IT, ANY OTHER ADDITIONAL RESPONSIBILITY OR LIABILITY FOR THE USE OF, OR RELIANCE UPON, THIS INFORMATION.

HAZARD WARNING INFORMATION

HEALTH	FIRE	REACTIVE	OTHER	DEGREE OF HAZARD	COLOR CODING	OTHER CODES
1	2	2		0 = Minimum Hazard 1 = Slight Hazard 2 = Moderate Hazard 3 = Serious Hazard 4 = Severe Hazard	HEALTH = BLUE FIRE = RED REACTIVITY = YELLOW OTHER = WHITE	OX = Oxidizer ACID = Acid ALK = Alkali COR = Corrosive W = Use No Water

SECTION I - GENERAL INFORMATION

Characterization	RCRA Number	EPA Class
Alicyclic Hydrocarbon	**None**	**Not Applicable**

DOT Proper Shipping Name	Chemical Abstract Service (CAS) Number
Cyclopentadiene	**542-92-7**
DOT Hazard Class and Label Requirements	**DOT Emergency Guide Code**
Not Found	**No Citation**
DOT Identification Number	**Chemical Formula**
Not Found	**C_5H_6**

Synonyms
1,3-Cyclopentadiene; pentole; pyropentylene; r-pentine.

SECTION II - HAZARDOUS INGREDIENTS AND IDENTITY INFORMATION

Hazardous Components (specific identity)	OSHA Exposure Criteria	NIOSH Exposure Criteria	Immediately Dangerous to Life and Health (IDLH)	ACGIH Exposure Criteria
Cyclopentadiene (derivation: From coal tar and cracked petroleum oils). **1 ppm = 2.75 mg/m³**	**PEL:** **75 ppm** **200 mg/m³** **STEL:** **Not Established**	**REL:** **75 ppm** **200 mg/m³** **STEL:** **Not Established**	**750 ppm**	**TLV:** **75 ppm** **200 mg/m³** **STEL:** **Not Established**

SECTION III - PHYSICAL AND CHEMICAL CHARACTERISTICS

Boiling Point	Specific Gravity (H_2O = 1)
109°F (43°C)	**0.809**
Vapor Pressure (gas)	Molecular Weight
400 at 69°F (20°C)	**66.1**
Vapor Density (Air = 1)	Melting Point
Not Determined	**-121°F (-85°C)**

Solubility
Insoluble in water. Soluble in most organic solvents.

Appearance and Odor
Colorless liquid with an irritating, terpene-like odor.

SECTION IV - FIRE AND EXPLOSION HAZARD DATA

Flash Point (method used)	Explosive Limits in Air % by Volume
77°F (25°C) opened cup	**LEL: 1.5% UEL: 9.4%**
NFPA Classification	Autoignition Temperature
Class 1C Flammable Liquid	**Not Determined**

Extinguishing Media
Use dry chemical, foam, or carbon dioxide. Water spray may be used to cool surrounding areas.

Special Fire Fighting Procedures
Wear full protective clothing and self-contained breathing apparatus (SCBA). Poisonous gases are produced in fire. Evacuate 1500 feet radius if fire becomes uncontrollable or container is exposed to flame.

Unusual Fire and Explosion Hazards
Dangerous fire and explosion hazard when exposed to heat or flame. Containers may explode in fire. Prolonged exposure to air may cause the formation of explosive peroxides. Moderate explosion hazard when in the form of vapor. Violent decomposition at high temperatures.

SECTION V - REACTIVITY DATA

Stability		Conditions to Avoid: Keep away from heat or flame and incompatible materials. Very reactive with incompatible materials. Reactions can be violent and even explosive.
Stable	Unstable	Incompatibility *(materials to avoid)*: Strong oxidizers, (such as permanganates, peroxides, nitrates, chlorates, and perchlorates), fuming nitric acid and sulfuric acid.
	X	
Hazardous Polymerization		Conditions to Avoid: Under normal conditions of temperature and pressure, hazardous polymerization of cyclopentadiene is not expected to occur.
May Occur	Will Not Occur	Hazardous Decomposition or By-products: Cyclopentadiene decomposes violently at high temperatures and pressures. Reaction is highly exothermic. When heated to decomposition, emits acrid smoke and fumes.
	X	

SECTION VI - HEALTH HAZARD DATA

Primary Route(s) of Entry:	Inhalation?	Absorption (skin)?	Ingestion?
	X		X

Health Hazards

INHALATION: Possible irritation of the eyes and upper respiratory tract with symptoms of headache and nausea. May depress the circulatory system and respiratory systems. High doses may cause narcosis, convulsions, liver and kidney damage, and death (human exposure data inconclusive).

SKIN & EYES: Mild to moderate irritation of the skin and eyes (vapor and liquid states).

INGESTION: No firm data reported. Possible gastrointestinal irritation, with diarrhea and nausea.

Carcinogenicity	NTP Listed?	IARC Cancer Review Group?	OSHA Regulated?	Target Organs?
No Citation	No	No	29 CFR 1910.1000 Table Z-1	Respiratory system, eyes.

Medical Conditions Generally Aggravated by Exposure

None reported.

Emergency and First-aid Procedures

Eye contact: Flush immediately with water for 15 minutes (minimum), seek medical attention. Skin contact: Remove all contaminated clothing. Immediately wash area with large amounts of soap and water. Seek medical assistance, if necessary. For inhalation: Remove the person from exposure. Provide respiratory assistance and CPR, if required. Transfer to medical facility. If swallowed: Never give an unconscious or convulsing person anything by mouth. Seek medical attention immediately.

SECTION VII - PRECAUTIONS FOR SAFE HANDLING AND USE

Steps to be Taken in Case Material is Released or Spilled

Absorb spills in vermiculite, dry sand, earth, or similar material and deposit in sealed containers using non-sparking tools. Ventilate area of spill or leak. Restrict those not involved in cleanup from entering area. Remove all sources of ignition. Contact state or local authorities, if required.

Preferred Waste Disposal Method

No Citation.

Precautions to be Taken in Handling and Storage

Store to avoid contact with incompatible materials in tightly closed containers in cool, well-ventilated area away from heat and flame. Use only non-sparking tools to open and close containers.

Other Precautions and Warnings

Sources of ignition are prohibited where cyclopentadiene is used, handled, or stored. Heat may cause containers to build up pressure and explode. Very reactive in presence of incompatible materials.

SECTION VIII - CONTROL MEASURES AND PERSONAL PROTECTIVE EQUIPMENT

Respiratory Protection *(specify type)*

For low exposures (up to PEL), use an organic vapor respirator. For higher exposures, use a supplied-air respirator with full facepiece, hood, or helmet in continuous flow mode, or a self-contained breathing apparatus (SCBA) operated in pressure demand or other positive pressure mode.

Ventilation

Local exhaust or general mechanical systems recommended.

Protective Gloves	Eye Protection	Other Protective Clothing
Impervious Rubber	Chemical Goggles or Face Mask	Rubber Apron

Work/Hygiene Practices

Always wash hands thoroughly after using chemical; never bring food, drink, or smoking materials into vicinity of chemicals.

CYCLOPENTADIENE

C_5H_6 CAS: 542-92-7

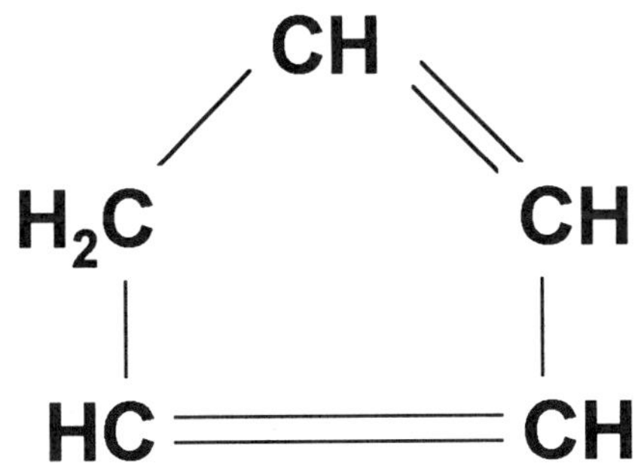

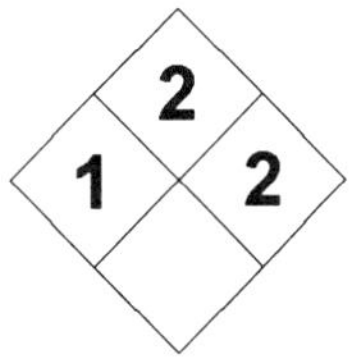

IDENTIFICATION AND TYPICAL USES

Cyclopentadiene is a colorless liquid with an irritating, terpene-like odor. It is used as a chemical intermediate, in organic synthesis, as a starting material for synthetic prostaglandin, in the chlorination of insecticides, and in the formation of sandwich compounds for chelation. It is also used in the manufacture of resins and as a ligand in the preparation of metal compounds.

RISK ASSESSMENT: HEALTH

General Assessment

Cyclopentadiene has exhibited low toxicity in test animals. Human exposure risk is primarily by *inhalation.* Skin and eye contact with the liquid or vapor can be mildly irritating.

Inhalation of cyclopentadiene vapors can be irritating to the eyes, nose, and upper respiratory system. There may be headache and nausea. High concentrations (3 mL) administered by injection to test animals produced narcosis, convulsions, and death. Prolonged exposures to vapors in animals tests led to liver and kidney injuries.

Acute Health Effects

The following acute (short-term) health effects may occur immediately or shortly after exposure to cyclopentadiene:

Skin: Mild irritation with possibility for rash.

Eye: Mild to moderate irritation (liquid or vapor contact) with tearing and redness. Prolonged contact can result in inflammation.

Lung: Irritation or the upper respiratory tract and depression of the respiratory system. There may be headache and nausea.

Other: Ingestion may cause irritation to the gastrointestinal system. There may be depression of the circulatory system.

Chronic Health Effects

The following chronic (long-term) health effects may occur at some time after exposure to cyclopentadiene and can last for months or even years:

Cancer Hazards: According to information presented in the references, cyclopentadiene has not been adequately tested for its ability to cause cancer in animals.

Reproductive Hazard: According to information presented in the references, cyclopentadiene has not been adequately tested for its ability to adversely affect reproduction.

Other Chronic Effects: Repeated or long-term exposures to cyclopentadiene may affect the liver and kidneys.

Recommended Risk-Reduction Measures

Personnel should avoid direct contact with cyclopentadiene. The long-term (chronic) toxicity of this chemical is not completely understood. Prudent risk management requires caution and a conservative approach when assessing the hazard of such chemicals.

If a less toxic chemical cannot be substituted for a hazardous substance, then *engineering controls* are the most effective method of reducing exposures. The best protection is to enclose operations and/or provide local exhaust ventilation at the site of chemical release. While not always operationally feasible, isolating operations can also reduce exposure. Using respiratory protection is less effective than the controls mentioned above, but is still advisable whenever working with or around cyclopentadiene. For exposures above the PEL (75 ppm), a full facepiece respirator equipped with an organic vapor cartridge is suggested. For high or prolonged exposures, a supplied-air respirator with full facepiece operated in continuous flow mode, or a self-contained breathing apparatus (SCBA) with full facepiece and pressure demand are the recommended respiratory protection methods of choice. If a full facepiece is not available, then chemical goggles should be worn to protect the eyes. Whenever a chemical splash hazard exists, a face shield and a rubber apron should be worn. To prevent hand and skin exposures, acid resistant gloves should be worn.

Administrative controls should also be in place to minimize the potential for human exposures. These may include written procedures or policies which specify the methods and techniques that will be practiced whenever personnel are to work with cyclopentadiene.

All personnel should receive *training* on the use, hazards, protective measures, emergency actions, and other precautions per 29 CFR 1910.1200 (Hazard Communication), prior to the first assignment in an area where cyclopentadiene is used or stored.

If symptoms develop or overexposure is suspected, the following tests are recommended:

☑ Kidney and liver function tests.

☑ Lung function tests (including Forced Vital Capacity, or "FVC").

Any evaluation should include a careful history of past and present symptoms with an examination. Medical tests that simply look for existing damage are not a substitute for controlling exposures.

Other methods to reduce exposure include:

☑ Where possible, enclose operations and use local exhaust ventilation at the site of chemical release. If local exhaust ventilation or enclosure is not used, respiratory protection should be mandatory. Also, if possible, liquids should be transferred automatically from storage containers to process containers.

☑ Always ensure that proper protective clothing is worn when using chemical substances and that personnel are trained on its use and care

☑ Wash thoroughly immediately after exposure to cyclopentadiene and at the end of the work shift or before eating, drinking, or smoking.

☑ Hazard warning information should be posted in the work area. In addition, as part of an on-going education and training program, all information on the health and safety hazards of cyclopentadiene should be communicated to all potentially exposed workers.

☑ Eye wash stations should be provided in the immediate work area for emergency use. If there is a possibility of skin exposure to cyclopentadiene, emergency shower facilities should also be provided.

☑ Personnel should never wear contaminated clothing home (family members can be exposed). Clothing should be laundered by personnel who have been trained on the hazards of cyclopentadiene.

RISK ASSESSMENT: ENVIRONMENT

General Assessment

The environment is at risk of exposure during transportation, storage, disposal, or destruction of cyclopentadiene. In almost every scenario, the threat of environmental exposure is contingent upon the proper handling of the chemical substance. Accidental spills, large or small, can result in fire, explosion, and possible contamination of the surrounding environmental mediums (water, soil, and air).

Cyclopentadiene is considered a class 1C flammable liquid (per OSHA 29 CFR 1910.106). Its relatively low flash point and boiling point make this chemical a moderate fire and explosion hazard concern. These characteristics require extreme caution in handling, storage, transportation, and disposal. It is a dangerous fire hazard when exposed to heat and flame. It is also a moderate explosion hazard in the form of a gas when exposed to heat or by chemical reaction. It decomposes violently at high temperatures and pressure. It is incompatible with oxidizers and will react explosively with fuming nitric acid. Mixed with oxygen, it forms a flame-sensitive explosive product. It will react vigorously in the presence of potassium hydroxide and is incompatible with acids (sulfuric), and oxides of nitrogen. Therefore, special consideration is required during any emergency situation involving a leak or spill of cyclopentadiene. Should cyclopentadiene ever come into contact with these incompatible substances during use, transportation, storage, or disposal, the formation of highly toxic and/or highly explosive commodities is extremely possible.

Cyclopentadiene may enter the environment through industrial discharges or spills.

Acute Ecological Effects

Acute (short-term) toxic effects may include the death of animals, birds, or fish and death or low growth rate in plants. Acute effects are seen 2 to 4 days after animals or plants are exposed to cyclopentadiene.

Insufficient data are available on the short-term effects of cyclopentadiene exposure to aquatic life, plants, birds or land animals.

Chronic Ecological Effects

Chronic (long-term) toxic effects may include shortened life span, reproductive problems, lower fertility, and changes in appearance or behavior in exposed animals. These effects can be seen long after first exposure(s) to toxic chemicals.

Insufficient data are available on the long-term effects of cyclopentadiene to aquatic life, plants, birds, or land animals.

Water Solubility

Cyclopentadiene is nearly insoluble in water. Concentrations of less than 100 milligrams may mix with a liter of water.

Persistence in the Environment

Cyclopentadiene is slightly persistent in water, with a half-life between 2 and 20 days. The half-life of a pollutant is the amount of time it takes for one half of the chemical to be degraded.

Bioaccumulation in Aquatic Organisms

Some substances increase in concentration, or *bioaccumulate,* in living organisms as they breathe contaminated air, drink contaminated water, or eat contaminated food. These chemicals can become concentrated in the tissues and internal organs of animals as well as humans.

The concentration of cyclopentadiene found in fish tissue is expected to be about the same as the average concentration of cyclopentadiene in the water from which the fish was taken.

Recommended Risk-Reduction Measures

Proper training of all transporters will reduce the likelihood of a mishap or accident resulting in a leak or spill to the environment. The correct use of DOT labeling (per the MSDS) on all containers, trucks, and rail cars will enable emergency responders to react properly and quickly to any disaster thereby reducing the potential risk to the environment and to personnel.

Storage of cyclopentadiene should be segregated from incompatible chemicals to minimize the risk of cross-contamination or contact. Buildings designated for storage should be equipped with appropriate fire protection systems (alarms, sprinklers, emergency lighting, portable extinguishers).

If a spill or leak to the environment has occurred, fire department, emergency response and/or hazardous materials spill personnel should be notified immediately. Cleanup should be attempted only by those trained in proper spill containment procedures. Contaminated soils should be removed for incineration and replaced with clean soil. If cyclopentadiene should contact the water table, aquifer, or navigable waterway, the local and/or state emergency response authorities must be notified. A comprehensive emergency response or disaster preparedness/recovery plan should be in place prior to any operations involving the use, transportation, storage, or disposal of cyclopentadiene. If cyclopentadiene is spilled or leaked, the following specific steps are recommended:

- ☑ Restrict persons not wearing protective clothing from area of spill or leak until cleanup is complete and area can be opened for normal work.
- ☑ Remove all ignition sources, ventilate area of leak.
- ☑ Absorb liquids in vermiculite, dry sand, earth, or similar material and deposit in sealed containers.
- ☑ It may be necessary to dispose of cyclopentadiene as a hazardous waste. The responsible state agency or the regional office of the federal Environmental Protection Agency (EPA) should be contacted for specific recommendations.

RISK ASSESSMENT: BUSINESS

General Assessment

Accidents or mishaps involving cyclopentadiene can present a significant threat to business operations. The loss or damage of equipment or facilities can significantly affect fiscal viability. Lawsuits that may result

from personnel injury/death, public exposures, and/or environmental contamination will also require a serious expenditure of resources. Media attention surrounding an injury, death, or environmental damage can also result in a loss of profits and loss of current as well as future business.

Recommended Risk-Reduction Measures

Company attorneys, safety and health professionals, and environmental specialists should be involved in the development of any procedures dealing with safety or emergency response. A company official should be pre-designated as a public relations officer with specific training in dealing with the press. Corporate plans and policies should be developed, approved, and implemented long before any need for such arises.

REFERENCES

Ahlbom, A. 1993. *Biostatistics for Epidemiologists.* New York: CRC Press/Lewis Publishers.

Asante-Duah, D. K. 1993. *Hazardous Waste Risk Assessment.* New York: CRC Press/Lewis Publishers.

Bloom, A. D. and F. J. DeSerres. 1995. *Ecotoxicity and Human Health.* New York: CRC Press/Lewis Publishers.

Brusick, D. J. 1994. *Methods for Genetic Risk Assessment.* New York: CRC Press/Lewis Publishers.

Calabrese, E. J., E. M. Kenyon. *Air Toxics and Risk Assessment.* New York: CRC Press/Lewis Publishers.

Calabrese, E. J. 1994. *Biological Effects of Low Level Exposures.* New York: CRC Press/Lewis Publishers.

Cockerman, L. G., B. S. Shane. 1994. *Basic Environmental Toxicology.* New York: CRC Press/Lewis Publishers.

Cothern, C. R., N. P. Ross. 1994. *Environmental Statistics, Assessment, and Forecasting.* New York: CRC Press/Lewis Publishers.

Cothern, R. R. 1993. *Comparative Environmental Risk Assessment.* New York: CRC Press/Lewis Publishers.

Cothern, C. R. 1995. *Handbook for Environmental Risk Decisionmaking.* New York: CRC Press/Lewis Publishers.

Gardner, R. W. 1994. *Chemical Intolerance.* New York: CRC Press/Lewis Publishers.

Hallenbeck, W. H. 1993. *Quantitative Risk Assessment for Environmental and Occupational Health,* 2nd Edition. New York: CRC Press/Lewis Publishers.

Howard, P. H. 1995. *Handbook of Environmental Fate and Exposures Data for Organic Chemicals,* Vol. I-IV. New York: CRC Press/Lewis Publishers.

Howard, P. H., et. al. 1991. *Handbook of Environmental Degradation Rates.* New York: CRC Press/Lewis Publishers.

Jones, F. E. 1994. *Toxic Organic Vapors in the Workplace.* New York: CRC Press/Lewis Publishers.

Keith, L. H. and M. M. Walker. 1995. *Handbook of Air Toxics.* New York: CRC Press/Lewis Publishers.

Landis, W. G., M. H. Yu. 1995. *Introduction to Environmental Toxicology.* New York: CRC Press/Lewis Publishers.

Lauwerys, R. R., P. Hoet. 1993. *Industrial Chemical Exposure.* New York: CRC Press/Lewis Publishers.

Lewis, R. J., Sr. 1992. *Sax's Dangerous Properties of Industrial Materials,* Volumes 1, 2, and 3, 8th Edition. New York: Van Nostrand Reinhold.

Lide, D. R. 1995. *Handbook of Chemistry and Physics,* 76th Edition. New York: CRC Press/Lewis Publishers.

Manahan, S. E. 1994. *Environmental Chemistry,* 6th Edition. New York: CRC Press/Lewis Publishers.

Manahan, S. E. 1992. *Toxicological Chemistry,* 2nd Edition. New York: CRC Press/Lewis Publishers.

Mellström, G. A., J. E. Wahlberg, and H. I. Maibach. 1994. *Protective Gloves for Occupational Use.* New York: CRC Press/Lewis Publishers.

Meyer, E. 1990. *Chemistry of Hazardous Materials.* Englewood Cliffs, NJ: Prentice Hall.

Mycock, J. C., J. D. McKenna, and L. Theodore. 1995. *Handbook of Air Pollution Control Engineering and Technology.* New York: CRC Press/Lewis Publishers.

Nelson, G. O. 1992. *Gas Mixtures.* New York: CRC Press/Lewis Publishers.

Ott, W. R. 1995. *Environmental Statistics and Data Analysis.* New York: CRC Press/Lewis Publishers.

Patnaik, P. 1992. *A Comprehensive Guide to the Hazardous Properties of Chemical Substances.* New York: Van Nostrand Reinhold.

Philp, R. B. 1995. *Environmental Hazards and Human Health.* New York: CRC Press/Lewis Publishers.

Quigley, D. R. 1994. *Handbook of Emergency Chemical Management.* New York: CRC Press/Lewis Publishers.

Rea, W. J. 1995. *Chemical Sensitivity,* Vol. I-III. New York: CRC Press/Lewis Publishers.

Richardson, M. 1995. *Dictionary of Substance and Their Effects,* Vol. I-VI. New York: CRC Press/Lewis Publishers.

Rylander, R., R. R. Jacobs. 1994. *Organic Dusts.* New York: CRC Press/Lewis Publishers.

Sacarello, H. L. A. 1994. *The Comprehensive Handbook of Hazardous Materials.* New York: CRC Press/Lewis Publishers.

Torkelson, T. R. V. K. Rowe. 1981. Halogenated aliphatic hydrocarbons. *Patty's Industrial Hygiene and Toxicology,* Vol. 2B, *Toxicology,* pp 3568-3572. New York: Wiley-Interscience.

U.S. Department of Health and Human Services, National Institute for Occupational Safety and Health. 1994. *NIOSH Pocket Guide to Chemical Hazards.* Washington, D.C.: U.S. Government Printing Office.

MATERIAL SAFETY DATA SHEET

CHEMICAL NAME

DDT
(Dichlorodiphenyltrichloroethane)

CRC PRESS/LEWIS PUBLISHERS DISCLAIMER: THE INFORMATION AND RECOMMENDATIONS PRESENTED HEREIN ARE BASED ON SOURCES BELIEVED TO BE RELIABLE. CRC MAKES NO REPRESENTATION ON ITS COMPLETENESS OR ACCURACY. IT IS THE RESPONSIBILITY OF THE USER TO DETERMINE THE CHEMICAL'S SUITABILITY FOR ITS INTENDED USE, THE CHEMICAL'S SAFE USE, AND THE CHEMICAL'S PROPER DISPOSAL. NO REPRESENTATIONS AND/OR WARRANTIES, EITHER EXPRESSED OR IMPLIED, OF THE MERCHANTABILITY OR FITNESS FOR A PARTICULAR PURPOSE, OR OF ANY OTHER NATURE, ARE MADE WITH RESPECT TO THE INFORMATION PROVIDED IN THIS MSDS OR TO THE CHEMICAL TO WHICH INFORMATION MAY REFER. CRC NEITHER ASSUMES NOR AUTHORIZES ANY OTHER PERSON TO ASSUME FOR IT, ANY OTHER ADDITIONAL RESPONSIBILITY OR LIABILITY FOR THE USE OF, OR RELIANCE UPON, THIS INFORMATION.

HAZARD WARNING INFORMATION

HEALTH	FIRE	REACTIVE	OTHER	DEGREE OF HAZARD	COLOR CODING	OTHER CODES
2	2	0		0 = Minimum Hazard 1 = Slight Hazard 2 = Moderate Hazard 3 = Serious Hazard 4 = Severe Hazard	HEALTH = BLUE FIRE = RED REACTIVITY = YELLOW OTHER = WHITE	OX = Oxidizer ACID = Acid ALK = Alkali COR = Corrosive ~~W~~ = Use No Water

SECTION I - GENERAL INFORMATION

Characterization	**Pesticide; Insecticide**	RCRA Number	**U061**
		EPA Class	**Toxic Waste**
DOT Proper Shipping Name	**Dichlorodiphenyltricloroethane**	Chemical Abstract Service (CAS) Number	**50-29-3**
DOT Hazard Class and Label Requirements	**ORM-A; Label: None**	DOT Emergency Guide Code	**55**
DOT Identification Number	**UN 2761**	Chemical Formula	**$(ClC_6H_8)_2CHCCl_3$**

Synonyms
***p,p'*-trichlorodiphenyltrichloroethane; 2,2-bis(*p*-chlorophenyl)-1,1,1-trichloroethane; 1,1'-(2,2,2-trichloroethylidene)bis(4-chlorobenzene); dicophane; chlorophenothane.**

SECTION II - HAZARDOUS INGREDIENTS AND IDENTITY INFORMATION

Hazardous Components (specific identity)	OSHA Exposure Criteria	NIOSH Exposure Criteria	Immediately Dangerous to Life and Health (IDLH)	ACGIH Exposure Criteria
DDT (derivation: Condensing chloral or chloral or chloral hydrate with chlorobenzene in presence of sulfuric acid).	**PEL (skin):** **1 mg/m³** **STEL: Not Established**	**REL (skin):** **0.5 mg/m³** **Suspected Human Carcinogen**	**500 mg/m³**	**TLV (skin):** **1 mg/m³** **STEL: Not Established**

SECTION III - PHYSICAL AND CHEMICAL CHARACTERISTICS

Boiling Point	**230°F (110°C) Decomposes**	Specific Gravity (H_2O = 1)	**0.99**
Vapor Pressure (mm Hg)	**<1 at 69°F (20°C)**	Molecular Weight	**354.5**
Vapor Density (Air = 1)	**12**	Melting Point	**227°F (108°C)**

Solubility
Insoluble in water, soluble in most organic solvents.

Appearance and Odor
Colorless crystals or white to slightly off-white powder, odorless or with slight aromatic odor. Odor Threshold = 0.2 ppm.

SECTION IV - FIRE AND EXPLOSION HAZARD DATA

Flash Point (method used)	**162-171°F (72-77°C) closed cup**	Explosive Limits in Air % by Volume	**LEL: N/A UEL: N/A**
NFPA Classification	**Combustible Solid**	Autoignition Temperature	**Not Applicable**

Extinguishing Media
Use dry chemical, carbon dioxide, foam, water, or water spray.

Special Fire Fighting Procedures
Poisonous gases are produced in fire, wear full protective clothing and self-contained breathing apparatus (SCBA). Move containers from fire if it can be done without risk.

Unusual Fire and Explosion Hazards
Containers may explode in fire due to buildup of excessive internal pressures. Exercise extreme caution when attempting to move or relocate containers that may have been exposed to high heat. DDT is a toxic waste; by-products of fire fighting operations should be handled as hazardous waste as well.

SECTION V - REACTIVITY DATA

Stability		Conditions to Avoid: Avoid contact with incompatible materials. DDT is sensitive to exposure to ultraviolet light. It is also sensitive to high temperatures. Solutions with water, ethanol or acetone should remain stable for 24 hours.
Stable	Unstable	Incompatibility *(materials to avoid)*: Strong oxidizers (such as chlorine, bromine, fluorine) and alkalis. May react with iron, aluminum and iron salts, ferric chloride and aluminum chloride.
X		
Hazardous Polymerization		Conditions to Avoid: Under normal conditions of temperature and pressure, hazardous polymerization of DDT is not expected to occur.
May Occur	Will Not Occur	Hazardous Decomposition or By-products: Toxic and irritating gases are produced when DDT is involved in fire.
	X	

SECTION VI - HEALTH HAZARD DATA

Primary Route(s) of Entry:	Inhalation?	Absorption (skin)?	Ingestion?
	X	X	X

Health Hazards

INHALATION: Symptoms include tremors of the head and neck muscles, tonic and clonic convulsions, cardiac or respiratory failure, and death. Prolonged exposures cause CNS depression and degeneration, weakness, convulsions, and coma.

ABSORPTION: Headache, dizziness, nausea, vomiting, tremor, ataxia, convulsions, CNS depression, respiratory failure, and may cause renal damage. May cause skin allergy.

INGESTION: May cause anesthesia, convulsions, headache, analgesia, cardiac arrhythmia, nausea, vomiting, sweating, and unspecified pulmonary changes.

Carcinogenicity	NTP Listed?	IARC Cancer Review Group?	OSHA Regulated?	Target Organs?
Suspected Human Confirmed Animal	5th Annual Report	Group 2B	29 CFR 1910.1000 Table Z-1 (skin)	CNS, liver, kidneys, skin, eyes, PNS, lymph glands.

Medical Conditions Generally Aggravated by Exposure

Renal and/or liver function problems; Skin conditions (dermatitis).

Emergency and First-aid Procedures

Eye contact: Flush immediately with water for 15 minutes (minimum), seek medical attention. Skin contact: Remove all contaminated clothing. Immediately wash area with large amounts of soap and water. Seek medical assistance. For inhalation: Remove the person from exposure. Provide respiratory assistance and CPR. Transfer to medical facility. If swallowed: If victim is conscious, give large amounts of water and milk. Do NOT induce vomiting. Never attempt to give an unconscious or convulsing person anything by mouth. Seek medical attention immediately.

SECTION VII - PRECAUTIONS FOR SAFE HANDLING AND USE

Steps to be Taken in Case Material is Released or Spilled

Collect powdered material in the most convenient (safe) way possible and deposit in DOT-approved drum. Ventilate area and remove ignition sources. Restrict those not involved in cleanup from entering area. Notify appropriate authorities, if applicable.

Preferred Waste Disposal Method

Chemical incinerator equipped with scrubber and afterburner.

Precautions to be Taken in Handling and Storage

Do not store DDT in the presence of incompatible chemicals or materials. Store in tightly closed containers in a cool, dark, well-ventilated area. Personnel working with DDT should be trained on its proper handing and storage *prior* to being assigned to such responsibilities.

Other Precautions and Warnings

The use of this compound was prohibited in the United States in 1973. It is still manufactured for use in other countries. Bulk storage is of DDT is not recommended.

SECTION VIII - CONTROL MEASURES AND PERSONAL PROTECTIVE EQUIPMENT

Respiratory Protection *(specify type)*

At any exposure level: Use a self-contained breathing apparatus (SCBA) with full facepiece and pressure demand or other positive pressure mode.

Ventilation

Local exhaust or general mechanical systems recommended.

Protective Gloves	Eye Protection	Other Protective Clothing
Rubber or Latex	Chemical/Dust Goggles or Face Mask	Apron

Work/Hygiene Practices

Always wash hands thoroughly after using chemical; never bring food, drink, or smoking materials into vicinity of chemicals.

DDT (dichlorodiphenyltrichloroethane)

$(ClC_6H_4)_2CHCCl_3$ CAS: 50-29-3

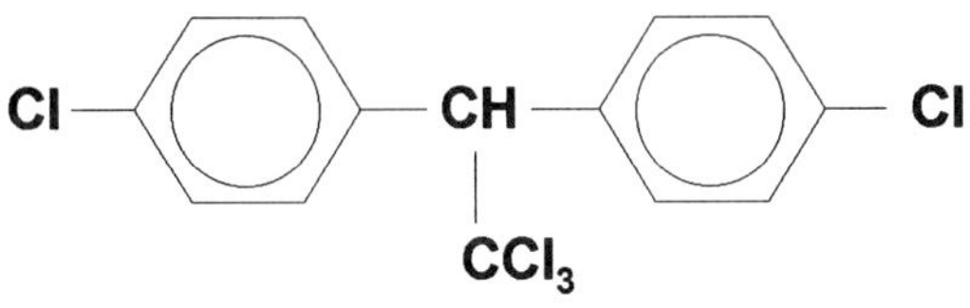

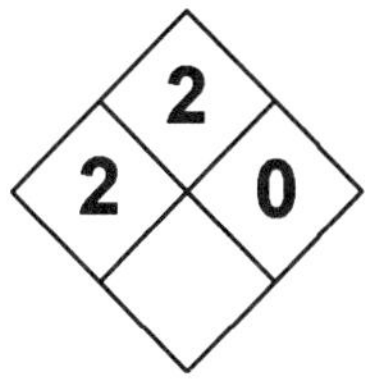

IDENTIFICATION AND TYPICAL USES

DDT appears as colorless crystals or off-white powder with a slight, aromatic odor. It is used as an insecticide, as a pesticide, and to control vectors of disease. It is used in veterinary medicine as an insecticide and pediculicide. It is used to kill head lice and disinfect clothing. It is also used as an ectoparasiticide. The use of this compound was prohibited in the United States in 1973. Food and Drug Administration (FDA) tolerance in foods is 5 ppm. It is not biodegradable.

RISK ASSESSMENT: HEALTH

General Assessment

DDT is a confirmed carcinogen in animals and is highly suspected to cause cancer in humans. Experimental neoplastigenic, tumorigenic, carcinogenic, and teratogenic data have been reported in numerous studies. Human mutation data have also been recorded. It is a human poison primarily by *ingestion,* but can also enter the body through *inhalation* and skin *absorption.*

Ingestion of large doses will result in a broad spectrum of toxic systemic effects as well as local gastric irritation. These include anesthesia, headache, analgesia, cardiac arrhythmia, nausea or vomiting, sweating, delayed emesis or diarrhea, and unspecified pulmonary changes. Other symptoms may appear 2-3 hours after small doses. These include tingling of the lips, tongue, and face. There may be malaise, headache, sore throat, fatigue, coarse tremors of the neck, head, and eyelids, apprehension, ataxia, and confusion. Convulsions may alternate with periods of coma and partial paralysis. Vital signs may appear as normal, but in severe poisonings, the pulse may be irregular or abnormally slow. Ventricular fibrillation and sudden death may occur.

Inhalation will irritate the eyes, nose, and upper respiratory tract. It can also produce many of the toxic systemic effects noted for ingestion. Pulmonary edema (fluid buildup in the lungs) does not normally occur. If it does, it may indicate solvent intoxication.

Skin contact produces mild irritation. If allowed to remain in contact for prolonged periods of time, may cause smarting and reddening of the skin. Absorption of the solid crystals is not likely. However, when DDT is mixed in solution, absorption through unbroken skin may occur on prolonged contact causing toxic systemic effects.

Acute Health Effects

The following acute (short-term) health effects may occur immediately or shortly after exposure to DDT:

Skin: Dermatitis associated with DDT is unusual. Most effects from skin contact are internal due to absorption of liquid solutions containing DDT. Symptoms include vomiting, nausea, malaise, and muscular twitching at high doses (above the PEL).

Eye: Possible irritation.

Lung: Irritation causing coughing and/or shortness of breath. Higher exposures can lead to respiratory problems.

CNS: Dizziness, headaches, nausea, vomiting, muscle jerks, severe seizures, and death.

Chronic Health Effects

The following chronic (long-term) health effects may occur at some time after exposure to DDT and can last for months or even years:

Cancer Hazards: DDT is a confirmed carcinogen in test animals and has been shown to effect animals in the wild as well. It is highly suspected to cause cancers in humans. It is a known human mutagen (causes genetic changes). The data on human carcinogenicity are conflicting and inconclusive.

Reproductive Hazard: DDT may be a teratogen in humans since it has been shown to be a teratogen in some test animals.

Other Chronic Effects: Prolonged exposures may damage the liver or kidney. In animals, chromic exposures have caused liver, lung, and lymphatic tumors. It is not known whether long-term human exposure will have the same effects.

Recommended Risk-Reduction Measures

Personnel should avoid direct contact with DDT. Even though it has not been permitted for use in the United States since 1973, it is still manufactured for export to other countries. Unless a less toxic chemical can be substituted for a hazardous substance, *engineering controls* are the most effective method of reducing exposures. The best protection is to enclose operations and/or provide local exhaust ventilation at the site of chemical release. This may not be feasible during use as a pesticide outdoors, but it may be possible during manufacture of DDT. While not always operationally feasible, isolating operations can also reduce exposure.

Using respiratory protection is less effective than the controls mentioned above, but is still advisable whenever working with or around DDT. A self-contained breathing apparatus (SCBA) with full facepiece and pressure demand is the recommended respiratory protection of choice. If a full facepiece is not available, then chemical/dust goggles should be worn to protect the eyes. Whenever DDT is being mixed with liquids, a chemical splash hazard exists and a face shield and rubber apron should be worn. To prevent hand and skin exposures, rubber gloves should be worn.

Administrative controls should also be in place to minimize the potential for human exposures. These may include written procedures or policies which specify the methods and techniques that will be practiced whenever personnel are to work with DDT.

All personnel should receive *training* on the use, hazards, protective measures, emergency actions, and other precautions per 29 CFR 1910.1200 (Hazard Communication), prior to the first assignment in an area where DDT is used or stored.

If symptoms develop or overexposure is suspected, the following medical tests may be useful:

- ☑ Lung function tests.
- ☑ Liver and kidney function tests.
- ☑ Evaluation by a qualified allergist, including careful exposure history and special testing (may help diagnose skin allergy).

Any evaluation should include a careful medical history of past and present symptoms with an examination. However, medical tests that evaluate existing damage are not a substitute for controlling exposure.

Other methods to reduce exposure include:

- ☑ Where possible, enclose operations and use local exhaust ventilation at the site of chemical release. If local exhaust ventilation or enclosure is not used, respiratory protection should be mandatory.
- ☑ Always ensure that proper protective clothing is worn when using chemical substances.
- ☑ Wash thoroughly immediately after exposure to DDT and at the end of the work shift or before eating, drinking, or smoking.
- ☑ Hazard warning information should be posted in the work area. In addition, as part of an on-going education and training program, all information on the health and safety hazards of DDT should be communicated to all potentially exposed workers.
- ☑ Eye wash stations should be provided in work areas. If the potential for whole body exposure exists, then safety showers should also be provided.

RISK ASSESSMENT: ENVIRONMENT

General Assessment

The environment is at risk of exposure during intended use, transportation, storage, disposal, or destruction of DDT. In almost every scenario, the threat of environmental exposure is contingent upon the proper handling of the chemical substance. Accidental spills, large or small, can result in fire, explosion, and possible contamination of the surrounding environmental mediums (water, soil, and air).

DDT is considered a combustible solid. While it is a minor fire hazard in and of itself, it is often mixed with liquids which are either flammable or combustible. These characteristics require special consideration during any emergency situation involving a leak or spill of DDT or DDT mixtures. Should DDT ever come into contact with incompatible substances such as strong oxidizers (chlorine, bromine, fluorine) or alkali materials either during use, transportation, or storage, violent reaction can occur.

The proper disposal/destruction method for DDT is to burn it in a chemical incinerator equipped with an afterburner and air scrubber.

DDT can enter the environment through its normal, intended use as a pesticide and through industrial discharges, urban runoff, and spills.

Acute Ecological Effects

Acute (short-term) toxic effects may include the death of animals, birds, or fish and death or low growth rate in plants. Acute effects are seen 2 to 4 days after animals or plants are exposed to DDT. This chemical has high acute toxicity to aquatic life and to birds. Insufficient data are available to evaluate or predict the short-term effects of DDT to plants or land animals.

Chronic Ecological Effects

Chronic toxic effects may include shortened life span, reproductive problems, lower fertility, and changes in appearance or behavior in exposed animals. These effects can be seen long after first exposure(s) to toxic chemicals.

DDT has high chronic toxicity to aquatic life and to birds. Insufficient data are available to evaluate or predict the long-term effects of DDT to plants or land animals.

Water Solubility

DDT is very slightly soluble in water. Concentrations of 1 milligram or less will mix with a liter of water.

Persistence in the Environment

DDT is highly persistent in water, with a half-greater than 200 days. The half-life of a pollutant is the amount of time it takes for one-half of the chemical to be degraded.

Bioaccumulation in Aquatic Organisms

Some substances increase in concentration, or *bioaccumulate,* in living organisms as they breathe contaminated air, drink contaminated water, or eat contaminated food. These chemicals can become concentrated in the tissues and internal organs of animals as well as humans.

The concentration of DDT found in fish tissues is expected to be much higher than the average concentration of DDT found in the water from which the fish was taken.

Recommended Risk-Reduction Measures

Proper training of all transporters will reduce the likelihood of a mishap or accident resulting in a leak or spill to the environment. The correct use of DOT labeling (per the MSDS) on all containers, trucks, and rail cars will enable emergency responders to react properly and quickly to any disaster thereby reducing the potential risk to the environment and to personnel.

Storage of DDT should be segregated from incompatible chemicals to minimize the risk of cross-contamination or contact.

If a spill or leak to the environment has occurred, fire department, emergency response and/or hazardous materials spill personnel should be notified immediately. Cleanup should be attempted only by those trained in proper spill containment procedures. Contaminated soils should be removed for incineration and replaced with clean soil. If DDT should contact the water table, aquifer, or navigable waterway, remediation activities should be prompt to ensure protection of aquatic sediments. When such spills occur, the local and/or state emergency response authorities must be notified. A comprehensive emergency response or disaster preparedness/recovery plan should be in place prior to any operations involving the use, transportation, storage, or disposal of DDT. If DDT is spilled or leaked, the following specific steps are recommended:

- ☑ Restrict persons not wearing protective clothing from area of spill or leak until cleanup is complete and area can be opened for normal work.
- ☑ Collect powdered material in the most convenient (safe) manner possible and deposit in sealed containers. If liquid solutions are spilled, use vermiculite, dry earth, or sand to absorb and place in a sealed drum for disposal.
- ☑ It may be necessary to dispose of DDT as a hazardous waste. The responsible state agency or the regional office of the federal Environmental Protection Agency (EPA) should be contacted for specific recommendations.

RISK ASSESSMENT: BUSINESS

General Assessment

Accidents or mishaps involving DDT can present a significant threat to business operations. The loss or damage of equipment or facilities can significantly affect fiscal viability. Lawsuits that may result from personnel injury/death, public exposures, and/or environmental contamination will also require a serious expenditure of resources. Media attention surrounding an injury, death, or environmental damage can also result in a loss of profits and loss of current as well as future business. The public is particularly aware of chemicals such as DDT because of the years of publicity concerning this chemical's hazardous nature. Risk managers should be aware of this when formulating any public relations policies.

Recommended Risk-Reduction Measures

Company attorneys, safety and health professionals, and environmental specialists should be involved in the development of any procedures dealing with safety or emergency response. A company official should be pre-designated as a public relations officer with specific training in dealing with the press. Corporate plans and policies should be developed, approved, and implemented long before any need for such arises.

REFERENCES

Ahlbom, A. 1993. *Biostatistics for Epidemiologists.* New York: CRC Press/Lewis Publishers.

American Conference of Governmental Industrial Hygienists. 1988. *Documentation of the Threshold Limit Values and Biological Exposure Indices,* 5th Edition (with updates). Cincinnati: ACGIH

Anthony, C. P., and N. J. Kolthoff. 1971. *Textbook of Anatomy and Physiology.* St. Louis: C. V. Mosby Company.

Asante-Duah, D. K. 1993. *Hazardous Waste Risk Assessment.* New York: CRC Press/Lewis Publishers.

Bloom, A. D. and F. J. DeSerres. 1995. *Ecotoxicity and Human Health.* New York: CRC Press/Lewis Publishers.

Brusick, D. J. 1994. *Methods for Genetic Risk Assessment.* New York: CRC Press/Lewis Publishers.

Bueche, J. R. 1972. *Principles of Physics,* 2nd Edition. New York: McGraw-Hill.

Burrell, R., D. K. Flaherty, and L. J. Sauers. 1992. *Toxicology of the Immune System.* New York: Van Nostrand Reinhold.

Calabrese, E. J. 1994. *Biological Effects of Low Level Exposures.* New York: CRC Press/Lewis Publishers.

Cockerman, L. G., B. S. Shane. 1994. *Basic Environmental Toxicology.* New York: CRC Press/Lewis Publishers.

Cothern, C. R., N. P. Ross. 1994. *Environmental Statistics, Assessment, and Forecasting.* New York: CRC Press/Lewis Publishers.

Gardner, R. W. 1994. *Chemical Intolerance.* New York: CRC Press/Lewis Publishers.

Hallenbeck, W. H. 1993. *Quantitative Risk Assessment for Environmental and Occupational Health,* 2nd Edition. New York: CRC Press/Lewis Publishers.

Hodgson, E., R. B. Mailman, and Janice E. Chambers. 1988. *Dictionary of Toxicology.* London: The MacMillian Company.

Howard, P. H. 1995. *Handbook of Environmental Fate and Exposures Data for Organic Chemicals,* Vol. I-IV. New York: CRC Press/Lewis Publishers.

Kamrin, M. A. 1988. *Toxicology: A Primer on Principles and Applications.* Chelsea, Michigan: Lewis Publishers.

Landis, W. G., M. H. Yu. 1995. *Introduction to Environmental Toxicology.* New York: CRC Press/Lewis Publishers.

Langerman, N. 1994 *Precautionary Labels for Chemical Containers.* New York: CRC Press/Lewis Publishers.

Lauwerys, R. R., P. Hoet. 1993. *Industrial Chemical Exposure.* New York: CRC Press/Lewis Publishers.

Lide, D. R. 1995. *Handbook of Chemistry and Physics,* 76th Edition. New York: CRC Press/Lewis Publishers.

Lewis, R. J., Sr. 1993. *Hawley's Condensed Chemical Dictionary,* 12th Edition. New York: Van Nostrand Reinhold.

Lewis, R. J., Sr. 1992. *Sax's Dangerous Properties of Industrial Materials,* Volumes 1, 2, and 3, 8th Edition. New York: Van Nostrand Reinhold.

Manahan, S. E. 1992. *Toxicological Chemistry,* 2nd Edition. New York: CRC Press/Lewis Publishers.

Mellström, G. A., J. E. Wahlberg, and H. I. Maibach. 1994. *Protective Gloves for Occupational Use.* New York: CRC Press/Lewis Publishers.

Meyer, E.. 1990. *Chemistry of Hazardous Materials.* Englewood Cliffs, NJ: Prentice Hall.

Nelson, G. O. 1992. *Gas Mixtures.* New York: CRC Press/Lewis Publishers.

Ott, W. R. 1995. *Environmental Statistics and Data Analysis.* New York: CRC Press/Lewis Publishers.

Patnaik, P. 1992. *A Comprehensive Guide to the Hazardous Properties of Chemical Substances.* New York: Van Nostrand Reinhold.

Philp, R. B. 1995. *Environmental Hazards and Human Health.* New York: CRC Press/Lewis Publishers.

Proctor, N. H., and James P. Hughes. 1991. *Chemical Hazards of the Workplace.* New York: Van Nostrand Reinhold.

Quigley, D. R. 1994. *Handbook of Emergency Chemical Management.* New York: CRC Press/Lewis Publishers.

Rea, W. J. 1995. *Chemical Sensitivity,* Vol. I-III. New York: CRC Press/Lewis Publishers.

Richardson, M. 1995. *Dictionary of Substance and Their Effects,* Vol. I-VI. New York: CRC Press/Lewis Publishers.

Rylander, R., R. R. Jacobs. 1994. *Organic Dusts.* New York: CRC Press/Lewis Publishers.

Sacarello, H. L. A. 1994. *The Comprehensive Handbook of Hazardous Materials.* New York: CRC Press/Lewis Publishers.

U.S. Department of Health and Human Services, National Institute for Occupational Safety and Health. 1994. *NIOSH Pocket Guide to Chemical Hazards.* Washington, D.C.: U.S. Government Printing Office.

MATERIAL SAFETY DATA SHEET

CHEMICAL NAME	
DECABORANE	**CRC PRESS/LEWIS PUBLISHERS DISCLAIMER:** THE INFORMATION AND RECOMMENDATIONS PRESENTED HEREIN ARE BASED ON SOURCES BELIEVED TO BE RELIABLE. CRC MAKES NO REPRESENTATION ON ITS COMPLETENESS OR ACCURACY. IT IS THE RESPONSIBILITY OF THE USER TO DETERMINE THE CHEMICAL'S SUITABILITY FOR ITS INTENDED USE, THE CHEMICAL'S SAFE USE, AND THE CHEMICAL'S PROPER DISPOSAL. NO REPRESENTATIONS AND/OR WARRANTIES, EITHER EXPRESSED OR IMPLIED, OF THE MERCHANTABILITY OR FITNESS FOR A PARTICULAR PURPOSE, OR OF ANY OTHER NATURE, ARE MADE WITH RESPECT TO THE INFORMATION PROVIDED IN THIS MSDS OR TO THE CHEMICAL TO WHICH INFORMATION MAY REFER. CRC NEITHER ASSUMES NOR AUTHORIZES ANY OTHER PERSON TO ASSUME FOR IT, ANY OTHER ADDITIONAL RESPONSIBILITY OR LIABILITY FOR THE USE OF, OR RELIANCE UPON, THIS INFORMATION.

HAZARD WARNING INFORMATION

HEALTH	FIRE	REACTIVE	OTHER	DEGREE OF HAZARD	COLOR CODING	OTHER CODES
3	2	1	~~W~~	0 = Minimum Hazard 1 = Slight Hazard 2 = Moderate Hazard 3 = Serious Hazard 4 = Severe Hazard	HEALTH = BLUE FIRE = RED REACTIVITY = YELLOW OTHER = WHITE	OX = Oxidizer ACID = Acid ALK = Alkali COR = Corrosive ~~W~~ = Use No Water

SECTION I - GENERAL INFORMATION

Characterization	RCRA Number	EPA Class
Hydride	**None**	**Not Applicable**

DOT Proper Shipping Name	Chemical Abstract Service (CAS) Number
Decaborane	**17702-41-9**
DOT Hazard Class and Label Requirements	DOT Emergency Guide Code
Flammable Solid and Poison	**34**
DOT Identification Number	Chemical Formula
UN 1868	**$B_{10}H_{14}$**

Synonyms
Decaboron tetradechydride.

SECTION II - HAZARDOUS INGREDIENTS AND IDENTITY INFORMATION

Hazardous Components (specific identity)	OSHA Exposure Criteria	NIOSH Exposure Criteria	Immediately Dangerous to Life and Health (IDLH)	ACGIH Exposure Criteria
Decaborane (derivation: A by-product of the pyrolysis of diborane). **1 ppm = 5.08 mg/m^3**	**PEL (skin):** **0.05 ppm** **0.3 mg/m^3** **STEL:** **0.15 ppm** **0.9 mg/m^3**	**REL (skin):** **0.05 ppm** **0.3 mg/m^3** **STEL:** **0.15 ppm** **0.9 mg/m^3**	**3 ppm** **(15 mg/m^3)**	**TLV (skin):** **0.05 ppm** **0.3 mg/m^3** **STEL:** **0.15 ppm** **0.9 mg/m^3**

SECTION III - PHYSICAL AND CHEMICAL CHARACTERISTICS

Boiling Point	Specific Gravity (H_2O = 1)
210°F (99°C) Decomposes	**0.94**
Vapor Pressure (mm Hg)	Molecular Weight
0.2 at 69°F (20°C)	**122.2**
Vapor Density (Air = 1)	Melting Point
4.2	**210°F (99°C)**

Solubility
Slightly soluble in cold water, decomposes in hot water. Dissolves in carbon disulfide, alcohol, benzene, ethyl acetate, and acetic acid.

Appearance and Odor
Colorless to white crystalline solid with an intense, bitter, chocolate-like odor.

SECTION IV - FIRE AND EXPLOSION HAZARD DATA

Flash Point (method used)	Explosive Limits in Air % by Volume
176°F (80°C) closed cup	**LEL: N/A UEL: N/A**
NFPA Classification	Autoignition Temperature
Combustible Solid	**300°F (149°C)**

Extinguishing Media
Use dry chemical, carbon dioxide, or water spray. Do not use halogenated extinguishing agents.

Special Fire Fighting Procedures
Poisonous gases are produced in fire, wear full protective clothing and self-contained breathing apparatus (SCBA). May give-off toxic fumes of unburned material. Move containers from fire area if it can be done without risk.

Unusual Fire and Explosion Hazards
Reacts slowly with water to form flammable hydrogen gas. Decaborane ignites spontaneously in oxygen. Reactions with halogen compounds yield products that may explode on impact. May explode when hot (burns with a green colored flame).

SECTION V - REACTIVITY DATA

Stability		Conditions to Avoid **Avoid contact with incompatible materials. Decaborane forms impact-sensitive explosive mixtures with ethers (e.g., dioxane) and halocarbons (carbon tetrachloride).**
Stable X	Unstable	Incompatibility *(materials to avoid)* **Oxidizers, water, halogenated compounds, dimethyl sulfoxide. Ignites spontaneously in oxygen and may ignite in air. It produces hydrogen when it hydrolyzes with water or on decomposition at 572°F (300°C).**
Hazardous Polymerization		Conditions to Avoid **Under normal conditions of temperature and pressure, hazardous polymerization of decaborane is not expected to occur.**
May Occur	Will Not Occur X	Hazardous Decomposition or By-products **When heated to decomposition, decaborane emits toxic fumes of boron oxides. Also produces flammable hydrogen gas at decomposition.**

SECTION VI - HEALTH HAZARD DATA

Primary Route(s) of Entry:	Inhalation?	Absorption (skin)?	Ingestion?
	X	X	X

Health Hazards

INHALATION: Headache, lightheadedness, nausea, vomiting, dizziness, muscle spasms, and possible convulsions. There may be dyspnea, weakness, and possible liver and kidney damage.

ABSORPTION: Relatively moderate systemic effects. Headache and mild nausea likely. Drowsiness and loss of coordination is possible. Rash may develop on prolonged contact.

INGESTION: May cause spasm, tremor, and convulsions.

Carcinogenicity	NTP Listed?	IARC Cancer Review Group?	OSHA Regulated?	Target Organs?
Unknown Human Unknown Animal	No	No	29 CFR 1910.1000 Table Z-1 (skin)	CNS, liver, kidneys, skin, eyes.

Medical Conditions Generally Aggravated by Exposure

None Reported.

Emergency and First-aid Procedures

Eye contact: Flush immediately with water for 15 minutes (minimum), seek medical attention. Skin contact: Remove all contaminated clothing. Wash area with large amounts of soap and water. Seek medical assistance. For inhalation: Remove the person from exposure. Provide respiratory assistance and CPR. Transfer to medical facility. If swallowed: If victim is conscious, give solution of one tablespoon of salt in 8 ounces of water. Repeat until vomitus is clear. Never attempt to give an unconscious or convulsing person anything by mouth. DOCTOR: Treat symptomatically. Administration of methocarbamol or other muscle relaxant may be helpful immediately following exposure.

SECTION VII - PRECAUTIONS FOR SAFE HANDLING AND USE

Steps to be Taken in Case Material is Released or Spilled

Collect powdered material in the most convenient (safe) way possible and deposit in DOT-approved drum. Decontaminate spill area with 3% aqueous ammonia solution. Ventilate area and remove ignition sources. Restrict those not involved in cleanup from entering area.

Preferred Waste Disposal Method

Carefully dissolve in water and treat with 1 *M* sulfuric acid to a pH of 1. Vent the hydrogen emissions. The solution is evaporated to dryness and the residue is buried in an approved landfill site.

Precautions to be Taken in Handling and Storage

Do not store decaborane in the presence of incompatible chemicals or materials. Store in tightly closed containers in a cool, dark, well-ventilated area. Personnel working with decaborane should be trained on its proper handing and storage *prior* to being assigned to such responsibilities.

Other Precautions and Warnings

Keep away from water and halogenated compounds. Bulk storage is of decaborane is not recommended. Use non-sparking tools, especially when opening and closing containers.

SECTION VIII - CONTROL MEASURES AND PERSONAL PROTECTIVE EQUIPMENT

Respiratory Protection *(specify type)*

PEL is extremely low (0.05 ppm). Use a self-contained breathing apparatus (SCBA) with full facepiece and pressure demand or other positive pressure mode.

Ventilation

Local exhaust or general mechanical systems recommended.

Protective Gloves	Eye Protection	Other Protective Clothing
Rubber or Latex	Chemical/Dust Goggles or Face Mask	Apron

Work/Hygiene Practices

Always wash hands thoroughly after using chemical; never bring food, drink, or smoking materials into vicinity of chemicals.

DECABORANE

$B_{10}H_{14}$ CAS: 17702-41-9

IDENTIFICATION AND TYPICAL USES

Decaborane appears as a colorless to white, orthorhombic, crystalline solid with an intense, bitter, chocolate-like odor. It is used in rocket propellants, as a catalyst in olefin polymerization, as a fuel additive, a corrosion inhibitor, and as a stabilizer. It is also used as a delustrant in the manufacture of rayon, as a mothproofing agent, a dye-stripping agent, a reducing agent, a fluxing agent, and as an oxygen scavenger.

RISK ASSESSMENT: HEALTH

General Assessment

Decaborane is a highly toxic compound by all routes of exposure (*inhalation, ingestion,* and skin *absorption*). It is a systemic poison in humans and acute high exposures or chronic low exposures can lead to death. Its carcinogenicity, teratogenicity, and mutagenicity are unknown in humans.

Inhalation of decaborane vapors or dusts can be irritating to the nose, throat, and upper respiratory tract. Primary effect of inhalation is its apparent action to depress the central nervous system (CNS). Symptoms may include headache, nausea, vomiting, dizziness, drowsiness, and lightheadedness. In severe cases of poisoning, there may be muscle spasm and convulsions. It is important to note that symptoms of exposure may be delayed for 24 to 48 hours following exposure, thereby creating a dangerous and risky false sense of security.

Ingestion can cause spasm, tremor, and convulsion. There may also be fatigue, weakness, muscle spasms, dypsnea, and liver and kidney damage.

Decaborane can be absorbed through unbroken skin to cause toxic systemic effects, including, but not limited to loss of coordination and drowsiness. Toxic systemic effects from absorption, however, are considered to be relatively moderate in comparison to other routes of entry.

Acute Health Effects

The following acute (short-term) health effects may occur immediately or shortly after exposure to decaborane:

Skin: Irritation and possible rash. If allowed to remain on skin, symptoms of headache, dizziness, nausea, and other systemic effects may occur.

Eye: Dusts will irritate the eyes. Liquid solutions may cause burns and damage.

Lung: Mild irritation of the nose and throat. May cause nausea and vomiting.

CNS: Dizziness, headaches, nausea, vomiting, muscle spasms, tremor, fatigue, convulsions, dypsnea, nervousness, drowsiness, and loss of coordination.

Chronic Health Effects

The following chronic (long-term) health effects may occur at some time after exposure to decaborane and can last for months or even years:

Cancer Hazards: According to the information presented in the references, decaborane has not been adequately tested for its ability to cause cancer in test animals.

Reproductive Hazard: According to the information presented in the references, decaborane has not been adequately tested for its ability to adversely effect reproduction.

Other Chronic Effects: Prolonged exposures may damage the liver or kidney.

Recommended Risk-Reduction Measures

Personnel should avoid direct contact with decaborane. Unless a less toxic chemical can be substituted for a hazardous substance, *engineering controls* are the most effective method of reducing exposures. The best protection is to enclose operations and/or provide local exhaust ventilation at the site of chemical release. While not always operationally feasible, isolating operations can also reduce exposure.

Using respiratory protection is less effective than the controls mentioned above, but is still advisable whenever working with or around decaborane. This chemical has an extremely low permissible exposure limit (0.05 ppm). Therefore, prudent risk management requires a conservative approach to personal protection. A self-contained breathing apparatus (SCBA) with full facepiece and pressure demand is the recommended respiratory protection of choice. If a full facepiece is not available, then chemical/dust goggles should be worn to protect the eyes. Whenever decaborane is being mixed with liquids, a chemical splash hazard exists and a face shield and rubber apron should be worn. To prevent hand and skin exposures, rubber gloves should be worn.

Administrative controls should also be in place to minimize the potential for human exposures. These may include written procedures or policies which specify the methods and techniques that will be practiced whenever personnel are to work with decaborane.

All personnel should receive *training* on the use, hazards, protective measures, emergency actions, and other precautions per 29 CFR 1910.1200 (Hazard Communication), prior to the first assignment in an area where decaborane is used or stored.

If symptoms develop or overexposure is suspected, the following medical tests may be useful:

- ☑ Lung function tests.
- ☑ Liver and kidney function tests.
- ☑ Evaluation by a qualified allergist, including careful exposure history and special testing (may help diagnose skin allergy).

Any evaluation should include a careful medical history of past and present symptoms with an examination. However, medical tests that evaluate existing damage are not a substitute for controlling exposure.

Physicians are advised to treat symptomatically. Administration of methocarbamol or other muscle relaxant may be helpful immediately following exposure and in the absence of symptoms (note that symptoms may be delayed up to 48 hours in some cases).

Other methods to reduce exposure include:

- ☑ Where possible, enclose operations and use local exhaust ventilation at the site of chemical release. If local exhaust ventilation or enclosure is not used, respiratory protection should be mandatory.
- ☑ Always ensure that proper protective clothing is worn when using chemical substances.
- ☑ Wash thoroughly immediately after exposure to decaborane and at the end of the work shift or before eating, drinking, or smoking.
- ☑ Hazard warning information should be posted in the work area. In addition, as part of an on-going education and training program, all information on the health and safety hazards of decaborane should be communicated to all potentially exposed workers.
- ☑ Eye wash stations should be provided in work areas. If the potential for whole body exposure exists, then safety showers should also be provided.

RISK ASSESSMENT: ENVIRONMENT

General Assessment

The environment is at risk of exposure during transportation, storage, disposal, or destruction of decaborane. In almost every scenario, the threat of environmental exposure is contingent upon the proper handling of the chemical substance. Accidental spills, large or small, can result in fire, explosion, and possible contamination of the surrounding environmental mediums (water, soil, and air).

Decaborane is considered a combustible solid. While it is a minor fire hazard in and of itself, it is often mixed with liquids which are either flammable or combustible. It ignites spontaneously in oxygen and may ignite in contact with air. In reaction to halogenated compounds, it yields products that may explode on impact. These characteristics require special consideration during any emergency situation involving a leak or spill of decaborane or decaborane mixtures. Should decaborane ever come into contact with incompatible substances such as strong oxidizers, water, halogenated compounds (especially carbon tetrachloride), or dimethyl sulfoxide either during use, transportation, or storage, violent reaction can occur.

Decaborane may be destroyed by carefully dissolving with water followed by treating with 1 *M* sulfuric acid to a pH of 1. Hydrogen gas will be emitted after addition of acid and should be vented to avoid accumulation of explosive atmospheres. The remaining solution should be allowed to evaporate to dryness. The residue should be buried in an approved landfill site. Decaborane can enter the environment through industrial discharges and from spills.

Acute Ecological Effects

Acute (short-term) toxic effects may include the death of animals, birds, or fish and death or low growth rate in plants. Acute effects are seen 2 to 4 days after animals or plants are exposed to decaborane.

Insufficient data are available to evaluate or predict the short-term effects of decaborane to aquatic life, plants, birds, or land animals.

Chronic Ecological Effects

Chronic toxic effects may include shortened life span, reproductive problems, lower fertility, and changes in appearance or behavior in exposed animals. These effects can be seen long after first exposure(s) to toxic chemicals.

Insufficient data are available to evaluate or predict the long-term effects of decaborane to aquatic life, plants, birds, or land animals.

Water Solubility

Decaborane is very slightly soluble in water. Concentrations of 1 milligram or less will mix with a liter of water.

Persistence in the Environment

Decaborane is highly persistent in water, with a half-life greater than 200 days. The half-life of a pollutant is the amount of time it takes for one-half of the chemical to be degraded.

Bioaccumulation in Aquatic Organisms

Some substances increase in concentration, or *bioaccumulate,* in living organisms as they breathe contaminated air, drink contaminated water, or eat contaminated food. These chemicals can become concentrated in the tissues and internal organs of animals as well as humans.

The concentration of decaborane found in fish tissues is expected to be much higher than the average concentration of decaborane in the water from which the fish was taken.

Recommended Risk-Reduction Measures

Proper training of all transporters will reduce the likelihood of a mishap or accident resulting in a leak or spill to the environment. The correct use of DOT labeling (per the MSDS) on all containers, trucks, and rail cars will enable emergency responders to react properly and quickly to any disaster thereby reducing the potential risk to the environment and to personnel.

Storage of decaborane should be segregated from incompatible chemicals to minimize the risk of cross-contamination or contact. A nitrogen purge is recommended inside all storage vessels and bulk storage containers.

If a spill or leak to the environment has occurred, fire department, emergency response and/or hazardous materials spill personnel should be notified immediately. Cleanup should be attempted only by those trained in proper spill containment procedures. Contaminated soils should be removed for incineration and replaced with clean soil. If decaborane should contact the water table, aquifer, or navigable waterway, remediation activities should be prompt to ensure protection of aquatic sediments. When such spills occur, the local and/or state emergency response authorities must be notified. A comprehensive emergency response or disaster preparedness/recovery plan should be in place prior to any operations involving the use, transportation, storage, or disposal of decaborane. If decaborane is spilled or leaked, the following specific steps are recommended:

- ☑ Restrict persons not wearing protective clothing from area of spill or leak until cleanup is complete and area can be opened for normal work.
- ☑ Collect powdered material in the most convenient (safe) manner possible and deposit in sealed containers. If liquid solutions are spilled, use vermiculite, dry earth, or sand to absorb and place in a sealed drum for disposal.
- ☑ It may be necessary to dispose of decaborane as a hazardous waste. The responsible state agency or the regional office of the federal Environmental Protection Agency (EPA) should be contacted for specific recommendations.

RISK ASSESSMENT: BUSINESS

General Assessment

Accidents or mishaps involving decaborane can present a significant threat to business operations. The loss or damage of equipment or facilities can significantly affect fiscal viability. Lawsuits that may result from personnel injury/death, public exposures, and/or environmental contamination will also require a serious expenditure of resources. Media attention surrounding an injury, death, or environmental damage can also result in a loss of profits and loss of current as well as future business.

ⓤ *Recommended Risk-Reduction Measures*

Company attorneys, safety and health professionals, and environmental specialists should be involved in the development of any procedures dealing with safety or emergency response. A company official should be pre-designated as a public relations officer with specific training in dealing with the press. Corporate plans and policies should be developed, approved, and implemented long before any need for such arises.

REFERENCES

Ahlbom, A. 1993. *Biostatistics for Epidemiologists.* New York: CRC Press/Lewis Publishers.

American Conference of Governmental Industrial Hygienists. 1988. *Documentation of the Threshold Limit Values and Biological Exposure Indices,* 5th Edition (with updates). Cincinnati: ACGIH

Anthony, C. P., and N. J. Kolthoff. 1971. *Textbook of Anatomy and Physiology.* St. Louis: C. V. Mosby Company.

Asante-Duah, D. K. 1993. *Hazardous Waste Risk Assessment.* New York: CRC Press/Lewis Publishers.

Bloom, A. D. and F. J. DeSerres. 1995. *Ecotoxicity and Human Health.* New York: CRC Press/Lewis Publishers.

Brusick, D. J. 1994. *Methods for Genetic Risk Assessment.* New York: CRC Press/Lewis Publishers.

Bueche, J. R. 1972. *Principles of Physics,* 2nd Edition. New York: McGraw-Hill.

Burrell, R., D. K. Flaherty, and L. J. Sauers. 1992. *Toxicology of the Immune System.* New York: Van Nostrand Reinhold.

Calabrese, E. J. 1994. *Biological Effects of Low Level Exposures.* New York: CRC Press/Lewis Publishers.

Cockerman, L. G., B. S. Shane. 1994. *Basic Environmental Toxicology.* New York: CRC Press/Lewis Publishers.

Cothern, C. R., N. P. Ross. 1994. *Environmental Statistics, Assessment, and Forecasting.* New York: CRC Press/Lewis Publishers.

Gardner, R. W. 1994. *Chemical Intolerance.* New York: CRC Press/Lewis Publishers.

Hallenbeck, W. H. 1993. *Quantitative Risk Assessment for Environmental and Occupational Health,* 2nd Edition. New York: CRC Press/Lewis Publishers.

Hodgson, E., R. B. Mailman, and Janice E. Chambers. 1988. *Dictionary of Toxicology.* London: The MacMillian Company.

Howard, P. H. 1995. *Handbook of Environmental Fate and Exposures Data for Organic Chemicals,* Vol. I-IV. New York: CRC Press/Lewis Publishers.

Kamrin, M. A. 1988. *Toxicology: A Primer on Principles and Applications.* Chelsea, Michigan: Lewis Publishers.

Landis, W. G., M. H. Yu. 1995. *Introduction to Environmental Toxicology.* New York: CRC Press/Lewis Publishers.

Langerman, N. 1994 *Precautionary Labels for Chemical Containers.* New York: CRC Press/Lewis Publishers.

Lauwerys, R. R., P. Hoet. 1993. *Industrial Chemical Exposure.* New York: CRC Press/Lewis Publishers.

Lide, D. R. 1995. *Handbook of Chemistry and Physics,* 76th Edition. New York: CRC Press/Lewis Publishers.

Lewis, R. J., Sr. 1993. *Hawley's Condensed Chemical Dictionary,* 12th Edition. New York: Van Nostrand Reinhold.

Lewis, R. J., Sr. 1992. *Sax's Dangerous Properties of Industrial Materials,* Volumes 1, 2, and 3, 8th Edition. New York: Van Nostrand Reinhold.

Manahan, S. E. 1992. *Toxicological Chemistry,* 2nd Edition. New York: CRC Press/Lewis Publishers.

Mellström, G. A., J. E. Wahlberg, and H. I. Maibach. 1994. *Protective Gloves for Occupational Use.* New York: CRC Press/Lewis Publishers.

Meyer, E. 1990. *Chemistry of Hazardous Materials.* Englewood Cliffs, NJ: Prentice Hall.

Nelson, G. O. 1992. *Gas Mixtures.* New York: CRC Press/Lewis Publishers.

Ott, W. R. 1995. *Environmental Statistics and Data Analysis.* New York: CRC Press/Lewis Publishers.

Patnaik, P. 1992. *A Comprehensive Guide to the Hazardous Properties of Chemical Substances.* New York: Van Nostrand Reinhold.

Philp, R. B. 1995. *Environmental Hazards and Human Health.* New York: CRC Press/Lewis Publishers.

Proctor, N. H., and James P. Hughes. 1991. *Chemical Hazards of the Workplace.* New York: Van Nostrand Reinhold.

Quigley, D. R. 1994. *Handbook of Emergency Chemical Management.* New York: CRC Press/Lewis Publishers.

Rea, W. J. 1995. *Chemical Sensitivity,* Vol. I-III. New York: CRC Press/Lewis Publishers.

Richardson, M. 1995. *Dictionary of Substance and Their Effects,* Vol. I-VI. New York: CRC Press/Lewis Publishers.

Rylander, R., R. R. Jacobs. 1994. *Organic Dusts.* New York: CRC Press/Lewis Publishers.

Sacarello, H. L. A. 1994. *The Comprehensive Handbook of Hazardous Materials.* New York: CRC Press/Lewis Publishers.

U.S. Department of Health and Human Services, National Institute for Occupational Safety and Health. 1994. *NIOSH Pocket Guide to Chemical Hazards.* Washington, D.C.: U.S. Government Printing Office.

MATERIAL SAFETY DATA SHEET

CHEMICAL NAME

DECAHYDRONAPHTHALENE

CRC PRESS/LEWIS PUBLISHERS DISCLAIMER: THE INFORMATION AND RECOMMENDATIONS PRESENTED HEREIN ARE BASED ON SOURCES BELIEVED TO BE RELIABLE. CRC MAKES NO REPRESENTATION ON ITS COMPLETENESS OR ACCURACY. IT IS THE RESPONSIBILITY OF THE USER TO DETERMINE THE CHEMICAL'S SUITABILITY FOR ITS INTENDED USE, THE CHEMICAL'S SAFE USE, AND THE CHEMICAL'S PROPER DISPOSAL. NO REPRESENTATIONS AND/OR WARRANTIES, EITHER EXPRESSED OR IMPLIED, OF THE MERCHANTABILITY OR FITNESS FOR A PARTICULAR PURPOSE, OR OF ANY OTHER NATURE, ARE MADE WITH RESPECT TO THE INFORMATION PROVIDED IN THIS MSDS OR TO THE CHEMICAL TO WHICH INFORMATION MAY REFER. CRC NEITHER ASSUMES NOR AUTHORIZES ANY OTHER PERSON TO ASSUME FOR IT, ANY OTHER ADDITIONAL RESPONSIBILITY OR LIABILITY FOR THE USE OF, OR RELIANCE UPON, THIS INFORMATION.

HAZARD WARNING INFORMATION

HEALTH	FIRE	REACTIVE	OTHER	DEGREE OF HAZARD	COLOR CODING	OTHER CODES
2	2	0		0 = Minimum Hazard 1 = Slight Hazard 2 = Moderate Hazard 3 = Serious Hazard 4 = Severe Hazard	HEALTH = BLUE FIRE = RED REACTIVITY = YELLOW OTHER = WHITE	OX = Oxidizer ACID = Acid ALK = Alkali COR = Corrosive W = Use No Water

SECTION I - GENERAL INFORMATION

Characterization	RCRA Number	EPA Class
Aromatic Hydrocarbon	**None**	**Not Applicable**

DOT Proper Shipping Name **Decalin**	Chemical Abstract Service (CAS) Number **91-17-8**
DOT Hazard Class and Label Requirements **Combustible Liquid**	DOT Emergency Guide Code **No Citation**
DOT Identification Number **UN 1147**	Molecular Formula **$C_{10}H_{18}$**

Synonyms
Bicyclo(4,4,0)decane; naphthalane; perhydronaphthalene; DEC; decalin; naphthane.

SECTION II - HAZARDOUS INGREDIENTS AND IDENTITY INFORMATION

Hazardous Components (specific identity)	OSHA Exposure Criteria	NIOSH Exposure Criteria	Immediately Dangerous to Life and Health (IDLH)	ACGIH Exposure Criteria
Decahydronaphthalene: (derivation: By treatment of naphthalene in a fused state (above 100°C) with hydrogen in the presence of a copper or nickel catalyst).	**PEL:** Not Established **STEL:** Not Established	**REL:** Not Established **STEL:** Not Established	**Not Determined**	**TLV:** Not Established **STEL:** Not Established

SECTION III - PHYSICAL AND CHEMICAL CHARACTERISTICS

Boiling Point **369°F (187°C)**	Specific Gravity (H_2O = 1) **0.9**
Vapor Pressure (mm Hg) **5 at 69°F (20°C)**	Molecular Weight (atomic weight) **138.3**
Vapor Density (Air = 1) **4.8**	Melting Point **-24°F (-31°C)**

Solubility
Insoluble in water. Soluble in alcohol and ether.

Appearance and Odor
Colorless, water-white liquid with an aromatic odor.

SECTION IV - FIRE AND EXPLOSION HAZARD DATA

Flash Point (method used) **136°F (58°C) closed cup**	Explosive Limits in Air % by Volume **LEL: 1.8%** **UEL: 6.9%**
NFPA Classification **Class IIA Combustible Liquid**	Autoignition Temperature **1118°F (604°C)**

Extinguishing Media
Use dry chemical, foam, or carbon dioxide or water.

Special Fire Fighting Procedures
Poisonous gases are produced in fire. Wear full protective clothing and self-contained breathing apparatus (SCBA) operated in pressure-demand mode. Move container from fire area if it can be done without risk. Isolate for ½ mile if rail or tank car is involved in fire.

Unusual Fire and Explosion Hazards
Decahydronaphthalene is a combustible liquid. Flammable and moderately explosive when exposed to heat or flame. Can react vigorously with oxidizing materials. Containers may explode in fire. Immediately evacuate if rising sound from venting device is heard or if fire is causing a discoloration of the tank.

SECTION V - REACTIVITY DATA

Stability		Conditions to Avoid Keep away from incompatible materials. Avoid contact with heat or flame. Do not allow containers to remain exposed to sunlight or in hot weather for extended periods of time.
Stable X	Unstable	Incompatibility *(materials to avoid)* Strong oxidizers (such as perchlorates, peroxides, chlorates, nitrates, and permanganates).
Hazardous Polymerization		Conditions to Avoid Under normal conditions of temperature and pressure, hazardous polymerization of decahydronaphthalene is not expected to occur.
May Occur	Will Not Occur X	Hazardous Decomposition or By-products When heated to decomposition, decahydronaphthalene will emit acrid and irritating smoke and fumes.

SECTION VI - HEALTH HAZARD DATA

Primary Route(s) of Entry:	Inhalation?	Absorption (skin/eye)?	Ingestion?
	X		X

Health Hazards

INHALATION: May cause severe irritation to the respiratory system. There may be conjunctiva irritation and unspecified olfactory and pulmonary changes. There may be numbness, headache, and vomiting. Urine may become blue and there may be kidney damage.

SKIN & EYES: Extremely irritating to the skin and eyes. Liquid contact will de-fat the skin. May cause cracking and secondary infection, eczema may also develop.

INGESTION: Gastrointestinal irritation. Possible kidney damage.

Carcinogenicity	NTP Listed?	IARC Cancer Review Group?	OSHA Regulated?	Target Organs?
Questioned Human Questioned Animal	No	No	No	Skin, eyes, respiratory system, kidney.

Medical Conditions Generally Aggravated by Exposure

None reported, however, skin conditions (dermatitis) and respiratory problems may be aggravated.

Emergency and First-aid Procedures

Eye contact: Do not allow victim to keep eyes tightly shut. Flush immediately with water for 15 minutes (minimum), seek medical attention. Skin contact: Remove all contaminated clothing, including shoes. Immediately wash area with flooding amounts of soap and water. Seek medical attention immediately. For inhalation: Remove the person from exposure. Provide respiratory assistance and CPR. Transfer to medical facility. Observe for shock. If swallowed: Seek medical attention immediately. If conscious, give warm salt water to drink. DOCTOR: Suggest a mild cathartic: Conserve liver and kidney functions.

SECTION VII - PRECAUTIONS FOR SAFE HANDLING AND USE

Steps to be Taken in Case Material is Released or Spilled

Wear self-contained breathing apparatus (SCBA) for respiratory protection. Restrict those not wearing not involved in cleanup from entering area. Avoid skin contact. Ventilate area of spill. Decontaminate spill area using soapy water. Take up with sand or vermiculite and deposit in sealed drum.

Preferred Waste Disposal Method

No Citation.

Precautions to be Taken in Handling and Storage

Avoid contact with heat, sparks, or flame. Can form peroxides on storage or on evaporation/distillation.

Other Precautions and Warnings

Use only non-sparking tools and equipment. Do not store in direct sunlight for extended periods of time. Bulk storage of decahydronaphthalene is permissible but not recommended.

SECTION VIII - CONTROL MEASURES AND PERSONAL PROTECTIVE EQUIPMENT

Respiratory Protection *(specify type)*

Exposure levels not established and chemical is moderately toxic by inhalation. Recommend using a supplied-air respirator with full facepiece operated in positive pressure mode or, use a self-contained breathing apparatus (SCBA) with full facepiece operated in positive pressure mode.

Ventilation

Local exhaust (preferred) at site of chemical work or general ventilation.

Protective Gloves	Eye Protection	Other Protective Clothing
Rubber Gloves	Chemical Goggles and Face Mask	Rubber Apron

Work/Hygiene Practices

Always wash hands thoroughly after using chemical; never bring food, drink, or smoking materials into vicinity of chemicals.

DECAHYDRONAPHTHALENE

$C_{10}H_{18}$ CAS: 91-17-8

IDENTIFICATION AND TYPICAL USES

Decahydronaphthalene is a colorless water-white liquid with a strong aromatic odor. It is used as a solvent for oils, fats, waxes, resins, and rubber. It is a substitute for turpentine and is used in cleaning machinery, as a stain remover, in shoe creams, in floor waxes, in other cleaning fluids, in lubricants, and as a motor fuel additive.

RISK ASSESSMENT: HEALTH

General Assessment

Decahydronaphthalene is moderately toxic to humans by *ingestion* and *inhalation*. It is a questionable carcinogen in humans and in animals. Mutation data have also been reported.

Inhalation and ingestion may cause nose and throat irritation with numbness, headache, and vomiting. There may also be conjunctiva irritation, and other unspecified olfactory and pulmonary system changes. It can cause kidney damage on high or prolonged exposures. Urine may become blue (indicating kidney effects).

The liquid will de-fat the skin and cause dryness, cracking, and secondary infections. Eczema may develop. Eye irritation will also occur on contact with the liquid or vapor, often followed by conjunctivitis.

Acute Health Effects

The following acute (short-term) health effects may occur immediately or shortly after exposure to decahydronaphthalene:

Skin: Irritation with possible eczema-like rash. There may be cracking of skin and secondary infections.

Eye: Mild to severe irritation with possibility for conjunctivitis.

Lung: Nose, throat, and respiratory irritation causing difficulty in breathing. There may be numbness, headache, vomiting, and possible kidney effects. The urine may become blue.

Other: Ingestion of decahydronaphthalene may cause nausea, vomiting, headaches, dizziness, and gastrointestinal irritation.

Chronic Health Effects

The following chronic (long-term) health effects may occur at some time after exposure to decahydronaphthalene and can last for months or even years:

Cancer Hazards: Decahydronaphthalene may cause mutations (genetic changes) in living cells. There is limited evidence in the references that exposure may cause cancer in animals. There is insufficient data to evaluate its carcinogenic potential in humans.

Reproductive Hazard: There is limited evidence in the references to indicate that decahydronaphthalene may affect the developing fetus.

Other Chronic Effects: Over the long-term, exposure to decahydronaphthalene can lead to skin allergy. If an allergy develops, very low future exposures can cause itching and rash. Very irritating substances such as decahydronaphthalene can affect the lungs. It is not known whether decahydronaphthalene causes lung damage. Chronic exposure may also affect the liver and cause kidney damage.

Recommended Risk-Reduction Measures

Personnel should avoid direct contact with decahydronaphthalene. Occupational exposure can occur primarily through inhalation and ingestion. Skin and eye contact also present risks to heath. If a less toxic material or compound cannot be substituted for decahydronaphthalene, then *engineering controls* are the most effective method of reducing exposure risk. The best protection is to enclose operations and/or provide local exhaust ventilation at the site of decahydronaphthalene dust release. While not always operationally feasible, isolating operations can also reduce exposure. Using respiratory protection is less effective than the controls mentioned above, but is still recommended whenever working with or around decahydronaphthalene. No exposure levels have been established for this chemical. However, this does not mean that unprotected exposures are without risk. Decahydronaphthalene has been shown to cause toxic systemic effects by at least two routes of entry (inhalation and ingestion), and skin contact also presents health risks. Therefore, for any exposure levels, a supplied-air respirator with full facepiece or an MSHA/NIOSH approved self-contained breathing apparatus (SCBA) with full facepiece and pressure demand are the appropriate level of respiratory protection. If a full facepiece is not available, then chemical goggles should be worn to protect the eyes, especially when a splash hazard exists. A face shield should also be considered. To prevent hand and skin exposures, rubber gloves should be worn.

Administrative controls should also be in place to minimize the potential for human exposures. These may include written procedures or policies which specify the methods and techniques that will be practiced whenever personnel are to work with decahydronaphthalene.

All personnel should receive *training* on the use, hazards, protective measures, emergency actions, and other precautions per 29 CFR 1910.1200 (Hazard Communication) prior to the first assignment in an area where decahydronaphthalene is used or stored.

Before beginning employment and at regular times thereafter, the following recommended medical tests should be considered for those with frequent or potentially high exposures:

☑ Lung function tests.

If symptoms develop or overexposure is suspected, the following may be useful:

☑ Skin testing with dilute decahydronaphthalene may help diagnose allergy, if performed by a qualified allergist.

☑ Consider chest X-ray following acute exposure.

☑ Liver and kidney function tests, including urinalysis.

It should be noted that medical tests that simply look for existing damage are not a substitute for controlling exposures. Medical histories are extremely important when assessing exposure risk.

Other methods to reduce exposure include:

☑ Decahydronaphthalene should be handled in an enclosed system. There should be local exhaust ventilation at the site of chemical release.

☑ Always ensure that proper protective clothing is worn when using chemical substances.

☑ Wash thoroughly immediately after exposure to decahydronaphthalene and at the end of the work shift or before eating, drinking, or smoking.

☑ Hazard warning information should be posted in the work area. In addition, as part of an on-going education and training program, all information on the health and safety hazards of decahydronaphthalene should be communicated to all potentially exposed workers.

☑ Eye wash stations should be provided in the immediate work area for emergency use. If there is a possibility of skin exposure to decahydronaphthalene, emergency shower facilities should also be provided.

☑ Workers whose clothing has been contaminated by decahydronaphthalene should change into clean clothes before leaving work. Contaminated work clothing should be laundered only by individuals who have been informed of the hazards of exposure to decahydronaphthalene.

RISK ASSESSMENT: ENVIRONMENT

General Assessment

The environment is at risk of exposure during transportation, storage, disposal, or destruction of decahydronaphthalene. In almost every scenario, the threat of environmental exposure is contingent upon the proper handling of the chemical substance. Accidental spills, large or small, can result in contamination of the surrounding environmental mediums (water, soil, and air).

Decahydronaphthalene is a combustible liquid. However, caution is always required in handling, storage, transportation, and disposal of decahydronaphthalene. It is flammable and moderately explosive when exposed to heat or flame. It can react vigorously with oxidizing materials (such as perchlorates, peroxides, chlorates, nitrates, and permanganates). Emergency responders should be made aware of the presence of decahydronaphthalene at any emergency response situation.

Decahydronaphthalene can enter the environment from industrial effluents and from spills.

Acute Ecological Effects

Acute (short-term) toxic effects may include the death of animals, birds, or fish and death or low growth rate in plants. Acute effects are seen 2 to 4 days after animals or plants are exposed to decahydronaphthalene. Decahydronaphthalene has moderate acute toxicity to aquatic life.

Insufficient data are available to evaluate the acute (short-term) effects of decahydronaphthalene on plants, birds, or land animals.

Chronic Ecological Effects

Chronic (long-term) toxic effects may include shortened life span, reproductive problems, lower fertility, and changes in appearance or behavior in exposed animals. These effects can be seen long after first exposure(s) to toxic chemicals.

Decahydronaphthalene has moderate chronic toxicity to aquatic life. Insufficient data are available to evaluate the chronic (long-term) effects of decahydronaphthalene on plants, birds, or land animals.

Water Solubility

Decahydronaphthalene is nearly insoluble in water. Concentrations up to 1 milligram will mix with a liter of water.

Persistence in the Environment

Decahydronaphthalene is slightly persistent in the environment, with a half-life of between 2 and 20 days. The half-life of a pollutant is the amount of time it takes for one-half of the chemical to be degraded.

The chemical properties of decahydronaphthalene indicate that approximately 93% will enter the air with the remaining 7% ending up in water.

Bioaccumulation in Aquatic Organisms

Some substances increase in concentration, or *bioaccumulate,* in living organisms as they breathe contaminated air, drink contaminated water, or eat contaminated food. These chemicals can become concentrated in the tissues and internal organs of animals as well as humans exposed to the chemical.

The concentration of decahydronaphthalene found in edible fish tissue is expected to be somewhat higher than the average concentration of decahydronaphthalene found in the water from which the fish was taken.

Recommended Risk-Reduction Measures

Proper training of all transporters will reduce the likelihood of a mishap or accident resulting in a leak or spill of decahydronaphthalene into the environment. Labels on all containers, trucks, and rail cars must meet DOT requirements and accurately reflect their contents to enable emergency responders to react properly and quickly to any disaster thereby reducing the potential risk to the environment and to personnel.

Storage of decahydronaphthalene should be segregated from other chemicals and materials to minimize the risk of cross-contamination and violent reactions.

If a spill or leak to the environment has occurred, fire department, emergency response and/or hazardous materials spill personnel should be notified immediately. Cleanup should be attempted only by those trained in proper spill containment procedures. Contaminated soils should be removed for incineration and replaced with clean soil.

If decahydronaphthalene should contact the water table, aquifer, or navigable waterway, time is of the essence. It is only slightly soluble and total remediation may be possible. The local and/or state emergency response authorities must be notified. A comprehensive emergency response or disaster preparedness/recovery plan should be in place prior to any operations involving the use, transportation, storage, or disposal of decahydronaphthalene.

If decahydronaphthalene is spilled or leaked, the following specific steps are recommended:

- ☑ Restrict persons not wearing protective clothing from area of spill or leak until cleanup is complete. Avoid skin and eye contact.
- ☑ Absorb liquids in vermiculite or other similar materials and deposit in sealed drum for disposal.
- ☑ Ventilate area of spill or leak.
- ☑ Remove all sources of heat and ignition from the immediate area.
- ☑ It may be necessary to dispose of decahydronaphthalene as a hazardous waste. The responsible state agency or the regional office of the federal Environmental Protection Agency (EPA) should be contacted for specific recommendations.

RISK ASSESSMENT: BUSINESS

General Assessment

Accidents or mishaps involving decahydronaphthalene can present a serious threat to business operations. The loss or damage of equipment or facilities can significantly affect fiscal viability. Lawsuits that may result from personnel illness, injury/death, public exposures, and/or environmental contamination will require a serious expenditure of resources. Media attention surrounding an injury, death, or environmental damage can also result in a loss of profits and loss of current as well as future business. Always remember that anytime the terms "carcinogen," "cancer," or "reproductive hazard" are used, public emotion and hysteria can run equally high. Risk managers must consider these phenomena when developing company policy.

Recommended Risk-Reduction Measures

Company attorneys, safety and health professionals, and environmental specialists should be involved in the development of any procedures for responding to chemical incidents. A company official should be predesignated as a public relations officer with specific training in dealing with the press. Corporate plans and policies should be developed, approved, and implemented long before any need for such arises.

REFERENCES

American Conference of Governmental Industrial Hygienists. 1988. *Documentation of the Threshold Limit Values and Biological Exposure Indices,* 5th Edition (with updates). Cincinnati: ACGIH

Anthony, C. P., and N. J. Kolthoff. 1971. *Textbook of Anatomy and Physiology.* St. Louis: C. V. Mosby Company.

Asante-Duah, D. K. 1993. *Hazardous Waste Risk Assessment.* New York: CRC Press/Lewis Publishers.

Bloom, A. D. and F. J. DeSerres. 1995. *Ecotoxicity and Human Health.* New York: CRC Press/Lewis Publishers.

Brusick, D. J. 1994. *Methods for Genetic Risk Assessment.* New York: CRC Press/Lewis Publishers.

Bueche, J. R. 1972. *Principles of Physics,* 2nd Edition. New York: McGraw-Hill.

Calabrese, E. J. 1994. *Biological Effects of Low Level Exposures.* New York: CRC Press/Lewis Publishers.

Cockerman, L. G., B. S. Shane. 1994. *Basic Environmental Toxicology.* New York: CRC Press/Lewis Publishers.

Danse, I. R. 1991. *Common Sense Toxics in the Workplace.* New York: Van Nostrand Reinhold.

Gardner, R. W. 1994. *Chemical Intolerance.* New York: CRC Press/Lewis Publishers.

Hallenbeck, W. H. 1993. *Quantitative Risk Assessment for Environmental and Occupational Health,* 2nd Edition. New York: CRC Press/Lewis Publishers.

Hodgson, E., R. B. Mailman, and J. E. Chambers. 1988. *Dictionary of Toxicology.* London: The MacMillian Company.

Howard, P. H., et. al. 1991. *Handbook of Environmental Degradation Rates.* New York: CRC Press/Lewis Publishers.

Jones, F. E. 1994. *Toxic Organic Vapors in the Workplace.* New York: CRC Press/Lewis Publishers.

Keith, L. H. and M. M. Walker. 1995. *Handbook of Air Toxics.* New York: CRC Press/Lewis Publishers.

Landis, W. G., M. H. Yu. 1995. *Introduction to Environmental Toxicology.* New York: CRC Press/Lewis Publishers.

Langerman, N. 1994 *Precautionary Labels for Chemical Containers.* New York: CRC Press/Lewis Publishers.

Lauwerys, R. R., P. Hoet. 1993. *Industrial Chemical Exposure.* New York: CRC Press/Lewis Publishers.

Lewis, R. J., Sr. 1993. *Hawley's Condensed Chemical Dictionary,* 12th Edition. New York: Van Nostrand Reinhold.

Lewis, R. J., Sr. 1992. *Sax's Dangerous Properties of Industrial Materials,* Volumes 1, 2, and 3, 8th Edition. New York: Van Nostrand Reinhold.

Mellström, G. A., J. E. Wahlberg, and H. I. Maibach. 1994. *Protective Gloves for Occupational Use.* New York: CRC Press/Lewis Publishers.

Meyer, E. 1990. *Chemistry of Hazardous Materials.* Englewood Cliffs, NJ: Prentice Hall.

Patnaik, P. 1992. *A Comprehensive Guide to the Hazardous Properties of Chemical Substances.* New York: Van Nostrand Reinhold.

Philp, R. B. 1995. *Environmental Hazards and Human Health.* New York: CRC Press/Lewis Publishers.

Proctor, N. H., J. P. Hughes. 1991. *Chemical Hazards of the Workplace.* New York: Van Nostrand Reinhold.

Quigley, D. R. 1994. *Handbook of Emergency Chemical Management.* New York: CRC Press/Lewis Publishers.

Richardson, M. 1995. *Dictionary of Substance and Their Effects,* Vol. I-VI. New York: CRC Press/Lewis Publishers.

Sacarello, H. L. A. 1994. *The Comprehensive Handbook of Hazardous Materials.* New York: CRC Press/Lewis Publishers.

Torkelson, T. R. V. K. Rowe. 1981. *Patty's Industrial Hygiene and Toxicology,* Vol. 2B. New York: Wiley-Interscience.

U.S. Department of Health and Human Services, National Institute for Occupational Safety and Health. 1994. *NIOSH Pocket Guide to Chemical Hazards.* Washington, D.C.: U.S. Government Printing Office.

MATERIAL SAFETY DATA SHEET

CHEMICAL NAME	CRC PRESS/LEWIS PUBLISHERS DISCLAIMER
DEMETON	

HAZARD WARNING INFORMATION

HEALTH	FIRE	REACTIVE	OTHER	DEGREE OF HAZARD	COLOR CODING	OTHER CODES
N/A	N/A	N/A		0 = Minimum Hazard 1 = Slight Hazard 2 = Moderate Hazard 3 = Serious Hazard 4 = Severe Hazard	HEALTH = BLUE FIRE = RED REACTIVITY = YELLOW OTHER = WHITE	OX = Oxidizer ACID = Acid ALK = Alkali COR = Corrosive W = Use No Water

SECTION I - GENERAL INFORMATION

Characterization	RCRA Number	EPA Class
Pesticide; Insecticide	**None**	**Not Applicable**

DOT Proper Shipping Name	**No Citation**	Chemical Abstract Service (CAS) Number	**8065-48-3**
DOT Hazard Class and Label Requirements	**No Citation**	DOT Emergency Guide Code	**No Citation**
DOT Identification Number	**No Cltation**	Chemical Formula	$(C_2H_6O)_2PSOC_2H_4SC_2H_6$

Synonyms

O-O-Diethyl-O(and S)-2-(ethylthio)ethyl phosphorothiolate mixture; Systox®; phosphorothioic acid-O-O-diethyl ester; demeton-O + demeton-S mixture.

SECTION II - HAZARDOUS INGREDIENTS AND IDENTITY INFORMATION

Hazardous Components (specific identity)	OSHA Exposure Criteria	NIOSH Exposure Criteria	Immediately Dangerous to Life and Health (IDLH)	ACGIH Exposure Criteria
Demeton (derivation: Mixture of O,O-diethyl-O-2-(ethylthio)ethyl phosphorothiolate [demeton-O] and O,O-diethyl-S-2-(ethylthio) ethyl phosphorothiolate [demeton-S]). **1 ppm = 10.74 mg/m^3**	**PEL (skin):** **0.9 ppm** **0.1 mg/m^3** **STEL:** **Not Established**	**REL (skin):** **0.9 ppm** **0.1 mg/m^3** **STEL:** **Not Established**	**10 mg/m^3**	**TLV (skin):** **0.01 ppm** **0.1 mg/m^3** **STEL:** **Not Established**

SECTION III - PHYSICAL AND CHEMICAL CHARACTERISTICS

Boiling Point	**273°F (134°C) Decomposes**	Specific Gravity (H_2O = 1)	**1.12**
Vapor Pressure (mm Hg)	**0.0003 at 69°F (20°C)**	Molecular Weight	**258.3**
Vapor Density (Air = 1)	**1**	Freezing Point	**<-13°F (<-25°C)**

Solubility

Slightly soluble in water (0.01%). Soluble in most organic solvents.

Appearance and Odor

Pale yellow or amber, oily liquid with a sulfur-like odor.

SECTION IV - FIRE AND EXPLOSION HAZARD DATA

Flash Point (method used)	**113°F (45°C) closed cup**	Explosive Limits in Air % by Volume	**LEL: 1.0% UEL: 5.3%**
NFPA Classification	**Class II Combustible Liquid**	Autoignition Temperature	**867°F (464°C)**

Extinguishing Media

Use dry chemical, carbon dioxide, or foam. Water spray may be used to cool surrounding areas.

Special Fire Fighting Procedures

Poisonous gases (sulfur dioxide and phosphoric acid) are produced in fire, wear full protective clothing and self-contained breathing apparatus (SCBA). Vapors are heavier than air and may travel for some distance to ignition source and flashback. Move containers from fire area if it can be done safely.

Unusual Fire and Explosion Hazards

Containers may explode in fire due to buildup of excessive internal pressures. Exercise extreme caution when attempting to move or relocate containers that may have been exposed to high heat.

SECTION V - REACTIVITY DATA

Stability		Conditions to Avoid
		Avoid contact with incompatible materials. Keep away from water and moisture.
Stable	Unstable	Incompatibility *(materials to avoid)*
X		Strong oxidizers (such as chlorine, bromine, fluorine) , water, and alkalis. May volatilize and form toxic fumes and vapors.
Hazardous Polymerization		Conditions to Avoid Under normal conditions of temperature and pressure, hazardous polymerization of demeton is not expected to occur.
May Occur	Will Not Occur	Hazardous Decomposition or By-products
	X	Toxic and irritating gases are produced when demeton is involved in fire, including fumes of sulfur dioxide and phosphoric acid.

SECTION VI - HEALTH HAZARD DATA

Primary Route(s) of Entry:	Inhalation?	Absorption (skin)?	Ingestion?
	X	X	X

Health Hazards

INHALATION: Symptoms include headache, vertigo, blurred vision, lachrymation, salivation, muscular weakness and ataxia, dypsnea, diarrhea, abdominal cramps, vomiting, coma, pulmonary edema, tightness in chest, wheezing, convulsions, and death.

ABSORPTION: Headache, dizziness, nausea, vomiting, tremor, ataxia, convulsions, CNS depression, respiratory failure, and may cause cardiac irregularities. May cause skin allergy.

INGESTION: May cause, nausea, vomiting, muscle twitching, and coma.

Carcinogenicity	NTP Listed?	IARC Cancer Review Group?	OSHA Regulated?	Target Organs?
Unknown Human Unknown Animal	No	No	29 CFR 1910.1000 Table Z-1 (skin)	Eyes, skin, respiratory system, CVS, CNS, blood.

Medical Conditions Generally Aggravated by Exposure

None Reported.

Emergency and First-aid Procedures

Eye contact: Flush immediately with water for 15 minutes (minimum), seek medical attention. Skin contact: Remove all contaminated clothing. Immediately wash area with large amounts of soap and water. Seek medical assistance. For inhalation: Remove the person from exposure. Provide respiratory assistance and CPR. Transfer to medical facility. Observe for 24-48 hours for lung effects. If swallowed: If victim is conscious, give milk, water, or salt water solution and induce vomiting repeatedly. Never attempt to give an unconscious or convulsing person anything by mouth.

SECTION VII - PRECAUTIONS FOR SAFE HANDLING AND USE

Steps to be Taken in Case Material is Released or Spilled

Absorb liquids in vermiculite or other absorbent and deposit in DOT-approved drum. Ventilate area and remove ignition sources. Restrict those not involved in cleanup from entering area. Notify appropriate authorities, if applicable.

Preferred Waste Disposal Method

No Citation.

Precautions to be Taken in Handling and Storage

Do not store demeton in the presence of incompatible chemicals or materials. Store in tightly closed containers in a cool, dark, well-ventilated area. Personnel working with demeton should be trained on its proper handing and storage *prior* to being assigned to such responsibilities.

Other Precautions and Warnings

Bulk storage of demeton is not recommended.

SECTION VIII - CONTROL MEASURES AND PERSONAL PROTECTIVE EQUIPMENT

Respiratory Protection *(specify type)*

PEL is extremely low. At any exposure level: Use a supplied-air respirator with full facepiece operated in pressure demand mode or a self-contained breathing apparatus (SCBA) with full facepiece operated in pressure demand or other positive pressure mode.

Ventilation

Local exhaust or general mechanical systems recommended.

Protective Gloves	Eye Protection	Other Protective Clothing
Rubber	Chemical Goggles or Face Mask	Apron

Work/Hygiene Practices

Always wash hands thoroughly after using chemical; never bring food, drink, or smoking materials into vicinity of chemicals.

DEMETON

$(C_2H_6O)_2PSOC_2H_4SC_2H_6$ CAS: 8065-48-3

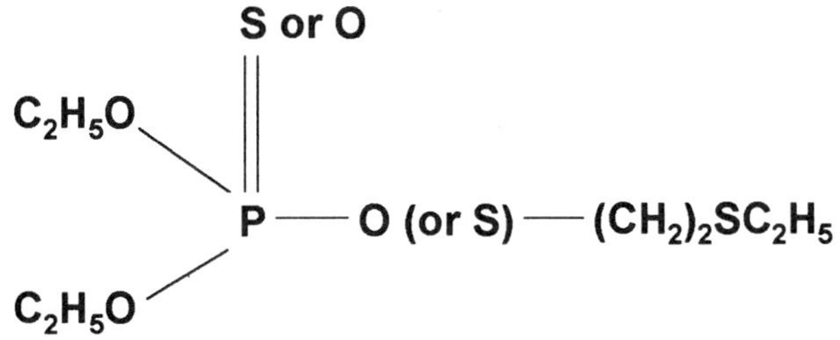

IDENTIFICATION AND TYPICAL USES

Demeton is a pale yellow or amber colored oily liquid with a sulfur-like odor. It is a mixture of demeton-O and demeton-S. It is used as a systemic insecticide. This means that it is absorbed by a plant, which then becomes toxic to sucking and chewing insects.

RISK ASSESSMENT: HEALTH

General Assessment

Demeton is a deadly human poison by *ingestion*. It is also extremely toxic by *inhalation* and skin contact (*absorption*). Experimental teratogenic data have been reported and human mutagenic data have been confirmed. Its primary systemic effect is to inhibit cholinesterase in humans and animals and, thus, causes a buildup of acetylcholine. Doses are cumulative in humans.

Inhalation appears to cause depression of the central nervous system (CNS) and adverse effects to the cardiovascular system (CVS). Symptoms of headache, vertigo, blurred vision, lachrymation, salivation, sweating, muscular weakness and ataxia, dypsnea, diarrhea, abdominal cramps, vomiting, coma, pulmonary edema, and death. Vapors are irritating to the eyes and skin. Other symptoms of exposure include rhinitis, tightness in chest, wheezing, spasms, giddiness, confusion, convulsions, low blood pressure, and irregular heartbeat.

Ingestion causes nausea, vomiting, muscle twitching, coma, CNS depression, and adverse cardiac effects. Death is likely on acute, high exposures.

Skin or eye contact results in severe irritation and discomfort. Demeton may be absorbed through unbroken skin to cause toxic systemic effects.

☠ *Acute Health Effects*

The following acute (short-term) health effects may occur immediately or shortly after exposure to demeton:

Skin: Irritation and possible allergy. Most effects from skin contact are internal due to absorption of demeton. Symptoms include vomiting, nausea, malaise, and muscular twitching at high doses.

Eye: Possible irritation. Possibility of damage if not removed immediately.

Lung: Irritation causing coughing and/or shortness of breath. Higher exposures can lead to respiratory problems, including pulmonary edema (fluid buildup in lungs) which is a medical emergency and can be fatal.

CNS: Dizziness, headaches, nausea, vomiting, muscle jerks, convulsions, coma, and death.

CVS: Abnormal cardiac rhythms, changes in blood pressure, and possible increase in blood cholesterol levels.

Chronic Health Effects

The following chronic (long-term) health effects may occur at some time after exposure to demeton and can last for months or even years:

Cancer Hazards: According to the information presented in the references, demeton has not been adequately tested for its ability to cause cancer in test animals. However, it is a known human mutagen (causes genetic changes) and some scientists believe that mutagenic chemicals may also pose a carcinogenic threat. More research is required on the carcinogenic properties of this chemical.

Reproductive Hazard: Demeton may be a teratogen in humans since it has been shown to be a teratogen in some test animals.

Other Chronic Effects: Demeton will accumulate with each dose. It will inhibit cholinesterase in humans and animals causing a buildup of acetylcholine.

Recommended Risk-Reduction Measures

Unless a less toxic chemical can be substituted for a hazardous substance, *engineering controls* are the most effective method of reducing exposures. The best protection is to enclose operations and/or provide local exhaust ventilation at the site of chemical release. This may not be feasible during use as a pesticide outdoors, but it may be possible during manufacture of demeton. While not always operationally feasible, isolating operations can also reduce exposure.

Using respiratory protection is less effective than the controls mentioned above, but is still advisable whenever working with or around demeton. A self-contained breathing apparatus (SCBA) with full facepiece and pressure demand is the recommended respiratory protection of choice. If a full facepiece is not available, then chemical goggles should be worn to protect the eyes. Whenever a chemical splash hazard exists a face shield and rubber apron should be worn. To prevent hand and skin exposures, rubber gloves should be worn.

Administrative controls should also be in place to minimize the potential for human exposures. These may include written procedures or policies which specify the methods and techniques that will be practiced whenever personnel are to work with demeton.

All personnel should receive *training* on the use, hazards, protective measures, emergency actions, and other precautions per 29 CFR 1910.1200 (Hazard Communication), prior to the first assignment in an area where demeton is used or stored.

If symptoms develop or overexposure is suspected, the following medical tests may be useful:

- ☑ Lung function tests.
- ☑ Electrocardiogram.
- ☑ Evaluation by a qualified allergist with careful evaluation of exposure history and administration of special tests (may help diagnose skin allergy).

Any evaluation should include a careful medical history of past and present symptoms with an examination. However, medical tests that evaluate existing damage are not a substitute for controlling exposure.

Other methods to reduce exposure include:

- ☑ Where possible, enclose operations and use local exhaust ventilation at the site of chemical release. If local exhaust ventilation or enclosure is not used, respiratory protection should be mandatory.
- ☑ Always ensure that proper protective clothing is worn when using chemical substances.
- ☑ Wash thoroughly immediately after exposure to demeton and at the end of the work shift or before eating, drinking, or smoking.
- ☑ Hazard warning information should be posted in the work area. In addition, as part of an on-going education and training program, all information on the health and safety hazards of demeton should be communicated to all exposed or potentially exposed workers.
- ☑ Eye wash stations should be provided in work areas. If the potential for whole body exposure exists, then safety showers should also be provided.

RISK ASSESSMENT: ENVIRONMENT

General Assessment

The environment is at risk of exposure during intended use, transportation, storage, disposal, or destruction of demeton. In almost every scenario, the threat of environmental exposure is contingent upon the proper handling of the chemical substance. Accidental spills, large or small, can result in fire, explosion, and possible contamination of the surrounding environmental mediums (water, soil, and air).

Demeton is considered a class II combustible liquid (per OSHA 29 CFR 1910.106). These characteristics require special consideration during any emergency situation involving a leak or spill of demeton or demeton mixtures. Should demeton ever come into contact with incompatible substances such as strong oxidizers (chlorine, bromine, fluorine), alkali materials, or even water either during use, transportation, or storage, violent reactions can occur.

Demeton can enter the environment through its normal, intended use as a pesticide and through industrial discharges, urban runoff, and spills.

Acute Ecological Effects

Acute (short-term) toxic effects may include the death of animals, birds, or fish and death or low growth rate in plants. Acute effects are seen 2 to 4 days after ani-

mals or plants are exposed to demeton. This chemical has not been shown to be acutely toxic to plants. Insufficient data are available to evaluate or predict the short-term effects of demeton to aquatic life, birds, or land animals.

Chronic Ecological Effects

Chronic toxic effects may include shortened life span, reproductive problems, lower fertility, and changes in appearance or behavior in exposed animals. These effects can be seen long after first exposure(s) to toxic chemicals. Demeton has not been shown to present chronic toxicity to plant life. Insufficient data are available to evaluate or predict the long-term effects of demeton to aquatic life, birds, or land animals.

Water Solubility

Demeton is very slightly soluble in water. Concentrations of 1 milligram or less will mix with a liter of water.

Persistence in the Environment

Demeton is highly persistent in water, with a half-greater than 200 days. The half-life of a pollutant is the amount of time it takes for one-half of the chemical to be degraded.

Bioaccumulation in Aquatic Organisms

Some substances increase in concentration, or *bioaccumulate,* in living organisms as they breathe contaminated air, drink contaminated water, or eat contaminated food. These chemicals can become concentrated in the tissues and internal organs of animals as well as humans.

The concentration of demeton found in fish tissues is expected to be much higher than the average concentration of demeton in the water from which the fish was taken.

Recommended Risk-Reduction Measures

Proper training of all transporters will reduce the likelihood of a mishap or accident resulting in a leak or spill to the environment. The correct use of DOT labeling (per the MSDS) on all containers, trucks, and rail cars will enable emergency responders to react properly and quickly to any disaster thereby reducing the potential risk to the environment and to personnel.

Storage of demeton should be segregated from incompatible chemicals to minimize the risk of cross-contamination or contact.

If a spill or leak to the environment has occurred, fire department, emergency response and/or hazardous materials spill personnel should be notified immediately. Cleanup should be attempted only by those trained in proper spill containment procedures. Contaminated soils should be removed for incineration and replaced with clean soil. If demeton should contact the water table, aquifer, or navigable waterway, remediation activities should be prompt to ensure protection of aquatic sediments. When such spills occur, the local and/or state emergency response authorities must be notified. A comprehensive emergency response or disaster preparedness/recovery plan should be in place prior to any operations involving the use, transportation, storage, or disposal of demeton. If demeton is spilled or leaked, the following specific steps are recommended:

- ☑ Restrict persons not wearing protective clothing from area of spill or leak until cleanup is complete and area can be opened for normal work.
- ☑ Absorb liquid spills using vermiculite, dry earth, or sand to absorb and place in a sealed drum for disposal.
- ☑ It may be necessary to dispose of demeton as a hazardous waste. The responsible state agency or the regional office of the federal Environmental Protection Agency (EPA) should be contacted for specific recommendations.

RISK ASSESSMENT: BUSINESS

General Assessment

Accidents or mishaps involving demeton can present a significant threat to business operations. The loss or damage of equipment or facilities can significantly affect fiscal viability. Lawsuits that may result from personnel injury/death, public exposures, and/or environmental contamination will also require a serious expenditure of resources. Media attention surrounding an injury, death, or environmental damage can also result in a loss of profits and loss of current as well as future business.

Recommended Risk-Reduction Measures

Company attorneys, safety and health professionals, and environmental specialists should be involved in the

development of any procedures dealing with safety or emergency response. A company official should be pre-designated as a public relations officer with specific training in dealing with the press. Corporate plans and policies should be developed, approved, and implemented long before any need for such arises.

REFERENCES

Ahlbom, A. 1993. *Biostatistics for Epidemiologists.* New York: CRC Press/Lewis Publishers.

American Conference of Governmental Industrial Hygienists. 1988. *Documentation of the Threshold Limit Values and Biological Exposure Indices,* 5th Edition (with updates). Cincinnati: ACGIH

Anthony, C. P., and N. J. Kolthoff. 1971. *Textbook of Anatomy and Physiology.* St. Louis: C. V. Mosby Company.

Asante-Duah, D. K. 1993. *Hazardous Waste Risk Assessment.* New York: CRC Press/Lewis Publishers.

Bloom, A. D. and F. J. DeSerres. 1995. *Ecotoxicity and Human Health.* New York: CRC Press/Lewis Publishers.

Brusick, D. J. 1994. *Methods for Genetic Risk Assessment.* New York: CRC Press/Lewis Publishers.

Bueche, J. R. 1972. *Principles of Physics,* 2nd Edition. New York: McGraw-Hill.

Burrell, R., D. K. Flaherty, and L. J. Sauers. 1992. *Toxicology of the Immune System.* New York: Van Nostrand Reinhold.

Calabrese, E. J., E. M. Kenyon. *Air Toxics and Risk Assessment.* New York: CRC Press/Lewis Publishers.

Calabrese, E. J. 1994. *Biological Effects of Low Level Exposures.* New York: CRC Press/Lewis Publishers.

Cockerman, L. G., B. S. Shane. 1994. *Basic Environmental Toxicology.* New York: CRC Press/Lewis Publishers.

Cothern, C. R., N. P. Ross. 1994. *Environmental Statistics, Assessment, and Forecasting.* New York: CRC Press/Lewis Publishers.

Cothern, R. R. 1993. *Comparative Environmental Risk Assessment.* New York: CRC Press/Lewis Publishers.

Cothern, C. R. 1995. *Handbook for Environmental Risk Decisionmaking.* New York: CRC Press/Lewis Publishers.

Gardner, R. W. 1994. *Chemical Intolerance.* New York: CRC Press/Lewis Publishers.

Hallenbeck, W. H. 1993. *Quantitative Risk Assessment for Environmental and Occupational Health,* 2nd Edition. New York: CRC Press/Lewis Publishers.

Hodgson, E., R. B. Mailman, and Janice E. Chambers. 1988. *Dictionary of Toxicology.* London: The MacMillian Company.

Howard, P. H. 1995. *Handbook of Environmental Fate and Exposures Data for Organic Chemicals,* Vol. I-IV. New York: CRC Press/Lewis Publishers.

Howard, P. H., et. al. 1991. *Handbook of Environmental Degradation Rates.* New York: CRC Press/Lewis Publishers.

Jones, F. E. 1994. *Toxic Organic Vapors in the Workplace.* New York: CRC Press/Lewis Publishers.

Kamrin, M. A. 1988. *Toxicology: A Primer on Principles and Applications.* Chelsea, Michigan: Lewis Publishers.

Keith, L. H. and M. M. Walker. 1995. *Handbook of Air Toxics.* New York: CRC Press/Lewis Publishers.

Landis, W. G., M. H. Yu. 1995. *Introduction to Environmental Toxicology.* New York: CRC Press/Lewis Publishers.

Langerman, N. 1994 *Precautionary Labels for Chemical Containers.* New York: CRC Press/Lewis Publishers.

Lauwerys, R. R., P. Hoet. 1993. *Industrial Chemical Exposure.* New York: CRC Press/Lewis Publishers.

Lide, D. R. 1995. *Handbook of Chemistry and Physics,* 76th Edition. New York: CRC Press/Lewis Publishers.

Lewis, R. J., Sr. 1993. *Hawley's Condensed Chemical Dictionary,* 12th Edition. New York: Van Nostrand Reinhold.

Lewis, R. J., Sr. 1992. *Sax's Dangerous Properties of Industrial Materials,* Volumes 1, 2, and 3, 8th Edition. New York: Van Nostrand Reinhold.

Manahan, S. E. 1992. *Toxicological Chemistry,* 2nd Edition. New York: CRC Press/Lewis Publishers.

Mellström, G. A., J. E. Wahlberg, and H. I. Maibach. 1994. *Protective Gloves for Occupational Use.* New York: CRC Press/Lewis Publishers.

Meyer, E. 1990. *Chemistry of Hazardous Materials.* Englewood Cliffs, NJ: Prentice Hall.

Nelson, G. O. 1992. *Gas Mixtures.* New York: CRC Press/Lewis Publishers.

Ott, W. R. 1995. *Environmental Statistics and Data Analysis.* New York: CRC Press/Lewis Publishers.

Patnaik, P. 1992. *A Comprehensive Guide to the Hazardous Properties of Chemical Substances.* New York: Van Nostrand Reinhold.

Philp, R. B. 1995. *Environmental Hazards and Human Health.* New York: CRC Press/Lewis Publishers.

Proctor, N. H., and James P. Hughes. 1991. *Chemical Hazards of the Workplace.* New York: Van Nostrand Reinhold.

Quigley, D. R. 1994. *Handbook of Emergency Chemical Management.* New York: CRC Press/Lewis Publishers.

Rea, W. J. 1995. *Chemical Sensitivity,* Vol. I-III. New York: CRC Press/Lewis Publishers.

Richardson, M. 1995. *Dictionary of Substance and Their Effects,* Vol. I-VI. New York: CRC Press/Lewis Publishers.

Rylander, R., R. R. Jacobs. 1994. *Organic Dusts.* New York: CRC Press/Lewis Publishers.

Sacarello, H. L. A. 1994. *The Comprehensive Handbook of Hazardous Materials.* New York: CRC Press/Lewis Publishers.

U.S. Department of Health and Human Services, National Institute for Occupational Safety and Health. 1994. *NIOSH Pocket Guide to Chemical Hazards.* Washington, D.C.: U.S. Government Printing Office.

MATERIAL SAFETY DATA SHEET

CHEMICAL NAME	
DIACETONE ALCOHOL	**CRC PRESS/LEWIS PUBLISHERS DISCLAIMER:** THE INFORMATION AND RECOMMENDATIONS PRESENTED HEREIN ARE BASED ON SOURCES BELIEVED TO BE RELIABLE. CRC MAKES NO REPRESENTATION ON ITS COMPLETENESS OR ACCURACY. IT IS THE RESPONSIBILITY OF THE USER TO DETERMINE THE CHEMICAL'S SUITABILITY FOR ITS INTENDED USE, THE CHEMICAL'S SAFE USE, AND THE CHEMICAL'S PROPER DISPOSAL. NO REPRESENTATIONS AND/OR WARRANTIES, EITHER EXPRESSED OR IMPLIED, OF THE MERCHANTABILITY OR FITNESS FOR A PARTICULAR PURPOSE, OR OF ANY OTHER NATURE, ARE MADE WITH RESPECT TO THE INFORMATION PROVIDED IN THIS MSDS OR TO THE CHEMICAL TO WHICH INFORMATION MAY REFER. CRC NEITHER ASSUMES NOR AUTHORIZES ANY OTHER PERSON TO ASSUME FOR IT, ANY OTHER ADDITIONAL RESPONSIBILITY OR LIABILITY FOR THE USE OF, OR RELIANCE UPON, THIS INFORMATION.

HAZARD WARNING INFORMATION

HEALTH	FIRE	REACTIVE	OTHER	DEGREE OF HAZARD	COLOR CODING	OTHER CODES
1	2	0		0 = Minimum Hazard 1 = Slight Hazard 2 = Moderate Hazard 3 = Serious Hazard 4 = Severe Hazard	HEALTH = BLUE FIRE = RED REACTIVITY = YELLOW OTHER = WHITE	OX = Oxidizer ACID = Acid ALK = Alkali COR = Corrosive W = Use No Water

SECTION I - GENERAL INFORMATION

Characterization	**Ketone**	RCRA Number	**None**
		EPA Class	**Not Applicable**
DOT Proper Shipping Name	**Diacetone Alcohol**	Chemical Abstract Service (CAS) Number	**123-42-2**
DOT Hazard Class and Label Requirements	**Flammable Liquid**	DOT Emergency Guide Code	**26**
DOT Identification Number	**UN 1148**	Chemical Formula	**$CH_3COCH_2C(CH_3)_2OH$**

Synonyms

Diacetone; 4-hydroxy-4-methylpentanone-2; 4-hydroxy-2-keto-4-methylpentane; 2-methyl-2-pentanol-4-one; pyranton; tyranton.

SECTION II - HAZARDOUS INGREDIENTS AND IDENTITY INFORMATION

Hazardous Components (specific identity)	OSHA Exposure Criteria	NIOSH Exposure Criteria	Immediately Dangerous to Life and Health (IDLH)	ACGIH Exposure Criteria
Diacetone alcohol (derivation: By the condensation of alcohol). **1 ppm = 4.83 mg/m^3**	**PEL:** **50 ppm** **240 mg/m^3** **STEL:** **Not Established**	**REL:** **50 ppm** **240 mg/m^3** **STEL:** **Not Established**	**1800 mg/m^3**	**TLV (skin):** **50 ppm** **240 mg/m^3** **STEL:** **Not Established**

SECTION III - PHYSICAL AND CHEMICAL CHARACTERISTICS

Boiling Point	**334°F (168°C) Decomposes**	Specific Gravity (H_2O = 1)	**0.9**
Vapor Pressure (mm Hg)	**<1 at 69°F (20°C)**	Molecular Weight	**116.2**
Vapor Density (Air = 1)	**4**	Melting Point	**-58°F (-50°C)**

Solubility

Miscible in water, alcohols, halogenated hydrocarbons, and esters.

Appearance and Odor

Colorless liquid with a faint but pleasant mint-like odor.

SECTION IV - FIRE AND EXPLOSION HAZARD DATA

Flash Point (method used)	**125°F (52°C) closed cup**	Explosive Limits in Air % by Volume	**LEL: 1.8% UEL: 6.9%**
NFPA Classification	**Class IIA Combustible Liquid**	Autoignition Temperature	**1118°F (604°C)**

Extinguishing Media

Use dry chemical, carbon dioxide, alcohol foam, or water.

Special Fire Fighting Procedures

Poisonous gases are produced in fire, wear full protective clothing and self-contained breathing apparatus (SCBA). Vapors are heavier than air and may travel for some distance to ignition source and flashback. Move containers from fire area if it can be done safely.

Unusual Fire and Explosion Hazards

Containers may explode in fire due to buildup of excessive internal pressures. Exercise extreme caution when attempting to move or relocate containers that may have been exposed to high heat.

SECTION V - REACTIVITY DATA

Stability		Conditions to Avoid **Avoid contact with incompatible materials. Keep away from heat and flame.**
Stable	Unstable	Incompatibility *(materials to avoid)*
X		**Strong oxidizers (such as chlorine, bromine, fluorine) and strong alkalis.**
Hazardous Polymerization		Conditions to Avoid **Under normal conditions of temperature and pressure, hazardous polymerization of diacetone alcohol is not expected to occur.**
May Occur	Will Not Occur	Hazardous Decomposition or By-products
	X	**Toxic and irritating gases are produced when diacetone alcohol is involved in fire.**

SECTION VI - HEALTH HAZARD DATA

Primary Route(s) of Entry:	Inhalation?	Absorption (skin)?	Ingestion?
	X		X

Health Hazards

INHALATION: Symptoms include headache, nausea or vomiting, eye and pulmonary changes. Can cause anemia and damage to liver and kidneys. Narcotic in high concentrations. Will irritate the mucous membranes of the respiratory tract.

SKIN & EYES: Severe irritant to the skin and eyes. May cause corneal damage and dermatitis.

INGESTION: Moderately toxic by ingestion. A human poison. Gastrointestinal irritation.

Carcinogenicity	NTP Listed?	IARC Cancer Review Group?	OSHA Regulated?	Target Organs?
Unknown Human Unknown Animal	No	No	29 CFR 1910.1000 Table Z-1	Eyes, skin, respiratory system, CNS, liver.

Medical Conditions Generally Aggravated by Exposure

None Reported.

Emergency and First-aid Procedures

Eye contact: Flush immediately with water for 15 minutes (minimum), seek medical attention. **Skin contact:** Remove all contaminated clothing. Immediately wash area with large amounts of soap and water. Seek medical assistance. For **inhalation**: Remove the person from exposure. Provide respiratory assistance and CPR. Transfer to medical facility. Observe for 24-48 hours for lung effects. If **swallowed**: Call the poison control center and seek medical attention immediately. Never attempt to give an unconscious or convulsing person anything by mouth.

SECTION VII - PRECAUTIONS FOR SAFE HANDLING AND USE

Steps to be Taken in Case Material is Released or Spilled

Absorb liquids in vermiculite or other absorbent and deposit in DOT-approved drum. Ventilate area and remove ignition sources. Restrict those not involved in cleanup from entering area. Notify appropriate authorities, if applicable.

Preferred Waste Disposal Method

No Citation.

Precautions to be Taken in Handling and Storage

Do not store diacetone alcohol in the presence of incompatible chemicals or materials. Store in tightly closed containers in a cool, well-ventilated area. Personnel working with diacetone alcohol should be trained on its proper handing and storage *prior* to being assigned to such responsibilities.

Other Precautions and Warnings

Bulk storage is of diacetone alcohol is not recommended.

SECTION VIII - CONTROL MEASURES AND PERSONAL PROTECTIVE EQUIPMENT

Respiratory Protection *(specify type)*

For exposures near or above the PEL (50 ppm), use a full facepiece respirator equipped with an organic vapor cartridge. For greater protection or high exposures, use a supplied-air respirator with full facepiece operated in pressure demand mode or a self-contained breathing apparatus (SCBA) with full facepiece operated in pressure demand or other positive pressure mode.

Ventilation

Local exhaust or general mechanical systems recommended.

Protective Gloves	Eye Protection	Other Protective Clothing
Rubber	Chemical Goggles or Face Mask	Apron

Work/Hygiene Practices

Always wash hands thoroughly after using chemical; never bring food, drink, or smoking materials into vicinity of chemicals.

DIACETONE ALCOHOL

$CH_3COCH_2C(CH_3)_2OH$ CAS: 123-42-2

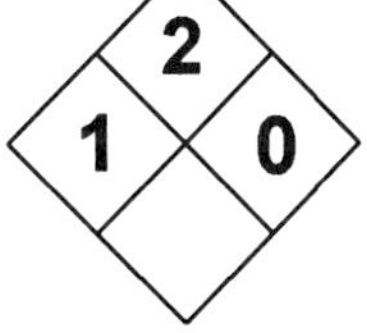

IDENTIFICATION AND TYPICAL USES

Diacetone alcohol is a colorless liquid with a faint but pleasant mint-like odor. It is used in the manufacture of nitrocellulose, cellulose acetate, various oils, resins, waxes, fats, dyes, tars, lacquers, dopes, coating compositions, wood preservatives, stains, rayon, and artificial leather. It is also used in the production of imitation gold leaf, in dyeing mixtures, in antifreeze mixtures, in the extraction of resins and waxes, as a preservative for animal tissue, in metal-cleaning compounds, in hydraulic fluids, in stripping agents (textiles), and as a laboratory reagent. The technical grade containing acetone has greater solvent power.

RISK ASSESSMENT: HEALTH

General Assessment

Diacetone alcohol is a human poison and is toxic by *ingestion, inhalation,* and also poses a risk to health on skin contact (absorption, however, has not been proven). Exposure appears to cause systemic effects by both routes of entry.

Inhalation and ingestion cause headache, nausea or vomiting, eye, and pulmonary changes. Its vapors are extremely irritating to the mucous membranes in the nose and throat. High or prolonged exposures can cause anemia and damage the liver and kidneys. There is also evidence of central nervous system (CNS) effects at high concentrations. These may include narcosis, dizziness, and anesthesia.

Skin contact results in severe irritation and may produce dermatitis with rash and itching. Eye contact can result in corneal damage.

Acute Health Effects

The following acute (short-term) health effects may occur immediately or shortly after exposure to diacetone alcohol:

Skin: Irritation and possible allergy. May cause dermatitis.

Eye: Severe irritation. Possibility of damage to the cornea.

Lung: Irritation causing coughing and/or shortness of breath. Higher exposures can lead to respiratory problems, including pulmonary edema (fluid buildup in lungs) which is a medical emergency and can be fatal.

CNS: Dizziness, headaches, nausea, vomiting, some anesthesia, narcosis (on high concentrations).

Chronic Health Effects

The following chronic (long-term) health effects may occur at some time after exposure to diacetone alcohol and can last for months or even years:

Cancer Hazards: According to the information presented in the references, diacetone alcohol has not been adequately tested for its ability to cause cancer in test animals.

Reproductive Hazard: According to the information presented in the references, diacetone alcohol has not been adequately tested for its ability to adversely affect reproduction.

Other Chronic Effects: Low, chronic exposures to skin may cause allergy or dermatitis. Once an allergy develops, even small future exposures will cause a recurrence of symptoms.

Recommended Risk-Reduction Measures

Unless a less toxic chemical can be substituted for a hazardous substance, *engineering controls* are the most effective method of reducing exposures. The best protection is to enclose operations and/or provide local exhaust ventilation at the site of chemical release. While not always operationally feasible, isolating operations can also reduce exposure.

Using respiratory protection is less effective than the controls mentioned above, but is still advisable whenever working with or around diacetone alcohol. At exposure levels near or slightly above the PEL (50 ppm), use a full facepiece respirator equipped with an organic vapor cartridge. For high concentrations, a self-contained breathing apparatus (SCBA) with full facepiece and pressure demand is the recommended respiratory protection of choice. If a full facepiece is not available, then chemical goggles should be worn to protect the eyes. Whenever a chemical splash hazard exists, a face shield and rubber apron should be worn. To prevent hand and skin exposures, rubber gloves should be worn.

Administrative controls should also be in place to minimize the potential for human exposures. These may include written procedures or policies which specify the methods and techniques that will be practiced whenever personnel are to work with diacetone alcohol.

All personnel should receive *training* on the use, hazards, protective measures, emergency actions, and other precautions per 29 CFR 1910.1200 (Hazard Communication), prior to the first assignment in an area where diacetone alcohol is used or stored.

If symptoms develop or overexposure is suspected, the following medical tests may be useful:

- ☑ Lung function tests.
- ☑ Liver and kidney function tests.
- ☑ Evaluation by a qualified allergist with careful evaluation of exposure history and administration of special tests (may help diagnose skin allergy).

Any evaluation should include a careful medical history of past and present symptoms with an examination. However, medical tests that evaluate existing damage are not a substitute for controlling exposure.
Other methods to reduce exposure include:

- ☑ Where possible, enclose operations and use local exhaust ventilation at the site of chemical release. If local exhaust ventilation or enclosure is not used, respiratory protection should be mandatory.
- ☑ Always ensure that proper protective clothing is worn when using chemical substances.
- ☑ Wash thoroughly immediately after exposure to diacetone alcohol and at the end of the work shift or before eating, drinking, or smoking.
- ☑ Hazard warning information should be posted in the work area. In addition, as part of an on-going education and training program, all information on the health and safety hazards of diacetone alcohol should be communicated to all potentially exposed workers.
- ☑ Eye wash stations should be provided in work areas. If the potential for whole body exposure exists, then safety showers should also be provided.

RISK ASSESSMENT: ENVIRONMENT

General Assessment

The environment is at risk of exposure during intended use, transportation, storage, disposal, or destruction of diacetone alcohol. In almost every scenario, the threat of environmental exposure is contingent upon the proper handling of the chemical substance. Accidental spills, large or small, can result in fire, explosion, and possible contamination of the surrounding environmental mediums (water, soil, and air).

Diacetone alcohol is considered a class II combustible liquid (per OSHA 29 CFR 1910.106). These characteristics require special consideration during any emergency situation involving a leak or spill of diacetone alcohol or diacetone alcohol mixtures. Should diacetone alcohol ever come into contact with incompatible substances such as strong oxidizers (chlorine, bromine, fluorine) or strong alkali materials either during use, transportation, or storage, violent reactions can occur.

Diacetone alcohol can enter the environment through industrial discharges and spills.

☠ *Acute Ecological Effects*

Acute (short-term) toxic effects may include the death of animals, birds, or fish and death or low growth rate in plants. Acute effects are seen 2 to 4 days after animals or plants are exposed to diacetone alcohol. Insufficient data are available to evaluate or predict the short-term effects of diacetone alcohol to aquatic life, plants, birds, or land animals.

Chronic Ecological Effects

Chronic toxic effects may include shortened life span, reproductive problems, lower fertility, and changes in appearance or behavior in exposed animals. These effects can be seen long after first exposure(s) to toxic chemicals. Diacetone alcohol has slight chronic toxicity to aquatic life. Insufficient data are available to evaluate or predict the long-term effects of diacetone alcohol to plants, birds, or land animals.

♦ *Water Solubility*

Diacetone alcohol is moderately soluble in water. Concentrations of 100 to 1000 milligrams will mix with a liter of water.

⌛ *Persistence in the Environment*

Diacetone alcohol is slightly persistent in water, with a half-life of between 2 to 20 days. The half-life of a pollutant is the amount of time it takes for one half of the chemical to be degraded. Approximately 50% of diacetone alcohol will eventually end up in air; the rest will end up in water.

Bioaccumulation in Aquatic Organisms

Some substances increase in concentration, or *bioaccumulate,* in living organisms as they breathe contaminated air, drink contaminated water, or eat contaminated food. These chemicals can become concentrated in the tissues and internal organs of animals as well as humans.

The concentration of diacetone alcohol found in fish tissues is expected to be about the same as the average concentration of diacetone alcohol in the water from which the fish was taken.

Recommended Risk-Reduction Measures

Proper training of all transporters will reduce the likelihood of a mishap or accident resulting in a leak or spill to the environment. The correct use of DOT labeling (per the MSDS) on all containers, trucks, and rail cars will enable emergency responders to react properly and quickly to any disaster thereby reducing the potential risk to the environment and to personnel.

Storage of diacetone alcohol should be segregated from incompatible chemicals to minimize the risk of cross-contamination or contact.

If a spill or leak to the environment has occurred, fire department, emergency response and/or hazardous materials spill personnel should be notified immediately. Cleanup should be attempted only by those trained in proper spill containment procedures. Contaminated soils should be removed for incineration and replaced with clean soil. If diacetone alcohol should contact the water table, aquifer, or navigable waterway, remediation activities should be prompt to ensure protection of aquatic sediments. When such spills occur, the local and/or state emergency response authorities must be notified. A comprehensive emergency response or disaster preparedness/recovery plan should be in place prior to any operations involving the use, transportation, storage, or disposal of diacetone alcohol. If diacetone alcohol is spilled or leaked, the following specific steps are recommended:

- ☑ Restrict persons not wearing protective clothing from area of spill or leak until cleanup is complete and area can be opened for normal work.
- ☑ Ventilate area of spill or leak.
- ☑ Remove all ignition sources.
- ☑ If source of leak is a cylinder and it cannot be repaired in place, remove to safe location (e.g., outdoors) and allow to vent until empty.
- ☑ Absorb liquid spills using vermiculite, dry earth, or sand to absorb and place in a sealed drum for disposal.
- ☑ It may be necessary to dispose of diacetone alcohol as a hazardous waste. The responsible state agency or the regional office of the federal Environmental Protection Agency (EPA) should be contacted for specific recommendations.

RISK ASSESSMENT: BUSINESS

General Assessment

Accidents or mishaps involving diacetone alcohol can present a significant threat to business operations. The loss or damage of equipment or facilities can significantly affect fiscal viability. Lawsuits that may result from personnel injury/death, public exposures, and/or environmental contamination will also require a serious expenditure of resources. Media attention surrounding an injury, death, or environmental damage can also result in a loss of profits and loss of current as well as future business.

Recommended Risk-Reduction Measures

Company attorneys, safety and health professionals, and environmental specialists should be involved in the development of any procedures dealing with safety or emergency response. A company official should be pre-designated as a public relations officer with specific training in dealing with the press. Corporate plans and policies should be developed, approved, and implemented long before any need for such arises.

REFERENCES

Ahlbom, A. 1993. *Biostatistics for Epidemiologists.* New York: CRC Press/Lewis Publishers.

American Conference of Governmental Industrial Hygienists. 1988. *Documentation of the Threshold Limit Values and Biological Exposure Indices,* 5th Edition (with updates). Cincinnati: ACGIH

Anthony, C. P., and N. J. Kolthoff. 1971. *Textbook of Anatomy and Physiology.* St. Louis: C. V. Mosby Company.

Asante-Duah, D. K. 1993. *Hazardous Waste Risk Assessment.* New York: CRC Press/Lewis Publishers.

Bloom, A. D. and F. J. DeSerres. 1995. *Ecotoxicity and Human Health.* New York: CRC Press/Lewis Publishers.

Brusick, D. J. 1994. *Methods for Genetic Risk Assessment.* New York: CRC Press/Lewis Publishers.

Bueche, J. R. 1972. *Principles of Physics,* 2nd Edition. New York: McGraw-Hill.

Burrell, R., D. K. Flaherty, and L. J. Sauers. 1992. *Toxicology of the Immune System.* New York: Van Nostrand Reinhold.

Calabrese, E. J., E. M. Kenyon. *Air Toxics and Risk Assessment.* New York: CRC Press/Lewis Publishers.

Calabrese, E. J. 1994. *Biological Effects of Low Level Exposures.* New York: CRC Press/Lewis Publishers.

Cockerman, L. G., B. S. Shane. 1994. *Basic Environmental Toxicology.* New York: CRC Press/Lewis Publishers.

Cothern, C. R., N. P. Ross. 1994. *Environmental Statistics, Assessment, and Forecasting.* New York: CRC Press/Lewis Publishers.

Cothern, R. R. 1993. *Comparative Environmental Risk Assessment.* New York: CRC Press/Lewis Publishers.

Cothern, C. R. 1995. *Handbook for Environmental Risk Decisionmaking.* New York: CRC Press/Lewis Publishers.

Gardner, R. W. 1994. *Chemical Intolerance.* New York: CRC Press/Lewis Publishers.

Hallenbeck, W. H. 1993. *Quantitative Risk Assessment for Environmental and Occupational Health,* 2nd Edition. New York: CRC Press/Lewis Publishers.

Hodgson, E., R. B. Mailman, and Janice E. Chambers. 1988. *Dictionary of Toxicology.* London: The MacMillian Company.

Howard, P. H. 1995. *Handbook of Environmental Fate and Exposures Data for Organic Chemicals,* Vol. I-IV. New York: CRC Press/Lewis Publishers.

Howard, P. H., et. al. 1991. *Handbook of Environmental Degradation Rates.* New York: CRC Press/Lewis Publishers.

Jones, F. E. 1994. *Toxic Organic Vapors in the Workplace.* New York: CRC Press/Lewis Publishers.

Kamrin, M. A. 1988. *Toxicology: A Primer on Principles and Applications.* Chelsea, Michigan: Lewis Publishers.

Keith, L. H. and M. M. Walker. 1995. *Handbook of Air Toxics.* New York: CRC Press/Lewis Publishers.

Landis, W. G., M. H. Yu. 1995. *Introduction to Environmental Toxicology.* New York: CRC Press/Lewis Publishers.

Langerman, N. 1994 *Precautionary Labels for Chemical Containers.* New York: CRC Press/Lewis Publishers.

Lauwerys, R. R., P. Hoet. 1993. *Industrial Chemical Exposure.* New York: CRC Press/Lewis Publishers.

Lide, D. R. 1995. *Handbook of Chemistry and Physics,* 76th Edition. New York: CRC Press/Lewis Publishers.

Lewis, R. J., Sr. 1993. *Hawley's Condensed Chemical Dictionary,* 12th Edition. New York: Van Nostrand Reinhold.

Lewis, R. J., Sr. 1992. *Sax's Dangerous Properties of Industrial Materials,* Volumes 1, 2, and 3, 8th Edition. New York: Van Nostrand Reinhold.

Manahan, S. E. 1994. *Environmental Chemistry,* 6th Edition. New York: CRC Press/Lewis Publishers.

Manahan, S. E. 1992. *Toxicological Chemistry,* 2nd Edition. New York: CRC Press/Lewis Publishers.

Mellström, G. A., J. E. Wahlberg, and H. I. Maibach. 1994. *Protective Gloves for Occupational Use.* New York: CRC Press/Lewis Publishers.

Meyer, E. 1990. *Chemistry of Hazardous Materials.* Englewood Cliffs, NJ: Prentice Hall.

Mycock, J. C., J. D. McKenna, and L. Theodore. 1995. *Handbook of Air Pollution Control Engineering and Technology.* New York: CRC Press/Lewis Publishers.

Nelson, G. O. 1992. *Gas Mixtures.* New York: CRC Press/Lewis Publishers.

Ott, W. R. 1995. *Environmental Statistics and Data Analysis.* New York: CRC Press/Lewis Publishers.

Patnaik, P. 1992. *A Comprehensive Guide to the Hazardous Properties of Chemical Substances.* New York: Van Nostrand Reinhold.

Philp, R. B. 1995. *Environmental Hazards and Human Health.* New York: CRC Press/Lewis Publishers.

Proctor, N. H., and James P. Hughes. 1991. *Chemical Hazards of the Workplace.* New York: Van Nostrand Reinhold.

Quigley, D. R. 1994. *Handbook of Emergency Chemical Management.* New York: CRC Press/Lewis Publishers.

Rea, W. J. 1995. *Chemical Sensitivity,* Vol. I-III. New York: CRC Press/Lewis Publishers.

Richardson, M. 1995. *Dictionary of Substance and Their Effects,* Vol. I-VI. New York: CRC Press/Lewis Publishers.

Rylander, R., R. R. Jacobs. 1994. *Organic Dusts.* New York: CRC Press/Lewis Publishers.

Sacarello, H. L. A. 1994. *The Comprehensive Handbook of Hazardous Materials.* New York: CRC Press/Lewis Publishers.

Sheftel, V. O. 1995. *Handbook of Toxic Properties of Monomers and Additives.* New York: CRC Press/Lewis Publishers.

U.S. Department of Health and Human Services, National Institute for Occupational Safety and Health. 1994. *NIOSH Pocket Guide to Chemical Hazards.* Washington, D.C.: U.S. Government Printing Office.

MATERIAL SAFETY DATA SHEET

CHEMICAL NAME	CRC PRESS/LEWIS PUBLISHERS DISCLAIMER
m-DIAMINOBENZENE	

HAZARD WARNING INFORMATION

HEALTH	FIRE	REACTIVE	OTHER	DEGREE OF HAZARD	COLOR CODING	OTHER CODES
4	1	0		0 = Minimum Hazard 1 = Slight Hazard 2 = Moderate Hazard 3 = Serious Hazard 4 = Severe Hazard	HEALTH = BLUE FIRE = RED REACTIVITY = YELLOW OTHER = WHITE	OX = Oxidizer ACID = Acid ALK = Alkali COR = Corrosive ~~W~~ = Use No Water

SECTION I - GENERAL INFORMATION

Characteristic	**Aromatic Amine**	RCRA Number	**None**
		EPA Class	**Not Applicable**
DOT Proper Shipping Name	***m*-Phenylenediamine, Solid**	Chemical Abstract Service (CAS) Number	**95-54-5**
DOT Hazard Class and Label Requirements	**ORM-A; St. Andrew's Cross**	DOT Emergency Guide Code	**53**
DOT Identification Number	**UN 1673**	Chemical Formula	**$C_64_3(NH_2)_2$**

Synonyms
***m*-Phenylenediamine; *m*-aminoaniline; *m*-benzenediamine; 3-aminoaniline; 1,3-benzendiamine; 1,3-diaminobenzene; 1,3-phenylenediamine.**

SECTION II - HAZARDOUS INGREDIENTS AND IDENTITY INFORMATION

Hazardous Components (specific identity)	OSHA Exposure Criteria	NIOSH Exposure Criteria	Immediately Dangerous to Life and Health (IDLH)	ACGIH Exposure Criteria
m-Diaminobenzene (derivation: By reduction of m-dinitrobenzene or nitroaniline with iron and hydrochloric acid. Purified by crystallization).	**PEL:** Not Established **STEL:** Not Established	**REL:** Not Established **STEL:** Not Established	**Not Determined**	**TLV:** Not Established **STEL:** Not Established

SECTION III - PHYSICAL AND CHEMICAL CHARACTERISTICS

Boiling Point	**547°F (286°C)**	Specific Gravity (H_2O = 1)	**1.13**
Vapor Pressure (mm Hg)	**1 mm at 212°F (100°C)**	Molecular Weight	**108.2**
Vapor Density (Air = 1)	**3.71**	Melting Point	**145°F (63°C)**

Solubility
Soluble in water, methanol, ethanol, chloroform, dimethylformaldehyde, and acetone. Slightly soluble in ether and carbon tetrachloride. Very slightly soluble in toluene and benzene.

Appearance and Odor
Colorless to white crystalline solid. Turns red on exposure to air. Mild phenolic odor.

SECTION IV - FIRE AND EXPLOSION HAZARD DATA

Flash Point (method used)	**312°F (155°C) closed cup**	Explosive Limits in Air % by Volume	**LEL: 1.4% UEL: Not Found**
NFPA Classification	**Combustible Solid**	Autoignition Temperature	**Not Determined**

Extinguishing Media
Use dry chemical, carbon dioxide, foam, or water fog. Water or foam may cause frothing.

Special Fire Fighting Procedures
Poisonous gases are produced in fire. Structural protective clothing is permeable and may not provide adequate protection. Wear full protective clothing and self-contained breathing apparatus (SCBA). Move containers from fire area if it can be done safely. Cool exposed container with water.

Unusual Fire and Explosion Hazards
Dusts can form explosive mixtures in air. Exercise extreme caution when attempting to move or relocate containers that may have been exposed to high heat.

SECTION V - REACTIVITY DATA

Stability		Conditions to Avoid
		Normally stable chemical. Keep away from heat, sparks, flame, and other ignition sources. May be affected by exposure to light.
Stable	Unstable	Incompatibility *(materials to avoid)*
X		Strong oxidizers (such as chlorine, bromine, fluorine). Reactions can be very violent.
Hazardous Polymerization		Conditions to Avoid Under normal conditions of temperature and pressure, hazardous polymerization of m-diaminobenzene is not expected to occur.
May Occur	Will Not Occur	Hazardous Decomposition or By-products
	X	When heated to decomposition, m-diaminobenzene will produce toxic and irritating gases and oxides of nitrogen and carbon dioxide and/or carbon monoxide.

SECTION VI - HEALTH HAZARD DATA

Primary Route(s) of Entry:	Inhalation?	Absorption (skin)?	Ingestion?
	X	X	X

Health Hazards

INHALATION: Irritation of the upper respiratory tract. Causes headache, cough, and may result in the development of allergic hypersensitivity asthma with inflammatory reactions of the larynx and pharynx. Systemic effects include lachrymation, salivation, ataxia, tremor, lower body temperature, increased pulse rate, and respiratory depression.

ABSORPTION: Can cause allergic skin reaction, hives, and dermatitis. Will pass through unbroken skin to cause toxic systemic effects. These include liver and spleen enlargement, vertigo, gastritis, jaundice, atrophy of the liver, and allergic asthma. Severe eye irritant. Can cause corneal ulcer, burning, and redness. Possible presbyopia and conjunctivitis.

INGESTION: Toxic systemic effects similar to those for inhalation and absorption.

Carcinogenicity	NTP Listed?	IARC Cancer Review Group?	OSHA Regulated?	Target Organs?
Questioned Human Questioned Animal	No	Group 3	No	Respiratory system, skin, eyes, liver, spleen.

Medical Conditions Generally Aggravated by Exposure

Respiratory conditions and skin disorders may be aggravated by exposure.

Emergency and First-aid Procedures

Eye contact: Flush immediately with water for 15 minutes (minimum) while holding upper and lower eyelids open. Seek medical attention. Skin contact: Remove all contaminated clothing. Immediately wash area with large amounts of soap and water. For inhalation: Remove the person from exposure. Provide respiratory assistance and CPR. Transfer to medical facility. If swallowed: Call the poison control center, give water to induce vomiting and seek medical attention immediately.

SECTION VII - PRECAUTIONS FOR SAFE HANDLING AND USE

Steps to be Taken in Case Material is Released or Spilled

Powders should be collected in safest manner possible and disposed of in sealed drums. Do NOT dry sweep. Use a high efficiency particulate air (HEPA) vacuum instead. Ventilate area and remove ignition sources. Restrict those not involved in cleanup from entering area. Notify appropriate authorities.

Preferred Waste Disposal Method

Consider reclamation, recycling, or destruction rather than landfill.

Precautions to be Taken in Handling and Storage

Store in tightly closed containers in a cool, dark, well-ventilated area. Personnel should be trained on the hazardous properties of m-diaminobenzene *prior* to working with the material.

Other Precautions and Warnings

Empty containers may still contain product residues and should be handled with caution.

SECTION VIII - CONTROL MEASURES AND PERSONAL PROTECTIVE EQUIPMENT

Respiratory Protection *(specify type)*

For low exposures, use a full facepiece respirator equipped with HEPA filter and dust pre-filter. For higher exposures, use a self-contained breathing apparatus (SCBA) with full facepiece operated in pressure demand or other positive pressure mode.

Ventilation

Local exhaust or general mechanical systems recommended.

Protective Gloves	Eye Protection	Other Protective Clothing
Impervious Gloves	Chemical/Dust Goggles or Face Mask	Protective Apron

Work/Hygiene Practices

Always wash hands thoroughly after using chemical; never bring food, drink, or smoking materials into vicinity of chemicals.

m-DIAMINOBENZENE

$C_6H_4(NH_2)_2$ CAS: 95-54-5

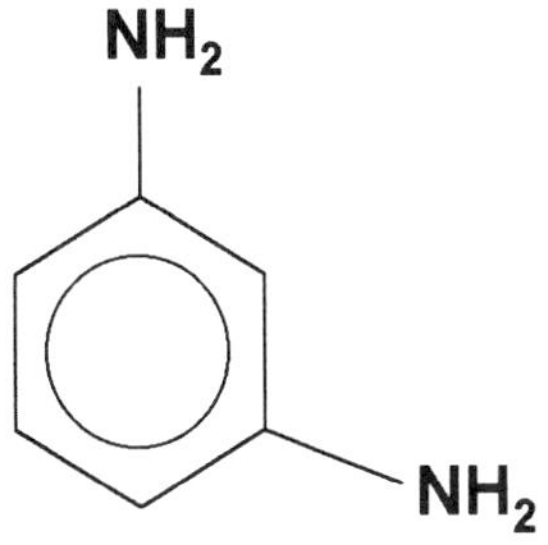

IDENTIFICATION AND TYPICAL USES

m-Diaminobenzene is a colorless to white crystalline solid. It will turn red on exposure to air. It has a mild, phenolic odor. It is used in the manufacture of a variety of dyes, in hair dye formulations, as a rubber curing agent, in petroleum additives, as a photographic developing agent, and as an analytical reagent.

RISK ASSESSMENT: HEALTH

General Assessment

m-Diaminobenzene is a moderate to highly toxic compound in humans. Of the three diamionbenzene isomers, the acute and chronic toxicity of the *"m"* or *meta*-amine is not as great as that of the *"p"* or *para*-isomers. It is a poison by *ingestion* and is mildly toxic by skin contact. *Absorption* through unbroken skin is possible. *Inhalation* of fine dusts will also result in toxic systemic effects.

Inhalation and ingestion will cause irritation of the larynx, pharynx, and may cause bronchial asthma. Reports of allergic hypersensitivity have been recorded in the references. Systemic effects may include lachrymation, salivation, ataxia, tremor, decreased body temperature, increased pulse rate, and respiratory depression. It may also cause tremor, excitement, convulsions, cyanosis, and collapse at fatal dose.

Skin contact will cause irritation and possible dermatitis with inflammation and redness. Absorption will result in systemic effects similar to those caused by inhalation and may also result in the enlargement of the liver, kidney, and possibly the spleen, vertigo, gastritis, jaundice, and atrophy of the liver. Eye contact results in burning, redness, corneal ulceration, and a possible presbyopia-like loss of elasticity in the lens of the eye.

Acute Health Effects

The following acute (short-term) health effects may occur immediately or shortly after exposure to m-diaminobenzene:

Skin: Mild to moderate irritation and possible dermatitis. Skin absorption is likely with subsequent toxic systemic effects.

Eye: Mild to severe irritation with possible damage to the cornea.

Lung: Irritation of the upper respiratory tract causing coughing and/or shortness of breath and may lead to secondary infections. Allergic hypersensitivity, bronchial asthma, and throat problems likely to occur.

CNS: Unconfirmed reaction, but may cause vertigo, ataxia, dizziness, tremor, excitement, convulsions, and other unspecified systemic effects by all routes of entry.

Chronic Health Effects

The following chronic (long-term) health effects may occur at some time after exposure to m-diaminobenzene and can last for months or even years:

Cancer Hazards: According to the information presented in the references, m-diaminobenzene has been shown to cause tumors in test animals. Its carcinogenicity in humans is questioned. However, the data in the references are conflicting and inconclusive in this regard. It is a mutagen. Until further studies are performed, it is suggested that m-diaminobenzene be handled with extreme caution.

Reproductive Hazard: According to the information presented in the references, m-diaminobenzene has not been adequately tested for its ability to adversely affect reproduction.

ⓘ *Recommended Risk-Reduction Measures*

Unless a less toxic chemical can be substituted for a hazardous substance, *engineering controls* are the most effective method of reducing exposures. The best protection is to enclose operations and/or provide local exhaust ventilation at the site of chemical release. While not always operationally feasible, isolating operations can also reduce exposure. All metal equipment used with this chemical should be grounded and bonded. Explosion-proof design of equipment is recommended.

Using respiratory protection is less effective than the controls mentioned above, but is still advisable whenever working with or around m-diaminobenzene. For low or relatively infrequent exposures, use a full facepiece respirator equipped with a high efficiency particulate air (HEPA) filter and a dust pre-filter. For higher concentrations, use of a self-contained breathing apparatus (SCBA) with full facepiece and pressure demand is the recommended respiratory protection of choice. If a full facepiece is not available, then chemical goggles should be worn to protect the eyes. Whenever a chemical splash hazard exists, a face shield and apron should be worn. To prevent hand and skin exposures, impervious gloves should be worn.

Administrative controls should also be in place to minimize the potential for human exposures. These may include written procedures or policies which specify the methods and techniques that will be practiced whenever personnel are to work with or around m-diaminobenzene.

All personnel should receive *training* on the use, hazards, protective measures, emergency actions, and other precautions per 29 CFR 1910.1200 (Hazard Communication), prior to the first assignment in an area where m-diaminobenzene is used or stored.

Before beginning employment and at regular intervals thereafter (e.g. annually), the following medical tests are recommended:

☑ Lung function tests (establish a baseline).

If symptoms develop or overexposure is suspected, the following medical tests may be useful:

☑ Lung function tests.

☑ Liver function tests.

☑ Consider chest X-ray after acute overexposure.

☑ Evaluation by a qualified allergist with careful consideration of exposure history and special tests (may help diagnose allergy).

Any evaluation should include a careful medical history of past and present symptoms with an examination. However, medical tests that evaluate existing damage are not a substitute for controlling exposure. Also, since smoking can cause heart diseases, emphysema, and numerous other respiratory disorders, smokers may be affected more rapidly than non-smokers to the same exposure conditions. Prudent risk assessment and management requires careful consideration of *all* possible risk factors that may be causing the appearance of symptoms in exposed workers.

Other methods to reduce exposure include:

☑ Where possible, enclose operations and use local exhaust ventilation at the site of chemical release. If local exhaust ventilation or enclosure is not used, respiratory protection should be mandatory.

☑ Always ensure that proper protective clothing is worn when using chemical substances.

☑ Wash thoroughly immediately after exposure to m-diaminobenzene and at the end of the work shift or before eating, drinking, or smoking.

☑ Hazard warning information should be posted in the work area. In addition, as part of an on-going education and training program, all information on the health and safety hazards of m-diaminobenzene should be communicated to all potentially exposed workers.

☑ Eye wash stations should be provided in work areas. If the potential for whole body exposure exists, then safety showers should also be provided.

RISK ASSESSMENT: ENVIRONMENT

General Assessment

The environment is at risk of exposure during intended use, transportation, storage, disposal, or destruction of m-diaminobenzene. In almost every scenario, the threat of environmental exposure is contingent upon the proper handling of the chemical substance. Accidental spills, large or small, can result in fire, explosion, and possible contamination of the surrounding environmental mediums (water, soil, and air).

m-Diaminobenzene should always be handled with extreme caution. It is incompatible with strong oxidizers and its dusts can form explosive mixtures in air.

These characteristics require special consideration during any emergency situation involving a leak or spill of m-diaminobenzene.

m-Diaminobenzene can enter the environment through industrial discharges and spills.

Acute Ecological Effects

Acute (short-term) toxic effects may include the death of animals, birds, or fish and death or low growth rate in plants. Acute effects are seen 2 to 4 days after animals or plants are exposed to m-diaminobenzene. Insufficient data are available to evaluate or predict the short-term effects of m-diaminobenzene to aquatic life, plants, birds, or land animals.

Chronic Ecological Effects

Chronic toxic effects may include shortened life span, reproductive problems, lower fertility, and changes in appearance or behavior in exposed animals. These effects can be seen long after first exposure(s) to toxic chemicals. Insufficient data are available to evaluate or predict the long-term effects of m-diaminobenzene to aquatic life, plants, birds, or land animals.

Water Solubility

m-Diaminobenzene is moderately soluble in water. Concentrations of 100 to 1000 milligrams will mix with a liter of water.

Persistence in the Environment

m-Diaminobenzene is highly persistent in water, with a half-life of greater than 200 days. The half-life of a pollutant is the amount of time it takes for one half of the chemical to be degraded.

Bioaccumulation in Aquatic Organisms

Some substances increase in concentration, or *bioaccumulate,* in living organisms as they breathe contaminated air, drink contaminated water, or eat contaminated food. These chemicals can become concentrated in the tissues and internal organs of animals as well as humans.

The concentration of m-diaminobenzene found in fish tissues is expected to be somewhat higher than the average concentration of m-diaminobenzene in the water from which the fish was taken.

Recommended Risk-Reduction Measures

Proper training of all transporters will reduce the likelihood of a mishap or accident resulting in a leak or spill to the environment. The correct use of labeling on all containers, trucks, and rail cars will enable emergency responders to react properly and quickly to any disaster thereby reducing the potential risk to the environment and to personnel.

Storage of m-diaminobenzene should be segregated from incompatible chemicals to minimize the risk of cross-contamination or contact. Containers should be protected from physical damage. Storage buildings should be equipped with the proper fire protection equipment (alarms, sprinklers, extinguishers, etc.).

If a spill or leak to the environment has occurred, fire department, emergency response and/or hazardous materials spill personnel should be notified immediately. Cleanup should be attempted only by those trained in proper spill containment procedures. Contaminated soils should be removed for incineration and replaced with clean soil. If m-diaminobenzene should contact the water table, aquifer, or navigable waterway, remediation activities should be prompt to ensure protection of aquatic sediments. When such spills occur, the local and/or state emergency response authorities must be notified. A comprehensive emergency response or disaster preparedness/recovery plan should be in place prior to any operations involving the use, transportation, storage, or disposal of m-diaminobenzene. If m-diaminobenzene is spilled or leaked, the following specific steps are recommended:

- ☑ Restrict persons not wearing protective clothing from area of spill or leak until cleanup is complete and area can be opened for normal work.
- ☑ Ventilate area of spill or leak.
- ☑ Remove all ignition sources.
- ☑ Collect powders in safest manner possible and dispose of in sealed drums. Do NOT dry sweep (creates airborne dusts). Use a vacuum equipped with a high efficiency particulate air (HEPA) filter cartridge or attachment.
- ☑ It may be necessary to dispose of m-diaminobenzene as a hazardous waste. The responsible state agency or the regional office of the federal Environmental Protection Agency (EPA) should be contacted for specific recommendations.

RISK ASSESSMENT: BUSINESS

General Assessment

Accidents or mishaps involving m-diaminobenzene can present a significant threat to business operations. The loss or damage of equipment or facilities can significantly affect fiscal viability. Lawsuits that may result from personnel injury/death, public exposures, and/or environmental contamination will also require a serious expenditure of resources. Media attention surrounding an injury, death, or environmental damage can also result in a loss of profits and loss of current as well as future business. Always remember that anytime the term "carcinogen" or "cancer" are used, public emotion, hysteria, and ignorance can run equally high. This should be considered during the development of any public relations policies.

Ʊ Recommended Risk-Reduction Measures

Company attorneys, safety and health professionals, and environmental specialists should be involved in the development of any procedures dealing with safety or emergency response. A company official should be pre-designated as a public relations officer with specific training in dealing with the press. Corporate plans and policies should be developed, approved, and implemented long before any need for such arises.

REFERENCES

Ahlbom, A. 1993. *Biostatistics for Epidemiologists.* New York: CRC Press/Lewis Publishers.

American Conference of Governmental Industrial Hygienists. 1988. *Documentation of the Threshold Limit Values and Biological Exposure Indices,* 5th Edition (with updates). Cincinnati: ACGIH

Anthony, C. P., and N. J. Kolthoff. 1971. *Textbook of Anatomy and Physiology.* St. Louis: C. V. Mosby Company.

Asante-Duah, D. K. 1993. *Hazardous Waste Risk Assessment.* New York: CRC Press/Lewis Publishers.

Bloom, A. D. and F. J. DeSerres. 1995. *Ecotoxicity and Human Health.* New York: CRC Press/Lewis Publishers.

Brusick, D. J. 1994. *Methods for Genetic Risk Assessment.* New York: CRC Press/Lewis Publishers.

Bueche, J. R. 1972. *Principles of Physics,* 2nd Edition. New York: McGraw-Hill.

Burrell, R., D. K. Flaherty, and L. J. Sauers. 1992. *Toxicology of the Immune System.* New York: Van Nostrand Reinhold.

Calabrese, E. J. 1994. *Biological Effects of Low Level Exposures.* New York: CRC Press/Lewis Publishers.

Cockerman, L. G., B. S. Shane. 1994. *Basic Environmental Toxicology.* New York: CRC Press/Lewis Publishers.

Cothern, C. R. 1995. *Handbook for Environmental Risk Decisionmaking.* New York: CRC Press/Lewis Publishers.

Gardner, R. W. 1994. *Chemical Intolerance.* New York: CRC Press/Lewis Publishers.

Hallenbeck, W. H. 1993. *Quantitative Risk Assessment for Environmental and Occupational Health,* 2nd Edition. New York: CRC Press/Lewis Publishers.

Hodgson, E., R. B. Mailman, and Janice E. Chambers. 1988. *Dictionary of Toxicology.* London: The MacMillian Company.

Howard, P. H. 1995. *Handbook of Environmental Fate and Exposures Data for Organic Chemicals,* Vol. I-IV. New York: CRC Press/Lewis Publishers.

Jones, F. E. 1994. *Toxic Organic Vapors in the Workplace.* New York: CRC Press/Lewis Publishers.

Keith, L. H. and M. M. Walker. 1995. *Handbook of Air Toxics.* New York: CRC Press/Lewis Publishers.

Landis, W. G., M. H. Yu. 1995. *Introduction to Environmental Toxicology.* New York: CRC Press/Lewis Publishers.

Langerman, N. 1994 *Precautionary Labels for Chemical Containers.* New York: CRC Press/Lewis Publishers.

Lauwerys, R. R., P. Hoet. 1993. *Industrial Chemical Exposure.* New York: CRC Press/Lewis Publishers.

Lewis, R. J., Sr. 1993. *Hawley's Condensed Chemical Dictionary,* 12th Edition. New York: Van Nostrand Reinhold.

Lewis, R. J., Sr. 1992. *Sax's Dangerous Properties of Industrial Materials,* Volumes 1, 2, and 3, 8th Edition. New York: Van Nostrand Reinhold.

Manahan, S. E. 1992. *Toxicological Chemistry,* 2nd Edition. New York: CRC Press/Lewis Publishers.

Mellström, G. A., J. E. Wahlberg, and H. I. Maibach. 1994. *Protective Gloves for Occupational Use.* New York: CRC Press/Lewis Publishers.

Meyer, E. 1990. *Chemistry of Hazardous Materials.* Englewood Cliffs, NJ: Prentice Hall.

Nelson, G. O. 1992. *Gas Mixtures.* New York: CRC Press/Lewis Publishers.

Ott, W. R. 1995. *Environmental Statistics and Data Analysis.* New York: CRC Press/Lewis Publishers.

Rea, W. J. 1995. *Chemical Sensitivity,* Vol. I-III. New York: CRC Press/Lewis Publishers.

Richardson, M. 1995. *Dictionary of Substance and Their Effects,* Vol. I-VI. New York: CRC Press/Lewis Publishers.

Sacarello, H. L. A. 1994. *The Comprehensive Handbook of Hazardous Materials.* New York: CRC Press/Lewis Publishers.

U.S. Department of Health and Human Services, National Institute for Occupational Safety and Health. 1994. *NIOSH Pocket Guide to Chemical Hazards.* Washington, D.C.: U.S. Government Printing Office.

MATERIAL SAFETY DATA SHEET

CHEMICAL NAME	
o-DIAMINOBENZENE	**CRC PRESS/LEWIS PUBLISHERS DISCLAIMER:** THE INFORMATION AND RECOMMENDATIONS PRESENTED HEREIN ARE BASED ON SOURCES BELIEVED TO BE RELIABLE. CRC MAKES NO REPRESENTATION ON ITS COMPLETENESS OR ACCURACY. IT IS THE RESPONSIBILITY OF THE USER TO DETERMINE THE CHEMICAL'S SUITABILITY FOR ITS INTENDED USE, THE CHEMICAL'S SAFE USE, AND THE CHEMICAL'S PROPER DISPOSAL. NO REPRESENTATIONS AND/OR WARRANTIES, EITHER EXPRESSED OR IMPLIED, OF THE MERCHANTABILITY OR FITNESS FOR A PARTICULAR PURPOSE, OR OF ANY OTHER NATURE, ARE MADE WITH RESPECT TO THE INFORMATION PROVIDED IN THIS MSDS OR TO THE CHEMICAL TO WHICH INFORMATION MAY REFER. CRC NEITHER ASSUMES NOR AUTHORIZES ANY OTHER PERSON TO ASSUME FOR IT, ANY OTHER ADDITIONAL RESPONSIBILITY OR LIABILITY FOR THE USE OF, OR RELIANCE UPON, THIS INFORMATION.

HAZARD WARNING INFORMATION

HEALTH	FIRE	REACTIVE	OTHER	DEGREE OF HAZARD	COLOR CODING	OTHER CODES
4	1	0		0 = Minimum Hazard 1 = Slight Hazard 2 = Moderate Hazard 3 = Serious Hazard 4 = Severe Hazard	HEALTH = BLUE FIRE = RED REACTIVITY = YELLOW OTHER = WHITE	OX = Oxidizer ACID = Acid ALK = Alkali COR = Corrosive ~~W~~ = Use No Water

SECTION I - GENERAL INFORMATION

Characterization	**Aromatic Amine**	RCRA Number	**None**
		EPA Class	**Not Applicable**
DOT Proper Shipping Name	***o*-Phenylenediamine, Solid**	Chemical Abstract Service (CAS) Number	**95-54-5**
DOT Hazard Class and Label Requirements	**ORM-A; St. Andrew's Cross**	DOT Emergency Guide Code	**53**
DOT Identification Number	**UN 1673**	Chemical Formula	**$C_64_3(NH_2)_2$**

Synonyms

***o*-Phenylenediamine; *o*-aminoaniline; *o*-benzenediamine; 2-aminoaniline; 1,2-benzendiamine; 1,2-diaminobenzene; 1,2-phenylenediamine.**

SECTION II - HAZARDOUS INGREDIENTS AND IDENTITY INFORMATION

Hazardous Components (specific identity)	OSHA Exposure Criteria	NIOSH Exposure Criteria	Immediately Dangerous to Life and Health (IDLH)	ACGIH Exposure Criteria
o-Diaminobenzene (derivation: By reduction of o-dinitrobenzene or nitroaniline with iron and hydrochloric acid. Purified by crystallization).	**PEL: Not Established** **STEL: Not Established**	**REL: Not Established** **STEL: Not Established**	**Not Determined**	**TLV (skin): 0.1 mg/m^3** **Suspected Human Carcinogen**

SECTION III - PHYSICAL AND CHEMICAL CHARACTERISTICS

Boiling Point	**495°F (257°C)**	Specific Gravity (H_2O = 1)	**>1**
Vapor Pressure (mm Hg)	**< 1 mm at 69°F (20°C)**	Molecular Weight	**108.2**
Vapor Density (Air = 1)	**3.71**	Melting Point	**217°F (103°C)**

Solubility

Slightly soluble in water. Very soluble in alcohol, chloroform, ether, and most organic solvents.

Appearance and Odor

Colorless, needle-like crystals that turn brownish-yellow or tan on exposure to air. Mild phenolic odor.

SECTION IV - FIRE AND EXPLOSION HAZARD DATA

Flash Point (method used)	**312°F (155°C) closed cup**	Explosive Limits in Air % by Volume	**LEL: 1.4% UEL: Not Found**
NFPA Classification	**Combustible Solid**	Autoignition Temperature	**Not Determined**

Extinguishing Media

Use dry chemical, carbon dioxide, foam, or water fog. Water or foam may cause frothing.

Special Fire Fighting Procedures

Poisonous gases are produced in fire. Structural protective clothing is permeable and may not provide adequate protection. Wear full protective clothing and self-contained breathing apparatus (SCBA). Move containers from fire area if it can be done safely. Cool exposed container with water.

Unusual Fire and Explosion Hazards

Dusts can form explosive mixtures in air. Exercise extreme caution when attempting to move or relocate containers that may have been exposed to high heat.

SECTION V - REACTIVITY DATA

Stability		Conditions to Avoid: **Normally stable chemical. Keep away from heat, sparks, flame, and other ignition sources. May be affected by exposure to light.**
Stable	Unstable	Incompatibility *(materials to avoid)*
X		**Strong oxidizers (such as chlorine, bromine, fluorine).**
Hazardous Polymerization		Conditions to Avoid: **Under normal conditions of temperature and pressure, hazardous polymerization of o-diaminobenzene is not expected to occur.**
May Occur	Will Not Occur	Hazardous Decomposition or By-products
	X	**When heated to decomposition, o-diaminobenzene will produce toxic and irritating gases and oxides of nitrogen and carbon dioxide and/or carbon monoxide.**

SECTION VI - HEALTH HAZARD DATA

Primary Route(s) of Entry:	Inhalation?	Absorption (skin)?	Ingestion?
	X	X	X

Health Hazards

INHALATION: **Irritation of the upper respiratory tract. Causes headache, cough, and may result in the development of allergic hypersensitivity asthma with inflammatory reactions of the larynx and pharynx. Systemic effects include lachrymation, salivation, ataxia, tremor, lower body temperature, increased pulse rate, and respiratory depression.**

ABSORPTION: **Can cause allergic skin reaction, hives, and dermatitis. Will pass through unbroken skin to cause toxic systemic effects. These include liver and spleen enlargement, vertigo, gastritis, jaundice, atrophy of the liver, and allergic asthma. Severe eye irritant. Can cause corneal ulcer, burning, and redness. Possible presbyopia and conjunctivitis.**

INGESTION: **Toxic systemic effects similar to those for inhalation and absorption.**

Carcinogenicity	NTP Listed?	IARC Cancer Review Group?	OSHA Regulated?	Target Organs?
Suspected Human Questioned Animal	**No**	**No**	**No**	**Respiratory system, skin, eyes, liver, spleen.**

Medical Conditions Generally Aggravated by Exposure

Respiratory conditions and skin disorders may be aggravated by exposure.

Emergency and First-aid Procedures

Eye contact: Flush immediately with water for 15 minutes (minimum) while holding upper and lower eyelids open. Seek medical attention. Skin contact: Remove all contaminated clothing. Immediately wash area with large amounts of soap and water. For inhalation: Remove the person from exposure. Provide respiratory assistance and CPR. Transfer to medical facility. If swallowed: Call the poison control center, give water to induce vomiting and seek medical attention immediately.

SECTION VII - PRECAUTIONS FOR SAFE HANDLING AND USE

Steps to be Taken in Case Material is Released or Spilled

Powders should be collected in safest manner possible and disposed of in sealed drums. Do NOT dry sweep. Use a high efficiency particulate air (HEPA) vacuum instead. Ventilate area and remove ignition sources. Restrict those not involved in cleanup from entering area. Notify appropriate authorities.

Preferred Waste Disposal Method

Consider reclamation, recycling, or destruction rather than landfill.

Precautions to be Taken in Handling and Storage

Store in tightly closed containers in a cool, dark, well-ventilated area. Personnel should be trained on the hazardous properties of o-diaminobenzene *prior* to working with the material.

Other Precautions and Warnings

Empty containers may still contain product residues and should be handled with caution.

SECTION VIII - CONTROL MEASURES AND PERSONAL PROTECTIVE EQUIPMENT

Respiratory Protection *(specify type)*

For low exposures, use a full facepiece respirator equipped with HEPA filter and dust pre-filter. For higher exposures, use a self-contained breathing apparatus (SCBA) with full facepiece operated in pressure demand or other positive pressure mode.

Ventilation

Local exhaust or general mechanical systems recommended.

Protective Gloves	Eye Protection	Other Protective Clothing
Impervious Gloves	**Chemical/Dust Goggles or Face Mask**	**Protective Apron**

Work/Hygiene Practices

Always wash hands thoroughly after using chemical; never bring food, drink, or smoking materials into vicinity of chemicals.

o-DIAMINOBENZENE

$C_6H_4(NH_2)_2$ CAS: 95-54-5

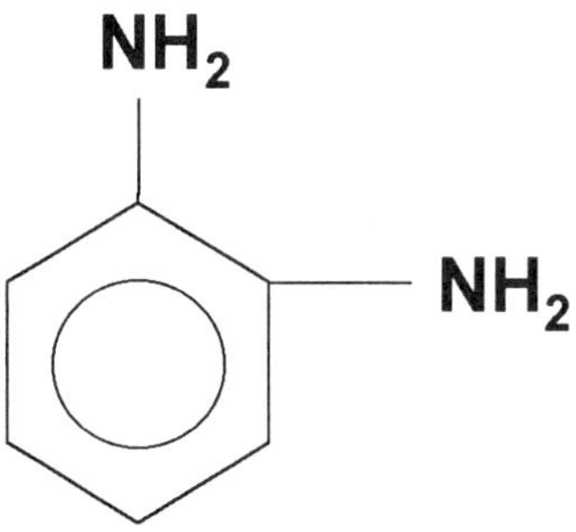

IDENTIFICATION AND TYPICAL USES

o-Diaminobenzene is a colorless to white crystalline, needle-like solid. It will turn brownish-yellow or tan on exposure to air. It has a mild, phenolic odor. It is used in the manufacture of dyes, as a photographic developing agent, in organic synthesis, and as a laboratory reagent.

RISK ASSESSMENT: HEALTH

General Assessment

o-Diaminobenzene is a moderate to highly toxic compound in humans. Of the three diamionbenzene isomers, the acute and chronic toxicity of the *"o"* or *ortho*-amine and the *"m"* or *meta*-amine is not as great as that of the *"p"* or *para*-isomer. However, it has been classified by the ACGIH as a suspected human carcinogen. It is a poison by *ingestion* and is mildly toxic by skin contact. *Absorption* through unbroken skin is possible. *Inhalation* of fine dusts will also result in toxic systemic effects.

Inhalation and ingestion will cause irritation of the throat and upper respiratory tract. Systemic effects may include lachrymation, salivation, ataxia, tremor, decreased body temperature, increased pulse rate, and respiratory depression. Ingestion will additionally result in irritation of the stomach.

Skin contact will cause irritation and possible dermatitis with inflammation and redness. Absorption will result in systemic effects similar to those caused by inhalation.

Eye contact results in burning, redness, corneal ulceration, and a possible presbyopia-like loss of elasticity in the crystalline lens of the eye.

In test animals (male rats and male and female mice), exposure produced liver tumors. Evidence of human carcinogenicity, however, is lacking.

Acute Health Effects

The following acute (short-term) health effects may occur immediately or shortly after exposure to o-diaminobenzene:

Skin: Mild to moderate irritation and possible dermatitis. Skin absorption is likely with subsequent toxic systemic effects.

Eye: Mild to severe irritation with possible damage to the cornea.

Lung: Irritation of the upper respiratory tract causing coughing and/or shortness of breath and may lead to secondary infections. There may be salivation and respiratory depression.

CNS: Unconfirmed reaction, but may cause excitement, tremor, convulsions, and other unspecified systemic effects by all routes of entry.

Chronic Health Effects

The following chronic (long-term) health effects may occur at some time after exposure to o-diaminobenzene and can last for months or even years:

Cancer Hazards: According to the information presented in the references, o-diaminobenzene has been shown to cause tumors in test animals. Its carcinogenicity in humans is questioned. Mutation data have been reported for this compound. Until further studies are performed, it is suggested that o-diaminobenzene be handled with extreme caution.

Reproductive Hazard: According to the information presented in the references, o-diaminobenzene has not been adequately tested for its ability to adversely affect reproduction.

ʊ *Recommended Risk-Reduction Measures*

Unless a less toxic chemical can be substituted for a hazardous substance, *engineering controls* are the most effective method of reducing exposures. The best protection is to enclose operations and/or provide local exhaust ventilation at the site of chemical release. While not always operationally feasible, isolating operations can also reduce exposure. All metal equipment used with this chemical should be grounded and bonded. Explosion-proof design of equipment is recommended.

Using respiratory protection is less effective than the controls mentioned above, but is still advisable whenever working with or around o-diaminobenzene. For low or relatively infrequent exposures, use a full facepiece respirator equipped with a high efficiency particulate air (HEPA) filter and a dust pre-filter. For higher concentrations, use of a self-contained breathing apparatus (SCBA) with full facepiece and pressure demand is the recommended respiratory protection of choice. If a full facepiece is not available, then chemical goggles should be worn to protect the eyes. Whenever a chemical splash hazard exists, a face shield and apron should be worn. To prevent hand and skin exposures, impervious gloves should be used. Glove manufacturers should be contacted and permeation studies obtained *before* final glove selection are made.

Administrative controls should also be in place to minimize the potential for human exposures. These may include written procedures or policies which specify the methods and techniques that will be practiced whenever personnel are to work with or around o-diaminobenzene.

All personnel should receive *training* on the use, hazards, protective measures, emergency actions, and other precautions per 29 CFR 1910.1200 (Hazard Communication), prior to the first assignment in an area where o-diaminobenzene is used or stored.

Before beginning employment and at regular intervals thereafter (e.g. annually), the following medical tests are recommended:

☑ Lung function tests (establish a baseline).

If symptoms develop or overexposure is suspected, the following medical tests may be useful:

☑ Lung function tests.

☑ Liver function tests.

☑ Consider chest X-ray after acute overexposure.

☑ Evaluation by a qualified allergist with careful consideration of exposure history and special tests (may help diagnose allergy).

Any evaluation should include a careful medical history of past and present symptoms with an examination. However, medical tests that evaluate existing damage are not a substitute for controlling exposure. Also, since smoking can cause heart diseases, emphysema, and numerous other respiratory disorders, smokers may be affected more rapidly than non-smokers to the same exposure conditions. Prudent risk assessment and management requires careful consideration of *all* possible risk factors that may be causing the appearance of symptoms in exposed workers.

Other methods to reduce exposure include:

☑ Where possible, enclose operations and use local exhaust ventilation at the site of chemical release. If local exhaust ventilation or enclosure is not used, respiratory protection should be mandatory.

☑ Always ensure that proper protective clothing is worn when using chemical substances.

☑ Wash thoroughly immediately after exposure to o-diaminobenzene and at the end of the work shift or before eating, drinking, or smoking.

☑ Hazard warning information should be posted in the work area. In addition, as part of an on-going education and training program, all information on the health and safety hazards of o-diaminobenzene should be communicated to all exposed and potentially exposed workers.

☑ Eye wash stations should be provided in work areas. If the potential for whole body exposure exists, then safety showers should also be provided.

RISK ASSESSMENT: ENVIRONMENT

General Assessment

The environment is at risk of exposure during intended use, transportation, storage, disposal, or destruction of o-diaminobenzene. In almost every scenario, the threat of environmental exposure is contingent upon the proper handling of the chemical substance. Accidental spills, large or small, can result in fire, explosion, and

possible contamination of the surrounding environmental mediums (water, soil, and air).

o-Diaminobenzene should always be handled with extreme caution. It is incompatible with strong oxidizers and its dusts can form explosive mixtures in air. These characteristics require special consideration during any emergency situation involving a leak or spill of o-diaminobenzene.

o-Diaminobenzene can enter the environment through its use as a fungicide, as well as through industrial discharges and spills.

Acute Ecological Effects

Acute (short-term) toxic effects may include the death of animals, birds, or fish and death or low growth rate in plants. Acute effects are seen 2 to 4 days after animals or plants are exposed to o-diaminobenzene. Insufficient data are available to evaluate or predict the short-term effects of o-diaminobenzene to aquatic life, plants, birds, or land animals.

Chronic Ecological Effects

Chronic toxic effects may include shortened life span, reproductive problems, lower fertility, and changes in appearance or behavior in exposed animals. These effects can be seen long after first exposure(s) to toxic chemicals. Insufficient data are available to evaluate or predict the long-term effects of o-diaminobenzene to aquatic life, plants, birds, or land animals.

Water Solubility

o-Diaminobenzene is slightly soluble in water. Concentrations of 1 to 100 milligrams will mix with a liter of water.

Persistence in the Environment

o-Diaminobenzene is moderately persistent in water, with a half-life of between 20 to 200 days. The half-life of a pollutant is the amount of time it takes for one half of the chemical to be degraded.

Bioaccumulation in Aquatic Organisms

Some substances increase in concentration, or *bioaccumulate,* in living organisms as they breathe contaminated air, drink contaminated water, or eat contaminated food. These chemicals can become concentrated in the tissues and organs of animals as well as humans.

The concentration of o-diaminobenzene found in fish tissues is expected to be about the same as the average concentration of o-diaminobenzene in the water from which the fish was taken.

Recommended Risk-Reduction Measures

Proper training of all transporters will reduce the likelihood of a mishap or accident resulting in a leak or spill to the environment. The correct use of labeling on all containers, trucks, and rail cars will enable emergency responders to react properly and quickly to any disaster thereby reducing the potential risk to the environment and to personnel.

Storage of o-diaminobenzene should be segregated from incompatible chemicals to minimize the risk of cross-contamination or contact. Containers should be protected from physical damage. Storage buildings should be equipped with the proper fire protection equipment (alarms, sprinklers, extinguishers, etc.).

If a spill or leak to the environment has occurred, fire department, emergency response and/or hazardous materials spill personnel should be notified immediately. Cleanup should be attempted only by those trained in proper spill containment procedures. Contaminated soils should be removed for incineration and replaced with clean soil. If o-diaminobenzene should contact the water table, aquifer, or navigable waterway, remediation activities should be prompt to ensure protection of aquatic sediments. When such spills occur, the local and/or state emergency response authorities must be notified. A comprehensive emergency response or disaster preparedness/recovery plan should be in place prior to any operations involving the use, transportation, storage, or disposal of o-diaminobenzene. If o-diaminobenzene is spilled or leaked, the following specific steps are recommended:

- ☑ Restrict persons not wearing protective clothing from area of spill or leak until cleanup of the area is complete.
- ☑ Ventilate area and remove all ignition sources.
- ☑ Collect powders in safest manner possible and dispose of in sealed drums. Do NOT dry sweep (creates airborne dusts). Use a vacuum equipped with a high efficiency particulate air (HEPA) filter attachment.
- ☑ It may be necessary to dispose of o-diaminobenzene as a hazardous waste. The responsible state agency or the regional office of the federal

Environmental Protection Agency (EPA) should be contacted for specific recommendations.

RISK ASSESSMENT: BUSINESS

General Assessment

Accidents or mishaps involving o-diaminobenzene can present a significant threat to business operations. The loss or damage of equipment or facilities can significantly affect fiscal viability. Lawsuits that may result from personnel injury/death, public exposures, and/or environmental contamination will also require a serious expenditure of resources. Media attention surrounding an injury, death, or environmental damage can also result in a loss of profits and loss of current as well as future business. Always remember that anytime the term "carcinogen" or "cancer" are used, public emotion, hysteria, and ignorance can run equally high. This should be considered during the development of any public relations policies.

Recommended Risk-Reduction Measures

Company attorneys, safety and health professionals, and environmental specialists should be involved in the development of any such procedures. A company official should be pre-designated as a public relations officer with specific training in dealing with the press. Corporate plans and policies should be developed, approved, and implemented long before any need for such arises.

REFERENCES

Ahlbom, A. 1993. *Biostatistics for Epidemiologists.* New York: CRC Press/Lewis Publishers.

American Conference of Governmental Industrial Hygienists. 1988. *Documentation of the Threshold Limit Values and Biological Exposure Indices,* 5th Edition (with updates). Cincinnati: ACGIH

Anthony, C. P., and N. J. Kolthoff. 1971. *Textbook of Anatomy and Physiology.* St. Louis: C. V. Mosby Company.

Asante-Duah, D. K. 1993. *Hazardous Waste Risk Assessment.* New York: CRC Press/Lewis Publishers.

Bloom, A. D. and F. J. DeSerres. 1995. *Ecotoxicity and Human Health.* New York: CRC Press/Lewis Publishers.

Brusick, D. J. 1994. *Methods for Genetic Risk Assessment.* New York: CRC Press/Lewis Publishers.

Bueche, J. R. 1972. *Principles of Physics,* 2nd Edition. New York: McGraw-Hill.

Burrell, R., D. K. Flaherty, and L. J. Sauers. 1992. *Toxicology of the Immune System.* New York: Van Nostrand Reinhold.

Calabrese, E. J., E. M. Kenyon. *Air Toxics and Risk Assessment.* New York: CRC Press/Lewis Publishers.

Cockerman, L. G., B. S. Shane. 1994. *Basic Environmental Toxicology.* New York: CRC Press/Lewis Publishers.

Cothern, C. R., N. P. Ross. 1994. *Environmental Statistics, Assessment, and Forecasting.* New York: CRC Press/Lewis Publishers.

Cothern, C. R. 1995. *Handbook for Environmental Risk Decisionmaking.* New York: CRC Press/Lewis Publishers.

Gardner, R. W. 1994. *Chemical Intolerance.* New York: CRC Press/Lewis Publishers.

Hallenbeck, W. H. 1993. *Quantitative Risk Assessment for Environmental and Occupational Health,* 2nd Edition. New York: CRC Press/Lewis Publishers.

Hodgson, E., R. B. Mailman, and Janice E. Chambers. 1988. *Dictionary of Toxicology.* London: The MacMillian Company.

Howard, P. H. 1995. *Handbook of Environmental Fate and Exposures Data for Organic Chemicals,* Vol. I-IV. New York: CRC Press/Lewis Publishers.

Jones, F. E. 1994. *Toxic Organic Vapors in the Workplace.* New York: CRC Press/Lewis Publishers.

Keith, L. H. and M. M. Walker. 1995. *Handbook of Air Toxics.* New York: CRC Press/Lewis Publishers.

Landis, W. G., M. H. Yu. 1995. *Introduction to Environmental Toxicology.* New York: CRC Press/Lewis Publishers.

Langerman, N. 1994 *Precautionary Labels for Chemical Containers.* New York: CRC Press/Lewis Publishers.

Lauwerys, R. R., P. Hoet. 1993. *Industrial Chemical Exposure.* New York: CRC Press/Lewis Publishers.

Lewis, R. J., Sr. 1993. *Hawley's Condensed Chemical Dictionary,* 12th Edition. New York: Van Nostrand Reinhold.

Lewis, R. J., Sr. 1992. *Sax's Dangerous Properties of Industrial Materials,* Volumes 1, 2, and 3, 8th Edition. New York: Van Nostrand Reinhold.

Manahan, S. E. 1992. *Toxicological Chemistry,* 2nd Edition. New York: CRC Press/Lewis Publishers.

Mellström, G. A., J. E. Wahlberg, and H. I. Maibach. 1994. *Protective Gloves for Occupational Use.* New York: CRC Press/Lewis Publishers.

Meyer, E. 1990. *Chemistry of Hazardous Materials.* Englewood Cliffs, NJ: Prentice Hall.

Ott, W. R. 1995. *Environmental Statistics and Data Analysis.* New York: CRC Press/Lewis Publishers.

Rea, W. J. 1995. *Chemical Sensitivity,* Vol. I-III. New York: CRC Press/Lewis Publishers.

Richardson, M. 1995. *Dictionary of Substance and Their Effects,* Vol. I-VI. New York: CRC Press/Lewis Publishers.

Sacarello, H. L. A. 1994. *The Comprehensive Handbook of Hazardous Materials.* New York: CRC Press/Lewis Publishers.

U.S. Department of Health and Human Services, National Institute for Occupational Safety and Health. 1994. *NIOSH Pocket Guide to Chemical Hazards.* Washington, D.C.: U.S. Government Printing Office.

MATERIAL SAFETY DATA SHEET

CHEMICAL NAME	
p-DIAMINOBENZENE	**CRC PRESS/LEWIS PUBLISHERS DISCLAIMER:** THE INFORMATION AND RECOMMENDATIONS PRESENTED HEREIN ARE BASED ON SOURCES BELIEVED TO BE RELIABLE. CRC MAKES NO REPRESENTATION ON ITS COMPLETENESS OR ACCURACY. IT IS THE RESPONSIBILITY OF THE USER TO DETERMINE THE CHEMICAL'S SUITABILITY FOR ITS INTENDED USE, THE CHEMICAL'S SAFE USE, AND THE CHEMICAL'S PROPER DISPOSAL. NO REPRESENTATIONS AND/OR WARRANTIES, EITHER EXPRESSED OR IMPLIED, OF THE MERCHANTABILITY OR FITNESS FOR A PARTICULAR PURPOSE, OR OF ANY OTHER NATURE, ARE MADE WITH RESPECT TO THE INFORMATION PROVIDED IN THIS MSDS OR TO THE CHEMICAL TO WHICH INFORMATION MAY REFER. CRC NEITHER ASSUMES NOR AUTHORIZES ANY OTHER PERSON TO ASSUME FOR IT, ANY OTHER ADDITIONAL RESPONSIBILITY OR LIABILITY FOR THE USE OF, OR RELIANCE UPON, THIS INFORMATION.

HAZARD WARNING INFORMATION

HEALTH	FIRE	REACTIVE	OTHER	DEGREE OF HAZARD	COLOR CODING	OTHER CODES
4	1	0		0 = Minimum Hazard 1 = Slight Hazard 2 = Moderate Hazard 3 = Serious Hazard 4 = Severe Hazard	HEALTH = BLUE FIRE = RED REACTIVITY = YELLOW OTHER = WHITE	OX = Oxidizer ACID = Acid ALK = Alkali COR = Corrosive W = Use No Water

SECTION I - GENERAL INFORMATION

Characterization	RCRA Number	EPA Class
Aromatic Amine	**None**	**Not Applicable**

DOT Proper Shipping Name ***p*-Phenylenediamine, Solid**	Chemical Abstract Service (CAS) Number **106-50-3**
DOT Hazard Class and Label Requirements **ORM-A; St. Andrew's Cross**	DOT Emergency Guide Code **53**
DOT Identification Number **UN 1673**	Chemical Formula **$C_64_3(NH_2)_2$**

Synonyms

***p*-Phenylenediamine; *p*-aminoaniline; *p*-benzenediamine; 4-aminoaniline; 1,4-benzendiamine; 1,4-diaminobenzene; 1,4-phenylenediamine.**

SECTION II - HAZARDOUS INGREDIENTS AND IDENTITY INFORMATION

Hazardous Components (specific identity)	OSHA Exposure Criteria	NIOSH Exposure Criteria	Immediately Dangerous to Life and Health (IDLH)	ACGIH Exposure Criteria
p-Diaminobenzene (derivation: By reduction of p-dinitrobenzene or nitroaniline with iron and hydrochloric acid. Purified by crystallization).	**PEL (skin): 0.1 mg/m^3** **STEL: Not Established**	**REL (skin): 0.1 mg/m^3** **STEL: Not Established**	**25 mg/m^3**	**TLV (skin): 0.1 mg/m^3** **STEL: Not Established**

SECTION III - PHYSICAL AND CHEMICAL CHARACTERISTICS

Boiling Point **512°F (267°C)**	Specific Gravity (H_2O = 1) **>1**
Vapor Pressure (mm Hg) **< 1 mm at 69°F (20°C)**	Molecular Weight **108.2**
Vapor Density (Air = 1) **3.71**	Melting Point **268°F (141°C)**

Solubility

Soluble in water (4.7% at 75°F). Soluble in alcohol, ether and chloroform.

Appearance and Odor

Colorless to white crystalline (sand-like) solid. Turns red, brown, or faintly black on exposure to air. Mild phenolic odor.

SECTION IV - FIRE AND EXPLOSION HAZARD DATA

Flash Point (method used) **312°F (155°C) closed cup**	Explosive Limits in Air % by Volume **LEL: 1.4% UEL: Not Found**
NFPA Classification **Combustible Solid**	Autoignition Temperature **Not Determined**

Extinguishing Media

Use dry chemical, carbon dioxide, foam, or water fog. Water or foam may cause frothing.

Special Fire Fighting Procedures

Poisonous gases are produced in fire. Structural protective clothing is permeable and may not provide adequate protection. Wear full protective clothing and self-contained breathing apparatus (SCBA). Move containers from fire area if it can be done safely. Cool exposed container with water.

Unusual Fire and Explosion Hazards

Dusts can form explosive mixtures in air. Exercise extreme caution when attempting to move or relocate containers that may have been exposed to high heat.

SECTION V - REACTIVITY DATA

Stability		Conditions to Avoid Normally stable chemical. Keep away from heat, sparks, flame, and other ignition sources. May be affected be exposure to light.
Stable X	Unstable	Incompatibility *(materials to avoid)* Strong oxidizers (such as chlorine, bromine, fluorine).
Hazardous Polymerization		Conditions to Avoid Under normal conditions of temperature and pressure, hazardous polymerization of p-diaminobenzene is not expected to occur.
May Occur	Will Not Occur X	Hazardous Decomposition or By-products When heated to decomposition, p-diaminobenzene will produce toxic and irritating gases and oxides of nitrogen and carbon dioxide and/or carbon monoxide.

SECTION VI - HEALTH HAZARD DATA

Primary Route(s) of Entry:	Inhalation?	Absorption (skin)?	Ingestion?
	X	X	X

Health Hazards

INHALATION: Irritation of the upper respiratory tract. Causes headache, cough, and may result in the development of allergic hypersensitivity asthma with inflammatory reactions of the larynx and pharynx. Systemic effects include lachrymation, salivation, ataxia, tremor, lower body temperature, increased pulse rate, and respiratory depression.

ABSORPTION: Can cause allergic skin reaction, hives, and dermatitis. Will pass through unbroken skin to cause toxic systemic effects. These include liver and spleen enlargement, vertigo, gastritis, jaundice, atrophy of the liver, and allergic asthma. Severe eye irritant. Can cause corneal ulcer, burning, and redness. Possible presbyopia and conjunctivitis.

INGESTION: Toxic systemic effects similar to those for inhalation and absorption.

Carcinogenicity	NTP Listed?	IARC Cancer Review Group?	OSHA Regulated?	Target Organs?
Questioned Human Questioned Animal	5th Annual Report	Group 3	29 CFR 1910.1000 Table Z-1	Respiratory system, skin, eyes, liver, spleen.

Medical Conditions Generally Aggravated by Exposure

Respiratory conditions and skin disorders may be aggravated by exposure.

Emergency and First-aid Procedures

Eye contact: Flush immediately with water for 15 minutes (minimum) while holding upper and lower eyelids open. Seek medical attention. Skin contact: Remove all contaminated clothing. Immediately wash area with large amounts of soap and water. For inhalation: Remove the person from exposure. Provide respiratory assistance and CPR. Transfer to medical facility. If swallowed: Call the poison control center, give water to induce vomiting and seek medical attention immediately.

SECTION VII - PRECAUTIONS FOR SAFE HANDLING AND USE

Steps to be Taken in Case Material is Released or Spilled

Powders should be collected in safest manner possible and disposed of in sealed drums. Do NOT dry sweep. Use a high efficiency particulate air (HEPA) vacuum instead. Ventilate area and remove ignition sources. Restrict those not involved in cleanup from entering area. Notify appropriate authorities.

Preferred Waste Disposal Method

Consider reclamation, recycling, or destruction rather than landfill.

Precautions to be Taken in Handling and Storage

Store in tightly closed containers in a cool, dark, well-ventilated area. Personnel should be trained on the hazardous properties of p-diaminobenzene *prior* to working with the material.

Other Precautions and Warnings

Empty containers may still contain product residues and should be handled with caution.

SECTION VIII - CONTROL MEASURES AND PERSONAL PROTECTIVE EQUIPMENT

Respiratory Protection *(specify type)*

For low exposures, use a full facepiece respirator equipped with HEPA filter and dust pre-filter. For higher exposures, use a self-contained breathing apparatus (SCBA) with full facepiece operated in pressure demand or other positive pressure mode.

Ventilation

Local exhaust or general mechanical systems recommended.

Protective Gloves	Eye Protection	Other Protective Clothing
Impervious Gloves	Chemical/Dust Goggles or Face Mask	Protective Apron

Work/Hygiene Practices

Always wash hands thoroughly after using chemical; never bring food, drink, or smoking materials into vicinity of chemicals.

p-DIAMINOBENZENE

$C_6H_4(NH_2)_2$ CAS: 106-50-3

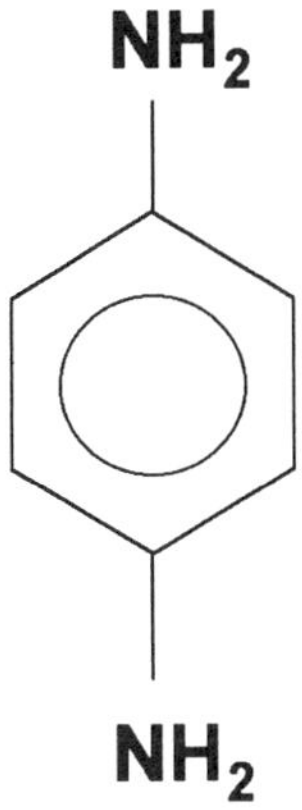

IDENTIFICATION AND TYPICAL USES

p-Diaminobenzene is a colorless to white crystalline solid. It will turn red, brown, or faintly black on exposure to air. It has a mild, phenolic odor. It is used in the dyeing of hair and fur, in the manufacture of azo dyes, in accelerating vulcanization of rubber, and in antioxidants.

RISK ASSESSMENT: HEALTH

General Assessment

p-Diaminobenzene is a moderate to highly toxic compound in humans. Of the three diamionbenzene isomers, the acute and chronic toxicity of the "p" or *para*-amine is greater than that of the *ortho*- or *meta*-isomers. It is a poison by *ingestion* and is mildly toxic by skin contact. *Absorption* through unbroken skin is possible. *Inhalation* of fine dusts will also result in toxic systemic effects.

Inhalation and ingestion will cause irritation of the larynx, pharynx, and may cause bronchial asthma. Reports of allergic hypersensitivity have been recorded in the references. Systemic effects may include lachrymation, salivation, ataxia, tremor, decreased body temperature, increased pulse rate, and respiratory depression.

Skin contact will cause irritation and possible dermatitis with inflammation and redness. Dermal absorption will result in systemic effects similar to those caused by inhalation and may also result in the enlargement of the liver and spleen, vertigo, gastritis, jaundice, and atrophy of the liver. Eye contact results in burning, redness, corneal ulceration, and a possible presbyopia-like loss of elasticity in the crystalline lens of the eye.

Acute Health Effects

The following acute (short-term) health effects may occur immediately or shortly after exposure to p-diaminobenzene:

Skin: Mild to moderate irritation and possible dermatitis. Skin absorption is likely with subsequent toxic systemic effects.

Eye: Mild to severe irritation with possible damage to the cornea.

Lung: Irritation of the upper respiratory tract causing coughing and/or shortness of breath and may lead to secondary infections. Allergic hypersensitivity, bronchial asthma, and throat problems likely to occur.

CNS: Unconfirmed reaction, but may cause vertigo, ataxia, dizziness, and other unspecified systemic effects by all routes of entry.

Chronic Health Effects

The following chronic (long-term) health effects may occur at some time after exposure to p-diaminobenzene and can last for months or even years:

Cancer Hazards: According to the information presented in the references, p-diaminobenzene has been shown to cause tumors in test animals. Its carcinogenicity in humans is questioned. However, the data in the references are conflicting and inconclusive in this regard. It is a mutagen. Until further studies are performed, it is suggested that p-diaminobenzene be handled with extreme caution.

Reproductive Hazard: According to the information presented in the references, p-diaminobenzene has not been adequately tested for its ability to adversely affect reproduction.

Other Chronic Effects: Repeated low level exposures may damage the liver and can be fatal.

Recommended Risk-Reduction Measures

Unless a less toxic chemical can be substituted for a hazardous substance, *engineering controls* are the most effective method of reducing exposures. The best protection is to enclose operations and/or provide local exhaust ventilation at the site of chemical release. While not always operationally feasible, isolating operations can also reduce exposure. All metal equipment used with this chemical should be grounded and bonded. Explosion-proof design of equipment is recommended.

Using respiratory protection is less effective than the controls mentioned above, but is still advisable whenever working with or around p-diaminobenzene. For low or relatively infrequent exposures, use a full facepiece respirator equipped with a high efficiency particulate air (HEPA) filter and a dust pre-filter. For higher concentrations, use of a self-contained breathing apparatus (SCBA) with full facepiece and pressure demand is the recommended respiratory protection of choice. If a full facepiece is not available, then chemical goggles should be worn to protect the eyes. Whenever a chemical splash hazard exists, a face shield and apron should be worn. To prevent hand and skin exposures, impervious gloves should be worn.

Administrative controls should also be in place to minimize the potential for human exposures. These may include written procedures or policies which specify the methods and techniques that will be practiced whenever personnel are to work with or around p-diaminobenzene.

All personnel should receive *training* on the use, hazards, protective measures, emergency actions, and other precautions per 29 CFR 1910.1200 (Hazard Communication), prior to the first assignment in an area where p-diaminobenzene is used or stored. Before beginning employment and at regular intervals thereafter (e.g. annually), the following medical tests are recommended:

☑ Lung function tests (establish a baseline).

If symptoms develop or overexposure is suspected, the following medical tests may be useful:

☑ Lung function tests.

☑ Liver function tests.

☑ Consider chest X-ray after acute overexposure.

☑ Evaluation by a qualified allergist with careful consideration of exposure history and special tests (may help diagnose allergy).

Any evaluation should include a careful medical history of past and present symptoms with an examination. However, medical tests that evaluate existing damage are not a substitute for controlling exposure. Also, since smoking can cause heart disease, emphysema, and numerous other respiratory disorders, smokers may be affected more rapidly than non-smokers to the same exposure conditions. Prudent risk assessment and management requires careful consideration of *all* possible risk factors that may be causing the appearance of symptoms in exposed workers.

Other methods to reduce exposure include:

☑ Where possible, enclose operations and use local exhaust ventilation at the site of chemical release. If local exhaust ventilation or enclosure is not used, respiratory protection should be mandatory.

☑ Always ensure that proper protective clothing is worn when using chemical substances.

☑ Wash thoroughly immediately after exposure to p-diaminobenzene and at the end of the work shift or before eating, drinking, or smoking.

☑ Hazard warning information should be posted in the work area. In addition, as part of an on-going education and training program, all information on the health and safety hazards of p-diaminobenzene should be communicated to all exposed workers.

☑ Eye wash stations should be provided in work areas. If the potential for whole body exposure exists, then safety showers should also be provided.

RISK ASSESSMENT: ENVIRONMENT

General Assessment

The environment is at risk of exposure during intended use, transportation, storage, disposal, or destruction of p-diaminobenzene. In almost every scenario, the threat of environmental exposure is contingent upon the proper handling of the chemical substance. Accidental spills, large or small, can result in fire, explosion, and possible contamination of the surrounding environmental mediums (water, soil, and air).

p-Diaminobenzene should always be handled with extreme caution. It is incompatible with strong oxidizers and its dusts can form explosive mixtures in air. These characteristics require special consideration during any emergency situation involving a leak or spill of p-diaminobenzene.

p-Diaminobenzene can enter the environment through industrial discharges and spills.

Acute Ecological Effects

Acute (short-term) toxic effects may include the death of animals, birds, or fish and death or low growth rate in plants. Acute effects are seen 2 to 4 days after animals or plants are exposed to p-diaminobenzene. Insufficient data are available to evaluate or predict the short-term effects of p-diaminobenzene to aquatic life, plants, birds, or land animals.

Chronic Ecological Effects

Chronic toxic effects may include shortened life span, reproductive problems, lower fertility, and changes in appearance or behavior in exposed animals. These effects can be seen long after first exposure(s) to toxic chemicals. Insufficient data are available to evaluate or predict the long-term effects of p-diaminobenzene to aquatic life, plants, birds, or land animals.

Water Solubility

p-Diaminobenzene is slightly soluble in water. Concentrations of 1 to 100 milligrams will mix with a liter of water.

Persistence in the Environment

p-Diaminobenzene is moderately persistent in water, with a half-life of between 20 to 200 days. The half-life of a pollutant is the amount of time it takes for one half of the chemical to be degraded.

Bioaccumulation in Aquatic Organisms

Some substances increase in concentration, or *bioaccumulate,* in living organisms as they breathe contaminated air, drink contaminated water, or eat contaminated food. These chemicals can become concentrated in the tissues and internal organs of animals as well as humans.

The concentration of p-diaminobenzene found in fish tissues is expected to be about the same as the average concentration of p-diaminobenzene in the water from which the fish was taken.

Recommended Risk-Reduction Measures

Proper training of all transporters will reduce the likelihood of a mishap or accident resulting in a leak or spill to the environment. The correct use of labeling on all containers, trucks, and rail cars will enable emergency responders to react properly and quickly to any disaster thereby reducing the potential risk to the environment and to personnel.

Storage of p-diaminobenzene should be segregated from incompatible chemicals to minimize the risk of cross-contamination or contact. Containers should be protected from physical damage. Storage buildings should be equipped with the proper fire protection equipment (alarms, sprinklers, extinguishers, etc.).

If a spill or leak to the environment has occurred, fire department, emergency response and/or hazardous materials spill personnel should be notified immediately. Cleanup should be attempted only by those trained in proper spill containment procedures. Contaminated soils should be removed for incineration and replaced with clean soil. If p-diaminobenzene should contact the water table, aquifer, or navigable waterway, remediation activities should be prompt to ensure protection of aquatic sediments. When such spills occur, the local and/or state emergency response authorities must be notified. A comprehensive emergency response or disaster preparedness/recovery plan should be in place prior to any operations involving the use, transportation, storage, or disposal of p-diaminobenzene. If p-diaminobenzene is spilled or leaked, the following specific steps are recommended:

- ☑ Restrict persons not wearing protective clothing from area of spill or leak until cleanup is complete.
- ☑ Ventilate area of spill or leak.
- ☑ Remove all ignition sources.
- ☑ Collect powders in safest manner possible and dispose of in sealed drums. Do NOT dry sweep (creates airborne dusts). Use a vacuum equipped with a high efficiency particulate air (HEPA) filter cartridge or attachment.

☑ It may be necessary to dispose of p-diaminobenzene as a hazardous waste. The responsible state agency or the regional office of the federal Environmental Protection Agency (EPA) should be contacted for specific recommendations.

RISK ASSESSMENT: BUSINESS

General Assessment

Accidents or mishaps involving p-diaminobenzene can present a significant threat to business operations. The loss or damage of equipment or facilities can significantly affect fiscal viability. Lawsuits that may result from personnel injury/death, public exposures, and/or environmental contamination will also require a serious expenditure of resources. Media attention surrounding an injury, death, or environmental damage can also result in a loss of profits and loss of current as well as future business. Always remember that anytime the term "carcinogen" or "cancer" are used, public emotion, hysteria, and ignorance can run equally high. This should be considered during the development of any public relations policies.

Recommended Risk-Reduction Measures

Company attorneys, safety and health professionals, and environmental specialists should be involved in the development of any procedures dealing with safety or emergency response. A company official should be pre-designated as a public relations officer with specific training in dealing with the press. Corporate plans and policies should be developed, approved, and implemented long before any need for such arises.

REFERENCES

Ahlbom, A. 1993. *Biostatistics for Epidemiologists.* New York: CRC Press/Lewis Publishers.

American Conference of Governmental Industrial Hygienists. 1988. *Documentation of the Threshold Limit Values and Biological Exposure Indices,* 5th Edition (with updates). Cincinnati: ACGIH

Anthony, C. P., and N. J. Kolthoff. 1971. *Textbook of Anatomy and Physiology.* St. Louis: C. V. Mosby Company.

Asante-Duah, D. K. 1993. *Hazardous Waste Risk Assessment.* New York: CRC Press/Lewis Publishers.

Bloom, A. D. and F. J. DeSerres. 1995. *Ecotoxicity and Human Health.* New York: CRC Press/Lewis Publishers.

Brusick, D. J. 1994. *Methods for Generic Risk Assessment.* New York: CRC Press/Lewis Publishers.

Bueche, J. R. 1972. *Principles of Physics,* 2nd Edition. New York: McGraw-Hill.

Burrell, R., D. K. Flaherty, and L. J. Sauers. 1992. *Toxicology of the Immune System.* New York: Van Nostrand Reinhold.

Cockerman, L. G., B. S. Shane. 1994. *Basic Environmental Toxicology.* New York: CRC Press/Lewis Publishers.

Cothern, C. R. 1995. *Handbook for Environmental Risk Decisionmaking.* New York: CRC Press/Lewis Publishers.

Hallenbeck, W. H. 1993. *Quantitative Risk Assessment for Environmental and Occupational Health,* 2nd Edition. New York: CRC Press/Lewis Publishers.

Hodgson, E., R. B. Mailman, and Janice E. Chambers. 1988. *Dictionary of Toxicology.* London: The MacMillian Company.

Howard, P. H. 1995. *Handbook of Environmental Fate and Exposures Data for Organic Chemicals,* Vol. I-IV. New York: CRC Press/Lewis Publishers.

Jones, F. E. 1994. *Toxic Organic Vapors in the Workplace.* New York: CRC Press/Lewis Publishers.

Keith, L. H. and M. M. Walker. 1995. *Handbook of Air Toxics.* New York: CRC Press/Lewis Publishers.

Landis, W. G., M. H. Yu. 1995. *Introduction to Environmental Toxicology.* New York: CRC Press/Lewis Publishers.

Langerman, N. 1994 *Precautionary Labels for Chemical Containers.* New York: CRC Press/Lewis Publishers.

Lauwerys, R. R., P. Hoet. 1993. *Industrial Chemical Exposure.* New York: CRC Press/Lewis Publishers.

Lewis, R. J., Sr. 1993. *Hawley's Condensed Chemical Dictionary,* 12th Edition. New York: Van Nostrand Reinhold.

Lewis, R. J., Sr. 1992. *Sax's Dangerous Properties of Industrial Materials,* Volumes 1, 2, and 3, 8th Edition. New York: Van Nostrand Reinhold.

Manahan, S. E. 1992. *Toxicological Chemistry,* 2nd Edition. New York: CRC Press/Lewis Publishers.

Mellström, G. A., J. E. Wahlberg, and H. I. Maibach. 1994. *Protective Gloves for Occupational Use.* New York: CRC Press/Lewis Publishers.

Meyer, E. 1990. *Chemistry of Hazardous Materials.* Englewood Cliffs, NJ: Prentice Hall.

Nelson, G. O. 1992. *Gas Mixtures.* New York: CRC Press/Lewis Publishers.

Ott, W. R. 1995. *Environmental Statistics and Data Analysis.* New York: CRC Press/Lewis Publishers.

Rea, W. J. 1995. *Chemical Sensitivity,* Vol. I-III. New York: CRC Press/Lewis Publishers.

Richardson, M. 1995. *Dictionary of Substance and Their Effects,* Vol. I-VI. New York: CRC Press/Lewis Publishers.

Sacarello, H. L. A. 1994. *The Comprehensive Handbook of Hazardous Materials.* New York: CRC Press/Lewis Publishers.

U.S. Department of Health and Human Services, National Institute for Occupational Safety and Health. 1994. *NIOSH Pocket Guide to Chemical Hazards.* Washington, D.C.: U.S. Government Printing Office.

MATERIAL SAFETY DATA SHEET

CHEMICAL NAME	
DIAZINON®	**CRC PRESS/LEWIS PUBLISHERS DISCLAIMER:** THE INFORMATION AND RECOMMENDATIONS PRESENTED HEREIN ARE BASED ON SOURCES BELIEVED TO BE RELIABLE. CRC MAKES NO REPRESENTATION ON ITS COMPLETENESS OR ACCURACY. IT IS THE RESPONSIBILITY OF THE USER TO DETERMINE THE CHEMICAL'S SUITABILITY FOR ITS INTENDED USE, THE CHEMICAL'S SAFE USE, AND THE CHEMICAL'S PROPER DISPOSAL. NO REPRESENTATIONS AND/OR WARRANTIES, EITHER EXPRESSED OR IMPLIED, OF THE MERCHANTABILITY OR FITNESS FOR A PARTICULAR PURPOSE, OR OF ANY OTHER NATURE, ARE MADE WITH RESPECT TO THE INFORMATION PROVIDED IN THIS MSDS OR TO THE CHEMICAL TO WHICH INFORMATION MAY REFER. CRC NEITHER ASSUMES NOR AUTHORIZES ANY OTHER PERSON TO ASSUME FOR IT, ANY OTHER ADDITIONAL RESPONSIBILITY OR LIABILITY FOR THE USE OF, OR RELIANCE UPON, THIS INFORMATION.

HAZARD WARNING INFORMATION

HEALTH	FIRE	REACTIVE	OTHER	DEGREE OF HAZARD	COLOR CODING	OTHER CODES
N/A	N/A	N/A		0 = Minimum Hazard 1 = Slight Hazard 2 = Moderate Hazard 3 = Serious Hazard 4 = Severe Hazard	HEALTH = BLUE FIRE = RED REACTIVITY = YELLOW OTHER = WHITE	OX = Oxidizer ACID = Acid ALK = Alkali COR = Corrosive ~~W~~ = Use No Water

SECTION I - GENERAL INFORMATION

Characterization **Pesticide**	RCRA Number **None**	EPA Class **Not Applicable**
DOT Proper Shipping Name **Diazinon**	Chemical Abstract Service (CAS) Number **333-41-5**	
DOT Hazard Class and Label Requirements **ORM-A**	DOT Emergency Guide Code **55**	
DOT Identification Number **NA 2783**	Chemical Formula $[C(CH_3)_2CHC_4N_2H(CH_3)O]PS(OC_2H_5)_2$	

Synonyms
O,O-diethyl-O-(2-isopropyl-4-methyl-6-pyrimidinyl)phosphorothioate; Dimpylate®; Basudin®; Diazide®; Spectracide®; Diazitol®; Niocidol; Kayazol.

SECTION II - HAZARDOUS INGREDIENTS AND IDENTITY INFORMATION

Hazardous Components (specific identity)	OSHA Exposure Criteria	NIOSH Exposure Criteria	Immediately Dangerous to Life and Health (IDLH)	ACGIH Exposure Criteria
O,O-diethyl-O-(2-isopropyl-4-methyl-6-pyrimidinyl)phosphorothioate (derivation: By condensing isobutyramidine with ethyl-3-oxobutryrate and treating the resulting pyrimidine with o,o-diethylphosphorochloriodothioate). **1 ppm = 12.65 mg/m³**	**PEL:** 0.1 mg/m³ (skin) **STEL:** Not Established	**REL:** 0.1 mg/m³ (skin) **STEL:** Not Established	Not Determined	**TLV:** 0.1 mg/m³ **STEL:** Not Established

SECTION III - PHYSICAL AND CHEMICAL CHARACTERISTICS

Boiling Point **Decomposes 183°F (84°C) at 0.002 mm**	Specific Gravity (H_2O = 1) **1.12**
Vapor Pressure (mm Hg) **<1 at 69°F (20°C)**	Molecular Weight **304.4**
Vapor Density (Air = 1) **10**	Melting Point **248°F (120°C)**

Solubility
Slightly soluble in water (0.004%). Soluble in petroleum solvents, alcohol, and ketones.

Appearance and Odor
Colorless liquid with a faint, ester-like odor. Technical grade Diazinon® appears as a pale to dark brown liquid. Commercial products also available as solid, tan or brown crystals or powder.

SECTION IV - FIRE AND EXPLOSION HAZARD DATA

Flash Point (method used) **82-105°F (28-40°C) closed cup**	Explosive Limits in Air % by Volume **LEL: Not Determined** **UEL: Not Determined**
NFPA Classification **Class IIIA Combustible Liquid**	Autoignition Temperature **Not Determined**

Extinguishing Media
Use dry chemical, water, or foam.

Special Fire Fighting Procedures
Poisonous gases are produced in fire. Wear full protective clothing and self-contained breathing apparatus (SCBA). If possible, fight fire from a distance and avoid smoke (consider down-wind conditions). Move containers from fire area if it can be done safely.

Unusual Fire and Explosion Hazards
If water is used on fire, use large amounts. With insufficient water, highly toxic tetraethyl monothiopyrophosphate is formed.

SECTION V - REACTIVITY DATA

Stability		
Stability		Conditions to Avoid **Avoid contact with incompatible materials. Keep away from heat and flame.**
Stable X	Unstable	Incompatibility *(materials to avoid)* **Strong acids (hydrofluoric, sulfuric, nitric), strong alkalis (sodium hydroxide, potassium hydroxide, lithium hydroxide), and copper-containing compounds.**
Hazardous Polymerization		Conditions to Avoid **Under normal conditions of temperature and pressure, hazardous polymerization of Diazinon is not expected to occur.**
May Occur	Will Not Occur X	Hazardous Decomposition or By-products **Acrid and irritating gases and fumes are produced when Diazinon is involved in fire, including oxides of nitrogen, potassium, and sulfur. Slowly hydrolyzes in water and dilute acid.**

SECTION VI - HEALTH HAZARD DATA

Primary Route(s) of Entry:	Inhalation?	Absorption (skin)?	Ingestion?
	X	X	X

Health Hazards

INHALATION: Headache, irritation of the mucous membranes, giddiness, blurred vision, nervousness, weakness, cramps, diarrhea, discomfort in chest, sweating, miosis, tearing, salivation, vomiting, cyanosis, papilledema, muscular twitching, convulsions, coma, loss of some reflexes, and loss of sphincter control. Also inhibits cholinesterase.

ABSORPTION: Irritation to the skin and eyes. May absorb to cause toxic systemic effects noted above.

INGESTION: Will result in systemic effects noted for inhalation.

Carcinogenicity	NTP Listed?	IARC Cancer Review Group?	OSHA Regulated?	Target Organs?
Unknown Human Unknown Animal	5th Annual Report	No	No	Eyes, respiratory system, CNS, CVS, blood.

Medical Conditions Generally Aggravated by Exposure

None Reported.

Emergency and First-aid Procedures

Eye contact: Flush immediately with water for 15 minutes (minimum), seek medical attention. **Skin contact:** Remove all contaminated clothing. Immediately wash area with large amounts of soap and water. Seek medical assistance. For **inhalation:** Remove the person from exposure. Provide respiratory assistance and CPR. Transfer to medical facility. If **swallowed:** If victim is conscious, give wet slurry of charcoal. Do NOT give milk or alcohol. Seek medical attention immediately.

SECTION VII - PRECAUTIONS FOR SAFE HANDLING AND USE

Steps to be Taken in Case Material is Released or Spilled

Absorb liquids in vermiculite or other absorbent and deposit in DOT-approved drum. Ventilate area and remove ignition sources. Restrict those not involved in cleanup from entering area. Collect any powdered materials in safest manner possible for reuse or disposal in sealed drums.

Preferred Waste Disposal Method

No Citation.

Precautions to be Taken in Handling and Storage

Do not store Diazinon in the presence of incompatible chemicals or materials. Store in tightly closed containers in a cool, well-ventilated, moisture-free area. Personnel working with Diazinon should be trained on its proper handling and storage *prior* to being assigned to such responsibilities.

Other Precautions and Warnings

Do not mix with other insecticides or pesticides, unless recommended by the manufacturer.

SECTION VIII - CONTROL MEASURES AND PERSONAL PROTECTIVE EQUIPMENT

Respiratory Protection *(specify type)*

For relatively low or infrequent exposure to powders, use a full facepiece respirator equipped with a high efficiency particulate air (HEPA) filter and dust/mist pre-filter. For greater protection, use a supplied-air respirator with full facepiece operated in pressure demand mode or a self-contained breathing apparatus (SCBA) with full facepiece operated in pressure demand or other positive pressure mode.

Ventilation

Local exhaust or general mechanical systems recommended.

Protective Gloves	Eye Protection	Other Protective Clothing
Rubber	Chemical Goggles or Face Mask	Apron

Work/Hygiene Practices

Always wash hands thoroughly after using chemical; never bring food, drink, or smoking materials into vicinity of chemicals.

DIAZINON

$[C(CH_3)_2CHC_4N_2H(CH_3)O]PS(OC_2H_5)_2$ CAS: 333-41-5

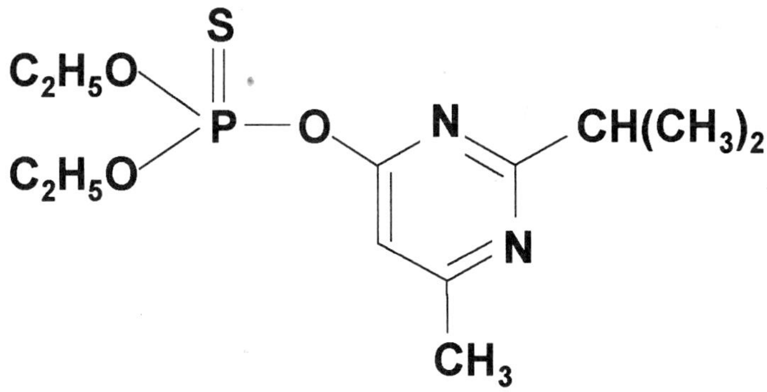

IDENTIFICATION AND TYPICAL USES

Diazinon is a colorless liquid with a faint ester-like odor. Its technical grade may appear as a pale to dark brown liquid and its commercial grade is often a dark brown liquid, tan to dark brown solid crystals, or a powder. It is used primarily and principally as a non-systemic insecticide and acaricide. It is used on a wide variety of agricultural crops, ornamentals, domestic animals, lawns, gardens, and to control household pests. It is permitted by the U.S. Environmental Protection Agency (EPA) for use against fire ants.

RISK ASSESSMENT: HEALTH

General Assessment

Diazinon is primarily toxic to humans by *ingestion*. However, its vapors can also cause adverse systemic effects by *inhalation,* and skin contact may result in *absorption* which will also cause adverse systemic effects. Contact with the eyes or skin will result in irritation. Human mutation data have been reported and it is an experimental teratogen with positive evidence of reproductive effects in animals. There are no reports of carcinogenic reactions subsequent to long-term exposure to Diazinon. However, the fact that mutation data have been reported requires some consideration.

Ingestion and inhalation of Diazinon will result in systemic effects with symptoms of headache, giddiness, blurred vision, nervousness, weakness, cramps, diarrhea, discomfort in the chest, sweating, miosis, lachrymation, salivation and other excessive respiratory tract secretions, vomiting, cyanosis, papilledema, uncontrollable muscle twitching, convulsions, loss of reflexes and motor functions, and loss of sphincter control. This compound is also an inhibitor of cholinesterase, an enzyme that is essential to the proper functioning of the muscles.

Acute Health Effects

The following acute (short-term) health effects may occur immediately or shortly after exposure to Diazinon:

Skin: Irritation and possible inflammation. May absorb through unbroken skin to cause toxic systemic effects.

Eye: Severe irritation with redness, lachrymation, and possible conjunctivitis.

Lung: Irritation of the mucosa of the nose, throat, and upper respiratory tract. Effects are more systemic than localized, with action on the blood, central nervous system, and (possibly) the cardiovascular system.

CNS: A depressant, causing headache, blurred vision, giddiness, nervousness, and other systemic effects.

Other: Ingestion of Diazinon will result in essentially the same symptoms as those reported for inhalation in terms of its action to depress the central nervous system.

Chronic Health Effects

The following chronic (long-term) health effects may occur at some time after exposure to Diazinon and can last for months or even years:

Cancer Hazards: According to the information presented in the references, Diazinon has not been adequately tested for its ability to cause cancer in test animals. It has been shown to cause mutations (genetic changes) and may pose a cancer risk as well.

Reproductive Hazard: According to the information presented in the references, Diazinon may damage the developing fetus in humans since it has been shown to be fetotoxic in animals.

Other Chronic Effects: Chronic exposure to Diazinon has not been sufficiently studied. However, there is some indication that long-term exposure or frequent high exposures may inhibit the essential enzyme cholinesterase.

Recommended Risk-Reduction Measures

Unless a less toxic chemical can be substituted for a hazardous substance, *engineering controls* are the most effective method of reducing exposures. The best protection is to enclose operations and/or provide local exhaust ventilation at the site of chemical release. While not always operationally feasible, isolating operations can also reduce exposure. These controls certainly may not be practical or applicable when using Diazinon as an insecticide, especially outdoors. However, during manufacture, transportation, or storage, certain engineering controls may be effective in reducing exposure risk.

Using respiratory protection is less effective than the controls mentioned above, but is still advisable whenever working with or around Diazinon. The exposure level for this compound is extremely low and may be difficult to measure and even more difficult to protect against. At low or infrequent exposures, a full facepiece respirator equipped with a high efficiency particulate air (HEPA) filter and a dust/mist pre-filter may suffice. However, for maximum exposure risk reduction or for exposures to high concentrations, a self-contained breathing apparatus (SCBA) with full facepiece and pressure demand is the recommended respiratory protection of choice. If a full facepiece is not available, then chemical goggles should be worn to protect the eyes. Whenever a chemical splash hazard exists, a face shield and rubber apron should be worn. To prevent hand and skin exposures, rubber gloves should be worn.

Administrative controls should also be in place to minimize the potential for human exposures. These may include written procedures or policies which specify the methods and techniques that will be practiced whenever personnel are to work with Diazinon.

All personnel should receive *training* on the use, hazards, protective measures, emergency actions, and other precautions per 29 CFR 1910.1200 (Hazard Communication), prior to the first assignment in an area where Diazinon is used or stored.

If symptoms develop or overexposure is suspected, the following medical tests may be useful:

☑ Complete blood count (CBC).

Any evaluation should include a careful medical history of past and present symptoms with an examination. However, medical tests that evaluate existing damage are not a substitute for controlling exposure.

Other methods to reduce exposure include:

☑ Where possible, enclose operations and use local exhaust ventilation at the site of chemical release. If local exhaust ventilation or enclosure is not used, respiratory protection should be mandatory.

☑ Always ensure that proper protective clothing is worn when using chemical substances.

☑ Wash thoroughly immediately after exposure to Diazinon and at the end of the work shift or before eating, drinking, or smoking.

☑ Hazard warning information should be posted in the work area. In addition, as part of an on-going education and training program, all information on the health and safety hazards of Diazinon should be communicated to all exposed and potentially exposed workers.

☑ Eye wash stations should be provided in work areas. If the potential for whole body exposure exists, then safety showers should also be provided.

RISK ASSESSMENT: ENVIRONMENT

General Assessment

The environment is at risk of exposure during the intended use as an insecticide, as well as during transportation, storage, disposal, or destruction of Diazinon. In almost every scenario, the threat of environmental exposure is contingent upon the proper handling of the chemical substance. Accidental spills, large or small, can result in fire, explosion, and possible contamination of the surrounding environmental mediums (water, soil, and, to a lesser extent, air).

Diazinon is considered a Class IIIA combustible liquid (per OSHA 29 CFR 1910.106). These characteristics require special consideration during any emergency situation involving a leak or spill of Diazinon. Should Diazinon ever come into contact with incompatible substances such as strong acids (hydrofluoric, sulfuric, nitric, hydrochloric), strong alkali materials (sodium hydroxide, potassium hydroxide, and lithium hydroxide), and/or any copper-containing compounds either during use, transportation, or storage, violent reaction can occur. Diazinon will slowly hydrolyze in water and dilute acid. If insufficient water is mixed, it will form highly toxic tetraethyl monothiopyrophosphate.

Diazinon can enter the environment through its intended use as an insecticide, through industrial discharges, and from spills.

☠ *Acute Ecological Effects*

Acute (short-term) toxic effects may include the death of animals, birds, or fish and death or low growth rate in plants. Acute effects are seen 2 to 4 days after animals or plants are exposed to Diazinon.

This chemical has high acute toxicity to aquatic life. Insufficient data are available to evaluate or predict the short-term effects of Diazinon to plants, birds, or land animals.

Chronic Ecological Effects

Chronic toxic effects may include shortened life span, reproductive problems, lower fertility, and changes in appearance or behavior in exposed animals. These effects can be seen long after first exposure(s) to toxic chemicals.

Diazinon has high chronic toxicity to aquatic life. Insufficient data are available to evaluate or predict the long-term effects of Diazinon to plants, birds, or land animals.

Water Solubility

Diazinon is very slightly soluble in water. Concentrations of 1 milligram or less will mix with a liter of water.

⌛ *Persistence in the Environment*

Diazinon is slightly persistent in water, with a half-life of between 2 to 20 days. The half-life of a pollutant is the amount of time it takes for one half of the chemical to be degraded.

Bioaccumulation in Aquatic Organisms

Some substances increase in concentration, or *bioaccumulate,* in living organisms as they breathe contaminated air, drink contaminated water, or eat contaminated food. These chemicals can become concentrated in the tissues and internal organs of animals as well as humans.

The concentration of Diazinon found in fish tissues is expected to be much higher than the average concentration of Diazinon in the water from which the fish was taken.

Recommended Risk-Reduction Measures

Proper training of all transporters will reduce the likelihood of a mishap or accident resulting in a leak or spill to the environment. The correct use of DOT labeling (per the MSDS) on all containers, trucks, and rail cars will enable emergency responders to react properly and quickly to any disaster thereby reducing the potential risk to the environment and to personnel.

Storage of Diazinon should be segregated from incompatible chemicals to minimize the risk of cross-contamination or contact.

If a spill or leak to the environment has occurred, fire department, emergency response and/or hazardous materials spill personnel should be notified immediately. Cleanup should be attempted only by those trained in proper spill containment procedures. Contaminated soils should be removed for incineration and replaced with clean soil. If Diazinon should contact the water table, aquifer, or navigable waterway, remediation activities should be prompt to ensure protection of aquatic sediments. When such spills occur, the local and/or state emergency response authorities must be notified. A comprehensive emergency response or disaster preparedness/recovery plan should be in place prior to any operations involving the use, transportation, storage, or disposal of Diazinon. If Diazinon is spilled or leaked, the following specific steps are recommended:

- ☑ Restrict persons not wearing protective clothing from area of spill or leak until cleanup is complete and area can be opened for normal work.
- ☑ Ventilate area and remove ignition sources.
- ☑ Absorb liquid spills using vermiculite, dry earth, or sand to absorb and place in a sealed drum for disposal. Collect any powder materials for reuse or dispose of in an approved, sealed drum.

☑ It may be necessary to dispose of Diazinon as a hazardous waste. The responsible state agency or the regional office of the federal Environmental Protection Agency (EPA) should be contacted for specific recommendations.

RISK ASSESSMENT: BUSINESS

General Assessment

Accidents or mishaps involving Diazinon can present a significant threat to business operations. The loss or damage of equipment or facilities can significantly affect fiscal viability. Lawsuits that may result from personnel injury/death, public exposures, and/or environmental contamination will also require a serious expenditure of resources. Media attention surrounding an injury, death, or environmental damage can also result in a loss of profits and loss of current as well as future business.

Recommended Risk-Reduction Measures

Company attorneys, safety and health professionals, and environmental specialists should be involved in the development of any procedures dealing with safety or emergency response. A company official should be pre-designated as a public relations officer with specific training in dealing with the press. Corporate plans and policies should be developed, approved, and implemented long before any need for such arises.

REFERENCES

Ahlbom, A. 1993. *Biostatistics for Epidemiologists.* New York: CRC Press/Lewis Publishers.

American Conference of Governmental Industrial Hygienists. 1988. *Documentation of the Threshold Limit Values and Biological Exposure Indices,* 5th Edition (with updates). Cincinnati: ACGIH

Asante-Duah, D. K. 1993. *Hazardous Waste Risk Assessment.* New York: CRC Press/Lewis Publishers.

Bloom, A. D. and F. J. DeSerres. 1995. *Ecotoxicity and Human Health.* New York: CRC Press/Lewis Publishers.

Bueche, J. R. 1972. *Principles of Physics,* 2nd Edition. New York: McGraw-Hill.

Burrell, R., D. K. Flaherty, and L. J. Sauers. 1992. *Toxicology of the Immune System.* New York: Van Nostrand Reinhold.

Calabrese, E. J., E. M. Kenyon. *Air Toxics and Risk Assessment.* New York: CRC Press/Lewis Publishers.

Cockerman, L. G., B. S. Shane. 1994. *Basic Environmental Toxicology.* New York: CRC Press/Lewis Publishers.

Cothern, C. R. 1995. *Handbook for Environmental Risk Decisionmaking.* New York: CRC Press/Lewis Publishers.

Gardner, R. W. 1994. *Chemical Intolerance.* New York: CRC Press/Lewis Publishers.

Hallenbeck, W. H. 1993. *Quantitative Risk Assessment for Environmental and Occupational Health,* 2nd Edition. New York: CRC Press/Lewis Publishers.

Howard, P. H. 1995. *Handbook of Environmental Fate and Exposures Data for Organic Chemicals,* Vol. I-IV. New York: CRC Press/Lewis Publishers.

Keith, L. H. and M. M. Walker. 1995. *Handbook of Air Toxics.* New York: CRC Press/Lewis Publishers.

Landis, W. G., M. H. Yu. 1995. *Introduction to Environmental Toxicology.* New York: CRC Press/Lewis Publishers.

Langerman, N. 1994 *Precautionary Labels for Chemical Containers.* New York: CRC Press/Lewis Publishers.

Lauwerys, R. R., P. Hoet. 1993. *Industrial Chemical Exposure.* New York: CRC Press/Lewis Publishers.

Lewis, R. J., Sr. 1993. *Hawley's Condensed Chemical Dictionary,* 12th Edition. New York: Van Nostrand Reinhold.

Lewis, R. J., Sr. 1992. *Sax's Dangerous Properties of Industrial Materials,* Volumes 1, 2, and 3, 8th Edition. New York: Van Nostrand Reinhold.

Manahan, S. E. 1992. *Toxicological Chemistry,* 2nd Edition. New York: CRC Press/Lewis Publishers.

Meyer, E. 1990. *Chemistry of Hazardous Materials.* Englewood Cliffs, NJ: Prentice Hall.

Mycock, J. C., J. D. McKenna, and L. Theodore. 1995. *Handbook of Air Pollution Control Engineering and Technology.* New York: CRC Press/Lewis Publishers.

Ott, W. R. 1995. *Environmental Statistics and Data Analysis.* New York: CRC Press/Lewis Publishers.

Philp, R. B. 1995. *Environmental Hazards and Human Health.* New York: CRC Press/Lewis Publishers.

Quigley, D. R. 1994. *Handbook of Emergency Chemical Management.* New York: CRC Press/Lewis Publishers.

Rea, W. J. 1995. *Chemical Sensitivity,* Vol. I-III. New York: CRC Press/Lewis Publishers.

Richardson, M. 1995. *Dictionary of Substance and Their Effects,* Vol. I-VI. New York: CRC Press/Lewis Publishers.

Rylander, R., R. R. Jacobs. 1994. *Organic Dusts.* New York: CRC Press/Lewis Publishers.

Sacarello, H. L. A. 1994. *The Comprehensive Handbook of Hazardous Materials.* New York: CRC Press/Lewis Publishers.

Sheftel, V. O. 1995. *Handbook of Toxic Properties of Monomers and Additives.* New York: CRC Press/Lewis Publishers.

U.S. Department of Health and Human Services, National Institute for Occupational Safety and Health. 1994. *NIOSH Pocket Guide to Chemical Hazards.* Washington, D.C.: U.S. Government Printing Office.

MATERIAL SAFETY DATA SHEET

CHEMICAL NAME

DIAZOMETHANE

CRC PRESS/LEWIS PUBLISHERS DISCLAIMER: THE INFORMATION AND RECOMMENDATIONS PRESENTED HEREIN ARE BASED ON SOURCES BELIEVED TO BE RELIABLE. CRC MAKES NO REPRESENTATION ON ITS COMPLETENESS OR ACCURACY. IT IS THE RESPONSIBILITY OF THE USER TO DETERMINE THE CHEMICAL'S SUITABILITY FOR ITS INTENDED USE, THE CHEMICAL'S SAFE USE, AND THE CHEMICAL'S PROPER DISPOSAL. NO REPRESENTATIONS AND/OR WARRANTIES, EITHER EXPRESSED OR IMPLIED, OF THE MERCHANTABILITY OR FITNESS FOR A PARTICULAR PURPOSE, OR OF ANY OTHER NATURE, ARE MADE WITH RESPECT TO THE INFORMATION PROVIDED IN THIS MSDS OR TO THE CHEMICAL TO WHICH INFORMATION MAY REFER. CRC NEITHER ASSUMES NOR AUTHORIZES ANY OTHER PERSON TO ASSUME FOR IT, ANY OTHER ADDITIONAL RESPONSIBILITY OR LIABILITY FOR THE USE OF, OR RELIANCE UPON, THIS INFORMATION.

HAZARD WARNING INFORMATION

HEALTH	FIRE	REACTIVE	OTHER	DEGREE OF HAZARD	COLOR CODING	OTHER CODES
4	3	3	W	0 = Minimum Hazard 1 = Slight Hazard 2 = Moderate Hazard 3 = Serious Hazard 4 = Severe Hazard	HEALTH = BLUE FIRE = RED REACTIVITY = YELLOW OTHER = WHITE	OX = Oxidizer ACID = Acid ALK = Alkali COR = Corrosive W = Use No Water

SECTION I - GENERAL INFORMATION

Characterization	Gas	RCRA Number	None
		EPA Class	Not Applicable
DOT Proper Shipping Name	Diazomethane	Chemical Abstract Service (CAS) Number	334-88-3
DOT Hazard Class and Label Requirements	No Citation	DOT Emergency Guide Code	No Citation
DOT Identification Number	None	Chemical Formula	CH_2N_2

Synonyms
Azimethylene; azomethylene; Diazirine.

SECTION II - HAZARDOUS INGREDIENTS AND IDENTITY INFORMATION

Hazardous Components (specific identity)	OSHA Exposure Criteria	NIOSH Exposure Criteria	Immediately Dangerous to Life and Health (IDLH)	ACGIH Exposure Criteria
Diazomethane 1 ppm = 1.75 mg/m^3	PEL: 0.2 ppm 0.4 mg/m^3 STEL: Not Established	REL: 0.2 ppm 0.4 mg/m^3 STEL: Not Established	2 ppm	TLV: 0.2 ppm 0.4 mg/m^3 STEL: Not Established

SECTION III - PHYSICAL AND CHEMICAL CHARACTERISTICS

Boiling Point	-9°F (-23°C)	Specific Gravity (H_2O = 1)	1.45
Vapor Pressure (atmospheres)	>1 at 69°F (20°C)	Molecular Weight	42.05
Vapor Density (Air = 1)	Not Found	Freezing Point	-229°F (-145°C)

Solubility
Reacts in water.

Appearance and Odor
Yellow gas with a musty odor. Normally shipped as a liquefied compressed gas.

SECTION IV - FIRE AND EXPLOSION HAZARD DATA

Flash Point (method used)	Not Applicable (gas)	Explosive Limits in Air % by Volume	LEL: Not Determined UEL: Not Determined
NFPA Classification	Flammable Gas (Explosive)	Autoignition Temperature	Not Applicable

Extinguishing Media
Diazomethane will explode in fire.

Special Fire Fighting Procedures
Diazomethane gas presents a serious explosion hazard when exposed to flame or involved in fire. Gas is extremely irritating: wear full protective clothing and self-contained breathing apparatus (SCBA). In the event of a fire, immediately evacuate the area.

Unusual Fire and Explosion Hazards
Emits toxic fumes of NOx when exposed to heat. To stop fire, stop the flow of gas. Heat, sunlight, or other bright lights can cause it to explode. Extremely reactive, shock sensitive, and impact sensitive. Explosions likely when exposed to rough surfaces.

SECTION V - REACTIVITY DATA

Stability		
Stability		Conditions to Avoid **Diazomethane is extremely reactive in the presence of heat, sunlight, bright lights, rough surfaces (such as ground glass), and incompatible materials. Explosions are likely to be violent and severe.**
Stable	Unstable **X**	Incompatibility *(materials to avoid)* **Diazomethane is incompatible with numerous commodities, including alkali metals, acid, acid fumes, water, drying agents (such as calcium arsenate) and calcium sulfate.**
Hazardous Polymerization		Conditions to Avoid **Under normal conditions of temperature and pressure, hazardous polymerization of diazomethane may occur.**
May Occur **X**	Will Not Occur	Hazardous Decomposition or By-products **Diazomethane decomposes upon heating or in the presence of acid or acid fumes to emit highly toxic fumes of Nox.**

SECTION VI - HEALTH HAZARD DATA

Primary Route(s) of Entry:	Inhalation?	Absorption (skin)?	Ingestion?
	X		

Health Hazards

INHALATION: **Extremely irritating. Exposure can cause severe lung damage with symptoms of coughing, chest pain, shortness of breath, fever and fatigue. Symptoms may not be felt until hours or days after exposure and this can cause death (pulmonary edema).**

SKIN & EYES: **Liquid or gas exposure to skin and eyes may cause severe burning and frostbite.**

INGESTION: **Not likely since diazomethane is a gas.**

Carcinogenicity	NTP Listed?	IARC Cancer Review Group?	OSHA Regulated?	Target Organs?
Suspected Human Confirmed Animal	**No**	**No**	**29 CFR 1910.1000 (Table Z-1)**	**Respiratory system; skin; eyes.**

Medical Conditions Generally Aggravated by Exposure

Respiratory function impairments (bronchitis, asthma, etc.).

Emergency and First-aid Procedures

Eye contact: Immediately flush with large amounts of water for 15 minutes (minimum), occasionally lifting eyelids, seek medical attention. Skin contact: Remove all contaminated clothing. Immediately wash area with large amounts of soap and water. For inhalation: Remove the person from exposure. Provide respiratory assistance and CPR. Transfer promptly to medical facility. Observation for 24 - 48 hours since pulmonary edema is possible. If swallowed: Not likely since this product is a gas.

SECTION VII - PRECAUTIONS FOR SAFE HANDLING AND USE

Steps to be Taken in Case Material is Released or Spilled

Remove all ignition sources. Ventilate area. Stop flow of gas. If the source of leak is a cylinder and cannot be stopped, remove to safe place in open air; repair leak or let cylinder empty. It may be necessary to evacuate area and wait for cylinder to empty.

Preferred Waste Disposal Method

No Citation.

Precautions to be Taken in Handling and Storage

Do not store with incompatible chemicals since violent reactions can occur. Store in tightly closed containers in a cool, dark, well-ventilated area away from sunlight, heat, and bright lights. Keep away from ignition sources such as fire, sparks, and flame.

Other Precautions and Warnings

Bulk storage of diazomethane is not recommended. Containers may explode in fire or under conditions of extreme heat (do not store outdoors or in direct sunlight).

SECTION VIII - CONTROL MEASURES AND PERSONAL PROTECTIVE EQUIPMENT

Respiratory Protection *(specify type)*

At any exposure level, use a powered supplied-air respirator set in positive pressure mode or use an MSHA/NIOSH-approved self-contained breathing apparatus (SCBA) with full facepiece and pressure demand or other positive pressure mode.

Ventilation

Local exhaust or general mechanical systems recommended.

Protective Gloves	Eye Protection	Other Protective Clothing
Butyl Rubber	**Chemical Goggles and/or Face Mask**	**Rubber Apron**

Work/Hygiene Practices

Always wash hands thoroughly after using chemical; never bring food, drink, or smoking materials into vicinity of chemicals.

DIAZOMETHANE

CH_2N_2 CAS: 334-88-3

IDENTIFICATION AND TYPICAL USES

Diazomethane is a yellow gas with a musty odor at room temperatures. It can also exist as a liquid under pressure. It is used as a methylating agent for acidic compounds such as carboxylic acids, phenols, and enols.

RISK ASSESSMENT: HEALTH

General Assessment

Diazomethane is a severe pulmonary irritant and a powerful allergen. As a gas, primary entry is by *inhalation.* It is a confirmed animal carcinogen, creating lung tumors in rats and mice. Human mutagenic data have also been reported.

Inhalation of diazomethane gas is extremely dangerous, causing irritation of the eyes, chest pain, cough, fever, and severe asthmatic attacks. It will cause immediate respiratory distress leading to pneumonitis and death. The warning properties of diazomethane are poor and symptoms may be delayed for several hours or even days after exposure. This presents a serious health risk. Other symptoms may include severe headache, generalized aching of muscles, a sensation of overwhelming tiredness, stupor, fatigue, flushed skin, and shortness of breath.

Skin or eye contact may cause severe damage to tissue. Liquid may cause frostbite, scaling, and burns.

Acute Health Effects

The following acute (short-term) health effects may occur immediately or shortly after exposure to diazomethane:

Skin: Irritation, burning, scaling, or frostbite on contact with liquid and possibly the gas.

Eye: Burns which can lead to permanent damage.

Lung: Severe lung damage. Symptoms include coughing, wheezing, shortness of breath, chest pain, dizziness, fatigue, headache, and irritation of the mouth, nose, and throat. Nausea, vomiting, and fever may also occur. Higher exposures can lead to pulmonary edema, a medical emergency which can be fatal.

Chronic Health Effects

The following chronic (long-term) health effects may occur at some time after exposure to diazomethane and can last for months or even years:

Cancer Hazards: Diazomethane has been shown to cause cancer (lung tumors) in test animals and its carcinogenic properties in humans is currently suspect. It has caused mutations (genetic changes) in humans and, therefore, should be treated with extreme caution.

Reproductive Hazards: According to information presented in the references, diazomethane has not been tested for its ability to affect reproduction.

Other Chronic Effects: Repeated exposure to diazomethane can cause an asthma-like lung allergy. Symptoms may include coughing, wheezing, and fatigue. The symptoms may become progressively worse with repeated attacks, and permanent lung damage (emphysema) may result.

Recommended Risk-Reduction Measures

Personnel should avoid direct contact with diazomethane. It is extremely toxic and exposures can cause death or serious physical injury or illness. Unless a less toxic chemical can be substituted for a hazardous substance, *engineering controls* are the most effective method of reducing exposures. The best protection is to enclose operations and/or provide local exhaust ventilation at the site of chemical release. While not always operationally feasible, isolating operations can also reduce exposure. Using respiratory protection is less effective than the controls mentioned above, but is still advisable whenever working with diazomethane. A NIOSH-approved self-contained breathing apparatus

(SCBA) with full facepiece and pressure demand is the recommended respiratory protection of choice. If a full facepiece is not available, then chemical goggles should be worn to protect the eyes. Whenever a chemical splash hazard exists, a face shield and a rubber apron should be worn. To prevent hand and skin exposures, butyl rubber gloves should be worn.

Administrative controls should also be in place to minimize the potential for human exposures. These may include written procedures or policies which specify the methods and techniques that will be practiced whenever personnel are to work with diazomethane.

All personnel should receive *training* on the use, hazards, protective measures, emergency actions, and other precautions per 29 CFR 1910.1200 (Hazard Communication), prior to the first assignment in an area where diazomethane is used or stored. Before beginning employment and at regular intervals thereafter (e.g., annually), the following medical tests are recommended:

- ☑ Lung function tests. These may be normal at first if the person is not having an attack at the time.

If symptoms develop or overexposure is suspected, the following may be useful:

- ☑ Consider chest X-ray after acute overexposure (may be negative if taken immediately after exposure due to delayed onset of pulmonary edema.
- ☑ Evaluation by a qualified allergist, including careful exposure history and special testing (may help diagnose allergy).

Other methods to reduce exposure to chemicals include:

- ☑ Where possible, enclose operations and use local exhaust ventilation at the site of chemical release. If local exhaust ventilation or enclosure is not used, respiratory protection should be mandatory.
- ☑ Always ensure that proper protective clothing is worn when using chemical substances.
- ☑ Wash thoroughly immediately after exposure to diazomethane and at the end of the work shift or before eating, drinking, or smoking. Work clothing contaminated with diazomethane should never be taken home for laundering. Only personnel trained in the hazards of exposure should launder contaminated clothing.
- ☑ Safety shower and eyewash stations should be readily available in work areas where diazomethane is used or stored.
- ☑ Hazard warning information should be posted in the work area. In addition, as part of an on-going education and training program, all information on the health and safety hazards of diazomethane should be communicated to all exposed and potentially exposed workers.

RISK ASSESSMENT: ENVIRONMENT

General Assessment

The environment is at risk of exposure during transportation, storage, disposal, or destruction of diazomethane. In almost every scenario, the threat of environmental exposure is contingent upon the proper handling of the chemical substance. Accidental releases, large or small, can result in fire, explosion, and possible contamination of the surrounding environmental mediums (water, soil, and air).

Diazomethane is considered to be a flammable gas. It is explosive by itself and, because it is incompatible with many other commodities, contact with any of them can result in violent explosion and fire. It is highly explosive when shocked, exposed to heat, or by chemical reaction. Undiluted liquid or gas may explode on contact with alkali metals, rough surfaces, heat, and high intensity light (especially sunlight). Therefore, diazomethane requires special consideration during any emergency involving a leak or release of diazomethane gas or liquid.

Diazomethane can enter the environment through industrial discharges, unchecked venting, or through spills.

☠ *Acute Ecological Effects*

Acute (short-term) toxic effects may include the death of animals, birds, or fish, and death or low growth rate in plants. Acute effects are seen 2 to 4 days after animals or plants are exposed to diazomethane.

Insufficient data are available to evaluate or predict the short-term effects of diazomethane on aquatic life, plants, birds, or land animals.

💣 *Chronic Ecological Effects*

Chronic (long-term) toxic effects may include shortened life span, reproductive problems, lower fertility, and changes in appearance or behavior in exposed animals. These effects can be seen long after first exposure(s) to toxic chemicals. Insufficient data are available to

evaluate the long-term effects of diazomethane on aquatic life, plants, birds, or land animals.

💧 *Water Solubility*

Diazomethane decomposes in water.

⌛ *Persistence in the Environment*

Diazomethane will react rapidly with water and therefore will not accumulate in the environment.

Bioaccumulation in Aquatic Organisms

Some substances increase in concentration, or *bioaccumulate,* in living organisms as they breathe contaminated air, drink contaminated water, or eat contaminated food. These chemicals can become concentrated in the tissues and internal organs of animals as well as humans.

Diazomethane will react rapidly with water and therefore is not expected to accumulate in the tissues of fish.

Recommended Risk-Reduction Measures

Proper training of all transporters will reduce the likelihood of a mishap or accident resulting in a leak or release to the environment. The correct use of labeling on all containers, trucks, and rail cars will enable emergency responders to react properly and quickly to any disaster thereby reducing the potential risk to the environment and to personnel.

Storage of diazomethane should be segregated from incompatible chemicals to minimize the risk of cross-contamination or contact. Under all circumstances, it should be protected from exposure to light, heat, and impacts. Buildings designated for storage should be equipped with appropriate fire protection systems (alarms, sprinklers, emergency lighting, portable extinguishers).

If a spill or release to the environment has occurred, fire department, emergency response, and/or hazardous materials spill personnel should be notified immediately. Cleanup should be attempted only by those trained in proper containment procedures. Contaminated soils should be removed for incineration and replaced with clean soil. If diazomethane should contact the water table, aquifer, or navigable waterway, remediation may not be possible. It will readily react with water and decompose and, therefore, total containment will not be possible. When such spills occur, the local and/or state emergency response authorities must be notified. A comprehensive emergency response or disaster preparedness/recovery plan should be in place prior to any operations involving the use, transportation, storage, or disposal of diazomethane. If diazomethane is spilled or leaked, the following specific steps are recommended:

- ☑ Restrict persons not wearing protective clothing from area of spill or leak until cleanup is complete.
- ☑ Remove all ignition sources.
- ☑ Ventilate area of spill or leak.
- ☑ If applicable, stop flow of leaking liquid or gas. If leak source is a cylinder and the leak cannot be stopped in place, carefully remove leaking cylinder to a safe place in the open air, and allow cylinder to empty. Repair when completely empty. Attempting to repair while venting can be extremely dangerous (explosions from shock or impact likely).
- ☑ Keep diazomethane out of a confined space, such as a sewer, because of the possibility of explosion (unless the sewer is designed to prevent the buildup of explosive concentrations).
- ☑ It may be necessary to dispose of diazomethane as a hazardous waste. The responsible state agency or the regional office of the federal Environmental Protection Agency (EPA) should be contacted for specific recommendations.

RISK ASSESSMENT: BUSINESS

General Assessment

Accidents or mishaps involving diazomethane can present a significant threat to business operations. The loss or damage of equipment or facilities can significantly affect fiscal viability. Lawsuits that may result from personnel injury, public exposures, and environmental contamination will also require a serious expenditure of resources. Media attention surrounding an injury, illness, or environmental damage can also result in a loss of profits and loss of current as well as future business.

Recommended Risk-Reduction Measures

Company attorneys, safety and health professionals, and environmental specialists should be involved in the development of any procedures dealing with safety or emergency response. A company official should be pre-designated as a public relations officer with specific

training in dealing with the press. Corporate plans and policies should be developed, approved, and implemented long before any need for such arises.

REFERENCES

Ahlbom, A. 1993. *Biostatistics for Epidemiologists.* New York: CRC Press/Lewis Publishers.

American Conference of Governmental Industrial Hygienists. 1988. *Documentation of the Threshold Limit Values and Biological Exposure Indices,* 5th Edition (with updates). Cincinnati: ACGIH

Anthony, C. P., and N. J. Kolthoff. 1971. *Textbook of Anatomy and Physiology.* St. Louis: C. V. Mosby Company.

Asante-Duah, D. K. 1993. *Hazardous Waste Risk Assessment.* New York: CRC Press/Lewis Publishers.

Bloom, A. D. and F. J. DeSerres. 1995. *Ecotoxicity and Human Health.* New York: CRC Press/Lewis Publishers.

Brusick, D. J. 1994. *Methods for Genetic Risk Assessment.* New York: CRC Press/Lewis Publishers.

Bueche, J. R. 1972. *Principles of Physics,* 2nd Edition. New York: McGraw-Hill.

Burrell, R., D. K. Flaherty, and L. J. Sauers. 1992. *Toxicology of the Immune System.* New York: Van Nostrand Reinhold.

Calabrese, E. J., E. M. Kenyon. *Air Toxics and Risk Assessment.* New York: CRC Press/Lewis Publishers.

Calabrese, E. J. 1994. *Biological Effects of Low Level Exposures.* New York: CRC Press/Lewis Publishers.

Cockerman, L. G., B. S. Shane. 1994. *Basic Environmental Toxicology.* New York: CRC Press/Lewis Publishers.

Cothern, R. R. 1993. *Comparative Environmental Risk Assessment.* New York: CRC Press/Lewis Publishers.

Cothern, C. R. 1995. *Handbook for Environmental Risk Decisionmaking.* New York: CRC Press/Lewis Publishers.

Gardner, R. W. 1994. *Chemical Intolerance.* New York: CRC Press/Lewis Publishers.

Hallenbeck, W. H. 1993. *Quantitative Risk Assessment for Environmental and Occupational Health,* 2nd Edition. New York: CRC Press/Lewis Publishers.

Hodgson, E., R. B. Mailman, and J. E. Chambers. 1988. *Dictionary of Toxicology.* London: The MacMillian Company.

Hosty, J. W., and P. Foster. 1990. *A Practical Guide to Chemical Spill Response.* New York: Van Nostrand Reinhold.

Howard, P. H. 1995. *Handbook of Environmental Fate and Exposures Data for Organic Chemicals,* Vol. I-IV. New York: CRC Press/Lewis Publishers.

Howard, P. H., et. al. 1991. *Handbook of Environmental Degradation Rates.* New York: CRC Press/Lewis Publishers.

International Agency for Research on Cancer (IARC). 1974. *IARC Monographs on the Evaluation of the Carcinogenic Risk of Chemicals to Man.* Volume 7. Some anti-thyroid and related substances, nitrofurans, and industrial chemicals. Pp 223-230. Lyon: IARC.

James, A., and M. Lord. 1992. *Index of Chemical and Physical Data.* New York: Van Nostrand Reinhold.

Jones, F. E. 1994. *Toxic Organic Vapors in the Workplace.* New York: CRC Press/Lewis Publishers.

Kamrin, M. A. 1988. *Toxicology: A Primer on Principles and Applications.* Chelsea, Michigan: Lewis Publishers.

Keith, L. H. and M. M. Walker. 1995. *Handbook of Air Toxics.* New York: CRC Press/Lewis Publishers.

Landis, W. G., M. H. Yu. 1995. *Introduction to Environmental Toxicology.* New York: CRC Press/Lewis Publishers.

Langerman, N. 1994 *Precautionary Labels for Chemical Containers.* New York: CRC Press/Lewis Publishers.

Lauwerys, R. R., P. Hoet. 1993. *Industrial Chemical Exposure.* New York: CRC Press/Lewis Publishers.

Lewis, C. E. 1964. Diazomethane poisoning, report of a case suggesting sensitization reaction. *Journal of Occupational Medicine.* 6:91-93.

Lewis, R. J., Sr. 1993. *Hawley's Condensed Chemical Dictionary,* 12th Edition. New York: Van Nostrand Reinhold.

Lewis, R. J., Sr. 1992. *Sax's Dangerous Properties of Industrial Materials,* Volumes 1, 2, and 3, 8th Edition. New York: Van Nostrand Reinhold.

Lide, D. R. 1995. *Handbook of Chemistry and Physics,* 76th Edition. New York: CRC Press/Lewis Publishers.

Manahan, S. E. 1994. *Environmental Chemistry,* 6th Edition. New York: CRC Press/Lewis Publishers.

Manahan, S. E. 1992. *Toxicological Chemistry,* 2nd Edition. New York: CRC Press/Lewis Publishers.

Mellström, G. A., J. E. Wahlberg, and H. I. Maibach. 1994. *Protective Gloves for Occupational Use.* New York: CRC Press/Lewis Publishers.

Meyer, E. 1990. *Chemistry of Hazardous Materials.* Englewood Cliffs, NJ: Prentice Hall.

Patnaik, P. 1992. *A Comprehensive Guide to the Hazardous Properties of Chemical Substances.* New York: Van Nostrand Reinhold.

Philp, R. B. 1995. *Environmental Hazards and Human Health.* New York: CRC Press/Lewis Publishers.

Proctor, N. H., and J. P. Hughes. 1991. *Chemical Hazards of the Workplace.* New York: Van Nostrand Reinhold.

Quigley, D. R. 1994. *Handbook of Emergency Chemical Management.* New York: CRC Press/Lewis Publishers.

Rea, W. J. 1995. *Chemical Sensitivity,* Vol. I-III. New York: CRC Press/Lewis Publishers.

Richardson, M. 1995. *Dictionary of Substance and Their Effects,* Vol. I-VI. New York: CRC Press/Lewis Publishers.

U.S. Department of Health and Human Services, National Institute for Occupational Safety and Health. 1994. *NIOSH Pocket Guide to Chemical Hazards.* Washington, D.C.: U.S. Government Printing Office.

MATERIAL SAFETY DATA SHEET

CHEMICAL NAME	
DIBORANE	**CRC PRESS/LEWIS PUBLISHERS DISCLAIMER:** THE INFORMATION AND RECOMMENDATIONS PRESENTED HEREIN ARE BASED ON SOURCES BELIEVED TO BE RELIABLE. CRC MAKES NO REPRESENTATION ON ITS COMPLETENESS OR ACCURACY. IT IS THE RESPONSIBILITY OF THE USER TO DETERMINE THE CHEMICAL'S SUITABILITY FOR ITS INTENDED USE, THE CHEMICAL'S SAFE USE, AND THE CHEMICAL'S PROPER DISPOSAL. NO REPRESENTATIONS AND/OR WARRANTIES, EITHER EXPRESSED OR IMPLIED, OF THE MERCHANTABILITY OR FITNESS FOR A PARTICULAR PURPOSE, OR OF ANY OTHER NATURE, ARE MADE WITH RESPECT TO THE INFORMATION PROVIDED IN THIS MSDS OR TO THE CHEMICAL TO WHICH INFORMATION MAY REFER. CRC NEITHER ASSUMES NOR AUTHORIZES ANY OTHER PERSON TO ASSUME FOR IT, ANY OTHER ADDITIONAL RESPONSIBILITY OR LIABILITY FOR THE USE OF, OR RELIANCE UPON, THIS INFORMATION.

HAZARD WARNING INFORMATION

HEALTH	FIRE	REACTIVE	OTHER	DEGREE OF HAZARD	COLOR CODING	OTHER CODES
3	4	3	~~W~~	0 = Minimum Hazard 1 = Slight Hazard 2 = Moderate Hazard 3 = Serious Hazard 4 = Severe Hazard	HEALTH = BLUE FIRE = RED REACTIVITY = YELLOW OTHER = WHITE	OX = Oxidizer ACID = Acid ALK = Alkali COR = Corrosive ~~W~~ = Use No Water

SECTION I - GENERAL INFORMATION

Characterization: **Hydride Gas**	RCRA Number: **None**	EPA Class: **Not Applicable**
DOT Proper Shipping Name: **Diborane**	Chemical Abstract Service (CAS) Number: **19287-45-7**	
DOT Hazard Class and Label Requirements: **Flammable Gas and Poison**	DOT Emergency Guide Code: **18**	
DOT Identification Number: **UN 1911**	Chemical Formula: **B_2H_6**	

Synonyms
Boroethane; boron hydride; diboron hexahydride.

SECTION II - HAZARDOUS INGREDIENTS AND IDENTITY INFORMATION

Hazardous Components (specific identity)	OSHA Exposure Criteria	NIOSH Exposure Criteria	Immediately Dangerous to Life and Health (IDLH)	ACGIH Exposure Criteria
Diborane (derivation: Reaction of lithium hydride and boron trifluoride catalyzed by ether at 25°C). **1 ppm = 1.15 mg/m³**	**PEL:** 0.1 ppm 0.1 mg/m³ **STEL:** Not Established	**REL:** 0.1 ppm 0.1 mg/m³ **STEL:** Not Established	**15 ppm**	**TLV:** 0.1 ppm 0.1 mg/m³ **STEL:** Not Established

SECTION III - PHYSICAL AND CHEMICAL CHARACTERISTICS

Boiling Point: **-135°F (-93°C)**	Specific Gravity (H_2O = 1): **0.97**
Vapor Pressure (atmospheres): **39.9 at 62°F (16°C)**	Molecular Weight: **27.7**
Vapor Density (Air = 1): **0.95**	Freezing Point: **-267°F (-166°C)**

Solubility
Reacts and decomposes in water. Soluble in carbon disulfide.

Appearance and Odor
Colorless gas with a repulsive, sweet odor. Usually shipped in pressurized cylinders diluted with hydrogen, argon, nitrogen, or helium. Odor Threshold = 3.3 ppm.

SECTION IV - FIRE AND EXPLOSION HAZARD DATA

Flash Point (method used): **-90°F (-68°C)**	Explosive Limits in Air % by Volume: **LEL: 0.8% UEL: 88%**
NFPA Classification: **Flammable Gas**	Autoignition Temperature: **104°F (40°C)**

Extinguishing Media
Stop flow of gas. Use water spray to keep fire-exposed containers cool. Do NOT use Halon.

Special Fire Fighting Procedures
Diborane gas presents a serious explosion hazard when exposed to flame or involved in fire. Gas is poisonous: wear full protective clothing and self-contained breathing apparatus (SCBA).

Unusual Fire and Explosion Hazards
To stop fire, stop the flow of gas. Extremely reactive substance. Can react explosive in air. In contact with water or when heated, will form explosive hydrogen gas. May re-ignite explosively after fire has been extinguished.

SECTION V - REACTIVITY DATA

Stability		Conditions to Avoid
		Diborane is extremely reactive and highly explosive in the presence of heat or flame. It is unstable under normal conditions of temperature and pressure. Explosions are likely to be violent and severe.
Stable	Unstable	Incompatibility *(materials to avoid)*
	X	Diborane is incompatible with hydrocarbons, amines, aluminum, neoprene rubber, air, water, oxidizing agents, halogens, halogenated compounds, nitric acid, and liquefied nitrogen trifluoride.
Hazardous Polymerization		Conditions to Avoid
		Under normal conditions of temperature and pressure, hazardous polymerization of diborane may occur. Avoid elevated temperatures.
May Occur	Will Not Occur	Hazardous Decomposition or By-products
X		Diborane decomposes upon heating or when exposed to water to evolve highly explosive hydrogen gas and boric acid. Combustion yields boron dust.

SECTION VI - HEALTH HAZARD DATA

Primary Route(s) of Entry:	Inhalation?	Absorption (skin)?	Ingestion?
	X		

Health Hazards

INHALATION: Extremely irritating, may cause infection in the respiratory passages. May deaden the sense of smell, cause headache, cough, nausea, tightness in chest, shortness of breath, chills, fever, weakness, and pulmonary edema (fluid in lungs), which can cause death.

SKIN & EYES: Liquid or gas exposure to skin and eyes may cause severe burning and frostbite.

INGESTION: Not likely since diborane is a gas.

Carcinogenicity	NTP Listed?	IARC Cancer Review Group?	OSHA Regulated?	Target Organs?
Unknown Human Unknown Animal	No	No	29 CFR 1910.1000 (Table Z-1)	Respiratory system; skin; eyes, CNS, liver, kidneys.

Medical Conditions Generally Aggravated by Exposure

Respiratory function impairments (bronchitis, asthma, etc.).

Emergency and First-aid Procedures

Eye contact: Immediately flush with large amounts of water for 15 minutes (minimum), occasionally lifting eyelids, seek medical attention. Skin contact: Remove all contaminated clothing. Immediately wash area with large amounts of soap and water. For inhalation: Remove the person from exposure. Provide respiratory assistance and CPR. Transfer promptly to medical facility. Observation for 24 - 48 hours since pulmonary edema is possible. If swallowed: Not likely since this product is a gas.

SECTION VII - PRECAUTIONS FOR SAFE HANDLING AND USE

Steps to be Taken in Case Material is Released or Spilled

Remove all ignition sources. Ventilate area. Stop flow of gas. If the source of leak is a cylinder and cannot be stopped, remove to safe place in open air; repair leak or let cylinder empty. It may be necessary to evacuate area and wait for cylinder to empty.

Preferred Waste Disposal Method

Burn by any suitable method, under controlled conditions.

Precautions to be Taken in Handling and Storage

Do not store with incompatible chemicals since violent reactions can occur. Store in tightly closed containers in a cool, dark, well-ventilated area away from heat and moisture. Suggest a nitrogen purge inside storage containers. Keep away from ignition sources such as fire, sparks, and flame.

Other Precautions and Warnings

Bulk storage of diborane is not recommended. Containers may explode in fire or under conditions of extreme heat (do not store outdoors or in direct sunlight).

SECTION VIII - CONTROL MEASURES AND PERSONAL PROTECTIVE EQUIPMENT

Respiratory Protection *(specify type)*

At any exposure level, use a powered supplied-air respirator set in positive pressure mode or use an MSHA/NIOSH approved self-contained breathing apparatus (SCBA) with full facepiece and pressure demand or other positive pressure mode.

Ventilation

Local exhaust or general mechanical systems recommended.

Protective Gloves	Eye Protection	Other Protective Clothing
Synthetic Impervious Material	Chemical Goggles and/or Face Mask	Protective Suit

Work/Hygiene Practices

Always wash hands thoroughly after using chemical; never bring food, drink, or smoking materials into vicinity of chemicals.

DIBORANE

B_2H_6 CAS: 19287-45-7

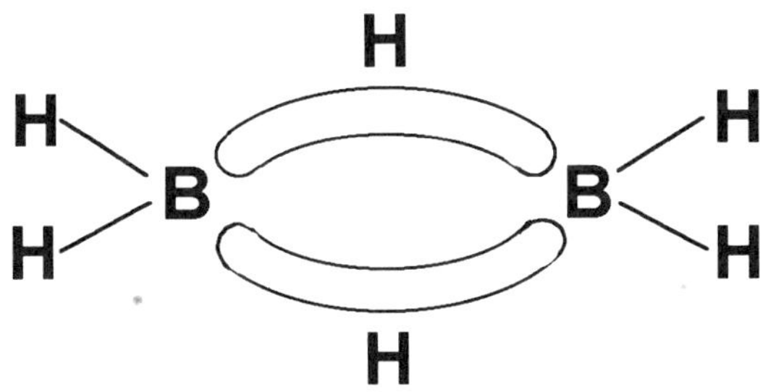

IDENTIFICATION AND TYPICAL USES

Diborane is a colorless gas with a repulsive, sweet odor. It is normally shipped in pressurized cylinders diluted with hydrogen, argon, nitrogen, or helium. It is used in the synthesis of organic boron compounds and metal borohydrides, as a polymerization catalyst for ethylene, a fuel for air-breathing engines and rockets, as a reducing agent, a doping agent for p-type semiconductors, and in the vulcanization of rubber.

RISK ASSESSMENT: HEALTH

General Assessment

Diborane is a severe pulmonary irritant that can cause secondary infections in the respiratory passageways (may precipitate bronchial pneumonia). As a gas, primary entry is by *inhalation*. Its toxicity is comparable to that of chlorine, fluorine, phosgene, and arsine. Its carcinogenic properties are unknown.

Inhalation of diborane gas is extremely dangerous, causing irritation of the eyes, and symptoms similar to that of metal fume fever. These include headache, nausea, tremors, convulsions, dypsnea, chest pain and tightness, shortness of breath, cough, chills, fever and weakness, delayed pulmonary edema and hemorrhage, loss of consciousness and death. It will deaden the sense of smell on short exposures. It may also cause immediate respiratory distress leading to pneumonitis and death. The warning properties of diborane are poor and symptoms may be delayed for several hours or even days after exposure. This presents a serious health risk. There may also be central nervous system effects (intoxication), liver, and kidney damage.

The liquid causes local inflammation, irritation, and burn. There may be blisters, redness, and painful swelling. Frostbite may also occur. Eye contact will result in similar effects, with redness, irritation, and swelling of the conjunctiva.

☠ *Acute Health Effects*

The following acute (short-term) health effects may occur immediately or shortly after exposure to diborane:

Skin: Irritation, burning, scaling, redness, blistering, swelling, or frostbite on contact with liquid and possibly the gas.

Eye: Burns which can lead to permanent damage. Inflammation and swelling of the conjunctiva and surrounding tissue.

Lung: Severe lung damage with symptoms of coughing, wheezing, shortness of breath, chest pain, dizziness, fatigue, headache, and irritation of the mouth, nose, and throat. Nausea, vomiting, and fever may also occur. Higher exposures can lead to pulmonary edema, a medical emergency which can be fatal.

CNS: High exposures may cause severe CNS irritation such as drowsiness, dizziness, visual disturbances, muscle twitching, and, in severe cases, painful muscle spasm.

Chronic Health Effects

The following chronic (long-term) health effects may occur at some time after exposure to diborane and can last for months or even years:

Cancer Hazards: According to information presented in the references, diborane has not been adequately tested for its ability to cause cancer in test animals.

Reproductive Hazards: According to information presented in the references, diborane has not been tested for its ability to affect reproduction.

Other Chronic Effects: Prolonged exposures may cause pulmonary irritation, headache, dizziness, muscle fatigue, and occasional transient tremors. Secondary infection may precipitate bronchial pneumonia. May also cause injuries to the central nervous system, liver, and kidneys.

Recommended Risk-Reduction Measures

Personnel should avoid direct contact with diborane. It is extremely toxic and exposures can cause death or serious physical injury or illness. Unless a less toxic chemical can be substituted for a hazardous substance, *engineering controls* are the most effective method of reducing exposures. The best protection is to enclose operations and/or provide local exhaust ventilation at the site of chemical release. While not always operationally feasible, isolating operations can also reduce exposure. Using respiratory protection is less effective than the controls mentioned above, but is always advisable whenever working with diborane. A NIOSH-approved self-contained breathing apparatus (SCBA) with full facepiece and pressure demand is the recommended respiratory protection of choice. If a full facepiece is not available, then chemical goggles should be worn to protect the eyes. Whenever a chemical splash hazard exists, a face shield and a rubber apron should be worn. To prevent hand and skin exposures, butyl rubber gloves should be worn.

Administrative controls should also be in place to minimize the potential for human exposures. These may include written procedures or policies which specify the methods and techniques that will be practiced whenever personnel are to work with diborane.

All personnel should receive *training* on the use, hazards, protective measures, emergency actions, and other precautions per 29 CFR 1910.1200 (Hazard Communication), prior to the first assignment in an area where diborane is used or stored. Before beginning employment and at regular intervals thereafter (e.g., annually), the following medical tests are recommended:

☑ Lung function tests. These may be normal at first if the person is not having an attack at the time.

If symptoms develop or overexposure is suspected, the following may be useful:

☑ Consider chest X-ray after acute overexposure.

☑ Evaluation by a qualified allergist, including careful exposure history and special testing (may help diagnose allergy).

☑ Liver and kidney function tests.

Other methods to reduce exposure to chemicals include:

☑ Where possible, enclose operations and use local exhaust ventilation at the site of chemical release. If local exhaust ventilation or enclosure is not used, respiratory protection should be mandatory.

☑ Always ensure that proper protective clothing is worn when using chemical substances.

☑ Wash thoroughly immediately after exposure to diborane and at the end of the work shift or before eating, drinking, or smoking. Work clothing contaminated with diborane should never be taken home for laundering. Only personnel trained in the hazards of exposure should launder contaminated clothing.

☑ Safety shower and eyewash stations should be readily available in work areas where diborane is used or stored.

☑ Hazard warning information should be posted in the work area. In addition, as part of an on-going education and training program, all information on the health and safety hazards of diborane should be communicated to all exposed and potentially exposed workers.

RISK ASSESSMENT: ENVIRONMENT

General Assessment

The environment is at risk of exposure during transportation, storage, disposal, or destruction of diborane. In almost every scenario, the threat of environmental exposure is contingent upon the proper handling of the chemical substance. Accidental releases, large or small, can result in fire, explosion, and possible contamination of the surrounding environmental mediums (water, soil, and air).

Diborane is considered to be a flammable gas. It is explosive by itself and, because it is incompatible with many other commodities, contact with any of them can result in violent explosion and fire. It is highly reactive and will ignite with moist air at room temperature. Diborane forms explosive mixtures in air over a wide range (0.8% - 88%) which makes it a serious fire and explosion risk. It will explode in contact with many oxidizing agents (especially fluorine, bromine, and chlorine). It is highly reactive in the presence of halogens and halogenated hydrocarbons. Therefore, diborane requires special consideration during any emergency involving a leak or release of diborane gas or liquid.

Diborane can enter the environment through industrial discharges, unchecked venting, or through spills.

Acute Ecological Effects

Acute (short-term) toxic effects may include the death of animals, birds, or fish, and death or low growth rate in plants. Acute effects are seen 2 to 4 days after animals or plants are exposed to diborane.

Insufficient data are available to evaluate or predict the short-term effects of diborane on aquatic life, plants, birds, or land animals.

Chronic Ecological Effects

Chronic (long-term) toxic effects may include shortened life span, reproductive problems, lower fertility, and changes in appearance or behavior in exposed animals. These effects can be seen long after first exposure(s) to toxic chemicals. Insufficient data are available to evaluate the long-term effects of diborane on aquatic life, plants, birds, or land animals.

Water Solubility

Diborane decomposes in water.

Persistence in the Environment

Diborane will react rapidly with water and therefore will not accumulate in the environment.

Bioaccumulation in Aquatic Organisms

Some substances increase in concentration, or *bioaccumulate,* in living organisms as they breathe contaminated air, drink contaminated water, or eat contaminated food. These chemicals can become concentrated in the tissues and internal organs of animals as well as humans.

Diborane will react rapidly with water and therefore is not expected to accumulate in the tissues of fish.

Recommended Risk-Reduction Measures

Proper training of all transporters will reduce the likelihood of a mishap or accident resulting in a leak or release to the environment. The correct use of DOT labeling (per the MSDS) on all containers, trucks, and rail cars will enable emergency responders to react properly and quickly to any disaster thereby reducing the potential risk to the environment and to personnel.

Storage of diborane should be segregated from incompatible chemicals to minimize the risk of cross-contamination or contact. Under all circumstances, it should be protected from exposure to moisture. Buildings designated for storage should be equipped with appropriate fire protection systems (alarms, sprinklers, emergency lighting, portable extinguishers).

If a spill or release to the environment has occurred, fire department, emergency response, and/or hazardous materials spill personnel should be notified immediately. Cleanup should be attempted only by those trained in proper containment procedures. Contaminated soils should be removed for incineration and replaced with clean soil. If diborane should contact the water table, aquifer, or navigable waterway, remediation may not be possible. It will readily react with water and decompose and, therefore, total containment will not be possible. When such spills occur, the local and/or state emergency response authorities must be notified. A comprehensive emergency response or disaster preparedness/recovery plan should be in place prior to any operations involving the use, transportation, storage, or disposal of diborane. If diborane is spilled or leaked, the following specific steps are recommended:

- ☑ Restrict persons not wearing protective clothing from area of spill or leak until cleanup is complete and area can be opened for normal work.
- ☑ Remove all ignition sources.
- ☑ Ventilate area of spill or leak.
- ☑ If applicable, stop flow of leaking liquid or gas. If leak source is a cylinder and the leak cannot be stopped in place, carefully remove leaking cylinder to a safe place in the open air, and allow cylinder to empty. Repair when completely empty. Attempting to repair while venting can be extremely dangerous.
- ☑ Keep diborane out of a confined space, such as a sewer, because of the possibility of explosion (unless the sewer is designed to prevent the buildup of explosive concentrations).
- ☑ It may be necessary to dispose of diborane as a hazardous waste. The responsible state agency or the regional office of the federal Environmental Protection Agency (EPA) should be contacted for specific recommendations.

RISK ASSESSMENT: BUSINESS

General Assessment

Accidents or mishaps involving diborane can present a significant threat to business operations. The loss or damage of equipment or facilities can significantly affect fiscal viability. Lawsuits that may result from personnel injury, public exposures, and environmental contamination will also require a serious expenditure of resources. Media attention surrounding an injury, illness, or environmental damage can also result in a loss of profits and loss of current as well as future business.

Recommended Risk-Reduction Measures

Company attorneys, safety and health professionals, and environmental specialists should be involved in the development of any procedures dealing with safety or emergency response. A company official should be pre-designated as a public relations officer with specific training in dealing with the press. Corporate plans and policies should be developed, approved, and implemented long before any need for such arises.

REFERENCES

American Conference of Governmental Industrial Hygienists. 1988. *Documentation of the Threshold Limit Values and Biological Exposure Indices,* 5th Edition (with updates). Cincinnati: ACGIH

Anthony, C. P., and N. J. Kolthoff. 1971. *Textbook of Anatomy and Physiology.* St. Louis: C. V. Mosby Company.

Asante-Duah, D. K. 1993. *Hazardous Waste Risk Assessment.* New York: CRC Press/Lewis Publishers.

Bloom, A. D. and F. J. DeSerres. 1995. *Ecotoxicity and Human Health.* New York: CRC Press/Lewis Publishers.

Brusick, D. J. 1994. *Methods for Genetic Risk Assessment.* New York: CRC Press/Lewis Publishers.

Bueche, J. R. 1972. *Principles of Physics,* 2nd Edition. New York: McGraw-Hill.

Burrell, R., D. K. Flaherty, and L. J. Sauers. 1992. *Toxicology of the Immune System.* New York: Van Nostrand Reinhold.

Calabrese, E. J., E. M. Kenyon. *Air Toxics and Risk Assessment.* New York: CRC Press/Lewis Publishers.

Cockerman, L. G., B. S. Shane. 1994. *Basic Environmental Toxicology.* New York: CRC Press/Lewis Publishers.

Cothern, R. R. 1993. *Comparative Environmental Risk Assessment.* New York: CRC Press/Lewis Publishers.

Gardner, R. W. 1994. *Chemical Intolerance.* New York: CRC Press/Lewis Publishers.

Hallenbeck, W. H. 1993. *Quantitative Risk Assessment for Environmental and Occupational Health,* 2nd Edition. New York: CRC Press/Lewis Publishers.

Hodgson, E., R. B. Mailman, and J. E. Chambers. 1988. *Dictionary of Toxicology.* London: The MacMillian Company.

Hosty, J. W., and P. Foster. 1990. *A Practical Guide to Chemical Spill Response.* New York: Van Nostrand Reinhold.

Howard, P. H. 1995. *Handbook of Environmental Fate and Exposures Data for Organic Chemicals,* Vol. I-IV. New York: CRC Press/Lewis Publishers.

Howard, P. H., et. al. 1991. *Handbook of Environmental Degradation Rates.* New York: CRC Press/Lewis Publishers.

James, A., and M. Lord. 1992. *Index of Chemical and Physical Data.* New York: Van Nostrand Reinhold.

Jones, F. E. 1994. *Toxic Organic Vapors in the Workplace.* New York: CRC Press/Lewis Publishers.

Kamrin, M. A. 1988. *Toxicology: A Primer on Principles and Applications.* Chelsea, Michigan: Lewis Publishers.

Keith, L. H. and M. M. Walker. 1995. *Handbook of Air Toxics.* New York: CRC Press/Lewis Publishers.

Landis, W. G., M. H. Yu. 1995. *Introduction to Environmental Toxicology.* New York: CRC Press/Lewis Publishers.

Langerman, N. 1994 *Precautionary Labels for Chemical Containers.* New York: CRC Press/Lewis Publishers.

Lauwerys, R. R., P. Hoet. 1993. *Industrial Chemical Exposure.* New York: CRC Press/Lewis Publishers.

Lewis, R. J., Sr. 1993. *Hawley's Condensed Chemical Dictionary,* 12th Edition. New York: Van Nostrand Reinhold.

Lewis, R. J., Sr. 1992. *Sax's Dangerous Properties of Industrial Materials,* Volumes 1, 2, and 3, 8th Edition. New York: Van Nostrand Reinhold.

Lide, D. R. 1995. *Handbook of Chemistry and Physics,* 76th Edition. New York: CRC Press/Lewis Publishers.

Manahan, S. E. 1992. *Toxicological Chemistry,* 2nd Edition. New York: CRC Press/Lewis Publishers.

Meyer, E. 1990. *Chemistry of Hazardous Materials.* Englewood Cliffs, NJ: Prentice Hall.

Patnaik, P. 1992. *A Comprehensive Guide to the Hazardous Properties of Chemical Substances.* New York: Van Nostrand Reinhold.

Philp, R. B. 1995. *Environmental Hazards and Human Health.* New York: CRC Press/Lewis Publishers.

Proctor, N. H., and J. P. Hughes. 1991. *Chemical Hazards of the Workplace.* New York: Van Nostrand Reinhold.

Quigley, D. R. 1994. *Handbook of Emergency Chemical Management.* New York: CRC Press/Lewis Publishers.

Rea, W. J. 1995. *Chemical Sensitivity,* Vol. I-III. New York: CRC Press/Lewis Publishers.

Richardson, M. 1995. *Dictionary of Substance and Their Effects,* Vol. I-VI. New York: CRC Press/Lewis Publishers.

U.S. Department of Health and Human Services, National Institute for Occupational Safety and Health. 1994. *NIOSH Pocket Guide to Chemical Hazards.* Washington, D.C.: U.S. Government Printing Office.

MATERIAL SAFETY DATA SHEET

CHEMICAL NAME	
DIBROMO-CHLOROPROPANE	**CRC PRESS/LEWIS PUBLISHERS DISCLAIMER:** THE INFORMATION AND RECOMMENDATIONS PRESENTED HEREIN ARE BASED ON SOURCES BELIEVED TO BE RELIABLE. CRC MAKES NO REPRESENTATION ON ITS COMPLETENESS OR ACCURACY. IT IS THE RESPONSIBILITY OF THE USER TO DETERMINE THE CHEMICAL'S SUITABILITY FOR ITS INTENDED USE, THE CHEMICAL'S SAFE USE, AND THE CHEMICAL'S PROPER DISPOSAL. NO REPRESENTATIONS AND/OR WARRANTIES, EITHER EXPRESSED OR IMPLIED, OF THE MERCHANTABILITY OR FITNESS FOR A PARTICULAR PURPOSE, OR OF ANY OTHER NATURE, ARE MADE WITH RESPECT TO THE INFORMATION PROVIDED IN THIS MSDS OR TO THE CHEMICAL TO WHICH INFORMATION MAY REFER. CRC NEITHER ASSUMES NOR AUTHORIZES ANY OTHER PERSON TO ASSUME FOR IT, ANY OTHER ADDITIONAL RESPONSIBILITY OR LIABILITY FOR THE USE OF, OR RELIANCE UPON, THIS INFORMATION.

HAZARD WARNING INFORMATION

HEALTH	FIRE	REACTIVE	OTHER	DEGREE OF HAZARD	COLOR CODING	OTHER CODES
3	1	0		0 = Minimum Hazard 1 = Slight Hazard 2 = Moderate Hazard 3 = Serious Hazard 4 = Severe Hazard	HEALTH = BLUE FIRE = RED REACTIVITY = YELLOW OTHER = WHITE	OX = Oxidizer ACID = Acid ALK = Alkali COR = Corrosive W = Use No Water

SECTION I - GENERAL INFORMATION

Characterization	RCRA Number	EPA Class
Pesticide (nematocide)	U066	Toxic Waste

DOT Proper Shipping Name Dibromochloropropane	Chemical Abstract Service (CAS) Number 96-12-8
DOT Hazard Class and Label Requirements Poison B	DOT Emergency Guide Code 58
DOT Identification Number UN 2872	Chemical Formula $CH_2BrCHBrCH_2Cl$

Synonyms
1,2-Dibromo-3-chloropropane; 1-chloro-2,3-dibromopropane, DBCP; Nematox; Nematocide; Nemanax; Nemazon; Nemagone.

SECTION II - HAZARDOUS INGREDIENTS AND IDENTITY INFORMATION

Hazardous Components (specific identity)	OSHA Exposure Criteria	NIOSH Exposure Criteria	Immediately Dangerous to Life and Health (IDLH)	ACGIH Exposure Criteria
Dibromochloropropane 1 ppm = 9.83 mg/m^3	PEL: 0.001 ppm 0.01 mg/m^3 Suspected Cancer Agent	REL (ceiling): 0.01 ppm 0.1 mg/m^3 Possible Human Carcinogen	Not Determined	TLV: Not Established STEL: Not Established

SECTION III - PHYSICAL AND CHEMICAL CHARACTERISTICS

Boiling Point 385°F (196°C)	Specific Gravity (H_2O = 1) 2.05
Vapor Pressure (mm Hg) 0.8 at 69°F (20°C)	Molecular Weight 236.4
Vapor Density (Air = 1) Not Available	Melting Point 43°F (6°C)

Solubility
Slightly soluble in water (0.1%), miscible in oils.

Appearance and Odor
Dense yellow or amber liquid with a pungent odor at high concentrations (a solid below 43°F).

SECTION IV - FIRE AND EXPLOSION HAZARD DATA

Flash Point (method used) 170°F (77°C) opened cup	Explosive Limits in Air % by Volume LEL: Not Determined UEL: Not Determined
NFPA Classification Class IIIA Combustible Liquid	Autoignition Temperature Not Determined

Extinguishing Media
Use dry chemical, carbon dioxide, foam, or water.

Special Fire Fighting Procedures
Poisonous gases are produced in fire, wear full protective clothing and self-contained breathing apparatus (SCBA). Vapors are heavier than air and may travel for some distance to ignition source and flashback. Move containers from fire area if it can be done safely.

Unusual Fire and Explosion Hazards
Containers may explode in fire due to buildup of excessive internal pressures. Exercise extreme caution when attempting to move or relocate containers that may have been exposed to high heat.

SECTION V - REACTIVITY DATA

Stability		Conditions to Avoid: Avoid contact with incompatible materials. Keep away from heat and flame.
Stable	Unstable	Incompatibility *(materials to avoid)*: Chemically active metals such as aluminum, magnesium, and tin alloys. Corrosive to metals.
X		
Hazardous Polymerization		Conditions to Avoid: Under normal conditions of temperature and pressure, hazardous polymerization of dibromochloropropane is not expected to occur.
May Occur	Will Not Occur	Hazardous Decomposition or By-products: Toxic and irritating gases of bromides and chlorides are produced when dibromochloropropane is heated to decomposition.
	X	

SECTION VI - HEALTH HAZARD DATA

Primary Route(s) of Entry:	Inhalation?	Absorption (skin)?	Ingestion?
	X	X	X

Health Hazards

INHALATION: Symptoms include headache, nausea or vomiting, eye and pulmonary changes. Can cause drowsiness and may act as a narcotic in high concentrations. May also affect male reproductive organs and cause sterility. A lung irritant that may cause fluid to build up in the lungs (pulmonary edema), which can be fatal.

ABSORPTION: Severe irritant. May pass through unbroken skin to cause toxic systemic effects.

INGESTION: Moderately toxic by ingestion. A human poison. Gastrointestinal irritation.

Carcinogenicity	NTP Listed?	IARC Cancer Review Group?	OSHA Regulated?	Target Organs?
Suspected Human Confirmed Animal	5th Annual Report	Group 2B	29 CFR 1910.1044	Eyes, skin, CNS, respiratory system, liver, kidneys, spleen, reproductive system, digestive system.

Medical Conditions Generally Aggravated by Exposure

None Reported.

Emergency and First-aid Procedures

Eye contact: Flush immediately with water for 15 minutes (minimum), seek medical attention. **Skin contact:** Remove all contaminated clothing. Immediately wash area with large amounts of soap and water. Seek medical assistance. For **inhalation:** Remove the person from exposure. Provide respiratory assistance and CPR. Transfer to medical facility. Observe for 24-48 hours for lung effects. If **swallowed:** Call the poison control center and seek medical attention immediately. Never attempt to give an unconscious or convulsing person anything by mouth.

SECTION VII - PRECAUTIONS FOR SAFE HANDLING AND USE

Steps to be Taken in Case Material is Released or Spilled

Absorb liquids in vermiculite or other absorbent and deposit in DOT-approved drum. Ventilate area and remove ignition sources. Restrict those not involved in cleanup from entering area. Notify appropriate authorities, if applicable.

Preferred Waste Disposal Method

No Citation.

Precautions to be Taken in Handling and Storage

Do not store dibromochloropropane in the presence of incompatible chemicals or materials. Store in tightly closed containers in a cool, well-ventilated area. Personnel working with dibromochloropropane should be trained on its proper handing and storage *prior* to being assigned to such responsibilities.

Other Precautions and Warnings

A carcinogen. Bulk storage is of dibromochloropropane is not recommended.

SECTION VIII - CONTROL MEASURES AND PERSONAL PROTECTIVE EQUIPMENT

Respiratory Protection *(specify type)*

For any exposure to a suspected human carcinogen, use MSHA/NIOSH-approved supplied-air respirator with full facepiece operated in continuous flow or other positive pressure mode (pressure demand), or use a self-contained breathing apparatus (SCBA) with full facepiece operated in pressure demand or other positive pressure mode.

Ventilation

Local exhaust or general mechanical systems recommended.

Protective Gloves	Eye Protection	Other Protective Clothing
Polyvinyl Chloride (Do Not Use Rubber)	Chemical Goggles or Face Mask	Apron

Work/Hygiene Practices

Always wash hands thoroughly after using chemical; never bring food, drink, or smoking materials into vicinity of chemicals.

DIBROMOCHLOROPROPANE

$CH_2BrCHBrCH_2CL$ CAS: 96-12-8

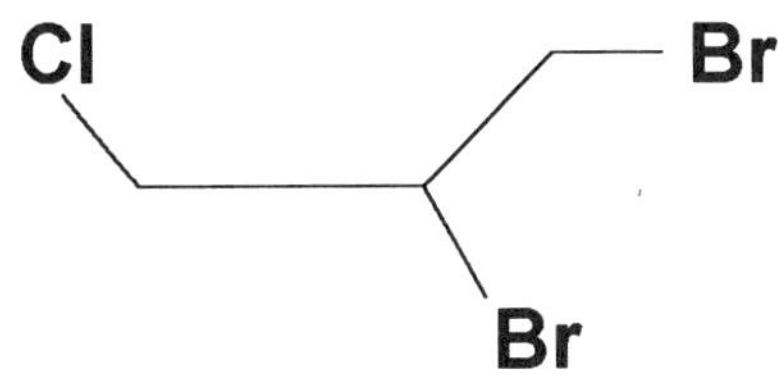

IDENTIFICATION AND TYPICAL USES

Dibromochloropropane (DBCP) is a dense yellow or amber liquid with a pungent odor at high concentrations. It will appear as a solid at temperatures below 43°F (6°C). It is primarily used as an insecticide, nematocide, and soil fumigant.

RISK ASSESSMENT: HEALTH

General Assessment

DBCP is a human poison and is toxic by *ingestion, inhalation,* and skin contact (*absorption*). Exposure appears to cause systemic effects by all routes of entry. It will irritate the nose and throat (mucous membranes) and is a severe eye and skin irritant. It also appears to have a narcotic effect at high concentrations and has also been implicated in male sterility. It is highly suspected as a human carcinogen. Human mutation data have also been reported. It may cause diminished renal function and degeneration and cirrhosis of the liver.

Inhalation will cause nausea, vomiting, conjunctivitis, respiratory irritation and distress, pulmonary congestion, pulmonary edema, testicular atrophy, spleen necrosis, liver and kidney damage, and sperm count depression. May also cause central nervous system (CNS) depression and subsequent symptoms of apathy, drowsiness, sluggishness, and ataxia.

Ingestion will result in digestive system irritation, unspecified but adverse gastrointestinal effects, nausea, vomiting, cramps, headache, and many of the same symptoms associated with inhalation exposures, including CNS depression.

Skin contact results in severe irritation and may produce dermatitis with rash, erythema, inflammation, and itching. Eye contact can result in severe irritation and possible corneal damage.

Acute Health Effects

The following acute (short-term) health effects may occur immediately or shortly after exposure to DBCP:

Skin: Irritation, inflammation, redness, itching, erythema, and possible dermatitis. May absorb through unbroken skin to cause toxic systemic effects.

Eye: Severe irritation, conjunctivitis, and possible damage to the cornea.

Lung: Irritation causing coughing and/or shortness of breath. Higher exposures can lead to respiratory problems, including pulmonary edema (fluid buildup in lungs) which is a medical emergency and can be fatal.

CNS: Dizziness, headaches, nausea, vomiting, some anesthesia, narcosis (on high concentrations).

Other: High concentrations may cause sperm count depression and changes in the liver, kidneys, and/or spleen.

Chronic Health Effects

The following chronic (long-term) health effects may occur at some time after exposure to DBCP and can last for months or even years:

Cancer Hazards: According to the information presented in the references, DBCP has been shown to cause cancer in test animals. Reports of human carcinogenicity in the references are inconclusive and conflicting. Extreme caution is warranted. Human mutation data have been confirmed.

Reproductive Hazard: According to the information presented in the references, DBCP may have the ability to adversely affect the male reproductive system and cause sperm count depression and testicular atrophy.

Other Chronic Effects: Low, chronic exposures to skin may cause allergy or dermatitis. Once an allergy develops, even small future exposures will cause a recurrence of symptoms. Repeated low exposures can also lead to liver, kidney, and spleen dysfunction.

Recommended Risk-Reduction Measures

Unless a less toxic chemical can be substituted for a hazardous substance, *engineering controls* are the most effective method of reducing exposures. The best protection is to enclose operations and/or provide local exhaust ventilation at the site of chemical release. While not always operationally feasible, isolating operations can also reduce exposure.

Using respiratory protection is less effective than the controls mentioned above, but is still advisable whenever working with or around DBCP. At any exposure to a suspected human carcinogen, a self-contained breathing apparatus (SCBA) with full facepiece and pressure demand is the recommended respiratory protection of choice. If a full facepiece is not available, then chemical goggles should be worn to protect the eyes. Whenever a chemical splash hazard exists, a face shield and apron should be worn. Rubber materials should be avoided since this chemical has been known to react with some types of natural and synthetic rubbers. To prevent hand and skin exposures, impervious gloves, such as polyvinyl chloride, should be worn.

Administrative controls should also be in place to minimize the potential for human exposures. These may include written procedures or policies which specify the methods and techniques that will be practiced whenever personnel are to work with DBCP.

All personnel should receive *training* on the use, hazards, protective measures, emergency actions, and other precautions per OSHA 29 CFR 1910.1044 (DBCP) and 29 CFR 1910.1200 (Hazard Communication), prior to the first assignment in an area where DBCP is used or stored.

Before beginning employment and at regular intervals thereafter (e.g. annually), the following medical tests are recommended:

☑ Lung function tests (establish a baseline).

If symptoms develop or overexposure is suspected, the following medical tests may be useful:

☑ Lung function tests.

☑ Liver and kidney function tests.

☑ Evaluation by a qualified allergist with careful evaluation of exposure history and administration of special tests (may help diagnose skin allergy).

Any evaluation should include a careful medical history of past and present symptoms with an examination. However, medical tests that evaluate existing damage are not a substitute for controlling exposure. Also, since smoking can cause heart disease, emphysema, and numerous other respiratory disorders, smokers may be affected more rapidly than non-smokers to the same exposure conditions. Prudent risk assessment and management requires careful consideration of *all* possible risk factors that may be causing the appearance of symptoms in exposed workers.

Other methods to reduce exposure include:

☑ Where possible, enclose operations and use local exhaust ventilation at the site of chemical release. If local exhaust ventilation or enclosure is not used, respiratory protection should be mandatory.

☑ Always ensure that proper protective clothing is worn when using chemical substances.

☑ Wash thoroughly immediately after exposure to DBCP and at the end of the work shift or before eating, drinking, or smoking.

☑ Hazard warning information should be posted in the work area. In addition, as part of an on-going education and training program, all information on the health and safety hazards of DBCP should be communicated to all potentially exposed workers.

☑ Eye wash stations should be provided in work areas. If the potential for whole body exposure exists, then safety showers should also be provided.

RISK ASSESSMENT: ENVIRONMENT

General Assessment

The environment is at risk of exposure during intended use, transportation, storage, disposal, or destruction of DBCP. In almost every scenario, the threat of environmental exposure is contingent upon the proper handling of the chemical substance. Accidental spills, large or small, can result in fire, explosion, and possible contamination of the surrounding environmental mediums (water, soil, and air).

DBCP is considered a Class IIIA combustible liquid (per OSHA 29 CFR 1910.106). It can react with common metals such as aluminum, magnesium, and tin alloys to cause a fire or explosion. These characteristics require special consideration during any emergency situation involving a leak or spill of DBCP or DBCP mixtures. Should DBCP ever come into contact with incompatible substances either during use, transportation, or storage, violent reactions can occur.

DBCP can enter the environment through its intended use as a soil fumigant and pesticide. However, its use in the United States has been severely curtailed and, in some instances, prohibited and, therefore, little release of DBCP presently occurs. It can reach the environment through industrial discharges and spills.

Acute Ecological Effects

Acute (short-term) toxic effects may include the death of animals, birds, or fish and death or low growth rate in plants. Acute effects are seen 2 to 4 days after animals or plants are exposed to DBCP.

Insufficient data are available to evaluate or predict the short-term effects of DBCP to aquatic life, plants, birds, or land animals.

Chronic Ecological Effects

Chronic toxic effects may include shortened life span, reproductive problems, lower fertility, and changes in appearance or behavior in exposed animals. These effects can be seen long after first exposure(s) to toxic chemicals.

DBCP has slight chronic toxicity to aquatic life. Insufficient data are available to evaluate or predict the long-term effects of DBCP to plants, birds, or land animals.

Water Solubility

DBCP is nearly insoluble in water. Concentrations of less than 1 milligram will mix with a liter of water.

Persistence in the Environment

DBCP is moderately persistent in water, with a half-life of between 20 to 200 days. The half-life of a pollutant is the amount of time it takes for one half of the chemical to be degraded.

Bioaccumulation in Aquatic Organisms

Some substances increase in concentration, or *bioaccumulate,* in living organisms as they breathe contaminated air, drink contaminated water, or eat contaminated food. These chemicals can become concentrated in the tissues and internal organs of animals as well as humans who have been exposed to the chemical.

The concentration of DBCP found in fish tissues is expected to be somewhat higher than the average concentration of DBCP in the water from which the fish was taken.

Recommended Risk-Reduction Measures

Proper training of all transporters will reduce the likelihood of a mishap or accident resulting in a leak or spill to the environment. The correct use of DOT labeling (per the MSDS) on all containers, trucks, and rail cars will enable emergency responders to react properly and quickly to any disaster thereby reducing the potential risk to the environment and to personnel.

Storage of DBCP should be segregated from incompatible chemicals to minimize the risk of cross-contamination or contact. Storage buildings should be equipped with the proper fire protection equipment (alarms, sprinklers, extinguishers, etc.).

If a spill or leak to the environment has occurred, fire department, emergency response and/or hazardous materials spill personnel should be notified immediately. Cleanup should be attempted only by those trained in proper spill containment procedures. This chemical will remain in soil for extended periods and may leach to underground aquifers. Contaminated soils should therefore be removed for incineration and replaced with clean soil. If DBCP should contact the water table, aquifer, or navigable waterway, remediation activities should be prompt to ensure protection of aquatic sediments. When such spills occur, the local and/or state emergency response authorities must be notified. A comprehensive emergency response or disaster preparedness/recovery plan should be in place prior to any operations involving the use, transportation, storage, or disposal of DBCP. If DBCP is spilled or leaked, the following specific steps are recommended:

- ☑ Restrict persons not wearing protective clothing from area of spill or leak until cleanup is complete and area can be opened for normal work.
- ☑ Ventilate area of spill or leak.
- ☑ Remove all ignition sources.
- ☑ Absorb liquid spills using vermiculite, dry earth, or sand to absorb and place in a sealed drum for disposal.
- ☑ It may be necessary to dispose of DBCP as a hazardous waste. The responsible state agency or the

regional office of the federal Environmental Protection Agency (EPA) should be contacted for specific recommendations.

RISK ASSESSMENT: BUSINESS

General Assessment

Accidents or mishaps involving DBCP can present a significant threat to business operations. The loss or damage of equipment or facilities can significantly affect fiscal viability. Lawsuits that may result from personnel injury/death, public exposures, and/or environmental contamination will also require a serious expenditure of resources. Media attention surrounding an injury, death, or environmental damage can also result in a loss of profits and loss of current as well as future business. Always remember that anytime the terms "carcinogen," "cancer," and "reproductive hazard" are used, public emotion and ignorance can run equally high. This should be a consideration in any policy planning activities.

Recommended Risk-Reduction Measures

Company attorneys, safety and health professionals, and environmental specialists should be involved in the development of any procedures dealing with safety or emergency response. A company official should be pre-designated as a public relations officer with specific training in dealing with the press. Corporate plans and policies should be developed, approved, and implemented long before any need for such arises.

REFERENCES

American Conference of Governmental Industrial Hygienists. 1988. *Documentation of the Threshold Limit Values and Biological Exposure Indices,* 5th Edition (with updates). Cincinnati: ACGIH

Anthony, C. P., and N. J. Kolthoff. 1971. *Textbook of Anatomy and Physiology.* St. Louis: C. V. Mosby Company.

Asante-Duah, D. K. 1993. *Hazardous Waste Risk Assessment.* New York: CRC Press/Lewis Publishers.

Bloom, A. D. and F. J. DeSerres. 1995. *Ecotoxicity and Human Health.* New York: CRC Press/Lewis Publishers.

Bueche, J. R. 1972. *Principles of Physics,* 2nd Edition. New York: McGraw-Hill.

Burrell, R., D. K. Flaherty, and L. J. Sauers. 1992. *Toxicology of the Immune System.* New York: Van Nostrand Reinhold.

Calabrese, E. J., E. M. Kenyon. *Air Toxics and Risk Assessment.* New York: CRC Press/Lewis Publishers.

Cockerman, L. G., B. S. Shane. 1994. *Basic Environmental Toxicology.* New York: CRC Press/Lewis Publishers.

Cothern, C. R. 1995. *Handbook for Environmental Risk Decisionmaking.* New York: CRC Press/Lewis Publishers.

Gardner, R. W. 1994. *Chemical Intolerance.* New York: CRC Press/Lewis Publishers.

Hodgson, E., R. B. Mailman, and Janice E. Chambers. 1988. *Dictionary of Toxicology.* London: The MacMillian Company.

Howard, P. H. 1995. *Handbook of Environmental Fate and Exposures Data for Organic Chemicals,* Vol. I-IV. New York: CRC Press/Lewis Publishers.

Jones, F. E. 1994. *Toxic Organic Vapors in the Workplace.* New York: CRC Press/Lewis Publishers.

Landis, W. G., M. H. Yu. 1995. *Introduction to Environmental Toxicology.* New York: CRC Press/Lewis Publishers.

Langerman, N. 1994 *Precautionary Labels for Chemical Containers.* New York: CRC Press/Lewis Publishers.

Lauwerys, R. R., P. Hoet. 1993. *Industrial Chemical Exposure.* New York: CRC Press/Lewis Publishers.

Lewis, R. J., Sr. 1993. *Hawley's Condensed Chemical Dictionary,* 12th Edition. New York: Van Nostrand Reinhold.

Lewis, R. J., Sr. 1992. *Sax's Dangerous Properties of Industrial Materials,* Volumes 1, 2, and 3, 8th Edition. New York: Van Nostrand Reinhold.

Manahan, S. E. 1992. *Toxicological Chemistry,* 2nd Edition. New York: CRC Press/Lewis Publishers.

Mellström, G. A., J. E. Wahlberg, and H. I. Maibach. 1994. *Protective Gloves for Occupational Use.* New York: CRC Press/Lewis Publishers.

Meyer, E. 1990. *Chemistry of Hazardous Materials.* Englewood Cliffs, NJ: Prentice Hall.

Nelson, G. O. 1992. *Gas Mixtures.* New York: CRC Press/Lewis Publishers.

Ott, W. R. 1995. *Environmental Statistics and Data Analysis.* New York: CRC Press/Lewis Publishers.

Philp, R. B. 1995. *Environmental Hazards and Human Health.* New York: CRC Press/Lewis Publishers.

Proctor, N. H., and James P. Hughes. 1991. *Chemical Hazards of the Workplace.* New York: Van Nostrand Reinhold.

Quigley, D. R. 1994. *Handbook of Emergency Chemical Management.* New York: CRC Press/Lewis Publishers.

Rea, W. J. 1995. *Chemical Sensitivity,* Vol. I-III. New York: CRC Press/Lewis Publishers.

Richardson, M. 1995. *Dictionary of Substance and Their Effects,* Vol. I-VI. New York: CRC Press/Lewis Publishers.

Sacarello, H. L. A. 1994. *The Comprehensive Handbook of Hazardous Materials.* New York: CRC Press/Lewis Publishers.

U.S. Department of Health and Human Services, National Institute for Occupational Safety and Health. 1994. *NIOSH Pocket Guide to Chemical Hazards.* Washington, D.C.: U.S. Government Printing Office.

MATERIAL SAFETY DATA SHEET

CHEMICAL NAME

DIBUTYLAMINE

HAZARD WARNING INFORMATION

HEALTH	FIRE	REACTIVE	OTHER	DEGREE OF HAZARD	COLOR CODING	OTHER CODES
3	2	0		0 = Minimum Hazard 1 = Slight Hazard 2 = Moderate Hazard 3 = Serious Hazard 4 = Severe Hazard	HEALTH = BLUE FIRE = RED REACTIVITY = YELLOW OTHER = WHITE	OX = Oxidizer ACID = Acid ALK = Alkali COR = Corrosive W = Use No Water

SECTION I - GENERAL INFORMATION

Characterization	**Aliphatic Amine**
RCRA Number	**None**
EPA Class	**Not Applicable**
DOT Proper Shipping Name	**Di(n-butyl)amine**
Chemical Abstract Service (CAS) Number	**111-92-2**
DOT Hazard Class and Label Requirements	**IMO; Corrosive Material**
DOT Emergency Guide Code	**68**
DOT Identification Number	**UN 2248**
Chemical Formula	**$(C_4H_9)_2NH$**

Synonyms

Di-*n*-butylamine; 1-butanamine-*n*-butyl; *N*-butyl-1-butamine.

SECTION II - HAZARDOUS INGREDIENTS AND IDENTITY INFORMATION

Hazardous Components (specific identity)	OSHA Exposure Criteria	NIOSH Exposure Criteria	Immediately Dangerous to Life and Health (IDLH)	ACGIH Exposure Criteria
Dibutylamine (derivation: By reaction of butanol or butyl chloride with ammonia).	**PEL:** Not Established **STEL:** Not Established	**REL:** Not Established **STEL:** Not Established	**Not Determined**	**TLV:** Not Established **STEL:** Not Established

SECTION III - PHYSICAL AND CHEMICAL CHARACTERISTICS

Boiling Point	**318°F (159°C)**
Specific Gravity (H_2O = 1)	**0.8**
Vapor Pressure (mm Hg)	**2 at 69°F (20°C)**
Molecular Weight	**129.3**
Vapor Density (Air = 1)	**4.5**
Melting Point	**-74°F (-59°C)**

Solubility

Nearly insoluble in water. Soluble in alcohol and ether and most organic solvents. Miscible with hydrocarbons.

Appearance and Odor

Colorless or clear liquid with a strong, amine-like odor.

SECTION IV - FIRE AND EXPLOSION HAZARD DATA

Flash Point (method used)	**125°F (51.6°C) opened cup**
Explosive Limits in Air % by Volume	**LEL: 1.1% UEL: Not Determined**
NFPA Classification	**Class II Combustible Liquid**
Autoignition Temperature	**Not Determined**

Extinguishing Media

Use dry chemical, carbon dioxide, foam, or water (water may not be effective by itself).

Special Fire Fighting Procedures

Poisonous gases are produced in fire, wear full protective clothing and self-contained breathing apparatus (SCBA). Vapors are heavier than air and may travel for some distance to ignition source and flashback. Move containers from fire area if it can be done safely. Cool exposed container with water.

Unusual Fire and Explosion Hazards

Containers may explode in fire due to buildup of excessive internal pressures. Exercise extreme caution when attempting to move or relocate containers that may have been exposed to high heat.

SECTION V - REACTIVITY DATA

Stability		Conditions to Avoid
		Avoid contact with incompatible materials. Keep away from heat and flame.
Stable	Unstable	Incompatibility *(materials to avoid)*
X		Oxidizing agents (such as chlorine, bromine, and oxygen) and other oxidizable materials.
Hazardous Polymerization		Conditions to Avoid Under normal conditions of temperature and pressure, hazardous polymerization of dibutylamine is not expected to occur.
May Occur	Will Not Occur	Hazardous Decomposition or By-products
	X	Toxic and irritating oxides of nitrogen are produced when dibutylamine is heated to decomposition or involved in a fire.

SECTION VI - HEALTH HAZARD DATA

Primary Route(s) of Entry:	Inhalation?	Absorption (skin)?	Ingestion?
	X		X

Health Hazards

INHALATION: Irritation of the nose, throat, and lungs. Will cause coughing, nausea, headache, and may cause vomiting. May cause fluid buildup in lungs (pulmonary edema) and death.

SKIN & EYES: A corrosive. Eye contact will cause irritation and inflammation. Skin contact causes severe irritation and possible dermatitis.

INGESTION: Poison by ingestion. Gastrointestinal irritation and irritation of the mouth and stomach.

Carcinogenicity	NTP Listed?	IARC Cancer Review Group?	OSHA Regulated?	Target Organs?
Unknown Human Unknown Animal	No	No	No	Eyes, skin, respiratory system (especially lungs).

Medical Conditions Generally Aggravated by Exposure

None Reported.

Emergency and First-aid Procedures

Eye contact: Flush immediately with water for 15 minutes (minimum), seek medical attention. Skin contact: Remove all contaminated clothing. Immediately wash area with large amounts of soap and water. Seek medical assistance. For inhalation: Remove the person from exposure. Provide respiratory assistance and CPR. Transfer to medical facility. Observe for 24-48 hours for lung effects. If swallowed: Call the poison control center and seek medical attention immediately. Never attempt to give an unconscious or convulsing person anything by mouth.

SECTION VII - PRECAUTIONS FOR SAFE HANDLING AND USE

Steps to be Taken in Case Material is Released or Spilled

Absorb liquids in vermiculite or other absorbent and deposit in DOT-approved drum. Ventilate area and remove ignition sources. Restrict those not involved in cleanup from entering area. Notify appropriate authorities, if applicable.

Preferred Waste Disposal Method

No Citation.

Precautions to be Taken in Handling and Storage

Do not store dibutylamine in the presence of incompatible chemicals or materials. Store in tightly closed containers in a cool, well-ventilated area. Personnel working with dibutylamine should be trained on its proper handing and storage *prior* to being assigned to such responsibilities.

Other Precautions and Warnings

A corrosive material. Long-term storage should be refrigeration under inert atmosphere.

SECTION VIII - CONTROL MEASURES AND PERSONAL PROTECTIVE EQUIPMENT

Respiratory Protection *(specify type)*

No exposure levels have been established. This does not mean exposures are without risk to heath. For low or infrequent exposures, use an organic vapor respirator. Greater protection at higher exposures is provided with an MSHA/NIOSH-approved supplied-air respirator with full facepiece operated in continuos flow or other positive pressure mode (pressure demand), or, a self-contained breathing apparatus (SCBA) with full facepiece operated in pressure demand or other positive pressure mode.

Ventilation

Local exhaust or general mechanical systems recommended.

Protective Gloves	Eye Protection	Other Protective Clothing
Polyvinyl Alcohol, nitrile, or Viton Gloves	Chemical Goggles or Face Mask	Protective Apron

Work/Hygiene Practices

Always wash hands thoroughly after using chemical; never bring food, drink, or smoking materials into vicinity of chemicals.

DIBUTYLAMINE

$(C_4H_9)_2NH$ CAS: 111-92-2

$CH_3 — CH_2 — CH_2 — CH_2 — NH — CH_2 — CH_2 — CH_2 — CH_3$

IDENTIFICATION AND TYPICAL USES

Dibutylamine is a colorless or clear liquid with an amine-like odor. It is used primarily as a corrosion inhibitor but is also used as an intermediate for emulsifiers, in rubber accelerators, in dyes, insecticides, and flotation agents. It is also used as an inhibitor for butadiene.

RISK ASSESSMENT: HEALTH

General Assessment

Dibutylamine is a human poison and is toxic by *ingestion* and *inhalation.* It is also moderately toxic by skin contact (corrosive). It will irritate the nose and throat (mucous membranes) and is a severe eye and skin irritant. Human mutation data have also been reported.

Inhalation will cause irritation to the nose, throat, and especially the lungs. It will cause cough, nausea, and headache. Its respiratory toxicity is primarily due to pulmonary effects. On high exposures, it may cause a fluid buildup in the lungs (pulmonary edema) which is a medical emergency and can be fatal. Symptoms may be delayed up to 48 hours, thereby creating a false sense of security relative to exposure risk.

Ingestion will result in digestive system irritation, unspecified but adverse gastrointestinal effects with irritation of the mouth and stomach. There may be nausea, vomiting, cramps, and headache.

Skin contact results in severe irritation and may produce dermatitis with rash and itching. Eye contact can result in severe irritation.

Acute Health Effects

The following acute (short-term) health effects may occur either immediately or shortly after (within hours or days) exposure to dibutylamine:

Skin: Irritation, inflammation, redness, itching, and possible dermatitis.

Eye: Severe irritation, inflammation, and possible damage to the cornea.

Lung: Irritation causing coughing and/or shortness of breath. Higher exposures can lead to respiratory problems, including pulmonary edema (fluid buildup in lungs) which is a medical emergency and can be fatal.

Chronic Health Effects

The following chronic (long-term) health effects may occur at some time after exposure to dibutylamine and can last for months or even years:

Cancer Hazards: According to the information presented in the references, dibutylamine has been shown to cause structural changes in the cells of test animals. It is a mutagen. There is insufficient evidence, however, to support or substantiate any claims of carcinogenicity in humans or animals.

Reproductive Hazard: According to the information presented in the references, dibutylamine has not been adequately tested for its ability to adversely affect reproduction.

Other Chronic Effects: Low, chronic exposures to skin may cause allergy or dermatitis. Once an allergy develops, even small future exposures will cause a recurrence of symptoms.

Recommended Risk-Reduction Measures

Unless a less toxic chemical can be substituted for a hazardous substance, *engineering controls* are the most effective method of reducing exposures. The best protection is to enclose operations and/or provide local exhaust ventilation at the site of chemical release.

While not always operationally feasible, isolating operations can also reduce exposure.

Using respiratory protection is less effective than the controls mentioned above, but is still advisable whenever working with or around dibutylamine. No exposure levels have been established for this chemical. This does not mean, however, that human exposures present no health risks. For relatively low, infrequent, or transient exposures, an organic vapor respirator may suffice. For higher or prolonged exposure, a self-contained breathing apparatus (SCBA) with full facepiece and pressure demand is the recommended respiratory protection of choice. If a full facepiece is not available, then chemical goggles should be worn to protect the eyes. Whenever a chemical splash hazard exists, a face shield and apron should be worn. To prevent hand and skin exposures, impervious nitrile, polyvinyl alcohol (PVA) or Viton gloves should be worn.

Administrative controls should also be in place to minimize the potential for human exposures. These may include written procedures or policies which specify the methods and techniques that will be practiced whenever personnel are to work with dibutylamine.

All personnel should receive *training* on the use, hazards, protective measures, emergency actions, and other precautions per 29 CFR 1910.1200 (Hazard Communication), prior to the first assignment in an area where dibutylamine is used or stored.

Before beginning employment and at regular intervals thereafter (e.g. annually), the following medical tests are recommended:

☑ Lung function tests (establish a baseline).

If symptoms develop or overexposure is suspected, the following medical tests may be useful:

☑ Lung function tests.

☑ Consider chest X-ray after acute overexposure (may be negative if performed immediately after exposure or if symptoms have not yet appeared).

☑ Evaluation by a qualified allergist with careful evaluation of exposure history and administration of special tests (may help diagnose skin allergy).

Any evaluation should include a careful medical history of past and present symptoms with an examination. However, medical tests that evaluate existing damage are not a substitute for controlling exposure. Also, since smoking can cause heart disease, emphysema, and numerous other respiratory disorders, smokers may be affected more rapidly than non-smokers to the same exposure conditions. Prudent risk assessment and management requires careful consideration of *all* possible risk factors that may be causing the appearance of symptoms in exposed workers.

Other methods to reduce exposure include:

☑ Where possible, enclose operations and use local exhaust ventilation at the site of chemical release. If local exhaust ventilation or enclosure is not used, respiratory protection should be mandatory. Also, if possible, automatically transfer liquids from storage containers to process containers using non-sparking tools.

☑ Always ensure that proper protective clothing is worn when using chemical substances.

☑ Wash thoroughly immediately after exposure to dibutylamine and at the end of the work shift or before eating, drinking, or smoking.

☑ Hazard warning information should be posted in the work area. In addition, as part of an on-going education and training program, all information on the health and safety hazards of dibutylamine should be communicated to all potentially exposed workers.

☑ Eye wash stations should be provided in work areas. If the potential for whole body exposure exists, then safety showers should also be provided.

RISK ASSESSMENT: ENVIRONMENT

General Assessment

The environment is at risk of exposure during intended use, transportation, storage, disposal, or destruction of dibutylamine. In almost every scenario, the threat of environmental exposure is contingent upon the proper handling of the chemical substance. Accidental spills, large or small, can result in fire, explosion, and possible contamination of the surrounding environmental mediums (water, soil, and air).

Dibutylamine is considered a Class II combustible liquid (per OSHA 29 CFR 1910.106). It can react with many common oxidizing materials to cause a fire or explosion. These characteristics require special consideration during any emergency situation involving a leak or spill of dibutylamine. Should dibutylamine ever come into contact with incompatible substances either during use, transportation, or storage, violent reactions can occur.

Dibutylamine can enter the environment through industrial discharges and spills.

☠ *Acute Ecological Effects*

Acute (short-term) toxic effects may include the death of animals, birds, or fish and death or low growth rate in plants. Acute effects are seen 2 to 4 days after animals or plants are exposed to dibutylamine. Insufficient data are available to evaluate or predict the short-term effects of dibutylamine to aquatic life, plants, birds, or land animals.

Chronic Ecological Effects

Chronic toxic effects may include shortened life span, reproductive problems, lower fertility, and changes in appearance or behavior in exposed animals. These effects can be seen long after first exposure(s) to toxic chemicals. Insufficient data are available to evaluate or predict the long-term effects of dibutylamine to aquatic life, plants, birds, or land animals.

Water Solubility

Dibutylamine is nearly insoluble in water. Concentrations of less than 1 milligram will mix with a liter of water.

⌛ *Persistence in the Environment*

Dibutylamine is moderately persistent in water, with a half-life of between 20 to 200 days. The half-life of a pollutant is the amount of time it takes for one half of the chemical to be degraded.

Bioaccumulation in Aquatic Organisms

Some substances increase in concentration, or *bioaccumulate,* in living organisms as they breathe contaminated air, drink contaminated water, or eat contaminated food. These chemicals can become concentrated in the tissues and internal organs of animals as well as humans.

The concentration of dibutylamine found in fish tissues is expected to be somewhat higher than the average concentration of dibutylamine in the water from which the fish was taken.

Recommended Risk-Reduction Measures

Proper training of all transporters will reduce the likelihood of a mishap or accident resulting in a leak or spill to the environment. The correct use of DOT labeling (per the MSDS) on all containers, trucks, and rail cars will enable emergency responders to react properly and quickly to any disaster thereby reducing the potential risk to the environment and to personnel.

Storage of dibutylamine should be segregated from incompatible chemicals to minimize the risk of cross-contamination or contact. Storage buildings should be equipped with the proper fire protection equipment (alarms, sprinklers, extinguishers, etc.).

If a spill or leak to the environment has occurred, fire department, emergency response and/or hazardous materials spill personnel should be notified immediately. Cleanup should be attempted only by those trained in proper spill containment procedures. This chemical will remain in soil for extended periods and may leach to underground aquifers. Contaminated soils should therefore be removed for incineration and replaced with clean soil. If dibutylamine should contact the water table, aquifer, or navigable waterway, remediation activities should be prompt to ensure protection of aquatic sediments. When such spills occur, the local and/or state emergency response authorities must be notified. A comprehensive emergency response or disaster preparedness/recovery plan should be in place prior to any operations involving the use, transportation, storage, or disposal of dibutylamine. If dibutylamine is spilled or leaked, the following specific steps are recommended:

- ☑ Restrict persons not wearing protective clothing from area of spill or leak until cleanup is complete and area can be opened for normal work.
- ☑ Ventilate area of spill or leak.
- ☑ Remove all ignition sources.
- ☑ Absorb liquid spills using vermiculite, dry earth, or sand to absorb and place in a sealed drum for disposal.
- ☑ It may be necessary to dispose of dibutylamine as a hazardous waste. The responsible state agency or the regional office of the federal Environmental Protection Agency (EPA) should be contacted for specific recommendations.

RISK ASSESSMENT: BUSINESS

General Assessment

Accidents or mishaps involving dibutylamine can present a significant threat to business operations. The loss or damage of equipment or facilities can significantly affect fiscal viability. Lawsuits that may result from personnel injury/death, public exposures, and/or environmental contamination will also require a serious expenditure of resources. Media attention surrounding an injury, death, or environmental damage can also result in a loss of profits and loss of current as well as future business.

Recommended Risk-Reduction Measures

Company attorneys, safety and health professionals, and environmental specialists should be involved in the development of any procedures dealing with safety or emergency response. A company official should be pre-designated as a public relations officer with specific training in dealing with the press. Corporate plans and policies should be developed, approved, and implemented long before any need for such arises.

REFERENCES

Ahlbom, A. 1993. *Biostatistics for Epidemiologists.* New York: CRC Press/Lewis Publishers.

American Conference of Governmental Industrial Hygienists. 1988. *Documentation of the Threshold Limit Values and Biological Exposure Indices,* 5th Edition (with updates). Cincinnati: ACGIH

Anthony, C. P., and N. J. Kolthoff. 1971. *Textbook of Anatomy and Physiology.* St. Louis: C. V. Mosby Company.

Asante-Duah, D. K. 1993. *Hazardous Waste Risk Assessment.* New York: CRC Press/Lewis Publishers.

Bloom, A. D. and F. J. DeSerres. 1995. *Ecotoxicity and Human Health.* New York: CRC Press/Lewis Publishers.

Bueche, J. R. 1972. *Principles of Physics,* 2nd Edition. New York: McGraw-Hill.

Burrell, R., D. K. Flaherty, and L. J. Sauers. 1992. *Toxicology of the Immune System.* New York: Van Nostrand Reinhold.

Calabrese, E. J., E. M. Kenyon. *Air Toxics and Risk Assessment.* New York: CRC Press/Lewis Publishers.

Calabrese, E. J. 1994. *Biological Effects of Low Level Exposures.* New York: CRC Press/Lewis Publishers.

Cockerman, L. G., B. S. Shane. 1994. *Basic Environmental Toxicology.* New York: CRC Press/Lewis Publishers.

Cothern, C. R., N. P. Ross. 1994. *Environmental Statistics, Assessment, and Forecasting.* New York: CRC Press/Lewis Publishers.

Cothern, R. R. 1993. *Comparative Environmental Risk Assessment.* New York: CRC Press/Lewis Publishers.

Cothern, C. R. 1995. *Handbook for Environmental Risk Decisionmaking.* New York: CRC Press/Lewis Publishers.

Gardner, R. W. 1994. *Chemical Intolerance.* New York: CRC Press/Lewis Publishers.

Hallenbeck, W. H. 1993. *Quantitative Risk Assessment for Environmental and Occupational Health,* 2nd Edition. New York: CRC Press/Lewis Publishers.

Hodgson, E., R. B. Mailman, and Janice E. Chambers. 1988. *Dictionary of Toxicology.* London: The MacMillian Company.

Howard, P. H. 1995. *Handbook of Environmental Fate and Exposures Data for Organic Chemicals,* Vol. I-IV. New York: CRC Press/Lewis Publishers.

Howard, P. H., et. al. 1991. *Handbook of Environmental Degradation Rates.* New York: CRC Press/Lewis Publishers.

Jones, F. E. 1994. *Toxic Organic Vapors in the Workplace.* New York: CRC Press/Lewis Publishers.

Landis, W. G., M. H. Yu. 1995. *Introduction to Environmental Toxicology.* New York: CRC Press/Lewis Publishers.

Langerman, N. 1994 *Precautionary Labels for Chemical Containers.* New York: CRC Press/Lewis Publishers.

Lauwerys, R. R., P. Hoet. 1993. *Industrial Chemical Exposure.* New York: CRC Press/Lewis Publishers.

Lewis, R. J., Sr. 1993. *Hawley's Condensed Chemical Dictionary,* 12th Edition. New York: Van Nostrand Reinhold.

Lewis, R. J., Sr. 1992. *Sax's Dangerous Properties of Industrial Materials,* Volumes 1, 2, and 3, 8th Edition. New York: Van Nostrand Reinhold.

Manahan, S. E. 1992. *Toxicological Chemistry,* 2nd Edition. New York: CRC Press/Lewis Publishers.

Mellström, G. A., J. E. Wahlberg, and H. I. Maibach. 1994. *Protective Gloves for Occupational Use.* New York: CRC Press/Lewis Publishers.

Meyer, E. 1990. *Chemistry of Hazardous Materials.* Englewood Cliffs, NJ: Prentice Hall.

Ott, W. R. 1995. *Environmental Statistics and Data Analysis.* New York: CRC Press/Lewis Publishers.

Philp, R. B. 1995. *Environmental Hazards and Human Health.* New York: CRC Press/Lewis Publishers.

Proctor, N. H., and James P. Hughes. 1991. *Chemical Hazards of the Workplace.* New York: Van Nostrand Reinhold.

Quigley, D. R. 1994. *Handbook of Emergency Chemical Management.* New York: CRC Press/Lewis Publishers.

Rea, W. J. 1995. *Chemical Sensitivity,* Vol. I-III. New York: CRC Press/Lewis Publishers.

Richardson, M. 1995. *Dictionary of Substance and Their Effects,* Vol. I-VI. New York: CRC Press/Lewis Publishers.

Rylander, R., R. R. Jacobs. 1994. *Organic Dusts.* New York: CRC Press/Lewis Publishers.

Sacarello, H. L. A. 1994. *The Comprehensive Handbook of Hazardous Materials.* New York: CRC Press/Lewis Publishers.

U.S. Department of Health and Human Services, National Institute for Occupational Safety and Health. 1994. *NIOSH Pocket Guide to Chemical Hazards.* Washington, D.C.: U.S. Government Printing Office.

MATERIAL SAFETY DATA SHEET

CHEMICAL NAME

DICHLOROACETYLENE

HAZARD WARNING INFORMATION

HEALTH	FIRE	REACTIVE	OTHER	DEGREE OF HAZARD	COLOR CODING	OTHER CODES
2	4	4		0 = Minimum Hazard 1 = Slight Hazard 2 = Moderate Hazard 3 = Serious Hazard 4 = Severe Hazard	HEALTH = BLUE FIRE = RED REACTIVITY = YELLOW OTHER = WHITE	OX = Oxidizer ACID = Acid ALK = Alkali COR = Corrosive ~~W~~ = Use No Water

SECTION I - GENERAL INFORMATION

Characterization: **Chlorinated Hydrocarbon**

RCRA Number: **D001, D003**

EPA Class: **Characteristic Waste (I) and (R)**

DOT Proper Shipping Name: **Dichloroacetylene**

Chemical Abstract Service (CAS) Number: **7572-29-4**

DOT Hazard Class and Label Requirements: **Forbidden**

DOT Emergency Guide Code: **No Citation**

DOT Identification Number: **None**

Chemical Formula: **C_2Cl_2**

Synonyms: **Dichloroethyne; DCA.**

SECTION II - HAZARDOUS INGREDIENTS AND IDENTITY INFORMATION

Hazardous Components (specific identity)	OSHA Exposure Criteria	NIOSH Exposure Criteria	Immediately Dangerous to Life and Health (IDLH)	ACGIH Exposure Criteria
Dichloroacetylene (derivation: An undesired decomposition product of trichloroethylene or trichloroethane; dehydrochlorination; or obtained by degradation of trichloroethylene, vinylidene chloride, and other chlorinated hydrocarbons by pyrolisis above 158°F, photolysis, or contact with alkaline materials). **1 ppm = 3.95 mg/m^3**	**PEL (ceiling):** **0.1 ppm** **0.4 mg/m^3** **STEL:** **Not Established**	**REL (ceiling):** **0.1 ppm** **0.4 mg/m^3** **Possible Cancer Agent**	**Not Determined**	**TLV (ceiling):** **0.1 ppm** **0.39 mg/m^3** **STEL:** **Not Established**

SECTION III - PHYSICAL AND CHEMICAL CHARACTERISTICS

Boiling Point: **90°F (32°C) Explodes**

Specific Gravity (H_2O = 1): **1.26**

Vapor Pressure (mm Hg): **Not Determined**

Molecular Weight: **94.9**

Vapor Density (Air = 1): **Not Determined**

Melting Point: **-86.8 to -83.6°F (-66 to - 64.2°C)**

Solubility: **Insoluble in water, soluble in alcohol, acetone, and ether.**

Appearance and Odor: **Can exist as a volatile oily liquid with a disagreeable, sweetish odor. Normally found in the workplace as a vapor or gas by-product of heated trichloroethylene (above 90°F). Not produced commercially.**

SECTION IV - FIRE AND EXPLOSION HAZARD DATA

Flash Point (method used): **Spontaneously Combustible**

Explosive Limits in Air % by Volume: **LEL: Not Determined** **UEL: Not Determined**

NFPA Classification: **Combustible Liquid**

Autoignition Temperature: **Not Determined**

Extinguishing Media: **For small fires, use carbon dioxide, water spray, or foam. On large fires, use water spray, foam, or fog.**

Special Fire Fighting Procedures: **Poisonous gases are produced in fire, wear full protective clothing and self-contained breathing apparatus (SCBA). Move containers from fire area if it can be done safely. Cool fire-exposed containers using water spray long after fire has been extinguished. Do not release to sewers or waterways.**

Unusual Fire and Explosion Hazards: **Containers may explode in fire due to buildup of excessive internal pressures. Vapors may explode when they reach or exceed 90°F (32.2°C). Do NOT use dry chemical on fires (may react explosively).**

SECTION V - REACTIVITY DATA

Stability		Conditions to Avoid: Dichloroacetylene is unstable, self-reactive, and shock-sensitive. In its gaseous state, it will spontaneously ignite on exposure to air. Avoid contact with incompatible materials. Keep away from heat and flame.
Stable	Unstable: X	Incompatibility *(materials to avoid)*: Dichloroacetylene will form toxic chloride fumes upon exposure to acids or acid fumes. It will react violently with strong oxidizers (such as chlorine, bromine, fluorine). Also incompatible with heat and shock.
Hazardous Polymerization		Conditions to Avoid: Avoid contact with acids and oxidizing materials. Do not allow exposure of vapors to air. Avoid situations where friction or shock might occur.
May Occur: X	Will Not Occur	Hazardous Decomposition or By-products: Toxic and irritating fumes of carbon dioxide and chloride are produced when dichloroacetylene is involved in fire.

SECTION VI - HEALTH HAZARD DATA

Primary Route(s) of Entry:	Inhalation?	Absorption (skin)?	Ingestion?
	X		X

Health Hazards

INHALATION: Headache, facial herpes, loss of appetite, intense jaw pain, vomiting, cranial nerve palsy causing uncontrollable tremor of the facial nerves, and nausea that can be extreme and disabling. High concentration can cause pulmonary or brain edema (a fluid buildup), which is a medical emergency and can be fatal. Long-term exposures can lead to CNS damage and behavioral changes, as well as kidney and liver injury.

SKIN & EYES: May cause reddened and blistered skin and injury or burning to the eyes.

INGESTION: Ingestion not likely. Insufficient data on results of ingesting liquid.

Carcinogenicity	NTP Listed?	IARC Cancer Review Group?	OSHA Regulated?	Target Organs?
Questioned Human Confirnmed Animal	No	Group 3	No	Respiratory system, CNS, kidneys, liver.

Medical Conditions Generally Aggravated by Exposure

None Reported.

Emergency and First-aid Procedures

Eye contact: Flush immediately with water for 15 minutes (minimum), seek medical attention. Skin contact: Remove all contaminated clothing. Immediately wash area with large amounts of soap and water. Seek medical assistance. For inhalation: Remove the person from exposure. Provide respiratory assistance and CPR. Transfer to medical facility. Observe for 24-48 hours for lung effects. If swallowed: Call the poison control center and seek medical attention immediately. Never attempt to give an unconscious or convulsing person anything by mouth.

SECTION VII - PRECAUTIONS FOR SAFE HANDLING AND USE

Steps to be Taken in Case Material is Released or Spilled

Evacuate area, deny entry, stay upwind. Dichloroacetylene is explosive as a liquid or gas. Cleanup personnel should wear fully encapsulated protective suit. If it can be done without risk, shut-off leak. For small spills, use absorbent material to collect liquids. For large spills, dike area far ahead and await for disposal or emergency response team. Follow OSHA, EPA, or other applicable guidelines.

Preferred Waste Disposal Method

No Citation.

Precautions to be Taken in Handling and Storage

Do not store dichloroacetylene in the presence of incompatible chemicals or materials. Store in tightly closed containers in a cool, well-ventilated area.

Other Precautions and Warnings

Bulk storage of dichloroacetylene is not recommended.

SECTION VIII - CONTROL MEASURES AND PERSONAL PROTECTIVE EQUIPMENT

Respiratory Protection *(specify type)*

Exposure level is low (0.01 ppm). Best protection is provided by a supplied-air respirator with full facepiece operated in pressure demand mode or a self-contained breathing apparatus (SCBA) with full facepiece operated in pressure demand or other positive pressure mode.

Ventilation

Local exhaust or general mechanical systems recommended.

Protective Gloves	Eye Protection	Other Protective Clothing
Chemically Impervious	Chemical Goggles or Face Mask	Apron and Gauntlets

Work/Hygiene Practices

Always wash hands thoroughly after using chemical; never bring food, drink, or smoking materials into vicinity of chemicals.

DICHLOROACETYLENE

C_2Cl_2 CAS: 7572-29-4

IDENTIFICATION AND TYPICAL USES

Dichloroacetylene is a volatile, oily liquid with a disagreeable, sweetish odor. It is usually found in the work environment as a vapor or gas by-product of heated trichloroethylene or, possibly, trichloroethane. It is not produced commercially for any known purpose.

RISK ASSESSMENT: HEALTH

General Assessment

Dichloroacetylene is a human poison and is toxic primarily by *inhalation,* and also poses a risk to health on skin contact (absorption, however, is not likely).

Inhalation of the normally heated vapors of dichloroacetylene can cause a wide range of health effects with symptoms such as headache, facial herpes, loss of appetite, intense jaw pain, vomiting, cranial nerve palsy with uncontrollable tremors of the facial nerves, and extreme nausea that may be disabling on prolonged exposures. Exposure to high concentrations may cause pulmonary (lung) or brain edema, which is a dangerous buildup of fluids in the respective organs, and can be fatal.

Skin contact may cause irritation and may produce redness and even blistering (prolonged contact). Eye contact can result in mild to severe irritation.

Acute Health Effects

The following acute (short-term) health effects may occur immediately or shortly after exposure to dichloroacetylene:

Skin: Irritation and possible redness and blistering.

Eye: Mild to severe irritation.

Lung: Irritation causing coughing and/or shortness of breath. Higher exposures can lead to respiratory problems, including pulmonary edema (fluid buildup in lungs) which is a medical emergency and can be fatal.

CNS: Symptoms of dizziness, headache, loss of appetite, jaw pain, cranial nerve palsy (may be permanent), and possible damage to the nervous system.

Chronic Health Effects

The following chronic (long-term) health effects may occur at some time after exposure to dichloroacetylene and can last for months or even years:

Cancer Hazards: According to the information presented in the references, dichloroacetylene has been shown to cause cancer in test animals. Human data are insufficient to support any claims of human carcinogenicity. Extreme caution is warranted.

Reproductive Hazard: According to the information presented in the references, dichloroacetylene has not been adequately tested for its ability to adversely affect reproduction.

Other Chronic Effects: Low, chronic exposures may lead to liver and kidney damage and possible injury to the central nervous system. There may be symptoms of behavioral changes and general weakness.

Recommended Risk-Reduction Measures

Unless a less toxic chemical can be substituted for a hazardous substance, *engineering controls* are the most effective method of reducing exposures. The best protection is to enclose operations and/or provide local exhaust ventilation at the site of chemical release. While not always operationally feasible, isolating operations can also reduce exposure.

Using respiratory protection is less effective than the controls mentioned above, but is still advisable whenever working with or around dichloroacetylene. The exposure level for this chemical is extremely low

(0.01 ppm), and difficult to measure accurately. Therefore, a self-contained breathing apparatus (SCBA) with full facepiece and pressure demand is the recommended respiratory protection of choice. If a full facepiece is not available, then chemical goggles should be worn to protect the eyes. Whenever a chemical splash hazard exists, a face shield and rubber apron should be worn. To prevent hand and skin exposures, rubber gloves should be worn.

Administrative controls should also be in place to minimize the potential for human exposures. These may include written procedures or policies which specify the methods and techniques that will be practiced whenever personnel are to work with dichloroacetylene.

All personnel should receive *training* on the use, hazards, protective measures, emergency actions, and other precautions per 29 CFR 1910.1200 (Hazard Communication), prior to the first assignment in an area where dichloroacetylene is used or stored.

If symptoms develop or overexposure is suspected, the following medical tests may be useful:

- ☑ Lung function tests.
- ☑ Liver and kidney function tests.
- ☑ Nervous system function tests, including interviews by a qualified neurologist to evaluate any changes in personality or behavior.

Any evaluation should include a careful medical history of past and present symptoms with an examination. However, medical tests that evaluate existing damage are not a substitute for controlling exposure.

Other methods to reduce exposure include:

- ☑ Where possible, enclose operations and use local exhaust ventilation at the site of chemical release. If local exhaust ventilation or enclosure is not used, respiratory protection should be mandatory. Automatic transfer of liquids from storage containers to process containers is recommended wherever possible.
- ☑ Always ensure that proper protective clothing is worn when using chemical substances.
- ☑ Wash thoroughly immediately after exposure to dichloroacetylene and at the end of the work shift or before eating, drinking, or smoking.
- ☑ Hazard warning information should be posted in the work area. In addition, as part of an on-going education and training program, all information on the health and safety hazards of dichloroacetylene should be communicated to all potentially exposed workers.
- ☑ Eye wash stations should be provided in work areas. If the potential for whole body exposure exists, then safety showers should also be provided.

RISK ASSESSMENT: ENVIRONMENT

General Assessment

The environment is at risk of exposure during intended use, transportation, storage, disposal, or destruction of dichloroacetylene. In almost every scenario, the threat of environmental exposure is contingent upon the proper handling of the chemical substance. Accidental spills, large or small, can result in fire, explosion, and possible contamination of the surrounding environmental mediums (water, soil, and air).

Dichloroacetylene is considered a combustible liquid. However, it is self-reactive, extremely sensitive to shock and impact, and can explode violently in contact with air at temperatures above 90°F (32.2°C). These characteristics require special consideration during any emergency situation involving a leak or spill of dichloroacetylene. Should dichloroacetylene ever come into contact with incompatible substances such as strong oxidizers (chlorine, bromine, fluorine), acids, acid fumes, or even air either during use, transportation, or storage, violent and explosive reactions can occur.

Dichloroacetylene can enter the environment through industrial discharges and spills.

☠ *Acute Ecological Effects*

Acute (short-term) toxic effects may include the death of animals, birds, or fish and death or low growth rate in plants. Acute effects are seen 2 to 4 days after animals or plants are exposed to dichloroacetylene. Insufficient data are available to evaluate or predict the short-term effects of dichloroacetylene to aquatic life, plants, birds, or land animals.

💣 *Chronic Ecological Effects*

Chronic toxic effects may include shortened life span, reproductive problems, lower fertility, and changes in appearance or behavior in exposed animals. These effects can be seen long after first exposure(s) to toxic chemicals. Insufficient data are available to evaluate or predict the long-term effects of dichloroacetylene to aquatic life, plants, birds, or land animals.

♦ *Water Solubility*

Dichloroacetylene is insoluble in water. Even concentrations of less than 1 milligram may not mix with a liter of water.

Persistence in the Environment

Dichloroacetylene is highly persistent in water, with a half-life of greater than 200 days. The half-life of a pollutant is the amount of time it takes for one half of the chemical to be degraded. Approximately 50% of dichloroacetylene will eventually end up in air; the rest will end up in water.

Bioaccumulation in Aquatic Organisms

Some substances increase in concentration, or *bioaccumulate,* in living organisms as they breathe contaminated air, drink contaminated water, or eat contaminated food. These chemicals can become concentrated in the tissues and internal organs of animals as well as humans.

The concentration of dichloroacetylene found in fish tissues is expected to be somewhat higher than the average concentration of dichloroacetylene in the water from which the fish was taken.

Recommended Risk-Reduction Measures

Proper training of all transporters will reduce the likelihood of a mishap or accident resulting in a leak or spill to the environment. The correct use of DOT labeling (per the MSDS) on all containers, trucks, and rail cars will enable emergency responders to react properly and quickly to any disaster thereby reducing the potential risk to the environment and to personnel.

Storage of dichloroacetylene should be segregated from incompatible chemicals to minimize the risk of cross-contamination or contact. Containers must be protected from shock and damage.

If a spill or leak to the environment has occurred, fire department, emergency response and/or hazardous materials spill personnel should be notified immediately. Cleanup should be attempted only by those trained in proper spill containment procedures. Contaminated soils should be removed for incineration and replaced with clean soil. If dichloroacetylene should contact the water table, aquifer, or navigable waterway, remediation activities should be prompt to ensure protection of aquatic sediments. When such spills occur, the local and/or state emergency response authorities must be notified. A comprehensive emergency response or disaster preparedness/recovery plan should be in place prior to any operations involving the use, transportation, storage, or disposal of dichloroacetylene. If dichloroacetylene is spilled or leaked, the following specific steps are recommended:

- ☑ Evacuate area and deny entry until cleanup is completed. Restrict persons not wearing protective clothing from hazard area.
- ☑ Ventilate area of spill or leak.
- ☑ Remove all ignition sources.
- ☑ If source of leak is a cylinder and it cannot be repaired in place, remove to safe location (e.g., outdoors) and allow to vent until empty. Exercise extreme caution. Chemical is shock and impact sensitive.
- ☑ Absorb liquid spills using vermiculite, dry earth, or sand to absorb and place in a sealed drum for disposal.
- ☑ It may be necessary to dispose of dichloroacetylene as a hazardous waste. The responsible state agency or the regional office of the federal Environmental Protection Agency (EPA) should be contacted for specific recommendations.

RISK ASSESSMENT: BUSINESS

General Assessment

Accidents or mishaps involving dichloroacetylene can present a significant threat to business operations. The loss or damage of equipment or facilities can significantly affect fiscal viability. Lawsuits that may result from personnel injury/death, public exposures, and/or environmental contamination will also require a serious expenditure of resources. Media attention surrounding an injury, death, or environmental damage can also result in a loss of profits and loss of current as well as future business. Always remember that anytime the terms "carcinogen," or "cancer" are used, public emotion, hysteria, and ignorance can run equally high. This should be considered during development of any public relations policies.

Recommended Risk-Reduction Measures

Company attorneys, safety and health professionals, and environmental specialists should be involved in the development of any procedures dealing with safety or emergency response. A company official should be pre-designated as a public relations officer with specific

training in dealing with the press. Corporate plans and policies should be developed, approved, and implemented long before any need for such arises.

REFERENCES

Ahlbom, A. 1993. *Biostatistics for Epidemiologists.* New York: CRC Press/Lewis Publishers.

American Conference of Governmental Industrial Hygienists. 1988. *Documentation of the Threshold Limit Values and Biological Exposure Indices,* 5th Edition (with updates). Cincinnati: ACGIH

Anthony, C. P., and N. J. Kolthoff. 1971. *Textbook of Anatomy and Physiology.* St. Louis: C. V. Mosby Company.

Asante-Duah, D. K. 1993. *Hazardous Waste Risk Assessment.* New York: CRC Press/Lewis Publishers.

Bloom, A. D. and F. J. DeSerres. 1995. *Ecotoxicity and Human Health.* New York: CRC Press/Lewis Publishers.

Brusick, D. J. 1994. *Methods for Genetic Risk Assessment.* New York: CRC Press/Lewis Publishers.

Bueche, J. R. 1972. *Principles of Physics,* 2nd Edition. New York: McGraw-Hill.

Burrell, R., D. K. Flaherty, and L. J. Sauers. 1992. *Toxicology of the Immune System.* New York: Van Nostrand Reinhold.

Calabrese, E. J., E. M. Kenyon. *Air Toxics and Risk Assessment.* New York: CRC Press/Lewis Publishers.

Calabrese, E. J. 1994. *Biological Effects of Low Level Exposures.* New York: CRC Press/Lewis Publishers.

Cockerman, L. G., B. S. Shane. 1994. *Basic Environmental Toxicology.* New York: CRC Press/Lewis Publishers.

Cothern, C. R., N. P. Ross. 1994. *Environmental Statistics, Assessment, and Forecasting.* New York: CRC Press/Lewis Publishers.

Cothern, R. R. 1993. *Comparative Environmental Risk Assessment.* New York: CRC Press/Lewis Publishers.

Cothern, C. R. 1995. *Handbook for Environmental Risk Decisionmaking.* New York: CRC Press/Lewis Publishers.

Gardner, R. W. 1994. *Chemical Intolerance.* New York: CRC Press/Lewis Publishers.

Hallenbeck, W. H. 1993. *Quantitative Risk Assessment for Environmental and Occupational Health,* 2nd Edition. New York: CRC Press/Lewis Publishers.

Hodgson, E., R. B. Mailman, and Janice E. Chambers. 1988. *Dictionary of Toxicology.* London: The MacMillian Company.

Howard, P. H. 1995. *Handbook of Environmental Fate and Exposures Data for Organic Chemicals,* Vol. I-IV. New York: CRC Press/Lewis Publishers.

Jones, F. E. 1994. *Toxic Organic Vapors in the Workplace.* New York: CRC Press/Lewis Publishers.

Kamrin, M. A. 1988. *Toxicology: A Primer on Principles and Applications.* Chelsea, Michigan: Lewis Publishers.

Keith, L. H. and M. M. Walker. 1995. *Handbook of Air Toxics.* New York: CRC Press/Lewis Publishers.

Landis, W. G., M. H. Yu. 1995. *Introduction to Environmental Toxicology.* New York: CRC Press/Lewis Publishers.

Langerman, N. 1994 *Precautionary Labels for Chemical Containers.* New York: CRC Press/Lewis Publishers.

Lauwerys, R. R., P. Hoet. 1993. *Industrial Chemical Exposure.* New York: CRC Press/Lewis Publishers.

Lide, D. R. 1995. *Handbook of Chemistry and Physics,* 76th Edition. New York: CRC Press/Lewis Publishers.

Lewis, R. J., Sr. 1993. *Hawley's Condensed Chemical Dictionary,* 12th Edition. New York: Van Nostrand Reinhold.

Lewis, R. J., Sr. 1992. *Sax's Dangerous Properties of Industrial Materials,* Volumes 1, 2, and 3, 8th Edition. New York: Van Nostrand Reinhold.

Manahan, S. E. 1994. *Environmental Chemistry,* 6th Edition. New York: CRC Press/Lewis Publishers.

Manahan, S. E. 1992. *Toxicological Chemistry,* 2nd Edition. New York: CRC Press/Lewis Publishers.

Mellström, G. A., J. E. Wahlberg, and H. I. Maibach. 1994. *Protective Gloves for Occupational Use.* New York: CRC Press/Lewis Publishers.

Meyer, E. 1990. *Chemistry of Hazardous Materials.* Englewood Cliffs, NJ: Prentice Hall.

Mycock, J. C., J. D. McKenna, and L. Theodore. 1995. *Handbook of Air Pollution Control Engineering and Technology.* New York: CRC Press/Lewis Publishers.

Nelson, G. O. 1992. *Gas Mixtures.* New York: CRC Press/Lewis Publishers.

Ott, W. R. 1995. *Environmental Statistics and Data Analysis.* New York: CRC Press/Lewis Publishers.

Patnaik, P. 1992. *A Comprehensive Guide to the Hazardous Properties of Chemical Substances.* New York: Van Nostrand Reinhold.

Philp, R. B. 1995. *Environmental Hazards and Human Health.* New York: CRC Press/Lewis Publishers.

Proctor, N. H., and James P. Hughes. 1991. *Chemical Hazards of the Workplace.* New York: Van Nostrand Reinhold.

Quigley, D. R. 1994. *Handbook of Emergency Chemical Management.* New York: CRC Press/Lewis Publishers.

Rea, W. J. 1995. *Chemical Sensitivity,* Vol. I-III. New York: CRC Press/Lewis Publishers.

Richardson, M. 1995. *Dictionary of Substance and Their Effects,* Vol. I-VI. New York: CRC Press/Lewis Publishers.

Rylander, R., R. R. Jacobs. 1994. *Organic Dusts.* New York: CRC Press/Lewis Publishers.

Sacarello, H. L. A. 1994. *The Comprehensive Handbook of Hazardous Materials.* New York: CRC Press/Lewis Publishers.

U.S. Department of Health and Human Services, National Institute for Occupational Safety and Health. 1994. *NIOSH Pocket Guide to Chemical Hazards.* Washington, D.C.: U.S. Government Printing Office.

MATERIAL SAFETY DATA SHEET

CHEMICAL NAME	
O-DICHLOROBENZENE	**CRC PRESS/LEWIS PUBLISHERS DISCLAIMER:** THE INFORMATION AND RECOMMENDATIONS PRESENTED HEREIN ARE BASED ON SOURCES BELIEVED TO BE RELIABLE. CRC MAKES NO REPRESENTATION ON ITS COMPLETENESS OR ACCURACY. IT IS THE RESPONSIBILITY OF THE USER TO DETERMINE THE CHEMICAL'S SUITABILITY FOR ITS INTENDED USE, THE CHEMICAL'S SAFE USE, AND THE CHEMICAL'S PROPER DISPOSAL. NO REPRESENTATIONS AND/OR WARRANTIES, EITHER EXPRESSED OR IMPLIED, OF THE MERCHANTABILITY OR FITNESS FOR A PARTICULAR PURPOSE, OR OF ANY OTHER NATURE, ARE MADE WITH RESPECT TO THE INFORMATION PROVIDED IN THIS MSDS OR TO THE CHEMICAL TO WHICH INFORMATION MAY REFER. CRC NEITHER ASSUMES NOR AUTHORIZES ANY OTHER PERSON TO ASSUME FOR IT, ANY OTHER ADDITIONAL RESPONSIBILITY OR LIABILITY FOR THE USE OF, OR RELIANCE UPON, THIS INFORMATION.

HAZARD WARNING INFORMATION

HEALTH	FIRE	REACTIVE	OTHER	DEGREE OF HAZARD	COLOR CODING	OTHER CODES
2	2	0		0 = Minimum Hazard 1 = Slight Hazard 2 = Moderate Hazard 3 = Serious Hazard 4 = Severe Hazard	HEALTH = BLUE FIRE = RED REACTIVITY = YELLOW OTHER = WHITE	OX = Oxidizer ACID = Acid ALK = Alkali COR = Corrosive W = Use No Water

SECTION I - GENERAL INFORMATION

Characterization	RCRA Number	EPA Class
Halogenated Hydrocarbon	**U070**	**Toxic Waste**

DOT Proper Shipping Name **Dichlorobenzene, Ortho, Liquid**	Chemical Abstract Service (CAS) Number **95-50-1**
DOT Hazard Class and Label Requirements **ORM-A**	DOT Emergency Guide Code **58**
DOT Identification Number **UN 1591**	Chemical Formula $C_6H_4Cl_2$

Synonyms
1,2-Dichlorobenzne; ortho dichlorobenzne; o-DCB; o-dichlorobenzol; chloroben; chloroden; DCB.

SECTION II - HAZARDOUS INGREDIENTS AND IDENTITY INFORMATION

Hazardous Components (specific identity)	OSHA Exposure Criteria	NIOSH Exposure Criteria	Immediately Dangerous to Life and Health (IDLH)	ACGIH Exposure Criteria
1,2-dichlorobenzene (derivation: By chlorination of monochlorobenzene). **1 ppm = 6.11 mg/m^3**	**PEL (ceiling):** **50 ppm** **300 mg/m^3** **STEL:** **Not Established**	**REL (ceiling):** **50 ppm** **300 mg/m^3** **STEL:** **Not Established**	**200 ppm**	**TLV (ceiling):** **50 ppm** **300 mg/m^3** **PROPOSED:** **25 ppm** **STEL:** **50 ppm**

SECTION III - PHYSICAL AND CHEMICAL CHARACTERISTICS

Boiling Point **357°F (180°C)**	Specific Gravity (H_2O = 1) **1.3**
Vapor Pressure (mm Hg) **1 at 69°F (20°C)**	Molecular Weight **147.0**
Vapor Density (Air = 1) **5.1**	Melting Point **0°F (-18°C)**

Solubility
Nearly insoluble in water (0.01%). Miscible with most organic solvents.

Appearance and Odor
Colorless to pale yellow liquid with a faint, pleasant, aromatic odor.

SECTION IV - FIRE AND EXPLOSION HAZARD DATA

Flash Point (method used) **151°F (66°C) closed cup**	Explosive Limits in Air % by Volume **LEL: 2.2% UEL: 9.2%**
NFPA Classification **Class IIIA Combustible Liquid**	Autoignition Temperature **1198°F (648°C)**

Extinguishing Media
Use dry chemical, carbon dioxide, foam, or water.

Special Fire Fighting Procedures
Poisonous gases are produced in fire, wear full protective clothing and self-contained breathing apparatus (SCBA). Vapors are heavier than air and may travel for some distance to ignition source and flashback. Move containers from fire area if it can be done safely. Cool exposed container with water.

Unusual Fire and Explosion Hazards
Containers may explode in fire due to buildup of excessive internal pressures. Exercise extreme caution when attempting to move or relocate containers that may have been exposed to high heat.

SECTION V - REACTIVITY DATA

Stability		Conditions to Avoid
		Avoid contact with incompatible materials. Keep away from heat and flame.
Stable	Unstable	Incompatibility *(materials to avoid)*
X		Strong oxidizers (such as chlorine, bromine, and oxygen) and other oxidizable materials, hot aluminum, and hot aluminum alloys, chlorides, acids, and acid fumes.
Hazardous Polymerization		Conditions to Avoid Under normal conditions of temperature and pressure, hazardous polymerization of o-dichlorobenzene is not expected to occur.
May Occur	Will Not Occur	Hazardous Decomposition or By-products
	X	When heated to decomposition, o-dichlorobenzene will produce toxic and irritating gases and vapors of hydrogen chloride and carbon monoxide.

SECTION VI - HEALTH HAZARD DATA

Primary Route(s) of Entry:	Inhalation?	Absorption (skin)?	Ingestion?
	X		X

Health Hazards

INHALATION: Damage to the lungs, liver, and kidneys. Coughing, central nervous system depression with transient anesthesia. Also causes lachrymation, irritation of the nose and throat, spasm of glottis and possible suffocation. Pulmonary edema and/or pneumonitis likely.

SKIN & EYES: Irritating to the skin, eyes, and membrane tissue. May cause dermatitis and blisters.

INGESTION: Poison by ingestion. Gastrointestinal irritation and irritation of the mouth and stomach.

Carcinogenicity	NTP Listed?	IARC Cancer Review Group?	OSHA Regulated?	Target Organs?
Questioned Human Questioned Animal	5th Annual Report	Group 3	29 CFR 1910.1000 Table Z-1	Eyes, skin, respiratory system, liver, kidneys.

Medical Conditions Generally Aggravated by Exposure

None Reported.

Emergency and First-aid Procedures

Eye contact: Flush immediately with water for 15 minutes (minimum), seek medical attention. Skin contact: Remove all contaminated clothing. Immediately wash area with large amounts of soap and water. Seek medical assistance. For inhalation: Remove the person from exposure. Provide respiratory assistance and CPR. Transfer to medical facility. Observe for 24-48 hours for lung effects. If swallowed: Call the poison control center and seek medical attention immediately. If conscious, induce vomiting. Never attempt to give an unconscious or convulsing person anything by mouth. DOCTOR: There is no known antidote. Treat symptomatically.

SECTION VII - PRECAUTIONS FOR SAFE HANDLING AND USE

Steps to be Taken in Case Material is Released or Spilled

Absorb liquids in vermiculite or other absorbent and deposit in DOT-approved drum. Ventilate area and remove ignition sources. Restrict those not involved in cleanup from entering area. Notify appropriate authorities, if applicable.

Preferred Waste Disposal Method

Burn in a chemical incinerator equipped with an afterburner and scrubber.

Precautions to be Taken in Handling and Storage

Do not store o-dichlorobenzene in the presence of incompatible chemicals or materials. Store in tightly closed containers in a cool, well-ventilated area. Do NOT store o-dichlorobenzene in aluminum containers (slow reaction may lead to explosion during storage).

Other Precautions and Warnings

Automatic transfer of liquids from storage to process containers is recommended.

SECTION VIII - CONTROL MEASURES AND PERSONAL PROTECTIVE EQUIPMENT

Respiratory Protection *(specify type)*

For low or infrequent exposures, use an organic vapor respirator. Greater protection at higher exposures is provided with an MSHA/NIOSH-approved supplied-air respirator with full facepiece operated in continuos flow or other positive pressure mode (pressure demand), or, a self-contained breathing apparatus (SCBA) with full facepiece operated in pressure demand or other positive pressure mode.

Ventilation

Local exhaust or general mechanical systems recommended.

Protective Gloves	Eye Protection	Other Protective Clothing
Butyl Rubber Gloves	Chemical Goggles or Face Mask	Protective Apron

Work/Hygiene Practices

Always wash hands thoroughly after using chemical; never bring food, drink, or smoking materials into vicinity of chemicals.

O-DICHLOROBENZENE

$C_6H_4Cl_2$ CAS: 95-50-1

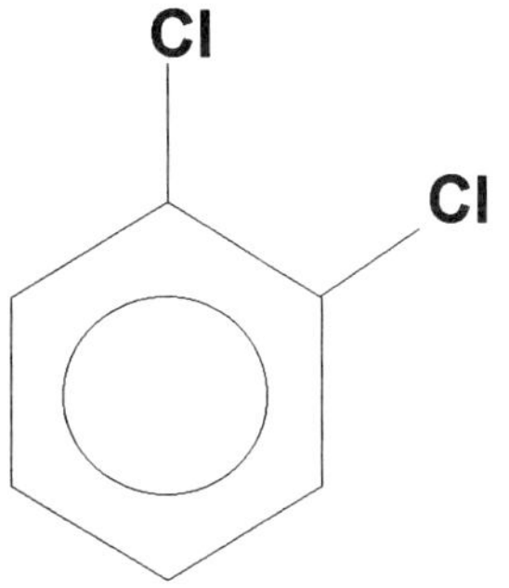

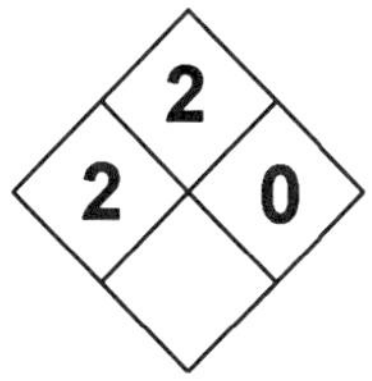

IDENTIFICATION AND TYPICAL USES

o-Dichlorobenzene is a colorless to pale yellow liquid with a faint, pleasant, aromatic odor. It is used in the manufacture of 3,4-dichloroaniline, as a solvent for a wide range of organic materials, and for oxides of nonferrous metals. It is a solvent carrier in the production of toluene diisocyanate. It is also used in dye manufacturing; as an insecticide, herbicide, and fumigant; for degreasing hides and wool; in metal polishes; for industrial odor control; and as a heat transfer medium.

RISK ASSESSMENT: HEALTH

General Assessment

o-Dichlorobenzene is a poison and is primarily toxic by *ingestion*. It is also moderately toxic by *inhalation* and will irritate the nose and throat (mucous membranes). It is a severe eye and skin irritant. It may cause liver and kidney injury and is also a questioned carcinogen and teratogen in animals. Human mutation data have also been reported.

Inhalation will cause irritation to the nose, throat, and especially the lungs. Symptoms range from cough and lachrymation to central nervous system depression and transient anesthesia. Its acute respiratory toxicity is primarily due to pulmonary effects. On high exposures, it may cause a fluid buildup in the lungs (pulmonary edema) which is a medical emergency and can be fatal. Symptoms may be delayed up to 48 hours, thereby creating a false sense of security relative to exposure risk.

Ingestion will result in digestive system irritation, unspecified but adverse gastrointestinal effects with irritation of the mouth and stomach. There may be nausea, vomiting, cramps, and headache.

Skin contact results in severe irritation and may produce dermatitis with rash and itching with blisters. Eye contact can result in severe irritation, inflammation, and possible damage to vision.

Acute Health Effects

The following acute (short-term) health effects may occur immediately or shortly after exposure to o-dichlorobenzene:

Skin: Irritation, inflammation, redness, itching, blistering, smarting, and possible dermatitis.

Eye: Severe irritation, inflammation, and possible damage to vision.

Lung: Irritation causing coughing and/or shortness of breath. Higher exposures can lead to respiratory problems, including pulmonary edema (fluid buildup in lungs) which is a medical emergency and can be fatal.

CNS: A central nervous system depressant causing narcotic effects with transient anesthesia.

Chronic Health Effects

The following chronic (long-term) health effects may occur at some time after exposure to o-dichlorobenzene and can last for months or even years:

Cancer Hazards: According to the information presented in the references, o-dichlorobenzene has been shown to cause liver and kidney changes in animals. Its carcinogenicity in humans is questioned. It is a mutagen. There is insufficient evidence, however, to support or substantiate any claims of carcinogenicity in humans or animals.

Reproductive Hazard: According to the information presented in the references, o-dichlorobenzene is suspected to adversely affect reproduction.

Other Chronic Effects: Low, chronic exposures to skin may cause allergy or dermatitis. Once an allergy develops, even small future exposures will cause a recurrence of symptoms. This chemical may cause damage to the liver or kidneys.

Recommended Risk-Reduction Measures

Unless a less toxic chemical can be substituted for a hazardous substance, *engineering controls* are the most effective method of reducing exposures. The best protection is to enclose operations and/or provide local exhaust ventilation at the site of chemical release. While not always operationally feasible, isolating operations can also reduce exposure.

Using respiratory protection is less effective than the controls mentioned above, but is still advisable whenever working with or around o-dichlorobenzene. For relatively low, infrequent, or transient exposures, an organic vapor respirator may suffice. For higher or prolonged, a self-contained breathing apparatus (SCBA) with full facepiece and pressure demand is the recommended respiratory protection of choice. If a full facepiece is not available, then chemical goggles should be worn to protect the eyes. Whenever a chemical splash hazard exists, a face shield and apron should be worn. To prevent hand and skin exposures, impervious butyl rubber gloves should be worn.

Administrative controls should also be in place to minimize the potential for human exposures. These may include written procedures or policies which specify the methods and techniques that will be practiced whenever personnel are to work with or around o-dichlorobenzene.

All personnel should receive *training* on the use, hazards, protective measures, emergency actions, and other precautions per 29 CFR 1910.1200 (Hazard Communication), prior to the first assignment in an area where o-dichlorobenzene is used or stored.

Before beginning employment and at regular intervals thereafter (e.g. annually), the following medical tests are recommended:

- ☑ Lung function tests (establish a baseline).

If symptoms develop or overexposure is suspected, the following medical tests may be useful:

- ☑ Lung function tests.
- ☑ Liver and kidney function tests.
- ☑ Consider chest X-ray after acute overexposure (may be negative if performed immediately after exposure or if symptoms have not yet appeared).
- ☑ Evaluation by a qualified allergist with careful evaluation of exposure history and administration of special tests (may help diagnose skin allergy).

Any evaluation should include a careful medical history of past and present symptoms with an examination. However, medical tests that evaluate existing damage are not a substitute for controlling exposure. Also, since smoking can cause heart disease, emphysema, and numerous other respiratory disorders, smokers may be affected more rapidly than non-smokers to the same exposure conditions. Prudent risk assessment and management requires careful consideration of *all* possible risk factors that may be causing the appearance of symptoms in exposed workers.

Other methods to reduce exposure include:

- ☑ Where possible, enclose operations and use local exhaust ventilation at the site of chemical release. If local exhaust ventilation or enclosure is not used, respiratory protection should be mandatory.
- ☑ Always ensure that proper protective clothing is worn when using chemical substances.
- ☑ Wash thoroughly immediately after exposure to o-dichlorobenzene and at the end of the work shift or before eating, drinking, or smoking.
- ☑ Hazard warning information should be posted in the work area. In addition, as part of an on-going education and training program, all information on the health and safety hazards of o-dichlorobenzene should be communicated to all potentially exposed workers.
- ☑ Eye wash stations should be provided in work areas. If the potential for whole body exposure exists, then safety showers should also be provided.

RISK ASSESSMENT: ENVIRONMENT

General Assessment

The environment is at risk of exposure during intended use, transportation, storage, disposal, or destruction of o-dichlorobenzene. In almost every scenario, the threat of environmental exposure is contingent upon the proper handling of the chemical substance. Accidental spills, large or small, can result in fire, explosion, and possible contamination of the surrounding environmental mediums (water, soil, and air).

o-Dichlorobenzene is considered a Class IIIA combustible liquid (per OSHA 29 CFR 1910.106). It can react with many strong oxidizers to cause a fire or explosion. In contact with aluminum, especially in heated conditions, it may react slowly and explode. Storage containers should not be made of aluminum. These characteristics require special consideration during any emergency situation involving a leak or spill of o-dichlorobenzene. Should o-dichlorobenzene ever come into contact with incompatible substances either during use, transportation, or storage, violent reactions can occur.

o-Dichlorobenzene can enter the environment through its intended use as an insecticide, herbicide, and fumigant or through industrial discharges and spills.

Acute Ecological Effects

Acute (short-term) toxic effects may include the death of animals, birds, or fish and death or low growth rate in plants. Acute effects are seen 2 to 4 days after animals or plants are exposed to o-dichlorobenzene.

As an herbicide, it does not appear that this chemical will cause damage to plants. However, insufficient data are available to evaluate or predict the short-term effects of o-dichlorobenzene to aquatic life, plants, birds, or land animals.

Chronic Ecological Effects

Chronic toxic effects may include shortened life span, reproductive problems, lower fertility, and changes in appearance or behavior in exposed animals. These effects can be seen long after first exposure(s) to toxic chemicals.

As an herbicide, it does not appear that this chemical will cause damage to plants. However, insufficient data are available to evaluate or predict the long-term effects of o-dichlorobenzene to aquatic life, plants, birds, or land animals.

Water Solubility

o-Dichlorobenzene is nearly insoluble in water (0.01%). Concentrations of less than 1 milligram will mix with a liter of water.

Persistence in the Environment

o-Dichlorobenzene is moderately persistent in water, with a half-life of between 20 to 200 days. The half-life of a pollutant is the amount of time it takes for one half of the chemical to be degraded.

Bioaccumulation in Aquatic Organisms

Some substances increase in concentration, or *bioaccumulate,* in living organisms as they breathe contaminated air, drink contaminated water, or eat contaminated food. These chemicals can become concentrated in the tissues and internal organs of animals as well as humans.

The concentration of o-dichlorobenzene found in fish tissues is expected to be somewhat higher than the average concentration of o-dichlorobenzene in the water from which the fish was taken.

Recommended Risk-Reduction Measures

Proper training of all transporters will reduce the likelihood of a mishap or accident resulting in a leak or spill to the environment. The correct use of DOT labeling (per the MSDS) on all containers, trucks, and rail cars will enable emergency responders to react properly and quickly to any disaster thereby reducing the potential risk to the environment and to personnel.

Storage of o-dichlorobenzene should be segregated from incompatible chemicals to minimize the risk of cross contamination or contact. Storage buildings should be equipped with the proper fire protection equipment (alarms, sprinklers, extinguishers, etc.).

If a spill or leak to the environment has occurred, fire department, emergency response and/or hazardous materials spill personnel should be notified immediately. Cleanup should be attempted only by those trained in proper spill containment procedures. This chemical will remain in soil for extended periods and may leach to underground aquifers. Contaminated soils should therefore be removed for incineration and replaced with clean soil. If o-dichlorobenzene should contact the water table, aquifer, or navigable waterway, remediation activities should be prompt to ensure protection of aquatic sediments. When such spills occur, the local and/or state emergency response authorities must be notified. A comprehensive emergency response or disaster preparedness/recovery plan should be in place prior to any operations involving the use, transportation, storage, or disposal of o-dichlorobenzene. If o-dichlorobenzene is spilled or leaked, the following specific steps are recommended:

- ☑ Restrict persons not wearing protective clothing from area of spill or leak until cleanup is complete and area can be opened for normal work.
- ☑ Ventilate area of spill or leak.
- ☑ Remove all ignition sources.
- ☑ Absorb liquid spills using vermiculite, dry earth, or sand to absorb and place in a sealed drum for disposal.
- ☑ It may be necessary to dispose of o-dichlorobenzene as a hazardous waste. The responsible state agency or the regional office of the federal Environmental Protection Agency (EPA) should be contacted for specific recommendations.

RISK ASSESSMENT: BUSINESS

General Assessment

Accidents or mishaps involving o-dichlorobenzene can present a significant threat to business operations. The loss or damage of equipment or facilities can significantly affect fiscal viability. Lawsuits that may result from personnel injury/death, public exposures, and/or environmental contamination will also require a serious expenditure of resources. Media attention surrounding an injury, death, or environmental damage can also result in a loss of profits and loss of current as well as future business.

Recommended Risk-Reduction Measures

Company attorneys, safety and health professionals, and environmental specialists should be involved in the development of any procedures dealing with safety or emergency response. A company official should be pre-designated as a public relations officer with specific training in dealing with the press. Corporate plans and policies should be developed, approved, and implemented long before any need for such arises.

REFERENCES

American Conference of Governmental Industrial Hygienists. 1988. *Documentation of the Threshold Limit Values and Biological Exposure Indices,* 5th Edition (with updates). Cincinnati: ACGIH

Anthony, C. P., and N. J. Kolthoff. 1971. *Textbook of Anatomy and Physiology.* St. Louis: C. V. Mosby Company.

Bloom, A. D. and F. J. DeSerres. 1995. *Ecotoxicity and Human Health.* New York: CRC Press/Lewis Publishers.

Bueche, J. R. 1972. *Principles of Physics,* 2nd Edition. New York: McGraw-Hill.

Burrell, R., D. K. Flaherty, and L. J. Sauers. 1992. *Toxicology of the Immune System.* New York: Van Nostrand Reinhold.

Calabrese, E. J., E. M. Kenyon. *Air Toxics and Risk Assessment.* New York: CRC Press/Lewis Publishers.

Cockerman, L. G., B. S. Shane. 1994. *Basic Environmental Toxicology.* New York: CRC Press/Lewis Publishers.

Cothern, C. R., N. P. Ross. 1994. *Environmental Statistics, Assessment, and Forecasting.* New York: CRC Press/Lewis Publishers.

Gardner, R. W. 1994. *Chemical Intolerance.* New York: CRC Press/Lewis Publishers.

Hallenbeck, W. H. 1993. *Quantitative Risk Assessment for Environmental and Occupational Health,* 2nd Edition. New York: CRC Press/Lewis Publishers.

Hodgson, E., R. B. Mailman, and Janice E. Chambers. 1988. *Dictionary of Toxicology.* London: The MacMillian Company.

Howard, P. H. 1995. *Handbook of Environmental Fate and Exposures Data for Organic Chemicals,* Vol. I-IV. New York: CRC Press/Lewis Publishers.

Langerman, N. 1994 *Precautionary Labels for Chemical Containers.* New York: CRC Press/Lewis Publishers.

Lauwerys, R. R., P. Hoet. 1993. *Industrial Chemical Exposure.* New York: CRC Press/Lewis Publishers.

Lewis, R. J., Sr. 1993. *Hawley's Condensed Chemical Dictionary,* 12th Edition. New York: Van Nostrand Reinhold.

Lewis, R. J., Sr. 1992. *Sax's Dangerous Properties of Industrial Materials,* Volumes 1, 2, and 3, 8th Edition. New York: Van Nostrand Reinhold.

Manahan, S. E. 1992. *Toxicological Chemistry,* 2nd Edition. New York: CRC Press/Lewis Publishers.

Meyer, E. 1990. *Chemistry of Hazardous Materials.* Englewood Cliffs, NJ: Prentice Hall.

Nelson, G. O. 1992. *Gas Mixtures.* New York: CRC Press/Lewis Publishers.

Ott, W. R. 1995. *Environmental Statistics and Data Analysis.* New York: CRC Press/Lewis Publishers.

Quigley, D. R. 1994. *Handbook of Emergency Chemical Management.* New York: CRC Press/Lewis Publishers.

Rea, W. J. 1995. *Chemical Sensitivity,* Vol. I-III. New York: CRC Press/Lewis Publishers.

Richardson, M. 1995. *Dictionary of Substance and Their Effects,* Vol. I-VI. New York: CRC Press/Lewis Publishers.

Sacarello, H. L. A. 1994. *The Comprehensive Handbook of Hazardous Materials.* New York: CRC Press/Lewis Publishers.

U.S. Department of Health and Human Services, National Institute for Occupational Safety and Health. 1994. *NIOSH Pocket Guide to Chemical Hazards.* Washington, D.C.: U.S. Government Printing Office.

MATERIAL SAFETY DATA SHEET

CHEMICAL NAME

P-DICHLOROBENZENE

HAZARD WARNING INFORMATION

HEALTH	FIRE	REACTIVE	OTHER	DEGREE OF HAZARD	COLOR CODING	OTHER CODES
2	2	0		0 = Minimum Hazard 1 = Slight Hazard 2 = Moderate Hazard 3 = Serious Hazard 4 = Severe Hazard	HEALTH = BLUE FIRE = RED REACTIVITY = YELLOW OTHER = WHITE	OX = Oxidizer ACID = Acid ALK = Alkali COR = Corrosive W = Use No Water

SECTION I - GENERAL INFORMATION

Characterization	**Halogenated Hydrocarbon**	RCRA Number / EPA Class	**U072** / **Toxic Waste**
DOT Proper Shipping Name	**Dichlorobenzene, Para, Solid**	Chemical Abstract Service (CAS) Number	**106-46-7**
DOT Hazard Class and Label Requirements	**ORM-A**	DOT Emergency Guide Code	**58**
DOT Identification Number	**UN 1591**	Chemical Formula	$C_6H_4Cl_2$

Synonyms

1,4-Dichlorobenzne; para dichlorobenzne; p-DCB; p-dichlorobenzol.

SECTION II - HAZARDOUS INGREDIENTS AND IDENTITY INFORMATION

Hazardous Components (specific identity)	OSHA Exposure Criteria	NIOSH Exposure Criteria	Immediately Dangerous to Life and Health (IDLH)	ACGIH Exposure Criteria
1,4-dichlorobenzene (derivation: By chlorination of monochlorobenzene). **1 ppm = 6.11 mg/m^3**	**PEL:** **75 ppm** **450 mg/m^3** **STEL:** **110 ppm** **670 mg/m^3**	**REL:** **Possible Cancer Agent** **STEL:** **Not Established**	**150 ppm**	**TLV:** **75 ppm** **450 mg/m^3** **Suspected Human Carcinogen** **STEL:** **110 ppm**

SECTION III - PHYSICAL AND CHEMICAL CHARACTERISTICS

Boiling Point	**345°F (174°C)**	Specific Gravity (H_2O = 1)	**1.25**
Vapor Pressure (mm Hg)	**1.3 at 69°F (20°C)**	Molecular Weight	**147.0**
Vapor Density (Air = 1)	**5.1**	Melting Point	**128°F (53°C)**

Solubility

Nearly insoluble in water (0.008%). Miscible with most organic solvents.

Appearance and Odor

Colorless or white crystalline solid with a characteristic mothball-like odor.

SECTION IV - FIRE AND EXPLOSION HAZARD DATA

Flash Point (method used)	**150°F (65°C) closed cup**	Explosive Limits in Air % by Volume	**LEL: 2.5% UEL: Not Determined**
NFPA Classification	**Combustible Solid**	Autoignition Temperature	**Not Determined**

Extinguishing Media

Use dry chemical, carbon dioxide, foam, or water.

Special Fire Fighting Procedures

Poisonous gases are produced in fire, wear full protective clothing and self-contained breathing apparatus (SCBA). Vapors are heavier than air and may travel for some distance to ignition source and flashback. Move containers from fire area if it can be done safely. Cool exposed container with water.

Unusual Fire and Explosion Hazards

Containers may explode in fire due to buildup of excessive internal pressures. Exercise extreme caution when attempting to move or relocate containers that may have been exposed to high heat.

SECTION V - REACTIVITY DATA

Stability		Conditions to Avoid: Avoid contact with incompatible materials. Keep away from heat and flame.
Stable	Unstable	Incompatibility *(materials to avoid)*: Strong oxidizers (such as chlorine, bromine, and oxygen) and other oxidizable materials.
X		
Hazardous Polymerization		Conditions to Avoid: Under normal conditions of temperature and pressure, hazardous polymerization of p-dichlorobenzene is not expected to occur. Becomes flammable when exposed to heat, flame, or oxidizers.
May Occur	Will Not Occur	Hazardous Decomposition or By-products: When heated to decomposition, p-dichlorobenzene will produce toxic and irritating gases and vapors of chlorine, hydrogen chloride, phosgene, and carbon monoxide.
	X	

SECTION VI - HEALTH HAZARD DATA

Primary Route(s) of Entry:	Inhalation?	Absorption (skin)?	Ingestion?
	X		X

Health Hazards

INHALATION: Irritation of the upper respiratory tract. Causes headache, weakness, dizziness, nausea, vomiting, diarrhea, loss of weight, and depression of the liver and kidneys. May cause jaundice and cirrhosis.

SKIN & EYES: Pain and mild irritation to the skin and eyes. Permanent problems not likely.

INGESTION: Moderately toxic. Causes unspecified changes in the eyes, lungs, thorax, and in respiration, with decreased motility or constipation. Will cause same symptoms as inhalation, with liver and kidney effects.

Carcinogenicity	NTP Listed?	IARC Cancer Review Group?	OSHA Regulated?	Target Organs?
Suspected Human Confirmed Animal	5th Annual Report	Group 2B	29 CFR 1910.1000 Table Z-1	Eyes, skin, respiratory system, liver, kidneys.

Medical Conditions Generally Aggravated by Exposure

None Reported.

Emergency and First-aid Procedures

Eye contact: Flush immediately with water for 15 minutes (minimum), seek medical attention. Skin contact: Remove all contaminated clothing. Immediately wash area with large amounts of soap and water. For inhalation: Remove the person from exposure. Provide respiratory assistance and CPR. Transfer to medical facility. Observe for 24-48 hours for lung effects (pulmonary edema). If swallowed: Call the poison control center and seek medical attention immediately.

SECTION VII - PRECAUTIONS FOR SAFE HANDLING AND USE

Steps to be Taken in Case Material is Released or Spilled

Absorb liquids in vermiculite or other absorbent and deposit in DOT-approved drum. Powders should be collected in safest manner possible and disposed of in sealed drums. Do NOT dry sweep. Use a HEPA vacuum instead. Ventilate area and remove ignition sources. Restrict those not involved in cleanup from entering area. Notify appropriate authorities, if applicable.

Preferred Waste Disposal Method

Dissolve in a combustible solvent and burn in a chemical incinerator equipped with an afterburner and scrubber.

Precautions to be Taken in Handling and Storage

Store in tightly closed containers in a cool, well-ventilated area. p-Dichlorobenzene may attack some forms of plastics, rubber, and coatings.

Other Precautions and Warnings

Automatic transfer of liquids from storage to process containers is recommended.

SECTION VIII - CONTROL MEASURES AND PERSONAL PROTECTIVE EQUIPMENT

Respiratory Protection *(specify type)*

For any exposure to a suspected carcinogen, use a supplied-air respirator with full facepiece operated in continuos flow or other positive pressure mode (pressure demand), or a self-contained breathing apparatus (SCBA) with full facepiece operated in pressure demand or other positive pressure mode.

Ventilation

Local exhaust or general mechanical systems recommended.

Protective Gloves	Eye Protection	Other Protective Clothing
Impervious Gloves	Chemical/Dust Goggles or Face Mask	Protective Apron

Work/Hygiene Practices

Always wash hands thoroughly after using chemical; never bring food, drink, or smoking materials into vicinity of chemicals.

p-DICHLOROBENZENE

$C_6H_4Cl_2$ CAS: 106-46-7

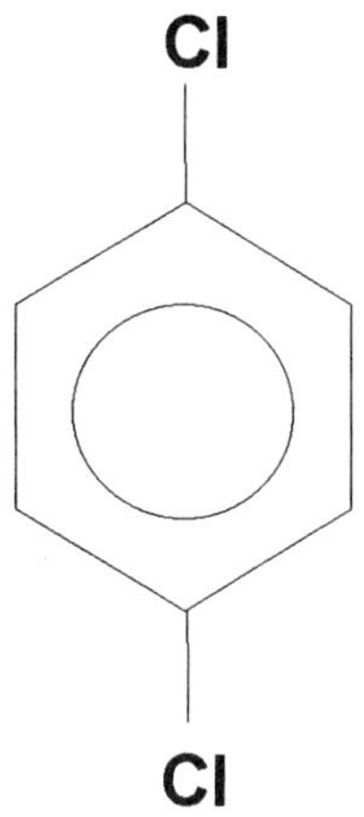

IDENTIFICATION AND TYPICAL USES

p-Dichlorobenzene is a colorless or white crystalline solid with a characteristic mothball-like odor. It is used as a moth repellent, a general insecticide, a germicide, a space odorant, in the manufacture of 2,5-dichloroaniline, in the manufacture of dyes, intermediates, in pharmaceuticals, and in the agricultural industry (soil fumigant).

RISK ASSESSMENT: HEALTH

General Assessment

p-Dichlorobenzene is moderately toxic to humans by *ingestion.* It is also moderately toxic by *inhalation* and will irritate the nose and throat (mucous membranes). It may cause liver and kidney injury and is also a confirmed carcinogen and suspected teratogen in animals. Human mutation data have also been reported.

Inhalation and ingestion will cause toxic symptoms of cough, headache, weakness, dizziness, nausea, vomiting, loss of weight, and injury to the liver and kidney. These symptoms occur from repeated inhalation of high concentrations of vapors or from ingestion. In addition, human systemic effects by ingestion also include unspecified changes in the eyes, lungs, thorax, and in respiration, and decreased motility or constipation. Chronic exposure may cause jaundice and cirrhosis. The vapors may also irritate the skin.

Skin and eye contact results in pain and mild irritation. Absorption is not likely.

☠ *Acute Health Effects*

The following acute (short-term) health effects may occur immediately or shortly after exposure to p-dichlorobenzene:

Skin: Mild irritation that may be painful on long exposures.

Eye: Mild to severe irritation with possible inflammation and conjunctivitis.

Lung: Irritation of the upper respiratory tract causing coughing and/or shortness of breath. Higher exposures can lead to respiratory problems, including pulmonary edema (fluid buildup in lungs) which is a medical emergency and can be fatal.

Other: Ingestion can cause symptoms ranging from no effect to headache, vomiting, weakness, loss of weight, and injury to the liver and kidneys.

Chronic Health Effects

The following chronic (long-term) health effects may occur at some time after exposure to p-dichlorobenzene and can last for months or even years:

Cancer Hazards: According to the information presented in the references, p-dichlorobenzene has been shown to cause liver and kidney changes in animals. Its carcinogenicity in humans is questioned. However, the data in the references are conflicting and inconclusive in this regard. It is a mutagen. Until further studies are performed, it is suggested that p-dichlorobenzene be handled with extreme caution.

Reproductive Hazard: According to the information presented in the references, p-dichlorobenzene is suspected to adversely affect reproduction in animals.

Other Chronic Effects: This chemical may cause unspecified damage to the eyes, lungs, thorax, liver, and kidneys. Repeated exposures may result in jaundice and cirrhosis.

Recommended Risk-Reduction Measures

Unless a less toxic chemical can be substituted for a hazardous substance, *engineering controls* are the most effective method of reducing exposures. The best protection is to enclose operations and/or provide local exhaust ventilation at the site of chemical release. While not always operationally feasible, isolating operations can also reduce exposure.

Using respiratory protection is less effective than the controls mentioned above, but is still advisable whenever working with or around p-dichlorobenzene. For any exposure to a suspected human carcinogen, a self-contained breathing apparatus (SCBA) with full facepiece and pressure demand is the recommended respiratory protection of choice. If a full facepiece is not available, then chemical goggles should be worn to protect the eyes. Whenever a chemical splash hazard exists, a face shield and apron should be worn. To prevent hand and skin exposures, impervious gloves should be worn. Rubber materials should be avoided since this chemical may attack some forms of rubber (as well as some plastics and coatings).

Administrative controls should also be in place to minimize the potential for human exposures. These may include written procedures or policies which specify the methods and techniques that will be practiced whenever personnel are to work with or around p-dichlorobenzene.

All personnel should receive *training* on the use, hazards, protective measures, emergency actions, and other precautions per 29 CFR 1910.1200 (Hazard Communication), prior to the first assignment in an area where p-dichlorobenzene is used or stored.

Before beginning employment and at regular intervals thereafter (e.g. annually), the following medical tests are recommended:

☑ Lung function tests (establish a baseline).

If symptoms develop or overexposure is suspected, the following medical tests may be useful:

☑ Lung function tests.

☑ Liver and kidney function tests.

☑ Consider chest X-ray after acute overexposure (may be negative if performed immediately after exposure or if symptoms have not yet appeared).

Any evaluation should include a careful medical history of past and present symptoms with an examination. However, medical tests that evaluate existing damage are not a substitute for controlling exposure. Also, since smoking can cause heart disease, emphysema, and numerous other respiratory disorders, smokers may be affected more rapidly than non-smokers to the same exposure conditions. Prudent risk assessment and management requires careful consideration of *all* possible risk factors that may be causing the appearance of symptoms in exposed workers.

Other methods to reduce exposure include:

☑ Where possible, enclose operations and use local exhaust ventilation at the site of chemical release. If local exhaust ventilation or enclosure is not used, respiratory protection should be mandatory.

☑ Always ensure that proper protective clothing is worn when using chemical substances.

☑ Wash thoroughly immediately after exposure to p-dichlorobenzene and at the end of the work shift or before eating, drinking, or smoking.

☑ Hazard warning information should be posted in the work area. In addition, as part of an on-going education and training program, all information on the health and safety hazards of p-dichlorobenzene should be communicated to all potentially exposed workers.

☑ Eye wash stations should be provided in work areas. If the potential for whole body exposure exists, then safety showers should also be provided.

RISK ASSESSMENT: ENVIRONMENT

General Assessment

The environment is at risk of exposure during intended use, transportation, storage, disposal, or destruction of p-dichlorobenzene. In almost every scenario, the threat of environmental exposure is contingent upon the proper handling of the chemical substance. Accidental spills, large or small, can result in fire, explosion, and possible contamination of the surrounding environmental mediums (water, soil, and air).

p-Dichlorobenzene is considered a combustible solid. It is often mixed in solutions with flammable materials. It can also react with many strong oxidizers to cause a fire or explosion. These characteristics require special consideration during any emergency situation involving a leak or spill of p-dichlorobenzene. Should p-dichlorobenzene ever come into contact with incompatible substances either during use, transportation, or storage, violent reactions can occur.

p-Dichlorobenzene can enter the environment through its intended use as an insecticide and fumigant or through industrial discharges and spills.

Acute Ecological Effects

Acute (short-term) toxic effects may include the death of animals, birds, or fish and death or low growth rate in plants. Acute effects are seen 2 to 4 days after animals or plants are exposed to p-dichlorobenzene. Insufficient data are available to evaluate or predict the short-term effects of p-dichlorobenzene to aquatic life, plants, birds, or land animals.

Chronic Ecological Effects

Chronic toxic effects may include shortened life span, reproductive problems, lower fertility, and changes in appearance or behavior in exposed animals. These effects can be seen long after first exposure(s) to toxic chemicals. Insufficient data are available to evaluate or predict the long-term effects of p-dichlorobenzene to aquatic life, plants, birds, or land animals.

Water Solubility

p-Dichlorobenzene is nearly insoluble in water (0.008%). Concentrations of less than 1 milligram will mix with a liter of water.

Persistence in the Environment

p-Dichlorobenzene is moderately persistent in water, with a half-life of between 20 to 200 days. The half-life of a pollutant is the amount of time it takes for one half of the chemical to be degraded.

Bioaccumulation in Aquatic Organisms

Some substances increase in concentration, or *bioaccumulate,* in living organisms as they breathe contaminated air, drink contaminated water, or eat contaminated food. These chemicals can become concentrated in the tissues and internal organs of animals as well as humans.

The concentration of p-dichlorobenzene found in fish tissues is expected to be somewhat higher than the average concentration of p-dichlorobenzene in the water from which the fish was taken.

Recommended Risk-Reduction Measures

Proper training of all transporters will reduce the likelihood of a mishap or accident resulting in a leak or spill to the environment. The correct use of DOT labeling (per the MSDS) on all containers, trucks, and rail cars will enable emergency responders to react properly and quickly to any disaster thereby reducing the potential risk to the environment and to personnel.

Storage of p-dichlorobenzene should be segregated from incompatible chemicals to minimize the risk of cross-contamination or contact. Storage buildings should be equipped with the proper fire protection equipment (alarms, sprinklers, extinguishers, etc.).

If a spill or leak to the environment has occurred, fire department, emergency response and/or hazardous materials spill personnel should be notified immediately. Cleanup should be attempted only by those trained in proper spill containment procedures. This chemical will remain in soil for extended periods and may leach to underground aquifers. Contaminated soils should therefore be removed for incineration and replaced with clean soil. If p-dichlorobenzene should contact the water table, aquifer, or navigable waterway, remediation activities should be prompt to ensure protection of aquatic sediments. When such spills occur, the local and/or state emergency response authorities must be notified. A comprehensive emergency response or disaster preparedness/recovery plan should be in place prior to any operations involving the use, transportation, storage, or disposal of p-dichlorobenzene. If p-dichlorobenzene is spilled or leaked, the following specific steps are recommended:

- ☑ Restrict persons not wearing protective clothing from area of spill or leak until cleanup is complete and area can be opened for normal work.
- ☑ Ventilate area of spill or leak.
- ☑ Remove all ignition sources.
- ☑ Absorb liquid spills using vermiculite, dry earth, or sand to absorb and place in a sealed drum for disposal. Collect powders in safest manner possible and dispose of in sealed drums. Do NOT dry

sweep (creates dusts). Use a vacuum with a high efficiency particulate air (HEPA) filter.

☑ It may be necessary to dispose of p-dichlorobenzene as a hazardous waste. The responsible state agency or the regional office of the federal Environmental Protection Agency (EPA) should be contacted for specific recommendations.

RISK ASSESSMENT: BUSINESS

General Assessment

Accidents or mishaps involving p-dichlorobenzene can present a significant threat to business operations. The loss or damage of equipment or facilities can significantly affect fiscal viability. Lawsuits that may result from personnel injury/death, public exposures, and/or environmental contamination will also require a serious expenditure of resources. Media attention surrounding an injury, death, or environmental damage can also result in a loss of profits and loss of current as well as future business. Always remember, anytime the terms "carcinogen," "cancer," or "reproductive" hazard is used, public emotion, hysteria, and ignorance can run equally high. This should be a consideration during the development of any corporate policies regarding public relations.

Recommended Risk-Reduction Measures

Company attorneys, safety and health professionals, and environmental specialists should be involved in the development of any procedures dealing with safety or emergency response. A company official should be pre-designated as a public relations officer with specific training in dealing with the press. Corporate plans and policies should be developed, approved, and implemented long before any need for such arises.

REFERENCES

American Conference of Governmental Industrial Hygienists. 1988. *Documentation of the Threshold Limit Values and Biological Exposure Indices,* 5th Edition (with updates). Cincinnati: ACGIH

Anthony, C. P., and N. J. Kolthoff. 1971. *Textbook of Anatomy and Physiology.* St. Louis: C. V. Mosby Company.

Bloom, A. D. and F. J. DeSerres. 1995. *Ecotoxicity and Human Health.* New York: CRC Press/Lewis Publishers.

Bueche, J. R. 1972. *Principles of Physics,* 2nd Edition. New York: McGraw-Hill.

Burrell, R., D. K. Flaherty, and L. J. Sauers. 1992. *Toxicology of the Immune System.* New York: Van Nostrand Reinhold.

Calabrese, E. J., E. M. Kenyon. *Air Toxics and Risk Assessment.* New York: CRC Press/Lewis Publishers.

Cockerman, L. G., B. S. Shane. 1994. *Basic Environmental Toxicology.* New York: CRC Press/Lewis Publishers.

Cothern, C. R., N. P. Ross. 1994. *Environmental Statistics, Assessment, and Forecasting.* New York: CRC Press/Lewis Publishers.

Gardner, R. W. 1994. *Chemical Intolerance.* New York: CRC Press/Lewis Publishers.

Hallenbeck, W. H. 1993. *Quantitative Risk Assessment for Environmental and Occupational Health,* 2nd Edition. New York: CRC Press/Lewis Publishers.

Hodgson, E., R. B. Mailman, and Janice E. Chambers. 1988. *Dictionary of Toxicology.* London: The MacMillian Company.

Howard, P. H. 1995. *Handbook of Environmental Fate and Exposures Data for Organic Chemicals,* Vol. I-IV. New York: CRC Press/Lewis Publishers.

Landis, W. G., M. H. Yu. 1995. *Introduction to Environmental Toxicology.* New York: CRC Press/Lewis Publishers.

Langerman, N. 1994 *Precautionary Labels for Chemical Containers.* New York: CRC Press/Lewis Publishers.

Lauwerys, R. R., P. Hoet. 1993. *Industrial Chemical Exposure.* New York: CRC Press/Lewis Publishers.

Lewis, R. J., Sr. 1993. *Hawley's Condensed Chemical Dictionary,* 12th Edition. New York: Van Nostrand Reinhold.

Lewis, R. J., Sr. 1992. *Sax's Dangerous Properties of Industrial Materials,* Volumes 1, 2, and 3, 8th Edition. New York: Van Nostrand Reinhold.

Manahan, S. E. 1992. *Toxicological Chemistry,* 2nd Edition. New York: CRC Press/Lewis Publishers.

Meyer, E. 1990. *Chemistry of Hazardous Materials.* Englewood Cliffs, NJ: Prentice Hall.

Nelson, G. O. 1992. *Gas Mixtures.* New York: CRC Press/Lewis Publishers.

Ott, W. R. 1995. *Environmental Statistics and Data Analysis.* New York: CRC Press/Lewis Publishers.

Quigley, D. R. 1994. *Handbook of Emergency Chemical Management.* New York: CRC Press/Lewis Publishers.

Rea, W. J. 1995. *Chemical Sensitivity,* Vol. I-III. New York: CRC Press/Lewis Publishers.

Richardson, M. 1995. *Dictionary of Substance and Their Effects,* Vol. I-VI. New York: CRC Press/Lewis Publishers.

Sacarello, H. L. A. 1994. *The Comprehensive Handbook of Hazardous Materials.* New York: CRC Press/Lewis Publishers.

U.S. Department of Health and Human Services, National Institute for Occupational Safety and Health. 1994. *NIOSH Pocket Guide to Chemical Hazards.* Washington, D.C.: U.S. Government Printing Office.

MATERIAL SAFETY DATA SHEET

CHEMICAL NAME

3,3'-DICHLOROBENZIDINE

CRC PRESS/LEWIS PUBLISHERS DISCLAIMER: THE INFORMATION AND RECOMMENDATIONS PRESENTED HEREIN ARE BASED ON SOURCES BELIEVED TO BE RELIABLE. CRC MAKES NO REPRESENTATION ON ITS COMPLETENESS OR ACCURACY. IT IS THE RESPONSIBILITY OF THE USER TO DETERMINE THE CHEMICAL'S SUITABILITY FOR ITS INTENDED USE, THE CHEMICAL'S SAFE USE, AND THE CHEMICAL'S PROPER DISPOSAL. NO REPRESENTATIONS AND/OR WARRANTIES, EITHER EXPRESSED OR IMPLIED, OF THE MERCHANTABILITY OR FITNESS FOR A PARTICULAR PURPOSE, OR OF ANY OTHER NATURE, ARE MADE WITH RESPECT TO THE INFORMATION PROVIDED IN THIS MSDS OR TO THE CHEMICAL TO WHICH INFORMATION MAY REFER. CRC NEITHER ASSUMES NOR AUTHORIZES ANY OTHER PERSON TO ASSUME FOR IT, ANY OTHER ADDITIONAL RESPONSIBILITY OR LIABILITY FOR THE USE OF, OR RELIANCE UPON, THIS INFORMATION.

HAZARD WARNING INFORMATION

HEALTH	FIRE	REACTIVE	OTHER	DEGREE OF HAZARD	COLOR CODING	OTHER CODES
4	0	0		0 = Minimum Hazard 1 = Slight Hazard 2 = Moderate Hazard 3 = Serious Hazard 4 = Severe Hazard	HEALTH = BLUE FIRE = RED REACTIVITY = YELLOW OTHER = WHITE	OX = Oxidizer ACID = Acid ALK = Alkali COR = Corrosive W = Use No Water

SECTION I - GENERAL INFORMATION

Characrterization	**Aromatic Amine**	RCRA Number	**U073**
		EPA Class	**Toxic Waste**
DOT Proper Shipping Name	**No Citation**	Chemical Abstract Service (CAS) Number	**91-94-1**
DOT Hazard Class and Label Requirements	**No Citation**	DOT Emergency Guide Code	**No Citation**
DOT Identification Number	**None**	Chemical Formula	**$NH_2ClC_6H_3C_6H_3ClNH_2$**

Synonyms

4,4'-Diamino-3,3'-dichlorobiphenyl; Dichlorobenzidine base; o,o'-dichlorobenzidine; 3,3'-dichlorobiphenyl-4,4'-diamine; 3,3'-dichloro-4,4'-biphenyldiamine; 3,3'-dichloro-4,4'-diaminobiphenyl.

SECTION II - HAZARDOUS INGREDIENTS AND IDENTITY INFORMATION

Hazardous Components (specific identity)	OSHA Exposure Criteria	NIOSH Exposure Criteria	Immediately Dangerous to Life and Health (IDLH)	ACGIH Exposure Criteria
3,3'-dichlorobenzidine	**PEL: Suspected Cancer Agent**	**REL: Possible Cancer Agent**	**Not Determined**	**TLV: Suspected Human Carcinogen**

SECTION III - PHYSICAL AND CHEMICAL CHARACTERISTICS

Boiling Point	**788°F (420°C)**	Specific Gravity (H_2O = 1)	**Not Determined**
Vapor Pressure (mm Hg)	**Not Found**	Molecular Weight	**253.1**
Vapor Density (Air = 1)	**Not Found**	Melting Point	**271°F (133°C)**

Solubility

Nearly insoluble in water (0.07% at 59°F). Soluble in alcohol and ether.

Appearance and Odor

Gray to purple crystalline solid.

SECTION IV - FIRE AND EXPLOSION HAZARD DATA

Flash Point (method used)	**Not Established**	Explosive Limits in Air % by Volume	**LEL: Not Determined UEL: Not Determined**
NFPA Classification	**No Citation**	Autoignition Temperature	**Not Determined**

Extinguishing Media

Use dry chemical, carbon dioxide, foam, or water.

Special Fire Fighting Procedures

Poisonous gases are produced in fire. Structural protective clothing is permeable and may not provide adequate protection. Wear full protective clothing and self-contained breathing apparatus (SCBA). Move containers from fire area if it can be done safely. Cool exposed container with water.

Unusual Fire and Explosion Hazards

Containers may explode in fire due to buildup of excessive internal pressures. Exercise extreme caution when attempting to move or relocate containers that may have been exposed to high heat.

SECTION V - REACTIVITY DATA

Stability		Conditions to Avoid **Keep away from heat and flame.**
Stable	Unstable	Incompatibility *(materials to avoid)* **None reported.**
X		
Hazardous Polymerization		Conditions to Avoid **Under normal conditions of temperature and pressure, hazardous polymerization of 3,3'-dichlorobenzidine is not expected to occur.**
May Occur	Will Not Occur	Hazardous Decomposition or By-products **When heated to decomposition, 3,3'-dichlorobenzidine will produce toxic and irritating gases and oxides of chlorine and nitrogen.**
	X	

SECTION VI - HEALTH HAZARD DATA

Primary Route(s) of Entry:	Inhalation?	Absorption (skin)?	Ingestion?
	X	X	X

Health Hazards

INHALATION: Irritation of the upper respiratory tract. Causes headache and dizziness. May cause dysuria, blood in the urine and painful, difficult, or frequent urination. Confirmed bladder cancer agent in animals. May cause upper respiratory infections.

ABSORPTION: Can cause allergic skin reaction, caustic burns, and dermatitis. Will pass through unbroken skin to cause toxic systemic effects. Severe eye irritant.

INGESTION: Mildly toxic. Can cause gastritis and affect the bladder (may cause cancer).

Carcinogenicity	NTP Listed?	IARC Cancer Review Group?	OSHA Regulated?	Target Organs?
Confirmed Human Confirmed Animal	5th Annual Report	No	29 CFR 1910.1007	Bladder, liver, lungs, skin, GI tract, eyes.

Medical Conditions Generally Aggravated by Exposure

None Reported.

Emergency and First-aid Procedures

Eye contact: Flush immediately with water for 15 minutes (minimum) while holding upper and lower eyelids open. Seek medical attention. **Skin contact:** Remove all contaminated clothing. Immediately wash area with large amounts of soap and water. For **inhalation:** Remove the person from exposure. Provide respiratory assistance and CPR. Transfer to medical facility. Keep person warm and at rest. If **swallowed:** Call the poison control center, keep person warm and at rest and seek medical attention immediately.

SECTION VII - PRECAUTIONS FOR SAFE HANDLING AND USE

Steps to be Taken in Case Material is Released or Spilled

Powders should be collected in safest manner possible and disposed of in sealed drums. Do NOT dry sweep. Use a high efficiency particulate air (HEPA) vacuum instead. Ventilate area and remove ignition sources. Restrict those not involved in cleanup from entering area. Notify appropriate authorities, if applicable.

Preferred Waste Disposal Method

Dissolve in a combustible solvent and burn in a chemical incinerator equipped with an afterburner and scrubber.

Precautions to be Taken in Handling and Storage

Store in tightly closed containers in a cool, well-ventilated area. A regulated and controlled work area should be established wherever 3,3'-dichlorobenzidine is used, handled, or stored.

Other Precautions and Warnings

Bulk storage of 3,3'-dichlorobenzidine should be avoided.

SECTION VIII - CONTROL MEASURES AND PERSONAL PROTECTIVE EQUIPMENT

Respiratory Protection *(specify type)*

For any exposure to a suspected carcinogen, use a supplied-air respirator with full facepiece operated in continuous flow or other positive pressure mode (pressure demand), or, a self-contained breathing apparatus (SCBA) with full facepiece operated in pressure demand or other positive pressure mode.

Ventilation

Local exhaust or general mechanical systems recommended.

Protective Gloves	Eye Protection	Other Protective Clothing
Impervious Gloves	Chemical/Dust Goggles or Face Mask	Protective Apron

Work/Hygiene Practices

Always wash hands thoroughly after using chemical; never bring food, drink, or smoking materials into vicinity of chemicals.

3,3'-DICHLOROBENZIDINE

$NH_2ClC_6H_3C_6H_3ClNH_2$

CAS: 91-94-1

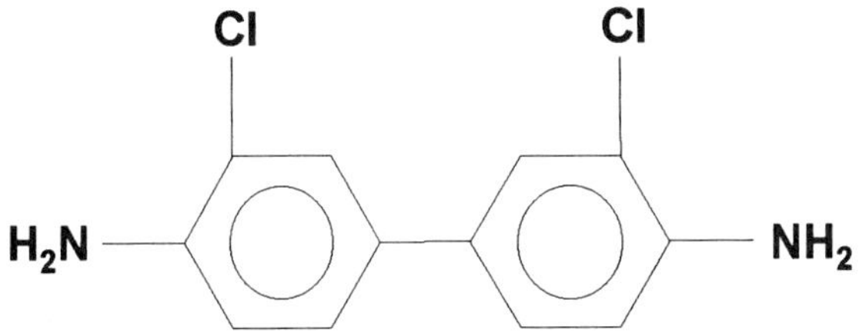

IDENTIFICATION AND TYPICAL USES

3,3'-Dichlorobenzidine is a gray to purple crystalline solid. It is used as an intermediate for dyes and pigments, and as a curing agent for isocyanate-terminated resins for urethane plastics.

RISK ASSESSMENT: HEALTH

General Assessment

3,3'-Dichlorobenzidine is mildly toxic to humans by *ingestion*. It is also toxic by *inhalation* and will irritate the nose and throat (mucous membranes). It has been shown to cause bladder cancer in test animals. Human mutation data have also been reported.

Inhalation can cause irritation of the upper respiratory tract with a potential for secondary infections. Inhalation and ingestion will cause toxic symptoms of cough, headache, and dizziness. Exposure can also cause blood in the urine and painful, difficult, or frequent urination. It is known to cause bladder cancer in test animals and its carcinogenic properties in humans are highly suspect. There may also be injuries to the liver. These symptoms occur from repeated inhalation of high concentrations of vapors or from ingestion.

This chemical can cause allergic skin reaction, dermatitis, and caustic burns. It can absorb through unbroken skin and cause toxic systemic effects, including a possibility of bladder cancer.

Acute Health Effects

The following acute (short-term) health effects may occur immediately or shortly after exposure to 3,3'-dichlorobenzidine:

Skin: Mild to moderate irritation and possible caustic burns. Skin absorption is likely.

Eye: Mild to severe irritation and inflammation.

Lung: Irritation of the upper respiratory tract causing coughing and/or shortness of breath and may lead to secondary infections. Can cause blood to appear in the urine. Urination may be painful, difficult, and frequent.

Other: Ingestion can cause symptoms ranging from no effect to headache, dizziness, and painful bloody urination.

Chronic Health Effects

The following chronic (long-term) health effects may occur at some time after exposure to 3,3'-dichlorobenzidine and can last for months or even years:

Cancer Hazards: According to the information presented in the references, 3,3'-dichlorobenzidine has been shown to cause liver, breast, bladder, intestine, and skin cancer in test animals. Its carcinogenicity in humans is questioned. However, the data in the references are conflicting and inconclusive in this regard. It is a mutagen. Until further studies are performed, it is suggested that 3,3'-dichlorobenzidine be handled with extreme caution.

Reproductive Hazard: According to the information presented in the references, 3,3'-dichlorobenzidine is suspected to adversely affect reproduction in animals. It has caused cancer in the offspring of test animals.

Other Chronic Effects: Repeated low-level exposures may cause dermatitis.

Recommended Risk-Reduction Measures

Unless a less toxic chemical can be substituted for a hazardous substance, *engineering controls* are the most effective method of reducing exposures. The best protection is to enclose operations and/or provide local exhaust ventilation at the site of chemical release.

While not always operationally feasible, isolating operations can also reduce exposure.

Using respiratory protection is less effective than the controls mentioned above, but is still advisable whenever working with or around 3,3'-dichlorobenzidine. For any exposure to a suspected human carcinogen, a self-contained breathing apparatus (SCBA) with full facepiece and pressure demand is the recommended respiratory protection of choice. If a full facepiece is not available, then chemical goggles should be worn to protect the eyes. Whenever a chemical splash hazard exists, a face shield and apron should be worn. To prevent hand and skin exposures, impervious gloves should be used. Glove manufacturers should be contacted and permeation studies obtained *before* final gloves selections have been made.

Administrative controls should also be in place to minimize the potential for human exposures. These may include written procedures or policies which specify the methods and techniques that will be practiced whenever personnel are to work with or around 3,3'-dichlorobenzidine.

All personnel should receive *training* on the use, hazards, protective measures, emergency actions, and other precautions per 29 CFR 1910.1200 (Hazard Communication), prior to the first assignment in an area where 3,3'-dichlorobenzidine is used or stored.

Before beginning employment and at regular intervals thereafter (e.g. annually), the following medical tests are recommended:

☑ Lung function tests (establish a baseline).

If symptoms develop or overexposure is suspected, the following medical tests may be useful:

☑ Lung function tests.

☑ Examination of the liver, breast, bladder, intestine, and skin.

☑ Evaluation by a qualified allergist, including careful exposure history and special testing (may help diagnose skin allergy).

Any evaluation should include a careful medical history of past and present symptoms with an examination. However, medical tests that evaluate existing damage are not a substitute for controlling exposure. Also, since smoking can cause heart disease, emphysema, and numerous other respiratory disorders, smokers may be affected more rapidly than non-smokers to the same exposure conditions. Prudent risk assessment and management requires careful consideration of *all* possible risk factors that may be causing the appearance of symptoms in exposed workers.

Other methods to reduce exposure include:

☑ Where possible, enclose operations and use local exhaust ventilation at the site of chemical release. If local exhaust ventilation or enclosure is not used, respiratory protection should be mandatory.

☑ Always ensure that proper protective clothing is worn when using chemical substances.

☑ Wash thoroughly immediately after exposure to 3,3'-dichlorobenzidine and at the end of the work shift or before eating, drinking, or smoking.

☑ Hazard warning information should be posted in the work area. In addition, as part of an on-going education and training program, all information on the health and safety hazards of 3,3'-dichlorobenzidine should be communicated to all potentially exposed workers.

☑ Eye wash stations should be provided in work areas. If the potential for whole body exposure exists, then safety showers should also be provided.

☑ Specific engineering controls are recommended for this chemical by NIOSH. Refer to NIOSH Criteria Document: Special Occupational Hazard Review for Benzidine Based Dyes (publication Number 80-109).

RISK ASSESSMENT: ENVIRONMENT

General Assessment

The environment is at risk of exposure during intended use, transportation, storage, disposal, or destruction of 3,3'-dichlorobenzidine. In almost every scenario, the threat of environmental exposure is contingent upon the proper handling of the chemical substance. Accidental spills, large or small, can result in fire, explosion, and possible contamination of the surrounding environmental mediums (water, soil, and air).

3,3'-Dichlorobenzidine should always be handled with extreme caution. While it is not reported to be incompatible with other materials, by itself it presents serious health exposure risk. These characteristics require special consideration during any emergency situation involving a leak or spill of 3,3'-dichlorobenzidine.

3,3'-Dichlorobenzidine can enter the environment through industrial discharges and spills.

Acute Ecological Effects

Acute (short-term) toxic effects may include the death of animals, birds, or fish and death or low growth rate in plants. Acute effects are seen 2 to 4 days after animals or plants are exposed to 3,3'-dichlorobenzidine.

This chemical has high acute toxicity to aquatic life. Insufficient data are available to evaluate or predict the short-term effects of 3,3'-dichlorobenzidine on plants, birds, or land animals.

Chronic Ecological Effects

Chronic toxic effects may include shortened life span, reproductive problems, lower fertility, and changes in appearance or behavior in exposed animals. These effects can be seen long after first exposure(s) to toxic chemicals.

3,3'-Dichlorobenzidine has high chronic toxicity to aquatic life. Insufficient data are available to evaluate or predict the long-term effects of 3,3'-dichlorobenzidine on plants, birds, or land animals.

Water Solubility

3,3'-Dichlorobenzidine is nearly insoluble in water (0.07%). Concentrations of less than 1 milligram will mix with a liter of water.

Persistence in the Environment

3,3'-Dichlorobenzidine is moderately persistent in water, with a half-life of between 20 to 200 days. The half-life of a pollutant is the amount of time it takes for one half of the chemical to be degraded. Approximately 60% will eventually end up in water, the rest will be divided about equally between terrestrial soils and aquatic sediments.

Bioaccumulation in Aquatic Organisms

Some substances increase in concentration, or *bioaccumulate,* in living organisms as they breathe contaminated air, drink contaminated water, or eat contaminated food. These chemicals can become concentrated in the tissues and internal organs of animals as well as humans.

The concentration of 3,3'-dichlorobenzidine found in fish tissues is expected to be somewhat higher than the average concentration of 3,3'-dichlorobenzidine in the water from which the fish was taken.

Recommended Risk-Reduction Measures

Proper training of all transporters will reduce the likelihood of a mishap or accident resulting in a leak or spill to the environment. The correct use of labeling on all containers, trucks, and rail cars will enable emergency responders to react properly and quickly to any disaster thereby reducing the potential risk to the environment and to personnel.

Storage of 3,3'-dichlorobenzidine should be segregated from incompatible chemicals to minimize the risk of cross-contamination or contact. Storage buildings should be equipped with the proper fire protection equipment (alarms, sprinklers, extinguishers, etc.).

If a spill or leak to the environment has occurred, fire department, emergency response and/or hazardous materials spill personnel should be notified immediately. Cleanup should be attempted only by those trained in proper spill containment procedures. Contaminated soils should be removed for incineration and replaced with clean soil. If 3,3'-dichlorobenzidine should contact the water table, aquifer, or navigable waterway, remediation activities should be prompt to ensure protection of aquatic sediments. When such spills occur, the local and/or state emergency response authorities must be notified. A comprehensive emergency response or disaster preparedness/recovery plan should be in place prior to any operations involving the use, transportation, storage, or disposal of 3,3'-dichlorobenzidine. If 3,3'-dichlorobenzidine is spilled or leaked, the following specific steps are recommended:

- ☑ Restrict persons not wearing protective clothing from area of spill or leak until cleanup is complete and area can be opened for normal work.
- ☑ Ventilate area of spill or leak.
- ☑ Remove all ignition sources.
- ☑ Collect powders in safest manner possible and dispose of in sealed drums. Do NOT dry sweep (creates airborne dusts). Use a vacuum equipped with a high efficiency particulate air (HEPA) filter.
- ☑ It may be necessary to dispose of p-dichlorobenzene as a hazardous waste. The responsible state agency or the regional office of the federal Environmental Protection Agency (EPA) should be contacted for specific recommendations.

RISK ASSESSMENT: BUSINESS

General Assessment

Accidents or mishaps involving 3,3'-dichlorobenzidine can present a significant threat to business operations. The loss or damage of equipment or facilities can significantly affect fiscal viability. Lawsuits that may result from personnel injury/death, public exposures, and/or environmental contamination will also require a serious expenditure of resources. Media attention surrounding an injury, death, or environmental damage can also result in a loss of profits and loss of current as well as future business. Always remember, anytime the terms "carcinogen," "cancer," or "reproductive" hazard is used, public emotion, hysteria, and ignorance can run equally high. This should be a consideration during the development of any corporate policies regarding public relations.

Recommended Risk-Reduction Measures

Company attorneys, safety and health professionals, and environmental specialists should be involved in the development of any procedures dealing with safety or emergency response. A company official should be pre-designated as a public relations officer with specific training in dealing with the press. Corporate plans and policies should be developed, approved, and implemented long before any need for such arises.

REFERENCES

Ahlbom, A. 1993. *Biostatistics for Epidemiologists.* New York: CRC Press/Lewis Publishers.

American Conference of Governmental Industrial Hygienists. 1988. *Documentation of the Threshold Limit Values and Biological Exposure Indices,* 5th Edition (with updates). Cincinnati: ACGIH

Anthony, C. P., and N. J. Kolthoff. 1971. *Textbook of Anatomy and Physiology.* St. Louis: C. V. Mosby Company.

Asante-Duah, D. K. 1993. *Hazardous Waste Risk Assessment.* New York: CRC Press/Lewis Publishers.

Bloom, A. D. and F. J. DeSerres. 1995. *Ecotoxicity and Human Health.* New York: CRC Press/Lewis Publishers.

Bueche, J. R. 1972. *Principles of Physics,* 2nd Edition. New York: McGraw-Hill.

Burrell, R., D. K. Flaherty, and L. J. Sauers. 1992. *Toxicology of the Immune System.* New York: Van Nostrand Reinhold.

Calabrese, E. J., E. M. Kenyon. *Air Toxics and Risk Assessment.* New York: CRC Press/Lewis Publishers.

Cockerman, L. G., B. S. Shane. 1994. *Basic Environmental Toxicology.* New York: CRC Press/Lewis Publishers.

Cothern, C. R. 1995. *Handbook for Environmental Risk Decisionmaking.* New York: CRC Press/Lewis Publishers.

Gardner, R. W. 1994. *Chemical Intolerance.* New York: CRC Press/Lewis Publishers.

Hallenbeck, W. H. 1993. *Quantitative Risk Assessment for Environmental and Occupational Health,* 2nd Edition. New York: CRC Press/Lewis Publishers.

Hodgson, E., R. B. Mailman, and Janice E. Chambers. 1988. *Dictionary of Toxicology.* London: The MacMillian Company.

Howard, P. H. 1995. *Handbook of Environmental Fate and Exposures Data for Organic Chemicals,* Vol. I-IV. New York: CRC Press/Lewis Publishers.

Jones, F. E. 1994. *Toxic Organic Vapors in the Workplace.* New York: CRC Press/Lewis Publishers.

Kamrin, M. A. 1988. *Toxicology: A Primer on Principles and Applications.* Chelsea, Michigan: Lewis Publishers.

Keith, L. H. and M. M. Walker. 1995. *Handbook of Air Toxics.* New York: CRC Press/Lewis Publishers.

Landis, W. G., M. H. Yu. 1995. *Introduction to Environmental Toxicology.* New York: CRC Press/Lewis Publishers.

Langerman, N. 1994 *Precautionary Labels for Chemical Containers.* New York: CRC Press/Lewis Publishers.

Lauwerys, R. R., P. Hoet. 1993. *Industrial Chemical Exposure.* New York: CRC Press/Lewis Publishers.

Lewis, R. J., Sr. 1993. *Hawley's Condensed Chemical Dictionary,* 12th Edition. New York: Van Nostrand Reinhold.

Lewis, R. J., Sr. 1992. *Sax's Dangerous Properties of Industrial Materials,* Volumes 1, 2, and 3, 8th Edition. New York: Van Nostrand Reinhold.

Manahan, S. E. 1992. *Toxicological Chemistry,* 2nd Edition. New York: CRC Press/Lewis Publishers.

Meyer, E. 1990. *Chemistry of Hazardous Materials.* Englewood Cliffs, NJ: Prentice Hall.

Nelson, G. O. 1992. *Gas Mixtures.* New York: CRC Press/Lewis Publishers.

Ott, W. R. 1995. *Environmental Statistics and Data Analysis.* New York: CRC Press/Lewis Publishers.

Richardson, M. 1995. *Dictionary of Substance and Their Effects,* Vol. I-VI. New York: CRC Press/Lewis Publishers.

Sacarello, H. L. A. 1994. *The Comprehensive Handbook of Hazardous Materials.* New York: CRC Press/Lewis Publishers.

U.S. Department of Health and Human Services, National Institute for Occupational Safety and Health (NIOSH). 1980. *Criteria Document: Special Occupational Hazard Review for Benzidine Based Dyes.* Publication Number: 80-109. Washington, D.C.: U.S. Government Printing Office.

U.S. Department of Health and Human Services, National Institute for Occupational Safety and Health. 1994. *NIOSH Pocket Guide to Chemical Hazards.* Washington, D.C.: U.S. Government Printing Office.

MATERIAL SAFETY DATA SHEET

CHEMICAL NAME	CRC PRESS/LEWIS PUBLISHERS DISCLAIMER
DICHLORODIFLUORO-METHANE	

HAZARD WARNING INFORMATION

HEALTH	FIRE	REACTIVE	OTHER	DEGREE OF HAZARD	COLOR CODING	OTHER CODES
1	0	0		0 = Minimum Hazard 1 = Slight Hazard 2 = Moderate Hazard 3 = Serious Hazard 4 = Severe Hazard	HEALTH = BLUE FIRE = RED REACTIVITY = YELLOW OTHER = WHITE	OX = Oxidizer ACID = Acid ALK = Alkali COR = Corrosive W = Use No Water

SECTION I - GENERAL INFORMATION

Characterization	RCRA Number	EPA Class
Halogenated Hydrocarbon	**U075**	**Toxic Waste**

DOT Proper Shipping Name **Dichlorodifluoromethane**	Chemical Abstract Service (CAS) Number **75-71-8**
DOT Hazard Class and Label Requirements **Non Flammable; Non Flammable Gas**	DOT Emergency Guide Code **12**
DOT Identification Number **UN 1028**	Chemical Formula **CCl_2F_2**

Synonyms
Difluorodichloromethane; fluorocarbon 12; Freon® 12; Genetron® 12; Halon® 122; Isotron® 12; propellant 12; refrigerant 12; Algofrene® type 2; Eskimon® 12; Ucon® 12.

SECTION II - HAZARDOUS INGREDIENTS AND IDENTITY INFORMATION

Hazardous Components (specific identity)	OSHA Exposure Criteria	NIOSH Exposure Criteria	Immediately Dangerous to Life and Health (IDLH)	ACGIH Exposure Criteria
Dichlorodifluoromethane (derivation: By reaction of carbon tetrachloride and anhydrous hydrogen fluoride, in the presence of an antimony halide catalyst; high-temperature chlorination of vinylidene fluoride, which is made by adding hydrogen fluoride to acetylene). **1 ppm = 5.03 mg/m^3**	**PEL (8-hour):** **1000 ppm** **4950 mg/m^3** **STEL:** **Not Established**	**REL (10 hour):** **1000 ppm** **4950 mg/m^3** **Possible Cancer Agent**	**50,000 ppm**	**TLV (8-hour):** **1000 ppm** **4950 mg/m^3** **STEL:** **Not Established**

SECTION III - PHYSICAL AND CHEMICAL CHARACTERISTICS

Boiling Point **-22°F (-30°C)**	Specific Gravity (H_2O = 1) **1.35**
Vapor Pressure (mm Hg) **5.7 at 69°F (20°C)**	Molecular Weight **120.9**
Vapor Density (Air = 1) **4.16**	Freezing Point **-252°F (-158°C)**

Solubility
Insoluble (0.03% at 77°F) in water. Soluble in most organic solvents (alcohol, ether, benzene, amyl chloride, bromobenzene, bromoform, *n*-butyl alcohol, butyl butyrate, carbon tetrachloride, and chloroform).

Appearance and Odor
Colorless gas with an ether-like odor at extremely high concentrations. Shipped as compressed gas.

SECTION IV - FIRE AND EXPLOSION HAZARD DATA

Flash Point (method used) **Non Flammable**	Explosive Limits in Air % by Volume **LEL: Not Determined** **UEL: Not Determined**
NFPA Classification **Non Flammable Gas**	Autoignition Temperature **Not Determined**

Extinguishing Media
Use extinguishing media suitable for surrounding fire. Dichlorodifluoromethane does not burn.

Special Fire Fighting Procedures
Poisonous gases are produced in fire, wear full protective clothing and self-contained breathing apparatus. Move containers from fire area if it can be done safely. Cool fire-exposed containers using water spray long after fire has been extinguished. May attack some rubbers, plastics, and coatings.

Unusual Fire and Explosion Hazards
Containers may explode in fire due to buildup of excessive internal pressures. Evacuate immediately if rising sound is heard from venting device or if there is any noticeable discoloration of tank or vessel.

SECTION V - REACTIVITY DATA

Stability		Conditions to Avoid Dichlorodifluoromethane is stable up to 1022°F (550°C), above which it will begin to decompose. Avoid exposures to excessive heat and contact with incompatible materials.
Stable X	Unstable	Incompatibility *(materials to avoid)* Dichlorodifluoromethane will react in the presence of chemically active metals, such as potassium, sodium, calcium, powdered aluminum, magnesium, or zinc.
Hazardous Polymerization		Conditions to Avoid Under normal conditions of temperature and pressure, hazardous polymerization of dichlorodifluoromethane can not occur.
May Occur	Will Not Occur X	Hazardous Decomposition or By-products Toxic and irritating fumes and gases are produced when dichlorodifluoromethane is involved in fire, including carbon dioxide, phosgene, chlorine, hydrogen chloride, hydrogen fluoride, and fluorine.

SECTION VI - HEALTH HAZARD DATA

Primary Route(s) of Entry:	Inhalation?	Absorption (skin)?	Ingestion?
	X		

Health Hazards

INHALATION: Low-level toxicity. A simple asphyxiant when oxygen levels are 15 to 16%. Produces unconsciousness at 6 to 8% oxygen. High concentrations may cause conjunctiva irritation, fibrosing alveolitis, and liver changes. Narcotic at high concentrations.

SKIN & EYES: Eye irritation at high concentrations. The liquid can cause frostbite burns to tissues.

INGESTION: Ingestion not likely.

Carcinogenicity	NTP Listed?	IARC Cancer Review Group?	OSHA Regulated?	Target Organs?
No Citation (Human) No Citation (Animal)	No	No	29 CFR 1910.1000 Table Z-1	Cardiovascular system, CNS, PNS

Medical Conditions Generally Aggravated by Exposure

Pre-existing ventricular ectopy (an abnormal displacement or situation in the ventricles of the heart).

Emergency and First-aid Procedures

Eye contact: Flush immediately with water for 15 minutes (minimum), seek medical attention. Skin contact: For frostbite, rapidly re-warm using water at 107°F (42°C) until flush returns. Do NOT use dry heat! Seek medical assistance. For inhalation: Remove the person from exposure. Provide respiratory assistance and CPR. Transfer to medical facility. If swallowed: Not likely. Liquid state only below temperatures of -22°F (-30°C). DOCTOR: Monitor for ventricular ectopy according to ACLS guidelines.

SECTION VII - PRECAUTIONS FOR SAFE HANDLING AND USE

Steps to be Taken in Case Material is Released or Spilled

Evacuate area, deny entry, stay upwind. If it can be done without risk, shut off leak. There may be a visible vapor cloud. If leak cannot be repaired, empty into a combustion chamber with a combustible fuel and burn (ensure complete combustion to prevent formation of phosgene).

Preferred Waste Disposal Method

Rotary kiln incineration at 1508 to 2912°F or fluidized bed incineration at 842 to 1796°F.

Precautions to be Taken in Handling and Storage

Do not store dichlorodifluoromethane in the presence of incompatible materials. Store in tightly closed containers in a cool (below 125°F), dry, well-ventilated area. Ship as a liquefied compressed gas.

Other Precautions and Warnings

Prevent physical damage to compressed gas storage cylinders.

SECTION VIII - CONTROL MEASURES AND PERSONAL PROTECTIVE EQUIPMENT

Respiratory Protection *(specify type)*

Best protection is provided by a supplied-air respirator with full facepiece operated in pressure demand mode or a self-contained breathing apparatus (SCBA) with full facepiece operated in pressure demand or other positive pressure mode.

Ventilation

Local exhaust or general mechanical systems recommended.

Protective Gloves	Eye Protection	Other Protective Clothing
Neoprene Rubber	Chemical Goggles or Face Mask	Neoprene Rubber Apron

Work/Hygiene Practices

Always wash hands thoroughly after using chemical; never bring food, drink, or smoking materials into vicinity of chemicals.

DICHLORODIFLUOROMETHANE

CCl_2F_2 CAS: 75-71-8

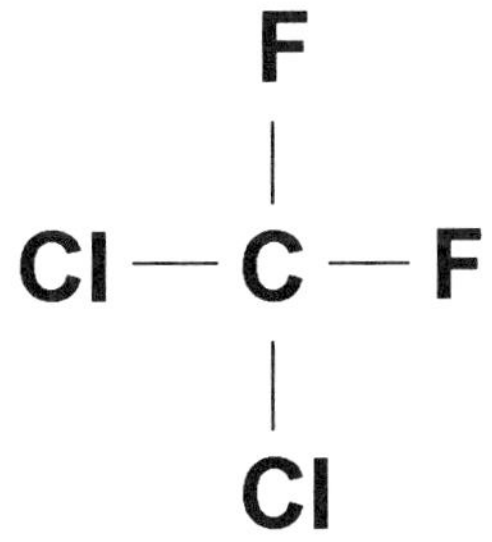

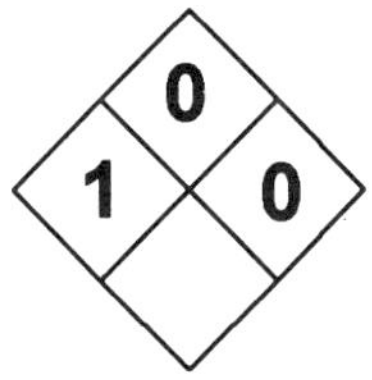

IDENTIFICATION AND TYPICAL USES

Dichlorodifluoromethane is colorless gas with an ether-like odor at extremely high concentrations. It is normally shipped as a liquefied compressed gas. It has been used primarily as a refrigerant in air conditioners. It has also been used in the manufacture of plastics, as a blowing agent (a propellant), as a low-temperature solvent, as a leak-detecting agent, and for freezing foods (by direct contact) and for chilling cocktail glasses.

RISK ASSESSMENT: HEALTH

General Assessment

Dichlorodifluoromethane is considered to have low toxicity by the *inhalation* route. Its is a simple asphyxiant and can produce some irritation and mild narcotic effects in high concentrations. Its greatest risk to health is its ability to induce serious cardiac arrhythmia at high concentrations or in personnel with pre-exiting ventricular ectopy (abnormalities).

Inhalation produces conjunctiva irritation, fibrosing alveolitis, and possible liver changes. It can also produce narcotic effects at high concentrations with primary action on the central nervous system (CNS) and the peripheral nervous system (PNS). Indications of asphyxia include an overwhelming feeling of "air starvation" or air hunger, and rapid, irregular breathing. There may be headache, fatigue, mental confusion, nausea and vomiting, giddiness and poor judgment, exhaustion, numbness in the legs and arms, convulsions, and death. Prolonged concentrations can lead to generalized paresthesia, ringing in the ears, apprehension, and slurred speech. These may be followed by a marked decrease in consciousness, amnesia, and cardiac arrhythmia.

Acute Health Effects

The following acute (short-term) health effects may occur immediately or shortly after exposure to dichlorodifluoromethane:

Skin: Irritation and possible frostbite (liquid contact).

Eye: Mild irritation and conjunctivitis. Liquid will burn tissue.

Lung: Irritation of the mucosa. Will cause asphyxiation and associated symptoms (up to and including death) at high concentrations.

Other: A narcotic at high concentrations. Will effect the central nervous system (CNS) and the peripheral nervous system (PNS). Also, can induce irregular or abnormal heart arrhythmia.

Chronic Health Effects

The following chronic (long-term) health effects may occur at some time after exposure to dichlorodifluoromethane and can last for months or even years:

Cancer Hazards: According to the information presented in the references, dichlorodifluoromethane has not been shown to cause cancer in test animals.

Reproductive Hazard: According to the information presented in the references, dichlorodifluoromethane has not been shown to adversely affect reproduction in test animals.

Other Chronic Effects: Low, chronic exposures may lead to liver and kidney damage and possible injury to the central nervous system. There may be symptoms of behavioral changes and general weakness. Ventricular ectopy may also be induced on high or prolonged exposures.

Recommended Risk-Reduction Measures

While dichlorodifluoromethane is considered to be of a low-order toxicity by inhalation, it can still pose some degree of health hazard risk. Unless a less toxic chemical can be substituted for a hazardous substance, *engineering controls* are the most effective method of reducing exposures. The best protection is to enclose operations and/or provide local exhaust ventilation at the site of chemical release. While not always operationally feasible, isolating operations can also reduce exposure.

Using respiratory protection is less effective than the controls mentioned above, but is still advisable whenever working with or around dichlorodifluoromethane. The exposure level for this chemical is extremely high (1000 ppm), and, therefore, respiratory protection may not always be necessary in well-ventilated work areas. However, when high exposure levels are predicted or prolonged exposures are anticipated, a supplied air respirator or a self-contained breathing apparatus (SCBA) with full facepiece and pressure demand are the recommended respiratory protection of choice. If a full facepiece is not available, then chemical goggles should be worn to protect the eyes. Whenever a chemical splash hazard exists, a face shield and rubber apron should be worn. To prevent hand and skin exposures, neoprene rubber gloves should be worn. Glove manufacturers should be contacted and permeation studies obtained *before* final glove selections are made.

Administrative controls should also be in place to minimize the potential for human exposures. These may include written procedures or policies which specify the methods and techniques that will be practiced whenever personnel are to work with dichlorodifluoromethane.

All personnel should receive *training* on the use, hazards, protective measures, emergency actions, and other precautions per 29 CFR 1910.1200 (Hazard Communication), prior to the first assignment in an area where dichlorodifluoromethane is used or stored.

Before beginning employment and at regular intervals thereafter (e.g., annually), the following medical protocol is suggested for those assigned to work with dichlorodifluoromethane:

- ☑ Electrocardiogram (EKG), to test for pre-existing ventricular ectopy and establish baseline.

If symptoms develop or overexposure is suspected, the following medical tests may be useful:

- ☑ Electrocardiogram (EKG), to compare with baseline.
- ☑ Liver function tests.
- ☑ Nervous system function tests, including interviews by a qualified neurologist to evaluate any changes in personality or behavior.

Any evaluation should include a careful medical history of past and present symptoms with an examination. However, medical tests that evaluate existing damage are not a substitute for controlling exposure.

Other methods to reduce exposure include:

- ☑ Where possible, enclose operations and use local exhaust ventilation at the site of chemical release. If local exhaust ventilation or enclosure is not used, respiratory protection should be mandatory.
- ☑ Always ensure that proper protective clothing is worn when using chemical substances.
- ☑ Wash thoroughly immediately after exposure to dichlorodifluoromethane and at the end of the work shift or before eating, drinking, or smoking.
- ☑ Hazard warning information should be posted in the work area. In addition, as part of an on-going education and training program, all information on the health and safety hazards of dichlorodifluoromethane should be communicated to all potentially exposed workers.
- ☑ Eye wash stations should be provided in work areas. If the potential for whole body exposure exists, then safety showers should also be provided.

RISK ASSESSMENT: ENVIRONMENT

General Assessment

The environment is at risk of exposure during intended use, transportation, storage, disposal, or destruction of dichlorodifluoromethane. In almost every scenario, the threat of environmental exposure is contingent upon the proper handling of the chemical substance. Accidental spills, large or small, can result in fire, explosion, and possible contamination of the surrounding environmental mediums (water, soil, and air).

Dichlorodifluoromethane is considered a non-flammable gas. However, it is incompatible with a number of common, chemically active metals and caution is required to ensure contact is prevented. It is also a chlorofluorocarbon (CFC) which is a group of halogenated hydrocarbons alleged to cause damage to the earth's protective ozone layer. Therefore, accidental release to the environment must not occur. These characteristics require special consideration during any emergency situation involving a leak or spill of dichlorodifluoromethane. Should dichlorodifluoromethane ever come into contact with incompatible substances either during use, transportation, or storage, violent and explosive reactions can occur.

Dichlorodifluoromethane is normally a gas. But it is shipped as a liquefied compressed gas. It can enter the environment through industrial discharges and spills.

Acute Ecological Effects

Acute (short-term) toxic effects may include the death of animals, birds, or fish and death or low growth rate in plants. Acute effects are seen 2 to 4 days after animals or plants are exposed to dichlorodifluoromethane.

This chemical is not expected to cause adverse short-term effects on aquatic life, plants, birds, or terrestrial animals.

Chronic Ecological Effects

Chronic toxic effects may include shortened life span, reproductive problems, lower fertility, and changes in appearance or behavior in exposed animals. These effects can be seen long after first exposure(s) to toxic chemicals.

Insufficient data are available to evaluate or predict the long-term effects of dichlorodifluoromethane to aquatic life, plants, birds, or land animals.

Water Solubility

Dichlorodifluoromethane is nearly insoluble in water (0.03% at 77°F). It is normally a gas.

Persistence in the Environment

Dichlorodifluoromethane is virtually non-persistent in water, with a half-life of less than 2 days. The half-life of a pollutant is the amount of time it takes for one half of the chemical to be degraded. Approximately 100% of dichlorodifluoromethane will end up in air.

Bioaccumulation in Aquatic Organisms

Some substances increase in concentration, or *bioaccumulate,* in living organisms as they breathe contaminated air, drink contaminated water, or eat contaminated food. These chemicals can become concentrated in the tissues and internal organs of animals as well as humans.

The concentration of dichlorodifluoromethane found in fish tissues is expected to much lower than the average concentration of dichlorodifluoromethane in the water from which the fish was taken.

Recommended Risk-Reduction Measures

Proper training of all transporters will reduce the likelihood of a mishap or accident resulting in a leak or spill to the environment. The correct use of DOT labeling (per the MSDS) on all containers, trucks, and rail cars will enable emergency responders to react properly and quickly to any disaster thereby reducing the potential risk to the environment and to personnel.

Storage of dichlorodifluoromethane should be segregated from incompatible chemicals to minimize the risk of cross-contamination or contact. Containers must be protected from shock, excessive heat (including the sun's rays), and damage.

If a spill or leak to the environment has occurred, fire department, emergency response and/or hazardous materials spill personnel should be notified immediately. Cleanup should be attempted only by those trained in proper spill containment procedures. If any liquid has spilled, contaminated soils may need to be removed for incineration and replaced with clean soil. If dichlorodifluoromethane should contact the water table, aquifer, or navigable waterway, remediation activities should be prompt to ensure protection of the aquatic environment. When such spills occur, the local and/or state emergency response authorities must be notified. A comprehensive emergency response or disaster preparedness/recovery plan should be in place prior to any operations involving the use, transportation, storage, or disposal of dichlorodifluoromethane. If dichlorodifluoromethane is spilled or leaked, the following specific steps are recommended:

- ☑ Evacuate area and deny entry until cleanup is completed. Restrict persons not wearing protective clothing from hazard area.
- ☑ Ventilate area of spill or leak.
- ☑ Remove all ignition sources.

☑ If source of leak is a cylinder and it cannot be repaired in place, remove to safe location (e.g., outdoors).

☑ Absorb liquid spills using vermiculite, dry earth, or sand to absorb and place in a sealed drum for disposal.

☑ It may be necessary to dispose of dichlorodifluoromethane as a hazardous waste. The responsible state agency or the regional office of the federal Environmental Protection Agency (EPA) should be contacted for specific recommendations.

RISK ASSESSMENT: BUSINESS

General Assessment

Accidents or mishaps involving dichlorodifluoromethane can present a significant threat to business operations. The loss or damage of equipment or facilities can significantly affect fiscal viability. Lawsuits that may result from personnel injury/death, public exposures, and/or environmental contamination will also require a serious expenditure of resources. Media attention surrounding an injury, death, or environmental damage can also result in a loss of profits and loss of current as well as future business.

Recommended Risk-Reduction Measures

Company attorneys, safety and health professionals, and environmental specialists should be involved in the development of any procedures dealing with safety or emergency response. A company official should be pre-designated as a public relations officer with specific training in dealing with the press. Corporate plans and policies should be developed, approved, and implemented long before any need for such arises.

REFERENCES

American Conference of Governmental Industrial Hygienists. 1988. *Documentation of the Threshold Limit Values and Biological Exposure Indices,* 5th Edition (with updates). Cincinnati: ACGIH

Anthony, C. P., and N. J. Kolthoff. 1971. *Textbook of Anatomy and Physiology.* St. Louis: C. V. Mosby Company.

Bueche, J. R. 1972. *Principles of Physics,* 2nd Edition. New York: McGraw-Hill.

Brusick, D. J. 1994. *Methods for Genetic Risk Assessment.* New York: CRC Press/Lewis Publishers.

Burrell, R., D. K. Flaherty, and L. J. Sauers. 1992. *Toxicology of the Immune System.* New York: Van Nostrand Reinhold.

Calabrese, E. J. 1994. *Biological Effects of Low Level Exposures.* New York: CRC Press/Lewis Publishers.

Cothern, C. R. 1995. *Handbook for Environmental Risk Decisionmaking.* New York: CRC Press/Lewis Publishers.

Gardner, R. W. 1994. *Chemical Intolerance.* New York: CRC Press/Lewis Publishers.

Hodgson, E., R. B. Mailman, and Janice E. Chambers. 1988. *Dictionary of Toxicology.* London: The MacMillian Company.

Jones, F. E. 1994. *Toxic Organic Vapors in the Workplace.* New York: CRC Press/Lewis Publishers.

Keith, L. H. and M. M. Walker. 1995. *Handbook of Air Toxics.* New York: CRC Press/Lewis Publishers.

Landis, W. G., M. H. Yu. 1995. *Introduction to Environmental Toxicology.* New York: CRC Press/Lewis Publishers.

Langerman, N. 1994 *Precautionary Labels for Chemical Containers.* New York: CRC Press/Lewis Publishers.

Lauwerys, R. R., P. Hoet. 1993. *Industrial Chemical Exposure.* New York: CRC Press/Lewis Publishers.

Lewis, R. J., Sr. 1993. *Hawley's Condensed Chemical Dictionary,* 12th Edition. New York: Van Nostrand Reinhold.

Lewis, R. J., Sr. 1992. *Sax's Dangerous Properties of Industrial Materials,* Volumes 1, 2, and 3, 8th Edition. New York: Van Nostrand Reinhold.

Mackay, D., W. Y. Shiu, and K. C. Ma. 1992. *Illustrated Handbook of Chemical-Physical Properties and Environmental Fate for Organic Chemicals.* New York: CRC Press/Lewis Publishers.

Manahan, S. E. 1992. *Toxicological Chemistry,* 2nd Edition. New York: CRC Press/Lewis Publishers.

Meyer, E. 1990. *Chemistry of Hazardous Materials.* Englewood Cliffs, NJ: Prentice Hall.

Mycock, J. C., J. D. McKenna, and L. Theodore. 1995. *Handbook of Air Pollution Control Engineering and Technology.* New York: CRC Press/Lewis Publishers.

Nelson, G. O. 1992. *Gas Mixtures.* New York: CRC Press/Lewis Publishers.

Ott, W. R. 1995. *Environmental Statistics and Data Analysis.* New York: CRC Press/Lewis Publishers.

Patnaik, P. 1992. *A Comprehensive Guide to the Hazardous Properties of Chemical Substances.* New York: Van Nostrand Reinhold.

Quigley, D. R. 1994. *Handbook of Emergency Chemical Management.* New York: CRC Press/Lewis Publishers.

Rea, W. J. 1995. *Chemical Sensitivity,* Vol. I-III. New York: CRC Press/Lewis Publishers.

Richardson, M. 1995. *Dictionary of Substance and Their Effects,* Vol. I-VI. New York: CRC Press/Lewis Publishers.

Sacarello, H. L. A. 1994. *The Comprehensive Handbook of Hazardous Materials.* New York: CRC Press/Lewis Publishers.

U.S. Department of Health and Human Services, National Institute for Occupational Safety and Health. 1994. *NIOSH Pocket Guide to Chemical Hazards.* Washington, D.C.: U.S. Government Printing Office.

MATERIAL SAFETY DATA SHEET

CHEMICAL NAME	
DIELDRIN	**CRC PRESS/LEWIS PUBLISHERS DISCLAIMER:** THE INFORMATION AND RECOMMENDATIONS PRESENTED HEREIN ARE BASED ON SOURCES BELIEVED TO BE RELIABLE. CRC MAKES NO REPRESENTATION ON ITS COMPLETENESS OR ACCURACY. IT IS THE RESPONSIBILITY OF THE USER TO DETERMINE THE CHEMICAL'S SUITABILITY FOR ITS INTENDED USE, THE CHEMICAL'S SAFE USE, AND THE CHEMICAL'S PROPER DISPOSAL. NO REPRESENTATIONS AND/OR WARRANTIES, EITHER EXPRESSED OR IMPLIED, OF THE MERCHANTABILITY OR FITNESS FOR A PARTICULAR PURPOSE, OR OF ANY OTHER NATURE, ARE MADE WITH RESPECT TO THE INFORMATION PROVIDED IN THIS MSDS OR TO THE CHEMICAL TO WHICH INFORMATION MAY REFER. CRC NEITHER ASSUMES NOR AUTHORIZES ANY OTHER PERSON TO ASSUME FOR IT, ANY OTHER ADDITIONAL RESPONSIBILITY OR LIABILITY FOR THE USE OF, OR RELIANCE UPON, THIS INFORMATION.

HAZARD WARNING INFORMATION

HEALTH	FIRE	REACTIVE	OTHER	DEGREE OF HAZARD	COLOR CODING	OTHER CODES
3	0	0		0 = Minimum Hazard 1 = Slight Hazard 2 = Moderate Hazard 3 = Serious Hazard 4 = Severe Hazard	HEALTH = BLUE FIRE = RED REACTIVITY = YELLOW OTHER = WHITE	OX = Oxidizer ACID = Acid ALK = Alkali COR = Corrosive W = Use No Water

SECTION I - GENERAL INFORMATION

Characterization	**Pesticide**	RCRA Number	**P037**
		EPA Class	**Acute Hazardous Waste**
DOT Proper Shipping Name	**Dieldrin**	Chemical Abstract Service (CAS) Number	**60-57-1**
DOT Hazard Class and Label Requirements	**Poison**	DOT Emergency Guide Code	**55**
DOT Identification Number	**UN 2761**	Chemical Formula	$C_{12}H_8Cl_6O$

Synonyms

1,2,3,4,10,10-Hexachloro-6,7-epoxy-1,4,4a,5,6,7,8,8a-octahydro-1,4-endo, exo-5,8-dimethanonaphthalene; HEOD; Octalox®; Compound 497; Alvit®; Dieldrex; Panoram D-31; Quintox.

SECTION II - HAZARDOUS INGREDIENTS AND IDENTITY INFORMATION

Hazardous Components (specific identity)	OSHA Exposure Criteria	NIOSH Exposure Criteria	Immediately Dangerous to Life and Health (IDLH)	ACGIH Exposure Criteria
Dieldrin: (derivation: By oxidation of aldrin with peracids; as a degradation product of aldrin).	PEL (8-hour): 0.25 mg/m^3 (skin) STEL: Not Established	REL (10-hour): 0.25 mg/m^3 (skin) Possible Cancer Agent	50 mg/m^3	TLV: 0.25 mg/m^3 (skin) STEL: Not Established

SECTION III - PHYSICAL AND CHEMICAL CHARACTERISTICS

Boiling Point	**Decomposes**	Specific Gravity (H_2O = 1)	**1.75**
Vapor Pressure (mm Hg)	**3.1 X 10^{-6} at 68°F (20°C)**	Molecular Weight (atomic weight)	**380.9**
Vapor Density (Air = 1)	**13.2**	Melting Point	**349°F (176°C)**

Solubility

Nearly insoluble (0.02%) in water. Soluble in most organic solvents, mineral oil, and aliphatic hydrocarbons. Moderately soluble in aromatic hydrocarbons, esters, and ketones.

Appearance and Odor

Pure dieldrin is a colorless to white crystalline solid. Technical grade is tan. Mild, chemical odor. Odor Threshold = 0.041 ppm

SECTION IV - FIRE AND EXPLOSION HAZARD DATA

Flash Point (method used)	**Non Combustible Solid**	Explosive Limits in Air % by Volume	**LEL: Not Applicable UEL: Not Applicable**
NFPA Classification	**Non Combustible Solid**	Autoignition Temperature	**Not Determined**

Extinguishing Media

Use agent suitable to surrounding fire (dieldrin itself does not burn).

Special Fire Fighting Procedures

Wear full protective clothing and self-contained breathing apparatus (SCBA). Structural fire-fighting equipment is permeable and may not provide sufficient protection. Remain clear of smoke, water fall-out and water run-off. Move containers from fire area if it can be done without risk. Cool exposed containers.

Unusual Fire and Explosion Hazards

Dieldrin may become molten when involved in fire and emit hydrogen chloride and other chloride gases.

SECTION V - REACTIVITY DATA

Stability		Conditions to Avoid: Dieldrin is normally stable at room temperature and pressure. It will degrade on exposure to sunlight to form photodieldrin. Avoid exposure to high temperatures or incompatible materials.
Stable	Unstable	Incompatibility (*materials to avoid*)
X		Reacts on contact with strong oxidizers (perchlorates, peroxides, permanganates, chlorates, and nitrates), active metals (sodium) mineral acids, strong acids (sulfuric, nitric), and phenols.
Hazardous Polymerization		Conditions to Avoid: Under normal conditions of temperature and pressure, hazardous polymerization of dieldrin is not known to occur.
May Occur	Will Not Occur	Hazardous Decomposition or By-products
	X	At high temperatures, dieldrin decomposes and may produce poisonous gases, including toxic hydrogen chloride and other chloride gases.

SECTION VI - HEALTH HAZARD DATA

Primary Route(s) of Entry:	Inhalation?	Absorption (skin/eye)?	Ingestion?
	X	X	X

Health Hazards

INHALATION: Breathing dusts can cause irritation of the nose and throat. Repeated contact will lead to toxic systemic effects. These include headache, dizziness, nausea, vomiting, malaise, profuse sweating, fatigue, clonic or tonic convulsions, myoclonic limb jerks, coma, and respiratory failure. May cause tremors, fainting, muscle spasms, weight loss, and damage to the liver and kidneys.

ABSORPTION: Irritation. Solution can pass through intact skin causing serious systemic effects.

INGESTION: Nausea, vomiting, diarrhea, and systemic effects (see inhalation).

Carcinogenicity	NTP Listed?	IARC Cancer Review Group?	OSHA Regulated?	Target Organs?
Questioned Human Questioned Animal	5th Annual Report	Group 3	29 CFR 1910.1000 Table Z-1	Central nervous system (CNS), liver, kidneys, skin.

Medical Conditions Generally Aggravated by Exposure

Central nervous system disorders and kidney or liver diseases may be aggravated by exposure.

Emergency and First-aid Procedures

Eye contact: Do not allow victim to keep eyes tightly shut. Flush immediately with water for 15 minutes, seek medical attention. Skin contact: Remove all contaminated clothing. Immediately wash area with flooding amounts of water followed by an alcohol wash and then a second wash with soap and water. If red skin or blisters appear, seek medical assistance. Inhalation: Remove the person from exposure. Provide respiratory assistance and CPR. Transfer to medical facility. If swallowed: Contact poison control center. Unless advised otherwise, do NOT induce vomiting. Seek medical attention immediately. Never give anything by mouth to an unconscious or convulsing person.

SECTION VII - PRECAUTIONS FOR SAFE HANDLING AND USE

Steps to be Taken in Case Material is Released or Spilled

Remove all sources of ignition. Ventilate area of spill. Restrict those not involved in cleanup from entering area. Notify appropriate authorities. Carefully scoop or collect powdered materials with a vacuum equipped with HEPA filter (do NOT generate dusts). Dispose of in sealed drums.

Preferred Waste Disposal Method

Rotary kiln incineration is recommended.

Precautions to be Taken in Handling and Storage

Store in tightly closed containers in cool, well-ventilated area away from heat and incompatibilities.

Other Precautions and Warnings

Since use of dieldrin is prohibited in the U.S., storage should only occur while awaiting disposal.

SECTION VIII - CONTROL MEASURES AND PERSONAL PROTECTIVE EQUIPMENT

Respiratory Protection (*specify type*)

Odor threshold is low and difficult to monitor. For exposures above 0.25 mg/m^3, use a NIOSH-approved supplied air respirator or a SCBA with full facepiece operated in positive pressure mode.

Ventilation

Local exhaust (preferred) at site of chemical work or general ventilation.

Protective Gloves	Eye Protection	Other Protective Clothing
Rubber	Chemical/Dust Goggles or Face Mask	Protective Apron, Boots (avoid skin contact)

Work/Hygiene Practices

Always wash hands thoroughly after using chemical; never bring food, drink, or smoking materials into vicinity of chemicals.

DIELDRIN

$C_{12}H_8Cl_6O$ CAS: 60-57-1

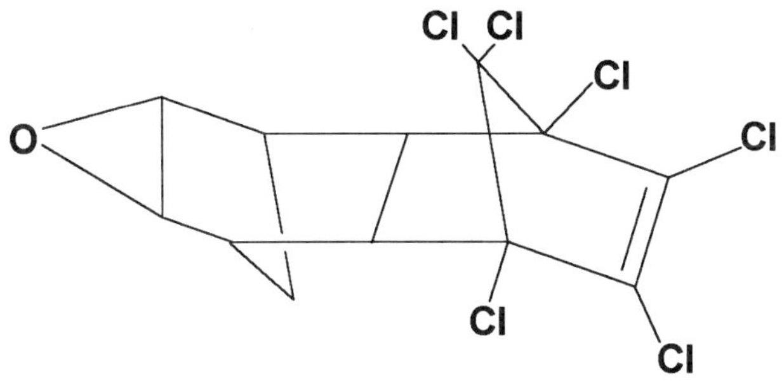

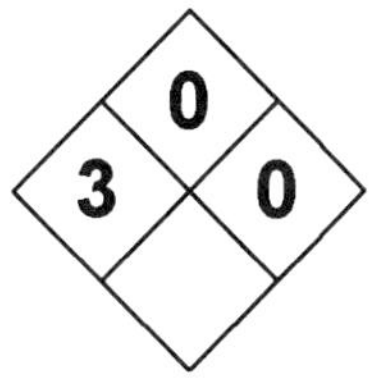

IDENTIFICATION AND TYPICAL USES

Dieldrin is a colorless to white crystalline solid in its pure form. The technical grade appears as a tan crystalline solid or powder. It has a mild chemical odor. It was formerly used widely as an insecticide until agricultural use was banned in 1974. Its use to control moth proliferation in woolen goods and for termite control continued until 1992. It is no longer produced or used in the United States. It is still used in other countries for the control of malaria mosquitoes and other related vector-borne diseases.

RISK ASSESSMENT: HEALTH

General Assessment

Dieldrin is highly toxic to humans by all routes of exposure (*inhalation, ingestion,* and skin *absorption*). It is a questionable carcinogen in animals and in humans, with experimental teratogenic and mutagenic data also reported. This chemical is no longer manufactured in the United States and, therefore, exposure risk is thought to be relatively low. However, because of the chemical's long persistence in the environment, exposure can still occur through contact with contaminated soils, water, and vegetation, at hazardous waste sites, or contaminated manufacturing sites. Should human exposure ever occur, dieldrin can cause serious toxic and poisonous systemic effects up to and including death.

Dieldrin is a central nervous system stimulant. Symptoms of exposure by both inhalation and skin contact (absorption) include respiratory depression with convulsions, hypersalivation, hyperirritability, profuse sweating, myoclonic limb jerks, clonic/tonic convulsions, and coma. There may also be headache, dizziness, fatigue, blurred vision, tremors, and respiratory failure. Inhalation also causes irritation of the eyes, nose, and throat. Skin absorption may produce greater toxicity than inhalation. Combined exposures (e.g., inhalation *and* absorption) are extremely dangerous and can lead to death in some cases. Caution is therefore warranted when inhalation and skin contact exposures are possible. Even though air levels may be within limits, personnel may be overexposed if dieldrin should contact the skin.

In addition to absorption and subsequent systemic effects, skin contact will result in moderate to severe irritation. Eye contact may also result in irritation. The chemical may even be absorbed through direct eye contact. Ingestion causes nausea, vomiting, diarrhea, and systemic effects of inhalation and skin absorption.

Acute Health Effects

The following acute (short-term) health effects may occur immediately or shortly after exposure to dieldrin:

Skin: Liquid can pass through unbroken skin. Contact may cause irritation and rash. Symptoms of exposure include headache, nausea, abdominal pain, vomiting, and CNS stimulation.

Eye: Mild to severe irritation. Absorption through direct eye contact is possible.

Lung: Irritation of the nose and throat. Breathing vapors can cause cough, convulsions, and adverse effects on the liver.

CNS: A central nervous system stimulant. Symptoms include, irritability, blurred vision, tremors, convulsions, and mental confusion.

Chronic Health Effects

The following chronic (long-term) health effects may occur at some time after exposure to dieldrin and can last for months or even years:

Cancer Hazards: Dieldrin may be a carcinogen in humans since it has been shown to cause liver cancer in animals.

Reproductive Hazard: There is some evidence that exposure to dieldrin causes damage to the developing fetus in test animals. Dieldrin may decrease fertility in males and females.

Other Chronic Effects: Repeated exposures may lead to chronic fainting, muscle spasms, tremors, and weight loss. Dieldrin is stored primarily in the fatty tissues and remains relatively unchanged.

Recommended Risk-Reduction Measures

Dieldrin is no longer manufactured in the United States. Its use is no longer permitted. However, in the unlikely event that personnel should be occupationally exposed to dieldrin, the risks to human health can be severe. Personnel should avoid direct contact with dieldrin. If a less toxic material or compound cannot be substituted for dieldrin, then *engineering controls* are the most effective method of reducing exposures. The best protection is to enclose operations and/or provide local exhaust ventilation at the site of dieldrin dust release. This may be practical in manufacturing operations, but is not feasible in application since it is used outdoors. While not always operationally possible, isolating operations involving dieldrin manufacturing can also reduce exposure. Using respiratory protection is less effective than the controls mentioned above, but is still advisable whenever working with or around dieldrin. The PEL is extremely low (0.25 mg/m^3) and difficult to accurately monitor. At exposures above the PEL, an MSHA or NIOSH-approved self-contained breathing apparatus (SCBA) with full facepiece and pressure demand is recommended for the greatest possible respiratory protection. If a full facepiece is not available, then chemical goggles should be worn to protect the eyes. A face shield and protective apron should also be worn. To prevent hand and skin exposures, impervious gloves should be worn.

Administrative controls should also be in place to minimize the potential for human exposures. These may include written procedures or policies which specify the methods and techniques that will be practiced whenever personnel are to work with dieldrin.

All personnel should receive *training* on the use, hazards, protective measures, emergency actions, and other precautions per 29 CFR 1910.1200 (Hazard Communication), prior to the first assignment in an area where dieldrin is used or stored.

If symptoms develop or overexposure is suspected, the following medical testing may be useful:

- ☑ Liver and kidney function tests.
- ☑ Urinalysis for halogenated compounds.
- ☑ Neurological evaluation of central nervous system.
- ☑ Gastric lavage may be indicated if substantial ingestion has occurred. Activated charcoal and sodium sulfate may be beneficial.

Any evaluation should include a careful history of past and present symptoms with an examination. Medical tests that simply look for existing damage are not a substitute for controlling exposures.

Other methods to reduce exposure include:

- ☑ Where possible, enclose operations and use local exhaust ventilation at the site of chemical release. If local exhaust ventilation or enclosure is not used, respiratory protection should be mandatory.
- ☑ Always ensure that proper protective clothing is worn when using chemical substances.
- ☑ Wash thoroughly immediately after exposure to dieldrin and at the end of the work shift or before eating, drinking, or smoking.
- ☑ Hazard warning information should be posted in the work area. In addition, as part of an on-going education and training program, all information on the health and safety hazards of dieldrin should be communicated to all potentially exposed workers.
- ☑ Eye wash stations should be provided in the immediate work area for emergency use. If there is a possibility of skin exposure to dieldrin, emergency shower facilities should also be provided.
- ☑ Workers whose clothing has been contaminated by dieldrin should change into clean clothes before leaving work. Contaminated work clothing should be laundered only by individuals who have been informed of the hazards of exposure to dieldrin.
- ☑ Specific engineering controls are recommended by NIOSH. Refer to NIOSH Criteria Document: Occupational Exposure During the Manufacture and Formulation of Pesticides (Publication Number 78-174).

RISK ASSESSMENT: ENVIRONMENT

General Assessment

Because of its former use as an insecticide and pesticide, dieldrin may be present in the environment. The environment is also at risk of exposure during transportation, storage, disposal, or destruction of dieldrin. In almost every scenario, the threat of environmental exposure is contingent upon the proper handling of the chemical substance. Accidental spills, large or small, where dieldrin contacts incompatible commodities can result in violent and even explosive reactions (depending upon conditions of spill), and possible contamination of the surrounding environmental mediums (water, soil, and air).

Dieldrin is considered a non-combustible solid. However, it is common practice to mix it with some flammable liquids such as benzene or xylene. Also, because it is incompatible with many common oxidizers, such as chlorine, extreme caution is required in handling, storage, transportation, and disposal of dieldrin. These characteristics also require special consideration during any emergency situation involving a leak or spill of dieldrin powder or dust.

Dieldrin use in the U.S. is no longer permitted. However, because of its once widespread, general usage in the past and its persistence, dieldrin can enter the environment in agricultural run-off.

Acute Ecological Effects

Acute (short-term) toxic effects may include the death of animals, birds, or fish and death or low growth rate in plants. Acute effects are seen 2 to 4 days after animals or plants are exposed to dieldrin. Dieldrin has high acute toxicity to aquatic life, land animals, and to birds. Insufficient data are available on the short-term effects of dieldrin to plants.

Chronic Ecological Effects

Chronic (long-term) toxic effects may include shortened life span, reproductive problems, lower fertility, and changes in appearance or behavior in exposed animals. These effects can be seen long after first exposure(s) to toxic chemicals. Dieldrin has high chronic toxicity to aquatic life, birds, and land animals. Insufficient data are available to evaluate or predict the long-term effects of dieldrin to plants.

Water Solubility

Dieldrin is nearly insoluble in water (0.02%). Concentrations of less than 1 milligram may not mix with a liter of water.

Persistence in the Environment

Dieldrin is highly persistent in the water, with a half-life of greater than 200 days. The half-life of a pollutant is the amount of time it takes for one-half of the chemical to be degraded. It will persist in soil for extremely long periods (7 years or more). It can leach to ground water and absorb in terrestrial and aquatic sediments. About 80% of dieldrin will eventually end up in terrestrial soil; approximately 19% will end up and aquatic sediments; the remainder will end up in water or air.

Bioaccumulation in Aquatic Organisms

Some substances increase in concentration, or *bioaccumulate,* in living organisms as they breathe contaminated air, drink contaminated water, or eat contaminated food. These chemicals can become concentrated in the tissues and internal organs of animals as well as humans.

The concentration of dieldrin found in fish tissues is expected to be considerably higher that the average concentration of dieldrin in the water from which the fish was taken.

Recommended Risk-Reduction Measures

Although dieldrin is no longer manufactured or used in the United States and it is unlikely that it will be transported, stored, handled, or destroyed, environmental contamination is still possible. This is based upon its extremely long persistence in the environment and the fact that it can be present in contaminated soils, at hazardous waste sites, and in some stockpiles awaiting destruction.

Storage of dieldrin should be segregated from incompatible materials to minimize the risk of cross-contamination or contact. Buildings or storage areas should be equipped with appropriate fire protection systems (alarms, sprinklers). Proper ventilation and protective equipment should be used while handing the powders or while preparing aqueous solutions.

If a spill or leak to the environment has occurred, fire department, emergency response and/or hazardous materials spill personnel should be notified immediately. Cleanup should be attempted only by those trained in proper spill containment procedures. If dieldrin should contact the water table, aquifer, or navigable waterway, reclamation procedures should be initiated as soon as practical. It is nearly insoluble in water and total remediation may be possible. The local and/or state emergency response authorities must be notified. A comprehensive emergency response or disaster preparedness/recovery plan should be in place prior to any operations involving the use, transportation, storage, or disposal of dieldrin. If dieldrin is spilled or leaked, the following specific steps are recommended:

- ☑ Restrict persons not wearing protective clothing from area of spill or leak until cleanup is complete and area can be opened for normal work.
- ☑ Remove all ignition sources.
- ☑ Collect solids using a HEPA vacuum to avoid the generation of dusts.
- ☑ It may be necessary to dispose of dieldrin as a hazardous waste. The responsible state agency or the regional office of the federal Environmental Protection Agency (EPA) should be contacted for specific recommendations.

RISK ASSESSMENT: BUSINESS

General Assessment

Accidents or mishaps involving dieldrin can present a significant threat to business operations. The loss or damage of equipment or facilities can significantly affect fiscal viability. Lawsuits that may result from personnel injury/death, public exposures, and/or environmental contamination will also require a serious expenditure of resources. Media attention surrounding an injury, death, or environmental damage can also result in a loss of profits and loss of current as well as future business. Always remember that anytime the terms "cancer," "carcinogen," or "reproductive hazard" are used, public emotion, hysteria, and ignorance can run equally high. This must be a consideration when developing public relations policies.

Recommended Risk-Reduction Measures

Company attorneys, safety and health professionals, and environmental specialists should be involved in the development of any procedures for responding to chemical incidents. A company official should be predesignated as a public relations officer with specific training in dealing with the press. Corporate plans and policies should be developed, approved, and implemented long before any need for such arises.

REFERENCES

American Conference of Governmental Industrial Hygienists. 1988. *Documentation of the Threshold Limit Values and Biological Exposure Indices,* 5th Edition (with updates). Cincinnati: ACGIH

Brusick, D. J. 1994. *Methods for Genetic Risk Assessment.* New York: CRC Press/Lewis Publishers.

Calabrese, E. J. 1994. *Biological Effects of Low Level Exposures.* New York: CRC Press/Lewis Publishers.

Cockerman, L. G., B. S. Shane. 1994. *Basic Environmental Toxicology.* New York: CRC Press/Lewis Publishers.

Cothern, C. R. 1995. *Handbook for Environmental Risk Decisionmaking.* New York: CRC Press/Lewis Publishers.

Howard, P. H., et. al. 1991. *Handbook of Environmental Degradation Rates.* New York: CRC Press/Lewis Publishers.

Lauwerys, R. R., P. Hoet. 1993. *Industrial Chemical Exposure.* New York: CRC Press/Lewis Publishers.

Lewis, R. J., Sr. 1992. *Sax's Dangerous Properties of Industrial Materials,* Volumes 1, 2, and 3, 8th Edition. New York: Van Nostrand Reinhold.

Manahan, S. E. 1992. *Toxicological Chemistry,* 2nd Edition. New York: CRC Press/Lewis Publishers.

Meyer, E. 1990. *Chemistry of Hazardous Materials.* Englewood Cliffs, NJ: Prentice Hall.

Ott, W. R. 1995. *Environmental Statistics and Data Analysis.* New York: CRC Press/Lewis Publishers.

Patnaik, P. 1992. *A Comprehensive Guide to the Hazardous Properties of Chemical Substances.* New York: Van Nostrand Reinhold.

Quigley, D. R. 1994. *Handbook of Emergency Chemical Management.* New York: CRC Press/Lewis Publishers.

Richardson, M. 1995. *Dictionary of Substance and Their Effects,* Vol. I-VI. New York: CRC Press/Lewis Publishers.

U.S. Department of Health and Human Services, National Institute for Occupational Health (NIOSH). 1978. *NIOS Criteria Document: Occupational Exposure During the Manufacture and Formulation of Pesticides.* Publication Number 78-174. Washington, DC: U.S. Government Printing Office.

MATERIAL SAFETY DATA SHEET

CHEMICAL NAME

DIETHANOLAMINE

HAZARD WARNING INFORMATION

HEALTH	FIRE	REACTIVE	OTHER	DEGREE OF HAZARD	COLOR CODING	OTHER CODES
1	1	0		0 = Minimum Hazard 1 = Slight Hazard 2 = Moderate Hazard 3 = Serious Hazard 4 = Severe Hazard	HEALTH = BLUE FIRE = RED REACTIVITY = YELLOW OTHER = WHITE	OX = Oxidizer ACID = Acid ALK = Alkali COR = Corrosive W = Use No Water

SECTION I - GENERAL INFORMATION

Characterization	RCRA Number	EPA Class
Aliphatic Amine	**None**	**Not Applicable**

DOT Proper Shipping Name **Not Listed**	Chemical Abstract Service (CAS) Number **111-42-2**
DOT Hazard Class and Label Requirements **Not Listed**	DOT Emergency Guide Code **No Citation**
DOT Identification Number **Not Listed**	Chemical Formula **$(HOCH_2CH_2)_2NH$**

Synonyms

DEA; di(2-hydroxyethyl)amine; 2,2'-dihydroxydiethylamine; diolamine; bis92-hydroxyethyl)amine; 2,2'-iminodiethanol.

SECTION II - HAZARDOUS INGREDIENTS AND IDENTITY INFORMATION

Hazardous Components (specific identity)	OSHA Exposure Criteria	NIOSH Exposure Criteria	Immediately Dangerous to Life and Health (IDLH)	ACGIH Exposure Criteria
Diethanolamine (derivation: By reaction of ammonia with ethylene oxide to form mono, di, and tri-ethanolamines and separated by fractionation). **1 ppm = 4.37 mg/m^3**	**PEL:** **Not Established** **STEL:** **Not Established**	**REL (10-hour):** **3 ppm** **15 mg/m^3** **STEL:** **Not Established**	**Not Determined**	**TLV (skin):** **0.46 ppm** **2 mg/m^3** **STEL:** **Not Established**

SECTION III - PHYSICAL AND CHEMICAL CHARACTERISTICS

Boiling Point **514°F (268°C)**	Specific Gravity (H_2O = 1) **1.10**
Vapor Pressure (mm Hg) **<0.01 at 69°F (20°C)**	Molecular Weight **105.2**
Vapor Density (Air = 1) **3.65**	Melting Point **82°F (28°C)**

Solubility

Very soluble in water (95%), acetone, methanol, and alcohol. Insoluble in ether and benzene.

Appearance and Odor

Colorless hygroscopic crystals or a syrupy, white liquid (above 82°F) with a mild, ammonia-like odor. Odor Threshold = 2.6 ppm.

SECTION IV - FIRE AND EXPLOSION HAZARD DATA

Flash Point (method used) **294°F (148°C) closed cup; 306°F (152°C) op. cup**	Explosive Limits in Air % by Volume **LEL: 1.6% UEL: 9.8%**
NFPA Classification **Class IIIB Combustible Liquid**	Autoignition Temperature **594°F (312°C)**

Extinguishing Media

Use dry chemical, carbon dioxide, foam, water spray, or fog. Water may cause frothing.

Special Fire Fighting Procedures

Poisonous gases are produced in fire, wear full protective clothing and self-contained breathing apparatus (SCBA). Move containers from fire area if it can be done safely. Cool exposed containers. Do not allow runoff from fire control methods to reach sewers or waterways.

Unusual Fire and Explosion Hazards

None Reported.

SECTION V - REACTIVITY DATA

Stability		Conditions to Avoid Diethanolamine is normally stable at room temperature in closed containers under normal conditions of storage and handling. Avoid contact with incompatible materials. Keep dry and away from heat and flame.
Stable X	Unstable	Incompatibility *(materials to avoid)* Oxidizing agents (chlorine, bromine, fluorine), strong acids (nitric, sulfuric), acid anhydrides, and halides. Chemical is hygroscopic (absorbs moisture from air). Corrosive to copper, zinc, and galvanized iron.
Hazardous Polymerization		Conditions to Avoid Under normal conditions of temperature and pressure, hazardous polymerization of diethanolamine is not expected to occur.
May Occur	Will Not Occur X	Hazardous Decomposition or By-products Toxic and irritating fumes and vapors are produced when diethanolamine is heated to decomposition. These include toxic oxides of nitrogen.

SECTION VI - HEALTH HAZARD DATA

Primary Route(s) of Entry:	Inhalation?	Absorption (skin)?	Ingestion?
	X		X

Health Hazards

INHALATION: Dusts can cause irritation of the respiratory tract with sneezing, coughing, and tearing.

SKIN & EYES: A corrosive. Severe irritation and possible corrosive burns to the eye with corneal damage likely and permanent loss of visual acuity. Skin irritation may be mild to moderate.

INGESTION: Severe irritation of the gastrointestinal tract with nausea, vomiting, and abdominal pain. There may also be somnolence, excitement, and muscular contraction. May damage the liver and kidneys (animal confirmation only).

Carcinogenicity	NTP Listed?	IARC Cancer Review Group?	OSHA Regulated?	Target Organs?
Unknown Human Unknown Animal	No	No	No	Eyes, skin, respiratory system.

Medical Conditions Generally Aggravated by Exposure

None Reported.

Emergency and First-aid Procedures

Eye contact: Flush immediately with water for 15 minutes (minimum), seek medical attention. Skin contact: Remove all contaminated clothing. Immediately wash area with large amounts of soap and water. Seek medical assistance. For inhalation: Remove the person from exposure. Provide respiratory assistance and CPR. Transfer to medical facility. If swallowed: Call the poison control center and seek medical attention immediately. Unless advised otherwise, give 1-2 glasses of water to dilute. Do NOT induce vomiting. Never attempt to give an unconscious or convulsing person anything by mouth. DOCTOR: Treatment is symptomatic and supportive.

SECTION VII - PRECAUTIONS FOR SAFE HANDLING AND USE

Steps to be Taken in Case Material is Released or Spilled

Absorb liquids in vermiculite or other absorbent and deposit in DOT-approved drum. Do NOT dry sweep solids. Use HEPA vacuum instead. Ventilate area and remove ignition sources. Restrict those not involved in cleanup from entering area. Notify appropriate authorities, if applicable.

Preferred Waste Disposal Method

No Citation.

Precautions to be Taken in Handling and Storage

Do not store diethanolamine in the presence of incompatible chemicals or materials. Store in tightly closed containers in a cool, dry well-ventilated area. Protect from light. If stored for prolonged periods, keep in stainless steel containers with nitrogen purge to prevent absorption of moisture.

Other Precautions and Warnings

None

SECTION VIII - CONTROL MEASURES AND PERSONAL PROTECTIVE EQUIPMENT

Respiratory Protection *(specify type)*

For low exposures, use a respirator equipped with HEPA filter and dust pre-filter. Otherwise, use a supplied-air respirator with full facepiece operated in positive pressure mode or a self-contained breathing apparatus with full facepiece operated in pressure demand or other positive pressure mode.

Ventilation

Local exhaust or general mechanical systems recommended.

Protective Gloves	Eye Protection	Other Protective Clothing
Impervious Rubber Gloves	Chemical/Dust Goggles or Face Mask	Protective Apron or Clothing

Work/Hygiene Practices

Always wash hands thoroughly after using chemical; never bring food, drink, or smoking materials into vicinity of chemicals.

DIETHANOLAMINE

$(HOCH_2CH_2)_2NH$ CAS: 111-42-2

$$HO - CH_2 - CH_2 - NH - CH_2 - CH_2 - OH$$

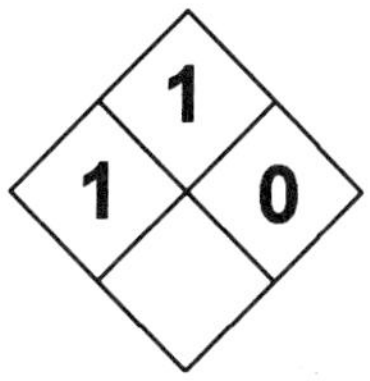

IDENTIFICATION AND TYPICAL USES

Diethanolamine is a colorless or clear crystalline solid or may appear as a syrupy, white, hygroscopic liquid with a mild, ammonia-like odor. It is used in the production of surface-active agents and lubricants for the textile industry. It is also used as an intermediate for rubber chemicals, as an emulsifier, as a humectant and softening agent, as a detergent in paints, shampoos, and other cleaners, and as an intermediate in resins and plasticizers.

RISK ASSESSMENT: HEALTH

General Assessment

Diethanolamine is a severe irritant and corrosive on contact with human tissue. Primary route of exposure is *inhalation* of the dusts or powders. It can also cause serious heath effects on *ingestion* and by skin contact (although absorption has not been a confirmed as a route of entry). There is no information in the references to indicate any carcinogenic, mutagenic, or teratogenic properties.

Inhalation will result in irritation of the mucosa, especially in the nose, throat, and upper respiratory tract. Symptoms may include coughing, sneezing, and tearing. There may also be nausea and headache. Inhalation of vapors or mists may be fatal due to the potential for spasm of the glottis which can lead to suffocation.

Skin contact with the liquid can cause mild to moderate irritation. Eye contact can result in corneal injury and permanent damage or loss of vision.

Ingestion of an alkaline material, such as diethanolamine, will cause irritation of the gastrointestinal tract with nausea, vomiting, and abdominal pain. Some animal studies have shown liver and kidney damage on ingestion. Human data are not conclusive in this area of study and more research is required.

Acute Health Effects

The following acute (short-term) health effects may occur immediately or shortly after exposure to diethanolamine:

Skin: Mild to moderate irritation. May cause allergy to develop.

Eye: Severe irritation, inflammation, with possible damage to the cornea and loss of vision.

Lung: Inhalation of the dusts or heated mists causes coughing, sneezing, and tearing. May cause spasm of the glottis and death by suffocation.

Other: Ingestion causes severe irritation of the digestive tract with nausea, vomiting, possible somnolence, excitement, and muscular contractions and abdominal pains.

Chronic Health Effects

The following chronic (long-term) health effects may occur at some time after exposure to diethanolamine and can last for months or even years:

Cancer Hazards: According to the information presented in the references, diethanolamine has not been adequately tested for its ability to cause cancer in test animals. More research is required.

Reproductive Hazard: According to the information presented in the references, diethanolamine has not been adequately tested for its ability to adversely affect reproduction.

Other Chronic Effects: Prolonged exposures may lead to liver damage and possible allergy.

Recommended Risk-Reduction Measures

Unless a less toxic chemical can be substituted for a hazardous substance, *engineering controls* are the most effective method of reducing exposures. The best protection is to enclose operations and/or provide local exhaust ventilation at the site of chemical release. To prevent static sparks, all containers and equipment should be electrically grounded and bonded. While not always operationally feasible, isolating operations can also reduce exposure.

Using respiratory protection is less effective than the controls mentioned above, but is still advisable whenever working with or around diethanolamine. For relatively low, infrequent, or transient exposures, an organic vapor respirator equipped with a high efficiency particulate air (HEPA) filter and dust pre-filter may suffice. For higher or prolonged exposure, a self-contained breathing apparatus (SCBA) with full facepiece and pressure demand is the recommended respiratory protection of choice. If a full facepiece is not available, then chemical goggles should be worn to protect the eyes. Whenever a chemical splash hazard exists, a face shield and apron should be worn. To prevent hand and skin exposures, impervious chemical gloves should be worn. Butyl, Neoprene, and nitrile rubber, as well as polyvinyl chloride and Viton are all recommended glove materials. Glove manufacturers should be contacted and permeation studies obtained *before* final glove selections have been made.

Administrative controls should also be in place to minimize the potential for human exposures. These may include written procedures or policies which specify the methods and techniques that will be practiced whenever personnel are to work with diethanolamine. Diethanolamine should not be used or mixed with N-nitosating agents since it may form N-nitrosodiethanolamine, which is a liver and nasal cavity carcinogen.

All personnel should receive *training* on the use, hazards, protective measures, emergency actions, and other precautions per 29 CFR 1910.1200 (Hazard Communication), prior to the first assignment in an area where diethanolamine is used or stored.

Before beginning employment and at regular intervals thereafter (e.g. annually), the following medical tests are recommended:

☑ Lung function tests (establish a baseline).

If symptoms develop or overexposure is suspected, the following medical tests may be useful:

☑ Lung function tests.

☑ Liver function tests.

Any evaluation should include a careful medical history of past and present symptoms with an examination. However, medical tests that evaluate existing damage are not a substitute for controlling exposure.

Other methods to reduce exposure include:

☑ Where possible, enclose operations and use local exhaust ventilation at the site of chemical release. If local exhaust ventilation or enclosure is not used, respiratory protection should be mandatory. Also, if possible, automatically transfer liquids from storage containers to process containers using non-sparking tools.

☑ Always ensure that proper protective clothing is worn when using chemical substances.

☑ Wash thoroughly immediately after exposure to diethanolamine and at the end of the work shift or before eating, drinking, or smoking.

☑ Hazard warning information should be posted in the work area. In addition, as part of an on-going education and training program, all information on the health and safety hazards of diethanolamine should be communicated to all potentially exposed workers.

☑ Eye wash stations should be provided in work areas. If the potential for whole body exposure exists, then safety showers should also be provided.

RISK ASSESSMENT: ENVIRONMENT

General Assessment

The environment is at risk of exposure during intended use, transportation, storage, disposal, or destruction of diethanolamine. In almost every scenario, the threat of environmental exposure is contingent upon the proper handling of the chemical substance. Accidental spills, large or small, can result in fire, explosion, and possible contamination of the surrounding environmental mediums (water, soil, and air).

Diethanolamine is considered a Class IIIB combustible liquid (per OSHA 29 CFR 1910.106). While this means it poses a minimum fire risk by itself, mixture or contact with incompatible materials can cause fire or explosion. It can react with many common oxidizing materials and strong acids. It is corrosive to copper, zinc, and galvanized iron. These characteristics require special consideration during any emergency situation involving a leak or spill of diethanolamine. Should diethanolamine ever come into contact with incompatible substances either during use, transportation, or storage, violent reactions can occur.

Diethanolamine can enter the environment through industrial discharges and spills.

Acute Ecological Effects

Acute (short-term) toxic effects may include the death of animals, birds, or fish and death or low growth rate in plants. Acute effects are seen 2 to 4 days after animals or plants are exposed to diethanolamine.

This chemical has low to moderate acute toxicity to aquatic life. Insufficient data are available to evaluate or predict the short-term effects of diethanolamine to plants, birds, or land animals.

Chronic Ecological Effects

Chronic toxic effects may include shortened life span, reproductive problems, lower fertility, and changes in appearance or behavior in exposed animals. These effects can be seen long after first exposure(s) to toxic chemicals.

Diethanolamine has low chronic toxicity to aquatic life. Insufficient data are available to evaluate or predict the long-term effects of diethanolamine to plants, birds, or land animals.

Water Solubility

Diethanolamine is highly soluble in water. Concentrations of 1000 milligrams and more will mix with a liter of water.

Persistence in the Environment

Diethanolamine is slightly persistent in water, with a half-life of between 2 to 20 days. The half-life of a pollutant is the amount of time it takes for one half of the chemical to be degraded.

Bioaccumulation in Aquatic Organisms

Some substances increase in concentration, or *bioaccumulate,* in living organisms as they breathe contaminated air, drink contaminated water, or eat contaminated food. These chemicals can become concentrated in the tissues and internal organs of animals as well as humans.

Diethanolamine is not expected to accumulate in fish tissue.

Recommended Risk-Reduction Measures

Proper training of all transporters will reduce the likelihood of a mishap or accident resulting in a leak or spill to the environment. The correct use of DOT labeling (per the MSDS) on all containers, trucks, and rail cars will enable emergency responders to react properly and quickly to any disaster thereby reducing the potential risk to the environment and to personnel.

Storage of diethanolamine should be segregated from incompatible chemicals to minimize the risk of cross-contamination or contact. Storage buildings should be equipped with the proper fire protection equipment (alarms, sprinklers, extinguishers, etc.).

If a spill or leak to the environment has occurred, fire department, emergency response and/or hazardous materials spill personnel should be notified immediately. Cleanup should be attempted only by those trained in proper spill containment procedures. Contaminated soils should therefore be removed for incineration and replaced with clean soil. If diethanolamine should contact the water table, aquifer, or navigable waterway, remediation activities should be prompt. Time is of the essence since it is highly soluble and total containment and remediation may not be possible. When such spills occur, the local and/or state emergency response authorities must be notified. A comprehensive emergency response or disaster preparedness/recovery plan should be in place prior to any operations involving the use, transportation, storage, or disposal of diethanolamine.

If diethanolamine is spilled or leaked, the following specific steps are recommended:

- ☑ Evacuate area and deny entry to those not involved in cleanup activities. Restrict persons not wearing protective clothing from area of spill or leak until cleanup is complete.
- ☑ Ventilate area of spill or leak.
- ☑ Remove all ignition sources.
- ☑ Collect dusts in safest and most convenient manner possible and dispose in sealed drums. Do not dry sweep (creates airborne dusts). Use a vacuum equipped with a high efficiency particulate air (HEPA) filter instead. Use a damp mop to collect residue and perform final cleanup of area.
- ☑ If liquids are spilled, absorb using vermiculite, dry earth, or sand and place in a sealed drum for disposal.
- ☑ It may be necessary to dispose of diethanolamine as a hazardous waste. The responsible state agency or the regional office of the federal Environmental Protection Agency (EPA) should be contacted for specific recommendations.

RISK ASSESSMENT: BUSINESS

General Assessment

Accidents or mishaps involving diethanolamine can present a significant threat to business operations. The loss or damage of equipment or facilities can significantly affect fiscal viability. Lawsuits that may result from personnel injury/death, public exposures, and/or environmental contamination will also require a serious expenditure of resources. Media attention surrounding an injury, death, or environmental damage can also result in a loss of profits and loss of current as well as future business.

Recommended Risk-Reduction Measures

Company attorneys, safety and health professionals, and environmental specialists should be involved in the development of any procedures dealing with safety or emergency response. A company official should be pre-designated as a public relations officer with specific training in dealing with the press. Corporate plans and policies should be developed, approved, and implemented long before any need for such arises.

REFERENCES

American Conference of Governmental Industrial Hygienists. 1988. *Documentation of the Threshold Limit Values and Biological Exposure Indices,* 5th Edition (with updates). Cincinnati: ACGIH

Anthony, C. P., and N. J. Kolthoff. 1971. *Textbook of Anatomy and Physiology.* St. Louis: C. V. Mosby Company.

Bloom, A. D. and F. J. DeSerres. 1995. *Ecotoxicity and Human Health.* New York: CRC Press/Lewis Publishers.

Bueche, J. R. 1972. *Principles of Physics,* 2nd Edition. New York: McGraw-Hill.

Calabrese, E. J., E. M. Kenyon. *Air Toxics and Risk Assessment.* New York: CRC Press/Lewis Publishers.

Calabrese, E. J. 1994. *Biological Effects of Low Level Exposures.* New York: CRC Press/Lewis Publishers.

Cockerman, L. G., B. S. Shane. 1994. *Basic Environmental Toxicology.* New York: CRC Press/Lewis Publishers.

Cothern, C. R., N. P. Ross. 1994. *Environmental Statistics, Assessment, and Forecasting.* New York: CRC Press/Lewis Publishers.

Cothern, R. R. 1993. *Comparative Environmental Risk Assessment.* New York: CRC Press/Lewis Publishers.

Cothern, C. R. 1995. *Handbook for Environmental Risk Decisionmaking.* New York: CRC Press/Lewis Publishers.

Gardner, R. W. 1994. *Chemical Intolerance.* New York: CRC Press/Lewis Publishers.

Hallenbeck, W. H. 1993. *Quantitative Risk Assessment for Environmental and Occupational Health,* 2nd Edition. New York: CRC Press/Lewis Publishers.

Hodgson, E., R. B. Mailman, and Janice E. Chambers. 1988. *Dictionary of Toxicology.* London: The MacMillian Company.

Howard, P. H. 1995. *Handbook of Environmental Fate and Exposures Data for Organic Chemicals,* Vol. I-IV. New York: CRC Press/Lewis Publishers.

Howard, P. H., et. al. 1991. *Handbook of Environmental Degradation Rates.* New York: CRC Press/Lewis Publishers.

Jones, F. E. 1994. *Toxic Organic Vapors in the Workplace.* New York: CRC Press/Lewis Publishers.

Landis, W. G., M. H. Yu. 1995. *Introduction to Environmental Toxicology.* New York: CRC Press/Lewis Publishers.

Langerman, N. 1994 *Precautionary Labels for Chemical Containers.* New York: CRC Press/Lewis Publishers.

Lauwerys, R. R., P. Hoet. 1993. *Industrial Chemical Exposure.* New York: CRC Press/Lewis Publishers.

Lewis, R. J., Sr. 1993. *Hawley's Condensed Chemical Dictionary,* 12th Edition. New York: Van Nostrand Reinhold.

Lewis, R. J., Sr. 1992. *Sax's Dangerous Properties of Industrial Materials,* Volumes 1, 2, and 3, 8th Edition. New York: Van Nostrand Reinhold.

Manahan, S. E. 1992. *Toxicological Chemistry,* 2nd Edition. New York: CRC Press/Lewis Publishers.

Mellström, G. A., J. E. Wahlberg, and H. I. Maibach. 1994. *Protective Gloves for Occupational Use.* New York: CRC Press/Lewis Publishers.

Meyer, E.. 1990. *Chemistry of Hazardous Materials.* Englewood Cliffs, NJ: Prentice Hall.

Quigley, D. R. 1994. *Handbook of Emergency Chemical Management.* New York: CRC Press/Lewis Publishers.

Rea, W. J. 1995. *Chemical Sensitivity,* Vol. I-III. New York: CRC Press/Lewis Publishers.

Richardson, M. 1995. *Dictionary of Substance and Their Effects,* Vol. I-VI. New York: CRC Press/Lewis Publishers.

Sacarello, H. L. A. 1994. *The Comprehensive Handbook of Hazardous Materials.* New York: CRC Press/Lewis Publishers.

U.S. Department of Health and Human Services, National Institute for Occupational Safety and Health. 1994. *NIOSH Pocket Guide to Chemical Hazards.* Washington, D.C.: U.S. Government Printing Office.

MATERIAL SAFETY DATA SHEET

CHEMICAL NAME	
DIETHYLAMINE	**CRC PRESS/LEWIS PUBLISHERS DISCLAIMER:** THE INFORMATION AND RECOMMENDATIONS PRESENTED HEREIN ARE BASED ON SOURCES BELIEVED TO BE RELIABLE. CRC MAKES NO REPRESENTATION ON ITS COMPLETENESS OR ACCURACY. IT IS THE RESPONSIBILITY OF THE USER TO DETERMINE THE CHEMICAL'S SUITABILITY FOR ITS INTENDED USE, THE CHEMICAL'S SAFE USE, AND THE CHEMICAL'S PROPER DISPOSAL. NO REPRESENTATIONS AND/OR WARRANTIES, EITHER EXPRESSED OR IMPLIED, OF THE MERCHANTABILITY OR FITNESS FOR A PARTICULAR PURPOSE, OR OF ANY OTHER NATURE, ARE MADE WITH RESPECT TO THE INFORMATION PROVIDED IN THIS MSDS OR TO THE CHEMICAL TO WHICH INFORMATION MAY REFER. CRC NEITHER ASSUMES NOR AUTHORIZES ANY OTHER PERSON TO ASSUME FOR IT, ANY OTHER ADDITIONAL RESPONSIBILITY OR LIABILITY FOR THE USE OF, OR RELIANCE UPON, THIS INFORMATION.

HAZARD WARNING INFORMATION

HEALTH	FIRE	REACTIVE	OTHER	DEGREE OF HAZARD	COLOR CODING	OTHER CODES
3	3	0		0 = Minimum Hazard 1 = Slight Hazard 2 = Moderate Hazard 3 = Serious Hazard 4 = Severe Hazard	HEALTH = BLUE FIRE = RED REACTIVITY = YELLOW OTHER = WHITE	OX = Oxidizer ACID = Acid ALK = Alkali COR = Corrosive W = Use No Water

SECTION I - GENERAL INFORMATION

Characterization	RCRA Number	EPA Class
Aliphatic Amine	**D001**	**Characteristic (I)**

DOT Proper Shipping Name **Diethylamine**	Chemical Abstract Service (CAS) Number **109-89-7**
DOT Hazard Class and Label Requirements **Flammable Liquid**	DOT Emergency Guide Code **68**
DOT Identification Number **UN 1154**	Chemical Formula **$(C_2H_5)_2NH$**

Synonyms

Diethamine; N,N-diethylamine; N-ethylethanamine.

SECTION II - HAZARDOUS INGREDIENTS AND IDENTITY INFORMATION

Hazardous Components (specific identity)	OSHA Exposure Criteria	NIOSH Exposure Criteria	Immediately Dangerous to Life and Health (IDLH)	ACGIH Exposure Criteria
Diethylamine (derivation: From ethyl chloride and ammonia under high heat and pressure; by continuously passing ammonia, aldehyde or ketone, and hydrogen over a catalyst under careful monitoring). **1 ppm = 3.04 mg/m^3**	**PEL (8-hour):** **25 ppm** **75 mg/m^3** **STEL:** **Not Established**	**REL (10-hour):** **10 ppm** **30 mg/m^3** **STEL:** **25 ppm** **75 mg/m^3**	**200 ppm**	**TLV (8-hour):** **5 ppm** **15 mg/m^3** **STEL:** **15 ppm** **45 mg/m^3**

SECTION III - PHYSICAL AND CHEMICAL CHARACTERISTICS

Boiling Point **132°F (55°C)**	Specific Gravity (H_2O = 1) **0.71**
Vapor Pressure (mm Hg) **194 at 69°F (20°C)**	Molecular Weight **73.1**
Vapor Density (Air = 1) **2.5**	Melting Point **-58°F (-50°C)**

Solubility

Soluble in water, alcohol, ether, chloroform, paraffin, hydrocarbons, aromatic and aliphatic hydrocarbons, fixed oils, mineral oils, oleic and stearic acids.

Appearance and Odor

Colorless liquid with a fishy, ammonia-like odor.

SECTION IV - FIRE AND EXPLOSION HAZARD DATA

Flash Point (method used) **-18°F (-28°C) closed cup**	Explosive Limits in Air % by Volume **LEL: 1.8% UEL: 10.1%**
NFPA Classification **Class IB Flammable Liquid**	Autoignition Temperature **594°F (312°C)**

Extinguishing Media

Use dry chemical, carbon dioxide, foam, water spray, or fog. Cool exposed container with water.

Special Fire Fighting Procedures

Use full protective clothing and self-contained breathing apparatus (SCBA). Vapors are heavier than air and may travel for some distance to ignition source and flashback.

Unusual Fire and Explosion Hazards

Containers may explode in fire due to buildup of excessive internal pressures. Do not attempt to extinguish fire unless liquid flow can be stopped first. Aqueous solutions are flammable unless diluted extensively. This chemical presents a vapor explosion hazard indoors, outdoors, and in sewers.

SECTION V - REACTIVITY DATA

Stability		Conditions to Avoid **Diethylamine is normally stable at room temperature in closed containers under normal conditions of storage and handling. Avoid contact with incompatible materials. Keep away from heat and flame.**
Stable **X**	Unstable	Incompatibility *(materials to avoid)* **Oxidizing materials, strong acids, halogens, cellulose nitrate, reactive organic compounds, and some metals. Also, diethylamine will attack some forms of rubber, plastics, and coatings.**
Hazardous Polymerization		Conditions to Avoid **Under normal conditions of temperature and pressure, hazardous polymerization of diethylamine is not expected to occur.**
May Occur	Will Not Occur **X**	Hazardous Decomposition or By-products **Toxic and irritating fumes and vapors are produced when diethylamine is heated to decomposition. These include carbon monoxide, carbon dioxide, hydrocarbons, nitrogen oxides, and amine vapors.**

SECTION VI - HEALTH HAZARD DATA

Primary Route(s) of Entry:	Inhalation?	Absorption (skin)?	Ingestion?
	X		X

Health Hazards

INHALATION: **Symptoms of nausea, coughing, shortness of breath and difficulty breathing, severe irritation of the nose, throat, and lungs. May cause pulmonary edema (fluid buildup in the lungs), which can be fatal. May also cause degenerative changes in the heart.**

SKIN & EYES: **A corrosive. Severe burns to the eye with corneal damage and permanent vision impairment. Vapor may cause fogging of vision. Liquid will burn the skin and damage tissue.**

INGESTION: **Severe pain with vomiting, diarrhea, collapse, breathing difficulties, and chest pain.**

Carcinogenicity	NTP Listed?	IARC Cancer Review Group?	OSHA Regulated?	Target Organs?
Unknown Human **Unknown Animal**	**No**	**No**	**29 CFR 1910.1000** **Table Z-1**	**Eyes, skin, respiratory system, CVS.**

Medical Conditions Generally Aggravated by Exposure

Chronic skin, eye, or respiratory diseases may be aggravated. Heart patients may also be affected.

Emergency and First-aid Procedures

Eye contact: Flush immediately with water for 15 minutes (minimum), seek medical attention. Skin contact: Remove all contaminated clothing. Immediately wash area with large amounts of soap and water. Seek medical assistance. For inhalation: Remove the person from exposure. Provide respiratory assistance and CPR. Transfer to medical facility. Observe for 24-48 hours for lung effects. If swallowed: Call the poison control center and seek medical attention immediately. Unless advised otherwise, give 8 ounces of milk. Do NOT induce vomiting. Never attempt to give an unconscious or convulsing person anything by mouth.

SECTION VII - PRECAUTIONS FOR SAFE HANDLING AND USE

Steps to be Taken in Case Material is Released or Spilled

Neutralize with sodium bisulfate and small amounts of water. Absorb liquids in vermiculite or other absorbent and deposit in DOT-approved drum. Ventilate area and remove ignition sources. Restrict those not involved in cleanup from entering area. Notify appropriate authorities, if applicable.

Preferred Waste Disposal Method

No Citation.

Precautions to be Taken in Handling and Storage

Do not store diethylamine in the presence of incompatible chemicals or materials. Store in tightly closed containers in a cool, well-ventilated area. Personnel working with diethylamine should be trained on its proper handing and storage *prior* to being assigned to such responsibilities.

Other Precautions and Warnings

Recommend storage in external (detached) facility away from main work areas.

SECTION VIII - CONTROL MEASURES AND PERSONAL PROTECTIVE EQUIPMENT

Respiratory Protection *(specify type)*

For low exposures, use an organic vapor respirator. Otherwise, use an MSHA/NIOSH-approved supplied-air respirator with full facepiece operated in positive pressure mode or a self-contained breathing apparatus (SCBA) with full facepiece operated in pressure demand or other positive pressure mode.

Ventilation

Local exhaust or general mechanical systems recommended.

Protective Gloves	Eye Protection	Other Protective Clothing
Impervious Gloves	**Chemical Goggles or Face Mask**	**Protective Apron**

Work/Hygiene Practices

Always wash hands thoroughly after using chemical; never bring food, drink, or smoking materials into vicinity of chemicals.

DIETHYLAMINE

$(C_2H_5)_2NH$ CAS: 109-89-7

$$CH_3 — CH_2 — NH — CH_2 — CH_3$$

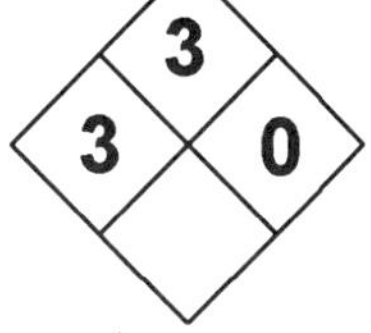

IDENTIFICATION AND TYPICAL USES

Diethylamine is a colorless or clear liquid with a strong, fishy, ammonia-like odor. It is used as a flotation agent, in the manufacture of dyes and pharmaceuticals, and in resins. It also finds applications in the rubber and petroleum industries, in the manufacture of textile specialties, in electroplating, as a corrosion inhibitor, and is used as a solvent.

RISK ASSESSMENT: HEALTH

General Assessment

Diethylamine is a severe irritant and corrosive on contact with human tissue. Primary route of exposure is *inhalation*. It can also cause serious heath effects on *ingestion* and by skin contact (although absorption has not been a confirmed as a route of entry). There is no information in the references to indicate any carcinogenic, mutagenic, or teratogenic properties.

Inhalation will result in irritation of the mucous membranes, especially in the nose, throat, and upper respiratory tract. Symptoms may include coughing, nausea, severe chest pains, difficult or labored breathing, and pulmonary edema (fluid buildup in the lungs), which is a medical emergency and can be fatal. There is also some evidence to indicate that high exposures may cause severe tissue degeneration of the heart (animal experiments only).

Skin contact with the liquid can cause severe irritation with necrosis and vesiculation (the appearance of tiny, fluid-filled lesions on the skin surface called vesicles). Eye contact can result in corneal injury and permanent damage or loss of vision. The vapor can cause foggy vision and tissue irritation.

Ingestion causes severe pain, vomiting, diarrhea, collapse, labored breathing, and chest pains. It may lead to liver or heart injuries (confirmed in animals studies only).

Acute Health Effects

The following acute (short-term) health effects may occur immediately or shortly after exposure to diethylamine:

Skin: Irritation, inflammation, redness, possible blistering, and tissue damage. May cause necrosis and vesiculation.

Eye: Severe irritation, inflammation, with possible damage to the cornea and loss of vision. Vapor can cause foggy vision, the appearance of halos around lights, and inflammation of tissues.

Lung: Irritation causing coughing and/or shortness of breath. Higher exposures can lead to respiratory problems, including pulmonary edema (fluid buildup in lungs) which is a medical emergency and can be fatal.

Other: Animal studies have shown a degeneration of heart tissue and liver damage on high exposures. It is not known if these effects will occur in humans.

Chronic Health Effects

The following chronic (long-term) health effects may occur at some time after exposure to diethylamine and can last for months or even years:

Cancer Hazards: According to the information presented in the references, diethylamine has not been adequately tested for its ability to cause cancer in test animals. More research is required.

Reproductive Hazard: According to the information presented in the references, diethylamine has not been adequately tested for its ability to adversely affect reproduction.

Other Chronic Effects: Low, chronic exposures may cause corneal damage, foggy vision, halos around lights, and irritation of the lungs, bronchi, and throat.

There is also a possibility that chronic exposures may cause injury to the heart and liver. However, the data are inconclusive in this regard.

Recommended Risk-Reduction Measures

Unless a less toxic chemical can be substituted for a hazardous substance, *engineering controls* are the most effective method of reducing exposures. The best protection is to enclose operations and/or provide local exhaust ventilation at the site of chemical release. To prevent static sparks, all containers and equipment should be electrically grounded and bonded. While not always operationally feasible, isolating operations can also reduce exposure.

Using respiratory protection is less effective than the controls mentioned above, but is still advisable whenever working with or around diethylamine. For relatively low, infrequent, or transient exposures, an organic vapor respirator may suffice. For higher or prolonged exposure, a self-contained breathing apparatus (SCBA) with full facepiece and pressure demand is the recommended respiratory protection of choice. If a full facepiece is not available, then chemical goggles should be worn to protect the eyes. Whenever a chemical splash hazard exists, a face shield and apron should be worn. To prevent hand and skin exposures, impervious chemical gloves should be worn. Butyl, natural, and nitrile rubber, as well as polyvinylchloride and Neoprene are not recommended.

Administrative controls should also be in place to minimize the potential for human exposures. These may include written procedures or policies which specify the methods and techniques that will be practiced whenever personnel are to work with diethylamine.

All personnel should receive *training* on the use, hazards, protective measures, emergency actions, and other precautions per 29 CFR 1910.1200 (Hazard Communication), prior to the first assignment in an area where diethylamine is used or stored.

Before beginning employment and at regular intervals thereafter (e.g. annually), the following medical tests are recommended:

- ☑ Lung function tests (establish a baseline).

If symptoms develop or overexposure is suspected, the following medical tests may be useful:

- ☑ Lung function tests. (Doctor: If pulmonary edema is present, consider PEEP. Steroids may be effective, but antibiotics are useful only if there is evidence of infection).
- ☑ Consider chest X-ray after acute overexposure.
- ☑ Complete blood count and blood gases.
- ☑ For oral exposures, evaluate for esophageal burns.
- ☑ For alkali burns, consider irrigating skin frequently for 24 hours with normal saline. (Note: Hydrotherapy may not be effective if not begun within 1 hour of the burn).

Any evaluation should include a careful medical history of past and present symptoms with an examination. However, medical tests that evaluate existing damage are not a substitute for controlling exposure. Also, since smoking can cause heart disease, emphysema, and numerous other respiratory disorders, smokers may be affected more rapidly than non-smokers to the same exposure conditions. Prudent risk assessment and management requires careful consideration of *all* possible risk factors that may be causing the appearance of symptoms in exposed workers.

Other methods to reduce exposure include:

- ☑ Where possible, enclose operations and use local exhaust ventilation at the site of chemical release. If local exhaust ventilation or enclosure is not used, respiratory protection should be mandatory. Also, if possible, automatically transfer liquids from storage containers to process containers using non-sparking tools.
- ☑ Always ensure that proper protective clothing is worn when using chemical substances.
- ☑ Wash thoroughly immediately after exposure to diethylamine and at the end of the work shift or before eating, drinking, or smoking.
- ☑ Hazard warning information should be posted in the work area. In addition, as part of an on-going education and training program, all information on the health and safety hazards of diethylamine should be communicated to all potentially exposed workers.
- ☑ Eye wash stations should be provided in work areas. If the potential for whole body exposure exists, then safety showers should also be provided.

RISK ASSESSMENT: ENVIRONMENT

General Assessment

The environment is at risk of exposure during intended use, transportation, storage, disposal, or destruction of diethylamine. In almost every scenario, the threat of

environmental exposure is contingent upon the proper handling of the chemical substance. Accidental spills, large or small, can result in fire, explosion, and possible contamination of the surrounding environmental mediums (water, soil, and air).

Diethylamine is considered a Class IB flammable liquid (per OSHA 29 CFR 1910.106). Its extremely low flash point and its relatively low boiling point make it a serious and dangerous fire and explosion risk. It can react with many common oxidizing materials to cause a fire or explosion. These characteristics require special consideration during any emergency situation involving a leak or spill of diethylamine. Should diethylamine ever come into contact with incompatible substances, such as oxidizers or acids, either during use, transportation, or storage, violent reactions can occur.

Diethylamine can enter the environment through industrial discharges and spills.

Acute Ecological Effects

Acute (short-term) toxic effects may include the death of animals, birds, or fish and death or low growth rate in plants. Acute effects are seen 2 to 4 days after animals or plants are exposed to diethylamine. This chemical has low to moderate acute toxicity to aquatic life. Insufficient data are available to evaluate or predict the short-term effects of diethylamine to plants, birds, or land animals.

Chronic Ecological Effects

Chronic toxic effects may include shortened life span, reproductive problems, lower fertility, and changes in appearance or behavior in exposed animals. These effects can be seen long after first exposure(s) to toxic chemicals. Insufficient data are available to evaluate or predict the long-term effects of diethylamine to aquatic life, plants, birds, or land animals.

Water Solubility

Diethylamine is highly soluble in water. Concentrations of 1000 milligrams and more will mix with a liter of water.

Persistence in the Environment

Diethylamine is slightly persistent in water, with a half-life of between 2 to 20 days. The half-life of a pollutant is the amount of time it takes for one half of the chemical to be degraded. It is not expected to accumulate in aquatic sediments or terrestrial soils.

Bioaccumulation in Aquatic Organisms

Some substances increase in concentration, or *bioaccumulate,* in living organisms as they breathe contaminated air, drink contaminated water, or eat contaminated food. These chemicals can become concentrated in the tissues and internal organs of animals as well as humans.

Diethylamine is not expected to accumulate in fish tissue.

Recommended Risk-Reduction Measures

Proper training of all transporters will reduce the likelihood of a mishap or accident resulting in a leak or spill to the environment. The correct use of DOT labeling (per the MSDS) on all containers, trucks, and rail cars will enable emergency responders to react properly and quickly to any disaster thereby reducing the potential risk to the environment and to personnel.

Storage of diethylamine should be segregated from incompatible chemicals to minimize the risk of cross-contamination or contact. Storage buildings should be equipped with the proper fire protection equipment (alarms, sprinklers, extinguishers, etc.).

If a spill or leak to the environment has occurred, fire department, emergency response and/or hazardous materials spill personnel should be notified immediately. Cleanup should be attempted only by those trained in proper spill containment procedures. This chemical will remain in soil for extended periods and may leach to underground aquifers. Contaminated soils should therefore be removed for incineration and replaced with clean soil. If diethylamine should contact the water table, aquifer, or navigable waterway, remediation activities should be prompt. When such spills occur, the local and/or state emergency response authorities must be notified. A comprehensive emergency response or disaster preparedness/recovery plan should be in place prior to any operations involving the use, transportation, storage, or disposal of diethylamine. If diethylamine is spilled or leaked, the following specific steps are recommended:

- ☑ Restrict persons not wearing protective clothing from area of spill or leak until cleanup is complete and area can be opened for normal work.
- ☑ Ventilate area of spill or leak.
- ☑ Remove all ignition sources.
- ☑ Absorb liquid spills using vermiculite, dry earth, or sand and place in a sealed drum for disposal.

☑ Use water spray to disperse vapors and dilute to non-flammable mixtures. Dispose of accordingly.

☑ It may be necessary to dispose of diethylamine as a hazardous waste. The responsible state agency or the regional office of the federal Environmental Protection Agency (EPA) should be contacted for specific recommendations.

RISK ASSESSMENT: BUSINESS

General Assessment

Accidents or mishaps involving diethylamine can present a significant threat to business operations. The loss or damage of equipment or facilities can significantly affect fiscal viability. Lawsuits that may result from personnel injury/death, public exposures, and/or environmental contamination will also require a serious expenditure of resources. Media attention surrounding an injury, death, or environmental damage can also result in a loss of profits and loss of current as well as future business.

Recommended Risk-Reduction Measures

Company attorneys, safety and health professionals, and environmental specialists should be involved in the development of any such procedures. A company official should be pre-designated as a public relations officer with specific training in dealing with the press. Corporate plans and policies should be developed, approved, and implemented long before any need for such arises.

REFERENCES

American Conference of Governmental Industrial Hygienists. 1988. *Documentation of the Threshold Limit Values and Biological Exposure Indices,* 5th Edition (with updates). Cincinnati: ACGIH

Anthony, C. P., and N. J. Kolthoff. 1971. *Textbook of Anatomy and Physiology.* St. Louis: C. V. Mosby Company.

Bloom, A. D. and F. J. DeSerres. 1995. *Ecotoxicity and Human Health.* New York: CRC Press/Lewis Publishers.

Bueche, J. R. 1972. *Principles of Physics,* 2nd Edition. New York: McGraw-Hill.

Calabrese, E. J., E. M. Kenyon. *Air Toxics and Risk Assessment.* New York: CRC Press/Lewis Publishers.

Cockerman, L. G., B. S. Shane. 1994. *Basic Environmental Toxicology.* New York: CRC Press/Lewis Publishers.

Cothern, C. R. 1995. *Handbook for Environmental Risk Decisionmaking.* New York: CRC Press/Lewis Publishers.

Gardner, R. W. 1994. *Chemical Intolerance.* New York: CRC Press/Lewis Publishers.

Hallenbeck, W. H. 1993. *Quantitative Risk Assessment for Environmental and Occupational Health,* 2nd Edition. New York: CRC Press/Lewis Publishers.

Hodgson, E., R. B. Mailman, and Janice E. Chambers. 1988. *Dictionary of Toxicology.* London: The MacMillian Company.

Howard, P. H. 1995. *Handbook of Environmental Fate and Exposures Data for Organic Chemicals,* Vol. I-IV. New York: CRC Press/Lewis Publishers.

Howard, P. H., et. al. 1991. *Handbook of Environmental Degradation Rates.* New York: CRC Press/Lewis Publishers.

Jones, F. E. 1994. *Toxic Organic Vapors in the Workplace.* New York: CRC Press/Lewis Publishers.

Landis, W. G., M. H. Yu. 1995. *Introduction to Environmental Toxicology.* New York: CRC Press/Lewis Publishers.

Langerman, N. 1994 *Precautionary Labels for Chemical Containers.* New York: CRC Press/Lewis Publishers.

Lauwerys, R. R., P. Hoet. 1993. *Industrial Chemical Exposure.* New York: CRC Press/Lewis Publishers.

Lewis, R. J., Sr. 1993. *Hawley's Condensed Chemical Dictionary,* 12th Edition. New York: Van Nostrand Reinhold.

Lewis, R. J., Sr. 1992. *Sax's Dangerous Properties of Industrial Materials,* Volumes 1, 2, and 3, 8th Edition. New York: Van Nostrand Reinhold.

Manahan, S. E. 1992. *Toxicological Chemistry,* 2nd Edition. New York: CRC Press/Lewis Publishers.

Mellström, G. A., J. E. Wahlberg, and H. I. Maibach. 1994. *Protective Gloves for Occupational Use.* New York: CRC Press/Lewis Publishers.

Meyer, E. 1990. *Chemistry of Hazardous Materials.* Englewood Cliffs, NJ: Prentice Hall.

Nelson, G. O. 1992. *Gas Mixtures.* New York: CRC Press/Lewis Publishers.

Quigley, D. R. 1994. *Handbook of Emergency Chemical Management.* New York: CRC Press/Lewis Publishers.

Rea, W. J. 1995. *Chemical Sensitivity,* Vol. I-III. New York: CRC Press/Lewis Publishers.

Richardson, M. 1995. *Dictionary of Substance and Their Effects,* Vol. I-VI. New York: CRC Press/Lewis Publishers.

Sacarello, H. L. A. 1994. *The Comprehensive Handbook of Hazardous Materials.* New York: CRC Press/Lewis Publishers.

U.S. Department of Health and Human Services, National Institute for Occupational Safety and Health. 1994. *NIOSH Pocket Guide to Chemical Hazards.* Washington, D.C.: U.S. Government Printing Office.

MATERIAL SAFETY DATA SHEET

CHEMICAL NAME

DIETHYLENE GLYCOL MONOBUTYL ETHER

HAZARD WARNING INFORMATION

HEALTH	FIRE	REACTIVE	OTHER	DEGREE OF HAZARD	COLOR CODING	OTHER CODES
1	2	0		0 = Minimum Hazard 1 = Slight Hazard 2 = Moderate Hazard 3 = Serious Hazard 4 = Severe Hazard	HEALTH = BLUE FIRE = RED REACTIVITY = YELLOW OTHER = WHITE	OX = Oxidizer ACID = Acid ALK = Alkali COR = Corrosive W = Use No Water

SECTION I - GENERAL INFORMATION

Characterization: **Glycol Ether**

RCRA Number: **None**

EPA Class: **Not Applicable**

DOT Proper Shipping Name: **Combustible Liquid, N.O.S.**

Chemical Abstract Service (CAS) Number: **112-34-5**

DOT Hazard Class and Label Requirements: **Combustible Liquid; Label: None**

DOT Emergency Guide Code: **29**

DOT Identification Number: **NA 1993**

Chemical Formula: **$C_8H_{18}O_3$**

Synonyms

Butyl carbitol; butoxy diglycol; butoxy diethylene glycol; 2-(2-butoxyethoxy) ethanol; butyl digol; diglycol monobutyl ether; Ektasolve DB; Jeffersol DB; Poly-Solve DB.

SECTION II - HAZARDOUS INGREDIENTS AND IDENTITY INFORMATION

Hazardous Components (specific identity)	OSHA Exposure Criteria	NIOSH Exposure Criteria	Immediately Dangerous to Life and Health (IDLH)	ACGIH Exposure Criteria
Diethylene glycol monobutyl ether (derivation: By reaction of ethylene oxide and *n*-butyl alcohol).	**PEL:** Not Established **STEL:** Not Established	**REL:** Not Established **STEL:** Not Established	**Not Determined**	**TLV:** Not Established **STEL:** Not Established

SECTION III - PHYSICAL AND CHEMICAL CHARACTERISTICS

Boiling Point: **448°F (231°C)**

Specific Gravity (H_2O = 1): **0.945**

Vapor Pressure (mm Hg): **0.02 at 69°F (20°C)**

Molecular Weight: **162.26**

Vapor Density (Air = 1): **5.58**

Melting Point: **-90.6°F (-68°C)**

Solubility

Completely soluble in water, oils, alcohol, acetone, benzene, and ether.

Appearance and Odor

Colorless liquid with a faint, pleasant odor.

SECTION IV - FIRE AND EXPLOSION HAZARD DATA

Flash Point (method used): **172°F (78°C) closed cup**

Explosive Limits in Air % by Volume: **LEL: 0.85% UEL: 24.6%**

NFPA Classification: **Class IIIB Combustible Liquid**

Autoignition Temperature: **400°F (204°C)**

Extinguishing Media

Use water, dry chemical, carbon dioxide, or alcohol resistant foam. Water may cause frothing.

Special Fire Fighting Procedures

Acrid and irritating fumes and smoke are produced in fire, wear full protective clothing and self-contained breathing apparatus (SCBA). Vapors are heavier than air and may travel for some distance to ignition source and flashback. Fight fire from distance if possible. Move containers from fire area if it can be done safely. Exercise extreme caution when moving containers.

Unusual Fire and Explosion Hazards

Containers may explode in fire due to buildup of excessive internal pressures. Apply cooling water to sides of container until long after fire has been extinguished.

SECTION V - REACTIVITY DATA

Stability		Conditions to Avoid: Under normal conditions of temperature and pressure, diethylene glycol monobutyl ether is stable. Avoid contact with incompatible materials. Keep away from heat and flame.
Stable	Unstable	Incompatibility (*materials to avoid*)
X		Strong oxidizers, such as chlorine, bromine, fluorine, and peroxides.
Hazardous Polymerization		Conditions to Avoid: Under normal conditions of temperature and pressure, hazardous polymerization of diethylene glycol monobutyl ether is not expected to occur.
May Occur	Will Not Occur	Hazardous Decomposition or By-products
	X	Acrid and irritated gases and fumes are produced when diethylene glycol monobutyl ether is involved in fire, including carbon dioxide.

SECTION VI - HEALTH HAZARD DATA

Primary Route(s) of Entry:	Inhalation?	Absorption (skin)?	Ingestion?
	X	X	X

Health Hazards

INHALATION: Essentially non-toxic on single exposures. Repeated exposures produce toxic effects with action on the CNS and the kidneys. May cause irritation to the eyes, nose, throat, and upper respiratory tract. Repeated exposures can cause cyanosis, acidosis, and tachypnea. Kidney function may be reduced. Tissue damage to the liver, kidney, spleen.

ABSORPTION: Low to no toxicity on single exposure. Repeated or heavy contact will pass through unbroken skin to cause toxic systemic effects. Eye contact with liquid can cause burns.

INGESTION: Moderately toxic with systemic effects as those noted for inhalation.

Carcinogenicity	NTP Listed?	IARC Cancer Review Group?	OSHA Regulated?	Target Organs?
Unknown Human Unknown Animal	No	No	No	CNS, kidneys, liver, eyes, skin.

Medical Conditions Generally Aggravated by Exposure

None reported.

Emergency and First-aid Procedures

Eye contact: Flush immediately with water for 15 minutes (minimum), seek medical attention. Skin contact: Remove all contaminated clothing. Immediately wash area with large amounts of soap and water. Seek medical assistance. For inhalation: Remove the person from exposure. Provide respiratory assistance and CPR. Transfer to medical facility. Administer 100% supplemental, humidified oxygen, as needed. If swallowed: Call the poison control center and seek medical attention immediately. Unless advised otherwise, give 1 - 2 glasses of water and induce vomiting. Never give an unconscious or convulsing person anything by mouth and never try to make such a person vomit.

SECTION VII - PRECAUTIONS FOR SAFE HANDLING AND USE

Steps to be Taken in Case Material is Released or Spilled

Absorb liquids in vermiculite or other absorbent and deposit in DOT-approved drum. Do not allow flow into sewers or waterways. Ventilate area. Restrict those not involved in cleanup from entering area.

Preferred Waste Disposal Method

Mix with a combustible solvent, burn in a chemical incinerator quipped with afterburner and scrubber.

Precautions to be Taken in Handling and Storage

Do not store diethylene glycol monobutyl ether in the presence of incompatible chemicals or materials. Store in tightly closed containers in a cool, dry, well-ventilated area away from ignition sources.

Other Precautions and Warnings

Use non-sparking tools for all operations involving this chemical compound.

SECTION VIII - CONTROL MEASURES AND PERSONAL PROTECTIVE EQUIPMENT

Respiratory Protection (*specify type*)

No exposure levels have been established. This does not mean exposure is without risk. For the best protection, use a supplied-air respirator with full facepiece operated in pressure demand mode or a self-contained breathing apparatus (SCBA) with full facepiece operated in pressure demand mode.

Ventilation

Local exhaust or general mechanical systems recommended.

Protective Gloves	Eye Protection	Other Protective Clothing
Butyl Rubber	Chemical Goggles or Face Mask	Protective Clothing or Apron

Work/Hygiene Practices

Always wash hands thoroughly after using chemical; never bring food, drink, or smoking materials into vicinity of chemicals.

DIETHYLENE GLYCOL MONOBUTYL ETHER

$C_8H_{18}O_3$ CAS: 112-34-5

$HO-CH_2-CH_2-O-CH_2-CH_2-O-CH_2-CH_2-CH_2-CH_3$

IDENTIFICATION AND TYPICAL USES

Diethylene glycol monobutyl ether (DGBE) is a colorless liquid with a faint, pleasant odor. It is used as a solvent for cellulose ester, oils, gums, soaps, polymers, lacquers, varnishes, and dyes. It is a plasticizer intermediate, a melting-out solution in the textile industry, and a diluting agent for hydraulic brake fluids. It is also used as a primary component of the aqueous film-forming foam that is used by the U.S. Navy in shipboard fire fighting systems.

RISK ASSESSMENT: HEALTH

General Assessment

DGBE is considered to be of low-order toxicity on single exposures. However, prolonged low exposures or acute high exposures, primarily by *ingestion* or *inhalation,* can cause serious risks to human health. It is also known for its ability to pass through unbroken skin (*absorption*) to cause toxic systemic effects. There is no evidence to indicate that it is carcinogenic or mutagenic in humans or test animals. There is some data to show it can damage the male reproductive organs in test animals. Human data in this regard are inconclusive.

In high concentrations, vapor inhalation may cause smarting of the eyes and respiratory tract. Ingestion and inhalation, as well as skin absorption, of repeated small exposures can cause cyanosis (bluish discoloration of the skin due to reduced hemoglobin in the blood), acidosis (accumulation of acid or depletion of the alkaline reserve in the blood and body tissue), and tachypnea (a respiratory neurosis marked by rapid respiration). Kidney function may be reduced with subsequent uremia (accumulation in the blood of constituents that are normally excreted in the urine), presenting a potential for severe systemic toxicity. Tissue damage to the liver, kidney, spleen, and testes were seen in test animals.

Skin contact is irritating and may cause rash to develop. Liquid contact with the eyes can result in severe irritation and possible damage to the cornea.

Acute Health Effects

The following acute (short-term) health effects may occur immediately or shortly after exposure to DGBE:

<u>Skin</u>: Low, single exposures are not likely to cause any toxic effects. Higher exposures can result in irritation, burns, inflammation, and possible absorption causing toxic systemic effects.

<u>Eye</u>: Severe irritation with burns, redness, and damage to the corneal tissues.

<u>Lung</u>: Single exposures are not likely to be toxic. High concentrations can cause irritation of the mucosa of the nose, throat, and upper respiratory tract. Can cause a number of toxic systemic effects, including cyanosis, acidosis, tachypnea, and uremia.

<u>Other</u>: Ingestion can cause the same symptoms as inhalation. Adverse action on the central nervous system is likely on high acute exposures.

Chronic Health Effects

The following chronic (long-term) health effects may occur at some time after exposure to DGBE and can last for months or even years:

<u>Cancer Hazards</u>: According to the information presented in the references, DGBE has not been shown to cause cancer in test animals.

<u>Reproductive Hazard</u>: According to the information presented in the references, DGBE may have teratogenic effects in test animals. It may cause tissue dam-

age to the testes. Human studies in this regard are inconclusive.

Other Chronic Effects: There is some indication that long-term exposure or frequent high exposures may cause tissue damage in the liver, kidney, and spleen. Kidney function may be reduced resulting in a toxic buildup of constituents normally excreted in urine (uremia).

Recommended Risk-Reduction Measures

Unless a less toxic chemical can be substituted for a hazardous substance, *engineering controls* are the most effective method of reducing exposures. The best protection is to enclose operations and/or provide local exhaust ventilation at the site of chemical release. While not always operationally feasible, isolating operations can also reduce exposure, especially is DGBE is to be agitated or heated.

Using respiratory protection is less effective than the controls mentioned above, but is still advisable whenever working with or around DGBE. No exposure levels have been established for this compound. However, this does not mean there is no heath hazard risk. For the best protection, a self-contained breathing apparatus (SCBA) with full facepiece and pressure demand or a supplied-air respirator operated in pressure demand or continuous flow mode are the recommended respiratory protection methods of choice. If a full facepiece is not available, then chemical goggles should be worn to protect the eyes. Whenever a chemical splash hazard exists, a face shield and rubber apron should be worn. To prevent hand and skin exposures, butyl rubber gloves should be worn.

Administrative controls should also be in place to minimize the potential for human exposures. These may include written procedures or policies which specify the methods and techniques that will be practiced whenever personnel are to work with DGBE.

All personnel should receive *training* on the use, hazards, protective measures, emergency actions, and other precautions per 29 CFR 1910.1200 (Hazard Communication), prior to the first assignment in an area where DGBE is used or stored.

If symptoms develop or overexposure is suspected, the following medical tests may be useful:

☑ Kidney and liver function tests.

☑ Urinalysis and blood gas tests.

Any evaluation should include a careful medical history of past and present symptoms with an examination. However, medical tests that evaluate existing damage are not a substitute for controlling exposure. Also, since smoking can cause heart disease, emphysema, and other respiratory disorders, exposure to DGBE can have a more pronounced effect on smokers than on non-smokers exposed to the same levels. Prudent risk assessment and management requires consideration of *all* possible risk factors when exposure symptoms appear in the workforce.

Other methods to reduce exposure include:

☑ Where possible, enclose operations and use local exhaust ventilation at the site of chemical release. If local exhaust ventilation or enclosure is not used, respiratory protection should be mandatory. Also, automatic transfer of liquids from storage containers to process containers is recommended.

☑ Always ensure that proper protective clothing is worn when using chemical substances.

☑ Wash thoroughly immediately after exposure to DGBE and at the end of the work shift or before eating, drinking, or smoking.

☑ Hazard warning information should be posted in the work area. In addition, as part of an on-going education and training program, all information on the health and safety hazards of DGBE should be communicated to all potentially exposed workers.

☑ Eye wash stations should be provided in work areas. If the potential for whole body exposure exists, then safety showers should also be provided.

RISK ASSESSMENT: ENVIRONMENT

General Assessment

The environment is at risk of exposure during use, transportation, storage, disposal, or destruction of DGBE. In almost every scenario, the threat of environmental exposure is contingent upon the proper handling of the chemical substance. Accidental spills, large or small, can result in fire, explosion, and possible contamination of the surrounding environmental mediums (water, soil, and air).

DGBE is considered a Class IIIB combustible liquid (per OSHA 29 CFR 1910.106). These characteristics require special consideration during any emergency situation involving a leak or spill of DGBE. Should DGBE ever come into contact with incompatible substances such as strong oxidizers (chlorine, bromine, fluorine, chlorine dioxide, and peroxides) either during use, transportation, or storage, violent reaction can occur.

The acceptable disposal method for DGBE is to mix it with a combustible solvent (higher alcohol, benzene, or petroleum ether) and burn in a chemical incinerator equipped with an afterburner and scrubber.

DGBE can enter the environment through industrial discharges and from spills.

Acute Ecological Effects

Acute (short-term) toxic effects may include the death of animals, birds, or fish and death or low growth rate in plants. Acute effects are seen 2 to 4 days after animals or plants are exposed to DGBE. This chemical is expected to have low to moderate acute toxicity to aquatic life. Insufficient data are available to evaluate or predict the short-term effects of DGBE to plants, birds, or land animals.

Chronic Ecological Effects

Chronic toxic effects may include shortened life span, reproductive problems, lower fertility, and changes in appearance or behavior in exposed animals. These effects can be seen long after first exposure(s) to toxic chemicals. DGBE is not expected to have chronic toxicity to aquatic life. Insufficient data are available to evaluate or predict the long-term effects of DGBE to plants, birds, or land animals.

Water Solubility

DGBE is highly soluble in water. Concentrations of 1000 milligrams and more will mix with a liter of water.

Persistence in the Environment

DGBE is highly persistent in the aquatic environment with a half-life of 200 days and more. The half-life of a pollutant is the amount of time it takes for one half of the chemical to be degraded. It is highly mobile in both water and soil and undergoes aerobic biodegeneration.

Bioaccumulation in Aquatic Organisms

Some substances increase in concentration, or *bioaccumulate,* in living organisms as they breathe contaminated air, drink contaminated water, or eat contaminated food. These chemicals can become concentrated in the tissues and internal organs of animals as well as humans who have been exposed.

The concentration of DGBE found in fish tissues is expected to be about the same as that found in the water from which the fish was taken.

Recommended Risk-Reduction Measures

Proper training of all transporters will reduce the likelihood of a mishap or accident resulting in a leak or spill to the environment. The correct use of DOT labeling (per the MSDS) on all containers, trucks, and rail cars will enable emergency responders to react properly and quickly to any disaster thereby reducing the potential risk to the environment and to personnel.

Storage of DGBE should be segregated from incompatible chemicals to minimize the risk of cross-contamination or contact. Buildings should be properly constructed for storage of hazardous chemicals (fire suppression equipment, alarms, separation, etc.).

If a spill or leak to the environment has occurred, fire department, emergency response and/or hazardous materials spill personnel should be notified immediately. Cleanup should be attempted only by those trained in proper spill containment procedures. Contaminated soils should be removed for incineration and replaced with clean soil. If DGBE should contact the water table, aquifer, or navigable waterway, remediation activities should be prompt. It is highly soluble in water and total containment and remediation may not be possible. When such spills occur, the local and/or state emergency response authorities must be notified. A comprehensive emergency response or disaster preparedness/recovery plan should be in place prior to any operations involving the use, transportation, storage, or disposal of DGBE. If DGBE is spilled or leaked, the following specific steps are recommended:

- ☑ Restrict persons not wearing protective clothing from area of spill or leak until cleanup is complete and area can be opened for normal work.
- ☑ Ventilate area of spill or leak.
- ☑ Remove all ignition sources.
- ☑ Collect liquid using vermiculite, dry earth, or sand to absorb and place in a sealed drum for disposal. Damp mop residue and collect. Contain any water from cleanup efforts and do NOT let it reach sewers or open waterways.
- ☑ It may be necessary to dispose of DGBE as a hazardous waste. The responsible state agency or the regional office of the federal Environmental Protection Agency (EPA) should be contacted for specific recommendations.

RISK ASSESSMENT: BUSINESS

General Assessment

Accidents or mishaps involving DGBE can present a significant threat to business operations. The loss or damage of equipment or facilities can significantly affect fiscal viability. Lawsuits that may result from personnel injury/death, public exposures, and/or environmental contamination will also require a serious expenditure of resources. Media attention surrounding an injury, death, or environmental damage can also result in a loss of profits and loss of current as well as future business. Always remember that anytime the term "reproductive hazard" is used, public emotion, hysteria, and ignorance can run equally high. This should be properly considered during the development of public relations policies.

ʊ *Recommended Risk-Reduction Measures*

Company attorneys, safety and health professionals, and environmental specialists should be involved in the development of any such procedures. A company official should be pre-designated as a public relations officer with specific training in dealing with the press. Corporate plans and policies should be developed, approved, and implemented long before any need for such arises.

REFERENCES

Ahlbom, A. 1993. *Biostatistics for Epidemiologists.* New York: CRC Press/Lewis Publishers.

American Conference of Governmental Industrial Hygienists. 1988. *Documentation of the Threshold Limit Values and Biological Exposure Indices,* 5th Edition (with updates). Cincinnati: ACGIH

Anthony, C. P., and N. J. Kolthoff. 1971. *Textbook of Anatomy and Physiology.* St. Louis: C. V. Mosby Company.

Asante-Duah, D. K. 1993. *Hazardous Waste Risk Assessment.* New York: CRC Press/Lewis Publishers.

Bloom, A. D. and F. J. DeSerres. 1995. *Ecotoxicity and Human Health.* New York: CRC Press/Lewis Publishers.

Brusick, D. J. 1994. *Methods for Genetic Risk Assessment.* New York: CRC Press/Lewis Publishers.

Bueche, J. R. 1972. *Principles of Physics,* 2nd Edition. New York: McGraw-Hill.

Burrell, R., D. K. Flaherty, and L. J. Sauers. 1992. *Toxicology of the Immune System.* New York: Van Nostrand Reinhold.

Calabrese, E. J., E. M. Kenyon. *Air Toxics and Risk Assessment.* New York: CRC Press/Lewis Publishers.

Cockerman, L. G., B. S. Shane. 1994. *Basic Environmental Toxicology.* New York: CRC Press/Lewis Publishers.

Cothern, C. R. 1995. *Handbook for Environmental Risk Decisionmaking.* New York: CRC Press/Lewis Publishers.

Gardner, R. W. 1994. *Chemical Intolerance.* New York: CRC Press/Lewis Publishers.

Hodgson, E., R. B. Mailman, and Janice E. Chambers. 1988. *Dictionary of Toxicology.* London: The MacMillian Company.

Howard, P. H. 1995. *Handbook of Environmental Fate and Exposures Data for Organic Chemicals,* Vol. I-IV. New York: CRC Press/Lewis Publishers.

Keith, L. H. and M. M. Walker. 1995. *Handbook of Air Toxics.* New York: CRC Press/Lewis Publishers.

Landis, W. G., M. H. Yu. 1995. *Introduction to Environmental Toxicology.* New York: CRC Press/Lewis Publishers.

Langerman, N. 1994 *Precautionary Labels for Chemical Containers.* New York: CRC Press/Lewis Publishers.

Lauwerys, R. R., P. Hoet. 1993. *Industrial Chemical Exposure.* New York: CRC Press/Lewis Publishers.

Lide, D. R. 1995. *Handbook of Chemistry and Physics,* 76th Edition. New York: CRC Press/Lewis Publishers.

Lewis, R. J., Sr. 1993. *Hawley's Condensed Chemical Dictionary,* 12th Edition. New York: Van Nostrand Reinhold.

Lewis, R. J., Sr. 1992. *Sax's Dangerous Properties of Industrial Materials,* Volumes 1, 2, and 3, 8th Edition. New York: Van Nostrand Reinhold.

Manahan, S. E. 1992. *Toxicological Chemistry,* 2nd Edition. New York: CRC Press/Lewis Publishers.

Mellström, G. A., J. E. Wahlberg, and H. I. Maibach. 1994. *Protective Gloves for Occupational Use.* New York: CRC Press/Lewis Publishers.

Meyer, E. 1990. *Chemistry of Hazardous Materials.* Englewood Cliffs, NJ: Prentice Hall.

Mycock, J. C., J. D. McKenna, and L. Theodore. 1995. *Handbook of Air Pollution Control Engineering and Technology.* New York: CRC Press/Lewis Publishers.

Nelson, G. O. 1992. *Gas Mixtures.* New York: CRC Press/Lewis Publishers.

Ott, W. R. 1995. *Environmental Statistics and Data Analysis.* New York: CRC Press/Lewis Publishers.

Patnaik, P. 1992. *A Comprehensive Guide to the Hazardous Properties of Chemical Substances.* New York: Van Nostrand Reinhold.

Philp, R. B. 1995. *Environmental Hazards and Human Health.* New York: CRC Press/Lewis Publishers.

Quigley, D. R. 1994. *Handbook of Emergency Chemical Management.* New York: CRC Press/Lewis Publishers.

Rea, W. J. 1995. *Chemical Sensitivity,* Vol. I-III. New York: CRC Press/Lewis Publishers.

Richardson, M. 1995. *Dictionary of Substance and Their Effects,* Vol. I-VI. New York: CRC Press/Lewis Publishers.

Sacarello, H. L. A. 1994. *The Comprehensive Handbook of Hazardous Materials.* New York: CRC Press/Lewis Publishers.

MATERIAL SAFETY DATA SHEET

CHEMICAL NAME

DIETHYLENETRIAMINE

HAZARD WARNING INFORMATION

HEALTH	FIRE	REACTIVE	OTHER	DEGREE OF HAZARD	COLOR CODING	OTHER CODES
3	1	0		0 = Minimum Hazard 1 = Slight Hazard 2 = Moderate Hazard 3 = Serious Hazard 4 = Severe Hazard	HEALTH = BLUE FIRE = RED REACTIVITY = YELLOW OTHER = WHITE	OX = Oxidizer ACID = Acid ALK = Alkali COR = Corrosive W̶ = Use No Water

SECTION I - GENERAL INFORMATION

Characterization	**Aliphatic Amine**	RCRA Number	**None**
		EPA Class	**Not Applicable**
DOT Proper Shipping Name	**Diethylenetriamine**	Chemical Abstract Service (CAS) Number	**111-40-0**
DOT Hazard Class and Label Requirements	**Corrosive**	DOT Emergency Guide Code	**29**
DOT Identification Number	**UN 2079**	Chemical Formula	**$(NH_2CH_2CH_2)_2NH$**

Synonyms

N-(2-Aminoethyl)1,2-ethanediamine; bis(2-Aminoethyl)amine; DETA; 2,2'-diaminoethylamine; aminoethylethandiamine.

SECTION II - HAZARDOUS INGREDIENTS AND IDENTITY INFORMATION

Hazardous Components (specific identity)	OSHA Exposure Criteria	NIOSH Exposure Criteria	Immediately Dangerous to Life and Health (IDLH)	ACGIH Exposure Criteria
Diethylenetriamine (derivation: By reaction between ethylene dichloride and ammonia). **1 ppm = 4.29 mg/m^3**	**PEL (8-hour):** **1 ppm** **4 mg/m^3** **STEL:** **Not Established**	**REL (skin):** **1 ppm** **4 mg/m^3** **STEL:** **Not Established**	**Not Determined**	**TLV (skin)):** **1 ppm** **4.2 mg/m^3** **STEL:** **Not Established**

SECTION III - PHYSICAL AND CHEMICAL CHARACTERISTICS

Boiling Point	**405°F (207°C)**	Specific Gravity (H_2O = 1)	**0.96**
Vapor Pressure (mm Hg)	**0.37 at 69°F (20°C)**	Molecular Weight	**103.2**
Vapor Density (Air = 1)	**3.56**	Freezing Point	**-38°F (-35°C)**

Solubility

Soluble in water, alcohol, and hydrocarbons. Insoluble in ether.

Appearance and Odor

Colorless to yellow hygroscopic liquid with a mild, ammonia-like odor.

SECTION IV - FIRE AND EXPLOSION HAZARD DATA

Flash Point (method used)	**208°F (98°C) opened cup**	Explosive Limits in Air % by Volume	**LEL: 2% at 302°F (150°C) UEL: 6.7% at 302°F (150°C)**
NFPA Classification	**Class IIIB Combustible Liquid**	Autoignition Temperature	**676°F (358°C)**

Extinguishing Media

Use dry chemical, carbon dioxide, foam, or water spray.

Special Fire Fighting Procedures

Poisonous gases are produced in fire, wear full protective clothing and self-contained breathing apparatus (SCBA). Move containers from fire area if it can be done safely. Cool exposed containers. Do not allow runoff from fire control methods to reach sewers or waterways.

Unusual Fire and Explosion Hazards

Combustion products may be more hazardous than the material itself. Vapors are heavier than air and may travel for some distance to an ignition source and flashback. Containers may explode in fire or when exposed to heat. Vapor explosions are possible indoors, outdoors, and in sewers.

SECTION V - REACTIVITY DATA

Stability		Conditions to Avoid Normally stable. Ignites on contact with cellulose nitrate of high surface area. Forms a shock-sensitive explosive with nitromethane. Avoid contact with incompatible materials. Keep away from heat and flame.
Stable X	Unstable	Incompatibility *(materials to avoid)* Corrosive to copper and its alloys. Incompatible with acids, oxidizers, halogenated compounds, oxidizing materials, and reactive organic compounds. Chemical is hygroscopic (absorbs moisture from air).
Hazardous Polymerization		Conditions to Avoid Under normal conditions of temperature and pressure, hazardous polymerization of diethylenetriamine is not expected to occur.
May Occur	Will Not Occur X	Hazardous Decomposition or By-products Toxic and irritating fumes and vapors are produced when diethylenetriamine is heated to decomposition, including carbon dioxide, carbon monoxide, hydrocarbons, amine vapors, and toxic nitrogen oxide fumes.

SECTION VI - HEALTH HAZARD DATA

Primary Route(s) of Entry:	Inhalation?	Absorption (skin)?	Ingestion?
	X		X

Health Hazards

INHALATION: Irritation to the mucosa of the respiratory tract producing cough, nausea, and vomiting. May cause asthmatic response and sensitization on prolonged exposures.

SKIN & EYES: A corrosive. Severe irritation and possible corrosive burns to the eye with corneal damage likely and permanent loss of visual acuity. Skin irritation may be mild to moderate.

INGESTION: Extremely painful burning and corrosion of the mouth, lips, and esophagus.

Carcinogenicity	NTP Listed?	IARC Cancer Review Group?	OSHA Regulated?	Target Organs?
Unknown Human Unknown Animal	No	No	29 CFR 1910.1000 Table Z-1	Eyes, skin, respiratory system.

Medical Conditions Generally Aggravated by Exposure

None Reported.

Emergency and First-aid Procedures

Eye contact: Flush immediately with water for 15 minutes (minimum), seek medical attention. Skin contact: Remove all contaminated clothing. Immediately wash area with large amounts of soap and water. For inhalation: Remove the person from exposure. Provide respiratory assistance and CPR. If swallowed: Call the poison control center and seek medical attention immediately. Unless advised otherwise, give 1-2 glasses of water to dilute. Do NOT give fruit juices or anything acidic. Do NOT induce vomiting. Never attempt to give an unconscious or convulsing person anything by mouth. DOCTOR: Early treatment of esophageal burns should include IV therapy, broad spectrum antibodies, sedation, parenteral hydrocortisone and prevention of esophageal obstruction, followed by dilation.

SECTION VII - PRECAUTIONS FOR SAFE HANDLING AND USE

Steps to be Taken in Case Material is Released or Spilled

Absorb liquids in vermiculite or other absorbent and deposit in DOT-approved drum. Ventilate area and remove ignition sources. Restrict those not involved in cleanup from entering area. Notify appropriate authorities, if applicable.

Preferred Waste Disposal Method

No Citation.

Precautions to be Taken in Handling and Storage

Do not store diethylenetriamine in the presence of incompatible chemicals or materials. Store in tightly closed containers in a cool, dry, well-ventilated area. Protect from physical damage. Outside or detached storage is preferred.

Other Precautions and Warnings

Electrically bond and ground all equipment used in manufacturing, use, and storage.

SECTION VIII - CONTROL MEASURES AND PERSONAL PROTECTIVE EQUIPMENT

Respiratory Protection *(specify type)*

Exposure level is low (1 ppm) and difficult to monitor. Best protection is provided using a supplied-air respirator with full facepiece operated in positive pressure mode or a self-contained breathing apparatus (SCBA) with full facepiece operated in pressure demand or other positive pressure mode.

Ventilation

Local exhaust or general mechanical systems recommended.

Protective Gloves	Eye Protection	Other Protective Clothing
Impervious Rubber Gloves	Chemical/Dust Goggles or Face Mask	Protective Apron or Clothing

Work/Hygiene Practices

Always wash hands thoroughly after using chemical; never bring food, drink, or smoking materials into vicinity of chemicals.

DIETHYLENETRIAMINE

$(NH_2CH_2CH_2)_2NH$ CAS: 111-40-0

$NH_2 — CH_2 — CH_2 — NH — CH_2 — CH_2 — NH_2$

IDENTIFICATION AND TYPICAL USES

Diethylenetriamine is a colorless to yellow liquid with a strong, ammonia-like odor. It is hygroscopic (absorbs moisture from the air). It is used as a solvent for sulfur, acid gases, various resins, and dyes. Also used as a saponification agent for acidic materials, as a fuel component, and in zylene or toluene as a hardener for epoxy resins.

RISK ASSESSMENT: HEALTH

General Assessment

Diethylenetriamine is a severe irritant and corrosive on contact with human tissue. Primary route of exposure is *inhalation* of its vapors. It can also cause serious heath effects on *ingestion* and by skin contact (although absorption has not been confirmed as a route of entry). There is no information in the references to indicate any carcinogenic, mutagenic, or teratogenic properties.

Inhalation will result in irritation of the mucosa, especially in the nose, throat, and upper respiratory tract. Symptoms may include coughing, dyspnea, nausea, and vomiting.

Skin contact with the liquid can cause mild to severe irritation. Dermatitis may develop with rash and painful itching and possible necrosis. Skin sensitization is also likely to occur over a period of prolonged or repeated contact. Eye contact can result in corneal injury and permanent damage or loss of vision. Tissue necrosis of the eye is also possible.

Ingestion can cause painful burning and corrosion of the mouth, lips, and esophagus.

Acute Health Effects

The following acute (short-term) health effects may occur immediately or shortly after exposure to diethylenetriamine:

Skin: Mild to severe irritation. May cause dermatitis and/or sensitization of the skin. Necrosis of skin tissue is possible.

Eye: Severe irritation, inflammation, with possible necrosis of eye tissue, damage to the cornea and loss of vision.

Lung: Coughing, nausea, and vomiting. May cause asthma-like symptoms and hypersensitivity.

Other: Ingestion causes severe irritation and corrosion of the mouth, lips, and esophagus.

Chronic Health Effects

The following chronic (long-term) health effects may occur at some time after exposure to diethylenetriamine and can last for months or even years:

Cancer Hazards: According to the information presented in the references, diethylenetriamine has not been adequately tested for its ability to cause cancer in test animals. More research is required.

Reproductive Hazard: According to the information presented in the references, diethylenetriamine has not been adequately tested for its ability to adversely affect reproduction.

Other Chronic Effects: Prolonged exposures may cause dermatitis and skin sensitization. Once allergy develops, even slight future exposures will cause a recurrence of the symptoms.

Recommended Risk-Reduction Measures

Unless a less toxic chemical can be substituted for a hazardous substance, *engineering controls* are the most effective method of reducing exposures. The best protection is to enclose operations and/or provide local exhaust ventilation at the site of chemical release. To prevent static sparks, all containers and equipment should be electrically grounded and bonded. While not

always operationally feasible, isolating operations can also reduce exposure.

Using respiratory protection is less effective than the controls mentioned above, but is still advisable whenever working with or around diethylenetriamine. The exposure level is extremely low (1 ppm) and difficult to monitor. An organic vapor respirator may suffice for low or transient exposures. However, greater protection is provided by a self-contained breathing apparatus (SCBA) with full facepiece and pressure demand or other positive pressure mode. If a full facepiece is not available, then chemical goggles should be worn to protect the eyes. Whenever a chemical splash hazard exists, a face shield and apron should be worn. To prevent hand and skin exposures, impervious chemical gloves should be used. Glove manufacturers should be contacted and permeation studies obtained *before* final glove selections are made.

Administrative controls should also be in place to minimize the potential for human exposures. These may include written procedures or policies which specify the methods and techniques that will be practiced whenever personnel are to work with diethylenetriamine.

All personnel should receive *training* on the use, hazards, protective measures, emergency actions, and other precautions per 29 CFR 1910.1200 (Hazard Communication), prior to the first assignment in an area where diethylenetriamine is used or stored.

Physicians should treat early for esophageal burns with IV therapy, broad spectrum antibodies, sedation, parenteral hydrocortisone, and, most importantly, prevention of esophageal obstruction, followed by dilation.

Any evaluation should include a careful medical history of past and present symptoms with an examination. However, medical tests that evaluate existing damage are not a substitute for controlling exposure.

Other methods to reduce exposure include:

- ☑ Where possible, enclose operations and use local exhaust ventilation at the site of chemical release. If local exhaust ventilation or enclosure is not used, respiratory protection should be mandatory. Also, if possible, automatically transfer liquids from storage containers to process containers using non-sparking tools.
- ☑ Always ensure that proper protective clothing is worn when using chemical substances.
- ☑ Wash thoroughly immediately after exposure to diethylenetriamine and at the end of the work shift or before eating, drinking, or smoking.
- ☑ Hazard warning information should be posted in the work area. In addition, as part of an on-going education and training program, all information on the health and safety hazards of diethylenetriamine should be communicated to all potentially exposed workers.
- ☑ Eye wash stations should be provided in work areas. If the potential for whole body exposure exists, then safety showers should also be provided.

RISK ASSESSMENT: ENVIRONMENT

General Assessment

The environment is at risk of exposure during intended use, transportation, storage, disposal, or destruction of diethylenetriamine. In almost every scenario, the threat of environmental exposure is contingent upon the proper handling of the chemical substance. Accidental spills, large or small, can result in fire, explosion, and possible contamination of the surrounding environmental mediums (water, soil, and air).

Diethylenetriamine is considered a Class IIIB combustible liquid (per OSHA 29 CFR 1910.106). While this means it poses a minimum fire risk by itself, mixture or contact with incompatible materials can cause fire or explosion. It can react with many common oxidizing materials, cellulose nitrate, and strong acids. It is corrosive to copper and its alloys. These characteristics require special consideration during any emergency situation involving a leak or spill of diethylenetriamine. Should diethylenetriamine ever come into contact with incompatible substances either during use, transportation, or storage, violent reactions can occur.

Diethylenetriamine can enter the environment through industrial discharges and spills.

☠ *Acute Ecological Effects*

Acute (short-term) toxic effects may include the death of animals, birds, or fish and death or low growth rate in plants. Acute effects are seen 2 to 4 days after animals or plants are exposed to diethylenetriamine. Insufficient data are available to evaluate or predict the short-term effects of diethylenetriamine to aquatic life, plants, birds, or land animals.

Chronic Ecological Effects

Chronic toxic effects may include shortened life span, reproductive problems, lower fertility, and changes in appearance or behavior in exposed animals. These effects can be seen long after first exposure(s) to toxic chemicals. Insufficient data are available to evaluate or predict the long-term effects of diethylenetriamine to aquatic life, plants, birds, or land animals.

Water Solubility

Diethylenetriamine is highly soluble in water. Concentrations of 1000 milligrams and more will mix with a liter of water.

Persistence in the Environment

Diethylenetriamine is slightly persistent in water, with a half-life of between 2 to 20 days. The half-life of a pollutant is the amount of time it takes for one half of the chemical to be degraded.

Bioaccumulation in Aquatic Organisms

Some substances increase in concentration, or *bioaccumulate,* in living organisms as they breathe contaminated air, drink contaminated water, or eat contaminated food. These chemicals can become concentrated in the tissues and internal organs of animals as well as humans.

The concentration of diethylenetriamine found in fish tissues is expected to be lower than the levels found in the water from which the fish was taken.

Recommended Risk-Reduction Measures

Proper training of all transporters will reduce the likelihood of a mishap or accident resulting in a leak or spill to the environment. The correct use of DOT labeling (per the MSDS) on all containers, trucks, and rail cars will enable emergency responders to react properly and quickly to any disaster thereby reducing the potential risk to the environment and to personnel.

Storage of diethylenetriamine should be segregated from incompatible chemicals to minimize the risk of cross-contamination or contact. Storage buildings should be equipped with the proper fire protection equipment (alarms, sprinklers, extinguishers, etc.).

If a spill or leak to the environment has occurred, fire department, emergency response and/or hazardous materials spill personnel should be notified immediately. Cleanup should be attempted only by those trained in proper spill containment procedures. Contaminated soils should therefore be removed for incineration and replaced with clean soil. If diethylenetriamine should contact the water table, aquifer, or navigable waterway, remediation activities should be prompt. Time is of the essence since it is highly soluble and total containment and remediation may not be possible. When such spills occur, the local and/or state emergency response authorities must be notified. A comprehensive emergency response or disaster preparedness/recovery plan should be in place prior to any operations involving the use, transportation, storage, or disposal of diethylenetriamine. If diethylenetriamine is spilled or leaked, the following specific steps are recommended:

- ☑ Evacuate area and deny entry to those not involved in cleanup activities. Restrict persons not wearing protective clothing from area of spill or leak until cleanup is complete.
- ☑ Ventilate area of spill or leak.
- ☑ Remove all ignition sources.
- ☑ Absorb liquids using vermiculite, dry earth, or sand and place in a sealed drum for disposal.
- ☑ It may be necessary to dispose of diethylenetriamine as a hazardous waste. The responsible state agency or the regional office of the federal Environmental Protection Agency (EPA) should be contacted for specific recommendations.

RISK ASSESSMENT: BUSINESS

General Assessment

Accidents or mishaps involving diethylenetriamine can present a significant threat to business operations. The loss or damage of equipment or facilities can significantly affect fiscal viability. Lawsuits that may result from personnel injury/death, public exposures, and/or environmental contamination will also require a serious expenditure of resources. Media attention surrounding an injury, death, or environmental damage can also result in a loss of profits and loss of current as well as future business.

Recommended Risk-Reduction Measures

Company attorneys, safety and health professionals, and environmental specialists should be involved in the development of any procedures dealing with safety or emergency response.

A company official should be pre-designated as a public relations officer with specific training in dealing with the press. Corporate plans and policies should be developed, approved, and implemented long before any need for such arises.

REFERENCES

Ahlbom, A. 1993. *Biostatistics for Epidemiologists.* New York: CRC Press/Lewis Publishers.

American Conference of Governmental Industrial Hygienists. 1988. *Documentation of the Threshold Limit Values and Biological Exposure Indices,* 5th Edition (with updates). Cincinnati: ACGIH

Anthony, C. P., and N. J. Kolthoff. 1971. *Textbook of Anatomy and Physiology.* St. Louis: C. V. Mosby Company.

Asante-Duah, D. K. 1993. *Hazardous Waste Risk Assessment.* New York: CRC Press/Lewis Publishers.

Bloom, A. D. and F. J. DeSerres. 1995. *Ecotoxicity and Human Health.* New York: CRC Press/Lewis Publishers.

Bueche, J. R. 1972. *Principles of Physics,* 2nd Edition. New York: McGraw-Hill.

Brusick, D. J. 1994. *Methods for Genetic Risk Assessment.* New York: CRC Press/Lewis Publishers.

Calabrese, E. J., E. M. Kenyon. *Air Toxics and Risk Assessment.* New York: CRC Press/Lewis Publishers.

Calabrese, E. J. 1994. *Biological Effects of Low Level Exposures.* New York: CRC Press/Lewis Publishers.

Cockerman, L. G., B. S. Shane. 1994. *Basic Environmental Toxicology.* New York: CRC Press/Lewis Publishers.

Cothern, C. R., N. P. Ross. 1994. *Environmental Statistics, Assessment, and Forecasting.* New York: CRC Press/Lewis Publishers.

Cothern, R. R. 1993. *Comparative Environmental Risk Assessment.* New York: CRC Press/Lewis Publishers.

Cothern, C. R. 1995. *Handbook for Environmental Risk Decisionmaking.* New York: CRC Press/Lewis Publishers.

Gardner, R. W. 1994. *Chemical Intolerance.* New York: CRC Press/Lewis Publishers.

Hallenbeck, W. H. 1993. *Quantitative Risk Assessment for Environmental and Occupational Health,* 2nd Edition. New York: CRC Press/Lewis Publishers.

Hodgson, E., R. B. Mailman, and Janice E. Chambers. 1988. *Dictionary of Toxicology.* London: The MacMillian Company.

Howard, P. H. 1995. *Handbook of Environmental Fate and Exposures Data for Organic Chemicals,* Vol. I-IV. New York: CRC Press/Lewis Publishers.

Howard, P. H., et. al. 1991. *Handbook of Environmental Degradation Rates.* New York: CRC Press/Lewis Publishers.

Jones, F. E. 1994. *Toxic Organic Vapors in the Workplace.* New York: CRC Press/Lewis Publishers.

Kamrin, M. A. 1988. *Toxicology: A Primer on Principles and Applications.* Chelsea, Michigan: Lewis Publishers.

Keith, L. H. and M. M. Walker. 1995. *Handbook of Air Toxics.* New York: CRC Press/Lewis Publishers.

Landis, W. G., M. H. Yu. 1995. *Introduction to Environmental Toxicology.* New York: CRC Press/Lewis Publishers.

Langerman, N. 1994 *Precautionary Labels for Chemical Containers.* New York: CRC Press/Lewis Publishers.

Lauwerys, R. R., P. Hoet. 1993. *Industrial Chemical Exposure.* New York: CRC Press/Lewis Publishers.

Lewis, R. J., Sr. 1993. *Hawley's Condensed Chemical Dictionary,* 12th Edition. New York: Van Nostrand Reinhold.

Lewis, R. J., Sr. 1992. *Sax's Dangerous Properties of Industrial Materials,* Volumes 1, 2, and 3, 8th Edition. New York: Van Nostrand Reinhold.

Lide, D. R. 1995. *Handbook of Chemistry and Physics,* 76th Edition. New York: CRC Press/Lewis Publishers.

Manahan, S. E. 1994. *Environmental Chemistry,* 6th Edition. New York: CRC Press/Lewis Publishers.

Manahan, S. E. 1992. *Toxicological Chemistry,* 2nd Edition. New York: CRC Press/Lewis Publishers.

Mellström, G. A., J. E. Wahlberg, and H. I. Maibach. 1994. *Protective Gloves for Occupational Use.* New York: CRC Press/Lewis Publishers.

Meyer, E. 1990. *Chemistry of Hazardous Materials.* Englewood Cliffs, NJ: Prentice Hall.

Mycock, J. C., J. D. McKenna, and L. Theodore. 1995. *Handbook of Air Pollution Control Engineering and Technology.* New York: CRC Press/Lewis Publishers.

Ott, W. R. 1995. *Environmental Statistics and Data Analysis.* New York: CRC Press/Lewis Publishers.

Philp, R. B. 1995. *Environmental Hazards and Human Health.* New York: CRC Press/Lewis Publishers.

Quigley, D. R. 1994. *Handbook of Emergency Chemical Management.* New York: CRC Press/Lewis Publishers.

Rea, W. J. 1995. *Chemical Sensitivity,* Vol. I-III. New York: CRC Press/Lewis Publishers.

Richardson, M. 1995. *Dictionary of Substance and Their Effects,* Vol. I-VI. New York: CRC Press/Lewis Publishers.

Rylander, R., R. R. Jacobs. 1994. *Organic Dusts.* New York: CRC Press/Lewis Publishers.

Sacarello, H. L. A. 1994. *The Comprehensive Handbook of Hazardous Materials.* New York: CRC Press/Lewis Publishers.

Sheftel, V. O. 1995. *Handbook of Toxic Properties of Monomers and Additives.* New York: CRC Press/Lewis Publishers.

Talbott, E., G. Craun. 1995. *An Introduction to Environmental Epidemiology.* New York: CRC Press/Lewis Publishers.

U.S. Department of Health and Human Services, National Institute for Occupational Safety and Health. 1994. *NIOSH Pocket Guide to Chemical Hazards.* Washington, D.C.: U.S. Government Printing Office.

MATERIAL SAFETY DATA SHEET

CHEMICAL NAME

DIETHYL KETONE

HAZARD WARNING INFORMATION

HEALTH	FIRE	REACTIVE	OTHER	DEGREE OF HAZARD	COLOR CODING	OTHER CODES
1	3	0		0 = Minimum Hazard 1 = Slight Hazard 2 = Moderate Hazard 3 = Serious Hazard 4 = Severe Hazard	HEALTH = BLUE FIRE = RED REACTIVITY = YELLOW OTHER = WHITE	OX = Oxidizer ACID = Acid ALK = Alkali COR = Corrosive W = Use No Water

SECTION I - GENERAL INFORMATION

Characterization: **Ketone**	RCRA Number: **None**	EPA Class: **Not Listed**
DOT Proper Shipping Name: **Diethyl ketone**	Chemical Abstract Service (CAS) Number: **96-22-0**	
DOT Hazard Class and Label Requirements: **Flammable Liquid**	DOT Emergency Guide Code: **26**	
DOT Identification Number: **UN 1156**	Chemical Formula: **$CH_3CH_3COCH_2CH_3$**	

Synonyms
DEK; dimethylacetone; ethyl ketone; metacetone; 3-pentanone; propione; ethyl propionyl; propione.

SECTION II - HAZARDOUS INGREDIENTS AND IDENTITY INFORMATION

Hazardous Components (specific identity)	OSHA Exposure Criteria	NIOSH Exposure Criteria	Immediately Dangerous to Life and Health (IDLH)	ACGIH Exposure Criteria
Diethyl ketone (derivation: By distilling sugar with an excess of lime). **1 ppm = 3.58 mg/m^3**	**PEL:** 200 ppm 705 mg/m^3 **STEL:** Not Established	**REL:** 200 ppm 705 mg/m^3 **STEL:** Not Established	**Not Determined**	**TLV:** 200 ppm 705 mg/m^3 **STEL:** Not Established

SECTION III - PHYSICAL AND CHEMICAL CHARACTERISTICS

Boiling Point: **215°F (101°C)**	Specific Gravity (H_2O = 1): **0.81**
Vapor Pressure (mm Hg): **35 at 77°F (25°C)**	Molecular Weight: **86.2**
Vapor Density (Air = 1): **3**	Melting Point: **-44°F (-42°C)**

Solubility
Slightly soluble (5%) in water. Soluble in alcohol and acetone.

Appearance and Odor
Colorless liquid with a characteristic, acetone-like odor.

SECTION IV - FIRE AND EXPLOSION HAZARD DATA

Flash Point (method used): **55°F (13°C) opened cup**	Explosive Limits in Air % by Volume: **LEL: 1.6% UEL: 6.4%**
NFPA Classification: **Class 1B Flammable Liquid**	Autoignition Temperature: **846°F (453°C)**

Extinguishing Media
Carbon dioxide, dry chemical, alcohol foam.

Special Fire Fighting Procedures
Wear full protective clothing and self-contained breathing apparatus (SCBA). Heat will build pressure and may rupture closed storage containers. Fight fire from a distance if possible. Because of low flash point, water may be ineffective on fire. Keep fire-exposed containers cool with water spray. If it can be done safely, move fire-exposed containers from fire area. Exercise extreme caution.

Unusual Fire and Explosion Hazards
Vapors are heavier than air; they may travel a distance to flashback and cause fire. Poisonous gases are produced in fire.

SECTION V - REACTIVITY DATA

Stability		Conditions to Avoid Normally stable. Do not allow diethyl ketone to come into contact with incompatible materials. Avoid contact with fire or high heat sources. Diethyl ketone is highly volatile. Do not use in confined spaces.
Stable X	Unstable	Incompatibility *(materials to avoid)* Strong oxidizers (chlorine, fluorine, bromine), alkalis, and mineral acids. Diethyl ketone reacts with hydrogen peroxide and nitric acid to form a heat-sensitive explosive peroxide.
Hazardous Polymerization		Conditions to Avoid Under normal conditions of temperature and pressure, hazardous polymerization of diethyl ketone is not expected to occur.
May Occur	Will Not Occur X	Hazardous Decomposition or By-products When heated to decomposition, diethyl ketone emits acrid, irritating smoke and fumes, including oxides of carbon.

SECTION VI - HEALTH HAZARD DATA

Primary Route(s) of Entry:	Inhalation? X	Absorption (skin)?	Ingestion? X

Health Hazards

INHALATION: Irritation of the conjunctiva of the eyes and mucosa of the nose and throat. Symptoms include dizziness, headache, nausea, vomiting, ataxia, and unconsciousness. May also cause potentially fatal heart rhythm disturbances. Narcotic effects also possible.

SKIN & EYES: Skin contact can result in irritation and may cause dermatitis. Severe eye irritation likely.

INGESTION: Symptoms of nausea and vomiting. Abdominal cramps.

Carcinogenicity	NTP Listed?	IARC Cancer Review Group?	OSHA Regulated?	Target Organs?
Unknown Human Unknown Animal	No	No	29 CFR 1910.1000 Table Z-1	Respiratory system; skin; eyes; CNS.

Medical Conditions Generally Aggravated by Exposure

None Reported. However, respiratory impairments, skin conditions (dermatitis) may be aggravated.

Emergency and First-aid Procedures

Eye contact: Immediately flush with large amounts of water for 15 minutes (minimum), occasionally lifting eyelids. Skin contact: Remove all contaminated clothing. Immediately wash area with large amounts of soap and water. For inhalation: Remove the person from exposure. Provide respiratory assistance and CPR. Transfer promptly to medical facility. If swallowed: Seek medical attention immediately. Never try to give an unconscious person anything by mouth. Do NOT induce vomiting.

SECTION VII - PRECAUTIONS FOR SAFE HANDLING AND USE

Steps to be Taken in Case Material is Released or Spilled

Remove all ignition sources. Restrict persons not wearing protective equipment from entering area of spill or leak until cleanup is complete. Ventilate area. Absorb liquid in dry sand or other material and deposit in sealed containers. Alert appropriate state or federal response agencies.

Preferred Waste Disposal Method

Incineration (with afterburner and scrubber).

Precautions to be Taken in Handling and Storage

Do not store with incompatible chemicals since violent reactions can occur. Store in tightly closed containers in a cool, dark, well-ventilated area away from fire, sparks, and flame. Use non-sparking tools on containers. Ground and bond metal containers during transfer operations.

Other Precautions and Warnings

Containers may explode in fire or under conditions of extreme heat (do not store outdoors or in direct sunlight). Protect containers from physical damage.

SECTION VIII - CONTROL MEASURES AND PERSONAL PROTECTIVE EQUIPMENT

Respiratory Protection *(specify type)*

Exposures above 200ppm: MSHA/NIOSH-approved full facepiece respirator with organic vapor cartridge. Greater protection is obtained from a self-contained breathing apparatus (SCBA) with full facepiece and pressure demand or other positive pressure mode.

Ventilation

Local exhaust or general mechanical systems recommended.

Protective Gloves	Eye Protection	Other Protective Clothing
Solvent Resistant Butyl Rubber	Chemical Goggles and/or Face Mask	Rubber Apron

Work/Hygiene Practices

Always wash hands thoroughly after using chemical; never bring food, drink, or smoking materials into vicinity of chemicals.

DIETHYL KETONE

$CH_2CII_2COCH_2CH_3$ CAS: 96-22-0

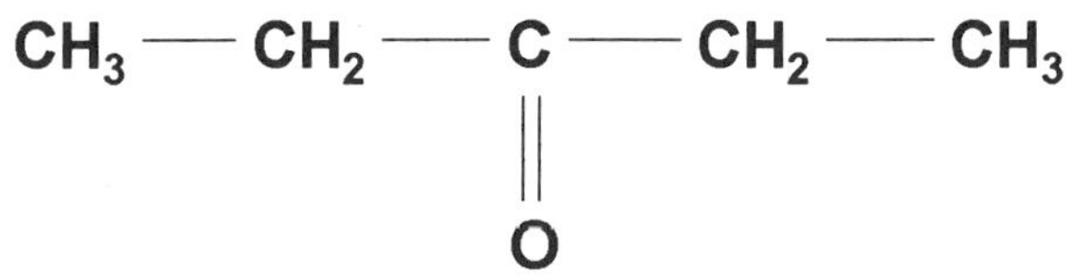

IDENTIFICATION AND TYPICAL USES

Diethyl ketone is a colorless liquid with a characteristic, acetone-like odor. It is used as a solvent. Also used in the manufacture of medicines and in organic synthesis.

RISK ASSESSMENT: HEALTH

General Assessment

Mutagenic effects in animals have been reported for diethyl ketone. However, there are no reports of carcinogenic or mutagenic effects in human. Exposure to diethyl ketone can result in harmful effects. Such exposures can occur primarily by *inhalation* of its vapors. Direct *ingestion* will also cause harmful effects. There are no reports in the references that indicate absorption is a risk factor. However, skin or eye contact can cause damage.

Inhalation of diethyl ketone vapors in high concentrations produces dryness of the mouth and throat, dizziness, headache, nausea, vomiting, ataxia, and unconsciousness. Inhalation may also produce mild narcotic effects. Inhalation of other organic solvents of similar structure are known to cause potentially fatal heart rhythm disturbances. It is not known if exposure to diethyl ketone will have this effect.

Ingestion can result in nausea and vomiting. Skin contact will defat the surface of the skin and may lead to dermatitis. Eye contact produces severe irritation and possible damage.

Acute Health Effects

The following acute (short-term) health effects may occur immediately or shortly after exposure to diethyl ketone:

Skin: Irritation/rash and possible dermatitis.

Eye: Irritation which can lead to permanent damage.

Lung: Irritation causing coughing and/or shortness of breath. Higher exposures can cause dizziness, headache, and possibly, coma.

CNS: A mild narcotic compound. Inhalation of high concentrations of the vapor may cause dizziness, lightheadedness, and possible loss of consciousness.

Chronic Health Effects

The following chronic (long-term) health effects may occur at some time after exposure to diethyl ketone and can last for months or even years:

Cancer Hazards: According to information presented in the references, exposure to diethyl ketone has not been shown to cause carcinogenic effects. Animal mutation data have been reported.

Reproduction: There are no reports to support any claims of reproductive hazards in humans or animals.

Sensitivity: Exposure to the skin may result in allergy (sensitization). If allergy develops, very low future exposure can result in skin irritation and rash. Repeated exposure can cause chronic irritation of the eyes and lungs.

Other Chronic Effects: Long-term exposures can also cause chronic nose and throat irritation.

Recommended Risk-Reduction Measures

Personnel should avoid direct contact with diethyl ketone. Unless a less toxic chemical can be substituted for a hazardous substance, *engineering controls* are the most effective method of reducing exposures. The best protection is to enclose operations and/or provide local exhaust ventilation at the site of chemical release. While not always operationally feasible, isolating operations can also reduce exposure. Using respiratory protection is less effective than the controls mentioned above, but is still advisable whenever working with diethyl ketone. For exposures to the PEL (200 ppm), use a NIOSH-approved, full facepiece organic vapor respirator. For higher or prolonged exposures, a self-contained breathing apparatus (SCBA) with full facepiece and pressure demand is the recommended respiratory protection of choice. If a full facepiece is not available, then chemical goggles should be worn to protect the eyes. Whenever a chemical splash hazard exists, a face shield and a rubber apron should be worn. To prevent hand and skin exposures, butyl rubber gloves should be worn.

Administrative controls should also be in place to minimize the potential for human exposures. These may include written procedures or policies which specify the methods and techniques that will be practiced whenever personnel are to work with diethyl ketone.

All personnel should receive *training* on the use, hazards, protective measures, emergency actions, and other precautions per 29 CFR 1910.1200 (Hazard Communication), prior to the first assignment in an area where diethyl ketone is used or stored.

Other methods to reduce exposure include:

- ☑ Where possible, enclose operations and use local exhaust ventilation at the site of chemical release. If local exhaust ventilation or enclosure is not used, respiratory protection should be mandatory.
- ☑ Always ensure that proper protective clothing is worn when using chemical substances. Clothing should be properly cared for and maintained by the responsible employee.
- ☑ Wash thoroughly immediately after exposure to diethyl ketone and at the end of the work shift or before eating, drinking, or smoking. Never wear clothes contaminated with diethyl ketone home. Family members can be exposed.
- ☑ Hazard warning information should be posted in the work area. As part of an on-going education and training program, all information on the health and safety hazards of diethyl ketone should be communicated to all potentially exposed workers.

RISK ASSESSMENT: ENVIRONMENT

General Assessment

The environment is at risk of exposure during transportation, storage, disposal, or destruction of diethyl ketone. In almost every scenario, the threat of environmental exposure is contingent upon the proper handling of the chemical substance. Accidental spills, large or small, can result in fire, explosion, and possible contamination of the surrounding environmental mediums (water, soil, and air).

Diethyl ketone is considered a highly flammable liquid. Because of its low flash point and relatively high boiling point, it is classified as a Class 1B flammable liquid (per OSHA 29 CFR 1910.106). Its low auto-ignition temperature and low flash point, together with its high vapor pressure make it a dangerous fire and explosion hazard. It is also incompatible with a number of commodities. Potentially explosive reactions can occur if diethyl ketone is mixed or comes into contact with oxidizers and/or mineral acids. These characteristics require special consideration during any emergency involving a leak or spill of diethyl ketone.

The proper disposal/destruction method for diethyl ketone is to burn it in a chemical incinerator (with molten salt treatment) equipped with an afterburner and air scrubber.

Diethyl ketone can enter the environment through manufacturing, unchecked discharge into effluents from industrial or municipal waste treatment plants, and through spills.

☠ *Acute Ecological Effects*

Acute (short-term) toxic effects may include the death of animals, birds, or fish, and death or low growth rate in plants. Acute effects are seen 2 to 4 days after animals or plants are exposed to diethyl ketone.

This chemical has a slight acute toxicity to aquatic life. No data are available on the short-term effects of diethyl ketone on plants, birds, or land animals.

Chronic Ecological Effects

Chronic (long-term) toxic effects may include shortened life span, reproductive problems, lower fertility, and changes in appearance or behavior in exposed animals. These effects can be seen long after first exposure(s) to toxic chemicals.

Diethyl ketone has slight chronic toxicity in aquatic life. No data are available to evaluate the long-term effects of diethyl ketone to plants, birds, or land animals.

Water Solubility

Diethyl ketone is slightly soluble in water. Concentrations between 1 to 100 milligrams will mix with a liter of water.

⌛ *Persistence in the Environment*

Diethyl ketone is slightly persistent in water, with a half-life of between 2 to 20 days. The half-life of a pollutant is the amount of time it takes for one half of the chemical to be degraded. Approximately 50% of diethyl ketone will eventually end up in air; the rest will end up in water.

Bioaccumulation in Aquatic Organisms

Some substances increase in concentration, or *bioaccumulate,* in living organisms as they breathe contaminated air, drink contaminated water, or eat contaminated food. These chemicals can become concentrated in the tissues and internal organs of animals as well as humans.

The concentration of diethyl ketone found in fish tissues is expected to be about the same as the average concentration of diethyl ketone in water from which the fish was taken.

Recommended Risk-Reduction Measures

Proper training of all transporters will reduce the likelihood of a mishap or accident resulting in a leak or spill to the environment. The correct use of DOT labeling (per the MSDS) on all containers, trucks, and rail cars will enable emergency responders to react properly and quickly to any disaster thereby reducing the potential risk to the environment and to personnel.

Storage of diethyl ketone should be segregated from incompatible chemicals to minimize the risk of cross-contamination or contact. Buildings designated for storage should be equipped with appropriate fire protection systems (alarms, sprinklers, emergency lighting, portable extinguishers).

If a spill or leak to the environment has occurred, fire department, emergency response, and/or hazardous materials spill personnel should be notified immediately. Cleanup should be attempted only by those trained in proper spill containment procedures. Contaminated soils should be removed for incineration and replaced with clean soil. If diethyl ketone should contact the water table, aquifer, or navigable waterway, time is of the essence. It is slightly soluble in water and, therefore, total containment and remediation may not be entirely possible. When such spills occur, the local and/or state emergency response authorities must be notified. A comprehensive emergency response or disaster preparedness/recovery plan should be in place prior to any operations involving the use, transportation, storage, or disposal of diethyl ketone. If diethyl ketone is spilled or leaked, the following specific steps are recommended:

- ☑ Restrict persons not wearing protective clothing from area of spill or leak until cleanup is complete.
- ☑ Remove all ignition sources.
- ☑ Ventilate area of spill or leak.
- ☑ Absorb liquids in vermiculite, dry sand, earth, or a similar material and deposit in sealed containers.
- ☑ If applicable, stop flow of leaking liquid or gas. If leak source is a cylinder and the leak cannot be stopped in place, remove leaking cylinder to a safe place in the open air, and repair or allow cylinder to empty.
- ☑ Keep diethyl ketone out of a confined space, such as a sewer, because of the possibility of explosion (unless the sewer is designed to prevent the build-up of explosive concentrations).
- ☑ It may be necessary to dispose of diethyl ketone as a hazardous waste. The responsible state agency or the regional office of the federal Environmental Protection Agency (EPA) should be contacted for specific recommendations.

RISK ASSESSMENT: BUSINESS

General Assessment

Accidents or mishaps involving diethyl ketone can present a significant threat to business operations. The loss or damage of equipment or facilities can significantly affect fiscal viability. Lawsuits that may result from personnel injury, public exposures, and/or environmental contamination will also require a serious expenditure of resources. Media attention surrounding an injury, illness, or environmental damage can also result in a loss of profits and loss of current as well as future business.

Recommended Risk-Reduction Measures

Company attorneys, safety and health professionals, and environmental specialists should be involved in the development of any procedures dealing with safety or emergency response. A company official should be pre-designated as a public relations officer with specific training in dealing with the press. Corporate plans and policies should be developed, approved, and implemented long before any need for such arises.

REFERENCES

Ahlbom, A. 1993. *Biostatistics for Epidemiologists.* New York: CRC Press/Lewis Publishers.

American Conference of Governmental Industrial Hygienists. 1988. *Documentation of the Threshold Limit Values and Biological Exposure Indices,* 5th Edition (with updates). Cincinnati: ACGIH

Anthony, C. P., and N. J. Kolthoff. 1971. *Textbook of Anatomy and Physiology.* St. Louis: C. V. Mosby Company.

Asante-Duah, D. K. 1993. *Hazardous Waste Risk Assessment.* New York: CRC Press/Lewis Publishers.

Bloom, A. D. and F. J. DeSerres. 1995. *Ecotoxicity and Human Health.* New York: CRC Press/Lewis Publishers.

Brusick, D. J. 1994. *Methods for Generic Risk Assessment.* New York: CRC Press/Lewis Publishers.

Bueche, J. R. 1972. *Principles of Physics,* 2nd Edition. New York: McGraw-Hill.

Burrell, R., D. K. Flaherty, and L. J. Sauers. 1992. *Toxicology of the Immune System.* New York: Van Nostrand Reinhold.

Calabrese, E. J., E. M. Kenyon. *Air Toxics and Risk Assessment.* New York: CRC Press/Lewis Publishers.

Calabrese, E. J. 1994. *Biological Effects of Low Level Exposures.* New York: CRC Press/Lewis Publishers.

Cockerman, L. G., B. S. Shane. 1994. *Basic Environmental Toxicology.* New York: CRC Press/Lewis Publishers.

Cothern, R. R. 1993. *Comparative Environmental Risk Assessment.* New York: CRC Press/Lewis Publishers.

Cothern, C. R. 1995. *Handbook for Environmental Risk Decisionmaking.* New York: CRC Press/Lewis Publishers.

DiVincenzo, G. O., F. J. Yanno, and B. D. Astill. 1973. Exposure of man and dog to low concentrations of acetone vapor. *American Industrial Hygiene Journal.* 34:329-336.

Gardner, R. W. 1994. *Chemical Intolerance.* New York: CRC Press/Lewis Publishers.

Hallenbeck, W. H. 1993. *Quantitative Risk Assessment for Environmental and Occupational Health,* 2nd Edition. New York: CRC Press/Lewis Publishers.

Hodgson, E., R. B. Mailman, and J. E. Chambers. 1988. *Dictionary of Toxicology.* London: The MacMillian Company.

Hosty, J. W., and P. Foster. 1990. *A Practical Guide to Chemical Spill Response.* New York: Van Nostrand Reinhold.

Howard, P. H. 1995. *Handbook of Environmental Fate and Exposures Data for Organic Chemicals,* Vol. I-IV. New York: CRC Press/Lewis Publishers.

Howard, P. H., et. al. 1991. *Handbook of Environmental Degradation Rates.* New York: CRC Press/Lewis Publishers.

James, A., and M. Lord. 1992. *Index of Chemical and Physical Data.* New York: Van Nostrand Reinhold.

Jones, F. E. 1994. *Toxic Organic Vapors in the Workplace.* New York: CRC Press/Lewis Publishers.

Kamrin, M. A. 1988. *Toxicology: A Primer on Principles and Applications.* Chelsea, Michigan: Lewis Publishers.

Keith, L. H. and M. M. Walker. 1995. *Handbook of Air Toxics.* New York: CRC Press/Lewis Publishers.

Landis, W. G., M. H. Yu. 1995. *Introduction to Environmental Toxicology.* New York: CRC Press/Lewis Publishers.

Langerman, N. 1994 *Precautionary Labels for Chemical Containers.* New York: CRC Press/Lewis Publishers.

Lauwerys, R. R., P. Hoet. 1993. *Industrial Chemical Exposure.* New York: CRC Press/Lewis Publishers.

Lewis, R. J., Sr. 1993. *Hawley's Condensed Chemical Dictionary,* 12th Edition. New York: Van Nostrand Reinhold.

Lewis, R. J., Sr. 1992. *Sax's Dangerous Properties of Industrial Materials,* Volumes 1, 2, and 3, 8th Edition. New York: Van Nostrand Reinhold.

Lide, D. R. 1995. *Handbook of Chemistry and Physics,* 76th Edition. New York: CRC Press/Lewis Publishers.

Manahan, S. E. 1994. *Environmental Chemistry,* 6th Edition. New York: CRC Press/Lewis Publishers.

Manahan, S. E. 1992. *Toxicological Chemistry,* 2nd Edition. New York: CRC Press/Lewis Publishers.

Mellström, G. A., J. E. Wahlberg, and H. I. Maibach. 1994. *Protective Gloves for Occupational Use.* New York: CRC Press/Lewis Publishers.

Meyer, E. 1990. *Chemistry of Hazardous Materials.* Englewood Cliffs, NJ: Prentice Hall.

Patnaik, P. 1992. *A Comprehensive Guide to the Hazardous Properties of Chemical Substances.* New York: Van Nostrand Reinhold.

Philp, R. B. 1995. *Environmental Hazards and Human Health.* New York: CRC Press/Lewis Publishers.

Proctor, N. H., and J. P. Hughes. 1991. *Chemical Hazards of the Workplace.* New York: Van Nostrand Reinhold.

Quigley, D. R. 1994. *Handbook of Emergency Chemical Management.* New York: CRC Press/Lewis Publishers.

Raleigh, R. L. and W. A. McGee. 1972. Effects of short, high concentration exposures to acetone as determined by observation in the work area. *Journal of Occupational Medicine.* 14:607-610.

Rea, W. J. 1995. *Chemical Sensitivity,* Vol. I-III. New York: CRC Press/Lewis Publishers.

Richardson, M. 1995. *Dictionary of Substance and Their Effects,* Vol. I-VI. New York: CRC Press/Lewis Publishers.

U.S. Department of Health and Human Services, National Institute for Occupational Safety and Health. 1994. *NIOSH Pocket Guide to Chemical Hazards.* Washington, D.C.: U.S. Government Printing Office.

U.S. Department of Health and Human Services, National Institute for Occupational Safety and Health. 1978. *Criteria for a Recommended Standard - Occupational Exposure to Ketones.* DHEW Publication Number: 78-173. Washington, D.C.: U.S. Government Printing Office.

MATERIAL SAFETY DATA SHEET

CHEMICAL NAME	
DIETHYL PHTHALATE	**CRC PRESS/LEWIS PUBLISHERS DISCLAIMER:** THE INFORMATION AND RECOMMENDATIONS PRESENTED HEREIN ARE BASED ON SOURCES BELIEVED TO BE RELIABLE. CRC MAKES NO REPRESENTATION ON ITS COMPLETENESS OR ACCURACY. IT IS THE RESPONSIBILITY OF THE USER TO DETERMINE THE CHEMICAL'S SUITABILITY FOR ITS INTENDED USE, THE CHEMICAL'S SAFE USE, AND THE CHEMICAL'S PROPER DISPOSAL. NO REPRESENTATIONS AND/OR WARRANTIES, EITHER EXPRESSED OR IMPLIED, OF THE MERCHANTABILITY OR FITNESS FOR A PARTICULAR PURPOSE, OR OF ANY OTHER NATURE, ARE MADE WITH RESPECT TO THE INFORMATION PROVIDED IN THIS MSDS OR TO THE CHEMICAL TO WHICH INFORMATION MAY REFER. CRC NEITHER ASSUMES NOR AUTHORIZES ANY OTHER PERSON TO ASSUME FOR IT, ANY OTHER ADDITIONAL RESPONSIBILITY OR LIABILITY FOR THE USE OF, OR RELIANCE UPON, THIS INFORMATION.

HAZARD WARNING INFORMATION

HEALTH	FIRE	REACTIVE	OTHER	DEGREE OF HAZARD	COLOR CODING	OTHER CODES
0	1	0		0 = Minimum Hazard 1 = Slight Hazard 2 = Moderate Hazard 3 = Serious Hazard 4 = Severe Hazard	HEALTH = BLUE FIRE = RED REACTIVITY = YELLOW OTHER = WHITE	OX = Oxidizer ACID = Acid ALK = Alkali COR = Corrosive W = Use No Water

SECTION I - GENERAL INFORMATION

Characterization: **Phthalate Ester**	RCRA Number: **U088**	EPA Class: **Toxic Waste**
DOT Proper Shipping Name: **Diethyl Ketone**	Chemical Abstract Service (CAS) Number: **84-66-2**	
DOT Hazard Class and Label Requirements: **Not Listed**	DOT Emergency Guide Code: **No Citation**	
DOT Identification Number: **Not Listed**	Chemical Formula: **$C_6H_4(COOC_2H_6)_2$**	

Synonyms

DEP; Diethyl ester of phthalic acid; ethyl phthalate; 1,2-benzenedicarboxylic acid, diethyl ester; diethyl-o-phthalate; phthalic acid, diethyl ester.

SECTION II - HAZARDOUS INGREDIENTS AND IDENTITY INFORMATION

Hazardous Components (specific identity)	OSHA Exposure Criteria	NIOSH Exposure Criteria	Immediately Dangerous to Life and Health (IDLH)	ACGIH Exposure Criteria
Diethyl phthalate (derivation: By reacting phthalic anhydride with ethanol followed by careful purification). **1 ppm = 9.24 mg/m³**	**PEL:** Not Established **STEL:** Not Established	**REL:** 0.5 ppm, 5 mg/m³ **STEL:** Not Established	**Not Determined**	**TLV:** 0.5 ppm, 5 mg/m³ **STEL:** 1.08 ppm, 10 mg/m³

SECTION III - PHYSICAL AND CHEMICAL CHARACTERISTICS

Boiling Point: **563°F (298°C)**	Specific Gravity (H_2O = 1): **1.12**
Vapor Pressure (mm Hg): **0.002 at 77°F (25°C)**	Molecular Weight: **222.3**
Vapor Density (Air = 1): **7.66**	Freezing Point: **-40°F (-40°C)**

Solubility

Insoluble in water. Miscible in alcohols, ketones, esters, aromatic hydrocarbons. Partly miscible with aliphatic solvents.

Appearance and Odor

Colorless to water-white, stable, oily liquid with a very slight, aromatic odor and a bitter taste.

SECTION IV - FIRE AND EXPLOSION HAZARD DATA

Flash Point (method used): **325°F (163°C) closed cup**	Explosive Limits in Air % by Volume: **LEL: 0.75% (at 368°F) UEL: Not Determined**
NFPA Classification: **Class IIIB Combustible Liquid**	Autoignition Temperature: **855°F (458°C)**

Extinguishing Media

Use water spray, dry chemical, carbon dioxide, or foam. Water may cause frothing.

Special Fire Fighting Procedures

Acrid and irritating fumes and smoke are produced in fire, wear full protective clothing and self-contained breathing apparatus (SCBA). Vapors are heavier than air and may travel for some distance to ignition source and flashback. Move containers from fire area if it can be done safely.

Unusual Fire and Explosion Hazards

Containers may explode in fire due to buildup of excessive internal pressures. Exercise extreme caution when attempting to move or relocate containers that may have been exposed to high heat.

SECTION V - REACTIVITY DATA

Stability		Conditions to Avoid
		Under normal conditions of temperature and pressure, diethyl phthalate is stable. Avoid contact with incompatible materials. Keep away from heat and flame.
Stable	Unstable	Incompatibility *(materials to avoid)*
X		Strong oxidizers, such as chlorine, bromine, and fluorine, strong acids (sulfuric, nitric, hydrochloric), permanganates, and water.
Hazardous Polymerization		Conditions to Avoid
		Under normal conditions of temperature and pressure, hazardous polymerization of diethyl phthalate is not expected to occur.
May Occur	Will Not Occur	Hazardous Decomposition or By-products
	X	Acrid and irritated gases and fumes are produced when diethyl phthalate is involved in fire.

SECTION VI - HEALTH HAZARD DATA

Primary Route(s) of Entry:	Inhalation?	Absorption (skin)?	Ingestion?
	X		X

Health Hazards

INHALATION: Symptoms include irritation to the eyes, nose, and throat. Causes headache, dizziness, nausea, lachrymation, pain, numbness, weakness, spasms in the arms and legs. Also, possible polyneuritis (inflammation of peripheral nerves) and respiratory obstruction. Narcotic at high concentrations.

SKIN & EYES: Contact may irritate the skin and eyes. May cause conjunctivitis.

INGESTION: Burning irritation of the mucous membranes. Gastrointestinal irritation.

Carcinogenicity	NTP Listed?	IARC Cancer Review Group?	OSHA Regulated?	Target Organs?
Unknown Human Unknown Animal	No	No	No	Eyes, skin, resp. system, CNS, PNS, repro. system.

Medical Conditions Generally Aggravated by Exposure

None Reported.

Emergency and First-aid Procedures

<u>Eye contact</u>: Flush immediately with water for 15 minutes (minimum), seek medical attention. <u>Skin contact</u>: Remove all contaminated clothing. Immediately wash area with large amounts of soap and water. Seek medical assistance. For <u>inhalation</u>: Remove the person from exposure. Provide respiratory assistance and CPR. Transfer to medical facility. If <u>swallowed</u>: Call the poison control center and seek medical attention immediately. Give 2-3 glasses of water and induce vomiting.

SECTION VII - PRECAUTIONS FOR SAFE HANDLING AND USE

Steps to be Taken in Case Material is Released or Spilled

Small spills can be collected using a paper towel or other absorbent for disposal. For larger spills, dike area ahead of spill and absorb liquids in vermiculite or other absorbent and deposit in DOT-approved drum. Do not allow flow into sewers or waterways. Ventilate area and remove ignition sources. Restrict those not involved in cleanup from entering area. Notify appropriate authorities, if applicable.

Preferred Waste Disposal Method

Burn in chemical incinerator equipped with an afterburner and scrubber.

Precautions to be Taken in Handling and Storage

Do not store diethyl phthalate in the presence of incompatible chemicals or materials. Store in tightly closed containers in a cool, well-ventilated area. Personnel working with diethyl phthalate should be trained on its proper handing and storage *prior* to being assigned to such responsibilities.

Other Precautions and Warnings

Bulk storage is of diethyl phthalate should be avoided.

SECTION VIII - CONTROL MEASURES AND PERSONAL PROTECTIVE EQUIPMENT

Respiratory Protection *(specify type)*

Exposure level is extremely low and may be difficult to monitor. For the best protection, use a supplied-air respirator with full facepiece operated in pressure demand mode or a self-contained breathing apparatus (SCBA) with full facepiece operated in pressure demand or other positive pressure mode.

Ventilation

Local exhaust or general mechanical systems recommended.

Protective Gloves	Eye Protection	Other Protective Clothing
Rubber	Chemical Goggles or Face Mask	Apron

Work/Hygiene Practices

Always wash hands thoroughly after using chemical; never bring food, drink, or smoking materials into vicinity of chemicals.

DIETHYL PHTHALATE

$C_6H_4(COOC_2H_6)_2$ CAS: 84-66-2

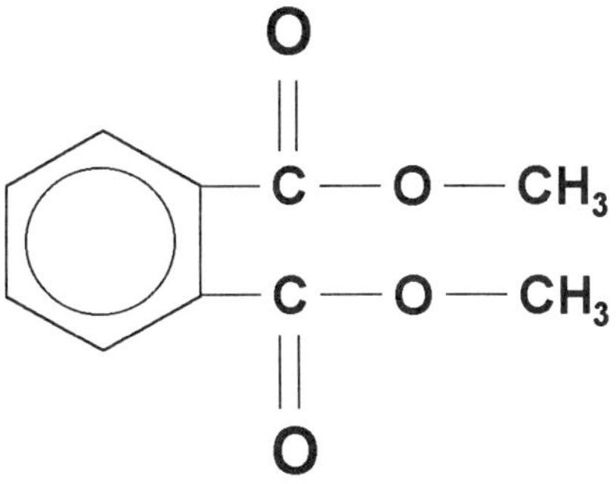

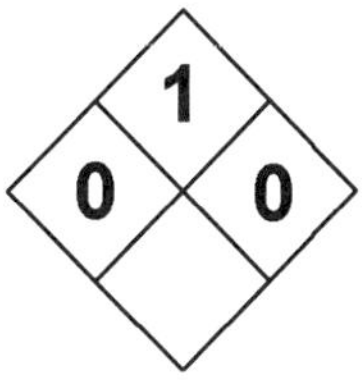

IDENTIFICATION AND TYPICAL USES

Diethyl phthalate is a colorless to water-white, oily liquid with a very slight, aromatic odor and a bitter taste. It is used as a solvent for nitrocellulose and cellulose acetate, as a plasticizer, a wetting agent, in the manufacture of insecticide sprays, as a camphor substitute, in plastics, in perfumery (as a fixative and solvent), as an alcohol denaturant, in mosquito repellents, and as a plasticizer in solid rocket propellants.

RISK ASSESSMENT: HEALTH

General Assessment

Diethyl phthalate is of low-order toxicity in animal studies. It is toxic in humans primarily through *inhalation* and can also cause adverse health effects on *ingestion.* Eye and skin contact also pose some degree of health hazard risk to humans. While it has not been shown to cause carcinogenic effects in humans, there are reports of teratogenicity in animals studies.

Inhalation causes toxic systemic effects, including lachrymation, respiratory obstruction, and other unspecified respiratory system effects. Also causes irritation of the eyes, nose, and throat. It is a narcotic on high concentrations. May also cause polyneuritis (inflammation of the peripheral nervous system) with symptoms of dizziness, nausea, pain, weakness, and spasms in the arms and legs.

Ingestion causes irritation and burning of the mucosa of the digestive tract. Causes somnolence, hypotension, and coma.

Skin and eye contact result in irritation, burning, and possible painful rash (on prolonged contact).

Acute Health Effects

The following acute (short-term) health effects may occur immediately or shortly after exposure to diethyl phthalate:

Skin: Irritation and possible inflammation.

Eye: Severe irritation with redness, lachrymation, and possible conjunctivitis.

Lung: Irritation of the mucosa of the nose, throat, and upper respiratory tract. Causes toxic systemic effects with action on the peripheral and central nervous systems.

Other: Narcotic effects, with a potential for inflammation of many of the peripheral nerves. Ingestion can cause gastritis, somnolence, hypotension, and coma.

Chronic Health Effects

The following chronic (long-term) health effects may occur at some time after exposure to diethyl phthalate and can last for months or even years:

Cancer Hazards: According to the information presented in the references, diethyl phthalate has not been adequately tested for its ability to cause cancer in test animals.

Reproductive Hazard: According to the information presented in the references, diethyl phthalate may damage the developing fetus in humans since it has been shown to be fetotoxic in animals. It may also damage fertility.

Other Chronic Effects: Chronic exposure to diethyl phthalate has not been sufficiently studied. However, there is some indication that long-term exposure or frequent high exposures may cause changes in the function of the nervous system, causing pain or numbness in the legs and arms. A mild decrease in white blood cells and platelets may occur.

❶ *Recommended Risk-Reduction Measures*

Unless a less toxic chemical can be substituted for a hazardous substance, *engineering controls* are the most effective method of reducing exposures. The best protection is to enclose operations and/or provide local exhaust ventilation at the site of chemical release. While not always operationally feasible, isolating operations can also reduce exposure.

Using respiratory protection is less effective than the controls mentioned above, but is still advisable whenever working with or around diethyl phthalate. The exposure level for this compound is extremely low and may be difficult to monitor. Cartridge-type respirators may not suffice, especially at high concentrations. For the best protection, a self-contained breathing apparatus (SCBA) with full facepiece and pressure demand or a supplied-air respirator operated in pressure demand or continuous flow mode are the recommended respiratory protection methods of choice. If a full facepiece is not available, then chemical goggles should be worn to protect the eyes. Whenever a chemical splash hazard exists, a face shield and rubber apron should be worn. To prevent hand and skin exposures, rubber gloves should be worn.

Administrative controls should also be in place to minimize the potential for human exposures. These may include written procedures or policies which specify the methods and techniques that will be practiced whenever personnel are to work with diethyl phthalate.

All personnel should receive *training* on the use, hazards, protective measures, emergency actions, and other precautions per 29 CFR 1910.1200 (Hazard Communication), prior to the first assignment in an area where diethyl phthalate is used or stored.

If symptoms develop or overexposure is suspected, the following medical tests may be useful:

- ☑ Nervous system examination with special testing by a qualified neurologist.
- ☑ Complete blood count and differential.

Any evaluation should include a careful medical history of past and present symptoms with an examination. However, medical tests that evaluate existing damage are not a substitute for controlling exposure.

Other methods to reduce exposure include:

- ☑ Where possible, enclose operations and use local exhaust ventilation at the site of chemical release. If local exhaust ventilation or enclosure is not used, respiratory protection should be mandatory. Also, automatic transfer of liquids from storage containers to process containers is recommended.
- ☑ Always ensure that proper protective clothing is worn when using chemical substances.
- ☑ Wash thoroughly immediately after exposure to diethyl phthalate and at the end of the work shift or before eating, drinking, or smoking.
- ☑ Hazard warning information should be posted in the work area. In addition, as part of an on-going education and training program, all information on the health and safety hazards of diethyl phthalate should be communicated to all potentially exposed workers.
- ☑ Eye wash stations should be provided in work areas. If the potential for whole body exposure exists, then safety showers should also be provided.

RISK ASSESSMENT: ENVIRONMENT

General Assessment

The environment is at risk of exposure during use, transportation, storage, disposal, or destruction of diethyl phthalate. In almost every scenario, the threat of environmental exposure is contingent upon the proper handling of the chemical substance. Accidental spills, large or small, can result in fire, explosion, and possible contamination of the surrounding environmental mediums (water, soil, and air).

Diethyl phthalate is considered a Class IIIB combustible liquid (per OSHA 29 CFR 1910.106). These characteristics require special consideration during any emergency situation involving a leak or spill of diethyl phthalate. Should diethyl phthalate ever come into contact with incompatible substances such as strong oxidizers (chlorine, bromine, fluorine, and chlorine dioxide), strong acids (sulfuric acid, hydrochloric acid, and nitric acid), permanganates, and water either during use, transportation, or storage, violent reaction can occur. Diethyl phthalate is especially reactive in contact with liquid chlorine.

Diethyl phthalate can enter the environment through industrial discharges and from spills.

Acute Ecological Effects

Acute (short-term) toxic effects may include the death of animals, birds, or fish and death or low growth rate in plants. Acute effects are seen 2 to 4 days after animals or plants are exposed to diethyl phthalate. This chemical has moderate acute toxicity to aquatic life. Insufficient data are available to evaluate or predict the short-term effects of diethyl phthalate to plants, birds, or land animals.

Chronic Ecological Effects

Chronic toxic effects may include shortened life span, reproductive problems, lower fertility, and changes in appearance or behavior in exposed animals. These effects can be seen long after first exposure(s) to toxic chemicals. Diethyl phthalate has moderate chronic toxicity to aquatic life. Insufficient data are available to evaluate or predict the long-term effects of diethyl phthalate to plants, birds, or land animals.

Water Solubility

Diethyl phthalate is nearly insoluble in water. Concentrations of 1 milligram or less will mix with a liter of water.

Persistence in the Environment

Diethyl phthalate is slightly persistent in water, with a half-life between 2 and 20 days. The half-life of a pollutant is the amount of time it takes for one half of the chemical to be degraded. About 83% of diethyl phthalate will eventually end up in water. Approximately 11.6% will end up in the air, and the rest will end up equally divided in terrestrial soils and aquatic sediments.

Bioaccumulation in Aquatic Organisms

Some substances increase in concentration, or *bioaccumulate,* in living organisms as they breathe contaminated air, drink contaminated water, or eat contaminated food. These chemicals can become concentrated in the tissues and internal organs of animals as well as humans.

The concentration of diethyl phthalate found in fish tissues is expected to somewhat higher than the average concentration of diethyl phthalate in the water from which the fish was taken.

Recommended Risk-Reduction Measures

Proper training of all transporters will reduce the likelihood of a mishap or accident resulting in a leak or spill to the environment. The correct use of DOT labeling (per the MSDS) on all containers, trucks, and rail cars will enable emergency responders to react properly and quickly to any disaster thereby reducing the potential risk to the environment and to personnel.

Storage of diethyl phthalate should be segregated from incompatible chemicals to minimize the risk of cross-contamination or contact.

Diethyl phthalate is destroyed by burning in a chemical incinerator equipped with an afterburner and scrubber.

If a spill or leak to the environment has occurred, fire department, emergency response and/or hazardous materials spill personnel should be notified immediately. Cleanup should be attempted only by those trained in proper spill containment procedures. Contaminated soils should be removed for incineration and replaced with clean soil. If diethyl phthalate should contact the water table, aquifer, or navigable waterway, remediation activities should be prompt to ensure protection of aquatic sediments. When such spills occur, the local and/or state emergency response authorities must be notified. A comprehensive emergency response or disaster preparedness/recovery plan should be in place prior to any operations involving the use, transportation, storage, or disposal of diethyl phthalate. If diethyl phthalate is spilled or leaked, the following specific steps are recommended:

- ☑ Restrict persons not wearing protective clothing from area of spill or leak until cleanup is complete and area can be opened for normal work.
- ☑ Ventilate area of spill or leak.
- ☑ Remove all ignition sources.
- ☑ Absorb small spills with paper towel or other absorbent material and dispose of as hazardous waste. For larger spills, collect liquid using vermiculite, dry earth, or sand to absorb and place in a sealed drum for disposal.
- ☑ It may be necessary to dispose of diethyl phthalate as a hazardous waste. The responsible state agency or the regional office of the federal Environmental Protection Agency (EPA) should be contacted for specific recommendations.

RISK ASSESSMENT: BUSINESS

General Assessment

Accidents or mishaps involving diethyl phthalate can present a significant threat to business operations. The loss or damage of equipment or facilities can significantly affect fiscal viability. Lawsuits that may result from personnel injury/death, public exposures, and/or environmental contamination will also require a serious expenditure of resources. Media attention surrounding an injury, death, or environmental damage can also result in a loss of profits and loss of current as well as future business.

Recommended Risk-Reduction Measures

Company attorneys, safety and health professionals, and environmental specialists should be involved in the development of any procedures dealing with safety or emergency response. A company official should be pre-designated as a public relations officer with specific training in dealing with the press. Corporate plans and policies should be developed, approved, and implemented long before any need for such arises.

REFERENCES

Ahlbom, A. 1993. *Biostatistics for Epidemiologists.* New York: CRC Press/Lewis Publishers.

American Conference of Governmental Industrial Hygienists. 1988. *Documentation of the Threshold Limit Values and Biological Exposure Indices,* 5th Edition (with updates). Cincinnati: ACGIH

Anthony, C. P., and N. J. Kolthoff. 1971. *Textbook of Anatomy and Physiology.* St. Louis: C. V. Mosby Company.

Asante-Duah, D. K. 1993. *Hazardous Waste Risk Assessment.* New York: CRC Press/Lewis Publishers.

Bloom, A. D. and F. J. DeSerres. 1995. *Ecotoxicity and Human Health.* New York: CRC Press/Lewis Publishers.

Bueche, J. R. 1972. *Principles of Physics,* 2nd Edition. New York: McGraw-Hill.

Burrell, R., D. K. Flaherty, and L. J. Sauers. 1992. *Toxicology of the Immune System.* New York: Van Nostrand Reinhold.

Calabrese, E. J., E. M. Kenyon. *Air Toxics and Risk Assessment.* New York: CRC Press/Lewis Publishers.

Cockerman, L. G., B. S. Shane. 1994. *Basic Environmental Toxicology.* New York: CRC Press/Lewis Publishers.

Cothern, C. R. 1995. *Handbook for Environmental Risk Decisionmaking.* New York: CRC Press/Lewis Publishers.

Gardner, R. W. 1994. *Chemical Intolerance.* New York: CRC Press/Lewis Publishers.

Hodgson, E., R. B. Mailman, and Janice E. Chambers. 1988. *Dictionary of Toxicology.* London: The MacMillian Company.

Howard, P. H. 1995. *Handbook of Environmental Fate and Exposures Data for Organic Chemicals,* Vol. I-IV. New York: CRC Press/Lewis Publishers.

Keith, L. H. and M. M. Walker. 1995. *Handbook of Air Toxics.* New York: CRC Press/Lewis Publishers.

Landis, W. G., M. H. Yu. 1995. *Introduction to Environmental Toxicology.* New York: CRC Press/Lewis Publishers.

Langerman, N. 1994 *Precautionary Labels for Chemical Containers.* New York: CRC Press/Lewis Publishers.

Lauwerys, R. R., P. Hoet. 1993. *Industrial Chemical Exposure.* New York: CRC Press/Lewis Publishers.

Lide, D. R. 1995. *Handbook of Chemistry and Physics,* 76th Edition. New York: CRC Press/Lewis Publishers.

Lewis, R. J., Sr. 1993. *Hawley's Condensed Chemical Dictionary,* 12th Edition. New York: Van Nostrand Reinhold.

Lewis, R. J., Sr. 1992. *Sax's Dangerous Properties of Industrial Materials,* Volumes 1, 2, and 3, 8th Edition. New York: Van Nostrand Reinhold.

Manahan, S. E. 1992. *Toxicological Chemistry,* 2nd Edition. New York: CRC Press/Lewis Publishers.

Mellström, G. A., J. E. Wahlberg, and H. I. Maibach. 1994. *Protective Gloves for Occupational Use.* New York: CRC Press/Lewis Publishers.

Meyer, E. 1990. *Chemistry of Hazardous Materials.* Englewood Cliffs, NJ: Prentice Hall.

Mycock, J. C., J. D. McKenna, and L. Theodore. 1995. *Handbook of Air Pollution Control Engineering and Technology.* New York: CRC Press/Lewis Publishers.

Nelson, G. O. 1992. *Gas Mixtures.* New York: CRC Press/Lewis Publishers.

Ott, W. R. 1995. *Environmental Statistics and Data Analysis.* New York: CRC Press/Lewis Publishers.

Patnaik, P. 1992. *A Comprehensive Guide to the Hazardous Properties of Chemical Substances.* New York: Van Nostrand Reinhold.

Philp, R. B. 1995. *Environmental Hazards and Human Health.* New York: CRC Press/Lewis Publishers.

Quigley, D. R. 1994. *Handbook of Emergency Chemical Management.* New York: CRC Press/Lewis Publishers.

Rea, W. J. 1995. *Chemical Sensitivity,* Vol. I-III. New York: CRC Press/Lewis Publishers.

Richardson, M. 1995. *Dictionary of Substance and Their Effects,* Vol. I-VI. New York: CRC Press/Lewis Publishers.

Rylander, R., R. R. Jacobs. 1994. *Organic Dusts.* New York: CRC Press/Lewis Publishers.

Sacarello, H. L. A. 1994. *The Comprehensive Handbook of Hazardous Materials.* New York: CRC Press/Lewis Publishers.

U.S. Department of Health and Human Services, National Institute for Occupational Safety and Health. 1994. *NIOSH Pocket Guide to Chemical Hazards.* Washington, D.C.: U.S. Government Printing Office.

MATERIAL SAFETY DATA SHEET

CHEMICAL NAME	CRC PRESS/LEWIS PUBLISHERS DISCLAIMER:
DIETHYL SULFATE	THE INFORMATION AND RECOMMENDATIONS PRESENTED HEREIN ARE BASED ON SOURCES BELIEVED TO BE RELIABLE. CRC MAKES NO REPRESENTATION ON ITS COMPLETENESS OR ACCURACY. IT IS THE RESPONSIBILITY OF THE USER TO DETERMINE THE CHEMICAL'S SUITABILITY FOR ITS INTENDED USE, THE CHEMICAL'S SAFE USE, AND THE CHEMICAL'S PROPER DISPOSAL. NO REPRESENTATIONS AND/OR WARRANTIES, EITHER EXPRESSED OR IMPLIED, OF THE MERCHANTABILITY OR FITNESS FOR A PARTICULAR PURPOSE, OR OF ANY OTHER NATURE, ARE MADE WITH RESPECT TO THE INFORMATION PROVIDED IN THIS MSDS OR TO THE CHEMICAL TO WHICH INFORMATION MAY REFER. CRC NEITHER ASSUMES NOR AUTHORIZES ANY OTHER PERSON TO ASSUME FOR IT, ANY OTHER ADDITIONAL RESPONSIBILITY OR LIABILITY FOR THE USE OF, OR RELIANCE UPON, THIS INFORMATION.

HAZARD WARNING INFORMATION

HEALTH	FIRE	REACTIVE	OTHER	DEGREE OF HAZARD	COLOR CODING	OTHER CODES
3	1	1		0 = Minimum Hazard 1 = Slight Hazard 2 = Moderate Hazard 3 = Serious Hazard 4 = Severe Hazard	HEALTH = BLUE FIRE = RED REACTIVITY = YELLOW OTHER = WHITE	OX = Oxidizer ACID = Acid ALK = Alkali COR = Corrosive W = Use No Water

SECTION I - GENERAL INFORMATION

Characterization	**Sulfate**	RCRA Number	**None**
		EPA Class	**Not Applicable**
DOT Proper Shipping Name	**Diethyl Sulfate**	Chemical Abstract Service (CAS) Number	**64-67-5**
DOT Hazard Class and Label Requirements	**Poison B**	DOT Emergency Guide Code	**15**
DOT Identification Number	**UN 1594**	Chemical Formula	$(CH_3CH_2)_2SO_4$

Synonyms
Diethyl monosulfate; diethyl sulphate; DS; ethyl sulfate; sulfuric acid diethyl ester.

SECTION II - HAZARDOUS INGREDIENTS AND IDENTITY INFORMATION

Hazardous Components (specific identity)	OSHA Exposure Criteria	NIOSH Exposure Criteria	Immediately Dangerous to Life and Health (IDLH)	ACGIH Exposure Criteria
Diethyl sulfate (derivation: From ethanol and fuming sulfuric acid or directly from ethanol and ethylhydrosulfuric acid).	**PEL:** Not Established **STEL:** Not Established	**REL:** Not Established **STEL:** Not Established	**Not Determined**	**TLV:** Not Established **STEL:** Not Established

SECTION III - PHYSICAL AND CHEMICAL CHARACTERISTICS

Boiling Point	**205°F (96°C)**	Specific Gravity (H_2O = 1)	**1.17**
Vapor Pressure (mm Hg)	**0.19 at 77°F (25°C)**	Molecular Weight	**154.2**
Vapor Density (Air = 1)	**5.31**	Melting Point	**-25°F (-13°C)**

Solubility
Insoluble in water (decomposes to form sulfuric acid. Rapidly decomposed by hot water into mono-ethyl sulfate and alcohol. Soluble in alcohol and ether.

Appearance and Odor
Colorless, oily liquid, darkens with age. Has a faint, ethereal or peppermint-like odor.

SECTION IV - FIRE AND EXPLOSION HAZARD DATA

Flash Point (method used)	**220°F (104°C) closed cup**	Explosive Limits in Air % by Volume	**LEL: 4.1% UEL: Not Determined**
NFPA Classification	**Class IIIB Combustible Liquid**	Autoignition Temperature	**817°F (436°C)**

Extinguishing Media
Use water spray, dry chemical, carbon dioxide, or foam. Water may cause frothing.

Special Fire Fighting Procedures
Acrid and irritating fumes and smoke are produced in fire, wear full protective clothing and self-contained breathing apparatus (SCBA). Vapors are heavier than air and may travel for some distance to ignition source and flashback. Fight fire from distance if possible. Move containers from fire area if it can be done safely. Exercise extreme caution when moving containers.

Unusual Fire and Explosion Hazards
Containers may explode in fire due to buildup of excessive internal pressures. Decomposes to ethyl ether in fire which is more flammable than diethyl sulfate.

SECTION V - REACTIVITY DATA

Stability		Conditions to Avoid Under normal conditions of temperature and pressure, diethyl sulfate is stable. Avoid contact with incompatible materials. Keep away from heat and flame.
Stable X	Unstable	Incompatibility *(materials to avoid)* Strong oxidizers, such as chlorine, bromine, fluorine, and peroxides; strong acids (especially concentrated nitric acid), moisture and water, aqueous alkaline solutions, and iron.
Hazardous Polymerization		Conditions to Avoid Under normal conditions of temperature and pressure, hazardous polymerization of diethyl sulfate is not expected to occur.
May Occur	Will Not Occur X	Hazardous Decomposition or By-products Acrid and irritated gases and fumes are produced when diethyl sulfate is involved in fire. These include carbon oxides, ethylene and sulfur oxides, and ethyl ether.

SECTION VI - HEALTH HAZARD DATA

Primary Route(s) of Entry:	Inhalation?	Absorption (skin)?	Ingestion?
	X	X	X

Health Hazards

INHALATION: Symptoms include coughing, wheezing, inflammation of the upper airway, throat, and upper respiratory tract. There may be headache, nausea, and vomiting. May cause also cause pulmonary edema (a fluid buildup in the lungs) which can be fatal.

SKIN & EYES: Contact may irritate and burn the skin and eyes. May pass through unbroken skin to cause toxic systemic effects.

INGESTION: Moderately toxic. Nausea, vomiting, abdominal pain, and collapse.

Carcinogenicity	NTP Listed?	IARC Cancer Review Group?	OSHA Regulated?	Target Organs?
Suspected Human Confirmed Animal	5th Annual Report	Group 2A	No	Respiratory system, eyes, skin.

Medical Conditions Generally Aggravated by Exposure

Chronic respiratory disease.

Emergency and First-aid Procedures

Eye contact: Flush immediately with water for 15 minutes (minimum), seek medical attention. Skin contact: Remove all contaminated clothing. Immediately wash area with large amounts of soap and water. Seek medical assistance. For inhalation: Remove the person from exposure. Provide respiratory assistance and CPR. Transfer to medical facility. Observe 24 - 48 hours for lung effects (pulmonary edema). If swallowed: Call the poison control center and seek medical attention immediately. Unless advised otherwise, do NOT induce vomiting due to the corrosive nature of diethyl sulfate.

SECTION VII - PRECAUTIONS FOR SAFE HANDLING AND USE

Steps to be Taken in Case Material is Released or Spilled

Absorb liquids in vermiculite or other absorbent and deposit in DOT-approved drum. Do not allow flow into sewers or waterways. Ventilate area and remove ignition sources. Restrict those not involved in cleanup from entering area. Notify appropriate authorities, if applicable.

Preferred Waste Disposal Method

No Citation.

Precautions to be Taken in Handling and Storage

Do not store diethyl sulfate in the presence of incompatible chemicals or materials. Store in tightly closed containers in a cool, dry, well-ventilated area away from ignition sources. Do not allow moisture to accumulate inside metal containers (hydrogen gas may build up and cause explosion).

Other Precautions and Warnings

Bulk storage should be avoided. Outside or external facility storage is recommended.

SECTION VIII - CONTROL MEASURES AND PERSONAL PROTECTIVE EQUIPMENT

Respiratory Protection *(specify type)*

No exposure levels have been established. This does not mean exposure is without risk. For the best protection, use a supplied-air respirator with full facepiece operated in pressure demand mode or a self-contained breathing apparatus (SCBA) with full facepiece operated in pressure demand mode.

Ventilation

Local exhaust or general mechanical systems recommended.

Protective Gloves	Eye Protection	Other Protective Clothing
Rubber	Chemical Goggles or Face Mask	Apron

Work/Hygiene Practices

Always wash hands thoroughly after using chemical; never bring food, drink, or smoking materials into vicinity of chemicals.

DIETHYL SULFATE

$(CH_3CH_2)_2SO_4$ CAS: 64-67-5

IDENTIFICATION AND TYPICAL USES

Diethyl sulfate is a colorless, oily liquid that darkens with age. It has a faint, ethereal or peppermint-like odor. It is primarily used as an ethylating agent in organic synthesis, as an alkylating agent, an accelerator in sulfation of ethylene, a chemical intermediate for ethyl derivatives of phenols, amines, and thiols, and in some sulfonations.

RISK ASSESSMENT: HEALTH

General Assessment

Diethyl sulfate is a confirmed animal carcinogen and is highly suspected to cause cancer in humans as well. It has been shown to cause reproductive problems in animals. It is a human poison by *inhalation, ingestion,* and skin *absorption.* Mutation data have also been recorded.

Inhalation results in headache, coughing, wheezing, inflammation of the upper respiratory system, nausea, and vomiting. Exposure may also result in severe chemical pneumonitis or pulmonary edema (fluid buildup in the lungs), which is a medical emergency and can lead to death. Symptoms may be delayed up to 48 hours creating a false sense of security with regard to exposure risk. Spasm of the glottis may occur causing death by suffocation.

Ingestion causes nausea, vomiting, abdominal pain, and collapse. Gastritis and other systemic effects are also possible.

Skin contact will cause burns and irritation. Absorption through unbroken skin is likely resulting in toxic systemic effects. Eye contact with the vapor causes irritation. The liquid will produce burns to the tissue on contact and subsequent inflammation of the surrounding area. If not removed immediately, there is a chance that permanent damage (up to and including blindness) may occur.

Acute Health Effects

The following acute (short-term) health effects may occur immediately or shortly after exposure to diethyl sulfate:

Skin: Irritation, burns, inflammation, and possible absorption causing toxic systemic effects.

Eye: Severe irritation with burns, redness, lachrymation, and possible conjunctivitis.

Lung: Irritation of the mucosa of the nose, throat, and upper respiratory tract. Can cause pulmonary edema (fluid buildup in the lungs) which can be fatal.

Other: Ingestion causes nausea, vomiting, abdominal pain, and possible collapse.

Chronic Health Effects

The following chronic (long-term) health effects may occur at some time after exposure to diethyl sulfate and can last for months or even years:

Cancer Hazards: According to the information presented in the references, diethyl sulfate has been shown to cause cancer in test animals and its carcinogenicity in humans is highly suspect.

Reproductive Hazard: According to the information presented in the references, diethyl sulfate may have teratogenic effects in test animals. It has caused cancer of the nervous system in the offspring of test animals.

Other Chronic Effects: There is some indication that long-term exposure or frequent high exposures may cause changes in cell structure (mutations) and damage the lungs.

Recommended Risk-Reduction Measures

Unless a less toxic chemical can be substituted for a hazardous substance, *engineering controls* are the most

effective method of reducing exposures. The best protection is to enclose operations and/or provide local exhaust ventilation at the site of chemical release. While not always operationally feasible, isolating operations can also reduce exposure.

Using respiratory protection is less effective than the controls mentioned above, but is still advisable whenever working with or around diethyl sulfate. No exposure levels have been established for this compound. However, this does not mean there is no health hazard risk. For the best protection, a self-contained breathing apparatus (SCBA) with full facepiece and pressure demand or a supplied-air respirator operated in pressure demand or continuous flow mode are the recommended respiratory protection methods of choice. If a full facepiece is not available, then chemical goggles should be worn to protect the eyes. Whenever a chemical splash hazard exists, a face shield and rubber apron should be worn. To prevent hand and skin exposures, rubber gloves should be worn.

Administrative controls should also be in place to minimize the potential for human exposures. These may include written procedures or policies which specify the methods and techniques that will be practiced whenever personnel are to work with diethyl sulfate.

All personnel should receive *training* on the use, hazards, protective measures, emergency actions, and other precautions per 29 CFR 1910.1200 (Hazard Communication), prior to the first assignment in an area where diethyl sulfate is used or stored.

If symptoms develop or overexposure is suspected, the following medical tests may be useful:

☑ Lung function test.

☑ Consider chest X-ray after acute overexposure (may be negative if done within the first 24 to 48 hours following exposure due to delayed onset of pulmonary edema).

Any evaluation should include a careful medical history of past and present symptoms with an examination. However, medical tests that evaluate existing damage are not a substitute for controlling exposure.

Also, since smoking can cause heart disease, emphysema, and other respiratory disorders, exposure to diethyl sulfate can have a more pronounced effect on smokers than on non-smokers exposed to the same levels. Prudent risk assessment and management requires consideration of *all* possible risk factors when exposure symptoms appear in the workforce.

Other methods to reduce exposure include:

☑ Where possible, enclose operations and use local exhaust ventilation at the site of chemical release. If local exhaust ventilation or enclosure is not used, respiratory protection should be mandatory. Also, automatic transfer of liquids from storage containers to process containers is recommended.

☑ Always ensure that proper protective clothing is worn when using chemical substances.

☑ Wash thoroughly immediately after exposure to diethyl sulfate and at the end of the work shift or before eating, drinking, or smoking.

☑ Hazard warning information should be posted in the work area. In addition, as part of an on-going education and training program, all information on the health and safety hazards of diethyl sulfate should be communicated to all potentially exposed workers.

☑ Eye wash stations should be provided in work areas. If the potential for whole body exposure exists, then safety showers should also be provided.

RISK ASSESSMENT: ENVIRONMENT

General Assessment

The environment is at risk of exposure during use, transportation, storage, disposal, or destruction of diethyl sulfate. In almost every scenario, the threat of environmental exposure is contingent upon the proper handling of the chemical substance. Accidental spills, large or small, can result in fire, explosion, and possible contamination of the surrounding environmental mediums (water, soil, and air).

Diethyl sulfate is considered a Class IIIB combustible liquid (per OSHA 29 CFR 1910.106). These characteristics require special consideration during any emergency situation involving a leak or spill of diethyl sulfate. Should diethyl sulfate ever come into contact with incompatible substances such as strong oxidizers (chlorine, bromine, fluorine, chlorine dioxide, and peroxides), strong acids (sulfuric acid, hydrochloric acid, and nitric acid), either during use, transportation, or storage, violent reaction can occur. Diethyl sulfate is especially reactive in contact with nitric acid. Diethyl sulfate may also react with moisture to yield sulfuric acid and subsequently react with metal to liberate flammable hydrogen gas.

Diethyl sulfate can enter the environment through industrial discharges and from spills.

Acute Ecological Effects

Acute (short-term) toxic effects may include the death of animals, birds, or fish and death or low growth rate in plants. Acute effects are seen 2 to 4 days after animals or plants are exposed to diethyl sulfate. This chemical is not expected to have acute toxicity to aquatic life. Insufficient data are available to evaluate or predict the short-term effects of diethyl sulfate to plants, birds, or land animals.

Chronic Ecological Effects

Chronic toxic effects may include shortened life span, reproductive problems, lower fertility, and changes in appearance or behavior in exposed animals. These effects can be seen long after first exposure(s) to toxic chemicals. Diethyl sulfate is not expected to have chronic toxicity to aquatic life. Insufficient data are available to evaluate or predict the long-term effects of diethyl sulfate to plants, birds, or land animals.

Water Solubility

Diethyl sulfate is known to rapidly hydrolyze in water to decompose and form sulfuric acid. The rate of hydrolysis will increase in both acidic and basic waters.

Persistence in the Environment

Diethyl sulfate is probably slightly persistent in the aquatic environment. It is not expected to adsorb to sediment and suspended organic matter. Approximately 93% will eventually end up in the water.

Bioaccumulation in Aquatic Organisms

Some substances increase in concentration, or *bioaccumulate,* in living organisms as they breathe contaminated air, drink contaminated water, or eat contaminated food. These chemicals can become concentrated in the tissues and internal organs of animals as well as humans.

Diethyl sulfate is not expected to accumulate in fish tissue or aquatic organisms.

Recommended Risk-Reduction Measures

Proper training of all transporters will reduce the likelihood of a mishap or accident resulting in a leak or spill to the environment. The correct use of DOT labeling (per the MSDS) on all containers, trucks, and rail cars will enable emergency responders to react properly and quickly to any disaster thereby reducing the potential risk to the environment and to personnel.

Storage of diethyl sulfate should be segregated from incompatible chemicals to minimize the risk of cross-contamination or contact. A blanket purge of inert gas inside containers is recommended to keep moisture from condensing since hydrogen gas may be formed presenting a dangerous explosion hazard.

If a spill or leak to the environment has occurred, fire department, emergency response and/or hazardous materials spill personnel should be notified immediately. Cleanup should be attempted only by those trained in proper spill containment procedures. Contaminated soils should be removed for incineration and replaced with clean soil. If diethyl sulfate should contact the water table, aquifer, or navigable waterway, remediation activities should be prompt. When such spills occur, the local and/or state emergency response authorities must be notified. A comprehensive emergency response or disaster preparedness/recovery plan should be in place prior to any operations involving the use, transportation, storage, or disposal of diethyl sulfate. If diethyl sulfate is spilled or leaked, the following specific steps are recommended:

- ☑ Restrict persons not wearing protective clothing from area of spill or leak until cleanup is complete and area can be opened for normal work.
- ☑ Ventilate area of spill or leak.
- ☑ Remove all ignition sources.
- ☑ Collect liquid using vermiculite, dry earth, or sand to absorb and place in a sealed drum for disposal.
- ☑ It may be necessary to dispose of diethyl sulfate as a hazardous waste. The responsible state agency or the regional office of the federal Environmental Protection Agency (EPA) should be contacted for specific recommendations.

RISK ASSESSMENT: BUSINESS

General Assessment

Accidents or mishaps involving diethyl sulfate can present a significant threat to business operations. The loss or damage of equipment or facilities can significantly affect fiscal viability. Lawsuits that may result from personnel injury/death, public exposures, and/or environmental contamination will also require a serious expenditure of resources. Media attention surrounding an injury, death, or environmental damage can also result in a loss of profits and loss of current as well as future business. Always remember that anytime the terms "cancer," "carcinogen," or "reproductive hazard" are used, public emotion, hysteria, and ignorance can run equally high. This should be properly considered during the development of public relations policies.

Recommended Risk-Reduction Measures

Company attorneys, safety and health professionals, and environmental specialists should be involved in the development of any procedures dealing with safety or emergency response. A company official should be pre-designated as a public relations officer with specific training in dealing with the press. Corporate plans and policies should be developed, approved, and implemented long before any need for such arises.

REFERENCES

Ahlbom, A. 1993. *Biostatistics for Epidemiologists.* New York: CRC Press/Lewis Publishers.

American Conference of Governmental Industrial Hygienists. 1988. *Documentation of the Threshold Limit Values and Biological Exposure Indices,* 5th Edition (with updates). Cincinnati: ACGIH

Anthony, C. P., and N. J. Kolthoff. 1971. *Textbook of Anatomy and Physiology.* St. Louis: C. V. Mosby Company.

Asante-Duah, D. K. 1993. *Hazardous Waste Risk Assessment.* New York: CRC Press/Lewis Publishers.

Bloom, A. D. and F. J. DeSerres. 1995. *Ecotoxicity and Human Health.* New York: CRC Press/Lewis Publishers.

Brusick, D. J. 1994. *Methods for Genetic Risk Assessment.* New York: CRC Press/Lewis Publishers.

Bueche, J. R. 1972. *Principles of Physics,* 2nd Edition. New York: McGraw-Hill.

Burrell, R., D. K. Flaherty, and L. J. Sauers. 1992. *Toxicology of the Immune System.* New York: Van Nostrand Reinhold.

Calabrese, E. J., E. M. Kenyon. *Air Toxics and Risk Assessment.* New York: CRC Press/Lewis Publishers.

Cockerman, L. G., B. S. Shane. 1994. *Basic Environmental Toxicology.* New York: CRC Press/Lewis Publishers.

Cothern, C. R. 1995. *Handbook for Environmental Risk Decisionmaking.* New York: CRC Press/Lewis Publishers.

Gardner, R. W. 1994. *Chemical Intolerance.* New York: CRC Press/Lewis Publishers.

Hodgson, E., R. B. Mailman, and Janice E. Chambers. 1988. *Dictionary of Toxicology.* London: The MacMillian Company.

Howard, P. H. 1995. *Handbook of Environmental Fate and Exposures Data for Organic Chemicals,* Vol. I-IV. New York: CRC Press/Lewis Publishers.

Kamrin, M. A. 1988. *Toxicology: A Primer on Principles and Applications.* Chelsea, Michigan: Lewis Publishers.

Keith, L. H. and M. M. Walker. 1995. *Handbook of Air Toxics.* New York: CRC Press/Lewis Publishers.

Landis, W. G., M. H. Yu. 1995. *Introduction to Environmental Toxicology.* New York: CRC Press/Lewis Publishers.

Langerman, N. 1994 *Precautionary Labels for Chemical Containers.* New York: CRC Press/Lewis Publishers.

Lauwerys, R. R., P. Hoet. 1993. *Industrial Chemical Exposure.* New York: CRC Press/Lewis Publishers.

Lide, D. R. 1995. *Handbook of Chemistry and Physics,* 76th Edition. New York: CRC Press/Lewis Publishers.

Lewis, R. J., Sr. 1993. *Hawley's Condensed Chemical Dictionary,* 12th Edition. New York: Van Nostrand Reinhold.

Lewis, R. J., Sr. 1992. *Sax's Dangerous Properties of Industrial Materials,* Volumes 1, 2, and 3, 8th Edition. New York: Van Nostrand Reinhold.

Manahan, S. E. 1992. *Toxicological Chemistry,* 2nd Edition. New York: CRC Press/Lewis Publishers.

Mellström, G. A., J. E. Wahlberg, and H. I. Maibach. 1994. *Protective Gloves for Occupational Use.* New York: CRC Press/Lewis Publishers.

Meyer, E. 1990. *Chemistry of Hazardous Materials.* Englewood Cliffs, NJ: Prentice Hall.

Mycock, J. C., J. D. McKenna, and L. Theodore. 1995. *Handbook of Air Pollution Control Engineering and Technology.* New York: CRC Press/Lewis Publishers.

Nelson, G. O. 1992. *Gas Mixtures.* New York: CRC Press/Lewis Publishers.

Ott, W. R. 1995. *Environmental Statistics and Data Analysis.* New York: CRC Press/Lewis Publishers.

Patnaik, P. 1992. *A Comprehensive Guide to the Hazardous Properties of Chemical Substances.* New York: Van Nostrand Reinhold.

Philp, R. B. 1995. *Environmental Hazards and Human Health.* New York: CRC Press/Lewis Publishers.

Quigley, D. R. 1994. *Handbook of Emergency Chemical Management.* New York: CRC Press/Lewis Publishers.

Rea, W. J. 1995. *Chemical Sensitivity,* Vol. I-III. New York: CRC Press/Lewis Publishers.

Richardson, M. 1995. *Dictionary of Substance and Their Effects,* Vol. I-VI. New York: CRC Press/Lewis Publishers.

Rylander, R., R. R. Jacobs. 1994. *Organic Dusts.* New York: CRC Press/Lewis Publishers.

Sacarello, H. L. A. 1994. *The Comprehensive Handbook of Hazardous Materials.* New York: CRC Press/Lewis Publishers.

Sheftel, V. O. 1995. *Handbook of Toxic Properties of Monomers and Additives.* New York: CRC Press/Lewis Publishers.

U.S. Department of Health and Human Services, National Institute for Occupational Safety and Health. 1994. *NIOSH Pocket Guide to Chemical Hazards.* Washington, D.C.: U.S. Government Printing Office.

MATERIAL SAFETY DATA SHEET

CHEMICAL NAME	
DIFLUORODIBROMO-METHANE	**CRC PRESS/LEWIS PUBLISHERS DISCLAIMER:** THE INFORMATION AND RECOMMENDATIONS PRESENTED HEREIN ARE BASED ON SOURCES BELIEVED TO BE RELIABLE. CRC MAKES NO REPRESENTATION ON ITS COMPLETENESS OR ACCURACY. IT IS THE RESPONSIBILITY OF THE USER TO DETERMINE THE CHEMICAL'S SUITABILITY FOR ITS INTENDED USE, THE CHEMICAL'S SAFE USE, AND THE CHEMICAL'S PROPER DISPOSAL. NO REPRESENTATIONS AND/OR WARRANTIES, EITHER EXPRESSED OR IMPLIED, OF THE MERCHANTABILITY OR FITNESS FOR A PARTICULAR PURPOSE, OR OF ANY OTHER NATURE, ARE MADE WITH RESPECT TO THE INFORMATION PROVIDED IN THIS MSDS OR TO THE CHEMICAL TO WHICH INFORMATION MAY REFER. CRC NEITHER ASSUMES NOR AUTHORIZES ANY OTHER PERSON TO ASSUME FOR IT, ANY OTHER ADDITIONAL RESPONSIBILITY OR LIABILITY FOR THE USE OF, OR RELIANCE UPON, THIS INFORMATION.

HAZARD WARNING INFORMATION

HEALTH	FIRE	REACTIVE	OTHER	DEGREE OF HAZARD	COLOR CODING	OTHER CODES
3	0	0		0 = Minimum Hazard 1 = Slight Hazard 2 = Moderate Hazard 3 = Serious Hazard 4 = Severe Hazard	HEALTH = BLUE FIRE = RED REACTIVITY = YELLOW OTHER = WHITE	OX = Oxidizer ACID = Acid ALK = Alkali COR = Corrosive W = Use No Water

SECTION I - GENERAL INFORMATION

Characterization	RCRA Number	EPA Class
Halogenated Hydrocarbon	**None**	**Not Applicable**

DOT Proper Shipping Name **Dibromodifluoromethane**	Chemical Abstract Service (CAS) Number **75-61-6**
DOT Hazard Class and Label Requirements **ORM-A; Label: None**	DOT Emergency Guide Code **58**
DOT Identification Number **UN 1941**	Chemical Formula **CBr_2F_2**

Synonyms
Dibromondifluoromethane; Freon® 12-2B; Halon® 1202.

SECTION II - HAZARDOUS INGREDIENTS AND IDENTITY INFORMATION

Hazardous Components (specific identity)	OSHA Exposure Criteria	NIOSH Exposure Criteria	Immediately Dangerous to Life and Health (IDLH)	ACGIH Exposure Criteria
Difluorodibromomethane (derivation: By vapor phase bromination of difluoromethane). **1 ppm = 8.72 mg/m^3**	**PEL (8-hour):** 100 ppm 860 mg/m^3 **STEL:** Not Established	**REL (10 hour):** 100 ppm 860 mg/m^3 **STEL:** Not Established	**2000 ppm**	**TLV (8-hour):** 1000 ppm 4950 mg/m^3 **STEL:** Not Established

SECTION III - PHYSICAL AND CHEMICAL CHARACTERISTICS

Boiling Point **76°F (24°C)**	Specific Gravity (H_2O = 1) **2.29**
Vapor Pressure (mm Hg) **620 at 69°F (20°C)**	Molecular Weight **209.83**
Vapor Density (Air = 1) **7.2**	Freezing Point **-231°F (-146°C)**

Solubility
Insoluble in water. Soluble in ether, acetone, benzene, and methanol.

Appearance and Odor
Colorless heavy liquid or gas (above 76°F) with a characteristic odor.

SECTION IV - FIRE AND EXPLOSION HAZARD DATA

Flash Point (method used) **None Reported**	Explosive Limits in Air % by Volume **LEL: Not Determined UEL: Not Determined**
NFPA Classification **Non Combustible Liquid; Non Flammable Gas**	Autoignition Temperature **Not Determined**

Extinguishing Media
Use extinguishing media suitable for surrounding fire. Difluorodibromomethane does not burn.

Special Fire Fighting Procedures
Poisonous and extremely irritating gases are produced in fire, wear full protective clothing and self-contained breathing apparatus (SCBA). Move containers from fire area if it can be done safely. Isolate area and deny entry to those not involved in fire fighting effort. Keep out of low areas and stay upwind. Be aware of runoff from fire control methods. Do not release to sewers or waterways.

Unusual Fire and Explosion Hazards
Containers may explode in fire due to buildup of excessive internal pressures. High vapor density and volatility pose potential for serious physical hazards.

SECTION V - REACTIVITY DATA

Stability		Conditions to Avoid: Difluorodibromomethane is stable at room temperature and pressure. Avoid exposures to excessive heat and contact with incompatible materials.
Stable	Unstable	Incompatibility *(materials to avoid)*: Difluorodibromomethane will react in the presence of chemically active metals, such as potassium, sodium, calcium, powdered aluminum, magnesium, or zinc.
X		
Hazardous Polymerization		Conditions to Avoid: Under normal conditions of temperature and pressure, hazardous polymerization of difluorodibromomethane cannot occur.
May Occur	Will Not Occur	Hazardous Decomposition or By-products: Toxic and irritating fumes and gases are produced when difluorodibromomethane is involved in fire, including bromine and fluorine.
	X	

SECTION VI - HEALTH HAZARD DATA

Primary Route(s) of Entry:	Inhalation?	Absorption (skin)?	Ingestion?
	X		X

Health Hazards

INHALATION: Low-level toxicity. May cause excitement, headache, and drowsiness. Also, possible sensitization of the heart's muscle cells to chemicals produced in the body (such as adrenaline), causing potentially fatal disturbances in the heart rhythm. Low-level exposures can cause irritation of the nose, eyes, or throat, palpitations, and lightheadedness. Pulmonary edema is also possible, with bronchial constriction and lung irritation. Damage to the liver may occur on prolonged exposures.

SKIN & EYES: Eye irritation at high concentrations. Contact dermatitis may also occur. Compressed gas may cause frostbite.

INGESTION: Ingestion not likely in the workplace.

Carcinogenicity	NTP Listed?	IARC Cancer Review Group?	OSHA Regulated?	Target Organs?
No Citation (Human) No Citation (Animal)	No	No	29 CFR 1910.1000 Table Z-1	Respiratory system, CNS, liver, and CVS.

Medical Conditions Generally Aggravated by Exposure

Chronic respiratory disorders.

Emergency and First-aid Procedures

Eye contact: Flush immediately with water for 15 minutes (minimum), seek medical attention. Skin contact: For frostbite, rapidly re-warm using water at 107°F (42°C) until flush returns. Do NOT use dry heat! Seek medical assistance. For inhalation: Remove the person from exposure. Provide respiratory assistance and CPR. Transfer to medical facility. Observe for 24 - 48 hours for lung effects. If swallowed: Not likely. If large amount is ingested, consider emesis and/or activated charcoal solution.

SECTION VII - PRECAUTIONS FOR SAFE HANDLING AND USE

Steps to be Taken in Case Material is Released or Spilled

Evacuate area, deny entry, stay upwind. If it can be done without risk, shut off leak. Vapor can accumulate in low areas. If possible, stop leak. If not, remove leaking cylinder to an open or safe area (fume hood) and allow to vent. Position so gas, not liquid, escapes.

Preferred Waste Disposal Method

No Citation.

Precautions to be Taken in Handling and Storage

Do not store difluorodibromomethane in the presence of incompatible materials. Store in tightly closed containers in a cool, well-ventilated area away from heat and the sun's rays.

Other Precautions and Warnings

Prevent physical damage to storage cylinders.

SECTION VIII - CONTROL MEASURES AND PERSONAL PROTECTIVE EQUIPMENT

Respiratory Protection *(specify type)*

A NIOSH-approved respirator will suffice for exposures to the PEL. Best protection is provided by a supplied-air respirator with full facepiece operated in pressure demand mode or a self-contained breathing apparatus with full facepiece operated in pressure demand or other positive pressure mode.

Ventilation

Local exhaust or general mechanical systems recommended.

Protective Gloves	Eye Protection	Other Protective Clothing
Impervious, Insulated Rubber	Chemical Goggles or Face Mask	Apron

Work/Hygiene Practices

Always wash hands thoroughly after using chemical; never bring food, drink, or smoking materials into vicinity of chemicals.

DIFLUORODIBROMOMETHANE

CBr_2F_2

CAS: 75-61-6

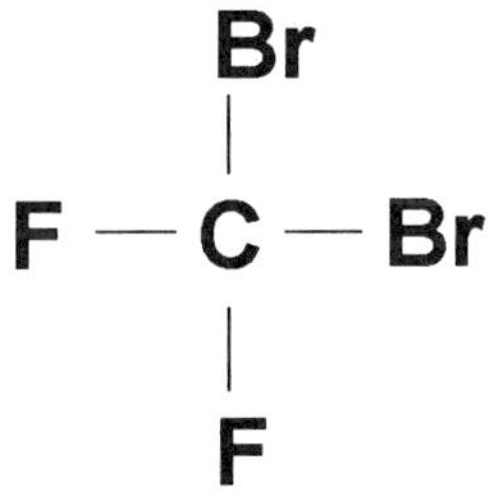

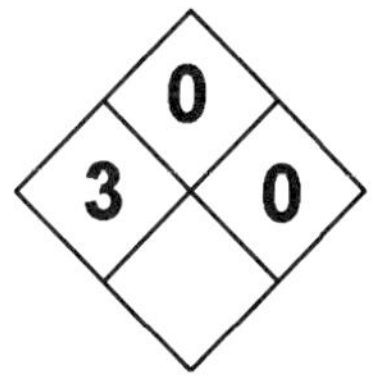

IDENTIFICATION AND TYPICAL USES

Difluorodibromomethane is colorless, heavy liquid or gas (above 76°F) with a characteristic odor. It is used as a fire extinguishing agent, in the synthesis of dyes, in the manufacture of pharmaceuticals, and as an intermediate in quaternary ammonium compounds.

RISK ASSESSMENT: HEALTH

General Assessment

Difluorodibromomethane is considered to be moderately toxic by *inhalation*. Skin contact also presents some level of health risk, although absorption is not likely to occur. Ingestion is also not likely in the industrial setting. Fluorinated hydrocarbons present their greatest toxicity when inhaled at high concentrations. This occurs most often as a result of intentional abuse rather than occupational exposure. There is no evidence in the references regarding the real or potential carcinogenicity, mutagenicity, or teratogenicity of this chemical in humans or test animals.

Inhalation can cause transient irritation of the eyes, nose, and throat. Can cause fluid to build up in the lungs (pulmonary edema), which is a medical emergency and can be fatal. Symptoms of cough, congestion, and tightness in chest may be delayed up to 48 hours, thereby creating a false sense of security with regard to heath risk. There may also be palpitations of the heart and lightheadedness. It may also act on the central nervous system as a depressant, although the references are not clear on the precise action. At high concentrations, muscle cells of the heart may become sensitized to essential chemicals normally produced by the body, such as adrenaline. This can lead to potentially fatal disturbances in the natural rhythm of the heart. Long-term exposures can also adversely affect the liver and cause damage.

Skin contact with the liquid can cause frostbite and irritation. It will defat the skin and may result in chronic contact dermatitis and erythema.

Acute Health Effects

The following acute (short-term) health effects may occur immediately or shortly after exposure to difluorodibromomethane:

Skin: Irritation, contact dermatitis, and possible frostbite (liquid contact).

Eye: Mild irritation. Liquid will burn tissue.

Lung: Irritation of the eyes, nose, and throat. Possible fatal disturbance in heart rhythm (palpitations), and lightheadedness. May cause pulmonary edema (fluid in the lungs), which can be fatal. Can cause lung irritation and bronchial constriction.

Other: Difluorodibromomethane is a narcotic at high concentrations with unspecified effect on the CNS.

Chronic Health Effects

The following chronic (long-term) health effects may occur at some time after exposure to dichlorodifluoromethane and can last for months or even years:

Cancer Hazards: According to the information presented in the references, difluorodibromomethane has not been adequately tested for its ability to cause cancer in test animals.

Reproductive Hazard: According to the information presented in the references, difluorodibromomethane has not been adequately tested for its ability to adversely affect reproduction in test animals.

Other Chronic Effects: Low, chronic exposures may lead to liver damage and possible adverse effect on the central nervous system. Chronic exposures can also lead to heart palpitations and lightheadedness.

Recommended Risk-Reduction Measures

While difluorodibromomethane is considered to be of a low-order toxicity by inhalation, it can still pose some degree of health hazard risk. Unless a less toxic chemical can be substituted for a hazardous substance, *engineering controls* are the most effective method of reducing exposures. The best protection is to enclose operations and/or provide local exhaust ventilation at the site of chemical release. While not always operationally feasible, isolating operations can also reduce exposure.

Using respiratory protection is less effective than the controls mentioned above, but is still advisable whenever working with or around dichlorodifluoromethane. For exposures to the PEL (100 ppm), a full facepiece respirator may suffice. However, when high exposure levels are predicted or prolonged exposures are anticipated, a supplied air respirator or a self-contained breathing apparatus (SCBA) with full facepiece and pressure demand are the recommended respiratory protection of choice. If a full facepiece is not available, then chemical goggles should be worn to protect the eyes. Whenever a chemical splash hazard exists, a face shield and rubber apron should be worn. To prevent hand and skin exposures, impervious and insulated rubber gloves should be worn.

Administrative controls should also be in place to minimize the potential for human exposures. These may include written procedures or policies which specify the methods and techniques that will be practiced whenever personnel are to work with difluorodibromomethane.

All personnel should receive *training* on the use, hazards, protective measures, emergency actions, and other precautions per 29 CFR 1910.1200 (Hazard Communication), prior to the first assignment in an area where difluorodibromomethane is used or stored.

Before beginning employment and at regular intervals thereafter (e.g., annually), the following medical protocol is suggested for those assigned to work with difluorodibromomethane:

- ☑ Electrocardiogram (EKG), to establish baseline and test for pre-existing heart rhythm abnormalities.

If symptoms develop or overexposure is suspected, the following medical tests may be useful:

- ☑ Electrocardiogram (EKG), to compare to baseline and test for heart rhythm disorders.
- ☑ Liver function tests.
- ☑ Lung function tests.
- ☑ Consider chest X-ray following acute overexposure (may be negative if taken within first 24 to 48 hours of exposure).

Any evaluation should include a careful medical history of past and present symptoms with an examination. However, medical tests that evaluate existing damage are not a substitute for controlling exposure. Also, since smoking can cause heart disease, emphysema, and other serious respiratory disorders, symptoms of exposure may be more pronounced in smokers than in non smokers under the same conditions. Prudent risk management requires proper consideration of *all* possible factors which may be causing the appearance of exposure symptoms among workers.

Other methods to reduce exposure include:

- ☑ Where possible, enclose operations and use local exhaust ventilation at the site of chemical release. If local exhaust ventilation or enclosure is not used, respiratory protection should be mandatory.
- ☑ Always ensure that proper protective clothing is worn when using chemical substances and that personnel are trained in its use and care.
- ☑ Wash thoroughly immediately after exposure to difluorodibromomethane and at the end of the work shift or before eating, drinking, or smoking.
- ☑ Hazard warning information should be posted in the work area. In addition, as part of an on-going education and training program, all information on the health and safety hazards of difluorodibromomethane should be communicated to all potentially exposed workers.
- ☑ Clothes that may be contaminated with chemical substances should never be worn home (family members may be exposed). Clothes should be removed at work and laundered only by personnel who have been trained on the hazards associated with the particular chemical contaminant.
- ☑ Eye wash stations should be provided in work areas. If the potential for whole body exposure exists, then safety showers should also be provided.

RISK ASSESSMENT: ENVIRONMENT

General Assessment

The environment is at risk of exposure during use, transportation, storage, disposal, or destruction of difluorodibromomethane. In almost every scenario, the threat of environmental exposure is contingent upon the proper handling of the chemical substance. Accidental spills, large or small, can result in fire, explosion, and possible contamination of the surrounding environmental mediums (water, soil, and air).

Difluorodibromomethane is considered a nonflammable liquid (or gas). However, it is incompatible with a number of common, chemically active metals and caution is required to ensure contact is prevented. It is also a chlorofluorocarbon (CFC) which is a group of halogenated hydrocarbons alleged to cause damage to the earth's protective ozone layer. Therefore, accidental release to the environment must not occur. These characteristics require special consideration during any emergency situation involving a leak or spill of difluorodibromomethane. Should difluorodibromomethane ever come into contact with incompatible substances either during use, transportation, or storage, violent and explosive reactions can occur.

Difluorodibromomethane can exist as a gas or liquid. It is usually shipped as a liquefied compressed gas. It can enter the environment through industrial discharges and spills.

Acute Ecological Effects

Acute (short-term) toxic effects may include the death of animals, birds, or fish and death or low growth rate in plants. Acute effects are seen 2 to 4 days after animals or plants are exposed to difluorodibromomethane.

This chemical is not expected to cause adverse short-term effects on aquatic life, plants, birds, or terrestrial animals.

Chronic Ecological Effects

Chronic toxic effects may include shortened life span, reproductive problems, lower fertility, and changes in appearance or behavior in exposed animals. These effects can be seen long after first exposure(s) to toxic chemicals.

Insufficient data are available to evaluate or predict the long-term effects of difluorodibromomethane to aquatic life, plants, birds, or land animals.

Water Solubility

Difluorodibromomethane is insoluble in water. It is a gas above 76°F. As a liquid, it will not mix with water.

Persistence in the Environment

Difluorodibromomethane is virtually non-persistent in water, with a half-life of less than 2 days. The half-life of a pollutant is the amount of time it takes for one half of the chemical to be degraded. Approximately 100% of difluorodibromomethane will eventually end up in air.

Bioaccumulation in Aquatic Organisms

Some substances increase in concentration, or *bioaccumulate,* in living organisms as they breathe contaminated air, drink contaminated water, or eat contaminated food. These chemicals can become concentrated in the tissues and internal organs of animals as well as humans.

The concentration of difluorodibromomethane found in fish tissues is expected to be much lower than the average concentration of difluorodibromomethane in the water from which the fish was taken.

Recommended Risk-Reduction Measures

Proper training of all transporters will reduce the likelihood of a mishap or accident resulting in a leak or spill to the environment. The correct use of DOT labeling (per the MSDS) on all containers, trucks, and rail cars will enable emergency responders to react properly and quickly to any disaster thereby reducing the potential risk to the environment and to personnel.

Storage of difluorodibromomethane should be segregated from incompatible chemicals to minimize the risk of cross-contamination or contact. Containers must be protected from shock, excessive heat (including the sun's rays), and damage.

If a spill or leak to the environment has occurred, fire department, emergency response and/or hazardous materials spill personnel should be notified immediately. Cleanup should be attempted only by those trained in proper spill containment procedures. If any liquid has spilled, contaminated soils may need to be removed for incineration and replaced with clean soil. If difluorodibromomethane should contact the water table, aquifer, or navigable waterway, remediation activities should be prompt. It is not soluble in water and total containment may be possible. When such spills occur, the local and/or state emergency response authorities must be notified. A comprehensive emergency response or disaster preparedness/recovery plan should be in place prior to any operations involving the use, transportation, storage, or disposal of difluorodibromomethane. If difluorodibromomethane is spilled or leaked, the following specific steps are recommended:

- ☑ Evacuate area and deny entry until cleanup is completed. Restrict persons not wearing protective clothing from hazard area.
- ☑ Ventilate area of spill or leak.
- ☑ If source of leak is a cylinder and it cannot be repaired in place, remove to safe location (e.g., outdoors).
- ☑ Absorb liquid spills using vermiculite, dry earth, or sand to absorb and place in a sealed drum for disposal.
- ☑ It may be necessary to dispose of dichlorodifluoromethane as a hazardous waste. The responsible state agency or the regional office of the federal Environmental Protection Agency (EPA) should be contacted for specific recommendations.

RISK ASSESSMENT: BUSINESS

General Assessment

Accidents or mishaps involving difluorodibromomethane can present a significant threat to business operations. The loss or damage of equipment or facilities can significantly affect fiscal viability. Lawsuits that may result from personnel injury/death, public exposures, and/or environmental contamination will also require a serious expenditure of resources. Media attention surrounding an injury, death, or environmental damage can also result in a loss of profits and loss of current as well as future business.

Recommended Risk-Reduction Measures

Company attorneys, safety and health professionals, and environmental specialists should be involved in the development of any procedures. A company official should be pre-designated as a public relations officer with specific training in dealing with the press. Corporate plans and policies should be developed, approved, and implemented long before any need for such arises.

REFERENCES

American Conference of Governmental Industrial Hygienists. 1988. *Documentation of the Threshold Limit Values and Biological Exposure Indices,* 5th Edition (with updates). Cincinnati: ACGIH

Anthony, C. P., and N. J. Kolthoff. 1971. *Textbook of Anatomy and Physiology.* St. Louis: C. V. Mosby Company.

Bueche, J. R. 1972. *Principles of Physics,* 2nd Edition. New York: McGraw-Hill.

Burrell, R., D. K. Flaherty, and L. J. Sauers. 1992. *Toxicology of the Immune System.* New York: Van Nostrand Reinhold.

Calabrese, E. J. 1994. *Biological Effects of Low Level Exposures.* New York: CRC Press/Lewis Publishers.

Cothern, C. R. 1995. *Handbook for Environmental Risk Decisionmaking.* New York: CRC Press/Lewis Publishers.

Hodgson, E., R. B. Mailman, and Janice E. Chambers. 1988. *Dictionary of Toxicology.* London: The MacMillian Company.

Jones, F. E. 1994. *Toxic Organic Vapors in the Workplace.* New York: CRC Press/Lewis Publishers.

Keith, L. H. and M. M. Walker. 1995. *Handbook of Air Toxics.* New York: CRC Press/Lewis Publishers.

Langerman, N. 1994 *Precautionary Labels for Chemical Containers.* New York: CRC Press/Lewis Publishers.

Lauwerys, R. R., P. Hoet. 1993. *Industrial Chemical Exposure.* New York: CRC Press/Lewis Publishers.

Lewis, R. J., Sr. 1993. *Hawley's Condensed Chemical Dictionary,* 12th Edition. New York: Van Nostrand Reinhold.

Lewis, R. J., Sr. 1992. *Sax's Dangerous Properties of Industrial Materials,* Volumes 1, 2, and 3, 8th Edition. New York: Van Nostrand Reinhold.

Manahan, S. E. 1992. *Toxicological Chemistry,* 2nd Edition. New York: CRC Press/Lewis Publishers.

Meyer, E. 1990. *Chemistry of Hazardous Materials.* Englewood Cliffs, NJ: Prentice Hall.

Nelson, G. O. 1992. *Gas Mixtures.* New York: CRC Press/Lewis Publishers.

Patnaik, P. 1992. *A Comprehensive Guide to the Hazardous Properties of Chemical Substances.* New York: Van Nostrand Reinhold.

Quigley, D. R. 1994. *Handbook of Emergency Chemical Management.* New York: CRC Press/Lewis Publishers.

Richardson, M. 1995. *Dictionary of Substance and Their Effects,* Vol. I-VI. New York: CRC Press/Lewis Publishers.

Sacarello, H. L. A. 1994. *The Comprehensive Handbook of Hazardous Materials.* New York: CRC Press/Lewis Publishers.

U.S. Department of Health and Human Services, National Institute for Occupational Safety and Health. 1994. *NIOSH Pocket Guide to Chemical Hazards.* Washington, D.C.: U.S. Government Printing Office.

MATERIAL SAFETY DATA SHEET

CHEMICAL NAME	
DIFLUOROETHANE	**CRC PRESS/LEWIS PUBLISHERS DISCLAIMER:** THE INFORMATION AND RECOMMENDATIONS PRESENTED HEREIN ARE BASED ON SOURCES BELIEVED TO BE RELIABLE. CRC MAKES NO REPRESENTATION ON ITS COMPLETENESS OR ACCURACY. IT IS THE RESPONSIBILITY OF THE USER TO DETERMINE THE CHEMICAL'S SUITABILITY FOR ITS INTENDED USE, THE CHEMICAL'S SAFE USE, AND THE CHEMICAL'S PROPER DISPOSAL. NO REPRESENTATIONS AND/OR WARRANTIES, EITHER EXPRESSED OR IMPLIED, OF THE MERCHANTABILITY OR FITNESS FOR A PARTICULAR PURPOSE, OR OF ANY OTHER NATURE, ARE MADE WITH RESPECT TO THE INFORMATION PROVIDED IN THIS MSDS OR TO THE CHEMICAL TO WHICH INFORMATION MAY REFER. CRC NEITHER ASSUMES NOR AUTHORIZES ANY OTHER PERSON TO ASSUME FOR IT, ANY OTHER ADDITIONAL RESPONSIBILITY OR LIABILITY FOR THE USE OF, OR RELIANCE UPON, THIS INFORMATION.

HAZARD WARNING INFORMATION

HEALTH	FIRE	REACTIVE	OTHER	DEGREE OF HAZARD	COLOR CODING	OTHER CODES
1	4	0		0 = Minimum Hazard 1 = Slight Hazard 2 = Moderate Hazard 3 = Serious Hazard 4 = Severe Hazard	HEALTH = BLUE FIRE = RED REACTIVITY = YELLOW OTHER = WHITE	OX = Oxidizer ACID = Acid ALK = Alkali COR = Corrosive W = Use No Water

SECTION I - GENERAL INFORMATION

Characterization	RCRA Number	EPA Class
Halogenated Hydrocarbon	**D001**	**Characteristic Waste (I)**

DOT Proper Shipping Name **Difluoroethane, R152a**	Chemical Abstract Service (CAS) Number **75-37-6**
DOT Hazard Class and Label Requirements **Flammable Gas**	DOT Emergency Guide Code **18**
DOT Identification Number **UN 1030**	Chemical Formula $C_2H_4F_2$

Synonyms

Algofrene type 67; Freon® 152a; Genetron® 152a; Halocarbon 152a; 1,1-difluoroethane; ethylene fluoride; ethylene difluoride; ethylidiene fluoride.

SECTION II - HAZARDOUS INGREDIENTS AND IDENTITY INFORMATION

Hazardous Components (specific identity)	OSHA Exposure Criteria	NIOSH Exposure Criteria	Immediately Dangerous to Life and Health (IDLH)	ACGIH Exposure Criteria
Difluoroethane (derivation: By adding hydrogen fluoride to acetylene).	**PEL:** Not Established **STEL:** Not Established	**REL:** Not Established **STEL:** Not Established	**Not Determined**	**TLV:** Not Established **STEL:** Not Established

SECTION III - PHYSICAL AND CHEMICAL CHARACTERISTICS

Boiling Point **-12°F (-25°C)**	Specific Gravity (H_2O = 1) **0.95**
Vapor Pressure (mm Hg) **4437 at 77°F (25°C)**	Molecular Weight **66.06**
Vapor Density (Air = 1) **2.27**	Freezing Point **-179°F (-117°C)**

Solubility

Insoluble in water.

Appearance and Odor

Colorless and odorless gas.

SECTION IV - FIRE AND EXPLOSION HAZARD DATA

Flash Point (method used) **Flammable Gas**	Explosive Limits in Air % by Volume **LEL: 3.7% UEL: 18%**
NFPA Classification **Flammable Gas**	Autoignition Temperature **Flammable Gas**

Extinguishing Media

Use dry chemical or carbon dioxide on small fires. For larger fires, use water spray of fog.

Special Fire Fighting Procedures

Poisonous and extremely irritating gases are produced in fire, wear full protective clothing and self-contained breathing apparatus (SCBA). Do not attempt to fight fire unless flow of gas can first be stopped. Apply cooling water to sides of containers until well after fire is out.

Unusual Fire and Explosion Hazards

Containers may explode in fire due to buildup of excessive internal pressures. BLEVE (boiling liquid expanding vapor explosion) is possible. Evacuate area if rising sound is heard from venting device or there is a noticeable discoloration of the storage tank or vessel.

SECTION V - REACTIVITY DATA

Stability		Conditions to Avoid **Difluoroethane is stable at room temperature and pressure. Avoid exposures to excessive heat, ignition sources, and contact with incompatible materials.**
Stable **X**	Unstable	Incompatibility *(materials to avoid)* **Difluoroethane will react in the presence of strong oxidizers (such as chlorine, bromine , and fluorine), and powdered or heated aluminum and other light, divalent metals (will form reactive materials).**
Hazardous Polymerization		Conditions to Avoid **Under normal conditions of temperature and pressure, hazardous polymerization of difluoroethane cannot occur.**
May Occur	Will Not Occur **X**	Hazardous Decomposition or By-products **Toxic and irritating fumes and gases are produced when difluoroethane is involved in fire, including extremely toxic hydrogen fluoride (hydrofluoric acid) and carbonyl fluoride.**

SECTION VI - HEALTH HAZARD DATA

Primary Route(s) of Entry:	Inhalation?	Absorption (skin)?	Ingestion?
	X		

Health Hazards

INHALATION: Can act as a simple asphyxiant and may produce narcosis at high concentrations. May also cause irritation of the respiratory tract, especially at high temperatures. Death can occur as a result of pulmonary edema (fluid in the lungs), oxygen displacement (asphyxiation), or ventricular fibrillation (heart muscle fluttering).

SKIN & EYES: Direct contact with the eyes or skin can cause frostbite.

INGESTION: Ingestion not likely since difluoroethane is a gas at room temperatures.

Carcinogenicity	NTP Listed?	IARC Cancer Review Group?	OSHA Regulated?	Target Organs?
No Citation (Human) No Citation (Animal)	No	No	No	Respiratory system, skin, eyes, and CVS.

Medical Conditions Generally Aggravated by Exposure

None reported.

Emergency and First-aid Procedures

Eye contact: Flush immediately with tepid water for 15 minutes (minimum), seek medical attention. Skin contact: For frostbite, rapidly re-warm using water at 107°F (42°C) until flush returns, Do NOT use dry heat! Do NOT rub affected area. Seek medical assistance. For inhalation: Remove the person from exposure. Provide respiratory assistance and CPR. Transfer to medical facility. Observe for 24 - 48 hours for lung effects. If swallowed: Not likely.

SECTION VII - PRECAUTIONS FOR SAFE HANDLING AND USE

Steps to be Taken in Case Material is Released or Spilled

Evacuate area, deny entry, stay upwind. If it can be done without risk, shut off leak. If possible, stop gas flow and remove all ignition sources. If leak from cylinder cannot be stopped, remove to safe area (outdoors) and allow to vent until empty. Keep area clear of personnel.

Preferred Waste Disposal Method

No Citation.

Precautions to be Taken in Handling and Storage

Avoid inhaling vapors. Do not store difluoroethane in the presence of incompatible materials. Store in tightly closed containers in a cool, well-ventilated area away from heat, flame, ignition sources, and even the sun's rays.

Other Precautions and Warnings

Prevent physical damage to storage cylinders.

SECTION VIII - CONTROL MEASURES AND PERSONAL PROTECTIVE EQUIPMENT

Respiratory Protection *(specify type)*

No exposure levels have been established and this compound is an asphyxiant. For best protection, use a supplied-air respirator with full facepiece operated in pressure demand mode or a self-contained breathing apparatus with full facepiece operated in pressure demand or other positive pressure mode.

Ventilation

Local exhaust or general mechanical systems recommended.

Protective Gloves	Eye Protection	Other Protective Clothing
Impervious, Insulated Rubber	Chemical Goggles or Face Mask	Apron

Work/Hygiene Practices

Always wash hands thoroughly after using chemical; never bring food, drink, or smoking materials into vicinity of chemicals.

DIFLUOROETHANE

$C_2H_4F_2$ CAS: 75-37-6

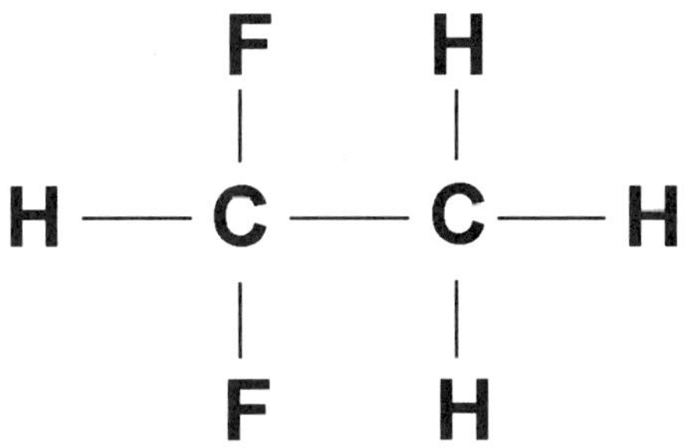

IDENTIFICATION AND TYPICAL USES

Difluoroethane is acolorless and odorless gas. It is used primarily as a chemical intermediate, a refrigerant, and as an aerosol propellant.

RISK ASSESSMENT: HEALTH

General Assessment

Difluoroethane is considered mildly toxic by *inhalation*. Skin and eye contact with the compressed gas or the super-cooled liquid also presents some level of health risk, although absorption is not likely to occur. Ingestion is also not likely since this compound is normally a gas at room temperatures. There is no evidence in the references regarding the real or potential carcinogenicity or teratogenicity of this chemical in humans or test animals. There is some evidence to indicate that exposure to difluoroethane may cause mutations (genetic changes) in animals. There are no human data in this area of study.

Inhalation can cause asphyxiation by displacing oxygen content in the cells. This may produce narcosis at high concentrations with numbness and tingling in the extremities, headache, loss of orientation, discomfort, loss of coordination, ataxia, loss of consciousness, and death. Inhalation may also cause respiratory tract irritation, especially at high temperatures due to the formation of irritating hydrogen chloride gas. Death can occur as a result of laryngeal spasm or edema (fluid buildup), pulmonary edema (fluid in the lungs), oxygen displacement (asphyxiation), or ventricular fibrillation (a fluttering of the heart muscle causing a loss of the necessary synchronization between pulse and heartbeat). It should be noted that symptoms of pulmonary edema (congestion, labored breathing, cough with phlegm, nausea) may be delayed up to 48 hours following exposure. This can create a false sense of security in terms of health risk.

Skin and eye contact can result in frostbite and subsequent damage to tissues.

Acute Health Effects

The following acute (short-term) health effects may occur immediately or shortly after exposure to difluoroethane:

Skin: Irritation and possible frostbite (liquid or gas contact).

Eye: Mild to severe irritation. Liquid or gas bursts will burn tissue (frostbite on contact).

Lung: Irritation of the eyes, nose, and throat. Possible fatal disturbance in heartbeat synchronization (ventricular fibrillation), and lightheadedness. May cause asphyxiation or pulmonary edema (fluid in the lungs), both of which can be fatal.

Other: A narcotic at high concentrations with unspecified effect on the CNS.

Chronic Health Effects

The following chronic (long-term) health effects may occur at some time after exposure to dichlorodifluoromethane and can last for months or even years:

Cancer Hazards: According to the information presented in the references, difluoroethane has not been adequately tested for its ability to cause cancer in test animals. Additional research is required.

Reproductive Hazard: According to the information presented in the references, difluoroethane has not been adequately tested for its ability to adversely affect reproduction in test animals.

❶ *Recommended Risk-Reduction Measures*

Unless a less toxic chemical can be substituted for a hazardous substance, *engineering controls* are the most effective method of reducing exposures. The best protection is to enclose operations and/or provide local exhaust ventilation at the site of chemical release. While not always operationally feasible, isolating operations can also reduce exposure.

Using respiratory protection is less effective than the controls mentioned above, but is still advisable whenever working with or around dichlorodifluoromethane. No exposure levels have yet been established for difluoroethane. However, this does not mean that it presents no risks to human health. The best protection is provided using a supplied air respirator or a self-contained breathing apparatus (SCBA) with full facepiece and pressure demand. If a full facepiece is not available, then chemical goggles should be worn to protect the eyes. Whenever a chemical splash hazard exists, a face shield and rubber apron should be worn. To prevent hand and skin exposures, impervious and insulated rubber gloves should be worn.

Administrative controls should also be in place to minimize the potential for human exposures. These may include written procedures or policies which specify the methods and techniques that will be practiced whenever personnel are to work with difluoroethane.

All personnel should receive *training* on the use, hazards, protective measures, emergency actions, and other precautions per 29 CFR 1910.1200 (Hazard Communication), prior to the first assignment in an area where difluoroethane is used or stored.

Before beginning employment and at regular intervals thereafter (e.g., annually), the following medical protocol is suggested for those assigned to work with difluoroethane:

☑ Electrocardiogram (EKG), to establish baseline and test for pre-existing heart rhythm abnormalities.

If symptoms develop or overexposure is suspected, the following medical tests may be useful:

☑ Electrocardiogram (EKG), to compare to baseline and test for heart rhythm disorders.

☑ Lung function tests.

☑ Consider chest X-ray following acute overexposure (may be negative if taken within first 24 to 48 hours of exposure).

Any evaluation should include a careful medical history of past and present symptoms with an examination. However, medical tests that evaluate existing damage are not a substitute for controlling exposure. Also, since smoking can cause heart disease, emphysema, and other serious respiratory disorders, symptoms of exposure may be more pronounced in smokers than in non smokers under the same conditions. Prudent risk management requires proper consideration of *all* possible factors which may be causing the appearance of exposure symptoms among workers.

Other methods to reduce exposure include:

☑ Where possible, enclose operations and use local exhaust ventilation at the site of chemical release. If local exhaust ventilation or enclosure is not used, respiratory protection should be mandatory.

☑ Always ensure that proper protective clothing is worn when using chemical substances.

☑ Wash thoroughly immediately after exposure to difluoroethane and at the end of the work shift or before eating, drinking, or smoking.

☑ Hazard warning information should be posted in the work area. In addition, as part of an on-going education and training program, all information on the health and safety hazards of difluoroethane should be communicated to all exposed and potentially exposed workers.

☑ Eye wash stations should be provided in work areas. If the potential for whole body exposure exists, then safety showers should also be provided.

RISK ASSESSMENT: ENVIRONMENT

General Assessment

The environment is at risk of exposure during use, transportation, storage, disposal, or destruction of difluoroethane. In almost every scenario, the threat of environmental exposure is contingent upon the proper handling of the chemical substance. Accidental spills, large or small, can result in fire, explosion, and possible contamination of the surrounding environmental mediums (water, soil, and air).

Difluoroethane is considered a flammable gas. It presents an extremely dangerous fire and explosion hazard. It is incompatible with a number of common, chemically active metals and caution is required to ensure contact is prevented. It is also incompatible with commonly encountered oxidizers, such as chlorine and bromine. These characteristics require special consideration during any emergency situation involving a leak or spill of difluoroethane. Should difluoroethane ever come into contact with incompatible substances either during use, transportation, or storage, violent and explosive reactions can occur.

Although difluoroethane can exist as a gas or liquid, it is normally present as a gas at room temperature. It can enter the environment through industrial discharges and leaks.

Acute Ecological Effects

Acute (short-term) toxic effects may include the death of animals, birds, or fish and death or low growth rate in plants. Acute effects are seen 2 to 4 days after animals or plants are exposed to difluoroethane. This chemical is not expected to cause adverse short-term effects on aquatic life, plants, birds, or terrestrial animals.

Chronic Ecological Effects

Chronic toxic effects may include shortened life span, reproductive problems, lower fertility, and changes in appearance or behavior in exposed animals. These effects can be seen long after first exposure(s) to toxic chemicals. Insufficient data are available to evaluate or predict the long-term effects of difluoroethane to aquatic life, plants, birds, or land animals.

Water Solubility

Difluoroethane is insoluble in water. It is normally present as a gas above -178°F.

Persistence in the Environment

Difluoroethane is virtually non-persistent in water, with a half-life of less than 2 days. The half-life of a pollutant is the amount of time it takes for one half of the chemical to be degraded. Approximately 100% of difluoroethane will eventually end up in air where it exists entirely in the vapor phase. It will react with photochemical-produced hydroxly radicals with a half life of 472 days in the troposphere. May gradually diffuse to the stratosphere where it will slowly photolyze. Its half-life there is expected to be equal to 20 years. Its contribution to ozone depletion requires more study. However, as a halogenated fluorocarbon (HFC), it is believed to be less of a threat to the stratospheric ozone layer than that posed by the chlorofluorocarbons (CFC) and halogenated chlorofluorocarbons (HCFCs).

Bioaccumulation in Aquatic Organisms

Some substances increase in concentration, or *bioaccumulate,* in living organisms as they breathe contaminated air, drink contaminated water, or eat contaminated food. These chemicals can become concentrated in the tissues and internal organs of animals as well as humans.

Bioaccumulation of difluoroethane is not expected to occur in aquatic organisms.

Recommended Risk-Reduction Measures

Proper training of all transporters will reduce the likelihood of a mishap or accident resulting in a leak or spill to the environment. The correct use of DOT labeling (per the MSDS) on all containers, trucks, and rail cars will enable emergency responders to react properly and quickly to any disaster thereby reducing the potential risk to the environment and to personnel.

Storage of difluoroethane should be segregated from incompatible chemicals to minimize the risk of cross-contamination or contact. Containers must be protected from shock, excessive heat (including the sun's rays), and damage.

If a spill or leak to the environment has occurred, fire department, emergency response and/or hazardous materials spill personnel should be notified immediately. Cleanup should be attempted only by those trained in proper spill containment procedures. If any liquid has spilled, it will rapidly volatilize although some may leach through soil to groundwater. Contaminated soils may need to be removed for incineration and replaced with clean soil. If difluoroethane should contact the water table, aquifer, or navigable waterway, remediation activities should be prompt. When such spills occur, the local and/or state emergency response authorities must be notified. A comprehensive emergency response or disaster preparedness/recovery plan should be in place prior to any operations involving the use, transportation, storage, or disposal of difluoroethane. If difluoroethane is spilled or leaked, the following specific steps are recommended:

- ☑ Evacuate area and deny entry until cleanup is completed. Restrict persons not wearing protective clothing from hazard area.
- ☑ Ventilate area of spill or leak.

☑ Remove all sources of ignition.

☑ If source of leak is a cylinder and it cannot be repaired in place, remove to safe location (e.g., outdoors). Keep personnel clear.

☑ It may be necessary to dispose of dichlorodifluoromethane as a hazardous waste. The responsible state agency or the regional office of the federal Environmental Protection Agency (EPA) should be contacted for specific recommendations.

RISK ASSESSMENT: BUSINESS

General Assessment

Accidents or mishaps involving difluoroethane can present a significant threat to business operations. The loss or damage of equipment or facilities can significantly affect fiscal viability. Lawsuits that may result from personnel injury/death, public exposures, and/or environmental contamination will also require a serious expenditure of resources. Media attention surrounding an injury, death, or environmental damage can also result in a loss of profits and loss of current as well as future business.

Recommended Risk-Reduction Measures

Company attorneys, safety and health professionals, and environmental specialists should be involved in the development of any procedures dealing with safety or emergency response. A company official should be pre-designated as a public relations officer with specific training in dealing with the press. Corporate plans and policies should be developed, approved, and implemented long before any need for such arises.

REFERENCES

Ahlbom, A. 1993. *Biostatistics for Epidemiologists.* New York: CRC Press/Lewis Publishers.

American Conference of Governmental Industrial Hygienists. 1988. *Documentation of the Threshold Limit Values and Biological Exposure Indices,* 5th Edition (with updates). Cincinnati: ACGIH

Anthony, C. P., and N. J. Kolthoff. 1971. *Textbook of Anatomy and Physiology.* St. Louis: C. V. Mosby Company.

Asante-Duah, D. K. 1993. *Hazardous Waste Risk Assessment.* New York: CRC Press/Lewis Publishers.

Bloom, A. D. and F. J. DeSerres. 1995. *Ecotoxicity and Human Health.* New York: CRC Press/Lewis Publishers.

Brusick, D. J. 1994. *Methods for Genetic Risk Assessment.* New York: CRC Press/Lewis Publishers.

Bueche, J. R. 1972. *Principles of Physics,* 2nd Edition. New York: McGraw-Hill.

Burrell, R., D. K. Flaherty, and L. J. Sauers. 1992. *Toxicology of the Immune System.* New York: Van Nostrand Reinhold.

Calabrese, E. J., E. M. Kenyon. *Air Toxics and Risk Assessment.* New York: CRC Press/Lewis Publishers.

Calabrese, E. J. 1994. *Biological Effects of Low Level Exposures.* New York: CRC Press/Lewis Publishers.

Cockerman, L. G., B. S. Shane. 1994. *Basic Environmental Toxicology.* New York: CRC Press/Lewis Publishers.

Cothern, C. R., N. P. Ross. 1994. *Environmental Statistics, Assessment, and Forecasting.* New York: CRC Press/Lewis Publishers.

Cothern, C. R. 1995. *Handbook for Environmental Risk Decisionmaking.* New York: CRC Press/Lewis Publishers.

Hodgson, E., R. B. Mailman, and Janice E. Chambers. 1988. *Dictionary of Toxicology.* London: The MacMillian Company.

Howard, P. H., et. al. 1991. *Handbook of Environmental Degradation Rates.* New York: CRC Press/Lewis Publishers.

Hosty, J. W., and P. Foster. 1990. *A Practical Guide to Chemical Spill Response.* New York: Van Nostrand Reinhold.

Jones, F. E. 1994. *Toxic Organic Vapors in the Workplace.* New York: CRC Press/Lewis Publishers.

Keith, L. H. and M. M. Walker. 1995. *Handbook of Air Toxics.* New York: CRC Press/Lewis Publishers.

Langerman, N. 1994 *Precautionary Labels for Chemical Containers.* New York: CRC Press/Lewis Publishers.

Lauwerys, R. R., P. Hoet. 1993. *Industrial Chemical Exposure.* New York: CRC Press/Lewis Publishers.

Lewis, R. J., Sr. 1993. *Hawley's Condensed Chemical Dictionary,* 12th Edition. New York: Van Nostrand Reinhold.

Lewis, R. J., Sr. 1992. *Sax's Dangerous Properties of Industrial Materials,* Volumes 1, 2, and 3, 8th Edition. New York: Van Nostrand Reinhold.

Manahan, S. E. 1992. *Toxicological Chemistry,* 2nd Edition. New York: CRC Press/Lewis Publishers.

Meyer, E.. 1990. *Chemistry of Hazardous Materials.* Englewood Cliffs, NJ: Prentice Hall.

Nelson, G. O. 1992. *Gas Mixtures.* New York: CRC Press/Lewis Publishers.

Patnaik, P. 1992. *A Comprehensive Guide to the Hazardous Properties of Chemical Substances.* New York: Van Nostrand Reinhold.

Quigley, D. R. 1994. *Handbook of Emergency Chemical Management.* New York: CRC Press/Lewis Publishers.

Richardson, M. 1995. *Dictionary of Substance and Their Effects,* Vol. I-VI. New York: CRC Press/Lewis Publishers.

Sacarello, H. L. A. 1994. *The Comprehensive Handbook of Hazardous Materials.* New York: CRC Press/Lewis Publishers.

U.S. Department of Health and Human Services, National Institute for Occupational Safety and Health. 1994. *NIOSH Pocket Guide to Chemical Hazards.* Washington, D.C.: U.S. Government Printing Office.

MATERIAL SAFETY DATA SHEET

CHEMICAL NAME	
DIGLYCIDYL ETHER	**CRC PRESS/LEWIS PUBLISHERS DISCLAIMER:** THE INFORMATION AND RECOMMENDATIONS PRESENTED HEREIN ARE BASED ON SOURCES BELIEVED TO BE RELIABLE. CRC MAKES NO REPRESENTATION ON ITS COMPLETENESS OR ACCURACY. IT IS THE RESPONSIBILITY OF THE USER TO DETERMINE THE CHEMICAL'S SUITABILITY FOR ITS INTENDED USE, THE CHEMICAL'S SAFE USE, AND THE CHEMICAL'S PROPER DISPOSAL. NO REPRESENTATIONS AND/OR WARRANTIES, EITHER EXPRESSED OR IMPLIED, OF THE MERCHANTABILITY OR FITNESS FOR A PARTICULAR PURPOSE, OR OF ANY OTHER NATURE, ARE MADE WITH RESPECT TO THE INFORMATION PROVIDED IN THIS MSDS OR TO THE CHEMICAL TO WHICH INFORMATION MAY REFER. CRC NEITHER ASSUMES NOR AUTHORIZES ANY OTHER PERSON TO ASSUME FOR IT, ANY OTHER ADDITIONAL RESPONSIBILITY OR LIABILITY FOR THE USE OF, OR RELIANCE UPON, THIS INFORMATION.

HAZARD WARNING INFORMATION

HEALTH	FIRE	REACTIVE	OTHER	DEGREE OF HAZARD	COLOR CODING	OTHER CODES
3	2	1		0 = Minimum Hazard 1 = Slight Hazard 2 = Moderate Hazard 3 = Serious Hazard 4 = Severe Hazard	HEALTH = BLUE FIRE = RED REACTIVITY = YELLOW OTHER = WHITE	OX = Oxidizer ACID = Acid ALK = Alkali COR = Corrosive W = Use No Water

SECTION I - GENERAL INFORMATION

Characterization: **Ether**	RCRA Number: **None**	EPA Class: **Not Applicable**
DOT Proper Shipping Name: **Not Listed**	Chemical Abstract Service (CAS) Number: **2238-07-5**	
DOT Hazard Class and Label Requirements: **Not Listed**	DOT Emergency Guide Code: **No Citation**	
DOT Identification Number: **Not Listed**	Molecular Formula: **$C_6H_{10}O_3$**	

Synonyms

DGE; diallyl ether dioxide; di92,3-epoxypropyl) ether; 2-epoxypropyl ether; bis(2,3-epoxypropyl) ether; glycidyl ether; 2,2-(oxybis(methylene))bis-.

SECTION II - HAZARDOUS INGREDIENTS AND IDENTITY INFORMATION

Hazardous Components (specific identity)	OSHA Exposure Criteria	NIOSH Exposure Criteria	Immediately Dangerous to Life and Health (IDLH)	ACGIH Exposure Criteria
Diglycidyl ether: (derivation: Appears as a terminal group of epoxy resin structures resulting from the reaction of epichlorohydrin and bisphenol A with alkaline catalyst). 1 ppm = 5.41 mg/m^3	PEL (8-hour): 0.1 ppm 0.5 mg/m^3 CEILING: 0.5 ppm 2.8 mg/m^3	REL: 0.1 ppm 0.53 mg/m^3 CEILING (15 min) 1.0 mg/m^3 Possible Cancer Agent	10 ppm	TLV: 0.1 ppm 0.5 mg/m^3 STEL: Not Established

SECTION III - PHYSICAL AND CHEMICAL CHARACTERISTICS

Boiling Point: **500°F (260°C)**	Specific Gravity (H_2O = 1): **1.12**
Vapor Pressure (mm Hg): **0.09 at 77°F (25°C)**	Molecular Weight (atomic weight): **130.2**
Vapor Density (Air = 1): **3.78**	Melting Point: **Not Found**

Solubility

Solubility not established.

Appearance and Odor

Colorless, clear liquid with a strong, pungent, irritating odor. Odor Threshold = 5 ppm (approximately).

SECTION IV - FIRE AND EXPLOSION HAZARD DATA

Flash Point (method used): **147°F (64°C) closed cup**	Explosive Limits in Air % by Volume: **LEL: Not Determined** **UEL: Not Determined**
NFPA Classification: **Class IIIA Combustible Liquid**	Autoignition Temperature: **Not Determined**

Extinguishing Media

Dry chemical, regular foam, water spray, or carbon dioxide.

Special Fire Fighting Procedures

Acrid and irritating smoke and fumes are produced in fire. Structural fire-fighting protective clothing will not provide adequate protection. Wear full protective clothing and self-contained breathing apparatus (SCBA). Move container from fire area if it can be done without risk.

Unusual Fire and Explosion Hazards

Containers may explode in fire. Presents a vapor explosion hazard indoors, outdoors, and in sewers.

SECTION V - REACTIVITY DATA

Stability		Conditions to Avoid: Normally stable at room temperature. Avoid contact with incompatible materials since violent reactions can occur. Also avoid contact with heat and ignition sources.
Stable	Unstable	Incompatibility (materials to avoid)
X		Incompatible with strong oxidizers, such as chlorine, bromine, and chlorine dioxide.
Hazardous Polymerization		Conditions to Avoid: In the presence of curing agents, diglycidyl ether may polymerize at room temperature to form hard epoxy resins.
May Occur	Will Not Occur	Hazardous Decomposition or By-products
X		When heated to decomposition, diglycidyl ether emits acrid smoke and fumes.

SECTION VI - HEALTH HAZARD DATA

Primary Route(s) of Entry:	Inhalation?	Absorption (skin/eye)?	Ingestion?
	X	X	X

Health Hazards

INHALATION: Severe irritant. Overexposure not likely due to the extreme irritating properties of this chemical. Vapors can cause severe irritation of the eyes, nose, throat, and skin at very low concentrations. May cause CNS depression, ataxia, loss of coordination, coma, and liver and kidney damage. Pulmonary edema (fluid in lungs) is also possible.

ABSORPTION: A severe eye and skin irritant. Can pass through unbroken skin causing toxic systemic effects. Causes skin burns and eye irritation.

INGESTION: May cause gastrointestinal tract irritation.

Carcinogenicity	NTP Listed?	IARC Cancer Review Group?	OSHA Regulated?	Target Organs?
Questioned Human Questioned Animal	No	No	29 CFR 1910.1000 Table Z-1	Skin, eyes, respiratory system, repro. system, CNS.

Medical Conditions Generally Aggravated by Exposure

None reported. However, existing respiratory problems (bronchitis, asthma) may be aggravated.

Emergency and First-aid Procedures

Eye contact: Do not allow victim to keep eyes tightly shut. Flush immediately with water for 15 minutes (minimum), seek medical attention. Skin contact: Wash with large amounts of soap and water for a minimum of 15 minutes. For inhalation: Remove the person from exposure. Provide respiratory assistance and CPR. Transfer to medical facility. Observe for 24-48 hours for lung effects. If swallowed: Seek medical attention immediately. Give 1 to 2 glasses of milk or water. Do not induce vomiting or attempt to give an unconscious or convulsing person anything by mouth.

SECTION VII - PRECAUTIONS FOR SAFE HANDLING AND USE

Steps to be Taken in Case Material is Released or Spilled

Wear self-contained breathing apparatus (SCBA) for respiratory protection. Restrict those not wearing protective equipment and who are not involved in cleanup from entering area. Avoid skin contact. Ventilate area of spill. Absorb liquids in vermiculite or other absorbent and deposit in sealed drum.

Preferred Waste Disposal Method

No Citation.

Precautions to be Taken in Handling and Storage

Store in tightly closed containers in a cool, well-ventilated area away from heat. Protect containers from physical damage.

Other Precautions and Warnings

Diglycidyl ether must be stored to avoid contact with strong oxidizers since violent reactions can occur.

SECTION VIII - CONTROL MEASURES AND PERSONAL PROTECTIVE EQUIPMENT

Respiratory Protection (specify type)

NIOSH suggests that diglycidyl ether is a possible human carcinogen. Use a NIOSH/MSHA-approved self-contained breathing apparatus (SCBA) with full facepiece operated in positive pressure mode.

Ventilation

Local exhaust (preferred) at site of chemical work or general ventilation.

Protective Gloves	Eye Protection	Other Protective Clothing
Butyl Rubber (or other non-permeable)	Chemical Goggles and Face Mask	Protective Uniform or Apron

Work/Hygiene Practices

Always wash hands thoroughly after using chemical; never bring food, drink, or smoking materials into vicinity of chemicals.

DIGLYCIDYL ETHER

$C_6H_{10}O_3$ CAS: 2238-07-5

IDENTIFICATION AND TYPICAL USES

Diglycidyl ether appears as a colorless, clear liquid with a strong, pungent, irritating odor. It is used as a chemical intermediate, a stabilizer of chlorinated organic compounds, a textile-treating agent, and a reactive diluting agent for epoxy resins. It may also be a trace contaminant in epoxy derivatives of epychlorohydrin.

RISK ASSESSMENT: HEALTH

General Assessment

Diglycidyl ether is a possible human carcinogen (has cancer-causing properties), according to NIOSH, with experimental mutagenic data reported. It is poisonous by *ingestion, inhalation,* and by *absorption* through the skin. A number of toxic systemic effects will result from either route of exposure.

Exposure to its vapors can cause severe irritation to the eyes, nose, and respiratory tract. Exposure to as little as 10 ppm will cause instant irritation of the respiratory tract. Higher exposures can cause death within days as a result of lung damage, chemical pneumonitis, or possibly pulmonary edema. In addition, animal studies have shown that exposure to diglycidyl ether causes significant changes in blood cell elements (up to and including possible bone marrow depression), depression of the central nervous system, ataxia, loss of coordination, coma, and liver and kidney damage. It has also demonstrated mutagenic properties and produces hyperkeratosis, epithelial hyperplasia, and skin papillomas.

Skin contact is quite irritating, causing redness, swelling, and burns. It will pass through unbroken skin to cause toxic systemic effects. Eye contact can cause conjunctival irritation and injury to the cornea.

Ingestion of this compound in small amounts may produce nausea, vomiting and gastrointestinal irritation.

Acute Health Effects

The following acute (short-term) health effects may occur immediately or shortly after exposure to diglycidyl ether:

Skin: Severe irritation and possible burns. On prolonged exposure, can be absorbed though intact skin causing toxic systemic effects.

Eye: Severe conjunctival irritation and possible damage to the cornea.

Lung: Severe irritation to respiratory tract causing shortness of breath and cough. High acute exposures can be fatal due to a buildup of fluid in the lungs (pulmonary edema).

CNS: Possible depression of the central nervous system causing a variety of toxic effects including loss of coordination and consciousness.

Other: Ingestion will result in gastritis and possible damage to the liver and kidneys.

Chronic Health Effects

The following chronic (long-term) health effects may occur at some time after exposure to diglycidyl ether and can last for months or even years:

Cancer Hazards: Diglycidyl ether may be a cancer causing agent in humans. NIOSH suggests it be treated as a potential human carcinogen. It is a mutagen (causes cell changes) in animals.

Reproductive Hazard: According to the references, diglycidyl ether has not been adequately tested for its ability to adversely affect reproduction.

Other Chronic Effects: Repeated low-dose exposures can irritate the lungs and result in bronchitis with symptoms of cough, phlegm, and shortness of breath. It may damage the liver and kidneys. Chronic exposure may produce bone marrow depression.

Recommended Risk-Reduction Measures

Personnel should avoid direct contact with diglycidyl ether. If a less toxic material or compound cannot be substituted for diglycidyl ether, then *engineering controls* are the most effective method of reducing exposure risk. The best protection is to enclose operations and/or provide local exhaust ventilation at the site of diglycidyl ether release. While not always operationally feasible, isolating operations can also reduce exposure. Using respiratory protection is less effective than the controls mentioned above, but is still required whenever working with or around diglycidyl ether. Exposure levels are extremely low (0.01 ppm) and difficult to accurately monitor. It is also suggested that diglycidyl ether is a possible human carcinogen. Therefore, for the best protection at any exposure level, a MSHA/NIOSH-approved self-contained breathing apparatus (SCBA) with full facepiece and pressure demand is the appropriate level of respiratory protection. If a full facepiece is not available, then chemical goggles should be worn to protect the eyes, especially when a splash hazard exists. A face shield should also be considered. To prevent hand and skin exposures, rubber or other non-permeable gloves should be worn.

Administrative controls should also be in place to minimize the potential for human exposures. These may include written procedures or policies which specify the methods and techniques that will be practiced whenever personnel are to work with diglycidyl ether.

All personnel should receive *training* on the use, hazards, protective measures, emergency actions, and other precautions per 29 CFR 1910.1200 (Hazard Communication) prior to the first assignment in an area where diglycidyl ether is used or stored.

Before beginning employment and at regular intervals thereafter (e.g., annually), the following medical test is recommended:

☑ Lung function tests.

If symptoms develop or overexposure is suspected, the following may be useful:

☑ Liver, kidney, and lung function tests.

☑ Consider chest X-ray after acute exposure.

It should be noted that medical tests that simply look for existing damage are not a substitute for controlling exposures. Medical histories are extremely important when assessing exposure risk. Also, since smoking can cause heart disease as well as lung cancer, emphysema, and other respiratory ailments, any exposure to diglycidyl ether can result in quicker and more devastating symptoms. Smokers should therefore avoid unprotected contact with this chemical. Prudent risk management requires proper consideration of *all* possible factors that may be causing the appearance of exposure symptoms among workers.

Other methods to reduce exposure include:

☑ Always ensure that proper protective clothing is worn when using chemical substances.

☑ Wash thoroughly immediately after exposure to diglycidyl ether and at the end of the work shift or before eating, drinking, or smoking.

☑ Hazard warning information should be posted in the work area. In addition, as part of an on-going education and training program, all information on the health and safety hazards of diglycidyl ether should be communicated to all exposed and potentially exposed workers.

☑ Eye wash stations should be provided in the immediate work area for emergency use. If there is a possibility of skin exposure to diglycidyl ether, emergency shower facilities should be provided.

☑ Workers whose clothing has been contaminated by diglycidyl ether should change into clean clothes before leaving work. Contaminated work clothing should be laundered only by individuals who have been informed of the hazards of exposure to diglycidyl ether.

RISK ASSESSMENT: ENVIRONMENT

General Assessment

The environment is at risk of exposure during transportation, storage, disposal, or destruction of diglycidyl ether. In almost every scenario, the threat of environmental exposure is contingent upon the proper handling of the chemical substance. Accidental spills, large or small, can result in contamination of the surrounding environmental mediums (water, soil, and air).

Diglycidyl ether is Class IIIA combustible liquid. It will support combustion as well as intensify any fire. It is incompatible with oxidizers and caution is always required in its handling, storage, transportation, and disposal. When heated to decomposition it emits very acrid and irritating smoke and fumes. Emergency responders should be made aware of the presence of diglycidyl ether at any emergency response situation.

Diglycidyl ether may enter the environment through industrial effluents and spills.

Acute Ecological Effects

Acute (short-term) toxic effects may include the death of animals, birds, or fish and death or low growth rate in plants. Acute effects are seen 2 to 4 days after animals or plants are exposed to diglycidyl ether. Insufficient data are available to evaluate or predict the acute (short-term) effects of diglycidyl ether to aquatic life, plants, birds, or land animals.

Chronic Ecological Effects

Chronic (long-term) toxic effects may include shortened life span, reproductive problems, lower fertility, and changes in appearance or behavior in exposed animals. These effects can be seen long after first exposure(s) to toxic chemicals. Insufficient data are available to evaluate the chronic (long-term) effects of diglycidyl ether on aquatic life, plants, birds, or land animals.

Water Solubility

The solubility of diglycidyl ether has not been established in the references.

Persistence in the Environment

There is no information in the references concerning the persistence of diglycidyl ether in the environment.

Bioaccumulation in Aquatic Organisms

Some substances increase in concentration, or *bioaccumulate,* in living organisms as they breathe contaminated air, drink contaminated water, or eat contaminated food. These chemicals can become concentrated in the tissues and internal organs of animals as well as humans.

The references do not address the bioaccumulation capabilities of diglycidyl ether in aquatic organisms.

Recommended Risk-Reduction Measures

Proper training of all transporters will reduce the likelihood of a mishap or accident resulting in a leak or spill of diglycidyl ether into the environment. Labels on all containers, trucks, and rail cars must meet DOT requirements and accurately reflect their contents to enable emergency responders to react properly and quickly to any disaster thereby reducing the potential risk to the environment and to personnel.

Storage of diglycidyl ether should be segregated from other chemicals to minimize the risk of cross-contamination. Diglycidyl ether must be stored to avoid contact with strong oxidizers, such as chlorine and bromine, since violent reactions can occur. Containers should be protected from physical damage and stored to avoid contact with heat.

If a spill or leak to the environment has occurred, fire department, emergency response and/or hazardous materials spill personnel should be notified immediately. Cleanup should be attempted only by those trained in proper spill containment procedures. Contaminated soils should be removed for incineration and replaced with clean soil.

If diglycidyl ether should contact the water table, aquifer, or navigable waterway, time is of the essence. Its specific solubility in water has not been established and total remediation may or may not be possible. The local and/or state emergency response authorities must be notified. A comprehensive emergency response or disaster preparedness/recovery plan should be in place prior to any operations involving the use, transportation, storage, or disposal of diglycidyl ether.

If diglycidyl ether is spilled or leaked, the following specific steps are recommended:

- ☑ Restrict persons not wearing protective clothing from area of spill or leak until cleanup is complete. Avoid skin and eye contact.
- ☑ Absorb spilled liquids with vermiculite, dry sand, or a similar material and deposit in sealed drum for disposal.
- ☑ Ventilate area of spill or leak.
- ☑ Remove all ignition sources.
- ☑ It may be necessary to dispose of diglycidyl ether as a hazardous waste. The responsible state agency or the regional office of the federal Environmental Protection Agency (EPA) should be contacted for specific recommendations.

RISK ASSESSMENT: BUSINESS

General Assessment

Accidents or mishaps involving diglycidyl ether can present a serious threat to business operations. The loss or damage of equipment or facilities can significantly affect fiscal viability. Lawsuits that may result from personnel illness, injury/death, public exposures, and/or environmental contamination will require a serious expenditure of resources. Media attention surrounding an injury, death, or environmental damage can also result in a loss of profits and loss of current as well as future business. Always remember that anytime the terms "cancer," or "carcinogen" are used, public emotion, hysteria, and ignorance can run equally high. This requires careful consideration during the development of public relations policies.

❶ *Recommended Risk-Reduction Measures*

Company attorneys, safety and health professionals, and environmental specialists should be involved in the development of any procedures for responding to chemical incidents. A company official should be predesignated as a public relations officer with specific training in dealing with the press. Corporate plans and policies should be developed, approved, and implemented long before any need for such arises.

REFERENCES

Ahlbom, A. 1993. *Biostatistics for Epidemiologists.* New York: CRC Press/Lewis Publishers.

American Conference of Governmental Industrial Hygienists. 1988. *Documentation of the Threshold Limit Values and Biological Exposure Indices,* 5th Edition (with updates). Cincinnati: ACGIH

Anthony, C. P., and N. J. Kolthoff. 1971. *Textbook of Anatomy and Physiology.* St. Louis: C. V. Mosby Company.

Asante-Duah, D. K. 1993. *Hazardous Waste Risk Assessment.* New York: CRC Press/Lewis Publishers.

Bloom, A. D. and F. J. DeSerres. 1995. *Ecotoxicity and Human Health.* New York: CRC Press/Lewis Publishers.

Brusick, D. J. 1994. *Methods for Genetic Risk Assessment.* New York: CRC Press/Lewis Publishers.

Calabrese, E. J., E. M. Kenyon. *Air Toxics and Risk Assessment.* New York: CRC Press/Lewis Publishers.

Calabrese, E. J. 1994. *Biological Effects of Low Level Exposures.* New York: CRC Press/Lewis Publishers.

Cockerman, L. G., B. S. Shane. 1994. *Basic Environmental Toxicology.* New York: CRC Press/Lewis Publishers.

Cothern, C. R., N. P. Ross. 1994. *Environmental Statistics, Assessment, and Forecasting.* New York: CRC Press/Lewis Publishers.

Cothern, R. R. 1993. *Comparative Environmental Risk Assessment.* New York: CRC Press/Lewis Publishers.

Cothern, C. R. 1995. *Handbook for Environmental Risk Decisionmaking.* New York: CRC Press/Lewis Publishers.

Danse, I. R. 1991. *Common Sense Toxics in the Workplace.* New York: Van Nostrand Reinhold.

Fodden, C. R., and J. L. Weddell. 1992. *Hazardous Materials Emergency Action Data.* New York: CRC Press/Lewis Publishers.

Gardner, R. W. 1994. *Chemical Intolerance.* New York: CRC Press/Lewis Publishers.

Hallenbeck, W. H. 1993. *Quantitative Risk Assessment for Environmental and Occupational Health,* 2nd Edition. New York: CRC Press/Lewis Publishers.

Hodgson, E., R. B. Mailman, and J. E. Chambers. 1988. *Dictionary of Toxicology.* London: The MacMillian Company.

Howard, P. H. 1995. *Handbook of Environmental Fate and Exposures Data for Organic Chemicals,* Vol. I-IV. New York: CRC Press/Lewis Publishers.

Howard, P. H., et. al. 1991. *Handbook of Environmental Degradation Rates.* New York: CRC Press/Lewis Publishers.

Hosty, J. W., and P. Foster. 1990. *A Practical Guide to Chemical Spill Response.* New York: Van Nostrand Reinhold.

Jones, F. E. 1994. *Toxic Organic Vapors in the Workplace.* New York: CRC Press/Lewis Publishers.

Kamrin, M. A. 1988. *Toxicology: A Primer on Principles and Applications.* Chelsea, Michigan: Lewis Publishers.

Keith, L. H. and M. M. Walker. 1995. *Handbook of Air Toxics.* New York: CRC Press/Lewis Publishers.

Landis, W. G., M. H. Yu. 1995. *Introduction to Environmental Toxicology.* New York: CRC Press/Lewis Publishers.

Langerman, N. 1994 *Precautionary Labels for Chemical Containers.* New York: CRC Press/Lewis Publishers.

Lauwerys, R. R., P. Hoet. 1993. *Industrial Chemical Exposure.* New York: CRC Press/Lewis Publishers.

Lewis, R. J., Sr. 1993. *Hawley's Condensed Chemical Dictionary,* 12th Edition. New York: Van Nostrand Reinhold.

Lewis, R. J., Sr. 1992. *Sax's Dangerous Properties of Industrial Materials,* Volumes 1, 2, and 3, 8th Edition. New York: Van Nostrand Reinhold.

Lide, D. R. 1995. *Handbook of Chemistry and Physics,* 76th Edition. New York: CRC Press/Lewis Publishers.

Manahan, S. E. 1992. *Toxicological Chemistry,* 2nd Edition. New York: CRC Press/Lewis Publishers.

Mellström, G. A., J. E. Wahlberg, and H. I. Maibach. 1994. *Protective Gloves for Occupational Use.* New York: CRC Press/Lewis Publishers.

Meyer, E. 1990. *Chemistry of Hazardous Materials.* Englewood Cliffs, NJ: Prentice Hall.

Patnaik, P. 1992. *A Comprehensive Guide to the Hazardous Properties of Chemical Substances.* New York: Van Nostrand Reinhold.

Philp, R. B. 1995. *Environmental Hazards and Human Health.* New York: CRC Press/Lewis Publishers.

Quigley, D. R. 1994. *Handbook of Emergency Chemical Management.* New York: CRC Press/Lewis Publishers.

Richardson, M. 1995. *Dictionary of Substance and Their Effects,* Vol. I-VI. New York: CRC Press/Lewis Publishers.

Torkelson, T. R. V. K. Rowe. 1981. *Patty's Industrial Hygiene and Toxicology,* Vol. 2B. New York: Wiley-Interscience.

MATERIAL SAFETY DATA SHEET

CHEMICAL NAME	
DIISOBUTYL KETONE	**CRC PRESS/LEWIS PUBLISHERS DISCLAIMER:** THE INFORMATION AND RECOMMENDATIONS PRESENTED HEREIN ARE BASED ON SOURCES BELIEVED TO BE RELIABLE. CRC MAKES NO REPRESENTATION ON ITS COMPLETENESS OR ACCURACY. IT IS THE RESPONSIBILITY OF THE USER TO DETERMINE THE CHEMICAL'S SUITABILITY FOR ITS INTENDED USE, THE CHEMICAL'S SAFE USE, AND THE CHEMICAL'S PROPER DISPOSAL. NO REPRESENTATIONS AND/OR WARRANTIES, EITHER EXPRESSED OR IMPLIED, OF THE MERCHANTABILITY OR FITNESS FOR A PARTICULAR PURPOSE, OR OF ANY OTHER NATURE, ARE MADE WITH RESPECT TO THE INFORMATION PROVIDED IN THIS MSDS OR TO THE CHEMICAL TO WHICH INFORMATION MAY REFER. CRC NEITHER ASSUMES NOR AUTHORIZES ANY OTHER PERSON TO ASSUME FOR IT, ANY OTHER ADDITIONAL RESPONSIBILITY OR LIABILITY FOR THE USE OF, OR RELIANCE UPON, THIS INFORMATION.

HAZARD WARNING INFORMATION

HEALTH	FIRE	REACTIVE	OTHER	DEGREE OF HAZARD	COLOR CODING	OTHER CODES
1	2	0		0 = Minimum Hazard 1 = Slight Hazard 2 = Moderate Hazard 3 = Serious Hazard 4 = Severe Hazard	HEALTH = BLUE FIRE = RED REACTIVITY = YELLOW OTHER = WHITE	OX = Oxidizer ACID = Acid ALK = Alkali COR = Corrosive ~~W~~ = Use No Water

SECTION I - GENERAL INFORMATION

Characterization **Ketone**	RCRA Number **None**	EPA Class **Not Listed**
DOT Proper Shipping Name **Diisobutyl ketone**	Chemical Abstract Service (CAS) Number **108-83-8**	
DOT Hazard Class and Label Requirements **Flammable or Combustible Liquid**	DOT Emergency Guide Code **26**	
DOT Identification Number **UN 1157**	Chemical Formula **$[(CH_3)_2CHCH_2]_2CO$**	

Synonyms
DIBK; sym-diisopropyl acetone; 2,6-dimethyl-4-heptanone; Isovalerone; Valerone.

SECTION II - HAZARDOUS INGREDIENTS AND IDENTITY INFORMATION

Hazardous Components (specific identity)	OSHA Exposure Criteria	NIOSH Exposure Criteria	Immediately Dangerous to Life and Health (IDLH)	ACGIH Exposure Criteria
Diisobutyl ketone **1 ppm = 5.92 mg/m^3**	**PEL:** **50 ppm** **290 mg/m^3** **STEL:** **Not Established**	**REL:** **25 ppm** **150 mg/m^3** **STEL:** **Not Established**	**500 ppm**	**TLV:** **25 ppm** **150 mg/m^3** **STEL:** **Not Established**

SECTION III - PHYSICAL AND CHEMICAL CHARACTERISTICS

Boiling Point **335°F (168°C)**	Specific Gravity (H_2O = 1) **0.806**
Vapor Pressure (mm Hg) **1.7 at 77°F (25°C)**	Molecular Weight **142.2**
Vapor Density (Air = 1) **4.9**	Freezing Point **108°F (-43°C)**

Solubility
Slightly soluble (0.05%) in water. Soluble in alcohol, ether, and chloroform, and most organic liquids.

Appearance and Odor
Colorless liquid with a mild, sweet odor. Odor Threshold = 0.31 ppm.

SECTION IV - FIRE AND EXPLOSION HAZARD DATA

Flash Point (method used) **120°F (49°C) opened cup**	Explosive Limits in Air % by Volume **LEL: 0.8% UEL: 7.1%**
NFPA Classification **Class II Combustible Liquid**	Autoignition Temperature **745°F (396°C)**

Extinguishing Media
Carbon dioxide, dry chemical, alcohol foam. Water may be ineffective.

Special Fire Fighting Procedures
Wear full protective clothing and self-contained breathing apparatus (SCBA). Fight fire from a distance if possible. Keep fire-exposed containers cool with water spray. Using a solid stream of water may spread fire. If it can be done safely, move fire-exposed containers from fire area.

Unusual Fire and Explosion Hazards
Containers may explode in fire due to the buildup of internal pressures. BLEVE (boiling liquid expanding vapor explosion) is possible. Immediately evacuate if rising sound is heard from venting device or a noticeable discoloration is observed on the sides of the tank or storage vessel.

SECTION V - REACTIVITY DATA

Stability		Conditions to Avoid Normally stable. Do not allow diisobutyl ketone to come into contact with incompatible materials. Avoid contact with fire or high heat sources. Do not use in confined spaces.
Stable X	Unstable	Incompatibility *(materials to avoid)* Incompatible with strong oxidizers (bromine, chlorine, fluorine).
Hazardous Polymerization		Conditions to Avoid Under normal conditions of temperature and pressure, hazardous polymerization of diisobutyl ketone is not expected to occur.
May Occur	Will Not Occur X	Hazardous Decomposition or By-products When heated to decomposition, diisobutyl ketone emits acrid, irritating smoke and fumes, including oxides of carbon.

SECTION VI - HEALTH HAZARD DATA

Primary Route(s) of Entry:	Inhalation?	Absorption (skin)?	Ingestion?
	X		X

Health Hazards

INHALATION: Irritation of the conjunctiva of the eyes and mucosa of the nose and throat. Symptoms include dizziness, headache, nausea, vomiting, ataxia, and unconsciousness. Narcotic effects also possible at high concentrations. Vapors can cause dermatitis.

SKIN & EYES: Skin contact can result in irritation and may cause dermatitis. Severe eye irritation likely.

INGESTION: Symptoms of nausea and vomiting. Abdominal cramps.

Carcinogenicity	NTP Listed?	IARC Cancer Review Group?	OSHA Regulated?	Target Organs?
Unknown Human Unknown Animal	No	No	29 CFR 1910.1000 Table Z-1	Respiratory system; skin; eyes; CNS, liver, kidneys.

Medical Conditions Generally Aggravated by Exposure

None Reported. However, respiratory impairments, skin conditions (dermatitis) may be aggravated.

Emergency and First-aid Procedures

Eye contact: Immediately flush large amounts of water for 15 minutes (minimum), occasionally lifting eyelids. Skin contact: Remove all contaminated clothing. Immediately wash area with large amounts of soap and water. For inhalation: Remove the person from exposure. Provide respiratory assistance and CPR. Transfer promptly to medical facility. If swallowed: Seek medical attention immediately. Give victim several glasses of water and induce vomiting. Never try to give an unconscious or convulsing person anything by mouth and never attempt to make such a person vomit.

SECTION VII - PRECAUTIONS FOR SAFE HANDLING AND USE

Steps to be Taken in Case Material is Released or Spilled

Remove all ignition sources. Restrict persons not wearing protective equipment from entering area of spill or leak until cleanup is complete. Ventilate area. Absorb liquid in dry sand or other material and deposit in sealed containers. Alert appropriate state or federal response agencies.

Preferred Waste Disposal Method

Incineration (with afterburner and scrubber), molten metal salt destruction.

Precautions to be Taken in Handling and Storage

Do not store with incompatible chemicals since violent reactions can occur. Store in tightly closed containers in a cool, dark, well-ventilated area away from fire, sparks, and flame. Use non-sparking tools on containers. Ground and bond metal containers during transfer operations.

Other Precautions and Warnings

Containers may explode in fire or under conditions of extreme heat (do not store outdoors or in direct sunlight). Protect containers from physical damage.

SECTION VIII - CONTROL MEASURES AND PERSONAL PROTECTIVE EQUIPMENT

Respiratory Protection *(specify type)*

Exposures above 50 ppm: MSHA/NIOSH-approved full facepiece respirator with organic vapor cartridge. Greater protection is obtained from a self-contained breathing apparatus (SCBA) with full facepiece and pressure demand or other positive pressure mode.

Ventilation

Local exhaust or general mechanical systems recommended.

Protective Gloves	Eye Protection	Other Protective Clothing
Solvent Resistant Butyl Rubber	Chemical Goggles and/or Face Mask	Rubber Apron

Work/Hygiene Practices

Always wash hands thoroughly after using chemical; never bring food, drink, or smoking materials into vicinity of chemicals.

DIISOBUTYL KETONE

$[(CH_3)_2CHCH_2]_2CO$ CAS: 108-83-8

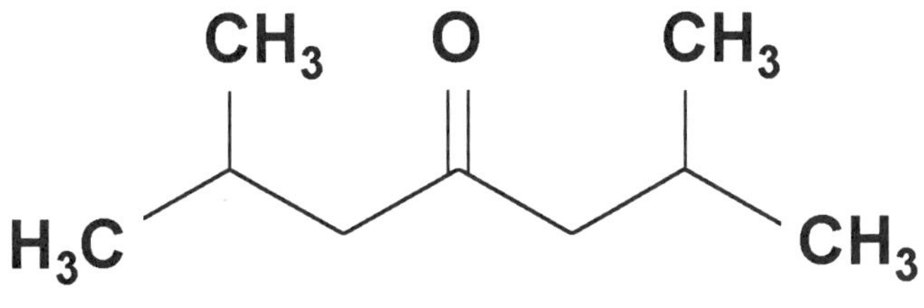

IDENTIFICATION AND TYPICAL USES

Diisobutyl ketone (DIBK) is a colorless liquid with a mild, sweet odor. It is used as a solvent for nitrocellulose, lacquers, and synthetic resins. It is also used in organic synthesis.

RISK ASSESSMENT: HEALTH

General Assessment

DIBK is moderately toxic to humans by *inhalation* and *ingestion.* There are no reports in the references that indicate skin absorption is a risk factor. However, skin or eye contact can cause damage.

Inhalation of DIBK vapors produces dryness and irritation of the eyes, nose, mouth, and throat. It will cause headache, vertigo, nausea, vomiting, coughing, and difficulty breathing. Inhalation may also produce mild narcotic effects at high concentrations. DIBK has also caused increased liver and kidney weights in test animals. It is not known if this chemical will have this same effect in humans.

Skin contact will result in mild to severe irritation. Because DIBK will defat the surface of the skin, cracking, scaling, and dermatitis can develop. Even contact with the vapors can cause skin reaction. Eye contact will also cause severe tissue irritation. May lead to damage of the cornea on prolonged contact.

Ingestion can result in headache, dizziness, and dermatitis.

Acute Health Effects

The following acute (short-term) health effects may occur immediately or shortly after exposure to DIBK:

Skin: Irritation/rash, cracking, scaling, and possible dermatitis.

Eye: Irritation which can lead to permanent damage.

Lung: Irritation causing coughing and/or shortness of breath. Higher exposures can cause dizziness, headache, and, possibly, coma.

CNS: A mild narcotic compound. Inhalation of high concentrations of the vapor may cause dizziness, lightheadedness, and possible loss of consciousness.

Chronic Health Effects

The following chronic (long-term) health effects may occur at some time after exposure to DIBK and can last for months or even years:

Cancer Hazards: According to information presented in the references, exposure to DIBK has not been shown to cause carcinogenic effects. Animal mutation data have been reported.

Reproduction: There are no reports to support any claims of reproductive hazards in humans or animals.

Sensitivity: Exposure to the skin may result in allergy (sensitization). If allergy develops, very low future exposure can result in skin irritation and rash. Repeated exposure can cause chronic irritation of the eyes and lungs.

Other Chronic Effects: Long-term exposures may cause damage to the liver and kidneys.

Recommended Risk-Reduction Measures

Personnel should avoid direct contact with DIBK. Unless a less toxic chemical can be substituted for a hazardous substance, *engineering controls* are the most effective method of reducing exposures. The best protection is to enclose operations and/or provide local exhaust ventilation at the site of chemical release. While not always operationally feasible, isolating operations can also reduce exposure. Using respiratory protection is less effective than the controls mentioned above, but is still advisable whenever working with DIBK. For exposures to the PEL (50 ppm), use a NIOSH-approved, full facepiece organic vapor respirator. For higher or prolonged exposures, a self-contained breathing apparatus (SCBA) with full facepiece and pressure demand is the recommended respiratory protection of choice. If a full facepiece is not available, then chemical goggles should be worn to protect the eyes. Whenever a chemical splash hazard exists, a face shield and a rubber apron should be worn. To prevent hand and skin exposures, butyl rubber gloves should be worn.

Administrative controls should also be in place to minimize the potential for human exposures. These may include written procedures or policies which specify the methods and techniques that will be practiced whenever personnel are to work with DIBK.

All personnel should receive *training* on the use, hazards, protective measures, emergency actions, and other precautions per 29 CFR 1910.1200 (Hazard Communication), prior to the first assignment in an area where DIBK is used or stored.

Other methods to reduce exposure include:

- ☑ Where possible, enclose operations and use local exhaust ventilation at the site of chemical release. If local exhaust ventilation or enclosure is not used, respiratory protection should be mandatory.
- ☑ Always ensure that proper protective clothing is worn when using chemical substances and that personnel that trained on its use and care.
- ☑ Wash thoroughly immediately after exposure to DIBK and at the end of the work shift or before eating, drinking, or smoking. Never wear clothes contaminated with DIBK home. Family members can be exposed.
- ☑ Hazard warning information should be posted in the work area. As part of an on-going education and training program, all information on the health and safety hazards of DIBK should be communicated to all potentially exposed workers.

RISK ASSESSMENT: ENVIRONMENT

General Assessment

The environment is at risk of exposure during transportation, storage, disposal, or destruction of DIBK. In almost every scenario, the threat of environmental exposure is contingent upon the proper handling of the chemical substance. Accidental spills, large or small, can result in fire, explosion, and possible contamination of the surrounding environmental mediums (water, soil, and air).

DIBK is considered a Class II combustible liquid (per OSHA 29 CFR 1910.106). It can react with strong oxidizers to cause a serious fire and explosion hazard. When involved in a fire, it can release acrid and irritating oxides of carbon. A boiling liquid expanding vapor explosion (BLEVE) is also a possibility. These characteristics require special consideration during any emergency involving a leak or spill of DIBK.

The proper disposal/destruction method for DIBK is to burn it in a chemical incinerator (with molten salt treatment) equipped with an afterburner and air scrubber.

DIBK can enter the environment through manufacturing, unchecked discharge into effluents from industrial or municipal waste treatment plants, and through spills.

Acute Ecological Effects

Acute (short-term) toxic effects may include the death of animals, birds, or fish, and death or low growth rate in plants. Acute effects are seen 2 to 4 days after animals or plants are exposed to DIBK. This chemical has a slight acute toxicity to aquatic life. No data are available on the short-term effects of DIBK on plants, birds, or land animals.

Chronic Ecological Effects

Chronic (long-term) toxic effects may include shortened life span, reproductive problems, lower fertility, and changes in appearance or behavior in exposed animals. These effects can be seen long after first exposure(s) to toxic chemicals. DIBK has slight chronic toxicity in aquatic life. No data are available to evaluate the long-term effects of DIBK to plants, birds, or land animals.

Water Solubility

DIBK is slightly soluble in water. Concentrations of less than 100 milligrams will mix with a liter of water.

Persistence in the Environment

DIBK is slightly persistent in water, with a half-life of between 2 to 20 days. The half-life of a pollutant is the amount of time it takes for one half of the chemical to be degraded. Approximately 75% of DIBK will eventually end up in air; the rest will end up in water.

Bioaccumulation in Aquatic Organisms

Some substances increase in concentration, or *bioaccumulate,* in living organisms as they breathe contaminated air, drink contaminated water, or eat contaminated food. These chemicals can become concentrated in the tissues and internal organs of animals as well as humans.

The concentration of DIBK found in fish tissues is expected to be about the same as the average concentration of DIBK in water from which the fish was taken.

Recommended Risk-Reduction Measures

Proper training of all transporters will reduce the likelihood of a mishap or accident resulting in a leak or spill to the environment. The correct use of DOT labeling (per the MSDS) on all containers, trucks, and rail cars will enable emergency responders to react properly and quickly to any disaster thereby reducing the potential risk to the environment and to personnel.

Storage of DIBK should be segregated from incompatible chemicals to minimize the risk of cross-contamination or contact. Buildings designated for storage should be equipped with appropriate fire protection systems (alarms, sprinklers, emergency lighting, portable extinguishers).

If a spill or leak to the environment has occurred, fire department, emergency response, and/or hazardous materials spill personnel should be notified immediately. Cleanup should be attempted only by those trained in proper spill containment procedures. Contaminated soils should be removed for incineration and replaced with clean soil. If DIBK should contact the water table, aquifer, or navigable waterway, time is of the essence. It is slightly soluble in water and, therefore, total containment and remediation may not be entirely possible. When such spills occur, the local and/or state emergency response authorities must be notified. A comprehensive emergency response or disaster preparedness/recovery plan should be in place prior to any operations involving the use, transportation, storage, or disposal of DIBK. If DIBK is spilled or leaked, the following specific steps are recommended:

- ☑ Restrict persons not wearing protective clothing from area of spill or leak until cleanup is complete and area can be opened for normal work.
- ☑ Remove ignition sources and ventilate area of spill or leak.
- ☑ Absorb liquids in vermiculite, dry sand, earth, or a similar material and deposit in sealed containers.
- ☑ If applicable, stop flow of leaking liquid or gas. If leak source is a cylinder and the leak cannot be stopped in place, remove leaking cylinder to a safe place in the open air, and repair or allow cylinder to empty.
- ☑ Keep DIBK out of a confined space, such as a sewer, because of the possibility of explosion (unless the sewer is designed to prevent the build-up of explosive concentrations).
- ☑ It may be necessary to dispose of DIBK as a hazardous waste. The responsible state agency or the regional office of the federal Environmental Protection Agency (EPA) should be contacted for specific recommendations.

RISK ASSESSMENT: BUSINESS

General Assessment

Accidents or mishaps involving DIBK can present a significant threat to business operations. The loss or damage of equipment or facilities can significantly affect fiscal viability. Lawsuits that may result from personnel injury, public exposures, and/or environmental contamination will also require a serious expenditure of resources. Media attention surrounding an injury, illness, or environmental damage can also result in a loss of profits and loss of current as well as future business.

Recommended Risk-Reduction Measures

Company attorneys, safety and health professionals, and environmental specialists should be involved in the development of any procedures dealing with safety or emergency response. A company official should be pre-designated as a public relations officer with specific training in dealing with the press. Corporate plans and policies should be developed, approved, and implemented long before any need for such arises.

REFERENCES

Ahlbom, A. 1993. *Biostatistics for Epidemiologists.* New York: CRC Press/Lewis Publishers.

American Conference of Governmental Industrial Hygienists. 1988. *Documentation of the Threshold Limit Values and Biological Exposure Indices,* 5th Edition (with updates). Cincinnati: ACGIH

Anthony, C. P., and N. J. Kolthoff. 1971. *Textbook of Anatomy and Physiology.* St. Louis: C. V. Mosby Company.

Asante-Duah, D. K. 1993. *Hazardous Waste Risk Assessment.* New York: CRC Press/Lewis Publishers.

Bloom, A. D. and F. J. DeSerres. 1995. *Ecotoxicity and Human Health.* New York: CRC Press/Lewis Publishers.

Brusick, D. J. 1994. *Methods for Geneic Risk Assessment.* New York: CRC Press/Lewis Publishers.

Bueche, J. R. 1972. *Principles of Physics,* 2nd Edition. New York: McGraw-Hill.

Burrell, R., D. K. Flaherty, and L. J. Sauers. 1992. *Toxicology of the Immune System.* New York: Van Nostrand Reinhold.

Calabrese, E. J., E. M. Kenyon. *Air Toxics and Risk Assessment.* New York: CRC Press/Lewis Publishers.

Calabrese, E. J. 1994. *Biological Effects of Low Level Exposures.* New York: CRC Press/Lewis Publishers.

Cockerman, L. G., B. S. Shane. 1994. *Basic Environmental Toxicology.* New York: CRC Press/Lewis Publishers.

Cothern, R. R. 1993. *Comparative Environmental Risk Assessment.* New York: CRC Press/Lewis Publishers.

Cothern, C. R. 1995. *Handbook for Environmental Risk Decisionmaking.* New York: CRC Press/Lewis Publishers.

DiVincenzo, G. O., F. J. Yanno, and B. D. Astill. 1973. Exposure of man and dog to low concentrations of acetone vapor. *American Industrial Hygiene Journal.* 34:329-336.

Gardner, R. W. 1994. *Chemical Intolerance.* New York: CRC Press/Lewis Publishers.

Hallenbeck, W. H. 1993. *Quantitative Risk Assessment for Environmental and Occupational Health,* 2nd Edition. New York: CRC Press/Lewis Publishers.

Hodgson, E., R. B. Mailman, and J. E. Chambers. 1988. *Dictionary of Toxicology.* London: The MacMillian Company.

Hosty, J. W., and P. Foster. 1990. *A Practical Guide to Chemical Spill Response.* New York: Van Nostrand Reinhold.

Howard, P. H. 1995. *Handbook of Environmental Fate and Exposures Data for Organic Chemicals,* Vol. I-IV. New York: CRC Press/Lewis Publishers.

Howard, P. H., et. al. 1991. *Handbook of Environmental Degradation Rates.* New York: CRC Press/Lewis Publishers.

James, A., and M. Lord. 1992. *Index of Chemical and Physical Data.* New York: Van Nostrand Reinhold.

Jones, F. E. 1994. *Toxic Organic Vapors in the Workplace.* New York: CRC Press/Lewis Publishers.

Kamrin, M. A. 1988. *Toxicology: A Primer on Principles and Applications.* Chelsea, Michigan: Lewis Publishers.

Keith, L. H. and M. M. Walker. 1995. *Handbook of Air Toxics.* New York: CRC Press/Lewis Publishers.

Landis, W. G., M. H. Yu. 1995. *Introduction to Environmental Toxicology.* New York: CRC Press/Lewis Publishers.

Langerman, N. 1994 *Precautionary Labels for Chemical Containers.* New York: CRC Press/Lewis Publishers.

Lauwerys, R. R., P. Hoet. 1993. *Industrial Chemical Exposure.* New York: CRC Press/Lewis Publishers.

Lewis, R. J., Sr. 1993. *Hawley's Condensed Chemical Dictionary,* 12th Edition. New York: Van Nostrand Reinhold.

Lewis, R. J., Sr. 1992. *Sax's Dangerous Properties of Industrial Materials,* Volumes 1, 2, and 3, 8th Edition. New York: Van Nostrand Reinhold.

Lide, D. R. 1995. *Handbook of Chemistry and Physics,* 76th Edition. New York: CRC Press/Lewis Publishers.

Manahan, S. E. 1994. *Environmental Chemistry,* 6th Edition. New York: CRC Press/Lewis Publishers.

Manahan, S. E. 1992. *Toxicological Chemistry,* 2nd Edition. New York: CRC Press/Lewis Publishers.

Mellström, G. A., J. E. Wahlberg, and H. I. Maibach. 1994. *Protective Gloves for Occupational Use.* New York: CRC Press/Lewis Publishers.

Meyer, E. 1990. *Chemistry of Hazardous Materials.* Englewood Cliffs, NJ: Prentice Hall.

Patnaik, P. 1992. *A Comprehensive Guide to the Hazardous Properties of Chemical Substances.* New York: Van Nostrand Reinhold.

Philp, R. B. 1995. *Environmental Hazards and Human Health.* New York: CRC Press/Lewis Publishers.

Proctor, N. H., and J. P. Hughes. 1991. *Chemical Hazards of the Workplace.* New York: Van Nostrand Reinhold.

Quigley, D. R. 1994. *Handbook of Emergency Chemical Management.* New York: CRC Press/Lewis Publishers.

Raleigh, R. L. and W. A. McGee. 1972. Effects of short, high concentration exposures to acetone as determined by observation in the work area. *Journal of Occupational Medicine.* 14:607-610.

Rea, W. J. 1995. *Chemical Sensitivity,* Vol. I-III. New York: CRC Press/Lewis Publishers.

Richardson, M. 1995. *Dictionary of Substance and Their Effects,* Vol. I-VI. New York: CRC Press/Lewis Publishers.

U.S. Department of Health and Human Services, National Institute for Occupational Safety and Health. 1994. *NIOSH Pocket Guide to Chemical Hazards.* Washington, D.C.: U.S. Government Printing Office.

U.S. Department of Health and Human Services, National Institute for Occupational Safety and Health. 1978. *Criteria for a Recommended Standard - Occupational Exposure to Ketones.* DHEW Publication Number: 78-173. Washington, D.C.: U.S. Government Printing Office.

MATERIAL SAFETY DATA SHEET

CHEMICAL NAME

DIISOPROPYLAMINE

CRC PRESS/LEWIS PUBLISHERS DISCLAIMER: THE INFORMATION AND RECOMMENDATIONS PRESENTED HEREIN ARE BASED ON SOURCES BELIEVED TO BE RELIABLE. CRC MAKES NO REPRESENTATION ON ITS COMPLETENESS OR ACCURACY. IT IS THE RESPONSIBILITY OF THE USER TO DETERMINE THE CHEMICAL'S SUITABILITY FOR ITS INTENDED USE, THE CHEMICAL'S SAFE USE, AND THE CHEMICAL'S PROPER DISPOSAL. NO REPRESENTATIONS AND/OR WARRANTIES, EITHER EXPRESSED OR IMPLIED, OF THE MERCHANTABILITY OR FITNESS FOR A PARTICULAR PURPOSE, OR OF ANY OTHER NATURE, ARE MADE WITH RESPECT TO THE INFORMATION PROVIDED IN THIS MSDS OR TO THE CHEMICAL TO WHICH INFORMATION MAY REFER. CRC NEITHER ASSUMES NOR AUTHORIZES ANY OTHER PERSON TO ASSUME FOR IT, ANY OTHER ADDITIONAL RESPONSIBILITY OR LIABILITY FOR THE USE OF, OR RELIANCE UPON, THIS INFORMATION.

HAZARD WARNING INFORMATION

HEALTH	FIRE	REACTIVE	OTHER	DEGREE OF HAZARD	COLOR CODING	OTHER CODES
3	3	0		0 = Minimum Hazard 1 = Slight Hazard 2 = Moderate Hazard 3 = Serious Hazard 4 = Severe Hazard	HEALTH = BLUE FIRE = RED REACTIVITY = YELLOW OTHER = WHITE	OX = Oxidizer ACID = Acid ALK = Alkali COR = Corrosive W = Use No Water

SECTION I - GENERAL INFORMATION

Characterization	RCRA Number	EPA Class
Aliphatic Amine	**D001**	**Characteristic Waste**

DOT Proper Shipping Name **Diisopropylamine**	Chemical Abstract Service (CAS) Number **108-18-9**
DOT Hazard Class and Label Requirements **Flammable Liquid**	DOT Emergency Guide Code **68**
DOT Identification Number **UN 1158**	Chemical Formula **$[(CH_3)_2CH]_2NH$**

Synonyms

DIPA; N-(1-methylethyl)-2-propanamine.

SECTION II - HAZARDOUS INGREDIENTS AND IDENTITY INFORMATION

Hazardous Components (specific identity)	OSHA Exposure Criteria	NIOSH Exposure Criteria	Immediately Dangerous to Life and Health (IDLH)	ACGIH Exposure Criteria
Diisopropylamine (derivation: From isopropyl chloride and ammonia). **1 ppm = 4.21 mg/m^3**	**PEL (skin):** **5 ppm** **20 mg/m^3** **STEL:** **Not Established**	**REL (skin):** **5 ppm** **20 mg/m^3** **STEL:** **Not Established**	**200 ppm**	**TLV (skin):** **5 ppm** **20 mg/m^3** **STEL:** **Not Established**

SECTION III - PHYSICAL AND CHEMICAL CHARACTERISTICS

Boiling Point **183°F (84°C)**	Specific Gravity (H_2O = 1) **0.72**
Vapor Pressure (mm Hg) **70 at 69°F (20°C)**	Molecular Weight **101.2**
Vapor Density (Air = 1) **3.5**	Melting Point **-141°F (-96°C)**

Solubility

Slightly soluble in water. Soluble in most organic solvents.

Appearance and Odor

Colorless liquid with a fishy, ammonia-like odor.

SECTION IV - FIRE AND EXPLOSION HAZARD DATA

Flash Point (method used) **21°F (-6°C) opened cup**	Explosive Limits in Air % by Volume **LEL: 1.1% UEL: 7.1%**
NFPA Classification **Class IB Flammable Liquid**	Autoignition Temperature **600°F (316°C)**

Extinguishing Media

Use dry chemical, carbon dioxide, alcohol foam, water spray. Water or foam may cause frothing.

Special Fire Fighting Procedures

Poisonous gases are produced in fire, wear full protective clothing and self-contained breathing apparatus (SCBA). Vapors are heavier than air and may travel for some distance to ignition source and flashback. Move containers from fire area if it can be done safely. Cool exposed container with water. Use water spray to disperse vapor of non-burning leak and to flush spills away from exposures.

Unusual Fire and Explosion Hazards

Containers may explode in fire due to buildup of excessive internal pressures. Aqueous solutions are flammable unless diluted extensively. This chemical presents a vapor explosion hazard indoors, outdoors, and in sewers.

SECTION V - REACTIVITY DATA

Stability		Conditions to Avoid **Diisopropylamine is normally stable at room temperature in closed containers under normal conditions of storage and handling. Avoid contact with incompatible materials. Keep away from heat and flame.**
Stable **X**	Unstable	Incompatibility *(materials to avoid)* **Strong oxidizers (such as chlorine, bromine, fluorine), and strong acids (such as sulfuric, nitric, and hydrochloric).**
Hazardous Polymerization		Conditions to Avoid **Under normal conditions of temperature and pressure, hazardous polymerization of diisopropylamine is not expected to occur.**
May Occur	Will Not Occur **X**	Hazardous Decomposition or By-products **Toxic and irritating fumes and vapors are produced when diisopropylamine is heated to decomposition. These include carbon monoxide, carbon dioxide, hydrocarbons, nitrogen oxides, and amine vapors.**

SECTION VI - HEALTH HAZARD DATA

Primary Route(s) of Entry:	Inhalation?	Absorption (skin)?	Ingestion?
	X	X	

Health Hazards

INHALATION: **Symptoms of nausea, headache, vomiting, visual disturbance (haziness), pulmonary and eye irritation. There may be cloudy swelling of the cornea with total or partial loss of vision. Pulmonary edema (fluid in lungs) is possible on high exposures.**

ABSORPTION: **A corrosive. Severe burns to the skin and eye on contact, with corneal damage and permanent vision impairment. Vapor may cause fogging of vision. Liquid will absorb through skin and cause systemic effects.**

INGESTION: **Severe pain and irritation to the digestive tract.**

Carcinogenicity	NTP Listed?	IARC Cancer Review Group?	OSHA Regulated?	Target Organs?
Unknown Human Unknown Animal	**No**	**No**	**29 CFR 1910.1000 Table Z-1**	**Eyes, skin, respiratory system.**

Medical Conditions Generally Aggravated by Exposure

None reported.

Emergency and First-aid Procedures

<u>Eye contact</u>: Flush immediately with water for 15 minutes (minimum), seek medical attention. <u>Skin contact</u>: Remove all contaminated clothing. Immediately wash area with large amounts of soap and water. Seek medical assistance. For <u>inhalation</u>: Remove the person from exposure. Provide respiratory assistance and CPR. Transfer to medical facility. Observe for 24-48 hours for lung effects. If <u>swallowed</u>: Unlikely. Call the poison control center and seek medical attention immediately. Unless advised otherwise, give 1-2 glasses of milk to dilute. Do NOT induce vomiting.

SECTION VII - PRECAUTIONS FOR SAFE HANDLING AND USE

Steps to be Taken in Case Material is Released or Spilled

Neutralize with sodium bisulfate and small amounts of water. Absorb liquids in vermiculite or other absorbent and deposit in DOT-approved drum. Ventilate area and remove ignition sources. Restrict those not involved in cleanup from entering area. Notify appropriate authorities, if applicable.

Preferred Waste Disposal Method

No Citation.

Precautions to be Taken in Handling and Storage

Do not store diisopropylamine in the presence of incompatible chemicals or materials. Store in tightly closed containers in a cool, well-ventilated area. Personnel working with diisopropylamine should be trained on its proper handing and storage *prior* to being assigned to such responsibilities.

Other Precautions and Warnings

Recommend storage in external (detached) facility away from main work areas.

SECTION VIII - CONTROL MEASURES AND PERSONAL PROTECTIVE EQUIPMENT

Respiratory Protection *(specify type)*

For low exposures, use an organic vapor respirator. Otherwise, use an MSHA/NIOSH-approved supplied-air respirator with full facepiece operated in positive pressure mode or a self-contained breathing apparatus (SCBA) with full facepiece operated in pressure demand or other positive pressure mode.

Ventilation

Local exhaust or general mechanical systems recommended.

Protective Gloves	Eye Protection	Other Protective Clothing
Impervious Butyl Rubber Gloves	**Chemical Goggles or Face Mask**	**Protective Apron**

Work/Hygiene Practices

Always wash hands thoroughly after using chemical; never bring food, drink, or smoking materials into vicinity of chemicals.

DIISOPROPYLAMINE

$[(CH_3)_2CH]_2NH$ CAS: 108-18-9

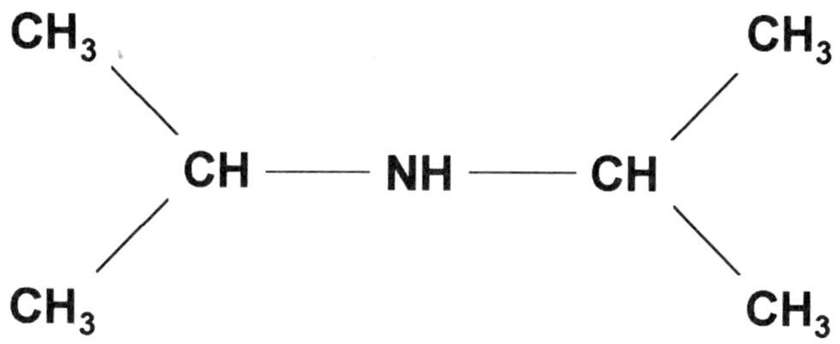

IDENTIFICATION AND TYPICAL USES

Diisopropylamine is a colorless or clear liquid with a strong, fishy, ammonia-like odor. It is used as a catalyst in organic synthesis and as an intermediate in chemical reactions.

RISK ASSESSMENT: HEALTH

General Assessment

Diisopropylamine is a severe irritant and corrosive on contact with human tissue. Primary route of exposure is *inhalation*. It can also cause serious heath effects on *ingestion* and by skin contact (*absorption*). There is no information in the references to indicate any carcinogenic or teratogenic risks. Mutation data have been reported. Some scientists believe that mutagenic chemicals may pose a cancer risk in the long-term. Additional research is required in this regard.

Inhalation will result in irritation of the mucosa, especially in the nose, throat, and upper respiratory tract. Symptoms may include headache, nausea, laryngitis, and difficulty or labored breathing. Exposure may also result in chemical pneumonitis and pulmonary edema (fluid buildup in the lungs), which is a medical emergency and can be fatal. Vapors will also cause eye irritation leading to possible visual disturbances (such as haziness) and cloudy swelling of the cornea with total or partial loss of vision.

Skin contact with the liquid can cause severe irritation with burns on contact. Absorption is likely to occur on prolonged contact to cause toxic systemic effects. Eye contact can result in corneal injury and permanent damage or loss of vision.

Ingestion toxicity is relatively low. Symptoms include irritation, nausea, and headache.

Acute Health Effects

The following acute (short-term) health effects may occur immediately or shortly after exposure to diisopropylamine:

Skin: Irritation, inflammation, redness, possible burning, and tissue damage. Absorption is likely on prolonged contact.

Eye: Severe irritation, inflammation, with possible damage to the cornea and loss of vision. Vapor can cause foggy or hazy vision, and inflammation of tissues.

Lung: Irritation causing coughing and/or shortness of breath. Higher exposures can lead to respiratory problems, including pulmonary edema (fluid buildup in lungs).

Chronic Health Effects

The following chronic (long-term) health effects may occur at some time after exposure to diisopropylamine and can last for months or even years:

Cancer Hazards: According to the information presented in the references, diisopropylamine has not been adequately tested for its ability to cause cancer in test animals. More research is required. Mutation data have been reported.

Reproductive Hazard: According to the information presented in the references, diisopropylamine has not been adequately tested for its ability to adversely affect reproduction.

Other Chronic Effects: Low, chronic exposures may cause corneal damage, foggy vision, and irritation of the lungs, bronchi, and throat.

☞ *Recommended Risk-Reduction Measures*

Unless a less toxic chemical can be substituted for a hazardous substance, *engineering controls* are the most effective method of reducing exposures. The best protection is to enclose operations and/or provide local exhaust ventilation at the site of chemical release. To prevent static sparks, all containers and equipment should be electrically grounded and bonded. While not always operationally feasible, isolating operations can also reduce exposure.

Using respiratory protection is less effective than the controls mentioned above, but is still advisable whenever working with or around diisopropylamine. For relatively low, infrequent, or transient exposures, an organic vapor respirator may suffice. For higher or prolonged exposure, a self-contained breathing apparatus (SCBA) with full facepiece and pressure demand is the recommended respiratory protection of choice. If a full facepiece is not available, then chemical goggles should be worn to protect the eyes. Whenever a chemical splash hazard exists, a face shield and apron should be worn. To prevent hand and skin exposures, impervious butyl rubber gloves should be worn.

Administrative controls should also be in place to minimize the potential for human exposures. These may include written procedures or policies which specify the methods and techniques that will be practiced whenever personnel are to work with diisopropylamine.

All personnel should receive *training* on the use, hazards, protective measures, emergency actions, and other precautions per 29 CFR 1910.1200 (Hazard Communication), prior to the first assignment in an area where diisopropylamine is used or stored.

Before beginning employment and at regular intervals thereafter (e.g. annually), the following medical tests are recommended:

☑ Lung function tests (establish a baseline).

If symptoms develop or overexposure is suspected, the following medical tests may be useful:

☑ Lung function tests. Doctor: If pulmonary edema is present, consider Positive Expiration End Pressure (PEEP) test. Steroids may be effective, but antibiotics are useful only if there is evidence of infection.

☑ Consider chest X-ray after acute overexposure.

☑ Complete blood count and blood gases.

Any evaluation should include a careful medical history of past and present symptoms with an examination. However, medical tests that evaluate existing damage are not a substitute for controlling exposure. Also, since smoking can cause heart disease, emphysema, and numerous other respiratory disorders, smokers may be affected more rapidly than non-smokers to the same exposure conditions. Prudent risk assessment and management requires careful consideration of *all* possible risk factors that may be causing the appearance of symptoms in exposed workers.

Other methods to reduce exposure include:

☑ Where possible, enclose operations and use local exhaust ventilation at the site of chemical release. If local exhaust ventilation or enclosure is not used, respiratory protection should be mandatory. Also, if possible, automatically transfer liquids from storage containers to process containers using non-sparking tools.

☑ Always ensure that proper protective clothing is worn when using chemical substances.

☑ Wash thoroughly immediately after exposure to diisopropylamine and at the end of the work shift or before eating, drinking, or smoking.

☑ Hazard warning information should be posted in the work area. In addition, as part of an on-going education and training program, all information on the health and safety hazards of diisopropylamine should be communicated to all potentially exposed workers.

☑ Eye wash stations should be provided in work areas. If the potential for whole body exposure exists, then safety showers should also be provided.

RISK ASSESSMENT: ENVIRONMENT

General Assessment

The environment is at risk of exposure during intended use, transportation, storage, disposal, or destruction of diisopropylamine. In almost every scenario, the threat of environmental exposure is contingent upon the proper handling of the chemical substance. Accidental spills, large or small, can result in fire, explosion, and possible contamination of the surrounding environmental mediums (water, soil, and air).

Diisopropylamine is considered a Class IB flammable liquid (per OSHA 29 CFR 1910.106). Its extremely low flash point and its relatively low boiling point make it a serious and dangerous fire and explosion risk. It can react with many common oxidizing agents and strong acids to cause a fire or explosion. These characteristics require special consideration during any emergency situation involving a leak or spill of diisopropylamine. Should diisopropylamine ever come into contact with incompatible substances, such as oxidizers or acids, either during use, transportation, or storage, violent reactions can occur.

Diisopropylamine can enter the environment through industrial discharges and spills.

Acute Ecological Effects

Acute (short-term) toxic effects may include the death of animals, birds, or fish and death or low growth rate in plants. Acute effects are seen 2 to 4 days after animals or plants are exposed to diisopropylamine.

Insufficient data are available to evaluate or predict the short-term effects of diisopropylamine to aquatic life, plants, birds, or land animals.

Chronic Ecological Effects

Chronic toxic effects may include shortened life span, reproductive problems, lower fertility, and changes in appearance or behavior in exposed animals. These effects can be seen long after first exposure(s) to toxic chemicals.

Insufficient data are available to evaluate or predict the long-term effects of diisopropylamine to aquatic life, plants, birds, or land animals.

Water Solubility

Diisopropylamine is highly soluble in water. Concentrations of 1000 milligrams and more will mix with a liter of water.

Persistence in the Environment

Diisopropylamine is slightly persistent in water, with a half-life of between 2 to 20 days. The half-life of a pollutant is the amount of time it takes for one half of the chemical to be degraded.

Bioaccumulation in Aquatic Organisms

Some substances increase in concentration, or *bioaccumulate,* in living organisms as they breathe contaminated air, drink contaminated water, or eat contaminated food. These chemicals can become concentrated in the tissues and internal organs of animals as well as humans who may ingest contaminated animals tissue.

The amount of diisopropylamine in edible fish tissue is expected to be about the same as that found in the water from which the fish was taken.

Recommended Risk-Reduction Measures

Proper training of all transporters will reduce the likelihood of a mishap or accident resulting in a leak or spill to the environment. The correct use of DOT labeling (per the MSDS) on all containers, trucks, and rail cars will enable emergency responders to react properly and quickly to any disaster thereby reducing the potential risk to the environment and to personnel.

Storage of diisopropylamine should be segregated from incompatible chemicals to minimize the risk of cross-contamination or contact. Storage buildings should be equipped with the proper fire protection equipment (alarms, sprinklers, extinguishers, etc.).

If a spill or leak to the environment has occurred, fire department, emergency response and/or hazardous materials spill personnel should be notified immediately. Cleanup should be attempted only by those trained in proper spill containment procedures. This chemical will remain in soil for extended periods and may leach to underground aquifers. Contaminated soils should therefore be removed for incineration and replaced with clean soil. If diisopropylamine should contact the water table, aquifer, or navigable waterway, remediation activities should be prompt. When such spills occur, the local and/or state emergency response authorities must be notified. A comprehensive emergency response or disaster preparedness/recovery plan should be in place prior to any operations involving the use, transportation, storage, or disposal of diisopropylamine. If diisopropylamine is spilled or leaked, the following specific steps are recommended:

- ☑ Restrict persons not wearing protective clothing from area of spill or leak until cleanup is complete and area can be opened for normal work.
- ☑ Ventilate area of spill or leak.
- ☑ Remove all ignition sources.
- ☑ Absorb liquid spills using vermiculite, dry earth, or sand to absorb and place in a sealed drum for disposal.
- ☑ Use water spray to disperse vapors and dilute to non-flammable mixtures. Dispose of accordingly.
- ☑ It may be necessary to dispose of diisopropylamine as a hazardous waste. The responsible state agency or the regional office of the federal

Environmental Protection Agency (EPA) should be contacted for specific recommendations.

RISK ASSESSMENT: BUSINESS

General Assessment

Accidents or mishaps involving diisopropylamine can present a significant threat to business operations. The loss or damage of equipment or facilities can significantly affect fiscal viability. Lawsuits that may result from personnel injury/death, public exposures, and/or environmental contamination will also require a serious expenditure of resources. Media attention surrounding an injury, death, or environmental damage can also result in a loss of profits and loss of current as well as future business.

⓾ Recommended Risk-Reduction Measures

Company attorneys, safety and health professionals, and environmental specialists should be involved in the development of any procedures dealing with safety and emergency response. A company official should be pre-designated as a public relations officer with specific training in dealing with the press. Corporate plans and policies should be developed, approved, and implemented long before any need for such arises.

REFERENCES

American Conference of Governmental Industrial Hygienists. 1988. *Documentation of the Threshold Limit Values and Biological Exposure Indices,* 5th Edition (with updates). Cincinnati: ACGIH

Anthony, C. P., and N. J. Kolthoff. 1971. *Textbook of Anatomy and Physiology.* St. Louis: C. V. Mosby Company.

Bloom, A. D. and F. J. DeSerres. 1995. *Ecotoxicity and Human Health.* New York: CRC Press/Lewis Publishers.

Bueche, J. R. 1972. *Principles of Physics,* 2nd Edition. New York: McGraw-Hill.

Calabrese, E. J., E. M. Kenyon. *Air Toxics and Risk Assessment.* New York: CRC Press/Lewis Publishers.

Cockerman, L. G., B. S. Shane. 1994. *Basic Environmental Toxicology.* New York: CRC Press/Lewis Publishers.

Cothern, C. R., N. P. Ross. 1994. *Environmental Statistics, Assessment, and Forecasting.* New York: CRC Press/Lewis Publishers.

Cothern, C. R. 1995. *Handbook for Environmental Risk Decisionmaking.* New York: CRC Press/Lewis Publishers.

Gardner, R. W. 1994. *Chemical Intolerance.* New York: CRC Press/Lewis Publishers.

Hallenbeck, W. H. 1993. *Quantitative Risk Assessment for Environmental and Occupational Health,* 2nd Edition. New York: CRC Press/Lewis Publishers.

Hodgson, E., R. B. Mailman, and Janice E. Chambers. 1988. *Dictionary of Toxicology.* London: The MacMillian Company.

Howard, P. H. 1995. *Handbook of Environmental Fate and Exposures Data for Organic Chemicals,* Vol. I-IV. New York: CRC Press/Lewis Publishers.

Howard, P. H., et. al. 1991. *Handbook of Environmental Degradation Rates.* New York: CRC Press/Lewis Publishers.

Jones, F. E. 1994. *Toxic Organic Vapors in the Workplace.* New York: CRC Press/Lewis Publishers.

Landis, W. G., M. H. Yu. 1995. *Introduction to Environmental Toxicology.* New York: CRC Press/Lewis Publishers.

Langerman, N. 1994 *Precautionary Labels for Chemical Containers.* New York: CRC Press/Lewis Publishers.

Lauwerys, R. R., P. Hoet. 1993. *Industrial Chemical Exposure.* New York: CRC Press/Lewis Publishers.

Lewis, R. J., Sr. 1993. *Hawley's Condensed Chemical Dictionary,* 12th Edition. New York: Van Nostrand Reinhold.

Lewis, R. J., Sr. 1992. *Sax's Dangerous Properties of Industrial Materials,* Volumes 1, 2, and 3, 8th Edition. New York: Van Nostrand Reinhold.

Manahan, S. E. 1992. *Toxicological Chemistry,* 2nd Edition. New York: CRC Press/Lewis Publishers.

Mellström, G. A., J. E. Wahlberg, and H. I. Maibach. 1994. *Protective Gloves for Occupational Use.* New York: CRC Press/Lewis Publishers.

Meyer, E. 1990. *Chemistry of Hazardous Materials.* Englewood Cliffs, NJ: Prentice Hall.

Nelson, G. O. 1992. *Gas Mixtures.* New York: CRC Press/Lewis Publishers.

Quigley, D. R. 1994. *Handbook of Emergency Chemical Management.* New York: CRC Press/Lewis Publishers.

Rea, W. J. 1995. *Chemical Sensitivity,* Vol. I-III. New York: CRC Press/Lewis Publishers.

Richardson, M. 1995. *Dictionary of Substance and Their Effects,* Vol. I-VI. New York: CRC Press/Lewis Publishers.

Sacarello, H. L. A. 1994. *The Comprehensive Handbook of Hazardous Materials.* New York: CRC Press/Lewis Publishers.

U.S. Department of Health and Human Services, National Institute for Occupational Safety and Health. 1994. *NIOSH Pocket Guide to Chemical Hazards.* Washington, D.C.: U.S. Government Printing Office.

MATERIAL SAFETY DATA SHEET

CHEMICAL NAME	CRC PRESS/LEWIS PUBLISHERS DISCLAIMER
DIMETHYLACETAMIDE	**CRC PRESS/LEWIS PUBLISHERS DISCLAIMER:** THE INFORMATION AND RECOMMENDATIONS PRESENTED HEREIN ARE BASED ON SOURCES BELIEVED TO BE RELIABLE. CRC MAKES NO REPRESENTATION ON ITS COMPLETENESS OR ACCURACY. IT IS THE RESPONSIBILITY OF THE USER TO DETERMINE THE CHEMICAL'S SUITABILITY FOR ITS INTENDED USE, THE CHEMICAL'S SAFE USE, AND THE CHEMICAL'S PROPER DISPOSAL. NO REPRESENTATIONS AND/OR WARRANTIES, EITHER EXPRESSED OR IMPLIED, OF THE MERCHANTABILITY OR FITNESS FOR A PARTICULAR PURPOSE, OR OF ANY OTHER NATURE, ARE MADE WITH RESPECT TO THE INFORMATION PROVIDED IN THIS MSDS OR TO THE CHEMICAL TO WHICH INFORMATION MAY REFER. CRC NEITHER ASSUMES NOR AUTHORIZES ANY OTHER PERSON TO ASSUME FOR IT, ANY OTHER ADDITIONAL RESPONSIBILITY OR LIABILITY FOR THE USE OF, OR RELIANCE UPON, THIS INFORMATION.

HAZARD WARNING INFORMATION

HEALTH	FIRE	REACTIVE	OTHER	DEGREE OF HAZARD	COLOR CODING	OTHER CODES
2	2	0		0 = Minimum Hazard 1 = Slight Hazard 2 = Moderate Hazard 3 = Serious Hazard 4 = Severe Hazard	HEALTH = BLUE FIRE = RED REACTIVITY = YELLOW OTHER = WHITE	OX = Oxidizer ACID = Acid ALK = Alkali COR = Corrosive W = Use No Water

SECTION I - GENERAL INFORMATION

Characterization **Acetic Anhydride**	RCRA Number **None**	EPA Class **Not Applicable**
DOT Proper Shipping Name **Dimethylacetamide**	Chemical Abstract Service (CAS) Number **127-19-5**	
DOT Hazard Class and Label Requirements **Combustible Liquid, N.O.S.**	DOT Emergency Guide Code **No Citation**	
DOT Identification Number **NA 1993**	Chemical Formula **$CH_3CON(CH_3)_2$**	

Synonyms
N,N-dimethyl acetamide; DMAC; acetic acid dimethylamide; dimethylamide acetate.

SECTION II - HAZARDOUS INGREDIENTS AND IDENTITY INFORMATION

Hazardous Components (specific identity)	OSHA Exposure Criteria	NIOSH Exposure Criteria	Immediately Dangerous to Life and Health (IDLH)	ACGIH Exposure Criteria
Dimethylacetamide (derivation: From acetic anhydride and dimethylformamide or from tris(dimethylamido) phosphate and acetic anhydride). **1 ppm = 3.62 mg/m^3**	**PEL:** **10 ppm** **35 mg/m^3** **STEL:** **Not Established**	**REL:** **10 ppm** **35 mg/m^3** **STEL:** **Not Established**	**300 ppm**	**TLV (skin):** **10 ppm** **35 mg/m^3** **STEL:** **Not Established**

SECTION III - PHYSICAL AND CHEMICAL CHARACTERISTICS

Boiling Point **329°F (165°C)**	Specific Gravity (H_2O = 1) **0.94**
Vapor Pressure (mm Hg) **2 at 69°F (20°C)**	Molecular Weight **87.1**
Vapor Density (Air = 1) **3**	Melting Point **–4°F (-20°C)**

Solubility
Miscible in water, aromatics, esters, ketones, and ethers.

Appearance and Odor
Colorless liquid with a weak fishy or ammonia-like odor. Odor Threshold between 21 and 48 ppm.

SECTION IV - FIRE AND EXPLOSION HAZARD DATA

Flash Point (method used) **158°F (70°C) opened cup**	Explosive Limits in Air % by Volume **LEL: 1.8% UEL: 11.5%**
NFPA Classification **Class IIIA Combustible Liquid**	Autoignition Temperature **914°F (490°C)**

Extinguishing Media
Use dry chemical, carbon dioxide, alcohol foam, water spray. Water or foam may cause frothing.

Special Fire Fighting Procedures
Poisonous gases are produced in fire, wear full protective clothing and self-contained breathing apparatus (SCBA). Move containers from fire area if it can be done safely. Cool exposed container with water. Do not direct stream of water into burning liquid (will scatter fire).

Unusual Fire and Explosion Hazards
Containers may explode in fire due to buildup of excessive internal pressures. Vapors are heavier than air and may travel for some distance to ignition source and flashback. This chemical presents a vapor explosion hazard indoors, outdoors, and in sewers.

SECTION V - REACTIVITY DATA

Stability		
Stability		Conditions to Avoid **Dimethylacetamide is normally stable at room temperature in closed containers under normal conditions of storage and handling. Avoid contact with incompatible materials. Keep away from heat and flame.**
Stable X	Unstable	Incompatibility *(materials to avoid)* **Strong oxidizers (such as chlorine, bromine, fluorine), and halogenated compounds (such as carbon tetrachloride, benzene) in the presence or in contact with iron.**
Hazardous Polymerization		Conditions to Avoid **Under normal conditions of temperature and pressure, hazardous polymerization of dimethylacetamide is not expected to occur.**
May Occur	Will Not Occur X	Hazardous Decomposition or By-products **Toxic and irritating fumes and vapors are produced when dimethylacetamide is heated to decomposition. These include ammonia, carbon monoxide, and carbon dioxide.**

SECTION VI - HEALTH HAZARD DATA

Primary Route(s) of Entry:	Inhalation?	Absorption (skin)?	Ingestion?
	X	X	

Health Hazards

INHALATION: Symptoms of headache, nausea, drowsiness, delusions, auditory and visual hallucinations, disorientation, depression of the CNS, lethargy, sweating, weakness, and moderate irritation of the respiratory system. May cause liver enlargement and affect reproduction. Cumulative liver damage is possible with elevated cholesterol levels.

ABSORPTION: Will pass through unbroken skin to cause toxic system effects (see inhalation above).

INGESTION: Depression, lethargy, confusion, visual and auditory hallucinations, disorientation, perceptual distortion, delusions, emotional detachment, and effective blunting.

Carcinogenicity	NTP Listed?	IARC Cancer Review Group?	OSHA Regulated?	Target Organs?
Unknown Human Unknown Animal	No	No	29 CFR 1910.1000 Table Z-1	Eyes, skin, respiratory system, liver, kidney, CNS.

Medical Conditions Generally Aggravated by Exposure

Liver disorders.

Emergency and First-aid Procedures

Eye contact: Flush immediately with water for 15 minutes (minimum), seek medical attention. **Skin contact:** Remove all contaminated clothing. Immediately wash area with large amounts of soap and water. Seek medical assistance. For **inhalation:** Remove the person from exposure. Provide respiratory assistance and CPR. Transfer to medical facility. If **swallowed:** Unlikely. Call the poison control center and seek medical attention immediately. Unless advised otherwise, give 1-2 glasses of milk and induce vomiting. DOCTOR: Suggest follow vomiting with gastric lavage and saline cathartics. Treatment for liver and kidney injury is supportive and symptomatic. Recommend Liver Function Test (LFT).

SECTION VII - PRECAUTIONS FOR SAFE HANDLING AND USE

Steps to be Taken in Case Material is Released or Spilled

Absorb liquids in vermiculite or other absorbent and deposit in DOT-approved drum. Ventilate area and remove ignition sources. Restrict those not involved in cleanup from entering area.

Preferred Waste Disposal Method

No Citation.

Precautions to be Taken in Handling and Storage

Do not store dimethylacetamide in the presence of incompatible chemicals or materials. Store in tightly closed containers in a cool, well-ventilated area away from heat and ignition sources. Shield from exposure to direct sunlight. Protect containers from physical damage.

Other Precautions and Warnings

If color or iron contamination is a concern, use aluminum or stainless steel containers.

SECTION VIII - CONTROL MEASURES AND PERSONAL PROTECTIVE EQUIPMENT

Respiratory Protection *(specify type)*

For low exposures, use an organic vapor respirator. Otherwise, use an MSHA/NIOSH-approved supplied-air respirator with full facepiece operated in positive pressure mode or a self-contained breathing apparatus (SCBA) with full facepiece operated in pressure demand or other positive pressure mode.

Ventilation

Local exhaust or general mechanical systems recommended.

Protective Gloves	Eye Protection	Other Protective Clothing
Impervious Gloves	Chemical Goggles or Face Mask	Protective Apron

Work/Hygiene Practices

Always wash hands thoroughly after using chemical; never bring food, drink, or smoking materials into vicinity of chemicals.

DIMETHYLACETAMIDE

$(CH_3CON(CH_3)_2$ CAS: 127-19-5

IDENTIFICATION AND TYPICAL USES

Dimethylacetamide is a colorless or clear liquid with a weak fishy, ammonia-like odor. It is used as a solvent for plastics, resins, gums, and electrolytes. It is an intermediate, a catalyst, a paint remover, and a high purity solvent for crystallization and purification.

RISK ASSESSMENT: HEALTH

General Assessment

Dimethylacetamide is an irritant to the skin and eyes. Primary routes of exposure are *inhalation* and skin contact (*absorption*). Although unlikely in the occupational setting, *ingestion* can also cause serious toxic systemic effects. There is no information in the references to indicate any carcinogenic properties. However, it has been shown to cause teratogenic and mutagenic effects in test animals. There is also some limited evidence that exposure will adversely affect human reproduction.

Inhalation, absorption, and even ingestion will result in a wide range of toxic systemic effects with noted action on the central nervous system (CNS). Symptoms include headache, dizziness, drowsiness, delusions, auditory and visual hallucinations, disorientation and confusion, perceptual distortion, emotional detachment, CNS depression, lethargy, sweating, weakness, and irritation of the skin, eyes, and respiratory system. There may also be liver enlargement and damage, elevated cholesterol, and jaundice.

Acute Health Effects

The following acute (short-term) health effects may occur immediately or shortly after exposure to dimethylacetamide:

Skin: Irritation, inflammation, and redness. Absorption is likely causing toxic systemic effects.

Eye: Mild irritation leading to severe and permanent damage if allowed to remain in contact with the liquid.

Lung: Irritation of the respiratory system causing coughing and/or shortness of breath, headache, and dizziness.

CNS: A known CNS depressant. Can cause hallucinations (both auditory and visual), with confusion, disorientation, emotional detachment, dizziness, drowsiness, lethargy, general weakness, and loss of consciousness.

Chronic Health Effects

The following chronic (long-term) health effects may occur at some time after exposure to dimethylacetamide and can last for months or even years:

Cancer Hazards: According to the information presented in the references, dimethylacetamide has not been adequately tested for its ability to cause cancer in test animals. More research is required. Mutation data have been reported.

Reproductive Hazard: According to the information presented in the references, dimethylacetamide may damage the liver in the developing fetus. It is a known teratogen in test animals.

Other Chronic Effects: Low, chronic exposures may cause liver enlargement, elevated cholesterol, cumulative liver damage, and jaundice.

Recommended Risk-Reduction Measures

Unless a less toxic chemical can be substituted for a hazardous substance, *engineering controls* are the most effective method of reducing exposures. The best protection is to enclose operations and/or provide local exhaust ventilation at the site of chemical release. To prevent static sparks, all containers and equipment should be electrically grounded and bonded. While not always operationally feasible, isolating operations can also reduce exposure.

Using respiratory protection is less effective than the controls mentioned above, but is still advisable whenever working with or around dimethylacetamide. For relatively low, infrequent, or transient exposures, an organic vapor respirator may suffice. For higher or prolonged exposure, a self-contained breathing apparatus (SCBA) with full facepiece and pressure demand is the recommended respiratory protection of choice. If a full facepiece is not available, then chemical goggles should be worn to protect the eyes. Whenever a chemical splash hazard exists, a face shield and apron should be worn. To prevent hand and skin exposures, impervious butyl rubber gloves should be worn.

Administrative controls should also be in place to minimize the potential for human exposures. These may include written procedures or policies which specify the methods and techniques that will be practiced whenever personnel are to work with dimethylacetamide.

All personnel should receive *training* on the use, hazards, protective measures, emergency actions, and other precautions per 29 CFR 1910.1200 (Hazard Communication), prior to the first assignment in an area where dimethylacetamide is used or stored.

If symptoms develop or overexposure is suspected, the following medical tests may be useful:

- ☑ Complete blood count and blood gases.
- ☑ Liver function tests.
- ☑ Examination of the nervous system by a qualified neurologist, with careful evaluation of exposure history and special tests to determine CNS effects.

Any evaluation should include a careful medical history of past and present symptoms with an examination. However, medical tests that evaluate existing damage are not a substitute for controlling exposure.

Other methods to reduce exposure include:

- ☑ Where possible, enclose operations and use local exhaust ventilation at the site of chemical release. If local exhaust ventilation or enclosure is not used, respiratory protection should be mandatory. Also, if possible, automatically transfer liquids from storage containers to process containers using non-sparking tools.
- ☑ Always ensure that proper protective clothing is worn when using chemical substances.
- ☑ Electrically ground and bond all containers and equipment used in shipping, handling, or storage.
- ☑ Containers or drums labeled "EMPTY" may still contain potentially explosive residues. Handle with extreme caution and never use pressurized air to empty drums that have contained dimethylacetamide.
- ☑ Wash thoroughly immediately after exposure to dimethylacetamide and at the end of the work shift or before eating, drinking, or smoking.
- ☑ Hazard warning information should be posted in the work area. In addition, as part of an on-going education and training program, all information on the health and safety hazards of dimethylacetamide should be communicated to all potentially exposed workers.
- ☑ Eye wash stations should be provided in work areas. If the potential for whole body exposure exists, then safety showers should also be provided.

RISK ASSESSMENT: ENVIRONMENT

General Assessment

The environment is at risk of exposure during intended use, transportation, storage, disposal, or destruction of dimethylacetamide. In almost every scenario, the threat of environmental exposure is contingent upon the proper handling of the chemical substance. Accidental spills, large or small, can result in fire, explosion, and possible contamination of the surrounding environmental mediums (water, soil, and air).

Dimethylacetamide is considered a Class IIIA combustible liquid (per OSHA 29 CFR 1910.106). It can react with many common oxidizing agents to cause a fire or explosion. These characteristics require special consideration during any emergency situation involving a leak or spill of dimethylacetamide. Should dimethylacetamide ever come into contact with incompatible substances either during use, transportation, or storage, violent reactions can occur.

Dimethylacetamide can enter the environment through industrial discharges and spills.

☠ *Acute Ecological Effects*

Acute (short-term) toxic effects may include the death of animals, birds, or fish and death or low growth rate in plants. Acute effects are seen 2 to 4 days after animals or plants are exposed to dimethylacetamide. Insufficient data are available to evaluate or predict the short-term effects of dimethylacetamide to aquatic life, plants, birds, or land animals.

Chronic Ecological Effects

Chronic toxic effects may include shortened life span, reproductive problems, lower fertility, and changes in appearance or behavior in exposed animals. These effects can be seen long after first exposure(s) to toxic chemicals. Insufficient data are available to evaluate or predict the long-term effects of dimethylacetamide to aquatic life, plants, birds, or land animals.

Water Solubility

Dimethylacetamide is highly soluble in water. Concentrations of 1000 milligrams and more will mix with a liter of water.

Persistence in the Environment

Dimethylacetamide is slightly persistent in water, with a half-life of between 2 to 20 days. The half-life of a pollutant is the amount of time it takes for one half of the chemical to be degraded.

Bioaccumulation in Aquatic Organisms

Some substances increase in concentration, or *bioaccumulate,* in living organisms as they breathe contaminated air, drink contaminated water, or eat contaminated food. These chemicals can become concentrated in the tissues and internal organs of animals as well as humans.

The amount of dimethylacetamide in edible fish tissue is expected to be about the same as that found in the water from which the fish was taken.

Recommended Risk-Reduction Measures

Proper training of all transporters will reduce the likelihood of a mishap or accident resulting in a leak or spill to the environment. The correct use of DOT labeling (per the MSDS) on all containers, trucks, and rail cars will enable emergency responders to react properly and quickly to any disaster thereby reducing the potential risk to the environment and to personnel.

Storage of dimethylacetamide should be segregated from incompatible chemicals to minimize the risk of cross-contamination or contact. Storage buildings should be equipped with the proper fire protection equipment (alarms, sprinklers, extinguishers, etc.).

If a spill or leak to the environment has occurred, fire department, emergency response and/or hazardous materials spill personnel should be notified immediately. Cleanup should be attempted only by those trained in proper spill containment procedures. This chemical will remain in soil for extended periods and may leach to underground aquifers. Contaminated soils should therefore be removed for incineration and replaced with clean soil. If dimethylacetamide should contact the water table, aquifer, or navigable waterway, remediation activities should be prompt. When such spills occur, the local and/or state emergency response authorities must be notified. A comprehensive emergency response or disaster preparedness/recovery plan should be in place prior to any operations involving the use, transportation, storage, or disposal of dimethylacetamide. If dimethylacetamide is spilled or leaked, the following specific steps are recommended:

- ☑ Restrict persons not wearing protective clothing from area of spill or leak until cleanup is complete and area can be opened for normal work.
- ☑ Ventilate area of spill or leak.
- ☑ Remove all ignition sources.
- ☑ Absorb liquid spills using vermiculite, dry earth, or sand to absorb and place in a sealed drum for disposal.
- ☑ Use water spray to disperse vapors and dilute to non-flammable mixtures. Dispose of accordingly.
- ☑ It may be necessary to dispose of dimethylacetamide as a hazardous waste. The responsible state agency or the regional office of the federal Environmental Protection Agency (EPA) should be contacted for specific recommendations.

RISK ASSESSMENT: BUSINESS

General Assessment

Accidents or mishaps involving dimethylacetamide can present a significant threat to business operations. The loss or damage of equipment or facilities can significantly affect fiscal viability. Lawsuits that may result from personnel injury/death, public exposures, and/or environmental contamination will also require a serious expenditure of resources. Media attention surrounding an injury, death, or environmental damage can also result in a loss of profits and loss of current as well as future business. Always remember that anytime the terms "reproductive hazard" or "teratogen" are used, public emotion, hysteria, and ignorance can run equally high. This must be carefully considered during the development of public relation policies.

Recommended Risk-Reduction Measures

Company attorneys, safety and health professionals, and environmental specialists should be involved in the development of any procedures dealing with safety or emergency response. A company official should be pre-designated as a public relations officer with specific training in dealing with the press. Corporate plans and policies should be developed, approved, and implemented long before any need for such arises.

REFERENCES

American Conference of Governmental Industrial Hygienists. 1988. *Documentation of the Threshold Limit Values and Biological Exposure Indices,* 5th Edition (with updates). Cincinnati: ACGIH

Anthony, C. P., and N. J. Kolthoff. 1971. *Textbook of Anatomy and Physiology.* St. Louis: C. V. Mosby Company.

Bloom, A. D. and F. J. DeSerres. 1995. *Ecotoxicity and Human Health.* New York: CRC Press/Lewis Publishers.

Bueche, J. R. 1972. *Principles of Physics,* 2nd Edition. New York: McGraw-Hill.

Calabrese, E. J., E. M. Kenyon. *Air Toxics and Risk Assessment.* New York: CRC Press/Lewis Publishers.

Calabrese, E. J. 1994. *Biological Effects of Low Level Exposures.* New York: CRC Press/Lewis Publishers.

Cockerman, L. G., B. S. Shane. 1994. *Basic Environmental Toxicology.* New York: CRC Press/Lewis Publishers.

Cothern, C. R., N. P. Ross. 1994. *Environmental Statistics, Assessment, and Forecasting.* New York: CRC Press/Lewis Publishers.

Cothern, R. R. 1993. *Comparative Environmental Risk Assessment.* New York: CRC Press/Lewis Publishers.

Cothern, C. R. 1995. *Handbook for Environmental Risk Decisionmaking.* New York: CRC Press/Lewis Publishers.

Fodden, C. R., and J. L. Weddell. 1992. *Hazardous Materials Emergency Action Data.* New York: CRC Press/Lewis Publishers.

Gardner, R. W. 1994. *Chemical Intolerance.* New York: CRC Press/Lewis Publishers.

Hallenbeck, W. H. 1993. *Quantitative Risk Assessment for Environmental and Occupational Health,* 2nd Edition. New York: CRC Press/Lewis Publishers.

Hodgson, E., R. B. Mailman, and Janice E. Chambers. 1988. *Dictionary of Toxicology.* London: The MacMillian Company.

Hosty, J. W., and P. Foster. 1990. *A Practical Guide to Chemical Spill Response.* New York: Van Nostrand Reinhold.

Howard, P. H. 1995. *Handbook of Environmental Fate and Exposures Data for Organic Chemicals,* Vol. I-IV. New York: CRC Press/Lewis Publishers.

Howard, P. H., et. al. 1991. *Handbook of Environmental Degradation Rates.* New York: CRC Press/Lewis Publishers.

Jones, F. E. 1994. *Toxic Organic Vapors in the Workplace.* New York: CRC Press/Lewis Publishers.

Landis, W. G., M. H. Yu. 1995. *Introduction to Environmental Toxicology.* New York: CRC Press/Lewis Publishers.

Langerman, N. 1994 *Precautionary Labels for Chemical Containers.* New York: CRC Press/Lewis Publishers.

Lauwerys, R. R., P. Hoet. 1993. *Industrial Chemical Exposure.* New York: CRC Press/Lewis Publishers.

Lewis, R. J., Sr. 1993. *Hawley's Condensed Chemical Dictionary,* 12th Edition. New York: Van Nostrand Reinhold.

Lewis, R. J., Sr. 1992. *Sax's Dangerous Properties of Industrial Materials,* Volumes 1, 2, and 3, 8th Edition. New York: Van Nostrand Reinhold.

Manahan, S. E. 1992. *Toxicological Chemistry,* 2nd Edition. New York: CRC Press/Lewis Publishers.

Mellström, G. A., J. E. Wahlberg, and H. I. Maibach. 1994. *Protective Gloves for Occupational Use.* New York: CRC Press/Lewis Publishers.

Meyer, E. 1990. *Chemistry of Hazardous Materials.* Englewood Cliffs, NJ: Prentice Hall.

Nelson, G. O. 1992. *Gas Mixtures.* New York: CRC Press/Lewis Publishers.

Quigley, D. R. 1994. *Handbook of Emergency Chemical Management.* New York: CRC Press/Lewis Publishers.

Rea, W. J. 1995. *Chemical Sensitivity,* Vol. I-III. New York: CRC Press/Lewis Publishers.

Richardson, M. 1995. *Dictionary of Substance and Their Effects,* Vol. I-VI. New York: CRC Press/Lewis Publishers.

Sacarello, H. L. A. 1994. *The Comprehensive Handbook of Hazardous Materials.* New York: CRC Press/Lewis Publishers.

Talbott, E., G. Craun. 1995. *An Introduction to Environmental Epidemiology.* New York: CRC Press/Lewis Publishers.

U.S. Department of Health and Human Services, National Institute for Occupational Safety and Health. 1994. *NIOSH Pocket Guide to Chemical Hazards.* Washington, D.C.: U.S. Government Printing Office.

Witorsch, P. and S. V. Spagnolo. 1994. *Air Pollution and Lung Disease in Adults.* New York: CRC Press/Lewis Publishers.

MATERIAL SAFETY DATA SHEET

CHEMICAL NAME	
DIMETHYLAMINE	**CRC PRESS/LEWIS PUBLISHERS DISCLAIMER:** THE INFORMATION AND RECOMMENDATIONS PRESENTED HEREIN ARE BASED ON SOURCES BELIEVED TO BE RELIABLE. CRC MAKES NO REPRESENTATION ON ITS COMPLETENESS OR ACCURACY. IT IS THE RESPONSIBILITY OF THE USER TO DETERMINE THE CHEMICAL'S SUITABILITY FOR ITS INTENDED USE, THE CHEMICAL'S SAFE USE, AND THE CHEMICAL'S PROPER DISPOSAL. NO REPRESENTATIONS AND/OR WARRANTIES, EITHER EXPRESSED OR IMPLIED, OF THE MERCHANTABILITY OR FITNESS FOR A PARTICULAR PURPOSE, OR OF ANY OTHER NATURE, ARE MADE WITH RESPECT TO THE INFORMATION PROVIDED IN THIS MSDS OR TO THE CHEMICAL TO WHICH INFORMATION MAY REFER. CRC NEITHER ASSUMES NOR AUTHORIZES ANY OTHER PERSON TO ASSUME FOR IT, ANY OTHER ADDITIONAL RESPONSIBILITY OR LIABILITY FOR THE USE OF, OR RELIANCE UPON, THIS INFORMATION.

HAZARD WARNING INFORMATION

HEALTH	FIRE	REACTIVE	OTHER	DEGREE OF HAZARD	COLOR CODING	OTHER CODES
3	4	0		0 = Minimum Hazard 1 = Slight Hazard 2 = Moderate Hazard 3 = Serious Hazard 4 = Severe Hazard	HEALTH = BLUE FIRE = RED REACTIVITY = YELLOW OTHER = WHITE	OX = Oxidizer ACID = Acid ALK = Alkali COR = Corrosive W = Use No Water

SECTION I - GENERAL INFORMATION

Characterization **Aliphatic Amine**	RCRA Number **U092**	EPA Class **Toxic Waste**
DOT Proper Shipping Name **Dimethylamine (aqueous solution)**	Chemical Abstract Service (CAS) Number **124-40-3**	
DOT Hazard Class and Label Requirements **Flammable Gas/Flammable Liquid (aqueous)**	DOT Emergency Guide Code **19 (gas) 26 (solution)**	
DOT Identification Number **UN 1032 (gas) UN 1160 (solution)**	Chemical Formula **$(CH_3)_2NH$**	

Synonyms
Diethamine; N,N-dimethylamine, N-ethylethanamine.

SECTION II - HAZARDOUS INGREDIENTS AND IDENTITY INFORMATION

Hazardous Components (specific identity)	OSHA Exposure Criteria	NIOSH Exposure Criteria	Immediately Dangerous to Life and Health (IDLH)	ACGIH Exposure Criteria
Dimethylamine (derivation: From the interaction of methanol and ammonia over a catalyst at high temperatures (572-932°F). The mono-, di-, and trimethylamines are all produced). **1 ppm = 1.87 mg/m^3**	**PEL (8-hour):** **10 ppm** **18 mg/m^3** **STEL:** **Not Established**	**REL (10-hour):** **10 ppm** **18 mg/m^3** **STEL:** **Not Established**	**500 ppm**	**TLV:** **5 ppm** **9.2 mg/m^3** **STEL:** **15 ppm** **27.6 mg/m^3**

SECTION III - PHYSICAL AND CHEMICAL CHARACTERISTICS

Boiling Point **45.3°F (7.4°C)**	Specific Gravity (H_2O = 1) **0.94**
Vapor Pressure (atmospheres) **1.7 at 69°F (20°C)**	Molecular Weight **45.1**
Vapor Density (Air = 1) **1.6**	Freezing Point **-134°F (-92°C)**

Solubility
Soluble in water, alcohol, and ether. Aqueous solution is strongly alkaline.

Appearance and Odor
Colorless gas with a fishy, ammonia-like odor. A liquid below 44°F. Normally shipped as a liquefied compressed gas. Odor Threshold = 0.005 - 0.016 ppm.

SECTION IV - FIRE AND EXPLOSION HAZARD DATA

Flash Point (method used) **0°F (-17.7°C) for gas; 20°F (-7°C) for liquid, (CC)**	Explosive Limits in Air % by Volume **LEL: 2.8% UEL: 14.4%**
NFPA Classification **Class IA Flammable Liquid**	Autoignition Temperature **806°F (430°C)**

Extinguishing Media
Stop flow of gas before fighting fire. Use water spray, dry chemical, or alcohol resistant foam.

Special Fire Fighting Procedures
Poisonous gases are produced in fire, wear full protective clothing and self-contained breathing apparatus (SCBA). Vapors are heavier than air and may travel for some distance to ignition source and flashback. Move containers from fire area if it can be done safely. Cool exposed container with water.

Unusual Fire and Explosion Hazards
Containers may explode in fire due to buildup of excessive internal pressures Aqueous solutions are flammable unless diluted extensively.

SECTION V - REACTIVITY DATA

Stability		Conditions to Avoid
		Dimethylamine is normally stable at room temperature in closed containers under normal conditions of storage and handling. Avoid contact with incompatible materials. Keep away from heat and flame.
Stable	Unstable	Incompatibility *(materials to avoid)*
X		Strong oxidizers and materials (fluorides, chlorine), acids, sources of halogens, mercury (will explode), hypochlorite, acraldehyde, maleic anhydride, aluminum, brass, copper, zinc, and 4-chloroacetophenone.
Hazardous Polymerization		Conditions to Avoid
		Under normal conditions of temperature and pressure, hazardous polymerization of dimethylamine is not expected to occur. Avoid contact with plastics, rubber, and coatings.
May Occur	Will Not Occur	Hazardous Decomposition or By-products
	X	Toxic and irritating fumes and vapors are produced when dimethylamine is heated to decomposition, including carbon monoxide, carbon dioxide, hydrocarbons, toxic oxides of nitrogen, and amine vapors.

SECTION VI - HEALTH HAZARD DATA

Primary Route(s) of Entry:	Inhalation?	Absorption (skin)?	Ingestion?
	X		X

Health Hazards

INHALATION: Symptoms of coughing, sneezing, labored breathing, nausea, severe irritation of the eyes, nose, and throat. Pulmonary edema (fluid in the lungs) is possible and can lead to death. May cause chemical bronchitis and possible liver damage in the long-term.

SKIN & EYES: Vapor contact causes severe irritation to skin and eyes, conjunctivitis, and corneal damage. Liquid will burn skin and eyes (frostbite) and damage tissue (necrosis).

INGESTION: Unlikely (dimethylamine is a gas at room temperature). The liquid will burn the mouth lips, throat, and gastrointestinal tract.

Carcinogenicity	NTP Listed?	IARC Cancer Review Group?	OSHA Regulated?	Target Organs?
Unknown Human Unknown Animal	No	No	29 CFR 1910.1000 Table Z-1	Eyes, skin, respiratory system, liver.

Medical Conditions Generally Aggravated by Exposure

Respiratory disorders (asthma, bronchitis); skin disease (eczema, dermatitis); eye or vision problems.

Emergency and First-aid Procedures

Eye contact: Flush immediately with water for 30 minutes (minimum), seek medical attention. Skin contact: Remove all contaminated clothing. Immediately wash area with large amounts of soap and water. For inhalation: Remove the person from exposure. Provide respiratory assistance and CPR. Transfer to medical facility. Observe for 24-48 hours for lung effects. If swallowed: Seek medical attention immediately. Unless advised otherwise, give 1-2 glasses of water. Do NOT induce vomiting. Never attempt to give an unconscious or convulsing person anything by mouth.

SECTION VII - PRECAUTIONS FOR SAFE HANDLING AND USE

Steps to be Taken in Case Material is Released or Spilled

Absorb liquids in vermiculite or other absorbent and deposit in DOT-approved drum. Ventilate area and remove ignition sources. Restrict those not involved in cleanup from entering area. Notify appropriate authorities, if applicable.

Preferred Waste Disposal Method

No Citation.

Precautions to be Taken in Handling and Storage

Do not store dimethylamine in the presence of incompatible chemicals or materials. Store in tightly closed containers in a cool, well-ventilated area. Protect containers from physical damage.

Other Precautions and Warnings

Store and handle cylinders in accordance with provisions of 29 CFR 1910.101.

SECTION VIII - CONTROL MEASURES AND PERSONAL PROTECTIVE EQUIPMENT

Respiratory Protection *(specify type)*

Use an MSHA/NIOSH-approved supplied-air respirator with full facepiece operated in positive pressure mode or a self-contained breathing apparatus (SCBA) with full facepiece operated in pressure demand or other positive pressure mode.

Ventilation

Local exhaust or general mechanical systems recommended.

Protective Gloves	Eye Protection	Other Protective Clothing
Impervious Butyl Rubber Gloves	Chemical/Gas Goggles or Face Mask	Protective Apron

Work/Hygiene Practices

Always wash hands thoroughly after using chemical; never bring food, drink, or smoking materials into vicinity of chemicals.

DIMETHYLAMINE

$(CH_3)_2NH$ CAS: 124-40-3

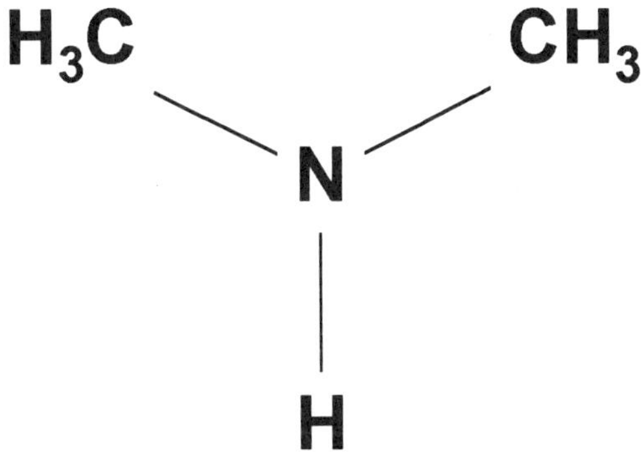

IDENTIFICATION AND TYPICAL USES

Dimethylamine is a colorless gas with an ammonia or fish-like odor. It is a liquid below 44°F (6.6°C) and is normally shipped as a liquefied compressed gas. It is used as an acid-gas absorbent, a solvent antioxidant, in the manufacture of dimethylformamide and dimethlacetamide, in dyes, as a flotation agent, a gasoline stabilizer, in pharmaceuticals, in the manufacture of textile chemicals, as a rubber accelerator, in electroplating, as a dehairing agent, in missile fuels, as a pesticide propellant, in rocket propellants, as a surfactant, and a reagent for magnesium.

RISK ASSESSMENT: HEALTH

General Assessment

Dimethylamine is a moderately toxic gas (or liquid) by *inhalation* and skin contact (absorption, however, has not been reported). Ingestion can occur, but is not likely since dimethlyamine is normally a gas at room temperatures. Mutation data have been reported. There is no information in the references to indicate that this compound has carcinogenic or teratogenic properties.

Inhalation of the vapors of dimethylamine is corrosive to the mucosa of the respiratory tract. It causes coughing, sneezing, difficulty breathing, nausea, severe irritation of the eyes, nose, and throat. It can also cause respiratory distress, bronchitis, pneumonitis, and pulmonary edema (fluid buildup in the lungs), which is a medical emergency and can be fatal. Symptoms of pulmonary edema may be delayed up to 48 hours creating a false sense of security with regard to exposure risk.

Skin contact with the vapors causes severe irritation. Liquid contact will cause frostbite, burns, and tissue necrosis. Eye contact with the vapors causes severe irritation and conjunctivitis. Corneal damage is quite likely to occur, either in contact with the liquid or the gas. Liquid contact with the eyes will cause intense pain with a potential for permanent corneal opacity and even blindness.

Ingestion is not likely. However, the liquid can burn the lips, mouth, and digestive tract.

☠ *Acute Health Effects*

The following acute (short-term) health effects may occur immediately or shortly after exposure to dimethylamine:

Skin: Irritation, inflammation, redness, and possible dermatitis. May cause necrosis and frostbite.

Eye: Severe irritation, inflammation, with possible damage to the cornea and loss of vision. Can cause permanent clouding of vision.

Lung: Irritation causing coughing and/or shortness of breath. Higher exposures can lead to respiratory problems, including life-threatening pulmonary edema (fluid buildup in lungs).

Chronic Health Effects

The following chronic (long-term) health effects may occur at some time after exposure to dimethylamine and can last for months or even years:

Cancer Hazards: According to the information presented in the references, dimethylamine has not been adequately tested for its ability to cause cancer in test animals. Mutation data have been reported and more research is required.

Reproductive Hazard: According to the information presented in the references, dimethylamine has not been adequately tested for its ability to adversely affect reproduction.

Other Chronic Effects: Animal studies have shown a propensity for liver damage. It is not known if these effects will occur in humans. Long-term exposures to low concentrations can cause dermatitis, chronic conjunctivitis, and chemical bronchitis. Also, severely irritating compounds are known to cause lung damage. It is not known if exposure to dimethylamine will have this effect in humans.

Recommended Risk-Reduction Measures

Unless a less toxic chemical can be substituted for a hazardous substance, *engineering controls* are the most effective method of reducing exposures. The best protection is to enclose operations and/or provide local exhaust ventilation at the site of chemical release. To prevent static sparks, all containers and equipment should be electrically grounded and bonded. Work areas should be equipped with Class I, Group C electrical equipment. Mercury-containing equipment or instruments must not be used. While not always operationally feasible, isolating operations can also reduce exposure.

Using respiratory protection is less effective than the controls mentioned above, but is still advisable whenever working with or around dimethylamine. For the best protection, a supplied-air respirator operated in the continuous flow mode or a self-contained breathing apparatus (SCBA) with full facepiece and pressure demand is the recommended respiratory protection of choice. If a full facepiece is not available, then chemical goggles should be worn to protect the eyes. Whenever a chemical splash hazard exists, a face shield and apron should be worn. To prevent hand and skin exposures, impervious butyl rubber chemical gloves should be worn.

Administrative controls should also be in place to minimize the potential for human exposures. These may include written procedures or policies which specify the methods and techniques that will be practiced whenever personnel are to work with dimethylamine.

All personnel should receive *training* on the use, hazards, protective measures, emergency actions, and other precautions per 29 CFR 1910.1200 (Hazard Communication), prior to the first assignment in an area where dimethylamine is used or stored.

Before beginning employment and at regular intervals thereafter (e.g. annually), the following medical tests are recommended:

☑ Lung function tests (establish a baseline).

If symptoms develop or overexposure is suspected, the following medical tests may be useful:

☑ Lung function tests.

☑ Consider chest X-ray after acute overexposure (may be negative if taken within first 48 hours of exposure).

Any evaluation should include a careful medical history of past and present symptoms with an examination. However, medical tests that evaluate existing damage are not a substitute for controlling exposure. Also, since smoking can cause heart disease, emphysema, and numerous other respiratory disorders, smokers may be affected more rapidly than non-smokers to the same exposure conditions. Prudent risk assessment and management requires careful consideration of *all* possible risk factors that may be causing the appearance of symptoms in exposed workers.

Other methods to reduce exposure include:

☑ Where possible, enclose operations and use local exhaust ventilation at the site of chemical release. If local exhaust ventilation or enclosure is not used, respiratory protection should be mandatory.

☑ Always ensure that proper protective clothing is worn when using chemical substances and that they are trained on its use and care.

☑ Wash thoroughly immediately after exposure to dimethylamine and at the end of the work shift or before eating, drinking, or smoking.

☑ Hazard warning information should be posted in the work area. In addition, as part of an on-going education and training program, all information on the health and safety hazards of dimethylamine should be communicated to all exposed and potentially exposed workers.

☑ Eye wash stations should be provided in work areas. If the potential for whole body exposure exists, then safety showers should also be provided.

RISK ASSESSMENT: ENVIRONMENT

General Assessment

The environment is at risk of exposure during intended use, transportation, storage, disposal, or destruction of dimethylamine. In almost every scenario, the threat of environmental exposure is contingent upon the proper handling of the chemical substance. Accidental spills, large or small, can result in fire, explosion, and possible contamination of the surrounding environmental mediums (water, soil, and air).

Dimethylamine is considered a Class IA flammable liquid (per OSHA 29 CFR 1910.106). Its extremely low flash point and its relatively low boiling point make it a serious and dangerous fire and explosion risk. Dimethylamine can react with many common oxidizing materials to cause a fire or explosion. Contact with acids, halogens, many metals (aluminum, brass, copper, zinc) can cause fire or violent explosion. In contact with mercury, dimethylamine explodes instantly. It can attack many plastic, rubber, and coating products (butyl rubber appears to hold up for longer than 8 hours). These characteristics require special consideration during any emergency situation involving a leak or spill of dimethylamine. Should dimethylamine ever come into contact with incompatible substances either during use, transportation, or storage, violent reactions can occur.

Dimethylamine can enter the environment through industrial discharges and spills.

Acute Ecological Effects

Acute (short-term) toxic effects may include the death of animals, birds, or fish and death or low growth rate in plants. Acute effects are seen 2 to 4 days after animals or plants are exposed to dimethylamine.

This chemical has low acute toxicity to aquatic life. Insufficient data are available to evaluate or predict the short-term effects of dimethylamine to plants, birds, or land animals.

Chronic Ecological Effects

Chronic toxic effects may include shortened life span, reproductive problems, lower fertility, and changes in appearance or behavior in exposed animals. These effects can be seen long after first exposure(s) to toxic chemicals.

Insufficient data are available to evaluate or predict the long-term effects of dimethylamine to aquatic life, plants, birds, or land animals.

Water Solubility

Dimethylamine is highly soluble in water. Concentrations of 1000 milligrams and more will mix with a liter of water.

Persistence in the Environment

Dimethylamine is slightly persistent in water, with a half-life of between 2 to 20 days. The half-life of a pollutant is the amount of time it takes for one half of the chemical to be degraded. It will volatilize slowly and its accumulation in aquatic or terrestrial sediments is not expected to be high.

Bioaccumulation in Aquatic Organisms

Some substances increase in concentration, or *bioaccumulate,* in living organisms as they breathe contaminated air, drink contaminated water, or eat contaminated food. These chemicals can become concentrated in the tissues and internal organs of animals as well as humans.

The concentration of dimethylamine found in fish tissues is expected to be somewhat lower than that found in the water from which the fish was taken.

Recommended Risk-Reduction Measures

Proper training of all transporters will reduce the likelihood of a mishap or accident resulting in a leak or spill to the environment. The correct use of DOT labeling (per the MSDS) on all containers, trucks, and rail cars will enable emergency responders to react properly and quickly to any disaster thereby reducing the potential risk to the environment and to personnel.

Storage of dimethylamine should be segregated from incompatible chemicals to minimize the risk of cross-contamination or contact. Storage buildings should be equipped with the proper fire protection equipment (alarms, sprinklers, extinguishers, etc.).

If a spill or leak to the environment has occurred, fire department, emergency response and/or hazardous materials spill personnel should be notified immediately. Cleanup should be attempted only by those trained in proper spill containment procedures. Contaminated soils should be removed for incineration and replaced with clean soil. If dimethylamine should contact the water table, aquifer, or navigable waterway, remediation activities should be prompt. When such spills occur, the local and/or state emergency response authorities must be notified. A comprehensive emergency response or disaster preparedness/recovery plan should be in place prior to any operations involving the use, transportation, storage, or disposal of dimethylamine. If dimethylamine is spilled or leaked, the following specific steps are recommended:

☑ Restrict persons not wearing protective clothing from area of spill or leak until cleanup is complete and area can be opened for normal work.

☑ Ventilate area of spill or leak.

☑ Remove all ignition sources.

☑ Gas leaks should be stopped and repaired. If repair cannot be done safety, cylinder should be removed to an open area and allowed to vent until empty. Liquids should be absorbed with vermiculite or sand and disposed of in a sealed container or drum.

☑ It may be necessary to dispose of dimethylamine as a hazardous waste. The responsible state agency or the regional office of the federal Environmental Protection Agency (EPA) should be contacted for specific recommendations.

RISK ASSESSMENT: BUSINESS

General Assessment

Accidents or mishaps involving dimethylamine can present a significant threat to business operations. The loss or damage of equipment or facilities can significantly affect fiscal viability. Lawsuits that may result from personnel injury/death, public exposures, and/or environmental contamination will also require a serious expenditure of resources. Media attention surrounding an injury, death, or environmental damage can also result in a loss of profits and loss of current as well as future business.

Recommended Risk-Reduction Measures

Company attorneys, safety and health professionals, and environmental specialists should be involved in the development of any procedures dealing with safety or emergency response. A company official should be pre-designated as a public relations officer with specific training in dealing with the press. Corporate plans and policies should be developed, approved, and implemented long before any need for such arises.

REFERENCES

American Conference of Governmental Industrial Hygienists. 1988. *Documentation of the Threshold Limit Values and Biological Exposure Indices,* 5th Edition (with updates). Cincinnati: ACGIH

Anthony, C. P., and N. J. Kolthoff. 1971. *Textbook of Anatomy and Physiology.* St. Louis: C. V. Mosby Company.

Bueche, J. R. 1972. *Principles of Physics,* 2nd Edition. New York: McGraw-Hill.

Calabrese, E. J., E. M. Kenyon. *Air Toxics and Risk Assessment.* New York: CRC Press/Lewis Publishers.

Cothern, C. R. 1995. *Handbook for Environmental Risk Decisionmaking.* New York: CRC Press/Lewis Publishers.

Gardner, R. W. 1994. *Chemical Intolerance.* New York: CRC Press/Lewis Publishers.

Hodgson, E., R. B. Mailman, and Janice E. Chambers. 1988. *Dictionary of Toxicology.* London: The MacMillian Company.

Howard, P. H., et. al. 1991. *Handbook of Environmental Degradation Rates.* New York: CRC Press/Lewis Publishers.

Jones, F. E. 1994. *Toxic Organic Vapors in the Workplace.* New York: CRC Press/Lewis Publishers.

Landis, W. G., M. H. Yu. 1995. *Introduction to Environmental Toxicology.* New York: CRC Press/Lewis Publishers.

Langerman, N. 1994 *Precautionary Labels for Chemical Containers.* New York: CRC Press/Lewis Publishers.

Lauwerys, R. R., P. Hoet. 1993. *Industrial Chemical Exposure.* New York: CRC Press/Lewis Publishers.

Lewis, R. J., Sr. 1993. *Hawley's Condensed Chemical Dictionary,* 12th Edition. New York: Van Nostrand Reinhold.

Lewis, R. J., Sr. 1992. *Sax's Dangerous Properties of Industrial Materials,* Volumes 1, 2, and 3, 8th Edition. New York: Van Nostrand Reinhold.

Manahan, S. E. 1992. *Toxicological Chemistry,* 2nd Edition. New York: CRC Press/Lewis Publishers.

Meyer, E. 1990. *Chemistry of Hazardous Materials.* Englewood Cliffs, NJ: Prentice Hall.

Nelson, G. O. 1992. *Gas Mixtures.* New York: CRC Press/Lewis Publishers.

Quigley, D. R. 1994. *Handbook of Emergency Chemical Management.* New York: CRC Press/Lewis Publishers.

Rea, W. J. 1995. *Chemical Sensitivity,* Vol. I-III. New York: CRC Press/Lewis Publishers.

Richardson, M. 1995. *Dictionary of Substance and Their Effects,* Vol. I-VI. New York: CRC Press/Lewis Publishers.

Sacarello, H. L. A. 1994. *The Comprehensive Handbook of Hazardous Materials.* New York: CRC Press/Lewis Publishers.

U.S. Department of Health and Human Services, National Institute for Occupational Safety and Health. 1994. *NIOSH Pocket Guide to Chemical Hazards.* Washington, D.C.: U.S. Government Printing Office.

MATERIAL SAFETY DATA SHEET

CHEMICAL NAME	
4-DIMETHYLAMINOAZO-BENZENE	**CRC PRESS/LEWIS PUBLISHERS DISCLAIMER:** THE INFORMATION AND RECOMMENDATIONS PRESENTED HEREIN ARE BASED ON SOURCES BELIEVED TO BE RELIABLE. CRC MAKES NO REPRESENTATION ON ITS COMPLETENESS OR ACCURACY. IT IS THE RESPONSIBILITY OF THE USER TO DETERMINE THE CHEMICAL'S SUITABILITY FOR ITS INTENDED USE, THE CHEMICAL'S SAFE USE, AND THE CHEMICAL'S PROPER DISPOSAL. NO REPRESENTATIONS AND/OR WARRANTIES, EITHER EXPRESSED OR IMPLIED, OF THE MERCHANTABILITY OR FITNESS FOR A PARTICULAR PURPOSE, OR OF ANY OTHER NATURE, ARE MADE WITH RESPECT TO THE INFORMATION PROVIDED IN THIS MSDS OR TO THE CHEMICAL TO WHICH INFORMATION MAY REFER. CRC NEITHER ASSUMES NOR AUTHORIZES ANY OTHER PERSON TO ASSUME FOR IT, ANY OTHER ADDITIONAL RESPONSIBILITY OR LIABILITY FOR THE USE OF, OR RELIANCE UPON, THIS INFORMATION.

HAZARD WARNING INFORMATION

HEALTH	FIRE	REACTIVE	OTHER	DEGREE OF HAZARD	COLOR CODING	OTHER CODES
2	0	0		0 = Minimum Hazard 1 = Slight Hazard 2 = Moderate Hazard 3 = Serious Hazard 4 = Severe Hazard	HEALTH = BLUE FIRE = RED REACTIVITY = YELLOW OTHER = WHITE	OX = Oxidizer ACID = Acid ALK = Alkali COR = Corrosive W = Use No Water

SECTION I - GENERAL INFORMATION

Characterization **Aromatic Amine**	RCRA Number **U093**	EPA Class **Toxic Waste**
DOT Proper Shipping Name **Environmentally Hazardous Substance,** N.O.S.	Chemical Abstract Service (CAS) Number **60-11-7**	
DOT Hazard Class and Label Requirements **Class 9**	DOT Emergency Guide Code **No Citation**	
DOT Identification Number **UN 3077**	Chemical Formula **$C_6H_5NNC_6H_4N(CH_3)_2$**	

Synonyms
Methyl yellow; butter yellow; DAB; p-dimethylaminoazobenzene; N,N-dimethyl-4-aminoazobenzene.

SECTION II - HAZARDOUS INGREDIENTS AND IDENTITY INFORMATION

Hazardous Components (specific identity)	OSHA Exposure Criteria	NIOSH Exposure Criteria	Immediately Dangerous to Life and Health (IDLH)	ACGIH Exposure Criteria
4-dimethylaminoazobenzene (derivation: By action of benzenediazonium chloride on dimethyl aniline).	**PEL:** Suspected Cancer Agent **STEL:** Not Applicable	**REL:** Possible Cancer Agent **STEL:** Not Applicable	**Not Determined**	**TLV:** Not Established **STEL:** Not Established

SECTION III - PHYSICAL AND CHEMICAL CHARACTERISTICS

Boiling Point **Sublimes**	Specific Gravity (H_2O = 1) **Not Established**
Vapor Pressure (mm Hg) **<1 mm at 69°F (20°C)**	Molecular Weight **225.3**
Vapor Density (Air = 1) **7.78**	Melting Point **237°F (114°C)**

Solubility
Slightly soluble in water (0.001%). Soluble in alcohol, benzene, chloroform, ether, petroleum ether, mineral acids, oils, and pyridine.

Appearance and Odor
Yellow, crystalline solid, appearing as leaflets.

SECTION IV - FIRE AND EXPLOSION HAZARD DATA

Flash Point (method used) **Non Combustible**	Explosive Limits in Air % by Volume **LEL: Not Determined UEL: Not Determined**
NFPA Classification **Non Combustible Solid**	Autoignition Temperature **Not Determined**

Extinguishing Media
Use an extinguishing agent suitable for surrounding fire.

Special Fire Fighting Procedures
Structural protective clothing is permeable. Remain clear of smoke, water fall-out, and water run-off. Poisonous gases are produced in fire. Move containers if it can be done without risk. Dry sweeping and mopping are prohibited by OSHA. Cool exposed containers. Evacuate non-essential personnel 2500 feet from the fire area (consider wind conditions).

Unusual Fire and Explosion Hazards
None Reported.

SECTION V - REACTIVITY DATA

Stability		Conditions to Avoid
		4-Dimethylaminoazobenzene is stable at room temperature in closed containers under normal conditions of storage and handling. Avoid generating dusts.
Stable	Unstable	Incompatibility *(materials to avoid)*
X		**None reported.**
Hazardous Polymerization		Conditions to Avoid
		Under normal conditions of temperature and pressure, hazardous polymerization of 4-dimethylamino-azobenzene is not expected to occur.
May Occur	Will Not Occur	Hazardous Decomposition or By-products
	X	**When heated to decomposition, 4-dimethylaminoazobenzene will produce toxic and irritating gases and oxides of nitrogen.**

SECTION VI - HEALTH HAZARD DATA

Primary Route(s) of Entry:	Inhalation?	Absorption (skin)?	Ingestion?
	X	X	X

Health Hazards

INHALATION: Irritation of the upper respiratory tract. Causes headache, wheezing, weakness, feeling of euphoria, shortness of breath, and irregular muscle action. Can also cause bloody sputum, dyspnea, bronchial secretions, frequent urination, cyanosis, hematuria (blood in urine), enlarged liver, liver and kidney dysfunction, possible liver and bladder tumors.

ABSORPTION: Can cause allergic skin reaction and dermatitis. Will pass through unbroken skin to cause toxic systemic effects noted above for inhalation.

INGESTION: A poison by ingestion. Systemic effects can occur.

Carcinogenicity	NTP Listed?	IARC Cancer Review Group?	OSHA Regulated?	Target Organs?
Confirmed Human Confirmed Animal	**5th Annual Report**	**Group 2B**	**29 CFR 1910.1015**	**Resp. system, skin, eyes, liver, kidney, bladder, CNS**

Medical Conditions Generally Aggravated by Exposure

Skin disorders.

Emergency and First-aid Procedures

Eye contact: Flush immediately with water for 15 minutes (minimum) while holding upper and lower eyelids open. Seek medical attention. Skin contact: Remove all contaminated clothing. Immediately wash area with large amounts of soap and water. For inhalation: Remove the person from exposure. Provide respiratory assistance and CPR. Transfer to medical facility. If swallowed: Call the poison control center, give 1-2 glasses of water to induce vomiting and seek medical attention immediately.

SECTION VII - PRECAUTIONS FOR SAFE HANDLING AND USE

Steps to be Taken in Case Material is Released or Spilled

Powders should be collected in safest manner possible and disposed of in sealed drums. Do NOT dry sweep. Use a high efficiency particulate air (HEPA) vacuum instead. Ventilate area and remove ignition sources. Restrict those not involved in cleanup from entering area. Notify appropriate authorities.

Preferred Waste Disposal Method

Consider rotary kiln or fluidized bed incineration (check with manufacturer first).

Precautions to be Taken in Handling and Storage

Store in tightly closed containers in a cool, dark, well-ventilated area. Personnel should be trained on the hazardous properties of 4-dimethylaminoazobenzene *prior* to working with the material.

Other Precautions and Warnings

Work should be performed in an enclosed area (glove box). A regulated, controlled work area should be established wherever this chemical is used. All containers must be labeled *cancer-suspect agent.*

SECTION VIII - CONTROL MEASURES AND PERSONAL PROTECTIVE EQUIPMENT

Respiratory Protection *(specify type)*

A carcinogenic agent. For any detectable concentrations, use a supplied-air respirator set in positive pressure or continuos flow mode or a self-contained breathing apparatus (SCBA) with full facepiece operated in pressure demand or other positive pressure mode.

Ventilation

Local exhaust or general mechanical systems recommended.

Protective Gloves	Eye Protection	Other Protective Clothing
Impervious Gloves	**Chemical/Dust Goggles or Face Mask**	**Protective Apron**

Work/Hygiene Practices

Always wash hands thoroughly after using chemical; never bring food, drink, or smoking materials into vicinity of chemicals.

4-DIMETHYLAMINOAZOBENZENE

$C_6H_5NNC_6H_4N(CH_3)_2$

CAS: 60-11-7

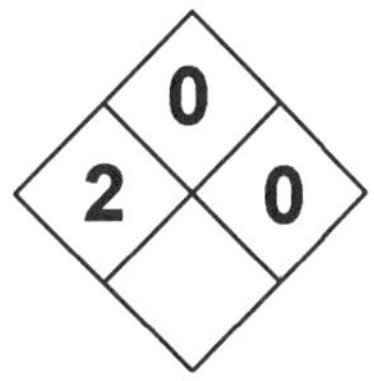

IDENTIFICATION AND TYPICAL USES

4-Dimethylaminoazobenzene appears as yellow, leaf-shaped crystals. It is no longer produced commercially in the United States and is only currently used for research in the determination of hydrochloric acid in gastric juice, in spot-test identification of peroxidized fats, as a pH indicator, and in laboratory testing for carcinogenicity. It was formally used for coloring polishes, in wax products, in butter, and in margarine.

RISK ASSESSMENT: HEALTH

General Assessment

4-Dimethylaminoazobenzene is a confirmed human and animal carcinogen. Human mutation data have been reported and there are animal studies to show some ability to produce adverse reproductive effects. It is primarily toxic by *inhalation,* but skin contact (*absorption*) and *ingestion* can also occur.

Inhalation, ingestion, and absorption can cause weakness, dizziness, euphoria, shortness of breath, and irregular muscular action. There can also be wheezing, dyspnea, bloody sputum, bronchial secretions, frequent urination, hematuria (blood in urine), and dysuria (painful urination). Systemic effects may lead to enlarged liver, kidney and liver dysfunction and, in test animals, it has been shown to cause liver, lung, and bladder tumors. It can also lead to chronic contact dermatitis, irritation, and rash.

Acute Health Effects

The following acute (short-term) health effects may occur immediately or shortly after exposure to 4-dimethylaminoazobenzene:

Skin: Mild to moderate irritation and possible dermatitis. Skin absorption is likely with subsequent toxic systemic effects.

Eye: Mild to severe irritation.

Lung: Irritation of the upper respiratory tract causing coughing and/or shortness of breath, wheezing, bronchial secretions, bloody sputum, and possible damage to lung tissue up to an including cancer (confirmed in laboratory test animals).

CNS: Dizziness, euphoria, weakness, and irregular muscle action.

Chronic Health Effects

The following chronic (long-term) health effects may occur at some time after exposure to 4-dimethylaminoazobenzene and can last for months or even years:

Cancer Hazards: According to the information presented in the references, 4-dimethylaminoazobenzene has been shown to cause tumors in the liver, bladder, and lungs of test animals. Its carcinogenicity in humans is expected to be at least equal to that of animals. Human mutation data have been reported for this compound.

Reproductive Hazard: According to the information presented in the references, this chemical has been shown to adversely affect reproduction in test animals.

Other Chronic Effects: Repeated skin contact can cause contact dermatitis. Chronic inhalation may cause cancer of the liver, bladder, and lung cancer.

Recommended Risk-Reduction Measures

4-Dimethylaminoazobenzene is listed by IARC, NTP, NIOSH, and OSHA as a carcinogen. Unless a less toxic chemical can be substituted for a hazardous substance, *engineering controls* are the most effective method of reducing exposures. The best protection is to enclose operations and/or provide local exhaust ventilation at the site of chemical release. Dusts must be prevented from entering the work area. Containers must be labeled "cancer-suspect agent." While not always operationally feasible, isolating operations can also reduce exposure.

Using respiratory protection is less effective than the controls mentioned above, but is still advisable whenever working with or around 4-dimethylaminoazobenzene. For any detectable exposures to a carcinogenic agent, use a supplied-air respirator set in positive pressure or continuous flow mode, or use self-contained breathing apparatus (SCBA) with full facepiece and pressure demand. If a full facepiece is not available, then chemical goggles should be worn to protect the eyes. Whenever an eye or face contact hazard exists, a face shield and apron should be worn. To prevent hand and skin exposures, impervious gloves should be worn.

Administrative controls should also be in place to minimize the potential for human exposures. These may include written procedures or policies which specify the methods and techniques that will be practiced whenever personnel are to work with or around 4-dimethylaminoazobenzene.

All personnel should receive *training* on the use, hazards, protective measures, emergency actions, and other precautions per 29 CFR 1910.1200 (Hazard Communication), prior to the first assignment in an area where 4-dimethylaminoazobenzene is used or stored.

Before beginning employment and at regular intervals thereafter (e.g. annually), the following medical tests are recommended:

☑ Lung function tests (establish a baseline).

If symptoms develop or overexposure is suspected, the following medical tests may be useful:

☑ Lung function tests.

☑ Liver and kidney function tests.

☑ Urinalysis.

☑ Consider chest X-ray after acute overexposure.

☑ Evaluation by a qualified allergist with careful consideration of exposure history and special tests (may help diagnose allergy).

Any evaluation should include a careful medical history of past and present symptoms with an examination. However, medical tests that evaluate existing damage are not a substitute for controlling exposure. Also, since smoking can cause heart disease, emphysema, and numerous other respiratory disorders, smokers may be affected more rapidly than non-smokers to the same exposure conditions. Prudent risk assessment and management requires careful consideration of *all* possible risk factors that may be causing the appearance of symptoms in exposed workers.

Other methods to reduce exposure include:

☑ Where possible, enclose operations and use local exhaust ventilation at the site of chemical release. If local exhaust ventilation or enclosure is not used, respiratory protection should be mandatory.

☑ Always ensure that proper protective clothing is worn when using chemical substances.

☑ Wash thoroughly immediately after exposure to 4-dimethylaminoazobenzene and at the end of the work shift or before eating, drinking, or smoking.

☑ Hazard warning information must be posted in the work area. All containers and work areas should be labeled as "cancer-suspect agent." A regulated, controlled work area should be established wherever this chemical is used, handled, or stored. In addition, as part of an on-going education and training program, all information on the health and safety hazards of 4-dimethyl-aminoazobenzene should be communicated to all exposed and potentially exposed workers.

☑ Eye wash stations should be provided in work areas. If the potential for whole body exposure exists, then safety showers should also be provided.

RISK ASSESSMENT: ENVIRONMENT

General Assessment

The environment is at risk of exposure during intended use, transportation, storage, disposal, or destruction of 4-dimethylaminoazobenzene. In almost every scenario, the threat of environmental exposure is contingent upon the proper handling of the chemical substance. Accidental spills, large or small, can result in fire, explosion, and possible contamination of the surrounding environmental mediums (water, soil, and air).

4-Dimethylaminoazobenzene should always be handled with extreme caution. While it is not flammable or combustible and is not reported to be incompatible with other commodities, it is an extremely hazardous and highly toxic substance. These characteristics require special consideration during any emergency situation involving a leak or spill of 4-dimethylaminoazobenzene.

4-Dimethylaminoazobenzene is no longer commercially produced and its use in laboratory research is extremely controlled such that release to the environment is unlikely. However, it may enter the environment through spills.

Acute Ecological Effects

Acute (short-term) toxic effects may include the death of animals, birds, or fish and death or low growth rate in plants. Acute effects are seen 2 to 4 days after animals or plants are exposed to 4-dimethylaminoazobenzene. Insufficient data are available to evaluate or predict the short-term effects of 4-dimethylaminoazobenzene to aquatic life, plants, birds, or land animals.

Chronic Ecological Effects

Chronic toxic effects may include shortened life span, reproductive problems, lower fertility, and changes in appearance or behavior in exposed animals. These effects can be seen long after first exposure(s) to toxic chemicals. Insufficient data are available to evaluate or predict the long-term effects of 4-dimethylaminoazobenzene to aquatic life, plants, birds, or land animals.

Water Solubility

4-Dimethylaminoazobenzene is slightly soluble in water. Concentrations of 1 to 100 milligrams will mix with a liter of water.

Persistence in the Environment

4-Dmethylaminoazobenzene is moderately persistent in water, with a half-life of between 20 to 200 days. The half-life of a pollutant is the amount of time it takes for one half of the chemical to be degraded. It is expected to absorb to sediment. In water, it should not hydrolyze.

Bioaccumulation in Aquatic Organisms

Some substances increase in concentration, or *bioaccumulate,* in living organisms as they breathe contaminated air, drink contaminated water, or eat contaminated food. These chemicals can become concentrated in the tissues and internal organs of animals as well as humans.

The concentration of 4-dimethylaminoazobenzene found in fish tissues is expected to be about the same as the average concentration found in the water from which the fish was taken.

Recommended Risk-Reduction Measures

Proper training of all transporters will reduce the likelihood of a mishap or accident resulting in a leak or spill to the environment. The correct use of labeling on all containers, trucks, and rail cars will enable emergency responders to react properly and quickly to any disaster thereby reducing the potential risk to the environment and to personnel.

Storage of 4-dimethylaminoazobenzene should be segregated from other chemicals to minimize the risk of cross-contamination or contact. Containers should be protected from physical damage and labeled as "cancer-suspect agent."

If a spill or leak to the environment has occurred, fire department, emergency response and/or hazardous materials spill personnel should be notified immediately. Cleanup should be attempted only by those trained in proper spill containment procedures and the health hazards associated with carcinogenic materials. Contaminated soils should be removed for incineration and replaced with clean soil. If 4-dimethylaminoazobenzene should contact the water table, aquifer, or navigable waterway, remediation activities should be prompt to ensure protection of aquatic sediments. When such spills occur, the local and/or state emergency response authorities must be notified. A comprehensive emergency response or disaster preparedness/recovery plan should be in place prior to any operations involving the use, transportation, storage, or disposal of 4-dimethylaminoazobenzene. If 4-dimethylaminoazobenzene is spilled or leaked, the following specific steps are recommended:

☑ Restrict persons not wearing protective clothing from area of spill or leak until cleanup of the area is complete.

☑ Ventilate area of leak or spill.

☑ Collect powders in safest manner possible and dispose of in sealed drums. Do NOT dry sweep (creates airborne dusts). Wet mopping is also prohibited (by OSHA). Use a vacuum equipped with a high efficiency particulate air (HEPA) filter attachment.

☑ It may be necessary to dispose of 4-dimethylaminoazobenzene as a hazardous waste. The responsible state agency or the regional office of the federal Environmental Protection Agency (EPA) should be contacted for specific recommendations.

RISK ASSESSMENT: BUSINESS

General Assessment

Accidents or mishaps involving 4-dimethylaminoazobenzene can present a significant threat to business operations. The loss or damage of equipment or facilities can significantly affect fiscal viability. Lawsuits that may result from personnel injury/death, public exposures, and/or environmental contamination will also require a serious expenditure of resources. Media attention surrounding an injury, death, or environmental damage can also result in a loss of profits and loss of current as well as future business. Always remember that anytime the term "carcinogen" or "cancer" are used, public emotion, hysteria, and ignorance can run equally high. This should be considered during the development of any public relations policies.

Recommended Risk-Reduction Measures

Company attorneys, safety and health professionals, and environmental specialists should be involved in the development of any procedures dealing with safety or emergency response. A company official should be pre-designated as a public relations officer with specific training in dealing with the press. Corporate plans and policies should be developed, approved, and implemented long before any need for such arises.

REFERENCES

Ahlbom, A. 1993. *Biostatistics for Epidemiologists.* New York: CRC Press/Lewis Publishers.

American Conference of Governmental Industrial Hygienists. 1988. *Documentation of the Threshold Limit Values and Biological Exposure Indices,* 5th Edition (with updates). Cincinnati: ACGIH

Anthony, C. P., and N. J. Kolthoff. 1971. *Textbook of Anatomy and Physiology.* St. Louis: C. V. Mosby Company.

Asante-Duah, D. K. 1993. *Hazardous Waste Risk Assessment.* New York: CRC Press/Lewis Publishers.

Bloom, A. D. and F. J. DeSerres. 1995. *Ecotoxicity and Human Health.* New York: CRC Press/Lewis Publishers.

Brusick, D. J. 1994. *Methods for Genetic Risk Assessment.* New York: CRC Press/Lewis Publishers.

Bueche, J. R. 1972. *Principles of Physics,* 2nd Edition. New York: McGraw-Hill.

Burrell, R., D. K. Flaherty, and L. J. Sauers. 1992. *Toxicology of the Immune System.* New York: Van Nostrand Reinhold.

Calabrese, E. J. 1994. *Biological Effects of Low Level Exposures.* New York: CRC Press/Lewis Publishers.

Cockerman, L. G., B. S. Shane. 1994. *Basic Environmental Toxicology.* New York: CRC Press/Lewis Publishers.

Cothern, C. R. 1995. *Handbook for Environmental Risk Decisionmaking.* New York: CRC Press/Lewis Publishers.

Gardner, R. W. 1994. *Chemical Intolerance.* New York: CRC Press/Lewis Publishers.

Hallenbeck, W. H. 1993. *Quantitative Risk Assessment for Environmental and Occupational Health,* 2nd Edition. New York: CRC Press/Lewis Publishers.

Hodgson, E., R. B. Mailman, and Janice E. Chambers. 1988. *Dictionary of Toxicology.* London: The MacMillian Company.

Howard, P. H. 1995. *Handbook of Environmental Fate and Exposures Data for Organic Chemicals,* Vol. I-IV. New York: CRC Press/Lewis Publishers.

Keith, L. H. and M. M. Walker. 1995. *Handbook of Air Toxics.* New York: CRC Press/Lewis Publishers.

Landis, W. G., M. H. Yu. 1995. *Introduction to Environmental Toxicology.* New York: CRC Press/Lewis Publishers.

Langerman, N. 1994 *Precautionary Labels for Chemical Containers.* New York: CRC Press/Lewis Publishers.

Lauwerys, R. R., P. Hoet. 1993. *Industrial Chemical Exposure.* New York: CRC Press/Lewis Publishers.

Lewis, R. J., Sr. 1993. *Hawley's Condensed Chemical Dictionary,* 12th Edition. New York: Van Nostrand Reinhold.

Lewis, R. J., Sr. 1992. *Sax's Dangerous Properties of Industrial Materials,* Volumes 1, 2, and 3, 8th Edition. New York: Van Nostrand Reinhold.

Manahan, S. E. 1992. *Toxicological Chemistry,* 2nd Edition. New York: CRC Press/Lewis Publishers.

Mellström, G. A., J. E. Wahlberg, and H. I. Maibach. 1994. *Protective Gloves for Occupational Use.* New York: CRC Press/Lewis Publishers.

Meyer, E. 1990. *Chemistry of Hazardous Materials.* Englewood Cliffs, NJ: Prentice Hall.

Ott, W. R. 1995. *Environmental Statistics and Data Analysis.* New York: CRC Press/Lewis Publishers.

Rea, W. J. 1995. *Chemical Sensitivity,* Vol. I-III. New York: CRC Press/Lewis Publishers.

Richardson, M. 1995. *Dictionary of Substance and Their Effects,* Vol. I-VI. New York: CRC Press/Lewis Publishers.

Sacarello, H. L. A. 1994. *The Comprehensive Handbook of Hazardous Materials.* New York: CRC Press/Lewis Publishers.

U.S. Department of Health and Human Services, National Institute for Occupational Safety and Health. 1994. *NIOSH Pocket Guide to Chemical Hazards.* Washington, D.C.: U.S. Government Printing Office.

MATERIAL SAFETY DATA SHEET

CHEMICAL NAME	
N,N-DIMETHYLANILINE	**CRC PRESS/LEWIS PUBLISHERS DISCLAIMER:** THE INFORMATION AND RECOMMENDATIONS PRESENTED HEREIN ARE BASED ON SOURCES BELIEVED TO BE RELIABLE. CRC MAKES NO REPRESENTATION ON ITS COMPLETENESS OR ACCURACY. IT IS THE RESPONSIBILITY OF THE USER TO DETERMINE THE CHEMICAL'S SUITABILITY FOR ITS INTENDED USE, THE CHEMICAL'S SAFE USE, AND THE CHEMICAL'S PROPER DISPOSAL. NO REPRESENTATIONS AND/OR WARRANTIES, EITHER EXPRESSED OR IMPLIED, OF THE MERCHANTABILITY OR FITNESS FOR A PARTICULAR PURPOSE, OR OF ANY OTHER NATURE, ARE MADE WITH RESPECT TO THE INFORMATION PROVIDED IN THIS MSDS OR TO THE CHEMICAL TO WHICH INFORMATION MAY REFER. CRC NEITHER ASSUMES NOR AUTHORIZES ANY OTHER PERSON TO ASSUME FOR IT, ANY OTHER ADDITIONAL RESPONSIBILITY OR LIABILITY FOR THE USE OF, OR RELIANCE UPON, THIS INFORMATION.

HAZARD WARNING INFORMATION

HEALTH	FIRE	REACTIVE	OTHER	DEGREE OF HAZARD	COLOR CODING	OTHER CODES
3	2	0		0 = Minimum Hazard 1 = Slight Hazard 2 = Moderate Hazard 3 = Serious Hazard 4 = Severe Hazard	HEALTH = BLUE FIRE = RED REACTIVITY = YELLOW OTHER = WHITE	OX = Oxidizer ACID = Acid ALK = Alkali COR = Corrosive W = Use No Water

SECTION I - GENERAL INFORMATION

Characterization	RCRA Number	EPA Class
Aromatic Amine	**None**	**Not Applicable**

DOT Proper Shipping Name **N,N-dimethylaniline**	Chemical Abstract Service (CAS) Number **121-69-7**
DOT Hazard Class and Label Requirements **Poison**	DOT Emergency Guide Code **57**
DOT Identification Number **UN 2253**	Chemical Formula $C_6H_5N(CH_3)_2$

Synonyms

Aniline, N,N-dimethyl; N,N-dimethylbenzeneamine; N,N-dimethylphenylamine; (dimethylamino) benzene.

SECTION II - HAZARDOUS INGREDIENTS AND IDENTITY INFORMATION

Hazardous Components (specific identity)	OSHA Exposure Criteria	NIOSH Exposure Criteria	Immediately Dangerous to Life and Health (IDLH)	ACGIH Exposure Criteria
N,N-dimethylaniline (derivation: By heating a mixture of aniline, aniline hydrochloride, and methyl alcohol (free from acetone) in an autoclave and distilling). **1 ppm = 5.04 mg/m³**	**PEL (skin):** **5 ppm** **25 mg/m³** **STEL:** **10 ppm** **50 mg/m³**	**REL (skin):** **5 ppm** **25 mg/m³** **STEL:** **10 ppm** **50 mg/m³**	**100 ppm**	**TLV (skin):** **5 ppm** **25 mg/m³** **STEL:** **10 ppm** **50 mg/m³**

SECTION III - PHYSICAL AND CHEMICAL CHARACTERISTICS

Boiling Point **379°F (193°C)**	Specific Gravity (H_2O = 1) **0.96**
Vapor Pressure (mm Hg) **1 at 69°F (20°C)**	Molecular Weight **121.2**
Vapor Density (Air = 1) **4.2**	Freezing Point **36°F (2°C)**

Solubility

Insoluble in water. Soluble in alcohol, ether, and chloroform.

Appearance and Odor

Pale-yellow, oily liquid with an amine-like odor. A solid below 36°F (2°C). Odor Threshold - 0.013 ppm.

SECTION IV - FIRE AND EXPLOSION HAZARD DATA

Flash Point (method used) **145°F (62°C) closed cup**	Explosive Limits in Air % by Volume **LEL: Not Determined UEL: Not Determined**
NFPA Classification **Class IIIA Combustible Liquid**	Autoignition Temperature **700°F (371°C)**

Extinguishing Media

High or low expansion foam, water fog, dry chemical, or carbon dioxide.

Special Fire Fighting Procedures

Wear full protective clothing and self-contained breathing apparatus (SCBA). Use a water spray to keep fire-exposed containers cool. Poisonous gases are produced in fire.

Unusual Fire and Explosion Hazards

Moderately flammable when exposed to heat. Containers may explode in fire. Firefighters should avoid contact with vapors produced during fire. Explodes on contact with benzoyl peroxide. Vapors are heavier than air and can travel for some distance to an ignition source and flashback to cause fire.

SECTION V - REACTIVITY DATA

Stability		Conditions to Avoid Under normal conditions of handling and storage, N,N-dimethylamine is stable at room temperature and pressure. Avoid exposure to incompatible chemicals, heat, and ignition sources.
Stable X	Unstable	Incompatibility *(materials to avoid)* Strong oxidizers (such as perchlorates, peroxides, permanganates, chlorates and nitrates); strong acids (such as hydrochloric, sulfuric, and nitric). Explodes on contract with benzoyl peroxide.
Hazardous Polymerization		Conditions to Avoid Under normal conditions of temperature and pressure, hazardous polymerization of N,N-dimethylaniline is not expected to occur.
May Occur	Will Not Occur X	Hazardous Decomposition or By-products When heated to decomposition, N,N-dimethylaniline can emit highly toxic/poisonous gases, including carbon monoxide and toxic aniline vapors.

SECTION VI - HEALTH HAZARD DATA

Primary Route(s) of Entry:	Inhalation?	Absorption (skin)?	Ingestion?
	X	X	

Health Hazards

INHALATION: Causes anoxia due to the formation of methemoglobin. Symptoms: headache, dizziness, ataxia, and cyanosis. Over exposure may lead to death due to respiratory paralysis. May also result in depression of the CNS, decreases in blood pressure, and tachycardia.

ABSORPTION: Absorbs freely through intact skin. Symptoms include headache and cyanosis, over-exposure can result in death. Skin contact with pure liquid can produce mild irritation.

INGESTION: Unspecified effects resulting from ingestion. Liver damage is suspect.

Carcinogenicity	NTP Listed?	IARC Cancer Review Group?	OSHA Regulated?	Target Organs?
Unknown Human Unknown Animal	No	No	29 CFR 1910.1000 Table Z-1	Eyes, skin, blood, kidneys, liver, CNS, CVS.

Medical Conditions Generally Aggravated by Exposure

None reported.

Emergency and First-aid Procedures

Eye contact: Flush immediately with water for 15 minutes (minimum), seek medical attention. Skin contact: Remove all contaminated clothing. Immediately wash area with large amounts of soap and water. Seek medical assistance. For inhalation: Remove the person from exposure. Provide respiratory assistance and CPR. Transfer to medical facility. If swallowed, seek medical attention immediately. Unless advised otherwise, give 2-3 glasses of water to a conscious person and induce vomiting.

SECTION VII - PRECAUTIONS FOR SAFE HANDLING AND USE

Steps to be Taken in Case Material is Released or Spilled

Cover spill with a 9:1 mixture of sand and soda ash and deposit in sealed containers. Remove all ignition sources. Ventilate area of spill or leak. Restrict those not involved in cleanup from entering area. Notify appropriate authorities, as required by SARA III, if applicable.

Preferred Waste Disposal Method

No Citation.

Precautions to be Taken in Handling and Storage

Store in tightly closed containers in cool, well-ventilated area away from heat. Ground and bond all containers prior to transferring liquids. Protect containers from physical damage and protect from exposure to direct sunlight.

Other Precautions and Warnings

Sources of ignition are prohibited where N,N-dimethylaniline is used, handled, or stored. Bulk storage of N,N-dimethylaniline should be kept in isolated or detached buildings or facilities.

SECTION VIII - CONTROL MEASURES AND PERSONAL PROTECTIVE EQUIPMENT

Respiratory Protection *(specify type)*

For low exposures (half the PEL), use an organic vapor respirator. Otherwise, use a self-contained breathing apparatus (SCBA) with full facepiece operated in positive pressure mode.

Ventilation

Local exhaust at site of chemical work.

Protective Gloves	Eye Protection	Other Protective Clothing
Impervious Rubber	Chemical Goggles or Face Mask	Rubber Apron

Work/Hygiene Practices

Always wash hands thoroughly after using chemical; never bring food, drink, or smoking materials into vicinity of chemicals.

N,N-DIMETHYLANILINE

$C_6H_5N(CH_3)_2$ CAS: 121-69-7

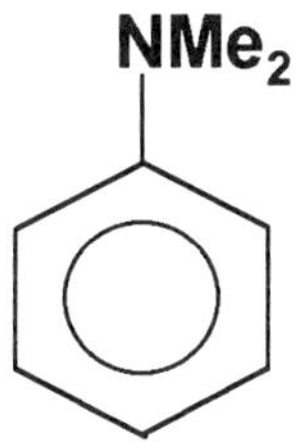

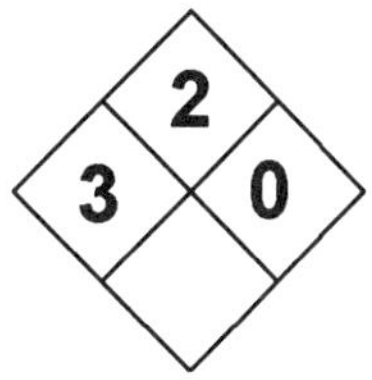

IDENTIFICATION AND TYPICAL USES

N,N-Dimethylaniline is a pale yellow to brown oily liquid with an amine-like odor. It is used in the manufacture of vanillin, Michler's Ketone, methyl violet, and other dyes. It is also used as a reagent in determination of ferric and ferrous ions for formaldehyde, methanol, methyl furfural, and hydrogen peroxide.

RISK ASSESSMENT: HEALTH

General Assessment

N,N-Dimethylaniline is a human poison by *ingestion* and is moderately toxic by *inhalation* and skin contact (*absorption*). There is insufficient evidence in the references to support any claims of human carcinogenicity, mutagenicity, or teratogenicity for this chemical. However, some related compounds (e.g., aniline) have demonstrated an ability to adversely affect female reproductive cycles disorders.

Inhalation is likely to cause systemic poisoning and depression of the central nervous system (CNS). Systemic poisoning is also characterized by methemoglobinemia, a dangerous and possibly fatal reduction in the red blood cell's capability to carry oxygen to the body's cells. An additional and completely separate effect (from the methemoglobinemia), is a dangerous reduction in blood pressure and tachycardia (rapid heartbeat). Symptoms include non-specific discomfort, nausea, headache, weakness, shortness of breath, anemia, and cyanosis. These symptoms may be delayed for several hours after exposure, creating a false sense of security with regard to health exposure risk.

Eye and skin contact will result in irritation. More importantly, however, is the fact that this compound is easily absorbed through intact skin causing the same toxic systemic effects as that noted for inhalation (especially depression of the CNS).

Acute Health Effects

The following acute (short-term) health effects may occur immediately or shortly after exposure to N,N-dimethylaniline:

Skin: Irritation and easy absorption through unbroken skin.

Eye: Severe irritation with possible corneal necrosis (tissue damage).

Lung: Irritation of the nose, throat, and upper respiratory tract. Easily absorbed and distributed through the respiration process to cause toxic systemic effects.

CVS: N,N-Dimethylaniline affects the blood's ability to transport oxygen and high exposure can cause death. It may also cause an unrelated drop in blood pressure and/or tachycardia.

CNS: A depressant of the central nervous system. Symptoms include headaches, weakness, irritability, drowsiness, and shortness of breath.

Chronic Health Effects

The following chronic (long-term) health effects may occur at some time after exposure to N,N-dimethylaniline and can last for months or even years:

Cancer Hazards: N,N-dimethylaniline has been tested for its ability to cause cancer in test animals and the results are inconclusive and conflicting. More research and testing is required in this regard.

Reproductive Hazard: There is limited evidence to show that exposure to N,N-dimethylaniline may adversely affect the female reproductive cycle and may lead to fetal abortion in exposed workers. Additional study is required in this regard.

Other Chronic Effects: Continuous exposure to small doses of N,N-dimethylaniline may produce anemia, loss of energy, digestive disturbance, and headache. Repeated exposures may cause anemia and methemoglobinemia to occur gradually over time rather than all at once.

Recommended Risk-Reduction Measures

Personnel should avoid direct contact with N,N-dimethylaniline. If a less toxic chemical cannot be substituted for a hazardous substance, then *engineering controls* are the most effective method of reducing exposures. The best protection is to enclose operations and/or provide local exhaust ventilation at the site of chemical release. While not always operationally feasible, isolating operations can also reduce exposure. Using respiratory protection is less effective than the controls mentioned above, but is still advisable whenever working with or around N,N-dimethylaniline. For relatively low exposures (half the PEL, or 2.5 ppm), a NIOSH-approved full facepiece respirator equipped with an organic vapor respirator will suffice. For higher exposures, a self-contained breathing apparatus (SCBA) with full facepiece and pressure demand is the recommended respiratory protection of choice. If a full facepiece is not available, then chemical goggles should be worn to protect the eyes. Whenever a chemical splash hazard exists, a face shield and a rubber apron should be worn. Skin contact must be prevented. To prevent hand and skin exposures, impervious rubber gloves should be worn.

Administrative controls should also be in place to minimize the potential for human exposures. These may include written procedures or policies which specify the methods and techniques that will be practiced whenever personnel are to work with N,N-dimethylaniline.

All personnel should receive *training* on the use, hazards, protective measures, emergency actions, and other precautions per 29 CFR 1910.1200 (Hazard Communication), prior to the first assignment in an area where N,N-dimethylaniline is used or stored.

If symptoms develop or overexposure is suspected, the following medical tests may be useful:

- ☑ Methemoglobin tests.

Any evaluation should include a careful history of past and present symptoms with an examination. Medical tests that simply look for existing damage are not a substitute for controlling exposures.

Other methods to reduce exposure include:

- ☑ Where possible, enclose operations and use local exhaust ventilation at the site of chemical release. If local exhaust ventilation or enclosure is not used, respiratory protection should be mandatory.
- ☑ Always ensure that proper protective clothing is worn when using chemical substances.
- ☑ Safety showers should be provided in areas where work is performed around N,N-dimethylaniline.
- ☑ Wash thoroughly immediately after exposure to N,N-dimethylaniline and at the end of the work shift or before eating, drinking, or smoking.
- ☑ Hazard warning information should be posted in the work area. In addition, as part of an on-going education and training program, all information on the health and safety hazards of N,N-dimethylaniline should be communicated to all potentially exposed workers.
- ☑ Eye wash stations should be provided in the immediate work area for emergency use. If there is a possibility of skin exposure to N,N-dimethylaniline, emergency shower facilities should also be provided.

RISK ASSESSMENT: ENVIRONMENT

General Assessment

The environment is at risk of exposure during transportation, storage, disposal, or destruction of N,N-dimethylaniline. In almost every scenario, the threat of environmental exposure is contingent upon the proper handling of the chemical substance. Accidental spills, large or small, can result in fire, explosion, and possible contamination of the surrounding environmental mediums (water, soil, and air).

N,N-Dimethylaniline is considered a Class IIIA combustible liquid. Alone, its high flash point and autoignition temperature do not present a serious fire or explosion hazard. However, because it is incompatible with so very many other commodities, extreme caution is required in handling, storage, transportation, and disposal of N,N-dimethylaniline. These characteristics also require special consideration during any emergency situation involving a leak or spill of N,N-dimethylaniline. Should N,N-dimethylaniline ever come into contact with incompatible substances such as oxidizers and acids, either during use, transportation, storage, or disposal, the formation of highly toxic and/or highly explosive commodities is extremely possible.

N,N-Dimethylaniline can enter the environment through industrial and municipal waste treatment plant discharges, or from spills.

Acute Ecological Effects

Acute (short-term) toxic effects may include the death of animals, birds, or fish and death or low growth rate in plants. Acute effects are seen 2 to 4 days after animals or plants are exposed to N,N-dimethylaniline.

N,N-Dimethylaniline has moderate acute toxicity to aquatic life. Insufficient data are available to evaluate or predict the short-term effects of N,N-dimethylaniline to plants, birds, or land animals.

Chronic Ecological Effects

Chronic (long-term) toxic effects may include shortened life span, reproductive problems, lower fertility, and changes in appearance or behavior in exposed animals. These effects can be seen long after first exposure(s) to toxic chemicals.

N,N-Dimethylaniline has moderate chronic toxicity in aquatic life. Insufficient data are available on the long-term effects of N,N-dimethylaniline to plants, birds, or land animals.

Water Solubility

N,N-Dimethylaniline is highly soluble in water. Concentrations of 1000 milligrams and more will mix with a liter of water.

Persistence in the Environment

N,N-Dimethylaniline is slightly persistent in water, with a half-life between 2 and 20 days. The half-life of a pollutant is the amount of time it takes for one half of the chemical to be degraded. Approximately 83% of N,N-dimethylaniline will eventually end up in air; about 16.5% will end up in water; and less than 1% will end up in about equally divided between terrestrial soil and aquatic sediments.

Bioaccumulation in Aquatic Organisms

Some substances increase in concentration, or *bioaccumulate,* in living organisms as they breathe contaminated air, drink contaminated water, or eat contaminated food. These chemicals can become concentrated in the tissues and internal organs of animals as well as humans.

The concentration of N,N-dimethylaniline found in fish tissues is expected to be somewhat higher than the average concentration of N,N-dimethylaniline in the water from which the fish was taken.

Recommended Risk-Reduction Measures

Proper training of all transporters will reduce the likelihood of a mishap or accident resulting in a leak or spill to the environment. The correct use of DOT labeling (per the MSDS) on all containers, trucks, and rail cars will enable emergency responders to react properly and quickly to any disaster thereby reducing the potential risk to the environment and to personnel.

Storage of N,N-dimethylaniline should be segregated from incompatible chemicals to minimize the risk of cross-contamination or contact.

If a spill or leak to the environment has occurred, fire department, emergency response and/or hazardous materials spill personnel should be notified immediately. Cleanup should be attempted only by those trained in proper spill containment procedures. Contaminated soils should be removed for incineration and replaced with clean soil. If N,N-dimethylaniline should contact the water table, aquifer, or navigable waterway, the local and/or state emergency response authorities must be notified. A comprehensive emergency response or disaster preparedness/recovery plan should be in place prior to any operations involving the use, transportation, storage, or disposal of N,N-dimethylaniline. If N,N-dimethylaniline is spilled or leaked, the following specific steps are recommended:

- ☑ Restrict persons not wearing protective clothing from area of spill or leak until cleanup is complete and area can be opened for normal.
- ☑ Remove all ignition sources.
- ☑ Cover liquid spills with a 9:1 mixture of sand and soda ash and mix. Collect neutralized product in safest manner possible using a scoop or shovel (non-sparking). Deposit in sealed drums form proper disposal.
- ☑ It may be necessary to dispose of N,N-dimethylaniline as a hazardous waste. The responsible state agency or the regional office of the federal Environmental Protection Agency (EPA) should be contacted for specific recommendations.

RISK ASSESSMENT: BUSINESS

General Assessment

Accidents or mishaps involving N,N-dimethylaniline can present a significant threat to business operations. The loss or damage of equipment or facilities can significantly affect fiscal viability. Lawsuits that may result from personnel injury/death, public exposures, and/or environmental contamination will also require a serious expenditure of resources. Media attention surrounding an injury, death, or environmental damage can also result in a loss of profits and loss of current as well as future business.

Recommended Risk-Reduction Measures

Company attorneys, safety and health professionals, and environmental specialists should be involved in the development of any procedures dealing with safety or emergency response. A company official should be pre-designated as a public relations officer with specific training in dealing with the press. Corporate plans and policies should be developed, approved, and implemented long before any need for such arises.

REFERENCES

Ahlbom, A. 1993. *Biostatistics for Epidemiologists.* New York: CRC Press/Lewis Publishers.

American Conference of Governmental Industrial Hygienists. 1988. *Documentation of the Threshold Limit Values and Biological Exposure Indices,* 5th Edition (with updates). Cincinnati: ACGIH

Asante-Duah, D. K. 1993. *Hazardous Waste Risk Assessment.* New York: CRC Press/Lewis Publishers.

Bloom, A. D. and F. J. DeSerres. 1995. *Ecotoxicity and Human Health.* New York: CRC Press/Lewis Publishers.

Brusick, D. J. 1994. *Methods for Genetic Risk Assessment.* New York: CRC Press/Lewis Publishers.

Bueche, J. R. 1972. *Principles of Physics,* 2nd Edition. New York: McGraw-Hill.

Calabrese, E. J. 1994. *Biological Effects of Low Level Exposures.* New York: CRC Press/Lewis Publishers.

Cockerman, L. G., B. S. Shane. 1994. *Basic Environmental Toxicology.* New York: CRC Press/Lewis Publishers.

Cothern, R. R. 1993. *Comparative Environmental Risk Assessment.* New York: CRC Press/Lewis Publishers.

Gardner, R. W. 1994. *Chemical Intolerance.* New York: CRC Press/Lewis Publishers.

Hallenbeck, W. H. 1993. *Quantitative Risk Assessment for Environmental and Occupational Health,* 2nd Edition. New York: CRC Press/Lewis Publishers.

Howard, P. H. 1995. *Handbook of Environmental Fate and Exposures Data for Organic Chemicals,* Vol. I-IV. New York: CRC Press/Lewis Publishers.

Jones, F. E. 1994. *Toxic Organic Vapors in the Workplace.* New York: CRC Press/Lewis Publishers.

Keith, L. H. and M. M. Walker. 1995. *Handbook of Air Toxics.* New York: CRC Press/Lewis Publishers.

Landis, W. G., M. H. Yu. 1995. *Introduction to Environmental Toxicology.* New York: CRC Press/Lewis Publishers.

Langerman, N. 1994 *Precautionary Labels for Chemical Containers.* New York: CRC Press/Lewis Publishers.

Lauwerys, R. R., P. Hoet. 1993. *Industrial Chemical Exposure.* New York: CRC Press/Lewis Publishers.

Lewis, R. J., Sr. 1993. *Hawley's Condensed Chemical Dictionary,* 12th Edition. New York: Van Nostrand Reinhold.

Lewis, R. J., Sr. 1992. *Sax's Dangerous Properties of Industrial Materials,* Volumes 1, 2, and 3, 8th Edition. New York: Van Nostrand Reinhold.

Lide, D. R. 1995. *Handbook of Chemistry and Physics,* 76th Edition. New York: CRC Press/Lewis Publishers.

Manahan, S. E. 1992. *Toxicological Chemistry,* 2nd Edition. New York: CRC Press/Lewis Publishers.

Mellström, G. A., J. E. Wahlberg, and H. I. Maibach. 1994. *Protective Gloves for Occupational Use.* New York: CRC Press/Lewis Publishers.

Meyer, E. 1990. *Chemistry of Hazardous Materials.* Englewood Cliffs, NJ: Prentice Hall.

National Cancer Institute. 1978. *Carcinogenesis Technical Report Series. Bioassay of Aniline for Possible Carcinogenicity.* DEHW (NIH) Publication Number: 78-1323, p 53. Washington, DC: U.S. Government Printing Office.

Ott, W. R. 1995. *Environmental Statistics and Data Analysis.* New York: CRC Press/Lewis Publishers.

Philp, R. B. 1995. *Environmental Hazards and Human Health.* New York: CRC Press/Lewis Publishers.

Quigley, D. R. 1994. *Handbook of Emergency Chemical Management.* New York: CRC Press/Lewis Publishers.

Rea, W. J. 1995. *Chemical Sensitivity,* Vol. I-III. New York: CRC Press/Lewis Publishers.

Richardson, M. 1995. *Dictionary of Substance and Their Effects,* Vol. I-VI. New York: CRC Press/Lewis Publishers.

Sacarello, H. L. A. 1994. *The Comprehensive Handbook of Hazardous Materials.* New York: CRC Press/Lewis Publishers.

Sheftel, V. O. 1995. *Handbook of Toxic Properties of Monomers and Additives.* New York: CRC Press/Lewis Publishers.

Talbott, E., G. Craun. 1995. *An Introduction to Environmental Epidemiology.* New York: CRC Press/Lewis Publishers.

U.S. Department of Health and Human Services, National Institute for Occupational Safety and Health. 1994. *NIOSH Pocket Guide to Chemical Hazards.* Washington, D.C.: U.S. Government Printing Office.

MATERIAL SAFETY DATA SHEET

CHEMICAL NAME	
DIMETHYLFORMAMIDE	**CRC PRESS/LEWIS PUBLISHERS DISCLAIMER:** THE INFORMATION AND RECOMMENDATIONS PRESENTED HEREIN ARE BASED ON SOURCES BELIEVED TO BE RELIABLE. CRC MAKES NO REPRESENTATION ON ITS COMPLETENESS OR ACCURACY. IT IS THE RESPONSIBILITY OF THE USER TO DETERMINE THE CHEMICAL'S SUITABILITY FOR ITS INTENDED USE, THE CHEMICAL'S SAFE USE, AND THE CHEMICAL'S PROPER DISPOSAL. NO REPRESENTATIONS AND/OR WARRANTIES, EITHER EXPRESSED OR IMPLIED, OF THE MERCHANTABILITY OR FITNESS FOR A PARTICULAR PURPOSE, OR OF ANY OTHER NATURE, ARE MADE WITH RESPECT TO THE INFORMATION PROVIDED IN THIS MSDS OR TO THE CHEMICAL TO WHICH INFORMATION MAY REFER. CRC NEITHER ASSUMES NOR AUTHORIZES ANY OTHER PERSON TO ASSUME FOR IT, ANY OTHER ADDITIONAL RESPONSIBILITY OR LIABILITY FOR THE USE OF, OR RELIANCE UPON, THIS INFORMATION.

HAZARD WARNING INFORMATION

HEALTH	FIRE	REACTIVE	OTHER	DEGREE OF HAZARD	COLOR CODING	OTHER CODES
1	2	0		0 = Minimum Hazard 1 = Slight Hazard 2 = Moderate Hazard 3 = Serious Hazard 4 = Severe Hazard	HEALTH = BLUE FIRE = RED REACTIVITY = YELLOW OTHER = WHITE	OX = Oxidizer ACID = Acid ALK = Alkali COR = Corrosive W = Use No Water

SECTION I - GENERAL INFORMATION

Characterization	**Aliphatic Amine**	RCRA Number	**None**
		EPA Class	**Not Applicable**
DOT Proper Shipping Name	**N,N-Dimethylformamide**	Chemical Abstract Service (CAS) Number	**68-12-2**
DOT Hazard Class and Label Requirements	**Flammable or Combustible Liquid**	DOT Emergency Guide Code	**57**
DOT Identification Number	**UN 2265**	Chemical Formula	**$HCON(CH_3)_2$**

Synonyms
Dimethyl formamide; N,N-dimethylformamide; DMF; N-formyldiemthylamine.

SECTION II - HAZARDOUS INGREDIENTS AND IDENTITY INFORMATION

Hazardous Components (specific identity)	OSHA Exposure Criteria	NIOSH Exposure Criteria	Immediately Dangerous to Life and Health (IDLH)	ACGIH Exposure Criteria
Dimethylformamide (derivation: By reaction of dimethylamine and formic acid; or from dimethylamine and hydrocyanic acid). **1 ppm = 3.04 mg/m³**	**PEL (skin):** **10 ppm** **30 mg/m³** **STEL:** **Not Established**	**REL (skin):** **10 ppm** **30 mg/m³** **STEL:** **Not Established**	**500 ppm**	**TLV (skin):** **10 ppm** **30 mg/m³** **STEL:** **Not Established**

SECTION III - PHYSICAL AND CHEMICAL CHARACTERISTICS

Boiling Point	**307°F (153°C)**	Specific Gravity (H_2O = 1)	**0.95**
Vapor Pressure (mm Hg)	**3 at 69°F (20°C)**	Molecular Weight	**73.1**
Vapor Density (Air = 1)	**2.51**	Freezing Point	**-78°F (-61°C)**

Solubility
Miscible in water and most organic solvents (except halogenated hydrocarbons).

Appearance and Odor
Colorless to pale yellow, high boiling liquid with a faint, amine-like odor.

SECTION IV - FIRE AND EXPLOSION HAZARD DATA

Flash Point (method used)	**136°F (58°C) closed cup**	Explosive Limits in Air % by Volume	**LEL: 2.2% (at 212°F) UEL: 15.2**
NFPA Classification	**Class II Combustible Liquid**	Autoignition Temperature	**833°F (445°C)**

Extinguishing Media
Use carbon dioxide, dry chemical, alcohol foam. Use water spray to cool fire-exposed containers, to subdue a leak that has not yet ignited, to protect personnel attempting to shut off leak, to disperse vapors, to dilute spills to nonflammable mixtures, and to flush spills away from incompatible chemicals.

Special Fire Fighting Procedures
Wear full protective clothing and self-contained breathing apparatus (SCBA). Move containers from fire area if it can be done without risk. Do NOT use Halon extinguishing agents to fight fire.

Unusual Fire and Explosion Hazards
Moderately flammable when exposed to heat. Containers may explode in fire. Vapors are heavier than air and can travel for some distance to an ignition source and flashback to cause fire.

SECTION V - REACTIVITY DATA

Stability		Conditions to Avoid **Under normal conditions of handling and storage, N,N-dimethylamine is stable at room temperature and pressure. Avoid exposure to incompatible chemicals, heat, and ignition sources.**
Stable X	Unstable	Incompatibility *(materials to avoid)* **Bromine, carbon tetrachloride, chromic anhydride, phosphorus oxychloride, hexachlorobenzene, magnesium nitrate, methylene diisocyanate, phosphorus trioxide, triethyl aluminum, organic nitrates.**
Hazardous Polymerization		Conditions to Avoid **Under normal conditions of temperature and pressure, hazardous polymerization of dimethylformamide is not expected to occur. However, methylene diisocyante can polymerize violently on contact.**
May Occur	Will Not Occur X	Hazardous Decomposition or By-products **When heated to decomposition, dimethylformamide can emit highly toxic/poisonous gases, including carbon monoxide, dimethyl amine, and oxides of nitrogen.**

SECTION VI - HEALTH HAZARD DATA

Primary Route(s) of Entry:	Inhalation?	Absorption (skin)?	Ingestion?
	X	X	

Health Hazards

INHALATION: Irritation of the eyes, mucosa, and upper respiratory tract, with dizziness, headache, liver and kidney damage, and intolerance to alcohol. May also cause abdominal distress, colicky abdominal pain, loss of appetite, nausea, vomiting, diarrhea or constipation, facial flushing, agitation, increased blood pressure, weakness, anorexia, abnormal liver function, dermatitis, hallucinations, hepatomegaly, and digestive tract disturbance.

ABSORPTION: Absorbs freely through intact skin. Systemic effects same as irritation. Eye contact will produce corneal damage.

INGESTION: Unspecified effects, possibly the same as inhalation. Liver damage is possible.

Carcinogenicity	NTP Listed?	IARC Cancer Review Group?	OSHA Regulated?	Target Organs?
Questioned Human Suspected Animal	5th Annual Report	Group 2B	29 CFR 1910.1000 Table Z-1	Eyes, skin, resp. system, kidneys, liver, CNS, CVS.

Medical Conditions Generally Aggravated by Exposure

Liver, kidney, and CNS disorders may be aggravated upon exposure to dimethylformamide.

Emergency and First-aid Procedures

Eye contact: Flush immediately with water for 15 minutes (minimum), seek medical attention. Skin contact: Remove all contaminated clothing. Immediately wash area with large amounts of soap and water. Seek medical assistance. For inhalation: Remove the person from exposure. Provide respiratory assistance and CPR. Transfer to medical facility. If swallowed, seek medical attention immediately. Unless advised otherwise, give 2-3 glasses of water or milk to a conscious person. Gastric lavage is preferable to induced vomiting due to the potential for additional injury to the mucosa.

SECTION VII - PRECAUTIONS FOR SAFE HANDLING AND USE

Steps to be Taken in Case Material is Released or Spilled

Absorb liquids in vermiculite or other material and deposit in sealed drums for disposal. Remove all ignition sources. Ventilate area of spill or leak. Restrict those not involved in cleanup from area.

Preferred Waste Disposal Method

No Citation.

Precautions to be Taken in Handling and Storage

Store in tightly closed containers in cool, well-ventilated area away from heat and sources of ignition. Protect containers from physical damage and from exposure to direct sunlight.

Other Precautions and Warnings

Ground and bond all equipment used with this chemical to prevent static sparks.

SECTION VIII - CONTROL MEASURES AND PERSONAL PROTECTIVE EQUIPMENT

Respiratory Protection *(specify type)*

For low exposures (half the PEL), use an organic vapor respirator. Otherwise, use a self-contained breathing apparatus (SCBA) with full facepiece operated in positive pressure mode.

Ventilation

Local exhaust at site of chemical work.

Protective Gloves	Eye Protection	Other Protective Clothing
Impervious Rubber	Chemical Goggles or Face Mask	Rubber Apron

Work/Hygiene Practices

Always wash hands thoroughly after using chemical; never bring food, drink, or smoking materials into vicinity of chemicals.

DIMETHYLFORMAMIDE

$HCON(CH_3)_2$ CAS: 68-12-2

IDENTIFICATION AND TYPICAL USES

Dimethylformamide is a colorless to pale yellow liquid with a faint, amine-like odor. It is used as a solvent for vinyl resins and acetylene, butadiene, acid gases, polyacrylic fibers, as a catalyst in carboxylation reactions, organic synthesis, and as a carrier for gases. It is also used as a booster or co-solvent in protective coatings, adhesives, films, printing inks, and whenever a solvent with a slow evaporation rate is required.

RISK ASSESSMENT: HEALTH

General Assessment

Dimethylformamide is moderately toxic by *ingestion* and mildly toxic by *inhalation* and *absorption*. It is a suspected animal carcinogen and a suspect human carcinogen. Human mutation data have also been reported and caution in use, handling, and storage is therefore warranted.

Inhalation, ingestion, and absorption will result in a variety of toxic systemic effects with action on the central nervous system (CNS), the cardiovascular system (CVS), and the respiratory system. Symptoms of exposure include irritation to the eyes, mucous membranes and upper respiratory tract. There may be dizziness, headache, liver and kidney damage, and a marked intolerance to alcoholic beverages that may last up to four days after exposure. This may cause facial flushing and blotchy redness of the neck, face, and upper extremities after consuming alcohol. These may be accompanied dizziness, nausea, dyspnea, and chest palpitations. Aside from the alcohol reaction, there may also be symptoms of abdominal distress, colicky abdominal pain, loss of appetite, nausea, vomiting, diarrhea or constipation, agitation, increased blood pressure, skin irritation, and rash. There may be other symptoms of weakness, abnormal liver function, dermatitis, anorexia, and hepatomegaly (enlarged liver). Further CNS effects include loss of coordination, hallucinations, and digestive tract disturbances with gastric cramps. The liquid can burn the skin and eyes and cause corneal damage and necrosis on contact.

Acute Health Effects

The following acute (short-term) health effects may occur immediately or shortly after exposure to dimethylformamide:

Skin: Irritation, dermatitis, rash, and inflammation with, easy absorption through unbroken skin.

Eye: Severe irritation with possible corneal necrosis (tissue damage).

Lung: Irritation of the nose, throat, and upper respiratory tract. Easily absorbed and distributed through the respiration process to cause a wide range of toxic systemic effects with action on the CNS and CVS pronounced.

CVS: Effects an increase in blood pressure and may cause potentially threatening heart palpitations.

CNS: A depressant of the central nervous system. Symptoms include headaches, weakness, irritability, drowsiness, and shortness of breath.

Other: Causes a marked intolerance to alcohol that can last up to four days after exposure. Consumption of amounts that may normally be tolerated under non-exposure circumstances can cause serious adverse health reactions.

Chronic Health Effects

The following chronic (long-term) health effects may occur at some time after exposure to dimethylformamide and can last for months or even years:

Cancer Hazards: Dimethylformamide has been tested for its ability to cause cancer in test animals and the results are inconclusive and conflicting. More research and testing is required in this regard. Its carcinogenicity is highly suspect in animals and questioned in humans.

Reproductive Hazard: There is limited evidence to show that exposure to dimethylformamide may adversely affect reproduction in test animals. Additional study is required in this regard.

Other Chronic Effects: Continuous exposure to small doses of dimethylformamide may produce chronic dermatitis. Liver damage may also occur in the long-term.

Recommended Risk-Reduction Measures

Personnel should avoid direct contact with dimethylformamide. If a less toxic chemical can not be substituted for a hazardous substance, then *engineering controls* are the most effective method of reducing exposures. The best protection is to enclose operations and/or provide local exhaust ventilation at the site of chemical release. While not always operationally feasible, isolating operations can also reduce exposure. Using respiratory protection is less effective than the controls mentioned above, but is still advisable whenever working with or around dimethylformamide. For relatively low exposures (half the PEL, or 5 ppm), a NIOSH-approved full facepiece respirator equipped with an organic vapor respirator will suffice. For higher exposures, a self-contained breathing apparatus (SCBA) with full facepiece and pressure demand is the recommended respiratory protection of choice. If a full facepiece is not available, then chemical goggles should be worn to protect the eyes. Whenever a chemical splash hazard exists, a face shield and a rubber apron should be worn. This chemical will easily pass through unbroken skin. Skin contact must therefore be avoided. To prevent hand and skin exposures, impervious rubber gloves should be worn.

Administrative controls should also be in place to minimize the potential for human exposures. These may include written procedures or policies which specify the methods and techniques that will be practiced whenever personnel are to work with dimethylformamide.

All personnel should receive *training* on the use, hazards, protective measures, emergency actions, and other precautions per 29 CFR 1910.1200 (Hazard Communication), prior to the first assignment in an area where dimethylformamide is used or stored.

If symptoms develop or overexposure is suspected, the following medical tests may be useful:

- ☑ Lung function tests.
- ☑ Liver function tests.

Any evaluation should include a careful history of past and present symptoms with an examination. Medical tests that simply look for existing damage are not a substitute for controlling exposures.

Other methods to reduce exposure include:

- ☑ Where possible, enclose operations and use local exhaust ventilation at the site of chemical release. If local exhaust ventilation or enclosure is not used, respiratory protection should be mandatory.
- ☑ Always ensure that proper protective clothing is worn when using chemical substances.
- ☑ All equipment used with dimethylformamide should be electrically grounded and bonded to prevent static spark and ignition.
- ☑ Wash thoroughly immediately after exposure to dimethylformamide and at the end of the work shift or before eating, drinking, or smoking.
- ☑ Hazard warning information should be posted in the work area. In addition, as part of an on-going education and training program, all information on the health and safety hazards dimethylformamide. should be communicated to all exposed and potentially exposed workers.
- ☑ Eye wash stations should be provided in the immediate work area for emergency use. If there is a possibility of skin exposure to dimethylformamide, emergency shower facilities should be provided.

RISK ASSESSMENT: ENVIRONMENT

General Assessment

The environment is at risk of exposure during transportation, storage, disposal, or destruction of dimethylformamide. In almost every scenario, the threat of environmental exposure is contingent upon the proper handling of the chemical substance. Accidental spills, large or small, can result in fire, explosion, and possible contamination of the surrounding environmental mediums (water, soil, and air).

Dimethylformamide is considered a Class II combustible liquid. Alone, its high flash point and autoignition temperature do not present a serious fire or explosion hazard. However, because it is incompatible with so very many other commodities, extreme caution is required in handling, storage, transportation, and disposal of dimethylformamide. These characteristics also require special consideration during any emergency situation involving a leak or spill of dimethylformamide. Should dimethylformamide ever come into contact with incompatible substances, such as bromine and other oxidizing materials, either during use, transportation, storage, or disposal, the formation of highly toxic and/or highly explosive commodities is extremely possible.

Dimethylformamide can enter the environment through industrial and municipal waste treatment plant discharges, or from spills.

Acute Ecological Effects

Acute (short-term) toxic effects may include the death of animals, birds, or fish and death or low growth rate in plants. Acute effects are seen 2 to 4 days after animals or plants are exposed to dimethylformamide. Insufficient data are available to evaluate or predict the short-term effects of dimethylformamide to aquatic life, plants, birds, or land animals.

Chronic Ecological Effects

Chronic (long-term) toxic effects may include shortened life span, reproductive problems, lower fertility, and changes in appearance or behavior in exposed animals. These effects can be seen long after first exposure(s) to toxic chemicals. Insufficient data are available on the long-term effects of dimethylformamide to aquatic life plants, birds, or land animals.

Water Solubility

Dimethylformamide is highly soluble in water. Concentrations of 1000 milligrams and more will mix with a liter of water.

Persistence in the Environment

Dimethylformamide is slightly persistent in water, with a half-life between 2 and 20 days. The half-life of a pollutant is the amount of time it takes for one half of the chemical to be degraded.

Bioaccumulation in Aquatic Organisms

Some substances increase in concentration, or *bioaccumulate,* in living organisms as they breathe contaminated air, drink contaminated water, or eat contaminated food. These chemicals can become concentrated in the tissues and internal organs of animals as well as humans.

The concentration of dimethylformamide found in fish tissues is expected to be about the same as the average concentration of dimethylformamide in the water from which the fish was taken.

Recommended Risk-Reduction Measures

Proper training of all transporters will reduce the likelihood of a mishap or accident resulting in a leak or spill to the environment. The correct use of DOT labeling (per the MSDS) on all containers, trucks, and rail cars will enable emergency responders to react properly and quickly to any disaster thereby reducing the potential risk to the environment and to personnel.

Storage of dimethylformamide should be segregated from incompatible chemicals to minimize the risk of cross-contamination or contact. Storage buildings should be equipped with proper fire protection and prevention equipment.

If a spill or leak to the environment has occurred, fire department, emergency response and/or hazardous materials spill personnel should be notified immediately. Cleanup should be attempted only by those trained in proper spill containment procedures. Contaminated soils should be removed for incineration and replaced with clean soil. If dimethylformamide should contact the water table, aquifer, or navigable waterway, the local and/or state emergency response authorities must be notified. A comprehensive emergency response or disaster preparedness/recovery plan should be in place prior to any operations involving the use, transportation, storage, or disposal of dimethylformamide. If dimethylformamide is spilled or leaked, the following specific steps are recommended:

- ☑ Restrict persons not wearing protective clothing from area of spill or leak until cleanup is complete and area can be opened for normal work.
- ☑ Remove all ignition sources and ventilate area of spill or leak.
- ☑ Liquids should be absorbed in vermiculite, sand, or other suitable, non-reactive material and disposed of is a sealed drum.
- ☑ It may be necessary to dispose of dimethylformamide as a hazardous waste. The responsible state agency or the regional office of the federal Environmental Protection Agency (EPA) should be contacted for specific recommendations.

RISK ASSESSMENT: BUSINESS

General Assessment

Accidents or mishaps involving dimethylformamide can present a significant threat to business operations. The loss or damage of equipment or facilities can significantly affect fiscal viability. Lawsuits that may result from personnel injury/death, public exposures, and/or environmental contamination will also require a serious expenditure of resources. Media attention surrounding an injury, death, or environmental damage can also result in a loss of profits and loss of current as well as future business. Always remember that anytime the terms "cancer" or "carcinogen" are used, pubic emotion, hysteria, and ignorance can run equally high. This should be a consideration during the development of public relation policies.

Recommended Risk-Reduction Measures

Company attorneys, safety and health professionals, and environmental specialists should be involved in the development of any procedures dealing with safety or emergency response. A company official should be pre-designated as a public relations officer with specific training in dealing with the press. Corporate plans and policies should be developed, approved, and implemented long before any need for such arises.

REFERENCES

Ahlbom, A. 1993. *Biostatistics for Epidemiologists.* New York: CRC Press/Lewis Publishers.

American Conference of Governmental Industrial Hygienists. 1988. *Documentation of the Threshold Limit Values and Biological Exposure Indices,* 5th Edition (with updates). Cincinnati: ACGIH

Asante-Duah, D. K. 1993. *Hazardous Waste Risk Assessment.* New York: CRC Press/Lewis Publishers.

Bloom, A. D. and F. J. DeSerres. 1995. *Ecotoxicity and Human Health.* New York: CRC Press/Lewis Publishers.

Brusick, D. J. 1994. *Methods for Genetic Risk Assessment.* New York: CRC Press/Lewis Publishers.

Bueche, J. R. 1972. *Principles of Physics,* 2nd Edition. New York: McGraw-Hill.

Calabrese, E. J. 1994. *Biological Effects of Low Level Exposures.* New York: CRC Press/Lewis Publishers.

Cockerman, L. G., B. S. Shane. 1994. *Basic Environmental Toxicology.* New York: CRC Press/Lewis Publishers.

Cothern, R. R. 1993. *Comparative Environmental Risk Assessment.* New York: CRC Press/Lewis Publishers.

Gardner, R. W. 1994. *Chemical Intolerance.* New York: CRC Press/Lewis Publishers.

Hallenbeck, W. H. 1993. *Quantitative Risk Assessment for Environmental and Occupational Health,* 2nd Edition. New York: CRC Press/Lewis Publishers.

Howard, P. H. 1995. *Handbook of Environmental Fate and Exposures Data for Organic Chemicals,* Vol. I-IV. New York: CRC Press/Lewis Publishers.

Jones, F. E. 1994. *Toxic Organic Vapors in the Workplace.* New York: CRC Press/Lewis Publishers.

Keith, L. H. and M. M. Walker. 1995. *Handbook of Air Toxics.* New York: CRC Press/Lewis Publishers.

Landis, W. G., M. H. Yu. 1995. *Introduction to Environmental Toxicology.* New York: CRC Press/Lewis Publishers.

Langerman, N. 1994 *Precautionary Labels for Chemical Containers.* New York: CRC Press/Lewis Publishers.

Lauwerys, R. R., P. Hoet. 1993. *Industrial Chemical Exposure.* New York: CRC Press/Lewis Publishers.

Lewis, R. J., Sr. 1993. *Hawley's Condensed Chemical Dictionary,* 12th Edition. New York: Van Nostrand Reinhold.

Lewis, R. J., Sr. 1992. *Sax's Dangerous Properties of Industrial Materials,* Volumes 1, 2, and 3, 8th Edition. New York: Van Nostrand Reinhold.

Manahan, S. E. 1992. *Toxicological Chemistry,* 2nd Edition. New York: CRC Press/Lewis Publishers.

Meyer, E. 1990. *Chemistry of Hazardous Materials.* Englewood Cliffs, NJ: Prentice Hall.

Ott, W. R. 1995. *Environmental Statistics and Data Analysis.* New York: CRC Press/Lewis Publishers.

Philp, R. B. 1995. *Environmental Hazards and Human Health.* New York: CRC Press/Lewis Publishers.

Quigley, D. R. 1994. *Handbook of Emergency Chemical Management.* New York: CRC Press/Lewis Publishers.

Rea, W. J. 1995. *Chemical Sensitivity,* Vol. I-III. New York: CRC Press/Lewis Publishers.

Richardson, M. 1995. *Dictionary of Substance and Their Effects,* Vol. I-VI. New York: CRC Press/Lewis Publishers.

Sacarello, H. L. A. 1994. *The Comprehensive Handbook of Hazardous Materials.* New York: CRC Press/Lewis Publishers.

Torkelson, T. R. V. K. Rowe. 1981. *Patty's Industrial Hygiene and Toxicology,* Vol. 2B. New York: Wiley-Interscience.

U.S. Department of Health and Human Services, National Institute for Occupational Safety and Health. 1994. *NIOSH Pocket Guide to Chemical Hazards.* Washington, D.C.: U.S. Government Printing Office.

MATERIAL SAFETY DATA SHEET

CHEMICAL NAME	CRC PRESS/LEWIS PUBLISHERS DISCLAIMER
1,1-DIMETHYLHYDRAZINE	**CRC PRESS/LEWIS PUBLISHERS DISCLAIMER:** THE INFORMATION AND RECOMMENDATIONS PRESENTED HEREIN ARE BASED ON SOURCES BELIEVED TO BE RELIABLE. CRC MAKES NO REPRESENTATION ON ITS COMPLETENESS OR ACCURACY. IT IS THE RESPONSIBILITY OF THE USER TO DETERMINE THE CHEMICAL'S SUITABILITY FOR ITS INTENDED USE, THE CHEMICAL'S SAFE USE, AND THE CHEMICAL'S PROPER DISPOSAL. NO REPRESENTATIONS AND/OR WARRANTIES, EITHER EXPRESSED OR IMPLIED, OF THE MERCHANTABILITY OR FITNESS FOR A PARTICULAR PURPOSE, OR OF ANY OTHER NATURE, ARE MADE WITH RESPECT TO THE INFORMATION PROVIDED IN THIS MSDS OR TO THE CHEMICAL TO WHICH INFORMATION MAY REFER. CRC NEITHER ASSUMES NOR AUTHORIZES ANY OTHER PERSON TO ASSUME FOR IT, ANY OTHER ADDITIONAL RESPONSIBILITY OR LIABILITY FOR THE USE OF, OR RELIANCE UPON, THIS INFORMATION.

HAZARD WARNING INFORMATION

HEALTH	FIRE	REACTIVE	OTHER	DEGREE OF HAZARD	COLOR CODING	OTHER CODES
3	3	1		0 = Minimum Hazard 1 = Slight Hazard 2 = Moderate Hazard 3 = Serious Hazard 4 = Severe Hazard	HEALTH = BLUE FIRE = RED REACTIVITY = YELLOW OTHER = WHITE	OX = Oxidizer ACID = Acid ALK = Alkali COR = Corrosive W = Use No Water

SECTION I - GENERAL INFORMATION

Characterization	**Unsymmetrical Compound**	RCRA Number	**U098**
		EPA Class	**Toxic Waste**
DOT Proper Shipping Name	**Dimethylhydrazine (unsymmetrical)**	Chemical Abstract Service (CAS) Number	**57-14-7**
DOT Hazard Class and Label Requirements	**Flammable Liquid**	DOT Emergency Guide Code	**57**
DOT Identification Number	**UN 1163**	Chemical Formula	$(CH_3)_2NNH_2$

Synonyms
Dimazine; DMH; UDMH; unsymmetrical dimethylhydrazine; N,N-dimethylhydrazine.

SECTION II - HAZARDOUS INGREDIENTS AND IDENTITY INFORMATION

Hazardous Components (specific identity)	OSHA Exposure Criteria	NIOSH Exposure Criteria	Immediately Dangerous to Life and Health (IDLH)	ACGIH Exposure Criteria
1,1-dimethylhydrazine (derivation: Reaction of diethylamine and chloramine; catalytic oxidation of dimethylamine and ammonia). **1 ppm = 2.50 mg/m³**	**PEL (skin):** **0.5 ppm** **1 mg/m³** **STEL:** **Not Applicable**	**REL:** **0.6 ppm** **0.15 mg/m³** **(2-hr ceiling)** **Possible Cancer Agent**	**15 ppm**	**TLV (skin):** **0.5 ppm** **1 mg/m³** **Suspected Human Carcinogen**

SECTION III - PHYSICAL AND CHEMICAL CHARACTERISTICS

Boiling Point	**147°F (64°C)**	Specific Gravity (H_2O = 1)	**0.79**
Vapor Pressure (mm Hg)	**103 mm at 69°F (20°C)**	Molecular Weight	**60.1**
Vapor Density (Air = 1)	**1.94**	Melting Point	**-72°F (-58°C)**

Solubility
Miscible in water, ether, and alcohol.

Appearance and Odor
Colorless liquid with an ammonia or fish-like odor.

SECTION IV - FIRE AND EXPLOSION HAZARD DATA

Flash Point (method used)	**5°F (-15°C) closed cup**	Explosive Limits in Air % by Volume	**LEL: 2% UEL: 90%**
NFPA Classification	**Class IB Flammable Liquid**	Autoignition Temperature	**485°F (252°C)**

Extinguishing Media
Use water spray, alcohol foam, or carbon dioxide. Water may be ineffective by itself (may spread fire).

Special Fire Fighting Procedures
Structural protective clothing is permeable. Remain clear of smoke, water fall-out, and water run-off. Poisonous gases are produced in fire. Move containers if it can be done without risk. Cool fire-exposed containers with water spray and use water spray to disperse vapors. Evacuate non-essential personnel 2500 feet from the fire area (consider wind conditions).

Unusual Fire and Explosion Hazards
Contact with oxidizing agents can cause violent and explosive ignition. A volatile liquid that is flammable over a wide vapor-to-air range. Vapors are heavier than air and may travel a considerable distance to a source of ignition and flashback to cause fire or explosion. Burns with clear flame.

SECTION V - REACTIVITY DATA

Stability		Conditions to Avoid 1,1-dimethylhydrazine is stable at room temperature in closed containers under normal conditions of storage and handling. Avoid contact with incompatible chemicals and heat.
Stable X	Unstable	Incompatibility *(materials to avoid)* Highly reactive with oxidizers (chlorine, bromine, nitrogen dioxide), halogens, ferric oxide (iron rust), metallic mercury, fuming nitric acid, hydrogen peroxide (ignites spontaneously with oxidizers).
Hazardous Polymerization		Conditions to Avoid Under normal conditions of temperature and pressure, hazardous polymerization of 1,1-dimethylhydrazine is not expected to occur.
May Occur	Will Not Occur X	Hazardous Decomposition or By-products When heated to decomposition, 1,1-dimethylhydrazine will produce toxic and irritating gases, including explosive hydrogen gas, ammonia, dimethylamine, and hydrazoic acid.

SECTION VI - HEALTH HAZARD DATA

Primary Route(s) of Entry:	Inhalation?	Absorption (skin)?	Ingestion?
	X	X	X

Health Hazards

INHALATION: Highly corrosive to the skin, eyes, and mucosa. Causes choking, chest pain, dyspnea, lethargy, nausea, convulsions, tremors, pulmonary irritation, liver injury, and hypoxia.

ABSORPTION: Will pass through unbroken skin to cause toxic systemic effects as noted above. Extremely corrosive to skin tissue causing chemical burns (2nd and 3rd degree) on short contact. Very injurious to the eyes, causing corneal damage and permanent blindness.

INGESTION: A poison by ingestion. Delayed gastrointestinal irritation. Systemic effects can occur.

Carcinogenicity	NTP Listed?	IARC Cancer Review Group?	OSHA Regulated?	Target Organs?
Suspected Human Confirmed Animal	5th Annual Report	Group 2B	29 CFR 1910.1000 Table Z-1	CNS, skin, liver, eyes, GI tract, blood, resp. system.

Medical Conditions Generally Aggravated by Exposure

Existing deficiencies in CNS function, liver, kidneys, and the circulatory system may be aggravated.

Emergency and First-aid Procedures

Eye contact: Flush immediately with water for 15 minutes (minimum) while holding upper and lower eyelids open. Seek medical attention. Skin contact: Remove all contaminated clothing. Immediately wash area with large amounts of water (15 minutes minimum). For inhalation: Remove the person from exposure. Provide respiratory assistance and CPR. Transfer to medical facility. Observe for 24-48 hours for lung effects (pulmonary edema). If swallowed: Call the poison control center. Give 1-2 glasses of water or milk to a conscious and alert person. Do NOT induce vomiting and seek medical attention immediately. Never attempt to give anything by mouth to an unconscious person.

SECTION VII - PRECAUTIONS FOR SAFE HANDLING AND USE

Steps to be Taken in Case Material is Released or Spilled

Evacuate area and deny entry to those not involved in cleanup activities. Protective clothing is mandatory. Remove all sources of ignition. Ventilate area of spill. Absorb liquids with toweling, vermiculite, or sand. Deposit in sealed drum for disposal. Notify appropriate authorities.

Preferred Waste Disposal Method

Incineration in a chemical incinerator equipped with afterburner and scrubber.

Precautions to be Taken in Handling and Storage

Store in tightly closed containers in a cool, dark, dry, well-ventilated area. Personnel should be trained on the hazardous properties of 1,1-dimethylhydrazine *prior* to working with the material.

Other Precautions and Warnings

Containers should be protected from physical damage. Use extreme caution when handling this compound. Use explosion-proof electrical equipment in areas where this chemical is used or stored.

SECTION VIII - CONTROL MEASURES AND PERSONAL PROTECTIVE EQUIPMENT

Respiratory Protection *(specify type)*

A carcinogenic agent. Exposure level is extremely low and difficult to monitor. For any exposures, use a supplied-air respirator set in positive pressure or continuous flow mode or a self-contained breathing apparatus (SCBA) with full facepiece operated in pressure demand or other positive pressure mode.

Ventilation

Local exhaust or general mechanical systems recommended.

Protective Gloves	Eye Protection	Other Protective Clothing
Impervious Gloves	Chemical Goggles or Face Mask	Full Protective Clothing

Work/Hygiene Practices

Always wash hands thoroughly after using chemical; never bring food, drink, or smoking materials into vicinity of chemicals.

1,1-DIMETHYLHYDRAZINE

$(CH_3)_2NNH_2$ CAS: 57-14-7

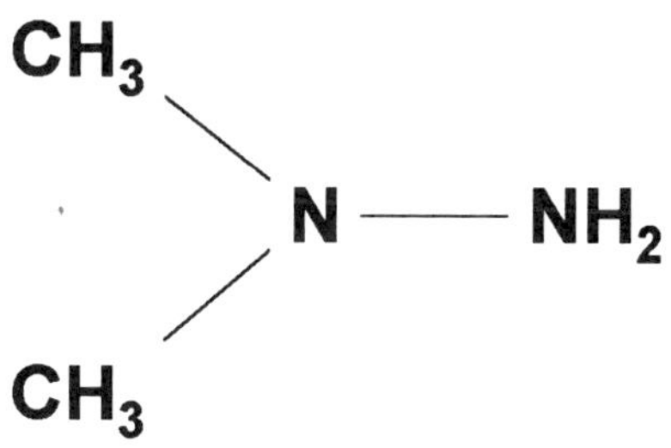

IDENTIFICATION AND TYPICAL USES

1,1-Dimethylhydrazine is a colorless liquid with an ammonia or fish-like odor. It is used as a component of jet and rocket fuel, as an absorbent for acid gases, and as an agent to control the growth of plants.

RISK ASSESSMENT: HEALTH

General Assessment

1,1-Dimethylhydrazine is a confirmed human and animal carcinogen. Human mutation data have also been reported. It is primarily toxic by *inhalation* and skin contact (*absorption*). It can also cause toxic and damaging effects by *ingestion*.

Inhalation will cause pulmonary irritation up to and including pulmonary edema (fluid buildup in the lungs), which is a medical emergency and can lead to death. Symptoms may be delayed up to 48 hours creating a false sense of security with regard to health exposure risk. Symptoms include choking, extreme chest pain, dyspnea (difficulty in breathing), lethargy, nausea, convulsions, tremors, liver injury, and hypoxia (reduced oxygen levels in the blood) leading to hemolytic anemia. Systemic intoxication, depression of the central nervous system (CNS), and delayed gastrointestinal irritation may also occur.

Aside from its ability to pass through intact skin (absorption), 1,1-dimethylhydrazine is also extremely corrosive to skin tissue and can cause 2nd and 3rd degree burns on contact. Damage to the eyes and cornea can be severe causing partial or complete loss of vision.

Ingestion is very unusual since the vapors of this chemical are so irritating, personnel are not likely to get close enough to ingest it. However, should it be swallowed, there can be burning of the mouth, throat, and esophagus, followed by gastrointestinal irritation.

Acute Health Effects

The following acute (short-term) health effects may occur immediately or shortly after exposure to 1,1-dimethylhydrazine:

Skin: Severe irritation and burning. Skin absorption will occur causing toxic systemic effects.

Eye: Severe irritation and tissue damage up to and including loss of vision.

Lung: Irritation of the upper respiratory tract causing choking, coughing, dypsnea, and pulmonary irritation leading to pulmonary edema (fluid in the lungs), which can be fatal.

CNS: Dizziness, convulsions, tremors, nausea, headaches, possible injury to the central nervous system.

Chronic Health Effects

The following chronic (long-term) health effects may occur at some time after exposure to 1,1-dimethylhydrazine and can last for months or even years:

Cancer Hazards: According to the information presented in the references, 1,1-dimethylhydrazine has been shown to cause cancer in test animals. Its carcinogenicity in humans is expected to be at least equal to that of animals. Human mutation data have been reported for this compound.

Reproductive Hazard: According to the information presented in the references, this chemical has been shown to adversely affect reproduction in test animals.

Other Chronic Effects: Repeated exposures can lead to hemolytic anemia and convulsions. Chronic exposures may lead to cancer.

Recommended Risk-Reduction Measures

1,1-Dimethylhydrazine is listed by the IARC as a confirmed animal carcinogen and the ACGIH as a suspected human carcinogen. Unless a less toxic chemical can be substituted for a hazardous substance, *engineering controls* are the most effective method of reducing exposures. The best protection is to enclose operations and/or provide local exhaust ventilation at the site of chemical release. While not always operationally feasible, isolating operations can also reduce exposure.

Using respiratory protection is less effective than the controls mentioned above, but is still advisable whenever working with or around 1,1-dimethylhydrazine. The exposure level is extremely low (0.5 ppm) and difficult to detect and monitor. For any detectable exposures to a suspected carcinogenic agent, use a supplied-air respirator set in positive pressure or continuous flow mode, or use a self-contained breathing apparatus (SCBA) with full facepiece and pressure demand. If a full facepiece is not available, then chemical goggles should be worn to protect the eyes. Whenever an eye or face contact hazard exists, a face shield and apron should be worn. To prevent hand and skin exposures, impervious gloves should be worn.

Administrative controls should also be in place to minimize the potential for human exposures. These may include written procedures or policies which specify the methods and techniques that will be practiced whenever personnel are to work with or around 1,1-dimethylhydrazine.

All personnel should receive *training* on the use, hazards, protective measures, emergency actions, and other precautions per 29 CFR 1910.1200 (Hazard Communication), prior to the first assignment in an area where 1,1-dimethylhydrazine is used or stored.

Before beginning employment and at regular intervals thereafter (e.g. annually), the following medical tests are recommended:

- ☑ Lung function tests (establish a baseline).

If symptoms develop or overexposure is suspected, the following medical tests may be useful:

- ☑ Lung function tests.
- ☑ Liver and kidney function tests.
- ☑ Consider chest X-ray after acute overexposure.
- ☑ Evaluation by a qualified allergist with careful consideration of exposure history and special tests (may help diagnose allergy).

Any evaluation should include a careful medical history of past and present symptoms with an examination. However, medical tests that evaluate existing damage are not a substitute for controlling exposure. Also, since smoking can cause heart disease, emphysema, and numerous other respiratory disorders, smokers may be affected more rapidly than non-smokers to the same exposure conditions. Prudent risk assessment and management requires careful consideration of *all* possible risk factors that may be causing the appearance of symptoms in exposed workers.

Other methods to reduce exposure include:

- ☑ Where possible, enclose operations and use local exhaust ventilation at the site of chemical release. If local exhaust ventilation or enclosure is not used, respiratory protection should be mandatory.
- ☑ Always ensure that proper protective clothing is worn when using chemical substances.
- ☑ Wash thoroughly immediately after exposure to 1,1-dimethylhydrazine and at the end of the work shift or before eating, drinking, or smoking.
- ☑ Hazard warning information must be posted in the work area. It is recommended that a regulated, controlled work area should be established wherever this chemical is used, handled, or stored. In addition, as part of an on-going education and training program, all information on the health and safety hazards of 1,1-dimethylhydrazine should be communicated to all potentially exposed workers.
- ☑ Eye wash stations should be provided in work areas. If the potential for whole body exposure exists, then safety showers should also be provided.

RISK ASSESSMENT: ENVIRONMENT

General Assessment

The environment is at risk of exposure during intended use, transportation, storage, disposal, or destruction of 1,1-dimethylhydrazine. In almost every scenario, the threat of environmental exposure is contingent upon the proper handling of the chemical substance. Accidental spills, large or small, can result in fire, explosion, and possible contamination of the surrounding environmental mediums (water, soil, and air).

1,1-Dimethylhydrazine should always be handled with extreme caution. It is a Class1B flammable liquid. Its extremely low flash point and relatively low boiling point make this chemical a dangerous fire and explosion hazard. Its vapors form explosive mixtures in air over a wide range of concentrations. It forms shock-sensitive explosive products when combined with metal perchlorates, alkali metals, or their dichromates. It is highly reactive with a number of substances, especially oxidizers and oxidizing agents. These characteristics require special consideration during any emergency situation involving a leak or spill of 1,1-dimethylhydrazine.

1,1-Dimethylhydrazine can enter the environment through industrial discharges or through spills.

Acute Ecological Effects

Acute (short-term) toxic effects may include the death of animals, birds, or fish and death or low growth rate in plants. Acute effects are seen 2 to 4 days after animals or plants are exposed to 1,1-dimethylhydrazine. This chemical has moderate acute toxicity to aquatic life and plant life. It has been used to control the growth rate of some ornamental and agricultural crops. Insufficient data are available to evaluate or predict the short-term effects of 1,1-dimethylhydrazine to birds or land animals.

Chronic Ecological Effects

Chronic toxic effects may include shortened life span, reproductive problems, lower fertility, and changes in appearance or behavior in exposed animals. These effects can be seen long after first exposure(s) to toxic chemicals. 1,1-Dimethylhydrazine has low chronic toxicity to aquatic life and moderate chronic toxicity to plant life. Insufficient data are available to evaluate or predict the long-term effects of 1,1-dimethylhydrazine to birds or land animals.

Water Solubility

1,1-Dimethylhydrazine is highly soluble in water. Concentrations of 1000 milligrams and more will mix with a liter of water.

Persistence in the Environment

1,1-Dimethylhydrazine is moderately persistent in water, with a half-life of between 20 to 200 days. The half-life of a pollutant is the amount of time it takes for one half of the chemical to be degraded. It is expected to absorb to sediment. No information is available in the references on the expected percentages of 1,1-dimethylhydrazine that will be divided through the environmental mediums (primarily air and water).

Bioaccumulation in Aquatic Organisms

Some substances increase in concentration, or *bioaccumulate,* in living organisms as they breathe contaminated air, drink contaminated water, or eat contaminated food. These chemicals can become concentrated in the tissues and internal organs of animals as well as humans.

The concentration of 1,1-dimethylhydrazine found in fish tissues is expected to be somewhat higher than the average concentration found in the water from which the fish was taken.

Recommended Risk-Reduction Measures

Proper training of all transporters will reduce the likelihood of a mishap or accident resulting in a leak or spill to the environment. The correct use of labeling on all containers, trucks, and rail cars will enable emergency responders to react properly and quickly to any disaster thereby reducing the potential risk to the environment and to personnel.

Storage of 1,1-dimethylhydrazine should be segregated from other chemicals to minimize the risk of cross-contamination or contact. Storage buildings should be equipped with proper fire protection and prevention equipment and explosion-proof electrical components and equipment. Bulk storage of 1,1-dimethylhydrazine is not recommended.

If a spill or leak to the environment has occurred, fire department, emergency response and/or hazardous materials spill personnel should be notified immediately. Cleanup should be attempted only by those trained in proper spill containment procedures and the health hazards associated with carcinogenic materials. Contaminated soils should be removed for incineration and replaced with clean soil.

If 1,1-dimethylhydrazine should contact the water table, aquifer, or navigable waterway, remediation activities should be prompt to ensure protection of aquatic sediments. It is highly soluble in water and total containment and remediation may not be possible. When such spills occur, the local and/or state emergency response authorities must be notified. A comprehensive emergency response or disaster preparedness/recovery plan should be in place prior to any operations involving the use, transportation, storage, or disposal of 1,1-dimethylhydrazine. Burning 1,1-dimethylhydrazine in a chemical incinerator equipped with an afterburner and an air scrubber is an acceptable disposal method.

If 1,1-dimethylhydrazine is spilled or leaked, the following specific steps are recommended:

☑ Restrict persons not wearing protective clothing from area of spill or leak until cleanup of the area is complete.

☑ Ventilate area and remove all ignition sources.

☑ Absorb liquids in vermiculite or sand and place in a sealed drums for disposal.

☑ It may be necessary to dispose of 1,1-dimethylhydrazine as a hazardous waste. The responsible state agency or the regional office of the federal Environmental Protection Agency (EPA) should be contacted for specific recommendations.

RISK ASSESSMENT: BUSINESS

General Assessment

Accidents or mishaps involving 1,1-dimethylhydrazine can present a significant threat to business operations. The loss or damage of equipment or facilities can significantly affect fiscal viability. Lawsuits that may result from personnel injury/death, public exposures, and/or environmental contamination will also require a serious expenditure of resources. Media attention surrounding an injury, death, or environmental damage can also result in a loss of profits and loss of current as well as future business. Always remember that anytime the term "carcinogen" or "cancer" are used, public emotion, hysteria, and ignorance can run equally high. This should be considered during the development of any public relations policies.

Recommended Risk-Reduction Measures

Company attorneys, safety and health professionals, and environmental specialists should be involved in the development of any procedures dealing with safety or emergency response. A company official should be pre-designated as a public relations officer with specific training in dealing with the press. Corporate plans and policies should be developed, approved, and implemented long before any need for such arises.

REFERENCES

Ahlbom, A. 1993. *Biostatistics for Epidemiologists.* New York: CRC Press/Lewis Publishers.

American Conference of Governmental Industrial Hygienists. 1988. *Documentation of the Threshold Limit Values and Biological Exposure Indices,* 5th Edition (with updates). Cincinnati: ACGIH

Anthony, C. P., and N. J. Kolthoff. 1971. *Textbook of Anatomy and Physiology.* St. Louis: C. V. Mosby Company.

Asante-Duah, D. K. 1993. *Hazardous Waste Risk Assessment.* New York: CRC Press/Lewis Publishers.

Bloom, A. D. and F. J. DeSerres. 1995. *Ecotoxicity and Human Health.* New York: CRC Press/Lewis Publishers.

Brusick, D. J. 1994. *Methods for Genetic Risk Assessment.* New York: CRC Press/Lewis Publishers.

Bueche, J. R. 1972. *Principles of Physics,* 2nd Edition. New York: McGraw-Hill.

Burrell, R., D. K. Flaherty, and L. J. Sauers. 1992. *Toxicology of the Immune System.* New York: Van Nostrand Reinhold.

Calabrese, E. J. 1994. *Biological Effects of Low Level Exposures.* New York: CRC Press/Lewis Publishers.

Cockerman, L. G., B. S. Shane. 1994. *Basic Environmental Toxicology.* New York: CRC Press/Lewis Publishers.

Cothern, C. R. 1995. *Handbook for Environmental Risk Decisionmaking.* New York: CRC Press/Lewis Publishers.

Gardner, R. W. 1994. *Chemical Intolerance.* New York: CRC Press/Lewis Publishers.

Hodgson, E., R. B. Mailman, and Janice E. Chambers. 1988. *Dictionary of Toxicology.* London: The MacMillian Company.

Howard, P. H. 1995. *Handbook of Environmental Fate and Exposures Data for Organic Chemicals,* Vol. I-IV. New York: CRC Press/Lewis Publishers.

Keith, L. H. and M. M. Walker. 1995. *Handbook of Air Toxics.* New York: CRC Press/Lewis Publishers.

Landis, W. G., M. H. Yu. 1995. *Introduction to Environmental Toxicology.* New York: CRC Press/Lewis Publishers.

Langerman, N. 1994 *Precautionary Labels for Chemical Containers.* New York: CRC Press/Lewis Publishers.

Lauwerys, R. R., P. Hoet. 1993. *Industrial Chemical Exposure.* New York: CRC Press/Lewis Publishers.

Lewis, R. J., Sr. 1993. *Hawley's Condensed Chemical Dictionary,* 12th Edition. New York: Van Nostrand Reinhold.

Lewis, R. J., Sr. 1992. *Sax's Dangerous Properties of Industrial Materials,* Volumes 1, 2, and 3, 8th Edition. New York: Van Nostrand Reinhold.

Manahan, S. E. 1992. *Toxicological Chemistry,* 2nd Edition. New York: CRC Press/Lewis Publishers.

Meyer, E. 1990. *Chemistry of Hazardous Materials.* Englewood Cliffs, NJ: Prentice Hall.

Ott, W. R. 1995. *Environmental Statistics and Data Analysis.* New York: CRC Press/Lewis Publishers.

Rea, W. J. 1995. *Chemical Sensitivity,* Vol. I-III. New York: CRC Press/Lewis Publishers.

Sacarello, H. L. A. 1994. *The Comprehensive Handbook of Hazardous Materials.* New York: CRC Press/Lewis Publishers.

U.S. Department of Health and Human Services, National Institute for Occupational Safety and Health. 1994. *NIOSH Pocket Guide to Chemical Hazards.* Washington, D.C.: U.S. Government Printing Office.

MATERIAL SAFETY DATA SHEET

CHEMICAL NAME	
1,2-DIMETHYLHYDRAZINE	**CRC PRESS/LEWIS PUBLISHERS DISCLAIMER:** THE INFORMATION AND RECOMMENDATIONS PRESENTED HEREIN ARE BASED ON SOURCES BELIEVED TO BE RELIABLE. CRC MAKES NO REPRESENTATION ON ITS COMPLETENESS OR ACCURACY. IT IS THE RESPONSIBILITY OF THE USER TO DETERMINE THE CHEMICAL'S SUITABILITY FOR ITS INTENDED USE, THE CHEMICAL'S SAFE USE, AND THE CHEMICAL'S PROPER DISPOSAL. NO REPRESENTATIONS AND/OR WARRANTIES, EITHER EXPRESSED OR IMPLIED, OF THE MERCHANTABILITY OR FITNESS FOR A PARTICULAR PURPOSE, OR OF ANY OTHER NATURE, ARE MADE WITH RESPECT TO THE INFORMATION PROVIDED IN THIS MSDS OR TO THE CHEMICAL TO WHICH INFORMATION MAY REFER. CRC NEITHER ASSUMES NOR AUTHORIZES ANY OTHER PERSON TO ASSUME FOR IT, ANY OTHER ADDITIONAL RESPONSIBILITY OR LIABILITY FOR THE USE OF, OR RELIANCE UPON, THIS INFORMATION.

HAZARD WARNING INFORMATION

HEALTH	FIRE	REACTIVE	OTHER	DEGREE OF HAZARD	COLOR CODING	OTHER CODES
3	3	1		0 = Minimum Hazard 1 = Slight Hazard 2 = Moderate Hazard 3 = Serious Hazard 4 = Severe Hazard	HEALTH = BLUE FIRE = RED REACTIVITY = YELLOW OTHER = WHITE	OX = Oxidizer ACID = Acid ALK = Alkali COR = Corrosive W = Use No Water

SECTION I - GENERAL INFORMATION

Characterization **Symmetrical Compound**	RCRA Number **U099**	EPA Class **Toxic Waste**
DOT Proper Shipping Name **Dimethylhydrazine (symmetrical)**	Chemical Abstract Service (CAS) Number **540-73-8**	
DOT Hazard Class and Label Requirements **Flammable Liquid**	DOT Emergency Guide Code **57**	
DOT Identification Number **UN 2382**	Chemical Formula **$C_2H_8N_2$**	

Synonyms
Hydrazomethane; DMH; symmetrical dimethylhydrazine; N,N-dimethylhydrazine.

SECTION II - HAZARDOUS INGREDIENTS AND IDENTITY INFORMATION

Hazardous Components (specific identity)	OSHA Exposure Criteria	NIOSH Exposure Criteria	Immediately Dangerous to Life and Health (IDLH)	ACGIH Exposure Criteria
1,2-dimethylhydrazine (derivation: Reaction of diethylamine and chloramine; catalytic oxidation of dimethylamine and ammonia). **1 ppm = 2.50 mg/m^3**	**PEL (skin): 0.5 ppm 1 mg/m^3** **STEL: Not Applicable**	**REL: 0.6 ppm 0.15 mg/m^3 (2-hr ceiling)** **Possible Cancer Agent**	**15 ppm**	**TLV (skin): 0.5 ppm 1 mg/m^3** **Suspected Human Carcinogen**

SECTION III - PHYSICAL AND CHEMICAL CHARACTERISTICS

Boiling Point **81°F (27°C)**	Specific Gravity (H_2O = 1) **0.82**
Vapor Pressure (mm Hg) **103 mm at 69°F (20°C)**	Molecular Weight **60.12**
Vapor Density (Air = 1) **1.94**	Melting Point **16°F (-9°C)**

Solubility
Miscible in water, ether, and alcohol.

Appearance and Odor
Colorless liquid with an ammonia or fish-like odor.

SECTION IV - FIRE AND EXPLOSION HAZARD DATA

Flash Point (method used) **73°F (23°C) closed cup**	Explosive Limits in Air % by Volume **LEL: 2% UEL: 90%**
NFPA Classification **Class IB Flammable Liquid**	Autoignition Temperature **485°F (252°C)**

Extinguishing Media
Use water spray, alcohol foam, or carbon dioxide. Water may be ineffective by itself (may spread fire).

Special Fire Fighting Procedures
Structural protective clothing is permeable. Remain clear of smoke, water fall-out, and water run-off. Poisonous gases are produced in fire. Move containers if it can be done without risk. Cool fire-exposed containers with water spray and use water spray to disperse vapors. Evacuate non-essential personnel 2500 feet from the fire area (consider wind conditions).

Unusual Fire and Explosion Hazards
Contact with oxidizing agents can cause violent and explosive ignition. A volatile liquid that is flammable over a wide vapor-to-air range. Vapors are heavier than air and may travel a considerable distance to a source of ignition and flashback to cause fire or explosion. Burns with clear flame.

SECTION V - REACTIVITY DATA

Stability		Conditions to Avoid 1,2-dimethylhydrazine is stable at room temperature in closed containers under normal conditions of storage and handling. Avoid contact with incompatible chemicals and heat.
Stable X	Unstable	Incompatibility (*materials to avoid*) Highly reactive with oxidizers (chlorine, bromine, nitrogen dioxide), halogens, ferric oxide (iron rust), metallic mercury, fuming nitric acid, hydrogen peroxide (ignites spontaneously with oxidizers).
Hazardous Polymerization		Conditions to Avoid Under normal conditions of temperature and pressure, hazardous polymerization of 1,2-dimethylhydrazine is not expected to occur.
May Occur	Will Not Occur X	Hazardous Decomposition or By-products When heated to decomposition, 1,2-dimethylhydrazine will produce toxic and irritating gases, including explosive hydrogen gas, ammonia, dimethylamine, and hydrazoic acid.

SECTION VI - HEALTH HAZARD DATA

Primary Route(s) of Entry:	Inhalation?	Absorption (skin)?	Ingestion?
	X	X	X

Health Hazards

INHALATION: Highly corrosive to the skin, eyes, and mucosa. Causes choking, chest pain, dyspnea, lethargy, nausea, convulsions, tremors, pulmonary irritation, liver injury, and hypoxia.

ABSORPTION: Will pass through unbroken skin to cause toxic systemic effects as noted above. Extremely corrosive to skin tissue causing chemical burns (2nd and 3rd degree) on short contact. Very injurious to the eyes, causing corneal damage and permanent blindness.

INGESTION: A poison by ingestion. Delayed gastrointestinal irritation. Systemic effects can occur.

Carcinogenicity	NTP Listed?	IARC Cancer Review Group?	OSHA Regulated?	Target Organs?
Suspected Human Confirmed Animal	5th Annual Report	Group 2B	29 CFR 1910.1000 Table Z-1	CNS, skin, liver, eyes, G.I. tract, blood, resp. system.

Medical Conditions Generally Aggravated by Exposure

Existing deficiencies in CNS function, liver, kidneys, and the circulatory system may be aggravated.

Emergency and First-aid Procedures

<u>Eye contact</u>: Flush immediately with water for 15 minutes (minimum) while holding upper and lower eyelids open. Seek medical attention. <u>Skin contact</u>: Remove all contaminated clothing. Immediately wash area with large amounts of water (15 minutes minimum). For <u>inhalation</u>: Remove the person from exposure. Provide respiratory assistance and CPR. Transfer to medical facility. Observe for 24-48 hours for lung effects (pulmonary edema). If <u>swallowed</u>: Call the poison control center. Give 1-2 glasses of water or milk to a conscious and alert person. Do NOT induce vomiting and seek medical attention immediately. Never attempt to give anything by mouth to an unconscious person.

SECTION VII - PRECAUTIONS FOR SAFE HANDLING AND USE

Steps to be Taken in Case Material is Released or Spilled

Evacuate area and deny entry to those not involved in cleanup activities. Protective clothing is mandatory. Remove all sources of ignition. Ventilate area of spill. Absorb liquids with toweling, vermiculite, or sand. Deposit in sealed drum for disposal. Notify appropriate authorities.

Preferred Waste Disposal Method

Incineration in a chemical incinerator equipped with afterburner and scrubber.

Precautions to be Taken in Handling and Storage

Store in tightly closed containers in a cool, dark, dry, well-ventilated area. Personnel should be trained on the hazardous properties of 1,2-dimethylhydrazine *prior* to working with the material.

Other Precautions and Warnings

Containers should be protected from physical damage. Use extreme caution when handling this compound. Use explosion-proof electrical equipment in areas where this chemical is used or stored.

SECTION VIII - CONTROL MEASURES AND PERSONAL PROTECTIVE EQUIPMENT

Respiratory Protection (*specify type*)

A carcinogenic agent. Exposure level is extremely low and difficult to monitor. For any exposures, use a supplied-air respirator set in positive pressure or continuous flow mode or a self-contained breathing apparatus (SCBA) with full facepiece operated in pressure demand or other positive pressure mode.

Ventilation

Local exhaust or general mechanical systems recommended.

Protective Gloves	Eye Protection	Other Protective Clothing
Impervious Gloves	Chemical Goggles or Face Mask	Full Protective Clothing

Work/Hygiene Practices

Always wash hands thoroughly after using chemical; never bring food, drink, or smoking materials into vicinity of chemicals.

1,2-DIMETHYLHYDRAZINE

$C_2H_8N_2$

CAS: 540-73-8

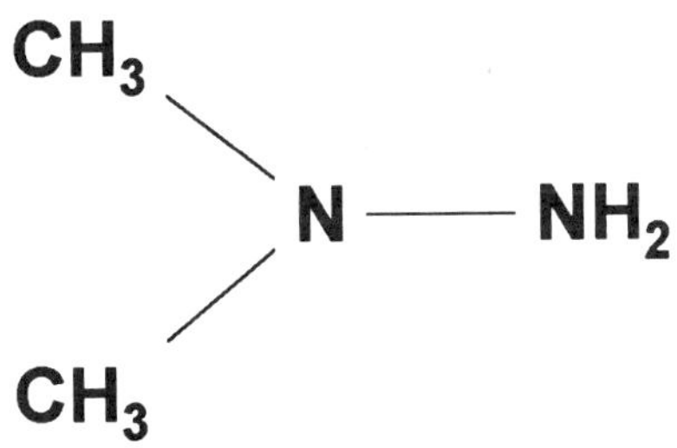

IDENTIFICATION AND TYPICAL USES

1,2-Dimethylhydrazine is a colorless liquid with an ammonia or fish-like odor. It is used as a component of jet and rocket fuel, as an absorbent for acid gases, in photography, and as an agent to control the growth of plants.

RISK ASSESSMENT: HEALTH

General Assessment

1,2-Dimethylhydrazine is a confirmed human and animal carcinogen. Human mutation data have also been reported. It is primarily toxic by *inhalation* and skin contact (*absorption*). It can also cause toxic and damaging effects by *ingestion*.

Inhalation will cause pulmonary irritation up to and including pulmonary edema (fluid buildup in the lungs), which is a medical emergency and can lead to death. Symptoms may be delayed up to 48 hours creating a false sense of security with regard to health exposure risk. Symptoms include choking, extreme chest pain, dyspnea (difficulty in breathing), lethargy, nausea, convulsions, tremors, liver injury, and hypoxia (reduced oxygen levels in the blood) leading to hemolytic anemia. Systemic intoxication, depression of the central nervous system (CNS), and delayed gastrointestinal irritation may also occur.

Aside from its ability to pass through intact skin (absorption), 1,2-dimethylhydrazine is also extremely corrosive to skin tissue and can cause 2nd and 3rd degree burns on contact. Damage to the eyes and cornea can be severe causing partial or complete loss of vision.

Ingestion is very unusual since the vapors of this chemical are so irritating, personnel are not likely to get close enough to ingest it. However, should it be swallowed, there can be burning of the mouth, throat, and esophagus, followed by gastrointestinal irritation.

Acute Health Effects

The following acute (short-term) health effects may occur immediately or shortly after exposure to 1,2-dimethylhydrazine:

Skin: Severe irritation and burning. Skin absorption will occur causing toxic systemic effects.

Eye: Severe irritation and tissue damage up to and including loss of vision.

Lung: Irritation of the upper respiratory tract causing choking, coughing, dypsnea, and pulmonary irritation leading to pulmonary edema (fluid in the lungs), which can be fatal.

CNS: Dizziness, convulsions, tremors, nausea, headaches, possible injury to the central nervous system.

Chronic Health Effects

The following chronic (long-term) health effects may occur at some time after exposure to 1,2-dimethylhydrazine and can last for months or even years:

Cancer Hazards: According to the information presented in the references, 1,2-dimethylhydrazine has been shown to cause cancer in test animals. Its carcinogenicity in humans is expected to be at least equal to that of animals. Human mutation data have been reported for this compound.

Reproductive Hazard: According to the information presented in the references, this chemical has been shown to adversely affect reproduction in test animals.

Other Chronic Effects: Repeated exposures can lead to hemolytic anemia and convulsions. Chronic exposures may lead to cancer.

Recommended Risk-Reduction Measures

1,2-Dimethylhydrazine is listed by the IARC as a confirmed animal carcinogen and the ACGIH as a suspected human carcinogen. Unless a less toxic chemical can be substituted for a hazardous substance, *engineering controls* are the most effective method of reducing exposures. The best protection is to enclose operations and/or provide local exhaust ventilation at the site of chemical release. While not always operationally feasible, isolating operations can also reduce exposure.

Using respiratory protection is less effective than the controls mentioned above, but is still advisable whenever working with or around 1,2-dimethylhydrazine. The exposure level is extremely low (0.5 ppm) and difficult to detect and monitor. For any detectable exposures to a suspected carcinogenic agent, use a supplied-air respirator set in positive pressure or continuous flow mode, or use a self-contained breathing apparatus (SCBA) with full facepiece and pressure demand. If a full facepiece is not available, then chemical goggles should be worn to protect the eyes. Whenever an eye or face contact hazard exists, a face shield and apron should be worn. To prevent hand and skin exposures, impervious gloves should be worn.

Administrative controls should also be in place to minimize the potential for human exposures. These may include written procedures or policies which specify the methods and techniques that will be practiced whenever personnel are to work with or around 1,2-dimethylhydrazine.

All personnel should receive *training* on the use, hazards, protective measures, emergency actions, and other precautions per 29 CFR 1910.1200 (Hazard Communication), prior to the first assignment in an area where 1,2-dimethylhydrazine is used or stored.

Before beginning employment and at regular intervals thereafter (e.g. annually), the following medical tests are recommended:

☑ Lung function tests (establish a baseline).

If symptoms develop or overexposure is suspected, the following medical tests may be useful:

☑ Lung function tests.

☑ Liver and kidney function tests.

☑ Consider chest X-ray after acute overexposure.

☑ Evaluation by a qualified allergist with careful consideration of exposure history and special tests (may help diagnose allergy).

Any evaluation should include a careful medical history of past and present symptoms with an examination. However, medical tests that evaluate existing damage are not a substitute for controlling exposure. Also, since smoking can cause heart disease, emphysema, and numerous other respiratory disorders, smokers may be affected more rapidly than non-smokers to the same exposure conditions. Prudent risk assessment and management requires careful consideration of *all* possible risk factors that may be causing the appearance of symptoms in exposed workers.

Other methods to reduce exposure include:

☑ Where possible, enclose operations and use local exhaust ventilation at the site of chemical release. If local exhaust ventilation or enclosure is not used, respiratory protection should be mandatory.

☑ Always ensure that proper protective clothing is worn when using chemical substances.

☑ Wash thoroughly immediately after exposure to 1,2-dimethylhydrazine and at the end of the work shift or before eating, drinking, or smoking.

☑ Hazard warning information must be posted in the work area. It is recommended that a regulated, controlled work area should be established wherever this chemical is used, handled, or stored. In addition, as part of an on-going education and training program, all information on the health and safety hazards of 1,2-dimethylhydrazine should be communicated to all potentially exposed workers.

☑ Eye wash stations should be provided in work areas. If the potential for whole body exposure exists, then safety showers should also be provided.

RISK ASSESSMENT: ENVIRONMENT

General Assessment

The environment is at risk of exposure during intended use, transportation, storage, disposal, or destruction of 1,2-dimethylhydrazine. In almost every scenario, the threat of environmental exposure is contingent upon the proper handling of the chemical substance. Accidental spills, large or small, can result in fire, explosion, and

possible contamination of the surrounding environmental mediums (water, soil, and air).

1,2-Dimethylhydrazine should always be handled with extreme caution. It is a Class1B flammable liquid. Its extremely low flash point and relatively low boiling point make this chemical a dangerous fire and explosion hazard. Its vapors form explosive mixtures in air over a wide range of concentrations. It forms shock-sensitive explosive products when combined with metal perchlorates, alkali metals, or their dichromates. It is highly reactive with a number of substances, especially oxidizers and oxidizing agents. These characteristics require special consideration during any emergency situation involving a leak or spill of 1,2-dimethylhydrazine.

1,2-Dimethylhydrazine can enter the environment through industrial discharges or through spills.

Acute Ecological Effects

Acute (short-term) toxic effects may include the death of animals, birds, or fish and death or low growth rate in plants. Acute effects are seen 2 to 4 days after animals or plants are exposed to 1,2-dimethylhydrazine. This chemical has moderate acute toxicity to aquatic life and plant life. It has been used to control the growth rate of some ornamental and agricultural crops. Insufficient data are available to evaluate or predict the short-term effects of 1,2-dimethylhydrazine to birds or land animals.

Chronic Ecological Effects

Chronic toxic effects may include shortened life span, reproductive problems, lower fertility, and changes in appearance or behavior in exposed animals. These effects can be seen long after first exposure(s) to toxic chemicals. 1,2-Dimethylhydrazine has low chronic toxicity to aquatic life and moderate chronic toxicity to plant life. Insufficient data are available to evaluate or predict the long-term effects of 1,2-dimethylhydrazine to birds or land animals.

Water Solubility

1,2-Dimethylhydrazine is highly soluble in water. Concentrations of 1000 milligrams and more will mix with a liter of water.

Persistence in the Environment

1,2-Dimethylhydrazine is moderately persistent in water, with a half-life of between 20 to 200 days. The half-life of a pollutant is the amount of time it takes for one half of the chemical to be degraded. It is expected to absorb to sediment. No information is available in the references on the expected percentages of 1,2-dimethylhydrazine that will be divided through the environmental mediums (primarily air and water).

Bioaccumulation in Aquatic Organisms

Some substances increase in concentration, or *bioaccumulate,* in living organisms as they breathe contaminated air, drink contaminated water, or eat contaminated food. These chemicals can become concentrated in the tissues and internal organs of animals as well as humans.

The concentration of 1,2-dimethylhydrazine found in fish tissues is expected to be somewhat higher than the average concentration found in the water from which the fish was taken.

Recommended Risk-Reduction Measures

Proper training of all transporters will reduce the likelihood of a mishap or accident resulting in a leak or spill to the environment. The correct use of labeling on all containers, trucks, and rail cars will enable emergency responders to react properly and quickly to any disaster thereby reducing the potential risk to the environment and to personnel.

Storage of 1,2-dimethylhydrazine should be segregated from other chemicals to minimize the risk of cross-contamination or contact. Storage buildings should be equipped with proper fire protection and prevention equipment and explosion-proof electrical components and equipment. Bulk storage of 1,2-dimethylhydrazine is not recommended.

If a spill or leak to the environment has occurred, fire department, emergency response and/or hazardous materials spill personnel should be notified immediately. Cleanup should be attempted only by those trained in proper spill containment procedures and the health hazards associated with carcinogenic materials. Contaminated soils should be removed for incineration and replaced with clean soil. If 1,2-dimethylhydrazine should contact the water table, aquifer, or navigable waterway, remediation activities should be prompt to ensure protection of aquatic sediments. It is highly soluble in water and total containment and remediation may not be possible. When such spills occur, the local and/or state emergency response authorities must be notified. A comprehensive emergency response or disaster preparedness/recovery plan should be in place prior to any operations involving the use, transportation, storage, or disposal of 1,2-dimethylhydrazine. If 1,2-dimethylhydrazine is spilled or leaked, the following specific steps are recommended:

☑ Restrict persons not wearing protective clothing from area of spill or leak until cleanup of the area is complete.

☑ Ventilate area and remove ignition sources.

☑ Absorb liquids in vermiculite or sand and place in a sealed drums for disposal.

☑ It may be necessary to dispose of 1,2-dimethylhydrazine as a hazardous waste. The responsible state agency or the regional office of the federal Environmental Protection Agency (EPA) should be contacted for specific recommendations.

RISK ASSESSMENT: BUSINESS

General Assessment

Accidents or mishaps involving 1,2-dimethylhydrazine can present a significant threat to business operations. The loss or damage of equipment or facilities can significantly affect fiscal viability. Lawsuits that may result from personnel injury/death, public exposures, and/or environmental contamination will also require a serious expenditure of resources. Media attention surrounding an injury, death, or environmental damage can also result in a loss of profits and loss of current as well as future business. Always remember that anytime the term "carcinogen" or "cancer" are used, public emotion, hysteria, and ignorance can run equally high. This should be considered during the development of any public relations policies.

Recommended Risk-Reduction Measures

Company attorneys, safety and health professionals, and environmental specialists should be involved in the development of any such procedures. A company official should be pre-designated as a public relations officer with specific training in dealing with the press. Corporate plans and policies should be developed, approved, and implemented long before any need for such arises.

REFERENCES

Ahlbom, A. 1993. *Biostatistics for Epidemiologists.* New York: CRC Press/Lewis Publishers.

Anthony, C. P., and N. J. Kolthoff. 1971. *Textbook of Anatomy and Physiology.* St. Louis: C. V. Mosby Company.

Asante-Duah, D. K. 1993. *Hazardous Waste Risk Assessment.* New York: CRC Press/Lewis Publishers.

Bloom, A. D. and F. J. DeSerres. 1995. *Ecotoxicity and Human Health.* New York: CRC Press/Lewis Publishers.

Brusick, D. J. 1994. *Methods for Genetic Risk Assessment.* New York: CRC Press/Lewis Publishers.

Bueche, J. R. 1972. *Principles of Physics,* 2nd Edition. New York: McGraw-Hill.

Burrell, R., D. K. Flaherty, and L. J. Sauers. 1992. *Toxicology of the Immune System.* New York: Van Nostrand Reinhold.

Calabrese, E. J. 1994. *Biological Effects of Low Level Exposures.* New York: CRC Press/Lewis Publishers.

Cockerman, L. G., B. S. Shane. 1994. *Basic Environmental Toxicology.* New York: CRC Press/Lewis Publishers.

Cothern, C. R. 1995. *Handbook for Environmental Risk Decisionmaking.* New York: CRC Press/Lewis Publishers.

Gardner, R. W. 1994. *Chemical Intolerance.* New York: CRC Press/Lewis Publishers.

Hodgson, E., R. B. Mailman, and Janice E. Chambers. 1988. *Dictionary of Toxicology.* London: The MacMillian Company.

Howard, P. H. 1995. *Handbook of Environmental Fate and Exposures Data for Organic Chemicals,* Vol. I-IV. New York: CRC Press/Lewis Publishers.

Keith, L. H. and M. M. Walker. 1995. *Handbook of Air Toxics.* New York: CRC Press/Lewis Publishers.

Landis, W. G., M. H. Yu. 1995. *Introduction to Environmental Toxicology.* New York: CRC Press/Lewis Publishers.

Langerman, N. 1994 *Precautionary Labels for Chemical Containers.* New York: CRC Press/Lewis Publishers.

Lauwerys, R. R., P. Hoet. 1993. *Industrial Chemical Exposure.* New York: CRC Press/Lewis Publishers.

Lewis, R. J., Sr. 1993. *Hawley's Condensed Chemical Dictionary,* 12th Edition. New York: Van Nostrand Reinhold.

Lewis, R. J., Sr. 1992. *Sax's Dangerous Properties of Industrial Materials,* Volumes 1, 2, and 3, 8th Edition. New York: Van Nostrand Reinhold.

Manahan, S. E. 1992. *Toxicological Chemistry,* 2nd Edition. New York: CRC Press/Lewis Publishers.

Meyer, E. 1990. *Chemistry of Hazardous Materials.* Englewood Cliffs, NJ: Prentice Hall.

Rea, W. J. 1995. *Chemical Sensitivity,* Vol. I-III. New York: CRC Press/Lewis Publishers.

Sacarello, H. L. A. 1994. *The Comprehensive Handbook of Hazardous Materials.* New York: CRC Press/Lewis Publishers.

U.S. Department of Health and Human Services, National Institute for Occupational Safety and Health. 1994. *NIOSH Pocket Guide to Chemical Hazards.* Washington, D.C.: U.S. Government Printing Office.

MATERIAL SAFETY DATA SHEET

CHEMICAL NAME	
DI-N-BUTYL PHTHALATE	**CRC PRESS/LEWIS PUBLISHERS DISCLAIMER:** THE INFORMATION AND RECOMMENDATIONS PRESENTED HEREIN ARE BASED ON SOURCES BELIEVED TO BE RELIABLE. CRC MAKES NO REPRESENTATION ON ITS COMPLETENESS OR ACCURACY. IT IS THE RESPONSIBILITY OF THE USER TO DETERMINE THE CHEMICAL'S SUITABILITY FOR ITS INTENDED USE, THE CHEMICAL'S SAFE USE, AND THE CHEMICAL'S PROPER DISPOSAL. NO REPRESENTATIONS AND/OR WARRANTIES, EITHER EXPRESSED OR IMPLIED, OF THE MERCHANTABILITY OR FITNESS FOR A PARTICULAR PURPOSE, OR OF ANY OTHER NATURE, ARE MADE WITH RESPECT TO THE INFORMATION PROVIDED IN THIS MSDS OR TO THE CHEMICAL TO WHICH INFORMATION MAY REFER. CRC NEITHER ASSUMES NOR AUTHORIZES ANY OTHER PERSON TO ASSUME FOR IT, ANY OTHER ADDITIONAL RESPONSIBILITY OR LIABILITY FOR THE USE OF, OR RELIANCE UPON, THIS INFORMATION.

HAZARD WARNING INFORMATION

HEALTH	FIRE	REACTIVE	OTHER	DEGREE OF HAZARD	COLOR CODING	OTHER CODES
2	1	0		0 = Minimum Hazard 1 = Slight Hazard 2 = Moderate Hazard 3 = Serious Hazard 4 = Severe Hazard	HEALTH = BLUE FIRE = RED REACTIVITY = YELLOW OTHER = WHITE	OX = Oxidizer ACID = Acid ALK = Alkali COR = Corrosive W = Use No Water

SECTION I - GENERAL INFORMATION

Characterization	RCRA Number	EPA Class
Phthalate Ester	**U069**	**Toxic Waste**

DOT Proper Shipping Name **n-Butyl Phthalate**	Chemical Abstract Service (CAS) Number **84-74-2**
DOT Hazard Class and Label Requirements **Combustible Liquid**	DOT Emergency Guide Code **31**
DOT Identification Number **NA 9095**	Chemical Formula **$C_6H_4(COOC_4H_9)_2$**

Synonyms

Dibutyl phthalate; o-benzenedicarboxylic acid, dibutyl ester; Celluflex DPB; dibutyl-1,2-benzenedicarboxylate; n-butyl phthalate.

SECTION II - HAZARDOUS INGREDIENTS AND IDENTITY INFORMATION

Hazardous Components (specific identity)	OSHA Exposure Criteria	NIOSH Exposure Criteria	Immediately Dangerous to Life and Health (IDLH)	ACGIH Exposure Criteria
Di-n-butyl phthalate (derivation: By treating n-butyl alcohol with phthalic anhydride followed by purification, which results in a product unusually free of odor and color). **1 ppm = 11.57 mg/m^3**	**PEL:** **0.4 ppm** **5 mg/m^3** **STEL:** **Not Established**	**REL:** **0.4 ppm** **5 mg/m^3** **STEL:** **Not Established**	**4000 mg/m^3**	**TLV:** **0.4 ppm** **5 mg/m^3** **STEL:** **Not Established**

SECTION III - PHYSICAL AND CHEMICAL CHARACTERISTICS

Boiling Point **644°F (340°C)**	Specific Gravity (H_2O = 1) **1.05**
Vapor Pressure (mm Hg) **<1 at 69°F (20°C)**	Molecular Weight **278.3**
Vapor Density (Air = 1) **9.58**	Freezing Point **-31°F (-35°C)**

Solubility

Miscible in common organic solvents, insoluble (0.001%) in water.

Appearance and Odor

Colorless to faint yellow oily liquid. Depending on the grade, there may be no noticeable odor or a slight aromatic odor.

SECTION IV - FIRE AND EXPLOSION HAZARD DATA

Flash Point (method used) **315°F (157°C) closed cup**	Explosive Limits in Air % by Volume **LEL: 0.5% (at 415°F)** **UEL: Not Determined**
NFPA Classification **Class IIIB Combustible Liquid**	Autoignition Temperature **757°F (403°C)**

Extinguishing Media

Use dry chemical or carbon dioxide.

Special Fire Fighting Procedures

Poisonous gases are produced in fire, wear full protective clothing and self-contained breathing apparatus (SCBA). Vapors are heavier than air and may travel for some distance to ignition source and flashback. Move containers from fire area if it can be done safely.

Unusual Fire and Explosion Hazards

Containers may explode in fire due to buildup of excessive internal pressures. Exercise extreme caution when attempting to move or relocate containers that may have been exposed to high heat.

SECTION V - REACTIVITY DATA

Stability		Conditions to Avoid **Avoid contact with incompatible materials. Keep away from heat and flame.**
Stable	Unstable	Incompatibility *(materials to avoid)* **Strong oxidizers, such as chlorine (especially liquid chlorine), bromine, and fluorine, strong alkalis (sodium hydroxide, potassium hydroxide, lithium hydroxide), strong acids, and nitrates.**
X		
Hazardous Polymerization		Conditions to Avoid **Under normal conditions of temperature and pressure, hazardous polymerization of di-n-butyl phthalate is not expected to occur.**
May Occur	Will Not Occur	Hazardous Decomposition or By-products **Acrid and irritated gases and fumes are produced when di-n-butyl phthalate is involved in fire.**
	X	

SECTION VI - HEALTH HAZARD DATA

Primary Route(s) of Entry:	Inhalation?	Absorption (skin)?	Ingestion?
	X		X

Health Hazards

INHALATION: Symptoms include irritation to the eyes and mucosa (especially in the upper respiratory tract). There may be nausea and vomiting.

SKIN & EYES: Contact may irritate the skin and eyes. May cause conjunctivitis with quick recovery.

INGESTION: Mildly toxic by ingestion. Gastrointestinal irritation. Systemic effects include blurred vision, hallucinations, distorted perceptions, nausea or vomiting, and kidney, ureter, or bladder changes.

Carcinogenicity	NTP Listed?	IARC Cancer Review Group?	OSHA Regulated?	Target Organs?
Unknown Human Unknown Animal	No	No	29 CFR 1910.1000 Table Z-1	Eyes, respiratory system, GI tract, kidney, bladder.

Medical Conditions Generally Aggravated by Exposure

None Reported.

Emergency and First-aid Procedures

Eye contact: Flush immediately with water for 15 minutes (minimum), seek medical attention. **Skin contact:** Remove all contaminated clothing. Immediately wash area with large amounts of soap and water. Seek medical assistance. For **inhalation:** Remove the person from exposure. Provide respiratory assistance and CPR. Transfer to medical facility. If **swallowed:** Call the poison control center and seek medical attention immediately.

SECTION VII - PRECAUTIONS FOR SAFE HANDLING AND USE

Steps to be Taken in Case Material is Released or Spilled

Absorb liquids in vermiculite or other absorbent and deposit in DOT-approved drum. Ventilate area and remove ignition sources. Restrict those not involved in cleanup from entering area. Notify appropriate authorities, if applicable.

Preferred Waste Disposal Method

Burn in chemical incinerator equipped with an afterburner and scrubber.

Precautions to be Taken in Handling and Storage

Do not store di-n-butyl phthalate in the presence of incompatible chemicals or materials. Store in tightly closed containers in a cool, well-ventilated area. Personnel working with di-n-butyl phthalate should be trained on its proper handling and storage *prior* to being assigned to such responsibilities.

Other Precautions and Warnings

Bulk storage is of di-n-butyl phthalate should be avoided.

SECTION VIII - CONTROL MEASURES AND PERSONAL PROTECTIVE EQUIPMENT

Respiratory Protection *(specify type)*

For relatively low or infrequent exposure, use a full facepiece respirator equipped with a high efficiency particulate air (HEPA) filter and dust/mist pre-filter. For greater protection, use a supplied-air respirator with full facepiece operated in pressure demand mode or a self-contained breathing apparatus (SCBA) with full facepiece operated in pressure demand or other positive pressure mode.

Ventilation

Local exhaust or general mechanical systems recommended.

Protective Gloves	Eye Protection	Other Protective Clothing
Rubber	Chemical Goggles or Face Mask	Apron

Work/Hygiene Practices

Always wash hands thoroughly after using chemical; never bring food, drink, or smoking materials into vicinity of chemicals.

DI-N-BUTYL PHTHALATE

$C_6H_4(COOC_4H_9)_2$ CAS: 84-74-2

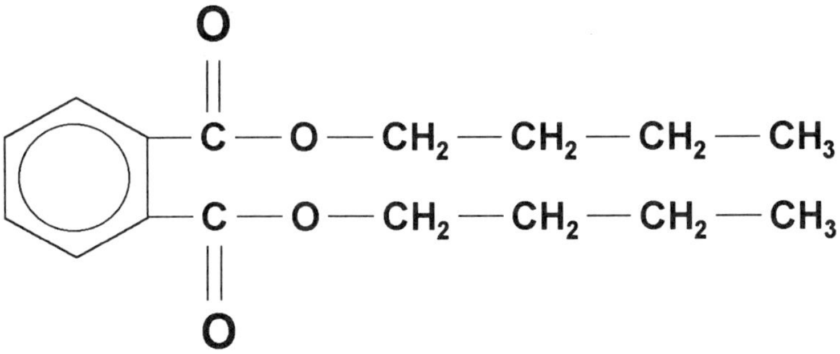

IDENTIFICATION AND TYPICAL USES

Di-n-butyl phthalate is a colorless to faint yellow oily, stable, liquid. Depending upon the grade, it may be odorless or have a slight, aromatic odor. It is used as a plasticizer in nitrocellulose lacquers, elastomers, nail polish, and solid rocket propellants. It is also used as a solvent for perfume oils, as a perfume fixative, a textile lubricating agent, in the manufacture of safety glass, and in the production of insecticides. It may also be found in certain printing inks, is resin solvents, in paper coatings, adhesives, and as an insect repellent for textiles.

RISK ASSESSMENT: HEALTH

General Assessment

Di-n-butyl phthalate is primarily toxic to humans by *ingestion*. However, its vapors can also cause adverse systemic effects by *inhalation*. Contact with the eyes or skin will result in irritation. It is known to cause damage to male reproductive organs in test animals. Human mutation data have been reported for this chemical.

Ingestion of di-n-butyl phthalate will result in systemic effects with symptoms of vomiting or nausea, and numerous vision-related disorders. Symptoms include hallucinations and distorted perceptions. There may also be kidney, ureter, or bladder changes.

Inhalation will irritate the mucous membranes of the nose and throat. There may also be systemic effects similar to those caused by ingestion, but this has not been confirmed by the references (the data are inconclusive and conflicting).

Acute Health Effects

The following acute (short-term) health effects may occur immediately or shortly after exposure to di-n-butyl phthalate:

Skin: Irritation and possible inflammation.

Eye: Severe irritation with redness, lachrymation, and possible conjunctivitis with prompt recovery.

Lung: Irritation of the mucosa of the nose, throat, and upper respiratory tract.

Other: Ingestion may cause hallucinations, blurred vision, and distorted perceptions. There may be vomiting or nausea.

Chronic Health Effects

The following chronic (long-term) health effects may occur at some time after exposure to di-n-butyl phthalate and can last for months or even years:

Cancer Hazards: According to the information presented in the references, di-n-butyl phthalate has not been adequately tested for its ability to cause cancer in test animals.

Reproductive Hazard: According to the information presented in the references, di-n-butyl phthalate may damage the developing fetus in humans since it has been shown to be fetotoxic in animals. It may also damage the testes (male reproductive glands).

Other Chronic Effects: Chronic exposure to di-n-butyl phthalate has not been sufficiently studied. However, there is some indication that long-term exposure or frequent high exposures may cause changes in the kidney, ureter, or bladder.

Recommended Risk-Reduction Measures

Unless a less toxic chemical can be substituted for a hazardous substance, *engineering controls* are the most effective method of reducing exposures. The best protection is to enclose operations and/or provide local exhaust ventilation at the site of chemical release. While not always operationally feasible, isolating operations can also reduce exposure.

Using respiratory protection is less effective than the controls mentioned above, but is still advisable whenever working with or around di-n-butyl phthalate. At low or infrequent exposures, use a full facepiece respirator equipped with a high efficiency particulate air (HEPA) filter and a dust/mist pre-filter. For high concentrations, a self-contained breathing apparatus (SCBA) with full facepiece and pressure demand is the recommended respiratory protection of choice. If a full facepiece is not available, then chemical goggles should be worn to protect the eyes. Whenever a chemical splash hazard exists, a face shield and rubber apron should be worn. To prevent hand and skin exposures, rubber gloves should be used. Glove manufacturers should be contacted for assistance with selection.

Administrative controls should also be in place to minimize the potential for human exposures. These may include written procedures or policies which specify the methods and techniques that will be practiced whenever personnel are to work with di-n-butyl phthalate.

All personnel should receive *training* on the use, hazards, protective measures, emergency actions, and other precautions per 29 CFR 1910.1200 (Hazard Communication), prior to the first assignment in an area where di-n-butyl phthalate is used or stored.

If symptoms develop or overexposure is suspected, the following medical tests may be useful:

☑ Kidney function tests.

Any evaluation should include a careful medical history of past and present symptoms with an examination. However, medical tests that evaluate existing damage are not a substitute for controlling exposure.

Other methods to reduce exposure include:

☑ Where possible, enclose operations and use local exhaust ventilation at the site of chemical release. If local exhaust ventilation or enclosure is not used, respiratory protection should be mandatory.

☑ Always ensure that proper protective clothing is worn when using chemical substances.

☑ Wash thoroughly immediately after exposure to di-n-butyl phthalate and at the end of the work shift or before eating, drinking, or smoking.

☑ Hazard warning information should be posted in the work area. In addition, as part of an on-going education and training program, all information on the health and safety hazards of di-n-butyl phthalate should be communicated to all potentially exposed workers.

☑ Eye wash stations should be provided in work areas. If the potential for whole body exposure exists, then safety showers should also be provided.

RISK ASSESSMENT: ENVIRONMENT

General Assessment

The environment is at risk of exposure during use, transportation, storage, disposal, or destruction of di-n-butyl phthalate. In almost every scenario, the threat of environmental exposure is contingent upon the proper handling of the chemical substance. Accidental spills, large or small, can result in fire, explosion, and possible contamination of the surrounding environmental mediums (water, soil, and air).

Di-n-butyl phthalate is considered a Class IIIB combustible liquid (per OSHA 29 CFR 1910.106). These characteristics require special consideration during any emergency situation involving a leak or spill of di-n-butyl phthalate. Should di-n-butyl phthalate ever come into contact with incompatible substances such as strong oxidizers (chlorine, bromine, fluorine, and chlorine dioxide), strong alkali materials (sodium hydroxide, potassium hydroxide, and lithium hydroxide), and strong acids (sulfuric acid, hydrochloric acid, and nitric acid), and other nitrates either during use, transportation, or storage, violent reaction can occur. Di-n-butyl phthalate is especially reactive in contact with liquid chlorine.

Di-n-butyl phthalate can enter the environment through industrial discharges, in non-point discharges from the wide and general use of this chemical, and from spills.

Acute Ecological Effects

Acute (short-term) toxic effects may include the death of animals, birds, or fish and death or low growth rate in plants. Acute effects are seen 2 to 4 days after ani-

mals or plants are exposed to di-n-butyl phthalate. This chemical has high acute toxicity to aquatic life. Insufficient data are available to evaluate or predict the short-term effects of di-n-butyl phthalate to plants, birds, or land animals.

Chronic Ecological Effects

Chronic toxic effects may include shortened life span, reproductive problems, lower fertility, and changes in appearance or behavior in exposed animals. These effects can be seen long after first exposure(s) to toxic chemicals. Di-n-butyl phthalate has high chronic toxicity to aquatic life. Insufficient data are available to evaluate or predict the long-term effects of di-n-butyl phthalate to plants, birds, or land animals.

Water Solubility

Di-n-butyl phthalate is very slightly soluble in water. Concentrations of 1 milligram or less will mix with a liter of water.

Persistence in the Environment

Di-n-butyl phthalate is slightly persistent in water, with a half-life of between 2 to 20 days. The half-life of a pollutant is the amount of time it takes for one half of the chemical to be degraded. Approximately 33% of di-n-butyl phthalate will eventually end up in terrestrial soil; about 31% will end up in aquatic sediments, about 26% will end up in air, and about 8% will end up in water.

Bioaccumulation in Aquatic Organisms

Some substances increase in concentration, or *bioaccumulate,* in living organisms as they breathe contaminated air, drink contaminated water, or eat contaminated food. These chemicals can become concentrated in the tissues and internal organs of animals as well as humans.

The concentration of di-n-butyl phthalate found in fish tissues is expected to be much higher than the average concentration of di-n-butyl phthalate in the water from which the fish was taken.

Recommended Risk-Reduction Measures

Proper training of all transporters will reduce the likelihood of a mishap or accident resulting in a leak or spill to the environment. The correct use of DOT labeling (per the MSDS) on all containers, trucks, and rail cars will enable emergency responders to react properly and quickly to any disaster thereby reducing the potential risk to the environment and to personnel.

Storage of di-n-butyl phthalate should be segregated from incompatible chemicals to minimize the risk of cross-contamination or contact.

Di-n-butyl phthalate is destroyed by burning in a chemical incinerator equipped with an afterburner and scrubber.

If a spill or leak to the environment has occurred, fire department, emergency response and/or hazardous materials spill personnel should be notified immediately. Cleanup should be attempted only by those trained in proper spill containment procedures. Contaminated soils should be removed for incineration and replaced with clean soil. If di-n-butyl phthalate should contact the water table, aquifer, or navigable waterway, remediation activities should be prompt to ensure protection of aquatic sediments. When such spills occur, the local and/or state emergency response authorities must be notified. A comprehensive emergency response or disaster preparedness/recovery plan should be in place prior to any operations involving the use, transportation, storage, or disposal of di-n-butyl phthalate. If di-n-butyl phthalate is spilled or leaked, the following specific steps are recommended:

- ☑ Restrict persons not wearing protective clothing from area of spill or leak until cleanup is complete and area can be opened for normal work.
- ☑ Ventilate area of spill or leak.
- ☑ Remove all ignition sources.
- ☑ Absorb liquid spills using vermiculite, dry earth, or sand to absorb and place in a sealed drum for disposal.
- ☑ It may be necessary to dispose of di-n-butyl phthalate as a hazardous waste. The responsible state agency or the regional office of the federal Environmental Protection Agency (EPA) should be contacted for specific recommendations.

RISK ASSESSMENT: BUSINESS

General Assessment

Accidents or mishaps involving di-n-butyl phthalate can present a significant threat to business operations. The loss or damage of equipment or facilities can significantly affect fiscal viability. Lawsuits that may result from personnel injury/death, public exposures, and/or environmental contamination will also require a serious expenditure of resources. Media attention surrounding an injury, death, or environmental damage can also result in a loss of profits and loss of current as well as future business.

Recommended Risk-Reduction Measures

Company attorneys, safety and health professionals, and environmental specialists should be involved in the development of any procedures dealing with safety or emergency response. A company official should be pre-designated as a public relations officer with specific training in dealing with the press. Corporate plans and policies should be developed, approved, and implemented long before any need for such arises.

REFERENCES

Ahlbom, A. 1993. *Biostatistics for Epidemiologists.* New York: CRC Press/Lewis Publishers.

American Conference of Governmental Industrial Hygienists. 1988. *Documentation of the Threshold Limit Values and Biological Exposure Indices,* 5th Edition (with updates). Cincinnati: ACGIH

Anthony, C. P., and N. J. Kolthoff. 1971. *Textbook of Anatomy and Physiology.* St. Louis: C. V. Mosby Company.

Asante-Duah, D. K. 1993. *Hazardous Waste Risk Assessment.* New York: CRC Press/Lewis Publishers.

Bloom, A. D. and F. J. DeSerres. 1995. *Ecotoxicity and Human Health.* New York: CRC Press/Lewis Publishers.

Brusick, D. J. 1994. *Methods for Genetic Risk Assessment.* New York: CRC Press/Lewis Publishers.

Bueche, J. R. 1972. *Principles of Physics,* 2nd Edition. New York: McGraw-Hill.

Burrell, R., D. K. Flaherty, and L. J. Sauers. 1992. *Toxicology of the Immune System.* New York: Van Nostrand Reinhold.

Calabrese, E. J., E. M. Kenyon. *Air Toxics and Risk Assessment.* New York: CRC Press/Lewis Publishers.

Cockerman, L. G., B. S. Shane. 1994. *Basic Environmental Toxicology.* New York: CRC Press/Lewis Publishers.

Cothern, C. R. 1995. *Handbook for Environmental Risk Decisionmaking.* New York: CRC Press/Lewis Publishers.

Gardner, R. W. 1994. *Chemical Intolerance.* New York: CRC Press/Lewis Publishers.

Hallenbeck, W. H. 1993. *Quantitative Risk Assessment for Environmental and Occupational Health,* 2nd Edition. New York: CRC Press/Lewis Publishers.

Hodgson, E., R. B. Mailman, and Janice E. Chambers. 1988. *Dictionary of Toxicology.* London: The MacMillian Company.

Kamrin, M. A. 1988. *Toxicology: A Primer on Principles and Applications.* Chelsea, Michigan: Lewis Publishers.

Keith, L. H. and M. M. Walker. 1995. *Handbook of Air Toxics.* New York: CRC Press/Lewis Publishers.

Landis, W. G., M. H. Yu. 1995. *Introduction to Environmental Toxicology.* New York: CRC Press/Lewis Publishers.

Langerman, N. 1994 *Precautionary Labels for Chemical Containers.* New York: CRC Press/Lewis Publishers.

Lauwerys, R. R., P. Hoet. 1993. *Industrial Chemical Exposure.* New York: CRC Press/Lewis Publishers.

Lide, D. R. 1995. *Handbook of Chemistry and Physics,* 76th Edition. New York: CRC Press/Lewis Publishers.

Lewis, R. J., Sr. 1993. *Hawley's Condensed Chemical Dictionary,* 12th Edition. New York: Van Nostrand Reinhold.

Lewis, R. J., Sr. 1992. *Sax's Dangerous Properties of Industrial Materials,* Volumes 1, 2, and 3, 8th Edition. New York: Van Nostrand Reinhold.

Manahan, S. E. 1992. *Toxicological Chemistry,* 2nd Edition. New York: CRC Press/Lewis Publishers.

Mellström, G. A., J. E. Wahlberg, and H. I. Maibach. 1994. *Protective Gloves for Occupational Use.* New York: CRC Press/Lewis Publishers.

Meyer, E. 1990. *Chemistry of Hazardous Materials.* Englewood Cliffs, NJ: Prentice Hall.

Mycock, J. C., J. D. McKenna, and L. Theodore. 1995. *Handbook of Air Pollution Control Engineering and Technology.* New York: CRC Press/Lewis Publishers.

Nelson, G. O. 1992. *Gas Mixtures.* New York: CRC Press/Lewis Publishers.

Ott, W. R. 1995. *Environmental Statistics and Data Analysis.* New York: CRC Press/Lewis Publishers.

Patnaik, P. 1992. *A Comprehensive Guide to the Hazardous Properties of Chemical Substances.* New York: Van Nostrand Reinhold.

Philp, R. B. 1995. *Environmental Hazards and Human Health.* New York: CRC Press/Lewis Publishers.

Quigley, D. R. 1994. *Handbook of Emergency Chemical Management.* New York: CRC Press/Lewis Publishers.

Rea, W. J. 1995. *Chemical Sensitivity,* Vol. I-III. New York: CRC Press/Lewis Publishers.

Richardson, M. 1995. *Dictionary of Substance and Their Effects,* Vol. I-VI. New York: CRC Press/Lewis Publishers.

Rylander, R., R. R. Jacobs. 1994. *Organic Dusts.* New York: CRC Press/Lewis Publishers.

Sacarello, H. L. A. 1994. *The Comprehensive Handbook of Hazardous Materials.* New York: CRC Press/Lewis Publishers.

U.S. Department of Health and Human Services, National Institute for Occupational Safety and Health. 1994. *NIOSH Pocket Guide to Chemical Hazards.* Washington, D.C.: U.S. Government Printing Office.

MATERIAL SAFETY DATA SHEET

CHEMICAL NAME	CRC PRESS/LEWIS PUBLISHERS DISCLAIMER:
DINITROBENZENE	

HAZARD WARNING INFORMATION

HEALTH	FIRE	REACTIVE	OTHER	DEGREE OF HAZARD	COLOR CODING	OTHER CODES
3	1	4		0 = Minimum Hazard 1 = Slight Hazard 2 = Moderate Hazard 3 = Serious Hazard 4 = Severe Hazard	HEALTH = BLUE FIRE = RED REACTIVITY = YELLOW OTHER = WHITE	OX = Oxidizer ACID = Acid ALK = Alkali COR = Corrosive W = Use No Water

SECTION I - GENERAL INFORMATION

Characterization **Phenol**	RCRA Number **D001**	EPA Class **Characteristic (I) Waste**
DOT Proper Shipping Name **Dinitrobenzene, solid, or Dinitrobenzol**	Chemical Abstract Service (CAS) Number **25154-54-5**	
DOT Hazard Class and Label Requirements **Poison B**	DOT Emergency Guide Code **56**	
DOT Identification Number **UN 1597**	Chemical Formula $C_6H_4(NO_2)_2$	

Synonyms
Dinitrobenzene (solution); dinitrobenzene (solid).

SECTION II - HAZARDOUS INGREDIENTS AND IDENTITY INFORMATION

Hazardous Components (specific identity)	OSHA Exposure Criteria	NIOSH Exposure Criteria	Immediately Dangerous to Life and Health (IDLH)	ACGIH Exposure Criteria
Dinitrobenzene (derivation: From the nitration of nitrobenzene with hot mixed acid). **1 ppm = 6.6 mg/m³**	**PEL (skin):** **1 ppm** **6.6 mg/m³** **STEL:** **Not Established**	**REL (skin):** **1 ppm** **6.6 mg/m³** **STEL:** **Not Established**	**50 mg/m³**	**TLV (skin):** **0.15 ppm** **1 mg/m³** **STEL:** **Not Established**

SECTION III - PHYSICAL AND CHEMICAL CHARACTERISTICS

Boiling Point **572°F (300°C)**	Specific Gravity (H_2O = 1) **1.58**
Vapor Pressure (gas) **<1 at 69°F (20°C)**	Molecular Weight **168.1**
Vapor Density (Air = 1) **5.8**	Melting Point **192°F (89°C)**

Solubility
Slightly soluble in water. Soluble in chloroform and ethyl acetate and sparingly soluble in benzene.

Appearance and Odor
Pale white or yellow crystalline solid.

SECTION IV - FIRE AND EXPLOSION HAZARD DATA

Flash Point (method used) **302°F (150°C) closed cup**	Explosive Limits in Air % by Volume **LEL: Not Determined UEL: Not Determined**
NFPA Classification **Combustible Solid**	Autoignition Temperature **Not Determined**

Extinguishing Media
For small fires, use dry chemical, carbon dioxide, water spray, or regular foam. For large fires, use water spray, fog, or regular foam. Apply water as fog in flooding amounts (solid stream may be ineffective).

Special Fire Fighting Procedures
Wear full protective clothing and self-contained breathing apparatus (SCBA). Poisonous gases are produced in fire. Use water spray to keep fire-exposed containers cool.

Unusual Fire and Explosion Hazards
Containers may explode in fire. Highly friction- and shock-sensitive. Dusts form explosive concentrations in air. May spontaneously ignite or explode at high temperatures due to spontaneous combustion.

SECTION V - REACTIVITY DATA

Stability		Conditions to Avoid
		Dinitrobenzene is unstable at high temperatures and is extremely shock- and friction-sensitive. Keep containers away from areas where shock, heat, and/or friction are likely to occur.
Stable	Unstable	Incompatibility *(materials to avoid)*
	X	Strong oxidizers, (such as chlorine, bromine, and fluorine), caustics, and many metals (such as zinc and tin). Explosive when mixed with nitric acid.
Hazardous Polymerization		Conditions to Avoid
		Under normal conditions of temperature and pressure, hazardous polymerization of dinitrobenzene is not expected to occur.
May Occur	Will Not Occur	Hazardous Decomposition or By-products
	X	When heated to decomposition, dinitrobenzene can emit highly toxic/poisonous gases, including carbon dioxide and oxides of nitrogen. Dinitrobenzene can explode when heated.

SECTION VI - HEALTH HAZARD DATA

Primary Route(s) of Entry:	Inhalation?	Absorption (skin)?	Ingestion?
	X	X	X

Health Hazards

INHALATION: Anoxia; visual disturbances; central scotomas; yellowing of hair, eyes, and skin; anemia; and liver, kidney, and central nervous system injury is possible. Forms methemoglobin.

ABSORPTION: Skin irritant. Will pass through skin to cause toxic systemic effects (same as inhalation).

INGESTION: Gastrointestinal disorders; tremor; convulsions; bad taste and burning feeling in mouth; dry throat; and thirst.

Carcinogenicity	NTP Listed?	IARC Cancer Review Group?	OSHA Regulated?	Target Organs?
Suspected Human Confirmed Animal	No	No	29 CFR 1910.1000 Table Z-1	Eyes, skin, blood, CVS, CNS, liver, kidney.

Medical Conditions Generally Aggravated by Exposure

Blood and liver disorders.

Emergency and First-aid Procedures

Eye contact: Flush immediately with water for 15 minutes (minimum), seek medical attention. Skin contact: Remove all contaminated clothing. Immediately wash area with large amounts of soap and water. Seek medical assistance. For inhalation: Remove the person from exposure. Provide respiratory assistance and CPR, if required. Transfer to medical facility. If swallowed: Unless advised otherwise, give 2 tablespoons Ipecac (adult dose) and 1-2 glasses of water. Induce vomiting. After vomiting, give 2 tablespoons of activated charcoal in 8 ounces of water. Seek medical attention immediately.

SECTION VII - PRECAUTIONS FOR SAFE HANDLING AND USE

Steps to be Taken in Case Material is Released or Spilled

Absorb liquids in vermiculite, dry sand, earth, or other non-reactive material and deposit is sealed drum. Ventilate area of spill or leak. For dry spills, collect for reuse or disposal. Do NOT dry sweep. Use water spray to knock-down vapors (for solution spills) and wet mop residue of dry spills.

Preferred Waste Disposal Method

No Citation.

Precautions to be Taken in Handling and Storage

Store to avoid contact with incompatible materials in tightly closed containers in cool, well-ventilated area away from heat and flame. Store in detached facility away from main plant or facility.

Other Precautions and Warnings

Sources of ignition are prohibited where dinitrobenzene is used, handled, or stored. Heat may cause solids to liquefy and containers to build up pressure and explode.

SECTION VIII - CONTROL MEASURES AND PERSONAL PROTECTIVE EQUIPMENT

Respiratory Protection *(specify type)*

PEL is low and difficult to monitor. For exposures above the PEL, use an MSHA/NIOSH-approved supplied air respirator with full facepiece, hood, or helmet in continuous flow mode, or use a self-contained breathing apparatus (SCBA) operated in pressure demand or other positive pressure mode.

Ventilation

Local exhaust or general mechanical systems recommended.

Protective Gloves	Eye Protection	Other Protective Clothing
Butyl Rubber	Chemical/Dust Goggles or Face Mask	Butyl Rubber Apron

Work/Hygiene Practices

Always wash hands thoroughly after using chemical; never bring food, drink, or smoking materials into vicinity of chemicals.

DINITROBENZENE

$C_6H_4(NO_2)_2$ CAS: 25154-54-5

ortho- isomer rating

IDENTIFICATION AND TYPICAL USES

Dinitrobenzene is a mixture of the three isomeric dinitrobenzenes, *m- (meta), o (ortho),* and *p (para)*. It is derived through the nitration of nitrobenzene with hot mixed acid. It is used organic synthesis, in the manufacture of dyes, as a camphor substitute in cellulose nitrate, and in explosives (the *ortho-* polymer is used in blasting charges and to fill artillery shells). It is extremely important to know which isomer is being assessed since their individual physical and toxicological properties differ.

RISK ASSESSMENT: HEALTH

General Assessment

Dinitrobenzene is a human poison by *inhalation, ingestion,* and skin *absorption.* It is an irritant to the eyes, skin, and respiratory system. It is a confirmed carcinogen in animals. Human carcinogenicity is suspect, although the data are inconclusive in this area of study. The *"m"* or *meta*-isomer is somewhat more toxicologically important than the other two isomers (*"p"* or *para-* and *"o"* or *ortho-*).

Inhalation can cause irritation to the respiratory system and associated mucosa. It can cause cough, congestion, with narcotic effects associated with the depression of the central nervous system (CNS). These include weakness, dyspnea (difficult or labored breathing), nausea, personality changes (irritability, mood changes), collapse, coma, and death (from respiratory failure). It also results in the formation of methemoglobin (a pigment similar to hemoglobin that is unable to perform the normal oxygenation process). Symptoms include headache, nausea, vomiting, dizziness, and jaundice (yellow coloring of the eyes, hair, and skin). There may be difficulty in breathing, visual disturbances (scotomas), general weakness, cyanosis, with possible progression to convulsions, coma, and even death.

Skin contact can produce the same systemic systems as inhalation as well as irritation, the appearance of small vesicles or blisters, redness or swelling, ulceration, and necrosis (issue death). Eye contact with the liquid solution or dusts can cause irritation, redness and swelling of the eyelids, painful sensitivity to light, and can cause severe eye damage and possible loss of vision.

Ingestion of dinitrobenzene can also cause systemic effects similar to those of inhalation. There can also be painful irritation of the mouth and stomach, stomach cramps, and diarrhea. A bitter almond taste or burning sensation in the mouth, dry throat, and thirst can occur.

Alcohol consumption, some medications, exposure to sunlight, or hot baths can aggravate and enhance symptoms.

Acute Health Effects

The following acute (short-term) health effects may occur immediately or shortly after exposure to dinitrobenzene:

Skin: Irritation and possible blistering, with ulceration. Will pass through unbroken skin to cause toxic systemic effects, including weakness, nausea, collapse, and death.

Eye: Severe irritation with damage and possible loss of vision likely.

Lung: Irritation to the respiratory tract and associated mucosa. Can cause weakness, nausea, confusion, collapse, coma, and death (respiratory failure).

CNS: A CNS depressant. Weakness, confusion, vertigo, mood changes and other personality changes, nausea, collapse.

Other: Ingestion will cause irritation to the gastrointestinal system with nausea, vomiting, convulsions, dypsnea, jaundice, tremor, confusion, and nervousness. It has also been found to be hepatotoxic (toxic to the liver) and nephrotoxic (toxic to the kidneys).

Chronic Health Effects

The following chronic (long-term) health effects may occur at some time after exposure to dinitrobenzene and can last for months or even years:

Cancer Hazards: Some studies have suggested that exposure to dinitrobenzene causes cancer in some test animals. While there is no evidence of carcinogenicity of this chemical in humans or animals, it is still considered a *suspected* carcinogen by some authorities.

Reproductive Hazard: According to information presented in the references, dinitrobenzene has not been adequately tested for its ability to adversely affect reproductively in animals.

Other Chronic Effects: High or repeated lower skin or inhalation exposures can cause liver or kidney disorders. Symptoms of late toxicity include secondary anemia, liver damage, irritability, weakness, headache, anorexia, weight loss, nausea, vomiting, cyanosis, dyspnea, and skin disorders. There may also be paresthesis in the feet, ankles, hands, and feet.

Recommended Risk-Reduction Measures

Personnel should avoid direct contact with dinitrobenzene. If a less toxic chemical cannot be substituted for a hazardous substance, then *engineering controls* are the most effective method of reducing exposures. The best protection is to enclose operations and/or provide local exhaust ventilation at the site of chemical release. While not always operationally feasible, isolating operations can also reduce exposure. Using respiratory protection is less effective than the controls mentioned above, but is still advisable whenever working with or around dinitrobenzene. Exposure level is low (1 ppm) and difficult to monitor. For exposures above the PEL, use an MSHA/NIOSH-approved full facepiece respirator equipped with an organic vapor cartridge equipped with a dust pre-filter. Increased protection is obtained from a supplied-air respirator with full facepiece operated in continuous flow mode, or a self-contained breathing apparatus (SCBA) with full facepiece and pressure demand. If a full facepiece is not available, then chemical goggles should be worn to protect the eyes. Whenever a chemical splash hazard exists, a face shield and a rubber apron should be worn. To prevent hand and skin exposures, butyl rubber gloves should be worn.

Administrative controls should also be in place to minimize the potential for human exposures. These may include written procedures or policies which specify the methods and techniques that will be practiced whenever personnel are to work with dinitrobenzene.

All personnel should receive *training* on the use, hazards, protective measures, emergency actions, and other precautions per 29 CFR 1910.1200 (Hazard Communication), prior to the first assignment in an area where dinitrobenzene is used or stored.

Before beginning employment with dinitrobenzene and at regular intervals thereafter (e.g., annually), those persons with high or frequent exposures should be provided the following recommended medical tests:

☑ Kidney and liver function tests.

☑ Complete blood count.

If symptoms develop or overexposure is suspected, the following tests are recommended:

☑ Kidney and liver function tests.

☑ Complete blood count.

☑ Consider possible reactions caused by alcohol consumption and/or medications.

Any evaluation should include a careful history of past and present symptoms with an examination. Medical tests that simply look for existing damage are not a substitute for controlling exposures. Physicians should note that symptoms may be delayed, so victim should be kept under observation. The methemoglobin concentration should be determined at regular intervals until indications of full reduction to hemoglobin. For ingestion, gastric lavage is suggested followed by saline catharsis.

Other methods to reduce exposure include:

☑ Where possible, enclose operations and use local exhaust ventilation at the site of chemical release. If local exhaust ventilation or enclosure is not used, respiratory protection should be mandatory.

☑ Always ensure that proper protective clothing is worn when using chemical substances and that personnel are trained on its care and use.

☑ Wash thoroughly immediately after exposure to dinitrobenzene and at the end of the work shift or before eating, drinking, or smoking.

☑ Hazard warning information should be posted in the work area. In addition, as part of an on-going education and training program, all information on the health and safety hazards of dinitrobenzene should be communicated to all potentially exposed workers.

☑ Eye wash stations should be provided in the immediate work area for emergency use. If there is a possibility of skin exposure to dinitrobenzene, emergency shower facilities should also be provided.

☑ Personnel should never wear contaminated clothing home (family members can be exposed). Clothing should be laundered by personnel who have been trained on the hazards of dinitrobenzene.

RISK ASSESSMENT: ENVIRONMENT

General Assessment

The environment is at risk of exposure during transportation, storage, disposal, or destruction of dinitrobenzene. In almost every scenario, the threat of environmental exposure is contingent upon the proper handling of the chemical substance. Accidental spills, large or small, can result in fire, explosion, and possible contamination of the surrounding environmental mediums (water, soil, and air).

Dinitrobenzene is normally considered a combustible solid, but it can also appear in mixed solutions. Its dusts can form explosive mixtures in air. These characteristics require caution in handling, storage, transportation, and disposal. It is incompatible with strong oxidizers, caustics, and some metals (such as zinc and tin). Contact with either can cause fire or explosion. Prolonged contact with heat may result in an explosion due to spontaneous decomposition. Therefore, special consideration is required during any emergency situation involving a leak or spill of dinitrobenzene. Should dinitrobenzene ever come into contact with these incompatible substances during use, transportation, storage, or disposal, the formation of highly toxic and/or highly explosive commodities is extremely possible.

Dinitrobenzene may enter the environment through industrial effluents or spills.

Acute Ecological Effects

Acute (short-term) toxic effects may include the death of animals, birds, or fish and death or low growth rate in plants. Acute effects are seen 2 to 4 days after animals or plants are exposed to dinitrobenzene. Insufficient data are available on the short-term effects of dinitrobenzene exposure to aquatic life, plants, birds or land animals.

Chronic Ecological Effects

Chronic (long-term) toxic effects may include shortened life span, reproductive problems, lower fertility, and changes in appearance or behavior in exposed animals. These effects can be seen long after first exposure(s) to toxic chemicals. Insufficient data are available on the long-term effects of dinitrobenzene to aquatic life, plants, birds, or land animals.

Water Solubility

Dinitrobenzene is slightly soluble in water. Concentrations of 1 to 100 milligrams will mix with a liter of water.

Persistence in the Environment

Dinitrobenzene is slightly persistent in water, with a half-life between 2 and 20 days. The half-life of a pollutant is the amount of time it takes for one half of the chemical to be degraded.

Bioaccumulation in Aquatic Organisms

Some substances increase in concentration, or *bioaccumulate,* in living organisms as they breathe contaminated air, drink contaminated water, or eat contaminated food. These chemicals can become concentrated in the tissues and internal organs of animals as well as humans.

The concentration of dinitrobenzene found in fish tissue is expected to be somewhat higher than the average concentration of dinitrobenzene in the water from which the fish was taken.

Recommended Risk-Reduction Measures

Proper training of all transporters will reduce the likelihood of a mishap or accident resulting in a leak or spill to the environment. The correct use of DOT labeling (per the MSDS) on all containers, trucks, and rail cars will enable emergency responders to react properly and quickly to any disaster thereby reducing the potential risk to the environment and to personnel.

Storage of dinitrobenzene should be segregated from incompatible chemicals to minimize the risk of

cross-contamination or contact. Buildings designated for storage should be equipped with appropriate fire protection systems (alarms, sprinklers, emergency lighting, portable extinguishers).

If a spill or leak to the environment has occurred, fire department, emergency response and/or hazardous materials spill personnel should be notified immediately. Cleanup should be attempted only by those trained in proper spill containment procedures. Contaminated soils should be removed for incineration and replaced with clean soil. If dinitrobenzene should contact the water table, aquifer, or navigable waterway, time is of the essence. It is slightly soluble in the aquatic environment and total remediation may not be possible. The local and/or state emergency response authorities must be notified. A comprehensive emergency response or disaster preparedness/recovery plan should be in place prior to any operations involving the use, transportation, storage, or disposal of dinitrobenzene. If dinitrobenzene is spilled or leaked, the following specific steps are recommended:

☑ Restrict persons not wearing protective clothing from area of spill or leak until cleanup is complete and area can be opened for normal work.

☑ Remove all ignition sources, ventilate area of leak.

☑ Absorb liquid spills in vermiculite, dry sand, earth, or similar material and deposit in sealed containers. Collect solid spills for reuse of disposal. Wet mop spill area to collect residue.

☑ It may be necessary to dispose of dinitrobenzene as a hazardous waste. The responsible state agency or the regional office of the federal Environmental Protection Agency (EPA) should be contacted for specific recommendations.

RISK ASSESSMENT: BUSINESS

General Assessment

Accidents or mishaps involving dinitrobenzene can present a significant threat to business operations. The loss or damage of equipment or facilities can significantly affect fiscal viability. Lawsuits that may result from personnel injury/death, public exposures, and/or environmental contamination will also require a serious expenditure of resources. Media attention surrounding an injury, death, or environmental damage can also result in a loss of profits and loss of current as well as future business. Always remember that any time the terms "carcinogen" or "cancer" are use, public emotion, hysteria, and ignorance can run equally high. This should be a consideration during the development of public relation policies.

Recommended Risk-Reduction Measures

Company attorneys, safety and health professionals, and environmental specialists should be involved in the development of any procedures dealing with safety or emergency response. A company official should be pre-designated as a public relations officer with specific training in dealing with the press. Corporate plans and policies should be developed, approved, and implemented long before any need for such arises.

REFERENCES

Asante-Duah, D. K. 1993. *Hazardous Waste Risk Assessment.* New York: CRC Press/Lewis Publishers.

Bloom, A. D. and F. J. DeSerres. 1995. *Ecotoxicity and Human Health.* New York: CRC Press/Lewis Publishers.

Calabrese, E. J. 1994. *Biological Effects of Low Level Exposures.* New York: CRC Press/Lewis Publishers.

Cockerman, L. G., B. S. Shane. 1994. *Basic Environmental Toxicology.* New York: CRC Press/Lewis Publishers.

Cothern, C. R., N. P. Ross. 1994. *Environmental Statistics, Assessment, and Forecasting.* New York: CRC Press/Lewis Publishers.

Hallenbeck, W. H. 1993. *Quantitative Risk Assessment for Environmental and Occupational Health,* 2nd Edition. New York: CRC Press/Lewis Publishers.

Howard, P. H. 1995. *Handbook of Environmental Fate and Exposures Data for Organic Chemicals,* Vol. I-IV. New York: CRC Press/Lewis Publishers.

Lauwerys, R. R., P. Hoet. 1993. *Industrial Chemical Exposure.* New York: CRC Press/Lewis Publishers.

Lewis, R. J., Sr. 1992. *Sax's Dangerous Properties of Industrial Materials,* Volumes 1, 2, and 3, 8th Edition. New York: Van Nostrand Reinhold.

Meyer, E. 1990. *Chemistry of Hazardous Materials.* Englewood Cliffs, NJ: Prentice Hall.

Quigley, D. R. 1994. *Handbook of Emergency Chemical Management.* New York: CRC Press/Lewis Publishers.

Torkelson, T. R. V. K. Rowe. 1981. Halogenated aliphatic hydrocarbons. *Patty's Industrial Hygiene and Toxicology,* Vol. 2B, *Toxicology,* pp 3568-3572. New York: Wiley-Interscience.

U.S. Department of Health and Human Services, National Institute for Occupational Safety and Health. 1994. *NIOSH Pocket Guide to Chemical Hazards.* Washington, D.C.: U.S. Government Printing Office.

MATERIAL SAFETY DATA SHEET

CHEMICAL NAME	CRC PRESS/LEWIS PUBLISHERS DISCLAIMER
***m*-DINITROBENZENE**	CRC PRESS/LEWIS PUBLISHERS DISCLAIMER: THE INFORMATION AND RECOMMENDATIONS PRESENTED HEREIN ARE BASED ON SOURCES BELIEVED TO BE RELIABLE. CRC MAKES NO REPRESENTATION ON ITS COMPLETENESS OR ACCURACY. IT IS THE RESPONSIBILITY OF THE USER TO DETERMINE THE CHEMICAL'S SUITABILITY FOR ITS INTENDED USE, THE CHEMICAL'S SAFE USE, AND THE CHEMICAL'S PROPER DISPOSAL. NO REPRESENTATIONS AND/OR WARRANTIES, EITHER EXPRESSED OR IMPLIED, OF THE MERCHANTABILITY OR FITNESS FOR A PARTICULAR PURPOSE, OR OF ANY OTHER NATURE, ARE MADE WITH RESPECT TO THE INFORMATION PROVIDED IN THIS MSDS OR TO THE CHEMICAL TO WHICH INFORMATION MAY REFER. CRC NEITHER ASSUMES NOR AUTHORIZES ANY OTHER PERSON TO ASSUME FOR IT, ANY OTHER ADDITIONAL RESPONSIBILITY OR LIABILITY FOR THE USE OF, OR RELIANCE UPON, THIS INFORMATION.

HAZARD WARNING INFORMATION

HEALTH	FIRE	REACTIVE	OTHER	DEGREE OF HAZARD	COLOR CODING	OTHER CODES
3	1	4		0 = Minimum Hazard 1 = Slight Hazard 2 = Moderate Hazard 3 = Serious Hazard 4 = Severe Hazard	HEALTH = BLUE FIRE = RED REACTIVITY = YELLOW OTHER = WHITE	OX = Oxidizer ACID = Acid ALK = Alkali COR = Corrosive W = Use No Water

SECTION I - GENERAL INFORMATION

Characterization	**Phenol**	RCRA Number / EPA Class	**D001** / **Characteristic (I) Waste**
DOT Proper Shipping Name	***m*-Dinitrobenzene, solid**	Chemical Abstract Service (CAS) Number	**99-65-0**
DOT Hazard Class and Label Requirements	**Poison B**	DOT Emergency Guide Code	**56**
DOT Identification Number	**UN 1597**	Chemical Formula	$C_6H_4(NO_2)_2$

Synonyms
1,3-Dinitrobenzene; *meta*-dinitrobenzene; binitrobenzene; 2,4-dinitrobenzene; 1,3-dinitobenzol

SECTION II - HAZARDOUS INGREDIENTS AND IDENTITY INFORMATION

Hazardous Components (specific identity)	OSHA Exposure Criteria	NIOSH Exposure Criteria	Immediately Dangerous to Life and Health (IDLH)	ACGIH Exposure Criteria
***m*-Dinitrobenzene** (derivation: From the nitration of nitrobenzene with hot mixed acid). **1 ppm = 6.6 mg/m³**	**PEL (skin):** 1 ppm 6.6 mg/m³ **STEL:** Not Established	**REL (skin):** 1 ppm 6.6 mg/m³ **STEL:** Not Established	**50 mg/m³**	**TLV (skin):** 0.15 ppm 1 mg/m³ **STEL:** Not Established

SECTION III - PHYSICAL AND CHEMICAL CHARACTERISTICS

Boiling Point	**572°F (300°C)**	Specific Gravity (H_2O = 1)	**1.58**
Vapor Pressure (gas)	**<1 at 69°F (20°C)**	Molecular Weight	**168.1**
Vapor Density (Air = 1)	**5.8**	Melting Point	**192°F (89°C)**

Solubility
Slightly soluble (0.02%). Soluble in chloroform and ethyl acetate and sparingly soluble in benzene.

Appearance and Odor
Pale white or yellow crystalline solid.

SECTION IV - FIRE AND EXPLOSION HAZARD DATA

Flash Point (method used)	**302°F (150°C) closed cup**	Explosive Limits in Air % by Volume	**LEL: Not Determined UEL: Not Determined**
NFPA Classification	**Combustible Solid**	Autoignition Temperature	**Not Determined**

Extinguishing Media
For small fires, use dry chemical, carbon dioxide, water spray, or regular foam. For large fires, use water spray, fog, or regular foam. Apply water as fog in flooding amounts (solid stream may be ineffective).

Special Fire Fighting Procedures
Wear full protective clothing and self-contained breathing apparatus (SCBA). Poisonous gases are produced in fire. Use water spray to keep fire-exposed containers cool.

Unusual Fire and Explosion Hazards
Containers may explode in fire. Highly friction- and shock-sensitive. Dusts form explosive concentrations in air. May spontaneously ignite or explode at high temperatures due to spontaneous combustion.

SECTION V - REACTIVITY DATA

Stability		Conditions to Avoid **m-Dinitrobenzene is unstable at high temperatures and is extremely shock- and friction-sensitive. Keep containers away from areas where shock, heat, and/or friction are likely to occur.**
Stable	Unstable **X**	Incompatibility *(materials to avoid)* **Strong oxidizers, (such as chlorine, bromine, and fluorine), caustics, tetranitromethane, and many metals (such as zinc and tin). Explosive when mixed with nitric acid.**
Hazardous Polymerization		Conditions to Avoid **Under normal conditions of temperature and pressure, hazardous polymerization of m-dinitrobenzene is not expected to occur.**
May Occur	Will Not Occur **X**	Hazardous Decomposition or By-products **When heated to decomposition, m-dinitrobenzene can emit highly toxic/poisonous gases, including carbon dioxide and oxides of nitrogen. m-Dinitrobenzene can explode when heated.**

SECTION VI - HEALTH HAZARD DATA

Primary Route(s) of Entry:	Inhalation?	Absorption (skin)?	Ingestion?
	X	X	X

Health Hazards

INHALATION: Anoxia; visual disturbances; central scotomas; yellowing of hair, eyes, and skin; anemia; and liver, kidney, and central nervous system injury is possible. Forms methemoglobin.

ABSORPTION: Skin irritant. Will pass through skin to cause toxic systemic effects (same as inhalation).

INGESTION: Gastrointestinal disorders; tremor; convulsions; bad taste and burning feeling in mouth; dry throat; and thirst.

Carcinogenicity	NTP Listed?	IARC Cancer Review Group?	OSHA Regulated?	Target Organs?
Suspected Human Confirmed Animal	No	No	29 CFR 1910.1000 Table Z-1	Eyes, skin, blood, CVS, CNS, liver, kidney.

Medical Conditions Generally Aggravated by Exposure

Blood and liver disorders.

Emergency and First-aid Procedures

Eye contact: Flush immediately with water for 15 minutes (minimum), seek medical attention. **Skin contact**: Remove all contaminated clothing. Immediately wash area with large amounts of soap and water. Seek medical assistance. For **inhalation**: Remove the person from exposure. Provide respiratory assistance and CPR, if required. Transfer to medical facility. If **swallowed**: Unless advised otherwise, give 2 tablespoons Ipecac (adult dose) and 1-2 glasses of water. Induce vomiting. After vomiting, give 2 tablespoons of activated charcoal in 8 ounces of water. Seek medical attention immediately.

SECTION VII - PRECAUTIONS FOR SAFE HANDLING AND USE

Steps to be Taken in Case Material is Released or Spilled

Absorb liquids in vermiculite, dry sand, earth, or other non-reactive material and deposit is sealed drum. Ventilate area of spill or leak. For dry spills, collect for reuse or disposal. Do NOT dry sweep. Use water spray to knock-down vapors (for solution spills) and wet mop residue of dry spills.

Preferred Waste Disposal Method

No Citation.

Precautions to be Taken in Handling and Storage

Store to avoid contact with incompatible materials in tightly closed containers in cool, well-ventilated area away from heat and flame. Store in detached facility away from main plant or facility.

Other Precautions and Warnings

Sources of ignition are prohibited where *m*-dinitrobenzene is used, handled, or stored. Heat may cause solids to liquefy and containers to build up pressure and explode.

SECTION VIII - CONTROL MEASURES AND PERSONAL PROTECTIVE EQUIPMENT

Respiratory Protection *(specify type)*

PEL is low and difficult to monitor. For exposures above the PEL, use a MSHA/NIOSH-approved supplied air respirator with full facepiece, hood, or helmet in continuous flow mode, or use a self-contained breathing apparatus (SCBA) operated in pressure demand or other positive pressure mode.

Ventilation

Local exhaust or general mechanical systems recommended.

Protective Gloves	Eye Protection	Other Protective Clothing
Butyl Rubber	**Chemical/Dust Goggles or Face Mask**	**Butyl Rubber Apron**

Work/Hygiene Practices

Always wash hands thoroughly after using chemical; never bring food, drink, or smoking materials into vicinity of chemicals.

m-DINITROBENZENE

$C_6H_4(NO_2)_2$ CAS: 99-65-0

ortho- isomer rating

IDENTIFICATION AND TYPICAL USES

m-Dinitrobenzene, or *meta*-dinitrobenzene, is one of three dinitrobenzene isomers, *o- (ortho),* and *p- (para)* being the other two. It is derived through the nitration of nitrobenzene with hot mixed acid. It is used organic synthesis, in the manufacture of dyes, as a camphor substitute in cellulose nitrate, and in explosives (the *ortho-* polymer is used in blasting charges and to fill artillery shells).

RISK ASSESSMENT: HEALTH

General Assessment

m-Dinitrobenzene is a human poison by *inhalation, ingestion,* and skin *absorption.* In contrast to the other isomers of dinitrobenzene, the *meta*-isomer is considered the most important from a health risk perspective. While exposure symptoms may be the same, it appears to cause toxic systemic effects more readily that other dinitrobenzene isomers of similar exposure levels and conditions. It is an irritant to the eyes, skin and respiratory system. It is a confirmed carcinogen in animals. Human carcinogenicity is suspect, although the data are inconclusive in this area of study.

Inhalation can cause irritation to the respiratory system and associated mucosa. It can cause cough, congestion, with narcotic effects associated with the depression of the central nervous system (CNS). These include weakness, dyspnea, nausea, personality changes (irritability, mood changes), collapse, coma, and death (from respiratory failure). It also results in the formation of methemoglobin (a pigment similar to hemoglobin that is unable to perform the normal oxygenation process). Symptoms include headache, nausea, vomiting, dizziness, and jaundice (yellow coloring of the eyes, hair, and skin). There may be difficulty in breathing, visual disturbances (scotomas), general weakness, cyanosis, with possible progression to convulsions, coma, and even death.

Skin contact can produce the same systemic systems as inhalation as well as irritation, the appearance of small vesicles or blisters, redness or swelling, ulceration, and necrosis (issue death). Eye contact with the liquid solution or dusts can cause irritation, redness and swelling of the eyelids, painful sensitivity to light, and can cause severe eye damage and possible loss of vision.

Ingestion of *m*-dinitrobenzene can also cause systemic effects similar to those of inhalation. There can also be painful irritation of the mouth and stomach, stomach cramps, and diarrhea. A bitter almond taste or burning sensation in the mouth, dry throat, and thirst can occur.

Alcohol consumption, some medications, exposure to sunlight, or hot baths can aggravate and enhance symptoms.

☠ *Acute Health Effects*

The following acute (short-term) health effects may occur immediately or shortly after exposure to *m*-dinitrobenzene:

Skin: Irritation and possible blistering, with ulceration. Will pass through unbroken skin to cause toxic systemic effects, including weakness, nausea, collapse, and death.

Eye: Severe irritation with damage and possible loss of vision likely.

Lung: Irritation to the respiratory tract and associated mucosa with weakness, nausea, confusion, collapse, coma, and death (respiratory failure).

CNS: A CNS depressant. Weakness, confusion, vertigo, mood changes and other personality changes, nausea, collapse.

Other: Ingestion will cause irritation to the gastrointestinal system with nausea, vomiting, convulsions, dypsnea, jaundice, tremor, confusion, and nervousness. It has also been found to be hepatotoxic (toxic to the liver) and nephrotoxic (toxic to the kidneys).

Chronic Health Effects

The following chronic (long-term) health effects may occur at some time after exposure to *m*-dinitrobenzene and can last for months or even years:

Cancer Hazards: Some studies have suggested that exposure to *m*-dinitrobenzene causes cancer in some test animals. While there is no evidence of carcinogenicity of this chemical in humans or animals, it is still considered a *suspected* carcinogen by some authorities.

Reproductive Hazard: According to information presented in the references, *m*-dinitrobenzene has not been adequately tested for its ability to adversely affect reproductively in animals.

Other Chronic Effects: High or repeated lower skin or inhalation exposures can cause liver or kidney disorders. Symptoms of late toxicity include secondary anemia, liver damage, irritability, weakness, headache, anorexia, weight loss, nausea, vomiting, cyanosis, dyspnea, and skin disorders. There may also be paresthesis in the feet, ankles, hands, and feet.

Recommended Risk-Reduction Measures

Personnel should avoid direct contact with *m*-dinitrobenzene. If a less toxic chemical cannot be substituted for a hazardous substance, then *engineering controls* are the most effective method of reducing exposures. The best protection is to enclose operations and/or provide local exhaust ventilation at the site of chemical release. While not always operationally feasible, isolating operations can also reduce exposure. Using respiratory protection is less effective than the controls mentioned above, but is still advisable whenever working with or around *m*-dinitrobenzene. Exposure level is low (1 ppm) and difficult to monitor. For exposures above the PEL, use an MSHA/NIOSH-approved full facepiece respirator equipped with an organic vapor cartridge equipped with a dust pre-filter. Increased protection is obtained from a supplied-air respirator with full facepiece operated in continuous flow mode, or a self-contained breathing apparatus (SCBA) with full facepiece and pressure demand. If a full facepiece is not available, then chemical goggles should be worn to protect the eyes. Whenever a chemical splash hazard exists, a face shield and a rubber apron should be worn. To prevent hand and skin exposures, butyl rubber gloves should be worn.

Administrative controls should also be in place to minimize the potential for human exposures. These may include written procedures or policies which specify the methods and techniques that will be practiced whenever personnel are to work with *m*-dinitrobenzene.

All personnel should receive *training* on the use, hazards, protective measures, emergency actions, and other precautions per 29 CFR 1910.1200 (Hazard Communication), prior to the first assignment in an area where *m*-dinitrobenzene is used or stored.

Before beginning employment with *m*-dinitrobenzene and at regular intervals thereafter (e.g., annually), those persons with high or frequent exposures should be provided the following recommended medical tests:

- ☑ Kidney and liver function tests.
- ☑ Complete blood count.

If symptoms develop or overexposure is suspected, the following tests are recommended:

- ☑ Kidney and liver function tests.
- ☑ Complete blood count.
- ☑ Consider possible reactions caused by alcohol consumption and/or medications.

Any evaluation should include a careful history of past and present symptoms with an examination. Medical tests that simply look for existing damage are not a substitute for controlling exposures. Physicians should note that symptoms may be delayed, so victim should be kept under observation. The methemoglobin concentration should be determined at regular intervals until indications of full reduction to hemoglobin. For ingestion, gastric lavage is suggested followed by saline catharsis.

Other methods to reduce exposure include:

- ☑ Where possible, enclose operations and use local exhaust ventilation at the site of chemical release. If local exhaust ventilation or enclosure is not used, respiratory protection should be mandatory.
- ☑ Always ensure that proper protective clothing is worn when using chemical substances.
- ☑ Wash thoroughly immediately after exposure to *m*-dinitrobenzene and at the end of the work shift or before eating, drinking, or smoking.

- ☑ Hazard warning information should be posted in the work area. In addition, as part of an on-going education and training program, all information on the health and safety hazards of *m*-dinitrobenzene should be communicated to all potentially exposed workers.
- ☑ Eye wash stations should be provided in the immediate work area for emergency use. If there is a possibility of skin exposure to *m*-dinitrobenzene, emergency shower facilities should also be provided.
- ☑ Personnel should never wear contaminated clothing home (family members can be exposed). Clothing should be laundered by personnel who have been trained on the hazards of *m*-dinitrobenzene.

RISK ASSESSMENT: ENVIRONMENT

General Assessment

The environment is at risk of exposure during transportation, storage, disposal, or destruction of *m*-dinitrobenzene. In almost every scenario, the threat of environmental exposure is contingent upon the proper handling of the chemical substance. Accidental spills, large or small, can result in fire, explosion, and possible contamination of the surrounding environmental mediums (water, soil, and air).

m-Dinitrobenzene is normally considered a combustible solid, but it can also appear in mixed solutions. Its dusts can form explosive mixtures in air. These characteristics require caution in handling, storage, transportation, and disposal. It is incompatible with strong oxidizers, caustics, and some metals (such as zinc and tin). Contact with either can cause fire or explosion. Prolonged contact with heat may result in an explosion due to spontaneous decomposition. Therefore, special consideration is required during any emergency situation involving a leak or spill of *m*-dinitrobenzene. Should *m*-dinitrobenzene ever come into contact with these incompatible substances during use, transportation, storage, or disposal, the formation of highly toxic and/or highly explosive commodities is extremely possible. *m*-Dinitrobenzene may enter the environment through industrial effluents or spills.

Acute Ecological Effects

Acute (short-term) toxic effects may include the death of animals, birds, or fish and death or low growth rate in plants. Acute effects are seen 2 to 4 days after animals or plants are exposed to *m*-dinitrobenzene. Insufficient data are available on the short-term effects of *m*-dinitrobenzene exposure to aquatic life, plants, birds or land animals.

Chronic Ecological Effects

Chronic (long-term) toxic effects may include shortened life span, reproductive problems, lower fertility, and changes in appearance or behavior in exposed animals. These effects can be seen long after first exposure(s) to toxic chemicals. Insufficient data are available on the long-term effects of *m*-dinitrobenzene to aquatic life, plants, birds, or land animals.

Water Solubility

m-Dinitrobenzene is slightly soluble in water. Concentrations of 1 to 100 milligrams will mix with a liter of water.

Persistence in the Environment

m-Dinitrobenzene is slightly persistent in water, with a half-life between 2 and 20 days. The half-life of a pollutant is the amount of time it takes for one half of the chemical to be degraded.

Bioaccumulation in Aquatic Organisms

Some substances increase in concentration, or *bioaccumulate,* in living organisms as they breathe contaminated air, drink contaminated water, or eat contaminated food. These chemicals can become concentrated in the tissues and internal organs of animals as well as humans.

The concentration of *m*-dinitrobenzene found in fish tissue is expected to be somewhat higher than the average concentration of *m*-dinitrobenzene in the water from which the fish was taken.

Recommended Risk-Reduction Measures

Proper training of all transporters will reduce the likelihood of a mishap or accident resulting in a leak or spill to the environment. The correct use of DOT labeling (per the MSDS) on all containers, trucks, and rail cars will enable emergency responders to react properly and quickly to any disaster thereby reducing the potential risk to the environment and to personnel.

Storage of *m*-dinitrobenzene should be segregated from incompatible chemicals to minimize the risk of

cross-contamination or contact. Buildings designated for storage should be equipped with appropriate fire protection systems (alarms, sprinklers, emergency lighting, portable extinguishers).

If a spill or leak to the environment has occurred, fire department, emergency response and/or hazardous materials spill personnel should be notified immediately. Cleanup should be attempted only by those trained in proper spill containment procedures. Contaminated soils should be removed for incineration and replaced with clean soil. If *m*-dinitrobenzene should contact the water table, aquifer, or navigable waterway, time is of the essence. It is slightly soluble in the aquatic environment and total remediation may not be possible. The local and/or state emergency response authorities must be notified. A comprehensive emergency response or disaster preparedness/recovery plan should be in place prior to any operations involving the use, transportation, storage, or disposal of *m*-dinitrobenzene. If *m*-dinitrobenzene is spilled or leaked, the following specific steps are recommended:

- ☑ Restrict persons not wearing protective clothing from area of spill or leak until cleanup is complete and area can be opened for normal work.
- ☑ Remove all ignition sources, ventilate area of leak.
- ☑ Absorb liquid spills in vermiculite, dry sand, earth, or similar material and deposit in sealed containers. Collect solid spills for reuse of disposal. Wet mop spill area to collect residue.
- ☑ It may be necessary to dispose of *m*-dinitrobenzene as a hazardous waste. The responsible state agency or the regional office of the federal Environmental Protection Agency (EPA) should be contacted for specific recommendations.

RISK ASSESSMENT: BUSINESS

General Assessment

Accidents or mishaps involving *m*-dinitrobenzene can present a significant threat to business operations. The loss or damage of equipment or facilities can significantly affect fiscal viability. Lawsuits that may result from personnel injury/death, public exposures, and/or environmental contamination will also require a serious expenditure of resources. Media attention surrounding an injury, death, or environmental damage can also result in a loss of profits and loss of current as well as future business. Always remember that any time the terms "carcinogen" or "cancer" are use, public emotion, hysteria, and ignorance can run equally high. This should be a consideration during the development of public relation policies.

Recommended Risk-Reduction Measures

Company attorneys, safety and health professionals, and environmental specialists should be involved in the development of any procedures dealing with safety or emergency response. A company official should be pre-designated as a public relations officer with specific training in dealing with the press. Corporate plans and policies should be developed, approved, and implemented long before any need for such arises.

REFERENCES

Asante-Duah, D. K. 1993. *Hazardous Waste Risk Assessment.* New York: CRC Press/Lewis Publishers.

Bloom, A. D. and F. J. DeSerres. 1995. *Ecotoxicity and Human Health.* New York: CRC Press/Lewis Publishers.

Calabrese, E. J. 1994. *Biological Effects of Low Level Exposures.* New York: CRC Press/Lewis Publishers.

Cockerman, L. G., B. S. Shane. 1994. *Basic Environmental Toxicology.* New York: CRC Press/Lewis Publishers.

Cothern, C. R., N. P. Ross. 1994. *Environmental Statistics, Assessment, and Forecasting.* New York: CRC Press/Lewis Publishers.

Hallenbeck, W. H. 1993. *Quantitative Risk Assessment for Environmental and Occupational Health,* 2nd Edition. New York: CRC Press/Lewis Publishers.

Howard, P. H. 1995. *Handbook of Environmental Fate and Exposures Data for Organic Chemicals,* Vol. I-IV. New York: CRC Press/Lewis Publishers.

Lauwerys, R. R., P. Hoet. 1993. *Industrial Chemical Exposure.* New York: CRC Press/Lewis Publishers.

Lewis, R. J., Sr. 1992. *Sax's Dangerous Properties of Industrial Materials,* Volumes 1, 2, and 3, 8th Edition. New York: Van Nostrand Reinhold.

Meyer, E. 1990. *Chemistry of Hazardous Materials.* Englewood Cliffs, NJ: Prentice Hall.

Quigley, D. R. 1994. *Handbook of Emergency Chemical Management.* New York: CRC Press/Lewis Publishers.

Torkelson, T. R. V. K. Rowe. 1981. Halogenated aliphatic hydrocarbons. *Patty's Industrial Hygiene and Toxicology,* Vol. 2B, *Toxicology,* pp 3568-3572. New York: Wiley-Interscience.

U.S. Department of Health and Human Services, National Institute for Occupational Safety and Health. 1994. *NIOSH Pocket Guide to Chemical Hazards.* Washington, D.C.: U.S. Government Printing Office.

MATERIAL SAFETY DATA SHEET

CHEMICAL NAME	CRC PRESS/LEWIS PUBLISHERS DISCLAIMER
***o*-DINITROBENZENE**	**CRC PRESS/LEWIS PUBLISHERS DISCLAIMER:** THE INFORMATION AND RECOMMENDATIONS PRESENTED HEREIN ARE BASED ON SOURCES BELIEVED TO BE RELIABLE. CRC MAKES NO REPRESENTATION ON ITS COMPLETENESS OR ACCURACY. IT IS THE RESPONSIBILITY OF THE USER TO DETERMINE THE CHEMICAL'S SUITABILITY FOR ITS INTENDED USE, THE CHEMICAL'S SAFE USE, AND THE CHEMICAL'S PROPER DISPOSAL. NO REPRESENTATIONS AND/OR WARRANTIES, EITHER EXPRESSED OR IMPLIED, OF THE MERCHANTABILITY OR FITNESS FOR A PARTICULAR PURPOSE, OR OF ANY OTHER NATURE, ARE MADE WITH RESPECT TO THE INFORMATION PROVIDED IN THIS MSDS OR TO THE CHEMICAL TO WHICH INFORMATION MAY REFER. CRC NEITHER ASSUMES NOR AUTHORIZES ANY OTHER PERSON TO ASSUME FOR IT, ANY OTHER ADDITIONAL RESPONSIBILITY OR LIABILITY FOR THE USE OF, OR RELIANCE UPON, THIS INFORMATION.

HAZARD WARNING INFORMATION

HEALTH	FIRE	REACTIVE	OTHER	DEGREE OF HAZARD	COLOR CODING	OTHER CODES
3	1	4		0 = Minimum Hazard 1 = Slight Hazard 2 = Moderate Hazard 3 = Serious Hazard 4 = Severe Hazard	HEALTH = BLUE FIRE = RED REACTIVITY = YELLOW OTHER = WHITE	OX = Oxidizer ACID = Acid ALK = Alkali COR = Corrosive W = Use No Water

SECTION I - GENERAL INFORMATION

Characterization	**Phenol**
RCRA Number	**D001**
EPA Class	**Characteristic (I) Waste**
DOT Proper Shipping Name	***o*-Dinitrobenzene, solid**
Chemical Abstract Service (CAS) Number	**528-29-0**
DOT Hazard Class and Label Requirements	**Poison B**
DOT Emergency Guide Code	**56**
DOT Identification Number	**UN 1597**
Chemical Formula	$C_6H_4(NO_2)_2$
Synonyms	**1,2-Dinitrobenzene; *ortho*-dinitrobenzene.**

SECTION II - HAZARDOUS INGREDIENTS AND IDENTITY INFORMATION

Hazardous Components (specific identity)	OSHA Exposure Criteria	NIOSH Exposure Criteria	Immediately Dangerous to Life and Health (IDLH)	ACGIH Exposure Criteria
***o*-Dinitrobenzene** (derivation: From the nitration of nitrobenzene with hot mixed acid). **1 ppm = 6.6 mg/m³**	**PEL (skin):** **1 ppm** **6.6 mg/m³** **STEL:** **Not Established**	**REL (skin):** **1 ppm** **6.6 mg/m³** **STEL:** **Not Established**	**50 mg/m³**	**TLV (skin):** **0.15 ppm** **1 mg/m³** **STEL:** **Not Established**

SECTION III - PHYSICAL AND CHEMICAL CHARACTERISTICS

Boiling Point	**606°F (319°C)**	Specific Gravity (H_2O = 1)	**1.57**
Vapor Pressure (gas)	**<1 at 69°F (20°C)**	Molecular Weight	**168.1**
Vapor Density (Air = 1)	**5.8**	Melting Point	**245°F (119°C)**

Solubility
Slightly soluble (0.05%). Soluble in chloroform and ethyl acetate and sparingly soluble in benzene.

Appearance and Odor
Yellowish needles or plates (a crystalline solid).

SECTION IV - FIRE AND EXPLOSION HAZARD DATA

Flash Point (method used)	**302°F (150°C) closed cup**	Explosive Limits in Air % by Volume	**LEL: Not Determined UEL: Not Determined**
NFPA Classification	**Combustible Solid**	Autoignition Temperature	**Not Determined**

Extinguishing Media
For small fires, use dry chemical, carbon dioxide, water spray, or regular foam. For large fires, use water spray, fog, or regular foam. Apply water as fog in flooding amounts (solid stream may be ineffective).

Special Fire Fighting Procedures
Wear full protective clothing and self-contained breathing apparatus (SCBA). Poisonous gases are produced in fire. Use water spray to keep fire-exposed containers cool.

Unusual Fire and Explosion Hazards
Containers may explode in fire. Highly friction- and shock-sensitive. Dusts form explosive concentrations in air. May spontaneously ignite or explode at high temperatures due to spontaneous combustion.

SECTION V - REACTIVITY DATA

Stability		Conditions to Avoid **o-Dinitrobenzene is unstable at high temperatures and is extremely shock- and friction-sensitive. Keep containers away from areas where shock, heat, and/or friction are likely to occur.**
Stable	Unstable X	Incompatibility *(materials to avoid)* **Strong oxidizers, (such as chlorine, bromine, and fluorine), caustics, and many metals (such as zinc and tin). Explosive when mixed with nitric acid.**
Hazardous Polymerization		Conditions to Avoid **Under normal conditions of temperature and pressure, hazardous polymerization of o-dinitrobenzene is not expected to occur.**
May Occur	Will Not Occur X	Hazardous Decomposition or By-products **When heated to decomposition, o-dinitrobenzene can emit highly toxic/poisonous gases, including carbon dioxide and oxides of nitrogen. o-Dinitrobenzene can explode when heated.**

SECTION VI - HEALTH HAZARD DATA

Primary Route(s) of Entry:	Inhalation?	Absorption (skin)?	Ingestion?
	X	X	X

Health Hazards

INHALATION: Anoxia; visual disturbances; central scotomas; yellowing of hair, eyes, and skin; anemia; and liver, kidney, and central nervous system injury is possible. Forms methemoglobin.

ABSORPTION: Skin irritant. Will pass through skin to cause toxic systemic effects (same as inhalation).

INGESTION: Gastrointestinal disorders; tremor; convulsions; bad taste and burning feeling in mouth; dry throat; and thirst.

Carcinogenicity	NTP Listed?	IARC Cancer Review Group?	OSHA Regulated?	Target Organs?
Suspected Human Confirmed Animal	No	No	29 CFR 1910.1000 Table Z-1	Eyes, skin, blood, CVS, CNS, liver, kidney.

Medical Conditions Generally Aggravated by Exposure

Blood and liver disorders.

Emergency and First-aid Procedures

Eye contact: Flush immediately with water for 15 minutes (minimum), seek medical attention. Skin contact: Remove all contaminated clothing. Immediately wash area with large amounts of soap and water. Seek medical assistance. For inhalation: Remove the person from exposure. Provide respiratory assistance and CPR, if required. Transfer to medical facility. If swallowed: Unless advised otherwise, give 2 tablespoons Ipecac (adult dose) and 1-2 glasses of water. Induce vomiting. After vomiting, give 2 tablespoons of activated charcoal in 8 ounces of water. Seek medical attention immediately.

SECTION VII - PRECAUTIONS FOR SAFE HANDLING AND USE

Steps to be Taken in Case Material is Released or Spilled

Absorb liquids in vermiculite, dry sand, earth, or other non-reactive material and deposit is sealed drum. Ventilate area of spill or leak. For dry spills, collect for reuse or disposal. Do NOT dry sweep. Use water spray to knock-down vapors (for solution spills) and wet mop residue of dry spills.

Preferred Waste Disposal Method

No Citation.

Precautions to be Taken in Handling and Storage

Store to avoid contact with incompatible materials in tightly closed containers in cool, well-ventilated area away from heat and flame. Store in detached facility away from main plant or facility.

Other Precautions and Warnings

Sources of ignition are prohibited where *o*-dinitrobenzene is used, handled, or stored. Heat may cause solids to liquefy and containers to build up pressure and explode.

SECTION VIII - CONTROL MEASURES AND PERSONAL PROTECTIVE EQUIPMENT

Respiratory Protection *(specify type)*

PEL is low and difficult to monitor. For exposures above the PEL, use an MSHA/NIOSH-approved supplied air respirator with full facepiece, hood, or helmet in continuous flow mode, or use a self-contained breathing apparatus (SCBA) operated in pressure demand or other positive pressure mode.

Ventilation

Local exhaust or general mechanical systems recommended.

Protective Gloves	Eye Protection	Other Protective Clothing
Butyl Rubber	Chemical/Dust Goggles or Face Mask	Butyl Rubber Apron

Work/Hygiene Practices

Always wash hands thoroughly after using chemical; never bring food, drink, or smoking materials into vicinity of chemicals.

o-DINITROBENZENE

$C_6H_4(NO_2)_2$ CAS: 528-29-0

ortho- isomer rating

IDENTIFICATION AND TYPICAL USES

o-Dinitrobenzene, or *ortho*-dinitrobenzene, is one of three dinitrobenzene isomers, *m- (meta),* and *p- (para)* being the other two. It is derived through the nitration of nitrobenzene with hot mixed acid. It is used in organic synthesis, in the manufacture of dyes, as a camphor substitute in cellulose nitrate, and in explosives (the *ortho-* polymer is used in blasting charges and to fill artillery shells).

RISK ASSESSMENT: HEALTH

General Assessment

o-Dinitrobenzene is a human poison by *inhalation, ingestion,* and skin *absorption.* It is an irritant to the eyes, skin and respiratory system. It is a confirmed carcinogen in animals. Human carcinogenicity is suspect, although the data are inconclusive in this area of study.

Inhalation can cause irritation to the respiratory system and associated mucosa. It can cause cough, congestion, with narcotic effects associated with the depression of the central nervous system (CNS). These include weakness, dyspnea, nausea, personality changes (irritability, mood changes), collapse, coma, and death (from respiratory failure). It also results in the formation of methemoglobin (a pigment similar to hemoglobin that is unable to perform the normal oxygenation process). Symptoms include headache, nausea, vomiting, dizziness, and jaundice (yellow coloring of the eyes, hair, and skin). There may be difficulty in breathing, visual disturbances (scotomas), general weakness, cyanosis, with possible progression to convulsions, coma, and even death.

Skin contact can produce the same systemic systems as inhalation as well as irritation, the appearance of small vesicles or blisters, redness or swelling, ulceration, and necrosis (issue death). Eye contact with the liquid solution or dusts can cause irritation, redness and swelling of the eyelids, painful sensitivity to light, and can cause severe eye damage and possible loss of vision.

Ingestion of *o*-dinitrobenzene can also cause systemic effects similar to those of inhalation. There can also be painful irritation of the mouth and stomach, stomach cramps, and diarrhea. A bitter almond taste or burning sensation in the mouth, dry throat, and thirst can occur.

Alcohol consumption, some medications, exposure to sunlight, or hot baths can aggravate and enhance symptoms.

☠ *Acute Health Effects*

The following acute (short-term) health effects may occur immediately or shortly after exposure to *o*-dinitrobenzene:

Skin: Irritation and possible blistering, with ulceration. Will pass through unbroken skin to cause toxic systemic effects, including weakness, nausea, collapse, and death.

Eye: Severe irritation with damage and possible loss of vision likely.

Lung: Irritation to the respiratory tract and associated mucosa. Can cause weakness, nausea, confusion, collapse, coma, and death (respiratory failure).

CNS: A CNS depressant. Weakness, confusion, vertigo, mood changes and other personality changes, nausea, collapse.

Other: Ingestion will cause irritation to the gastrointestinal system with nausea, vomiting, convulsions, dypsnea, jaundice, tremor, confusion, and nervousness. It has also been found to be hepatotoxic (toxic to the liver) and nephrotoxic (toxic to the kidneys).

Chronic Health Effects

The following chronic (long-term) health effects may occur at some time after exposure to *o*-dinitrobenzene and can last for months or even years:

Cancer Hazards: Some studies have suggested that exposure to *o*-dinitrobenzene causes cancer in some test animals. While there is no evidence of carcinogenicity of this chemical in humans or animals, it is still considered a *suspected* carcinogen by some authorities.

Reproductive Hazard: According to information presented in the references, *o*-dinitrobenzene has not been adequately tested for its ability to adversely affect reproductively in animals.

Other Chronic Effects: High or repeated lower skin or inhalation exposures can cause liver or kidney disorders. Symptoms of late toxicity include secondary anemia, liver damage, irritability, weakness, headache, anorexia, weight loss, nausea, vomiting, cyanosis, dyspnea, and skin disorders. There may also be paresthesis in the feet, ankles, hands, and feet.

Recommended Risk-Reduction Measures

Personnel should avoid direct contact with *o*-dinitrobenzene. If a less toxic chemical cannot be substituted for a hazardous substance, then *engineering controls* are the most effective method of reducing exposures. The best protection is to enclose operations and/or provide local exhaust ventilation at the site of chemical release. While not always operationally feasible, isolating operations can also reduce exposure. Using respiratory protection is less effective than the controls mentioned above, but is still advisable whenever working with or around *o*-dinitrobenzene. Exposure level is low (1 ppm) and difficult to monitor. For exposures above the PEL, use an MSHA/NIOSH-approved full facepiece respirator equipped with an organic vapor cartridge equipped with a dust pre-filter. Increased protection is obtained from a supplied-air respirator with full facepiece operated in continuous flow mode, or a self-contained breathing apparatus (SCBA) with full facepiece and pressure demand. If a full facepiece is not available, then chemical goggles should be worn to protect the eyes. Whenever a chemical splash hazard exists, a face shield and a rubber apron should be worn. To prevent hand and skin exposures, butyl rubber gloves should be worn.

Administrative controls should also be in place to minimize the potential for human exposures. These may include written procedures or policies which specify the methods and techniques that will be practiced whenever personnel are to work with *o*-dinitrobenzene.

All personnel should receive *training* on the use, hazards, protective measures, emergency actions, and other precautions per 29 CFR 1910.1200 (Hazard Communication), prior to the first assignment in an area where *o*-dinitrobenzene is used or stored.

Before beginning employment with *o*-dinitrobenzene and at regular intervals thereafter (e.g., annually), those persons with high or frequent exposures should be provided the following recommended medical tests:

- ☑ Kidney and liver function tests.
- ☑ Complete blood count.

If symptoms develop or overexposure is suspected, the following tests are recommended:

- ☑ Kidney and liver function tests.
- ☑ Complete blood count.
- ☑ Consider possible reactions caused by alcohol consumption and/or medications.

Any evaluation should include a careful history of past and present symptoms with an examination. Medical tests that simply look for existing damage are not a substitute for controlling exposures. Physicians should note that symptoms may be delayed, so victim should be kept under observation. The methemoglobin concentration should be determined at regular intervals until indications of full reduction to hemoglobin. For ingestion, gastric lavage is suggested followed by saline catharsis.

Other methods to reduce exposure include:

- ☑ Where possible, enclose operations and use local exhaust ventilation at the site of chemical release. If local exhaust ventilation or enclosure is not used, respiratory protection should be mandatory.
- ☑ Always ensure that proper protective clothing is worn when using chemical substances.
- ☑ Wash thoroughly immediately after exposure to *o*-dinitrobenzene and at the end of the work shift or before eating, drinking, or smoking.
- ☑ Hazard warning information should be posted in the work area. In addition, as part of an on-going education and training program, all information on

the health and safety hazards of *o*-dinitrobenzene should be communicated to all potentially exposed workers.

☑ Eye wash stations should be provided in the immediate work area for emergency use. If there is a possibility of skin exposure to *o*-dinitrobenzene, emergency shower facilities should also be provided.

☑ Personnel should never wear contaminated clothing home (family members can be exposed). Clothing should be laundered by personnel who have been trained on the hazards of *o*-dinitrobenzene.

RISK ASSESSMENT: ENVIRONMENT

General Assessment

The environment is at risk of exposure during transportation, storage, disposal, or destruction of *o*-dinitrobenzene. In almost every scenario, the threat of environmental exposure is contingent upon the proper handling of the chemical substance. Accidental spills, large or small, can result in fire, explosion, and possible contamination of the surrounding environmental mediums (water, soil, and air).

o-Dinitrobenzene is normally considered a combustible solid, but it can also appear in mixed solutions. Its dusts can form explosive mixtures in air. These characteristics require caution in handling, storage, transportation, and disposal. It is incompatible with strong oxidizers, caustics, and some metals (such as zinc and tin). Contact with either can cause fire or explosion. Prolonged contact with heat may result in an explosion due to spontaneous decomposition. Therefore, special consideration is required during any emergency situation involving a leak or spill of *o*-dinitrobenzene. Should *o*-dinitrobenzene ever come into contact with these incompatible substances during use, transportation, storage, or disposal, the formation of highly toxic and/or highly explosive commodities is extremely possible.

o-Dinitrobenzene may enter the environment through industrial effluents or spills.

Acute Ecological Effects

Acute (short-term) toxic effects may include the death of animals, birds, or fish and death or low growth rate in plants. Acute effects are seen 2 to 4 days after animals or plants are exposed to *o*-dinitrobenzene. Insufficient data are available on the short-term effects of *o*-dinitrobenzene exposure to aquatic life, plants, birds or land animals.

Chronic Ecological Effects

Chronic (long-term) toxic effects may include shortened life span, reproductive problems, lower fertility, and changes in appearance or behavior in exposed animals. These effects can be seen long after first exposure(s) to toxic chemicals. Insufficient data are available on the long-term effects of *o*-dinitrobenzene to aquatic life, plants, birds, or land animals.

Water Solubility

o-Dinitrobenzene is slightly soluble in water. Concentrations of 1 to 100 milligrams will mix with a liter of water.

Persistence in the Environment

o-Dinitrobenzene is slightly persistent in water, with a half-life between 2 and 20 days. The half-life of a pollutant is the amount of time it takes for one half of the chemical to be degraded.

Bioaccumulation in Aquatic Organisms

Some substances increase in concentration, or *bioaccumulate,* in living organisms as they breathe contaminated air, drink contaminated water, or eat contaminated food. These chemicals can become concentrated in the tissues and internal organs of animals as well as humans.

The concentration of *o*-dinitrobenzene found in fish tissue is expected to be somewhat higher than the average concentration of *o*-dinitrobenzene in the water from which the fish was taken.

Recommended Risk-Reduction Measures

Proper training of all transporters will reduce the likelihood of a mishap or accident resulting in a leak or spill to the environment. The correct use of DOT labeling (per the MSDS) on all containers, trucks, and rail cars will enable emergency responders to react properly and quickly to any disaster thereby reducing the potential risk to the environment and to personnel.

Storage of *o*-dinitrobenzene should be segregated from incompatible chemicals to minimize the risk of cross-contamination or contact. Buildings designated for storage should be equipped with appropriate fire protection systems (alarms, sprinklers, emergency lighting, portable extinguishers).

If a spill or leak to the environment has occurred, fire department, emergency response and/or hazardous materials spill personnel should be notified immediately. Cleanup should be attempted only by those trained in proper spill containment procedures. Contaminated soils should be removed for incineration and replaced with clean soil. If *o*-dinitrobenzene should contact the water table, aquifer, or navigable waterway, time is of the essence. It is slightly soluble in the aquatic environment and total remediation may not be possible. The local and/or state emergency response authorities must be notified. A comprehensive emergency response or disaster preparedness/recovery plan should be in place prior to any operations involving the use, transportation, storage, or disposal of *o*-dinitrobenzene. If *o*-dinitrobenzene is spilled or leaked, the following specific steps are recommended:

- ☑ Restrict persons not wearing protective clothing from area of spill or leak until cleanup is complete and area can be opened for normal work.
- ☑ Remove all ignition sources, ventilate area of leak.
- ☑ Absorb liquid spills in vermiculite, dry sand, earth, or similar material and deposit in sealed containers. Collect solid spills for reuse of disposal. Wet mop spill area to collect residue.
- ☑ It may be necessary to dispose of *o*-dinitrobenzene as a hazardous waste. The responsible state agency or the regional office of the federal Environmental Protection Agency (EPA) should be contacted for specific recommendations.

RISK ASSESSMENT: BUSINESS

General Assessment

Accidents or mishaps involving *o*-dinitrobenzene can present a significant threat to business operations. The loss or damage of equipment or facilities can significantly affect fiscal viability. Lawsuits that may result from personnel injury/death, public exposures, and/or environmental contamination will also require a serious expenditure of resources. Media attention surrounding an injury, death, or environmental damage can also result in a loss of profits and loss of current as well as future business. Always remember that any time the terms "carcinogen" or "cancer" are use, public emotion, hysteria, and ignorance can run equally high. This should be a consideration during the development of public relation policies.

Recommended Risk-Reduction Measures

Company attorneys, safety and health professionals, and environmental specialists should be involved in the development of any procedures dealing with safety or emergency response. A company official should be pre-designated as a public relations officer with specific training in dealing with the press. Corporate plans and policies should be developed, approved, and implemented long before any need for such arises.

REFERENCES

Asante-Duah, D. K. 1993. *Hazardous Waste Risk Assessment.* New York: CRC Press/Lewis Publishers.

Bloom, A. D. and F. J. DeSerres. 1995. *Ecotoxicity and Human Health.* New York: CRC Press/Lewis Publishers.

Calabrese, E. J. 1994. *Biological Effects of Low Level Exposures.* New York: CRC Press/Lewis Publishers.

Cockerman, L. G., B. S. Shane. 1994. *Basic Environmental Toxicology.* New York: CRC Press/Lewis Publishers.

Cothern, C. R., N. P. Ross. 1994. *Environmental Statistics, Assessment, and Forecasting.* New York: CRC Press/Lewis Publishers.

Hallenbeck, W. H. 1993. *Quantitative Risk Assessment for Environmental and Occupational Health,* 2nd Edition. New York: CRC Press/Lewis Publishers.

Howard, P. H. 1995. *Handbook of Environmental Fate and Exposures Data for Organic Chemicals,* Vol. I-IV. New York: CRC Press/Lewis Publishers.

Lauwerys, R. R., P. Hoet. 1993. *Industrial Chemical Exposure.* New York: CRC Press/Lewis Publishers.

Lewis, R. J., Sr. 1992. *Sax's Dangerous Properties of Industrial Materials,* Volumes 1, 2, and 3, 8th Edition. New York: Van Nostrand Reinhold.

Meyer, E. 1990. *Chemistry of Hazardous Materials.* Englewood Cliffs, NJ: Prentice Hall.

Quigley, D. R. 1994. *Handbook of Emergency Chemical Management.* New York: CRC Press/Lewis Publishers.

Torkelson, T. R. V. K. Rowe. 1981. Halogenated aliphatic hydrocarbons. *Patty's Industrial Hygiene and Toxicology,* Vol. 2B, *Toxicology,* pp 3568-3572. New York: Wiley-Interscience.

U.S. Department of Health and Human Services, National Institute for Occupational Safety and Health. 1994. *NIOSH Pocket Guide to Chemical Hazards.* Washington, D.C.: U.S. Government Printing Office.

MATERIAL SAFETY DATA SHEET

CHEMICAL NAME	CRC PRESS/LEWIS PUBLISHERS DISCLAIMER:
p-DINITROBENZENE	THE INFORMATION AND RECOMMENDATIONS PRESENTED HEREIN ARE BASED ON SOURCES BELIEVED TO BE RELIABLE. CRC MAKES NO REPRESENTATION ON ITS COMPLETENESS OR ACCURACY. IT IS THE RESPONSIBILITY OF THE USER TO DETERMINE THE CHEMICAL'S SUITABILITY FOR ITS INTENDED USE, THE CHEMICAL'S SAFE USE, AND THE CHEMICAL'S PROPER DISPOSAL. NO REPRESENTATIONS AND/OR WARRANTIES, EITHER EXPRESSED OR IMPLIED, OF THE MERCHANTABILITY OR FITNESS FOR A PARTICULAR PURPOSE, OR OF ANY OTHER NATURE, ARE MADE WITH RESPECT TO THE INFORMATION PROVIDED IN THIS MSDS OR TO THE CHEMICAL TO WHICH INFORMATION MAY REFER. CRC NEITHER ASSUMES NOR AUTHORIZES ANY OTHER PERSON TO ASSUME FOR IT, ANY OTHER ADDITIONAL RESPONSIBILITY OR LIABILITY FOR THE USE OF, OR RELIANCE UPON, THIS INFORMATION.

HAZARD WARNING INFORMATION

HEALTH	FIRE	REACTIVE	OTHER	DEGREE OF HAZARD	COLOR CODING	OTHER CODES
3	1	4		0 = Minimum Hazard 1 = Slight Hazard 2 = Moderate Hazard 3 = Serious Hazard 4 = Severe Hazard	HEALTH = BLUE FIRE = RED REACTIVITY = YELLOW OTHER = WHITE	OX = Oxidizer ACID = Acid ALK = Alkali COR = Corrosive W = Use No Water

SECTION I - GENERAL INFORMATION

Characterization	Phenol	RCRA Number: D001 / EPA Class	Characteristic (I) Waste
DOT Proper Shipping Name	*p*-Dinitrobenzene, solid	Chemical Abstract Service (CAS) Number	100-25-4
DOT Hazard Class and Label Requirements	Poison B	DOT Emergency Guide Code	56
DOT Identification Number	UN 1597	Chemical Formula	$C_6H_4(NO_2)_2$

Synonyms
1,4-Dinitrobenzene; *para*-dinitrobenzene; Dithane A-4.

SECTION II - HAZARDOUS INGREDIENTS AND IDENTITY INFORMATION

Hazardous Components (specific identity)	OSHA Exposure Criteria	NIOSH Exposure Criteria	Immediately Dangerous to Life and Health (IDLH)	ACGIH Exposure Criteria
p-Dinitrobenzene (derivation: From the nitration of nitrobenzene with hot mixed acid). 1 ppm = 6.6 mg/m³	PEL (skin): 1 ppm 6.6 mg/m³ STEL: Not Established	REL (skin): 1 ppm 6.6 mg/m³ STEL: Not Established	50 mg/m³	TLV (skin): 0.15 ppm 1 mg/m³ STEL: Not Established

SECTION III - PHYSICAL AND CHEMICAL CHARACTERISTICS

Boiling Point	570°F (299°C)	Specific Gravity (H_2O = 1)	1.63
Vapor Pressure (gas)	<1 at 69°F (20°C)	Molecular Weight	168.1
Vapor Density (Air = 1)	5.8	Melting Point	343°F (173°C)

Solubility
Slightly soluble (0.01%). Soluble in chloroform and ethyl acetate and sparingly soluble in benzene.

Appearance and Odor
Pale white or yellow crystalline solid.

SECTION IV - FIRE AND EXPLOSION HAZARD DATA

Flash Point (method used)	302°F (150°C) closed cup	Explosive Limits in Air % by Volume	LEL: Not Determined UEL: Not Determined
NFPA Classification	Combustible Solid	Autoignition Temperature	Not Determined

Extinguishing Media
For small fires, use dry chemical, carbon dioxide, water spray, or regular foam. For large fires, use water spray, fog, or regular foam. Apply water as fog in flooding amounts (solid stream may be ineffective).

Special Fire Fighting Procedures
Wear full protective clothing and self-contained breathing apparatus (SCBA). Poisonous gases are produced in fire. Use water spray to keep fire-exposed containers cool.

Unusual Fire and Explosion Hazards
Containers may explode in fire. Highly friction- and shock-sensitive. Dusts form explosive concentrations in air. May spontaneously ignite or explode at high temperatures due to spontaneous combustion.

SECTION V - REACTIVITY DATA

Stability		Conditions to Avoid **p-Dinitrobenzene is unstable at high temperatures and is extremely shock- and friction-sensitive. Keep containers away from areas where shock, heat, and/or friction are likely to occur.**
Stable	Unstable X	Incompatibility *(materials to avoid)* **Strong oxidizers, (such as chlorine, bromine, and fluorine), caustics, and many metals (such as zinc and tin). Explosive when mixed with nitric acid or when in contact with steam.**
Hazardous Polymerization		Conditions to Avoid **Under normal conditions of temperature and pressure, hazardous polymerization of p-dinitrobenzene is not expected to occur.**
May Occur	Will Not Occur X	Hazardous Decomposition or By-products **When heated to decomposition, p-dinitrobenzene can emit highly toxic/poisonous gases, including carbon dioxide and oxides of nitrogen. p-Dinitrobenzene can explode when heated.**

SECTION VI - HEALTH HAZARD DATA

Primary Route(s) of Entry:	Inhalation?	Absorption (skin)?	Ingestion?
	X	X	X

Health Hazards

INHALATION: Anoxia; visual disturbances; central scotomas; yellowing of hair, eyes, and skin; anemia; and liver, kidney, and central nervous system injury is possible. Forms methemoglobin.

ABSORPTION: Skin irritant. Will pass through skin to cause toxic systemic effects (same as inhalation).

INGESTION: Gastrointestinal disorders; tremor; convulsions; bad taste and burning feeling in mouth; dry throat; and thirst.

Carcinogenicity	NTP Listed?	IARC Cancer Review Group?	OSHA Regulated?	Target Organs?
Suspected Human Confirmed Animal	No	No	29 CFR 1910.1000 Table Z-1	Eyes, skin, blood, CVS, CNS, liver, kidney.

Medical Conditions Generally Aggravated by Exposure

Blood and liver disorders.

Emergency and First-aid Procedures

Eye contact: Flush immediately with water for 15 minutes (minimum), seek medical attention. Skin contact: Remove all contaminated clothing. Immediately wash area with large amounts of soap and water. Seek medical assistance. For inhalation: Remove the person from exposure. Provide respiratory assistance and CPR, if required. Transfer to medical facility. If swallowed: Unless advised otherwise, give 2 tablespoons Ipecac (adult dose) and 1-2 glasses of water. Induce vomiting. After vomiting, give 2 tablespoons of activated charcoal in 8 ounces of water. Seek medical attention immediately.

SECTION VII - PRECAUTIONS FOR SAFE HANDLING AND USE

Steps to be Taken in Case Material is Released or Spilled

Absorb liquids in vermiculite, dry sand, earth, or other non-reactive material and deposit is sealed drum. Ventilate area of spill or leak. For dry spills, collect for reuse or disposal. Do NOT dry sweep. Use water spray to knock-down vapors (for solution spills) and wet mop residue of dry spills.

Preferred Waste Disposal Method

No Citation.

Precautions to be Taken in Handling and Storage

Store to avoid contact with incompatible materials in tightly closed containers in cool, well-ventilated area away from heat and flame. Store in detached facility away from main plant or facility.

Other Precautions and Warnings

Sources of ignition are prohibited where *p*-dinitrobenzene is used, handled, or stored. Heat may cause solids to liquefy and containers to build up pressure and explode.

SECTION VIII - CONTROL MEASURES AND PERSONAL PROTECTIVE EQUIPMENT

Respiratory Protection *(specify type)*

PEL is low and difficult to monitor. For exposures above the PEL, use an MSHA/NIOSH-approved supplied air respirator with full facepiece, hood, or helmet in continuous flow mode, or use a self-contained breathing apparatus (SCBA) operated in pressure demand or other positive pressure mode.

Ventilation

Local exhaust or general mechanical systems recommended.

Protective Gloves	Eye Protection	Other Protective Clothing
Butyl Rubber	**Chemical/Dust Goggles or Face Mask**	**Butyl Rubber Apron**

Work/Hygiene Practices

Always wash hands thoroughly after using chemical; never bring food, drink, or smoking materials into vicinity of chemicals.

p-DINITROBENZENE

$C_6H_4(NO_2)_2$ CAS: 100-25-4

ortho- isomer rating

IDENTIFICATION AND TYPICAL USES

p-Dinitrobenzene, or *para*-dinitrobenzene, is one of three dinitrobenzene isomers, *o- (ortho),* and *m- (meta)* being the other two. It is derived through the nitration of nitrobenzene with hot mixed acid. It is used in organic synthesis, in the manufacture of dyes, as a camphor substitute in cellulose nitrate, and in explosives (the *ortho-* polymer is used in blasting charges and to fill artillery shells).

RISK ASSESSMENT: HEALTH

General Assessment

p-Dinitrobenzene is a human poison by *inhalation, ingestion,* and skin *absorption.* It is an irritant to the eyes, skin and respiratory system. It is a confirmed carcinogen in animals. Human carcinogenicity is suspect, although the data are inconclusive in this area of study.

Inhalation can cause irritation to the respiratory system and associated mucosa. It can cause cough, congestion, with narcotic effects associated with the depression of the central nervous system (CNS). These include weakness, dyspnea, nausea, personality changes (irritability, mood changes), collapse, coma, and death (from respiratory failure). It also results in the formation of methemoglobin (a pigment similar to hemoglobin that is unable to perform the normal oxygenation process). Symptoms include headache, nausea, vomiting, dizziness, and jaundice (yellow coloring of the eyes, hair, and skin). There may be difficulty in breathing, visual disturbances (scotomas), general weakness, cyanosis, with possible progression to convulsions, coma, and even death.

Skin contact can produce the same systemic systems as inhalation as well as irritation, the appearance of small vesicles or blisters, redness or swelling, ulceration, and necrosis (issue death). Eye contact with the liquid solution or dusts can cause irritation, redness and swelling of the eyelids, painful sensitivity to light, and can cause severe eye damage and possible loss of vision.

Ingestion of *p*-dinitrobenzene can also cause systemic effects similar to those of inhalation. There can also be painful irritation of the mouth and stomach, stomach cramps, and diarrhea. A bitter almond taste or burning sensation in the mouth, dry throat, and thirst can occur.

Alcohol consumption, some medications, exposure to sunlight, or hot baths can aggravate and enhance symptoms.

☠ *Acute Health Effects*

The following acute (short-term) health effects may occur immediately or shortly after exposure to *p*-dinitrobenzene:

Skin: Irritation and possible blistering, with ulceration. Will pass through unbroken skin to cause toxic systemic effects, including weakness, nausea, collapse, and death.

Eye: Severe irritation with damage and possible loss of vision likely.

Lung: Irritation to the respiratory tract and associated mucosa. Can cause weakness, nausea, confusion, collapse, coma, and death (respiratory failure).

CNS: A CNS depressant. Weakness, confusion, vertigo, mood changes and other personality changes, nausea, collapse.

Other: Ingestion will cause irritation to the gastrointestinal system with nausea, vomiting, convulsions, dypsnea, jaundice, tremor, confusion, and nervousness. It has also been found to be hepatotoxic (toxic to the liver) and nephrotoxic (toxic to the kidneys).

Chronic Health Effects

The following chronic (long-term) health effects may occur at some time after exposure to *p*-dinitrobenzene and can last for months or even years:

Cancer Hazards: Some studies have suggested that exposure to *p*-dinitrobenzene causes cancer in some test animals. While there is no evidence of carcinogenicity of this chemical in humans or animals, it is still considered a *suspected* carcinogen by some authorities.

Reproductive Hazard: According to information presented in the references, *p*-dinitrobenzene has not been adequately tested for its ability to adversely affect reproductively in animals.

Other Chronic Effects: High or repeated lower skin or inhalation exposures can cause liver or kidney disorders. Symptoms of late toxicity include secondary anemia, liver damage, irritability, weakness, headache, anorexia, weight loss, nausea, vomiting, cyanosis, dyspnea, and skin disorders. There may also be paresthesis in the feet, ankles, hands, and feet.

Recommended Risk-Reduction Measures

Personnel should avoid direct contact with *p*-dinitrobenzene. If a less toxic chemical cannot be substituted for a hazardous substance, then *engineering controls* are the most effective method of reducing exposures. The best protection is to enclose operations and/or provide local exhaust ventilation at the site of chemical release. While not always operationally feasible, isolating operations can also reduce exposure. Using respiratory protection is less effective than the controls mentioned above, but is still advisable whenever working with or around *p*-dinitrobenzene. Exposure level is low (1 ppm) and difficult to monitor. For exposures above the PEL, use an MSHA/NIOSH-approved full facepiece respirator equipped with an organic vapor cartridge equipped with a dust pre-filter. Increased protection is obtained from a supplied-air respirator with full facepiece operated in continuous flow mode, or a self-contained breathing apparatus (SCBA) with full facepiece and pressure demand. If a full facepiece is not available, then chemical goggles should be worn to protect the eyes. Whenever a chemical splash hazard exists, a face shield and a rubber apron should be worn. To prevent hand and skin exposures, butyl rubber gloves should be worn.

Administrative controls should also be in place to minimize the potential for human exposures. These may include written procedures or policies which specify the methods and techniques that will be practiced whenever personnel are to work with *p*-dinitrobenzene.

All personnel should receive *training* on the use, hazards, protective measures, emergency actions, and other precautions per 29 CFR 1910.1200 (Hazard Communication), prior to the first assignment in an area where *p*-dinitrobenzene is used or stored.

Before beginning employment with *p*-dinitrobenzene and at regular intervals thereafter (e.g., annually), those persons with high or frequent exposures should be provided the following recommended medical tests:

- ☑ Kidney and liver function tests.
- ☑ Complete blood count.

If symptoms develop or overexposure is suspected, the following tests are recommended:

- ☑ Kidney and liver function tests.
- ☑ Complete blood count.
- ☑ Consider possible reactions caused by alcohol consumption and/or medications.

Any evaluation should include a careful history of past and present symptoms with an examination. Medical tests that simply look for existing damage are not a substitute for controlling exposures. Physicians should note that symptoms may be delayed, so victim should be kept under observation. The methemoglobin concentration should be determined at regular intervals until indications of full reduction to hemoglobin. For ingestion, gastric lavage is suggested followed by saline catharsis.

Other methods to reduce exposure include:

- ☑ Where possible, enclose operations and use local exhaust ventilation at the site of chemical release. If local exhaust ventilation or enclosure is not used, respiratory protection should be mandatory.
- ☑ Always ensure that proper protective clothing is worn when using chemical substances.
- ☑ Wash thoroughly immediately after exposure to *p*-dinitrobenzene and at the end of the work shift or before eating, drinking, or smoking.
- ☑ Hazard warning information should be posted in the work area. In addition, as part of an on-going education and training program, all information on

the health and safety hazards of *p*-dinitrobenzene should be communicated to all potentially exposed workers.

☑ Eye wash stations should be provided in the immediate work area for emergency use. If there is a possibility of skin exposure to *p*-dinitrobenzene, emergency shower facilities should also be provided.

☑ Personnel should never wear contaminated clothing home (family members can be exposed). Clothing should be laundered by personnel who have been trained on the hazards of *p*-dinitrobenzene.

RISK ASSESSMENT: ENVIRONMENT

General Assessment

The environment is at risk of exposure during transportation, storage, disposal, or destruction of *p*-dinitrobenzene. In almost every scenario, the threat of environmental exposure is contingent upon the proper handling of the chemical substance. Accidental spills, large or small, can result in fire, explosion, and possible contamination of the surrounding environmental mediums (water, soil, and air).

p-Dinitrobenzene is normally considered a combustible solid, but it can also appear in mixed solutions. Its dusts can form explosive mixtures in air. These characteristics require caution in handling, storage, transportation, and disposal. It is incompatible with strong oxidizers, caustics, and some metals (such as zinc and tin). Contact with either can cause fire or explosion. Prolonged contact with heat may result in an explosion due to spontaneous decomposition. Therefore, special consideration is required during any emergency situation involving a leak or spill of *p*-dinitrobenzene. Should *p*-dinitrobenzene ever come into contact with these incompatible substances during use, transportation, storage, or disposal, the formation of highly toxic and/or highly explosive commodities is extremely possible.

p-Dinitrobenzene may enter the environment through industrial effluents or spills.

Acute Ecological Effects

Acute (short-term) toxic effects may include the death of animals, birds, or fish and death or low growth rate in plants. Acute effects are seen 2 to 4 days after animals or plants are exposed to *p*-dinitrobenzene.

Insufficient data are available on the short-term effects of *p*-dinitrobenzene exposure to aquatic life, plants, birds or land animals.

Chronic Ecological Effects

Chronic (long-term) toxic effects may include shortened life span, reproductive problems, lower fertility, and changes in appearance or behavior in exposed animals. These effects can be seen long after first exposure(s) to toxic chemicals.

Insufficient data are available on the long-term effects of *p*-dinitrobenzene to aquatic life, plants, birds, or land animals.

Water Solubility

p-Dinitrobenzene is slightly soluble in water. Concentrations of 1 to 100 milligrams will mix with a liter of water.

Persistence in the Environment

p-Dinitrobenzene is slightly persistent in water, with a half-life between 2 and 20 days. The half-life of a pollutant is the amount of time it takes for one half of the chemical to be degraded.

Bioaccumulation in Aquatic Organisms

Some substances increase in concentration, or *bioaccumulate,* in living organisms as they breathe contaminated air, drink contaminated water, or eat contaminated food. These chemicals can become concentrated in the tissues and internal organs of animals as well as humans.

The concentration of *p*-dinitrobenzene found in fish tissue is expected to be somewhat higher than the average concentration of *p*-dinitrobenzene in the water from which the fish was taken.

Recommended Risk-Reduction Measures

Proper training of all transporters will reduce the likelihood of a mishap or accident resulting in a leak or spill to the environment. The correct use of DOT labeling (per the MSDS) on all containers, trucks, and rail cars will enable emergency responders to react properly and quickly to any disaster thereby reducing the potential risk to the environment and to personnel.

Storage of *p*-dinitrobenzene should be segregated from incompatible chemicals to minimize the risk of cross-contamination or contact. Buildings designated for storage should be equipped with appropriate fire protection systems (alarms, sprinklers, emergency lighting, portable extinguishers).

If a spill or leak to the environment has occurred, fire department, emergency response and/or hazardous materials spill personnel should be notified immediately. Cleanup should be attempted only by those trained in proper spill containment procedures. Contaminated soils should be removed for incineration and replaced with clean soil. If *p*-dinitrobenzene should contact the water table, aquifer, or navigable waterway, time is of the essence. It is slightly soluble in the aquatic environment and total remediation may not be possible. The local and/or state emergency response authorities must be notified. A comprehensive emergency response or disaster preparedness/recovery plan should be in place prior to any operations involving the use, transportation, storage, or disposal of *p*-dinitrobenzene. If *p*-dinitrobenzene is spilled or leaked, the following specific steps are recommended:

- ☑ Restrict persons not wearing protective clothing from area of spill or leak until cleanup is complete and area can be opened for normal work.
- ☑ Remove all ignition sources, ventilate area of leak.
- ☑ Absorb liquid spills in vermiculite, dry sand, earth, or similar material and deposit in sealed containers. Collect solid spills for reuse of disposal. Wet mop spill area to collect residue.
- ☑ It may be necessary to dispose of *p*-dinitrobenzene as a hazardous waste. The responsible state agency or the regional office of the federal Environmental Protection Agency (EPA) should be contacted for specific recommendations.

RISK ASSESSMENT: BUSINESS

General Assessment

Accidents or mishaps involving *p*-dinitrobenzene can present a significant threat to business operations. The loss or damage of equipment or facilities can significantly affect fiscal viability. Lawsuits that may result from personnel injury/death, public exposures, and/or environmental contamination will also require a serious expenditure of resources. Media attention surrounding an injury, death, or environmental damage can also result in a loss of profits and loss of current as well as future business. Always remember that any time the terms "carcinogen" or "cancer" are use, public emotion, hysteria, and ignorance can run equally high. This should be a consideration during the development of public relation policies.

Recommended Risk-Reduction Measures

Company attorneys, safety and health professionals, and environmental specialists should be involved in the development of any such procedures dealing with safety or emergency response. A company official should be pre-designated as a public relations officer with specific training in dealing with the press. Corporate plans and policies should be developed, approved, and implemented long before any need for such arises.

REFERENCES

Asante-Duah, D. K. 1993. *Hazardous Waste Risk Assessment.* New York: CRC Press/Lewis Publishers.

Bloom, A. D. and F. J. DeSerres. 1995. *Ecotoxicity and Human Health.* New York: CRC Press/Lewis Publishers.

Calabrese, E. J. 1994. *Biological Effects of Low Level Exposures.* New York: CRC Press/Lewis Publishers.

Cockerman, L. G., B. S. Shane. 1994. *Basic Environmental Toxicology.* New York: CRC Press/Lewis Publishers.

Cothern, C. R., N. P. Ross. 1994. *Environmental Statistics, Assessment, and Forecasting.* New York: CRC Press/Lewis Publishers.

Hallenbeck, W. H. 1993. *Quantitative Risk Assessment for Environmental and Occupational Health,* 2nd Edition. New York: CRC Press/Lewis Publishers.

Howard, P. H. 1995. *Handbook of Environmental Fate and Exposures Data for Organic Chemicals,* Vol. I-IV. New York: CRC Press/Lewis Publishers.

Lauwerys, R. R., P. Hoet. 1993. *Industrial Chemical Exposure.* New York: CRC Press/Lewis Publishers.

Lewis, R. J., Sr. 1992. *Sax's Dangerous Properties of Industrial Materials,* Volumes 1, 2, and 3, 8th Edition. New York: Van Nostrand Reinhold.

Meyer, E. 1990. *Chemistry of Hazardous Materials.* Englewood Cliffs, NJ: Prentice Hall.

Quigley, D. R. 1994. *Handbook of Emergency Chemical Management.* New York: CRC Press/Lewis Publishers.

Torkelson, T. R. V. K. Rowe. 1981. Halogenated aliphatic hydrocarbons. *Patty's Industrial Hygiene and Toxicology,* Vol. 2B, *Toxicology,* pp 3568-3572. New York: Wiley-Interscience.

U.S. Department of Health and Human Services, National Institute for Occupational Safety and Health. 1994. *NIOSH Pocket Guide to Chemical Hazards.* Washington, D.C.: U.S. Government Printing Office.

MATERIAL SAFETY DATA SHEET

CHEMICAL NAME	
DINITRO-O-CRESOL	**CRC PRESS/LEWIS PUBLISHERS DISCLAIMER:** THE INFORMATION AND RECOMMENDATIONS PRESENTED HEREIN ARE BASED ON SOURCES BELIEVED TO BE RELIABLE. CRC MAKES NO REPRESENTATION ON ITS COMPLETENESS OR ACCURACY. IT IS THE RESPONSIBILITY OF THE USER TO DETERMINE THE CHEMICAL'S SUITABILITY FOR ITS INTENDED USE, THE CHEMICAL'S SAFE USE, AND THE CHEMICAL'S PROPER DISPOSAL. NO REPRESENTATIONS AND/OR WARRANTIES, EITHER EXPRESSED OR IMPLIED, OF THE MERCHANTABILITY OR FITNESS FOR A PARTICULAR PURPOSE, OR OF ANY OTHER NATURE, ARE MADE WITH RESPECT TO THE INFORMATION PROVIDED IN THIS MSDS OR TO THE CHEMICAL TO WHICH INFORMATION MAY REFER. CRC NEITHER ASSUMES NOR AUTHORIZES ANY OTHER PERSON TO ASSUME FOR IT, ANY OTHER ADDITIONAL RESPONSIBILITY OR LIABILITY FOR THE USE OF, OR RELIANCE UPON, THIS INFORMATION.

HAZARD WARNING INFORMATION

HEALTH	FIRE	REACTIVE	OTHER	DEGREE OF HAZARD	COLOR CODING	OTHER CODES
N/A	N/A	N/A		0 = Minimum Hazard 1 = Slight Hazard 2 = Moderate Hazard 3 = Serious Hazard 4 = Severe Hazard	HEALTH = BLUE FIRE = RED REACTIVITY = YELLOW OTHER = WHITE	OX = Oxidizer ACID = Acid ALK = Alkali COR = Corrosive W = Use No Water

SECTION I - GENERAL INFORMATION

Characterization: **Phenol**	RCRA Number: **P047**	EPA Class: **Acute Hazardous Waste**
DOT Proper Shipping Name: **Dinitro-Ortho-Cresol**	Chemical Abstract Service (CAS) Number: **534-52-1**	
DOT Hazard Class and Label Requirements: **Not Citation**	DOT Emergency Guide Code: **53**	
DOT Identification Number: **UN 1598**	Chemical Formula: **$CH_3C_6H_2OH(NO_2)_2$**	

Synonyms
4,6-Dinitro-o-cresol; 3,5-dinitro-2-hydroxytoluene; 4,6-dinitro-2-methyl phenol; DNC; DNOC.

SECTION II - HAZARDOUS INGREDIENTS AND IDENTITY INFORMATION

Hazardous Components (specific identity)	OSHA Exposure Criteria	NIOSH Exposure Criteria	Immediately Dangerous to Life and Health (IDLH)	ACGIH Exposure Criteria
Dinitro-o-cresol	**PEL (skin): 0.2 mg/m³** **STEL: Not Established**	**REL (skin): 0.2 mg/m³** **STEL: Not Established**	**5 mg/m³**	**TLV (skin): 0.2 mg/m³** **STEL: Not Established**

SECTION III - PHYSICAL AND CHEMICAL CHARACTERISTICS

Boiling Point: **594°F (312°C)**	Specific Gravity (H_2O = 1): **>1.1 (estimated)**
Vapor Pressure (gas): **<1 at 69°F (20°C)**	Molecular Weight: **198.1**
Vapor Density (Air = 1): **6.8**	Melting Point: **190°F (88°C)**

Solubility
Slightly soluble (0.01%). Soluble in alcohol, acetone, and ether.

Appearance and Odor
Yellow, odorless solid (powder).

SECTION IV - FIRE AND EXPLOSION HAZARD DATA

Flash Point (method used): **Not Applicable**	Explosive Limits in Air % by Volume: **LEL: Not Determined UEL: Not Determined**
NFPA Classification: **Non-Combustible Solid**	Autoignition Temperature: **Not Determined**

Extinguishing Media
Use carbon dioxide, water spray, dry chemical, foam, or fog.

Special Fire Fighting Procedures
Wear full protective clothing and self-contained breathing apparatus (SCBA). Poisonous gases are produced in fire. Use water spray to keep fire-exposed containers cool. Move containers from fire if it can be done without risk. Fight fire from a distance or protected position, if possible.

Unusual Fire and Explosion Hazards
Containers may explode in fire. Highly friction- and shock-sensitive (explosions can occur) when moving). Avoid generating airborne dusts or powders. May spontaneously ignite or explode at high temperatures.

SECTION V - REACTIVITY DATA

Stability		Conditions to Avoid: Dinitro-o-cresol is unstable at high temperatures and is extremely shock- and friction-sensitive. Keep containers away from areas where shock, heat, and/or friction are likely to occur.
Stable	Unstable	Incompatibility *(materials to avoid)*
	X	Strong oxidizers, (such as chlorine, bromine, and fluorine).
Hazardous Polymerization		Conditions to Avoid: Under normal conditions of temperature and pressure, hazardous polymerization of dinitro-o-cresol is not expected to occur.
May Occur	Will Not Occur	Hazardous Decomposition or By-products
	X	When heated to decomposition, dinitro-o-cresol can emit highly toxic/poisonous gases, including carbon dioxide and oxides of nitrogen. Dinitro-o-cresol can explode when heated.

SECTION VI - HEALTH HAZARD DATA

Primary Route(s) of Entry:	Inhalation?	Absorption (skin)?	Ingestion?
	X	X	X

Health Hazards

INHALATION: Can cause extremely high fever as well as the symptoms of ingestion.

ABSORPTION: Skin irritant. Will pass through skin to cause toxic systemic effects (same as ingestion).

INGESTION: Initial symptom is a false sense of well-being followed by headache, fever, lassitude; profuse sweating, excess thirst, tachycardia, hyperpnea, cough, shortness of breath, yellow urine, increased basal metabolism, vomiting, convulsions, coma, and death.

Carcinogenicity	NTP Listed?	IARC Cancer Review Group?	OSHA Regulated?	Target Organs?
Unknown Human Unknown Animal	No	No	29 CFR 1910.1000 Table Z-1	CNS, CVS, endocrine system

Medical Conditions Generally Aggravated by Exposure

None reported.

Emergency and First-aid Procedures

Eye contact: Flush immediately with water for 15 minutes (minimum), seek medical attention. Skin contact: Remove all contaminated clothing. Immediately wash area with large amounts of soap and water. Seek medical assistance. For inhalation: Remove the person from exposure. Provide respiratory assistance and CPR, if required. Transfer to medical facility. If swallowed: Unless advised otherwise, give large amounts of water and induce vomiting. Seek medical attention immediately. DOCTOR: Suggest applying ice packs to promote heat loss; replace fluids and electrolytes; allay anxiety.

SECTION VII - PRECAUTIONS FOR SAFE HANDLING AND USE

Steps to be Taken in Case Material is Released or Spilled

Minimize dust generation. Do NOT dry sweep. Use a vacuum equipped with a high efficiency particulate air (HEPA) filter instead. Collect and dispose of in sealed drum. If in run-off, dike area to contain, cover with absorbent (vermiculite or other) and dispose of in sealed drum.

Preferred Waste Disposal Method

No Citation.

Precautions to be Taken in Handling and Storage

Store to avoid contact with incompatible materials in tightly closed containers in cool, well-ventilated area away from heat and flame. Store in detached facility away from main plant or facility.

Other Precautions and Warnings

Sources of ignition are prohibited where dinitro-o-cresol is used, handled, or stored. Heat may cause containers to build up pressure and explode.

SECTION VIII - CONTROL MEASURES AND PERSONAL PROTECTIVE EQUIPMENT

Respiratory Protection *(specify type)*

PEL is low and difficult to monitor. For any exposure level, recommend an MSHA/NIOSH-approved supplied air respirator with full facepiece, hood, or helmet in continuous flow mode, or use a self-contained breathing apparatus (SCBA) operated in pressure demand or other positive pressure mode.

Ventilation

Local exhaust or general mechanical systems recommended.

Protective Gloves	Eye Protection	Other Protective Clothing
Butyl Rubber	Chemical/Dust Goggles or Face Mask	Butyl Rubber Apron

Work/Hygiene Practices

Always wash hands thoroughly after using chemical; never bring food, drink, or smoking materials into vicinity of chemicals.

DINITRO-O-CRESOL

$CH_3C_6H_2OH(NO_2)_2$

CAS: 534-52-1

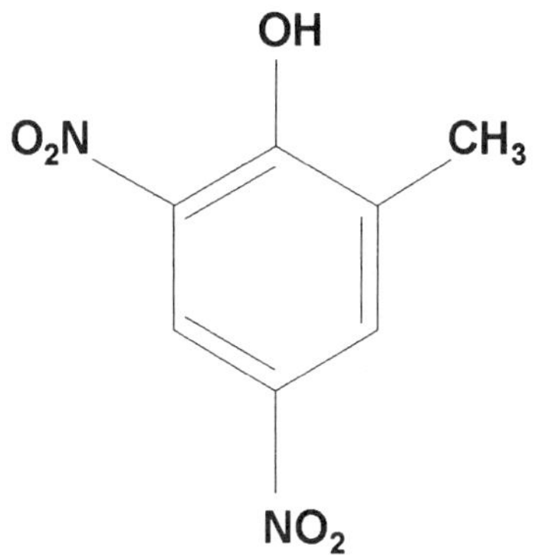

IDENTIFICATION AND TYPICAL USES

Dinitro-o-cresol is a yellow, odorless solid that appears as a powder. A dormant ovicidal spray for fruit tress (highly phototoxic and cannot be used successfully on actively growing plants), a selective herbicide as well as an insecticide.

RISK ASSESSMENT: HEALTH

General Assessment

Dinitro-o-cresol exhibits cumulative toxicity in humans. Although the specific route of exposure has not been accurately determined, it is believed to be a human poison by *inhalation, ingestion,* and skin *absorption.* It is an irritant to the eyes and skin. It is a confirmed carcinogen in animals. Mutation data (genetic changes) have been reported in test animals. The human carcinogenicity of this compound has not been adequately investigated.

Inhalation, ingestion, and absorption can cause a variety of toxic systemic effects. Symptoms of exposure include an initial false sense of well-being followed by somnolence (sleepiness or unnatural drowsiness), headache, fever, lassitude (weakness or exhaustion), profuse sweating, excessive thirst, tachycardia (abnormally rapid heart rate), hyperpnea (abnormal increase in depth and rate of respiration), cough, shortness of breath, coma, and death. On examination, brain recordings from specific areas of the central nervous, cardiac, and gastrointestinal systems may show functional changes.

Skin or eye contact will result in irritation and possible local necrosis (tissue damage) at point of contact.

Acute Health Effects

The following acute (short-term) health effects may occur immediately or shortly after exposure to dinitro-o-cresol:

Skin: Irritation and possible tissue damage. Will pass through unbroken skin to cause dangerous systemic effects.

Eye: Irritation with damage to the surrounding tissue. If not removed immediately, damage may be more severe and permanent.

Lung: Mild irritation to the respiratory tract and associated mucosa. Not particularly damaging to the lungs on acute exposures. Primary effects of inhalation are the subsequent absorption into the body leading to toxic systemic effects.

CNS: Causes unspecified changes in the function of the central nervous system with euphoria, headache, fever, lassitude, sweating, thirst, hyperpnea, coma, and even death.

Other: Ingestion will cause a variety of systemic effects, including euphoria, sweating, increased basal metabolism, yellow urine, marked thirst, vomiting, convulsions, coma, and death. These may be due to this compound's apparent ability to affect the function of the central nervous system (CNS) and the endocrine system.

Chronic Health Effects

The following chronic (long-term) health effects may occur at some time after exposure to dinitro-o-cresol and can last for months or even years:

Cancer Hazards: According to the information available in the references, dinitro-o-cresol has not been adequately tested for its ability to cause cancer in test animals. Experimental mutation data have been reported and many scientists believe that such chemicals may have the ability to cause cancer. This, however, requires much more study and research before any claims of carcinogenicity can be accurately asserted.

Reproductive Hazard: According to information presented in the references, dinitro-o-cresol has not been adequately tested for its ability to adversely affect reproductively in animals.

Recommended Risk-Reduction Measures

Personnel should avoid direct contact with dinitro-o-cresol. If a less toxic chemical cannot be substituted for a hazardous substance, then *engineering controls* are the most effective method of reducing exposures. The best protection is to enclose operations and/or provide local exhaust ventilation at the site of chemical release. While not always operationally feasible, isolating operations can also reduce exposure. Using respiratory protection is less effective than the controls mentioned above, but is still advisable whenever working with or around dinitro-o-cresol. The exposure level is low (0.2 mg/m^3) and difficult to monitor. Therefore, an MSHA/NIOSH-approved full facepiece respirator equipped with an organic vapor cartridge equipped with a dust pre-filter should be used if any exposure is anticipated or known. Increased protection is obtained from a supplied-air respirator with full facepiece operated in continuous flow mode, or a self-contained breathing apparatus (SCBA) with full facepiece and pressure demand. If a full facepiece is not available, then chemical dust goggles should be worn to protect the eyes. Whenever an airborne chemical dust hazard exists, a face shield and a rubber apron should be worn. To prevent hand and skin exposures, butyl rubber gloves should be used. Glove manufacturers should be contacted and permeation studies obtained *before* final glove selections are made.

Administrative controls should also be in place to minimize the potential for human exposures. These may include written procedures or policies which specify the methods and techniques that will be practiced whenever personnel are to work with dinitro-o-cresol.

All personnel should receive *training* on the use, hazards, protective measures, emergency actions, and other precautions per 29 CFR 1910.1200 (Hazard Communication), prior to the first assignment in an area where dinitro-o-cresol is used or stored.

Before beginning employment with dinitro-o-cresol and at regular intervals thereafter (e.g., annually), those persons with high or frequent exposures should be provided the following recommended medical tests:

☑ Complete blood count (establish baseline).

If symptoms develop or overexposure is suspected, the following tests are recommended:

☑ Complete blood count (compare to baseline). Some authorities suggest all exposed workers have blood tests regularly to determine the level of the substance present. Further contact with dinitro-o-cresol should be avoided if the blood levels exceed 20 micrograms per gram.

Any evaluation should include a careful history of past and present symptoms with an examination. Medical tests that simply look for existing damage are not a substitute for controlling exposures. It is suggested that physicians try application of ice packs to promote heat loss. Fluids and electrolytes should be replaced.

Other methods to reduce exposure include:

☑ Where possible, enclose operations and use local exhaust ventilation at the site of chemical release. If local exhaust ventilation or enclosure is not used, respiratory protection should be mandatory.

☑ Always ensure that proper protective clothing is worn when using chemical substances.

☑ Wash thoroughly immediately after exposure to dinitro-o-cresol and at the end of the work shift or before eating, drinking, or smoking.

☑ Hazard warning information should be posted in the work area. In addition, as part of an on-going education and training program, all information on the health and safety hazards of dinitro-o-cresol should be communicated to all potentially exposed workers.

☑ Eye wash stations should be provided in the immediate work area for emergency use. If there is a possibility of skin exposure to dinitro-o-cresol, emergency shower facilities should also be provided.

☑ Personnel should never wear contaminated clothing home (family members can be exposed). Clothing should be laundered only by personnel who have been specifically trained on the hazards of dinitro-o-cresol.

RISK ASSESSMENT: ENVIRONMENT

General Assessment

The environment is at risk of exposure during transportation, storage, disposal, or destruction of dinitro-o-cresol. In almost every scenario, the threat of environmental exposure is contingent upon the proper handling of the chemical substance. Accidental spills, large or small, can result in fire, explosion, and possible contamination of the surrounding environmental mediums (water, soil, and air).

Dinitro-o-cresol is normally considered a noncombustible solid, but it can often appear in mixtures with combustible or flammable solutions. Its dusts may form explosive mixtures in air, especially in confined areas. These characteristics require caution in handling, storage, transportation, and disposal. It is incompatible with strong oxidizers (such as chlorine, fluorine, and bromine) and contact can cause fire or explosion. Prolonged contact with heat may result in an explosion due to the potential for spontaneous decomposition. Therefore, special consideration is required during any emergency situation involving a leak or spill of dinitro-o-cresol. Should dinitro-o-cresol ever come into contact with these incompatible substances during use, transportation, storage, or disposal, the formation of highly toxic and/or highly explosive commodities is extremely possible.

Dinitro-o-cresol may enter the environment through industrial effluents or spills.

Acute Ecological Effects

Acute (short-term) toxic effects may include the death of animals, birds, or fish and death or low growth rate in plants. Acute effects are seen 2 to 4 days after animals or plants are exposed to dinitro-o-cresol.

Insufficient data are available on the short-term effects of dinitro-o-cresol exposure to aquatic life, plants, birds or land animals.

Chronic Ecological Effects

Chronic (long-term) toxic effects may include shortened life span, reproductive problems, lower fertility, and changes in appearance or behavior in exposed animals. These effects can be seen long after first exposure(s) to toxic chemicals.

Insufficient data are available on the long-term effects of dinitro-o-cresol to aquatic life, plants, birds, or land animals.

Water Solubility

Dinitro-o-cresol is slightly soluble in water. Concentrations of 1 to 100 milligrams will mix with a liter of water.

Persistence in the Environment

Dinitro-o-cresol is slightly persistent in water, with a half-life between 2 and 20 days. The half-life of a pollutant is the amount of time it takes for one half of the chemical to be degraded.

Bioaccumulation in Aquatic Organisms

Some substances increase in concentration, or *bioaccumulate,* in living organisms as they breathe contaminated air, drink contaminated water, or eat contaminated food. These chemicals can become concentrated in the tissues and internal organs of animals as well as humans.

The concentration of dinitro-o-cresol found in fish tissue is expected to be somewhat higher than the average concentration of dinitro-o-cresol in the water from which the fish was taken.

Recommended Risk-Reduction Measures

Proper training of all transporters will reduce the likelihood of a mishap or accident resulting in a leak or spill to the environment. The correct use of DOT labeling (per the MSDS) on all containers, trucks, and rail cars will enable emergency responders to react properly and quickly to any disaster thereby reducing the potential risk to the environment and to personnel.

Storage of dinitro-o-cresol should be segregated from incompatible chemicals to minimize the risk of cross-contamination or contact. Containers must be protected from heat, damage, and shock. Buildings designated for storage should be equipped with appropriate fire protection systems (alarms, sprinklers, emergency lighting, portable extinguishers).

If a spill or leak to the environment has occurred, fire department, emergency response and/or hazardous materials spill personnel should be notified immediately. Cleanup should be attempted only by those trained in proper spill containment procedures. Contaminated soils should be removed for incineration and replaced with clean soil. If dinitro-o-cresol should contact the water table, aquifer, or navigable waterway, time is of the essence. It is slightly soluble in the aquatic environment and total remediation may not be possible. The local and/or state emergency response authorities must be notified. A comprehensive emergency response or disaster preparedness/recovery plan should be in place prior to any operations involving the use, transportation, storage, or disposal of dinitro-o-cresol. If dinitro-o-cresol is spilled or leaked, the following specific steps are recommended:

☑ Restrict persons not wearing protective clothing from area of spill or leak until cleanup is complete and area can be opened for normal work.

☑ Remove all ignition sources, ventilate area of leak.

☑ Absorb liquid spills in vermiculite, dry sand, earth, or similar material and deposit in sealed containers. Collect solid spills for reuse of disposal. Wet-mop spill area to collect residue.

☑ It may be necessary to dispose of dinitro-o-cresol as a hazardous waste. The responsible state agency or the regional office of the federal Environmental Protection Agency (EPA) should be contacted for specific recommendations.

RISK ASSESSMENT: BUSINESS

General Assessment

Accidents or mishaps involving dinitro-o-cresol can present a significant threat to business operations. The loss or damage of equipment or facilities can significantly affect fiscal viability. Lawsuits that may result from personnel injury/death, public exposures, and/or environmental contamination will also require a serious expenditure of resources. Media attention surrounding an injury, death, or environmental damage can also result in a loss of profits and loss of current as well as future business.

Recommended Risk-Reduction Measures

Company attorneys, safety and health professionals, and environmental specialists should be involved in the development of any procedures dealing with safety or emergency response. A company official should be pre-designated as a public relations officer with specific training in dealing with the press. Corporate plans and policies should be developed, approved, and implemented long before any need for such arises.

REFERENCES

Ahlbom, A. 1993. *Biostatistics for Epidemiologists.* New York: CRC Press/Lewis Publishers.

American Conference of Governmental Industrial Hygienists. 1988. *Documentation of the Threshold Limit Values and Biological Exposure Indices,* 5th Edition (with updates). Cincinnati: ACGIH

Anthony, C. P., and N. J. Kolthoff. 1971. *Textbook of Anatomy and Physiology.* St. Louis: C. V. Mosby Company.

Asante-Duah, D. K. 1993. *Hazardous Waste Risk Assessment.* New York: CRC Press/Lewis Publishers.

Bloom, A. D. and F. J. DeSerres. 1995. *Ecotoxicity and Human Health.* New York: CRC Press/Lewis Publishers.

Brusick, D. J. 1994. *Methods for Genetic Risk Assessment.* New York: CRC Press/Lewis Publishers.

Bueche, J. R. 1972. *Principles of Physics,* 2nd Edition. New York: McGraw-Hill.

Burrell, R., D. K. Flaherty, and L. J. Sauers. 1992. *Toxicology of the Immune System.* New York: Van Nostrand Reinhold.

Calabrese, E. J. 1994. *Biological Effects of Low Level Exposures.* New York: CRC Press/Lewis Publishers.

Cockerman, L. G., B. S. Shane. 1994. *Basic Environmental Toxicology.* New York: CRC Press/Lewis Publishers.

Cothern, C. R., N. P. Ross. 1994. *Environmental Statistics, Assessment, and Forecasting.* New York: CRC Press/Lewis Publishers.

Danse, I. R. 1991. *Common Sense Toxics in the Workplace.* New York: Van Nostrand Reinhold.

Environmental Protection Agency. 1983. *Chemical Hazard Information Profiles.* Washington, D.C.: U.S. EPA.

Fodden, C. R., and J. L. Weddell. 1992. *Hazardous Materials Emergency Action Data.* New York: CRC Press/Lewis Publishers.

Gardner, R. W. 1994. *Chemical Intolerance.* New York: CRC Press/Lewis Publishers.

Hallenbeck, W. H. 1993. *Quantitative Risk Assessment for Environmental and Occupational Health,* 2nd Edition. New York: CRC Press/Lewis Publishers.

Howard, P. H. 1995. *Handbook of Environmental Fate and Exposures Data for Organic Chemicals,* Vol. I-IV. New York: CRC Press/Lewis Publishers.

Lewis, R. J., Sr. 1992. *Sax's Dangerous Properties of Industrial Materials,* Volumes 1, 2, and 3, 8th Edition. New York: Van Nostrand Reinhold.

Meyer, E. 1990. *Chemistry of Hazardous Materials.* Englewood Cliffs, NJ: Prentice Hall.

Quigley, D. R. 1994. *Handbook of Emergency Chemical Management.* New York: CRC Press/Lewis Publishers.

Torkelson, T. R. V. K. Rowe. 1981. Halogenated aliphatic hydrocarbons. *Patty's Industrial Hygiene and Toxicology,* Vol. 2B, *Toxicology,* pp 3568-3572. New York: Wiley-Interscience.

U.S. Department of Health and Human Services, National Institute for Occupational Safety and Health. 1994. *NIOSH Pocket Guide to Chemical Hazards.* Washington, D.C.: U.S. Government Printing Office.

MATERIAL SAFETY DATA SHEET

CHEMICAL NAME	CRC PRESS/LEWIS PUBLISHERS DISCLAIMER
DINITROPHENOL	

HAZARD WARNING INFORMATION

HEALTH	FIRE	REACTIVE	OTHER	DEGREE OF HAZARD	COLOR CODING	OTHER CODES
DRY: 3 WET: 3	DRY: 3 WET: 1	DRY: 3 WET: 0		0 = Minimum Hazard 1 = Slight Hazard 2 = Moderate Hazard 3 = Serious Hazard 4 = Severe Hazard	HEALTH = BLUE FIRE = RED REACTIVITY = YELLOW OTHER = WHITE	OX = Oxidizer ACID = Acid ALK = Alkali COR = Corrosive W = Use No Water

SECTION I - GENERAL INFORMATION

Characterization	**Phenol**	RCRA Number	**None**
		EPA Class	**Not Applicable**
DOT Proper Shipping Name	**Dinitrophenol** (solution)	Chemical Abstract Service (CAS) Number	**25550-58-7**
DOT Hazard Class and Label Requirements	**Poison B** (solution), **Flammable Solid**	DOT Emergency Guide Code	**53**
DOT Identification Number	**UN 1320, 1321** (solid); **UN 1599** (solution)	Chemical Formula	$C_6H_4N_2O_5$

Synonyms

None reported (a mixture of all isomers of dinitrophenol).

SECTION II - HAZARDOUS INGREDIENTS AND IDENTITY INFORMATION

Hazardous Components (specific identity)	OSHA Exposure Criteria	NIOSH Exposure Criteria	Immediately Dangerous to Life and Health (IDLH)	ACGIH Exposure Criteria
Dinitrophenol (derivation: By heating phenol with dilute sulfuric acid, cooling the product, and then nitrating while keeping the temperature approximately 50°C; or by nitrating with mixed acid under careful temperature control).	**PEL (skin):** 0.2 mg/m³ **STEL:** Not Established	**REL:** Not Established **STEL:** Not Established	**Not Determined**	**TLV:** Not Established **STEL:** Not Established

SECTION III - PHYSICAL AND CHEMICAL CHARACTERISTICS

Boiling Point	**Sublimes (when heated)**	Specific Gravity (H_2O = 1)	**1.68**
Vapor Pressure (gas)	**5-10 at 69°F (20°C)**	Molecular Weight	**184.1**
Vapor Density (Air = 1)	**6.3**	Melting Point	**234°F (112°C)**

Solubility

Slightly soluble in cold water; freely soluble in hot water. Soluble in alcohol, ether, acetone, pyridine, benzene, chloroform, and most organic solvents. Soluble in aqueous alkaline solutions.

Appearance and Odor

Yellow, odorless solid crystals (powder).

SECTION IV - FIRE AND EXPLOSION HAZARD DATA

Flash Point (method used)	**Not Applicable**	Explosive Limits in Air % by Volume	**LEL: Not Determined UEL: Not Determined**
NFPA Classification	**Flammable Solid**	Autoignition Temperature	**Not Determined**

Extinguishing Media

Use carbon dioxide, water spray, dry chemical, foam, or fog.

Special Fire Fighting Procedures

Wear full protective clothing and self-contained breathing apparatus (SCBA). Poisonous gases are produced in fire. Use water spray to keep fire-exposed containers cool. Move containers from fire if it can be done without risk. Fight fire from a distance or protected position, if possible.

Unusual Fire and Explosion Hazards

Containers may explode in fire. Fires involving this material should be considered an explosion hazard. Extremely explosive in dry form when exposed to heat, flame, or shock.

SECTION V - REACTIVITY DATA

Stability		
Stability		Conditions to Avoid **Dinitrophenol is normally stable. It will remain stable in mixtures up to 15% by weight with water. It becomes explosive when dry and subjected to shock, heat, or flame.**
Stable **X**	Unstable	Incompatibility *(materials to avoid)* **Strong oxidizers, (such as chlorine, bromine, and fluorine), and steam. It will form explosive salts with alkalis and/or ammonia.**
Hazardous Polymerization		Conditions to Avoid **Under normal conditions of temperature and pressure, hazardous polymerization of dinitrophenol is not expected to occur.**
May Occur	Will Not Occur **X**	Hazardous Decomposition or By-products **When heated to decomposition, dinitrophenol can emit highly toxic/poisonous gases, including carbon dioxide and oxides of nitrogen.**

SECTION VI - HEALTH HAZARD DATA

Primary Route(s) of Entry:	Inhalation?	Absorption (skin)?	Ingestion?
	X	X	X

Health Hazards

INHALATION: Can cause headache, coughing, sweating, thirst, intense fatigue, high fever, rapid pulse, warm and flushed skin, rapid and difficult breathing, anxiety, confusion, convulsions, loss of consciousness, pulmonary edema (fluid in lungs), and death.

ABSORPTION: Irritation with yellow staining, redness, swelling, blisters, and delayed symptoms like those of inhalation (will pass through skin to cause same toxic systemic effects).

INGESTION: Burning of mouth and throat, salivation, dizziness, nausea, vomiting, bright yellow stool, and symptoms of inhalation.

Carcinogenicity	NTP Listed?	IARC Cancer Review Group?	OSHA Regulated?	Target Organs?
Unknown Human Unknown Animal	No	No	No	Skin, respiratory system, CNS, CVS, liver, kidney

Medical Conditions Generally Aggravated by Exposure

None reported. However, existing thyroid, liver, kidney, and cardiovascular diseases may be aggravated.

Emergency and First-aid Procedures

Eye contact: Flush immediately with water for 15 minutes (minimum), seek medical attention. Skin contact: Remove all contaminated clothing. Immediately wash area with large amounts of soap and water. Seek medical assistance. Yellow discoloration need not be totally removed to prevent absorption. For inhalation: Remove the person from exposure. Provide CPR if required. Transfer to medical facility. Observe for 24-48 hours for lung effects. If swallowed: Unless advised otherwise, give 1-2 glasses of water and induce vomiting, then give 2 tablespoons of activated charcoal in 8 ounces of water.

SECTION VII - PRECAUTIONS FOR SAFE HANDLING AND USE

Steps to be Taken in Case Material is Released or Spilled

Minimize dust generation. Do NOT dry sweep. Use a vacuum equipped with a high efficiency particulate air (HEPA) filter instead. Collect and dispose of in sealed drum. If in run-off, dike area to contain, cover with absorbent (vermiculite or other) and dispose of in sealed drum.

Preferred Waste Disposal Method

No Citation.

Precautions to be Taken in Handling and Storage

Store in tightly closed containers in cool, well-ventilated area away from heat or flame. Heat may cause containers to build up pressure and explode. Store in detached facility away from main plant or facility.

Other Precautions and Warnings

Sources of ignition are prohibited where dinitrophenol is used, handled, or stored.

SECTION VIII - CONTROL MEASURES AND PERSONAL PROTECTIVE EQUIPMENT

Respiratory Protection *(specify type)*

PEL is low and difficult to monitor. For any exposure level, recommend a MSHA/NIOSH-approved supplied air respirator with full facepiece, hood, or helmet in continuous flow mode, or use a self-contained breathing apparatus (SCBA) operated in pressure demand or other positive pressure mode.

Ventilation

Local exhaust or general mechanical systems recommended.

Protective Gloves	Eye Protection	Other Protective Clothing
Butyl Rubber	**Chemical/Dust Goggles or Face Mask**	**Butyl Rubber Apron**

Work/Hygiene Practices

Always wash hands thoroughly after using chemical; never bring food, drink, or smoking materials into vicinity of chemicals.

DINITROPHENOL

$C_6H_4N_2O_5$ CAS: 25550-58-7

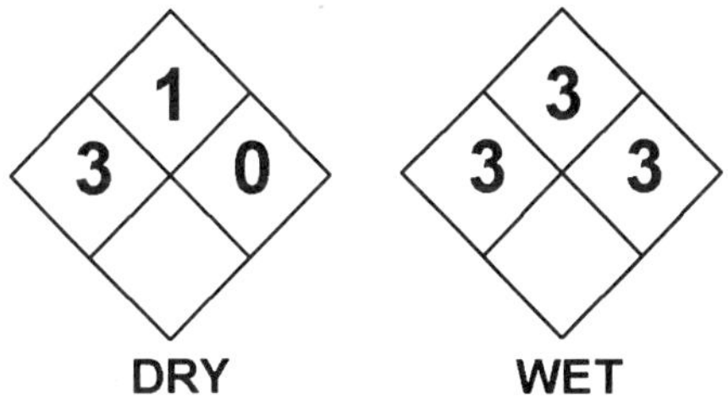

IDENTIFICATION AND TYPICAL USES

Dinitrophenol is a yellow, odorless solid that appears as a powder. It is used in the manufacture of dyes, especially sulfur colors, in picric acid picramic acid, in the preservation of lumber, in the manufacture of the photographic developer diaminophenol hydrochloride, in explosives manufacture, as an indicator, and a reagent for potassium and ammonium ions.

RISK ASSESSMENT: HEALTH

General Assessment

Dinitrophenol is a deadly human poison by *ingestion,* with equally toxic systemic consequences occurring by way of *inhalation,* and skin contact (*absorption*). It has been shown to cause teratogenic effects in experimental tests and mutation data have also been reported. Its carcinogenic properties are unknown. Liver or kidney damage can occur up to 3 days following exposure (in test animals).

Inhalation can result in headache, coughing, profuse sweating, marked thirst, intense fatigue, high fever, rapid pulse or heart rate (tachycardia), warm and flushed skin, rapid respiration (tachypnea), anxiety and confusion, convulsions, loss of consciousness, pulmonary edema (fluid in the lungs), and death due to circulatory or respiratory collapse. Exposure may also lead to a deadly buildup of fluids in the lungs in a condition known as pulmonary edema. This is a medical emergency requiring immediate attention. Symptoms of cough, tightness in chest, congestion, and dypsnea can be delayed up to 48 hours, creating a false sense of security with regard to health exposure risk.

Skin contact will result in nearly immediate yellow staining, irritation, swelling, redness, blisters, and delayed systemic effects from absorption with symptoms nearly identical to those of inhalation. Eye contact causes burning pain, tearing, and inflammation of the eyelids. If not removed immediately, damage to vision may occur.

Ingestion can cause painful burning of the mouth and throat, salivation, dizziness, nausea and vomiting, the appearance of bright yellow stools, and the symptoms of systemic poisoning similar to those of inhalation.

Acute Health Effects

The following acute (short-term) health effects may occur immediately or shortly after exposure to dinitrophenol:

Skin: Irritation with yellowing of skin and possible tissue damage. Will pass through unbroken skin to cause dangerous systemic effects.

Eye: Irritation with damage to the surrounding tissue. If not removed immediately, damage may be more severe and permanent.

Lung: Severe irritation to the respiratory tract and associated mucosa. Can cause coughing, congestion, and possibly pulmonary edema (fluid in the lungs), which can be fatal.

CNS: Can cause confusion, anxiety, headache, changes in the function of the respiratory and circulatory systems, convulsions, coma.

Other: Can cause a disruption in cellular metabolism in all tissues resulting in high fever, elevated blood pressure, pulse, and respiration. Can also cause death by hyperthermia. Effects are more severe in hot work environments.

Chronic Health Effects

The following chronic (long-term) health effects may occur at some time after exposure to dinitrophenol and can last for months or even years:

Cancer Hazards: According to the information available in the references, dinitrophenol has not been adequately tested for its ability to cause cancer in test animals. Experimental mutation data have been reported and many scientists believe that such chemicals may have the ability to cause cancer. This, however, requires much more study and research before any claims of carcinogenicity can be accurately asserted.

Reproductive Hazard: According to information presented in the references, dinitrophenol has been shown to cause adverse effects in reproduction in animals. It is considered an experimental teratogen. Human effects in this regard are inconclusive and require additional research.

Other Chronic Effects: Long-term exposures can result in damage to the liver or kidneys. Hypersensitivity may also result from repeated exposures.

Recommended Risk-Reduction Measures

Personnel should avoid direct contact with dinitrophenol. If a less toxic chemical cannot be substituted for a hazardous substance, then *engineering controls* are the most effective method of reducing exposures. The best protection is to enclose operations and/or provide local exhaust ventilation at the site of chemical release. While not always operationally feasible, isolating operations can also reduce exposure. Using respiratory protection is less effective than the controls mentioned above, but is still advisable whenever working with or around dinitrophenol. The exposure level is low (0.2 mg/m^3) and difficult to monitor. Therefore, a MSHA/NIOSH approved full facepiece respirator equipped with an organic vapor cartridge equipped with a dust pre-filter should be used if any exposure is anticipated or known. Increased protection is obtained from a supplied-air respirator with full facepiece operated in continuous flow mode, or a self-contained breathing apparatus (SCBA) with full facepiece and pressure demand. If a full facepiece is not available, then chemical dust goggles should be worn to protect the eyes. Whenever an airborne chemical dust hazard exists, a face shield and a rubber apron should be worn. To prevent hand and skin exposures, butyl rubber gloves should be worn.

Administrative controls should also be in place to minimize the potential for human exposures. These may include written procedures or policies which specify the methods and techniques that will be practiced whenever personnel are to work with dinitrophenol.

All personnel should receive *training* on the use, hazards, protective measures, emergency actions, and other precautions per 29 CFR 1910.1200 (Hazard Communication), prior to the first assignment in an area where dinitrophenol is used or stored.

Before beginning employment with dinitrophenol and at regular intervals thereafter (e.g., annually), those persons with high or frequent exposures should be provided the following recommended medical tests:

- ☑ Complete blood count (establish baseline).
- ☑ Urinalysis (establish baseline nitrophenol levels).

If symptoms develop or overexposure is suspected, the following tests are recommended:

- ☑ Complete blood count (compare to baseline).
- ☑ Urinalysis (compare to baseline).
- ☑ Consider chest X-ray following acute overexposure (may be negative if taken immediately after exposure due to delayed onset of pulmonary edema).
- ☑ Lung function tests.
- ☑ Liver and kidney function tests.

Any evaluation should include a careful history of past and present symptoms with an examination. Medical tests that simply look for existing damage are not a substitute for controlling exposures. Physician should monitor fluids and electrolytes, control hyperthermia, and test for dinitrophenol or its metabolite (aminophenol) in the urine using Derrian's Test. The biological half-life of dinitrophenol is 5-14 days.

Other methods to reduce exposure include:

- ☑ Where possible, enclose operations and use local exhaust ventilation at the site of chemical release. If local exhaust ventilation or enclosure is not used, respiratory protection should be mandatory.
- ☑ Always ensure that proper protective clothing is worn when using chemical substances.
- ☑ Wash thoroughly immediately after exposure to dinitrophenol and at the end of the work shift or before eating, drinking, or smoking.
- ☑ Hazard warning information should be posted in the work area. In addition, as part of an on-going

education and training program, all information on the health and safety hazards of dinitrophenol should be communicated to all potentially exposed workers.

☑ Eye wash stations should be provided in the immediate work area for emergency use. If there is a possibility of skin exposure to dinitrophenol, emergency shower facilities should also be provided.

☑ Personnel should never wear contaminated clothing home (family members can be exposed). Clothing should be laundered by personnel who have been trained on the hazards of dinitrophenol.

RISK ASSESSMENT: ENVIRONMENT

General Assessment

The environment is at risk of exposure during transportation, storage, disposal, or destruction of dinitrophenol. In almost every scenario, the threat of environmental exposure is contingent upon the proper handling of the chemical substance. Accidental spills, large or small, can result in fire, explosion, and possible contamination of the surrounding environmental mediums (water, soil, and air).

Dinitrophenol is normally considered a flammable solid. In solution, it can often appear in mixtures with other combustible or flammable commodities. Its dusts may form explosive mixtures in air. These characteristics require caution in handling, storage, transportation, and disposal. It is incompatible with strong oxidizers and contact can cause fire or explosion. It forms explosive salts in contact with alkalis or with ammonia. Prolonged contact with heat may result in an explosion due to spontaneous decomposition. Therefore, special consideration is required during any emergency situation involving a leak or spill of dinitrophenol. Should dinitrophenol ever come into contact with these incompatible substances during use, transportation, storage, or disposal, the formation of highly toxic and/or highly explosive commodities is extremely possible.

Dinitrophenol may enter the environment through its intended use as an insecticide, as well as through industrial effluents or spills.

Acute Ecological Effects

Acute (short-term) toxic effects may include the death of animals, birds, or fish and death or low growth rate in plants. Acute effects are seen 2 to 4 days after animals or plants are exposed to dinitrophenol.

Dinitrophenol has moderate acute toxicity to aquatic life and mild acute toxicity to plant life. Insufficient data are available on the short-term effects of dinitrophenol exposure to birds or land animals.

Chronic Ecological Effects

Chronic (long-term) toxic effects may include shortened life span, reproductive problems, lower fertility, and changes in appearance or behavior in exposed animals. These effects can be seen long after first exposure(s) to toxic chemicals.

Dinitrophenol has moderate chronic toxicity to aquatic life. Insufficient data are available on the long-term effects of dinitrophenol to plants, birds, or land animals.

Water Solubility

Dinitrophenol is slightly soluble in cold water. Concentrations of 1 to 100 milligrams will mix with a liter of water.

Persistence in the Environment

Dinitrophenol is slightly persistent in water, with an ecological half-life between 2 and 20 days. The half-life of a pollutant is the amount of time it takes for one half of the chemical to be degraded.

Bioaccumulation in Aquatic Organisms

Some substances increase in concentration, or *bioaccumulate,* in living organisms as they breathe contaminated air, drink contaminated water, or eat contaminated food. These chemicals can become concentrated in the tissues and internal organs of animals as well as humans.

The concentration of dinitrophenol found in fish tissue is expected to be somewhat higher than the average concentration of dinitrophenol in the water from which the fish was taken.

Recommended Risk-Reduction Measures

Proper training of all transporters will reduce the likelihood of a mishap or accident resulting in a leak or spill to the environment. The correct use of DOT labeling (per the MSDS) on all containers, trucks, and rail cars will enable emergency responders to react properly and

quickly to any disaster thereby reducing the potential risk to the environment and to personnel.

Storage of dinitrophenol should be segregated from incompatible chemicals to minimize the risk of cross-contamination or contact. Containers must be protected from heat, damage, and shock. Buildings designated for storage should be equipped with appropriate fire protection systems (alarms, sprinklers, emergency lighting, portable extinguishers).

If a spill or leak to the environment has occurred, fire department, emergency response and/or hazardous materials spill personnel should be notified immediately. Cleanup should be attempted only by those trained in proper spill containment procedures. Contaminated soils should be removed for incineration and replaced with clean soil. If dinitrophenol should contact the water table, aquifer, or navigable waterway, time is of the essence. It is slightly soluble in the aquatic environment and total remediation may not be possible. The local and/or state emergency response authorities must be notified. A comprehensive emergency response or disaster preparedness/recovery plan should be in place prior to any operations involving the use, transportation, storage, or disposal of dinitrophenol. If dinitrophenol is spilled or leaked, the following specific steps are recommended:

- ☑ Restrict persons not wearing protective clothing from area of spill or leak until cleanup is complete and area can be opened for normal work.
- ☑ Remove all ignition sources, ventilate area of leak.
- ☑ Absorb liquid spills in vermiculite, dry sand, earth, or similar material and deposit in sealed containers. Collect solid spills for reuse of disposal. Wet-mop spill area to collect residue.
- ☑ It may be necessary to dispose of dinitrophenol as a hazardous waste. The responsible state agency or the regional office of the federal Environmental Protection Agency (EPA) should be contacted for specific recommendations.

RISK ASSESSMENT: BUSINESS

General Assessment

Accidents or mishaps involving dinitrophenol can present a significant threat to business operations. The loss or damage of equipment or facilities can significantly affect fiscal viability. Lawsuits that may result from personnel injury/death, public exposures, and/or environmental contamination will also require a serious expenditure of resources. Media attention surrounding an injury, death, or environmental damage can also result in a loss of profits and loss of current as well as future business.

❶ *Recommended Risk-Reduction Measures*

Company attorneys, safety and health professionals, and environmental specialists should be involved in the development of any procedures dealing with safety or emergency response. A company official should be pre-designated as a public relations officer with specific training in dealing with the press. Corporate plans and policies should be developed, approved, and implemented long before any need for such arises.

REFERENCES

Ahlbom, A. 1993. *Biostatistics for Epidemiologists.* New York: CRC Press/Lewis Publishers.

Anthony, C. P., and N. J. Kolthoff. 1971. *Textbook of Anatomy and Physiology.* St. Louis: C. V. Mosby Company.

Bloom, A. D. and F. J. DeSerres. 1995. *Ecotoxicity and Human Health.* New York: CRC Press/Lewis Publishers.

Bueche, J. R. 1972. *Principles of Physics,* 2nd Edition. New York: McGraw-Hill.

Burrell, R., D. K. Flaherty, and L. J. Sauers. 1992. *Toxicology of the Immune System.* New York: Van Nostrand Reinhold.

Calabrese, E. J. 1994. *Biological Effects of Low Level Exposures.* New York: CRC Press/Lewis Publishers.

Cockerman, L. G., B. S. Shane. 1994. *Basic Environmental Toxicology.* New York: CRC Press/Lewis Publishers.

Environmental Protection Agency. 1983. *Chemical Hazard Information Profiles.* Washington, D.C.: U.S. EPA.

Fodden, C. R., and J. L. Weddell. 1992. *Hazardous Materials Emergency Action Data.* New York: CRC Press/Lewis Publishers.

Hallenbeck, W. H. 1993. *Quantitative Risk Assessment for Environmental and Occupational Health,* 2nd Edition. New York: CRC Press/Lewis Publishers.

Howard, P. H. 1995. *Handbook of Environmental Fate and Exposures Data for Organic Chemicals,* Vol. I-IV. New York: CRC Press/Lewis Publishers.

Lewis, R. J., Sr. 1992. *Sax's Dangerous Properties of Industrial Materials,* Volumes 1, 2, and 3, 8th Edition. New York: Van Nostrand Reinhold.

Meyer, E. 1990. *Chemistry of Hazardous Materials.* Englewood Cliffs, NJ: Prentice Hall.

Quigley, D. R. 1994. *Handbook of Emergency Chemical Management.* New York: CRC Press/Lewis Publishers.

MATERIAL SAFETY DATA SHEET

CHEMICAL NAME	
2,3-DINITROPHENOL	**CRC PRESS/LEWIS PUBLISHERS DISCLAIMER:** THE INFORMATION AND RECOMMENDATIONS PRESENTED HEREIN ARE BASED ON SOURCES BELIEVED TO BE RELIABLE. CRC MAKES NO REPRESENTATION ON ITS COMPLETENESS OR ACCURACY. IT IS THE RESPONSIBILITY OF THE USER TO DETERMINE THE CHEMICAL'S SUITABILITY FOR ITS INTENDED USE, THE CHEMICAL'S SAFE USE, AND THE CHEMICAL'S PROPER DISPOSAL. NO REPRESENTATIONS AND/OR WARRANTIES, EITHER EXPRESSED OR IMPLIED, OF THE MERCHANTABILITY OR FITNESS FOR A PARTICULAR PURPOSE, OR OF ANY OTHER NATURE, ARE MADE WITH RESPECT TO THE INFORMATION PROVIDED IN THIS MSDS OR TO THE CHEMICAL TO WHICH INFORMATION MAY REFER. CRC NEITHER ASSUMES NOR AUTHORIZES ANY OTHER PERSON TO ASSUME FOR IT, ANY OTHER ADDITIONAL RESPONSIBILITY OR LIABILITY FOR THE USE OF, OR RELIANCE UPON, THIS INFORMATION.

HAZARD WARNING INFORMATION

HEALTH	FIRE	REACTIVE	OTHER	DEGREE OF HAZARD	COLOR CODING	OTHER CODES
DRY: 3 WET: 3	DRY: 3 WET: 1	DRY: 3 WET: 0		0 = Minimum Hazard 1 = Slight Hazard 2 = Moderate Hazard 3 = Serious Hazard 4 = Severe Hazard	HEALTH = BLUE FIRE = RED REACTIVITY = YELLOW OTHER = WHITE	OX = Oxidizer ACID = Acid ALK = Alkali COR = Corrosive W = Use No Water

SECTION I - GENERAL INFORMATION

Characterization	**Phenol**	RCRA Number	**None**
		EPA Class	**Not Applicable**
DOT Proper Shipping Name	**Dinitrophenol (solution)**	Chemical Abstract Service (CAS) Number	**66-56-8**
DOT Hazard Class and Label Requirements	**Poison B (solution), Flammable Solid**	DOT Emergency Guide Code	**53**
DOT Identification Number	**UN 1320, 1321 (solid); UN 1599 (solution)**	Chemical Formula	$C_6H_4N_2O_5$

Synonyms
None reported.

SECTION II - HAZARDOUS INGREDIENTS AND IDENTITY INFORMATION

Hazardous Components (specific identity)	OSHA Exposure Criteria	NIOSH Exposure Criteria	Immediately Dangerous to Life and Health (IDLH)	ACGIH Exposure Criteria
2,3-dinitrophenol (derivation: By heating phenol with dilute sulfuric acid, cooling the product, and then nitrating while keeping the temperature approximately 50°C; or by nitrating with mixed acid under careful temperature control).	**PEL (skin):** 0.2 mg/m³ **STEL:** Not Established	**REL:** Not Established **STEL:** Not Established	**Not Determined**	**TLV:** Not Established **STEL:** Not Established

SECTION III - PHYSICAL AND CHEMICAL CHARACTERISTICS

Boiling Point	**Sublimes (when heated)**	Specific Gravity (H_2O = 1)	**1.681**
Vapor Pressure (gas)	**5-10 at 69°F (20°C)**	Molecular Weight	**184.12**
Vapor Density (Air = 1)	**6.35**	Melting Point	**291°F (144°C)**

Solubility
Slightly soluble in cold water; freely soluble in hot water. Soluble in alcohol, ether, acetone, pyridine, benzene, chloroform, and most organic solvents. Soluble in aqueous alkaline solutions.

Appearance and Odor
Yellow, odorless solid crystals (powder).

SECTION IV - FIRE AND EXPLOSION HAZARD DATA

Flash Point (method used)	**Not Applicable**	Explosive Limits in Air % by Volume	**LEL: Not Determined** **UEL: Not Determined**
NFPA Classification	**Flammable Solid**	Autoignition Temperature	**Not Determined**

Extinguishing Media
Use carbon dioxide, water spray, dry chemical, foam, or fog.

Special Fire Fighting Procedures
Wear full protective clothing and self-contained breathing apparatus (SCBA). Poisonous gases are produced in fire. Use water spray to keep fire-exposed containers cool. Move containers from fire if it can be done without risk. Fight fire from a distance or protected position, if possible.

Unusual Fire and Explosion Hazards
Containers may explode in fire. Fires involving this material should be considered an explosion hazard. Extremely explosive in dry form when exposed to heat, flame, or shock.

SECTION V - REACTIVITY DATA

Stability		Conditions to Avoid 2,3-Dinitrophenol is normally stable. It will remain stable in mixtures up to 15% by weight with water. It becomes explosive when dry and subjected to shock, heat, or flame.
Stable X	Unstable	Incompatibility *(materials to avoid)* Strong oxidizers, (such as chlorine, bromine, and fluorine), and steam. It will form explosive salts with alkalis and/or ammonia.
Hazardous Polymerization		Conditions to Avoid Under normal conditions of temperature and pressure, hazardous polymerization of 2,3-dinitrophenol is not expected to occur.
May Occur	Will Not Occur X	Hazardous Decomposition or By-products When heated to decomposition, 2,3-dinitrophenol can emit highly toxic/poisonous gases, including carbon dioxide and oxides of nitrogen.

SECTION VI - HEALTH HAZARD DATA

Primary Route(s) of Entry:	Inhalation?	Absorption (skin)?	Ingestion?
	X	X	X

Health Hazards

INHALATION: Can cause headache, coughing, sweating, thirst, intense fatigue, high fever, rapid pulse, warm and flushed skin, rapid and difficult breathing, anxiety, confusion, convulsions, loss of consciousness, pulmonary edema (fluid in lungs), and death.

ABSORPTION: Irritation with yellow staining, redness, swelling, blisters, and delayed symptoms like those of inhalation (will pass through skin to cause same toxic systemic effects).

INGESTION: Burning of mouth and throat, salivation, dizziness, nausea, vomiting, bright yellow stool, and symptoms of inhalation.

Carcinogenicity	NTP Listed?	IARC Cancer Review Group?	OSHA Regulated?	Target Organs?
Unknown Human Unknown Animal	No	No	No	Skin, respiratory system, CNS, CVS, liver, kidney

Medical Conditions Generally Aggravated by Exposure

None reported. However, existing thyroid, liver, kidney, and cardiovascular diseases may be aggravated.

Emergency and First-aid Procedures

Eye contact: Flush immediately with water for 15 minutes (minimum), seek medical attention. Skin contact: Remove all contaminated clothing. Immediately wash area with large amounts of soap and water. Seek medical assistance. Yellow discoloration need not be totally removed to prevent absorption. For inhalation: Remove the person from exposure. Provide CPR if required. Transfer to medical facility. Observe for 24-48 hours for lung effects. If swallowed: Unless advised otherwise, give 1-2 glasses of water and induce vomiting, then give 2 tablespoons of activated charcoal in 8 ounces of water.

SECTION VII - PRECAUTIONS FOR SAFE HANDLING AND USE

Steps to be Taken in Case Material is Released or Spilled

Minimize dust generation. Do NOT dry sweep. Use a vacuum equipped with a high efficiency particulate air (HEPA) filter instead. Collect and dispose of in sealed drum. If in run-off, dike area to contain, cover with absorbent (vermiculite or other) and dispose of in sealed drum.

Preferred Waste Disposal Method

No Citation.

Precautions to be Taken in Handling and Storage

Store in tightly closed containers in cool, well-ventilated area away from heat or flame. Heat may cause containers to build up pressure and explode. Store in detached facility away from main plant or facility.

Other Precautions and Warnings

Sources of ignition are prohibited where 2,3-dinitrophenol is used, handled, or stored.

SECTION VIII - CONTROL MEASURES AND PERSONAL PROTECTIVE EQUIPMENT

Respiratory Protection *(specify type)*

PEL is low and difficult to monitor. For any exposure level, recommend a MSHA/NIOSH-approved supplied air respirator with full facepiece, hood, or helmet in continuous flow mode, or use a self-contained breathing apparatus (SCBA) operated in pressure demand or other positive pressure mode.

Ventilation

Local exhaust or general mechanical systems recommended.

Protective Gloves	Eye Protection	Other Protective Clothing
Butyl Rubber	Chemical/Dust Goggles or Face Mask	Butyl Rubber Apron

Work/Hygiene Practices

Always wash hands thoroughly after using chemical; never bring food, drink, or smoking materials into vicinity of chemicals.

2,3-DINITROPHENOL

$C_6H_4N_2O_5$ CAS: 66-56-8

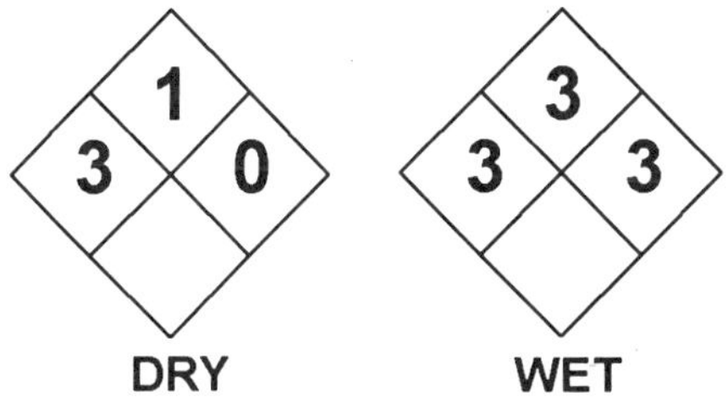

IDENTIFICATION AND TYPICAL USES

2,3-Dinitrophenol is a yellow, odorless solid that appears as a powder. It is used in the manufacture of dyes, especially sulfur colors, in picric acid picramic acid, in the preservation of lumber, in the manufacture of the photographic developer diaminophenol hydrochloride, in explosives manufacture as a component in shell and bomb charges, as an indicator, and a reagent for potassium and ammonium ions.

RISK ASSESSMENT: HEALTH

General Assessment

2,3-Dinitrophenol is a deadly human poison by *ingestion,* with equally toxic systemic consequences occurring by way of *inhalation,* and skin contact (*absorption*). It has been shown to cause teratogenic effects in experimental tests and mutation data have also been reported. Its carcinogenic properties are unknown. Liver or kidney damage can occur up to 3 days following exposure. It is a powerful stimulant of the metabolism by excessive oxidation.

Inhalation can result in headache, coughing, profuse sweating, marked thirst, intense fatigue, high fever, rapid pulse or heart rate (tachycardia), warm and flushed skin, rapid respiration (tachypnea), anxiety and confusion, convulsions, loss of consciousness, pulmonary edema (fluid in the lungs), and death due to circulatory or respiratory collapse. Exposure may also lead to a deadly buildup of fluids in the lungs in a condition known as pulmonary edema. This is a medical emergency requiring immediate attention. Symptoms can be delayed up to 48 hours, creating a false sense of security with regard to health exposure risk.

Skin contact will result in nearly immediate yellow staining, irritation, swelling, redness, blisters, and delayed systemic effects from absorption with symptoms identical to those of inhalation. It is also a skin allergen. Eye contact causes burning pain, tearing, and inflammation of the eyelids.

Ingestion can cause painful burning of the mouth and throat, salivation, dizziness, nausea and vomiting, the appearance of bright yellow stools, and the symptoms of systemic poisoning similar to those of inhalation.

☠ *Acute Health Effects*

The following acute (short-term) health effects may occur immediately or shortly after exposure to 2,3-dinitrophenol:

Skin: Irritation with yellowing of skin and possible tissue damage. Can cause skin allergies. Will pass through unbroken skin to cause dangerous systemic effects.

Eye: Irritation with damage to the surrounding tissue. If not removed immediately, damage may be more severe and permanent.

Lung: Severe irritation to the respiratory tract and associated mucosa. Can cause coughing, congestion, and possibly pulmonary edema (fluid in the lungs), which can be fatal.

CNS: Can cause confusion, anxiety, headache, changes in the function of the respiratory and circulatory systems, convulsions, coma.

Other: Can cause a disruption in cellular metabolism in all tissues resulting in high fever, elevated blood pressure, pulse, and respiration. Can also cause death by hyperthermia. Effects are more severe in hot work environments.

Chronic Health Effects

The following chronic (long-term) health effects may occur at some time after exposure to 2,3-dinitrophenol and can last for months or even years:

Cancer Hazards: According to the information available in the references, 2,3-dinitrophenol has not been adequately tested for its ability to cause cancer in test animals. Experimental mutation data have been reported.

Reproductive Hazard: According to information presented in the references, 2,3-dinitrophenol has not been adequately tested for its ability to adversely affect reproduction. However, a similar compound, 2,4-dinitrophenol, has demonstrated teratogenic effects in test animals. Human effects in this regard are inconclusive and require additional research.

Other Chronic Effects: Long-term exposures can result in damage to the liver or kidneys. Hypersensitivity may also result from repeated exposures.

Recommended Risk-Reduction Measures

Personnel should avoid direct contact with 2,3-dinitrophenol. If a less toxic chemical cannot be substituted for a hazardous substance, then *engineering controls* are the most effective method of reducing exposures. The best protection is to enclose operations and/or provide local exhaust ventilation at the site of chemical release. While not always operationally feasible, isolating operations can also reduce exposure. Using respiratory protection is less effective than the controls mentioned above, but is still advisable whenever working with or around 2,3-dinitrophenol. The exposure level is low (0.2 mg/m^3) and difficult to monitor. Therefore, a MSHA/NIOSH approved full facepiece respirator equipped with an organic vapor cartridge equipped with a dust pre-filter should be used if any exposure is anticipated or known. Increased protection is obtained from a supplied-air respirator with full facepiece operated in continuous flow mode, or a self-contained breathing apparatus (SCBA) with full facepiece and pressure demand. If a full facepiece is not available, then chemical dust goggles should be worn to protect the eyes. Whenever an airborne chemical dust hazard exists, a face shield and a rubber apron should be worn. To prevent hand and skin exposures, butyl rubber gloves should be worn.

Administrative controls should also be in place to minimize the potential for human exposures. These may include written procedures or policies which specify the methods and techniques that will be practiced whenever personnel are to work with 2,3-dinitrophenol.

All personnel should receive *training* on the use, hazards, protective measures, emergency actions, and other precautions per 29 CFR 1910.1200 (Hazard Communication), prior to the first assignment in an area where 2,3-dinitrophenol is used or stored.

Before beginning employment with 2,3-dinitrophenol and at regular intervals thereafter (e.g., annually), those persons with high or frequent exposures should be provided the following recommended medical tests:

☑ Complete blood count (establish baseline).

☑ Urinalysis (establish baseline nitrophenol levels).

If symptoms develop or overexposure is suspected, the following tests are recommended:

☑ Complete blood count (compare to baseline).

☑ Urinalysis (compare to baseline).

☑ Consider chest X-ray following acute overexposure (may be negative if taken immediately after exposure).

☑ Lung function tests.

☑ Liver and kidney function tests.

Any evaluation should include a careful history of past and present symptoms with an examination. Medical tests that simply look for existing damage are not a substitute for controlling exposures. Physician should monitor fluids and electrolytes, control hyperthermia, and test for dinitrophenol or its metabolite (aminophenol) in the urine using Derrian's Test. The biological half-life of 2,3-dinitrophenol is 5-14 days.

Other methods to reduce exposure include:

☑ Where possible, enclose operations and use local exhaust ventilation at the site of chemical release. If local exhaust ventilation or enclosure is not used, respiratory protection should be mandatory.

☑ Always ensure that proper protective clothing is worn when using chemical substances.

☑ Wash thoroughly immediately after exposure to 2,3-dinitrophenol and at the end of the work shift or before eating, drinking, or smoking.

☑ Hazard warning information should be posted in the work area. In addition, as part of an on-going

education and training program, all information on the health and safety hazards of 2,3-dinitrophenol should be communicated to all potentially exposed workers.

☑ Eye wash stations should be provided in the immediate work area for emergency use. If there is a possibility of skin exposure to 2,3-dinitrophenol, emergency shower facilities should also be provided.

☑ Personnel should never wear contaminated clothing home (family members can be exposed). Clothing should be laundered by personnel who have been trained on the hazards of 2,3-dinitrophenol.

RISK ASSESSMENT: ENVIRONMENT

General Assessment

The environment is at risk of exposure during transportation, storage, disposal, or destruction of 2,3-dinitrophenol. In almost every scenario, the threat of environmental exposure is contingent upon the proper handling of the chemical substance. Accidental spills, large or small, can result in fire, explosion, and possible contamination of the surrounding environmental mediums (water, soil, and air).

2,3-Dinitrophenol is normally considered a flammable solid. In solution, it can often appear in mixtures with other combustible or flammable commodities. Its dusts may form explosive mixtures in air. These characteristics require caution in handling, storage, transportation, and disposal. It is incompatible with strong oxidizers and contact can cause fire or explosion. It forms explosive salts in contact with alkalis or with ammonia. Prolonged contact with heat may result in an explosion due to spontaneous decomposition. Therefore, special consideration is required during any emergency situation involving a leak or spill of 2,3-dinitrophenol. Should 2,3-dinitrophenol ever come into contact with these incompatible substances during use, transportation, storage, or disposal, the formation of highly toxic and/or highly explosive commodities is extremely possible.

2,3-Dinitrophenol may enter the environment through industrial effluents or spills.

Acute Ecological Effects

Acute (short-term) toxic effects may include the death of animals, birds, or fish and death or low growth rate in plants. Acute effects are seen 2 to 4 days after animals or plants are exposed to 2,3-dinitrophenol. Insufficient data are available on the short-term effects of 2,3-dinitrophenol exposure to aquatic life, plants, birds or land animals.

Chronic Ecological Effects

Chronic (long-term) toxic effects may include shortened life span, reproductive problems, lower fertility, and changes in appearance or behavior in exposed animals. These effects can be seen long after first exposure(s) to toxic chemicals. Insufficient data are available on the long-term effects of 2,3-dinitrophenol to aquatic life, plants, birds, or land animals.

Water Solubility

2,3-Dinitrophenol is slightly soluble in cold water. Concentrations of 1 to 100 milligrams will mix with a liter of water.

Persistence in the Environment

2,3-Dinitrophenol is slightly persistent in water, with an ecological half-life between 2 and 20 days. The half-life of a pollutant is the amount of time it takes for one half of the chemical to be degraded.

Bioaccumulation in Aquatic Organisms

Some substances increase in concentration, or *bioaccumulate,* in living organisms as they breathe contaminated air, drink contaminated water, or eat contaminated food. These chemicals can become concentrated in the tissues and internal organs of animals as well as humans.

It is not known if 2,3-dinitrophenol will accumulate in the edible tissues of fish.

Recommended Risk-Reduction Measures

Proper training of all transporters will reduce the likelihood of a mishap or accident resulting in a leak or spill to the environment. The correct use of DOT labeling (per the MSDS) on all containers, trucks, and rail cars will enable emergency responders to react properly and quickly to any disaster thereby reducing the potential risk to the environment and to personnel.

Storage of 2,3-dinitrophenol should be segregated from incompatible chemicals to minimize the risk of cross-contamination or contact. Containers must be protected from heat, damage, and shock. Buildings designated for storage should be equipped with appropriate fire protection systems (alarms, sprinklers, emergency lighting, portable extinguishers).

If a spill or leak to the environment has occurred, fire department, emergency response and/or hazardous materials spill personnel should be notified immediately. Cleanup should be attempted only by those trained in proper spill containment procedures. Contaminated soils should be removed for incineration and replaced with clean soil. If 2,3-dinitrophenol should contact the water table, aquifer, or navigable waterway, time is of the essence. It is slightly soluble in the aquatic environment and total remediation may not be possible. The local and/or state emergency response authorities must be notified. A comprehensive emergency response or disaster preparedness/recovery plan should be in place prior to any operations involving the use, transportation, storage, or disposal of 2,3-dinitrophenol. If 2,3-dinitrophenol is spilled or leaked, the following specific steps are recommended:

- ☑ Restrict persons not wearing protective clothing from area of spill or leak until cleanup is complete and area can be opened for normal work.
- ☑ Remove all ignition sources, ventilate area of leak.
- ☑ Absorb liquid spills in vermiculite, dry sand, earth, or similar material and deposit in sealed containers. Collect solid spills for reuse of disposal. Wet mop spill area to collect residue.
- ☑ It may be necessary to dispose of 2,3-dinitrophenol as a hazardous waste. The responsible state agency or the regional office of the federal Environmental Protection Agency (EPA) should be contacted for specific recommendations.

RISK ASSESSMENT: BUSINESS

General Assessment

Accidents or mishaps involving 2,3-dinitrophenol can present a significant threat to business operations. The loss or damage of equipment or facilities can significantly affect fiscal viability. Lawsuits that may result from personnel injury/death, public exposures, and/or environmental contamination will also require a serious expenditure of resources. Media attention surrounding an injury, death, or environmental damage can also result in a loss of profits and loss of current as well as future business.

Recommended Risk-Reduction Measures

Company attorneys, safety and health professionals, and environmental specialists should be involved in the development of any procedures dealing with safety and emergency response. A company official should be pre-designated as a public relations officer with specific training in dealing with the press. Corporate plans and policies should be developed, approved, and implemented long before any need for such arises.

REFERENCES

Ahlbom, A. 1993. *Biostatistics for Epidemiologists.* New York: CRC Press/Lewis Publishers.

Anthony, C. P., and N. J. Kolthoff. 1971. *Textbook of Anatomy and Physiology.* St. Louis: C. V. Mosby Company.

Bloom, A. D. and F. J. DeSerres. 1995. *Ecotoxicity and Human Health.* New York: CRC Press/Lewis Publishers.

Bueche, J. R. 1972. *Principles of Physics,* 2nd Edition. New York: McGraw-Hill.

Burrell, R., D. K. Flaherty, and L. J. Sauers. 1992. *Toxicology of the Immune System.* New York: Van Nostrand Reinhold.

Calabrese, E. J. 1994. *Biological Effects of Low Level Exposures.* New York: CRC Press/Lewis Publishers.

Cockerman, L. G., B. S. Shane. 1994. *Basic Environmental Toxicology.* New York: CRC Press/Lewis Publishers.

Environmental Protection Agency. 1983. *Chemical Hazard Information Profiles.* Washington, D.C.: U.S. EPA.

Fodden, C. R., and J. L. Weddell. 1992. *Hazardous Materials Emergency Action Data.* New York: CRC Press/Lewis Publishers.

Hallenbeck, W. H. 1993. *Quantitative Risk Assessment for Environmental and Occupational Health,* 2nd Edition. New York: CRC Press/Lewis Publishers.

Howard, P. H. 1995. *Handbook of Environmental Fate and Exposures Data for Organic Chemicals,* Vol. I-IV. New York: CRC Press/Lewis Publishers.

Lewis, R. J., Sr. 1992. *Sax's Dangerous Properties of Industrial Materials,* Volumes 1, 2, and 3, 8th Edition. New York: Van Nostrand Reinhold.

Meyer, E. 1990. *Chemistry of Hazardous Materials.* Englewood Cliffs, NJ: Prentice Hall.

Quigley, D. R. 1994. *Handbook of Emergency Chemical Management.* New York: CRC Press/Lewis Publishers.

U.S. Department of Health and Human Services, National Institute for Occupational Safety and Health. 1994. *NIOSH Pocket Guide to Chemical Hazards.* Washington, D.C.: U.S. Government Printing Office.

MATERIAL SAFETY DATA SHEET

CHEMICAL NAME	
2,4-DINITROPHENOL	**CRC PRESS/LEWIS PUBLISHERS DISCLAIMER:** THE INFORMATION AND RECOMMENDATIONS PRESENTED HEREIN ARE BASED ON SOURCES BELIEVED TO BE RELIABLE. CRC MAKES NO REPRESENTATION ON ITS COMPLETENESS OR ACCURACY. IT IS THE RESPONSIBILITY OF THE USER TO DETERMINE THE CHEMICAL'S SUITABILITY FOR ITS INTENDED USE, THE CHEMICAL'S SAFE USE, AND THE CHEMICAL'S PROPER DISPOSAL. NO REPRESENTATIONS AND/OR WARRANTIES, EITHER EXPRESSED OR IMPLIED, OF THE MERCHANTABILITY OR FITNESS FOR A PARTICULAR PURPOSE, OR OF ANY OTHER NATURE, ARE MADE WITH RESPECT TO THE INFORMATION PROVIDED IN THIS MSDS OR TO THE CHEMICAL TO WHICH INFORMATION MAY REFER. CRC NEITHER ASSUMES NOR AUTHORIZES ANY OTHER PERSON TO ASSUME FOR IT, ANY OTHER ADDITIONAL RESPONSIBILITY OR LIABILITY FOR THE USE OF, OR RELIANCE UPON, THIS INFORMATION.

HAZARD WARNING INFORMATION

HEALTH	FIRE	REACTIVE	OTHER	DEGREE OF HAZARD	COLOR CODING	OTHER CODES
DRY: 3 WET: 3	DRY: 3 WET: 1	DRY: 3 WET: 0		0 = Minimum Hazard 1 = Slight Hazard 2 = Moderate Hazard 3 = Serious Hazard 4 = Severe Hazard	HEALTH = BLUE FIRE = RED REACTIVITY = YELLOW OTHER = WHITE	OX = Oxidizer ACID = Acid ALK = Alkali COR = Corrosive W = Use No Water

SECTION I - GENERAL INFORMATION

Characterization **Phenol**	RCRA Number **P048**	EPA Class **Acute Hazardous Waste**
DOT Proper Shipping Name **Dinitrophenol (solution)**	Chemical Abstract Service (CAS) Number **51-28-5**	
DOT Hazard Class and Label Requirements **Poison B (solution), Flammable Solid**	DOT Emergency Guide Code **53**	
DOT Identification Number **UN 1320, 1321 (solid); UN 1599 (solution)**	Chemical Formula $C_6H_4N_2O_5$	

Synonyms
Aldifen; Chemox PE; α-dinitrophenol; 2,4-DNP; Solfo Black BB; Tertrosulfur Black PB.

SECTION II - HAZARDOUS INGREDIENTS AND IDENTITY INFORMATION

Hazardous Components (specific identity)	OSHA Exposure Criteria	NIOSH Exposure Criteria	Immediately Dangerous to Life and Health (IDLH)	ACGIH Exposure Criteria
2,4-dinitrophenol (derivation: By heating phenol with dilute sulfuric acid, cooling the product, and then nitrating while keeping the temperature approximately 50°C; or by nitrating with mixed acid under careful temperature control).	**PEL (skin):** 0.2 mg/m³ **STEL:** Not Established	**REL:** Not Established **STEL:** Not Established	Not Determined	**TLV:** Not Established **STEL:** Not Established

SECTION III - PHYSICAL AND CHEMICAL CHARACTERISTICS

Boiling Point **Sublimes (when heated)**	Specific Gravity (H_2O = 1) **1.68**
Vapor Pressure (gas) **5-10 at 69°F (20°C)**	Molecular Weight **184.1**
Vapor Density (Air = 1) **6.3**	Melting Point **234°F (112°C)**

Solubility
Slightly soluble in cold water; freely soluble in hot water. Soluble in alcohol, ether, acetone, pyridine, benzene, chloroform, and most organic solvents. Soluble in aqueous alkaline solutions.

Appearance and Odor
Yellow, odorless solid crystals (powder).

SECTION IV - FIRE AND EXPLOSION HAZARD DATA

Flash Point (method used) **Not Applicable**	Explosive Limits in Air % by Volume **LEL: Not Determined UEL: Not Determined**
NFPA Classification **Flammable Solid**	Autoignition Temperature **Not Determined**

Extinguishing Media
Use carbon dioxide, water spray, dry chemical, foam, or fog.

Special Fire Fighting Procedures
Wear full protective clothing and self-contained breathing apparatus (SCBA). Poisonous gases are produced in fire. Use water spray to keep fire-exposed containers cool. Move containers from fire if it can be done without risk. Fight fire from a distance or protected position, if possible.

Unusual Fire and Explosion Hazards
Containers may explode in fire. Fires involving this material should be considered an explosion hazard. Extremely explosive in dry form when exposed to heat, flame, or shock.

SECTION V - REACTIVITY DATA

Stability		Conditions to Avoid **2,4-Dinitrophenol is normally stable. It will remain stable in mixtures up to 15% by weight with water. It becomes explosive when dry and subjected to shock, heat, or flame.**
Stable **X**	Unstable	Incompatibility *(materials to avoid)* **Strong oxidizers, (such as chlorine, bromine, and fluorine), and steam. It will form explosive salts with alkalis and/or ammonia.**
Hazardous Polymerization		Conditions to Avoid **Under normal conditions of temperature and pressure, hazardous polymerization of 2,4-dinitrophenol is not expected to occur.**
May Occur	Will Not Occur **X**	Hazardous Decomposition or By-products **When heated to decomposition, 2,4-dinitrophenol can emit highly toxic/poisonous gases, including carbon dioxide and oxides of nitrogen.**

SECTION VI - HEALTH HAZARD DATA

Primary Route(s) of Entry:	Inhalation?	Absorption (skin)?	Ingestion?
	X	X	X

Health Hazards

INHALATION: **Can cause headache, coughing, sweating, thirst, intense fatigue, high fever, rapid pulse, warm and flushed skin, rapid and difficult breathing, anxiety, confusion, convulsions, loss of consciousness, pulmonary edema (fluid in lungs), and death.**

ABSORPTION: **Irritation with yellow staining, redness, swelling, blisters, and delayed symptoms like those of inhalation (will pass through skin to cause same toxic systemic effects).**

INGESTION: **Burning of mouth and throat, salivation, dizziness, nausea, vomiting, bright yellow stool, and symptoms of inhalation.**

Carcinogenicity	NTP Listed?	IARC Cancer Review Group?	OSHA Regulated?	Target Organs?
Unknown Human Unknown Animal	**No**	**No**	**No**	**Skin, respiratory system, CNS, CVS, liver, kidney**

Medical Conditions Generally Aggravated by Exposure

None reported. However, existing thyroid, liver, kidney, and cardiovascular diseases may be aggravated.

Emergency and First-aid Procedures

Eye contact: Flush immediately with water for 15 minutes (minimum), seek medical attention. Skin contact: Remove all contaminated clothing. Immediately wash area with large amounts of soap and water. Seek medical assistance. Yellow discoloration need not be totally removed to prevent absorption. For inhalation: Remove the person from exposure. Provide CPR if required. Transfer to medical facility. Observe for 24-48 hours for lung effects. If swallowed: Unless advised otherwise, give 1-2 glasses of water and induce vomiting, then give 2 tablespoons of activated charcoal in 8 ounces of water.

SECTION VII - PRECAUTIONS FOR SAFE HANDLING AND USE

Steps to be Taken in Case Material is Released or Spilled

Minimize dust generation. Do NOT dry sweep. Use a vacuum equipped with a high efficiency particulate air (HEPA) filter instead. Collect and dispose of in sealed drum. If in run-off, dike area to contain, cover with absorbent (vermiculite or other) and dispose of in sealed drum.

Preferred Waste Disposal Method

No Citation.

Precautions to be Taken in Handling and Storage

Store in tightly closed containers in cool, well-ventilated area away from heat or flame. Heat may cause containers to build up pressure and explode. Store in detached facility away from main plant or facility.

Other Precautions and Warnings

Sources of ignition are prohibited where 2,4-dinitrophenol is used, handled, or stored.

SECTION VIII - CONTROL MEASURES AND PERSONAL PROTECTIVE EQUIPMENT

Respiratory Protection *(specify type)*

PEL is low and difficult to monitor. For any exposure level, recommend a MSHA/NIOSH-approved supplied air respirator with full facepiece, hood, or helmet in continuous flow mode, or use a self-contained breathing apparatus (SCBA) operated in pressure demand or other positive pressure mode.

Ventilation

Local exhaust or general mechanical systems recommended.

Protective Gloves	Eye Protection	Other Protective Clothing
Butyl Rubber	**Chemical/Dust Goggles or Face Mask**	**Butyl Rubber Apron**

Work/Hygiene Practices

Always wash hands thoroughly after using chemical; never bring food, drink, or smoking materials into vicinity of chemicals.

2,4-DINITROPHENOL

$C_6H_4N_2O_5$ CAS: 51-28-5

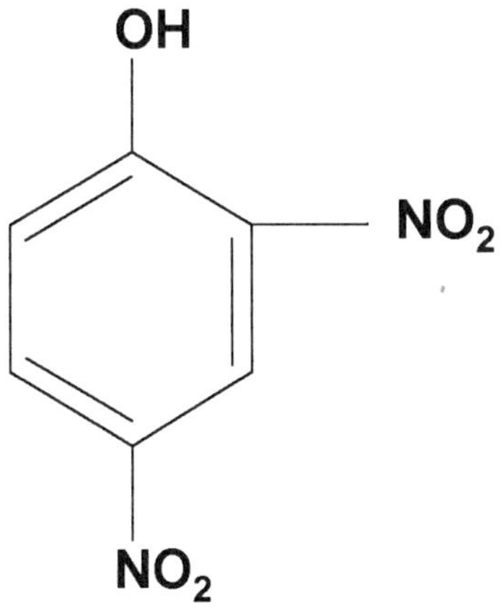

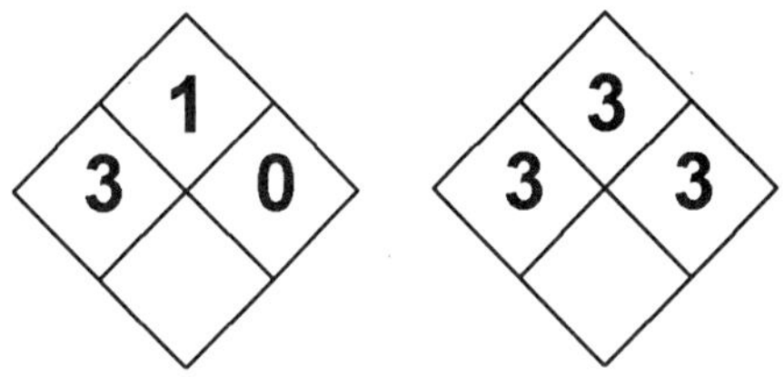

DRY WET

IDENTIFICATION AND TYPICAL USES

2,4-Dinitrophenol is a yellow, odorless solid that appears as a powder. It is used in the manufacture of dyes, especially sulfur colors, in picric acid picramic acid, in the preservation of lumber, in the manufacture of the photographic developer diaminophenol hydrochloride, in explosives manufacture, as an indicator, and a reagent for potassium and ammonium ions.

RISK ASSESSMENT: HEALTH

General Assessment

2,4-Dinitrophenol is a human poison by *ingestion,* with equally toxic systemic consequences occurring by way of *inhalation,* and skin contact (*absorption*). It has been shown to cause teratogenic effects in experimental tests and mutation data have also been reported. Its carcinogenic properties are unknown. Liver or kidney damage can occur up to 3 days following exposure.

Inhalation can result in headache, coughing, profuse sweating, marked thirst, intense fatigue, high fever, rapid pulse or heart rate (tachycardia), warm and flushed skin, difficulty in breathing (tachypnea), anxiety and confusion, convulsions, loss of consciousness, pulmonary edema (fluid in the lungs), and death due to circulatory or respiratory collapse. Exposure may also lead to a deadly buildup of fluids in the lungs in a condition known as pulmonary edema. This is a medical emergency requiring immediate attention. Symptoms can be delayed up to 48 hours, creating a false sense of security with regard to health exposure risk.

Skin contact will result in nearly immediate yellow staining, irritation, swelling, redness, blisters, and delayed systemic effects from absorption with symptoms identical to those of inhalation. Eye contact causes burning pain, tearing, and inflammation of the eyelids.

Ingestion can cause painful burning of the mouth and throat, salivation, dizziness, nausea and vomiting, the appearance of bright yellow stools, and the symptoms of systemic poisoning similar to those of inhalation.

Acute Health Effects

The following acute (short-term) health effects may occur immediately or shortly after exposure to 2,4-dinitrophenol:

Skin: Irritation with yellowing of skin and possible tissue damage. Will pass through unbroken skin to cause dangerous systemic effects.

Eye: Irritation with damage to the surrounding tissue. If not removed immediately, damage may be more severe and permanent.

Lung: Severe irritation to the respiratory tract and associated mucosa. Can cause coughing, congestion, and possibly pulmonary edema (fluid in the lungs), which can be fatal.

CNS: Can cause confusion, anxiety, headache, changes in the function of the respiratory and circulatory systems, convulsions, coma.

Other: Can cause a disruption in cellular metabolism in all tissues resulting in high fever, elevated blood pressure, pulse, and respiration. Can also cause death by hyperthermia. Effects are more severe in hot work environments.

Chronic Health Effects

The following chronic (long-term) health effects may occur at some time after exposure to 2,4-dinitrophenol and can last for months or even years:

Cancer Hazards: According to the information available in the references, 2,4-dinitrophenol has not been adequately tested for its ability to cause cancer in test animals. Experimental mutation data have been reported and many scientists believe that such chemicals may have the ability to cause cancer. This, however, requires much more study and research before any claims of carcinogenicity can be accurately asserted.

Reproductive Hazard: According to information presented in the references, 2,4-dinitrophenol has been shown to cause adverse effects in reproduction in animals. It is considered an experimental teratogen. Human effects in this regard are inconclusive and require additional research.

Other Chronic Effects: Long-term exposures can result in damage to the liver or kidneys. Hypersensitivity may also result from repeated exposures.

Recommended Risk-Reduction Measures

Personnel should avoid direct contact with 2,4-dinitrophenol. If a less toxic chemical cannot be substituted for a hazardous substance, then *engineering controls* are the most effective method of reducing exposures. The best protection is to enclose operations and/or provide local exhaust ventilation at the site of chemical release. While not always operationally feasible, isolating operations can also reduce exposure. Using respiratory protection is less effective than the controls mentioned above, but is still advisable whenever working with or around 2,4-dinitrophenol. The exposure level is low (0.2 mg/m^3) and difficult to monitor. Therefore, a MSHA/NIOSH approved full facepiece respirator equipped with an organic vapor cartridge equipped with a dust pre-filter should be used if any exposure is anticipated or known. Increased protection is obtained from a supplied-air respirator with full facepiece operated in continuous flow mode, or a self-contained breathing apparatus (SCBA) with full facepiece and pressure demand. If a full facepiece is not available, then chemical dust goggles should be worn to protect the eyes. Whenever an airborne chemical dust hazard exists, a face shield and a rubber apron should be worn. To prevent hand and skin exposures, butyl rubber gloves should be worn.

Administrative controls should also be in place to minimize the potential for human exposures. These may include written procedures or policies which specify the methods and techniques that will be practiced whenever personnel are to work with 2,4-dinitrophenol.

All personnel should receive *training* on the use, hazards, protective measures, emergency actions, and other precautions per 29 CFR 1910.1200 (Hazard Communication), prior to the first assignment in an area where 2,4-dinitrophenol is used or stored.

Before beginning employment with 2,4-dinitrophenol and at regular intervals thereafter (e.g., annually), those persons with high or frequent exposures should be provided the following recommended medical tests:

- ☑ Complete blood count (establish baseline).
- ☑ Urinalysis (establish baseline nitrophenol levels).

If symptoms develop or overexposure is suspected, the following tests are recommended:

- ☑ Complete blood count (compare to baseline).
- ☑ Urinalysis (compare to baseline).
- ☑ Consider chest X-ray following acute overexposure (may be negative if taken immediately after exposure).
- ☑ Lung function tests.
- ☑ Liver and kidney function tests.

Any evaluation should include a careful history of past and present symptoms with an examination. Medical tests that simply look for existing damage are not a substitute for controlling exposures. Physician should monitor fluids and electrolytes, control hyperthermia, and test for dinitrophenol or its metabolite (aminophenol) in the urine using Derrian's Test. The biological half-life of 2,4-dinitrophenol is 5-14 days.

Other methods to reduce exposure include:

- ☑ Where possible, enclose operations and use local exhaust ventilation at the site of chemical release. If local exhaust ventilation or enclosure is not used, respiratory protection should be mandatory.
- ☑ Always ensure that proper protective clothing is worn when using chemical substances.
- ☑ Wash thoroughly immediately after exposure to 2,4-dinitrophenol and at the end of the work shift or before eating, drinking, or smoking.

☑ Hazard warning information should be posted in the work area. In addition, as part of an on-going education and training program, all information on the health and safety hazards of 2,4-dinitrophenol should be communicated to all exposed and potentially exposed workers.

☑ Eye wash stations should be provided in the immediate work area for emergency use. If there is a possibility of skin exposure to 2,4-dinitrophenol, emergency shower facilities should also be provided in the work area.

☑ Personnel should never wear contaminated clothing home (family members can be exposed). Clothing should be laundered by personnel who have been trained on the hazards of working with and around 2,4-dinitrophenol.

RISK ASSESSMENT: ENVIRONMENT

General Assessment

The environment is at risk of exposure during transportation, storage, disposal, or destruction of 2,4-dinitrophenol. In almost every scenario, the threat of environmental exposure is contingent upon the proper handling of the chemical substance. Accidental spills, large or small, can result in fire, explosion, and possible contamination of the surrounding environmental mediums (water, soil, and air).

2,4-Dinitrophenol is normally considered a flammable solid. In solution, it can often appear in mixtures with other combustible or flammable commodities. Its dusts may form explosive mixtures in air. These characteristics require caution in handling, storage, transportation, and disposal. It is incompatible with strong oxidizers and contact can cause fire or explosion. It forms explosive salts in contact with alkalis or with ammonia. Prolonged contact with heat may result in an explosion due to spontaneous decomposition. Therefore, special consideration is required during any emergency situation involving a leak or spill of 2,4-dinitrophenol. Should 2,4-dinitrophenol ever come into contact with these incompatible substances during use, transportation, storage, or disposal, the formation of highly toxic and/or highly explosive commodities is extremely possible.

2,4-Dinitrophenol may enter the environment through its intended use as an insecticide, through agricultural use, as well as through municipal waste discharges, industrial effluents, or spills.

Acute Ecological Effects

Acute (short-term) toxic effects may include the death of animals, birds, or fish and death or low growth rate in plants. Acute effects are seen 2 to 4 days after animals or plants are exposed to 2,4-dinitrophenol.

2,4-Dinitrophenol has moderate acute toxicity to aquatic life and mild acute toxicity to plant life. Insufficient data are available on the short-term effects of 2,4-dinitrophenol exposure to birds or land animals.

Chronic Ecological Effects

Chronic (long-term) toxic effects may include shortened life span, reproductive problems, lower fertility, and changes in appearance or behavior in exposed animals. These effects can be seen long after first exposure(s) to toxic chemicals.

2,4-Dinitrophenol has moderate chronic toxicity to aquatic life. Insufficient data are available on the long-term effects of 2,4-dinitrophenol to plants, birds, or land animals.

Water Solubility

2,4-Dinitrophenol is slightly soluble in cold water. Concentrations of 1 to 100 milligrams will mix with a liter of water.

Persistence in the Environment

2,4-Dinitrophenol is slightly persistent in water, with a half-life between 2 and 20 days. The ecological half-life of a pollutant is the amount of time it takes for one half of the chemical to be degraded.

Bioaccumulation in Aquatic Organisms

Some substances increase in concentration, or *bioaccumulate,* in living organisms as they breathe contaminated air, drink contaminated water, or eat contaminated food. These chemicals can become concentrated in the tissues and internal organs of animals as well as humans.

The concentration of 2,4-dinitrophenol found in fish tissue is expected to be somewhat higher than the average concentration of 2,4-dinitrophenol in the water from which the fish was taken.

Recommended Risk-Reduction Measures

Proper training of all transporters will reduce the likelihood of a mishap or accident resulting in a leak or spill to the environment. The correct use of DOT labeling (per the MSDS) on all containers, trucks, and rail cars will enable emergency responders to react properly and quickly to any disaster thereby reducing the potential risk to the environment and to personnel.

Storage of 2,4-dinitrophenol should be segregated from incompatible chemicals to minimize the risk of cross-contamination or contact. Containers must be protected from heat, damage, and shock. Buildings designated for storage should be equipped with appropriate fire protection systems (alarms, sprinklers, emergency lighting, portable extinguishers).

If a spill or leak to the environment has occurred, fire department, emergency response and/or hazardous materials spill personnel should be notified immediately. Cleanup should be attempted only by those trained in proper spill containment procedures. Contaminated soils should be removed for incineration and replaced with clean soil. If 2,4-dinitrophenol should contact the water table, aquifer, or navigable waterway, time is of the essence. It is slightly soluble in the aquatic environment and total remediation may not be possible. The local and/or state emergency response authorities must be notified. A comprehensive emergency response or disaster preparedness/recovery plan should be in place prior to any operations involving the use, transportation, storage, or disposal of 2,4-dinitrophenol. If 2,4-dinitrophenol is spilled or leaked, the following specific steps are recommended:

- ☑ Restrict persons not wearing protective clothing from area of spill or leak until cleanup is complete and area can be opened for normal work.
- ☑ Remove all ignition sources, ventilate area of leak.
- ☑ Absorb liquid spills in vermiculite, dry sand, earth, or similar material and deposit in sealed containers. Collect solid spills for reuse of disposal. Wet mop spill area to collect residue.
- ☑ It may be necessary to dispose of 2,4-dinitrophenol as a hazardous waste. The responsible state agency or the regional office of the federal Environmental Protection Agency (EPA) should be contacted for specific recommendations.

RISK ASSESSMENT: BUSINESS

General Assessment

Accidents or mishaps involving 2,4-dinitrophenol can present a significant threat to business operations. The loss or damage of equipment or facilities can significantly affect fiscal viability. Lawsuits that may result from personnel injury/death, public exposures, and/or environmental contamination will also require a serious expenditure of resources. Media attention surrounding an injury, death, or environmental damage can also result in a loss of profits and loss of current as well as future business.

Recommended Risk-Reduction Measures

Company attorneys, safety and health professionals, and environmental specialists should be involved in the development of any procedures dealing with safety or emergency response. A company official should be pre-designated as a public relations officer with specific training in dealing with the press. Corporate plans and policies should be developed, approved, and implemented long before any need for such arises.

REFERENCES

Ahlbom, A. 1993. *Biostatistics for Epidemiologists.* New York: CRC Press/Lewis Publishers.

Anthony, C. P., and N. J. Kolthoff. 1971. *Textbook of Anatomy and Physiology.* St. Louis: C. V. Mosby Company.

Bloom, A. D. and F. J. DeSerres. 1995. *Ecotoxicity and Human Health.* New York: CRC Press/Lewis Publishers.

Bueche, J. R. 1972. *Principles of Physics,* 2nd Edition. New York: McGraw-Hill.

Burrell, R., D. K. Flaherty, and L. J. Sauers. 1992. *Toxicology of the Immune System.* New York: Van Nostrand Reinhold.

Calabrese, E. J. 1994. *Biological Effects of Low Level Exposures.* New York: CRC Press/Lewis Publishers.

Cockerman, L. G., B. S. Shane. 1994. *Basic Environmental Toxicology.* New York: CRC Press/Lewis Publishers.

Environmental Protection Agency. 1983. *Chemical Hazard Information Profiles.* Washington, D.C.: U.S. EPA.

Fodden, C. R., and J. L. Weddell. 1992. *Hazardous Materials Emergency Action Data.* New York: CRC Press/Lewis Publishers.

Hallenbeck, W. H. 1993. *Quantitative Risk Assessment for Environmental and Occupational Health,* 2nd Edition. New York: CRC Press/Lewis Publishers.

Howard, P. H. 1995. *Handbook of Environmental Fate and Exposures Data for Organic Chemicals,* Vol. I-IV. New York: CRC Press/Lewis Publishers.

Lewis, R. J., Sr. 1992. *Sax's Dangerous Properties of Industrial Materials,* Volumes 1, 2, and 3, 8th Edition. New York: Van Nostrand Reinhold.

Meyer, E. 1990. *Chemistry of Hazardous Materials.* Englewood Cliffs, NJ: Prentice Hall.

Quigley, D. R. 1994. *Handbook of Emergency Chemical Management.* New York: CRC Press/Lewis Publishers.

MATERIAL SAFETY DATA SHEET

CHEMICAL NAME	
2,5-DINITROPHENOL	CRC PRESS/LEWIS PUBLISHERS DISCLAIMER: THE INFORMATION AND RECOMMENDATIONS PRESENTED HEREIN ARE BASED ON SOURCES BELIEVED TO BE RELIABLE. CRC MAKES NO REPRESENTATION ON ITS COMPLETENESS OR ACCURACY. IT IS THE RESPONSIBILITY OF THE USER TO DETERMINE THE CHEMICAL'S SUITABILITY FOR ITS INTENDED USE, THE CHEMICAL'S SAFE USE, AND THE CHEMICAL'S PROPER DISPOSAL. NO REPRESENTATIONS AND/OR WARRANTIES, EITHER EXPRESSED OR IMPLIED, OF THE MERCHANTABILITY OR FITNESS FOR A PARTICULAR PURPOSE, OR OF ANY OTHER NATURE, ARE MADE WITH RESPECT TO THE INFORMATION PROVIDED IN THIS MSDS OR TO THE CHEMICAL TO WHICH INFORMATION MAY REFER. CRC NEITHER ASSUMES NOR AUTHORIZES ANY OTHER PERSON TO ASSUME FOR IT, ANY OTHER ADDITIONAL RESPONSIBILITY OR LIABILITY FOR THE USE OF, OR RELIANCE UPON, THIS INFORMATION.

HAZARD WARNING INFORMATION

HEALTH	FIRE	REACTIVE	OTHER	DEGREE OF HAZARD	COLOR CODING	OTHER CODES
DRY: 3 WET: 3	DRY: 3 WET: 1	DRY: 3 WET: 0		0 = Minimum Hazard 1 = Slight Hazard 2 = Moderate Hazard 3 = Serious Hazard 4 = Severe Hazard	HEALTH = BLUE FIRE = RED REACTIVITY = YELLOW OTHER = WHITE	OX = Oxidizer ACID = Acid ALK = Alkali COR = Corrosive W = Use No Water

SECTION I - GENERAL INFORMATION

Characterization	Phenol	RCRA Number	None
		EPA Class	Not Applicable
DOT Proper Shipping Name	Dinitrophenol (solution)	Chemical Abstract Service (CAS) Number	329-71-5
DOT Hazard Class and Label Requirements	Poison B (solution), Flammable Solid	DOT Emergency Guide Code	53
DOT Identification Number	UN 1320, 1321 (solid); UN 1599 (solution)	Chemical Formula	$C_6H_4N_2O_5$
Synonyms	Γ-Dinitrophenol; 2,5-DNP.		

SECTION II - HAZARDOUS INGREDIENTS AND IDENTITY INFORMATION

Hazardous Components (specific identity)	OSHA Exposure Criteria	NIOSH Exposure Criteria	Immediately Dangerous to Life and Health (IDLH)	ACGIH Exposure Criteria
2,5-dinitrophenol (derivation: By heating phenol with dilute sulfuric acid, cooling the product, and then nitrating while keeping the temperature approximately 50°C; or by nitrating with mixed acid under careful temperature control).	PEL (skin): 0.2 mg/m³ STEL: Not Established	REL: Not Established STEL: Not Established	Not Determined	TLV: Not Established STEL: Not Established

SECTION III - PHYSICAL AND CHEMICAL CHARACTERISTICS

Boiling Point	Sublimes (when heated)	Specific Gravity (H_2O = 1)	1.68
Vapor Pressure (gas)	5-10 at 69°F (20°C)	Molecular Weight	184.1
Vapor Density (Air = 1)	6.3	Melting Point	2264°F (108°C)

Solubility

Slightly soluble in cold water; freely soluble in hot water. Soluble in hot alcohol, ether, acetone, pyridine, benzene, chloroform, and most organic solvents, and alkali hydroxides.

Appearance and Odor

Yellow, odorless solid crystals (powder).

SECTION IV - FIRE AND EXPLOSION HAZARD DATA

Flash Point (method used)	Not Applicable	Explosive Limits in Air % by Volume	LEL: Not Determined UEL: Not Determined
NFPA Classification	Flammable Solid	Autoignition Temperature	Not Determined

Extinguishing Media

Use carbon dioxide, water spray, dry chemical, foam, or fog.

Special Fire Fighting Procedures

Wear full protective clothing and self-contained breathing apparatus (SCBA). Poisonous gases are produced in fire. Use water spray to keep fire-exposed containers cool. Move containers from fire if it can be done without risk. Fight fire from a distance or protected position, if possible.

Unusual Fire and Explosion Hazards

Containers may explode in fire. Fires involving this material should be considered an explosion hazard. Extremely explosive in dry form when exposed to heat, flame, or shock.

SECTION V - REACTIVITY DATA

Stability		Conditions to Avoid 2,5-Dinitrophenol is normally stable. It will remain stable in mixtures up to 15% by weight with water. It becomes explosive when dry and subjected to shock, heat, or flame.
Stable X	Unstable	Incompatibility *(materials to avoid)* Strong oxidizers, (such as chlorine, bromine, and fluorine), and steam. It will form explosive salts with alkalis and/or ammonia.
Hazardous Polymerization		Conditions to Avoid Under normal conditions of temperature and pressure, hazardous polymerization of 2,5-dinitrophenol is not expected to occur.
May Occur	Will Not Occur X	Hazardous Decomposition or By-products When heated to decomposition, 2,5-dinitrophenol can emit highly toxic/poisonous gases, including carbon dioxide and oxides of nitrogen.

SECTION VI - HEALTH HAZARD DATA

Primary Route(s) of Entry:	Inhalation?	Absorption (skin)?	Ingestion?
	X	X	X

Health Hazards

INHALATION: Can cause headache, coughing, sweating, thirst, intense fatigue, high fever, rapid pulse, warm and flushed skin, rapid and difficult breathing, anxiety, confusion, convulsions, loss of consciousness, pulmonary edema (fluid in lungs), and death.

ABSORPTION: Irritation with yellow staining, redness, swelling, blisters, and delayed symptoms like those of inhalation (will pass through skin to cause same toxic systemic effects).

INGESTION: Burning of mouth and throat, salivation, dizziness, nausea, vomiting, bright yellow stool, and symptoms of inhalation.

Carcinogenicity	NTP Listed?	IARC Cancer Review Group?	OSHA Regulated?	Target Organs?
Unknown Human Unknown Animal	No	No	No	Skin, respiratory system, CNS, CVS, liver, kidney

Medical Conditions Generally Aggravated by Exposure

None reported. However, existing thyroid, liver, kidney, and cardiovascular diseases may be aggravated.

Emergency and First-aid Procedures

Eye contact: Flush immediately with water for 15 minutes (minimum), seek medical attention. Skin contact: Remove all contaminated clothing. Immediately wash area with large amounts of soap and water. Seek medical assistance. Yellow discoloration need not be totally removed to prevent absorption. For inhalation: Remove the person from exposure. Provide CPR if required. Transfer to medical facility. Observe for 24-48 hours for lung effects. If swallowed: Unless advised otherwise, give 1-2 glasses of water and induce vomiting, then give 2 tablespoons of activated charcoal in 8 ounces of water.

SECTION VII - PRECAUTIONS FOR SAFE HANDLING AND USE

Steps to be Taken in Case Material is Released or Spilled

Minimize dust generation. Do NOT dry sweep. Use a vacuum equipped with a high efficiency particulate air (HEPA) filter instead. Collect and dispose of in sealed drum. If in run-off, dike area to contain, cover with absorbent (vermiculite or other) and dispose of in sealed drum.

Preferred Waste Disposal Method

No Citation.

Precautions to be Taken in Handling and Storage

Store in tightly closed containers in cool, well-ventilated area away from heat or flame. Heat may cause containers to build up pressure and explode. Store in detached facility away from main plant or facility.

Other Precautions and Warnings

Sources of ignition are prohibited where 2,5-dinitrophenol is used, handled, or stored.

SECTION VIII - CONTROL MEASURES AND PERSONAL PROTECTIVE EQUIPMENT

Respiratory Protection *(specify type)*

PEL is low and difficult to monitor. For any exposure level, recommend a MSHA/NIOSH-approved supplied-air respirator with full facepiece, hood, or helmet in continuous flow mode, or use a self-contained breathing apparatus (SCBA) operated in pressure demand or other positive pressure mode.

Ventilation

Local exhaust or general mechanical systems recommended.

Protective Gloves	Eye Protection	Other Protective Clothing
Butyl Rubber	Chemical/Dust Goggles or Face Mask	Butyl Rubber Apron

Work/Hygiene Practices

Always wash hands thoroughly after using chemical; never bring food, drink, or smoking materials into vicinity of chemicals.

2,5-DINITROPHENOL

$C_6H_4N_2O_5$ CAS: 329-71-5

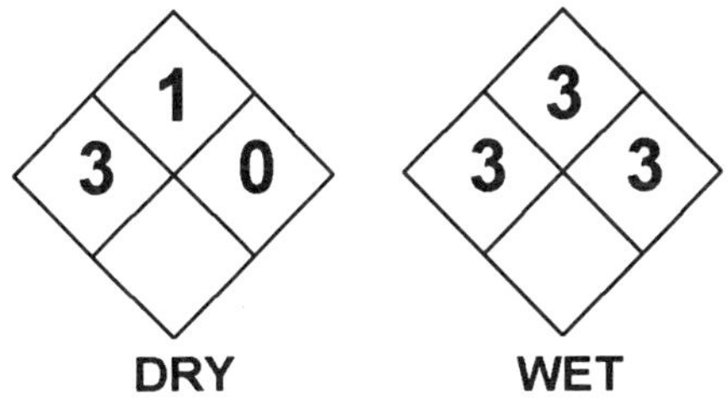

IDENTIFICATION AND TYPICAL USES

2,5-Dinitrophenol is a yellow, odorless solid that appears as a powder. It is used in the manufacture of dyes, especially sulfur colors, in picric acid picramic acid, in the preservation of lumber, in the manufacture of the photographic developer diaminophenol hydrochloride, in explosives manufacture, as an indicator, and a reagent for potassium and ammonium ions.

RISK ASSESSMENT: HEALTH

General Assessment

2,5-Dinitrophenol is a deadly human poison by *ingestion,* with equally toxic systemic consequences occurring by way of *inhalation,* and skin contact (*absorption*). It has been shown to cause teratogenic effects in experimental tests and mutation data have also been reported. Its carcinogenic properties are unknown. Liver or kidney damage can occur up to 3 days following exposure (in test animals).

Inhalation can result in headache, coughing, profuse sweating, marked thirst, intense fatigue, high fever, rapid pulse or heart rate (tachycardia), warm and flushed skin, rapid respiration (tachypnea), anxiety and confusion, convulsions, loss of consciousness, pulmonary edema (fluid in the lungs), and death due to circulatory or respiratory collapse. Exposure may also lead to a deadly buildup of fluids in the lungs in a condition known as pulmonary edema. This is a medical emergency requiring immediate attention. Symptoms of cough, tightness in chest, congestion, and dypsnea can be delayed up to 48 hours, creating a false sense of security with regard to health exposure risk.

Skin contact will result in nearly immediate yellow staining, irritation, swelling, redness, blisters, and delayed systemic effects from absorption with symptoms nearly identical to those of inhalation. Eye contact causes burning pain, tearing, and inflammation of the eyelids. If not removed immediately, damage to vision may occur.

Ingestion can cause painful burning of the mouth and throat, salivation, dizziness, nausea and vomiting, the appearance of bright yellow stools, and the symptoms of systemic poisoning similar to those of inhalation.

Acute Health Effects

The following acute (short-term) health effects may occur immediately or shortly after exposure to 2,5-dinitrophenol:

<u>Skin</u>: Irritation with yellowing of skin and possible tissue damage. Will pass through unbroken skin to cause dangerous systemic effects.

<u>Eye</u>: Irritation with damage to the surrounding tissue. If not removed immediately, damage may be more severe and permanent.

<u>Lung</u>: Severe irritation to the respiratory tract and associated mucosa. Can cause coughing, congestion, and possibly pulmonary edema (fluid in the lungs), which can be fatal.

<u>CNS</u>: Can cause confusion, anxiety, headache, changes in the function of the respiratory and circulatory systems, convulsions, coma.

<u>Other</u>: Can cause a disruption in cellular metabolism in all tissues resulting in high fever, elevated blood pressure, pulse, and respiration. Can also cause death by hyperthermia. Effects are more severe in hot work environments.

Chronic Health Effects

The following chronic (long-term) health effects may occur at some time after exposure to 2,5-dinitrophenol and can last for months or even years:

Cancer Hazards: According to the information available in the references, 2,5-dinitrophenol has not been adequately tested for its ability to cause cancer in test animals. Experimental mutation data have been reported and many scientists believe that such chemicals may have the ability to cause cancer. This, however, requires much more study and research before any claims of carcinogenicity can be accurately asserted.

Reproductive Hazard: According to information presented in the references, 2,5-dinitrophenol has been shown to cause adverse effects in reproduction in animals. It is considered an experimental teratogen. Human effects in this regard are inconclusive and require additional research.

Other Chronic Effects: Long-term exposures can result in damage to the liver or kidneys. Hypersensitivity may also result from repeated exposures.

Recommended Risk-Reduction Measures

Personnel should avoid direct contact with 2,5-dinitrophenol. If a less toxic chemical cannot be substituted for a hazardous substance, then *engineering controls* are the most effective method of reducing exposures. The best protection is to enclose operations and/or provide local exhaust ventilation at the site of chemical release. While not always operationally feasible, isolating operations can also reduce exposure. Using respiratory protection is less effective than the controls mentioned above, but is still advisable whenever working with or around 2,5-dinitrophenol. The exposure level is low (0.2 mg/m^3) and difficult to monitor. Therefore, a MSHA/NIOSH approved full facepiece respirator equipped with an organic vapor cartridge equipped with a dust pre-filter should be used if any exposure is anticipated or known. Increased protection is obtained from a supplied-air respirator with full facepiece operated in continuous flow mode, or a self-contained breathing apparatus (SCBA) with full facepiece and pressure demand. If a full facepiece is not available, then chemical dust goggles should be worn to protect the eyes. Whenever an airborne chemical dust hazard exists, a face shield and a rubber apron should be worn. To prevent hand and skin exposures, butyl rubber gloves should be worn.

Administrative controls should also be in place to minimize the potential for human exposures. These may include written procedures or policies which specify the methods and techniques that will be practiced whenever personnel are to work with 2,5-dinitrophenol.

All personnel should receive *training* on the use, hazards, protective measures, emergency actions, and other precautions per 29 CFR 1910.1200 (Hazard Communication), prior to the first assignment in an area where 2,5-dinitrophenol is used or stored.

Before beginning employment with 2,5-dinitrophenol and at regular intervals thereafter (e.g., annually), those persons with high or frequent exposures should be provided the following recommended medical tests:

- ☑ Complete blood count (establish baseline).
- ☑ Urinalysis (establish baseline nitrophenol levels).

If symptoms develop or overexposure is suspected, the following tests are recommended:

- ☑ Complete blood count (compare to baseline).
- ☑ Urinalysis (compare to baseline).
- ☑ Consider chest X-ray following acute overexposure (may be negative if taken immediately after exposure).
- ☑ Lung function tests.
- ☑ Liver and kidney function tests.

Any evaluation should include a careful history of past and present symptoms with an examination. Medical tests that simply look for existing damage are not a substitute for controlling exposures. Physician should monitor fluids and electrolytes, control hyperthermia, and test for dinitrophenol or its metabolite (aminophenol) in the urine using Derrian's Test. The biological half-life of 2,5-dinitrophenol is 5-14 days.

Other methods to reduce exposure include:

- ☑ Where possible, enclose operations and use local exhaust ventilation at the site of chemical release. If local exhaust ventilation or enclosure is not used, respiratory protection should be mandatory.
- ☑ Always ensure that proper protective clothing is worn when using chemical substances.
- ☑ Wash thoroughly immediately after exposure to 2,5-dinitrophenol and at the end of the work shift or before eating, drinking, or smoking.
- ☑ Hazard warning information should be posted in the work area. In addition, as part of an on-going

education and training program, all information on the health and safety hazards of 2,5-dinitrophenol should be communicated to all exposed workers.

☑ Eye wash stations should be provided in the immediate work area for emergency use. If there is a possibility of skin exposure to 2,5-dinitrophenol, emergency shower facilities should also be provided.

☑ Personnel should never wear contaminated clothing home (family members can be exposed). Clothing should be laundered by personnel who have been trained on the hazards of 2,5-dinitrophenol.

RISK ASSESSMENT: ENVIRONMENT

General Assessment

The environment is at risk of exposure during transportation, storage, disposal, or destruction of 2,5-dinitrophenol. In almost every scenario, the threat of environmental exposure is contingent upon the proper handling of the chemical substance. Accidental spills, large or small, can result in fire, explosion, and possible contamination of the surrounding environmental mediums (water, soil, and air).

2,5-Dinitrophenol is normally considered a flammable solid. In solution, it can often appear in mixtures with other combustible or flammable commodities. Its dusts may form explosive mixtures in air. These characteristics require caution in handling, storage, transportation, and disposal. It is incompatible with strong oxidizers and contact can cause fire or explosion. It forms explosive salts in contact with alkalies or with ammonia. Prolonged contact with heat may result in an explosion due to spontaneous decomposition. Therefore, special consideration is required during any emergency situation involving a leak or spill of 2,5-dinitrophenol. Should 2,5-dinitrophenol ever come into contact with these incompatible substances during use, transportation, storage, or disposal, the formation of highly toxic and/or highly explosive commodities is extremely possible.

2,5-Dinitrophenol may enter the environment through its intended use as an insecticide, as well as through industrial effluents or spills.

☠ *Acute Ecological Effects*

Acute (short-term) toxic effects may include the death of animals, birds, or fish and death or low growth rate in plants. Acute effects are seen 2 to 4 days after animals or plants are exposed to 2,5-dinitrophenol.

2,5-Dinitrophenol has moderate acute toxicity to aquatic life and mild acute toxicity to plant life. Insufficient data are available on the short-term effects of 2,5-dinitrophenol exposure to birds or land animals.

Chronic Ecological Effects

Chronic (long-term) toxic effects may include shortened life span, reproductive problems, lower fertility, and changes in appearance or behavior in exposed animals. These effects can be seen long after first exposure(s) to toxic chemicals.

2,5-Dinitrophenol has moderate chronic toxicity to aquatic life. Insufficient data are available on the long-term effects of 2,5-dinitrophenol to plants, birds, or land animals.

Water Solubility

2,5-Dinitrophenol is slightly soluble in cold water. Concentrations of 1 to 100 milligrams will mix with a liter of water.

⌛ *Persistence in the Environment*

2,5-Dinitrophenol is slightly persistent in water, with a half-life between 2 and 20 days. The half-life of a pollutant is the amount of time it takes for one half of the chemical to be degraded.

Bioaccumulation in Aquatic Organisms

Some substances increase in concentration, or *bioaccumulate,* in living organisms as they breathe contaminated air, drink contaminated water, or eat contaminated food. These chemicals can become concentrated in the tissues and internal organs of animals as well as humans.

The concentration of 2,5-dinitrophenol found in fish tissue is expected to be somewhat higher than the average concentration of 2,5-dinitrophenol in the water from which the fish was taken.

Recommended Risk-Reduction Measures

Proper training of all transporters will reduce the likelihood of a mishap or accident resulting in a leak or spill to the environment. The correct use of DOT labeling (per the MSDS) on all containers, trucks, and rail cars will enable emergency responders to react properly and

quickly to any disaster thereby reducing the potential risk to the environment and to personnel.

Storage of 2,5-dinitrophenol should be segregated from incompatible chemicals to minimize the risk of cross-contamination or contact. Containers must be protected from heat, damage, and shock. Buildings designated for storage should be equipped with appropriate fire protection systems (alarms, sprinklers, emergency lighting, portable extinguishers).

If a spill or leak to the environment has occurred, fire department, emergency response and/or hazardous materials spill personnel should be notified immediately. Cleanup should be attempted only by those trained in proper spill containment procedures. Contaminated soils should be removed for incineration and replaced with clean soil. If 2,5-dinitrophenol should contact the water table, aquifer, or navigable waterway, time is of the essence. It is slightly soluble in the aquatic environment and total remediation may not be possible. The local and/or state emergency response authorities must be notified. A comprehensive emergency response or disaster preparedness/recovery plan should be in place prior to any operations involving the use, transportation, storage, or disposal of 2,5-dinitrophenol. If 2,5-dinitrophenol is spilled or leaked, the following specific steps are recommended:

☑ Restrict persons not wearing protective clothing from area of spill or leak until cleanup is complete and area can be opened for normal work.

☑ Remove all ignition sources, ventilate area of leak.

☑ Absorb liquid spills in vermiculite, dry sand, earth, or similar material and deposit in sealed containers. Collect solid spills for reuse of disposal. Wet-mop spill area to collect residue.

☑ It may be necessary to dispose of 2,5-dinitrophenol as a hazardous waste. The responsible state agency or the regional office of the federal Environmental Protection Agency (EPA) should be contacted for specific recommendations.

RISK ASSESSMENT: BUSINESS

General Assessment

Accidents or mishaps involving 2,5-dinitrophenol can present a significant threat to business operations. The loss or damage of equipment or facilities can significantly affect fiscal viability. Lawsuits that may result from personnel injury/death, public exposures, and/or environmental contamination will also require a serious expenditure of resources. Media attention surrounding an injury, death, or environmental damage can also result in a loss of profits and loss of current as well as future business.

Recommended Risk-Reduction Measures

Company attorneys, safety and health professionals, and environmental specialists should be involved in the development of any procedures dealing with safety or emergency response. A company official should be pre-designated as a public relations officer with specific training in dealing with the press. Corporate plans and policies should be developed, approved, and implemented long before any need for such arises.

REFERENCES

Ahlbom, A. 1993. *Biostatistics for Epidemiologists.* New York: CRC Press/Lewis Publishers.

Anthony, C. P., and N. J. Kolthoff. 1971. *Textbook of Anatomy and Physiology.* St. Louis: C. V. Mosby Company.

Bloom, A. D. and F. J. DeSerres. 1995. *Ecotoxicity and Human Health.* New York: CRC Press/Lewis Publishers.

Bueche, J. R. 1972. *Principles of Physics,* 2nd Edition. New York: McGraw-Hill.

Burrell, R., D. K. Flaherty, and L. J. Sauers. 1992. *Toxicology of the Immune System.* New York: Van Nostrand Reinhold.

Calabrese, E. J. 1994. *Biological Effects of Low Level Exposures.* New York: CRC Press/Lewis Publishers.

Cockerman, L. G., B. S. Shane. 1994. *Basic Environmental Toxicology.* New York: CRC Press/Lewis Publishers.

Environmental Protection Agency. 1983. *Chemical Hazard Information Profiles.* Washington, D.C.: U.S. EPA.

Fodden, C. R., and J. L. Weddell. 1992. *Hazardous Materials Emergency Action Data.* New York: CRC Press/Lewis Publishers.

Hallenbeck, W. H. 1993. *Quantitative Risk Assessment for Environmental and Occupational Health,* 2nd Edition. New York: CRC Press/Lewis Publishers.

Howard, P. H. 1995. *Handbook of Environmental Fate and Exposures Data for Organic Chemicals,* Vol. I-IV. New York: CRC Press/Lewis Publishers.

Lewis, R. J., Sr. 1992. *Sax's Dangerous Properties of Industrial Materials,* Volumes 1, 2, and 3, 8th Edition. New York: Van Nostrand Reinhold.

Meyer, E. 1990. *Chemistry of Hazardous Materials.* Englewood Cliffs, NJ: Prentice Hall.

Quigley, D. R. 1994. *Handbook of Emergency Chemical Management.* New York: CRC Press/Lewis Publishers.

MATERIAL SAFETY DATA SHEET

CHEMICAL NAME	
3,4-DINITROPHENOL	**CRC PRESS/LEWIS PUBLISHERS DISCLAIMER:** THE INFORMATION AND RECOMMENDATIONS PRESENTED HEREIN ARE BASED ON SOURCES BELIEVED TO BE RELIABLE. CRC MAKES NO REPRESENTATION ON ITS COMPLETENESS OR ACCURACY. IT IS THE RESPONSIBILITY OF THE USER TO DETERMINE THE CHEMICAL'S SUITABILITY FOR ITS INTENDED USE, THE CHEMICAL'S SAFE USE, AND THE CHEMICAL'S PROPER DISPOSAL. NO REPRESENTATIONS AND/OR WARRANTIES, EITHER EXPRESSED OR IMPLIED, OF THE MERCHANTABILITY OR FITNESS FOR A PARTICULAR PURPOSE, OR OF ANY OTHER NATURE, ARE MADE WITH RESPECT TO THE INFORMATION PROVIDED IN THIS MSDS OR TO THE CHEMICAL TO WHICH INFORMATION MAY REFER. CRC NEITHER ASSUMES NOR AUTHORIZES ANY OTHER PERSON TO ASSUME FOR IT, ANY OTHER ADDITIONAL RESPONSIBILITY OR LIABILITY FOR THE USE OF, OR RELIANCE UPON, THIS INFORMATION.

HAZARD WARNING INFORMATION

HEALTH	FIRE	REACTIVE	OTHER	DEGREE OF HAZARD	COLOR CODING	OTHER CODES
DRY: 3 WET: 3	DRY: 3 WET: 1	DRY: 3 WET: 0		0 = Minimum Hazard 1 = Slight Hazard 2 = Moderate Hazard 3 = Serious Hazard 4 = Severe Hazard	HEALTH = BLUE FIRE = RED REACTIVITY = YELLOW OTHER = WHITE	OX = Oxidizer ACID = Acid ALK = Alkali COR = Corrosive W = Use No Water

SECTION I - GENERAL INFORMATION

Characterization **Phenol**	RCRA Number **None**	EPA Class **Not Applicable**
DOT Proper Shipping Name **Dinitrophenol (solution)**	Chemical Abstract Service (CAS) Number **577-71-9**	
DOT Hazard Class and Label Requirements **Poison B (solution), Flammable Solid**	DOT Emergency Guide Code **53**	
DOT Identification Number **UN 1320, 1321 (solid); UN 1599 (solution)**	Chemical Formula $C_6H_4N_2O_5$	

Synonyms
3,4-DNP.

SECTION II - HAZARDOUS INGREDIENTS AND IDENTITY INFORMATION

Hazardous Components (specific identity)	OSHA Exposure Criteria	NIOSH Exposure Criteria	Immediately Dangerous to Life and Health (IDLH)	ACGIH Exposure Criteria
3,4-dinitrophenol (derivation: By heating phenol with dilute sulfuric acid, cooling the product, and then nitrating while keeping the temperature approximately 50°C; or by nitrating with mixed acid under careful temperature control).	**PEL (skin):** 0.2 mg/m³ **STEL:** Not Established	**REL:** Not Established **STEL:** Not Established	**Not Determined**	**TLV:** Not Established **STEL:** Not Established

SECTION III - PHYSICAL AND CHEMICAL CHARACTERISTICS

Boiling Point **Sublimes (when heated)**	Specific Gravity (H_2O = 1) **1.68**
Vapor Pressure (gas) **5-10 at 69°F (20°C)**	Molecular Weight **184.1**
Vapor Density (Air = 1) **6.3**	Melting Point **234°F (112°C)**

Solubility
Slightly soluble in cold water; freely soluble in hot water. Soluble in alcohol, ether, acetone, pyridine, benzene, chloroform, and most organic solvents. Soluble in aqueous alkaline solutions.

Appearance and Odor
Yellow, odorless solid crystals (powder).

SECTION IV - FIRE AND EXPLOSION HAZARD DATA

Flash Point (method used) **Not Applicable**	Explosive Limits in Air % by Volume **LEL: Not Determined UEL: Not Determined**
NFPA Classification **Flammable Solid**	Autoignition Temperature **Not Determined**

Extinguishing Media
Use carbon dioxide, water spray, dry chemical, foam, or fog.

Special Fire Fighting Procedures
Wear full protective clothing and self-contained breathing apparatus (SCBA). Poisonous gases are produced in fire. Use water spray to keep fire-exposed containers cool. Move containers from fire if it can be done without risk. Fight fire from a distance or protected position, if possible.

Unusual Fire and Explosion Hazards
Containers may explode in fire. Fires involving this material should be considered an explosion hazard. Extremely explosive in dry form when exposed to heat, flame, or shock.

SECTION V - REACTIVITY DATA

Stability		Conditions to Avoid 3,4-Dinitrophenol is normally stable. It will remain stable in mixtures up to 15% by weight with water. It becomes explosive when dry and subjected to shock, heat, or flame.
Stable X	Unstable	Incompatibility *(materials to avoid)* Strong oxidizers, (such as chlorine, bromine, and fluorine), and steam. It will form explosive salts with alkalis and/or ammonia.
Hazardous Polymerization		Conditions to Avoid Under normal conditions of temperature and pressure, hazardous polymerization of 3,4-dinitrophenol is not expected to occur.
May Occur	Will Not Occur X	Hazardous Decomposition or By-products When heated to decomposition, 3,4-dinitrophenol can emit highly toxic/poisonous gases, including carbon dioxide and oxides of nitrogen.

SECTION VI - HEALTH HAZARD DATA

Primary Route(s) of Entry:	Inhalation?	Absorption (skin)?	Ingestion?
	X	X	X

Health Hazards

INHALATION: Can cause headache, coughing, sweating, thirst, intense fatigue, high fever, rapid pulse, warm and flushed skin, rapid and difficult breathing, anxiety, confusion, convulsions, loss of consciousness, pulmonary edema (fluid in lungs), and death.

ABSORPTION: Irritation with yellow staining, redness, swelling, blisters, and delayed symptoms like those of inhalation (will pass through skin to cause same toxic systemic effects).

INGESTION: Burning of mouth and throat, salivation, dizziness, nausea, vomiting, bright yellow stool, and symptoms of inhalation.

Carcinogenicity	NTP Listed?	IARC Cancer Review Group?	OSHA Regulated?	Target Organs?
Unknown Human Unknown Animal	No	No	No	Skin, respiratory system, CNS, CVS, liver, kidney

Medical Conditions Generally Aggravated by Exposure

None reported. However, existing thyroid, liver, kidney, and cardiovascular diseases may be aggravated.

Emergency and First-aid Procedures

Eye contact: Flush immediately with water for 15 minutes (minimum), seek medical attention. Skin contact: Remove all contaminated clothing. Immediately wash area with large amounts of soap and water. Seek medical assistance. Yellow discoloration need not be totally removed to prevent absorption. For inhalation: Remove the person from exposure. Provide CPR if required. Transfer to medical facility. Observe for 24-48 hours for lung effects. If swallowed: Unless advised otherwise, give 1-2 glasses of water and induce vomiting, then give 2 tablespoons of activated charcoal in 8 ounces of water.

SECTION VII - PRECAUTIONS FOR SAFE HANDLING AND USE

Steps to be Taken in Case Material is Released or Spilled

Minimize dust generation. Do NOT dry sweep. Use a vacuum equipped with a high efficiency particulate air (HEPA) filter instead. Collect and dispose of in sealed drum. If in run-off, dike area to contain, cover with absorbent (vermiculite or other) and dispose of in sealed drum.

Preferred Waste Disposal Method

No Citation.

Precautions to be Taken in Handling and Storage

Store in tightly closed containers in cool, well-ventilated area away from heat or flame. Heat may cause containers to build up pressure and explode. Store in detached facility away from main plant or facility.

Other Precautions and Warnings

Sources of ignition are prohibited where 3,4-dinitrophenol is used, handled, or stored.

SECTION VIII - CONTROL MEASURES AND PERSONAL PROTECTIVE EQUIPMENT

Respiratory Protection *(specify type)*

PEL is low and difficult to monitor. For any exposure level, recommend a MSHA/NIOSH-approved supplied air respirator with full facepiece, hood, or helmet in continuous flow mode, or use a self-contained breathing apparatus (SCBA) operated in pressure demand or other positive pressure mode.

Ventilation

Local exhaust or general mechanical systems recommended.

Protective Gloves	Eye Protection	Other Protective Clothing
Butyl Rubber	Chemical/Dust Goggles or Face Mask	Butyl Rubber Apron

Work/Hygiene Practices

Always wash hands thoroughly after using chemical; never bring food, drink, or smoking materials into vicinity of chemicals.

3,4-DINITROPHENOL

$C_6H_4N_2O_5$ CAS: 577-71-9

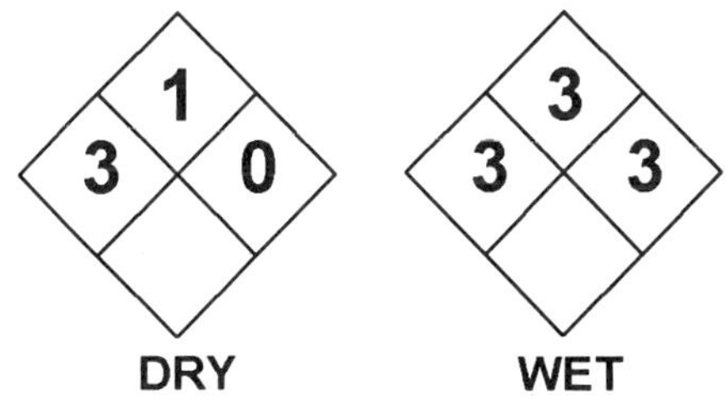

IDENTIFICATION AND TYPICAL USES

3,4-Dinitrophenol is a yellow, odorless solid that appears as a powder. It is used in the manufacture of dyes, especially sulfur colors, in picric acid picramic acid, in the preservation of lumber, in the manufacture of the photographic developer diaminophenol hydrochloride, in explosives manufacture, as an indicator, and a reagent for potassium and ammonium ions.

RISK ASSESSMENT: HEALTH

General Assessment

3,4-Dinitrophenol is a deadly human poison by unspecified routes of exposure. Therefore, all routes of exposure (*ingestion, inhalation,* and skin *absorption*) are suspect in this regard. While the 3,4-isomer of dinitrophenol has not been specifically tested for its ability to adversely affect reproduction, the very closely related compound 2,4-dinitrophenol, has been shown to cause teratogenic effects in experimental tests and mutation data have also been reported. Its carcinogenic properties are unknown. Liver or kidney damage can occur up to 3 days following exposure.

Inhalation can result in headache, coughing, profuse sweating, marked thirst, intense fatigue, high fever, rapid pulse or heart rate (tachycardia), warm and flushed skin, difficulty in breathing (tachypnea), anxiety and confusion, convulsions, loss of consciousness, pulmonary edema (fluid in the lungs), and death due to circulatory or respiratory collapse. Exposure may also lead to a deadly buildup of fluids in the lungs in a condition known as pulmonary edema. This is a medical emergency requiring immediate attention. Symptoms can be delayed up to 48 hours, creating a false sense of security with regard to health exposure risk.

Skin contact will result in nearly immediate yellow staining, irritation, swelling, redness, blisters, and delayed systemic effects from absorption with symptoms identical to those of inhalation. Eye contact causes burning pain, tearing, and inflammation of the eyelids.

Ingestion can cause painful burning of the mouth and throat, salivation, dizziness, nausea and vomiting, the appearance of bright yellow stools, and the symptoms of systemic poisoning similar to those of inhalation.

☠ *Acute Health Effects*

The following acute (short-term) health effects may occur immediately or shortly after exposure to 3,4-dinitrophenol:

Skin: Irritation with yellowing of skin and possible tissue damage. Will pass through unbroken skin to cause dangerous systemic effects.

Eye: Irritation with damage to the surrounding tissue. If not removed immediately, damage may be more severe and permanent.

Lung: Severe irritation to the respiratory tract and associated mucosa. Can cause coughing, congestion, and possibly pulmonary edema (fluid in the lungs), which can be fatal.

CNS: Can cause confusion, anxiety, headache, changes in the function of the respiratory and circulatory systems, convulsions, coma.

Other: Can cause a disruption in cellular metabolism in all tissues resulting in high fever, elevated blood pressure, pulse, and respiration. Can also cause death by hyperthermia. Effects are more severe in hot work environments.

Chronic Health Effects

The following chronic (long-term) health effects may occur at some time after exposure to 3,4-dinitrophenol and can last for months or even years:

Cancer Hazards: According to the information available in the references, 3,4-dinitrophenol has not been adequately tested for its ability to cause cancer in test animals. A closely related compound, 2,4-dinitrophenol, has been shown to produce mutagenic effects in test animals. Data on human effects are inconclusive.

Reproductive Hazard: According to information presented in the references, 3,4-dinitrophenol not been adequately tested for its ability to produce reproductive effects in test animals. The 2,4-isomer of dinitrophenol, which is closely related to the 3,4-isomer, has been shown to adversely affect reproduction in animals. It is considered an experimental teratogen.

Other Chronic Effects: Long-term exposures can result in damage to the liver or kidneys. Hypersensitivity may also result from repeated exposures.

Recommended Risk-Reduction Measures

Personnel should avoid direct contact with 3,4-dinitrophenol. If a less toxic chemical cannot be substituted for a hazardous substance, then *engineering controls* are the most effective method of reducing exposures. The best protection is to enclose operations and/or provide local exhaust ventilation at the site of chemical release. While not always operationally feasible, isolating operations can also reduce exposure. Using respiratory protection is less effective than the controls mentioned above, but is still advisable whenever working with or around 3,4-dinitrophenol. The exposure level is low (0.2 mg/m^3) and difficult to monitor. Therefore, a MSHA/NIOSH approved full facepiece respirator equipped with an organic vapor cartridge equipped with a dust pre-filter should be used if any exposure is anticipated or known. Increased protection is obtained from a supplied-air respirator with full facepiece operated in continuous flow mode, or a self-contained breathing apparatus (SCBA) with full facepiece and pressure demand. If a full facepiece is not available, then chemical dust goggles should be worn to protect the eyes. Whenever an airborne chemical dust hazard exists, a face shield and a rubber apron should be worn. To prevent hand and skin exposures, butyl rubber gloves should be worn.

Administrative controls should also be in place to minimize the potential for human exposures. These may include written procedures or policies which specify the methods and techniques that will be practiced whenever personnel are to work with 3,4-dinitrophenol.

All personnel should receive *training* on the use, hazards, protective measures, emergency actions, and other precautions per 29 CFR 1910.1200 (Hazard Communication), prior to the first assignment in an area where 3,4-dinitrophenol is used or stored.

Before beginning employment with 3,4-dinitrophenol and at regular intervals thereafter (e.g., annually), those persons with high or frequent exposures should be provided the following recommended medical tests:

- ☑ Complete blood count (establish baseline).
- ☑ Urinalysis (establish baseline nitrophenol levels).

If symptoms develop or overexposure is suspected, the following tests are recommended:

- ☑ Complete blood count (compare to baseline).
- ☑ Urinalysis (compare to baseline).
- ☑ Consider chest X-ray following acute overexposure (may be negative if taken immediately after exposure).
- ☑ Lung function tests.
- ☑ Liver and kidney function tests.

Any evaluation should include a careful history of past and present symptoms with an examination. Medical tests that simply look for existing damage are not a substitute for controlling exposures. Physician should monitor fluids and electrolytes, control hyperthermia, and test for dinitrophenol or its metabolite (aminophenol) in the urine using Derrian's Test. The biological half-life of 3,4-dinitrophenol is 5-14 days.

Other methods to reduce exposure include:

- ☑ Where possible, enclose operations and use local exhaust ventilation at the site of chemical release. If local exhaust ventilation or enclosure is not used, respiratory protection should be mandatory.
- ☑ Always ensure that proper protective clothing is worn when using chemical substances.
- ☑ Wash thoroughly immediately after exposure to 3,4-dinitrophenol and at the end of the work shift or before eating, drinking, or smoking.
- ☑ Hazard warning information should be posted in the work area. In addition, as part of an on-going education and training program, all information on the health and safety hazards of 3,4-dinitrophenol

should be communicated to all potentially exposed workers.

☑ Eye wash stations should be provided in the immediate work area for emergency use. If there is a possibility of skin exposure to 3,4-dinitrophenol, emergency shower facilities should also be provided.

☑ Personnel should never wear contaminated clothing home (family members can be exposed). Clothing should be laundered by personnel who have been trained on the hazards of 3,4-dinitrophenol.

RISK ASSESSMENT: ENVIRONMENT

General Assessment

The environment is at risk of exposure during transportation, storage, disposal, or destruction of 3,4-dinitrophenol. In almost every scenario, the threat of environmental exposure is contingent upon the proper handling of the chemical substance. Accidental spills, large or small, can result in fire, explosion, and possible contamination of the surrounding environmental mediums (water, soil, and air).

3,4-Dinitrophenol is normally considered a flammable solid. In solution, it can often appear in mixtures with other combustible or flammable commodities. Its dusts may form explosive mixtures in air. These characteristics require caution in handling, storage, transportation, and disposal. It is incompatible with strong oxidizers and contact can cause fire or explosion. It forms explosive salts in contact with alkalis or with ammonia. Prolonged contact with heat may result in an explosion due to spontaneous decomposition. Therefore, special consideration is required during any emergency situation involving a leak or spill of 3,4-dinitrophenol. Should 3,4-dinitrophenol ever come into contact with these incompatible substances during use, transportation, storage, or disposal, the formation of highly toxic and/or highly explosive commodities is extremely possible.

3,4-Dinitrophenol may enter the environment through its intended use as an insecticide, as well as through industrial effluents or spills.

Acute Ecological Effects

Acute (short-term) toxic effects may include the death of animals, birds, or fish and death or low growth rate in plants. Acute effects are seen 2 to 4 days after animals or plants are exposed to 3,4-dinitrophenol. Insufficient data are available on the short-term effects of 3,4-dinitrophenol exposure to aquatic life, plants, birds or land animals.

Chronic Ecological Effects

Chronic (long-term) toxic effects may include shortened life span, reproductive problems, lower fertility, and changes in appearance or behavior in exposed animals. These effects can be seen long after first exposure(s) to toxic chemicals. Insufficient data are available on the long-term effects of 3,4-dinitrophenol to aquatic life, plants, birds, or land animals.

Water Solubility

3,4-Dinitrophenol is slightly soluble in cold water. Concentrations of 1 to 100 milligrams will mix with a liter of water.

Persistence in the Environment

3,4-Dinitrophenol is slightly persistent in water, with an ecological half-life between 2 and 20 days. The half-life of a pollutant is the amount of time it takes for one half of the chemical to be degraded.

Bioaccumulation in Aquatic Organisms

Some substances increase in concentration, or *bioaccumulate,* in living organisms as they breathe contaminated air, drink contaminated water, or eat contaminated food. These chemicals can become concentrated in the tissues and internal organs of animals as well as humans.

It is not known if 3,4-dinitrophenol will accumulate in the edible tissues of fish.

Recommended Risk-Reduction Measures

Proper training of all transporters will reduce the likelihood of a mishap or accident resulting in a leak or spill to the environment. The correct use of DOT labeling (per the MSDS) on all containers, trucks, and rail cars will enable emergency responders to react properly and quickly to any disaster thereby reducing the potential risk to the environment and to personnel.

Storage of 3,4-dinitrophenol should be segregated from incompatible chemicals to minimize the risk of cross-contamination or contact. Containers must be protected from heat, damage, and shock. Buildings designated for storage should be equipped with appropriate fire protection systems (alarms, sprinklers, emergency lighting, portable extinguishers).

If a spill or leak to the environment has occurred, fire department, emergency response and/or hazardous materials spill personnel should be notified immediately. Cleanup should be attempted only by those trained in proper spill containment procedures. Contaminated soils should be removed for incineration and replaced with clean soil. If 3,4-dinitrophenol should contact the water table, aquifer, or navigable waterway, time is of the essence. It is slightly soluble in the aquatic environment and total remediation may not be possible. The local and/or state emergency response authorities must be notified. A comprehensive emergency response or disaster preparedness/recovery plan should be in place prior to any operations involving the use, transportation, storage, or disposal of 3,4-dinitrophenol. If 3,4-dinitrophenol is spilled or leaked, the following specific steps are recommended:

☑ Restrict persons not wearing protective clothing from area of spill or leak until cleanup is complete and area can be opened for normal work.

☑ Remove all ignition sources, ventilate area of leak.

☑ Absorb liquid spills in vermiculite, dry sand, earth, or similar material and deposit in sealed containers. Collect solid spills for reuse of disposal. Wet-mop spill area to collect residue.

☑ It may be necessary to dispose of 3,4-dinitrophenol as a hazardous waste. The responsible state agency or the regional office of the federal Environmental Protection Agency (EPA) should be contacted for specific recommendations.

RISK ASSESSMENT: BUSINESS

General Assessment

Accidents or mishaps involving 3,4-dinitrophenol can present a significant threat to business operations. The loss or damage of equipment or facilities can significantly affect fiscal viability. Lawsuits that may result from personnel injury/death, public exposures, and/or environmental contamination will also require a serious expenditure of resources. Media attention surrounding an injury, death, or environmental damage can also result in a loss of profits and loss of current as well as future business.

Recommended Risk-Reduction Measures

Company attorneys, safety and health professionals, and environmental specialists should be involved in the development of any procedures dealing with safety and emergency response. A company official should be pre-designated as a public relations officer with specific training in dealing with the press. Corporate plans and policies should be developed, approved, and implemented long before any need for such arises.

REFERENCES

Ahlbom, A. 1993. *Biostatistics for Epidemiologists.* New York: CRC Press/Lewis Publishers.

Anthony, C. P., and N. J. Kolthoff. 1971. *Textbook of Anatomy and Physiology.* St. Louis: C. V. Mosby Company.

Bloom, A. D. and F. J. DeSerres. 1995. *Ecotoxicity and Human Health.* New York: CRC Press/Lewis Publishers.

Bueche, J. R. 1972. *Principles of Physics,* 2nd Edition. New York: McGraw-Hill.

Burrell, R., D. K. Flaherty, and L. J. Sauers. 1992. *Toxicology of the Immune System.* New York: Van Nostrand Reinhold.

Calabrese, E. J. 1994. *Biological Effects of Low Level Exposures.* New York: CRC Press/Lewis Publishers.

Cockerman, L. G., B. S. Shane. 1994. *Basic Environmental Toxicology.* New York: CRC Press/Lewis Publishers.

Environmental Protection Agency. 1983. *Chemical Hazard Information Profiles.* Washington, D.C.: U.S. EPA.

Hallenbeck, W. H. 1993. *Quantitative Risk Assessment for Environmental and Occupational Health,* 2nd Edition. New York: CRC Press/Lewis Publishers.

Howard, P. H. 1995. *Handbook of Environmental Fate and Exposures Data for Organic Chemicals,* Vol. I-IV. New York: CRC Press/Lewis Publishers.

Lewis, R. J., Sr. 1992. *Sax's Dangerous Properties of Industrial Materials,* Volumes 1, 2, and 3, 8th Edition. New York: Van Nostrand Reinhold.

Meyer, E. 1990. *Chemistry of Hazardous Materials.* Englewood Cliffs, NJ: Prentice Hall.

Quigley, D. R. 1994. *Handbook of Emergency Chemical Management.* New York: CRC Press/Lewis Publishers.

U.S. Department of Health and Human Services, National Institute for Occupational Safety and Health. 1994. *NIOSH Pocket Guide to Chemical Hazards.* Washington, D.C.: U.S. Government Printing Office.

MATERIAL SAFETY DATA SHEET

CHEMICAL NAME	
3,5-DINITROPHENOL	**CRC PRESS/LEWIS PUBLISHERS DISCLAIMER:** THE INFORMATION AND RECOMMENDATIONS PRESENTED HEREIN ARE BASED ON SOURCES BELIEVED TO BE RELIABLE. CRC MAKES NO REPRESENTATION ON ITS COMPLETENESS OR ACCURACY. IT IS THE RESPONSIBILITY OF THE USER TO DETERMINE THE CHEMICAL'S SUITABILITY FOR ITS INTENDED USE, THE CHEMICAL'S SAFE USE, AND THE CHEMICAL'S PROPER DISPOSAL. NO REPRESENTATIONS AND/OR WARRANTIES, EITHER EXPRESSED OR IMPLIED, OF THE MERCHANTABILITY OR FITNESS FOR A PARTICULAR PURPOSE, OR OF ANY OTHER NATURE, ARE MADE WITH RESPECT TO THE INFORMATION PROVIDED IN THIS MSDS OR TO THE CHEMICAL TO WHICH INFORMATION MAY REFER. CRC NEITHER ASSUMES NOR AUTHORIZES ANY OTHER PERSON TO ASSUME FOR IT, ANY OTHER ADDITIONAL RESPONSIBILITY OR LIABILITY FOR THE USE OF, OR RELIANCE UPON, THIS INFORMATION.

HAZARD WARNING INFORMATION

HEALTH	FIRE	REACTIVE	OTHER	DEGREE OF HAZARD	COLOR CODING	OTHER CODES
DRY: 3 WET: 3	DRY: 3 WET: 1	DRY: 3 WET: 0		0 = Minimum Hazard 1 = Slight Hazard 2 = Moderate Hazard 3 = Serious Hazard 4 = Severe Hazard	HEALTH = BLUE FIRE = RED REACTIVITY = YELLOW OTHER = WHITE	OX = Oxidizer ACID = Acid ALK = Alkali COR = Corrosive W = Use No Water

SECTION I - GENERAL INFORMATION

Characterization	**Phenol**	RCRA Number	**None**
		EPA Class	**Not Applicable**
DOT Proper Shipping Name	**Dinitrophenol (solution)**	Chemical Abstract Service (CAS) Number	**586-11-8**
DOT Hazard Class and Label Requirements	**Poison B (solution), Flammable Solid**	DOT Emergency Guide Code	**53**
DOT Identification Number	**UN 1320, 1321 (solid); UN 1599 (solution)**	Chemical Formula	**$C_6H_4N_2O_5$**
Synonyms	**3,5-DNP.**		

SECTION II - HAZARDOUS INGREDIENTS AND IDENTITY INFORMATION

Hazardous Components (specific identity)	OSHA Exposure Criteria	NIOSH Exposure Criteria	Immediately Dangerous to Life and Health (IDLH)	ACGIH Exposure Criteria
3,5-dinitrophenol (derivation: By heating phenol with dilute sulfuric acid, cooling the product, and then nitrating while keeping the temperature approximately 50°C; or by nitrating with mixed acid under careful temperature control).	**PEL (skin):** 0.2 mg/m^3 **STEL:** Not Established	**REL:** Not Established **STEL:** Not Established	**Not Determined**	**TLV:** Not Established **STEL:** Not Established

SECTION III - PHYSICAL AND CHEMICAL CHARACTERISTICS

Boiling Point	**Sublimes (when heated)**	Specific Gravity (H_2O = 1)	**1.68**
Vapor Pressure (gas)	**5-10 at 69°F (20°C)**	Molecular Weight	**184.1**
Vapor Density (Air = 1)	**6.3**	Melting Point	**234°F (112°C)**

Solubility

Slightly soluble in cold water; freely soluble in hot water. Soluble in alcohol, ether, acetone, pyridine, benzene, chloroform, and most organic solvents. Soluble in aqueous alkaline solutions.

Appearance and Odor

Yellow, odorless solid crystals (powder).

SECTION IV - FIRE AND EXPLOSION HAZARD DATA

Flash Point (method used)	**Not Applicable**	Explosive Limits in Air % by Volume	**LEL: Not Determined UEL: Not Determined**
NFPA Classification	**Flammable Solid**	Autoignition Temperature	**Not Determined**

Extinguishing Media

Use carbon dioxide, water spray, dry chemical, foam, or fog.

Special Fire Fighting Procedures

Wear full protective clothing and self-contained breathing apparatus (SCBA). Poisonous gases are produced in fire. Use water spray to keep fire-exposed containers cool. Move containers from fire if it can be done without risk. Fight fire from a distance or protected position, if possible.

Unusual Fire and Explosion Hazards

Containers may explode in fire. Fires involving this material should be considered an explosion hazard. Extremely explosive in dry form when exposed to heat, flame, or shock.

SECTION V - REACTIVITY DATA

Stability		Conditions to Avoid 3,5-Dinitrophenol is normally stable. It will remain stable in mixtures up to 15% by weight with water. It becomes explosive when dry and subjected to shock, heat, or flame.
Stable X	Unstable	Incompatibility *(materials to avoid)* Strong oxidizers, (such as chlorine, bromine, and fluorine), and steam. It will form explosive salts with alkalis and/or ammonia.
Hazardous Polymerization		Conditions to Avoid Under normal conditions of temperature and pressure, hazardous polymerization of 3,5-dinitrophenol is not expected to occur.
May Occur	Will Not Occur X	Hazardous Decomposition or By-products When heated to decomposition, 3,5-dinitrophenol can emit highly toxic/poisonous gases, including carbon dioxide and oxides of nitrogen.

SECTION VI - HEALTH HAZARD DATA

Primary Route(s) of Entry:	Inhalation?	Absorption (skin)?	Ingestion?
	X	X	X

Health Hazards

INHALATION: Can cause headache, coughing, sweating, thirst, intense fatigue, high fever, rapid pulse, warm and flushed skin, rapid and difficult breathing, anxiety, confusion, convulsions, loss of consciousness, pulmonary edema (fluid in lungs), and death.

ABSORPTION: Irritation with yellow staining, redness, swelling, blisters, and delayed symptoms like those of inhalation (will pass through skin to cause same toxic systemic effects).

INGESTION: Burning of mouth and throat, salivation, dizziness, nausea, vomiting, bright yellow stool, and symptoms of inhalation.

Carcinogenicity	NTP Listed?	IARC Cancer Review Group?	OSHA Regulated?	Target Organs?
Unknown Human Unknown Animal	No	No	No	Skin, respiratory system, CNS, CVS, liver, kidney

Medical Conditions Generally Aggravated by Exposure

None reported. However, existing thyroid, liver, kidney, and cardiovascular diseases may be aggravated.

Emergency and First-aid Procedures

Eye contact: Flush immediately with water for 15 minutes (minimum), seek medical attention. Skin contact: Remove all contaminated clothing. Immediately wash area with large amounts of soap and water. Seek medical assistance. Yellow discoloration need not be totally removed to prevent absorption. For inhalation: Remove the person from exposure. Provide CPR if required. Transfer to medical facility. Observe for 24-48 hours for lung effects. If swallowed: Unless advised otherwise, give 1-2 glasses of water and induce vomiting, then give 2 tablespoons of activated charcoal in 8 ounces of water.

SECTION VII - PRECAUTIONS FOR SAFE HANDLING AND USE

Steps to be Taken in Case Material is Released or Spilled

Minimize dust generation. Do NOT dry sweep. Use a vacuum equipped with a high efficiency particulate air (HEPA) filter instead. Collect and dispose of in sealed drum. If in run-off, dike area to contain, cover with absorbent (vermiculite or other) and dispose of in sealed drum.

Preferred Waste Disposal Method

No Citation.

Precautions to be Taken in Handling and Storage

Store in tightly closed containers in cool, well-ventilated area away from heat or flame. Heat may cause containers to build up pressure and explode. Store in detached facility away from main plant or facility.

Other Precautions and Warnings

Sources of ignition are prohibited where 3,5-dinitrophenol is used, handled, or stored.

SECTION VIII - CONTROL MEASURES AND PERSONAL PROTECTIVE EQUIPMENT

Respiratory Protection *(specify type)*

PEL is low and difficult to monitor. For any exposure level, recommend a MSHA/NIOSH-approved supplied air respirator with full facepiece, hood, or helmet in continuous flow mode, or use a self-contained breathing apparatus (SCBA) operated in pressure demand or other positive pressure mode.

Ventilation

Local exhaust or general mechanical systems recommended.

Protective Gloves	Eye Protection	Other Protective Clothing
Butyl Rubber	Chemical/Dust Goggles or Face Mask	Butyl Rubber Apron

Work/Hygiene Practices

Always wash hands thoroughly after using chemical; never bring food, drink, or smoking materials into vicinity of chemicals.

3,5-DINITROPHENOL

$C_6H_4N_2O_5$ CAS: 586-11-8

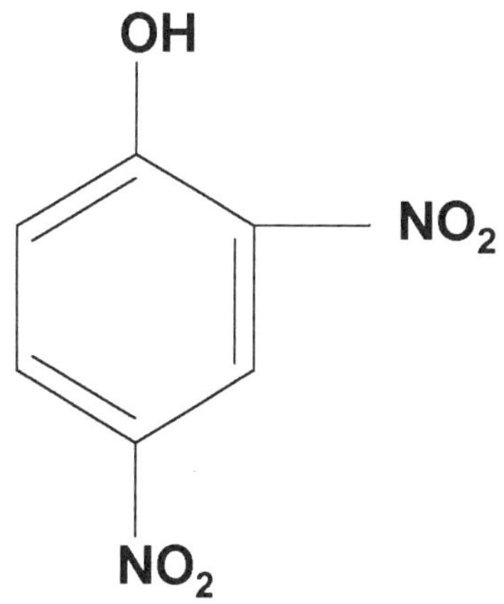

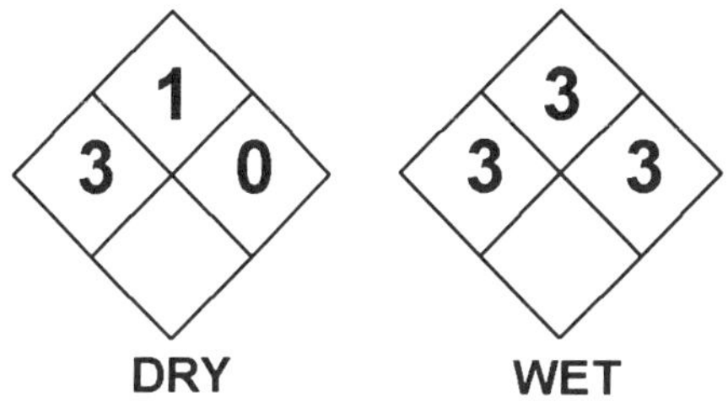

IDENTIFICATION AND TYPICAL USES

3,5-Dinitrophenol is a yellow, odorless solid that appears as a powder. It is used in the manufacture of dyes, especially sulfur colors, in picric acid picramic acid, in the preservation of lumber, in the manufacture of the photographic developer diaminophenol hydrochloride, in explosives manufacture, as an indicator, and a reagent for potassium and ammonium ions.

RISK ASSESSMENT: HEALTH

General Assessment

3,5-Dinitrophenol is a deadly human poison by unspecified routes of exposure. Therefore, all routes of exposure (*ingestion, inhalation,* and skin *absorption*) are suspect in this regard. While the 3,5-isomer of dinitrophenol has not been specifically tested for its ability to adversely affect reproduction, the very closely related compound 2,4-dinitrophenol, has been shown to cause teratogenic effects in experimental tests and mutation data have also been reported. Its carcinogenic properties are unknown. Liver or kidney damage can occur up to 3 days following exposure.

Inhalation can result in headache, coughing, profuse sweating, marked thirst, intense fatigue, high fever, rapid pulse or heart rate (tachycardia), warm and flushed skin, difficulty in breathing (tachypnea), anxiety and confusion, convulsions, loss of consciousness, pulmonary edema (fluid in the lungs), and death due to circulatory or respiratory collapse. Exposure may also lead to a deadly buildup of fluids in the lungs in a condition known as pulmonary edema. This is a medical emergency requiring immediate attention. Symptoms can be delayed up to 48-hours, creating a false sense of security with regard to health exposure risk.

Skin contact will result in nearly immediate yellow staining, irritation, swelling, redness, blisters, and delayed systemic effects from absorption with symptoms nearly identical to those of inhalation. Eye contact causes burning pain, tearing, and inflammation of the eyelids.

Ingestion can cause painful burning of the mouth and throat, salivation, dizziness, nausea and vomiting, the appearance of bright yellow stools, and the symptoms of systemic poisoning similar to those of inhalation.

☠ *Acute Health Effects*

The following acute (short-term) health effects may occur immediately or shortly after exposure to 3,5-dinitrophenol:

<u>Skin</u>: Irritation with yellowing of skin and possible tissue damage. Will pass through unbroken skin to cause dangerous systemic effects.

<u>Eye</u>: Irritation with damage to the surrounding tissue. If not removed immediately, damage may be more severe and permanent.

<u>Lung</u>: Severe irritation to the respiratory tract and associated mucosa. Can cause coughing, congestion, and possibly pulmonary edema (fluid in the lungs), which can be fatal.

CNS: Can cause confusion, anxiety, headache, changes in the function of the respiratory and circulatory systems, convulsions, coma.

Other: Can cause a disruption in cellular metabolism in all tissues resulting in high fever, elevated blood pressure, pulse, and respiration. Can also cause death by hyperthermia. Effects are more severe in hot work environments.

Chronic Health Effects

The following chronic (long-term) health effects may occur at some time after exposure to 3,5-dinitrophenol and can last for months or even years:

Cancer Hazards: According to the information available in the references, 3,5-dinitrophenol has not been adequately tested for its ability to cause cancer in test animals. A closely related compound, 2,4-dinitrophenol, has been shown to produce mutagenic effects in test animals. Data on human effects are inconclusive.

Reproductive Hazard: According to information presented in the references, 3,5-dinitrophenol not been adequately tested for its ability to produce reproductive effects in test animals. The 2,4-isomer of dinitrophenol, which is closely related to the 3,5-isomer, has been shown to adversely affect reproduction in animals. It is considered an experimental teratogen.

Other Chronic Effects: Long-term exposures can result in damage to the liver or kidneys. Hypersensitivity may also result from repeated exposures.

Recommended Risk-Reduction Measures

Personnel should avoid direct contact with 3,5-dinitrophenol. If a less toxic chemical cannot be substituted for a hazardous substance, then *engineering controls* are the most effective method of reducing exposures. The best protection is to enclose operations and/or provide local exhaust ventilation at the site of chemical release. While not always operationally feasible, isolating operations can also reduce exposure. Using respiratory protection is less effective than the controls mentioned above, but is still advisable whenever working with or around 3,5-dinitrophenol. The exposure level is low (0.2 mg/m^3) and difficult to monitor. Therefore, a MSHA/NIOSH approved full facepiece respirator equipped with an organic vapor cartridge equipped with a dust pre-filter should be used if any exposure is anticipated or known. Increased protection is obtained from a supplied-air respirator with full facepiece operated in continuous flow mode, or a self-contained breathing apparatus (SCBA) with full facepiece and pressure demand. If a full facepiece is not available, then chemical dust goggles should be worn to protect the eyes. Whenever an airborne chemical dust hazard exists, a face shield and a rubber apron should be worn. To prevent hand and skin exposures, butyl rubber gloves should be worn.

Administrative controls should also be in place to minimize the potential for human exposures. These may include written procedures or policies which specify the methods and techniques that will be practiced whenever personnel are to work with 3,5-dinitrophenol.

All personnel should receive *training* on the use, hazards, protective measures, emergency actions, and other precautions per 29 CFR 1910.1200 (Hazard Communication), prior to the first assignment in an area where 3,5-dinitrophenol is used or stored.

Before beginning employment with 3,5-dinitrophenol and at regular intervals thereafter (e.g., annually), those persons with high or frequent exposures should be provided the following recommended medical tests:

☑ Complete blood count (establish baseline).

☑ Urinalysis (establish baseline nitrophenol levels).

If symptoms develop or overexposure is suspected, the following tests are recommended:

☑ Complete blood count (compare to baseline).

☑ Urinalysis (compare to baseline).

☑ Consider chest X-ray following acute overexposure (may be negative if taken immediately after exposure).

☑ Lung function tests.

☑ Liver and kidney function tests.

Any evaluation should include a careful history of past and present symptoms with an examination. Medical tests that simply look for existing damage are not a substitute for controlling exposures. Physician should monitor fluids and electrolytes, control hyperthermia, and test for dinitrophenol or its metabolite (aminophenol) in the urine using Derrian's Test. The biological half-life of 3,5-dinitrophenol is 5-14 days.

Other methods to reduce exposure include:

☑ Where possible, enclose operations and use local exhaust ventilation at the site of chemical release. If local exhaust ventilation or enclosure is not used, respiratory protection should be mandatory.

☑ Always ensure that proper protective clothing is worn when using chemical substances.

☑ Wash thoroughly immediately after exposure to 3,5-dinitrophenol and at the end of the work shift or before eating, drinking, or smoking.

☑ Hazard warning information should be posted in the work area. In addition, as part of an on-going education and training program, all information on the health and safety hazards of 3,5-dinitrophenol should be communicated to all potentially exposed workers.

☑ Eye wash stations should be provided in the immediate work area for emergency use. If there is a possibility of skin exposure to 3,5-dinitrophenol, emergency shower facilities should also be provided.

☑ Personnel should never wear contaminated clothing home (family members can be exposed). Clothing should be laundered by personnel who have been trained on the hazards of 3,5-dinitrophenol.

RISK ASSESSMENT: ENVIRONMENT

General Assessment

The environment is at risk of exposure during transportation, storage, disposal, or destruction of 3,5-dinitrophenol. In almost every scenario, the threat of environmental exposure is contingent upon the proper handling of the chemical substance. Accidental spills, large or small, can result in fire, explosion, and possible contamination of the surrounding environmental mediums (water, soil, and air).

3,5-Dinitrophenol is normally considered a flammable solid. In solution, it can often appear in mixtures with other combustible or flammable commodities. Its dusts may form explosive mixtures in air. These characteristics require caution in handling, storage, transportation, and disposal. It is incompatible with strong oxidizers and contact can cause fire or explosion. It forms explosive salts in contact with alkalis or with ammonia. Prolonged contact with heat may result in an explosion due to spontaneous decomposition. Therefore, special consideration is required during any emergency situation involving a leak or spill of 3,5-dinitrophenol. Should 3,5-dinitrophenol ever come into contact with these incompatible substances during use, transportation, storage, or disposal, the formation of highly toxic and/or highly explosive commodities is extremely possible. 3,5-Dinitrophenol may enter the environment through its intended use as an insecticide, as well as through industrial effluents or spills.

Acute Ecological Effects

Acute (short-term) toxic effects may include the death of animals, birds, or fish and death or low growth rate in plants. Acute effects are seen 2 to 4 days after animals or plants are exposed to 3,5-dinitrophenol. Insufficient data are available on the short-term effects of 3,5-dinitrophenol exposure to aquatic life, plants, birds or land animals.

Chronic Ecological Effects

Chronic (long-term) toxic effects may include shortened life span, reproductive problems, lower fertility, and changes in appearance or behavior in exposed animals. These effects can be seen long after first exposure(s) to toxic chemicals. Insufficient data are available on the long-term effects of 3,5-dinitrophenol to aquatic life, plants, birds, or land animals.

Water Solubility

3,5-Dinitrophenol is slightly soluble in cold water. Concentrations of 1 to 100 milligrams will mix with a liter of water.

Persistence in the Environment

3,5-Dinitrophenol is slightly persistent in water, with a half-life between 2 and 20 days. The ecological half-life of a pollutant is the amount of time it takes for one half of the chemical to be degraded.

Bioaccumulation in Aquatic Organisms

Some substances increase in concentration, or *bioaccumulate,* in living organisms as they breathe contaminated air, drink contaminated water, or eat contaminated food. These chemicals can become concentrated in the tissues and internal organs of animals as well as humans.

It is not known if 3,5-dinitrophenol will accumulate in the edible tissues of fish.

Recommended Risk-Reduction Measures

Proper training of all transporters will reduce the likelihood of a mishap or accident resulting in a leak or spill to the environment. The correct use of DOT labeling (per the MSDS) on all containers, trucks, and rail cars will enable emergency responders to react properly and quickly to any disaster thereby reducing the potential risk to the environment and to personnel.

Storage of 3,5-dinitrophenol should be segregated from incompatible chemicals to minimize the risk of cross-contamination or contact. Containers must be protected from heat, damage, and shock. Buildings designated for storage should be equipped with appropriate fire protection systems (alarms, sprinklers, emergency lighting, portable extinguishers).

If a spill or leak to the environment has occurred, fire department, emergency response and/or hazardous materials spill personnel should be notified immediately. Cleanup should be attempted only by those trained in proper spill containment procedures. Contaminated soils should be removed for incineration and replaced with clean soil. If 3,5-dinitrophenol should contact the water table, aquifer, or navigable waterway, time is of the essence. It is slightly soluble in the aquatic environment and total remediation may not be possible. The local and/or state emergency response authorities must be notified. A comprehensive emergency response or disaster preparedness/recovery plan should be in place prior to any operations involving the use, transportation, storage, or disposal of 3,5-dinitrophenol. If 3,5-dinitrophenol is spilled or leaked, the following specific steps are recommended:

- ☑ Restrict persons not wearing protective clothing from area of spill or leak until cleanup is complete and area can be opened for normal work.
- ☑ Remove all ignition sources, ventilate area of leak.
- ☑ Absorb liquid spills in vermiculite, dry sand, earth, or similar material and deposit in sealed containers. Collect solid spills for reuse of disposal. Wet mop spill area to collect residue.
- ☑ It may be necessary to dispose of 3,5-dinitrophenol as a hazardous waste. The responsible state agency or the regional office of the federal Environmental Protection Agency (EPA) should be contacted for specific recommendations.

RISK ASSESSMENT: BUSINESS

General Assessment

Accidents or mishaps involving 3,5-dinitrophenol can present a significant threat to business operations. The loss or damage of equipment or facilities can significantly affect fiscal viability. Lawsuits that may result from personnel injury/death, public exposures, and/or environmental contamination will also require a serious expenditure of resources. Media attention surrounding an injury, death, or environmental damage can also result in a loss of profits and loss of current as well as future business.

Recommended Risk-Reduction Measures

Company attorneys, safety and health professionals, and environmental specialists should be involved in the development of any procedures dealing with safety and emergency response. A company official should be pre-designated as a public relations officer with specific training in dealing with the press. Corporate plans and policies should be developed, approved, and implemented long before any need for such arises.

REFERENCES

Ahlbom, A. 1993. *Biostatistics for Epidemiologists.* New York: CRC Press/Lewis Publishers.

Anthony, C. P., and N. J. Kolthoff. 1971. *Textbook of Anatomy and Physiology.* St. Louis: C. V. Mosby Company.

Bloom, A. D. and F. J. DeSerres. 1995. *Ecotoxicity and Human Health.* New York: CRC Press/Lewis Publishers.

Bueche, J. R. 1972. *Principles of Physics,* 2nd Edition. New York: McGraw-Hill.

Burrell, R., D. K. Flaherty, and L. J. Sauers. 1992. *Toxicology of the Immune System.* New York: Van Nostrand Reinhold.

Calabrese, E. J. 1994. *Biological Effects of Low Level Exposures.* New York: CRC Press/Lewis Publishers.

Cockerman, L. G., B. S. Shane. 1994. *Basic Environmental Toxicology.* New York: CRC Press/Lewis Publishers.

Environmental Protection Agency. 1983. *Chemical Hazard Information Profiles.* Washington, D.C.: U.S. EPA.

Hallenbeck, W. H. 1993. *Quantitative Risk Assessment for Environmental and Occupational Health,* 2nd Edition. New York: CRC Press/Lewis Publishers.

Howard, P. H. 1995. *Handbook of Environmental Fate and Exposures Data for Organic Chemicals,* Vol. I-IV. New York: CRC Press/Lewis Publishers.

Lewis, R. J., Sr. 1992. *Sax's Dangerous Properties of Industrial Materials,* Volumes 1, 2, and 3, 8th Edition. New York: Van Nostrand Reinhold.

Meyer, E. 1990. *Chemistry of Hazardous Materials.* Englewood Cliffs, NJ: Prentice Hall.

Quigley, D. R. 1994. *Handbook of Emergency Chemical Management.* New York: CRC Press/Lewis Publishers.

MATERIAL SAFETY DATA SHEET

CHEMICAL NAME	CRC PRESS/LEWIS PUBLISHERS DISCLAIMER
DINITROTOLUENE	

HAZARD WARNING INFORMATION

HEALTH	FIRE	REACTIVE	OTHER	DEGREE OF HAZARD	COLOR CODING	OTHER CODES
3	1	3		0 = Minimum Hazard 1 = Slight Hazard 2 = Moderate Hazard 3 = Serious Hazard 4 = Severe Hazard	HEALTH = BLUE FIRE = RED REACTIVITY = YELLOW OTHER = WHITE	OX = Oxidizer ACID = Acid ALK = Alkali COR = Corrosive W = Use No Water

SECTION I - GENERAL INFORMATION

Characterization: **Aromatic Hydrocarbon**	RCRA Number: **None**	EPA Class: **Not Applicable**
DOT Proper Shipping Name: **Dinitrotoluene (solid)**	Chemical Abstract Service (CAS) Number: **25321-14-6**	
DOT Hazard Class and Label Requirements: **Poison B; Label: Poison**	DOT Emergency Guide Code: **56**	
DOT Identification Number: **UN 1600 (molten); UN 2038 (solid)**	Chemical Formula: $C_6H_3CH_3(NO_2)_2$	

Synonyms
Dinitrophenylmethane; ar,ar-dinitrotoluene; methyldinitrobenzene.

SECTION II - HAZARDOUS INGREDIENTS AND IDENTITY INFORMATION

Hazardous Components (specific identity)	OSHA Exposure Criteria	NIOSH Exposure Criteria	Immediately Dangerous to Life and Health (IDLH)	ACGIH Exposure Criteria
Dinitrotoluene (derivation: By nitration of nitrotoluene with nitric acid in the presence of concentrated sulfuric acid).	**PEL (skin):** 1.5 mg/m³ **STEL:** Not Established	**REL (skin):** 1.5 mg/m³ Possible Cancer Agent	50 mg/m³	**TLV (skin):** 1.5 mg/m³ **STEL:** Not Established

SECTION III - PHYSICAL AND CHEMICAL CHARACTERISTICS

Boiling Point: **482°F (250°C)**	Specific Gravity (H_2O = 1): **1.32 (liquid) 1.5 (solid)**
Vapor Pressure (gas): **1 at 69°F (20°C)**	Molecular Weight: **182.14**
Vapor Density (Air = 1): **6.27**	Melting Point: **158°F (70°C)**

Solubility
Nearly insoluble in water. Soluble in alcohol, acetone, benzene, carbon disulfide, ether, and pyridine.

Appearance and Odor
Orange to yellow crystalline solid with a slight characteristic odor. Usually shipped as molten.

SECTION IV - FIRE AND EXPLOSION HAZARD DATA

Flash Point (method used): **404°F (207°C) closed cup**	Explosive Limits in Air % by Volume: **LEL: 0.3% at 474°F (245°C) UEL: Not Determined**
NFPA Classification: **Combustible Solid**	Autoignition Temperature: **Not Determined**

Extinguishing Media
Use carbon dioxide, water, or dry chemical from a protected location.

Special Fire Fighting Procedures
A carcinogenic. Wear full protective clothing and self-contained breathing apparatus (SCBA). Poisonous gases are produced in fire. Use water spray to keep fire-exposed containers cool. Move containers from fire if it can be done without risk. Fight fire from a distance or protected position, if possible.

Unusual Fire and Explosion Hazards
Containers may explode in fire. Fires involving this material should be considered an explosion hazard. Extremely explosive in dry form when exposed to heat, flame, or shock. Decomposition is self-sustaining at 280°F (138°C).

SECTION V - REACTIVITY DATA

Stability		Conditions to Avoid **Dinitrotoluene is explosive when exposed to heat, friction, or contamination. It can be detonated using a very strong initiating device. It is slightly sensitive to impact shock, especially in solid form.**
Stable	Unstable **X**	Incompatibility *(materials to avoid)* **Strong oxidizers, (such as chlorine, bromine, and fluorine), organic materials, strong reducing agents (such as sodium sulfide, zinc powder, sodium hyposulfite, and metallic hydrides), caustics, and metals.**
Hazardous Polymerization		Conditions to Avoid **Under normal conditions of temperature and pressure, hazardous polymerization of dinitrotoluene is not expected to occur.**
May Occur	Will Not Occur **X**	Hazardous Decomposition or By-products **When heated to decomposition, dinitrotoluene can emit highly toxic/poisonous gases, including carbon dioxide and oxides of nitrogen.**

SECTION VI - HEALTH HAZARD DATA

Primary Route(s) of Entry:	Inhalation?	Absorption (skin)?	Ingestion?
	X	X	X

Health Hazards

INHALATION: Can cause cyanosis, headache, irritability, anoxia, anemia, dizziness, weakness, nausea, vomiting, dypsnea, drowsiness, collapse, loss of consciousness, and death by cardiovascular collapse.

ABSORPTION: Irritation, small blisters, redness, swelling, ulceration, and necrosis (will pass through skin to cause same toxic systemic effects as inhalation). Eye contact will result in burns, irritation of the lids, redness, swelling, pain on exposure to light, and tissue damage.

INGESTION: Irritation of the mouth, throat, and stomach, stomach cramps, and diarrhea.

Carcinogenicity	NTP Listed?	IARC Cancer Review Group?	OSHA Regulated?	Target Organs?
Suspected Human Confirmed Animal	5th Annual Report	No	29 CFR 1910-1000 Table Z-1	Blood, liver, CVS, reproductive system, liver, skin

Medical Conditions Generally Aggravated by Exposure

Persons with blood disorders may be at an increased risk if exposed to dinitrotoluene.

Emergency and First-aid Procedures

Eye contact: Flush immediately with water for 15 minutes (minimum), seek medical attention. Skin contact: Remove all contaminated clothing. Immediately wash area with large amounts of soap and water. Seek medical assistance. For inhalation: Remove the person from exposure. Provide CPR if required. Transfer to medical facility. If swallowed: Give large amounts of water or milk and seek medical attention. Never attempt to give an unconscious or convulsing person anything by mouth.

SECTION VII - PRECAUTIONS FOR SAFE HANDLING AND USE

Steps to be Taken in Case Material is Released or Spilled

Remove all ignition sources and ventilate area. Evacuate personnel not involved in cleanup. Stay upwind. Cleanup crew should wear fully-encapsulating suits. Use water spray to reduce vapors. Collect spilled powders and place in sealed steel drums for disposal.

Preferred Waste Disposal Method

No Citation.

Precautions to be Taken in Handling and Storage

Store in tightly closed containers in cool, well-ventilated area away from heat or flame. Heat may cause containers to build up pressure and explode. Usually stored as a molten liquid.

Other Precautions and Warnings

Sources of ignition are prohibited where dinitrotoluene is used, handled, or stored.

SECTION VIII - CONTROL MEASURES AND PERSONAL PROTECTIVE EQUIPMENT

Respiratory Protection *(specify type)*

PEL is low and difficult to monitor. For any exposure to a suspected carcinogen, recommend a supplied-air respirator with full facepiece, hood, or helmet in continuous flow mode, or use a self-contained breathing apparatus (SCBA) operated in pressure demand or other positive pressure mode.

Ventilation

Local exhaust or general mechanical systems recommended.

Protective Gloves	Eye Protection	Other Protective Clothing
Butyl Rubber	Chemical/Dust Goggles or Face Mask	Butyl Rubber Apron

Work/Hygiene Practices

Always wash hands thoroughly after using chemical; never bring food, drink, or smoking materials into vicinity of chemicals.

DINITROTOLUENE

$C_7H_6N_2O_4$ CAS: 25321-14-6

IDENTIFICATION AND TYPICAL USES

Dinitrotoluene is an orange to yellow crystalline solid with a characteristic odor. It is normally shipped as a molten liquid. It is a mixture of numerous isomers of dinitrotoluene. It is used primarily in the manufacture of explosives and dyes, in organic synthesis, as a chemical intermediate, a plasticizer, and as a modifier for smokeless powders in the munitions industry.

RISK ASSESSMENT: HEALTH

General Assessment

Dinitrotoluene is a human poison by *ingestion,* with equally toxic systemic consequences occurring by way of *inhalation,* and skin contact (*absorption*). It has been shown to cause teratogenic effects in experimental tests and mutation data have also been reported. It is a confirmed animal carcinogen and a suspected human carcinogen. Liver or kidney damage can occur up to 3 days following exposure.

Inhalation and skin absorption can result in cyanosis, methemoglobinemia, headache, irritability, dizziness, weakness, nausea, vomiting, dyspnea, drowsiness, and loss of consciousness. If allowed to go untreated, the result can be death due to cardiovascular collapse. Repeated exposures can lead to anemia, and liver damage. It has caused mutations and liver tumors in test animals.

Ingestion will result in irritation of the mouth and stomach, stomach cramps, and diarrhea. Possible systemic effects similar to inhalation and absorption may also occur. Ingestion of alcohol shortly before or after exposure may increase individual susceptibility to toxicity.

Skin contact will not only result in absorption, it will also produce irritation, small blisters, redness, swelling, ulceration, and localized necrosis. Eye contact causes burning pain, irritation, redness, and swelling of the eyelids. There may be pain on exposure to light and severe eye damage. Contact with the molten liquid will cause thermal burns.

Acute Health Effects

The following acute (short-term) health effects may occur immediately or shortly after exposure to dinitrotoluene:

Skin: Irritation and possible tissue damage. Will pass through unbroken skin to cause serious systemic effects.

Eye: Irritation, swelling, redness, pain on exposure to light, and potential damage to the surrounding tissue.

Lung: Irritation to the respiratory tract and associated mucosa. Absorption through the respiration process causes toxic systemic effects (including anemia and methemoglobinemia).

Chronic Health Effects

The following chronic (long-term) health effects may occur at some time after exposure to dinitrotoluene and can last for months or even years:

Cancer Hazards: According to the information available in the references, dinitrotoluene is a confirmed carcinogen in test animals. Its carcinogenic properties in humans is highly suspect. Experimental mutation data have been reported and many scientists believe that such chemicals may have the ability to cause cancer.

Reproductive Hazard: According to information presented in the references, dinitrotoluene has been shown to cause adverse effects in reproduction in animals. It is considered an experimental teratogen. Human effects in this regard are inconclusive and require additional research.

Other Chronic Effects: Long-term exposures can result in anemia and possible liver damage. It is also known to cause mutations and tumors in the livers of test animals.

ʘ *Recommended Risk-Reduction Measures*

Personnel should avoid direct contact with dinitrotoluene. If a less toxic chemical cannot be substituted for a hazardous substance, then *engineering controls* are the most effective method of reducing exposures. The best protection is to enclose operations and/or provide local exhaust ventilation at the site of chemical release. While not always operationally feasible, isolating operations can also reduce exposure. Using respiratory protection is less effective than the controls mentioned above, but is still advisable whenever working with or around dinitrotoluene. The exposure level is low (1.5 mg/m^3) and difficult to monitor. Therefore, a MSHA/NIOSH approved full facepiece respirator equipped with an organic vapor cartridge equipped with a dust pre-filter should be used if any exposure is anticipated or known. Increased protection is obtained from a supplied-air respirator with full facepiece operated in continuous flow mode, or a self-contained breathing apparatus (SCBA) with full facepiece and pressure demand. If a full facepiece is not available, then chemical dust goggles should be worn to protect the eyes. Whenever an airborne chemical dust hazard exists, a face shield and a rubber apron should be worn. To prevent hand and skin exposures, butyl rubber gloves should be worn.

Administrative controls should also be in place to minimize the potential for human exposures. These may include written procedures or policies which specify the methods and techniques that will be practiced whenever personnel are to work with dinitrotoluene.

All personnel should receive *training* on the use, hazards, protective measures, emergency actions, and other precautions per 29 CFR 1910.1200 (Hazard Communication), prior to the first assignment in an area where dinitrotoluene is used or stored.

Before beginning employment with dinitrotoluene and at regular intervals thereafter (e.g., annually), those persons with high or frequent exposures should be provided the following recommended medical tests:

☑ Liver and kidney function tests.

☑ Urinalysis (establish baseline).

If symptoms develop or overexposure is suspected, the following tests are recommended:

☑ Urinalysis (compare to baseline).

☑ Complete blood count.

☑ Liver and kidney function tests.

Any evaluation should include a careful history of past and present symptoms with an examination. Medical tests that simply look for existing damage are not a substitute for controlling exposures.

Other methods to reduce exposure include:

☑ Where possible, enclose operations and use local exhaust ventilation at the site of chemical release. If local exhaust ventilation or enclosure is not used, respiratory protection should be mandatory.

☑ Always ensure that proper protective clothing is worn when using chemical substances.

☑ Wash thoroughly immediately after exposure to dinitrotoluene and at the end of the work shift or before eating, drinking, or smoking.

☑ Hazard warning information should be posted in the work area. In addition, as part of an on-going education and training program, all information on the health and safety hazards of dinitrotoluene should be communicated to all potentially exposed workers.

☑ Eye wash stations should be provided in the immediate work area for emergency use. If there is a possibility of skin exposure to dinitrotoluene, emergency shower facilities should also be provided.

☑ Personnel should never wear contaminated clothing home (family members can be exposed). Clothing should be laundered by personnel who have been trained on the hazards of dinitrotoluene.

RISK ASSESSMENT: ENVIRONMENT

General Assessment

The environment is at risk of exposure during transportation, storage, disposal, or destruction of dinitrotoluene. In almost every scenario, the threat of environmental exposure is contingent upon the proper handling of the chemical substance. Accidental spills, large or small, can result in fire, explosion, and possi-

ble contamination of the surrounding environmental mediums (water, soil, and air).

Dinitrotoluene is normally considered a combustible solid. Its dusts may form explosive mixtures in air. These characteristics require caution in handling, storage, transportation, and disposal. It is incompatible with strong oxidizers, caustics, and metals (such as tin and zinc). Contact with either of these can cause fire or explosion. Therefore, special consideration is required during any emergency situation involving a leak or spill of dinitrotoluene. Should dinitrotoluene ever come into contact with any incompatible substances during use, transportation, storage, or disposal, the formation of highly toxic and/or highly explosive commodities is extremely possible.

Dinitrotoluene may enter the environment through its intended use as an insecticide, as well as through industrial effluents or spills.

Acute Ecological Effects

Acute (short-term) toxic effects may include the death of animals, birds, or fish and death or low growth rate in plants. Acute effects are seen 2 to 4 days after animals or plants are exposed to dinitrotoluene.

Dinitrotoluene has high acute toxicity to aquatic life. Insufficient data are available on the short-term effects of dinitrotoluene exposure to plants, birds, or land animals.

Chronic Ecological Effects

Chronic (long-term) toxic effects may include shortened life span, reproductive problems, lower fertility, and changes in appearance or behavior in exposed animals. These effects can be seen long after first exposure(s) to toxic chemicals. Dinitrotoluene has moderate chronic toxicity to aquatic life. Insufficient data are available on the long-term effects of dinitrotoluene to plants, birds, or land animals.

Water Solubility

Dinitrotoluene is nearly insoluble in water. Concentrations of less than 1 milligram may mix with a liter of water.

Persistence in the Environment

Dinitrotoluene is highly persistent in the aquatic environment, with a half-life of greater than 200 days. The half-life of a pollutant is the amount of time it takes for one half of the chemical to be degraded.

Bioaccumulation in Aquatic Organisms

Some substances increase in concentration, or *bioaccumulate,* in living organisms as they breathe contaminated air, drink contaminated water, or eat contaminated food. These chemicals can become concentrated in the tissues and internal organs of animals as well as humans.

The concentration of dinitrotoluene found in fish tissue is expected to be somewhat higher than the average concentration of dinitrotoluene in the water from which the fish was taken.

Recommended Risk-Reduction Measures

Proper training of all transporters will reduce the likelihood of a mishap or accident resulting in a leak or spill to the environment. The correct use of DOT labeling (per the MSDS) on all containers, trucks, and rail cars will enable emergency responders to react properly and quickly to any disaster thereby reducing the potential risk to the environment and to personnel.

Storage of dinitrotoluene should be segregated from incompatible chemicals to minimize the risk of cross-contamination or contact. Containers must be protected from heat, damage, and shock. Buildings designated for storage should be equipped with appropriate fire protection systems (alarms, sprinklers, emergency lighting, portable extinguishers).

If a spill or leak to the environment has occurred, fire department, emergency response and/or hazardous materials spill personnel should be notified immediately. Cleanup should be attempted only by those trained in proper spill containment procedures. Contaminated soils should be removed for incineration and replaced with clean soil. If dinitrotoluene should contact the water table, aquifer, or navigable waterway, time is of the essence. It is slightly soluble in the aquatic environment and total remediation may not be possible. The local and/or state emergency response authorities must be notified. A comprehensive emergency response or disaster preparedness/recovery plan should be in place prior to any operations involving the use, transportation, storage, or disposal of dinitrotoluene.

If dinitrotoluene is spilled or leaked, the following specific steps are recommended:

- ☑ Restrict persons not wearing protective clothing from area of spill or leak until cleanup is complete and area can be opened for normal work.
- ☑ Remove all ignition sources, ventilate area of leak.

☑ Use a water spray to knock-down vapors. Spills should be collected and placed in steel drums. Area should be flushed with hot water to remove solid particulates. Any liquids should be contained, as well as any liquid run-off.

☑ It may be necessary to dispose of dinitrotoluene as a hazardous waste. The responsible state agency or the regional office of the federal Environmental Protection Agency (EPA) should be contacted for specific recommendations.

RISK ASSESSMENT: BUSINESS

General Assessment

Accidents or mishaps involving dinitrotoluene can present a significant threat to business operations. The loss or damage of equipment or facilities can significantly affect fiscal viability. Lawsuits that may result from personnel injury/death, public exposures, and/or environmental contamination will also require a serious expenditure of resources. Media attention surrounding an injury, death, or environmental damage can also result in a loss of profits and loss of current as well as future business. Always remember that anytime the terms "carcinogen," "cancer," or "reproductive hazard" are used, public emotion, hysteria, and ignorance can run equally high. This must be properly considered whenever developing public relations policies.

Recommended Risk-Reduction Measures

Company attorneys, safety and health professionals, and environmental specialists should be involved in the development of any procedures dealing with safety and emergency response. A company official should be pre-designated as a public relations officer with specific training in dealing with the press. Corporate plans and policies should be developed, approved, and implemented long before any need for such arises.

REFERENCES

Ahlbom, A. 1993. *Biostatistics for Epidemiologists.* New York: CRC Press/Lewis Publishers.

American Conference of Governmental Industrial Hygienists. 1988. *Documentation of the Threshold Limit Values and Biological Exposure Indices,* 5th Edition (with updates). Cincinnati: ACGIH

Anthony, C. P., and N. J. Kolthoff. 1971. *Textbook of Anatomy and Physiology.* St. Louis: C. V. Mosby Company.

Asante-Duah, D. K. 1993. *Hazardous Waste Risk Assessment.* New York: CRC Press/Lewis Publishers.

Bloom, A. D. and F. J. DeSerres. 1995. *Ecotoxicity and Human Health.* New York: CRC Press/Lewis Publishers.

Bueche, J. R. 1972. *Principles of Physics,* 2nd Edition. New York: McGraw-Hill.

Burrell, R., D. K. Flaherty, and L. J. Sauers. 1992. *Toxicology of the Immune System.* New York: Van Nostrand Reinhold.

Calabrese, E. J. 1994. *Biological Effects of Low Level Exposures.* New York: CRC Press/Lewis Publishers.

Cockerman, L. G., B. S. Shane. 1994. *Basic Environmental Toxicology.* New York: CRC Press/Lewis Publishers.

Environmental Protection Agency. 1983. *Chemical Hazard Information Profiles.* Washington, D.C.: U.S. EPA.

Fodden, C. R., and J. L. Weddell. 1992. *Hazardous Materials Emergency Action Data.* New York: CRC Press/Lewis Publishers.

Hallenbeck, W. H. 1993. *Quantitative Risk Assessment for Environmental and Occupational Health,* 2nd Edition. New York: CRC Press/Lewis Publishers.

Howard, P. H. 1995. *Handbook of Environmental Fate and Exposures Data for Organic Chemicals,* Vol. I-IV. New York: CRC Press/Lewis Publishers.

International Agency for Research on Cancer (IARC). 1987. *IARC Monographs on the Evaluation of Carcinogenic Risk to Humans, Overall Evaluations of Carcinogenicity: An updating of IARC Monographs Volumes 1-42.* Lyon: IARC.

Keith, L. H. (Editor). 1992. *Integrated Risk Information System (IRIS) Chemical Information Database.* Prepared by the U.S. Environmental Protection Agency. Boca Raton, FL: CRC Press/Lewis Publishers.

Keith, L. H. and D. B. Walters. (Editors) 1992. *National Toxicology Program's Chemical Database, Volumes I-VII.* Chelsea, MI: Lewis Publishers.

Lewis, R. J., Sr. 1992. *Sax's Dangerous Properties of Industrial Materials,* Volumes 1, 2, and 3, 8th Edition. New York: Van Nostrand Reinhold.

Mackay, D., W. Y. Shiu, and K. C. Ma. 1992. *Illustrated Handbook of Chemical-Physical Properties and Environmental Fate for Organic Chemicals.* New York: CRC Press/Lewis Publishers.

Manahan, S. E. 1992. *Toxicological Chemistry,* 2nd Edition. New York: CRC Press/Lewis Publishers.

Meyer, E. 1990. *Chemistry of Hazardous Materials.* Englewood Cliffs, NJ: Prentice Hall.

Quigley, D. R. 1994. *Handbook of Emergency Chemical Management.* New York: CRC Press/Lewis Publishers.

U.S. Department of Health and Human Services, National Institute for Occupational Safety and Health. 1994. *NIOSH Pocket Guide to Chemical Hazards.* Washington, D.C.: U.S. Government Printing Office.

MATERIAL SAFETY DATA SHEET

CHEMICAL NAME	
2,3-DINITROTOLUENE	**CRC PRESS/LEWIS PUBLISHERS DISCLAIMER:** THE INFORMATION AND RECOMMENDATIONS PRESENTED HEREIN ARE BASED ON SOURCES BELIEVED TO BE RELIABLE. CRC MAKES NO REPRESENTATION ON ITS COMPLETENESS OR ACCURACY. IT IS THE RESPONSIBILITY OF THE USER TO DETERMINE THE CHEMICAL'S SUITABILITY FOR ITS INTENDED USE, THE CHEMICAL'S SAFE USE, AND THE CHEMICAL'S PROPER DISPOSAL. NO REPRESENTATIONS AND/OR WARRANTIES, EITHER EXPRESSED OR IMPLIED, OF THE MERCHANTABILITY OR FITNESS FOR A PARTICULAR PURPOSE, OR OF ANY OTHER NATURE, ARE MADE WITH RESPECT TO THE INFORMATION PROVIDED IN THIS MSDS OR TO THE CHEMICAL TO WHICH INFORMATION MAY REFER. CRC NEITHER ASSUMES NOR AUTHORIZES ANY OTHER PERSON TO ASSUME FOR IT, ANY OTHER ADDITIONAL RESPONSIBILITY OR LIABILITY FOR THE USE OF, OR RELIANCE UPON, THIS INFORMATION.

HAZARD WARNING INFORMATION

HEALTH	FIRE	REACTIVE	OTHER	DEGREE OF HAZARD	COLOR CODING	OTHER CODES
3	1	3		0 = Minimum Hazard 1 = Slight Hazard 2 = Moderate Hazard 3 = Serious Hazard 4 = Severe Hazard	HEALTH = BLUE FIRE = RED REACTIVITY = YELLOW OTHER = WHITE	OX = Oxidizer ACID = Acid ALK = Alkali COR = Corrosive W̶ = Use No Water

SECTION I - GENERAL INFORMATION

Characterization	**Aromatic Hydrocarbon**	RCRA Number	**None**
		EPA Class	**Not Applicable**
DOT Proper Shipping Name	**2,3-Dinitrotoluene (solid)**	Chemical Abstract Service (CAS) Number	**602-01-7**
DOT Hazard Class and Label Requirements	**Poison B; Label: Poison**	DOT Emergency Guide Code	**56**
DOT Identification Number	**UN 1600 (molten); UN 2038 (solid)**	Chemical Formula	$C_6H_3CH_3(NO_2)_2$

Synonyms
2,3-DNT; 1-methyl-2,3-dinitrobenzene.

SECTION II - HAZARDOUS INGREDIENTS AND IDENTITY INFORMATION

Hazardous Components (specific identity)	OSHA Exposure Criteria	NIOSH Exposure Criteria	Immediately Dangerous to Life and Health (IDLH)	ACGIH Exposure Criteria
2,3-Dinitrotoluene (derivation: By nitration of toluene with nitric acid in the presence of concentrated sulfuric acid).	**PEL (skin):** 1.5 mg/m³ **STEL:** Not Established	**REL (skin):** 1.5 mg/m³ **Possible Cancer Agent**	**50 mg/m³**	**TLV (skin):** 1.5 mg/m³ **STEL:** Not Established

SECTION III - PHYSICAL AND CHEMICAL CHARACTERISTICS

Boiling Point	**482°F (250°C)**	Specific Gravity (H_2O = 1)	**1.32 (liquid) 1.5 (solid)**
Vapor Pressure (gas)	**1 at 69°F (20°C)**	Molecular Weight	**182.14**
Vapor Density (Air = 1)	**6.27**	Melting Point	**158°F (70°C)**

Solubility
Nearly insoluble in water. Soluble in alcohol, acetone, benzene, carbon disulfide, ether, and pyridine.

Appearance and Odor
Orange to yellow crystalline solid with a slight characteristic odor. Usually shipped as molten.

SECTION IV - FIRE AND EXPLOSION HAZARD DATA

Flash Point (method used)	**404°F (207°C) closed cup**	Explosive Limits in Air % by Volume	**LEL: 0.3% at 474°F (245°C) UEL: Not Determined**
NFPA Classification	**Combustible Solid**	Autoignition Temperature	**Not Determined**

Extinguishing Media
Use carbon dioxide, water, or dry chemical from a protected location.

Special Fire Fighting Procedures
A carcinogenic. Wear full protective clothing and self-contained breathing apparatus (SCBA). Poisonous gases are produced in fire. Use water spray to keep fire-exposed containers cool. Move containers from fire if it can be done without risk. Fight fire from a distance or protected position, if possible.

Unusual Fire and Explosion Hazards
Containers may explode in fire. Fires involving this material should be considered an explosion hazard. Extremely explosive in dry form when exposed to heat, flame, or shock. Decomposition is self-sustaining at 280°F (138°C).

SECTION V - REACTIVITY DATA

Stability		Conditions to Avoid **2,3-Dinitrotoluene is explosive when exposed to heat, friction, or contamination. It can be detonated using a very strong initiating device. It is slightly sensitive to impact shock, especially in solid form.**
Stable	Unstable **X**	Incompatibility *(materials to avoid)* **Strong oxidizers, (such as chlorine, bromine, and fluorine), organic materials, strong reducing agents (such as sodium sulfide, zinc powder, sodium hyposulfite, and metallic hydrides), caustics, and metals.**
Hazardous Polymerization		Conditions to Avoid **Under normal conditions of temperature and pressure, hazardous polymerization of 2,3-dinitrotoluene is not expected to occur.**
May Occur	Will Not Occur **X**	Hazardous Decomposition or By-products **When heated to decomposition, 2,3-dinitrotoluene can emit highly toxic/poisonous gases, including carbon dioxide and oxides of nitrogen.**

SECTION VI - HEALTH HAZARD DATA

Primary Route(s) of Entry:	Inhalation?	Absorption (skin)?	Ingestion?
	X	X	X

Health Hazards

INHALATION: **Can cause cyanosis, headache, irritability, anoxia, anemia, dizziness, weakness, nausea, vomiting, dypsnea, drowsiness, collapse, loss of consciousness, and death by cardio-vascular collapse.**

ABSORPTION: **Irritation, small blisters, redness, swelling, ulceration, and necrosis (will pass through skin to cause same toxic systemic effects as inhalation). Eye contact will result in burns, irritation of the lids, redness, swelling, pain on exposure to light, and tissue damage.**

INGESTION: **Irritation of the mouth, throat, and stomach, stomach cramps, and diarrhea.**

Carcinogenicity	NTP Listed?	IARC Cancer Review Group?	OSHA Regulated?	Target Organs?
Suspected Human Confirmed Animal	**5th Annual Report**	**No**	**29 CFR 1910-1000 Table Z-1**	**Blood, liver, CVS, reproductive system, liver, skin**

Medical Conditions Generally Aggravated by Exposure

Persons with blood disorders may be at an increased risk if exposed to 2,3-dinitrotoluene.

Emergency and First-aid Procedures

<u>Eye contact</u>: Flush immediately with water for 15 minutes (minimum), seek medical attention. <u>Skin contact</u>: Remove all contaminated clothing. Immediately wash area with large amounts of soap and water. Seek medical assistance. For <u>inhalation</u>: Remove the person from exposure. Provide CPR if required. Transfer to medical facility. If <u>swallowed</u>: Give large amounts of water or milk and seek medical attention. Never attempt to give an unconscious or convulsing person anything by mouth.

SECTION VII - PRECAUTIONS FOR SAFE HANDLING AND USE

Steps to be Taken in Case Material is Released or Spilled

Remove all ignition sources and ventilate area. Evacuate personnel not involved in cleanup. Stay upwind. Cleanup crew should wear fully-encapsulating suits. Use water spray to reduce vapors. Collect spilled powders and place in sealed steel drums for disposal.

Preferred Waste Disposal Method

No Citation.

Precautions to be Taken in Handling and Storage

Store in tightly closed containers in cool, well-ventilated area away from heat or flame. Heat may cause containers to build up pressure and explode. Usually stored as a molten liquid.

Other Precautions and Warnings

Sources of ignition are prohibited where 2,3-dinitrotoluene is used, handled, or stored.

SECTION VIII - CONTROL MEASURES AND PERSONAL PROTECTIVE EQUIPMENT

Respiratory Protection *(specify type)*

PEL is low and difficult to monitor. For any exposure to a suspected carcinogen, recommend a supplied air respirator with full facepiece, hood, or helmet in continuous flow mode, or use a self-contained breathing apparatus (SCBA) operated in pressure demand or other positive pressure mode.

Ventilation

Local exhaust or general mechanical systems recommended.

Protective Gloves	Eye Protection	Other Protective Clothing
Butyl Rubber	**Chemical/Dust Goggles or Face Mask**	**Butyl Rubber Apron**

Work/Hygiene Practices

Always wash hands thoroughly after using chemical; never bring food, drink, or smoking materials into vicinity of chemicals.

2,3-DINITROTOLUENE

$C_7H_6N_2O_4$ CAS: 602-01-7

IDENTIFICATION AND TYPICAL USES

2,3-Dinitrotoluene is an orange to yellow crystalline solid with a characteristic odor. It is normally shipped as a molten liquid. It is one of numerous isomers of dinitrotoluene. It is used primarily in the manufacture of explosives and dyes, in organic synthesis, as a chemical intermediate, a plasticizer, and as a modifier for smokeless powders in the munitions industry.

RISK ASSESSMENT: HEALTH

General Assessment

2,3-Dinitrotoluene is a human poison by *ingestion,* with equally toxic systemic consequences occurring by way of *inhalation,* and skin contact (*absorption*). The closely-related 2,4-isomer has been shown to cause teratogenic effects in experimental tests and mutation data have also been reported. It is also a confirmed animal carcinogen and a suspected human carcinogen. Liver or kidney damage can occur up to 3 days following exposure. While these effects have not been confirmed for the 2,3-isomer, extreme caution is recommended in use, handling, storage, or disposal.

Inhalation and skin absorption can result in cyanosis, methemoglobinemia, headache, irritability, dizziness, weakness, nausea, vomiting, dyspnea, drowsiness, and loss of consciousness. If allowed to go untreated, the result can be death due to cardiovascular collapse. Repeated exposures can lead to anemia and liver damage. It has caused mutations and liver tumors in test animals.

Ingestion will result in irritation of the mouth and stomach, stomach cramps, and diarrhea. Possible systemic effects similar to inhalation and absorption may also occur. Ingestion of alcohol shortly before or after exposure may increase individual susceptibility to toxicity.

Skin contact will not only result in absorption, it will also produce irritation, small blisters, redness, swelling, ulceration, and localized necrosis. Eye contact causes burning pain, irritation, redness, and swelling of the eyelids. There may be pain on exposure to light and severe eye damage. Contact with the molten liquid will cause thermal burns.

Acute Health Effects

The following acute (short-term) health effects may occur immediately or shortly after exposure to 2,3-dinitrotoluene:

Skin: Irritation and possible tissue damage. Will pass through unbroken skin to cause serious systemic effects.

Eye: Irritation, swelling, redness, pain on exposure to light, and potential damage to the surrounding tissue.

Lung: Irritation to the respiratory tract and associated mucosa. Absorption through the respiration process causes toxic systemic effects (including anemia and methemoglobinemia).

Chronic Health Effects

The following chronic (long-term) health effects may occur at some time after exposure to 2,3-dinitrotoluene and can last for months or even years:

Cancer Hazards: According to the information available in the references, the 2,4-isomer of dinitrotoluene is a confirmed carcinogen in test animals. Its carcinogenic properties in humans is highly suspect. Experimental mutation data have been reported and many sci-

entists believe that such chemicals may have the ability to cause cancer. It is not known whether 2,3-isomer of dinitrotoluene will cause these effects.

Reproductive Hazard: According to information presented in the references, 2,4-dinitrotoluene has been shown to cause adverse effects in reproduction in animals. It is considered an experimental teratogen. It is not known whether 2,3-isomer of dinitrotoluene will cause these effects. Human effects in this regard are inconclusive for both isomers.

Other Chronic Effects: Long-term exposures can result in anemia and possible liver damage. It is also known to cause mutations and tumors in the livers of animals.

Recommended Risk-Reduction Measures

Personnel should avoid direct contact with 2,3-dinitrotoluene. If a less toxic chemical cannot be substituted for a hazardous substance, then *engineering controls* are the most effective method of reducing exposures. The best protection is to enclose operations and/or provide local exhaust ventilation at the site of chemical release. While not always operationally feasible, isolating operations can also reduce exposure. Using respiratory protection is less effective than the controls mentioned above, but is still advisable whenever working with or around 2,3-dinitrotoluene. The exposure level is low (1.5 mg/m^3) and difficult to monitor. Therefore, a MSHA/NIOSH approved full facepiece respirator equipped with an organic vapor cartridge equipped with a dust pre-filter should be used if any exposure is anticipated or known. Increased protection is obtained from a supplied-air respirator with full facepiece operated in continuous flow mode, or a self-contained breathing apparatus (SCBA) with full facepiece and pressure demand. If a full facepiece is not available, then chemical dust goggles should be worn to protect the eyes. Whenever an airborne chemical dust hazard exists, a face shield and a rubber apron should be worn. To prevent hand and skin exposures, butyl rubber gloves should be worn.

Administrative controls should also be in place to minimize the potential for human exposures. These may include written procedures or policies which specify the methods and techniques that will be practiced whenever personnel are to work with 2,3-dinitrotoluene.

All personnel should receive *training* on the use, hazards, protective measures, emergency actions, and other precautions per 29 CFR 1910.1200 (Hazard Communication), prior to the first assignment in an area where 2,3-dinitrotoluene is used or stored.

Before beginning employment with 2,3-dinitrotoluene and at regular intervals thereafter (e.g., annually), those persons with high or frequent exposures should be provided the following recommended medical tests:

- ☑ Liver and kidney function tests.
- ☑ Urinalysis (establish baseline).

If symptoms develop or overexposure is suspected, the following tests are recommended:

- ☑ Urinalysis (compare to baseline).
- ☑ Complete blood count.
- ☑ Liver and kidney function tests.

Any evaluation should include a careful history of past and present symptoms with an examination. Medical tests that simply look for existing damage are not a substitute for controlling exposures.

Other methods to reduce exposure include:

- ☑ Where possible, enclose operations and use local exhaust ventilation at the site of chemical release. If local exhaust ventilation or enclosure is not used, respiratory protection should be mandatory.
- ☑ Always ensure that proper protective clothing is worn when using chemical substances.
- ☑ Wash thoroughly immediately after exposure to 2,3-dinitrotoluene and at the end of the work shift or before eating, drinking, or smoking.
- ☑ Hazard warning information should be posted in the work area. In addition, as part of an on-going education and training program, all information on the health and safety hazards of 2,3-dinitrotoluene should be communicated to all exposed workers.
- ☑ Eye wash stations should be provided in the immediate work area for emergency use. If there is a possibility of skin exposure to 2,3-dinitrotoluene, emergency shower facilities should be provided.
- ☑ Personnel should not wear contaminated clothing home (family members can be exposed). Clothing should be laundered by persons who have been trained on the hazards of 2,3-dinitrotoluene.

RISK ASSESSMENT: ENVIRONMENT

General Assessment

The environment is at risk of exposure during transportation, storage, disposal, or destruction of 2,3-dinitrotoluene. In almost every scenario, the threat of

environmental exposure is contingent upon the proper handling of the chemical substance. Accidental spills, large or small, can result in fire, explosion, and possible contamination of the surrounding environmental mediums (water, soil, and air).

2,3-Dinitrotoluene is normally considered a combustible solid. Its dusts may form explosive mixtures in air. These characteristics require caution in handling, storage, transportation, and disposal. It is incompatible with strong oxidizers, caustics, and metals (such as tin and zinc). Contact with either of these can cause fire or explosion. Also, commercial grades will decompose at 482°F (250°C) with sustaining decomposition at 536°F (280°C). Therefore, special consideration is required during any emergency situation involving a leak or spill of 2,3-dinitrotoluene. Should 2,3-dinitrotoluene ever come into contact with any incompatible substances during use, transportation, storage, or disposal, the formation of highly toxic and/or highly explosive commodities is extremely possible.

2,3-Dinitrotoluene may enter the environment through its intended use as an insecticide, as well as through industrial effluents or spills.

Acute Ecological Effects

Acute (short-term) toxic effects may include the death of animals, birds, or fish and death or low growth rate in plants. Acute effects are seen 2 to 4 days after animals or plants are exposed to 2,3-dinitrotoluene.

2,3-Dinitrotoluene has high acute toxicity to aquatic life. Insufficient data are available on the short-term effects of 2,3-dinitrotoluene exposure to plants, birds, or land animals.

Chronic Ecological Effects

Chronic (long-term) toxic effects may include shortened life span, reproductive problems, lower fertility, and changes in appearance or behavior in exposed animals. These effects can be seen long after first exposure(s) to toxic chemicals.

2,3-Dinitrotoluene has moderate chronic toxicity to aquatic life. Insufficient data are available on the long-term effects of 2,3-dinitrotoluene to plants, birds, or land animals.

Water Solubility

2,3-Dinitrotoluene is nearly insoluble in water. Concentrations of less than 1 milligram may mix with a liter of water.

Persistence in the Environment

2,3-Dinitrotoluene is highly persistent in the aquatic environment, with a half-life of greater than 200 days. The half-life of a pollutant is the amount of time it takes for one half of the chemical to be degraded.

Bioaccumulation in Aquatic Organisms

Some substances increase in concentration, or *bioaccumulate,* in living organisms as they breathe contaminated air, drink contaminated water, or eat contaminated food. These chemicals can become concentrated in the tissues and internal organs of animals as well as humans.

The concentration of 2,3-dinitrotoluene found in fish tissue is expected to be somewhat higher than the average concentration of 2,3-dinitrotoluene in the water from which the fish was taken.

Recommended Risk-Reduction Measures

Proper training of all transporters will reduce the likelihood of a mishap or accident resulting in a leak or spill to the environment. The correct use of DOT labeling (per the MSDS) on all containers, trucks, and rail cars will enable emergency responders to react properly and quickly to any disaster thereby reducing the potential risk to the environment and to personnel.

Storage of 2,3-dinitrotoluene should be segregated from incompatible chemicals to minimize the risk of cross-contamination or contact. Containers must be protected from heat, damage, and shock. Buildings designated for storage should be equipped with appropriate fire protection systems (alarms, sprinklers, emergency lighting, portable extinguishers).

If a spill or leak to the environment has occurred, fire department, emergency response and/or hazardous materials spill personnel should be notified immediately. Cleanup should be attempted only by those trained in proper spill containment procedures. Contaminated soils should be removed for incineration and replaced with clean soil. If 2,3-dinitrotoluene should contact the water table, aquifer, or navigable waterway, time is of the essence. It is slightly soluble in the aquatic environment and total remediation may not be possible. The local and/or state emergency response authorities must be notified. A comprehensive emergency response or disaster preparedness/recovery plan should be in place prior to any operations involving the use, transportation, storage, or disposal of 2,3-dinitrotoluene. If 2,3-dinitrotoluene is spilled or leaked, the following specific steps are recommended:

☑ Restrict persons not wearing protective clothing from area of spill or leak until cleanup is complete and area can be opened for normal work.

☑ Remove all ignition sources, ventilate area of leak.

☑ Use a water spray to knock-down vapors. Spills should be collected and placed in steel drums. Area should be flushed with hot water to remove solid particulates. Any liquids should be contained, as well as any liquid run-off.

☑ It may be necessary to dispose of 2,3-dinitrotoluene as a hazardous waste. The responsible state agency or the regional office of the federal Environmental Protection Agency (EPA) should be contacted for specific recommendations.

RISK ASSESSMENT: BUSINESS

General Assessment

Accidents or mishaps involving 2,3-dinitrotoluene can present a significant threat to business operations. The loss or damage of equipment or facilities can significantly affect fiscal viability. Lawsuits that may result from personnel injury/death, public exposures, and/or environmental contamination will also require a serious expenditure of resources. Media attention surrounding an injury, death, or environmental damage can also result in a loss of profits and loss of current as well as future business. Always remember that anytime the terms "carcinogen," "cancer," or "reproductive hazard" are used, public emotion, hysteria, and ignorance can run equally high. This must be properly considered whenever developing public relations policies.

Recommended Risk-Reduction Measures

Company attorneys, safety and health professionals, and environmental specialists should be involved in the development of any procedures dealing with safety and emergency response. A company official should be pre-designated as a public relations officer with specific training in dealing with the press. Corporate plans and policies should be developed, approved, and implemented long before any need for such arises.

REFERENCES

American Conference of Governmental Industrial Hygienists. 1988. *Documentation of the Threshold Limit Values and Biological Exposure Indices,* 5th Edition (with updates). Cincinnati: ACGIH

Anthony, C. P., and N. J. Kolthoff. 1971. *Textbook of Anatomy and Physiology.* St. Louis: C. V. Mosby Company.

Bloom, A. D. and F. J. DeSerres. 1995. *Ecotoxicity and Human Health.* New York: CRC Press/Lewis Publishers.

Bueche, J. R. 1972. *Principles of Physics,* 2nd Edition. New York: McGraw-Hill.

Burrell, R., D. K. Flaherty, and L. J. Sauers. 1992. *Toxicology of the Immune System.* New York: Van Nostrand Reinhold.

Calabrese, E. J. 1994. *Biological Effects of Low Level Exposures.* New York: CRC Press/Lewis Publishers.

Cockerman, L. G., B. S. Shane. 1994. *Basic Environmental Toxicology.* New York: CRC Press/Lewis Publishers.

Environmental Protection Agency. 1983. *Chemical Hazard Information Profiles.* Washington, D.C.: U.S. EPA.

Fodden, C. R., and J. L. Weddell. 1992. *Hazardous Materials Emergency Action Data.* New York: CRC Press/Lewis Publishers.

Hallenbeck, W. H. 1993. *Quantitative Risk Assessment for Environmental and Occupational Health,* 2nd Edition. New York: CRC Press/Lewis Publishers.

Howard, P. H. 1995. *Handbook of Environmental Fate and Exposures Data for Organic Chemicals,* Vol. I-IV. New York: CRC Press/Lewis Publishers.

International Agency for Research on Cancer (IARC). 1987. *IARC Monographs on the Evaluation of Carcinogenic Risk to Humans, Overall Evaluations of Carcinogenicity: An updating of IARC Monographs Volumes 1-42.* Lyon: IARC.

Keith, L. H. (Editor). 1992. *Integrated Risk Information System (IRIS) Chemical Information Database.* Prepared by the U.S. Environmental Protection Agency. Boca Raton, FL: CRC Press/Lewis Publishers.

Keith, L. H. and D. B. Walters. (Editors) 1992. *National Toxicology Program's Chemical Database, Volumes I-VII.* Chelsea, MI: Lewis Publishers.

Lewis, R. J., Sr. 1992. *Sax's Dangerous Properties of Industrial Materials,* Volumes 1, 2, and 3, 8th Edition. New York: Van Nostrand Reinhold.

Meyer, E. 1990. *Chemistry of Hazardous Materials.* Englewood Cliffs, NJ: Prentice Hall.

Torkelson, T. R. V. K. Rowe. 1981. *Patty's Industrial Hygiene and Toxicology,* Vol. 2B. New York: Wiley-Interscience.

U.S. Department of Health and Human Services, National Institute for Occupational Safety and Health. 1994. *NIOSH Pocket Guide to Chemical Hazards.* Washington, D.C.: U.S. Government Printing Office.

MATERIAL SAFETY DATA SHEET

CHEMICAL NAME	
2,4-DINITROTOLUENE	**CRC PRESS/LEWIS PUBLISHERS DISCLAIMER:** THE INFORMATION AND RECOMMENDATIONS PRESENTED HEREIN ARE BASED ON SOURCES BELIEVED TO BE RELIABLE. CRC MAKES NO REPRESENTATION ON ITS COMPLETENESS OR ACCURACY. IT IS THE RESPONSIBILITY OF THE USER TO DETERMINE THE CHEMICAL'S SUITABILITY FOR ITS INTENDED USE, THE CHEMICAL'S SAFE USE, AND THE CHEMICAL'S PROPER DISPOSAL. NO REPRESENTATIONS AND/OR WARRANTIES, EITHER EXPRESSED OR IMPLIED, OF THE MERCHANTABILITY OR FITNESS FOR A PARTICULAR PURPOSE, OR OF ANY OTHER NATURE, ARE MADE WITH RESPECT TO THE INFORMATION PROVIDED IN THIS MSDS OR TO THE CHEMICAL TO WHICH INFORMATION MAY REFER. CRC NEITHER ASSUMES NOR AUTHORIZES ANY OTHER PERSON TO ASSUME FOR IT, ANY OTHER ADDITIONAL RESPONSIBILITY OR LIABILITY FOR THE USE OF, OR RELIANCE UPON, THIS INFORMATION.

HAZARD WARNING INFORMATION

HEALTH	FIRE	REACTIVE	OTHER	DEGREE OF HAZARD	COLOR CODING	OTHER CODES
3	1	3		0 = Minimum Hazard 1 = Slight Hazard 2 = Moderate Hazard 3 = Serious Hazard 4 = Severe Hazard	HEALTH = BLUE FIRE = RED REACTIVITY = YELLOW OTHER = WHITE	OX = Oxidizer ACID = Acid ALK = Alkali COR = Corrosive W = Use No Water

SECTION I - GENERAL INFORMATION

Characterization: **Aromatic Hydrocarbon**	RCRA Number: **U105**	EPA Class: **Toxic Waste**
DOT Proper Shipping Name: **2,4-Dinitrotoluene (solid)**	Chemical Abstract Service (CAS) Number: **121-14-2**	
DOT Hazard Class and Label Requirements: **Poison B; Label: Poison**	DOT Emergency Guide Code: **56**	
DOT Identification Number: **UN 1600 (molten); UN 2038 (solid)**	Chemical Formula: **$C_6H_3CH_3(NO_2)_2$**	

Synonyms
2,4-Dinitrotoluol; 2,4-DNT; 1-methyl-2,4-dinitrobenzene.

SECTION II - HAZARDOUS INGREDIENTS AND IDENTITY INFORMATION

Hazardous Components (specific identity)	OSHA Exposure Criteria	NIOSH Exposure Criteria	Immediately Dangerous to Life and Health (IDLH)	ACGIH Exposure Criteria
2,4-Dinitrotoluene (derivation: By nitration of toluene with nitric acid in the presence of concentrated sulfuric acid to yield approximately 80% 2,4- and 20% 2,6-isomers from which the 2,4-isomer is not usually separated).	**PEL (skin): 1.5 mg/m³** **STEL: Not Established**	**REL (skin): 1.5 mg/m³** **Possible Cancer Agent**	**50 mg/m³**	**TLV (skin): 1.5 mg/m³** **STEL: Not Established**

SECTION III - PHYSICAL AND CHEMICAL CHARACTERISTICS

Boiling Point: **482°F (250°C)**	Specific Gravity (H_2O = 1): **1.32 (liquid) 1.5 (solid)**
Vapor Pressure (gas): **1 at 69°F (20°C)**	Molecular Weight: **182.14**
Vapor Density (Air = 1): **6.27**	Melting Point: **158°F (70°C)**

Solubility
Nearly insoluble in water. Soluble in alcohol, acetone, benzene, carbon disulfide, ether, and pyridine.

Appearance and Odor
Orange to yellow crystalline solid with a slight characteristic odor. Usually shipped as molten.

SECTION IV - FIRE AND EXPLOSION HAZARD DATA

Flash Point (method used): **404°F (207°C) closed cup**	Explosive Limits in Air % by Volume: **LEL: 0.3% at 474°F (245°C) UEL: Not Determined**
NFPA Classification: **Combustible Solid**	Autoignition Temperature: **Not Determined**

Extinguishing Media
Use carbon dioxide, water, or dry chemical from a protected location.

Special Fire Fighting Procedures
A carcinogenic. Wear full protective clothing and self-contained breathing apparatus (SCBA). Poisonous gases are produced in fire. Use water spray to keep fire-exposed containers cool. Move containers from fire if it can be done without risk. Fight fire from a distance or protected position, if possible.

Unusual Fire and Explosion Hazards
Containers may explode in fire. Fires involving this material should be considered an explosion hazard. Extremely explosive in dry form when exposed to heat, flame, or shock. Decomposition is self-sustaining at 280°F (138°C).

SECTION V - REACTIVITY DATA

Stability		Conditions to Avoid
		2,4-Dinitrotoluene is explosive when exposed to heat, friction, or contamination. It can be detonated using a very strong initiating device. It is slightly sensitive to impact shock, especially in solid form.
Stable	Unstable	Incompatibility (*materials to avoid*)
	X	Strong oxidizers, (such as chlorine, bromine, and fluorine), organic materials, strong reducing agents (such as sodium sulfide, zinc powder, sodium hyposulfite, and metallic hydrides), caustics, and metals.
Hazardous Polymerization		Conditions to Avoid
		Under normal conditions of temperature and pressure, hazardous polymerization of 2,4-dinitrotoluene is not expected to occur.
May Occur	Will Not Occur	Hazardous Decomposition or By-products
	X	When heated to decomposition, 2,4-dinitrotoluene can emit highly toxic/poisonous gases, including carbon dioxide and oxides of nitrogen.

SECTION VI - HEALTH HAZARD DATA

Primary Route(s) of Entry:	Inhalation?	Absorption (skin)?	Ingestion?
	X	X	X

Health Hazards

INHALATION: Can cause cyanosis, headache, irritability, anoxia, anemia, dizziness, weakness, nausea, vomiting, dypsnea, drowsiness, collapse, loss of consciousness, and death by cardiovascular collapse.

ABSORPTION: Irritation, small blisters, redness, swelling, ulceration, and necrosis (will pass through skin to cause same toxic systemic effects as inhalation). Eye contact will result in burns, irritation of the lids, redness, swelling, pain on exposure to light, and tissue damage.

INGESTION: Irritation of the mouth, throat, and stomach, stomach cramps, and diarrhea.

Carcinogenicity	NTP Listed?	IARC Cancer Review Group?	OSHA Regulated?	Target Organs?
Suspected Human Confirmed Animal	5th Annual Report	No	29 CFR 1910-1000 Table Z-1	Blood, liver, CVS, reproductive system, liver, skin

Medical Conditions Generally Aggravated by Exposure

Persons with blood disorders may be at an increased risk if exposed to 2,4-dinitrotoluene.

Emergency and First-aid Procedures

Eye contact: Flush immediately with water for 15 minutes (minimum), seek medical attention. Skin contact: Remove all contaminated clothing. Immediately wash area with large amounts of soap and water. Seek medical assistance. For inhalation: Remove the person from exposure. Provide CPR if required. Transfer to medical facility. If swallowed: Give large amounts of water or milk and seek medical attention. Never attempt to give an unconscious or convulsing person anything by mouth.

SECTION VII - PRECAUTIONS FOR SAFE HANDLING AND USE

Steps to be Taken in Case Material is Released or Spilled

Remove all ignition sources and ventilate area. Evacuate personnel not involved in cleanup. Stay upwind. Cleanup crew should wear fully-encapsulating suits. Use water spray to reduce vapors. Collect spilled powders and place in sealed steel drums for disposal.

Preferred Waste Disposal Method

No Citation.

Precautions to be Taken in Handling and Storage

Store in tightly closed containers in cool, well-ventilated area away from heat or flame. Heat may cause containers to build up pressure and explode. Usually stored as a molten liquid.

Other Precautions and Warnings

Sources of ignition are prohibited where 2,4-dinitrotoluene is used, handled, or stored.

SECTION VIII - CONTROL MEASURES AND PERSONAL PROTECTIVE EQUIPMENT

Respiratory Protection (*specify type*)

PEL is low and difficult to monitor. For any exposure to a suspected carcinogen, recommend a supplied air respirator with full facepiece, hood, or helmet in continuous flow mode, or use a self-contained breathing apparatus (SCBA) operated in pressure demand or other positive pressure mode.

Ventilation

Local exhaust or general mechanical systems recommended.

Protective Gloves	Eye Protection	Other Protective Clothing
Butyl Rubber	Chemical/Dust Goggles or Face Mask	Butyl Rubber Apron

Work/Hygiene Practices

Always wash hands thoroughly after using chemical; never bring food, drink, or smoking materials into vicinity of chemicals.

2,4-DINITROTOLUENE

$C_7H_6N_2O_4$ CAS: 121-14-2

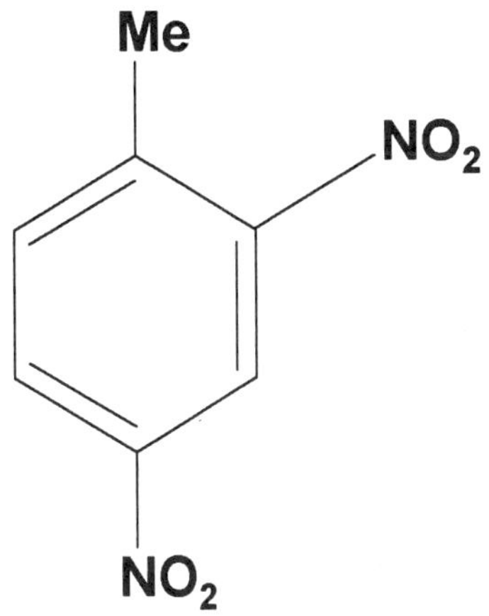

IDENTIFICATION AND TYPICAL USES

2,4-Dinitrotoluene is an orange to yellow crystalline solid with a characteristic odor. It is normally shipped as a molten liquid. It is one of numerous isomers of dinitrotoluene and is generally considered the most commonly encountered in industry. It is used primarily in the manufacture of explosives and dyes, in organic synthesis, as a chemical intermediate, a plasticizer, and as a modifier for smokeless powders in the munitions industry.

RISK ASSESSMENT: HEALTH

General Assessment

2,4-Dinitrotoluene is a human poison by *ingestion,* with equally toxic systemic consequences occurring by way of *inhalation,* and skin contact (*absorption*). It has been shown to cause teratogenic effects in experimental tests and mutation data have also been reported. It is a confirmed animal carcinogen and a suspected human carcinogen. Liver or kidney damage can occur up to 3 days following exposure.

Inhalation and skin absorption can result in cyanosis, methemoglobinemia, headache, irritability, dizziness, weakness, nausea, vomiting, dyspnea, drowsiness, and loss of consciousness. If allowed to go untreated, death due to cardiovascular collapse can occur. Repeated exposures can lead to anemia and liver damage. It has also been known to cause mutations and liver tumors in tests animals.

Ingestion will result in irritation of the mouth and stomach, stomach cramps, and diarrhea. Possible systemic effects similar to inhalation and absorption may also occur. Ingestion of alcohol shortly before or after exposure may increase individual susceptibility to toxicity.

Skin contact will not only result in absorption, it will also produce irritation, small blisters, redness, swelling, ulceration, and localized necrosis. Eye contact causes burning pain, irritation, redness, and swelling of the eyelids. There may be intense pain on exposure to light and severe eye damage (may be permanent). Contact with the molten liquid will cause thermal burns.

Acute Health Effects

The following acute (short-term) health effects may occur immediately or shortly after exposure to 2,4-dinitrotoluene:

<u>Skin</u>: Irritation and possible tissue damage. Will pass through unbroken skin to cause serious systemic effects.

<u>Eye</u>: Irritation, swelling, redness, pain on exposure to light, and potential damage to the surrounding tissue.

<u>Lung</u>: Irritation to the respiratory tract and associated mucosa. Absorption through the respiration process causes toxic systemic effects (including anemia and methemoglobinemia).

Chronic Health Effects

The following chronic (long-term) health effects may occur at some time after exposure to 2,4-dinitrotoluene and can last for months or even years:

Cancer Hazards: According to the information available in the references, 2,4-dinitrotoluene is a confirmed carcinogen in test animals. Its carcinogenic properties in humans is highly suspect. Experimental mutation data have been reported and many scientists believe that such chemicals may have the ability to cause cancer.

Reproductive Hazard: According to information presented in the references, 2,4-dinitrotoluene has been shown to cause adverse effects in reproduction in animals. It is considered an experimental teratogen. Human effects in this regard are inconclusive and require additional research.

Other Chronic Effects: Long-term exposures can result in anemia and possible liver damage. It is also known to cause mutations and tumors in the livers of test animals.

Recommended Risk-Reduction Measures

Personnel should avoid direct contact with 2,4-dinitrotoluene. If a less toxic chemical cannot be substituted for a hazardous substance, then *engineering controls* are the most effective method of reducing exposures. The best protection is to enclose operations and/or provide local exhaust ventilation at the site of chemical release. While not always operationally feasible, isolating operations can also reduce exposure. Using respiratory protection is less effective than the controls mentioned above, but is still advisable whenever working with or around 2,4-dinitrotoluene. The exposure level is low (1.5 mg/m^3) and difficult to monitor. Therefore, a MSHA/NIOSH approved full facepiece respirator equipped with an organic vapor cartridge equipped with a dust pre-filter should be used if any exposure is anticipated or known. Increased protection is obtained from a supplied-air respirator with full facepiece operated in continuous flow mode, or a self-contained breathing apparatus (SCBA) with full facepiece and pressure demand. If a full facepiece is not available, then chemical dust goggles should be worn to protect the eyes. Whenever an airborne chemical dust hazard exists, a face shield and a rubber apron should be worn. To prevent hand and skin exposures, butyl rubber gloves should be worn.

Administrative controls should also be in place to minimize the potential for human exposures. These may include written procedures or policies which specify the methods and techniques that will be practiced whenever personnel are to work with 2,4-dinitrotoluene.

All personnel should receive *training* on the use, hazards, protective measures, emergency actions, and other precautions per 29 CFR 1910.1200 (Hazard Communication), prior to the first assignment in an area where 2,4-dinitrotoluene is used or stored.

Before beginning employment with 2,4-dinitrotoluene and at regular intervals thereafter (e.g., annually), those persons with high or frequent exposures should be provided the following recommended medical tests:

☑ Liver and kidney function tests.

☑ Urinalysis (establish baseline).

If symptoms develop or overexposure is suspected, the following tests are recommended:

☑ Urinalysis (compare to baseline).

☑ Complete blood count.

☑ Liver and kidney function tests.

Any evaluation should include a careful history of past and present symptoms with an examination. Medical tests that simply look for existing damage are not a substitute for controlling exposures.

Other methods to reduce exposure include:

☑ Where possible, enclose operations and use local exhaust ventilation at the site of chemical release. If local exhaust ventilation or enclosure is not used, respiratory protection should be mandatory.

☑ Always ensure that proper protective clothing is worn when using chemical substances.

☑ Wash thoroughly immediately after exposure to 2,4-dinitrotoluene and at the end of the work shift or before eating, drinking, or smoking.

☑ Hazard warning information should be posted in the work area. In addition, as part of an on-going education and training program, all information on the health and safety hazards of 2,4-dinitrotoluene should be communicated to all exposed and potentially exposed workers.

☑ Eye wash stations should be provided in the immediate work area for emergency use. If there is a possibility of skin exposure to 2,4-dinitrotoluene, emergency shower facilities should be provided.

☑ Personnel should never wear contaminated clothing home (family members can be exposed). Clothing should be laundered by personnel who have been trained on the hazards of 2,4-dinitrotoluene.

RISK ASSESSMENT: ENVIRONMENT

General Assessment

The environment is at risk of exposure during transportation, storage, disposal, or destruction of 2,4-dinitrotoluene. In almost every scenario, the threat of environmental exposure is contingent upon the proper handling of the chemical substance. Accidental spills, large or small, can result in fire, explosion, and possible contamination of the surrounding environmental mediums (water, soil, and air).

2,4-Dinitrotoluene is normally considered a combustible solid. Its dusts may form explosive mixtures in air. These characteristics require caution in handling, storage, transportation, and disposal. It is incompatible with strong oxidizers, caustics, and metals (such as tin and zinc). Contact with either of these can cause fire or explosion. Also, commercial grades will decompose at 482°F (250°C) with sustaining decomposition at 536°F (280°C). Therefore, special consideration is required during any emergency situation involving a leak or spill of 2,4-dinitrotoluene. Should 2,4-dinitrotoluene ever come into contact with any incompatible substances during use, transportation, storage, or disposal, the formation of highly toxic and/or highly explosive commodities is extremely possible.

2,4-Dinitrotoluene may enter the environment through its intended use as an insecticide, as well as through industrial effluents or spills.

Acute Ecological Effects

Acute (short-term) toxic effects may include the death of animals, birds, or fish and death or low growth rate in plants. Acute effects are seen 2 to 4 days after animals or plants are exposed to 2,4-dinitrotoluene.

2,4-Dinitrotoluene has high acute toxicity to aquatic life. Insufficient data are available on the short-term effects of 2,4-dinitrotoluene exposure to plants, birds, or land animals.

Chronic Ecological Effects

Chronic (long-term) toxic effects may include shortened life span, reproductive problems, lower fertility, and changes in appearance or behavior in exposed animals. These effects can be seen long after first exposure(s) to toxic chemicals.

2,4-Dinitrotoluene has moderate chronic toxicity to aquatic life. Insufficient data are available on the long-term effects of 2,4-dinitrotoluene to plants, birds, or land animals.

Water Solubility

2,4-Dinitrotoluene is nearly insoluble in water. Concentrations of less than 1 milligram may mix with a liter of water.

Persistence in the Environment

2,4-Dinitrotoluene is highly persistent in the aquatic environment, with a half-life of greater than 200 days. The half-life of a pollutant is the amount of time it takes for one half of the chemical to be degraded.

Bioaccumulation in Aquatic Organisms

Some substances increase in concentration, or *bioaccumulate,* in living organisms as they breathe contaminated air, drink contaminated water, or eat contaminated food. These chemicals can become concentrated in the tissues and internal organs of animals as well as humans.

The concentration of 2,4-dinitrotoluene found in fish tissue is expected to be somewhat higher than the average concentration of 2,4-dinitrotoluene in the water from which the fish was taken.

Recommended Risk-Reduction Measures

Proper training of all transporters will reduce the likelihood of a mishap or accident resulting in a leak or spill to the environment. The correct use of DOT labeling (per the MSDS) on all containers, trucks, and rail cars will enable emergency responders to react properly and quickly to any disaster thereby reducing the potential risk to the environment and to personnel.

Storage of 2,4-dinitrotoluene should be segregated from incompatible chemicals to minimize the risk of cross-contamination or contact. Containers must be protected from heat, damage, and shock. Buildings designated for storage should be equipped with appropriate fire protection systems (alarms, sprinklers, emergency lighting, portable extinguishers).

If a spill or leak to the environment has occurred, fire department, emergency response and/or hazardous materials spill personnel should be notified immedi-

ately. Cleanup should be attempted only by those trained in proper spill containment procedures. Contaminated soils should be removed for incineration and replaced with clean soil. If 2,4-dinitrotoluene should contact the water table, aquifer, or navigable waterway, time is of the essence. It is slightly soluble in the aquatic environment and total remediation may not be possible. The local and/or state emergency response authorities must be notified. A comprehensive emergency response or disaster preparedness/recovery plan should be in place prior to any operations involving the use, transportation, storage, or disposal of 2,4-dinitrotoluene. If 2,4-dinitrotoluene is spilled or leaked, the following specific steps are recommended:

- ☑ Restrict persons not wearing protective clothing from area of spill or leak until cleanup is complete and area can be opened for normal work.
- ☑ Remove all ignition sources, ventilate area of leak.
- ☑ Use a water spray to knock-down vapors. Spills should be collected and placed in steel drums. Area should be flushed with hot water to remove solid particulates. Any liquids should be contained, as well as any liquid run-off.
- ☑ It may be necessary to dispose of 2,4-dinitrotoluene as a hazardous waste. The responsible state agency or the regional office of the federal Environmental Protection Agency (EPA) should be contacted for specific recommendations.

RISK ASSESSMENT: BUSINESS

General Assessment

Accidents or mishaps involving 2,4-dinitrotoluene can present a significant threat to business operations. The loss or damage of equipment or facilities can significantly affect fiscal viability. Lawsuits that may result from personnel injury/death, public exposures, and/or environmental contamination will also require a serious expenditure of resources.

Media attention surrounding an injury, death, or environmental damage can also result in a loss of profits and loss of current as well as future business. Always remember that anytime the terms "carcinogen," "cancer," or "reproductive hazard" are used, public emotion, hysteria, and ignorance can run equally high. This must be properly considered whenever developing public relations policies.

Recommended Risk-Reduction Measures

Company attorneys, safety and health professionals, and environmental specialists should be involved in the development of any procedures dealing with safety or emergency response. A company official should be pre-designated as a public relations officer with specific training in dealing with the press. Corporate plans and policies should be developed, approved, and implemented long before any need for such arises.

REFERENCES

Ahlbom, A. 1993. *Biostatistics for Epidemiologists.* New York: CRC Press/Lewis Publishers.

Anthony, C. P., and N. J. Kolthoff. 1971. *Textbook of Anatomy and Physiology.* St. Louis: C. V. Mosby Company.

Bloom, A. D. and F. J. DeSerres. 1995. *Ecotoxicity and Human Health.* New York: CRC Press/Lewis Publishers.

Bueche, J. R. 1972. *Principles of Physics,* 2nd Edition. New York: McGraw-Hill.

Burrell, R., D. K. Flaherty, and L. J. Sauers. 1992. *Toxicology of the Immune System.* New York: Van Nostrand Reinhold.

Calabrese, E. J. 1994. *Biological Effects of Low Level Exposures.* New York: CRC Press/Lewis Publishers.

Cockerman, L. G., B. S. Shane. 1994. *Basic Environmental Toxicology.* New York: CRC Press/Lewis Publishers.

Environmental Protection Agency. 1983. *Chemical Hazard Information Profiles.* Washington, D.C.: U.S. EPA.

Fodden, C. R., and J. L. Weddell. 1992. *Hazardous Materials Emergency Action Data.* New York: CRC Press/Lewis Publishers.

Hallenbeck, W. H. 1993. *Quantitative Risk Assessment for Environmental and Occupational Health,* 2nd Edition. New York: CRC Press/Lewis Publishers.

Howard, P. H. 1995. *Handbook of Environmental Fate and Exposures Data for Organic Chemicals,* Vol. I-IV. New York: CRC Press/Lewis Publishers.

Lewis, R. J., Sr. 1992. *Sax's Dangerous Properties of Industrial Materials,* Volumes 1, 2, and 3, 8th Edition. New York: Van Nostrand Reinhold.

Meyer, E. 1990. *Chemistry of Hazardous Materials.* Englewood Cliffs, NJ: Prentice Hall.

Quigley, D. R. 1994. *Handbook of Emergency Chemical Management.* New York: CRC Press/Lewis Publishers.

U.S. Department of Health and Human Services, National Institute for Occupational Safety and Health. 1994. *NIOSH Pocket Guide to Chemical Hazards.* Washington, D.C.: U.S. Government Printing Office.

MATERIAL SAFETY DATA SHEET

CHEMICAL NAME	
2,6-DINITROTOLUENE	**CRC PRESS/LEWIS PUBLISHERS DISCLAIMER:** THE INFORMATION AND RECOMMENDATIONS PRESENTED HEREIN ARE BASED ON SOURCES BELIEVED TO BE RELIABLE. CRC MAKES NO REPRESENTATION ON ITS COMPLETENESS OR ACCURACY. IT IS THE RESPONSIBILITY OF THE USER TO DETERMINE THE CHEMICAL'S SUITABILITY FOR ITS INTENDED USE, THE CHEMICAL'S SAFE USE, AND THE CHEMICAL'S PROPER DISPOSAL. NO REPRESENTATIONS AND/OR WARRANTIES, EITHER EXPRESSED OR IMPLIED, OF THE MERCHANTABILITY OR FITNESS FOR A PARTICULAR PURPOSE, OR OF ANY OTHER NATURE, ARE MADE WITH RESPECT TO THE INFORMATION PROVIDED IN THIS MSDS OR TO THE CHEMICAL TO WHICH INFORMATION MAY REFER. CRC NEITHER ASSUMES NOR AUTHORIZES ANY OTHER PERSON TO ASSUME FOR IT, ANY OTHER ADDITIONAL RESPONSIBILITY OR LIABILITY FOR THE USE OF, OR RELIANCE UPON, THIS INFORMATION.

HAZARD WARNING INFORMATION

HEALTH	FIRE	REACTIVE	OTHER	DEGREE OF HAZARD	COLOR CODING	OTHER CODES
3	1	3		0 = Minimum Hazard 1 = Slight Hazard 2 = Moderate Hazard 3 = Serious Hazard 4 = Severe Hazard	HEALTH = BLUE FIRE = RED REACTIVITY = YELLOW OTHER = WHITE	OX = Oxidizer ACID = Acid ALK = Alkali COR = Corrosive W = Use No Water

SECTION I - GENERAL INFORMATION

Characterization	RCRA Number	EPA Class
Aromatic Hydrocarbon	**U106**	**Toxic Waste**

DOT Proper Shipping Name **2,6-Dinitrotoluene (solid)**	Chemical Abstract Service (CAS) Number **606-20-2**
DOT Hazard Class and Label Requirements **Poison B; Label: Poison**	DOT Emergency Guide Code **56**
DOT Identification Number **UN 1600 (molten); UN 2038 (solid)**	Chemical Formula $C_6H_3CH_3(NO_2)_2$

Synonyms
2,6-DNT; 2-methyl-1,3-dinitrobenzene.

SECTION II - HAZARDOUS INGREDIENTS AND IDENTITY INFORMATION

Hazardous Components (specific identity)	OSHA Exposure Criteria	NIOSH Exposure Criteria	Immediately Dangerous to Life and Health (IDLH)	ACGIH Exposure Criteria
2,6-Dinitrotoluene (derivation: By nitration of toluene with nitric acid in the presence of concentrated sulfuric acid to yield approximately 80% 2,4- and 20% 2,6-isomers from which the 2,4-isomer is not usually separated).	**PEL (skin):** 1.5 mg/m³ **STEL:** Not Established	**REL (skin):** 1.5 mg/m³ **Possible Cancer Agent**	**50 mg/m³**	**TLV (skin):** 1.5 mg/m³ **STEL:** Not Established

SECTION III - PHYSICAL AND CHEMICAL CHARACTERISTICS

Boiling Point **482°F (250°C)**	Specific Gravity (H_2O = 1) **1.32 (liquid) 1.5 (solid)**
Vapor Pressure (gas) **1 at 69°F (20°C)**	Molecular Weight **182.14**
Vapor Density (Air = 1) **6.27**	Melting Point **158°F (70°C)**

Solubility
Nearly insoluble in water. Soluble in alcohol, acetone, benzene, carbon disulfide, ether, and pyridine.

Appearance and Odor
Orange to yellow crystalline solid with a slight characteristic odor. Usually shipped as molten.

SECTION IV - FIRE AND EXPLOSION HAZARD DATA

Flash Point (method used) **404°F (207°C) closed cup**	Explosive Limits in Air % by Volume **LEL: 0.3% at 474°F (245°C) UEL: Not Determined**
NFPA Classification **Combustible Solid**	Autoignition Temperature **Not Determined**

Extinguishing Media
Use carbon dioxide, water, or dry chemical from a protected location.

Special Fire Fighting Procedures
A carcinogenic. Wear full protective clothing and self-contained breathing apparatus (SCBA). Poisonous gases are produced in fire. Use water spray to keep fire-exposed containers cool. Move containers from fire if it can be done without risk. Fight fire from a distance or protected position, if possible.

Unusual Fire and Explosion Hazards
Containers may explode in fire. Fires involving this material should be considered an explosion hazard. Extremely explosive in dry form when exposed to heat, flame, or shock. Decomposition is self-sustaining at 280°F (138°C).

SECTION V - REACTIVITY DATA

Stability		Conditions to Avoid 2,6-Dinitrotoluene is explosive when exposed to heat, friction, or contamination. It can be detonated using a very strong initiating device. It is slightly sensitive to impact shock, especially in solid form.
Stable	Unstable X	Incompatibility *(materials to avoid)* Strong oxidizers, (such as chlorine, bromine, and fluorine), organic materials, strong reducing agents (such as sodium sulfide, zinc powder, sodium hyposulfite, and metallic hydrides), caustics, and metals.
Hazardous Polymerization		Conditions to Avoid Under normal conditions of temperature and pressure, hazardous polymerization of 2,6-dinitrotoluene is not expected to occur.
May Occur	Will Not Occur X	Hazardous Decomposition or By-products When heated to decomposition, 2,6-dinitrotoluene can emit highly toxic/poisonous gases, including carbon dioxide and oxides of nitrogen.

SECTION VI - HEALTH HAZARD DATA

Primary Route(s) of Entry:	Inhalation?	Absorption (skin)?	Ingestion?
	X	X	X

Health Hazards

INHALATION: Can cause cyanosis, headache, irritability, anoxia, anemia, dizziness, weakness, nausea, vomiting, dypsnea, drowsiness, collapse, loss of consciousness, and death by cardiovascular collapse.

ABSORPTION: Irritation, small blisters, redness, swelling, ulceration, and necrosis (will pass through skin to cause same toxic systemic effects as inhalation). Eye contact will result in burns, irritation of the lids, redness, swelling, pain on exposure to light, and tissue damage.

INGESTION: Irritation of the mouth, throat, and stomach, stomach cramps, and diarrhea.

Carcinogenicity	NTP Listed?	IARC Cancer Review Group?	OSHA Regulated?	Target Organs?
Suspected Human Confirmed Animal	5th Annual Report	No	29 CFR 1910-1000 Table Z-1	Blood, liver, CVS, reproductive system, liver, skin

Medical Conditions Generally Aggravated by Exposure

Persons with blood disorders may be at an increased risk if exposed to 2,6-dinitrotoluene.

Emergency and First-aid Procedures

Eye contact: Flush immediately with water for 15 minutes (minimum), seek medical attention. Skin contact: Remove all contaminated clothing. Immediately wash area with large amounts of soap and water. Seek medical assistance. For inhalation: Remove the person from exposure. Provide CPR if required. Transfer to medical facility. If swallowed: Give large amounts of water or milk and seek medical attention. Never attempt to give an unconscious or convulsing person anything by mouth.

SECTION VII - PRECAUTIONS FOR SAFE HANDLING AND USE

Steps to be Taken in Case Material is Released or Spilled

Remove all ignition sources and ventilate area. Evacuate personnel not involved in cleanup. Stay upwind. Cleanup crew should wear fully-encapsulating suits. Use water spray to reduce vapors. Collect spilled powders and place in sealed steel drums for disposal.

Preferred Waste Disposal Method

No Citation.

Precautions to be Taken in Handling and Storage

Store in tightly closed containers in cool, well-ventilated area away from heat or flame. Heat may cause containers to build up pressure and explode. Usually stored as a molten liquid.

Other Precautions and Warnings

Sources of ignition are prohibited where 2,6-dinitrotoluene is used, handled, or stored.

SECTION VIII - CONTROL MEASURES AND PERSONAL PROTECTIVE EQUIPMENT

Respiratory Protection *(specify type)*

PEL is low and difficult to monitor. For any exposure to a suspected carcinogen, recommend a supplied-air respirator with full facepiece, hood, or helmet in continuous flow mode, or use a self-contained breathing apparatus (SCBA) operated in pressure demand or other positive pressure mode.

Ventilation

Local exhaust or general mechanical systems recommended.

Protective Gloves	Eye Protection	Other Protective Clothing
Butyl Rubber	Chemical/Dust Goggles or Face Mask	Butyl Rubber Apron

Work/Hygiene Practices

Always wash hands thoroughly after using chemical; never bring food, drink, or smoking materials into vicinity of chemicals.

2,6-DINITROTOLUENE

$C_7H_6N_2O_4$ CAS: 606-20-2

IDENTIFICATION AND TYPICAL USES

2,6-Dinitrotoluene is an orange to yellow crystalline solid with a characteristic odor. It is normally shipped as a molten liquid. It is one of numerous isomers of dinitrotoluene. It is used in the manufacture of explosives and dyes, in organic synthesis, as a chemical intermediate, a plasticizer, and as a modifier for smokeless powders in the munitions industry.

RISK ASSESSMENT: HEALTH

General Assessment

2,6-Dinitrotoluene is a human poison by *ingestion,* with equally toxic systemic consequences occurring by way of *inhalation,* and skin contact (*absorption*). The closely-related 2,4-isomer has been shown to cause teratogenic effects in experimental tests and mutation data have also been reported. It is also a confirmed animal carcinogen and a suspected human carcinogen. Liver or kidney damage can occur up to 3 days following exposure. While these effects have not been confirmed for the 2,6-isomer, extreme caution is recommended in use, handling, storage, or disposal.

Inhalation and skin absorption can result in cyanosis, methemoglobinemia, headache, irritability, dizziness, weakness, nausea, vomiting, dyspnea, drowsiness, and loss of consciousness. If allowed to go untreated, the result can be death due to cardiovascular collapse. Repeated exposures can lead to anemia and liver damage. It has caused mutations and liver tumors in test animals.

Ingestion will result in irritation of the mouth and stomach, stomach cramps, and diarrhea. Possible systemic effects similar to inhalation and absorption may also occur. Ingestion of alcohol shortly before or after exposure may increase individual susceptibility to toxicity.

Skin contact will not only result in absorption, it will also produce irritation, small blisters, redness, swelling, ulceration, and localized necrosis. Eye contact causes burning pain, irritation, redness, and swelling of the eyelids. There may be pain on exposure to light and severe eye damage. Contact with the molten liquid will cause thermal burns.

Acute Health Effects

The following acute (short-term) health effects may occur immediately or shortly after exposure to 2,6-dinitrotoluene:

Skin: Irritation and possible tissue damage. Will pass through unbroken skin to cause serious systemic effects.

Eye: Irritation, swelling, redness, pain on exposure to light, and potential damage to the surrounding tissue.

Lung: Irritation to the respiratory tract and associated mucosa. Absorption through the respiration process causes toxic systemic effects (including anemia and methemoglobinemia).

Chronic Health Effects

The following chronic (long-term) health effects may occur at some time after exposure to 2,6-dinitrotoluene and can last for months or even years:

Cancer Hazards: According to the information available in the references, the 2,4-isomer of dinitrotoluene is a confirmed carcinogen in test animals. Its carcinogenic properties in humans is highly suspect. Experimental mutation data have been reported and many sci-

entists believe that such chemicals may have the ability to cause cancer. It is not known whether 2,6-isomer of dinitrotoluene will cause these effects.

Reproductive Hazard: According to information presented in the references, 2,4-dinitrotoluene has been shown to cause adverse effects in reproduction in animals. It is considered an experimental teratogen. It is not known whether 2,6-isomer of dinitrotoluene will cause these effects. Human effects in this regard are inconclusive for both isomers and additional research is required.

Other Chronic Effects: Long-term exposures can result in anemia and possible liver damage. It is also known to cause mutations and tumors in the livers of test animals.

Recommended Risk-Reduction Measures

Personnel should avoid direct contact with 2,6-dinitrotoluene. If a less toxic chemical cannot be substituted for a hazardous substance, then *engineering controls* are the most effective method of reducing exposures. The best protection is to enclose operations and/or provide local exhaust ventilation at the site of chemical release. While not always operationally feasible, isolating operations can also reduce exposure. Using respiratory protection is less effective than the controls mentioned above, but is still advisable whenever working with or around 2,6-dinitrotoluene. The exposure level is low (1.5 mg/m^3) and difficult to monitor. Therefore, a MSHA/NIOSH approved full facepiece respirator equipped with an organic vapor cartridge equipped with a dust pre-filter should be used if any exposure is anticipated or known. Increased protection is obtained from a supplied-air respirator with full facepiece operated in continuous flow mode, or a self-contained breathing apparatus (SCBA) with full facepiece and pressure demand. If a full facepiece is not available, then chemical dust goggles should be worn to protect the eyes. Whenever an airborne chemical dust hazard exists, a face shield and a rubber apron should be worn. To prevent hand and skin exposures, butyl rubber gloves should be worn. Glove manufactures should be contacted and permeation studies obtained *before* final glove selections are made.

Administrative controls should also be in place to minimize the potential for human exposures. These may include written procedures or policies which specify the methods and techniques that will be practiced whenever personnel are to work with 2,6-dinitrotoluene.

All personnel should receive *training* on the use, hazards, protective measures, emergency actions, and other precautions per 29 CFR 1910.1200 (Hazard Communication), prior to the first assignment in an area where 2,6-dinitrotoluene is used or stored.

Before beginning employment with 2,6-dinitrotoluene and at regular intervals thereafter (e.g., annually), those persons with high or frequent exposures should be provided the following recommended medical tests:

☑ Liver and kidney function tests.

☑ Urinalysis (establish baseline).

If symptoms develop or overexposure is suspected, the following tests are recommended:

☑ Urinalysis (compare to baseline).

☑ Complete blood count.

☑ Liver and kidney function tests.

Any evaluation should include a careful history of past and present symptoms with an examination. Medical tests that simply look for existing damage are not a substitute for controlling exposures.

Other methods to reduce exposure include:

☑ Where possible, enclose operations and use local exhaust ventilation at the site of chemical release. If local exhaust ventilation or enclosure is not used, respiratory protection should be mandatory.

☑ Always ensure that proper protective clothing is worn when using chemical substances and that personnel are trained in its use and care.

☑ Wash thoroughly immediately after exposure to 2,6-dinitrotoluene and at the end of the work shift or before eating, drinking, or smoking.

☑ Hazard warning information should be posted in the work area. In addition, as part of an on-going education and training program, all information on the health and safety hazards of 2,6-dinitrotoluene should be communicated to all potentially exposed workers.

☑ Eye wash stations should be provided in the immediate work area for emergency use. If there is a possibility of skin exposure to 2,6-dinitrotoluene, emergency shower facilities should also be provided.

☑ Personnel should never wear contaminated clothing home (family members can be exposed). Clothing should be laundered by personnel who have been trained on the hazards of 2,6-dinitrotoluene.

RISK ASSESSMENT: ENVIRONMENT

General Assessment

The environment is at risk of exposure during transportation, storage, disposal, or destruction of 2,6-dinitrotoluene. In almost every scenario, the threat of environmental exposure is contingent upon the proper handling of the chemical substance. Accidental spills, large or small, can result in fire, explosion, and possible contamination of the surrounding environmental mediums (water, soil, and air).

2,6-Dinitrotoluene is normally considered a combustible solid. Its dusts may form explosive mixtures in air. These characteristics require caution in handling, storage, transportation, and disposal. It is incompatible with strong oxidizers, caustics, and metals (such as tin and zinc). Contact with either of these can cause fire or explosion. Also, commercial grades will decompose at 482°F (250°C) with sustaining decomposition at 536°F (280°C). Therefore, special consideration is required during any emergency situation involving a leak or spill of 2,6-dinitrotoluene. Should 2,6-dinitrotoluene ever come into contact with any incompatible substances during use, transportation, storage, or disposal, the formation of highly toxic and/or highly explosive commodities is extremely possible.

2,6-Dinitrotoluene may enter the environment through its intended use as an insecticide, as well as through industrial effluents or spills.

Acute Ecological Effects

Acute (short-term) toxic effects may include the death of animals, birds, or fish and death or low growth rate in plants. Acute effects are seen 2 to 4 days after animals or plants are exposed to 2,6-dinitrotoluene.

2,6-Dinitrotoluene has high acute toxicity to aquatic life. Insufficient data are available on the short-term effects of 2,6-dinitrotoluene exposure to plants, birds, or land animals.

Chronic Ecological Effects

Chronic (long-term) toxic effects may include shortened life span, reproductive problems, lower fertility, and changes in appearance or behavior in exposed animals. These effects can be seen long after first exposure(s) to toxic chemicals.

2,6-Dinitrotoluene has moderate chronic toxicity to aquatic life. Insufficient data are available on the long-term effects of 2,6-dinitrotoluene to plants, birds, or land animals.

Water Solubility

2,6-Dinitrotoluene is nearly insoluble in water. Concentrations of less than 1 milligram may mix with a liter of water.

Persistence in the Environment

2,6-Dinitrotoluene is highly persistent in the aquatic environment, with a half-life of greater than 200 days. The half-life of a pollutant is the amount of time it takes for one half of the chemical to be degraded.

Bioaccumulation in Aquatic Organisms

Some substances increase in concentration, or *bioaccumulate*, in living organisms as they breathe contaminated air, drink contaminated water, or eat contaminated food. These chemicals can become concentrated in the tissues and internal organs of animals as well as humans.

The concentration of 2,6-dinitrotoluene found in fish tissue is expected to be somewhat higher than the average concentration of 2,6-dinitrotoluene in the water from which the fish was taken.

Recommended Risk-Reduction Measures

Proper training of all transporters will reduce the likelihood of a mishap or accident resulting in a leak or spill to the environment. The correct use of DOT labeling (per the MSDS) on all containers, trucks, and rail cars will enable emergency responders to react properly and quickly to any disaster thereby reducing the potential risk to the environment and to personnel.

Storage of 2,6-dinitrotoluene should be segregated from incompatible chemicals to minimize the risk of cross-contamination or contact. Containers must be protected from heat, damage, and shock. Buildings designated for storage should be equipped with appropriate fire protection systems (alarms, sprinklers, emergency lighting, portable extinguishers).

If a spill or leak to the environment has occurred, fire department, emergency response and/or hazardous materials spill personnel should be notified immediately. Cleanup should be attempted only by those trained in proper spill containment procedures. Contaminated soils should be removed for incineration and replaced with clean soil. If 2,6-dinitrotoluene should contact the water table, aquifer, or navigable waterway, time is of the essence. It is slightly soluble in the aquatic environment and total remediation may not be possible. The local and/or state emergency response authorities must be notified. A comprehensive emergency response or disaster preparedness/recovery plan should be in place prior to any operations involving the use, transportation, storage, or disposal of 2,6-dinitrotoluene. If 2,6-dinitrotoluene is spilled or leaked, the following specific steps are recommended:

- ☑ Restrict persons not wearing protective clothing from area of spill or leak until cleanup is complete and area can be opened for normal work.
- ☑ Remove all ignition sources, ventilate area of leak.
- ☑ Use a water spray to knock-down vapors. Spills should be collected and placed in steel drums. Area should be flushed with hot water to remove solid particulates. Any liquids should be contained, as well as any liquid run-off.
- ☑ It may be necessary to dispose of 2,6-dinitrotoluene as a hazardous waste. The responsible state agency or the regional office of the federal Environmental Protection Agency (EPA) should be contacted for specific recommendations.

RISK ASSESSMENT: BUSINESS

General Assessment

Accidents or mishaps involving 2,6-dinitrotoluene can present a significant threat to business operations. The loss or damage of equipment or facilities can significantly affect fiscal viability. Lawsuits that may result from personnel injury/death, public exposures, and/or environmental contamination will also require a serious expenditure of resources. Media attention surrounding an injury, death, or environmental damage can also result in a loss of profits and loss of current as well as future business. Always remember that anytime the terms "carcinogen," "cancer," or "reproductive hazard" are used, public emotion, hysteria, and ignorance can run equally high. This must be properly considered whenever developing public relations policies.

Recommended Risk-Reduction Measures

Company attorneys, safety and health professionals, and environmental specialists should be involved in the development of any procedures dealing with safety and emergency response. A company official should be pre-designated as a public relations officer with specific training in dealing with the press. Corporate plans and policies should be developed, approved, and implemented long before any need for such arises.

REFERENCES

Ahlbom, A. 1993. *Biostatistics for Epidemiologists.* New York: CRC Press/Lewis Publishers.

American Conference of Governmental Industrial Hygienists. 1988. *Documentation of the Threshold Limit Values and Biological Exposure Indices,* 5th Edition (with updates). Cincinnati: ACGIH

Anthony, C. P., and N. J. Kolthoff. 1971. *Textbook of Anatomy and Physiology.* St. Louis: C. V. Mosby Company.

Asante-Duah, D. K. 1993. *Hazardous Waste Risk Assessment.* New York: CRC Press/Lewis Publishers.

Bloom, A. D. and F. J. DeSerres. 1995. *Ecotoxicity and Human Health.* New York: CRC Press/Lewis Publishers.

Bueche, J. R. 1972. *Principles of Physics,* 2nd Edition. New York: McGraw-Hill.

Burrell, R., D. K. Flaherty, and L. J. Sauers. 1992. *Toxicology of the Immune System.* New York: Van Nostrand Reinhold.

Calabrese, E. J. 1994. *Biological Effects of Low Level Exposures.* New York: CRC Press/Lewis Publishers.

Cockerman, L. G., B. S. Shane. 1994. *Basic Environmental Toxicology.* New York: CRC Press/Lewis Publishers.

Environmental Protection Agency. 1983. *Chemical Hazard Information Profiles.* Washington, D.C.: U.S. EPA.

Fodden, C. R., and J. L. Weddell. 1992. *Hazardous Materials Emergency Action Data.* New York: CRC Press/Lewis Publishers.

Hallenbeck, W. H. 1993. *Quantitative Risk Assessment for Environmental and Occupational Health,* 2nd Edition. New York: CRC Press/Lewis Publishers.

Howard, P. H. 1995. *Handbook of Environmental Fate and Exposures Data for Organic Chemicals,* Vol. I-IV. New York: CRC Press/Lewis Publishers.

Keith, L. H. (Editor). 1992. *Integrated Risk Information System (IRIS) Chemical Information Database.* Prepared by the U.S. Environmental Protection Agency. Boca Raton, FL: CRC Press/Lewis Publishers.

Lewis, R. J., Sr. 1992. *Sax's Dangerous Properties of Industrial Materials,* Volumes 1, 2, and 3, 8th Edition. New York: Van Nostrand Reinhold.

Meyer, E. 1990. *Chemistry of Hazardous Materials.* Englewood Cliffs, NJ: Prentice Hall.

Quigley, D. R. 1994. *Handbook of Emergency Chemical Management.* New York: CRC Press/Lewis Publishers.

U.S. Department of Health and Human Services, National Institute for Occupational Safety and Health. 1994. *NIOSH Pocket Guide to Chemical Hazards.* Washington, D.C.: U.S. Government Printing Office.

MATERIAL SAFETY DATA SHEET

CHEMICAL NAME	
3,4-DINITROTOLUENE	**CRC PRESS/LEWIS PUBLISHERS DISCLAIMER:** THE INFORMATION AND RECOMMENDATIONS PRESENTED HEREIN ARE BASED ON SOURCES BELIEVED TO BE RELIABLE. CRC MAKES NO REPRESENTATION ON ITS COMPLETENESS OR ACCURACY. IT IS THE RESPONSIBILITY OF THE USER TO DETERMINE THE CHEMICAL'S SUITABILITY FOR ITS INTENDED USE, THE CHEMICAL'S SAFE USE, AND THE CHEMICAL'S PROPER DISPOSAL. NO REPRESENTATIONS AND/OR WARRANTIES, EITHER EXPRESSED OR IMPLIED, OF THE MERCHANTABILITY OR FITNESS FOR A PARTICULAR PURPOSE, OR OF ANY OTHER NATURE, ARE MADE WITH RESPECT TO THE INFORMATION PROVIDED IN THIS MSDS OR TO THE CHEMICAL TO WHICH INFORMATION MAY REFER. CRC NEITHER ASSUMES NOR AUTHORIZES ANY OTHER PERSON TO ASSUME FOR IT, ANY OTHER ADDITIONAL RESPONSIBILITY OR LIABILITY FOR THE USE OF, OR RELIANCE UPON, THIS INFORMATION.

HAZARD WARNING INFORMATION

HEALTH	FIRE	REACTIVE	OTHER	DEGREE OF HAZARD	COLOR CODING	OTHER CODES
3	1	3		0 = Minimum Hazard 1 = Slight Hazard 2 = Moderate Hazard 3 = Serious Hazard 4 = Severe Hazard	HEALTH = BLUE FIRE = RED REACTIVITY = YELLOW OTHER = WHITE	OX = Oxidizer ACID = Acid ALK = Alkali COR = Corrosive W = Use No Water

SECTION I - GENERAL INFORMATION

Characterization	RCRA Number	EPA Class
Aromatic Hydrocarbon	**None**	**Not Applicable**

DOT Proper Shipping Name **3,4-Dinitrotoluene (solid)**	Chemical Abstract Service (CAS) Number **610-39-9**
DOT Hazard Class and Label Requirements **Poison B; Label: Poison**	DOT Emergency Guide Code **56**
DOT Identification Number **UN 1600 (molten); UN 2038 (solid)**	Chemical Formula $C_6H_3CH_3(NO_2)_2$

Synonyms
3,4-DNT; 4-methyl-1,2-dinitrobenzene.

SECTION II - HAZARDOUS INGREDIENTS AND IDENTITY INFORMATION

Hazardous Components (specific identity)	OSHA Exposure Criteria	NIOSH Exposure Criteria	Immediately Dangerous to Life and Health (IDLH)	ACGIH Exposure Criteria
3,4-Dinitrotoluene (derivation: By nitration of toluene with nitric acid in the presence of concentrated sulfuric acid).	**PEL (skin):** 1.5 mg/m³ **STEL:** Not Established	**REL (skin):** 1.5 mg/m³ **Possible Cancer Agent**	**50 mg/m³**	**TLV (skin):** 1.5 mg/m³ **STEL:** Not Established

SECTION III - PHYSICAL AND CHEMICAL CHARACTERISTICS

Boiling Point **482°F (250°C)**	Specific Gravity (H_2O = 1) **1.32 (liquid) 1.5 (solid)**
Vapor Pressure (gas) **1 at 69°F (20°C)**	Molecular Weight **182.14**
Vapor Density (Air = 1) **6.27**	Melting Point **158°F (70°C)**

Solubility
Nearly insoluble in water. Soluble in alcohol, acetone, benzene, carbon disulfide, ether, and pyridine.

Appearance and Odor
Orange to yellow crystalline solid with a slight characteristic odor. Usually shipped as molten.

SECTION IV - FIRE AND EXPLOSION HAZARD DATA

Flash Point (method used) **404°F (207°C) closed cup**	Explosive Limits in Air % by Volume **LEL: 0.3% at 474°F (245°C) UEL: Not Determined**
NFPA Classification **Combustible Solid**	Autoignition Temperature **Not Determined**

Extinguishing Media
Use carbon dioxide, water, or dry chemical from a protected location.

Special Fire Fighting Procedures
A carcinogenic. Wear full protective clothing and self-contained breathing apparatus (SCBA). Poisonous gases are produced in fire. Use water spray to keep fire-exposed containers cool. Move containers from fire if it can be done without risk. Fight fire from a distance or protected position, if possible.

Unusual Fire and Explosion Hazards
Containers may explode in fire. Fires involving this material should be considered an explosion hazard. Extremely explosive in dry form when exposed to heat, flame, or shock. Decomposition is self-sustaining at 280°F (138°C).

SECTION V - REACTIVITY DATA

Stability		Conditions to Avoid
		3,4-Dinitrotoluene is explosive when exposed to heat, friction, or contamination. It can be detonated using a very strong initiating device. It is slightly sensitive to impact shock, especially in solid form.
Stable	Unstable	Incompatibility *(materials to avoid)*
	X	Strong oxidizers, (such as chlorine, bromine, and fluorine), organic materials, strong reducing agents (such as sodium sulfide, zinc powder, sodium hyposulfite, and metallic hydrides), caustics, and metals.
Hazardous Polymerization		Conditions to Avoid
		Under normal conditions of temperature and pressure, hazardous polymerization of 3,4-dinitrotoluene is not expected to occur.
May Occur	Will Not Occur	Hazardous Decomposition or By-products
	X	When heated to decomposition, 3,4-dinitrotoluene can emit highly toxic/poisonous gases, including carbon dioxide and oxides of nitrogen.

SECTION VI - HEALTH HAZARD DATA

Primary Route(s) of Entry:	Inhalation?	Absorption (skin)?	Ingestion?
	X	X	X

Health Hazards

INHALATION: Can cause cyanosis, headache, irritability, anoxia, anemia, dizziness, weakness, nausea, vomiting, dypsnea, drowsiness, collapse, loss of consciousness, and death by cardiovascular collapse.

ABSORPTION: Irritation, small blisters, redness, swelling, ulceration, and necrosis (will pass through skin to cause same toxic systemic effects as inhalation). Eye contact will result in burns, irritation of the lids, redness, swelling, pain on exposure to light, and tissue damage.

INGESTION: Irritation of the mouth, throat, and stomach, stomach cramps, and diarrhea.

Carcinogenicity	NTP Listed?	IARC Cancer Review Group?	OSHA Regulated?	Target Organs?
Suspected Human Confirmed Animal	No	No	29 CFR 1910-1000 Table Z-1	Blood, liver, CVS, reproductive system, liver, skin

Medical Conditions Generally Aggravated by Exposure

Persons with blood disorders may be at an increased risk if exposed to 3,4-dinitrotoluene.

Emergency and First-aid Procedures

Eye contact: Flush immediately with water for 15 minutes (minimum), seek medical attention. Skin contact: Remove all contaminated clothing. Immediately wash area with large amounts of soap and water. Seek medical assistance. For inhalation: Remove the person from exposure. Provide CPR if required. Transfer to medical facility. If swallowed: Give large amounts of water or milk and seek medical attention. Never attempt to give an unconscious or convulsing person anything by mouth.

SECTION VII - PRECAUTIONS FOR SAFE HANDLING AND USE

Steps to be Taken in Case Material is Released or Spilled

Remove all ignition sources and ventilate area. Evacuate personnel not involved in cleanup. Stay upwind. Cleanup crew should wear fully-encapsulating suits. Use water spray to reduce vapors. Collect spilled powders and place in sealed steel drums for disposal.

Preferred Waste Disposal Method

No Citation.

Precautions to be Taken in Handling and Storage

Store in tightly closed containers in cool, well-ventilated area away from heat or flame. Heat may cause containers to build up pressure and explode. Usually stored as a molten liquid.

Other Precautions and Warnings

Sources of ignition are prohibited where 3,4-dinitrotoluene is used, handled, or stored.

SECTION VIII - CONTROL MEASURES AND PERSONAL PROTECTIVE EQUIPMENT

Respiratory Protection *(specify type)*

PEL is low and difficult to monitor. For any exposure to a suspected carcinogen, recommend a supplied air respirator with full facepiece, hood, or helmet in continuous flow mode, or use a self-contained breathing apparatus (SCBA) operated in pressure demand or other positive pressure mode.

Ventilation

Local exhaust or general mechanical systems recommended.

Protective Gloves	Eye Protection	Other Protective Clothing
Butyl Rubber	Chemical/Dust Goggles or Face Mask	Butyl Rubber Apron

Work/Hygiene Practices

Always wash hands thoroughly after using chemical; never bring food, drink, or smoking materials into vicinity of chemicals.

3,4-DINITROTOLUENE

$C_7H_6N_2O_4$ CAS: 610-39-9

IDENTIFICATION AND TYPICAL USES

3,4-Dinitrotoluene is an orange to yellow crystalline solid with a characteristic odor. It is normally shipped as a molten liquid. It is one of numerous isomers of dinitrotoluene. It is used in the manufacture of explosives and dyes, in organic synthesis, as a chemical intermediate, a plasticizer, and as a modifier for smokeless powders in the munitions industry.

RISK ASSESSMENT: HEALTH

General Assessment

3,4-Dinitrotoluene is a human poison by *ingestion*, with equally toxic systemic consequences occurring by way of *inhalation*, and skin contact (*absorption*). The closely-related 2,4-isomer has been shown to cause teratogenic effects in experimental tests and mutation data have also been reported. It is also a confirmed animal carcinogen and a suspected human carcinogen. Liver or kidney damage can occur up to 3 days following exposure. While these effects have not been confirmed for the 3,4-isomer, extreme caution is recommended in use, handling, storage, or disposal.

Inhalation and skin absorption can result in cyanosis, methemoglobinemia, headache, irritability, dizziness, weakness, nausea, vomiting, dyspnea, drowsiness, and loss of consciousness. If allowed to go untreated, the result can be death due to cardiovascular collapse. Repeated exposures can lead to anemia and liver damage. It has caused mutations and liver tumors in test animals.

Ingestion will result in irritation of the mouth and stomach, stomach cramps, and diarrhea. Possible systemic effects similar to inhalation and absorption may also occur. Ingestion of alcohol shortly before or after exposure may increase individual susceptibility to the chemical's toxicity.

Skin contact will not only result in absorption, it will also produce irritation, small blisters, redness, swelling, ulceration, and localized necrosis. Eye contact causes burning pain, irritation, redness, and swelling of the eyelids. There may be pain on exposure to light and severe eye damage. Contact with the molten liquid will cause thermal burns.

☠ *Acute Health Effects*

The following acute (short-term) health effects may occur immediately or shortly after exposure to 3,4-dinitrotoluene:

Skin: Irritation and possible tissue damage. Will pass through unbroken skin to cause serious systemic effects.

Eye: Irritation, swelling, redness, pain on exposure to light, and potential damage to the surrounding tissue.

Lung: Irritation to the respiratory tract and associated mucosa. Absorption through the respiration process causes toxic systemic effects (including anemia and methemoglobinemia).

Chronic Health Effects

The following chronic (long-term) health effects may occur at some time after exposure to 3,4-dinitrotoluene and can last for months or even years:

Cancer Hazards: According to the information available in the references, the 2,4-isomer of dinitrotoluene is a confirmed carcinogen in test animals. Its carcinogenic properties in humans is highly suspect. Experimental mutation data have been reported and many sci-

entists believe that such chemicals may have the ability to cause cancer. It is not known whether 3,4-isomer of dinitrotoluene will cause these effects.

Reproductive Hazard: According to information presented in the references, 2,4-dinitrotoluene has been shown to cause adverse effects in reproduction in animals. It is considered an experimental teratogen. It is not known whether 3,4-isomer of dinitrotoluene will cause these effects. Human effects in this regard are inconclusive for both isomers and additional research is required.

Other Chronic Effects: Long-term exposures can result in anemia and possible liver damage. It is also known to cause mutations and tumors in the livers of test animals.

❶ *Recommended Risk-Reduction Measures*

Personnel should avoid direct contact with 3,4-dinitrotoluene. If a less toxic chemical cannot be substituted for a hazardous substance, then *engineering controls* are the most effective method of reducing exposures. The best protection is to enclose operations and/or provide local exhaust ventilation at the site of chemical release. While not always operationally feasible, isolating operations can also reduce exposure. Using respiratory protection is less effective than the controls mentioned above, but is still advisable whenever working with or around 3,4-dinitrotoluene. The exposure level is low (1.5 mg/m^3) and difficult to monitor. Therefore, a MSHA/NIOSH approved full facepiece respirator equipped with an organic vapor cartridge equipped with a dust pre-filter should be used if any exposure is anticipated or known. Increased protection is obtained from a supplied-air respirator with full facepiece operated in continuous flow mode, or a self-contained breathing apparatus (SCBA) with full facepiece and pressure demand. If a full facepiece is not available, then chemical dust goggles should be worn to protect the eyes. Whenever an airborne chemical dust hazard exists, a face shield and a rubber apron should be worn. To prevent hand and skin exposures, butyl rubber gloves should be worn. Glove manufactures should be contacted and permeation studies obtained *before* final glove selections are made.

Administrative controls should also be in place to minimize the potential for human exposures. These may include written procedures or policies which specify the methods and techniques that will be practiced whenever personnel are to work with 3,4-dinitrotoluene.

All personnel should receive *training* on the use, hazards, protective measures, emergency actions, and other precautions per 29 CFR 1910.1200 (Hazard Communication), prior to the first assignment in an area where 3,4-dinitrotoluene is used or stored.

Before beginning employment with 3,4-dinitrotoluene and at regular intervals thereafter (e.g., annually), those persons with high or frequent exposures should be provided the following recommended medical tests:

- ☑ Liver and kidney function tests.
- ☑ Urinalysis (establish baseline).

If symptoms develop or overexposure is suspected, the following tests are recommended:

- ☑ Urinalysis (compare to baseline).
- ☑ Complete blood count.
- ☑ Liver and kidney function tests.

Any evaluation should include a careful history of past and present symptoms with an examination. Medical tests that simply look for existing damage are not a substitute for controlling exposures.

Other methods to reduce exposure include:

- ☑ Where possible, enclose operations and use local exhaust ventilation at the site of chemical release. If local exhaust ventilation or enclosure is not used, respiratory protection should be mandatory.
- ☑ Always ensure that proper protective clothing is worn when using chemical substances.
- ☑ Wash thoroughly immediately after exposure to 3,4-dinitrotoluene and at the end of the work shift or before eating, drinking, or smoking.
- ☑ Hazard warning information should be posted in the work area. In addition, as part of an on-going education and training program, all information on the health and safety hazards of 3,4-dinitrotoluene should be communicated to all potentially exposed workers.
- ☑ Eye wash stations should be provided in the immediate work area for emergency use. If there is a possibility of skin exposure to 3,4-dinitrotoluene, emergency shower facilities should also be provided.
- ☑ Personnel should not wear contaminated clothing home (family members can be exposed). Clothing should be laundered by personnel who have been trained on the hazards of 3,4-dinitrotoluene.

RISK ASSESSMENT: ENVIRONMENT

General Assessment

The environment is at risk of exposure during transportation, storage, disposal, or destruction of 3,4-dinitrotoluene. In almost every scenario, the threat of environmental exposure is contingent upon the proper handling of the chemical substance. Accidental spills, large or small, can result in fire, explosion, and possible contamination of the surrounding environmental mediums (water, soil, and air).

3,4-Dinitrotoluene is normally considered a combustible solid. Its dusts may form explosive mixtures in air. These characteristics require caution in handling, storage, transportation, and disposal. It is incompatible with strong oxidizers, caustics, and metals (such as tin and zinc). Contact with either of these can cause fire or explosion. Also, commercial grades will decompose at 482°F (250°C) with sustaining decomposition at 536°F (280°C). Therefore, special consideration is required during any emergency situation involving a leak or spill of 3,4-dinitrotoluene. Should 3,4-dinitrotoluene ever come into contact with any incompatible substances during use, transportation, storage, or disposal, the formation of highly toxic and/or highly explosive commodities is extremely possible.

3,4-Dinitrotoluene may enter the environment through its intended use as an insecticide, as well as through industrial effluents or spills.

Acute Ecological Effects

Acute (short-term) toxic effects may include the death of animals, birds, or fish and death or low growth rate in plants. Acute effects are seen 2 to 4 days after animals or plants are exposed to 3,4-dinitrotoluene.

3,4-Dinitrotoluene has high acute toxicity to aquatic life. Insufficient data are available on the short-term effects of 3,4-dinitrotoluene exposure to plants, birds, or land animals.

Chronic Ecological Effects

Chronic (long-term) toxic effects may include shortened life span, reproductive problems, lower fertility, and changes in appearance or behavior in exposed animals. These effects can be seen long after first exposure(s) to toxic chemicals.

3,4-Dinitrotoluene has moderate chronic toxicity to aquatic life. Insufficient data are available on the long-term effects of 3,4-dinitrotoluene to plants, birds, or land animals.

Water Solubility

3,4-Dinitrotoluene is nearly insoluble in water. Concentrations of less than 1 milligram may mix with a liter of water.

Persistence in the Environment

3,4-Dinitrotoluene is highly persistent in the aquatic environment, with a half-life of greater than 200 days. The half-life of a pollutant is the amount of time it takes for one half of the chemical to be degraded.

Bioaccumulation in Aquatic Organisms

Some substances increase in concentration, or *bioaccumulate,* in living organisms as they breathe contaminated air, drink contaminated water, or eat contaminated food. These chemicals can become concentrated in the tissues and internal organs of animals as well as humans.

The concentration of 3,4-dinitrotoluene found in fish tissue is expected to be somewhat higher than the average concentration of 3,4-dinitrotoluene in the water from which the fish was taken.

Recommended Risk-Reduction Measures

Proper training of all transporters will reduce the likelihood of a mishap or accident resulting in a leak or spill to the environment. The correct use of DOT labeling (per the MSDS) on all containers, trucks, and rail cars will enable emergency responders to react properly and quickly to any disaster thereby reducing the potential risk to the environment and to personnel.

Storage of 3,4-dinitrotoluene should be segregated from incompatible chemicals to minimize the risk of cross-contamination or contact. Containers must be protected from heat, damage, and shock. Buildings designated for storage should be equipped with appropriate fire protection systems (alarms, sprinklers, emergency lighting, portable extinguishers).

If a spill or leak to the environment has occurred, fire department, emergency response and/or hazardous materials spill personnel should be notified immediately. Cleanup should be attempted only by those trained in proper spill containment procedures. Contaminated soils should be removed for incineration and replaced with clean soil. If 3,4-dinitrotoluene should contact the water table, aquifer, or navigable waterway, time is of the essence. It is slightly soluble in the aquatic environment and total remediation may not be possible. The local and/or state emergency response authorities must be notified. A comprehensive emergency response or disaster preparedness/recovery plan should be in place prior to any operations involving the use, transportation, storage, or disposal of 3,4-dinitrotoluene. If 3,4-dinitrotoluene is spilled or leaked, the following specific steps are recommended:

- ☑ Restrict persons not wearing protective clothing from area of spill or leak until cleanup is complete and area can be opened for normal work.
- ☑ Remove all ignition sources, ventilate area of leak.
- ☑ Use a water spray to knock-down vapors. Spills should be collected and placed in steel drums. Area should be flushed with hot water to remove solid particulates. Any liquids should be contained, as well as any liquid run-off.
- ☑ It may be necessary to dispose of 3,4-dinitrotoluene as a hazardous waste. The responsible state agency or the regional office of the federal Environmental Protection Agency (EPA) should be contacted for specific recommendations.

RISK ASSESSMENT: BUSINESS

General Assessment

Accidents or mishaps involving 3,4-dinitrotoluene can present a significant threat to business operations. The loss or damage of equipment or facilities can significantly affect fiscal viability. Lawsuits that may result from personnel injury/death, public exposures, and/or environmental contamination will also require a serious expenditure of resources. Media attention surrounding an injury, death, or environmental damage can also result in a loss of profits and loss of current as well as future business. Always remember that anytime the terms "carcinogen," "cancer," or "reproductive hazard" are used, public emotion, hysteria, and ignorance can run equally high. This must be properly considered whenever developing public relations policies.

Recommended Risk-Reduction Measures

Company attorneys, safety and health professionals, and environmental specialists should be involved in the development of any procedures dealing with safety and emergency response. A company official should be pre-designated as a public relations officer with specific training in dealing with the press. Corporate plans and policies should be developed, approved, and implemented long before any need for such arises.

REFERENCES

Ahlbom, A. 1993. *Biostatistics for Epidemiologists.* New York: CRC Press/Lewis Publishers.

American Conference of Governmental Industrial Hygienists. 1988. *Documentation of the Threshold Limit Values and Biological Exposure Indices,* 5th Edition (with updates). Cincinnati: ACGIH

Anthony, C. P., and N. J. Kolthoff. 1971. *Textbook of Anatomy and Physiology.* St. Louis: C. V. Mosby Company.

Asante-Duah, D. K. 1993. *Hazardous Waste Risk Assessment.* New York: CRC Press/Lewis Publishers.

Bloom, A. D. and F. J. DeSerres. 1995. *Ecotoxicity and Human Health.* New York: CRC Press/Lewis Publishers.

Bueche, J. R. 1972. *Principles of Physics,* 2nd Edition. New York: McGraw-Hill.

Burrell, R., D. K. Flaherty, and L. J. Sauers. 1992. *Toxicology of the Immune System.* New York: Van Nostrand Reinhold.

Calabrese, E. J. 1994. *Biological Effects of Low Level Exposures.* New York: CRC Press/Lewis Publishers.

Cockerman, L. G., B. S. Shane. 1994. *Basic Environmental Toxicology.* New York: CRC Press/Lewis Publishers.

Environmental Protection Agency. 1983. *Chemical Hazard Information Profiles.* Washington, D.C.: U.S. EPA.

Fodden, C. R., and J. L. Weddell. 1992. *Hazardous Materials Emergency Action Data.* New York: CRC Press/Lewis Publishers.

Hallenbeck, W. H. 1993. *Quantitative Risk Assessment for Environmental and Occupational Health,* 2nd Edition. New York: CRC Press/Lewis Publishers.

Howard, P. H. 1995. *Handbook of Environmental Fate and Exposures Data for Organic Chemicals,* Vol. I-IV. New York: CRC Press/Lewis Publishers.

Keith, L. H. (Editor). 1992. *Integrated Risk Information System (IRIS) Chemical Information Database.* Prepared by the U.S. Environmental Protection Agency. Boca Raton, FL: CRC Press/Lewis Publishers.

Lewis, R. J., Sr. 1992. *Sax's Dangerous Properties of Industrial Materials,* Volumes 1, 2, and 3, 8th Edition. New York: Van Nostrand Reinhold.

Meyer, E. 1990. *Chemistry of Hazardous Materials.* Englewood Cliffs, NJ: Prentice Hall.

Quigley, D. R. 1994. *Handbook of Emergency Chemical Management.* New York: CRC Press/Lewis Publishers.

U.S. Department of Health and Human Services, National Institute for Occupational Safety and Health. 1994. *NIOSH Pocket Guide to Chemical Hazards.* Washington, D.C.: U.S. Government Printing Office.

MATERIAL SAFETY DATA SHEET

CHEMICAL NAME

3,5-DINITROTOLUENE

HAZARD WARNING INFORMATION

HEALTH	FIRE	REACTIVE	OTHER	DEGREE OF HAZARD	COLOR CODING	OTHER CODES
3	1	3		0 = Minimum Hazard 1 = Slight Hazard 2 = Moderate Hazard 3 = Serious Hazard 4 = Severe Hazard	HEALTH = BLUE FIRE = RED REACTIVITY = YELLOW OTHER = WHITE	OX = Oxidizer ACID = Acid ALK = Alkali COR = Corrosive W = Use No Water

SECTION I - GENERAL INFORMATION

Characterization **Aromatic Hydrocarbon**	RCRA Number **None**	EPA Class **Not Applicable**
DOT Proper Shipping Name **3,5-Dinitrotoluene (solid)**	Chemical Abstract Service (CAS) Number **618-85-9**	
DOT Hazard Class and Label Requirements **Poison B; Label: Poison**	DOT Emergency Guide Code **56**	
DOT Identification Number **UN 1600 (molten); UN 2038 (solid)**	Chemical Formula $C_6H_3CH_3(NO_2)_2$	

Synonyms
3,5-DNT; 1-methyl-3,5-dinitro-benzene.

SECTION II - HAZARDOUS INGREDIENTS AND IDENTITY INFORMATION

Hazardous Components (specific identity)	OSHA Exposure Criteria	NIOSH Exposure Criteria	Immediately Dangerous to Life and Health (IDLH)	ACGIH Exposure Criteria
3,5-Dinitrotoluene (derivation: By nitration of toluene with nitric acid in the presence of concentrated sulfuric acid).	**PEL (skin): 1.5 mg/m^3** **STEL: Not Established**	**REL (skin): 1.5 mg/m^3** **Possible Cancer Agent**	**50 mg/m^3**	**TLV (skin): 1.5 mg/m^3** **STEL: Not Established**

SECTION III - PHYSICAL AND CHEMICAL CHARACTERISTICS

Boiling Point **482°F (250°C)**	Specific Gravity (H_2O = 1) **1.32 (liquid) 1.5 (solid)**
Vapor Pressure (gas) **1 at 69°F (20°C)**	Molecular Weight **182.14**
Vapor Density (Air = 1) **6.27**	Melting Point **158°F (70°C)**

Solubility
Nearly insoluble in water. Soluble in alcohol, acetone, benzene, carbon disulfide, ether, and pyridine.

Appearance and Odor
Orange to yellow crystalline solid with a slight characteristic odor. Usually shipped as molten.

SECTION IV - FIRE AND EXPLOSION HAZARD DATA

Flash Point (method used) **404°F (207°C) closed cup**	Explosive Limits in Air % by Volume **LEL: 0.3% at 474°F (245°C) UEL: Not Determined**
NFPA Classification **Combustible Solid**	Autoignition Temperature **Not Determined**

Extinguishing Media
Use carbon dioxide, water, or dry chemical from a protected location.

Special Fire Fighting Procedures
A carcinogenic. Wear full protective clothing and self-contained breathing apparatus (SCBA). Poisonous gases are produced in fire. Use water spray to keep fire-exposed containers cool. Move containers from fire if it can be done without risk. Fight fire from a distance or protected position, if possible.

Unusual Fire and Explosion Hazards
Containers may explode in fire. Fires involving this material should be considered an explosion hazard. Extremely explosive in dry form when exposed to heat, flame, or shock. Decomposition is self-sustaining at 280°F (138°C).

SECTION V - REACTIVITY DATA

Stability		Conditions to Avoid **3,5-Dinitrotoluene is explosive when exposed to heat, friction, or contamination. It can be detonated using a very strong initiating device. It is slightly sensitive to impact shock, especially in solid form.**
Stable	Unstable	Incompatibility *(materials to avoid)*
	X	**Strong oxidizers, (such as chlorine, bromine, and fluorine), organic materials, strong reducing agents (such as sodium sulfide, zinc powder, sodium hyposulfite, and metallic hydrides), caustics, and metals.**
Hazardous Polymerization		Conditions to Avoid **Under normal conditions of temperature and pressure, hazardous polymerization of 3,5-dinitrotoluene is not expected to occur.**
May Occur	Will Not Occur	Hazardous Decomposition or By-products
	X	**When heated to decomposition, 3,5-dinitrotoluene can emit highly toxic/poisonous gases, including carbon dioxide and oxides of nitrogen.**

SECTION VI - HEALTH HAZARD DATA

Primary Route(s) of Entry:	Inhalation?	Absorption (skin)?	Ingestion?
	X	X	X

Health Hazards

INHALATION: Can cause cyanosis, headache, irritability, anoxia, anemia, dizziness, weakness, nausea, vomiting, dypsnea, drowsiness, collapse, loss of consciousness, and death by cardiovascular collapse.

ABSORPTION: Irritation, small blisters, redness, swelling, ulceration, and necrosis (will pass through skin to cause same toxic systemic effects as inhalation). Eye contact will result in burns, irritation of the lids, redness, swelling, pain on exposure to light, and tissue damage.

INGESTION: Irritation of the mouth, throat, and stomach, stomach cramps, and diarrhea.

Carcinogenicity	NTP Listed?	IARC Cancer Review Group?	OSHA Regulated?	Target Organs?
Suspected Human Confirmed Animal	No	No	29 CFR 1910-1000 Table Z-1	Blood, liver, CVS, reproductive system, liver, skin

Medical Conditions Generally Aggravated by Exposure

Persons with blood disorders may be at an increased risk if exposed to 3,5-dinitrotoluene.

Emergency and First-aid Procedures

Eye contact: Flush immediately with water for 15 minutes (minimum), seek medical attention. Skin contact: Remove all contaminated clothing. Immediately wash area with large amounts of soap and water. Seek medical assistance. For inhalation: Remove the person from exposure. Provide CPR if required. Transfer to medical facility. If swallowed: Give large amounts of water or milk and seek medical attention. Never attempt to give an unconscious or convulsing person anything by mouth.

SECTION VII - PRECAUTIONS FOR SAFE HANDLING AND USE

Steps to be Taken in Case Material is Released or Spilled

Remove all ignition sources and ventilate area. Evacuate personnel not involved in cleanup. Stay upwind. Cleanup crew should wear fully-encapsulating suits. Use water spray to reduce vapors. Collect spilled powders and place in sealed steel drums for disposal.

Preferred Waste Disposal Method

No Citation.

Precautions to be Taken in Handling and Storage

Store in tightly closed containers in cool, well-ventilated area away from heat or flame. Heat may cause containers to build up pressure and explode. Usually stored as a molten liquid.

Other Precautions and Warnings

Sources of ignition are prohibited where 3,5-dinitrotoluene is used, handled, or stored.

SECTION VIII - CONTROL MEASURES AND PERSONAL PROTECTIVE EQUIPMENT

Respiratory Protection *(specify type)*

PEL is low and difficult to monitor. For any exposure to a suspected carcinogen, recommend a supplied air respirator with full facepiece, hood, or helmet in continuous flow mode, or use a self-contained breathing apparatus (SCBA) operated in pressure demand or other positive pressure mode.

Ventilation

Local exhaust or general mechanical systems recommended.

Protective Gloves	Eye Protection	Other Protective Clothing
Butyl Rubber	Chemical/Dust Goggles or Face Mask	Butyl Rubber Apron

Work/Hygiene Practices

Always wash hands thoroughly after using chemical; never bring food, drink, or smoking materials into vicinity of chemicals.

3,5-DINITROTOLUENE

$C_7H_6N_2O_4$ CAS: 618-85-9

IDENTIFICATION AND TYPICAL USES

3,5-Dinitrotoluene is an orange to yellow crystalline solid with a characteristic odor. It is normally shipped as a molten liquid. It is one of numerous isomers of dinitrotoluene. It is used in the manufacture of explosives and dyes, in organic synthesis, as a chemical intermediate, a plasticizer, and as a modifier for smokeless powders in the munitions industry.

RISK ASSESSMENT: HEALTH

General Assessment

3,5-Dinitrotoluene is a human poison by *ingestion,* with equally toxic systemic consequences occurring by way of *inhalation,* and skin contact (*absorption*). The 2,4-isomer has been shown to cause teratogenic effects in experimental tests and mutation data have also been reported. It is also a confirmed animal carcinogen and a suspected human carcinogen. Liver or kidney damage can occur up to 3 days following exposure. While these effects have not been confirmed for the 3,5-isomer, extreme caution is recommended in use, handling, storage, or disposal.

Inhalation and skin absorption can result in cyanosis, methemoglobinemia, headache, irritability, dizziness, weakness, nausea, vomiting, dyspnea, drowsiness, and loss of consciousness. If allowed to go untreated, the result can be death due to cardiovascular collapse. Repeated exposures can lead to anemia and liver damage. It has caused mutations and liver tumors in test animals.

Ingestion will result in irritation of the mouth and stomach, stomach cramps, and diarrhea. Possible systemic effects similar to inhalation and absorption may also occur. Ingestion of alcohol shortly before or after exposure to 3,5-dinitrotoluene may increase individual susceptibility to toxicity.

Skin contact will not only result in absorption, it will also produce irritation, small blisters, redness, swelling, ulceration, and localized necrosis. Eye contact causes burning pain, irritation, redness, and swelling of the eyelids. There may be pain on exposure to light and severe eye damage. Contact with the molten liquid will cause thermal burns.

Acute Health Effects

The following acute (short-term) health effects may occur immediately or shortly after exposure to 3,5-dinitrotoluene:

Skin: Irritation and possible tissue damage. Will pass through unbroken skin to cause serious systemic effects.

Eye: Irritation, swelling, redness, pain on exposure to light, and potential damage to the surrounding tissue.

Lung: Irritation to the respiratory tract and associated mucosa. Absorption through the respiration process causes toxic systemic effects (including anemia and methemoglobinemia).

Chronic Health Effects

The following chronic (long-term) health effects may occur at some time after exposure to 3,5-dinitrotoluene and can last for months or even years:

Cancer Hazards: According to the information available in the references, the 2,4-isomer of dinitrotoluene is a confirmed carcinogen in test animals. Its carcinogenic properties in humans is highly suspect. Experimental mutation data have been reported and many scientists believe that such chemicals may have the ability to cause cancer. It is not known whether 3,5-isomer of dinitrotoluene will cause these effects.

Reproductive Hazard: According to information presented in the references, 2,4-dinitrotoluene has been shown to cause adverse effects in reproduction in animals. It is considered an experimental teratogen. It is not known whether 3,5-isomer of dinitrotoluene will cause these effects. Human effects in this regard are inconclusive for both isomers and additional research is required.

Other Chronic Effects: Long-term exposures can result in anemia and possible liver damage. It is also known to cause mutations and tumors in the livers of test animals.

Recommended Risk-Reduction Measures

Personnel should avoid direct contact with 3,5-dinitrotoluene. If a less toxic chemical cannot be substituted for a hazardous substance, then *engineering controls* are the most effective method of reducing exposures. The best protection is to enclose operations and/or provide local exhaust ventilation at the site of chemical release. While not always operationally feasible, isolating operations can also reduce exposure. Using respiratory protection is less effective than the controls mentioned above, but is still advisable whenever working with or around 3,5-dinitrotoluene. The exposure level is low (1.5 mg/m^3) and difficult to monitor. Therefore, a MSHA/NIOSH approved full facepiece respirator equipped with an organic vapor cartridge equipped with a dust pre-filter should be used if any exposure is anticipated or known. Increased protection is obtained from a supplied-air respirator with full facepiece operated in continuous flow mode, or a self-contained breathing apparatus (SCBA) with full facepiece and pressure demand. If a full facepiece is not available, then chemical dust goggles should be worn to protect the eyes. Whenever an airborne chemical dust hazard exists, a face shield and a rubber apron should be worn. To prevent hand and skin exposures, butyl rubber gloves should be used. Glove manufacturers should be contacted for glove selection.

Administrative controls should also be in place to minimize the potential for human exposures. These may include written procedures or policies which specify the methods and techniques that will be practiced whenever personnel are to work with 3,5-dinitrotoluene.

All personnel should receive *training* on the use, hazards, protective measures, emergency actions, and other precautions per 29 CFR 1910.1200 (Hazard Communication), prior to the first assignment in an area where 3,5-dinitrotoluene is used or stored.

Before beginning employment with 3,5-dinitrotoluene and at regular intervals thereafter (e.g., annually), those persons with high or frequent exposures should be provided the following recommended medical tests:

- ☑ Liver and kidney function tests.
- ☑ Urinalysis (establish baseline).

If symptoms develop or overexposure is suspected, the following tests are recommended:

- ☑ Urinalysis (compare to baseline).
- ☑ Complete blood count.
- ☑ Liver and kidney function tests.

Any evaluation should include a careful history of past and present symptoms with an examination. Medical tests that simply look for existing damage are not a substitute for controlling exposures.

Other methods to reduce exposure include:

- ☑ Where possible, enclose operations and use local exhaust ventilation at the site of chemical release. If local exhaust ventilation or enclosure is not used, respiratory protection should be mandatory.
- ☑ Always ensure that proper protective clothing is worn when using chemical substances.
- ☑ Wash thoroughly immediately after exposure to 3,5-dinitrotoluene and at the end of the work shift or before eating, drinking, or smoking.
- ☑ Hazard warning information should be posted in the work area. In addition, as part of an on-going education and training program, all information on the health and safety hazards of 3,5-dinitrotoluene should be communicated to all exposed workers.
- ☑ Eye wash stations should be provided in the immediate work area for emergency use. If there is a possibility of skin exposure to 3,5-dinitrotoluene, emergency shower facilities should be provided.
- ☑ Personnel should not wear contaminated clothing home (family members can be exposed). Clothing should be laundered by personnel who have been trained on the hazards of 3,5-dinitrotoluene.

RISK ASSESSMENT: ENVIRONMENT

General Assessment

The environment is at risk of exposure during transportation, storage, disposal, or destruction of 3,5-dinitrotoluene. In almost every scenario, the threat of

environmental exposure is contingent upon the proper handling of the chemical substance. Accidental spills, large or small, can result in fire, explosion, and possible contamination of the surrounding environmental mediums (water, soil, and air).

3,5-Dinitrotoluene is normally considered a combustible solid. Its dusts may form explosive mixtures in air. These characteristics require caution in handling, storage, transportation, and disposal. It is incompatible with strong oxidizers, caustics, and metals (such as tin and zinc). Contact with either of these can cause fire or explosion. Also, commercial grades will decompose at 482°F (250°C) with sustaining decomposition at 536°F (280°C). Therefore, special consideration is required during any emergency situation involving a leak or spill of 3,5-dinitrotoluene. Should 3,5-dinitrotoluene ever come into contact with any incompatible substances during use, transportation, storage, or disposal, the formation of highly toxic and/or highly explosive commodities is extremely possible.

3,5-Dinitrotoluene may enter the environment through its intended use as an insecticide, as well as through industrial effluents or spills.

Acute Ecological Effects

Acute (short-term) toxic effects may include the death of animals, birds, or fish and death or low growth rate in plants. Acute effects are seen 2 to 4 days after animals or plants are exposed to 3,5-dinitrotoluene.

3,5-Dinitrotoluene has high acute toxicity to aquatic life. Insufficient data are available on the short-term effects of 3,5-dinitrotoluene exposure to plants, birds, or land animals.

Chronic Ecological Effects

Chronic (long-term) toxic effects may include shortened life span, reproductive problems, lower fertility, and changes in appearance or behavior in exposed animals. These effects can be seen long after first exposure(s) to toxic chemicals.

3,5-Dinitrotoluene has moderate chronic toxicity to aquatic life. Insufficient data are available on the long-term effects of 3,5-dinitrotoluene to plants, birds, or land animals.

Water Solubility

3,5-Dinitrotoluene is nearly insoluble in water. Concentrations of less than 1 milligram may mix with a liter of water.

Persistence in the Environment

3,5-Dinitrotoluene is highly persistent in the aquatic environment, with a half-life of greater than 200 days. The half-life of a pollutant is the amount of time it takes for one half of the chemical to be degraded.

Bioaccumulation in Aquatic Organisms

Some substances increase in concentration, or *bioaccumulate,* in living organisms as they breathe contaminated air, drink contaminated water, or eat contaminated food. These chemicals can become concentrated in the tissues and internal organs of animals as well as humans.

The concentration of 3,5-dinitrotoluene found in fish tissue is expected to be somewhat higher than the average concentration of 3,5-dinitrotoluene in the water from which the fish was taken.

Recommended Risk-Reduction Measures

Proper training of all transporters will reduce the likelihood of a mishap or accident resulting in a leak or spill to the environment. The correct use of DOT labeling (per the MSDS) on all containers, trucks, and rail cars will enable emergency responders to react properly and quickly to any disaster thereby reducing the potential risk to the environment and to personnel.

Storage of 3,5-dinitrotoluene should be segregated from incompatible chemicals to minimize the risk of cross-contamination or contact. Containers must be protected from heat, damage, and shock. Buildings designated for storage should be equipped with appropriate fire protection systems (alarms, sprinklers, emergency lighting, portable extinguishers). Personnel must be trained on the storage of hazardous chemicals.

If a spill or leak to the environment has occurred, fire department, emergency response and/or hazardous materials spill personnel should be notified immediately. Cleanup should be attempted only by those trained in proper spill containment procedures. Contaminated soils should be removed for incineration and replaced with clean soil. If 3,5-dinitrotoluene should contact the water table, aquifer, or navigable waterway, time is of the essence. It is slightly soluble in the aquatic environment and total remediation may not be possible. The local and/or state emergency response authorities must be notified. A comprehensive emergency response or disaster preparedness/recovery plan should be in place prior to any operations involving the use, transportation, storage, or disposal of 3,5-

dinitrotoluene. If 3,5-dinitrotoluene is spilled or leaked, the following specific steps are recommended:

- ☑ Restrict persons not wearing protective clothing from area of spill or leak until cleanup is complete and area can be opened for normal work.
- ☑ Remove all ignition sources.
- ☑ Ventilate area of leak or spill.
- ☑ Use a water spray to knock-down vapors. Spills should be collected and placed in steel drums. Area should be flushed with hot water to remove solid particulates. Any liquids should be contained, as well as any liquid run-off.
- ☑ It may be necessary to dispose of 3,5-dinitrotoluene as a hazardous waste. The responsible state agency or the regional office of the federal Environmental Protection Agency (EPA) should be contacted for specific recommendations.

RISK ASSESSMENT: BUSINESS

General Assessment

Accidents or mishaps involving 3,5-dinitrotoluene can present a significant threat to business operations. The loss or damage of equipment or facilities can significantly affect fiscal viability. Lawsuits that may result from personnel injury/death, public exposures, and/or environmental contamination will also require a serious expenditure of resources. Media attention surrounding an injury, death, or environmental damage can also result in a loss of profits and loss of current as well as future business. Always remember that anytime the terms "carcinogen," "cancer," or "reproductive hazard" are used, public emotion, hysteria, and ignorance can run equally high. This must be properly considered whenever developing public relations policies.

Recommended Risk-Reduction Measures

Company attorneys, safety and health professionals, and environmental specialists should be involved in the development of any procedures dealing with safety and emergency response. A company official should be pre-designated as a public relations officer with specific training in dealing with the press. Corporate plans and policies should be developed, approved, and implemented long before any need for such arises.

REFERENCES

Ahlbom, A. 1993. *Biostatistics for Epidemiologists.* New York: CRC Press/Lewis Publishers.

American Conference of Governmental Industrial Hygienists. 1988. *Documentation of the Threshold Limit Values and Biological Exposure Indices,* 5th Edition (with updates). Cincinnati: ACGIH

Anthony, C. P., and N. J. Kolthoff. 1971. *Textbook of Anatomy and Physiology.* St. Louis: C. V. Mosby Company.

Asante-Duah, D. K. 1993. *Hazardous Waste Risk Assessment.* New York: CRC Press/Lewis Publishers.

Bloom, A. D. and F. J. DeSerres. 1995. *Ecotoxicity and Human Health.* New York: CRC Press/Lewis Publishers.

Bueche, J. R. 1972. *Principles of Physics,* 2nd Edition. New York: McGraw-Hill.

Burrell, R., D. K. Flaherty, and L. J. Sauers. 1992. *Toxicology of the Immune System.* New York: Van Nostrand Reinhold.

Calabrese, E. J. 1994. *Biological Effects of Low Level Exposures.* New York: CRC Press/Lewis Publishers.

Cockerman, L. G., B. S. Shane. 1994. *Basic Environmental Toxicology.* New York: CRC Press/Lewis Publishers.

Environmental Protection Agency. 1983. *Chemical Hazard Information Profiles.* Washington, D.C.: U.S. EPA.

Fodden, C. R., and J. L. Weddell. 1992. *Hazardous Materials Emergency Action Data.* New York: CRC Press/Lewis Publishers.

Hallenbeck, W. H. 1993. *Quantitative Risk Assessment for Environmental and Occupational Health,* 2nd Edition. New York: CRC Press/Lewis Publishers.

Howard, P. H. 1995. *Handbook of Environmental Fate and Exposures Data for Organic Chemicals,* Vol. I-IV. New York: CRC Press/Lewis Publishers.

Keith, L. H. (Editor). 1992. *Integrated Risk Information System (IRIS) Chemical Information Database.* Prepared by the U.S. Environmental Protection Agency. Boca Raton, FL: CRC Press/Lewis Publishers.

Lewis, R. J., Sr. 1992. *Sax's Dangerous Properties of Industrial Materials,* Volumes 1, 2, and 3, 8th Edition. New York: Van Nostrand Reinhold.

Meyer, E. 1990. *Chemistry of Hazardous Materials.* Englewood Cliffs, NJ: Prentice Hall.

Quigley, D. R. 1994. *Handbook of Emergency Chemical Management.* New York: CRC Press/Lewis Publishers.

U.S. Department of Health and Human Services, National Institute for Occupational Safety and Health. 1994. *NIOSH Pocket Guide to Chemical Hazards.* Washington, D.C.: U.S. Government Printing Office.

MATERIAL SAFETY DATA SHEET

CHEMICAL NAME

DI-N-OCTYL PHTHALATE

HAZARD WARNING INFORMATION

HEALTH	FIRE	REACTIVE	OTHER	DEGREE OF HAZARD	COLOR CODING	OTHER CODES
0	1	0		0 = Minimum Hazard 1 = Slight Hazard 2 = Moderate Hazard 3 = Serious Hazard 4 = Severe Hazard	HEALTH = BLUE FIRE = RED REACTIVITY = YELLOW OTHER = WHITE	OX = Oxidizer ACID = Acid ALK = Alkali COR = Corrosive W = Use No Water

SECTION I - GENERAL INFORMATION

Characterization	Phthalate Ester	RCRA Number	U107
		EPA Class	Toxic Waste
DOT Proper Shipping Name	None	Chemical Abstract Service (CAS) Number	117-84-0
DOT Hazard Class and Label Requirements	Not Applicable	DOT Emergency Guide Code	No Citation
DOT Identification Number	None	Chemical Formula	$C_{24}H_{38}O_4$

Synonyms

Dioctyl phthalate; 1,2-benzenedicarboxylic acid dioctyl ester; dioctyl *o*-benzenedicarboxylate; phthalic acid dioctyl ester; bis(2-ethylhexyl) phthalate; di(2-ethylhexyl) phthalate; n-octyl phthalate.

SECTION II - HAZARDOUS INGREDIENTS AND IDENTITY INFORMATION

Hazardous Components (specific identity)	OSHA Exposure Criteria	NIOSH Exposure Criteria	Immediately Dangerous to Life and Health (IDLH)	ACGIH Exposure Criteria
Di-n-octyl phthalate (derivation: By reaction of 2-ethylhexanol and phthalic anhydride). **1 ppm = 16.23 mg/m³**	**PEL:** Not Established **STEL:** Not Established	**REL:** Not Established **STEL:** Not Established	**Not Determined**	**TLV:** Not Established **STEL:** Not Established

SECTION III - PHYSICAL AND CHEMICAL CHARACTERISTICS

Boiling Point	727°F (386°C)	Specific Gravity (H_2O = 1)	0.99
Vapor Pressure (mm Hg)	<0.01 at 69°F (20°C)	Molecular Weight	390.62
Vapor Density (Air = 1)	Not Found	Freezing Point	-58°F (-42°C)

Solubility

Insoluble in water (0.00003%). Miscible in mineral oil.

Appearance and Odor

Colorless, oily liquid with a slight ester-like odor.

SECTION IV - FIRE AND EXPLOSION HAZARD DATA

Flash Point (method used)	425°F (218°C) closed cup	Explosive Limits in Air % by Volume	LEL: 0.3% (at 474°F) UEL: Not Determined
NFPA Classification	Class IIIB Combustible Liquid	Autoignition Temperature	Not Determined

Extinguishing Media

Use dry chemical, foam, carbon dioxide, or water.

Special Fire Fighting Procedures

Acrid and irritating gases and fumes are produced in fire, wear full protective clothing and self-contained breathing apparatus (SCBA). Water and foam may cause frothing. Move containers from fire area if it can be done safely.

Unusual Fire and Explosion Hazards

Exercise extreme caution when attempting to move or relocate containers that may have been exposed to high heat. Cool exposed containers. Containers may explode at high temperatures.

SECTION V - REACTIVITY DATA

Stability		Conditions to Avoid Avoid contact with incompatible materials. Keep away from heat and flame.
Stable X	Unstable	Incompatibility *(materials to avoid)* Strong oxidizers, such as chlorine, bromine, and fluorine, strong alkalis (sodium hydroxide, potassium hydroxide, lithium hydroxide), strong acids, and nitrates.
Hazardous Polymerization		Conditions to Avoid Under normal conditions of temperature and pressure, hazardous polymerization of di-n-octyl phthalate is not expected to occur.
May Occur	Will Not Occur X	Hazardous Decomposition or By-products Acrid and irritated gases and fumes are produced when di-n-octyl phthalate is involved in fire.

SECTION VI - HEALTH HAZARD DATA

Primary Route(s) of Entry:	Inhalation?	Absorption (skin)?	Ingestion?
	X	X	X

Health Hazards

INHALATION: Symptoms include irritation to the eyes and mucosa (especially in the upper respiratory tract). High exposures can irritate the lungs and may cause death by pulmonary edema.

ABSORPTION: Contact may irritate the skin and eyes. Absorption is possible. May cause drying and cracking of skin and secondary infections.

INGESTION: Mildly toxic by ingestion. Gastrointestinal irritation. Systemic effects and possibility of liver and/or kidney damage on high exposures.

Carcinogenicity	NTP Listed?	IARC Cancer Review Group?	OSHA Regulated?	Target Organs?
Unknown Human Unknown Animal	No	No	No	Eyes, respiratory system, GI tract, kidney, liver.

Medical Conditions Generally Aggravated by Exposure

None Reported.

Emergency and First-aid Procedures

Eye contact: Flush immediately with water for 15 minutes (minimum), seek medical attention. Skin contact: Remove all contaminated clothing. Immediately wash area with large amounts of soap and water. Seek medical assistance. For inhalation: Remove the person from exposure. Provide respiratory assistance and CPR. Transfer to medical facility. If swallowed: Call the poison control center and seek medical attention immediately.

SECTION VII - PRECAUTIONS FOR SAFE HANDLING AND USE

Steps to be Taken in Case Material is Released or Spilled

Absorb liquids in vermiculite or other absorbent and deposit in DOT-approved drum. Ventilate area and remove ignition sources. Restrict those not involved in cleanup from entering area. Notify appropriate authorities, if applicable.

Preferred Waste Disposal Method

Burn in chemical incinerator equipped with an afterburner and scrubber.

Precautions to be Taken in Handling and Storage

Do not store di-n-octyl phthalate in the presence of incompatible chemicals or materials. Store in tightly closed containers in a cool, well-ventilated area. Personnel working with di-n-octyl phthalate should be trained on its proper handing and storage *prior* to being assigned to such responsibilities.

Other Precautions and Warnings

Bulk storage is of di-n-octyl phthalate should be avoided.

SECTION VIII - CONTROL MEASURES AND PERSONAL PROTECTIVE EQUIPMENT

Respiratory Protection *(specify type)*

For relatively low or infrequent exposure, use a full facepiece respirator equipped with a high efficiency particulate air (HEPA) filter and dust/mist pre-filter. For greater protection, use a supplied-air respirator with full facepiece operated in pressure demand mode or a self-contained breathing apparatus (SCBA) with full facepiece operated in pressure demand or other positive pressure mode.

Ventilation

Local exhaust or general mechanical systems recommended.

Protective Gloves	Eye Protection	Other Protective Clothing
Rubber	Chemical Goggles or Face Mask	Apron

Work/Hygiene Practices

Always wash hands thoroughly after using chemical; never bring food, drink, or smoking materials into vicinity of chemicals.

DI-N-OCTYL PHTHALATE

$C_{24}H_{38}O_4$ CAS: 117-84-0

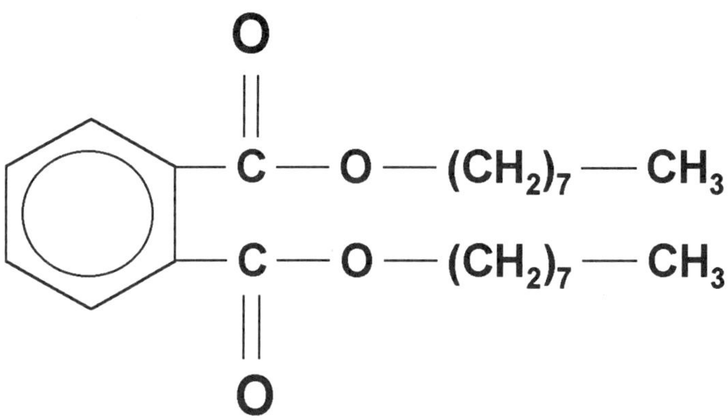

IDENTIFICATION AND TYPICAL USES

Di-n-octyl phthalate is a colorless oily liquid with a slight, ester-like odor. It is used as a plasticizer in nitrocellulose lacquers, elastomers, nail polish, and solid rocket propellants. It is also used as a solvent for perfume oils, as a perfume fixative, a textile lubricating agent, in the manufacture of safety glass, and in the production of insecticides. It may also be found in certain printing inks, is resin solvents, in paper coatings, adhesives, and as an insect repellent for textiles.

RISK ASSESSMENT: HEALTH

General Assessment

Di-n-octyl phthalate is primarily toxic to humans by *ingestion*. However, its vapors can also cause adverse systemic effects by *inhalation*. Contact with the eyes or skin will result in irritation and, if in contact for long periods, *absorption*. It is known to be a teratogen in test animals. Human mutation data have been reported for this chemical.

Inhalation of di-n-octyl phthalate will directly irritate the mucous membranes of the respiratory tract (nose, throat, and bronchial tubes). Inhalation will also result in systemic effects with symptoms of vomiting or nausea, and numerous indications of depression of the central nervous system (CNS). These include hallucinations and distorted perceptions. There may be somnolence and lachrymation. In high exposures, lung irritation can lead to pulmonary edema (fluid in the lungs) and death. In animals, exposure has caused liver tumors. It is not known whether human exposure will have this effect.

Ingestion toxicity is considered relatively low compared to other phthalates. There may also be systemic effects similar to those caused by inhalation.

Acute Health Effects

The following acute (short-term) health effects may occur immediately or shortly after exposure to di-n-octyl phthalate:

Skin: Irritation and possible inflammation. Absorption on prolonged exposure is likely, leading to toxic systemic effects.

Eye: Severe irritation with redness, lachrymation, and possible conjunctivitis.

Lung: Irritation of the mucosa of the nose, throat, and respiratory tract (especially the bronchial tubes). High exposure can lead to pulmonary edema and death.

Other: Ingestion may cause hallucinations, blurred vision, and distorted perceptions. There may be vomiting or nausea.

Chronic Health Effects

The following chronic (long-term) health effects may occur at some time after exposure to di-n-octyl phthalate and can last for months or even years:

Cancer Hazards: According to the information presented in the references, di-n-octyl phthalate has not been adequately tested for its ability to cause cancer in test animals.

Reproductive Hazard: There is limited evidence that di-n-octyl phthalate is a teratogen in test animals. Until further testing has been performed, it is suggested that it be treated as a possible teratogen in humans.

Other Chronic Effects: Repeated skin contact causes dryness, cracking, and rash. Frequent or repeated high exposures may damage the kidneys and liver. Very irritating substances may affect the lungs. It has not been confirmed whether di-n-octyl phthalate causes lung damage (pulmonary edema usually occurs before damage can be assessed).

Recommended Risk-Reduction Measures

Unless a less toxic chemical can be substituted for a hazardous substance, *engineering controls* are the most effective method of reducing exposures. The best protection is to enclose operations and/or provide local exhaust ventilation at the site of chemical release. While not always operationally feasible, isolating operations can also reduce exposure.

Using respiratory protection is less effective than the controls mentioned above, but is still advisable whenever working with or around di-n-octyl phthalate. At low or infrequent exposures, use a full facepiece respirator equipped with a high efficiency particulate air (HEPA) filter and a dust/mist pre-filter. For high concentrations, a self-contained breathing apparatus (SCBA) with full facepiece and pressure demand is the recommended respiratory protection of choice. If a full facepiece is not available, then chemical goggles should be worn to protect the eyes. Whenever a chemical splash hazard exists, a face shield and rubber apron should be worn. To prevent hand and skin exposures, rubber gloves should be worn.

Administrative controls should also be in place to minimize the potential for human exposures. These may include written procedures or policies which specify the methods and techniques that will be practiced whenever personnel are to work with di-n-octyl phthalate.

All personnel should receive *training* on the use, hazards, protective measures, emergency actions, and other precautions per 29 CFR 1910.1200 (Hazard Communication), prior to the first assignment in an area where di-n-octyl phthalate is used or stored.

Before beginning employment and at regular times thereafter (e.g., annually), for those with frequent or potentially high exposures, the following medical test is recommended:

☑ Lung function tests.

If symptoms develop or overexposure is suspected, the following medical tests may be useful:

☑ Liver and kidney function tests.

☑ Consider chest X-ray after acute overexposure.

Any evaluation should include a careful medical history of past and present symptoms with an examination. However, medical tests that evaluate existing damage are not a substitute for controlling exposure. Also, because smoking can cause heart disease as well as lung cancer, emphysema, and other respiratory problems, smokers exposed to di-n-octyl phthalate may have a more severe reaction than non-smokers. Prudent risk assessment and management requires consideration of *all* possible causes of health related symptoms.

Other methods to reduce exposure include:

☑ Where possible, enclose operations and use local exhaust ventilation at the site of chemical release. If local exhaust ventilation or enclosure is not used, respiratory protection should be mandatory.

☑ Always ensure that proper protective clothing is worn when using chemical substances.

☑ Wash thoroughly immediately after exposure to di-n-octyl phthalate and at the end of the work shift or before eating, drinking, or smoking.

☑ Hazard warning information should be posted in the work area. In addition, as part of an on-going education and training program, all information on the health and safety hazards of di-n-octyl phthalate should be communicated to all potentially exposed workers.

☑ Eye wash stations should be provided in work areas. If the potential for whole body exposure exists, then safety showers should also be provided.

RISK ASSESSMENT: ENVIRONMENT

General Assessment

The environment is at risk of exposure during intended use, transportation, storage, disposal, or destruction of di-n-octyl phthalate. In almost every scenario, the threat of environmental exposure is contingent upon the proper handling of the chemical substance. Accidental spills, large or small, can result in fire, explosion, and

possible contamination of the surrounding environmental mediums (water, soil, and air).

Di-n-octyl phthalate is considered a class IIIB combustible liquid (per OSHA 29 CFR 1910.106). These characteristics require special consideration during any emergency situation involving a leak or spill of di-n-octyl phthalate. Should di-n-octyl phthalate ever come into contact with incompatible substances such as strong oxidizers (chlorine, bromine, fluorine, and chlorine dioxide), strong alkali materials (sodium hydroxide, potassium hydroxide, and lithium hydroxide), and strong acids (sulfuric acid, hydrochloric acid, and nitric acid), and other nitrates either during use, transportation, or storage, violent reaction can occur. Di-n-octyl phthalate is especially reactive in contact with liquid chlorine.

Di-n-octyl phthalate can enter the environment through industrial discharges, in non-point discharges from the wide and general use of this chemical, and from spills.

Acute Ecological Effects

Acute (short-term) toxic effects may include the death of animals, birds, or fish and death or low growth rate in plants. Acute effects are seen 2 to 4 days after animals or plants are exposed to di-n-octyl phthalate.

This chemical has moderate acute toxicity to aquatic life. Insufficient data are available to evaluate or predict the short-term effects of di-n-octyl phthalate to plants, birds, or land animals.

Chronic Ecological Effects

Chronic toxic effects may include shortened life span, reproductive problems, lower fertility, and changes in appearance or behavior in exposed animals. These effects can be seen long after first exposure(s) to toxic chemicals. Di-n-octyl phthalate has moderate chronic toxicity to aquatic life. Insufficient data are available to evaluate or predict the long-term effects of di-n-octyl phthalate to plants, birds, or land animals.

Water Solubility

Di-n-octyl phthalate is very slightly soluble in water. Concentrations of 1 milligram or less will mix with a liter of water.

Persistence in the Environment

Di-n-octyl phthalate is slightly persistent in water, with a half-life of between 2 to 20 days. The half-life of a pollutant is the amount of time it takes for one half of the chemical to be degraded. Approximately 51% of di-n-octyl phthalate will eventually end up in terrestrial soil; about 47% will end up in aquatic sediments, and about 2% will end up in air.

Bioaccumulation in Aquatic Organisms

Some substances increase in concentration, or *bioaccumulate,* in living organisms as they breathe contaminated air, drink contaminated water, or eat contaminated food. These chemicals can become concentrated in the tissues and internal organs of animals as well as humans.

The concentration of di-n-octyl phthalate found in fish tissues is expected to be considerably higher than the average concentration of di-n-octyl phthalate in the water from which the fish was taken.

Recommended Risk-Reduction Measures

Proper training of all transporters will reduce the likelihood of a mishap or accident resulting in a leak or spill to the environment. The correct use of DOT labeling (per the MSDS) on all containers, trucks, and rail cars will enable emergency responders to react properly and quickly to any disaster thereby reducing the potential risk to the environment and to personnel.

Storage of di-n-octyl phthalate should be segregated from incompatible chemicals to minimize the risk of cross-contamination or contact.

Di-n-octyl phthalate is destroyed by burning in a chemical incinerator equipped with an afterburner and scrubber. Destruction of chemicals must be performed by a properly permitted facility.

If a spill or leak to the environment has occurred, fire department, emergency response and/or hazardous materials spill personnel should be notified immediately. Cleanup should be attempted only by those trained in proper spill containment procedures. Contaminated soils should be removed for incineration and replaced with clean soil. If di-n-octyl phthalate should contact the water table, aquifer, or navigable waterway, remediation activities should be prompt to ensure protection of aquatic sediments. When such spills occur, the local and/or state emergency response authorities must be notified. A comprehensive emergency response or disaster preparedness/recovery plan should be in place prior to any operations involving the use, transportation, storage, or disposal of di-n-octyl phthalate.

If di-n-octyl phthalate is spilled or leaked, the following specific steps are recommended:

☑ Restrict persons not wearing protective clothing from area of spill or leak until cleanup is complete and area can be opened for normal work.

☑ Ventilate area of spill or leak.

☑ Remove all ignition sources.

☑ Absorb liquid spills using vermiculite, dry earth, or sand to absorb and place in a sealed drum for disposal.

☑ It may be necessary to dispose of di-n-octyl phthalate as a hazardous waste. The responsible state agency or the regional office of the federal Environmental Protection Agency (EPA) should be contacted for specific recommendations.

RISK ASSESSMENT: BUSINESS

General Assessment

Accidents or mishaps involving di-n-octyl phthalate can present a significant threat to business operations. The loss or damage of equipment or facilities can significantly affect fiscal viability. Lawsuits that may result from personnel injury/death, public exposures, and/or environmental contamination will also require a serious expenditure of resources. Media attention surrounding an injury, death, or environmental damage can also result in a loss of profits and loss of current as well as future business.

Recommended Risk-Reduction Measures

Company attorneys, safety and health professionals, and environmental specialists should be involved in the development of any procedures dealing with safety or emergency response. A company official should be pre-designated as a public relations officer with specific training in dealing with the press. Corporate plans and policies should be developed, approved, and implemented long before any need for such arises.

REFERENCES

American Conference of Governmental Industrial Hygienists. 1988. *Documentation of the Threshold Limit Values and Biological Exposure Indices,* 5th Edition (with updates). Cincinnati: ACGIH

Anthony, C. P., and N. J. Kolthoff. 1971. *Textbook of Anatomy and Physiology.* St. Louis: C. V. Mosby Company.

Bloom, A. D. and F. J. DeSerres. 1995. *Ecotoxicity and Human Health.* New York: CRC Press/Lewis Publishers.

Bueche, J. R. 1972. *Principles of Physics,* 2nd Edition. New York: McGraw-Hill.

Burrell, R., D. K. Flaherty, and L. J. Sauers. 1992. *Toxicology of the Immune System.* New York: Van Nostrand Reinhold.

Calabrese, E. J., E. M. Kenyon. *Air Toxics and Risk Assessment.* New York: CRC Press/Lewis Publishers.

Cockerman, L. G., B. S. Shane. 1994. *Basic Environmental Toxicology.* New York: CRC Press/Lewis Publishers.

Cothern, C. R. 1995. *Handbook for Environmental Risk Decisionmaking.* New York: CRC Press/Lewis Publishers.

Gardner, R. W. 1994. *Chemical Intolerance.* New York: CRC Press/Lewis Publishers.

Hallenbeck, W. H. 1993. *Quantitative Risk Assessment for Environmental and Occupational Health,* 2nd Edition. New York: CRC Press/Lewis Publishers.

Hodgson, E., R. B. Mailman, and Janice E. Chambers. 1988. *Dictionary of Toxicology.* London: The MacMillian Company.

Landis, W. G., M. H. Yu. 1995. *Introduction to Environmental Toxicology.* New York: CRC Press/Lewis Publishers.

Lauwerys, R. R., P. Hoet. 1993. *Industrial Chemical Exposure.* New York: CRC Press/Lewis Publishers.

Lewis, R. J., Sr. 1993. *Hawley's Condensed Chemical Dictionary,* 12th Edition. New York: Van Nostrand Reinhold.

Lewis, R. J., Sr. 1992. *Sax's Dangerous Properties of Industrial Materials,* Volumes 1, 2, and 3, 8th Edition. New York: Van Nostrand Reinhold.

Manahan, S. E. 1992. *Toxicological Chemistry,* 2nd Edition. New York: CRC Press/Lewis Publishers.

Meyer, E. 1990. *Chemistry of Hazardous Materials.* Englewood Cliffs, NJ: Prentice Hall.

Mycock, J. C., J. D. McKenna, and L. Theodore. 1995. *Handbook of Air Pollution Control Engineering and Technology.* New York: CRC Press/Lewis Publishers.

Nelson, G. O. 1992. *Gas Mixtures.* New York: CRC Press/Lewis Publishers.

Ott, W. R. 1995. *Environmental Statistics and Data Analysis.* New York: CRC Press/Lewis Publishers.

Patnaik, P. 1992. *A Comprehensive Guide to the Hazardous Properties of Chemical Substances.* New York: Van Nostrand Reinhold.

Quigley, D. R. 1994. *Handbook of Emergency Chemical Management.* New York: CRC Press/Lewis Publishers.

Rea, W. J. 1995. *Chemical Sensitivity,* Vol. I-III. New York: CRC Press/Lewis Publishers.

Richardson, M. 1995. *Dictionary of Substance and Their Effects,* Vol. I-VI. New York: CRC Press/Lewis Publishers.

Rylander, R., R. R. Jacobs. 1994. *Organic Dusts.* New York: CRC Press/Lewis Publishers.

Sacarello, H. L. A. 1994. *The Comprehensive Handbook of Hazardous Materials.* New York: CRC Press/Lewis Publishers.

U.S. Department of Health and Human Services, National Institute for Occupational Safety and Health. 1994. *NIOSH Pocket Guide to Chemical Hazards.* Washington, D.C.: U.S. Government Printing Office.

MATERIAL SAFETY DATA SHEET

CHEMICAL NAME	
DIOXANE	**CRC PRESS/LEWIS PUBLISHERS DISCLAIMER:** THE INFORMATION AND RECOMMENDATIONS PRESENTED HEREIN ARE BASED ON SOURCES BELIEVED TO BE RELIABLE. CRC MAKES NO REPRESENTATION ON ITS COMPLETENESS OR ACCURACY. IT IS THE RESPONSIBILITY OF THE USER TO DETERMINE THE CHEMICAL'S SUITABILITY FOR ITS INTENDED USE, THE CHEMICAL'S SAFE USE, AND THE CHEMICAL'S PROPER DISPOSAL. NO REPRESENTATIONS AND/OR WARRANTIES, EITHER EXPRESSED OR IMPLIED, OF THE MERCHANTABILITY OR FITNESS FOR A PARTICULAR PURPOSE, OR OF ANY OTHER NATURE, ARE MADE WITH RESPECT TO THE INFORMATION PROVIDED IN THIS MSDS OR TO THE CHEMICAL TO WHICH INFORMATION MAY REFER. CRC NEITHER ASSUMES NOR AUTHORIZES ANY OTHER PERSON TO ASSUME FOR IT, ANY OTHER ADDITIONAL RESPONSIBILITY OR LIABILITY FOR THE USE OF, OR RELIANCE UPON, THIS INFORMATION.

HAZARD WARNING INFORMATION

HEALTH	FIRE	REACTIVE	OTHER	DEGREE OF HAZARD	COLOR CODING	OTHER CODES
2	3	1		0 = Minimum Hazard 1 = Slight Hazard 2 = Moderate Hazard 3 = Serious Hazard 4 = Severe Hazard	HEALTH = BLUE FIRE = RED REACTIVITY = YELLOW OTHER = WHITE	OX = Oxidizer ACID = Acid ALK = Alkali COR = Corrosive W = Use No Water

SECTION I - GENERAL INFORMATION

Characterization **Glycol Ether**	RCRA Number **U108**	EPA Class **Toxic Waste**
DOT Proper Shipping Name **Dioxane**	Chemical Abstract Service (CAS) Number **123-91-1**	
DOT Hazard Class and Label Requirements **Flammable Liquid**	DOT Emergency Guide Code **26**	
DOT Identification Number **UN 1165**	Molecular Formula $C_4H_8O_2$	

Synonyms

Diethylene dioxide; diethylene ether; *p*-Dioxane; 1,4-dioxane; 1,4-dioxacyclohexane; 1,4-diethylene dioxide; di(ethylene oxide).

SECTION II - HAZARDOUS INGREDIENTS AND IDENTITY INFORMATION

Hazardous Components (specific identity)	OSHA Exposure Criteria	NIOSH Exposure Criteria	Immediately Dangerous to Life and Health (IDLH)	ACGIH Exposure Criteria
Dioxane: (derivation: From ethylene glycol by treatment with acid; from β,β-dichloroethyl ether by treatment with alkali). **1 ppm = 3.66 mg/m^3**	**PEL (skin):** 25 ppm 90 mg/m^3 **STEL:** Not Established	**REL (ceiling):** 1 ppm 3.6 mg/m^3 (30 minutes) Possible Cancer Agent	500 ppm	**TLV (skin):** 25 ppm 90 mg/m^3 **STEL:** Not Established

SECTION III - PHYSICAL AND CHEMICAL CHARACTERISTICS

Boiling Point **214°F (103°C)**	Specific Gravity (H_2O = 1) **1.03**
Vapor Pressure (mm Hg) **29 at 68°F (20°C)**	Molecular Weight (atomic weight) **88.1**
Vapor Density (Air = 1) **3**	Melting Point **14°F (-10°C)**

Solubility

Miscible with water and most organic solvents.

Appearance and Odor

Colorless liquid (or solid, below 53°F) with a mild, ether-like odor.

SECTION IV - FIRE AND EXPLOSION HAZARD DATA

Flash Point (method used) **55°F (13°C) closed cup**	Explosive Limits in Air % by Volume **LEL: 2% UEL: 22%**
NFPA Classification **Class IB Combustible Liquid**	Autoignition Temperature **356°F (180°C)**

Extinguishing Media

Dry chemical, carbon dioxide, or alcohol foam.

Special Fire Fighting Procedures

Acrid and irritating smoke and fumes are produced in fire. Structural fire-fighting protective clothing will not provide adequate protection. Wear full protective clothing and self-contained breathing apparatus (SCBA). Move container from fire area if it can be done without risk. Use water spray to cool containers.

Unusual Fire and Explosion Hazards

Containers may explode in fire. Presents a vapor explosion hazard indoors, outdoors, and in sewers.

SECTION V - REACTIVITY DATA

Stability		Conditions to Avoid Normally stable at room temperature, if kept under a nitrogen blanket purge. Dioxane is extremely volatile and very flammable. Keep away from ALL sources of heat or ignition and incompatible materials.
Stable	Unstable	Incompatibility *(materials to avoid)*
X		A dangerous fire and explosion hazard when exposed to heat or flame. Reacts vigorously with oxidizing materials. Forms explosive peroxides in air. Incompatible with acids and some metals.
Hazardous Polymerization		Conditions to Avoid Under normal conditions of temperature and pressure, hazardous polymerization of dioxane is not expected to occur. Avoid using bulk quantities of dioxane in work operations.
May Occur	Will Not Occur	Hazardous Decomposition or By-products
	X	When heated to decomposition, dioxane emits acrid smoke and irritating fumes (such as carbon monoxide).

SECTION VI - HEALTH HAZARD DATA

Primary Route(s) of Entry:	Inhalation?	Absorption (skin/eye)?	Ingestion?
	X	X	X

Health Hazards

INHALATION: Extremely dangerous to humans. Systemic poisoning. Causes drowsiness, headache, respiratory distress, nausea, anorexia, and vomiting. Depression of the CNS, coma, and death may occur (from kidney failure, liver necrosis, pulmonary edema, or brain edema).

ABSORPTION: Can pass through unbroken skin causing toxic systemic effects. May cause mild to no surface irritation on initial contact. Toxic systemic effects can equal that of inhalation.

INGESTION: Moderately toxic. May cause unspecified gastrointestinal effects.

Carcinogenicity	NTP Listed?	IARC Cancer Review Group?	OSHA Regulated?	Target Organs?
Suspected Human Confirmed Animal	5th Annual Report	Group 2B	29 CFR 1910.1000 Table Z-1	Skin, eyes, liver, kidneys, CNS, brain, resp. system

Medical Conditions Generally Aggravated by Exposure

Kidney, liver, and skin disorders may be aggravated by exposure.

Emergency and First-aid Procedures

Eye contact: Do not allow victim to keep eyes tightly shut. Flush immediately with water for 15 minutes (minimum), seek medical attention. Skin contact: Wash with large amounts of soap and water for a minimum of 15 minutes. For inhalation: Remove the person from exposure. Provide respiratory assistance and CPR. Transfer to medical facility. Observe for 24-48 hours for lung effects. If swallowed: Seek medical attention immediately. Give large amounts of milk or water and induce vomiting. Do NOT attempt to give an unconscious or convulsing person anything by mouth.

SECTION VII - PRECAUTIONS FOR SAFE HANDLING AND USE

Steps to be Taken in Case Material is Released or Spilled

Spills of this material constitute a serious emergency. Wear self-contained breathing apparatus (SCBA) for respiratory protection. Restrict those not involved in cleanup from entering area. Avoid skin contact. Ventilate area of spill. Absorb in vermiculite or other material and dispose of in sealed drums.

Preferred Waste Disposal Method

Burn in chemical incinerator equipped with and afterburner and scrubber.

Precautions to be Taken in Handling and Storage

Store in tightly closed containers (preferably under a nitrogen purge) in a cool, well ventilated area away from heat (including direct sunlight). Protect containers from physical damage.

Other Precautions and Warnings

Dioxane must be stored to avoid contact with strong oxidizers since violent reactions can occur.

SECTION VIII - CONTROL MEASURES AND PERSONAL PROTECTIVE EQUIPMENT

Respiratory Protection *(specify type)*

Dioxane is a possible human carcinogen. Use a NIOSH/MSHA approved self-contained breathing apparatus (SCBA) with full facepiece operated in positive pressure mode.

Ventilation

Local exhaust (preferred) at site of chemical work or general ventilation.

Protective Gloves	Eye Protection	Other Protective Clothing
Butyl Rubber (or other non-permeable)	Chemical Goggles and Face Mask	Protective Uniform or Apron

Work/Hygiene Practices

Always wash hands thoroughly after using chemical; never bring food, drink, or smoking materials into vicinity of chemicals.

DIOXANE

$C_4H_8O_2$

CAS: 123-91-1

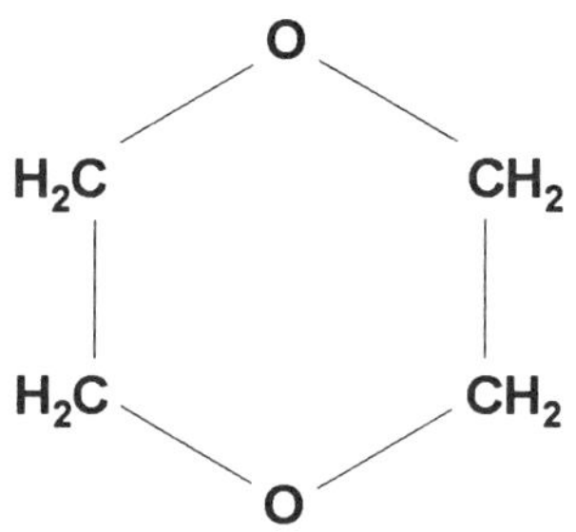

IDENTIFICATION AND TYPICAL USES

Dioxane appears as a colorless liquid (or a solid below 55°F) with a mild, ether-like odor. It is used as a solvent for cellulose esters and a wide range of organic products. It is found in lacquers, varnishes, and paints, as well as paint and varnish removers. It is a wetting and dispersing agent in textile processing, in dye baths, and in stain and printing compositions. It has uses in cleaning and detergent preparations, in cements, in cosmetics and deodorants, as a fumigant, in emulsions, in polishing compositions, as a stabilizer for chlorinated solvents, and in scintillation counters.

RISK ASSESSMENT: HEALTH

General Assessment

Dioxane presents a serious human health risk. It is a confirmed animal carcinogen and its carcinogenicity in humans is highly suspect. In addition, mutation data have been reported and it has been shown to adversely affect reproduction in test animals. It can enter the body and cause dangerous, life-threatening toxic systemic effects through *inhalation, ingestion,* and by passing through intact skin (*absorption*).

Inhalation and absorption will cause symptoms of systemic poisoning with epigastric (intestinal) pain, anorexia, headache, drowsiness, vomiting, nausea, coma, and death. There may be lachrymation and respiratory distress with irritation of the eyes, nose, and throat. It may lead to pulmonary edema (fluid in the lungs), which can be fatal. Symptoms of cough, congestion, and shortness of breath may be delayed for 24-48 hours, creating a false sense of security with regard to health exposure risk. Dioxane can also act to depress the central nervous system (CNS) and lead to damage of the liver (tissue destruction, or "necrosis"), kidneys, lungs (edema), and brain (edema).

Skin contact may produce no significant or apparent irritation on brief exposure. However, prolonged or repeated contact may cause a rash or even a burn to tissues.

Ingestion may result in unspecified gastrointestinal and respiratory effects. Systemic poisoning is extremely possible with symptoms similar to that of inhalation and absorption.

Acute Health Effects

The following acute (short-term) health effects may occur immediately or shortly after exposure to dioxane:

Skin: Mild to severe irritation and possible burns. On prolonged exposure, can be absorbed though intact skin causing toxic systemic effects.

Eye: Severe conjunctival irritation and possible damage to the cornea.

Lung: Severe irritation to respiratory tract causing shortness of breath and cough. High acute exposures can be fatal due to a buildup of fluid in the lungs (pulmonary edema), a medical emergency.

CNS: A depressant of the central nervous system causing a variety of toxic effects including drowsiness, coma, and death.

Other: May cause fluid to build up in the skull (brain edema), which can lead to unconsciousness, coma, and death. Ingestion can cause systemic poisoning along with gastrointestinal distress and symptoms of vomiting and nausea. There may be oliguria (reduce urinary function), or even anuria (almost no urinary function) depending upon the nature and extent of kidney damage.

Chronic Health Effects

The following chronic (long-term) health effects may occur at some time after exposure to dioxane and can last for months or even years:

Cancer Hazards: Dioxane may be a cancer causing agent in humans. Both the National Toxicology Program (NTP) and the International Agency for Research on Cancer (IARC) list dioxane as an anticipated or probable human carcinogen. It is a mutagen (causes cell changes) in animals.

Reproductive Hazard: According to the references, dioxane has been shown to cause adverse effects in reproduction in test animals. Its teratogenic effects in humans have not been adequately studies.

Other Chronic Effects: Repeated low-dose exposures can irritate the lungs and result in bronchitis with symptoms of cough, phlegm, and shortness of breath. It may damage the liver and kidneys. Chronic exposure can lead to jaundice and renal failure.

Recommended Risk-Reduction Measures

Dioxane is extremely hazardous and poses a serious (toxic) risk to human health. Personnel should avoid direct contact with dioxane. If a less toxic material or compound cannot be substituted for dioxane, then *engineering controls* are the most effective method of reducing exposure risk. The best protection is to enclose operations and/or provide local exhaust ventilation at the site of dioxane release. While not always operationally feasible, isolating operations can also reduce exposure. Using respiratory protection is less effective than the controls mentioned above, but is still required whenever working with or around dioxane. Exposure levels are relatively low (25 ppm) and may be difficult to accurately monitor. A NIOSH/MSHA-approved organic vapor respirator may be used for low-level exposures (at or near the PEL). However, the warning properties of dioxane are low and cartridge-type respirators may not suffice. Also, it has been suggested that dioxane is a possible human carcinogen. Therefore, for the best protection at any exposure level, a MSHA/NIOSH approved self-contained breathing apparatus (SCBA) with full facepiece and pressure demand is the appropriate level of respiratory protection. If a full facepiece is not available, then chemical goggles should be worn to protect the eyes, especially when a splash hazard exists. A face shield should also be considered. Skin contract must be avoided. To prevent hand and skin exposures, rubber or other non-permeable gloves should be worn.

Administrative controls should also be in place to minimize the potential for human exposures. These may include written procedures or policies which specify the methods and techniques that will be practiced whenever personnel are to work with dioxane.

All personnel should receive *training* on the use, hazards, protective measures, emergency actions, and other precautions per 29 CFR 1910.1200 (Hazard Communication) prior to the first assignment in an area where dioxane is used or stored.

Before beginning employment and at regular intervals thereafter (e.g., annually), the following medical test is recommended:

☑ Lung function tests.

If symptoms develop or overexposure is suspected, the following may be useful:

☑ Liver, kidney, and lung function tests.

☑ Consider chest X-ray after acute exposure (may be negative if taken immediately following exposure).

☑ Blood urea nitrogen test with serum creatinine and urine sediment.

It should be noted that medical tests that simply look for existing damage are not a substitute for controlling exposures. Medical histories are extremely important when assessing exposure risk. Also, since smoking can cause heart disease as well as lung cancer, emphysema, and other respiratory ailments, any exposure to dioxane can result in quicker and more devastating symptoms. Smokers should therefore avoid unprotected contact with this chemical. Prudent risk management requires proper consideration of *all* possible factors that may be causing the appearance of exposure symptoms among workers.

Other methods to reduce exposure include:

☑ Always ensure that proper protective clothing is worn when using chemical substances.

☑ Wash thoroughly immediately after exposure to dioxane and at the end of the work shift or before eating, drinking, or smoking.

☑ Hazard warning information should be posted in the work area. In addition, as part of an on-going education and training program, all information on

the health and safety hazards of dioxane should be communicated to all potentially exposed workers.

☑ Eye wash stations should be provided in the immediate work area for emergency use. If there is a possibility of skin exposure to dioxane, emergency shower facilities should also be provided.

☑ Workers whose clothing has been contaminated by dioxane should change into clean clothes before leaving work. Contaminated work clothing should be laundered only by individuals who have been informed of the hazards of exposure to dioxane.

RISK ASSESSMENT: ENVIRONMENT

General Assessment

The environment is at risk of exposure during transportation, storage, disposal, or destruction of dioxane. In almost every scenario, the threat of environmental exposure is contingent upon the proper handling of the chemical substance. Accidental spills, large or small, can result in contamination of the surrounding environmental mediums (water, soil, and air).

Dioxane is class IB combustible liquid (per OSHA 29 CFR 1910.106). It is extremely volatile and presents a serious risk of fire and or explosion. It will form dangerous (explosive) peroxides in air. It also forms explosive complexes with the perchlorates of many metals, such as silver or mercury. It is incompatible with oxidizers and some acids (nitric, for example), reacting vigorously and violently on contact. Therefore, caution is always required in handling, storage, transportation, and disposal of dioxane. When heated to decomposition it emits very acrid and irritating smoke and fumes. Emergency responders, hazardous materials personnel, and firefighters should be made aware of the presence of dioxane at any emergency response situation.

Dioxane may enter the environment through industrial effluents and spills.

Acute Ecological Effects

Acute (short-term) toxic effects may include the death of animals, birds, or fish and death or low growth rate in plants. Acute effects are seen 2 to 4 days after animals or plants are exposed to dioxane.

Insufficient data are available to evaluate or predict the acute (short-term) effects of dioxane to aquatic life, plants, birds, or land animals.

Chronic Ecological Effects

Chronic (long-term) toxic effects may include shortened life span, reproductive problems, lower fertility, and changes in appearance or behavior in exposed animals. These effects can be seen long after first exposure(s) to toxic chemicals.

Insufficient data are available to evaluate the chronic (long-term) effects of dioxane on aquatic life, plants, birds, or land animals.

Water Solubility

Dioxane is highly soluble in water. Concentrations of 1000 milligrams and more will mix with a liter of water.

Persistence in the Environment

Dioxane is slightly persistent in water, with a half-life of between 2 to 20 days. The half-life of a pollutant is the amount of time it takes for one half of the chemical to be degraded. Approximately 50% of dioxane will eventually end up in air; the rest will end up in water.

Bioaccumulation in Aquatic Organisms

Some substances increase in concentration, or *bioaccumulate,* in living organisms as they breathe contaminated air, drink contaminated water, or eat contaminated food. These chemicals can become concentrated in the tissues and internal organs of animals as well as humans.

Dioxane is not expected to accumulate in aquatic organisms. The concentration of dioxane in edible fish tissues is expected to be negligible (trace levels) as compared to the levels found the water from which the fish was taken.

Recommended Risk-Reduction Measures

Proper training of all transporters will reduce the likelihood of a mishap or accident resulting in a leak or spill of dioxane into the environment. Labels on all containers, trucks, and rail cars must meet DOT requirements and accurately reflect their contents to enable emergency responders to react properly and quickly to any disaster thereby reducing the potential risk to the environment and to personnel.

Storage of dioxane should be segregated from other chemicals to minimize the risk of cross-contamination. Dioxane must be stored to avoid contact with strong oxidizers, such as chlorine and bromine, since violent reactions can occur. Containers should be protected from physical damage and stored to avoid contact with heat and direct sunlight.

If a spill or leak to the environment has occurred, fire department, emergency response and/or hazardous materials spill personnel should be notified immediately. Cleanup should be attempted only by those trained in proper spill containment procedures. Contaminated soils should be removed for incineration and replaced with clean soil.

If dioxane should contact the water table, aquifer, or navigable waterway, time is of the essence. Its specific solubility in water has not been established and total remediation may or may not be possible. The local and/or state emergency response authorities must be notified. A comprehensive emergency response or disaster preparedness/recovery plan should be in place prior to any operations involving the use, transportation, storage, or disposal of dioxane.

If dioxane is spilled or leaked, the following specific steps are recommended:

- ☑ Restrict persons not wearing protective clothing from area of spill or leak until cleanup is complete. Avoid skin and eye contact.
- ☑ Absorb spilled liquids with vermiculite, dry sand, or a similar material and deposit in sealed drum for disposal.
- ☑ Ventilate area of spill or leak.
- ☑ Remove all ignition sources.
- ☑ It may be necessary to dispose of dioxane as a hazardous waste. The responsible state agency or the regional office of the federal Environmental Protection Agency (EPA) should be contacted for specific recommendations.

RISK ASSESSMENT: BUSINESS

General Assessment

Accidents or mishaps involving dioxane can present a serious threat to business operations. The loss or damage of equipment or facilities can significantly affect fiscal viability. Lawsuits that may result from personnel illness, injury/death, public exposures, and/or environmental contamination will require a serious expenditure of resources. Media attention surrounding an injury, death, or environmental damage can also result in a loss of profits and loss of current as well as future business. Always remember that anytime the terms "cancer," or "carcinogen" are used, public emotion, hysteria, and ignorance can run equally high. This requires careful consideration during the development of public relations policies.

Recommended Risk-Reduction Measures

Company attorneys, safety and health professionals, and environmental specialists should be involved in the development of any procedures for responding to chemical incidents. A company official should be predesignated as a public relations officer with specific training in dealing with the press. Corporate plans and policies should be developed, approved, and implemented long before any need for such arises.

REFERENCES

Anthony, C. P., and N. J. Kolthoff. 1971. *Textbook of Anatomy and Physiology.* St. Louis: C. V. Mosby Company.

Brusick, D. J. 1994. *Methods for Genetic Risk Assessment.* New York: CRC Press/Lewis Publishers.

Calabrese, E. J. 1994. *Biological Effects of Low Level Exposures.* New York: CRC Press/Lewis Publishers.

Cothern, C. R. 1995. *Handbook for Environmental Risk Decisionmaking.* New York: CRC Press/Lewis Publishers.

Fodden, C. R., and J. L. Weddell. 1992. *Hazardous Materials Emergency Action Data.* New York: CRC Press/Lewis Publishers.

Gardner, R. W. 1994. *Chemical Intolerance.* New York: CRC Press/Lewis Publishers.

Hodgson, E., R. B. Mailman, and J. E. Chambers. 1988. *Dictionary of Toxicology.* London: The MacMillian Company.

International Agency for Research on Cancer (IARC). 1987. *IARC Monographs on the Evaluation of Carcinogenic Risk to Humans, Overall Evaluations of Carcinogenicity: An updating of IARC Monographs Volumes 1-42.* Lyon: IARC.

Langerman, N. 1994 *Precautionary Labels for Chemical Containers.* New York: CRC Press/Lewis Publishers.

Lauwerys, R. R., P. Hoet. 1993. *Industrial Chemical Exposure.* New York: CRC Press/Lewis Publishers.

Lewis, R. J., Sr. 1992. *Sax's Dangerous Properties of Industrial Materials,* Volumes 1, 2, and 3, 8th Edition. New York: Van Nostrand Reinhold.

Manahan, S. E. 1992. *Toxicological Chemistry,* 2nd Edition. New York: CRC Press/Lewis Publishers.

Meyer, E. 1990. *Chemistry of Hazardous Materials.* Englewood Cliffs, NJ: Prentice Hall.

Patnaik, P. 1992. *A Comprehensive Guide to the Hazardous Properties of Chemical Substances.* New York: Van Nostrand Reinhold.

Torkelson, T. R. V. K. Rowe. 1981. *Patty's Industrial Hygiene and Toxicology,* Vol. 2B. New York: Wiley-Interscience.

MATERIAL SAFETY DATA SHEET

CHEMICAL NAME	
DIPHENYL	**CRC PRESS/LEWIS PUBLISHERS DISCLAIMER:** THE INFORMATION AND RECOMMENDATIONS PRESENTED HEREIN ARE BASED ON SOURCES BELIEVED TO BE RELIABLE. CRC MAKES NO REPRESENTATION ON ITS COMPLETENESS OR ACCURACY. IT IS THE RESPONSIBILITY OF THE USER TO DETERMINE THE CHEMICAL'S SUITABILITY FOR ITS INTENDED USE, THE CHEMICAL'S SAFE USE, AND THE CHEMICAL'S PROPER DISPOSAL. NO REPRESENTATIONS AND/OR WARRANTIES, EITHER EXPRESSED OR IMPLIED, OF THE MERCHANTABILITY OR FITNESS FOR A PARTICULAR PURPOSE, OR OF ANY OTHER NATURE, ARE MADE WITH RESPECT TO THE INFORMATION PROVIDED IN THIS MSDS OR TO THE CHEMICAL TO WHICH INFORMATION MAY REFER. CRC NEITHER ASSUMES NOR AUTHORIZES ANY OTHER PERSON TO ASSUME FOR IT, ANY OTHER ADDITIONAL RESPONSIBILITY OR LIABILITY FOR THE USE OF, OR RELIANCE UPON, THIS INFORMATION.

HAZARD WARNING INFORMATION

HEALTH	FIRE	REACTIVE	OTHER	DEGREE OF HAZARD	COLOR CODING	OTHER CODES
2	1	0		0 = Minimum Hazard 1 = Slight Hazard 2 = Moderate Hazard 3 = Serious Hazard 4 = Severe Hazard	HEALTH = BLUE FIRE = RED REACTIVITY = YELLOW OTHER = WHITE	OX = Oxidizer ACID = Acid ALK = Alkali COR = Corrosive W = Use No Water

SECTION I - GENERAL INFORMATION

Characterization: **Aromatic Hydrocarbon**	RCRA Number: **None**	EPA Class: **Not Applicable**
DOT Proper Shipping Name: **Not Listed**	Chemical Abstract Service (CAS) Number: **92-52-4**	
DOT Hazard Class and Label Requirements: **Not Listed**	DOT Emergency Guide Code: **No Citation**	
DOT Identification Number: **Not Listed**	Molecular Formula: $C_6H_5C_6H_5$	

Synonyms

Biphenyl; phenyl benzene; bibenzene; xenene; 1,1'-biphenyl; Phenador-X.

SECTION II - HAZARDOUS INGREDIENTS AND IDENTITY INFORMATION

Hazardous Components (specific identity)	OSHA Exposure Criteria	NIOSH Exposure Criteria	Immediately Dangerous to Life and Health (IDLH)	ACGIH Exposure Criteria
Diphenyl (derivation: By slowly passing benzene through a red-hot iron tube; by heating bromobenzene and sodium with subsequent distillation). 1 ppm = 6.41 mg/m^3	**PEL:** 0.2 ppm 1.2 mg/m^3 **STEL:** Not Established	**REL:** 0.2 ppm 1.2 mg/m^3 **STEL:** Not Established	100 mg/m^3	**TLV:** 0.2 ppm 1.2 mg/m^3 **STEL:** Not Established

SECTION III - PHYSICAL AND CHEMICAL CHARACTERISTICS

Boiling Point: **491°F (255°C)**	Specific Gravity (H_2O = 1): **1.04**
Vapor Pressure (mm Hg): **0.005 at 69°F (20°C)**	Molecular Weight (atomic weight): **154.2**
Vapor Density (Air = 1): **5.31**	Melting Point: **158°F (70°C)**

Solubility

Insoluble in water. Soluble in alcohol and ether.

Appearance and Odor

Colorless to pale yellow solid (scale or leaf-like crystals) with a pleasant, peculiar, characteristic (fungicide-like) odor.

SECTION IV - FIRE AND EXPLOSION HAZARD DATA

Flash Point (method used): **235°F (113°C) closed cup**	Explosive Limits in Air % by Volume: **LEL: 0.6% at 232°F (111°C) UEL: 5.8% at 311°F (155°C)**
NFPA Classification: **Combustible Solid**	Autoignition Temperature: **1004°F (540°C)**

Extinguishing Media

Use dry chemical, carbon dioxide, foam, or water spray, mist or fog.

Special Fire Fighting Procedures

Poisonous gases are produced in fire. Structural fire-fighting clothing is permeable and may not provide adequate protection. Wear self-contained breathing apparatus (SCBA) operated in positive pressure mode. Move fire-exposed containers from fire if it can be done without risk. Cool with water spray.

Unusual Fire and Explosion Hazards

Slight fire hazard when exposed to heat or flame. Dust explosions can be significant.

SECTION V - REACTIVITY DATA

Stability		Conditions to Avoid
		Diphenyl is stable at room temperature under normal conditions of handling and storage. Avoid the formation of dust clouds of diphenyl and prevent contact with ignition sources or oxidizers.
Stable	Unstable	Incompatibility *(materials to avoid)*
X		Strong oxidizers, such as chlorine, bromine, and fluorine.
Hazardous Polymerization		Conditions to Avoid
		Under normal conditions of temperature and pressure, hazardous polymerization of diphenyl is not expected to occur.
May Occur	Will Not Occur	Hazardous Decomposition or By-products
	X	Poisonous gases, including oxides of carbon (carbon monoxide) and aromatic compounds are produced when diphenyl is involved in a fire.

SECTION VI - HEALTH HAZARD DATA

Primary Route(s) of Entry:	Inhalation?	Absorption (skin/eye)?	Ingestion?
	X	X	X

Health Hazards

INHALATION: Irritation of the eyes, nose, throat, and upper respiratory tract. Symptoms of headache, nausea, fatigue, malaise, numbness in limbs, gastrointestinal pain, indigestion, possible liver damage. Also toxic to the central and peripheral nervous systems (CNS and PNS).

ABSORPTION: Irritation and possible sensitization. Also, absorbs through the skin to cause toxic systemic effects. Eye contact may result in severe irritation and swelling, with redness.

INGESTION: Moderately toxic by ingestion. Can result in same systemic effects as inhalation, but not as quickly. Diffuse gastrointestinal pain, stomach cramps, and nausea likely.

Carcinogenicity	NTP Listed?	IARC Cancer Review Group?	OSHA Regulated?	Target Organs?
Unknown Human Unknown Animal	No	No	29 CFR 1910.1000 Table Z-1	Respiratory system, eyes, skin, CNS, PNS, liver.

Medical Conditions Generally Aggravated by Exposure

None identified.

Emergency and First-aid Procedures

Eye contact: Do not allow victim to keep eyes tightly shut. Flush immediately with water for 15 minutes (minimum), seek medical attention. Skin contact: Remove all contaminated clothing, including shoes. Immediately wash area with flooding amounts of soap and water. Seek medical attention immediately. For inhalation: Remove the person from exposure. Provide respiratory assistance and CPR. Transfer to medical facility. If swallowed: Seek medical attention immediately. Do NOT induce vomiting.

SECTION VII - PRECAUTIONS FOR SAFE HANDLING AND USE

Steps to be Taken in Case Material is Released or Spilled

Remove all ignition sources. Wear SCBA respiratory protection. Restrict those not wearing protective equipment and who are not involved in cleanup from entering area. Avoid skin contact. Do not create dust cloud. Ventilate area of spill. Do NOT dry sweep. Use HEPA vacuum and deposit in sealed drums.

Preferred Waste Disposal Method

No Citation.

Precautions to be Taken in Handling and Storage

Store in tightly closed containers in cool, well ventilated location. Protect containers from physical damage. Prevent dust buildup or similar conditions in storage and use areas.

Other Precautions and Warnings

Sources of ignition (open flame or smoking), are prohibited where diphenyl is used or stored. Metal containers should be bonded and grounded. Use explosion-proof equipment in bulk storage facilities.

SECTION VIII - CONTROL MEASURES AND PERSONAL PROTECTIVE EQUIPMENT

Respiratory Protection *(specify type)*

Use a NIOSH approved respirator equipped with HEPA filter, or a powered air purifying respirator, or use a self-contained breathing apparatus (SCBA) with full facepiece operated in positive pressure mode.

Ventilation

Local exhaust (preferred) at site of chemical work or general ventilation.

Protective Gloves	Eye Protection	Other Protective Clothing
Impervious Gloves	Chemical/Dust Goggles or Face Mask	Protective Clothing (prevent skin contact)

Work/Hygiene Practices

Always wash hands thoroughly after using chemical; never bring food, drink, or smoking materials into vicinity of chemicals.

DIPHENYL

$C_6H_5C_6H_5$ CAS: 92-52-4

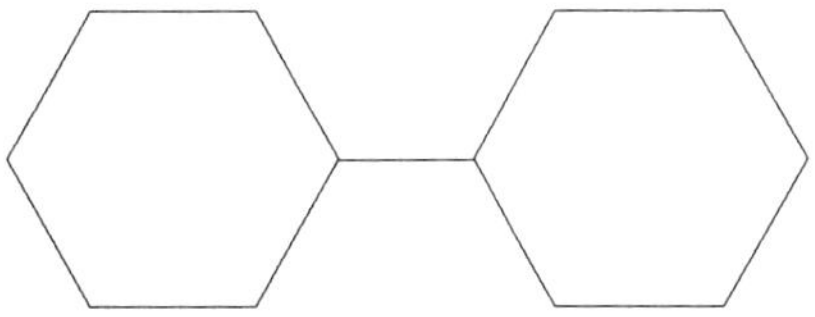

IDENTIFICATION AND TYPICAL USES

Diphenyl is a colorless to pale yellow solid (scales or leaf-like crystals) with a pleasant, peculiar, aromatic odor characteristic of fungicides. It is used in organic synthesis, as a heat transfer agent (either with or without diphenyl oxide) because of its high thermal stability. It is also used as a fungicide in the packaging of citrus fruits and to prevent the growth of blue mold and other plant diseases. It has also been used as a dyeing assistant for polyester fibers.

RISK ASSESSMENT: HEALTH

General Assessment

Diphenyl is a human poison by *inhalation, ingestion,* and *absorption* through intact skin. While its human carcinogenicity is unknown, there are some studies which suspect cancer-causing capabilities in laboratory test animals. However, the data in the references are conflicting and inconclusive in this regard. Mutation data have been reported and some scientists believe that mutagenic chemicals may also have carcinogenic properties.

Inhalation of the dusts, as well as skin absorption, causes toxic systemic effects with action on the central nervous system (CNS) and the peripheral nervous system (PNS). There may be headaches with nausea, general fatigue, malaise, gastrointestinal pain, indigestion, numbness and aching pain in the legs and arms, and possible damage to the liver (necrosis or atrophy). In addition, inhalation will irritate the mucosa of the nose, throat, and upper respiratory tract.

Skin contact may result in mild to severe irritation. In some individuals, sensitization may develop with dermatitis and itching rash. Once sensitization occurs, even the slightest future exposures will cause a recurrence of the symptoms.

Ingestion is not likely in the industrial setting but is still a possible route of entry. Swallowing diphenyl dusts can result in unspecified but diffuse gastrointestinal irritation and possible indigestion. The symptoms of inhalation and absorption may also appear on ingestion, although not as pronounced or as quick.

☠ *Acute Health Effects*

The following acute (short-term) health effects may occur immediately or shortly after exposure to diphenyl:

Skin: Irritation with rash and localized inflammation. Absorption through intact skin is extremely likely on short exposures.

Eye: Severe irritation; pain and inflammation.

Lung: Nose, throat, and respiratory tract irritation causing difficulty in breathing.

CNS: Confusion, dizziness, numbness in the leg and arm muscles, cranial pressure (headaches), possible coma and death.

💣 *Chronic Health Effects*

The following chronic (long-term) health effects may occur at some time after exposure to diphenyl and can last for months or even years:

Cancer Hazards: According to the information presented in the references, diphenyl has not been adequately tested for its ability to cause cancer in test animals.

Reproductive Hazard: There is no information in the references to indicate that diphenyl causes adverse effects in the reproduction process of test animals. Additional research is required in this area of study.

Other Chronic Effects: Chronic dermatitis sensitization is possible in certain individuals. Once sensitization occurs, even slight future exposures will cause a recurrence of symptoms. Liver damage (necrosis and/or atrophy) can occur on long-term exposures.

Recommended Risk-Reduction Measures

Personnel should avoid direct contact with diphenyl. Occupational poisoning occurs most commonly by inhalation. If a less toxic material or compound cannot be substituted for diphenyl, then *engineering controls* are the most effective method of reducing exposures. The best protection is to enclose operations and/or provide local exhaust ventilation at the site of diphenyl release. While not always operationally feasible, isolating operations can also reduce exposure. Using respiratory protection is less effective than the controls mentioned above, but is still advisable whenever working with or around diphenyl. Personnel should wear an MSHA/NIOSH-approved respirator equipped with a high efficiency particulate air (HEPA) filter for low or infrequent exposures. An MSHA/NIOSH-approved self-contained breathing apparatus (SCBA) with full facepiece and pressure demand is a more appropriate level of respiratory protection when high exposures to diphenyl are anticipated. If a full facepiece is not available, then chemical/dust goggles should be worn to protect the eyes. A face shield should also be considered. To prevent hand and skin exposures, polyvinyl alcohol gloves should be worn.

Administrative controls should also be in place to minimize the potential for human exposures. These may include written procedures or policies which specify the methods and techniques that will be practiced whenever personnel are to work with diphenyl.

All personnel should receive *training* on the use, hazards, protective measures, emergency actions, and other precautions per 29 CFR 1910.1200 (Hazard Communication), prior to the first assignment in an area where diphenyl is used or stored.

Before beginning employment and at regular times thereafter, the following recommended medical tests should be considered:

- ☑ Complete blood count (CBC).
- ☑ Urinary phenol (a test to see if diphenyl is already present in the body).

If symptoms should develop or overexposure is suspected, the following additional tests may be helpful:

- ☑ Complete blood count.
- ☑ Urinary phenol.
- ☑ Evaluation by a qualified allergist, with careful review of exposure history and special tests (may help diagnose skin allergy).
- ☑ Liver function tests.
- ☑ Neurological tests of the CNS and PNS.

Any evaluation should include a careful history of past and present symptoms with an examination. Medical tests that simply look for existing damage are not a substitute for controlling exposures.

Other methods to reduce exposure include:

- ☑ Where possible, enclose operations and use local exhaust ventilation at the site of chemical release. If local exhaust ventilation or enclosure is not used, respiratory protection should be mandatory.
- ☑ Always ensure that proper protective clothing is worn when using chemical substances.
- ☑ Wash thoroughly immediately after exposure to diphenyl and at the end of the work shift or before eating, drinking, or smoking.
- ☑ Hazard warning information should be posted in the work area. In addition, as part of an on-going education and training program, all information on the health and safety hazards of diphenyl should be communicated to all exposed and potentially exposed workers.
- ☑ Eye wash stations should be provided in the immediate work area for emergency use. If there is a possibility of skin exposure to diphenyl, emergency shower facilities should also be provided.
- ☑ Workers whose clothing has been contaminated by diphenyl should change into clean clothes before leaving work. Contaminated work clothing should be laundered only by individuals who have been informed of the hazards of exposure to diphenyl.

RISK ASSESSMENT: ENVIRONMENT

General Assessment

The environment is at risk of exposure during transportation, storage, disposal, or destruction of diphenyl. In almost every scenario, the threat of environmental exposure is contingent upon the proper handling of the chemical substance. Accidental spills, large or small, can result in contamination of the surrounding environmental mediums (water, soil, and air).

Diphenyl is a combustible solid. It can present a dangerous explosion hazard if its dusts form a cloud and come into contact with an ignition source. It is also incompatible with oxidizing materials and contact can result in violent and explosive reactions. Caution is always required in handling, storage, transportation, and disposal of diphenyl. Emergency responders should be made aware of the presence of diphenyl at any emergency response situation.

Diphenyl can enter the environment from industrial effluents and from spills.

Acute Ecological Effects

Acute (short-term) toxic effects may include the death of animals, birds, or fish and death or low growth rate in plants. Acute effects are seen 2 to 4 days after animals or plants are exposed to diphenyl.

Diphenyl has moderate acute toxicity to aquatic life, moderate-to-high acute toxicity to land animals, and high acute toxicity to birds. While it has not been shown to cause damage or injury to agricultural crops or plants in its use as a fungicide, insufficient data are available to properly evaluate the short-term effects of diphenyl on plants.

Chronic Ecological Effects

Chronic (long-term) toxic effects may include shortened life span, reproductive problems, lower fertility, and changes in appearance or behavior in exposed animals. These effects can be seen long after first exposure(s) to toxic chemicals.

Diphenyl has high chronic toxicity to aquatic life. Insufficient data are available to evaluate or predict the long-term effects of diphenyl to plants, birds, or terrestrial animals.

Water Solubility

Diphenyl is insoluble in water. Concentrations of less than 1 milligram will not mix with a liter of water.

Persistence in the Environment

Diphenyl is slightly persistent in water, with a half-life between 2 and 20 days. The half-life of a pollutant is the amount of time it takes for one half of the chemical to be degraded.

Approximately 59% of diphenyl will eventually end up in air; about 15.6% will end up in water; and the rest will be divided nearly equally between terrestrial soils and aquatic sediments.

Bioaccumulation in Aquatic Organisms

Some substances increase in concentration, or *bioaccumulate,* in living organisms as they breathe contaminated air, drink contaminated water, or eat contaminated food. These chemicals can become concentrated in the tissues and internal organs of animals as well as humans.

The concentration of diphenyl found in fish tissues is expected to be somewhat higher than the average concentration of diphenyl in the water from which the fish was taken.

Recommended Risk-Reduction Measures

Proper training of all transporters will reduce the likelihood of a mishap or accident resulting in a leak or spill of diphenyl into the environment. Labels on all containers, trucks, and rail cars must meet DOT requirements and accurately reflect their contents to enable emergency responders to react properly and quickly to any disaster thereby reducing the potential risk to the environment and to personnel.

Storage of diphenyl should be segregated from other chemicals and materials to minimize the risk of cross-contamination. Storage facilities should be designed with explosion-proof electrical equipment.

If a spill or leak to the environment has occurred, fire department, emergency response and/or hazardous materials spill personnel should be notified immediately. Cleanup should be attempted only by those trained in proper spill containment procedures. Contaminated soils should be removed for incineration and replaced with clean soil. If diphenyl should contact the water table, aquifer, or navigable waterway, time is of the essence. It is insoluble and total remediation may be possible. The local and/or state emergency response authorities must be notified. A comprehensive emergency response or disaster preparedness/recovery plan should be in place prior to any operations involving the use, transportation, storage, or disposal of diphenyl.

For disposal, diphenyl may be burned in a chemical incinerator equipped with an afterburner and scrubber.

If diphenyl is spilled or leaked, the following specific steps are recommended:

- ☑ Restrict persons not wearing protective clothing from area of spill or leak until cleanup is complete.
- ☑ Do NOT generate dusts. Use a scoop of a vacuum equipped with a high efficiency particulate

air filter (HEPA) to collect spills. Place materials in drum for recovery or disposal, as required.

☑ Ventilate area of spill or leak. Container should be removed to safer area if it can be done without increasing the risk.

☑ It may be necessary to dispose of diphenyl as a hazardous waste. The responsible state agency or the regional office of the federal Environmental Protection Agency (EPA) should be contacted for specific recommendations.

RISK ASSESSMENT: BUSINESS

General Assessment

Accidents or mishaps involving diphenyl can present a serious threat to business operations. The loss or damage of equipment or facilities can significantly affect fiscal viability. Lawsuits that may result from personnel illness (leukemia), injury/death, public exposures, and/or environmental contamination will require a serious expenditure of resources. Media attention surrounding an injury, death, or environmental damage can also result in a loss of profits and loss of current as well as future business.

Recommended Risk-Reduction Measures

Company attorneys, safety and health professionals, and environmental specialists should be involved in the development of any procedures for responding to chemical incidents. A company official should be predesignated as a public relations officer with specific training in dealing with the press. Corporate plans and policies should be developed, approved, and implemented long before any need for such arises.

REFERENCES

American Conference of Governmental Industrial Hygienists. 1988. *Documentation of the Threshold Limit Values and Biological Exposure Indices,* 5th Edition (with updates). Cincinnati: ACGIH.

Bloom, A. D. and F. J. DeSerres. 1995. *Ecotoxicity and Human Health.* New York: CRC Press/Lewis Publishers.

Brusick, D. J. 1994. *Methods for Genetic Risk Assessment.* New York: CRC Press/Lewis Publishers.

Bueche, J. R. 1972. *Principles of Physics,* 2nd Edition. New York: McGraw-Hill.

Calabrese, E. J., E. M. Kenyon. *Air Toxics and Risk Assessment.* New York: CRC Press/Lewis Publishers.

Cockerman, L. G., B. S. Shane. 1994. *Basic Environmental Toxicology.* New York: CRC Press/Lewis Publishers.

Cothern, C. R., N. P. Ross. 1994. *Environmental Statistics, Assessment, and Forecasting.* New York: CRC Press/Lewis Publishers.

Danse, I. R. 1991. *Common Sense Toxics in the Workplace.* New York: Van Nostrand Reinhold.

Gerarde, H. W. 1960. *Toxicology and Biochemistry of Aromatic Hydrocarbons.* New York: Elsevier.

Hallenbeck, W. H. 1993. *Quantitative Risk Assessment for Environmental and Occupational Health,* 2nd Edition. New York: CRC Press/Lewis Publishers.

Howard, P. H., et. al. 1991. *Handbook of Environmental Degradation Rates.* New York: CRC Press/Lewis Publishers.

Jones, F. E. 1994. *Toxic Organic Vapors in the Workplace.* New York: CRC Press/Lewis Publishers.

Keith, L. H. and M. M. Walker. 1995. *Handbook of Air Toxics.* New York: CRC Press/Lewis Publishers.

Landis, W. G., M. H. Yu. 1995. *Introduction to Environmental Toxicology.* New York: CRC Press/Lewis Publishers.

Langerman, N. 1994 *Precautionary Labels for Chemical Containers.* New York: CRC Press/Lewis Publishers.

Lauwerys, R. R., P. Hoet. 1993. *Industrial Chemical Exposure.* New York: CRC Press/Lewis Publishers.

Lewis, R. J., Sr. 1993. *Hawley's Condensed Chemical Dictionary,* 12th Edition. New York: Van Nostrand Reinhold.

Lewis, R. J., Sr. 1992. *Sax's Dangerous Properties of Industrial Materials,* Volumes 1, 2, and 3, 8th Edition. New York: Van Nostrand Reinhold.

Manahan, S. E. 1994. *Environmental Chemistry,* 6th Edition. New York: CRC Press/Lewis Publishers.

Meyer, E. 1990. *Chemistry of Hazardous Materials.* Englewood Cliffs, NJ: Prentice Hall.

Ott, W. R. 1995. *Environmental Statistics and Data Analysis.* New York: CRC Press/Lewis Publishers.

Patnaik, P. 1992. *A Comprehensive Guide to the Hazardous Properties of Chemical Substances.* New York: Van Nostrand Reinhold.

Philp, R. B. 1995. *Environmental Hazards and Human Health.* New York: CRC Press/Lewis Publishers.

Proctor, N. H., J. P. Hughes. 1991. *Chemical Hazards of the Workplace.* New York: Van Nostrand Reinhold.

Quigley, D. R. 1994. *Handbook of Emergency Chemical Management.* New York: CRC Press/Lewis Publishers.

Sacarello, H. L. A. 1994. *The Comprehensive Handbook of Hazardous Materials.* New York: CRC Press/Lewis Publishers.

Torkelson, T. R. V. K. Rowe. 1981. *Patty's Industrial Hygiene and Toxicology,* Vol. 2B. New York: Wiley-Interscience.

U.S. Department of Health and Human Services, National Institute for Occupational Safety and Health. 1994. *NIOSH Pocket Guide to Chemical Hazards.* Washington, D.C.: U.S. Government Printing Office.

MATERIAL SAFETY DATA SHEET

CHEMICAL NAME	
DIPHENYLAMINE	**CRC PRESS/LEWIS PUBLISHERS DISCLAIMER:** THE INFORMATION AND RECOMMENDATIONS PRESENTED HEREIN ARE BASED ON SOURCES BELIEVED TO BE RELIABLE. CRC MAKES NO REPRESENTATION ON ITS COMPLETENESS OR ACCURACY. IT IS THE RESPONSIBILITY OF THE USER TO DETERMINE THE CHEMICAL'S SUITABILITY FOR ITS INTENDED USE, THE CHEMICAL'S SAFE USE, AND THE CHEMICAL'S PROPER DISPOSAL. NO REPRESENTATIONS AND/OR WARRANTIES, EITHER EXPRESSED OR IMPLIED, OF THE MERCHANTABILITY OR FITNESS FOR A PARTICULAR PURPOSE, OR OF ANY OTHER NATURE, ARE MADE WITH RESPECT TO THE INFORMATION PROVIDED IN THIS MSDS OR TO THE CHEMICAL TO WHICH INFORMATION MAY REFER. CRC NEITHER ASSUMES NOR AUTHORIZES ANY OTHER PERSON TO ASSUME FOR IT, ANY OTHER ADDITIONAL RESPONSIBILITY OR LIABILITY FOR THE USE OF, OR RELIANCE UPON, THIS INFORMATION.

HAZARD WARNING INFORMATION

HEALTH	FIRE	REACTIVE	OTHER	DEGREE OF HAZARD	COLOR CODING	OTHER CODES
3	1	0		0 = Minimum Hazard 1 = Slight Hazard 2 = Moderate Hazard 3 = Serious Hazard 4 = Severe Hazard	HEALTH = BLUE FIRE = RED REACTIVITY = YELLOW OTHER = WHITE	OX = Oxidizer ACID = Acid ALK = Alkali COR = Corrosive W = Use No Water

SECTION I - GENERAL INFORMATION

Characterization	**Aromatic Amine**	RCRA Number	**None**
		EPA Class	**Not Applicable**
DOT Proper Shipping Name	**Not Listed**	Chemical Abstract Service (CAS) Number	**122-39-4**
DOT Hazard Class and Label Requirements	**Not Listed**	DOT Emergency Guide Code	**No Citation**
DOT Identification Number	**Not Listed**	Molecular Formula	**$(C_6H_5)_2NH$**

Synonyms
Anilinobenzene; DPA; Phenylaniline; N-phenylaniline; N-phenylbenzamine.

SECTION II - HAZARDOUS INGREDIENTS AND IDENTITY INFORMATION

Hazardous Components (specific identity)	OSHA Exposure Criteria	NIOSH Exposure Criteria	Immediately Dangerous to Life and Health (IDLH)	ACGIH Exposure Criteria
Diphenylamine (derivation: By heating equal formula weights of aniline and aniline hydrochloride in an autoclave. The product is boiled with dilute hydrochloric acid to remove the unaltered aniline, and the residue is distilled).	**PEL:** Not Established **STEL:** Not Established	**REL:** 10 mg/m^3 **STEL:** Not Established	**Not Determined**	**TLV:** 10 mg/m^3 **STEL:** Not Established

SECTION III - PHYSICAL AND CHEMICAL CHARACTERISTICS

Boiling Point	**572°F (300°C)**	Specific Gravity (H_2O = 1)	**1.04**
Vapor Pressure (mm Hg)	**1 at 227°F (108°C)**	Molecular Weight (atomic weight)	**1.16**
Vapor Density (Air = 1)	**5.82**	Melting Point	**127°F (53°C)**

Solubility
Nearly insoluble in water (0.03%). Soluble in alcohol, benzene, ether, and carbon disulfide.

Appearance and Odor
Colorless, tan, amber, or brown crystalline solid (discolors when exposed to light) with a floral odor.

SECTION IV - FIRE AND EXPLOSION HAZARD DATA

Flash Point (method used)	**307°F (153°C) closed cup**	Explosive Limits in Air % by Volume	**LEL: Not Determined UEL: Not Determined**
NFPA Classification	**Combustible Solid**	Autoignition Temperature	**1173°F (634°C)**

Extinguishing Media
Use dry chemical, carbon dioxide, alcohol foam, or water (spray, mist or fog).

Special Fire Fighting Procedures
Poisonous gases are produced in fire. Structural fire-fighting clothing is permeable and may not provide adequate protection. Wear self-contained breathing apparatus (SCBA) operated in positive pressure mode. Move fire-exposed containers from fire if it can be done without risk. Cool with water spray.

Unusual Fire and Explosion Hazards
Dusts can form extremely explosive mixtures if dispersed through the air in proper proportions and exposed to an ignition source. Do not direct a solid stream of water at bulk quantities of diphenylamine.

SECTION V - REACTIVITY DATA

Stability		Conditions to Avoid **Diphenylamine is stable at room temperature under normal conditions of handling and storage. Avoid the formation of dust clouds of diphenylamine and prevent contact with ignition sources or oxidizers.**
Stable **X**	Unstable	Incompatibility *(materials to avoid)* **Violent reactions in contact with strong oxidizers, hexachloromelamine, and trichloromelamine.**
Hazardous Polymerization		Conditions to Avoid **Under normal conditions of temperature and pressure, hazardous polymerization of diphenylamine is not expected to occur.**
May Occur	Will Not Occur **X**	Hazardous Decomposition or By-products **Poisonous gases, including oxides of carbon (carbon monoxide) and oxides of nitrogen are produced when diphenylamine is involved in a fire.**

SECTION VI - HEALTH HAZARD DATA

Primary Route(s) of Entry:	Inhalation?	Absorption (skin/eye)?	Ingestion?
	X	X	X

Health Hazards

INHALATION: **Irritation of the nose, throat, and respiratory tract. Symptoms include cough, sneezing, tachycardia, hypertension, cyanosis, headache, fatigue, anorexia, vomiting, emaciation, and bladder problems causing hematuria (blood in urine), proteinuria (excessive proteins in urine), and bladder injury. Will accumulate in body organs (liver, kidney, brain). Can also cause methemoglobinemia (decreased oxygen-carrying capabilities of the blood).**

ABSORPTION: **Irritation of the skin and eyes. Absorbs through the skin (see effects of inhalation).**

INGESTION: **Not likely. Can cause similar systemic effects as noted for inhalation.**

Carcinogenicity	NTP Listed?	IARC Cancer Review Group?	OSHA Regulated?	Target Organs?
Unknown Human Unknown Animal	**No**	**No**	**No**	**Resp. sys. Bladder, eyes, skin, blood, repro. sys.**

Medical Conditions Generally Aggravated by Exposure

None identified.

Emergency and First-aid Procedures

Eye contact: Do not allow victim to keep eyes tightly shut. Flush immediately with water for 15 minutes (minimum), seek medical attention. Skin contact: Remove all contaminated clothing, including shoes. Immediately wash area with flooding amounts of soap and water. Seek medical attention immediately. For inhalation: Remove the person from exposure. Provide respiratory assistance and CPR. Transfer to medical facility. If swallowed: Seek medical attention immediately.

SECTION VII - PRECAUTIONS FOR SAFE HANDLING AND USE

Steps to be Taken in Case Material is Released or Spilled

Remove all ignition sources. Wear SCBA respiratory protection. Restrict those not wearing protective equipment and who are not involved in cleanup from entering area. Avoid skin contact. Do not create dust cloud. Ventilate area of spill. Do NOT dry sweep. Use HEPA vacuum and deposit in sealed drums.

Preferred Waste Disposal Method

No Citation.

Precautions to be Taken in Handling and Storage

Store in tightly closed containers in cool, well ventilated location away from heat (including direct sunlight). Protect containers from physical damage. Prevent dust buildup in storage and use areas.

Other Precautions and Warnings

Sources of ignition (open flame or smoking), are prohibited where diphenylamine is used or stored. Metal containers should be used. Do NOT generate dust clouds or dusty conditions when working.

SECTION VIII - CONTROL MEASURES AND PERSONAL PROTECTIVE EQUIPMENT

Respiratory Protection *(specify type)*

Use a NIOSH-approved respirator equipped with HEPA filter, or a powered air purifying respirator, or use a self-contained breathing apparatus (SCBA) with full facepiece operated in positive pressure mode.

Ventilation

Local exhaust (preferred) at site of chemical work or general ventilation.

Protective Gloves	Eye Protection	Other Protective Clothing
Impervious Gloves	**Chemical/Dust Goggles or Face Mask**	**Protective Clothing (prevent skin contact)**

Work/Hygiene Practices

Always wash hands thoroughly after using chemical; never bring food, drink, or smoking materials into vicinity of chemicals.

DIPHENYLAMINE

$(C_6H_5)_2NH$ CAS: 122-39-4

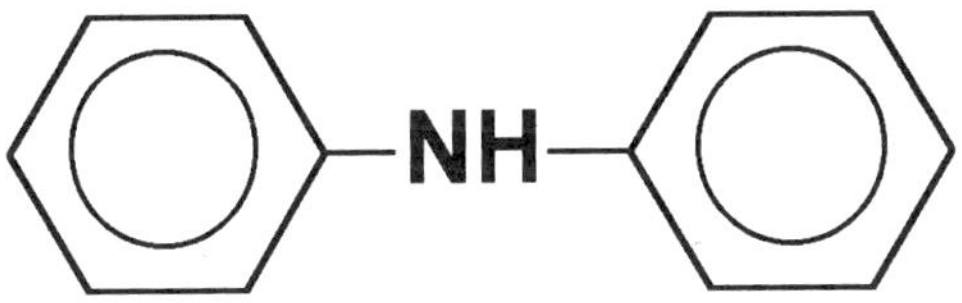

IDENTIFICATION AND TYPICAL USES

Diphenylamine is a colorless, tan, amber, or brown crystalline solid with a pleasant, floral odor. It is used in rubber antioxidants and accelerators, in solid rocket propellants, in pesticides, dyes, pharmaceuticals, veterinary medicines, as a storage preservative for apples, as a stabilizer for nitrocellulose, and applications in analytical chemistry as a reagent for colorimetric tests for nitrate and chlorate.

RISK ASSESSMENT: HEALTH

General Assessment

Diphenylamine is a human poison by *inhalation* and *absorption* through intact skin. *Ingestion* is unlikely but may also be a route of entry into the body. Its human carcinogenicity is unknown. It has caused adverse reproductive effects in animals causing developmental abnormalities. Its human teratogenicity is unknown. Diphenylamine is soluble in fats and lipids. Once absorbed, it can accumulate in body organs, such as the brain, liver, and kidneys.

Inhalation of the dusts, as well as skin absorption, causes irritation of the skin, eyes, nose, throat, and the mucous membranes lining the respiratory tract. There can be eczema, tachycardia (rapid heartbeat) hypertension (increased blood pressure) coughing, sneezing, headache, fatigue, anorexia, painful diarrhea, loss of appetite with subsequent weight loss, low body temperature, cyanosis (blue discoloration of the skin, especially on face, hands, and feet due to oxygen deficiency), vomiting, and emaciation. There can also be serious bladder disorders, including hematuria (blood in urine) and proteinuria (excessive proteins in urine), as well as bladder damage.

Prolonged exposures can lead to methemoglobinemia (decrease in blood's oxygen-carrying capabilities).

Acute Health Effects

The following acute (short-term) health effects may occur immediately or shortly after exposure to diphenylamine:

Skin: Irritation with rash and localized inflammation. Absorption through intact skin is likely on short exposures.

Eye: Severe irritation; pain and inflammation.

Lung: Nose, throat, and respiratory tract irritation causing coughing, sneezing, and toxic systemic effects.

Chronic Health Effects

The following chronic (long-term) health effects may occur at some time after exposure to diphenylamine and can last for months or even years:

Cancer Hazards: According to the information presented in the references, diphenylamine has not been adequately tested for its ability to cause cancer in test animals.

Reproductive Hazard: According to information presented in the references, diphenylamine has been shown to adversely affect reproduction in test animals causing developmental abnormalities. Its human teratogenicity is unknown and requires additional study.

Other Chronic Effects: Chronic exposure to skin can cause eczema. Long-term or repeated contact may result in loss of appetite and weight loss, diarrhea, low body temperature, and accumulation of diphenylamine in the liver, kidney, and brain with unspecified but potentially detrimental effects.

Recommended Risk-Reduction Measures

Personnel should avoid direct contact with diphenylamine. Occupational poisoning occurs most commonly by inhalation and skin absorption. If a less toxic material or compound cannot be substituted for diphenylamine, then *engineering controls* are the most effective method of reducing exposures. The best protection is to enclose operations and/or provide local exhaust ventilation at the site of diphenylamine release. While not always operationally feasible, isolating operations can also reduce exposure. Using respiratory protection is less effective than the controls mentioned above, but is still advisable whenever working with or around diphenylamine. Personnel should be required to wear an MSHA/NIOSH-approved respirator equipped with a high efficiency particulate air (HEPA) filter for low or infrequent exposures. An MSHA/NIOSH-approved self-contained breathing apparatus (SCBA) with full facepiece and pressure demand is a more appropriate level of respiratory protection when high exposures to diphenylamine are anticipated. If a full facepiece is not available, then chemical/dust goggles should be worn to protect the eyes. A face shield should also be considered. To prevent hand and skin exposures, polyvinyl alcohol gloves should be worn.

Administrative controls should also be in place to minimize the potential for human exposures. These may include written procedures or policies which specify the methods and techniques that will be practiced whenever personnel are to work with diphenylamine.

All personnel should receive *training* on the use, hazards, protective measures, emergency actions, and other precautions per 29 CFR 1910.1200 (Hazard Communication), prior to the first assignment in an area where diphenylamine is used or stored.

Before beginning employment and at regular times thereafter, the following recommended medical tests should be considered:

☑ Complete blood count (CBC).

☑ Urinalysis to test for blood and protein levels in urine (establish baseline).

If symptoms should develop or overexposure is suspected, the following additional tests may be helpful:

☑ Complete blood count.

☑ Urinalysis (compare to baseline).

☑ Liver and kidney function tests.

Any evaluation should include a careful history of past and present symptoms with an examination. Medical tests that simply look for existing damage are not a substitute for controlling exposures.

Other methods to reduce exposure include:

☑ Where possible, enclose operations and use local exhaust ventilation at the site of chemical release. If local exhaust ventilation or enclosure is not used, respiratory protection should be mandatory.

☑ Always ensure that proper protective clothing is worn when using chemical substances and that personnel are trained on its use and care.

☑ Wash thoroughly immediately after exposure to diphenylamine and at the end of the work shift or before eating, drinking, or smoking.

☑ Hazard warning information should be posted in the work area. In addition, as part of an on-going education and training program, all information on the health and safety hazards of diphenylamine should be communicated to all potentially exposed workers.

☑ Eye wash stations should be provided in the immediate work area for emergency use. If there is a possibility of skin exposure to diphenylamine, emergency shower facilities should be provided.

☑ Workers whose clothing has been contaminated by diphenylamine should change into clean clothes before leaving work. Contaminated work clothing should be laundered only by individuals who have been informed of the hazards of exposure to diphenylamine.

RISK ASSESSMENT: ENVIRONMENT

General Assessment

The environment is at risk of exposure during transportation, storage, disposal, or destruction of diphenylamine. In almost every scenario, the threat of environmental exposure is contingent upon the proper handling of the chemical substance. Accidental spills, large or small, can result in contamination of the surrounding environmental mediums (water, soil, and air).

Diphenylamine is a combustible solid. It can present a dangerous explosion hazard if its dusts form a cloud and come into contact with an ignition source. It is also incompatible with oxidizing materials and contact can result in violent and explosive reactions. Caution is always required in handling, storage, transportation, and disposal of diphenylamine. Emergency responders should be made aware of the presence of diphenylamine at any emergency response situation.

Diphenylamine can enter the environment from industrial effluents and from spills.

Acute Ecological Effects

Acute (short-term) toxic effects may include the death of animals, birds, or fish and death or low growth rate in plants. Acute effects are seen 2 to 4 days after animals or plants are exposed to diphenylamine. Diphenylamine has high acute toxicity to aquatic life. Insufficient data are available to evaluate the short-term effects of diphenylamine on plants, birds, or terrestrial animals.

Chronic Ecological Effects

Chronic (long-term) toxic effects may include shortened life span, reproductive problems, lower fertility, and changes in appearance or behavior in exposed animals. These effects can be seen long after first exposure(s) to toxic chemicals. Diphenylamine has high chronic toxicity to aquatic life. Insufficient data are available to evaluate or predict the long-term effects of diphenylamine to plants, birds, or terrestrial animals.

Water Solubility

Diphenylamine is nearly insoluble in water. Concentrations of less than 1 milligram will mix with a liter of water.

Persistence in the Environment

Diphenylamine is highly persistent in water, with a half-life of greater than 200 days. The half-life of a pollutant is the amount of time it takes for one half of the chemical to be degraded.

Bioaccumulation in Aquatic Organisms

Some substances increase in concentration, or *bioaccumulate,* in living organisms as they breathe contaminated air, drink contaminated water, or eat contaminated food. These chemicals can become concentrated in the tissues and internal organs of animals as well as humans.

The concentration of diphenylamine found in fish tissues is expected to be somewhat higher than the average concentration of diphenylamine in the water from which the fish was taken.

Recommended Risk-Reduction Measures

Proper training of all transporters will reduce the likelihood of a mishap or accident resulting in a leak or spill of diphenylamine into the environment. Labels on all containers, trucks, and rail cars must meet DOT requirements and accurately reflect their contents to enable emergency responders to react properly and quickly to any disaster thereby reducing the potential risk to the environment and to personnel.

Storage of diphenylamine should be segregated from other chemicals and materials to minimize the risk of cross-contamination.

If a spill or leak to the environment has occurred, fire department, emergency response and/or hazardous materials spill personnel should be notified immediately. Cleanup should be attempted only by those trained in proper spill containment procedures. Contaminated soils should be removed for incineration and replaced with clean soil. If diphenylamine should contact the water table, aquifer, or navigable waterway, time is of the essence. It is practically insoluble and total containment and remediation may be possible. The local and/or state emergency response authorities must be notified. A comprehensive emergency response or disaster preparedness/recovery plan should be in place prior to any operations involving the use, transportation, storage, or disposal of diphenylamine.

If diphenylamine is spilled or leaked, the following specific steps are recommended:

- ☑ Restrict persons not wearing protective clothing from area of spill or leak until cleanup is complete and area can be opened for normal work.
- ☑ Do NOT generate dusts. Dust clouds of the right proportions in air present a dangerous explosion hazard. Use a scoop or a vacuum equipped with a high efficiency particulate air filter (HEPA) to collect spills. Place materials in drum for recovery or disposal, as required.
- ☑ Ventilate area and remove ignition sources.
- ☑ Containers should be removed to safer area if it can be done without increasing the risk.
- ☑ It may be necessary to dispose of diphenylamine as a hazardous waste. The responsible state agency or the regional office of the federal Environmental Protection Agency (EPA) should be contacted for specific recommendations.

RISK ASSESSMENT: BUSINESS

General Assessment

Accidents or mishaps involving diphenylamine can present a serious threat to business operations. The loss or damage of equipment or facilities can significantly affect fiscal viability. Lawsuits that may result from personnel illness, injury/death, public exposures, and/or environmental contamination will require a serious expenditure of resources. Media attention surrounding an injury, death, or environmental damage can also result in a loss of profits and loss of current as well as future business.

Recommended Risk-Reduction Measures

Company attorneys, safety and health professionals, and environmental specialists should be involved in the development of any procedures for responding to chemical incidents. A company official should be predesignated as a public relations officer with specific training in dealing with the press. Corporate plans and policies should be developed, approved, and implemented long before any need for such arises.

REFERENCES

American Conference of Governmental Industrial Hygienists. 1988. *Documentation of the Threshold Limit Values and Biological Exposure Indices,* 5th Edition (with updates). Cincinnati: ACGIH

Bloom, A. D. and F. J. DeSerres. 1995. *Ecotoxicity and Human Health.* New York: CRC Press/Lewis Publishers.

Brusick, D. J. 1994. *Methods for Genetic Risk Assessment.* New York: CRC Press/Lewis Publishers.

Bueche, J. R. 1972. *Principles of Physics,* 2nd Edition. New York: McGraw-Hill.

Calabrese, E. J., E. M. Kenyon. *Air Toxics and Risk Assessment.* New York: CRC Press/Lewis Publishers.

Calabrese, E. J. 1994. *Biological Effects of Low Level Exposures.* New York: CRC Press/Lewis Publishers.

Cockerman, L. G., B. S. Shane. 1994. *Basic Environmental Toxicology.* New York: CRC Press/Lewis Publishers.

Cothern, C. R., N. P. Ross. 1994. *Environmental Statistics, Assessment, and Forecasting.* New York: CRC Press/Lewis Publishers.

Cothern, R. R. 1993. *Comparative Environmental Risk Assessment.* New York: CRC Press/Lewis Publishers.

Danse, I. R. 1991. *Common Sense Toxics in the Workplace.* New York: Van Nostrand Reinhold.

Gerarde, H. W. 1960. *Toxicology and Biochemistry of Aromatic Hydrocarbons.* New York: Elsevier.

Hallenbeck, W. H. 1993. *Quantitative Risk Assessment for Environmental and Occupational Health,* 2nd Edition. New York: CRC Press/Lewis Publishers.

Howard, P. H., et. al. 1991. *Handbook of Environmental Degradation Rates.* New York: CRC Press/Lewis Publishers.

Jones, F. E. 1994. *Toxic Organic Vapors in the Workplace.* New York: CRC Press/Lewis Publishers.

Keith, L. H. and M. M. Walker. 1995. *Handbook of Air Toxics.* New York: CRC Press/Lewis Publishers.

Landis, W. G., M. H. Yu. 1995. *Introduction to Environmental Toxicology.* New York: CRC Press/Lewis Publishers.

Langerman, N. 1994 *Precautionary Labels for Chemical Containers.* New York: CRC Press/Lewis Publishers.

Lauwerys, R. R., P. Hoet. 1993. *Industrial Chemical Exposure.* New York: CRC Press/Lewis Publishers.

Lewis, R. J., Sr. 1993. *Hawley's Condensed Chemical Dictionary,* 12th Edition. New York: Van Nostrand Reinhold.

Lewis, R. J., Sr. 1992. *Sax's Dangerous Properties of Industrial Materials,* Volumes 1, 2, and 3, 8th Edition. New York: Van Nostrand Reinhold.

Manahan, S. E. 1994. *Environmental Chemistry,* 6th Edition. New York: CRC Press/Lewis Publishers.

Mellström, G. A., J. E. Wahlberg, and H. I. Maibach. 1994. *Protective Gloves for Occupational Use.* New York: CRC Press/Lewis Publishers.

Meyer, E. 1990. *Chemistry of Hazardous Materials.* Englewood Cliffs, NJ: Prentice Hall.

Ott, W. R. 1995. *Environmental Statistics and Data Analysis.* New York: CRC Press/Lewis Publishers.

Patnaik, P. 1992. *A Comprehensive Guide to the Hazardous Properties of Chemical Substances.* New York: Van Nostrand Reinhold.

Philp, R. B. 1995. *Environmental Hazards and Human Health.* New York: CRC Press/Lewis Publishers.

Proctor, N. H., J. P. Hughes. 1991. *Chemical Hazards of the Workplace.* New York: Van Nostrand Reinhold.

Quigley, D. R. 1994. *Handbook of Emergency Chemical Management.* New York: CRC Press/Lewis Publishers.

Sacarello, H. L. A. 1994. *The Comprehensive Handbook of Hazardous Materials.* New York: CRC Press/Lewis Publishers.

Torkelson, T. R. V. K. Rowe. 1981. *Patty's Industrial Hygiene and Toxicology,* Vol. 2B. New York: Wiley-Interscience.

U.S. Department of Health and Human Services, National Institute for Occupational Safety and Health. 1994. *NIOSH Pocket Guide to Chemical Hazards.* Washington, D.C.: U.S. Government Printing Office.

MATERIAL SAFETY DATA SHEET

CHEMICAL NAME	CRC PRESS/LEWIS PUBLISHERS DISCLAIMER
DIPROPYLENE GLYCOL METHYL ETHER	THE INFORMATION AND RECOMMENDATIONS PRESENTED HEREIN ARE BASED ON SOURCES BELIEVED TO BE RELIABLE. CRC MAKES NO REPRESENTATION ON ITS COMPLETENESS OR ACCURACY. IT IS THE RESPONSIBILITY OF THE USER TO DETERMINE THE CHEMICAL'S SUITABILITY FOR ITS INTENDED USE, THE CHEMICAL'S SAFE USE, AND THE CHEMICAL'S PROPER DISPOSAL. NO REPRESENTATIONS AND/OR WARRANTIES, EITHER EXPRESSED OR IMPLIED, OF THE MERCHANTABILITY OR FITNESS FOR A PARTICULAR PURPOSE, OR OF ANY OTHER NATURE, ARE MADE WITH RESPECT TO THE INFORMATION PROVIDED IN THIS MSDS OR TO THE CHEMICAL TO WHICH INFORMATION MAY REFER. CRC NEITHER ASSUMES NOR AUTHORIZES ANY OTHER PERSON TO ASSUME FOR IT, ANY OTHER ADDITIONAL RESPONSIBILITY OR LIABILITY FOR THE USE OF, OR RELIANCE UPON, THIS INFORMATION.

HAZARD WARNING INFORMATION

HEALTH	FIRE	REACTIVE	OTHER	DEGREE OF HAZARD	COLOR CODING	OTHER CODES
0	2	0		0 = Minimum Hazard 1 = Slight Hazard 2 = Moderate Hazard 3 = Serious Hazard 4 = Severe Hazard	HEALTH = BLUE FIRE = RED REACTIVITY = YELLOW OTHER = WHITE	OX = Oxidizer ACID = Acid ALK = Alkali COR = Corrosive W = Use No Water

SECTION I - GENERAL INFORMATION

Characterization	**Glycol Ether**	RCRA Number	**None**
		EPA Class	**Not Applicable**
DOT Proper Shipping Name	**Not Listed**	Chemical Abstract Service (CAS) Number	**34590-94-8**
DOT Hazard Class and Label Requirements	**Not Listed**	DOT Emergency Guide Code	**No Citation**
DOT Identification Number	**Not Listed**	Chemical Formula	**$CH_3OC_3H_6OC_3H_6OH$**

Synonyms

Dipropylene glycol monoethyl ether; Dowanol® 50B; DPGME; Arcosolv®; UCAR Solvent 2LM.

SECTION II - HAZARDOUS INGREDIENTS AND IDENTITY INFORMATION

Hazardous Components (specific identity)	OSHA Exposure Criteria	NIOSH Exposure Criteria	Immediately Dangerous to Life and Health (IDLH)	ACGIH Exposure Criteria
Dipropylene glycol methyl ether (derivation: By reaction of ethylene oxide and *n*-butyl alcohol). **1 ppm = 6.16 mg/m^3**	**PEL (skin):** 100 ppm 600 mg/m^3 **STEL:** Not Established	**REL (skin):** 100 ppm 600 mg/m^3 **STEL:** 150 ppm 900 mg/m^3	**600 ppm**	**TLV (skin):** 100 ppm 600 mg/m^3 **STEL:** 150 ppm 900 mg/m^3

SECTION III - PHYSICAL AND CHEMICAL CHARACTERISTICS

Boiling Point	**372°F (189°C)**	Specific Gravity (H_2O = 1)	**0.95**
Vapor Pressure (mm Hg)	**0.5 at 69°F (20°C)**	Molecular Weight	**148.2**
Vapor Density (Air = 1)	**5.1**	Melting Point	**-112°F (-80°C)**

Solubility

Completely miscible in water, VM&P naphtha, acetone, ethanol, benzene, carbon tetrachloride, ether, methanol, monochlorobenzene, and petroleum ether.

Appearance and Odor

Colorless liquid with a mild, pleasant, ether-like odor with a bitter taste.

SECTION IV - FIRE AND EXPLOSION HAZARD DATA

Flash Point (method used)	**185°F (85°C) opened cup**	Explosive Limits in Air % by Volume	**LEL: 1.1% at 392°F (200°C) UEL: 3.0%**
NFPA Classification	**Class IIIA Combustible Liquid**	Autoignition Temperature	**Not Determined**

Extinguishing Media

Use dry chemical, alcohol foam, carbon dioxide, or water fog. Water and foam may cause frothing.

Special Fire Fighting Procedures

Acrid and irritating fumes and smoke are produced in fire, wear full protective clothing and self-contained breathing apparatus (SCBA). Use water spray to cool fire-exposed containers. Fight fire from distance if possible. Move containers from fire area if it can be done safely. Exercise extreme caution when moving containers.

Unusual Fire and Explosion Hazards

A combustible liquid. A moderate fire hazard when exposed to heat, sparks, or open flame. Apply cooling water to sides of container until long after fire has been extinguished.

SECTION V - REACTIVITY DATA

Stability		Conditions to Avoid
		Under normal conditions of temperature and pressure, dipropylene glycol methyl ether is stable. Avoid contact with incompatible materials. Keep away from heat, sparks, and flame.
Stable	Unstable	Incompatibility *(materials to avoid)*
X		**Strong oxidizers (such as chlorine, bromine, fluorine, and peroxides), and oxidizing agents.**
Hazardous Polymerization		Conditions to Avoid
		Under normal conditions of temperature and pressure, hazardous polymerization of dipropylene glycol methyl ether is not expected to occur.
May Occur	Will Not Occur	Hazardous Decomposition or By-products
	X	**Acrid and irritated gases and fumes are produced when dipropylene glycol methyl ether is involved in fire, including carbon dioxide and carbon monoxide.**

SECTION VI - HEALTH HAZARD DATA

	Inhalation?	Absorption (skin)?	Ingestion?
Primary Route(s) of Entry:	X	X	X

Health Hazards

INHALATION: Causes irritation of the eyes, nose, throat, weakness, lightheadedness, headache, and possible effects on the central nervous system (CNS). Overexposure to hazardous concentrations is not likely due to the irritating and disagreeable odor and cannot normally be tolerated.

ABSORPTION: Repeated or heavy contact will pass through unbroken skin to cause toxic systemic effects. Eye contact with liquid can cause burns.

INGESTION: Mildly toxic with systemic effects as those noted for inhalation.

Carcinogenicity	NTP Listed?	IARC Cancer Review Group?	OSHA Regulated?	Target Organs?
Unknown Human **Unknown Animal**	**No**	**No**	**29 CFR 1910.1000 Table Z-1**	**CNS, respiratory system, skin.**

Medical Conditions Generally Aggravated by Exposure

Pre-existing disorders of the CNS may be aggravated by exposure.

Emergency and First-aid Procedures

Eye contact: Flush immediately with water for 15 minutes (minimum), seek medical attention. Skin contact: Remove all contaminated clothing. Immediately wash area with large amounts of soap and water to prevent absorption. Seek medical assistance. For inhalation: Remove the person from exposure. Provide respiratory assistance and CPR. Transfer to medical facility. If swallowed: Call the poison control center and seek medical attention immediately. Unless advised otherwise, give a conscious and alert person 1-2 glasses of water or milk to drink. Do NOT induce vomiting.

SECTION VII - PRECAUTIONS FOR SAFE HANDLING AND USE

Steps to be Taken in Case Material is Released or Spilled

Evacuate area of spill and remove all ignition sources. Remove leaking containers to a safe place (outdoors), if it can be done without risk. Absorb liquids with vermiculite or other material and place in sealed drum for disposal. Use non-sparking tools. Use water spray to flush spills away from exposures and dilute to nonflammable mixtures (keep out of sewers).

Preferred Waste Disposal Method

Consider reclamation, recycling before disposal.

Precautions to be Taken in Handling and Storage

Do not store dipropylene glycol methyl ether in the presence of incompatible chemicals or materials. Store in tightly closed containers in a cool, dry, well-ventilated area away from ignition sources.

Other Precautions and Warnings

Use non-sparking tools and electrically ground and bond all containers, especially during transfer.

SECTION VIII - CONTROL MEASURES AND PERSONAL PROTECTIVE EQUIPMENT

Respiratory Protection *(specify type)*

Use an air-purifying respirator or a supplied-air respirator with full facepiece operated in pressure demand mode or a self-contained breathing apparatus with operated in pressure demand mode.

Ventilation

Local exhaust or general mechanical systems recommended.

Protective Gloves	Eye Protection	Other Protective Clothing
Impervious Rubber	**Chemical Goggles or Face Mask**	**Protective Clothing or Apron**

Work/Hygiene Practices

Always wash hands thoroughly after using chemical; never bring food, drink, or smoking materials into vicinity of chemicals.

DIPROPYLENE GLYCOL METHYL ETHER

$CH_3OC_3H_6OC_3H_6OH$ CAS: 34590-94-8

IDENTIFICATION AND TYPICAL USES

Dipropylene glycol methyl ether (DPGME) is a colorless liquid with a mild, pleasant, ether-like odor and a bitter taste. It is used primarily in hydraulic brake fluids and as a high-boiling point solvent.

RISK ASSESSMENT: HEALTH

General Assessment

DPGME is mildly toxic by *ingestion* and *absorption* through the skin. *Inhalation* of its vapors at high concentrations (above 300 ppm) can will also present a significant health hazard risk. There is no information in the references on the carcinogenic, teratogenic, or mutagenic properties of DPGME. Because of its extremely disagreeable (at high concentrations) and irritating (intolerable) odor, overexposure is unlikely under normal operating conditions.

Inhalation can cause irritation of the eyes, nose, throat, and the mucosa of the upper respiratory tract. Inhalation and skin absorption can lead to headaches, lightheadedness, weakness, and action to depress the function of the central nervous system (CNS). On the skin, there may be rash with irritation and localized inflammation. DPGME will absorb rapidly through intact skin.

Ingestion will cause unspecified gastrointestinal effects and possibly lead to toxic systemic effects with action primarily to depress the CNS.

Acute Health Effects

The following acute (short-term) health effects may occur immediately or shortly after exposure to DPGME:

Skin: Irritation with possible rash and quick absorption causing toxic systemic effects.

Eye: Severe irritation with localized inflammation.

Lung: Irritation of the mucosa of the nose, throat, and upper respiratory tract. Can cause a number of toxic systemic effects with symptoms of headaches, lightheadedness, weakness, and CNS depression.

Other: Ingestion can cause the same symptoms as inhalation. Adverse action on the central nervous system is likely on high acute exposures.

Chronic Health Effects

The following chronic (long-term) health effects may occur at some time after exposure to DPGME and can last for months or even years:

Cancer Hazards: According to the information presented in the references, DPGME has not been shown to cause cancer in test animals.

Reproductive Hazard: According to the information presented in the references, DPGME has not been adequately tested for its ability to cause adverse effects on reproduction in test animals.

Other Chronic Effects: Low, repeated exposures may lead to damage to the central nervous system (CNS).

Recommended Risk-Reduction Measures

Unless a less toxic chemical can be substituted for a hazardous substance, *engineering controls* are the most effective method of reducing exposures. The best protection is to enclose operations and/or provide local exhaust ventilation at the site of chemical release. While not always operationally feasible, isolating operations can also reduce exposure, especially if DPGME is to be agitated or heated.

Using respiratory protection is less effective than the controls mentioned above, but is still advisable whenever working with or around DPGME. For low exposures up to or over the PEL (100 ppm), use a NIOSH-approved organic vapor respirator with full facepiece. For the best protection, especially at high exposures or during emergency situations, a self-contained breathing apparatus (SCBA) with full facepiece operated in pressure demand mode, or a supplied-air respirator operated in pressure demand or continuos flow mode are the recommended respiratory protection methods of choice. If a full facepiece is not available, then chemical goggles should be worn to protect the eyes. Whenever a chemical splash hazard exists, a face shield and rubber apron should be worn. Skin contact must be avoided. To prevent hand and skin exposures, impervious rubber gloves should be worn.

Administrative controls should also be in place to minimize the potential for human exposures. These may include written procedures or policies which specify the methods and techniques that will be practiced whenever personnel are to work with DPGME.

All personnel should receive *training* on the use, hazards, protective measures, emergency actions, and other precautions per 29 CFR 1910.1200 (Hazard Communication), prior to the first assignment in an area where DPGME is used or stored.

If symptoms develop or overexposure is suspected, the following medical tests may be useful:

- ☑ Evaluation by a neurologist with special tests for central nervous system (CNS) functions.
- ☑ Lung function tests.

Any evaluation should include a careful medical history of past and present symptoms with an examination. However, medical tests that evaluate existing damage are not a substitute for controlling exposure. Also, since smoking can cause heart disease, emphysema, and other respiratory disorders, exposure to DPGME can have a more pronounced effect on smokers than on non-smokers exposed to the same levels. Prudent risk assessment and management requires consideration of *all* possible risk factors when exposure symptoms appear in the workforce.

Other methods to reduce exposure include:

- ☑ Where possible, enclose operations and use local exhaust ventilation at the site of chemical release. If local exhaust ventilation or enclosure is not used, respiratory protection should be mandatory. Also, automatic transfer of liquids from storage containers to process containers is recommended.
- ☑ Always ensure that proper protective clothing is worn when using chemical substances and that personnel are trained in its use and care.
- ☑ Wash thoroughly immediately after exposure to DPGME and at the end of the work shift or before eating, drinking, or smoking.
- ☑ Hazard warning information should be posted in the work area. In addition, as part of an on-going education and training program, all information on the health and safety hazards of DPGME should be communicated to all exposed and potentially exposed workers.
- ☑ Eye wash stations should be provided in work areas. If the potential for whole body exposure exists, then safety showers should also be provided.

RISK ASSESSMENT: ENVIRONMENT

General Assessment

The environment is at risk of exposure during use, transportation, storage, disposal, or destruction of DPGME. In almost every scenario, the threat of environmental exposure is contingent upon the proper handling of the chemical substance. Accidental spills, large or small, can result in fire, explosion, and possible contamination of the surrounding environmental mediums (water, soil, and air).

DPGME is considered a class IIIA combustible liquid (per OSHA 29 CFR 1910.106). These characteristics require special consideration during any emergency situation involving a leak or spill of DPGME. Should DPGME ever come into contact with incompatible substances such as strong oxidizers (chlorine, bromine, fluorine, chlorine dioxide, and peroxides) either during use, transportation, or storage, violent reaction can occur.

Wastes should be considered for recovery, reclamation, or recycling rather than land-fill disposal. Destruction in a chemical incinerator equipped with an afterburner and scrubber is also acceptable.

DPGME can enter the environment through industrial discharges and from spills.

☠ *Acute Ecological Effects*

Acute (short-term) toxic effects may include the death of animals, birds, or fish and death or low growth rate in plants. Acute effects are seen 2 to 4 days after animals or plants are exposed to DPGME. Insufficient data are available to evaluate or predict the acute (short-term) effects of DPGME to aquatic life, plants, birds, or land animals.

Chronic Ecological Effects

Chronic toxic effects may include shortened life span, reproductive problems, lower fertility, and changes in appearance or behavior in exposed animals. These effects can be seen long after first exposure(s) to toxic chemicals. Insufficient data are available to evaluate or predict the long-term (chronic) effects of DPGME to aquatic life, plants, birds, or land animals.

Water Solubility

DPGME is highly soluble in water. Concentrations of 1000 milligrams and more will mix with a liter of water.

Persistence in the Environment

DPGME is highly persistent in the aquatic environment with a half-life of 200 days and more. The half-life of a pollutant is the amount of time it takes for one half of the chemical to be degraded. It is highly mobile in both water and soil and undergoes aerobic biodegeneration.

Bioaccumulation in Aquatic Organisms

Some substances increase in concentration, or *bioaccumulate,* in living organisms as they breathe contaminated air, drink contaminated water, or eat contaminated food. These chemicals can become concentrated in the tissues and internal organs of animals as well as humans.

The concentration of DPGME found in fish tissues is expected to be about the same as that found in the water from which the fish was taken.

Recommended Risk-Reduction Measures

Proper training of all transporters will reduce the likelihood of a mishap or accident resulting in a leak or spill to the environment. The correct use of DOT labeling (per the MSDS) on all containers, trucks, and rail cars will enable emergency responders to react properly and quickly to any disaster thereby reducing the potential risk to the environment and to personnel.

Storage of DPGME should be segregated from incompatible chemicals to minimize the risk of cross-contamination or contact. All containers used in shipping or transfer operations must be electrically grounded to prevent static sparks. Buildings should be properly constructed for storage of hazardous chemicals (fire suppression equipment, alarms, separation and distance requirements, etc.).

If a spill or leak to the environment has occurred, fire department, emergency response and/or hazardous materials spill personnel should be notified immediately. Cleanup should be attempted only by those trained in proper spill containment procedures. Contaminated soils should be removed for incineration and replaced with clean soil. If DPGME should contact the water table, aquifer, or navigable waterway, remediation activities should be prompt. It is highly soluble in water and total containment and remediation may not be possible. When such spills occur, the local and/or state emergency response authorities must be notified. A comprehensive emergency response or disaster preparedness/recovery plan should be in place prior to any operations involving the use, transportation, storage, or disposal of DPGME. If DPGME is spilled or leaked, the following specific steps are recommended:

- ☑ Restrict persons not wearing protective clothing from area of spill or leak until cleanup is complete and area can be opened for normal work.
- ☑ Ventilate area of spill or leak.
- ☑ Remove all ignition sources.
- ☑ Collect liquid using vermiculite, dry earth, or sand to absorb and place in a sealed drum for disposal.
- ☑ It may be necessary to dispose of DPGME as a hazardous waste. The responsible state agency or the regional office of the federal Environmental Protection Agency (EPA) should be contacted for specific recommendations.

RISK ASSESSMENT: BUSINESS

General Assessment

Accidents or mishaps involving DPGME can present a significant threat to business operations. The loss or damage of equipment or facilities can significantly affect fiscal viability. Lawsuits that may result from personnel injury/death, public exposures, and/or environmental contamination will also require a serious expenditure of resources. Media attention surrounding an injury, death, or environmental damage can also result in a loss of profits and loss of current as well as future business.

Recommended Risk-Reduction Measures

Company attorneys, safety and health professionals, and environmental specialists should be involved in the development of any procedures dealing with safety and emergency response. A company official should be pre-designated as a public relations officer with specific training in dealing with the press. Corporate plans and policies should be developed, approved, and implemented long before any need for such arises.

REFERENCES

Ahlbom, A. 1993. *Biostatistics for Epidemiologists.* New York: CRC Press/Lewis Publishers.

American Conference of Governmental Industrial Hygienists. 1988. *Documentation of the Threshold Limit Values and Biological Exposure Indices,* 5th Edition (with updates). Cincinnati: ACGIH

Anthony, C. P., and N. J. Kolthoff. 1971. *Textbook of Anatomy and Physiology.* St. Louis: C. V. Mosby Company.

Asante-Duah, D. K. 1993. *Hazardous Waste Risk Assessment.* New York: CRC Press/Lewis Publishers.

Bloom, A. D. and F. J. DeSerres. 1995. *Ecotoxicity and Human Health.* New York: CRC Press/Lewis Publishers.

Brusick, D. J. 1994. *Methods for Genetic Risk Assessment.* New York: CRC Press/Lewis Publishers.

Bueche, J. R. 1972. *Principles of Physics,* 2nd Edition. New York: McGraw-Hill.

Burrell, R., D. K. Flaherty, and L. J. Sauers. 1992. *Toxicology of the Immune System.* New York: Van Nostrand Reinhold.

Calabrese, E. J., E. M. Kenyon. *Air Toxics and Risk Assessment.* New York: CRC Press/Lewis Publishers.

Cockerman, L. G., B. S. Shane. 1994. *Basic Environmental Toxicology.* New York: CRC Press/Lewis Publishers.

Cothern, C. R., N. P. Ross. 1994. *Environmental Statistics, Assessment, and Forecasting.* New York: CRC Press/Lewis Publishers.

Cothern, R. R. 1993. *Comparative Environmental Risk Assessment.* New York: CRC Press/Lewis Publishers.

Cothern, C. R. 1995. *Handbook for Environmental Risk Decisionmaking.* New York: CRC Press/Lewis Publishers.

Danse, I. R. 1991. *Common Sense Toxics in the Workplace.* New York: Van Nostrand Reinhold.

Environmental Protection Agency. 1983. *Chemical Hazard Information Profiles.* Washington, D.C.: U.S. EPA.

Fodden, C. R., and J. L. Weddell. 1992. *Hazardous Materials Emergency Action Data.* New York: CRC Press/Lewis Publishers.

Forseberg, K., and L. H. Keith. 1993. *Chemical; Protective Clothing Permeation and Degradation Database.* Chelsea, MI: Lewis Publishers.

Gardner, R. W. 1994. *Chemical Intolerance.* New York: CRC Press/Lewis Publishers.

Hallenbeck, W. H. 1993. *Quantitative Risk Assessment for Environmental and Occupational Health,* 2nd Edition. New York: CRC Press/Lewis Publishers.

Hodgson, E., R. B. Mailman, and Janice E. Chambers. 1988. *Dictionary of Toxicology.* London: The MacMillian Company.

Howard, P. H. 1995. *Handbook of Environmental Fate and Exposures Data for Organic Chemicals,* Vol. I-IV. New York: CRC Press/Lewis Publishers.

Howard, P. H., et. al. 1991. *Handbook of Environmental Degradation Rates.* New York: CRC Press/Lewis Publishers.

Hosty, J. W., and P. Foster. 1990. *A Practical Guide to Chemical Spill Response.* New York: Van Nostrand Reinhold.

Jones, F. E. 1994. *Toxic Organic Vapors in the Workplace.* New York: CRC Press/Lewis Publishers.

Keith, L. H. and M. M. Walker. 1995. *Handbook of Air Toxics.* New York: CRC Press/Lewis Publishers.

Landis, W. G., M. H. Yu. 1995. *Introduction to Environmental Toxicology.* New York: CRC Press/Lewis Publishers.

Langerman, N. 1994 *Precautionary Labels for Chemical Containers.* New York: CRC Press/Lewis Publishers.

Lauwerys, R. R., P. Hoet. 1993. *Industrial Chemical Exposure.* New York: CRC Press/Lewis Publishers.

Lide, D. R. 1995. *Handbook of Chemistry and Physics,* 76th Edition. New York: CRC Press/Lewis Publishers.

Lewis, R. J., Sr. 1993. *Hawley's Condensed Chemical Dictionary,* 12th Edition. New York: Van Nostrand Reinhold.

Lewis, R. J., Sr. 1992. *Sax's Dangerous Properties of Industrial Materials,* Volumes 1, 2, and 3, 8th Edition. New York: Van Nostrand Reinhold.

Manahan, S. E. 1992. *Toxicological Chemistry,* 2nd Edition. New York: CRC Press/Lewis Publishers.

Mellström, G. A., J. E. Wahlberg, and H. I. Maibach. 1994. *Protective Gloves for Occupational Use.* New York: CRC Press/Lewis Publishers.

Meyer, E.. 1990. *Chemistry of Hazardous Materials.* Englewood Cliffs, NJ: Prentice Hall.

Mycock, J. C., J. D. McKenna, and L. Theodore. 1995. *Handbook of Air Pollution Control Engineering and Technology.* New York: CRC Press/Lewis Publishers.

Ott, W. R. 1995. *Environmental Statistics and Data Analysis.* New York: CRC Press/Lewis Publishers.

Patnaik, P. 1992. *A Comprehensive Guide to the Hazardous Properties of Chemical Substances.* New York: Van Nostrand Reinhold.

Philp, R. B. 1995. *Environmental Hazards and Human Health.* New York: CRC Press/Lewis Publishers.

Quigley, D. R. 1994. *Handbook of Emergency Chemical Management.* New York: CRC Press/Lewis Publishers.

Rea, W. J. 1995. *Chemical Sensitivity,* Vol. I-III. New York: CRC Press/Lewis Publishers.

Richardson, M. 1995. *Dictionary of Substance and Their Effects,* Vol. I-VI. New York: CRC Press/Lewis Publishers.

Sacarello, H. L. A. 1994. *The Comprehensive Handbook of Hazardous Materials.* New York: CRC Press/Lewis Publishers.

MATERIAL SAFETY DATA SHEET

CHEMICAL NAME	
DIPROPYL KETONE	**CRC PRESS/LEWIS PUBLISHERS DISCLAIMER:** THE INFORMATION AND RECOMMENDATIONS PRESENTED HEREIN ARE BASED ON SOURCES BELIEVED TO BE RELIABLE. CRC MAKES NO REPRESENTATION ON ITS COMPLETENESS OR ACCURACY. IT IS THE RESPONSIBILITY OF THE USER TO DETERMINE THE CHEMICAL'S SUITABILITY FOR ITS INTENDED USE, THE CHEMICAL'S SAFE USE, AND THE CHEMICAL'S PROPER DISPOSAL. NO REPRESENTATIONS AND/OR WARRANTIES, EITHER EXPRESSED OR IMPLIED, OF THE MERCHANTABILITY OR FITNESS FOR A PARTICULAR PURPOSE, OR OF ANY OTHER NATURE, ARE MADE WITH RESPECT TO THE INFORMATION PROVIDED IN THIS MSDS OR TO THE CHEMICAL TO WHICH INFORMATION MAY REFER. CRC NEITHER ASSUMES NOR AUTHORIZES ANY OTHER PERSON TO ASSUME FOR IT, ANY OTHER ADDITIONAL RESPONSIBILITY OR LIABILITY FOR THE USE OF, OR RELIANCE UPON, THIS INFORMATION.

HAZARD WARNING INFORMATION

HEALTH	FIRE	REACTIVE	OTHER	DEGREE OF HAZARD	COLOR CODING	OTHER CODES
2	2	0		0 = Minimum Hazard 1 = Slight Hazard 2 = Moderate Hazard 3 = Serious Hazard 4 = Severe Hazard	HEALTH = BLUE FIRE = RED REACTIVITY = YELLOW OTHER = WHITE	OX = Oxidizer ACID = Acid ALK = Alkali COR = Corrosive W = Use No Water

SECTION I - GENERAL INFORMATION

Characterization	RCRA Number	EPA Class
Ketone	**D001**	**Characteristic (I) Waste**

DOT Proper Shipping Name **Butyrone**	Chemical Abstract Service (CAS) Number **123-19-3**
DOT Hazard Class and Label Requirements **Flammable or Combustible Liquid**	DOT Emergency Guide Code **26**
DOT Identification Number **UN 2710**	Chemical Formula **$(CH_3CH_2CH_2)_2CO$**

Synonyms
Butyrone; DPK; 4-heptanone; heptan-4-one; propyl ketone.

SECTION II - HAZARDOUS INGREDIENTS AND IDENTITY INFORMATION

Hazardous Components (specific identity)	OSHA Exposure Criteria	NIOSH Exposure Criteria	Immediately Dangerous to Life and Health (IDLH)	ACGIH Exposure Criteria
Dipropyl ketone (derivation: By passing butyric acid over precipitated calcium carbonate at 842°F or 450°C) 1 ppm = 4.75 mg/m^3	**PEL:** 50 ppm 235 mg/m^3 **STEL:** Not Established	**REL:** 50 ppm 235 mg/m^3 **STEL:** Not Established	Not Determined	**TLV:** 50 ppm 235 mg/m^3 **STEL:** Not Established

SECTION III - PHYSICAL AND CHEMICAL CHARACTERISTICS

Boiling Point **290°F (144°C)**	Specific Gravity (H_2O = 1) **0.82**
Vapor Pressure (mm Hg) **5.2 at 69°F (20°C)**	Molecular Weight **142.2**
Vapor Density (Air = 1) **3.93**	Freezing Point **-27°F (-33°C)**

Solubility
Insoluble in water. Soluble in alcohol and ether.

Appearance and Odor
Colorless liquid with a strong, penetrating, but pleasant odor and burning taste.

SECTION IV - FIRE AND EXPLOSION HAZARD DATA

Flash Point (method used) **120°F (49°C) closed cup**	Explosive Limits in Air % by Volume **LEL: Not Determined UEL: Not Determined**
NFPA Classification **Class II Combustible Liquid**	Autoignition Temperature **Not Found**

Extinguishing Media
Large fires: Alcohol foam or water spray. Small fires: Dry chemical, carbon dioxide, water spray, foam.

Special Fire Fighting Procedures
Poisonous gases are produced in fire. Wear full protective clothing and self-contained breathing apparatus. Fight fire from a distance if possible. Keep fire-exposed containers cool with water spray. A solid stream of water may spread fire. If it can be done safely, move fire-exposed containers from area.

Unusual Fire and Explosion Hazards
Containers may explode in fire due to the buildup of internal pressures. BLEVE (boiling liquid expanding vapor explosion) is possible. Immediately evacuate if rising sound is heard from venting device or a noticeable discoloration is observed on the sides of the tank or storage vessel.

SECTION V - REACTIVITY DATA

Stability		Conditions to Avoid Normally stable. Do not allow dipropyl ketone to come into contact with incompatible materials. Avoid contact with fire or high heat sources. Do not use in confined spaces.
Stable X	Unstable	Incompatibility *(materials to avoid)* Incompatible with strong oxidizers (bromine, chlorine, fluorine).
Hazardous Polymerization		Conditions to Avoid Under normal conditions of temperature and pressure, hazardous polymerization of dipropyl ketone is not expected to occur.
May Occur	Will Not Occur X	Hazardous Decomposition or By-products When heated to decomposition, dipropyl ketone emits acrid, irritating smoke and fumes, including oxides of carbon.

SECTION VI - HEALTH HAZARD DATA

Primary Route(s) of Entry:	Inhalation?	Absorption (skin)?	Ingestion?
	X		X

Health Hazards

INHALATION: Irritation of the mucosa of the respiratory tract causing respiratory depression causing gasping, choking, and dypsnea. Higher concentrations causes CNS depression with headache, dizziness, fainting, tremor, chemical pneumonitis, coma, and possibly death.

SKIN & EYES: Skin contact with liquid or vapor can result in irritation and may cause dermatitis and paresthesia (tingling). Eye contact causes pain, irritation, and corneal damage, tearing.

INGESTION: Unlikely. Symptoms of irritation, sore throat, nausea, coughing, salivation, vomiting.

Carcinogenicity	NTP Listed?	IARC Cancer Review Group?	OSHA Regulated?	Target Organs?
Unknown Human Unknown Animal	No	No	29 CFR 1910.1000 Table Z-1	Respiratory system; skin; eyes; CNS, liver.

Medical Conditions Generally Aggravated by Exposure

Respiratory impairments, skin conditions (dermatitis), and liver disorders may be aggravated.

Emergency and First-aid Procedures

Eye contact: Immediately flush with large amounts of water for 15 minutes (minimum), occasionally lifting eyelids. Skin contact: Remove all contaminated clothing. Immediately wash area with large amounts of soap and water. For inhalation: Remove the person from exposure. Provide respiratory assistance and CPR. Transfer promptly to medical facility. If swallowed: Seek medical attention immediately. Give a conscious and alert victim several glasses of milk or water. Vomiting is not recommended (causes direct aspiration of the liquid which may cause chemical pneumonitis).

SECTION VII - PRECAUTIONS FOR SAFE HANDLING AND USE

Steps to be Taken in Case Material is Released or Spilled

Remove all ignition sources. Restrict persons not wearing protective equipment from entering area of spill or leak until cleanup is complete. Ventilate area. Absorb liquid in dry sand or other material and deposit in sealed containers. Alert appropriate state or federal response agencies.

Preferred Waste Disposal Method

Burn in a chemical incinerator equipped with an afterburner and scrubber.

Precautions to be Taken in Handling and Storage

Do not store with incompatible chemicals since violent reactions can occur. Store in tightly closed containers in a cool, dark, well-ventilated area away from fire, sparks, and flame. Use non-sparking tools on containers. Ground and bond metal containers during transfer operations.

Other Precautions and Warnings

Containers may explode in fire or under conditions of extreme heat (do not store outdoors or in direct sunlight). Protect containers from physical damage.

SECTION VIII - CONTROL MEASURES AND PERSONAL PROTECTIVE EQUIPMENT

Respiratory Protection *(specify type)*

Exposures above 50 ppm: MSHA/NIOSH approved full facepiece respirator with organic vapor cartridge. Greater protection is obtained from a self-contained breathing apparatus (SCBA) with full facepiece and pressure demand or other positive pressure mode.

Ventilation

Local exhaust or general mechanical systems recommended.

Protective Gloves	Eye Protection	Other Protective Clothing
Solvent Resistant Butyl Rubber	Chemical Goggles and/or Face Mask	Rubber Apron

Work/Hygiene Practices

Always wash hands thoroughly after using chemical; never bring food, drink, or smoking materials into vicinity of chemicals.

DIPROPYL KETONE

$(CH_3CH_2CH_2)_2CO$ CAS: 123-19-3

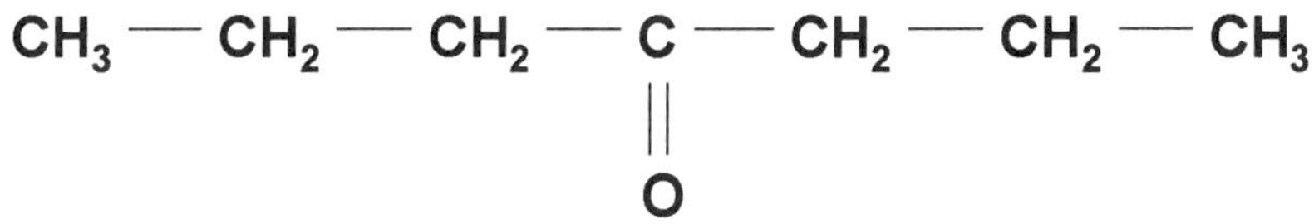

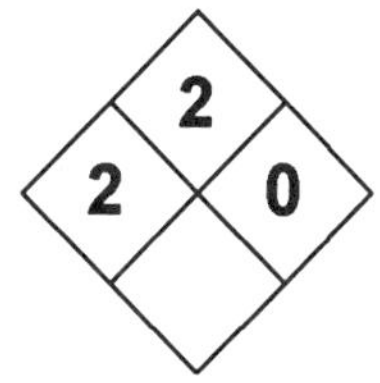

IDENTIFICATION AND TYPICAL USES

Dipropyl ketone is a colorless liquid with a strong, penetrating, but pleasant odor and a burning taste. It is used in lacquers and as a solvent for raw and blown oils, resins, nitrocellulose, and polymers. It is also used as a flavoring compound.

RISK ASSESSMENT: HEALTH

General Assessment

Dipropyl ketone is moderately toxic to humans by *inhalation* and *ingestion.* There are no reports in the references that indicate skin absorption is a risk factor. However, skin or eye contact can cause surface tissue damage and inflammation. The references contain no information on the carcinogenic, mutagenic, or teratogenic properties of dipropyl ketone. However, long-term exposures have caused damage to the liver in test animals.

Inhalation of dipropyl ketone causes irritation of the mucosa of the respiratory tract leading to respiratory depression with symptoms of gasping, choking, and dypsnea. Exposure to higher concentrations can cause depression of the central nervous system (CNS) with symptoms of headache, lachrymation, somnolence, decreased breathing, dizziness, fainting, tremor, coma, and possibly death (rare but possible). If the liquid should be aspirated, it can lead to chemical pneumonitis.

Skin contact with the liquid or vapor can lead to chemical dermatitis and localized paresthesia with symptoms of prickling or tingling with no apparent stimulus and associated sensory nerve damage. Eye contact causes ocular pain, corneal damage, lachrymation, and possible loss of vision.

Although ingestion is unlikely in the occupational setting, it can cause irritation of the mouth, sore throat, coughing, salivation, nausea, and vomiting.

☠ *Acute Health Effects*

The following acute (short-term) health effects may occur immediately or shortly after exposure to dipropyl ketone:

Skin: Irritation, dermatitis, and possible localized paresthesia.

Eye: Irritation with pain and inflammation that can lead to corneal damage and possible loss of vision.

Lung: Irritation causing coughing and/or shortness of breath. Higher exposures can cause dizziness, headache, and, possibly, coma and death.

CNS: A mild narcotic compound. Inhalation of high concentrations of the vapor may cause dizziness, lightheadedness, and possible loss of consciousness.

Chronic Health Effects

The following chronic (long-term) health effects may occur at some time after exposure to dipropyl ketone and can last for months or even years:

Cancer Hazards: According to information presented in the references, dipropyl ketone has not been adequately tested for its ability to cause cancer in laboratory test animals.

Reproduction: There are no reports to support any claims of reproductive hazards in humans or animals.

Sensitivity: Exposure to the skin may result in dermatitis and allergy (sensitization). If allergy develops, very low future exposure can result in skin irritation and rash. Repeated exposure can cause chronic irritation of the eyes and lungs.

Other Chronic Effects: Long-term exposures may cause damage to the liver and/or injury to the central nervous system.

Recommended Risk-Reduction Measures

Personnel should avoid direct contact with dipropyl ketone. Unless a less toxic chemical can be substituted for a hazardous substance, *engineering controls* are the most effective method of reducing exposures. The best protection is to enclose operations and/or provide local exhaust ventilation at the site of chemical release. While not always operationally feasible, isolating operations can also reduce exposure. Using respiratory protection is less effective than the controls mentioned above, but is still advisable whenever working with dipropyl ketone. For exposures to the PEL (50 ppm), use a NIOSH-approved, full facepiece organic vapor respirator. For higher or prolonged exposures, a self-contained breathing apparatus (SCBA) with full facepiece operated in pressure demand mode, or a supplied-air respirator operated in continuous flow mode are the recommended respiratory protection methods of choice. If a full facepiece is not available, then chemical goggles should be worn to protect the eyes. Whenever a chemical splash hazard exists, a face shield and a rubber apron should be worn. To prevent hand and skin exposures, butyl rubber gloves should be worn. Glove manufacturers should be contacted and permeation studies obtained *before* final glove selections are made.

Administrative controls should also be in place to minimize the potential for human exposures. These may include written procedures or policies which specify the methods and techniques that will be practiced whenever personnel are to work with dipropyl ketone.

All personnel should receive *training* on the use, hazards, protective measures, emergency actions, and other precautions per 29 CFR 1910.1200 (Hazard Communication), prior to the first assignment in an area where dipropyl ketone is used or stored.

If symptoms develop or overexposure is suspected, the following medical tests may be helpful:

☑ Liver function tests.

☑ Neurological evaluation of the central nervous system.

Other methods to reduce exposure include:

☑ Where possible, enclose operations and use local exhaust ventilation at the site of chemical release. If local exhaust ventilation or enclosure is not used, respiratory protection should be mandatory.

☑ Always ensure that proper protective clothing is worn when using chemical substances.

☑ Wash thoroughly immediately after exposure to dipropyl ketone and at the end of the work shift or before eating, drinking, or smoking. Never wear clothes contaminated with dipropyl ketone home. Family members can be exposed.

☑ Hazard warning information should be posted in the work area. As part of an on-going education and training program, all information on the health and safety hazards of dipropyl ketone should be communicated to all potentially exposed workers.

RISK ASSESSMENT: ENVIRONMENT

General Assessment

The environment is at risk of exposure during transportation, storage, disposal, or destruction of dipropyl ketone. In almost every scenario, the threat of environmental exposure is contingent upon the proper handling of the chemical substance. Accidental spills, large or small, can result in fire, explosion, and possible contamination of the surrounding environmental mediums (water, soil, and air).

Dipropyl ketone is considered a class II combustible liquid (per OSHA 29 CFR 1910.106). It can react with strong oxidizers to cause a serious fire and explosion hazard. When involved in a fire, it can release acrid and irritating oxides of carbon. A boiling liquid expanding vapor explosion (BLEVE) is also a possibility. These characteristics require special consideration during any emergency involving a leak or spill of dipropyl ketone.

The proper disposal/destruction method for dipropyl ketone is to burn it in a chemical incinerator (with molten salt treatment) equipped with an afterburner and air scrubber.

Dipropyl ketone can enter the environment through manufacturing, unchecked discharge into effluents from industrial or municipal waste treatment plants, and through spills.

Acute Ecological Effects

Acute (short-term) toxic effects may include the death of animals, birds, or fish, and death or low growth rate in plants. Acute effects are seen 2 to 4 days after animals or plants are exposed to dipropyl ketone. This chemical has a slight acute toxicity to aquatic life. No data are available on the short-term effects of dipropyl ketone on plants, birds, or land animals.

Chronic Ecological Effects

Chronic (long-term) toxic effects may include shortened life span, reproductive problems, lower fertility, and changes in appearance or behavior in exposed animals. These effects can be seen long after first exposure(s) to toxic chemicals. Dipropyl ketone has slight chronic toxicity in aquatic life. No data are available to evaluate the long-term effects of dipropyl ketone to plants, birds, or land animals.

Water Solubility

Dipropyl ketone is insoluble in water. Concentrations of less than 1 milligram may mix with a liter of water.

Persistence in the Environment

Dipropyl ketone is moderately persistent in water, with a half-life of between 20 to 200 days. The half-life of a pollutant is the amount of time it takes for one half of the chemical to be degraded.

Bioaccumulation in Aquatic Organisms

Some substances increase in concentration, or *bioaccumulate,* in living organisms as they breathe contaminated air, drink contaminated water, or eat contaminated food. These chemicals can become concentrated in the tissues and internal organs of animals as well as humans.

The concentration of dipropyl ketone found in fish tissues is expected to be about the same as the average concentration of dipropyl ketone in water from which the fish was taken.

Recommended Risk-Reduction Measures

Proper training of all transporters will reduce the likelihood of a mishap or accident resulting in a leak or spill to the environment. The correct use of DOT labeling (per the MSDS) on all containers, trucks, and rail cars will enable emergency responders to react properly and quickly to any disaster thereby reducing the potential risk to the environment and to personnel.

Storage of dipropyl ketone should be segregated from incompatible chemicals to minimize the risk of cross-contamination or contact. Buildings designated for storage should be equipped with appropriate fire protection systems (alarms, sprinklers, emergency lighting, portable extinguishers).

If a spill or leak to the environment has occurred, fire department, emergency response, and/or hazardous materials spill personnel should be notified immediately. Cleanup should be attempted only by those trained in proper spill containment procedures. Contaminated soils should be removed for incineration and replaced with clean soil. If dipropyl ketone should contact the water table, aquifer, or navigable waterway, time is of the essence. It is insoluble in water and, therefore, total containment and remediation may be possible. When such spills occur, the local and/or state emergency response authorities must be notified. A comprehensive emergency response or disaster preparedness/recovery plan should be in place prior to any operations involving the use, transportation, storage, or disposal of dipropyl ketone.

If dipropyl ketone is spilled or leaked, the following specific steps are recommended:

- ☑ Restrict persons not wearing protective clothing from area of spill or leak until cleanup is complete and area can be opened for normal work.
- ☑ Remove all ignition sources.
- ☑ Ventilate area of spill or leak.
- ☑ Absorb liquids in vermiculite, dry sand, earth, or a similar material and deposit in sealed containers for disposal.
- ☑ Keep dipropyl ketone out of a confined space, such as a sewer, because of the possibility of explosion (unless the sewer is designed to prevent the buildup of explosive concentrations).
- ☑ It may be necessary to dispose of dipropyl ketone as a hazardous waste. The responsible state agency or the regional office of the federal Environmental Protection Agency (EPA) should be contacted for specific recommendations.

RISK ASSESSMENT: BUSINESS

General Assessment

Accidents or mishaps involving dipropyl ketone can present a significant threat to business operations. The loss or damage of equipment or facilities can significantly affect fiscal viability. Lawsuits that may result from personnel injury, public exposures, and/or environmental contamination will also require a serious expenditure of resources.

Media attention surrounding an injury, illness, or environmental damage can also result in a loss of profits and loss of current as well as future business prospects and contacts.

Recommended Risk-Reduction Measures

Company attorneys, safety and health professionals, and environmental specialists should be involved in the development of any procedures dealing with safety or emergency response. A company official should be pre-designated as a public relations officer with specific training in dealing with the press. Corporate plans and policies should be developed, approved, and implemented long before any need for such arises.

REFERENCES

Ahlbom, A. 1993. *Biostatistics for Epidemiologists.* New York: CRC Press/Lewis Publishers.

American Conference of Governmental Industrial Hygienists. 1988. *Documentation of the Threshold Limit Values and Biological Exposure Indices,* 5th Edition (with updates). Cincinnati: ACGIH.

Anthony, C. P., and N. J. Kolthoff. 1971. *Textbook of Anatomy and Physiology.* St. Louis: C. V. Mosby Company.

Asante-Duah, D. K. 1993. *Hazardous Waste Risk Assessment.* New York: CRC Press/Lewis Publishers.

Bloom, A. D. and F. J. DeSerres. 1995. *Ecotoxicity and Human Health.* New York: CRC Press/Lewis Publishers.

Brusick, D. J. 1994. *Methods for Genetic Risk Assessment.* New York: CRC Press/Lewis Publishers.

Bueche, J. R. 1972. *Principles of Physics,* 2nd Edition. New York: McGraw-Hill.

Burrell, R., D. K. Flaherty, and L. J. Sauers. 1992. *Toxicology of the Immune System.* New York: Van Nostrand Reinhold.

Calabrese, E. J., E. M. Kenyon. *Air Toxics and Risk Assessment.* New York: CRC Press/Lewis Publishers.

Calabrese, E. J. 1994. *Biological Effects of Low Level Exposures.* New York: CRC Press/Lewis Publishers.

Cockerman, L. G., B. S. Shane. 1994. *Basic Environmental Toxicology.* New York: CRC Press/Lewis Publishers.

Cothern, R. R. 1993. *Comparative Environmental Risk Assessment.* New York: CRC Press/Lewis Publishers.

Cothern, C. R. 1995. *Handbook for Environmental Risk Decisionmaking.* New York: CRC Press/Lewis Publishers.

Gardner, R. W. 1994. *Chemical Intolerance.* New York: CRC Press/Lewis Publishers.

Hodgson, E., R. B. Mailman, and J. E. Chambers. 1988. *Dictionary of Toxicology.* London: The MacMillian Company.

Hosty, J. W., and P. Foster. 1990. *A Practical Guide to Chemical Spill Response.* New York: Van Nostrand Reinhold.

Howard, P. H. 1995. *Handbook of Environmental Fate and Exposures Data for Organic Chemicals,* Vol. I-IV. New York: CRC Press/Lewis Publishers.

Howard, P. H., et. al. 1991. *Handbook of Environmental Degradation Rates.* New York: CRC Press/Lewis Publishers.

James, A., and M. Lord. 1992. *Index of Chemical and Physical Data.* New York: Van Nostrand Reinhold.

Jones, F. E. 1994. *Toxic Organic Vapors in the Workplace.* New York: CRC Press/Lewis Publishers.

Kamrin, M. A. 1988. *Toxicology: A Primer on Principles and Applications.* Chelsea, Michigan: Lewis Publishers.

Keith, L. H. and M. M. Walker. 1995. *Handbook of Air Toxics.* New York: CRC Press/Lewis Publishers.

Landis, W. G., M. H. Yu. 1995. *Introduction to Environmental Toxicology.* New York: CRC Press/Lewis Publishers.

Langerman, N. 1994 *Precautionary Labels for Chemical Containers.* New York: CRC Press/Lewis Publishers.

Lauwerys, R. R., P. Hoet. 1993. *Industrial Chemical Exposure.* New York: CRC Press/Lewis Publishers.

Lewis, R. J., Sr. 1993. *Hawley's Condensed Chemical Dictionary,* 12th Edition. New York: Van Nostrand Reinhold.

Lewis, R. J., Sr. 1992. *Sax's Dangerous Properties of Industrial Materials,* Volumes 1, 2, and 3, 8th Edition. New York: Van Nostrand Reinhold.

Lide, D. R. 1995. *Handbook of Chemistry and Physics,* 76th Edition. New York: CRC Press/Lewis Publishers.

Manahan, S. E. 1992. *Toxicological Chemistry,* 2nd Edition. New York: CRC Press/Lewis Publishers.

Mellström, G. A., J. E. Wahlberg, and H. I. Maibach. 1994. *Protective Gloves for Occupational Use.* New York: CRC Press/Lewis Publishers.

Meyer, E. 1990. *Chemistry of Hazardous Materials.* Englewood Cliffs, NJ: Prentice Hall.

Patnaik, P. 1992. *A Comprehensive Guide to the Hazardous Properties of Chemical Substances.* New York: Van Nostrand Reinhold.

Philp, R. B. 1995. *Environmental Hazards and Human Health.* New York: CRC Press/Lewis Publishers.

Quigley, D. R. 1994. *Handbook of Emergency Chemical Management.* New York: CRC Press/Lewis Publishers.

Raleigh, R. L. and W. A. McGee. 1972. Effects of short, high concentration exposures to acetone as determined by observation in the work area. *Journal of Occupational Medicine.* 14:607-610.

Rea, W. J. 1995. *Chemical Sensitivity,* Vol. I-III. New York: CRC Press/Lewis Publishers.

Richardson, M. 1995. *Dictionary of Substance and Their Effects,* Vol. I-VI. New York: CRC Press/Lewis Publishers.

U.S. Department of Health and Human Services, National Institute for Occupational Safety and Health. 1994. *NIOSH Pocket Guide to Chemical Hazards.* Washington, D.C.: U.S. Government Printing Office.

U.S. Department of Health and Human Services, National Institute for Occupational Safety and Health. 1978. *Criteria for a Recommended Standard - Occupational Exposure to Ketones.* DHEW Publication Number: 78-173. Washington, D.C.: U.S. Government Printing Office.

MATERIAL SAFETY DATA SHEET

CHEMICAL NAME	
EMERY	**CRC PRESS/LEWIS PUBLISHERS DISCLAIMER:** THE INFORMATION AND RECOMMENDATIONS PRESENTED HEREIN ARE BASED ON SOURCES BELIEVED TO BE RELIABLE. CRC MAKES NO REPRESENTATION ON ITS COMPLETENESS OR ACCURACY. IT IS THE RESPONSIBILITY OF THE USER TO DETERMINE THE CHEMICAL'S SUITABILITY FOR ITS INTENDED USE, THE CHEMICAL'S SAFE USE, AND THE CHEMICAL'S PROPER DISPOSAL. NO REPRESENTATIONS AND/OR WARRANTIES, EITHER EXPRESSED OR IMPLIED, OF THE MERCHANTABILITY OR FITNESS FOR A PARTICULAR PURPOSE, OR OF ANY OTHER NATURE, ARE MADE WITH RESPECT TO THE INFORMATION PROVIDED IN THIS MSDS OR TO THE CHEMICAL TO WHICH INFORMATION MAY REFER. CRC NEITHER ASSUMES NOR AUTHORIZES ANY OTHER PERSON TO ASSUME FOR IT, ANY OTHER ADDITIONAL RESPONSIBILITY OR LIABILITY FOR THE USE OF, OR RELIANCE UPON, THIS INFORMATION.

HAZARD WARNING INFORMATION

HEALTH	FIRE	REACTIVE	OTHER	DEGREE OF HAZARD	COLOR CODING	OTHER CODES
1	0	0		0 = Minimum Hazard 1 = Slight Hazard 2 = Moderate Hazard 3 = Serious Hazard 4 = Severe Hazard	HEALTH = BLUE FIRE = RED REACTIVITY = YELLOW OTHER = WHITE	OX = Oxidizer ACID = Acid ALK = Alkali COR = Corrosive W = Use No Water

SECTION I - GENERAL INFORMATION

Characterization	**Metal Oxide**	RCRA Number	**Not Listed**
		EPA Class	**Not Listed**
DOT Proper Shipping Name	**Chemicals, N.O.S.**	Chemical Abstract Service (CAS) Number	**12415-34-8 (emery) 1302-74-5 (corundum)**
DOT Hazard Class and Label Requirements	**Non-Regulated**	DOT Emergency Guide Code	**No Citation**
DOT Identification Number	**None Established**	Molecular Formula	**Al_2O_3**

Synonyms

Aluminum oxide; aluminum trioxide; corundum; impure corundum; natural aluminum oxide. Often confused with methyl oleate (CAS: 112-62-9) which is also called emery but is a liquid ($C_{10}H_{36}O_2$).

SECTION II - HAZARDOUS INGREDIENTS AND IDENTITY INFORMATION

Hazardous Components (specific identity)	OSHA Exposure Criteria	NIOSH Exposure Criteria	Immediately Dangerous to Life and Health (IDLH)	ACGIH Exposure Criteria
Emery (derivation: Naturally occurring in Greece, Asia Minor, and New York as an impure variety of natural aluminum oxide containing minute amounts of iron, magnesium, and silica. Corundum is natural aluminum oxide).	**PEL:** 15 mg/m³ (total) 5 mg/m³ (respiratory)	**REL:** 10 mg/m³ (proposed)	**Not Determined**	**TLV:** 10 mg/m³ **STEL:** Not Established

SECTION III - PHYSICAL AND CHEMICAL CHARACTERISTICS

Boiling Point	**5396°F (2977°C)**	Specific Gravity (H_2O = 1)	**4.0**
Vapor Pressure (mm Hg)	**1 (approx.) at 2158°C**	Molecular Weight	**101.96**
Vapor Density (Air = 1)	**Not Applicable**	Melting Point	**3686°F (2030°C)**

Solubility

Negligible solubility (less than 0.1%). Difficult, but soluble in mineral acids and strong alkali.

Appearance and Odor

White, odorless, crystalline powder. Depending upon method of preparation, appearance can also be as white balls, or lumps of various mesh.

SECTION IV - FIRE AND EXPLOSION HAZARD DATA

Flash Point (method used)	**Not Applicable**	Explosive Limits in Air % by Volume	**LEL: N/A UEL: N/A**
NFPA Classification	**Noncombustible Solid**	Autoignition Temperature	**Not Applicable**

Extinguishing Media

Use extinguishing media suitable to surrounding fires.

Special Fire Fighting Procedures

None noted. But, emery dust is toxic by inhalation. Fire fighters should take care to avoid dusty situations or wear appropriate respiratory protection where emery may be present.

Unusual Fire and Explosion Hazards

Emery is a noncombustible solid. But, its dusts may form explosive mixtures in air. Caution is warranted when approaching situations that may involve emery dusts.

SECTION V - REACTIVITY DATA

Stability		Conditions to Avoid: Under normal conditions of temperature and pressure, emery is considered to be stable. However, emery should be kept away from contact with hot chlorinated rubber.
Stable	Unstable	Incompatibility *(materials to avoid)*: Emery is not compatible with hot chlorinated rubber; acids; oxidizers (especially ethylene oxide and chlorine trifluoride).
X		
Hazardous Polymerization		Conditions to Avoid: Explosive hydrogen gas may be formed when finely divided iron contacts moisture during crushing and milling operations.
May Occur	Will Not Occur	Hazardous Decomposition or By-products: Exothermic reaction above 392°F (200°C) with halocarbon vapors produces toxic hydrochloric acid and phosgene. Also, toxic aluminum fumes can be produced in fire.
	X	

SECTION VI - HEALTH HAZARD DATA

Primary Route(s) of Entry:	Inhalation?	Absorption (skin/eye)?	Ingestion?
	X		

Health Hazards

INHALATION: Primarily a "nuisance dust." Inhaling finely divided particles may cause lung damage (Shaver's disease) and possibly lead to pneumoconiosis (more probable because of the silica content of emery). In general, dust inhalation may cause tightness and pain in chest, coughing, difficulty breathing.

SKIN & EYES: Contact with the skin or eyes may cause irritation.

Carcinogenicity	NTP Listed?	IARC Cancer Review Group?	OSHA Regulated?	Target Organs?
Unknown Human Questioned Animal	No	No	29 CFR 1910.1000 Table Z-1-A & Z-3	Respiratory system

Medical Conditions Generally Aggravated by Exposure

Pre-existing lung disease or dysfunction may be aggravated by exposure to ammonium oxide dusts.

Emergency and First-aid Procedures

Eye contact: Do not allow victim to keep eyes tightly shut. Flush immediately with water for 15 minutes (minimum), seek medical attention. Skin contact: Remove all contaminated clothing. Immediately wash area with large amounts of soap and water. For inhalation: Remove the person from exposure. Provide respiratory assistance and CPR. Transfer to medical facility. If swallowed: Remove person to fresh air and seek medical attention immediately. Begin rescue breathing if breathing has stopped and CPR if heart action has stopped. If conscious, give 1-2 glasses of water. Do NOT attempt to give an unconscious or convulsing person anything by mouth.

SECTION VII - PRECAUTIONS FOR SAFE HANDLING AND USE

Steps to be Taken in Case Material is Released or Spilled

Ventilate area of spill or leak. Restrict those not involved in cleanup from entering area. Collect spilled powdered materials in safest manner possible and deposit in sealed drum for disposal.

Preferred Waste Disposal Method

No Citation. Check applicable federal, state, and/or local environmental regulations.

Precautions to be Taken in Handling and Storage

Emery is a noncombustible solid. Store to avoid contact with strong acids and oxidizers. Store in tightly closed containers in cool, well-ventilated area.

Other Precautions and Warnings

Avoid damage to containers. Emery is suitable for storage in general chemical storage areas.

SECTION VIII - CONTROL MEASURES AND PERSONAL PROTECTIVE EQUIPMENT

Respiratory Protection *(specify type)*

None required where adequate ventilation conditions exist. If airborne concentrations exceed the PEL, a dust/mist respirator is recommended. For high exposures, or if concentrations exceed the capacity of the respirator, use a self-contained breathing apparatus with full facepiece operated in positive pressure mode.

Ventilation

Local exhaust (preferred) at site of chemical work or general ventilation.

Protective Gloves	Eye Protection	Other Protective Clothing
Impervious Rubber	Chemical/Dust Goggles	Not Applicable

Work/Hygiene Practices

Always wash hands thoroughly after using chemical; never bring food, drink, or smoking materials into vicinity of chemicals.

EMERY

Al_2O_3 CAS: 12415-34-8 (for emery)
CAS: 1302-74-5 (for corundum)

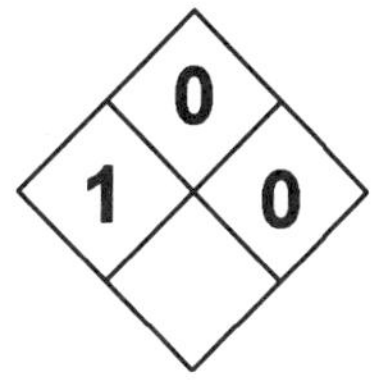

IDENTIFICATION AND TYPICAL USES

Emery is a white crystalline sand-like powder with no odor. The mineral corundum is natural emery. Emery, ruby, and sapphire are impure crystalline varieties. The mixed mineral bauxite is a hydrated emery. It is used in many polishing and abrasive operations in the form of emery cloth, grinding wheels, and grinding belts. Fine emery is used primarily as a lens grinder in the gem-cutting process.

RISK ASSESSMENT: HEALTH

General Assessment

The aluminas are generally considered nuisance dusts and under some conditions of human exposure may cause a minimal pulmonary nodular response. However, excessive *inhalation* of finely divided particles may lead to lung damage (Shaver's disease) characterized by scarred tissue ("pulmonary fibrosis") with shortness of breath, tightness and pain in chest, and cough. The condition can be fatal.

It should be noted that population studies of potentially exposed workers have shown minimal evidence of pulmonary fibrosis or pneumoconiosis. However, such diseases may develop more as a result of the silica content in emery. Also, animal experiments with emery have shown that the type of reaction in lung tissue varies with the form of alumina used and its particle size, the species of animal tested, and the route of administration. However, the data from both human and animal studies seem to indicate that alumina is efficiently eliminated from the lung and has a low degree of fibrogenicity.

This does not mean there is no health risk associated with emery. There is no way of estimating exact human response under all possible types of exposure conditions. Prudent risk management requires caution when working with or around emery.

Acute Health Effects

The following acute (short-term) health effects may occur immediately or shortly after exposure to emery:

Skin: Irritation.

Eye: Severe eye irritation likely.

Lung: Can irritate the respiratory tract. In large doses it may cause pulmonary fibrosis.

Chronic Health Effects

The following chronic (long-term) health effects may occur at some time after exposure to emery and can last for months or even years:

Cancer Hazards: According to the information presented in the references, emery has been tested and may lead to cancer in test animals. No data were found on its human carcinogenic capabilities.

Reproductive Hazard: According to the information presented in the references, emery has been tested and has not been shown to affect reproduction.

Other Chronic Effects: Repeated exposure to large amounts of emery dust may damage the lungs, leading to scarring of the tissue (pulmonary fibrosis), with shortness of breath and possible death.

Recommended Risk-Reduction Measures

Personnel should avoid direct contact with emery. The exact nature of its harmful capabilities is not clearly understood in the references. Therefore, it is recommended that it be handled with extreme. If a less toxic material cannot be substituted for emery, then *engineering controls* are the most effective method of reducing exposures. The best protection is to enclose operations and/or provide local exhaust ventilation at the site of emery release. While not always operationally feasible, isolating operations can reduce exposure.

Using respiratory protection and other personal protective equipment (PPE) is less effective than the engineering controls mentioned above, but is still advisable whenever working with or around emery. Under normal operating conditions of low exposure potential, a dust/mist respirator will suffice. However, for maximum protection in large concentrations or when the capacity of the dust/mist respirator is exceeded, personnel should wear a self-contained breathing apparatus (SCBA) with full facepiece and pressure demand. If a full facepiece is not available, then chemical/dust goggles, safety glass with side shields, or a face shield should be worn to protect the eyes. To prevent hand and skin exposures, impervious rubber gloves should be used.

Administrative controls should also be in place to minimize the potential for human exposures. These may include written procedures or policies which specify the methods and techniques that will be practiced whenever personnel are to work with emery.

All personnel should receive *training* on the use, hazards, protective measures, emergency actions, and other precautions per 29 CFR 1910.1200 (Hazard Communication), prior to the first assignment in an area where emery is used or stored.

If there has been a long history of high exposure, or if lung symptoms develop, the following medical tests are recommended:

☑ Chest X-ray.

☑ Pulmonary function tests.

Any evaluation should include a careful history of past and present symptoms with an examination. Medical tests that simply look for existing damage are not a substitute for controlling exposures. Also, because smoking can cause lung cancer, emphysema, and a host of other medical problems, working with or around emery may aggravate these conditions further.

Other methods to reduce exposure to emery include:

☑ Where possible, enclose operations and use local exhaust ventilation at the site of chemical release. If local exhaust ventilation or enclosure is not used, then respiratory protection should be required.

☑ Always ensure that proper protective clothing is worn when using chemical substances. Never allow contaminated clothing to be taken home.

☑ Hazard warning information should be posted in the work area. In addition, as part of an on-going education and training program, all information on the health and safety hazards of emery should be communicated to all potentially exposed workers.

☑ Eye wash stations should be provided in the immediate work area for emergency use.

RISK ASSESSMENT: ENVIRONMENT

General Assessment

The environment is at risk of exposure during transportation, storage, disposal, or destruction of emery. In almost every scenario, the threat of environmental exposure is contingent upon the proper handling of the chemical substance. Accidental spills, large or small, where emery contacts incompatible commodities can result in violent reactions as well as the possible contamination of the surrounding environmental mediums (water, soil, and air).

Emery is considered a noncombustible solid. However, its dusts may form explosive mixtures in air. Also, because it is incompatible with strong oxidizers and strong acids, extreme caution is required in handling, storage, transportation, and disposal of emery. These characteristics also require special consideration during any emergency situation involving a leak or spill of emery. Should emery ever come into contact with incompatible substances either during use, transportation, storage, or disposal, dangerous reactions are possible.

Emery occurs naturally in the environment in the minerals bauxite, baerites, boehmite, corundum, diaspore, and gibbsite. It therefore occurs naturally in the aquatic environment but can also enter it through industrial discharges and/or spills.

☠ *Acute Ecological Effects*

Acute (short-term) toxic effects may include the death of animals, birds, or fish and death or low growth rate in plants. Acute effects are seen 2 to 4 days after animals or plants are exposed to emery.

Emery has slight acute toxicity to aquatic life, but its toxicity will be increased under alkaline conditions. Insufficient data are available to evaluate or predict the short-term toxic effects of emery to plants, birds, and terrestrial animals.

Chronic Ecological Effects

Chronic (long-term) toxic effects may include shortened life span, reproductive problems, lower fertility, and changes in appearance or behavior in exposed animals. These effects can be seen long after first exposure to toxic chemicals. Emery has slight chronic toxicity to aquatic life, but this toxicity will increase under alkaline conditions. Insufficient data are available to evaluate or predict the long-term toxic effects of emery to plants, birds, or land animals.

Water Solubility

Emery is slightly solubility in water. Concentrations of 1 milligram or less will mix with a liter of water.

Persistence in the Environment

Emery is highly persistent in water, with a half-life greater than 200 days. The half-life of a pollutant is the amount of time it takes for one-half of the chemical to be degraded.

Bioaccumulation in Aquatic Organisms

Some substances increase in concentration, or *bioaccumulate,* in living organisms as they breathe contaminated air, drink contaminated water, or eat contaminated food. These chemicals can become concentrated in the tissues and internal organs of animals as well as humans. Emery will not accumulate in the edible tissues of aquatic life that are consumed by humans.

Recommended Risk-Reduction Measures

Proper training of all transporters will reduce the likelihood of a mishap or accident resulting in a leak or spill of emery into the environment. The correct use of labeling on all containers, trucks, and rail cars will enable emergency responders to react properly and quickly to any disaster thereby reducing the potential risk to the environment and to personnel.

Storage of emery should be segregated from incompatible chemicals to minimize the risk of cross-contamination or contact.

If a spill or leak to the environment has occurred, fire department, emergency response and/or hazardous materials spill personnel should be notified immediately. Cleanup should be attempted only by those trained in proper spill containment procedures. If emery should contact the water table, aquifer, or navigable waterway, the local and/or state emergency response authorities must be notified. A comprehensive emergency response or disaster preparedness/recovery plan should be in place prior to any operations involving the use, transportation, storage, or disposal of emery.

If emery is spilled or leaked, the following specific steps are recommended:

- ☑ Restrict persons not wearing protective clothing from area of spill or leak until cleanup is complete and area can be opened for normal work.
- ☑ Do not allow exposure to incompatible materials.
- ☑ Collect powdered materials in the safest and most efficient manner possible. Do NOT dry sweep and avoid the generation of airborne dusts. Use a wet mop method or a vacuum equipped with a high efficiency particulate air (HEPA) filter. Materials should be placed in a sealed drum for disposal.

RISK ASSESSMENT: BUSINESS

General Assessment

Accidents or mishaps involving emery can present a moderate threat to business operations. Although not likely, equipment and facilities may be lost or their use disrupted as a result of an incident involving emery which can significantly affect fiscal viability. Lawsuits that may result from personnel injury, illness, and/or death, public exposures, and/or environmental contamination will also require a serious expenditure of resources. Media attention surrounding an injury, death, or environmental damage can also result in a loss of profits and loss of current as well as future business. Also, some studies indicate that emery may cause cancer in test animals. Always remember that whenever the terms "cancer" or "carcinogen" are used, public emotion, hysteria, and ignorance can run equally high. This must be carefully considered whenever developing public relations policies.

Recommended Risk-Reduction Measures

Company attorneys, safety and health professionals, and environmental specialists should be involved in the development of any procedures that focus on emergency response, disaster preparedness, and disaster recovery. A company official should be pre-designated as a public relations officer with specific training in dealing with the press. Corporate plans and policies should be developed, approved, and implemented long before any need for such arises.

REFERENCES

Ahlbom, A. 1993. *Biostatistics for Epidemiologists.* New York: CRC Press/Lewis Publishers.

American Conference of Governmental Industrial Hygienists. 1988. *Documentation of the Threshold Limit Values and Biological Exposure Indices,* 5th Edition (with updates). Cincinnati: ACGIH.

Bloom, A. D. and F. J. DeSerres. 1995. *Ecotoxicity and Human Health.* New York: CRC Press/Lewis Publishers.

Brusick, D. J. 1994. *Methods for Genetic Risk Assessment.* New York: CRC Press/Lewis Publishers.

Bueche, J. R. 1972. *Principles of Physics,* 2nd Edition. New York: McGraw-Hill.

Calabrese, E. J., E. M. Kenyon. *Air Toxics and Risk Assessment.* New York: CRC Press/Lewis Publishers.

Calabrese, E. J. 1994. *Biological Effects of Low Level Exposures.* New York: CRC Press/Lewis Publishers.

Cockerman, L. G., B. S. Shane. 1994. *Basic Environmental Toxicology.* New York: CRC Press/Lewis Publishers.

Cothern, R. R. 1993. *Comparative Environmental Risk Assessment.* New York: CRC Press/Lewis Publishers.

Danse, I. R. 1991. *Common Sense Toxics in the Workplace.* New York: Van Nostrand Reinhold.

Dinman, B. D. 1988. Alumina-related pulmonary disease. *Journal of Occupational Medicine.* 30:328-335.

Environmental Protection Agency. 1983. *Chemical Hazard Information Profiles.* Washington, D.C.: U.S. EPA.

Fodden, C. R., and J. L. Weddell. 1992. *Hazardous Materials Emergency Action Data.* New York: CRC Press/Lewis Publishers.

Forseberg, K., and L. H. Keith. 1993. *Chemical; Protective Clothing Permeation and Degradation Database.* Chelsea, MI: Lewis Publishers.

Gardner, R. W. 1994. *Chemical Intolerance.* New York: CRC Press/Lewis Publishers.

Gibbs, G. W. 1985. Mortality of aluminum reduction plant workers, 1950 through 1977. *Journal of Occupational Medicine.* 27:761-770.

Hallenbeck, W. H. 1993. *Quantitative Risk Assessment for Environmental and Occupational Health,* 2nd Edition. New York: CRC Press/Lewis Publishers.

Howard, P. H. 1995. *Handbook of Environmental Fate and Exposures Data for Organic Chemicals,* Vol. I-IV. New York: CRC Press/Lewis Publishers.

Howard, P. H., et. al. 1991. *Handbook of Environmental Degradation Rates.* New York: CRC Press/Lewis Publishers.

Hosty, J. W., and P. Foster. 1990. *A Practical Guide to Chemical Spill Response.* New York: Van Nostrand Reinhold.

International Agency for Research on Cancer (IARC). 1987. *IARC Monographs on the Evaluation of Carcinogenic Risk to Humans, Overall Evaluations of Carcinogenicity: An updating of IARC Monographs Volumes 1-42.* Lyon: IARC.

Jones, F. E. 1994. *Toxic Organic Vapors in the Workplace.* New York: CRC Press/Lewis Publishers.

Kamrin, M. A. 1988. *Toxicology: A Primer on Principles and Applications.* Chelsea, Michigan: Lewis Publishers.

Keith, L. H. , M. T. Johnston, C. E. Hudak, and M. Conoley. 1992. *National Toxicology Program's Gloves+ Database.* Chelsea, MI: Lewis Publishers.

Keith, L. H. and M. M. Walker. 1995. *Handbook of Air Toxics.* New York: CRC Press/Lewis Publishers.

Landis, W. G., M. H. Yu. 1995. *Introduction to Environmental Toxicology.* New York: CRC Press/Lewis Publishers.

Langerman, N. 1994 *Precautionary Labels for Chemical Containers.* New York: CRC Press/Lewis Publishers.

Lauwerys, R. R., P. Hoet. 1993. *Industrial Chemical Exposure.* New York: CRC Press/Lewis Publishers.

Lewis, R. J., Sr. 1993. *Hawley's Condensed Chemical Dictionary,* 12th Edition. New York: Van Nostrand Reinhold.

Lewis, R. J., Sr. 1992. *Sax's Dangerous Properties of Industrial Materials,* Volumes 1, 2, and 3, 8th Edition. New York: Van Nostrand Reinhold.

Lide, D. R. 1995. *Handbook of Chemistry and Physics,* 76th Edition. New York: CRC Press/Lewis Publishers.

Manahan, S. E. 1994. *Environmental Chemistry,* 6th Edition. New York: CRC Press/Lewis Publishers.

Mellström, G. A., J. E. Wahlberg, and H. I. Maibach. 1994. *Protective Gloves for Occupational Use.* New York: CRC Press/Lewis Publishers.

Meyer, E. 1990. *Chemistry of Hazardous Materials.* Englewood Cliffs, NJ: Prentice Hall.

Mycock, J. C., J. D. McKenna, and L. Theodore. 1995. *Handbook of Air Pollution Control Engineering and Technology.* New York: CRC Press/Lewis Publishers.

Patnaik, P. 1992. *A Comprehensive Guide to the Hazardous Properties of Chemical Substances.* New York: Van Nostrand Reinhold.

Proctor, N. H., J. P. Hughes. 1991. *Chemical Hazards of the Workplace.* New York: Van Nostrand Reinhold.

Quigley, D. R. 1994. *Handbook of Emergency Chemical Management.* New York: CRC Press/Lewis Publishers.

Rea, W. J. 1995. *Chemical Sensitivity,* Vol. I-III. New York: CRC Press/Lewis Publishers.

Richardson, M. 1995. *Dictionary of Substance and Their Effects,* Vol. I-VI. New York: CRC Press/Lewis Publishers.

Sacarello, H. L. A. 1994. *The Comprehensive Handbook of Hazardous Materials.* New York: CRC Press/Lewis Publishers.

Sheftel, V. O. 1995. *Handbook of Toxic Properties of Monomers and Additives.* New York: CRC Press/Lewis Publishers.

Talbott, E., G. Craun. 1995. *An Introduction to Environmental Epidemiology.* New York: CRC Press/Lewis Publishers.

Torkelson, T. R. V. K. Rowe. 1981. *Patty's Industrial Hygiene and Toxicology,* Vol. 2B. New York: Wiley-Interscience.

MATERIAL SAFETY DATA SHEET

CHEMICAL NAME	
ENDRIN	**CRC PRESS/LEWIS PUBLISHERS DISCLAIMER:** THE INFORMATION AND RECOMMENDATIONS PRESENTED HEREIN ARE BASED ON SOURCES BELIEVED TO BE RELIABLE. CRC MAKES NO REPRESENTATION ON ITS COMPLETENESS OR ACCURACY. IT IS THE RESPONSIBILITY OF THE USER TO DETERMINE THE CHEMICAL'S SUITABILITY FOR ITS INTENDED USE, THE CHEMICAL'S SAFE USE, AND THE CHEMICAL'S PROPER DISPOSAL. NO REPRESENTATIONS AND/OR WARRANTIES, EITHER EXPRESSED OR IMPLIED, OF THE MERCHANTABILITY OR FITNESS FOR A PARTICULAR PURPOSE, OR OF ANY OTHER NATURE, ARE MADE WITH RESPECT TO THE INFORMATION PROVIDED IN THIS MSDS OR TO THE CHEMICAL TO WHICH INFORMATION MAY REFER. CRC NEITHER ASSUMES NOR AUTHORIZES ANY OTHER PERSON TO ASSUME FOR IT, ANY OTHER ADDITIONAL RESPONSIBILITY OR LIABILITY FOR THE USE OF, OR RELIANCE UPON, THIS INFORMATION.

HAZARD WARNING INFORMATION

HEALTH	FIRE	REACTIVE	OTHER	DEGREE OF HAZARD	COLOR CODING	OTHER CODES
4	0	0		0 = Minimum Hazard 1 = Slight Hazard 2 = Moderate Hazard 3 = Serious Hazard 4 = Severe Hazard	HEALTH = BLUE FIRE = RED REACTIVITY = YELLOW OTHER = WHITE	OX = Oxidizer ACID = Acid ALK = Alkali COR = Corrosive W = Use No Water

SECTION I - GENERAL INFORMATION

Characterization **Pesticide**	RCRA Number **P051**	EPA Class **Acute Hazardous Waste**
DOT Proper Shipping Name **Organochlorine pesticides, solid, toxic**	Chemical Abstract Service (CAS) Number **72-20-8**	
DOT Hazard Class and Label Requirements **Poison B; Poison Liquid**	DOT Emergency Guide Code **55**	
DOT Identification Number **UN 2761**	Chemical Formula $C_{12}H_8Cl_6O$	

Synonyms
1,2,3,4,10,10-Hexachloro-6,7-epoxy-1,4,4a,5,6,7,8,8a-octahydro-1,4-endo; endo-5,8-dimethanonaphthalene; Hexadrin®; Compound 269; 1,4:5,8-dimethanonaphthalene; Nendrin; Mendrin; OMS 197.

SECTION II - HAZARDOUS INGREDIENTS AND IDENTITY INFORMATION

Hazardous Components (specific identity)	OSHA Exposure Criteria	NIOSH Exposure Criteria	Immediately Dangerous to Life and Health (IDLH)	ACGIH Exposure Criteria
Endrin: (derivation: From isodrin or by condensing hexachlorocyclopentadiene with vinyl chloride to produce heptachlorbicyclo-(2.2.1)-2-heptane, followed by condensation with cyclopentadiene to form isodrin which is then oxidized to the 6,7-epoxide with peracetic or perbenzoic acid).	PEL (8-hour): 0.1 mg/m^3 (skin) STEL: Not Established	REL (10-hour): 0.1 mg/m3 (skin) STEL: Not Established	2 mg/m^3	TLV: 0.1 mg/m^3 (skin) STEL: Not Established

SECTION III - PHYSICAL AND CHEMICAL CHARACTERISTICS

Boiling Point **473°F (245°C) Decomposes**	Specific Gravity (H_2O = 1) **1.70**
Vapor Pressure (mm Hg) **2 X 10^{-7} at 77°F (25°C)**	Molecular Weight (atomic weight) **380.9**
Vapor Density (Air = 1) **Not Determined**	Melting Point **392°F (200°C)**

Solubility
Insoluble in water. Soluble in most organic solvents.

Appearance and Odor
Pure endrin is a colorless to white crystalline solid. Technical grade is tan. Mild, chemical odor. Odor Threshold = <.1 ppm (1.80 x 10^{-2}).

SECTION IV - FIRE AND EXPLOSION HAZARD DATA

Flash Point (method used) **Non-Combustible Solid**	Explosive Limits in Air % by Volume **LEL: Not Applicable UEL: Not Applicable**
NFPA Classification **Non-Combustible Solid**	Autoignition Temperature **Not Determined**

Extinguishing Media
Dry chemical, carbon dioxide, water (spray or fog), or regular foam.

Special Fire Fighting Procedures
Wear full protective clothing and self-contained breathing apparatus (SCBA). Structural fire-fighting clothing is permeable and may not provide sufficient protection. Remain clear of smoke, water fall-out and water run-off. Move containers from fire area if it can be done without risk. Cool exposed containers. Fight fire from distance, if possible. Poisonous gases are emitted (hydrogen chloride and phosgene).

Unusual Fire and Explosion Hazards
Endrin is a noncombustible solid, but it may be dissolved in flammable liquids, such as benzene.

SECTION V - REACTIVITY DATA

Stability		Conditions to Avoid Endrin is normally stable at room temperature and pressure. It becomes unstable in the presence of strong acids, sunlight, or heat. Avoid contact with incompatible materials and do not store near heat or heat source.
Stable X	Unstable	Incompatibility *(materials to avoid)* Reacts on contact with strong oxidizers (perchlorates, peroxides, permanganates, chlorates, and nitrates), acids (sulfuric, nitric), parathion, and diethyl 4-nitrophenyl thionophosphate. Slightly corrosive to metals.
Hazardous Polymerization		Conditions to Avoid Under normal conditions of temperature and pressure, hazardous polymerization of endrin is not known to occur.
May Occur	Will Not Occur X	Hazardous Decomposition or By-products At high temperatures, endrin decomposes and may produce poisonous gases, including toxic hydrogen chloride and phosgene.

SECTION VI - HEALTH HAZARD DATA

Primary Route(s) of Entry:	Inhalation?	Absorption (skin/eye)?	Ingestion?
	X	X	X

Health Hazards

INHALATION: Breathing dusts can cause irritation of the nose and throat. Repeated contact will lead to toxic systemic effects. Initial symptoms include epileptic convulsions lasting several minutes. There may be respiratory depression, hypersalivation, frothing (from the nose and mouth), cardiac arrhythmia, deafness, kidney dysfunction, headache, dizziness, weakness in the legs, abdominal discomfort, nausea, vomiting, insomnia, agitation, and confusion.

ABSORPTION: Irritation. Solution can pass through intact skin causing systemic effects (see inhalation).

INGESTION: Nausea, vomiting, diarrhea, and systemic effects (see inhalation). Can be fatal.

Carcinogenicity	NTP Listed?	IARC Cancer Review Group?	OSHA Regulated?	Target Organs?
Questioned Human Questioned Animal	5th Annual Report	Group 3	29 CFR 1910.1000 Table Z-1	Central nervous system (CNS), liver.

Medical Conditions Generally Aggravated by Exposure

Convulsive disorders (e.g., epilepsy) and liver disease may be aggravated by exposure to endrin.

Emergency and First-aid Procedures

Eye contact: Do not allow victim to keep eyes tightly shut. Flush immediately with water for 15 minutes, seek medical attention. Skin contact: Remove all contaminated clothing. Immediately wash area with flooding amounts of water. Do NOT scrub. Inhalation: Remove the person from exposure. Provide respiratory assistance and CPR. Transfer to medical facility. If swallowed: Contact poison control center. Unless advised otherwise, if victim is conscious, give 1-2 glasses of water. Due to potential for sudden seizures, do NOT induce vomiting. Do NOT give any oils by mouth. Seek medical attention immediately. Never give anything by mouth to an unconscious or convulsing person.

SECTION VII - PRECAUTIONS FOR SAFE HANDLING AND USE

Steps to be Taken in Case Material is Released or Spilled

Remove all sources of ignition. Ventilate area of spill. Restrict those not involved in cleanup from entering area. Notify appropriate authorities. Collect powdered materials with scoop or use a vacuum equipped with HEPA filter (do NOT generate dusts). Dispose of in sealed drums.

Preferred Waste Disposal Method

Chemical treatment by reductive chlorination, or burn in a chemical incinerator (with afterburner).

Precautions to be Taken in Handling and Storage

Store in tightly closed containers in cool, well ventilated area away from heat and incompatibilities.

Other Precautions and Warnings

Store in detached or external facility, if possible. Bulk storage is not recommended in plant.

SECTION VIII - CONTROL MEASURES AND PERSONAL PROTECTIVE EQUIPMENT

Respiratory Protection *(specify type)*

Odor threshold is low and difficult to monitor. For exposures above 0.1 mg/m^3, use a NIOSH-approved supplied air respirator or a SCBA with full facepiece operated in positive pressure mode.

Ventilation

Local exhaust (preferred) at site of chemical work or general ventilation.

Protective Gloves	Eye Protection	Other Protective Clothing
Rubber	Chemical/Dust Goggles or Face Mask	Protective Apron, Boots (avoid skin contact)

Work/Hygiene Practices

Always wash hands thoroughly after using chemical; never bring food, drink, or smoking materials into vicinity of chemicals.

ENDRIN

$C_{12}H_8Cl_6O$ CAS: 72-20-8

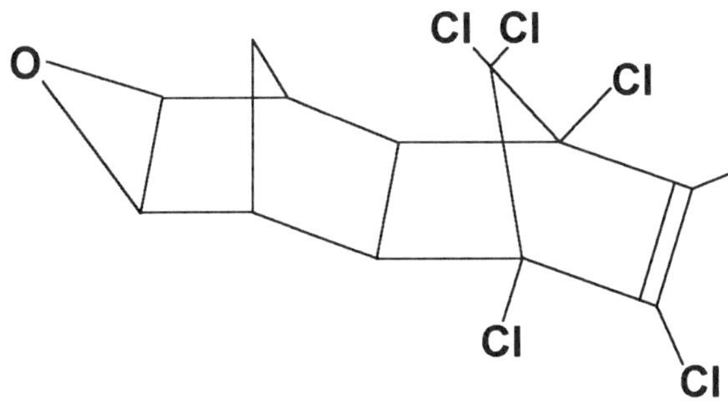

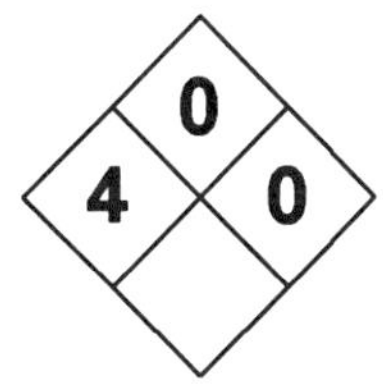

IDENTIFICATION AND TYPICAL USES

Endrin is a colorless to white crystalline solid in its pure form. The technical grade appears as a tan crystalline solid or powder. It has a mild chemical odor. It was formerly used as an insecticide or pesticide to control insects, rodents, and birds. However, because of its persistence in the environment, its manufacture has been discontinued in the United States since 1986. It was commonly used in field crops (cotton, grains, etc.). Also used to control grasshoppers in non-cropland and to control voles and mice in orchards.

RISK ASSESSMENT: HEALTH

General Assessment

Endrin is highly toxic to humans by all routes of exposure (*inhalation, ingestion,* and skin *absorption*). It is a questionable carcinogen in animals and in humans, with experimental teratogenic and mutagenic data also reported. This chemical is no longer manufactured in the United States and, therefore, exposure risk is thought to be relatively low. However, should human exposure ever occur, endrin can cause serious toxic and poisonous systemic effects up to and including death.

Endrin is a central nervous system stimulant. The initial symptom of exposure is the appearance of epileptic convulsions occurring up to 10 hours following an overexposure and lasting for several minutes.

Symptoms of exposure by both inhalation and skin contact (absorption) include respiratory depression with convulsions, hypersalivation, frothing from the nose and mouth, cardiac arrhythmia, deafness, renal dysfunction (kidney problems), and acidosis (a decease in the alkaline balance in the blood and tissues). Recovery is usually rapid but headache, dizziness, lethargy, weakness, and loss of appetite may persist. There may also be weakness in the limbs (primarily legs), abdominal discomfort, nausea, vomiting, insomnia, agitation and aggressive behavior, and mental confusion. Inhalation also causes irritation of the eyes, nose, and throat.

In addition to absorption and subsequent systemic effects, skin contact will result in moderate to severe irritation. Eye contact may also result in irritation. Ingestion causes nausea, vomiting, diarrhea, and systemic effects of inhalation and skin absorption.

Caution is warranted when inhalation and skin contact exposures are possible. Even though air levels may be within limits, personnel may be overexposed if endrin should contact the skin.

☠ *Acute Health Effects*

The following acute (short-term) health effects may occur immediately or shortly after exposure to endrin:

Skin: Liquid can pass through unbroken skin. Contact may cause irritation and rash. Symptoms of exposure include headache, nausea, abdominal pain, vomiting, and CNS stimulation.

Eye: Severe irritation.

Lung: Irritation of the nose and throat. Breathing vapors can cause cough, convulsions, and adverse effects on the liver.

CNS: A powerful central nervous system stimulant. Symptoms of exposure may include agitation, aggressiveness, excitement, tremors, convulsions, blurred vision, mental confusion, delirium, and irritability.

Chronic Health Effects

The following chronic (long-term) health effects may occur at some time after exposure to endrin and can last for months or even years:

Cancer Hazards: Endrin may be a carcinogen in humans since it has been shown to cause liver cancer in animals.

Reproductive Hazard: There is some evidence that exposure to endrin cause damage to the developing fetus in test animals. Endrin may decrease fertility in males and females.

Other Chronic Effects: Endrin may damage the liver causing enlargement (hepatomegaly). It may also cause damage to the central nervous system (CNS) causing loss of appetite, fatigue, and malaise.

Recommended Risk-Reduction Measures

Endrin is no longer manufactured in the United States. Its use is no longer permitted. However, in the unlikely event that personnel should be occupationally exposed to endrin, the risks to human health can be severe. Personnel should avoid direct contact with endrin. If a less toxic material or compound cannot be substituted for endrin, then *engineering controls* are the most effective method of reducing exposures. The best protection is to enclose operations and/or provide local exhaust ventilation at the site of endrin dust release. This may be practical in manufacturing operations, but is not feasible in application since it is used outdoors. While not always operationally possible, isolating operations involving endrin manufacturing can also reduce exposure. Using respiratory protection is less effective than the controls mentioned above, but is still advisable whenever working with or around endrin. The PEL is extremely low (0.1 mg/m^3) and difficult to accurately monitor. At exposures above the PEL, an MSHA or NIOSH-approved self-contained breathing apparatus (SCBA) with full facepiece and pressure demand is recommended for the greatest possible respiratory protection. If a full facepiece is not available, then chemical goggles should be worn to protect the eyes. A face shield and protective apron should also be worn. To prevent hand and skin exposures, impervious gloves should be worn.

Administrative controls should also be in place to minimize the potential for human exposures. These may include written procedures or policies which specify the methods and techniques that will be practiced whenever personnel are to work with endrin.

All personnel should receive *training* on the use, hazards, protective measures, emergency actions, and other precautions per 29 CFR 1910.1200 (Hazard Communication), prior to the first assignment in an area where endrin is used or stored.

If symptoms develop or overexposure is suspected, the following medical testing may be useful:

- ☑ Liver and kidney function tests.
- ☑ Electroencephalogram (EEG).
- ☑ Blood endrin levels.
- ☑ Recommended anticonvulsants include pentobarbital, phenobarbital, and benzodiazepines. Administration of adrenergic amines should be avoided (may increase myocardial irritability).

Any evaluation should include a careful history of past and present symptoms with an examination. Medical tests that simply look for existing damage are not a substitute for controlling exposures.

Other methods to reduce exposure include:

- ☑ Where possible, enclose operations and use local exhaust ventilation at the site of chemical release. If local exhaust ventilation or enclosure is not used, respiratory protection should be mandatory.
- ☑ Always ensure that proper protective clothing is worn when using chemical substances and that personnel are trained on its use and care.
- ☑ Wash thoroughly immediately after exposure to endrin and at the end of the work shift or before eating, drinking, or smoking.
- ☑ Hazard warning information should be posted in the work area. In addition, as part of an on-going education and training program, all information on the health and safety hazards of endrin should be communicated to all potentially exposed workers.
- ☑ Eye wash stations should be provided in the immediate work area for emergency use. If there is a possibility of skin exposure to endrin, emergency shower facilities should also be provided.
- ☑ Workers whose clothing has been contaminated by endrin should change into clean clothes before leaving work. Contaminated work clothing should be laundered only by individuals who have been informed of the hazards of exposure to endrin.
- ☑ Specific engineering controls are recommended by NIOSH. Refer to NIOSH Criteria Document: Occupational Exposure During the Manufacture

and Formulation of Pesticides (Publication Number 78-174)

RISK ASSESSMENT: ENVIRONMENT

General Assessment

Because of its former use as an insecticide and pesticide, endrin may be present in the environment. The environment is also at risk of exposure during transportation, storage, disposal, or destruction of endrin. In almost every scenario, the threat of environmental exposure is contingent upon the proper handling of the chemical substance. Accidental spills, large or small, where endrin contacts incompatible commodities can result in violent and even explosive reactions (depending upon conditions of spill), and possible contamination of the surrounding environmental mediums (water, soil, and air).

Endrin is considered a non-combustible solid. However, it was common practice to mix it with some flammable liquids such as benzene or xylene. Also, because it is incompatible with many common oxidizers, such as chlorine, extreme caution is required in handling, storage, transportation, and disposal of endrin. These characteristics also require special consideration during any emergency situation involving a leak or spill of endrin powder or dust.

Endrin use in the U.S. is no longer permitted. However, because of its once widespread, general usage in the past and its persistence in the environment, endrin may still enter the environment in agricultural run-off.

Acute Ecological Effects

Acute (short-term) toxic effects may include the death of animals, birds, or fish and death or low growth rate in plants. Acute effects are seen 2 to 4 days after animals or plants are exposed to endrin.

Endrin has high acute toxicity to aquatic life, land animals, and to birds. Insufficient data are available on the short-term effects of endrin to plants.

Chronic Ecological Effects

Chronic (long-term) toxic effects may include shortened life span, reproductive problems, lower fertility, and changes in appearance or behavior in exposed animals. These effects can be seen long after first exposure(s) to toxic chemicals.

Endrin has high chronic toxicity to aquatic life, birds, and land animals. Insufficient data are available to evaluate or predict the long-term effects of endrin to plants.

Water Solubility

Endrin is insoluble in water. Concentrations of less than 1 milligram may not mix with a liter of water.

Persistence in the Environment

Endrin is highly persistent in the water, with a half-life of greater than 200 days. The half-life of a pollutant is the amount of time it takes for one-half of the chemical to be degraded. It will persist in soil for extremely long periods (15 years or more). It can leach to ground water and absorb in terrestrial and aquatic sediments. About 75% of endrin will eventually end up in terrestrial soil; approximately 24% will end up in aquatic sediments; the remainder will end up in water or air.

Bioaccumulation in Aquatic Organisms

Some substances increase in concentration, or *bioaccumulate,* in living organisms as they breathe contaminated air, drink contaminated water, or eat contaminated food. These chemicals can become concentrated in the tissues and internal organs of animals as well as humans.

The concentration of endrin found in fish tissues is expected to be considerably higher than the average concentration of endrin in the water from which the fish was taken.

Recommended Risk-Reduction Measures

Although endrin is no longer manufactured or used in the United States and it is unlikely that it will be transported, stored, handled, or destroyed, environmental contamination is still possible. This is based upon endrin's extremely long persistence in the environment and the fact that some private enterprises may still have a supply of endrin in use on private croplands.

Storage of endrin should be segregated from incompatible materials to minimize the risk of cross-contamination or contact. Buildings or storage areas should be equipped with appropriate fire protection systems (alarms, sprinklers). Proper ventilation and protective equipment should be used while handing the powders or while preparing aqueous solutions.

If a spill or leak to the environment has occurred, fire department, emergency response and/or hazardous materials spill personnel should be notified immediately. Cleanup should be attempted only by those trained in proper spill containment procedures. If endrin should contact the water table, aquifer, or navigable waterway, reclamation procedures should be initiated as soon as practical. It is only slightly soluble in water and total remediation may be possible. The local and/or state emergency response authorities must be notified. A comprehensive emergency response or disaster preparedness/recovery plan should be in place prior to any operations involving the use, transportation, storage, or disposal of endrin. If endrin is spilled or leaked, the following specific steps are recommended:

- ☑ Restrict persons not wearing protective clothing from area of spill or leak until cleanup is complete and area can be opened for normal work.
- ☑ Remove all ignition sources.
- ☑ Collect solids using a HEPA vacuum to avoid the generation of dusts.
- ☑ It may be necessary to dispose of endrin as a hazardous waste. The responsible state agency or the regional office of the federal Environmental Protection Agency (EPA) should be contacted for specific recommendations.

RISK ASSESSMENT: BUSINESS

General Assessment

Accidents or mishaps involving endrin can present a significant threat to business operations. The loss or damage of equipment or facilities can significantly affect fiscal viability. Lawsuits that may result from personnel injury/death, public exposures, and/or environmental contamination will also require a serious expenditure of resources. Media attention surrounding an injury, death, or environmental damage can also result in a loss of profits and loss of current as well as future business. Always remember that anytime the terms "cancer," "carcinogen," or "reproductive hazard" are used, public emotion, hysteria, and ignorance can run equally high. This must be a consideration when developing public relations policies.

Recommended Risk-Reduction Measures

Company attorneys, safety and health professionals, and environmental specialists should be involved in the development of any procedures for responding to chemical incidents. A company official should be predesignated as a public relations officer with specific training in dealing with the press. Corporate plans and policies should be developed, approved, and implemented long before any need for such arises.

REFERENCES

American Conference of Governmental Industrial Hygienists. 1988. *Documentation of the Threshold Limit Values and Biological Exposure Indices,* 5th Edition (with updates). Cincinnati: ACGIH.

Brusick, D. J. 1994. *Methods for Genetic Risk Assessment.* New York: CRC Press/Lewis Publishers.

Calabrese, E. J. 1994. *Biological Effects of Low Level Exposures.* New York: CRC Press/Lewis Publishers.

Cockerman, L. G., B. S. Shane. 1994. *Basic Environmental Toxicology.* New York: CRC Press/Lewis Publishers.

Cothern, C. R. 1995. *Handbook for Environmental Risk Decisionmaking.* New York: CRC Press/Lewis Publishers.

Howard, P. H., et. al. 1991. *Handbook of Environmental Degradation Rates.* New York: CRC Press/Lewis Publishers.

Langerman, N. 1994 *Precautionary Labels for Chemical Containers.* New York: CRC Press/Lewis Publishers.

Lauwerys, R. R., P. Hoet. 1993. *Industrial Chemical Exposure.* New York: CRC Press/Lewis Publishers.

Lewis, R. J., Sr. 1992. *Sax's Dangerous Properties of Industrial Materials,* Volumes 1, 2, and 3, 8th Edition. New York: Van Nostrand Reinhold.

Manahan, S. E. 1992. *Toxicological Chemistry,* 2nd Edition. New York: CRC Press/Lewis Publishers.

Meyer, E. 1990. *Chemistry of Hazardous Materials.* Englewood Cliffs, NJ: Prentice Hall.

Quigley, D. R. 1994. *Handbook of Emergency Chemical Management.* New York: CRC Press/Lewis Publishers.

Rea, W. J. 1995. *Chemical Sensitivity,* Vol. I-III. New York: CRC Press/Lewis Publishers.

Richardson, M. 1995. *Dictionary of Substance and Their Effects,* Vol. I-VI. New York: CRC Press/Lewis Publishers.

Sacarello, H. L. A. 1994. *The Comprehensive Handbook of Hazardous Materials.* New York: CRC Press/Lewis Publishers.

U.S. Department of Health and Human Services, National Institute for Occupational Health (NIOSH). 1978. *NIOS Criteria Document: Occupational Exposure During the Manufacture and Formulation of Pesticides.* Publication Number 78-174. Washington, DC: U.S. Government Printing Office.

MATERIAL SAFETY DATA SHEET

CHEMICAL NAME	CRC PRESS/LEWIS PUBLISHERS DISCLAIMER
EPICHLOROHYDRIN	**CRC PRESS/LEWIS PUBLISHERS DISCLAIMER:** THE INFORMATION AND RECOMMENDATIONS PRESENTED HEREIN ARE BASED ON SOURCES BELIEVED TO BE RELIABLE. CRC MAKES NO REPRESENTATION ON ITS COMPLETENESS OR ACCURACY. IT IS THE RESPONSIBILITY OF THE USER TO DETERMINE THE CHEMICAL'S SUITABILITY FOR ITS INTENDED USE, THE CHEMICAL'S SAFE USE, AND THE CHEMICAL'S PROPER DISPOSAL. NO REPRESENTATIONS AND/OR WARRANTIES, EITHER EXPRESSED OR IMPLIED, OF THE MERCHANTABILITY OR FITNESS FOR A PARTICULAR PURPOSE, OR OF ANY OTHER NATURE, ARE MADE WITH RESPECT TO THE INFORMATION PROVIDED IN THIS MSDS OR TO THE CHEMICAL TO WHICH INFORMATION MAY REFER. CRC NEITHER ASSUMES NOR AUTHORIZES ANY OTHER PERSON TO ASSUME FOR IT, ANY OTHER ADDITIONAL RESPONSIBILITY OR LIABILITY FOR THE USE OF, OR RELIANCE UPON, THIS INFORMATION.

HAZARD WARNING INFORMATION

HEALTH	FIRE	REACTIVE	OTHER	DEGREE OF HAZARD	COLOR CODING	OTHER CODES
4	3	2		0 = Minimum Hazard 1 = Slight Hazard 2 = Moderate Hazard 3 = Serious Hazard 4 = Severe Hazard	HEALTH = BLUE FIRE = RED REACTIVITY = YELLOW OTHER = WHITE	OX = Oxidizer ACID = Acid ALK = Alkali COR = Corrosive W = Use No Water

SECTION I - GENERAL INFORMATION

Characterization	Chlorohydrin (Epoxy Resin)	RCRA Number	U041
		EPA Class	Toxic Waste
DOT Proper Shipping Name	Epichlorohydrin	Chemical Abstract Service (CAS) Number	106-89-8
DOT Hazard Class and Label Requirements	Flammable Liquid	DOT Emergency Guide Code	30
DOT Identification Number	UN 2023	Chemical Formula	C_3H_5ClO

Synonyms

1-Chloro-2,3-epoxypropane; 2-chloropropylene oxide; gamma-chloropropylene oxide; glycerol epichlorohydrin; alpha-epichlorohydrin; (chloromethyl)ethylene oxide.

SECTION II - HAZARDOUS INGREDIENTS AND IDENTITY INFORMATION

Hazardous Components (specific identity)	OSHA Exposure Criteria	NIOSH Exposure Criteria	Immediately Dangerous to Life and Health (IDLH)	ACGIH Exposure Criteria
Epichlorohydrin: (derivation: By removing hydrogen chloride from dichlorohydrin). 1 ppm = 3.85 mg/m^3	PEL (skin): 2 ppm 8 mg/m^3 STEL: Not Established	REL: Minimum Possible Exposure Possible Cancer Agent	75 ppm	TLV (skin): 2 ppm 8 mg/m^3 Suspected Human Carcinogen

SECTION III - PHYSICAL AND CHEMICAL CHARACTERISTICS

Boiling Point	242°F (116°C) Decomposes	Specific Gravity (H_2O = 1)	1.18
Vapor Pressure (mm Hg)	13 at 68°F (20°C)	Molecular Weight (atomic weight)	92.5
Vapor Density (Air = 1)	3.29	Melting Point	-54°F (-48°C)

Solubility

Slightly soluble in water (7%). Miscible in most organic solvents.

Appearance and Odor

A colorless, mobile liquid with a slightly irritating, chloroform-like odor. Odor Threshold = 10 ppm.

SECTION IV - FIRE AND EXPLOSION HAZARD DATA

Flash Point (method used)	93°F (34°C) closed cup	Explosive Limits in Air % by Volume	LEL: 3.8% UEL: 21%
NFPA Classification	Class 1C Flammable Liquid	Autoignition Temperature	772°F (411°C)

Extinguishing Media

Alcohol foam is recommended. Also use carbon dioxide, dry chemical, or water spray.

Special Fire Fighting Procedures

Poisonous gases are emitted (including phosgene). Wear full protective clothing and self-contained breathing apparatus (SCBA). Remain clear of smoke, water fall-out and water run-off. Move containers from fire area if it can be done without risk. Cool exposed containers. Fight fire from distance, if possible.

Unusual Fire and Explosion Hazards

Vapors form explosive mixtures in air. Vapors are heavier than air and may travel from some distance to an ignition source and flashback.

SECTION V - REACTIVITY DATA

Stability		Conditions to Avoid: Epichlorohydrin is a highly volatile and unstable liquid. Avoid contact with incompatible materials, heat, ignition sources, and even direct sunlight.
Stable	Unstable: X	Incompatibility (*materials to avoid*): Reacts violently with liberation of heat with strong oxidizers (chlorine, bromine, fluorine), strong acids, (sulfuric, nitric), and chemically active metals (aluminum, caustics, chlorides or iron, aluminum, and zinc).
Hazardous Polymerization		Conditions to Avoid: Highly exothermic polymerization of epichlorohydrin can be expected to occur in contact with incompatible materials.
May Occur: X	Will Not Occur	Hazardous Decomposition or By-products: At high temperatures, epichlorohydrin decomposes and may produce poisonous gases, including toxic hydrogen chloride, carbon dioxide, carbon monoxide, chlorine, and phosgene.

SECTION VI - HEALTH HAZARD DATA

Primary Route(s) of Entry:	Inhalation?	Absorption (skin/eye)?	Ingestion?
	X	X	

Health Hazards

INHALATION: Irritation of the eyes, nose, and respiratory tract. May cause nausea, vomiting, abdominal pain and discomfort, headache, and pain in the area of the liver. There may be difficulty breathing, cyanosis, coughing, nasal and throat irritation, and chemical pneumonitis. High exposures can depress the central nervous system and lead to death. Pulmonary edema (fluid in the lungs) is also possible. There can be liver and kidney damage as well.

ABSORPTION: Irritation and burns. Will pass through skin causing systemic effects (see inhalation).

INGESTION: Tremor, somnolence, ataxia, and systemic effects (see inhalation).

Carcinogenicity	NTP Listed?	IARC Cancer Review Group?	OSHA Regulated?	Target Organs?
Suspected Human Confirmed Animal	5th Annual Report	Group 2A	29 CFR 1910.1000 Table Z-1	Resp. sys., skin, eyes, CNS, liver, kidney, repro. sys.

Medical Conditions Generally Aggravated by Exposure

Liver, kidney, lung, or nervous system disorders may be aggravated by exposure to epichlorohydrin.

Emergency and First-aid Procedures

Eye contact: Do not allow victim to keep eyes tightly shut. Flush immediately with water for 15 minutes, seek medical attention. Skin contact: Remove all contaminated clothing. Immediately wash area with flooding amounts of water. Inhalation: Remove the person from exposure. Provide respiratory assistance and CPR. Transfer to medical facility. Observe for 1-2 days for lung effects. If swallowed: Contact poison control center. Unless advised otherwise, if victim is conscious, give 1-2 glasses of water and induce vomiting repeatedly until vomit is clear. Seek medical attention immediately. Never give anything by mouth to an unconscious or convulsing person.

SECTION VII - PRECAUTIONS FOR SAFE HANDLING AND USE

Steps to be Taken in Case Material is Released or Spilled

Remove all sources of ignition. Ventilate area of spill. Restrict those not involved in cleanup from entering area. Notify appropriate authorities. Absorb in vermiculite or other suitable material and dispose of in sealed drums. Do NOT release to sewers, storm drains, surface waters, or soils.

Preferred Waste Disposal Method

Mix with vermiculite, caustic soda and slaked lime, wrap in paper and burn in chemical incinerator.

Precautions to be Taken in Handling and Storage

Store in tightly closed containers in cool, well ventilated area away from heat and incompatibilities.

Other Precautions and Warnings

Prevent static sparks by grounding and bonding all containers. Treat “empty” containers as if full.

SECTION VIII - CONTROL MEASURES AND PERSONAL PROTECTIVE EQUIPMENT

Respiratory Protection (*specify type*)

For any exposure to a suspected carcinogen, use a NIOSH-approved supplied air respirator operated in positive pressure mode, or a SCBA with full facepiece operated in positive pressure mode.

Ventilation

Local exhaust (preferred) at site of chemical work or general ventilation.

Protective Gloves	Eye Protection	Other Protective Clothing
Impervious Rubber (test for adequacy)	Chemical Goggles or Face Mask	Protective Apron, Boots (avoid skin contact)

Work/Hygiene Practices

Always wash hands thoroughly after using chemical; never bring food, drink, or smoking materials into vicinity of chemicals.

EPICHLOROHYDRIN

C_3H_5ClO CAS: 106-89-8

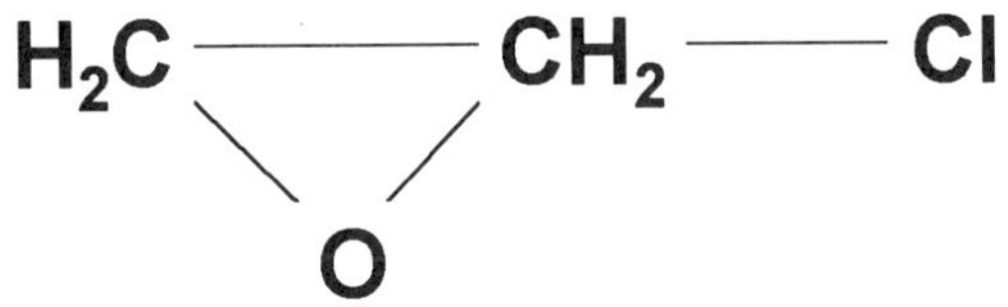

IDENTIFICATION AND TYPICAL USES

Epichlorohydrin is a colorless liquid with a slightly irritating chloroform-like odor. It is a major raw material for epoxy and phenoxy resins. It is also used in the manufacture of glycerol, adhesives, and castings, as a derivative in producing dyes, pharmaceuticals, surfactants, and plasticizers, in curing propylene-based rubber products, as a solvent for resins, gums, varnishes, and cellulose esters and ethers, and as a high wet-strength resin in the paper industry.

RISK ASSESSMENT: HEALTH

General Assessment

Epichlorohydrin is highly toxic to humans by *inhalation* and skin *absorption*. If *ingestion* occurs, it can also result in extremely toxic systemic effects. It is a confirmed carcinogen, a suspected teratogen (causes reproductive effects), and a known mutagen (causes genetic changes). Symptoms of exposure through all routes can be delayed for several hours, creating a false sense of security with regard to health hazard risk.

Inhalation results in moderate to severe irritation of the eyes, nose, throat, and respiratory tract. High concentrations can cause depression of and even injury to the central nervous system (CNS), leading the death. It is a known sensitizer by inhalation and absorption. Prolonged exposures can injure the liver, kidney, lungs, and cause changes in lymphocytes (cells in the blood or lymph). Symptoms of exposure through inhalation and absorption can also include headache, nausea, vomiting, abdominal discomfort and pain in the liver region, difficulty breathing, cyanosis, coughing, followed by nasal irritation, and chemical pneumonitis or pulmonary edema (fluid buildup in the lungs), which is a medical emergency and can be fatal.

In addition to absorption and subsequent systemic effects, skin and eye contact can produce severe burning in the area of contact, with blistering and deep pain occurring several hours later. There may also be allergic reactions (eczema or dermatitis) in some individuals.

Ingestion can produce tremors, somnolence, ataxia, and potentially serious damage to the liver and kidneys that may result in death.

☠ *Acute Health Effects*

The following acute (short-term) health effects may occur immediately or shortly after exposure to epichlorohydrin:

Skin: Liquid can pass through unbroken skin. Contact can cause severe irritation, burns, and deep-seated pain. Symptoms of exposure include headache, nausea, abdominal pain, vomiting, and CNS depression.

Eye: Severe irritation and burns that can lead to permanent damage to vision.

Lung: Irritation of the nose and throat. Breathing vapors can damage the lungs and cause chemical pneumonitis and even pulmonary edema (fluid in the lungs).

CNS: A central nervous system depressant. Symptoms include tremors, somnolence, ataxia, and possible death.

Chronic Health Effects

The following chronic (long-term) health effects may occur at some time after exposure to epichlorohydrin and can last for months or even years:

Cancer Hazards: Epichlorohydrin is highly suspected to be a human carcinogen. There is some evidence in the references that exposure can cause lung cancers in humans and it has been shown to cause stomach, liver, skin, and nasal cavity cancers in animals.

Reproductive Hazard: Exposure by all routes has lead to impaired male fertility and/or spermatogenesis. It is unclear at this time whether this is a reversible effect. In animals, there is evidence of fetotoxic effects.

Other Chronic Effects: Long-term exposure may cause skin allergies and sensitization to develop, with red rash and itching. If this occurs, even low future exposures could cause a recurrence of symptoms. Repeated exposures could cause lung damage and also may affect the liver and kidneys. Lung allergy (asthma) with wheezing, cough, and chest tightness may develop.

Recommended Risk-Reduction Measures

Epichlorohydrin presents a serious health hazard risk. Personnel should avoid direct contact with epichlorohydrin. If a less toxic material or compound cannot be substituted for epichlorohydrin, then *engineering controls* are the most effective method of reducing exposures. The best protection is to enclose operations and/or provide local exhaust ventilation at the site of epichlorohydrin release. While not always operationally possible, isolating operations involving epichlorohydrin manufacturing or use can also reduce exposure. Personnel must follow proper industrial hygiene and good housekeeping work practices. Using respiratory protection is less effective than the controls mentioned above, but is still advisable whenever working with or around epichlorohydrin. The PEL is relatively low (5 ppm) and difficult to accurately monitor. At exposures above the PEL, an MSHA or NIOSH-approved powered air-supplied respirator or a self-contained breathing apparatus (SCBA) with full facepiece and operated in pressure demand are recommended for the greatest possible respiratory protection. If a full facepiece is not available, then chemical goggles should be worn to protect the eyes. A face shield and protective apron should also be worn. Skin contact must not occur. To prevent hand and skin exposures, impervious gloves should be worn.

Administrative controls should also be in place to minimize the potential for human exposures. These may include written procedures or policies which specify the methods and techniques that will be practiced whenever personnel are to work with epichlorohydrin.

All personnel should receive *training* on the use, hazards, protective measures, emergency actions, and other precautions per 29 CFR 1910.1200 (Hazard Communication), prior to the first assignment in an area where epichlorohydrin is used or stored.

Before beginning initial assignments with this chemical and at regular intervals thereafter (e.g., annually), the following medical protocol are recommended:

- ☑ Lung function tests (establish initial baseline and compare thereafter).
- ☑ Liver and kidney function tests (establish initial baseline and compare thereafter).

If symptoms develop or overexposure is suspected, the following medical testing may be useful:

- ☑ Lung function tests.
- ☑ Exposure may cause renal or hepatic failure. Liver and kidney function tests are recommended. Monitor for serial liver enzymes for several days following exposure and urine should be tested for casts, protein, and abnormal cells.
- ☑ Consider chest X-ray after acute overexposure (may be negative if taken immediately following exposure).
- ☑ Evaluation by a qualified allergist with careful evaluation of exposure history and special testing (may help diagnose allergies).

Any evaluation should include a careful history of past and present symptoms with an examination. Medical tests that simply look for existing damage are not a substitute for controlling exposures. Also, since smoking can cause heart disease, emphysema, and other respiratory problems, personnel who smoke may show symptoms of exposure much quicker and more pronounced than those who do not. Prudent risk management requires careful consideration of *all* possible and potential exposure risk factors to ensure proper implementation of control measures and management of the exposure risk.

Other methods to reduce exposure include:

- ☑ Where possible, enclose operations and use local exhaust ventilation at the site of chemical release. If local exhaust ventilation or enclosure is not used, respiratory protection should be mandatory.
- ☑ Always ensure that proper protective clothing is worn when using chemical substances.
- ☑ Wash thoroughly immediately after exposure to epichlorohydrin and at the end of the work shift or before eating, drinking, or smoking.
- ☑ Hazard warning information should be posted in the work area. In addition, as part of an on-going

education and training program, all information on the health and safety hazards of epichlorohydrin should be communicated to all potentially exposed workers.

☑ Eye wash stations should be provided in the immediate work area for emergency use. If there is a possibility of skin exposure to epichlorohydrin, emergency shower facilities should also be provided.

☑ Workers whose clothing has been contaminated by epichlorohydrin should change into clean clothes *before* leaving work. Contaminated work clothing should be laundered only by individuals who have been informed of the hazards of exposure to epichlorohydrin.

☑ Specific engineering controls are recommended by NIOSH. Refer to NIOSH Criteria Document: Epichlorohydrin (Publication Number 76-206).

RISK ASSESSMENT: ENVIRONMENT

General Assessment

The environment is at risk of exposure during transportation, storage, disposal, or destruction of epichlorohydrin. In almost every scenario, the threat of environmental exposure is contingent upon the proper handling of the chemical substance. Accidental spills, large or small, where epichlorohydrin contacts incompatible commodities can result in violent and even explosive reactions (depending upon conditions of spill), and possible contamination of the surrounding environmental mediums (water, soil, and air).

Epichlorohydrin is considered a class 1C flammable liquid (per OSHA 29 CFR 1910.106). Its relatively low flash point and high boiling point make it a moderate risk in the presence of an ignition source, heat, or flame. Also, because it is incompatible with many common oxidizers (such as chlorine, bromine, and fluorine), acids (such as hydrochloric, nitric, and sulfuric), solvents (forms explosive dichloroacetylene in contact with trichloroethylene), and aniline (reacts explosively), extreme caution is required in handling, storage, transportation, and disposal of epichlorohydrin. In fact, epichlorohydrin has been known to polymerize in the presence of strong acids, especially when hot. These characteristics also require special consideration during any emergency situation involving a leak or spill of epichlorohydrin.

The proper disposal method is to mix epichlorohydrin with vermiculite and with caustic soda and slaked lime. The resultant mixture is then wrapped in paper and burned in a chemical incinerator equipped with an afterburner and air scrubber.

Epichlorohydrin may enter the environment through industrial effluents and from spills.

Acute Ecological Effects

Acute (short-term) toxic effects may include the death of animals, birds, or fish and death or low growth rate in plants. Acute effects are seen 2 to 4 days after animals or plants are exposed to epichlorohydrin.

Epichlorohydrin has moderate acute toxicity to aquatic life. Insufficient data are available on the short-term effects of epichlorohydrin to plants, land animals, and birds.

Chronic Ecological Effects

Chronic (long-term) toxic effects may include shortened life span, reproductive problems, lower fertility, and changes in appearance or behavior in exposed animals. These effects can be seen long after first exposure(s) to toxic chemicals.

Epichlorohydrin has moderate chronic toxicity to aquatic life. Insufficient data are available to evaluate or predict the long-term effects of epichlorohydrin to plants, birds, and land animals.

Water Solubility

Epichlorohydrin is highly soluble in water. Concentrations of 1000 milligrams and more will mix with a liter of water.

Persistence in the Environment

Epichlorohydrin is slightly persistent in water, with a half-life between 2 and 20 days. The half-life of a pollutant is the amount of time it takes for one-half of the chemical to be degraded. About 90% of epichlorohydrin will eventually end up in water; the remainder will end up in the air.

Bioaccumulation in Aquatic Organisms

Some substances increase in concentration, or *bioaccumulate,* in living organisms as they breathe contaminated air, drink contaminated water, or eat contaminated food. These chemicals can become concentrated in the tissues and internal organs of animals as well as humans.

The concentration of epichlorohydrin found in fish tissues is expected to be about the same as the average concentration of epichlorohydrin in the water from which the fish was taken.

Recommended Risk-Reduction Measures

Proper training of all transporters will reduce the likelihood of a mishap or accident resulting in a leak or spill to the environment. The correct use of DOT labeling (per the MSDS) on all containers, trucks, and rail cars will enable emergency responders to react properly and quickly to any disaster thereby reducing the potential risk to the environment and to personnel.

Storage of epichlorohydrin should be segregated from incompatible materials to minimize the risk of cross-contamination or contact. Storage facilities should be equipped with the proper fire protection equipment (sprinklers, alarms, extinguishers). Two-stage storage with water should be avoided since a slow exothermic reaction may result. Proper ventilation and protective equipment should be used while handling the powders or while preparing aqueous solutions. Non-sparking tools should be provided and used to ensure the prevention of ignition. Also, all equipment and containers used in shipping, receiving, or transferring should be electrically grounded and bonded.

If a spill or leak to the environment has occurred, fire department, emergency response and/or hazardous materials spill personnel should be notified immediately. Cleanup should be attempted only by those trained in proper spill containment procedures. If epichlorohydrin should contact the water table, aquifer, or navigable waterway, reclamation procedures should be initiated as soon as practical. It is highly soluble in water and total remediation may not be possible. The local and/or state emergency response authorities must be notified. A comprehensive emergency response or disaster preparedness/recovery plan should be in place prior to any operations involving the use, transportation, storage, or disposal of epichlorohydrin. If epichlorohydrin is spilled or leaked, the following specific steps are recommended:

- ☑ Restrict persons not wearing protective clothing from area of spill or leak until cleanup is complete and area can be opened for normal work.
- ☑ Remove all ignition sources and ventilate area of spill or leak.
- ☑ Absorb liquids in vermiculite or other compatible material and dispose of in sealed drums.
- ☑ It may be necessary to dispose of epichlorohydrin as a hazardous waste. The responsible state agency or the regional office of the federal Environmental Protection Agency (EPA) should be contacted for specific recommendations.

RISK ASSESSMENT: BUSINESS

General Assessment

Accidents or mishaps involving epichlorohydrin can present a significant threat to business operations. The loss or damage of equipment or facilities can significantly affect fiscal viability. Lawsuits that may result from personnel injury/death, public exposures, and/or environmental contamination will also require a serious expenditure of resources. Media attention surrounding an injury, death, or environmental damage can also result in a loss of profits and loss of current as well as future business. Always remember that anytime the terms "cancer," "carcinogen," or "reproductive hazard" are used, public emotion, hysteria, and ignorance can run equally high. This must be a consideration when developing public relations policies.

Recommended Risk-Reduction Measures

Company attorneys, safety and health professionals, and environmental specialists should be involved in the development of any procedures for responding to chemical incidents. A company official should be predesignated as a public relations officer with specific training in dealing with the press. Corporate plans and policies should be developed, approved, and implemented long before any need for such arises.

REFERENCES

American Conference of Governmental Industrial Hygienists. 1988. *Documentation of the Threshold Limit Values and Biological Exposure Indices,* 5th Edition (with updates). Cincinnati: ACGIH

Lewis, R. J., Sr. 1992. *Sax's Dangerous Properties of Industrial Materials,* Volumes 1, 2, and 3, 8th Edition. New York: Van Nostrand Reinhold.

Richardson, M. 1995. *Dictionary of Substance and Their Effects,* Vol. I-VI. New York: CRC Press/Lewis Publishers.

U.S. Department of Health and Human Services, National Institute for Occupational Health (NIOSH). 1978. *NIOS Criteria Document: Epichlorohydrin.* Publication Number 76-206. Washington, DC: U.S. Government Printing Office.

MATERIAL SAFETY DATA SHEET

CHEMICAL NAME	
ETHANE	**CRC PRESS/LEWIS PUBLISHERS DISCLAIMER:** THE INFORMATION AND RECOMMENDATIONS PRESENTED HEREIN ARE BASED ON SOURCES BELIEVED TO BE RELIABLE. CRC MAKES NO REPRESENTATION ON ITS COMPLETENESS OR ACCURACY. IT IS THE RESPONSIBILITY OF THE USER TO DETERMINE THE CHEMICAL'S SUITABILITY FOR ITS INTENDED USE, THE CHEMICAL'S SAFE USE, AND THE CHEMICAL'S PROPER DISPOSAL. NO REPRESENTATIONS AND/OR WARRANTIES, EITHER EXPRESSED OR IMPLIED, OF THE MERCHANTABILITY OR FITNESS FOR A PARTICULAR PURPOSE, OR OF ANY OTHER NATURE, ARE MADE WITH RESPECT TO THE INFORMATION PROVIDED IN THIS MSDS OR TO THE CHEMICAL TO WHICH INFORMATION MAY REFER. CRC NEITHER ASSUMES NOR AUTHORIZES ANY OTHER PERSON TO ASSUME FOR IT, ANY OTHER ADDITIONAL RESPONSIBILITY OR LIABILITY FOR THE USE OF, OR RELIANCE UPON, THIS INFORMATION.

HAZARD WARNING INFORMATION

HEALTH	FIRE	REACTIVE	OTHER	DEGREE OF HAZARD	COLOR CODING	OTHER CODES
1	4	0		0 = Minimum Hazard 1 = Slight Hazard 2 = Moderate Hazard 3 = Serious Hazard 4 = Severe Hazard	HEALTH = BLUE FIRE = RED REACTIVITY = YELLOW OTHER = WHITE	OX = Oxidizer ACID = Acid ALK = Alkali COR = Corrosive W = Use No Water

SECTION I - GENERAL INFORMATION

Characterization **Gas, Hydrocarbon**	RCRA Number **None**	EPA Class **Not Applicable**
DOT Proper Shipping Name **Ethane, Compressed or Ethane, Refrigerated Liquid**	Chemical Abstract Service (CAS) Number **74-84-0**	
DOT Hazard Class and Label Requirements **Flammable Gas (both forms)**	DOT Emergency Guide Code **15**	
DOT Identification Number **UN 1035 (compressed); UN 1961 (liquid)**	Chemical Formula **C_2H_6**	

Synonyms
Bimethyl; dimethyl; methylmethane; ethyl hydride.

SECTION II - HAZARDOUS INGREDIENTS AND IDENTITY INFORMATION

Hazardous Components (specific identity)	OSHA Exposure Criteria	NIOSH Exposure Criteria	Immediately Dangerous to Life and Health (IDLH)	ACGIH Exposure Criteria
Ethane (derivation from: Fractionation of natural gas). **1 ppm = 0.71 mg/m^3**	**PEL:** Not Established **STEL:** Not Established	**REL:** Not Established **STEL:** Not Established	**Not Determined**	**TLV:** Not Established **STEL:** Not Established

SECTION III - PHYSICAL AND CHEMICAL CHARACTERISTICS

Boiling Point **-128°F (-89°C)**	Specific Gravity (H_2O = 1) **Not Found**
Vapor Pressure (atmospheres) **5.44 (psig) at 70°F (21°C)**	Molecular Weight **30.08**
Vapor Density (Air = 1) **1.04 at 32°F (0°C)**	Freezing Point **-297°F (-183.2°C)**

Solubility
Only slightly soluble in water. More soluble in organic solvents.

Appearance and Odor
Colorless, odorless gas.

SECTION IV - FIRE AND EXPLOSION HAZARD DATA

Flash Point (method used) **-211°F (-153°C) closed cup**	Explosive Limits in Air % by Volume **LEL: 3% UEL: 12.5%**
NFPA Classification **Flammable Gas**	Autoignition Temperature **959°F (515°C)**

Extinguishing Media
Stop flow of gas first (if possible). Use foam, carbon dioxide, dry chemical, or water spray on fire.

Special Fire Fighting Procedures
Ethane is an extremely flammable, very dangerous fire and explosion hazard. Poisonous gases are evolved from fire. Wear a self-contained breathing apparatus (SCBA) with full facepiece. Ensure ALL sources of escaping gas are controlled. Monitor area long after fire is extinguished. Move containers from fire area if it can be done without risk.

Unusual Fire and Explosion Hazards
Gas is explosive. Containers may explode in flame. Even if fire is extinguished, escaping gas can form explosive mixtures in air quickly and ignite without warning.

SECTION V - REACTIVITY DATA

Stability		Conditions to Avoid Ethane is normally stable in its closed containers, during routine operations (those that do not expose it to ignition sources). Avoid contact with heat, sparks, flame, lighted tobacco products, and steam lines.
Stable X	Unstable	Incompatibility *(materials to avoid)* Ethane is reactive in contact with chlorine or chlorine dioxide, other oxidizers, heat, sparks, and open flame.
Hazardous Polymerization		Conditions to Avoid Under normal conditions of temperature and pressure, hazardous polymerization of ethane is not expected to occur.
May Occur	Will Not Occur X	Hazardous Decomposition or By-products If involved in fire, ethane can produce toxic carbon monoxide.

SECTION VI - HEALTH HAZARD DATA

Primary Route(s) of Entry:	Inhalation? X	Absorption (skin)?	Ingestion?

Health Hazards

INHALATION: An asphyxiant gas. Initial symptoms are those of oxygen starvation. Feeling of "air hunger," diminished mental alertness, headache, dullness, shortness of breath, and an impairment in muscle coordination. Continued exposures lead to faulty judgment, depression of senses, rapid fatigue, unconsciousness, convulsions, coma, and death.

SKIN & EYES: Contact with the compressed gas or refrigerated liquid can cause frostbite and burns.

INGESTION: Not likely since ethane is a gas at normal working (room) temperatures.

Carcinogenicity	NTP Listed?	IARC Cancer Review Group?	OSHA Regulated?	Target Organs?
Insufficient data	No	No	No	None Reported

Medical Conditions Generally Aggravated by Exposure

None Reported.

Emergency and First-aid Procedures

Eye contact: Immediately flush with large amounts of water for 15 minutes (minimum), occasionally lifting eyelids, seek medical attention. Skin contact: Remove all contaminated clothing. Immediately wash area with large amounts of soap and water. For inhalation: Remove the person from exposure. Provide respiratory assistance and CPR. Transfer promptly to medical facility. If swallowed: Not likely to occur under normal or even emergency occupational conditions.

SECTION VII - PRECAUTIONS FOR SAFE HANDLING AND USE

Steps to be Taken in Case Material is Released or Spilled

Remove all ignition sources. Ventilate area. Stop flow of gas. If the source of leak is a cylinder and cannot be stopped, remove to safe place in open air; repair leak or let cylinder empty. If gas has not yet ignited, use a water spray to knock down vapors and direct them from ignition source.

Preferred Waste Disposal Method

None specified in the references (incineration of a gas may be the only alternative).

Precautions to be Taken in Handling and Storage

Do not store with incompatible materials or in the presence of heat since violent reactions or explosions can occur. Store in tightly closed pressurized containers in a cool, dark, well-ventilated area away from sunlight. Keep away from ignition sources such as fire, sparks, and flame.

Other Precautions and Warnings

Bulk storage of ethane is not recommended. Containers may explode in fire or under conditions of extreme heat (do not store outdoors or in direct sunlight).

SECTION VIII - CONTROL MEASURES AND PERSONAL PROTECTIVE EQUIPMENT

Respiratory Protection *(specify type)*

No exposure limits have been established. Ethane is a simple asphyxiant. Do NOT use cartridge-type respirators (ineffective for this type of gas). Use a powered supplied-air respirator or a self-contained breathing apparatus with full facepiece operated in pressure demand or other positive pressure mode.

Ventilation

Local exhaust or general mechanical systems recommended.

Protective Gloves	Eye Protection	Other Protective Clothing
Insulated Rubber	Goggles and/or Face Mask	Leather Apron

Work/Hygiene Practices

Always wash hands thoroughly after using chemical; never bring food, drink, or smoking materials into vicinity of chemicals.

ETHANE

C_2H_6 CAS: 74-84-0

IDENTIFICATION AND TYPICAL USES

Ethane is a colorless and odorless gas. It is used in the manufacture of chlorinated derivatives, as a minor ingredient in bottled fuel gas, in the petrochemical industry (as a source of ethylene and halogenated ethanes), and as a refrigerant.

RISK ASSESSMENT: HEALTH

General Assessment

Ethane is a simple asphyxiant gas. As such, its primary route of entry into the body is through *inhalation.* While it does not pose a significant physiological effect (i.e., no systemic poisoning or toxicity), it can displace oxygen levels to below that which is necessary to sustain life. Skin contact with the liquid can be destructive to tissues due to the extremely cold temperatures but absorption is not a likely exposure route. Since ethane usually exists as a gas at room temperatures, ingestion is not likely to occur either.

Inhalation of ethane gas can cause headache, dullness, shortness of breath, and the symptoms of asphyxiation. These include rapid respiration and a feeling of "air hunger" or air starvation followed by a decrease in mental alertness, and impaired muscular coordination. If exposure continues, there can be nausea, vomiting, prostration, unconsciousness, coma, and death.

Skin or eye contact with the vapor poses no serious threat. The vapors have not been shown to cause any physiological response. However, if the liquid should contact either the skin or the eyes, there can be serious irritation, burns, tissue damage, and frostbite.

Acute Health Effects

The following acute (short-term) health effects may occur immediately or shortly after exposure to ethane:

Skin: Liquid threat only. Redness, pain, swelling, blisters, and severe frostbite.

Eye: Liquid threat only. Causes frostbite, redness, irritation, severe pain, and blurred vision.

Lung: Vapor is non-irritating and therefore poses no threat of damage to the lungs. However, in high concentrations, ethane can lead to symptoms of asphyxiation and oxygen depravation up to and including death.

Chronic Health Effects

The following chronic (long-term) health effects may occur at some time after exposure to ethane and can last for months or even years:

Cancer Hazards: According to information presented in the references, exposure to ethane does not lead to carcinogenic effects (it is a simple asphyxiant).

Reproductive Hazards: There are no reports to support any claims of reproductive hazards.

Other Chronic Effects: None reported.

Recommended Risk-Reduction Measures

Personnel should avoid direct contact with ethane. If another gas cannot be substituted for a ethane, then *engineering controls* are the most effective method of reducing exposures. The best protection is to separate operations and provide local exhaust ventilation at the site of chemical release. While not always operationally feasible, isolating operations can also reduce exposure. Using equipment with explosion-proof designs and ensuring all metal containers and tools are electrically grounded and bonded can further reduce the potential for injury or damage. Using respiratory protection is less effective than the controls mentioned above, but is still advisable whenever working with ethane. Air-purifying respirators (e.g., the cartridge-type or gas masks) cannot be used with ethane. These type respirators only purify the air and can do nothing to protect in oxygen deficient atmospheres. A powered air-supplied respirator operated in pressure demand or other positive pressure mode, or a self-contained breathing apparatus (SCBA) with full facepiece operated in pressure demand mode are the recommended respiratory protection methods of choice. If a full facepiece is not available, then chemical goggles should be worn to protect the eyes. Whenever a chemical splash hazard exists, a face shield and a rubber or leather apron should be worn. To prevent hand and skin exposures with the liquid, thermal gloves should be worn.

Administrative controls should also be in place to minimize the potential for human exposures. These may include written procedures or policies which specify the methods and techniques that will be practiced whenever personnel are to work with ethane.

All personnel should receive *training* on the use, hazards, protective measures, emergency actions, and other precautions per 29 CFR 1910.1200 (Hazard Communication), prior to the first assignment in an area where ethane is used or stored.

Other methods to reduce exposure to chemicals include:

- ☑ Where possible, consider installing automatic sensors that warn personnel of dropping oxygen levels and use local exhaust ventilation at the site of chemical release. If local exhaust ventilation or enclosure is not used, respiratory protection should be mandatory.
- ☑ Always ensure that proper protective clothing is worn when using chemical substances.
- ☑ Wash thoroughly immediately after exposure to ethane and at the end of the work shift or before eating, drinking, or smoking.
- ☑ Safety shower and eyewash stations should be readily available in work areas where ethane is used or stored.
- ☑ Hazard warning information should be posted in the work area. In addition, as part of an on-going education and training program, all information on the health and safety hazards of ethane should be communicated to all potentially exposed workers.

RISK ASSESSMENT: ENVIRONMENT

General Assessment

The environment is at risk of exposure during transportation, storage, disposal, or destruction of ethane. In almost every scenario, the threat of environmental exposure is contingent upon the proper handling of the chemical substance. Accidental releases, large or small, can result in fire, explosion, and possible contamination of the surrounding environmental mediums (water, soil, and air). Ethane is an extremely flammable and highly explosive gas. Its extremely low flash point and boiling point present a serious fire hazard. Also, since it can react violently in the presence of oxidizers, such as chlorine, ethane requires special consideration during any emergency involving a leak or release of ethane gas or liquid.

Ethane can enter the environment through industrial discharges, unchecked venting to air, or through spills of the liquid. Since liquids will most likely vaporize rapidly, water and soil contamination are possible but improbable.

Acute Ecological Effects

Acute (short-term) toxic effects may include the death of animals, birds, or fish, and death or low growth rate in plants. Acute effects are seen 2 to 4 days after animals or plants are exposed to ethane. Insufficient data are available on the short-term effects of ethane on aquatic life, plants, birds, or land animals. However, as a relatively non-toxic, simple asphyxiant, it might be assumed that ecological toxicity will be relatively low.

Chronic Ecological Effects

Chronic (long-term) toxic effects may include shortened life span, reproductive problems, lower fertility, and changes in appearance or behavior in exposed animals. These effects can be seen long after first exposure(s) to toxic chemicals. Insufficient data are available to evaluate the long-term effects of ethane to aquatic life, plants, birds, or land animals. However, as a relatively non-toxic, simple asphyxiant, it might be assumed that any chronic ecological toxicity will be relatively low.

Water Solubility

Ethane is only slightly soluble in water. Concentrations of 1 milligram or less will mix with a liter of water.

Persistence in the Environment

Ethane is non-persistent in water, with a half-life of less than 2 days. The half-life of a pollutant is the amount of time it takes for one half of the chemical to be degraded.

Bioaccumulation in Aquatic Organisms

Some substances increase in concentration, or *bioaccumulate,* in living organisms as they breathe contami-

nated air, drink contaminated water, or eat contaminated food. These chemicals can become concentrated in the tissues and internal organs of animals as well as humans.

Ethane is not expected to accumulate in the tissues of fish.

ʊ *Recommended Risk-Reduction Measures*

Proper training of all transporters will reduce the likelihood of a mishap or accident resulting in a leak or release to the environment. The correct use of DOT labeling (per the MSDS) on all containers, trucks, and rail cars will enable emergency responders to react properly and quickly to any disaster thereby reducing the potential risk to the environment and to personnel.

Storage of ethane should be segregated from incompatible chemicals to minimize the risk of cross-contamination or contact. Buildings designated for storage should be equipped with appropriate fire protection systems (alarms, sprinklers, emergency lighting, portable extinguishers).

If a spill or release to the environment has occurred, fire department, emergency response, and/or hazardous materials spill personnel should be notified immediately. Cleanup should be attempted only by those trained in proper containment procedures. Contaminated soils should be removed for incineration and replaced with clean soil. When spills or leaks occur, it may be necessary to contact the local and/or state emergency response authorities. A comprehensive emergency response or disaster preparedness/recovery plan should be in place prior to any operations involving the use, transportation, storage, or disposal of ethane. If ethane is spilled or leaked, the following specific steps are recommended:

☑ Restrict persons not wearing protective clothing from area of spill or leak until cleanup is complete.

☑ Remove all ignition sources.

☑ Ventilate area of spill or leak.

☑ If applicable, stop flow of leaking liquid or gas. If leak source is a cylinder and the leak cannot be stopped in place, remove leaking cylinder to a safe place in the open air, and repair or allow cylinder to empty.

☑ Keep ethane out of a confined space, such as a sewer, because of the possibility of explosion (unless the sewer is designed to prevent the build-up of explosive concentrations).

☑ It may be necessary to dispose of ethane as a hazardous waste. The responsible state agency or the regional office of the federal Environmental Protection Agency (EPA) should be contacted for specific recommendations.

RISK ASSESSMENT: BUSINESS

General Assessment

Accidents or mishaps involving ethane can present a significant threat to business operations. The loss or damage of equipment or facilities can significantly affect fiscal viability. Lawsuits that may result from personnel injury, public exposures, and environmental contamination will also require a serious expenditure of resources. Media attention surrounding an injury, illness, or environmental damage can also result in a loss of profits and loss of current as well as future business.

ʊ *Recommended Risk-Reduction Measures*

Company attorneys, safety and health professionals, and environmental specialists should be involved in the development of any procedures dealing with safety and emergency response. A company official should be pre-designated as a public relations officer with specific training in dealing with the press. Corporate plans and policies should be developed, approved, and implemented long before any need for such arises.

REFERENCES

Ahlbom, A. 1993. *Biostatistics for Epidemiologists.* New York: CRC Press/Lewis Publishers.

American Conference of Governmental Industrial Hygienists. 1988. *Documentation of the Threshold Limit Values and Biological Exposure Indices,* 5th Edition (with updates). Cincinnati: ACGIH

Anthony, C. P., and N. J. Kolthoff. 1971. *Textbook of Anatomy and Physiology.* St. Louis: C. V. Mosby Company.

Asante-Duah, D. K. 1993. *Hazardous Waste Risk Assessment.* New York: CRC Press/Lewis Publishers.

Bloom, A. D. and F. J. DeSerres. 1995. *Ecotoxicity and Human Health.* New York: CRC Press/Lewis Publishers.

Brusick, D. J. 1994. *Methods for Genetic Risk Assessment.* New York: CRC Press/Lewis Publishers.

Bueche, J. R. 1972. *Principles of Physics,* 2nd Edition. New York: McGraw-Hill.

Burrell, R., D. K. Flaherty, and L. J. Sauers. 1992. *Toxicology of the Immune System.* New York: Van Nostrand Reinhold.

Calabrese, E. J., E. M. Kenyon. *Air Toxics and Risk Assessment.* New York: CRC Press/Lewis Publishers.

Calabrese, E. J. 1994. *Biological Effects of Low Level Exposures.* New York: CRC Press/Lewis Publishers.

Cockerman, L. G., B. S. Shane. 1994. *Basic Environmental Toxicology.* New York: CRC Press/Lewis Publishers.

Cothern, R. R. 1993. *Comparative Environmental Risk Assessment.* New York: CRC Press/Lewis Publishers.

Cothern, C. R. 1995. *Handbook for Environmental Risk Decisionmaking.* New York: CRC Press/Lewis Publishers.

Danse, I. R. 1991. *Common Sense Toxics in the Workplace.* New York: Van Nostrand Reinhold.

Environmental Protection Agency. 1983. *Chemical Hazard Information Profiles.* Washington, D.C.: U.S. EPA.

Fodden, C. R., and J. L. Weddell. 1992. *Hazardous Materials Emergency Action Data.* New York: CRC Press/Lewis Publishers.

Forseberg, K., and L. H. Keith. 1993. *Chemical; Protective Clothing Permeation and Degradation Database.* Chelsea, MI: Lewis Publishers.

Gardner, R. W. 1994. *Chemical Intolerance.* New York: CRC Press/Lewis Publishers.

Hallenbeck, W. H. 1993. *Quantitative Risk Assessment for Environmental and Occupational Health,* 2nd Edition. New York: CRC Press/Lewis Publishers.

Hodgson, E., R. B. Mailman, and J. E. Chambers. 1988. *Dictionary of Toxicology.* London: The MacMillian Company.

Hosty, J. W., and P. Foster. 1990. *A Practical Guide to Chemical Spill Response.* New York: Van Nostrand Reinhold.

Howard, P. H. 1995. *Handbook of Environmental Fate and Exposures Data for Organic Chemicals,* Vol. I-IV. New York: CRC Press/Lewis Publishers.

Howard, P. H., et. al. 1991. *Handbook of Environmental Degradation Rates.* New York: CRC Press/Lewis Publishers.

James, A., and M. Lord. 1992. *Index of Chemical and Physical Data.* New York: Van Nostrand Reinhold.

Jones, F. E. 1994. *Toxic Organic Vapors in the Workplace.* New York: CRC Press/Lewis Publishers.

Kamrin, M. A. 1988. *Toxicology: A Primer on Principles and Applications.* Chelsea, Michigan: Lewis Publishers.

Keith, L. H. , M. T. Johnston, C. E. Hudak, and M. Conoley. 1992. *National Toxicology Program's Gloves+ Database.* Chelsea, MI: Lewis Publishers.

Keith, L. H. and M. M. Walker. 1995. *Handbook of Air Toxics.* New York: CRC Press/Lewis Publishers.

Keith, L. H. (Editor). 1992. *Integrated Risk Information System (IRIS) Chemical Information Database.* Prepared by the U.S. Environmental Protection Agency. Boca Raton, FL: CRC Press/Lewis Publishers.

Keith, L. H. and D. B. Walters. (Editors) 1992. *National Toxicology Program's Chemical Database, Volumes I-VII.* Chelsea, MI: Lewis Publishers.

Landis, W. G., M. H. Yu. 1995. *Introduction to Environmental Toxicology.* New York: CRC Press/Lewis Publishers.

Langerman, N. 1994 *Precautionary Labels for Chemical Containers.* New York: CRC Press/Lewis Publishers.

Lauwerys, R. R., P. Hoet. 1993. *Industrial Chemical Exposure.* New York: CRC Press/Lewis Publishers.

Lewis, R. J., Sr. 1993. *Hawley's Condensed Chemical Dictionary,* 12th Edition. New York: Van Nostrand Reinhold.

Lewis, R. J., Sr. 1992. *Sax's Dangerous Properties of Industrial Materials,* Volumes 1, 2, and 3, 8th Edition. New York: Van Nostrand Reinhold.

Lide, D. R. 1995. *Handbook of Chemistry and Physics,* 76th Edition. New York: CRC Press/Lewis Publishers.

Manahan, S. E. 1994. *Environmental Chemistry,* 6th Edition. New York: CRC Press/Lewis Publishers.

Manahan, S. E. 1992. *Toxicological Chemistry,* 2nd Edition. New York: CRC Press/Lewis Publishers.

Mellström, G. A., J. E. Wahlberg, and H. I. Maibach. 1994. *Protective Gloves for Occupational Use.* New York: CRC Press/Lewis Publishers.

Meyer, E. 1990. *Chemistry of Hazardous Materials.* Englewood Cliffs, NJ: Prentice Hall.

Patnaik, P. 1992. *A Comprehensive Guide to the Hazardous Properties of Chemical Substances.* New York: Van Nostrand Reinhold.

Philp, R. B. 1995. *Environmental Hazards and Human Health.* New York: CRC Press/Lewis Publishers.

Proctor, N. H., and J. P. Hughes. 1991. *Chemical Hazards of the Workplace.* New York: Van Nostrand Reinhold.

Quigley, D. R. 1994. *Handbook of Emergency Chemical Management.* New York: CRC Press/Lewis Publishers.

Rea, W. J. 1995. *Chemical Sensitivity,* Vol. I-III. New York: CRC Press/Lewis Publishers.

Richardson, M. 1995. *Dictionary of Substance and Their Effects,* Vol. I-VI. New York: CRC Press/Lewis Publishers.

Sacarello, H. L. A. 1994. *The Comprehensive Handbook of Hazardous Materials.* New York: CRC Press/Lewis Publishers.

Sheftel, V. O. 1995. *Handbook of Toxic Properties of Monomers and Additives.* New York: CRC Press/Lewis Publishers.

Sherman, J. D. 1988. *Chemical Exposure and Disease.* New York: Van Nostrand Reinhold.

Talbott, E., G. Craun. 1995. *An Introduction to Environmental Epidemiology.* New York: CRC Press/Lewis Publishers.

The Merck Index, 10th edition. 1983. Rahway, NJ: Merck & Company.

Torkelson, T. R. V. K. Rowe. 1981. *Patty's Industrial Hygiene and Toxicology,* Vol. 2B. New York: Wiley-Interscience.

U.S. Department of Health and Human Services, National Institute for Occupational Safety and Health. 1994. *NIOSH Pocket Guide to Chemical Hazards.* Washington, D.C.: U.S. Government Printing Office.

Witorsch, P. and S. V. Spagnolo. 1994. *Air Pollution and Lung Disease in Adults.* New York: CRC Press/Lewis Publishers.

MATERIAL SAFETY DATA SHEET

CHEMICAL NAME	
ETHANOLAMINE	**CRC PRESS/LEWIS PUBLISHERS DISCLAIMER:** THE INFORMATION AND RECOMMENDATIONS PRESENTED HEREIN ARE BASED ON SOURCES BELIEVED TO BE RELIABLE. CRC MAKES NO REPRESENTATION ON ITS COMPLETENESS OR ACCURACY. IT IS THE RESPONSIBILITY OF THE USER TO DETERMINE THE CHEMICAL'S SUITABILITY FOR ITS INTENDED USE, THE CHEMICAL'S SAFE USE, AND THE CHEMICAL'S PROPER DISPOSAL. NO REPRESENTATIONS AND/OR WARRANTIES, EITHER EXPRESSED OR IMPLIED, OF THE MERCHANTABILITY OR FITNESS FOR A PARTICULAR PURPOSE, OR OF ANY OTHER NATURE, ARE MADE WITH RESPECT TO THE INFORMATION PROVIDED IN THIS MSDS OR TO THE CHEMICAL TO WHICH INFORMATION MAY REFER. CRC NEITHER ASSUMES NOR AUTHORIZES ANY OTHER PERSON TO ASSUME FOR IT, ANY OTHER ADDITIONAL RESPONSIBILITY OR LIABILITY FOR THE USE OF, OR RELIANCE UPON, THIS INFORMATION.

HAZARD WARNING INFORMATION

HEALTH	FIRE	REACTIVE	OTHER	DEGREE OF HAZARD	COLOR CODING	OTHER CODES
2	2	0		0 = Minimum Hazard 1 = Slight Hazard 2 = Moderate Hazard 3 = Serious Hazard 4 = Severe Hazard	HEALTH = BLUE FIRE = RED REACTIVITY = YELLOW OTHER = WHITE	OX = Oxidizer ACID = Acid ALK = Alkali COR = Corrosive W = Use No Water

SECTION I - GENERAL INFORMATION

Characterization	RCRA Number	EPA Class
Aliphatic Amine	**None**	**Not Applicable**

DOT Proper Shipping Name **Ethanolamine (solution)**	Chemical Abstract Service (CAS) Number **141-43-5**
DOT Hazard Class and Label Requirements **Corrosive Material**	DOT Emergency Guide Code **60**
DOT Identification Number **UN 2491**	Chemical Formula **$NH_2CH_2CH_2OH$**

Synonyms
2-Aminoethanol; β-aminoethyl alcohol; ethylolamine; 2-hydroxyethylamine; monoethanolamine.

SECTION II - HAZARDOUS INGREDIENTS AND IDENTITY INFORMATION

Hazardous Components (specific identity)	OSHA Exposure Criteria	NIOSH Exposure Criteria	Immediately Dangerous to Life and Health (IDLH)	ACGIH Exposure Criteria
Ethanolamine (derivation: By reaction of ammonia with ethylene oxide to form mono, di, and tri-ethanolamines and separated by fractionation). **1 ppm = 2.54 mg/m^3**	**PEL (8-hour):** **3 ppm** **8 mg/m^3** **STEL:** **6 ppm** **15 mg/m^3**	**REL (10-hour):** **3 ppm** **8 mg/m^3** **STEL:** **6 ppm** **15 mg/m^3**	**30 ppm**	**TLV:** **3 ppm** **8 mg/m^3** **STEL:** **6 ppm** **15 mg/m^3**

SECTION III - PHYSICAL AND CHEMICAL CHARACTERISTICS

Boiling Point **339°F (170°C)**	Specific Gravity (H_2O – 1) **1.02**
Vapor Pressure (mm Hg) **0.4 at 69°F (20°C)**	Molecular Weight **61.1**
Vapor Density (Air = 1) **2.1**	Melting Point **51°F (10°C)**

Solubility
Completely soluble in water. Soluble in alcohol and acetone.

Appearance and Odor
Colorless, viscous, hygroscopic liquid or solid (below 51°F) with an unpleasant, ammonia-like odor.

SECTION IV - FIRE AND EXPLOSION HAZARD DATA

Flash Point (method used) **185°F (85°C) closed cup**	Explosive Limits in Air % by Volume **LEL: 3.0% UEL: 23.5%**
NFPA Classification **Class IIIA Combustible Liquid**	Autoignition Temperature **1436°F (780°C)**

Extinguishing Media
Use dry chemical, carbon dioxide, or "alcohol" foam. Do NOT use direct stream of water (may scatter or spread the fire).

Special Fire Fighting Procedures
Poisonous gases are produced in fire, wear full protective clothing and self-contained breathing apparatus (SCBA). Move containers from fire area if it can be done safely. Use water spray to cool fire-exposed containers. Do not allow run-off from fire control methods to reach sewers or waterways.

Unusual Fire and Explosion Hazards
Moderate fire and explosion hazard. Vapors form explosive mixtures in air. Vapors are slightly heavier than air and can travel for great distances to an ignition source to flashback causing fire or explosion.

SECTION V - REACTIVITY DATA

Stability		Conditions to Avoid Ethanolamine is normally stable at room temperature in closed containers under normal conditions of storage and handling. Avoid contact with incompatible materials. Keep dry and away from heat and flame.
Stable X	Unstable	Incompatibility *(materials to avoid)* Strong oxidizers (chlorine, bromine, fluorine), strong acids (sulfuric, nitric). It will dissolve and strip some paints and varnishes. It will attack some forms of plastics, rubber, coatings, copper, and brass.
Hazardous Polymerization		Conditions to Avoid Under normal conditions of temperature and pressure, hazardous polymerization of ethanolamine is not expected to occur.
May Occur	Will Not Occur X	Hazardous Decomposition or By-products Toxic and irritating fumes and vapors are produced when ethanolamine is heated to decomposition. These include toxic oxides of nitrogen and carbon.

SECTION VI - HEALTH HAZARD DATA

Primary Route(s) of Entry:	Inhalation?	Absorption (skin)?	Ingestion?
	possible but not likely		X

Health Hazards

INHALATION: Vapors are irritating to the eyes, nose, and upper respiratory tract. It can also cause depression of the central nervous system (CNS) with symptoms of lethargy (drowsiness or indifference. The extremely unpleasant odor makes this an unlikely exposure route.

SKIN & EYES: Moderately irritating to the skin causing redness and swelling. Severely irritating to the eyes causing redness and possible burns.

INGESTION: May damage the gastrointestinal tract.

Carcinogenicity	NTP Listed?	IARC Cancer Review Group?	OSHA Regulated?	Target Organs?
Unknown Human Unknown Animal	No	No	29 CFR 1910.1000 Table Z-1	Eyes, skin, respiratory system, CNS

Medical Conditions Generally Aggravated by Exposure

None Reported.

Emergency and First-aid Procedures

Eye contact: Flush immediately with water for 15 minutes (minimum), seek medical attention. Skin contact: Remove all contaminated clothing. Immediately wash area with large amounts of soap and water. Seek medical assistance. For inhalation: Remove the person from exposure. Provide respiratory assistance and CPR. Transfer to medical facility. If swallowed: Call the poison control center and seek medical attention immediately. Do NOT induce vomiting. Never attempt to give an unconscious or convulsing person anything by mouth.

SECTION VII - PRECAUTIONS FOR SAFE HANDLING AND USE

Steps to be Taken in Case Material is Released or Spilled

Absorb liquids in vermiculite or other absorbent and deposit in DOT-approved drum. Ventilate area and remove ignition sources. Restrict those not involved in cleanup from entering area. Neutralize residues or small spills with sodium bisulfite and flush to a drain with copious amounts of water. Notify appropriate authorities, if applicable.

Preferred Waste Disposal Method

Mix with a flammable solvent and burn in a chemical incinerator (with afterburner and scrubber).

Precautions to be Taken in Handling and Storage

Store in tightly closed containers in a cool, dry well-ventilated area (preferably at 65-95°F/18-35°C). Protect containers from physical damage. Keep away from incompatible materials. Ground and bond metal containers and equipment when making transfers (prevent static sparks).

Other Precautions and Warnings

Empty containers may still contain hazardous residues and should be handled accordingly.

SECTION VIII - CONTROL MEASURES AND PERSONAL PROTECTIVE EQUIPMENT

Respiratory Protection *(specify type)*

Corrosive to lung tissue and mucus membranes. Use a supplied-air respirator with full facepiece operated in positive pressure mode or a self-contained breathing apparatus (SCBA) with full facepiece operated in pressure demand or other positive pressure mode.

Ventilation

Local exhaust or general mechanical systems recommended.

Protective Gloves	Eye Protection	Other Protective Clothing
Impervious Gloves (not rubber)	Chemical Goggles or Face Mask	Protective Apron or Clothing

Work/Hygiene Practices

Always wash hands thoroughly after using chemical; never bring food, drink, or smoking materials into vicinity of chemicals.

ETHANOLAMINE

$NH_2CH_2CH_2OH$ CAS: 141-43-5

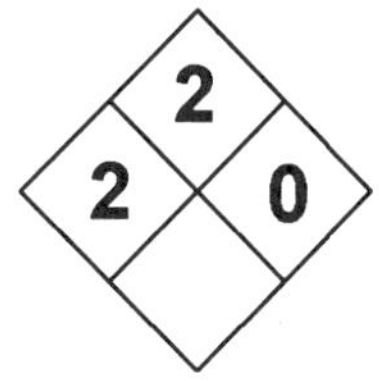

IDENTIFICATION AND TYPICAL USES

Ethanolamine is a colorless, viscous, hygroscopic liquid (or solid below 51°F) with an extremely unpleasant, ammonia-like odor. It is used as a dispersing agent for agricultural chemicals, in the synthesis of surface active agents, as a softening agent for hides, and in emulsifiers, polishes, and hair solutions.

RISK ASSESSMENT: HEALTH

General Assessment

Ethanolamine is moderately toxic by *ingestion* and skin contact (*absorption*) has been reported as a route of entry into the body. While *inhalation* is certainly a probable exposure risk, the extremely unpleasant odor of ethanolamine makes overexposure by inhalation an unlikely event. There is no information in the references on the carcinogenic or teratogenic properties of ethanolamine. However, human mutation data have been reported. Some scientists believe that mutagenic chemicals may have carcinogenic effects in the long-term. Additional study is required in this area. Lung, kidney, and liver injuries can also occur following prolonged or repeated exposures.

Inhalation of ethanolamine vapors causes irritation to the eyes, nose, throat, and upper respiratory tract. Excessive exposure to the vapor lead to depression of the central nervous system (CNS) with symptoms of lethargy (drowsiness, indifference).

Skin contact will cause moderate irritation with redness and swelling. Ethanolamine can be absorbed through the skin in toxic amounts. Eye contact results in severe irritation and may result in burns and tissue damage.

Ingestion can occur through careless working conditions, accidents, and, improper hygienic practices. It causes damage to the gastrointestinal tract and may lead to systemic poisoning.

Acute Health Effects

The following acute (short-term) health effects may occur immediately or shortly after exposure to ethanolamine:

Skin: Mild to moderate irritation. May cause burns with redness and swelling. Absorption can also occur.

Eye: Severe irritation, inflammation, with possible damage to vision.

Lung: Inhalation causes irritation of the eyes, nose, and upper respiratory tract with distressed breathing, coughing, congestion, and tearing.

CNS: Depression of the central nervous system can occur with lethargic symptoms, fatigue, somnolence, and dizziness.

Chronic Health Effects

The following chronic (long-term) health effects may occur at some time after exposure to ethanolamine and can last for months or even years:

Cancer Hazards: According to the information presented in the references, ethanolamine has not been adequately tested for its ability to cause cancer in test animals. Human mutation data have been reported. More research is required.

Reproductive Hazard: According to the information presented in the references, ethanolamine has not been adequately tested for its ability to adversely affect reproduction.

Other Chronic Effects: Prolonged exposures may lead to lung, liver, and/or kidney damage.

Recommended Risk-Reduction Measures

Unless a less toxic chemical can be substituted for a hazardous substance, *engineering controls* are the most

effective method of reducing exposures. The best protection is to enclose operations and/or provide local exhaust ventilation at the site of chemical release. To prevent static sparks, all containers and equipment should be electrically grounded and bonded. While not always operationally feasible, isolating operations can also reduce exposure.

Using respiratory protection is less effective than the controls mentioned above, but is still advisable whenever working with or around ethanolamine. For relatively low, infrequent, or transient exposures, an organic vapor respirator equipped with a cartridge protecting against amines may suffice. For higher or prolonged exposure, a supplied-air respirator with full facepiece operated in positive pressure mode or a self-contained breathing apparatus (SCBA) with full facepiece and operated in pressure demand are the recommended respiratory protection methods of choice. If a full facepiece is not available, then chemical goggles should be worn to protect the eyes. Whenever a chemical splash hazard exists, a face shield and apron should be worn. To prevent hand and skin exposures, impervious chemical gloves should be worn. Caution should be used when selecting gloves. Ethanolamine will attack some forms of rubber. Glove manufacturers should be contacted for permeation studies before selecting a protective glove.

Administrative controls should also be in place to minimize the potential for human exposures. These may include written procedures or policies which specify the methods and techniques that will be practiced whenever personnel are to work with ethanolamine.

All personnel should receive *training* on the use, hazards, protective measures, emergency actions, and other precautions per 29 CFR 1910.1200 (Hazard Communication), prior to the first assignment in an area where ethanolamine is used or stored.

Before beginning employment and at regular intervals thereafter (e.g. annually), the following medical tests are recommended:

☑ Lung function tests (establish a baseline).

If symptoms develop or overexposure is suspected, the following medical tests may be useful:

☑ Lung function tests.

☑ Liver and kidney function tests.

Any evaluation should include a careful medical history of past and present symptoms with an examination. However, medical tests that evaluate existing damage are not a substitute for controlling exposure.

Other methods to reduce exposure include:

☑ Where possible, enclose operations and use local exhaust ventilation at the site of chemical release. If local exhaust ventilation or enclosure is not used, respiratory protection should be mandatory. Also, if possible, automatically transfer liquids from storage containers to process containers using non-sparking tools.

☑ Always ensure that proper protective clothing is worn when using chemical substances.

☑ Wash thoroughly immediately after exposure to ethanolamine and at the end of the work shift or before eating, drinking, or smoking.

☑ Hazard warning information should be posted in the work area. In addition, as part of an on-going education and training program, all information on the health and safety hazards of ethanolamine should be communicated to all exposed workers.

☑ Eye wash stations should be provided in work areas. If the potential for whole body exposure exists, then safety showers should also be provided.

RISK ASSESSMENT: ENVIRONMENT

General Assessment

The environment is at risk of exposure during intended use, transportation, storage, disposal, or destruction of ethanolamine. In almost every scenario, the threat of environmental exposure is contingent upon the proper handling of the chemical substance. Accidental spills, large or small, can result in fire, explosion, and possible contamination of the surrounding environmental mediums (water, soil, and air).

Ethanolamine is considered a class IIIA combustible liquid (per OSHA 29 CFR 1910.106). While this means it poses a minimum fire risk by itself, mixture or contact with incompatible materials can cause fire or explosion. It can react with many common oxidizing materials and strong acids. It is corrosive to copper, brass, galvanized iron, and rubber. Its vapors can form explosive mixtures in air. These characteristics require special consideration during any emergency situation involving a leak or spill of ethanolamine. Should ethanolamine ever come into contact with incompatible substances either during use, transportation, or storage, violent reactions can occur.

Ethanolamine can enter the environment through its intended use as a dispersing agent for agricultural chemicals, as well as through industrial discharges and spills.

Acute Ecological Effects

Acute (short-term) toxic effects may include the death of animals, birds, or fish and death or low growth rate in plants. Acute effects are seen 2 to 4 days after animals or plants are exposed to ethanolamine. Insufficient data are available to evaluate or predict the short-term effects of ethanolamine to aquatic life, plants, birds, or land animals.

Chronic Ecological Effects

Chronic toxic effects may include shortened life span, reproductive problems, lower fertility, and changes in appearance or behavior in exposed animals. These effects can be seen long after first exposure(s) to toxic chemicals. Insufficient data are available to evaluate or predict the long-term effects of ethanolamine to aquatic life, plants, birds, or land animals.

Water Solubility

Ethanolamine is highly soluble in water. Concentrations of 1000 milligrams and more will mix with a liter of water.

Persistence in the Environment

Ethanolamine is slightly persistent in water, with a half-life of between 2 to 20 days. The half-life of a pollutant is the amount of time it takes for one half of the chemical to be degraded.

Bioaccumulation in Aquatic Organisms

Some substances increase in concentration, or *bioaccumulate,* in living organisms as they breathe contaminated air, drink contaminated water, or eat contaminated food. These chemicals can become concentrated in the tissues and internal organs of animals as well as humans.

Ethanolamine is not expected to accumulate in fish tissue.

Recommended Risk-Reduction Measures

Proper training of all transporters will reduce the likelihood of a mishap or accident resulting in a leak or spill to the environment. The correct use of DOT labeling (per the MSDS) on all containers, trucks, and rail cars will enable emergency responders to react properly and quickly to any disaster thereby reducing the potential risk to the environment and to personnel.

Storage of ethanolamine should be segregated from incompatible chemicals to minimize the risk of cross-contamination or contact. Storage buildings should be equipped with the proper fire protection equipment (alarms, sprinklers, extinguishers, etc.).

If a spill or leak to the environment has occurred, fire department, emergency response and/or hazardous materials spill personnel should be notified immediately. Cleanup should be attempted only by those trained in proper spill containment procedures. Contaminated soils should therefore be removed for incineration and replaced with clean soil. If ethanolamine should contact the water table, aquifer, or navigable waterway, remediation activities should be prompt. Time is of the essence since it is highly soluble and total containment and remediation may not be possible. When such spills occur, the local and/or state emergency response authorities must be notified. A comprehensive emergency response or disaster preparedness/recovery plan should be in place prior to any operations involving the use, transportation, storage, or disposal of ethanolamine. Ethanolamine may be disposed of by placing it in a chemical incinerator equipped with an afterburner and scrubber.

If ethanolamine is spilled or leaked, the following specific steps are recommended:

- ☑ Evacuate area and deny entry to those not involved in cleanup activities. Restrict persons not wearing protective clothing from area of spill or leak until cleanup is complete.
- ☑ Ventilate area of leak, remove ignition sources.
- ☑ Absorb liquids using vermiculite, dry earth, or sand and place in a sealed drum for disposal. Trace residues or small spills should be neutralized with sodium bisulfite and then flushed to a drain using copious amounts of water.
- ☑ It may be necessary to dispose of ethanolamine as a hazardous waste. The responsible state agency or the regional office of the federal Environmental Protection Agency (EPA) should be contacted for specific recommendations.

RISK ASSESSMENT: BUSINESS

General Assessment

Accidents or mishaps involving ethanolamine can present a significant threat to business operations. The loss or damage of equipment or facilities can significantly affect fiscal viability. Lawsuits that may result from personnel injury/death, public exposures, and/or environmental contamination will also require a serious expenditure of resources. Media attention surrounding an injury, death, or environmental damage can also result in a loss of profits and loss of current as well as future business.

Recommended Risk-Reduction Measures

Company attorneys, safety and health professionals, and environmental specialists should be involved in the development of any procedures dealing with safety or emergency response. A company official should be pre-designated as a public relations officer with specific training in dealing with the press. Corporate plans and policies should be developed, approved, and implemented long before any need for such arises.

REFERENCES

American Conference of Governmental Industrial Hygienists. 1988. *Documentation of the Threshold Limit Values and Biological Exposure Indices,* 5th Edition (with updates). Cincinnati: ACGIH.

Anthony, C. P., and N. J. Kolthoff. 1971. *Textbook of Anatomy and Physiology.* St. Louis: C. V. Mosby Company.

Bloom, A. D. and F. J. DeSerres. 1995. *Ecotoxicity and Human Health.* New York: CRC Press/Lewis Publishers.

Bueche, J. R. 1972. *Principles of Physics,* 2nd Edition. New York: McGraw-Hill.

Calabrese, E. J., E. M. Kenyon. *Air Toxics and Risk Assessment.* New York: CRC Press/Lewis Publishers.

Calabrese, E. J. 1994. *Biological Effects of Low Level Exposures.* New York: CRC Press/Lewis Publishers.

Cockerman, L. G., B. S. Shane. 1994. *Basic Environmental Toxicology.* New York: CRC Press/Lewis Publishers.

Cothern, C. R., N. P. Ross. 1994. *Environmental Statistics, Assessment, and Forecasting.* New York: CRC Press/Lewis Publishers.

Cothern, R. R. 1993. *Comparative Environmental Risk Assessment.* New York: CRC Press/Lewis Publishers.

Cothern, C. R. 1995. *Handbook for Environmental Risk Decisionmaking.* New York: CRC Press/Lewis Publishers.

Danse, I. R. 1991. *Common Sense Toxics in the Workplace.* New York: Van Nostrand Reinhold.

Environmental Protection Agency. 1983. *Chemical Hazard Information Profiles.* Washington, D.C.: U.S. EPA.

Fodden, C. R., and J. L. Weddell. 1992. *Hazardous Materials Emergency Action Data.* New York: CRC Press/Lewis Publishers.

Forseberg, K., and L. H. Keith. 1993. *Chemical; Protective Clothing Permeation and Degradation Database.* Chelsea, MI: Lewis Publishers.

Gardner, R. W. 1994. *Chemical Intolerance.* New York: CRC Press/Lewis Publishers.

Hallenbeck, W. H. 1993. *Quantitative Risk Assessment for Environmental and Occupational Health,* 2nd Edition. New York: CRC Press/Lewis Publishers.

Hodgson, E., R. B. Mailman, and Janice E. Chambers. 1988. *Dictionary of Toxicology.* London: The MacMillian Company.

Howard, P. H. 1995. *Handbook of Environmental Fate and Exposures Data for Organic Chemicals,* Vol. I-IV. New York: CRC Press/Lewis Publishers.

Howard, P. H., et. al. 1991. *Handbook of Environmental Degradation Rates.* New York: CRC Press/Lewis Publishers.

Jones, F. E. 1994. *Toxic Organic Vapors in the Workplace.* New York: CRC Press/Lewis Publishers.

Landis, W. G., M. H. Yu. 1995. *Introduction to Environmental Toxicology.* New York: CRC Press/Lewis Publishers.

Langerman, N. 1994 *Precautionary Labels for Chemical Containers.* New York: CRC Press/Lewis Publishers.

Lauwerys, R. R., P. Hoet. 1993. *Industrial Chemical Exposure.* New York: CRC Press/Lewis Publishers.

Lewis, R. J., Sr. 1993. *Hawley's Condensed Chemical Dictionary,* 12th Edition. New York: Van Nostrand Reinhold.

Lewis, R. J., Sr. 1992. *Sax's Dangerous Properties of Industrial Materials,* Volumes 1, 2, and 3, 8th Edition. New York: Van Nostrand Reinhold.

Manahan, S. E. 1992. *Toxicological Chemistry,* 2nd Edition. New York: CRC Press/Lewis Publishers.

Mellström, G. A., J. E. Wahlberg, and H. I. Maibach. 1994. *Protective Gloves for Occupational Use.* New York: CRC Press/Lewis Publishers.

Meyer, E. 1990. *Chemistry of Hazardous Materials.* Englewood Cliffs, NJ: Prentice Hall.

Nelson, G. O. 1992. *Gas Mixtures.* New York: CRC Press/Lewis Publishers.

Quigley, D. R. 1994. *Handbook of Emergency Chemical Management.* New York: CRC Press/Lewis Publishers.

Rea, W. J. 1995. *Chemical Sensitivity,* Vol. I-III. New York: CRC Press/Lewis Publishers.

Richardson, M. 1995. *Dictionary of Substance and Their Effects,* Vol. I-VI. New York: CRC Press/Lewis Publishers.

Sacarello, H. L. A. 1994. *The Comprehensive Handbook of Hazardous Materials.* New York: CRC Press/Lewis Publishers.

U.S. Department of Health and Human Services, National Institute for Occupational Safety and Health. 1994. *NIOSH Pocket Guide to Chemical Hazards.* Washington, D.C.: U.S. Government Printing Office.

MATERIAL SAFETY DATA SHEET

CHEMICAL NAME	
ETHYL ACETATE	**CRC PRESS/LEWIS PUBLISHERS DISCLAIMER:** THE INFORMATION AND RECOMMENDATIONS PRESENTED HEREIN ARE BASED ON SOURCES BELIEVED TO BE RELIABLE. CRC MAKES NO REPRESENTATION ON ITS COMPLETENESS OR ACCURACY. IT IS THE RESPONSIBILITY OF THE USER TO DETERMINE THE CHEMICAL'S SUITABILITY FOR ITS INTENDED USE, THE CHEMICAL'S SAFE USE, AND THE CHEMICAL'S PROPER DISPOSAL. NO REPRESENTATIONS AND/OR WARRANTIES, EITHER EXPRESSED OR IMPLIED, OF THE MERCHANTABILITY OR FITNESS FOR A PARTICULAR PURPOSE, OR OF ANY OTHER NATURE, ARE MADE WITH RESPECT TO THE INFORMATION PROVIDED IN THIS MSDS OR TO THE CHEMICAL TO WHICH INFORMATION MAY REFER. CRC NEITHER ASSUMES NOR AUTHORIZES ANY OTHER PERSON TO ASSUME FOR IT, ANY OTHER ADDITIONAL RESPONSIBILITY OR LIABILITY FOR THE USE OF, OR RELIANCE UPON, THIS INFORMATION.

HAZARD WARNING INFORMATION

HEALTH	FIRE	REACTIVE	OTHER	DEGREE OF HAZARD	COLOR CODING	OTHER CODES
1	3	0		0 = Minimum Hazard 1 = Slight Hazard 2 = Moderate Hazard 3 = Serious Hazard 4 = Severe Hazard	HEALTH = BLUE FIRE = RED REACTIVITY = YELLOW OTHER = WHITE	OX = Oxidizer ACID = Acid ALK = Alkali COR = Corrosive W = Use No Water

SECTION I - GENERAL INFORMATION

Characterization	RCRA Number	EPA Class
Aliphatic Ester	**U112**	**Toxic Waste**

DOT Proper Shipping Name **Ethyl Acetate**	Chemical Abstract Service (CAS) Number **141-78-6**
DOT Hazard Class and Label Requirements **Flammable Liquid**	DOT Emergency Guide Code **26**
DOT Identification Number **UN 1173**	Chemical Formula $CH_3COOC_2H_5$

Synonyms

Acetic ester; acetic ether; ethyl ester of acetic acid; ethyl ethanoate; vinegar naphtha.

SECTION II - HAZARDOUS INGREDIENTS AND IDENTITY INFORMATION

Hazardous Components (specific identity)	OSHA Exposure Criteria	NIOSH Exposure Criteria	Immediately Dangerous to Life and Health (IDLH)	ACGIH Exposure Criteria
Ethyl acetate (derivation: By heating acetic acid and ethyl alcohol in the presence of sulfuric acid and distilling). **1 ppm = 3.66 mg/m^3**	**PEL:** **400 ppm** **1400 mg/m^3** **STEL:** **Not Determined**	**REL:** **400 ppm** **1400 mg/m^3** **STEL:** **Not Determined**	**2000 ppm**	**TLV:** **400 ppm** **1400 mg/m^3** **STEL:** **Not Determined**

SECTION III - PHYSICAL AND CHEMICAL CHARACTERISTICS

Boiling Point **171°F (77°C)**	Specific Gravity (H_2O = 1) **0.902**
Vapor Pressure (mm Hg) **73 at 96°F (35.5°C)**	Molecular Weight **88.1**
Vapor Density (Air = 1) **3.0**	Melting Point **-117°F (-83°C)**

Solubility

Somewhat soluble in water (10% at 20°C). Soluble in chloroform, alcohol, and ether.

Appearance and Odor

Clear colorless liquid with an strong ether-like, fruity odor. Odor Threshold = 50 ppm.

SECTION IV - FIRE AND EXPLOSION HAZARD DATA

Flash Point (method used) **24°F (-4.4°C) closed cup**	Explosive Limits in Air % by Volume **LEL: 2.0%** **UEL: 11.5%**
NFPA Classification **Class IB Flammable Liquid**	Autoignition Temperature **800°F (427°C)**

Extinguishing Media

Dry chemical, alcohol foam, carbon dioxide, water fog. Use water to keep fire-exposed containers cool.

Special Fire Fighting Procedures

Poisonous gases are produced in fire. Wear full protective clothing and self-contained breathing apparatus (SCBA). Use blanketing effect with foam to smother large fires. Continue to cool containers with water after fire is extinguished. Water on fire itself may be ineffective and actually spread fire further.

Unusual Fire and Explosion Hazards

Immediately withdraw if rising sound from venting device is heard or if fire is causing discoloration to the tank. Evacuate 1500 feet radius if fire becomes uncontrollable or container is exposed to direct flame. Vapors are heavier than air and may travel a distances to ignition source to flashback.

SECTION V - REACTIVITY DATA

Stability		Conditions to Avoid **Ethyl acetate is stable under normal conditions of storage and operation. Avoid contact with incompatible materials. Keep away from heat, flame, and other sources of ignition.**
Stable **X**	Unstable	Incompatibility *(materials to avoid)* **Reacts violently in contact with nitrates, strong oxidizers, alkalis, and acids. Particular reactive with chlorosulfonic acid, lithium aluminum hydride and 2-chloromethyl furan, oleum, and potassium t-butoxide.**
Hazardous Polymerization		Conditions to Avoid **Under normal conditions of temperature and pressure, hazardous polymerization of ethyl acetate is not expected to occur. It will attack some forms of plastics, rubber, and coatings.**
May Occur	Will Not Occur **X**	Hazardous Decomposition or By-products **When heated to decomposition, ethyl acetate emits acrid and irritating fumes.**

SECTION VI - HEALTH HAZARD DATA

Primary Route(s) of Entry:	Inhalation?	Absorption (skin)?	Ingestion?
	X		X

Health Hazards

INHALATION: Irritation of the eyes, nose, throat, and respiratory tract. Exposures above 400 ppm may cause narcosis and headache. Systemic effects include olfactory changes, conjunctiva irritation, and pulmonary changes. It is irritating to the mucosa, particularly of the eyes, gums, and respiratory passages. Can also cause corneal clouding.

SKIN & EYES: Will irritate the skin. May cause contact dermatitis. Eye contact may cause corneal damage with cloudy swelling, and fatty degeneration of the viscera.

INGESTION: May cause headache, drowsiness, and unconsciousness.

Carcinogenicity	NTP Listed?	IARC Cancer Review Group?	OSHA Regulated?	Target Organs?
Unknown Human Unknown Animal	No	No	29 CFR 1910.1000 Table Z-1	Respiratory system; skin; eyes, CNS.

Medical Conditions Generally Aggravated by Exposure

None reported.

Emergency and First-aid Procedures

Eye contact: Flush immediately with water for 15 minutes (minimum); seek medical attention. **Skin contact:** Remove all contaminated clothing. Immediately wash area with large amounts of soap and water. Seek medical assistance. For **inhalation:** Remove the person from exposure. Provide respiratory assistance and CPR. Transfer to medical facility. If **swallowed:** Seek medical attention immediately. Call poison control center for advice. Give 1-2 glasses of milk or water. Do NOT induce vomiting unless advised otherwise. Never give an unconscious or convulsing person anything by mouth.

SECTION VII - PRECAUTIONS FOR SAFE HANDLING AND USE

Steps to be Taken in Case Material is Released or Spilled

Contain spills using absorbent material (vermiculite, or other). Ventilate area of spill. Have water source available in case of fire. Store materials in DOT-approved containers. Restrict those not involved in cleanup from entering area. Notify appropriate authorities, as required.

Preferred Waste Disposal Method

Incineration in chemical incinerator equipped with scrubber and afterburner.

Precautions to be Taken in Handling and Storage

Store in tightly closed containers in a cool, dark, well-ventilated area. Protect containers from physical damage. Keep heat, fire and flame away. If possible, automatically transfer liquids between containers.

Other Precautions and Warnings

Bulk storage of ethyl acetate is not recommended. Ground and bond all metal containers.

SECTION VIII - CONTROL MEASURES AND PERSONAL PROTECTIVE EQUIPMENT

Respiratory Protection *(specify type)*

Where potential exists for exposure over 400 ppm, use an MSHA/NIOSH-approved full facepiece respirator with an organic vapor cartridge. For high exposures, use a self-contained breathing apparatus (SCBA) with full facepiece operated in pressure demand or other positive pressure mode.

Ventilation

Local exhaust or general mechanical systems recommended.

Protective Gloves	Eye Protection	Other Protective Clothing
Impervious Gloves (avoid rubber)	Chemical Goggles or Face Mask	Impermeable Apron

Work/Hygiene Practices

Always wash hands thoroughly after using chemical; never bring food, drink, or smoking materials into vicinity of chemicals.

ETHYL ACETATE

$CH_3COOC_2H_5$ CAS: 141-78-6

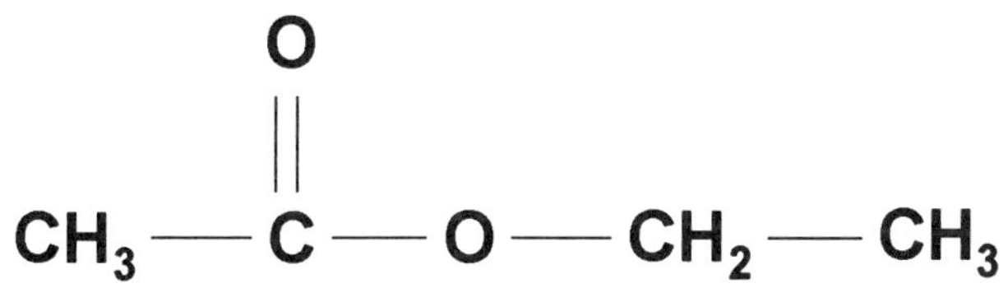

IDENTIFICATION AND TYPICAL USES

Ethyl acetate is a colorless liquid with a characteristic, fruity, ether-like odor. It is used as a solvent for varnishes, lacquers, and nitrocellulose. It is also used in cleaning textiles, in the leather industry, and in the manufacture of artificial silk. It is found in some perfumes, is often used as an artificial fruit flavoring, and is used in photographic films and plates.

RISK ASSESSMENT: HEALTH

General Assessment

Ethyl acetate is a human poison by *inhalation* causing toxic systemic effects. It is mildly toxic by *ingestion*. Skin contact can produce localized effects but dermal absorption has not been reported. While its specific carcinogenic and teratogenic properties are unknown, mutation data in animal studies have been documented for ethyl acetate.

Inhalation of ethyl acetate vapors is not likely to occur under normal operating conditions. The extremely irritating properties of the vapor will cause personnel to vacate the area before toxic exposure levels occur. However, should an exposure occur by inhalation, it will likely result in irritation of the eyes, nose, throat, and upper respiratory tract, including the mucosa (particularly the eyes, gums, and respiratory passages). Inhalation can cause changes in the sense of smell (olfactory changes), conjunctiva irritation, and pulmonary changes. In high concentrations, there can be depression of the central nervous system with mild narcotic effects.

Skin contact results in the removal of the skin's natural protective oily layer on its surface with subsequent drying, cracking, and possible secondary infections and dermatitis. Eye contact can cause serious damage to the cornea with a possibility for permanent effects to vision.

Ingestion of ethyl acetate may cause headache, drowsiness, unconsciousness, and unspecified gastrointestinal effects.

Acute Health Effects

The following acute (short-term) health effects may occur immediately or shortly after exposure to ethyl acetate:

Skin: Mild to moderate irritation. Will remove the skin's protective oily layer to cause drying, cracking, and possible secondary infection.

Eye: Severe irritation which can lead to injury and permanent damage if allowed to remain in contact with the eyes.

Lung: Irritation of the eyes, nose, throat and lungs (especially the mucus membranes). May cause changes in the sense of smell and in pulmonary function.

CNS: High concentrations can depress the central nervous system leading to narcoses and possible loss of consciousness.

Chronic Health Effects

The following chronic (long-term) health effects may occur at some time after exposure to ethyl acetate and can last for months or even years:

Cancer Hazards: According to the references, ethyl acetate has not been adequately tested for its ability to cause cancer in test animals. However, mutation data have been reported. This does not mean that ethyl acetate is carcinogenic. It does mean that further research is required and caution is warranted in use, handling, and storage.

Reproduction: According to the references, ethyl acetate has not been adequately tested for its ability to adversely affect reproduction in test animals.

Other Chronic Effects: Very high or prolonged exposure may lead to chronic poisoning with congestion of the liver and kidneys. There may be anemia, leukocytosis (transient increase in the white blood cell count), and cloudy swelling and fatty degeneration of the viscera.

Recommended Risk-Reduction Measures

Personnel should avoid direct contact with ethyl acetate. The best risk reduction measure is to use a less toxic chemical as a substitute for ethyl acetate. If this is not possible or feasible, then *engineering controls* are the most effective method of reducing exposures. The best protection is to enclose operations and/or provide local exhaust ventilation at the site of chemical release. While not always operationally feasible, isolating operations can also reduce exposure risk.

Using respiratory protection is less effective than the controls mentioned above, but is still advisable whenever working with or around ethyl acetate. For exposures over 400 ppm (8 hours), an MSHA/NIOSH-approved chemical respirator with an organic vapor cartridge should be used. For higher exposures, a self-contained breathing apparatus (SCBA) with full facepiece and operated in pressure demand mode or a supplied-air respirator with full facepiece operated in pressure demand are the recommended respiratory protection methods of choice. If a full facepiece is not available, then chemical goggles should be worn to protect the eyes. Whenever a chemical splash hazard exists, a face shield and a protective apron should be worn. To prevent hand and skin exposures, impervious gloves should be used. Caution is warranted in glove selection since ethyl acetate will attack some rubber products. Glove manufacturers should be contacted to obtain permeation studies before gloves are selected.

Administrative controls should also be in place to minimize the potential for human exposures. These may include written procedures or policies which specify the methods and techniques that will be practiced whenever personnel are to work with ethyl acetate.

All personnel should receive *training* on the use, hazards, protective measures, emergency actions, and other precautions per 29 CFR 1910.1200 (Hazard Communication), prior to the first assignment in an area where ethyl acetate is used or stored.

If symptoms develop or overexposure is suspected, the following medical tests are recommended:

- ☑ Liver and kidney function tests.
- ☑ Lung function tests.
- ☑ Skin testing with dilute ethyl acetate to help diagnose allergy (performed by a qualified allergist).
- ☑ Complete blood count (CBC) with particular attention on leukocyte levels.

Any medical evaluation should include a careful history of past and present symptoms with an examination. Medical tests that look for existing damage are not a substitute for controlling exposures.

Other methods to reduce exposure include:

- ☑ Where possible, enclose operations and use local exhaust ventilation at the site of chemical release. If local exhaust ventilation or enclosure is not used, respiratory protection should be mandatory.
- ☑ Always ensure that proper protective clothing is worn when using chemical substances.
- ☑ Wash thoroughly immediately after exposure to ethyl acetate and at the end of the work shift or before eating, drinking, or smoking.
- ☑ Hazard warning information should be posted in the work area. In addition, as part of an on-going education and training program, all information on the health and safety hazards of ethyl acetate should be communicated to all potentially exposed workers.
- ☑ Never eat, drink, or smoke in areas were ethyl acetate is used, handled or stored.

RISK ASSESSMENT: ENVIRONMENT

General Assessment

The environment is at risk of exposure during transportation, storage, disposal, or destruction of ethyl acetate. In almost every scenario, the threat of environmental exposure is contingent upon the proper handling of the chemical substance.

Accidental spills, large or small, can result in fire, explosion, and possible contamination of the surrounding environmental mediums (water, soil, and air). Ethyl acetate is considered a class IB flammable liquid (according to OSHA 29 CFR 1910.106). Its low flash point and relatively low boiling point present a serious fire and explosion hazard concern. Also, because it is incompatible with a number of common materials, especially strong oxidizers and acids, contact can result in violent and explosive reactions. It can also form explosive mixtures in air and can ignite on contact with heat, fire, or sparks. These characteristics require special consideration during any emergency situation involving a leak or spill of ethyl acetate.

The proper disposal/destruction method for ethyl acetate is to burn it in a chemical incinerator equipped with an afterburner and air scrubber. Ethyl acetate can enter the environment through unchecked industrial discharges into effluents and through spills.

Acute Ecological Effects

Acute (short-term) toxic effects may include the death of animals, birds, or fish, and death or low growth rate in plants. Acute effects are seen 2 to 4 days after animals or plants are exposed to ethyl acetate. This chemical has moderate acute toxicity to aquatic life. Insufficient data are available to evaluate or predict the short-term effects of ethyl acetate to plants, birds, or terrestrial animals.

Chronic Ecological Effects

Chronic toxic effects may include shortened life span, reproductive problems, lower fertility, and changes in appearance or behavior in exposed animals. These effects can be seen long after first exposure(s) to toxic chemicals. Ethyl acetate has moderate chronic toxicity to aquatic life. Insufficient data are available to evaluate or predict the long-term effects of ethyl acetate to plants, birds, or land animals.

Water Solubility

Ethyl acetate is moderately soluble in water. Concentrations of 1 to 1000 milligrams will mix with a liter of water.

Persistence in the Environment

Ethyl acetate is slightly persistent in water, with a half-life of between 2 to 20 days. The half-life of a pollutant is the amount of time it takes for one-half of the chemical to be degraded.

Bioaccumulation in Aquatic Organisms

Some substances increase in concentration, or *bioaccumulate,* in living organisms as they breathe contaminated air, drink contaminated water, or eat contaminated food. These chemicals can become concentrated in the tissues and internal organs of animals as well as humans.

The concentration of ethyl acetate found in fish tissues is expected to be somewhat higher then the average concentration of ethyl acetate in water from which the fish was taken.

Recommended Risk-Reduction Measures

Proper training of all transporters will reduce the likelihood of a mishap or accident resulting in a leak or spill to the environment. The correct use of DOT labeling (per the MSDS) on all containers, trucks, and rail cars will enable emergency responders to react properly and quickly to any disaster thereby reducing the potential risk to the environment and to personnel.

Storage of ethyl acetate should be segregated from incompatible chemicals to minimize the risk of cross-contamination or contact. Buildings designated for storage should be equipped with appropriate fire protection systems (alarms, sprinklers, emergency lighting, portable extinguishers).

If a spill or leak to the environment has occurred, fire department, emergency response, and/or hazardous materials spill personnel should be notified immediately. Cleanup should be attempted only by those trained in proper spill containment procedures. Contaminated soils should be removed for incineration and replaced with clean soil.

If ethyl acetate should contact the water table, aquifer, or navigable waterway, time is of the essence. It is moderately soluble in water and, therefore, total containment and remediation may not be entirely possible. When such spills occur, the local and/or state emergency response authorities must be notified. A comprehensive emergency response or disaster preparedness/recovery plan should be in place prior to any operations involving the use, transportation, storage, or disposal of ethyl acetate. If ethyl acetate is spilled or leaked, the following specific steps are recommended:

- ☑ Restrict persons not wearing protective clothing from area of spill or leak until cleanup is complete and area can be opened for normal work.
- ☑ Ventilate area of spill or leak.
- ☑ Remove all ignition sources from area immediately.
- ☑ Absorb liquids in vermiculite, dry sand, earth, or a similar material and deposit in sealed containers. Use non-sparking tools. Avoid brass since ethyl acetate has been known to attack brass.
- ☑ It may be necessary to dispose of ethyl acetate as a hazardous waste. The responsible state agency or the regional office of the federal Environmental

Protection Agency (EPA) should be contacted for specific recommendations.

RISK ASSESSMENT: BUSINESS

General Assessment

Accidents or mishaps involving ethyl acetate can present a moderate threat to business operations. The loss or damage of equipment or facilities can significantly affect fiscal viability. Lawsuits that may result from personnel injury/death, public exposures, and/or environmental contamination will also require a serious expenditure of resources.

Media attention surrounding an injury, death, or environmental damage can also result in a loss of profits and loss of current as well as future business.

Recommended Risk-Reduction Measures

Company attorneys, safety and health professionals, and environmental specialists should be involved in the development of any procedures or policies intended to manage the use of chemicals in the workplace. A company official should be pre-designated as a public relations officer with specific training in dealing with the press. Corporate plans and policies should be developed, approved, and implemented long before any need for such arises.

REFERENCES

American Conference of Governmental Industrial Hygienists. 1988. *Documentation of the Threshold Limit Values and Biological Exposure Indices,* 5th Edition (with updates). Cincinnati: ACGIH.

Anthony, C. P., and N. J. Kolthoff. 1971. *Textbook of Anatomy and Physiology.* St. Louis: C. V. Mosby Company.

Bloom, A. D. and F. J. DeSerres. 1995. *Ecotoxicity and Human Health.* New York: CRC Press/Lewis Publishers.

Brusick, D. J. 1994. *Methods for Genetic Risk Assessment.* New York: CRC Press/Lewis Publishers.

Bueche, J. R. 1972. *Principles of Physics,* 2nd Edition. New York: McGraw-Hill.

Burrell, R., D. K. Flaherty, and L. J. Sauers. 1992. *Toxicology of the Immune System.* New York: Van Nostrand Reinhold.

Calabrese, E. J. 1994. *Biological Effects of Low Level Exposures.* New York: CRC Press/Lewis Publishers.

Cockerman, L. G., B. S. Shane. 1994. *Basic Environmental Toxicology.* New York: CRC Press/Lewis Publishers.

Cothern, C. R., N. P. Ross. 1994. *Environmental Statistics, Assessment, and Forecasting.* New York: CRC Press/Lewis Publishers.

Danse, I. R. 1991. *Common Sense Toxics in the Workplace.* New York: Van Nostrand Reinhold.

Environmental Protection Agency. 1983. *Chemical Hazard Information Profile: Ethyl acetate.* Washington, D.C.: U.S. EPA.

Gardner, R. W. 1994. *Chemical Intolerance.* New York: CRC Press/Lewis Publishers.

Hallenbeck, W. H. 1993. *Quantitative Risk Assessment for Environmental and Occupational Health,* 2nd Edition. New York: CRC Press/Lewis Publishers.

Hodgson, E., R. B. Mailman, and J. E. Chambers. 1988. *Dictionary of Toxicology.* London: The MacMillian Company.

Hosty, J. W., and P. Foster. 1990. *A Practical Guide to Chemical Spill Response.* New York: Van Nostrand Reinhold.

Howard, P. H., et. al. 1991. *Handbook of Environmental Degradation Rates.* New York: CRC Press/Lewis Publishers.

James, A., and M. Lord. 1992. *Index of Chemical and Physical Data.* New York: Van Nostrand Reinhold.

Jones, F. E. 1994. *Toxic Organic Vapors in the Workplace.* New York: CRC Press/Lewis Publishers.

Kamrin, M. 1 A. 1988. *Toxicology: A Primer on Principles and Applications.* Chelsea, Michigan: Lewis Publishers.

Keith, L. H. and M. M. Walker. 1995. *Handbook of Air Toxics.* New York: CRC Press/Lewis Publishers.

Landis, W. G., M. H. Yu. 1995. *Introduction to Environmental Toxicology.* New York: CRC Press/Lewis Publishers.

Langerman, N. 1994 *Precautionary Labels for Chemical Containers.* New York: CRC Press/Lewis Publishers.

Lauwerys, R. R., P. Hoet. 1993. *Industrial Chemical Exposure.* New York: CRC Press/Lewis Publishers.

Lide, D. R. 1995. *Handbook of Chemistry and Physics,* 76th Edition. New York: CRC Press/Lewis Publishers.

Lewis, Richard J., Sr. 1992. *Sax's Dangerous Properties of Industrial Materials,* Volumes 1, 2, and 3, 8th Edition. New York: Van Nostrand Reinhold.

Lisella, Frank S. 1994. *The VNR Dictionary of Environmental Health and Safety.* New York: Van Nostrand Reinhold.

Manahan, S. E. 1994. *Environmental Chemistry,* 6th Edition. New York: CRC Press/Lewis Publishers.

Manahan, S. E. 1992. *Toxicological Chemistry,* 2nd Edition. New York: CRC Press/Lewis Publishers.

Meyer, Eugene. 1990. *Chemistry of Hazardous Materials.* Englewood Cliffs, NJ: Prentice Hall.

Patnaik, P. 1992. *A Comprehensive Guide to the Hazardous Properties of Chemical Substances.* New York: Van Nostrand Reinhold.

Proctor, N. H., and J. P. Hughes. 1991. *Chemical Hazards of the Workplace.* New York: Van Nostrand Reinhold.

MATERIAL SAFETY DATA SHEET

CHEMICAL NAME	
ETHYL ACRYLATE	**CRC PRESS/LEWIS PUBLISHERS DISCLAIMER:** THE INFORMATION AND RECOMMENDATIONS PRESENTED HEREIN ARE BASED ON SOURCES BELIEVED TO BE RELIABLE. CRC MAKES NO REPRESENTATION ON ITS COMPLETENESS OR ACCURACY. IT IS THE RESPONSIBILITY OF THE USER TO DETERMINE THE CHEMICAL'S SUITABILITY FOR ITS INTENDED USE, THE CHEMICAL'S SAFE USE, AND THE CHEMICAL'S PROPER DISPOSAL. NO REPRESENTATIONS AND/OR WARRANTIES, EITHER EXPRESSED OR IMPLIED, OF THE MERCHANTABILITY OR FITNESS FOR A PARTICULAR PURPOSE, OR OF ANY OTHER NATURE, ARE MADE WITH RESPECT TO THE INFORMATION PROVIDED IN THIS MSDS OR TO THE CHEMICAL TO WHICH INFORMATION MAY REFER. CRC NEITHER ASSUMES NOR AUTHORIZES ANY OTHER PERSON TO ASSUME FOR IT, ANY OTHER ADDITIONAL RESPONSIBILITY OR LIABILITY FOR THE USE OF, OR RELIANCE UPON, THIS INFORMATION.

HAZARD WARNING INFORMATION

HEALTH	FIRE	REACTIVE	OTHER	DEGREE OF HAZARD	COLOR CODING	OTHER CODES
2	3	2		0 = Minimum Hazard 1 = Slight Hazard 2 = Moderate Hazard 3 = Serious Hazard 4 = Severe Hazard	HEALTH = BLUE FIRE = RED REACTIVITY = YELLOW OTHER = WHITE	OX = Oxidizer ACID = Acid ALK = Alkali COR = Corrosive ~~W~~ = Use No Water

SECTION I - GENERAL INFORMATION

Characterization	**Aliphatic Ester**	RCRA Number	**U113**
		EPA Class	**Toxic Waste**
DOT Proper Shipping Name	**Ethyl acrylate, Inhibited**	Chemical Abstract Service (CAS) Number	**140-88-5**
DOT Hazard Class and Label Requirements	**Flammable Liquid**	DOT Emergency Guide Code	**27 (Inhibited)**
DOT Identification Number	**UN 1917**	Chemical Formula	**CH_2:$CHCOOC_2H_5$**

Synonyms
Ethyl acrylate (inhibited); ethyl ester of acrylic acid; ethyl propenoate; acrylic acid ethyl ester.

SECTION II - HAZARDOUS INGREDIENTS AND IDENTITY INFORMATION

Hazardous Components (specific identity)	OSHA Exposure Criteria	NIOSH Exposure Criteria	Immediately Dangerous to Life and Health (IDLH)	ACGIH Exposure Criteria
Ethyl acrylate (derivation: From ethylene cyanohydrin, ethyl alcohol, and dilute sulfuric acid; Oxo reaction of acetylene, carbon monoxide, and ethyl alcohol in the presence of nickel or cobalt catalyst). **1 ppm = 4.16 mg/m^3**	**PEL (skin):** **25 ppm** **100 mg/m^3** **STEL:** **Not Determined**	**REL:** **Lowest Possible Level** **Possible Human Carcinogen**	**300 ppm**	**TLV:** **5 ppm** **20 mg/m^3** **15 ppm (STEL)** **Suspected Human Carcinogen**

SECTION III - PHYSICAL AND CHEMICAL CHARACTERISTICS

Boiling Point	**211°F (99.6°C)**	Specific Gravity (H_2O = 1)	**0.92**
Vapor Pressure (mm Hg)	**29 at 96°F (35.5°C)**	Molecular Weight	**100.1**
Vapor Density (Air = 1)	**3.45**	Melting Point	**-98°F (-72°C)**

Solubility
Slightly soluble in water (2% at 20°C). Soluble in chloroform, alcohol, and ether.

Appearance and Odor
Clear, colorless liquid with a sharp, acrid odor. Odor Threshold = 0.0012 ppm.

SECTION IV - FIRE AND EXPLOSION HAZARD DATA

Flash Point (method used)	**48°F (9°C) closed cup**	Explosive Limits in Air % by Volume	**LEL: 1.4%** **UEL: 14%**
NFPA Classification	**Class IB Flammable Liquid**	Autoignition Temperature	**721°F (383°C)**

Extinguishing Media
Dry chemical, alcohol foam, or carbon dioxide. Water may be ineffective and may spread fire.

Special Fire Fighting Procedures
Poisonous gases are produced in fire. Wear full protective clothing and self-contained breathing apparatus (SCBA). Use blanketing effect with foam to smother large fires. Continue to cool containers with water after fire is extinguished. Move containers from fire area if it can be done without risk.

Unusual Fire and Explosion Hazards
Evacuate 2500 feet radius if fire becomes uncontrollable or container is exposed to direct flame. Vapors are heavier than air and may travel a distances to ignition source to flashback. Heat and/or an insufficient concentration of inhibitor can cause violent polymerization with the rupture of container.

SECTION V - REACTIVITY DATA

Stability		Conditions to Avoid Ethyl acrylate is kept stable under normal conditions of storage and operation with an inhibitor. Avoid contact with incompatible materials. Keep away from heat, flame, and other sources of ignition.
Stable X	Unstable	Incompatibility *(materials to avoid)* Reacts violently in contact with oxidizers (peroxides, nitrates, perchlorates, chlorates, permanganates) polymerizers, strong alkalis (sodium hydroxide, potassium hydroxide), moisture, chlorosulfonic acid.
Hazardous Polymerization		Conditions to Avoid Hazardous polymerization can occur if the inhibitor levels become low or if ethyl acrylate is exposed to heat, sparks, or open flames. It can react by itself in closed containers with explosive polymerization.
May Occur X	Will Not Occur	Hazardous Decomposition or By-products When heated to decomposition, ethyl acrylate emits acrid and irritating fumes, including oxides of carbon (monoxide and dioxide).

SECTION VI - HEALTH HAZARD DATA

Primary Route(s) of Entry:	Inhalation?	Absorption (skin)?	Ingestion?
	X	X	X

Health Hazards

INHALATION: Severe irritation to the eyes, nose, throat, and respiratory system. Exposure can cause drowsiness, headache, vomiting, and nausea. Extreme exposures can cause the development of pulmonary edema (fluid in the lungs), which can be fatal. There may be dypsnea, cyanosis, convulsions, lachrymation, and laryngitis.

ABSORPTION: Severe irritation of the skin and eyes. Will pass through unbroken skin. May cause skin sensitization. Systemic effects include dypsnea, cyanosis, and convulsive movements.

INGESTION: Severe local irritation of the gastrointestinal tract and toxic degenerative changes of the cardiac (heart), hepatic (liver), renal (kidney, and splenic (spleen) tissues.

Carcinogenicity	NTP Listed?	IARC Cancer Review Group?	OSHA Regulated?	Target Organs?
Suspected Human Confirmed Animal	5th Annual Report	Group 2B	29 CFR 1910.1000 Table Z-1	Respiratory system; skin; eyes.

Medical Conditions Generally Aggravated by Exposure

None reported.

Emergency and First-aid Procedures

Eye contact: Flush immediately with water for 15 minutes (minimum); seek medical attention. Skin contact: Remove all contaminated clothing. Immediately wash area with large amounts of soap and water. Seek medical assistance. For inhalation: Remove the person from exposure. Provide respiratory assistance and CPR. Transfer to medical facility. Observe for 48 hours for lung effects. If swallowed: Seek medical attention immediately. Do NOT induce vomiting.

SECTION VII - PRECAUTIONS FOR SAFE HANDLING AND USE

Steps to be Taken in Case Material is Released or Spilled

Contain spills using absorbent material (vermiculite, or other). Ventilate area of spill. Have extinguishing agent available in case of fire. Store materials in DOT-approved containers. Restrict those not involved in cleanup from entering area. Notify appropriate authorities, as required.

Preferred Waste Disposal Method

Incineration in chemical incinerator equipped with scrubber and afterburner.

Precautions to be Taken in Handling and Storage

Store in tightly closed containers in a cool, dark, well-ventilated area. Protect containers from physical damage. Keep heat, fire and flame away. Outside building storage is recommended.

Other Precautions and Warnings

Bulk storage of ethyl acrylate is not recommended. Ground and bond all metal containers.

SECTION VIII - CONTROL MEASURES AND PERSONAL PROTECTIVE EQUIPMENT

Respiratory Protection *(specify type)*

For exposures to suspected carcinogens, use a supplied-air respirator or a self-contained breathing apparatus (SCBA) with full facepiece operated in pressure demand or other positive pressure mode.

Ventilation

Local exhaust or general mechanical systems recommended.

Protective Gloves	Eye Protection	Other Protective Clothing
Impervious Gloves	Chemical Goggles or Face Mask	Impermeable Apron or Protective Suit

Work/Hygiene Practices

Always wash hands thoroughly after using chemical; never bring food, drink, or smoking materials into vicinity of chemicals.

ETHYL ACRYLATE

$CH_2{:}CHCOOC_2H_5$ CAS: 140-88-5

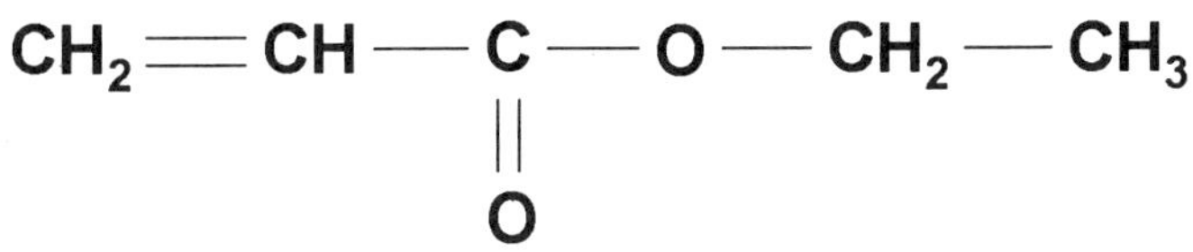

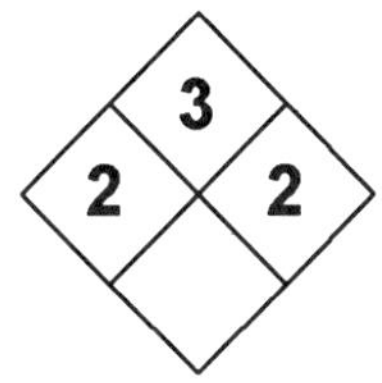

IDENTIFICATION AND TYPICAL USES

Ethyl acrylate is a colorless liquid with an acrid odor. It is used in the manufacture of acrylic resins, acrylic fibers, textile and paper coatings, adhesives, and leather finish resins. It is also sometimes used as a flavoring agent.

RISK ASSESSMENT: HEALTH

General Assessment

Ethyl acrylate is toxic to humans by *inhalation, ingestion,* and by skin contact (*absorption*). It is a confirmed animal carcinogen and its carcinogenicity in humans is considered highly suspect by the National Toxicology Program (NTP). Prolonged exposures may lead to congestion of the lungs and changes in the heart, kidney, and liver.

Inhalation results in irritation of the mucus membranes of the eyes, nose, throat, and respiratory system. It can cause changes in the sense of smell and in pulmonary function. In addition, there may be symptoms such as lethargy, cyanosis, and convulsive movements. High concentrations can cause dizziness with dypsnea, nausea, headache, vomiting, and can lead to death as a result of chemical pneumonitis or pulmonary edema (fluid in the lungs). Symptoms of edema can be delayed up to 48 hours following exposure which may create a false sense of security with regard to health exposure risk.

Skin contact can cause serious localized irritation at the site of contact with marked erythema, edema, thickening, and vascular damage. There may be 1st degree burns on short exposures and 2nd degree burns on prolonged exposures. Repeated exposures in some individuals may develop sensitivity. It will pass through intact skin to cause toxic systemic effects.

Ingestion of ethyl acrylate causes irritation of the gastrointestinal tract with subsequent absorption through the digestive process leading to toxic systemic effects (especially changes to liver tissues).

Acute Health Effects

The following acute (short-term) health effects may occur immediately or shortly after exposure to ethyl acrylate:

Skin: Serious irritation and possible 1st or 2nd degree burns. There can be marked localized irritation, erythema, edema, thickening, and vascular damage at the site of contact. It will pass through intact skin to cause toxic systemic effects.

Eye: Severe irritation which can lead to injury, burns, painful irritation, and permanent damage if allowed to remain in contact with the eyes.

Lung: Irritation of the eyes, nose, throat and lungs (especially the mucus membranes). May cause changes in the sense of smell and in pulmonary function. Can lead to chemical pneumonitis and/or pulmonary edema (fluid in the lungs).

CNS: High concentrations may cause dizziness, headache, nausea, vomiting, dypsnea, lethargy, and convulsive movements.

Chronic Health Effects

The following chronic (long-term) health effects may occur at some time after exposure to ethyl acrylate and can last for months or even years:

Cancer Hazards: According to the references, ethyl acrylate has been shown to cause cancer of the forestomach in test animals. Because of this and because of its ability to effect changes in the function of the liver, kidney, spleen, heart, and lungs, human carcinogenicity is considered highly suspect.

Reproduction: According to the references, ethyl acrylate has not been adequately tested for its ability to adversely affect reproduction in test animals.

Other Chronic Effects: Very high or prolonged exposure may lead to congestion of the lungs and degenerative changes in hepatic (liver), renal (kidney), cardiac (heart), and splenic (spleen) tissues. Prolonged contact may cause sensitization.

Recommended Risk-Reduction Measures

Because of its suspected human carcinogenicity, as well as its known ability to cause serious topical and systemic effects, personnel should avoid direct contact with ethyl acrylate. The best risk reduction measure is to use a less toxic chemical as a substitute for a ethyl acrylate. If this is not possible or feasible, then *engineering controls* are the most effective method of reducing exposures. The best protection is to enclose operations and/or provide local exhaust ventilation at the site of chemical release. While not always operationally feasible, isolating operations can also reduce exposure risk.

Using respiratory protection is less effective than the controls mentioned above, but is still advisable whenever working with or around ethyl acrylate. For any exposure to a suspected human carcinogen, a supplied-air respirator with full facepiece operated in positive pressure mode, or a self-contained breathing apparatus (SCBA) with full facepiece and operated in pressure demand mode are the recommended respiratory protection methods of choice. If a full facepiece is not available, then chemical goggles should be worn to protect the eyes. Whenever a chemical splash hazard exists, a face shield and a protective apron should be worn. To prevent hand and skin exposures, impervious gloves should be used. Glove manufacturers should be contacted to obtain permeation studies before gloves are selected.

Administrative controls should also be in place to minimize the potential for human exposures. These may include written procedures or policies which specify the methods and techniques that will be practiced whenever personnel are to work with ethyl acrylate.

All personnel should receive *training* on the use, hazards, protective measures, emergency actions, and other precautions per 29 CFR 1910.1200 (Hazard Communication), prior to the first assignment in an area where ethyl acrylate is used or stored.

Before beginning employment and at regular intervals thereafter (e.g., annually), those persons having frequent or potentially high exposures should be given the following medical tests:

- ☑ Lung function tests.

If symptoms develop or overexposure is suspected, the following medical tests are recommended:

- ☑ Liver and kidney function tests.
- ☑ Lung function tests.
- ☑ Skin testing with dilute ethyl acrylate to help diagnose allergy (performed by a qualified allergist).
- ☑ Consider chest X-ray following acute overexposure (may be negative if taken immediately after exposure).

Any medical evaluation should include a careful history of past and present symptoms with an examination. Medical tests that look for existing damage are not a substitute for controlling exposures. Also, since smoking can cause heart disease, emphysema, and other respiratory disorders, smokers exposed to ethyl acrylate may develop symptoms much quicker and with greater intensity than non-smokers under identical exposure conditions. Prudent risk management requires careful consideration of *all* possible factors which may be causing the appearance of exposure symptoms.

Other methods to reduce exposure include:

- ☑ Where possible, enclose operations and use local exhaust ventilation at the site of chemical release. If local exhaust ventilation or enclosure is not used, respiratory protection should be mandatory.
- ☑ Always ensure that proper protective clothing is worn when using chemical substances.
- ☑ Wash thoroughly immediately after exposure to ethyl acrylate and at the end of the work shift or before eating, drinking, or smoking.
- ☑ Hazard warning information should be posted in the work area. In addition, as part of an on-going education and training program, all information on the health and safety hazards of ethyl acrylate should be communicated to all potentially exposed workers.
- ☑ Never eat, drink, or smoke in areas were ethyl acrylate is used, handled or stored.
- ☑ Work clothing contaminated with ethyl acrylate should never be worn home where family members can be exposed. It should be laundered only by those who have been trained on the health risks associated with ethyl acrylate.

RISK ASSESSMENT: ENVIRONMENT

General Assessment

The environment is at risk of exposure during transportation, storage, disposal, or destruction of ethyl acrylate. In almost every scenario, the threat of environmental exposure is contingent upon the proper handling of the chemical substance. Accidental spills, large or small, can result in fire, explosion, and possible contamination of the surrounding environmental mediums (water, soil, and air).

Ethyl acrylate is considered a class IB flammable liquid (according to OSHA 29 CFR 1910.106). Its low flash point and relatively low boiling point present a serious fire and explosion hazard concern. Also, because it is incompatible with a number of common materials, especially strong oxidizers and acids, contact can result in violent and explosive reactions. It can also form explosive mixtures in air and can ignite on contact with heat, fire, or sparks. Heat can cause ethyl acrylate to react by itself. If this happens in a closed container, an explosion can occur. Ethyl acrylate usually contains an inhibitor such as hydroquinone or its methyl ether to prevent a self reaction. If the inhibitor is absent or present in too low a quantity, the explosive reaction may occur without the application of heat. These characteristics require special consideration during any emergency situation involving a leak or spill of ethyl acrylate.

Ethyl acrylate can enter the environment through unchecked industrial discharges into effluents and through spills.

Acute Ecological Effects

Acute (short-term) toxic effects may include the death of animals, birds, or fish, and death or low growth rate in plants. Acute effects are seen 2 to 4 days after animals or plants are exposed to ethyl acrylate. This chemical has moderate acute toxicity to aquatic life. Insufficient data are available to evaluate or predict the short-term effects of ethyl acrylate to plants, birds, or terrestrial animals.

Chronic Ecological Effects

Chronic toxic effects may include shortened life span, reproductive problems, lower fertility, and changes in appearance or behavior in exposed animals. These effects can be seen long after first exposure(s) to toxic chemicals. Ethyl acrylate has moderate chronic toxicity to aquatic life. Insufficient data are available to evaluate or predict the long-term effects of ethyl acrylate to plants, birds, or land animals.

Water Solubility

Ethyl acrylate is moderately soluble in water. Concentrations of 1 to 1000 milligrams will mix with a liter of water.

Persistence in the Environment

Ethyl acrylate is slightly persistent in water, with a half-life of between 2 to 20 days. The half-life of a pollutant is the amount of time it takes for one-half of the chemical to be degraded. About 91% of ethyl acrylate will eventually end up in the air; the remainder will end up in water.

Bioaccumulation in Aquatic Organisms

Some substances increase in concentration, or *bioaccumulate,* in living organisms as they breathe contaminated air, drink contaminated water, or eat contaminated food. These chemicals can become concentrated in the tissues and internal organs of animals as well as humans.

The concentration of ethyl acrylate found in fish tissues is expected to be about the same as the average concentration of ethyl acrylate in water from which the fish was taken.

Recommended Risk-Reduction Measures

Proper training of all transporters will reduce the likelihood of a mishap or accident resulting in a leak or spill to the environment. The correct use of DOT labeling (per the MSDS) on all containers, trucks, and rail cars will enable emergency responders to react properly and quickly to any disaster thereby reducing the potential risk to the environment and to personnel.

Storage of ethyl acrylate should be segregated from incompatible chemicals to minimize the risk of cross-contamination or contact. Buildings designated for storage should be equipped with appropriate fire protection systems (alarms, sprinklers, emergency lighting, portable extinguishers). Equipment should be designed to meet explosion-proof standards.

If a spill or leak to the environment has occurred, fire department, emergency response, and/or hazardous materials spill personnel should be notified immediately. Cleanup should be attempted only by those trained in proper spill containment procedures. Contaminated soils should be removed for incineration and replaced with clean soil. If ethyl acrylate should contact the water table, aquifer, or navigable waterway, time is of the essence. It is moderately soluble in water and, therefore, total containment and remediation may not be entirely possible. When such spills occur, the local and/or state emergency response authorities must be notified. A comprehensive emergency response or disaster preparedness/recovery plan should be in place prior to any operations involving the use, transportation, storage, or disposal of ethyl acrylate. If ethyl acrylate is spilled or leaked, the following specific steps are recommended:

- ☑ Restrict persons not wearing protective clothing from area of spill or leak until cleanup is complete.
- ☑ Ventilate area and remove ignition sources.
- ☑ Absorb liquids in vermiculite, dry sand, earth, or a similar material and deposit in sealed containers. Use non-sparking tools.
- ☑ It may be necessary to dispose of ethyl acrylate as a hazardous waste. The responsible state agency or the regional office of the federal Environmental Protection Agency (EPA) should be contacted for specific recommendations.

RISK ASSESSMENT: BUSINESS

General Assessment

Accidents or mishaps involving ethyl acrylate can present a moderate threat to business operations. The loss or damage of equipment or facilities can significantly affect fiscal viability. Lawsuits that may result from personnel injury/death, public exposures, and/or environmental contamination will also require a serious expenditure of resources. Media attention surrounding an injury, death, or environmental damage can also result in a loss of profits and loss of current as well as future business. Always remember that anytime the terms "cancer" or "carcinogen" are used, public emotion, hysteria, and ignorance can run equally high. This should be carefully considered when constructing a public relations policy.

❶ *Recommended Risk-Reduction Measures*

Company attorneys, safety and health professionals, and environmental specialists should be involved in the development of any procedures or policies intended to manage the use of chemicals in the workplace. A company official should be pre-designated as a public relations officer with specific training in dealing with the press. Corporate plans and policies should be developed, approved, and implemented long before any need for such arises.

REFERENCES

American Conference of Governmental Industrial Hygienists. 1988. *Documentation of the Threshold Limit Values and Biological Exposure Indices,* 5th Edition (with updates). Cincinnati: ACGIH

Anthony, C. P., and N. J. Kolthoff. 1971. *Textbook of Anatomy and Physiology.* St. Louis: C. V. Mosby Company.

Bloom, A. D. and F. J. DeSerres. 1995. *Ecotoxicity and Human Health.* New York: CRC Press/Lewis Publishers.

Brusick, D. J. 1994. *Methods for Genetic Risk Assessment.* New York: CRC Press/Lewis Publishers.

Bueche, J. R. 1972. *Principles of Physics,* 2nd Edition. New York: McGraw-Hill.

Burrell, R., D. K. Flaherty, and L. J. Sauers. 1992. *Toxicology of the Immune System.* New York: Van Nostrand Reinhold.

Calabrese, E. J. 1994. *Biological Effects of Low Level Exposures.* New York: CRC Press/Lewis Publishers.

Cothern, C. R., N. P. Ross. 1994. *Environmental Statistics, Assessment, and Forecasting.* New York: CRC Press/Lewis Publishers.

Gardner, R. W. 1994. *Chemical Intolerance.* New York: CRC Press/Lewis Publishers.

Hodgson, E., R. B. Mailman, and J. E. Chambers. 1988. *Dictionary of Toxicology.* London: The MacMillian Company.

Hosty, J. W., and P. Foster. 1990. *A Practical Guide to Chemical Spill Response.* New York: Van Nostrand Reinhold.

Howard, P. H., et. al. 1991. *Handbook of Environmental Degradation Rates.* New York: CRC Press/Lewis Publishers.

Keith, L. H. and D. B. Walters. (Editors) 1992. *National Toxicology Program's Chemical Database, Volumes I-VII.* Chelsea, MI: Lewis Publishers.

Landis, W. G., M. H. Yu. 1995. *Introduction to Environmental Toxicology.* New York: CRC Press/Lewis Publishers.

Lauwerys, R. R., P. Hoet. 1993. *Industrial Chemical Exposure.* New York: CRC Press/Lewis Publishers.

Lewis, Richard J., Sr. 1992. *Sax's Dangerous Properties of Industrial Materials,* Volumes 1, 2, and 3, 8th Edition. New York: Van Nostrand Reinhold.

Manahan, S. E. 1992. *Toxicological Chemistry,* 2nd Edition. New York: CRC Press/Lewis Publishers.

Meyer, Eugene. 1990. *Chemistry of Hazardous Materials.* Englewood Cliffs, NJ: Prentice Hall.

Patnaik, P. 1992. *A Comprehensive Guide to the Hazardous Properties of Chemical Substances.* New York: Van Nostrand Reinhold.

MATERIAL SAFETY DATA SHEET

CHEMICAL NAME	
ETHYL ALCOHOL	**CRC PRESS/LEWIS PUBLISHERS DISCLAIMER:** THE INFORMATION AND RECOMMENDATIONS PRESENTED HEREIN ARE BASED ON SOURCES BELIEVED TO BE RELIABLE. CRC MAKES NO REPRESENTATION ON ITS COMPLETENESS OR ACCURACY. IT IS THE RESPONSIBILITY OF THE USER TO DETERMINE THE CHEMICAL'S SUITABILITY FOR ITS INTENDED USE, THE CHEMICAL'S SAFE USE, AND THE CHEMICAL'S PROPER DISPOSAL. NO REPRESENTATIONS AND/OR WARRANTIES, EITHER EXPRESSED OR IMPLIED, OF THE MERCHANTABILITY OR FITNESS FOR A PARTICULAR PURPOSE, OR OF ANY OTHER NATURE, ARE MADE WITH RESPECT TO THE INFORMATION PROVIDED IN THIS MSDS OR TO THE CHEMICAL TO WHICH INFORMATION MAY REFER. CRC NEITHER ASSUMES NOR AUTHORIZES ANY OTHER PERSON TO ASSUME FOR IT, ANY OTHER ADDITIONAL RESPONSIBILITY OR LIABILITY FOR THE USE OF, OR RELIANCE UPON, THIS INFORMATION.

HAZARD WARNING INFORMATION

HEALTH	FIRE	REACTIVE	OTHER	DEGREE OF HAZARD	COLOR CODING	OTHER CODES
0	3	0		0 = Minimum Hazard 1 = Slight Hazard 2 = Moderate Hazard 3 = Serious Hazard 4 = Severe Hazard	HEALTH = BLUE FIRE = RED REACTIVITY = YELLOW OTHER = WHITE	OX = Oxidizer ACID = Acid ALK = Alkali COR = Corrosive W = Use No Water

SECTION I - GENERAL INFORMATION

Characterization	RCRA Number	EPA Class
Alcohol	**D001**	**Characteristic (I) Waste**

DOT Proper Shipping Name **Ethanol or Ethyl Alcohol**	Chemical Abstract Service (CAS) Number **64-17-5**
DOT Hazard Class and Label Requirements **Flammable Liquid**	DOT Emergency Guide Code **26**
DOT Identification Number **UN 1170**	Chemical Formula **CH_3CH_2OH**

Synonyms
Alcohol; ethanol; cologne spirits; EtOH; grain alcohol; ethyl hydroxide; ethyl hydrate; rectified spirits.

SECTION II - HAZARDOUS INGREDIENTS AND IDENTITY INFORMATION

Hazardous Components (specific identity)	OSHA Exposure Criteria	NIOSH Exposure Criteria	Immediately Dangerous to Life and Health (IDLH)	ACGIH Exposure Criteria
Ethyl alcohol (derivation: From ethylene by direct catalytic hydration or with ethyl sulfate as intermediate; fermentation of biomass, especially agricultural wastes; enzymatic hydrolysis of cellulose). **1 ppm = 1.92 mg/m^3**	**PEL:** **1000 ppm** **1900 mg/m^3** **STEL:** **Not Established**	**REL:** **1000 ppm** **1900 mg/m^3** **STEL:** **Not Established**	**3300 ppm**	**TLV:** **1000 ppm** **1880 mg/m^3** **STEL:** **Not Established**

SECTION III - PHYSICAL AND CHEMICAL CHARACTERISTICS

Boiling Point **173°F (78°C)**	Specific Gravity (H_2O = 1) **0.79**
Vapor Pressure (mm Hg) **44 at 68°F (20°C)**	Molecular Weight **46.1**
Vapor Density (Air = 1) **1.59**	Melting Point **-173°F (-114°C)**

Solubility
Completely soluble in water, chloroform, acetone, ether, benzene, and methanol.

Appearance and Odor
Colorless volatile liquid with a weak, vinous, alcohol odor and bitter taste. Odor Threshold = 84 ppm.

SECTION IV - FIRE AND EXPLOSION HAZARD DATA

Flash Point (method used) **55°F (13°C) closed cup**	Explosive Limits in Air % by Volume **LEL: 3.3%** **UEL: 19%**
NFPA Classification **Class IB Flammable Liquid**	Autoignition Temperature **685°F (363°C)**

Extinguishing Media
Dry chemical, carbon dioxide, or alcohol resistant foam. Use water as fog in flooding quantities.

Special Fire Fighting Procedures
Poisonous gases are produced in fire. Wear full protective clothing and self-contained breathing apparatus (SCBA). Continue to cool containers with water after fire is extinguished. Move containers from fire area if it can be done without risk. For large fires, use unmanned hose apparatus, if possible. Evacuate non-essential personnel 1500 feet radius if fire. Consider down-wind conditions.

Unusual Fire and Explosion Hazards
Vapors are heavier than air and may travel a distances to ignition source to flashback. Vapors may explode if ignited in an enclosed area.

SECTION V - REACTIVITY DATA

Stability		Conditions to Avoid Ethyl alcohol is stable under normal conditions of storage and operation. Avoid contact with incompatible materials. Keep away from heat, flame, and other ignition sources.
Stable X	Unstable	Incompatibility *(materials to avoid)* Reacts violently in contact with oxidizers, potassium dioxide, bromine pentafluoride, acetyl bromide, acetyl chloride, platinum, sodium. Forms explosive products in reaction with many metal nitrates.
Hazardous Polymerization		Conditions to Avoid Under normal conditions of temperature and pressure, ethyl alcohol is not expected to undergo hazardous polymerization.
May Occur	Will Not Occur X	Hazardous Decomposition or By-products When heated to decomposition, ethyl alcohol emits acrid and irritating fumes, including oxides of carbon (monoxide and dioxide).

SECTION VI - HEALTH HAZARD DATA

Primary Route(s) of Entry:	Inhalation?	Absorption (skin)?	Ingestion?
	X		X

Health Hazards

INHALATION: There may be slight irritation of the eyes, nose, and mucosa of the respiratory tract. Exposure can also cause stupor, fatigue, and sleepiness. There are no reports of liver damage (cirrhosis) occurring as a result of inhalation (consumption causes cancer).

SKIN & EYES: As a solvent, cause skin dryness, cracking, and possible dermatitis. Eye contact can result in burning, stinging, tearing, corneal injury, and hyperemia of the conjunctiva.

INGESTION: Sleep disorders, hallucinations, distorted perceptions, convulsions, ataxia and motor activity changes, coma, headache, pulmonary changes, nausea, vomiting, menstrual cycle changes, glandular changes, and decrease of body temperature.

Carcinogenicity	NTP Listed?	IARC Cancer Review Group?	OSHA Regulated?	Target Organs?
Confirmed Human Confirmed Animal	No	Sufficient Evidence	29 CFR 1910.1000 Table Z-1	Skin, eyes, resp. sys., CNS, liver, blood, repro.

Medical Conditions Generally Aggravated by Exposure

Chronic liver disease.

Emergency and First-aid Procedures

Eye contact: Flush immediately with water for 15 minutes (minimum); seek medical attention. Skin contact: Remove all contaminated clothing. Immediately wash area with large amounts of soap and water. Seek medical assistance. For inhalation: Remove the person from exposure. Provide respiratory assistance and CPR. Transfer to medical facility. If swallowed: Seek medical attention immediately. Give 1-2 glasses of water and induce vomiting.

SECTION VII - PRECAUTIONS FOR SAFE HANDLING AND USE

Steps to be Taken in Case Material is Released or Spilled

Contain spills using absorbent material (vermiculite or other). Ventilate area of spill. Have extinguishing agent available in case of fire. Store materials in DOT-approved containers.

Preferred Waste Disposal Method

Incineration in chemical incinerator equipped with scrubber and afterburner.

Precautions to be Taken in Handling and Storage

Store in tightly closed containers in a cool, dark, well-ventilated area. Protect containers from physical damage. Keep heat, fire and flame away. Outside building or underground storage is recommended.

Other Precautions and Warnings

Ground and bond all metal containers. Ensure Class 1, Group D electrical installations.

SECTION VIII - CONTROL MEASURES AND PERSONAL PROTECTIVE EQUIPMENT

Respiratory Protection *(specify type)*

For vapor exposures, use an organic vapor air-purifying respirator. Greater protection is obtained using a supplied-air respirator or a self-contained breathing apparatus (SCBA) with full facepiece operated in pressure demand or other positive pressure mode.

Ventilation

Local exhaust or general mechanical systems recommended.

Protective Gloves	Eye Protection	Other Protective Clothing
Impervious Butyl Rubber Gloves	Chemical Goggles or Face Mask	Impermeable Apron or Protective Suit

Work/Hygiene Practices

Always wash hands thoroughly after using chemical; never bring food, drink, or smoking materials into vicinity of chemicals.

ETHYL ALCOHOL

CH_3CH_2OH CAS: 64-17-5

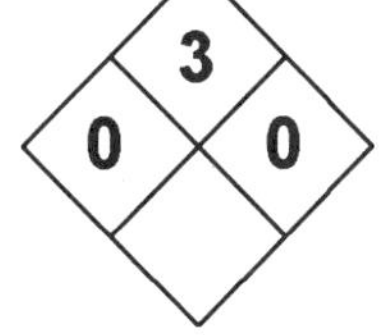

$$CH_3 — CH_2 — OH$$

IDENTIFICATION AND TYPICAL USES

Ethyl alcohol is a colorless or clear liquid with a weak, ethereal, vinous, or alcohol-like odor and bitter taste. It is primarily used as a solvent. It is an important industrial solvent for resins, lacquers, pharmaceuticals, toilet preparations, and cleaning agents. It is also used in the production of raw materials for cosmetics, perfumes, drugs, and plasticizers. It is a major component of antifreeze solutions and is also used as an additive in automobile fuel. Its most known application is in the making of alcoholic beverages.

RISK ASSESSMENT: HEALTH

General Assessment

The toxicity of ethyl alcohol is much lower in comparison to methanol or propanol. Ethyl alcohol is primarily toxic to humans by *ingestion.* While inhalation of its vapors can produce some toxic effects, its ability to enhance the effects of other chemicals poses a greater health risk for inhalation. Skin contact can cause topical damage and absorption is, therefore, not likely. It should be noted that most manufacturers of ethyl alcohol for use in industrial applications will normally mix it with a denaturant (a substance added to make it undesirable to drink). These include gasoline, acetone, formaldehyde, or methyl alcohol. Therefore, industrial exposures resulting from ingestion are very unlikely.

Inhalation can cause irritation of the eyes, nose, throat, upper respiratory tract, and associated mucosa. There may be headache, nervousness, tremors, dizziness, tearing, fatigue, nausea, somnolence, and narcosis with stupor and loss of consciousness. There are no reports of cirrhosis occurring as a result of inhalation exposures. However, chronic exposure to ethyl alcohol vapors caused brain damage in mice. Vapor exposure can also increase the toxic effects of other chemicals being inhaled. Also, the toxicity of ethyl alcohol is enhanced with the presence of compounds such as barbiturates, carbon monoxide, and methyl mercury.

Liquid contact with the eyes causes immediate burning and stinging with lachrymation and reflex closure of the lids. There may be injury to the corneal epithelium and possible hyperemia (excessive blood) of the conjunctiva. Skin contact results in drying and cracking which can lead to secondary infections and dermatitis.

Ingestion of ethyl alcohol is not likely to occur in the industrial environment. However, if it does, symptoms can include sleep disorders, hallucinations, distorted perceptions, ataxia, motor function changes, convulsions and tremors, coma, headaches, pulmonary changes, alteration of gastric secretions, menstrual cycle changes, glandular changes, nausea or vomiting, and decrease in body temperature.

Acute Health Effects

The following acute (short-term) health effects may occur immediately or shortly after exposure to ethyl alcohol:

Skin: Causes dryness and cracking leading to dermatitis and possible infection.

Eye: Severe irritation with burning and stinging with possible damage to the cornea and conjunctiva.

Lung: Irritation of the eyes, nose, throat, and respiratory tract.

CNS: High concentrations can cause depression of the CNS with symptoms of sleepiness and lack of concentration.

Chronic Health Effects

The following chronic (long-term) health effects may occur at some time after exposure to ethyl alcohol and can last for months or even years:

Cancer Hazards: Ethyl alcohol is known to cause liver cancer in humans, primarily due to ingestion. Industrial exposures through ingestion are not likely but are certainly possible.

Reproduction: According to the references, ethyl alcohol can affect human reproduction by ingestion. It causes changes in the female fertility index. Effects on newborns include changes in the apgar score, neonatal measures or effects, and drug dependence.

Other Chronic Effects: Very high or prolonged exposure may result in mucous membrane irritation, headache, and depression of the CNS with symptoms of somnolence and lack of concentration. Prolonged skin contact can cause dermatitis.

Recommended Risk-Reduction Measures

Even though ethyl alcohol is a known carcinogen, this effect is primarily the result of ingesting large amounts of alcoholic beverages. Industrial exposures by this route are not likely to occur. The best risk reduction measure is to use a less toxic chemical as a substitute for a ethyl alcohol. However, based upon the fact that ethyl alcohol is one of the most widely used industrial solvents, substitution is usually not an alternative. Therefore, *engineering controls* are the most effective methods of reducing exposures. The best protection is to enclose operations and/or provide local exhaust ventilation at the site of chemical release. While not always operationally feasible, isolating operations can also reduce exposure risk.

Using respiratory protection is less effective than the controls mentioned above, but is still advisable whenever working with or around ethyl alcohol. For concentrations over the PEL (1000 ppm), an air-purifying respirator with an organic vapor cartridge will suffice. For higher exposures, a supplied-air respirator with full facepiece operated in positive pressure mode, or a self-contained breathing apparatus (SCBA) with full facepiece and operated in pressure demand mode are the recommended respiratory protection methods of choice. If a full facepiece is not available, then chemical goggles should be worn to protect the eyes. Whenever a chemical splash hazard exists, a face shield and a protective apron should be worn. To prevent hand and skin exposures, impervious gloves should be used.

Administrative controls should also be in place to minimize the potential for human exposures. These may include written procedures or policies which specify the methods and techniques that will be practiced whenever personnel are to work with ethyl alcohol.

All personnel should receive *training* on the use, hazards, protective measures, emergency actions, and other precautions per 29 CFR 1910.1200 (Hazard Communication), prior to the first assignment in an area where ethyl alcohol is used or stored.

If symptoms develop or overexposure is suspected, the following medical tests are recommended:

- ☑ Liver function tests.
- ☑ Skin testing with dilute ethyl alcohol to help diagnose allergy (performed by a qualified allergist).

Any medical evaluation should include a careful history of past and present symptoms with an examination. Medical tests that look for existing damage are not a substitute for controlling exposures. Also, since consuming large quantities of alcoholic beverages can lead to liver dysfunction and even cancer, persons with alcohol addiction who are exposed to ethyl alcohol on the job may develop symptoms much quicker and with greater intensity than those who do not drink under identical exposure conditions. Prudent risk management requires careful consideration of *all* possible factors which may be causing the appearance of exposure symptoms.

Other methods to reduce exposure include:

- ☑ Where possible, enclose operations and use local exhaust ventilation at the site of chemical release. If local exhaust ventilation or enclosure is not used, respiratory protection should be mandatory.
- ☑ Always ensure that proper protective clothing is worn when using chemical substances.
- ☑ Wash thoroughly immediately after exposure to ethyl alcohol and at the end of the work shift or before eating, drinking, or smoking.
- ☑ Hazard warning information should be posted in the work area. In addition, as part of an on-going education and training program, all information on the health and safety hazards of ethyl alcohol should be communicated to all potentially exposed workers.

RISK ASSESSMENT: ENVIRONMENT

General Assessment

The environment is at risk of exposure during transportation, storage, disposal, or destruction of ethyl alcohol. In almost every scenario, the threat of environmental exposure is contingent upon the proper handling of the chemical substance. Accidental spills, large or small, can result in fire, explosion, and possible contamination of the surrounding environmental mediums (water, soil, and air).

Ethyl alcohol is considered a class IB flammable liquid (according to OSHA 29 CFR 1910.106). Its low flash point and relatively low boiling point present a serious fire and explosion hazard concern. Also, because it is incompatible with a number of common materials, especially strong oxidizers and many metal nitrates, contact can result in violent and explosive reactions. It can form explosive mixtures in air and can ignite on contact with heat, fire, or sparks. It will react and then explode in contact with acetic anhydride + sodium hydrogen sulfate. It also reacts violently with acetyl bromide (evolves hydrogen bromide). These characteristics require special consideration during any emergency situation involving a leak or spill of ethyl alcohol.

Ethyl alcohol can enter the environment through unchecked industrial discharges into effluents and through spills.

☠ *Acute Ecological Effects*

Acute (short-term) toxic effects may include the death of animals, birds, or fish, and death or low growth rate in plants. Acute effects are seen 2 to 4 days after animals or plants are exposed to ethyl alcohol. This chemical has moderate acute toxicity to aquatic life. Insufficient data are available to evaluate or predict the short-term effects of ethyl alcohol to plants, birds, or terrestrial animals.

Chronic Ecological Effects

Chronic toxic effects may include shortened life span, reproductive problems, lower fertility, and changes in appearance or behavior in exposed animals. These effects can be seen long after first exposure(s) to toxic chemicals. Ethyl alcohol has moderate chronic toxicity to aquatic life. Insufficient data are available to evaluate or predict the long-term effects of ethyl alcohol to plants, birds, or land animals.

Water Solubility

Ethyl alcohol is highly soluble in water. Concentrations of 1000 milligrams and more can be expected to mix with a liter of water.

⌛ *Persistence in the Environment*

Ethyl alcohol is slightly persistent in water, with a half-life of between 2 to 20 days. The half-life of a pollutant is the amount of time it takes for one-half of the chemical to be degraded. About 90% of ethyl alcohol will eventually end up in the air; the remainder will end up in water.

Bioaccumulation in Aquatic Organisms

Some substances increase in concentration, or *bioaccumulate,* in living organisms as they breathe contaminated air, drink contaminated water, or eat contaminated food. These chemicals can become concentrated in the tissues and internal organs of animals as well as humans.

The concentration of ethyl alcohol found in fish tissues is expected to be about the same as the average concentration of ethyl alcohol in water from which the fish was taken.

Recommended Risk-Reduction Measures

Proper training of all transporters will reduce the likelihood of a mishap or accident resulting in a leak or spill to the environment. The correct use of DOT labeling (per the MSDS) on all containers, trucks, and rail cars will enable emergency responders to react properly and quickly to any disaster thereby reducing the potential risk to the environment and to personnel.

Storage of ethyl alcohol should be segregated from incompatible chemicals to minimize the risk of cross-contamination or contact. Buildings designated for storage should be equipped with appropriate fire protection systems (alarms, sprinklers, emergency lighting, portable extinguishers). Equipment should be designed to meet explosion-proof standards.

If a spill or leak to the environment has occurred, fire department, emergency response, and/or hazardous materials spill personnel should be notified immediately. Cleanup should be attempted only by those trained in proper spill containment procedures. Contaminated soils should be removed for incineration and replaced with clean soil. If ethyl alcohol should contact the water table, aquifer, or navigable waterway, time is of the essence. It is highly soluble in water and, therefore, total containment and remediation may not be entirely possible. When such spills occur, the local and/or state emergency response authorities must be notified. A comprehensive emergency response or disaster preparedness/recovery plan should be in place prior to any operations involving the use, transportation, storage, or disposal of ethyl alcohol. If ethyl alcohol is spilled or leaked, the following specific steps are recommended:

- ☑ Restrict persons not wearing protective clothing from area of spill or leak until cleanup is complete and area can be opened for normal work.
- ☑ Ventilate area and remove ignition sources.
- ☑ Absorb liquids in vermiculite, dry sand, earth, or a similar material and deposit in sealed containers. Use non-sparking tools.
- ☑ It may be necessary to dispose of ethyl alcohol as a hazardous waste. The responsible state agency or the regional office of the federal Environmental Protection Agency (EPA) should be contacted for specific recommendations.

RISK ASSESSMENT: BUSINESS

General Assessment

Accidents or mishaps involving ethyl alcohol can present a moderate threat to business operations. The loss or damage of equipment or facilities can significantly affect fiscal viability. Lawsuits that may result from personnel injury/death, public exposures, and/or environmental contamination will also require a serious expenditure of resources. Media attention surrounding an injury, death, or environmental damage can also result in a loss of profits and loss of current as well as future business.

Recommended Risk-Reduction Measures

Company attorneys, safety and health professionals, and environmental specialists should be involved in the development of any procedures or policies intended to manage the use of chemicals in the workplace. A company official should be pre-designated as a public relations officer with specific training in dealing with the press. Corporate plans and policies should be developed, approved, and implemented long before any need for such arises.

REFERENCES

American Conference of Governmental Industrial Hygienists. 1988. *Documentation of the Threshold Limit Values and Biological Exposure Indices,* 5th Edition (with updates). Cincinnati: ACGIH

Anthony, C. P., and N. J. Kolthoff. 1971. *Textbook of Anatomy and Physiology.* St. Louis: C. V. Mosby Company.

Bloom, A. D. and F. J. DeSerres. 1995. *Ecotoxicity and Human Health.* New York: CRC Press/Lewis Publishers.

Brusick, D. J. 1994. *Methods for Genetic Risk Assessment.* New York: CRC Press/Lewis Publishers.

Bueche, J. R. 1972. *Principles of Physics,* 2nd Edition. New York: McGraw-Hill.

Burrell, R., D. K. Flaherty, and L. J. Sauers. 1992. *Toxicology of the Immune System.* New York: Van Nostrand Reinhold.

Calabrese, E. J. 1994. *Biological Effects of Low Level Exposures.* New York: CRC Press/Lewis Publishers.

Cothern, C. R., N. P. Ross. 1994. *Environmental Statistics, Assessment, and Forecasting.* New York: CRC Press/Lewis Publishers.

Danse, I. R. 1991. *Common Sense Toxics in the Workplace.* New York: Van Nostrand Reinhold.

Environmental Protection Agency. 1983. *Chemical Hazard Information Profiles.* Washington, D.C.: U.S. EPA.

Fodden, C. R., and J. L. Weddell. 1992. *Hazardous Materials Emergency Action Data.* New York: CRC Press/Lewis Publishers.

Forseberg, K., and L. H. Keith. 1993. *Chemical; Protective Clothing Permeation and Degradation Database.* Chelsea, MI: Lewis Publishers.

Gardner, R. W. 1994. *Chemical Intolerance.* New York: CRC Press/Lewis Publishers.

Hodgson, E., R. B. Mailman, and J. E. Chambers. 1988. *Dictionary of Toxicology.* London: The MacMillian Company.

Hosty, J. W., and P. Foster. 1990. *A Practical Guide to Chemical Spill Response.* New York: Van Nostrand Reinhold.

Howard, P. H., et. al. 1991. *Handbook of Environmental Degradation Rates.* New York: CRC Press/Lewis Publishers.

International Agency for Research on Cancer (IARC). 1987. *IARC Monographs on the Evaluation of Carcinogenic Risk to Humans, Overall Evaluations of Carcinogenicity: An updating of IARC Monographs Volumes 1-42.* Lyon: IARC.

Keith, L. H. and D. B. Walters. (Editors) 1992. *National Toxicology Program's Chemical Database, Volumes I-VII.* Chelsea, MI: Lewis Publishers.

Landis, W. G., M. H. Yu. 1995. *Introduction to Environmental Toxicology.* New York: CRC Press/Lewis Publishers.

Lauwerys, R. R., P. Hoet. 1993. *Industrial Chemical Exposure.* New York: CRC Press/Lewis Publishers.

Lewis, Richard J., Sr. 1992. *Sax's Dangerous Properties of Industrial Materials,* Volumes 1, 2, and 3, 8th Edition. New York: Van Nostrand Reinhold.

Manahan, S. E. 1992. *Toxicological Chemistry,* 2nd Edition. New York: CRC Press/Lewis Publishers.

Meyer, Eugene. 1990. *Chemistry of Hazardous Materials.* Englewood Cliffs, NJ: Prentice Hall.

Patnaik, P. 1992. *A Comprehensive Guide to the Hazardous Properties of Chemical Substances.* New York: Van Nostrand Reinhold.

Philp, R. B. 1995. *Environmental Hazards and Human Health.* New York: CRC Press/Lewis Publishers.

MATERIAL SAFETY DATA SHEET

CHEMICAL NAME

ETHYLAMINE

CRC PRESS/LEWIS PUBLISHERS DISCLAIMER: THE INFORMATION AND RECOMMENDATIONS PRESENTED HEREIN ARE BASED ON SOURCES BELIEVED TO BE RELIABLE. CRC MAKES NO REPRESENTATION ON ITS COMPLETENESS OR ACCURACY. IT IS THE RESPONSIBILITY OF THE USER TO DETERMINE THE CHEMICAL'S SUITABILITY FOR ITS INTENDED USE, THE CHEMICAL'S SAFE USE, AND THE CHEMICAL'S PROPER DISPOSAL. NO REPRESENTATIONS AND/OR WARRANTIES, EITHER EXPRESSED OR IMPLIED, OF THE MERCHANTABILITY OR FITNESS FOR A PARTICULAR PURPOSE, OR OF ANY OTHER NATURE, ARE MADE WITH RESPECT TO THE INFORMATION PROVIDED IN THIS MSDS OR TO THE CHEMICAL TO WHICH INFORMATION MAY REFER. CRC NEITHER ASSUMES NOR AUTHORIZES ANY OTHER PERSON TO ASSUME FOR IT, ANY OTHER ADDITIONAL RESPONSIBILITY OR LIABILITY FOR THE USE OF, OR RELIANCE UPON, THIS INFORMATION.

HAZARD WARNING INFORMATION

HEALTH	FIRE	REACTIVE	OTHER	DEGREE OF HAZARD	COLOR CODING	OTHER CODES
3	4	0		0 = Minimum Hazard 1 = Slight Hazard 2 = Moderate Hazard 3 = Serious Hazard 4 = Severe Hazard	HEALTH = BLUE FIRE = RED REACTIVITY = YELLOW OTHER = WHITE	OX = Oxidizer ACID = Acid ALK = Alkali COR = Corrosive W = Use No Water

SECTION I - GENERAL INFORMATION

Characterization **Aliphatic Amine**	RCRA Number **D001**	EPA Class **Characteristic (I) Waste**
DOT Proper Shipping Name **Ethylamine or Monoethylamine**	Chemical Abstract Service (CAS) Number **75-04-7**	
DOT Hazard Class and Label Requirements **Flammable Gas (or Flammable Liquid)**	DOT Emergency Guide Code **68**	
DOT Identification Number **UN 1036 (compressed); UN 2270 (liquid)**	Chemical Formula **$CH_3CH_2NH_2$**	

Synonyms
Aminoethane; ethylamine (anhydrous), monoethylamine; ethanamine.

SECTION II - HAZARDOUS INGREDIENTS AND IDENTITY INFORMATION

Hazardous Components (specific identity)	OSHA Exposure Criteria	NIOSH Exposure Criteria	Immediately Dangerous to Life and Health (IDLH)	ACGIH Exposure Criteria
Ethylamine (derivation from: Chloroethane heated with alcoholic ammonia; from ethyl chloride and alcoholic ammonia under heat and pressure; by hydrogenation of nitroethane). **1 ppm = 1.87 mg/m^3**	**PEL:** **10 ppm** **18 mg/m^3** **STEL:** **Not Established**	**REL:** **10 ppm** **18 mg/m^3** **STEL:** **Not Established**	**600 ppm**	**TLV:** **5 ppm** **9 mg/m^3** **STEL:** **15 ppm** **28 mg/m^3**

SECTION III - PHYSICAL AND CHEMICAL CHARACTERISTICS

Boiling Point **62°F (16°C)**	Specific Gravity (H_2O = 1) **0.68 (as liquid) at 68°F (20°C)**
Vapor Pressure (mmHg) **873 at 68°F (20°C)**	Molecular Weight **30.08**
Vapor Density (Air = 1) **1.55**	Freezing Point **-114°F (-81°C)**

Solubility
Completely soluble in water. Soluble in alcohol and ether.

Appearance and Odor
Colorless gas or a volatile, water-white liquid with an ammonia-like odor.

SECTION IV - FIRE AND EXPLOSION HAZARD DATA

Flash Point (method used) **0°F (-17°C) closed cup**	Explosive Limits in Air % by Volume **LEL: 3.5% UEL: 14%**
NFPA Classification **Flammable Gas or Class 1A Flammable Liquid**	Autoignition Temperature **725°F (385°C)**

Extinguishing Media
Stop flow of gas first (if possible). If liquid, use foam, carbon dioxide, or dry chemical.

Special Fire Fighting Procedures
Ethylamine is an extremely flammable, very dangerous fire and explosion hazard. Poisonous gases are evolved from fire. Wear a self-contained breathing apparatus with full facepiece. Ensure ALL sources of escaping gas are controlled. Move containers from fire area if it can be done without risk. Water may be ineffective on fire. Keep cooling containers with water long after fire is out.

Unusual Fire and Explosion Hazards
Gas is explosive. Containers may explode in flame. Vapors are heavier than air and will travel for a great distance to an ignition source and flashback to cause fire or explosion.

SECTION V - REACTIVITY DATA

Stability		Conditions to Avoid Ethylamine is normally stable in its closed containers, during routine operations (those that do not expose it to ignition sources). Avoid contact with heat, sparks, flame, and other sources of ignition.
Stable X	Unstable	Incompatibility *(materials to avoid)* Strong oxidizing materials, strong acids, which causes splattering, cellulose nitrate, and copper, tin, or zinc with moisture. Will strip and dissolve paint, most plastics, and will swell most rubber products.
Hazardous Polymerization		Conditions to Avoid Under normal conditions of temperature and pressure, hazardous polymerization of ethylamine is not expected to occur.
May Occur	Will Not Occur X	Hazardous Decomposition or By-products If involved in fire, ethylamine can produce toxic oxides of carbon and nitrogen.

SECTION VI - HEALTH HAZARD DATA

Primary Route(s) of Entry:	Inhalation?	Absorption (skin)?	Ingestion?
	X		

Health Hazards

INHALATION: Vapors are irritating to the eyes, skin, and respiratory tract. Causes pulmonary edema (fluid in the lungs), which is a medical emergency and can lead to death. There may also be blurred or hazy vision (temporary) resulting from damage to the corneal epithelium.

SKIN & EYES: Contact with the compressed gas or liquid can cause frostbite and 1st degree burns.

INGESTION: Not likely since ethylamine is a gas at normal working (room) temperatures. If it occurs, there can be severe burns of the mouth and stomach that can be fatal.

Carcinogenicity	NTP Listed?	IARC Cancer Review Group?	OSHA Regulated?	Target Organs?
Insufficient data	No	No	29 CFR 1910.1000 Table Z-1	Eyes, skin, respiratory tract

Medical Conditions Generally Aggravated by Exposure

None Reported.

Emergency and First-aid Procedures

Eye contact: For frostbite, rinse with warm (104°F/40°C) water. Skin contact: Remove all contaminated clothing. Immediately wash area with large amounts warm (104°F/40°C) water. For inhalation: Remove the person from exposure. Provide respiratory assistance and CPR. Transfer promptly to medical facility. Observe for 24-48 hours for lung effects (pulmonary edema). If swallowed: Contact poison control center. Unless advised otherwise, give 1-2 glasses of water to dilute. Do NOT induce vomiting.

SECTION VII - PRECAUTIONS FOR SAFE HANDLING AND USE

Steps to be Taken in Case Material is Released or Spilled

Remove all ignition sources. Ventilate area. Stop flow of gas. If the source of leak is a cylinder and cannot be stopped, remove to safe place in open air; repair leak or let cylinder empty. If gas has not yet ignited, use a water spray to knock down vapors and direct them from ignition source.

Preferred Waste Disposal Method

None specified in the references (incineration of a gas may be the only alternative).

Precautions to be Taken in Handling and Storage

Do not store with incompatible materials or in the presence of heat since violent reactions or explosions can occur. Store in tightly closed pressurized containers in a cool, dark, well-ventilated area away from sunlight. Keep away from ignition sources such as fire, sparks, and flame.

Other Precautions and Warnings

Bulk storage of ethylamine is not recommended. Containers may explode in fire or under conditions of extreme heat (do not store outdoors or in direct sunlight).

SECTION VIII - CONTROL MEASURES AND PERSONAL PROTECTIVE EQUIPMENT

Respiratory Protection *(specify type)*

For exposures to the PEL, use a cartridge-type respirator For higher exposures, use a powered supplied-air respirator or a self-contained breathing apparatus with full facepiece operated in pressure demand or other positive pressure mode.

Ventilation

Local exhaust or general mechanical systems recommended.

Protective Gloves	Eye Protection	Other Protective Clothing
Insulated Butyl Rubber	Goggles and/or Face Mask	Leather or Butyl Rubber Apron

Work/Hygiene Practices

Always wash hands thoroughly after using chemical; never bring food, drink, or smoking materials into vicinity of chemicals.

ETHYLAMINE

$CH_3CH_2NH_2$ CAS: 75-04-7

$CH_3 — CH_2 — NH_2$

IDENTIFICATION AND TYPICAL USES

Ethylamine is a colorless gas with an ammonia odor. It can also appear as a water-white liquid. It is used in organic synthesis, in resin chemistry, in solvent extraction, in petroleum refining, a stabilizer for rubber latex, in detergents, and as an intermediate for dyestuffs.

RISK ASSESSMENT: HEALTH

General Assessment

Ethylamine normally appears as a gas. As such, its primary route of entry into the body is through *inhalation.* Skin contact with the liquid can be destructive to tissues due to the extremely cold temperatures and irritating properties of ethylamine, but absorption is not a likely exposure route. Since ethylamine usually exists as a gas at room temperatures, ingestion is not likely to occur under normal conditions. It is not known if exposure to ethylamine will lead to carcinogenic, mutagenic, or teratogenic effects in humans or animals.

Inhalation of ethylamine vapors causes irritation of the eyes, nose, throat, and respiratory tract. Exposure can lead to pulmonary edema (fluid buildup in the lungs) which is a medical emergency and can be fatal. Symptoms of congestion, cough, phlegm, tightness and possibly pain in chest, and difficulty breathing may be delayed up to 48 hours thereby creating a false sense of security with regard to health hazard/exposure risk. There is also a possibility that exposure can cause blurred or hazy vision, apparently as a result of corneal epithelium disturbances.

Skin or eye contact with the vapor causes irritation. Liquid contact with either the skin or the eyes causes serious irritation and even first degree burns with subsequent tissue damage and frostbite.

Although ingestion is unlikely to occur under normal conditions (ethylamine is a gas at room temperature), swallowing it will cause severe burns to the mouth and stomach and can lead to death.

Acute Health Effects

The following acute (short-term) health effects may occur immediately or shortly after exposure to ethylamine:

Skin: Redness, smarting, pain, possible dermatitis, first degree burns, and severe frostbite.

Eye: Causes frostbite, redness, irritation, severe pain, and blurred vision.

Lung: Vapor is seriously irritating and therefore poses a significant threat of damage to the lungs. Irritation can lead to pulmonary edema (fluid buildup in the lungs).

Other: Ethylamine solutions are strongly alkaline and can cause deep tissue destruction.

Chronic Health Effects

The following chronic (long-term) health effects may occur at some time after exposure to ethylamine and can last for months or even years:

Cancer Hazards: According to information presented in the references, it is not known if exposure to ethylamine will lead to carcinogenic or mutagenic effects.

Reproductive Hazards: There are no reports to support any claims of reproductive hazards.

Other Chronic Effects: None reported.

Recommended Risk-Reduction Measures

Personnel should avoid direct contact with ethylamine. Unless a less toxic chemical can be substituted for a hazardous substance, *engineering controls* are the most effective method of reducing exposures. The best protection is to separate operations and provide local exhaust ventilation at the site of chemical release. While not always operationally feasible, isolating operations can also reduce exposure. Using equipment with explosion-proof designs and ensuring all metal containers and tools are electrically grounded and bonded can further reduce the potential for injury or damage. Using respiratory protection is less effective than the controls mentioned above, but is still advisable whenever working with ethylamine. For exposures above the PEL (10 ppm), use an MSHA/NIOSH-approved chemical cartridge respirator with a full facepiece and cartridges suitable for protection against amines. For higher or more frequent exposures, use a powered supplied-air respirator operated in pressure demand or other positive pressure mode, or a self-contained breathing apparatus (SCBA) with full facepiece operated in pressure demand mode. If a full facepiece is not available, then chemical goggles should be worn to protect the eyes. Whenever a chemical splash hazard exists, a face shield and a rubber or leather apron should be worn. To prevent hand and skin exposures with the liquid, thermal gloves should be worn.

Administrative controls should also be in place to minimize the potential for human exposures. These may include written procedures or policies which specify the methods and techniques that will be practiced whenever personnel are to work with ethylamine.

All personnel should receive *training* on the use, hazards, protective measures, emergency actions, and other precautions per 29 CFR 1910.1200 (Hazard Communication), prior to the first assignment in an area where ethylamine is used or stored. Before beginning employment and at regular intervals thereafter (e.g., annually), the following medical tests are recommended:

☑ Lung function tests (establish baseline).

If symptoms develop or overexposure is suspected, the following additional tests may be helpful:

☑ Lung function tests (compare to baseline).

☑ Consider chest X-ray after acute overexposure (may be negative if taken immediately following exposure).

Other methods to reduce exposure to chemicals include:

☑ Where possible, use local exhaust ventilation at the site of chemical release. If local exhaust ventilation or enclosure is not used, respiratory protection should be mandatory.

☑ Prevent static sparks by electrically grounding and bonding all equipment used with or around ethylamine.

☑ Always ensure that proper protective clothing is worn when using chemical substances.

☑ Wash thoroughly immediately after exposure to ethylamine and at the end of the work shift or before eating, drinking, or smoking.

☑ Safety shower and eyewash stations should be readily available in work areas where ethylamine is used or stored.

☑ Hazard warning information should be posted in the work area. In addition, as part of an on-going education and training program, all information on the health and safety hazards of ethylamine should be communicated to all exposed and potentially exposed workers.

RISK ASSESSMENT: ENVIRONMENT

General Assessment

The environment is at risk of exposure during transportation, storage, disposal, or destruction of ethylamine. In almost every scenario, the threat of environmental exposure is contingent upon the proper handling of the chemical substance. Accidental releases, large or small, can result in fire, explosion, and possible contamination of the surrounding environmental mediums (water, soil, and air).

Ethylamine is an extremely flammable and highly explosive gas. As a liquid, it is considered a class 1A flammable liquid (per OSHA 29 CFR 1910.106). Its extremely low flash point and boiling point present a serious fire hazard. Also, since it can react violently in the presence of oxidizers, such as chlorine, ethylamine requires special consideration during any emergency involving a leak or release of ethylamine gas or liquid.

Ethylamine can enter the environment through industrial discharges, unchecked venting to air, or through spills of the liquid.

Acute Ecological Effects

Acute (short-term) toxic effects may include the death of animals, birds, or fish, and death or low growth rate in plants. Acute effects are seen 2 to 4 days after animals or plants are exposed to ethylamine. Insufficient data are available on the short-term effects of ethylamine on aquatic life, plants, birds, or land animals.

Chronic Ecological Effects

Chronic (long-term) toxic effects may include shortened life span, reproductive problems, lower fertility, and changes in appearance or behavior in exposed animals.

These effects can be seen long after first exposure(s) to toxic chemicals. Insufficient data are available to evaluate the long-term effects of ethylamine to aquatic life, plants, birds, or land animals.

Water Solubility

Ethylamine is highly soluble in water. Concentrations of 1000 milligrams and more will mix with a liter of water.

Persistence in the Environment

Ethylamine is slightly persistent in the aquatic environment with a half-life of 2 to 20 days. The half-life of a pollutant is the amount of time it takes for one half of the chemical to be degraded. In moist soil, ethylamine should readily degrade with a half-life of 2 months. In dry soil, it should volatilize rapidly.

Bioaccumulation in Aquatic Organisms

Some substances increase in concentration, or *bioaccumulate,* in living organisms as they breathe contaminated air, drink contaminated water, or eat contaminated food. These chemicals can become concentrated in the tissues and internal organs of animals as well as humans.

Ethylamine is not expected to accumulate in the tissues of fish.

Recommended Risk-Reduction Measures

Proper training of all transporters will reduce the likelihood of a mishap or accident resulting in a leak or release to the environment. The correct use of DOT labeling (per the MSDS) on all containers, trucks, and rail cars will enable emergency responders to react properly and quickly to any disaster thereby reducing the potential risk to the environment and to personnel.

Storage of ethylamine should be segregated from incompatible chemicals to minimize the risk of cross-contamination or contact. Buildings designated for storage should be equipped with appropriate fire protection systems (alarms, sprinklers, emergency lighting, portable extinguishers).

If a spill or release to the environment has occurred, fire department, emergency response, and/or hazardous materials spill personnel should be notified immediately. Cleanup should be attempted only by those trained in proper containment procedures. Contaminated soils should be removed for incineration and replaced with clean soil. If ethylamine should contact the water table, aquifer, or navigable waterway, time is of the essence. It is readily miscible in water and total containment and remediation may not be entirely possible. When such spills occur, it may be necessary to contact the local and/or state emergency response authorities. A comprehensive emergency response or disaster preparedness/recovery plan should be in place prior to any operations involving the use, transportation, storage, or disposal of ethylamine. If ethylamine is spilled or leaked, the following specific steps are recommended:

- ☑ Restrict persons not wearing protective clothing from area of spill or leak until cleanup is complete and area can be opened for normal work.
- ☑ Remove all ignition sources.
- ☑ Ventilate area of spill or leak.
- ☑ If applicable, stop flow of leaking liquid or gas. If leak source is a cylinder and the leak cannot be stopped in place, remove leaking cylinder to a safe place in the open air, and repair or allow cylinder to empty.
- ☑ Keep ethylamine out of a confined space, such as a sewer, because of the possibility of explosion (unless the sewer is designed to prevent the build-up of explosive concentrations).
- ☑ It may be necessary to dispose of ethylamine as a hazardous waste. The responsible state agency or the regional office of the federal Environmental Protection Agency (EPA) should be contacted for specific recommendations.

RISK ASSESSMENT: BUSINESS

General Assessment

Accidents or mishaps involving ethylamine can present a significant threat to business operations. The loss or damage of equipment or facilities can significantly affect fiscal viability. Lawsuits that may result from personnel injury, public exposures, and environmental contamination will also require a serious expenditure of resources. Media attention surrounding an injury, illness, or environmental damage can also result in a loss of profits and loss of current as well as future business.

Recommended Risk-Reduction Measures

Company attorneys, safety and health professionals, and environmental specialists should be involved in the development of any such procedures. A company official should be pre-designated as a public relations officer with specific training in dealing with the press. Corporate plans and policies should be developed, approved, and implemented long before any need for such arises.

REFERENCES

Ahlbom, A. 1993. *Biostatistics for Epidemiologists.* New York: CRC Press/Lewis Publishers.

American Conference of Governmental Industrial Hygienists. 1988. *Documentation of the Threshold Limit Values and Biological Exposure Indices,* 5th Edition (with updates). Cincinnati: ACGIH

Anthony, C. P., and N. J. Kolthoff. 1971. *Textbook of Anatomy and Physiology.* St. Louis: C. V. Mosby Company.

Asante-Duah, D. K. 1993. *Hazardous Waste Risk Assessment.* New York: CRC Press/Lewis Publishers.

Bloom, A. D. and F. J. DeSerres. 1995. *Ecotoxicity and Human Health.* New York: CRC Press/Lewis Publishers.

Brusick, D. J. 1994. *Methods for Genetic Risk Assessment.* New York: CRC Press/Lewis Publishers.

Bueche, J. R. 1972. *Principles of Physics,* 2nd Edition. New York: McGraw-Hill.

Burrell, R., D. K. Flaherty, and L. J. Sauers. 1992. *Toxicology of the Immune System.* New York: Van Nostrand Reinhold.

Calabrese, E. J. 1994. *Biological Effects of Low Level Exposures.* New York: CRC Press/Lewis Publishers.

Cockerman, L. G., B. S. Shane. 1994. *Basic Environmental Toxicology.* New York: CRC Press/Lewis Publishers.

Cothern, R. R. 1993. *Comparative Environmental Risk Assessment.* New York: CRC Press/Lewis Publishers.

Danse, I. R. 1991. *Common Sense Toxics in the Workplace.* New York: Van Nostrand Reinhold.

Environmental Protection Agency. 1983. *Chemical Hazard Information Profiles.* Washington, D.C.: U.S. EPA.

Fodden, C. R., and J. L. Weddell. 1992. *Hazardous Materials Emergency Action Data.* New York: CRC Press/Lewis Publishers.

Forseberg, K., and L. H. Keith. 1993. *Chemical; Protective Clothing Permeation and Degradation Database.* Chelsea, MI: Lewis Publishers.

Gardner, R. W. 1994. *Chemical Intolerance.* New York: CRC Press/Lewis Publishers.

Hallenbeck, W. H. 1993. *Quantitative Risk Assessment for Environmental and Occupational Health,* 2nd Edition. New York: CRC Press/Lewis Publishers.

Hodgson, E., R. B. Mailman, and J. E. Chambers. 1988. *Dictionary of Toxicology.* London: The MacMillian Company.

Hosty, J. W., and P. Foster. 1990. *A Practical Guide to Chemical Spill Response.* New York: Van Nostrand Reinhold.

Howard, P. H., et. al. 1991. *Handbook of Environmental Degradation Rates.* New York: CRC Press/Lewis Publishers.

James, A., and M. Lord. 1992. *Index of Chemical and Physical Data.* New York: Van Nostrand Reinhold.

Jones, F. E. 1994. *Toxic Organic Vapors in the Workplace.* New York: CRC Press/Lewis Publishers.

Kamrin, M. A. 1988. *Toxicology: A Primer on Principles and Applications.* Chelsea, Michigan: Lewis Publishers.

Keith, L. H. and M. M. Walker. 1995. *Handbook of Air Toxics.* New York: CRC Press/Lewis Publishers.

Keith, L. H. (Editor). 1992. *Integrated Risk Information System (IRIS) Chemical Information Database.* Prepared by the U.S. Environmental Protection Agency. Boca Raton, FL: CRC Press/Lewis Publishers.

Landis, W. G., M. H. Yu. 1995. *Introduction to Environmental Toxicology.* New York: CRC Press/Lewis Publishers.

Langerman, N. 1994 *Precautionary Labels for Chemical Containers.* New York: CRC Press/Lewis Publishers.

Lauwerys, R. R., P. Hoet. 1993. *Industrial Chemical Exposure.* New York: CRC Press/Lewis Publishers.

Lewis, R. J., Sr. 1993. *Hawley's Condensed Chemical Dictionary,* 12th Edition. New York: Van Nostrand Reinhold.

Lewis, R. J., Sr. 1992. *Sax's Dangerous Properties of Industrial Materials,* Volumes 1, 2, and 3, 8th Edition. New York: Van Nostrand Reinhold.

Lide, D. R. 1995. *Handbook of Chemistry and Physics,* 76th Edition. New York: CRC Press/Lewis Publishers.

Manahan, S. E. 1992. *Toxicological Chemistry,* 2nd Edition. New York: CRC Press/Lewis Publishers.

Meyer, E. 1990. *Chemistry of Hazardous Materials.* Englewood Cliffs, NJ: Prentice Hall.

Patnaik, P. 1992. *A Comprehensive Guide to the Hazardous Properties of Chemical Substances.* New York: Van Nostrand Reinhold.

Philp, R. B. 1995. *Environmental Hazards and Human Health.* New York: CRC Press/Lewis Publishers.

Quigley, D. R. 1994. *Handbook of Emergency Chemical Management.* New York: CRC Press/Lewis Publishers.

Rea, W. J. 1995. *Chemical Sensitivity,* Vol. I-III. New York: CRC Press/Lewis Publishers.

Richardson, M. 1995. *Dictionary of Substance and Their Effects,* Vol. I-VI. New York: CRC Press/Lewis Publishers.

Sacarello, H. L. A. 1994. *The Comprehensive Handbook of Hazardous Materials.* New York: CRC Press/Lewis Publishers.

Sherman, J. D. 1988. *Chemical Exposure and Disease.* New York: Van Nostrand Reinhold.

Talbott, E., G. Craun. 1995. *An Introduction to Environmental Epidemiology.* New York: CRC Press/Lewis Publishers.

Torkelson, T. R. V. K. Rowe. 1981. *Patty's Industrial Hygiene and Toxicology,* Vol. 2B. New York: Wiley-Interscience.

U.S. Department of Health and Human Services, National Institute for Occupational Safety and Health. 1994. *NIOSH Pocket Guide to Chemical Hazards.* Washington, D.C.: U.S. Government Printing Office.

MATERIAL SAFETY DATA SHEET

CHEMICAL NAME	
N-ETHYLANILINE	**CRC PRESS/LEWIS PUBLISHERS DISCLAIMER:** THE INFORMATION AND RECOMMENDATIONS PRESENTED HEREIN ARE BASED ON SOURCES BELIEVED TO BE RELIABLE. CRC MAKES NO REPRESENTATION ON ITS COMPLETENESS OR ACCURACY. IT IS THE RESPONSIBILITY OF THE USER TO DETERMINE THE CHEMICAL'S SUITABILITY FOR ITS INTENDED USE, THE CHEMICAL'S SAFE USE, AND THE CHEMICAL'S PROPER DISPOSAL. NO REPRESENTATIONS AND/OR WARRANTIES, EITHER EXPRESSED OR IMPLIED, OF THE MERCHANTABILITY OR FITNESS FOR A PARTICULAR PURPOSE, OR OF ANY OTHER NATURE, ARE MADE WITH RESPECT TO THE INFORMATION PROVIDED IN THIS MSDS OR TO THE CHEMICAL TO WHICH INFORMATION MAY REFER. CRC NEITHER ASSUMES NOR AUTHORIZES ANY OTHER PERSON TO ASSUME FOR IT, ANY OTHER ADDITIONAL RESPONSIBILITY OR LIABILITY FOR THE USE OF, OR RELIANCE UPON, THIS INFORMATION.

HAZARD WARNING INFORMATION

HEALTH	FIRE	REACTIVE	OTHER	DEGREE OF HAZARD	COLOR CODING	OTHER CODES
3	2	0		0 = Minimum Hazard 1 = Slight Hazard 2 = Moderate Hazard 3 = Serious Hazard 4 = Severe Hazard	HEALTH = BLUE FIRE = RED REACTIVITY = YELLOW OTHER = WHITE	OX = Oxidizer ACID = Acid ALK = Alkali COR = Corrosive ~~W~~ = Use No Water

SECTION I - GENERAL INFORMATION

Characterization	**Aromatic Amine**	RCRA Number	**None**
		EPA Class	**Not Applicable**
DOT Proper Shipping Name	***n*-Ethylaniline**	Chemical Abstract Service (CAS) Number	**103-69-5**
DOT Hazard Class and Label Requirements	**Poison B**	DOT Emergency Guide Code	**57**
DOT Identification Number	**UN 2272**	Chemical Formula	**$C_2H_5NHC_6H_5$**

Synonyms

***n*-Ethylaminobenzene; ethylaniline; n-ethylbenzenamine; *n*-ethylbenzenamino; ethylphenylamine.**

SECTION II - HAZARDOUS INGREDIENTS AND IDENTITY INFORMATION

Hazardous Components (specific identity)	OSHA Exposure Criteria	NIOSH Exposure Criteria	Immediately Dangerous to Life and Health (IDLH)	ACGIH Exposure Criteria
***n*-Ethylaniline** (derivation: By heating aniline and ethyl alcohol in presence of sulfuric acid with subsequent distillation). Ingredients: *n*-Ethylaniline (CAS: 103-69-5): 97.5% *n,n*-Diethylaniline (CAS: 91-66-7) 2.0% Aniline (CAS: 62-53-3) .5%	**PEL:** Not Established **STEL:** Not Established	**REL:** Not Established **STEL:** Not Established	**Not Determined**	**TLV:** Not Established **STEL:** Not Established

SECTION III - PHYSICAL AND CHEMICAL CHARACTERISTICS

Boiling Point	**400°F (204°C)**	Specific Gravity (H_2O = 1)	**0.9**
Vapor Pressure (mm Hg)	**<1 at 69°F (20°C)**	Molecular Weight	**121.2**
Vapor Density (Air = 1)	**4.2**	Melting Point	**-83°F (-64°C)**

Solubility

Insoluble in water. Soluble in alcohol.

Appearance and Odor

Oily, colorless or pale yellow liquid which darkens to brown on exposure to light; aromatic, amine-like odor and burning taste.

SECTION IV - FIRE AND EXPLOSION HAZARD DATA

Flash Point (method used)	**185°F (85°C) opened cup**	Explosive Limits in Air % by Volume	**LEL: 1.6% UEL: 9.5%**
NFPA Classification	**Class IIIA Combustible Liquid**	Autoignition Temperature	**Not Determined**

Extinguishing Media

Carbon dioxide, dry chemical, alcohol foam, water spray or fog.

Special Fire Fighting Procedures

Wear full protective clothing and self-contained breathing apparatus (SCBA). Use a water spray to keep fire-exposed containers cool and remove from fire if possible. Poisonous gases are produced in fire.

Unusual Fire and Explosion Hazards

Moderately flammable when exposed to heat. Containers may explode in fire. Firefighters should avoid contact with vapors produced during fire. Hypergolic reaction with red fuming nitric acid. Can react with oxidizers to cause fire or explosion.

SECTION V - REACTIVITY DATA

Stability		Conditions to Avoid *n*-Ethylaniline is considered stable in closed containers under normal conditions of temperature and pressure. Avoid contact with incompatible materials and keep away from heat or flame.
Stable X	Unstable	Incompatibility *(materials to avoid)* Strong oxidizers (such as perchlorates, peroxides, permanganates, chlorates and nitrates), and strong acids (such as hydrochloric, sulfuric, and nitric). Reacts hypergolically with red fuming nitric acid.
Hazardous Polymerization		Conditions to Avoid Under normal conditions of temperature and pressure, hazardous polymerization of *n*-ethylaniline is not expected to occur.
May Occur	Will Not Occur X	Hazardous Decomposition or By-products When heated to decomposition, *n*-ethylaniline can emit highly toxic/poisonous fumes of aniline and oxides of nitrogen.

SECTION VI - HEALTH HAZARD DATA

Primary Route(s) of Entry:	Inhalation?	Absorption (skin)?	Ingestion?
	X	X	X

Health Hazards

INHALATION: Causes anoxia due to the formation of methemoglobin. Symptoms: headache, dizziness, ataxia, and cyanosis. Overexposure may lead to death due to respiratory paralysis.

ABSORPTION: Absorbs freely through intact skin. Symptoms include headache and cyanosis; overexposure can result in death. Skin contact with pure liquid can produce mild irritation.

INGESTION: Unspecified effects resulting from ingestion. Liver damage is suspect.

Carcinogenicity	NTP Listed?	IARC Cancer Review Group?	OSHA Regulated?	Target Organs?
Unknown Human Unknown Animal	No	No	No	Respiratory sys., eyes, skin, blood, kidney, liver.

Medical Conditions Generally Aggravated by Exposure

Respiratory impairments (bronchitis, asthma); Skin conditions (dermatitis); Renal or liver disorders.

Emergency and First-aid Procedures

Eye contact: Flush immediately with water for 15 minutes (minimum), seek medical attention. Skin contact: Remove all contaminated clothing. Immediately wash area with large amounts of soap and water. Seek medical assistance. For inhalation: Remove the person from exposure. Provide respiratory assistance and CPR. Transfer to medical facility. If swallowed: Seek medical attention immediately. Unless advised otherwise, give 1-2 glasses of water and induce vomiting.

SECTION VII - PRECAUTIONS FOR SAFE HANDLING AND USE

Steps to be Taken in Case Material is Released or Spilled

Absorb liquids in vermiculite, dry sand, earth, or similar material and deposit in sealed containers. Ventilate area of spill or leak and remove ignition sources. Restrict those not involved in cleanup from entering area. Notify appropriate authorities, if applicable.

Preferred Waste Disposal Method

Dissolve in combustible solvent; burn in chemical incinerator equipped with scrubber and afterburner.

Precautions to be Taken in Handling and Storage

n-Ethylaniline is a combustible liquid. Store to avoid contact with strong oxidizers and acids, since violent reactions can occur. Store in tightly closed containers in cool, well-ventilated area away from heat. Ground and bond all containers prior to transferring liquids.

Other Precautions and Warnings

Sources of ignition are prohibited where *n*-ethylaniline is used, handled, or stored. Bulk storage of *n*-ethylaniline is not recommended.

SECTION VIII - CONTROL MEASURES AND PERSONAL PROTECTIVE EQUIPMENT

Respiratory Protection *(specify type)*

Exposure limits have not been established. For best protection, use a supplied air respirator with full facepiece, hood, or helmet in continuous flow mode, or a self-contained breathing apparatus (SCBA) with full facepiece operated in positive pressure mode.

Ventilation

Local exhaust at site of chemical work.

Protective Gloves	Eye Protection	Other Protective Clothing
Natural Rubber	Chemical Goggles or Face Mask	Rubber Apron

Work/Hygiene Practices

Always wash hands thoroughly after using chemical; never bring food, drink, or smoking materials into vicinity of chemicals.

N-ETHYLANILINE

$C_2H_5NHC_6H_5$ CAS: 103-69-5

IDENTIFICATION AND TYPICAL USES

n-Ethylaniline is an oily, colorless or pale yellow liquid with characteristic odor and burning taste; color darkens to brown or tan upon exposure to air. *n*-Ethylaniline is an intermediate in chemical synthesis and, as an aniline compound, it is among one of the most important of the organic bases.

RISK ASSESSMENT: HEALTH

General Assessment

n-Ethylaniline is moderately toxic through *inhalation, ingestion,* and *absorption* through the skin. While *n*-ethylaniline has not been adequately tested for its ability to cause cancer or for its effects on reproduction, it should be noted that aniline itself causes mutations (genetic changes) in living cells.

Toxic symptoms of *n*-ethylaniline exposure by all routes include headache, weakness, dizziness, ataxia, and cyanosis (blue tinting of the lips, skin, and nail beds due to oxygen deprivation). Cyanosis is due to the formation of methemoglobin which restricts the blood's oxygen carrying capabilities. Cyanosis occurs when the methemoglobin concentration is 15% or more. Blueness develops first in the lips, nose, and earlobes. At 40% concentration, there is weakness and dizziness. At up to 70% concentration, symptoms may include ataxia, dyspnea on mild exertion, and tachycardia. Coma may ensue at concentrations over 70%. Overexposure (between 85% and 90% concentrations) may lead to death from respiratory paralysis.

Contact with the pure liquid *n*-ethylaniline on the skin can produce moderate irritation, while the effect on the eyes can be quite severe.

Acute Health Effects

The following acute (short-term) health effects may occur immediately or shortly after exposure to *n*-ethylaniline:

Skin: Irritation and possible color change around nails and fingers. Absorbs freely through the skin.

Eye: Severe irritation; may cause corneal damage.

Lung: Not known to be acutely irritating to lung tissue. However, overexposure can lead to death due to respiratory arrest.

Other: *n*-Ethylaniline affects the blood's ability to transport oxygen and high exposure can cause death. Symptoms include headaches, weakness, irritability, drowsiness, and shortness of breath.

Chronic Health Effects

The following chronic (long-term) health effects may occur at some time after exposure to *n*-ethylaniline and can last for months or even years:

Cancer Hazards: *n*-Ethylaniline has not been adequately tested for its ability to cause cancer in test animals. Aniline, however, may cause mutations (genetic changes) in living cells. The data on its carcinogenicity are inconclusive.

Reproductive Hazard: There is no evidence to show that exposure to *n*-ethylaniline causes damage to the developing fetus. Aniline, however, may cause such effects (more study is required in this area).

Other Chronic Effects: Continuous exposure to small doses of *n*-ethylaniline may produce anemia, loss of energy, digestive disturbance, and headache. Repeated exposures may cause anemia and brownish staining of the eyes can occur.

Recommended Risk-Reduction Measures

Personnel should avoid direct contact with *n*-ethylaniline. If a less toxic chemical cannot be substituted for a hazardous substance, then *engineering controls* are the most effective method of reducing exposures. The best protection is to enclose operations and/or provide local exhaust ventilation at the site of chemical release. While not always operationally feasible, isolating operations can also reduce exposure. Using respiratory protection is less effective than the controls mentioned above, but is still advisable whenever working with or around *n*-ethylaniline. Since no exposure limits have been established, an air-purifying (cartridge-type) respirator may not provide adequate protection. A supplied-air respirator operated in positive pressure mode or a self-contained breathing apparatus (SCBA) with full facepiece and operated in pressure demand are the recommended respiratory protection methods of choice. If a full facepiece is not available, then chemical goggles should be worn to protect the eyes. Whenever a chemical splash hazard exists, a face shield and a rubber apron should be worn. To prevent hand and skin exposures, natural rubber gloves should be worn.

Administrative controls should also be in place to minimize the potential for human exposures. These may include written procedures or policies which specify the methods and techniques that will be practiced whenever personnel are to work with *n*-ethylaniline.

All personnel should receive *training* on the use, hazards, protective measures, emergency actions, and other precautions per 29 CFR 1910.1200 (Hazard Communication), prior to the first assignment in an area where *n*-ethylaniline is used or stored.

If symptoms develop or overexposure is suspected, the following medical tests may be useful:

☑ Methemoglobin tests.

☑ Urinalysis (for blood content).

Any evaluation should include a careful history of past and present symptoms with an examination. Medical tests that simply look for existing damage are not a substitute for controlling exposures.

Other methods to reduce exposure include:

☑ Where possible, enclose operations and use local exhaust ventilation at the site of chemical release. If local exhaust ventilation or enclosure is not used, respiratory protection should be mandatory.

☑ Always ensure that proper protective clothing is worn when using chemical substances.

☑ Wash thoroughly immediately after exposure to *n*-ethylaniline and at the end of the work shift or before eating, drinking, or smoking.

☑ Hazard warning information should be posted in the work area. In addition, as part of an on-going education and training program, all information on the health and safety hazards of *n*-ethylaniline should be communicated to all potentially exposed workers.

☑ Contaminated clothing (including shoes and undergarments) should never worn home where family members can be exposed. Work clothing should be laundered only by personnel who have been briefed on the hazards of *n*-ethylaniline.

☑ Eye wash stations should be provided in the immediate work area for emergency use. If there is a possibility of skin exposure to *n*-ethylaniline, emergency shower facilities should also be provided.

RISK ASSESSMENT: ENVIRONMENT

General Assessment

The environment is at risk of exposure during transportation, storage, disposal, or destruction of *n*-ethylaniline. In almost every scenario, the threat of environmental exposure is contingent upon the proper handling of the chemical substance. Accidental spills, large or small, can result in fire, explosion, and possible contamination of the surrounding environmental mediums (water, soil, and air).

n-Ethylaniline is considered a Class IIIA combustible liquid. Alone, its high flash point and autoignition temperature do not present a serious fire or explosion hazard. However, because it is incompatible with so very many other commodities, extreme caution is required in handling, storage, transportation, and disposal of *n*-ethylaniline. These characteristics also require special consideration during any emergency situation involving a leak or spill of *n*-ethylaniline. Should *n*-ethylaniline ever come into contact with incompatible substances such as oxidizers, acids, or alkalis either during use, transportation, storage, or disposal, the formation of highly toxic and/or highly explosive commodities is extremely possible.

The proper disposal/destruction method for *n*-ethylaniline waste is to dissolve it in a combustible solvent and burn it in a chemical incinerator equipped with an afterburner and air scrubber. *n*-Ethylaniline can enter the environment through direct disposal and indirect use, such as industrial discharges and non-point sources from agricultural use.

Acute Ecological Effects

Acute (short-term) toxic effects may include the death of animals, birds, or fish and death or low growth rate in plants. Acute effects are seen 2 to 4 days after animals or plants are exposed to *n*-ethylaniline. This chemical has high acute toxicity to aquatic life and moderate acute toxicity to birds. No data are available on the short-term effects of *n*-ethylaniline to aquatic life, plants, or land animals. However, aniline causes germination decrease, stunting and size decreases on numerous agricultural crops.

Chronic Ecological Effects

Chronic (long-term) toxic effects may include shortened life span, reproductive problems, lower fertility, and changes in appearance or behavior in exposed animals. These effects can be seen long after first exposure(s) to toxic chemicals. No data are available on the long-term effects of *n*-ethylaniline to aquatic life, plants, birds, or land animals. However, aniline has high chronic toxicity in aquatic life.

Water Solubility

n-Ethylaniline is insoluble in water. Concentrations of less than 1 milligram will not mix with a liter of water.

Persistence in the Environment

n-Ethylaniline is slightly persistent in water, with a half-life between 2 and 20 days. The half-life of a pollutant is the amount of time it takes for one half of the chemical to be degraded.

Bioaccumulation in Aquatic Organisms

Some substances increase in concentration, or *bioaccumulate,* in living organisms as they breathe contaminated air, drink contaminated water, or eat contaminated food. These chemicals can become concentrated in the tissues and internal organs of animals as well as humans.

The concentration of *n*-ethylaniline found in fish tissues is expected to be somewhat higher than the average concentration of *n*-ethylaniline in the water from which the fish was taken.

Recommended Risk-Reduction Measures

Proper training of all transporters will reduce the likelihood of a mishap or accident resulting in a leak or spill to the environment. The correct use of DOT labeling (per the MSDS) on all containers, trucks, and rail cars will enable emergency responders to react properly and quickly to any disaster thereby reducing the potential risk to the environment and to personnel.

Storage of *n*-ethylaniline should be segregated from incompatible chemicals to minimize the risk of cross-contamination or contact.

If a spill or leak to the environment has occurred, fire department, emergency response and/or hazardous materials spill personnel should be notified immediately. Cleanup should be attempted only by those trained in proper spill containment procedures. Contaminated soils should be removed for incineration and replaced with clean soil. If *n*-ethylaniline should contact the water table, aquifer, or navigable waterway, the local and/or state emergency response authorities must be notified. A comprehensive emergency response or disaster preparedness/recovery plan should be in place prior to any operations involving the use, transportation, storage, or disposal of *n*-ethylaniline. If *n*-ethylaniline is spilled or leaked, the following specific steps are recommended:

- ☑ Restrict persons not wearing protective clothing from area of spill or leak until cleanup is complete and area can be opened for normal work.
- ☑ Remove all ignition sources.
- ☑ Ventilate area of spill or leak.
- ☑ Absorb liquids in vermiculite, dry sand, earth, or similar material and deposit in sealed containers.
- ☑ It may be necessary to dispose of *n*-ethylaniline as a hazardous waste. The responsible state agency or the regional office of the federal Environmental Protection Agency (EPA) should be contacted for specific recommendations.

RISK ASSESSMENT: BUSINESS

General Assessment

Accidents or mishaps involving *n*-ethylaniline can present a significant threat to business operations. The loss or damage of equipment or facilities can significantly affect fiscal viability. Lawsuits that may result from personnel injury/death, public exposures, and/or environmental contamination will also require a serious expenditure of resources. Media attention surrounding an injury, death, or environmental damage can also result in a loss of profits and loss of current as well as future business.

Recommended Risk-Reduction Measures

Company attorneys, safety and health professionals, and environmental specialists should be involved in the development of any procedures dealing with safety or emergency response. A company official should be pre-designated as a public relations officer with specific training in dealing with the press. Corporate plans and policies should be developed, approved, and implemented long before any need for such arises.

REFERENCES

Ahlbom, A. 1993. *Biostatistics for Epidemiologists.* New York: CRC Press/Lewis Publishers.

American Conference of Governmental Industrial Hygienists. 1988. *Documentation of the Threshold Limit Values and Biological Exposure Indices,* 5th Edition (with updates). Cincinnati: ACGIH

Asante-Duah, D. K. 1993. *Hazardous Waste Risk Assessment.* New York: CRC Press/Lewis Publishers.

Bloom, A. D. and F. J. DeSerres. 1995. *Ecotoxicity and Human Health.* New York: CRC Press/Lewis Publishers.

Brusick, D. J. 1994. *Methods for Genetic Risk Assessment.* New York: CRC Press/Lewis Publishers.

Bueche, J. R. 1972. *Principles of Physics,* 2nd Edition. New York: McGraw-Hill.

Calabrese, E. J. 1994. *Biological Effects of Low Level Exposures.* New York: CRC Press/Lewis Publishers.

Cockerman, L. G., B. S. Shane. 1994. *Basic Environmental Toxicology.* New York: CRC Press/Lewis Publishers.

Cothern, R. R. 1993. *Comparative Environmental Risk Assessment.* New York: CRC Press/Lewis Publishers.

Gardner, R. W. 1994. *Chemical Intolerance.* New York: CRC Press/Lewis Publishers.

Hallenbeck, W. H. 1993. *Quantitative Risk Assessment for Environmental and Occupational Health,* 2nd Edition. New York: CRC Press/Lewis Publishers.

Howard, P. H. 1995. *Handbook of Environmental Fate and Exposures Data for Organic Chemicals,* Vol. I-IV. New York: CRC Press/Lewis Publishers.

Jones, F. E. 1994. *Toxic Organic Vapors in the Workplace.* New York: CRC Press/Lewis Publishers.

Landis, W. G., M. H. Yu. 1995. *Introduction to Environmental Toxicology.* New York: CRC Press/Lewis Publishers.

Langerman, N. 1994 *Precautionary Labels for Chemical Containers.* New York: CRC Press/Lewis Publishers.

Lauwerys, R. R., P. Hoet. 1993. *Industrial Chemical Exposure.* New York: CRC Press/Lewis Publishers.

Lewis, R. J., Sr. 1993. *Hawley's Condensed Chemical Dictionary,* 12th Edition. New York: Van Nostrand Reinhold.

Lewis, R. J., Sr. 1992. *Sax's Dangerous Properties of Industrial Materials,* Volumes 1, 2, and 3, 8th Edition. New York: Van Nostrand Reinhold.

Manahan, S. E. 1994. *Environmental Chemistry,* 6th Edition. New York: CRC Press/Lewis Publishers.

Manahan, S. E. 1992. *Toxicological Chemistry,* 2nd Edition. New York: CRC Press/Lewis Publishers.

Mellström, G. A., J. E. Wahlberg, and H. I. Maibach. 1994. *Protective Gloves for Occupational Use.* New York: CRC Press/Lewis Publishers.

Meyer, E. 1990. *Chemistry of Hazardous Materials.* Englewood Cliffs, NJ: Prentice Hall.

Mycock, J. C., J. D. McKenna, and L. Theodore. 1995. *Handbook of Air Pollution Control Engineering and Technology.* New York: CRC Press/Lewis Publishers.

National Cancer Institute. 1978. *Carcinogenesis Technical Report Series. Bioassay of Aniline for Possible Carcinogenicity.* DEHW (NIH) Publication Number: 78-1323, p 53. Washington, DC: U.S. Government Printing Office.

Ott, W. R. 1995. *Environmental Statistics and Data Analysis.* New York: CRC Press/Lewis Publishers.

Patnaik, P. 1992. *A Comprehensive Guide to the Hazardous Properties of Chemical Substances.* New York: Van Nostrand Reinhold.

Philp, R. B. 1995. *Environmental Hazards and Human Health.* New York: CRC Press/Lewis Publishers.

Proctor, N. H., J. P. Hughes. 1991. *Chemical Hazards of the Workplace.* New York: Van Nostrand Reinhold.

Quigley, D. R. 1994. *Handbook of Emergency Chemical Management.* New York: CRC Press/Lewis Publishers.

Rea, W. J. 1995. *Chemical Sensitivity,* Vol. I-III. New York: CRC Press/Lewis Publishers.

Richardson, M. 1995. *Dictionary of Substance and Their Effects,* Vol. I-VI. New York: CRC Press/Lewis Publishers.

Sacarello, H. L. A. 1994. *The Comprehensive Handbook of Hazardous Materials.* New York: CRC Press/Lewis Publishers.

Sheftel, V. O. 1995. *Handbook of Toxic Properties of Monomers and Additives.* New York: CRC Press/Lewis Publishers.

Talbott, E., G. Craun. 1995. *An Introduction to Environmental Epidemiology.* New York: CRC Press/Lewis Publishers.

Torkelson, T. R. V. K. Rowe. 1981. *Patty's Industrial Hygiene and Toxicology,* Vol. 2B. New York: Wiley-Interscience.

U.S. Department of Health and Human Services, National Institute for Occupational Safety and Health. 1994. *NIOSH Pocket Guide to Chemical Hazards.* Washington, D.C.: U.S. Government Printing Office.

MATERIAL SAFETY DATA SHEET

CHEMICAL NAME	
O-ETHYLANILINE	**CRC PRESS/LEWIS PUBLISHERS DISCLAIMER:** THE INFORMATION AND RECOMMENDATIONS PRESENTED HEREIN ARE BASED ON SOURCES BELIEVED TO BE RELIABLE. CRC MAKES NO REPRESENTATION ON ITS COMPLETENESS OR ACCURACY. IT IS THE RESPONSIBILITY OF THE USER TO DETERMINE THE CHEMICAL'S SUITABILITY FOR ITS INTENDED USE, THE CHEMICAL'S SAFE USE, AND THE CHEMICAL'S PROPER DISPOSAL. NO REPRESENTATIONS AND/OR WARRANTIES, EITHER EXPRESSED OR IMPLIED, OF THE MERCHANTABILITY OR FITNESS FOR A PARTICULAR PURPOSE, OR OF ANY OTHER NATURE, ARE MADE WITH RESPECT TO THE INFORMATION PROVIDED IN THIS MSDS OR TO THE CHEMICAL TO WHICH INFORMATION MAY REFER. CRC NEITHER ASSUMES NOR AUTHORIZES ANY OTHER PERSON TO ASSUME FOR IT, ANY OTHER ADDITIONAL RESPONSIBILITY OR LIABILITY FOR THE USE OF, OR RELIANCE UPON, THIS INFORMATION.

HAZARD WARNING INFORMATION

HEALTH	FIRE	REACTIVE	OTHER	DEGREE OF HAZARD	COLOR CODING	OTHER CODES
3	2	0		0 = Minimum Hazard 1 = Slight Hazard 2 = Moderate Hazard 3 = Serious Hazard 4 = Severe Hazard	HEALTH = BLUE FIRE = RED REACTIVITY = YELLOW OTHER = WHITE	OX = Oxidizer ACID = Acid ALK = Alkali COR = Corrosive W = Use No Water

SECTION I - GENERAL INFORMATION

Characterization	**Aromatic Amine**	RCRA Number	**None**
		EPA Class	**Not Applicable**
DOT Proper Shipping Name	***o*-Ethylaniline**	Chemical Abstract Service (CAS) Number	**578-54-1**
DOT Hazard Class and Label Requirements	**Poison B**	DOT Emergency Guide Code	**57**
DOT Identification Number	**UN 2273**	Chemical Formula	$C_2H_5NHC_6H_5$

Synonyms

***o*-Aminoethylbenzene; *o*-ethylaniline; 2-ethylbenzenamine.**

SECTION II - HAZARDOUS INGREDIENTS AND IDENTITY INFORMATION

Hazardous Components (specific identity)	OSHA Exposure Criteria	NIOSH Exposure Criteria	Immediately Dangerous to Life and Health (IDLH)	ACGIH Exposure Criteria
***o*-Ethylaniline** (derivation: By heating aniline and ethyl alcohol in presence of sulfuric acid with subsequent distillation).	**PEL:** Not Established **STEL:** Not Established	**REL:** Not Established **STEL:** Not Established	**Not Determined**	**TLV:** Not Established **STEL:** Not Established

SECTION III - PHYSICAL AND CHEMICAL CHARACTERISTICS

Boiling Point	**419°F (215°C)**	Specific Gravity (H_2O = 1)	**1.0**
Vapor Pressure (mm Hg)	**<1 at 69°F (20°C)**	Molecular Weight	**121.2**
Vapor Density (Air = 1)	**4.2**	Melting Point	**-47°F (-44°C)**

Solubility

Slightly soluble in water. Soluble in alcohol and toluene.

Appearance and Odor

Oily, pale yellow, red, or brown liquid with an aromatic, amine-like odor and burning taste.

SECTION IV - FIRE AND EXPLOSION HAZARD DATA

Flash Point (method used)	**196°F (91°C) opened cup**	Explosive Limits in Air % by Volume	**LEL: Not Determined UEL: Not Determined**
NFPA Classification	**Class IIIA Combustible Liquid**	Autoignition Temperature	**Not Determined**

Extinguishing Media

Carbon dioxide, dry chemical, alcohol foam, water spray or fog.

Special Fire Fighting Procedures

Wear full protective clothing and self-contained breathing apparatus (SCBA). Use a water spray to keep fire-exposed containers cool and remove from fire area if possible. Poisonous gases are produced in fire. Avoid contact with skin. Fight fire from distance if possible.

Unusual Fire and Explosion Hazards

Moderately flammable when exposed to heat. Containers may explode in fire. Firefighters should avoid contact with vapors produced during fire. Hypergolic reaction with red fuming nitric acid. Can react with oxidizers to cause fire or explosion.

SECTION V - REACTIVITY DATA

Stability		Conditions to Avoid **o-Ethylaniline is considered stable in closed containers under normal conditions of temperature and pressure. Avoid contact with incompatible materials and keep away from heat or flame.**
Stable	Unstable	Incompatibility *(materials to avoid)* **Strong oxidizers (such as perchlorates, peroxides, permanganates, chlorates and nitrates), and strong acids (such as hydrochloric, sulfuric, and nitric). Reacts hypergolically with red fuming nitric acid.**
X		
Hazardous Polymerization		Conditions to Avoid **Under normal conditions of temperature and pressure, hazardous polymerization of *o*-ethylaniline is not expected to occur.**
May Occur	Will Not Occur	Hazardous Decomposition or By-products **When heated to decomposition, *o*-ethylaniline can emit highly toxic/poisonous fumes of aniline and oxides of nitrogen.**
	X	

SECTION VI - HEALTH HAZARD DATA

	Inhalation?	Absorption (skin)?	Ingestion?
Primary Route(s) of Entry:	X	X	X

Health Hazards

INHALATION: Causes anoxia due to the formation of methemoglobin. Symptoms: headache, dizziness, ataxia, and cyanosis. Overexposure may lead to death due to respiratory paralysis.

ABSORPTION: Absorbs freely through intact skin. Symptoms include headache and cyanosis; overexposure can result in death. Skin contact with pure liquid can produce mild irritation.

INGESTION: Unspecified effects resulting from ingestion. Liver damage is suspect.

Carcinogenicity	NTP Listed?	IARC Cancer Review Group?	OSHA Regulated?	Target Organs?
Unknown Human Unknown Animal	No	No	No	Respiratory sys., eyes, skin, blood, kidney, liver.

Medical Conditions Generally Aggravated by Exposure

Respiratory impairments (bronchitis, asthma); Skin conditions (dermatitis); Renal or liver disorders.

Emergency and First-aid Procedures

Eye contact: Flush immediately with water for 15 minutes (minimum), seek medical attention. Skin contact: Remove all contaminated clothing. Immediately wash area with large amounts of soap and water. Seek medical assistance. For inhalation: Remove the person from exposure. Provide respiratory assistance and CPR. Transfer to medical facility. If swallowed: Seek medical attention immediately. Unless advised otherwise, give 1-2 glasses of water and induce vomiting.

SECTION VII - PRECAUTIONS FOR SAFE HANDLING AND USE

Steps to be Taken in Case Material is Released or Spilled

Absorb liquids in vermiculite, dry sand, earth, or similar material and deposit in sealed containers. Ventilate area of spill or leak and remove ignition sources. Restrict those not involved in cleanup from entering area. Notify appropriate authorities, if applicable.

Preferred Waste Disposal Method

Dissolve in combustible solvent; burn in chemical incinerator equipped with scrubber and afterburner.

Precautions to be Taken in Handling and Storage

***o*-Ethylaniline is a combustible liquid. Store to avoid contact with strong oxidizers and acids, since violent reactions can occur. Store in tightly closed containers in cool, well-ventilated area away from heat. Ground and bond all containers prior to transferring liquids.**

Other Precautions and Warnings

Sources of ignition are prohibited where *o*-ethylaniline is used, handled, or stored. Bulk storage of *o*-ethylaniline is not recommended.

SECTION VIII - CONTROL MEASURES AND PERSONAL PROTECTIVE EQUIPMENT

Respiratory Protection *(specify type)*

Exposure limits have not been established. For best protection, use a supplied air respirator with full facepiece, hood, or helmet in continuous flow mode, or a self-contained breathing apparatus (SCBA) with full facepiece operated in positive pressure mode.

Ventilation

Local exhaust at site of chemical work.

Protective Gloves	Eye Protection	Other Protective Clothing
Natural Rubber	**Chemical Goggles or Face Mask**	**Rubber Apron**

Work/Hygiene Practices

Always wash hands thoroughly after using chemical; never bring food, drink, or smoking materials into vicinity of chemicals.

O-ETHYLANILINE

$C_2H_5NHC_6H_5$ CAS: 578-54-1

IDENTIFICATION AND TYPICAL USES

o-Ethylaniline is an oily, colorless or pale yellow, red, or brown liquid with characteristic odor and burning taste. It is used as an intermediate for pharmaceuticals, dyestuffs, pesticides, and other products. It may also be used in chemical synthesis and, as an aniline compound, it is among one of the most important of the organic bases.

RISK ASSESSMENT: HEALTH

General Assessment

o-Ethylaniline is moderately toxic through *inhalation, ingestion,* and *absorption* through the skin. While *o*-ethylaniline has not been adequately tested for its ability to cause cancer or for its effects on reproduction, it should be noted that aniline itself causes mutations (genetic changes) in living cells. Many scientists believe that mutagenic chemicals may also have carcinogenic capabilities in the long-term. However, the data in the references are conflicting and inconclusive in this regard.

Toxic symptoms of *o*-ethylaniline exposure by all routes include headache, weakness, dizziness, ataxia, and cyanosis (blue tinting of the lips, skin, and nail beds due to oxygen deprivation). Cyanosis is due to the formation of methemoglobin which restricts the blood's oxygen carrying capabilities. Cyanosis occurs when the methemoglobin concentration is 15% or more. Blueness develops first in the lips, nose, and earlobes. At 40% concentration, there is weakness and dizziness. At up to 70% concentration, symptoms may include ataxia, dyspnea on mild exertion, and tachycardia. Coma may ensue at concentrations over 70%. Overexposure (between 85% and 90% concentrations) may lead to death from respiratory paralysis.

Contact with the pure liquid *o*-ethylaniline on the skin can produce moderate irritation, while the effect on the eyes can be quite severe.

Acute Health Effects

The following acute (short-term) health effects may occur immediately or shortly after exposure to *o*-ethylaniline:

Skin: Irritation and possible color change around nails and fingers. Absorbs freely through the skin.

Eye: Severe irritation; may cause corneal damage.

Lung: Not known to be acutely irritating to lung tissue. However, overexposure can lead to death due to respiratory arrest.

Other: *o*-Ethylaniline affects the blood's ability to transport oxygen and high exposure can cause death. Symptoms include headaches, weakness, irritability, drowsiness, and shortness of breath.

Chronic Health Effects

The following chronic (long-term) health effects may occur at some time after exposure to *o*-ethylaniline and can last for months or even years:

Cancer Hazards: *o*-Ethylaniline has not been adequately tested for its ability to cause cancer in test animals. Aniline, however, may cause mutations (genetic changes) in living cells. The data on its carcinogenicity are inconclusive.

Reproductive Hazard: There is no evidence to show that exposure to *o*-ethylaniline causes damage to the developing fetus. Aniline, however, may cause such effects (more study is required in this area).

Other Chronic Effects: Continuous exposure to small doses of *o*-ethylaniline may produce anemia, loss of energy, digestive disturbance, and headache. Repeated exposures may cause anemia and brownish staining of the eyes can occur.

ʊ *Recommended Risk-Reduction Measures*

Personnel should avoid direct contact with *o*-ethylaniline. If a less toxic chemical cannot be substituted for a hazardous substance, then *engineering controls* are the most effective method of reducing exposures. The best protection is to enclose operations and/or provide local exhaust ventilation at the site of chemical release. While not always operationally feasible, isolating operations can also reduce exposure. Using respiratory protection is less effective than the controls mentioned above, but is still advisable whenever working with or around *o*-ethylaniline. Since no exposure limits have been established, an air-purifying (cartridge-type) respirator may not provide adequate protection. A supplied-air respirator operated in positive pressure mode or a self-contained breathing apparatus (SCBA) with full facepiece and operated in pressure demand are the recommended respiratory protection methods of choice. If a full facepiece is not available, then chemical goggles should be worn to protect the eyes. Whenever a chemical splash hazard exists, a face shield and a rubber apron should be worn. To prevent hand and skin exposures, natural rubber gloves should be used. Glove manufacturers should be contacted and permeation studies obtained *before* final glove selections are made.

Administrative controls should also be in place to minimize the potential for human exposures. These may include written procedures or policies which specify the methods and techniques that will be practiced whenever personnel are to work with *o*-ethylaniline.

All personnel should receive *training* on the use, hazards, protective measures, emergency actions, and other precautions per 29 CFR 1910.1200 (Hazard Communication), prior to the first assignment in an area where *o*-ethylaniline is used or stored.

If symptoms develop or overexposure is suspected, the following medical tests may be useful:

☑ Methemoglobin tests.

☑ Urinalysis (for blood content).

Any evaluation should include a careful history of past and present symptoms with an examination. Medical tests that simply look for existing damage are not a substitute for controlling exposures.

Other methods to reduce exposure include:

☑ Where possible, enclose operations and use local exhaust ventilation at the site of chemical release. If local exhaust ventilation or enclosure is not used, respiratory protection should be mandatory.

☑ Always ensure that proper protective clothing is worn when using chemical substances.

☑ Wash thoroughly immediately after exposure to *o*-ethylaniline and at the end of the work shift or before eating, drinking, or smoking.

☑ Hazard warning information should be posted in the work area. In addition, as part of an on-going education and training program, all information on the health and safety hazards of *o*-ethylaniline should be communicated to all exposed workers.

☑ Contaminated clothing (including shoes and undergarments) should never worn home where family members can be exposed. Work clothing should be laundered only by personnel who have been briefed on the hazards of *o*-ethylaniline.

☑ Eye wash stations should be provided in the immediate work area for emergency use. If there is a possibility of skin exposure to *o*-ethylaniline, emergency shower facilities should also be provided in the immediate work area.

RISK ASSESSMENT: ENVIRONMENT

General Assessment

The environment is at risk of exposure during transportation, storage, disposal, or destruction of *o*-ethylaniline. In almost every scenario, the threat of environmental exposure is contingent upon the proper handling of the chemical substance. Accidental spills, large or small, can result in fire, explosion, and possible contamination of the surrounding environmental mediums (water, soil, and air).

o-Ethylaniline is considered a class IIIA combustible liquid. Alone, its high flash point and autoignition temperature do not present a serious fire or explosion hazard. However, because it is incompatible with so very many other commodities, extreme caution is required in handling, storage, transportation, and disposal of *o*-ethylaniline. These characteristics also require special consideration during any emergency situation involving a leak or spill of *o*-ethylaniline. Should *o*-ethylaniline ever come into contact with incompatible substances such as oxidizers, acids, or alkalis either during use, transportation, storage, or disposal, the formation of highly toxic and/or highly explosive commodities is extremely possible.

The proper disposal/destruction method for *o*-ethylaniline waste is to dissolve it in a combustible solvent and burn it in a chemical incinerator equipped with an afterburner and air scrubber. *o*-Ethylaniline can enter the environment through direct disposal and indirect use, such as industrial discharges and non-point sources from agricultural use.

Acute Ecological Effects

Acute (short-term) toxic effects may include the death of animals, birds, or fish and death or low growth rate in plants. Acute effects are seen 2 to 4 days after animals or plants are exposed to *o*-ethylaniline. This chemical has high acute toxicity to aquatic life and moderate acute toxicity to birds. No data are available on the short-term effects of *o*-ethylaniline to aquatic life, plants, or land animals. However, aniline causes germination decrease, stunting and size decreases on numerous agricultural crops.

Chronic Ecological Effects

Chronic (long-term) toxic effects may include shortened life span, reproductive problems, lower fertility, and changes in appearance or behavior in exposed animals. These effects can be seen long after first exposure(s) to toxic chemicals. No data are available on the long-term effects of *o*-ethylaniline to aquatic life, plants, birds, or land animals. However, aniline has high chronic toxicity in aquatic life.

Water Solubility

o-Ethylaniline is insoluble in water. Concentrations of less than 1 milligram will not mix with a liter of water.

Persistence in the Environment

o-Ethylaniline is slightly persistent in water, with a half-life between 2 and 20 days. The half-life of a pollutant is the amount of time it takes for one half of the chemical to be degraded.

Bioaccumulation in Aquatic Organisms

Some substances increase in concentration, or *bioaccumulate,* in living organisms as they breathe contaminated air, drink contaminated water, or eat contaminated food. These chemicals can become concentrated in the tissues and internal organs of animals as well as humans.

The concentration of *o*-ethylaniline found in fish tissues is expected to be somewhat higher than the average concentration of *o*-ethylaniline in the water from which the fish was taken.

Recommended Risk-Reduction Measures

Proper training of all transporters will reduce the likelihood of a mishap or accident resulting in a leak or spill to the environment. The correct use of DOT labeling (per the MSDS) on all containers, trucks, and rail cars will enable emergency responders to react properly and quickly to any disaster thereby reducing the potential risk to the environment and to personnel.

Storage of *o*-ethylaniline should be segregated from incompatible chemicals to minimize the risk of cross-contamination or contact.

If a spill or leak to the environment has occurred, fire department, emergency response and/or hazardous materials spill personnel should be notified immediately. Cleanup should be attempted only by those trained in proper spill containment procedures. Contaminated soils should be removed for incineration and replaced with clean soil. If *o*-ethylaniline should contact the water table, aquifer, or navigable waterway, the local and/or state emergency response authorities must be notified. A comprehensive emergency response or disaster preparedness/recovery plan should be in place prior to any operations involving the use, transportation, storage, or disposal of *o*-ethylaniline. If *o*-ethylaniline is spilled or leaked, the following specific steps are recommended:

- ☑ Restrict persons not wearing protective clothing from area of spill or leak until cleanup is complete and area can be opened for normal work.
- ☑ Remove all ignition sources.
- ☑ Ventilate are of spill or leak.
- ☑ Absorb liquids in vermiculite, dry sand, earth, or similar material and deposit in sealed containers. Wash area with detergent and water. Cover washed area with absorbent, sweep up, and place in sealed containers for disposal.
- ☑ It may be necessary to dispose of *o*-ethylaniline as a hazardous waste. The responsible state agency or the regional office of the federal Environmental Protection Agency (EPA) should be contacted for specific recommendations.

RISK ASSESSMENT: BUSINESS

General Assessment

Accidents or mishaps involving *o*-ethylaniline can present a significant threat to business operations. The loss or damage of equipment or facilities can significantly

affect fiscal viability. Lawsuits that may result from personnel injury/death, public exposures, and/or environmental contamination will also require a serious expenditure of resources. Media attention surrounding an injury, death, or environmental damage can also result in a loss of profits and loss of current as well as future business.

❶ *Recommended Risk-Reduction Measures*

Company attorneys, safety and health professionals, and environmental specialists should be involved in the development of any procedures dealing with safety or emergency response. A company official should be pre-designated as a public relations officer with specific training in dealing with the press. Corporate plans and policies should be developed, approved, and implemented long before any need for such arises.

REFERENCES

Ahlbom, A. 1993. *Biostatistics for Epidemiologists.* New York: CRC Press/Lewis Publishers.

American Conference of Governmental Industrial Hygienists. 1988. *Documentation of the Threshold Limit Values and Biological Exposure Indices,* 5th Edition (with updates). Cincinnati: ACGIH

Asante-Duah, D. K. 1993. *Hazardous Waste Risk Assessment.* New York: CRC Press/Lewis Publishers.

Bloom, A. D. and F. J. DeSerres. 1995. *Ecotoxicity and Human Health.* New York: CRC Press/Lewis Publishers.

Brusick, D. J. 1994. *Methods for Genetic Risk Assessment.* New York: CRC Press/Lewis Publishers.

Bueche, J. R. 1972. *Principles of Physics,* 2nd Edition. New York: McGraw-Hill.

Calabrese, E. J. 1994. *Biological Effects of Low Level Exposures.* New York: CRC Press/Lewis Publishers.

Cockerman, L. G., B. S. Shane. 1994. *Basic Environmental Toxicology.* New York: CRC Press/Lewis Publishers.

Cothern, R. R. 1993. *Comparative Environmental Risk Assessment.* New York: CRC Press/Lewis Publishers.

Gardner, R. W. 1994. *Chemical Intolerance.* New York: CRC Press/Lewis Publishers.

Hallenbeck, W. H. 1993. *Quantitative Risk Assessment for Environmental and Occupational Health,* 2nd Edition. New York: CRC Press/Lewis Publishers.

Howard, P. H. 1995. *Handbook of Environmental Fate and Exposures Data for Organic Chemicals,* Vol. I-IV. New York: CRC Press/Lewis Publishers.

Jones, F. E. 1994. *Toxic Organic Vapors in the Workplace.* New York: CRC Press/Lewis Publishers.

Keith, L. H. and M. M. Walker. 1995. *Handbook of Air Toxics.* New York: CRC Press/Lewis Publishers.

Landis, W. G., M. H. Yu. 1995. *Introduction to Environmental Toxicology.* New York: CRC Press/Lewis Publishers.

Langerman, N. 1994 *Precautionary Labels for Chemical Containers.* New York: CRC Press/Lewis Publishers.

Lauwerys, R. R., P. Hoet. 1993. *Industrial Chemical Exposure.* New York: CRC Press/Lewis Publishers.

Lewis, R. J., Sr. 1993. *Hawley's Condensed Chemical Dictionary,* 12th Edition. New York: Van Nostrand Reinhold.

Lewis, R. J., Sr. 1992. *Sax's Dangerous Properties of Industrial Materials,* Volumes 1, 2, and 3, 8th Edition. New York: Van Nostrand Reinhold.

Manahan, S. E. 1992. *Toxicological Chemistry,* 2nd Edition. New York: CRC Press/Lewis Publishers.

Mellström, G. A., J. E. Wahlberg, and H. I. Maibach. 1994. *Protective Gloves for Occupational Use.* New York: CRC Press/Lewis Publishers.

Meyer, E. 1990. *Chemistry of Hazardous Materials.* Englewood Cliffs, NJ: Prentice Hall.

National Cancer Institute. 1978. *Carcinogenesis Technical Report Series. Bioassay of Aniline for Possible Carcinogenicity.* DEHW (NIH) Publication Number: 78-1323, p 53. Washington, DC: U.S. Government Printing Office.

Ott, W. R. 1995. *Environmental Statistics and Data Analysis.* New York: CRC Press/Lewis Publishers.

Patnaik, P. 1992. *A Comprehensive Guide to the Hazardous Properties of Chemical Substances.* New York: Van Nostrand Reinhold.

Philp, R. B. 1995. *Environmental Hazards and Human Health.* New York: CRC Press/Lewis Publishers.

Proctor, N. H., J. P. Hughes. 1991. *Chemical Hazards of the Workplace.* New York: Van Nostrand Reinhold.

Quigley, D. R. 1994. *Handbook of Emergency Chemical Management.* New York: CRC Press/Lewis Publishers.

Rea, W. J. 1995. *Chemical Sensitivity,* Vol. I-III. New York: CRC Press/Lewis Publishers.

Richardson, M. 1995. *Dictionary of Substance and Their Effects,* Vol. I-VI. New York: CRC Press/Lewis Publishers.

Sacarello, H. L. A. 1994. *The Comprehensive Handbook of Hazardous Materials.* New York: CRC Press/Lewis Publishers.

Sheftel, V. O. 1995. *Handbook of Toxic Properties of Monomers and Additives.* New York: CRC Press/Lewis Publishers.

Talbott, E., G. Craun. 1995. *An Introduction to Environmental Epidemiology.* New York: CRC Press/Lewis Publishers.

Torkelson, T. R. V. K. Rowe. 1981. *Patty's Industrial Hygiene and Toxicology,* Vol. 2B. New York: Wiley-Interscience.

MATERIAL SAFETY DATA SHEET

CHEMICAL NAME	
P-ETHYLANILINE	**CRC PRESS/LEWIS PUBLISHERS DISCLAIMER:** THE INFORMATION AND RECOMMENDATIONS PRESENTED HEREIN ARE BASED ON SOURCES BELIEVED TO BE RELIABLE. CRC MAKES NO REPRESENTATION ON ITS COMPLETENESS OR ACCURACY. IT IS THE RESPONSIBILITY OF THE USER TO DETERMINE THE CHEMICAL'S SUITABILITY FOR ITS INTENDED USE, THE CHEMICAL'S SAFE USE, AND THE CHEMICAL'S PROPER DISPOSAL. NO REPRESENTATIONS AND/OR WARRANTIES, EITHER EXPRESSED OR IMPLIED, OF THE MERCHANTABILITY OR FITNESS FOR A PARTICULAR PURPOSE, OR OF ANY OTHER NATURE, ARE MADE WITH RESPECT TO THE INFORMATION PROVIDED IN THIS MSDS OR TO THE CHEMICAL TO WHICH INFORMATION MAY REFER. CRC NEITHER ASSUMES NOR AUTHORIZES ANY OTHER PERSON TO ASSUME FOR IT, ANY OTHER ADDITIONAL RESPONSIBILITY OR LIABILITY FOR THE USE OF, OR RELIANCE UPON, THIS INFORMATION.

HAZARD WARNING INFORMATION

HEALTH	FIRE	REACTIVE	OTHER	DEGREE OF HAZARD	COLOR CODING	OTHER CODES
3	2	0		0 = Minimum Hazard 1 = Slight Hazard 2 = Moderate Hazard 3 = Serious Hazard 4 = Severe Hazard	HEALTH = BLUE FIRE = RED REACTIVITY = YELLOW OTHER = WHITE	OX = Oxidizer ACID = Acid ALK = Alkali COR = Corrosive W = Use No Water

SECTION I - GENERAL INFORMATION

Characterization **Aromatic Amine**	RCRA Number **None**	EPA Class **Not Applicable**
DOT Proper Shipping Name ***p*-Ethylaniline**	Chemical Abstract Service (CAS) Number **589-16-2**	
DOT Hazard Class and Label Requirements **Poison B**	DOT Emergency Guide Code **57**	
DOT Identification Number **UN 2272**	Chemical Formula $C_2H_5NHC_6H_5$	

Synonyms
1-amino-4-ethylbenzene; 4-ethylaniline.

SECTION II - HAZARDOUS INGREDIENTS AND IDENTITY INFORMATION

Hazardous Components (specific identity)	OSHA Exposure Criteria	NIOSH Exposure Criteria	Immediately Dangerous to Life and Health (IDLH)	ACGIH Exposure Criteria
***p*-Ethylaniline** (derivation: By heating aniline and ethyl alcohol in presence of sulfuric acid with subsequent distillation).	**PEL:** Not Established **STEL:** Not Established	**REL:** Not Established **STEL:** Not Established	**Not Determined**	**TLV:** Not Established **STEL:** Not Established

SECTION III - PHYSICAL AND CHEMICAL CHARACTERISTICS

Boiling Point **401°F (205°C)**	Specific Gravity (H_2O = 1) **0.963**
Vapor Pressure (mm Hg) **<1 at 69°F (20°C)**	Molecular Weight **121.2**
Vapor Density (Air = 1) **4.2**	Melting Point **-87°F (-66°C)**

Solubility
Insoluble in water. Soluble in alcohol and ether.

Appearance and Odor
Oily, colorless or pale yellow liquid which darkens to brown on exposure to light; aromatic, amine-like odor and burning taste.

SECTION IV - FIRE AND EXPLOSION HAZARD DATA

Flash Point (method used) **185°F (85°C) opened cup**	Explosive Limits in Air % by Volume **LEL: 1.6% UEL: 9.5%**
NFPA Classification **Class IIIA Combustible Liquid**	Autoignition Temperature **Not Determined**

Extinguishing Media
Carbon dioxide, dry chemical, alcohol foam, water spray or fog.

Special Fire Fighting Procedures
Wear full protective clothing and self-contained breathing apparatus (SCBA). Use a water spray to keep fire-exposed containers cool and remove from fire if possible. Poisonous gases are produced in fire.

Unusual Fire and Explosion Hazards
Moderately flammable when exposed to heat. Containers may explode in fire. Firefighters should avoid contact with vapors produced during fire. Hypergolic reaction with red fuming nitric acid. Can react with oxidizers to cause fire or explosion.

SECTION V - REACTIVITY DATA

Stability		Conditions to Avoid *p*-Ethylaniline is considered stable in closed containers under normal conditions of temperature and pressure. Avoid contact with incompatible materials and keep away from heat or flame.
Stable X	Unstable	Incompatibility *(materials to avoid)* Strong oxidizers (such as perchlorates, peroxides, permanganates, chlorates and nitrates), and strong acids (such as hydrochloric, sulfuric, and nitric). Reacts hypergolically with red fuming nitric acid.
Hazardous Polymerization		Conditions to Avoid Under normal conditions of temperature and pressure, hazardous polymerization of *p*-ethylaniline is not expected to occur.
May Occur	Will Not Occur X	Hazardous Decomposition or By-products When heated to decomposition, *p*-ethylaniline can emit highly toxic/poisonous fumes of aniline and oxides of nitrogen.

SECTION VI - HEALTH HAZARD DATA

Primary Route(s) of Entry:	Inhalation?	Absorption (skin)?	Ingestion?
	X	X	X

Health Hazards

INHALATION: Causes anoxia due to the formation of methemoglobin. Symptoms: headache, dizziness, ataxia, and cyanosis. Overexposure may lead to death due to respiratory paralysis.

ABSORPTION: Absorbs freely through intact skin. Symptoms include headache and cyanosis; overexposure can result in death. Skin contact with pure liquid can produce mild irritation.

INGESTION: Unspecified effects resulting from ingestion. Liver damage is suspect.

Carcinogenicity	NTP Listed?	IARC Cancer Review Group?	OSHA Regulated?	Target Organs?
Unknown Human Unknown Animal	No	No	No	Respiratory system, eyes, skin, blood, kidney, liver.

Medical Conditions Generally Aggravated by Exposure

Respiratory impairments (bronchitis, asthma); Skin conditions (dermatitis); Renal or liver disorders.

Emergency and First-aid Procedures

Eye contact: Flush immediately with water for 15 minutes (minimum), seek medical attention. Skin contact: Remove all contaminated clothing. Immediately wash area with large amounts of soap and water. Seek medical assistance. For inhalation: Remove the person from exposure. Provide respiratory assistance and CPR. Transfer to medical facility. If swallowed: Seek medical attention immediately. Unless advised otherwise, give 1-2 glasses of water and induce vomiting.

SECTION VII - PRECAUTIONS FOR SAFE HANDLING AND USE

Steps to be Taken in Case Material is Released or Spilled

Absorb liquids in vermiculite, dry sand, earth, or similar material and deposit in sealed containers. Ventilate area of spill or leak and remove ignition sources. Restrict those not involved in cleanup from entering area. Notify appropriate authorities, if applicable.

Preferred Waste Disposal Method

Dissolve in combustible solvent; burn in chemical incinerator equipped with scrubber and afterburner.

Precautions to be Taken in Handling and Storage

p-Ethylaniline is a combustible liquid. Store to avoid contact with strong oxidizers and acids, since violent reactions can occur. Store in tightly closed containers in cool, well-ventilated area away from heat. Ground and bond all containers prior to transferring liquids.

Other Precautions and Warnings

Sources of ignition are prohibited where *p*-Ethylaniline is used, handled, or stored. Bulk storage of *p*-Ethylaniline is not recommended.

SECTION VIII - CONTROL MEASURES AND PERSONAL PROTECTIVE EQUIPMENT

Respiratory Protection *(specify type)*

Exposure limits have not been established. For best protection, use a supplied air respirator with full facepiece, hood, or helmet in continuous flow mode, or a self-contained breathing apparatus (SCBA) with full facepiece operated in positive pressure mode.

Ventilation

Local exhaust at site of chemical work.

Protective Gloves	Eye Protection	Other Protective Clothing
Natural Rubber	Chemical Goggles or Face Mask	Rubber Apron

Work/Hygiene Practices

Always wash hands thoroughly after using chemical; never bring food, drink, or smoking materials into vicinity of chemicals.

P-ETHYLANILINE

$C_2H_5NHC_6H_5$ CAS: 589-16-2

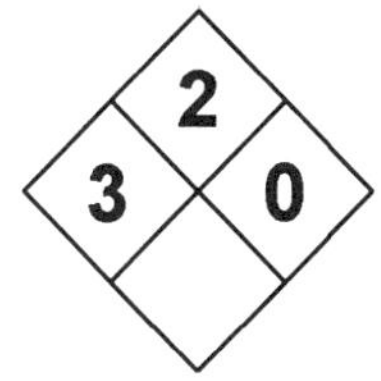

IDENTIFICATION AND TYPICAL USES

p-Ethylaniline is an oily, colorless or pale yellow liquid with characteristic odor and burning taste; color darkens to brown or tan upon exposure to air. *p*-Ethylaniline is an intermediate in chemical synthesis and, as an aniline compound, it is among one of the most important of the organic bases.

RISK ASSESSMENT: HEALTH

General Assessment

p-Ethylaniline is moderately toxic through *inhalation, ingestion,* and *absorption* through the skin. While *p*-ethylaniline has not been adequately tested for its ability to cause cancer or for its effects on reproduction, it should be noted that aniline itself causes mutations (genetic changes) in living cells.

Toxic symptoms of *p*-ethylaniline exposure by all routes include headache, weakness, dizziness, ataxia, and cyanosis (blue tinting of the lips, skin, and nail beds due to oxygen deprivation). Cyanosis is due to the formation of methemoglobin which restricts the blood's oxygen carrying capabilities. Cyanosis occurs when the methemoglobin concentration is 15% or more. Blueness develops first in the lips, nose, and earlobes. At 40% concentration, there is weakness and dizziness. At up to 70% concentration, symptoms may include ataxia, dyspnea on mild exertion, and tachycardia. Coma may ensue at concentrations over 70%. Overexposure (between 85% and 90% concentrations) may lead to death from respiratory paralysis.

Contact with the pure liquid *p*-ethylaniline on the skin can produce moderate irritation, while the effect on the eyes can be quite severe.

Acute Health Effects

The following acute (short-term) health effects may occur immediately or shortly after exposure to *p*-ethylaniline:

Skin: Irritation and possible color change around nails and fingers. Absorbs freely through the skin.

Eye: Severe irritation; may cause corneal damage.

Lung: Not known to be acutely irritating to lung tissue. However, overexposure can lead to death due to respiratory arrest.

Other: *p*-Ethylaniline affects the blood's ability to transport oxygen and high exposure can cause death. Symptoms include headaches, weakness, irritability, drowsiness, and shortness of breath.

Chronic Health Effects

The following chronic (long-term) health effects may occur at some time after exposure to *p*-ethylaniline and can last for months or even years:

Cancer Hazards: *p*-Ethylaniline has not been adequately tested for its ability to cause cancer in test animals. Aniline, however, may cause mutations (genetic changes) in living cells. The data on its carcinogenicity are inconclusive.

Reproductive Hazard: There is no evidence to show that exposure to *p*-ethylaniline causes damage to the developing fetus. Aniline, however, may cause such effects (more study is required in this area).

Other Chronic Effects: Continuous exposure to small doses of *p*-ethylaniline may produce anemia, loss of energy, digestive disturbance, and headache. Repeated exposures may cause anemia and brownish staining of the eyes can occur.

Recommended Risk-Reduction Measures

Personnel should avoid direct contact with *p*-ethylaniline. If a less toxic chemical cannot be substituted for a hazardous substance, then *engineering controls* are the most effective method of reducing exposures. The best protection is to enclose operations and/or provide local exhaust ventilation at the site of chemical release. While not always operationally feasible, isolating operations can also reduce exposure. Using respiratory protection is less effective than the controls mentioned above, but is still advisable whenever working with or around *p*-ethylaniline. Since no exposure limits have been established, an air-purifying (cartridge-type) respirator may not provide adequate protection. A supplied-air respirator operated in positive pressure mode or a self-contained breathing apparatus (SCBA) with full facepiece and operated in pressure demand are the recommended respiratory protection methods of choice. If a full facepiece is not available, then chemical goggles should be worn to protect the eyes. Whenever a chemical splash hazard exists, a face shield and a rubber apron should be worn. To prevent hand and skin exposures, natural rubber gloves should be worn.

Administrative controls should also be in place to minimize the potential for human exposures. These may include written procedures or policies which specify the methods and techniques that will be practiced whenever personnel are to work with *p*-ethylaniline.

All personnel should receive *training* on the use, hazards, protective measures, emergency actions, and other precautions per 29 CFR 1910.1200 (Hazard Communication), prior to the first assignment in an area where *p*-ethylaniline is used or stored.

If symptoms develop or overexposure is suspected, the following medical tests may be useful:

☑ Methemoglobin tests.

☑ Urinalysis (for blood content).

Any evaluation should include a careful history of past and present symptoms with an examination. Medical tests that simply look for existing damage are not a substitute for controlling exposures.

Other methods to reduce exposure include:

☑ Where possible, enclose operations and use local exhaust ventilation at the site of chemical release. If local exhaust ventilation or enclosure is not used, respiratory protection should be mandatory.

☑ Always ensure that proper protective clothing is worn when using chemical substances.

☑ Wash thoroughly immediately after exposure to *p*-ethylaniline and at the end of the work shift or before eating, drinking, or smoking.

☑ Hazard warning information should be posted in the work area. In addition, as part of an on-going education and training program, all information on the health and safety hazards of *p*-ethylaniline should be communicated to all potentially exposed workers.

☑ Contaminated clothing (including shoes and undergarments) should never worn home where family members can be exposed. Work clothing should be laundered only by personnel who have been briefed on the hazards of *p*-ethylaniline.

☑ Eye wash stations should be provided in the immediate work area for emergency use. If there is a possibility of skin exposure to *p*-ethylaniline, emergency shower facilities should be provided.

RISK ASSESSMENT: ENVIRONMENT

General Assessment

The environment is at risk of exposure during transportation, storage, disposal, or destruction of *p*-ethylaniline. In almost every scenario, the threat of environmental exposure is contingent upon the proper handling of the chemical substance. Accidental spills, large or small, can result in fire, explosion, and possible contamination of the surrounding environmental mediums (water, soil, and air).

p-Ethylaniline is considered a class IIIA combustible liquid. Alone, its high flash point and autoignition temperature do not present a serious fire or explosion hazard. However, because it is incompatible with so very many other commodities, extreme caution is required in handling, storage, transportation, and disposal of *p*-ethylaniline. These characteristics also require special consideration during any emergency situation involving a leak or spill of *p*-ethylaniline. Should *p*-ethylaniline ever come into contact with incompatible substances such as oxidizers, acids, or alkalis either during use, transportation, storage, or disposal, the formation of highly toxic and/or highly explosive commodities is extremely possible.

The proper disposal/destruction method for *p*-ethylaniline waste is to dissolve it in a combustible solvent and burn it in a chemical incinerator equipped with an afterburner and air scrubber. *p*-Ethylaniline can enter the environment through direct disposal and indirect use, such as industrial discharges and non-point sources from agricultural use.

Acute Ecological Effects

Acute (short-term) toxic effects may include the death of animals, birds, or fish and death or low growth rate in plants. Acute effects are seen 2 to 4 days after animals or plants are exposed to *p*-ethylaniline. This chemical has high acute toxicity to aquatic life and moderate acute toxicity to birds. No data are available on the short-term effects of *p*-ethylaniline to aquatic life, plants, or land animals. However, aniline causes germination decrease, stunting and size decreases on numerous agricultural crops.

Chronic Ecological Effects

Chronic (long-term) toxic effects may include shortened life span, reproductive problems, lower fertility, and changes in appearance or behavior in exposed animals. These effects can be seen long after first exposure(s) to toxic chemicals. No data are available on the long-term effects of *p*-ethylaniline to aquatic life, plants, birds, or land animals. However, aniline has high chronic toxicity in aquatic life.

Water Solubility

p-Ethylaniline is slightly soluble in water. Concentrations of 1 to 100 milligrams will not mix with a liter of water.

Persistence in the Environment

p-Ethylaniline is slightly persistent in water, with a half-life between 2 and 20 days. The half-life of a pollutant is the amount of time it takes for one half of the chemical to be degraded.

Bioaccumulation in Aquatic Organisms

Some substances increase in concentration, or *bioaccumulate,* in living organisms as they breathe contaminated air, drink contaminated water, or eat contaminated food. These chemicals can become concentrated in the tissues and internal organs of animals as well as humans.

The concentration of *p*-ethylaniline found in fish tissues is expected to be somewhat higher than the average concentration of *p*-ethylaniline in the water from which the fish was taken.

Recommended Risk-Reduction Measures

Proper training of all transporters will reduce the likelihood of a mishap or accident resulting in a leak or spill to the environment. The correct use of DOT labeling (per the MSDS) on all containers, trucks, and rail cars will enable emergency responders to react properly and quickly to any disaster thereby reducing the potential risk to the environment and to personnel.

Storage of *p*-ethylaniline should be segregated from incompatible chemicals to minimize the risk of cross-contamination or contact.

If a spill or leak to the environment has occurred, fire department, emergency response and/or hazardous materials spill personnel should be notified immediately. Cleanup should be attempted only by those trained in proper spill containment procedures. Contaminated soils should be removed for incineration and replaced with clean soil. If *p*-ethylaniline should contact the water table, aquifer, or navigable waterway, the local and/or state emergency response authorities must be notified. A comprehensive emergency response or disaster preparedness/recovery plan should be in place prior to any operations involving the use, transportation, storage, or disposal of *p*-ethylaniline. If *p*-ethylaniline is spilled or leaked, the following specific steps are recommended:

- ☑ Restrict persons not wearing protective clothing from area of spill or leak until cleanup is complete.
- ☑ Remove all ignition sources.
- ☑ Ventilate area of spill or leak.
- ☑ Absorb liquids in vermiculite, dry sand, earth, or similar material and deposit in sealed containers.
- ☑ It may be necessary to dispose of *p*-ethylaniline as a hazardous waste. The responsible state agency or the regional office of the federal Environmental Protection Agency (EPA) should be contacted for specific recommendations.

RISK ASSESSMENT: BUSINESS

General Assessment

Accidents or mishaps involving *p*-ethylaniline can present a significant threat to business operations. The loss or damage of equipment or facilities can significantly affect fiscal viability. Lawsuits that may result from personnel injury/death, public exposures, and/or environmental contamination will also require a serious expenditure of resources. Media attention surrounding an injury, death, or environmental damage can also result in a loss of profits and loss of current as well as future business.

⊍ *Recommended Risk-Reduction Measures*

Company attorneys, safety and health professionals, and environmental specialists should be involved in the development of any procedures dealing with safety or emergency response. A company official should be pre-designated as a public relations officer with specific training in dealing with the press. Corporate plans and policies should be developed, approved, and implemented long before any need for such arises.

REFERENCES

Ahlbom, A. 1993. *Biostatistics for Epidemiologists.* New York: CRC Press/Lewis Publishers.

American Conference of Governmental Industrial Hygienists. 1988. *Documentation of the Threshold Limit Values and Biological Exposure Indices,* 5th Edition (with updates). Cincinnati: ACGIH

Asante-Duah, D. K. 1993. *Hazardous Waste Risk Assessment.* New York: CRC Press/Lewis Publishers.

Bloom, A. D. and F. J. DeSerres. 1995. *Ecotoxicity and Human Health.* New York: CRC Press/Lewis Publishers.

Brusick, D. J. 1994. *Methods for Genetic Risk Assessment.* New York: CRC Press/Lewis Publishers.

Bueche, J. R. 1972. *Principles of Physics,* 2nd Edition. New York: McGraw-Hill.

Calabrese, E. J. 1994. *Biological Effects of Low Level Exposures.* New York: CRC Press/Lewis Publishers.

Cockerman, L. G., B. S. Shane. 1994. *Basic Environmental Toxicology.* New York: CRC Press/Lewis Publishers.

Cothern, R. R. 1993. *Comparative Environmental Risk Assessment.* New York: CRC Press/Lewis Publishers.

Gardner, R. W. 1994. *Chemical Intolerance.* New York: CRC Press/Lewis Publishers.

Hallenbeck, W. H. 1993. *Quantitative Risk Assessment for Environmental and Occupational Health,* 2nd Edition. New York: CRC Press/Lewis Publishers.

Howard, P. H. 1995. *Handbook of Environmental Fate and Exposures Data for Organic Chemicals,* Vol. I-IV. New York: CRC Press/Lewis Publishers.

Jones, F. E. 1994. *Toxic Organic Vapors in the Workplace.* New York: CRC Press/Lewis Publishers.

Keith, L. H. and M. M. Walker. 1995. *Handbook of Air Toxics.* New York: CRC Press/Lewis Publishers.

Landis, W. G., M. H. Yu. 1995. *Introduction to Environmental Toxicology.* New York: CRC Press/Lewis Publishers.

Langerman, N. 1994 *Precautionary Labels for Chemical Containers.* New York: CRC Press/Lewis Publishers.

Lauwerys, R. R., P. Hoet. 1993. *Industrial Chemical Exposure.* New York: CRC Press/Lewis Publishers.

Lewis, R. J., Sr. 1993. *Hawley's Condensed Chemical Dictionary,* 12th Edition. New York: Van Nostrand Reinhold.

Lewis, R. J., Sr. 1992. *Sax's Dangerous Properties of Industrial Materials,* Volumes 1, 2, and 3, 8th Edition. New York: Van Nostrand Reinhold.

Manahan, S. E. 1994. *Environmental Chemistry,* 6th Edition. New York: CRC Press/Lewis Publishers.

Manahan, S. E. 1992. *Toxicological Chemistry,* 2nd Edition. New York: CRC Press/Lewis Publishers.

Mellström, G. A., J. E. Wahlberg, and H. I. Maibach. 1994. *Protective Gloves for Occupational Use.* New York: CRC Press/Lewis Publishers.

Meyer, E. 1990. *Chemistry of Hazardous Materials.* Englewood Cliffs, NJ: Prentice Hall.

Mycock, J. C., J. D. McKenna, and L. Theodore. 1995. *Handbook of Air Pollution Control Engineering and Technology.* New York: CRC Press/Lewis Publishers.

National Cancer Institute. 1978. *Carcinogenesis Technical Report Series. Bioassay of Aniline for Possible Carcinogenicity.* DEHW (NIH) Publication Number: 78-1323, p 53. Washington, DC: U.S. Government Printing Office.

Ott, W. R. 1995. *Environmental Statistics and Data Analysis.* New York: CRC Press/Lewis Publishers.

Patnaik, P. 1992. *A Comprehensive Guide to the Hazardous Properties of Chemical Substances.* New York: Van Nostrand Reinhold.

Philp, R. B. 1995. *Environmental Hazards and Human Health.* New York: CRC Press/Lewis Publishers.

Proctor, N. H., J. P. Hughes. 1991. *Chemical Hazards of the Workplace.* New York: Van Nostrand Reinhold.

Quigley, D. R. 1994. *Handbook of Emergency Chemical Management.* New York: CRC Press/Lewis Publishers.

Rea, W. J. 1995. *Chemical Sensitivity,* Vol. I-III. New York: CRC Press/Lewis Publishers.

Richardson, M. 1995. *Dictionary of Substance and Their Effects,* Vol. I-VI. New York: CRC Press/Lewis Publishers.

Sacarello, H. L. A. 1994. *The Comprehensive Handbook of Hazardous Materials.* New York: CRC Press/Lewis Publishers.

Sheftel, V. O. 1995. *Handbook of Toxic Properties of Monomers and Additives.* New York: CRC Press/Lewis Publishers.

Talbott, E., G. Craun. 1995. *An Introduction to Environmental Epidemiology.* New York: CRC Press/Lewis Publishers.

Torkelson, T. R. V. K. Rowe. 1981. *Patty's Industrial Hygiene and Toxicology,* Vol. 2B. New York: Wiley-Interscience.

U.S. Department of Health and Human Services, National Institute for Occupational Safety and Health. 1994. *NIOSH Pocket Guide to Chemical Hazards.* Washington, D.C.: U.S. Government Printing Office.

MATERIAL SAFETY DATA SHEET

CHEMICAL NAME	CRC PRESS/LEWIS PUBLISHERS DISCLAIMER
ETHYL BENZENE	

HAZARD WARNING INFORMATION

HEALTH	FIRE	REACTIVE	OTHER	DEGREE OF HAZARD	COLOR CODING	OTHER CODES
2	3	0		0 = Minimum Hazard 1 = Slight Hazard 2 = Moderate Hazard 3 = Serious Hazard 4 = Severe Hazard	HEALTH = BLUE FIRE = RED REACTIVITY = YELLOW OTHER = WHITE	OX = Oxidizer ACID = Acid ALK = Alkali COR = Corrosive ~~W~~ = Use No Water

SECTION I - GENERAL INFORMATION

Characterization	RCRA Number	EPA Class
Aromatic Hydrocarbon	**D001**	**Characteristic (I) Waste**

DOT Proper Shipping Name	**Ethylbenzene**	Chemical Abstract Service (CAS) Number	**100-41-4**
DOT Hazard Class and Label Requirements	**Flammable Liquid**	DOT Emergency Guide Code	**26**
DOT Identification Number	**UN 1175**	Molecular Formula	$CH_3CH_2C_6H_5$

Synonyms
Phenylethane; EB; ethyl benzol.

SECTION II - HAZARDOUS INGREDIENTS AND IDENTITY INFORMATION

Hazardous Components (specific identity)	OSHA Exposure Criteria	NIOSH Exposure Criteria	Immediately Dangerous to Life and Health (IDLH)	ACGIH Exposure Criteria
Ethyl benzene (derivation: By heating benzene and ethylene in the presence of aluminum chloride with subsequent distillation; by fractionation directly from the mixed xylene stream in petroleum refining). **1 ppm = 4.41 mg/m^3**	**PEL:** **100 ppm** **435 mg/m^3** **STEL (15 min):** **125 ppm** **545 mg/m^3**	**REL:** **100 ppm** **435 mg/m^3** **STEL:** **125 ppm** **545 mg/m^3**	**800 ppm**	**TLV:** **100 ppm** **435 mg/m^3** **STEL (15 min):** **125 ppm** **545 mg/m^3**

SECTION III - PHYSICAL AND CHEMICAL CHARACTERISTICS

Boiling Point	**227°F (136°C)**	Specific Gravity (H_2O = 1)	**0.87**
Vapor Pressure (mm Hg)	**7 at 69°F (20°C)**	Molecular Weight (atomic weight)	**106.2**
Vapor Density (Air = 1)	**3.7**	Melting Point	**-139°F (-95°C)**

Solubility
Slightly soluble in water (0.01%). Miscible in alcohol and ether. Soluble in carbon tetrachloride, benzene, sulfur dioxide, and most organic solvents. Insoluble in ammonia.

Appearance and Odor
Colorless liquid with an aromatic odor.

SECTION IV - FIRE AND EXPLOSION HAZARD DATA

Flash Point (method used)	**64°F (18°C) closed cup**	Explosive Limits in Air % by Volume	**LEL: 1.0%** **UEL: 6.7%**
NFPA Classification	**Class 1B Flammable Liquid**	Autoignition Temperature	**810°F (432°C)**

Extinguishing Media
Use dry chemical, carbon dioxide, or alcohol resistant foam. Water may be ineffective (ethyl benzene will float on water and may travel to an ignition source).

Special Fire Fighting Procedures
Wear self-contained breathing apparatus (SCBA). Poisonous gases are produced in fire. Move containers from fire if it can be done without risk. Cool fire-exposed containers with water even after fire is out.

Unusual Fire and Explosion Hazards
Dangerous fire and explosion hazard. Containers may explode in fire. Immediately withdraw if rising sound is heard from venting device or there is a discoloration on the sides of tanks or containers.

SECTION V - REACTIVITY DATA

Stability		Conditions to Avoid
		Under normal conditions of handling and storage, ethyl benzene is considered stable. Avoid heat, flame, other sources of ignition, and contact with incompatible materials and chemicals.
Stable	Unstable	Incompatibility *(materials to avoid)*
X		Strong oxidizers (chlorine, bromine, fluorine).
Hazardous Polymerization		Conditions to Avoid
		Hazardous polymerization has not been reported to occur under normal conditions of temperature and pressure.
May Occur	Will Not Occur	Hazardous Decomposition or By-products
	X	Poisonous gases and acrid fumes are produced in fire.

SECTION VI - HEALTH HAZARD DATA

Primary Route(s) of Entry:	Inhalation?	Absorption (skin/eye)?	Ingestion?
	X	X	X

Health Hazards

INHALATION: May cause nausea, headache, vomiting, dizziness, narcosis, suffocation, lower blood pressure, CNS depression. Vapors may cause severe irritation or burns of the respiratory system, pulmonary edema, or lung inflammation. Causes narcosis, coma, and death.

ABSORPTION: Irritation with possible redness and pain to the skin and eyes. Also, the liquid will absorb through the skin but the vapor will not. May cause dermatitis. Eye contact may result in severe irritation and possible inflammation of the cornea.

INGESTION: Unspecified gastrointestinal effects.

Carcinogenicity	NTP Listed?	IARC Cancer Review Group?	OSHA Regulated?	Target Organs?
Unknown Human Unknown Animal	No	No	29 CFR 1910.1000 Table Z-1	Eyes, skin, respiratory sys. blood, CNS

Medical Conditions Generally Aggravated by Exposure

Skin and CNS diseases and respiratory dysfunction (obstructive airway disease).

Emergency and First-aid Procedures

Eye contact: Do not allow victim to keep eyes tightly shut. Flush immediately with water for 15 minutes (minimum), seek medical attention. Skin contact: Remove all contaminated clothing, including shoes. Immediately wash area with flooding amounts of soap and water. Seek medical attention immediately. For inhalation: Remove the person from exposure. Provide respiratory assistance and CPR. Transfer to medical facility. Observe for 24-48 hours for lung effects. If swallowed: Seek medical attention immediately. Do not induce vomiting (aspiration of vomitus can cause serious damage to lung tissue).

SECTION VII - PRECAUTIONS FOR SAFE HANDLING AND USE

Steps to be Taken in Case Material is Released or Spilled

Remove all ignition sources, wear SCBA respiratory protection. Restrict those not wearing protective equipment and who are not involved in cleanup from entering area. Stop leak if it can be done without risk. Ventilate area of spill. Absorb with vermiculite and deposit in sealed containers.

Preferred Waste Disposal Method

Burn in chemical incinerator equipped with an afterburner and scrubber.

Precautions to be Taken in Handling and Storage

Store in tightly closed containers in cool, well ventilated location. Keep away from sources of heat.

Other Precautions and Warnings

Sources of ignition such as open flame or smoking, are prohibited where ethyl benzene is used or stored. Metal containers should be bonded and grounded. Bulk storage not recommended.

SECTION VIII - CONTROL MEASURES AND PERSONAL PROTECTIVE EQUIPMENT

Respiratory Protection *(specify type)*

Use an MSHA/NIOSH-approved air purifying (cartridge type) respirator or use a supplied-air respirator or a self-contained breathing apparatus (SCBA) with full facepiece operated in positive pressure mode.

Ventilation

Local exhaust (preferred) at site of chemical work or general ventilation.

Protective Gloves	Eye Protection	Other Protective Clothing
Polyvinylchloride or Viton	Chemical Goggles or Face Mask	Protective Apron if Splash is Likely

Work/Hygiene Practices

Always wash hands thoroughly after using chemical; never bring food, drink, or smoking materials into vicinity of chemicals.

ETHYL BENZENE

$CH_3CH_2C_6H_5$ CAS: 100-41-4

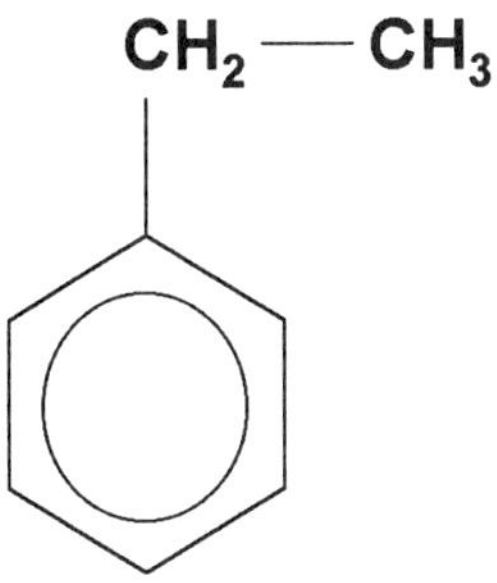

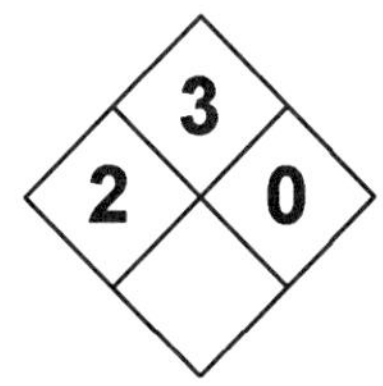

IDENTIFICATION AND TYPICAL USES

Ethyl benzene is a clear, colorless liquid with a pungent, aromatic odor. It is used as a solvent, as an antiknock agent in gasoline, and as an intermediate to produce styrene monomer and synthetic rubber.

RISK ASSESSMENT: HEALTH

General Assessment

Industrial exposures to ethyl benzene itself are unlikely under normal operating conditions. This is because ethyl benzene is usually present in mixtures with other solvents. However, if exposure does occur, the health risks can be serious. Ethyl benzene is moderately toxic by *ingestion* and mildly toxic by *inhalation* and skin contact (*absorption*). While it is not listed by any sources as a carcinogen, human mutation data have been reported and it has caused teratogenic effects in laboratory test animals.

Inhalation of ethyl benzene vapors causes eye irritation with profuse watering of the eyes, nasal irritation, chest constriction, and vertigo. Symptoms worsen as vapor concentrations increase. As levels approach 5000 ppm, exposure becomes intolerable and causes nasal and eye irritation. There may be changes in sleeping patterns and changes in pulmonary function. Irritating properties may cause pulmonary edema (fluid buildup in the lungs), which is a medical emergency. Symptoms can be delayed up to 48 hours, thereby creating a false sense of security with regard to health hazard risk. Higher or prolonged concentrations will result central nervous system (CNS) effects with narcosis, cramps, lightheadedness, dizziness, and even death due to respiratory paralysis.

Contact with the skin causes drying, cracking, scaling, and even blistering. There may be 1st degree burns on short exposures and 2nd degree burns on prolonged contact. Ethyl benzene will pass through unbroken skin to cause toxic systemic effects.

Acute Health Effects

The following acute (short-term) health effects may occur immediately or shortly after exposure to ethyl benzene:

Skin: Irritation with potential for redness, blistering, and serious burning. Chemical will pass through intact skin to cause toxic systemic effects, including narcosis.

Eye: Severe irritation, burning, pain, and permanent damage possible.

Lung: Nose, throat, and respiratory tract irritation causing difficulty in breathing, possible pulmonary edema (fluid in lungs) on high exposure.

CNS: Narcosis, headaches, lightheadedness, dizziness, vertigo, coma, and death to due respiratory paralysis.

Chronic Health Effects

The following chronic (long-term) health effects may occur at some time after exposure to ethyl benzene and can last for months or even years:

Cancer Hazards: Ethyl benzene has not been adequately tested for its ability to cause cancer in test animals. However, human mutation (genetic changes) data have been reported. Many scientists believe that mutagenic chemicals may have the ability to cause cancer in the long-term. Further research is required in this regard.

Reproductive Hazard: There is limited evidence that ethyl benzene is a teratogen in animals. Until further tests are performed, it should be treated as a possible human teratogen.

Other Chronic Effects: Repeated skin contact can cause chronic dryness and scaling of the skin's surface. Prolonged contact may cause blistering. Repeated inhalation of ethyl benzene vapors may result in liver damage and blood disorders, particularly leukopenia (deficiency in white blood cell count) and lymphocytosis (the formation of cysts containing lymph). This chemical has not been adequately evaluated to determine whether brain or other nerve damage could occur with repeated exposure. However, many other solvents and petroleum-based chemicals have been known to cause such damage. Effects include reduced memory and concentration, personality changes (irritability, withdrawal), fatigue, sleep disturbances, reduced coordination, and/or effects on the nerves supplying the internal organs (autonomic nerves) and/or nerves to the arms and legs with weakness and a feeling of "pins and needles."

Recommended Risk-Reduction Measures

Personnel should avoid direct contact with ethyl benzene. Occupational poisoning occurs most commonly by inhalation and absorption. If a less toxic material or compound cannot be substituted for ethyl benzene, then *engineering controls* are the most effective method of reducing exposures. The best protection is to enclose operations and/or provide local exhaust ventilation at the site of ethyl benzene release. While not always operationally feasible, isolating operations can also reduce exposure. Using respiratory protection is less effective than the controls mentioned above, but is still advisable whenever working with or around ethyl benzene. For exposure over the PEL (100 ppm), an MSHA/NIOSH-approved air-purifying respiratory with organic vapor cartridge(s) will suffice. For exposures to higher concentrations, or when the concentration is unknown, use a supplied air respirator with full facepiece and mask operated in positive pressure mode or a self-contained breathing apparatus (SCBA) with full facepiece operated in pressure demand. If a full facepiece is not available, then chemical goggles should be worn to protect the eyes. A face shield should also be considered. To prevent hand and skin exposures, polyvinylchloride or Viton gloves should be worn.

Administrative controls should also be in place to minimize the potential for human exposures. These may include written procedures or policies which specify the methods and techniques that will be practiced whenever personnel are to work with ethyl benzene.

All personnel should receive *training* on the use, hazards, protective measures, emergency actions, and other precautions per 29 CFR 1910.1200 (Hazard Communication), prior to the first assignment in an area where ethyl benzene is used or stored.

If symptoms develop or overexposure is suspected, the following medical tests should be considered:

☑ Liver function tests.

☑ Lung function tests.

☑ Consider chest X-ray following acute overexposure (may be negative if taken immediately after exposure).

☑ Urinalysis (mandelic acid in urine, 1.5 g/g of creatinine), sample at end of shift, last day of workweek. Also test for expired ethyl benzene in air for confirmation.

☑ Interview for brain effects, including recent changes in memory, mood, or concentration, headaches, malaise, and altered sleeping patterns. Consider cerebellar, autonomic, and peripheral nervous system evaluation. Positive and borderline cases should be referred for neuropsychological testing.

Any evaluation should include a careful history of past and present symptoms with an examination. Medical tests that simply look for existing damage are not a substitute for controlling exposures.

Other methods to reduce exposure include:

☑ Where possible, enclose operations and use local exhaust ventilation at the site of chemical release. If local exhaust ventilation or enclosure is not used, respiratory protection should be mandatory.

☑ Always ensure that proper protective clothing is worn when using chemical substances.

☑ Wash thoroughly immediately after exposure to ethyl benzene and at the end of the work shift or before eating, drinking, or smoking.

☑ Hazard warning information should be posted in the work area. In addition, as part of an on-going education and training program, all information on the health and safety hazards of ethyl benzene should be communicated to all exposed and potentially exposed workers.

☑ Eye wash stations should be provided in the immediate work area for emergency use. If there is a possibility of skin exposure to ethyl benzene, emergency shower facilities should be provided.

☑ Workers whose clothing has been contaminated by ethyl benzene should change into clean clothes before leaving work. Contaminated work clothing should be laundered only by individuals who have been informed of the hazards of ethyl benzene.

☑ Where possible, automatically transfer liquids containing ethyl benzene from drums or other containers to process containers.

RISK ASSESSMENT: ENVIRONMENT

General Assessment

The environment is at risk of exposure during transportation, storage, disposal, or destruction of ethyl benzene. In almost every scenario, the threat of environmental exposure is contingent upon the proper handling of the chemical substance. Accidental spills, large or small, can result in contamination of the surrounding environmental mediums (water, soil, and air).

Ethyl benzene is an moderately flammable liquid. Because of its low flash point and high boiling point, it is classified as a Class 1B flammable liquid (per OSHA 29 CFR 1910.106). It is a dangerous fire hazard and moderate explosion hazard when exposed to heat or flame. Caution is always required in handling, storage, transportation, and disposal of ethyl benzene. Emergency responders should be made aware of the presence of ethyl benzene at any emergency response situation.

Ethyl benzene can enter the environment from industrial effluents and from spills.

Acute Ecological Effects

Acute (short-term) toxic effects may include the death of animals, birds, or fish and death or low growth rate in plants. Acute effects are seen 2 to 4 days after animals or plants are exposed to ethyl benzene. Ethyl benzene has high acute toxicity to aquatic life. It has caused injury to various agricultural crops. Insufficient data are available to evaluate the short-term effects of ethyl benzene on birds or land animals.

Chronic Ecological Effects

Chronic (long-term) toxic effects may include shortened life span, reproductive problems, lower fertility, and changes in appearance or behavior in exposed animals. These effects can be seen long after first exposure(s) to toxic chemicals. Ethyl benzene has high chronic toxicity to aquatic life. Insufficient data are available to evaluate or predict the long-term effects of ethyl benzene to plants, birds, or terrestrial animals.

Water Solubility

Ethyl benzene is slightly to moderately soluble in water. Concentrations between 1 and 100 milligrams will mix with a liter of water.

Persistence in the Environment

Ethyl benzene is non-persistent in water (due to volatilization), with a half-life of less than 2 days. The half-life of a pollutant is the amount of time it takes for one half of the chemical to be degraded. About 99.5% of ethyl benzene will eventually end up in air; the rest will end up in water.

Bioaccumulation in Aquatic Organisms

Some substances increase in concentration, or *bioaccumulate,* in living organisms as they breathe contaminated air, drink contaminated water, or eat contaminated food. These chemicals can become concentrated in the tissues and internal organs of animals as well as humans.

The concentration of ethyl benzene found in fish tissues is expected to be somewhat higher than the average concentration of ethyl benzene in the water from which the fish was taken.

Recommended Risk-Reduction Measures

Proper training of all transporters will reduce the likelihood of a mishap or accident resulting in a leak or spill of ethyl benzene into the environment. Labels on all containers, trucks, and rail cars must meet DOT requirements and accurately reflect their contents to enable emergency responders to react properly and quickly to any disaster thereby reducing the potential risk to the environment and to personnel.

Storage of ethyl benzene should be segregated from other chemicals and materials to minimize the risk of cross-contamination.

If a spill or leak to the environment has occurred, fire department, emergency response and/or hazardous materials spill personnel should be notified immediately. Cleanup should be attempted only by those trained in proper spill containment procedures. Contaminated soils should be removed for incineration and replaced with clean soil. If ethyl benzene should contact the water table, aquifer, or navigable waterway, time is of the essence. It is slightly to moderately soluble and total remediation may not be possible. The local and/or state emergency response authorities must be notified. A comprehensive emergency response or disaster preparedness/recovery plan should be in place prior to any operations involving the use, transportation, storage, or disposal of ethyl benzene.

For disposal, ethyl benzene may be burned in a chemical incinerator equipped with an afterburner and scrubber.

If ethyl benzene is spilled or leaked, the following specific steps are recommended:

☑ Restrict persons not wearing protective clothing from area of spill or leak until cleanup is complete. Use non-sparking tools.

☑ Absorb liquids in vermiculite, sand, earth, or similar material and deposit in sealed drum.

☑ Ventilate area of spill or leak. Container should be removed to safer area if it can be done without increasing the risk.

☑ Use water spray to knock-down vapors.

☑ It may be necessary to dispose of ethyl benzene as a hazardous waste. The responsible state agency or the regional office of the federal Environmental Protection Agency (EPA) should be contacted for specific recommendations.

RISK ASSESSMENT: BUSINESS

General Assessment

Accidents or mishaps involving ethyl benzene can present a serious threat to business operations. The loss or damage of equipment or facilities can significantly affect fiscal viability. Lawsuits that may result from personnel illness, injury/death, public exposures, and/or environmental contamination will require a serious expenditure of resources. Media attention surrounding an injury, death, or environmental damage can also result in a loss of profits and loss of current as well as future business.

Recommended Risk-Reduction Measures

Company attorneys, safety and health professionals, and environmental specialists should be involved in the development of any procedures for responding to chemical incidents. A company official should be predesignated as a public relations officer with specific training in dealing with the press. Corporate plans and policies should be developed, approved, and implemented long before any need for such arises.

REFERENCES

American Conference of Governmental Industrial Hygienists. 1988. *Documentation of the Threshold Limit Values and Biological Exposure Indices,* 5th Edition (with updates). Cincinnati: ACGIH

Bloom, A. D. and F. J. DeSerres. 1995. *Ecotoxicity and Human Health.* New York: CRC Press/Lewis Publishers.

Bueche, J. R. 1972. *Principles of Physics,* 2nd Edition. New York: McGraw-Hill.

Cockerman, L. G., B. S. Shane. 1994. *Basic Environmental Toxicology.* New York: CRC Press/Lewis Publishers.

Gerarde, H. W. 1960. *Toxicology and Biochemistry of Aromatic Hydrocarbons.* New York: Elsevier.

Hallenbeck, W. H. 1993. *Quantitative Risk Assessment for Environmental and Occupational Health,* 2nd Edition. New York: CRC Press/Lewis Publishers.

Howard, P. H., et. al. 1991. *Handbook of Environmental Degradation Rates.* New York: CRC Press/Lewis Publishers.

Lewis, R. J., Sr. 1993. *Hawley's Condensed Chemical Dictionary,* 12th Edition. New York: Van Nostrand Reinhold.

Lewis, R. J., Sr. 1992. *Sax's Dangerous Properties of Industrial Materials,* Volumes 1, 2, and 3, 8th Edition. New York: Van Nostrand Reinhold.

Manahan, S. E. 1994. *Environmental Chemistry,* 6th Edition. New York: CRC Press/Lewis Publishers.

Meyer, E. 1990. *Chemistry of Hazardous Materials.* Englewood Cliffs, NJ: Prentice Hall.

Patnaik, P. 1992. *A Comprehensive Guide to the Hazardous Properties of Chemical Substances.* New York: Van Nostrand Reinhold.

Quigley, D. R. 1994. *Handbook of Emergency Chemical Management.* New York: CRC Press/Lewis Publishers.

Sacarello, H. L. A. 1994. *The Comprehensive Handbook of Hazardous Materials.* New York: CRC Press/Lewis Publishers.

Torkelson, T. R. V. K. Rowe. 1981. *Patty's Industrial Hygiene and Toxicology,* Vol. 2B. New York: Wiley-Interscience.

U.S. Department of Health and Human Services, National Institute for Occupational Safety and Health. 1994. *NIOSH Pocket Guide to Chemical Hazards.* Washington, D.C.: U.S. Government Printing Office.

Yin, S. N., et. al. 1987. Leukemia in benzene workers: A retrospective cohort study. *British Journal of Industrial Medicine.* 44:124-128.

MATERIAL SAFETY DATA SHEET

CHEMICAL NAME	
ETHYL BROMIDE	**CRC PRESS/LEWIS PUBLISHERS DISCLAIMER:** THE INFORMATION AND RECOMMENDATIONS PRESENTED HEREIN ARE BASED ON SOURCES BELIEVED TO BE RELIABLE. CRC MAKES NO REPRESENTATION ON ITS COMPLETENESS OR ACCURACY. IT IS THE RESPONSIBILITY OF THE USER TO DETERMINE THE CHEMICAL'S SUITABILITY FOR ITS INTENDED USE, THE CHEMICAL'S SAFE USE, AND THE CHEMICAL'S PROPER DISPOSAL. NO REPRESENTATIONS AND/OR WARRANTIES, EITHER EXPRESSED OR IMPLIED, OF THE MERCHANTABILITY OR FITNESS FOR A PARTICULAR PURPOSE, OR OF ANY OTHER NATURE, ARE MADE WITH RESPECT TO THE INFORMATION PROVIDED IN THIS MSDS OR TO THE CHEMICAL TO WHICH INFORMATION MAY REFER. CRC NEITHER ASSUMES NOR AUTHORIZES ANY OTHER PERSON TO ASSUME FOR IT, ANY OTHER ADDITIONAL RESPONSIBILITY OR LIABILITY FOR THE USE OF, OR RELIANCE UPON, THIS INFORMATION.

HAZARD WARNING INFORMATION

HEALTH	FIRE	REACTIVE	OTHER	DEGREE OF HAZARD	COLOR CODING	OTHER CODES
2	1	0		0 = Minimum Hazard 1 = Slight Hazard 2 = Moderate Hazard 3 = Serious Hazard 4 = Severe Hazard	HEALTH = BLUE FIRE = RED REACTIVITY = YELLOW OTHER = WHITE	OX = Oxidizer ACID = Acid ALK = Alkali COR = Corrosive W = Use No Water

SECTION I - GENERAL INFORMATION

Characterization: **Halogenated Hydrocarbon**	RCRA Number: **None**	EPA Class: **Not Applicable**
DOT Proper Shipping Name: **Ethyl Bromide**	Chemical Abstract Service (CAS) Number: **74-96-4**	
DOT Hazard Class and Label Requirements: **Poison B**	DOT Emergency Guide Code: **58**	
DOT Identification Number: **UN 1891**	Molecular Formula: **CH_3CH_2Br**	

Synonyms
Bromoethane; monobromoethane; Halon 2001; hydrobromic ether.

SECTION II - HAZARDOUS INGREDIENTS AND IDENTITY INFORMATION

Hazardous Components (specific identity)	OSHA Exposure Criteria	NIOSH Exposure Criteria	Immediately Dangerous to Life and Health (IDLH)	ACGIH Exposure Criteria
Ethyl bromide (derivation: From ethanol or ethylene and hydrobromic acid. One process uses γ-radiation to initiate the combination). **1 ppm = 4.53 mg/m^3**	**PEL:** 200 ppm, 890 mg/m^3 **STEL:** 250 ppm, 1110 mg/m^3	**REL:** Not Established **STEL:** Not Established	**2000 ppm**	**TLV:** 200 ppm, 890 mg/m^3 **STEL:** 250 ppm, 1110 mg/m^3

SECTION III - PHYSICAL AND CHEMICAL CHARACTERISTICS

Boiling Point: **101°F (38°C)**	Specific Gravity (H_2O = 1): **1.46**
Vapor Pressure (mm Hg): **375 at 69°F (20°C)**	Molecular Weight (atomic weight): **109.0**
Vapor Density (Air = 1): **3.8**	Melting Point: **-182°F (-119°C)**

Solubility
Slightly soluble in water (0.9%). Soluble in alcohol, ether, and most organic solvents.

Appearance and Odor
Colorless liquid, turns yellow on exposure to light and air with a strong ether-like odor. (Note: Exists as a gas above 101°F).

SECTION IV - FIRE AND EXPLOSION HAZARD DATA

Flash Point (method used): **<-4°F (<-20°C) closed cup**	Explosive Limits in Air % by Volume: **LEL: 6.8% UEL: 11.3%**
NFPA Classification: **Class 1B Flammable Liquid**	Autoignition Temperature: **952°F (511°C)**

Extinguishing Media
Use dry chemical, carbon dioxide, or foam. Water may be ineffective as an extinguishing agent by itself.

Special Fire Fighting Procedures
Poisonous gases are produced in fire. Wear self-contained breathing apparatus (SCBA) and full protective gear. Move containers from fire if it can be done without risk. Keep cooling sides of fire-exposed containers with water long after fire is out.

Unusual Fire and Explosion Hazards
Dangerous fire and explosion hazard. Containers may explode in fire. Its vapors are heavier than air and can travel for great distances to an ignition source to flashback and cause fire or explosion.

SECTION V - REACTIVITY DATA

Stability		Conditions to Avoid
		Under normal conditions of handling and storage, ethyl bromide is considered stable. Avoid heat, flame, other sources of ignition, and contact with incompatible materials and chemicals.
Stable	Unstable	Incompatibility *(materials to avoid)*
X		Strong oxidizers (chlorine, bromine, fluorine), chemically active metals (such as sodium, potassium, calcium, powdered aluminum, zinc, and magnesium).
Hazardous Polymerization		Conditions to Avoid
		Hazardous polymerization of ethyl bromide has not been reported to occur under normal conditions of temperature and pressure.
May Occur	Will Not Occur	Hazardous Decomposition or By-products
	X	Poisonous gases and acrid fumes are produced in fire, including carbon dioxide, carbon monoxide, and toxic fumes of bromine.

SECTION VI - HEALTH HAZARD DATA

Primary Route(s) of Entry:	Inhalation?	Absorption (skin/eye)?	Ingestion?
	X		

Health Hazards

INHALATION: Causes depression of the central nervous system (CNS) and fatty degeneration of the liver heart, and renal tissues. May damage liver, kidneys, and intestines. Vapors may cause severe irritation of the respiratory system with acute congestion and pulmonary edema. Can cause narcosis, coma, and death.

SKIN & EYES: Irritation with possible redness and pain to the skin and eyes. Eye contact may result in severe irritation and possible inflammation of the cornea.

INGESTION: Unspecified gastrointestinal effects. An unlikely exposure route, but still possible.

Carcinogenicity	NTP Listed?	IARC Cancer Review Group?	OSHA Regulated?	Target Organs?
Suspected Human Suspected Animal	5th Annual Report	No	29 CFR 1910.1000 Table Z-1	Eyes, skin, respiratory sys., liver, kidney, CVS, CNS.

Medical Conditions Generally Aggravated by Exposure

Liver, kidney, heart, and respiratory problems may be aggravated by exposure.

Emergency and First-aid Procedures

Eye contact: Do not allow victim to keep eyes tightly shut. Flush immediately with water for 15 minutes (minimum), seek medical attention. Skin contact: Remove all contaminated clothing, including shoes. Immediately wash area with flooding amounts of soap and water. Seek medical attention immediately. For inhalation: Remove the person from exposure. Provide respiratory assistance and CPR. Transfer to medical facility. Observe for 24-48 hours for lung effects. If swallowed: Seek medical attention immediately. Give conscious and alert person 1-2 glasses of milk or water. Do not induce vomiting.

SECTION VII - PRECAUTIONS FOR SAFE HANDLING AND USE

Steps to be Taken in Case Material is Released or Spilled

Remove all ignition sources, wear SCBA respiratory protection. Restrict those not wearing protective equipment and who are not involved in cleanup from entering area. Stop leak if it can be done without risk. Provide explosion-proof ventilation. Absorb with vermiculite and deposit in sealed containers.

Preferred Waste Disposal Method

Mix with combustible solvent; burn in chemical incinerator equipped with an afterburner and scrubber.

Precautions to be Taken in Handling and Storage

Store in tightly closed containers in cool, well ventilated location. Keep away from sources of heat.

Other Precautions and Warnings

Sources of ignition such as open flame or smoking, are prohibited where ethyl bromide is used or stored. Metal containers should be bonded and grounded. Empty containers contain residue, handle with care.

SECTION VIII - CONTROL MEASURES AND PERSONAL PROTECTIVE EQUIPMENT

Respiratory Protection *(specify type)*

Use an MSHA/NIOSH-approved air purifying respirator (cartridge type) or use a supplied-air respirator or a self-contained breathing apparatus (SCBA) with full facepiece operated in positive pressure mode.

Ventilation

Local exhaust (preferred) at site of chemical work or general ventilation.

Protective Gloves	Eye Protection	Other Protective Clothing
Impervious Gloves	Chemical Goggles or Face Mask	Protective Apron if Splash is Likely

Work/Hygiene Practices

Always wash hands thoroughly after using chemical; never bring food, drink, or smoking materials into vicinity of chemicals.

ETHYL BROMIDE

CH_3CH_2Br CAS: 74-96-4

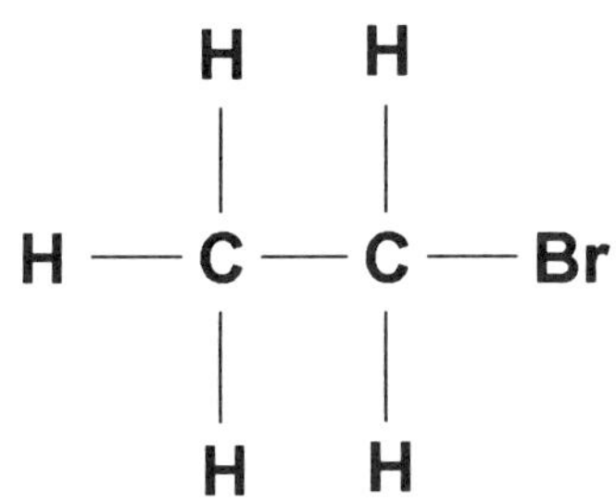

IDENTIFICATION AND TYPICAL USES

Ethyl bromide is a colorless to yellow-colored liquid with a strong ether-like odor. It can also exist as a gas above 101°F (38°C). It is used as a refrigerant and an ethylating agent. It has uses as a solvent, and as a fruit and grain fumigant. It was formerly used as a topical anesthetic.

RISK ASSESSMENT: HEALTH

General Assessment

Ethyl bromide is a human poison. It is moderately toxic by *ingestion* and mildly toxic by *inhalation.* While absorption through the skin has not been known to occur, skin contact can cause localized damage to tissues. Ethyl bromide has been listed as a suspected human carcinogen by the National Toxicology Program (NTP). Physiologically, it is an anesthetic and narcotic. The primary human response to ethyl bromide is depression of the central nervous system (CNS).

Inhalation causes irritation of the lungs, marked congestion and fatty degeneration of the liver, kidneys, heart, and intestine tissues. Symptoms of exposure and CNS depression include drowsiness, slurred speech, lack of coordination, and possible loss of consciousness. Irritation to the eyes and respiratory system is also know to occur on inhalation. In fact, pulmonary edema (fluid buildup in the lungs) has been reported. This is a medical emergency that can be fatal. Symptoms of congestion, difficulty breathing, cough with phlegm, and chest pain may be delayed for up to 48 hours following exposure. This may create a false sense of security with regard to health risk. Other possible symptoms of exposure include liver and kidney disease, cardiac arrhythmia, and cardiac arrest.

Skin contact may cause localized burning at site of contact with irritation and possible rash. Absorption is not know to occur. Eye contact can result in severe irritation to the cornea, with redness and swelling. Damage can be serious if not immediately flushed from eye surface.

Acute Health Effects

The following acute (short-term) health effects may occur immediately or shortly after exposure to ethyl bromide:

Skin: Irritation with potential for redness, pain, and inflammation.

Eye: Severe irritation, burning, pain, and permanent damage possible.

Lung: Nose, throat, and respiratory tract irritation causing difficulty in breathing, possible pulmonary edema (fluid in lungs) on high exposure.

CNS: Narcosis, headaches, lightheadedness, dizziness, vertigo, coma, and death due to respiratory paralysis or cardiac arrest.

Chronic Health Effects

The following chronic (long-term) health effects may occur at some time after exposure to ethyl bromide and can last for months or even years:

Cancer Hazards: According to the National Toxicology Program (NTP), there is sufficient evidence that exposure to ethyl bromide may cause cancer in test animals. Human data are lacking but human carcinogenicity is considered suspect. More research is required since the data in the references are inconclusive and conflicting in this regard.

Reproductive Hazard: According to information presented in the references, ethyl bromide has not been adequately tested for its ability to cause adverse reproductive effects in test animals.

Other Chronic Effects: Long-term exposure may lead to liver and kidney damage. Extremely irritating substances have been know to damage lung tissues. It is not known whether exposure to ethyl bromide will have this effect on humans.

Recommended Risk-Reduction Measures

Personnel should avoid direct contact with ethyl bromide. Occupational poisoning occurs most commonly by inhalation. If a less toxic material or compound cannot be substituted for ethyl bromide, then *engineering controls* are the most effective method of reducing exposures. The best protection is to enclose operations and/or provide local exhaust ventilation at the site of ethyl bromide release. While not always operationally feasible, isolating operations can also reduce exposure. Using respiratory protection is less effective than the controls mentioned above, but is still advisable whenever working with or around ethyl bromide. Caution should be exercised when selecting respiratory protection since it is not known if exposure to ethyl bromide will cause cancer in humans. For exposure over the PEL (200 ppm), an MSHA/NIOSH-approved air-purifying respiratory with organic vapor cartridge(s) will suffice. For exposures to higher concentrations, or when the concentration is unknown, use a supplied air respirator with full facepiece and mask operated in positive pressure mode or a self-contained breathing apparatus (SCBA) with full facepiece operated in pressure demand. If a full facepiece is not available, then chemical goggles should be worn to protect the eyes. A face shield should also be considered. To prevent hand and skin exposures, impervious rubber gloves should be worn.

Administrative controls should also be in place to minimize the potential for human exposures. These may include written procedures or policies which specify the methods and techniques that will be practiced whenever personnel are to work with ethyl bromide.

All personnel should receive *training* on the use, hazards, protective measures, emergency actions, and other precautions per 29 CFR 1910.1200 (Hazard Communication), prior to the first assignment in an area where ethyl bromide is used or stored.

If symptoms develop or overexposure is suspected, the following recommended medical tests should be considered:

☑ Liver and kidney function tests.

☑ Lung function tests.

☑ Consider chest X-ray following acute overexposure (may be negative if taken immediately after exposure).

Any evaluation should include a careful history of past and present symptoms with an examination. Medical tests that simply look for existing damage are not a substitute for controlling exposures. Also, since smoking causes heart disease, emphysema, and other respiratory problems, smokers exposed to ethyl bromide may experience symptoms quicker and more pronounced than non-smokers under the same conditions of exposure. Prudent risk management requires careful consideration of *all* possible factors that may be causing the appearance of exposures symptoms in the work place.

Other methods to reduce exposure include:

☑ Where possible, enclose operations and use local exhaust ventilation at the site of chemical release. If local exhaust ventilation or enclosure is not used, respiratory protection should be mandatory.

☑ Always ensure that proper protective clothing is worn when using chemical substances.

☑ Wash thoroughly immediately after exposure to ethyl bromide and at the end of the work shift or before eating, drinking, or smoking.

☑ Hazard warning information should be posted in the work area. In addition, as part of an on-going education and training program, all information on the health and safety hazards of ethyl bromide should be communicated to all potentially exposed workers.

☑ Eye wash stations should be provided in the immediate work area for emergency use. If there is a possibility of skin exposure to ethyl bromide, emergency shower facilities should be provided.

☑ Workers whose clothing has been contaminated by ethyl bromide should change into clean clothes before leaving work. Contaminated work clothing should be laundered only by individuals who have been informed of the hazards of ethyl bromide.

☑ Where possible, automatically transfer liquids containing ethyl bromide from drums or other containers to process containers.

RISK ASSESSMENT: ENVIRONMENT

General Assessment

The environment is at risk of exposure during transportation, storage, disposal, or destruction of ethyl bromide. In almost every scenario, the threat of environmental exposure is contingent upon the proper handling of the chemical substance. Accidental spills, large or small, can result in contamination of the surrounding environmental mediums (water, soil, and air).

Ethyl bromide is a moderately flammable liquid. Because of its low flash point and high boiling point, it is classified as a Class 1B flammable liquid (per OSHA 29 CFR 1910.106). It is a dangerous fire hazard and moderate explosion hazard when exposed to heat or flame. It is incompatible with strong oxidizers, such as chlorine, fluorine, and bromine, and many chemically-active metals, such as sodium, potassium, calcium, powdered aluminum, zinc, and magnesium. Caution is always required in handling, storage, transportation, and disposal of ethyl bromide. Emergency responders should be made aware of the presence of ethyl bromide at any emergency response situation.

Ethyl bromide can enter the environment from industrial effluents and from spills.

Acute Ecological Effects

Acute (short-term) toxic effects may include the death of animals, birds, or fish and death or low growth rate in plants. Acute effects are seen 2 to 4 days after animals or plants are exposed to ethyl bromide. Ethyl bromide has high acute toxicity to aquatic life. It has caused injury to various agricultural crops. Insufficient data are available to evaluate the short-term effects of ethyl bromide on birds or land animals.

Chronic Ecological Effects

Chronic (long-term) toxic effects may include shortened life span, reproductive problems, lower fertility, and changes in appearance or behavior in exposed animals. These effects can be seen long after first exposure(s) to toxic chemicals. Ethyl bromide has high chronic toxicity to aquatic life. Insufficient data are available to evaluate or predict the long-term effects of ethyl bromide to plants, birds, or terrestrial animals.

Water Solubility

Ethyl bromide is slightly to moderately soluble in water. Concentrations between 1 and 100 milligrams will mix with a liter of water.

Persistence in the Environment

Ethyl bromide is slightly persistent in water, with a half-life between 2 and 20 days. The half-life of a pollutant is the amount of time it takes for one half of the chemical to be degraded.

Bioaccumulation in Aquatic Organisms

Some substances increase in concentration, or *bioaccumulate,* in living organisms as they breathe contaminated air, drink contaminated water, or eat contaminated food. These chemicals can become concentrated in the tissues and internal organs of animals as well as humans.

The concentration of ethyl bromide found in fish tissues is expected to be about the same as the average concentration of ethyl bromide in the water from which the fish was taken.

Recommended Risk-Reduction Measures

Proper training of all transporters will reduce the likelihood of a mishap or accident resulting in a leak or spill of ethyl bromide into the environment. Labels on all containers, trucks, and rail cars must meet DOT requirements and accurately reflect their contents to enable emergency responders to react properly and quickly to any disaster thereby reducing the potential risk to the environment and to personnel.

Storage of ethyl bromide should be segregated from other chemicals and materials to minimize the risk of cross-contamination.

If a spill or leak to the environment has occurred, fire department, emergency response and/or hazardous materials spill personnel should be notified immediately. Cleanup should be attempted only by those trained in proper spill containment procedures. Contaminated soils should be removed for incineration and replaced with clean soil. If ethyl bromide should contact the water table, aquifer, or navigable waterway, time is of the essence. It is slightly soluble and total remediation may not be possible. The local and/or state emergency response authorities must be notified. A comprehensive emergency response or disaster preparedness/recovery plan should be in place prior to any operations involving the use, transportation, storage, or disposal of ethyl bromide.

For disposal, ethyl bromide may be mixed with a combustible solvent and burned in a chemical incinerator equipped with an afterburner and scrubber.

If ethyl bromide is spilled or leaked, the following specific steps are recommended:

- ☑ Restrict persons not wearing protective clothing from area of spill or leak until cleanup is complete. Use non-sparking tools.
- ☑ Absorb liquids in vermiculite, sand, earth, or similar material and deposit in sealed drum.
- ☑ Ventilate area of spill or leak. Container should be removed to safer area if it can be done without increasing the risk.
- ☑ Use water spray to knock-down vapors.
- ☑ It may be necessary to dispose of ethyl bromide as a hazardous waste. The responsible state agency or the regional office of the federal Environmental Protection Agency (EPA) should be contacted for specific recommendations.

RISK ASSESSMENT: BUSINESS

General Assessment

Accidents or mishaps involving ethyl bromide can present a serious threat to business operations. The loss or damage of equipment or facilities can significantly affect fiscal viability. Lawsuits that may result from personnel illness, injury/death, public exposures, and/or environmental contamination will require a serious expenditure of resources. Media attention surrounding an injury, death, or environmental damage can also result in a loss of profits and loss of current as well as future business. Always remember that anytime terms such as "cancer" or "carcinogen" are used, public emotion, hysteria, and ignorance can run equally high. This must be carefully considered whenever developing or implementing a public relations policy.

Recommended Risk-Reduction Measures

Company attorneys, safety and health professionals, and environmental specialists should be involved in the development of any procedures for responding to chemical incidents. A company official should be predesignated as a public relations officer with specific training in dealing with the press. Corporate plans and policies should be developed, approved, and implemented long before any need for such arises.

REFERENCES

Ahlbom, A. 1993. *Biostatistics for Epidemiologists.* New York: CRC Press/Lewis Publishers.

American Conference of Governmental Industrial Hygienists. 1988. *Documentation of the Threshold Limit Values and Biological Exposure Indices,* 5th Edition (with updates). Cincinnati: ACGIH

Anthony, C. P., and N. J. Kolthoff. 1971. *Textbook of Anatomy and Physiology.* St. Louis: C. V. Mosby Company.

Asante-Duah, D. K. 1993. *Hazardous Waste Risk Assessment.* New York: CRC Press/Lewis Publishers.

Bloom, A. D. and F. J. DeSerres. 1995. *Ecotoxicity and Human Health.* New York: CRC Press/Lewis Publishers.

Bueche, J. R. 1972. *Principles of Physics,* 2nd Edition. New York: McGraw-Hill.

Calabrese, E. J. 1994. *Biological Effects of Low Level Exposures.* New York: CRC Press/Lewis Publishers.

Cockerman, L. G., B. S. Shane. 1994. *Basic Environmental Toxicology.* New York: CRC Press/Lewis Publishers.

Cothern, C. R., N. P. Ross. 1994. *Environmental Statistics, Assessment, and Forecasting.* New York: CRC Press/Lewis Publishers.

Hallenbeck, W. H. 1993. *Quantitative Risk Assessment for Environmental and Occupational Health,* 2nd Edition. New York: CRC Press/Lewis Publishers.

Howard, P. H., et. al. 1991. *Handbook of Environmental Degradation Rates.* New York: CRC Press/Lewis Publishers.

Lewis, R. J., Sr. 1993. *Hawley's Condensed Chemical Dictionary,* 12th Edition. New York: Van Nostrand Reinhold.

Lewis, R. J., Sr. 1992. *Sax's Dangerous Properties of Industrial Materials,* Volumes 1, 2, and 3, 8th Edition. New York: Van Nostrand Reinhold.

Manahan, S. E. 1994. *Environmental Chemistry,* 6th Edition. New York: CRC Press/Lewis Publishers.

Meyer, E. 1990. *Chemistry of Hazardous Materials.* Englewood Cliffs, NJ: Prentice Hall.

Patnaik, P. 1992. *A Comprehensive Guide to the Hazardous Properties of Chemical Substances.* New York: Van Nostrand Reinhold.

Quigley, D. R. 1994. *Handbook of Emergency Chemical Management.* New York: CRC Press/Lewis Publishers.

Sacarello, H. L. A. 1994. *The Comprehensive Handbook of Hazardous Materials.* New York: CRC Press/Lewis Publishers.

Torkelson, T. R. V. K. Rowe. 1981. *Patty's Industrial Hygiene and Toxicology,* Vol. 2B. New York: Wiley-Interscience.

U.S. Department of Health and Human Services, National Institute for Occupational Safety and Health. 1994. *NIOSH Pocket Guide to Chemical Hazards.* Washington, D.C.: U.S. Government Printing Office.

MATERIAL SAFETY DATA SHEET

CHEMICAL NAME

ETHYL BUTYL KETONE

CRC PRESS/LEWIS PUBLISHERS DISCLAIMER: THE INFORMATION AND RECOMMENDATIONS PRESENTED HEREIN ARE BASED ON SOURCES BELIEVED TO BE RELIABLE. CRC MAKES NO REPRESENTATION ON ITS COMPLETENESS OR ACCURACY. IT IS THE RESPONSIBILITY OF THE USER TO DETERMINE THE CHEMICAL'S SUITABILITY FOR ITS INTENDED USE, THE CHEMICAL'S SAFE USE, AND THE CHEMICAL'S PROPER DISPOSAL. NO REPRESENTATIONS AND/OR WARRANTIES, EITHER EXPRESSED OR IMPLIED, OF THE MERCHANTABILITY OR FITNESS FOR A PARTICULAR PURPOSE, OR OF ANY OTHER NATURE, ARE MADE WITH RESPECT TO THE INFORMATION PROVIDED IN THIS MSDS OR TO THE CHEMICAL TO WHICH INFORMATION MAY REFER. CRC NEITHER ASSUMES NOR AUTHORIZES ANY OTHER PERSON TO ASSUME FOR IT, ANY OTHER ADDITIONAL RESPONSIBILITY OR LIABILITY FOR THE USE OF, OR RELIANCE UPON, THIS INFORMATION.

HAZARD WARNING INFORMATION

HEALTH	FIRE	REACTIVE	OTHER	DEGREE OF HAZARD	COLOR CODING	OTHER CODES
1	2	0		0 = Minimum Hazard 1 = Slight Hazard 2 = Moderate Hazard 3 = Serious Hazard 4 = Severe Hazard	HEALTH = BLUE FIRE = RED REACTIVITY = YELLOW OTHER = WHITE	OX = Oxidizer ACID = Acid ALK = Alkali COR = Corrosive W = Use No Water

SECTION I - GENERAL INFORMATION

Characterization: **Ketone**	RCRA Number: **D001**	EPA Class: **Characteristic (I) Waste**
DOT Proper Shipping Name: **Ketones, Liquid, N.O.S.**	Chemical Abstract Service (CAS) Number: **106-35-4**	
DOT Hazard Class and Label Requirements: **Flammable Liquid**	DOT Emergency Guide Code: **26**	
DOT Identification Number: **UN 1224**	Chemical Formula: **$CH_3CH_2CO[CH_2]_3CH_3$**	

Synonyms

Butyl ethyl ketone; 3-heptanone; *n*-butyl ethyl ketone; ethyl *n*-butyl ketone; heptan-3-one.

SECTION II - HAZARDOUS INGREDIENTS AND IDENTITY INFORMATION

Hazardous Components (specific identity)	OSHA Exposure Criteria	NIOSH Exposure Criteria	Immediately Dangerous to Life and Health (IDLH)	ACGIH Exposure Criteria
Ethyl *n*-butyl ketone (derivation: By catalytic dehydrogenation of 3-heptanol; by hydrogenation of the mixed alcohol condensation product of propionaldehyde and methyl ethyl ketone). May Contain Trace Impurities **1 ppm = 4.75 mg/m^3**	**PEL (8 hour):** 50 ppm 230 mg/m^3 **STEL:** Not Established	**REL (10 hour):** 50 ppm 230 mg/m^3 **STEL:** Not Established	1000 ppm	**TLV:** 50 ppm 234 mg/m^3 **STEL:** Not Established

SECTION III - PHYSICAL AND CHEMICAL CHARACTERISTICS

Boiling Point: **298°F (148°C)**	Specific Gravity ($H_2O = 1$): **0.82**
Vapor Pressure (mm Hg): **1.4 at 77°F (25°C)**	Molecular Weight: **114.2**
Vapor Density (Air = 1): **3.93**	Freezing Point: **-34°F (-37°C)**

Solubility

Nearly insoluble in cold water (1.0%). Soluble in hot water at 300°F (149°C), alcohol, and ether.

Appearance and Odor

Clear, colorless liquid with a powerful fruity odor.

SECTION IV - FIRE AND EXPLOSION HAZARD DATA

Flash Point (method used): **115°F (46°C) opened cup**	Explosive Limits in Air % by Volume: **LEL: 1.4% UEL: 8.8%**
NFPA Classification: **Class II Combustible Liquid**	Autoignition Temperature: **Not Determined**

Extinguishing Media

Carbon dioxide, dry chemical, alcohol foam, water spray.

Special Fire Fighting Procedures

Wear full protective clothing and self-contained breathing apparatus (SCBA). Fight fire from a distance if possible. Keep fire-exposed containers cool with water spray. Using a solid stream of water may spread fire. If it can be done safely, move fire-exposed containers from fire area.

Unusual Fire and Explosion Hazards

Containers may explode in fire due to the buildup of internal pressures. BLEVE (boiling liquid expanding vapor explosion) is possible. Immediately evacuate if rising sound is heard from venting device or a noticeable discoloration is observed on the sides of the tank or storage vessel.

SECTION V - REACTIVITY DATA

Stability		Conditions to Avoid
		Normally stable. Do not allow ethyl butyl ketone to come into contact with incompatible materials. Avoid contact with fire or high heat sources. Do not use in confined spaces.
Stable	Unstable	Incompatibility *(materials to avoid)*
X		Incompatible with strong oxidizers (bromine, chlorine, fluorine), acetaldehyde, and perchloric acid. Also note that ketones + nitric acid and hydrogen peroxide form highly explosive solids or oils.
Hazardous Polymerization		Conditions to Avoid
		Under normal conditions of temperature and pressure, hazardous polymerization of ethyl butyl ketone is not expected to occur.
May Occur	Will Not Occur	Hazardous Decomposition or By-products
	X	When heated to decomposition, ethyl butyl ketone emits acrid, irritating smoke and fumes, including oxides of carbon.

SECTION VI - HEALTH HAZARD DATA

Primary Route(s) of Entry:	Inhalation?	Absorption (skin)?	Ingestion?
	X		X

Health Hazards

INHALATION: Irritation of the conjunctiva of the eyes and mucosa of the nose and throat. Symptoms of CNS depression include dizziness, headache and unconsciousness. Neurotoxic effects to the peripheral nervous system may also occur.

SKIN & EYES: Mild irritation on contact with the skin and eyes. Can lead to dermatitis in the long-term.

INGESTION: Symptoms of gastrointestinal irritation with nausea and vomiting.

Carcinogenicity	NTP Listed?	IARC Cancer Review Group?	OSHA Regulated?	Target Organs?
Unknown Human Unknown Animal	No	No	29 CFR 1910.1000 Table Z-1	Respiratory system; skin; eyes; CNS.

Medical Conditions Generally Aggravated by Exposure

Skin conditions (dermatitis) may be aggravated.

Emergency and First-aid Procedures

Eye contact: Immediately flush large amounts of water for 15 minutes (minimum), occasionally lifting eyelids. Skin contact: Remove all contaminated clothing. Immediately wash area with large amounts of soap and water. For inhalation: Remove the person from exposure. Provide respiratory assistance and CPR. Transfer promptly to medical facility. If swallowed: Seek medical attention immediately. Give victim several glasses of water. Do NOT induce vomiting (possibility of liquid aspiration to lungs). Never try to give an unconscious or convulsing person anything by mouth.

SECTION VII - PRECAUTIONS FOR SAFE HANDLING AND USE

Steps to be Taken in Case Material is Released or Spilled

Remove all ignition sources. Restrict persons not wearing protective equipment from entering area of spill or leak until cleanup is complete. Ventilate area. Absorb liquid in dry sand or other material and deposit in sealed containers. Alert appropriate state or federal response agencies.

Preferred Waste Disposal Method

Incineration (with afterburner and scrubber), molten metal salt destruction.

Precautions to be Taken in Handling and Storage

Do not store with incompatible chemicals since violent reactions can occur. Store in tightly closed containers in a cool, dark, well-ventilated area away from fire, sparks, and flame. Use non-sparking tools on containers. Ground and bond metal containers during transfer operations.

Other Precautions and Warnings

Containers may explode in fire or under conditions of extreme heat (do not store outdoors or in direct sunlight). Protect containers from physical damage. Dike around outside storage areas.

SECTION VIII - CONTROL MEASURES AND PERSONAL PROTECTIVE EQUIPMENT

Respiratory Protection *(specify type)*

Exposures above 50 ppm: MSHA/NIOSH-approved full facepiece respirator with organic vapor cartridge. Greater protection is obtained from a self-contained breathing apparatus (SCBA) with full facepiece and operated in pressure demand or other positive pressure mode.

Ventilation

Local exhaust or general mechanical systems recommended.

Protective Gloves	Eye Protection	Other Protective Clothing
Solvent Resistant Butyl Rubber	Chemical Goggles and/or Face Mask	Rubber Apron

Work/Hygiene Practices

Always wash hands thoroughly after using chemical; never bring food, drink, or smoking materials into vicinity of chemicals.

ETHYL BUTYL KETONE

$CH_3CH_2CO[CH_2]_3CH_3$ CAS: 106-35-4

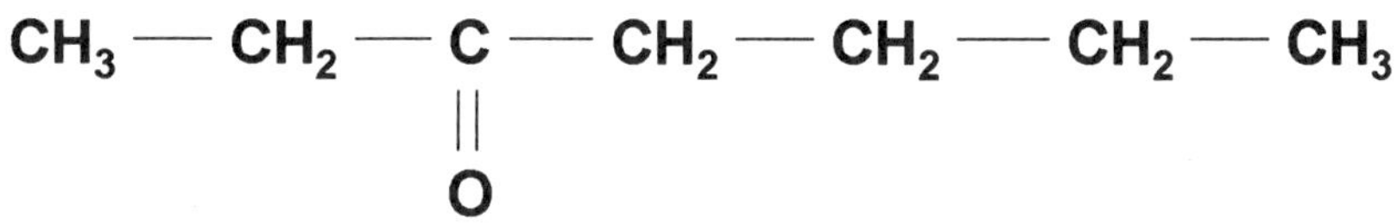

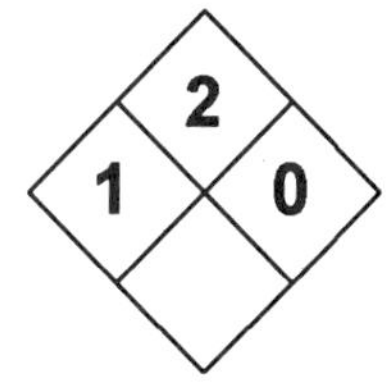

IDENTIFICATION AND TYPICAL USES

Ethyl butyl ketone is a colorless liquid with a powerful, fruity odor. It is used as a solvent for nitrocellulose and polyvinyl resins, and as an intermediate in organic synthesis.

RISK ASSESSMENT: HEALTH

General Assessment

Ethyl butyl ketone is mildly toxic to humans by *inhalation* and skin contact, although absorption is not likely. *Ingestion* is also a possible exposure route. There is relatively little human exposure data available in the references. However, based upon animals studies, the toxicity of ethyl butyl ketones appears to be relatively low. There are no reports in the references of carcinogenic, mutagenic, or teratogenic effects of this chemical in humans or animals.

Inhalation of ethyl butyl ketone vapors causes mild irritation to the skin, eyes, and mucous membranes of the respiratory tract with little evidence of toxic systemic effects at low concentrations. However, exposures to extremely high levels of vapor concentration can cause depression of the central nervous system (CNS) leading to symptoms of narcosis with headache, dizziness, nausea, loss of consciousness, and possible coma. Neurotoxic effects also appear to focus on the actions of the peripheral nervous system, apparently due to the metabolites of ethyl butyl ketone.

Skin contact will result in mild irritation. Prolonged exposures can lead to dryness, cracking, contact dermatitis, and a possibility for secondary infection at the point of contact. Eye contact causes mild irritation with very little potential for permanent injury or damage.

Ingestion of ethyl butyl ketone can produce gastrointestinal irritation, nausea, and vomiting.

Acute Health Effects

The following acute (short-term) health effects may occur immediately or shortly after exposure to ethyl butyl ketone:

Skin: Mild irritation and possible dermatitis.

Eye: Mild irritation.

Lung: Irritation of the mucous membranes of the respiratory tract.

CNS: A mild narcotic compound. Inhalation of high concentrations of the vapor may cause dizziness, lightheadedness, and possible loss of consciousness.

Chronic Health Effects

The following chronic (long-term) health effects may occur at some time after exposure to ethyl butyl ketone and can last for months or even years:

Cancer Hazards: According to information presented in the references, exposure to ethyl butyl ketone has not been shown to cause carcinogenic effects.

Reproduction: There are no reports to support any claims of reproductive hazards in humans or animals.

Sensitivity: Repeated exposure to the skin may result in dermatitis.

Recommended Risk-Reduction Measures

Personnel should avoid direct contact with ethyl butyl ketone. Unless a less toxic chemical can be substituted for a hazardous substance, *engineering controls* are the most effective method of reducing exposures. The best protection is to enclose operations and/or provide local exhaust ventilation at the site of chemical release. While not always operationally feasible, isolating operations can also reduce exposure. Using respiratory protection is less effective than the controls mentioned above, but is still advisable whenever working with ethyl butyl ketone. For exposures to the PEL (50 ppm), use a NIOSH-approved, full facepiece organic vapor respirator. For higher or prolonged exposures, a supplied-air respirator with full facepiece operated in a positive pressure mode or a self-contained breathing apparatus (SCBA) with full facepiece and pressure demand are the recommended respiratory protection methods of choice. If a full facepiece is not available, then chemical goggles should be worn to protect the eyes. Whenever a chemical splash hazard exists, a face shield and a rubber apron should be worn. Natural rubber and polyvinylchloride (PVC) gloves will NOT protect against ketones. To prevent hand and skin exposures, butyl rubber gloves should be used.

Administrative controls should also be in place to minimize the potential for human exposures. These may include written procedures or policies which specify the methods and techniques that will be practiced whenever personnel are to work with ethyl butyl ketone.

All personnel should receive *training* on the use, hazards, protective measures, emergency actions, and other precautions per 29 CFR 1910.1200 (Hazard Communication), prior to the first assignment in an area where ethyl butyl ketone is used or stored.

Other methods to reduce exposure include:

☑ Where possible, enclose operations and use local exhaust ventilation at the site of chemical release. If local exhaust ventilation or enclosure is not used, respiratory protection should be mandatory.

☑ Always ensure that proper protective clothing is worn when using chemical substances.

☑ Wash thoroughly immediately after exposure to ethyl butyl ketone and at the end of the work shift or before eating, drinking, or smoking.

☑ Hazard warning information should be posted in the work area. As part of an on-going education and training program, all information on the health and safety hazards of ethyl butyl ketone should be communicated to all potentially exposed workers.

RISK ASSESSMENT: ENVIRONMENT

General Assessment

The environment is at risk of exposure during transportation, storage, disposal, or destruction of ethyl butyl ketone. In almost every scenario, the threat of environmental exposure is contingent upon the proper handling of the chemical substance. Accidental spills, large or small, can result in fire, explosion, and possible contamination of the surrounding environmental mediums (water, soil, and air).

Ethyl butyl ketone is considered a Class II combustible liquid (per OSHA 29 CFR 1910.106). It can react with strong oxidizers to cause a serious fire and explosion hazard. It is also incompatible with acetaldehyde and perchloric acid. Contact with either will result in fire or explosion. When mixed with nitric acid and hydrogen peroxide, it will form solids or oils that are highly explosive. When involved in a fire, it can release acrid and irritating oxides of carbon. A boiling liquid expanding vapor explosion (BLEVE) is also a possibility. These characteristics require special consideration during any emergency involving a leak or spill of ethyl butyl ketone.

The proper disposal/destruction method for ethyl butyl ketone is to burn it in a chemical incinerator equipped with an afterburner and air scrubber. Waste treatment methods include steam stripping and high-temperature catalytic hydrogenation.

Ethyl butyl ketone can enter the environment through manufacturing, unchecked discharge into effluents from industrial or municipal waste treatment plants, and through spills.

☠ *Acute Ecological Effects*

Acute (short-term) toxic effects may include the death of animals, birds, or fish, and death or low growth rate in plants. Acute effects are seen 2 to 4 days after animals or plants are exposed to ethyl butyl ketone. Insufficient data are available to evaluate or predict the short-term effects of ethyl butyl ketone on aquatic life, plants, birds, or land animals.

Chronic Ecological Effects

Chronic (long-term) toxic effects may include shortened life span, reproductive problems, lower fertility, and changes in appearance or behavior in exposed animals. These effects can be seen long after first exposure(s) to toxic chemicals. Insufficient data are available to evaluate or predict the long-term effects of ethyl butyl ketone on aquatic life, plants, birds, or land animals.

Water Solubility

Ethyl butyl ketone is slightly soluble in water. Concentrations of less than 100 milligrams will mix with a liter of water.

Persistence in the Environment

Ethyl butyl ketone is slightly persistent in water, with a half-life of between 2 to 20 days. The half-life of a pollutant is the amount of time it takes for one half of the chemical to be degraded. There is no information in the references on the specific persistence percentages for this chemical in the various environmental mediums (air, soil, water).

Bioaccumulation in Aquatic Organisms

Some substances increase in concentration, or *bioaccumulate,* in living organisms as they breathe contaminated air, drink contaminated water, or eat contaminated food. These chemicals can become concentrated in the tissues and internal organs of animals as well as humans.

The concentration of ethyl butyl ketone found in fish tissues is expected to be about the same as the average concentration of ethyl butyl ketone in water from which the fish was taken.

Recommended Risk-Reduction Measures

Proper training of all transporters will reduce the likelihood of a mishap or accident resulting in a leak or spill to the environment. The correct use of DOT labeling (per the MSDS) on all containers, trucks, and rail cars will enable emergency responders to react properly and quickly to any disaster thereby reducing the potential risk to the environment and to personnel.

Storage of ethyl butyl ketone should be segregated from incompatible chemicals to minimize the risk of cross-contamination or contact. Buildings designated for storage should be equipped with appropriate fire protection systems (alarms, sprinklers, emergency lighting, portable extinguishers).

If a spill or leak to the environment has occurred, fire department, emergency response, and/or hazardous materials spill personnel should be notified immediately. Cleanup should be attempted only by those trained in proper spill containment procedures. Contaminated soils should be removed for incineration and replaced with clean soil. If ethyl butyl ketone should contact the water table, aquifer, or navigable waterway, time is of the essence. It is slightly soluble in water and, therefore, total containment and remediation may not be entirely possible. When such spills occur, the local and/or state emergency response authorities must be notified. A comprehensive emergency response or disaster preparedness/recovery plan should be in place prior to any operations involving the use, transportation, storage, or disposal of ethyl butyl ketone.

If ethyl butyl ketone is spilled or leaked, the following specific steps are recommended:

- ☑ Restrict persons not wearing protective clothing from area of spill or leak until cleanup is complete and area can be opened for normal work.
- ☑ Remove all ignition sources.
- ☑ Ventilate area of spill or leak.
- ☑ Absorb liquids in vermiculite, dry sand, earth, or a similar material and deposit in sealed containers.
- ☑ For large spills, dike area far ahead of spill to ensure total containment. Do not allow spilled materials or run-off to reach waterways.
- ☑ Keep ethyl butyl ketone out of a confined space, such as a sewer, because of the possibility of explosion (unless the sewer is designed to prevent the buildup of explosive concentrations).
- ☑ It may be necessary to dispose of ethyl butyl ketone as a hazardous waste. The responsible state agency or the regional office of the federal Environmental Protection Agency (EPA) should be contacted for specific recommendations.

RISK ASSESSMENT: BUSINESS

General Assessment

Accidents or mishaps involving ethyl butyl ketone can present a moderate threat to business operations. The loss or damage of equipment or facilities can significantly affect fiscal viability. Lawsuits that may result from personnel injury, public exposures, and/or environmental contamination will also require a serious expenditure of resources. Media attention surrounding an injury, illness, or environmental damage can also result in a loss of profits and loss of current as well as future business.

Recommended Risk-Reduction Measures

Company attorneys, safety and health professionals, and environmental specialists should be involved in the development of any procedures dealing with safety, health, or emergency response. A company official should be pre-designated as a public relations officer with specific training in dealing with the press. Corporate plans and policies should be developed, approved, and implemented long before any need for such arises.

REFERENCES

Ahlbom, A. 1993. *Biostatistics for Epidemiologists.* New York: CRC Press/Lewis Publishers.

American Conference of Governmental Industrial Hygienists. 1988. *Documentation of the Threshold Limit Values and Biological Exposure Indices,* 5th Edition (with updates). Cincinnati: ACGIH

Anthony, C. P., and N. J. Kolthoff. 1971. *Textbook of Anatomy and Physiology.* St. Louis: C. V. Mosby Company.

Asante-Duah, D. K. 1993. *Hazardous Waste Risk Assessment.* New York: CRC Press/Lewis Publishers.

Bloom, A. D. and F. J. DeSerres. 1995. *Ecotoxicity and Human Health.* New York: CRC Press/Lewis Publishers.

Brusick, D. J. 1994. *Methods for Genetic Risk Assessment.* New York: CRC Press/Lewis Publishers.

Bueche, J. R. 1972. *Principles of Physics,* 2nd Edition. New York: McGraw-Hill.

Burrell, R., D. K. Flaherty, and L. J. Sauers. 1992. *Toxicology of the Immune System.* New York: Van Nostrand Reinhold.

Calabrese, E. J., E. M. Kenyon. *Air Toxics and Risk Assessment.* New York: CRC Press/Lewis Publishers.

Calabrese, E. J. 1994. *Biological Effects of Low Level Exposures.* New York: CRC Press/Lewis Publishers.

Cockerman, L. G., B. S. Shane. 1994. *Basic Environmental Toxicology.* New York: CRC Press/Lewis Publishers.

Cothern, R. R. 1993. *Comparative Environmental Risk Assessment.* New York: CRC Press/Lewis Publishers.

Cothern, C. R. 1995. *Handbook for Environmental Risk Decisionmaking.* New York: CRC Press/Lewis Publishers.

DiVincenzo, G. O., F. J. Yanno, and B. D. Astill. 1973. Exposure of man and dog to low concentrations of acetone vapor. *American Industrial Hygiene Journal.* 34:329-336.

Gardner, R. W. 1994. *Chemical Intolerance.* New York: CRC Press/Lewis Publishers.

Hallenbeck, W. H. 1993. *Quantitative Risk Assessment for Environmental and Occupational Health,* 2nd Edition. New York: CRC Press/Lewis Publishers.

Hodgson, E., R. B. Mailman, and J. E. Chambers. 1988. *Dictionary of Toxicology.* London: The MacMillian Company.

Hosty, J. W., and P. Foster. 1990. *A Practical Guide to Chemical Spill Response.* New York: Van Nostrand Reinhold.

Howard, P. H. 1995. *Handbook of Environmental Fate and Exposures Data for Organic Chemicals,* Vol. I-IV. New York: CRC Press/Lewis Publishers.

Howard, P. H., et. al. 1991. *Handbook of Environmental Degradation Rates.* New York: CRC Press/Lewis Publishers.

James, A., and M. Lord. 1992. *Index of Chemical and Physical Data.* New York: Van Nostrand Reinhold.

Jones, F. E. 1994. *Toxic Organic Vapors in the Workplace.* New York: CRC Press/Lewis Publishers.

Kamrin, M. A. 1988. *Toxicology: A Primer on Principles and Applications.* Chelsea, Michigan: Lewis Publishers.

Keith, L. H. and M. M. Walker. 1995. *Handbook of Air Toxics.* New York: CRC Press/Lewis Publishers.

Landis, W. G., M. H. Yu. 1995. *Introduction to Environmental Toxicology.* New York: CRC Press/Lewis Publishers.

Langerman, N. 1994 *Precautionary Labels for Chemical Containers.* New York: CRC Press/Lewis Publishers.

Lauwerys, R. R., P. Hoet. 1993. *Industrial Chemical Exposure.* New York: CRC Press/Lewis Publishers.

Lewis, R. J., Sr. 1993. *Hawley's Condensed Chemical Dictionary,* 12th Edition. New York: Van Nostrand Reinhold.

Lewis, R. J., Sr. 1992. *Sax's Dangerous Properties of Industrial Materials,* Volumes 1, 2, and 3, 8th Edition. New York: Van Nostrand Reinhold.

Lide, D. R. 1995. *Handbook of Chemistry and Physics,* 76th Edition. New York: CRC Press/Lewis Publishers.

Manahan, S. E. 1994. *Environmental Chemistry,* 6th Edition. New York: CRC Press/Lewis Publishers.

Manahan, S. E. 1992. *Toxicological Chemistry,* 2nd Edition. New York: CRC Press/Lewis Publishers.

Mellström, G. A., J. E. Wahlberg, and H. I. Maibach. 1994. *Protective Gloves for Occupational Use.* New York: CRC Press/Lewis Publishers.

Meyer, E. 1990. *Chemistry of Hazardous Materials.* Englewood Cliffs, NJ: Prentice Hall.

Patnaik, P. 1992. *A Comprehensive Guide to the Hazardous Properties of Chemical Substances.* New York: Van Nostrand Reinhold.

Philp, R. B. 1995. *Environmental Hazards and Human Health.* New York: CRC Press/Lewis Publishers.

Proctor, N. H., and J. P. Hughes. 1991. *Chemical Hazards of the Workplace.* New York: Van Nostrand Reinhold.

Quigley, D. R. 1994. *Handbook of Emergency Chemical Management.* New York: CRC Press/Lewis Publishers.

Raleigh, R. L. and W. A. McGee. 1972. Effects of short, high concentration exposures to acetone as determined by observation in the work area. *Journal of Occupational Medicine.* 14:607-610.

Rea, W. J. 1995. *Chemical Sensitivity,* Vol. I-III. New York: CRC Press/Lewis Publishers.

Richardson, M. 1995. *Dictionary of Substance and Their Effects,* Vol. I-VI. New York: CRC Press/Lewis Publishers.

U.S. Department of Health and Human Services, National Institute for Occupational Safety and Health. 1994. *NIOSH Pocket Guide to Chemical Hazards.* Washington, D.C.: U.S. Government Printing Office.

U.S. Department of Health and Human Services, National Institute for Occupational Safety and Health. 1978. *Criteria for a Recommended Standard - Occupational Exposure to Ketones.* DHEW Publication Number: 78-173. Washington, D.C.: U.S. Government Printing Office.

MATERIAL SAFETY DATA SHEET

CHEMICAL NAME	
ETHYL CHLORIDE	**CRC PRESS/LEWIS PUBLISHERS DISCLAIMER:** THE INFORMATION AND RECOMMENDATIONS PRESENTED HEREIN ARE BASED ON SOURCES BELIEVED TO BE RELIABLE. CRC MAKES NO REPRESENTATION ON ITS COMPLETENESS OR ACCURACY. IT IS THE RESPONSIBILITY OF THE USER TO DETERMINE THE CHEMICAL'S SUITABILITY FOR ITS INTENDED USE, THE CHEMICAL'S SAFE USE, AND THE CHEMICAL'S PROPER DISPOSAL. NO REPRESENTATIONS AND/OR WARRANTIES, EITHER EXPRESSED OR IMPLIED, OF THE MERCHANTABILITY OR FITNESS FOR A PARTICULAR PURPOSE, OR OF ANY OTHER NATURE, ARE MADE WITH RESPECT TO THE INFORMATION PROVIDED IN THIS MSDS OR TO THE CHEMICAL TO WHICH INFORMATION MAY REFER. CRC NEITHER ASSUMES NOR AUTHORIZES ANY OTHER PERSON TO ASSUME FOR IT, ANY OTHER ADDITIONAL RESPONSIBILITY OR LIABILITY FOR THE USE OF, OR RELIANCE UPON, THIS INFORMATION.

HAZARD WARNING INFORMATION

HEALTH	FIRE	REACTIVE	OTHER	DEGREE OF HAZARD	COLOR CODING	OTHER CODES
2	4	0		0 = Minimum Hazard 1 = Slight Hazard 2 = Moderate Hazard 3 = Serious Hazard 4 = Severe Hazard	HEALTH = BLUE FIRE = RED REACTIVITY = YELLOW OTHER = WHITE	OX = Oxidizer ACID = Acid ALK = Alkali COR = Corrosive W = Use No Water

SECTION I - GENERAL INFORMATION

Characterization	RCRA Number	EPA Class
Halogenated Hydrocarbon	**D001**	**Characteristic (I) Waste**

DOT Proper Shipping Name **Ethyl chloride**	Chemical Abstract Service (CAS) Number **75-00-3**
DOT Hazard Class and Label Requirements **Flammable Gas**	DOT Emergency Guide Code **27**
DOT Identification Number **UN 1037**	Molecular Formula **CH_3CH_2Cl**

Synonyms
Chloroethane; monochloroethane; muriatic ether; hydrochloric ether.

SECTION II - HAZARDOUS INGREDIENTS AND IDENTITY INFORMATION

Hazardous Components (specific identity)	OSHA Exposure Criteria	NIOSH Exposure Criteria	Immediately Dangerous to Life and Health (IDLH)	ACGIH Exposure Criteria
Ethyl chloride (derivation: By passing hydrogen chloride into a solution of zinc chloride and ethanol or by action of chlorine on ethylene in the presence of the chlorides of copper, iron, antimony, and calcium). **1 ppm = 2.68 mg/m^3**	**PEL:** **1000 ppm** **2600 mg/m^3** **STEL:** **Not Established**	**REL:** **Reduce to Lowest Possible Level**	**20,000 ppm**	**TLV:** **1000 ppm** **2600 mg/m^3** **STEL:** **Not Established**

SECTION III - PHYSICAL AND CHEMICAL CHARACTERISTICS

Boiling Point **54°F (12°C)**	Specific Gravity (H_2O = 1) **0.92 at 32°F (0°C)**
Vapor Pressure (mm Hg) **1060 at 69°F (20°C)**	Molecular Weight (atomic weight) **64.5**
Vapor Density (Air = 1) **2.22**	Freezing Point **-218°F (-139°C)**

Solubility
Slightly soluble in water (0.6%). Soluble in alcohol, ether, and most organic solvents.

Appearance and Odor
Colorless gas, with a strong, pungent ether-like odor. (Note: Exists as a liquid below 54°F). Normally shipped as a liquefied compressed gas. Odor Threshold = 4.2 ppm

SECTION IV - FIRE AND EXPLOSION HAZARD DATA

Flash Point (method used) **-58°F (-50°C) closed cup (as a liquid)**	Explosive Limits in Air % by Volume **LEL: 3.8%** **UEL: 15.4%**
NFPA Classification **Flammable Gas; Class 1A Flammable Liquid**	Autoignition Temperature **966°F (519°C)**

Extinguishing Media
Stop flow of gas first, if possible. Use dry chemical, carbon dioxide, water spray, or foam.

Special Fire Fighting Procedures
Poisonous gases are produced in fire. Wear self-contained breathing apparatus (SCBA) and full protective gear. Move containers from fire if it can be done without risk. Cool sides of fire-exposed containers with water until long after fire is out.

Unusual Fire and Explosion Hazards
Dangerous fire and explosion hazard. Containers may explode in fire. Its vapors are heavier than air and can travel for great distances to an ignition source to flashback and cause fire or explosion.

SECTION V - REACTIVITY DATA

Stability		Conditions to Avoid Ethyl chloride is stable up to 752°F (400°C). Avoid heat, flame, other sources of ignition, and contact with incompatible materials and chemicals.
Stable X	Unstable	Incompatibility *(materials to avoid)* Strong oxidizers (chlorine, bromine, fluorine), chemically active metals (such as sodium, potassium, calcium, powdered aluminum, zinc, and magnesium). Water or steam (reacts to form hydrochloric acid).
Hazardous Polymerization		Conditions to Avoid Under normal conditions of temperature and pressure, hazardous polymerization of ethyl chloride cannot occur.
May Occur	Will Not Occur X	Hazardous Decomposition or By-products Poisonous gases and acrid fumes are produced in fire, including extremely toxic phosgene and hydrogen chloride gas.

SECTION VI - HEALTH HAZARD DATA

Primary Route(s) of Entry:	Inhalation?	Absorption (skin/eye)?	Ingestion?
	X	X	

Health Hazards

INHALATION: Causes depression of the central nervous system (CNS) with stupor, eye irritation, lack of coordination, abdominal cramps, anesthetic effects, unconsciousness, and cardiac arrest. May damage liver and kidneys. Vapors may cause irritation to the eyes, nose, and throat.

ABSORPTION: Irritation, redness and pain from frostbite to the skin. Eye contact may result in severe irritation and possible permanent damage due to frostbite. Will pass through intact skin.

INGESTION: Unspecified gastrointestinal effects. An unlikely exposure route, but still possible.

Carcinogenicity	NTP Listed?	IARC Cancer Review Group?	OSHA Regulated?	Target Organs?
Unknown Human Questioned Animal	Yes	No	29 CFR 1910.1000 Table Z-1	Eyes, skin, respiratory sys., liver, kidney, CVS, CNS.

Medical Conditions Generally Aggravated by Exposure

None reported. However, liver, kidney, heart, and respiratory problems may be aggravated by exposure.

Emergency and First-aid Procedures

Eye contact: Do not allow victim to keep eyes tightly shut. Flush immediately with tepid water (104°F), for 15 minutes (minimum), seek medical attention. Skin contact: Remove all contaminated clothing, including shoes. Immediately wash area with flooding amounts of warm water (104°F). Do NOT rub affected area or use dry heat. Seek medical attention immediately. For inhalation: Remove the person from exposure. Provide respiratory assistance and CPR. Transfer to medical facility. If swallowed: Seek medical attention immediately. Give conscious and alert person 1-2 glasses of water and induce vomiting. Never give an unconscious or convulsing person anything by mouth.

SECTION VII - PRECAUTIONS FOR SAFE HANDLING AND USE

Steps to be Taken in Case Material is Released or Spilled

Remove all ignition sources, wear SCBA respiratory protection. Restrict those not wearing protective equipment and who are not involved in cleanup from entering area. Stop leak if it can be done without risk. Spilled liquids are likely to evaporate quickly. Absorb with vermiculite and place in sealed container.

Preferred Waste Disposal Method

Burn in chemical incinerator equipped with an afterburner and scrubber.

Precautions to be Taken in Handling and Storage

Store in tightly closed containers in cool, well ventilated location. Keep away from sources of heat.

Other Precautions and Warnings

Sources of ignition such as open flame or smoking, are prohibited where ethyl chloride is used or stored. Metal containers should be bonded and grounded. Empty cylinders contain residue, handle with care.

SECTION VIII - CONTROL MEASURES AND PERSONAL PROTECTIVE EQUIPMENT

Respiratory Protection *(specify type)*

Use an MSHA/NIOSH-approved air purifying respirator (cartridge type) or use a supplied-air respirator or a self-contained breathing apparatus (SCBA) with full facepiece operated in positive pressure mode.

Ventilation

Local exhaust (preferred) at site of chemical work or general ventilation.

Protective Gloves	Eye Protection	Other Protective Clothing
Viton, Nitrile, or Butyl Rubber Gloves	Chemical Goggles or Face Mask	Protective Apron if Splash is Likely

Work/Hygiene Practices

Always wash hands thoroughly after using chemical; never bring food, drink, or smoking materials into vicinity of chemicals.

ETHYL CHLORIDE

CH_3CH_2Cl CAS: 75-00-3

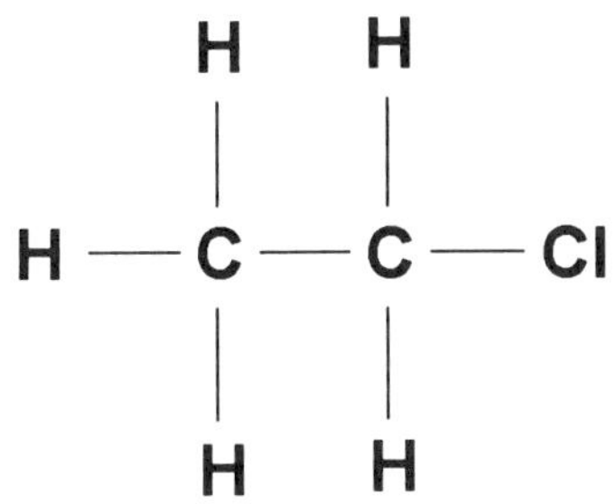

IDENTIFICATION AND TYPICAL USES

Ethyl chloride is a colorless gas with a strong ether-like odor. It can also exist as a liquid below 54°F (12°C). It is used as a refrigerant, as a solvent, in the manufacture of tetraethyl lead, and as an alkylating agent. It is also used as a topical anesthetic.

RISK ASSESSMENT: HEALTH

General Assessment

Ethyl chloride is mildly toxic by *inhalation.* Although it is normally a gas at room temperature, it can also exist as a liquid at temperatures below 54°F (12°C) and, therefore, ethyl chloride can enter the body through *ingestion* of the liquid. The liquid is also known the pass through the skin (*absorption*) to cause toxic systemic effects. Skin contact with the liquid can cause localized tissue damage due to the extremely cold temperatures of the compressed or liquefied gas.

Inhalation is irritating to the eyes, nose, and mucosa of the respiratory tract. High concentrations of ethyl chloride causes depression of the central nervous system and cardiac arrhythmia. Subsequent symptoms include dizziness, lightheadedness, headache, stupor, drunkenness and inebriation, loss of coordination, abdominal cramps, unconsciousness, and possible cardiac arrest, liver, and kidney damage. It should be noted that such symptoms as inebriation and drunkenness can also lead to accidents. Personnel may injure themselves or others, or damage equipment and property, as a result of the intoxicating effects of exposure.

Aside from the absorption potential of ethyl chloride, skin contact can also produce allergic eczema, and severe tissue damage due to frostbite. Eye contact will also cause serious injury to tissues as a result of the extremely cold temperatures of both the liquid and the compressed gas.

☠ *Acute Health Effects*

The following acute (short-term) health effects may occur immediately or shortly after exposure to ethyl chloride:

Skin: Irritation with potential for redness, pain, frostbite, and tissue damage. Possible allergic eczema can develop.

Eye: Severe irritation, burning (frostbite), pain, and smarting of the conjunctiva.

Lung: Nose, throat, and respiratory tract irritation. Absorption through the lungs is quick.

CNS: Headaches, lightheadedness, dizziness, inebriation, loss of coordination, stupor, and drunkenness. Death can occur due to cardiac arrhythmia and possible cardiac arrest.

Chronic Health Effects

The following chronic (long-term) health effects may occur at some time after exposure to ethyl chloride and can last for months or even years:

Cancer Hazards: The National Institute for Occupational Safety and Health (NIOSH) considers ethyl chloride to be a carcinogen and the National Toxicology Program (NTP) has conducted studies indicated evidence that ethyl chloride is a tumorigenic agent. However, no carcinogenic designations have been assigned to this chemical. Further study and confirmation are required. It is recommended that users proceed with due caution until more is known about the carcinogenic properties of ethyl chloride.

Reproductive Hazard: According to information presented in the references, ethyl chloride has not been adequately tested for its ability to cause adverse reproductive effects in test animals.

Other Chronic Effects: Long-term exposure may lead to liver and kidney damage. Such damage has been demonstrated in animals and is possible, but not probable, in humans.

Recommended Risk-Reduction Measures

Exposure to ethyl chloride does not appear to pose a serious threat in the industrial environment since exposure to very high concentrations (13,000 - 40,000 ppm and higher) are required to cause toxic effects. However, exposure is still possible under certain operating conditions and during emergency situations. Personnel should therefore avoid direct contact with ethyl chloride. Occupational poisoning occurs most commonly by inhalation. If a less toxic material or compound cannot be substituted for ethyl chloride, then *engineering controls* are the most effective method of reducing exposures. The best protection is to enclose operations and/or provide local exhaust ventilation at the site of ethyl chloride release. While not always operationally feasible, isolating operations can also reduce exposure. Using respiratory protection is less effective than the controls mentioned above, but is still advisable whenever working with or around ethyl chloride. Caution should be exercised when selecting respiratory protection since it is not known if exposure to ethyl chloride will cause cancer in humans. For exposure over the PEL (1000 ppm), an MSHA/NIOSH-approved air-purifying respiratory with organic vapor cartridge(s) will suffice. For exposures to higher concentrations, or when the concentration is unknown, use a supplied air respirator with full facepiece and mask operated in positive pressure mode or a self-contained breathing apparatus (SCBA) with full facepiece operated in pressure demand. If a full facepiece is not available, then chemical goggles should be worn to protect the eyes. A face shield should also be considered. To prevent hand and skin exposures, nitrile, Viton, or butyl rubber gloves should be worn.

Administrative controls should also be in place to minimize the potential for human exposures. These may include written procedures or policies which specify the methods and techniques that will be practiced whenever personnel are to work with ethyl chloride.

All personnel should receive *training* on the use, hazards, protective measures, emergency actions, and other precautions per 29 CFR 1910.1200 (Hazard Communication), prior to the first assignment in an area where ethyl chloride is used or stored.

If symptoms develop or overexposure is suspected, the following recommended medical tests should be considered:

- ☑ Liver and kidney function tests.
- ☑ Special 24-hour electrocardiogram (EKG) using the Holter Monitor to evaluate for irregular heartbeat.
- ☑ Interview for brain effects, including recent changes in memory, mood (irritability, withdrawal), and concentration. Look for headaches, malaise, and altered sleeping patterns. Cerebellar, autonomic, and peripheral nervous system evaluation should be considered. Positive or borderline cases should be referred for further neuropsychological testing.

Any evaluation should include a careful history of past and present symptoms with an examination. Medical tests that simply look for existing damage are not a substitute for controlling exposures.

Other methods to reduce exposure include:

- ☑ Where possible, enclose operations and use local exhaust ventilation at the site of chemical release. If local exhaust ventilation or enclosure is not used, respiratory protection should be mandatory.
- ☑ Always ensure that proper protective clothing is worn when using chemical substances.
- ☑ Wash thoroughly immediately after exposure to ethyl chloride and at the end of the work shift or before eating, drinking, or smoking.
- ☑ Hazard warning information should be posted in the work area. In addition, as part of an on-going education and training program, all information on the health and safety hazards of ethyl chloride should be communicated to all potentially exposed workers.
- ☑ Eye wash stations should be provided in the immediate work area for emergency use. If there is a possibility of skin exposure to ethyl chloride, emergency shower facilities should also be provided in the immediate area.
- ☑ Workers whose clothing has been contaminated by ethyl chloride should change into clean clothes before leaving work. Contaminated work clothing should be laundered only by individuals who have

been informed of the hazards of exposure to ethyl chloride.

☑ Where possible, automatically transfer liquids containing ethyl chloride from drums or other containers to process containers.

RISK ASSESSMENT: ENVIRONMENT

General Assessment

Although ethyl chloride normally exists as a gas, it is usually shipped as a liquefied compressed gas. The environment is at risk of exposure during transportation, storage, disposal, or destruction of ethyl chloride. In almost every scenario, the threat of environmental exposure is contingent upon the proper handling of the chemical substance. Accidental spills, large or small, can result in contamination of the surrounding environmental mediums (water, soil, and air).

Ethyl chloride is very dangerous flammable liquid. Because of its low flash point and low boiling point, it is classified as a Class 1A flammable liquid or a flammable gas, depending on its state (per OSHA 29 CFR 1910.106). It is a serious fire hazard and moderate explosion hazard when exposed to heat or flame. It is incompatible with strong oxidizers, such as chlorine, fluorine, and bromine, and many chemically-active metals, such as sodium, potassium, calcium, powdered aluminum, zinc, and magnesium. Caution is always required in handling, storage, transportation, and disposal of ethyl chloride. Emergency responders should be made aware of the presence of ethyl chloride at any emergency response situation.

Ethyl chloride can enter the environment from industrial effluents and from spills.

Acute Ecological Effects

Acute (short-term) toxic effects may include the death of animals, birds, or fish and death or low growth rate in plants. Acute effects are seen 2 to 4 days after animals or plants are exposed to ethyl chloride.

Ethyl chloride has slight acute toxicity to aquatic life. It has caused respiration and germination increases in potatoes. No data are available on the short-term effects of ethyl chloride on birds or land animals.

Chronic Ecological Effects

Chronic (long-term) toxic effects may include shortened life span, reproductive problems, lower fertility, and changes in appearance or behavior in exposed animals. These effects can be seen long after first exposure(s) to toxic chemicals.

Ethyl chloride has slight chronic toxicity to aquatic life. Insufficient data are available to evaluate or predict the long-term effects of ethyl chloride to plants, birds, or terrestrial animals.

Water Solubility

Ethyl chloride is moderately soluble in water. Concentrations between 1 and 1000 milligrams will mix with a liter of water.

Persistence in the Environment

Ethyl chloride is non-persistent in water, with a half-life of less than 2 days. The half-life of a pollutant is the amount of time it takes for one half of the chemical to be degraded. About 99.8% of ethyl chloride will eventually end up in air; the remainder will end up in water.

Bioaccumulation in Aquatic Organisms

Some substances increase in concentration, or *bioaccumulate,* in living organisms as they breathe contaminated air, drink contaminated water, or eat contaminated food. These chemicals can become concentrated in the tissues and internal organs of animals as well as humans.

The concentration of ethyl chloride found in fish tissues is expected to be about the same as the average concentration of ethyl chloride in the water from which the fish was taken.

Recommended Risk-Reduction Measures

Proper training of all transporters will reduce the likelihood of a mishap or accident resulting in a leak or spill of ethyl chloride into the environment. Labels on all containers, trucks, and rail cars must meet DOT requirements and accurately reflect their contents to enable emergency responders to react properly and quickly to any disaster thereby reducing the potential risk to the environment and to personnel.

Storage of ethyl chloride should be segregated from other chemicals and materials to minimize the risk of cross-contamination.

If a spill or leak to the environment has occurred, fire department, emergency response and/or hazardous materials spill personnel should be notified immediately. Cleanup should be attempted only by those trained in proper spill containment procedures. Contaminated soils should be removed for incineration and

replaced with clean soil. If ethyl chloride should contact the water table, aquifer, or navigable waterway, time is of the essence. It is moderately soluble and total remediation may not be possible. The local and/or state emergency response authorities must be notified. A comprehensive emergency response or disaster preparedness/recovery plan should be in place prior to any operations involving the use, transportation, storage, or disposal of ethyl chloride.

For disposal, ethyl chloride may be burned in a chemical incinerator equipped with an afterburner and scrubber to remove the haloacids. It should be noted that complete destruction by incineration is difficult unless a direct gas feed to the incinerator is used. These are normally available in permit-approved facilities only. Complete combustion is necessary to prevent the formation of phosgene gas.

If ethyl chloride is spilled or leaked, the following specific steps are recommended:

- ☑ Restrict persons not wearing protective clothing from area of spill or leak until cleanup is complete. Use only non-sparking tools.
- ☑ If source of leak is a cylinder, attempt to shut off leak if it can be done without risk. If not, try to move cylinder to a safe place (outdoors) and allow to vent. If fire occurs, use water spray to knock down vapors and direct away from personnel. Absorb liquids in vermiculite, sand, earth, or similar material and deposit in sealed drum.
- ☑ Ventilate area of spill or leak and remove all ignition sources.
- ☑ It may be necessary to dispose of ethyl chloride as a hazardous waste. The responsible state agency or the regional office of the federal Environmental Protection Agency (EPA) should be contacted for specific recommendations.

RISK ASSESSMENT: BUSINESS

General Assessment

Accidents or mishaps involving ethyl chloride can present a serious threat to business operations. The loss or damage of equipment or facilities can significantly affect fiscal viability. Lawsuits that may result from personnel illness, injury/death, public exposures, and/or environmental contamination will require a serious expenditure of resources. Media attention surrounding an injury, death, or environmental damage can also result in a loss of profits and loss of current as well as future business.

Recommended Risk-Reduction Measures

Company attorneys, safety and health professionals, and environmental specialists should be involved in the development of any procedures for responding to chemical incidents. A company official should be predesignated as a public relations officer with specific training in dealing with the press. Corporate plans and policies should be developed, approved, and implemented long before any need for such arises.

REFERENCES

American Conference of Governmental Industrial Hygienists. 1988. *Documentation of the Threshold Limit Values and Biological Exposure Indices,* 5th Edition (with updates). Cincinnati: ACGIH

Anthony, C. P., and N. J. Kolthoff. 1971. *Textbook of Anatomy and Physiology.* St. Louis: C. V. Mosby Company.

Bueche, J. R. 1972. *Principles of Physics,* 2nd Edition. New York: McGraw-Hill.

Calabrese, E. J. 1994. *Biological Effects of Low Level Exposures.* New York: CRC Press/Lewis Publishers.

Cockerman, L. G., B. S. Shane. 1994. *Basic Environmental Toxicology.* New York: CRC Press/Lewis Publishers.

Cothern, C. R., N. P. Ross. 1994. *Environmental Statistics, Assessment, and Forecasting.* New York: CRC Press/Lewis Publishers.

Hallenbeck, W. H. 1993. *Quantitative Risk Assessment for Environmental and Occupational Health,* 2nd Edition. New York: CRC Press/Lewis Publishers.

Lewis, R. J., Sr. 1993. *Hawley's Condensed Chemical Dictionary,* 12th Edition. New York: Van Nostrand Reinhold.

Lewis, R. J., Sr. 1992. *Sax's Dangerous Properties of Industrial Materials,* Volumes 1, 2, and 3, 8th Edition. New York: Van Nostrand Reinhold.

Meyer, E. 1990. *Chemistry of Hazardous Materials.* Englewood Cliffs, NJ: Prentice Hall.

Patnaik, P. 1992. *A Comprehensive Guide to the Hazardous Properties of Chemical Substances.* New York: Van Nostrand Reinhold.

Quigley, D. R. 1994. *Handbook of Emergency Chemical Management.* New York: CRC Press/Lewis Publishers.

U.S. Department of Health and Human Services, National Institute for Occupational Safety and Health. 1994. *NIOSH Pocket Guide to Chemical Hazards.* Washington, D.C.: U.S. Government Printing Office.

MATERIAL SAFETY DATA SHEET

CHEMICAL NAME	
ETHYL CHLOROFORMATE	**CRC PRESS/LEWIS PUBLISHERS DISCLAIMER:** THE INFORMATION AND RECOMMENDATIONS PRESENTED HEREIN ARE BASED ON SOURCES BELIEVED TO BE RELIABLE. CRC MAKES NO REPRESENTATION ON ITS COMPLETENESS OR ACCURACY. IT IS THE RESPONSIBILITY OF THE USER TO DETERMINE THE CHEMICAL'S SUITABILITY FOR ITS INTENDED USE, THE CHEMICAL'S SAFE USE, AND THE CHEMICAL'S PROPER DISPOSAL. NO REPRESENTATIONS AND/OR WARRANTIES, EITHER EXPRESSED OR IMPLIED, OF THE MERCHANTABILITY OR FITNESS FOR A PARTICULAR PURPOSE, OR OF ANY OTHER NATURE, ARE MADE WITH RESPECT TO THE INFORMATION PROVIDED IN THIS MSDS OR TO THE CHEMICAL TO WHICH INFORMATION MAY REFER. CRC NEITHER ASSUMES NOR AUTHORIZES ANY OTHER PERSON TO ASSUME FOR IT, ANY OTHER ADDITIONAL RESPONSIBILITY OR LIABILITY FOR THE USE OF, OR RELIANCE UPON, THIS INFORMATION.

HAZARD WARNING INFORMATION

HEALTH	FIRE	REACTIVE	OTHER	DEGREE OF HAZARD	COLOR CODING	OTHER CODES
3	3	1		0 = Minimum Hazard 1 = Slight Hazard 2 = Moderate Hazard 3 = Serious Hazard 4 = Severe Hazard	HEALTH = BLUE FIRE = RED REACTIVITY = YELLOW OTHER = WHITE	OX = Oxidizer ACID = Acid ALK = Alkali COR = Corrosive W = Use No Water

SECTION I - GENERAL INFORMATION

Characterization **Chlorinated hydrocarbon**	RCRA Number **None**	EPA Class **Not Applicable**
DOT Proper Shipping Name **Ethyl Chlorocarbonate**	Chemical Abstract Service (CAS) Number **541-41-3**	
DOT Hazard Class and Label Requirements **Flammable Liquid; Poison**	DOT Emergency Guide Code **57**	
DOT Identification Number **UN 1182**	Chemical Formula **$ClCOOOC_2H_5$**	

Synonyms
Ethyl chlorocarbonate; chloroformic acid ethyl ether.

SECTION II - HAZARDOUS INGREDIENTS AND IDENTITY INFORMATION

Hazardous Components (specific identity)	OSHA Exposure Criteria	NIOSH Exposure Criteria	Immediately Dangerous to Life and Health (IDLH)	ACGIH Exposure Criteria
Ethyl chloroformate (derivation: Reaction of carbon monoxide with gaseous chlorine, producing phosgene ($COCL_2$) which is then treated with anhydrous ethanol, giving ethyl chlorocarbonate and splitting of hydrogen chloride).	**PEL:** Not Established **STEL:** Not Established	**REL:** Not Established **STEL:** Not Established	**Not Determined**	**TLV:** Not Established **STEL:** Not Established

SECTION III - PHYSICAL AND CHEMICAL CHARACTERISTICS

Boiling Point **203°F (95°C)**	Specific Gravity (H_2O = 1) **1.35**
Vapor Pressure (gas) **53 at 69°F (20°C)**	Molecular Weight **108.5**
Vapor Density (Air = 1) **3.74**	Melting Point **-113°F (-81°C)**

Solubility
Decomposes in water and alcohol; soluble in benzene, chloroform, and ether.

Appearance and Odor
Colorless, water-white liquid with an irritating odor.

SECTION IV - FIRE AND EXPLOSION HAZARD DATA

Flash Point (method used) **61°F (16°C) closed cup**	Explosive Limits in Air % by Volume **LEL: Not Determined UEL: Not Determined**
NFPA Classification **Class 1A Flammable Liquid**	Autoignition Temperature **932°F (500°C)**

Extinguishing Media
Alcohol resistant foam, carbon dioxide, dry chemical, water spray, or fog.

Special Fire Fighting Procedures
Highly flammable and corrosive liquid. Wear full protective clothing and self-contained breathing apparatus (SCBA). Poisonous gases are produced in fire. Use water spray to cool containers.

Unusual Fire and Explosion Hazards
Dangerous fire and explosion hazard when exposed to heat or flame. Containers may explode in fire. Immediately withdraw if rising sound is heard from venting device or fire is causing a discoloration of tank, a boiling liquid expanding vapor explosion (BLEVE) may be imminent. Isolate ½ mile distance.

SECTION V - REACTIVITY DATA

Stability		Conditions to Avoid **Under normal conditions of temperature, storage, pressure, and handling, ethyl chloroformate is stable. Keep away from heat or flame. Keep away from incompatible materials. Corrosive to metals.**
Stable **X**	Unstable	Incompatibility *(materials to avoid)* **Acids, strong bases, alcohols, oxidizers (such as perchlorates, peroxides, permanganates, chlorates, and nitrates). Decomposes by water, moist air, or steam to release toxic and corrosive gases.**
Hazardous Polymerization		Conditions to Avoid **Under normal conditions of temperature and pressure, hazardous polymerization of ethyl chloroformate is not expected to occur.**
May Occur	Will Not Occur **X**	Hazardous Decomposition or By-products **When heated to decomposition, ethyl chloroformate can emit highly toxic/poisonous gases, including chloride fumes.**

SECTION VI - HEALTH HAZARD DATA

Primary Route(s) of Entry:	Inhalation?	Absorption (skin)?	Ingestion?
	X	X	X

Health Hazards

INHALATION: **Irritation of the eyes, nose, throat, and lungs causing coughing or shortness of breath. Higher exposures may cause severe irritation leading to a dangerous (life-threatening) condition known as pulmonary edema (fluid buildup in the lungs). Repeated exposure may cause liver and kidney damage.**

ABSORPTION: **Skin or eye contact results in acid-type burns (1st or 2nd degree) possible. It will also pass through intact skin to cause toxic systemic effects similar to that of inhalation.**

INGESTION: **Severe burns to the mouth and stomach.**

Carcinogenicity	NTP Listed?	IARC Cancer Review Group?	OSHA Regulated?	Target Organs?
Unknown Human **Questioned Animal**	**No**	**No**	**No**	**Respiratory system, eyes, skin, liver, kidney.**

Medical Conditions Generally Aggravated by Exposure

None reported.

Emergency and First-aid Procedures

<u>Eye contact</u>: Flush immediately with water for 15 minutes (minimum), seek medical attention. <u>Skin contact</u>: Remove all contaminated clothing. Immediately wash area with large amounts of water (15 minutes minimum) then apply a dilute solution of sodium bicarbonate. For <u>inhalation</u>: Remove the person from exposure. Provide respiratory assistance and CPR. Transfer to medical facility. Observe for 48 hours for lung effects. If <u>swallowed</u>: If conscious, give large amounts of water, do NOT induce vomiting. Seek medical attention immediately.

SECTION VII - PRECAUTIONS FOR SAFE HANDLING AND USE

Steps to be Taken in Case Material is Released or Spilled

Absorb liquids in vermiculite, dry sand or similar material and deposit in sealed containers. Ventilate area of spill or leak. Restrict those not involved in cleanup from area. Remove all sources of ignition.

Preferred Waste Disposal Method

No Citation.

Precautions to be Taken in Handling and Storage

Store to avoid contact with incompatible materials since violent reactions can occur. Store in tightly closed containers in cool, well-ventilated area away from heat and flame.

Other Precautions and Warnings

Heat may cause containers to build up pressure and explode. Use only non-sparking tools. Drums must be equipped with self-closing valves, pressure vacuum bungs, and flame arresters.

SECTION VIII - CONTROL MEASURES AND PERSONAL PROTECTIVE EQUIPMENT

Respiratory Protection *(specify type)*

Exposure levels have not been established. Recommend using a supplied-air respirator with full facepiece, hood, or helmet in continuous flow mode, or use a self-contained breathing apparatus (SCBA) operated in pressure demand or other positive pressure mode.

Ventilation

Local exhaust or general mechanical systems recommended.

Protective Gloves	Eye Protection	Other Protective Clothing
Solvent-resistant Rubber	**Chemical Goggles or Face Mask**	**Rubber Apron**

Work/Hygiene Practices

Always wash hands thoroughly after using chemical; never bring food, drink, or smoking materials into vicinity of chemicals.

ETHYL CHLOROFORMATE

$ClCOOC_2H_5$ CAS: 541-41-3

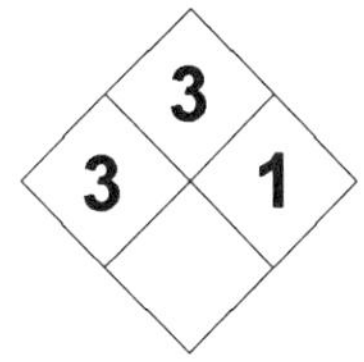

IDENTIFICATION AND TYPICAL USES

Ethyl chloroformate is a colorless, water-white liquid with an irritating odor. It is used in organic synthesis, as an intermediate in making diethyl carbonate, in flotation agents, polymers, and isocyanates.

RISK ASSESSMENT: HEALTH

General Assessment

Ethyl chloroformate is a human poison by *inhalation, ingestion,* and skin *absorption.* Its carcinogenicity is questioned in animals, although human carcinogenicity studies have proven inconclusive in this regard. It has been confirmed that exposure to ethyl chloroformate can cause nasal cancers in some test animals (rats) by inhalation. A conservative approach to exposure risk management is recommended for ethyl chloroformate since the exact nature of its cancer risk is unresolved.

Inhalation of ethyl chloroformate vapors can cause irritation of the eyes, nose, throat, and respiratory tract with symptoms of burning sensation, coughing, sneezing, wheezing, laryngitis, severe lachrymation, headache, vomiting, nausea, shortness of breath. Higher concentrations may be fatal as a result of spasm, inflammation, and edema of the bronchi and larynx, chemical pneumonitis, or pulmonary edema (fluid in the lungs). Pulmonary edema can be delayed up to 48 hours following exposure, thereby creating a false sense of security with regard to health exposure risk. Repeated or long-term exposures may cause liver or kidney damage.

Skin contact results in serious burns and corrosive tissue destruction leading to permanent damage. Short exposures can cause 1st degree burns and longer exposures may result in 2nd degree burns. It may also cause skin sensitization. Eye contact with the vapor produces severe lachrymation (tearing) and the liquid will cause smarting, irritation, and painful burns with tissue damage and possible loss of vision (depending upon the length of time the liquid remains in contract with the eye surface.

Acute Health Effects

The following acute (short-term) health effects may occur immediately or shortly after exposure to ethyl chloroformate:

Skin: Immediate irritation and burn on contact. Burns may be severe, leading to permanent damage to the skin.

Eye: Acid-type burns with possibility for permanent tissue damage.

Lung: Irritation of the respiratory tract. High level exposures can lead to death as a result of spasm or edema of the bronchi or larynx. Death can also result from chemical pneumonitis and/or pulmonary edema (fluid in the lungs).

Chronic Health Effects

The following chronic (long-term) health effects may occur at some time after exposure to ethyl chloroformate and can last for months or even years:

Cancer Hazards: There is limited evidence in the references that exposure to ethyl chloroformate may cause nasal cancer (positive studies with rats). Human carcinogenic research is limited and inconclusive.

Reproductive Hazard: According to information presented in the references, ethyl chloroformate has not been adequately tested for its ability to cause cancer in test animals.

Other Chronic Effects: Repeated exposures may damage the liver and kidneys. There are no data regarding any other possible or potential long-term (chronic) effects of exposure to ethyl chloroformate.

☋ *Recommended Risk-Reduction Measures*

Personnel should avoid direct contact with ethyl chloroformate or any other chemical with suspected or unconfirmed human carcinogenic characteristics. If a less toxic chemical cannot be substituted for a hazardous substance, then *engineering controls* are the most effective method of reducing exposures. The best protection is to enclose operations and/or provide local exhaust ventilation at the site of chemical release. While not always operationally feasible, isolating operations can also reduce exposure. Using respiratory protection is less effective than the controls mentioned above, but is still advisable whenever working with or around ethyl chloroformate. No exposure levels have been established for this substance. For any concentration, a supplied-air respirator operated in continuous flow or other positive pressure mode or a self-contained breathing apparatus (SCBA) with full facepiece operated in pressure demand are the recommended respiratory protection methods of choice. If a full facepiece is not available, then chemical goggles should be worn to protect the eyes. Whenever a chemical splash hazard exists, a face shield and a rubber apron should be worn. To prevent hand and skin exposures, acid resistant gloves should be worn.

Administrative controls should also be in place to minimize the potential for human exposures. These may include written procedures or policies which specify the methods and techniques that will be practiced whenever personnel are to work with ethyl chloroformate.

All personnel should receive *training* on the use, hazards, protective measures, emergency actions, and other precautions per 29 CFR 1910.1200 (Hazard Communication), prior to the first assignment in an area where ethyl chloroformate is used or stored.

Before beginning employment and at regular intervals thereafter (e.g., annually), the following medical tests are recommended:

☑ Lung function tests.

☑ Kidney and liver function tests.

☑ Consider chest X-ray after acute overexposure (may be negative if taken immediately following exposure as pulmonary edema may be delayed up to 48 hours).

Any evaluation should include a careful history of past and present symptoms with an examination. Medical tests that simply look for existing damage are not a substitute for controlling exposures. Also, since smoking can cause heart disease, emphysema, and other respiratory disorders, smokers exposed to ethyl chloroformate may experience symptoms more quickly and more pronounced than non-smokers under the same exposure conditions. Prudent risk assessment and management requires careful consideration of *all* possible factors which may be causing the appearance of exposure symptoms among the work force.

Other methods to reduce exposure include:

☑ Where possible, enclose operations and use local exhaust ventilation at the site of chemical release. If local exhaust ventilation or enclosure is not used, respiratory protection should be mandatory.

☑ Always ensure that proper protective clothing is worn when using chemical substances.

☑ Wash thoroughly immediately after exposure to ethyl chloroformate and at the end of the work shift or before eating, drinking, or smoking.

☑ Hazard warning information should be posted in the work area. In addition, as part of an on-going education and training program, all information on the health and safety hazards of ethyl chloroformate should be communicated to all potentially exposed workers.

☑ Eye wash stations should be provided in the immediate work area for emergency use. If there is a possibility of skin exposure to ethyl chloroformate, emergency shower facilities should also be provided.

☑ Personnel should never wear contaminated clothing home (family members can be exposed). Clothing should be laundered by personnel who have been trained on the hazards of ethyl chloroformate.

RISK ASSESSMENT: ENVIRONMENT

General Assessment

The environment is at risk of exposure during transportation, storage, disposal, or destruction of ethyl chloroformate. In almost every scenario, the threat of environmental exposure is contingent upon the proper handling of the chemical substance. Accidental spills, large or small, can result in fire, explosion, and possible contamination of the surrounding environmental mediums (water, soil, and air).

Ethyl chloroformate is considered a Class 1A flammable liquid (per OSHA 29 CFR 1910.106). Its extremely low flash point and relatively low boiling point make this chemical a serious fire and explosion hazard concern. These characteristics require extreme caution in handling, storage, transportation, and disposal. It is incompatible with a number of materials, including many common acids and oxidizers. Contact with either can cause fire or explosion. Therefore, special consideration is required during any emergency situation involving a leak or spill of ethyl chloroformate. Should ethyl chloroformate ever come into contact with incompatible substances during use, transportation, storage, or disposal, the formation of highly toxic and/or highly explosive commodities is possible.

Ethyl chloroformate may enter the environment through industrial discharges or spills.

Acute Ecological Effects

Acute (short-term) toxic effects may include the death of animals, birds, or fish and death or low growth rate in plants. Acute effects are seen 2 to 4 days after animals or plants are exposed to ethyl chloroformate.

Insufficient data are available on the short-term effects of ethyl chloroformate exposure to aquatic life, plants, birds, or land animals.

Chronic Ecological Effects

Chronic (long-term) toxic effects may include shortened life span, reproductive problems, lower fertility, and changes in appearance or behavior in exposed animals. These effects can be seen long after first exposure(s) to toxic chemicals.

Insufficient data are available on the long-term effects of ethyl chloroformate to aquatic life, plants, birds, or land animals.

Water Solubility

Ethyl chloroformate decomposes in water to form toxic gases and fumes (including chloride fumes and phosgene gas).

Persistence in the Environment

Ethyl chloroformate is non-persistent in water. It will react rapidly and, therefore, will not persist in the aquatic environment.

Bioaccumulation in Aquatic Organisms

Some substances increase in concentration, or *bioaccumulate,* in living organisms as they breathe contaminated air, drink contaminated water, or eat contaminated food. These chemicals can become concentrated in the tissues and internal organs of animals as well as humans.

Ethyl chloroformate will react rapidly in water and, therefore, will not accumulate in the edible tissues of aquatic organisms that are consumed by humans.

Recommended Risk-Reduction Measures

Proper training of all transporters will reduce the likelihood of a mishap or accident resulting in a leak or spill to the environment. The correct use of DOT labeling (per the MSDS) on all containers, trucks, and rail cars will enable emergency responders to react properly and quickly to any disaster thereby reducing the potential risk to the environment and to personnel.

Storage of ethyl chloroformate should be segregated from incompatible chemicals to minimize the risk of cross-contamination or contact. Buildings designated for storage should be equipped with appropriate fire protection systems (alarms, sprinklers, emergency lighting, portable extinguishers).

If a spill or leak to the environment has occurred, fire department, emergency response and/or hazardous materials spill personnel should be notified immediately. Cleanup should be attempted only by those trained in proper spill containment procedures using non-sparking tools. Contaminated soils should be removed for incineration and replaced with clean soil. If ethyl chloroformate should contact the water table, aquifer, or navigable waterway, the local and/or state emergency response authorities must be notified. A comprehensive emergency response or disaster preparedness/recovery plan should be in place prior to any operations involving the use, transportation, storage, or disposal of ethyl chloroformate. If ethyl chloroformate is spilled or leaked, the following specific steps are recommended:

- ☑ Restrict persons not wearing protective clothing from area of spill or leak until cleanup is complete and area can be opened for normal work.
- ☑ Remove all ignition sources.
- ☑ Ventilate area of leak.
- ☑ Absorb liquids in vermiculite, dry sand, earth, or similar material and deposit in sealed containers.
- ☑ It may be necessary to dispose of ethyl chloroformate as a hazardous waste. The responsible state agency or the regional office of the federal Envi-

ronmental Protection Agency (EPA) should be contacted for specific recommendations.

RISK ASSESSMENT: BUSINESS

General Assessment

Accidents or mishaps involving ethyl chloroformate can present a significant threat to business operations. The loss or damage of equipment or facilities can significantly affect fiscal viability. Lawsuits that may result from personnel injury/death, public exposures, and/or environmental contamination will also require a serious expenditure of resources. Media attention surrounding an injury, death, or environmental damage can also result in a loss of profits and loss of current as well as future business.

Recommended Risk-Reduction Measures

Company attorneys, safety and health professionals, and environmental specialists should be involved in the development of any procedures dealing with safety and emergency response. A company official should be pre-designated as a public relations officer with specific training in dealing with the press. Corporate plans and policies should be developed, approved, and implemented long before any need for such arises.

REFERENCES

Ahlbom, A. 1993. *Biostatistics for Epidemiologists.* New York: CRC Press/Lewis Publishers.

American Conference of Governmental Industrial Hygienists. 1988. *Documentation of the Threshold Limit Values and Biological Exposure Indices,* 5th Edition (with updates). Cincinnati: ACGIH

Anthony, C. P., and N. J. Kolthoff. 1971. *Textbook of Anatomy and Physiology.* St. Louis: C. V. Mosby Company.

Asante-Duah, D. K. 1993. *Hazardous Waste Risk Assessment.* New York: CRC Press/Lewis Publishers.

Bloom, A. D. and F. J. DeSerres. 1995. *Ecotoxicity and Human Health.* New York: CRC Press/Lewis Publishers.

Brusick, D. J. 1994. *Methods for Genetic Risk Assessment.* New York: CRC Press/Lewis Publishers.

Bueche, J. R. 1972. *Principles of Physics,* 2nd Edition. New York: McGraw-Hill.

Burrell, R., D. K. Flaherty, and L. J. Sauers. 1992. *Toxicology of the Immune System.* New York: Van Nostrand Reinhold.

Calabrese, E. J. 1994. *Biological Effects of Low Level Exposures.* New York: CRC Press/Lewis Publishers.

Cockerman, L. G., B. S. Shane. 1994. *Basic Environmental Toxicology.* New York: CRC Press/Lewis Publishers.

Gardner, R. W. 1994. *Chemical Intolerance.* New York: CRC Press/Lewis Publishers.

Hallenbeck, W. H. 1993. *Quantitative Risk Assessment for Environmental and Occupational Health,* 2nd Edition. New York: CRC Press/Lewis Publishers.

Jones, F. E. 1994. *Toxic Organic Vapors in the Workplace.* New York: CRC Press/Lewis Publishers.\

Keith, L. H. and M. M. Walker. 1995. *Handbook of Air Toxics.* New York: CRC Press/Lewis Publishers.

Landis, W. G., M. H. Yu. 1995. *Introduction to Environmental Toxicology.* New York: CRC Press/Lewis Publishers.

Langerman, N. 1994 *Precautionary Labels for Chemical Containers.* New York: CRC Press/Lewis Publishers.

Lauwerys, R. R., P. Hoet. 1993. *Industrial Chemical Exposure.* New York: CRC Press/Lewis Publishers.

Lewis, R. J., Sr. 1992. *Sax's Dangerous Properties of Industrial Materials,* Volumes 1, 2, and 3, 8th Edition. New York: Van Nostrand Reinhold.

Lide, D. R. 1995. *Handbook of Chemistry and Physics,* 76th Edition. New York: CRC Press/Lewis Publishers.

Manahan, S. E. 1994. *Environmental Chemistry,* 6th Edition. New York: CRC Press/Lewis Publishers.

Mellström, G. A., J. E. Wahlberg, and H. I. Maibach. 1994. *Protective Gloves for Occupational Use.* New York: CRC Press/Lewis Publishers.

Meyer, E. 1990. *Chemistry of Hazardous Materials.* Englewood Cliffs, NJ: Prentice Hall.

Patnaik, P. 1992. *A Comprehensive Guide to the Hazardous Properties of Chemical Substances.* New York: Van Nostrand Reinhold.

Proctor, N. H., J. P. Hughes. 1991. *Chemical Hazards of the Workplace.* New York: Van Nostrand Reinhold.

Quigley, D. R. 1994. *Handbook of Emergency Chemical Management.* New York: CRC Press/Lewis Publishers.

Rea, W. J. 1995. *Chemical Sensitivity,* Vol. I-III. New York: CRC Press/Lewis Publishers.

Richardson, M. 1995. *Dictionary of Substance and Their Effects,* Vol. I-VI. New York: CRC Press/Lewis Publishers.

Sacarello, H. L. A. 1994. *The Comprehensive Handbook of Hazardous Materials.* New York: CRC Press/Lewis Publishers.

Sheftel, V. O. 1995. *Handbook of Toxic Properties of Monomers and Additives.* New York: CRC Press/Lewis Publishers.

Torkelson, T. R. V. K. Rowe. 1981. Halogenated aliphatic hydrocarbons. *Patty's Industrial Hygiene and Toxicology,* Vol. 2B, *Toxicology,* pp 3568-3572. New York: Wiley-Interscience.

U.S. Department of Health and Human Services, National Institute for Occupational Safety and Health. 1994. *NIOSH Pocket Guide to Chemical Hazards.* Washington, D.C.: U.S. Government Printing Office.

MATERIAL SAFETY DATA SHEET

CHEMICAL NAME	
ETHYLENE	**CRC PRESS/LEWIS PUBLISHERS DISCLAIMER:** THE INFORMATION AND RECOMMENDATIONS PRESENTED HEREIN ARE BASED ON SOURCES BELIEVED TO BE RELIABLE. CRC MAKES NO REPRESENTATION ON ITS COMPLETENESS OR ACCURACY. IT IS THE RESPONSIBILITY OF THE USER TO DETERMINE THE CHEMICAL'S SUITABILITY FOR ITS INTENDED USE, THE CHEMICAL'S SAFE USE, AND THE CHEMICAL'S PROPER DISPOSAL. NO REPRESENTATIONS AND/OR WARRANTIES, EITHER EXPRESSED OR IMPLIED, OF THE MERCHANTABILITY OR FITNESS FOR A PARTICULAR PURPOSE, OR OF ANY OTHER NATURE, ARE MADE WITH RESPECT TO THE INFORMATION PROVIDED IN THIS MSDS OR TO THE CHEMICAL TO WHICH INFORMATION MAY REFER. CRC NEITHER ASSUMES NOR AUTHORIZES ANY OTHER PERSON TO ASSUME FOR IT, ANY OTHER ADDITIONAL RESPONSIBILITY OR LIABILITY FOR THE USE OF, OR RELIANCE UPON, THIS INFORMATION.

HAZARD WARNING INFORMATION

HEALTH	FIRE	REACTIVE	OTHER	DEGREE OF HAZARD	COLOR CODING	OTHER CODES
1	4	2		0 = Minimum Hazard 1 = Slight Hazard 2 = Moderate Hazard 3 = Serious Hazard 4 = Severe Hazard	HEALTH = BLUE FIRE = RED REACTIVITY = YELLOW OTHER = WHITE	OX = Oxidizer ACID = Acid ALK = Alkali COR = Corrosive W = Use No Water

SECTION I - GENERAL INFORMATION

Characterization **Gas**	RCRA Number **None**	EPA Class **Not Applicable**
DOT Proper Shipping Name **Ethylene, compressed or Ethylene, refrigerated liquid**	Chemical Abstract Service (CAS) Number **74-85-1**	
DOT Hazard Class and Label Requirements **Flammable Gas (both forms)**	DOT Emergency Guide Code **22**	
DOT Identification Number **UN 1038 (compressed gas); UN 1962 (liquid)**	Chemical Formula **C_2H_4**	

Synonyms
Bicarburetted hydrogen; ethene; liquid ethylene; olefiant gas.

SECTION II - HAZARDOUS INGREDIENTS AND IDENTITY INFORMATION

Hazardous Components (specific identity)	OSHA Exposure Criteria	NIOSH Exposure Criteria	Immediately Dangerous to Life and Health (IDLH)	ACGIH Exposure Criteria
Ethylene (derivation from: By thermal cracking of hydrocarbon gases at temperatures between 800°-900°C; by dehydration of ethanol; or from synthesis gas with Ru as catalysts. Also found naturally in ripening fruit).	**PEL:** Not Established **STEL:** Not Established	**REL:** Not Established **STEL:** Not Established	**Not Determined**	**TLV:** Not Established **STEL:** Not Established

SECTION III - PHYSICAL AND CHEMICAL CHARACTERISTICS

Boiling Point **-152°F (-102°C)**	Specific Gravity (H_2O = 1) **Not Found**
Vapor Pressure (atmospheres) **Gas**	Molecular Weight **28.06**
Vapor Density (Air = 1) **0.98**	Freezing Point **-272°F (-169°C)**

Solubility
Soluble in water, alcohol, and ether.

Appearance and Odor
A colorless gas with a sweet odor and taste.

SECTION IV - FIRE AND EXPLOSION HAZARD DATA

Flash Point (method used) **-213°F (-136°C) closed cup**	Explosive Limits in Air % by Volume **LEL: 3% UEL: 34%**
NFPA Classification **Flammable Gas**	Autoignition Temperature **1009°F (543°C)**

Extinguishing Media
Stop flow of gas first (if possible). Use foam, carbon dioxide, dry chemical, or water spray on fire.

Special Fire Fighting Procedures
Ethylene is an extremely flammable, very dangerous fire and explosion hazard. Poisonous gases are evolved from fire. Wear a self-contained breathing apparatus (SCBA) with full facepiece. Ensure ALL sources of escaping gas are controlled. Monitor area for long after fire is extinguished. Move containers from fire area if it can be done without risk. Use water spray to knock down vapors.

Unusual Fire and Explosion Hazards
Gas is explosive. Containers may explode in flame. Even if fire is extinguished, escaping gas can form explosive mixtures in air quickly and ignite without warning.

SECTION V - REACTIVITY DATA

Stability		Conditions to Avoid
		Ethylene is normally stable in its closed containers, during routine operations (those that do not expose it to ignition sources). Avoid contact with heat, sparks, flame, lighted tobacco products, and steam lines.
Stable	Unstable	Incompatibility *(materials to avoid)*
X		Ethylene is reactive in contact with chlorine or chlorine dioxide, other oxidizers, heat, sparks, and open flame.
Hazardous Polymerization		Conditions to Avoid
		Under normal conditions of temperature and pressure, hazardous polymerization of ethylene is not expected to occur.
May Occur	Will Not Occur	Hazardous Decomposition or By-products
	X	If involved in fire, ethylene can produce toxic carbon monoxide.

SECTION VI - HEALTH HAZARD DATA

Primary Route(s) of Entry:	Inhalation?	Absorption (skin)?	Ingestion?
	X		

Health Hazards

INHALATION: An asphyxiant gas. Initial symptoms are those of oxygen starvation. Feeling of "air hunger," diminished mental alertness, headache, dullness, shortness of breath, and an impairment in muscle coordination. Continued exposures lead to faulty judgment, depression of senses, rapid fatigue, unconsciousness, convulsions, coma, and death.

SKIN & EYES: Contact with the compressed gas or refrigerated liquid can cause frostbite and burns.

INGESTION: Not likely since ethylene is a gas at normal working (room) temperatures.

Carcinogenicity	NTP Listed?	IARC Cancer Review Group?	OSHA Regulated?	Target Organs?
Insufficient data	No	No	No	None Reported

Medical Conditions Generally Aggravated by Exposure

None Reported.

Emergency and First-aid Procedures

Eye contact: Immediately flush large amounts of water for 15 minutes (minimum), occasionally lifting eyelids, seek medical attention. Skin contact: Remove all contaminated clothing. Immediately wash area with large amounts of soap and water. For inhalation: Remove the person from exposure. Provide respiratory assistance and CPR. Transfer promptly to medical facility. If swallowed: Not likely to occur under normal or even emergency occupational conditions.

SECTION VII - PRECAUTIONS FOR SAFE HANDLING AND USE

Steps to be Taken in Case Material is Released or Spilled

Remove all ignition sources. Ventilate area. Stop flow of gas. If the source of leak is a cylinder and cannot be stopped, remove to safe place in open air; repair leak or let cylinder empty. If gas has not yet ignited, use a water spray to knock down vapors and direct them from ignition source.

Preferred Waste Disposal Method

None specified in the references (incineration of a gas may be the only alternative).

Precautions to be Taken in Handling and Storage

Do not store with incompatible materials or in the presence of heat since violent reactions or explosions can occur. Store in tightly closed pressurized containers in a cool, dark, well-ventilated area away from sunlight. Keep away from ignition sources such as fire, sparks, and flame.

Other Precautions and Warnings

Bulk storage of ethylene is not recommended. Containers may explode in fire or under conditions of extreme heat (do not store outdoors or in direct sunlight).

SECTION VIII - CONTROL MEASURES AND PERSONAL PROTECTIVE EQUIPMENT

Respiratory Protection *(specify type)*

No exposure limits have been established. Ethylene is a simple asphyxiant. Do NOT use cartridge-type respirators (ineffective for this type of gas). Use a powered supplied-air respirator or a self-contained breathing apparatus with full facepiece operated in pressure demand or other positive pressure mode.

Ventilation

Local exhaust or general mechanical systems recommended.

Protective Gloves	Eye Protection	Other Protective Clothing
Insulated Rubber	Goggles and/or Face Mask	Leather Apron

Work/Hygiene Practices

Always wash hands thoroughly after using chemical; never bring food, drink, or smoking materials into vicinity of chemicals.

ETHYLENE

C_2H_4 CAS: 74-85-1

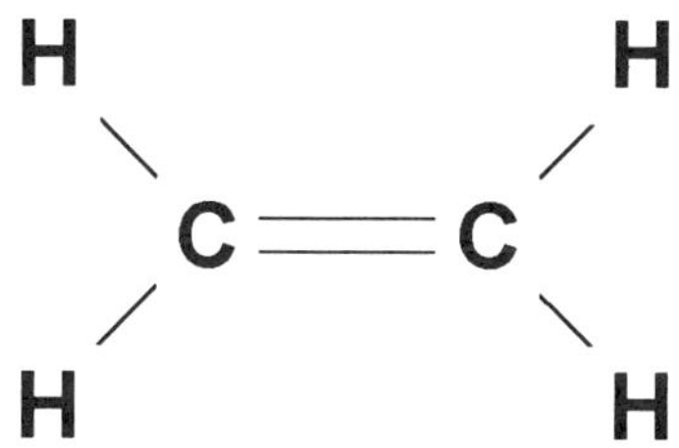

IDENTIFICATION AND TYPICAL USES

Ethylene is a colorless gas with a sweet odor and taste. While it can be produced synthetically, it occurs naturally in ripening fruit as a result of the dehydration of alcohol. It is found in illuminating gases and in petroleum gases. It is used in oxyethylene flame for welding and cutting metals, in the manufacture of polyethylene, polystyrene, and other plastics. It is also used in the manufacture of ethylene oxide and in orchard sprays to accelerate fruit ripening. It has even been used as a inhalation anesthetic.

RISK ASSESSMENT: HEALTH

General Assessment

Ethylene is a simple asphyxiant gas. As such, its primary route of entry into the body is through *inhalation.* It can displace oxygen levels to below that which is necessary to sustain life. At high concentrations, it is a narcotic and can cause anesthesia and unconsciousness. Skin contact with the liquid can be destructive to tissues due to the extremely cold temperatures but absorption is not a likely exposure route. Since ethylene usually exists as a gas at room temperatures, ingestion is not likely to occur either.

Inhalation of ethylene gas can cause headache, dullness, shortness of breath, and the symptoms of asphyxiation. These include rapid respiration and a feeling of "air hunger" or air starvation followed by a decrease in mental alertness, and impaired muscular coordination. If exposure continues, there can be narcosis, anesthesia, faulty judgment, depression of the senses, emotional instability, nausea, vomiting, prostration, unconsciousness, convulsions, coma, and death.

Skin or eye contact with the vapor poses no serious threat. The vapors have not been shown to cause any physiological response. However, if the liquid should contact either the skin or the eyes, there can be serious irritation, burns, tissue damage, and frostbite.

Acute Health Effects

The following acute (short-term) health effects may occur immediately or shortly after exposure to ethylene:

Skin: Liquid threat only. Redness, pain, swelling, blisters, and severe frostbite.

Eye: Liquid threat only. Causes frostbite, redness, irritation, severe pain, and blurred vision.

Lung: Vapor is non-irritating and therefore poses no threat of damage to the lungs. However, in high concentrations, ethylene can lead to symptoms of asphyxiation and oxygen depravation up to and including loss of consciousness and death.

Chronic Health Effects

The following chronic (long-term) health effects may occur at some time after exposure to ethylene and can last for months or even years:

Cancer Hazards: There is insufficient evidence in the references to evaluate ethylene for its ability to cause cancer in animals. It is a simple asphyxiant.

Reproductive Hazards: There are no reports to support any claims of reproductive hazards.

Other Chronic Effects: Ethylene has not been tested for other chronic (long-term) health effects.

Recommended Risk-Reduction Measures

Personnel should avoid direct contact with ethylene. If another gas cannot be substituted for a ethylene, then *engineering controls* are the most effective method of reducing exposures. The best protection is to separate operations and provide local exhaust ventilation at the site of chemical release. While not always operationally feasible, isolating operations can also reduce exposure. Using equipment with explosion-proof designs and ensuring all metal containers and tools are electrically grounded and bonded can further reduce the potential for injury or damage. Using respiratory protection is less effective than the controls mentioned above, but is still advisable whenever working with ethylene. Air-purifying respirators (e.g., the cartridge-type or gas masks) cannot be used with ethylene. These type respirators only purify the air and can do nothing to protect in oxygen deficient atmospheres. A powered supplied-air respirator operated in pressure demand or other positive pressure mode, or a self-contained breathing apparatus (SCBA) with full facepiece operated in pressure demand mode are the recommended respiratory protection methods of choice. If a full facepiece is not available, then chemical goggles should be worn to protect the eyes. Whenever a chemical splash hazard exists, a face shield and a rubber or leather apron should be worn. To prevent hand and skin exposures with the liquid, thermal gloves should be worn.

Administrative controls should also be in place to minimize the potential for human exposures. These may include written procedures or policies which specify the methods and techniques that will be practiced whenever personnel are to work with ethylene. Also, hot-work operations that use ethylene (welding, cutting, soldering) can present other health hazards, such as the generation of toxic carbon and metal oxide fumes. Administrative controls as well as engineering controls must consider the effects of working with ethylene.

All personnel should receive *training* on the use, hazards, protective measures, emergency actions, and other precautions per 29 CFR 1910.1200 (Hazard Communication), prior to the first assignment in an area where ethylene is used or stored.

Other methods to reduce exposure to chemicals include:

- ☑ Where possible, consider installing automatic sensors that warn personnel of dropping oxygen levels and use local exhaust ventilation at the site of chemical release. If local exhaust ventilation or enclosure is not used, respiratory protection should be mandatory.
- ☑ Always ensure that proper protective clothing is worn when using chemical substances.
- ☑ Wash thoroughly immediately after exposure to ethylene and at the end of the work shift or before eating, drinking, or smoking.
- ☑ Safety shower and eyewash stations should be readily available in work areas where ethylene is used or stored.
- ☑ Hazard warning information should be posted in the work area. In addition, as part of an on-going education and training program, all information on the health and safety hazards of ethylene should be communicated to all exposed and potentially exposed workers.

RISK ASSESSMENT: ENVIRONMENT

General Assessment

The environment is at risk of exposure during transportation, storage, disposal, or destruction of ethylene. In almost every scenario, the threat of environmental exposure is contingent upon the proper handling of the chemical substance. Accidental releases, large or small, can result in fire, explosion, and possible contamination of the surrounding environmental mediums (water, soil, and air). Ethylene is an extremely flammable and highly explosive gas. Its extremely low flash point and boiling point present a serious fire hazard. It can react with ozone to form ethylene ozonide which is unstable and explodes on mechanical shock. Also, since it can react violently in the presence of oxidizers, such as chlorine, ethylene requires special consideration during any emergency involving a leak or release of ethylene gas or liquid.

Ethylene can enter the environment through industrial discharges, unchecked venting to air, or through spills of the liquid. Since liquids will most likely vaporize rapidly, water and soil contamination are possible but improbable.

☠ *Acute Ecological Effects*

Acute (short-term) toxic effects may include the death of animals, birds, or fish, and death or low growth rate in plants. Acute effects are seen 2 to 4 days after animals or plants are exposed to ethylene.

Ethylene has slight acute toxicity to aquatic life. It has caused various types of injuries to a number of agricultural crops. Insufficient data are available on the acute (short-term) effects of ethylene on birds or terrestrial animals.

Chronic Ecological Effects

Chronic (long-term) toxic effects may include shortened life span, reproductive problems, lower fertility, and changes in appearance or behavior in exposed animals. These effects can be seen long after first exposure(s) to toxic chemicals. Ethylene has slight chronic toxicity to aquatic life. Insufficient data are available to evaluate the long-term effects of ethylene to plants, birds, or land animals.

Water Solubility

Ethylene is highly soluble in water. Concentrations of 1000 milligrams and more will mix with a liter of water.

Persistence in the Environment

Ethylene is non-persistent in water, with a half-life of less than 2 days. The half-life of a pollutant is the amount of time it takes for one half of the chemical to be degraded. About 99.9% of ethylene will eventually end up in air; the rest will end up in the water.

Bioaccumulation in Aquatic Organisms

Some substances increase in concentration, or *bioaccumulate,* in living organisms as they breathe contaminated air, drink contaminated water, or eat contaminated food. These chemicals can become concentrated in the tissues and internal organs of animals as well as humans.

Ethylene is not expected to accumulate in the tissues of fish.

Recommended Risk-Reduction Measures

Proper training of all transporters will reduce the likelihood of a mishap or accident resulting in a leak or release to the environment. The correct use of DOT labeling (per the MSDS) on all containers, trucks, and rail cars will enable emergency responders to react properly and quickly to any disaster thereby reducing the potential risk to the environment and to personnel.

Storage of ethylene should be segregated from incompatible chemicals to minimize the risk of cross-contamination or contact. Buildings designated for storage should be equipped with appropriate fire protection systems (alarms, sprinklers, emergency lighting, portable extinguishers).

If a spill or release to the environment has occurred, fire department, emergency response, and/or hazardous materials spill personnel should be notified immediately. Cleanup should be attempted only by those trained in proper containment procedures. Contaminated soils should be removed for incineration and replaced with clean soil. When spills or leaks occur, it may be necessary to contact the local and/or state emergency response authorities. A comprehensive emergency response or disaster preparedness/recovery plan should be in place prior to any operations involving the use, transportation, storage, or disposal of ethylene. If ethylene is spilled or leaked, the following specific steps are recommended:

- ☑ Restrict persons not wearing protective clothing from area of spill or leak until cleanup is complete.
- ☑ Remove all ignition sources.
- ☑ Ventilate area of spill or leak.
- ☑ If applicable, stop flow of leaking liquid or gas. If leak source is a cylinder and the leak cannot be stopped in place, remove leaking cylinder to a safe place in the open air, and repair or allow cylinder to empty.
- ☑ Keep ethylene out of a confined space, such as a sewer, because of the possibility of explosion (unless the sewer is designed to prevent the build-up of explosive concentrations).
- ☑ It may be necessary to dispose of ethylene as a hazardous waste. The responsible state agency or the regional office of the federal Environmental Protection Agency (EPA) should be contacted for specific recommendations.

RISK ASSESSMENT: BUSINESS

General Assessment

Accidents or mishaps involving ethylene can present a significant threat to business operations. The loss or damage of equipment or facilities can significantly affect fiscal viability. Lawsuits that may result from personnel injury, public exposures, and environmental contamination will also require a serious expenditure of resources. Media attention surrounding an injury, illness, or environmental damage can also result in a loss of profits and loss of current as well as future business.

Recommended Risk-Reduction Measures

Company attorneys, safety and health professionals, and environmental specialists should be involved in the

development of any such procedures. A company official should be pre-designated as a public relations officer with specific training in dealing with the press. Corporate plans and policies should be developed, approved, and implemented long before any need for such arises.

REFERENCES

Ahlbom, A. 1993. *Biostatistics for Epidemiologists.* New York: CRC Press/Lewis Publishers.

American Conference of Governmental Industrial Hygienists. 1988. *Documentation of the Threshold Limit Values and Biological Exposure Indices,* 5th Edition (with updates). Cincinnati: ACGIH

Anthony, C. P., and N. J. Kolthoff. 1971. *Textbook of Anatomy and Physiology.* St. Louis: C. V. Mosby Company.

Asante-Duah, D. K. 1993. *Hazardous Waste Risk Assessment.* New York: CRC Press/Lewis Publishers.

Bloom, A. D. and F. J. DeSerres. 1995. *Ecotoxicity and Human Health.* New York: CRC Press/Lewis Publishers.

Brusick, D. J. 1994. *Methods for Genetic Risk Assessment.* New York: CRC Press/Lewis Publishers.

Bueche, J. R. 1972. *Principles of Physics,* 2nd Edition. New York: McGraw-Hill.

Burrell, R., D. K. Flaherty, and L. J. Sauers. 1992. *Toxicology of the Immune System.* New York: Van Nostrand Reinhold.

Calabrese, E. J. 1994. *Biological Effects of Low Level Exposures.* New York: CRC Press/Lewis Publishers.

Cockerman, L. G., B. S. Shane. 1994. *Basic Environmental Toxicology.* New York: CRC Press/Lewis Publishers.

Cothern, C. R. 1995. *Handbook for Environmental Risk Decisionmaking.* New York: CRC Press/Lewis Publishers.

Danse, I. R. 1991. *Common Sense Toxics in the Workplace.* New York: Van Nostrand Reinhold.

Environmental Protection Agency. 1983. *Chemical Hazard Information Profiles.* Washington, D.C.: U.S. EPA.

Fodden, C. R., and J. L. Weddell. 1992. *Hazardous Materials Emergency Action Data.* New York: CRC Press/Lewis Publishers.

Forseberg, K., and L. H. Keith. 1993. *Chemical; Protective Clothing Permeation and Degradation Database.* Chelsea, MI: Lewis Publishers.

Gardner, R. W. 1994. *Chemical Intolerance.* New York: CRC Press/Lewis Publishers.

Hallenbeck, W. H. 1993. *Quantitative Risk Assessment for Environmental and Occupational Health,* 2nd Edition. New York: CRC Press/Lewis Publishers.

Hodgson, E., R. B. Mailman, and J. E. Chambers. 1988. *Dictionary of Toxicology.* London: The MacMillian Company.

Howard, P. H. 1995. *Handbook of Environmental Fate and Exposures Data for Organic Chemicals,* Vol. I-IV. New York: CRC Press/Lewis Publishers.

James, A., and M. Lord. 1992. *Index of Chemical and Physical Data.* New York: Van Nostrand Reinhold.

Jones, F. E. 1994. *Toxic Organic Vapors in the Workplace.* New York: CRC Press/Lewis Publishers.

Keith, L. H. and M. M. Walker. 1995. *Handbook of Air Toxics.* New York: CRC Press/Lewis Publishers.

Keith, L. H. (Editor). 1992. *Integrated Risk Information System (IRIS) Chemical Information Database.* Prepared by the U.S. Environmental Protection Agency. Boca Raton, FL: CRC Press/Lewis Publishers.

Keith, L. H. and D. B. Walters. (Editors) 1992. *National Toxicology Program's Chemical Database, Volumes I-VII.* Chelsea, MI: Lewis Publishers.

Landis, W. G., M. H. Yu. 1995. *Introduction to Environmental Toxicology.* New York: CRC Press/Lewis Publishers.

Langerman, N. 1994 *Precautionary Labels for Chemical Containers.* New York: CRC Press/Lewis Publishers.

Lauwerys, R. R., P. Hoet. 1993. *Industrial Chemical Exposure.* New York: CRC Press/Lewis Publishers.

Lewis, R. J., Sr. 1993. *Hawley's Condensed Chemical Dictionary,* 12th Edition. New York: Van Nostrand Reinhold.

Lewis, R. J., Sr. 1992. *Sax's Dangerous Properties of Industrial Materials,* Volumes 1, 2, and 3, 8th Edition. New York: Van Nostrand Reinhold.

Manahan, S. E. 1992. *Toxicological Chemistry,* 2nd Edition. New York: CRC Press/Lewis Publishers.

Meyer, E. 1990. *Chemistry of Hazardous Materials.* Englewood Cliffs, NJ: Prentice Hall.

Patnaik, P. 1992. *A Comprehensive Guide to the Hazardous Properties of Chemical Substances.* New York: Van Nostrand Reinhold.

Philp, R. B. 1995. *Environmental Hazards and Human Health.* New York: CRC Press/Lewis Publishers.

Proctor, N. H., and J. P. Hughes. 1991. *Chemical Hazards of the Workplace.* New York: Van Nostrand Reinhold.

Quigley, D. R. 1994. *Handbook of Emergency Chemical Management.* New York: CRC Press/Lewis Publishers.

Rea, W. J. 1995. *Chemical Sensitivity,* Vol. I-III. New York: CRC Press/Lewis Publishers.

Richardson, M. 1995. *Dictionary of Substance and Their Effects,* Vol. I-VI. New York: CRC Press/Lewis Publishers.

Sacarello, H. L. A. 1994. *The Comprehensive Handbook of Hazardous Materials.* New York: CRC Press/Lewis Publishers.

Sherman, J. D. 1988. *Chemical Exposure and Disease.* New York: Van Nostrand Reinhold.

Torkelson, T. R. V. K. Rowe. 1981. *Patty's Industrial Hygiene and Toxicology,* Vol. 2B. New York: Wiley-Interscience.

U.S. Department of Health and Human Services, National Institute for Occupational Safety and Health. 1994. *NIOSH Pocket Guide to Chemical Hazards.* Washington, D.C.: U.S. Government Printing Office.

MATERIAL SAFETY DATA SHEET

CHEMICAL NAME

ETHYLENE CHLOROHYDRIN

HAZARD WARNING INFORMATION

HEALTH	FIRE	REACTIVE	OTHER	DEGREE OF HAZARD	COLOR CODING	OTHER CODES
4	2	0		0 = Minimum Hazard 1 = Slight Hazard 2 = Moderate Hazard 3 = Serious Hazard 4 = Severe Hazard	HEALTH = BLUE FIRE = RED REACTIVITY = YELLOW OTHER = WHITE	OX = Oxidizer ACID = Acid ALK = Alkali COR = Corrosive W = Use No Water

SECTION I - GENERAL INFORMATION

Characterization	**Chlorohydrin**	RCRA Number / EPA Class	**None** / **Not Applicable**
DOT Proper Shipping Name	**Ethylene Chlorohydrin**	Chemical Abstract Service (CAS) Number	**107-07-3**
DOT Hazard Class and Label Requirements	**Poison B**	DOT Emergency Guide Code	**55**
DOT Identification Number	**UN 1135**	Molecular Formula	**$ClCH_2CH_2OH$**

Synonyms
2-chloroethanol; 2-chlorethyl alcohol; glycol monochlorohydrin; ethylene glycol chlorohydrin; β-chloroethy alcohol.

SECTION II - HAZARDOUS INGREDIENTS AND IDENTITY INFORMATION

Hazardous Components (specific identity)	OSHA Exposure Criteria	NIOSH Exposure Criteria	Immediately Dangerous to Life and Health (IDLH)	ACGIH Exposure Criteria
Ethylene chlorohydrin: (derivation: By the action of hypochlorous acid on ethylene. It is also formed during the sterilization of grains and spices, drugs, and surgical and ophthalmic supplies with ethylene oxide). 1 ppm = 3.35 mg/m^3	**PEL (skin):** 5 ppm 16 mg/m^3 **CEILING:** Not Established	**REL (skin): CEILING** (15 min) 1 ppm 3.35 mg/m^3	7 ppm	**TLV (skin): CEILING** (15 min) 0.1 ppm 0.5 mg/m^3

SECTION III - PHYSICAL AND CHEMICAL CHARACTERISTICS

Boiling Point	**262°F (128°C)**	Specific Gravity (H_2O = 1)	**1.197**
Vapor Pressure (mm Hg)	**4.9 at 68°F (20°C)**	Molecular Weight (atomic weight)	**80.5**
Vapor Density (Air = 1)	**2.78**	Melting Point	**-90°F (-68°C)**

Solubility
Soluble in water, alcohol, and ether.

Appearance and Odor
Colorless, clear liquid with a faint smell of ether.

SECTION IV - FIRE AND EXPLOSION HAZARD DATA

Flash Point (method used)	**140°F (60°C) opened cup**	Explosive Limits in Air % by Volume	**LEL: 4.9% UEL: 15.9**
NFPA Classification	**Class IIIA Combustible Liquid**	Autoignition Temperature	**797°F (425°C)**

Extinguishing Media
Stop liquid flow first. Use dry chemical, regular foam, water spray, or carbon dioxide.

Special Fire Fighting Procedures
Acrid and irritating smoke and fumes are produced in fire. Structural fire-fighting protective clothing will not provide adequate protection. Wear full protective clothing and self-contained breathing apparatus (SCBA). Move container from fire area if it can be done without risk.

Unusual Fire and Explosion Hazards
Containers may explode in fire. Presents a vapor explosion hazard indoors, outdoors, and in sewers. Vapors form explosive mixtures in air. Fire may produce toxic hydrogen chloride and phosgene fumes.

SECTION V - REACTIVITY DATA

Stability		Conditions to Avoid: Normally stable at room temperature. Avoid contact with incompatible materials since violent reactions can occur. Also avoid contact with heat and ignition sources.
Stable	Unstable	Incompatibility *(materials to avoid)*: Incompatible with strong oxidizers, such as chlorine, bromine, and chlorine dioxide. Also incompatible with acids, bases, ethylene diamine, chlorosulfonic acid, and sodium hydroxide. Reacts with water or steam.
X		
Hazardous Polymerization		Conditions to Avoid: In the presence of curing agents, ethylene chlorohydrin may polymerize at room temperature to form hard epoxy resins.
May Occur	Will Not Occur	Hazardous Decomposition or By-products: When heated to decomposition, ethylene chlorohydrin emits toxic and irritating fumes, including hydrogen chloride and phosgene fumes.
	X	

SECTION VI - HEALTH HAZARD DATA

Primary Route(s) of Entry:	Inhalation?	Absorption (skin/eye)?	Ingestion?
	X	X	X

Health Hazards

INHALATION: Extremely irritating to the mucosa of the upper respiratory tract. Vapors can cause severe irritation of the eyes, nose, throat, and skin at very low concentrations. May cause CNS depression, ataxia, loss of coordination, coma, and liver and kidney damage. Pulmonary edema (fluid in lungs) is also possible. Death can occur by respiratory paralysis.

ABSORPTION: Can pass through unbroken skin, without immediate irritation, causing toxic systemic effects with action on the CNS.

INGESTION: Causes nausea, headache, delirium, coma, and collapse.

Carcinogenicity	NTP Listed?	IARC Cancer Review Group?	OSHA Regulated?	Target Organs?
No Evidence Human No Evidence Animal	NTP-TR-275,85	No	29 CFR 1910.1000 Table Z-1	CNS, resp. sys., kidneys, liver, digestive sys., CVS.

Medical Conditions Generally Aggravated by Exposure

Liver diseases and central nervous system disorders may be aggravated by exposure.

Emergency and First-aid Procedures

Eye contact: Do not allow victim to keep eyes tightly shut. Flush immediately with water for 15 minutes (minimum), seek medical attention. Skin contact: Quickly remove contaminated clothing. Wash with large amounts of soap and water for a minimum of 15 minutes. For inhalation: Remove the person from exposure. Provide respiratory assistance and CPR. Transfer to medical facility. Observe for 24- 48 hours for lung effects. If swallowed: Seek medical attention immediately. Give 1 to 2 glasses of milk or water. Do NOT induce vomiting or attempt to give an unconscious or convulsing person anything by mouth.

SECTION VII - PRECAUTIONS FOR SAFE HANDLING AND USE

Steps to be Taken in Case Material is Released or Spilled

Wear self-contained breathing apparatus (SCBA) for respiratory protection. Restrict those not wearing protective equipment and who are not involved in cleanup from entering area. Avoid skin contact. Ventilate area of spill. Absorb liquids in vermiculite or other material and deposit in sealed drums.

Preferred Waste Disposal Method

Mix with a combustible solvent and burn in a chemical incinerator (with afterburner and scrubber).

Precautions to be Taken in Handling and Storage

Store in tightly closed containers in a cool, well ventilated area away from heat. Protect containers from physical damage.

Other Precautions and Warnings

Store to avoid contact with strong oxidizers since violent reactions can occur.

SECTION VIII - CONTROL MEASURES AND PERSONAL PROTECTIVE EQUIPMENT

Respiratory Protection *(specify type)*

Use a NIOSH/MSHA-approved air purifying respirator or a supplied air respirator, or use a self-contained breathing apparatus (SCBA) with full facepiece operated in positive pressure mode.

Ventilation

Local exhaust (preferred) at site of chemical work or general ventilation.

Protective Gloves	Eye Protection	Other Protective Clothing
Butyl Rubber (or other non-permeable)	Chemical Goggles and Face Mask	Protective Uniform or Apron

Work/Hygiene Practices

Always wash hands thoroughly after using chemical; never bring food, drink, or smoking materials into vicinity of chemicals.

ETHYLENE CHLOROHYDRIN

$ClCH_2CH_2OH$ CAS: 107-07-3

Cl — CH$_2$ — CH$_2$ — OH

IDENTIFICATION AND TYPICAL USES

Ethylene chlorohydrin appears as a colorless, clear liquid with a faint odor of ether. It is used in the manufacture of insecticides, as a solvent for cellulose esters, in treating sweet potato plants before planting, and in making ethylene glycol and ethylene oxide. Exposure risks to this compound can occur when ethylene chlorohydrin is formed from ethylene oxide during sterilization of grain and spices, drugs, and surgical supplies.

RISK ASSESSMENT: HEALTH

General Assessment

Ethylene chlorohydrin is a severe acute poison by *inhalation, ingestion,* and skin *absorption.* There is no evidence that exposure to ethylene chlorohydrin causes cancer in test animals. However, mutation data have been reported and there is some indication that ethylene chlorohydrin causes teratogenic effects in test animals.

Inhalation of ethylene chlorohydrin vapors causes severe irritation to the mucous membranes of the eyes, nose, and throat. It causes central nervous system (CNS) and respiratory system depression and exerts biological effects on the cardiovascular system, liver, kidneys, and gastrointestinal tract, skin, and eyes. Pulmonary edema (fluid buildup in the lungs) can also occur. This is a medical emergency and can be fatal. Respiratory failure may result. Symptoms may be delayed up to 48 hours, creating a false sense of security with regard to exposure risk. Other symptoms include loss of coordination, confusion, vertigo, weakness, nausea, headache, delirium, vomiting, seizures, low blood pressure, collapse, and loss of consciousness. There is also a possibility that exposure will damage the liver and/or kidneys.

Skin contact results in no immediate indications of irritation or pain. However, it will be absorbed freely through the skin to cause toxic systemic effects similar to that of inhalation (it is more toxic than ingestion). Prolonged contact may cause mild irritation to the skin and corneal burns to the eyes.

Ingestion causes nausea, headache, delirium, coma, and collapse. It can also cause symptoms similar to that of inhalation.

Acute Health Effects

The following acute (short-term) health effects may occur immediately or shortly after exposure to ethylene chlorohydrin:

Skin: Mild irritation with rapid absorption through intact skin.

Eye: Possible cornea burns with recovery within 48 hours.

Lung: Severe irritation to respiratory tract causing shortness of breath and cough. High acute exposures can be fatal due to a buildup of fluid in the lungs (pulmonary edema).

CNS: Possible depression of the central nervous system causing a variety of toxic effects including loss of coordination and consciousness.

Other: Exposure can cause liver and/or kidney damage. There may also be adverse effects on the cardiovascular system (CVS).

Chronic Health Effects

The following chronic (long-term) health effects may occur at some time after exposure to ethylene chlorohydrin and can last for months or even years:

Cancer Hazards: There is insufficient evidence in the references to indicate that exposure to ethylene chlorohydrin causes cancer in test animals. However, limited data suggests that it is a mutagen (causes cell changes) in animals. Some scientists believe that such chemicals may cause cancer in the long-term. More research is required in this area of study.

Reproductive Hazard: According to the references, there is limited evidence that ethylene chlorohydrin has the ability to adversely affect reproduction in experimental test animals.

Other Chronic Effects: Repeated low-dose exposures may damage the liver and kidneys. Prolonged exposure may depress the CNS and the respiratory system.

Recommended Risk-Reduction Measures

Personnel should avoid direct contact with ethylene chlorohydrin. If a less toxic material or compound cannot be substituted for ethylene chlorohydrin, then *engineering controls* are the most effective method of reducing exposure risk. The best protection is to enclose operations and/or provide local exhaust ventilation at the site of ethylene chlorohydrin release. While not always operationally feasible, isolating operations can also reduce exposure. Using respiratory protection is less effective than the controls mentioned above, but is still required whenever working with or around ethylene chlorohydrin. Air purifying respirators may provide some protection at low or infrequent exposures. For the best protection, an MSHA/NIOSH-approved supplied-air respirator or a self-contained breathing apparatus (SCBA) with full facepiece and pressure demand are the recommended methods of respiratory protection. If a full facepiece is not available, then chemical goggles should be worn to protect the eyes, especially when a splash hazard exists. A face shield should also be considered. To prevent hand and skin exposures, rubber or other non-permeable gloves should be worn.

Administrative controls should also be in place to minimize the potential for human exposures. These may include written procedures or policies which specify the methods and techniques that will be practiced whenever personnel are to work with ethylene chlorohydrin.

All personnel should receive *training* on the use, hazards, protective measures, emergency actions, and other precautions per 29 CFR 1910.1200 (Hazard Communication) prior to the first assignment in an area where ethylene chlorohydrin is used or stored.

Before beginning employment and at regular intervals thereafter (e.g., annually), the following medical test is recommended:

- ☑ Lung function tests (establish baseline).

If symptoms develop or overexposure is suspected, the following may be useful:

- ☑ Liver, kidney, and lung function tests.
- ☑ Consider chest X-ray after acute exposure (may be negative if taken immediately after exposure, as indications of pulmonary edema may be delayed).
- ☑ Urinalysis to look for increase in red blood cells, albumin, and casts (a mass of plastic-like matter formed by diseased organs and discharged from the body).
- ☑ Evaluation of the central nervous system by a qualified neurologist for evidence of CNS depression or other indications of exposure effects.

It should be noted that medical tests that simply look for existing damage are not a substitute for controlling exposures. Medical histories are extremely important when assessing exposure risk. Also, since smoking can cause heart disease as well as lung cancer, emphysema, and other respiratory ailments, any exposure to ethylene chlorohydrin can result in quicker and more devastating symptoms. Smokers should therefore avoid unprotected contact with this chemical. Prudent risk management requires proper consideration of *all* possible factors that may be causing the appearance of exposure symptoms among workers.

Other methods to reduce exposure include:

- ☑ Always ensure that proper protective clothing is worn when using chemical substances.
- ☑ Wash thoroughly immediately after exposure to ethylene chlorohydrin and at the end of the work shift or before eating, drinking, or smoking.
- ☑ Hazard warning information should be posted in the work area. In addition, as part of an on-going education and training program, all information on the health and safety hazards of ethylene chlorohydrin should be communicated to all potentially exposed workers.
- ☑ Eye wash stations should be provided in the immediate work area for emergency use. If there is a possibility of skin exposure to ethylene chlorohydrin, emergency shower facilities should also be provided.
- ☑ Workers whose clothing has been contaminated by ethylene chlorohydrin should change into clean clothes before leaving work. Contaminated work clothing should be laundered only by individuals who have been informed of the hazards of exposure to ethylene chlorohydrin.

RISK ASSESSMENT: ENVIRONMENT

General Assessment

The environment is at risk of exposure during transportation, storage, disposal, or destruction of ethylene chlorohydrin. In almost every scenario, the threat of environmental exposure is contingent upon the proper handling of the chemical substance. Accidental spills, large or small, can result in contamination of the surrounding environmental mediums (water, soil, and air).

Ethylene chlorohydrin is class IIIA combustible liquid. It will support combustion as well as intensify any fire. It is incompatible with oxidizers and caution is always required in handling, storage, transportation, and disposal of ethylene chlorohydrin. When heated to decomposition it emits very acrid and irritating smoke and fumes, including toxic hydrogen chloride and phosgene. In the presence of water or steam, it will produce toxic and corrosive fumes. Emergency responders should be made aware of the presence of ethylene chlorohydrin at any emergency response situation.

Ethylene chlorohydrin may enter the environment through industrial effluents and discharges, and through spills.

Acute Ecological Effects

Acute (short-term) toxic effects may include the death of animals, birds, or fish and death or low growth rate in plants. Acute effects are seen 2 to 4 days after animals or plants are exposed to ethylene chlorohydrin.

Insufficient data are available to evaluate or predict the acute (short-term) effects of ethylene chlorohydrin to aquatic life, plants, birds, or land animals.

Chronic Ecological Effects

Chronic (long-term) toxic effects may include shortened life span, reproductive problems, lower fertility, and changes in appearance or behavior in exposed animals. These effects can be seen long after first exposure(s) to toxic chemicals.

Insufficient data are available to evaluate the chronic (long-term) effects of ethylene chlorohydrin on aquatic life, plants, birds, or land animals.

Water Solubility

Ethylene chlorohydrin highly is soluble in water. Concentrations of 1000 milligrams and more will mix with a liter of water.

Persistence in the Environment

There is no information in the references concerning the persistence of ethylene chlorohydrin in the environment (air, water, or soil).

Bioaccumulation in Aquatic Organisms

Some substances increase in concentration, or *bioaccumulate,* in living organisms as they breathe contaminated air, drink contaminated water, or eat contaminated food. These chemicals can become concentrated in the tissues and internal organs of animals as well as humans.

The references do not address the bioaccumulation capabilities of ethylene chlorohydrin in the edible tissues of aquatic organisms.

Recommended Risk-Reduction Measures

Proper training of all transporters will reduce the likelihood of a mishap or accident resulting in a leak or spill of ethylene chlorohydrin into the environment. Labels on all containers, trucks, and rail cars must meet DOT requirements and accurately reflect their contents to enable emergency responders to react properly and quickly to any disaster thereby reducing the potential risk to the environment and to personnel.

Storage of ethylene chlorohydrin should be segregated from other chemicals to minimize the risk of cross-contamination. Ethylene chlorohydrin must be stored to avoid contact with strong oxidizers, such as chlorine and bromine, since violent reactions can occur. Containers should be protected from physical damage and stored to avoid contact with heat.

If a spill or leak to the environment has occurred, fire department, emergency response and/or hazardous materials spill personnel should be notified immediately. Cleanup should be attempted only by those trained in proper spill containment procedures. Contaminated soils should be removed for incineration and replaced with clean soil.

If ethylene chlorohydrin should contact the water table, aquifer, or navigable waterway, time is of the essence. It is highly soluble in water and total remediation may or may not be possible. The local and/or state emergency response authorities must be notified. A comprehensive emergency response or disaster preparedness/recovery plan should be in place prior to any operations involving the use, transportation, storage, or disposal of ethylene chlorohydrin.

To dispose of ethylene chlorohydrin, it should be mixed with a combustible solvent and burned in a chemical incinerator equipped with an afterburner and a scrubber.

If ethylene chlorohydrin is spilled or leaked, the following specific steps are recommended:

☑ Restrict persons not wearing protective clothing from area of spill or leak until cleanup is complete. Avoid skin and eye contact.

☑ Absorb spilled liquids with vermiculite, dry sand, or a similar material and deposit in sealed drum for disposal. Use only non-sparking tools.

☑ Ventilate area and remove ignition sources.

☑ It may be necessary to dispose of ethylene chlorohydrin as a hazardous waste. The responsible state agency or the regional office of the federal Environmental Protection Agency (EPA) should be contacted for specific recommendations.

RISK ASSESSMENT: BUSINESS

General Assessment

Accidents or mishaps involving ethylene chlorohydrin can present a serious threat to business operations. The loss or damage of equipment or facilities can significantly affect fiscal viability. Lawsuits that may result from personnel illness, injury/death, public exposures, and/or environmental contamination will require a serious expenditure of resources. Media attention surrounding an injury, death, or environmental damage can also result in a loss of profits and loss of current as well as future business.

Recommended Risk-Reduction Measures

Company attorneys, safety and health professionals, and environmental specialists should be involved in the development of any procedures for responding to chemical incidents. A company official should be predesignated as a public relations officer with specific training in dealing with the press. Corporate plans and policies should be developed, approved, and implemented long before any need for such arises.

REFERENCES

American Conference of Governmental Industrial Hygienists. 1988. *Documentation of the Threshold Limit Values and Biological Exposure Indices,* 5th Edition (with updates). Cincinnati: ACGIH

Anthony, C. P., and N. J. Kolthoff. 1971. *Textbook of Anatomy and Physiology.* St. Louis: C. V. Mosby Company.

Bloom, A. D. and F. J. DeSerres. 1995. *Ecotoxicity and Human Health.* New York: CRC Press/Lewis Publishers.

Calabrese, E. J., E. M. Kenyon. *Air Toxics and Risk Assessment.* New York: CRC Press/Lewis Publishers.

Cockerman, L. G., B. S. Shane. 1994. *Basic Environmental Toxicology.* New York: CRC Press/Lewis Publishers.

Cothern, C. R. 1995. *Handbook for Environmental Risk Decisionmaking.* New York: CRC Press/Lewis Publishers.

Danse, I. R. 1991. *Common Sense Toxics in the Workplace.* New York: Van Nostrand Reinhold.

Fodden, C. R., and J. L. Weddell. 1992. *Hazardous Materials Emergency Action Data.* New York: CRC Press/Lewis Publishers.

Gardner, R. W. 1994. *Chemical Intolerance.* New York: CRC Press/Lewis Publishers.

Hodgson, E., R. B. Mailman, and J. E. Chambers. 1988. *Dictionary of Toxicology.* London: The MacMillian Company.

Howard, P. H., et. al. 1991. *Handbook of Environmental Degradation Rates.* New York: CRC Press/Lewis Publishers.

Hosty, J. W., and P. Foster. 1990. *A Practical Guide to Chemical Spill Response.* New York: Van Nostrand Reinhold.

Jones, F. E. 1994. *Toxic Organic Vapors in the Workplace.* New York: CRC Press/Lewis Publishers.

Kamrin, M. A. 1988. *Toxicology: A Primer on Principles and Applications.* Chelsea, Michigan: Lewis Publishers.

Keith, L. H. and M. M. Walker. 1995. *Handbook of Air Toxics.* New York: CRC Press/Lewis Publishers.

Landis, W. G., M. H. Yu. 1995. *Introduction to Environmental Toxicology.* New York: CRC Press/Lewis Publishers.

Langerman, N. 1994 *Precautionary Labels for Chemical Containers.* New York: CRC Press/Lewis Publishers.

Lauwerys, R. R., P. Hoet. 1993. *Industrial Chemical Exposure.* New York: CRC Press/Lewis Publishers.

Lewis, R. J., Sr. 1993. *Hawley's Condensed Chemical Dictionary,* 12th Edition. New York: Van Nostrand Reinhold.

Lewis, R. J., Sr. 1992. *Sax's Dangerous Properties of Industrial Materials,* Volumes 1, 2, and 3, 8th Edition. New York: Van Nostrand Reinhold.

Manahan, S. E. 1992. *Toxicological Chemistry,* 2nd Edition. New York: CRC Press/Lewis Publishers.

Meyer, E. 1990. *Chemistry of Hazardous Materials.* Englewood Cliffs, NJ: Prentice Hall.

Quigley, D. R. 1994. *Handbook of Emergency Chemical Management.* New York: CRC Press/Lewis Publishers.

Richardson, M. 1995. *Dictionary of Substance and Their Effects,* Vol. I-VI. New York: CRC Press/Lewis Publishers.

Torkelson, T. R. V. K. Rowe. 1981. *Patty's Industrial Hygiene and Toxicology,* Vol. 2B. New York: Wiley-Interscience.

MATERIAL SAFETY DATA SHEET

CHEMICAL NAME	CRC PRESS/LEWIS PUBLISHERS DISCLAIMER
ETHYLENEDIAMINE	

HAZARD WARNING INFORMATION

HEALTH	FIRE	REACTIVE	OTHER	DEGREE OF HAZARD	COLOR CODING	OTHER CODES
3	3	0		0 = Minimum Hazard 1 = Slight Hazard 2 = Moderate Hazard 3 = Serious Hazard 4 = Severe Hazard	HEALTH = BLUE FIRE = RED REACTIVITY = YELLOW OTHER = WHITE	OX = Oxidizer ACID = Acid ALK = Alkali COR = Corrosive W = Use No Water

SECTION I - GENERAL INFORMATION

Characterization: **Aliphatic Amine**	RCRA Number: **D001**	EPA Class: **Characteristic (I) Waste**
DOT Proper Shipping Name: **Ethylenediamine**	Chemical Abstract Service (CAS) Number: **107-15-3**	
DOT Hazard Class and Label Requirements: **Corrosive Material**	DOT Emergency Guide Code: **29**	
DOT Identification Number: **UN 1604**	Chemical Formula: **$NH_2CH_2CH_2NH_2$**	

Synonyms
1,2-Ethanediamine; 1,2-diaminoethane; 1,2-ethylenediamine; Algicode 106 L; Amerstat 274; β-amino-ethylamine; Caswell No. 437; dimethylenediamine.

SECTION II - HAZARDOUS INGREDIENTS AND IDENTITY INFORMATION

Hazardous Components (specific identity)	OSHA Exposure Criteria	NIOSH Exposure Criteria	Immediately Dangerous to Life and Health (IDLH)	ACGIH Exposure Criteria
Ethylenediamine (derivation: Catalytic reaction of ethylene dichloride and ammonia. Nickel or copper catalysts are normally used in the process. Also, may be derived by catalytic animation of monoethanolamine). **1 ppm = 2.50 mg/m^3**	**PEL (8-hour):** 10 ppm 25 mg/m^3 **STEL:** Not Established	**REL (10-hour):** 10 ppm 25 mg/m^3 **STEL:** Not Established	**1000 ppm**	**TLV:** 10 ppm 25 mg/m^3 **STEL:** Not Established

SECTION III - PHYSICAL AND CHEMICAL CHARACTERISTICS

Boiling Point: **244°F (118°C)**	Specific Gravity (H_2O = 1): **0.91**
Vapor Pressure (mm Hg): **10.7 at 69°F (20°C)**	Molecular Weight: **60.1**
Vapor Density (Air = 1): **2.07**	Melting Point: **47°F (8.3°C)**

Solubility
Soluble in water and alcohol, slightly soluble in ether. Insoluble in benzene (unless dried first).

Appearance and Odor
Colorless, viscous liquid (or solid below 47°F) with an unpleasant, ammonia-like odor.

SECTION IV - FIRE AND EXPLOSION HAZARD DATA

Flash Point (method used): **93°F (34°C) closed cup**	Explosive Limits in Air % by Volume: **LEL: 2.5% (at 212°F) UEL: 12% (at 212°F)**
NFPA Classification: **Class IC Flammable Liquid**	Autoignition Temperature: **716°F (380°C)**

Extinguishing Media
Use dry chemical, carbon dioxide, "alcohol-resistant" foam, or water spray. Do NOT use direct stream of water (may scatter or spread the fire).

Special Fire Fighting Procedures
Poisonous gases are produced in fire, wear full protective clothing and self-contained breathing apparatus (SCBA). Move containers from fire area if it can be done safely. Use water spray to cool fire-exposed containers. Do not allow run-off from fire control methods to reach sewers or waterways.

Unusual Fire and Explosion Hazards
Moderate fire and explosion hazard. Vapors form explosive mixtures in air. Vapors are slightly heavier than air and can travel for great distances to an ignition source to flashback causing fire or explosion.

SECTION V - REACTIVITY DATA

Stability		Conditions to Avoid
		Ethylenediamine is normally stable. It does absorb carbon dioxide from air to form a nonvolatile carbonate. Avoid contact with incompatible materials. Keep dry and away from heat and flame.
Stable	Unstable	Incompatibility *(materials to avoid)*
X		Corrosive to metals. Incompatible with strong oxidizers, strong acids, carbon tetrachloride and other chlorinated compounds, carbon disulfide, acrylonitrile, epichlorohydrin, oleum, nitromethane, vinyl acetate.
Hazardous Polymerization		Conditions to Avoid
		Under normal conditions of temperature and pressure, hazardous polymerization of ethylenediamine is not expected to occur.
May Occur	Will Not Occur	Hazardous Decomposition or By-products
	X	Toxic and irritating fumes and vapors are produced when ethylenediamine is heated to decomposition. These include toxic oxides of nitrogen and carbon, as well as ammonia.

SECTION VI - HEALTH HAZARD DATA

Primary Route(s) of Entry:	Inhalation?	Absorption (skin)?	Ingestion?
	X	X	X

Health Hazards

INHALATION: Coughing, wheezing, laryngitis, shortness of breath, headache, nausea and vomiting, dizziness, abdominal cramps. Death can occur due to spasm, inflammation and edema of the larynx and bronchi, chemical pneumonia, or pulmonary edema (fluid in lungs). Possible allergic sensitization (asthma) may develop. Liver or kidney damage possible.

ABSORPTION: Skin contact may cause dermatitis with irritation. Eye contact may cause swelling (painless) of the cornea epithelium causing visual disturbances. Will pass through skin.

INGESTION: May damage the gastrointestinal tract and cause toxic systemic effects.

Carcinogenicity	NTP Listed?	IARC Cancer Review Group?	OSHA Regulated?	Target Organs?
Unknown Human Unknown Animal	No	No	29 CFR 1910.1000 Table Z-1	Eyes, skin, resp. sys., CNS, liver, kidney, spleen.

Medical Conditions Generally Aggravated by Exposure

Skin (dermatitis), respiratory (asthma), or kidney problems may be aggravated by exposure.

Emergency and First-aid Procedures

Eye contact: Flush immediately with water for 15 minutes (minimum), seek medical attention. Skin contact: Quickly remove all contaminated clothing. Immediately wash area with large amounts of soap and water. Seek medical assistance. For inhalation: Remove the person from exposure. Provide respiratory assistance and CPR. Transfer to medical facility. Observe for 48 hours for lung effects. If swallowed: Call the poison control center and seek medical attention immediately. Do NOT induce vomiting. Never attempt to give an unconscious or convulsing person anything by mouth.

SECTION VII - PRECAUTIONS FOR SAFE HANDLING AND USE

Steps to be Taken in Case Material is Released or Spilled

Absorb liquids in vermiculite or other absorbent and deposit in sealed drum. Ventilate area and remove ignition sources. Restrict those not involved in cleanup from entering area. Neutralize residues or small spills with sodium bicarbonate or soda ash and flush to a drain with copious amounts of water.

Preferred Waste Disposal Method

Large amounts: Collect and atomize in a combustion chamber with an effluent gas cleaning device.

Precautions to be Taken in Handling and Storage

Store in tightly closed containers in a cool, dry well-ventilated area (preferably above 50°F/18°C). Protect containers from physical damage. Use explosion proof (Class 1, Group D) electrical equipment. Ground and bond metal containers and equipment when making transfers (prevent static sparks).

Other Precautions and Warnings

Empty containers may still contain hazardous residues and should be handled accordingly.

SECTION VIII - CONTROL MEASURES AND PERSONAL PROTECTIVE EQUIPMENT

Respiratory Protection *(specify type)*

For low concentrations (<250 ppm), use a chemical cartridge respirator. A supplied-air respirator with full facepiece operated in positive pressure mode or a self-contained breathing apparatus with full facepiece operated in pressure demand or other positive pressure mode provide greater protection.

Ventilation

Local exhaust or general mechanical systems recommended.

Protective Gloves	Eye Protection	Other Protective Clothing
Butyl Rubber or Teflon Gloves	Chemical Goggles or Face Mask	Protective Apron or Clothing

Work/Hygiene Practices

Always wash hands thoroughly after using chemical; never bring food, drink, or smoking materials into vicinity of chemicals.

ETHYLENEDIAMINE

$NH_2CH_2CH_2NH_2$ CAS: 107-15-3

IDENTIFICATION AND TYPICAL USES

Ethylenediamine is a colorless, viscous liquid (or solid below 47°F) with an extremely unpleasant, ammonia-like odor. It is used as a stabilizer for latex rubber, as an emulsifier, as an inhibitor in antifreeze solutions, and in textile lubricants. It is also used as a solvent for albumin, shellac, sulfur, and other substances.

RISK ASSESSMENT: HEALTH

General Assessment

Ethylenediamine is an irritant poison in humans and is toxic by *inhalation, ingestion,* and skin contact (*absorption*). The extremely unpleasant odor of ethylenediamine makes overexposure by inhalation an uncommon event since personnel will not usually tolerate moderate or high vapor concentrations. There is no information in the references on the carcinogenic or teratogenic properties of ethylenediamine. However, mutation data have been reported. Some scientists believe that mutagenic chemicals may have carcinogenic effects in the long-term. Additional study is required in this area. Lung, kidney, and liver injuries can also occur following prolonged or repeated exposures.

Inhalation of ethylenediamine vapors causes severe and corrosive irritation to the eyes, nose, throat, and upper respiratory tract. Excessive exposure to the vapor leads to neurotoxic effects on the central nervous system (CNS) due to its ability to inhibit neuronal firing. Symptoms of exposure include coughing, wheezing, laryngitis, shortness of breath, nausea and vomiting, dizziness, allergic sensitization (asthma), and abdominal cramps. Liver, lung, and/or kidney damage may occur. Exposure to ethylenediamine may be fatal as a result on spasms and inflammation with edema of the larynx and bronchi. Chemical pneumonia and/or pulmonary edema (fluid in the lungs) are also possible, and each can lead to death. Symptoms of pulmonary edema may be delayed up to 48 hours creating a false sense of security with regard to exposure risk.

Ethylenediamine is a severe skin irritant, producing sensitization, blistering, and 2nd degree burns on prolonged contact. Because ethylenediamine tends to be fat soluble, it can be easily absorbed through the skin in toxic amounts. Eye contact results in severe irritation and may cause painless swelling of the corneal epithelium causing the appearance of "halos" around lights. Eye contact may also cause burns and tissue damage.

Ingestion can occur through careless working conditions, accidents, and improper hygienic practices. It causes damage to the gastrointestinal tract and may lead to systemic poisoning.

Acute Health Effects

The following acute (short-term) health effects may occur immediately or shortly after exposure to ethylenediamine:

- Skin: Severe, burning irritation. If not removed immediately, absorption through intact skin can also occur.
- Eye: Irritation, inflammation, redness, and swelling with damage to vision possible if the chemical is not removed immediately.
- Lung: Inhalation causes irritation of the eyes, nose, and upper respiratory tract with distressed breathing, coughing, congestion, and tearing.
- CNS: Action on the peripheral nervous system causing headache, nausea, dizziness, and interference or disruption in neuronal firing with subsequent paresthesia of the peripheral nerves.

Chronic Health Effects

The following chronic (long-term) health effects may occur at some time after exposure to ethylenediamine and can last for months or even years:

Cancer Hazards: According to the information presented in the references, ethylenediamine has not been adequately tested for its ability to cause cancer in test animals. Mutation data have been reported, but human results are conflicting and inconclusive. More research is required.

Reproductive Hazard: According to the information presented in the references, ethylenediamine has not been adequately tested for its ability to adversely affect reproduction.

Other Chronic Effects: Prolonged exposures may lead to lung, liver, and/or kidney damage. Ethylenediamine is an allergen and may cause sensitization resulting in asthma and/or dermatitis (depending on the route of prolonged exposure).

Recommended Risk-Reduction Measures

Unless a less toxic chemical can be substituted for a hazardous substance, *engineering controls* are the most effective method of reducing exposures. The best protection is to enclose operations and/or provide local exhaust ventilation at the site of chemical release. To prevent static sparks, all containers and equipment should be electrically grounded and bonded. While not always operationally feasible, isolating operations can also reduce exposure.

Using respiratory protection is less effective than the controls mentioned above, but is still advisable whenever working with or around ethylenediamine. For relatively low, infrequent, or transient exposures, an organic vapor respirator equipped with a cartridge protecting against amines may suffice. For higher or prolonged exposure, a supplied-air respirator with full facepiece operated in positive pressure mode or a self-contained breathing apparatus (SCBA) with full facepiece and operated in pressure demand are the recommended respiratory protection methods of choice. If a full facepiece is not available, then chemical goggles should be worn to protect the eyes. Whenever a chemical splash hazard exists, a face shield and apron should be worn. To prevent hand and skin exposures, butyl rubber gloves should be worn.

Administrative controls should also be in place to minimize the potential for human exposures. These may include written procedures or policies which specify the methods and techniques that will be practiced whenever personnel are to work with ethylenediamine.

All personnel should receive *training* on the use, hazards, protective measures, emergency actions, and other precautions per 29 CFR 1910.1200 (Hazard Communication), prior to the first assignment in an area where ethylenediamine is used or stored.

Before beginning employment and at regular intervals thereafter (e.g. annually), the following medical tests are recommended:

☑ Lung function tests (establish a baseline).

If symptoms develop or overexposure is suspected, the following medical tests may be useful:

☑ Lung function tests.

☑ Liver and kidney function tests.

☑ Consider chest X-ray following acute overexposure (may be negative if taken immediately after exposure since indication of pulmonary edema may be delayed).

☑ Evaluation by a qualified allergist, with careful consideration of exposure symptoms and special testing (may help diagnose allergy).

Any evaluation should include a careful medical history of past and present symptoms with an examination. However, medical tests that evaluate existing damage are not a substitute for controlling exposure. Also, since smoking causes heart disease, emphysema, and other respiratory disorders, and can decrease the time between first exposure an the onset of respiratory symptoms, smokers exposed to ethylenediamine may experience symptoms much quicker and more pronounced than non-smokers under the same exposure conditions. Prudent risk assessment and management requires careful consideration of *all* possible factors which may be causing the appearance of symptoms in the work force.

Other methods to reduce exposure include:

☑ Where possible, enclose operations and use local exhaust ventilation at the site of chemical release. If local exhaust ventilation or enclosure is not used, respiratory protection should be mandatory. Also, if possible, automatically transfer liquids from storage containers to process containers using non-sparking tools.

☑ Always ensure that proper protective clothing is worn when using chemical substances.

☑ Wash thoroughly immediately after exposure to ethylenediamine and at the end of the work shift or before eating, drinking, or smoking.

☑ Hazard warning information should be posted in the work area. In addition, as part of an on-going

education and training program, all information on the health and safety hazards of ethylenediamine should be communicated to all potentially exposed workers.

☑ Eye wash stations should be provided in work areas. If the potential for whole body exposure exists, then safety showers should also be provided.

RISK ASSESSMENT: ENVIRONMENT

General Assessment

The environment is at risk of exposure during intended use, transportation, storage, disposal, or destruction of ethylenediamine. In almost every scenario, the threat of environmental exposure is contingent upon the proper handling of the chemical substance. Accidental spills, large or small, can result in fire, explosion, and possible contamination of the surrounding environmental mediums (water, soil, and air).

Ethylenediamine is considered a Class IC flammable liquid (per OSHA 29 CFR 1910.106). This means it poses a serious fire risk by itself or in contact with incompatible materials (can cause fire or explosion). It can react with many common oxidizing materials and strong acids. It is corrosive to copper, brass, galvanized iron, and many other metals compounds. Its vapors can form explosive mixtures in air. These characteristics require special consideration during any emergency situation involving a leak or spill of ethylenediamine. Should ethylenediamine ever come into contact with incompatible substances either during use, transportation, or storage, violent reactions can occur.

Ethylenediamine can enter the environment through its intended use as a fungicide, as well as through industrial discharges and spills.

Acute Ecological Effects

Acute (short-term) toxic effects may include the death of animals, birds, or fish and death or low growth rate in plants. Acute effects are seen 2 to 4 days after animals or plants are exposed to ethylenediamine. Ethylenediamine has moderate acute toxicity to aquatic life. Insufficient data are available to evaluate or predict the short-term effects of ethylenediamine on plants, birds, or land animals.

Chronic Ecological Effects

Chronic toxic effects may include shortened life span, reproductive problems, lower fertility, and changes in appearance or behavior in exposed animals. These effects can be seen long after first exposure(s) to toxic chemicals. Ethylenediamine has slight chronic toxicity to aquatic life. Insufficient data are available to evaluate or predict the long-term effects of ethylenediamine on plants, birds, or land animals.

Water Solubility

Ethylenediamine is highly soluble in water. Concentrations of 1000 milligrams and more will mix with a liter of water.

Persistence in the Environment

Ethylenediamine is slightly persistent in water, with a half-life of between 2 to 20 days. The half-life of a pollutant is the amount of time it takes for one half of the chemical to be degraded.

Bioaccumulation in Aquatic Organisms

Some substances increase in concentration, or *bioaccumulate,* in living organisms as they breathe contaminated air, drink contaminated water, or eat contaminated food. These chemicals can become concentrated in the tissues and internal organs of animals as well as humans.

The concentration of ethylenediamine found in fish tissues is expected to be about the same as the average concentration of ethylenediamine in the water from which the fish was taken.

Recommended Risk-Reduction Measures

Proper training of all transporters will reduce the likelihood of a mishap or accident resulting in a leak or spill to the environment. The correct use of DOT labeling (per the MSDS) on all containers, trucks, and rail cars will enable emergency responders to react properly and quickly to any disaster thereby reducing the potential risk to the environment and to personnel.

Storage of ethylenediamine should be segregated from incompatible chemicals to minimize the risk of cross-contamination or contact. Storage buildings should be equipped with the proper fire protection equipment (alarms, sprinklers, extinguishers, etc.). Electrical equipment in storage areas should be rated for Class 1, Group D environments.

If a spill or leak to the environment has occurred, fire department, emergency response and/or hazardous materials spill personnel should be notified immediately. Cleanup should be attempted only by those

trained in proper spill containment procedures. Contaminated soils should therefore be removed for incineration and replaced with clean soil. If ethylenediamine should contact the water table, aquifer, or navigable waterway, remediation activities should be prompt. Time is of the essence since it is highly soluble and total containment and remediation may not be possible. When such spills occur, the local and/or state emergency response authorities must be notified. A comprehensive emergency response or disaster preparedness/recovery plan should be in place prior to any operations involving the use, transportation, storage, or disposal of ethylenediamine.

For disposal of small quantities, add the ethylenediamine to a layer of sodium bisulfate in a large evaporating dish and spray with water. Neutralize and then flush to the sanitary sewer. Large amounts should be collected and atomized in a combustion chamber equipped with an effluent gas cleaning device.

If ethylenediamine is spilled or leaked, the following specific steps are recommended:

☑ Evacuate area and deny entry to those not involved in cleanup activities. Restrict persons not wearing protective clothing from area of spill or leak until cleanup is complete.

☑ Ventilate area and remove all ignition sources.

☑ Absorb liquids using vermiculite, dry earth, or sand and place in a sealed drum for disposal. Trace residues or small spills should be neutralized with sodium bicarbonate or soda ash and then flushed to a drain using copious amounts of water.

☑ It may be necessary to dispose of ethylenediamine as a hazardous waste. The responsible state agency or the regional office of the federal Environmental Protection Agency (EPA) should be contacted for specific recommendations.

RISK ASSESSMENT: BUSINESS

General Assessment

Accidents or mishaps involving ethylenediamine can present a significant threat to business operations. The loss or damage of equipment or facilities can significantly affect fiscal viability. Lawsuits that may result from personnel injury/death, public exposures, and/or environmental contamination will also require a serious expenditure of resources. Media attention surrounding an injury, death, or environmental damage can also result in a loss of profits and loss of current as well as future business.

Recommended Risk-Reduction Measures

Company attorneys, safety and health professionals, and environmental specialists should be involved in the development of any procedures dealing with safety or emergency response. A company official should be pre-designated as a public relations officer with specific training in dealing with the press. Corporate plans and policies should be developed, approved, and implemented long before any need for such arises.

REFERENCES

American Conference of Governmental Industrial Hygienists. 1988. *Documentation of the Threshold Limit Values and Biological Exposure Indices,* 5th Edition (with updates). Cincinnati: ACGIH

Bueche, J. R. 1972. *Principles of Physics,* 2nd Edition. New York: McGraw-Hill.

Calabrese, E. J. 1994. *Biological Effects of Low Level Exposures.* New York: CRC Press/Lewis Publishers.

Danse, I. R. 1991. *Common Sense Toxics in the Workplace.* New York: Van Nostrand Reinhold.

Environmental Protection Agency. 1983. *Chemical Hazard Information Profiles.* Washington, D.C.: U.S. EPA.

Fodden, C. R., and J. L. Weddell. 1992. *Hazardous Materials Emergency Action Data.* New York: CRC Press/Lewis Publishers.

Hodgson, E., R. B. Mailman, and Janice E. Chambers. 1988. *Dictionary of Toxicology.* London: The MacMillian Company.

Lauwerys, R. R., P. Hoet. 1993. *Industrial Chemical Exposure.* New York: CRC Press/Lewis Publishers.

Lewis, R. J., Sr. 1993. *Hawley's Condensed Chemical Dictionary,* 12th Edition. New York: Van Nostrand Reinhold.

Lewis, R. J., Sr. 1992. *Sax's Dangerous Properties of Industrial Materials,* Volumes 1, 2, and 3, 8th Edition. New York: Van Nostrand Reinhold.

Manahan, S. E. 1992. *Toxicological Chemistry,* 2nd Edition. New York: CRC Press/Lewis Publishers.

Meyer, E. 1990. *Chemistry of Hazardous Materials.* Englewood Cliffs, NJ: Prentice Hall.

Quigley, D. R. 1994. *Handbook of Emergency Chemical Management.* New York: CRC Press/Lewis Publishers.

Richardson, M. 1995. *Dictionary of Substance and Their Effects,* Vol. I-VI. New York: CRC Press/Lewis Publishers.

Sacarello, H. L. A. 1994. *The Comprehensive Handbook of Hazardous Materials.* New York: CRC Press/Lewis Publishers.

U.S. Department of Health and Human Services, National Institute for Occupational Safety and Health. 1994. *NIOSH Pocket Guide to Chemical Hazards.* Washington, D.C.: U.S. Government Printing Office.

MATERIAL SAFETY DATA SHEET

CHEMICAL NAME	
ETHYLENE DIBROMIDE	**CRC PRESS/LEWIS PUBLISHERS DISCLAIMER:** THE INFORMATION AND RECOMMENDATIONS PRESENTED HEREIN ARE BASED ON SOURCES BELIEVED TO BE RELIABLE. CRC MAKES NO REPRESENTATION ON ITS COMPLETENESS OR ACCURACY. IT IS THE RESPONSIBILITY OF THE USER TO DETERMINE THE CHEMICAL'S SUITABILITY FOR ITS INTENDED USE, THE CHEMICAL'S SAFE USE, AND THE CHEMICAL'S PROPER DISPOSAL. NO REPRESENTATIONS AND/OR WARRANTIES, EITHER EXPRESSED OR IMPLIED, OF THE MERCHANTABILITY OR FITNESS FOR A PARTICULAR PURPOSE, OR OF ANY OTHER NATURE, ARE MADE WITH RESPECT TO THE INFORMATION PROVIDED IN THIS MSDS OR TO THE CHEMICAL TO WHICH INFORMATION MAY REFER. CRC NEITHER ASSUMES NOR AUTHORIZES ANY OTHER PERSON TO ASSUME FOR IT, ANY OTHER ADDITIONAL RESPONSIBILITY OR LIABILITY FOR THE USE OF, OR RELIANCE UPON, THIS INFORMATION.

HAZARD WARNING INFORMATION

HEALTH	FIRE	REACTIVE	OTHER	DEGREE OF HAZARD	COLOR CODING	OTHER CODES
3	0	0		0 = Minimum Hazard 1 = Slight Hazard 2 = Moderate Hazard 3 = Serious Hazard 4 = Severe Hazard	HEALTH = BLUE FIRE = RED REACTIVITY = YELLOW OTHER = WHITE	OX = Oxidizer ACID = Acid ALK = Alkali COR = Corrosive W = Use No Water

SECTION I - GENERAL INFORMATION

Characterization	**Halogenated Hydrocarbon**	RCRA Number	**U067**
		EPA Class	**Toxic Waste**
DOT Proper Shipping Name	**Ethylene Dibromide**	Chemical Abstract Service (CAS) Number	**106-93-4**
DOT Hazard Class and Label Requirements	**ORM-A; Poison B**	DOT Emergency Guide Code	**55**
DOT Identification Number	**UN 1605**	Molecular Formula	**$BrCH_2CH_2Br$**

Synonyms
1,2-Dibromoethane; ethylene bromide; glycol dibromide.

SECTION II - HAZARDOUS INGREDIENTS AND IDENTITY INFORMATION

Hazardous Components (specific identity)	OSHA Exposure Criteria	NIOSH Exposure Criteria	Immediately Dangerous to Life and Health (IDLH)	ACGIH Exposure Criteria
Ethylene dibromide (derivation: By action of bromine on ethylene; from acetylene and hydrogen bromide). **1 ppm = 7.81 mg/m^3**	**PEL:** **20 ppm** **156 mg/m^3** **CEILING:** **30 ppm** **PEAK (5min/8hr):** **50 ppm**	**REL:** **0.45 ppm** **0.38 mg/m^3** **CEILING:** **0.13 ppm** **1 mg/m^3**	**100 ppm**	**TLV:** **Suspected Human Carcinogen**

SECTION III - PHYSICAL AND CHEMICAL CHARACTERISTICS

Boiling Point	**268°F (131°C)**	Specific Gravity (H_2O = 1)	**2.17 at 77°F (25°C)**
Vapor Pressure (mm Hg)	**12 at 69°F (20°C)**	Molecular Weight (atomic weight)	**187.9**
Vapor Density (Air = 1)	**6.48**	Freezing Point	**50°F (10°C)**

Solubility
Slightly soluble in water (0.4%). Miscible with most organic solvents.

Appearance and Odor
Colorless, heavy, viscous liquid (or solid below 50°F) with a sweet, pleasant, and agreeable chloroform-like odor.

SECTION IV - FIRE AND EXPLOSION HAZARD DATA

Flash Point (method used)	**Not Applicable**	Explosive Limits in Air % by Volume	**LEL: Not Applicable UEL: Not Applicable**
NFPA Classification	**Not Applicable**	Autoignition Temperature	**Not Applicable**

Extinguishing Media
Ethylene dibromide will not burn. Use extinguishing agent suitable to surrounding fire.

Special Fire Fighting Procedures
Poisonous gases are produced in fire. Wear self-contained breathing apparatus (SCBA) and full protective gear. Move containers from fire if it can be done without risk. Cool sides of fire-exposed containers with water until long after fire is out. Reacts with hot aluminum and magnesium.

Unusual Fire and Explosion Hazards
None reported. Ethylene dibromide is a non-flammable and non-combustible liquid.

SECTION V - REACTIVITY DATA

Stability		Conditions to Avoid
		Ethylene dibromide is stable under normal conditions of handling and storage. Avoid exposing ethylene dibromide to heat or light in the presence of moisture (it may slowly hydrolyze to produce hydrogen bromide).
Stable	Unstable	Incompatibility *(materials to avoid)*
X		Strong oxidizers (chlorine, bromine, fluorine), chemically active metals (such as sodium, potassium, calcium, zinc, hot aluminum, and hot magnesium), liquid ammonia, and heat or light in the presence of moisture.
Hazardous Polymerization		Conditions to Avoid
		Under normal conditions of temperature and pressure, hazardous polymerization of ethylene dibromide cannot occur.
May Occur	Will Not Occur	Hazardous Decomposition or By-products
	X	Poisonous gases and acrid fumes are produced in fire, including extremely toxic oxides of carbon (carbon monoxide) and hydrogen bromide.

SECTION VI - HEALTH HAZARD DATA

Primary Route(s) of Entry:	Inhalation?	Absorption (skin/eye)?	Ingestion?
	X	X	X

Health Hazards

INHALATION: Depression of the central nervous system (CNS) with general weakness, headache, loss of appetite and subsequent weight loss, depression, vomiting, diarrhea, chest pain, cough, shortness of breath, abdominal pain, damage to major organs, and pulmonary edema.

ABSORPTION: Irritation and blistering of skin with vesicles. Eye contact may result in severe irritation. Will rapidly pass through intact skin. Together with inhalation, effects can be serious.

INGESTION: Possible necrosis of the liver and kidneys that can lead to death.

Carcinogenicity	NTP Listed?	IARC Cancer Review Group?	OSHA Regulated?	Target Organs?
Suspected Human Confirmed Animal	5th Annual Report	Group 2A	29 CFR 1910.1000 Table Z-1	Eyes, skin, resp. sys., liver, kidney, CNS, repro. sys.

Medical Conditions Generally Aggravated by Exposure

None reported.

Emergency and First-aid Procedures

Eye contact: Do not allow victim to keep eyes tightly shut. Flush immediately with water for 15 minutes (minimum), seek medical attention. Skin contact: Remove all contaminated clothing, including shoes. Immediately wash area with flooding amounts of soap and water. Seek medical attention immediately. For inhalation: Remove the person from exposure. Provide respiratory assistance and CPR. Transfer to medical facility. Observe for 48 hours for lung effects. If swallowed: Seek medical attention immediately. Give conscious and alert person 1-2 glasses of water and induce vomiting.

SECTION VII - PRECAUTIONS FOR SAFE HANDLING AND USE

Steps to be Taken in Case Material is Released or Spilled

Wear SCBA respiratory protection. Restrict those not wearing protective equipment and who are not involved in cleanup from entering area. Absorb with vermiculite and place in sealed container. Vapors are likely to collect in low lying areas in high concentrations. Exercise caution.

Preferred Waste Disposal Method

Mix with a combustible solvent, burn in chemical incinerator equipped with an afterburner and scrubber.

Precautions to be Taken in Handling and Storage

Store in tightly closed containers in cool, well ventilated location. Keep away from sources of heat.

Other Precautions and Warnings

Recommend storage under a nitrogen purge blanket to prevent condensation of moisture inside containers. Workers must avoid contact since ethylene dibromide is a cancer and reproductive hazard.

SECTION VIII - CONTROL MEASURES AND PERSONAL PROTECTIVE EQUIPMENT

Respiratory Protection *(specify type)*

An MSHA/NIOSH-approved air purifying respirator (cartridge type), is permissible. However, for best protection against a suspected carcinogen use a supplied-air respirator or a self-contained breathing apparatus (SCBA) with full facepiece operated in positive pressure mode.

Ventilation

Local exhaust (preferred) at site of chemical work or general ventilation.

Protective Gloves	Eye Protection	Other Protective Clothing
Impervious Rubber Gloves	Chemical Goggles or Face Mask	Protective Apron if Splash is Likely

Work/Hygiene Practices

Always wash hands thoroughly after using chemical; never bring food, drink, or smoking materials into vicinity of chemicals.

ETHYLENE DIBROMIDE

$BrCH_2CH_2Br$ CAS: 106-93-4

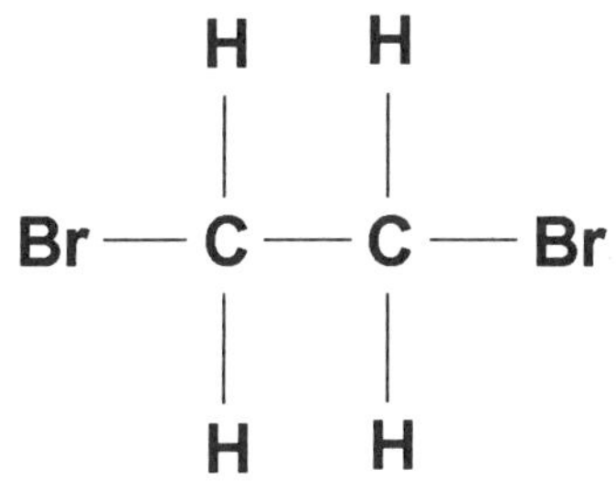

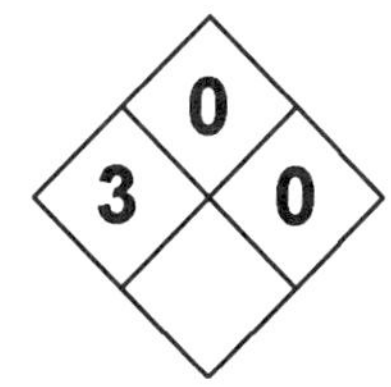

IDENTIFICATION AND TYPICAL USES

Ethylene dibromide is a colorless, heavy, viscous liquid with a sweet, pleasant, chloroform-like odor. It can also exist as a solid below 50°F (10°C). It is widely used as a fumigant for grains. Because of its carcinogenic properties, this use is under constant review. It is also used as a scavenger of lead in leaded gasoline (antiknock agent), as a solvent, in paints, varnish, and finish removers, in metal degreasing, in soaps and scouring agents, in ore flotation, in water purifying preparations, and in organic synthesis.

RISK ASSESSMENT: HEALTH

General Assessment

Ethylene dibromide is moderately toxic by *inhalation*. The liquid is also known the pass through the skin (*absorption*) to cause toxic systemic effects. Although unlikely under normal operating conditions, *ingestion* is also a possible exposure route. This chemical is known to cause cancer in test animals (skin and lung tumors) and this effect is suspected to occur in humans as well. Human mutation data have been reported. In addition, ethylene dibromide has been shown to adversely affect reproduction in test animals.

Inhalation is irritating to the eyes, nose, and mucosa of the respiratory tract. High concentrations of ethylene dibromide cause depression of the central nervous system. Subsequent symptoms include dizziness, lightheadedness, headache, loss of appetite and subsequent weight loss, abdominal pain with progression to major organ system damage (primarily the liver and kidneys), unconsciousness, and death due to liver and kidney damage (necrosis). Other symptoms include general weakness and fatigue, drowsiness, nausea and vomiting, diarrhea, chest pains, cough, and shortness of breath. Irritation to lung tissues can lead to a dangerous buildup of fluids in the lungs (pulmonary edema), which is a medical emergency and can be fatal. Symptoms may be delayed up to 48 hours creating a false sense of security with regard to health risk.

Skin contact can produce severe localized irritation and blistering with vesicles. Absorption will occur rapidly causing an increase in toxicity. If coupled with inhalation, the systemic effects can be quite severe. Eye contact will also cause irritation to the conjunctiva and surrounding tissues.

Acute Health Effects

The following acute (short-term) health effects may occur immediately or shortly after exposure to ethylene dibromide:

Skin: Irritation with potential for redness, pain, smarting, burns, blistering with vesicles, and tissue damage.

Eye: Irritation and smarting of the conjunctiva. No evidence of permanent damage.

Lung: Irritation of the mucous membranes of the nose, throat, and respiratory tract causing coughing, chest pains, difficulty in breathing, and possible pulmonary edema.

CNS: Depression of the central nervous system with symptoms of headaches, lightheadedness, dizziness, nausea, vomiting, diarrhea, loss of appetite and weight.

Other: High exposures can lead to rapid progressive damage to the kidneys, liver, heart, and lungs which can lead to death.

Chronic Health Effects

The following chronic (long-term) health effects may occur at some time after exposure to ethylene dibromide and can last for months or even years:

Cancer Hazards: According to the references, several agencies consider ethylene dibromide to be a potential or suspect human carcinogen. These include the IARC, the NTP, and the ACGIH. Human mutation data have also been reported.

Reproductive Hazard: According to information presented in the references, ethylene dibromide is a possible teratogen in humans and has been shown to adversely affect reproduction in test animals. It may damage the reproductive system causing abnormal sperm in males and decreased fertility in females.

Other Chronic Effects: Exposure will irritate the lungs, If pulmonary edema does not develop, there may be bronchitis with cough, phlegm, and/or shortness of breath. Long-term exposure may lead to liver and kidney damage.

Recommended Risk-Reduction Measures

Personnel should avoid direct contact with ethylene dibromide. Occupational poisoning occurs most commonly by inhalation and skin absorption. If a less toxic material or compound cannot be substituted for ethylene dibromide, then *engineering controls* are the most effective method of reducing exposures. The best protection is to enclose operations and/or provide local exhaust ventilation at the site of ethylene dibromide release. While not always operationally feasible, isolating operations can also reduce exposure. Using respiratory protection is less effective than the controls mentioned above, but is still advisable whenever working with or around ethylene dibromide. Caution should be exercised when selecting respiratory protection since it is not known if exposure to ethylene dibromide will cause cancer in humans. For exposure over the PEL (20 ppm), an MSHA/NIOSH-approved air-purifying respiratory with organic vapor cartridge(s) may suffice. However, for the best protection against a *possible carcinogen,* especially when the concentration is unknown, use a supplied air respirator with full facepiece and mask operated in positive pressure mode or a self-contained breathing apparatus (SCBA) with full facepiece operated in pressure demand. If a full facepiece is not available, then chemical goggles should be worn to protect the eyes. A face shield should also be considered. To prevent hand and skin exposures, impervious rubber gloves should be worn.

Administrative controls should also be in place to minimize the potential for human exposures. These may include written procedures or policies which specify the methods and techniques that will be practiced whenever personnel are to work with ethylene dibromide.

All personnel should receive *training* on the use, hazards, protective measures, emergency actions, and other precautions per 29 CFR 1910.1200 (Hazard Communication), prior to the first assignment in an area where ethylene dibromide is used or stored.

Before beginning employment with ethylene dibromide and at regular intervals thereafter (e.g., annually), the following medical tests are recommended:

☑ Liver and kidney function tests (establish baseline).

☑ Lung function tests (establish baseline).

If symptoms develop or overexposure is suspected, the following tests should be considered:

☑ Liver and kidney function tests.

☑ Consider chest X-ray after acute overexposure (may be negative if taken immediately after exposure due to the delayed effects of pulmonary edema).

Any evaluation should include a careful history of past and present symptoms with an examination. Medical tests that simply look for existing damage are not a substitute for controlling exposures. Also, since smoking can cause heart disease, lung cancer, emphysema, and other respiratory problems, smokers may be effected more rapidly and with more pronounced symptoms than non-smokers. In addition, those who are taking the drug Disulfiram (antabuse) and are exposed to ethylene dibromide at work may experience more severe toxic effects. Prudent risk management requires careful consideration of *all* risk factors that might be causing the appearance of symptoms in the work force.

Other methods to reduce exposure include:

☑ Where possible, enclose operations and use local exhaust ventilation at the site of chemical release. If local exhaust ventilation or enclosure is not used, respiratory protection should be mandatory.

☑ Always ensure that proper protective clothing is worn when using chemical substances.

- ☑ Wash thoroughly immediately after exposure to ethylene dibromide and at the end of the work shift or before eating, drinking, or smoking.
- ☑ Hazard warning information should be posted in the work area. In addition, as part of an on-going education and training program, all information on the health and safety hazards of ethylene dibromide should be communicated to all potentially exposed workers.
- ☑ Eye wash stations should be provided in the immediate work area for emergency use. If there is a possibility of skin exposure to ethylene dibromide, emergency shower facilities should also be provided.
- ☑ Workers whose clothing has been contaminated by ethylene dibromide should change into clean clothes before leaving work. Contaminated work clothing should be laundered only by individuals who have been informed of the hazards of exposure to ethylene dibromide.

RISK ASSESSMENT: ENVIRONMENT

General Assessment

Although ethylene dibromide normally exists as a gas, it is usually shipped as a liquefied compressed gas. The environment is at risk of exposure during transportation, storage, disposal, or destruction of ethylene dibromide. In almost every scenario, the threat of environmental exposure is contingent upon the proper handling of the chemical substance. Accidental spills, large or small, can result in contamination of the surrounding environmental mediums (water, soil, and air).

Ethylene dibromide is a non-flammable liquid. However, it can react violently in the presence of chemically-active metals such as sodium, potassium, calcium, and strong oxidizing agents. In contact with water or moisture in the presence of heat or light it can produce toxic hydrogen bromide. Caution is always required in handling, storage, transportation, and disposal of ethylene dibromide. Emergency responders should be made aware of the presence of ethylene dibromide at any emergency response situation.

Ethylene dibromide can enter the environment from industrial effluents and from spills.

Acute Ecological Effects

Acute (short-term) toxic effects may include the death of animals, birds, or fish and death or low growth rate in plants. Acute effects are seen 2 to 4 days after animals or plants are exposed to ethylene dibromide.

Ethylene dibromide has moderate acute toxicity to aquatic life. Insufficient data are available to evaluate the short-term effects of ethylene dibromide on plants, birds, or land animals.

Chronic Ecological Effects

Chronic (long-term) toxic effects may include shortened life span, reproductive problems, lower fertility, and changes in appearance or behavior in exposed animals. These effects can be seen long after first exposure(s) to toxic chemicals.

Ethylene dibromide has moderate chronic toxicity to aquatic life. Insufficient data are available to evaluate or predict the long-term effects of ethylene dibromide to plants, birds, or terrestrial animals.

Water Solubility

Ethylene dibromide is highly soluble in water. Concentrations of 1000 milligrams and more will mix with a liter of water.

Persistence in the Environment

Ethylene dibromide is slightly persistent in water, with a half-life between 2 and 20 days. The half-life of a pollutant is the amount of time it takes for one half of the chemical to be degraded. About 91.6% of ethylene dibromide will eventually end up in air; the remainder will end up in water.

Bioaccumulation in Aquatic Organisms

Some substances increase in concentration, or *bioaccumulate,* in living organisms as they breathe contaminated air, drink contaminated water, or eat contaminated food. These chemicals can become concentrated in the tissues and internal organs of animals as well as humans.

The concentration of ethylene dibromide found in fish tissues is expected to be about the same as the average concentration of ethylene dibromide in the water from which the fish was taken.

Recommended Risk-Reduction Measures

Proper training of all transporters will reduce the likelihood of a mishap or accident resulting in a leak or spill of ethylene dibromide into the environment. Labels on all containers, trucks, and rail cars must meet DOT requirements and accurately reflect their contents to enable emergency responders to react properly and quickly to any disaster thereby reducing the potential risk to the environment and to personnel.

Storage of ethylene dibromide should be segregated from other chemicals and materials to minimize the risk of cross-contamination.

If a spill or leak to the environment has occurred, fire department, emergency response and/or hazardous materials spill personnel should be notified immediately. Cleanup should be attempted only by those trained in proper spill containment procedures. Safety and handling equipment should be selected carefully since ethylene dibromide attacks certain forms of plastics and resins that can be found in some pipes, pumps, and respirators. Contaminated soils should be removed for incineration and replaced with clean soil. If ethylene dibromide should contact the water table, aquifer, or navigable waterway, time is of the essence. It is highly soluble and total remediation may not be possible. The local and/or state emergency response authorities must be notified. A comprehensive emergency response or disaster preparedness/recovery plan should be in place prior to any operations involving the use, transportation, storage, or disposal of ethylene dibromide.

For disposal, ethylene dibromide may be burned in a chemical incinerator equipped with an afterburner and scrubber.

If ethylene dibromide is spilled or leaked, the following specific steps are recommended:

☑ Restrict persons not wearing protective clothing from area of spill until cleanup is complete.

☑ Absorb liquids in vermiculite, sand, earth, or similar material and deposit in sealed drum.

☑ Ventilate area of spill or leak.

☑ It may be necessary to dispose of ethylene dibromide as a hazardous waste. The responsible state agency or the regional office of the federal Environmental Protection Agency (EPA) should be contacted for specific recommendations.

RISK ASSESSMENT: BUSINESS

General Assessment

Accidents or mishaps involving ethylene dibromide can present a serious threat to business operations. The loss or damage of equipment or facilities can significantly affect fiscal viability. Lawsuits that may result from personnel illness, injury/death, public exposures, and/or environmental contamination will require a serious expenditure of resources. Media attention surrounding an injury, death, or environmental damage can also result in a loss of profits and loss of current as well as future business. Always remember that anytime the terms "cancer," "carcinogen," or "reproductive hazard" are used, public emotion, hysteria, and ignorance can run equally high. This must be carefully considered when developing or implementing any public relation policies.

Recommended Risk-Reduction Measures

Company attorneys, safety and health professionals, and environmental specialists should be involved in the development of any procedures for responding to chemical incidents. A company official should be predesignated as a public relations officer with specific training in dealing with the press. Corporate plans and policies should be developed, approved, and implemented long before any need for such arises.

REFERENCES

American Conference of Governmental Industrial Hygienists. 1988. *Documentation of the Threshold Limit Values and Biological Exposure Indices,* 5th Edition (with updates). Cincinnati: ACGIH

Anthony, C. P., and N. J. Kolthoff. 1971. *Textbook of Anatomy and Physiology.* St. Louis: C. V. Mosby Company.

Bueche, J. R. 1972. *Principles of Physics,* 2nd Edition. New York: McGraw-Hill.

Calabrese, E. J. 1994. *Biological Effects of Low Level Exposures.* New York: CRC Press/Lewis Publishers.

Cockerman, L. G., B. S. Shane. 1994. *Basic Environmental Toxicology.* New York: CRC Press/Lewis Publishers.

Lewis, R. J., Sr. 1993. *Hawley's Condensed Chemical Dictionary,* 12th Edition. New York: Van Nostrand Reinhold.

Lewis, R. J., Sr. 1992. *Sax's Dangerous Properties of Industrial Materials,* Volumes 1, 2, and 3, 8th Edition. New York: Van Nostrand Reinhold.

Meyer, E. 1990. *Chemistry of Hazardous Materials.* Englewood Cliffs, NJ: Prentice Hall.

Patnaik, P. 1992. *A Comprehensive Guide to the Hazardous Properties of Chemical Substances.* New York: Van Nostrand Reinhold.

Quigley, D. R. 1994. *Handbook of Emergency Chemical Management.* New York: CRC Press/Lewis Publishers.

U.S. Department of Health and Human Services, National Institute for Occupational Safety and Health. 1994. *NIOSH Pocket Guide to Chemical Hazards.* Washington, D.C.: U.S. Government Printing Office.

MATERIAL SAFETY DATA SHEET

CHEMICAL NAME	
ETHYLENE DICHLORIDE	**CRC PRESS/LEWIS PUBLISHERS DISCLAIMER:** THE INFORMATION AND RECOMMENDATIONS PRESENTED HEREIN ARE BASED ON SOURCES BELIEVED TO BE RELIABLE. CRC MAKES NO REPRESENTATION ON ITS COMPLETENESS OR ACCURACY. IT IS THE RESPONSIBILITY OF THE USER TO DETERMINE THE CHEMICAL'S SUITABILITY FOR ITS INTENDED USE, THE CHEMICAL'S SAFE USE, AND THE CHEMICAL'S PROPER DISPOSAL. NO REPRESENTATIONS AND/OR WARRANTIES, EITHER EXPRESSED OR IMPLIED, OF THE MERCHANTABILITY OR FITNESS FOR A PARTICULAR PURPOSE, OR OF ANY OTHER NATURE, ARE MADE WITH RESPECT TO THE INFORMATION PROVIDED IN THIS MSDS OR TO THE CHEMICAL TO WHICH INFORMATION MAY REFER. CRC NEITHER ASSUMES NOR AUTHORIZES ANY OTHER PERSON TO ASSUME FOR IT, ANY OTHER ADDITIONAL RESPONSIBILITY OR LIABILITY FOR THE USE OF, OR RELIANCE UPON, THIS INFORMATION.

HAZARD WARNING INFORMATION

HEALTH	FIRE	REACTIVE	OTHER	DEGREE OF HAZARD	COLOR CODING	OTHER CODES
2	3	0		0 = Minimum Hazard 1 = Slight Hazard 2 = Moderate Hazard 3 = Serious Hazard 4 = Severe Hazard	HEALTH = BLUE FIRE = RED REACTIVITY = YELLOW OTHER = WHITE	OX = Oxidizer ACID = Acid ALK = Alkali COR = Corrosive W = Use No Water

SECTION I - GENERAL INFORMATION

Characterization	RCRA Number	EPA Class
Halogenated Hydrocarbon	**U077**	**Toxic Waste**

DOT Proper Shipping Name **Ethylene Dichloride**	Chemical Abstract Service (CAS) Number **107-06-2**
DOT Hazard Class and Label Requirements **Flammable Liquid**	DOT Emergency Guide Code **26**
DOT Identification Number **UN 1184**	Molecular Formula **$ClCH_2CH_2Cl$**

Synonyms
1,2-Dichloroethane; ethylene chloride; glycol dichloride; *sym*-dichloroethane; 1,2-ethylene dichloride.

SECTION II - HAZARDOUS INGREDIENTS AND IDENTITY INFORMATION

Hazardous Components (specific identity)	OSHA Exposure Criteria	NIOSH Exposure Criteria	Immediately Dangerous to Life and Health (IDLH)	ACGIH Exposure Criteria
Ethylene dichloride (derivation: By action of chlorine on ethylene with subsequent distillation with a metallic catalyst; by reaction of acetylene and hydrochloric acid). 1 ppm = 4.11 mg/m^3	**PEL:** 50 ppm 205 mg/m^3 **CEILING:** 100 ppm **PEAK** (5min/3hr): 200 ppm	**REL:** 1 ppm 4 mg/m^3 **CEILING:** 2 ppm 8 mg/m^3 Cancer Agent	50 ppm	**TLV:** 10 ppm 40 mg/m^3 **STEL:** Not Established

SECTION III - PHYSICAL AND CHEMICAL CHARACTERISTICS

Boiling Point **182°F (83°C)**	Specific Gravity (H_2O = 1) **1.24**
Vapor Pressure (mm Hg) **64 at 69°F (20°C)**	Molecular Weight (atomic weight) **99.0**
Vapor Density (Air = 1) **3.4**	Freezing Point **-32°F (-35°C)**

Solubility
Slightly soluble in water (0.9%). Miscible in alcohol, chloroform, ether, and most organic solvents. It decomposes slowly, becomes acidic and darkens in color.

Appearance and Odor
Colorless or clear liquid with a sweet, pleasant, chloroform-like odor. Odor Threshold = 40 ppm

SECTION IV - FIRE AND EXPLOSION HAZARD DATA

Flash Point (method used) **56°F (13°C) closed cup**	Explosive Limits in Air % by Volume **LEL: 6.2% UEL: 16%**
NFPA Classification **Class 1B Flammable Liquid**	Autoignition Temperature **775°F (413°C)**

Extinguishing Media
Dry chemical, carbon dioxide, alcohol foam, water spray/fog, or dry sand. Direct water may be ineffective.

Special Fire Fighting Procedures
Poisonous gases are produced in fire. Wear self-contained breathing apparatus (SCBA) and full protective gear. Move containers from fire if it can be done without risk. Cool sides of fire-exposed containers with water until long after fire is out.

Unusual Fire and Explosion Hazards
Dangerous fire and explosion hazard. Containers may explode in fire. Its vapors are heavier than air and can travel for great distances to an ignition source to flashback and cause fire or explosion.

SECTION V - REACTIVITY DATA

Stability		Conditions to Avoid **Ethylene dichloride is stable. Avoid heat, flame, other sources of ignition, and contact with incompatible materials and chemicals.**
Stable X	Unstable	Incompatibility *(materials to avoid)* **Strong oxidizers (chlorine, bromine, fluorine), chemically active metals (such as sodium, potassium, calcium, powdered aluminum, zinc, and magnesium), caustic agents, and liquid ammonia.**
Hazardous Polymerization		Conditions to Avoid **Under normal conditions of temperature and pressure, hazardous polymerization of ethylene dichloride cannot occur.**
May Occur	Will Not Occur X	Hazardous Decomposition or By-products **Poisonous gases and acrid fumes are produced in fire, including extremely toxic phosgene, vinyl chloride, and chloride fumes.**

SECTION VI - HEALTH HAZARD DATA

Primary Route(s) of Entry:	Inhalation?	Absorption (skin/eye)?	Ingestion?
	X	X	X

Health Hazards

INHALATION: Respiratory irritation, intoxication, narcotic and anesthetic effects, vomiting, dizziness, depression, diarrhea, CNS depression, eye irritation, corneal opacity, ulceration of the respiratory tract, somnolence, cyanosis, pulmonary edema, coma, and death. The liver, kidneys, heart, adrenal glands, and skin may all be adversely affected.

ABSORPTION: Irritation, de-fatting, and burning (1st or 2nd degree). Eye contact may result in severe irritation and possible permanent damage. Will pass through intact skin.

INGESTION: Ingestion of the liquid can cause death.

Carcinogenicity	NTP Listed?	IARC Cancer Review Group?	OSHA Regulated?	Target Organs?
Suspected Human Confirmed Animal	5th Annual Report	Group 2A	29 CFR 1910.1000 Table Z-1	Eyes, skin, respiratory system, liver, kidney, CVS, and CNS.

Medical Conditions Generally Aggravated by Exposure

Those on anticoagulants may show increased bleeding; those on insulin may show high blood sugars.

Emergency and First-aid Procedures

Eye contact: Do not allow victim to keep eyes shut. Flush immediately with water for 15 minutes minimum, seek medical attention. Skin contact: Remove all contaminated clothing. Immediately wash area with flooding amounts of soap and water. Seek medical attention immediately. For inhalation: Remove the person from exposure. Provide respiratory assistance and CPR. Transfer to medical facility. Observe for 48 hours for lung effects. If swallowed: Seek medical attention immediately. Induce vomiting in conscious and alert person. Never give an unconscious or convulsing person anything by mouth.

SECTION VII - PRECAUTIONS FOR SAFE HANDLING AND USE

Steps to be Taken in Case Material is Released or Spilled

Remove all ignition sources, wear SCBA respiratory protection. Restrict those not wearing protective equipment who are not involved in cleanup from entering area. Absorb with vermiculite and place in sealed container. Use non-sparking tools. Flush residues with water (keep away from sewers).

Preferred Waste Disposal Method

Burn in chemical incinerator equipped with an afterburner and scrubber.

Precautions to be Taken in Handling and Storage

Store in tightly closed containers in cool, well ventilated location. Keep away from sources of heat.

Other Precautions and Warnings

Sources of ignition are prohibited where ethylene dichloride is used or stored. Metal containers should be bonded and grounded. Empty cylinders contain residue, handle with care.

SECTION VIII - CONTROL MEASURES AND PERSONAL PROTECTIVE EQUIPMENT

Respiratory Protection *(specify type)*

Use an MSHA/NIOSH-approved air purifying respirator (cartridge type) or use a supplied-air respirator or a self-contained breathing apparatus (SCBA) with full facepiece operated in positive pressure mode.

Ventilation

Local exhaust (preferred) at site of chemical work or general ventilation.

Protective Gloves	Eye Protection	Other Protective Clothing
Impervious Rubber Gloves	**Chemical Goggles or Face Mask**	**Protective Apron if Splash is Likely**

Work/Hygiene Practices

Always wash hands thoroughly after using chemical; never bring food, drink, or smoking materials into vicinity of chemicals.

ETHYLENE DICHLORIDE

$ClCH_2CH_2Cl$ CAS: 107-06-2

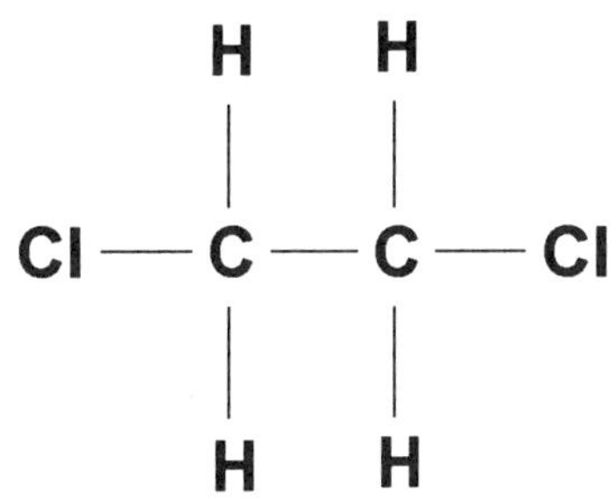

IDENTIFICATION AND TYPICAL USES

Ethylene dichloride is a colorless liquid with a sweet, pleasant, chloroform-like odor. It decomposes slowly, becomes acidic and darkens in color. It is used in the manufacture of acetyl cellulose and vinyl chloride, in paint removers, as a fumigant, as a degreaser, as a wetting and penetration agent, as an intermediate for pharmaceuticals, and as a solvent for oils, fats, waxes, gums, resins, and rubber.

RISK ASSESSMENT: HEALTH

General Assessment

Ethylene dichloride is moderately toxic by *inhalation*. The liquid is also known the pass through the skin (*absorption*) to cause toxic systemic effects, and it is considered a dangerous, life-threatening human poison by *ingestion*. This chemical is known to cause cancer in test animals (fore-stomach, mammary glands, and circulatory system) and this effect is suspected to occur in humans as well. Human mutation data have been reported. In addition, ethylene dichloride has been shown to adversely affect reproduction in test animals. The systemic effects from overexposure can be seen in the liver, kidneys, digestive tract, blood, lungs, adrenal glands, and in the central nervous system.

Inhalation is irritating to the eyes, nose, and mucosa of the respiratory tract and may produce ulceration of the mucous membrane. High concentrations of ethylene dichloride causes depression of the central nervous system (CNS). Symptoms include dizziness, intoxication, narcotic and anesthetic effects, lightheadedness, headache, somnolence, cyanosis, unconsciousness, coma, and death.

This chemical poses a significant hepatoxic (injurious to the liver) threat. Irritation to lung tissues can lead to a dangerous buildup of fluids in the lungs (pulmonary edema), which is a medical emergency and can be fatal. Symptoms may be delayed up to 48 hours creating a false sense of security with regard to health risk.

Skin contact can produce severe localized irritation, burns, de-fatting of the surface leading to possible cracking, drying, and secondary infections. Absorption will occur rapidly causing an increase in toxicity. If coupled with inhalation, the systemic effects can be quite severe. Eye contact will also cause irritation and, if not removed immediately, injury (corneal opacity).

Ingestion of ethylene dichloride can result in death. A fatal dose in humans may range between 30 and 50 milliliters.

Acute Health Effects

The following acute (short-term) health effects may occur immediately or shortly after exposure to ethylene dichloride:

Skin: Irritation with potential for redness, pain, burns, and tissue damage.

Eye: Irritation and smarting of the conjunctiva. Can cause clouding of the cornea (corneal opacity).

Lung: Irritation of the mucous membranes of the nose, throat, and respiratory tract causing possible ulceration, coughing, chest pains, difficulty breathing, and pulmonary edema.

CNS: Depression with headaches, lightheadedness, dizziness, nausea, vomiting, diarrhea, narcotic and anesthetic effects.

Chronic Health Effects

The following chronic (long-term) health effects may occur at some time after exposure to ethylene dichloride and can last for months or even years:

Cancer Hazards: According to the references, ethylene dichloride is a confirmed animal carcinogen and several agencies consider ethylene dichloride to be a potential or suspect human carcinogen. These include the IARC (International Association for Research on Cancer), the NTP (National Toxicology Program), and the NCI (National Cancer Institute). It has caused cancer of the lung, digestive system (fore-stomach), skin, mammary glands, and circulatory system in animals. Human mutation data have also been reported.

Reproductive Hazard: According to information presented in the references, ethylene dichloride is a possible teratogen in humans and has been shown to adversely affect reproduction in test animals. Animal testing has shown reproductive failure and fetal abortion. There may also be an increased risk to nursing infants of exposed mothers.

Other Chronic Effects: Prolonged exposure will cause injuries to the liver and kidneys, with weight loss, decreased blood pressure, jaundice, difficult urination, or anemia.

Recommended Risk-Reduction Measures

Personnel should avoid direct contact with ethylene dichloride. Occupational poisoning occurs most commonly by inhalation and skin absorption. If a less toxic material or compound cannot be substituted for ethylene dichloride, then *engineering controls* are the most effective method of reducing exposures. The best protection is to enclose operations and/or provide local exhaust ventilation at the site of ethylene dichloride release. While not always operationally feasible, isolating operations can also reduce exposure. Using respiratory protection is less effective than the controls mentioned above, but is still advisable whenever working with or around ethylene dichloride. Caution should be exercised when selecting respiratory protection since it is not known if exposure to ethylene dichloride will cause cancer in humans. For exposure over the PEL (50 ppm), an MSHA/NIOSH-approved air-purifying respirator with organic vapor cartridge(s) may suffice. However, for the best protection against a *possible carcinogen,* especially when the concentration is unknown, use a supplied air respirator with full facepiece and mask operated in positive pressure mode or a self-contained breathing apparatus (SCBA) with full facepiece operated in pressure demand. If a full facepiece is not available, then chemical goggles should be worn to protect the eyes. A face shield should also be considered. To prevent hand and skin exposures, impervious rubber gloves should be worn.

Administrative controls should also be in place to minimize the potential for human exposures. These may include written procedures or policies which specify the methods and techniques that will be practiced whenever personnel are to work with ethylene dichloride.

All personnel should receive *training* on the use, hazards, protective measures, emergency actions, and other precautions per 29 CFR 1910.1200 (Hazard Communication), prior to the first assignment in an area where ethylene dichloride is used or stored.

Before beginning employment with ethylene dichloride and at regular intervals thereafter (e.g., annually), the following medical tests are recommended:

☑ Liver and kidney function tests (establish baseline).

☑ Lung function tests (establish baseline).

If symptoms develop or overexposure is suspected, the following tests should be considered:

☑ Liver and kidney function tests.

☑ Consider chest X-ray after acute overexposure (may be negative if taken immediately after exposure due to the delayed effects of pulmonary edema).

☑ Lung function tests (compare to baseline).

Any evaluation should include a careful history of past and present symptoms with an examination. Medical tests that simply look for existing damage are not a substitute for controlling exposures.

Also, since smoking can cause heart disease, lung cancer, emphysema, and other respiratory problems, smokers may be affected more rapidly and with more pronounced symptoms than non-smokers. Prudent risk management requires careful consideration of *all* risk factors that might be causing the appearance of symptoms in the work force.

Other methods to reduce exposure include:

☑ Where possible, enclose operations and use local exhaust ventilation at the site of chemical release. If local exhaust ventilation or enclosure is not feasible, then respiratory protection should be provided and its use mandatory.

- ☑ Always ensure that proper protective clothing is worn when using chemical substances.
- ☑ Wash thoroughly immediately after exposure to ethylene dichloride and at the end of the work shift or before eating, drinking, or smoking.
- ☑ Hazard warning information should be posted in the work area. In addition, as part of an on-going education and training program, all information on the health and safety hazards of ethylene dichloride should be communicated to all potentially exposed workers.
- ☑ Eye wash stations should be provided in the immediate work area for emergency use. If there is a possibility of skin exposure to ethylene dichloride, emergency shower facilities should also be provided in the immediate work area.
- ☑ Workers whose clothing has been contaminated by ethylene dichloride should change into clean clothes before leaving work. Contaminated work clothing should be laundered only by individuals who have been informed of the hazards of exposure to ethylene dichloride.

RISK ASSESSMENT: ENVIRONMENT

General Assessment

Although ethylene dichloride normally exists as a gas, it is usually shipped as a liquefied compressed gas. The environment is at risk of exposure during transportation, storage, disposal, or destruction of ethylene dichloride. In almost every scenario, the threat of environmental exposure is contingent upon the proper handling of the chemical substance. Accidental spills, large or small, can result in contamination of the surrounding environmental mediums (water, soil, and air).

Ethylene dichloride is very dangerous flammable liquid. Because of its low flash point and low boiling point, it is classified as a Class 1B flammable liquid (per OSHA 29 CFR 1910.106). It is a serious fire hazard and moderate explosion hazard when exposed to heat or flame. It can also react violently in the presence of chemically-active metals such as sodium, potassium, calcium, and strong oxidizing agents. Caution is always required in handling, storage, transportation, and disposal of ethylene dichloride. Emergency responders should be made aware of the presence of ethylene dichloride at any emergency response situation.

Ethylene dichloride can enter the environment from industrial effluents and from spills.

Acute Ecological Effects

Acute (short-term) toxic effects may include the death of animals, birds, or fish and death or low growth rate in plants. Acute effects are seen 2 to 4 days after animals or plants are exposed to ethylene dichloride.

Insufficient data are available to evaluate the short-term effects of ethylene dichloride on aquatic life, plants, birds, or land animals.

Chronic Ecological Effects

Chronic (long-term) toxic effects may include shortened life span, reproductive problems, lower fertility, and changes in appearance or behavior in exposed animals. These effects can be seen long after first exposure(s) to toxic chemicals.

Insufficient data are available to evaluate or predict the long-term effects of ethylene dichloride to aquatic life, plants, birds, or terrestrial animals.

Water Solubility

Ethylene dichloride is slightly soluble in water. Concentrations of 1 to 100 milligrams will mix with a liter of water.

Persistence in the Environment

Ethylene dichloride is slightly persistent in water, with a half-life between 2 and 20 days. The half-life of a pollutant is the amount of time it takes for one half of the chemical to be degraded.

Bioaccumulation in Aquatic Organisms

Some substances increase in concentration, or *bioaccumulate,* in living organisms as they breathe contaminated air, drink contaminated water, or eat contaminated food. These chemicals can become concentrated in the tissues and internal organs of animals as well as humans.

The concentration of ethylene dichloride found in fish tissues is expected to be about the same as the average concentration of ethylene dichloride in the water from which the fish was taken.

Recommended Risk-Reduction Measures

Proper training of all transporters will reduce the likelihood of a mishap or accident resulting in a leak or spill of ethylene dichloride into the environment. Labels on all containers, trucks, and rail cars must meet DOT requirements and accurately reflect their contents to enable emergency responders to react properly and quickly to any disaster thereby reducing the potential risk to the environment and to personnel.

Storage of ethylene dichloride should be segregated from other chemicals and materials to minimize the risk of cross-contamination.

If a spill or leak to the environment has occurred, fire department, emergency response and/or hazardous materials spill personnel should be notified immediately. Cleanup should be attempted only by those trained in proper spill containment procedures using non-sparking tools. Contaminated soils should be removed for incineration and replaced with clean soil. If ethylene dichloride should contact the water table, aquifer, or navigable waterway, time is of the essence. It is only slightly soluble and total remediation may be possible. The local and/or state emergency response authorities must be notified. A comprehensive emergency response or disaster preparedness/recovery plan should be in place prior to any operations involving the use, transportation, storage, or disposal of ethylene dichloride.

For disposal, ethylene dichloride may be burned in a chemical incinerator equipped with an afterburner and scrubber.

If ethylene dichloride is spilled or leaked, the following specific steps are recommended:

- ☑ Restrict persons not wearing protective clothing from area of spill until cleanup is complete.
- ☑ Absorb liquids in vermiculite, sand, earth, or similar material and deposit in sealed drum.
- ☑ Remove ignition sources and ventilate area of spill or leak.
- ☑ It may be necessary to dispose of ethylene dichloride as a hazardous waste. The responsible state agency or the regional office of the federal Environmental Protection Agency (EPA) should be contacted for specific recommendations.

RISK ASSESSMENT: BUSINESS

General Assessment

Accidents or mishaps involving ethylene dichloride can present a serious threat to business operations. The loss or damage of equipment or facilities can significantly affect fiscal viability. Lawsuits that may result from personnel illness, injury/death, public exposures, and/or environmental contamination will require a serious expenditure of resources. Media attention surrounding an injury, death, or environmental damage can also result in a loss of profits and loss of current as well as future business. Always remember that anytime the terms "cancer," "carcinogen," or "reproductive hazard" are used, public emotion, hysteria, and ignorance can run equally high. This must be carefully considered when developing any public relation policies.

Recommended Risk-Reduction Measures

Company attorneys, safety and health professionals, and environmental specialists should be involved in the development of any procedures for responding to chemical incidents. A company official should be predesignated as a public relations officer with specific training in dealing with the press. Corporate plans and policies should be developed, approved, and implemented long before any need for such arises.

REFERENCES

American Conference of Governmental Industrial Hygienists. 1988. *Documentation of the Threshold Limit Values and Biological Exposure Indices,* 5th Edition (with updates). Cincinnati: ACGIH

Anthony, C. P., and N. J. Kolthoff. 1971. *Textbook of Anatomy and Physiology.* St. Louis: C. V. Mosby Company.

Bueche, J. R. 1972. *Principles of Physics,* 2nd Edition. New York: McGraw-Hill.

Calabrese, E. J. 1994. *Biological Effects of Low Level Exposures.* New York: CRC Press/Lewis Publishers.

Cockerman, L. G., B. S. Shane. 1994. *Basic Environmental Toxicology.* New York: CRC Press/Lewis Publishers.

Lewis, R. J., Sr. 1993. *Hawley's Condensed Chemical Dictionary,* 12th Edition. New York: Van Nostrand Reinhold.

Lewis, R. J., Sr. 1992. *Sax's Dangerous Properties of Industrial Materials,* Volumes 1, 2, and 3, 8th Edition. New York: Van Nostrand Reinhold.

Meyer, E. 1990. *Chemistry of Hazardous Materials.* Englewood Cliffs, NJ: Prentice Hall.

Patnaik, P. 1992. *A Comprehensive Guide to the Hazardous Properties of Chemical Substances.* New York: Van Nostrand Reinhold.

Quigley, D. R. 1994. *Handbook of Emergency Chemical Management.* New York: CRC Press/Lewis Publishers.

U.S. Department of Health and Human Services, National Institute for Occupational Safety and Health. 1994. *NIOSH Pocket Guide to Chemical Hazards.* Washington, D.C.: U.S. Government Printing Office.

MATERIAL SAFETY DATA SHEET

CHEMICAL NAME	CRC PRESS/LEWIS PUBLISHERS DISCLAIMER
ETHYLENE GLYCOL	**CRC PRESS/LEWIS PUBLISHERS DISCLAIMER:** THE INFORMATION AND RECOMMENDATIONS PRESENTED HEREIN ARE BASED ON SOURCES BELIEVED TO BE RELIABLE. CRC MAKES NO REPRESENTATION ON ITS COMPLETENESS OR ACCURACY. IT IS THE RESPONSIBILITY OF THE USER TO DETERMINE THE CHEMICAL'S SUITABILITY FOR ITS INTENDED USE, THE CHEMICAL'S SAFE USE, AND THE CHEMICAL'S PROPER DISPOSAL. NO REPRESENTATIONS AND/OR WARRANTIES, EITHER EXPRESSED OR IMPLIED, OF THE MERCHANTABILITY OR FITNESS FOR A PARTICULAR PURPOSE, OR OF ANY OTHER NATURE, ARE MADE WITH RESPECT TO THE INFORMATION PROVIDED IN THIS MSDS OR TO THE CHEMICAL TO WHICH INFORMATION MAY REFER. CRC NEITHER ASSUMES NOR AUTHORIZES ANY OTHER PERSON TO ASSUME FOR IT, ANY OTHER ADDITIONAL RESPONSIBILITY OR LIABILITY FOR THE USE OF, OR RELIANCE UPON, THIS INFORMATION.

HAZARD WARNING INFORMATION

HEALTH	FIRE	REACTIVE	OTHER	DEGREE OF HAZARD	COLOR CODING	OTHER CODES
1	1	0		0 = Minimum Hazard 1 = Slight Hazard 2 = Moderate Hazard 3 = Serious Hazard 4 = Severe Hazard	HEALTH = BLUE FIRE = RED REACTIVITY = YELLOW OTHER = WHITE	OX = Oxidizer ACID = Acid ALK = Alkali COR = Corrosive W = Use No Water

SECTION I - GENERAL INFORMATION

Characterization	**Glycol Ether**	RCRA Number	**None**
		EPA Class	**Not Applicable**
DOT Proper Shipping Name	**Combustible Liquid, N.O.S.**	Chemical Abstract Service (CAS) Number	**107-21-1**
DOT Hazard Class and Label Requirements	**Combustible Liquid**	DOT Emergency Guide Code	**No Citation**
DOT Identification Number	**NA 1993**	Chemical Formula	**$HOCH_2CH_2OH$**

Synonyms

1,2-Dihydroxyethane; 1,2-ethanediol; glycol' glycol alcohol; monoethylene glycol; ethylene alcohol.

SECTION II - HAZARDOUS INGREDIENTS AND IDENTITY INFORMATION

Hazardous Components (specific identity)	OSHA Exposure Criteria	NIOSH Exposure Criteria	Immediately Dangerous to Life and Health (IDLH)	ACGIH Exposure Criteria
Ethylene glycol (derivation: By air oxidation of ethylene followed by hydration of the ethylene oxide formed; acetoxylation; from carbon monoxide and hydrogen (synthesis gas) from coal gasification; or from the Oxirane process). **1 ppm = 2.58 mg/m³**	**PEL (ceiling):** **50 ppm** **125 mg/m³** **STEL:** **Not Established**	**REL:** **Not Established** **STEL:** **Not Established**	**Not Determined**	**TLV (ceiling):** **50 ppm** **127 mg/m³** **STEL:** **Not Established**

SECTION III - PHYSICAL AND CHEMICAL CHARACTERISTICS

Boiling Point	**387°F (197°C)**	Specific Gravity (H_2O = 1)	**1.11**
Vapor Pressure (gas)	**0.06 at 69°F (20°C)**	Molecular Weight	**62.1**
Vapor Density (Air = 1)	**2.14**	Freezing Point	**9°F (-13°C)**

Solubility

Soluble in water, alcohol, acetone, acetic acid, and pyridine. Slightly soluble in ether. Insoluble in benzene, chloroform, and petroleum ether.

Appearance and Odor

Clear, colorless, hygroscopic, syrupy, odorless liquid (a solid below 9°F) with a bittersweet taste.

SECTION IV - FIRE AND EXPLOSION HAZARD DATA

Flash Point (method used)	**292°F (111°C) closed cup**	Explosive Limits in Air % by Volume	**LEL: 3.2% UEL: 15.3%**
NFPA Classification	**Class IIIB Combustible Liquid**	Autoignition Temperature	**748°F (398°C)**

Extinguishing Media

Use dry chemical, carbon dioxide, water, or alcohol-resistant foam (water or foam may cause frothing).

Special Fire Fighting Procedures

Wear full protective clothing and self-contained breathing apparatus (SCBA). Poisonous gases are produced in fire. Use water spray to keep fire-exposed containers cool. Containers may explode in fire.

Unusual Fire and Explosion Hazards

Dangerous fire hazard when exposed to heat. Chemical will react on contact with oxidizable materials. Moderate explosion hazard when in the form of vapor. Vapors are heavier than air.

SECTION V - REACTIVITY DATA

Stability		Conditions to Avoid **Ethylene glycol is stable under normal conditions. Keep away from sources of heat or flame. Keep away from incompatible materials.**
Stable **X**	Unstable	Incompatibility *(materials to avoid)* **Strong oxidizers, such as potassium permanganate, chromium trioxide, sodium peroxide, nitrates, chlorates, and perchlorates. Hygroscopic (i.e., absorbs moisture from air).**
Hazardous Polymerization		Conditions to Avoid **Under normal conditions of temperature and pressure, hazardous polymerization of ethylene glycol is not expected to occur.**
May Occur	Will Not Occur **X**	Hazardous Decomposition or By-products **When heated to decomposition, ethylene glycol can emit highly toxic/poisonous and acrid fumes and gases, including carbon dioxide.**

SECTION VI - HEALTH HAZARD DATA

Primary Route(s) of Entry:	Inhalation?	Absorption (skin)?	Ingestion?
	X	X	X

Health Hazards

INHALATION: **Symptoms include lachrymation, irritation of the throat and upper respiratory tract, general anesthesia, headache, burning cough, respiratory stimulation, nausea or vomiting, liver, kidney, or pulmonary changes, and coma (especially if vapors are heated).**

ABSORPTION: **Will pass through intact skin, but toxic systemic effects are rarely observed.**

INGESTION: **Same symptoms as inhalation, only more severe. May also cause CNS depression with excitement, nausea, vomiting, drowsiness, coma, respiratory or kidney failure, death.**

Carcinogenicity	NTP Listed?	IARC Cancer Review Group?	OSHA Regulated?	Target Organs?
No Evidence Human **No Evidence Animal**	**No**	**No**	**29 CFR 1910.1000 Table Z-1-A**	**Resp. system, eyes, skin, liver, kidneys, CNS, CVS.**

Medical Conditions Generally Aggravated by Exposure

Any liver, kidney, cardiovascular, or nervous system disorders may be affected.

Emergency and First-aid Procedures

Eye contact: Flush immediately with water for 15 minutes (minimum), seek medical attention. Skin contact: Remove all contaminated clothing. Immediately wash area with large amounts of soap and water. Seek medical assistance. For inhalation: Remove the person from exposure. Provide respiratory assistance and CPR, if required. Transfer to medical facility. If swallowed: Seek medical attention immediately. Unless told otherwise, give a conscious victim 1-2 glasses of water and induce vomiting.

SECTION VII - PRECAUTIONS FOR SAFE HANDLING AND USE

Steps to be Taken in Case Material is Released or Spilled

Absorb large spills in vermiculite or similar material and deposit in sealed containers. Ventilate area of spill or leak. Restrict those not involved in cleanup from entering area. Remove all sources of ignition.

Preferred Waste Disposal Method

Burn in chemical incinerator equipped with an afterburner and scrubber.

Precautions to be Taken in Handling and Storage

Store to avoid contact with incompatible materials in tightly closed containers in cool, well-ventilated area away from heat and flame. Use only non-sparking tools to open and close containers.

Other Precautions and Warnings

Sources of ignition are prohibited where ethylene glycol is used, handled, or stored. Heat may cause containers to build up pressure and explode. Containers must be grounded and bonded during transfer.

SECTION VIII - CONTROL MEASURES AND PERSONAL PROTECTIVE EQUIPMENT

Respiratory Protection *(specify type)*

For low exposures, uses a chemical cartridge respirator. Otherwise, use an MSHA/NIOSH-approved supplied air respirator with full facepiece, hood, or helmet in continuous flow mode, or use a self-contained breathing apparatus (SCBA) operated in pressure demand or other positive pressure mode.

Ventilation

Local exhaust or general mechanical systems recommended.

Protective Gloves	Eye Protection	Other Protective Clothing
Nitrile Rubber or Neoprene	**Chemical Goggles or Face Mask**	**Rubber Apron**

Work/Hygiene Practices

Always wash hands thoroughly after using chemical; never bring food, drink, or smoking materials into vicinity of chemicals.

ETHYLENE GLYCOL

$HOCH_2CH_2OH$ CAS: 107-21-1

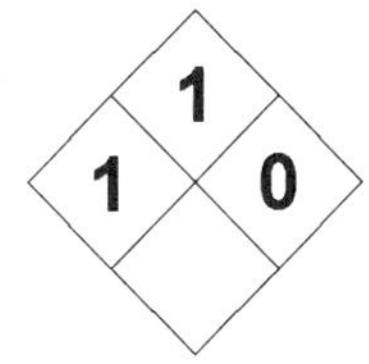

IDENTIFICATION AND TYPICAL USES

Ethylene glycol is a clear, colorless, hygroscopic, syrupy, odorless liquid with a bittersweet taste. It is used primarily as an antifreeze in heating and cooling systems (automobile radiators and as a coolant for aircraft engines). It is also used in the manufacture of hydraulic brake fluids, as a solvent for paints, plastics, and inks, as a softening agent for cellophane, and in the manufacture of plasticizers, elastomers, alkyd resins, and in synthetic fibers and waxes.

RISK ASSESSMENT: HEALTH

General Assessment

Ethylene glycol is a dangerous human poison by *ingestion* and is moderately toxic by *inhalation*. It will also pass through unbroken skin (*absorption*); however reports of toxic systemic effects by this route of exposure are lacking in the references. There are no reports in the references to indicate that exposure causes cancer in humans or animals. However, human mutation data (genetic changes) have been reported and many scientists believe that such chemicals may pose a carcinogenic threat in the long-term. Data are conflicting and inconsistent in this regard. It has been shown to cause teratogenic effects (damage to the developing fetus) in animal studies.

Inhalation is rare because of this chemical's low vapor pressure. However, when it is heated or misted, it presents a significant health hazard risk. There will be some initial irritation of the eyes, nose, and throat followed by lachrymation, feeling of intoxication or "drunkenness" (inebriation), burning cough, headache, nystagmus (rapid eye movements), and coma.

Skin and eye contact may produce mild to moderate irritation on contact and may cause an allergic reaction and rash. It will pass through unbroken skin, but its ability to produce toxic systemic effects by this route has not been confirmed.

On ingestion, ethylene glycol poses a serious poison threat. Initial symptoms are stimulation or excitement, followed by depression of the central nervous system (CNS). Symptoms include nystagmus, partial or complete paralysis of the eye muscles, swelling and protrusion of the eyes from their sockets, appearance of a "blind spot" in visual range, depressed reflexes, and convulsions. Symptoms can progress to include cardiovascular and pulmonary system effects, mild hypertension, tachycardia (rapid heartbeat), tachypnea (rapid respiration), and systemic capillary damage. Finally, death may occur due to kidney complications leading to necrosis and renal failure (24-72 hours post-ingestion). Other effects can include acidosis and meningoencephalitis (inflammation of the brain and meninges).

Acute Health Effects

The following acute (short-term) health effects may occur immediately or shortly after exposure to ethylene glycol:

Skin: Some irritation at site of contact. Allergic rash may develop. Absorption is possible.

Eye: Mild irritation (liquid only).

Lung: Nose and throat irritation. Produces intoxicating effects and may cause lung effects.

CNS: Stimulation and depression with symptoms of dizziness, headache, lightheadedness, nausea, vomiting, and coma.

Chronic Health Effects

The following chronic (long-term) health effects may occur at some time after exposure to ethylene glycol and can last for months or even years:

Cancer Hazards: According to information presented in the references, ethylene glycol has not been adequately tested for its ability to cause cancer in animals. Some studies indicate that it causes mutations in humans, while others are inconclusive in this regard.

Reproductive Hazard: According to information presented in the references, ethylene glycol may be a teratogen in humans since it has been shown to be a teratogen in animals. It may cause damage to the developing fetus.

Other Chronic Effects: Repeated or high exposures may cause kidney damage and/or the development of kidney stones. Brain damage may also occur. Skin allergy can develop in the long-term. If this happens, even small future exposures can cause recurrence of symptoms (rash and itching).

Recommended Risk-Reduction Measures

Personnel should avoid direct contact with ethylene glycol. This chemical is a dangerous human poison by two exposure routes (inhalation and ingestion). If a less toxic chemical cannot be substituted for a hazardous substance, then *engineering controls* are the most effective method of reducing exposures. The best protection is to enclose operations and/or provide local exhaust ventilation at the site of chemical release. While not always operationally feasible, isolating operations can also reduce exposure. Using respiratory protection is less effective than the controls mentioned above, but is still advisable whenever working with or around ethylene glycol. Chemical cartridge respirators may provide adequate protection up to the PEL. However, the best protection is provided by a supplied-air respirator with full facepiece operated in continuous flow mode, or a self-contained breathing apparatus (SCBA) with full facepiece and pressure demand. If a full facepiece is not available, then chemical goggles should be worn to protect the eyes. Whenever a chemical splash hazard exists, a face shield and a rubber apron should be worn. To prevent hand and skin exposures, impervious rubber gloves should be worn.

Administrative controls should also be in place to minimize the potential for human exposures. These may include written procedures or policies which specify the methods and techniques that will be practiced whenever personnel are to work with ethylene glycol.

All personnel should receive *training* on the use, hazards, protective measures, emergency actions, and other precautions per 29 CFR 1910.1200 (Hazard Communication), prior to the first assignment in an area where ethylene glycol is used or stored.

If symptoms develop or overexposure is suspected, the following tests are recommended:

- ☑ Kidney function tests.
- ☑ Urine oxalate level. Urinalysis may show low specific gravity, proteinuria, pyuria, cylinduria, hematuria, calcium oxide, and hippuric crystals.

Any evaluation should include a careful history of past and present symptoms with an examination. Medical tests that simply look for existing damage are not a substitute for controlling exposures. Sodium bicarbonate may be used intravenously to treat acidosis. Ethanol may be used in antedotal treatment. However, blood glucose levels must be monitored for hypoglycemia. A diuretic, such as mannitol, can be infused to help in the prevention of brain edema and hemodialysis (to remove ethylene glycol from circulation).

Other methods to reduce exposure include:

- ☑ Where possible, enclose operations and use local exhaust ventilation at the site of chemical release. If local exhaust ventilation or enclosure is not used, respiratory protection should be mandatory.
- ☑ Where possible, automatically pump liquid ethylene glycol from drums to process containers.
- ☑ Always ensure that proper protective clothing is worn when using chemical substances.
- ☑ Wash thoroughly immediately after exposure to ethylene glycol and at the end of the work shift or before eating, drinking, or smoking.
- ☑ Hazard warning information should be posted in the work area. In addition, as part of an on-going education and training program, all information on the health and safety hazards of ethylene glycol should be communicated to all potentially exposed workers.
- ☑ Eye wash stations should be provided in the immediate work area for emergency use. If there is a possibility of skin exposure to ethylene glycol, emergency shower facilities should also be provided.
- ☑ Personnel should never wear contaminated clothing home (family members can be exposed). Clothing should be laundered only by personnel who have been properly trained on the hazards of exposure to ethylene glycol.

RISK ASSESSMENT: ENVIRONMENT

General Assessment

The environment is at risk of exposure during transportation, storage, disposal, or destruction of ethylene glycol. In almost every scenario, the threat of environmental exposure is contingent upon the proper handling of the chemical substance. Accidental spills, large or small, can result in fire, explosion, and possible contamination of the surrounding environmental mediums (water, soil, and air).

Ethylene glycol is considered a Class IIIB combustible liquid (per OSHA 29 CFR 1910.106). Under normal operating conditions, it presents a mild fire hazard. However, when heated or misted, it becomes a serious fire and explosion hazard concern. These characteristics require extreme caution in handling, storage, transportation, and disposal. It is incompatible with many common peroxides and other oxidizers and contact can cause fire or explosion. It will ignite on contact with chromium trioxide, potassium permanganate, and sodium peroxide. Therefore, special consideration is required during any emergency situation involving a leak or spill of ethylene glycol. Should ethylene glycol ever come into contact with these incompatible substances during use, transportation, storage, or disposal, the formation of highly toxic and/or highly explosive commodities is extremely possible.

The proper disposal method for ethylene glycol is to burn it in a chemical incinerator equipped with an afterburner and scrubber.

Ethylene glycol may enter the environment through industrial discharges, municipal waste treatment plant discharges, agricultural runoff or disposal, or spills.

Acute Ecological Effects

Acute (short-term) toxic effects may include the death of animals, birds, or fish and death or low growth rate in plants. Acute effects are seen 2 to 4 days after animals or plants are exposed to ethylene glycol.

Ethylene glycol has moderate acute toxicity to aquatic life. It has caused chromosomal damage to agricultural crops. Insufficient data are available on the short-term effects of ethylene glycol exposure to birds or land animals.

Chronic Ecological Effects

Chronic (long-term) toxic effects may include shortened life span, reproductive problems, lower fertility, and changes in appearance or behavior in exposed animals. These effects can be seen long after first exposure(s) to toxic chemicals.

Ethylene glycol has moderate chronic toxicity to aquatic life. Insufficient data are available on the long-term effects of ethylene glycol exposure on birds, plants, or land animals.

Water Solubility

Ethylene glycol is highly soluble in water. Concentrations of 1000 milligrams and more will mix with a liter of water.

Persistence in the Environment

Ethylene glycol is slightly persistent in water, with a half-life between 2 and 20 days. The half-life of a pollutant is the amount of time it takes for one half of the chemical to be degraded. Virtually 100% of ethylene glycol will eventually end up in water.

Bioaccumulation in Aquatic Organisms

Some substances increase in concentration, or *bioaccumulate,* in living organisms as they breathe contaminated air, drink contaminated water, or eat contaminated food. These chemicals can become concentrated in the tissues and internal organs of animals as well as humans.

The concentration of ethylene glycol found in fish tissue is expected to be about the same as the average concentration of ethylene glycol in the water from which the fish was taken.

Recommended Risk-Reduction Measures

Proper training of all transporters will reduce the likelihood of a mishap or accident resulting in a leak or spill to the environment. The correct use of DOT labeling (per the MSDS) on all containers, trucks, and rail cars will enable emergency responders to react properly and quickly to any disaster thereby reducing the potential risk to the environment and to personnel.

Storage of ethylene glycol should be segregated from incompatible chemicals to minimize the risk of cross-contamination or contact. Buildings designated for storage should be equipped with appropriate fire protection systems (alarms, sprinklers, emergency lighting, portable extinguishers).

If a spill or leak to the environment has occurred, fire department, emergency response and/or hazardous materials spill personnel should be notified immediately. Cleanup should be attempted only by those

trained in proper spill containment procedures using non-sparking tools. Contaminated soils should be removed for incineration and replaced with clean soil. If ethylene glycol should contact the water table, aquifer, or navigable waterway, cleanup should be started immediately. It is highly soluble in water and total remediation may not be possible. The local and/or state emergency response authorities must be notified. A comprehensive emergency response or disaster preparedness/recovery plan should be in place prior to any operations involving the use, transportation, storage, or disposal of ethylene glycol. If ethylene glycol is spilled or leaked, the following specific steps are recommended:

- ☑ Restrict persons not wearing protective clothing from area of spill or leak until cleanup is complete and area can be opened for normal work.
- ☑ Remove all ignition sources, ventilate area of leak.
- ☑ Absorb liquids in vermiculite, dry sand, earth, or similar material and deposit in sealed containers.
- ☑ It may be necessary to dispose of ethylene glycol as a hazardous waste. The responsible state agency or the regional office of the federal Environmental Protection Agency (EPA) should be contacted for specific recommendations.

RISK ASSESSMENT: BUSINESS

General Assessment

Accidents or mishaps involving ethylene glycol can present a significant threat to business operations. The loss or damage of equipment or facilities can significantly affect fiscal viability. Lawsuits that may result from personnel injury/death, public exposures, and/or environmental contamination will also require a serious expenditure of resources. Media attention surrounding an injury, death, or environmental damage can also result in a loss of profits and loss of current as well as future business.

Recommended Risk-Reduction Measures

Company attorneys, safety and health professionals, and environmental specialists should be involved in the development of any procedures dealing with safety or emergency response. A company official should be pre-designated as a public relations officer with specific training in dealing with the press. Corporate plans and policies should be developed, approved, and implemented long before any need for such arises.

REFERENCES

American Conference of Governmental Industrial Hygienists. 1988. *Documentation of the Threshold Limit Values and Biological Exposure Indices,* 5th Edition (with updates). Cincinnati: ACGIH

Bueche, J. R. 1972. *Principles of Physics,* 2nd Edition. New York: McGraw-Hill.

Calabrese, E. J. 1994. *Biological Effects of Low Level Exposures.* New York: CRC Press/Lewis Publishers.

Cockerman, L. G., B. S. Shane. 1994. *Basic Environmental Toxicology.* New York: CRC Press/Lewis Publishers.

Gardner, R. W. 1994. *Chemical Intolerance.* New York: CRC Press/Lewis Publishers.

Hallenbeck, W. H. 1993. *Quantitative Risk Assessment for Environmental and Occupational Health,* 2nd Edition. New York: CRC Press/Lewis Publishers.

Keith, L. H. and M. M. Walker. 1995. *Handbook of Air Toxics.* New York: CRC Press/Lewis Publishers.

Landis, W. G., M. H. Yu. 1995. *Introduction to Environmental Toxicology.* New York: CRC Press/Lewis Publishers.

Langerman, N. 1994 *Precautionary Labels for Chemical Containers.* New York: CRC Press/Lewis Publishers.

Lauwerys, R. R., P. Hoet. 1993. *Industrial Chemical Exposure.* New York: CRC Press/Lewis Publishers.

Lewis, R. J., Sr. 1992. *Sax's Dangerous Properties of Industrial Materials,* Volumes 1, 2, and 3, 8th Edition. New York: Van Nostrand Reinhold.

Manahan, S. E. 1994. *Environmental Chemistry,* 6th Edition. New York: CRC Press/Lewis Publishers.

Meyer, E. 1990. *Chemistry of Hazardous Materials.* Englewood Cliffs, NJ: Prentice Hall.

Patnaik, P. 1992. *A Comprehensive Guide to the Hazardous Properties of Chemical Substances.* New York: Van Nostrand Reinhold.

Quigley, D. R. 1994. *Handbook of Emergency Chemical Management.* New York: CRC Press/Lewis Publishers.

Rea, W. J. 1995. *Chemical Sensitivity,* Vol. I-III. New York: CRC Press/Lewis Publishers.

Richardson, M. 1995. *Dictionary of Substance and Their Effects,* Vol. I-VI. New York: CRC Press/Lewis Publishers.

Sacarello, H. L. A. 1994. *The Comprehensive Handbook of Hazardous Materials.* New York: CRC Press/Lewis Publishers.

Torkelson, T. R. V. K. Rowe. 1981. Halogenated aliphatic hydrocarbons. *Patty's Industrial Hygiene and Toxicology,* Vol. 2B, *Toxicology,* pp 3568-3572. New York: Wiley-Interscience.

U.S. Department of Health and Human Services, National Institute for Occupational Safety and Health. 1994. *NIOSH Pocket Guide to Chemical Hazards.* Washington, D.C.: U.S. Government Printing Office.

MATERIAL SAFETY DATA SHEET

CHEMICAL NAME

ETHYLENE GLYCOL DINITRATE

CRC PRESS/LEWIS PUBLISHERS DISCLAIMER: THE INFORMATION AND RECOMMENDATIONS PRESENTED HEREIN ARE BASED ON SOURCES BELIEVED TO BE RELIABLE. CRC MAKES NO REPRESENTATION ON ITS COMPLETENESS OR ACCURACY. IT IS THE RESPONSIBILITY OF THE USER TO DETERMINE THE CHEMICAL'S SUITABILITY FOR ITS INTENDED USE, THE CHEMICAL'S SAFE USE, AND THE CHEMICAL'S PROPER DISPOSAL. NO REPRESENTATIONS AND/OR WARRANTIES, EITHER EXPRESSED OR IMPLIED, OF THE MERCHANTABILITY OR FITNESS FOR A PARTICULAR PURPOSE, OR OF ANY OTHER NATURE, ARE MADE WITH RESPECT TO THE INFORMATION PROVIDED IN THIS MSDS OR TO THE CHEMICAL TO WHICH INFORMATION MAY REFER. CRC NEITHER ASSUMES NOR AUTHORIZES ANY OTHER PERSON TO ASSUME FOR IT, ANY OTHER ADDITIONAL RESPONSIBILITY OR LIABILITY FOR THE USE OF, OR RELIANCE UPON, THIS INFORMATION.

HAZARD WARNING INFORMATION

HEALTH	FIRE	REACTIVE	OTHER	DEGREE OF HAZARD	COLOR CODING	OTHER CODES
3	3	2		0 = Minimum Hazard 1 = Slight Hazard 2 = Moderate Hazard 3 = Serious Hazard 4 = Severe Hazard	HEALTH = BLUE FIRE = RED REACTIVITY = YELLOW OTHER = WHITE	OX = Oxidizer ACID = Acid ALK = Alkali COR = Corrosive W = Use No Water

SECTION I - GENERAL INFORMATION

Characterization: **Nitrate**	RCRA Number: **None** — EPA Class: **Not Applicable**
DOT Proper Shipping Name: **No Citation**	Chemical Abstract Service (CAS) Number: **628-96-6**
DOT Hazard Class and Label Requirements: **No Citation**	DOT Emergency Guide Code: **No Citation**
DOT Identification Number: **No Citation**	Chemical Formula: **$O_2NOCH_2CH_2ONO_2$**

Synonyms

EGDN; 1,2-ethanediol dinitrate; ethanediol dinitrate; ethylene dinitrate; ethylene nitrate; glycol dinitrate, nitroglycol; dinitroglycol.

SECTION II - HAZARDOUS INGREDIENTS AND IDENTITY INFORMATION

Hazardous Components (specific identity)	OSHA Exposure Criteria	NIOSH Exposure Criteria	Immediately Dangerous to Life and Health (IDLH)	ACGIH Exposure Criteria
Ethylene glycol dinitrate (derivation: An explosive mixed with nitroglycerin to lower its melting point). **1 ppm = 6.12 mg/m^3**	**PEL (ceiling):** **0.2 ppm** **1 mg/m^3** **(skin)** **STEL:** **0.1 mg/m^3** **(skin)**	**REL (ceiling):** **0.1 mg/m^3** **(20 minutes)** **(skin)** **STEL:** **Not Established**	**75 mg/m^3**	**TLV:** **0.05 ppm** **0.3 mg/m^3** **(skin)** **STEL:** **Not Established**

SECTION III - PHYSICAL AND CHEMICAL CHARACTERISTICS

Boiling Point: **Explodes at 237°F (114°C)**	Specific Gravity (H_2O = 1): **1.49**
Vapor Pressure (gas): **0.05 at 69°F (20°C)**	Molecular Weight: **152.1**
Vapor Density (Air = 1): **5.25**	Freezing Point: **-8°F (-22°C)**

Solubility

Insoluble in water.

Appearance and Odor

Colorless to yellow, oily, odorless liquid.

SECTION IV - FIRE AND EXPLOSION HAZARD DATA

Flash Point (method used): **419°F (215°C) closed cup**	Explosive Limits in Air % by Volume: **LEL: Not Determined** **UEL: Not Determined**
NFPA Classification: **Explosive Liquid**	Autoignition Temperature: **Not Determined**

Extinguishing Media

Water spray, alcohol foam, dry chemical, or carbon dioxide (fight fire from remote location, if possible).

Special Fire Fighting Procedures

Wear full protective clothing and self-contained breathing apparatus (SCBA). Poisonous gases are produced in fire. Use water spray to keep fire-exposed containers cool. Containers may explode in fire.

Unusual Fire and Explosion Hazards

A component of dynamite. Very explosive. Fight fires from an explosion-proof location, if possible. Vapors are heavier than air and can travel for great distance to ignition source and explode.

SECTION V - REACTIVITY DATA

Stability		Conditions to Avoid Ethylene glycol dinitrate is stable under normal conditions. Keep away from sources of heat or flame. Keep away from incompatible materials.
Stable	Unstable	Incompatibility *(materials to avoid)*
X		Reacts in contact with numerous materials, including acids (sulfuric, nitric, and hydrochloric), alkali materials, and even by itself in the presence of an ignition source.
Hazardous Polymerization		Conditions to Avoid Under normal conditions of temperature and pressure, hazardous polymerization of ethylene glycol dinitrate cannot occur.
May Occur	Will Not Occur	Hazardous Decomposition or By-products
	X	When heated to decomposition, ethylene glycol dinitrate can emit highly toxic/poisonous and acrid fumes and gases, including oxides of nitrogen and carbon.

SECTION VI - HEALTH HAZARD DATA

Primary Route(s) of Entry:	Inhalation?	Absorption (skin)?	Ingestion?
	X	X	X

Health Hazards

INHALATION: Vasodilation (expansion of the blood vessel walls), decreased blood pressure, intense and throbbing headache (cerebral dilation), flushed skin, dizziness, nausea, vomiting, abdominal pain, heart palpitation, angina, methemoglobinemia, delirium, depression of the CNS, tachycardia, diastolic hypertension, reduced pulse, liver and kidney damage.

ABSORPTION: Will pass through intact skin and cause toxic systemic effects (see inhalation).

INGESTION: Systemic effects with many of the same symptoms as inhalation.

Carcinogenicity	NTP Listed?	IARC Cancer Review Group?	OSHA Regulated?	Target Organs?
No Evidence Human No Evidence Animal	No	No	29 CFR 1910.1000 Table Z-1-A	Skin, CNS, liver, kidneys, blood, CVS.

Medical Conditions Generally Aggravated by Exposure

None reported. However, heart conditions or kidney/liver disorders may be affected.

Emergency and First-aid Procedures

Eye contact: Flush immediately with water for 15 minutes (minimum), seek medical attention. Skin contact: Remove all contaminated clothing. Immediately wash area with large amounts of soap and water. Seek medical assistance. For inhalation: Remove the person from exposure. Provide respiratory assistance and CPR, if required. Transfer to medical facility. If swallowed: Seek medical attention immediately. Induce vomiting in conscious persons (within 30 minutes of exposure) using an emetic.

SECTION VII - PRECAUTIONS FOR SAFE HANDLING AND USE

Steps to be Taken in Case Material is Released or Spilled

Remove all sources of ignition. Ventilate area of spill or leak. Neutralize using sodium bisulfate. Collect mixture with non-sparking tools. Restrict those not involved in cleanup from entering area.

Preferred Waste Disposal Method

No citation (check with chemical manufacturer before attempting to dispose of explosive wastes).

Precautions to be Taken in Handling and Storage

Store to avoid contact with incompatible materials in tightly closed containers in cool, well-ventilated area away from heat (including radiant heat). Use only non-sparking tools to open and close containers.

Other Precautions and Warnings

Sources of ignition are prohibited where ethylene glycol dinitrate is used, handled, or stored. Heat may cause containers to build up pressure and explode. Containers must be electrically grounded and bonded during transfer procedures. All electrical equipment should be explosion proof.

SECTION VIII - CONTROL MEASURES AND PERSONAL PROTECTIVE EQUIPMENT

Respiratory Protection *(specify type)*

For low exposures, uses a chemical cartridge respirator. Otherwise, use an MSHA/NIOSH-approved supplied air respirator with full facepiece, hood, or helmet in continuous flow mode, or use a self-contained breathing apparatus (SCBA) operated in pressure demand or other positive pressure mode.

Ventilation

Local exhaust or general mechanical systems recommended.

Protective Gloves	Eye Protection	Other Protective Clothing
Butyl Rubber	Chemical Goggles or Face Mask	Butyl Rubber Apron

Work/Hygiene Practices

Always wash hands thoroughly after using chemical; never bring food, drink, or smoking materials into vicinity of chemicals.

ETHYLENE GLYCOL DINITRATE

$O_2NOCH_2CH_2ONO_2$ CAS: 628-96-6

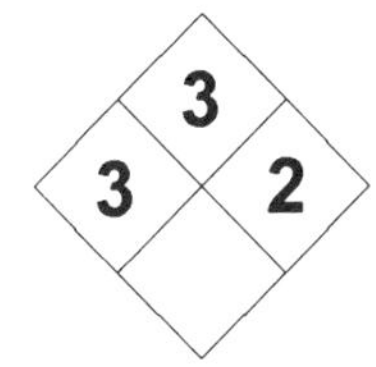

IDENTIFICATION AND TYPICAL USES

Ethylene glycol dinitrate is a colorless to yellow, oily, odorless liquid. It is used as an explosive ingredient (60-80%) in dynamite along with nitroglycerin (40-20%). It is mixed with nitroglycerin to lower its melting point and to reduce the hazards associated with frozen dynamite.

RISK ASSESSMENT: HEALTH

General Assessment

Ethylene glycol dinitrate is very poisonous and highly toxic by *inhalation, ingestion,* and skin *absorption* (i.e., it will also pass through unbroken skin). There are no reports in the references to indicate that exposure causes cancer in humans or animals. There is no evidence of teratogenicity or mutagenicity caused by exposure to this chemical.

Effects of exposure by all routes include vasodilatation (expansion of the blood vessel walls), decrease in blood pressure (hypotension), intense and throbbing headaches (caused by cerebral vasodilatation); flushed skin (especially the face), dizziness, nausea, vomiting, abdominal pain, heart palpitations and persistent tachycardia (rapid heartbeat), methemoglobinemia, angina, delirium, depression of the central nervous system, diastolic hypertension, and a reduction in pulse pressure. Death due to cardiac arrest is also possible. In animals, exposure also resulted in anemia and liver and kidney damage.

Skin and eye contact may produce mild to moderate irritation on contact. As stated above, it will pass through unbroken skin to cause toxic systemic effects.

Acute Health Effects

The following acute (short-term) health effects may occur immediately or shortly after exposure to ethylene glycol dinitrate:

Skin: Some irritation at site of contact. Absorption very likely to occur.

Eye: Mild irritation.

Lung: Nose and throat irritation. Absorption through lungs is rapid and toxic systemic effects can be severe.

CNS: Depression with symptoms of dizziness, headache, nausea, and vomiting.

CVS: Numerous adverse actions on the cardiovascular system with effects on blood pressure, blood vessels, and heart function (beat, pulse, and pressure).

Chronic Health Effects

The following chronic (long-term) health effects may occur at some time after exposure to ethylene glycol dinitrate and can last for months or even years:

Cancer Hazards: According to information presented in the references, ethylene glycol dinitrate has not been adequately tested for its ability to cause cancer in test animals.

Reproductive Hazard: According to information presented in the references, ethylene glycol dinitrate has not been adequately tested for its ability to adversely affect reproduction in test animals.

Other Chronic Effects: Repeated or high exposures may cause kidney or liver damage.

Recommended Risk-Reduction Measures

Personnel should avoid direct contact with ethylene glycol dinitrate. This chemical is a dangerous human poison by all exposure routes (inhalation, absorption, and ingestion). If a less toxic chemical cannot be substituted for a hazardous substance, then *engineering controls* are the most effective method of reducing exposures. The best protection is to enclose operations and/or provide explosion-proof, local exhaust ventilation at the site of chemical release. While not always operationally feasible, isolating operations can also reduce exposure. Using respiratory protection is less effective than the controls mentioned above, but is still advisable whenever working with or around ethylene glycol dinitrate. Chemical cartridge respirators may provide adequate protection up to the permissible exposure ceiling. However, this level is extremely low (0.2 ppm) and the best protection may be provided by a supplied-air respirator with full facepiece operated in continuous flow mode, or a self-contained breathing apparatus (SCBA) with full facepiece and pressure demand. If a full facepiece is not available, then chemical goggles should be worn to protect the eyes. Whenever a chemical splash hazard exists, a face shield and a rubber apron should be worn. To prevent hand and skin exposures, butyl rubber gloves should be worn.

Administrative controls should also be in place to minimize the potential for human exposures. These may include written procedures or policies which specify the methods and techniques that will be practiced whenever personnel are to work with ethylene glycol dinitrate.

All personnel should receive *training* on the use, hazards, protective measures, emergency actions, and other precautions per 29 CFR 1910.1200 (Hazard Communication), prior to the first assignment in an area where ethylene glycol dinitrate is used or stored.

If symptoms develop or overexposure is suspected, the following tests are recommended:

☑ Kidney function tests.

☑ Evaluation by a qualified neurologist for central nervous system effects.

☑ Electrocardiogram (EKG).

Any evaluation should include a careful history of past and present symptoms with an examination. Medical tests that simply look for existing damage are not a substitute for controlling exposures.

Other methods to reduce exposure include:

☑ Where possible, enclose operations and use local exhaust ventilation at the site of chemical release. If local exhaust ventilation or enclosure is not used, respiratory protection should be mandatory.

☑ Where possible, automatically pump liquid ethylene glycol dinitrate from storage drums or containers to process containers.

☑ Always ensure that proper protective clothing is worn when using chemical substances.

☑ Wash thoroughly immediately after exposure to ethylene glycol dinitrate and at the end of the work shift or before eating, drinking, or smoking.

☑ Hazard warning information should be posted in the work area. In addition, as part of an on-going education and training program, all information on the health and safety hazards of ethylene glycol dinitrate should be communicated to all potentially exposed workers.

☑ Eye wash stations should be provided in the immediate work area for emergency use. If there is a possibility of skin exposure to ethylene glycol dinitrate, emergency shower facilities should also be provided.

☑ Personnel should never wear contaminated clothing home (family members can be exposed). Clothing should be laundered only by personnel who have been properly trained on the hazards of exposure to ethylene glycol dinitrate.

RISK ASSESSMENT: ENVIRONMENT

General Assessment

The environment is at risk of exposure during transportation, storage, disposal, or destruction of ethylene glycol dinitrate. In almost every scenario, the threat of environmental exposure is contingent upon the proper handling of the chemical substance. Accidental spills, large or small, can result in fire, explosion, and possible contamination of the surrounding environmental mediums (water, soil, and air).

Ethylene glycol dinitrate is considered an explosive liquid. Under normal operating conditions, it presents a dangerous explosion hazard. These characteristics require extreme caution in handling, storage, transportation, and disposal. It is incompatible with many substances, especially acids and alkalis, and contact can cause fire or explosion. It will even react explosively by itself if an ignition source is introduced. Therefore, special consideration is required during any emergency situation involving a leak or spill of ethylene glycol dinitrate. Should ethylene glycol dinitrate ever come into contact with these incompatible substances during use, transportation, storage, or disposal, the formation of highly toxic and/or highly explosive commodities is extremely possible.

Ethylene glycol dinitrate may enter the environment through industrial discharges, municipal waste treatment plant discharges, or spills.

Acute Ecological Effects

Acute (short-term) toxic effects may include the death of animals, birds, or fish and death or low growth rate in plants. Acute effects are seen 2 to 4 days after animals or plants are exposed to ethylene glycol dinitrate.

Insufficient data are available to evaluate or predict the short-term effects of ethylene glycol dinitrate exposure to aquatic life, plants, birds, or land animals.

Chronic Ecological Effects

Chronic (long-term) toxic effects may include shortened life span, reproductive problems, lower fertility, and changes in appearance or behavior in exposed animals. These effects can be seen long after first exposure(s) to toxic chemicals.

Insufficient data are available to evaluate or predict the long-term effects of ethylene glycol dinitrate exposure to aquatic life, plants, birds, or land animals.

Water Solubility

Ethylene glycol dinitrate is insoluble in water. Concentrations of less than 1 milligram will not mix with a liter of water.

Persistence in the Environment

Ethylene glycol dinitrate is highly persistent in water, with a half-life greater than 200 days. The half-life of a pollutant is the amount of time it takes for one half of the chemical to be degraded.

Bioaccumulation in Aquatic Organisms

Some substances increase in concentration, or *bioaccumulate,* in living organisms as they breathe contaminated air, drink contaminated water, or eat contaminated food. These chemicals can become concentrated in the tissues and internal organs of animals as well as humans.

The data are insufficient to evaluate or predict the concentration of ethylene glycol dinitrate that may be found in edible fish tissue.

Recommended Risk-Reduction Measures

Proper training of all transporters will reduce the likelihood of a mishap or accident resulting in a leak or spill to the environment. The correct use of DOT labeling (per the MSDS) on all containers, trucks, and rail cars will enable emergency responders to react properly and quickly to any disaster thereby reducing the potential risk to the environment and to personnel.

Storage of ethylene glycol dinitrate should be segregated from incompatible chemicals to minimize the risk of cross-contamination or contact. Buildings designated for storage should be equipped with appropriate fire protection systems (alarms, sprinklers, emergency lighting, portable extinguishers).

If a spill or leak to the environment has occurred, fire department, emergency response and/or hazardous materials spill personnel should be notified immediately. Cleanup should be attempted only by those trained in proper spill containment procedures using non-sparking tools. Contaminated soils should be carefully removed and replaced with clean soil. If ethylene glycol dinitrate should contact the water table, aquifer, or navigable waterway, cleanup should be started immediately. It is insoluble in water and total containment and remediation may be possible. The local and/or state emergency response authorities must be notified. A comprehensive emergency response or disaster preparedness/recovery plan, with consideration for explosive commodities, should be in place prior to any operations involving the use, transportation, storage, or disposal of ethylene glycol dinitrate. If ethylene glycol dinitrate is spilled or leaked, the following specific steps are recommended:

- ☑ Restrict persons not wearing protective clothing from area of spill or leak until cleanup is complete and area can be opened for normal work.
- ☑ Remove all ignition sources, ventilate area of leak.
- ☑ Neutralize spilled liquids by mixing with sodium bisulfate. Mixture should be collected using non-sparking tools and placed in sealed containers.
- ☑ It may be necessary to dispose of ethylene glycol dinitrate as a hazardous waste. The responsible state agency or the regional office of the federal Environmental Protection Agency (EPA) should be contacted for specific recommendations.

RISK ASSESSMENT: BUSINESS

General Assessment

Accidents or mishaps involving ethylene glycol dinitrate can present a serious threat to business operations. The loss or damage of equipment or facilities can significantly affect fiscal viability. Lawsuits that may result from personnel injury/death, public exposures, and/or environmental contamination/destruction will also require a serious expenditure of resources. Media attention surrounding an injury, death, or environmental damage can also result in a loss of profits and loss of current as well as future business.

Recommended Risk-Reduction Measures

Company attorneys, safety and health professionals, and environmental specialists should be involved in the development of any procedures dealing with safety or emergency response. A company official should be pre-designated as a public relations officer with specific training in dealing with the press. Corporate plans and policies should be developed, approved, and implemented long before any need for such arises.

REFERENCES

Ahlbom, A. 1993. *Biostatistics for Epidemiologists.* New York: CRC Press/Lewis Publishers.

American Conference of Governmental Industrial Hygienists. 1988. *Documentation of the Threshold Limit Values and Biological Exposure Indices,* 5th Edition (with updates). Cincinnati: ACGIH

Anthony, C. P., and N. J. Kolthoff. 1971. *Textbook of Anatomy and Physiology.* St. Louis: C. V. Mosby Company.

Asante-Duah, D. K. 1993. *Hazardous Waste Risk Assessment.* New York: CRC Press/Lewis Publishers.

Bloom, A. D. and F. J. DeSerres. 1995. *Ecotoxicity and Human Health.* New York: CRC Press/Lewis Publishers.

Brusick, D. J. 1994. *Methods for Genetic Risk Assessment.* New York: CRC Press/Lewis Publishers.

Bueche, J. R. 1972. *Principles of Physics,* 2nd Edition. New York: McGraw-Hill.

Burrell, R., D. K. Flaherty, and L. J. Sauers. 1992. *Toxicology of the Immune System.* New York: Van Nostrand Reinhold.

Calabrese, E. J. 1994. *Biological Effects of Low Level Exposures.* New York: CRC Press/Lewis Publishers.

Cockerman, L. G., B. S. Shane. 1994. *Basic Environmental Toxicology.* New York: CRC Press/Lewis Publishers.

Cothern, R. R. 1993. *Comparative Environmental Risk Assessment.* New York: CRC Press/Lewis Publishers.

Danse, I. R. 1991. *Common Sense Toxics in the Workplace.* New York: Van Nostrand Reinhold.

Environmental Protection Agency. 1983. *Chemical Hazard Information Profiles.* Washington, D.C.: U.S. EPA.

Fodden, C. R., and J. L. Weddell. 1992. *Hazardous Materials Emergency Action Data.* New York: CRC Press/Lewis Publishers.

Forseberg, K., and L. H. Keith. 1993. *Chemical; Protective Clothing Permeation and Degradation Database.* Chelsea, MI: Lewis Publishers.

Gardner, R. W. 1994. *Chemical Intolerance.* New York: CRC Press/Lewis Publishers.

Hallenbeck, W. H. 1993. *Quantitative Risk Assessment for Environmental and Occupational Health,* 2nd Edition. New York: CRC Press/Lewis Publishers.

Hodgson, E., R. B. Mailman, and Janice E. Chambers. 1988. *Dictionary of Toxicology.* London: The MacMillian Company.

Howard, P. H. 1995. *Handbook of Environmental Fate and Exposures Data for Organic Chemicals,* Vol. I-IV. New York: CRC Press/Lewis Publishers.

Hosty, J. W., and P. Foster. 1990. *A Practical Guide to Chemical Spill Response.* New York: Van Nostrand Reinhold.

Keith, L. H. and M. M. Walker. 1995. *Handbook of Air Toxics.* New York: CRC Press/Lewis Publishers.

Landis, W. G., M. H. Yu. 1995. *Introduction to Environmental Toxicology.* New York: CRC Press/Lewis Publishers.

Langerman, N. 1994 *Precautionary Labels for Chemical Containers.* New York: CRC Press/Lewis Publishers.

Lauwerys, R. R., P. Hoet. 1993. *Industrial Chemical Exposure.* New York: CRC Press/Lewis Publishers.

Lewis, R. J., Sr. 1992. *Sax's Dangerous Properties of Industrial Materials,* Volumes 1, 2, and 3, 8th Edition. New York: Van Nostrand Reinhold.

Manahan, S. E. 1994. *Environmental Chemistry,* 6th Edition. New York: CRC Press/Lewis Publishers.

Mackay, D., W. Y. Shiu, and K. C. Ma. 1992. *Illustrated Handbook of Chemical-Physical Properties and Environmental Fate for Organic Chemicals.* New York: CRC Press/Lewis Publishers.

Meyer, E. 1990. *Chemistry of Hazardous Materials.* Englewood Cliffs, NJ: Prentice Hall.

Miller, B. F., M.D., and C. B. Keane, R.N. 1972. *Encyclopedia and Dictionary of Medicine and Nursing.* Philadelphia: W.B. Sanders Company.

Patnaik, P. 1992. *A Comprehensive Guide to the Hazardous Properties of Chemical Substances.* New York: Van Nostrand Reinhold.

Quigley, D. R. 1994. *Handbook of Emergency Chemical Management.* New York: CRC Press/Lewis Publishers.

Rea, W. J. 1995. *Chemical Sensitivity,* Vol. I-III. New York: CRC Press/Lewis Publishers.

Richardson, M. 1995. *Dictionary of Substance and Their Effects,* Vol. I-VI. New York: CRC Press/Lewis Publishers.

Sacarello, H. L. A. 1994. *The Comprehensive Handbook of Hazardous Materials.* New York: CRC Press/Lewis Publishers.

Torkelson, T. R. V. K. Rowe. 1981. Halogenated aliphatic hydrocarbons. *Patty's Industrial Hygiene and Toxicology,* Vol. 2B, *Toxicology,* pp 3568-3572. New York: Wiley-Interscience.

U.S. Department of Health and Human Services, National Institute for Occupational Safety and Health. 1994. *NIOSH Pocket Guide to Chemical Hazards.* Washington, D.C.: U.S. Government Printing Office.

MATERIAL SAFETY DATA SHEET

CHEMICAL NAME

ETHYLENE GLYCOL MONOETHYL ETHER

HAZARD WARNING INFORMATION

HEALTH	FIRE	REACTIVE	OTHER	DEGREE OF HAZARD	COLOR CODING	OTHER CODES
2	2	0		0 = Minimum Hazard 1 = Slight Hazard 2 = Moderate Hazard 3 = Serious Hazard 4 = Severe Hazard	HEALTH = BLUE FIRE = RED REACTIVITY = YELLOW OTHER = WHITE	OX = Oxidizer ACID = Acid ALK = Alkali COR = Corrosive W = Use No Water

SECTION I - GENERAL INFORMATION

Characterization **Glycol Ether**	RCRA Number **U359**	EPA Class **Toxic Waste**
DOT Proper Shipping Name **Ethylene Glycol Monoethyl Ether**	Chemical Abstract Service (CAS) Number **110-80-5**	
DOT Hazard Class and Label Requirements **Combustible Liquid**	DOT Emergency Guide Code **55**	
DOT Identification Number **UN 1171**	Chemical Formula $C_4H_{10}O_2$	

Synonyms
2-Ethoxyethanol; Cellosolve; Dowanol EE; EGEE; glycol ethyl ether; hydroxyether, Oxitol.

SECTION II - HAZARDOUS INGREDIENTS AND IDENTITY INFORMATION

Hazardous Components (specific identity)	OSHA Exposure Criteria	NIOSH Exposure Criteria	Immediately Dangerous to Life and Health (IDLH)	ACGIH Exposure Criteria
Ethylene glycol monoethyl ether (derivation: By reaction of ethylene oxide and ethanol). **1 ppm = 3.7 mg/m^3**	**PEL (skin):** **200 ppm** **740 mg/m^3** **STEL:** **Not Established**	**REL (skin):** **0.5 ppm** **1.8 mg/m^3** **STEL:** **Not Established**	**6000 ppm**	**TLV (skin):** **5 ppm** **18.50 mg/m^3** **STEL:** **Not Established**

SECTION III - PHYSICAL AND CHEMICAL CHARACTERISTICS

Boiling Point **275°F (135°C)**	Specific Gravity (H_2O = 1) **0.931**
Vapor Pressure (mm Hg) **6.2 at 69°F (20°C)**	Molecular Weight **90.14**
Vapor Density (Air = 1) **3.10**	Melting Point **-94°F (-70°C)**

Solubility
Soluble in water, alcohol, acetone, ether, liquid esters, and hydrocarbons.

Appearance and Odor
Colorless liquid with a sweet, pleasant, ether-like odor and slightly bitter taste.

SECTION IV - FIRE AND EXPLOSION HAZARD DATA

Flash Point (method used) **111°F (44°C) closed cup**	Explosive Limits in Air % by Volume **LEL: 1.7% at 200°F (93°C) UEL: 15.6%**
NFPA Classification **Class II Combustible Liquid**	Autoignition Temperature **455°F (235°C)**

Extinguishing Media
Use dry chemical, carbon dioxide, water spray, fog, or alcohol-resistant foam.

Special Fire Fighting Procedures
Acrid and irritating fumes and smoke are produced in fire, wear full protective clothing and self-contained breathing apparatus (SCBA). Use water spray to cool fire-exposed containers. Fight fire from distance if possible. Move containers from fire area if it can be done safely. Exercise extreme caution when moving containers. Stay clear of ends of tanks and vessels.

Unusual Fire and Explosion Hazards
A combustible liquid. A moderate fire hazard when exposed to heat, sparks, or open flame. Boiling liquid expanding vapor explosion (BLEVE) is possible. Immediately withdraw if rising sound is hear from venting device or fire is causing discoloration on sides of tank. Stay clear one-half mile.

SECTION V - REACTIVITY DATA

Stability		Conditions to Avoid **Ethylene glycol monoethyl ether will form highly explosive peroxides when exposed to air (this will NOT occur if it is mixed in solution). Avoid contact with incompatible materials. Keep away from heat and flame.**
Stable X	Unstable	Incompatibility *(materials to avoid)* **Strong oxidizers (such as chlorine, bromine, fluorine, and, especially hydrogen peroxide), and other oxidizing agents.**
Hazardous Polymerization		Conditions to Avoid **Under normal conditions of temperature and pressure, hazardous polymerization of ethylene glycol monoethyl ether is not expected to occur.**
May Occur	Will Not Occur X	Hazardous Decomposition or By-products **Acrid and irritated gases and fumes are produced when ethylene glycol monoethyl ether is involved in fire, including toxic oxides of carbon.**

SECTION VI - HEALTH HAZARD DATA

Primary Route(s) of Entry:	Inhalation?	Absorption (skin)?	Ingestion?
	X	X	X

Health Hazards

INHALATION: Causes dizziness and cyanosis. High concentrations can lead to potentially fatal pulmonary edema (fluid in the lungs), albuminuria (excess albumin proteins in urine), hyperbilirubinemia (excess bilirubin in blood), leukopenia (low white cells), hematuria (blood in urine), oligospermia (low sperm count), unconsciousness, gynecological disorders, fetal abnormalities, possible anemia, and liver or kidney damage.

ABSORPTION: Repeated or heavy contact will pass through unbroken skin to cause toxic systemic effects. Eye contact causes irritation to the cornea and conjunctiva.

INGESTION: Cyanosis, pulmonary edema, tonic-clonic seizures, unconsciousness, acidosis.

Carcinogenicity	NTP Listed?	IARC Cancer Review Group?	OSHA Regulated?	Target Organs?
Unknown Human Unknown Animal	No	No	29 CFR 1910.1000 Table Z-1	CNS, respiratory system, skin.

Medical Conditions Generally Aggravated by Exposure

Pre-existing disorders of the CNS, reproductive system, and the blood may be aggravated by exposure.

Emergency and First-aid Procedures

Eye contact: Flush immediately with water for 15 minutes (minimum), seek medical attention. Skin contact: Remove all contaminated clothing. Immediately wash area with large amounts of soap and water to prevent absorption. Seek medical assistance. For inhalation: Remove the person from exposure. Provide respiratory assistance and CPR. Transfer to medical facility. If swallowed: Call the poison control center and seek medical attention immediately. Unless advised otherwise, give a conscious and alert person 1-2 glasses of water or milk, then induce vomiting (within 30 minutes).

SECTION VII - PRECAUTIONS FOR SAFE HANDLING AND USE

Steps to be Taken in Case Material is Released or Spilled

Evacuate area of spill and remove all ignition sources. Remove leaking containers to a safe place (outdoors), if it can be done without risk. Absorb liquids with vermiculite or other material and place in sealed drum for disposal. Use non-sparking tools. Use water spray to flush spills away from exposures and dilute to nonflammable mixtures (keep out of sewers and away from waterways).

Preferred Waste Disposal Method

Burn in a chemical incinerator equipped with an afterburner and scrubber.

Precautions to be Taken in Handling and Storage

Do not store ethylene glycol monoethyl ether in the presence of incompatible chemicals or materials. Store in tightly closed containers in a cool, dry, well-ventilated area away from ignition sources.

Other Precautions and Warnings

Use non-sparking tools and electrically ground and bond all containers, especially during transfer.

SECTION VIII - CONTROL MEASURES AND PERSONAL PROTECTIVE EQUIPMENT

Respiratory Protection *(specify type)*

Use an air-purifying respirator or a supplied-air respirator with full facepiece operated in pressure demand mode or a self-contained breathing apparatus with operated in pressure demand mode.

Ventilation

Local exhaust or general mechanical systems recommended.

Protective Gloves	Eye Protection	Other Protective Clothing
Butyl Rubber	**Chemical Goggles or Face Mask**	**Protective Clothing or Apron**

Work/Hygiene Practices

Always wash hands thoroughly after using chemical; never bring food, drink, or smoking materials into vicinity of chemicals.

ETHYLENE GLYCOL MONOETHYL ETHER

$C_4H_{10}O_2$ CAS: 110-80-5

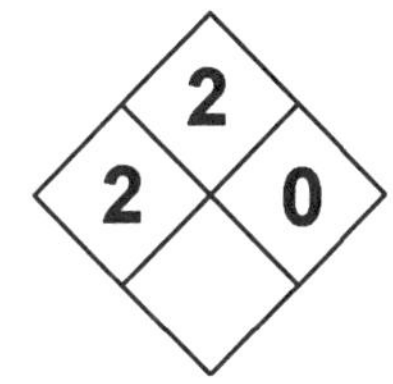

$$HO—CH_2—CH_2—O—CH_2—CH_3$$

IDENTIFICATION AND TYPICAL USES

Ethylene glycol monoethyl ether (EGEE) is a colorless liquid with a mild, pleasant (sweet), ether-like odor and a slightly bitter taste. It is used as a solvent for nitrocellulose and natural and synthetic resins, as an anti-icing additive in aviation fuels, as a chemical intermediate for 2-ethoxyethyl acetate, in lacquers, in the dyeing and printing of textiles, in varnish removers, cleaning solutions, in products for treating leather goods, and as an emulsion stabilizer.

RISK ASSESSMENT: HEALTH

General Assessment

EGEE is mildly to moderately toxic by *ingestion, inhalation,* and *absorption* through the skin. There is no information in the references on the carcinogenic or mutagenic properties of EGEE. However, it poses a serious teratogenic threat to humans (both male and female) and exposure must therefore be kept to the lowest possible levels.

Inhalation can cause some (mild) irritation of the eyes, nose, throat, and the mucosa of the upper respiratory tract. Inhalation and skin absorption can lead to serious systemic effects. Symptoms include dizziness and cyanosis (bluish color of the lips and appendages). Exposure to high concentrations can cause a number of life-threatening illnesses and disorders. These include pulmonary edema (fluid buildup in the lungs) with symptoms of cough with phlegm, pain in chest, and difficulty breathing. These symptoms may be delayed up to 48 hours creating a false sense of security with regard to health risk. Exposure can also lead to albuminuria (excessive levels of albumin protein in urine), hyperbilirubinemia (excessive levels of bilirubin in blood), oligospermia (low sperm count), loss of consciousness, and a number of varied gynecological disorders and possible fetal abnormalities.

In addition, animal studies have shown a potential for reversible liver and kidney changes, lesions of the liver, kidney, pituitary gland, spleen, and testis, enlargement of the spleen, gastrointestinal hemorrhages, and blood disorders (e.g., anemia, leukopenia, and hematuria).

Ingestion can cause unspecified gastrointestinal effects, cyanosis, pulmonary edema, tonic-clonic seizures, acetone odor on breath, acidosis, and unconsciousness.

Acute Health Effects

The following acute (short-term) health effects may occur immediately or shortly after exposure to EGEE:

Skin: Mild irritation with absorption through the skin (if contact is prolonged) causing toxic systemic effects.

Eye: Irritation of the cornea and conjunctiva.

Lung: Mild irritation of the mucosa of the nose, throat, and upper respiratory tract. Can cause a number of toxic systemic effects with symptoms of headaches, lightheadedness, weakness, and CNS depression.

Other: Ingestion can cause the same symptoms as inhalation. Adverse action on the central nervous system is likely on high acute exposures (may cause seizures and loss of consciousness).

Chronic Health Effects

The following chronic (long-term) health effects may occur at some time after exposure to EGEE and can last for months or even years:

Cancer Hazards: According to the information presented in the references, EGEE has not been shown to cause cancer in test animals.

Reproductive Hazard: According to the information presented in the references, EGEE is a known teratogen, especially at high concentrations. Male sperm count reductions have been reported and female gynecological effects have been documented. It may also cause birth defects, primarily involving the heart and skeletal system.

Other Chronic Effects: Low, repeated exposures may lead to damage to the liver and kidneys. There may be albuminuria and hyperbilirubinemia resulting from long-term exposures.

Recommended Risk-Reduction Measures

Unless a less toxic chemical can be substituted for a hazardous substance, *engineering controls* are the most effective method of reducing exposures. The best protection is to enclose operations and/or provide local exhaust ventilation at the site of chemical release. While not always operationally feasible, isolating operations can also reduce exposure.

Using respiratory protection is less effective than the controls mentioned above, but is still advisable whenever working with or around EGEE. For low exposures up to or over the PEL (200 ppm), use a NIOSH-approved organic vapor respirator with full facepiece. For the best protection, especially at high exposures or during emergency situations, use a self-contained breathing apparatus (SCBA) with full facepiece operated in pressure demand mode, or a supplied-air respirator operated in pressure demand or continuous flow mode. If a full facepiece is not available, then chemical goggles should be worn to protect the eyes. Whenever a chemical splash hazard exists, a face shield and rubber apron should be worn. Skin contact must be avoided. To prevent hand and skin exposures, butyl rubber gloves should be worn. Natural rubber, polyvinylchloride, or polyvinyl alcohol gloves are not appropriate. Glove manufacturers should be contacted and permeation studies obtained *before* making final glove selection.

Administrative controls should also be in place to minimize the potential for human exposures. These may include written procedures or policies which specify the methods and techniques that will be practiced whenever personnel are to work with EGEE.

All personnel should receive *training* on the use, hazards, protective measures, emergency actions, and other precautions per 29 CFR 1910.1200 (Hazard Communication), prior to the first assignment in an area where EGEE is used or stored.

If symptoms develop or overexposure is suspected, the following medical tests may be useful:

- ☑ Urinalysis (check for albuminuria and hematuria).
- ☑ Complete blood count.
- ☑ Liver and kidney function tests.
- ☑ Lung function tests.
- ☑ Consider chest X-ray following acute overexposure (may be negative if taken immediately after exposure due to the delayed development of pulmonary edema).

Any evaluation should include a careful medical history of past and present symptoms with an examination. However, medical tests that evaluate existing damage are not a substitute for controlling exposure. Also, since smoking can cause heart disease, emphysema, and other respiratory disorders, exposure to EGEE can have a more pronounced effect on smokers than on non-smokers exposed to the same levels. Prudent risk assessment and management requires consideration of *all* possible risk factors when exposure symptoms appear in the workforce.

Other methods to reduce exposure include:

- ☑ Where possible, enclose operations and use local exhaust ventilation at the site of chemical release. If local exhaust ventilation or enclosure is not used, respiratory protection should be mandatory. Also, automatic transfer of liquids from storage containers to process containers is recommended.
- ☑ Always ensure that proper protective clothing is worn when using chemical substances.
- ☑ Wash thoroughly immediately after exposure to EGEE and at the end of the work shift or before eating, drinking, or smoking.
- ☑ Hazard warning information should be posted in the work area. In addition, as part of an on-going education and training program, all information on the health and safety hazards of EGEE should be communicated to all potentially exposed workers.
- ☑ Eye wash stations should be provided in work areas. If the potential for whole body exposure exists, then safety showers should also be provided.

RISK ASSESSMENT: ENVIRONMENT

General Assessment

The environment is at risk of exposure during use, transportation, storage, disposal, or destruction of EGEE. In almost every scenario, the threat of environmental exposure is contingent upon the proper handling of the chemical substance. Accidental spills, large or small, can result in fire, explosion, and possible contamination of the surrounding environmental mediums (water, soil, and air).

EGEE is considered a Class II combustible liquid (per OSHA 29 CFR 1910.106). These characteristics require special consideration during any emergency situation involving a leak or spill of EGEE. Should EGEE ever come into contact with incompatible substances such as strong oxidizers (chlorine, bromine, fluorine, chlorine dioxide, and peroxides) either during use, transportation, or storage, violent reaction can occur.

Wastes should be considered for recovery, reclamation, or recycling rather than land-fill disposal. Destruction in a chemical incinerator equipped with an afterburner and scrubber is also acceptable.

EGEE can enter the environment through industrial discharges and from spills.

☠ *Acute Ecological Effects*

Acute (short-term) toxic effects may include the death of animals, birds, or fish and death or low growth rate in plants. Acute effects are seen 2 to 4 days after animals or plants are exposed to EGEE. Insufficient data are available to evaluate or predict the acute (short-term) effects of EGEE to aquatic life, plants, birds, or land animals.

Chronic Ecological Effects

Chronic toxic effects may include shortened life span, reproductive problems, lower fertility, and changes in appearance or behavior in exposed animals. These effects can be seen long after first exposure(s) to toxic chemicals. Insufficient data are available to evaluate or predict the long-term (chronic) effects of EGEE to aquatic life, plants, birds, or land animals.

Water Solubility

EGEE is highly soluble in water. Concentrations of 1000 milligrams and more will mix with a liter of water.

⌛ *Persistence in the Environment*

EGEE is highly persistent in the aquatic environment with a half-life of 200 days and more. The half-life of a pollutant is the amount of time it takes for one half of the chemical to be degraded. It is highly mobile in both water and soil and undergoes aerobic biodegeneration.

Bioaccumulation in Aquatic Organisms

Some substances increase in concentration, or *bioaccumulate,* in living organisms as they breathe contaminated air, drink contaminated water, or eat contaminated food. These chemicals can become concentrated in the tissues and internal organs of animals as well as humans.

The concentration of EGEE found in fish tissues is expected to be somewhat higher than the average concentration found in the water from which the fish was taken.

Recommended Risk-Reduction Measures

Proper training of all transporters will reduce the likelihood of a mishap or accident resulting in a leak or spill to the environment. The correct use of DOT labeling (per the MSDS) on all containers, trucks, and rail cars will enable emergency responders to react properly and quickly to any disaster thereby reducing the potential risk to the environment and to personnel.

Storage of EGEE should be segregated from incompatible chemicals to minimize the risk of cross-contamination or contact. All containers used in shipping or transfer operations must be electrically grounded to prevent static sparks. Buildings should be properly constructed for storage of hazardous chemicals (fire suppression equipment, alarms, separation). Personnel should be trained on the hazards associated with the storage of hazardous chemicals.

If a spill or leak to the environment has occurred, fire department, emergency response and/or hazardous materials spill personnel should be notified immediately. Cleanup should be attempted only by those trained in proper spill containment procedures. Contaminated soils should be removed for incineration and replaced with clean soil. If EGEE should contact the water table, aquifer, or navigable waterway, remediation activities should be prompt. It is highly soluble in water and total containment and remediation may not be possible. When such spills occur, the local and/or state emergency response authorities must be notified. A comprehensive emergency response or disaster preparedness/recovery plan should be in place prior to any operations involving the use, transportation, storage, or disposal of EGEE.

If EGEE is spilled or leaked, the following specific steps are recommended:

☑ Restrict persons not wearing protective clothing from area of spill or leak until cleanup is complete and area can be opened for normal work.

☑ Ventilate area and remove all ignition sources.

☑ Collect liquid using vermiculite, dry earth, or sand to absorb and place in a sealed drum for disposal.

☑ It may be necessary to dispose of EGEE as a hazardous waste. The responsible state agency or the regional office of the federal Environmental Protection Agency (EPA) should be contacted for specific recommendations.

RISK ASSESSMENT: BUSINESS

General Assessment

Accidents or mishaps involving EGEE can present a significant threat to business operations. The loss or damage of equipment or facilities can significantly affect fiscal viability. Lawsuits that may result from personnel injury/death, public exposures, and/or environmental contamination will also require a serious expenditure of resources. Media attention surrounding an injury, death, or environmental damage can also result in a loss of profits and loss of current as well as future business. Always remember that anytime the term "reproductive hazard" is used, public hysteria, emotion, and ignorance can run equally high. This should be carefully considered during the development of public relations policies.

Recommended Risk-Reduction Measures

Company attorneys, safety and health professionals, and environmental specialists should be involved in the development of any procedures dealing with safety or emergency response. A company official should be pre-designated as a public relations officer with specific training in dealing with the press. Corporate plans and policies should be developed, approved, and implemented long before any need for such arises.

REFERENCES

Ahlbom, A. 1993. *Biostatistics for Epidemiologists.* New York: CRC Press/Lewis Publishers.

American Conference of Governmental Industrial Hygienists. 1988. *Documentation of the Threshold Limit Values and Biological Exposure Indices,* 5th Edition (with updates). Cincinnati: ACGIH

Anthony, C. P., and N. J. Kolthoff. 1971. *Textbook of Anatomy and Physiology.* St. Louis: C. V. Mosby Company.

Bloom, A. D. and F. J. DeSerres. 1995. *Ecotoxicity and Human Health.* New York: CRC Press/Lewis Publishers.

Bueche, J. R. 1972. *Principles of Physics,* 2nd Edition. New York: McGraw-Hill.

Burrell, R., D. K. Flaherty, and L. J. Sauers. 1992. *Toxicology of the Immune System.* New York: Van Nostrand Reinhold.

Calabrese, E. J., E. M. Kenyon. *Air Toxics and Risk Assessment.* New York: CRC Press/Lewis Publishers.

Cockerman, L. G., B. S. Shane. 1994. *Basic Environmental Toxicology.* New York: CRC Press/Lewis Publishers.

Danse, I. R. 1991. *Common Sense Toxics in the Workplace.* New York: Van Nostrand Reinhold.

Environmental Protection Agency. 1983. *Chemical Hazard Information Profiles.* Washington, D.C.: U.S. EPA.

Fodden, C. R., and J. L. Weddell. 1992. *Hazardous Materials Emergency Action Data.* New York: CRC Press/Lewis Publishers.

Gardner, R. W. 1994. *Chemical Intolerance.* New York: CRC Press/Lewis Publishers.

Hodgson, E., R. B. Mailman, and Janice E. Chambers. 1988. *Dictionary of Toxicology.* London: The MacMillian Company.

Howard, P. H., et. al. 1991. *Handbook of Environmental Degradation Rates.* New York: CRC Press/Lewis Publishers.

Hosty, J. W., and P. Foster. 1990. *A Practical Guide to Chemical Spill Response.* New York: Van Nostrand Reinhold.

Keith, L. H. and M. M. Walker. 1995. *Handbook of Air Toxics.* New York: CRC Press/Lewis Publishers.

Landis, W. G., M. H. Yu. 1995. *Introduction to Environmental Toxicology.* New York: CRC Press/Lewis Publishers.

Lauwerys, R. R., P. Hoet. 1993. *Industrial Chemical Exposure.* New York: CRC Press/Lewis Publishers.

Lewis, R. J., Sr. 1993. *Hawley's Condensed Chemical Dictionary,* 12th Edition. New York: Van Nostrand Reinhold.

Lewis, R. J., Sr. 1992. *Sax's Dangerous Properties of Industrial Materials,* Volumes 1, 2, and 3, 8th Edition. New York: Van Nostrand Reinhold.

Manahan, S. E. 1992. *Toxicological Chemistry,* 2nd Edition. New York: CRC Press/Lewis Publishers.

Meyer, E. 1990. *Chemistry of Hazardous Materials.* Englewood Cliffs, NJ: Prentice Hall.

Ott, W. R. 1995. *Environmental Statistics and Data Analysis.* New York: CRC Press/Lewis Publishers.

Patnaik, P. 1992. *A Comprehensive Guide to the Hazardous Properties of Chemical Substances.* New York: Van Nostrand Reinhold.

Quigley, D. R. 1994. *Handbook of Emergency Chemical Management.* New York: CRC Press/Lewis Publishers.

Rea, W. J. 1995. *Chemical Sensitivity,* Vol. I-III. New York: CRC Press/Lewis Publishers.

Sacarello, H. L. A. 1994. *The Comprehensive Handbook of Hazardous Materials.* New York: CRC Press/Lewis Publishers.

MATERIAL SAFETY DATA SHEET

CHEMICAL NAME	CRC PRESS/LEWIS PUBLISHERS DISCLAIMER
ETHYLENE GLYCOL MONOMETHYL ETHER	

HAZARD WARNING INFORMATION

HEALTH	FIRE	REACTIVE	OTHER	DEGREE OF HAZARD	COLOR CODING	OTHER CODES
2	2	0		0 = Minimum Hazard 1 = Slight Hazard 2 = Moderate Hazard 3 = Serious Hazard 4 = Severe Hazard	HEALTH = BLUE FIRE = RED REACTIVITY = YELLOW OTHER = WHITE	OX = Oxidizer ACID = Acid ALK = Alkali COR = Corrosive W = Use No Water

SECTION I - GENERAL INFORMATION

Characterization	**Glycol Ether**	RCRA Number	**None**
		EPA Class	**Not Applicable**
DOT Proper Shipping Name	**Ethylene Glycol Monomethyl Ether**	Chemical Abstract Service (CAS) Number	**109-86-4**
DOT Hazard Class and Label Requirements	**Combustible Liquid**	DOT Emergency Guide Code	**55**
DOT Identification Number	**UN 1188**	Chemical Formula	$C_3H_8O_2$

Synonyms
2-Methoxyethanol; Methyl Cellosolve; Dowanol EM; EGME; glycol methyl ether; methyl glycol; methyl oxitol; ethylene glycol methyl ether; Poly-Solve EM

SECTION II - HAZARDOUS INGREDIENTS AND IDENTITY INFORMATION

Hazardous Components (specific identity)	OSHA Exposure Criteria	NIOSH Exposure Criteria	Immediately Dangerous to Life and Health (IDLH)	ACGIH Exposure Criteria
Ethylene glycol monomethyl ether (derivation: By reaction of ethylene oxide and ethanol). **1 ppm = 3.2 mg/m^3**	**PEL (skin):** **25 ppm** **80 mg/m^3** **STEL:** **Not Established**	**REL:** **Lowest Feasible Concentration**	**Not Determined**	**TLV (skin):** **5 ppm** **16 mg/m^3** **STEL:** **Not Established**

SECTION III - PHYSICAL AND CHEMICAL CHARACTERISTICS

Boiling Point	**255°F (124°C)**	Specific Gravity (H_2O = 1)	**0.966**
Vapor Pressure (mm Hg)	**6 at 69°F (20°C)**	Molecular Weight	**76.11**
Vapor Density (Air = 1)	**2.64**	Melting Point	**-121°F (-85°C)**

Solubility
Miscible with water, alcohol, acetone, ether, liquid esters, and hydrocarbons.

Appearance and Odor
Colorless liquid with a mild, agreeable odor. Odor Threshold = 0.4 ppm.

SECTION IV - FIRE AND EXPLOSION HAZARD DATA

Flash Point (method used)	**107°F (41°C) closed cup**	Explosive Limits in Air % by Volume	**LEL: 2.5% UEL: 14%**
NFPA Classification	**Class II Combustible Liquid**	Autoignition Temperature	**545°F (285°C)**

Extinguishing Media
Use dry chemical, carbon dioxide, water spray, fog, or alcohol-resistant foam (smothering effect).

Special Fire Fighting Procedures
Acrid and irritating fumes and smoke are produced in fire, wear full protective clothing and self-contained breathing apparatus (SCBA). Use water spray to cool fire-exposed containers. Fight fire from distance if possible. Move containers from fire area if it can be done safely.

Unusual Fire and Explosion Hazards
A combustible liquid. A moderate fire hazard when exposed to heat, sparks, or open flame. Reaction with oxidizers forms explosive peroxides that are heat and shock sensitive. Vapors are heavier than air and can travel for great distances to an ignition source to flashback and cause fire or explosion.

SECTION V - REACTIVITY DATA

Stability		Conditions to Avoid Ethylene glycol monomethyl ether will form highly explosive peroxides when exposed to air (this will NOT occur if it is mixed in solution). Avoid contact with incompatible materials. Keep away from heat and flame.
Stable X	Unstable	Incompatibility *(materials to avoid)* Strong oxidizers (such as chlorine, bromine, fluorine, and, especially hydrogen peroxide), and other oxidizing agents.
Hazardous Polymerization		Conditions to Avoid Under normal conditions of temperature and pressure, hazardous polymerization of ethylene glycol monomethyl ether is not expected to occur.
May Occur	Will Not Occur X	Hazardous Decomposition or By-products Acrid and irritated gases and fumes are produced when ethylene glycol monomethyl ether is involved in fire, including toxic oxides of carbon.

SECTION VI - HEALTH HAZARD DATA

Primary Route(s) of Entry:	Inhalation?	Absorption (skin)?	Ingestion?
	X	X	X

Health Hazards

INHALATION: A teratogen. Also causes systemic poisoning with anemia, macrocytosis (enlarged red blood cells), other blood changes, conjunctivitis, blurred vision, irritation of the respiratory tract with dypsnea and coughing), nausea, vomiting, headache, and depression of the central nervous system (CNS) causing slurred speech, lethargy, confusion, tremors, disorientation, ataxia, and stupor. May cause hematuria, proteinuria, testicular atrophy, and birth defects. Kidney damage is also possible on long-term exposures.

ABSORPTION: Will pass through unbroken skin to cause same toxic systemic effects as inhalation. Eye contact causes irritation to the cornea and conjunctiva.

INGESTION: Macrocytotic anemia, anesthesia, and many of the systemic effects of inhalation.

Carcinogenicity	NTP Listed?	IARC Cancer Review Group?	OSHA Regulated?	Target Organs?
No Evidence Human No Evidence Animal	No	No	29 CFR 1910.1000 Table Z-1	Skin, eyes, blood, CNS, kidneys.

Medical Conditions Generally Aggravated by Exposure

Pre-existing disorders of the CNS, reproductive system, and the blood may be aggravated by exposure.

Emergency and First-aid Procedures

Eye contact: Flush immediately with water for 15 minutes (minimum), seek medical attention. Skin contact: Remove all contaminated clothing. Immediately wash area with large amounts of soap and water to prevent absorption. Seek medical assistance. For inhalation: Remove the person from exposure. Provide respiratory assistance and CPR. Transfer to medical facility. If swallowed: Call the poison control center and seek medical attention immediately. Unless advised otherwise, give a conscious and alert person 3-4 glasses of water, then induce vomiting (best if within 30 minutes).

SECTION VII - PRECAUTIONS FOR SAFE HANDLING AND USE

Steps to be Taken in Case Material is Released or Spilled

Evacuate area of spill and remove all ignition sources. Remove leaking containers to a safe place (outdoors), if it can be done without risk. Absorb liquids with vermiculite or other material and place in sealed drum for disposal. Use non-sparking tools. Use water spray to flush spills away from exposures and dilute to nonflammable mixtures (keep out of sewers and away from waterways).

Preferred Waste Disposal Method

Burn in a chemical incinerator equipped with an afterburner and scrubber.

Precautions to be Taken in Handling and Storage

Do not store ethylene glycol monomethyl ether in the presence of incompatible chemicals or materials. Store in tightly closed containers in a cool, dry, well-ventilated area away from ignition sources.

Other Precautions and Warnings

Use non-sparking tools and electrically ground and bond all containers, especially during transfer.

SECTION VIII - CONTROL MEASURES AND PERSONAL PROTECTIVE EQUIPMENT

Respiratory Protection *(specify type)*

Use an air-purifying respirator or a supplied-air respirator with full facepiece operated in pressure demand mode or a self-contained breathing apparatus with operated in pressure demand mode.

Ventilation

Local exhaust or general mechanical systems recommended.

Protective Gloves	Eye Protection	Other Protective Clothing
Buna-N or Neoprene Rubber	Chemical Goggles or Face Mask	Protective Clothing or Apron

Work/Hygiene Practices

Always wash hands thoroughly after using chemical; never bring food, drink, or smoking materials into vicinity of chemicals.

ETHYLENE GLYCOL MONOMETHYL ETHER

$C_3H_8O_2$ CAS: 109-86-4

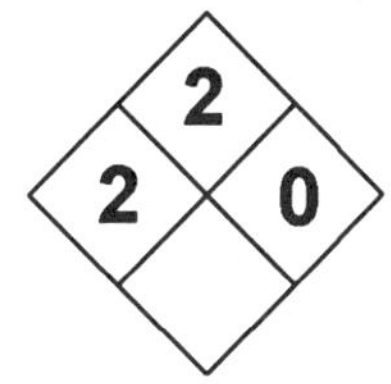

$$HO—CH_2—CH_2—O—CH_3$$

IDENTIFICATION AND TYPICAL USES

Ethylene glycol monomethyl ether (EGME) is a colorless liquid with a mild, agreeable odor. It is primarily used as a solvent for cellulose acetate, certain synthetic and natural resins, and dyes. Also used in jet fuel dicing agents, sealing and moisture-proofing cellophane, as a perfume fixative, and in dyeing leather. Its is used in nail polishes, varnishes, lacquers, and enamels.

RISK ASSESSMENT: HEALTH

General Assessment

EGME is mildly to moderately toxic by *ingestion, inhalation,* and *absorption* through the skin. There is no information in the references on the carcinogenic properties of EGME. However, mutation data have been reported and it is known to pose a serious teratogenic threat to humans (both male and female) and exposure must therefore be kept to the lowest possible levels. It is a systemic poison with target effects on the blood and the central nervous system (CNS).

Inhalation can cause some (mild) irritation of the eyes, nose, throat, and the mucosa of the upper respiratory tract with coughing and dypsnea (difficulty breathing). Inhalation and skin absorption can lead to serious systemic effects. Symptoms include conjunctivitis, blurred vision, nausea, vomiting, headache, and depression of the CNS causing stupor, slurred speech, weakness, drowsiness, lethargy, confusion, disorientation, tremors, and ataxia. Exposure can also lead to hematuria (blood in the urine) and proteinuria (protein in the urine) due to the chemical's damaging effects on the kidneys. High exposures can lead to testicular atrophy and birth defects.

Ingestion can cause unspecified gastrointestinal effects, anesthesia and other narcotic effects, macrocytotic anemia with macrocytosis (enlargement of the red blood cells), and possible death.

☠ *Acute Health Effects*

The following acute (short-term) health effects may occur immediately or shortly after exposure to EGME:

Skin: Mild irritation with absorption through the skin (if contact is prolonged) causing toxic systemic effects.

Eye: Irritation of the cornea and conjunctiva. May cause blurred vision.

Lung: Mild irritation of the mucosa of the nose, throat, and upper respiratory tract. Can cause a number of toxic systemic effects with symptoms of headaches, lightheadedness, weakness, and CNS depression.

Other: Ingestion can cause the same symptoms as inhalation. Adverse action on the central nervous system is likely on high acute exposures (may cause seizures and loss of consciousness). Serious blood effects and kidney damage also possible.

Chronic Health Effects

The following chronic (long-term) health effects may occur at some time after exposure to EGME and can last for months or even years:

Cancer Hazards: According to the information presented in the references, EGME has not been shown to cause cancer in test animals.

Reproductive Hazard: According to the information presented in the references, EGME is a known teratogen, especially at high concentrations. It causes testicular atrophy and may lead to birth defects (developmental abnormalities in the urogenital and musculoskeletal systems).

Other Chronic Effects: Low, repeated exposures may lead to damage to the kidneys, blood disturbances, chronic fatigue and sleepiness, and chronic CNS depression leading to personality changes.

Recommended Risk-Reduction Measures

Unless a less toxic chemical can be substituted for a hazardous substance, *engineering controls* are the most effective method of reducing exposures. The best protection is to enclose operations and/or provide local exhaust ventilation at the site of chemical release. While not always operationally feasible, isolating operations can also reduce exposure.

Using respiratory protection is less effective than the controls mentioned above, but is still advisable whenever working with or around EGME. For low exposures up to or over the PEL (25 ppm), use a NIOSH-approved organic vapor respirator with full facepiece. For the best protection, especially at high exposures or during emergency situations, a self-contained breathing apparatus (SCBA) with full facepiece operated in pressure demand mode, or a supplied-air respirator operated in pressure demand or continuous flow mode are the recommended respiratory protection methods of choice. If a full facepiece is not available, then chemical goggles should be worn to protect the eyes. Whenever a chemical splash hazard exists, a face shield and rubber apron should be worn. Skin contact must be avoided. To prevent hand and skin exposures, butyl rubber gloves should be worn. Buna-N or neoprene rubber should provide appropriate protection. Contact glove manufacturers to obtain permeation studies *before* making glove selection.

Administrative controls should also be in place to minimize the potential for human exposures. These may include written procedures or policies which specify the methods and techniques that will be practiced whenever personnel are to work with EGME.

All personnel should receive *training* on the use, hazards, protective measures, emergency actions, and other precautions per 29 CFR 1910.1200 (Hazard Communication), prior to the first assignment in an area where EGME is used or stored.

If symptoms develop or overexposure is suspected, the following medical tests may be useful:

- ☑ Urinalysis (check for proteinuria and hematuria).
- ☑ Complete blood count (check for macrocytosis).
- ☑ Kidney function tests.
- ☑ Evaluation of the central nervous system by a qualified neurologist to check for personality changes (moodiness, irritability, sleep pattern changes, stupor, fatigue, weakness).

Any evaluation should include a careful medical history of past and present symptoms with an examination. However, medical tests that evaluate existing damage are not a substitute for controlling exposure.

Other methods to reduce exposure include:

- ☑ Where possible, enclose operations and use local exhaust ventilation at the site of chemical release. If local exhaust ventilation or enclosure is not used, respiratory protection should be mandatory. Also, automatic transfer of liquids from storage containers to process containers is recommended.
- ☑ Always ensure that proper protective clothing is worn when using chemical substances.
- ☑ Wash thoroughly immediately after exposure to EGME and at the end of the work shift or before eating, drinking, or smoking.
- ☑ Hazard warning information should be posted in the work area. In addition, as part of an on-going education and training program, all information on the health and safety hazards of EGME should be communicated to all potentially exposed workers.
- ☑ Eye wash stations should be provided in work areas. If the potential for whole body exposure exists, then safety showers should also be provided.

RISK ASSESSMENT: ENVIRONMENT

General Assessment

The environment is at risk of exposure during use, transportation, storage, disposal, or destruction of EGME. In almost every scenario, the threat of environmental exposure is contingent upon the proper handling of the chemical substance. Accidental spills, large or small, can result in fire, explosion, and possible contamination of the surrounding environmental mediums (water, soil, and air).

EGME is considered a Class II combustible liquid (per OSHA 29 CFR 1910.106). These characteristics require special consideration during any emergency situation involving a leak or spill of EGME. Should EGME ever come into contact with incompatible substances such as strong oxidizers (chlorine, bromine, fluorine, chlorine dioxide, and peroxides) either during use, transportation, or storage, violent reaction can occur. Prolonged contact with air can cause the formation of dangerous peroxides which are both shock and heat sensitive.

Wastes should be considered for recovery, reclamation, or recycling rather than land-fill disposal. Destruction in a chemical incinerator equipped with an afterburner and scrubber is also acceptable.

EGME can enter the environment through industrial discharges and from spills.

Acute Ecological Effects

Acute (short-term) toxic effects may include the death of animals, birds, or fish and death or low growth rate in plants. Acute effects are seen 2 to 4 days after animals or plants are exposed to EGME. Insufficient data are available to evaluate or predict the acute (short-term) effects of EGME to aquatic life, plants, birds, or land animals.

Chronic Ecological Effects

Chronic toxic effects may include shortened life span, reproductive problems, lower fertility, and changes in appearance or behavior in exposed animals. These effects can be seen long after first exposure(s) to toxic chemicals. Insufficient data are available to evaluate or predict the long-term (chronic) effects of EGME to aquatic life, plants, birds, or land animals.

Water Solubility

EGME is highly soluble in water. Concentrations of 1000 milligrams and more will easily mix with a liter of water.

Persistence in the Environment

EGME is highly persistent in the aquatic environment with a half-life of 200 days and more. The half-life of a pollutant is the amount of time it takes for one half of the chemical to be degraded.

Bioaccumulation in Aquatic Organisms

Some substances increase in concentration, or *bioaccumulate,* in living organisms as they breathe contaminated air, drink contaminated water, or eat contaminated food. These chemicals can become concentrated in the tissues and internal organs of animals as well as humans.

The concentration of EGME found in fish tissues is expected to be somewhat higher than the average concentration found in the water from which the fish was taken.

Recommended Risk-Reduction Measures

Proper training of all transporters will reduce the likelihood of a mishap or accident resulting in a leak or spill to the environment. The correct use of DOT labeling (per the MSDS) on all containers, trucks, and rail cars will enable emergency responders to react properly and quickly to any disaster thereby reducing the potential risk to the environment and to personnel.

Storage of EGME should be segregated from incompatible chemicals to minimize the risk of cross-contamination or contact. All containers used in shipping or transfer operations must be electrically grounded to prevent static sparks. Buildings should be properly constructed for storage of hazardous chemicals (fire suppression equipment, alarms, separation). Electrical equipment should be designed to meet explosion-proof criteria.

If a spill or leak to the environment has occurred, fire department, emergency response and/or hazardous materials spill personnel should be notified immediately. Cleanup should be attempted only by those trained in proper spill containment procedures. Contaminated soils should be removed for incineration and replaced with clean soil. If EGME should contact the water table, aquifer, or navigable waterway, remediation activities should be prompt. It is highly soluble in water and total containment and remediation may not be possible. When such spills occur, the local and/or state emergency response authorities must be notified. A comprehensive emergency response or disaster preparedness/recovery plan should be in place prior to any operations involving the use, transportation, storage, or disposal of EGME. If EGME is spilled or leaked, the following specific steps are recommended:

- ☑ Ventilate area and remove all ignition sources.
- ☑ Restrict persons not wearing protective clothing from area of spill or leak until cleanup is complete.
- ☑ Absorb using vermiculite, dry earth, or sand and place in a sealed drum for disposal.
- ☑ It may be necessary to dispose of EGME as a hazardous waste. The responsible state agency or the regional office of the federal Environmental Protection Agency (EPA) should be contacted for specific recommendations.

RISK ASSESSMENT: BUSINESS

General Assessment

Accidents or mishaps involving EGME can present a significant threat to business operations. The loss or damage of equipment or facilities can significantly affect fiscal viability. Lawsuits that may result from personnel injury/death, public exposures, and/or environmental contamination will also require a serious expenditure of resources. Media attention surrounding an injury, death, or environmental damage can also result in a loss of profits and loss of current as well as future business. Always remember that anytime the term "reproductive hazard" is used, public hysteria, emotion, and ignorance can run equally high. This should be carefully considered during the development of public relations policies.

Recommended Risk-Reduction Measures

Company attorneys, safety and health professionals, and environmental specialists should be involved in the development of any procedures dealing with safety or emergency response. A company official should be pre-designated as a public relations officer with specific training in dealing with the press. Corporate plans and policies should be developed, approved, and implemented long before any need for such arises.

REFERENCES

Ahlbom, A. 1993. *Biostatistics for Epidemiologists.* New York: CRC Press/Lewis Publishers.

American Conference of Governmental Industrial Hygienists. 1988. *Documentation of the Threshold Limit Values and Biological Exposure Indices,* 5th Edition (with updates). Cincinnati: ACGIH

Anthony, C. P., and N. J. Kolthoff. 1971. *Textbook of Anatomy and Physiology.* St. Louis: C. V. Mosby Company.

Bloom, A. D. and F. J. DeSerres. 1995. *Ecotoxicity and Human Health.* New York: CRC Press/Lewis Publishers.

Bueche, J. R. 1972. *Principles of Physics,* 2nd Edition. New York: McGraw-Hill.

Burrell, R., D. K. Flaherty, and L. J. Sauers. 1992. *Toxicology of the Immune System.* New York: Van Nostrand Reinhold.

Calabrese, E. J., E. M. Kenyon. *Air Toxics and Risk Assessment.* New York: CRC Press/Lewis Publishers.

Cockerman, L. G., B. S. Shane. 1994. *Basic Environmental Toxicology.* New York: CRC Press/Lewis Publishers.

Danse, I. R. 1991. *Common Sense Toxics in the Workplace.* New York: Van Nostrand Reinhold.

Environmental Protection Agency. 1983. *Chemical Hazard Information Profiles.* Washington, D.C.: U.S. EPA.

Fodden, C. R., and J. L. Weddell. 1992. *Hazardous Materials Emergency Action Data.* New York: CRC Press/Lewis Publishers.

Gardner, R. W. 1994. *Chemical Intolerance.* New York: CRC Press/Lewis Publishers.

Hodgson, E., R. B. Mailman, and Janice E. Chambers. 1988. *Dictionary of Toxicology.* London: The MacMillian Company.

Howard, P. H., et. al. 1991. *Handbook of Environmental Degradation Rates.* New York: CRC Press/Lewis Publishers.

Hosty, J. W., and P. Foster. 1990. *A Practical Guide to Chemical Spill Response.* New York: Van Nostrand Reinhold.

Keith, L. H. and M. M. Walker. 1995. *Handbook of Air Toxics.* New York: CRC Press/Lewis Publishers.

Landis, W. G., M. H. Yu. 1995. *Introduction to Environmental Toxicology.* New York: CRC Press/Lewis Publishers.

Lauwerys, R. R., P. Hoet. 1993. *Industrial Chemical Exposure.* New York: CRC Press/Lewis Publishers.

Lewis, R. J., Sr. 1993. *Hawley's Condensed Chemical Dictionary,* 12th Edition. New York: Van Nostrand Reinhold.

Lewis, R. J., Sr. 1992. *Sax's Dangerous Properties of Industrial Materials,* Volumes 1, 2, and 3, 8th Edition. New York: Van Nostrand Reinhold.

Manahan, S. E. 1992. *Toxicological Chemistry,* 2nd Edition. New York: CRC Press/Lewis Publishers.

Meyer, E. 1990. *Chemistry of Hazardous Materials.* Englewood Cliffs, NJ: Prentice Hall.

Ott, W. R. 1995. *Environmental Statistics and Data Analysis.* New York: CRC Press/Lewis Publishers.

Patnaik, P. 1992. *A Comprehensive Guide to the Hazardous Properties of Chemical Substances.* New York: Van Nostrand Reinhold.

Quigley, D. R. 1994. *Handbook of Emergency Chemical Management.* New York: CRC Press/Lewis Publishers.

Rea, W. J. 1995. *Chemical Sensitivity,* Vol. I-III. New York: CRC Press/Lewis Publishers.

Sacarello, H. L. A. 1994. *The Comprehensive Handbook of Hazardous Materials.* New York: CRC Press/Lewis Publishers.

Sheftel, V. O. 1995. *Handbook of Toxic Properties of Monomers and Additives.* New York: CRC Press/Lewis Publishers.

Sherman, J. D. 1988. *Chemical Exposure and Disease.* New York: Van Nostrand Reinhold.

Talbott, E., G. Craun. 1995. *An Introduction to Environmental Epidemiology.* New York: CRC Press/Lewis Publishers.

The Merck Index, 10th edition. 1983. Rahway, NJ: Merck & Company.

Torkelson, T. R. V. K. Rowe. 1981. *Patty's Industrial Hygiene and Toxicology,* Vol. 2B. New York: Wiley-Interscience.

U.S. Department of Labor. 1995. *Code of Federal Regulations, Title 29, Part 1910. Regulations Relating to Labor: Safety for General Industry.* Washington, D.C.: U.S. Government Printing Office.

MATERIAL SAFETY DATA SHEET

CHEMICAL NAME	
ETHYLENEIMINE	**CRC PRESS/LEWIS PUBLISHERS DISCLAIMER:** THE INFORMATION AND RECOMMENDATIONS PRESENTED HEREIN ARE BASED ON SOURCES BELIEVED TO BE RELIABLE. CRC MAKES NO REPRESENTATION ON ITS COMPLETENESS OR ACCURACY. IT IS THE RESPONSIBILITY OF THE USER TO DETERMINE THE CHEMICAL'S SUITABILITY FOR ITS INTENDED USE, THE CHEMICAL'S SAFE USE, AND THE CHEMICAL'S PROPER DISPOSAL. NO REPRESENTATIONS AND/OR WARRANTIES, EITHER EXPRESSED OR IMPLIED, OF THE MERCHANTABILITY OR FITNESS FOR A PARTICULAR PURPOSE, OR OF ANY OTHER NATURE, ARE MADE WITH RESPECT TO THE INFORMATION PROVIDED IN THIS MSDS OR TO THE CHEMICAL TO WHICH INFORMATION MAY REFER. CRC NEITHER ASSUMES NOR AUTHORIZES ANY OTHER PERSON TO ASSUME FOR IT, ANY OTHER ADDITIONAL RESPONSIBILITY OR LIABILITY FOR THE USE OF, OR RELIANCE UPON, THIS INFORMATION.

HAZARD WARNING INFORMATION

HEALTH	FIRE	REACTIVE	OTHER	DEGREE OF HAZARD	COLOR CODING	OTHER CODES
3	3	2		0 = Minimum Hazard 1 = Slight Hazard 2 = Moderate Hazard 3 = Serious Hazard 4 = Severe Hazard	HEALTH = BLUE FIRE = RED REACTIVITY = YELLOW OTHER = WHITE	OX = Oxidizer ACID = Acid ALK = Alkali COR = Corrosive W = Use No Water

SECTION I - GENERAL INFORMATION

Characterization	RCRA Number	EPA Class
Aliphatic Amine	**P054**	**Acute Hazardous Waste**

DOT Proper Shipping Name **Ethyleneimine, Inhibited**	Chemical Abstract Service (CAS) Number **151-56-4**
DOT Hazard Class and Label Requirements **Flammable Liquid and Poison**	DOT Emergency Guide Code **30**
DOT Identification Number **UN 1185**	Chemical Formula **C_2H_5N**

Synonyms
Dimethyleneimine; aminothylene; azirane; aziridine; ethyleneimine; ethylimine.

SECTION II - HAZARDOUS INGREDIENTS AND IDENTITY INFORMATION

Hazardous Components (specific identity)	OSHA Exposure Criteria	NIOSH Exposure Criteria	Immediately Dangerous to Life and Health (IDLH)	ACGIH Exposure Criteria
Ethyleneimine (derivation: From ethyl chloride and ammonia by use of an acid acceptor; by treating 2-chloroethylamine hydrochloride with sodium hydroxide). **1 ppm = 1.79 mg/m^3**	**PEL (8-hour):** **0.5 ppm** **1 mg/m^3** **STEL:** **Not Established**	**REL:** **Lowest Possible Level** **Possible Cancer Agent**	**100 ppm**	**TLV (8-hour):** **0.5 ppm** **1 mg/m^3** **STEL:** **Not Established**

SECTION III - PHYSICAL AND CHEMICAL CHARACTERISTICS

Boiling Point **131°F (55°C)**	Specific Gravity (H_2O = 1) **0.83**
Vapor Pressure (mm Hg) **160 at 69°F (20°C)**	Molecular Weight **43.1**
Vapor Density (Air = 1) **1.5**	Melting Point **-97°F (-71.5°C)**

Solubility
Completely soluble in water. Soluble in most organic solvents.

Appearance and Odor
Clear, colorless liquid with an intense, fishy, ammonia-like odor. Odor Threshold = (ranges) 1.3 - 2 ppm in air.

SECTION IV - FIRE AND EXPLOSION HAZARD DATA

Flash Point (method used) **12°F (-11°C) closed cup**	Explosive Limits in Air % by Volume **LEL: 3.3% UEL: 54.8%**
NFPA Classification **Class IB Flammable Liquid**	Autoignition Temperature **612°F (322°C)**

Extinguishing Media
Use dry chemical, carbon dioxide, alcohol foam, or water spray. Cool exposed container with water.

Special Fire Fighting Procedures
Use full protective clothing and self-contained breathing apparatus (SCBA). Fight fire from distance if possible. Move containers from fire area if it can be done without risk.

Unusual Fire and Explosion Hazards
Containers may explode in fire due to buildup of excessive internal pressures. Do not attempt to extinguish fire unless liquid flow can be stopped first. Aqueous solutions are flammable unless diluted extensively. This chemical presents a vapor explosion hazard indoors, outdoors, and in sewers.

SECTION V - REACTIVITY DATA

Stability		Conditions to Avoid **Ethyleneimine can be kept stable in closed containers containing a lump of caustic sodium hydroxide (NaOH) to inhibit polymerization. Avoid contact with incompatible materials. Keep away from heat.**
Stable X	Unstable	Incompatibility *(materials to avoid)* **Reacts with acetic anhydride, acrolein, allyl chloride, carbon disulfide, chlorine, epichlorohydrin, glyoxal, oleum, β-propiolacetone, silver, sodium hypochlorite, vinyl acetate, and acids.**
Hazardous Polymerization		Conditions to Avoid **Will polymerize explosively in the presence of acids. Can also polymerize if there is an insufficient concentration of inhibitor, at elevated temperatures, or in contact with a catalyst.**
May Occur X	Will Not Occur	Hazardous Decomposition or By-products **Toxic and irritating fumes and vapors are produced when ethyleneimine is heated to decomposition. These include oxides of carbon and nitrogen.**

SECTION VI - HEALTH HAZARD DATA

Primary Route(s) of Entry:	Inhalation?	Absorption (skin)?	Ingestion?
	X	X	X

Health Hazards

INHALATION: Symptoms of nausea, coughing, shortness of breath and difficulty breathing, severe irritation of the nose, throat, and lungs. May cause pulmonary edema (fluid buildup in the lungs) and kidney damage. May also cause degenerative changes in the heart.

ABSORPTION: A skin sensitizer. Will pass through intact skin to cause toxic systemic effects similar to those of inhalation. Causes opaque cornea, keratoconus and necrosis of the cornea.

INGESTION: Severe pain with vomiting, diarrhea, collapse, breathing difficulties, and chest pain.

Carcinogenicity	NTP Listed?	IARC Cancer Review Group?	OSHA Regulated?	Target Organs?
Suspected Human Suspected Animal	No	Group 3	29 CFR 1910.1000 Table Z-1	Eyes, skin, lungs, resp. sys., kidneys, repro., CNS.

Medical Conditions Generally Aggravated by Exposure

Liver, kidney, and pulmonary diseases may be aggravated by exposure.

Emergency and First-aid Procedures

Eye contact: Flush immediately with water for 15 minutes (minimum), seek medical attention. Skin contact: Remove all contaminated clothing. Immediately wash area with large amounts of soap and water. Seek medical assistance. For inhalation: Remove the person from exposure. Provide respiratory assistance and CPR. Transfer to medical facility. Observe for 24-48 hours for lung effects. If swallowed: Call the poison control center and seek medical attention immediately. Unless advised otherwise, give 1-2 glasses of water or milk. Do NOT induce vomiting. Never attempt to give an unconscious or convulsing person anything by mouth.

SECTION VII - PRECAUTIONS FOR SAFE HANDLING AND USE

Steps to be Taken in Case Material is Released or Spilled

Neutralize with sand and soda ash at a ratio of 9 to 1. Collect materials using non-sparking tools and deposit in DOT-approved drum. Ventilate area and remove ignition sources. Restrict those not involved in cleanup from entering area. Notify appropriate authorities, if applicable.

Preferred Waste Disposal Method

No Citation.

Precautions to be Taken in Handling and Storage

Do not store ethyleneimine in the presence of incompatible chemicals or materials. Store in tightly closed containers in a cool, well-ventilated area. Protect containers from physical damage and provide shade from radiant heat (including direct sunlight).

Other Precautions and Warnings

Recommend explosion-proof design for bulk storage facilities. Electrically ground and bond equipment, especially during transfer operations. Detached facility storage is recommended.

SECTION VIII - CONTROL MEASURES AND PERSONAL PROTECTIVE EQUIPMENT

Respiratory Protection *(specify type)*

For low exposures, use an organic vapor respirator. Otherwise, use an MSHA/NIOSH-approved supplied-air respirator with full facepiece operated in positive pressure mode or a self-contained breathing apparatus (SCBA) with full facepiece operated in pressure demand or other positive pressure mode.

Ventilation

Local exhaust or general mechanical systems recommended.

Protective Gloves	Eye Protection	Other Protective Clothing
Impervious Butyl Rubber Gloves	**Chemical Goggles or Face Mask**	**Protective Apron**

Work/Hygiene Practices

Always wash hands thoroughly after using chemical; never bring food, drink, or smoking materials into vicinity of chemicals.

ETHYLENEIMINE

C_2H_5N CAS: 151-56-4

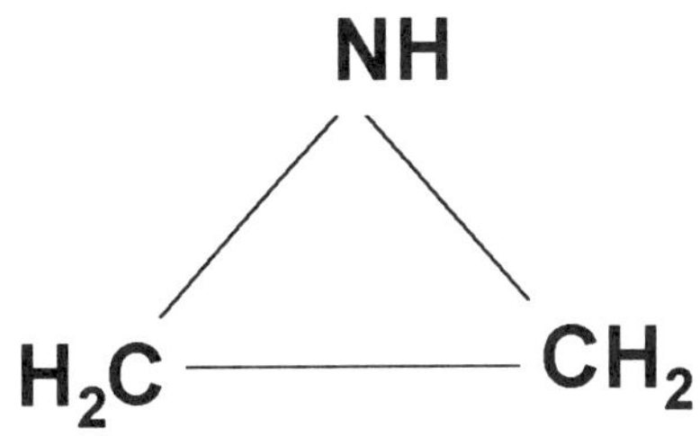

IDENTIFICATION AND TYPICAL USES

Ethyleneimine is a colorless or clear liquid with a strong, fishy, ammonia-like odor. It is used in the manufacture of triethylenemelamine in the paper industry and as a flocculation agent in the clarification of effluents. It is also an intermediate and monomer for fuel-oil and lubricant refining. It is used in ion exchange, in protective coatings, in the manufacture of pharmaceuticals, adhesives, as a polymer stabilizer, and in surfactants. Alkyl-substituted forms (known as alkyl aziranes) are used as intermediates and for microbial control.

RISK ASSESSMENT: HEALTH

General Assessment

Ethyleneimine is a severe irritant and is corrosive on contact with human tissue. It is extremely toxic by all routes of exposure (*inhalation*, *ingestion*, and by skin *absorption*). There is sufficient evidence in the references to indicate that exposure to ethyleneimine poses a significant cancer risk. Animal carcinogenicity is considered highly suspect. Human mutation data have also been reported and it is suggested that this chemical should be treated as though it were a human carcinogen. There is also sufficient evidence to indicate that this chemical causes teratogenic effects in test animals.

Inhalation will result in irritation of the mucosa, especially in the nose, throat, and upper respiratory tract. The toxic effects of exposure include dizziness, headache, congestion, coughing, nausea, vomiting (may be delayed), severe chest pains, difficulty or labored breathing, hemorrhaging, kidney damage, proteinuria (excessive protein content in urine), increased BUN (blood urea nitrogen), and delayed lung injury or pulmonary edema (fluid buildup in the lungs), which is a medical emergency and can be fatal. Symptoms of pulmonary edema can be delayed up to 48 hours, thereby creating a false sense of security with regard to health exposure risk. There is also some evidence to indicate that high exposures may cause severe tissue degeneration of the heart.

Skin contact with the liquid can cause severe irritation with a possibility for 1st and even 2nd degree burns on prolonged contact. Eye contact can result in permanent corneal injury, conjunctival scarring, and permanent damage or loss of vision. Vapor can cause foggy vision, keratoconus, and cornea necrosis.

Ingestion causes vomiting, dizziness, headache, nausea and vomiting (possibly delayed), possible scarring of the esophagus, and pulmonary edema. It may also lead to liver and heart injuries (confirmed in animals studies only).

☠ *Acute Health Effects*

The following acute (short-term) health effects may occur immediately or shortly after exposure to ethyleneimine:

Skin: Irritation, inflammation, redness, possible blistering, serious burns, and tissue damage. May cause necrosis and vesiculation.

Eye: Severe irritation, inflammation, with possible damage to the cornea and loss of vision. Vapor can cause foggy vision, the appearance of halos around lights, and inflammation and scarring of conjunctival tissues.

Lung: Irritation causing coughing and/or shortness of breath. Higher exposures can lead to respiratory problems, including pulmonary edema (fluid buildup in lungs) which is a medical emergency and can be fatal.

Other: Animal studies have shown a degeneration of heart tissue and liver damage on high exposures to amines. It is not known if these effects will occur in humans.

Chronic Health Effects

The following chronic (long-term) health effects may occur at some time after exposure to ethyleneimine and can last for months or even years:

Cancer Hazards: According to the information presented in the references, ethyleneimine may cause cancer in humans since there is sufficient evidence to show that it will cause cancer (lung and liver) in animals. Human mutation (genetic changes) data have also been reported.

Reproductive Hazard: According to the information presented in the references, ethyleneimine may cause reproductive toxicity, indicating specific developmental abnormalities in the central nervous system, eyes, and ears.

Other Chronic Effects: Low, chronic exposures may cause corneal damage, foggy vision, and irritation of the lungs, bronchi, and throat. There is also a possibility that chronic exposures may cause injury to the heart and liver. However, the data are inconclusive in this regard.

Recommended Risk-Reduction Measures

Ethyleneimine is a suspected human carcinogen. Unless a less toxic chemical can be substituted for a hazardous substance, *engineering controls* are the most effective method of reducing exposures. The best protection is to enclose operations and/or provide local exhaust ventilation at the site of chemical release. To prevent static sparks, all containers and equipment should be electrically grounded and bonded. While not always operationally feasible, isolating operations can also reduce exposure.

Using respiratory protection is less effective than the controls mentioned above, but is still advisable whenever working with or around ethyleneimine. For relatively low, infrequent, or transient exposures, an organic vapor respirator may suffice. For higher or prolonged exposure, a self-contained breathing apparatus (SCBA) with full facepiece and pressure demand is the recommended respiratory protection of choice. If a full facepiece is not available, then chemical goggles should be worn to protect the eyes. Whenever a chemical splash hazard exists, a face shield and apron should be worn. To prevent hand and skin exposures, impervious chemical gloves should be worn.

Administrative controls should also be in place to minimize the potential for human exposures. These may include written procedures or policies which specify the methods and techniques that will be practiced whenever personnel are to work with ethyleneimine.

All personnel should receive *training* on the use, hazards, protective measures, emergency actions, and other precautions per 29 CFR 1910.1200 (Hazard Communication), prior to the first assignment in an area where ethyleneimine is used or stored. OSHA 20 CFR 1910.1012 ("Ethyleneimine") contains specific administrative and engineering controls that are required whenever using this chemical.

Before beginning employment and at regular intervals thereafter (e.g. annually), the following medical tests are recommended:

- ☑ Lung function tests (establish a baseline).

If symptoms develop or overexposure is suspected, the following medical tests may be useful:

- ☑ Lung function tests.
- ☑ Consider chest X-ray after acute overexposure (may be negative if taken immediately after exposure due to delayed development of pulmonary edema).
- ☑ Complete blood count and blood gases (check BUN levels).
- ☑ For oral exposures, evaluate for esophageal burns.
- ☑ Evaluation by a qualified allergist, including careful consideration of exposure history and special testing (may help diagnose skin allergy).

Any evaluation should include a careful medical history of past and present symptoms with an examination. However, medical tests that evaluate existing damage are not a substitute for controlling exposure. Also, since smoking can cause heart disease, emphysema, and numerous other respiratory disorders, smokers may be affected more rapidly than non-smokers under the same conditions of exposure. Prudent risk assessment and management requires careful consideration of *all* possible risk factors that may be causing the appearance of symptoms in exposed workers.

Other methods to reduce exposure include:

- ☑ Where possible, enclose operations and use local exhaust ventilation at the site of chemical release. If local exhaust ventilation or enclosure is not used, respiratory protection should be mandatory. Also, if possible, automatically transfer liquids from storage containers to process containers using non-sparking tools.

- ☑ A marked, regulated control area must be established wherever ethyleneimine is used, handled, or stored (refer to OSHA 29 CFR 1910.1012 for specific control measures).
- ☑ Always ensure that proper protective clothing is worn when using chemical substances.
- ☑ Wash thoroughly immediately after exposure to ethyleneimine and at the end of the work shift or before eating, drinking, or smoking.
- ☑ Hazard warning information should be posted in the work area. In addition, as part of an on-going education and training program, all information on the health and safety hazards of ethyleneimine should be communicated to all potentially exposed workers.
- ☑ Eye wash stations should be provided in work areas. If the potential for whole body exposure exists, then safety showers should also be provided.

RISK ASSESSMENT: ENVIRONMENT

General Assessment

The environment is at risk of exposure during intended use, transportation, storage, disposal, or destruction of ethyleneimine. In almost every scenario, the threat of environmental exposure is contingent upon the proper handling of the chemical substance. Accidental spills, large or small, can result in fire, explosion, and possible contamination of the surrounding environmental mediums (water, soil, and air).

Ethyleneimine is considered a Class IB flammable liquid (per OSHA 29 CFR 1910.106). Its extremely low flash point and its relatively low boiling point make it a serious and dangerous fire and explosion risk. It can react with many common acids and many oxidizers to cause a fire or explosion. These characteristics require special consideration during any emergency situation involving a leak or spill of ethyleneimine. Should ethyleneimine ever come into contact with these incompatible substances either during use, transportation, or storage, violent reactions can occur.

Ethyleneimine can enter the environment through industrial discharges, municipal waste treatment discharges, and spills.

Acute Ecological Effects

Acute (short-term) toxic effects may include the death of animals, birds, or fish and death or low growth rate in plants. Acute effects are seen 2 to 4 days after animals or plants are exposed to ethyleneimine.

This chemical has moderate acute toxicity to aquatic life. It has caused germination decrease, sterility increase, and mutation increases in various crops. Insufficient data are available to evaluate or predict the short-term effects of ethyleneimine to birds or land animals.

Chronic Ecological Effects

Chronic toxic effects may include shortened life span, reproductive problems, lower fertility, and changes in appearance or behavior in exposed animals. These effects can be seen long after first exposure(s) to toxic chemicals.

Ethyleneimine has moderate chronic toxicity to aquatic life. Insufficient data are available to evaluate or predict the long-term effects of ethyleneimine to plants, birds, or land animals.

Water Solubility

Ethyleneimine is highly soluble in water. Concentrations of 1000 milligrams and more will mix with a liter of water.

Persistence in the Environment

Ethyleneimine is moderately persistent in water, with a half-life of between 20 to 200 days. The half-life of a pollutant is the amount of time it takes for one half of the chemical to be degraded. About 93% of ethyleneimine will eventually end up in the water; the rest will end up in the air. It is not expected to accumulate in aquatic sediments or terrestrial soils.

Bioaccumulation in Aquatic Organisms

Some substances increase in concentration, or *bioaccumulate,* in living organisms as they breathe contaminated air, drink contaminated water, or eat contaminated food. These chemicals can become concentrated in the tissues and internal organs of animals as well as humans.

The concentration of ethyleneimine found in fish tissues is expected to be about the same as the average concentration of ethyleneimine in the water from which the fish was taken.

ⓤ Recommended Risk-Reduction Measures

Proper training of all transporters will reduce the likelihood of a mishap or accident resulting in a leak or spill to the environment. The correct use of DOT labeling (per the MSDS) on all containers, trucks, and rail cars will enable emergency responders to react properly and quickly to any disaster thereby reducing the potential risk to the environment and to personnel.

Storage of ethyleneimine should be segregated from incompatible chemicals to minimize the risk of cross-contamination or contact. Storage buildings should be equipped with the proper fire protection equipment (alarms, sprinklers, extinguishers, etc.). Explosion-proof design is highly recommended for all storage facilities and operational equipment.

If a spill or leak to the environment has occurred, fire department, emergency response and/or hazardous materials spill personnel should be notified immediately. Cleanup should be attempted only by those trained in proper spill containment procedures. Contaminated soils should be removed for incineration and replaced with clean soil. If ethyleneimine should contact the water table, aquifer, or navigable waterway, remediation activities should be prompt. When such spills occur, the local and/or state emergency response authorities must be notified. A comprehensive emergency response or disaster preparedness/recovery plan should be in place prior to any operations involving the use, transportation, storage, or disposal of ethyleneimine. If ethyleneimine is spilled or leaked, the following specific steps are recommended:

- ☑ Restrict persons not wearing protective clothing from area of spill or leak until cleanup is complete and area can be opened for normal work
- ☑ Ventilate area and remove all ignition sources.
- ☑ Neutralize with a 9 to 1 solution of soda ash and sand and place in a sealed drum for disposal. Use non-sparking tools during cleanup.
- ☑ Use water spray to disperse vapors and dilute spills to non-flammable mixtures. Flush spills away from exposures. Dispose of accordingly.
- ☑ It may be necessary to dispose of ethyleneimine as a hazardous waste. The responsible state agency or the regional office of the federal Environmental Protection Agency (EPA) should be contacted for specific recommendations.

RISK ASSESSMENT: BUSINESS

General Assessment

Accidents or mishaps involving ethyleneimine can present a significant threat to business operations. The loss or damage of equipment or facilities can significantly affect fiscal viability. Lawsuits that may result from personnel injury/death, public exposures, and/or environmental contamination will also require a serious expenditure of resources. Media attention surrounding an injury, death, or environmental damage can also result in a loss of profits and loss of current as well as future business. Always remember that anytime the terms "cancer," carcinogen," or reproductive hazard" are used, public emotion, hysteria, and ignorance can run equally high. This requires careful consideration during the development of any public relations policy.

ⓤ Recommended Risk-Reduction Measures

Company attorneys, safety and health professionals, and environmental specialists should be involved in the development of any procedures dealing with safety or emergency response. A company official should be pre-designated as a public relations officer with specific training in dealing with the press. Corporate plans and policies should be developed, approved, and implemented long before any need for such arises.

REFERENCES

Anthony, C. P., and N. J. Kolthoff. 1971. *Textbook of Anatomy and Physiology.* St. Louis: C. V. Mosby Company.

Bueche, J. R. 1972. *Principles of Physics,* 2nd Edition. New York: McGraw-Hill.

Hodgson, E., R. B. Mailman, and Janice E. Chambers. 1988. *Dictionary of Toxicology.* London: The MacMillian Company.

Lewis, R. J., Sr. 1993. *Hawley's Condensed Chemical Dictionary,* 12th Edition. New York: Van Nostrand Reinhold.

Lewis, R. J., Sr. 1992. *Sax's Dangerous Properties of Industrial Materials,* Volumes 1, 2, and 3, 8th Edition. New York: Van Nostrand Reinhold.

Meyer, E. 1990. *Chemistry of Hazardous Materials.* Englewood Cliffs, NJ: Prentice Hall.

Richardson, M. 1995. *Dictionary of Substance and Their Effects,* Vol. I-VI. New York: CRC Press/Lewis Publishers.

Sacarello, H. L. A. 1994. *The Comprehensive Handbook of Hazardous Materials.* New York: CRC Press/Lewis Publishers.

U.S. Department of Health and Human Services, National Institute for Occupational Safety and Health. 1994. *NIOSH Pocket Guide to Chemical Hazards.* Washington, D.C.: U.S. Government Printing Office.

MATERIAL SAFETY DATA SHEET

CHEMICAL NAME	
ETHYLENE OXIDE	**CRC PRESS/LEWIS PUBLISHERS DISCLAIMER:** THE INFORMATION AND RECOMMENDATIONS PRESENTED HEREIN ARE BASED ON SOURCES BELIEVED TO BE RELIABLE. CRC MAKES NO REPRESENTATION ON ITS COMPLETENESS OR ACCURACY. IT IS THE RESPONSIBILITY OF THE USER TO DETERMINE THE CHEMICAL'S SUITABILITY FOR ITS INTENDED USE, THE CHEMICAL'S SAFE USE, AND THE CHEMICAL'S PROPER DISPOSAL. NO REPRESENTATIONS AND/OR WARRANTIES, EITHER EXPRESSED OR IMPLIED, OF THE MERCHANTABILITY OR FITNESS FOR A PARTICULAR PURPOSE, OR OF ANY OTHER NATURE, ARE MADE WITH RESPECT TO THE INFORMATION PROVIDED IN THIS MSDS OR TO THE CHEMICAL TO WHICH INFORMATION MAY REFER. CRC NEITHER ASSUMES NOR AUTHORIZES ANY OTHER PERSON TO ASSUME FOR IT, ANY OTHER ADDITIONAL RESPONSIBILITY OR LIABILITY FOR THE USE OF, OR RELIANCE UPON, THIS INFORMATION.

HAZARD WARNING INFORMATION

HEALTH	FIRE	REACTIVE	OTHER	DEGREE OF HAZARD	COLOR CODING	OTHER CODES
2	4	3		0 = Minimum Hazard 1 = Slight Hazard 2 = Moderate Hazard 3 = Serious Hazard 4 = Severe Hazard	HEALTH = BLUE FIRE = RED REACTIVITY = YELLOW OTHER = WHITE	OX = Oxidizer ACID = Acid ALK = Alkali COR = Corrosive W = Use No Water

SECTION I - GENERAL INFORMATION

Characterization	**Epoxide Ether**	RCRA Number	**U115**
		EPA Class	**Toxic Waste**
DOT Proper Shipping Name	**Ethylene Oxide**	Chemical Abstract Service (CAS) Number	**75-21-8**
DOT Hazard Class and Label Requirements	**Flammable Liquid; Flammable Gas**	DOT Emergency Guide Code	**69**
DOT Identification Number	**UN 1040**	Molecular Formula	C_2H_4O

Synonyms

Dimethylene oxide; 1,2-epoxy ethane; oxirane; amprolene; anprolene; dihydroxirene; ETO; oxyfume; oxyfume 12; dimethyl oxide; carboxyfume.

SECTION II - HAZARDOUS INGREDIENTS AND IDENTITY INFORMATION

Hazardous Components (specific identity)	OSHA Exposure Criteria	NIOSH Exposure Criteria	Immediately Dangerous to Life and Health (IDLH)	ACGIH Exposure Criteria
Ethylene oxide: (derivation: By oxidation of ethylene in air or oxygen with silver catalyst; or by action of an alkali on ethylene chlorohydrin). 1 ppm = 1.83 mg/m^3	**PEL (8-hour):** 1 ppm 1.8 mg/m^3 **CEILING:** 5 ppm 9 mg/m^3	**REL:** <0.1 ppm <0.18 mg/m^3 **CEILING** (10 min) 9 mg/m^3 Possible Cancer Agent	800 ppm	**TLV:** 1 ppm 1.8 mg/m^3 Suspected Cancer Agent

SECTION III - PHYSICAL AND CHEMICAL CHARACTERISTICS

Boiling Point	**51°F (11°C)**	Specific Gravity (H_2O = 1)	**0.9**
Vapor Pressure (mm Hg)	**1.46 at 77°F (25°C)**	Molecular Weight (atomic weight)	**44.1**
Vapor Density (Air = 1)	**1.52**	Melting Point	**-170°F (-112°C)**

Solubility

Miscible in water, alcohol, acetone, benzene, ether, carbon tetrachloride, and most organic solvents.

Appearance and Odor

Colorless gas (or liquid below 51°F) with an ether-like odor. Odor Threshold = 430 ppm.

SECTION IV - FIRE AND EXPLOSION HAZARD DATA

Flash Point (method used)	**-4°F (-20°C) closed cup (liquid)**	Explosive Limits in Air % by Volume	**LEL: 3% UEL: 100%**
NFPA Classification	**Class IA Flammable Liquid**	Autoignition Temperature	**804°F (429°C)**

Extinguishing Media

Dry chemical, carbon dioxide, alcohol-resistant foam, water spray, or fog.

Special Fire Fighting Procedures

Poisonous gases are produced in fire. Wear full protective clothing and self-contained breathing apparatus (SCBA). Move container from fire area if it can be done without risk. Fight fire from maximum distance. Stay away from ends of tanks. If gas fire, stop flow before attempting to extinguish fire.

Unusual Fire and Explosion Hazards

Containers may explode in fire. Boiling liquid expanding vapor exposure (BLEVE) possible. Immediately withdraw if rising sound is heard from venting device. Vapors are heavier than air and explosive.

SECTION V - REACTIVITY DATA

Stability		Conditions to Avoid **Highly reactive and potentially unstable. Do not store for more than 60 days. It is relatively stable in aqueous solutions. Keep containers cool at all times. Avoid contact with heat and ignition sources.**
Stable	Unstable X	Incompatibility *(materials to avoid)* **Incompatible with oxidizers, alcohols, mercaptans, magnesium perchlorates, alkane thiols, trimethylamine, copper, and bromomethane. It is absorbed by leather, rubber, and many plastics.**
Hazardous Polymerization		Conditions to Avoid **Polymerization occurs with ammonia, alkali hydroxides, amines, metallic potassium, acids, covalent halides (aluminum chloride, iron chloride, tin chloride, aluminum oxide, iron oxide, rust), hot glycerol.**
May Occur X	Will Not Occur	Hazardous Decomposition or By-products **When heated to decomposition, ethylene oxide emits carbon dioxide, acrid smoke and irritating fumes.**

SECTION VI - HEALTH HAZARD DATA

Primary Route(s) of Entry:	Inhalation?	Absorption (skin/eye)?	Ingestion?
	X	X	X

Health Hazards

INHALATION: Severe irritation to the eyes, nose, throat, and lungs. Causes headache, nausea, vomiting, difficulty breathing, cyanosis, pulmonary edema (fluid in the lungs), depression of the CNS with drowsiness, weakness, loss of coordination, and convulsions.

ABSORPTION: A severe eye and skin irritant. Can pass through unbroken skin causing toxic systemic effects. Causes skin burns and eye irritation.

INGESTION: Unlikely. If liquid is swallowed, it may cause gastrointestinal tract irritation and burns.

Carcinogenicity	NTP Listed?	IARC Cancer Review Group?	OSHA Regulated?	Target Organs?
Suspected Human Confirmed Animal	5th Annual Report	Group 2A	29 CFR 1910.1047	Skin, eyes, resp., repro., liver, blood, CNS, kidneys.

Medical Conditions Generally Aggravated by Exposure

None reported. However, existing respiratory problems (bronchitis, asthma) may be aggravated.

Emergency and First-aid Procedures

Eye contact: Do not allow victim to keep eyes tightly shut. Flush immediately with water for 15 minutes (minimum). Skin contact: Wash with large amounts of water for a minimum of 15 minutes. For inhalation: Remove the person from exposure. Provide respiratory assistance and CPR. Transfer to medical facility. Observe for 24-48 hours for lung effects. If swallowed: Seek medical attention immediately. Give 1-2 glasses of water, induce vomiting then 2 tablespoons activated charcoal in 8 ounces of water.

SECTION VII - PRECAUTIONS FOR SAFE HANDLING AND USE

Steps to be Taken in Case Material is Released or Spilled

Wear self-contained breathing apparatus (SCBA). Stop flow of gas. Restrict those not wearing protective equipment and who are not involved in cleanup from entering area. Avoid skin contact. Ventilate area of spill. Absorb liquids in vermiculite or other absorbent and deposit in sealed drum.

Preferred Waste Disposal Method

No Citation.

Precautions to be Taken in Handling and Storage

Store in tightly closed, insulated, steel containers in a cool, well ventilated area away from heat. Protect containers from physical damage. Regularly inspect for leaks, cracks, or faulty valves.

Other Precautions and Warnings

Electrically ground and bond all electrical equipment. Monitor inventory; do not store more than 60 days.

SECTION VIII - CONTROL MEASURES AND PERSONAL PROTECTIVE EQUIPMENT

Respiratory Protection *(specify type)*

Ethylene oxide is a suspected human carcinogen. Use a NIOSH/MSHA-approved supplied-air respirator or a self-contained breathing apparatus (SCBA) with full facepiece operated in positive pressure mode.

Ventilation

Local exhaust (preferred) at site of chemical work or general ventilation.

Protective Gloves	Eye Protection	Other Protective Clothing
Polyvinylchloride	Chemical Goggles and Face Mask	Protective Uniform or Apron

Work/Hygiene Practices

Always wash hands thoroughly after using chemical; never bring food, drink, or smoking materials into vicinity of chemicals.

ETHYLENE OXIDE

C_2H_4O CAS: 75-21-8

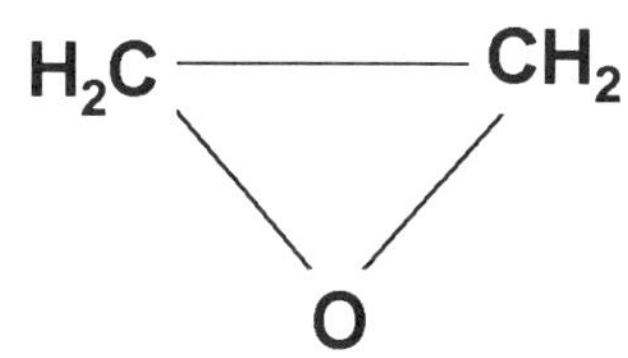

IDENTIFICATION AND TYPICAL USES

Ethylene oxide appears as a colorless gas (or liquid below 51°F) with an ether-like odor. It is one of the 25 most produced chemicals in the United States. It is used widely as a sterilization agent and as a fumigant. It is also used as a propellant, as a petroleum de-emulsifier, to accelerate the maturing of tobacco leaves, in the production of explosives, in the manufacture of ethylene glycol, polyethylene oxide, glycol ethers, crown ethers, ethanolamines, and other derivatives, and in organic synthesis. Also investigated for use as an agent to improve the durability of wood.

RISK ASSESSMENT: HEALTH

General Assessment

Ethylene oxide is a severe irritant as well as toxic, carcinogenic, mutagenic, and teratogenic compound. As a gas, it can enter the body through *inhalation.* However, it can also exist as a liquid (below 51°F) and, therefore, *absorption* through intact skin and even *ingestion* are also possible exposure routes.

Inhalation of ethylene oxide can cause severe irritation to the eyes, nose, throat, and respiratory system. Symptoms can be delayed for 1 to 5 hours and may include headache, nausea, vomiting, dypsnea (difficult breathing), diarrhea, cyanosis (bluish discoloration of the lips, fingers, and toes), olfactory changes (numbing of the sense of smell making detection more difficult), and possible CNS depression with drowsiness, weakness, loss of coordination, memory impairment, peripheral neuropathy, convulsions, coma, and death. Lung irritation can be quite severe and may lead to pulmonary edema (fluid buildup in the lungs), which is a medical emergency and can be fatal. The symptoms of pulmonary edema can be delayed up to 48 hours, thereby creating a false sense of security with regard to health exposure risk.

Skin contact with pure anhydrous ethylene oxide has very little noticeable effect on dry skin. However, contact with solutions cause characteristic burns due to frostbite resulting from evaporative cooling. Also, the liquid solution will cause edema (fluid-filled) blistering of the surface of the skin with redness and dryness. Healing takes approximately three weeks and may leave a residual discoloration (usually brown) on the skin's surface. Ethylene oxide will pass through intact skin to cause toxic systemic effects. If inhalation has also occurred, the compounded effects may be quite severe (life threatening). Eye contact causes irritation, watering, inflammation of the lids, chemical burns to the cornea, and (possibly) severe lesions.

Although ingestion is not likely under normal operating conditions (ethylene oxide is normally a gas), the liquid may be swallowed under conditions of emergency or other unplanned circumstance. If swallowed, symptoms of systemic toxicity similar to that of inhalation may result.

☠ *Acute Health Effects*

The following acute (short-term) health effects may occur immediately or shortly after exposure to ethylene oxide:

Skin: Severe irritation with localized tissue damage and possible burns. Can be absorbed through intact skin to cause toxic systemic effects.

Eye: Severe conjunctival irritation and possible damage to the cornea and loss of vision.

Lung: Severe irritation to respiratory tract causing shortness of breath and cough. High acute exposures can be fatal due to a buildup of fluid in the lungs (pulmonary edema).

CNS: Possible depression of the central nervous system causing a variety of toxic effects including loss of coordination and consciousness.

Other: Ingestion will result in gastritis and possible damage to the digestive tract. Kidney damage may also occur as a result of high/acute overexposures.

Chronic Health Effects

The following chronic (long-term) health effects may occur at some time after exposure to ethylene oxide and can last for months or even years:

Cancer Hazards: Ethylene oxide may be a cancer causing agent in humans since it has been shown to cause a number of cancers in animals, including brain, liver, gastrointestinal, and blood cancers (leukemia). It is a mutagen (causes cell changes) in animals.

Reproductive Hazard: According to the references, ethylene oxide has been shown to cause serious teratogenic effects. Laboratory tests on animals indicate that exposure could cause fetal deaths, specific development abnormalities, and paternal effects related to testes and sperm ducts. There are reports of birth defects caused by exposure to ethylene oxide and, there is evidence of increased spontaneous abortions in hospital workers.

Other Chronic Effects: Repeated low-dose exposures can cause olfactory changes (numbing of the sense of smell), memory impairment, peripheral neuropathy (sensory and reflex loss and nerve changes in the extremities), which may resolve in 1-7 months. There may be alkylated hemoglobin, systemic sensitization, increased possibility of leukemia and other cancers.

Recommended Risk-Reduction Measures

It is very possible that exposure to ethylene oxide may lead to the development of cancer in humans. Personnel should, therefore, avoid direct contact with ethylene oxide. If a less toxic material or compound cannot be substituted for ethylene oxide, then *engineering controls* are the most effective method of reducing exposure risk. OSHA has established a standard (29 CFR 1910.1047) to specifically regulate ethylene oxide. The best protection is to enclose operations and/or provide local exhaust ventilation at the site of ethylene oxide release. A regulated and controlled work area must be established wherever ethylene oxide is used, handled, or stored. While not always operationally feasible, isolating operations can also reduce exposure. Using respiratory protection is less effective than the controls mentioned above, but is still required whenever working with or around ethylene oxide. Exposure levels are extremely low (1 ppm) and difficult to accurately monitor. Since the odor threshold is very high (430 to 500 ppm or higher) and, since ethylene oxide will also numb the sense of smell, personnel will not have adequate warning of hazardous exposures. Therefore, for the best protection at any exposure level, a MSHA/NIOSH-approved supplied-air respirator or a self-contained breathing apparatus (SCBA) with full facepiece and operated in pressure demand mode will provide the appropriate level of respiratory protection. If a full facepiece is not available, then chemical goggles should be worn to protect the eyes, especially when a splash hazard exists. A face shield should also be considered. To prevent hand and skin exposures, polyvinylchloride gloves should be worn.

Administrative controls should also be in place to minimize the potential for human exposures. These may include written procedures or policies which specify the methods and techniques that will be practiced whenever personnel are to work with ethylene oxide.

All personnel should receive *training* on the use, hazards, protective measures, emergency actions, and other precautions per OSHA 29 CFR 1910.1047 and 29 CFR 1910.1200 (Hazard Communication) prior to the first assignment in an area where ethylene oxide is used or stored.

Before beginning employment and at regular intervals thereafter (e.g., annually), the following medical test is recommended:

☑ Lung function tests.

If symptoms develop or overexposure is suspected, the following may be useful:

☑ Liver, kidney, and lung function tests.

☑ Consider chest X-ray after acute exposure (may be negative if taken immediately following exposure due to delayed development of pulmonary edema).

☑ Complete blood count.

☑ Evaluation by a qualified allergist, including careful consideration of exposure history and special tests (may help diagnose skin allergy).

☑ Examination and evaluation of the reproductive system.

It should be noted that medical tests that simply look for existing damage are not a substitute for controlling exposures. Medical histories are extremely important when assessing exposure risk. Also, since smoking can cause heart disease as well as lung cancer, emphysema, and other respiratory ailments, any exposure to ethylene oxide can result in quicker and more devastating symptoms. Smokers should therefore avoid unprotected contact with this chemical. Prudent risk management requires proper consideration of *all* possible factors that may be causing the appearance of exposure symptoms among workers.

Other methods to reduce exposure include:

- ☑ Always ensure that proper protective clothing is worn when using chemical substances.
- ☑ Wash thoroughly immediately after exposure to ethylene oxide and at the end of the work shift or before eating, drinking, or smoking.
- ☑ Hazard warning information should be posted in the work area. In addition, as part of an on-going education and training program, all information on the health and safety hazards of ethylene oxide should be communicated to all potentially exposed workers.
- ☑ Eye wash stations should be provided in the immediate work area for emergency use. If there is a possibility of skin exposure to ethylene oxide, emergency shower facilities should also be provided.
- ☑ Workers whose clothing has been contaminated by ethylene oxide should change into clean clothes before leaving work. Contaminated work clothing should be laundered only by individuals who have been informed of the hazards of exposure to ethylene oxide.

RISK ASSESSMENT: ENVIRONMENT

General Assessment

The environment is at risk of exposure during transportation, storage, disposal, or destruction of ethylene oxide. In almost every scenario, the threat of environmental exposure is contingent upon the proper handling of the chemical substance. Accidental spills, large or small, can result in contamination of the surrounding environmental mediums (water, soil, and air).

Ethylene oxide is flammable gas and, when in the liquid state, a Class IA combustible liquid (per OSHA 29 CFR 1910.106). Its extremely low flash point and low boiling point together with the fact that it has a very wide flammability range (between 3 and 100% in air), create a dangerous fire and explosion risk. It is incompatible with oxidizers, acids, alkalis, and the chlorides of many metals. Caution is always required in handling, storage, transportation, and disposal of ethylene oxide. When heated to decomposition it emits very acrid and irritating smoke and fumes. Emergency responders should be made aware of the presence of ethylene oxide at any emergency response situation. A boiling liquid expanding vapor explosion (BLEVE) is extremely possible if a tank or rail car containing ethylene oxide is involved in fire.

Ethylene oxide may enter the environment through industrial discharges and from spills.

Acute Ecological Effects

Acute (short-term) toxic effects may include the death of animals, birds, or fish and death or low growth rate in plants. Acute effects are seen 2 to 4 days after animals or plants are exposed to ethylene oxide.

Ethylene oxide has moderate acute toxicity to aquatic life. Various rose varieties have shown delayed maturation, wilting, and size decrease from exposure to ethylene oxide. Insufficient data are available to evaluate or predict the acute (short-term) effects of ethylene oxide on birds or land animals.

Chronic Ecological Effects

Chronic (long-term) toxic effects may include shortened life span, reproductive problems, lower fertility, and changes in appearance or behavior in exposed animals. These effects can be seen long after first exposure(s) to toxic chemicals.

Ethylene oxide has moderate chronic toxicity to aquatic life. Insufficient data are available to evaluate the chronic (long-term) effects of ethylene oxide on plants, birds, or land animals.

Water Solubility

Ethylene oxide is highly soluble in water. Concentrations of 1000 milligrams and more will mix with a liter of water.

Persistence in the Environment

Ethylene oxide is slightly persistent in the aquatic environment, with a half-life between 2 and 20 days. The half-life of a pollutant is the amount of time it takes for one-half of the chemical to be degraded.

Bioaccumulation in Aquatic Organisms

Some substances increase in concentration, or *bioaccumulate,* in living organisms as they breathe contaminated air, drink contaminated water, or eat contaminated food. These chemicals can become concentrated in the tissues and internal organs of animals as well as humans.

The concentration of ethylene oxide found in fish tissues is expected to be about the same as the average concentration of ethylene oxide in the water from which the fish was taken.

Recommended Risk-Reduction Measures

Proper training of all transporters will reduce the likelihood of a mishap or accident resulting in a leak or spill of ethylene oxide into the environment. Labels on all containers, trucks, and rail cars must meet DOT requirements and accurately reflect their contents to enable emergency responders to react properly and quickly to any disaster thereby reducing the potential risk to the environment and to personnel.

Storage of ethylene oxide should be segregated from other chemicals to minimize the risk of cross-contamination. Ethylene oxide must be stored to avoid contact with strong oxidizers, such as chlorine and bromine, since violent reactions can occur. Containers should be protected from physical damage and stored to avoid contact with heat. If a spill or leak to the environment has occurred, fire department, emergency response and/or hazardous materials spill personnel should be notified immediately. Cleanup should be attempted only by those trained in proper spill containment procedures. Contaminated soils should be removed for incineration and replaced with clean soil.

If ethylene oxide should contact the water table, aquifer, or navigable waterway, time is of the essence. Its specific solubility in water has not been established and total remediation may or may not be possible. The local and/or state emergency response authorities must be notified. A comprehensive emergency response or disaster preparedness/recovery plan should be in place prior to any operations involving the use, transportation, storage, or disposal of ethylene oxide.

If ethylene oxide is spilled or leaked, the following specific steps are recommended:

- ☑ Restrict persons not wearing protective clothing from area of spill or leak until cleanup is complete. Avoid skin and eye contact.
- ☑ Dilute small spills with water to form non-flammable mixtures. Absorb spilled liquids with vermiculite or a similar material and deposit in sealed drum for disposal. Use non-sparking tools.
- ☑ Ventilate area and remove ignition sources.
- ☑ If source if leak is a cylinder, attempt to stop flow. If this cannot be done or cylinder is damaged, remove to safe place outdoors and allow to vent until empty.

RISK ASSESSMENT: BUSINESS

General Assessment

Accidents or mishaps involving ethylene oxide can present a serious threat to business operations. The loss or damage of equipment or facilities can significantly affect fiscal viability. Lawsuits that may result from personnel illness, injury/death, public exposures, and/or environmental contamination will require a serious expenditure of resources. Media attention surrounding an injury, death, or environmental damage can also result in a loss of profits and loss of current as well as future business. Always remember that anytime the terms "cancer," or "carcinogen" are used, public emotion, hysteria, and ignorance can run equally high. This requires careful consideration during the development of public relations policies.

Recommended Risk-Reduction Measures

Company attorneys, safety and health professionals, and environmental specialists should be involved in the development of any procedures for responding to chemical incidents. A company official should be predesignated as a public relations officer with specific training in dealing with the press. Corporate plans and policies should be developed, approved, and implemented long before any need for such arises.

REFERENCES

American Conference of Governmental Industrial Hygienists. 1988. *Documentation of the Threshold Limit Values and Biological Exposure Indices,* 5th Edition (with updates). Cincinnati: ACGIH

Lewis, R. J., Sr. 1993. *Hawley's Condensed Chemical Dictionary,* 12th Edition. New York: Van Nostrand Reinhold.

Lewis, R. J., Sr. 1992. *Sax's Dangerous Properties of Industrial Materials,* Volumes 1, 2, and 3, 8th Edition. New York: Van Nostrand Reinhold.

Patnaik, P. 1992. *A Comprehensive Guide to the Hazardous Properties of Chemical Substances.* New York: Van Nostrand Reinhold.

Torkelson, T. R. V. K. Rowe. 1981. *Patty's Industrial Hygiene and Toxicology,* Vol. 2B. New York: Wiley-Interscience.

U.S. Department of Labor. 1995. *Code of Federal Regulations, Title 29, Part 1910.1047 "Ethylene Oxide."* Washington, D.C.: U.S. Government Printing Office.

MATERIAL SAFETY DATA SHEET

CHEMICAL NAME	
ETHYLENE THIOUREA	**CRC PRESS/LEWIS PUBLISHERS DISCLAIMER:** THE INFORMATION AND RECOMMENDATIONS PRESENTED HEREIN ARE BASED ON SOURCES BELIEVED TO BE RELIABLE. CRC MAKES NO REPRESENTATION ON ITS COMPLETENESS OR ACCURACY. IT IS THE RESPONSIBILITY OF THE USER TO DETERMINE THE CHEMICAL'S SUITABILITY FOR ITS INTENDED USE, THE CHEMICAL'S SAFE USE, AND THE CHEMICAL'S PROPER DISPOSAL. NO REPRESENTATIONS AND/OR WARRANTIES, EITHER EXPRESSED OR IMPLIED, OF THE MERCHANTABILITY OR FITNESS FOR A PARTICULAR PURPOSE, OR OF ANY OTHER NATURE, ARE MADE WITH RESPECT TO THE INFORMATION PROVIDED IN THIS MSDS OR TO THE CHEMICAL TO WHICH INFORMATION MAY REFER. CRC NEITHER ASSUMES NOR AUTHORIZES ANY OTHER PERSON TO ASSUME FOR IT, ANY OTHER ADDITIONAL RESPONSIBILITY OR LIABILITY FOR THE USE OF, OR RELIANCE UPON, THIS INFORMATION.

HAZARD WARNING INFORMATION

HEALTH	FIRE	REACTIVE	OTHER	DEGREE OF HAZARD	COLOR CODING	OTHER CODES
2	1	0		0 = Minimum Hazard 1 = Slight Hazard 2 = Moderate Hazard 3 = Serious Hazard 4 = Severe Hazard	HEALTH = BLUE FIRE = RED REACTIVITY = YELLOW OTHER = WHITE	OX = Oxidizer ACID = Acid ALK = Alkali COR = Corrosive W = Use No Water

SECTION I - GENERAL INFORMATION

Characterization	**Amine**	RCRA Number	**U116**
		EPA Class	**Toxic Waste**
DOT Proper Shipping Name	**No Citation**	Chemical Abstract Service (CAS) Number	**96-45-7**
DOT Hazard Class and Label Requirements	**No Citation**	DOT Emergency Guide Code	**No Citation**
DOT Identification Number	**No Citation**	Chemical Formula	**$C_3H_6N_2S$**

Synonyms

1,3-Ethylene-2-thiourea; N,N-ethylenethiourea; ETU; 2-imidazolidine-2-thione; 2-imidazolidinethione; 2-mercapto-imidazoline; 4,5-dihydroimidazole-2(3H)-thione; 2-thiol-dihydroglyoxaline.

SECTION II - HAZARDOUS INGREDIENTS AND IDENTITY INFORMATION

Hazardous Components (specific identity)	OSHA Exposure Criteria	NIOSH Exposure Criteria	Immediately Dangerous to Life and Health (IDLH)	ACGIH Exposure Criteria
Ethylene thiourea (derivation: By the action of ethylenediamine upon carbon disulfide in aqueous alcohol; also occurs as a main degradation product of the metal salts of ethylene bis-dithiocarbamic acid).	**PEL:** Not Established **STEL:** Not Established	**REL:** Lowest Possible Level (Potential Human Carcinogen)	**Not Determined**	**TLV:** Not Established **STEL:** Not Established

SECTION III - PHYSICAL AND CHEMICAL CHARACTERISTICS

Boiling Point	**446-595°F (230-313°C)**	Specific Gravity (H_2O = 1)	**Not Found**
Vapor Pressure (mm Hg)	**16 at 69°F (20°C)**	Molecular Weight	**102.2**
Vapor Density (Air = 1)	**Not Found**	Melting Point	**390-400°F (199-204°C)**

Solubility

Slightly soluble in cold water, very soluble in hot water. Slightly soluble at room temperature in methanol, ethanol, acetic acid, and naphtha.

Appearance and Odor

White to pale-green, crystalline solid with a faint, amine-like (fishy) odor and bitter taste.

SECTION IV - FIRE AND EXPLOSION HAZARD DATA

Flash Point (method used)	**486°F (252°C) closed cup**	Explosive Limits in Air % by Volume	**LEL: Not Determined UEL: Not Determined**
NFPA Classification	**Combustible Solid**	Autoignition Temperature	**Not Found**

Extinguishing Media

Use dry chemical, carbon dioxide, foam, water spray, or fog.

Special Fire Fighting Procedures

Acrid and toxic oxides of sulfur and nitrogen are produced in fire. Use full protective clothing and self-contained breathing apparatus (SCBA) with a full facepiece operated in the pressure-demand or positive pressure mode.

Unusual Fire and Explosion Hazards

None reported.

SECTION V - REACTIVITY DATA

Stability		Conditions to Avoid: Ethylene thiourea is normally stable at room temperature in closed containers under normal conditions of storage and handling. Avoid contact with incompatible materials. Avoid human exposures.
Stable	Unstable	Incompatibility *(materials to avoid)*
X		Ethylene thiourea will react in contact with acrolein to cause fire or explosion.
Hazardous Polymerization		Conditions to Avoid: Under normal conditions of temperature and pressure, hazardous polymerization of ethylene thiourea is not expected to occur.
May Occur	Will Not Occur	Hazardous Decomposition or By-products
	X	Toxic and irritating fumes and vapors are produced when ethylene thiourea is heated to decomposition. These include oxides of nitrogen and sulfur.

SECTION VI - HEALTH HAZARD DATA

Primary Route(s) of Entry:	Inhalation?	Absorption (skin)?	Ingestion?
	X	X	X

Health Hazards

INHALATION: Irritation of the eyes, nose, throat, and lungs. High exposure may cause sweating, nausea, shortness of breath, and increased heart rate and blood pressure. May also cause pulmonary edema (a dangerous buildup of fluid in the lungs). Low, repeated exposures may lead to enlarged thyroid (goiter), and altered thyroid function. A teratogen.

ABSORPTION: Causes burns and irritation to the skin and eyes. Will pass through intact skin, thereby increasing the effects of other exposures (inhalation or ingestion).

INGESTION: Its bitter taste prevents ingestion of acutely toxic amounts.

Carcinogenicity	NTP Listed?	IARC Cancer Review Group?	OSHA Regulated?	Target Organs?
Suspected Human Confirmed Animal	5th Annual Report	Group 2B	Not Listed	Eyes, skin, thyroid, resp. system, repro. sys., liver.

Medical Conditions Generally Aggravated by Exposure

Thyroid ailments and disorders may be aggravated by exposure.

Emergency and First-aid Procedures

Eye contact: Flush immediately with water for 15 minutes (minimum), seek medical attention. Skin contact: Remove all contaminated clothing. Immediately wash area with large amounts of soap and water. Seek medical assistance. For inhalation: Remove the person from exposure. Provide respiratory assistance and CPR. Transfer to medical facility. Observe for 24-48 hours for lung effects. If swallowed: Not likely. Call the poison control center and seek medical attention immediately. Unless advised otherwise, give 1-2 glasses of water. Never give an unconscious or convulsing person anything by mouth. DOCTOR: Examination of thyroid is suggested.

SECTION VII - PRECAUTIONS FOR SAFE HANDLING AND USE

Steps to be Taken in Case Material is Released or Spilled

Collect powdered materials in safest manner possible and deposit in DOT-approved drum. Do NOT generate dusts. Use vacuum equipped with HEPA filter instead. Ventilate area. Restrict those not involved in cleanup from entering area. Notify appropriate authorities, if applicable.

Preferred Waste Disposal Method

Decompose to less hazardous ethylene urea by adding sodium hypochlorite solution.

Precautions to be Taken in Handling and Storage

Store in tightly closed containers in a cool, well-ventilated area. Protect containers from physical damage and shield from direct sunlight or other ultraviolet light sources.

Other Precautions and Warnings

Personnel should be trained on proper handing and storage *prior* to working with ethylene thiourea.

SECTION VIII - CONTROL MEASURES AND PERSONAL PROTECTIVE EQUIPMENT

Respiratory Protection *(specify type)*

No exposure limits have been established. This chemical is a known teratogen and a suspected carcinogen. Use an MSHA/NIOSH-approved supplied-air respirator or a self-contained breathing apparatus (SCBA) with full facepiece operated in pressure demand or other positive pressure mode.

Ventilation

Local exhaust or general mechanical systems recommended.

Protective Gloves	Eye Protection	Other Protective Clothing
Impervious Gloves	Chemical Goggles or Face Mask	Protective Apron

Work/Hygiene Practices

Always wash hands thoroughly after using chemical; never bring food, drink, or smoking materials into vicinity of chemicals.

ETHYLENE THIOUREA

$C_3H_6N_2S$ CAS: 96-45-7

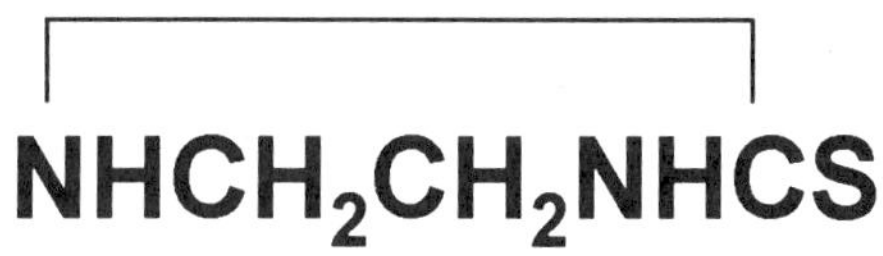

IDENTIFICATION AND TYPICAL USES

Ethylene thiourea is a white to pale-green, crystalline solid with a faint, amine-like (fishy) odor. It is used extensively as a vulcanization accelerator for curing polychloroprene (Neoprene®) and other elastomers. It is also used in electroplating baths, intermediate for antioxidants, insecticides, fungicides, dyes, pharmaceuticals, and synthetic resins.

RISK ASSESSMENT: HEALTH

General Assessment

Ethylene thiourea is human poison by *inhalation, ingestion,* and by skin contact (*absorption*). It is known to cause cancer in laboratory test animals (liver, thyroid, and lymphatic system tumors) and its carcinogenic potential in humans is highly suspect. It is also a known teratogen (causes birth defects) and may cause other adverse effects on the human reproductive system.

Inhalation of the dusts or, if mixed in aqueous solution, the vapors or mists, can cause irritation to the eyes, nose, throat, bronchial tubes, and lungs. This could cause soreness, hoarseness, cough, and phlegm. High exposure may cause sweating, nausea, increased heart rate, and blood pressure. It can cause a dangerous buildup of fluid in the lungs, a condition known as pulmonary edema, which is a medical emergency and can be fatal. Symptoms of difficulty breathing, tightness in chest, coughing, and congestion may be delayed up to 48 hours following exposure thereby creating a false sense of security with regard to health exposure risk. Low or repeated exposures can cause enlarged thyroid (goiter) and changes in thyroid function.

Skin contact will result in irritation. Ethylene thiourea will pass through the skin to cause toxic systemic effects. If coupled with other exposure routes (inhalation or ingestion), absorption can increase the toxic effects to dangerous levels.

Ingestion is not very likely since the bitter taste of ethylene thiourea will prevent ingestion of acutely toxic amounts. However, ingestion is still a possible exposure route. It can lead to a reduction of thyroid gland activity and result in symptoms such as a yellowish or ivory complexion, thickening and drying of the skin, an enlarged tongue, and dry, brittle hair.

Acute Health Effects

The following acute (short-term) health effects may occur immediately or shortly after exposure to ethylene thiourea:

Skin: Localized irritation at sight of contact. Will readily absorb through skin to cause toxic systemic effects.

Eye: Irritation and possible inflammation.

Lung: Irritation causing coughing and/or shortness of breath. Higher exposures can lead to respiratory problems, including pulmonary edema (fluid buildup in lungs) with cough, phlegm, dypsnea, nausea, and sweating.

Other: A related chemical (Ziram) causes brain swelling, hemorrhage, muscle weakness, and liver and kidney effects. It is not known whether ethylene thiourea can cause these changes.

Chronic Health Effects

The following chronic (long-term) health effects may occur at some time after exposure to ethylene thiourea and can last for months or even years:

Cancer Hazards: According to the information presented in the references, ethylene thiourea has been shown to cause cancers of the liver, thyroid, and lymphatic system in test animals. Its ability to cause cancer in humans is highly suspect.

Reproductive Hazard: According to the information presented in the references, ethylene thiourea has been shown to cause teratogenic effects in tests animals and it is thought to be a human teratogen as well.

Other Chronic Effects: Chronic ingestion of small quantities can cause a reduction of thyroid gland activity with symptoms of yellowish or ivory-colored skin, thickening of the skin, and enlarged tongue, and dry, brittle hair. A related chemical (maneb), an agricultural fungicide, can cause tremors, weakness, poor coordination, and other nerve effects. It is not known if ethylene thiourea will cause these effects.

Recommended Risk-Reduction Measures

Ethylene thiourea is a probable human carcinogen and a known teratogen. Unless a less toxic chemical can be substituted for a hazardous substance, *engineering controls* are the most effective method of reducing exposures. The best protection is to enclose operations and/or provide local exhaust ventilation at the site of chemical release. While not always operationally feasible, isolating operations can also reduce exposure.

Using respiratory protection is less effective than the controls mentioned above, but is still advisable whenever working with or around ethylene thiourea. No exposure levels have been established for this chemical. This does not mean that respiratory protection is not necessary. In fact, with ethylene thiourea, the opposite is true. It may pose a cancer and reproductive hazard and, therefore, *no* exposure level should be tolerated. A supplied-air respirator or a self-contained breathing apparatus (SCBA) with full facepiece and pressure demand are the recommended respiratory protection methods of choice. If a full facepiece is not available, then chemical goggles should be worn to protect the eyes. Whenever a chemical splash hazard exists, a face shield and apron should be worn. To prevent hand and skin exposures, impervious chemical gloves should be worn. Contact glove manufacturers and obtain permeation studies *before* selecting gloves.

Administrative controls should also be in place to minimize the potential for human exposures. These may include written procedures or policies which specify the methods and techniques that will be practiced whenever personnel are to work with ethylene thiourea.

All personnel should receive *training* on the use, hazards, protective measures, emergency actions, and other precautions per 29 CFR 1910.1200 (Hazard Communication), prior to the first assignment in an area where ethylene thiourea is used or stored.

If symptoms develop or overexposure is suspected, the following medical tests may be useful:

- ☑ Lung function tests.
- ☑ Consider chest X-ray after acute overexposure (may be negative if taken immediately after exposure due to the delayed development of pulmonary edema).
- ☑ Thyroid function tests.
- ☑ Examination of the nervous system.

Any evaluation should include a careful medical history of past and present symptoms with an examination. However, medical tests that evaluate existing damage are not a substitute for controlling exposure. Also, since smoking can cause heart diseases, emphysema, and numerous other respiratory disorders, smokers may be affected more rapidly than non-smokers to the same exposure conditions. Prudent risk assessment and management requires careful consideration of *all* possible risk factors that may be causing the appearance of symptoms in exposed workers.

Other methods to reduce exposure include:

- ☑ Where possible, enclose operations and use local exhaust ventilation at the site of chemical release. If local exhaust ventilation or enclosure is not used, respiratory protection should be mandatory.
- ☑ Always ensure that proper protective clothing is worn when using chemical substances and that personnel are trained on its use and care.
- ☑ Wash thoroughly immediately after exposure to ethylene thiourea and at the end of the work shift or before eating, drinking, or smoking.
- ☑ Hazard warning information should be posted in the work area. In addition, as part of an on-going education and training program, all information on the health and safety hazards of ethylene thiourea should be communicated to all exposed and potentially exposed workers.
- ☑ Eye wash stations should be provided in work areas. If the potential for whole body exposure exists, then safety showers should also be provided.
- ☑ Specific engineering controls are required for ethylene thiourea. Refer to NIOSH Current Intelligence Bulletin Number 22, Publication Number 78-144 ("Ethylene Thiourea").

RISK ASSESSMENT: ENVIRONMENT

General Assessment

The environment is at risk of exposure during intended use, transportation, storage, disposal, or destruction of ethylene thiourea. In almost every scenario, the threat of environmental exposure is contingent upon the proper handling of the chemical substance. Accidental spills, large or small, can result in possible contamination of the surrounding environmental mediums (water, soil, and air).

Ethylene thiourea is considered a combustible solid. It can react with acrolein and can decompose into ethylene urea if reacted with alkaline hypochlorite or exposed to ultraviolet radiation in the presence of a sensitizer such as acetone or riboflavin. These characteristics require special consideration during any emergency situation involving a leak or spill of ethylene thiourea.

Ethylene thiourea can enter the environment through industrial discharges and spills.

Acute Ecological Effects

Acute (short-term) toxic effects may include the death of animals, birds, or fish and death or low growth rate in plants. Acute effects are seen 2 to 4 days after animals or plants are exposed to ethylene thiourea.

This chemical has moderate acute toxicity to aquatic life and to land animals. At concentrations of 100 ppm, ethylene thiourea had no effect on the size of wheat plants. Insufficient data are available to evaluate or predict the short-term effects of ethylene thiourea on birds.

Chronic Ecological Effects

Chronic toxic effects may include shortened life span, reproductive problems, lower fertility, and changes in appearance or behavior in exposed animals. These effects can be seen long after first exposure(s) to toxic chemicals.

Ethylene thiourea has moderate chronic toxicity to aquatic life. Insufficient data are available to evaluate or predict the long-term effects of ethylene thiourea to plants, birds, or land animals.

Water Solubility

Ethylene thiourea is highly soluble in water. Concentrations of 1000 milligrams and more will mix with a liter of water.

Persistence in the Environment

Ethylene thiourea is moderately persistent in water, with a half-life of between 20 to 200 days. The half-life of a pollutant is the amount of time it takes for one half of the chemical to be degraded. Virtually 100% of ethylene thiourea will end up in water.

Bioaccumulation in Aquatic Organisms

Some substances increase in concentration, or *bioaccumulate,* in living organisms as they breathe contaminated air, drink contaminated water, or eat contaminated food. These chemicals can become concentrated in the tissues and internal organs of animals as well as humans.

The concentration of ethylene thiourea found in fish tissues is expected to be about the same as the average concentration of ethylene thiourea in the water from which the fish was taken.

Recommended Risk-Reduction Measures

Proper training of all transporters will reduce the likelihood of a mishap or accident resulting in a leak or spill to the environment. The correct use of DOT labeling (per the MSDS) on all containers, trucks, and rail cars will enable emergency responders to react properly and quickly to any disaster thereby reducing the potential risk to the environment and to personnel.

Storage of ethylene thiourea should be segregated from incompatible chemicals and shielded from radiant heat to minimize the risk of cross-contamination, contact, or reaction.

If a spill or leak to the environment has occurred, fire department, emergency response and/or hazardous materials spill personnel should be notified immediately. Cleanup should be attempted only by those trained in proper spill containment procedures. Contaminated soils should be removed for incineration and replaced with clean soil. If ethylene thiourea should contact the water table, aquifer, or navigable waterway, remediation activities should be prompt. When such spills occur, the local and/or state emergency response authorities must be notified. A comprehensive emergency response or disaster preparedness/recovery plan should be in place prior to any operations involving the use, transportation, storage, or disposal of ethylene thiourea. If ethylene thiourea is spilled or leaked, the following specific steps are recommended:

☑ Restrict persons not wearing protective clothing from area of spill or leak until cleanup is complete and area can be opened for normal work.

☑ Ventilate area of spill or leak.

☑ Collect powdered materials in safest manner possible using a vacuum and HEPA filter and place in a sealed drum for disposal.

☑ It may be necessary to dispose of ethylene thiourea as a hazardous waste. The responsible state agency or the regional office of the federal Environmental Protection Agency (EPA) should be contacted for specific recommendations.

RISK ASSESSMENT: BUSINESS

General Assessment

Accidents or mishaps involving ethylene thiourea can present a significant threat to business operations. The loss or damage of equipment or facilities can significantly affect fiscal viability. Lawsuits that may result from personnel injury/death, public exposures, and/or environmental contamination will also require a serious expenditure of resources. Media attention surrounding an injury, death, or environmental damage can also result in a loss of profits and loss of current as well as future business. Always remember that anytime the terms "cancer," "carcinogen," and "reproductive hazard" are used, public emotion, hysteria, and ignorance can run equally high. This should be carefully considered whenever developing public relations policies.

Recommended Risk-Reduction Measures

Company attorneys, safety and health professionals, and environmental specialists should be involved in the development of any such procedures. A company official should be pre-designated as a public relations officer with specific training in dealing with the press. Corporate plans and policies should be developed, approved, and implemented long before any need for such arises.

REFERENCES

American Conference of Governmental Industrial Hygienists. 1988. *Documentation of the Threshold Limit Values and Biological Exposure Indices,* 5th Edition (with updates). Cincinnati: ACGIH

Anthony, C. P., and N. J. Kolthoff. 1971. *Textbook of Anatomy and Physiology.* St. Louis: C. V. Mosby Company.

Bueche, J. R. 1972. *Principles of Physics,* 2nd Edition. New York: McGraw-Hill.

Calabrese, E. J., E. M. Kenyon. *Air Toxics and Risk Assessment.* New York: CRC Press/Lewis Publishers.

Cockerman, L. G., B. S. Shane. 1994. *Basic Environmental Toxicology.* New York: CRC Press/Lewis Publishers.

Cothern, C. R. 1995. *Handbook for Environmental Risk Decisionmaking.* New York: CRC Press/Lewis Publishers.

Gardner, R. W. 1994. *Chemical Intolerance.* New York: CRC Press/Lewis Publishers.

Hallenbeck, W. H. 1993. *Quantitative Risk Assessment for Environmental and Occupational Health,* 2nd Edition. New York: CRC Press/Lewis Publishers.

Hodgson, E., R. B. Mailman, and Janice E. Chambers. 1988. *Dictionary of Toxicology.* London: The MacMillian Company.

Howard, P. H. 1995. *Handbook of Environmental Fate and Exposures Data for Organic Chemicals,* Vol. I-IV. New York: CRC Press/Lewis Publishers.

Jones, F. E. 1994. *Toxic Organic Vapors in the Workplace.* New York: CRC Press/Lewis Publishers.

Landis, W. G., M. H. Yu. 1995. *Introduction to Environmental Toxicology.* New York: CRC Press/Lewis Publishers.

Langerman, N. 1994 *Precautionary Labels for Chemical Containers.* New York: CRC Press/Lewis Publishers.

Lauwerys, R. R., P. Hoet. 1993. *Industrial Chemical Exposure.* New York: CRC Press/Lewis Publishers.

Lewis, R. J., Sr. 1993. *Hawley's Condensed Chemical Dictionary,* 12th Edition. New York: Van Nostrand Reinhold.

Lewis, R. J., Sr. 1992. *Sax's Dangerous Properties of Industrial Materials,* Volumes 1, 2, and 3, 8th Edition. New York: Van Nostrand Reinhold.

Meyer, E. 1990. *Chemistry of Hazardous Materials.* Englewood Cliffs, NJ: Prentice Hall.

Nelson, G. O. 1992. *Gas Mixtures.* New York: CRC Press/Lewis Publishers.

Quigley, D. R. 1994. *Handbook of Emergency Chemical Management.* New York: CRC Press/Lewis Publishers.

Rea, W. J. 1995. *Chemical Sensitivity,* Vol. I-III. New York: CRC Press/Lewis Publishers.

Richardson, M. 1995. *Dictionary of Substance and Their Effects,* Vol. I-VI. New York: CRC Press/Lewis Publishers.

Sacarello, H. L. A. 1994. *The Comprehensive Handbook of Hazardous Materials.* New York: CRC Press/Lewis Publishers.

U.S. Department of Health and Human Services, National Institute for Occupational Safety and Health. 1994. *NIOSH Pocket Guide to Chemical Hazards.* Washington, D.C.: U.S. Government Printing Office.

U.S. Department of Health and Human Services, National Institute for Occupational Safety and Health. 1978. Current Intelligence Bulletin #22 (*Ethylene Thiourea*). Publication Number 78-144. Washington, D.C.: U.S. Government Printing Office.

MATERIAL SAFETY DATA SHEET

CHEMICAL NAME	CRC PRESS/LEWIS PUBLISHERS DISCLAIMER
ETHYL ETHER	

HAZARD WARNING INFORMATION

HEALTH	FIRE	REACTIVE	OTHER	DEGREE OF HAZARD	COLOR CODING	OTHER CODES
2	4	1		0 = Minimum Hazard 1 = Slight Hazard 2 = Moderate Hazard 3 = Serious Hazard 4 = Severe Hazard	HEALTH = BLUE FIRE = RED REACTIVITY = YELLOW OTHER = WHITE	OX = Oxidizer ACID = Acid ALK = Alkali COR = Corrosive ~~W~~ = Use No Water

SECTION I - GENERAL INFORMATION

Characterization: **Ether**	RCRA Number: **U117**	EPA Class: **Toxic Waste**
DOT Proper Shipping Name: **Ethyl Ether**	Chemical Abstract Service (CAS) Number: **60-29-7**	
DOT Hazard Class and Label Requirements: **Flammable Liquid**	DOT Emergency Guide Code: **26**	
DOT Identification Number: **UN 1155**	Molecular Formula: **$C_2H_5OC_2H_5$**	

Synonyms
Diethyl ether; diethyl oxide; ether; ethyl oxide; solvent ether; anesthesia ether; anesthetic ether; ethoxy ethane; sulfuric ether; 1,1'-oxybisethane.

SECTION II - HAZARDOUS INGREDIENTS AND IDENTITY INFORMATION

Hazardous Components (specific identity)	OSHA Exposure Criteria	NIOSH Exposure Criteria	Immediately Dangerous to Life and Health (IDLH)	ACGIH Exposure Criteria
Ethyl ether: (derivation: By the action of sulfuric acid on ethanol or ethylene followed by distillation). 1 ppm = 3.08 mg/m^3	PEL (8-hour): 400 ppm 1200 mg/m^3 STEL: 500 ppm 1500 mg/m^3	REL: Not Established STEL: Not Established	1900 ppm	TLV: 400 ppm 1200 mg/m^3 STEL: 500 ppm 1500 mg/m^3

SECTION III - PHYSICAL AND CHEMICAL CHARACTERISTICS

Boiling Point: **94°F (34°C)**	Specific Gravity (H_2O = 1): **0.71**
Vapor Pressure (mm Hg): **440 at 77°F (25°C)**	Molecular Weight (atomic weight): **74.1**
Vapor Density (Air = 1): **2.55**	Melting Point: **-177°F (-116°C)**

Solubility
Slightly soluble in water (8%). Soluble in alcohol, chloroform, benzene, solvent naphtha, and oils.

Appearance and Odor
A clear, colorless, mobile, water-white, extremely volatile liquid with a strong, sweet, aromatic odor and burning sweet taste.

SECTION IV - FIRE AND EXPLOSION HAZARD DATA

Flash Point (method used): **-49°F (-45°C) closed cup**	Explosive Limits in Air % by Volume: **LEL: 1.85% UEL: 36%**
NFPA Classification: **Class IA Flammable Liquid**	Autoignition Temperature: **356°F (180°C)**

Extinguishing Media
Dry chemical, alcohol foam, or carbon dioxide. Use water spray to keep fire-exposed containers cool.

Special Fire Fighting Procedures
Wear full protective clothing and self-contained breathing apparatus (SCBA). Move container from fire area if it can be done without risk. Use water spray to flush spills from ignition sources and to knock down vapors that have not yet ignited. Treat as a gas fire; use blanketing effect to smother flames.

Unusual Fire and Explosion Hazards
Containers may explode in fire. Presents a vapor explosion hazard indoors, outdoors, and in sewers.

SECTION V - REACTIVITY DATA

Stability		Conditions to Avoid **Ethyl ether is NOT stable under normal conditions of handing and storage. It can form unstable peroxides if exposed to air and sunlight for prolonged periods of time. Peroxide can explode if distilled or heated.**
Stable	Unstable X	Incompatibility *(materials to avoid)* **Incompatible with strong oxidizers, peroxides, permanganates, acids, anhydrides, metallic hydrides, air, liquid oxygen, and nitrates.**
Hazardous Polymerization		Conditions to Avoid **Under normal conditions of temperature and pressure, hazardous polymerization of ethyl ether is not expected to occur.**
May Occur	Will Not Occur X	Hazardous Decomposition or By-products **When heated to decomposition, ethyl ether emits acrid smoke and fumes, including toxic oxides of carbon.**

SECTION VI - HEALTH HAZARD DATA

Primary Route(s) of Entry:	Inhalation?	Absorption (skin/eye)?	Ingestion?
	X	X	

Health Hazards

INHALATION: Causes headache, nausea, vomiting, loss of consciousness, and at high concentrations, death. Depression of the central nervous system (CNS) causing anesthesia, exhaustion, loss of appetite, sleepiness, dizziness, excitation, anorexia, giddiness, narcotic effects.

ABSORPTION: An eye and skin irritant. Can pass through unbroken skin causing toxic systemic effects. May cause skin burns and eye irritation (smarting).

INGESTION: Not likely. The liquid evaporates rapidly and makes ingestion an unlikely exposure route.

Carcinogenicity	NTP Listed?	IARC Cancer Review Group?	OSHA Regulated?	Target Organs?
Unknown Human No Evidence Animal	**No**	**Not Listed But Reviewed**	**29 CFR 1910.1000 Table Z-1**	**Skin, eyes, respiratory system, CNS, blood, liver.**

Medical Conditions Generally Aggravated by Exposure

None reported.

Emergency and First-aid Procedures

Eye contact: Do not allow victim to keep eyes tightly shut. Flush immediately with water for 15 minutes (minimum), seek medical attention. Skin contact: No serious hazard since evaporation is quick. Prolonged contact may cause drying, cracking, or burning Contaminated clothing should be removed (fire hazard). Wash area with large amounts of soap and water. For inhalation: Remove the person from exposure. Provide respiratory assistance and CPR. Transfer to medical facility. If swallowed: Not very likely. Seek medical attention immediately. Unless advised otherwise, give several glasses of water. Do not induce vomiting because of danger of aspiration (breathing the liquid into the lungs). Never attempt to give an unconscious or convulsing person anything by mouth.

SECTION VII - PRECAUTIONS FOR SAFE HANDLING AND USE

Steps to be Taken in Case Material is Released or Spilled

Wear self-contained breathing apparatus (SCBA). Restrict those not wearing protective equipment and who are not involved in cleanup from area. Avoid skin contact. Ventilate area of spill and remove ignition sources. Absorb liquids in vermiculite, collect with non-sparking tools, and deposit in sealed drum.

Preferred Waste Disposal Method

Mix with a higher-boiling solvent and burn in chemical incinerator with afterburner and scrubber.

Precautions to be Taken in Handling and Storage

Store in tightly closed (air-tight), metal containers in a cool, dry, well-ventilated area away from heat and incompatible materials. Control inventories. Protect containers from physical damage.

Other Precautions and Warnings

Before using bulk material, test small quantity to ascertain quality and peroxide content.

SECTION VIII - CONTROL MEASURES AND PERSONAL PROTECTIVE EQUIPMENT

Respiratory Protection *(specify type)*

Use a NIOSH-approved air-purifying respirator for low levels. Best protection is obtained using a supplied-air respirator or a SCBA with full facepiece operated in positive pressure mode.

Ventilation

Local exhaust (preferred) at site of chemical work or general ventilation.

Protective Gloves	Eye Protection	Other Protective Clothing
Impervious Polyvinyl	**Chemical Goggles and Face Mask**	**Protective Uniform or Apron**

Work/Hygiene Practices

Always wash hands thoroughly after using chemical; never bring food, drink, or smoking materials into vicinity of chemicals.

ETHYL ETHER

$C_2H_5OC_2H_5$ CAS: 60-29-7

$$H_3C — CH_2 — O — CH_2 — CH_3$$

IDENTIFICATION AND TYPICAL USES

Ethyl ether appears as a clear, colorless, volatile, mobile, water-white liquid with a characteristic, sweet, pungent, aromatic odor and bittersweet taste. It is used as a solvent for fats, oils, waxes, gums, perfumes, and nitrocellulose. It is also used in the making of gun powder, as an anesthetic, and in organic synthesis.

RISK ASSESSMENT: HEALTH

General Assessment

Ethyl ether is a narcotic substance and is considered mildly to moderately toxic by *inhalation* and skin *absorption.* Although unlikely under normal operating conditions, *ingestion* may also be an exposure route. There are no reports to indicate that this chemical poses a carcinogenic or teratogenic effect in humans. However, mutation (genetic changes) data have been reported in animals. It is not known whether it will cause this effect in humans.

Inhalation of ethyl ether vapors causes anesthetic effects because of the chemical's ability to quickly and effectively depress the central nervous system. Symptoms of exposure include exhaustion, loss of appetite, sleepiness, dizziness, headache, giddiness, excitation, nausea, vomiting, anorexia, loss of consciousness, possible seizures and convulsions, psychic disturbances, and death due to respiratory arrest. There may also be an occurrence of polycythemia, a condition characterized by an abnormal increase in circulatory red blood cells. Exposure can cause irritation of the eyes, nose, throat, and respiratory system with subsequent changes (numbing) of the sense of smell.

Skin contact with the liquid poses no serious threat because ethyl ether will evaporate rather quickly. However, if allowed to remain in contact (in wet clothing, for example) for extended periods, there can be drying, cracking, even some burning of the skin's surface. The chemical will also pass through unbroken skin to cause toxic systemic effects similar to that caused by inhalation. The vapors will cause irritation and slight smarting of the eyes.

Ingestion is not likely since the liquid will evaporate so quickly leaving very little time for ingestion to occur. However, it is a remote possibility and the symptoms of inhalation can be expected to occur on ingestion of ethyl ether.

Acute Health Effects

The following acute (short-term) health effects may occur immediately or shortly after exposure to ethyl ether:

Skin: Mild irritation with drying, cracking, secondary infection, and possible burns. On prolonged exposure, can be absorbed through intact skin causing toxic systemic effects.

Eye: Irritation and smarting of eye tissues and associated skin surfaces (lids).

Lung: Irritation to respiratory tract causing cough and sore throat. May also cause olfactory changes (numbing of the sense of smell). Absorption through inhalation causes anesthetic effects and narcosis (deep unconsciousness).

CNS: Rapid depression of the central nervous system causing a variety of toxic effects including loss of coordination and consciousness, and even death due to respiratory arrest.

Other: May cause an abnormal increase in red blood cells of the circulatory system (polycythemia).

Chronic Health Effects

The following chronic (long-term) health effects may occur at some time after exposure to ethyl ether and can last for months or even years:

Cancer Hazards: There is no evidence in the references that exposure to ethyl ether causes cancer in animals or humans. However, mutation data have been reported in test animals. Some scientists believe that mutagenic chemicals may pose a cancer risk in the long-term. Additional research is required in this area of study before any conclusions can be made.

Reproductive Hazard: According to the references, ethyl ether has not been adequately tested for its ability to adversely affect reproduction.

Other Chronic Effects: Repeated low-dose exposures can cause anorexia, chronic fatigue or exhaustion, persistent headache, drowsiness, dizziness, excitation, and psychic disturbances due to its depression of the CNS.

Recommended Risk-Reduction Measures

Personnel should avoid direct contact with ethyl ether. If a less toxic material or compound cannot be substituted for ethyl ether, then *engineering controls* are the most effective method of reducing exposure risk. The best protection is to enclose operations and/or provide local exhaust ventilation at the site of ethyl ether release. While not always operationally feasible, isolating operations can also reduce exposure. Using respiratory protection is less effective than the controls mentioned above, but is still required whenever working with or around ethyl ether. For exposures to the PEL (400 ppm), an air purifying respirator should provide adequate protection. For high exposures or when the concentration is unknown, an MSHA/NIOSH-approved supplied air respirator or a self-contained breathing apparatus (SCBA) with full facepiece and operated in pressure demand mode are the recommended methods of respiratory protection. If a full facepiece is not available, then chemical goggles should be worn to protect the eyes, especially when a splash hazard exists. A face shield should also be considered. To prevent hand and skin exposures, polyvinyl or other non-permeable gloves should be worn. Glove manufacturers should be contacted and permeation studies obtained *before* selecting gloves.

Administrative controls should also be in place to minimize the potential for human exposures. These may include written procedures or policies which specify the methods and techniques that will be practiced whenever personnel are to work with ethyl ether.

All personnel should receive *training* on the use, hazards, protective measures, emergency actions, and other precautions per 29 CFR 1910.1200 (Hazard Communication) prior to the first assignment in an area where ethyl ether is used or stored.

If symptoms develop or overexposure is suspected, the following may be useful:

- ☑ Liver, kidney, and lung function tests.
- ☑ Evaluation of the central nervous system.

It should be noted that medical tests that simply look for existing damage are not a substitute for controlling exposures. Medical histories are extremely important when assessing exposure risk.

Other methods to reduce exposure include:

- ☑ Always ensure that proper protective clothing is worn when using chemical substances.
- ☑ Wash thoroughly immediately after exposure to ethyl ether and at the end of the work shift or before eating, drinking, or smoking.
- ☑ Hazard warning information should be posted in the work area. In addition, as part of an on-going education and training program, all information on the health and safety hazards of ethyl ether should be communicated to all exposed or potentially exposed workers.
- ☑ Eye wash stations should be provided in the immediate work area for emergency use. If there is a possibility of skin exposure to ethyl ether, emergency shower facilities should also be provided.
- ☑ Workers whose clothing has been contaminated by ethyl ether should change into clean clothes before leaving work. Contaminated work clothing presents a serious fire hazard and should be laundered only by individuals who have been informed of the hazards of exposure to ethyl ether.

RISK ASSESSMENT: ENVIRONMENT

General Assessment

The environment is at risk of exposure during transportation, storage, disposal, or destruction of ethyl ether. In almost every scenario, the threat of environmental exposure is contingent upon the proper handling of the chemical substance. Accidental spills, large or small, can result in contamination of the surrounding environmental mediums (primarily air, but water and soil contamination can also occur).

Ethyl ether is Class IA flammable liquid (per OSHA 29 CFR 1910.106). This presents a very dangerous fire and explosion hazard. Its vapors are heavier than air and can travel for some distance to an ignition source to flashback and cause fire or explosion. It can form explosive peroxides if distilled to near dryness or heated to 212°F (100°C). It is incompatible with many oxidizers and acids and caution is always required in handling, storage, transportation, and disposal of ethyl ether. When heated to decomposition it emits very acrid and irritating smoke and fumes. Emergency responders should be made aware of the presence of ethyl ether at any emergency response situation.

Ethyl ether may enter the environment through industrial effluents and spills.

Acute Ecological Effects

Acute (short-term) toxic effects may include the death of animals, birds, or fish and death or low growth rate in plants. Acute effects are seen 2 to 4 days after animals or plants are exposed to ethyl ether. Insufficient data are available to evaluate or predict the acute (short-term) effects of ethyl ether to aquatic life, plants, birds, or land animals.

Chronic Ecological Effects

Chronic (long-term) toxic effects may include shortened life span, reproductive problems, lower fertility, and changes in appearance or behavior in exposed animals. These effects can be seen long after first exposure(s) to toxic chemicals. Insufficient data are available to evaluate the chronic (long-term) effects of ethyl ether on aquatic life, plants, birds, or land animals.

Water Solubility

Ethyl ether is slightly soluble in water. Concentrations of 1 to 100 milligrams will mix with a liter of water.

Persistence in the Environment

Ethyl ether is non-persistent in water, with a half-life of less than 2 days. The half-life of a pollutant is the amount of time it takes for one half of the chemical to be degraded. Approximately 100% of ethyl ether will eventually end up in air.

Bioaccumulation in Aquatic Organisms

Some substances increase in concentration, or *bioaccumulate,* in living organisms as they breathe contaminated air, drink contaminated water, or eat contaminated food. These chemicals can become concentrated in the tissues and internal organs of animals as well as humans.

The references do not address the bioaccumulation capabilities of ethyl ether in aquatic organisms. However, considering the extremely volatile nature of this chemical, bioaccumulation may not be possible.

Recommended Risk-Reduction Measures

Proper training of all transporters will reduce the likelihood of a mishap or accident resulting in a leak or spill of ethyl ether into the environment. Labels on all containers, trucks, and rail cars must meet DOT requirements and accurately reflect their contents to enable emergency responders to react properly and quickly to any disaster thereby reducing the potential risk to the environment and to personnel.

Storage of ethyl ether should be segregated from other chemicals to minimize the risk of cross-contamination. Ethyl ether must be stored to avoid contact with strong oxidizers, such as chlorine and bromine, since violent reactions can occur. Containers should be protected from physical damage and stored to avoid contact with heat, including sunlight.

If a spill or leak to the environment has occurred, fire department, emergency response and/or hazardous materials spill personnel should be notified immediately. Cleanup should be attempted only by those trained in proper spill containment procedures. Contaminated soils should be removed for incineration and replaced with clean soil.

If ethyl ether should contact the water table, aquifer, or navigable waterway, time is of the essence. Its rapid evaporation may make total remediation impossible. The local and/or state emergency response authorities must be notified. A comprehensive emergency response or disaster preparedness/recovery plan should be in place prior to any operations involving the use, transportation, storage, or disposal of ethyl ether.

Ethyl ether may be disposed of by mixing with an excess of a higher-boiling solvent and then burning in a chemical incinerator equipped with an afterburner and scrubber. If ethyl ether is spilled or leaked, the following specific steps are recommended:

- ☑ Restrict persons not wearing protective clothing from area of spill or leak until cleanup is complete. Avoid skin and eye contact.
- ☑ Absorb spilled liquids with vermiculite, dry sand, or a similar material and deposit in sealed drum for disposal. Use non-sparking tools!

☑ Ventilate area and remove all ignition sources.

☑ It may be necessary to dispose of ethyl ether as a hazardous waste. The responsible state agency or the regional office of the federal Environmental Protection Agency (EPA) should be contacted for specific recommendations.

RISK ASSESSMENT: BUSINESS

General Assessment

Accidents or mishaps involving ethyl ether can present a serious threat to business operations. The loss or damage of equipment or facilities can significantly affect fiscal viability. Lawsuits that may result from personnel illness, injury/death, public exposures, and/or environmental contamination will require a serious expenditure of resources. Media attention surrounding an injury, death, or environmental damage can also result in a loss of profits and loss of current as well as future business.

Recommended Risk-Reduction Measures

Company attorneys, safety and health professionals, and environmental specialists should be involved in the development of any procedures for responding to chemical incidents. A company official should be predesignated as a public relations officer with specific training in dealing with the press. Corporate plans and policies should be developed, approved, and implemented long before any need for such arises.

REFERENCES

Ahlbom, A. 1993. *Biostatistics for Epidemiologists.* New York: CRC Press/Lewis Publishers.

American Conference of Governmental Industrial Hygienists. 1988. *Documentation of the Threshold Limit Values and Biological Exposure Indices,* 5th Edition (with updates). Cincinnati: ACGIH

Anthony, C. P., and N. J. Kolthoff. 1971. *Textbook of Anatomy and Physiology.* St. Louis: C. V. Mosby Company.

Asante-Duah, D. K. 1993. *Hazardous Waste Risk Assessment.* New York: CRC Press/Lewis Publishers.

Bloom, A. D. and F. J. DeSerres. 1995. *Ecotoxicity and Human Health.* New York: CRC Press/Lewis Publishers.

Brusick, D. J. 1994. *Methods for Genetic Risk Assessment.* New York: CRC Press/Lewis Publishers.

Calabrese, E. J. 1994. *Biological Effects of Low Level Exposures.* New York: CRC Press/Lewis Publishers.

Cothern, C. R., N. P. Ross. 1994. *Environmental Statistics, Assessment, and Forecasting.* New York: CRC Press/Lewis Publishers.

Danse, I. R. 1991. *Common Sense Toxics in the Workplace.* New York: Van Nostrand Reinhold.

Fodden, C. R., and J. L. Weddell. 1992. *Hazardous Materials Emergency Action Data.* New York: CRC Press/Lewis Publishers.

Gardner, R. W. 1994. *Chemical Intolerance.* New York: CRC Press/Lewis Publishers.

Hallenbeck, W. H. 1993. *Quantitative Risk Assessment for Environmental and Occupational Health,* 2nd Edition. New York: CRC Press/Lewis Publishers.

Hodgson, E., R. B. Mailman, and J. E. Chambers. 1988. *Dictionary of Toxicology.* London: The MacMillian Company.

Howard, P. H., et. al. 1991. *Handbook of Environmental Degradation Rates.* New York: CRC Press/Lewis Publishers.

Hosty, J. W., and P. Foster. 1990. *A Practical Guide to Chemical Spill Response.* New York: Van Nostrand Reinhold.

Jones, F. E. 1994. *Toxic Organic Vapors in the Workplace.* New York: CRC Press/Lewis Publishers.

Keith, L. H. and M. M. Walker. 1995. *Handbook of Air Toxics.* New York: CRC Press/Lewis Publishers.

Landis, W. G., M. H. Yu. 1995. *Introduction to Environmental Toxicology.* New York: CRC Press/Lewis Publishers.

Langerman, N. 1994 *Precautionary Labels for Chemical Containers.* New York: CRC Press/Lewis Publishers.

Lauwerys, R. R., P. Hoet. 1993. *Industrial Chemical Exposure.* New York: CRC Press/Lewis Publishers.

Lewis, R. J., Sr. 1993. *Hawley's Condensed Chemical Dictionary,* 12th Edition. New York: Van Nostrand Reinhold.

Lewis, R. J., Sr. 1992. *Sax's Dangerous Properties of Industrial Materials,* Volumes 1, 2, and 3, 8th Edition. New York: Van Nostrand Reinhold.

Manahan, S. E. 1992. *Toxicological Chemistry,* 2nd Edition. New York: CRC Press/Lewis Publishers.

Mellström, G. A., J. E. Wahlberg, and H. I. Maibach. 1994. *Protective Gloves for Occupational Use.* New York: CRC Press/Lewis Publishers.

Meyer, E. 1990. *Chemistry of Hazardous Materials.* Englewood Cliffs, NJ: Prentice Hall.

Patnaik, P. 1992. *A Comprehensive Guide to the Hazardous Properties of Chemical Substances.* New York: Van Nostrand Reinhold.

Philp, R. B. 1995. *Environmental Hazards and Human Health.* New York: CRC Press/Lewis Publishers.

Quigley, D. R. 1994. *Handbook of Emergency Chemical Management.* New York: CRC Press/Lewis Publishers.

Richardson, M. 1995. *Dictionary of Substance and Their Effects,* Vol. I-VI. New York: CRC Press/Lewis Publishers.

Torkelson, T. R. V. K. Rowe. 1981. *Patty's Industrial Hygiene and Toxicology,* Vol. 2B. New York: Wiley-Interscience.

MATERIAL SAFETY DATA SHEET

CHEMICAL NAME	CRC PRESS/LEWIS PUBLISHERS DISCLAIMER
ETHYL FORMATE	**CRC PRESS/LEWIS PUBLISHERS DISCLAIMER:** THE INFORMATION AND RECOMMENDATIONS PRESENTED HEREIN ARE BASED ON SOURCES BELIEVED TO BE RELIABLE. CRC MAKES NO REPRESENTATION ON ITS COMPLETENESS OR ACCURACY. IT IS THE RESPONSIBILITY OF THE USER TO DETERMINE THE CHEMICAL'S SUITABILITY FOR ITS INTENDED USE, THE CHEMICAL'S SAFE USE, AND THE CHEMICAL'S PROPER DISPOSAL. NO REPRESENTATIONS AND/OR WARRANTIES, EITHER EXPRESSED OR IMPLIED, OF THE MERCHANTABILITY OR FITNESS FOR A PARTICULAR PURPOSE, OR OF ANY OTHER NATURE, ARE MADE WITH RESPECT TO THE INFORMATION PROVIDED IN THIS MSDS OR TO THE CHEMICAL TO WHICH INFORMATION MAY REFER. CRC NEITHER ASSUMES NOR AUTHORIZES ANY OTHER PERSON TO ASSUME FOR IT, ANY OTHER ADDITIONAL RESPONSIBILITY OR LIABILITY FOR THE USE OF, OR RELIANCE UPON, THIS INFORMATION.

HAZARD WARNING INFORMATION

HEALTH	FIRE	REACTIVE	OTHER	DEGREE OF HAZARD	COLOR CODING	OTHER CODES
2	3	0		0 = Minimum Hazard 1 = Slight Hazard 2 = Moderate Hazard 3 = Serious Hazard 4 = Severe Hazard	HEALTH = BLUE FIRE = RED REACTIVITY = YELLOW OTHER = WHITE	OX = Oxidizer ACID = Acid ALK = Alkali COR = Corrosive W = Use No Water

SECTION I - GENERAL INFORMATION

Characterization	**Aliphatic Ester**	RCRA Number	**None**
		EPA Class	**Not Applicable**
DOT Proper Shipping Name	**Ethyl Formate**	Chemical Abstract Service (CAS) Number	**109-94-4**
DOT Hazard Class and Label Requirements	**Flammable Liquid**	DOT Emergency Guide Code	**26**
DOT Identification Number	**UN 1190**	Chemical Formula	**CH_3CH_2OCHO**

Synonyms
Ethyl ester of formic acid; ethyl methanoate; ethyl formic ester; formic ether; Areginal.

SECTION II - HAZARDOUS INGREDIENTS AND IDENTITY INFORMATION

Hazardous Components (specific identity)	OSHA Exposure Criteria	NIOSH Exposure Criteria	Immediately Dangerous to Life and Health (IDLH)	ACGIH Exposure Criteria
Ethyl formate (derivation: By heating ethanol with formic acid in the presence of sulfuric acid and then distilling). **1 ppm = 3.08 mg/m^3**	**PEL (8-hour):** **100 ppm** **300 mg/m^3** **STEL:** **Not Determined**	**REL (10-hour):** **100 ppm** **300 mg/m^3** **STEL:** **Not Determined**	**1500 ppm**	**TLV:** **100 ppm** **303 mg/m^3** **STEL:** **Not Determined**

SECTION III - PHYSICAL AND CHEMICAL CHARACTERISTICS

Boiling Point	**130°F (54°C)**	Specific Gravity (H_2O = 1)	**0.92**
Vapor Pressure (mm Hg)	**200 at 68°F (20°C)**	Molecular Weight	**74.1**
Vapor Density (Air = 1)	**2.6**	Melting Point	**-113°F (-80°C)**

Solubility
Moderately soluble in water (13.5% at 20°C). Soluble in alcohol, acetone, benzene, and most organic solvents. Insoluble in ethyl ether.

Appearance and Odor
Clear colorless liquid with a fruity odor and slightly bitter taste. Odor Threshold = 31 ppm.

SECTION IV - FIRE AND EXPLOSION HAZARD DATA

Flash Point (method used)	**-4°F (-20°C) closed cup**	Explosive Limits in Air % by Volume	**LEL: 2.8%** **UEL: 16%**
NFPA Classification	**Class IB Flammable Liquid**	Autoignition Temperature	**851°F (455°C)**

Extinguishing Media
Dry chemical, alcohol foam, carbon dioxide, fog. Use water spray to keep fire-exposed containers cool.

Special Fire Fighting Procedures
Wear full protective clothing and self-contained breathing apparatus (SCBA). Continue to cool containers with water after fire is extinguished. Move containers from fire if it can be done without risk.

Unusual Fire and Explosion Hazards
Immediately withdraw if rising sound from venting device is heard or if fire is causing discoloration to the tank. Evacuate 1500 feet radius if fire becomes uncontrollable or container is exposed to direct flame. Vapors are heavier than air and may travel a distances to ignition source to flashback.

SECTION V - REACTIVITY DATA

Stability		Conditions to Avoid Ethyl formate is stable under normal conditions of storage and operation. Avoid contact with incompatible materials. Keep away from heat, flame, and other sources of ignition.
Stable X	Unstable	Incompatibility *(materials to avoid)* Reacts violently in contact with nitrates, strong oxidizers, alkalis, and acids. It will slowly decompose in water to form ethyl alcohol and formic acid.
Hazardous Polymerization		Conditions to Avoid Under normal conditions of temperature and pressure, hazardous polymerization of ethyl formate is not expected to occur.
May Occur	Will Not Occur X	Hazardous Decomposition or By-products When heated to decomposition, ethyl formate emits acrid and irritating fumes.

SECTION VI - HEALTH HAZARD DATA

Primary Route(s) of Entry:	Inhalation?	Absorption (skin)?	Ingestion?
	X		X

Health Hazards

INHALATION: Irritation of the eyes, nose, and respiratory tract. Irritation of the eyes and mucous membranes may persist for hours. Inhalation of high concentrations may cause pulmonary edema (fluid in the lungs) and CNS depression with deep narcosis. Death can occur due to circulatory system or respiratory system failure.

SKIN & EYE: Vapors or liquid will irritate the eyes. The liquid will irritate the skin (redness, blistering).

INGESTION: Irritation of the mouth and possible esophageal burns, unconsciousness and death.

Carcinogenicity	NTP Listed?	IARC Cancer Review Group?	OSHA Regulated?	Target Organs?
Unknown Human Unknown Animal	No	No	29 CFR 1910.1000 Table Z-1	Respiratory system; skin; eyes, CNS.

Medical Conditions Generally Aggravated by Exposure

None reported.

Emergency and First-aid Procedures

Eye contact: Flush immediately with water for 15 minutes (minimum); seek medical attention. Skin contact: Remove all contaminated clothing. Immediately wash area with large amounts of soap and water. Seek medical assistance. For inhalation: Remove the person from exposure. Provide respiratory assistance and CPR. Transfer to medical facility. If swallowed: Seek medical attention immediately. Call poison control center for advice. Give 1-2 glasses of water. Do NOT induce vomiting unless advised otherwise. Never give an unconscious or convulsing person anything by mouth.

SECTION VII - PRECAUTIONS FOR SAFE HANDLING AND USE

Steps to be Taken in Case Material is Released or Spilled

Contain spills using absorbent material (vermiculite or other). Ventilate area of spill. Have water source available in case of fire. Store materials in DOT-approved containers. Restrict those not involved in cleanup from entering area. Notify appropriate authorities, as required.

Preferred Waste Disposal Method

Incineration in chemical incinerator equipped with scrubber and afterburner.

Precautions to be Taken in Handling and Storage

Store in tightly closed containers in a cool, dark, well-ventilated area. Protect containers from physical damage. Keep heat, fire and flame away. If possible, automatically transfer liquids between containers.

Other Precautions and Warnings

Bulk storage of ethyl formate is not recommended. Ground and bond all metal containers.

SECTION VIII - CONTROL MEASURES AND PERSONAL PROTECTIVE EQUIPMENT

Respiratory Protection *(specify type)*

Where potential exists for exposure over 100 ppm, use an MSHA/NIOSH-approved full facepiece respirator with an organic vapor cartridge. For high exposures, or when exposure levels are unknown (as in emergency response), use a supplied air respirator or a self-contained breathing apparatus (SCBA) with full facepiece operated in pressure demand or other positive pressure mode.

Ventilation

Local exhaust or general mechanical systems recommended.

Protective Gloves	Eye Protection	Other Protective Clothing
Butyl Rubber or Viton	Chemical Goggles or Face Mask	Impermeable Apron

Work/Hygiene Practices

Always wash hands thoroughly after using chemical; never bring food, drink, or smoking materials into vicinity of chemicals.

ETHYL FORMATE

CH_3CH_2OCHO CAS: 109-94-4

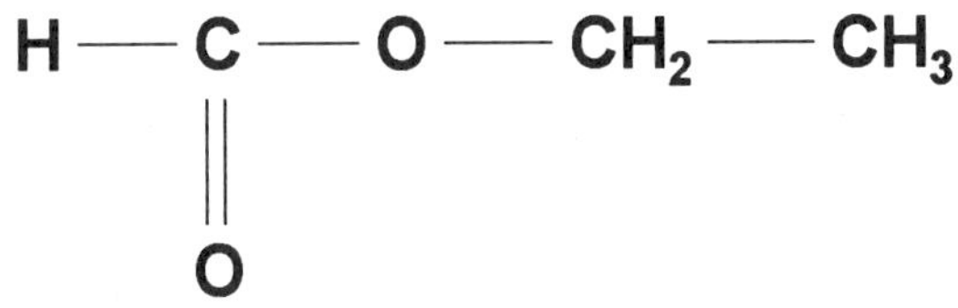

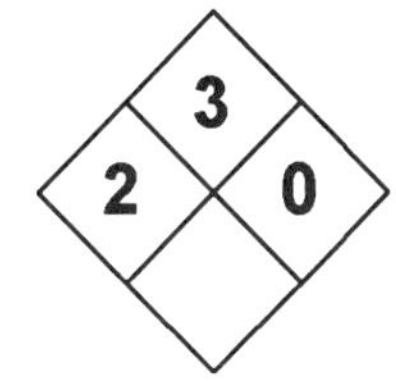

IDENTIFICATION AND TYPICAL USES

Ethyl formate is a colorless liquid with a characteristic, fruity odor and slightly bitter taste. It is used as a solvent for cellulose nitrate and acetate, as an acetone substitute, in organic synthesis, as a flavoring agent for lemonades and essences, and as a fungicide and larvicide for cereals, dry fruits, tobacco, and other similar products.

RISK ASSESSMENT: HEALTH

General Assessment

Ethyl formate is toxic by *inhalation* causing a variety of systemic effects. It is mildly toxic by *ingestion*. Skin contact can produce localized effects but absorption has not been reported. While its specific carcinogenic and mutagenic properties are unknown, mutation data have been reported.

Inhalation of ethyl formate vapors causes irritation of the respiratory tract and difficulty breathing. Irritation of the mucosa can last for several hours following the last exposure to ethyl formate. Inhalation of high concentrations can cause pulmonary edema, a dangerous buildup of fluid in the lungs, which can be fatal. Indications of pulmonary edema may be delayed up to 48 hours, thereby creating a false sense of security with regard to health exposure risk. Varying degrees of central nervous system (CNS) depression is also possible with symptoms of narcosis (deep unconsciousness) and death due to respiratory system or circulatory system failure.

Skin and eye contact can result in mild to moderate irritation. No reports of permanent damage have been recorded in the references.

Ingestion of ethyl formate may cause a variety of toxic effects, including somnolence (sleepiness or excessive drowsiness), narcosis, gastritis, dypsnea (difficulty breathing), irritation of the mouth and stomach with possible esophageal burns. If not treated promptly, there may be unconsciousness and death.

Acute Health Effects

The following acute (short-term) health effects may occur immediately or shortly after exposure to ethyl formate:

Skin: Mild to moderate irritation.

Eye: Mild to moderate irritation.

Lung: Irritation of the eyes, nose, throat and lungs (especially the mucus membranes). May cause pulmonary edema (fluid in the lungs).

CNS: High concentrations can depress the central nervous system leading to narcosis and possible loss of consciousness and even death.

Other: Ingestion can cause a serious irritation of the mouth and stomach with a possibility of esophageal burns, unconsciousness, and death.

Chronic Health Effects

The following chronic (long-term) health effects may occur at some time after exposure to ethyl formate and can last for months or even years:

Cancer Hazards: According to the references, ethyl formate has not been adequately tested for its ability to cause cancer in test animals. However, mutation data have been reported. This does not mean that ethyl formate is carcinogenic. It does mean that further research is required and caution is warranted in use, handling, and storage until further information is known about ethyl formate.

Reproduction: According to the references, ethyl formate has not been adequately tested for its ability to adversely affect reproduction in test animals.

Other Chronic Effects: None reported.

ʊ *Recommended Risk-Reduction Measures*

Personnel should avoid direct contact with ethyl formate. The best risk reduction measure is to use a less toxic chemical as a substitute for a ethyl formate. If this is not possible or feasible, then *engineering controls* are the most effective method of reducing exposures. The best protection is to enclose operations and/or provide local exhaust ventilation at the site of chemical release. While not always operationally feasible, isolating operations can also help reduce exposure risk.

Using respiratory protection is less effective than the controls mentioned above, but is still advisable whenever working with or around ethyl formate. For exposures over 100 ppm (8 hours), an MSHA/NIOSH-approved chemical respirator with an organic vapor cartridge should be used. For higher exposures, or when concentration levels are not known (as in emergency response), a self-contained breathing apparatus (SCBA) with full facepiece and operated in pressure demand mode or a supplied-air respirator with full facepiece operated in pressure demand are the recommended respiratory protection methods of choice. If a full facepiece is not available, then chemical goggles should be worn to protect the eyes. Whenever a chemical splash hazard exists, a face shield and a protective apron should be worn. To prevent hand and skin exposures, impervious butyl rubber, Viton, or chlorinated polyethylene gloves should be used. Glove manufacturers should be contacted to obtain permeation studies before gloves are selected.

Administrative controls should also be in place to minimize the potential for human exposures. These may include written procedures or policies which specify the methods and techniques that will be practiced whenever personnel are to work with ethyl formate.

All personnel should receive *training* on the use, hazards, protective measures, emergency actions, and other precautions per 29 CFR 1910.1200 (Hazard Communication), prior to the first assignment in an area where ethyl formate is used or stored.

If symptoms develop or overexposure is suspected, the following medical tests are recommended:

☑ Lung function tests.

☑ Consider chest X-ray following acute overexposure (may be negative if taken immediately after exposure due to the delayed development of pulmonary edema).

☑ Complete evaluation of the central nervous system.

Any medical evaluation should include a careful history of past and present symptoms with an examination. Medical tests that look for existing damage are not a substitute for controlling exposures. Also, since smoking can cause heart diseases, emphysema, and a variety of other respiratory disorders, smokers exposed to ethyl formate may experience symptoms much more rapidly and more pronounced than non-smokers under the same conditions of exposure. Prudent risk management requires the accurate assessment and consideration of *all* possible factors that may be causing the appearance of exposure symptoms in the work force.

Other methods to reduce exposure include:

☑ Where possible, enclose operations and use local exhaust ventilation at the site of chemical release. If local exhaust ventilation or enclosure is not used, respiratory protection should be mandatory.

☑ Always ensure that proper protective clothing is worn when using chemical substances.

☑ Wash thoroughly immediately after exposure to ethyl formate and at the end of the work shift or before eating, drinking, or smoking.

☑ Hazard warning information should be posted in the work area. In addition, as part of an on-going education and training program, all information on the health and safety hazards of ethyl formate should be communicated to all potentially exposed workers.

☑ Never eat, drink, or smoke in areas were ethyl formate is used, handled or stored.

RISK ASSESSMENT: ENVIRONMENT

General Assessment

The environment is at risk of exposure during transportation, storage, disposal, or destruction of ethyl formate. In almost every scenario, the threat of environmental exposure is contingent upon the proper handling of the chemical substance.

Accidental spills, large or small, can result in fire, explosion, and possible contamination of the surrounding environmental mediums (water, soil, and air). Ethyl formate is considered a Class IB flammable liquid (according to OSHA 29 CFR 1910.106). Its low flash point and relatively low boiling point present a serious fire and explosion hazard concern. Also, because it is incompatible with a number of common materials, especially strong oxidizers and acids, contact can result in violent and explosive reactions. It can also form explosive mixtures in air and can ignite on contact with heat, fire, or sparks. These characteristics require special consideration during any emergency situation involving a leak or spill of ethyl formate.

The proper disposal/destruction method for ethyl formate is to burn it in a chemical incinerator equipped with an afterburner and air scrubber.

Ethyl formate can enter the environment through unchecked industrial discharges into effluents and through spills.

Acute Ecological Effects

Acute (short-term) toxic effects may include the death of animals, birds, or fish, and death or low growth rate in plants. Acute effects are seen 2 to 4 days after animals or plants are exposed to ethyl formate. This chemical has moderate acute toxicity to aquatic life. Insufficient data are available to evaluate or predict the short-term effects of ethyl formate to plants, birds, or terrestrial animals.

Chronic Ecological Effects

Chronic toxic effects may include shortened life span, reproductive problems, lower fertility, and changes in appearance or behavior in exposed animals. These effects can be seen long after first exposure(s) to toxic chemicals. Insufficient data are available to evaluate or predict the long-term effects of ethyl formate to aquatic life, plants, birds, or land animals.

Water Solubility

Ethyl formate is moderately soluble in water. Concentrations of 1 to 1000 milligrams will mix with a liter of water. However, it will gradually decompose to form ethanol and formic acid.

Persistence in the Environment

Ethyl formate is non-persistent in water, with a half-life of less than 2 days. The half-life of a pollutant is the amount of time it takes for one-half of the chemical to be degraded. Since ethyl formate will decompose in water, it is not expected to be absorbed by aquatic sediments or terrestrial soils.

Bioaccumulation in Aquatic Organisms

Some substances increase in concentration, or *bioaccumulate,* in living organisms as they breathe contaminated air, drink contaminated water, or eat contaminated food. These chemicals can become concentrated in the tissues and internal organs of animals as well as humans.

Since ethyl formate will decompose in water, it is not expected to accumulate in fish tissues.

Recommended Risk-Reduction Measures

Proper training of all transporters will reduce the likelihood of a mishap or accident resulting in a leak or spill to the environment. The correct use of DOT labeling (per the MSDS) on all containers, trucks, and rail cars will enable emergency responders to react properly and quickly to any disaster thereby reducing the potential risk to the environment and to personnel.

Storage of ethyl formate should be segregated from incompatible chemicals to minimize the risk of cross-contamination or contact. Buildings designated for storage should be equipped with appropriate fire protection systems (alarms, sprinklers, emergency lighting, portable extinguishers).

If a spill or leak to the environment has occurred, fire department, emergency response, and/or hazardous materials spill personnel should be notified immediately. Cleanup should be attempted only by those trained in proper spill containment procedures. Contaminated soils should be removed for incineration and replaced with clean soil.

If ethyl formate should contact the water table, aquifer, or navigable waterway, time is of the essence. It will slowly decompose in water and, therefore, total containment and remediation may not be entirely possible. When such spills occur, the local and/or state emergency response authorities must be notified. A comprehensive emergency response or disaster preparedness/recovery plan should be in place prior to any operations involving the use, transportation, storage, or disposal of ethyl formate. If ethyl formate is spilled or leaked, the following specific steps are recommended:

- ☑ Restrict persons not wearing protective clothing from area of spill or leak until cleanup is complete and area can be opened for normal work.
- ☑ Ventilate area and remove all ignition sources.
- ☑ Absorb liquids in vermiculite, dry sand, earth, or a similar material and deposit in sealed containers. Use non-sparking tools!
- ☑ It may be necessary to dispose of ethyl formate as a hazardous waste. The responsible state agency or the regional office of the federal Environmental

Protection Agency (EPA) should be contacted for specific recommendations.

RISK ASSESSMENT: BUSINESS

General Assessment

Accidents or mishaps involving ethyl formate can present a moderate threat to business operations. The loss or damage of equipment or facilities can significantly affect fiscal viability. Lawsuits that may result from personnel injury/death, public exposures, and/or environmental contamination will also require a serious expenditure of resources.

Media attention surrounding an injury, death, or environmental damage can also result in a loss of profits and loss of current as well as future business.

Recommended Risk-Reduction Measures

Company attorneys, safety and health professionals, and environmental specialists should be involved in the development of any procedures or policies intended to manage the use of chemicals in the workplace. A company official should be pre-designated as a public relations officer with specific training in dealing with the press. Corporate plans and policies should be developed, approved, and implemented long before any need for such arises.

REFERENCES

American Conference of Governmental Industrial Hygienists. 1988. *Documentation of the Threshold Limit Values and Biological Exposure Indices,* 5th Edition (with updates). Cincinnati: ACGIH

Anthony, C. P., and N. J. Kolthoff. 1971. *Textbook of Anatomy and Physiology.* St. Louis: C. V. Mosby Company.

Bloom, A. D. and F. J. DeSerres. 1995. *Ecotoxicity and Human Health.* New York: CRC Press/Lewis Publishers.

Brusick, D. J. 1994. *Methods for Genetic Risk Assessment.* New York: CRC Press/Lewis Publishers.

Bueche, J. R. 1972. *Principles of Physics,* 2nd Edition. New York: McGraw-Hill.

Burrell, R., D. K. Flaherty, and L. J. Sauers. 1992. *Toxicology of the Immune System.* New York: Van Nostrand Reinhold.

Calabrese, E. J. 1994. *Biological Effects of Low Level Exposures.* New York: CRC Press/Lewis Publishers.

Cockerman, L. G., B. S. Shane. 1994. *Basic Environmental Toxicology.* New York: CRC Press/Lewis Publishers.

Cothern, C. R., N. P. Ross. 1994. *Environmental Statistics, Assessment, and Forecasting.* New York: CRC Press/Lewis Publishers.

Danse, I. R. 1991. *Common Sense Toxics in the Workplace.* New York: Van Nostrand Reinhold.

Environmental Protection Agency. 1983. *Chemical Hazard Information Profile: Ethyl formate.* Washington, D.C.: U.S. EPA.

Gardner, R. W. 1994. *Chemical Intolerance.* New York: CRC Press/Lewis Publishers.

Hallenbeck, W. H. 1993. *Quantitative Risk Assessment for Environmental and Occupational Health,* 2nd Edition. New York: CRC Press/Lewis Publishers.

Hodgson, E., R. B. Mailman, and J. E. Chambers. 1988. *Dictionary of Toxicology.* London: The MacMillian Company.

Hosty, J. W., and P. Foster. 1990. *A Practical Guide to Chemical Spill Response.* New York: Van Nostrand Reinhold.

Howard, P. H., et. al. 1991. *Handbook of Environmental Degradation Rates.* New York: CRC Press/Lewis Publishers.

James, A., and M. Lord. 1992. *Index of Chemical and Physical Data.* New York: Van Nostrand Reinhold.

Jones, F. E. 1994. *Toxic Organic Vapors in the Workplace.* New York: CRC Press/Lewis Publishers.

Kamrin, M. I A. 1988. *Toxicology: A Primer on Principles and Applications.* Chelsea, Michigan: Lewis Publishers.

Keith, L. H. and M. M. Walker. 1995. *Handbook of Air Toxics.* New York: CRC Press/Lewis Publishers.

Landis, W. G., M. H. Yu. 1995. *Introduction to Environmental Toxicology.* New York: CRC Press/Lewis Publishers.

Langerman, N. 1994 *Precautionary Labels for Chemical Containers.* New York: CRC Press/Lewis Publishers.

Lauwerys, R. R., P. Hoet. 1993. *Industrial Chemical Exposure.* New York: CRC Press/Lewis Publishers.

Lide, D. R. 1995. *Handbook of Chemistry and Physics,* 76th Edition. New York: CRC Press/Lewis Publishers.

Lewis, Richard J., Sr. 1992. *Sax's Dangerous Properties of Industrial Materials,* Volumes 1, 2, and 3, 8th Edition. New York: Van Nostrand Reinhold.

Manahan, S. E. 1994. *Environmental Chemistry,* 6th Edition. New York: CRC Press/Lewis Publishers.

Manahan, S. E. 1992. *Toxicological Chemistry,* 2nd Edition. New York: CRC Press/Lewis Publishers.

Meyer, Eugene. 1990. *Chemistry of Hazardous Materials.* Englewood Cliffs, NJ: Prentice Hall.

Patnaik, P. 1992. *A Comprehensive Guide to the Hazardous Properties of Chemical Substances.* New York: Van Nostrand Reinhold.

Proctor, N. H., and J. P. Hughes. 1991. *Chemical Hazards of the Workplace.* New York: Van Nostrand Reinhold.

U.S. Department of Health and Human Services, National Institute for Occupational Safety and Health. 1994. *NIOSH Pocket Guide to Chemical Hazards.* Washington, D.C.: U.S. Government Printing Office.

MATERIAL SAFETY DATA SHEET

CHEMICAL NAME	
2-ETHYLHEXANOL	**CRC PRESS/LEWIS PUBLISHERS DISCLAIMER:** THE INFORMATION AND RECOMMENDATIONS PRESENTED HEREIN ARE BASED ON SOURCES BELIEVED TO BE RELIABLE. CRC MAKES NO REPRESENTATION ON ITS COMPLETENESS OR ACCURACY. IT IS THE RESPONSIBILITY OF THE USER TO DETERMINE THE CHEMICAL'S SUITABILITY FOR ITS INTENDED USE, THE CHEMICAL'S SAFE USE, AND THE CHEMICAL'S PROPER DISPOSAL. NO REPRESENTATIONS AND/OR WARRANTIES, EITHER EXPRESSED OR IMPLIED, OF THE MERCHANTABILITY OR FITNESS FOR A PARTICULAR PURPOSE, OR OF ANY OTHER NATURE, ARE MADE WITH RESPECT TO THE INFORMATION PROVIDED IN THIS MSDS OR TO THE CHEMICAL TO WHICH INFORMATION MAY REFER. CRC NEITHER ASSUMES NOR AUTHORIZES ANY OTHER PERSON TO ASSUME FOR IT, ANY OTHER ADDITIONAL RESPONSIBILITY OR LIABILITY FOR THE USE OF, OR RELIANCE UPON, THIS INFORMATION.

HAZARD WARNING INFORMATION

HEALTH	FIRE	REACTIVE	OTHER	DEGREE OF HAZARD	COLOR CODING	OTHER CODES
2	2	0		0 = Minimum Hazard 1 = Slight Hazard 2 = Moderate Hazard 3 = Serious Hazard 4 = Severe Hazard	HEALTH = BLUE FIRE = RED REACTIVITY = YELLOW OTHER = WHITE	OX = Oxidizer ACID = Acid ALK = Alkali COR = Corrosive W = Use No Water

SECTION I - GENERAL INFORMATION

Characterization	**Alcohol**	RCRA Number	**None**
		EPA Class	**Not Applicable**
DOT Proper Shipping Name	**No Citation**	Chemical Abstract Service (CAS) Number	**104-76-7**
DOT Hazard Class and Label Requirements	**No Citation**	DOT Emergency Guide Code	**No Citation**
DOT Identification Number	**No Citation**	Chemical Formula	$C_8H_{18}O$

Synonyms

2-ethylhexyl alcohol; octyl alcohol; 2-ethyl-1-hexanol.

SECTION II - HAZARDOUS INGREDIENTS AND IDENTITY INFORMATION

Hazardous Components (specific identity)	OSHA Exposure Criteria	NIOSH Exposure Criteria	Immediately Dangerous to Life and Health (IDLH)	ACGIH Exposure Criteria
2-Ethylhexanol (derivation: By condensation of butyraldehyde followed by hydrogenation, the fermentation of alcohol, or by hydrogenation of 2-ethyl-1-hexanol at 392°F (200°C) using a copper or nickel catalyst).	**PEL:** Not Established **STEL:** Not Established	**REL:** Not Established **STEL:** Not Established	Not Determined	**TLV:** Not Established **STEL:** Not Established

SECTION III - PHYSICAL AND CHEMICAL CHARACTERISTICS

Boiling Point	**359°F (182°C)**	Specific Gravity (H_2O = 1)	**0.834**
Vapor Pressure (gas)	**0.2 at 69°F (20°C)**	Molecular Weight	**130.22**
Vapor Density (Air = 1)	**4.5**	Melting Point	**<-105°F (<-76°C)**

Solubility

Nearly insoluble in water (0.1%). Soluble in most organic solvents, including ethyl alcohol and ethyl ether.

Appearance and Odor

Clear or colorless liquid with a sweet, agreeable, characteristic floral (rose-like) odor.

SECTION IV - FIRE AND EXPLOSION HAZARD DATA

Flash Point (method used)	**178°F (81°C) closed cup**	Explosive Limits in Air % by Volume	**LEL: 0.8% UEL: 9.7%**
NFPA Classification	**Class IIIA Combustible Liquid**	Autoignition Temperature	**488°F (231°C)**

Extinguishing Media

Use dry chemical, carbon dioxide, water spray, or alcohol-resistant foam.

Special Fire Fighting Procedures

Wear full protective clothing and self-contained breathing apparatus (SCBA). Poisonous gases are produced in fire. Use water spray to keep fire-exposed containers cool. Containers may explode in fire.

Unusual Fire and Explosion Hazards

Dangerous fire hazard when exposed to heat. Chemical will react on contact with oxidizable materials. Moderate explosion hazard when in the form of vapor.

SECTION V - REACTIVITY DATA

Stability		Conditions to Avoid: Keep away from sources of heat (including direct sunlight) or flame. Large quantities should be stored in metal tanks or drums. Keep away from incompatible materials.
Stable	Unstable	Incompatibility *(materials to avoid)*
X		Strong oxidizers, such as permanganates, peroxides, nitrates, chlorates, and perchlorates.
Hazardous Polymerization		Conditions to Avoid: Under normal conditions of temperature and pressure, hazardous polymerization of 2-ethylhexanol is not expected to occur.
May Occur	Will Not Occur	Hazardous Decomposition or By-products
	X	When heated to decomposition, 2-ethylhexanol can emit highly toxic/poisonous and acrid fumes and gases, including carbon dioxide.

SECTION VI - HEALTH HAZARD DATA

Primary Route(s) of Entry:	Inhalation?	Absorption (skin)?	Ingestion?
	X	X	

Health Hazards

INHALATION: Can produce human systemic effects, including central nervous system (CNS) depression with symptoms of headache, muscle weakness, nausea, giddiness, and ataxia. Will also cause irritation to the eyes and respiratory tract.

ABSORPTION: Irritating to the skin. May be absorbed, but seldom in toxic amounts. There may be skin sensitization and redness. Eye contact will cause mild irritation.

INGESTION: Not likely to occur.

Carcinogenicity	NTP Listed?	IARC Cancer Review Group?	OSHA Regulated?	Target Organs?
Unknown Human Unknown Animal	No	No	Not Listed	Respiratory system, eyes, skin, CNS.

Medical Conditions Generally Aggravated by Exposure

None reported. Skin conditions (dermatitis) or respiratory problems (asthma) may be aggravated.

Emergency and First-aid Procedures

Eye contact: Flush immediately with water for 15 minutes (minimum), seek medical attention. Skin contact: Remove all contaminated clothing. Immediately wash area with large amounts of soap and water. Seek medical assistance. For inhalation: Remove the person from exposure. Provide respiratory assistance and CPR, if required. Transfer to medical facility. Keep victim warm and at rest. If swallowed: Not likely. Contact poison control center. Give 1-2 glasses of water and induce vomiting.

SECTION VII - PRECAUTIONS FOR SAFE HANDLING AND USE

Steps to be Taken in Case Material is Released or Spilled

Absorb large spills in vermiculite or similar material and deposit in sealed containers for disposal. Ventilate area of spill or leak. Restrict those not involved in cleanup from entering area. Remove all sources of ignition.

Preferred Waste Disposal Method

No citation.

Precautions to be Taken in Handling and Storage

Store to avoid contact with incompatible materials in tightly closed containers in cool, well-ventilated area away from heat and flame. Use only non-sparking tools to open and close containers.

Other Precautions and Warnings

Sources of ignition are prohibited where 2-ethylhexanol is used, handled, or stored. Heat may cause containers to build up pressure and explode. Containers must be grounded and bonded.

SECTION VIII - CONTROL MEASURES AND PERSONAL PROTECTIVE EQUIPMENT

Respiratory Protection *(specify type)*

No exposure limits have been established. For best protection use a supplied-air respirator with full facepiece or use a self-contained breathing apparatus (SCBA) in positive pressure or other pressure demand mode.

Ventilation

Local exhaust or general mechanical systems recommended.

Protective Gloves	Eye Protection	Other Protective Clothing
Impervious	Chemical Goggles or Face Mask	protective Apron or Suit

Work/Hygiene Practices

Always wash hands thoroughly after using chemical; never bring food, drink, or smoking materials into vicinity of chemicals.

2-ETHYLHEXANOL

$C_8H_{18}O$ CAS: 104-76-7

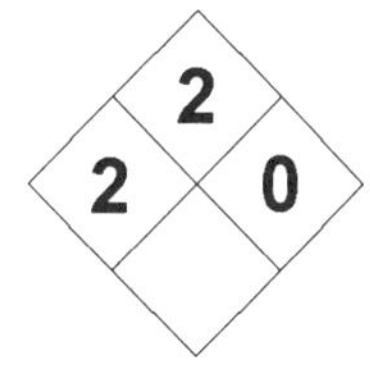

$$CH_3(CH_2)_3CHC_2H_5CH_2OH$$

IDENTIFICATION AND TYPICAL USES

2-Ethylhexanol is a clear, colorless, liquid with a sweet, agreeable, characteristic floral (rose-like) odor. It is used as a plasticizer for polyvinylchloride (PVC) resins, as a de-foaming agent, a wetting agent, in organic synthesis, as a solvent mix for nitrocellulose, in paints, lacquers, baking finishes, in cotton refinement and textile finishing compounds, in inks, rubber, paper, lubricants, photography, and in the dry cleaning industry.

RISK ASSESSMENT: HEALTH

General Assessment

2-Ethylhexanol is relatively non-toxic to humans at room temperature (due to its low vapor pressure). However, when heated or misted it can present mild to moderate toxicity by *inhalation*. It can also pass through intact skin (*absorption*), although seldom in toxic amounts from a single exposure. Its carcinogenicity in humans is not known but it has been shown to cause mutations in animal studies. Experimental teratogenic data have also been reported in animal tests.

Inhalation causes irritation to the eyes and respiratory tract. In high concentrations, exposure may cause depression of the central nervous system (CNS) resulting in headache, muscle weakness, nausea, giddiness, and ataxia (loss of muscle control or coordination).

Skin and eye contact may produce mild irritation on contact. It may cause redness and sensitization on prolonged contact. It will pass through unbroken skin to cause toxic systemic effects primarily associated with CNS depression. Repeated dermal exposures have resulted in liver and kidney damage in test animals.

Acute Health Effects

The following acute (short-term) health effects may occur immediately or shortly after exposure to 2-ethylhexanol:

Skin: Mild irritation with redness and possible sensitization. Will pass through unbroken skin.

Eye: Mild to moderate irritation and pain (smarting).

Lung: Nose and throat irritation. Produces intoxicating effects and may cause lung effects as well.

CNS: Serious depression of the CNS at high exposures. Can lead to loss of consciousness.

Chronic Health Effects

The following chronic (long-term) health effects may occur at some time after exposure to 2-ethylhexanol and can last for months or even years:

Cancer Hazards: According to information presented in the references, 2-ethylhexanol has not been adequately tested for its ability to cause cancer in animals. Some studies indicate that it causes mutations in animals, while others are inconclusive in this regard.

Reproductive Hazard: According to information presented in the references, 2-ethylhexanol has produced an increase in the mortality of the offspring of test animals (mice). The teratogenicity of 2-ethylhexanol in humans has not been adequately studied.

Other Chronic Effects: Prolonged skin contact may cause drying, cracking, and chapping of the affected area. Repeated or long-term exposures to 2-ethylhexanol may affect the liver and kidneys.

Recommended Risk-Reduction Measures

Personnel should avoid direct contact with 2-ethylhexanol. The exact nature of the toxicity of this chemical is not entirely understood in the references. Therefore, prudent risk management requires it be treated as though it posed a significant health risk in the event of overexposure. If a less toxic chemical cannot be substituted for a hazardous substance, then *engi-*

neering controls are the most effective method of reducing exposures. The best protection is to enclose operations and/or provide local exhaust ventilation at the site of chemical release. While not always operationally feasible, isolating operations can also reduce exposure. Using respiratory protection is less effective than the controls mentioned above, but is still advisable whenever working with or around 2-ethylhexanol. No exposure levels have been established for this chemical. This does not mean that exposure is without risk. A supplied-air respirator with full facepiece operated in continuous flow mode, or a self-contained breathing apparatus (SCBA) with full facepiece and pressure demand are the recommended respiratory protection methods of choice. If a full facepiece is not available, then chemical goggles should be worn to protect the eyes. Whenever a chemical splash hazard exists, a face shield and a rubber apron should be worn. To prevent hand and skin exposures, impervious rubber gloves should be worn. Glove manufacturers should be contacted and permeation studies obtained *before* glove selections are made.

Administrative controls should also be in place to minimize the potential for human exposures. These may include written procedures or policies which specify the methods and techniques that will be practiced whenever personnel are to work with 2-ethylhexanol.

All personnel should receive *training* on the use, hazards, protective measures, emergency actions, and other precautions per 29 CFR 1910.1200 (Hazard Communication), prior to the first assignment in an area where 2-ethylhexanol is used or stored.

If symptoms develop or overexposure is suspected, the following tests are recommended:

☑ Kidney and liver function tests.

☑ Evaluation of central nervous system function.

Any evaluation should include a careful history of past and present symptoms with an examination. Medical tests that simply look for existing damage are not a substitute for controlling exposures.

Other methods to reduce exposure include:

☑ Where possible, enclose operations and use local exhaust ventilation at the site of chemical release. If local exhaust ventilation or enclosure is not used, respiratory protection should be mandatory.

☑ Always ensure that proper protective clothing is worn when using chemical substances and that personnel are trained on its use and care.

☑ Wash thoroughly immediately after exposure to 2-ethylhexanol and at the end of the work shift or before eating, drinking, or smoking.

☑ Hazard warning information should be posted in the work area. In addition, as part of an on-going education and training program, all information on the health and safety hazards of 2-ethylhexanol should be communicated to all potentially exposed workers.

☑ Eye wash stations should be provided in the immediate work area for emergency use. If there is a possibility of skin exposure to 2-ethylhexanol, emergency shower facilities should also be provided.

☑ Personnel should never wear contaminated clothing home (family members can be exposed). Clothing should be laundered by personnel who have been trained on the hazards of 2-ethylhexanol.

RISK ASSESSMENT: ENVIRONMENT

General Assessment

The environment is at risk of exposure during transportation, storage, disposal, or destruction of 2-ethylhexanol. In almost every scenario, the threat of environmental exposure is contingent upon the proper handling of the chemical substance. Accidental spills, large or small, can result in fire, explosion, and possible contamination of the surrounding environmental mediums (water, soil, and air).

2-Ethylhexanol is considered a Class IIIA combustible liquid (per OSHA 29 CFR 1910.106). While this presents a moderate fire hazard, it can react violently in contact with incompatible materials and is an explosion hazard in vapor form. These characteristics require extreme caution in handling, storage, transportation, and disposal. It is incompatible with common oxidizers and contact can cause fire or explosion. It will ignite on contact with chromium trioxide. Therefore, special consideration is required during any emergency situation involving a leak or spill of 2-ethylhexanol. Should 2-ethylhexanol ever come into contact with these incompatible substances during use, transportation, storage, or disposal, the formation of highly toxic and/or highly explosive commodities is possible.

The proper disposal method for 2-ethylhexanol is to burn it in a chemical incinerator equipped with an afterburner and scrubber.

2-Ethylhexanol may enter the environment through industrial discharges or spills.

Acute Ecological Effects

Acute (short-term) toxic effects may include the death of animals, birds, or fish and death or low growth rate in plants. Acute effects are seen 2 to 4 days after animals or plants are exposed to 2-ethylhexanol.

Insufficient data are available on the short-term effects of 2-ethylhexanol exposure to aquatic life, birds, plants, or land animals.

Chronic Ecological Effects

Chronic (long-term) toxic effects may include shortened life span, reproductive problems, lower fertility, and changes in appearance or behavior in exposed animals. These effects can be seen long after first exposure(s) to toxic chemicals.

Insufficient data are available on the long-term effects of 2-ethylhexanol exposure to aquatic life, birds, plants, or land animals.

Water Solubility

2-Ethylhexanol is nearly insoluble in water. Concentrations of less than 1 milligram may mix with a liter of water.

Persistence in the Environment

2-Ethylhexanol is slightly persistent in water, with a half-life between 2 and 20 days. The half-life of a pollutant is the amount of time it takes for one half of the chemical to be degraded.

Bioaccumulation in Aquatic Organisms

Some substances increase in concentration, or *bioaccumulate*, in living organisms as they breathe contaminated air, drink contaminated water, or eat contaminated food. These chemicals can become concentrated in the tissues and internal organs of animals as well as humans.

The concentration of 2-ethylhexanol found in fish tissue is expected to be about the same as the average concentration of 2-ethylhexanol in the water from which the fish was taken.

Recommended Risk-Reduction Measures

Proper training of all transporters will reduce the likelihood of a mishap or accident resulting in a leak or spill to the environment. The correct use of DOT labeling (per the MSDS) on all containers, trucks, and rail cars will enable emergency responders to react properly and quickly to any disaster thereby reducing the potential risk to the environment and to personnel.

Storage of 2-ethylhexanol should be segregated from incompatible chemicals to minimize the risk of cross-contamination or contact. Buildings designated for storage should be equipped with appropriate fire protection systems (alarms, sprinklers, emergency lighting, portable extinguishers).

If a spill or leak to the environment has occurred, fire department, emergency response and/or hazardous materials spill personnel should be notified immediately. Cleanup should be attempted only by those trained in proper spill containment procedures. Contaminated soils should be removed for incineration and replaced with clean soil. If 2-ethylhexanol should contact the water table, aquifer, or navigable waterway, cleanup should be started immediately. It is only slightly soluble in water and total remediation may be possible. The local and/or state emergency response authorities must be notified. A comprehensive emergency response or disaster preparedness/recovery plan should be in place prior to any operations involving the use, transportation, storage, or disposal of 2-ethylhexanol.

If 2-ethylhexanol is spilled or leaked, the following specific steps are recommended:

☑ Restrict persons not wearing protective clothing from area of spill or leak until cleanup is complete and area is opened for normal work.

☑ Remove all ignition sources.

☑ Ventilate area of spill or leak.

☑ Absorb liquids in vermiculite, dry sand, earth, or similar material and deposit in sealed containers.

☑ It may be necessary to dispose of 2-ethylhexanol as a hazardous waste. The responsible state agency or the regional office of the federal Environmental Protection Agency (EPA) should be contacted for specific recommendations.

RISK ASSESSMENT: BUSINESS

General Assessment

Accidents or mishaps involving 2-ethylhexanol can present a significant threat to business operations. The loss or damage of equipment or facilities can significantly affect fiscal viability. Lawsuits that may result from personnel injury/death, public exposures, and/or environmental contamination will also require a serious expenditure of resources. Media attention surrounding an injury, death, or environmental damage can also result in a loss of profits and loss of current as well as future business. Always remember that anytime the term "reproductive hazard" is used, public emotion, hysteria, and ignorance can run equally high. This requires careful consideration during the development of any public relations policies.

Recommended Risk-Reduction Measures

Company attorneys, safety and health professionals, and environmental specialists should be involved in the development of any procedures dealing with safety or emergency response. A company official should be pre-designated as a public relations officer with specific training in dealing with the press. Corporate plans and policies should be developed, approved, and implemented long before any need for such arises.

REFERENCES

Ahlbom, A. 1993. *Biostatistics for Epidemiologists.* New York: CRC Press/Lewis Publishers.

American Conference of Governmental Industrial Hygienists. 1988. *Documentation of the Threshold Limit Values and Biological Exposure Indices,* 5th Edition (with updates). Cincinnati: ACGIH

Anthony, C. P., and N. J. Kolthoff. 1971. *Textbook of Anatomy and Physiology.* St. Louis: C. V. Mosby Company.

Asante-Duah, D. K. 1993. *Hazardous Waste Risk Assessment.* New York: CRC Press/Lewis Publishers.

Bloom, A. D. and F. J. DeSerres. 1995. *Ecotoxicity and Human Health.* New York: CRC Press/Lewis Publishers.

Brusick, D. J. 1994. *Methods for Genetic Risk Assessment.* New York: CRC Press/Lewis Publishers.

Bueche, J. R. 1972. *Principles of Physics,* 2nd Edition. New York: McGraw-Hill.

Burrell, R., D. K. Flaherty, and L. J. Sauers. 1992. *Toxicology of the Immune System.* New York: Van Nostrand Reinhold.

Calabrese, E. J. 1994. *Biological Effects of Low Level Exposures.* New York: CRC Press/Lewis Publishers.

Cockerman, L. G., B. S. Shane. 1994. *Basic Environmental Toxicology.* New York: CRC Press/Lewis Publishers.

Danse, I. R. 1991. *Common Sense Toxics in the Workplace.* New York: Van Nostrand Reinhold.

Environmental Protection Agency. 1983. *Chemical Hazard Information Profiles.* Washington, D.C.: U.S. EPA.

Fodden, C. R., and J. L. Weddell. 1992. *Hazardous Materials Emergency Action Data.* New York: CRC Press/Lewis Publishers.

Forseberg, K., and L. H. Keith. 1993. *Chemical; Protective Clothing Permeation and Degradation Database.* Chelsea, MI: Lewis Publishers.

Gardner, R. W. 1994. *Chemical Intolerance.* New York: CRC Press/Lewis Publishers.

Hallenbeck, W. H. 1993. *Quantitative Risk Assessment for Environmental and Occupational Health,* 2nd Edition. New York: CRC Press/Lewis Publishers.

Jones, F. E. 1994. *Toxic Organic Vapors in the Workplace.* New York: CRC Press/Lewis Publishers.

Keith, L. H. and M. M. Walker. 1995. *Handbook of Air Toxics.* New York: CRC Press/Lewis Publishers.

Landis, W. G., M. H. Yu. 1995. *Introduction to Environmental Toxicology.* New York: CRC Press/Lewis Publishers.

Langerman, N. 1994 *Precautionary Labels for Chemical Containers.* New York: CRC Press/Lewis Publishers.

Lauwerys, R. R., P. Hoet. 1993. *Industrial Chemical Exposure.* New York: CRC Press/Lewis Publishers.

Lewis, R. J., Sr. 1992. *Sax's Dangerous Properties of Industrial Materials,* Volumes 1, 2, and 3, 8th Edition. New York: Van Nostrand Reinhold.

Lide, D. R. 1995. *Handbook of Chemistry and Physics,* 76th Edition. New York: CRC Press/Lewis Publishers.

Manahan, S. E. 1994. *Environmental Chemistry,* 6th Edition. New York: CRC Press/Lewis Publishers.

Mellström, G. A., J. E. Wahlberg, and H. I. Maibach. 1994. *Protective Gloves for Occupational Use.* New York: CRC Press/Lewis Publishers.

Meyer, E. 1990. *Chemistry of Hazardous Materials.* Englewood Cliffs, NJ: Prentice Hall.

Patnaik, P. 1992. *A Comprehensive Guide to the Hazardous Properties of Chemical Substances.* New York: Van Nostrand Reinhold.

Proctor, N. H., J. P. Hughes. 1991. *Chemical Hazards of the Workplace.* New York: Van Nostrand Reinhold.

Quigley, D. R. 1994. *Handbook of Emergency Chemical Management.* New York: CRC Press/Lewis Publishers.

Rea, W. J. 1995. *Chemical Sensitivity,* Vol. I-III. New York: CRC Press/Lewis Publishers.

Richardson, M. 1995. *Dictionary of Substance and Their Effects,* Vol. I-VI. New York: CRC Press/Lewis Publishers.

Sacarello, H. L. A. 1994. *The Comprehensive Handbook of Hazardous Materials.* New York: CRC Press/Lewis Publishers.

Sheftel, V. O. 1995. *Handbook of Toxic Properties of Monomers and Additives.* New York: CRC Press/Lewis Publishers.

Torkelson, T. R. V. K. Rowe. 1981. Halogenated aliphatic hydrocarbons. *Patty's Industrial Hygiene and Toxicology,* Vol. 2B, *Toxicology,* pp 3568-3572. New York: Wiley-Interscience.

U.S. Department of Health and Human Services, National Institute for Occupational Safety and Health. 1994. *NIOSH Pocket Guide to Chemical Hazards.* Washington, D.C.: U.S. Government Printing Office.

MATERIAL SAFETY DATA SHEET

CHEMICAL NAME

ETHYLIDENE NORBORNENE

CRC PRESS/LEWIS PUBLISHERS DISCLAIMER: THE INFORMATION AND RECOMMENDATIONS PRESENTED HEREIN ARE BASED ON SOURCES BELIEVED TO BE RELIABLE. CRC MAKES NO REPRESENTATION ON ITS COMPLETENESS OR ACCURACY. IT IS THE RESPONSIBILITY OF THE USER TO DETERMINE THE CHEMICAL'S SUITABILITY FOR ITS INTENDED USE, THE CHEMICAL'S SAFE USE, AND THE CHEMICAL'S PROPER DISPOSAL. NO REPRESENTATIONS AND/OR WARRANTIES, EITHER EXPRESSED OR IMPLIED, OF THE MERCHANTABILITY OR FITNESS FOR A PARTICULAR PURPOSE, OR OF ANY OTHER NATURE, ARE MADE WITH RESPECT TO THE INFORMATION PROVIDED IN THIS MSDS OR TO THE CHEMICAL TO WHICH INFORMATION MAY REFER. CRC NEITHER ASSUMES NOR AUTHORIZES ANY OTHER PERSON TO ASSUME FOR IT, ANY OTHER ADDITIONAL RESPONSIBILITY OR LIABILITY FOR THE USE OF, OR RELIANCE UPON, THIS INFORMATION.

HAZARD WARNING INFORMATION

HEALTH	FIRE	REACTIVE	OTHER	DEGREE OF HAZARD	COLOR CODING	OTHER CODES
1	1	2		0 = Minimum Hazard 1 = Slight Hazard 2 = Moderate Hazard 3 = Serious Hazard 4 = Severe Hazard	HEALTH = BLUE FIRE = RED REACTIVITY = YELLOW OTHER = WHITE	OX = Oxidizer ACID = Acid ALK = Alkali COR = Corrosive W = Use No Water

SECTION I - GENERAL INFORMATION

Characterization: **Cyclic Olefin Hydrocarbon**

RCRA Number: **None**

EPA Class: **Not Applicable**

DOT Proper Shipping Name: **No Citation**

Chemical Abstract Service (CAS) Number: **16219-75-3**

DOT Hazard Class and Label Requirements: **No Citation**

DOT Emergency Guide Code: **No Citation**

DOT Identification Number: **No Citation**

Chemical Formula: **C_9H_{12}**

Synonyms: **ENB; ethylidenebicyclo(2,2,1)hep-2-ene; 5-ethylidene-2-norbornene.**

SECTION II - HAZARDOUS INGREDIENTS AND IDENTITY INFORMATION

Hazardous Components (specific identity)	OSHA Exposure Criteria	NIOSH Exposure Criteria	Immediately Dangerous to Life and Health (IDLH)	ACGIH Exposure Criteria
Ethylidene norbornene (derivation: A cyclic olefin hydrocarbon related to pinene used as the third monomer in EPDM elastomers). **1 ppm = 5.00 mg/m^3**	**PEL (ceiling):** 5 ppm 25 mg/m^3 **STEL:** Not Applicable	**REL (ceiling):** 5 ppm 25 mg/m^3 **STEL:** Not Applicable	**Not Determined**	**TLV (ceiling):** 5 ppm 25 mg/m^3 **STEL:** Not Applicable

SECTION III - PHYSICAL AND CHEMICAL CHARACTERISTICS

Boiling Point: **153°F (67°C)**

Specific Gravity (H_2O = 1): **0.90**

Vapor Pressure (gas): **4.2 at 69°F (20°C)**

Molecular Weight: **120.2**

Vapor Density (Air = 1): **4.1**

Melting Point: **-112°F (-80°C)**

Solubility: **Solubility not reported.**

Appearance and Odor: **A colorless to white liquid with a slight, turpentine-like odor. Odor Threshold = 0.007 to 0.014 ppm.**

SECTION IV - FIRE AND EXPLOSION HAZARD DATA

Flash Point (method used): **101°F (38°C)**

Explosive Limits in Air % by Volume: **LEL: Not Determined UEL: Not Determined**

NFPA Classification: **Class II Combustible Liquid**

Autoignition Temperature: **Not Determined**

Extinguishing Media: **Use dry chemical, carbon dioxide, or alcohol foam (best for large fires). Water may be ineffective by itself on fires. Direct streams may spread or scatter fire.**

Special Fire Fighting Procedures: **Wear full protective clothing and self-contained breathing apparatus (SCBA). Poisonous gases are produced in fire. Use water spray to keep fire-exposed containers cool.**

Unusual Fire and Explosion Hazards: **Combustible liquid but a dangerous fire hazard when exposed to heat. Violent reactions can occur with oxygen.**

SECTION V - REACTIVITY DATA

Stability		Conditions to Avoid **Normally stable but will react violently in contact with oxygen. Should be kept in tightly closed containers with a nitrogen purge. It may be stabilized with 100 ppm or *tert*-butyl catechol.**
Stable X	Unstable	Incompatibility *(materials to avoid)* **Ethylidene norbornene is incompatible with oxygen.**
Hazardous Polymerization		Conditions to Avoid **Under normal conditions of temperature and pressure, hazardous polymerization of ethylidene norbornene is not expected to occur.**
May Occur	Will Not Occur X	Hazardous Decomposition or By-products **When heated to decomposition, ethylidene norbornene can emit highly toxic/poisonous and acrid fumes and gases, including carbon dioxide.**

SECTION VI - HEALTH HAZARD DATA

Primary Route(s) of Entry:	Inhalation?	Absorption (skin)?	Ingestion?
	X	X	X

Health Hazards

INHALATION: **Irritating to the eyes, nose, throat and respiratory system causing cough, dypsnea, headache, nausea, vomiting, olfactory and taste changes, chemical pneumonia, confusion, and possible liver, kidney, and bone marrow effects. Animal studies showed testicular atrophy and lesions (no evidence in humans).**

ABSORPTION: **Smarting and reddening of the skin. Absorption is likely, causing toxic systemic effects.**

INGESTION: **Nausea, vomiting, and irritation of the entire digestive system.**

Carcinogenicity	NTP Listed?	IARC Cancer Review Group?	OSHA Regulated?	Target Organs?
No Evidence Human No Evidence Animal	**No**	**No**	**Not Listed**	**Eyes, skin, resp. sys., CNS, liver, kidneys, urogenital sys., bone marrow.**

Medical Conditions Generally Aggravated by Exposure

None reported.

Emergency and First-aid Procedures

Eye contact: Flush immediately with water for 15 minutes (minimum), seek medical attention. Skin contact: Remove all contaminated clothing. Immediately wash area with large amounts of soap and water. Seek medical assistance. For inhalation: Remove the person from exposure. Provide respiratory assistance and CPR, if required. Transfer to medical facility. If swallowed: Seek medical attention immediately. Give conscious and alert person 1-2 glasses of water. Do NOT induce vomiting.

SECTION VII - PRECAUTIONS FOR SAFE HANDLING AND USE

Steps to be Taken in Case Material is Released or Spilled

Absorb in vermiculite, earth, or similar material and deposit in sealed containers. Ventilate area of spill or leak. Restrict those not involved in cleanup from entering area. Remove all sources of ignition.

Preferred Waste Disposal Method

No Citation.

Precautions to be Taken in Handling and Storage

Store to avoid contact with incompatible materials in tightly closed containers in cool, well-ventilated area away from heat and flame under a nitrogen purge. Prevent physical damage to containers.

Other Precautions and Warnings

Sources of ignition are prohibited where ethylidene norbornene is used, handled, or stored.

SECTION VIII - CONTROL MEASURES AND PERSONAL PROTECTIVE EQUIPMENT

Respiratory Protection *(specify type)*

For exposures above the PEL, use an MSHA/NIOSH-approved supplied-air respirator with full facepiece, hood, or helmet in continuous flow mode, or use a self-contained breathing apparatus (SCBA) operated in pressure demand or other positive pressure mode.

Ventilation

Local exhaust or general mechanical systems recommended.

Protective Gloves	Eye Protection	Other Protective Clothing
Impervious Rubber	**Chemical Goggles or Face Mask**	**Protective Apron**

Work/Hygiene Practices

Always wash hands thoroughly after using chemical; never bring food, drink, or smoking materials into vicinity of chemicals.

ETHYLIDENE NORBORNENE

C_9H_{12}

CAS: 16219-75-3

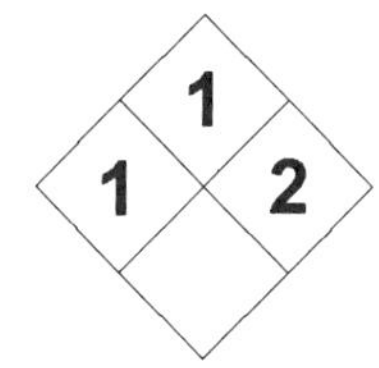

IDENTIFICATION AND TYPICAL USES

Ethylidene norbornene is a colorless to white liquid with a strong, turpentine-like odor. It is a cyclic olefin hydrocarbon related to pinene and is used as the third monomer in EPDM elastomers.

RISK ASSESSMENT: HEALTH

General Assessment

Ethylidene norbornene is moderately toxic by *ingestion* and mildly toxic by *inhalation*. It will also pass through unbroken skin (*absorption*) to cause toxic systemic effects. Its carcinogenicity in humans is not known but is questioned in some animal studies. It has been shown to cause changes in the reproductive system of experimental animals. Human teratogenicity is suspected but not confirmed.

Inhalation of ethylidene norbornene can cause irritation to the eyes, nose, throat, and respiratory system causing cough, dypsnea, nausea, vomiting, and possible chemical pneumonia. The latter occurs mainly from the aspiration of vomitus into the lungs. High concentrations may depress the central nervous system (CNS) causing symptoms of headache, nausea, and confusion. There may be changes in the senses of smell and taste. In animals studies, there have been reports of testicular atrophy, liver lesions, kidney injury, urogenital injury, and bone marrow effects. Although no such toxicity has been reported in humans, caution is still warranted with regard to health exposure risk.

Skin contact causes smarting and reddening of the skin. Absorption will occur if liquid remains in contact with the skin for prolonged periods. If absorption and inhalation occur simultaneously, the compounded effects can be quite severe (possibly life-threatening). Eye contact causes irritation and inflammation of the conjunctival tissues. If not removed immediately, more serious damage may occur.

Acute Health Effects

The following acute (short-term) health effects may occur immediately or shortly after exposure to ethylidene norbornene:

Skin: Smarting and reddening. Will pass through intact skin to cause toxic systemic effects.

Eye: Irritation and inflammation.

Lung: Nose and throat irritation. Causes coughing, dypsnea, nausea, and vomiting.

CNS: Possible depression causing headache and confusion.

Other: Ingestion may produce nausea, vomiting, and severe irritation of the entire digestive tract.

Chronic Health Effects

The following chronic (long-term) health effects may occur at some time after exposure to ethylidene norbornene and can last for months or even years:

Cancer Hazards: According to information presented in the references, ethylidene norbornene has not been adequately tested for its ability to cause cancer in animals.

Reproductive Hazard: According to information presented in the references, ethylidene norbornene may have the ability to adversely affect reproduction in animals. It has caused testicular atrophy and other urogenital injuries. The teratogenicity of ethylidene norbornene in humans has not been adequately studied.

Other Chronic Effects: Causes adverse effects to the kidneys, liver, and testes in laboratory animals. Humans data in this regard are insufficient to draw accurate conclusions.

Recommended Risk-Reduction Measures

Personnel should avoid direct contact with ethylidene norbornene. The exact nature of the toxicity of this chemical is not entirely understood in the references. Therefore, prudent risk management requires it be treated as though it posed a significant health risk in the event of overexposure. If a less toxic chemical cannot be substituted for a hazardous substance, then *engineering controls* are the most effective method of reducing exposures. The best protection is to enclose operations and/or provide local exhaust ventilation at the site of chemical release. While not always operationally feasible, isolating operations can also reduce exposure. Using respiratory protection is less effective than the controls mentioned above, but is still advisable whenever working with or around ethylidene norbornene. For exposures above the OSHA ceiling of 5 ppm, a supplied-air respirator with full facepiece operated in continuous flow mode, or a self-contained breathing apparatus (SCBA) with full facepiece and pressure demand are the recommended respiratory protection methods of choice. If a full facepiece is not available, then chemical goggles should be worn to protect the eyes. Whenever a chemical splash hazard exists, a face shield and a rubber apron should be worn. To prevent hand and skin exposures, impervious gloves should be worn. Glove manufacturers should be contacted to obtain permeation studies *before* glove selection has been made.

Administrative controls should also be in place to minimize the potential for human exposures. These may include written procedures or policies which specify the methods and techniques that will be practiced whenever personnel are to work with ethylidene norbornene.

All personnel should receive *training* on the use, hazards, protective measures, emergency actions, and other precautions per 29 CFR 1910.1200 (Hazard Communication), prior to the first assignment in an area where ethylidene norbornene is used or stored.

If symptoms develop or overexposure is suspected, the following tests are recommended:

☑ Kidney and liver function tests.

☑ Neurological examination.

Any evaluation should include a careful history of past and present symptoms with an examination. Medical tests that simply look for existing damage are not a substitute for controlling exposures.

Other methods to reduce exposure include:

☑ Where possible, enclose operations and use local exhaust ventilation at the site of chemical release. If local exhaust ventilation or enclosure is not used, respiratory protection should be mandatory.

☑ Always ensure that proper protective clothing is worn when using chemical substances.

☑ Wash thoroughly immediately after exposure to ethylidene norbornene and at the end of the work shift or before eating, drinking, or smoking.

☑ Hazard warning information should be posted in the work area. In addition, as part of an on-going education and training program, all information on the health and safety hazards of ethylidene norbornene should be communicated to all potentially exposed workers.

☑ Eye wash stations should be provided in the immediate work area for emergency use. If there is a possibility of skin exposure to ethylidene norbornene, emergency shower facilities should also be provided.

☑ Personnel should never wear contaminated clothing home (family members can be exposed). Clothing should be laundered by personnel who have been trained on the hazards of ethylidene norbornene.

RISK ASSESSMENT: ENVIRONMENT

General Assessment

The environment is at risk of exposure during transportation, storage, disposal, or destruction of ethylidene norbornene. In almost every scenario, the threat of environmental exposure is contingent upon the proper handling of the chemical substance. Accidental spills, large or small, can result in fire, explosion, and possible contamination of the surrounding environmental mediums (water, soil, and air).

Ethylidene norbornene is considered a Class II combustible liquid (per OSHA 29 CFR 1910.106). It presents a moderate fire and explosion hazard risk, especially in contact with oxygen. These characteristics require extreme caution in handling, storage, transportation, and disposal. Therefore, special consideration is required during any emergency situation involving a leak or spill of ethylidene norbornene. Should ethylidene norbornene ever come into contact with oxygen during use, transportation, storage, or disposal, explosion and fire are extremely possible.

Ethylidene norbornene may enter the environment through industrial discharges into the effluent and through spills.

Acute Ecological Effects

Acute (short-term) toxic effects may include the death of animals, birds, or fish and death or low growth rate in plants. Acute effects are seen 2 to 4 days after animals or plants are exposed to ethylidene norbornene.

Insufficient data are available on the short-term effects of ethylidene norbornene exposure to aquatic life, birds, plants, or land animals.

Chronic Ecological Effects

Chronic (long-term) toxic effects may include shortened life span, reproductive problems, lower fertility, and changes in appearance or behavior in exposed animals. These effects can be seen long after first exposure(s) to toxic chemicals.

Insufficient data are available on the long-term effects of ethylidene norbornene exposure to aquatic life, birds, plants, or land animals.

Water Solubility

The solubility of ethylidene norbornene has not been clearly established.

Persistence in the Environment

The persistence of ethylidene norbornene in the environment could not be determined in the references.

Bioaccumulation in Aquatic Organisms

Some substances increase in concentration, or *bioaccumulate,* in living organisms as they breathe contaminated air, drink contaminated water, or eat contaminated food. These chemicals can become concentrated in the tissues and internal organs of animals as well as humans.

No information on the bioaccumulation of ethylidene norbornene could be found in the references.

Recommended Risk-Reduction Measures

Proper training of all transporters will reduce the likelihood of a mishap or accident resulting in a leak or spill to the environment. The correct use of DOT labeling (per the MSDS) on all containers, trucks, and rail cars will enable emergency responders to react properly and quickly to any disaster thereby reducing the potential risk to the environment and to personnel.

Storage of ethylidene norbornene should be segregated from oxygen to minimize the risk of cross-contamination or contact. A nitrogen purge is recommended and 100 ppm of *tert*-butyl catechol is often used as a stabilizer. Buildings designated for storage should be equipped with appropriate fire protection systems (alarms, sprinklers, emergency lighting, portable extinguishers).

If a spill or leak to the environment has occurred, fire department, emergency response and/or hazardous materials spill personnel should be notified immediately. Cleanup should be attempted only by those trained in proper spill containment procedures. Contaminated soils should be removed for incineration and replaced with clean soil. If ethylidene norbornene should contact the water table, aquifer, or navigable waterway, cleanup should be started immediately. Its specific solubility could not be determined and total remediation may not be possible. The local and/or state emergency response authorities must be notified. A comprehensive emergency response or disaster preparedness/recovery plan should be in place prior to any operations involving the use, transportation, storage, or disposal of ethylidene norbornene. If ethylidene norbornene is spilled or leaked, the following specific steps are recommended:

- ☑ Restrict persons not wearing protective clothing from area of spill or leak until cleanup is complete and area can be opened for normal work.
- ☑ Remove all ignition sources, ventilate area of leak.
- ☑ Absorb liquids in vermiculite, dry sand, earth, or similar material and deposit in sealed containers.
- ☑ It may be necessary to dispose of ethylidene norbornene as a hazardous waste. The responsible state agency or the regional office of the federal Environmental Protection Agency (EPA) should be contacted for specific recommendations.

RISK ASSESSMENT: BUSINESS

General Assessment

Accidents or mishaps involving ethylidene norbornene can present a significant threat to business operations. The loss or damage of equipment or facilities can significantly affect fiscal viability. Lawsuits that may result from personnel injury/death, public exposures, and/or environmental contamination will also require a serious expenditure of resources. Media attention surrounding an injury, death, or environmental damage can also result in a loss of profits and loss of current as well as future business. Always remember that whenever the term "reproductive hazard" is used, public emotion, ignorance, and hysteria can run equally high. This must be carefully considered when developing any public relation policies.

Recommended Risk-Reduction Measures

Company attorneys, safety and health professionals, and environmental specialists should be involved in the development of any procedures dealing with safety or emergency response. A company official should be pre-designated as a public relations officer with specific training in dealing with the press. Corporate plans and policies should be developed, approved, and implemented long before any need for such arises.

REFERENCES

Ahlbom, A. 1993. *Biostatistics for Epidemiologists.* New York: CRC Press/Lewis Publishers.

American Conference of Governmental Industrial Hygienists. 1988. *Documentation of the Threshold Limit Values and Biological Exposure Indices,* 5th Edition (with updates). Cincinnati: ACGIH

Anthony, C. P., and N. J. Kolthoff. 1971. *Textbook of Anatomy and Physiology.* St. Louis: C. V. Mosby Company.

Asante-Duah, D. K. 1993. *Hazardous Waste Risk Assessment.* New York: CRC Press/Lewis Publishers.

Bloom, A. D. and F. J. DeSerres. 1995. *Ecotoxicity and Human Health.* New York: CRC Press/Lewis Publishers.

Brusick, D. J. 1994. *Methods for Genetic Risk Assessment.* New York: CRC Press/Lewis Publishers.

Bueche, J. R. 1972. *Principles of Physics,* 2nd Edition. New York: McGraw-Hill.

Burrell, R., D. K. Flaherty, and L. J. Sauers. 1992. *Toxicology of the Immune System.* New York: Van Nostrand Reinhold.

Calabrese, E. J. 1994. *Biological Effects of Low Level Exposures.* New York: CRC Press/Lewis Publishers.

Cockerman, L. G., B. S. Shane. 1994. *Basic Environmental Toxicology.* New York: CRC Press/Lewis Publishers.

Cothern, C. R. 1995. *Handbook for Environmental Risk Decisionmaking.* New York: CRC Press/Lewis Publishers.

Danse, I. R. 1991. *Common Sense Toxics in the Workplace.* New York: Van Nostrand Reinhold.

Environmental Protection Agency. 1983. *Chemical Hazard Information Profiles.* Washington, D.C.: U.S. EPA.

Fodden, C. R., and J. L. Weddell. 1992. *Hazardous Materials Emergency Action Data.* New York: CRC Press/Lewis Publishers.

Forseberg, K., and L. H. Keith. 1993. *Chemical; Protective Clothing Permeation and Degradation Database.* Chelsea, MI: Lewis Publishers.

Gardner, R. W. 1994. *Chemical Intolerance.* New York: CRC Press/Lewis Publishers.

Gerarde, H. W. 1960. *Toxicology and Biochemistry of Aromatic Hydrocarbons.* New York: Elsevier.

Hallenbeck, W. H. 1993. *Quantitative Risk Assessment for Environmental and Occupational Health,* 2nd Edition. New York: CRC Press/Lewis Publishers.

Hodgson, E., R. B. Mailman, and Janice E. Chambers. 1988. *Dictionary of Toxicology.* London: The MacMillian Company.

Howard, P. H. 1995. *Handbook of Environmental Fate and Exposures Data for Organic Chemicals,* Vol. I-IV. New York: CRC Press/Lewis Publishers.

Hosty, J. W., and P. Foster. 1990. *A Practical Guide to Chemical Spill Response.* New York: Van Nostrand Reinhold.

Keith, L. H. and M. M. Walker. 1995. *Handbook of Air Toxics.* New York: CRC Press/Lewis Publishers.

Langerman, N. 1994 *Precautionary Labels for Chemical Containers.* New York: CRC Press/Lewis Publishers.

Lauwerys, R. R., P. Hoet. 1993. *Industrial Chemical Exposure.* New York: CRC Press/Lewis Publishers.

Lewis, R. J., Sr. 1992. *Sax's Dangerous Properties of Industrial Materials,* Volumes 1, 2, and 3, 8th Edition. New York: Van Nostrand Reinhold.

Lide, D. R. 1995. *Handbook of Chemistry and Physics,* 76th Edition. New York: CRC Press/Lewis Publishers.

Mackay, D., W. Y. Shiu, and K. C. Ma. 1992. *Illustrated Handbook of Chemical-Physical Properties and Environmental Fate for Organic Chemicals.* New York: CRC Press/Lewis Publishers.

Manahan, S. E. 1994. *Environmental Chemistry,* 6th Edition. New York: CRC Press/Lewis Publishers.

Meyer, E. 1990. *Chemistry of Hazardous Materials.* Englewood Cliffs, NJ: Prentice Hall.

Patnaik, P. 1992. *A Comprehensive Guide to the Hazardous Properties of Chemical Substances.* New York: Van Nostrand Reinhold.

Proctor, N. H., J. P. Hughes. 1991. *Chemical Hazards of the Workplace.* New York: Van Nostrand Reinhold.

Quigley, D. R. 1994. *Handbook of Emergency Chemical Management.* New York: CRC Press/Lewis Publishers.

Rea, W. J. 1995. *Chemical Sensitivity,* Vol. I-III. New York: CRC Press/Lewis Publishers.

Richardson, M. 1995. *Dictionary of Substance and Their Effects,* Vol. I-VI. New York: CRC Press/Lewis Publishers.

Sacarello, H. L. A. 1994. *The Comprehensive Handbook of Hazardous Materials.* New York: CRC Press/Lewis Publishers.

Sheftel, V. O. 1995. *Handbook of Toxic Properties of Monomers and Additives.* New York: CRC Press/Lewis Publishers.

Torkelson, T. R. V. K. Rowe. 1981. Halogenated aliphatic hydrocarbons. *Patty's Industrial Hygiene and Toxicology,* Vol. 2B, *Toxicology,* pp 3568-3572. New York: Wiley-Interscience.

U.S. Department of Health and Human Services, National Institute for Occupational Safety and Health. 1994. *NIOSH Pocket Guide to Chemical Hazards.* Washington, D.C.: U.S. Government Printing Office.

MATERIAL SAFETY DATA SHEET

CHEMICAL NAME

ETHYL MERCAPTAN

HAZARD WARNING INFORMATION

HEALTH	FIRE	REACTIVE	OTHER	DEGREE OF HAZARD	COLOR CODING	OTHER CODES
2	4	0		0 = Minimum Hazard 1 = Slight Hazard 2 = Moderate Hazard 3 = Serious Hazard 4 = Severe Hazard	HEALTH = BLUE FIRE = RED REACTIVITY = YELLOW OTHER = WHITE	OX = Oxidizer ACID = Acid ALK = Alkali COR = Corrosive W = Use No Water

SECTION I - GENERAL INFORMATION

Characterization **Sulfur-Containing Organic**	RCRA Number **D001** — EPA Class **Characteristic (I) Waste**
DOT Proper Shipping Name **Ethyl Mercaptan**	Chemical Abstract Service (CAS) Number **75-08-1**
DOT Hazard Class and Label Requirements **Flammable Liquid**	DOT Emergency Guide Code **27**
DOT Identification Number **UN 2363**	Molecular Formula CH_3CH_2SH

Synonyms

Ethanethiol; ethyl sulfhydrate; mercaptoethane; ethyl hydrosulfide; ethyl thioalcohol; LPG ethyl mercaptan 1010®, thioethanol; thioethyl alcohol.

SECTION II - HAZARDOUS INGREDIENTS AND IDENTITY INFORMATION

Hazardous Components (specific identity)	OSHA Exposure Criteria	NIOSH Exposure Criteria	Immediately Dangerous to Life and Health (IDLH)	ACGIH Exposure Criteria
Ethyl mercaptan: (derivation: By saturating potassium hydroxide solution with hydrogen sulfide, mixing with calcium ethylsulfate solution, and distilling on a water bath. Also occurs in some vegetables from vinous fermentation, in illuminating gas, "sour" natural gas, coal tar, and various petroleum distillates). **1 ppm = 2.58 mg/m^3**	**PEL:** **0.5 ppm** **1 mg/m^3** **CEILING:** **10 ppm** **26 mg/m^3**	**REL (ceiling):** **0.5 ppm** **1.3 mg/m^3** **(15 minutes)**	**500 ppm**	**TLV (skin):** **0.5 ppm** **1.3 mg/m^3** **STEL:** **Not Established**

SECTION III - PHYSICAL AND CHEMICAL CHARACTERISTICS

Boiling Point **97°F (36°C)**	Specific Gravity (H_2O = 1) **0.84**
Vapor Pressure (mm Hg) **442 at 68°F (20°C)**	Molecular Weight (atomic weight) **62.1**
Vapor Density (Air = 1) **2.14**	Melting Point **-228°F (-144°C)**

Solubility

Slightly soluble in water (0.7%). Soluble in alcohol, ether, and petroleum naphtha.

Appearance and Odor

Colorless liquid (or gas, above 95°F) with strong, persistent, characteristic, disagreeable, pungent, "skunk-like," or garlic, or leek odor. Odor Threshold = 0.001 ppm.

SECTION IV - FIRE AND EXPLOSION HAZARD DATA

Flash Point (method used) **-55°F (-48°C) closed cup**	Explosive Limits in Air % by Volume **LEL: 2.8% UEL: 18%**
NFPA Classification **Class IA Flammable Liquid**	Autoignition Temperature **570°F (298°C)**

Extinguishing Media

Dry chemical, carbon dioxide, water spray, or regular foam. Direct water stream may be ineffective.

Special Fire Fighting Procedures

Wear full protective clothing and self-contained breathing apparatus (SCBA). Move container from fire area if it can be done without risk. Use water spray to cool containers.

Unusual Fire and Explosion Hazards

Containers may explode in fire. Presents a vapor explosion hazard indoors, outdoors, and in sewers.

SECTION V - REACTIVITY DATA

Stability		Conditions to Avoid
		Normally stable at room temperature, if kept under a nitrogen blanket purge. Ethyl mercaptan is extremely volatile and very flammable. Keep away from ALL sources of heat or ignition and incompatible materials.
Stable	Unstable	Incompatibility *(materials to avoid)*
X		Reacts vigorously with oxidizing materials. Reacts violently with dry bleach, calcium oxychloride, and water or steam to produce toxic flammable vapors. Dangerous when exposed to heat, flame, or sparks.
Hazardous Polymerization		Conditions to Avoid
		Under normal conditions of temperature and pressure, hazardous polymerization of ethyl mercaptan is not expected to occur.
May Occur	Will Not Occur	Hazardous Decomposition or By-products
	X	When heated to decomposition or in contact with acid or acrif fumes, ethyl mercaptan can produce highly toxic fumes of sulfur oxides.

SECTION VI - HEALTH HAZARD DATA

Primary Route(s) of Entry:	Inhalation?	Absorption (skin/eye)?	Ingestion?
	X		X

Health Hazards

INHALATION: Irritation of the mucous membranes, headache, nausea, and possible disturbances of the central nervous system (CNS). At high concentrations, causes narcosis with drowsiness, dizziness, restlessness, loss of muscle control, skeletal muscle paralysis. Cyanosis, respiratory depression, coma, and death. May produce changes in smell and taste.

SKIN & EYES: Irritation of the eyes (conjunctiva) and skin from liquid contact.

INGESTION: Moderately toxic. May cause toxic systemic effects similar to that of inhalation.

Carcinogenicity	NTP Listed?	IARC Cancer Review Group?	OSHA Regulated?	Target Organs?
Unknown Human Unknown Animal	No	No	29 CFR 1910.1000 Table Z-1	Skin, eyes, liver, kidneys, CNS, blood, resp. system

Medical Conditions Generally Aggravated by Exposure

None Reported.

Emergency and First-aid Procedures

Eye contact: Do not allow victim to keep eyes tightly shut. Flush immediately with water for 15 minutes (minimum), seek medical attention. Skin contact: Wash with large amounts of water for a minimum of 15 minutes. For inhalation: Remove the person from exposure. Provide respiratory assistance and CPR. Give 100% humidified supplemental O_2 with assisted ventilation. Transfer to medical facility. If swallowed: Seek medical attention immediately. Give 1-2 glasses of water, do NOT induce vomiting. Do NOT attempt to give an unconscious or convulsing person anything by mouth.

SECTION VII - PRECAUTIONS FOR SAFE HANDLING AND USE

Steps to be Taken in Case Material is Released or Spilled

Remove all ignition sources. Use water spray to knock down vapors. Wear self-contained breathing apparatus (SCBA) for respiratory protection. Restrict those not involved in cleanup from entering area. Avoid skin contact. Ventilate area of spill. Absorb in sand or other material and dispose of in sealed drums.

Preferred Waste Disposal Method

No Citation.

Precautions to be Taken in Handling and Storage

Store in tightly closed containers in a cool, dry, well ventilated area away from heat (including direct sunlight). Protect containers from physical damage. Inspect containers regularly.

Other Precautions and Warnings

Ethyl mercaptan must be stored to avoid contact with strong oxidizers since violent reactions can occur.

SECTION VIII - CONTROL MEASURES AND PERSONAL PROTECTIVE EQUIPMENT

Respiratory Protection *(specify type)*

PEL is low and difficult to monitor. Air-purifying respirators may suffice but best protection is provided by a self-contained breathing apparatus (SCBA) with full facepiece operated in positive pressure mode.

Ventilation

Local exhaust (preferred) at site of chemical work or general ventilation.

Protective Gloves	Eye Protection	Other Protective Clothing
Impervious Gloves	Chemical Goggles and Face Mask	Protective Uniform or Apron

Work/Hygiene Practices

Always wash hands thoroughly after using chemical; never bring food, drink, or smoking materials into vicinity of chemicals.

ETHYL MERCAPTAN

CH_3CH_2SH CAS: 75-08-1

$CH_3 — CH_2 — SH$

IDENTIFICATION AND TYPICAL USES

Ethyl mercaptan appears as a colorless liquid (or a gas above 95°F) with an extremely strong, penetrating, pungent, characteristic, disagreeable, "skunk-like," or garlic, or leek-like odor. It can be manufactured synthetically, but also occurs naturally in certain vegetables like cabbage, as a result of vinous fermentation, and in illuminating gas, "sour" natural gas from wells, coal tar, and various petroleum distillates. It is used as an intermediate in the manufacture of insecticides, plastics and antioxidants, and as an additive to natural gas, butane, liquefied petroleum gas, and propane to give odor.

RISK ASSESSMENT: HEALTH

General Assessment

Very little data are available in the references on the human toxicity of ethyl mercaptan. It is believed to present moderate toxicity by *ingestion* and *inhalation*. There is insufficient data on the carcinogenicity, mutagenicity, and teratogenicity of this compound in humans.

Inhalation may cause disturbances to the central nervous system (CNS) with headache, fatigue, narcosis, drowsiness, and dizziness. Animal studies confirm additional CNS effects with restlessness, loss of muscular coordination, skeletal muscle paralysis, severe to mild cyanosis, kidney and liver damage, respiratory depression, coma, and death. There can also be irritation of the eyes, nose, throat, and respiratory system and changes to the senses of smell and taste. Animal studies have also produced cardiovascular disorders and depression of the hematopoietic system.

Skin contact may produce no significant or apparent irritation on brief exposure. However, prolonged or repeated contact may cause mild to severe irritation.

Ingestion may result in unspecified gastrointestinal and respiratory effects. Systemic poisoning is extremely possible with symptoms similar to that of inhalation.

☠ *Acute Health Effects*

The following acute (short-term) health effects may occur immediately or shortly after exposure to ethyl mercaptan:

Skin: Mild to moderate irritation on prolonged exposure to the liquid.

Eye: Mild to severe conjunctival irritation in contact with the liquid or vapor.

Lung: Severe irritation to respiratory tract. Absorption through the respiration process appears to cause a variety of adverse effects to the central nervous system.

CNS: A depressant of the central nervous system causing a variety of toxic effects including drowsiness, coma, and death.

Chronic Health Effects

The following chronic (long-term) health effects may occur at some time after exposure to ethyl mercaptan and can last for months or even years:

Cancer Hazards: There is little human data on the effects of long-term exposure to ethyl mercaptan. There are no reports of carcinogenic or mutagenic occurrences in humans or test animals resulting from exposure to ethyl mercaptan.

Reproductive Hazard: There are no reports to substantiate claims of teratogenicity resulting from exposure to ethyl mercaptan.

Other Chronic Effects: Chronic exposures may produce changes in taste and smell. Animal studies have shown disorders of the cardiovascular system and depression of the hematopoietic (blood-forming) system.

ⓘ *Recommended Risk-Reduction Measures*

Information on the specific human toxicity of ethyl mercaptan is limited in the references. Personnel should avoid direct contact with ethyl mercaptan. If a less toxic material or compound cannot be substituted for ethyl mercaptan, then *engineering controls* are the most effective method of reducing exposure risk. The best protection is to enclose operations and/or provide local exhaust ventilation at the site of ethyl mercaptan release. While not always operationally feasible, isolating operations can also reduce exposure. Using respiratory protection is less effective than the controls mentioned above, but is still required whenever working with or around ethyl mercaptan. The OSHA permissible exposure ceiling for this chemical is relatively low (10 ppm) and the specific human toxicity is not clear. An air-purifying respirator may suffice for low-level exposures. However, best protection is provided using an MSHA/NIOSH-approved self-contained breathing apparatus (SCBA) with full facepiece and pressure demand is the appropriate level of respiratory protection. If a full facepiece is not available, then chemical goggles should be worn to protect the eyes, especially when a splash hazard exists. A face shield should also be considered. Skin contract must be avoided. To prevent hand and skin exposures, rubber or other non-permeable gloves should be worn.

Administrative controls should also be in place to minimize the potential for human exposures. These may include written procedures or policies which specify the methods and techniques that will be practiced whenever personnel are to work with ethyl mercaptan.

All personnel should receive *training* on the use, hazards, protective measures, emergency actions, and other precautions per 29 CFR 1910.1200 (Hazard Communication) prior to the first assignment in an area where ethyl mercaptan is used or stored.

Before beginning employment and at regular intervals thereafter (e.g., annually), the following medical test is recommended:

- ☑ Lung function tests.

If symptoms develop or overexposure is suspected, the following may be useful:

- ☑ Liver, kidney, and lung function tests.
- ☑ Complete evaluation of the central nervous system.

It should be noted that medical tests that simply look for existing damage are not a substitute for controlling exposures. Medical histories are extremely important when assessing exposure risk.

Other methods to reduce exposure include:

- ☑ Always ensure that proper protective clothing is worn when using chemical substances.
- ☑ Wash thoroughly immediately after exposure to ethyl mercaptan and at the end of the work shift or before eating, drinking, or smoking.
- ☑ Hazard warning information should be posted in the work area. In addition, as part of an on-going education and training program, all information on the health and safety hazards of ethyl mercaptan should be communicated to all potentially exposed workers.
- ☑ Eye wash stations should be provided in the immediate work area for emergency use. If there is a possibility of skin exposure to ethyl mercaptan, emergency shower facilities should also be provided.
- ☑ Workers whose clothing has been contaminated by ethyl mercaptan should change into clean clothes before leaving work. Contaminated work clothing should be laundered only by individuals who have been informed of the hazards of exposure to ethyl mercaptan.

RISK ASSESSMENT: ENVIRONMENT

General Assessment

The environment is at risk of exposure during transportation, storage, disposal, or destruction of ethyl mercaptan. In almost every scenario, the threat of environmental exposure is contingent upon the proper handling of the chemical substance. Accidental spills, large or small, can result in contamination of the surrounding environmental mediums (water, soil, and air).

Ethyl mercaptan is Class IA flammable liquid (per OSHA 29 CFR 1910.106). It is extremely volatile and presents a serious risk of fire and/or explosion. It will form dangerous (explosive) mixtures in air. It is incompatible with oxidizers and reacts violently with calcium hypochlorite on contact. Therefore, caution is always required in handling, storage, transportation, and disposal of ethyl mercaptan. When heated to decomposition it emits very acrid and irritating smoke and fumes of sulfur oxides. Emergency responders, hazardous materials personnel, and firefighters should be made aware of the presence of ethyl mercaptan at any emergency response situation.

Ethyl mercaptan may enter the environment through industrial effluents and spills.

Acute Ecological Effects

Acute (short-term) toxic effects may include the death of animals, birds, or fish and death or low growth rate in plants. Acute effects are seen 2 to 4 days after animals or plants are exposed to ethyl mercaptan.

This chemical has moderate acute toxicity to aquatic life. Insufficient data are available to evaluate or predict the acute (short-term) effects of ethyl mercaptan to plants, birds, or land animals.

Chronic Ecological Effects

Chronic (long-term) toxic effects may include shortened life span, reproductive problems, lower fertility, and changes in appearance or behavior in exposed animals. These effects can be seen long after first exposure(s) to toxic chemicals.

Insufficient data are available to evaluate the chronic (long-term) effects of ethyl mercaptan on aquatic life, plants, birds, or land animals.

Water Solubility

Ethyl mercaptan is slightly soluble in water. Concentrations of 1 to 100 milligrams will mix with a liter of water.

Persistence in the Environment

Ethyl mercaptan is moderately persistent in water, with a half-life of between 20 to 200 days. The half-life of a pollutant is the amount of time it takes for one half of the chemical to be degraded.

Bioaccumulation in Aquatic Organisms

Some substances increase in concentration, or *bioaccumulate,* in living organisms as they breathe contaminated air, drink contaminated water, or eat contaminated food. These chemicals can become concentrated in the tissues and internal organs of animals as well as humans.

Insufficient data are available to evaluate or predict the bioaccumulation characteristics of ethyl mercaptan. In the edible tissues of fish.

Recommended Risk-Reduction Measures

Proper training of all transporters will reduce the likelihood of a mishap or accident resulting in a leak or spill of ethyl mercaptan into the environment. Labels on all containers, trucks, and rail cars must meet DOT requirements and accurately reflect their contents to enable emergency responders to react properly and quickly to any disaster thereby reducing the potential risk to the environment and to personnel.

Storage of ethyl mercaptan should be segregated from other chemicals to minimize the risk of cross-contamination. Ethyl mercaptan must be stored to avoid contact with strong oxidizers, such as chlorine and bromine, since violent reactions can occur. Containers should be protected from physical damage and stored to avoid contact with heat and direct sunlight.

If a spill or leak to the environment has occurred, fire department, emergency response and/or hazardous materials spill personnel should be notified immediately. Cleanup should be attempted only by those trained in proper spill containment procedures. Contaminated soils should be removed for incineration and replaced with clean soil.

If ethyl mercaptan should contact the water table, aquifer, or navigable waterway, time is of the essence. It is only slightly soluble in water and total remediation may or may not be possible. The local and/or state emergency response authorities must be notified. A comprehensive emergency response or disaster preparedness/recovery plan should be in place prior to any operations involving the use, transportation, storage, or disposal of ethyl mercaptan.

If ethyl mercaptan is spilled or leaked, the following specific steps are recommended:

- ☑ Restrict persons not wearing protective clothing from area of spill or leak until cleanup is complete. Avoid skin and eye contact and use only non-sparking tools.
- ☑ Absorb spilled liquids with vermiculite, dry sand, or a similar material and deposit in sealed drum for disposal.
- ☑ Ventilate area of spill or leak.
- ☑ Remove all ignition sources.
- ☑ It may be necessary to dispose of ethyl mercaptan as a hazardous waste. The responsible state agency or the regional office of the federal Environmental Protection Agency (EPA) should be contacted for specific recommendations.

RISK ASSESSMENT: BUSINESS

General Assessment

Accidents or mishaps involving ethyl mercaptan can present a serious threat to business operations. The loss or damage of equipment or facilities can significantly affect fiscal viability. Lawsuits that may result from personnel illness, injury/death, public exposures, and/or environmental contamination will require a serious expenditure of resources. Media attention surrounding an injury, death, or environmental damage can also result in a loss of profits and loss of current as well as future business.

Recommended Risk-Reduction Measures

Company attorneys, safety and health professionals, and environmental specialists should be involved in the development of any procedures for responding to chemical incidents. A company official should be predesignated as a public relations officer with specific training in dealing with the press. Corporate plans and policies should be developed, approved, and implemented long before any need for such arises.

REFERENCES

Ahlbom, A. 1993. *Biostatistics for Epidemiologists.* New York: CRC Press/Lewis Publishers.

American Conference of Governmental Industrial Hygienists. 1988. *Documentation of the Threshold Limit Values and Biological Exposure Indices,* 5th Edition (with updates). Cincinnati: ACGIH

Anthony, C. P., and N. J. Kolthoff. 1971. *Textbook of Anatomy and Physiology.* St. Louis: C. V. Mosby Company.

Asante-Duah, D. K. 1993. *Hazardous Waste Risk Assessment.* New York: CRC Press/Lewis Publishers.

Bloom, A. D. and F. J. DeSerres. 1995. *Ecotoxicity and Human Health.* New York: CRC Press/Lewis Publishers.

Brusick, D. J. 1994. *Methods for Genetic Risk Assessment.* New York: CRC Press/Lewis Publishers.

Bueche, J. R. 1972. *Principles of Physics,* 2nd Edition. New York: McGraw-Hill.

Burrell, R., D. K. Flaherty, and L. J. Sauers. 1992. *Toxicology of the Immune System.* New York: Van Nostrand Reinhold.

Calabrese, E. J., E. M. Kenyon. *Air Toxics and Risk Assessment.* New York: CRC Press/Lewis Publishers.

Calabrese, E. J. 1994. *Biological Effects of Low Level Exposures.* New York: CRC Press/Lewis Publishers.

Cockerman, L. G., B. S. Shane. 1994. *Basic Environmental Toxicology.* New York: CRC Press/Lewis Publishers.

Cothern, C. R., N. P. Ross. 1994. *Environmental Statistics, Assessment, and Forecasting.* New York: CRC Press/Lewis Publishers.

Cothern, R. R. 1993. *Comparative Environmental Risk Assessment.* New York: CRC Press/Lewis Publishers.

Cothern, C. R. 1995. *Handbook for Environmental Risk Decisionmaking.* New York: CRC Press/Lewis Publishers.

Danse, I. R. 1991. *Common Sense Toxics in the Workplace.* New York: Van Nostrand Reinhold.

Environmental Protection Agency. 1983. *Chemical Hazard Information Profiles.* Washington, D.C.: U.S. EPA.

Fodden, C. R., and J. L. Weddell. 1992. *Hazardous Materials Emergency Action Data.* New York: CRC Press/Lewis Publishers.

Forseberg, K., and L. H. Keith. 1993. *Chemical; Protective Clothing Permeation and Degradation Database.* Chelsea, MI: Lewis Publishers.

Gardner, R. W. 1994. *Chemical Intolerance.* New York: CRC Press/Lewis Publishers.

Gerarde, H. W. 1960. *Toxicology and Biochemistry of Aromatic Hydrocarbons.* New York: Elsevier.

Hallenbeck, W. H. 1993. *Quantitative Risk Assessment for Environmental and Occupational Health,* 2nd Edition. New York: CRC Press/Lewis Publishers.

Hodgson, E., R. B. Mailman, and Janice E. Chambers. 1988. *Dictionary of Toxicology.* London: The MacMillian Company.

Howard, P. H. 1995. *Handbook of Environmental Fate and Exposures Data for Organic Chemicals,* Vol. I-IV. New York: CRC Press/Lewis Publishers.

Hosty, J. W., and P. Foster. 1990. *A Practical Guide to Chemical Spill Response.* New York: Van Nostrand Reinhold.

Jones, F. E. 1994. *Toxic Organic Vapors in the Workplace.* New York: CRC Press/Lewis Publishers.

Langerman, N. 1994 *Precautionary Labels for Chemical Containers.* New York: CRC Press/Lewis Publishers.

Lauwerys, R. R., P. Hoet. 1993. *Industrial Chemical Exposure.* New York: CRC Press/Lewis Publishers.

Lewis, R. J., Sr. 1992. *Sax's Dangerous Properties of Industrial Materials,* Volumes 1, 2, and 3, 8th Edition. New York: Van Nostrand Reinhold.

Manahan, S. E. 1992. *Toxicological Chemistry,* 2nd Edition. New York: CRC Press/Lewis Publishers.

Meyer, E. 1990. *Chemistry of Hazardous Materials.* Englewood Cliffs, NJ: Prentice Hall.

Patnaik, P. 1992. *A Comprehensive Guide to the Hazardous Properties of Chemical Substances.* New York: Van Nostrand Reinhold.

Torkelson, T. R. V. K. Rowe. 1981. *Patty's Industrial Hygiene and Toxicology,* Vol. 2B. New York: Wiley-Interscience.

U.S. Department of Health and Human Services, National Institute for Occupational Safety and Health. 1994. *NIOSH Pocket Guide to Chemical Hazards.* Washington, D.C.: U.S. Government Printing Office.

MATERIAL SAFETY DATA SHEET

CHEMICAL NAME	
ETHYLMORPHOLINE	**CRC PRESS/LEWIS PUBLISHERS DISCLAIMER:** THE INFORMATION AND RECOMMENDATIONS PRESENTED HEREIN ARE BASED ON SOURCES BELIEVED TO BE RELIABLE. CRC MAKES NO REPRESENTATION ON ITS COMPLETENESS OR ACCURACY. IT IS THE RESPONSIBILITY OF THE USER TO DETERMINE THE CHEMICAL'S SUITABILITY FOR ITS INTENDED USE, THE CHEMICAL'S SAFE USE, AND THE CHEMICAL'S PROPER DISPOSAL. NO REPRESENTATIONS AND/OR WARRANTIES, EITHER EXPRESSED OR IMPLIED, OF THE MERCHANTABILITY OR FITNESS FOR A PARTICULAR PURPOSE, OR OF ANY OTHER NATURE, ARE MADE WITH RESPECT TO THE INFORMATION PROVIDED IN THIS MSDS OR TO THE CHEMICAL TO WHICH INFORMATION MAY REFER. CRC NEITHER ASSUMES NOR AUTHORIZES ANY OTHER PERSON TO ASSUME FOR IT, ANY OTHER ADDITIONAL RESPONSIBILITY OR LIABILITY FOR THE USE OF, OR RELIANCE UPON, THIS INFORMATION.

HAZARD WARNING INFORMATION

HEALTH	FIRE	REACTIVE	OTHER	DEGREE OF HAZARD	COLOR CODING	OTHER CODES
2	3	0		0 = Minimum Hazard 1 = Slight Hazard 2 = Moderate Hazard 3 = Serious Hazard 4 = Severe Hazard	HEALTH = BLUE FIRE = RED REACTIVITY = YELLOW OTHER = WHITE	OX = Oxidizer ACID = Acid ALK = Alkali COR = Corrosive W = Use No Water

SECTION I - GENERAL INFORMATION

Characterization: **Amine**	RCRA Number: **D001**	EPA Class: **Characteristic (I) Waste**
DOT Proper Shipping Name: **Flammable Liquids, n.o.s.**	Chemical Abstract Service (CAS) Number: **100-74-3**	
DOT Hazard Class and Label Requirements: **Flammable Liquid**	DOT Emergency Guide Code: **No Citation**	
DOT Identification Number: **UN 1993**	Chemical Formula: **$C_4H_8ONCH_2CH_3$**	

Synonyms

4-Ethylmorpholine; n-ethylmorpholine.

SECTION II - HAZARDOUS INGREDIENTS AND IDENTITY INFORMATION

Hazardous Components (specific identity)	OSHA Exposure Criteria	NIOSH Exposure Criteria	Immediately Dangerous to Life and Health (IDLH)	ACGIH Exposure Criteria
Ethylmorpholine (derivation: Produced by reaction of morpholine with diethyl sulfate). **1 ppm = 4.79 mg/m^3**	**PEL (skin):** **20 ppm** **94 mg/m^3** **STEL:** **Not Established**	**REL (skin):** **5 ppm** **23 mg/m^3** **STEL:** **Not Established**	**100 ppm**	**TLV (skin):** **5 ppm** **24 mg/m^3** **STEL:** **Not Established**

SECTION III - PHYSICAL AND CHEMICAL CHARACTERISTICS

Boiling Point: **280°F (138°C)**	Specific Gravity (H_2O = 1): **0.90**
Vapor Pressure (mm Hg): **6.1 at 69°F (20°C)**	Molecular Weight: **115.2**
Vapor Density (Air = 1): **4.00**	Melting Point: **-81°F (-63°C)**

Solubility

Completely soluble in water. Soluble in alcohol, acetone, benzene, and ether.

Appearance and Odor

Clear, colorless liquid with an ammonia-like odor.

SECTION IV - FIRE AND EXPLOSION HAZARD DATA

Flash Point (method used): **90°F (32°C) opened cup**	Explosive Limits in Air % by Volume: **LEL: 1% UEL: 9.8%**
NFPA Classification: **Class IC Flammable Liquid**	Autoignition Temperature: **365°F (185°C)**

Extinguishing Media

Use alcohol foam, foam, carbon dioxide, water spray, or dry chemical.

Special Fire Fighting Procedures

Poisonous gases are produced in fire. Use full protective clothing and self-contained breathing apparatus (SCBA). Fight fire from maximum distance if possible. Move containers from fire area if it can be done without risk. Otherwise, use water spray to cool sides of exposed containers. Do NOT release runoff from fire control methods to sewers or waterways.

Unusual Fire and Explosion Hazards

Containers may explode in fire due to buildup of excessive internal pressures. Do not attempt to extinguish fire unless liquid flow can be stopped first. Aqueous solutions are flammable unless diluted extensively. This chemical presents a vapor explosion hazard indoors, outdoors, and in sewers.

SECTION V - REACTIVITY DATA

Stability		Conditions to Avoid **Ethylmorpholine is stable at room temperature in closed containers under normal conditions of handling and storage. Avoid contact with incompatible materials. Keep away from heat and ignition sources.**
Stable X	Unstable	Incompatibility *(materials to avoid)* **Reacts with strong oxidizers (such as chlorine, fluorine, and bromine), and strong acids (including nitric, sulfuric, and hydrochloric).**
Hazardous Polymerization		Conditions to Avoid **Under normal conditions of temperature and pressure, hazardous polymerization of ethylmorpholine is not expected to occur.**
May Occur	Will Not Occur X	Hazardous Decomposition or By-products **Toxic and irritating fumes and vapors are produced when ethylmorpholine is heated to decomposition. These include oxides of carbon and nitrogen, and ammonia gas.**

SECTION VI - HEALTH HAZARD DATA

Primary Route(s) of Entry:	Inhalation?	Absorption (skin)?	Ingestion?
	X	X	X

Health Hazards

INHALATION: Symptoms of respiratory tract irritation (also, eyes, nose, and throat) and drowsiness. There may be olfactory fatigue, corneal edema (eye swelling) with "halos" appearing around lights, and blurred vision or blue-gray vision.

ABSORPTION: Minor irritation on contact. Will pass through intact skin to cause toxic systemic effects similar to those of inhalation.

INGESTION: May cause gastrointestinal tract irritation.

Carcinogenicity	NTP Listed?	IARC Cancer Review Group?	OSHA Regulated?	Target Organs?
Unknown Human Unknown Animal	No	No	29 CFR 1910.1000 Table Z-1	Eyes, skin, respiratory system.

Medical Conditions Generally Aggravated by Exposure

Chronic respiratory system problems and/or eye disorders may be affected.

Emergency and First-aid Procedures

Eye contact: Flush immediately with water for 15 minutes (minimum), seek medical attention. Skin contact: Remove all contaminated clothing. Immediately wash area with large amounts of soap and water. Seek medical assistance. For inhalation: Remove the person from exposure. Provide respiratory assistance and CPR. Transfer to medical facility. If swallowed: Call the poison control center and seek medical attention immediately. Unless advised otherwise, give 1-2 glasses of water to dilute. Do NOT induce vomiting. Never attempt to give an unconscious or convulsing person anything by mouth.

SECTION VII - PRECAUTIONS FOR SAFE HANDLING AND USE

Steps to be Taken in Case Material is Released or Spilled

Use water spray to dilute spill to non-flammable mixture. Absorb in vermiculite or other material, collect using non-sparking tools, and deposit in DOT-approved drum. Ventilate area and remove ignition sources. Restrict those not involved in cleanup from entering area.

Preferred Waste Disposal Method

Incinerate in a combustible chamber equipped with an appropriate effluent gas cleaning device.

Precautions to be Taken in Handling and Storage

Do not store ethylmorpholine in the presence of incompatible chemicals or materials. Store in tightly closed containers in a cool (below room temperature), well-ventilated area. Protect containers from physical damage. Storage under an inert atmosphere (such as nitrogen) is highly recommended.

Other Precautions and Warnings

Recommend explosion proof design for bulk storage facilities. Electrically ground and bond equipment, especially during transfer operations. Detached facility storage is recommended.

SECTION VIII - CONTROL MEASURES AND PERSONAL PROTECTIVE EQUIPMENT

Respiratory Protection *(specify type)*

For low exposures, use an organic vapor respirator. Otherwise, use an MSHA/NIOSH-approved supplied-air respirator with full facepiece operated in positive pressure mode or a self-contained breathing apparatus (SCBA) with full facepiece operated in pressure demand or other positive pressure mode.

Ventilation

Local exhaust or general mechanical systems recommended.

Protective Gloves	Eye Protection	Other Protective Clothing
Impervious Butyl Rubber Gloves	Chemical Goggles or Face Mask	Protective Apron

Work/Hygiene Practices

Always wash hands thoroughly after using chemical; never bring food, drink, or smoking materials into vicinity of chemicals.

ETHYLMORPHOLINE

$C_4H_8ONCH_2CH_3$ CAS: 100-74-3

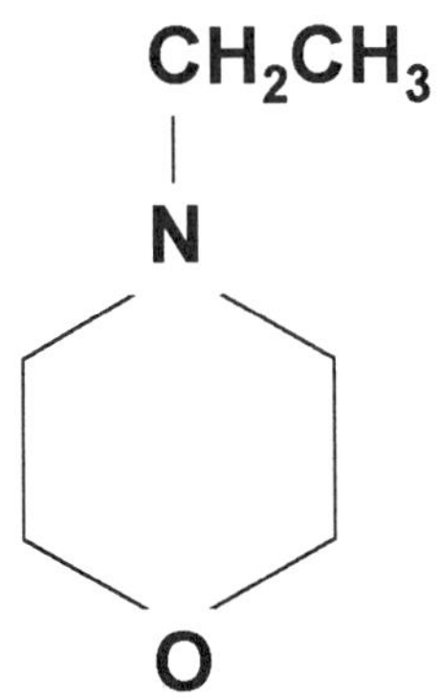

IDENTIFICATION AND TYPICAL USES

Ethylmorpholine is a colorless or clear liquid with a strong, ammonia-like odor. It is used as an intermediate for dyestuffs and in pharmaceuticals; in rubber accelerators and emulsifying agents; as a solvent for dyes resins, and oils; and as a catalyst in making polyurethane foams.

RISK ASSESSMENT: HEALTH

General Assessment

Ethylmorpholine is a mildly toxic compound by *inhalation, ingestion,* and by skin *absorption.* Its primary health risk is due its irritating properties and the fact that exposure can occur simultaneously by more than one route. When this occurs, it compounds the health effects of exposure. Its carcinogenicity, teratogenicity, or mutagenicity have not been established. Ethylmorpholine is primarily a skin and eye irritant.

Inhalation causes irritation of the eyes, nose, throat, and respiratory tract. Inhalation may cause olfactory fatigue making exposure awareness difficult. It causes corneal edema (swelling of the eyes), visual disturbances, including blurred vision, blue-gray vision, and/or colored halos appearing around lights.

Ethylmorpholine is a skin irritant and prolonged contact may cause a rash to develop. It will pass through intact skin to cause toxic systemic effects. Skin absorption can compound the effects of exposure by other routes (e.g., inhalation).

Ingestion of this compound causes irritation of the mouth and esophagus as well as other unspecified gastrointestinal effects. Ethylmorpholine is considered a poison.

Acute Health Effects

The following acute (short-term) health effects may occur immediately or shortly after exposure to ethylmorpholine:

Skin: Irritation with possible redness and rash. It will pass through intact skin.

Eye: Severe irritation, swelling (corneal edema), with blurred vision, blur-gray vision, and colored halos appearing around lights.

Lung: Irritation causing coughing and/or shortness of breath. May cause olfactory changes (numbing of the sense of smell).

Chronic Health Effects

The following chronic (long-term) health effects may occur at some time after exposure to ethylmorpholine and can last for months or even years:

Cancer Hazards: According to the information presented in the references, ethylmorpholine has not been adequately tested for its ability to cause cancer in test animals.

Reproductive Hazard: According to the information presented in the references, ethylmorpholine has not been adequately tested for its ability to adversely affect reproduction in test animals.

Sensitivity: Prolonged skin contact to low levels may result in chronic skin rash.

Recommended Risk-Reduction Measures

Unless a less toxic chemical can be substituted for a hazardous substance, *engineering controls* are the most

effective method of reducing exposures. The best protection is to enclose operations and/or provide local exhaust ventilation at the site of chemical release. To prevent static sparks, all containers and equipment should be electrically grounded and bonded. While not always operationally feasible, isolating operations can also reduce exposure.

Using respiratory protection is less effective than the controls mentioned above, but is still advisable whenever working with or around ethylmorpholine. For relatively low, infrequent, or transient exposures, an organic vapor respirator may suffice. For higher or prolonged exposure, a supplied-air respirator or a self-contained breathing apparatus (SCBA) with full facepiece operated in pressure demand mode are the recommended respiratory protection methods of choice. If a full facepiece is not available, then chemical goggles should be worn to protect the eyes. Whenever a chemical splash hazard exists, a face shield and apron should be worn. To prevent hand and skin exposures, impervious butyl rubber gloves should be worn.

Administrative controls should also be in place to minimize the potential for human exposures. These may include written procedures or policies which specify the methods and techniques that will be practiced whenever personnel are to work with ethylmorpholine.

All personnel should receive *training* on the use, hazards, protective measures, emergency actions, and other precautions per 29 CFR 1910.1200 (Hazard Communication), prior to the first assignment in an area where ethylmorpholine is used or stored.

Before beginning employment and at regular intervals thereafter (e.g. annually), the following medical tests are recommended:

☑ Lung function tests (establish a baseline).

☑ Visual acuity testing (establish a baseline).

If symptoms develop or overexposure is suspected, the following medical tests may be useful:

☑ Lung function tests (compare to baseline).

☑ Visual acuity testing (compare to baseline).

☑ Evaluation by a qualified allergist, including careful consideration of exposure history and special testing (may help diagnose skin allergy).

Any evaluation should include a careful medical history of past and present symptoms with an examination. However, medical tests that evaluate existing damage are not a substitute for controlling exposure. Also, since smoking can cause heart diseases, emphysema, and numerous other respiratory disorders, smokers may be affected more rapidly than non-smokers to the same exposure conditions. Prudent risk assessment and management requires careful consideration of *all* possible risk factors that may be causing the appearance of symptoms in exposed workers.

Other methods to reduce exposure include:

☑ Where possible, enclose operations and use local exhaust ventilation at the site of chemical release. If local exhaust ventilation or enclosure is not used, respiratory protection should be mandatory. Also, if possible, automatically transfer liquids from storage containers to process containers using non-sparking tools.

☑ Always ensure that proper protective clothing is worn when using chemical substances.

☑ Wash thoroughly immediately after exposure to ethylmorpholine and at the end of the work shift or before eating, drinking, or smoking.

☑ Hazard warning information should be posted in the work area. In addition, as part of an on-going education and training program, all information on the health and safety hazards of ethylmorpholine should be communicated to all potentially exposed workers.

☑ Work clothing contaminated with ethylmorpholine should never be worn home where family members can be exposed. Also, clothing soaked with ethylmorpholine presents a serious fire hazard. Contaminated clothing should be laundered by personnel who have been briefed on the hazards of exposure to ethylmorpholine.

☑ Eye wash stations should be provided in work areas. If the potential for whole body exposure exists, then safety showers should also be provided.

RISK ASSESSMENT: ENVIRONMENT

General Assessment

The environment is at risk of exposure during intended use, transportation, storage, disposal, or destruction of ethylmorpholine. In almost every scenario, the threat of environmental exposure is contingent upon the proper handling of the chemical substance. Accidental spills, large or small, can result in fire, explosion, and possible contamination of the surrounding environmental mediums (water, soil, and air).

Ethylmorpholine is considered a Class IC flammable liquid (per OSHA 29 CFR 1910.106). Its extremely low flash point and its boiling point make it a serious and dangerous fire and explosion risk. It can react with many common acids and many oxidizers to cause a fire or explosion. These characteristics require special consideration during any emergency situation involving a leak or spill of ethylmorpholine. Should ethylmorpholine ever come into contact with these incompatible substances either during use, transportation, or storage, violent reactions can occur.

Ethylmorpholine can enter the environment through industrial discharges, municipal waste treatment discharges, and spills.

Acute Ecological Effects

Acute (short-term) toxic effects may include the death of animals, birds, or fish and death or low growth rate in plants. Acute effects are seen 2 to 4 days after animals or plants are exposed to ethylmorpholine.

Insufficient data are available to evaluate or predict the short-term effects of ethylmorpholine to aquatic life, plants, birds, or land animals.

Chronic Ecological Effects

Chronic toxic effects may include shortened life span, reproductive problems, lower fertility, and changes in appearance or behavior in exposed animals. These effects can be seen long after first exposure(s) to toxic chemicals.

Insufficient data are available to evaluate or predict the long-term effects of ethylmorpholine to aquatic life, plants, birds, or land animals.

Water Solubility

Ethylmorpholine is highly soluble in water. Concentrations of 1000 milligrams and more will mix with a liter of water.

Persistence in the Environment

Insufficient data are available to evaluate or predict the ability of ethylmorpholine to persist in the environmental mediums.

Bioaccumulation in Aquatic Organisms

Some substances increase in concentration, or *bioaccumulate,* in living organisms as they breathe contaminated air, drink contaminated water, or eat contaminated food. These chemicals can become concentrated in the tissues and internal organs of animals as well as humans.

There is no data available on the bioaccumulation characteristics of ethylmorpholine.

Recommended Risk-Reduction Measures

Proper training of all transporters will reduce the likelihood of a mishap or accident resulting in a leak or spill to the environment. The correct use of DOT labeling (per the MSDS) on all containers, trucks, and rail cars will enable emergency responders to react properly and quickly to any disaster thereby reducing the potential risk to the environment and to personnel.

Storage of ethylmorpholine should be segregated from incompatible chemicals to minimize the risk of cross-contamination or contact. Storage buildings should be equipped with the proper fire protection equipment (alarms, sprinklers, extinguishers, etc.). Explosion proof design is highly recommended for all storage facility and operational equipment.

If a spill or leak to the environment has occurred, fire department, emergency response and/or hazardous materials spill personnel should be notified immediately. Cleanup should be attempted only by those trained in proper spill containment procedures using non-sparking tools. Contaminated soils should be removed for incineration and replaced with clean soil. If ethylmorpholine should contact the water table, aquifer, or navigable waterway, remediation activities should be prompt. It is highly soluble in water and total containment may not be possible. When such spills occur, the local and/or state emergency response authorities must be notified. A comprehensive emergency response or disaster preparedness/recovery plan should be in place prior to any operations involving the use, transportation, storage, or disposal of ethylmorpholine. If ethylmorpholine is spilled or leaked, the following specific steps are recommended:

- ☑ Restrict persons not wearing protective clothing from area of spill or leak until cleanup is complete and area can be opened for normal work.
- ☑ Ventilate area of spill or leak.
- ☑ Remove all ignition sources.
- ☑ Use water spray to disperse vapors and dilute spills to non-flammable mixtures. Flush spills away from exposures. Dispose of accordingly (do not allow runoff to enter sewers or waterways.

Absorb liquids with vermiculite and deposit in sealed drums for disposal.

☑ It may be necessary to dispose of ethylmorpholine as a hazardous waste. The responsible state agency or the regional office of the federal Environmental Protection Agency (EPA) should be contacted for specific recommendations.

RISK ASSESSMENT: BUSINESS

General Assessment

Accidents or mishaps involving ethylmorpholine can present a significant threat to business operations. The loss or damage of equipment or facilities can significantly affect fiscal viability. Lawsuits that may result from personnel injury/death, public exposures, and/or environmental contamination will also require a serious expenditure of resources. Media attention surrounding an injury, death, or environmental damage can also result in a loss of profits and loss of current as well as future business.

Recommended Risk-Reduction Measures

Company attorneys, safety and health professionals, and environmental specialists should be involved in the development of any procedures dealing with safety and emergency response. A company official should be pre-designated as a public relations officer with specific training in dealing with the press. Corporate plans and policies should be developed, approved, and implemented long before any need for such arises.

REFERENCES

Ahlbom, A. 1993. *Biostatistics for Epidemiologists.* New York: CRC Press/Lewis Publishers.

American Conference of Governmental Industrial Hygienists. 1988. *Documentation of the Threshold Limit Values and Biological Exposure Indices,* 5th Edition (with updates). Cincinnati: ACGIH

Anthony, C. P., and N. J. Kolthoff. 1971. *Textbook of Anatomy and Physiology.* St. Louis: C. V. Mosby Company.

Asante-Duah, D. K. 1993. *Hazardous Waste Risk Assessment.* New York: CRC Press/Lewis Publishers.

Bloom, A. D. and F. J. DeSerres. 1995. *Ecotoxicity and Human Health.* New York: CRC Press/Lewis Publishers.

Brusick, D. J. 1994. *Methods for Genetic Risk Assessment.* New York: CRC Press/Lewis Publishers.

Bueche, J. R. 1972. *Principles of Physics,* 2nd Edition. New York: McGraw-Hill.

Burrell, R., D. K. Flaherty, and L. J. Sauers. 1992. *Toxicology of the Immune System.* New York: Van Nostrand Reinhold.

Calabrese, E. J. 1994. *Biological Effects of Low Level Exposures.* New York: CRC Press/Lewis Publishers.

Cockerman, L. G., B. S. Shane. 1994. *Basic Environmental Toxicology.* New York: CRC Press/Lewis Publishers.

Cothern, C. R., N. P. Ross. 1994. *Environmental Statistics, Assessment, and Forecasting.* New York: CRC Press/Lewis Publishers.

Danse, I. R. 1991. *Common Sense Toxics in the Workplace.* New York: Van Nostrand Reinhold.

Environmental Protection Agency. 1983. *Chemical Hazard Information Profiles.* Washington, D.C.: U.S. EPA.

Fodden, C. R., and J. L. Weddell. 1992. *Hazardous Materials Emergency Action Data.* New York: CRC Press/Lewis Publishers.

Forseberg, K., and L. H. Keith. 1993. *Chemical; Protective Clothing Permeation and Degradation Database.* Chelsea, MI: Lewis Publishers.

Gardner, R. W. 1994. *Chemical Intolerance.* New York: CRC Press/Lewis Publishers.

Gerarde, H. W. 1960. *Toxicology and Biochemistry of Aromatic Hydrocarbons.* New York: Elsevier.

Hallenbeck, W. H. 1993. *Quantitative Risk Assessment for Environmental and Occupational Health,* 2nd Edition. New York: CRC Press/Lewis Publishers.

Hodgson, E., R. B. Mailman, and Janice E. Chambers. 1988. *Dictionary of Toxicology.* London: The MacMillian Company.

Howard, P. H. 1995. *Handbook of Environmental Fate and Exposures Data for Organic Chemicals,* Vol. I-IV. New York: CRC Press/Lewis Publishers.

Hosty, J. W., and P. Foster. 1990. *A Practical Guide to Chemical Spill Response.* New York: Van Nostrand Reinhold.

Kamrin, M. A. 1988. *Toxicology: A Primer on Principles and Applications.* Chelsea, Michigan: Lewis Publishers.

Lewis, R. J., Sr. 1993. *Hawley's Condensed Chemical Dictionary,* 12th Edition. New York: Van Nostrand Reinhold.

Lewis, R. J., Sr. 1992. *Sax's Dangerous Properties of Industrial Materials,* Volumes 1, 2, and 3, 8th Edition. New York: Van Nostrand Reinhold.

Mackay, D., W. Y. Shiu, and K. C. Ma. 1992. *Illustrated Handbook of Chemical-Physical Properties and Environmental Fate for Organic Chemicals.* New York: CRC Press/Lewis Publishers.

Meyer, E. 1990. *Chemistry of Hazardous Materials.* Englewood Cliffs, NJ: Prentice Hall.

Richardson, M. 1995. *Dictionary of Substance and Their Effects,* Vol. I-VI. New York: CRC Press/Lewis Publishers.

Sacarello, H. L. A. 1994. *The Comprehensive Handbook of Hazardous Materials.* New York: CRC Press/Lewis Publishers.

U.S. Department of Health and Human Services, National Institute for Occupational Safety and Health. 1994. *NIOSH Pocket Guide to Chemical Hazards.* Washington, D.C.: U.S. Government Printing Office.

MATERIAL SAFETY DATA SHEET

CHEMICAL NAME

ETHYL NITRITE

HAZARD WARNING INFORMATION

HEALTH	FIRE	REACTIVE	OTHER	DEGREE OF HAZARD	COLOR CODING	OTHER CODES
2	4	4		0 = Minimum Hazard 1 = Slight Hazard 2 = Moderate Hazard 3 = Serious Hazard 4 = Severe Hazard	HEALTH = BLUE FIRE = RED REACTIVITY = YELLOW OTHER = WHITE	OX = Oxidizer ACID = Acid ALK = Alkali COR = Corrosive W = Use No Water

SECTION I - GENERAL INFORMATION

Characterization	RCRA Number	EPA Class
Ester	**D001, D003**	**Characteristic (I) and (R) Waste**

DOT Proper Shipping Name **Ethyl nitrite**	Chemical Abstract Service (CAS) Number **109-95-5**
DOT Hazard Class and Label Requirements **Flammable Liquid**	DOT Emergency Guide Code **No Citation**
DOT Identification Number **UN 1194**	Chemical Formula $C_2H_5NO_2$

Synonyms

Ethyl nitrite solution; hyponitrous ether; nitrosyl ethoxide; nitrous acid ethyl ester; nitrous ethyl ether; spirit of ethyl nitrite; spirit of ether nitrite; sweet spirit of niter.

SECTION II - HAZARDOUS INGREDIENTS AND IDENTITY INFORMATION

Hazardous Components (specific identity)	OSHA Exposure Criteria	NIOSH Exposure Criteria	Immediately Dangerous to Life and Health (IDLH)	ACGIH Exposure Criteria
Ethyl nitrite (derivation: By reaction of nitrous acid and ethanol or by the action of sodium nitrile on a mixture of alcohol and sulfuric acid under cool conditions).	**PEL:** Not Established **STEL:** Not Established	**REL:** Not Established **STEL:** Not Established	**Not Determined**	**TLV:** Not Established **STEL:** Not Established

SECTION III - PHYSICAL AND CHEMICAL CHARACTERISTICS

Boiling Point **63°F (17°C)**	Specific Gravity (H_2O = 1) **0.90**
Vapor Pressure (mm Hg) **Gas above 63°F (17°C)**	Molecular Weight **75.1**
Vapor Density (Air = 1) **2.59**	Melting Point **-58°F (-50°C)**

Solubility

Slightly soluble in water (decomposes). Soluble in alcohol and ether.

Appearance and Odor

Colorless or yellowish, clear liquid with a sweet, characteristic, rum-like odor and fruity taste.

SECTION IV - FIRE AND EXPLOSION HAZARD DATA

Flash Point (method used) **-31°F (-35°C) closed cup**	Explosive Limits in Air % by Volume **LEL: 3% UEL: 50%**
NFPA Classification **Class IA Flammable Liquid**	Autoignition Temperature **194°F (90°C)**

Extinguishing Media

Use foam, carbon dioxide, water spray, or dry chemical. Use water spray to cool exposed containers.

Special Fire Fighting Procedures

Poisonous gases are produced in fire. Use full protective clothing and self-contained breathing apparatus (SCBA). Fight fire from maximum distance if possible. Move containers from fire area if it can be done without risk. Do NOT release runoff from fire control methods to sewers or waterways.

Unusual Fire and Explosion Hazards

Containers may explode in fire due to buildup of excessive internal pressures. It will explode when heated above 194°F (90°C). Oxygen is not required for decomposition. This chemical presents a vapor explosion hazard indoors, outdoors, and in sewers.

SECTION V - REACTIVITY DATA

Stability		Conditions to Avoid Ethyl nitrite is thermally unstable. It will decompose when exposed to air, light, or moisture. On standing, it slowly decomposes to acid and oxides of nitrogen. Avoid contact with incompatible materials.
Stable	Unstable X	Incompatibility *(materials to avoid)* Reacts violently with a broad range of materials, including acids, acid fumes, and oxidizers. Also extremely flammable and explosive. Reacts with heat, flame, and ignition sources.
Hazardous Polymerization		Conditions to Avoid Under normal conditions of temperature and pressure, hazardous polymerization of ethyl nitrite is not expected to occur.
May Occur	Will Not Occur X	Hazardous Decomposition or By-products Toxic and irritating fumes and vapors are produced when ethyl nitrite is heated to decomposition. These include oxides of nitrogen.

SECTION VI - HEALTH HAZARD DATA

Primary Route(s) of Entry:	Inhalation?	Absorption (skin)?	Ingestion?
	X		X

Health Hazards

INHALATION: Symptoms of respiratory tract irritation (also, eyes, nose, and throat) and possibly may be corrosive to nasal tissues. High exposures can cause methemoglobinemia to develop with cyanosis and respiratory depression. Narcotic in high concentrations. Lowers blood pressure and may cause tachycardia and loss of consciousness.

SKIN & EYES: Minor irritation on contact. May be corrosive depending upon the intensity and duration of skin contact/exposure. Mild to moderate eye irritation also possible.

INGESTION: May cause gastrointestinal tract irritation.

Carcinogenicity	NTP Listed?	IARC Cancer Review Group?	OSHA Regulated?	Target Organs?
Unknown Human Unknown Animal	No	No	No	Eyes, skin, and blood.

Medical Conditions Generally Aggravated by Exposure

None reported.

Emergency and First-aid Procedures

Eye contact: Flush immediately with water for 15 minutes (minimum), seek medical attention. Skin contact: Remove all contaminated clothing. Immediately wash area with large amounts water (15 minutes). Seek medical assistance. For inhalation: Remove the person from exposure. Provide respiratory assistance and CPR. Transfer to medical facility. If swallowed: Call the poison control center and seek medical attention immediately. Unless advised otherwise, give 1-2 glasses of water to dilute and induce vomiting. Never attempt to give an unconscious or convulsing person anything by mouth.

SECTION VII - PRECAUTIONS FOR SAFE HANDLING AND USE

Steps to be Taken in Case Material is Released or Spilled

Use regular foam to blanket the release and control vapors. Absorb in vermiculite or other material, collect using non-sparking tools, and deposit in DOT-approved drum. Ventilate area and remove ignition sources. Restrict those not involved in cleanup from entering area.

Preferred Waste Disposal Method

No citation.

Precautions to be Taken in Handling and Storage

Do not store ethyl nitrite in the presence of incompatible chemicals or materials. Store in tightly closed containers in a cool, well-ventilated area. Protect containers from physical damage away from heat (including sunlight). Automatic transfer of liquids between containers is recommended.

Other Precautions and Warnings

Recommend explosion proof design for bulk storage facilities. Electrically ground and bond equipment, especially during transfer operations. Detached facility storage is recommended.

SECTION VIII - CONTROL MEASURES AND PERSONAL PROTECTIVE EQUIPMENT

Respiratory Protection *(specify type)*

Exposure level not established. Recommend using an MSHA/NIOSH-approved supplied-air respirator with full facepiece operated in positive pressure mode or a self-contained breathing apparatus (SCBA) with full facepiece operated in pressure demand or other positive pressure mode.

Ventilation

Local exhaust or general mechanical systems recommended.

Protective Gloves	Eye Protection	Other Protective Clothing
Impervious Gloves	Chemical Goggles or Face Mask	Protective Apron

Work/Hygiene Practices

Always wash hands thoroughly after using chemical; never bring food, drink, or smoking materials into vicinity of chemicals.

ETHYL NITRITE

$C_2H_5NO_2$ CAS: 109-95-5

IDENTIFICATION AND TYPICAL USES

Ethyl nitrite is a colorless or yellowish, clear liquid with a characteristic, sweet, rum-like odor and fruity taste. It is no longer produced commercially in the United States. It is used in preparing spirit nitrous ether, as a chemical intermediate, in organic reactions, and for synthetic flavorings.

RISK ASSESSMENT: HEALTH

General Assessment

Ethyl nitrite is a mildly toxic compound by *inhalation* and *ingestion*. Although it is a skin irritant and can be corrosive, skin absorption has not been reported. There are no reports on its carcinogenicity, mutagenicity, or teratogenicity.

Inhalation causes irritation of the eyes, nose, throat, and respiratory tract. Inhalation may cause headache, tachycardia (rapid pulse rate), hypotension (lowered blood pressure), methemoglobinemia (lack of oxygen-carrying capacity of the blood due to the formation of methemoglobin from the oxygen-carrying pigment, hemoglobin) causing cyanosis (bluish discoloration of the skin), respiratory depression, and loss of consciousness.

Skin contact produces irritation and possible burns. Ingestion of this compound causes irritation with unspecified gastrointestinal effects. Ethyl nitrite is considered a poison.

Acute Health Effects

The following acute (short-term) health effects may occur immediately or shortly after exposure to ethyl nitrite:

Skin: Irritation with possible burns, redness, blistering, and rash.

Eye: Severe irritation with redness and burns.

Lung: Irritation of the respiratory tract and lung tissues resulting in cough and congestion.

Other: Acute overexposure may cause rapid pulse, low blood pressure, methemoglobinemia, and loss of consciousness.

Chronic Health Effects

The following chronic (long-term) health effects may occur at some time after exposure to ethyl nitrite and can last for months or even years:

Cancer Hazards: According to the information presented in the references, ethyl nitrite has not been adequately tested for its ability to cause cancer in test animals.

Reproductive Hazard: According to the information presented in the references, ethyl nitrite has not been adequately tested for its ability to adversely affect reproduction in test animals.

Other Chronic Effects: Prolonged exposure may cause chronic methemoglobinemia (decrease in oxygen-carrying capability of the blood).

Recommended Risk-Reduction Measures

Unless a less toxic chemical can be substituted for a hazardous substance, *engineering controls* are the most effective method of reducing exposures. The best protection is to enclose operations and/or provide local exhaust ventilation at the site of chemical release. To prevent static sparks, all containers and equipment should be electrically grounded and bonded. While not always operationally feasible, isolating operations can also reduce exposure.

Using respiratory protection is less effective than the controls mentioned above, but is still advisable whenever working with or around ethyl nitrite. No exposure levels have been established for this chemical.

However, this does not mean that exposure to ethyl nitrite is without risk. For the best possible protection, a supplied-air respirator or a self-contained breathing apparatus (SCBA) with full facepiece operated in pressure demand mode are the recommended respiratory protection methods of choice. If a full facepiece is not available, then chemical goggles should be worn to protect the eyes. Whenever a chemical splash hazard exists, a face shield and apron should be worn. To prevent hand and skin exposures, impervious rubber gloves should be worn. Glove manufacturers should be contacted and permeation studies obtained *before* glove selections are made.

Administrative controls should also be in place to minimize the potential for human exposures. These may include written procedures or policies which specify the methods and techniques that will be practiced whenever personnel are to work with ethyl nitrite.

All personnel should receive *training* on the use, hazards, protective measures, emergency actions, and other precautions per 29 CFR 1910.1200 (Hazard Communication), prior to the first assignment in an area where ethyl nitrite is used or stored.

If symptoms develop or overexposure is suspected, the following medical tests may be useful:

☑ Complete blood count and blood gas tests.

Any evaluation should include a careful medical history of past and present symptoms with an examination. However, medical tests that evaluate existing damage are not a substitute for controlling exposure. Also, since smoking can cause heart diseases, emphysema, numerous other respiratory disorders, and interfere with the blood's oxygen-carrying capabilities, smokers may be affected more rapidly than non-smokers to the same exposure conditions. Prudent risk assessment and management requires careful consideration of *all* possible risk factors that may be causing the appearance of symptoms in exposed workers.

Other methods to reduce exposure include:

☑ Where possible, enclose operations and use local exhaust ventilation at the site of chemical release. If local exhaust ventilation or enclosure is not used, respiratory protection should be mandatory. Also, if possible, automatically transfer liquids from storage containers to process containers using non-sparking tools.

☑ Always ensure that proper protective clothing is worn when using chemical substances.

☑ Wash thoroughly immediately after exposure to ethyl nitrite and at the end of the work shift or before eating, drinking, or smoking.

☑ Hazard warning information should be posted in the work area. In addition, as part of an on-going education and training program, all information on the health and safety hazards of ethyl nitrite should be communicated to all potentially exposed workers.

☑ Work clothing contaminated with ethyl nitrite should never be worn home where family members can be exposed. Also, clothing soaked with ethyl nitrite presents a serious fire hazard. Contaminated clothing should be laundered by personnel who have been briefed on the hazards of exposure to ethyl nitrite.

☑ Eye wash stations should be provided in work areas. If the potential for whole body exposure exists, then safety showers should also be provided.

RISK ASSESSMENT: ENVIRONMENT

General Assessment

The environment is at risk of exposure during intended use, transportation, storage, disposal, or destruction of ethyl nitrite. In almost every scenario, the threat of environmental exposure is contingent upon the proper handling of the chemical substance. Accidental spills, large or small, can result in fire, explosion, and possible contamination of the surrounding environmental mediums (water, soil, and air).

Ethyl nitrite is considered a Class IA flammable liquid (per OSHA 29 CFR 1910.106). Its extremely low flash point and relatively low boiling point and autoignition temperature make it a serious and dangerous fire and explosion risk. Its vapors are heavier than air and can travel for some distance to an ignition source to flashback and cause a fire or explosion. It can react with many common acids and many oxidizers to cause a fire or explosion. These characteristics require special consideration during any emergency situation involving a leak or spill of ethyl nitrite. Should ethyl nitrite ever come into contact with these incompatible substances either during use, transportation, or storage, violent reactions can occur.

Ethyl nitrite can enter the environment through industrial discharges, municipal waste treatment discharges, and spills.

☠ *Acute Ecological Effects*

Acute (short-term) toxic effects may include the death of animals, birds, or fish and death or low growth rate in plants. Acute effects are seen 2 to 4 days after animals or plants are exposed to ethyl nitrite.

Insufficient data are available to evaluate or predict the short-term effects of ethyl nitrite to aquatic life, plants, birds, or land animals.

Chronic Ecological Effects

Chronic toxic effects may include shortened life span, reproductive problems, lower fertility, and changes in appearance or behavior in exposed animals. These effects can be seen long after first exposure(s) to toxic chemicals.

Insufficient data are available to evaluate or predict the long-term effects of ethyl nitrite to aquatic life, plants, birds, or land animals.

Water Solubility

Ethyl nitrite is highly soluble in water. Concentrations of 1000 milligrams and more will mix with a liter of water.

⌛ *Persistence in the Environment*

Insufficient data are available to evaluate or predict the ability of ethyl nitrite to persist in the environmental mediums.

Bioaccumulation in Aquatic Organisms

Some substances increase in concentration, or *bioaccumulate,* in living organisms as they breathe contaminated air, drink contaminated water, or eat contaminated food. These chemicals can become concentrated in the tissues and internal organs of animals as well as humans.

There is no data available on the bioaccumulation characteristics of ethyl nitrite.

Recommended Risk-Reduction Measures

Proper training of all transporters will reduce the likelihood of a mishap or accident resulting in a leak or spill to the environment. The correct use of DOT labeling (per the MSDS) on all containers, trucks, and rail cars will enable emergency responders to react properly and quickly to any disaster thereby reducing the potential risk to the environment and to personnel.

Storage of ethyl nitrite should be segregated from incompatible chemicals to minimize the risk of cross-contamination or contact. Storage buildings should be equipped with the proper fire protection equipment (alarms, sprinklers, extinguishers, etc.). Explosion proof design is highly recommended for all storage facility and operational equipment.

If a spill or leak to the environment has occurred, fire department, emergency response and/or hazardous materials spill personnel should be notified immediately. Cleanup should be attempted only by those trained in proper spill containment procedures using non-sparking tools. Contaminated soils should be removed for incineration and replaced with clean soil. If ethyl nitrite should contact the water table, aquifer, or navigable waterway, remediation activities should be prompt. It is highly soluble in water and total containment may not be possible. When such spills occur, the local and/or state emergency response authorities must be notified. A comprehensive emergency response or disaster preparedness/recovery plan should be in place prior to any operations involving the use, transportation, storage, or disposal of ethyl nitrite. If ethyl nitrite is spilled or leaked, the following specific steps are recommended:

- ☑ Restrict persons not wearing protective clothing from area of spill or leak until cleanup is complete and area can be opened for normal work.
- ☑ Ventilate area of spill or leak.
- ☑ Remove all ignition sources.
- ☑ Use water spray to disperse vapors and dilute spills to non-flammable mixtures. Flush spills away from exposures. Dispose of accordingly (do not allow runoff to enter sewers or waterways). Absorb liquids with vermiculite and deposit in sealed drums for disposal.
- ☑ It may be necessary to dispose of ethyl nitrite as a hazardous waste. The responsible state agency or the regional office of the federal Environmental Protection Agency (EPA) should be contacted for specific recommendations.

RISK ASSESSMENT: BUSINESS

General Assessment

Accidents or mishaps involving ethyl nitrite can present a significant threat to business operations. The loss or damage of equipment or facilities can significantly affect fiscal viability. Lawsuits that may result from personnel injury/death, public exposures, and/or environmental contamination will also require a serious expenditure of resources. Media attention surrounding an injury, death, or environmental damage can also result in a loss of profits and loss of current as well as future business.

Recommended Risk-Reduction Measures

Company attorneys, safety and health professionals, and environmental specialists should be involved in the development of any procedures dealing with safety or emergency response. A company official should be pre-designated as a public relations officer with specific training in dealing with the press. Corporate plans and policies should be developed, approved, and implemented long before any need for such arises.

REFERENCES

Ahlbom, A. 1993. *Biostatistics for Epidemiologists.* New York: CRC Press/Lewis Publishers.

American Conference of Governmental Industrial Hygienists. 1988. *Documentation of the Threshold Limit Values and Biological Exposure Indices,* 5th Edition (with updates). Cincinnati: ACGIH

Anthony, C. P., and N. J. Kolthoff. 1971. *Textbook of Anatomy and Physiology.* St. Louis: C. V. Mosby Company.

Asante-Duah, D. K. 1993. *Hazardous Waste Risk Assessment.* New York: CRC Press/Lewis Publishers.

Bloom, A. D. and F. J. DeSerres. 1995. *Ecotoxicity and Human Health.* New York: CRC Press/Lewis Publishers.

Brusick, D. J. 1994. *Methods for Genetic Risk Assessment.* New York: CRC Press/Lewis Publishers.

Bueche, J. R. 1972. *Principles of Physics,* 2nd Edition. New York: McGraw-Hill.

Burrell, R., D. K. Flaherty, and L. J. Sauers. 1992. *Toxicology of the Immune System.* New York: Van Nostrand Reinhold.

Calabrese, E. J. 1994. *Biological Effects of Low Level Exposures.* New York: CRC Press/Lewis Publishers.

Cockerman, L. G., B. S. Shane. 1994. *Basic Environmental Toxicology.* New York: CRC Press/Lewis Publishers.

Cothern, C. R., N. P. Ross. 1994. *Environmental Statistics, Assessment, and Forecasting.* New York: CRC Press/Lewis Publishers.

Danse, I. R. 1991. *Common Sense Toxics in the Workplace.* New York: Van Nostrand Reinhold.

Environmental Protection Agency. 1983. *Chemical Hazard Information Profiles.* Washington, D.C.: U.S. EPA.

Fodden, C. R., and J. L. Weddell. 1992. *Hazardous Materials Emergency Action Data.* New York: CRC Press/Lewis Publishers.

Forseberg, K., and L. H. Keith. 1993. *Chemical; Protective Clothing Permeation and Degradation Database.* Chelsea, MI: Lewis Publishers.

Gardner, R. W. 1994. *Chemical Intolerance.* New York: CRC Press/Lewis Publishers.

Gerarde, H. W. 1960. *Toxicology and Biochemistry of Aromatic Hydrocarbons.* New York: Elsevier.

Hallenbeck, W. H. 1993. *Quantitative Risk Assessment for Environmental and Occupational Health,* 2nd Edition. New York: CRC Press/Lewis Publishers.

Hodgson, E., R. B. Mailman, and Janice E. Chambers. 1988. *Dictionary of Toxicology.* London: The MacMillian Company.

Howard, P. H. 1995. *Handbook of Environmental Fate and Exposures Data for Organic Chemicals,* Vol. I-IV. New York: CRC Press/Lewis Publishers.

Hosty, J. W., and P. Foster. 1990. *A Practical Guide to Chemical Spill Response.* New York: Van Nostrand Reinhold.

Kamrin, M. A. 1988. *Toxicology: A Primer on Principles and Applications.* Chelsea, Michigan: Lewis Publishers.

Lewis, R. J., Sr. 1993. *Hawley's Condensed Chemical Dictionary,* 12th Edition. New York: Van Nostrand Reinhold.

Lewis, R. J., Sr. 1992. *Sax's Dangerous Properties of Industrial Materials,* Volumes 1, 2, and 3, 8th Edition. New York: Van Nostrand Reinhold.

Mackay, D., W. Y. Shiu, and K. C. Ma. 1992. *Illustrated Handbook of Chemical-Physical Properties and Environmental Fate for Organic Chemicals.* New York: CRC Press/Lewis Publishers.

Manahan, S. E. 1992. *Toxicological Chemistry,* 2nd Edition. New York: CRC Press/Lewis Publishers.

Mellström, G. A., J. E. Wahlberg, and H. I. Maibach. 1994. *Protective Gloves for Occupational Use.* New York: CRC Press/Lewis Publishers.

Meyer, E. 1990. *Chemistry of Hazardous Materials.* Englewood Cliffs, NJ: Prentice Hall.

Miller, B. F., M.D., and C. B. Keane, R.N. 1972. *Encyclopedia and Dictionary of Medicine and Nursing.* Philadelphia: W.B. Sanders Company.

Mycock, J. C., J. D. McKenna, and L. Theodore. 1995. *Handbook of Air Pollution Control Engineering and Technology.* New York: CRC Press/Lewis Publishers.

Nelson, G. O. 1992. *Gas Mixtures.* New York: CRC Press/Lewis Publishers.

Ott, W. R. 1995. *Environmental Statistics and Data Analysis.* New York: CRC Press/Lewis Publishers.

Patnaik, P. 1992. *A Comprehensive Guide to the Hazardous Properties of Chemical Substances.* New York: Van Nostrand Reinhold.

Richardson, M. 1995. *Dictionary of Substance and Their Effects,* Vol. I-VI. New York: CRC Press/Lewis Publishers.

Sacarello, H. L. A. 1994. *The Comprehensive Handbook of Hazardous Materials.* New York: CRC Press/Lewis Publishers.

U.S. Department of Health and Human Services, National Institute for Occupational Safety and Health. 1994. *NIOSH Pocket Guide to Chemical Hazards.* Washington, D.C.: U.S. Government Printing Office.

MATERIAL SAFETY DATA SHEET

CHEMICAL NAME

ETHYL SILICATE

HAZARD WARNING INFORMATION

HEALTH	FIRE	REACTIVE	OTHER	DEGREE OF HAZARD	COLOR CODING	OTHER CODES
2	2	0		0 = Minimum Hazard 1 = Slight Hazard 2 = Moderate Hazard 3 = Serious Hazard 4 = Severe Hazard	HEALTH = BLUE FIRE = RED REACTIVITY = YELLOW OTHER = WHITE	OX = Oxidizer ACID = Acid ALK = Alkali COR = Corrosive ~~W~~ = Use No Water

SECTION I - GENERAL INFORMATION

Characterization	RCRA Number	EPA Class
Ester	**None**	**Not Applicable**

DOT Proper Shipping Name	Chemical Abstract Service (CAS) Number
Tetraethyl Silicate	**78-10-4**
DOT Hazard Class and Label Requirements	DOT Emergency Guide Code
Combustible Liquid	**29**
DOT Identification Number	Chemical Formula
UN 1292	**$(C_2H_5)_4SiO_4$**

Synonyms

Ethyl orthosilicate; ethyl silicate (condensed); tetraethoxysilane; tetraethyl orthosilicate (TEOS); tetratethyl silicate; silicic acid tetraethyl ester.

SECTION II - HAZARDOUS INGREDIENTS AND IDENTITY INFORMATION

Hazardous Components (specific identity)	OSHA Exposure Criteria	NIOSH Exposure Criteria	Immediately Dangerous to Life and Health (IDLH)	ACGIH Exposure Criteria
Ethyl silicate (derivation: From absolute alcohol and silicon tetrachloride). **1 ppm = 8.66 mg/m^3**	**PEL:** **100 ppm** **850 mg/m^3** **STEL:** **Not Established**	**REL:** **10 ppm** **85 mg/m^3** **STEL:** **Not Established**	**700 ppm**	**TLV:** **10 ppm** **85 mg/m^3** **STEL:** **Not Established**

SECTION III - PHYSICAL AND CHEMICAL CHARACTERISTICS

Boiling Point	Specific Gravity (H_2O = 1)
336°F (169°C)	**0.93**
Vapor Pressure (mm Hg)	Molecular Weight
1at 63°F (17°C)	**208.3**
Vapor Density (Air = 1)	Melting Point
7.22	**-107°F (-77°C)**

Solubility

Decomposes in water (hydrolyzed to an adhesive form of silica). Miscible with alcohol.

Appearance and Odor

Colorless liquid with a mild, sweet, alcohol-like odor. Odor Threshold = 85 ppm.

SECTION IV - FIRE AND EXPLOSION HAZARD DATA

Flash Point (method used)	Explosive Limits in Air % by Volume
99°F (37°C) closed cup	**LEL: 1.3% UEL: 23%**
NFPA Classification	Autoignition Temperature
Class IA Flammable Liquid	**Not Determined**

Extinguishing Media

Use foams, carbon dioxide, water spray, or dry chemical.

Special Fire Fighting Procedures

Poisonous gases are produced in fire. Use full protective clothing and self-contained breathing apparatus (SCBA). Fight fire from maximum distance if possible. Move containers from fire area if it can be done without risk. Do NOT release runoff from fire control methods to sewers or waterways.

Unusual Fire and Explosion Hazards

Ethyl silicate can form potentially explosive mixtures in air (note its wide flammability range). Vapors are heavier than air and can travel for some distance to an ignition source and flashback to cause fire or explosion. This chemical presents a vapor explosion hazard indoors, outdoors, and in sewers.

SECTION V - REACTIVITY DATA

Stability		Conditions to Avoid Ethyl silicate is normally stable at room temperature in closed containers under normal storage and handling conditions. Avoid contact with incompatible materials.
Stable X	Unstable	Incompatibility *(materials to avoid)* Reacts dangerously with strong oxidizers (such as chlorine, bromine, and fluorine), and will decompose to form a silicone adhesive (a milky-white mass) on contact with water.
Hazardous Polymerization		Conditions to Avoid Under normal conditions of temperature and pressure, hazardous polymerization of ethyl silicate is not expected to occur.
May Occur	Will Not Occur X	Hazardous Decomposition or By-products Toxic and irritating fumes and vapors are produced when ethyl silicate is heated to decomposition. These include oxides of carbon (dioxide and monoxide) and various silicon compounds.

SECTION VI - HEALTH HAZARD DATA

Primary Route(s) of Entry:	Inhalation?	Absorption (skin)?	Ingestion?
	X		X

Health Hazards

INHALATION: Symptoms of eyes, nose, and respiratory tract irritation and may cause pulmonary, liver, and kidney damage. There could be shortness of breath, wheezing, and coughing. It is narcotic in high concentrations causing tremors and other unspecified effects. Exposure may also cause anemia and pulmonary edema (fluid in lungs).

SKIN & EYES: Irritation on skin contact. Will cause contact dermatitis and redness. Severe irritation to the eyes causing tearing, redness, and swelling.

INGESTION: May cause gastrointestinal tract irritation.

Carcinogenicity	NTP Listed?	IARC Cancer Review Group?	OSHA Regulated?	Target Organs?
Unknown Human Unknown Animal	No	No	29 CFR 1910.1000 Table Z-1	Eyes, skin, liver, kidneys, respiratory system, blood.

Medical Conditions Generally Aggravated by Exposure

None reported.

Emergency and First-aid Procedures

Eye contact: Flush immediately with water for 15 minutes (minimum), seek medical attention. Skin contact: Remove all contaminated clothing. Immediately wash area with large amounts water (15 minutes). Seek medical assistance. For inhalation: Remove the person from exposure. Provide respiratory assistance and CPR. Transfer to medical facility. Observe 48 hours for lung effects. If swallowed: Give 1-2 glasses of water and induce vomiting until vomitus is clear. Never attempt to give an unconscious or convulsing person anything by mouth.

SECTION VII - PRECAUTIONS FOR SAFE HANDLING AND USE

Steps to be Taken in Case Material is Released or Spilled

Use self-contained breathing apparatus (SCBA). Absorb in vermiculite or other suitable material, collect using non-sparking tools, and deposit in DOT-approved drum. Ventilate area and remove ignition sources. Restrict those not involved in cleanup from entering area.

Preferred Waste Disposal Method

No Citation.

Precautions to be Taken in Handling and Storage

Do not store ethyl silicate in the presence of incompatible chemicals or materials. Store in tightly closed containers in a cool, well-ventilated area. Protect containers from physical damage away from heat. Automatic transfer of liquids between containers is recommended.

Other Precautions and Warnings

Recommend explosion proof design for bulk storage facilities. Electrically ground and bond equipment, especially during transfer operations. Detached facility storage is recommended.

SECTION VIII - CONTROL MEASURES AND PERSONAL PROTECTIVE EQUIPMENT

Respiratory Protection *(specify type)*

For low exposure, use an air-purifying respirator. For higher or frequent exposure, use a supplied-air respirator with full facepiece operated in positive pressure mode or a self-contained breathing apparatus (SCBA) with full facepiece operated in pressure demand or other positive pressure mode.

Ventilation

Local exhaust or general mechanical systems recommended.

Protective Gloves	Eye Protection	Other Protective Clothing
Impervious Gloves	Chemical Goggles or Face Mask	Protective Apron

Work/Hygiene Practices

Always wash hands thoroughly after using chemical; never bring food, drink, or smoking materials into vicinity of chemicals.

ETHYL SILICATE

$(C_2H_5)_4SiO_4$ CAS: 78-10-4

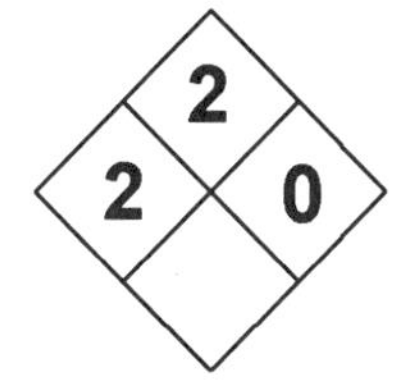

IDENTIFICATION AND TYPICAL USES

Ethyl silicate is a colorless liquid with a mild, sweet, characteristic, alcohol-like odor. It is used for weatherproofing and acid-proofing cements, mortar, refractory bricks, and other molded objects. Also used in the manufacture of heat-resistant paints, chemical resistant paints, protective coatings for industrial buildings and castings, in lacquers, as a bonding agent, and an intermediate. It has also been used to harden stone (to prevent decay and disintegration).

RISK ASSESSMENT: HEALTH

General Assessment

Ethyl silicate is a mildly toxic compound by *inhalation* and *ingestion*. Although it is a severe skin and eye irritant, skin absorption has not been reported. There are no reports on its carcinogenicity, mutagenicity, or teratogenicity.

Inhalation causes irritation of the eyes, nose, throat, and mucosa of the upper respiratory tract. Inhalation may cause headache, anemia, shortness of breath (dypsnea), tearing (lachrymation), coughing, and wheezing. It is considered narcotic in high concentrations causing tremors, unsteadiness, loss of consciousness, and other unspecified effects. Liver, kidney, and pulmonary damage is possible. Acute overexposure can result in the buildup of fluid in the lungs, a condition known as pulmonary edema. This is a medical emergency and can lead to death. Symptoms may be delayed up to 48 hours, creating a false sense of security with regard to health exposure risk.

Skin contact produces severe irritation, redness, drying, cracking, and possible contact dermatitis. Eye contact can cause severe irritation with redness, burning, and tearing. Ingestion nausea, vomiting, and cramps. Narcosis is also possible at high concentrations or frequent ingestion.

Acute Health Effects

The following acute (short-term) health effects may occur immediately or shortly after exposure to ethyl silicate:

Skin: Irritation with possible burns, redness, and contact dermatitis. Causes drying and cracking that may lead to secondary infection.

Eye: Severe irritation with redness and tearing.

Lung: Irritation of the respiratory tract and lung tissues resulting in coughing, wheezing, and congestion. May cause pulmonary edema.

Other: Acute overexposure may cause narcotic effects (tremor, unsteadiness, loss of consciousness) and may cause anemia.

Chronic Health Effects

The following chronic (long-term) health effects may occur at some time after exposure to ethyl silicate and can last for months or even years:

Cancer Hazards: According to the information presented in the references, ethyl silicate has not been adequately tested for its ability to cause cancer in test animals.

Reproductive Hazard: According to the information presented in the references, ethyl silicate has not been adequately tested for its ability to adversely affect reproduction in test animals.

Recommended Risk-Reduction Measures

Unless a less toxic chemical can be substituted for a hazardous substance, *engineering controls* are the most effective method of reducing exposures. The best protection is to enclose operations and/or provide local exhaust ventilation at the site of chemical release. To prevent static sparks, all containers and equipment should be electrically grounded and bonded. While not

always operationally feasible, isolating operations can also reduce exposure.

Using respiratory protection is less effective than the controls mentioned above, but is still advisable whenever working with or around ethyl silicate. For low or infrequent exposures above the PEL (100 ppm), an air purifying respirator is suggested. For the best possible protection, however, a supplied-air respirator or a self-contained breathing apparatus (SCBA) with full facepiece operated in pressure demand mode are the recommended respiratory protection methods of choice. If a full facepiece is not available, then chemical goggles should be worn to protect the eyes. Whenever a chemical splash hazard exists, a face shield and apron should be worn. To prevent hand and skin exposures, impervious rubber gloves should be used. Glove manufacturers should be contacted and permeation studies obtained *before* glove selections are made.

Administrative controls should also be in place to minimize the potential for human exposures. These may include written procedures or policies which specify the methods and techniques that will be practiced whenever personnel are to work with ethyl silicate.

All personnel should receive *training* on the use, hazards, protective measures, emergency actions, and other precautions per 29 CFR 1910.1200 (Hazard Communication), prior to the first assignment in an area where ethyl silicate is used or stored.

Toxicity data on human health effects are limited in the references. However, with symptoms of narcosis, dypsnea, wheezing, and anemia, animal studies have shown lung, kidney, and liver effects.

If symptoms develop or overexposure is suspected, the following medical tests may be useful:

☑ Lung function tests.

☑ Consider chest X-ray after acute overexposure (may be negative if taken immediately after exposure due to the delayed development of pulmonary edema).

☑ Evaluation by a qualified allergist with consideration of exposure history and special testing (may help diagnose skin allergy).

☑ Liver and kidney function tests.

Any evaluation should include a careful medical history of past and present symptoms with an examination. However, medical tests that evaluate existing damage are not a substitute for controlling exposure. Also, since smoking can cause heart diseases, emphysema, numerous other respiratory disorders, and interfere with the blood's oxygen-carrying capabilities, smokers may be affected more rapidly than non-smokers to the same exposure conditions. Prudent risk assessment and management requires careful consideration of *all* possible risk factors that may be causing the appearance of symptoms in exposed workers.

Other methods to reduce exposure include:

☑ Where possible, enclose operations and use local exhaust ventilation at the site of chemical release. If local exhaust ventilation or enclosure is not used, respiratory protection should be mandatory. Also, if possible, automatically transfer liquids from storage containers to process containers using non-sparking tools.

☑ Always ensure that proper protective clothing is worn when using chemical substances.

☑ Wash thoroughly immediately after exposure to ethyl silicate and at the end of the work shift or before eating, drinking, or smoking.

☑ Hazard warning information should be posted in the work area. In addition, as part of an on-going education and training program, all information on the health and safety hazards of ethyl silicate should be communicated to all potentially exposed workers.

☑ Work clothing contaminated with ethyl silicate should never be worn home where family members can be exposed. Also, clothing soaked with ethyl silicate presents a serious fire hazard. Contaminated clothing should be laundered by personnel who have been briefed on the hazards of exposure to ethyl silicate.

☑ Eye wash stations should be provided in work areas. If the potential for whole body exposure exists, then safety showers should also be provided in the immediate area.

RISK ASSESSMENT: ENVIRONMENT

General Assessment

The environment is at risk of exposure during intended use, transportation, storage, disposal, or destruction of ethyl silicate. In almost every scenario, the threat of environmental exposure is contingent upon the proper handling of the chemical substance. Accidental spills, large or small, can result in fire, explosion, and possible contamination of the surrounding environmental mediums (water, soil, and air).

Ethyl silicate is considered a Class IC flammable liquid (per OSHA 29 CFR 1910.106). Its low flash point and moderate boiling point temperatures make it a dangerous fire and explosion risk. Its vapors are heavier than air and can travel for some distance to an ignition source to flashback and cause a fire or explosion. It can react with many common oxidizers to cause a fire or explosion. These characteristics require special consideration during any emergency situation involving a leak or spill of ethyl silicate. Should ethyl silicate ever come into contact with these incompatible substances either during use, transportation, or storage, violent reactions can occur.

Ethyl silicate can enter the environment through industrial discharges, municipal waste treatment discharges, and spills.

Acute Ecological Effects

Acute (short-term) toxic effects may include the death of animals, birds, or fish and death or low growth rate in plants. Acute effects are seen 2 to 4 days after animals or plants are exposed to ethyl silicate.

Insufficient data are available to evaluate or predict the short-term effects of ethyl silicate to aquatic life, plants, birds, or land animals.

Chronic Ecological Effects

Chronic toxic effects may include shortened life span, reproductive problems, lower fertility, and changes in appearance or behavior in exposed animals. These effects can be seen long after first exposure(s) to toxic chemicals.

Insufficient data are available to evaluate or predict the long-term effects of ethyl silicate to aquatic life, plants, birds, or land animals.

Water Solubility

Ethyl silicate decomposes in water (hydrolyzes to an adhesive form of silica).

Persistence in the Environment

Insufficient data are available to evaluate or predict the ability of ethyl silicate to persist in the environmental mediums.

Bioaccumulation in Aquatic Organisms

Some substances increase in concentration, or *bioaccumulate,* in living organisms as they breathe contaminated air, drink contaminated water, or eat contaminated food. These chemicals can become concentrated in the tissues and internal organs of animals as well as humans.

There is no data available on the bioaccumulation characteristics of ethyl silicate.

Recommended Risk-Reduction Measures

Proper training of all transporters will reduce the likelihood of a mishap or accident resulting in a leak or spill to the environment. The correct use of DOT labeling (per the MSDS) on all containers, trucks, and rail cars will enable emergency responders to react properly and quickly to any disaster thereby reducing the potential risk to the environment and to personnel.

Storage of ethyl silicate should be segregated from incompatible chemicals to minimize the risk of cross-contamination or contact. Storage buildings should be equipped with the proper fire protection equipment (alarms, sprinklers, extinguishers, etc.). Explosion proof design is highly recommended for all storage facility and operational equipment.

If a spill or leak to the environment has occurred, fire department, emergency response and/or hazardous materials spill personnel should be notified immediately. Cleanup should be attempted only by those trained in proper spill containment procedures using non-sparking tools. Contaminated soils should be removed for incineration and replaced with clean soil. If ethyl silicate should contact the water table, aquifer, or navigable waterway, remediation activities should be prompt. It slowly decomposes in water and total containment may not be possible. When such spills occur, the local and/or state emergency response authorities must be notified. A comprehensive emergency response or disaster preparedness/recovery plan should be in place prior to any operations involving the use, transportation, storage, or disposal of ethyl silicate. If ethyl silicate is spilled or leaked, the following specific steps are recommended:

- ☑ Restrict persons not wearing protective clothing from area of spill or leak until cleanup is complete and area can be opened for normal work.
- ☑ Ventilate area of spill or leak.
- ☑ Remove all ignition sources.
- ☑ Absorb liquids with vermiculite and deposit in sealed drums for disposal.
- ☑ It may be necessary to dispose of ethyl silicate as a hazardous waste. The responsible state agency

or the regional office of the federal Environmental Protection Agency (EPA) should be contacted for specific recommendations.

RISK ASSESSMENT: BUSINESS

General Assessment

Accidents or mishaps involving ethyl silicate can present a significant threat to business operations. The loss or damage of equipment or facilities can significantly affect fiscal viability. Lawsuits that may result from personnel injury/death, public exposures, and/or environmental contamination will also require a serious expenditure of resources. Media attention surrounding an injury, death, or environmental damage can also result in a loss of profits and loss of current as well as future business.

⋃ *Recommended Risk-Reduction Measures*

Company attorneys, safety and health professionals, and environmental specialists should be involved in the development of any procedures dealing with safety or emergency response. A company official should be pre-designated as a public relations officer with specific training in dealing with the press. Corporate plans and policies should be developed, approved, and implemented long before any need for such arises.

REFERENCES

American Conference of Governmental Industrial Hygienists. 1988. *Documentation of the Threshold Limit Values and Biological Exposure Indices,* 5th Edition (with updates). Cincinnati: ACGIH

Anthony, C. P., and N. J. Kolthoff. 1971. *Textbook of Anatomy and Physiology.* St. Louis: C. V. Mosby Company.

Asante-Duah, D. K. 1993. *Hazardous Waste Risk Assessment.* New York: CRC Press/Lewis Publishers.

Bloom, A. D. and F. J. DeSerres. 1995. *Ecotoxicity and Human Health.* New York: CRC Press/Lewis Publishers.

Brusick, D. J. 1994. *Methods for Genetic Risk Assessment.* New York: CRC Press/Lewis Publishers.

Bueche, J. R. 1972. *Principles of Physics,* 2nd Edition. New York: McGraw-Hill.

Burrell, R., D. K. Flaherty, and L. J. Sauers. 1992. *Toxicology of the Immune System.* New York: Van Nostrand Reinhold.

Calabrese, E. J. 1994. *Biological Effects of Low Level Exposures.* New York: CRC Press/Lewis Publishers.

Cockerman, L. G., B. S. Shane. 1994. *Basic Environmental Toxicology.* New York: CRC Press/Lewis Publishers.

Cothern, C. R., N. P. Ross. 1994. *Environmental Statistics, Assessment, and Forecasting.* New York: CRC Press/Lewis Publishers.

Danse, I. R. 1991. *Common Sense Toxics in the Workplace.* New York: Van Nostrand Reinhold.

Environmental Protection Agency. 1983. *Chemical Hazard Information Profiles.* Washington, D.C.: U.S. EPA.

Fodden, C. R., and J. L. Weddell. 1992. *Hazardous Materials Emergency Action Data.* New York: CRC Press/Lewis Publishers.

Forseberg, K., and L. H. Keith. 1993. *Chemical; Protective Clothing Permeation and Degradation Database.* Chelsea, MI: Lewis Publishers.

Gardner, R. W. 1994. *Chemical Intolerance.* New York: CRC Press/Lewis Publishers.

Gerarde, H. W. 1960. *Toxicology and Biochemistry of Aromatic Hydrocarbons.* New York: Elsevier.

Hallenbeck, W. H. 1993. *Quantitative Risk Assessment for Environmental and Occupational Health,* 2nd Edition. New York: CRC Press/Lewis Publishers.

Hodgson, E., R. B. Mailman, and Janice E. Chambers. 1988. *Dictionary of Toxicology.* London: The MacMillian Company.

Howard, P. H. 1995. *Handbook of Environmental Fate and Exposures Data for Organic Chemicals,* Vol. I-IV. New York: CRC Press/Lewis Publishers.

Hosty, J. W., and P. Foster. 1990. *A Practical Guide to Chemical Spill Response.* New York: Van Nostrand Reinhold.

Kamrin, M. A. 1988. *Toxicology: A Primer on Principles and Applications.* Chelsea, Michigan: Lewis Publishers.

Lewis, R. J., Sr. 1993. *Hawley's Condensed Chemical Dictionary,* 12th Edition. New York: Van Nostrand Reinhold.

Lewis, R. J., Sr. 1992. *Sax's Dangerous Properties of Industrial Materials,* Volumes 1, 2, and 3, 8th Edition. New York: Van Nostrand Reinhold.

Manahan, S. E. 1992. *Toxicological Chemistry,* 2nd Edition. New York: CRC Press/Lewis Publishers.

Meyer, E. 1990. *Chemistry of Hazardous Materials.* Englewood Cliffs, NJ: Prentice Hall.

Miller, B. F., M.D., and C. B. Keane, R.N. 1972. *Encyclopedia and Dictionary of Medicine and Nursing.* Philadelphia: W.B. Sanders Company.

Nelson, G. O. 1992. *Gas Mixtures.* New York: CRC Press/Lewis Publishers.

Ott, W. R. 1995. *Environmental Statistics and Data Analysis.* New York: CRC Press/Lewis Publishers.

Patnaik, P. 1992. *A Comprehensive Guide to the Hazardous Properties of Chemical Substances.* New York: Van Nostrand Reinhold.

Richardson, M. 1995. *Dictionary of Substance and Their Effects,* Vol. I-VI. New York: CRC Press/Lewis Publishers.

Sacarello, H. L. A. 1994. *The Comprehensive Handbook of Hazardous Materials.* New York: CRC Press/Lewis Publishers.

U.S. Department of Health and Human Services, National Institute for Occupational Safety and Health. 1994. *NIOSH Pocket Guide to Chemical Hazards.* Washington, D.C.: U.S. Government Printing Office.

MATERIAL SAFETY DATA SHEET

CHEMICAL NAME	CRC PRESS/LEWIS PUBLISHERS DISCLAIMER
ETHYL TRICHLOROSILANE	**CRC PRESS/LEWIS PUBLISHERS DISCLAIMER:** THE INFORMATION AND RECOMMENDATIONS PRESENTED HEREIN ARE BASED ON SOURCES BELIEVED TO BE RELIABLE. CRC MAKES NO REPRESENTATION ON ITS COMPLETENESS OR ACCURACY. IT IS THE RESPONSIBILITY OF THE USER TO DETERMINE THE CHEMICAL'S SUITABILITY FOR ITS INTENDED USE, THE CHEMICAL'S SAFE USE, AND THE CHEMICAL'S PROPER DISPOSAL. NO REPRESENTATIONS AND/OR WARRANTIES, EITHER EXPRESSED OR IMPLIED, OF THE MERCHANTABILITY OR FITNESS FOR A PARTICULAR PURPOSE, OR OF ANY OTHER NATURE, ARE MADE WITH RESPECT TO THE INFORMATION PROVIDED IN THIS MSDS OR TO THE CHEMICAL TO WHICH INFORMATION MAY REFER. CRC NEITHER ASSUMES NOR AUTHORIZES ANY OTHER PERSON TO ASSUME FOR IT, ANY OTHER ADDITIONAL RESPONSIBILITY OR LIABILITY FOR THE USE OF, OR RELIANCE UPON, THIS INFORMATION.

HAZARD WARNING INFORMATION

HEALTH	FIRE	REACTIVE	OTHER	DEGREE OF HAZARD	COLOR CODING	OTHER CODES
3	3	2	~~W~~	0 = Minimum Hazard 1 = Slight Hazard 2 = Moderate Hazard 3 = Serious Hazard 4 = Severe Hazard	HEALTH = BLUE FIRE = RED REACTIVITY = YELLOW OTHER = WHITE	OX = Oxidizer ACID = Acid ALK = Alkali COR = Corrosive ~~W~~ = Use No Water

SECTION I - GENERAL INFORMATION

Characterization **Chlorosilanes**	RCRA Number **D001, D003**	EPA Class **Characteristic (I) and (R) Waste**
DOT Proper Shipping Name **Ethyl Trichlorosilane**	Chemical Abstract Service (CAS) Number **115-21-9**	
DOT Hazard Class and Label Requirements **Flammable Liquid**	DOT Emergency Guide Code **No Citation**	
DOT Identification Number **UN 1196**	Chemical Formula $C_2H_5Cl_3Si$	

Synonyms
Ethylsilicon trichloride; trichloroethylsilane; trichloroethylsilicane; ethyl silicon trichloride.

SECTION II - HAZARDOUS INGREDIENTS AND IDENTITY INFORMATION

Hazardous Components (specific identity)	OSHA Exposure Criteria	NIOSH Exposure Criteria	Immediately Dangerous to Life and Health (IDLH)	ACGIH Exposure Criteria
Ethyl trichlorosilane (derivation: By reaction of ethylene and trichlorosilane in the presence of a peroxide catalyst).	**PEL:** Not Established **STEL:** Not Established	**REL:** Not Established **STEL:** Not Established	Not Determined	**TLV:** Not Established **STEL:** Not Established

SECTION III - PHYSICAL AND CHEMICAL CHARACTERISTICS

Boiling Point **208°F (99°C)**	Specific Gravity (H_2O = 1) **1.24**
Vapor Pressure (mm Hg) **Not Found**	Molecular Weight **163.51**
Vapor Density (Air = 1) **5.6**	Melting Point **-158°F (-106°C)**

Solubility
Insoluble in water (decomposes). Soluble in benzene, ether, heptane, and perchloroethylene. Decomposes in alcohol.

Appearance and Odor
Colorless liquid with an acrid, irritating odor.

SECTION IV - FIRE AND EXPLOSION HAZARD DATA

Flash Point (method used) **64°F (18°C) opened cup**	Explosive Limits in Air % by Volume **LEL: Not Determined UEL: Not Determined**
NFPA Classification **Class IB Flammable Liquid**	Autoignition Temperature **Not Determined**

Extinguishing Media
Use dry chemical, carbon dioxide, regular foam or fog. Use water spray to cool fire-exposed containers.

Special Fire Fighting Procedures
Poisonous gases are produced in fire. Use full protective clothing and self-contained breathing apparatus (SCBA). Fight fire from maximum distance if possible. Move containers from fire area if it can be done without risk. Do NOT release runoff from fire control methods to sewers or waterways.

Unusual Fire and Explosion Hazards
Containers may explode in fire due to excessive internal pressures caused by intense heat. Boiling liquid expanding vapor explosion (BLEVE) is likely. Immediately withdraw if rising sound is heard from venting device or there is tank discoloration due to fire. Vapors are heavier than air and can flashback.

SECTION V - REACTIVITY DATA

Stability		Conditions to Avoid Ethyl trichlorosilane is normally stable at room temperatures in closed containers under normal conditions of storage and handling. Avoid ignition and heat sources, water, and contact with incompatible materials.
Stable X	Unstable	Incompatibility *(materials to avoid)* Reacts with water to release hydrochloric acid. It is also incompatible with oxidizing materials (chlorine, fluorine) and may explode on contact with some bases.
Hazardous Polymerization		Conditions to Avoid Under normal conditions of temperature and pressure, hazardous polymerization of ethyl trichlorosilane is not expected to occur.
May Occur	Will Not Occur X	Hazardous Decomposition or By-products Toxic and irritating fumes and vapors are produced when ethyl trichlorosilane is heated to decomposition. These include carbon dioxide, chloride, and phosgene fumes.

SECTION VI - HEALTH HAZARD DATA

Primary Route(s) of Entry:	Inhalation?	Absorption (skin)?	Ingestion?
	X		X

Health Hazards

INHALATION: Irritation of the eyes, nose, and throat. Severely irritant and corrosive. Causes serious dypsnea (difficulty breathing) with swelling of the larynx. There may be serious eye and lung injury. Destruction of red blood cells is also a possibility.

SKIN & EYES: Vapor and liquid can irritate skin and eyes. Liquid causes severe burns, possibly even 2nd and 3rd degree. Eye contact causes tearing, irritation, pain, and corneal damage.

INGESTION: Corrosive to mouth, throat, esophagus, and digestive tract. Causes abdominal pain, intense thirst, difficulty swallowing, nausea, vomiting, shock, rapid pulse, and collapse of the circulatory system and death. Possible kidney failure, lesions on the liver and heart.

Carcinogenicity	NTP Listed?	IARC Cancer Review Group?	OSHA Regulated?	Target Organs?
Unknown Human Unknown Animal	No	No	No	Eyes, skin, respiratory system, CVS.

Medical Conditions Generally Aggravated by Exposure

None reported.

Emergency and First-aid Procedures

Eye contact: Flush immediately with water for 15 minutes (minimum), seek medical attention. Skin contact: Remove all contaminated clothing. Immediately wash area with large amounts water (15 minutes). Seek medical assistance. For inhalation: Remove the person from exposure. Provide respiratory assistance and administer 100% humidified oxygen. Transfer to medical facility. If swallowed: Contact poison control center and seek medical attention immediately. Unless advised otherwise, give 1-2 glasses of water. Do NOT induce vomiting. Never give an unconscious or convulsing victim anything by mouth.

SECTION VII - PRECAUTIONS FOR SAFE HANDLING AND USE

Steps to be Taken in Case Material is Released or Spilled

Use water spray to disperse vapors (do not spray directly on spilled materials). Neutralize spill with crushed limestone, collect using non-sparking tools, and deposit in DOT-approved drum. Ventilate area and remove ignition sources. Restrict those not involved in cleanup from entering area.

Preferred Waste Disposal Method

No Citation.

Precautions to be Taken in Handling and Storage

Store in tightly closed containers in a cool, well-ventilated area. Protect containers from physical damage away from heat. Automatic transfer of liquids between containers is recommended.

Other Precautions and Warnings

Recommend explosion proof design for bulk storage facilities. Electrically ground and bond equipment, especially during transfer operations. Detached facility storage is recommended.

SECTION VIII - CONTROL MEASURES AND PERSONAL PROTECTIVE EQUIPMENT

Respiratory Protection *(specify type)*

Exposure level not established. Recommend using an MSHA/NIOSH-approved supplied-air respirator with full facepiece operated in positive pressure mode or a self-contained breathing apparatus (SCBA) with full facepiece operated in pressure demand or other positive pressure mode.

Ventilation

Local exhaust or general mechanical systems recommended.

Protective Gloves	Eye Protection	Other Protective Clothing
Butyl Rubber Gloves	Chemical Goggles or Face Mask	Protective Apron

Work/Hygiene Practices

Always wash hands thoroughly after using chemical; never bring food, drink, or smoking materials into vicinity of chemicals.

ETHYL TRICHLOROSILANE

$C_2H_5SiCl_3$ CAS: 115-21-9

IDENTIFICATION AND TYPICAL USES

Ethyl trichlorosilane is a clear, colorless liquid with an acrid odor. It is used primarily as an intermediate for silicones.

RISK ASSESSMENT: HEALTH

General Assessment

Ethyl trichlorosilane is a highly toxic and extremely dangerous corrosive compound. It is destructive to skin tissue on contact and can pose a serious health hazard risk through *inhalation* and *ingestion*. Its carcinogenic, mutagenic, and teratogenic potential has not been adequately studied. However, because of the extremely corrosive and tissue-destroying capability of ethyl trichlorosilane, extreme caution is warranted when handling or using this material.

Inhalation causes moderate to severe irritation of the eyes, nose, and throat. It causes serious dypsnea (difficulty in breathing) with swelling of the larynx. Vapor exposure may lead to serious eye and lung injury. It is also possible that exposure will cause red blood cell destruction. Because of their extremely corrosive nature, vapors are not tolerable even at low concentrations.

Skin contact with vapors or liquids can cause severe irritation and even 2nd and 3rd degree burns. Eye contact causes irritation, pain, tearing, and corneal damage and scars.

Ingestion poses serious, even life-threatening risks. There may be corrosion of the mucous membranes in the mouth, throat, and esophagus. There will be abdominal pain, difficulty in swallowing, nausea and vomiting, intense thirst, shock with cold skin, tachycardia (rapid pulse), shallow respiration, and possible circulatory system collapse, and death. Prior to death, circulatory collapse can cause kidney failure and the development of lesions on the liver and heart.

Acute Health Effects

The following acute (short-term) health effects may occur immediately or shortly after exposure to ethyl trichlorosilane:

Skin: Severe irritation with deep painful burns, redness, blistering, and rash.

Eye: Severe irritation with redness and burns. Corneal damage is likely on short contact.

Lung: Irritation of the respiratory tract and lung tissues resulting in possible injury.

Other: Acute overexposure, especially by ingestion, may cause rapid pulse, blood pressure changes, and circulatory system collapse.

Chronic Health Effects

The following chronic (long-term) health effects may occur at some time after exposure to ethyl trichlorosilane and can last for months or even years:

Cancer Hazards: According to the information presented in the references, ethyl trichlorosilane has not been adequately tested for its ability to cause cancer in test animals.

Reproductive Hazard: According to the information presented in the references, ethyl trichlorosilane has not been adequately tested for its ability to adversely affect reproduction in test animals.

Recommended Risk-Reduction Measures

Unless a less toxic chemical can be substituted for a hazardous substance, *engineering controls* are the most effective method of reducing exposures. The best protection is to enclose operations and/or provide local exhaust ventilation at the site of chemical release. To prevent static sparks, all containers and equipment should be electrically grounded and bonded. While not

always operationally feasible, isolating operations can also reduce exposure.

Using respiratory protection is less effective than the controls mentioned above, but is still advisable whenever working with or around ethyl trichlorosilane. No exposure levels have been established for this chemical. However, this does not mean that exposure to ethyl trichlorosilane is without risk. For the best possible protection, a supplied-air respirator or a self-contained breathing apparatus (SCBA) with full facepiece operated in pressure demand mode are the recommended respiratory protection methods of choice. If a full facepiece is not available, then chemical goggles should be worn to protect the eyes. Whenever a chemical splash hazard exists, a face shield and apron should be worn. To prevent hand and skin exposures, impervious butyl rubber gloves should be worn. Glove manufacturers should be contacted and permeation studies obtained *before* glove selections are made.

Administrative controls should also be in place to minimize the potential for human exposures. These may include written procedures or policies which specify the methods and techniques that will be practiced whenever personnel are to work with or around ethyl trichlorosilane.

All personnel should receive *training* on the use, hazards, protective measures, emergency actions, and other precautions per 29 CFR 1910.1200 (Hazard Communication), prior to the first assignment in an area where ethyl trichlorosilane is used or stored.

If symptoms develop or overexposure is suspected, the following medical tests may be useful:

- ☑ Complete blood count.
- ☑ Lung function tests.
- ☑ Consider chest X-ray after acute overexposure.
- ☑ Liver and kidney function tests.

Any evaluation should include a careful medical history of past and present symptoms with an examination. However, medical tests that evaluate existing damage are not a substitute for controlling exposure. Also, since smoking can cause heart diseases, emphysema, numerous other respiratory disorders, and interfere with the blood's oxygen-carrying capabilities, smokers may be affected more rapidly than non-smokers to the same exposure conditions. Prudent risk assessment and management requires careful consideration of *all* possible risk factors that may be causing the appearance of symptoms in exposed workers.

Other methods to reduce exposure include:

- ☑ Where possible, enclose operations and use local exhaust ventilation at the site of chemical release. If local exhaust ventilation or enclosure is not used, respiratory protection should be mandatory. Also, if possible, automatically transfer liquids from storage containers to process containers using non-sparking tools.
- ☑ Always ensure that proper protective clothing is worn when using chemical substances.
- ☑ Wash thoroughly immediately after exposure to ethyl trichlorosilane and at the end of the work shift or before eating, drinking, or smoking.
- ☑ Hazard warning information should be posted in the work area. In addition, as part of an on-going education and training program, all information on the health and safety hazards of ethyl trichlorosilane should be communicated to all potentially exposed workers.
- ☑ Work clothing contaminated with ethyl trichlorosilane should never be worn home where family members can be exposed. Also, clothing soaked with ethyl trichlorosilane presents a serious fire hazard. Contaminated clothing should be laundered by personnel who have been briefed on the hazards of exposure to ethyl trichlorosilane.
- ☑ Eye wash stations should be provided in work areas. If the potential for whole body exposure exists, then safety showers should also be provided.

RISK ASSESSMENT: ENVIRONMENT

General Assessment

The environment is at risk of exposure during intended use, transportation, storage, disposal, or destruction of ethyl trichlorosilane. In almost every scenario, the threat of environmental exposure is contingent upon the proper handling of the chemical substance. Accidental spills, large or small, can result in fire, explosion, and possible contamination of the surrounding environmental mediums (water, soil, and air).

Ethyl trichlorosilane is considered a class IB flammable liquid (per OSHA 29 CFR 1910.106). Its extremely low flash point and relatively low boiling point and autoignition temperature make it a serious and dangerous fire and explosion risk. Its vapors are heavier than air and can travel for some distance to an ignition source to flashback and cause a fire or explosion. It can react with many oxidizers to cause a fire or explosion. It will react with water to form hydrochloric acid. These characteristics require special consideration during any emergency situation involving a leak or spill of ethyl trichlorosilane. Should this chemical ever come into contact with incompatible substances either during use, transportation, or storage, violent reactions can occur.

Ethyl trichlorosilane can enter the environment through industrial discharges, municipal waste treatment discharges, and spills.

Acute Ecological Effects

Acute (short-term) toxic effects may include the death of animals, birds, or fish and death or low growth rate in plants. Acute effects are seen 2 to 4 days after animals or plants are exposed to ethyl trichlorosilane.

Insufficient data are available to evaluate or predict the short-term effects of ethyl trichlorosilane to aquatic life, plants, birds, or land animals.

Chronic Ecological Effects

Chronic toxic effects may include shortened life span, reproductive problems, lower fertility, and changes in appearance or behavior in exposed animals. These effects can be seen long after first exposure(s) to toxic chemicals.

Insufficient data are available to evaluate or predict the long-term effects of ethyl trichlorosilane to aquatic life, plants, birds, or land animals.

Water Solubility

Ethyl trichlorosilane decomposes in water to release hydrogen chloride gas.

Persistence in the Environment

Insufficient data are available to evaluate or predict the ability of ethyl trichlorosilane to persist in the environmental mediums.

Bioaccumulation in Aquatic Organisms

Some substances increase in concentration, or *bioaccumulate,* in living organisms as they breathe contaminated air, drink contaminated water, or eat contaminated food. These chemicals can become concentrated in the tissues and internal organs of animals as well as humans.

There is no data available on the bioaccumulation characteristics of ethyl trichlorosilane.

Recommended Risk-Reduction Measures

Proper training of all transporters will reduce the likelihood of a mishap or accident resulting in a leak or spill to the environment. The correct use of DOT labeling (per the MSDS) on all containers, trucks, and rail cars will enable emergency responders to react properly and quickly to any disaster thereby reducing the potential risk to the environment and to personnel.

Storage of ethyl trichlorosilane should be segregated from incompatible chemicals to minimize the risk of cross-contamination or contact. Storage buildings should be equipped with the proper fire protection equipment (alarms, sprinklers, extinguishers, etc.). Explosion proof design is highly recommended for all storage facility and operational equipment.

If a spill or leak to the environment has occurred, fire department, emergency response and/or hazardous materials spill personnel should be notified immediately. Cleanup should be attempted only by those trained in proper spill containment procedures using non-sparking tools. Contaminated soils should be removed for incineration and replaced with clean soil. If ethyl trichlorosilane should contact the water table, aquifer, or navigable waterway, remediation activities should be prompt. It decomposes to hydrogen chloride gas and total containment may not be possible. When such spills occur, the local and/or state emergency response authorities must be notified. A comprehensive emergency response or disaster preparedness/recovery plan should be in place prior to any operations involving the use, transportation, storage, or disposal of ethyl trichlorosilane. If ethyl trichlorosilane is spilled or leaked, the following specific steps are recommended:

- ☑ Restrict persons not wearing protective clothing from area of spill or leak until cleanup is complete and area can be opened for normal work.
- ☑ Ventilate area of spill or leak.
- ☑ Remove all ignition sources.
- ☑ Stop leak if it can be done without risk. Use water spray to disperse vapors but do not allow water to directly contact spilled materials. Neutralize with crushed limestone, soda ash, or lime. Absorb small spills with vermiculite and deposit in sealed drums for disposal.
- ☑ It may be necessary to dispose of ethyl trichlorosilane as a hazardous waste. The responsible state agency or the regional office of the federal Environmental Protection Agency (EPA) should be contacted for specific recommendations.

RISK ASSESSMENT: BUSINESS

General Assessment

Accidents or mishaps involving ethyl trichlorosilane can present a significant threat to business operations. The loss or damage of equipment or facilities can significantly affect fiscal viability. Lawsuits that may result from personnel injury/death, public exposures, and/or environmental contamination will also require a serious expenditure of resources. Media attention surrounding an injury, death, or environmental damage can also result in a loss of profits and loss of current as well as future business.

ʊ Recommended Risk-Reduction Measures

Company attorneys, safety and health professionals, and environmental specialists should be involved in the development of any procedures dealing with safety or emergency response. A company official should be pre-designated as a public relations officer with specific training in dealing with the press. Corporate plans and policies should be developed, approved, and implemented long before any need for such arises.

REFERENCES

Ahlbom, A. 1993. *Biostatistics for Epidemiologists.* New York: CRC Press/Lewis Publishers.

American Conference of Governmental Industrial Hygienists. 1988. *Documentation of the Threshold Limit Values and Biological Exposure Indices,* 5th Edition (with updates). Cincinnati: ACGIH

Anthony, C. P., and N. J. Kolthoff. 1971. *Textbook of Anatomy and Physiology.* St. Louis: C. V. Mosby Company.

Asante-Duah, D. K. 1993. *Hazardous Waste Risk Assessment.* New York: CRC Press/Lewis Publishers.

Bloom, A. D. and F. J. DeSerres. 1995. *Ecotoxicity and Human Health.* New York: CRC Press/Lewis Publishers.

Brusick, D. J. 1994. *Methods for Genetic Risk Assessment.* New York: CRC Press/Lewis Publishers.

Bueche, J. R. 1972. *Principles of Physics,* 2nd Edition. New York: McGraw-Hill.

Burrell, R., D. K. Flaherty, and L. J. Sauers. 1992. *Toxicology of the Immune System.* New York: Van Nostrand Reinhold.

Calabrese, E. J. 1994. *Biological Effects of Low Level Exposures.* New York: CRC Press/Lewis Publishers.

Cockerman, L. G., B. S. Shane. 1994. *Basic Environmental Toxicology.* New York: CRC Press/Lewis Publishers.

Cothern, C. R., N. P. Ross. 1994. *Environmental Statistics, Assessment, and Forecasting.* New York: CRC Press/Lewis Publishers.

Danse, I. R. 1991. *Common Sense Toxics in the Workplace.* New York: Van Nostrand Reinhold.

Environmental Protection Agency. 1983. *Chemical Hazard Information Profiles.* Washington, D.C.: U.S. EPA.

Fodden, C. R., and J. L. Weddell. 1992. *Hazardous Materials Emergency Action Data.* New York: CRC Press/Lewis Publishers.

Forseberg, K., and L. H. Keith. 1993. *Chemical; Protective Clothing Permeation and Degradation Database.* Chelsea, MI: Lewis Publishers.

Gardner, R. W. 1994. *Chemical Intolerance.* New York: CRC Press/Lewis Publishers.

Gerarde, H. W. 1960. *Toxicology and Biochemistry of Aromatic Hydrocarbons.* New York: Elsevier.

Hallenbeck, W. H. 1993. *Quantitative Risk Assessment for Environmental and Occupational Health,* 2nd Edition. New York: CRC Press/Lewis Publishers.

Hodgson, E., R. B. Mailman, and Janice E. Chambers. 1988. *Dictionary of Toxicology.* London: The MacMillian Company.

Howard, P. H. 1995. *Handbook of Environmental Fate and Exposures Data for Organic Chemicals,* Vol. I-IV. New York: CRC Press/Lewis Publishers.

Hosty, J. W., and P. Foster. 1990. *A Practical Guide to Chemical Spill Response.* New York: Van Nostrand Reinhold.

Kamrin, M. A. 1988. *Toxicology: A Primer on Principles and Applications.* Chelsea, Michigan: Lewis Publishers.

Lewis, R. J., Sr. 1993. *Hawley's Condensed Chemical Dictionary,* 12th Edition. New York: Van Nostrand Reinhold.

Lewis, R. J., Sr. 1992. *Sax's Dangerous Properties of Industrial Materials,* Volumes 1, 2, and 3, 8th Edition. New York: Van Nostrand Reinhold.

Mackay, D., W. Y. Shiu, and K. C. Ma. 1992. *Illustrated Handbook of Chemical-Physical Properties and Environmental Fate for Organic Chemicals.* New York: CRC Press/Lewis Publishers.

Manahan, S. E. 1992. *Toxicological Chemistry,* 2nd Edition. New York: CRC Press/Lewis Publishers.

Mellström, G. A., J. E. Wahlberg, and H. I. Maibach. 1994. *Protective Gloves for Occupational Use.* New York: CRC Press/Lewis Publishers.

Meyer, E. 1990. *Chemistry of Hazardous Materials.* Englewood Cliffs, NJ: Prentice Hall.

Miller, B. F., M.D., and C. B. Keane, R.N. 1972. *Encyclopedia and Dictionary of Medicine and Nursing.* Philadelphia: W.B. Sanders Company.

Mycock, J. C., J. D. McKenna, and L. Theodore. 1995. *Handbook of Air Pollution Control Engineering and Technology.* New York: CRC Press/Lewis Publishers.

Nelson, G. O. 1992. *Gas Mixtures.* New York: CRC Press/Lewis Publishers.

Ott, W. R. 1995. *Environmental Statistics and Data Analysis.* New York: CRC Press/Lewis Publishers.

Patnaik, P. 1992. *A Comprehensive Guide to the Hazardous Properties of Chemical Substances.* New York: Van Nostrand Reinhold.

Richardson, M. 1995. *Dictionary of Substance and Their Effects,* Vol. I-VI. New York: CRC Press/Lewis Publishers.

MATERIAL SAFETY DATA SHEET

CHEMICAL NAME	
FENTHION	**CRC PRESS/LEWIS PUBLISHERS DISCLAIMER:** THE INFORMATION AND RECOMMENDATIONS PRESENTED HEREIN ARE BASED ON SOURCES BELIEVED TO BE RELIABLE. CRC MAKES NO REPRESENTATION ON ITS COMPLETENESS OR ACCURACY. IT IS THE RESPONSIBILITY OF THE USER TO DETERMINE THE CHEMICAL'S SUITABILITY FOR ITS INTENDED USE, THE CHEMICAL'S SAFE USE, AND THE CHEMICAL'S PROPER DISPOSAL. NO REPRESENTATIONS AND/OR WARRANTIES, EITHER EXPRESSED OR IMPLIED, OF THE MERCHANTABILITY OR FITNESS FOR A PARTICULAR PURPOSE, OR OF ANY OTHER NATURE, ARE MADE WITH RESPECT TO THE INFORMATION PROVIDED IN THIS MSDS OR TO THE CHEMICAL TO WHICH INFORMATION MAY REFER. CRC NEITHER ASSUMES NOR AUTHORIZES ANY OTHER PERSON TO ASSUME FOR IT, ANY OTHER ADDITIONAL RESPONSIBILITY OR LIABILITY FOR THE USE OF, OR RELIANCE UPON, THIS INFORMATION.

HAZARD WARNING INFORMATION

HEALTH	FIRE	REACTIVE	OTHER	DEGREE OF HAZARD	COLOR CODING	OTHER CODES
2	0	0		0 = Minimum Hazard 1 = Slight Hazard 2 = Moderate Hazard 3 = Serious Hazard 4 = Severe Hazard	HEALTH = BLUE FIRE = RED REACTIVITY = YELLOW OTHER = WHITE	OX = Oxidizer ACID = Acid ALK = Alkali COR = Corrosive W = Use No Water

SECTION I - GENERAL INFORMATION

Characterization	**Pesticide, Insecticide**	RCRA Number	**Not Listed**
		EPA Class	**Not Applicable**
DOT Proper Shipping Name	**Pesticides, Liquid, Toxic, n.o.s.**	Chemical Abstract Service (CAS) Number	**55-38-9**
DOT Hazard Class and Label Requirements	**Keep Away From Food**	DOT Emergency Guide Code	**No Citation**
DOT Identification Number	**UN 3018**	Chemical Formula	$C_{10}H_{15}O_3PS_2$

Synonyms

Baytex®; Entex®; O,O-dimethyl O-3-methyl-4-methylthiophenyl phosphorothioate; mercaptophos.

SECTION II - HAZARDOUS INGREDIENTS AND IDENTITY INFORMATION

Hazardous Components (specific identity)	OSHA Exposure Criteria	NIOSH Exposure Criteria	Immediately Dangerous to Life and Health (IDLH)	ACGIH Exposure Criteria
Fenthion (derivation: By condensation of 4-methylmercapto-*m*-cresol and *o,o*-dimethyl phosphorochloridothioate). **1 ppm = 11.57 mg/m³**	**PEL:** Not Established **STEL:** Not Established	**REL:** Not Established **Possible Cancer Agent**	**Not Determined**	**TLV (skin):** 0.02 ppm 0.2 mg/m³ **STEL:** Not Established

SECTION III - PHYSICAL AND CHEMICAL CHARACTERISTICS

Boiling Point	**189°F (87°C) at 0.01 mmHg**	Specific Gravity (H_2O = 1)	**1.25**
Vapor Pressure (mm Hg)	**3 x 10^{-5} at 69°F (20°C)**	Molecular Weight (atomic weight)	**278.3**
Vapor Density (Air = 1)	**Not Found**	Melting Point	**43°F (6°C)**

Solubility

Practically insoluble in water (0.006% at 77°F). Soluble in acetone, glyceride, ethanol, methanol, ether, petroleum ether, many organic solvents, and especially the chlorinated hydrocarbons.

Appearance and Odor

A colorless to brown or tan oily liquid with a slight garlic-like odor.

SECTION IV - FIRE AND EXPLOSION HAZARD DATA

Flash Point (method used)	**Non-Combustible**	Explosive Limits in Air % by Volume	**LEL: N/A UEL: N/A**
NFPA Classification	**Non-Combustible Liquid**	Autoignition Temperature	**Not Found**

Extinguishing Media

Use agent suitable to surrounding fire. Fenthion itself will not burn.

Special Fire Fighting Procedures

Poisonous gases are produced in fire. Wear full protective clothing and self-contained breathing apparatus (SCBA). Structural fire-fighting equipment is permeable and may not provide sufficient protection. Move containers from fire area if it can be done without risk. Cool exposed containers with water spray.

Unusual Fire and Explosion Hazards

None reported.

SECTION V - REACTIVITY DATA

Stability		Conditions to Avoid Fenthion is thermally stable up to 410°F (210°C) and is resistant to light. Avoid contact with incompatible materials, and exposure to heat.
Stable X	Unstable	Incompatibility *(materials to avoid)* Incompatible with oxidizers, and alkalis. Will react when exposed to excessive heat.
Hazardous Polymerization		Conditions to Avoid Under normal conditions of temperature and pressure, hazardous polymerization of fenthion is not expected to occur.
May Occur	Will Not Occur X	Hazardous Decomposition or By-products At high temperatures, fenthion decomposes and may produce poisonous gases, including oxides of sulfur and phosphorus.

SECTION VI - HEALTH HAZARD DATA

Primary Route(s) of Entry:	Inhalation?	Absorption (skin/eye)?	Ingestion?
	X	X	X

Health Hazards

INHALATION: Symptoms of exposure include nausea and vomiting, abdominal pain, diarrhea, dizziness, fatigue, blurred vision, lachrymation (tearing), twitching and loss of muscle coordination, speech disturbances (slurring), confusion, labored breathing, convulsions, coma, and death from respiratory paralysis. There may also be tightness in chest, runny nose, and the development of pulmonary edema (fluid in the lungs).

ABSORPTION: Will pass through skin and cause many of the same symptoms as inhalation.

INGESTION: Causes the same systemic effects as inhalation.

Carcinogenicity	NTP Listed?	IARC Cancer Review Group?	OSHA Regulated?	Target Organs?
Unknown Human Questioned Animal	No	Group 3	29 CFR 1910.1000 Table Z-1-A	Respiratory system, CNS, CVS.

Medical Conditions Generally Aggravated by Exposure

None reported. However, persons with existing lung dysfunction may experience effects quicker.

Emergency and First-aid Procedures

Eye contact: Do not allow victim to keep eyes tightly shut. Flush immediately with water for 15 minutes (minimum), seek medical attention. Skin contact: Remove all contaminated clothing. Immediately wash area with flooding amounts of soap and water. For inhalation: Remove the person from exposure. Maintain an open airway. Provide respiratory assistance and CPR. Transfer to medical facility. If swallowed: Call poison control center for advice. Unless advised otherwise, give 1-2 glasses of water and induce vomiting or perform gastric lavage. Never give anything by mouth to an unconscious or convulsing person. DOCTOR: The primary antidote is atropine.

SECTION VII - PRECAUTIONS FOR SAFE HANDLING AND USE

Steps to be Taken in Case Material is Released or Spilled

Restrict those not involved in cleanup from entering area. Cleanup crew must avoid exposure by using appropriate personal protective equipment. Absorb any liquids in vermiculite or absorbent and deposit in sealed drum. Damp mop residue with a dilute alkaline solution. Ventilate area of spill.

Preferred Waste Disposal Method

Hydrolyze with excess calcium oxide or sodium hydroxide; or, pack in 17H epoxy-lined drum and dispose of at EPA waste site; or, burn in a chemical incinerator equipped with an afterburner and scrubber.

Precautions to be Taken in Handling and Storage

Store in tightly closed containers free from moisture. Keep in cool (below 86°F/30°C), well-ventilated area.

Other Precautions and Warnings

Keep away from heat, strong alkalis, or oxidizers. Use numerous small containers rather than large ones.

SECTION VIII - CONTROL MEASURES AND PERSONAL PROTECTIVE EQUIPMENT

Respiratory Protection *(specify type)*

PEL is low (0.2 mg/m^3) and difficult to monitor. Use a respirator with a pesticide cartridge. For best protection, use a supplied-air respirator or an SCBA with full facepiece operated in positive pressure mode.

Ventilation

Local exhaust (preferred) at site of chemical work or general ventilation.

Protective Gloves	Eye Protection	Other Protective Clothing
Polyvinylchloride Material	Chemical Goggles or Face Mask	Protective Apron, Boots (avoid skin contact)

Work/Hygiene Practices

Always wash hands thoroughly after using chemical; never bring food, drink, or smoking materials into vicinity of chemicals.

FENTHION

$C_{10}H_{15}O_3PS_2$ CAS: 55-38-9

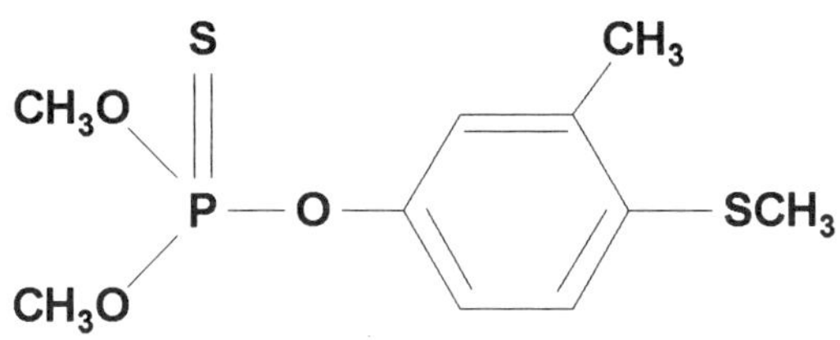

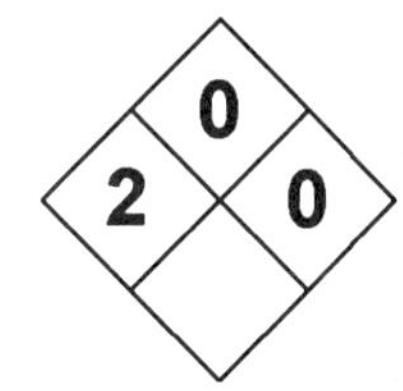

IDENTIFICATION AND TYPICAL USES

Fenthion is a colorless to tan or brown liquid with a slight, garlic-like odor. It is used as an acaricide (for mite control) and an insecticide primarily for use in controlling mosquitoes. It is also used as a chemical intermediate for pharmaceuticals.

RISK ASSESSMENT: HEALTH

General Assessment

Fenthion is a poison by *ingestion, inhalation,* and skin contact (*absorption*). It is moderately toxic to humans through all routes of exposure. Its carcinogenicity and teratogenicity have been tested and the results are inconclusive. Human mutagenic and tumorigenic data have been reported. It is known to inhibit the development of cholinesterase (an important enzyme for proper nerve function in muscle contraction and blood vessel dilation).

Symptoms from *all* routes of exposure appear to focus on action to the central nervous system. These include nausea and vomiting with abdominal cramps and pain, diarrhea, dizziness, fatigue, blurred vision, miosis, tearing (lachrymation), muscular twitching and loss of coordination, slurring of speech, confusion, salivation, headache, giddiness, vertigo, weakness, rhinitis, labored breathing (dypsnea), convulsions, coma, and death, usually from respiratory paralysis.

In addition, inhalation has been reported to cause tightness in chest, runny nose, and pulmonary edema (fluid in the lungs). The development of pulmonary edema can be delayed up to 48 hours creating a false sense of security with regard to health exposure risk.

Ingestion has been shown to cause additional symptoms of hypermotility, rapid pulse (tachycardia), vomiting, and diarrhea.

Acute Health Effects

The following acute (short-term) health effects may occur immediately or shortly after exposure to fenthion:

Skin: No reported effects.

Eye: Possible local irritation.

Lung: Irritation of the nose and throat. High exposures may cause a delayed buildup of pulmonary edema (fluid in the lungs), which is a medical emergency and can be fatal.

CNS: Numerous effects resulting from exposure including dizziness, giddiness, blurred vision, fatigue, muscular twitching, confusion, vertigo, weakness, convulsions, and coma. Death may occur from respiratory paralysis.

Other: A cholinesterase inhibitor causing interference with nerve function and other important systemic operations.

Chronic Health Effects

The following chronic (long-term) health effects may occur at some time after exposure to fenthion and can last for months or even years:

Cancer Hazards: Fenthion may be a mutagen (causes genetic changes). Such chemicals may have a cancer risk and there is limited evidence that fenthion does cause cancer in some test animals. Human carcinogenicity is unknown at this time.

Reproductive Hazard: There is some evidence that exposure to fenthion causes damage to the developing fetus in test animals. Human effects are unknown.

Other Chronic Effects: Fenthion may cause chronic cholinesterase depression. However, the data are inconclusive in this regard. Further study is required.

Recommended Risk-Reduction Measures

Personnel should avoid direct contact with fenthion. If a less toxic material or compound cannot be substituted for fenthion, then *engineering controls* are the most effective method of reducing exposures. While not feasible for outdoor operations, the best protection in manufacturing and production is to enclose operations and/or provide local exhaust ventilation at the site of fenthion release. While not always operationally possible, isolating operations involving fenthion manufacturing can also reduce exposure. Using respiratory protection is less effective than the controls mentioned above, but is still advisable whenever working with or around fenthion. For low or infrequent exposures, an MSHA/NIOSH-approved respirator equipped with a pesticide cartridge may be adequate. However, the permissible exposure level (0.2 mg/m^3) is extremely low and difficult to monitor. The best protection is obtained using a supplied-air respirator or a self-contained breathing apparatus (SCBA) with full facepiece operated in pressure demand mode. If a full facepiece is not available, then chemical/dust goggles should be worn to protect the eyes. A face shield and protective apron should also be worn. To prevent hand and skin exposures, polyvinylchloride gloves should be worn.

Administrative controls should also be in place to minimize the potential for human exposures. These may include written procedures or policies which specify the methods and techniques that will be practiced whenever personnel are to work with fenthion.

All personnel should receive *training* on the use, hazards, protective measures, emergency actions, and other precautions per 29 CFR 1910.1200 (Hazard Communication), prior to the first assignment in an area where fenthion is used or stored.

Before beginning employment and regular intervals thereafter (e.g., annually), the following medical tests are recommended:

☑ Lung function tests (establish baseline).

☑ Determine baseline cholinesterase levels.

If symptoms develop or overexposure is suspected, the following medical testing may be useful:

☑ Complete neurological evaluation.

☑ Lung function tests (compare to baseline).

☑ Cholinesterase level test. Personnel showing a 25% decrease should be prevented from using fenthion.

☑ Consider chest X-ray after acute over exposure (may be negative if taken immediately after exposure due to delayed development of pulmonary edema).

☑ If visual disturbances are presented, recommend using a 0.15 to 1% solution of atropine sulfate distilled in the eye.

☑ Cholinesterase reactivation agents, such as Toxogonin, may be effective treatment.

Any evaluation should include a careful history of past and present symptoms with an examination. Medical tests that simply look for existing damage are not a substitute for controlling exposures. Also, since smoking can cause heart disease, emphysema, and a number of other respiratory disorders, smokers may be effected more quickly and with more pronounced symptoms than non-smokers under the same conditions of exposure. Prudent risk management requires careful consideration of *all* factors which may be causing the development of exposure symptoms in the work force.

Other methods to reduce exposure include:

☑ Where possible, enclose operations and use local exhaust ventilation at the site of chemical release. If local exhaust ventilation or enclosure is not used, respiratory protection should be mandatory.

☑ Always ensure that proper protective clothing is worn when using chemical substances.

☑ Wash thoroughly immediately after exposure to fenthion and at the end of the work shift or before eating, drinking, or smoking.

☑ Hazard warning information should be posted in the work area. When taken for use outdoors, proper labeling and warning information should be sent with it. In addition, as part of an on-going education and training program, all information on the health and safety hazards of fenthion should be communicated to all potentially exposed workers.

☑ Eye wash stations should be provided in the immediate work area for emergency use. If there is a possibility of skin exposure to fenthion, emergency shower facilities should also be provided.

☑ Workers whose clothing has been contaminated by fenthion should change into clean clothes before leaving work. Contaminated work clothing should be laundered only by individuals who have been informed of the hazards of exposure to fenthion.

RISK ASSESSMENT: ENVIRONMENT

General Assessment

Because of its use as an insecticide and pesticide, fenthion can be present in the environment at any given time and at a variety of levels. The environment is also at risk of exposure during transportation, storage, disposal, or destruction of fenthion. In almost every scenario, the threat of environmental exposure is contingent upon the proper handling of the chemical substance. Accidental spills, large or small, where fenthion contacts incompatible commodities can result in the possible contamination of the surrounding environmental mediums (water, soil, and air).

Fenthion is considered a non-combustible solid. However, it is incompatible with alkalis and oxidizers and contact can cause violent reactions. Also, if involved in a fire, it will emit toxic oxides of sulfur and phosphorus. These characteristics require special consideration during any emergency situation involving a leak or spill of fenthion.

Fenthion can enter the environment from industrial and municipal discharges, through agricultural run-off, and from spills. Also, as an insecticide and pesticide, it will also be present in the environment as a result of its intended use.

Acute Ecological Effects

Acute (short-term) toxic effects may include the death of animals, birds, or fish and death or low growth rate in plants. Acute effects are seen 2 to 4 days after animals or plants are exposed to fenthion. Fenthion has moderate acute toxicity to aquatic life and birds. No data are available on the short-term effects of fenthion to plants and land animals.

Chronic Ecological Effects

Chronic (long-term) toxic effects may include shortened life span, reproductive problems, lower fertility, and changes in appearance or behavior in exposed animals. These effects can be seen long after first exposure(s) to toxic chemicals. Fenthion has moderate chronic toxicity to aquatic life. No data are available to evaluate or predict the long-term effects of fenthion to plants, birds, or land animals.

Water Solubility

Fenthion is very slightly soluble (practically insoluble) in water. Concentrations of less than 1 milligram will mix with a liter of water.

Persistence in the Environment

Fenthion is highly persistent in water, with a half-life of greater than 200 days. The half-life of a pollutant is the amount of time it takes for one-half of the chemical to be degraded. About 97% of fenthion will eventually end up in water; approximately 1.5% will end up in terrestrial soil and aquatic sediments. Hydrolysis is quicker in warm water containing large quantities of microorganisms.

Bioaccumulation in Aquatic Organisms

Some substances increase in concentration, or *bioaccumulate,* in living organisms as they breathe contaminated air, drink contaminated water, or eat contaminated food. These chemicals can become concentrated in the tissues and internal organs of animals as well as humans.

The concentration of fenthion found in fish tissues is expected to be somewhat higher that the average concentration of fenthion in the water from which the fish was taken.

Recommended Risk-Reduction Measures

Proper training of all transporters will reduce the likelihood of a mishap or accident resulting in a leak or spill of fenthion dusts or powders into the environment. The correct use of DOT labeling (per the MSDS) on all containers, trucks, and rail cars will enable emergency responders to react properly and quickly to any disaster thereby reducing the potential risk to the environment and to personnel.

Storage of fenthion should be segregated from incompatible materials to minimize the risk of cross-contamination or contact. Proper ventilation and protective equipment should be used while handing or using fenthion.

If a spill or leak to the environment has occurred, fire department, emergency response and/or hazardous materials spill personnel should be notified immediately. Cleanup should be attempted only by those trained in proper spill containment procedures. If fenthion should contact the water table, aquifer, or navigable waterway, reclamation procedures should be initiated as soon as practical. It is only slightly soluble in water and total remediation may be possible. The local and/or state emergency response authorities must be notified. A comprehensive emergency response or disaster preparedness/recovery plan should be in place prior to any operations involving the use, transportation, storage, or disposal of fenthion. If fenthion powder is spilled or leaked, the following specific steps are recommended:

☑ Restrict persons not wearing protective clothing from area of spill or leak until cleanup is complete and area can be opened for normal work.

☑ Absorb liquids in vermiculite or other material and place in sealed drum for disposal or reclamation. Damp mop any residue with a dilute alkaline solution.

☑ It may be necessary to dispose of fenthion as a hazardous waste. The responsible state agency or the regional office of the federal Environmental Protection Agency (EPA) should be contacted for specific recommendations.

RISK ASSESSMENT: BUSINESS

General Assessment

Accidents or mishaps involving fenthion can present a significant threat to business operations. The loss or damage of equipment or facilities can significantly affect fiscal viability. Lawsuits that may result from personnel injury/death, public exposures, and/or environmental contamination will also require a serious expenditure of resources. Media attention surrounding an injury, death, or environmental damage can also result in a loss of profits and loss of current as well as future business.

Recommended Risk-Reduction Measures

Company attorneys, safety and health professionals, and environmental specialists should be involved in the development of any procedures for responding to chemical incidents. A company official should be predesignated as a public relations officer with specific training in dealing with the press. Corporate plans and policies should be developed, approved, and implemented long before any need for such arises.

REFERENCES

Ahlbom, A. 1993. *Biostatistics for Epidemiologists.* New York: CRC Press/Lewis Publishers.

American Conference of Governmental Industrial Hygienists. 1988. *Documentation of the Threshold Limit Values and Biological Exposure Indices,* 5th Edition (with updates). Cincinnati: ACGIH

Brusick, D. J. 1994. *Methods for Genetic Risk Assessment.* New York: CRC Press/Lewis Publishers.

Bueche, J. R. 1972. *Principles of Physics,* 2nd Edition. New York: McGraw-Hill.

Calabrese, E. J. 1994. *Biological Effects of Low Level Exposures.* New York: CRC Press/Lewis Publishers.

Cockerman, L. G., B. S. Shane. 1994. *Basic Environmental Toxicology.* New York: CRC Press/Lewis Publishers.

Cothern, C. R., N. P. Ross. 1994. *Environmental Statistics, Assessment, and Forecasting.* New York: CRC Press/Lewis Publishers.

Gardner, R. W. 1994. *Chemical Intolerance.* New York: CRC Press/Lewis Publishers.

Howard, P. H., et. al. 1991. *Handbook of Environmental Degradation Rates.* New York: CRC Press/Lewis Publishers.

Jones, F. E. 1994. *Toxic Organic Vapors in the Workplace.* New York: CRC Press/Lewis Publishers.

Keith, L. H. and M. M. Walker. 1995. *Handbook of Air Toxics.* New York: CRC Press/Lewis Publishers.

Landis, W. G., M. H. Yu. 1995. *Introduction to Environmental Toxicology.* New York: CRC Press/Lewis Publishers.

Lauwerys, R. R., P. Hoet. 1993. *Industrial Chemical Exposure.* New York: CRC Press/Lewis Publishers.

Lewis, R. J., Sr. 1992. *Sax's Dangerous Properties of Industrial Materials,* Volumes 1, 2, and 3, 8th Edition. New York: Van Nostrand Reinhold.

Manahan, S. E. 1992. *Toxicological Chemistry,* 2nd Edition. New York: CRC Press/Lewis Publishers.

Meyer, E. 1990. *Chemistry of Hazardous Materials.* Englewood Cliffs, NJ: Prentice Hall.

Nelson, G. O. 1992. *Gas Mixtures.* New York: CRC Press/Lewis Publishers.

Ott, W. R. 1995. *Environmental Statistics and Data Analysis.* New York: CRC Press/Lewis Publishers.

Patnaik, P. 1992. *A Comprehensive Guide to the Hazardous Properties of Chemical Substances.* New York: Van Nostrand Reinhold.

Quigley, D. R. 1994. *Handbook of Emergency Chemical Management.* New York: CRC Press/Lewis Publishers.

Rea, W. J. 1995. *Chemical Sensitivity,* Vol. I-III. New York: CRC Press/Lewis Publishers.

Reeves, A. L. 1979. Fenthion. *Handbook on the Toxicology of Metals.* pp 321-328. New York: Elsevier North-Holland.

Richardson, M. 1995. *Dictionary of Substance and Their Effects,* Vol. I-VI. New York: CRC Press/Lewis Publishers.

Sacarello, H. L. A. 1994. *The Comprehensive Handbook of Hazardous Materials.* New York: CRC Press/Lewis Publishers.

Talbott, E., G. Craun. 1995. *An Introduction to Environmental Epidemiology.* New York: CRC Press/Lewis Publishers.

Torkelson, T. R. V. K. Rowe. 1981. *Patty's Industrial Hygiene and Toxicology,* Vol. 2B. New York: Wiley-Interscience.

MATERIAL SAFETY DATA SHEET

CHEMICAL NAME	
FERBAM	**CRC PRESS/LEWIS PUBLISHERS DISCLAIMER:** THE INFORMATION AND RECOMMENDATIONS PRESENTED HEREIN ARE BASED ON SOURCES BELIEVED TO BE RELIABLE. CRC MAKES NO REPRESENTATION ON ITS COMPLETENESS OR ACCURACY. IT IS THE RESPONSIBILITY OF THE USER TO DETERMINE THE CHEMICAL'S SUITABILITY FOR ITS INTENDED USE, THE CHEMICAL'S SAFE USE, AND THE CHEMICAL'S PROPER DISPOSAL. NO REPRESENTATIONS AND/OR WARRANTIES, EITHER EXPRESSED OR IMPLIED, OF THE MERCHANTABILITY OR FITNESS FOR A PARTICULAR PURPOSE, OR OF ANY OTHER NATURE, ARE MADE WITH RESPECT TO THE INFORMATION PROVIDED IN THIS MSDS OR TO THE CHEMICAL TO WHICH INFORMATION MAY REFER. CRC NEITHER ASSUMES NOR AUTHORIZES ANY OTHER PERSON TO ASSUME FOR IT, ANY OTHER ADDITIONAL RESPONSIBILITY OR LIABILITY FOR THE USE OF, OR RELIANCE UPON, THIS INFORMATION.

HAZARD WARNING INFORMATION

HEALTH	FIRE	REACTIVE	OTHER	DEGREE OF HAZARD	COLOR CODING	OTHER CODES
3	0	0		0 = Minimum Hazard 1 = Slight Hazard 2 = Moderate Hazard 3 = Serious Hazard 4 = Severe Hazard	HEALTH = BLUE FIRE = RED REACTIVITY = YELLOW OTHER = WHITE	OX = Oxidizer ACID = Acid ALK = Alkali COR = Corrosive W = Use No Water

SECTION I - GENERAL INFORMATION

Characterization **Carbamate (Fungicide)**	RCRA Number **None**	EPA Class **Not Applicable**
DOT Proper Shipping Name **No Citation**	Chemical Abstract Service (CAS) Number **14484-64-1**	
DOT Hazard Class and Label Requirements **No Citation**	DOT Emergency Guide Code **No Citation**	
DOT Identification Number **No Citation**	Chemical Formula **$[(CH_3)_2NCSS]_3Fe$**	

Synonyms
Generic name for ferric dimethyldithiocarbamate; tris(dimethyldithiocarbamato)iron.

SECTION II - HAZARDOUS INGREDIENTS AND IDENTITY INFORMATION

Hazardous Components (specific identity)	OSHA Exposure Criteria	NIOSH Exposure Criteria	Immediately Dangerous to Life and Health (IDLH)	ACGIH Exposure Criteria
Ferbam (derivation: By addition of carbon disulfide to an alcoholic solution of dimethylamine and precipitation with a ferric salt).	**PEL:** 5 mg/m^3 (respirable) 15 mg/m^3 (total dusts) **STEL:** Not Established	**REL:** 5 mg/m^3 (respirable) 15 mg/m^3 (total dusts) **STEL:** Not Established	800 mg/m^3	**TLV:** 10 mg/m^3 **STEL:** Not Established

SECTION III - PHYSICAL AND CHEMICAL CHARACTERISTICS

Boiling Point **Decomposes**	Specific Gravity (H_2O = 1) **Not Found**
Vapor Pressure (mm Hg) **0 (approx.) at 69°F (20°C)**	Molecular Weight (atomic weight) **416.51**
Vapor Density (Air = 1) **Not Found**	Melting Point **>356°F (180°C) Decomposes**

Solubility
Practically insoluble in water (0.01%).

Appearance and Odor
Black or dark-colored, fluffy powder (solid). It may be mixed in solution and appear as a dark liquid.

SECTION IV - FIRE AND EXPLOSION HAZARD DATA

Flash Point (method used) **Non-Combustible**	Explosive Limits in Air % by Volume **LEL: N/A UEL: N/A**
NFPA Classification **Combustible Liquid**	Autoignition Temperature **Not Found**

Extinguishing Media
Use dry powder, foam, carbon dioxide, and water.

Special Fire Fighting Procedures
Decomposes above 356°F (180°C). Poisonous gases are produced in fire. Wear full protective clothing and self-contained breathing apparatus (SCBA). Structural fire-fighting equipment is permeable and may not provide sufficient protection. Move containers from fire area if it can be done without risk. Cool exposed containers with water spray. Do not release run off to sewers or waterways.

Unusual Fire and Explosion Hazards
None reported.

SECTION V - REACTIVITY DATA

Stability		Conditions to Avoid Ferbam is normally stable under at room temperature and pressure under routine conditions of handling and storage. Avoid contact with incompatible materials, and exposure to heat.
Stable X	Unstable	Incompatibility *(materials to avoid)* Exposure to oxidizers may cause fire or explosion.
Hazardous Polymerization		Conditions to Avoid Under normal conditions of temperature and pressure, hazardous polymerization of ferbam is not expected to occur.
May Occur	Will Not Occur X	Hazardous Decomposition or By-products At high temperatures, ferbam decomposes and may produce poisonous gases, including oxides of sulfur, nitrogen, and carbon monoxide.

SECTION VI - HEALTH HAZARD DATA

Primary Route(s) of Entry:	Inhalation? X	Absorption (skin/eye)?	Ingestion? X

Health Hazards

INHALATION: Causes irritation of the eyes, nose, throat, and upper respiratory tract. There may be some mild or moderate coughing, sneezing, sore throat, and possibly nausea. Possible central nervous system (CNS) depression in animals and may cause this effect in humans.

EYES & SKIN: May cause skin rash or dermatitis (especially in persons sensitive to sulfur). Eye irritation can occur. If not removed immediately, may lead to inflammation, redness, and tearing.

INGESTION: Causes a variety of gastrointestinal disturbances.

Carcinogenicity	NTP Listed?	IARC Cancer Review Group?	OSHA Regulated?	Target Organs?
Unknown Human Questioned Animal	No	Group 3	29 CFR 1910.1000 Table Z-1	Respiratory system, eyes, skin, G.I. tract.

Medical Conditions Generally Aggravated by Exposure

None reported. However, persons with dermatitis or other skin disorders may be sensitive to exposure.

Emergency and First-aid Procedures

Eye contact: Do not allow victim to keep eyes tightly shut. Flush immediately with water for 15 minutes (minimum), seek medical attention. Skin contact: Remove all contaminated clothing. Immediately wash area with flooding amounts of soap and water. For inhalation: Remove the person from exposure. Maintain an open airway. Provide respiratory assistance and CPR. Transfer to medical facility. If swallowed: Call poison control center for advice. Unless advised otherwise, give large amounts of water and induce vomiting. Never give anything by mouth to an unconscious or convulsing person.

SECTION VII - PRECAUTIONS FOR SAFE HANDLING AND USE

Steps to be Taken in Case Material is Released or Spilled

Restrict those not involved in cleanup from entering area. Ventilate area of spill. Cleanup crew must avoid exposure by using appropriate personal protective equipment. Collect dusts using a vacuum equipped with a HEPA filter. Damp mop residue. Absorb any liquids or spilled solutions in vermiculite or absorbent and deposit in sealed drum.

Preferred Waste Disposal Method

No Citation.

Precautions to be Taken in Handling and Storage

Store in tightly closed containers in cool, dry, well-ventilated area. Protect containers from physical damage. Avoid creating dusty conditions in storage.

Other Precautions and Warnings

Caution is warranted when mixing ferbam in solution. Protective equipment is required.

SECTION VIII - CONTROL MEASURES AND PERSONAL PROTECTIVE EQUIPMENT

Respiratory Protection *(specify type)*

PEL is low (5 mg/m^3) and may be difficult to monitor. Use a respirator with a fungicide cartridge. For best protection, use a supplied-air respirator or a self-contained breathing apparatus (SCBA) with full facepiece operated in positive pressure mode.

Ventilation

Local exhaust (preferred) at site of chemical work or general ventilation.

Protective Gloves	Eye Protection	Other Protective Clothing
Impervious Material	Chemical/Dust Goggles or Face Mask	Protective Apron, Boots (avoid skin contact)

Work/Hygiene Practices

Always wash hands thoroughly after using chemical; never bring food, drink, or smoking materials into vicinity of chemicals.

FERBAM

$[(CH_3)_2NCSS]_3Fe$ CAS: 14484-64-1

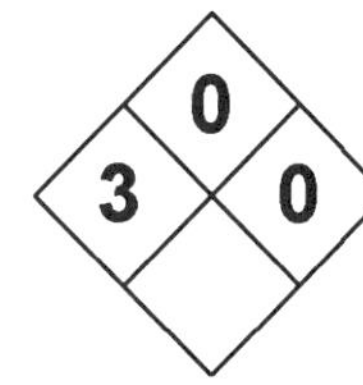

IDENTIFICATION AND TYPICAL USES

Ferbam is a dark brown or black, odorless solid. It may be mixed in solution and used as a spray. Used as a fungicide to control scab and rust diseases.

RISK ASSESSMENT: HEALTH

General Assessment

Ferbam is an irritant by *inhalation* and is considered moderately toxic by *ingestion*. Skin contact also presents a contact hazard. Its carcinogenicity and teratogenicity are not clearly understood in the references. Animal mutagenic and tumorigenic data have been reported.

Inhalation causes irritation of the eyes, nose, throat, and respiratory tract. There is some indication that exposure can cause central nervous system (CNS) depression in test animals and it is expected that severe exposure may have this effect in humans. Symptoms may include drowsiness, headache, fatigue, and other unspecified effects.

Skin contact is irritating and may cause contact dermatitis in some individuals. Eye contact can cause irritation and, if not removed immediately, inflammation and redness. Ingestion causes gastrointestinal irritation and other related effects (possible nausea, stomach cramps, vomiting).

Acute Health Effects

The following acute (short-term) health effects may occur immediately or shortly after exposure to ferbam:

Skin: Irritation, rash, and possible dermatitis in some individuals.

Eye: Irritation and inflammation.

Lung: Irritation of the nose, throat, and lungs.

CNS: A CNS depressant in test animals. Numerous effects resulting from exposure may include dizziness, fatigue, muscular twitching, confusion, vertigo, weakness, convulsions, and coma.

Chronic Health Effects

The following chronic (long-term) health effects may occur at some time after exposure to ferbam and can last for months or even years:

Cancer Hazards: Ferbam may be a mutagen (causes genetic changes). Such chemicals may have a cancer risk and there is limited evidence that ferbam does cause cancer in some test animals. Human carcinogenicity is unknown at this time.

Reproductive Hazard: There is some evidence that exposure to ferbam cause damage to the developing fetus in test animals. Human teratogenicity is unknown.

Other Chronic Effects: Long-term exposures may lead to chronic dermatitis in some individuals.

Recommended Risk-Reduction Measures

Personnel should avoid direct contact with ferbam. If a less toxic material or compound cannot be substituted for ferbam, then *engineering controls* are the most effective method of reducing exposures. While not feasible for outdoor operations, the best protection in manufacturing and production is to enclose operations and/or provide local exhaust ventilation at the site of ferbam release. While not always operationally possible, isolating operations involving ferbam manufacturing can also reduce exposure. Using respiratory protection is less effective than the controls mentioned above, but is still advisable whenever working with or around ferbam. For low or infrequent exposures over the PEL (15 mg/m^3 for total dusts or 5 mg/m^3 for respirable fraction), an organic vapor respirator equipped with a dust/mist pre-filter may be adequate. For higher exposures or when the exposure level is unknown (as in an emergency, for example), the best protection is obtained using a supplied-air respirator or a self-contained breathing apparatus (SCBA) with full facepiece operated in pressure demand mode. If a full facepiece is not available, then chemical/dust goggles should be worn to protect the eyes. A face shield and protective apron should also be worn. To prevent hand and skin exposures, polyvinylchloride gloves should be used.

Administrative controls should also be in place to minimize the potential for human exposures. These may include written procedures or policies which specify the methods and techniques that will be practiced whenever personnel are to work with ferbam.

All personnel should receive *training* on the use, hazards, protective measures, emergency actions, and other precautions per 29 CFR 1910.1200 (Hazard Communication), prior to the first assignment in an area where ferbam is used or stored.

Before beginning employment and regular intervals thereafter (e.g., annually), the following medical tests are recommended:

☑ Lung function tests (establish baseline).

If symptoms develop or overexposure is suspected, the following medical testing may be useful:

☑ Lung function tests (compare to baseline).

☑ Evaluation by a qualified allergist with careful consideration of exposure history and special testing (may help diagnose skin allergy).

Any evaluation should include a careful history of past and present symptoms with an examination. Medical tests that simply look for existing damage are not a substitute for controlling exposures. Also, since smoking can cause heart disease, emphysema, and a number of other respiratory disorders, smokers may be effected more quickly and with more pronounced symptoms than non-smokers under the same conditions of exposure. Prudent risk management requires careful consideration of *all* factors which may be causing the development of exposure symptoms in the work force.

Other methods to reduce exposure include:

☑ Where possible, enclose operations and use local exhaust ventilation at the site of chemical release. If local exhaust ventilation or enclosure is not used, respiratory protection should be mandatory.

☑ Always ensure that proper protective clothing is worn when using chemical substances.

☑ Wash thoroughly immediately after exposure to ferbam and at the end of the work shift or before eating, drinking, or smoking.

☑ Hazard warning information should be posted in the work area. When taken for use outdoors, proper labeling and warning information should be sent with it. In addition, as part of an on-going education and training program, all information on the health and safety hazards of ferbam should be communicated to all potentially exposed workers.

☑ Eye wash stations should be provided in the immediate work area for emergency use. If there is a possibility of skin exposure to ferbam, emergency shower facilities should also be provided.

☑ Workers whose clothing has been contaminated by ferbam should change into clean clothes before leaving work. Contaminated work clothing should be laundered only by individuals who have been informed of the hazards of exposure to ferbam.

RISK ASSESSMENT: ENVIRONMENT

General Assessment

Because of its use as a fungicide, ferbam can be present in the environment at any given time and at a variety of levels. However, the environment is also at risk of exposure during transportation, storage, disposal, or destruction of ferbam. In almost every scenario, the threat of environmental exposure is contingent upon the proper handling of the chemical substance. Accidental spills, large or small, where ferbam contacts incompatible commodities can result in the possible contamination of the surrounding environmental mediums (water, soil, and air).

Ferbam is considered a combustible solid and will also support the combustion of other combustible materials in a fire. In powdered form, it can form an explosive mixture in air. It is incompatible with oxidizers, such as chlorine, bromine, and fluorine, and contact can cause violent reactions. Also, if involved in a fire, it will emit toxic oxides of sulfur and nitrogen oxides and carbon monoxide. These characteristics require special consideration during any emergency situation involving a leak or spill of ferbam.

Ferbam can enter the environment from industrial and municipal discharges, through agricultural run-off, and from spills.

Acute Ecological Effects

Acute (short-term) toxic effects may include the death of animals, birds, or fish and death or low growth rate in plants. Acute effects are seen 2 to 4 days after animals or plants are exposed to ferbam.

Insufficient data are available to evaluate or predict the short-term (acute) effects of ferbam to aquatic life, birds, plants, and land animals.

Chronic Ecological Effects

Chronic (long-term) toxic effects may include shortened life span, reproductive problems, lower fertility, and changes in appearance or behavior in exposed animals. These effects can be seen long after first exposure(s) to toxic chemicals.

Insufficient data are available to evaluate or predict the long-term (chronic) effects of ferbam to aquatic life, birds, plants, and land animals.

Water Solubility

Ferbam is nearly insoluble in water. Concentrations of less than 1 milligram may mix with a liter of water.

Persistence in the Environment

Ferbam is highly persistent in water, with a half-life of greater than 200 days. The half-life of a pollutant is the amount of time it takes for one-half of the chemical to be degraded.

Bioaccumulation in Aquatic Organisms

Some substances increase in concentration, or *bioaccumulate,* in living organisms as they breathe contaminated air, drink contaminated water, or eat contaminated food. These chemicals can become concentrated in the tissues and internal organs of animals as well as humans.

There is no information on the bioaccumulation properties of ferbam in the edible tissues of fish.

Recommended Risk-Reduction Measures

Proper training of all transporters will reduce the likelihood of a mishap or accident resulting in a leak or spill of ferbam dusts or powders into the environment. The correct use labeling on all containers, trucks, and rail cars will enable emergency responders to react properly and quickly to any disaster thereby reducing the potential risk to the environment and to personnel.

Storage of ferbam should be segregated from incompatible materials to minimize the risk of cross-contamination or contact. Proper ventilation and protective equipment should be used while handing or using ferbam.

If a spill or leak to the environment has occurred, fire department, emergency response and/or hazardous materials spill personnel should be notified immediately. Cleanup should be attempted only by those trained in proper spill containment procedures. If ferbam should contact the water table, aquifer, or navigable waterway, reclamation procedures should be initiated as soon as practical. It is nearly insoluble in water and total remediation may be possible. The local and/or state emergency response authorities must be notified. A comprehensive emergency response or disaster preparedness/recovery plan should be in place prior to any operations involving the use, transportation, storage, or disposal of ferbam. If ferbam powder is spilled or leaked, the following specific steps are recommended:

- ☑ Restrict persons not wearing protective clothing from area of spill or leak until cleanup is complete and area can be opened for normal work.
- ☑ Collect powdered materials in safest and most convenient manner possible. Do NOT dry sweep (generates dusts). Use vacuum equipped with a high-efficiency particulate air filter instead. Damp mop residue. Absorb any liquid spills containing ferbam solution in vermiculite or other material and place in sealed drum for disposal or reclamation.
- ☑ It may be necessary to dispose of ferbam as a hazardous waste. The responsible state agency or the regional office of the federal Environmental Protection Agency (EPA) should be contacted for specific recommendations.

RISK ASSESSMENT: BUSINESS

General Assessment

Accidents or mishaps involving ferbam can present a significant threat to business operations. The loss or damage of equipment or facilities can significantly affect fiscal viability. Lawsuits that may result from personnel injury/death, public exposures, and/or environmental contamination will also require a serious expenditure of resources. Media attention surrounding an injury, death, or environmental damage can also result in a loss of profits as well as a loss of current and future business potential.

Recommended Risk-Reduction Measures

Company attorneys, safety and health professionals, and environmental specialists should be involved in the development of any procedures for responding to chemical incidents. A company official should be predesignated as a public relations officer with specific training in dealing with the press. Corporate plans and policies should be developed, approved, and implemented long before any need for such arises.

REFERENCES

Ahlbom, A. 1993. *Biostatistics for Epidemiologists.* New York: CRC Press/Lewis Publishers.

American Conference of Governmental Industrial Hygienists. 1988. *Documentation of the Threshold Limit Values and Biological Exposure Indices,* 5th Edition (with updates). Cincinnati: ACGIH

Anthony, C. P., and N. J. Kolthoff. 1971. *Textbook of Anatomy and Physiology.* St. Louis: C. V. Mosby Company.

Asante-Duah, D. K. 1993. *Hazardous Waste Risk Assessment.* New York: CRC Press/Lewis Publishers.

Bloom, A. D. and F. J. DeSerres. 1995. *Ecotoxicity and Human Health.* New York: CRC Press/Lewis Publishers.

Brusick, D. J. 1994. *Methods for Genetic Risk Assessment.* New York: CRC Press/Lewis Publishers.

Bueche, J. R. 1972. *Principles of Physics,* 2nd Edition. New York: McGraw-Hill.

Calabrese, E. J. 1994. *Biological Effects of Low Level Exposures.* New York: CRC Press/Lewis Publishers.

Cockerman, L. G., B. S. Shane. 1994. *Basic Environmental Toxicology.* New York: CRC Press/Lewis Publishers.

Cothern, C. R., N. P. Ross. 1994. *Environmental Statistics, Assessment, and Forecasting.* New York: CRC Press/Lewis Publishers.

Danse, I. R. 1991. *Common Sense Toxics in the Workplace.* New York: Van Nostrand Reinhold.

Environmental Protection Agency. 1983. *Chemical Hazard Information Profiles.* Washington, D.C.: U.S. EPA.

Fodden, C. R., and J. L. Weddell. 1992. *Hazardous Materials Emergency Action Data.* New York: CRC Press/Lewis Publishers.

Forseberg, K., and L. H. Keith. 1993. *Chemical; Protective Clothing Permeation and Degradation Database.* Chelsea, MI: Lewis Publishers.

Gardner, R. W. 1994. *Chemical Intolerance.* New York: CRC Press/Lewis Publishers.

Howard, P. H., et. al. 1991. *Handbook of Environmental Degradation Rates.* New York: CRC Press/Lewis Publishers.

Hosty, J. W., and P. Foster. 1990. *A Practical Guide to Chemical Spill Response.* New York: Van Nostrand Reinhold.

International Agency for Research on Cancer (IARC). 1987. *IARC Monographs on the Evaluation of Carcinogenic Risk to Humans, Overall Evaluations of Carcinogenicity: An updating of IARC Monographs Volumes 1-42.* Lyon: IARC.

Jones, F. E. 1994. *Toxic Organic Vapors in the Workplace.* New York: CRC Press/Lewis Publishers.

Keith, L. H. and M. M. Walker. 1995. *Handbook of Air Toxics.* New York: CRC Press/Lewis Publishers.

Landis, W. G., M. H. Yu. 1995. *Introduction to Environmental Toxicology.* New York: CRC Press/Lewis Publishers.

Lauwerys, R. R., P. Hoet. 1993. *Industrial Chemical Exposure.* New York: CRC Press/Lewis Publishers.

Lewis, .J., Sr. 1993. *Hawley's Condensed Chemical Dictionary,* 12th Edition. New York: Van Nostrand Reinhold.

Lewis, R. J., Sr. 1992. *Sax's Dangerous Properties of Industrial Materials,* Volumes 1, 2, and 3, 8th Edition. New York: Van Nostrand Reinhold.

Manahan, S. E. 1994. *Environmental Chemistry,* 6th Edition. New York: CRC Press/Lewis Publishers.

Mellström, G. A., J. E. Wahlberg, and H. I. Maibach. 1994. *Protective Gloves for Occupational Use.* New York: CRC Press/Lewis Publishers.

Meyer, E. 1990. *Chemistry of Hazardous Materials.* Englewood Cliffs, NJ: Prentice Hall.

Miller, B. F., M.D., and C. B. Keane, R.N. 1972. *Encyclopedia and Dictionary of Medicine and Nursing.* Philadelphia: W.B. Sanders Company.

Mycock, J. C., J. D. McKenna, and L. Theodore. 1995. *Handbook of Air Pollution Control Engineering and Technology.* New York: CRC Press/Lewis Publishers.

Ott, W. R. 1995. *Environmental Statistics and Data Analysis.* New York: CRC Press/Lewis Publishers.

Patnaik, P. 1992. *A Comprehensive Guide to the Hazardous Properties of Chemical Substances.* New York: Van Nostrand Reinhold.

Proctor, N. H., and J. P. Hughes. 1991. *Chemical Hazards of the Workplace.* New York: Van Nostrand Reinhold.

Quigley, D. R. 1994. *Handbook of Emergency Chemical Management.* New York: CRC Press/Lewis Publishers.

Rea, W. J. 1995. *Chemical Sensitivity,* Vol. I-III. New York: CRC Press/Lewis Publishers.

Richardson, M. 1995. *Dictionary of Substance and Their Effects,* Vol. I-VI. New York: CRC Press/Lewis Publishers.

Sacarello, H. L. A. 1994. *The Comprehensive Handbook of Hazardous Materials.* New York: CRC Press/Lewis Publishers.

Talbott, E., G. Craun. 1995. *An Introduction to Environmental Epidemiology.* New York: CRC Press/Lewis Publishers.

Torkelson, T. R. V. K. Rowe. 1981. *Patty's Industrial Hygiene and Toxicology,* Vol. 2B. New York: Wiley-Interscience.

MATERIAL SAFETY DATA SHEET

CHEMICAL NAME

FERRIC CHLORIDE (solution)

CRC PRESS/LEWIS PUBLISHERS DISCLAIMER: THE INFORMATION AND RECOMMENDATIONS PRESENTED HEREIN ARE BASED ON SOURCES BELIEVED TO BE RELIABLE. CRC MAKES NO REPRESENTATION ON ITS COMPLETENESS OR ACCURACY. IT IS THE RESPONSIBILITY OF THE USER TO DETERMINE THE CHEMICAL'S SUITABILITY FOR ITS INTENDED USE, THE CHEMICAL'S SAFE USE, AND THE CHEMICAL'S PROPER DISPOSAL. NO REPRESENTATIONS AND/OR WARRANTIES, EITHER EXPRESSED OR IMPLIED, OF THE MERCHANTABILITY OR FITNESS FOR A PARTICULAR PURPOSE, OR OF ANY OTHER NATURE, ARE MADE WITH RESPECT TO THE INFORMATION PROVIDED IN THIS MSDS OR TO THE CHEMICAL TO WHICH INFORMATION MAY REFER. CRC NEITHER ASSUMES NOR AUTHORIZES ANY OTHER PERSON TO ASSUME FOR IT, ANY OTHER ADDITIONAL RESPONSIBILITY OR LIABILITY FOR THE USE OF, OR RELIANCE UPON, THIS INFORMATION.

HAZARD WARNING INFORMATION

HEALTH	FIRE	REACTIVE	OTHER	DEGREE OF HAZARD	COLOR CODING	OTHER CODES
2	0	0		0 = Minimum Hazard 1 = Slight Hazard 2 = Moderate Hazard 3 = Serious Hazard 4 = Severe Hazard	HEALTH = BLUE FIRE = RED REACTIVITY = YELLOW OTHER = WHITE	OX = Oxidizer ACID = Acid ALK = Alkali COR = Corrosive W = Use No Water

SECTION I - GENERAL INFORMATION

Characterization: **Chloride**	RCRA Number: **None** — EPA Class: **Not Applicable**
DOT Proper Shipping Name: **Ferric chloride (solution)**	Chemical Abstract Service (CAS) Number: **7705-08-0**
DOT Hazard Class and Label Requirements: **Corrosive Material**	DOT Emergency Guide Code: **57**
DOT Identification Number: **UN 2582**	Chemical Formula: **$FeCl_3$**

Synonyms
Iron (III) chloride (solution); iron perchloride; iron sesquichloride; iron trichloride.

SECTION II - HAZARDOUS INGREDIENTS AND IDENTITY INFORMATION

Hazardous Components (specific identity)	OSHA Exposure Criteria	NIOSH Exposure Criteria	Immediately Dangerous to Life and Health (IDLH)	ACGIH Exposure Criteria
Ferric chloride solution (derivation: By the action of chlorine on ferrous sulfate or chloride; by reaction of chlorine gas on red hot iron, obtained from steel pickling liquors; or as a by-product of titanium dioxide manufactures by the chlorine process. Occurs in nature as the mineral molysite)	**PEL:** 1 mg(Fe)/m^3 (soluble iron salts) **STEL:** Not Established	**REL:** Not Established **STEL:** Not Established	**Not Determined**	**TLV:** 1 mg(Fe)/m^3 (soluble iron salts) **STEL:** Not Established

SECTION III - PHYSICAL AND CHEMICAL CHARACTERISTICS

Boiling Point: **606°F (319°C)**	Specific Gravity (H_2O = 1): **2.8**
Vapor Pressure (mm Hg): **40 at 69°F (20°C)**	Molecular Weight: **162.22**
Vapor Density (Air = 1): **5.6**	Melting Point: **-58°F (-50°C) at 30%**

Solubility
Soluble in water, alcohol, acetone, ether, and glycerol. Slightly soluble in carbon disulfide.

Appearance and Odor
A green-to-black colored solution with a slight hydrochloric acid odor.

SECTION IV - FIRE AND EXPLOSION HAZARD DATA

Flash Point (method used): **Non-Combustible**	Explosive Limits in Air % by Volume: **LEL: Not Applicable** **UEL: Not Applicable**
NFPA Classification: **Non-Combustible Solution**	Autoignition Temperature: **Non-Combustible**

Extinguishing Media
Use extinguishing media suitable to surrounding fire. Ferric chloride solution is non-combustible.

Special Fire Fighting Procedures
Toxic and poisonous gases and fumes are produced in fire. Structural fire-fighting clothing may not provide adequate protection. Wear full protective clothing and self-contained breathing apparatus (SCBA). Use a water spray to keep fire-exposed containers cool (to prevent crystallization of solution). Keep clear of ends of tanks. Do NOT release runoff to sewers or waterways.

Unusual Fire and Explosion Hazards
Irritating hydrogen chloride fumes and other gases are formed in fires. Firefighters should avoid contact with vapors produced during fire.

SECTION V - REACTIVITY DATA

Stability		Conditions to Avoid
Stable	Unstable	Ferric chloride solution is normally stable in closed containers under normal conditions of storage and handling. Avoid contact with incompatible materials and heat. Keep cool to avoid crystallization.
X		Incompatibility *(materials to avoid)*: Keep away from most metals (except titanium and tantalum) due to corrosive nature of ferric chloride solution. Incompatible with allyl chloride, potassium, sodium, ethylene oxide, nylon, alkalis, and most oxidizers.
Hazardous Polymerization		Conditions to Avoid: Under normal conditions of temperature and pressure, ferric chloride (solution) is not expected to undergo hazardous polymerization.
May Occur	Will Not Occur	Hazardous Decomposition or By-products
	X	When heated to decomposition, ferric chloride can emit highly toxic/poisonous gases, including hydrogen chloride and chloride gas.

SECTION VI - HEALTH HAZARD DATA

Primary Route(s) of Entry:	Inhalation?	Absorption (skin)?	Ingestion?
	X		X

Health Hazards

INHALATION: Irritation of the eyes, nose, throat, and mucous membrane of the respiratory tract. There may be death due to glottis spasms and/or suffocation. Can cause chemical pneumonitis and pulmonary edema (fluid in the lungs) which can cause death.

SKIN & EYES: Contact with the eyes can cause localized discoloration (brown) of the conjunctiva along with irritation and pain. Skin contact causes irritation and destruction of tissue.

INGESTION: Irritation of the mouth and stomach with abdominal pain, vomiting, nausea, diarrhea, rapid and weak pulse, low blood pressure, cyanosis, secondary shock, and coma.

Carcinogenicity	NTP Listed?	IARC Cancer Review Group?	OSHA Regulated?	Target Organs?
Unknown Human Unknown Animal	No	No	29 CFR 1910.1000 Table Z-1-A	Respiratory system, eyes, skin, kidneys, liver.

Medical Conditions Generally Aggravated by Exposure

None reported.

Emergency and First-aid Procedures

Eye contact: Flush immediately with water for 15 minutes (minimum), seek medical attention. Skin contact: Remove all contaminated clothing. Immediately wash area with large amounts of water. Seek medical assistance. For inhalation: Remove the person from exposure. Provide respiratory assistance and CPR. Transfer to medical facility. Observe for 24-48 hours for lung effects. If swallowed: Seek medical attention immediately. Unless advised otherwise, give 1-2 glasses of water and induce vomiting.

SECTION VII - PRECAUTIONS FOR SAFE HANDLING AND USE

Steps to be Taken in Case Material is Released or Spilled

Absorb liquids in vermiculite, or similar material and deposit in sealed containers. Ventilate area of spill or leak. Restrict those not involved in cleanup from entering area. Rinse area with sodium bicarbonate.

Preferred Waste Disposal Method

Precipitate as the sulfide, adjust the pH to neutral, filter insolubles, and dispose as hazardous waste.

Precautions to be Taken in Handling and Storage

Store in tightly closed containers in cool, well-ventilated area away from incompatibles. Protect against physical damage. Discard solutions that become cloudy. Store small quantities in polyethylene bottles. Keep tightly closed. If polyethylene is not available, use glass and refrigerate. The solution may leach alkali from the glass and form a yellow ferric oxide precipitate.

Other Precautions and Warnings

Store large quantities in rubber-lined steel tanks or fiberglass reinforced polyester tanks.

SECTION VIII - CONTROL MEASURES AND PERSONAL PROTECTIVE EQUIPMENT

Respiratory Protection *(specify type)*

For exposures below the PEL, use a full mask respirator with organic vapor cartridge. For all other exposures, use supplied-air respirator with full facepiece, hood, or helmet in continuous flow mode, or use a full facepiece operated in pressure demand or other positive pressure mode.

Ventilation

Local exhaust or general mechanical systems recommended.

Protective Gloves	Eye Protection	Other Protective Clothing
Polycarbonate or Butyl Rubber	Chemical Goggles or Face Mask	Rubber Apron

Work/Hygiene Practices

Always wash hands thoroughly after using chemical; never bring food, drink, or smoking materials into vicinity of chemicals.

FERRIC CHLORIDE SOLUTION

$FeCl_3$ CAS: 7705-08-0

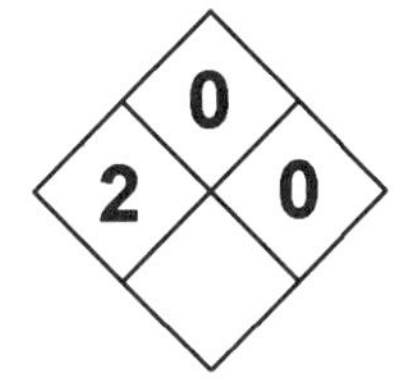

IDENTIFICATION AND TYPICAL USES

Ferric chloride solution is a greenish-black liquid with a slight hydrochloric acid odor. It is used primarily in the treatment of sewage and industrial wastes. It is also used as an etching agent for engraving, in photography, and in printed circuitry. It is a condensation catalyst in Friedel-Crafts reactions; a mordant, an oxidizing, chlorinating, and condensing agent; a disinfectant; a pigment; a feed additive; and is used in water purification.

RISK ASSESSMENT: HEALTH

General Assessment

Exposure to ferric chloride solution is possible through *inhalation* of its vapors or by direct *ingestion* of the liquid. The latter presents the most serious risks to human health. While skin absorption has not been reported, skin contact can result in localized irritation and should be avoided. There are no data on its carcinogenic or teratogenic properties. Mutation data (causes genetic changes) have been reported in test animals. This does not mean that it will cause the same reaction in humans. More study is required in this regard.

Inhalation can cause headache and nausea with irritation to the eyes, nose, and throat. Inhalation of vapors can be fatal as a result of spasm of the glottis and suffocation. High-level exposures may lead to chemical pneumonia and pulmonary edema (fluid in the lungs) which can also be fatal. Development of pulmonary edema may be delayed up to 48 hours thereby creating a false sense of security with regard to health exposure risk.

Skin and eye contact is irritating. Prolonged contact with the eyes can cause a brownish discoloration of the conjunctiva.

Ingestion of ferric chloride solution will cause moderate to severe irritation of the mouth and stomach. Symptoms may appear up to an hour after ingestion and include abdominal pain and discomfort, vomiting and nausea, diarrhea, weak and rapid pulse, hypotension (decreased blood pressure), cyanosis (bluish discoloration of the fingers, toes, lips, and skin surface due to lack of oxygen in blood), secondary shock, and coma. The exact type and degree of these symptoms are largely dependent upon the percentage of ferric chloride solution contained in the swallowed solution.

Acute Health Effects

The following acute (short-term) health effects may occur immediately or shortly after exposure to ferric chloride solution:

Skin: Localized surface irritation.

Eye: Severe irritation; may cause brownish discoloration of the conjunctival tissues.

Lung: Irritation of the nose, throat, and lungs following exposure. Pulmonary edema is possible (24 to 48 hours following a high exposure).

Other: Ingestion can cause a variety of gastrointestinal and systemic effects up to and including loss of consciousness and even death.

Chronic Health Effects

The following chronic (long-term) health effects may occur at some time after exposure to ferric chloride solution and can last for months or even years:

Cancer Hazards: Ferric chloride solution may cause mutations (genetic changes) in living cells. There are no data on its carcinogenicity.

Reproductive Hazard: According to the information available in the references, ferric chloride solution has not been adequately tested for its ability to adversely affect the reproductive process in test animals.

Other Chronic Effects: There are no reports on chronic effects resulting from exposure to ferric chloride solution in humans or in test animals.

ʊ *Recommended Risk-Reduction Measures*

Personnel should avoid direct contact with ferric chloride solution. If a less toxic chemical cannot be substituted for a hazardous substance, then *engineering controls* are the most effective method of reducing exposures. The best protection is to enclose operations and/or provide local exhaust ventilation at the site of chemical release. While not always operationally feasible, isolating operations can also reduce exposure.

Using respiratory protection is less effective than the controls mentioned above, but is still advisable whenever working with or around ferric chloride solution. The permissible exposure level (PEL) for this chemical is extremely low (1 mg/m^3) and difficult to monitor. An air purifying respirator may suffice. However, a supplied-air respirator or a self-contained breathing apparatus (SCBA) with full facepiece operated in pressure demand mode would provide optimum respiratory protection. If a full facepiece is not available, then chemical goggles should be worn to protect the eyes. Whenever a chemical splash hazard exists, a face shield and a rubber apron should be worn. To prevent hand and skin exposures, butyl rubber or polycarbonate gloves should be worn.

Administrative controls should also be in place to minimize the potential for human exposures. These may include written procedures or policies which specify the methods and techniques that will be practiced whenever personnel are to work with ferric chloride solution.

All personnel should receive *training* on the use, hazards, protective measures, emergency actions, and other precautions per 29 CFR 1910.1200 (Hazard Communication), prior to the first assignment in an area where ferric chloride solution is used or stored.

For those personnel with frequent or potentially high exposure (half the PEL or greater), the following are recommended medical tests to be provided before initial assignment and at regular intervals thereafter (e.g., annually):

- ☑ Lung function tests.
- ☑ Consider chest X-ray after acute overexposure (results may be negative if taken immediately after exposure due to delayed development of pulmonary edema).

Any evaluation should include a careful history of past and present symptoms with an examination. Medical tests that simply look for existing damage are not a substitute for controlling exposures. Also, because smoking can cause heart disease, as well as lung cancer, emphysema, and other respiratory problems, symptoms of exposure to ferric chloride solution may be more pronounced in smokers than in non-smokers under the same conditions of exposure. Prudent risk management requires careful consideration of *all* possible exposure factors.

Other methods to reduce exposure include:

- ☑ Where possible, enclose operations and use local exhaust ventilation at the site of chemical release. If local exhaust ventilation or enclosure is not used, respiratory protection should be mandatory.
- ☑ Always ensure that proper protective clothing is worn when using chemical substances.
- ☑ Wash thoroughly immediately after exposure to ferric chloride solution and at the end of the work shift or before eating, drinking, or smoking.
- ☑ Hazard warning information should be posted in the work area. In addition, as part of an on-going education and training program, all information on the health and safety hazards of ferric chloride solution should be communicated to all potentially exposed workers.
- ☑ Eye wash stations should be provided in the immediate work area for emergency use. If there is a possibility of skin exposure to ferric chloride solution, emergency shower facilities should also be provided.
- ☑ Personnel should never wear contaminated clothing home (family members can be exposed). Clothing should be laundered by personnel who have been trained on the hazards of ferric chloride solution.

RISK ASSESSMENT: ENVIRONMENT

General Assessment

The environment is at risk of exposure during transportation, storage, disposal, or destruction of ferric chloride solution. In almost every scenario, the threat of environmental exposure is contingent upon the proper handling of the chemical substance. Accidental spills, large or small, can result in fire, explosion, and possible contamination of the surrounding environmental mediums (water, soil, and air).

Ferric chloride solution is considered a noncombustible liquid. However, it is corrosive to many metals and can react vigorously with allyl chloride, potassium, ethylene oxide, nylon, alkalis, and many oxidizers and contact can cause fire or explosion. These characteristics require special consideration during any emergency situation involving a leak or spill of ferric chloride solution. Should ferric chloride solution ever come into contact with these incompatible either during use, transportation, storage, or disposal, the formation of highly toxic and/or highly explosive commodities is extremely possible.

The proper disposal/destruction method for ferric chloride solution waste is to precipitate as sulfide, adjust the pH to neutral, filter the insoluble materials and dispose of at a hazardous waste site. Excess sulfide is destroyed using sodium hypochlorite and neutralized before flushing to drain.

Ferric chloride solution can enter the environment mainly through industrial effluents or spills.

Acute Ecological Effects

Acute (short-term) toxic effects may include the death of animals, birds, or fish and death or low growth rate in plants. Acute effects are seen 2 to 4 days after animals or plants are exposed to ferric chloride.

There is insufficient evidence to evaluate or predict the short-term effects of ferric chloride solution to aquatic life, plants, birds, or land animals.

Chronic Ecological Effects

Chronic (long-term) toxic effects may include shortened life span, reproductive problems, lower fertility, and changes in appearance or behavior in exposed animals. These effects can be seen long after first exposure(s) to toxic chemicals.

There is insufficient evidence to evaluate or predict the long-term effects of ferric chloride solution to aquatic life, plants, birds, or land animals.

Water Solubility

Ferric chloride solution is highly soluble in water. Concentrations of 1000 milligrams and more will mix with a liter of water.

Persistence in the Environment

Ferric chloride solution is highly persistent in water, with a half-life of more than 200 days. The half-life of a pollutant is the amount of time it takes for one half of the chemical to be degraded.

Bioaccumulation in Aquatic Organisms

Some substances increase in concentration, or *bioaccumulate,* in living organisms as they breathe contaminated air, drink contaminated water, or eat contaminated food. These chemicals can become concentrated in the tissues and internal organs of animals as well as humans.

The concentration of ferric chloride found in fish tissues is expected to be about the same as the average concentration of ferric chloride in the water from which the fish was taken.

Recommended Risk-Reduction Measures

Proper training of all transporters will reduce the likelihood of a mishap or accident resulting in a leak or spill to the environment. The correct use of DOT labeling (per the MSDS) on all containers, trucks, and rail cars will enable emergency responders to react properly and quickly to any disaster thereby reducing the potential risk to the environment and to personnel.

Storage of ferric chloride solution should be segregated from incompatible chemicals to minimize the risk of cross-contamination or contact. Small quantities can be stored in polyethylene bottles. If polyethylene is not available, then glass bottles will suffice but they must be refrigerated. The solution may leach alkali from the glass to form a yellow ferric oxide precipitate. Large quantities should be kept in rubber-lined steel tanks or fiberglass reinforced polyester tanks.

If a spill or leak to the environment has occurred, fire department, emergency response and/or hazardous materials spill personnel should be notified immediately. Cleanup should be attempted only by those trained in proper spill containment procedures. Contaminated soils should be removed for incineration and replaced with clean soil. If ferric chloride solution should contact the water table, aquifer, or navigable waterway, the local and/or state emergency response authorities must be notified. A comprehensive emergency response or disaster preparedness/recovery plan should be in place prior to any operations involving the use, transportation, storage, or disposal of ferric chloride solution. If ferric chloride solution is spilled or leaked, the following specific steps are recommended:

- ☑ Restrict persons not wearing protective clothing from area of spill or leak until cleanup is complete and area can be opened for normal work.
- ☑ Ventilate are or spill or leak.

- ☑ Neutralize with agricultural lime or crushed limestone if landfill disposal is being considered. Rinse area with sodium bicarbonate or soda ash solution. Absorb liquids in vermiculite, dry sand, earth, or similar material and deposit in sealed containers.
- ☑ It may be necessary to dispose of ferric chloride solution as a hazardous waste. The responsible state agency or the regional office of the federal Environmental Protection Agency (EPA) should be contacted for specific recommendations.

RISK ASSESSMENT: BUSINESS

General Assessment

Accidents or mishaps involving ferric chloride solution can present a significant threat to business operations. The loss or damage of equipment or facilities can significantly affect fiscal viability. Lawsuits that may result from personnel injury/death, public exposures, and/or environmental contamination will also require a serious expenditure of resources. Media attention surrounding an injury, death, or environmental damage can also result in a loss of profits and loss of current as well as future business.

Recommended Risk-Reduction Measures

Company attorneys, safety and health professionals, and environmental specialists should be involved in the development of any safety or emergency response procedures. A company official should be pre-designated as a public relations officer with specific training in dealing with the press. Corporate plans and policies should be developed, approved, and implemented long before any need for such arises.

REFERENCES

American Conference of Governmental Industrial Hygienists. 1988. *Documentation of the Threshold Limit Values and Biological Exposure Indices,* 5th Edition (with updates). Cincinnati: ACGIH

Asante-Duah, D. K. 1993. *Hazardous Waste Risk Assessment.* New York: CRC Press/Lewis Publishers.

Brusick, D. J. 1994. *Methods for Genetic Risk Assessment.* New York: CRC Press/Lewis Publishers.

Bueche, J. R. 1972. *Principles of Physics,* 2nd Edition. New York: McGraw-Hill.

Cockerman, L. G., B. S. Shane. 1994. *Basic Environmental Toxicology.* New York: CRC Press/Lewis Publishers.

Cothern, R. R. 1993. *Comparative Environmental Risk Assessment.* New York: CRC Press/Lewis Publishers.

Gardner, R. W. 1994. *Chemical Intolerance.* New York: CRC Press/Lewis Publishers.

Hallenbeck, W. H. 1993. *Quantitative Risk Assessment for Environmental and Occupational Health,* 2nd Edition. New York: CRC Press/Lewis Publishers.

Howard, P. H. 1995. *Handbook of Environmental Fate and Exposures Data for Organic Chemicals,* Vol. I-IV. New York: CRC Press/Lewis Publishers.

Keith, L. H. and M. M. Walker. 1995. *Handbook of Air Toxics.* New York: CRC Press/Lewis Publishers.

Landis, W. G., M. H. Yu. 1995. *Introduction to Environmental Toxicology.* New York: CRC Press/Lewis Publishers.

Lauwerys, R. R., P. Hoet. 1993. *Industrial Chemical Exposure.* New York: CRC Press/Lewis Publishers.

Lewis, R. J., Sr. 1993. *Hawley's Condensed Chemical Dictionary,* 12th Edition. New York: Van Nostrand Reinhold.

Lewis, R. J., Sr. 1992. *Sax's Dangerous Properties of Industrial Materials,* Volumes 1, 2, and 3, 8th Edition. New York: Van Nostrand Reinhold.

Manahan, S. E. 1992. *Toxicological Chemistry,* 2nd Edition. New York: CRC Press/Lewis Publishers.

Meyer, E. 1990. *Chemistry of Hazardous Materials.* Englewood Cliffs, NJ: Prentice Hall.

Mycock, J. C., J. D. McKenna, and L. Theodore. 1995. *Handbook of Air Pollution Control Engineering and Technology.* New York: CRC Press/Lewis Publishers.

Ott, W. R. 1995. *Environmental Statistics and Data Analysis.* New York: CRC Press/Lewis Publishers.

Patnaik, P. 1992. *A Comprehensive Guide to the Hazardous Properties of Chemical Substances.* New York: Van Nostrand Reinhold.

Philp, R. B. 1995. *Environmental Hazards and Human Health.* New York: CRC Press/Lewis Publishers.

Quigley, D. R. 1994. *Handbook of Emergency Chemical Management.* New York: CRC Press/Lewis Publishers.

Rea, W. J. 1995. *Chemical Sensitivity,* Vol. I-III. New York: CRC Press/Lewis Publishers.

Richardson, M. 1995. *Dictionary of Substance and Their Effects,* Vol. I-VI. New York: CRC Press/Lewis Publishers.

Sacarello, H. L. A. 1994. *The Comprehensive Handbook of Hazardous Materials.* New York: CRC Press/Lewis Publishers.

Torkelson, T. R. V. K. Rowe. 1981. Halogenated aliphatic hydrocarbons. *Patty's Industrial Hygiene and Toxicology,* Vol. 2B, *Toxicology,* pp 3568-3572. New York: Wiley-Interscience.

MATERIAL SAFETY DATA SHEET

CHEMICAL NAME	
FERRIC CHLORIDE (anhydrous)	**CRC PRESS/LEWIS PUBLISHERS DISCLAIMER:** THE INFORMATION AND RECOMMENDATIONS PRESENTED HEREIN ARE BASED ON SOURCES BELIEVED TO BE RELIABLE. CRC MAKES NO REPRESENTATION ON ITS COMPLETENESS OR ACCURACY. IT IS THE RESPONSIBILITY OF THE USER TO DETERMINE THE CHEMICAL'S SUITABILITY FOR ITS INTENDED USE, THE CHEMICAL'S SAFE USE, AND THE CHEMICAL'S PROPER DISPOSAL. NO REPRESENTATIONS AND/OR WARRANTIES, EITHER EXPRESSED OR IMPLIED, OF THE MERCHANTABILITY OR FITNESS FOR A PARTICULAR PURPOSE, OR OF ANY OTHER NATURE, ARE MADE WITH RESPECT TO THE INFORMATION PROVIDED IN THIS MSDS OR TO THE CHEMICAL TO WHICH INFORMATION MAY REFER. CRC NEITHER ASSUMES NOR AUTHORIZES ANY OTHER PERSON TO ASSUME FOR IT, ANY OTHER ADDITIONAL RESPONSIBILITY OR LIABILITY FOR THE USE OF, OR RELIANCE UPON, THIS INFORMATION.

HAZARD WARNING INFORMATION

HEALTH	FIRE	REACTIVE	OTHER	DEGREE OF HAZARD	COLOR CODING	OTHER CODES
1	0	0		0 = Minimum Hazard 1 = Slight Hazard 2 = Moderate Hazard 3 = Serious Hazard 4 = Severe Hazard	HEALTH = BLUE FIRE = RED REACTIVITY = YELLOW OTHER = WHITE	OX = Oxidizer ACID = Acid ALK = Alkali COR = Corrosive W = Use No Water

SECTION I - GENERAL INFORMATION

Characterization	Chloride	RCRA Number	None
		EPA Class	Not Applicable
DOT Proper Shipping Name	Ferric Chloride (solid)	Chemical Abstract Service (CAS) Number	7705-08-0
DOT Hazard Class and Label Requirements	ORM-B	DOT Emergency Guide Code	No Citation
DOT Identification Number	UN 1773	Chemical Formula	Cl_3Fe

Synonyms
Iron (III) chloride; iron chloride; iron sesquichloride; iron trichloride; ferric trichloride; ferric perchloride.

SECTION II - HAZARDOUS INGREDIENTS AND IDENTITY INFORMATION

Hazardous Components (specific identity)	OSHA Exposure Criteria	NIOSH Exposure Criteria	Immediately Dangerous to Life and Health (IDLH)	ACGIH Exposure Criteria
Ferric chloride anhydrous (derivation: By the action of chlorine on ferrous sulfate or chloride).	**PEL:** 1 mg(Fe)/m^3 (soluble iron salts) **STEL:** Not Established	**REL:** Not Established **STEL:** Not Established	Not Determined	**TLV:** 1 mg(Fe)/m^3 (soluble iron salts) **STEL:** Not Established

SECTION III - PHYSICAL AND CHEMICAL CHARACTERISTICS

Boiling Point	606°F (319°C)	Specific Gravity (H_2O = 1)	2.8
Vapor Pressure (mm Hg)	40 at 69°F (20°C)	Molecular Weight	162.2
Vapor Density (Air = 1)	5.6	Melting Point	558°F (292°C)

Solubility
Soluble in water, alcohol, acetone, ether, and glycerol. Slightly soluble in carbon disulfide.

Appearance and Odor
Black-to-brown solid hexagonal leaflets or plates, turning red to green when exposed to light. Odorless.

SECTION IV - FIRE AND EXPLOSION HAZARD DATA

Flash Point (method used)	Non-Combustible	Explosive Limits in Air % by Volume	LEL: Not Applicable UEL: Not Applicable
NFPA Classification	Non-Combustible Solid	Autoignition Temperature	Non-Combustible

Extinguishing Media
Use extinguishing media suitable to surrounding fire. Ferric chloride is non-combustible.

Special Fire Fighting Procedures
Toxic and poisonous gases and fumes are produced in fire. Structural fire-fighting clothing may not provide adequate protection. Wear full protective clothing and self-contained breathing apparatus (SCBA). Use a water spray to keep fire-exposed containers cool. Very hygroscopic, prevent water used in fire-fighting activities from contacting material.

Unusual Fire and Explosion Hazards
Irritating hydrogen chloride fumes and other gases are formed in fires. Firefighters should avoid contact with vapors produced during fire.

SECTION V - REACTIVITY DATA

Stability		Conditions to Avoid
		Ferric chloride is normally stable in closed (airtight) containers under normal conditions of storage and handling. Avoid contact with incompatible materials and heat. Keep cool and dry.
Stable	Unstable	Incompatibility *(materials to avoid)*
X		Keep away water, moisture, and steam (may soften the solid and eventually dissolve it). Incompatible with allyl chloride, potassium, sodium, ethylene oxide, nylon, alkalis, and most oxidizers.
Hazardous Polymerization		Conditions to Avoid
		Under normal conditions of temperature and pressure, ferric chloride (anhydrous) is not expected to undergo hazardous polymerization.
May Occur	Will Not Occur	Hazardous Decomposition or By-products
	X	When heated to decomposition, ferric chloride can emit highly toxic/poisonous gases, including ferrous chloride and chloride gas. Solutions can emit toxic hydrogen chloride gas.

SECTION VI - HEALTH HAZARD DATA

Primary Route(s) of Entry:	Inhalation?	Absorption (skin)?	Ingestion?
	X		X

Health Hazards

INHALATION: Irritation of the eyes, nose, throat, and mucous membrane of the respiratory tract. There may death due to glottis spasms and/or suffocation. Can cause chemical pneumonitis and pulmonary edema (fluid in the lungs) which can cause death.

SKIN & EYES: Contact with the eyes can cause localized discoloration (brown) of the conjunctiva along with irritation and pain. Skin contact with dusts or solution causes irritation.

INGESTION: Irritation of the mouth and stomach with abdominal pain, vomiting of blood, nausea, diarrhea, rapid & weak pulse, low blood pressure, cyanosis, secondary shock, and coma.

Carcinogenicity	NTP Listed?	IARC Cancer Review Group?	OSHA Regulated?	Target Organs?
Unknown Human Unknown Animal	No	No	29 CFR 1910.1000 Table Z-1-A	Respiratory system, eyes, skin.

Medical Conditions Generally Aggravated by Exposure

None reported.

Emergency and First-aid Procedures

<u>Eye contact</u>: Flush immediately with water for 15 minutes (minimum), seek medical attention. <u>Skin contact</u>: Remove all contaminated clothing. Immediately wash area with large amounts of water. Seek medical assistance. For <u>inhalation</u>: Remove the person from exposure. Provide respiratory assistance and CPR. Transfer to medical facility. Observe for 24-48 hours for lung effects. If <u>swallowed</u>: Seek medical attention immediately. Unless advised otherwise, give 1-2 glasses of water and induce vomiting.

SECTION VII - PRECAUTIONS FOR SAFE HANDLING AND USE

Steps to be Taken in Case Material is Released or Spilled

Collect dusts in safest manner possible and deposit in sealed containers. Do NOT dry sweep, use HEPA vacuum instead. Ventilate area of spill or leak. Restrict those not involved in cleanup from entering area.

Preferred Waste Disposal Method

No Citation.

Precautions to be Taken in Handling and Storage

Store in tightly closed containers in cool, well-ventilated area away from incompatibles. Protect against physical damage. Keep away from all forms of water (moisture, steam) since ferric chloride will almost completely dissolve in water. Keep storage facilities airtight to prevent condensation inside containers.

Other Precautions and Warnings

Test drums containing ferric chloride for hydrogen chloride gas.

SECTION VIII - CONTROL MEASURES AND PERSONAL PROTECTIVE EQUIPMENT

Respiratory Protection *(specify type)*

For exposures below the PEL, use a full mask respirator with dust pre-filter. For all other exposures, use supplied-air respirator with full facepiece, hood, or helmet in continuous flow mode, or use a full facepiece operated in pressure demand or other positive pressure mode.

Ventilation

Local exhaust or general mechanical systems recommended.

Protective Gloves	Eye Protection	Other Protective Clothing
Polycarbonate or Butyl Rubber	Chemical/Dust Goggles or Face Mask	Rubber Apron

Work/Hygiene Practices

Always wash hands thoroughly after using chemical; never bring food, drink, or smoking materials into vicinity of chemicals.

FERRIC CHLORIDE ANHYDROUS

Cl_3Fe CAS: 7705-08-0

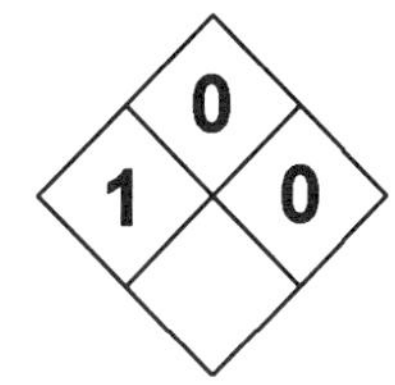

IDENTIFICATION AND TYPICAL USES

Ferric chloride (anhydrous) appears as black-to-brown hexagonal leaflets or plates that may turn red by transmitted light or green by reflected light. There is no distinguishable odor as a solid. When mixed in solution, it may have a slight hydrochloric acid odor. It is used in photography, in the manufacture of other iron salts, as a catalyst in organic synthesis reactions, as an oxidizing agent in dye manufacturing, in the chlorination of silver and copper ores, and in purifying factory effluents and deodorizing sewage. It should not be confused with ferric chloride solution or ferrous chloride.

RISK ASSESSMENT: HEALTH

General Assessment

Exposure to ferric chloride is possible through *inhalation* of its dusts or by direct *ingestion* of the solid material. The latter presents the most serious risks to human health. While skin absorption has not been reported, skin contact can result in localized irritation and should be avoided. There are no data on its carcinogenic or teratogenic properties. Mutation data (causes genetic changes) have been reported in test animals. This does not mean that it will cause the same reaction in humans. More study is required in this regard.

Inhalation of dusts can cause headache and nausea with irritation to the eyes, nose, and throat. Inhalation of vapors can be fatal as a result of spasm of the glottis and suffocation. High-level exposures to solutions containing ferric chloride may lead to chemical pneumonia and pulmonary edema (fluid in the lungs) which can also be fatal. Development of pulmonary edema may be delayed up to 48 hours thereby creating a false sense of security with regard to health exposure risk.

Skin contact is irritating and is aggravated by the presence of moisture (sweat). Prolonged contact with the eyes can cause irritation and a possible brownish discoloration of the conjunctiva.

Ingestion of ferric chloride will cause moderate to severe irritation of the mouth and stomach. Symptoms may appear up to an hour after ingestion and include abdominal pain and discomfort, vomiting of blood, nausea, diarrhea, weak and rapid pulse, hypotension (decreased blood pressure), cyanosis (bluish discoloration of the fingers, toes, lips, and skin surface due to lack of oxygen in blood), shock, and coma.

Acute Health Effects

The following acute (short-term) health effects may occur immediately or shortly after exposure to ferric chloride:

Skin: Localized surface irritation.

Eye: Severe irritation; may cause brownish discoloration of the conjunctival tissues.

Lung: Irritation of the nose, throat, and lungs following exposure. Pulmonary edema is possible (24 to 48 hours following a high exposure to ferric oxide solution).

Other: Ingestion can cause a variety of gastrointestinal and systemic effects up to and including loss of consciousness and even death. Excessive intake of ferric chloride may cause an increased accumulation of iron in body organs and systems such as the liver, spleen, and lymphatic system.

Chronic Health Effects

The following chronic (long-term) health effects may occur at some time after exposure to ferric chloride and can last for months or even years:

Cancer Hazards: Ferric chloride anhydrous may cause mutations (genetic changes) in living cells. There are no data on its carcinogenicity.

Reproductive Hazard: According to the information available in the references, ferric chloride anhydrous has not been adequately tested for its ability to adversely affect the reproductive process in test animals.

Other Chronic Effects: Pulmonary changes (spotting and blotching of the lungs) may occur from long-term exposure to ferric chloride dusts.

Recommended Risk-Reduction Measures

Personnel should avoid direct contact with ferric chloride. If a less toxic chemical cannot be substituted for a hazardous substance, then *engineering controls* are the most effective method of reducing exposures. The best protection is to enclose operations and/or provide local exhaust ventilation at the site of chemical release. While not always operationally feasible, isolating operations can also reduce exposure.

Using respiratory protection is less effective than the controls mentioned above, but is still advisable whenever working with or around ferric chloride. The permissible exposure level (PEL) for this chemical is extremely low (1 mg/m^3) and difficult to monitor. An air purifying respirator with a dust pre-filter may suffice. However, a supplied-air respirator or a self-contained breathing apparatus (SCBA) with full facepiece operated in pressure demand mode would provide optimum respiratory protection. If a full facepiece is not available, then chemical goggles should be worn to protect the eyes. Whenever a dust hazard exists, a face shield and a rubber apron should be worn. To prevent hand and skin exposures, butyl rubber or polycarbonate gloves should be worn.

Administrative controls should also be in place to minimize the potential for human exposures. These may include written procedures or policies which specify the methods and techniques that will be practiced whenever personnel are to work with ferric chloride.

All personnel should receive *training* on the use, hazards, protective measures, emergency actions, and other precautions per 29 CFR 1910.1200 (Hazard Communication), prior to the first assignment in an area where ferric chloride is used or stored.

For those personnel with frequent or potentially high exposure (half the PEL or greater), the following are recommended medical tests to be provided before initial assignment and at regular intervals thereafter (e.g., annually):

☑ Lung function tests.

☑ Consider chest X-ray after acute overexposure (results may be negative if taken immediately after exposure due to delayed development of pulmonary edema).

Any evaluation should include a careful history of past and present symptoms with an examination. Medical tests that simply look for existing damage are not a substitute for controlling exposures. Also, because smoking can cause heart disease, as well as lung cancer, emphysema, and other respiratory problems, symptoms of exposure to ferric chloride anhydrous may be more pronounced in smokers than in non-smokers under the same conditions of exposure. Prudent risk management requires careful consideration of *all* possible exposure factors.

Other methods to reduce exposure include:

☑ Where possible, enclose operations and use local exhaust ventilation at the site of chemical release. If local exhaust ventilation or enclosure is not used, respiratory protection should be mandatory.

☑ Always ensure that proper protective clothing is worn when using chemical substances.

☑ Wash thoroughly immediately after exposure to ferric chloride anhydrous and at the end of the work shift or before eating, drinking, or smoking.

☑ Hazard warning information should be posted in the work area. In addition, as part of an on-going education and training program, all information on the health and safety hazards of ferric chloride anhydrous should be communicated to all potentially exposed workers.

☑ Eye wash stations should be provided in the immediate work area for emergency use. If there is a possibility of skin exposure to ferric chloride, emergency shower facilities should also be provided in the immediate work area.

☑ Personnel should never wear contaminated clothing home (family members can be exposed). Clothing should be laundered by personnel who have been trained on the hazards of ferric chloride.

RISK ASSESSMENT: ENVIRONMENT

General Assessment

The environment is at risk of exposure during transportation, storage, disposal, or destruction of ferric chloride. In almost every scenario, the threat of environmental exposure is contingent upon the proper handling of the chemical substance. Accidental spills, large or small, can result in fire, explosion, and possible contamination of the surrounding environmental mediums (water, soil, and air).

Ferric chloride is considered a non-combustible solid. However, it is corrosive to many metals (especially in solution) and can react vigorously with allyl chloride, potassium, ethylene oxide, nylon, alkalis, and many oxidizers and contact can cause fire or explosion. In contact with water it will nearly dissolve completely. It may emit toxic hydrogen chloride gas in the process. These characteristics require special consideration during any emergency situation involving a leak or spill of ferric chloride. Should ferric chloride ever come into contact with these incompatible either during use, transportation, storage, or disposal, the formation of highly toxic and/or highly explosive commodities is extremely possible.

Ferric chloride can enter the environment mainly through industrial effluents or spills.

Acute Ecological Effects

Acute (short-term) toxic effects may include the death of animals, birds, or fish and death or low growth rate in plants. Acute effects are seen 2 to 4 days after animals or plants are exposed to ferric chloride.

There is insufficient evidence to evaluate or predict the short-term effects of ferric chloride to aquatic life, plants, birds, or land animals.

Chronic Ecological Effects

Chronic (long-term) toxic effects may include shortened life span, reproductive problems, lower fertility, and changes in appearance or behavior in exposed animals. These effects can be seen long after first exposure(s) to toxic chemicals.

There is insufficient evidence to evaluate or predict the long-term effects of ferric chloride to aquatic life, plants, birds, or land animals.

Water Solubility

Ferric chloride is highly soluble in water. Concentrations of 1000 milligrams and more will mix with a liter of water. It will be dissolved completely in contact with water or moisture.

Persistence in the Environment

Ferric chloride is non-persistent in water, with a half-life of less than 2 days. The half-life of a pollutant is the amount of time it takes for one half of the chemical to be degraded.

Bioaccumulation in Aquatic Organisms

Some substances increase in concentration, or *bioaccumulate,* in living organisms as they breathe contaminated air, drink contaminated water, or eat contaminated food. These chemicals can become concentrated in the tissues and internal organs of animals as well as humans.

The concentration of ferric chloride found in fish tissues is expected to be much lower than the average concentration of ferric chloride in the water from which the fish was taken.

Recommended Risk-Reduction Measures

Proper training of all transporters will reduce the likelihood of a mishap or accident resulting in a leak or spill to the environment. The correct use of DOT labeling (per the MSDS) on all containers, trucks, and rail cars will enable emergency responders to react properly and quickly to any disaster thereby reducing the potential risk to the environment and to personnel.

Storage of ferric chloride should be segregated from incompatible chemicals to minimize the risk of cross-contamination or contact. Exposure to any form of water (moisture, steam) can soften the material and may eventually dissolve ferric chloride.

If a spill or leak to the environment has occurred, fire department, emergency response and/or hazardous materials spill personnel should be notified immediately. Cleanup should be attempted only by those trained in proper spill containment procedures. Contaminated soils should be removed for incineration and replaced with clean soil. If ferric chloride should contact the water table, aquifer, or navigable waterway, the local and/or state emergency response authorities must be notified. A comprehensive emergency response or disaster preparedness/recovery plan should be in place prior to any operations involving the use, transportation, storage, or disposal of ferric chloride. If ferric chloride is spilled or leaked, the following specific steps are recommended:

- ☑ Restrict persons not wearing protective clothing from area of spill or leak until cleanup is complete and area can be opened for normal work.
- ☑ Ventilate area of spill or leak.
- ☑ Collect solid materials in safest and most efficient manner possible. Do NOT dry sweep (generates

airborne dusts). Use vacuum equipped with high efficiency particulate air (HEPA) filter instead. Absorbed spilled solutions in vermiculite or other material and deposit in sealed containers.

☑ It may be necessary to dispose of ferric chloride as a hazardous waste. The responsible state agency or the regional office of the federal Environmental Protection Agency (EPA) should be contacted for specific recommendations.

RISK ASSESSMENT: BUSINESS

General Assessment

Accidents or mishaps involving ferric chloride can present a significant threat to business operations. The loss or damage of equipment or facilities can significantly affect fiscal viability. Lawsuits that may result from personnel injury/death, public exposures, and/or environmental contamination will also require a serious expenditure of resources. Media attention surrounding an injury, death, or environmental damage can also result in a loss of profits and loss of current as well as future business.

Recommended Risk-Reduction Measures

Company attorneys, safety and health professionals, and environmental specialists should be involved in the development of any safety or emergency response procedures. A company official should be pre-designated as a public relations officer with specific training in dealing with the press. Corporate plans and policies should be developed, approved, and implemented long before any need for such arises.

REFERENCES

Ahlbom, A. 1993. *Biostatistics for Epidemiologists.* New York: CRC Press/Lewis Publishers.

American Conference of Governmental Industrial Hygienists. 1988. *Documentation of the Threshold Limit Values and Biological Exposure Indices,* 5th Edition (with updates). Cincinnati: ACGIH

Asante-Duah, D. K. 1993. *Hazardous Waste Risk Assessment.* New York: CRC Press/Lewis Publishers.

Brusick, D. J. 1994. *Methods for Genetic Risk Assessment.* New York: CRC Press/Lewis Publishers.

Bueche, J. R. 1972. *Principles of Physics,* 2nd Edition. New York: McGraw-Hill.

Cockerman, L. G., B. S. Shane. 1994. *Basic Environmental Toxicology.* New York: CRC Press/Lewis Publishers.

Cothern, R. R. 1993. *Comparative Environmental Risk Assessment.* New York: CRC Press/Lewis Publishers.

Gardner, R. W. 1994. *Chemical Intolerance.* New York: CRC Press/Lewis Publishers.

Hallenbeck, W. H. 1993. *Quantitative Risk Assessment for Environmental and Occupational Health,* 2nd Edition. New York: CRC Press/Lewis Publishers.

Howard, P. H. 1995. *Handbook of Environmental Fate and Exposures Data for Organic Chemicals,* Vol. I-IV. New York: CRC Press/Lewis Publishers.

Keith, L. H. and M. M. Walker. 1995. *Handbook of Air Toxics.* New York: CRC Press/Lewis Publishers.

Landis, W. G., M. H. Yu. 1995. *Introduction to Environmental Toxicology.* New York: CRC Press/Lewis Publishers.

Lauwerys, R. R., P. Hoet. 1993. *Industrial Chemical Exposure.* New York: CRC Press/Lewis Publishers.

Lewis, R. J., Sr. 1993. *Hawley's Condensed Chemical Dictionary,* 12th Edition. New York: Van Nostrand Reinhold.

Lewis, R. J., Sr. 1992. *Sax's Dangerous Properties of Industrial Materials,* Volumes 1, 2, and 3, 8th Edition. New York: Van Nostrand Reinhold.

Manahan, S. E. 1992. *Toxicological Chemistry,* 2nd Edition. New York: CRC Press/Lewis Publishers.

Meyer, E. 1990. *Chemistry of Hazardous Materials.* Englewood Cliffs, NJ: Prentice Hall.

Mycock, J. C., J. D. McKenna, and L. Theodore. 1995. *Handbook of Air Pollution Control Engineering and Technology.* New York: CRC Press/Lewis Publishers.

Ott, W. R. 1995. *Environmental Statistics and Data Analysis.* New York: CRC Press/Lewis Publishers.

Patnaik, P. 1992. *A Comprehensive Guide to the Hazardous Properties of Chemical Substances.* New York: Van Nostrand Reinhold.

Philp, R. B. 1995. *Environmental Hazards and Human Health.* New York: CRC Press/Lewis Publishers.

Quigley, D. R. 1994. *Handbook of Emergency Chemical Management.* New York: CRC Press/Lewis Publishers.

Rea, W. J. 1995. *Chemical Sensitivity,* Vol. I-III. New York: CRC Press/Lewis Publishers.

Richardson, M. 1995. *Dictionary of Substance and Their Effects,* Vol. I-VI. New York: CRC Press/Lewis Publishers.

Sacarello, H. L. A. 1994. *The Comprehensive Handbook of Hazardous Materials.* New York: CRC Press/Lewis Publishers.

Sherman, J. D. 1988. *Chemical Exposure and Disease.* New York: Van Nostrand Reinhold.

Talbott, E., G. Craun. 1995. *An Introduction to Environmental Epidemiology.* New York: CRC Press/Lewis Publishers.

Torkelson, T. R. V. K. Rowe. 1981. Halogenated aliphatic hydrocarbons. *Patty's Industrial Hygiene and Toxicology,* Vol. 2B, *Toxicology,* pp 3568-3572. New York: Wiley-Interscience.

MATERIAL SAFETY DATA SHEET

CHEMICAL NAME	CRC PRESS/LEWIS PUBLISHERS DISCLAIMER:
FERRIC NITRATE	THE INFORMATION AND RECOMMENDATIONS PRESENTED HEREIN ARE BASED ON SOURCES BELIEVED TO BE RELIABLE. CRC MAKES NO REPRESENTATION ON ITS COMPLETENESS OR ACCURACY. IT IS THE RESPONSIBILITY OF THE USER TO DETERMINE THE CHEMICAL'S SUITABILITY FOR ITS INTENDED USE, THE CHEMICAL'S SAFE USE, AND THE CHEMICAL'S PROPER DISPOSAL. NO REPRESENTATIONS AND/OR WARRANTIES, EITHER EXPRESSED OR IMPLIED, OF THE MERCHANTABILITY OR FITNESS FOR A PARTICULAR PURPOSE, OR OF ANY OTHER NATURE, ARE MADE WITH RESPECT TO THE INFORMATION PROVIDED IN THIS MSDS OR TO THE CHEMICAL TO WHICH INFORMATION MAY REFER. CRC NEITHER ASSUMES NOR AUTHORIZES ANY OTHER PERSON TO ASSUME FOR IT, ANY OTHER ADDITIONAL RESPONSIBILITY OR LIABILITY FOR THE USE OF, OR RELIANCE UPON, THIS INFORMATION.

HAZARD WARNING INFORMATION

HEALTH	FIRE	REACTIVE	OTHER	DEGREE OF HAZARD	COLOR CODING	OTHER CODES
2	2	0	OX	0 = Minimum Hazard 1 = Slight Hazard 2 = Moderate Hazard 3 = Serious Hazard 4 = Severe Hazard	HEALTH = BLUE FIRE = RED REACTIVITY = YELLOW OTHER = WHITE	OX = Oxidizer ACID = Acid ALK = Alkali COR = Corrosive W = Use No Water

SECTION I - GENERAL INFORMATION

Characterization	RCRA Number	EPA Class
Inorganic Salt	**None**	**Not Applicable**

DOT Proper Shipping Name **Ferric nitrate**	Chemical Abstract Service (CAS) Number **10421-48-4**
DOT Hazard Class and Label Requirements **Oxidizer**	DOT Emergency Guide Code **42**
DOT Identification Number **UN 1466**	Molecular Formula **$Fe(NO_3)_3 \cdot 9H_2O$**

Synonyms

Iron nitrate; iron (III) nitrate, anhydrous; iron trinitrate; nitric acid, iron (3+) salt.

SECTION II - HAZARDOUS INGREDIENTS AND IDENTITY INFORMATION

Hazardous Components (specific identity)	OSHA Exposure Criteria	NIOSH Exposure Criteria	Immediately Dangerous to Life and Health (IDLH)	ACGIH Exposure Criteria
Ferric nitrate (derivation: By action of concentrated nitric acid on scrap iron or iron oxide and crystallization).	**PEL:** 1 mg(Fe)/m^3 (as iron salt) **STEL:** Not Established	**REL:** 1 mg(Fe)/m^3 (as iron salt) **STEL:** Not Established	**Not Determined**	**TLV:** 1 mg(Fe)/m^3 (as iron salt) **STEL:** Not Established

SECTION III - PHYSICAL AND CHEMICAL CHARACTERISTICS

Boiling Point **257°F (125°C) Decomposes**	Specific Gravity (H_2O = 1) **1.68**
Vapor Pressure (mm Hg) **Not Found**	Molecular Weight (atomic weight) **241.88**
Vapor Density (Air = 1) **Not Found**	Melting Point **117°F (47°C)**

Solubility

Soluble in water. Very soluble in alcohol and acetone, slightly soluble in cold concentrated nitric acid.

Appearance and Odor

Pale violet or grayish-white deliquescent crystals.

SECTION IV - FIRE AND EXPLOSION HAZARD DATA

Flash Point (method used) **Not Determined**	Explosive Limits in Air % by Volume **LEL: Not Determined UEL: Not Determined**
NFPA Classification **Combustible Solid**	Autoignition Temperature **Not Found**

Extinguishing Media

Water only, do not use dry chemical, carbon dioxide, or halon. For larger fires, flood area with water from a distance. For massive fire in cargo areas, use unmanned hose holder or monitor nozzles.

Special Fire Fighting Procedures

Wear full protective clothing and self-contained breathing apparatus (SCBA). Poisonous gases are produced in fire. Move container from fire area if it can be done without risk. Apply cooling water to sides of containers exposed to flames until well after fire is out. Stay away from ends of tanks.

Unusual Fire and Explosion Hazards

Ferric nitrate is an oxidizer. By itself, it will not burn. However, contact with organic or combustible materials such as wood, paper, or oil, can cause fires or explosions. If fires are out of control, consider evacuation of 1/2 mile radius.

SECTION V - REACTIVITY DATA

Stability		Conditions to Avoid
Stability		Normally stable, but avoid contact with combustible materials. Ferric nitrate may increase the burning rate. Finely divided materials may ignite or explode on contact.
Stable	Unstable	Incompatibility *(materials to avoid)*
X		Organic and combustible materials (such as wood, paper, oils, and fuels), aluminum, cyanides, phosphorus, acetylene gas, reducing agents (stannous chloride, thiocyanates, sodium hypophosphite, tin, magnesium).
Hazardous Polymerization		Conditions to Avoid: Hazardous polymerization has not been reported to occur under normal conditions of temperature and pressure.
May Occur	Will Not Occur	Hazardous Decomposition or By-products
	X	Poisonous gases, including oxides of nitrogen and nitric acid vapor, are emitted during fire.

SECTION VI - HEALTH HAZARD DATA

Primary Route(s) of Entry:	Inhalation?	Absorption (skin/eye)?	Ingestion?
	X		X

Health Hazards

INHALATION: May cause irritation of the respiratory tract with symptoms of sore throat, coughing, and shortness of breath or labored breathing.

EYE & SKIN: Irritation with redness and pain. Repeated eye contact may cause conjunctivitis.

INGESTION: Severe gastrointestinal irritation with abdominal pain, nausea, vomiting (possible bloody), diarrhea, colic, headache, fatigue, excessive salivation, shock, liver injury, and death. Dryness and constriction of mouth and throat with metallic taste may also occur.

Carcinogenicity	NTP Listed?	IARC Cancer Review Group?	OSHA Regulated?	Target Organs?
Unknown Human Unknown Animal	No	No	29 CFR 1910.1000 Table Z-1	Resp. tract, liver, skin, kidneys, G.I., mucosa.

Medical Conditions Generally Aggravated by Exposure

None reported.

Emergency and First-aid Procedures

Eye contact: Do not allow victim to keep eyes tightly shut. Flush immediately with water for 15 minutes (minimum), seek medical attention. Skin contact: Remove all contaminated clothing, including shoes. Immediately wash area with flooding amounts of soap and water. Seek medical attention immediately. For inhalation: Remove the person from exposure. Provide respiratory assistance and CPR. Transfer to medical facility. If swallowed: Induce vomiting immediately, seek medical attention immediately.

SECTION VII - PRECAUTIONS FOR SAFE HANDLING AND USE

Steps to be Taken in Case Material is Released or Spilled

Keep combustibles away from spilled material. Do not touch spilled materials with bare skin. For small spills, use a clean shovel to place materials into clean, dry container and cover. For large spills, dike area far ahead of spill for later disposal. Restrict those not involved in cleanup from entering area.

Preferred Waste Disposal Method

Refer to local, state, and/or federal disposal regulations for characteristic waste disposal requirements.

Precautions to be Taken in Handling and Storage

Protect against physical damage. Separate from combustible, organic or other easily oxidized materials. Avoid storage on wood floors. Immediately remove and dispose of any spilled materials. Store in tightly closed containers in cool, well-ventilated location. Keep away from heat and direct sunlight.

Other Precautions and Warnings

Avoid storing aqueous solutions in metal containers because of ferric nitrate's extremely corrosive nature.

SECTION VIII - CONTROL MEASURES AND PERSONAL PROTECTIVE EQUIPMENT

Respiratory Protection *(specify type)*

For dust exposures below PEL, use a respirator with a high efficiency particulate air (HEPA) filter. Greater protection is provided by a powered air-purifying respirator or use a self-contained breathing apparatus (SCBA) with full facepiece operated in positive pressure mode.

Ventilation

Local exhaust (preferred) at site of chemical work or general ventilation.

Protective Gloves	Eye Protection	Other Protective Clothing
Impervious Rubber	Chemical/Dust Goggles or Face Mask	Protective Apron, Boots

Work/Hygiene Practices

Always wash hands thoroughly after using chemical; never bring food, drink, or smoking materials into vicinity of chemicals.

FERRIC NITRATE

$Fe(NO_3)_3 \cdot 9H_2O$ CAS: 10421-48-4

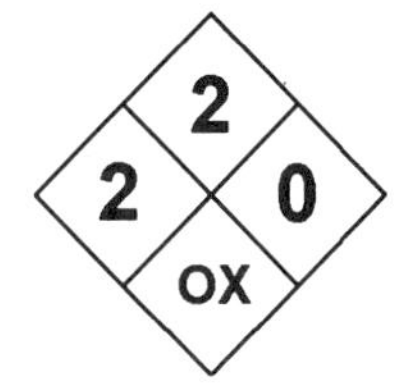

IDENTIFICATION AND TYPICAL USES

Ferric nitrate appears as pale violet or grayish-white, deliquescent crystals. It is used as a dyeing mordant for buffs and blacks and in weighting silk, in tanning, as a corrosion inhibitor, and in analytical chemistry.

RISK ASSESSMENT: HEALTH

General Assessment

Ferric nitrate is a human poison via *ingestion* and *inhalation.* It is also a strong irritant to the eyes and skin. Symptoms of inhalation include nose and throat irritation, bronchial tube inflammation causing cough and phlegm.

The most severe cases of poisoning from ferric nitrate exposure are those resulting from ingestion. The initial effects from soluble ferric compounds are those of severe gastrointestinal irritation, including nausea, vomiting and diarrhea that may be accompanied by abdominal colic, and excessive salivation. There may also be dryness, a sense of constriction of the mouth and throat, and a metallic taste. Systemic effects follow and may include repeated and often bloody vomiting; diarrhea with water and probably tar-colored feces indicating the presence of blood; dehydration, headache; fatigue and weakness; shock with pale and clammy skin; rapid and weak pulse; low blood pressure; and possible progression to acidosis, liver injury, and possible death.

Rarely, inorganic nitrates may be converted to nitrates by nitrate-reducing bacteria in the digestive tract, resulting in methemoglobinemia.

Acute Health Effects

The following acute (short-term) health effects may occur immediately or shortly after exposure to ferric nitrate:

Skin: Severe irritation and burns, redness, and pain.

Eye: Severe irritation, redness, and pain.

Lung: Respiratory tract irritation causing cough, labored breathing, and phlegm discharge.

Other: Ingestion can cause a systemic reaction with symptoms leading to potentially fatal shock.

Chronic Health Effects

The following chronic (long-term) health effects may occur at some time after exposure to ferric nitrate and can last for months or even years:

Cancer Hazards: According to the references, ferric nitrate has not been tested for its ability to cause cancer in animals.

Reproductive Hazard: According to the references, ferric nitrate has not been tested for its ability to affect reproduction.

Other Chronic Effects: Repeated, low-level ingestion may cause a progressive weakening or degeneration of the liver.

Recommended Risk-Reduction Measures

Personnel should avoid direct contact with ferric nitrate dusts, powders, or solutions. If a less toxic material or compound cannot be substituted for ferric nitrate, then *engineering controls* are the most effective method of reducing exposures. The best protection is to enclose operations and/or provide local exhaust ventilation at the site of ferric nitrate dust release. While not always operationally feasible, isolating operations can also reduce exposure. Using respiratory protection is less effective than the controls mentioned above, but is still advisable whenever working with or around ferric nitrate. A supplied-air respirator or a self-contained breathing apparatus (SCBA) with full facepiece operated in pressure demand or other positive pressure mode are recommended for the greatest possible respiratory protection, especially when exposure levels are unknown (as in an emergency). However, for low or infrequent exposure, an MSHA/NIOSH-approved full facepiece respirator with a high efficiency particulate air (HEPA) filter may be used. If a full facepiece is not available, then chemical/dust goggles should be worn to protect the eyes. A face shield and protective apron should also be worn. To prevent hand and skin exposures, gloves should be worn.

Administrative controls should also be in place to minimize the potential for human exposures. These may include written procedures or policies which specify the methods and techniques that will be practiced whenever personnel are to work with ferric nitrate.

All personnel should receive *training* on the use, hazards, protective measures, emergency actions, and other precautions per 29 CFR 1910.1200 (Hazard Communication), prior to the first assignment in an area where ferric nitrate is used or stored.

Before beginning employment working with ferric nitrate, and at regular intervals thereafter (e.g., annually), those personnel with frequent or potentially high exposures should be provided the following recommended medical tests:

☑ Lung function test.

☑ Liver function test.

If symptoms develop or overexposure is suspected, the following additional tests and procedures are recommended:

☑ Complete blood count (CBC).

☑ Monitor electrolytes, blood sugar, serum iron, and iron binding (TIBC).

☑ Consider abdominal X-ray.

Any evaluation should include a careful history of past and present symptoms with an examination. Medical tests that simply look for existing damage are not a substitute for controlling exposures.

Other methods to reduce exposure include:

☑ Where possible, enclose operations and use local exhaust ventilation at the site of chemical release. If local exhaust ventilation or enclosure is not used, respiratory protection should be mandatory.

☑ Always ensure that proper protective clothing is worn when using chemical substances.

☑ Wash thoroughly immediately after exposure to ferric nitrate and at the end of the work shift or before eating, drinking, or smoking.

☑ Hazard warning information should be posted in the work area. In addition, as part of an on-going education and training program, all information on the health and safety hazards of ferric nitrate should be communicated to all potentially exposed workers.

☑ Eye wash stations should be provided in the immediate work area for emergency use. If there is a possibility of skin exposure to ferric nitrate, emergency shower facilities should also be provided.

☑ Workers whose clothing has been contaminated by ferric nitrate should change into clean clothes before leaving work. Contaminated work clothing should be laundered only by individuals who have been informed of the hazards of exposure to ferric nitrate.

RISK ASSESSMENT: ENVIRONMENT

General Assessment

The environment is at risk of exposure during transportation, storage, disposal, or destruction of ferric nitrate. In almost every scenario, the threat of environmental exposure is contingent upon the proper handling of the chemical substance. Accidental spills, large or small, where ferric nitrate contacts incompatible commodities can result in fire, explosion (depending upon conditions of spill), and possible contamination of the surrounding environmental mediums (water, soil, and air).

Ferric nitrate is an oxidizer. By itself, it will not burn. However, it readily facilitates the combustion of other materials. It can therefore present a moderate fire and explosion hazard. Also, because it is incompatible so many commonly encountered materials such as wood and paper, as well as other commodities, extreme caution is required in handling, storage, transportation, and disposal of ferric nitrate. These characteristics also require special consideration during any emergency situation involving a leak or spill of ferric nitrate powder or dust. Should ferric nitrate ever come into contact with incompatible substances (such as combustible materials, aluminum, or acetylene gas) either during use, transportation, storage, or disposal, violent and even explosive reactions are possible.

Ferric nitrate enter the environment from industrial and municipal waste treatment plant discharges and from spills.

Acute Ecological Effects

Acute (short-term) toxic effects may include the death of animals, birds, or fish and death or low growth rate in plants. Acute effects are seen 2 to 4 days after animals or plants are exposed to ferric nitrate. Iron and its salts have moderate acute toxicity to aquatic life. Insufficient data are available to evaluate the short-term effects of iron or its salts to plants, birds, or land animals.

Chronic Ecological Effects

Chronic (long-term) toxic effects may include shortened life span, reproductive problems, lower fertility, and changes in appearance or behavior in exposed animals. These effects can be seen long after first exposure(s) to toxic chemicals. Iron and its salts have moderate chronic toxicity to aquatic life. Insufficient data are available to evaluate or predict the long-term effects of iron or its salts to plants, birds, or terrestrial animals.

Water Solubility

Most of the iron salts are either highly or moderately soluble in water. Concentrations of 1 to 1000 milligrams and more will mix with a liter of water.

Persistence in the Environment

Ferric nitrate is highly persistent in water, with a half-life greater than 200 days. The half-life of a pollutant is the amount of time it takes for one half of the chemical to be degraded. No data are available on the estimated percentages of ferric nitrate persistence in the aquatic or terrestrial environments.

Bioaccumulation in Aquatic Organisms

Some substances increase in concentration, or *bioaccumulate,* in living organisms as they breathe contaminated air, drink contaminated water, or eat contaminated food. These chemicals can become concentrated in the tissues and internal organs of animals as well as humans.

The concentration of ferric nitrate found in fish tissues is expected to be about the same as the average concentration of ferric nitrate in the water from which the fish was taken.

Recommended Risk-Reduction Measures

Proper training of all transporters will reduce the likelihood of a mishap or accident resulting in a leak or spill of ferric nitrate dusts or powders into the environment. The correct use of DOT labeling (per the MSDS) on all containers, trucks, and rail cars will enable emergency responders to react properly and quickly to any disaster thereby reducing the potential risk to the environment and to personnel.

Storage of ferric nitrate should be segregated from incompatible chemicals and materials to minimize the risk of cross-contamination or contact.

If a spill or leak to the environment has occurred, fire department, emergency response and/or hazardous materials spill personnel should be notified immediately. Cleanup should be attempted only by those trained in proper spill containment procedures. Contaminated soils should be removed for incineration and replaced with clean soil. If ferric nitrate should contact the water table, aquifer, or navigable waterway, time is of the essence. It is highly soluble and total remediation may not be possible. The local and/or state emergency response authorities must be notified. A comprehensive emergency response or disaster preparedness/recovery plan should be in place prior to any operations involving the use, transportation, storage, or disposal of ferric nitrate. If ferric nitrate is spilled or leaked, the following specific steps are recommended:

- ☑ Restrict persons not wearing protective clothing from area of spill or leak until cleanup is complete and area can be opened for normal work.
- ☑ Collect powdered materials in the safest manner possible and deposit in sealed containers. Recommend using clean shovel for small spills, depositing materials in clean container. For larger spills, use a vacuum with HEPA filter. Do not dry sweep (generates airborne dusts).
- ☑ It may be necessary to dispose of ferric nitrate as a hazardous waste. The responsible state agency or the regional office of the federal Environmental Protection Agency (EPA) should be contacted for specific recommendations.

RISK ASSESSMENT: BUSINESS

General Assessment

Accidents or mishaps involving ferric nitrate can present a significant threat to business operations. The loss or damage of equipment or facilities can significantly affect fiscal viability. Lawsuits that may result from personnel injury/death, public exposures, and/or environmental contamination will also require a serious expenditure of resources. Media attention surrounding an injury, death, or environmental damage can also result in a loss of profits and loss of current as well as future business.

Recommended Risk-Reduction Measures

Company attorneys, safety and health professionals, and environmental specialists should be involved in the development of any procedures for responding to chemical incidents. A company official should be predesignated as a public relations officer with specific training in dealing with the press. Corporate plans and policies should be developed, approved, and implemented long before any need for such arises.

REFERENCES

Ahlbom, A. 1993. *Biostatistics for Epidemiologists.* New York: CRC Press/Lewis Publishers.

American Conference of Governmental Industrial Hygienists. 1988. *Documentation of the Threshold Limit Values and Biological Exposure Indices,* 5th Edition (with updates). Cincinnati: ACGIH

Anthony, C. P., and N. J. Kolthoff. 1971. *Textbook of Anatomy and Physiology.* St. Louis: C. V. Mosby Company.

Asante-Duah, D. K. 1993. *Hazardous Waste Risk Assessment.* New York: CRC Press/Lewis Publishers.

Bloom, A. D. and F. J. DeSerres. 1995. *Ecotoxicity and Human Health.* New York: CRC Press/Lewis Publishers.

Brusick, D. J. 1994. *Methods for Genetic Risk Assessment.* New York: CRC Press/Lewis Publishers.

Bueche, J. R. 1972. *Principles of Physics,* 2nd Edition. New York: McGraw-Hill.

Calabrese, E. J. 1994. *Biological Effects of Low Level Exposures.* New York: CRC Press/Lewis Publishers.

Cockerman, L. G., B. S. Shane. 1994. *Basic Environmental Toxicology.* New York: CRC Press/Lewis Publishers.

Cothern, R. R. 1993. *Comparative Environmental Risk Assessment.* New York: CRC Press/Lewis Publishers.

Danse, I. R. 1991. *Common Sense Toxics in the Workplace.* New York: Van Nostrand Reinhold.

Environmental Protection Agency. 1983. *Chemical Hazard Information Profiles.* Washington, D.C.: U.S. EPA.

Fodden, C. R., and J. L. Weddell. 1992. *Hazardous Materials Emergency Action Data.* New York: CRC Press/Lewis Publishers.

Forseberg, K., and L. H. Keith. 1993. *Chemical; Protective Clothing Permeation and Degradation Database.* Chelsea, MI: Lewis Publishers.

Gardner, R. W. 1994. *Chemical Intolerance.* New York: CRC Press/Lewis Publishers.

Hallenbeck, W. H. 1993. *Quantitative Risk Assessment for Environmental and Occupational Health,* 2nd Edition. New York: CRC Press/Lewis Publishers.

Howard, P. H. 1995. *Handbook of Environmental Fate and Exposures Data for Organic Chemicals,* Vol. I-IV. New York: CRC Press/Lewis Publishers.

Kamrin, M. A. 1988. *Toxicology: A Primer on Principles and Applications.* Chelsea, Michigan: Lewis Publishers.

Keith, L. H. , M. T. Johnston, C. E. Hudak, and M. Conoley. 1992. *National Toxicology Program's Gloves+ Database.* Chelsea, MI: Lewis Publishers.

Keith, L. H. and M. M. Walker. 1995. *Handbook of Air Toxics.* New York: CRC Press/Lewis Publishers.

Landis, W. G., M. H. Yu. 1995. *Introduction to Environmental Toxicology.* New York: CRC Press/Lewis Publishers.

Langerman, N. 1994 *Precautionary Labels for Chemical Containers.* New York: CRC Press/Lewis Publishers.

Lauwerys, R. R., P. Hoet. 1993. *Industrial Chemical Exposure.* New York: CRC Press/Lewis Publishers.

Lewis, R. J., Sr. 1993. *Hawley's Condensed Chemical Dictionary,* 12th Edition. New York: Van Nostrand Reinhold.

Lewis, R. J., Sr. 1992. *Sax's Dangerous Properties of Industrial Materials,* Volumes 1, 2, and 3, 8th Edition. New York: Van Nostrand Reinhold.

Lide, D. R. 1995. *Handbook of Chemistry and Physics,* 76th Edition. New York: CRC Press/Lewis Publishers.

Lisella, F. S. 1994. *The VNR Dictionary of Environmental Health and Safety.* New York: Van Nostrand Reinhold.

Mackay, D., W. Y. Shiu, and K. C. Ma. 1992. *Illustrated Handbook of Chemical-Physical Properties and Environmental Fate for Organic Chemicals.* New York: CRC Press/Lewis Publishers.

Manahan, S. E. 1994. *Environmental Chemistry,* 6th Edition. New York: CRC Press/Lewis Publishers.

Meyer, E. 1990. *Chemistry of Hazardous Materials.* Englewood Cliffs, NJ: Prentice Hall.

Patnaik, P. 1992. *A Comprehensive Guide to the Hazardous Properties of Chemical Substances.* New York: Van Nostrand Reinhold.

Proctor, N. H., J. P. Hughes. 1991. *Chemical Hazards of the Workplace.* New York: Van Nostrand Reinhold.

Quigley, D. R. 1994. *Handbook of Emergency Chemical Management.* New York: CRC Press/Lewis Publishers.

Rea, W. J. 1995. *Chemical Sensitivity,* Vol. I-III. New York: CRC Press/Lewis Publishers.

Richardson, M. 1995. *Dictionary of Substance and Their Effects,* Vol. I-VI. New York: CRC Press/Lewis Publishers.

Sacarello, H. L. A. 1994. *The Comprehensive Handbook of Hazardous Materials.* New York: CRC Press/Lewis Publishers.

Talbott, E., G. Craun. 1995. *An Introduction to Environmental Epidemiology.* New York: CRC Press/Lewis Publishers.

Torkelson, T. R. V. K. Rowe. 1981. *Patty's Industrial Hygiene and Toxicology,* Vol. 2B. New York: Wiley-Interscience.

MATERIAL SAFETY DATA SHEET

CHEMICAL NAME	
FERROCENE	**CRC PRESS/LEWIS PUBLISHERS DISCLAIMER:** THE INFORMATION AND RECOMMENDATIONS PRESENTED HEREIN ARE BASED ON SOURCES BELIEVED TO BE RELIABLE. CRC MAKES NO REPRESENTATION ON ITS COMPLETENESS OR ACCURACY. IT IS THE RESPONSIBILITY OF THE USER TO DETERMINE THE CHEMICAL'S SUITABILITY FOR ITS INTENDED USE, THE CHEMICAL'S SAFE USE, AND THE CHEMICAL'S PROPER DISPOSAL. NO REPRESENTATIONS AND/OR WARRANTIES, EITHER EXPRESSED OR IMPLIED, OF THE MERCHANTABILITY OR FITNESS FOR A PARTICULAR PURPOSE, OR OF ANY OTHER NATURE, ARE MADE WITH RESPECT TO THE INFORMATION PROVIDED IN THIS MSDS OR TO THE CHEMICAL TO WHICH INFORMATION MAY REFER. CRC NEITHER ASSUMES NOR AUTHORIZES ANY OTHER PERSON TO ASSUME FOR IT, ANY OTHER ADDITIONAL RESPONSIBILITY OR LIABILITY FOR THE USE OF, OR RELIANCE UPON, THIS INFORMATION.

HAZARD WARNING INFORMATION

HEALTH	FIRE	REACTIVE	OTHER	DEGREE OF HAZARD	COLOR CODING	OTHER CODES
1	2	0		0 = Minimum Hazard 1 = Slight Hazard 2 = Moderate Hazard 3 = Serious Hazard 4 = Severe Hazard	HEALTH = BLUE FIRE = RED REACTIVITY = YELLOW OTHER = WHITE	OX = Oxidizer ACID = Acid ALK = Alkali COR = Corrosive W = Use No Water

SECTION I - GENERAL INFORMATION

Characterization	RCRA Number	EPA Class
Aromatic Hydrocarbon	**None**	**Not Applicable**

DOT Proper Shipping Name **Not Found**	Chemical Abstract Service (CAS) Number **102-54-5**
DOT Hazard Class and Label Requirements **Not Found**	DOT Emergency Guide Code **No Citation**
DOT Identification Number **Not Found**	Chemical Formula $C_{10}H_{10}Fe$

Synonyms

Biscyclopentadienyl iron; di-2,4-cycolpentadien-1-yl iron; di-pi-cyclopentadienyliron; ferrotsen; dicyclopentadienyl iron; iron bis(cyclopentadiene); iron dicycolpentadienyl.

SECTION II - HAZARDOUS INGREDIENTS AND IDENTITY INFORMATION

Hazardous Components (specific identity)	OSHA Exposure Criteria	NIOSH Exposure Criteria	Immediately Dangerous to Life and Health (IDLH)	ACGIH Exposure Criteria
Ferrocene (derivation: By reaction of cyclopentadiene with reduced iron in the presence of metal oxides; by reaction of cyclopentadiene sodium with ferrous chloride).	**PEL (8-hour):** 15 mg/m³ (total dust) 5 mg/m³ (respirable) **STEL:** Not Applicable	**REL:** Not Established **STEL:** Not Established	**Not Determined**	**TLV:** 10 mg/m³ **STEL:** Not Established

SECTION III - PHYSICAL AND CHEMICAL CHARACTERISTICS

Boiling Point **480°F (249°C)**	Specific Gravity (H_2O = 1) **Not Found**
Vapor Pressure (gas) **Not Found**	Molecular Weight **186.05**
Vapor Density (Air = 1) **Not Determined**	Melting Point **343°F (173°C)**

Solubility

Insoluble in water. Soluble in benzene, alcohol, and ether. Sublimes above 212°F (100°C).

Appearance and Odor

Orange, crystalline solid with a camphor-like odor.

SECTION IV - FIRE AND EXPLOSION HAZARD DATA

Flash Point (method used) **Not Determined**	Explosive Limits in Air % by Volume **LEL: Not Determined UEL: Not Determined**
NFPA Classification **Flammable Solid**	Autoignition Temperature **Not Determined**

Extinguishing Media

Use dry chemical, alcohol foam, water spray, or carbon dioxide.

Special Fire Fighting Procedures

Wear full protective clothing and self-contained breathing apparatus (SCBA). Poisonous gases are produced in fire. Evacuate 1500 feet radius if fire becomes uncontrollable or container is exposed to flame.

Unusual Fire and Explosion Hazards

Dangerous fire and explosion hazard when exposed to heat or flame. Containers may explode in fire. May present a moderate explosion hazard when powders become airborne in concentrated areas.

SECTION V - REACTIVITY DATA

Stability		Conditions to Avoid **Normally stable. Keep away from heat or flame and incompatible materials. Very reactive with incompatible materials. Reactions can be violent and even explosive.**
Stable **X**	Unstable	Incompatibility *(materials to avoid)* **Reacts violently with ammonium perchlorate. When mixed with mercury (II) nitrate, can become explosive and sensitive to electrostatic charges.**
Hazardous Polymerization		Conditions to Avoid **Under normal conditions of temperature and pressure, hazardous polymerization of ferrocene is not expected to occur.**
May Occur	Will Not Occur **X**	Hazardous Decomposition or By-products **Hazardous decomposition of ferrocene will yield toxic and irritating smoke and fumes.**

SECTION VI - HEALTH HAZARD DATA

Primary Route(s) of Entry:	Inhalation?	Absorption (skin)?	Ingestion?
	X		X

Health Hazards

INHALATION: Possible irritation of the eyes and upper respiratory tract with symptoms of headache and nausea. May depress the circulatory system and respiratory system. High doses may cause narcosis, convulsions, liver and kidney damage, and death (human exposure data inconclusive).

SKIN & EYES: Mild to moderate irritation of the skin and eyes. Contact dermatitis possible.

INGESTION: No firm data reported. Possible gastrointestinal irritation, with diarrhea and nausea.

Carcinogenicity	NTP Listed?	IARC Cancer Review Group?	OSHA Regulated?	Target Organs?
Unknown Human Questioned Animal	**No**	**No**	**29 CFR 1910.1000 Table Z-1**	**Respiratory system, eyes, CNS, blood, liver.**

Medical Conditions Generally Aggravated by Exposure

None reported.

Emergency and First-aid Procedures

Eye contact: Flush immediately with water for 15 minutes (minimum), seek medical attention. Skin contact: Remove all contaminated clothing. Immediately wash area with large amounts of soap and water. Seek medical assistance, if necessary. For inhalation: Remove the person from exposure. Provide respiratory assistance and CPR, if required. Transfer to medical facility. If swallowed: Never give an unconscious or convulsing person anything by mouth. Seek medical attention immediately.

SECTION VII - PRECAUTIONS FOR SAFE HANDLING AND USE

Steps to be Taken in Case Material is Released or Spilled

Remove all sources of ignition. Collect powdered materials in safest manner possible. Avoid generating dusts. Do NOT dry sweep. Use a HEPA vacuum and damp mop residue using non-sparking tools. Ventilate area of spill or leak. Restrict those not involved in cleanup from entering area.

Preferred Waste Disposal Method

No Citation.

Precautions to be Taken in Handling and Storage

Store to avoid contact with incompatible materials in tightly closed containers in cool, well-ventilated area away from heat and flame. Use only non-sparking tools to open and close containers.

Other Precautions and Warnings

Sources of ignition are prohibited where ferrocene is used, handled, or stored. Protect containers from physical damage.

SECTION VIII - CONTROL MEASURES AND PERSONAL PROTECTIVE EQUIPMENT

Respiratory Protection *(specify type)*

For low exposures (up to PEL), use an organic vapor respirator with HEPA filter and dust pre-filter. For higher exposures, use a supplied-air respirator with full facepiece in continuous flow mode, or a self-contained breathing apparatus (SCBA) operated in pressure demand or other positive pressure mode.

Ventilation

Local exhaust or general mechanical systems recommended.

Protective Gloves	Eye Protection	Other Protective Clothing
Impervious Rubber	**Chemical Goggles or Face Mask**	**Rubber Apron**

Work/Hygiene Practices

Always wash hands thoroughly after using chemical; never bring food, drink, or smoking materials into vicinity of chemicals.

FERROCENE

$C_{10}H_{10}Fe$

CAS: 102-54-5

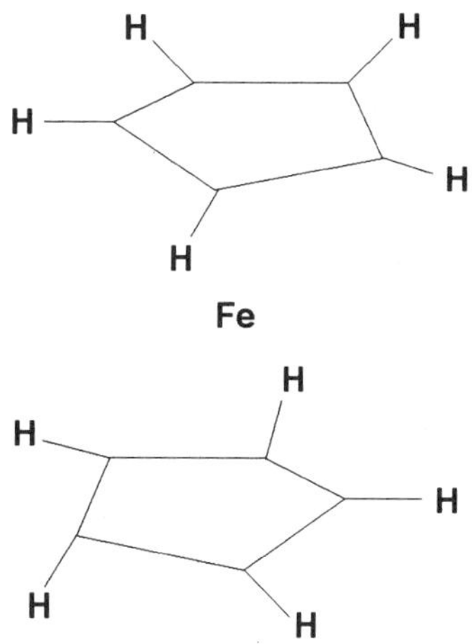

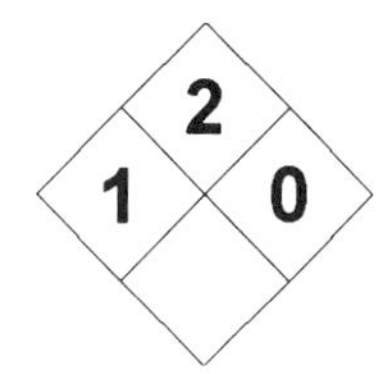

IDENTIFICATION AND TYPICAL USES

Ferrocene appears as an orange crystalline solid with a camphor-like odor. It exists as a "sandwich-like" structure with ferrous iron located at the center of two cyclopentadiene rings (see structural rendering above). This gives the molecule the properties of an aromatic hydrocarbon. It is used as an additive for fuel oils to improve efficiency of combustion (antiknock) and eliminate smoke. It is also a catalyst for accelerating vulcanization and polymerization. It is used as a coating for missiles and satellites, a high temperature lubricant, and an intermediate for high temperature polymers.

RISK ASSESSMENT: HEALTH

General Assessment

Ferrocene is considered moderately toxic by *ingestion* and *inhalation*. It is a questionable carcinogen in animals studies with experimental tumorigenic and mutagenic data reported. Human data are limited and insufficient to evaluate or predict human cancer potential.

Inhalation of ferrocene dusts or powders can be irritating to the eyes, nose, and upper respiratory system. There may be headache and nausea. High concentrations may produce depression of the central nervous system (CNS) with symptoms of narcosis, altered respiratory rates, convulsions, and death. Prolonged exposures may lead to liver and kidney injuries. Animal studies have shown ferrocene to stimulate blood cell formation or increase hemoglobin content in the blood. It is not known if human exposure will have this effect.

Acute Health Effects

The following acute (short-term) health effects may occur immediately or shortly after exposure to ferrocene:

Skin: Mild irritation with possibility for rash.

Eye: Mild to moderate irritation with tearing and redness.

Lung: Irritation of the upper respiratory tract and depression of the respiratory system. There may be headache and nausea.

Other: Ingestion may cause irritation to the gastrointestinal system. There may be depression of the circulatory system and other blood effects.

Chronic Health Effects

The following chronic (long-term) health effects may occur at some time after exposure to ferrocene and can last for months or even years:

Cancer Hazards: According to information presented in the references, ferrocene has been shown to produce tumors in test animals. Mutation data have also been reported. It is not known if human exposure to ferrocene will result in cancer.

Reproductive Hazard: According to information presented in the references, ferrocene has not been adequately tested for its ability to adversely affect reproduction.

Other Chronic Effects: Repeated or long-term exposures to ferrocene may cause dermatitis and possible sensitization. It may also adversely affect the liver and kidneys.

Recommended Risk-Reduction Measures

Personnel should avoid direct contact with ferrocene. The exact nature of its human toxicity is not clearly understood by the references. Prudent risk management requires caution and a conservative approach when assessing the hazard of such chemicals.

If a less toxic chemical cannot be substituted for a hazardous substance, then *engineering controls* are the most effective method of reducing exposures. The best protection is to enclose operations and/or provide local exhaust ventilation at the site of chemical release. While not always operationally feasible, isolating operations can also reduce exposure. Using respiratory protection is less effective than the controls mentioned above, but is still advisable whenever working with or around ferrocene. For exposures above the PEL (15 mg/m^3 total dust or 5 mgm^3 respirable fraction), a full facepiece air purifying respirator equipped with a HEPA filter and dust/mist pre-filter is suggested. For high or prolonged exposures, a supplied-air respirator with full facepiece operated in continuous flow mode, or a self-contained breathing apparatus (SCBA) with full facepiece and pressure demand are the recommended respiratory protection methods of choice. If a full facepiece is not available, then chemical goggles should be worn to protect the eyes. Whenever an airborne chemical dust hazard exists, a face shield and a rubber apron should be worn. To prevent hand and skin exposures, impervious gloves should be used.

Administrative controls should also be in place to minimize the potential for human exposures. These may include written procedures or policies which specify the methods and techniques that will be practiced whenever personnel are to work with ferrocene.

All personnel should receive *training* on the use, hazards, protective measures, emergency actions, and other precautions per 29 CFR 1910.1200 (Hazard Communication), prior to the first assignment in an area where ferrocene is used or stored.

If symptoms develop or overexposure is suspected, the following tests are recommended:

- ☑ Kidney and liver function tests.
- ☑ Nervous system evaluation.
- ☑ Evaluation by a qualified allergist with careful evaluation of exposure history and special testing (may help diagnose skin allergy).
- ☑ Complete blood count (CBC). Monitor hemoglobin levels, check for unusually high levels.

Any evaluation should include a careful history of past and present symptoms with an examination. Medical tests that simply look for existing damage are not a substitute for controlling exposures.

Other methods to reduce exposure include:

- ☑ Where possible, enclose operations and use local exhaust ventilation at the site of chemical release. If local exhaust ventilation or enclosure is not used, respiratory protection should be mandatory.
- ☑ Always ensure that proper protective clothing is worn when using chemical substances.
- ☑ Wash thoroughly immediately after exposure to ferrocene and at the end of the work shift or before eating, drinking, or smoking.
- ☑ Hazard warning information should be posted in the work area. In addition, as part of an on-going education and training program, all information on the health and safety hazards of ferrocene should be communicated to all exposed or potentially exposed workers.
- ☑ Eye wash stations should be provided in the immediate work area for emergency use. If there is a possibility of skin exposure to ferrocene, emergency shower facilities should also be provided.
- ☑ Personnel should never wear contaminated clothing home (family members can be exposed). Clothing should be laundered by personnel who have been trained on the hazards of ferrocene.

RISK ASSESSMENT: ENVIRONMENT

General Assessment

The environment is at risk of exposure during transportation, storage, disposal, or destruction of ferrocene. In almost every scenario, the threat of environmental exposure is contingent upon the proper handling of the chemical substance. Accidental spills, large or small, can result in fire, explosion, and possible contamination of the surrounding environmental mediums (water, soil, and, to a lesser extent, air).

Ferrocene is considered a flammable solid. It can ignite and burn vigorously. In the proper concentrations in air, its powders may ignite explosively. These characteristics require extreme caution in handling, storage, transportation, and disposal. It is a dangerous fire hazard when exposed to heat and flame. Therefore, special consideration is required during any emergency situation involving a leak or spill of ferrocene. Should ferrocene ever come into contact with incompatible substances, especially ammonium perchlorate, either during use, transportation, storage, or disposal, an explosive reaction is likely.

Ferrocene may enter the environment through industrial discharges or spills.

Acute Ecological Effects

Acute (short-term) toxic effects may include the death of animals, birds, or fish and death or low growth rate in plants. Acute effects are seen 2 to 4 days after animals or plants are exposed to ferrocene.

Insufficient data are available on the short-term effects of ferrocene exposure to aquatic life, plants, birds or land animals.

Chronic Ecological Effects

Chronic (long-term) toxic effects may include shortened life span, reproductive problems, lower fertility, and changes in appearance or behavior in exposed animals. These effects can be seen long after first exposure(s) to toxic chemicals.

Insufficient data are available on the long-term effects of ferrocene to aquatic life, plants, birds, or land animals.

Water Solubility

Ferrocene is nearly insoluble in water. Concentrations of less than 100 milligrams may mix with a liter of water.

Persistence in the Environment

Ferrocene is slightly persistent in water, with a half-life between 2 and 20 days. The half-life of a pollutant is the amount of time it takes for one half of the chemical to be degraded.

Bioaccumulation in Aquatic Organisms

Some substances increase in concentration, or *bioaccumulate,* in living organisms as they breathe contaminated air, drink contaminated water, or eat contaminated food. These chemicals can become concentrated in the tissues and internal organs of animals as well as humans.

The concentration of ferrocene found in fish tissue is expected to be about the same as the average concentration of ferrocene in the water from which the fish was taken.

Recommended Risk-Reduction Measures

Proper training of all transporters will reduce the likelihood of a mishap or accident resulting in a leak or spill to the environment. The correct use of labeling on all containers, trucks, and rail cars will enable emergency responders to react properly and quickly to any disaster thereby reducing the potential risk to the environment and to personnel.

Storage of ferrocene should be segregated from incompatible chemicals to minimize the risk of cross-contamination or contact. Buildings designated for storage should be equipped with appropriate fire protection systems (alarms, sprinklers, emergency lighting, portable extinguishers).

If a spill or leak to the environment has occurred, fire department, emergency response and/or hazardous materials spill personnel should be notified immediately. Cleanup should be attempted only by those trained in proper spill containment procedures. Contaminated soils should be removed for incineration and replaced with clean soil. If ferrocene should contact the water table, aquifer, or navigable waterway, the local and/or state emergency response authorities must be notified. A comprehensive emergency response or disaster preparedness/recovery plan should be in place prior to any operations involving the use, transportation, storage, or disposal of ferrocene. If ferrocene is spilled or leaked, the following specific steps are recommended:

- ☑ Restrict persons not wearing protective clothing from area of spill or leak until cleanup is complete and area can be opened for normal work.
- ☑ Remove all ignition sources, ventilate area of leak.
- ☑ Collect powdered materials in safest manner possible. Do NOT dry sweep (generates airborne dusts). Use HEPA vacuum instead. Damp mop residue and dispose of properly.
- ☑ It may be necessary to dispose of ferrocene as a hazardous waste. The responsible state agency or the regional office of the federal Environmental

Protection Agency (EPA) should be contacted for specific recommendations.

RISK ASSESSMENT: BUSINESS

General Assessment

Accidents or mishaps involving ferrocene can present a significant threat to business operations. The loss or damage of equipment or facilities can significantly affect fiscal viability. Lawsuits that may result from personnel injury/death, public exposures, and/or environmental contamination will also require a serious expenditure of resources. Media attention surrounding an injury, death, or environmental damage can also result in a loss of profits and loss of current as well as future business. Always remember that anytime the words "cancer" or "carcinogen" are used, public emotion, hysteria, and ignorance can run equally high. This should be carefully considered whenever planning, developing, or implementing public relations policies.

➊ *Recommended Risk-Reduction Measures*

Company attorneys, safety and health professionals, and environmental specialists should be involved in the development of any such policies or procedures. A company official should be pre-designated as a public relations officer with specific training in dealing with the press. Corporate plans and policies should be developed, approved, and implemented long before any need for such arises.

REFERENCES

Ahlbom, A. 1993. *Biostatistics for Epidemiologists.* New York: CRC Press/Lewis Publishers.

Asante-Duah, D. K. 1993. *Hazardous Waste Risk Assessment.* New York: CRC Press/Lewis Publishers.

Bloom, A. D. and F. J. DeSerres. 1995. *Ecotoxicity and Human Health.* New York: CRC Press/Lewis Publishers.

Brusick, D. J. 1994. *Methods for Genetic Risk Assessment.* New York: CRC Press/Lewis Publishers.

Calabrese, E. J. 1994. *Biological Effects of Low Level Exposures.* New York: CRC Press/Lewis Publishers.

Cockerman, L. G., B. S. Shane. 1994. *Basic Environmental Toxicology.* New York: CRC Press/Lewis Publishers.

Cothern, C. R. 1995. *Handbook for Environmental Risk Decisionmaking.* New York: CRC Press/Lewis Publishers.

Gardner, R. W. 1994. *Chemical Intolerance.* New York: CRC Press/Lewis Publishers.

Hallenbeck, W. H. 1993. *Quantitative Risk Assessment for Environmental and Occupational Health,* 2nd Edition. New York: CRC Press/Lewis Publishers.

Howard, P. H. 1995. *Handbook of Environmental Fate and Exposures Data for Organic Chemicals,* Vol. I-IV. New York: CRC Press/Lewis Publishers.

Jones, F. E. 1994. *Toxic Organic Vapors in the Workplace.* New York: CRC Press/Lewis Publishers.

Keith, L. H. and M. M. Walker. 1995. *Handbook of Air Toxics.* New York: CRC Press/Lewis Publishers.

Landis, W. G., M. H. Yu. 1995. *Introduction to Environmental Toxicology.* New York: CRC Press/Lewis Publishers.

Lauwerys, R. R., P. Hoet. 1993. *Industrial Chemical Exposure.* New York: CRC Press/Lewis Publishers.

Lewis, R. J., Sr. 1992. *Sax's Dangerous Properties of Industrial Materials,* Volumes 1, 2, and 3, 8th Edition. New York: Van Nostrand Reinhold.

Manahan, S. E. 1992. *Toxicological Chemistry,* 2nd Edition. New York: CRC Press/Lewis Publishers.

Mellström, G. A., J. E. Wahlberg, and H. I. Maibach. 1994. *Protective Gloves for Occupational Use.* New York: CRC Press/Lewis Publishers.

Meyer, E. 1990. *Chemistry of Hazardous Materials.* Englewood Cliffs, NJ: Prentice Hall.

Mycock, J. C., J. D. McKenna, and L. Theodore. 1995. *Handbook of Air Pollution Control Engineering and Technology.* New York: CRC Press/Lewis Publishers.

Nelson, G. O. 1992. *Gas Mixtures.* New York: CRC Press/Lewis Publishers.

Ott, W. R. 1995. *Environmental Statistics and Data Analysis.* New York: CRC Press/Lewis Publishers.

Patnaik, P. 1992. *A Comprehensive Guide to the Hazardous Properties of Chemical Substances.* New York: Van Nostrand Reinhold.

Philp, R. B. 1995. *Environmental Hazards and Human Health.* New York: CRC Press/Lewis Publishers.

Quigley, D. R. 1994. *Handbook of Emergency Chemical Management.* New York: CRC Press/Lewis Publishers.

Rea, W. J. 1995. *Chemical Sensitivity,* Vol. I-III. New York: CRC Press/Lewis Publishers.

Richardson, M. 1995. *Dictionary of Substance and Their Effects,* Vol. I-VI. New York: CRC Press/Lewis Publishers.

Rylander, R., R. R. Jacobs. 1994. *Organic Dusts.* New York: CRC Press/Lewis Publishers.

Sacarello, H. L. A. 1994. *The Comprehensive Handbook of Hazardous Materials.* New York: CRC Press/Lewis Publishers.

Torkelson, T. R. V. K. Rowe. 1981. Halogenated aliphatic hydrocarbons. *Patty's Industrial Hygiene and Toxicology,* Vol. 2B, *Toxicology,* pp 3568-3572. New York: Wiley-Interscience.

U.S. Department of Health and Human Services, National Institute for Occupational Safety and Health. 1994. *NIOSH Pocket Guide to Chemical Hazards.* Washington, D.C.: U.S. Government Printing Office.

MATERIAL SAFETY DATA SHEET

CHEMICAL NAME	
FERROUS CHLORIDE	**CRC PRESS/LEWIS PUBLISHERS DISCLAIMER:** THE INFORMATION AND RECOMMENDATIONS PRESENTED HEREIN ARE BASED ON SOURCES BELIEVED TO BE RELIABLE. CRC MAKES NO REPRESENTATION ON ITS COMPLETENESS OR ACCURACY. IT IS THE RESPONSIBILITY OF THE USER TO DETERMINE THE CHEMICAL'S SUITABILITY FOR ITS INTENDED USE, THE CHEMICAL'S SAFE USE, AND THE CHEMICAL'S PROPER DISPOSAL. NO REPRESENTATIONS AND/OR WARRANTIES, EITHER EXPRESSED OR IMPLIED, OF THE MERCHANTABILITY OR FITNESS FOR A PARTICULAR PURPOSE, OR OF ANY OTHER NATURE, ARE MADE WITH RESPECT TO THE INFORMATION PROVIDED IN THIS MSDS OR TO THE CHEMICAL TO WHICH INFORMATION MAY REFER. CRC NEITHER ASSUMES NOR AUTHORIZES ANY OTHER PERSON TO ASSUME FOR IT, ANY OTHER ADDITIONAL RESPONSIBILITY OR LIABILITY FOR THE USE OF, OR RELIANCE UPON, THIS INFORMATION.

HAZARD WARNING INFORMATION

HEALTH	FIRE	REACTIVE	OTHER	DEGREE OF HAZARD	COLOR CODING	OTHER CODES
1	0	0		0 = Minimum Hazard 1 = Slight Hazard 2 = Moderate Hazard 3 = Serious Hazard 4 = Severe Hazard	HEALTH = BLUE FIRE = RED REACTIVITY = YELLOW OTHER = WHITE	OX = Oxidizer ACID = Acid ALK = Alkali COR = Corrosive W = Use No Water

SECTION I - GENERAL INFORMATION

Characterization	**Chloride**	RCRA Number	**None**
		EPA Class	**Not Applicable**
DOT Proper Shipping Name	**Ferrous Chloride (solid or solution)**	Chemical Abstract Service (CAS) Number	**7758-94-3**
DOT Hazard Class and Label Requirements	**ORM-B (solid); Corrosive (solution)**	DOT Emergency Guide Code	**No Citation**
DOT Identification Number	**NA 1759 NA 1760 (solution)**	Chemical Formula	**$FeCl_2$**

Synonyms

Iron (II) chloride (1:2); iron dichloride; iron protochloride; iron chloride.

SECTION II - HAZARDOUS INGREDIENTS AND IDENTITY INFORMATION

Hazardous Components (specific identity)	OSHA Exposure Criteria	NIOSH Exposure Criteria	Immediately Dangerous to Life and Health (IDLH)	ACGIH Exposure Criteria
Ferrous chloride (derivation: By the action of hydrochloric acid on an excess of iron with subsequent crystallization).	**PEL:** 1 mg(Fe)/m^3 (soluble iron salts) **STEL:** Not Established	**REL:** Not Established **STEL:** Not Established	**Not Determined**	**TLV:** 1 mg(Fe)/m^3 (soluble iron salts) **STEL:** Not Established

SECTION III - PHYSICAL AND CHEMICAL CHARACTERISTICS

Boiling Point	**1873°F (1023°C)**	Specific Gravity (H_2O = 1)	**3.16**
Vapor Pressure (mm Hg)	**10 at 1292°F (700°C)**	Molecular Weight	**126.75**
Vapor Density (Air = 1)	**Not Found**	Melting Point	**1245°F (674°C)**

Solubility

Soluble in water (hygroscopic). Also soluble in alcohol.

Appearance and Odor

White or pale green, solid, deliquescent crystals with no odor.

SECTION IV - FIRE AND EXPLOSION HAZARD DATA

Flash Point (method used)	**Non-Combustible**	Explosive Limits in Air % by Volume	**LEL: Not Applicable UEL: Not Applicable**
NFPA Classification	**Non-Combustible Solid**	Autoignition Temperature	**Non-Combustible**

Extinguishing Media

Use extinguishing media suitable to surrounding fire. Ferrous chloride is non-combustible.

Special Fire Fighting Procedures

Toxic and poisonous gases and fumes are produced in fire. Structural fire-fighting clothing may not provide adequate protection. Wear full protective clothing and self-contained breathing apparatus (SCBA). Very hygroscopic, prevent water used in fire-fighting activities from contacting material.

Unusual Fire and Explosion Hazards

Firefighters should avoid contact with vapors produced during fire.

SECTION V - REACTIVITY DATA

Stability		Conditions to Avoid **Ferrous chloride is normally stable in closed (airtight) containers under normal conditions of storage and handling. Avoid contact with incompatible materials and heat. Keep cool and dry.**
Stable **X**	Unstable	Incompatibility *(materials to avoid)* **Keep away from water, moisture, and steam (may soften the solid and eventually dissolve it). Incompatible with ethylene oxide, potassium, and sodium.**
Hazardous Polymerization		Conditions to Avoid **Under normal conditions of temperature and pressure, ferrous chloride is not expected to undergo hazardous polymerization.**
May Occur	Will Not Occur **X**	Hazardous Decomposition or By-products **When heated to decomposition, ferrous chloride can emit highly toxic/poisonous gases, including highly toxic chloride fumes.**

SECTION VI - HEALTH HAZARD DATA

Primary Route(s) of Entry:	Inhalation?	Absorption (skin)?	Ingestion?
	X		X

Health Hazards

INHALATION: Dust causes irritation of the eyes, nose, throat, and mucous membrane of the respiratory tract. Its vapors may cause the glottis to spasm leading to suffocation. It can also cause chemical pneumonitis and pulmonary edema (fluid in the lungs) which can cause death.

SKIN & EYES: Mild irritation to the skin and eyes.

INGESTION: Gastrointestinal irritation.

Carcinogenicity	NTP Listed?	IARC Cancer Review Group?	OSHA Regulated?	Target Organs?
Unknown Human Unknown Animal	**No**	**No**	**29 CFR 1910.1000 Table Z-1-A**	**Respiratory system, eyes, skin.**

Medical Conditions Generally Aggravated by Exposure

None reported.

Emergency and First-aid Procedures

Eye contact: Flush immediately with water for 15 minutes (minimum), seek medical attention. Skin contact: Remove all contaminated clothing. Immediately wash area with large amounts of water. Seek medical assistance. For inhalation: Remove the person from exposure. Provide respiratory assistance and CPR. Transfer to medical facility. Observe for 24-48 hours for lung effects. If swallowed: Seek medical attention immediately. Unless advised otherwise, give 1-2 glasses of water and induce vomiting.

SECTION VII - PRECAUTIONS FOR SAFE HANDLING AND USE

Steps to be Taken in Case Material is Released or Spilled

Collect dusts in safest manner possible and deposit in sealed containers. Do NOT dry sweep, use HEPA vacuum instead. Ventilate area of spill or leak. Restrict those not involved in cleanup from entering area.

Preferred Waste Disposal Method

No Citation.

Precautions to be Taken in Handling and Storage

Store in tightly closed containers in cool, well-ventilated area away from incompatibles. Protect against physical damage. Keep away from all forms of water (moisture, steam) since ferrous chloride will almost completely dissolve in water. Keep storage facilities airtight to prevent condensation inside containers.

Other Precautions and Warnings

None reported.

SECTION VIII - CONTROL MEASURES AND PERSONAL PROTECTIVE EQUIPMENT

Respiratory Protection *(specify type)*

For exposures below the PEL, use a full mask respirator with dust pre-filter. For all other exposures, use supplied-air respirator with full facepiece, hood, or helmet in continuous flow mode, or use a full facepiece operated in pressure demand or other positive pressure mode.

Ventilation

Local exhaust or general mechanical systems recommended.

Protective Gloves	Eye Protection	Other Protective Clothing
Polycarbonate or Butyl Rubber	**Chemical/Dust Goggles or Face Mask**	**Rubber Apron**

Work/Hygiene Practices

Always wash hands thoroughly after using chemical; never bring food, drink, or smoking materials into vicinity of chemicals.

FERROUS CHLORIDE

$FeCl_2$

CAS: 7758-94-3

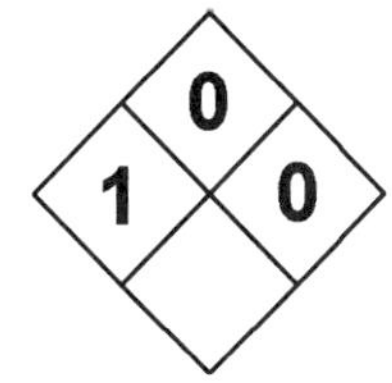

IDENTIFICATION AND TYPICAL USES

Ferrous chloride appears as white or pale green solid crystals with no noticeable odor. It is used as a mordant in fixing dyes, in metallurgy, in the making of some pharmaceuticals, in the manufacture of ferric chloride, and in sewage treatment. It should not be confused with ferric chloride solution or ferric chloride.

RISK ASSESSMENT: HEALTH

General Assessment

Exposure to ferrous chloride is possible through *inhalation* of its dusts or by direct *ingestion* of the solid material. The latter presents the most serious risks to human health, especially in solution form. While skin absorption has not been reported, skin contact can result in localized irritation and should be avoided. There are no data on its carcinogenic, mutagenic, or teratogenic properties. More study is required in this regard.

Inhalation of dusts can cause headache and nausea with irritation to the eyes, nose, and throat. Inhalation of vapors can be fatal as a result of spasm of the glottis and suffocation. High-level exposures to solutions containing ferrous chloride may lead to chemical pneumonia and pulmonary edema (fluid in the lungs) which can also be fatal. Development of pulmonary edema may be delayed up to 48 hours thereby creating a false sense of security with regard to health exposure risk.

Skin contact is irritating and is aggravated by the presence of moisture (sweat). Contact with the eyes can cause irritation.

Ingestion of ferrous chloride will cause moderate to severe irritation of the mouth, stomach, and gastrointestinal tract.

Acute Health Effects

The following acute (short-term) health effects may occur immediately or shortly after exposure to ferrous chloride:

Skin: Localized surface irritation.

Eye: Mild irritation.

Lung: Irritation of the nose, throat, and lungs following exposure to dusts or vapors (of solution). Pulmonary edema is possible (24 to 48 hours following a high exposure to ferrous oxide solution).

Other: Ingestion may cause a variety of gastrointestinal and systemic effects.

Chronic Health Effects

The following chronic (long-term) health effects may occur at some time after exposure to ferrous chloride and can last for months or even years:

Cancer Hazards: According to the information available in the references, ferrous chloride has not been adequately tested for its ability to cause cancer in test animals.

Reproductive Hazard: According to the information available in the references, ferrous chloride has not been adequately tested for its ability to adversely affect the reproductive process in test animals.

Other Chronic Effects: No other chronic effects have been reported as a result if long-term exposure to ferrous chloride.

Recommended Risk-Reduction Measures

Personnel should avoid direct contact with ferrous chloride. If a less toxic chemical cannot be substituted for a hazardous substance, then *engineering controls* are the most effective method of reducing exposures. The best protection is to enclose operations and/or provide local exhaust ventilation at the site of chemical release. While not always operationally feasible, isolating operations can also reduce exposure.

Using respiratory protection is less effective than the controls mentioned above, but is still advisable whenever working with or around ferrous chloride. The permissible exposure level (PEL) for this chemical is extremely low (1 mg/m^3) and difficult to monitor. An air purifying respirator with a dust pre-filter may suffice. However, a supplied-air respirator or a self-contained breathing apparatus (SCBA) with full face-piece operated in pressure demand mode would provide optimum respiratory protection. If a full facepiece is not available, then chemical goggles should be worn to protect the eyes. Whenever a dust hazard exists, a face shield and a rubber apron should be worn. To prevent hand and skin exposures, butyl rubber or polycarbonate gloves should be worn.

Administrative controls should also be in place to minimize the potential for human exposures. These may include written procedures or policies which specify the methods and techniques that will be practiced whenever personnel are to work with ferrous chloride.

All personnel should receive *training* on the use, hazards, protective measures, emergency actions, and other precautions per 29 CFR 1910.1200 (Hazard Communication), prior to the first assignment in an area where ferrous chloride is used or stored.

For those personnel with frequent or potentially high exposure (half the PEL or greater), the following are recommended medical tests to be provided before initial assignment and at regular intervals thereafter (e.g., annually):

☑ Lung function tests.

☑ Consider chest X-ray after acute overexposure (results may be negative if taken immediately after exposure due to delayed development of pulmonary edema).

Any evaluation should include a careful history of past and present symptoms with an examination. Medical tests that simply look for existing damage are not a substitute for controlling exposures. Also, because smoking can cause heart disease, as well as lung cancer, emphysema, and other respiratory problems, symptoms of exposure to ferrous chloride anhydrous may be more pronounced in smokers than in non-smokers under the same conditions of exposure. Prudent risk management requires careful consideration of *all* possible exposure factors.

Other methods to reduce exposure include:

☑ Where possible, enclose operations and use local exhaust ventilation at the site of chemical release. If local exhaust ventilation or enclosure is not used, respiratory protection should be mandatory.

☑ Always ensure that proper protective clothing is worn when using chemical substances and that personnel are trained on its use and care.

☑ Wash thoroughly immediately after exposure to ferrous chloride anhydrous and at the end of the work shift or before eating, drinking, or smoking.

☑ Hazard warning information should be posted in the work area. In addition, as part of an on-going education and training program, all information on the health and safety hazards of ferrous chloride anhydrous should be communicated to all potentially exposed workers.

☑ Eye wash stations should be provided in the immediate work area for emergency use. If there is a possibility of skin exposure to ferrous chloride, emergency shower facilities should also be provided.

☑ Personnel should never wear contaminated clothing home (family members can be exposed). Clothing should be laundered by personnel who have been trained on the hazards of ferrous chloride.

RISK ASSESSMENT: ENVIRONMENT

General Assessment

The environment is at risk of exposure during transportation, storage, disposal, or destruction of ferrous chloride. In almost every scenario, the threat of environmental exposure is contingent upon the proper handling of the chemical substance. Accidental spills, large or small, can result in fire, explosion, and possible contamination of the surrounding environmental mediums (water, soil, and air).

Ferrous chloride is considered a non-combustible solid. However, it is corrosive to many metals (especially in solution) and can react vigorously with potassium, ethylene oxide, sodium, and many oxidizers and contact can cause fire or explosion. In contact with water it will nearly dissolve completely. These characteristics require special consideration during any emergency situation involving a leak or spill of ferrous chloride. Should ferrous chloride ever come into contact with these incompatible either during use, transportation, storage, or disposal, the formation of highly toxic and/or highly explosive commodities is extremely possible.

Ferrous chloride can enter the environment mainly through industrial effluents or spills.

☠ *Acute Ecological Effects*

Acute (short-term) toxic effects may include the death of animals, birds, or fish and death or low growth rate in plants. Acute effects are seen 2 to 4 days after animals or plants are exposed to ferrous chloride.

There is insufficient evidence to evaluate or predict the short-term effects of ferrous chloride anhydrous to aquatic life, plants, birds, or land animals.

Chronic Ecological Effects

Chronic (long-term) toxic effects may include shortened life span, reproductive problems, lower fertility, and changes in appearance or behavior in exposed animals. These effects can be seen long after first exposure(s) to toxic chemicals.

There is insufficient evidence to evaluate or predict the long-term effects of ferrous chloride to aquatic life, plants, birds, or land animals.

Water Solubility

Ferrous chloride is highly soluble in water. Concentrations of 1000 milligrams and more will mix with a liter of water. It will be dissolved completely in contact with water or moisture.

⌛ *Persistence in the Environment*

Ferrous chloride is non-persistent in water, with a half-life of less than 2 days. The half-life of a pollutant is the amount of time it takes for one half of the chemical to be degraded.

Bioaccumulation in Aquatic Organisms

Some substances increase in concentration, or *bioaccumulate,* in living organisms as they breathe contaminated air, drink contaminated water, or eat contaminated food. These chemicals can become concentrated in the tissues and internal organs of animals as well as humans.

The concentration of ferrous chloride found in fish tissues is expected to be much lower than the average concentration of ferrous chloride in the water from which the fish was taken.

Recommended Risk-Reduction Measures

Proper training of all transporters will reduce the likelihood of a mishap or accident resulting in a leak or spill to the environment. The correct use of DOT labeling (per the MSDS) on all containers, trucks, and rail cars will enable emergency responders to react properly and quickly to any disaster thereby reducing the potential risk to the environment and to personnel.

Storage of ferrous chloride should be segregated from incompatible chemicals to minimize the risk of cross-contamination or contact. Exposure to any form of water (moisture, steam) can soften the material and may eventually dissolve ferrous chloride.

If a spill or leak to the environment has occurred, fire department, emergency response and/or hazardous materials spill personnel should be notified immediately. Cleanup should be attempted only by those trained in proper spill containment procedures. Contaminated soils should be removed for incineration and replaced with clean soil. If ferrous chloride should contact the water table, aquifer, or navigable waterway, the local and/or state emergency response authorities must be notified. A comprehensive emergency response or disaster preparedness/recovery plan should be in place prior to any operations involving the use, transportation, storage, or disposal of ferrous chloride. If ferrous chloride is spilled or leaked, the following specific steps are recommended:

- ☑ Restrict persons not wearing protective clothing from area of spill or leak until cleanup is complete and area can be opened for normal work.
- ☑ Ventilate are or spill or leak.
- ☑ Collect solid materials in safest and most efficient manner possible. Do NOT dry sweep (generates airborne dusts). Use vacuum equipped with high efficiency particulate air (HEPA) filter instead. Absorbed spilled solutions in vermiculite or other material and deposit in sealed containers.
- ☑ It may be necessary to dispose of ferrous chloride as a hazardous waste. The responsible state agency or the regional office of the federal Environmental Protection Agency (EPA) should be contacted for specific recommendations.

RISK ASSESSMENT: BUSINESS

General Assessment

Accidents or mishaps involving ferrous chloride can present a significant threat to business operations. The loss or damage of equipment or facilities can significantly affect fiscal viability. Lawsuits that may result from personnel injury/death, public exposures, and/or environmental contamination will also require a serious expenditure of resources. Media attention surrounding an injury, death, or environmental damage can also result in a loss of profits and loss of current as well as future business.

Recommended Risk-Reduction Measures

Company attorneys, safety and health professionals, and environmental specialists should be involved in the development of any safety or emergency response procedures. A company official should be pre-designated as a public relations officer with specific training in dealing with the press. Corporate plans and policies should be developed, approved, and implemented long before any need for such arises.

REFERENCES

Ahlbom, A. 1993. *Biostatistics for Epidemiologists.* New York: CRC Press/Lewis Publishers.

American Conference of Governmental Industrial Hygienists. 1988. *Documentation of the Threshold Limit Values and Biological Exposure Indices,* 5th Edition (with updates). Cincinnati: ACGIH

Anthony, C. P., and N. J. Kolthoff. 1971. *Textbook of Anatomy and Physiology.* St. Louis: C. V. Mosby Company.

Asante-Duah, D. K. 1993. *Hazardous Waste Risk Assessment.* New York: CRC Press/Lewis Publishers.

Brusick, D. J. 1994. *Methods for Genetic Risk Assessment.* New York: CRC Press/Lewis Publishers.

Bueche, J. R. 1972. *Principles of Physics,* 2nd Edition. New York: McGraw-Hill.

Calabrese, E. J., E. M. Kenyon. *Air Toxics and Risk Assessment.* New York: CRC Press/Lewis Publishers.

Calabrese, E. J. 1994. *Biological Effects of Low Level Exposures.* New York: CRC Press/Lewis Publishers.

Cockerman, L. G., B. S. Shane. 1994. *Basic Environmental Toxicology.* New York: CRC Press/Lewis Publishers.

Cothern, C. R., N. P. Ross. 1994. *Environmental Statistics, Assessment, and Forecasting.* New York: CRC Press/Lewis Publishers.

Cothern, R. R. 1993. *Comparative Environmental Risk Assessment.* New York: CRC Press/Lewis Publishers.

Cothern, C. R. 1995. *Handbook for Environmental Risk Decisionmaking.* New York: CRC Press/Lewis Publishers.

Danse, I. R. 1991. *Common Sense Toxics in the Workplace.* New York: Van Nostrand Reinhold.

Environmental Protection Agency. 1983. *Chemical Hazard Information Profiles.* Washington, D.C.: U.S. EPA.

Fodden, C. R., and J. L. Weddell. 1992. *Hazardous Materials Emergency Action Data.* New York: CRC Press/Lewis Publishers.

Forseberg, K., and L. H. Keith. 1993. *Chemical; Protective Clothing Permeation and Degradation Database.* Chelsea, MI: Lewis Publishers.

Gardner, R. W. 1994. *Chemical Intolerance.* New York: CRC Press/Lewis Publishers.

Gerarde, H. W. 1960. *Toxicology and Biochemistry of Aromatic Hydrocarbons.* New York: Elsevier.

Hallenbeck, W. H. 1993. *Quantitative Risk Assessment for Environmental and Occupational Health,* 2nd Edition. New York: CRC Press/Lewis Publishers.

Howard, P. H. 1995. *Handbook of Environmental Fate and Exposures Data for Organic Chemicals,* Vol. I-IV. New York: CRC Press/Lewis Publishers.

Keith, L. H. and M. M. Walker. 1995. *Handbook of Air Toxics.* New York: CRC Press/Lewis Publishers.

Landis, W. G., M. H. Yu. 1995. *Introduction to Environmental Toxicology.* New York: CRC Press/Lewis Publishers.

Lauwerys, R. R., P. Hoet. 1993. *Industrial Chemical Exposure.* New York: CRC Press/Lewis Publishers.

Lewis, R. J., Sr. 1993. *Hawley's Condensed Chemical Dictionary,* 12th Edition. New York: Van Nostrand Reinhold.

Lewis, R. J., Sr. 1992. *Sax's Dangerous Properties of Industrial Materials,* Volumes 1, 2, and 3, 8th Edition. New York: Van Nostrand Reinhold.

Manahan, S. E. 1992. *Toxicological Chemistry,* 2nd Edition. New York: CRC Press/Lewis Publishers.

Meyer, E. 1990. *Chemistry of Hazardous Materials.* Englewood Cliffs, NJ: Prentice Hall.

Mycock, J. C., J. D. McKenna, and L. Theodore. 1995. *Handbook of Air Pollution Control Engineering and Technology.* New York: CRC Press/Lewis Publishers.

Ott, W. R. 1995. *Environmental Statistics and Data Analysis.* New York: CRC Press/Lewis Publishers.

Patnaik, P. 1992. *A Comprehensive Guide to the Hazardous Properties of Chemical Substances.* New York: Van Nostrand Reinhold.

Philp, R. B. 1995. *Environmental Hazards and Human Health.* New York: CRC Press/Lewis Publishers.

Quigley, D. R. 1994. *Handbook of Emergency Chemical Management.* New York: CRC Press/Lewis Publishers.

Rea, W. J. 1995. *Chemical Sensitivity,* Vol. I-III. New York: CRC Press/Lewis Publishers.

Richardson, M. 1995. *Dictionary of Substance and Their Effects,* Vol. I-VI. New York: CRC Press/Lewis Publishers.

Sacarello, H. L. A. 1994. *The Comprehensive Handbook of Hazardous Materials.* New York: CRC Press/Lewis Publishers.

Sherman, J. D. 1988. *Chemical Exposure and Disease.* New York: Van Nostrand Reinhold.

Talbott, E., G. Craun. 1995. *An Introduction to Environmental Epidemiology.* New York: CRC Press/Lewis Publishers.

Torkelson, T. R. V. K. Rowe. 1981. Halogenated aliphatic hydrocarbons. *Patty's Industrial Hygiene and Toxicology,* Vol. 2B, *Toxicology,* pp 3568-3572. New York: Wiley-Interscience.

MATERIAL SAFETY DATA SHEET

CHEMICAL NAME

FERROVANADIUM DUST

HAZARD WARNING INFORMATION

HEALTH	FIRE	REACTIVE	OTHER	DEGREE OF HAZARD	COLOR CODING	OTHER CODES
1	1	0		0 = Minimum Hazard 1 = Slight Hazard 2 = Moderate Hazard 3 = Serious Hazard 4 = Severe Hazard	HEALTH = BLUE FIRE = RED REACTIVITY = YELLOW OTHER = WHITE	OX = Oxidizer ACID = Acid ALK = Alkali COR = Corrosive W = Use No Water

SECTION I - GENERAL INFORMATION

Characterization: **Vanadium Alloy**	RCRA Number: **None**	EPA Class: **Not Applicable**
DOT Proper Shipping Name: **Not Found**	Chemical Abstract Service (CAS) Number: **12604-58-9**	
DOT Hazard Class and Label Requirements: **Not Found**	DOT Emergency Guide Code: **No Citation**	
DOT Identification Number: **Not Found**	Chemical Formula: **FeV**	

Synonyms
Ferrovanadium.

SECTION II - HAZARDOUS INGREDIENTS AND IDENTITY INFORMATION

Hazardous Components (specific identity)	OSHA Exposure Criteria	NIOSH Exposure Criteria	Immediately Dangerous to Life and Health (IDLH)	ACGIH Exposure Criteria
Ferrovanadium dust (derivation: By reduction of the oxide with aluminum or silicon in presence of iron in an electric furnace). Impurities: 13% silicon 1.5% aluminum	**PEL (8-hour): 1 mg/m^3** **STEL: 3 mg/m^3 (15 minute)**	**REL (10-hour): 1 mg/m^3** **STEL: 3 mg/m^3 (15 minute)**	**500 mg/m^3**	**TLV: 1 mg/m^3** **STEL: 3 mg/m^3 (15 minute)**

SECTION III - PHYSICAL AND CHEMICAL CHARACTERISTICS

Boiling Point: **Not Found**	Specific Gravity (H_2O = 1): **Not Found**
Vapor Pressure (gas): **0 (approx.) at 68°F (20°C)**	Molecular Weight: **106.8**
Vapor Density (Air = 1): **Not Applicable**	Melting Point: **2696-2768°F (1480-1520°C)**

Solubility
Insoluble in water.

Appearance and Odor
A dark (grayish black), odorless particulate (dust) dispersed in air. May be furnished in a variety of lump, crushed, and ground sizes. Ferrovanadium metal is an alloy containing 50-80% vanadium.

SECTION IV - FIRE AND EXPLOSION HAZARD DATA

Flash Point (method used): **Not Determined**	Explosive Limits in Air % by Volume: **LEL: Not Determined** **UEL: Not Determined**
NFPA Classification: **Non-Combustible Solid**	Autoignition Temperature: **Not Determined**

Extinguishing Media
Use dry chemical, water spray, fog, regular foam, or carbon dioxide.

Special Fire Fighting Procedures
Wear full protective clothing and self-contained breathing apparatus (SCBA). Poisonous gases are produced in fire. Do NOT scatter fire by directing a high pressure stream at fire.

Unusual Fire and Explosion Hazards
A non-combustible solid, but the dusts can form explosive mixtures in air presenting a dangerous fire and explosion hazard when exposed to heat or flame. Air concentrations as low as 1.3 gr/34 ft^3 can cause an explosion.

SECTION V - REACTIVITY DATA

Stability		Conditions to Avoid: Normally stable. Keep away from heat or flame and incompatible materials. Reactions can be violent and even explosive. Avoid conditions that can generate a lot of dusts.
Stable	Unstable	Incompatibility *(materials to avoid)*
X		Reacts violently with strong oxidizers (such as fluorine, chlorine, bromine), heat, or flame.
Hazardous Polymerization		Conditions to Avoid: Under normal conditions of temperature and pressure, hazardous polymerization of ferrovanadium dust is not expected to occur.
May Occur	Will Not Occur	Hazardous Decomposition or By-products
	X	Hazardous decomposition of ferrovanadium dust will yield toxic and irritating smoke and fumes, including iron and vanadium oxides.

SECTION VI - HEALTH HAZARD DATA

Primary Route(s) of Entry:	Inhalation?	Absorption (skin)?	Ingestion?
	X		

Health Hazards

INHALATION: Possible irritation of the eyes, nose, and upper respiratory tract. Chronic exposure has caused pulmonary changes (bronchitis and perialveolitis) in test animals.

SKIN & EYES: Mild to moderate irritation of the eyes. Skin contact may cause no noticeable response.

INGESTION: Not likely, but possible in high dusty concentrations. Throat and digestive tract irritation is possible.

Carcinogenicity	NTP Listed?	IARC Cancer Review Group?	OSHA Regulated?	Target Organs?
Unknown Human Unknown Animal	No	No	29 CFR 1910.1000 Table Z-1	Respiratory system, eyes.

Medical Conditions Generally Aggravated by Exposure

None reported.

Emergency and First-aid Procedures

Eye contact: Flush immediately with water for 15 minutes (minimum), seek medical attention. Skin contact: Ferrovanadium dust is not known to be a skin irritant but it should still be washed off as soon as possible using soap and water. For inhalation: Remove the person from exposure. Provide respiratory assistance and CPR, if required. Transfer to medical facility. If swallowed: Contact poison control center. Unless advised otherwise, give 1-2 glasses of water and induce vomiting. Never give an unconscious or convulsing person anything by mouth or attempt to make such a person vomit.

SECTION VII - PRECAUTIONS FOR SAFE HANDLING AND USE

Steps to be Taken in Case Material is Released or Spilled

Remove all sources of ignition. Collect powdered materials in safest manner possible. Avoid generating dusts. Do NOT dry sweep. Use a HEPA vacuum and damp mop residue using non-sparking tools. Ventilate area of spill or leak. Restrict those not involved in cleanup from entering area.

Preferred Waste Disposal Method

No Citation.

Precautions to be Taken in Handling and Storage

Store to avoid contact with incompatible materials in tightly closed containers in cool, well-ventilated area away from heat and flame. Use only non-sparking tools to open and close containers.

Other Precautions and Warnings

Sources of ignition are prohibited where ferrovanadium dust is used, handled, or stored. Protect containers from physical damage. Open-air storage (ore piles) should be kept damp to prevent dust release.

SECTION VIII - CONTROL MEASURES AND PERSONAL PROTECTIVE EQUIPMENT

Respiratory Protection *(specify type)*

For low exposures (up to PEL), use and organic vapor respirator with HEPA filter and dust pre-filter. For higher exposures, use a supplied-air respirator with full facepiece in continuous flow mode, or a self-contained breathing apparatus (SCBA) operated in pressure demand or other positive pressure mode.

Ventilation

Local exhaust or general mechanical systems recommended.

Protective Gloves	Eye Protection	Other Protective Clothing
Chemically Protective Rubber	Chemical/Dust Goggles or Face Mask	Rubber Apron

Work/Hygiene Practices

Always wash hands thoroughly after using chemical; never bring food, drink, or smoking materials into vicinity of chemicals. Practice good housekeeping and ensure adequate dust control measures.

FERROVANADIUM DUST

FeV

CAS: 12604-58-9

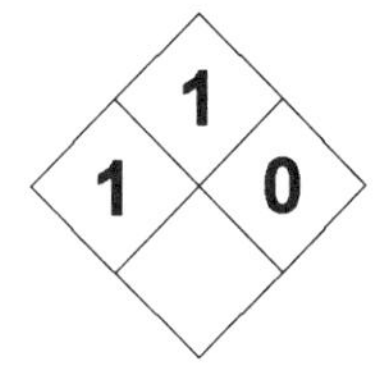

IDENTIFICATION AND TYPICAL USES

Ferrovanadium dust appears as a dark, grayish-black, particulate (dust) when dispersed in air. It can be found in a variety of lump, crushed, and ground sizes. Ferrovanadium metal is an typically contains between 50 and 80% vanadium. It is used to add vanadium to steel to produce fineness of grain, toughness, and resistance to high temperature and torsion.

RISK ASSESSMENT: HEALTH

General Assessment

Ferrovanadium dust is considered toxic by *inhalation*. There are no reports of carcinogenic, mutagenic, or teratogenic effects subsequent to exposure.

Inhalation of ferrovanadium dusts or powders can be irritating to the eyes, nose, and upper respiratory system. There may be headache and nausea. Animal studies have shown ferrovanadium dust to cause pulmonary damage, including bronchitis and inflammation of the lungs (perialveolitis).

Eye contact may cause irritation and, if not removed immediately may cause inflammation. Skin contact has not been shown to produce any indications of irritation or inflammation. Ingestion is not likely but can occur in extremely dusty work environments. The references contain no reports of adverse health effects resulting from ingestion, although high concentrations are likely to produce some localized irritation, especially of the mucous membrane in the throat and digestive tract.

Acute Health Effects

The following acute (short-term) health effects may occur immediately or shortly after exposure to ferrovanadium dust:

Skin: No reported effects.

Eye: Mild to moderate irritation with tearing and redness.

Lung: Irritation of the upper respiratory tract with coughing and sneezing. There may be headache and nausea.

Chronic Health Effects

The following chronic (long-term) health effects may occur at some time after exposure to ferrovanadium dust and can last for months or even years:

Cancer Hazards: According to information presented in the references, ferrovanadium dust has not been adequately tested for its ability to cause cancer in test animals.

Reproductive Hazard: According to information presented in the references, ferrovanadium dust has not been adequately tested for its ability to adversely affect reproduction.

Other Chronic Effects: Repeated or long-term exposures to ferrovanadium dust may cause pulmonary damage, including bronchitis and inflammation of the lungs (perialveolitis).

Recommended Risk-Reduction Measures

Personnel should avoid direct contact with ferrovanadium dust. The exact nature of its human toxicity is believed to be relatively low, but is not clearly understood by the references. Prudent risk management requires caution and a conservative approach when assessing the hazard of such chemicals.

If a less toxic chemical cannot be substituted for a hazardous substance, then *engineering controls* are the most effective method of reducing exposures. The best protection is to enclose operations and/or provide local exhaust ventilation at the site of chemical release. While not always operationally feasible, isolating operations can also reduce exposure. Positive measures should be instituted to ensure dust generation is controlled. These may include wet suppression control measures, artificial humidity modification, or other suitable techniques. Using respiratory protection is less effective than the controls mentioned above, but is still advisable whenever working with or around ferrovanadium dust. For exposures above the PEL (1 mg/m^3), a full facepiece air purifying respirator equipped with a HEPA filter and dust/mist pre-filter is suggested. For high or prolonged exposures, a supplied-air respirator with full facepiece operated in continuous flow mode, or a self-contained breathing apparatus (SCBA) with full facepiece and pressure demand are the recommended respiratory protection methods of choice. If a full facepiece is not available, then chemical/dust goggles should be worn to protect the eyes. Whenever an airborne chemical dust hazard exists, a face shield and a rubber apron should be worn. To prevent hand and skin exposures, impervious gloves should be used.

Administrative controls should also be in place to minimize the potential for human exposures. These may include written procedures or policies which specify the methods and techniques that will be practiced whenever personnel are to work with ferrovanadium dust.

All personnel should receive *training* on the use, hazards, protective measures, emergency actions, and other precautions per 29 CFR 1910.1200 (Hazard Communication), prior to the first assignment in an area where ferrovanadium dust is used or stored.

If symptoms develop or overexposure is suspected, the following tests are recommended:

☑ Lung function tests.

☑ Consider chest X-ray.

Any evaluation should include a careful history of past and present symptoms with an examination. Medical tests that simply look for existing damage are not a substitute for controlling exposures. Also, since smoking can cause emphysema and other lung disease, smokers exposed to irritating materials such as ferrovanadium dust may experience symptoms more quickly than non-smokers under the same conditions of exposure. Prudent risk management requires proper consideration of *all* health exposure risks.

Other methods to reduce exposure include:

☑ Where possible, enclose operations and use local exhaust ventilation at the site of chemical release. If local exhaust ventilation or enclosure is not used, respiratory protection should be mandatory.

☑ Always ensure that proper protective clothing is worn when using chemical substances and that personnel are trained in its use and care.

☑ Wash thoroughly immediately after exposure to ferrovanadium dust and at the end of the work shift or before eating, drinking, or smoking.

☑ Hazard warning information should be posted in the work area. In addition, as part of an on-going education and training program, all information on the health and safety hazards of ferrovanadium dust should be communicated to all exposed or potentially exposed workers.

☑ Eye wash stations should be provided in the immediate work area for emergency use. If there is a possibility of skin exposure to ferrovanadium dust, emergency shower facilities should also be provided.

☑ Personnel should never wear contaminated clothing home (family members can be exposed). Clothing should be laundered by personnel who have been trained on the hazards of ferrovanadium dust.

RISK ASSESSMENT: ENVIRONMENT

General Assessment

The environment is at risk of exposure during transportation, storage, disposal, or destruction of ferrovanadium dust. In almost every scenario, the threat of environmental exposure is contingent upon the proper handling of the chemical substance. Accidental spills, large or small, can result in fire, explosion, and possible contamination of the surrounding environmental mediums (water, soil, and, to a lesser extent, air).

Ferrovanadium dust is considered a noncombustible solid. However, airborne dust concentrations can ignite, burn vigorously, and even explode. It can form explosive concentrations as low as 1.3 gr/34 ft^3 in air. These characteristics require extreme caution in handling, storage, transportation, and disposal. It is also incompatible with strong oxidizers. Should ferrovanadium dust ever come into contact with incompatible substances either during use, transportation, storage, or disposal, an explosive reaction is likely. Therefore, special consideration is required during any emergency situation involving a leak or spill of ferrovanadium dust.

Ferrovanadium dust may enter the environment through industrial discharges or spills.

Acute Ecological Effects

Acute (short-term) toxic effects may include the death of animals, birds, or fish and death or low growth rate in plants. Acute effects are seen 2 to 4 days after animals or plants are exposed to ferrovanadium dust.

Insufficient data are available on the short-term effects of ferrovanadium dust exposure to aquatic life, plants, birds or land animals.

Chronic Ecological Effects

Chronic (long-term) toxic effects may include shortened life span, reproductive problems, lower fertility, and changes in appearance or behavior in exposed animals. These effects can be seen long after first exposure(s) to toxic chemicals.

Insufficient data are available on the long-term effects of ferrovanadium dust to aquatic life, plants, birds, or land animals.

Water Solubility

Ferrovanadium dust is insoluble in water. Concentrations of less than 1 milligram will not mix with a liter of water.

Persistence in the Environment

Ferrovanadium dust is highly persistent in water, with a half-life greater than 200 days. The half-life of a pollutant is the amount of time it takes for one half of the chemical to be degraded.

Bioaccumulation in Aquatic Organisms

Some substances increase in concentration, or *bioaccumulate,* in living organisms as they breathe contaminated air, drink contaminated water, or eat contaminated food. These chemicals can become concentrated in the tissues and internal organs of animals as well as humans.

Ferrovanadium dust is not known to accumulate in the edible tissues of fish.

Recommended Risk-Reduction Measures

Proper training of all transporters will reduce the likelihood of a mishap or accident resulting in a leak or spill to the environment. The correct use of labeling on all containers, trucks, and rail cars will enable emergency responders to react properly and quickly to any disaster thereby reducing the potential risk to the environment and to personnel.

Storage of ferrovanadium dust should be segregated from incompatible chemicals (oxidizers) to minimize the risk of cross-contamination or contact. Buildings designated for storage should be equipped with appropriate fire protection systems (alarms, sprinklers, emergency lighting, portable extinguishers).

If a spill or leak to the environment has occurred, fire department, emergency response and/or hazardous materials spill personnel should be notified immediately. Cleanup should be attempted only by those trained in proper spill containment procedures. Contaminated soils may need to be removed and replaced with clean soil. If ferrovanadium dust should contact the water table, aquifer, or navigable waterway, time is of the essence. It is insoluble and total remediation may be possible. The local and/or state emergency response authorities should be notified. A comprehensive emergency response or disaster preparedness/recovery plan should be in place prior to any operations involving the use, transportation, storage, or disposal of ferrovanadium dust. If ferrovanadium dust is spilled or leaked, the following specific steps are recommended:

- ☑ Restrict persons not wearing protective clothing from area of spill or leak until cleanup is complete and area can be opened for normal work.
- ☑ Remove all ignition sources, ventilate area of leak.
- ☑ Collect powdered materials in safest manner possible. Do NOT dry sweep (generates airborne dusts). Use HEPA vacuum instead. Damp mop residue and dispose of properly.
- ☑ It may be necessary to dispose of ferrovanadium dust as a hazardous waste. The responsible state agency or the regional office of the federal Environmental Protection Agency (EPA) should be contacted for specific recommendations.

RISK ASSESSMENT: BUSINESS

General Assessment

Accidents or mishaps involving ferrovanadium dust can present a significant threat to business operations. The loss or damage of equipment or facilities can significantly affect fiscal viability. Lawsuits that may result from personnel injury/death, public exposures, and/or environmental contamination will also require a serious expenditure of resources. Media attention surrounding an injury, death, or environmental damage can also result in a loss of profits and loss of current as well as future business.

Recommended Risk-Reduction Measures

Company attorneys, safety and health professionals, and environmental specialists should be involved in the development of any such policies or procedures. A company official should be pre-designated as a public relations officer with specific training in dealing with the press. Corporate plans and policies should be developed, approved, and implemented long before any need for such arises.

REFERENCES

Ahlbom, A. 1993. *Biostatistics for Epidemiologists.* New York: CRC Press/Lewis Publishers.

American Conference of Governmental Industrial Hygienists. 1988. *Documentation of the Threshold Limit Values and Biological Exposure Indices,* 5th Edition (with updates). Cincinnati: ACGIH

Anthony, C. P., and N. J. Kolthoff. 1971. *Textbook of Anatomy and Physiology.* St. Louis: C. V. Mosby Company.

Asante-Duah, D. K. 1993. *Hazardous Waste Risk Assessment.* New York: CRC Press/Lewis Publishers.

Bloom, A. D. and F. J. DeSerres. 1995. *Ecotoxicity and Human Health.* New York: CRC Press/Lewis Publishers.

Brusick, D. J. 1994. *Methods for Genetic Risk Assessment.* New York: CRC Press/Lewis Publishers.

Bueche, J. R. 1972. *Principles of Physics,* 2nd Edition. New York: McGraw-Hill.

Burrell, R., D. K. Flaherty, and L. J. Sauers. 1992. *Toxicology of the Immune System.* New York: Van Nostrand Reinhold.

Calabrese, E. J. 1994. *Biological Effects of Low Level Exposures.* New York: CRC Press/Lewis Publishers.

Cockerman, L. G., B. S. Shane. 1994. *Basic Environmental Toxicology.* New York: CRC Press/Lewis Publishers.

Cothern, C. R. 1995. *Handbook for Environmental Risk Decisionmaking.* New York: CRC Press/Lewis Publishers.

Danse, I. R. 1991. *Common Sense Toxics in the Workplace.* New York: Van Nostrand Reinhold.

Environmental Protection Agency. 1983. *Chemical Hazard Information Profiles.* Washington, D.C.: U.S. EPA.

Fodden, C. R., and J. L. Weddell. 1992. *Hazardous Materials Emergency Action Data.* New York: CRC Press/Lewis Publishers.

Forseberg, K., and L. H. Keith. 1993. *Chemical; Protective Clothing Permeation and Degradation Database.* Chelsea, MI: Lewis Publishers.

Gardner, R. W. 1994. *Chemical Intolerance.* New York: CRC Press/Lewis Publishers.

Hallenbeck, W. H. 1993. *Quantitative Risk Assessment for Environmental and Occupational Health,* 2nd Edition. New York: CRC Press/Lewis Publishers.

Howard, P. H. 1995. *Handbook of Environmental Fate and Exposures Data for Organic Chemicals,* Vol. I-IV. New York: CRC Press/Lewis Publishers.

Jones, F. E. 1994. *Toxic Organic Vapors in the Workplace.* New York: CRC Press/Lewis Publishers.

Keith, L. H. and M. M. Walker. 1995. *Handbook of Air Toxics.* New York: CRC Press/Lewis Publishers.

Landis, W. G., M. H. Yu. 1995. *Introduction to Environmental Toxicology.* New York: CRC Press/Lewis Publishers.

Lauwerys, R. R., P. Hoet. 1993. *Industrial Chemical Exposure.* New York: CRC Press/Lewis Publishers.

Lewis, R. J., Sr. 1992. *Sax's Dangerous Properties of Industrial Materials,* Volumes 1, 2, and 3, 8th Edition. New York: Van Nostrand Reinhold.

Manahan, S. E. 1992. *Toxicological Chemistry,* 2nd Edition. New York: CRC Press/Lewis Publishers.

Mellström, G. A., J. E. Wahlberg, and H. I. Maibach. 1994. *Protective Gloves for Occupational Use.* New York: CRC Press/Lewis Publishers.

Meyer, E. 1990. *Chemistry of Hazardous Materials.* Englewood Cliffs, NJ: Prentice Hall.

Mycock, J. C., J. D. McKenna, and L. Theodore. 1995. *Handbook of Air Pollution Control Engineering and Technology.* New York: CRC Press/Lewis Publishers.

Nelson, G. O. 1992. *Gas Mixtures.* New York: CRC Press/Lewis Publishers.

Ott, W. R. 1995. *Environmental Statistics and Data Analysis.* New York: CRC Press/Lewis Publishers.

Patnaik, P. 1992. *A Comprehensive Guide to the Hazardous Properties of Chemical Substances.* New York: Van Nostrand Reinhold.

Philp, R. B. 1995. *Environmental Hazards and Human Health.* New York: CRC Press/Lewis Publishers.

Quigley, D. R. 1994. *Handbook of Emergency Chemical Management.* New York: CRC Press/Lewis Publishers.

Rea, W. J. 1995. *Chemical Sensitivity,* Vol. I-III. New York: CRC Press/Lewis Publishers.

Richardson, M. 1995. *Dictionary of Substance and Their Effects,* Vol. I-VI. New York: CRC Press/Lewis Publishers.

Rylander, R., R. R. Jacobs. 1994. *Organic Dusts.* New York: CRC Press/Lewis Publishers.

Sacarello, H. L. A. 1994. *The Comprehensive Handbook of Hazardous Materials.* New York: CRC Press/Lewis Publishers.

Torkelson, T. R. V. K. Rowe. 1981. Halogenated aliphatic hydrocarbons. *Patty's Industrial Hygiene and Toxicology,* Vol. 2B, *Toxicology,* pp 3568-3572. New York: Wiley-Interscience.

U.S. Department of Health and Human Services, National Institute for Occupational Safety and Health. 1994. *NIOSH Pocket Guide to Chemical Hazards.* Washington, D.C.: U.S. Government Printing Office.

MATERIAL SAFETY DATA SHEET

CHEMICAL NAME

FLUOMETURON

CRC PRESS/LEWIS PUBLISHERS DISCLAIMER: THE INFORMATION AND RECOMMENDATIONS PRESENTED HEREIN ARE BASED ON SOURCES BELIEVED TO BE RELIABLE. CRC MAKES NO REPRESENTATION ON ITS COMPLETENESS OR ACCURACY. IT IS THE RESPONSIBILITY OF THE USER TO DETERMINE THE CHEMICAL'S SUITABILITY FOR ITS INTENDED USE, THE CHEMICAL'S SAFE USE, AND THE CHEMICAL'S PROPER DISPOSAL. NO REPRESENTATIONS AND/OR WARRANTIES, EITHER EXPRESSED OR IMPLIED, OF THE MERCHANTABILITY OR FITNESS FOR A PARTICULAR PURPOSE, OR OF ANY OTHER NATURE, ARE MADE WITH RESPECT TO THE INFORMATION PROVIDED IN THIS MSDS OR TO THE CHEMICAL TO WHICH INFORMATION MAY REFER. CRC NEITHER ASSUMES NOR AUTHORIZES ANY OTHER PERSON TO ASSUME FOR IT, ANY OTHER ADDITIONAL RESPONSIBILITY OR LIABILITY FOR THE USE OF, OR RELIANCE UPON, THIS INFORMATION.

HAZARD WARNING INFORMATION

HEALTH	FIRE	REACTIVE	OTHER	DEGREE OF HAZARD	COLOR CODING	OTHER CODES
3	0	0		0 = Minimum Hazard 1 = Slight Hazard 2 = Moderate Hazard 3 = Serious Hazard 4 = Severe Hazard	HEALTH = BLUE FIRE = RED REACTIVITY = YELLOW OTHER = WHITE	OX = Oxidizer ACID = Acid ALK = Alkali COR = Corrosive W = Use No Water

SECTION I - GENERAL INFORMATION

Characterization	RCRA Number	EPA Class
Herbicide	**None**	**Not Applicable**

DOT Proper Shipping Name	Chemical Abstract Service (CAS) Number
Not Listed	**2164-17-2**

DOT Hazard Class and Label Requirements	DOT Emergency Guide Code
Not Listed	**No Citation**

DOT Identification Number	Chemical Formula
Not Listed	$C_{10}H_{11}F_3N_2O$

Synonyms

[N-(3-trifluoromethylphenyl)-N',N'-dimethylurea]; Cotoran; Lanex; Herbicide C-2059; 1,1-dimethyl-3-(α, α, α-trifluoro-*m*-tolyl) urea.

SECTION II - HAZARDOUS INGREDIENTS AND IDENTITY INFORMATION

Hazardous Components (specific identity)	OSHA Exposure Criteria	NIOSH Exposure Criteria	Immediately Dangerous to Life and Health (IDLH)	ACGIH Exposure Criteria
Fluometuron (derivation: By reaction of dimethylamine with 3-trifluoromethylphenyl isocyanate).	**PEL:** Not Established **STEL:** Not Established	**REL:** Not Established **STEL:** Not Established	**Not Determined**	**TLV:** Not Established **STEL:** Not Established

SECTION III - PHYSICAL AND CHEMICAL CHARACTERISTICS

Boiling Point	Specific Gravity (H_2O = 1)
Not Found	**Not Found**

Vapor Pressure (mm Hg)	Molecular Weight (atomic weight)
0 (approx.) at 69°F (20°C)	**232.23**

Vapor Density (Air = 1)	Melting Point
Not Found	**Not Found**

Solubility

Partially to moderately soluble in water. Soluble in alcohol and acetone.

Appearance and Odor

Colorless or white, crystalline (sand-like) solid.

SECTION IV - FIRE AND EXPLOSION HAZARD DATA

Flash Point (method used)	Explosive Limits in Air % by Volume
Not Determined	**LEL: N/A UEL: N/A**

NFPA Classification	Autoignition Temperature
Non-Combustible Solid	**Not Found**

Extinguishing Media

Use extinguishing media suitable to surrounding fire.

Special Fire Fighting Procedures

Wear full protective clothing and self-contained breathing apparatus (SCBA). Structural fire-fighting equipment is permeable and may not provide sufficient protection. Evacuate personnel 2500 feet from fire area. Move containers from fire area if it can be done without risk. Cool exposed containers.

Unusual Fire and Explosion Hazards

Fluometuron is a non-combustible solid. However, it may be dissolved in some flammable liquids. In sufficient quantities in air, the dusts may form an explosive or easily ignited mixture.

SECTION V - REACTIVITY DATA

Stability		Conditions to Avoid: Fluometuron is stable in closed containers under routine conditions of storage and handling. Avoid contact with heat.
Stable	Unstable	Incompatibility *(materials to avoid)*
X		None reported.
Hazardous Polymerization		Conditions to Avoid: Under normal conditions of temperature and pressure, hazardous polymerization of fluometuron is not expected to occur.
May Occur	Will Not Occur	Hazardous Decomposition or By-products
	X	When heated to decomposition, fluometuron emits toxic and acrid fumes and gases, including toxic fluoride and oxides of nitrogen.

SECTION VI - HEALTH HAZARD DATA

Primary Route(s) of Entry:	Inhalation?	Absorption (skin/eye)?	Ingestion?
	X	X	X

Health Hazards

INHALATION: Irritation of the eyes, nose, throat, and respiratory tract. In high concentrations, exposure may cause changes in blood cell count (increased white, reduced red) and may lead to liver damage and altered thyroid function.

ABSORPTION: Will pass through unbroken skin and enter the bloodstream. Contact may severely irritate the skin and eyes. Prolonged contact can cause allergies (sensitization).

INGESTION: Moderately toxic to humans resulting in toxic systemic effects. May also cause severe gastroenteritis.

Carcinogenicity	NTP Listed?	IARC Cancer Review Group?	OSHA Regulated?	Target Organs?
Questioned Human Suspected Animal	No	Group 3	No	Respiratory system, skin, eyes, liver, thyroid.

Medical Conditions Generally Aggravated by Exposure

None reported. However, persons with existing lung dysfunction may experience effects quicker.

Emergency and First-aid Procedures

Eye contact: Do not allow victim to keep eyes tightly shut. Flush immediately with water for 15 minutes (minimum), seek medical attention. Skin contact: Remove all contaminated clothing. Immediately wash area with flooding amounts of soap and water. For inhalation: Remove the person from exposure. Maintain an open airway. Provide respiratory assistance and CPR. Transfer to medical facility. If swallowed: Seek medical attention immediately. Call poison control center for advice. Never give anything by mouth to an unconscious or convulsing person. Do not make an unconscious person vomit.

SECTION VII - PRECAUTIONS FOR SAFE HANDLING AND USE

Steps to be Taken in Case Material is Released or Spilled

Collect powdered materials in safest manner possible and deposit in sealed containers. Remove all sources of ignition. Ventilate area of spill. Use HEPA vacuum. Restrict those not involved in cleanup from entering area. Notify appropriate authorities. Absorb any liquids containing fluometuron in vermiculite, dry sand, earth, or similar material and deposit in sealed drum for disposal or reclamation.

Preferred Waste Disposal Method

No Citation.

Precautions to be Taken in Handling and Storage

Store in tightly closed containers in cool, dry, well-ventilated area.

Other Precautions and Warnings

Personnel required to work with or around fluometuron should be trained on its hazards.

SECTION VIII - CONTROL MEASURES AND PERSONAL PROTECTIVE EQUIPMENT

Respiratory Protection *(specify type)*

No exposure levels established and fluometuron is a questionable carcinogen. For best protection, use a supplied-air respirator of a SCBA with full facepiece operated in positive pressure mode.

Ventilation

Local exhaust (preferred) at site of chemical work or general ventilation.

Protective Gloves	Eye Protection	Other Protective Clothing
Impervious Material	Chemical/Dust Goggles or Face Mask	Protective Apron, Boots (avoid skin contact)

Work/Hygiene Practices

Always wash hands thoroughly after using chemical; never bring food, drink, or smoking materials into vicinity of chemicals.

FLUOMETURON

$C_{10}H_{11}F_3N_2O$ CAS: 2164-17-2

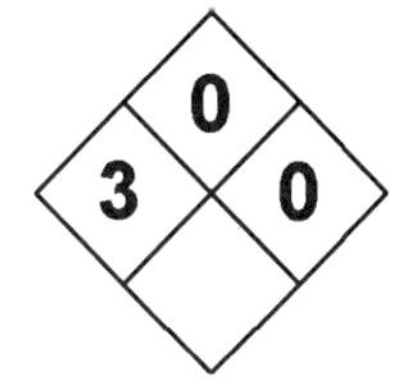

IDENTIFICATION AND TYPICAL USES

Fluometuron is a colorless to white crystalline (sand-like) solid material. It is used as an herbicide to control broadleaf and grass weeds of cotton and sugarcane crops.

RISK ASSESSMENT: HEALTH

General Assessment

Fluometuron is a moderately toxic compound by *ingestion.* It can also cause mild to serious health hazard risks by *inhalation,* and skin contact (*absorption*). It is a questionable carcinogen in animals and there is also some experimental mutagenic data reported.

Breathing its fine dusts can irritate the respiratory system, especially the nose and throat. There is evidence that exposure can lead to changes in blood cell count (increased white blood cells, decreased red blood cells). The liver and thyroid gland may also be damaged or injured.

Contact with skin and eyes can be severely irritating. Prolonged contact can cause an allergic reaction (sensitization). Any exposures after that will result in a return of allergic symptoms (rash, irritation). Ingestion may cause gastrointestinal irritation.

Acute Health Effects

The following acute (short-term) health effects may occur immediately or shortly after exposure to fluometuron:

Skin: Possible irritation and allergic reaction. Absorption is likely to occur, especially when chemical is in solution.

Eye: Severe irritation and possible damage on contact.

Lung: Irritation of the nose, throat, and respiratory tract.

Chronic Health Effects

The following chronic (long-term) health effects may occur at some time after exposure to fluometuron and can last for months or even years:

Cancer Hazards: Fluometuron is a mutagen (causes genetic changes). Such chemicals may have a cancer risk and there is limited evidence that fluometuron does cause cancer in test animals. Human carcinogenicity should be considered questionable.

Reproductive Hazard: According to information presented in the references, fluometuron has not been adequately tested for its ability to cause adverse reproductive effects in test animals.

Other Chronic Effects: Fluometuron causes skin sensitization and possible allergies. If allergy develops, very low future exposures can cause itching and skin rash. Prolonged or repeated exposures may lead to damaged liver and/or changes in thyroid function. Chronic exposure may cause blood cell changes with a reduction in red blood cell count and/or an increase in white blood cell count.

Recommended Risk-Reduction Measures

Personnel should avoid direct contact with fluometuron dusts or powders. Its carcinogenic properties in humans is not clearly understood and this alone represents a serious health hazard risk. If a less toxic material or compound cannot be substituted for fluometuron, then *engineering controls* are the most effective method of reducing exposures. The best protection is to enclose operations and/or provide local exhaust ventilation at the site of fluometuron dust release. This may be practical in manufacturing operations, but is not feasible in application since it is used outdoors. While not always operationally possible, isolating operations involving fluometuron manufacturing can also reduce exposure. Using respiratory protection is less effective than the controls mentioned above, but is still advisable whenever working with or around fluometuron. No exposure levels have been established for this compound. However, this does not mean that exposure to fluometuron is without health hazard risk. For the best protection, use a supplied-air respirator or a self-contained breathing apparatus (SCBA) with full facepiece operated in pressure demand or other continuous flow mode. If a full facepiece is not available, then chemical/dust goggles should be worn to protect the eyes. A face shield and protective apron should also be worn. To prevent hand and skin exposures, gloves should be worn. Glove manufacturers should be contacted and permeation studies obtained *before* final glove selections are made.

Administrative controls should also be in place to minimize the potential for human exposures. These may include written procedures or policies which specify the methods and techniques that will be practiced whenever personnel are to work with fluometuron.

All personnel should receive *training* on the use, hazards, protective measures, emergency actions, and other precautions per 29 CFR 1910.1200 (Hazard Communication), prior to the first assignment in an area where fluometuron is used or stored.

If symptoms develop or overexposure is suspected, the following medical testing may be useful:

☑ Lung function tests.

☑ Complete blood count (CBC) to monitor for white and red blood cell count changes.

☑ Liver and thyroid examination and testing.

☑ Evaluation by a qualified allergist, including careful consideration of exposure history and special testing. Skin testing with dilute fluometuron may help diagnose skin allergy, if done by a qualified allergist.

Any evaluation should include a careful history of past and present symptoms with an examination. Medical tests that simply look for existing damage are not a substitute for controlling exposures.

Other methods to reduce exposure include:

☑ Where possible, enclose operations and use local exhaust ventilation at the site of chemical release. If local exhaust ventilation or enclosure is not used, respiratory protection should be mandatory.

☑ Always ensure that proper protective clothing is worn when using chemical substances and that personnel are trained in its use and care.

☑ Wash thoroughly immediately after exposure to fluometuron and at the end of the work shift or before eating, drinking, or smoking.

☑ Hazard warning information should be posted in the work area. In addition, as part of an on-going education and training program, all information on the health and safety hazards of fluometuron should be communicated to all exposed and potentially exposed workers.

☑ Eye wash stations should be provided in the immediate work area for emergency use. If there is a possibility of skin exposure to fluometuron, emergency shower facilities should also be provided.

☑ Workers whose clothing has been contaminated by fluometuron should change into clean clothes before leaving work. Contaminated work clothing should be laundered only by individuals who have been informed of the hazards of exposure to fluometuron.

RISK ASSESSMENT: ENVIRONMENT

General Assessment

Because of its use as an herbicide, fluometuron can be present in the environment at any given time and at a variety of levels. The environment is also at risk of exposure during transportation, storage, disposal, or destruction of fluometuron. In almost every scenario, the threat of environmental exposure is contingent upon the proper handling of the chemical substance. Accidental spills, large or small, where fluometuron contacts incompatible commodities can result in the release of hazardous fumes and gases (depending upon conditions of spill), and possible contamination of the surrounding environmental mediums (water, soil, and air).

In its powder form, fluometuron is considered a non-combustible solid. However, in sufficient quantities mixed with air, it may present an explosion or fire hazard. Also, it may be dissolved in some flammable or combustible liquids. Therefore, extreme caution is required in handling, storage, transportation, and disposal of fluometuron, and special consideration is warranted during any emergency situation involving a leak or spill of fluometuron powder or dust.

Fluometuron can enter the environment from industrial and municipal discharges, through agricultural run-off, and from spills. Also, as an herbicide, it can be present in the environment due to its intended use.

Acute Ecological Effects

Acute (short-term) toxic effects may include the death of animals, birds, or fish and death or low growth rate in plants. Acute effects are seen 2 to 4 days after animals or plants are exposed to fluometuron.

Fluometuron has high acute toxicity to aquatic life and has caused injury to various types of agricultural and ornamental crops. Insufficient data are available to evaluate or predict the short-term effects of fluometuron to birds or land animals.

Chronic Ecological Effects

Chronic (long-term) toxic effects may include shortened life span, reproductive problems, lower fertility, and changes in appearance or behavior in exposed animals. These effects can be seen long after first exposure(s) to toxic chemicals.

Fluometuron has high chronic toxicity to aquatic life. Insufficient data are available to evaluate or predict the long-term effects of fluometuron to plants, birds, or land animals.

Water Solubility

Fluometuron is moderately soluble in water. Concentrations of 1 to 1000 milligrams will mix with a liter of water.

Persistence in the Environment

Fluometuron is moderately persistent in the water, with a half-life of 20 to 200 days. The half-life of a pollutant is the amount of time it takes for one-half of the chemical to be degraded. About 57% of fluometuron will eventually end up in water; approximately 41.3% will end up in air, and the remainder will be divided nearly equally between terrestrial soil and aquatic sediments.

Bioaccumulation in Aquatic Organisms

Some substances increase in concentration, or *bioaccumulate,* in living organisms as they breathe contaminated air, drink contaminated water, or eat contaminated food. These chemicals can become concentrated in the tissues and internal organs of animals as well as humans.

The concentration of fluometuron found in fish tissues is expected to be somewhat higher that the average concentration of fluometuron in the water from which the fish was taken.

Recommended Risk-Reduction Measures

Proper training of all transporters will reduce the likelihood of a mishap or accident resulting in a leak or spill of fluometuron dusts or powders into the environment. The correct use labeling on all containers, trucks, and rail cars will enable emergency responders to react properly and quickly to any disaster thereby reducing the potential risk to the environment and to personnel.

Storage of fluometuron should be controlled to minimize the risk of cross-contamination or contact with other materials. Proper ventilation and protective equipment should be used while handing the powders or while preparing aqueous solutions.

If a spill or leak to the environment has occurred, fire department, emergency response and/or hazardous materials spill personnel should be notified immediately. Cleanup should be attempted only by those trained in proper spill containment procedures. If fluometuron should contact the water table, aquifer, or navigable waterway, reclamation procedures should be initiated as soon as practical. It is moderately soluble in water and total remediation may not be possible. The local and/or state emergency response authorities must be notified. A comprehensive emergency response or disaster preparedness/recovery plan should be in place prior to any operations involving the use, transportation, storage, or disposal of fluometuron. If fluometuron powder is spilled or leaked, the following specific steps are recommended:

- ☑ Restrict persons not wearing protective clothing from area of spill or leak until cleanup is complete and area can be opened for normal work.
- ☑ Remove all ignition sources.
- ☑ Ventilate area of spill or leak.
- ☑ Collect powdered materials in the safest manner possible and deposit in sealed containers. Do NOT dry sweep (generates airborne dusts). Recommend using vacuum with a high-efficiency particulate air (HEPA) filter instead. Damp mop residue and contain for disposal. If liquid solutions are spilled, absorb in vermiculite or other material and place in sealed drum for disposal or reclamation.
- ☑ It may be necessary to dispose of fluometuron as a hazardous waste. The responsible state agency or the regional office of the federal Environmental Protection Agency (EPA) should be contacted for specific recommendations.

RISK ASSESSMENT: BUSINESS

General Assessment

Accidents or mishaps involving fluometuron can present a significant threat to business operations. The loss or damage of equipment or facilities can significantly affect fiscal viability. Lawsuits that may result from personnel injury/death, public exposures, and/or environmental contamination will also require a serious expenditure of resources. Media attention surrounding an injury, death, or environmental damage can also result in a loss of profits and loss of current as well as future business. Always remember that anytime the words "cancer" or "carcinogen" are used, public emotion, hysteria, and ignorance can run equally high. This is a matter for careful consideration whenever developing or implementing public relation policies.

Recommended Risk-Reduction Measures

Company attorneys, safety and health professionals, and environmental specialists should be involved in the development of any procedures for responding to chemical incidents. A company official should be predesignated as a public relations officer with specific training in dealing with the press. Corporate plans and policies should be developed, approved, and implemented long before any need for such arises.

REFERENCES

Ahlbom, A. 1993. *Biostatistics for Epidemiologists.* New York: CRC Press/Lewis Publishers.

American Conference of Governmental Industrial Hygienists. 1988. *Documentation of the Threshold Limit Values and Biological Exposure Indices,* 5th Edition (with updates). Cincinnati: ACGIH

Asante-Duah, D. K. 1993. *Hazardous Waste Risk Assessment.* New York: CRC Press/Lewis Publishers.

Brusick, D. J. 1994. *Methods for Genetic Risk Assessment.* New York: CRC Press/Lewis Publishers.

Bueche, J. R. 1972. *Principles of Physics,* 2nd Edition. New York: McGraw-Hill.

Calabrese, E. J. 1994. *Biological Effects of Low Level Exposures.* New York: CRC Press/Lewis Publishers.

Cockerman, L. G., B. S. Shane. 1994. *Basic Environmental Toxicology.* New York: CRC Press/Lewis Publishers.

Cothern, C. R. 1995. *Handbook for Environmental Risk Decisionmaking.* New York: CRC Press/Lewis Publishers.

Gardner, R. W. 1994. *Chemical Intolerance.* New York: CRC Press/Lewis Publishers.

Hallenbeck, W. H. 1993. *Quantitative Risk Assessment for Environmental and Occupational Health,* 2nd Edition. New York: CRC Press/Lewis Publishers.

Howard, P. H., et. al. 1991. *Handbook of Environmental Degradation Rates.* New York: CRC Press/Lewis Publishers.

Jones, F. E. 1994. *Toxic Organic Vapors in the Workplace.* New York: CRC Press/Lewis Publishers.

Keith, L. H. and M. M. Walker. 1995. *Handbook of Air Toxics.* New York: CRC Press/Lewis Publishers.

Landis, W. G., M. H. Yu. 1995. *Introduction to Environmental Toxicology.* New York: CRC Press/Lewis Publishers.

Langerman, N. 1994 *Precautionary Labels for Chemical Containers.* New York: CRC Press/Lewis Publishers.

Lauwerys, R. R., P. Hoet. 1993. *Industrial Chemical Exposure.* New York: CRC Press/Lewis Publishers.

Lewis, R. J., Sr. 1992. *Sax's Dangerous Properties of Industrial Materials,* Volumes 1, 2, and 3, 8th Edition. New York: Van Nostrand Reinhold.

Manahan, S. E. 1992. *Toxicological Chemistry,* 2nd Edition. New York: CRC Press/Lewis Publishers.

Mellström, G. A., J. E. Wahlberg, and H. I. Maibach. 1994. *Protective Gloves for Occupational Use.* New York: CRC Press/Lewis Publishers.

Meyer, E. 1990. *Chemistry of Hazardous Materials.* Englewood Cliffs, NJ: Prentice Hall.

Mycock, J. C., J. D. McKenna, and L. Theodore. 1995. *Handbook of Air Pollution Control Engineering and Technology.* New York: CRC Press/Lewis Publishers.

Nelson, G. O. 1992. *Gas Mixtures.* New York: CRC Press/Lewis Publishers.

Ott, W. R. 1995. *Environmental Statistics and Data Analysis.* New York: CRC Press/Lewis Publishers.

Patnaik, P. 1992. *A Comprehensive Guide to the Hazardous Properties of Chemical Substances.* New York: Van Nostrand Reinhold.

Proctor, N. H., J. P. Hughes. 1991. *Chemical Hazards of the Workplace.* New York: Van Nostrand Reinhold.

Quigley, D. R. 1994. *Handbook of Emergency Chemical Management.* New York: CRC Press/Lewis Publishers.

Rea, W. J. 1995. *Chemical Sensitivity,* Vol. I-III. New York: CRC Press/Lewis Publishers.

Reeves, A. L. 1979. Fluometuron. *Handbook on the Toxicology of Metals.* pp 321-328. New York: Elsevier North-Holland.

Richardson, M. 1995. *Dictionary of Substance and Their Effects,* Vol. I-VI. New York: CRC Press/Lewis Publishers.

Sacarello, H. L. A. 1994. *The Comprehensive Handbook of Hazardous Materials.* New York: CRC Press/Lewis Publishers.

Talbott, E., G. Craun. 1995. *An Introduction to Environmental Epidemiology.* New York: CRC Press/Lewis Publishers.

Torkelson, T. R. V. K. Rowe. 1981. *Patty's Industrial Hygiene and Toxicology,* Vol. 2B. New York: Wiley-Interscience.

MATERIAL SAFETY DATA SHEET

CHEMICAL NAME	
FLUORINE	**CRC PRESS/LEWIS PUBLISHERS DISCLAIMER:** THE INFORMATION AND RECOMMENDATIONS PRESENTED HEREIN ARE BASED ON SOURCES BELIEVED TO BE RELIABLE. CRC MAKES NO REPRESENTATION ON ITS COMPLETENESS OR ACCURACY. IT IS THE RESPONSIBILITY OF THE USER TO DETERMINE THE CHEMICAL'S SUITABILITY FOR ITS INTENDED USE, THE CHEMICAL'S SAFE USE, AND THE CHEMICAL'S PROPER DISPOSAL. NO REPRESENTATIONS AND/OR WARRANTIES, EITHER EXPRESSED OR IMPLIED, OF THE MERCHANTABILITY OR FITNESS FOR A PARTICULAR PURPOSE, OR OF ANY OTHER NATURE, ARE MADE WITH RESPECT TO THE INFORMATION PROVIDED IN THIS MSDS OR TO THE CHEMICAL TO WHICH INFORMATION MAY REFER. CRC NEITHER ASSUMES NOR AUTHORIZES ANY OTHER PERSON TO ASSUME FOR IT, ANY OTHER ADDITIONAL RESPONSIBILITY OR LIABILITY FOR THE USE OF, OR RELIANCE UPON, THIS INFORMATION.

HAZARD WARNING INFORMATION

HEALTH	FIRE	REACTIVE	OTHER	DEGREE OF HAZARD	COLOR CODING	OTHER CODES
4	0	4	~~W~~	0 = Minimum Hazard 1 = Slight Hazard 2 = Moderate Hazard 3 = Serious Hazard 4 = Severe Hazard	HEALTH = BLUE FIRE = RED REACTIVITY = YELLOW OTHER = WHITE	OX = Oxidizer ACID = Acid ALK = Alkali COR = Corrosive ~~W~~ = Use No Water

SECTION I - GENERAL INFORMATION

Characterization **Halogen**	RCRA Number **P056**	EPA Class **Acute Hazardous Waste**
DOT Proper Shipping Name **Fluorine**	Chemical Abstract Service (CAS) Number **7782-41-4**	
DOT Hazard Class and Label Requirements **Non-flammable Gas; Poison and Oxidizer**	DOT Emergency Guide Code **20** (compressed gas) **25** (cryogenic liquid)	
DOT Identification Number **UN 1045** (compressed gas) **UN 9192** (cryogenic liquid)	Chemical Formula F_2	
Synonyms **Fluorine-19**		

SECTION II - HAZARDOUS INGREDIENTS AND IDENTITY INFORMATION

Hazardous Components (specific identity)	OSHA Exposure Criteria	NIOSH Exposure Criteria	Immediately Dangerous to Life and Health (IDLH)	ACGIH Exposure Criteria
Fluorine (derivation: By electrolysis of molten anhydrous hydrofluoric acid and potassium fluoride solution melts with special copper-bearing carbon anodes, steel cathodes and containers, and monel screens). **1 ppm = 1.58 mg/m³**	**PEL (skin):** **0.1 ppm** **0.2 mg/m³** **STEL** **Not Established**	**REL:** **0.1 ppm** **0.2 mg/m³** **STEL:** **Not Established**	**25 ppm**	**TLV:** **1 ppm** **1.6 mg/m³** **STEL:** **2 ppm** **3.1 mg/m³**

SECTION III - PHYSICAL AND CHEMICAL CHARACTERISTICS

Boiling Point **-307°F (-188 °C)**	Specific Gravity (H_2O = 1) **1.31**
Vapor Pressure (atmospheres) **<1 at 69°F (20°C)**	Molecular Weight **38.0**
Vapor Density (Air = 1) **1.7**	Melting Point **-363°F (-220°C)**

Solubility
Reacts with water to emit hydrofluoric acid, oxygen fluoride, hydrogen peroxide, oxygen, and ozone. Also reacts vigorously with most oxidizable substances at room temperature, frequently with ignition.

Appearance and Odor
Pale-yellow to greenish colored gas with a pungent, irritating odor. Odor Threshold = 0.035 ppm.

SECTION IV - FIRE AND EXPLOSION HAZARD DATA

Flash Point (method used) **Not Applicable**	Explosive Limits in Air % by Volume **LEL: N/A UEL: N/A**
NFPA Classification **Non-flammable Gas**	Autoignition Temperature **Not Applicable**

Extinguishing Media
Extinguish fires only if flow of gas can be stopped first. Extinguishing agents may react. Allow to burn out if possible. Use extinguishing agent suitable to surrounding media. Fluorine itself does not burn.

Special Fire Fighting Procedures
Fluorine is a powerful oxidizer. Wear full protective clothing and self-contained breathing apparatus (SCBA). Use a water spray to keep fire-exposed containers cool. Poisonous gases are produced in fire.

Unusual Fire and Explosion Hazards
Dangerous fire and explosion hazard when exposed to heat or incompatible materials. Containers may explode in fire. Firefighters should avoid contact with vapors produced during fire.

SECTION V - REACTIVITY DATA

Stability		Conditions to Avoid Fluorine is the most reactive non-metal known. Avoid contact with incompatible materials, heat, and ignition sources.
Stable	Unstable X	Incompatibility *(materials to avoid)* Fluorine is incompatible with a large number of substances, both organic and inorganic compounds. It combines directly and indirectly to form fluorides with all elements except helium, neon, and argon.
Hazardous Polymerization		Conditions to Avoid Under normal conditions of temperature and pressure, hazardous polymerization of fluorine is not known to occur.
May Occur	Will Not Occur X	Hazardous Decomposition or By-products When heated to decomposition, fluorine can emit highly toxic/poisonous fluoride fumes.

SECTION VI - HEALTH HAZARD DATA

Primary Route(s) of Entry:	Inhalation?	Absorption (skin)?	Ingestion?
	X		

Health Hazards

INHALATION: Extremely irritating to the lungs, throat, nose and eyes. Coughing and shortness of breath that may persist for hours after exposure. Can cause permanent damage to the lungs. Fluid can build up in the lungs (pulmonary edema), which is a medical emergency. Gastrointestinal symptoms may also occur.

SKIN & EYES: Contact with the liquid can cause severe skin burns and persistent rash. The liquid can cause severe skin and eye burns (frostbite) leading to permanent damage.

Carcinogenicity	NTP Listed?	IARC Cancer Review Group?	OSHA Regulated?	Target Organs?
Unknown Human Unknown Animal	No	No	29 CFR 1910.1000 Table Z-1	Respiratory system, eyes, skin, liver, kidneys.

Medical Conditions Generally Aggravated by Exposure

Existing lung diseases may be aggravated by exposure.

Emergency and First-aid Procedures

Eye contact: Flush immediately with water for 15 minutes (minimum), seek medical attention. Skin contact: Remove all contaminated clothing. Immediately wash area with large amounts of soap and water. For red or blistered skin, seek medical assistance. For inhalation: Remove the person from exposure. Provide respiratory assistance and CPR. Transfer to medical facility. Observe for 24-48 hours since pulmonary edema may be a delayed reaction. If swallowed: Unlikely, since fluorine is a gas.

SECTION VII - PRECAUTIONS FOR SAFE HANDLING AND USE

Steps to be Taken in Case Material is Released or Spilled

Restrict those not involved in cleanup from entering area. Isolate area 150 feet in all directions. Ventilate area of spill or leak. If gas is leaked, stop source if possible or remove leaking cylinders to safe area outdoors and allow to empty. Report any leak in excess of 10 pounds.

Preferred Waste Disposal Method

No Citation.

Precautions to be Taken in Handling and Storage

Fluorine is a non-flammable gas but an extremely powerful oxidizer. Store to avoid contact with incompatible materials since violent reactions can occur. Store in tightly closed containers in cool, well-ventilated area away from heat and sunlight. Ship in steel pressure cylinders and tank cars.

Other Precautions and Warnings

Sources of ignition are prohibited where fluorine is used, handled, or stored. Heat may cause containers to build up pressure and explode. Detached facility storage is recommended.

SECTION VIII - CONTROL MEASURES AND PERSONAL PROTECTIVE EQUIPMENT

Respiratory Protection *(specify type)*

For exposures below the PEL, use a full facepiece respirator with fluorine gas cartridge. For higher exposures or when levels are not known, use a supplied-air respirator with full facepiece, hood, or helmet in continuous flow mode, or use a SCBA operated in pressure demand or other positive pressure mode.

Ventilation

Local exhaust or general mechanical systems recommended.

Protective Gloves	Eye Protection	Other Protective Clothing
Impervious and Insulated Material	Chemical Goggles or Face Mask	Protective Suit

Work/Hygiene Practices

Always wash hands thoroughly after using chemical; never bring food, drink, or smoking materials into vicinity of chemicals.

FLUORINE

F_2 CAS: 7782-41-4

IDENTIFICATION AND TYPICAL USES

Fluorine is a pale-yellow to greenish colored gas with a pungent, sharp, irritating odor. It may be shipped as a compressed gas or a cryogenic liquid (at temperatures below 363°F). It is one of the most reactive non-metallic substances known and rarely occurs in the elemental state in nature. It is found in fluorspar, cryolite, and fluorapetite. It is the most electronegative element and one of the most powerful oxidizing agents known. It is primarily used in conversion of uranium tetrafluroide to uranium hexafluroide. It is also used in the manufacture of fluorochemicals (both metallic and organic). It is used in rocket fuel systems, and as a chemical intermediate, as an active constituent of fluoridating compounds used in drinking water and toothpaste, and in insecticide sprays for fruits and vegetables.

RISK ASSESSMENT: HEALTH

General Assessment

Fluorine is a severely irritating gas, causing irritation of the eyes, nose, throat, and respiratory tract. It is considered moderately toxic to humans by *inhalation*. Human mutation data are also reported. It acts as a cellular poison, interfering with the metabolism of calcium and certain other enzyme mechanics. Its carcinogenic and teratogenic potential in humans is not known.

Inhalation causes eye, nose, throat, and respiratory tract irritation with severe dypsnea (difficulty breathing) and quick progression to respiratory damage. Pulmonary edema (fluid buildup in the lungs) may occur within 24-48 hours of exposure. This is a medical emergency and can lead to death. Because of its delayed onset, personnel often neglect seeking treatment or even reporting exposure. Inhalation of excessive concentration can also cause bronchospasm, laryngospasm, and suffocation.

Escaping gas can cause frostbite. Contact with the skin or eyes may produce deep, painful burns with possible bone involvement due to the fluoride ion's affinity for calcium.

Acute Health Effects

The following acute (short-term) health effects may occur immediately or shortly after exposure to fluorine:

Skin: Immediate irritation and burn on contact. May cause permanent damage to skin tissue and frostbite.

Eye: Severe irritation; may cause burns and permanent damage.

Lung: Irritation of the nose, throat, and lungs following exposure. Pulmonary edema is possible (24 to 48 hours following a high exposure). It can cause tearing, choking, coughing, sputum, bloody nose, and chest pain.

Chronic Health Effects

The following chronic (long-term) health effects may occur at some time after exposure to fluorine and can last for months or even years:

Cancer Hazards: Fluorine may cause mutations (genetic changes) in living cells. It has not been adequately tested for its ability to cause cancer in test animals.

Reproductive Hazard: According to information presented in the references, fluorine has not been adequately tested for its ability to adversely affect reproduction.

Other Chronic Effects: Fluorine can irritate the lungs. Repeated exposure may cause bronchitis to develop with cough, phlegm, and/or shortness of breath. Long-term exposure can cause corrosion of the teeth, inflammation of the mucous membranes, calcification of ligaments, respiratory ailments, and liver and kidney damage. Also, osteosclerosis (hardening of bone and/or bone marrow), especially in the pelvic bones and spine, has been reported by workers occupationally or environmentally exposed for periods of 20 to 30 years.

Recommended Risk-Reduction Measures

Personnel should avoid direct contact with fluorine. If a less toxic chemical cannot be substituted for a hazardous substance, then *engineering controls* are the most effective method of reducing exposures. The best protection is to enclose operations and/or provide local exhaust ventilation at the site of chemical release. While not always operationally feasible, isolating operations can also reduce exposure. Using respiratory protection is less effective than the controls mentioned above, but is still advisable whenever working with or around fluorine. For low concentrations (below the PEL), an air-purifying organic vapor respirator equipped with a fluorine gas cartridge is acceptable protection. For higher concentrations, a self-contained breathing apparatus (SCBA) with full facepiece operated in pressure demand mode or a supplied-air respirator will provide greater protection. If a full facepiece is not available, then chemical goggles should be worn to protect the eyes. Whenever a chemical splash hazard exists, a face shield and a rubber apron should be worn. To prevent hand and skin exposures, chemical resistant and thermally insulated gloves should be worn.

Administrative controls should also be in place to minimize the potential for human exposures. These may include written procedures or policies which specify the methods and techniques that will be practiced whenever personnel are to work with fluorine.

All personnel should receive *training* on the use, hazards, protective measures, emergency actions, and other precautions per 29 CFR 1910.1200 (Hazard Communication), prior to the first assignment in an area where fluorine is used or stored.

For those personnel with frequent or potentially high exposure (half the PEL or greater), the following are recommended medical tests to be provided before initial assignment and at regular intervals thereafter (e.g., annually):

- ☑ Dental examination to check teeth for indications of erosion.
- ☑ Lung function tests.

If symptoms develop or overexposure is suspected, the following tests are recommended:

- ☑ Lung function tests.
- ☑ Consider chest X-ray after acute overexposures (may be negative if taken immediately after exposure due to delayed development of pulmonary edema).

Any evaluation should include a careful history of past and present symptoms with an examination. Medical tests that simply look for existing damage are not a substitute for controlling exposures. Also, because smoking can cause heart disease, as well as lung cancer, emphysema, and other respiratory problems, smokers exposed to fluorine may experience symptoms more quickly and more defined than non-smokers under the same conditions of exposure. Prudent risk management requires careful consideration of *all* possible factors which may be causing the appearance of symptoms in the work place.

Other methods to reduce exposure include:

- ☑ Where possible, enclose operations and use local exhaust ventilation at the site of chemical release. If local exhaust ventilation or enclosure is not used, respiratory protection should be mandatory.
- ☑ Always ensure that proper protective clothing is worn when using chemical substances.
- ☑ Wash thoroughly immediately after exposure to fluorine and at the end of the work shift or before eating, drinking, or smoking.
- ☑ Hazard warning information should be posted in the work area. In addition, as part of an on-going education and training program, all information on the health and safety hazards of fluorine should be communicated to all potentially exposed workers.
- ☑ Eye wash stations should be provided in the immediate work area for emergency use. If there is a possibility of skin exposure to fluorine, emergency shower facilities should also be provided.
- ☑ Personnel should never wear contaminated clothing home (family members can be exposed). Clothing should be laundered by personnel who have been trained on the hazards of fluorine.

RISK ASSESSMENT: ENVIRONMENT

General Assessment

The environment is at risk of exposure during transportation, storage, disposal, or destruction of fluorine. In almost every scenario, the threat of environmental exposure is contingent upon the proper handling of the chemical substance. Accidental spills, large or small, can result in fire, explosion, and possible contamination of the surrounding environmental mediums (water, soil, and air).

Fluorine is considered a non-flammable gas. However, it is a powerful oxidizing agent and reactions with incompatible commodities can result in fire and serious explosions. These characteristics require extreme caution in handling, storage, transportation, and disposal. It is incompatible with a wide range of materials, including water, acids, oxidizers, organic compounds, and many combustible materials. These characteristics require special consideration during any emergency situation involving a leak or spill of fluorine. Should fluorine ever come into contact with incompatible substances during use, transportation, storage, or disposal, the formation of highly toxic and/or highly explosive commodities is extremely possible. It will react with water to form hydrofluoric acid which is extremely toxic and dangerously corrosive.

Fluorine is a natural element of common occurrence but, because it is so reactive, it is rarely found in its natural state. It can enter the environment through industrial and municipal discharges and spills.

Acute Ecological Effects

Acute (short-term) toxic effects may include the death of animals, birds, or fish and death or low growth rate in plants. Acute effects are seen 2 to 4 days after animals or plants are exposed to fluorine.

This chemical has high acute toxicity to aquatic life. No data are available on the short-term effects of fluorine exposure to plants, birds, or land animals.

Chronic Ecological Effects

Chronic (long-term) toxic effects may include shortened life span, reproductive problems, lower fertility, and changes in appearance or behavior in exposed animals. These effects can be seen long after first exposure(s) to toxic chemicals.

Fluorine has high chronic toxicity in aquatic life. No data are available on the long-term effects of fluorine to plants, birds, or land animals.

Water Solubility

Fluorine is reacts with water to form hydrofluoric acid which is very soluble in water. Concentrations of 1000 milligrams and more will mix with a liter of water.

Persistence in the Environment

Fluorine is non-persistent in water, with a half-life of less than 2 days. The half-life of a pollutant is the amount of time it takes for one half of the chemical to be degraded. No data are available on the amount of fluorine that will remain in air or water.

Bioaccumulation in Aquatic Organisms

Some substances increase in concentration, or *bioaccumulate,* in living organisms as they breathe contaminated air, drink contaminated water, or eat contaminated food. These chemicals can become concentrated in the tissues and internal organs of animals as well as humans.

Fluorine is not expected to accumulate in the edible tissues of fish since the fish will probably die first.

Recommended Risk-Reduction Measures

Proper training of all transporters will reduce the likelihood of a mishap or accident resulting in a leak or spill to the environment. The correct use of DOT labeling (per the MSDS) on all containers, trucks, and rail cars will enable emergency responders to react properly and quickly to any disaster thereby reducing the potential risk to the environment and to personnel.

Storage of fluorine should be segregated from incompatible chemicals to minimize the risk of cross-contamination or contact. Buildings designated for storage should be equipped with appropriate fire protection systems (alarms, sprinklers, emergency lighting, portable extinguishers).

If a spill or leak to the environment has occurred, fire department, emergency response and/or hazardous materials spill personnel should be notified immediately. Cleanup should be attempted only by those trained in proper spill containment procedures. Contaminated soils should be removed for incineration and replaced with clean soil. If fluorine should contact the water table, aquifer, or navigable waterway, time is of the essence. It reacts quickly in water to form hydrofluoric acid and total remediation and containment may not be possible. The local and/or state emergency response authorities should be notified. A comprehensive emergency response or disaster preparedness/recovery plan should be in place prior to any operations involving the use, transportation, storage, or disposal of fluorine.

If fluorine is spilled or leaked, the following specific steps are recommended:

☑ Restrict persons not wearing protective clothing from area of spill or leak until cleanup is complete and area can be opened for normal work.

☑ Ventilate area of leak.

☑ If cryogenic liquids are spilled, cleanup may not be possible before evaporation occurs.

☑ If gas is leaking, stop the flow of gas. If the source is a cylinder and leak cannot be stopped in place, remove cylinder to a safe place in the open air and repair or allow to empty.

☑ It may be necessary to dispose of fluorine as a hazardous waste. The responsible state agency or the regional office of the federal Environmental Protection Agency (EPA) should be contacted for specific recommendations.

RISK ASSESSMENT: BUSINESS

General Assessment

Accidents or mishaps involving fluorine can present a significant threat to business operations. The loss or damage of equipment or facilities can significantly effect fiscal viability. Lawsuits that may result from personnel injury/death, public exposures, and/or environmental contamination will also require a serious expenditure of resources. Media attention surrounding an injury, death, or environmental damage can also result in a loss of profits and loss of current as well as future business.

Recommended Risk-Reduction Measures

Company attorneys, safety and health professionals, and environmental specialists should be involved in the development of any procedures dealing with safety or emergency response. A company official should be pre-designated as a public relations officer with specific training in dealing with the press. Corporate plans and policies should be developed, approved, and implemented long before any need for such arises.

REFERENCES

American Conference of Governmental Industrial Hygienists. 1988. *Documentation of the Threshold Limit Values and Biological Exposure Indices,* 5th Edition (with updates). Cincinnati: ACGIH

Bueche, J. R. 1972. *Principles of Physics,* 2nd Edition. New York: McGraw-Hill.

Calabrese, E. J. 1994. *Biological Effects of Low Level Exposures.* New York: CRC Press/Lewis Publishers.

Cockerman, L. G., B. S. Shane. 1994. *Basic Environmental Toxicology.* New York: CRC Press/Lewis Publishers.

Gardner, R. W. 1994. *Chemical Intolerance.* New York: CRC Press/Lewis Publishers.

Hallenbeck, W. H. 1993. *Quantitative Risk Assessment for Environmental and Occupational Health,* 2nd Edition. New York: CRC Press/Lewis Publishers.

Jones, F. E. 1994. *Toxic Organic Vapors in the Workplace.* New York: CRC Press/Lewis Publishers.

Keith, L. H. and M. M. Walker. 1995. *Handbook of Air Toxics.* New York: CRC Press/Lewis Publishers.

Landis, W. G., M. H. Yu. 1995. *Introduction to Environmental Toxicology.* New York: CRC Press/Lewis Publishers.

Langerman, N. 1994 *Precautionary Labels for Chemical Containers.* New York: CRC Press/Lewis Publishers.

Lauwerys, R. R., P. Hoet. 1993. *Industrial Chemical Exposure.* New York: CRC Press/Lewis Publishers.

Lewis, R. J., Sr. 1992. *Sax's Dangerous Properties of Industrial Materials,* Volumes 1, 2, and 3, 8th Edition. New York: Van Nostrand Reinhold.

Lide, D. R. 1995. *Handbook of Chemistry and Physics,* 76th Edition. New York: CRC Press/Lewis Publishers.

Manahan, S. E. 1994. *Environmental Chemistry,* 6th Edition. New York: CRC Press/Lewis Publishers.

Mellström, G. A., J. E. Wahlberg, and H. I. Maibach. 1994. *Protective Gloves for Occupational Use.* New York: CRC Press/Lewis Publishers.

Meyer, E. 1990. *Chemistry of Hazardous Materials.* Englewood Cliffs, NJ: Prentice Hall.

Patnaik, P. 1992. *A Comprehensive Guide to the Hazardous Properties of Chemical Substances.* New York: Van Nostrand Reinhold.

Proctor, N. H., J. P. Hughes. 1991. *Chemical Hazards of the Workplace.* New York: Van Nostrand Reinhold.

Quigley, D. R. 1994. *Handbook of Emergency Chemical Management.* New York: CRC Press/Lewis Publishers.

Rea, W. J. 1995. *Chemical Sensitivity,* Vol. I-III. New York: CRC Press/Lewis Publishers.

Richardson, M. 1995. *Dictionary of Substance and Their Effects,* Vol. I-VI. New York: CRC Press/Lewis Publishers.

Sacarello, H. L. A. 1994. *The Comprehensive Handbook of Hazardous Materials.* New York: CRC Press/Lewis Publishers.

Sheftel, V. O. 1995. *Handbook of Toxic Properties of Monomers and Additives.* New York: CRC Press/Lewis Publishers.

Torkelson, T. R. V. K. Rowe. 1981. Halogenated aliphatic hydrocarbons. *Patty's Industrial Hygiene and Toxicology,* Vol. 2B, *Toxicology,* pp 3568-3572. New York: Wiley-Interscience.

U.S. Department of Health and Human Services, National Institute for Occupational Safety and Health. 1994. *NIOSH Pocket Guide to Chemical Hazards.* Washington, D.C.: U.S. Government Printing Office.

MATERIAL SAFETY DATA SHEET

CHEMICAL NAME	
FLUOROACETIC ACID	**CRC PRESS/LEWIS PUBLISHERS DISCLAIMER:** THE INFORMATION AND RECOMMENDATIONS PRESENTED HEREIN ARE BASED ON SOURCES BELIEVED TO BE RELIABLE. CRC MAKES NO REPRESENTATION ON ITS COMPLETENESS OR ACCURACY. IT IS THE RESPONSIBILITY OF THE USER TO DETERMINE THE CHEMICAL'S SUITABILITY FOR ITS INTENDED USE, THE CHEMICAL'S SAFE USE, AND THE CHEMICAL'S PROPER DISPOSAL. NO REPRESENTATIONS AND/OR WARRANTIES, EITHER EXPRESSED OR IMPLIED, OF THE MERCHANTABILITY OR FITNESS FOR A PARTICULAR PURPOSE, OR OF ANY OTHER NATURE, ARE MADE WITH RESPECT TO THE INFORMATION PROVIDED IN THIS MSDS OR TO THE CHEMICAL TO WHICH INFORMATION MAY REFER. CRC NEITHER ASSUMES NOR AUTHORIZES ANY OTHER PERSON TO ASSUME FOR IT, ANY OTHER ADDITIONAL RESPONSIBILITY OR LIABILITY FOR THE USE OF, OR RELIANCE UPON, THIS INFORMATION.

HAZARD WARNING INFORMATION

HEALTH	FIRE	REACTIVE	OTHER	DEGREE OF HAZARD	COLOR CODING	OTHER CODES
4	0	0		0 = Minimum Hazard 1 = Slight Hazard 2 = Moderate Hazard 3 = Serious Hazard 4 = Severe Hazard	HEALTH = BLUE FIRE = RED REACTIVITY = YELLOW OTHER = WHITE	OX = Oxidizer ACID = Acid ALK = Alkali COR = Corrosive W = Use No Water

SECTION I - GENERAL INFORMATION

Characterization	RCRA Number	EPA Class
Aliphatic Ester	**None**	**No Applicable**

DOT Proper Shipping Name **Fluoroacetic Acid**	Chemical Abstract Service (CAS) Number **144-49-0**
DOT Hazard Class and Label Requirements **Poison B; Poison**	DOT Emergency Guide Code **26**
DOT Identification Number **UN 2642**	Chemical Formula **$C_2H_3FO_2$**

Synonyms

Fluoroethanoic acid; cymonic acid; fluoroacetate; FAA; 2-fluoroacetic acid; gifblaar poison; monofluoroacetic acid.

SECTION II - HAZARDOUS INGREDIENTS AND IDENTITY INFORMATION

Hazardous Components (specific identity)	OSHA Exposure Criteria	NIOSH Exposure Criteria	Immediately Dangerous to Life and Health (IDLH)	ACGIH Exposure Criteria
Fluoroacetic acid (derivation: By heating methyl chloroacetate with potassium fluoride and siphoning the methyl ester with baryta; or by heating methyl iodoacetate with silver fluoride or mercurous fluoride).	**PEL:** Not Established **STEL:** Not Established	**REL:** Not Established **STEL:** Not Established	**Not Determined**	**TLV:** Not Established **STEL:** Not Established

SECTION III - PHYSICAL AND CHEMICAL CHARACTERISTICS

Boiling Point **329°F (165°C)**	Specific Gravity (H_2O = 1) **Not Listed**
Vapor Pressure (mm Hg) **Not Found**	Molecular Weight **78.05**
Vapor Density (Air = 1) **Not Found**	Melting Point **91°F (33°C)**

Solubility

Soluble in water and alcohol.

Appearance and Odor

Colorless crystalline solid.

SECTION IV - FIRE AND EXPLOSION HAZARD DATA

Flash Point (method used) **Not Determined**	Explosive Limits in Air % by Volume **LEL: Not Determined** **UEL: Not Determined**
NFPA Classification **Non-flammable Solid**	Autoignition Temperature **Not Determined**

Extinguishing Media

Use dry chemical, carbon dioxide, water spray, or foam.

Special Fire Fighting Procedures

Poisonous gases are produced in fire. Wear full protective clothing and self-contained breathing apparatus (SCBA). Use blanketing effect with foam to smother large fires. Continue to cool containers with water after fire is extinguished. Be aware of fire control measure by-products. Do NOT release runoff to sewers or waterways. Dike and control for later disposal.

Unusual Fire and Explosion Hazards

None reported.

SECTION V - REACTIVITY DATA

Stability		Conditions to Avoid **Fluoroacetic acid is stable under normal conditions of storage and operation. Avoid contact with incompatible materials. Keep away from heat, flame, and other sources of ignition.**
Stable	Unstable	Incompatibility *(materials to avoid)*
X		**Reacts violently in contact with most alkaline metals.**
Hazardous Polymerization		Conditions to Avoid **Under normal conditions of temperature and pressure, hazardous polymerization of fluoroacetic acid is not expected to occur.**
May Occur	Will Not Occur	Hazardous Decomposition or By-products
	X	**When heated to decomposition, fluoroacetic acid emits acrid, poisonous, and irritating fluoride fumes and oxides of sodium.**

SECTION VI - HEALTH HAZARD DATA

Primary Route(s) of Entry:	Inhalation?	Absorption (skin)?	Ingestion?
	X		X

Health Hazards

INHALATION: Irritation to the respiratory system. Effects of exposure are systemic with primary action on the central nervous system (CNS). Respiratory depression, ventricular fibrillation, and cardiac arrest are also possible.

SKIN & EYES: Irritation of the skin and eyes with possible burns.

INGESTION: Hyperactivity, muscular spasms, convulsions, respiratory depression, irregular heartbeat and ventricular fibrillation, cardiac arrest, and death.

Carcinogenicity	NTP Listed?	IARC Cancer Review Group?	OSHA Regulated?	Target Organs?
Unknown Human Unknown Animal	No	No	No	Respiratory system; skin; eyes, CNS, CVS.

Medical Conditions Generally Aggravated by Exposure

None reported.

Emergency and First-aid Procedures

Eye contact: Flush immediately with water for 15 minutes (minimum); seek medical attention. **Skin contact:** Remove all contaminated clothing. Immediately wash area with large amounts of soap and water. Seek medical assistance. For **inhalation**: Remove the person from exposure. Provide respiratory assistance and CPR. Transfer to medical facility. If **swallowed**: Seek medical attention immediately. Give 1-2 glasses of water and induce vomiting, repeat until vomitus is clear, unless advised otherwise. Never give an unconscious or convulsing person anything by mouth.

SECTION VII - PRECAUTIONS FOR SAFE HANDLING AND USE

Steps to be Taken in Case Material is Released or Spilled

Collect powdered materials using HEPA vacuum. Do NOT dry sweep (generates dusts). Damp mop residues and collect for disposal. Ventilate area of spill. Store materials in DOT-approved containers. Restrict those not involved in cleanup from entering area. Notify appropriate authorities, as required.

Preferred Waste Disposal Method

No Citation.

Precautions to be Taken in Handling and Storage

Store in tightly closed containers in a cool, dark, well-ventilated area. Protect containers from physical damage. Keep heat, fire and flame away. Keep away from alkaline metals.

Other Precautions and Warnings

A highly toxic irritant. Handle with care and use proper personal protective equipment.

SECTION VIII - CONTROL MEASURES AND PERSONAL PROTECTIVE EQUIPMENT

Respiratory Protection *(specify type)*

No exposure levels established. For best protection, recommend using an MSHA/NIOSH-approved supplied-air respirator or a self-contained breathing apparatus (SCBA) with full facepiece operated in pressure demand or other positive pressure mode.

Ventilation

Local exhaust or general mechanical systems recommended.

Protective Gloves	Eye Protection	Other Protective Clothing
Impervious Gloves (avoid rubber)	Chemical Goggles or Face Mask	Impermeable Apron

Work/Hygiene Practices

Always wash hands thoroughly after using chemical; never bring food, drink, or smoking materials into vicinity of chemicals.

FLUOROACETIC ACID

$C_2H_3FO_2$ CAS: 144-49-0

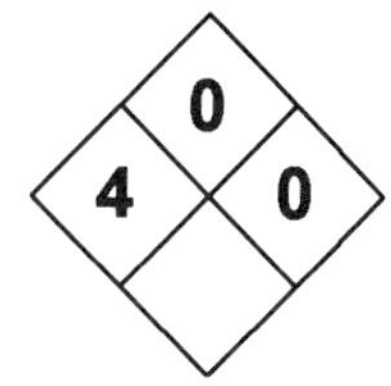

IDENTIFICATION AND TYPICAL USES

Fluoroacetic acid is a colorless, crystalline solid. It can occur naturally in an extremely poisonous plant indigenous to South Africa known as *Gifblaar*. Its sodium salt is mixed with water and used as a powerful rodenticide.

RISK ASSESSMENT: HEALTH

General Assessment

Fluoroacetic acid is a human poison by *ingestion* causing toxic systemic effects. It is mildly toxic by *inhalation*. Skin contact can produce localized effects (irritation and burns) but absorption has not been reported. Its carcinogenic, mutagenic, and teratogenic properties are not known.

Inhalation of fluoroacetic acid dusts or mists (from the mixed solution) can cause serious irritation to the mucosa of the eyes, nose, throat, and respiratory system. Once absorbed through the respiration process, the effects appear to target the central nervous system (CNS). There may be headache, dizziness, nausea, respiratory depression, irregular heartbeat and possible ventricular fibrillation leading to cardiac arrest and death.

Ingestion of fluoroacetic acid may cause gastrointestinal tract irritation with nausea and vomiting. Once absorbed through the digestive process, there can be adverse CNS effects with hyperactivity, muscular spasms in the hands and feet, and convulsions. This may lead to respiratory depression, rapid pulse, ventricular fibrillation, cardiac arrest, and death.

Acute Health Effects

The following acute (short-term) health effects may occur immediately or shortly after exposure to fluoroacetic acid:

Skin: Moderate to severe irritation and possible burns at area of contact.

Eye: Severe irritation and burns. If not removed immediately, may cause damage to vision.

Lung: Irritation of the eyes, nose, throat and lungs (especially the mucus membranes).

CNS: Adverse effects on the CNS through inhalation and ingestion. Can cause death if allowed to go untreated. Cardiac arrest and respiratory depression are possible.

Chronic Health Effects

The following chronic (long-term) health effects may occur at some time after exposure to fluoroacetic acid and can last for months or even years:

Cancer Hazards: According to the references, fluoroacetic acid has not been adequately tested for its ability to cause cancer in test animals. Its human carcinogenic or mutagenic properties are not known.

Reproduction: According to the references, fluoroacetic acid has not been adequately tested for its ability to adversely affect reproduction in test animals.

Other Known Effects: There are no other chronic effects of exposure to fluoroacetic acid reported in the references.

Recommended Risk-Reduction Measures

Personnel should avoid direct contact with fluoroacetic acid. The best risk reduction measure is to use a less toxic chemical as a substitute for fluoroacetic acid. If this is not possible or feasible, then *engineering controls* are the most effective method of reducing exposures. The best protection is to enclose operations and/or provide local exhaust ventilation at the site of chemical release. While not always operationally feasible, isolating operations can also reduce exposure risk. Using respiratory protection is less effective than the controls mentioned above, but is still advisable whenever working with or around fluoroacetic acid. No exposure levels have been established for fluoroacetic acid. This does not mean that exposure is without risk. It is a serious irritant and exposure can cause a number of toxic systemic effects up to and including death. For the best protection, use a self-contained breathing apparatus (SCBA) with full facepiece and operated in pressure demand mode or a supplied-air respirator with full facepiece operated in pressure demand or other positive pressure mode. If a full facepiece is not available, then chemical goggles should be worn to protect the eyes. Whenever a chemical splash or dust hazards exists, a face shield and a protective apron should be worn. To prevent hand and skin exposures, impervious gloves should be used. Glove manufacturers should be contacted to obtain permeation studies *before* gloves are selected.

Administrative controls should also be in place to minimize the potential for human exposures. These may include written procedures or policies which specify the methods and techniques that will be practiced whenever personnel are to work with fluoroacetic acid.

All personnel should receive *training* on the use, hazards, protective measures, emergency actions, and other precautions per 29 CFR 1910.1200 (Hazard Communication), prior to the first assignment in an area where fluoroacetic acid is used or stored.

If symptoms develop or overexposure is suspected, the following medical tests are recommended:

☑ Lung function tests.

☑ Skin testing with dilute fluoroacetic acid to help diagnose allergy (performed by a qualified allergist).

☑ Neurological evaluation and special testing.

Any medical evaluation should include a careful history of past and present symptoms with an examination. Medical tests that look for existing damage are not a substitute for controlling exposures. Also, since smoking can cause heart disease, emphysema, and other respiratory disorders, smokers exposed to irritating substances such as fluoroacetic acid may experience symptoms more quickly than non-smokers under the same conditions of exposure. Prudent risk management requires careful consideration of *all* possible factors that may be causing the appearance of exposure symptoms in the workplace.

Other methods to reduce exposure include:

☑ Where possible, enclose operations and use local exhaust ventilation at the site of chemical release. If local exhaust ventilation or enclosure is not used, respiratory protection should be mandatory.

☑ Always ensure that proper protective clothing is worn when using chemical substances.

☑ Wash thoroughly immediately after exposure to fluoroacetic acid and at the end of the work shift or before eating, drinking, or smoking.

☑ Hazard warning information should be posted in the work area. In addition, as part of an on-going education and training program, all information on the health and safety hazards of fluoroacetic acid should be communicated to all exposed and potentially exposed workers.

☑ Never eat, drink, or smoke in areas were fluoroacetic acid is used, handled or stored.

RISK ASSESSMENT: ENVIRONMENT

General Assessment

The environment is at risk of exposure during transportation, storage, disposal, or destruction of fluoroacetic acid. In almost every scenario, the threat of environmental exposure is contingent upon the proper handling of the chemical substance. Accidental spills, large or small, can result in fire, explosion, and possible contamination of the surrounding environmental mediums (water, soil, and air).

Fluoroacetic acid is non-flammable solid. However, if it is involved in a fire it can produce toxic and poisonous fumes and smoke. Its finely divided powders may create an explosive mixture in air, especially in confined areas. Also, because it is incompatible with a number of common alkaline metals, contact can result in violent and explosive reactions. These characteristics require special consideration during any emergency situation involving a leak or spill of fluoroacetic acid.

Fluoroacetic acid can enter the environment through its intended use as a rodenticide, from agricultural runoff, from unchecked industrial discharges into effluents, and through spills.

☠ *Acute Ecological Effects*

Acute (short-term) toxic effects may include the death of animals, birds, or fish, and death or low growth rate in plants. Acute effects are seen 2 to 4 days after animals or plants are exposed to fluoroacetic acid. Insufficient data are available to evaluate or predict the short-term effects of fluoroacetic acid to aquatic life, plants, birds, or terrestrial animals.

Chronic Ecological Effects

Chronic toxic effects may include shortened life span, reproductive problems, lower fertility, and changes in appearance or behavior in exposed animals. These effects can be seen long after first exposure(s) to toxic chemicals. Insufficient data are available to evaluate or predict the long-term effects of fluoroacetic acid to aquatic life, plants, birds, or land animals.

Water Solubility

Fluoroacetic acid is highly soluble in water. Concentrations of 1000 milligrams and more will mix with a liter of water.

Persistence in the Environment

Fluoroacetic acid is highly persistent in water, with a half-life of greater than 200 days. The half-life of a pollutant is the amount of time it takes for one-half of the chemical to be degraded. Over 90% of fluoroacetic acid will eventually end up in water; the remainder will end up in terrestrial and aquatic sediments. It is not expected to be persistent in the air.

Bioaccumulation in Aquatic Organisms

Some substances increase in concentration, or *bioaccumulate,* in living organisms as they breathe contaminated air, drink contaminated water, or eat contaminated food. These chemicals can become concentrated in the tissues and internal organs of animals as well as humans.

The concentration of fluoroacetic acid found in fish tissues is expected to be somewhat higher then the average concentration of fluoroacetic acid in water from which the fish was taken.

Recommended Risk-Reduction Measures

Proper training of all transporters will reduce the likelihood of a mishap or accident resulting in a leak or spill to the environment. The correct use of DOT labeling (per the MSDS) on all containers, trucks, and rail cars will enable emergency responders to react properly and quickly to any disaster thereby reducing the potential risk to the environment and to personnel.

Storage of fluoroacetic acid should be segregated from incompatible alkaline metals to minimize the risk of cross-contamination or contact. Buildings designated for storage should be equipped with appropriate fire protection systems (alarms, sprinklers, emergency lighting, portable extinguishers).

If a spill or leak to the environment has occurred, fire department, emergency response, and/or hazardous materials spill personnel should be notified immediately. Cleanup should be attempted only by those trained in proper spill containment procedures. Contaminated soils should be removed for incineration and replaced with clean soil. If fluoroacetic acid should contact the water table, aquifer, or navigable waterway, time is of the essence. It is highly soluble in water and, therefore, total containment and remediation may not be possible. When such spills occur, the local and/or state emergency response authorities must be notified. A comprehensive emergency response or disaster preparedness/recovery plan should be in place prior to any operations involving the use, transportation, storage, or disposal of fluoroacetic acid. If fluoroacetic acid is spilled or leaked, the following specific steps are recommended:

- ☑ Restrict persons not wearing protective clothing from area of spill or leak until cleanup is complete and area can be opened for normal work.
- ☑ Ventilate area of spill, remove ignition sources.
- ☑ Collect spilled powders using a HEPA vacuum. Do NOT dry sweep (generates airborne dusts). Damp mop residues and collect for disposal. Absorb liquid or solution spills in vermiculite, dry sand, earth, or a similar material and deposit in sealed containers. Use non-sparking tools.
- ☑ It may be necessary to dispose of fluoroacetic acid as a hazardous waste. The responsible state agency or the regional office of the federal Environmental Protection Agency (EPA) should be contacted for specific recommendations.

RISK ASSESSMENT: BUSINESS

General Assessment

Accidents or mishaps involving fluoroacetic acid can present a moderate threat to business operations. The loss or damage of equipment or facilities can significantly affect fiscal viability. Lawsuits that may result from personnel injury/death, public exposures, and/or environmental contamination will also require a serious expenditure of resources.

Media attention surrounding an injury, death, or environmental damage can also result in a loss of profits and loss of current as well as future business.

Recommended Risk-Reduction Measures

Company attorneys, safety and health professionals, and environmental specialists should be involved in the development of any procedures or policies intended to manage the use of chemicals in the workplace. A company official should be pre-designated as a public relations officer with specific training in dealing with the press. Corporate plans and policies should be developed, approved, and implemented long before any need for such arises.

REFERENCES

American Conference of Governmental Industrial Hygienists. 1988. *Documentation of the Threshold Limit Values and Biological Exposure Indices,* 5th Edition (with updates). Cincinnati: ACGIH

Anthony, C. P., and N. J. Kolthoff. 1971. *Textbook of Anatomy and Physiology.* St. Louis: C. V. Mosby Company.

Bloom, A. D. and F. J. DeSerres. 1995. *Ecotoxicity and Human Health.* New York: CRC Press/Lewis Publishers.

Brusick, D. J. 1994. *Methods for Genetic Risk Assessment.* New York: CRC Press/Lewis Publishers.

Bueche, J. R. 1972. *Principles of Physics,* 2nd Edition. New York: McGraw-Hill.

Burrell, R., D. K. Flaherty, and L. J. Sauers. 1992. *Toxicology of the Immune System.* New York: Van Nostrand Reinhold.

Calabrese, E. J. 1994. *Biological Effects of Low Level Exposures.* New York: CRC Press/Lewis Publishers.

Cockerman, L. G., B. S. Shane. 1994. *Basic Environmental Toxicology.* New York: CRC Press/Lewis Publishers.

Cothern, C. R., N. P. Ross. 1994. *Environmental Statistics, Assessment, and Forecasting.* New York: CRC Press/Lewis Publishers.

Danse, I. R. 1991. *Common Sense Toxics in the Workplace.* New York: Van Nostrand Reinhold.

Environmental Protection Agency. 1983. *Chemical Hazard Information Profile: Fluoroacetic acid.* Washington, D.C.: U.S. EPA.

Gardner, R. W. 1994. *Chemical Intolerance.* New York: CRC Press/Lewis Publishers.

Hallenbeck, W. H. 1993. *Quantitative Risk Assessment for Environmental and Occupational Health,* 2nd Edition. New York: CRC Press/Lewis Publishers.

Hodgson, E., R. B. Mailman, and J. E. Chambers. 1988. *Dictionary of Toxicology.* London: The MacMillian Company.

Hosty, J. W., and P. Foster. 1990. *A Practical Guide to Chemical Spill Response.* New York: Van Nostrand Reinhold.

Howard, P. H., et. al. 1991. *Handbook of Environmental Degradation Rates.* New York: CRC Press/Lewis Publishers.

International Agency for Research on Cancer (IARC). 1987. *IARC Monographs on the Evaluation of Carcinogenic Risk to Humans, Overall Evaluations of Carcinogenicity: An updating of IARC Monographs Volumes 1-42.* Lyon: IARC.

James, A., and M. Lord. 1992. *Index of Chemical and Physical Data.* New York: Van Nostrand Reinhold.

Jones, F. E. 1994. *Toxic Organic Vapors in the Workplace.* New York: CRC Press/Lewis Publishers.

Kamrin, M. 1 A. 1988. *Toxicology: A Primer on Principles and Applications.* Chelsea, Michigan: Lewis Publishers.

Keith, L. H. and M. M. Walker. 1995. *Handbook of Air Toxics.* New York: CRC Press/Lewis Publishers.

Landis, W. G., M. H. Yu. 1995. *Introduction to Environmental Toxicology.* New York: CRC Press/Lewis Publishers.

Langerman, N. 1994 *Precautionary Labels for Chemical Containers.* New York: CRC Press/Lewis Publishers.

Lauwerys, R. R., P. Hoet. 1993. *Industrial Chemical Exposure.* New York: CRC Press/Lewis Publishers.

Lide, D. R. 1995. *Handbook of Chemistry and Physics,* 76th Edition. New York: CRC Press/Lewis Publishers.

Lewis, Richard J., Sr. 1992. *Sax's Dangerous Properties of Industrial Materials,* Volumes 1, 2, and 3, 8th Edition. New York: Van Nostrand Reinhold.

Lisella, Frank S. 1994. *The VNR Dictionary of Environmental Health and Safety.* New York: Van Nostrand Reinhold.

Manahan, S. E. 1994. *Environmental Chemistry,* 6th Edition. New York: CRC Press/Lewis Publishers.

Manahan, S. E. 1992. *Toxicological Chemistry,* 2nd Edition. New York: CRC Press/Lewis Publishers.

Meyer, Eugene. 1990. *Chemistry of Hazardous Materials.* Englewood Cliffs, NJ: Prentice Hall.

Patnaik, P. 1992. *A Comprehensive Guide to the Hazardous Properties of Chemical Substances.* New York: Van Nostrand Reinhold.

Proctor, N. H., and J. P. Hughes. 1991. *Chemical Hazards of the Workplace.* New York: Van Nostrand Reinhold.

Reeves, A. L. 1979. *Handbook on the Toxicology of Metals.* New York: Elsevier North-Holland.

Richardson, M. 1995. *Dictionary of Substance and Their Effects,* Vol. I-VI. New York: CRC Press/Lewis Publishers.

Royal Society of Chemistry. 1992. *Chemical Safety Data Sheets, Volumes 1-5.* Newcastle-upon-Tyne, UK: Athenaeum Press.

Rylander, R., R. R. Jacobs. 1994. *Organic Dusts.* New York: CRC Press/Lewis Publishers.

MATERIAL SAFETY DATA SHEET

CHEMICAL NAME

FLUORODICHLORO-METHANE

CRC PRESS/LEWIS PUBLISHERS DISCLAIMER: THE INFORMATION AND RECOMMENDATIONS PRESENTED HEREIN ARE BASED ON SOURCES BELIEVED TO BE RELIABLE. CRC MAKES NO REPRESENTATION ON ITS COMPLETENESS OR ACCURACY. IT IS THE RESPONSIBILITY OF THE USER TO DETERMINE THE CHEMICAL'S SUITABILITY FOR ITS INTENDED USE, THE CHEMICAL'S SAFE USE, AND THE CHEMICAL'S PROPER DISPOSAL. NO REPRESENTATIONS AND/OR WARRANTIES, EITHER EXPRESSED OR IMPLIED, OF THE MERCHANTABILITY OR FITNESS FOR A PARTICULAR PURPOSE, OR OF ANY OTHER NATURE, ARE MADE WITH RESPECT TO THE INFORMATION PROVIDED IN THIS MSDS OR TO THE CHEMICAL TO WHICH INFORMATION MAY REFER. CRC NEITHER ASSUMES NOR AUTHORIZES ANY OTHER PERSON TO ASSUME FOR IT, ANY OTHER ADDITIONAL RESPONSIBILITY OR LIABILITY FOR THE USE OF, OR RELIANCE UPON, THIS INFORMATION.

HAZARD WARNING INFORMATION

HEALTH	FIRE	REACTIVE	OTHER	DEGREE OF HAZARD	COLOR CODING	OTHER CODES
2	1	0		0 = Minimum Hazard 1 = Slight Hazard 2 = Moderate Hazard 3 = Serious Hazard 4 = Severe Hazard	HEALTH = BLUE FIRE = RED REACTIVITY = YELLOW OTHER = WHITE	OX = Oxidizer ACID = Acid ALK = Alkali COR = Corrosive W = Use No Water

SECTION I - GENERAL INFORMATION

Characterization	**Halogenated Hydrocarbon**	RCRA Number	**None**
		EPA Class	**Not Applicable**
DOT Proper Shipping Name	**Dichloromonofluoromethane**	Chemical Abstract Service (CAS) Number	**75-43-4**
DOT Hazard Class and Label Requirements	**Non Flammable; Non Flammable Gas**	DOT Emergency Guide Code	**No Citation**
DOT Identification Number	**UN 1029**	Chemical Formula	**$CHCl_2F$**

Synonyms

Dichlorofluoromethane; dichloromonofluoromethane; Freon 21®; Algofrene Type 5®; Arcton 7®; F21; fluorocarbon 21.

SECTION II - HAZARDOUS INGREDIENTS AND IDENTITY INFORMATION

Hazardous Components (specific identity)	OSHA Exposure Criteria	NIOSH Exposure Criteria	Immediately Dangerous to Life and Health (IDLH)	ACGIH Exposure Criteria
Dichloromonofluoromethane (derivation: By reaction of chloroform and hydrogen fluoride). **1 ppm = 4.2 mg/m^3**	**PEL (8-hour):** **1000 ppm** **4200 mg/m^3** **STEL:** **Not Established**	**REL (10 hour):** **10 ppm** **40 mg/m^3** **STEL:** **Not Established**	**5000 ppm**	**TLV (8-hour):** **10 ppm** **45 mg/m^3** **STEL:** **12 ppm** **50 mg/m^3**

SECTION III - PHYSICAL AND CHEMICAL CHARACTERISTICS

Boiling Point	**48°F (9°C)**	Specific Gravity (H_2O = 1)	**1.40**
Vapor Pressure (atmospheres)	**2 at 83°F (28°C)**	Molecular Weight	**102.92**
Vapor Density (Air = 1)	**3.82**	Freezing Point	**-211°F (-135°C)**

Solubility

Insoluble in water. Soluble in alcohol and ether.

Appearance and Odor

Colorless, heavy gas with a slight ether-like odor, especially at extremely high concentrations.

SECTION IV - FIRE AND EXPLOSION HAZARD DATA

Flash Point (method used)	**Non-Flammable**	Explosive Limits in Air % by Volume	**LEL: Not Determined UEL: Not Determined**
NFPA Classification	**Non-Flammable Gas**	Autoignition Temperature	**972°F (522°C)**

Extinguishing Media

Fluorodichloromethane is a non-flammable gas. If it is involved in fire, use dry chemical, carbon dioxide, water spray, fog, or foam or extinguishing agent suitable to surrounding fire.

Special Fire Fighting Procedures

Poisonous gases are produced in fire, wear full protective clothing and self-contained breathing apparatus. Move containers from fire area if it can be done safely. Cool fire-exposed containers using water spray long after fire is out. Vapors may cause dizziness and suffocation. Liquid causes frostbite.

Unusual Fire and Explosion Hazards

Containers may explode in fire due to buildup of excessive internal pressures. Evacuate immediately if rising sound is heard from venting device or if there is any noticeable discoloration of tank or vessel.

SECTION V - REACTIVITY DATA

Stability		Conditions to Avoid **Fluorodichloromethane is stable in closed containers under normal conditions of handling and storage. Avoid exposures to excessive heat and contact with incompatible materials.**
Stable X	Unstable	Incompatibility *(materials to avoid)* **Fluorodichloromethane will react in the presence of chemically active metals, such as potassium, sodium, calcium, powdered aluminum, magnesium, or zinc. Contact with acid or acid fumes produces toxic gases.**
Hazardous Polymerization		Conditions to Avoid **Under normal conditions of temperature and pressure, hazardous polymerization of fluorodichloromethane cannot occur.**
May Occur	Will Not Occur X	Hazardous Decomposition or By-products **Toxic and irritating fumes and gases are produced when fluorodichloromethane is involved in fire, including, phosgene, chlorine, hydrogen chloride, hydrogen fluoride, and fluorine.**

SECTION VI - HEALTH HAZARD DATA

Primary Route(s) of Entry:	Inhalation?	Absorption (skin)?	Ingestion?
	X		

Health Hazards

INHALATION: Low-level toxicity. A simple asphyxiant when oxygen levels are 15 to 16%. Produces unconsciousness at 6 to 8% oxygen. High concentrations may cause giddiness, lightheadedness, disorientation, nausea, vomiting, narcosis, cardiac dysrhythmias, lowered blood pressure, and death.

SKIN & EYES: Eye irritation at high concentrations. The liquid can cause frostbite burns to tissues.

INGESTION: Ingestion not likely.

Carcinogenicity	NTP Listed?	IARC Cancer Review Group?	OSHA Regulated?	Target Organs?
No Citation (Human) No Citation (Animal)	No	No	29 CFR 1910.1000 Table Z-1	Respiratory system, skin, eyes, cardiovascular sys.

Medical Conditions Generally Aggravated by Exposure

None reported.

Emergency and First-aid Procedures

Eye contact: Flush immediately with water for 15 minutes (minimum), seek medical attention. Skin contact: For frostbite, rapidly re-warm using water at 107°F (42°C) until flush returns, Do NOT use dry heat! Seek medical assistance. For inhalation: Remove the person from exposure. Provide respiratory assistance, 100% supplemental humidified oxygen, and CPR if required. Transfer to medical facility. Observe for heart disturbances. If swallowed: Not likely since it exists in liquid state only below temperatures of -211°F (-135°C).

SECTION VII - PRECAUTIONS FOR SAFE HANDLING AND USE

Steps to be Taken in Case Material is Released or Spilled

Evacuate area, deny entry, stay upwind. Wear protective clothing. Contact with the liquid causes frostbite. If it can be done without risk, shut off leak (a thin film of soapy solution may help detect leak). Ventilate area to avoid displacement of oxygen. If leak cannot be repaired, move container to open area (if possible), preferably outdoors, and allow to empty. Keep away from heat.

Preferred Waste Disposal Method

No Citation.

Precautions to be Taken in Handling and Storage

Do not store fluorodichloromethane in the presence of incompatible materials. Store in tightly closed containers in a cool, dry, well-ventilated area. Separate empty from useable cylinders.

Other Precautions and Warnings

Prevent physical damage to compressed gas storage cylinders.

SECTION VIII - CONTROL MEASURES AND PERSONAL PROTECTIVE EQUIPMENT

Respiratory Protection *(specify type)*

Air purifying respirators are NOT advised for use with asphyxiating gases. Best protection is provided by a supplied-air respirator, or a self-contained breathing apparatus (SCBA) with full facepiece operated in pressure demand or other positive pressure mode.

Ventilation

Local exhaust or general mechanical systems recommended.

Protective Gloves	Eye Protection	Other Protective Clothing
Impervious Material (Insulated)	**Chemical Goggles or Face Mask**	**Protective Clothing or Apron**

Work/Hygiene Practices

Always wash hands thoroughly after using chemical; never bring food, drink, or smoking materials into vicinity of chemicals.

FLUORODICHLOROMETHANE

$CHCl_2F$ CAS: 75-43-4

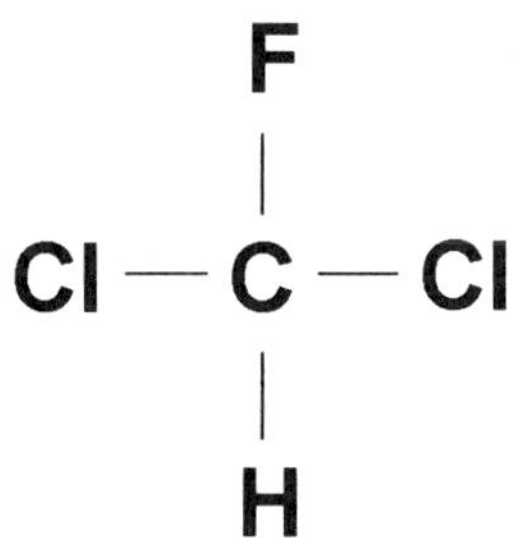

IDENTIFICATION AND TYPICAL USES

Fluorodichloromethane is a heavy, colorless gas with an ether-like odor, especially at extremely high concentrations. It is used primarily as a solvent and a refrigerant, especially in the aerospace industry.

RISK ASSESSMENT: HEALTH

General Assessment

Fluorodichloromethane is considered to have low toxicity by the *inhalation* route. It is a simple asphyxiant and can produce some irritation and mild to moderate narcotic effects in high concentrations. Its greatest risk to health is its ability to induce serious cardiac arrhythmia at high concentrations or in personnel with pre-exiting heart dysfunction. Acute and chronic exposure data for humans is limited in the references.

Inhalation produces some mild irritation of the eyes, nose, throat, and respiratory tract. At high concentrations (5000 ppm), there is a possibility for liver damage. It can also produce narcotic effects at high concentrations with primary action on the central nervous system (CNS). Symptoms include giddiness, disorientation, lightheadedness, nausea and vomiting, narcosis, hypotension, potentially dangerous changes in heart function, and death. Also, inhalation of fumes produced during fire can cause eye, nose, and throat irritation, chemical pneumonitis, or pulmonary edema (fluid in the lungs), which can lead to death. Indications of asphyxia include an overwhelming feeling of "air starvation" or air hunger, and rapid-irregular breathing. There may be headache, fatigue, mental confusion, nausea and vomiting, giddiness and poor judgment, exhaustion, numbness in the legs and arms, convulsions, and death. Prolonged concentrations can lead to generalized paresthesia, ringing in the ears, apprehension, and slurred speech. These may be followed by a marked decrease in consciousness, amnesia, and cardiac arrhythmia. Heart rhythm changes result from fluorodichloromethane's ability to increase the heart's sensitivity to adrenaline which is always present in the blood stream.

Skin contact with the compressed gas or cryogenic liquid will cause mild to severe frostbite burns. Contact with the eyes can cause extremely painful tissue destruction due to the cryogenic properties of this compound.

Acute Health Effects

The following acute (short-term) health effects may occur immediately or shortly after exposure to fluorodichloromethane:

Skin: Irritation and possible frostbite.

Eye: Moderate to severe irritation and possible burns and frostbite.

Lung: Irritation of the mucosa. Will cause asphyxiation and associated symptoms (up to and including death) at high concentrations.

Other: A narcotic at high concentrations. Will effect the central nervous system (CNS) with narcosis (drowsiness), giddiness, and disorientation. It can also induce irregular or abnormal heart rhythm (arrhythmia).

Chronic Health Effects

The following chronic (long-term) health effects may occur at some time after exposure to fluorodichloromethane and can last for months or even years:

Cancer Hazards: According to the information presented in the references, fluorodichloromethane has not been shown to cause cancer in test animals.

Reproductive Hazard: According to the information presented in the references, there is limited evidence that exposure to fluorodichloromethane may adversely affect reproduction in test animals.

Other Chronic Effects: Low, chronic exposures may lead to liver and kidney damage and possible injury to the central nervous system. Cardiac rhythm changes may also be induced on high or prolonged exposures.

Recommended Risk-Reduction Measures

While fluorodichloromethane is considered to be of a low-order toxicity by inhalation, it can still pose some degree of health hazard risk. Unless a less toxic chemical can be substituted for a hazardous substance, *engineering controls* are the most effective method of reducing exposures. The best protection is to enclose operations and/or provide local exhaust ventilation at the site of chemical release. While not always operationally feasible, isolating operations can also reduce exposure.

Using respiratory protection is less effective than the controls mentioned above, but is still advisable whenever working with or around fluorodichloromethane. The exposure level for this chemical is extremely high (1000 ppm), and, therefore, respiratory protection may not always be necessary in well-ventilated work areas. However, when high exposure levels are predicted or prolonged exposures are anticipated, a supplied-air respirator or a self-contained breathing apparatus (SCBA) with full facepiece operated in pressure demand mode are the recommended respiratory protection of choice. If a full facepiece is not available, then chemical goggles should be worn to protect the eyes. Whenever a chemical splash hazard exists, a face shield and rubber apron should be worn. To prevent hand and skin exposures, impervious, insulated gloves should be used.

Administrative controls should also be in place to minimize the potential for human exposures. These may include written procedures or policies which specify the methods and techniques that will be practiced whenever personnel are to work with fluorodichloromethane.

All personnel should receive *training* on the use, hazards, protective measures, emergency actions, and other precautions per 29 CFR 1910.1200 (Hazard Communication), prior to the first assignment in an area where fluorodichloromethane is used or stored.

Before beginning employment and at regular intervals thereafter (e.g., annually), the following medical protocol is suggested for those assigned to work with fluorodichloromethane:

☑ Electrocardiogram (EKG), to test for pre-existing cardiac rhythm dysfunction and establish baseline.

If symptoms develop or overexposure is suspected, the following medical tests may be useful:

☑ Electrocardiogram (EKG), to compare with baseline.

☑ Liver function tests.

☑ Nervous system function tests, including interviews by a qualified neurologist to evaluate any changes in personality or behavior.

Any evaluation should include a careful medical history of past and present symptoms with an examination. However, medical tests that evaluate existing damage are not a substitute for controlling exposure.

Other methods to reduce exposure include:

☑ Where possible, enclose operations and use local exhaust ventilation at the site of chemical release. If local exhaust ventilation or enclosure is not used, respiratory protection should be mandatory.

☑ Always ensure that proper protective clothing is worn when using chemical substances.

☑ Wash thoroughly immediately after exposure to fluorodichloromethane and at the end of the work shift or before eating, drinking, or smoking.

☑ Hazard warning information should be posted in the work area. In addition, as part of an on-going education and training program, all information on the health and safety hazards of fluorodichloromethane should be communicated to all potentially exposed workers.

☑ Eye wash stations should be provided in work areas. If the potential for whole body exposure exists, then safety showers should also be provided.

RISK ASSESSMENT: ENVIRONMENT

General Assessment

The environment is at risk of exposure during intended use, transportation, storage, disposal, or destruction of fluorodichloromethane. In almost every scenario, the threat of environmental exposure is contingent upon the proper handling of the chemical substance. Accidental spills, large or small, can result in fire, explosion, and possible contamination of the surrounding environmental mediums (water, soil, and air).

Fluorodichloromethane is considered a non-flammable gas. However, it is incompatible with a number of common, chemically active metals and caution is required to ensure contact is prevented. It is also a chlorofluorocarbon (CFC) which is a group of halogenated hydrocarbons alleged to cause damage to the earth's protective ozone layer. Therefore, accidental release to the environment must not occur. These characteristics require special consideration during any emergency situation involving a leak or spill of fluorodichloromethane. Should fluorodichloromethane ever come into contact with incompatible substances either during use, transportation, or storage, violent and explosive reactions can occur.

Fluorodichloromethane is normally a gas. But it can also exist as a cryogenic liquid. It can enter the environment through industrial discharges and spills.

☠ *Acute Ecological Effects*

Acute (short-term) toxic effects may include the death of animals, birds, or fish and death or low growth rate in plants. Acute effects are seen 2 to 4 days after animals or plants are exposed to fluorodichloromethane. This chemical is not expected to cause adverse short-term effects on aquatic life, plants, birds, or terrestrial animals.

Chronic Ecological Effects

Chronic toxic effects may include shortened life span, reproductive problems, lower fertility, and changes in appearance or behavior in exposed animals. These effects can be seen long after first exposure(s) to toxic chemicals. Insufficient data are available to evaluate or predict the long-term effects of fluorodichloromethane to aquatic life, plants, birds, or land animals.

Water Solubility

Fluorodichloromethane is insoluble in water. It is normally a gas.

⌛ *Persistence in the Environment*

Fluorodichloromethane is virtually non-persistent in water, with a half-life of less than 2 days. The half-life of a pollutant is the amount of time it takes for one half of the chemical to be degraded. Approximately 100% of fluorodichloromethane will eventually end up in air.

Bioaccumulation in Aquatic Organisms

Some substances increase in concentration, or *bioaccumulate,* in living organisms as they breathe contaminated air, drink contaminated water, or eat contaminated food. These chemicals can become concentrated in the tissues and internal organs of animals as well as humans.

The concentration of fluorodichloromethane found in fish tissues is expected to much lower than the average concentration of fluorodichloromethane in the water from which the fish was taken.

Recommended Risk-Reduction Measures

Proper training of all transporters will reduce the likelihood of a mishap or accident resulting in a leak or spill to the environment. The correct use of DOT labeling (per the MSDS) on all containers, trucks, and rail cars will enable emergency responders to react properly and quickly to any disaster thereby reducing the potential risk to the environment and to personnel.

Storage of fluorodichloromethane should be segregated from incompatible chemicals to minimize the risk of cross-contamination or contact. Containers must be protected from shock, excessive heat (including the sun's rays), and damage.

If a spill or leak to the environment has occurred, fire department, emergency response and/or hazardous materials spill personnel should be notified immediately. Cleanup should be attempted only by those trained in proper spill containment procedures. If any liquid has spilled, contaminated soils may need to be removed for incineration and replaced with clean soil. If fluorodichloromethane should contact the water table, aquifer, or navigable waterway, remediation activities should be prompt to ensure protection of the aquatic environment. When such spills occur, the local and/or state emergency response authorities should be notified. A comprehensive emergency response or disaster preparedness/recovery plan should be in place prior to any operations involving the use, transportation, storage, or disposal of fluorodichloromethane.

If fluorodichloromethane is spilled or leaked, the following specific steps are recommended:

☑ Evacuate area and deny entry until cleanup is completed. Restrict persons not wearing protective clothing from hazard area.

☑ Ventilate area of spill or leak.

☑ If source of leak is a cylinder and it cannot be repaired in place, remove to safe location (e.g., outdoors) and allow to empty. Stay clear of area.

☑ It may be necessary to dispose of dichlorodifluoromethane as a hazardous waste. The responsible state agency or the regional office of the federal Environmental Protection Agency (EPA) should be contacted for specific recommendations.

RISK ASSESSMENT: BUSINESS

General Assessment

Accidents or mishaps involving fluorodichloromethane can present a significant threat to business operations. The loss or damage of equipment or facilities can significantly affect fiscal viability. Lawsuits that may result from personnel injury/death, public exposures, and/or environmental contamination will also require a serious expenditure of resources. Media attention surrounding an injury, death, or environmental damage can also result in a loss of profits and loss of current as well as future business.

Recommended Risk-Reduction Measures

Company attorneys, safety and health professionals, and environmental specialists should be involved in the development of any procedures dealing with safety or emergency response. A company official should be pre-designated as a public relations officer with specific training in dealing with the press. Corporate plans and policies should be developed, approved, and implemented long before any need for such arises.

REFERENCES

American Conference of Governmental Industrial Hygienists. 1988. *Documentation of the Threshold Limit Values and Biological Exposure Indices,* 5th Edition (with updates). Cincinnati: ACGIH

Anthony, C. P., and N. J. Kolthoff. 1971. *Textbook of Anatomy and Physiology.* St. Louis: C. V. Mosby Company.

Bueche, J. R. 1972. *Principles of Physics,* 2nd Edition. New York: McGraw-Hill.

Brusick, D. J. 1994. *Methods for Genetic Risk Assessment.* New York: CRC Press/Lewis Publishers.

Burrell, R., D. K. Flaherty, and L. J. Sauers. 1992. *Toxicology of the Immune System.* New York: Van Nostrand Reinhold.

Calabrese, E. J. 1994. *Biological Effects of Low Level Exposures.* New York: CRC Press/Lewis Publishers.

Cothern, C. R. 1995. *Handbook for Environmental Risk Decisionmaking.* New York: CRC Press/Lewis Publishers.

Gardner, R. W. 1994. *Chemical Intolerance.* New York: CRC Press/Lewis Publishers.

Hodgson, E., R. B. Mailman, and Janice E. Chambers. 1988. *Dictionary of Toxicology.* London: The MacMillian Company.

Jones, F. E. 1994. *Toxic Organic Vapors in the Workplace.* New York: CRC Press/Lewis Publishers.

Keith, L. H. and M. M. Walker. 1995. *Handbook of Air Toxics.* New York: CRC Press/Lewis Publishers.

Landis, W. G., M. H. Yu. 1995. *Introduction to Environmental Toxicology.* New York: CRC Press/Lewis Publishers.

Langerman, N. 1994 *Precautionary Labels for Chemical Containers.* New York: CRC Press/Lewis Publishers.

Lauwerys, R. R., P. Hoet. 1993. *Industrial Chemical Exposure.* New York: CRC Press/Lewis Publishers.

Lewis, R. J., Sr. 1993. *Hawley's Condensed Chemical Dictionary,* 12th Edition. New York: Van Nostrand Reinhold.

Lewis, R. J., Sr. 1992. *Sax's Dangerous Properties of Industrial Materials,* Volumes 1, 2, and 3, 8th Edition. New York: Van Nostrand Reinhold.

Mackay, D., W. Y. Shiu, and K. C. Ma. 1992. *Illustrated Handbook of Chemical-Physical Properties and Environmental Fate for Organic Chemicals.* New York: CRC Press/Lewis Publishers.

Manahan, S. E. 1992. *Toxicological Chemistry,* 2nd Edition. New York: CRC Press/Lewis Publishers.

Meyer, E. 1990. *Chemistry of Hazardous Materials.* Englewood Cliffs, NJ: Prentice Hall.

Mycock, J. C., J. D. McKenna, and L. Theodore. 1995. *Handbook of Air Pollution Control Engineering and Technology.* New York: CRC Press/Lewis Publishers.

Nelson, G. O. 1992. *Gas Mixtures.* New York: CRC Press/Lewis Publishers.

Ott, W. R. 1995. *Environmental Statistics and Data Analysis.* New York: CRC Press/Lewis Publishers.

Patnaik, P. 1992. *A Comprehensive Guide to the Hazardous Properties of Chemical Substances.* New York: Van Nostrand Reinhold.

Quigley, D. R. 1994. *Handbook of Emergency Chemical Management.* New York: CRC Press/Lewis Publishers.

Rea, W. J. 1995. *Chemical Sensitivity,* Vol. I-III. New York: CRC Press/Lewis Publishers.

Richardson, M. 1995. *Dictionary of Substance and Their Effects,* Vol. I-VI. New York: CRC Press/Lewis Publishers.

Sacarello, H. L. A. 1994. *The Comprehensive Handbook of Hazardous Materials.* New York: CRC Press/Lewis Publishers.

Torkelson, T. R. V. K. Rowe. 1981. *Patty's Industrial Hygiene and Toxicology,* Vol. 2B. New York: Wiley-Interscience.

U.S. Department of Health and Human Services, National Institute for Occupational Safety and Health. 1994. *NIOSH Pocket Guide to Chemical Hazards.* Washington, D.C.: U.S. Government Printing Office.

MATERIAL SAFETY DATA SHEET

CHEMICAL NAME	
FLUOROTRICHLORO-METHANE	**CRC PRESS/LEWIS PUBLISHERS DISCLAIMER:** THE INFORMATION AND RECOMMENDATIONS PRESENTED HEREIN ARE BASED ON SOURCES BELIEVED TO BE RELIABLE. CRC MAKES NO REPRESENTATION ON ITS COMPLETENESS OR ACCURACY. IT IS THE RESPONSIBILITY OF THE USER TO DETERMINE THE CHEMICAL'S SUITABILITY FOR ITS INTENDED USE, THE CHEMICAL'S SAFE USE, AND THE CHEMICAL'S PROPER DISPOSAL. NO REPRESENTATIONS AND/OR WARRANTIES, EITHER EXPRESSED OR IMPLIED, OF THE MERCHANTABILITY OR FITNESS FOR A PARTICULAR PURPOSE, OR OF ANY OTHER NATURE, ARE MADE WITH RESPECT TO THE INFORMATION PROVIDED IN THIS MSDS OR TO THE CHEMICAL TO WHICH INFORMATION MAY REFER. CRC NEITHER ASSUMES NOR AUTHORIZES ANY OTHER PERSON TO ASSUME FOR IT, ANY OTHER ADDITIONAL RESPONSIBILITY OR LIABILITY FOR THE USE OF, OR RELIANCE UPON, THIS INFORMATION.

HAZARD WARNING INFORMATION

HEALTH	FIRE	REACTIVE	OTHER	DEGREE OF HAZARD	COLOR CODING	OTHER CODES
2	0	0		0 = Minimum Hazard 1 = Slight Hazard 2 = Moderate Hazard 3 = Serious Hazard 4 = Severe Hazard	HEALTH = BLUE FIRE = RED REACTIVITY = YELLOW OTHER = WHITE	OX = Oxidizer ACID = Acid ALK = Alkali COR = Corrosive ~~W~~ = Use No Water

SECTION I - GENERAL INFORMATION

Characterization	RCRA Number	EPA Class
Halogenated Hydrocarbon	**U121**	**Toxic Waste**

DOT Proper Shipping Name **Refrigerant gases, n.o.s.**	Chemical Abstract Service (CAS) Number **75-69-4**
DOT Hazard Class and Label Requirements **Non Flammable; Non Flammable Gas**	DOT Emergency Guide Code **No Citation**
DOT Identification Number **UN 1078**	Chemical Formula **ClF_3**

Synonyms

Trichlorofluoromethane; trichloromonofluoromethane; Freon 11®; Algofrene Type 1®; Arcton 11®; FC-11; monofluorotrichloromethane; refrigerant 11; fluorocarbon 11; Genetron 11®; Halocarbon 11®.

SECTION II - HAZARDOUS INGREDIENTS AND IDENTITY INFORMATION

Hazardous Components (specific identity)	OSHA Exposure Criteria	NIOSH Exposure Criteria	Immediately Dangerous to Life and Health (IDLH)	ACGIH Exposure Criteria
Trichlorofluoromethane (derivation: From carbon tetrachloride and hydrogen fluoride in the presence of fluorinating agents such as antimony *tri-* and *penta*-fluoride). **1 ppm = 5.71 mg/m^3**	**PEL (ceiling):** **1000 ppm** **5600 mg/m^3** **STEL:** **Not Established**	**REL (ceiling):** **1000 ppm** **5600 mg/m^3** **STEL:** **Not Established**	**2000 ppm**	**TLV (ceiling):** **1000 ppm** **5620 mg/m^3** **STEL:** **Not Established**

SECTION III - PHYSICAL AND CHEMICAL CHARACTERISTICS

Boiling Point **75°F (24°C)**	Specific Gravity (H_2O = 1) **1.47 (liquid at 75°F/24°C)**
Vapor Pressure (mmHg) **690 at 68°F (20°C)**	Molecular Weight **137.4**
Vapor Density (Air = 1) **5.04**	Freezing Point **-168°F (-111°C)**

Solubility

Slightly soluble in water (0.1% at 75°F). Soluble in alcohol, ether, and other organic solvents.

Appearance and Odor

Colorless or water-white liquid, nearly odorless. A gas above 75°F (24°C).

SECTION IV - FIRE AND EXPLOSION HAZARD DATA

Flash Point (method used) **Non-Flammable**	Explosive Limits in Air % by Volume **LEL: Not Determined UEL: Not Determined**
NFPA Classification **Non-Flammable Gas**	Autoignition Temperature **Not Determined**

Extinguishing Media

Fluorotrichloromethane is a non-flammable gas. If it is involved in fire, use dry chemical, carbon dioxide, water spray, fog, or foam or extinguishing agent suitable to surrounding fire.

Special Fire Fighting Procedures

Poisonous gases are produced in fire, wear full protective clothing and self-contained breathing apparatus. Move containers from fire area if it can be done safely. Cool fire-exposed containers using water spray long after fire is out. Vapors may cause dizziness and suffocation. Liquid causes frostbite.

Unusual Fire and Explosion Hazards

Containers may explode in fire due to buildup of excessive internal pressures. Evacuate immediately if rising sound is heard from venting device or if there is any noticeable discoloration of tank or vessel.

SECTION V - REACTIVITY DATA

Stability		Conditions to Avoid
		Fluorotrichloromethane is stable in closed containers under normal conditions of handling and storage. Avoid exposures to excessive heat and contact with incompatible materials.
Stable	Unstable	Incompatibility *(materials to avoid)*
X		Fluorotrichloromethane will react in the presence of chemically active metals, (potassium, sodium, calcium, powdered aluminum, magnesium, or zinc). Reacts violently with aluminum, barium, and lithium.
Hazardous Polymerization		Conditions to Avoid
		Under normal conditions of temperature and pressure, hazardous polymerization of fluorotrichloromethane cannot occur. It will attack some forms of plastic, rubber, and coatings.
May Occur	Will Not Occur	Hazardous Decomposition or By-products
	X	Toxic and irritating fumes and gases are produced when fluorotrichloromethane is involved in fire, including CO_2, phosgene, chlorine, carbonyl fluoride, hydrogen chloride, hydrogen fluoride, and fluorine.

SECTION VI - HEALTH HAZARD DATA

Primary Route(s) of Entry:	Inhalation?	Absorption (skin)?	Ingestion?
	X		X

Health Hazards

INHALATION: High concentrations cause narcosis and anesthesia. Systemic effects include dizziness, lightheadedness, drowsiness (narcosis), disorientation, conjunctival irritation, fibrosing alveolitis, bronchospasm, upper respiratory tract irritation, cardiac arrhythmia, and liver changes. Peripheral neuropathy may also occur.

SKIN & EYES: Eye irritation at high concentrations. The liquid causes frostbite burns to tissues. There may be dryness, cracking, redness, contact dermatitis, and secondary infection.

INGESTION: Ingestion is not likely but can occur. Causes necrosis and stomach perforations.

Carcinogenicity	NTP Listed?	IARC Cancer Review Group?	OSHA Regulated?	Target Organs?
No Citation (Human) No Citation (Animal)	No	No	29 CFR 1910.1000 Table Z-1	Respiratory system, skin, eyes, CNS, CVS.

Medical Conditions Generally Aggravated by Exposure

Cardiovascular diseases and skin diseases (dermatitis) may be aggravated by exposure.

Emergency and First-aid Procedures

Eye contact: Flush immediately with water for 15 minutes (minimum), seek medical attention. Skin contact: For frostbite, rapidly re-warm using water at 107°F (42°C) until flush returns, Do NOT use dry heat! Do NOT rub affected area. Seek medical assistance. For inhalation: Remove the person from exposure. Provide respiratory assistance, 100% supplemental humidified oxygen, and CPR if required. Transfer to medical facility. Observe for heart disturbances. If swallowed: Contact poison control center. Unless advised otherwise, give 1-2 glasses of water and induce vomiting.

SECTION VII - PRECAUTIONS FOR SAFE HANDLING AND USE

Steps to be Taken in Case Material is Released or Spilled

Evacuate area, deny entry, stay upwind. Wear protective clothing. Absorb any liquid spills with vermiculite and deposit in sealed drums. Ventilate area to avoid displacement of oxygen. Contact with the liquid causes frostbite. If it can be done without risk, shut off leak. If leak cannot be repaired, move container to open area (if possible), preferably outdoors, and allow to empty. Keep away from heat.

Preferred Waste Disposal Method

Rotary kiln and fluidized bed incineration with acid scrubber to remove halo acids.

Precautions to be Taken in Handling and Storage

Do not store fluorotrichloromethane in the presence of incompatible materials. Store in tightly closed containers in a cool, dry, well-ventilated area. Separate empty from useable cylinders.

Other Precautions and Warnings

Prevent physical damage to compressed gas storage cylinders.

SECTION VIII - CONTROL MEASURES AND PERSONAL PROTECTIVE EQUIPMENT

Respiratory Protection *(specify type)*

Air purifying respirators are NOT advised for use with asphyxiating gases. Best protection is provided by a supplied-air respirator, or a self-contained breathing apparatus (SCBA) with full facepiece operated in pressure demand or other positive pressure mode.

Ventilation

Local exhaust or general mechanical systems recommended.

Protective Gloves	Eye Protection	Other Protective Clothing
Polycarbonate, Viton, or Polyvinylchloride	Chemical Goggles or Face Mask	Protective Clothing or Apron

Work/Hygiene Practices

Always wash hands thoroughly after using chemical; never bring food, drink, or smoking materials into vicinity of chemicals.

FLUOROTRICHLOROMETHANE

Cl_3F

CAS: 75-69-4

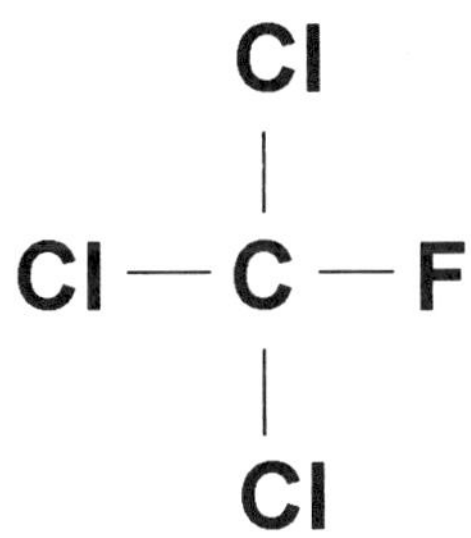

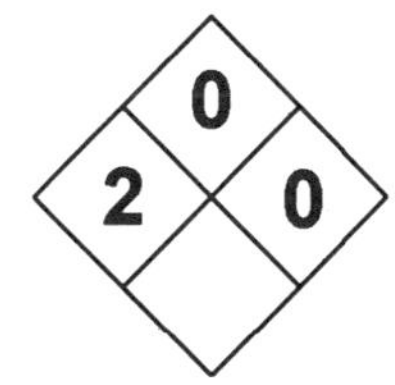

IDENTIFICATION AND TYPICAL USES

Fluorotrichloromethane is colorless to water-white liquid with a very faint, ether-like odor. It can also exist as a gas above 75°F (24°C). It is used as an industrial solvent and degreasing agent, in fire extinguishers, in electric insulation, in the production of polymeric resin, as a chemical intermediate, and as a blowing agent. It was formerly used in aerosol propellants but was banned from this use in 1978 in the United States. It is alleged to damage (deplete) the stratospheric ozone.

RISK ASSESSMENT: HEALTH

General Assessment

Fluorotrichloromethane is considered to have moderate toxicity by the *inhalation* route. Its is a simple asphyxiant and can produce some irritation and mild to moderate narcotic effects in high concentrations. It can also cause physical damage and irritation to tissues and mucosa. Its greatest risk to health is its ability to induce serious cardiac arrhythmia at high concentrations or in personnel with pre-exiting heart dysfunction.

Inhalation causes dizziness, narcosis (drowsiness), lightheadedness, disorientation, irritation of conjunctival tissues, bronchospasm, upper respiratory tract irritation (including the nose and throat), cardiac arrhythmia, and changes in the liver. Peripheral neuropathy has also been reported. Death can occur from respiratory depression or cardiac sensitization (this chemical has the ability to increase the heart's sensitivity to adrenaline which is always present in the blood stream).

Skin contact with the compressed gas or cryogenic liquid will cause mild to severe frostbite burns. Contact with the eyes can cause extremely painful tissue destruction due to the cryogenic properties of this compound.

Since fluorotrichloromethane can exist as a liquid at room temperatures, ingestion is possible. If swallowed, symptoms of necrosis (tissue destruction) and stomach perforations may occur.

Acute Health Effects

The following acute (short-term) health effects may occur immediately or shortly after exposure to fluorotrichloromethane:

Skin: Irritation and possible frostbite.

Eye: Moderate to severe irritation and possible burns and frostbite.

Lung: Irritation of the mucosa. May cause bronchospasm, cough, and congestion or tightness in chest.

Other: A narcotic at high concentrations. Will effect the central nervous system (CNS) with narcosis (drowsiness), giddiness, and disorientation. It can also induce irregular or abnormal heart arrhythmia.

Chronic Health Effects

The following chronic (long-term) health effects may occur at some time after exposure to fluorotrichloromethane and can last for months or even years:

Cancer Hazards: According to the information presented in the references, fluorotrichloromethane has not been shown to cause cancer in test animals.

Reproductive Hazard: According to the information presented in the references, fluorotrichloromethane has not been adequately tested for its ability to adversely affect reproduction in test animals.

Other Chronic Effects: Low, chronic exposures may lead to liver damage and possible injury to the eyes and lungs. Cardiac rhythm changes may also be induced on high or prolonged exposures, although this is conjecture in the references and cannot be substantiated with human or animal studies. Prolonged skin contact may cause dryness, cracking, redness, secondary infection, and contact dermatitis.

❶ *Recommended Risk-Reduction Measures*

If a less toxic chemical cannot be substituted for a hazardous substance, *engineering controls* are the most effective method of reducing exposures. The best protection is to enclose operations and/or provide local exhaust ventilation at the site of chemical release. While not always operationally feasible, isolating operations can also reduce exposure.

Using respiratory protection is less effective than the controls mentioned above, but is still advisable whenever working with or around a chemical such as fluorotrichloromethane. The exposure level for this chemical is extremely high (1000 ppm) and, therefore, respiratory protection may not always be necessary in well-ventilated work areas. However, when high exposure levels are predicted or prolonged exposures are anticipated, a supplied-air respirator or a self-contained breathing apparatus (SCBA) with full facepiece operated in pressure demand mode are the recommended respiratory protection of choice. If a full facepiece is not available, then chemical goggles should be worn to protect the eyes. Whenever a chemical splash hazard exists, a face shield and rubber apron should be worn. To prevent hand and skin exposures, impervious polyvinylchloride, Viton, or polycarbonate gloves should be used.

Administrative controls should also be in place to minimize the potential for human exposures. These may include written procedures or policies which specify the methods and techniques that will be practiced whenever personnel are to work with or around fluorotrichloromethane.

All personnel should receive *training* on the use, hazards, protective measures, emergency actions, and other precautions per 29 CFR 1910.1200 (Hazard Communication), prior to the first assignment in an area where fluorotrichloromethane is used or stored.

Before beginning employment and at regular intervals thereafter (e.g., annually), the following medical protocol is suggested for those assigned to work with fluorotrichloromethane:

- ☑ Electrocardiogram (EKG), to test for pre-existing cardiac rhythm dysfunction and establish baseline.
- ☑ Lung function tests (establish baseline).

If symptoms develop or overexposure is suspected, the following medical tests may be useful:

- ☑ Electrocardiogram (EKG), to compare with baseline.
- ☑ Liver function tests.
- ☑ Lung function tests (compare with baseline).
- ☑ Nervous system function tests, including interviews by a qualified neurologist to evaluate any changes in personality or behavior.
- ☑ Evaluation by a qualified allergist with careful consideration of exposure history and special testing (may help diagnose skin allergy).

Any evaluation should include a careful medical history of past and present symptoms with an examination. However, medical tests that evaluate existing damage are not a substitute for controlling exposure.

Other methods to reduce exposure include:

- ☑ Where possible, enclose operations and use local exhaust ventilation at the site of chemical release. If local exhaust ventilation or enclosure is not used, respiratory protection should be mandatory.
- ☑ Always ensure that proper protective clothing is worn when using chemical substances.
- ☑ Wash thoroughly immediately after exposure to fluorotrichloromethane and at the end of the work shift or before eating, drinking, or smoking.
- ☑ Hazard warning information should be posted in the work area. In addition, as part of an on-going education and training program, all information on the health and safety hazards of fluorotrichloromethane should be communicated to all potentially exposed workers.
- ☑ Eye wash stations should be provided in work areas. If the potential for whole body exposure exists, then safety showers should also be provided.
- ☑ Work clothing that has been contaminated with fluorotrichloromethane should never be worn home where family members can be exposed. It should laundered only by personnel who have been briefed on its hazards.

RISK ASSESSMENT: ENVIRONMENT

General Assessment

The environment is at risk of exposure during intended use, transportation, storage, disposal, or destruction of fluorotrichloromethane. In almost every scenario, the threat of environmental exposure is contingent upon the proper handling of the chemical substance. Accidental spills, large or small, can result in fire, explosion, and possible contamination of the surrounding environmental mediums (water, soil, and air).

Fluorotrichloromethane is considered to be a nonflammable gas (or liquid above 75°F). However, it is incompatible with a number of common, chemically active metals and caution is required to ensure contact is prevented. It is also a chlorofluorocarbon (CFC) which is a group of halogenated hydrocarbons alleged to cause damage to the earth's protective ozone layer. Therefore, accidental release to the environment must not occur. These characteristics require special consideration during any emergency situation involving a leak or spill of fluorotrichloromethane. Should fluorotrichloromethane ever come into contact with incompatible substances either during use, transportation, or storage, violent and explosive reactions can occur.

Fluorotrichloromethane can exist as a gas or a liquid. Both can damage the environment. It can enter the environment through industrial discharges and spills.

Acute Ecological Effects

Acute (short-term) toxic effects may include the death of animals, birds, or fish and death or low growth rate in plants. Acute effects are seen 2 to 4 days after animals or plants are exposed to fluorotrichloromethane.

Fluorotrichloromethane has moderate acute toxicity to aquatic life. Insufficient data are available to evaluate or predict the short-term effects of exposure to plants, birds, or terrestrial animals.

Chronic Ecological Effects

Chronic toxic effects may include shortened life span, reproductive problems, lower fertility, and changes in appearance or behavior in exposed animals. These effects can be seen long after first exposure(s) to toxic chemicals.

Fluorotrichloromethane has moderate acute toxicity to aquatic life. Insufficient data are available to evaluate or predict the long-term effects of exposure to fluorotrichloromethane on plants, birds, or terrestrial animals.

Water Solubility

Fluorotrichloromethane is nearly insoluble in water. Concentrations of 100 milligrams or less may mix with a liter of water.

Persistence in the Environment

Fluorotrichloromethane is highly persistent in the aquatic environment, with a half-life of greater than 200 days. The half-life of a pollutant is the amount of time it takes for one half of the chemical to be degraded. If the gas reaches the atmosphere, it is very stable and is expected to remain in the troposphere for a very long time (half-life in troposphere is between 50 and 200 years).

Bioaccumulation in Aquatic Organisms

Some substances increase in concentration, or *bioaccumulate,* in living organisms as they breathe contaminated air, drink contaminated water, or eat contaminated food. These chemicals can become concentrated in the tissues and internal organs of animals as well as humans.

Bioaccumulation of fluorotrichloromethane in the edible tissues of fish is not expected to occur.

Recommended Risk-Reduction Measures

Proper training of all transporters will reduce the likelihood of a mishap or accident resulting in a leak or spill to the environment. The correct use of DOT labeling (per the MSDS) on all containers, trucks, and rail cars will enable emergency responders to react properly and quickly to any disaster thereby reducing the potential risk to the environment and to personnel.

Storage of fluorotrichloromethane should be segregated from incompatible chemicals to minimize the risk of cross-contamination or contact. Containers must be protected from shock, excessive heat (including the sun's rays), and damage.

If a spill or leak to the environment has occurred, fire department, emergency response and/or hazardous materials spill personnel should be notified immediately. Cleanup should be attempted only by those trained in proper spill containment procedures. If any liquid has spilled, contaminated soils may need to be removed for incineration and replaced with clean soil. If fluorotrichloromethane should contact the water table, aquifer, or navigable waterway, remediation activities should be prompt to ensure protection of the aquatic environment. When such spills occur, the local and/or state emergency response authorities should be notified. A comprehensive emergency response or disaster preparedness/recovery plan should be in place prior to any operations involving the use, transportation, storage, or disposal of fluorotrichloromethane. If fluorotrichloromethane is spilled or leaked, the following specific steps are recommended:

- ☑ Evacuate area and deny entry until cleanup is completed. Restrict persons not wearing protective clothing from hazard area.
- ☑ Ventilate area of spill or leak and remove all ignition sources.
- ☑ If source of leak is a cylinder and it cannot be repaired in place, remove to safe location (e.g., outdoors) and allow to empty. Stay clear of area. If liquids are released, absorb in vermiculite or other material and dispose of in sealed drums.
- ☑ It may be necessary to dispose of dichlorodifluoromethane as a hazardous waste. The responsible state agency or the regional office of the federal Environmental Protection Agency (EPA) should be contacted for specific recommendations.

RISK ASSESSMENT: BUSINESS

General Assessment

Accidents or mishaps involving fluorotrichloromethane can present a significant threat to business operations. The loss or damage of equipment or facilities can significantly affect fiscal viability. Lawsuits that may result from personnel injury/death, public exposures, and/or environmental contamination will also require a serious expenditure of resources.

Media attention surrounding an injury, death, or environmental damage can also result in a loss of profits and loss of current as well as future business.

Recommended Risk-Reduction Measures

Company attorneys, safety and health professionals, and environmental specialists should be involved in the development of any procedures dealing with safety and emergency response. A company official should be pre-designated as a public relations officer with specific training in dealing with the press. Corporate plans and policies that deal with public relations should be developed, approved, and implemented long before any need for such arises.

REFERENCES

American Conference of Governmental Industrial Hygienists. 1988. *Documentation of the Threshold Limit Values and Biological Exposure Indices,* 5th Edition (with updates). Cincinnati: ACGIH

Ahlbom, A. 1993. *Biostatistics for Epidemiologists.* New York: CRC Press/Lewis Publishers.

Anthony, C. P., and N. J. Kolthoff. 1971. *Textbook of Anatomy and Physiology.* St. Louis: C. V. Mosby Company.

Burrell, R., D. K. Flaherty, and L. J. Sauers. 1992. *Toxicology of the Immune System.* New York: Van Nostrand Reinhold.

Calabrese, E. J. 1994. *Biological Effects of Low Level Exposures.* New York: CRC Press/Lewis Publishers.

Cothern, C. R. 1995. *Handbook for Environmental Risk Decisionmaking.* New York: CRC Press/Lewis Publishers.

Gardner, R. W. 1994. *Chemical Intolerance.* New York: CRC Press/Lewis Publishers.

Hodgson, E., R. B. Mailman, and Janice E. Chambers. 1988. *Dictionary of Toxicology.* London: The MacMillian Company.

Keith, L. H. and M. M. Walker. 1995. *Handbook of Air Toxics.* New York: CRC Press/Lewis Publishers.

Landis, W. G., M. H. Yu. 1995. *Introduction to Environmental Toxicology.* New York: CRC Press/Lewis Publishers.

Langerman, N. 1994 *Precautionary Labels for Chemical Containers.* New York: CRC Press/Lewis Publishers.

Lauwerys, R. R., P. Hoet. 1993. *Industrial Chemical Exposure.* New York: CRC Press/Lewis Publishers.

Lewis, R. J., Sr. 1993. *Hawley's Condensed Chemical Dictionary,* 12th Edition. New York: Van Nostrand Reinhold.

Lewis, R. J., Sr. 1992. *Sax's Dangerous Properties of Industrial Materials,* Volumes 1, 2, and 3, 8th Edition. New York: Van Nostrand Reinhold.

Manahan, S. E. 1992. *Toxicological Chemistry,* 2nd Edition. New York: CRC Press/Lewis Publishers.

Meyer, E. 1990. *Chemistry of Hazardous Materials.* Englewood Cliffs, NJ: Prentice Hall.

Ott, W. R. 1995. *Environmental Statistics and Data Analysis.* New York: CRC Press/Lewis Publishers.

Quigley, D. R. 1994. *Handbook of Emergency Chemical Management.* New York: CRC Press/Lewis Publishers.

Richardson, M. 1995. *Dictionary of Substance and Their Effects,* Vol. I-VI. New York: CRC Press/Lewis Publishers.

Sacarello, H. L. A. 1994. *The Comprehensive Handbook of Hazardous Materials.* New York: CRC Press/Lewis Publishers.

Torkelson, T. R. V. K. Rowe. 1981. *Patty's Industrial Hygiene and Toxicology,* Vol. 2B. New York: Wiley-Interscience.

U.S. Department of Health and Human Services, National Institute for Occupational Safety and Health. 1994. *NIOSH Pocket Guide to Chemical Hazards.* Washington, D.C.: U.S. Government Printing Office.

MATERIAL SAFETY DATA SHEET

CHEMICAL NAME	
FORMALDEHYDE	**CRC PRESS/LEWIS PUBLISHERS DISCLAIMER:** THE INFORMATION AND RECOMMENDATIONS PRESENTED HEREIN ARE BASED ON SOURCES BELIEVED TO BE RELIABLE. CRC MAKES NO REPRESENTATION ON ITS COMPLETENESS OR ACCURACY. IT IS THE RESPONSIBILITY OF THE USER TO DETERMINE THE CHEMICAL'S SUITABILITY FOR ITS INTENDED USE, THE CHEMICAL'S SAFE USE, AND THE CHEMICAL'S PROPER DISPOSAL. NO REPRESENTATIONS AND/OR WARRANTIES, EITHER EXPRESSED OR IMPLIED, OF THE MERCHANTABILITY OR FITNESS FOR A PARTICULAR PURPOSE, OR OF ANY OTHER NATURE, ARE MADE WITH RESPECT TO THE INFORMATION PROVIDED IN THIS MSDS OR TO THE CHEMICAL TO WHICH INFORMATION MAY REFER. CRC NEITHER ASSUMES NOR AUTHORIZES ANY OTHER PERSON TO ASSUME FOR IT, ANY OTHER ADDITIONAL RESPONSIBILITY OR LIABILITY FOR THE USE OF, OR RELIANCE UPON, THIS INFORMATION.

HAZARD WARNING INFORMATION

HEALTH	FIRE	REACTIVE	OTHER	DEGREE OF HAZARD	COLOR CODING	OTHER CODES
3	2	0		0 = Minimum Hazard 1 = Slight Hazard 2 = Moderate Hazard 3 = Serious Hazard 4 = Severe Hazard	HEALTH = BLUE FIRE = RED REACTIVITY = YELLOW OTHER = WHITE	OX = Oxidizer ACID = Acid ALK = Alkali COR = Corrosive ~~W~~ = Use No Water

SECTION I - GENERAL INFORMATION

Characterization	RCRA Number	EPA Class
Aldehyde	**U122**	**Toxic Waste**

DOT Proper Shipping Name **Formaldehyde (solutions)**	Chemical Abstract Service (CAS) Number **50-00-0**
DOT Hazard Class and Label Requirements **Flammable Gas**	DOT Emergency Guide Code **29**
DOT Identification Number **UN 1198 (gas) UN 2209 (solutions)**	Chemical Formula **HCHO**

Synonyms
Methanal; methyl aldehyde; methylene oxide; oxymethylene; formic aldehyde; formalin solution.

SECTION II - HAZARDOUS INGREDIENTS AND IDENTITY INFORMATION

Hazardous Components (specific identity)	OSHA Exposure Criteria	NIOSH Exposure Criteria	Immediately Dangerous to Life and Health (IDLH)	ACGIH Exposure Criteria
Formaldehyde (derivation: By oxidation of synthetic methanol or low-boiling petroleum gases such as propane and butane. Silver, copper, or an iron-molybdenum oxide are the most common catalysts). **1 ppm = 1.25 mg/m³**	**PEL:** **0.75 ppm** **.94 mg/m³** **STEL:** **2 ppm** **2.5 mg/m³**	**REL:** **0.016 ppm** **CEILING** **(15 minute)** **0.1 ppm** **Possible** **Cancer Agent**	**20 ppm**	**TLV:** **1 ppm** **1.2 mg/m³** **Suspected** **Carcinogen** **STEL:** **2 ppm**

SECTION III - PHYSICAL AND CHEMICAL CHARACTERISTICS

Boiling Point **-6°F (-21°C)**	Specific Gravity (H_2O = 1) **1.1**
Vapor Pressure (atmospheres) **>1 at 69°F (20°C)**	Molecular Weight **30**
Vapor Density (Air = 1) **1.0**	Freezing Point **-134°F (-92°C)**

Solubility
Miscible in water, acetone, benzene, chloroform, diethyl ether, ethanol, and ether.

Appearance and Odor
Nearly colorless gas with a pungent, suffocating odor. May also exist in aqueous solution. Odor Threshold = 0.83 ppm

SECTION IV - FIRE AND EXPLOSION HAZARD DATA

Flash Point (method used) **In solution: 182°F (83°C) closed cup**	Explosive Limits in Air % by Volume **LEL: 7.0% UEL: 73%**
NFPA Classification **Flammable Gas; Class IIIA Combustible Liquid**	Autoignition Temperature **572°F (300°C) as gas; 806°F (430°C) as solution**

Extinguishing Media
Dry chemical, carbon dioxide, water spray, or alcohol-resistant foam.

Special Fire Fighting Procedures
Wear full protective clothing and self-contained breathing apparatus (SCBA). Heat will build pressure and may rupture closed storage containers. Keep fire-exposed containers cool with water spray. Vapors are heavier than air; they may travel a distance to ignition source, flashback and cause fire.

Unusual Fire and Explosion Hazards
Stay away from ends of tanks since internal pressures can cause explosion; easily undergoes polymerization. Poisonous gases are produced in fire.

SECTION V - REACTIVITY DATA

Stability		Conditions to Avoid
		Slowly oxidizes in air to formic acid. On standing, especially cold, it may become cloudy. At very low temperatures, a small amount of trioxymethylene is formed. Avoid exposure to heat, incompatible materials.
Stable	Unstable	Incompatibility *(materials to avoid)*
	X	Liquid formaldehyde corrodes steel (the vapor will not). Will react with strong oxidizers, alkalis, acids, phenols, and urea. Reacts with hydrochloric acid to form bischloromethyl ether (a carcinogen).
Hazardous Polymerization		Conditions to Avoid
		Polymerizes readily in both the gaseous and liquid states. Do not expose to very cold or very warm temperatures.
May Occur	Will Not Occur	Hazardous Decomposition or By-products
X		When heated to decomposition, formaldehyde emits acrid, irritating smoke and fumes.

SECTION VI - HEALTH HAZARD DATA

Primary Route(s) of Entry:	Inhalation?	Absorption (skin)?	Ingestion?
	X	X	X

Health Hazards

INHALATION: Irritation and burning of the eyes, nose, throat, and respiratory tract with tearing, coughing, and severe, persistent lung irritation leading to pulmonary edema (fluid buildup in the lungs), pneumonitis, and death. Systemic effects include CNS depression, metabolic acidosis, vertigo, convulsions, stupor, coma, and respiratory failure.

ABSORPTION: Liquid irritates the skin causing rash or burning feeling. Can also cause severe burns and permanent damage. Itching, scaling, blistering occur.

INGESTION: Immediate burning of the mouth, esophagus, and stomach with pain, cramps, nausea, vomiting, bloody diarrhea, digestive hemorrhage, and stomach ulceration/perforation.

Carcinogenicity	NTP Listed?	IARC Cancer Review Group?	OSHA Regulated?	Target Organs?
Suspected Human Confirmed Animal	5th Annual Report	Group 2A	29 CFR 1910.1048 29 CFR 1910.1000	Respiratory system; skin; eyes; kidneys, CNS

Medical Conditions Generally Aggravated by Exposure

Respiratory impairments (asthma); Renal function abnormalities; Skin conditions (dermatitis).

Emergency and First-aid Procedures

Eye contact: Flush immediately with water for 30 minutes (minimum); seek medical attention. Skin contact: Remove all contaminated clothing. Immediately wash area with large amounts of soap and water. For inhalation: Remove the person from exposure. Provide respiratory assistance and CPR. Transfer to medical facility. Observe 24-48 hours for lung effects. If swallowed: Contact poison control. Unless advised otherwise, give 1-2 glasses of water with ammonium acetate, induce vomiting.

SECTION VII - PRECAUTIONS FOR SAFE HANDLING AND USE

Steps to be Taken in Case Material is Released or Spilled

Absorb spills using absorbent vermiculite or other. Ventilate area of spill. Remove ignition sources. Have water source available in case of fire. Store materials in DOT-approved containers. Restrict those not involved in cleanup from entering area. Notify appropriate authorities, as required.

Preferred Waste Disposal Method

Incineration.

Precautions to be Taken in Handling and Storage

Formaldehyde is highly volatile. Store in tightly closed containers in a cool, dark, well-ventilated area. Keep fire and flame away. Use non-sparking tools to open or close containers.

Other Precautions and Warnings

Install Class 1, Group B electrical equipment.

SECTION VIII - CONTROL MEASURES AND PERSONAL PROTECTIVE EQUIPMENT

Respiratory Protection *(specify type)*

For any exposure level, use a self-contained breathing apparatus (SCBA) with full facepiece and pressure demand or other positive pressure mode or supplied-air respirator.

Ventilation

Local exhaust or general mechanical systems recommended.

Protective Gloves	Eye Protection	Other Protective Clothing
Polyethylene, Nitrile, or Butyl Rubber	Chemical Goggles or Face Mask	Protective Clothing

Work/Hygiene Practices

Always wash hands thoroughly after using chemical; never bring food, drink, or smoking materials into vicinity of chemicals.

FORMALDEHYDE

HCHO

CAS: 50-00-0

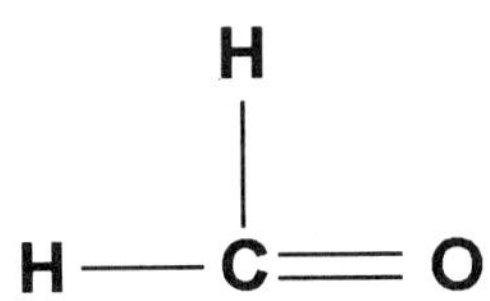

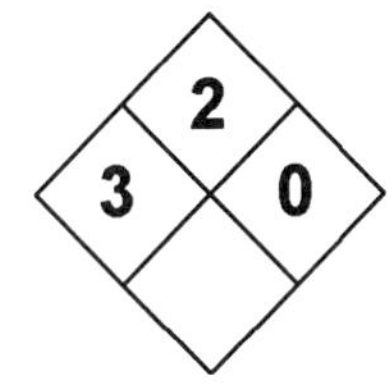

IDENTIFICATION AND TYPICAL USES

Formaldehyde is a nearly colorless gas with a pungent, suffocating odor. It is often used in an aqueous solution. It is used in the manufacture of phenolic resins, cellulose esters, artificial silk, dyes, explosives, and organic chemicals. Also used as a germicide, fungicide, and disinfectant. It has uses in tanning products, in adhesives, in waterproofing fabrics, for tonic and chrome printing in photography, in cosmetics, as embalming fluid, in foam insulation and wood preservatives, as a chemical intermediate, as a stabilizer in gasoline, as a corrosion inhibitor in the metal industry, a starch modifier, in medicines (for athletes foot medicine and cough drops), and treating skin diseases in animals.

RISK ASSESSMENT: HEALTH

General Assessment

Formaldehyde presents a moderate to severe health hazard risk. It is a primary irritant of the eyes, skin, and respiratory system. It is mutagenic (causes changes in cell structure), teratogenic (adversely affects fetal development), is carcinogenic (cancer-causing) in animals and is suspected to cause cancer in humans as well. Exposure occurs through all routes of exposure (*ingestion, inhalation,* and *absorption*).

Inhalation causes lachrymation (tearing), olfactory changes (numbing of the sense of smell), burning of eyes, nose, and throat, dry and sore throat, sneezing, coughing, headache, feeling of thirst, and disturbed sleep. There may be listlessness, dypsnea, loss of body weight, inflammation in the nasal tissues, and pathological changes in the nose, larynx, trachea, and bronchi. Respiratory irritation may progress to severe injury, including pulmonary edema (fluid in lungs), chemical pneumonitis, and death. Symptoms of pulmonary edema may be delayed up to 48 hours creating a false sense of security with regard to health exposure risk.

Skin contact with the liquid causes moderate to severe irritation with redness and a possibility for burning, drying, cracking, scaling, and blistering. The liquid may pass through intact skin to cause toxic systemic effects. When combined with inhalation or ingestion exposures, the results can be fatal. Skin allergy may develop. Eye contact will result injuries up to and including permanent corneal opacity and vision loss.

Ingestion causes immediate burns to the mouth, esophagus, and stomach. There may be abdominal pain and cramps, nausea and vomiting, bloody diarrhea, digestive hemorrhage, and ulceration or perforation of the stomach.

Acute Health Effects

The following acute (short-term) health effects may occur immediately or shortly after exposure to formaldehyde:

Skin: Irritation and rash or burning sensation on contact. Absorption of the liquid is possible.

Eye: Burns leading to permanent damage.

Lung: Irritation causing coughing and/or shortness of breath. Higher exposures can cause a buildup of fluid in the lungs (pulmonary edema), which is a medical emergency.

CNS: Systemic effects of exposure include CNS depression with symptoms of metabolic acidosis, vertigo, convulsions, stupor, and coma. Death may occur due to respiratory failure.

Chronic Health Effects

The following chronic (long-term) health effects may occur at some time after exposure to formaldehyde and can last for months or even years:

Cancer Hazards: Formaldehyde is a probable carcinogen in humans since it has been shown to cause cancer in animals. Formaldehyde also causes mutations (genetic changes). Such chemicals may have a cancer or reproductive risk. Formaldehyde may be a teratogen in humans since it has been shown to cause birth defects in animals.

Reproductive Hazards: According to information presented in the references, exposure to formaldehyde can adversely affect the reproductive process causing pathological changes in the ovaries and subsequent changes in the menstrual cycle. It may also cause low birth weight children (results of research in this area are conflicting and inconclusive).

Sensitivity: Exposure to the skin may result in allergy (sensitization). If allergy develops, very low future exposure can result in skin irritation and rash. Repeated exposure can cause chronic irritation of the eyes and lungs.

Recommended Risk-Reduction Measures

Formaldehyde is a suspected human carcinogen. Personnel should avoid direct contact with formaldehyde. Unless a less toxic chemical can be substituted for a hazardous substance, *engineering controls* are the most effective method of reducing exposures. The best protection is to enclose operations and/or provide local exhaust ventilation at the site of chemical release. While not always operationally feasible, isolating operations can also reduce exposure.

Using respiratory protection is less effective than the controls mentioned above, but is still advisable whenever working with or around formaldehyde. For exposures to any level, use a supplied-air respirator operated in positive pressure mode or use a self-contained breathing apparatus (SCBA) with full facepiece operated in pressure demand mode are the recommended respiratory protection methods of choice. If a full facepiece is not available, then chemical goggles should be worn to protect the eyes. Whenever a chemical splash hazard exists, a face shield and a rubber apron should be worn. To prevent hand and skin exposures, polyethylene, butyl, and nitrile rubber gloves should be worn. Glove manufacturers should be contacted and permeation studies obtained *before* glove selections are made.

Administrative controls should also be in place to minimize the potential for human exposures. These may include written procedures or policies which specify the methods and techniques that will be practiced whenever personnel are to work with formaldehyde.

All personnel should receive *training* on the use, hazards, protective measures, emergency actions, and other precautions per 29 CFR 1910.1200 (Hazard Communication), prior to the first assignment in an area where formaldehyde is used or stored. OSHA 29 CFR 1910.1048 (Formaldehyde) provides additional information on personnel training requirements.

Other methods to reduce exposure include:

- ☑ Where possible, enclose operations and use local exhaust ventilation at the site of chemical release. If local exhaust ventilation or enclosure is not used, respiratory protection should be mandatory. Also, if possible, automatically pump liquid formaldehyde from drums or other storage containers to process containers.
- ☑ Specific engineering controls are recommended for formaldehyde by NIOSH. Refer to the NIOSH criteria document: *Occupational Exposure to Formaldehyde* (Publication Number 77-126). Also refer to OSHA 29 CFR 1910.1048 (Formaldehyde) for additional engineering and administrative control requirements.
- ☑ Always ensure that proper protective clothing is worn when using chemical substances.
- ☑ Wash thoroughly immediately after exposure to formaldehyde and at the end of the work shift or before eating, drinking, or smoking.
- ☑ Hazard warning information should be posted in the work area. In addition, as part of an on-going education and training program, all information on the health and safety hazards of formaldehyde should be communicated to all potentially exposed workers.
- ☑ Never eat, drink, or smoke in areas were formaldehyde is used, handled or stored.

RISK ASSESSMENT: ENVIRONMENT

General Assessment

The environment is at risk of exposure during transportation, storage, disposal, or destruction of formaldehyde. In almost every scenario, the threat of environmental exposure is contingent upon the proper handling of the chemical substance. Accidental spills, large or small, can result in fire, explosion, and possible contamination of the surrounding environmental mediums (water, soil, and air).

Formaldehyde is considered a highly flammable gas. As a liquid, it is a Class III combustible. It is extremely volatile and presents a serious fire and explosion hazard when exposed to heat or in contact with incompatible materials. Rail cars and storage tanks are subject to boiling liquid expanding vapor explosion (BLEVE) if involved in fire. These characteristics require special consideration during any emergency situation involving a leak or spill of formaldehyde. Should formaldehyde ever come into contact with incompatible substances such as oxidizers, acids, alkalies, phenols, or permanganates, either during use, transportation, or storage, the formation of highly toxic and/or highly explosive commodities is extremely possible. It polymerizes readily in both the gaseous and liquid states and slowly oxidizes in air to formic acid.

The proper disposal/destruction method for formaldehyde is to burn it in a chemical incinerator equipped with an afterburner and air scrubber (permitted facilities only).

Formaldehyde can enter the environment through manufacturing processes, unchecked waste discharges into effluents, and through spills.

Acute Ecological Effects

Acute (short-term) toxic effects may include the death of animals, birds, or fish, and death or low growth rate in plants. Acute effects are seen 2 to 4 days after animals or plants are exposed to formaldehyde.

This chemical has high acute toxicity to aquatic life. No data are available on the short-term effects of formaldehyde to terrestrial animals.

Chronic Ecological Effects

Chronic toxic effects may include shortened life span, reproductive problems, lower fertility, and changes in appearance or behavior in exposed animals. These effects can be seen long after first exposure(s) to toxic chemicals.

Formaldehyde has high chronic toxicity in aquatic life. It has caused cancer and other chronic effects in laboratory animals (rodents). Birds and terrestrial animals exposed to formaldehyde may show similar effects. No data are available on the long-term effects of formaldehyde on plants, birds, or land animals.

Water Solubility

Formaldehyde is highly soluble in water. Concentrations of 1000 milligrams and more will mix with a liter of water.

Persistence in the Environment

Formaldehyde is slightly persistent in water, with a half-life of between 2 to 20 days. The half-life of a pollutant is the amount of time it takes for one-half of the chemical to be degraded. Approximately 99% of formaldehyde will eventually end up in air; the rest will end up in water.

Bioaccumulation in Aquatic Organisms

Some substances increase in concentration, or *bioaccumulate,* in living organisms as they breathe contaminated air, drink contaminated water, or eat contaminated food. These chemicals can become concentrated in the tissues and internal organs of animals as well as humans.

The concentration of formaldehyde found in fish tissues is expected to be about the same as the average concentration of formaldehyde in water from which the fish was taken.

Recommended Risk-Reduction Measures

Proper training of all transporters will reduce the likelihood of a mishap or accident resulting in a leak or spill to the environment. The correct use of DOT labeling (pcr thc MSDS) on all containers, trucks, and rail cars will enable emergency responders to react properly and quickly to any disaster thereby reducing the potential risk to the environment and to personnel.

Storage of formaldehyde should be segregated from incompatible chemicals to minimize the risk of cross-contamination or contact. Maximum storage temperatures should not exceed 83°F/28°C (30% solution with 0.05% methanol) or 29°F/-5°C (solutions 15% methanol) to inhibit polymerization. Buildings designated for storage should be equipped with appropriate fire protection systems (alarms, sprinklers, emergency lighting, portable extinguishers).

If a spill or leak to the environment has occurred, fire department, emergency response, and/or hazardous materials spill personnel should be notified immediately. Cleanup should be attempted only by those trained in proper spill containment procedures. Contaminated soils should be removed for incineration and replaced with clean soil. If formaldehyde should contact the water table, aquifer, or navigable waterway, time is of the essence. It is readily miscible in water and, therefore, total containment and remediation may not be entirely possible. When such spills occur, the local and/or state emergency response authorities must be notified. A comprehensive emergency response or disaster preparedness/recovery plan should be in place prior to any operations involving the use, transportation, storage, or disposal of formaldehyde. If formaldehyde is spilled or leaked, the following specific steps are recommended:

- ☑ Restrict persons not wearing protective clothing from area of spill or leak until cleanup is complete and area can be opened for normal work.
- ☑ Ventilate area and remove all ignition sources.
- ☑ Absorb liquids in vermiculite, dry sand, earth, or a similar material and deposit in sealed containers. Use non-sparking tools.
- ☑ If applicable, stop flow of leaking liquid or gas. If leak source is a cylinder and the leak cannot be stopped in place, remove leaking cylinder to a safe place in the open air and repair or allow cylinder to empty.
- ☑ Keep formaldehyde out of a confined space, such as a sewer, because of the possibility of explosion (unless the sewer is designed to prevent the build-up of explosive concentrations).
- ☑ It may be necessary to dispose of formaldehyde as a hazardous waste. The responsible state agency or the regional office of the federal Environmental Protection Agency (EPA) should be contacted for specific recommendations.

RISK ASSESSMENT: BUSINESS

General Assessment

Accidents or mishaps involving formaldehyde can present a significant threat to business operations. The loss or damage of equipment or facilities can significantly affect fiscal viability. Lawsuits that may result from personnel injury/death, public exposures, and/or environmental contamination will also require a serious expenditure of resources. Media attention surrounding an injury, death, or environmental damage can also result in a loss of profits and loss of current as well as future business.

Always remember that anytime to the terms "cancer," "carcinogen," or "reproductive hazard" are used, public hysteria, emotion, and ignorance can run equally high. This should be carefully considered whenever developing or implementing public relation policies.

❶ *Recommended Risk-Reduction Measures*

Company attorneys, safety and health professionals, and environmental specialists should be involved in the development of any procedures dealing with safety or emergency response. A company official should be pre-designated as a public relations officer with specific training in dealing with the press. Corporate plans and policies should be developed, approved, and implemented long before any need for such arises.

REFERENCES

American Conference of Governmental Industrial Hygienists. 1988. *Documentation of the Threshold Limit Values and Biological Exposure Indices,* 5th Edition (with updates). Cincinnati: ACGIH

Calabrese, E. J. 1994. *Biological Effects of Low Level Exposures.* New York: CRC Press/Lewis Publishers.

Hodgson, E., R. B. Mailman, and J. E. Chambers. 1988. *Dictionary of Toxicology.* London: The MacMillian Company.

Heck, H. 1985. Mechanisms of aldehyde toxicity: Structure activity Studies. *CIIT Activities,* 5(10):106.

International Agency for Research on Cancer (IARC). 1985. *IARC Monographs on the Evaluation of Carcinogenic Risk of Chemicals to Humans,* Vol. 36, *Allyl Compounds, Aldehydes, Epoxides, and Peroxides,* pp 101-136. Lyon: IARC.

Landis, W. G., M. H. Yu. 1995. *Introduction to Environmental Toxicology.* New York: CRC Press/Lewis Publishers.

Lewis, .J., Sr. 1993. *Hawley's Condensed Chemical Dictionary,* 12th Edition. New York: Van Nostrand Reinhold.

Lewis, . J., Sr. 1992. *Sax's Dangerous Properties of Industrial Materials,* Volumes 1, 2, and 3, 8th Edition. New York: Van Nostrand Reinhold.

Patnaik, P. 1992. *A Comprehensive Guide to the Hazardous Properties of Chemical Substances.* New York: Van Nostrand Reinhold.

U.S. Department of Health and Human Services, National Institute for Occupational Safety and Health. 1977. *NIOSH Criteria Document: Occupational Exposure to Formaldehyde.* (Pub. No. 77-126). Washington, D.C.: U.S. Government Printing Office.

U.S. Department of Health and Human Services, National Institute for Occupational Safety and Health. 1994. *NIOSH Pocket Guide to Chemical Hazards.* Washington, D.C.: U.S. Government Printing Office.

MATERIAL SAFETY DATA SHEET

CHEMICAL NAME	CRC PRESS/LEWIS PUBLISHERS DISCLAIMER
FORMAMIDE	

HAZARD WARNING INFORMATION

HEALTH	FIRE	REACTIVE	OTHER	DEGREE OF HAZARD	COLOR CODING	OTHER CODES
2	1	0		0 = Minimum Hazard 1 = Slight Hazard 2 = Moderate Hazard 3 = Serious Hazard 4 = Severe Hazard	HEALTH = BLUE FIRE = RED REACTIVITY = YELLOW OTHER = WHITE	OX = Oxidizer ACID = Acid ALK = Alkali COR = Corrosive W = Use No Water

SECTION I - GENERAL INFORMATION

Characterization	**Aldehyde**	RCRA Number	**None**
		EPA Class	**Not Applicable**
DOT Proper Shipping Name	**Not Listed**	Chemical Abstract Service (CAS) Number	**75-12-7**
DOT Hazard Class and Label Requirements	**No Citation**	DOT Emergency Guide Code	**No Citation**
DOT Identification Number	**No Citation**	Chemical Formula	**CH_3NO**

Synonyms
Carbamaldehyde; methanamide.

SECTION II - HAZARDOUS INGREDIENTS AND IDENTITY INFORMATION

Hazardous Components (specific identity)	OSHA Exposure Criteria	NIOSH Exposure Criteria	Immediately Dangerous to Life and Health (IDLH)	ACGIH Exposure Criteria
Formamide (derivation: By interaction of ethyl formate and ammonia at high temperature and pressure with subsequent distillation). **1 ppm = 1.87 mg/m³**	**PEL (skin):** **20 ppm** **30 mg/m³** **STEL:** **30 ppm** **45 mg/m³**	**REL (skin):** **10 ppm** **15 mg/m³** **Possible Cancer Agent**	**Not Determined**	**TLV (skin):** **10 ppm** **15 mg/m³** **STEL:** **Not Established**

SECTION III - PHYSICAL AND CHEMICAL CHARACTERISTICS

Boiling Point	**410°F (210°C) Decomposes**	Specific Gravity (H_2O = 1)	**1.13**
Vapor Pressure (mm Hg)	**0.1 at 86°F (30°C)**	Molecular Weight	**45.1**
Vapor Density (Air = 1)	**1.6**	Freezing Point	**37°F (3°C)**

Solubility
Completely soluble in water. Soluble in alcohol. Slightly soluble in ether. Formamide is hygroscopic.

Appearance and Odor
Colorless, clear, oily or viscous, hygroscopic liquid. Odorless or slight ammonia-like odor.

SECTION IV - FIRE AND EXPLOSION HAZARD DATA

Flash Point (method used)	**310°F (154°C) opened cup**	Explosive Limits in Air % by Volume	**LEL: Not Determined UEL: Not Determined**
NFPA Classification	**Class 111B Flammable Liquid**	Autoignition Temperature	**Not Determined**

Extinguishing Media
Use water spray or fog, alcohol foam, carbon dioxide, or dry chemical.

Special Fire Fighting Procedures
A combustible liquid but a moderate fire hazard when exposed to sources of ignition. Poisonous gases are produced in fire. Wear full protective clothing and self-contained breathing apparatus (SCBA). Keep fire-exposed containers cool with water spray. Vapors are slightly heavier than air; they may travel a distance to an ignition source and, in sufficient quantities, may flashback to cause fire.

Unusual Fire and Explosion Hazards
Remove containers from fire if it can be done without risk. Heat will build pressure and may rupture closed storage containers.

SECTION V - REACTIVITY DATA

Stability		Conditions to Avoid Formamide is stable in closed containers during normal conditions of handling and storage. Avoid exposure to heat. Thermal decomposition occurs at 410°F (210°C) yielding toxic fumes and gases.
Stable X	Unstable	Incompatibility *(materials to avoid)* Strong oxidizers, acids, bases, iodine, pyridine, sulfur trioxide, copper, brass, lead, and moisture. Formamide is hygroscopic and will absorb moisture from any source (air, steam, water).
Hazardous Polymerization		Conditions to Avoid Under normal conditions of temperature and pressure, hazardous polymerization of formamide is not expected to occur.
May Occur	Will Not Occur X	Hazardous Decomposition or By-products When heated to decomposition, formamide emits toxic gases and fumes, including fumes of ammonia, oxides of nitrogen, and oxides of carbon.

SECTION VI - HEALTH HAZARD DATA

Primary Route(s) of Entry:	Inhalation?	Absorption (skin)?	Ingestion?
	X	X	X

Health Hazards

INHALATION: Irritation of the eyes and mucosa of the upper respiratory tract. Excessive inhalation of vapor can cause headache, drowsiness, fatigue, nausea, acidosis, abdominal spasms, and other effects on the central nervous system (CNS) including loss of consciousness.

ABSORPTION: Skin absorption can compound effects of inhalation. Skin eruptions can occur.

INGESTION: Central nervous system (CNS) depression. Toxic to liver and kidney.

Carcinogenicity	NTP Listed?	IARC Cancer Review Group?	OSHA Regulated?	Target Organs?
Unknown Human Unknown Animal	No	No	29 CFR 1910.1000 (Table Z-1)	Respiratory system, skin, eyes, CNS, repro. system.

Medical Conditions Generally Aggravated by Exposure

None reported.

Emergency and First-aid Procedures

Eye contact: Flush immediately with water for 15 minutes (minimum); seek medical attention. Skin contact: Remove all contaminated clothing. Immediately wash area with large amounts of soap and water. Seek medical assistance. For inhalation: Remove the person from exposure. Provide respiratory assistance and CPR. Transfer to medical facility. Keep victim warm and at rest. Treat symptomatically and supportively. If swallowed: Unlikely to occur. Seek medical attention immediately.

SECTION VII - PRECAUTIONS FOR SAFE HANDLING AND USE

Steps to be Taken in Case Material is Released or Spilled

Contain spills using absorbent material (vermiculite, or other). Ventilate area of spill. Have water source available in case of fire. Store materials in DOT-approved containers. Restrict those not involved in cleanup from entering area. Notify appropriate authorities, as required.

Preferred Waste Disposal Method

No citation.

Precautions to be Taken in Handling and Storage

Do not store with incompatible chemicals. Store in tightly closed containers in a cool, dark, well-ventilated area. Keep away from heat, flame, and moisture. Aluminum or stainless steel containers are recommended for shipping and storing formamide.

Other Precautions and Warnings

Bulk storage facilities should be equipped with air vents and dryers to prevent moist air from prolonged contact with stored formamide. Copper and brass materials should be avoided.

SECTION VIII - CONTROL MEASURES AND PERSONAL PROTECTIVE EQUIPMENT

Respiratory Protection *(specify type)*

Exposures above the PEL (20 ppm): MSHA/NIOSH-approved full facepiece respirator with organic vapor cartridge. For higher exposures: Supplied-air respirator or a self-contained breathing apparatus (SCBA) with full facepiece and pressure demand or other positive pressure mode.

Ventilation

Local exhaust or general mechanical systems recommended.

Protective Gloves	Eye Protection	Other Protective Clothing
Butyl Rubber	Chemical Goggles or Face Mask	Rubber Apron

Work/Hygiene Practices

Always wash hands thoroughly after using chemical; never bring food, drink, or smoking materials into vicinity of chemicals.

FORMAMIDE

CH_3NO

CAS: 75-12-7

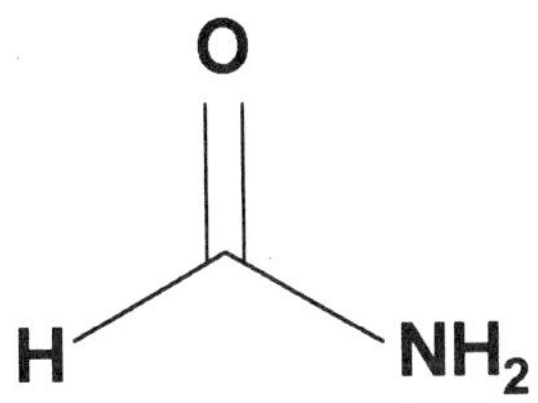

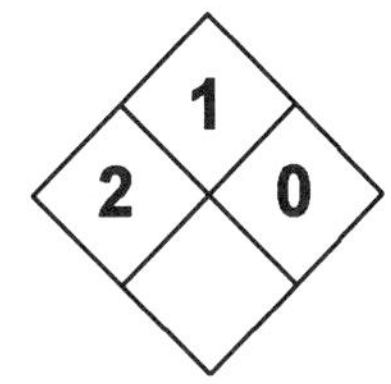

IDENTIFICATION AND TYPICAL USES

Formamide is a clear, colorless, viscous (oily), hygroscopic liquid with a slight ammonia-like odor. It is used in the manufacture of formic esters and hydrocyanic acid by catalytic dehydration. It is also used as an ionizing solvent, as a softener for animal glues, paper, and water-soluble gums.

RISK ASSESSMENT: HEALTH

General Assessment

Formamide is a human poison by skin contact (*absorption*) and by *inhalation*. Although ingestion is unlikely, it can occur with moderately toxic results. Its carcinogenic properties have not been adequately investigated. However, it has been shown to cause mutations (genetic changes) in laboratory experiments with test animals. It has also demonstrated experimental teratogenic (fetal changes) and reproductive effects. Human data in this regard are non-specific and inconclusive.

Inhalation causes immediate irritation to the eyes, nose, throat, and mucous membranes lining the respiratory tract. High concentrations can cause drowsiness, fatigue, nausea, vomiting, acidosis, loss of consciousness, and other indications of central nervous system depression.

Skin contact cause surface irritation, skin eruptions, redness, blistering, and associated pain. Because it is absorbed readily through intact skin, it can compound the effects of inhalation to life-threatening proportions.

Acute Health Effects

The following acute (short-term) health effects may occur immediately or shortly after exposure to formamide:

Skin: Irritation and rash or burning sensation on contact. Absorption is likely and the systemic effects are expected to be moderate to severe.

Eye: Irritation and redness. Possible inflammation of the conjunctiva.

Lung: Irritation of the nose, throat, and mucosa of the upper respiratory tract.

CNS: Inhalation of high concentrations of the vapor may cause drowsiness, fatigue, nausea, acidosis, and even loss of consciousness. These symptoms indicate CNS depression.

Chronic Health Effects

The following chronic (long-term) health effects may occur at some time after exposure to formamide and can last for months or even years:

Cancer Hazards: It has been shown that formamide causes mutations (genetic changes) in test animals. Such chemicals may have a cancer or reproductive risk. There are no reports that formamide causes cancer in animals. Additional study is required in this area with regard to human exposures.

Reproductive Hazards: Formamide may be a teratogen in humans since it has been shown to cause birth defects in animals. Reproductive effects have also been demonstrated.

Recommended Risk-Reduction Measures

Personnel should avoid direct contact with formamide. Unless a less toxic chemical can be substituted for a hazardous substance, *engineering controls* are the most effective method of reducing exposures. The best protection is to enclose operations and/or provide local exhaust ventilation at the site of chemical release. While not always operationally feasible, isolating operations can also reduce exposure.

Using respiratory protection is less effective than the controls mentioned above, but is still advisable whenever working with or around formamide. For low-level or infrequent exposures (over the PEL of 20 ppm), an organic vapor respirator may suffice. For higher or more frequent exposures, a self-contained breathing apparatus (SCBA) with full facepiece operated in pressure demand mode or a supplied-air respirator are the recommended respiratory protection methods of choice. If a full facepiece is not available, then chemical goggles should be worn to protect the eyes. Whenever a chemical splash hazard exists, a face shield and a rubber apron should be worn. To prevent hand and skin exposures, butyl rubber gloves should be worn. Glove manufacturers should be contacted and permeation studies obtained *before* final glove selections are made.

Administrative controls should also be in place to minimize the potential for human exposures. These may include written procedures or policies which specify the methods and techniques that will be practiced whenever personnel are to work with formamide.

All personnel should receive *training* on the use, hazards, protective measures, emergency actions, and other precautions per 29 CFR 1910.1200 (Hazard Communication), prior to the first assignment in an area where formamide is used or stored.

Other methods to reduce exposure include:

☑ Where possible, enclose operations and use local exhaust ventilation at the site of chemical release. If local exhaust ventilation or enclosure is not used, respiratory protection should be mandatory. Also, automatic transfer of liquids between containers is recommended.

☑ Always ensure that proper protective clothing is worn when using chemical substances.

☑ Wash thoroughly immediately after exposure to formamide and at the end of the work shift or before eating, drinking, or smoking.

☑ Hazard warning information should be posted in the work area. In addition, as part of an on-going education and training program, all information on the health and safety hazards of formamide should be communicated to all exposed or potentially exposed workers.

☑ Never eat, drink, or smoke in areas were formamide is used, handled or stored.

RISK ASSESSMENT: ENVIRONMENT

General Assessment

The environment is at risk of exposure during transportation, storage, disposal, or destruction of formamide. In almost every scenario, the threat of environmental exposure is contingent upon the proper handling of the chemical substance. Accidental spills, large or small, can result in fire, explosion, and possible contamination of the surrounding environmental mediums (water, soil, and air). Formamide is considered a Class 111B combustible liquid. It presents a moderate fire hazard when exposed to heat or sources of ignition. It is also hygroscopic (will absorb moisture). It will attack lead, copper, and brass materials. These characteristics require special consideration during any emergency situation involving a leak or spill of formamide. Should formamide ever come into contact with incompatible substances such as oxidizers, acids, bases, iodine, pyridine, or sulfur trioxide, the formation of highly toxic and/or highly explosive commodities is extremely possible.

Formamide can enter the environment through manufacturing, unchecked discharge into effluents, and through spills.

☠ *Acute Ecological Effects*

Acute (short-term) toxic effects may include the death of animals, birds, or fish, and death or low growth rate in plants. Acute effects are seen 2 to 4 days after animals or plants are exposed to formamide.

This chemical has high acute toxicity to aquatic life. No data are available on the short-term effects of formamide to terrestrial animals.

Chronic Ecological Effects

Chronic toxic effects may include shortened life span, reproductive problems, lower fertility, and changes in appearance or behavior in exposed animals. These effects can be seen long after first exposure(s) to toxic chemicals.

Formamide has high chronic toxicity in aquatic life. No data are available on the long-term effects of formamide to plants, birds, or land animals.

Water Solubility

Formamide is highly soluble in water. Concentrations of 1000 milligrams will mix with a liter of water.

⌛ *Persistence in the Environment*

Formamide is moderately persistent in water, with a half-life of between 2 to 20 days. The half-life of a pollutant is the amount of time it takes for one-half of the chemical to be degraded. It is expected to have high mobility in soil, especially dry soil.

Bioaccumulation in Aquatic Organisms

Some substances increase in concentration, or *bioaccumulate,* in living organisms as they breathe contaminated air, drink contaminated water, or eat contaminated food. These chemicals can become concentrated in the tissues and internal organs of animals as well as humans.

The concentration of formamide found in fish tissues is expected to be about the same or lower than the average concentration of formamide in water from which the fish was taken.

Recommended Risk-Reduction Measures

Proper training of all transporters will reduce the likelihood of a mishap or accident resulting in a leak or spill to the environment. The correct use of DOT labeling (per the MSDS) on all containers, trucks, and rail cars will enable emergency responders to react properly and quickly to any disaster thereby reducing the potential risk to the environment and to personnel.

Storage of formamide should be segregated from incompatible chemicals to minimize the risk of cross-contamination or contact. Buildings designated for storage should be equipped with appropriate fire protection systems (alarms, sprinklers, emergency lighting, portable extinguishers).

If a spill or leak to the environment has occurred, fire department, emergency response, and/or hazardous materials spill personnel should be notified immediately. Cleanup should be attempted only by those trained in proper spill containment procedures. Contaminated soils should be removed for incineration and replaced with clean soil.

If formamide should contact the water table, aquifer, or navigable waterway, time is of the essence. It is readily miscible in water and, therefore, total containment and remediation may not be entirely possible. When such spills occur, the local and/or state emergency response authorities must be notified. A comprehensive emergency response or disaster preparedness/recovery plan should be in place prior to any operations involving the use, transportation, storage, or disposal of formamide. If formamide is spilled or leaked, the following specific steps are recommended:

- ☑ Restrict persons not wearing protective clothing from area of spill or leak until cleanup is complete. Responders must wear appropriate personal protective equipment (PPE), including self-contained breathing apparatus (SCBA).
- ☑ Remove all ignition sources.
- ☑ Ventilate area of spill or leak.
- ☑ Absorb liquids in vermiculite, dry sand, earth, or a similar material and deposit in sealed containers. If non-sparking tools are used, avoid copper, brass, or lead since formamide will attack these metals. Suggest plastic or rubber materials.
- ☑ If applicable, stop flow of leaking liquid or gas. If leak source is a cylinder and the leak can not be stopped in place, remove leaking cylinder to a safe place in the open air and repair or allow cylinder to empty.
- ☑ Keep formamide out of a confined space, such as a sewer, because of the possibility of explosion (unless the sewer is designed to prevent the build-up of explosive concentrations).
- ☑ It may be necessary to dispose of formamide as a hazardous waste. The responsible state agency or the regional office of the federal Environmental Protection Agency (EPA) should be contacted for specific recommendations.

RISK ASSESSMENT: BUSINESS

General Assessment

Accidents or mishaps involving formamide can present a significant threat to business operations. The loss or damage of equipment or facilities can significantly affect fiscal viability. Lawsuits that may result from personnel injury/death, public exposures, and/or environmental contamination will also require a serious expenditure of resources. Media attention surrounding an injury, death, or environmental damage can also result in a loss of profits and loss of current as well as future business.

Always remember that anytime the term "reproductive hazard" is used, public emotion, hysteria, and ignorance can run equally high. This should be serious considered whenever developing or implementing public relations policies.

Recommended Risk-Reduction Measures

Company attorneys, safety and health professionals, and environmental specialists should be involved in the development of any procedures dealing with safety or emergency response. A company official should be pre-designated as a public relations officer with specific training in dealing with the press. Corporate plans and policies should be developed, approved, and implemented long before any need for such arises.

REFERENCES

Ahlbom, A. 1993. *Biostatistics for Epidemiologists.* New York: CRC Press/Lewis Publishers.

American Conference of Governmental Industrial Hygienists. 1988. *Documentation of the Threshold Limit Values and Biological Exposure Indices,* 5th Edition (with updates). Cincinnati: ACGIH

Anthony, C. P., and N. J. Kolthoff. 1971. *Textbook of Anatomy and Physiology.* St. Louis: C. V. Mosby Company.

Asante-Duah, D. K. 1993. *Hazardous Waste Risk Assessment.* New York: CRC Press/Lewis Publishers.

Bloom, A. D. and F. J. DeSerres. 1995. *Ecotoxicity and Human Health.* New York: CRC Press/Lewis Publishers.

Brusick, D. J. 1994. *Methods for Genetic Risk Assessment.* New York: CRC Press/Lewis Publishers.

Bueche, J. R. 1972. *Principles of Physics,* 2nd Edition. New York: McGraw-Hill.

Burrell, R., D. K. Flaherty, and L. J. Sauers. 1992. *Toxicology of the Immune System.* New York: Van Nostrand Reinhold.

Calabrese, E. J., E. M. Kenyon. *Air Toxics and Risk Assessment.* New York: CRC Press/Lewis Publishers.

Calabrese, E. J. 1994. *Biological Effects of Low Level Exposures.* New York: CRC Press/Lewis Publishers.

Cockerman, L. G., B. S. Shane. 1994. *Basic Environmental Toxicology.* New York: CRC Press/Lewis Publishers.

Cothern, C. R., N. P. Ross. 1994. *Environmental Statistics, Assessment, and Forecasting.* New York: CRC Press/Lewis Publishers.

Cothern, R. R. 1993. *Comparative Environmental Risk Assessment.* New York: CRC Press/Lewis Publishers.

Cothern, C. R. 1995. *Handbook for Environmental Risk Decisionmaking.* New York: CRC Press/Lewis Publishers.

Danse, I. R. 1991. *Common Sense Toxics in the Workplace.* New York: Van Nostrand Reinhold.

Environmental Protection Agency. 1983. *Chemical Hazard Information Profile: Formamide.* Washington, D.C.: U.S. EPA.

Gardner, R. W. 1994. *Chemical Intolerance.* New York: CRC Press/Lewis Publishers.

Hallenbeck, W. H. 1993. *Quantitative Risk Assessment for Environmental and Occupational Health,* 2nd Edition. New York: CRC Press/Lewis Publishers.

Hodgson, E., R. B. Mailman, and J. E. Chambers. 1988. *Dictionary of Toxicology.* London: The MacMillian Company.

Heck, H. 1985. Mechanisms of aldehyde toxicity: Structure activity Studies. *CIIT Activities,* 5(10):106.

Hosty, J. W., and P. Foster. 1990. *A Practical Guide to Chemical Spill Response.* New York: Van Nostrand Reinhold.

Howard, P. H. 1995. *Handbook of Environmental Fate and Exposures Data for Organic Chemicals,* Vol. I-IV. New York: CRC Press/Lewis Publishers.

International Agency for Research on Cancer (IARC). 1985. *IARC Monographs on the Evaluation of Carcinogenic Risk of Chemicals to Humans,* Vol. 36, *Allyl Compounds, Aldehydes, Epoxides, and Peroxides,* pp 101-136. Lyon: IARC.

James, A., and M. Lord. 1992. *Index of Chemical and Physical Data.* New York: Van Nostrand Reinhold.

Jones, F. E. 1994. *Toxic Organic Vapors in the Workplace.* New York: CRC Press/Lewis Publishers.

Kamrin, M. 1 A. 1988. *Toxicology: A Primer on Principles and Applications.* Chelsea, Michigan: Lewis Publishers.

Keith, L. H. and M. M. Walker. 1995. *Handbook of Air Toxics.* New York: CRC Press/Lewis Publishers.

Landis, W. G., M. H. Yu. 1995. *Introduction to Environmental Toxicology.* New York: CRC Press/Lewis Publishers.

Lewis, .J., Sr. 1993. *Hawley's Condensed Chemical Dictionary,* 12th Edition. New York: Van Nostrand Reinhold.

Lewis, . J., Sr. 1992. *Sax's Dangerous Properties of Industrial Materials,* Volumes 1, 2, and 3, 8th Edition. New York: Van Nostrand Reinhold.

Lisella, F. S. 1994. *The VNR Dictionary of Environmental Health and Safety.* New York: Van Nostrand Reinhold.

Meyer, E. 1990. *Chemistry of Hazardous Materials.* Englewood Cliffs, NJ: Prentice Hall.

Patnaik, P. 1992. *A Comprehensive Guide to the Hazardous Properties of Chemical Substances.* New York: Van Nostrand Reinhold.

Proctor, N. H., and J. P. Hughes. 1991. *Chemical Hazards of the Workplace.* New York: Van Nostrand Reinhold.

Silverman, L., H. F. Schuttle, and M. W. First. 1946. Further studies on sensory response to certain industrial solvent vapors. *Journal of Industrial Hygiene Toxicology,* 28:262-266.

Sherman, J. D. 1988. *Chemical Exposure and Disease.* New York: Van Nostrand Reinhold.

U.S. Department of Health and Human Services, National Institute for Occupational Safety and Health. 1994. *NIOSH Pocket Guide to Chemical Hazards.* Washington, D.C.: U.S. Government Printing Office.

MATERIAL SAFETY DATA SHEET

CHEMICAL NAME	
FORMIC ACID	**CRC PRESS/LEWIS PUBLISHERS DISCLAIMER:** THE INFORMATION AND RECOMMENDATIONS PRESENTED HEREIN ARE BASED ON SOURCES BELIEVED TO BE RELIABLE. CRC MAKES NO REPRESENTATION ON ITS COMPLETENESS OR ACCURACY. IT IS THE RESPONSIBILITY OF THE USER TO DETERMINE THE CHEMICAL'S SUITABILITY FOR ITS INTENDED USE, THE CHEMICAL'S SAFE USE, AND THE CHEMICAL'S PROPER DISPOSAL. NO REPRESENTATIONS AND/OR WARRANTIES, EITHER EXPRESSED OR IMPLIED, OF THE MERCHANTABILITY OR FITNESS FOR A PARTICULAR PURPOSE, OR OF ANY OTHER NATURE, ARE MADE WITH RESPECT TO THE INFORMATION PROVIDED IN THIS MSDS OR TO THE CHEMICAL TO WHICH INFORMATION MAY REFER. CRC NEITHER ASSUMES NOR AUTHORIZES ANY OTHER PERSON TO ASSUME FOR IT, ANY OTHER ADDITIONAL RESPONSIBILITY OR LIABILITY FOR THE USE OF, OR RELIANCE UPON, THIS INFORMATION.

HAZARD WARNING INFORMATION

HEALTH	FIRE	REACTIVE	OTHER	DEGREE OF HAZARD	COLOR CODING	OTHER CODES
3	2	0		0 = Minimum Hazard 1 = Slight Hazard 2 = Moderate Hazard 3 = Serious Hazard 4 = Severe Hazard	HEALTH = BLUE FIRE = RED REACTIVITY = YELLOW OTHER = WHITE	OX = Oxidizer ACID = Acid ALK = Alkali COR = Corrosive W = Use No Water

SECTION I - GENERAL INFORMATION

Characterization	**Acid, Carboxylic**	RCRA Number	**U123**
		EPA Class	**Toxic Waste**
DOT Proper Shipping Name	**Formic Acid Solution**	Chemical Abstract Service (CAS) Number	**64-18-6**
DOT Hazard Class and Label Requirements	**Corrosive Material**	DOT Emergency Guide Code	**60**
DOT Identification Number	**UN 1779**	Chemical Formula	**CH_2O_2**

Synonyms
Formic acid (85-95% in aqueous solution); hydrogen carboxylic acid; methanoic acid; aminic acid.

SECTION II - HAZARDOUS INGREDIENTS AND IDENTITY INFORMATION

Hazardous Components (specific identity)	OSHA Exposure Criteria	NIOSH Exposure Criteria	Immediately Dangerous to Life and Health (IDLH)	ACGIH Exposure Criteria
Formic acid (derivation: By treatment of sodium formate and sodium acid formate with sulfuric acid at low temperatures and distilling in vacuum; by acid hydrolysis of methyl formate; as a byproduct in manufacture of acetaldehyde and formaldehyde). **1 ppm = 1.91 mg/m^3**	**PEL:** 5 ppm 9 mg/m^3 **STEL:** Not Established	**REL:** 5 ppm 9 mg/m^3 **STEL:** Not Established	**30 ppm**	**TLV:** 5 ppm 9 mg/m^3 **STEL:** 10 ppm 19 mg/m^3

SECTION III - PHYSICAL AND CHEMICAL CHARACTERISTICS

Boiling Point	**221°F (105°C)**	Specific Gravity (H_2O = 1)	**1.22 (90% solution)**
Vapor Pressure (mm Hg)	**35 at 69°F (20°C)**	Molecular Weight	**46.0**
Vapor Density (Air = 1)	**1.59**	Melting Point	**37°F (3°C)**

Solubility
Very miscible. Also miscible in alcohol and ether.

Appearance and Odor
Colorless, fuming, hygroscopic liquid with an extremely pungent and penetrating odor. Odor Threshold Range = <1 ppm to 38 ppm.

SECTION IV - FIRE AND EXPLOSION HAZARD DATA

Flash Point (method used)	**310°F (154°C) opened cup**	Explosive Limits in Air % by Volume	**LEL: 18% UEL: 57%**
NFPA Classification	**Class II Combustible Liquid**	Autoignition Temperature	**813°F (434°C)**

Extinguishing Media
Carbon dioxide, dry chemical, water spray, or alcohol foam.

Special Fire Fighting Procedures
Wear full protective clothing and self-contained breathing apparatus (SCBA). Use water spray to flush and dilute a spill and to disperse the vapors. Also, use a water spray to keep fire-exposed containers cool. Do not release runoff to sewers and waterways. Poisonous gases are produced in fire.

Unusual Fire and Explosion Hazards
Potentially explosive reaction with strong oxidizers, strong caustics, and concentrated sulfuric acid. Containers may explode in fire. Firefighters should avoid contact with vapors produced during fire.

SECTION V - REACTIVITY DATA

Stability		Conditions to Avoid **Normally stable in closed containers under normal conditions of storage and handling. However, at room temperature, glacial formic acid slowly decomposes and builds up pressure in sealed containers.**
Stable X	Unstable	Incompatibility *(materials to avoid)* **Strong oxidizers (such as chlorine, fluorine, and bromine), strong caustics, and concentrated sulfuric acid. Explosive reactions with furfuryl alcohol, hydrogen peroxide, nitromethane, and phosphorus pentoxide.**
Hazardous Polymerization		Conditions to Avoid **Under normal conditions of temperature and pressure, hazardous polymerization of formic acid is not expected to occur.**
May Occur	Will Not Occur X	Hazardous Decomposition or By-products **When heated to decomposition, formic acid can emit highly toxic/poisonous gases.**

SECTION VI - HEALTH HAZARD DATA

Primary Route(s) of Entry:	Inhalation?	Absorption (skin)?	Ingestion?
	X	X	X

Health Hazards

INHALATION: Vapors cause irritation to the upper respiratory tract with coughing, dypsnea, headache, bronchitis, weakness, tearing, nasal discharge, and nausea.

SKIN & EYES: Eye contact causes scarring of the cornea. Skin contact causes burns, blistering, local necroses, and membrane ulceration. Sensitization may occur.

INGESTION: Severe poisoning, kidney damage, salivation, bloody vomiting, burning in mouth, pain, and diarrhea. Severe cases may cause shock, dypsnea, circulatory collapse, and death.

Carcinogenicity	NTP Listed?	IARC Cancer Review Group?	OSHA Regulated?	Target Organs?
Unknown Human Unknown Animal	No	No	29 CFR 1910.1000 Table Z-1	Respiratory system, eyes, skin, kidneys, liver.

Medical Conditions Generally Aggravated by Exposure

Respiratory impairments (bronchitis, asthma); Skin conditions (dermatitis); Renal or liver disorders.

Emergency and First-aid Procedures

Eye contact: Flush immediately with water for 15 minutes (minimum), seek medical attention. Skin contact: Remove all contaminated clothing. Immediately wash area with large amounts of water. Seek medical assistance. For inhalation: Remove the person from exposure. Provide respiratory assistance and CPR. Transfer to medical facility. Inhaling a 5% sodium bicarbonate in water aerosol may reduce effects. If swallowed: Seek medical attention, contact poison control center. Do NOT induce vomiting. Give milk or water to drink to dilute followed by milk of magnesia or aluminum hydroxide.

SECTION VII - PRECAUTIONS FOR SAFE HANDLING AND USE

Steps to be Taken in Case Material is Released or Spilled

Absorb liquids in vermiculite or other material, collect using non-sparking tools, and deposit in sealed containers. Ventilate area of spill or leak. Restrict those not involved in cleanup from entering area.

Preferred Waste Disposal Method

No Citation.

Precautions to be Taken in Handling and Storage

Formic acid is a combustible liquid. Avoid contact with strong oxidizers such as chlorine since violent reactions occur. Store in tightly closed containers in cool, well-ventilated area away from heat.

Other Precautions and Warnings

Sources of ignition are prohibited where formic acid is used, handled, or stored. Detached facility storage is preferred. Electrically ground and bond all containers.

SECTION VIII - CONTROL MEASURES AND PERSONAL PROTECTIVE EQUIPMENT

Respiratory Protection *(specify type)*

Use half-mask respirator with acid cartridge for low or transient exposures (cartridge should be designed to protect against acid vapor or mist). For higher exposures, use a supplied-air respirator or a self-contained breathing apparatus operated in pressure demand or other positive pressure mode.

Ventilation

Local exhaust or general mechanical systems recommended.

Protective Gloves	Eye Protection	Other Protective Clothing
Acid resistant synthetic rubber	Chemical Goggles or Face Mask	Rubber Apron

Work/Hygiene Practices

Always wash hands thoroughly after using chemical; never bring food, drink, or smoking materials into vicinity of chemicals.

FORMIC ACID

CH_2O_2 CAS: 64-18-6

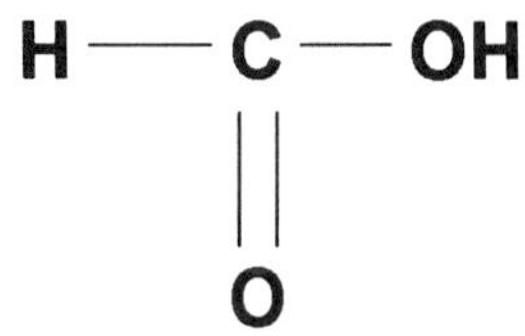

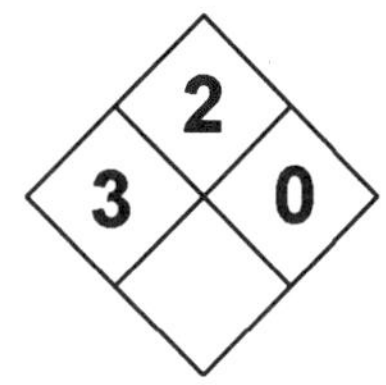

IDENTIFICATION AND TYPICAL USES

Formic acid is a colorless liquid with pungent, penetrating, and irritating odor. It is often used in aqueous solution of 90% formic acid or in a 98% (glacial) solution). It is used the dyeing and finishing of textiles, in the manufacture of fumigants, in the treatment of leathers, as a solvent for perfumes and lacquers, in electroplating operations, in the coagulation of rubber latex, as an aid in regenerating old rubber, in chemical analysis, as a decalcifier, as an antiseptic in brewing, in silvering glass products, in ore flotation, as an animal feed additive, a food preservative, and as a flavor adjunct. It is also used in vinyl resin plasticizers.

RISK ASSESSMENT: HEALTH

General Assessment

Exposure to formic acid is possible through *inhalation* of its vapors and by *ingestion.* Although it is considered to be of moderate to severe toxicity, it is a highly caustic and, therefore, corrosive compound. There are no reports on its carcinogenic, mutagenic, or teratogenic properties.

Inhalation produces severe irritation of the eyes, nose, and respiratory tract with coughing, dypsnea (labored breathing), bronchitis, headache, and total body weakness. There may be tearing, nasal discharge, throat irritation, and nausea.

Skin contact causes severe burning with possible vesiculation. Its extremely corrosive properties cause immediate burns, membrane ulceration, skin blistering, and local necrosis (tissue destruction) upon contact. Skin sensitization may also occur. Eye contact causes damage and the possibility of permanent corneal scarring.

Ingestion is extremely dangerous, even life-threatening. It causes severe poisoning, kidney damage, salivation, bloody vomiting, burning sensation on the mouth, acute tissue damage, diarrhea with pain, shock, breathing difficulties, circulatory collapse and death.

Acute Health Effects

The following acute (short-term) health effects may occur immediately or shortly after exposure to formic acid:

Skin: Irritation/rash or burning feeling on contact with redness, blistering and vesiculation, and localized tissue destruction.

Eye: Severe irritation; may cause burns and damage with permanent corneal scarring.

Lung: Severe irritation of the mucous membranes in the nose, throat, and lungs following exposure.

CNS: Inconclusive (possible lethargy, weight loss).

Chronic Health Effects

The following chronic (long-term) health effects may occur at some time after exposure to formic acid and can last for months or even years:

Cancer Hazards: According to information presented in the references, formic acid has not been adequately tested for its ability to cause cancer in test animals.

Reproductive Hazard: According to information presented in the references, formic acid has not been adequately tested for its ability to adversely affect reproduction in test animals.

Other Chronic Effects: Skin contact causes irritation and burns. There may be some lasting effects of chronic exposures such as sensitization. Once an individual becomes "sensitized," even the slightest future exposures can cause an allergic reaction.

Recommended Risk-Reduction Measures

Personnel should avoid direct contact with formic acid. It is extremely corrosive and moderately toxic to humans. If a less toxic chemical cannot be substituted for a hazardous substance, then *engineering controls* are the most effective method of reducing exposures.

The best protection is to enclose operations and/or provide local exhaust ventilation at the site of chemical release. While not always operationally feasible, isolating operations can also reduce exposure. Using respiratory protection is less effective than the controls mentioned above, but is still advisable whenever working with or around formic acid. A self-contained breathing apparatus (SCBA) or a supplied-air respirator with full facepiece operated in pressure demand mode are the recommended respiratory protection methods of choice. If a full facepiece is not available, then chemical goggles should be worn to protect the eyes. For low concentrations (near the 5 ppm exposure limit), an acid gas cartridge respirator (equipped with a mist/vapor filter) is acceptable protection. Whenever a chemical splash hazard exists, a face shield and a rubber apron should be worn. To prevent hand and skin exposures, acid resistant gloves should be worn.

Administrative controls should also be in place to minimize the potential for human exposures. These may include written procedures or policies which specify the methods and techniques that will be practiced whenever personnel are to work with formic acid.

All personnel should receive *training* on the use, hazards, protective measures, emergency actions, and other precautions per 29 CFR 1910.1200 (Hazard Communication), prior to the first assignment in an area where formic acid is used or stored.

Before beginning employment and at regular intervals thereafter (e.g., annually), the following tests are recommended:

☑ Lung function tests (look for chronic respiratory diseases).

☑ Kidney and liver function tests.

If symptoms occur or overexposure is suspected, the following additional tests are suggested:

☑ Evaluation by a qualified allergist with careful consideration of exposure history and special testing (may help diagnose skin allergy).

☑ If formic acid is ingested, contact poison control center. Unless advised otherwise and victim is conscious and alert, give 1-2 glasses of water or milk to dilute followed by milk of magnesia or aluminum hydroxide. Do NOT induce vomiting!

Other methods to reduce exposure include:

☑ Where possible, enclose operations and use local exhaust ventilation at the site of chemical release. If local exhaust ventilation or enclosure is not used, respiratory protection should be mandatory.

☑ Always ensure that proper protective clothing is worn when using chemical substances.

☑ Wash thoroughly immediately after exposure to formic acid and at the end of the work shift or before eating, drinking, or smoking.

☑ Hazard warning information should be posted in the work area. In addition, as part of an on-going education and training program, all information on the health and safety hazards of formic acid should be communicated to all exposed workers.

RISK ASSESSMENT: ENVIRONMENT

General Assessment

The environment is at risk of exposure during transportation, storage, disposal, or destruction of formic acid. In almost every scenario, the threat of environmental exposure is contingent upon the proper handling of the chemical substance. Accidental spills, large or small, can result in fire, explosion, and possible contamination of the surrounding environmental mediums (water, soil, and, to a lesser degree, air).

Formic acid is considered a combustible liquid. It is also extremely corrosive to many materials. It is incompatible with oxidizing agents and contact can cause fire or explosion. These characteristics require special consideration during any emergency situation involving a leak or spill of formic acid.

Should formic acid ever come into contact with incompatible substances such as oxidizers, caustics, or acids either during use, transportation, or storage, the formation of highly toxic and/or highly explosive commodities is extremely possible.

Formic acid can enter the environment mainly through industrial discharges or spills.

☠ *Acute Ecological Effects*

Acute (short-term) toxic effects may include the death of animals, birds, or fish and death or low growth rate in plants. Acute effects are seen 2 to 4 days after animals or plants are exposed to formic acid.

Insufficient data are available on the short-term effects of formic acid exposure to aquatic life, plants, birds, or land animals.

Chronic Ecological Effects

Chronic (long-term) toxic effects may include shortened life span, reproductive problems, lower fertility, and changes in appearance or behavior in exposed animals. These effects can be seen long after first exposure(s) to toxic chemicals. Insufficient data are available on the long-term effects of formic acid to aquatic life, plants, birds, or land animals.

Water Solubility

Formic acid is highly soluble in water. Concentrations of over 1000 milligrams will mix with a liter of water.

Persistence in the Environment

Formic acid is slightly persistent in the aquatic environment, with a half-life of between 2 to 20 days. The half-life of a pollutant is the amount of time it takes for one half of the chemical to be degraded.

Bioaccumulation in Aquatic Organisms

Some substances increase in concentration, or *bioaccumulate,* in living organisms as they breathe contaminated air, drink contaminated water, or eat contaminated food. These chemicals can become concentrated in the tissues and internal organs of animals as well as humans.

The concentration of formic acid found in fish tissues is expected to be about the same as the average concentration of formic acid in the water from which the fish was taken.

Recommended Risk-Reduction Measures

Proper training of all transporters will reduce the likelihood of a mishap or accident resulting in a leak or spill to the environment. The correct use of DOT labeling (per the MSDS) on all containers, trucks, and rail cars will enable emergency responders to react properly and quickly to any disaster thereby reducing the potential risk to the environment and to personnel.

Storage of formic acid should be segregated from incompatible chemicals to minimize the risk of cross-contamination or contact. Buildings designated for storage should be equipped with appropriate fire protection systems (alarms, sprinklers, emergency lighting, portable extinguishers).

If a spill or leak to the environment has occurred, fire department, emergency response and/or hazardous materials spill personnel should be notified immediately. Cleanup should be attempted only by those trained in proper spill containment procedures. Contaminated soils should be removed for incineration and replaced with clean soil. If formic acid should contact the water table, aquifer, or navigable waterway, time is of the essence. It is highly miscible in water and, therefore, total containment and remediation may not be entirely possible. When such spills occur, the local and/or state emergency response authorities must be notified. A comprehensive emergency response or disaster preparedness/recovery plan should be in place prior to any operations involving the use, transportation, storage, or disposal of formic acid. If formic acid is spilled or leaked, the following specific steps are recommended:

- ☑ Restrict persons not wearing protective clothing from area of spill or leak until cleanup is complete. Use non-sparking tools to cleanup area.
- ☑ Remove all ignition sources.
- ☑ Ventilate area or spill or leak.
- ☑ Absorb liquids in vermiculite, dry sand, earth, or similar material and deposit in sealed containers. Neutralize remaining residue with soda ash or sodium bicarbonate. Do not flush to drains, sewers, or waterways.
- ☑ It may be necessary to dispose of formic acid as a hazardous waste. The responsible state agency or the regional office of the federal Environmental Protection Agency (EPA) should be contacted for specific recommendations.

RISK ASSESSMENT: BUSINESS

General Assessment

Accidents or mishaps involving formic acid can present a significant threat to business operations. The loss or damage of equipment or facilities can significantly affect fiscal viability. Lawsuits that may result from personnel injury/death, public exposures, and/or environmental contamination will also require a serious expenditure of resources. Media attention surrounding an injury, death, or environmental damage can also result in a loss of profits and loss of current as well as future business.

Recommended Risk-Reduction Measures

Company attorneys, safety and health professionals, and environmental specialists should be involved in the

development of any procedures dealing with safety or emergency response. A company official should be pre-designated as a public relations officer with specific training in dealing with the press. Corporate plans and policies should be developed, approved, and implemented long before any need for such arises.

REFERENCES

Ahlbom, A. 1993. *Biostatistics for Epidemiologists.* New York: CRC Press/Lewis Publishers.

American Conference of Governmental Industrial Hygienists. 1988. *Documentation of the Threshold Limit Values and Biological Exposure Indices,* 5th Edition (with updates). Cincinnati: ACGIH

Asante-Duah, D. K. 1993. *Hazardous Waste Risk Assessment.* New York: CRC Press/Lewis Publishers.

Bloom, A. D. and F. J. DeSerres. 1995. *Ecotoxicity and Human Health.* New York: CRC Press/Lewis Publishers.

Bueche, J. R. 1972. *Principles of Physics,* 2nd Edition. New York: McGraw-Hill.

Burrell, R., D. K. Flaherty, and Leonard J. Sauers. 1992. *Toxicology of the Immune System.* New York: Van Nostrand Reinhold.

Calabrese, E. J., E. M. Kenyon. *Air Toxics and Risk Assessment.* New York: CRC Press/Lewis Publishers.

Cockerman, L. G., B. S. Shane. 1994. *Basic Environmental Toxicology.* New York: CRC Press/Lewis Publishers.

Cothern, C. R. 1995. *Handbook for Environmental Risk Decisionmaking.* New York: CRC Press/Lewis Publishers.

Danse, I. R. 1991. *Common Sense Toxics in the Workplace.* New York: Van Nostrand Reinhold.

Gardner, R. W. 1994. *Chemical Intolerance.* New York: CRC Press/Lewis Publishers.

Hallenbeck, W. H. 1993. *Quantitative Risk Assessment for Environmental and Occupational Health,* 2nd Edition. New York: CRC Press/Lewis Publishers.

Hosty, J. W., P. Foster. 1990. *A Practical Guide to Chemical Spill Response.* New York: Van Nostrand Reinhold.

Howard, P. H., et. al. 1991. *Handbook of Environmental Degradation Rates.* New York: CRC Press/Lewis Publishers.

Jones, F. E. 1994. *Toxic Organic Vapors in the Workplace.* New York: CRC Press/Lewis Publishers.

Keith, L. H. and M. M. Walker. 1995. *Handbook of Air Toxics.* New York: CRC Press/Lewis Publishers.

Landis, W. G., M. H. Yu. 1995. *Introduction to Environmental Toxicology.* New York: CRC Press/Lewis Publishers.

Langerman, N. 1994 *Precautionary Labels for Chemical Containers.* New York: CRC Press/Lewis Publishers.

Lauwerys, R. R., P. Hoet. 1993. *Industrial Chemical Exposure.* New York: CRC Press/Lewis Publishers.

Lewis, R. J., Sr. 1993. *Hawley's Condensed Chemical Dictionary,* 12th Edition. New York: Van Nostrand Reinhold.

Lewis, R. J., Sr. 1992. *Sax's Dangerous Properties of Industrial Materials,* Volumes 1, 2, and 3, 8th Edition. New York: Van Nostrand Reinhold.

Manahan, S. E. 1992. *Toxicological Chemistry,* 2nd Edition. New York: CRC Press/Lewis Publishers.

Mellström, G. A., J. E. Wahlberg, and H. I. Maibach. 1994. *Protective Gloves for Occupational Use.* New York: CRC Press/Lewis Publishers.

Meyer, E. 1990. *Chemistry of Hazardous Materials.* Englewood Cliffs, NJ: Prentice Hall.

Mycock, J. C., J. D. McKenna, and L. Theodore. 1995. *Handbook of Air Pollution Control Engineering and Technology.* New York: CRC Press/Lewis Publishers.

Nelson, G. O. 1992. *Gas Mixtures.* New York: CRC Press/Lewis Publishers.

Ott, W. R. 1995. *Environmental Statistics and Data Analysis.* New York: CRC Press/Lewis Publishers.

Patnaik, P. 1992. *A Comprehensive Guide to the Hazardous Properties of Chemical Substances.* New York: Van Nostrand Reinhold.

Philp, R. B. 1995. *Environmental Hazards and Human Health.* New York: CRC Press/Lewis Publishers.

Quigley, D. R. 1994. *Handbook of Emergency Chemical Management.* New York: CRC Press/Lewis Publishers.

Rea, W. J. 1995. *Chemical Sensitivity,* Vol. I-III. New York: CRC Press/Lewis Publishers.

Richardson, M. 1995. *Dictionary of Substance and Their Effects,* Vol. I-VI. New York: CRC Press/Lewis Publishers.

Rylander, R., R. R. Jacobs. 1994. *Organic Dusts.* New York: CRC Press/Lewis Publishers.

Sacarello, H. L. A. 1994. *The Comprehensive Handbook of Hazardous Materials.* New York: CRC Press/Lewis Publishers.

Sheftel, V. O. 1995. *Handbook of Toxic Properties of Monomers and Additives.* New York: CRC Press/Lewis Publishers.

Sherman, J. D. 1988. *Chemical Exposure and Disease.* New York: Van Nostrand Reinhold.

Talbott, E., G. Craun. 1995. *An Introduction to Environmental Epidemiology.* New York: CRC Press/Lewis Publishers.

U.S. Department of Health and Human Services, National Institute for Occupational Safety and Health. 1994. *NIOSH Pocket Guide to Chemical Hazards.* Washington, D.C.: U.S. Government Printing Office.

Witorsch, P. and S. V. Spagnolo. 1994. *Air Pollution and Lung Disease in Adults.* New York: CRC Press/Lewis Publishers.

MATERIAL SAFETY DATA SHEET

CHEMICAL NAME	
FURAN	**CRC PRESS/LEWIS PUBLISHERS DISCLAIMER:** THE INFORMATION AND RECOMMENDATIONS PRESENTED HEREIN ARE BASED ON SOURCES BELIEVED TO BE RELIABLE. CRC MAKES NO REPRESENTATION ON ITS COMPLETENESS OR ACCURACY. IT IS THE RESPONSIBILITY OF THE USER TO DETERMINE THE CHEMICAL'S SUITABILITY FOR ITS INTENDED USE, THE CHEMICAL'S SAFE USE, AND THE CHEMICAL'S PROPER DISPOSAL. NO REPRESENTATIONS AND/OR WARRANTIES, EITHER EXPRESSED OR IMPLIED, OF THE MERCHANTABILITY OR FITNESS FOR A PARTICULAR PURPOSE, OR OF ANY OTHER NATURE, ARE MADE WITH RESPECT TO THE INFORMATION PROVIDED IN THIS MSDS OR TO THE CHEMICAL TO WHICH INFORMATION MAY REFER. CRC NEITHER ASSUMES NOR AUTHORIZES ANY OTHER PERSON TO ASSUME FOR IT, ANY OTHER ADDITIONAL RESPONSIBILITY OR LIABILITY FOR THE USE OF, OR RELIANCE UPON, THIS INFORMATION.

NATIONAL FIRE PROTECTION ASSOCIATION (NFPA) LABELING INFORMATION

HEALTH	FIRE	REACTIVE	OTHER	DEGREE OF HAZARD	COLOR CODING	OTHER CODES
1	4	1		0 = Minimum Hazard 1 = Slight Hazard 2 = Moderate Hazard 3 = Serious Hazard 4 = Severe Hazard	HEALTH = BLUE FIRE = RED REACTIVITY = YELLOW OTHER = WHITE	OX = Oxidizer ACID = Acid ALK = Alkali COR = Corrosive W = Use No Water

SECTION I - GENERAL INFORMATION

Characterization **Heterocyclic Compound (of Oxygen)**	RCRA Number **U124**	EPA Class **Toxic Waste**
DOT Proper Shipping Name **Furan**	Chemical Abstract Service (CAS) Number **110-00-9**	
DOT Hazard Class and Label Requirements **Flammable Liquid**	DOT Emergency Guide Code **No Citation**	
DOT Identification Number **UN 2389**	Chemical Formula **C_4H_4O**	

Synonyms
Furfuran; oxole; divinyl oxide; tetrole; oxacyclopentadiene; 1,4-epoxy-1,3-butadiene.

SECTION II - HAZARDOUS INGREDIENTS AND IDENTITY INFORMATION

Hazardous Components (specific identity)	OSHA Exposure Criteria	NIOSH Exposure Criteria	Immediately Dangerous to Life and Health (IDLH)	ACGIH Exposure Criteria
Furan (derivation: By dry distillation of furoic acid from furfural. Also found in the gaseous state in cigarette and wood smoke, and gasoline and diesel engine exhaust).	**PEL:** Not Established **STEL:** Not Established	**REL:** Not Established **STEL:** Not Established	**Not Determined**	**TLV:** Not Established **STEL:** Not Established

SECTION III - PHYSICAL AND CHEMICAL CHARACTERISTICS

Boiling Point **90°F (32°C)**	Specific Gravity (H_2O = 1) **0.94**
Vapor Pressure (mm Hg) **600 at 77°F (25°C)**	Molecular Weight **68.08**
Vapor Density (Air = 1) **2.35**	Melting Point **-123°F (-86°C)**

Solubility
Insoluble in water, soluble in alcohol and ether.

Appearance and Odor
Colorless, hygroscopic liquid with a slight ether-like odor. Color may darken to brown on standing. Color change is retarded by the addition of a small amount of water.

SECTION IV - FIRE AND EXPLOSION HAZARD DATA

Flash Point (method used) **-32°F (-36°C) closed cup**	Explosive Limits in Air % by Volume **LEL: 2.3% UEL: 14.3%**
NFPA Classification **Class 1A Flammable Liquid**	Autoignition Temperature **Not Found**

Extinguishing Media
Carbon dioxide, dry chemical, or foam. Water may be ineffective and could actually spread fire further.

Special Fire Fighting Procedures
Wear full protective clothing and self-contained breathing apparatus (SCBA). Poisonous gases are produced in fire. Move containers from fire area if it can be done without risk.

Unusual Fire and Explosion Hazards
Furan forms explosive peroxides in air. Moderately explosive when exposed to flame. Vapors are heavier than air and may travel for some distance to an ignition source to flashback and cause fire.

SECTION V - REACTIVITY DATA

Stability		Conditions to Avoid
		Furan is stable in tightly closed containers under normal conditions of storage and handling. Exposure to air forms unstable peroxides. Furan can be stabilized using butylated hydroxytoluene.
Stable	Unstable	Incompatibility *(materials to avoid)*
X		Incompatible with strong oxidizers (such as chlorine, bromine, and fluorine), and strong acids (such as sulfuric, nitric, and hydrochloric). Contact with acids can cause violent exothermic reaction.
Hazardous Polymerization		Conditions to Avoid
		Under normal conditions of temperature and pressure, hazardous polymerization of furan is not expected to occur. Do not distill to dryness since the furan product may contain explosive peroxides.
May Occur	Will Not Occur	Hazardous Decomposition or By-products
	X	When heated to decomposition, furan emits acrid and irritating smoke.

SECTION VI - HEALTH HAZARD DATA

Primary Route(s) of Entry:	Inhalation?	Absorption (skin/eye)?	Ingestion?
	X	X	X

Health Hazards

INHALATION: Irritation to the eyes, nose, throat, and mucosa of the respiratory tract. Causes fatigue, dizziness, suffocation, headache, vomiting, gastrointestinal and bronchial congestion, and low blood pressure. High levels may cause CNS depression, lung damage, death.

ABSORPTION: Skin contact causes dryness and irritation. Possible sensitizer. Absorption is likely. Eye contact causes tearing and irritation.

INGESTION: Gastrointestinal tract irritation, vomiting, nausea, and congestion.

Carcinogenicity	NTP Listed?	IARC Cancer Review Group?	OSHA Regulated?	Target Organs?
Unknown Human Unknown Animal	No	No	No	CNS, skin, eyes, lungs, mucous membranes.

Medical Conditions Generally Aggravated by Exposure

Chronic pulmonary disease may be aggravated by exposure to furan.

Emergency and First-aid Procedures

Eye contact: Do not allow victim to keep eyes tightly shut. Flush immediately with water for 15 minutes (minimum), seek medical attention. Skin contact: Remove all contaminated clothing. Immediately wash area with large amounts of soap and water. For inhalation: Remove the person from exposure. Provide respiratory assistance and CPR. Transfer to medical facility. Observe for 24-48 hours for lung effects (pulmonary edema). If swallowed: Seek medical attention immediately. Give 1-2 glasses of water and induce vomiting. Never give anything by mouth to an unconscious or convulsing person.

SECTION VII - PRECAUTIONS FOR SAFE HANDLING AND USE

Steps to be Taken in Case Material is Released or Spilled

Absorb spills in vermiculite or other material and deposit in sealed containers. Ventilate area of spill or leak. Remove all ignition sources. Restrict those not involved in cleanup from entering area. Notify appropriate authorities, as required by applicable environmental regulations.

Preferred Waste Disposal Method

No Citation.

Precautions to be Taken in Handling and Storage

Store to avoid contact with strong oxidizers, since violent reactions can occur. Store in tightly closed containers in cool, well-ventilated area away from heat. Electrically ground and bond all equipment.

Other Precautions and Warnings

Sources of ignition (heat, flame, cigarettes, incompatibles) are prohibited where furan is used, handled, or stored. Bulk storage of furan is not recommended. Protect containers from physical damage.

SECTION VIII - CONTROL MEASURES AND PERSONAL PROTECTIVE EQUIPMENT

Respiratory Protection *(specify type)*

No exposure limits have been established. Use a supplied-air respirator with full facepiece operated in continuous flow mode, or use a self-contained breathing apparatus (SCBA) with full facepiece operated in positive pressure mode.

Ventilation

Local exhaust (preferred) at site of chemical work or general ventilation.

Protective Gloves	Eye Protection	Other Protective Clothing
Impervious Gloves	Chemical Goggles or Face Mask	Boots; Aprons; Gauntlets

Work/Hygiene Practices

Always wash hands thoroughly after using chemical; never bring food, drink, or smoking materials into vicinity of chemicals.

FURAN

C_4H_4O CAS: 110-00-9

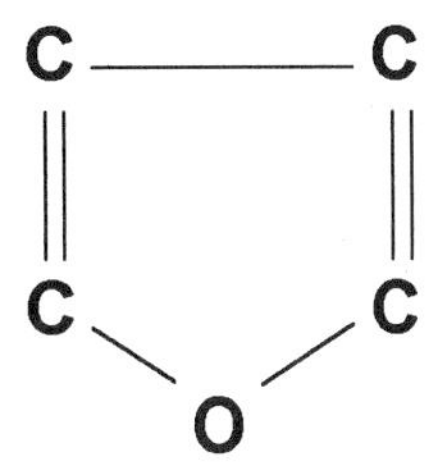

IDENTIFICATION AND TYPICAL USES

Furan is a clear, colorless liquid with an ethereal odor. The liquid turns brown on standing. The color change can be retarded if a small amount of water is added. Furan is used as an intermediate in organic synthesis, especially for pyrrole, tetrahydrofuran, and thiophene. Also in the formation of lacquers and solvent for resins. Furan is a common constituent of some foods.

RISK ASSESSMENT: HEALTH

General Assessment

Furan is a highly toxic compound. It is a poison by *inhalation, ingestion,* and is moderately toxic by skin contact (*absorption*). While its carcinogenic properties have not been adequately investigated, mutation data have been reported.

Inhalation of furan can cause dizziness, suffocation, fatigue, headache, vomiting, bronchial congestion, low blood pressure, suffocation, and a possibility of pulmonary edema (fluid in the lungs), which is a medical emergency and can lead to death. Symptoms may be delayed up to 48 hours, thereby creating a false sense of security with regard to health exposure risk. Furan is a central nervous system (CNS) depressant that may cause respiratory arrest and death.

Skin contact with the liquid or vapor can cause irritation and redness. Burning and watering of eyes have been reported on contact with the vapors of furan. If ingested, there can be gastrointestinal congestion along with many of the systemic (CNS) effects noted for inhalation.

Acute Health Effects

The following acute (short-term) health effects may occur immediately or shortly after exposure to furan:

Skin: Irritation with burning, itching, redness, and possible swelling.

Eye: Severe irritation of the conjunctiva with burning, itching, and watering.

Lung: Irritation of the bronchial tubes and other members of the respiratory tract. May cause pulmonary edema (fluid in the lungs) at high exposures, which can be fatal.

CNS: Narcotic effects at high concentrations with dizziness, fatigue, headache, weakness, and other symptoms of CNS depression.

Chronic Health Effects

The following chronic (long-term) health effects may occur at some time after exposure to furan and can last for months or even years:

Cancer Hazards: Furan may cause mutations (genetic changes) in living cells. The data on its carcinogenicity are inconclusive.

Reproductive Hazard: According to the references, furan has not been adequately tested for its ability to adversely affect reproduction.

Other Chronic Effects: Long-term exposures to low levels of furan may lead to liver damage and may result in permanent lung damage.

Recommended Risk-Reduction Measures

Personnel should avoid direct contact with furan. If a less toxic chemical cannot be substituted for a hazardous substance, then *engineering controls* are the most effective method of reducing exposures. The best protection is to enclose operations and/or provide local exhaust ventilation at the site of chemical release. While not always operationally feasible, isolating operations can also reduce exposure. Using respiratory protection is less effective than the controls mentioned above, but is still advisable whenever working with or around furan. No exposure levels have been established for this chemical. However, this does not mean that exposure is without risk. For the best protection, a supplied-air respirator or a self-contained breathing apparatus (SCBA) with full facepiece operated in continuos flow or positive pressure mode are recommended. If a full facepiece is not available, then chemical goggles should be worn to protect the eyes. A face shield and a rubber apron should also be worn. To prevent hand and skin exposures, impervious gloves should be worn. Glove manufacturers should be contact and permeation studies obtained *before* glove selections have been made.

Administrative controls should also be in place to minimize the potential for human exposures. These may include written procedures or policies which specify the methods and techniques that will be practiced whenever personnel are to work with furan.

All personnel should receive *training* on the use, hazards, protective measures, emergency actions, and other precautions per 29 CFR 1910.1200 (Hazard Communication), prior to the first assignment in an area where furan is used or stored.

Before beginning employment, and at regular intervals thereafter (e.g., annually), the following medical tests are recommended for personnel working with or around furan:

☑ Liver function tests.

☑ Checks for blood pressure and pulse rate changes.

If symptoms develop or overexposure is suspected, the following additional tests are recommended:

☑ Lung function tests.

☑ Consider chest X-ray after acute overexposure (may be negative if taken immediately after exposure due to the delayed onset of pulmonary edema).

Any evaluation should include a careful history of past and present symptoms with an examination. Medical tests that simply look for existing damage are not a substitute for controlling exposures. Also, because smoking can cause heart disease, emphysema, and other respiratory dysfunction, smokers may develop exposure symptoms more quickly than non-smokers under the same conditions of exposure. Prudent risk management requires careful consideration of *all* possible exposure factors when assessing health exposure risk.

Other methods to reduce exposure include:

☑ Where possible, enclose operations and use local exhaust ventilation at the site of chemical release. If local exhaust ventilation or enclosure is not used, respiratory protection should be mandatory.

☑ Always ensure that proper protective clothing is worn when using chemical substances.

☑ Wash thoroughly immediately after exposure to furan and at the end of the work shift or before eating, drinking, or smoking. Avoid contact with direct sunlight.

☑ Hazard warning information should be posted in the work area. In addition, as part of an on-going education and training program, all information on the health and safety hazards of furan should be communicated to all potentially exposed workers.

☑ Eye wash stations should be provided in the immediate work area for emergency use. If there is a possibility of skin exposure to furan, emergency shower facilities should also be provided.

RISK ASSESSMENT: ENVIRONMENT

General Assessment

The environment is at risk of exposure during transportation, storage, disposal, or destruction of furan. In almost every scenario, the threat of environmental exposure is contingent upon the proper handling of the chemical substance. Accidental spills, large or small, can result in fire, explosion (depending upon conditions of spill), and possible contamination of the surrounding environmental mediums (water, soil, and air).

Furan is considered a Class 1A flammable liquid (per OSHA 29 CFR 1910.106). Its extremely low flash point and low boiling point present a dangerous fire and explosion risk. Also, because it is incompatible with many common oxidizers and acids, extreme caution is required in handling, storage, transportation, and disposal of furan. If it becomes unstable, it may form shock-sensitive and explosive peroxides in air. These characteristics also require special consideration during any emergency situation involving a leak or spill of furan crystals. Should furan ever come into contact with incompatible substances either during use, transportation, storage, or disposal, the formation of highly toxic and/or highly explosive commodities is extremely possible.

Furan can enter the environment from non-point sources, as well as industrial and municipal discharges, and from accidental spills.

Acute Ecological Effects

Acute (short-term) toxic effects may include the death of animals, birds, or fish and death or low growth rate in plants. Acute effects are seen 2 to 4 days after animals or plants are exposed to furan.

Insufficient data are available to evaluate or predict the short-term effects of furan to aquatic life, plants, birds, or land animals.

Chronic Ecological Effects

Chronic (long-term) toxic effects may include shortened life span, reproductive problems, lower fertility, and changes in appearance or behavior in exposed animals. These effects can be seen long after first exposure(s) to toxic chemicals. Insufficient data are available to evaluate or predict the long-term effects of furan to aquatic life, plants, birds, or land animals.

Water Solubility

Furan is not soluble in water. Concentrations of 1 milligram or less will mix with a liter of water.

Persistence in the Environment

Furan is moderately persistent in water, with a half-life between 20 and 200 days. The half-life of a pollutant is the amount of time it takes for one half of the chemical to be degraded.

Bioaccumulation in Aquatic Organisms

Some substances increase in concentration, or *bioaccumulate,* in living organisms as they breathe contaminated air, drink contaminated water, or eat contaminated food. These chemicals can become concentrated in the tissues and internal organs of animals as well as humans.

The concentration of furan found in fish tissues is expected to be higher than the average concentration of furan in the water from which the fish was taken.

Recommended Risk-Reduction Measures

Proper training of all transporters will reduce the likelihood of a mishap or accident resulting in a leak or spill of furan into the environment. The correct use of DOT labeling (per the MSDS) on all containers, trucks, and rail cars will enable emergency responders to react properly and quickly to any disaster thereby reducing the potential risk to the environment and to personnel.

Storage of furan should be segregated from incompatible chemicals to minimize the risk of cross-contamination or contact. To prevent static sparks, all containers and equipment used in shipping, receiving, or transferring operations, should be grounded and bonded. Storage facilities should be designed with proper fire protection and prevention.

If a spill or leak to the environment has occurred, fire department, emergency response and/or hazardous materials spill personnel should be notified immediately. Cleanup should be attempted only by those trained in proper spill containment procedures. Contaminated soils should be removed for incineration and replaced with clean soil. If furan should contact the water table, aquifer, or navigable waterway, the local and/or state emergency response authorities must be notified. A comprehensive emergency response or disaster preparedness/recovery plan should be in place prior to any operations involving the use, transportation, storage, or disposal of furan. If furan is spilled or leaked, the following specific steps are recommended:

- ☑ Restrict persons not wearing protective clothing from area of spill or leak until cleanup is complete and area can be opened for normal work.
- ☑ Remove all ignition sources.
- ☑ Ventilate area of spill or leak.
- ☑ Absorb liquids in vermiculite or other similar material and place in sealed drums for disposal. Use non-sparking tools and avoid skin contact.
- ☑ It may be necessary to dispose of furan as a hazardous waste. The responsible state agency or the regional office of the federal Environmental Protection Agency (EPA) should be contacted for specific recommendations.

RISK ASSESSMENT: BUSINESS

General Assessment

Accidents or mishaps involving furan can present a significant threat to business operations. The loss or damage of equipment or facilities can significantly affect fiscal viability. Lawsuits that may result from personnel injury/death, public exposures, and/or environmental contamination will also require a serious expenditure of resources.

Media attention surrounding an injury, death, or environmental damage can also result in a loss of profits and loss of current as well as future business.

ʊ Recommended Risk-Reduction Measures

Company attorneys, safety and health professionals, and environmental specialists should be involved in the development of any procedures dealing with safety or emergency response. A company official should be pre-designated as a public relations officer with specific training in dealing with the press. Corporate plans and policies should be developed, approved, and implemented long before any need for such arises.

REFERENCES

Ahlbom, A. 1993. *Biostatistics for Epidemiologists.* New York: CRC Press/Lewis Publishers.

American Conference of Governmental Industrial Hygienists. 1988. *Documentation of the Threshold Limit Values and Biological Exposure Indices,* 5th Edition (with updates). Cincinnati: ACGIH

Asante-Duah, D. K. 1993. *Hazardous Waste Risk Assessment.* New York: CRC Press/Lewis Publishers.

Bloom, A. D. and F. J. DeSerres. 1995. *Ecotoxicity and Human Health.* New York: CRC Press/Lewis Publishers.

Brusick, D. J. 1994. *Methods for Genetic Risk Assessment.* New York: CRC Press/Lewis Publishers.

Bueche, J. R. 1972. *Principles of Physics,* 2nd Edition. New York: McGraw-Hill.

Burrell, R., D. K. Flaherty, and L. J. Sauers. 1992. *Toxicology of the Immune System.* New York: Van Nostrand Reinhold.

Calabrese, E. J. 1994. *Biological Effects of Low Level Exposures.* New York: CRC Press/Lewis Publishers.

Cockerman, L. G., B. S. Shane. 1994. *Basic Environmental Toxicology.* New York: CRC Press/Lewis Publishers.

Cothern, C. R., N. P. Ross. 1994. *Environmental Statistics, Assessment, and Forecasting.* New York: CRC Press/Lewis Publishers.

Danse, I. R. 1991. *Common Sense Toxics in the Workplace.* New York: Van Nostrand Reinhold.

Fodden, C. R., and J. L. Weddell. 1992. *Hazardous Materials Emergency Action Data.* New York: CRC Press/Lewis Publishers.

Forseberg, K., and L. H. Keith. 1993. *Chemical; Protective Clothing Permeation and Degradation Database.* Chelsea, MI: Lewis Publishers.

Gardner, R. W. 1994. *Chemical Intolerance.* New York: CRC Press/Lewis Publishers.

Gerarde, H. W. 1960. *Toxicology and Biochemistry of Aromatic Hydrocarbons.* New York: Elsevier.

Hallenbeck, W. H. 1993. *Quantitative Risk Assessment for Environmental and Occupational Health,* 2nd Edition. New York: CRC Press/Lewis Publishers.

Hodgson, E., R. B. Mailman, and Janice E. Chambers. 1988. *Dictionary of Toxicology.* London: The MacMillian Company.

Howard, P. H. 1995. *Handbook of Environmental Fate and Exposures Data for Organic Chemicals,* Vol. I-IV. New York: CRC Press/Lewis Publishers.

Jones, F. E. 1994. *Toxic Organic Vapors in the Workplace.* New York: CRC Press/Lewis Publishers.

Keith, L. H. and M. M. Walker. 1995. *Handbook of Air Toxics.* New York: CRC Press/Lewis Publishers.

Landis, W. G., M. H. Yu. 1995. *Introduction to Environmental Toxicology.* New York: CRC Press/Lewis Publishers.

Langerman, N. 1994 *Precautionary Labels for Chemical Containers.* New York: CRC Press/Lewis Publishers.

Lauwerys, R. R., P. Hoet. 1993. *Industrial Chemical Exposure.* New York: CRC Press/Lewis Publishers.

Lewis, R. J., Sr. 1993. *Hawley's Condensed Chemical Dictionary,* 12th Edition. New York: Van Nostrand Reinhold.

Lewis, R. J., Sr. 1992. *Sax's Dangerous Properties of Industrial Materials,* Volumes 1, 2, and 3, 8th Edition. New York: Van Nostrand Reinhold.

Manahan, S. E. 1994. *Environmental Chemistry,* 6th Edition. New York: CRC Press/Lewis Publishers.

Mellström, G. A., J. E. Wahlberg, and H. I. Maibach. 1994. *Protective Gloves for Occupational Use.* New York: CRC Press/Lewis Publishers.

Meyer, E. 1990. *Chemistry of Hazardous Materials.* Englewood Cliffs, NJ: Prentice Hall.

Mycock, J. C., J. D. McKenna, and L. Theodore. 1995. *Handbook of Air Pollution Control Engineering and Technology.* New York: CRC Press/Lewis Publishers.

Ott, W. R. 1995. *Environmental Statistics and Data Analysis.* New York: CRC Press/Lewis Publishers.

Patnaik, P. 1992. *A Comprehensive Guide to the Hazardous Properties of Chemical Substances.* New York: Van Nostrand Reinhold.

Philp, R. B. 1995. *Environmental Hazards and Human Health.* New York: CRC Press/Lewis Publishers.

Proctor, N. H., J. P. Hughes. 1991. *Chemical Hazards of the Workplace.* New York: Van Nostrand Reinhold.

Quigley, D. R. 1994. *Handbook of Emergency Chemical Management.* New York: CRC Press/Lewis Publishers.

Rea, W. J. 1995. *Chemical Sensitivity,* Vol. I-III. New York: CRC Press/Lewis Publishers.

Richardson, M. 1995. *Dictionary of Substance and Their Effects,* Vol. I-VI. New York: CRC Press/Lewis Publishers.

Sacarello, H. L. A. 1994. *The Comprehensive Handbook of Hazardous Materials.* New York: CRC Press/Lewis Publishers.

Sheftel, V. O. 1995. *Handbook of Toxic Properties of Monomers and Additives.* New York: CRC Press/Lewis Publishers.

Talbott, E., G. Craun. 1995. *An Introduction to Environmental Epidemiology.* New York: CRC Press/Lewis Publishers.

Torkelson, T. R. V. K. Rowe. 1981. *Patty's Industrial Hygiene and Toxicology,* Vol. 2B. New York: Wiley-Interscience.

MATERIAL SAFETY DATA SHEET

CHEMICAL NAME	
FURFURAL	**CRC PRESS/LEWIS PUBLISHERS DISCLAIMER:** THE INFORMATION AND RECOMMENDATIONS PRESENTED HEREIN ARE BASED ON SOURCES BELIEVED TO BE RELIABLE. CRC MAKES NO REPRESENTATION ON ITS COMPLETENESS OR ACCURACY. IT IS THE RESPONSIBILITY OF THE USER TO DETERMINE THE CHEMICAL'S SUITABILITY FOR ITS INTENDED USE, THE CHEMICAL'S SAFE USE, AND THE CHEMICAL'S PROPER DISPOSAL. NO REPRESENTATIONS AND/OR WARRANTIES, EITHER EXPRESSED OR IMPLIED, OF THE MERCHANTABILITY OR FITNESS FOR A PARTICULAR PURPOSE, OR OF ANY OTHER NATURE, ARE MADE WITH RESPECT TO THE INFORMATION PROVIDED IN THIS MSDS OR TO THE CHEMICAL TO WHICH INFORMATION MAY REFER. CRC NEITHER ASSUMES NOR AUTHORIZES ANY OTHER PERSON TO ASSUME FOR IT, ANY OTHER ADDITIONAL RESPONSIBILITY OR LIABILITY FOR THE USE OF, OR RELIANCE UPON, THIS INFORMATION.

NATIONAL FIRE PROTECTION ASSOCIATION (NFPA) LABELING INFORMATION

HEALTH	FIRE	REACTIVE	OTHER	DEGREE OF HAZARD	COLOR CODING	OTHER CODES
2	2	0		0 = Minimum Hazard 1 = Slight Hazard 2 = Moderate Hazard 3 = Serious Hazard 4 = Severe Hazard	HEALTH = BLUE FIRE = RED REACTIVITY = YELLOW OTHER = WHITE	OX = Oxidizer ACID = Acid ALK = Alkali COR = Corrosive W = Use No Water

SECTION I - GENERAL INFORMATION

Characterization **Aldehyde**	RCRA Number **U125**	EPA Class **Toxic Waste**
DOT Proper Shipping Name **Furfural**	Chemical Abstract Service (CAS) Number **98-01-1**	
DOT Hazard Class and Label Requirements **Combustible Liquid**	DOT Emergency Guide Code **29**	
DOT Identification Number **UN 1199**	Chemical Formula $C_5H_4O_2$	

Synonyms
Fural; 2-furancarboxaldehyde; furfuraldehyde; 2-furfuraldehyde; furfurol; artificial ant oil.

SECTION II - HAZARDOUS INGREDIENTS AND IDENTITY INFORMATION

Hazardous Components (specific identity)	OSHA Exposure Criteria	NIOSH Exposure Criteria	Immediately Dangerous to Life and Health (IDLH)	ACGIH Exposure Criteria
Furfural (derivation: From oat hulls, rice hulls, corn cobs, bagasse, and other cellulosic waste minerals by steam-acid digestion). **1 ppm = 3.99 mg/m^3**	**PEL (skin):** **5 ppm** **20 mg/m^3** **STEL:** **Not Established**	**REL (skin):** **Not Established** **(2 ppm proposed)** **STEL:** **Not Established**	**100 ppm**	**TLV (skin):** **2 ppm** **8 mg/m^3** **STEL:** **Not Established**

SECTION III - PHYSICAL AND CHEMICAL CHARACTERISTICS

Boiling Point **322°F (162°C)**	Specific Gravity (H_2O = 1) **1.16**
Vapor Pressure (mm Hg) **2 at 68°F (20°C)**	Molecular Weight **96.1**
Vapor Density (Air = 1) **3.3**	Melting Point **-34°F (-36°C)**

Solubility
Slightly soluble (8%) in water. Slightly soluble in alcohol, ether, and benzene.

Appearance and Odor
Colorless, oily liquid in pure state. Becomes reddish-brown on exposure to air and light. Odor is similar to benzaldehyde (almond-like). Odor Threshold = .25 ppm.

SECTION IV - FIRE AND EXPLOSION HAZARD DATA

Flash Point (method used) **140°F (60°C) closed cup**	Explosive Limits in Air % by Volume **LEL: 2.1% UEL: 19.3%**
NFPA Classification **Class 111A Flammable Liquid**	Autoignition Temperature **600°F (316°C)**

Extinguishing Media
Dry chemical, alcohol foam, carbon dioxide, or water spray. Use water spray to cool containers.

Special Fire Fighting Procedures
Wear full protective clothing and self-contained breathing apparatus (SCBA). Poisonous gases are produced in fire. Move containers from fire area if it can be done without risk.

Unusual Fire and Explosion Hazards
Violent polymerization occurs at high temperatures. Moderately explosive when exposed to flame. Vapors are heavier than air and may travel for some distance to an ignition source to flashback.

SECTION V - REACTIVITY DATA

Stability		Conditions to Avoid: Furfural is normally not stable. It can be fairly stable if kept pure and container in tightly sealed containers and protected from exposure to light or high temperatures. Avoid contact with incompatible materials.
Stable	Unstable: X	Incompatibility *(materials to avoid)*: Incompatible with strong oxidizers (such as chlorine, bromine, and fluorine), and strong mineral acids, and alkalis. At room temperature, reaction with atmospheric oxygen is likely.
Hazardous Polymerization		Conditions to Avoid: Explosive and violent exothermic polymerization occurs in contact with strong mineral acids, alkalis, or at elevated temperatures. Fire-exposed containers must be kept cool.
May Occur: X	Will Not Occur	Hazardous Decomposition or By-products: When heated to decomposition, furfural can emit acrid, irritating oxides of carbon.

SECTION VI - HEALTH HAZARD DATA

Primary Route(s) of Entry:	Inhalation?	Absorption (skin/eye)?	Ingestion?
	X	X	X

Health Hazards

INHALATION: Causes headache, eye irritation, and irritation of the upper respiratory system. There could be numbness of the tongue and mucous membranes of the mouth, absence of taste, itching throat, difficulty breathing, pulmonary edema, and eye irritation.

ABSORPTION: Skin contact causes dryness, irritation, possible skin staining (yellow-brown), and may cause dermatitis. Absorption is likely. Eye contact causes tearing and irritation.

INGESTION: Headache and possible cirrhosis of the liver (positive in rats).

Carcinogenicity	NTP Listed?	IARC Cancer Review Group?	OSHA Regulated?	Target Organs?
Unknown Human Unknown Animal	No	No	29 CFR 1910.1000 Table Z-1	Respiratory system, skin, CNS, eyes

Medical Conditions Generally Aggravated by Exposure

Chronic pulmonary disease and skin diseases may be aggravated by exposure to furfural.

Emergency and First-aid Procedures

Eye contact: Do not allow victim to keep eyes tightly shut. Flush immediately with water for 15 minutes (minimum), seek medical attention. Skin contact: Remove all contaminated clothing. Immediately wash area with large amounts of soap and water. For inhalation: Remove the person from exposure. Provide respiratory assistance and CPR. Transfer to medical facility. Observe for 24-48 hours for lung effects (pulmonary edema). If swallowed: Seek medical attention immediately. Give 1-2 glasses of water or milk and induce vomiting. Never give anything by mouth to an unconscious or convulsing person.

SECTION VII - PRECAUTIONS FOR SAFE HANDLING AND USE

Steps to be Taken in Case Material is Released or Spilled

Disperse unignited vapors with water spray. Absorb spills in vermiculite or other material and deposit in sealed containers. Ventilate area of spill or leak. Remove all ignition sources. Restrict those not involved in cleanup from entering area. Notify appropriate authorities, as required.

Preferred Waste Disposal Method

No citation.

Precautions to be Taken in Handling and Storage

Store to avoid contact with strong oxidizers, since violent reactions can occur. Store in tightly closed containers in cool, well-ventilated area away from heat. Electrically ground and bond all equipment.

Other Precautions and Warnings

Sources of ignition (heat, flame, cigarettes, incompatibles) are prohibited where furfural is used, handled, or stored. Storage under an inert gas is recommended. Protect containers from physical damage.

SECTION VIII - CONTROL MEASURES AND PERSONAL PROTECTIVE EQUIPMENT

Respiratory Protection *(specify type)*

Exposure limits are very low and difficult to monitor. An air purifying respirator may suffice. For best protection, use a supplied-air respirator with full facepiece operated in continuous flow mode, or use a self-contained breathing apparatus (SCBA) with full facepiece operated in positive pressure mode.

Ventilation

Local exhaust (preferred) at site of chemical work or general ventilation.

Protective Gloves	Eye Protection	Other Protective Clothing
Impervious Rubber Gloves	Chemical Goggles or Face Mask	Boots; Aprons; Gauntlets

Work/Hygiene Practices

Always wash hands thoroughly after using chemical; never bring food, drink, or smoking materials into vicinity of chemicals.

FURFURAL

$C_5H_4O_2$

CAS: 98-01-1

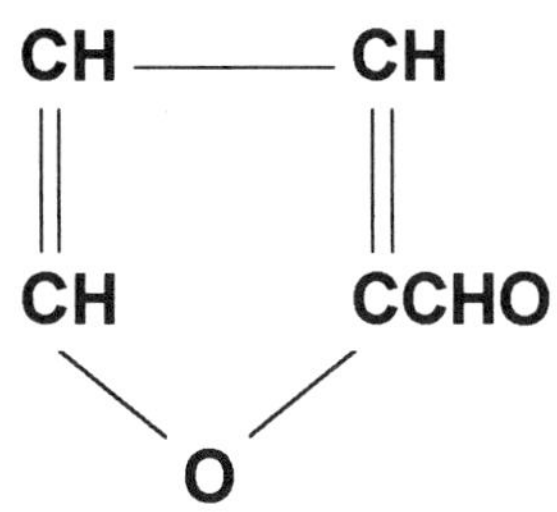

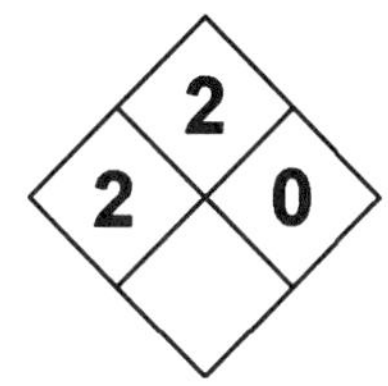

IDENTIFICATION AND TYPICAL USES

Furfural is a clear, colorless, oily liquid that turns reddish brown on exposure to air and light. It has an almond-like odor similar to benzaldehyde. It is used in solvent refining of lubricating oils, butadiene, rosin, and other organic materials, as a solvent for nitrocellulose, cellulose acetate, shoe dyes, as an intermediate for tertahydrofuran and furfuryl alcohol; in phenolic and furan polymers, as a wetting agent in the manufacture of abrasive wheels and brake linings, as a weed killer, a fungicide, in road construction materials, in the production of lysine, in the refining of rare earth and metals, in flavoring, and as an analytical reagent.

RISK ASSESSMENT: HEALTH

General Assessment

Furfural is a human poison by *ingestion* and is considered moderately toxic by *inhalation* and skin contact (*absorption*). While its carcinogenic and teratogenic properties have not been adequately investigated, mutation data have been reported.

Inhalation of furfural can cause dangerous irritation of the mucous membranes of the respiratory system. It can cause numbness of the tongue and loss of the sense of taste. There may be headache, itching throat, eye irritation, difficulty in breathing, and a possibility of pulmonary edema (fluid in the lungs), which is a medical emergency and can lead to death. Symptoms of pulmonary edema can be delayed up to 48 hours creating a false sense of security with regard to health exposure risk. Furan is also a central nervous system (CNS) depressant that may cause respiratory arrest and death.

Skin contact with the liquid or vapor can cause irritation and redness. Repeated contact may cause skin discoloration to a yellow-brown. It can absorb quickly through the skin to further compound the effects of exposure by other routes (e.g., inhalation). Eye contact causes burning and watering of eyes and may result in damage. If ingested, there can be severe headache and unspecified gastrointestinal effects.

Acute Health Effects

The following acute (short-term) health effects may occur immediately or shortly after exposure to furfural:

Skin: Irritation with itching, redness, and possible staining. Absorption is likely.

Eye: Severe irritation of the conjunctiva with burning, itching, and watering.

Lung: Irritation of the mucous membrane of the respiratory system. May cause pulmonary edema (fluid in the lungs) at high exposures, which can bc fatal.

CNS: Narcotic effects at high concentrations with dizziness, fatigue, headache, weakness, and other symptoms of CNS depression.

Chronic Health Effects

The following chronic (long-term) health effects may occur at some time after exposure to furfural and can last for months or even years.

Cancer Hazards: Furfural may cause mutations (genetic changes) in living cells. The data on its carcinogenicity are inconclusive.

Reproductive Hazard: According to the references, furfural has not been adequately tested for its ability to adversely affect reproduction.

Other Chronic Effects: Long-term exposure causes eczema, allergic sensitization, and photosensitization in some individuals. Once allergy develops, even the slightest future exposures can cause recurrence of symptoms.

✪ *Recommended Risk-Reduction Measures*

Personnel should avoid direct contact with furfural. If a less toxic chemical cannot be substituted for a hazardous substance, then *engineering controls* are the most effective method of reducing exposures. The best protection is to enclose operations and/or provide local exhaust ventilation at the site of chemical release. While not always operationally feasible, isolating operations can also reduce exposure. Using respiratory protection is less effective than the controls mentioned above, but is still advisable whenever working with or around furfural. The exposure level for this chemical is relatively low (5 ppm) and may be difficult to monitor. For the best protection, a supplied-air respirator or a self-contained breathing apparatus (SCBA) with full facepiece operated in continuous flow or positive pressure mode are recommended. If a full facepiece is not available, then chemical goggles should be worn to protect the eyes. A face shield and a rubber apron should also be worn. To prevent hand and skin exposures, impervious gloves should be worn. Glove manufacturers should be contacted and permeation studies obtained *before* glove selections have been made.

Administrative controls should also be in place to minimize the potential for human exposures. These may include written procedures or policies which specify the methods and techniques that will be practiced whenever personnel are to work with furfural.

All personnel should receive *training* on the use, hazards, protective measures, emergency actions, and other precautions per 29 CFR 1910.1200 (Hazard Communication), prior to the first assignment in an area where furfural is used or stored.

Before beginning employment, and at regular intervals thereafter (e.g., annually), the following medical tests are recommended for personnel working with or around furfural:

☑ Liver function tests.

☑ Checks for blood pressure and pulse rate changes.

If symptoms develop or overexposure is suspected, the following additional tests are recommended:

☑ Lung function tests.

☑ Consider chest X-ray after acute overexposure (may be negative if taken immediately after exposure due to the delayed pulmonary edema).

Any evaluation should include a careful history of past and present symptoms with an examination. Medical tests that simply look for existing damage are not a substitute for controlling exposures. Also, because smoking can cause heart disease, emphysema, and other respiratory dysfunction, smokers may develop exposure symptoms more quickly than non-smokers under the same conditions of exposure. Prudent risk management requires careful consideration of *all* possible exposure factors when assessing health exposure risk in the work place.

Other methods to reduce exposure include:

☑ Where possible, enclose operations and use local exhaust ventilation at the site of chemical release. If local exhaust ventilation or enclosure is not used, respiratory protection should be mandatory.

☑ Always ensure that proper protective clothing is worn when using chemical substances.

☑ Wash thoroughly immediately after exposure to furfural and at the end of the work shift or before eating, drinking, or smoking.

☑ Hazard warning information should be posted in the work area. In addition, as part of an on-going education and training program, all information on the health and safety hazards of furfural should be communicated to all potentially exposed workers.

☑ Eye wash stations should be provided in the immediate work area for emergency use. If there is a possibility of skin exposure to furfural, emergency shower facilities should also be provided.

RISK ASSESSMENT: ENVIRONMENT

General Assessment

The environment is at risk of exposure during transportation, storage, disposal, or destruction of furfural. In almost every scenario, the threat of environmental exposure is contingent upon the proper handling of the chemical substance. Accidental spills, large or small, can result in fire, explosion (depending upon conditions of spill), and possible contamination of the surrounding environmental mediums (water, soil, and air).

Furfural is considered a Class 111A flammable liquid (per OSHA 29 CFR 1910.106). Its moderately low flash point presents a fire and explosion hazard risk. Also, because it is incompatible with many common oxidizers and acids, extreme caution is required in handling, storage, transportation, and disposal of furfural. It is unstable in the presence of light, air, heat, and can polymerize violently. These characteristics also require special consideration during any emergency situation involving a leak or spill of furfural crystals. Should furfural ever come into contact with incompatible substances either during use, transportation, storage, or disposal, the formation of highly toxic and/or highly explosive commodities is highly possible.

Furfural can enter the environment from non-point sources, as well as industrial and municipal discharges, and from accidental spills.

Acute Ecological Effects

Acute (short-term) toxic effects may include the death of animals, birds, or fish and death or low growth rate in plants. Acute effects are seen 2 to 4 days after animals or plants are exposed to furfural.

Insufficient data are available to evaluate or predict the short-term effects of furfural to aquatic life, plants, birds, or land animals.

Chronic Ecological Effects

Chronic (long-term) toxic effects may include shortened life span, reproductive problems, lower fertility, and changes in appearance or behavior in exposed animals. These effects can be seen long after first exposure(s) to toxic chemicals.

Insufficient data are available to evaluate or predict the long-term effects of furfural to aquatic life, plants, birds, or land animals.

Water Solubility

Furfural is slightly soluble in water. Concentrations of 1 to 100 milligrams will mix with a liter of water.

Persistence in the Environment

Furfural is moderately persistent in water, with a half-life between 20 and 200 days. The half-life of a pollutant is the amount of time it takes for one half of the chemical to be degraded.

Bioaccumulation in Aquatic Organisms

Some substances increase in concentration, or *bioaccumulate,* in living organisms as they breathe contaminated air, drink contaminated water, or eat contaminated food. These chemicals can become concentrated in the tissues and internal organs of animals as well as humans who have consumed contaminated animal tissues.

The concentration of furfural found in fish tissues is expected to be higher than the average concentration of furfural in the water from which the fish was taken.

Recommended Risk-Reduction Measures

Proper training of all transporters will reduce the likelihood of a mishap or accident resulting in a leak or spill of furfural into the environment. The correct use of DOT labeling (per the MSDS) on all containers, trucks, and rail cars will enable emergency responders to react properly and quickly to any disaster thereby reducing the potential risk to the environment and to personnel.

Storage of furfural should be segregated from incompatible chemicals to minimize the risk of cross-contamination or contact. To prevent static sparks, all containers and equipment used in shipping, receiving, or transferring, should be grounded and bonded.

If a spill or leak to the environment has occurred, fire department, emergency response and/or hazardous materials spill personnel should be notified immediately. Cleanup should be attempted only by those trained in proper spill containment procedures. Contaminated soils should be removed for incineration and replaced with clean soil. If furfural should contact the water table, aquifer, or navigable waterway, the local and/or state emergency response authorities must be notified. A comprehensive emergency response or disaster preparedness/recovery plan should be in place prior to any operations involving the use, transportation, storage, or disposal of furfural. If furfural is spilled or leaked, the following specific steps are recommended:

- ☑ Restrict persons not wearing protective clothing from area of spill or leak until cleanup is complete and area can be opened for normal work.
- ☑ Remove all ignition sources.
- ☑ Ventilate area of spill or leak.
- ☑ Absorb liquids in vermiculite or other similar material and place in sealed drums for disposal. Use non-sparking tools and avoid skin contact.
- ☑ It may be necessary to dispose of furfural as a hazardous waste. The responsible state agency or the regional office of the federal Environmental Protection Agency (EPA) should be contacted for specific recommendations.

RISK ASSESSMENT: BUSINESS

General Assessment

Accidents or mishaps involving furfural can present a significant threat to business operations. The loss or damage of equipment or facilities can significantly affect fiscal viability. Lawsuits that may result from personnel injury/death, public exposures, and/or environmental contamination will also require a serious expenditure of resources. Media attention surrounding an injury, death, or environmental damage can also result in a loss of profits and loss of current as well as future business.

Recommended Risk-Reduction Measures

Company attorneys, safety and health professionals, and environmental specialists should be involved in the development of any procedures dealing with safety or emergency response. A company official should be pre-designated as a public relations officer with specific training in dealing with the press. Corporate plans and policies should be developed, approved, and implemented long before any need for such arises.

REFERENCES

American Conference of Governmental Industrial Hygienists. 1988. *Documentation of the Threshold Limit Values and Biological Exposure Indices*, 5th Edition (with updates). Cincinnati: ACGIH.

Asante-Duah, D. K. 1993. *Hazardous Waste Risk Assessment.* New York: CRC Press/Lewis Publishers.

Bloom, A. D. and F. J. DeSerres. 1995. *Ecotoxicity and Human Health.* New York: CRC Press/Lewis Publishers.

Brusick, D. J. 1994. *Methods for Genetic Risk Assessment.* New York: CRC Press/Lewis Publishers.

Bueche, J. R. 1972. *Principles of Physics*, 2nd Edition. New York: McGraw-Hill.

Burrell, R., D. K. Flaherty, and L. J. Sauers. 1992. *Toxicology of the Immune System.* New York: Van Nostrand Reinhold.

Calabrese, E. J. 1994. *Biological Effects of Low Level Exposures.* New York: CRC Press/Lewis Publishers.

Cockerman, L. G., B. S. Shane. 1994. *Basic Environmental Toxicology.* New York: CRC Press/Lewis Publishers.

Cothern, R. R. 1993. *Comparative Environmental Risk Assessment.* New York: CRC Press/Lewis Publishers.

Danse, I. R. 1991. *Common Sense Toxics in the Workplace.* New York: Van Nostrand Reinhold.

Fodden, C. R., and J. L. Weddell. 1992. *Hazardous Materials Emergency Action Data.* New York: CRC Press/Lewis Publishers.

Forseberg, K., and L. H. Keith. 1993. *Chemical; Protective Clothing Permeation and Degradation Database.* Chelsea, MI: Lewis Publishers.

Gardner, R. W. 1994. *Chemical Intolerance.* New York: CRC Press/Lewis Publishers.

Gerarde, H. W. 1960. *Toxicology and Biochemistry of Aromatic Hydrocarbons.* New York: Elsevier.

Hodgson, E., R. B. Mailman, and Janice E. Chambers. 1988. *Dictionary of Toxicology.* London: The MacMillian Company.

Howard, P. H. 1995. *Handbook of Environmental Fate and Exposures Data for Organic Chemicals*, Vol. I-IV. New York: CRC Press/Lewis Publishers.

Jones, F. E. 1994. *Toxic Organic Vapors in the Workplace.* New York: CRC Press/Lewis Publishers.

Keith, L. H. and M. M. Walker. 1995. *Handbook of Air Toxics.* New York: CRC Press/Lewis Publishers.

Langerman, N. 1994 *Precautionary Labels for Chemical Containers.* New York: CRC Press/Lewis Publishers.

Lauwerys, R. R., P. Hoet. 1993. *Industrial Chemical Exposure.* New York: CRC Press/Lewis Publishers.

Lewis, R. J., Sr. 1993. *Hawley's Condensed Chemical Dictionary*, 12th Edition. New York: Van Nostrand Reinhold.

Lewis, R. J., Sr. 1992. *Sax's Dangerous Properties of Industrial Materials*, Volumes 1, 2, and 3, 8th Edition. New York: Van Nostrand Reinhold.

Manahan, S. E. 1992. *Toxicological Chemistry*, 2nd Edition. New York: CRC Press/Lewis Publishers.

Meyer, E. 1990. *Chemistry of Hazardous Materials.* Englewood Cliffs, NJ: Prentice Hall.

Mycock, J. C., J. D. McKenna, and L. Theodore. 1995. *Handbook of Air Pollution Control Engineering and Technology.* New York: CRC Press/Lewis Publishers.

Patnaik, P. 1992. *A Comprehensive Guide to the Hazardous Properties of Chemical Substances.* New York: Van Nostrand Reinhold.

Quigley, D. R. 1994. *Handbook of Emergency Chemical Management.* New York: CRC Press/Lewis Publishers.

Rea, W. J. 1995. *Chemical Sensitivity*, Vol. I-III. New York: CRC Press/Lewis Publishers.

Richardson, M. 1995. *Dictionary of Substance and Their Effects*, Vol. I-VI. New York: CRC Press/Lewis Publishers.

Sacarello, H. L. A. 1994. *The Comprehensive Handbook of Hazardous Materials.* New York: CRC Press/Lewis Publishers.

Sheftel, V. O. 1995. *Handbook of Toxic Properties of Monomers and Additives.* New York: CRC Press/Lewis Publishers.

Talbott, E., G. Craun. 1995. *An Introduction to Environmental Epidemiology.* New York: CRC Press/Lewis Publishers.

Torkelson, T. R. V. K. Rowe. 1981. *Patty's Industrial Hygiene and Toxicology*, Vol. 2B. New York: Wiley-Interscience.

Witorsch, P. and S. V. Spagnolo. 1994. *Air Pollution and Lung Disease in Adults.* New York: CRC Press/Lewis Publishers.

MATERIAL SAFETY DATA SHEET

CHEMICAL NAME

FURFURYL ALCOHOL

CRC PRESS/LEWIS PUBLISHERS DISCLAIMER: THE INFORMATION AND RECOMMENDATIONS PRESENTED HEREIN ARE BASED ON SOURCES BELIEVED TO BE RELIABLE. CRC MAKES NO REPRESENTATION ON ITS COMPLETENESS OR ACCURACY. IT IS THE RESPONSIBILITY OF THE USER TO DETERMINE THE CHEMICAL'S SUITABILITY FOR ITS INTENDED USE, THE CHEMICAL'S SAFE USE, AND THE CHEMICAL'S PROPER DISPOSAL. NO REPRESENTATIONS AND/OR WARRANTIES, EITHER EXPRESSED OR IMPLIED, OF THE MERCHANTABILITY OR FITNESS FOR A PARTICULAR PURPOSE, OR OF ANY OTHER NATURE, ARE MADE WITH RESPECT TO THE INFORMATION PROVIDED IN THIS MSDS OR TO THE CHEMICAL TO WHICH INFORMATION MAY REFER. CRC NEITHER ASSUMES NOR AUTHORIZES ANY OTHER PERSON TO ASSUME FOR IT, ANY OTHER ADDITIONAL RESPONSIBILITY OR LIABILITY FOR THE USE OF, OR RELIANCE UPON, THIS INFORMATION.

NATIONAL FIRE PROTECTION ASSOCIATION (NFPA) LABELING INFORMATION

HEALTH	FIRE	REACTIVE	OTHER	DEGREE OF HAZARD	COLOR CODING	OTHER CODES
1	2	1		0 = Minimum Hazard 1 = Slight Hazard 2 = Moderate Hazard 3 = Serious Hazard 4 = Severe Hazard	HEALTH = BLUE FIRE = RED REACTIVITY = YELLOW OTHER = WHITE	OX = Oxidizer ACID = Acid ALK = Alkali COR = Corrosive W = Use No Water

SECTION I - GENERAL INFORMATION

Characterization **Alcohol**	RCRA Number **None**	EPA Class **Not Applicable**
DOT Proper Shipping Name **Furfuryl alcohol**	Chemical Abstract Service (CAS) Number **98-00-0**	
DOT Hazard Class and Label Requirements **Poison B**	DOT Emergency Guide Code **55**	
DOT Identification Number **UN 2874**	Chemical Formula **$C_5H_6O_2$**	

Synonyms
2-Furymethanol; 2-hydroxymethylfuran; furfural alcohol; 2-furylcarbinol.

SECTION II - HAZARDOUS INGREDIENTS AND IDENTITY INFORMATION

Hazardous Components (specific identity)	OSHA Exposure Criteria	NIOSH Exposure Criteria	Immediately Dangerous to Life and Health (IDLH)	ACGIH Exposure Criteria
Furfuryl alcohol (derivation: Continuos vapor phase hydrogenation of furfural). **1 ppm = 4.08 mg/m^3**	**PEL (skin):** **50 ppm** **200 mg/m^3** **STEL:** **Not Established**	**REL (skin):** **10 ppm** **40 mg/m^3** **STEL (15 min):** **15 ppm** **60 mg/m^3**	**75 ppm**	**TLV (skin):** **10 ppm** **40 mg/m^3** **STEL:** **15 ppm** **60 mg/m^3**

SECTION III - PHYSICAL AND CHEMICAL CHARACTERISTICS

Boiling Point **340°F (171°C)**	Specific Gravity (H_2O = 1) **1.13**
Vapor Pressure (mm Hg) **0.6 at 77°F (25°C)**	Molecular Weight **98.1**
Vapor Density (Air = 1) **3.37**	Melting Point **59°F (15°C)**

Solubility
Very soluble in water, alcohol, chloroform, and benzene. Insoluble in paraffin hydrocarbons.

Appearance and Odor
Colorless, mobile liquid with a burning, ether-like odor and bitter taste. Becomes brown to dark red on exposure to air and light. Odor Threshold = 8 ppm.

SECTION IV - FIRE AND EXPLOSION HAZARD DATA

Flash Point (method used) **167°F (75°C) opened cup**	Explosive Limits in Air % by Volume **LEL: 1.8% UEL: 16.3%**
NFPA Classification **Class 111A Combustible Liquid**	Autoignition Temperature **915°F (490°C)**

Extinguishing Media
Dry chemical, alcohol foam, carbon dioxide, water spray, or fog. Use water spray to cool containers.

Special Fire Fighting Procedures
Wear full protective clothing and self-contained breathing apparatus (SCBA). Poisonous gases are produced in fire. Move containers from fire area if it can be done without risk.

Unusual Fire and Explosion Hazards
Container may explode when exposed to heat. Moderately explosive when exposed to flame. Vapors are heavier than air and may travel for some distance to an ignition source to flashback.

SECTION V - REACTIVITY DATA

Stability		Conditions to Avoid **Furfuryl alcohol is normally stable in closed containers under routine conditions of handling and storage. Keep away from incompatible materials and heat.**
Stable X	Unstable	Incompatibility *(materials to avoid)* **Incompatible with strong oxidizers (such as chlorine, bromine, and fluorine), and strong acids, sulfur tetrafluoride, and 85% hydrogen peroxide.**
Hazardous Polymerization		Conditions to Avoid **Hazardous polymerization can occur in contact with many organic acids, such as cyanoacetic acid, formic acid, and nitric acid. Reactions can be explosive and violent.**
May Occur X	Will Not Occur	Hazardous Decomposition or By-products **When heated to decomposition, furfuryl alcohol can emit acrid, irritating and toxic fumes and smoke.**

SECTION VI - HEALTH HAZARD DATA

Primary Route(s) of Entry:	Inhalation?	Absorption (skin/eye)?	Ingestion?
	X	X	X

Health Hazards

INHALATION: Causes headache, eye irritation, and irritation of the upper respiratory system. There could be depression of the central nervous system (CNS) with dizziness, excitability, hot or "flushed" feeling, drunken stupor and behavior, staggering, loss of coordination, headache, mental confusion, visual disturbances, fatigue, nausea, vomiting, pale skin color, diarrhea, diuresis, low body temperature, unconsciousness, coma, and death.

ABSORPTION: Dryness, irritation, possible dermatitis. Absorption is likely with effects of inhalation.

INGESTION: Gastrointestinal irritation and the systemic effects of inhalation.

Carcinogenicity	NTP Listed?	IARC Cancer Review Group?	OSHA Regulated?	Target Organs?
Unknown Human Unknown Animal	No	No	29 CFR 1910.1000 Table Z-1	Respiratory system, skin, CNS, eyes.

Medical Conditions Generally Aggravated by Exposure

Chronic pulmonary disease and skin diseases may be aggravated by exposure to furfuryl alcohol.

Emergency and First-aid Procedures

Eye contact: Do not allow victim to keep eyes tightly shut. Flush immediately with water for 15 minutes (minimum), seek medical attention. Skin contact: Remove all contaminated clothing. Immediately wash area with large amounts of soap and water. For red or blistered skin, contact physician. For inhalation: Remove the person from exposure. Provide respiratory assistance and CPR. Transfer to medical facility. If swallowed: Seek medical attention immediately. Give glass of water. Do NOT induce vomiting. Never give anything by mouth to an unconscious or convulsing person.

SECTION VII - PRECAUTIONS FOR SAFE HANDLING AND USE

Steps to be Taken in Case Material is Released or Spilled

Disperse unignited vapors with water spray. Absorb spills in vermiculite or other material and deposit in sealed containers. Ventilate area of spill or leak. Remove all ignition sources. Restrict those not involved in cleanup from entering area. Notify appropriate authorities, as required.

Preferred Waste Disposal Method

No Citation.

Precautions to be Taken in Handling and Storage

Store in steel or baked phenolic-lined containers in cool, well-ventilated area away from heat. Do not keep more than six months supply. Storage under inert (nitrogen) atmosphere is recommended.

Other Precautions and Warnings

Protect containers from physical damage. Electrical grounding and bonding is recommended.

SECTION VIII - CONTROL MEASURES AND PERSONAL PROTECTIVE EQUIPMENT

Respiratory Protection *(specify type)*

For exposure over the PEL (50 ppm), use an air-purifying respirator. For higher exposures, use a supplied-air respirator with full facepiece operated in continuous flow mode, or use a self-contained breathing apparatus (SCBA) with full facepiece operated in positive pressure mode.

Ventilation

Local exhaust (preferred) at site of chemical work or general ventilation.

Protective Gloves	Eye Protection	Other Protective Clothing
Impervious Rubber Gloves	Chemical Goggles or Face Mask	Boots; Aprons; Gauntlets

Work/Hygiene Practices

Always wash hands thoroughly after using chemical; never bring food, drink, or smoking materials into vicinity of chemicals.

FURFURYL ALCOHOL

$C_5H_6O_2$ CAS: 98-00-0

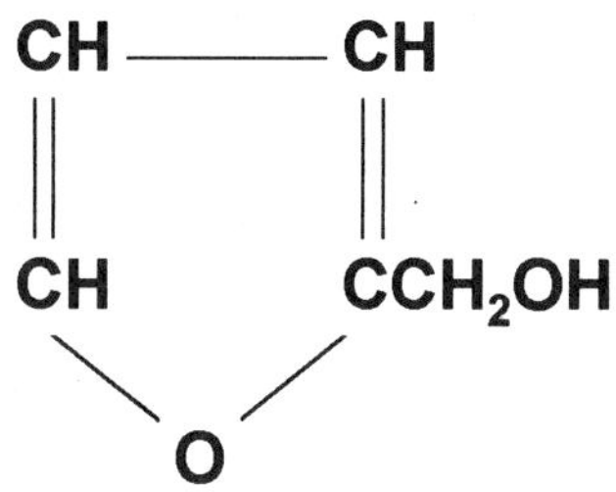

IDENTIFICATION AND TYPICAL USES

Furfuryl alcohol is a clear, colorless, mobile liquid that turns brownish-black on exposure to air. It has a strong ether-like odor and bitter taste. It is used as a wetting agent, in furan polymers, in corrosion-resistant sealants and cements, in foundry cores, in modified urea-formaldehyde polymers, as a penetrant, a solvent for dyes and resins, and as a flavoring agent (food and feed). The polymer is used as a mortar for bonding acid-proof brick and chemical masonry.

RISK ASSESSMENT: HEALTH

General Assessment

Furfuryl alcohol is a human poison by *ingestion* and is considered moderately toxic by *inhalation* and skin contact (*absorption*). While its carcinogenic and teratogenic properties have not been adequately investigated, mutation data have been reported.

Inhalation of furfuryl alcohol can cause dangerous irritation of the eyes, nose, and mucous membranes of the respiratory system. It can cause central nervous system (CNS) depression with excitability, hot-flushed skin, excessive talking, drunken behavior, staggering and loss of coordination, headache, mental confusion, visual disturbances, fatigue, nausea and vomiting, diarrhea, diuresis, lower body temperatures, pale skin color, and dizziness. High exposures can lead to stupor, loss of consciousness, coma, and death.

Skin contact results initially in a cooling sensation at the sight of contact. Since it will effectively remove the protective oily layer on the skin's surface, there will be drying, cracking, redness, and possible progression to dermatitis. Eye contact causes immediate stinging and burning with tearing, inflammation of surrounding tissues, and a painful sensitivity to light. If absorption occurs, it will compound the systemic effects caused by other simultaneous exposures (inhalation or ingestion).

Ingestion causes gastrointestinal irritation and many of the same systemic effects caused by inhalation, particularly those affecting the CNS.

Acute Health Effects

The following acute (short-term) health effects may occur immediately or shortly after exposure to furfuryl alcohol:

Skin: Irritation with dryness, cracking, and a possibility for secondary infection and dermatitis. Absorption is possible.

Eye: Severe irritation of the eye tissue with burning, itching, and watering.

Lung: Irritation of the mucosa of the respiratory system. Can cause cough and itchy throat.

CNS: A powerful depressant of the central nervous system with symptoms similar to that of intoxication. High exposures can be fatal.

Chronic Health Effects

The following chronic (long-term) health effects may occur at some time after exposure to furfuryl alcohol and can last for months or even years:

Cancer Hazards: Furfuryl alcohol may cause mutations (genetic changes) in living cells. The data on its carcinogenicity are inconclusive.

Reproductive Hazard: According to the references, furfuryl alcohol has not been adequately tested for its ability to adversely affect reproduction.

Other Chronic Effects: Prolonged or repeated skin contact may lead to dermatitis in some individuals.

Recommended Risk-Reduction Measures

Personnel should avoid direct contact with furfuryl alcohol. If a less toxic chemical cannot be substituted for a hazardous substance, then *engineering controls* are the most effective method of reducing exposures. The best protection is to enclose operations and/or provide local exhaust ventilation at the site of chemical release. While not always operationally feasible, isolating operations can also reduce exposure. Using respiratory protection is less effective than the controls mentioned above, but is still advisable whenever working with or around furfuryl alcohol. An air purifying respirator with an organic vapor cartridge may provide adequate protection for low or infrequent exposures above the PEL (50 ppm). For the best protection, a supplied-air respirator or a self-contained breathing apparatus (SCBA) with full facepiece operated in continuos flow or positive pressure mode are recommended. If a full facepiece is not available, then chemical goggles should be worn to protect the eyes. A face shield and a rubber apron should also be worn. To prevent hand and skin exposures, impervious gloves should be worn. Glove manufacturers should be contacted and permeation studies obtained *before* glove selections are made.

Administrative controls should also be in place to minimize the potential for human exposures. These may include written procedures or policies which specify the methods and techniques that will be practiced whenever personnel are to work with furfuryl alcohol.

All personnel should receive *training* on the use, hazards, protective measures, emergency actions, and other precautions per 29 CFR 1910.1200 (Hazard Communication), prior to the first assignment in an area where furfuryl alcohol is used or stored.

Before beginning employment, and at regular intervals thereafter (e.g., annually), the following medical tests are recommended for personnel working with or around furfuryl alcohol:

☑ Lung function test (establish baseline).

☑ Checks for blood pressure and pulse rate changes.

If symptoms develop or overexposure is suspected, the following additional tests are recommended:

☑ Lung function tests (compare to baseline).

☑ Complete neurological evaluation to check for nervous system disorders.

Any evaluation should include a careful history of past and present symptoms with an examination. Medical tests that simply look for existing damage are not a substitute for controlling exposures. Also, because smoking can cause heart disease, emphysema, and other respiratory dysfunction, smokers may develop exposure symptoms more quickly than non-smokers under the same conditions of exposure. Prudent risk management requires careful consideration of *all* possible exposure factors when assessing health exposure risk in the work place.

Other methods to reduce exposure include:

☑ Where possible, enclose operations and use local exhaust ventilation at the site of chemical release. If local exhaust ventilation or enclosure is not used, respiratory protection should be mandatory.

☑ Always ensure that proper protective clothing is worn when using chemical substances.

☑ Wash thoroughly immediately after exposure to furfuryl alcohol and at the end of the work shift or before eating, drinking, or smoking.

☑ Hazard warning information should be posted in the work area. In addition, as part of an on-going education and training program, all information on the health and safety hazards of furfuryl alcohol should be communicated to all potentially exposed workers.

☑ Eye wash stations should be provided in the immediate work area for emergency use. If there is a possibility of skin exposure to furfuryl alcohol, emergency shower facilities should also be provided.

RISK ASSESSMENT: ENVIRONMENT

General Assessment

The environment is at risk of exposure during transportation, storage, disposal, or destruction of furfuryl alcohol. In almost every scenario, the threat of environmental exposure is contingent upon the proper handling of the chemical substance. Accidental spills, large or small, can result in fire, explosion (depending upon conditions of spill), and possible contamination of the surrounding environmental mediums (water, soil, and air).

Furfuryl alcohol is considered a Class 111A flammable liquid (per OSHA 29 CFR 1910.106). Its moderately low flash point presents a fire and explosion hazard risk. Also, because it is incompatible with many common oxidizers and acids, extreme caution is required in handling, storage, transportation, and disposal of furfuryl alcohol. It can polymerize violently in contact with many organic acids (such as formic and nitric acid). These characteristics also require special consideration during any emergency situation involving a leak or spill of furfuryl alcohol crystals. Should furfuryl alcohol ever come into contact with incompatible substances either during use, transportation, storage, or disposal, the formation of highly toxic and/or highly explosive commodities is highly possible.

Furfuryl alcohol can enter the environment from industrial and municipal discharges, and from accidental spills.

Acute Ecological Effects

Acute (short-term) toxic effects may include the death of animals, birds, or fish and death or low growth rate in plants. Acute effects are seen 2 to 4 days after animals or plants are exposed to furfuryl alcohol.

Insufficient data are available to evaluate or predict the short-term effects of furfuryl alcohol to aquatic life, plants, birds, or land animals.

Chronic Ecological Effects

Chronic (long-term) toxic effects may include shortened life span, reproductive problems, lower fertility, and changes in appearance or behavior in exposed animals. These effects can be seen long after first exposure(s) to toxic chemicals.

Insufficient data are available to evaluate or predict the long-term effects of furfuryl alcohol to aquatic life, plants, birds, or land animals.

Water Solubility

Furfuryl alcohol is highly soluble in water. Concentrations of 1000 milligrams and more will mix with a liter of water.

Persistence in the Environment

Furfuryl alcohol is moderately persistent in water, with a half-life between 20 and 200 days. The half-life of a pollutant is the amount of time it takes for one half of the chemical to be degraded.

Bioaccumulation in Aquatic Organisms

Some substances increase in concentration, or *bioaccumulate,* in living organisms as they breathe contaminated air, drink contaminated water, or eat contaminated food. These chemicals can become concentrated in the tissues and internal organs of animals as well as humans.

The concentration of furfuryl alcohol found in fish tissues is expected to be higher than the average concentration of furfuryl alcohol in the water from which the fish was taken.

Recommended Risk-Reduction Measures

Proper training of all transporters will reduce the likelihood of a mishap or accident resulting in a leak or spill of furfuryl alcohol into the environment. The correct use of DOT labeling (per the MSDS) on all containers, trucks, and rail cars will enable emergency responders to react properly and quickly to any disaster thereby reducing the potential risk to the environment and to personnel.

Storage of furfuryl alcohol should be segregated from incompatible chemicals to minimize the risk of cross-contamination or contact. To prevent static sparks, all containers and equipment used in shipping, receiving, or transferring operations, should be electrically grounded and bonded.

If a spill or leak to the environment has occurred, fire department, emergency response and/or hazardous materials spill personnel should be notified immediately. Cleanup should be attempted only by those trained in proper spill containment procedures. Contaminated soils should be removed for incineration and replaced with clean soil. If furfuryl alcohol should contact the water table, aquifer, or navigable waterway, the local and/or state emergency response authorities must be notified. A comprehensive emergency response or disaster preparedness/recovery plan should be in place prior to any operations involving the use, transportation, storage, or disposal of furfuryl alcohol. If furfuryl alcohol is spilled or leaked, the following specific steps are recommended:

- ☑ Restrict persons not wearing protective clothing from area of spill or leak until cleanup is complete.
- ☑ Remove all ignition sources.
- ☑ Ventilate area of spill or leak.
- ☑ Absorb liquids in vermiculite or other similar material and place in sealed drums for disposal. Use non-sparking tools and avoid skin contact.
- ☑ It may be necessary to dispose of furfuryl alcohol as a hazardous waste. The responsible state agency or the regional office of the federal Envi-

ronmental Protection Agency (EPA) should be contacted for specific recommendations.

RISK ASSESSMENT: BUSINESS

General Assessment

Accidents or mishaps involving furfuryl alcohol can present a significant threat to business operations. The loss or damage of equipment or facilities can significantly affect fiscal viability. Lawsuits that may result from personnel injury/death, public exposures, and/or environmental contamination will also require a serious expenditure of resources. Media attention surrounding an injury, death, or environmental damage can also result in a loss of profits and loss of current as well as future business.

Recommended Risk-Reduction Measures

Company attorneys, safety and health professionals, and environmental specialists should be involved in the development of any procedures dealing with safety or emergency response. A company official should be pre-designated as a public relations officer with specific training in dealing with the press. Corporate plans and policies should be developed, approved, and implemented long before any need for such arises.

REFERENCES

American Conference of Governmental Industrial Hygienists. 1988. *Documentation of the Threshold Limit Values and Biological Exposure Indices,* 5th Edition (with updates). Cincinnati: ACGIH

Asante-Duah, D. K. 1993. *Hazardous Waste Risk Assessment.* New York: CRC Press/Lewis Publishers.

Bloom, A. D. and F. J. DeSerres. 1995. *Ecotoxicity and Human Health.* New York: CRC Press/Lewis Publishers.

Bueche, J. R. 1972. *Principles of Physics,* 2nd Edition. New York: McGraw-Hill.

Burrell, R., D. K. Flaherty, and L. J. Sauers. 1992. *Toxicology of the Immune System.* New York: Van Nostrand Reinhold.

Calabrese, E. J. 1994. *Biological Effects of Low Level Exposures.* New York: CRC Press/Lewis Publishers.

Cockerman, L. G., B. S. Shane. 1994. *Basic Environmental Toxicology.* New York: CRC Press/Lewis Publishers.

Cothern, R. R. 1993. *Comparative Environmental Risk Assessment.* New York: CRC Press/Lewis Publishers.

Danse, I. R. 1991. *Common Sense Toxics in the Workplace.* New York: Van Nostrand Reinhold.

Fodden, C. R., and J. L. Weddell. 1992. *Hazardous Materials Emergency Action Data.* New York: CRC Press/Lewis Publishers.

Forseberg, K., and L. H. Keith. 1993. *Chemical; Protective Clothing Permeation and Degradation Database.* Chelsea, MI: Lewis Publishers.

Gardner, R. W. 1994. *Chemical Intolerance.* New York: CRC Press/Lewis Publishers.

Gerarde, H. W. 1960. *Toxicology and Biochemistry of Aromatic Hydrocarbons.* New York: Elsevier.

Hodgson, E., R. B. Mailman, and Janice E. Chambers. 1988. *Dictionary of Toxicology.* London: The MacMillian Company.

Howard, P. H. 1995. *Handbook of Environmental Fate and Exposures Data for Organic Chemicals,* Vol. I-IV. New York: CRC Press/Lewis Publishers.

Jones, F. E. 1994. *Toxic Organic Vapors in the Workplace.* New York: CRC Press/Lewis Publishers.

Keith, L. H. and M. M. Walker. 1995. *Handbook of Air Toxics.* New York: CRC Press/Lewis Publishers.

Langerman, N. 1994 *Precautionary Labels for Chemical Containers.* New York: CRC Press/Lewis Publishers.

Lauwerys, R. R., P. Hoet. 1993. *Industrial Chemical Exposure.* New York: CRC Press/Lewis Publishers.

Lewis, R. J., Sr. 1993. *Hawley's Condensed Chemical Dictionary,* 12th Edition. New York: Van Nostrand Reinhold.

Lewis, R. J., Sr. 1992. *Sax's Dangerous Properties of Industrial Materials,* Volumes 1, 2, and 3, 8th Edition. New York: Van Nostrand Reinhold.

Manahan, S. E. 1992. *Toxicological Chemistry,* 2nd Edition. New York: CRC Press/Lewis Publishers.

Meyer, E. 1990. *Chemistry of Hazardous Materials.* Englewood Cliffs, NJ: Prentice Hall.

Mycock, J. C., J. D. McKenna, and L. Theodore. 1995. *Handbook of Air Pollution Control Engineering and Technology.* New York: CRC Press/Lewis Publishers.

Patnaik, P. 1992. *A Comprehensive Guide to the Hazardous Properties of Chemical Substances.* New York: Van Nostrand Reinhold.

Quigley, D. R. 1994. *Handbook of Emergency Chemical Management.* New York: CRC Press/Lewis Publishers.

Rea, W. J. 1995. *Chemical Sensitivity,* Vol. I-III. New York: CRC Press/Lewis Publishers.

Richardson, M. 1995. *Dictionary of Substance and Their Effects,* Vol. I-VI. New York: CRC Press/Lewis Publishers.

Sacarello, H. L. A. 1994. *The Comprehensive Handbook of Hazardous Materials.* New York: CRC Press/Lewis Publishers.

Sheftel, V. O. 1995. *Handbook of Toxic Properties of Monomers and Additives.* New York: CRC Press/Lewis Publishers.

Talbott, E., G. Craun. 1995. *An Introduction to Environmental Epidemiology.* New York: CRC Press/Lewis Publishers.

Torkelson, T. R. V. K. Rowe. 1981. *Patty's Industrial Hygiene and Toxicology,* Vol. 2B. New York: Wiley-Interscience.

Witorsch, P. and S. V. Spagnolo. 1994. *Air Pollution and Lung Disease in Adults.* New York: CRC Press/Lewis Publishers.

VOLUME II
G – Z

RISK MANAGEMENT FOR HAZARDOUS CHEMICALS

VOLUME II
G–Z

RISK MANAGEMENT FOR HAZARDOUS CHEMICALS

Jeffrey W. Vincoli, CSP

LEWIS PUBLISHERS

Boca Raton New York London Tokyo

Acquiring Editor:	Ken McCombs
Project Editor:	Joan Moscrop
Marketing Manager:	Greg Daurelle
Direct Marketing Manager:	Arline Massey
Cover design:	Jonathan Pennell
Manufacturing:	Sheri Schwartz

Library of Congress Cataloging-in-Publication Data

Vincoli, Jeffrey W.
Risk management for hazardous chemicals / Jeffrey W. Vincoli.
p. cm.
Includes bibliographical references and index.
ISBN 1-56670-200-3
1. Hazardous substances--Safety measures--Handbooks, manuals, etc.
2. Hazardous substances--Risk assessment--Handbooks, manuals, etc.
I. Title.
T55.3.H3V55 1996
604.7--DC20 96-35282
CIP

Direct all inquiries to CRC Press, Inc., 2000 Corporate Blvd. N.W., Boca Raton, Florida 33431.

International Standard Book Number 1-56670-200-3
Library of Congress Card Number 96-35282
Printed in the United States of America 1 2 3 4 5 6 7 8 9 0
Printed on acid-free paper

Table of Contents

MATERIAL SAFETY DATA SHEET

CHEMICAL NAME	CRC PRESS/LEWIS PUBLISHERS DISCLAIMER
GALLIC ACID	**CRC PRESS/LEWIS PUBLISHERS DISCLAIMER:** THE INFORMATION AND RECOMMENDATIONS PRESENTED HEREIN ARE BASED ON SOURCES BELIEVED TO BE RELIABLE. CRC MAKES NO REPRESENTATION ON ITS COMPLETENESS OR ACCURACY. IT IS THE RESPONSIBILITY OF THE USER TO DETERMINE THE CHEMICAL'S SUITABILITY FOR ITS INTENDED USE, THE CHEMICAL'S SAFE USE, AND THE CHEMICAL'S PROPER DISPOSAL. NO REPRESENTATIONS AND/OR WARRANTIES, EITHER EXPRESSED OR IMPLIED, OF THE MERCHANTABILITY OR FITNESS FOR A PARTICULAR PURPOSE, OR OF ANY OTHER NATURE, ARE MADE WITH RESPECT TO THE INFORMATION PROVIDED IN THIS MSDS OR TO THE CHEMICAL TO WHICH INFORMATION MAY REFER. CRC NEITHER ASSUMES NOR AUTHORIZES ANY OTHER PERSON TO ASSUME FOR IT, ANY OTHER ADDITIONAL RESPONSIBILITY OR LIABILITY FOR THE USE OF, OR RELIANCE UPON, THIS INFORMATION.

HAZARD WARNING INFORMATION

HEALTH	FIRE	REACTIVE	OTHER	DEGREE OF HAZARD	COLOR CODING	OTHER CODES
2	0	0		0 = Minimum Hazard 1 = Slight Hazard 2 = Moderate Hazard 3 = Serious Hazard 4 = Severe Hazard	HEALTH = BLUE FIRE = RED REACTIVITY = YELLOW OTHER = WHITE	OX = Oxidizer ACID = Acid ALK = Alkali COR = Corrosive W = Use No Water

SECTION I - GENERAL INFORMATION

Characterization	**Benzoic Acid**	RCRA Number	**None**
		EPA Class	**Not Applicable**
DOT Proper Shipping Name	**No Citation**	Chemical Abstract Service (CAS) Number	**149-91-7**
DOT Hazard Class and Label Requirements	**No Citation**	DOT Emergency Guide Code	**No Citation**
DOT Identification Number	**No Citation**	Chemical Formula	$C_6H_2(OH)_3CO_2H$

Synonyms
3,4,5-trihydroxybenzoic acid.

SECTION II - HAZARDOUS INGREDIENTS AND IDENTITY INFORMATION

Hazardous Components (specific identity)	OSHA Exposure Criteria	NIOSH Exposure Criteria	Immediately Dangerous to Life and Health (IDLH)	ACGIH Exposure Criteria
Gallic acid (derivation: By action of mold on solutions of tannin or by boiling the latter with strong acid or caustic soda).	**PEL:** Not Established **STEL:** Not Established	**REL:** Not Established **STEL:** Not Established	**Not Established**	**TLV:** Not Established **STEL:** Not Established

SECTION III - PHYSICAL AND CHEMICAL CHARACTERISTICS

Boiling Point	**Decomposes**	Specific Gravity (H_2O – 1)	**1.694**
Vapor Pressure (mm Hg)	**Not Found**	Molecular Weight	**170.13**
Vapor Density (Air = 1)	**Not Found**	Melting Point	**107-121°F (225-250°C) Decomposes**

Solubility
Sparingly soluble in water and ether. Soluble in alcohol and glycerol.

Appearance and Odor
Colorless or slightly yellow crystalline needles or prisms.

SECTION IV - FIRE AND EXPLOSION HAZARD DATA

Flash Point (method used)	**Not Determined**	Explosive Limits in Air % by Volume	**LEL: Not Determined UEL: Not Determined**
NFPA Classification	**Non-Combustible Solid**	Autoignition Temperature	**Not Determined**

Extinguishing Media
Use agent suitable to surrounding media. Gallic acid itself does not burn.

Special Fire Fighting Procedures
Poisonous gases are produced in fire. Wear full protective clothing and self-contained breathing apparatus (SCBA). Use water spray to flush and dilute a spill and to disperse the vapors. Also, use a water spray to keep fire-exposed containers cool. Do not release runoff from fire control measure to sewers or waterways. Firefighters should avoid contact with vapors produced during fire.

Unusual Fire and Explosion Hazards
None reported.

SECTION V - REACTIVITY DATA

Stability		Conditions to Avoid **Gallic acid is normally stable under routine conditions of handling and storage. Avoid contact with heat and ignition sources.**
Stable **X**	Unstable	Incompatibility *(materials to avoid)* **Violent reaction with potassium chlorite.**
Hazardous Polymerization		Conditions to Avoid **Under normal conditions of temperature and pressure, hazardous polymerization of gallic acid is not expected to occur.**
May Occur	Will Not Occur **X**	Hazardous Decomposition or By-products **When heated to decomposition, gallic acid can emit highly toxic and acrid smoke and irritating fumes.**

SECTION VI - HEALTH HAZARD DATA

Primary Route(s) of Entry:	Inhalation?	Absorption (skin)?	Ingestion?
	X		X

Health Hazards

INHALATION: Extremely irritating to the eyes, nose, throat, and respiratory system. Inhaled dusts may also cause bronchial irritation and inflammation of associated mucosa.

EYES & SKIN: Skin and eye contact can result in irritation. Absorption is not reported. Prolonged eye contact may cause corneal abrasions.

INGESTION: Poisonous by ingestion with experimental reproductive effects. Irritation of the gastrointestinal tract with abdominal cramps and nausea likely.

Carcinogenicity	NTP Listed?	IARC Cancer Review Group?	OSHA Regulated?	Target Organs?
Unknown Human Unknown Animal	**No**	**No**	**No**	**Respiratory system, eyes, skin.**

Medical Conditions Generally Aggravated by Exposure

None reported.

Emergency and First-aid Procedures

Eye contact: Flush immediately with water for 15 minutes (minimum), seek medical attention. Skin contact: Remove all contaminated clothing. Immediately flush area with large amounts of water. For red or blistered skin, wash with soap and water. Seek medical assistance. For inhalation: Remove the person from exposure. Provide respiratory assistance and CPR. Transfer to medical facility. If swallowed: Contact poison control center. Never attempt to give anything by mouth to an unconscious or convulsing person. Seek medical attention.

SECTION VII - PRECAUTIONS FOR SAFE HANDLING AND USE

Steps to be Taken in Case Material is Released or Spilled

Collect powdered materials in safest manner possible. Do NOT dry sweep (generates dusts). Use a HEPA vacuum instead. Absorb solution spills in vermiculite or similar materials and deposit in sealed containers. Ventilate area of spill or leak. Restrict those not involved in cleanup from entering area.

Preferred Waste Disposal Method

No Citation.

Precautions to be Taken in Handling and Storage

Avoid contact with potassium chlorite. Store in tightly closed containers in cool, well-ventilated area away from heat.

Other Precautions and Warnings

Avoid generating dusty conditions in storage areas.

SECTION VIII - CONTROL MEASURES AND PERSONAL PROTECTIVE EQUIPMENT

Respiratory Protection *(specify type)*

No exposure levels have been established for gallic acid. It is suspected to cause reproductive effects. An air purifying respirator with acid cartridge and a HEPA filter may suffice. However, better protection is provided by a supplied-air respirator with full facepiece operated in pressure demand or other positive pressure mode.

Ventilation

Local exhaust or general mechanical systems recommended.

Protective Gloves	Eye Protection	Other Protective Clothing
Acid resistant synthetic rubber	**Chemical/Dust Goggles or Face Mask**	**Rubber Apron**

Work/Hygiene Practices

Always wash hands thoroughly after using chemical; never bring food, drink, or smoking materials into vicinity of chemicals.

GALLIC ACID

$C_6H_2(OH)_3CO_2H$ CAS: 149-91-7

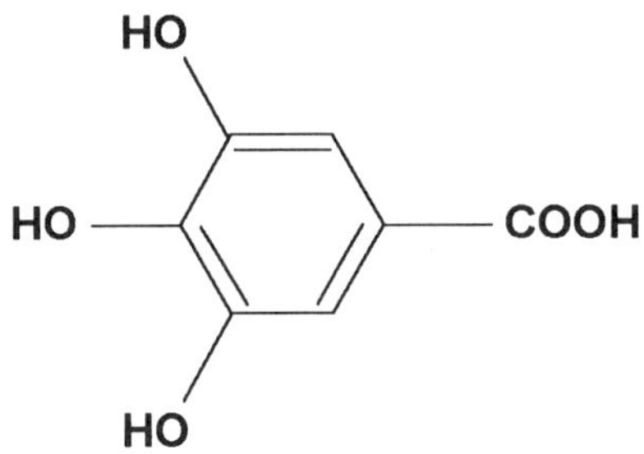

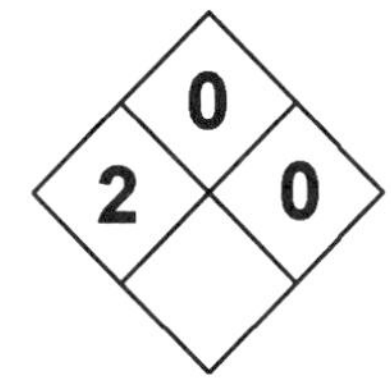

IDENTIFICATION AND TYPICAL USES

Gallic acid appears as colorless or slightly yellow crystalline needles or prisms. It is used in photography, in writing ink, in dyeing, in the manufacture of pyrogallol, as a tanning agent and in the manufacture of tannins, in the manufacture of paper, in pharmaceuticals, in engraving and lithography, and as an analytical reagent.

RISK ASSESSMENT: HEALTH

General Assessment

Exposure to gallic acid is possible through *inhalation* of its dusts and by *ingestion*. The data in the references are limited regarding its specific toxicity. However, reproductive effects have been reported in test animals and mutation data have also been reported. There are no reports of animal or human carcinogenicity.

Inhalation of gallic acid dusts will cause irritation of the eyes, nose, throat, and mucous membranes of the upper respiratory tract. Bronchial tube irritation may also occur with progression to bronchitis if exposures continue.

Skin and eye contact can result in mild to moderate irritation. If allowed to remain in contact with the eye for prolonged periods, corneal abrasions may occur. Ingestion may cause systemic poisoning with abdominal cramps and possible nausea and vomiting.

Acute Health Effects

The following acute (short-term) health effects may occur immediately or shortly after exposure to gallic acid:

Skin: Irritation/rash or burning feeling on contact.

Eye: Severe irritation; may cause burns and damage.

Lung: Irritation of the nose, throat, and lungs following exposure.

Other: Ingestion causes gastrointestinal irritation.

Chronic Health Effects

The following chronic (long-term) health effects may occur at some time after exposure to gallic acid and can last for months or even years:

Cancer Hazards: According to information presented in the references, gallic acid has not been adequately tested for its ability to cause cancer in animals. Mutation data have been reported and some scientists believe that long-term exposure to mutagens may pose a cancer risk There is no indication that exposure to gallic acid will have this effect. Additional research is required.

Reproductive Hazard: There is limited evidence to show that exposure to gallic acid causes adverse reproductive effects in laboratory animals.

Recommended Risk-Reduction Measures

Personnel should avoid direct contact with gallic acid. The exact nature of its toxicity is not clearly understood in the references. If a less toxic chemical cannot be substituted for a hazardous substance, then *engineering controls* are the most effective method of reducing exposures.

The best protection is to enclose operations and/or provide local exhaust ventilation at the site of chemical release. While not always operationally feasible, isolating operations can also reduce exposure. Using respiratory protection is less effective than the controls mentioned above, but is still advisable whenever working with or around gallic acid. No exposure levels have been established for gallic acid. This does not mean that exposure is without health risk. Prudent risk management may require a conservative approach to respiratory protection until further data are known about gallic acid. For transient exposures, an air-purifying respirator equipped with an acid cartridge, a HEPA filter, and a dust/mist pre-filter may suffice. Greater protection is obtained using a supplied-air respirator or a self-contained breathing apparatus (SCBA) with full facepiece operated in pressure demand or other positive pressure mode. If a full facepiece is not available, then chemical goggles should be worn to protect the eyes. Whenever a chemical splash/dust hazard exists, a face shield and a rubber apron should be worn. To prevent hand and skin exposures, acid resistant gloves should be used. Glove manufacturers should be contacted and permeation studies obtained *before* final glove selections have been made.

Administrative controls should also be in place to minimize the potential for human exposures. These may include written procedures or policies which specify the methods and techniques that will be practiced whenever personnel are to work with gallic acid.

All personnel should receive *training* on the use, hazards, protective measures, emergency actions, and other precautions per 29 CFR 1910.1200 (Hazard Communication), prior to the first assignment in an area where gallic acid is used or stored. Other methods to reduce exposure include:

- ☑ Where possible, enclose operations and use local exhaust ventilation at the site of chemical release. If local exhaust ventilation or enclosure is not used, respiratory protection should be mandatory.
- ☑ Always ensure that proper protective clothing is worn when using chemical substances.
- ☑ Wash thoroughly immediately after exposure to gallic acid and at the end of the work shift or before eating, drinking, or smoking.
- ☑ Hazard warning information should be posted in the work area. In addition, as part of an on-going education and training program, all information on the health and safety hazards of gallic acid should be communicated to all exposed or potentially exposed workers.

RISK ASSESSMENT: ENVIRONMENT

General Assessment

The environment is at risk of exposure during transportation, storage, disposal, or destruction of gallic acid. In almost every scenario, the threat of environmental exposure is contingent upon the proper handling of the chemical substance. Accidental spills, large or small, can result in fire, explosion, and possible contamination of the surrounding environmental mediums (water, soil, and, to a lesser degree, air).

Gallic acid is considered a non-combustible solid. It may be corrosive, especially when mixed in solution. It is incompatible with potassium chlorite and contact can cause violent reaction, including fire or explosion. These characteristics require special consideration during any emergency situation involving a leak or spill of gallic acid.

Gallic acid can enter the environment mainly industrial discharges or spills.

Acute Ecological Effects

Acute (short-term) toxic effects may include the death of animals, birds, or fish and death or low growth rate in plants. Acute effects are seen 2 to 4 days after animals or plants are exposed to gallic acid.

Insufficient data are available to evaluate or predict the short-term effects of gallic acid exposure to aquatic life, birds, plants, or land animals.

Chronic Ecological Effects

Chronic (long-term) toxic effects may include shortened life span, reproductive problems, lower fertility, and changes in appearance or behavior in exposed animals. These effects can be seen long after first exposure(s) to toxic chemicals. Insufficient data are available to evaluate or predict the long-term effects of gallic acid exposure to aquatic life, birds, plants, or land animals.

Water Solubility

Gallic acid is slightly soluble in water. Concentrations of 1 to 100 milligrams will mix with a liter of water.

Persistence in the Environment

Gallic acid is slightly persistent in the aquatic environment, with a half-life of between 2 to 20 days. The half-life of a pollutant is the amount of time it takes for one half of the chemical to be degraded.

Bioaccumulation in Aquatic Organisms

Some substances increase in concentration, or *bioaccumulate,* in living organisms as they breathe contaminated air, drink contaminated water, or eat contaminated food. These chemicals can become concentrated in the tissues and internal organs of animals as well as humans.

There is no information in the references on the expected bioaccumulation of gallic acid in the edible tissues of fish.

Recommended Risk-Reduction Measures

Proper training of all transporters will reduce the likelihood of a mishap or accident resulting in a leak or spill to the environment. The correct use of labeling on all containers, trucks, and rail cars will enable emergency responders to react properly and quickly to any disaster thereby reducing the potential risk to the environment and to personnel.

Storage of gallic acid should be segregated from incompatible chemicals to minimize the risk of cross-contamination or contact. Buildings designated for storage should be equipped with appropriate fire protection systems (alarms, sprinklers, emergency lighting, portable extinguishers).

If a spill or leak to the environment has occurred, fire department, emergency response and/or hazardous materials spill personnel should be notified immediately. Cleanup should be attempted only by those trained in proper spill containment procedures. Contaminated soils should be removed for incineration and replaced with clean soil. If gallic acid should contact the water table, aquifer, or navigable waterway, time is of the essence. It is slightly miscible in water and, therefore, total containment and remediation may not be entirely possible. When such spills occur, the local and/or state emergency response authorities must be notified. A comprehensive emergency response or disaster preparedness/recovery plan should be in place prior to any operations involving the use, transportation, storage, or disposal of gallic acid. If gallic acid is spilled or leaked, the following specific steps are recommended:

- ☑ Restrict persons not wearing protective clothing from area of spill or leak until cleanup is complete and area can be opened for normal work.
- ☑ Remove all ignition sources and ventilate area of spill.
- ☑ Collect all powdered materials in safest and most convenient manner possible and deposit in sealed drums for disposal. Do NOT dry sweep (generates airborne dusts). Use a vacuum equipped with a high efficiency particulate air (HEPA) filter instead. Damp mop residue and collect for disposal. Absorb any solution spills in vermiculite, dry sand, earth, or similar material and deposit in sealed containers.
- ☑ It may be necessary to dispose of gallic acid as a hazardous waste. The responsible state agency or the regional office of the federal Environmental Protection Agency (EPA) should be contacted for specific recommendations.

RISK ASSESSMENT: BUSINESS

General Assessment

Accidents or mishaps involving gallic acid can present a significant threat to business operations. The loss or damage of equipment or facilities can significantly affect fiscal viability. Lawsuits that may result from personnel injury/death, public exposures, and/or environmental contamination will also require a serious expenditure of resources. Media attention surrounding an injury, death, or environmental damage can also result in a loss of profits and loss of current as well as future business.

Recommended Risk-Reduction Measures

Company attorneys, safety and health professionals, and environmental specialists should be involved in the development of any such procedures. A company official should be pre-designated as a public relations officer with specific training in dealing with the press. Corporate plans and policies should be developed, approved, and implemented long before any need for such arises.

REFERENCES

Ahlbom, A. 1993. *Biostatistics for Epidemiologists.* New York: CRC Press/Lewis Publishers.

American Conference of Governmental Industrial Hygienists. 1988. *Documentation of the Threshold Limit Values and Biological Exposure Indices,* 5th Edition (with updates). Cincinnati: ACGIH

Asante-Duah, D. K. 1993. *Hazardous Waste Risk Assessment.* New York: CRC Press/Lewis Publishers.

Bloom, A. D. and F. J. DeSerres. 1995. *Ecotoxicity and Human Health.* New York: CRC Press/Lewis Publishers.

Brusick, D. J. 1994. *Methods for Genetic Risk Assessment.* New York: CRC Press/Lewis Publishers.

Bueche, J. R. 1972. *Principles of Physics,* 2nd Edition. New York: McGraw-Hill.

Burrell, R., D. K. Flaherty, and Leonard J. Sauers. 1992. *Toxicology of the Immune System.* New York: Van Nostrand Reinhold.

Calabrese, E. J., E. M. Kenyon. *Air Toxics and Risk Assessment.* New York: CRC Press/Lewis Publishers.

Calabrese, E. J. 1994. *Biological Effects of Low Level Exposures.* New York: CRC Press/Lewis Publishers.

Cockerman, L. G., B. S. Shane. 1994. *Basic Environmental Toxicology.* New York: CRC Press/Lewis Publishers.

Cothern, C. R., N. P. Ross. 1994. *Environmental Statistics, Assessment, and Forecasting.* New York: CRC Press/Lewis Publishers.

Cothern, R. R. 1993. *Comparative Environmental Risk Assessment.* New York: CRC Press/Lewis Publishers.

Cothern, C. R. 1995. *Handbook for Environmental Risk Decisionmaking.* New York: CRC Press/Lewis Publishers.

Danse, I. R. 1991. *Common Sense Toxics in the Workplace.* New York: Van Nostrand Reinhold.

Gardner, R. W. 1994. *Chemical Intolerance.* New York: CRC Press/Lewis Publishers.

Hallenbeck, W. H. 1993. *Quantitative Risk Assessment for Environmental and Occupational Health,* 2nd Edition. New York: CRC Press/Lewis Publishers.

Hosty, J. W., P. Foster. 1990. *A Practical Guide to Chemical Spill Response.* New York: Van Nostrand Reinhold.

Howard, P. H. 1995. *Handbook of Environmental Fate and Exposures Data for Organic Chemicals,* Vol. I-IV. New York: CRC Press/Lewis Publishers.

Howard, P. H., et. al. 1991. *Handbook of Environmental Degradation Rates.* New York: CRC Press/Lewis Publishers.

James, A., M. Lord. 1992. *Index of Chemical and Physical Data.* New York: Van Nostrand Reinhold.

Jones, F. E. 1994. *Toxic Organic Vapors in the Workplace.* New York: CRC Press/Lewis Publishers.

Keith, L. H. and M. M. Walker. 1995. *Handbook of Air Toxics.* New York: CRC Press/Lewis Publishers.

Landis, W. G., M. H. Yu. 1995. *Introduction to Environmental Toxicology.* New York: CRC Press/Lewis Publishers.

Langerman, N. 1994 *Precautionary Labels for Chemical Containers.* New York: CRC Press/Lewis Publishers.

Lauwerys, R. R., P. Hoet. 1993. *Industrial Chemical Exposure.* New York: CRC Press/Lewis Publishers.

Lewis, R. J., Sr. 1993. *Hawley's Condensed Chemical Dictionary,* 12th Edition. New York: Van Nostrand Reinhold.

Lewis, R. J., Sr. 1992. *Sax's Dangerous Properties of Industrial Materials,* Volumes 1, 2, and 3, 8th Edition. New York: Van Nostrand Reinhold.

Lide, D. R. 1995. *Handbook of Chemistry and Physics,* 76th Edition. New York: CRC Press/Lewis Publishers.

Manahan, S. E. 1994. *Environmental Chemistry,* 6th Edition. New York: CRC Press/Lewis Publishers.

Manahan, S. E. 1992. *Toxicological Chemistry,* 2nd Edition. New York: CRC Press/Lewis Publishers.

Mellström, G. A., J. E. Wahlberg, and H. I. Maibach. 1994. *Protective Gloves for Occupational Use.* New York: CRC Press/Lewis Publishers.

Meyer, E. 1990. *Chemistry of Hazardous Materials.* Englewood Cliffs, NJ: Prentice Hall.

Mycock, J. C., J. D. McKenna, and L. Theodore. 1995. *Handbook of Air Pollution Control Engineering and Technology.* New York: CRC Press/Lewis Publishers.

Nelson, G. O. 1992. *Gas Mixtures.* New York: CRC Press/Lewis Publishers.

Ott, W. R. 1995. *Environmental Statistics and Data Analysis.* New York: CRC Press/Lewis Publishers.

Patnaik, P. 1992. *A Comprehensive Guide to the Hazardous Properties of Chemical Substances.* New York: Van Nostrand Reinhold.

Philp, R. B. 1995. *Environmental Hazards and Human Health.* New York: CRC Press/Lewis Publishers.

Proctor, N. H., J. P. Hughes. 1991. *Chemical Hazards of the Workplace.* New York: Van Nostrand Reinhold.

Quigley, D. R. 1994. *Handbook of Emergency Chemical Management.* New York: CRC Press/Lewis Publishers.

Rea, W. J. 1995. *Chemical Sensitivity,* Vol. I-III. New York: CRC Press/Lewis Publishers.

Richardson, M. 1995. *Dictionary of Substance and Their Effects,* Vol. I-VI. New York: CRC Press/Lewis Publishers.

Rylander, R., R. R. Jacobs. 1994. *Organic Dusts.* New York: CRC Press/Lewis Publishers.

Sacarello, H. L. A. 1994. *The Comprehensive Handbook of Hazardous Materials.* New York: CRC Press/Lewis Publishers.

Sheftel, V. O. 1995. *Handbook of Toxic Properties of Monomers and Additives.* New York: CRC Press/Lewis Publishers.

Sherman, J. D. 1988. *Chemical Exposure and Disease.* New York: Van Nostrand Reinhold.

Talbott, E., G. Craun. 1995. *An Introduction to Environmental Epidemiology.* New York: CRC Press/Lewis Publishers.

The Merck Index, 10th edition. 1983. Rahway, NJ: Merck & Company.

U.S. Department of Health and Human Services, National Institute for Occupational Safety and Health. 1980. *Gallic acid. Registry of Toxic Effects of Chemical Substances,* Vol. 1. Washington, D.C.: U.S. Government Printing Office.

U.S. Department of Health and Human Services, National Institute for Occupational Safety and Health. 1994. *NIOSH Pocket Guide to Chemical Hazards.* Washington, D.C.: U.S. Government Printing Office.

Witorsch, P. and S. V. Spagnolo. 1994. *Air Pollution and Lung Disease in Adults.* New York: CRC Press/Lewis Publishers.

MATERIAL SAFETY DATA SHEET

CHEMICAL NAME	
GALLIUM ARSENIDE	**CRC PRESS/LEWIS PUBLISHERS DISCLAIMER:** THE INFORMATION AND RECOMMENDATIONS PRESENTED HEREIN ARE BASED ON SOURCES BELIEVED TO BE RELIABLE. CRC MAKES NO REPRESENTATION ON ITS COMPLETENESS OR ACCURACY. IT IS THE RESPONSIBILITY OF THE USER TO DETERMINE THE CHEMICAL'S SUITABILITY FOR ITS INTENDED USE, THE CHEMICAL'S SAFE USE, AND THE CHEMICAL'S PROPER DISPOSAL. NO REPRESENTATIONS AND/OR WARRANTIES, EITHER EXPRESSED OR IMPLIED, OF THE MERCHANTABILITY OR FITNESS FOR A PARTICULAR PURPOSE, OR OF ANY OTHER NATURE, ARE MADE WITH RESPECT TO THE INFORMATION PROVIDED IN THIS MSDS OR TO THE CHEMICAL TO WHICH INFORMATION MAY REFER. CRC NEITHER ASSUMES NOR AUTHORIZES ANY OTHER PERSON TO ASSUME FOR IT, ANY OTHER ADDITIONAL RESPONSIBILITY OR LIABILITY FOR THE USE OF, OR RELIANCE UPON, THIS INFORMATION.

HAZARD WARNING INFORMATION

HEALTH	FIRE	REACTIVE	OTHER	DEGREE OF HAZARD	COLOR CODING	OTHER CODES
3	1	2	W	0 = Minimum Hazard 1 = Slight Hazard 2 = Moderate Hazard 3 = Serious Hazard 4 = Severe Hazard	HEALTH = BLUE FIRE = RED REACTIVITY = YELLOW OTHER = WHITE	OX = Oxidizer ACID = Acid ALK = Alkali COR = Corrosive W = Use No Water

SECTION I - GENERAL INFORMATION

Characterization	**Gallium Compound; Arsenic Compound**
RCRA Number	**D004**
EPA Class	**Characteristic Waste**
DOT Proper Shipping Name	**No Citation**
Chemical Abstract Service (CAS) Number	**1303-00-0**
DOT Hazard Class and Label Requirements	**No Citation**
DOT Emergency Guide Code	**No Citation**
DOT Identification Number	**No Citation**
Molecular Formula	**AsGa**
Synonyms	**Gallium monoarsenide.**

SECTION II - HAZARDOUS INGREDIENTS AND IDENTITY INFORMATION

Hazardous Components (specific identity)	OSHA Exposure Criteria	NIOSH Exposure Criteria	Immediately Dangerous to Life and Health (IDLH)	ACGIH Exposure Criteria
Gallium arsenide (derivation: Passing a mixture of hydrogen and arsenic vapor over gallium (III) oxide at 1112°F/600°C. It is often alloyed with gallium phosphide or iridium arsenide).	**PEL:** **0.001 mg(As)/m³** **Possible Cancer Agent**	**REL (ceiling):** **0.002 mg(As)/m³** **(15 minutes)** **Possible Human Carcinogen**	**Not Determined**	**TLV:** **0.2 mg(As)/m³** **STEL:** **Not Established**

SECTION III - PHYSICAL AND CHEMICAL CHARACTERISTICS

Boiling Point	**Not Found**
Specific Gravity (H_2O = 1)	**5.31**
Vapor Pressure (mm Hg)	**Not Found**
Molecular Weight (atomic weight)	**144.64**
Vapor Density (Air = 1)	**Not Found**
Melting Point	**2260°F (1238°C)**
Solubility	**Insoluble (decomposes) in water.**
Appearance and Odor	**Dark gray cubic crystals with a metallic sheen. No odor when dry; garlic odor when moist.**

SECTION IV - FIRE AND EXPLOSION HAZARD DATA

Flash Point (method used)	**Not Found**
Explosive Limits in Air % by Volume	**LEL: N/A UEL: N/A**
NFPA Classification	**Flammable Solid (in powder form)**
Autoignition Temperature	**Not Found**
Extinguishing Media	**Use a Class D extinguishing agent or smother with dry sand, dry clay, or dry limestone. Do NOT use carbon dioxide, water, or halogenated compounds.**
Special Fire Fighting Procedures	**Wear full protective clothing and self-contained breathing apparatus (SCBA). Poisonous gases are produced in fire, including arsenic. Use water mist from unmanned device to keep fire-exposed containers cool. Keep cooling containers after fire is extinguished. Fight fire from maximum distance, if possible.**
Unusual Fire and Explosion Hazards	**Move containers from fire if it can be done without risk. Containers may build up pressure and explode violently in fire. Explosive decomposition may occur when gallium arsenide is involved in fire.**

SECTION V - REACTIVITY DATA

Stability		Conditions to Avoid Normally stable compound. Avoid heat and open flame, especially around dusts and powders of barium. Do not allow contact with heat since violent and explosive reaction can occur. Avoid incompatible agents.
Stable X	Unstable	Incompatibility *(materials to avoid)* Will react in the presence of steam, acids, and acid fumes to evolve poisonous arsine gas. Molten gallium arsenide attacks quartz.
Hazardous Polymerization		Conditions to Avoid Under normal conditions of temperature and pressure, hazardous polymerization of gallium arsenide is not expected to occur.
May Occur	Will Not Occur X	Hazardous Decomposition or By-products When heated to decomposition, gallium arsenide emits toxic and poisonous fumes and gases including arsine, arsenic, and gallium.

SECTION VI - HEALTH HAZARD DATA

Primary Route(s) of Entry:	Inhalation?	Absorption (skin/eye)?	Ingestion?
	X		X

Health Hazards

INHALATION: Breathing the dusts of gallium arsenide may cause the release of arsenic in the body. This may cause cardiovascular changes, pulmonary edema (fluid in the lungs), and serious effects on the nervous system (peripheral neuropathy), muscle cramps, nausea, vomiting, diarrhea, dehydration, abdominal pain, kidney failure, jaundice, delirium, headaches, convulsions, paresthesis, and death from cardiovascular collapse or pulmonary edema.

EYES & SKIN: Exposure to the eyes and/or skin may cause moderate to severe irritation and even burns.

INGESTION: Severe gastrointestinal effects along with many of the same symptoms as inhalation.

Carcinogenicity	NTP Listed?	IARC Cancer Review Group?	OSHA Regulated?	Target Organs?
Suspected Human Confirmed Animal	5th Annual Report	Group 1 (as As)	29 CFR 1910.1000 Table Z-1 (as As)	Liver, kidneys, blood, lungs, CNS

Medical Conditions Generally Aggravated by Exposure

Skin conditions (dermatitis); respiratory problems; renal conditions and hepatic disorders.

Emergency and First-aid Procedures

Eye contact: Do not allow victim to keep eyes tightly shut. Flush immediately with water for 15 minutes (minimum), seek medical attention. Skin contact: Remove all contaminated clothing. Immediately wash area with flooding amounts of water. For inhalation: Remove the person from exposure. Provide respiratory assistance or CPR. Transfer to medical facility. Observe 48 hours for lung effects. If swallowed: Seek medical attention immediately. Never give anything by mouth to an unconscious or convulsing person.

SECTION VII - PRECAUTIONS FOR SAFE HANDLING AND USE

Steps to be Taken in Case Material is Released or Spilled

Collect powdered materials in most convenient and safe manner and deposit in sealed containers. Dry sweep not recommended. Use HEPA vacuum. Remove all ignition sources. Restrict those not involved in cleanup from entering area. Notify appropriate authorities, as required by applicable regulations.

Preferred Waste Disposal Method

No citation.

Precautions to be Taken in Handling and Storage

Store powders in sealed containers in cool, dry, well-ventilated area away from incompatible materials, heat, and moisture. Protect containers from physical damage.

Other Precautions and Warnings

Detached or separate facility storage is recommended. Use only non-sparking tools or equipment, especially when opening or closing containers.

SECTION VIII - CONTROL MEASURES AND PERSONAL PROTECTIVE EQUIPMENT

Respiratory Protection *(specify type)*

For dust exposures below PEL, use a respirator with a high efficiency particulate air (HEPA) filter, otherwise use a NIOSH/MSHA-approved supplied-air respirator or a self-contained breathing apparatus with full facepiece operated in positive pressure mode.

Ventilation

Local exhaust (preferred) at site of chemical work or general ventilation.

Protective Gloves	Eye Protection	Other Protective Clothing
Impervious Rubber	Chemical/Dust Goggles or Face Mask	Protective Apron, Boots

Work/Hygiene Practices

Always wash hands thoroughly after using chemical; never bring food, drink, or smoking materials into vicinity of chemicals.

GALLIUM ARSENIDE

AsGa CAS: 1303-00-0

IDENTIFICATION AND TYPICAL USES

Gallium arsenide appears as a dark gray, cubic, crystalline solid. It has no noticeable odor when dry but may have a slight odor like garlic when moist. It is used as a semiconductor in light-emitting diodes for telephone dials, injection lasers, solar cells, magnetoresistance devices, thermistors, and microwave generation. Also used in medical lasers and in transistors for military electronics equipment (radar-jamming devices).

RISK ASSESSMENT: HEALTH

General Assessment

Gallium arsenide is toxic by *inhalation* and *ingestion* due to its arsenic content. Skin contact causes some topical risk but absorption has not been reported. Arsenic is a suspected human teratogen (cause birth defects), a known carcinogen (causes cancer), and a confirmed mutagen (causes genetic changes). Arsenic has also been shown to accumulate in the body over prolonged periods of exposure. Gallium itself is poorly absorbed in the body but may cause bone marrow depression and dermatitis.

Inhalation of gallium dusts result in the release of arsenic into the body. This can cause irregular heart contractions and serious (life-threatening) lung injury. Pulmonary edema (fluid in the lungs) can develop up to 48 hours after exposure. This is a medical emergency and can lead to death. Because of the delayed development, victims may experience a false sense of security with regard to heath exposure risk. Other symptoms of exposure include central nervous system (CNS) effects with peripheral neuropathy, delirium, headache, convulsions, muscle cramps, and paresthesis (tingling of the skin). Death can occur from cardiovascular collapse, kidney failure, or pulmonary edema. Gastrointestinal symptoms include vomiting, nausea, and abdominal pains. There may also be intravascular breakage of red blood cells resulting in reversible kidney failure. There may be jaundice and liver damage as well as weakness, anorexia, gastrointestinal disturbances, liver damage, and injury to the nervous system on prolonged contact.

Acute Health Effects

The following acute (short-term) health effects may occur immediately or shortly after exposure to gallium arsenide:

<u>Skin</u>: May cause severe irritation, redness, rash, and possible dermatitis.

<u>Eye</u>: Severe irritation and possible damage.

<u>Lung</u>: Dusts will irritate the lungs causing cough and phlegm. The may be a dangerous buildup of fluid in the lungs (pulmonary edema) which can be fatal.

<u>CNS</u>: Serious depression of the central nervous system with effects ranging from headache to delirium, convulsions, coma, and death.

Chronic Health Effects

The following chronic (long-term) health effects may occur at some time after exposure to gallium arsenide and can last for months or even years:

<u>Cancer Hazards</u>: According to the references, gallium arsenide has not been tested for its ability to cause cancer in animals. However arsenic and arsine, both of which are produced by gallium arsenide once it enters the body, are known carcinogens. It is highly probable that gallium arsenide is also a human carcinogen.

<u>Reproductive Hazard</u>: According to the references, gallium arsenide has not been tested for its ability to affect reproduction. Arsenic, a derivative of gallium arsenide, is a known teratogen.

<u>Other Chronic Effects</u>: Repeated exposure can cause nerve damage with "pins and needles" effect, burning,

numbness, and later weakness of arms and legs. Repeated skin contact can cause thickened skin and/or patchy areas of darkening and loss of pigment. Some persons develop white lines on the nails. Repeated exposures can also damage the liver, cause narrowing of the blood vessels, or interfere with the bone marrow's ability to make red blood cells. Chronic exposure may cause anorexia, general weakness, gastrointestinal disturbances, and peripheral neuropathy.

Recommended Risk-Reduction Measures

Personnel should avoid direct contact with gallium arsenide dusts or powders. If a less toxic material or compound cannot be substituted for gallium arsenide, then *engineering controls* are the most effective method of reducing exposures. The best protection is to enclose operations and/or provide local exhaust ventilation at the site of gallium arsenide dust release. While not always operationally feasible, isolating operations can also reduce exposure. Using respiratory protection is less effective than the controls mentioned above, but is still advisable whenever working with or around gallium arsenide. A self-contained breathing apparatus (SCBA) with full facepiece and pressure demand is recommended for the greatest possible respiratory protection, especially when exposures levels are unknown. However, for low or infrequent exposure, an MSHA/NIOSH-approved full facepiece respirator with a high efficiency particulate air (HEPA) filter can be used. If a full facepiece is not available, then chemical/dust goggles should be worn to protect the eyes. A face shield and protective apron should also be worn. To prevent hand and skin exposures, protective rubber gloves should be worn.

Administrative controls should also be in place to minimize the potential for human exposures. These may include written procedures or policies which specify the methods and techniques that will be practiced whenever personnel are to work with gallium arsenide.

All personnel should receive *training* on the use, hazards, protective measures, emergency actions, and other precautions per 29 CFR 1910.1200 (Hazard Communication), prior to the first assignment in an area where gallium arsenide is used or stored.

Before beginning employment working with gallium arsenide, and at regular intervals thereafter (e.g., annually), those personnel with frequent or potentially high exposures should be provided the following recommended medical tests:

☑ Lung function test.

☑ Obtain baseline arterial blood gases.

☑ Complete blood count.

☑ Test for urine arsenic (may not be accurate within 2 days of eating shellfish or fish; most accurate at the end of a work shift) should not be greater than 100 micrograms per gram creatinine in the urine.

☑ Liver and kidney function tests.

If symptoms develop or over exposure is suspected, the following additional tests may be helpful:

☑ Lung function tests.

☑ Consider chest X-ray after acute overexposure (may be negative if taken immediately after exposure due to delayed development of pulmonary edema).

☑ Evaluation by a qualified allergist with careful consideration of exposure history and special tests (may help diagnose allergy).

Any evaluation should include a careful history of past and present symptoms with an examination. Medical tests that simply look for existing damage are not a substitute for controlling exposures. Also, since smoking can cause heart disease as well as lung cancer, emphysema, and other known respiratory problems, smokers exposed to gallium arsenide may experience symptoms much more quickly and with more pronounced indications than non-smokers under the same conditions of exposure.

Other methods to reduce exposure include:

☑ Where possible, enclose operations and use local exhaust ventilation at the site of chemical release. If local exhaust ventilation or enclosure is not used, respiratory protection should be mandatory.

☑ Always ensure that proper protective clothing is worn when using chemical substances.

☑ Wash thoroughly immediately after exposure to gallium arsenide and at the end of the work shift or before eating, drinking, or smoking.

☑ Hazard warning information should be posted in the work area. In addition, as part of an on-going education and training program, all information on the health and safety hazards of gallium arsenide should be communicated to all potentially exposed workers.

☑ Eye wash stations should be provided in the immediate work area for emergency use. If there is a possibility of skin exposure to gallium arsenide, emergency shower facilities should also be provided.

☑ Workers whose clothing has been contaminated by gallium arsenide should change into clean clothes before leaving work. Contaminated work clothing should be laundered only by individuals who have been informed of the hazards of exposure to gallium arsenide.

RISK ASSESSMENT: ENVIRONMENT

General Assessment

The environment is at risk of exposure during transportation, storage, disposal, or destruction of gallium arsenide. In almost every scenario, the threat of environmental exposure is contingent upon the proper handling of the chemical substance. Accidental spills, large or small, where gallium arsenide contacts incompatible commodities can result in fire, explosion (depending upon conditions of spill), and possible contamination of the surrounding environmental mediums (water, soil, and air).

In its powdered form, gallium arsenide is considered a non-combustible solid. However, it can react explosively and present a serious fire hazard. Because it is incompatible with many commodities, including steam, acids, and acid fumes, extreme caution is required in handling, storage, transportation, and disposal of gallium arsenide. These characteristics also require special consideration during any emergency situation involving a leak or spill of gallium arsenide powder or dust. Should gallium arsenide ever come into contact with incompatible substances either during use, transportation, storage, or disposal, violent and even explosive reactions are possible.

Gallium arsenide can enter the environment from industrial and municipal waste treatment plant discharges and from spills.

Acute Ecological Effects

Acute (short-term) toxic effects may include the death of animals, birds, or fish and death or low growth rate in plants. Acute effects are seen 2 to 4 days after animals or plants are exposed to barium.

The main ecological hazard of gallium arsenide is due to its arsenic derivative.

Arsenic has high acute toxicity to aquatic life, birds, and land animals. Except where soil arsenic content is high (e.g., around smelters and where arsenic-based pesticides are known to have been used), arsenic does not accumulate in plants to toxic levels. Where soil content is high, growth of crops and crop yields can be decreased.

Chronic Ecological Effects

Chronic (long-term) toxic effects may include shortened life span, reproductive problems, lower fertility, and changes in appearance or behavior in exposed animals. These effects can be seen long after first exposure(s) to toxic chemicals.

Gallium arsenide has high chronic toxicity to aquatic life and moderate chronic toxicity to birds and land animals.

Water Solubility

Gallium arsenide is insoluble in water (decomposes). As arsenic, concentrations of less than 1 milligram will mix with a liter of water.

Persistence in the Environment

Gallium arsenide is highly persistent in water as arsenic, with a half-life greater than 200 days. The half-life of a pollutant is the amount of time it takes for one half of the chemical to be degraded.

Bioaccumulation in Aquatic Organisms

Some substances increase in concentration, or *bioaccumulate,* in living organisms as they breathe contaminated air, drink contaminated water, or eat contaminated food. These chemicals can become concentrated in the tissues and internal organs of animals as well as humans.

The concentration of gallium arsenide found in fish tissues as arsenic is expected to be much higher than the average concentration of arsenic in the water from which the fish was taken.

Recommended Risk-Reduction Measures

Proper training of all transporters will reduce the likelihood of a mishap or accident resulting in a leak or spill of gallium arsenide dusts or powders into the environment. The correct use of DOT labeling (per the MSDS) on all containers, trucks, and rail cars will enable emergency responders to react properly and quickly to any disaster thereby reducing the potential risk to the environment and to personnel.

Storage of gallium arsenide should be segregated from incompatible chemicals to minimize the risk of cross-contamination or contact.

If a spill or leak to the environment has occurred, fire department, emergency response and/or hazardous materials spill personnel should be notified immediately. Cleanup should be attempted only by those trained in proper spill containment procedures. Contaminated soils should be removed for incineration and replaced with clean soil. If gallium arsenide should contact the water table, aquifer, or navigable waterway, time is of the essence. It decomposes in water and total remediation may not be possible. The local and/or state emergency response authorities must be notified. A comprehensive emergency response or disaster preparedness/recovery plan should be in place prior to any operations involving the use, transportation, storage, or disposal of gallium arsenide. If gallium arsenide powder is spilled or leaked, the following specific steps are recommended:

- ☑ Restrict persons not wearing protective clothing from area of spill or leak until cleanup is complete and area can be opened for normal work.
- ☑ Remove all ignition sources, ventilate area.
- ☑ Collect powdered materials in the safest manner possible and deposit in sealed containers. Recommend using vacuum with HEPA filter. Do not dry sweep (generates airborne dusts).
- ☑ It may be necessary to dispose of gallium arsenide as a hazardous waste. The responsible state agency or the regional office of the federal Environmental Protection Agency (EPA) should be contacted for specific recommendations.

RISK ASSESSMENT: BUSINESS

General Assessment

Accidents or mishaps involving gallium arsenide can present a significant threat to business operations. The loss or damage of equipment or facilities can significantly affect fiscal viability. Lawsuits that may result from personnel injury/death, public exposures, and/or environmental contamination will also require a serious expenditure of resources. Media attention surrounding an injury, death, or environmental damage can also result in a loss of profits and loss of current as well as future business. Always remember that anytime the terms "cancer," "carcinogen," or "reproductive hazard" are used, public hysteria, emotion, and ignorance can run equally high. This must be carefully considered whenever developing or implementing public relations policies.

Recommended Risk-Reduction Measures

Company attorneys, safety and health professionals, and environmental specialists should be involved in the development of any procedures for responding to chemical incidents. A company official should be predesignated as a public relations officer with specific training in dealing with the press. Corporate plans and policies should be developed, approved, and implemented long before any need for such arises.

REFERENCES

American Conference of Governmental Industrial Hygienists. 1988. *Documentation of the Threshold Limit Values and Biological Exposure Indices,* 5th Edition (with updates). Cincinnati: ACGIH

Asante-Duah, D. K. 1993. *Hazardous Waste Risk Assessment.* New York: CRC Press/Lewis Publishers.

Brusick, D. J. 1994. *Methods for Genetic Risk Assessment.* New York: CRC Press/Lewis Publishers.

Bueche, J. R. 1972. *Principles of Physics,* 2nd Edition. New York: McGraw-Hill.

Cockerman, L. G., B. S. Shane. 1994. *Basic Environmental Toxicology.* New York: CRC Press/Lewis Publishers.

Cothern, R. R. 1993. *Comparative Environmental Risk Assessment.* New York: CRC Press/Lewis Publishers.

Landis, W. G., M. H. Yu. 1995. *Introduction to Environmental Toxicology.* New York: CRC Press/Lewis Publishers.

Lauwerys, R. R., P. Hoet. 1993. *Industrial Chemical Exposure.* New York: CRC Press/Lewis Publishers.

Lewis, R. J., Sr. 1993. *Hawley's Condensed Chemical Dictionary,* 12th Edition. New York: Van Nostrand Reinhold.

Lewis, R. J., Sr. 1992. *Sax's Dangerous Properties of Industrial Materials,* Volumes 1, 2, and 3, 8th Edition. New York: Van Nostrand Reinhold.

Meyer, E. 1990. *Chemistry of Hazardous Materials.* Englewood Cliffs, NJ: Prentice Hall.

Patnaik, P. 1992. *A Comprehensive Guide to the Hazardous Properties of Chemical Substances.* New York: Van Nostrand Reinhold.

Quigley, D. R. 1994. *Handbook of Emergency Chemical Management.* New York: CRC Press/Lewis Publishers.

Rea, W. J. 1995. *Chemical Sensitivity,* Vol. I-III. New York: CRC Press/Lewis Publishers.

Richardson, M. 1995. *Dictionary of Substance and Their Effects,* Vol. I-VI. New York: CRC Press/Lewis Publishers.

Torkelson, T. R. V. K. Rowe. 1981. *Patty's Industrial Hygiene and Toxicology,* Vol. 2B. New York: Wiley-Interscience.

MATERIAL SAFETY DATA SHEET

CHEMICAL NAME

GASOLINE

HAZARD WARNING INFORMATION

HEALTH	FIRE	REACTIVE	OTHER	DEGREE OF HAZARD	COLOR CODING	OTHER CODES
1	3	0		0 = Minimum Hazard 1 = Slight Hazard 2 = Moderate Hazard 3 = Serious Hazard 4 = Severe Hazard	HEALTH = BLUE FIRE = RED REACTIVITY = YELLOW OTHER = WHITE	OX = Oxidizer ACID = Acid ALK = Alkali COR = Corrosive W = Use No Water

SECTION I - GENERAL INFORMATION

Characterization	RCRA Number	EPA Class
Aliphatic Hydrocarbon (mixture)	**D001**	**Characteristic (I) Waste**

DOT Proper Shipping Name	Chemical Abstract Service (CAS) Number
Gasoline (casing head and natural)	**8006-61-9**
DOT Hazard Class and Label Requirements	DOT Emergency Guide Code
Flammable Liquid	**27**
DOT Identification Number	Molecular Formula
UN 1203	**Varies with grade.**

Synonyms

Motor spirit; petrol; casing head gasoline; motor fuel; natural gasoline; benzin; gasolene.

SECTION II - HAZARDOUS INGREDIENTS AND IDENTITY INFORMATION

Hazardous Components (specific identity)	OSHA Exposure Criteria	NIOSH Exposure Criteria	Immediately Dangerous to Life and Health (IDLH)	ACGIH Exposure Criteria
Gasoline (derivation: A mixture of volatile hydrocarbons of mainly branched-chain paraffins, cycloparaffins, olefins, naphthenes, and aromatics. Produced from petroleum, shale oil, Athabasca tar sands, and coal. The different grades are derived from a variety or processes). **1 ppm = 3 mg/m^3 (approx.)**	**PEL:** **300 ppm** **900 mg/m^3** **STEL (15 min):** **500 ppm** **1500 mg/m^3**	**REL:** **Reduce to Lowest Level** **Possible Human Carcinogen**	**Not Determined**	**TLV (proposed):** **300 ppm** **890 mg/m^3** **STEL:** **500 ppm** **1480 mg/m^3**

SECTION III - PHYSICAL AND CHEMICAL CHARACTERISTICS

Boiling Point	Specific Gravity (H_2O = 1)
102°F (39°C)	**0.72 to 0.76 at 69°F (20°C)**
Vapor Pressure (mm Hg)	Molecular Weight (atomic weight)
38 to 300 at 69°F (20°C)	**72 (approx.)**
Vapor Density (Air = 1)	Melting Point
3.0 to 4.0	**Not Found**

Solubility

Insoluble in water. Freely soluble in alcohol, ether, chloroform, and benzene.

Appearance and Odor

A clear or synthetically colored (with dye), mobile, volatile, aromatic liquid; a mixture of aliphatic hydrocarbons. Odor Threshold = 10 ppm.

SECTION IV - FIRE AND EXPLOSION HAZARD DATA

Flash Point (method used)	Explosive Limits in Air % by Volume
-45°F (-43°C) closed cup	**LEL: 1.4% UEL: 7.6%**
NFPA Classification	**Autoignition Temperature**
Class 1B Flammable Liquid	**536 to 853°F (280 to 456°C)**

Extinguishing Media

Use alcohol foam, dry chemical, or carbon dioxide. Water may be ineffective and may actually spread fire. A water spray may be used to knock down vapors.

Special Fire Fighting Procedures

Isolate fire area. Wear self-contained breathing apparatus (SCBA). Poisonous gases are produced in fire. Use water spray to cool fire-exposed containers until long after fire is out. Use non-sparking tools.

Unusual Fire and Explosion Hazards

Dangerous fire and explosion hazard when exposed to heat or flame. Containers may explode in fire. The vapor is heavier than air and may travel for some distance to cause fire to flashback to source.

SECTION V - REACTIVITY DATA

Stability		Conditions to Avoid
		Gasoline is normally stable under routine conditions of handling and storage. Avoid contact with heat, flame, other sources of ignition, and contact with incompatible materials and chemicals.
Stable	Unstable	Incompatibility *(materials to avoid)*
X		Strong oxidizers (such as peroxides), nitric acid, and perchlorates.
Hazardous Polymerization		Conditions to Avoid
		Hazardous polymerization has not been reported to occur under normal conditions of temperature and pressure.
May Occur	Will Not Occur	Hazardous Decomposition or By-products
	X	When heated to decomposition, gasoline emits toxic and acrid fumes, such as oxides of carbon and partially oxidized hydrocarbons.

SECTION VI - HEALTH HAZARD DATA

Primary Route(s) of Entry:	Inhalation?	Absorption (skin/eye)?	Ingestion?
	X	X	X

Health Hazards

INHALATION: Vapors are poisonous. Causes irritation of the respiratory system mucosa and depression of the central nervous system (CNS). High concentrations even for short periods may lead to fatal pulmonary edema (fluid in the lungs). Intoxicating and anesthetic effects.

SKIN & EYES: Irritation with drying, cracking, redness, blistering (prolonged contact). May lead to contact dermatitis and secondary infection.

INGESTION: May cause nausea, vomiting, headache, dizziness, gastrointestinal irritation; blurring of vision, and other effects of central nervous system depression.

Carcinogenicity	NTP Listed?	IARC Cancer Review Group?	OSHA Regulated?	Target Organs?
Questioned Human Suspected Animal	5th Annual Report	Group 2B	Not Listed	Eyes, skin, respiratory sys. liver, CNS, kidneys.

Medical Conditions Generally Aggravated by Exposure

None identified.

Emergency and First-aid Procedures

Eye contact: Do not allow victim to keep eyes tightly shut. Flush immediately with water for 15 minutes (minimum), seek medical attention. Skin contact: Remove all contaminated clothing, including shoes. Immediately wash area with flooding amounts of soap and water. Seek medical attention immediately. For inhalation: Remove the person from exposure. Provide respiratory assistance and CPR. Transfer to medical facility. Observe 48 hours for lung effects. If swallowed: Seek medical attention immediately. Do not induce vomiting due to possible aspiration of vomitus/chemical into the lungs.

SECTION VII - PRECAUTIONS FOR SAFE HANDLING AND USE

Steps to be Taken in Case Material is Released or Spilled

Remove all ignition sources. Wear SCBA respiratory protection. Restrict those not wearing protective equipment and who are not involved in cleanup from entering area. Avoid skin contact. Stop leak if it can be done without risk. Provide maximum explosion-proof ventilation. Absorb with vermiculite or other similar material and deposit in sealed containers. Use non-sparking tools during cleanup.

Preferred Waste Disposal Method

No Citation.

Precautions to be Taken in Handling and Storage

Store in tightly closed containers in cool, well ventilated location. Keep away from sources of heat.

Other Precautions and Warnings

Sources of ignition (open flame or smoking), are prohibited where gasoline is used or stored. Metal containers should be bonded and grounded during transfer operations. Bulk storage not recommended.

SECTION VIII - CONTROL MEASURES AND PERSONAL PROTECTIVE EQUIPMENT

Respiratory Protection *(specify type)*

Use a MSHA/NIOSH-approved air-purifying organic vapor respirator, a powered air purifying respirator, or use a self-contained breathing apparatus (SCBA) with full facepiece operated in positive pressure mode.

Ventilation

Local exhaust (preferred) at site of chemical work or general ventilation.

Protective Gloves	Eye Protection	Other Protective Clothing
Neoprene or Polyvinyl Alcohol Gloves	Chemical Goggles or Face Mask	Protective Apron if Splash is Likely

Work/Hygiene Practices

Always wash hands thoroughly after using chemical; never bring food, drink, or smoking materials into vicinity of chemicals.

GASOLINE

Formula varies with grade

CAS: 8006-61-9

IDENTIFICATION AND TYPICAL USES

Gasoline is a clear or colored (with dye), mobile, aromatic, volatile liquid with a characteristic odor. It is a mixture of volatile hydrocarbons suitable for use in a spark-ignited internal combustion engine. It has an octane number of at least 60. Octane numbers of 90 or slightly lower are almost chiefly used as automotive fuels and octane numbers higher than 100 are aviation fuels. Other variations and formulations exist for *anti-knock gasoline, cracked gasoline, lead-free gasoline, natural gasoline, polymer gasoline, pyrolysis gasoline, reformed gasoline, straight-run gasoline,* and *white gasoline.*

RISK ASSESSMENT: HEALTH

General Assessment

Gasoline is considered moderately poisonous by *inhalation* of its vapors or by *ingestion* of the liquid. Skin contact will also cause localized damage at site of contact. When large quantities are in contact with skin, toxic amounts may be absorbed. It is suspected to cause cancer in animals but there is inadequate evidence to rate its human carcinogenic potential. Its teratogenic and mutagenic properties have not been reported in the references.

Inhalation of the gasoline vapors can cause mild to severe irritation of the eyes, nose, throat, and respiratory tract mucosa. Systemic effects include cough, conjunctiva irritation, flushing of the face, headaches, and blurred vision. High concentrations even for short periods can cause potentially fatal pulmonary edema (fluid in the lungs). Symptoms of coughing, wheezing, congestion, tightness in chest, and dypsnea may be delayed up to 48 hours, thereby creating a false sense of security with regard to health exposure risk. Inhalation can also lead to depression of the central nervous system (CNS) with symptoms of intoxication, hallucination, mental confusion, staggering gait or stupor, slurred speech, loss of consciousness and possible convulsions.

Skin contact may cause dermatitis and absorption can occur with harmful effects. It is a comparatively strong irritant, producing erythema, and burning, and, in severe cases, edema and even blistering. Contact dermatitis is possible, as is secondary infection. Eye contact can cause irritation of the conjunctiva with redness, swelling, tearing, and localized inflammation of eye tissues.

Ingestion may cause nausea, vomiting, headaches, dizziness, gastrointestinal irritation, blurred vision, drunkenness, fever, drowsiness, confusion, cyanosis, and symptoms of CNS depression similar to that caused by inhalation. If vomiting occurs, aspiration of the liquid into the lungs can cause choking, difficulty breathing or shortness of breath (dypsnea), tachypnea (rapid respiration), tachycardia (rapid heartbeat), fever, bronchitis, and pneumonitis.

Acute Health Effects

The following acute (short-term) health effects may occur immediately or shortly after exposure to gasoline:

Skin: Irritation with potential for redness, blistering, burning, and absorption.

Eye: Severe irritation, conjunctivitis, redness, possible pain, and visual disturbances.

Lung: Nose, throat, and respiratory tract irritation causing difficulty in breathing, possible pulmonary edema on high exposure (a medical emergency) which can be fatal.

CNS: Inebriation, dizziness, stupor, hallucinations, slurred speech, mental confusion, loss of consciousness, and possible convulsions.

Other: Acute exposure can also lead to acute hemorrhage of the pancreas, fatty degeneration of the liver and kidneys, congestion of the spleen.

Chronic Health Effects

The following chronic (long-term) health effects may occur at some time after exposure to gasoline and can last for months or even years:

Cancer Hazards: Gasoline is a suspected carcinogen in animal studies. Its ability to cause cancer in humans has not been adequately determined. The IARC states insufficient evidence exists for a human carcinogenic determination.

Reproductive Hazard: There is no evidence in the references to indicate that exposure to gasoline causes any adverse effects to reproduction.

Other Chronic Effects: Chronic inhalation of gasoline vapors can cause loss of appetite, nausea, weight loss, insomnia, hyperesthesia, motor weakness, muscular degeneration, diminished tendon reflexes, and loss of coordination. Repeated skin contact causes blistering, drying, and lesions.

Recommended Risk-Reduction Measures

Personnel should avoid direct contact with or exposure to gasoline. Occupational poisoning occurs most commonly by inhalation. Substitution is not usually an option with gasoline since its use is quite specific. However, a less toxic material or compound should be used in place of gasoline whenever possible. If substitution cannot be done, *engineering controls* are the most effective method of reducing exposures. The best protection is to enclose operations and/or provide local exhaust ventilation at the site of gasoline release. While not always operationally feasible, isolating operations can also reduce exposure. Using respiratory protection is less effective than the controls mentioned above, but is still advisable whenever working with or around gasoline. An MSHA/NIOSH-approved air-purifying respirator with an organic vapor cartridge, a supplied-air respirator operated in continuos flow mode, or a self-contained breathing apparatus (SCBA) with full facepiece operated in pressure demand are the appropriate methods of respiratory protection when exposure to gasoline is anticipated. If a full facepiece is not available, then chemical goggles should be worn to protect the eyes. A face shield should also be considered. To prevent hand and skin exposures, neoprene or polyvinyl alcohol gloves should be worn. Glove manufacturers should be contacted and permeation studies obtained *before* final glove selections are made.

Administrative controls should also be in place to minimize the potential for human exposures. These may include written procedures or policies which specify the methods and techniques that will be practiced whenever personnel are to work with gasoline.

All personnel should receive *training* on the use, hazards, protective measures, emergency actions, and other precautions per 29 CFR 1910.1200 (Hazard Communication), prior to the first assignment in an area where gasoline is used or stored.

Before beginning employment and at regular times thereafter, the following recommended medical tests should be considered:

- ☑ Lung function tests (establish baseline).

If symptoms occur or overexposure is suspected, the following additional tests are suggested:

- ☑ Lung function tests (compare to baseline).
- ☑ Consider chest X-ray following acute overexposure (may be negative if taken immediately after exposure due to delayed development of pulmonary edema).
- ☑ Liver and kidney function tests.

Any evaluation should include a careful history of past and present symptoms with an examination. Medical tests that simply look for existing damage are not a substitute for controlling exposures. Also, since smoking can cause heart disease, lung cancer, emphysema, and other respiratory disorders, smokers exposed to gasoline may experience symptoms more rapidly than non-smokers under the same conditions of exposure. Prudent risk management requires careful consideration of *all* possible risk factors.

Other methods to reduce exposure include:

- ☑ Where possible, enclose operations and use local exhaust ventilation at the site of chemical release. If local exhaust ventilation or enclosure is not used, respiratory protection should be mandatory.
- ☑ Always ensure that proper protective clothing is worn when using chemical substances and that personnel have been trained on its use, care, and maintenance.
- ☑ Wash thoroughly immediately after exposure to gasoline and at the end of the work shift or before eating, drinking, or smoking.
- ☑ Hazard warning information should be posted in the work area. In addition, as part of an on-going education and training program, all information on the health and safety hazards of gasoline should be communicated to all potentially exposed workers.

- ☑ Eye wash stations should be provided in the immediate work area for emergency use. If there is a possibility of skin exposure to gasoline, emergency shower facilities should also be provided.
- ☑ Workers whose clothing has been contaminated by gasoline should change into clean clothes before leaving work. Contaminated work clothing should be laundered only by individuals who have been informed of the hazards of exposure to gasoline.
- ☑ Where possible, automatically transfer liquids containing gasoline from drums or other containers to process containers. Always ground and bond all containers and equipment. Use non-sparking tools.

RISK ASSESSMENT: ENVIRONMENT

General Assessment

The environment is at risk of exposure during transportation, storage, disposal, or destruction of gasoline. In almost every scenario, the threat of environmental exposure is contingent upon the proper handling of the chemical substance. Accidental spills, large or small, can result in contamination of the surrounding environmental mediums (water, soil, and air).

Gasoline is a moderately flammable liquid. Because of its extremely low flash point and high boiling point, it is classified as a Class 1B flammable liquid (per OSHA 29 CFR 1910.106). It is a dangerous fire hazard and moderate explosion hazard when exposed to heat or flame. It is incompatible with commonly encountered oxidizers and some acids (such as nitric). Caution is always required in handling, storage, transportation, and disposal of gasoline. Emergency responders should be made aware of the presence of gasoline at any emergency response situation.

Gasoline can enter the environment from industrial effluents and from spills.

Acute Ecological Effects

Acute (short-term) toxic effects may include the death of animals, birds, or fish and death or low growth rate in plants. Acute effects are seen 2 to 4 days after animals or plants are exposed to gasoline. Gasoline has high acute toxicity to aquatic life. It can cause death in plants and roots and membrane damage in leaves of various agricultural crops. Insufficient data are available to evaluate the short-term effects of gasoline on birds or land animals.

Chronic Ecological Effects

Chronic (long-term) toxic effects may include shortened life span, reproductive problems, lower fertility, and changes in appearance or behavior in exposed animals. These effects can be seen long after first exposure(s) to toxic chemicals. Gasoline has high chronic toxicity to aquatic life. Insufficient data are available to evaluate or predict the long-term effects of gasoline to plants, birds, or terrestrial animals.

Water Solubility

Gasoline is not soluble in water. Concentrations of 1 milligram will not mix with a liter of water.

Persistence in the Environment

Gasoline is slightly persistent in water, with a half-life of between 2 to 20 days. The half-life of a pollutant is the amount of time it takes for one half of the chemical to be degraded. About 99.5% of gasoline will eventually end up in air; the rest will end up in water.

Bioaccumulation in Aquatic Organisms

Some substances increase in concentration, or *bioaccumulate,* in living organisms as they breathe contaminated air, drink contaminated water, or eat contaminated food. These chemicals can become concentrated in the tissues and internal organs of animals as well as humans.

The concentration of gasoline found in fish tissues is expected to be somewhat higher than the average concentration of gasoline in the water from which the fish was taken.

Recommended Risk-Reduction Measures

Proper training of all transporters will reduce the likelihood of a mishap or accident resulting in a leak or spill of gasoline into the environment. Labels on all containers, trucks, and rail cars must meet DOT requirements and accurately reflect their contents to enable emergency responders to react properly and quickly to any disaster thereby reducing the potential risk to the environment and to personnel.

Storage of gasoline should be segregated from other chemicals and materials to minimize the risk of cross-contamination. Facilities should be designed to meet explosion-proof requirements.

If a spill or leak to the environment has occurred, fire department, emergency response and/or hazardous materials spill personnel should be notified immediately. Cleanup should be attempted only by those trained in proper spill containment procedures using non-sparking tools. Contaminated soils should be removed for incineration and replaced with clean soil. If gasoline should contact the water table, aquifer, or navigable waterway, time is of the essence. It is not soluble and total remediation may be possible. In general, gasoline will float on the surface and may be captured if response actions are prompt. The local and/or state emergency response authorities must be notified. A comprehensive emergency response or disaster preparedness/recovery plan should be in place prior to any operations involving the use, transportation, storage, or disposal of gasoline.

If gasoline is spilled or leaked, the following specific steps are recommended:

- ☑ Restrict persons not wearing protective clothing from area of spill or leak until cleanup is complete and area can be opened for normal work.
- ☑ Absorb liquids in vermiculite, sand, earth, or similar material and deposit in sealed drum. Use non-sparking tools during cleanup procedures.
- ☑ Ventilate area of spill or leak. Container should be removed to safer area if it can be done without increasing the risk.
- ☑ Use water spray to knock down vapors.
- ☑ It may be necessary to dispose of gasoline as a hazardous waste. The responsible state agency or the regional office of the federal Environmental Protection Agency (EPA) should be contacted for specific recommendations.

RISK ASSESSMENT: BUSINESS

General Assessment

Accidents or mishaps involving gasoline can present a serious threat to business operations. The loss or damage of equipment or facilities can significantly affect fiscal viability. Lawsuits that may result from personnel illness, injury/death, public exposures, and/or environmental contamination will require a serious expenditure of resources. Media attention surrounding an injury, death, or environmental damage can also result in a loss of profits and loss of current as well as future business. Always remember that anytime the terms "cancer," or "carcinogen" are used, public emotion, hysteria, and ignorance can run equally high. This must be carefully considered when developing or implementing any public relations policies.

Recommended Risk-Reduction Measures

Company attorneys, safety and health professionals, and environmental specialists should be involved in the development of any procedures for responding to chemical incidents. A company official should be predesignated as a public relations officer with specific training in dealing with the press. Corporate plans and policies should be developed, approved, and implemented long before any need for such arises.

REFERENCES

American Conference of Governmental Industrial Hygienists. 1988. *Documentation of the Threshold Limit Values and Biological Exposure Indices*, 5th Edition (with updates). Cincinnati: ACGIH

Bloom, A. D. and F. J. DeSerres. 1995. *Ecotoxicity and Human Health.* New York: CRC Press/Lewis Publishers.

Bueche, J. R. 1972. *Principles of Physics,* 2nd Edition. New York: McGraw-Hill.

Cockerman, L. G., B. S. Shane. 1994. *Basic Environmental Toxicology.* New York: CRC Press/Lewis Publishers.

Howard, P. H., et. al. 1991. *Handbook of Environmental Degradation Rates.* New York: CRC Press/Lewis Publishers.

Jones, F. E. 1994. *Toxic Organic Vapors in the Workplace.* New York: CRC Press/Lewis Publishers.

Lauwerys, R. R., P. Hoet. 1993. *Industrial Chemical Exposure.* New York: CRC Press/Lewis Publishers.

Lewis, R. J., Sr. 1993. *Hawley's Condensed Chemical Dictionary,* 12th Edition. New York: Van Nostrand Reinhold.

Lewis, R. J., Sr. 1992. *Sax's Dangerous Properties of Industrial Materials,* Volumes 1, 2, and 3, 8th Edition. New York: Van Nostrand Reinhold.

Meyer, E. 1990. *Chemistry of Hazardous Materials.* Englewood Cliffs, NJ: Prentice Hall.

Patnaik, P. 1992. *A Comprehensive Guide to the Hazardous Properties of Chemical Substances.* New York: Van Nostrand Reinhold.

Proctor, N. H., J. P. Hughes. 1991. *Chemical Hazards of the Workplace.* New York: Van Nostrand Reinhold.

Quigley, D. R. 1994. *Handbook of Emergency Chemical Management.* New York: CRC Press/Lewis Publishers.

Sacarello, H. L. A. 1994. *The Comprehensive Handbook of Hazardous Materials.* New York: CRC Press/Lewis Publishers.

Torkelson, T. R. V. K. Rowe. 1981. *Patty's Industrial Hygiene and Toxicology,* Vol. 2B. New York: Wiley-Interscience.

U.S. Department of Health and Human Services, National Institute for Occupational Safety and Health. 1994. *NIOSH Pocket Guide to Chemical Hazards.* Washington, D.C.: U.S. Government Printing Office.

MATERIAL SAFETY DATA SHEET

CHEMICAL NAME	
GERMANIUM	CRC PRESS/LEWIS PUBLISHERS DISCLAIMER: THE INFORMATION AND RECOMMENDATIONS PRESENTED HEREIN ARE BASED ON SOURCES BELIEVED TO BE RELIABLE. CRC MAKES NO REPRESENTATION ON ITS COMPLETENESS OR ACCURACY. IT IS THE RESPONSIBILITY OF THE USER TO DETERMINE THE CHEMICAL'S SUITABILITY FOR ITS INTENDED USE, THE CHEMICAL'S SAFE USE, AND THE CHEMICAL'S PROPER DISPOSAL. NO REPRESENTATIONS AND/OR WARRANTIES, EITHER EXPRESSED OR IMPLIED, OF THE MERCHANTABILITY OR FITNESS FOR A PARTICULAR PURPOSE, OR OF ANY OTHER NATURE, ARE MADE WITH RESPECT TO THE INFORMATION PROVIDED IN THIS MSDS OR TO THE CHEMICAL TO WHICH INFORMATION MAY REFER. CRC NEITHER ASSUMES NOR AUTHORIZES ANY OTHER PERSON TO ASSUME FOR IT, ANY OTHER ADDITIONAL RESPONSIBILITY OR LIABILITY FOR THE USE OF, OR RELIANCE UPON, THIS INFORMATION.

HAZARD WARNING INFORMATION

HEALTH	FIRE	REACTIVE	OTHER	DEGREE OF HAZARD	COLOR CODING	OTHER CODES
1	0	2		0 = Minimum Hazard 1 = Slight Hazard 2 = Moderate Hazard 3 = Serious Hazard 4 = Severe Hazard	HEALTH = BLUE FIRE = RED REACTIVITY = YELLOW OTHER = WHITE	OX = Oxidizer ACID = Acid ALK = Alkali COR = Corrosive W = Use No Water

SECTION I - GENERAL INFORMATION

Characterization	RCRA Number	EPA Class
Non-metallic Element	**None**	**Not Applicable**

DOT Proper Shipping Name **No Citation**	Chemical Abstract Service (CAS) Number **7440-56-4**
DOT Hazard Class and Label Requirements **No Citation**	DOT Emergency Guide Code **No Citation**
DOT Identification Number **No Citation**	Chemical Formula **Ge**

Synonyms
None Reported.

SECTION II - HAZARDOUS INGREDIENTS AND IDENTITY INFORMATION

Hazardous Components (specific identity)	OSHA Exposure Criteria	NIOSH Exposure Criteria	Immediately Dangerous to Life and Health (IDLH)	ACGIH Exposure Criteria
Germanium (derivation: Non-metallic element of atomic number 32. Germanium is recovered from residues of zinc and other sources, by heating in the presence of air and chlorine. Also present in some coals and can be recovered from their ash.).	**PEL:** Not Established **STEL:** Not Established	**REL:** Not Established **STEL:** Not Established	**Not Determined**	**TLV:** Not Established **STEL:** Not Established

SECTION III - PHYSICAL AND CHEMICAL CHARACTERISTICS

Boiling Point **5126°F (2830°C)**	Specific Gravity (H_2O = 1) **5.323**
Vapor Pressure (mmHg) **Not Found**	Molecular Weight **72.59**
Vapor Density (Air = 1) **Not Found**	Freezing Point **1719°F (932°C)**

Solubility
Insoluble in water, hydrochloric acid, dilute alcohol, and hydroxides. Attacked by nitric acid and aqua regia.

Appearance and Odor
A grayish-white metalloid, crystalline and brittle solid.

SECTION IV - FIRE AND EXPLOSION HAZARD DATA

Flash Point (method used) **Not Found**	Explosive Limits in Air % by Volume **LEL: Not Determined UEL: Not Determined**
NFPA Classification **Not Found**	Autoignition Temperature **Not Determined**

Extinguishing Media
In solid form, germanium will not burn. Its dusts can ignite and are reactive with a number of other compounds. Use extinguishing media suitable to surrounding fire.

Special Fire Fighting Procedures
Germanium can create a dangerous fire and explosion hazard in contact with incompatible materials. Poisonous gases are produced in fire. Wear self-contained breathing apparatus. Use water spray to keep containers cool. Fight fire from distance, if possible.

Unusual Fire and Explosion Hazards
None Reported.

SECTION V - REACTIVITY DATA

Stability		Conditions to Avoid: Germanium is normally stable in closed containers under normal conditions of temperature and pressure and routine conditions of handling and storage. Avoid contact with heat and incompatible materials.
Stable	Unstable	Incompatibility *(materials to avoid)*
X		Potassium chlorate, potassium nitrate, nitric acid, chlorine, bromine, potassium hydroxide, aqua regia, concentrated sulfuric acid, fused alkalies, nitrates, carbonates, halogens, and oxidants.
Hazardous Polymerization		Conditions to Avoid: Under normal conditions of temperature and pressure, hazardous polymerization of germanium is not expected to occur.
May Occur	Will Not Occur	Hazardous Decomposition or By-products
	X	None reported.

SECTION VI - HEALTH HAZARD DATA

Primary Route(s) of Entry:	Inhalation?	Absorption (skin)?	Ingestion?
	X		X

Health Hazards

INHALATION: Irritation of the upper respiratory tract. Conjunctivitis; visual disturbances, may cause ulceration or perforation of nasal septum similar to arsenic exposure. Causes irritation to the lungs. Inhaled dusts are rapidly absorbed. Systemic effects are not specified.

SKIN & EYES: Skin contact may cause irritation and possible rash with redness. Eye irritation is also likely to occur on contact with the dusts. Conjunctivitis may occur in some cases.

INGESTION: Unspecified gastrointestinal effects.

Carcinogenicity	NTP Listed?	IARC Cancer Review Group?	OSHA Regulated?	Target Organs?
Unknown Human Unknown Animal	No	No	No	Respiratory system; skin; eyes.

Medical Conditions Generally Aggravated by Exposure

None Reported.

Emergency and First-aid Procedures

Eye contact: Immediately flush large amounts of water for 15 minutes (minimum), occasionally lifting eyelids, seek medical attention. Skin contact: Remove all contaminated clothing. Immediately wash area with large amounts of soap and water. For inhalation: Remove the person from exposure. Provide respiratory assistance and CPR. Transfer promptly to medical facility. If swallowed: Seek medical attention immediately. Never attempt to give an unconscious and convulsing person anything by mouth.

SECTION VII - PRECAUTIONS FOR SAFE HANDLING AND USE

Steps to be Taken in Case Material is Released or Spilled

Remove all ignition sources. Ventilate area. Evacuate non-essential personnel. Collect powdered materials in safest manner possible. Do NOT dry sweep. Use HEPA vacuum instead. Have water source available to fight any surrounding fires. Emergency personnel must avoid contact with airborne dusts.

Preferred Waste Disposal Method

No Citation.

Precautions to be Taken in Handling and Storage

Store in closed containers in cool, dry, well-ventilated area away from heat. Avoid handling activities that create dusty conditions. Respiratory protection should be worn by all personnel involved in handling and storage operations.

Other Precautions and Warnings

Keep away from incompatible materials since violent reactions can easily occur.

SECTION VIII - CONTROL MEASURES AND PERSONAL PROTECTIVE EQUIPMENT

Respiratory Protection *(specify type)*

PEL has not been established. An air-purifying respirator with a HEPA filter and dust/mist pre-filter is the minimum suggested protection. Recommend using a powered supplied-air respirator with full facepiece operated in pressure demand mode. Greater protection is obtained from a self-contained breathing apparatus (SCBA) with full facepiece in pressure demand or other positive pressure mode.

Ventilation

Local exhaust or general mechanical systems recommended.

Protective Gloves	Eye Protection	Other Protective Clothing
Impervious Rubber	Chemical/Dust Goggles and/or Face Mask	Protective Apron

Work/Hygiene Practices

Always wash hands thoroughly after using chemical; never bring food, drink, or smoking materials into vicinity of chemicals.

GERMANIUM

Ge

CAS: 7440-56-4

IDENTIFICATION AND TYPICAL USES

Germanium is a non-metallic element of atomic number 32, atomic weight of 72.59, valances of 2,4, group IVA of the periodic system. It appears as a grayish-white metalloid, crystalline, and brittle solid. It is used in solid-state electronic devices such as transistors and diodes. Also used in semiconductor applications, in brazing alloys, in phosphors, in gold and beryllium alloys, and in infrared-transmitting glass.

RISK ASSESSMENT: HEALTH

General Assessment

Germanium is considered moderately toxic by *inhalation* and *ingestion*. Human and animal data are lacking in the references. There is no information on the carcinogenic, teratogenic, or mutagenic properties of germanium. However, some germanium compounds, such as tributyl germanium and germanium tetrachloride are known mutagens and dimethyl germanium is a confirmed teratogen.

Inhalation of large concentrations of germanium dusts will cause irritation to the eyes, nose, throat, and respiratory tract. Eye irritation can progress to conjunctivitis. Inhaled dusts are rapidly absorbed through the respiration process. Once in the body, the effects are considered toxic, systemic, but are not clearly specified in the references. It does have a close chemical relationship with arsenic and caution is suggested in its use and handling.

Acute Health Effects

The following acute (short-term) health effects may occur immediately or shortly after exposure to germanium:

Skin: Irritation and possible rash in some individuals.

Eye: Mild to moderate irritation that may progress to conjunctivitis.

Lung: Irritation of the mouth, nose, and throat causing coughing and/or shortness of breath.

Other: Ingestion of the dusts may cause throat and stomach irritation with unspecified gastrointestinal effects.

Chronic Health Effects

The following chronic (long-term) health effects may occur at some time after exposure to germanium and can last for months or even years:

Cancer Hazards: According to information presented in the references, germanium has not been adequately tested for its ability to cause cancer in test animals. Some compounds, such as germanium tetrachloride, are known mutagens.

Reproductive Hazards: There are no reports to support any claims of reproductive hazards following exposure to germanium. However, some germanium compounds (such as dimethyl germanium) are known teratogens.

Recommended Risk-Reduction Measures

Personnel should avoid direct contact with germanium dusts and powders. Although the toxicity of the base element is not clearly understood in the references, many germanium compounds are known to cause moderate to severe toxic health effects. Unless a less toxic chemical can be substituted for a hazardous substance, *engineering controls* are the most effective method of reducing exposures. The best protection is to enclose operations and/or provide local exhaust ventilation at the site of chemical release. Because germanium can be highly reactive, explosion-proof ventilation is recommended. While not always operationally feasible, isolating operations can also reduce exposure. Using respiratory protection is less effective than the controls mentioned above, but is still advisable whenever working with germanium. No exposure level has been established for elemental germanium. However, this does not mean that exposure is entirely risk free. An MSHA/NIOSH-approved air-purifying respirator equipped with a high-efficiency particulate air (HEPA) filter and a dust/mist pre-filter may suffice. The best protection is obtained using a powered supplied-air respirator operated in continuous flow mode or a self-contained breathing apparatus (SCBA) with full facepiece in pressure demand mode. If a full facepiece is not available, then chemical goggles should be worn to protect the eyes. Whenever an airborne dust hazard exists, a face shield and a protective apron should be worn. To prevent hand and skin exposures, impervious protective gloves should be worn.

Administrative controls should also be in place to minimize the potential for human exposures. These may include written procedures or policies which specify the methods and techniques that will be practiced whenever personnel are to work with germanium.

All personnel should receive *training* on the use, hazards, protective measures, emergency actions, and other precautions per 29 CFR 1910.1200 (Hazard Communication), prior to the first assignment in an area where germanium is used or stored.

Before beginning employment and at regular intervals thereafter (e.g., annually), the following medical tests are recommended:

☑ Lung function tests.

If symptoms develop or overexposure is suspected, the following may be useful:

☑ Consider chest X-ray after acute overexposure.

Any evaluation should include a careful history of past and present symptoms with an examination. Medical tests that simply look for existing damage are not a substitute for controlling exposures. Also, since smoking can cause heart disease, lung cancer, emphysema, and other respiratory disorders, smokers exposed to germanium may experience symptoms more rapidly than non-smokers under the same conditions of exposure. Prudent risk management requires careful consideration of *all* possible risk factors.

Other methods to reduce exposure to chemicals include:

☑ Where possible, enclose operations and use local exhaust ventilation at the site of chemical release. If local exhaust ventilation or enclosure is not used, respiratory protection should be mandatory.

☑ Always ensure that proper protective clothing is worn when using chemical substances and that personnel are trained in its use and care.

☑ Wash thoroughly immediately after exposure to germanium and at the end of the work shift or before eating, drinking, or smoking. Work clothing contaminated with germanium should never be taken home for laundering (fire hazard). Only personnel trained in the hazards of exposure should launder contaminated clothing.

☑ Safety shower and eyewash stations should be readily available in work areas where germanium is used or stored.

☑ Hazard warning information should be posted in the work area. In addition, as part of an on-going education and training program, all information on the health and safety hazards of germanium should be communicated to all exposed and potentially exposed workers.

RISK ASSESSMENT: ENVIRONMENT

General Assessment

The environment is at risk of exposure during transportation, storage, disposal, or destruction of germanium. In almost every scenario, the threat of environmental exposure is contingent upon the proper handling of the chemical substance. Accidental releases, large or small, can result in fire, explosion, and possible contamination of the surrounding environmental mediums (primarily air, but water and soil may also be contaminated depending upon the condition of the spill).

Germanium is considered to be a dangerous, reactive material. It can react explosively when heated with potassium chlorate or potassium nitrate. It can react violently with nitric acid and will ignite on contact with chlorine or bromine plus heat. It will react incandescently when heated with oxygen or potassium hydroxide. It is incompatible with aqua regia, concentrated sulfuric acid, fused alkalis, nitrates and carbonates, halogens, and oxidants. These characteristic require special consideration during any emergency involving a leak or release of germanium gas.

Germanium can be present naturally and can enter the environment through spills or leaks. Unchecked discharges from industrial sources are also possible.

☠ *Acute Ecological Effects*

Acute (short-term) toxic effects may include the death of animals, birds, or fish, and death or low growth rate in plants. Acute effects are seen 2 to 4 days after animals or plants are exposed to germanium.

Insufficient data are available to accurately evaluate or predict the short-term effects of germanium exposure to aquatic life, plants, birds, or land animals.

Chronic Ecological Effects

Chronic (long-term) toxic effects may include shortened life span, reproductive problems, lower fertility, and changes in appearance or behavior in exposed animals. These effects can be seen long after first exposure(s) to toxic chemicals.

Insufficient data are available to evaluate or predict the long-term (chronic) effects of germanium to aquatic life, plants, birds, or land animals.

Water Solubility

Germanium is insoluble in water. Concentrations of 1 milligram will not mix with a liter of water.

Persistence in the Environment

Germanium is highly persistent in water, with a half-life of greater than 200 days. The half-life of a pollutant is the amount of time it takes for one half of the chemical to be degraded. Virtually all of germanium will end up in terrestrial soil and aquatic sediments.

Bioaccumulation in Aquatic Organisms

Some substances increase in concentration, or *bioaccumulate,* in living organisms as they breathe contaminated air, drink contaminated water, or eat contaminated food. These chemicals can become concentrated in the tissues and internal organs of animals as well as humans.

The concentration of germanium found in fish tissues is expected to be somewhat higher than the average concentration of germanium in the water from which the fish was taken.

Recommended Risk-Reduction Measures

Proper training of all transporters will reduce the likelihood of a mishap or accident resulting in a leak or release to the environment. The correct use of labeling on all containers, trucks, and rail cars will enable emergency responders to react properly and quickly to any disaster thereby reducing the potential risk to the environment and to personnel.

Storage of germanium should be segregated from sources or heat or ignition and kept away from incompatible materials. Containers should be kept closed and in a cool, dry, well-ventilated area. Buildings designated for storage should be equipped with appropriate fire protection systems (alarms, sprinklers, emergency lighting, portable extinguishers).

If a spill or release to the environment has occurred, fire department, emergency response, and/or hazardous materials spill personnel should be notified immediately. Contaminated soils should be removed and replaced with clean soils. Cleanup should be attempted only by those trained in proper containment procedures. If germanium should contact the water table, aquifer, or navigable waterway, time is of the essence. It is insoluble in water and total containment and remediation may be possible. When such spills occur, the local and/or state emergency response authorities must be notified. A comprehensive emergency response or disaster preparedness/recovery plan should be in place prior to any operations involving the use, transportation, storage, or disposal of germanium. If germanium is spilled or leaked, the following steps are recommended:

- ☑ Restrict persons not wearing protective clothing from area of spill or leak until cleanup is complete and area can be opened for normal work.
- ☑ Remove all ignition sources.
- ☑ Ventilate area of spill or leak.
- ☑ Collect powdered materials in safest manner possible. Do NOT dry sweep (generates airborne dusts). Use a vacuum equipped with a high-efficiency particulate air (HEPA) filter instead. Dispose of in sealed containers.
- ☑ It may be necessary to dispose of germanium as a hazardous waste. The responsible state agency or the regional office of the federal Environmental Protection Agency (EPA) should be contacted for specific recommendations.

RISK ASSESSMENT: BUSINESS

General Assessment

Accidents or mishaps involving germanium can present a significant threat to business operations. The loss or damage of equipment or facilities can significantly affect fiscal viability. Lawsuits that may result from personnel injury, public exposures, and environmental contamination will also require a serious expenditure of resources. Media attention surrounding an injury, illness, or environmental damage can also result in a loss of profits and loss of current as well as future business.

Recommended Risk-Reduction Measures

Company attorneys, safety and health professionals, and environmental specialists should be involved in the development of any procedures dealing with safety or emergency response. A company official should be pre-designated as a public relations officer with specific training in dealing with the press. Corporate plans and policies should be developed, approved, and implemented long before any need for such arises.

REFERENCES

Ahlbom, A. 1993. *Biostatistics for Epidemiologists.* New York: CRC Press/Lewis Publishers.

American Conference of Governmental Industrial Hygienists. 1988. *Documentation of the Threshold Limit Values and Biological Exposure Indices,* 5th Edition (with updates). Cincinnati: ACGIH

Anthony, C. P., and N. J. Kolthoff. 1971. *Textbook of Anatomy and Physiology.* St. Louis: C. V. Mosby Company.

Asante-Duah, D. K. 1993. *Hazardous Waste Risk Assessment.* New York: CRC Press/Lewis Publishers.

Bloom, A. D. and F. J. DeSerres. 1995. *Ecotoxicity and Human Health.* New York: CRC Press/Lewis Publishers.

Brusick, D. J. 1994. *Methods for Genetic Risk Assessment.* New York: CRC Press/Lewis Publishers.

Bueche, J. R. 1972. *Principles of Physics,* 2nd Edition. New York: McGraw-Hill.

Burrell, R., D. K. Flaherty, and L. J. Sauers. 1992. *Toxicology of the Immune System.* New York: Van Nostrand Reinhold.

Calabrese, E. J. 1994. *Biological Effects of Low Level Exposures.* New York: CRC Press/Lewis Publishers.

Cockerman, L. G., B. S. Shane. 1994. *Basic Environmental Toxicology.* New York: CRC Press/Lewis Publishers.

Cothern, C. R. 1995. *Handbook for Environmental Risk Decisionmaking.* New York: CRC Press/Lewis Publishers.

Gardner, R. W. 1994. *Chemical Intolerance.* New York: CRC Press/Lewis Publishers.

Hallenbeck, W. H. 1993. *Quantitative Risk Assessment for Environmental and Occupational Health,* 2nd Edition. New York: CRC Press/Lewis Publishers.

Hodgson, E., R. B. Mailman, and J. E. Chambers. 1988. *Dictionary of Toxicology.* London: The MacMillian Company.

Hosty, J. W., and P. Foster. 1990. *A Practical Guide to Chemical Spill Response.* New York: Van Nostrand Reinhold.

Howard, P. H. 1995. *Handbook of Environmental Fate and Exposures Data for Organic Chemicals,* Vol. I-IV. New York: CRC Press/Lewis Publishers.

Howard, P. H., et. al. 1991. *Handbook of Environmental Degradation Rates.* New York: CRC Press/Lewis Publishers.

James, A., and M. Lord. 1992. *Index of Chemical and Physical Data.* New York: Van Nostrand Reinhold.

Jones, F. E. 1994. *Toxic Organic Vapors in the Workplace.* New York: CRC Press/Lewis Publishers.

Kamrin, M. A. 1988. *Toxicology: A Primer on Principles and Applications.* Chelsea, Michigan: Lewis Publishers.

Keith, L. H. and M. M. Walker. 1995. *Handbook of Air Toxics.* New York: CRC Press/Lewis Publishers.

Landis, W. G., M. H. Yu. 1995. *Introduction to Environmental Toxicology.* New York: CRC Press/Lewis Publishers.

Langerman, N. 1994 *Precautionary Labels for Chemical Containers.* New York: CRC Press/Lewis Publishers.

Lauwerys, R. R., P. Hoet. 1993. *Industrial Chemical Exposure.* New York: CRC Press/Lewis Publishers.

Lewis, R. J., Sr. 1993. *Hawley's Condensed Chemical Dictionary,* 12th Edition. New York: Van Nostrand Reinhold.

Lewis, R. J., Sr. 1992. *Sax's Dangerous Properties of Industrial Materials,* Volumes 1, 2, and 3, 8th Edition. New York: Van Nostrand Reinhold.

Lide, D. R. 1995. *Handbook of Chemistry and Physics,* 76th Edition. New York: CRC Press/Lewis Publishers.

Manahan, S. E. 1992. *Toxicological Chemistry,* 2nd Edition. New York: CRC Press/Lewis Publishers.

Meyer, E. 1990. *Chemistry of Hazardous Materials.* Englewood Cliffs, NJ: Prentice Hall.

Patnaik, P. 1992. *A Comprehensive Guide to the Hazardous Properties of Chemical Substances.* New York: Van Nostrand Reinhold.

Philp, R. B. 1995. *Environmental Hazards and Human Health.* New York: CRC Press/Lewis Publishers.

Proctor, N. H., and J. P. Hughes. 1991. *Chemical Hazards of the Workplace.* New York: Van Nostrand Reinhold.

Quigley, D. R. 1994. *Handbook of Emergency Chemical Management.* New York: CRC Press/Lewis Publishers.

Rea, W. J. 1995. *Chemical Sensitivity,* Vol. I-III. New York: CRC Press/Lewis Publishers.

Richardson, M. 1995. *Dictionary of Substance and Their Effects,* Vol. I-VI. New York: CRC Press/Lewis Publishers.

MATERIAL SAFETY DATA SHEET

CHEMICAL NAME	CRC PRESS/LEWIS PUBLISHERS DISCLAIMER
GERMANIUM CHLORIDE	THE INFORMATION AND RECOMMENDATIONS PRESENTED HEREIN ARE BASED ON SOURCES BELIEVED TO BE RELIABLE. CRC MAKES NO REPRESENTATION ON ITS COMPLETENESS OR ACCURACY. IT IS THE RESPONSIBILITY OF THE USER TO DETERMINE THE CHEMICAL'S SUITABILITY FOR ITS INTENDED USE, THE CHEMICAL'S SAFE USE, AND THE CHEMICAL'S PROPER DISPOSAL. NO REPRESENTATIONS AND/OR WARRANTIES, EITHER EXPRESSED OR IMPLIED, OF THE MERCHANTABILITY OR FITNESS FOR A PARTICULAR PURPOSE, OR OF ANY OTHER NATURE, ARE MADE WITH RESPECT TO THE INFORMATION PROVIDED IN THIS MSDS OR TO THE CHEMICAL TO WHICH INFORMATION MAY REFER. CRC NEITHER ASSUMES NOR AUTHORIZES ANY OTHER PERSON TO ASSUME FOR IT, ANY OTHER ADDITIONAL RESPONSIBILITY OR LIABILITY FOR THE USE OF, OR RELIANCE UPON, THIS INFORMATION.

HAZARD WARNING INFORMATION

HEALTH	FIRE	REACTIVE	OTHER	DEGREE OF HAZARD	COLOR CODING	OTHER CODES
Not Found	Not Found	Not Found		0 = Minimum Hazard 1 = Slight Hazard 2 = Moderate Hazard 3 = Serious Hazard 4 = Severe Hazard	HEALTH = BLUE FIRE = RED REACTIVITY = YELLOW OTHER = WHITE	OX = Oxidizer ACID = Acid ALK = Alkali COR = Corrosive W = Use No Water

SECTION I - GENERAL INFORMATION

Characterization	**Metallic Compound**	RCRA Number	**None**
		EPA Class	**Not Listed**
DOT Proper Shipping Name	**Germanium Compound**	Chemical Abstract Service (CAS) Number	**10038-98-9**
DOT Hazard Class and Label Requirements	**No Citation**	DOT Emergency Guide Code	**No Citation**
DOT Identification Number	**No Citation**	Atomic Formula	**Cl_4Ge**

Synonyms

Extrema; Germanium tetrachloride.

SECTION II - HAZARDOUS INGREDIENTS AND IDENTITY INFORMATION

Hazardous Components (specific identity)	OSHA Exposure Criteria	NIOSH Exposure Criteria	Immediately Dangerous to Life and Health (IDLH)	ACGIH Exposure Criteria
Germanium chloride: (derivation: Reaction of chlorine with elemental germanium).	**PEL:** Not Established **STEL:** Not Established	**REL:** Not Established **STEL:** Not Established	**Not Determined**	**TLV:** Not Established **STEL:** Not Established

SECTION III - PHYSICAL AND CHEMICAL CHARACTERISTICS

Boiling Point	**181°F (83°C)**	Specific Gravity (H_2O = 1)	**1.874**
Vapor Pressure (mm Hg)	**Volatile at room temperature**	Molecular Weight (atomic weight)	**214.39**
Vapor Density (Air = 1)	**Not Found**	Melting Point	**-56°F (-49°C)**

Solubility

Soluble in benzene, alcohol, ether, carbon disulfide, and other organic solvents. Insoluble in concentrated hydrochloric acid. Reacts violently in water.

Appearance and Odor

Colorless, mobile, liquid with a peculiar, acidic odor. Fumes in air.

SECTION IV - FIRE AND EXPLOSION HAZARD DATA

Flash Point (method used)	**Not Determined**	Explosive Limits in Air % by Volume	**LEL: Not Determined** **UEL: Not Determined**
NFPA Classification	**Non-Combustible Solid (bulk form)**	Autoignition Temperature	**Not Determined**

Extinguishing Media

Use extinguishing agent suitable for the type of surrounding fire. Germanium chloride itself does not burn. DO NOT USE WATER.

Special Fire Fighting Procedures

Poisonous gases and fumes are produced in fire. Wear full protective clothing, self-contained breathing apparatus (SCBA) operated in pressure demand mode. Move container from fire area if it can be done without risk. Remain clear of smoke, water fall out and water run off. Evacuate non-essential personnel 1500 feet from fire area.

Unusual Fire and Explosion Hazards

Germanium chloride will react violently in contact with water or steam.

SECTION V - REACTIVITY DATA

Stability		Conditions to Avoid Avoid contact with incompatible materials, water, and moisture.
Stable X	Unstable	Incompatibility *(materials to avoid)* Incompatible with water and steam. Will react to produce toxic and corrosive fumes.
Hazardous Polymerization		Conditions to Avoid Under normal conditions of temperature and pressure, germanium chloride is not known to undergo hazardous polymerization.
May Occur	Will Not Occur X	Hazardous Decomposition or By-products When heated to decomposition in air it emits highly toxic fumes of chloride.

SECTION VI - HEALTH HAZARD DATA

Primary Route(s) of Entry:	Inhalation?	Absorption (skin/eye)?	Ingestion?
	X		X

Health Hazards

INHALATION: May cause severe irritation to the respiratory system with symptoms of pneumonitis, nasopharyngitis, tracheobronchitis, dyspnea, and chronic cough. Sever bronchitis can occur days after high exposure with cough, shortness of breath, pulmonary edema, and death.

SKIN & EYES: Eye contact will result in conjunctival irritation. Reactions may be worse on moist skin. Allergic skin rash may occur.

INGESTION: Poison by ingestion. Causes irritation of the mouth and stomach.

Carcinogenicity	NTP Listed?	IARC Cancer Review Group?	OSHA Regulated?	Target Organs?
Unknown Human Unknown Animal	No	No	No	Skin, eyes, respiratory system.

Medical Conditions Generally Aggravated by Exposure

None reported. However, existing respiratory problems (bronchitis, asthma) may be aggravated.

Emergency and First-aid Procedures

Eye contact: Do not allow victim to keep eyes tightly shut. Flush immediately with water for 15 minutes (minimum), seek medical attention. Skin contact: Wash with flooding amounts of water for at least 15 minutes. If redness or blisters develop, seek medical attention. For inhalation: Remove the person from exposure. Provide respiratory assistance. Transfer to medical facility. Observe for 48 hours for lung effects. If swallowed: Seek medical attention immediately. Do not attempt to give an unconscious or convulsing person anything by mouth.

SECTION VII - PRECAUTIONS FOR SAFE HANDLING AND USE

Steps to be Taken in Case Material is Released or Spilled

Wear self-contained breathing apparatus (SCBA) for respiratory protection. Restrict those not wearing protective equipment and who are not involved in cleanup from entering area. Avoid skin contact. Ventilate area of spill. Do not use water. Absorb in vermiculite and deposit in sealed drums.

Preferred Waste Disposal Method

No Citation.

Precautions to be Taken in Handling and Storage

Store in tightly closed containers in a cool, well-ventilated area away from heat. Protect containers from physical damage. Keep dry and away from water and moisture.

Other Precautions and Warnings

An inert nitrogen blanket is recommended inside containers to prevent condensation buildup.

SECTION VIII - CONTROL MEASURES AND PERSONAL PROTECTIVE EQUIPMENT

Respiratory Protection *(specify type)*

Exposure levels not established. Recommend a NIOSH/MSHA-approved supplied-air respirator or a self-contained breathing apparatus (SCBA) with full facepiece operated in positive pressure mode.

Ventilation

Local exhaust (preferred) at site of chemical work or general ventilation.

Protective Gloves	Eye Protection	Other Protective Clothing
Impervious Material	Chemical Goggles and Face Mask	Protective Uniform or Apron

Work/Hygiene Practices

Always wash hands thoroughly after using chemical; never bring food, drink, or smoking materials into vicinity of chemicals.

GERMANIUM CHLORIDE

$GeCl_4$ CAS: 10038-98-9

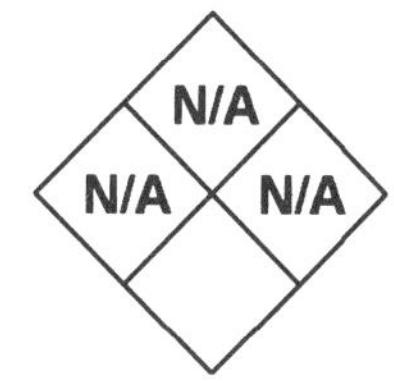

IDENTIFICATION AND TYPICAL USES

Germanium chloride appears as a colorless, mobile liquid that may fume in air. It has a peculiar acidic odor that can be distinguished from that of concentrated hydrochloric acid. It is used primarily in the semiconductor industry.

RISK ASSESSMENT: HEALTH

General Assessment

Germanium chloride is toxic by *ingestion* and can also cause toxic effects on *inhalation.* Skin contact will produced localized damage to tissues and may be destructive, depending upon the skin's surface moisture. There are no reports in the references of any carcinogenic, mutagenic, or teratogenic response following exposure to germanium chloride.

Inhalation may cause pneumonitis, nasopharyngitis, tracheobronchitis, dyspnea, chronic cough, and shortness of breath. Germanium chloride produces chronic systemic disease that primarily affects the lungs. High concentrations can also cause a dangerous (life-threatening) buildup of fluids in the lungs in a condition known as pulmonary edema. Symptoms may be delayed up to 48 hours, thereby creating a false sense of security with regard to health exposure risk.

Ingestion causes irritation of the mouth and stomach. Eye contact causes conjunctival irritation. Germanium chloride can react with the water content of the eyes to cause corrosive destruction of tissues.

Acute Health Effects

The following acute (short-term) health effects may occur immediately or shortly after exposure to germanium chloride:

Skin: Severe irritation and possible burns. Localized and corrosive skin destruction may also be possible, depending on moisture levels (sweat).

Eye: Severe irritation, itching, and burning on contact. Damage may be severe and permanent, depending on length of contact time.

Lung: Overexposure can severely irritate the airways and lungs, causing nasal discharge, tightness in the chest, cough, shortness of breath, and/or fever. Future exposures can cause further attacks. High acute exposures can lead to pneumonitis or life-threatening pulmonary edema. Symptoms may be delayed for hours or even days following exposure.

Chronic Health Effects

The following chronic (long-term) health effects may occur at some time after exposure to germanium chloride and can last for months or even years:

Cancer Hazards: According to information presented in the references, germanium chloride has not been adequately tested for its ability to cause cancer in test animals.

Reproductive Hazard: There is no evidence in the published studies to show that germanium chloride has the ability to adversely affect reproduction.

Other Chronic Effects: There are no other chronic effects reported in the references for germanium chloride.

Recommended Risk-Reduction Measures

The specific toxicity of germanium chloride is not clearly understood in the references. Personnel should avoid direct contact with germanium chloride. If a less toxic material or compound cannot be substituted for germanium, then *engineering controls* are the most effective method of reducing exposure risk. The best protection is to enclose operations and/or provide local exhaust ventilation at the site of germanium chloride release. While not always operationally feasible, isolating operations can also reduce exposure. Using respiratory protection is less effective than the controls mentioned above, but is still required whenever working with or around germanium chloride. No exposure levels have been established for this compound. This does not mean that exposure to germanium chloride is without health hazard risk. A MSHA/NIOSH approved supplied-air respirator operated in positive pressure mode, or a self-contained breathing apparatus (SCBA) with full facepiece operated in continuous flow or other positive pressure mode are the recommended methods of respiratory protection. If a full facepiece is not available, then chemical goggles should be worn to protect the eyes, especially when a splash hazard exists. A face shield should also be considered. To prevent hand and skin exposures, impervious gloves should be worn. Glove manufacturers should be contacted and permeation studies obtained *before* final glove selections are made.

Administrative controls should also be in place to minimize the potential for human exposures. These may include written procedures or policies which specify the methods and techniques that will be practiced whenever personnel are to work with germanium chloride.

All personnel should receive *training* on the use, hazards, protective measures, emergency actions, and other precautions per 29 CFR 1910.1200 (Hazard Communication) prior to the first assignment in an area where germanium is used or stored.

Before beginning employment and at regular times thereafter, the following recommended medical tests should be considered for those with frequent or potentially high exposures:

- ☑ Lung function tests (spirometry and gas transfer testing).
- ☑ Chest X-ray.
- ☑ Careful physical examination of the skin and lungs.

If symptoms develop or overexposure is suspected, the following may be useful:

- ☑ Consider chest X-ray after acute exposure.

It should be noted that medical tests that simply look for existing damage are not a substitute for controlling exposures. Medical histories are extremely important when assessing exposure risk. Also, since smoking can cause heart disease, emphysema, and other lung diseases, smokers exposed to irritating substances such as germanium chloride may experience symptoms more quickly than non-smokers under the same conditions of exposure. Prudent risk management requires careful consideration of *all* possible factors which may be causing the appearance of exposure symptoms in the workplace.

Other methods to reduce exposure include:

- ☑ Always ensure that proper protective clothing is worn when using chemical substances and that personnel are trained in its correct use and care.
- ☑ Wash thoroughly immediately after exposure to germanium chloride and at the end of the work shift or before eating, drinking, or smoking.
- ☑ Hazard warning information should be posted in the work area. In addition, as part of an on-going education and training program, all information on the health and safety hazards of germanium chloride should be communicated to all exposed and potentially exposed workers.
- ☑ Eye wash stations should be provided in the immediate work area for emergency use. If there is a possibility of skin exposure to germanium chloride, emergency shower facilities should also be provided in the immediate work area.
- ☑ Workers whose clothing has been contaminated by germanium chloride should change into clean clothes before leaving work. Contaminated work clothing should be laundered only by individuals who have been informed of the hazards of exposure to germanium chloride.

RISK ASSESSMENT: ENVIRONMENT

General Assessment

The environment is at risk of exposure during transportation, storage, disposal, or destruction of germanium chloride. In almost every scenario, the threat of environmental exposure is contingent upon the proper handling of the chemical substance. Accidental spills, large or small, can result in contamination of the surrounding environmental mediums (water, soil, and air).

Germanium chloride is generally non-combustible. However, it can react violently in contact with water or steam and caution is always required in handling, storage, transportation, and disposal of germanium chloride. When heated to decomposition in air it emits very toxic fumes of chloride. Emergency responders should be made aware of the presence of germanium chloride at any emergency response situation.

Germanium chloride may enter the environment through industrial discharge and municipal operations, and from accidental spills or leaks.

Acute Ecological Effects

Acute (short-term) toxic effects may include the death of animals, birds, or fish and death or low growth rate in plants. Acute effects are seen 2 to 4 days after animals or plants are exposed to germanium chloride.

Insufficient data are available to evaluate or predict the acute (short-term) effects of germanium chloride to aquatic life, birds, plants or land animals.

Chronic Ecological Effects

Chronic (long-term) toxic effects may include shortened life span, reproductive problems, lower fertility, and changes in appearance or behavior in exposed animals. These effects can be seen long after first exposure(s) to toxic chemicals.

Insufficient data are available to evaluate the chronic (long-term) effects of germanium to aquatic life, plants, birds, or land animals.

Water Solubility

Germanium chloride reacts violently in contact with water or moisture to produce toxic and corrosive fumes.

Persistence in the Environment

Germanium chloride will decompose in violent reaction with water and is virtually non-persistent in the aquatic environment, with a half-life of less than 2 days. The half-life of a pollutant is the amount of time it takes for one-half of the chemical to be degraded.

Bioaccumulation in Aquatic Organisms

Some substances increase in concentration, or *bioaccumulate,* in living organisms as they breathe contaminated air, drink contaminated water, or eat contaminated food. These chemicals can become concentrated in the tissues and internal organs of animals as well as humans.

There is no data on the bioaccumulation of germanium chloride in aquatic organisms. However, since it is known to react violently in water to form other toxic and corrosive materials, it is highly probable that bioaccumulation will not occur. More research is required.

Recommended Risk-Reduction Measures

Proper training of all transporters will reduce the likelihood of a mishap or accident resulting in a leak or spill of germanium chloride into the environment. Labels on all containers, trucks, and rail cars must accurately reflect their contents to enable emergency responders to react properly and quickly to any disaster thereby reducing the risks to the environment and personnel.

Storage of germanium chloride should be segregated from other chemicals and materials to minimize the risk of cross-contamination. Germanium chloride must be stored to avoid contact with water and moisture since violent reactions can occur. Containers should be protected from physical damage and stored to avoid contact with heat.

If a spill or leak to the environment has occurred, fire department, emergency response and/or hazardous materials spill personnel should be notified immediately. Cleanup should be attempted only by those trained in proper spill containment procedures. Contaminated soils should be removed for incineration and replaced with clean soil. If germanium chloride should contact the water table, aquifer, or navigable waterway, the local and/or state emergency response authorities must be notified. A comprehensive emergency response or disaster preparedness/recovery plan should be in place prior to any operations involving the use, transportation, storage, or disposal of germanium chloride.

If germanium chloride is spilled or leaked, the following specific steps are recommended:

- ☑ Restrict persons not wearing protective clothing from area of spill or leak until cleanup is complete. Avoid skin and eye contact.
- ☑ Absorb liquid spills with vermiculite or other material and place in sealed drums for disposal.
- ☑ Ventilate area of spill or leak.
- ☑ It may be necessary to dispose of germanium chloride as a hazardous waste. The responsible state agency or the regional office of the federal Environmental Protection Agency (EPA) should be contacted for specific recommendations.

RISK ASSESSMENT: BUSINESS

General Assessment

Accidents or mishaps involving germanium chloride can present a serious threat to business operations. The loss or damage of equipment or facilities can significantly affect fiscal viability. Lawsuits that may result from personnel illness, injury/death, public exposures, and/or environmental contamination will require a serious expenditure of resources. Media attention surrounding an injury, death, or environmental damage can also result in a loss of profits and loss of current as well as future business.

Recommended Risk-Reduction Measures

Company attorneys, safety and health professionals, and environmental specialists should be involved in the development of any procedures for responding to chemical incidents. A company official should be predesignated as a public relations officer with specific training in dealing with the press. Corporate plans and policies should be developed, approved, and implemented long before any need for such arises.

REFERENCES

American Conference of Governmental Industrial Hygienists. 1988. *Documentation of the Threshold Limit Values and Biological Exposure Indices,* 5th Edition (with updates). Cincinnati: ACGIH

Anthony, C. P., and N. J. Kolthoff. 1971. *Textbook of Anatomy and Physiology.* St. Louis: C. V. Mosby Company.

Asante-Duah, D. K. 1993. *Hazardous Waste Risk Assessment.* New York: CRC Press/Lewis Publishers.

Brusick, D. J. 1994. *Methods for Genetic Risk Assessment.* New York: CRC Press/Lewis Publishers.

Bueche, J. R. 1972. *Principles of Physics,* 2nd Edition. New York: McGraw-Hill.

Calabrese, E. J. 1994. *Biological Effects of Low Level Exposures.* New York: CRC Press/Lewis Publishers.

Cockerman, L. G., B. S. Shane. 1994. *Basic Environmental Toxicology.* New York: CRC Press/Lewis Publishers.

Danse, I. R. 1991. *Common Sense Toxics in the Workplace.* New York: Van Nostrand Reinhold.

Environmental Protection Agency. 1983. *Chemical Hazard Information Profiles.* Washington, D.C.: U.S. EPA.

Fodden, C. R., and J. L. Weddell. 1992. *Hazardous Materials Emergency Action Data.* New York: CRC Press/Lewis Publishers.

Forseberg, K., and L. H. Keith. 1993. *Chemical; Protective Clothing Permeation and Degradation Database.* Chelsea, MI: Lewis Publishers.

Gardner, R. W. 1994. *Chemical Intolerance.* New York: CRC Press/Lewis Publishers.

Hallenbeck, W. H. 1993. *Quantitative Risk Assessment for Environmental and Occupational Health,* 2nd Edition. New York: CRC Press/Lewis Publishers.

Hodgson, E., R. B. Mailman, and Janice E. Chambers. 1988. *Dictionary of Toxicology.* London: The MacMillian Company.

Howard, P. H. 1995. *Handbook of Environmental Fate and Exposures Data for Organic Chemicals,* Vol. I-IV. New York: CRC Press/Lewis Publishers.

Howard, P. H., et. al. 1991. *Handbook of Environmental Degradation Rates.* New York: CRC Press/Lewis Publishers.

Hosty, J. W., and P. Foster. 1990. *A Practical Guide to Chemical Spill Response.* New York: Van Nostrand Reinhold.

Jones, F. E. 1994. *Toxic Organic Vapors in the Workplace.* New York: CRC Press/Lewis Publishers.

Kamrin, M. A. 1988. *Toxicology: A Primer on Principles and Applications.* Chelsea, Michigan: Lewis Publishers.

Keith, L. H. and M. M. Walker. 1995. *Handbook of Air Toxics.* New York: CRC Press/Lewis Publishers.

Landis, W. G., M. H. Yu. 1995. *Introduction to Environmental Toxicology.* New York: CRC Press/Lewis Publishers.

Langerman, N. 1994 *Precautionary Labels for Chemical Containers.* New York: CRC Press/Lewis Publishers.

Lauwerys, R. R., P. Hoet. 1993. *Industrial Chemical Exposure.* New York: CRC Press/Lewis Publishers.

Lewis, R. J., Sr. 1993. *Hawley's Condensed Chemical Dictionary,* 12th Edition. New York: Van Nostrand Reinhold.

Lewis, R. J., Sr. 1992. *Sax's Dangerous Properties of Industrial Materials,* Volumes 1, 2, and 3, 8th Edition. New York: Van Nostrand Reinhold.

Mackay, D., W. Y. Shiu, and K. C. Ma. 1992. *Illustrated Handbook of Chemical-Physical Properties and Environmental Fate for Organic Chemicals.* New York: CRC Press/Lewis Publishers.

Manahan, S. E. 1994. *Environmental Chemistry,* 6th Edition. New York: CRC Press/Lewis Publishers.

Mellström, G. A., J. E. Wahlberg, and H. I. Maibach. 1994. *Protective Gloves for Occupational Use.* New York: CRC Press/Lewis Publishers.

Meyer, E. 1990. *Chemistry of Hazardous Materials.* Englewood Cliffs, NJ: Prentice Hall.

Miller, B. F., M.D., and C. B. Keane, R.N. 1972. *Encyclopedia and Dictionary of Medicine and Nursing.* Philadelphia: W.B. Sanders Company.

Mycock, J. C., J. D. McKenna, and L. Theodore. 1995. *Handbook of Air Pollution Control Engineering and Technology.* New York: CRC Press/Lewis Publishers.

Ott, W. R. 1995. *Environmental Statistics and Data Analysis.* New York: CRC Press/Lewis Publishers.

Patnaik, P. 1992. *A Comprehensive Guide to the Hazardous Properties of Chemical Substances.* New York: Van Nostrand Reinhold.

Philp, R. B. 1995. *Environmental Hazards and Human Health.* New York: CRC Press/Lewis Publishers.

Quigley, D. R. 1994. *Handbook of Emergency Chemical Management.* New York: CRC Press/Lewis Publishers.

Richardson, M. 1995. *Dictionary of Substance and Their Effects,* Vol. I-VI. New York: CRC Press/Lewis Publishers.

Torkelson, T. R. V. K. Rowe. 1981. *Patty's Industrial Hygiene and Toxicology,* Vol. 2B. New York: Wiley-Interscience.

MATERIAL SAFETY DATA SHEET

CHEMICAL NAME

GERMANIUM TETRAHYDRIDE

HAZARD WARNING INFORMATION

HEALTH	FIRE	REACTIVE	OTHER	DEGREE OF HAZARD	COLOR CODING	OTHER CODES
1	4	2		0 = Minimum Hazard 1 = Slight Hazard 2 = Moderate Hazard 3 = Serious Hazard 4 = Severe Hazard	HEALTH = BLUE FIRE = RED REACTIVITY = YELLOW OTHER = WHITE	OX = Oxidizer ACID = Acid ALK = Alkali COR = Corrosive W = Use No Water

SECTION I - GENERAL INFORMATION

Characterization: **Gas; Hydride**

RCRA Number: **None**

EPA Class: **Not Applicable**

DOT Proper Shipping Name: **Germanium tetrahydride**

Chemical Abstract Service (CAS) Number: **7782-65-2**

DOT Hazard Class and Label Requirements: **Poison Gas; Flammable Gas**

DOT Emergency Guide Code: **18**

DOT Identification Number: **UN 2192**

Chemical Formula: **GeH_4**

Synonyms
Germanium hydride; germane; germanomethane; monogermanium tetrahydride.

SECTION II - HAZARDOUS INGREDIENTS AND IDENTITY INFORMATION

Hazardous Components (specific identity)	OSHA Exposure Criteria	NIOSH Exposure Criteria	Immediately Dangerous to Life and Health (IDLH)	ACGIH Exposure Criteria
Germanium tetrahydride (derivation from: Germanium is derived from residues from refining of zinc and other sources, by heating in the presence of air and chlorine. It is also present in some coals and can be recovered from their ash). **1 ppm = 3.18 mg/m^3**	**PEL:** **Not Established** **STEL:** **Not Established**	**REL:** **0.2 ppm** **0.6 mg/m^3** **STEL:** **Not Established**	**Not Determined**	**TLV:** **0.2 ppm** **0.6 mg/m^3** **STEL:** **Not Established**

SECTION III - PHYSICAL AND CHEMICAL CHARACTERISTICS

Boiling Point: **-126°F (-88°C)**

Specific Gravity (H_2O = 1): **2.65**

Vapor Pressure (atmospheres): **>1 at 69°F (20°C)**

Molecular Weight: **76.6**

Vapor Density (Air = 1): **Not Found**

Freezing Point: **-267°F (-165°C)**

Solubility
Insoluble in water; slightly soluble in hot hydrochloric acid; soluble in liquid ammonia. Decomposes in nitric acid.

Appearance and Odor
Colorless gas with a pungent odor. Normally shipped as a compressed gas.

SECTION IV - FIRE AND EXPLOSION HAZARD DATA

Flash Point (method used): **Not Applicable (gas)**

Explosive Limits in Air % by Volume: **LEL: Not Determined UEL: Not Determined**

NFPA Classification: **Flammable Gas**

Autoignition Temperature: **Not Determined**

Extinguishing Media
Stop flow of gas, if possible. Allow to burn, do not attempt to extinguish since combustion of germanium tetrahydride reduces the possibility of explosions due to the formation of unstable air mixtures.

Special Fire Fighting Procedures
Germanium tetrahydride is a dangerous fire and explosion hazard. The gas is pyrophoric and may spontaneously ignite in air. Poisonous gases are produced in fire. Wear self-contained breathing apparatus. Use water spray to keep containers cool.

Unusual Fire and Explosion Hazards
Fire-exposed containers can explode violently in fire. They should be moved from fire areas if it can be done without risk. Cooling water should be applied from a safe distance if possible.

SECTION V - REACTIVITY DATA

Stability		Conditions to Avoid **Germanium tetrahydride is a highly reactive gas and can ignite spontaneously on exposure to air. Avoid contact with heat, air, ignition sources, and electrostatic sparks.**
Stable	Unstable	Incompatibility *(materials to avoid)*
	X	**None reported.**
Hazardous Polymerization		Conditions to Avoid **Under normal conditions of temperature and pressure, hazardous polymerization of germanium tetrahydride is not expected to occur.**
May Occur	Will Not Occur	Hazardous Decomposition or By-products
	X	**When germanium tetrahydride is involved in fire, it will emit toxic and acrid fumes and gases, including germanium and highly explosive hydrogen gas.**

SECTION VI - HEALTH HAZARD DATA

Primary Route(s) of Entry:	Inhalation?	Absorption (skin)?	Ingestion?
	X		

Health Hazards

INHALATION: Eye, nose, and throat irritant. Toxic effects include lachrymation, respiratory distress, chest pain, possible seizures, and pulmonary edema (fluid in the lungs).

SKIN & EYES: Skin contact may cause irritation and burns (frostbite from the compressed gas). Eye irritation is also likely to occur on contact with the gas.

INGESTION: Not likely.

Carcinogenicity	NTP Listed?	IARC Cancer Review Group?	OSHA Regulated?	Target Organs?
Unknown Human Unknown Animal	No	No	No	Respiratory system; liver; kidneys; CNS; skin; eyes.

Medical Conditions Generally Aggravated by Exposure

Respiratory function impairments (bronchitis, asthma, etc.); Skin conditions (dermatitis).

Emergency and First-aid Procedures

Eye contact: Immediately flush large amounts of water for 15 minutes (minimum), occasionally lifting eyelids, seek medical attention. Skin contact: Remove all contaminated clothing. Immediately wash area with large amounts of soap and water. For inhalation: Remove the person from exposure. Provide respiratory assistance and CPR. Transfer promptly to medical facility. Observation for 24 - 48 hours for lung effects and seizures. If swallowed: Unlikely.

SECTION VII - PRECAUTIONS FOR SAFE HANDLING AND USE

Steps to be Taken in Case Material is Released or Spilled

Remove all ignition sources. Ventilate area. Leaking gas can ignite spontaneously. Evacuate non-essential personnel. Stop flow of gas. If the source of leak is a cylinder and cannot be stopped, clear area and allow to empty. Have water source available to fight any surrounding fires. Stay clear and allow to burn until out. Emergency personnel must avoid contact with vapors and gases.

Preferred Waste Disposal Method

None specified in the references (incineration of a gas may be the only alternative).

Precautions to be Taken in Handling and Storage

Do not store in open air since violent and explosive fire may occur. Store in tightly closed containers in a cool, dark, well-ventilated area away from sunlight. Keep away from ignition sources such as fire, sparks, and flame.

Other Precautions and Warnings

Bulk storage of germanium tetrahydride is not recommended. Containers may explode in fire or under conditions of extreme heat (do not store outdoors or in direct sunlight).

SECTION VIII - CONTROL MEASURES AND PERSONAL PROTECTIVE EQUIPMENT

Respiratory Protection *(specify type)*

PEL is low and difficult to monitor (0.2 ppm). Recommend using a powered supplied-air respirator with full facepiece operated in pressure demand mode. Greater protection is obtained from a self-contained breathing apparatus (SCBA) with full facepiece in pressure demand or other positive pressure mode.

Ventilation

Local exhaust or general mechanical systems recommended.

Protective Gloves	Eye Protection	Other Protective Clothing
Impervious and Thermal Rubber	Chemical Goggles and/or Face Mask	Protective Apron

Work/Hygiene Practices

Always wash hands thoroughly after using chemical; never bring food, drink, or smoking materials into vicinity of chemicals.

GERMANIUM TETRAHYDRIDE

GeH_4 CAS: 7782-65-2

IDENTIFICATION AND TYPICAL USES

Germanium tetrahydride is a colorless gas with a pungent odor, normally shipped as a compressed gas. It is used primarily as a doping agent for solid-state electronic components.

RISK ASSESSMENT: HEALTH

General Assessment

Germanium tetrahydride gas is a severe irritant of the eyes, respiratory tract, and skin. As a gas, it is primarily toxic by *inhalation*. If the compressed gas should contact the skin or eyes, burning and irritation is likely to occur although absorption as an exposure route has not been reported. Ingestion is not likely to occur. There are no reports of carcinogenic, teratogenic, or mutagenic indications.

Inhalation of large concentrations causes dyspnea (difficulty breathing), bronchospasm, chest pain, congestion, and pulmonary edema, which may be fatal. Development of pulmonary edema may be delayed up to 48 hours after exposure, thereby creating a false sense of security with regard to health exposure risk. There may also be seizures following exposure to germanium tetrahydride gas. Germanium tetrahydride is a hemolytic agent which means it can induce hemolysis, a breakdown of the red blood cells with a liberation of hemoglobin. Liver and kidney damage are also possible (positive animal studies).

Acute Health Effects

The following acute (short-term) health effects may occur immediately or shortly after exposure to germanium tetrahydride:

Skin: Irritation and burning on contact with the cold compressed gas.

Eye: Burns which may cause damage to the eye and surrounding tissues.

Lung: Irritation of the mouth, nose, and throat causing coughing and/or shortness of breath. Higher exposures can lead to pulmonary edema.

CNS: Data on the acute effects of germanium tetrahydride on the central nervous system are inconclusive. However, animal studies have shown some indication of seizure activity at high concentrations.

Chronic Health Effects

The following chronic (long-term) health effects may occur at some time after exposure to germanium tetrahydride and can last for months or even years:

Cancer Hazards: According to information presented in the references, germanium tetrahydride has not been adequately tested for its ability to cause cancer in test animals.

Reproductive Hazards: There are no reports to support any claims of reproductive hazards following exposure to germanium tetrahydride.

Other Chronic Effects: Germanium tetrahydride is a hemolytic agent and repeated or prolonged exposure may damage the blood, liver, and kidneys.

Recommended Risk-Reduction Measures

Personnel should avoid direct contact with germanium tetrahydride gas. Unless a less toxic chemical can be substituted for a hazardous substance, *engineering controls* are the most effective method of reducing exposures. The best protection is to enclose operations and/or provide local exhaust ventilation at the site of chemical release. Because germanium tetrahydride is a highly flammable and reactive gas, explosion-proof ventilation is required. While not always operationally feasible, isolating operations can also reduce exposure. Using respiratory protection is less effective than the controls mentioned above, but is still advisable whenever working with germanium tetrahydride. The exposure level is extremely low (0.2 ppm) and difficult to accurately monitor. The best protection is obtained using a powered supplied-air respirator operated in continuous flow mode or a self-contained breathing apparatus (SCBA) with full facepiece in pressure demand mode. If a full facepiece is not available, then chemical goggles should be worn to protect the eyes. Whenever a chemical splash hazard exists, a face shield and a rubber apron should be worn. To prevent hand and skin exposures, impervious, thermal protective gloves should be worn. Glove manufacturers should be contacted and permeation studies obtained *before* final glove selections are made.

Administrative controls should also be in place to minimize the potential for human exposures. These may include written procedures or policies which specify the methods and techniques that will be practiced whenever personnel are to work with germanium tetrahydride.

All personnel should receive *training* on the use, hazards, protective measures, emergency actions, and other precautions per 29 CFR 1910.1200 (Hazard Communication), prior to the first assignment in an area where germanium tetrahydride is used or stored.

Before beginning employment and at regular intervals thereafter (e.g., annually), the following medical tests are recommended:

☑ Lung function tests.

If symptoms develop or overexposure is suspected, the following may be useful:

☑ Consider chest X-ray after acute overexposure (may be negative if taken immediately following exposure due to delayed development of pulmonary edema).

☑ Liver and kidney function tests.

Any evaluation should include a careful history of past and present symptoms with an examination. Medical tests that simply look for existing damage are not a substitute for controlling exposures. Also, since smoking can cause heart disease, lung cancer, emphysema, and other respiratory disorders, smokers exposed to germanium tetrahydride may experience symptoms more rapidly than non-smokers under the same conditions of exposure. Prudent risk management requires careful consideration of *all* possible risk factors.

Other methods to reduce exposure to chemicals include:

☑ Where possible, enclose operations and use local exhaust ventilation at the site of chemical release. If local exhaust ventilation or enclosure is not used, respiratory protection should be mandatory.

☑ Always ensure that proper protective clothing is worn when using chemical substances.

☑ Wash thoroughly immediately after exposure to germanium tetrahydride and at the end of the work shift or before eating, drinking, or smoking. Work clothing contaminated with germanium tetrahydride should never be taken home for laundering (fire hazard). Only personnel trained in the hazards of exposure should launder contaminated clothing.

☑ Safety shower and eyewash stations should be readily available in work areas where germanium tetrahydride is used or stored.

☑ Hazard warning information should be posted in the work area. In addition, as part of an on-going education and training program, all information on the health and safety hazards of germanium tetrahydride should be communicated to all exposed and potentially exposed workers.

RISK ASSESSMENT: ENVIRONMENT

General Assessment

The environment is at risk of exposure during transportation, storage, disposal, or destruction of germanium tetrahydride. In almost every scenario, the threat of environmental exposure is contingent upon the proper handling of the chemical substance. Accidental releases, large or small, can result in fire, explosion, and possible contamination of the surrounding environmental mediums (primarily air, but water and soil may also be contaminated depending upon the condition of the spill).

Germanium tetrahydride is considered to be a dangerous, reactive, and explosive flammable gas. It is pyrophoric, which means it can ignite spontaneously on exposure to air and burn vigorously. This characteristic requires special consideration during any emergency involving a leak or release of germanium tetrahydride gas.

Germanium tetrahydride can enter the environment through spills or leaks. Unchecked emissions from industrial sources are also possible.

Acute Ecological Effects

Acute (short-term) toxic effects may include the death of animals, birds, or fish, and death or low growth rate in plants. Acute effects are seen 2 to 4 days after animals or plants are exposed to germanium tetrahydride.

As a gas, germanium tetrahydride may not be expected to pose a serious threat to aquatic life. However, insufficient data are available to accurately evaluate or predict the short-term effects of germanium tetrahydride exposure to aquatic life, plants, birds, or land animals.

Chronic Ecological Effects

Chronic (long-term) toxic effects may include shortened life span, reproductive problems, lower fertility, and changes in appearance or behavior in exposed animals. These effects can be seen long after first exposure(s) to toxic chemicals.

Insufficient data are available to evaluate or predict the long-term (chronic) effects of germanium tetrahydride to aquatic life, plants, birds, or land animals.

Water Solubility

Germanium tetrahydride is insoluble in water. Concentrations of 1 milligram will not mix with a liter of water.

Persistence in the Environment

Germanium tetrahydride is non-persistent in water, with a half-life of less than 2 days. The half-life of a pollutant is the amount of time it takes for one half of the chemical to be degraded. Virtually 100% of germanium tetrahydride will eventually end up in the air.

Bioaccumulation in Aquatic Organisms

Some substances increase in concentration, or *bioaccumulate,* in living organisms as they breathe contaminated air, drink contaminated water, or eat contaminated food. These chemicals can become concentrated in the tissues and internal organs of animals as well as humans.

Germanium tetrahydride is not expected to accumulate in the edible tissues of fish.

Recommended Risk-Reduction Measures

Proper training of all transporters will reduce the likelihood of a mishap or accident resulting in a leak or release to the environment. The correct use of DOT labeling (per the MSDS) on all containers, trucks, and rail cars will enable emergency responders to react properly and quickly to any disaster thereby reducing the potential risk to the environment and to personnel.

Storage of germanium tetrahydride should be segregated from sources of heat or ignition. Containers should be air-tight and enclosed. Buildings designated for storage should be equipped with appropriate explosion-proof design as well as the proper fire protection systems (alarms, sprinklers, emergency lighting, portable extinguishers).

If a spill or release to the environment has occurred, fire department, emergency response, and/or hazardous materials spill personnel should be notified immediately. Cleanup should be attempted only by those trained in proper containment procedures. When such spills occur, the local and/or state emergency response authorities must be notified. A comprehensive emergency response or disaster preparedness/recovery plan should be in place prior to any operations involving the use, transportation, storage, or disposal of germanium tetrahydride. If germanium tetrahydride is spilled or leaked, the following steps are recommended:

- ☑ Restrict persons not wearing protective clothing from area of spill or leak until cleanup is complete and area can be opened for normal work.
- ☑ Remove all ignition sources.
- ☑ Ventilate area of spill or leak using maximum explosion-proof equipment.
- ☑ If possible, stop flow of leaking liquid or gas. Gas will spontaneously ignite in contact with air. Remove all combustible materials from the area and have extinguishing equipment available. Use water spray to knock down vapors. Once germanium tetrahydride gas is ignited, it may be best to let it burn since combustion will reduce the likelihood of explosions of unstable germanium tetrahydride mixtures in air.
- ☑ Keep germanium tetrahydride out of a confined space, such as a sewer, because of the possibility of explosion (unless the sewer is designed to prevent the buildup of explosive concentrations).

RISK ASSESSMENT: BUSINESS

General Assessment

Accidents or mishaps involving germanium tetrahydride can present a significant threat to business operations.

The loss or damage of equipment or facilities can significantly affect fiscal viability. Lawsuits that may result from personnel injury, public exposures, and environmental contamination will also require a serious expenditure of resources. Media attention surrounding an injury, illness, or environmental damage can also result in a loss of profits and loss of current as well as future business.

Recommended Risk-Reduction Measures

Company attorneys, safety and health professionals, and environmental specialists should be involved in the development of any procedures dealing with safety or emergency response. A company official should be pre-designated as a public relations officer with specific training in dealing with the press. Corporate plans and policies should be developed, approved, and implemented long before any need for such arises.

REFERENCES

Ahlbom, A. 1993. *Biostatistics for Epidemiologists.* New York: CRC Press/Lewis Publishers.

American Conference of Governmental Industrial Hygienists. 1988. *Documentation of the Threshold Limit Values and Biological Exposure Indices,* 5th Edition (with updates). Cincinnati: ACGIH

Anthony, C. P., and N. J. Kolthoff. 1971. *Textbook of Anatomy and Physiology.* St. Louis: C. V. Mosby Company.

Asante-Duah, D. K. 1993. *Hazardous Waste Risk Assessment.* New York: CRC Press/Lewis Publishers.

Bloom, A. D. and F. J. DeSerres. 1995. *Ecotoxicity and Human Health.* New York: CRC Press/Lewis Publishers.

Brusick, D. J. 1994. *Methods for Genetic Risk Assessment.* New York: CRC Press/Lewis Publishers.

Bueche, J. R. 1972. *Principles of Physics,* 2nd Edition. New York: McGraw-Hill.

Burrell, R., D. K. Flaherty, and L. J. Sauers. 1992. *Toxicology of the Immune System.* New York: Van Nostrand Reinhold.

Calabrese, E. J. 1994. *Biological Effects of Low Level Exposures.* New York: CRC Press/Lewis Publishers.

Cockerman, L. G., B. S. Shane. 1994. *Basic Environmental Toxicology.* New York: CRC Press/Lewis Publishers.

Cothern, C. R. 1995. *Handbook for Environmental Risk Decisionmaking.* New York: CRC Press/Lewis Publishers.

Gardner, R. W. 1994. *Chemical Intolerance.* New York: CRC Press/Lewis Publishers.

Hallenbeck, W. H. 1993. *Quantitative Risk Assessment for Environmental and Occupational Health,* 2nd Edition. New York: CRC Press/Lewis Publishers.

Hodgson, E., R. B. Mailman, and J. E. Chambers. 1988. *Dictionary of Toxicology.* London: The MacMillian Company.

Hosty, J. W., and P. Foster. 1990. *A Practical Guide to Chemical Spill Response.* New York: Van Nostrand Reinhold.

Howard, P. H. 1995. *Handbook of Environmental Fate and Exposures Data for Organic Chemicals,* Vol. I-IV. New York: CRC Press/Lewis Publishers.

Howard, P. H., et. al. 1991. *Handbook of Environmental Degradation Rates.* New York: CRC Press/Lewis Publishers.

James, A., and M. Lord. 1992. *Index of Chemical and Physical Data.* New York: Van Nostrand Reinhold.

Jones, F. E. 1994. *Toxic Organic Vapors in the Workplace.* New York: CRC Press/Lewis Publishers.

Kamrin, M. A. 1988. *Toxicology: A Primer on Principles and Applications.* Chelsea, Michigan: Lewis Publishers.

Keith, L. H. and M. M. Walker. 1995. *Handbook of Air Toxics.* New York: CRC Press/Lewis Publishers.

Landis, W. G., M. H. Yu. 1995. *Introduction to Environmental Toxicology.* New York: CRC Press/Lewis Publishers.

Langerman, N. 1994 *Precautionary Labels for Chemical Containers.* New York: CRC Press/Lewis Publishers.

Lauwerys, R. R., P. Hoet. 1993. *Industrial Chemical Exposure.* New York: CRC Press/Lewis Publishers.

Lewis, R. J., Sr. 1993. *Hawley's Condensed Chemical Dictionary,* 12th Edition. New York: Van Nostrand Reinhold.

Lewis, R. J., Sr. 1992. *Sax's Dangerous Properties of Industrial Materials,* Volumes 1, 2, and 3, 8th Edition. New York: Van Nostrand Reinhold.

Lide, D. R. 1995. *Handbook of Chemistry and Physics,* 76th Edition. New York: CRC Press/Lewis Publishers.

Manahan, S. E. 1992. *Toxicological Chemistry,* 2nd Edition. New York: CRC Press/Lewis Publishers.

Meyer, E. 1990. *Chemistry of Hazardous Materials.* Englewood Cliffs, NJ: Prentice Hall.

Patnaik, P. 1992. *A Comprehensive Guide to the Hazardous Properties of Chemical Substances.* New York: Van Nostrand Reinhold.

Philp, R. B. 1995. *Environmental Hazards and Human Health.* New York: CRC Press/Lewis Publishers.

Proctor, N. H., and J. P. Hughes. 1991. *Chemical Hazards of the Workplace.* New York: Van Nostrand Reinhold.

Quigley, D. R. 1994. *Handbook of Emergency Chemical Management.* New York: CRC Press/Lewis Publishers.

Rea, W. J. 1995. *Chemical Sensitivity,* Vol. I-III. New York: CRC Press/Lewis Publishers.

Richardson, M. 1995. *Dictionary of Substance and Their Effects,* Vol. I-VI. New York: CRC Press/Lewis Publishers.

U.S. Department of Health and Human Services, National Institute for Occupational Safety and Health. 1994. *NIOSH Pocket Guide to Chemical Hazards.* Washington, D.C.: U.S. Government Printing Office.

MATERIAL SAFETY DATA SHEET

CHEMICAL NAME	
GLUTARALDEHYDE	**CRC PRESS/LEWIS PUBLISHERS DISCLAIMER:** THE INFORMATION AND RECOMMENDATIONS PRESENTED HEREIN ARE BASED ON SOURCES BELIEVED TO BE RELIABLE. CRC MAKES NO REPRESENTATION ON ITS COMPLETENESS OR ACCURACY. IT IS THE RESPONSIBILITY OF THE USER TO DETERMINE THE CHEMICAL'S SUITABILITY FOR ITS INTENDED USE, THE CHEMICAL'S SAFE USE, AND THE CHEMICAL'S PROPER DISPOSAL. NO REPRESENTATIONS AND/OR WARRANTIES, EITHER EXPRESSED OR IMPLIED, OF THE MERCHANTABILITY OR FITNESS FOR A PARTICULAR PURPOSE, OR OF ANY OTHER NATURE, ARE MADE WITH RESPECT TO THE INFORMATION PROVIDED IN THIS MSDS OR TO THE CHEMICAL TO WHICH INFORMATION MAY REFER. CRC NEITHER ASSUMES NOR AUTHORIZES ANY OTHER PERSON TO ASSUME FOR IT, ANY OTHER ADDITIONAL RESPONSIBILITY OR LIABILITY FOR THE USE OF, OR RELIANCE UPON, THIS INFORMATION.

HAZARD WARNING INFORMATION

HEALTH	FIRE	REACTIVE	OTHER	DEGREE OF HAZARD	COLOR CODING	OTHER CODES
2	0	1		0 = Minimum Hazard 1 = Slight Hazard 2 = Moderate Hazard 3 = Serious Hazard 4 = Severe Hazard	HEALTH = BLUE FIRE = RED REACTIVITY = YELLOW OTHER = WHITE	OX = Oxidizer ACID = Acid ALK = Alkali COR = Corrosive W = Use No Water

SECTION I - GENERAL INFORMATION

Characterization: **Aldehyde**	RCRA Number: **None**	EPA Class: **Not Applicable**
DOT Proper Shipping Name: **Corrosive Liquids, N.O.S.**	Chemical Abstract Service (CAS) Number: **111-30-8**	
DOT Hazard Class and Label Requirements: **Corrosive**	DOT Emergency Guide Code: **No Citation**	
DOT Identification Number: **UN 1760**	Chemical Formula: **$OCH(CH_2)_3CHO$**	

Synonyms

Glutaric dialdehyde; 1,5-pentanedial; Cidex; 1,3-diformylpropane; Glutarol; glutardialdehyde; Glutarex; Sonacide; Sporicidin; Succin-dialdehyde; Verucasep.

SECTION II - HAZARDOUS INGREDIENTS AND IDENTITY INFORMATION

Hazardous Components (specific identity)	OSHA Exposure Criteria	NIOSH Exposure Criteria	Immediately Dangerous to Life and Health (IDLH)	ACGIH Exposure Criteria
Glutaraldehyde (derivation: By treating 2-ethoxy-3,4-dihydro-2H-pyran with aqueous hydrochloric acid; or by reaction of methyl vinyl ether with acrolein to form dihydropyrans, with hydrolysis to glutaraldehyde). Usually available as 2% or 50% solution. **1 ppm = 4.16 mg/m³**	**PEL:** Not Established **STEL:** Not Established	**REL (ceiling):** 0.2 ppm, 0.8 mg/m³ **STEL:** Not Applicable	**Not Determined**	**TLV (ceiling):** 0.2 ppm, 0.8 mg/m³ **STEL:** Not Applicable

SECTION III - PHYSICAL AND CHEMICAL CHARACTERISTICS

Boiling Point: **212°F (100°C)**	Specific Gravity (H_2O = 1): **1.10**
Vapor Pressure (mm Hg): **17 at 69°F (20°C)**	Molecular Weight: **100.1**
Vapor Density (Air = 1): **0.72**	Freezing Point: **7°F (-14°C)**

Solubility

Miscible in water. Soluble in benzene, ethanol, ether, and most other organic solvents.

Appearance and Odor

Colorless to pale-yellow liquid with a pungent, fruity (rotten apple) odor. It can appear as a colorless, crystalline solid at temperatures below 7°F (-14°C). Odor Threshold = 0.04 ppm.

SECTION IV - FIRE AND EXPLOSION HAZARD DATA

Flash Point (method used): **Not Applicable**	Explosive Limits in Air % by Volume: **LEL: Not Applicable** **UEL: Not Applicable**
NFPA Classification: **Non-Combustible Liquid**	Autoignition Temperature: **Not Determined**

Extinguishing Media

Use extinguishing media suitable to surrounding fires. Glutaraldehyde does not burn.

Special Fire Fighting Procedures

Wear full protective clothing and self-contained breathing apparatus (SCBA). Move fire-exposed containers from area if it can be done without risk. Dike area for control and containment of runoff.

Unusual Fire and Explosion Hazards

None Reported.

SECTION V - REACTIVITY DATA

Stability		Conditions to Avoid
		Glutaraldehyde is not normally stable. It readily oxidizes when exposed to air. It can be kept stable for up to 2 weeks when solutions buffered to neutral pH. Avoid contact with heat and incompatible materials.
Stable	Unstable	Incompatibility *(materials to avoid)*
	X	Strong oxidizers, acids, bases, alcohols, ammonia and amines, phenols, ketones, hydrogen cyanide, hydrogen sulfide, acid anhydrides, and halogens.
Hazardous Polymerization		Conditions to Avoid Glutaraldehyde polymerizes in alkaline solutions (above pH 9) or when heated. Avoid conditions which might expose glutaraldehyde to heat, including radiant heat and sunlight.
May Occur	Will Not Occur	Hazardous Decomposition or By-products
X		When heated to decomposition, glutaraldehyde emits acrid, irritating smoke and fumes.

SECTION VI - HEALTH HAZARD DATA

Primary Route(s) of Entry:	Inhalation?	Absorption (skin)?	Ingestion?
	X	X	X

Health Hazards

INHALATION: Eye, nose, and throat irritant. Causes headache, nasal inflammation and nosebleeds, tightness in chest, delayed asthma, nausea, vomiting, and possible liver damage.

ABSORPTION: Skin contact can result in irritating rash and skin staining. Can cause severe eye burns leading to permanent damage. Can result in delayed dermatitis. Absorption is likely.

INGESTION: Stomach irritation and possible hemorrhage, ulceration, and perforation of the stomach and esophagus.

Carcinogenicity	NTP Listed?	IARC Cancer Review Group?	OSHA Regulated?	Target Organs?
No Evidence Human Confirmed Animal	No	No	No	Respiratory system; skin; eyes; liver (possibly).

Medical Conditions Generally Aggravated by Exposure

Any conditions that deplete glutathione or inhibit aldehyde dehydrogenase may be aggravated.

Emergency and First-aid Procedures

Eye contact: Flush immediately with water for 30 minutes (minimum); seek medical attention. Skin contact: Remove all contaminated clothing. Immediately wash area with large amounts of soap and water. Seek medical assistance. For inhalation: Remove the person from exposure. Provide respiratory assistance and CPR. Transfer to medical facility. If swallowed: Seek medical attention immediately. Unless advised otherwise, give 1-2 glasses of water to dilute and induce vomiting. Never attempt to give an unconscious or convulsing person anything by mouth. Never attempt to make such a person vomit.

SECTION VII - PRECAUTIONS FOR SAFE HANDLING AND USE

Steps to be Taken in Case Material is Released or Spilled

Contain spills using absorbent material (vermiculite, or other). Ventilate area of spill. Have water source available in case of fire. Store materials in DOT-approved containers. Restrict those not involved in cleanup from entering area. Notify appropriate authorities, as required.

Preferred Waste Disposal Method

Dissolve in a combustible solvent and burn in a chemical incinerator (with afterburner and scrubber).

Precautions to be Taken in Handling and Storage

Do not store with incompatible chemicals. Store in tightly closed containers in a cool, dark, well-ventilated area. Keep fire and flame away.

Other Precautions and Warnings

Prevent physical damage to containers.

SECTION VIII - CONTROL MEASURES AND PERSONAL PROTECTIVE EQUIPMENT

Respiratory Protection *(specify type)*

The NIOSH exposure ceiling is low and difficult to monitor (0.2 ppm). For relatively low or infrequent exposures, use an MSHA/NIOSH-approved full-facepiece respirator with organic vapor cartridge. For best protection, especially at high exposures or during emergencies, use a supplied-air respirator or a self-contained breathing apparatus with full facepiece and operated in a positive pressure mode.

Ventilation

Local exhaust or general mechanical systems recommended.

Protective Gloves	Eye Protection	Other Protective Clothing
Impervious Material	Chemical Goggles or Face Mask	Protective Apron

Work/Hygiene Practices

Always wash hands thoroughly after using chemical; never bring food, drink, or smoking materials into vicinity of chemicals.

GLUTARALDEHYDE

$OCH(CH_2)_3CHO$ CAS: 111-30-8

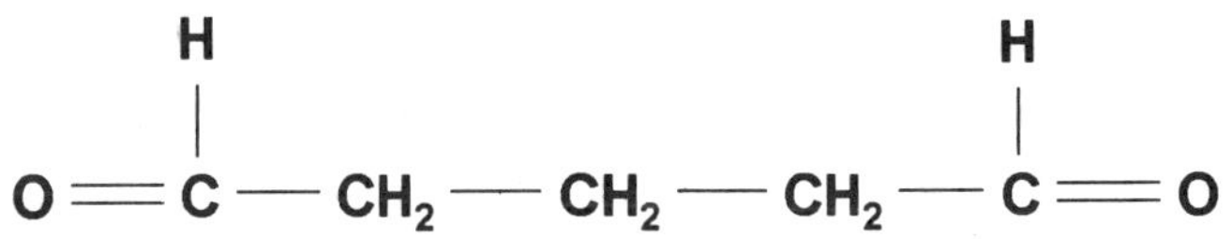

IDENTIFICATION AND TYPICAL USES

Glutaraldehyde is a colorless to pale-yellow liquid with a pungent, fruity, rotten apple-like odor. It can also appear as a colorless, crystalline solid at temperatures below 7°F (-14°C). It is used as a leather tanning agent, as a tissue fixative, a cross-linking agent for proteins, in X-ray processing solutions and film emulsions, as a disinfectant in hospitals (2% solution), in embalming fluids (25% solution), in pharmaceuticals as a therapeutic treatment for warts (up to 10% solution), and as a preservative in certain cosmetic products. It is also used as a chemical intermediate for adhesives, sealants, and electrical products.

RISK ASSESSMENT: HEALTH

General Assessment

Glutaraldehyde is a poison by *ingestion* and is considered moderately toxic by *inhalation* and skin contact (*absorption*). While it is not currently listed as a carcinogen, animal mutation data have been reported. It is also an experimental teratogen causing reproductive effects in some test animals.

Because of its relatively low vapor pressure, the potential for inhalation exposures is considered to be equally low. However, respiratory irritation and other effects have been recorded. Inhalation may also cause headache, nasal inflammation with accompanying nose bleeds, tightness in chest, difficulty breathing, nausea, and vomiting. Liver damage (hepatitis) has occurred following inhalation studies conducted with animals.

Skin contact results in severe irritation and may cause brown staining to develop on the skin's surface at the point of contact. If glutaraldehyde should contact the eye, severe irritation and possible ocular damage may occur.

Ingestion causes mouth and stomach irritation. High concentrations can lead to hemorrhage, ulceration, and perforation of the stomach and esophagus.

Acute Health Effects

The following acute (short-term) health effects may occur immediately or shortly after exposure to glutaraldehyde:

Skin: Irritation and rash or burning sensation on contact with subsequent brown staining of the skin's surface.

Eye: Burns leading to ocular damage that may be permanent.

Lung: Irritation causing coughing and/or shortness of breath. May cause nasal inflammation and nosebleeds.

Other: Ingestion causes serious irritation to the gastrointestinal tract with a possibility for hemorrhage and ulcerations in vital organs and organ members.

Chronic Health Effects

The following chronic (long-term) health effects may occur at some time after exposure to glutaraldehyde and can last for months or even years:

Cancer Hazards: Glutaraldehyde causes mutations (genetic changes). Such chemicals may have a cancer risk. There is limited evidence that glutaraldehyde causes cancer in animals. More research is required with regard to human carcinogenic potential.

Reproductive Hazards: According to information presented in the references, glutaraldehyde may be a teratogen in humans since it has been shown to cause birth defects and adversely affect the reproductive process in test animals.

Other Chronic Effects: Animal studies show that prolonged exposure to glutaraldehyde may cause liver damage (hepatitis). It is not known if it will have this effect on humans. Dermatitis and asthma may develop in some individuals (sensitization).

☞ *Recommended Risk-Reduction Measures*

Personnel should avoid direct contact with glutaraldehyde. Unless a less toxic chemical can be substituted for a hazardous substance, *engineering controls* are the most effective method of reducing exposures. The best protection is to enclose operations and/or provide local exhaust ventilation at the site of chemical release. While not always operationally feasible, isolating operations can also reduce exposure.

Using respiratory protection is less effective than the controls mentioned above, but is still advisable whenever working with or around glutaraldehyde. For low or infrequent exposures, a full facepiece respirator equipped with an organic vapor cartridge may provide appropriate protection. However, the NIOSH exposure ceiling is low and difficult to accurately monitor (0.2 ppm). The best protection is provided using a supplied-air respirator or a self-contained breathing apparatus (SCBA) with full facepiece operated in pressure demand or other positive pressure or continuous flow mode. If a full facepiece is not available, then chemical goggles should be worn to protect the eyes. Whenever a chemical splash hazard exists, a face shield and a rubber apron should be worn. To prevent hand and skin exposures, impervious gloves should be worn. Rubber products may not provide suitable protection. Glove manufacturers should be contacted and permeation studies obtained *before* final glove selections have been made.

Administrative controls should also be in place to minimize the potential for human exposures. These may include written procedures or policies which specify the methods and techniques that will be practiced whenever personnel are to work with glutaraldehyde.

All personnel should receive *training* on the use, hazards, protective measures, emergency actions, and other precautions per 29 CFR 1910.1200 (Hazard Communication), prior to the first assignment in an area where glutaraldehyde is used or stored.

If symptoms develop or overexposure is suspected, the following medical tests may be helpful:

- ☑ Lung function tests, to include forced vital capacity (FVC) and forced expiatory volume (FEV).
- ☑ Monitor liver function tests and blood gases for indications of acidosis.
- ☑ Evaluation by a qualified allergist with careful consideration of exposure history and special testing (may help diagnose skin allergy).

Any evaluation should include a careful history of past and present symptoms with an examination. Medical tests that simply look for existing damage are not a substitute for controlling exposures. Also, since smoking can cause heart disease, lung cancer, emphysema, and other respiratory disorders, smokers exposed to glutaraldehyde may experience symptoms more rapidly than non-smokers under the same conditions of exposure. Prudent risk management requires careful consideration of *all* possible risk factors.

Other methods to reduce exposure include:

- ☑ Where possible, enclose operations and use local exhaust ventilation at the site of chemical release. If local exhaust ventilation or enclosure is not used, respiratory protection should be mandatory.
- ☑ Always ensure that proper protective clothing is worn when using chemical substances.
- ☑ Wash thoroughly immediately after exposure to glutaraldehyde and at the end of the work shift or before eating, drinking, or smoking.
- ☑ Hazard warning information should be posted in the work area. In addition, as part of an on-going education and training program, all information on the health and safety hazards of glutaraldehyde should be communicated to all exposed workers.
- ☑ Never eat, drink, or smoke in areas were glutaraldehyde is used, handled or stored.

RISK ASSESSMENT: ENVIRONMENT

General Assessment

The environment is at risk of exposure during transportation, storage, disposal, or destruction of glutaraldehyde. In almost every scenario, the threat of environmental exposure is contingent upon the proper handling of the chemical substance. Accidental spills, large or small, can result in fire, explosion, and possible contamination of the surrounding environmental mediums (water, soil, and air).

Glutaraldehyde is considered a non-combustible liquid. However, it will readily oxidize when exposed to air and can polymerize on exposure to alkaline solutions or when heated. It is incompatible with a number of commonly encountered materials, such as strong oxidizers and strong bases. Alkaline solutions of glutaraldehyde react with alcohol, ketones, amines, hydrazines, and proteins. These characteristics require special consideration during any emergency situation involving a leak or spill of glutaraldehyde. Should glutaraldehyde ever come into contact with any of these incompatible substances either during use, transportation, or storage, the formation of highly toxic and/or highly explosive commodities is extremely possible.

The proper disposal/destruction method for glutaraldehyde is to combine it with a combustible solvent and burn it in a chemical incinerator equipped with an afterburner and air scrubber. Glutaraldehyde can enter the environment through manufacturing, unchecked discharge into effluents, and through spills.

Acute Ecological Effects

Acute (short-term) toxic effects may include the death of animals, birds, or fish, and death or low growth rate in plants. Acute effects are seen 2 to 4 days after animals or plants are exposed to glutaraldehyde. This chemical has high acute toxicity to aquatic life. No data are available on the short-term effects of glutaraldehyde to plants, birds, or terrestrial animals.

Chronic Ecological Effects

Chronic toxic effects may include shortened life span, reproductive problems, lower fertility, and changes in appearance or behavior in exposed animals. These effects can be seen long after first exposure(s) to toxic chemicals. Glutaraldehyde has high chronic toxicity in aquatic life. No data are available on the long-term effects of glutaraldehyde to plants, birds, or land animals.

Water Solubility

Glutaraldehyde is highly soluble in water. Concentrations of 1000 milligrams and more will mix with a liter of water.

Persistence in the Environment

Glutaraldehyde is moderately persistent in water, with a half-life of between 2 to 20 days. The half-life of a pollutant is the amount of time it takes for one-half of the chemical to be degraded. Approximately 73% of glutaraldehyde will eventually end up in air; the rest will end up in water.

Bioaccumulation in Aquatic Organisms

Some substances increase in concentration, or *bioaccumulate,* in living organisms as they breathe contaminated air, drink contaminated water, or eat contaminated food. These chemicals can become concentrated in the tissues and internal organs of animals as well as humans.

The concentration of glutaraldehyde found in fish tissues is expected to be about the same as the average concentration of glutaraldehyde in water from which the fish was taken.

Recommended Risk-Reduction Measures

Proper training of all transporters will reduce the likelihood of a mishap or accident resulting in a leak or spill to the environment. The correct use of DOT labeling (per the MSDS) on all containers, trucks, and rail cars will enable emergency responders to react properly and quickly to any disaster thereby reducing the potential risk to the environment and to personnel.

Storage of glutaraldehyde should be segregated from incompatible chemicals to minimize the risk of cross-contamination or contact. Buildings designated for storage should be equipped with appropriate fire protection systems (alarms, sprinklers, emergency lighting, portable extinguishers).

If a spill or leak to the environment has occurred, fire department, emergency response, and/or hazardous materials spill personnel should be notified immediately. Cleanup should be attempted only by those trained in proper spill containment procedures. Contaminated soils should be removed for incineration and replaced with clean soil. If glutaraldehyde should contact the water table, aquifer, or navigable waterway, time is of the essence. It is readily miscible in water and, therefore, total containment and remediation may not be entirely possible. When such spills occur, the local and/or state emergency response authorities must be notified. A comprehensive emergency response or disaster preparedness/recovery plan should be in place prior to any operations involving the use, transportation, storage, or disposal of glutaraldehyde. If glutaraldehyde is spilled or leaked, the following specific steps are recommended:

- ☑ Restrict persons not wearing protective clothing from area of spill or leak until cleanup is complete and area can be opened for normal work.
- ☑ Ventilate area of spill or leak.
- ☑ Absorb liquids in vermiculite, dry sand, earth, or a similar material and deposit in sealed containers.
- ☑ Keep glutaraldehyde out of a confined space, such as a sewer, because of the possibility of explosion

(unless the sewer is designed to prevent the buildup of explosive concentrations).

☑ It may be necessary to dispose of glutaraldehyde as a hazardous waste. The responsible state agency or the regional office of the federal Environmental Protection Agency (EPA) should be contacted for specific recommendations.

RISK ASSESSMENT: BUSINESS

General Assessment

Accidents or mishaps involving glutaraldehyde can present a significant threat to business operations. The loss or damage of equipment or facilities can significantly affect fiscal viability. Lawsuits that may result from personnel injury/death, public exposures, and/or environmental contamination will also require a serious expenditure of resources. Media attention surrounding an injury, death, or environmental damage can also result in a loss of profits and loss of current as well as future business. Always remember that anytime the term "reproductive hazard" is used, public emotion, hysteria, and ignorance can run equally high. This requires careful consideration during the development an implementation of public relations policies.

Recommended Risk-Reduction Measures

Company attorneys, safety and health professionals, and environmental specialists should be involved in the development of any procedures dealing with safety or emergency response. A company official should be pre-designated as a public relations officer with specific training in dealing with the press. Corporate plans and policies should be developed, approved, and implemented long before any need for such arises.

REFERENCES

American Conference of Governmental Industrial Hygienists. 1988. *Documentation of the Threshold Limit Values and Biological Exposure Indices,* 5th Edition (with updates). Cincinnati: ACGIH

Anthony, C. P., and N. J. Kolthoff. 1971. *Textbook of Anatomy and Physiology.* St. Louis: C. V. Mosby Company.

Bloom, A. D. and F. J. DeSerres. 1995. *Ecotoxicity and Human Health.* New York: CRC Press/Lewis Publishers.

Brusick, D. J. 1994. *Methods for Genetic Risk Assessment.* New York: CRC Press/Lewis Publishers.

Bueche, J. R. 1972. *Principles of Physics,* 2nd Edition. New York: McGraw-Hill.

Burrell, R., D. K. Flaherty, and L. J. Sauers. 1992. *Toxicology of the Immune System.* New York: Van Nostrand Reinhold.

Calabrese, E. J. 1994. *Biological Effects of Low Level Exposures.* New York: CRC Press/Lewis Publishers.

Cockerman, L. G., B. S. Shane. 1994. *Basic Environmental Toxicology.* New York: CRC Press/Lewis Publishers.

Cothern, C. R. 1995. *Handbook for Environmental Risk Decisionmaking.* New York: CRC Press/Lewis Publishers.

Danse, I. R. 1991. *Common Sense Toxics in the Workplace.* New York: Van Nostrand Reinhold.

Gardner, R. W. 1994. *Chemical Intolerance.* New York: CRC Press/Lewis Publishers.

Hallenbeck, W. H. 1993. *Quantitative Risk Assessment for Environmental and Occupational Health,* 2nd Edition. New York: CRC Press/Lewis Publishers.

Hodgson, E., R. B. Mailman, and J. E. Chambers. 1988. *Dictionary of Toxicology.* London: The MacMillian Company.

Heck, H. 1985. Mechanisms of aldehyde toxicity: Structure activity Studies. *CIIT Activities,* 5(10):106.

Hosty, J. W., and P. Foster. 1990. *A Practical Guide to Chemical Spill Response.* New York: Van Nostrand Reinhold.

Howard, P. H. 1995. *Handbook of Environmental Fate and Exposures Data for Organic Chemicals,* Vol. I-IV. New York: CRC Press/Lewis Publishers.

James, A., and M. Lord. 1992. *Index of Chemical and Physical Data.* New York: Van Nostrand Reinhold.

Jones, F. E. 1994. *Toxic Organic Vapors in the Workplace.* New York: CRC Press/Lewis Publishers.

Kamrin, M. 1 A. 1988. *Toxicology: A Primer on Principles and Applications.* Chelsea, Michigan: Lewis Publishers.

Keith, L. H. and M. M. Walker. 1995. *Handbook of Air Toxics.* New York: CRC Press/Lewis Publishers.

Landis, W. G., M. H. Yu. 1995. *Introduction to Environmental Toxicology.* New York: CRC Press/Lewis Publishers.

Lewis, .J., Sr. 1993. *Hawley's Condensed Chemical Dictionary,* 12th Edition. New York: Van Nostrand Reinhold.

Lewis, . J., Sr. 1992. *Sax's Dangerous Properties of Industrial Materials,* Volumes 1, 2, and 3, 8th Edition. New York: Van Nostrand Reinhold.

Meyer, E. 1990. *Chemistry of Hazardous Materials.* Englewood Cliffs, NJ: Prentice Hall.

Patnaik, P. 1992. *A Comprehensive Guide to the Hazardous Properties of Chemical Substances.* New York: Van Nostrand Reinhold.

Silverman, L., H. F. Schuttle, and M. W. First. 1946. Further studies on sensory response to certain industrial solvent vapors. *Journal of Industrial Hygiene Toxicology,* 28:262-266.

Sherman, J. D. 1988. *Chemical Exposure and Disease.* New York: Van Nostrand Reinhold.

U.S. Department of Health and Human Services, National Institute for Occupational Safety and Health. 1994. *NIOSH Pocket Guide to Chemical Hazards.* Washington, D.C.: U.S. Government Printing Office.

MATERIAL SAFETY DATA SHEET

CHEMICAL NAME	
GLYCERIN	**CRC PRESS/LEWIS PUBLISHERS DISCLAIMER:** THE INFORMATION AND RECOMMENDATIONS PRESENTED HEREIN ARE BASED ON SOURCES BELIEVED TO BE RELIABLE. CRC MAKES NO REPRESENTATION ON ITS COMPLETENESS OR ACCURACY. IT IS THE RESPONSIBILITY OF THE USER TO DETERMINE THE CHEMICAL'S SUITABILITY FOR ITS INTENDED USE, THE CHEMICAL'S SAFE USE, AND THE CHEMICAL'S PROPER DISPOSAL. NO REPRESENTATIONS AND/OR WARRANTIES, EITHER EXPRESSED OR IMPLIED, OF THE MERCHANTABILITY OR FITNESS FOR A PARTICULAR PURPOSE, OR OF ANY OTHER NATURE, ARE MADE WITH RESPECT TO THE INFORMATION PROVIDED IN THIS MSDS OR TO THE CHEMICAL TO WHICH INFORMATION MAY REFER. CRC NEITHER ASSUMES NOR AUTHORIZES ANY OTHER PERSON TO ASSUME FOR IT, ANY OTHER ADDITIONAL RESPONSIBILITY OR LIABILITY FOR THE USE OF, OR RELIANCE UPON, THIS INFORMATION.

HAZARD WARNING INFORMATION

HEALTH	FIRE	REACTIVE	OTHER	DEGREE OF HAZARD	COLOR CODING	OTHER CODES
1	1	0		0 = Minimum Hazard 1 = Slight Hazard 2 = Moderate Hazard 3 = Serious Hazard 4 = Severe Hazard	HEALTH = BLUE FIRE = RED REACTIVITY = YELLOW OTHER = WHITE	OX = Oxidizer ACID = Acid ALK = Alkali COR = Corrosive W = Use No Water

SECTION I - GENERAL INFORMATION

Characterization	RCRA Number	EPA Class
Chlorohydrin	**None**	**Not Applicable**

DOT Proper Shipping Name	Chemical Abstract Service (CAS) Number
No Citation	**56-81-5**
DOT Hazard Class and Label Requirements	DOT Emergency Guide Code
No Citation	**No Citation**
DOT Identification Number	Chemical Formula
No Citation	**$HOCH_2CH(OH)CH_2OH$**

Synonyms

Glycerin (anhydrous); glycerol; glycyl alcohol; 1,2,3-propanetriol; trihydroxypropane; Superol; synthetic glycerin; 90 technical glycerin; 1,2,3-trihydroxypropane.

SECTION II - HAZARDOUS INGREDIENTS AND IDENTITY INFORMATION

Hazardous Components (specific identity)	OSHA Exposure Criteria	NIOSH Exposure Criteria	Immediately Dangerous to Life and Health (IDLH)	ACGIH Exposure Criteria
Glycerin: (derivation: By-product of soap manufacture; from propylene and chlorine to form allyl chloride, converted to the dichlorohydrin with hypochlorous acid, then saponified to glucerol with caustic solution; isomerization of propylene oxide to allyl alcohol and reacted with peracetic acid, the resulting glycidol is hydrolyzed to glycerol). **1 ppm = 3.83 mg/m^3**	**PEL:** **(Respirable)** **1.3 ppm** **5 mg/m^3** **(Total Mist)** **3.9 ppm** **15 mg/m^3**	**REL:** **Not Established** **STEL:** **Not Established**	**Not Determined**	**TLV:** **(Vapor)** **2.6 ppm** **10 mg/m^3** **STEL:** **Not Established**

SECTION III - PHYSICAL AND CHEMICAL CHARACTERISTICS

Boiling Point	Specific Gravity (H_2O = 1)
290°F (544°C) Decomposes	**1.26**
Vapor Pressure (mm Hg)	Molecular Weight (atomic weight)
0.003 at 122°F (50°C)	**92.1**
Vapor Density (Air = 1)	Melting Point
3.2	**64°F (18°C)**

Solubility

Miscible in water (hygroscopic). Soluble in alcohol. Insoluble in ether, benzene, and chloroform in oils.

Appearance and Odor

Clear, colorless, or pale-yellow, odorless, syrupy liquid (or solid below 64°F) with a sweet, warm taste.

SECTION IV - FIRE AND EXPLOSION HAZARD DATA

Flash Point (method used)	Explosive Limits in Air % by Volume
320°F (160°C) closed cup	**LEL: 0.9% UEL: Not Found**
NFPA Classification	Autoignition Temperature
Class IIIB Combustible Liquid	**752°F (400°C)**

Extinguishing Media

Dry chemical, alcohol foam (recommended), carbon dioxide, water. Water or foam may cause frothing.

Special Fire Fighting Procedures

Poisonous gases are emitted (including phosgene). Wear full protective clothing and self-contained breathing apparatus (SCBA). Remain clear of smoke, water fallout and water run off. Move containers from fire area if it can be done without risk. Cool exposed containers. Fight fire from distance, if possible.

Unusual Fire and Explosion Hazards

Material is corrosive. Containers may explode in fire due to buildup of pressure.

SECTION V - REACTIVITY DATA

Stability		Conditions to Avoid Glycerin is stable in closed containers under routine conditions of handling and storage. Avoid contact with heat and incompatible materials. Material is hygroscopic (absorbs moisture from the air).
Stable X	Unstable	Incompatibility *(materials to avoid)* Strong oxidizers (chromium trioxide, potassium chlorate, potassium permanganate, hydrogen peroxide), calcium hypochlorite, nitric acid mixed with sulfuric acid, some metallic oxide mixtures (lead oxide).
Hazardous Polymerization		Conditions to Avoid Under normal conditions of temperature and pressure, hazardous polymerization of glycerin is not expected to occur.
May Occur	Will Not Occur X	Hazardous Decomposition or By-products At high temperatures, glycerin decomposes and may produce poisonous gases, including toxic oxides of carbon and corrosive acrolein gas.

SECTION VI - HEALTH HAZARD DATA

Primary Route(s) of Entry:	Inhalation?	Absorption (skin/eye)?	Ingestion?
			X

Health Hazards

INHALATION: In the form of mist, it is a nuisance particulate and an inhalation irritant. It is considered relatively non-toxic, causing very few noticeable effects.

SKIN & EYES: May cause mild irritation to the eyes and skin.

INGESTION: Causes headache, nausea, vomiting, and diarrhea. May cause kidney injury.

Carcinogenicity	NTP Listed?	IARC Cancer Review Group?	OSHA Regulated?	Target Organs?
No Evidence Human No Evidence Animal	No	No	29 CFR 1910.1000 Table Z-1	Respiratory system, eyes, skin, kidneys.

Medical Conditions Generally Aggravated by Exposure

None reported.

Emergency and First-aid Procedures

Eye contact: Do not allow victim to keep eyes tightly shut. Flush immediately with water for 15 minutes, seek medical attention. Skin contact: Remove all contaminated clothing. Immediately flood area with water for 15 minutes. Then wash affected areas with soap and water. Inhalation: Remove the person from exposure. Provide respiratory assistance and CPR. Transfer to medical facility, if required. If swallowed: Contact poison control center. Rinse victim's mouth with water. Unless advised otherwise by poison control center and victim is conscious and alert, give 2-3 glasses of water to drink to dilute and then induce vomiting repeatedly until vomitus is clear. Never give anything by mouth to an unconscious or convulsing person. Never try to make an unconscious or convulsing person vomit. Seek medical attention immediately. Keep victim warm and at rest.

SECTION VII - PRECAUTIONS FOR SAFE HANDLING AND USE

Steps to be Taken in Case Material is Released or Spilled

Remove all sources of ignition. Ventilate area of spill. Restrict those not involved in cleanup from entering area. Notify appropriate authorities. Absorb in vermiculite or other suitable material and dispose of in sealed drums. Do NOT release to sewers, storm drains, surface waters, or soils.

Preferred Waste Disposal Method

No Citation.

Precautions to be Taken in Handling and Storage

Store in tightly closed containers in cool, dry, well-ventilated area away from heat and incompatibilities. Material is hygroscopic and will absorb moisture from the air. Ensure air-tight seals on all containers.

Other Precautions and Warnings

Prevent static sparks by grounding and bonding all containers.

SECTION VIII - CONTROL MEASURES AND PERSONAL PROTECTIVE EQUIPMENT

Respiratory Protection *(specify type)*

For exposures above the permissible limit of 5 mg/m^3 (respirable fraction), use an MSHA/NIOSH-approved half-mask respirator with an organic vapor cartridge and equipped with a dust/mist pre-filter. For higher concentrations, or in emergencies, use a self-contained breathing apparatus in positive pressure mode.

Ventilation

Local exhaust (preferred) at site of chemical work or general ventilation.

Protective Gloves	Eye Protection	Other Protective Clothing
Impervious Rubber (test for adequacy)	Chemical Goggles or Face Mask	Protective Apron, Boots (avoid skin contact)

Work/Hygiene Practices

Always wash hands thoroughly after using chemical; never bring food, drink, or smoking materials into vicinity of chemicals.

GLYCERIN

$HOCH_3CH(OH)CH_2OH$

CAS: 56-81-5

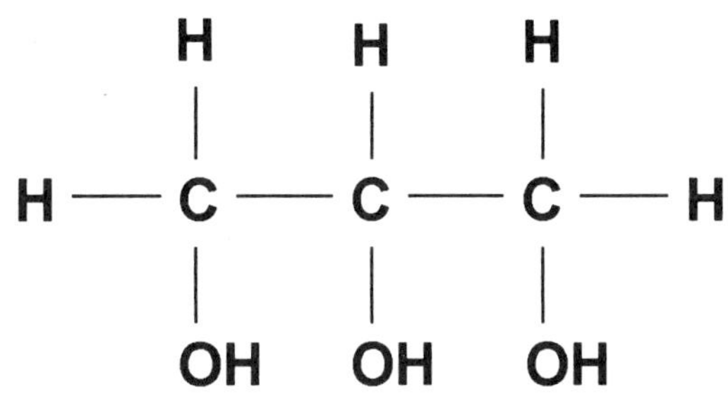

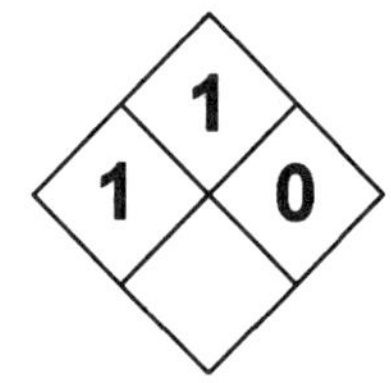

IDENTIFICATION AND TYPICAL USES

Glycerin is a clear, colorless, or pale-yellow odorless, syrupy liquid with a sweet, warm taste. It may appear as a colorless solid at temperatures below 64°F (18°C). It has a wide range of uses, including: used in alkyd resins, in dynamite, in ester gums, in pharmaceuticals, in perfumery, as a plasticizer for the regeneration of cellulose, in cosmetics, in foodstuffs, in conditioning of tobacco, in certain liquors, as a solvent, in printer's ink rolls, in polyurethane polyols, as an emulsifying agent, in rubber stamp and copying inks, as a binder for cements and mixes, in special soaps, as a lubricant and softener, as a bacteriostat, a penetrant, in hydraulic fluid, as a humectant, in the fermentation of nutrients, and in anti-freeze mixtures.

RISK ASSESSMENT: HEALTH

General Assessment

Glycerin is a relatively non-toxic material. However, it is considered mildly toxic by *ingestion*. Although inhalation and skin contact can also cause some reaction, they are not normally thought of as exposure routes. While there is no indication that exposure will cause any carcinogenic effects on long-term exposure, human mutation data have been reported. It has also been shown to adversely affect reproduction in experimental testing with animals.

In the form of mist, glycerin is a nuisance particulate and an inhalation irritant. It can cause mild irritation of the eyes, nose, throat, and respiratory tract. Eye and skin contact can result in irritation.

Ingestion causes the most noticeable and severe systemic effects. It can cause headache, nausea, and vomiting. It may also cause kidney damage.

Acute Health Effects

The following acute (short-term) health effects may occur immediately or shortly after exposure to glycerin:

Skin: Local irritation at contact site.

Eye: Irritation and possible inflammation.

Lung: Irritation of the nose, throat, and respiratory tract.

Other: Ingestion causes headache, nausea, vomiting, and diarrhea.

Chronic Health Effects

The following chronic (long-term) health effects may occur at some time after exposure to glycerin and can last for months or even years:

Cancer Hazards: According to information presented in the references, glycerin has not been shown to cause cancer in test animals. Human mutation data have been reported and many scientists believe that exposure to mutagenic substances may present a cancer hazard in the long-term. Additional study is required before any determinations can be made in this regard for glycerin.

Reproductive Hazard: Glycerin has been shown to cause adverse reproductive effects in laboratory experiments. It is not known if it will have this effect in humans exposed to glycerin.

Other Chronic Effects: Long-term exposure may cause kidney injury or damage.

Recommended Risk-Reduction Measures

Glycerin presents a relatively low to moderate health hazard risk. However, personnel should still avoid direct contact with glycerin. If a less toxic material or compound cannot be substituted for glycerin, then *engineering controls* are the most effective method of reducing exposures. The best protection is to enclose operations and/or provide local exhaust ventilation at the site of glycerin release. While not always operationally possible, isolating operations involving glycerin manufacturing or use can also reduce exposure. Personnel must follow proper industrial hygiene and good housekeeping work practices. Using respiratory protection is less effective than the controls mentioned above, but is still advisable whenever working with or around glycerin. At exposures above the PEL (5 mg/m^3, respirable fraction or 15 mg/m^3 total mist), an MSHA or NIOSH half-mask respirator equipped with an organic vapor cartridge and a dust/mist pre-filter may suffice. For higher exposures or when the exposure levels are unknown (as in an emergency), a powered supplied-air respirator or a self-contained breathing apparatus (SCBA) with full facepiece and operated in pressure demand mode are recommended for the greatest possible respiratory protection. If a full facepiece is not available, then chemical goggles should be worn to protect the eyes. A face shield and protective apron should also be worn. Skin contact must not occur. To prevent hand and skin exposures, impervious gloves should be worn. Glove manufacturers should be contacted and permeation studies obtained *before* final glove selections have been made.

Administrative controls should also be in place to minimize the potential for human exposures. These may include written procedures or policies which specify the methods and techniques that will be practiced whenever personnel are to work with glycerin.

All personnel should receive *training* on the use, hazards, protective measures, emergency actions, and other precautions per 29 CFR 1910.1200 (Hazard Communication), prior to the first assignment in an area where glycerin is used or stored.

Before beginning initial assignments with this chemical and at regular intervals thereafter (e.g., annually), the following medical protocol are recommended:

- ☑ Lung function tests (establish initial baseline and compare thereafter).
- ☑ Liver and kidney function tests (establish initial baseline and compare thereafter).

If symptoms develop or overexposure is suspected, the following medical testing may be useful:

- ☑ Lung function tests (compare to initial and subsequent baseline tests).

Any evaluation should include a careful history of past and present symptoms with an examination. Medical tests that simply look for existing damage are not a substitute for controlling exposures. Also, since smoking can cause heart disease, emphysema, and other respiratory problems, personnel who smoke may show symptoms of exposure much quicker and more pronounced than those who do not. Prudent risk management requires careful consideration of *all* possible and potential exposure risk factors to ensure proper implementation of control measures and management of the exposure risk.

Other methods to reduce exposure include:

- ☑ Where possible, enclose operations and use local exhaust ventilation at the site of chemical release. If local exhaust ventilation or enclosure is not used, respiratory protection should be mandatory.
- ☑ Always ensure that proper protective clothing is worn when using chemical substances and that personnel are trained in the proper use, care, and maintenance of their protective equipment.
- ☑ Wash thoroughly immediately after exposure to glycerin and at the end of the work shift or before eating, drinking, or smoking.
- ☑ Hazard warning information should be posted in the work area. In addition, as part of an on-going education and training program, all information on the health and safety hazards of glycerin should be communicated to all potentially exposed workers.
- ☑ Eye wash stations should be provided in the immediate work area for emergency use. If there is a possibility of skin exposure to glycerin, emergency shower facilities should also be provided in the immediate work area.
- ☑ Workers whose clothing has been contaminated by glycerin should change into clean clothes *before* leaving work. Contaminated work clothing should never be worn home where family members can be contaminated. The clothing should be laundered only by individuals who have been informed of the hazards of exposure to glycerin.

RISK ASSESSMENT: ENVIRONMENT

General Assessment

The environment is at risk of exposure during transportation, storage, disposal, or destruction of glycerin. In almost every scenario, the threat of environmental exposure is contingent upon the proper handling of the chemical substance. Accidental spills, large or small, where glycerin contacts incompatible commodities can result in violent and even explosive reactions (depending upon conditions of spill), and possible contamination of the surrounding environmental mediums (water, soil, and air).

Glycerin is considered a Class 111A combustible liquid (per OSHA 29 CFR 1910.106). It can ignite when exposed to heat and support the burning of other materials in fire. Mixtures with hydrogen peroxide are highly explosive. It will ignite on contact with potassium permanganate and calcium hypochlorite. When mixed with nitric acid plus sulfuric acid, it forms the highly explosive glycerol nitrate. Mixtures with other oxidizers, and some acids can result in the formation of highly toxic and explosive compounds. Therefore, extreme caution is required in handling, storage, transportation, and disposal of glycerin. It is hygroscopic (will absorb moisture from the air to form other materials. These characteristics also require special consideration during any emergency situation involving a leak or spill of glycerin.

Glycerin may enter the environment through industrial effluents and from spills.

Acute Ecological Effects

Acute (short-term) toxic effects may include the death of animals, birds, or fish and death or low growth rate in plants. Acute effects are seen 2 to 4 days after animals or plants are exposed to glycerin.

Insufficient data are available on the short-term effects of glycerin to aquatic life, plants, land animals, and birds.

Chronic Ecological Effects

Chronic (long-term) toxic effects may include shortened life span, reproductive problems, lower fertility, and changes in appearance or behavior in exposed animals. These effects can be seen long after first exposure(s) to toxic chemicals.

Insufficient data are available to evaluate or predict the long-term effects of glycerin to aquatic life, plants, birds, and land animals.

Water Solubility

Glycerin is highly soluble in water. Concentrations of 1000 milligrams and even more will readily mix with a liter of water.

Persistence in the Environment

Glycerin is moderately persistent in the water, with a half-life between 20 and 200 days. The half-life of a pollutant is the amount of time it takes for one-half of the chemical to be degraded.

Bioaccumulation in Aquatic Organisms

Some substances increase in concentration, or *bioaccumulate,* in living organisms as they breathe contaminated air, drink contaminated water, or eat contaminated food. These chemicals can become concentrated in the tissues and internal organs of animals as well as humans.

The concentration of glycerin found in fish tissues is expected to be about the same as the average concentration of glycerin in the water from which the fish was taken.

Recommended Risk-Reduction Measures

Proper training of all transporters will reduce the likelihood of a mishap or accident resulting in a leak or spill to the environment. The correct use of DOT labeling (per the MSDS) on all containers, trucks, and rail cars will enable emergency responders to react properly and quickly to any disaster thereby reducing the potential risk to the environment and to personnel.

Storage of glycerin should be segregated from incompatible materials to minimize the risk of cross-contamination or contact. Storage facilities should be equipped with the proper fire protection equipment (sprinklers, alarms, extinguishers). To prevent static sparks, all equipment and containers used in shipping, receiving, or transferring should be electrically grounded and bounded.

If a spill or leak to the environment has occurred, fire department, emergency response and/or hazardous materials spill personnel should be notified immediately. Contaminated soils should be removed for incineration and replaced with clean soils. Cleanup should be attempted only by those trained in proper spill containment procedures. If glycerin should contact the water table, aquifer, or navigable waterway, reclamation procedures should be initiated as soon as practical. It is highly soluble in water and total remediation may not be possible. The local and/or state emergency response authorities must be notified. A comprehensive emergency response or disaster preparedness/recovery plan should be in place prior to any operations involving the use, transportation, storage, or disposal of glycerin. If glycerin is spilled or leaked, the following specific steps are recommended:

- ☑ Restrict persons not wearing protective clothing from area of spill or leak until cleanup is complete and area can be opened for normal work.
- ☑ Remove all ignition sources.
- ☑ Ventilate area of leak or spill.
- ☑ Absorb liquids in vermiculite or other compatible material and dispose of in sealed drums for disposal. Do NOT allow material or run off to contact waterways or enter sewers.
- ☑ It may be necessary to dispose of glycerin as a hazardous waste. The responsible state agency or the regional office of the federal Environmental Protection Agency (EPA) should be contacted for specific recommendations.

RISK ASSESSMENT: BUSINESS

General Assessment

Accidents or mishaps involving glycerin can present a significant threat to business operations. The loss or damage of equipment or facilities can significantly affect fiscal viability. Lawsuits that may result from personnel injury/death, public exposures, and/or environmental contamination will also require a serious expenditure of resources. Media attention surrounding an injury, death, or environmental damage can also result in a loss of profits and loss of current as well as future business.

Recommended Risk-Reduction Measures

Company attorneys, safety and health professionals, and environmental specialists should be involved in the development of any procedures for responding to chemical incidents. A company official should be predesignated as a public relations officer with specific training in dealing with the press. Corporate plans and policies should be developed, approved, and implemented long before any need for such arises.

REFERENCES

Ahlbom, A. 1993. *Biostatistics for Epidemiologists.* New York: CRC Press/Lewis Publishers.

American Conference of Governmental Industrial Hygienists. 1988. *Documentation of the Threshold Limit Values and Biological Exposure Indices,* 5th Edition (with updates). Cincinnati: ACGIH

Anthony, C. P., and N. J. Kolthoff. 1971. *Textbook of Anatomy and Physiology.* St. Louis: C. V. Mosby Company.

Asante-Duah, D. K. 1993. *Hazardous Waste Risk Assessment.* New York: CRC Press/Lewis Publishers.

Bloom, A. D. and F. J. DeSerres. 1995. *Ecotoxicity and Human Health.* New York: CRC Press/Lewis Publishers.

Brusick, D. J. 1994. *Methods for Genetic Risk Assessment.* New York: CRC Press/Lewis Publishers.

Bueche, J. R. 1972. *Principles of Physics,* 2nd Edition. New York: McGraw-Hill.

Burrell, R., D. K. Flaherty, and L. J. Sauers. 1992. *Toxicology of the Immune System.* New York: Van Nostrand Reinhold.

Calabrese, E. J. 1994. *Biological Effects of Low Level Exposures.* New York: CRC Press/Lewis Publishers.

Cockerman, L. G., B. S. Shane. 1994. *Basic Environmental Toxicology.* New York: CRC Press/Lewis Publishers.

Cothern, C. R. 1995. *Handbook for Environmental Risk Decisionmaking.* New York: CRC Press/Lewis Publishers.

Danse, I. R. 1991. *Common Sense Toxics in the Workplace.* New York: Van Nostrand Reinhold.

Environmental Protection Agency. 1983. *Chemical Hazard Information Profiles.* Washington, D.C.: U.S. EPA.

Fodden, C. R., and J. L. Weddell. 1992. *Hazardous Materials Emergency Action Data.* New York: CRC Press/Lewis Publishers.

Forseberg, K., and L. H. Keith. 1993. *Chemical; Protective Clothing Permeation and Degradation Database.* Chelsea, MI: Lewis Publishers.

Gardner, R. W. 1994. *Chemical Intolerance.* New York: CRC Press/Lewis Publishers.

Hallenbeck, W. H. 1993. *Quantitative Risk Assessment for Environmental and Occupational Health,* 2nd Edition. New York: CRC Press/Lewis Publishers.

Hodgson, E., R. B. Mailman, and Janice E. Chambers. 1988. *Dictionary of Toxicology.* London: The MacMillian Company.

Howard, P. H. 1995. *Handbook of Environmental Fate and Exposures Data for Organic Chemicals,* Vol. I-IV. New York: CRC Press/Lewis Publishers.

Hosty, J. W., and P. Foster. 1990. *A Practical Guide to Chemical Spill Response.* New York: Van Nostrand Reinhold.

Jones, F. E. 1994. *Toxic Organic Vapors in the Workplace.* New York: CRC Press/Lewis Publishers.

Kamrin, M. A. 1988. *Toxicology: A Primer on Principles and Applications.* Chelsea, Michigan: Lewis Publishers.

Lewis, .J., Sr. 1993. *Hawley's Condensed Chemical Dictionary,* 12th Edition. New York: Van Nostrand Reinhold.

Lewis, . J., Sr. 1992. *Sax's Dangerous Properties of Industrial Materials,* Volumes 1, 2, and 3, 8th Edition. New York: Van Nostrand Reinhold.

Mackay, D., W. Y. Shiu, and K. C. Ma. 1992. *Illustrated Handbook of Chemical-Physical Properties and Environmental Fate for Organic Chemicals.* New York: CRC Press/Lewis Publishers.

MATERIAL SAFETY DATA SHEET

CHEMICAL NAME	
GLYCEROL α-MONOCHLOROHYDRIN	**CRC PRESS/LEWIS PUBLISHERS DISCLAIMER:** THE INFORMATION AND RECOMMENDATIONS PRESENTED HEREIN ARE BASED ON SOURCES BELIEVED TO BE RELIABLE. CRC MAKES NO REPRESENTATION ON ITS COMPLETENESS OR ACCURACY. IT IS THE RESPONSIBILITY OF THE USER TO DETERMINE THE CHEMICAL'S SUITABILITY FOR ITS INTENDED USE, THE CHEMICAL'S SAFE USE, AND THE CHEMICAL'S PROPER DISPOSAL. NO REPRESENTATIONS AND/OR WARRANTIES, EITHER EXPRESSED OR IMPLIED, OF THE MERCHANTABILITY OR FITNESS FOR A PARTICULAR PURPOSE, OR OF ANY OTHER NATURE, ARE MADE WITH RESPECT TO THE INFORMATION PROVIDED IN THIS MSDS OR TO THE CHEMICAL TO WHICH INFORMATION MAY REFER. CRC NEITHER ASSUMES NOR AUTHORIZES ANY OTHER PERSON TO ASSUME FOR IT, ANY OTHER ADDITIONAL RESPONSIBILITY OR LIABILITY FOR THE USE OF, OR RELIANCE UPON, THIS INFORMATION.

HAZARD WARNING INFORMATION

HEALTH	FIRE	REACTIVE	OTHER	DEGREE OF HAZARD	COLOR CODING	OTHER CODES
3	1	0		0 = Minimum Hazard 1 = Slight Hazard 2 = Moderate Hazard 3 = Serious Hazard 4 = Severe Hazard	HEALTH = BLUE FIRE = RED REACTIVITY = YELLOW OTHER = WHITE	OX = Oxidizer ACID = Acid ALK = Alkali COR = Corrosive W = Use No Water

SECTION I - GENERAL INFORMATION

Characterization	Chlorohydrin	RCRA Number	None
		EPA Class	Not Applicable
DOT Proper Shipping Name	Glycerol α-Monochlorohydrin	Chemical Abstract Service (CAS) Number	96-24-2
DOT Hazard Class and Label Requirements	Poison B; St. Andrews Cross	DOT Emergency Guide Code	30
DOT Identification Number	UN 2689	Chemical Formula	$CH_2ClCH_2OCH_2Cl$

Synonyms
α-Propenyldichlorohydrin; α-dichlorohydrin; 1,3-dichloro-2-propanol; sym-dichloroisopropyl alcohol; α-chlorohydrin; chlorodeoxyglycerol; 1-chloro-2,3-propanediol; chlorohydrin; 3-chloropropylene glycol.

SECTION II - HAZARDOUS INGREDIENTS AND IDENTITY INFORMATION

Hazardous Components (specific identity)	OSHA Exposure Criteria	NIOSH Exposure Criteria	Immediately Dangerous to Life and Health (IDLH)	ACGIH Exposure Criteria
Glycerol monochlorohydrin: (derivation: By interaction of glycerol and dry hydrogen chloride gas and subsequent distillation).	PEL: Not Established STEL: Not Established	REL: Not Established STEL: Not Established	Not Determined	TLV: Not Established STEL: Not Established

SECTION III - PHYSICAL AND CHEMICAL CHARACTERISTICS

Boiling Point	345°F (174°C) Decomposes	Specific Gravity (H_2O = 1)	1.36
Vapor Pressure (mm Hg)	7 at 122°F (50°C)	Molecular Weight (atomic weight)	110.55
Vapor Density (Air = 1)	Not Found	Melting Point	27°F (-4°C)

Solubility
Soluble in water, alcohol, and ether. Miscible in most organic solvents, vegetable oils.

Appearance and Odor
Colorless, slightly viscous liquid with a faint, pleasant, chloroform-like odor and sweet taste.

SECTION IV - FIRE AND EXPLOSION HAZARD DATA

Flash Point (method used)	165°F (74°C) closed cup	Explosive Limits in Air % by Volume	LEL: Not Determined UEL: Not Determined
NFPA Classification	Combustible Liquid	Autoignition Temperature	Not Determined

Extinguishing Media
Use extinguishing agents suitable to surrounding fire.

Special Fire Fighting Procedures
Poisonous gases are emitted during fire and structural fire-fighting clothing may not provide adequate protection. Wear full protective clothing and self-contained breathing apparatus (SCBA). Remain clear of smoke, water fallout and water run off. Move containers from fire area if it can be done without risk. Cool exposed containers with water spray until after fire is out. Fight fire from distance, if possible.

Unusual Fire and Explosion Hazards
Material is extremely toxic. Avoid contact. The vapor may form explosive mixtures in air.

SECTION V - REACTIVITY DATA

Stability		Conditions to Avoid
		Glycerol α- monochlorohydrin is an unstable liquid. However, the commercial product is often a stable mixture of two isomers. Keep away from incompatible materials, heat, and flame.
Stable	Unstable	Incompatibility *(materials to avoid)*
	X	Strong acids, such as nitric, sulfuric, and especially perchloric (forms a sensitive explosive product more powerful than glyceryl nitrate). In the presence of a catalyst, it can produce propylene oxide, a carcinogen.
Hazardous Polymerization		Conditions to Avoid Under normal conditions of temperature and pressure, hazardous polymerization of glycerol α- monochlorohydrin is not expected to occur.
May Occur	Will Not Occur	Hazardous Decomposition or By-products
	X	At high temperatures, glycerol α-monochlorohydrin decomposes and may produce poisonous gases, including toxic oxides of carbon and corrosive acrolein gas.

SECTION VI - HEALTH HAZARD DATA

Primary Route(s) of Entry:	Inhalation?	Absorption (skin/eye)?	Ingestion?
	X		X

Health Hazards

INHALATION: Irritation to the eyes, nose, throat, and respiratory system. Causes toxic systemic effects such as headache, nausea, vomiting, diarrhea, and symptoms of central nervous system (CNS) depression such as drowsiness (narcosis). Possible thyroid tumors.

SKIN & EYES: Irritation to skin with absorption likely. Causes toxic systemic effects with target on the liver, kidneys, and CNS. Eye contact may cause moderate to severe irritation.

INGESTION: Causes headache, nausea, vomiting, and diarrhea. Gastrointestinal irritation is likely.

Carcinogenicity	NTP Listed?	IARC Cancer Review Group?	OSHA Regulated?	Target Organs?
Questioned Human Suspected Animal	No	Not Found	No	Respiratory system, eyes, skin, CNS, thyroid.

Medical Conditions Generally Aggravated by Exposure

None reported.

Emergency and First-aid Procedures

Eye contact: Do not allow victim to keep eyes tightly shut. Flush immediately with water for 15 minutes, seek medical attention. Skin contact: Remove all contaminated clothing. Immediately flood area with water for 15 minutes. Then wash affected areas with soap and water. Inhalation: Remove the person from exposure. Provide respiratory assistance and CPR. Transfer to medical facility, if required. If swallowed: Contact poison control center. Never attempt to give anything by mouth to an unconscious or convulsing person. Never try to make an unconscious or convulsing person vomit. Seek medical attention immediately. Keep victim warm and at rest.

SECTION VII - PRECAUTIONS FOR SAFE HANDLING AND USE

Steps to be Taken in Case Material is Released or Spilled

Remove all sources of ignition. Ventilate area of spill. Restrict those not involved in cleanup from entering area. Notify appropriate authorities. Absorb in vermiculite or other suitable material and dispose of in sealed drums. Do NOT release to sewers, storm drains, surface waters, or soils.

Preferred Waste Disposal Method

Burn in chemical incinerator equipped with afterburner and scrubber.

Precautions to be Taken in Handling and Storage

Store in tightly closed containers in cool, dry, well ventilated area away from heat and incompatibilities. Avoid contact with material during storage and shipping operations. Use protective equipment.

Other Precautions and Warnings

Prevent static sparks by grounding and bonding all containers.

SECTION VIII - CONTROL MEASURES AND PERSONAL PROTECTIVE EQUIPMENT

Respiratory Protection *(specify type)*

No exposure levels have been established. For any exposure to a possible carcinogen, use a NIOSH-approved supplied-air respirator operated in continuous flow mode, or a self-contained breathing apparatus (SCBA) with full facepiece operated in pressure demand or some other positive pressure mode.

Ventilation

Local exhaust (preferred) at site of chemical work or general ventilation.

Protective Gloves	Eye Protection	Other Protective Clothing
Impervious Rubber (test for adequacy)	Chemical Goggles or Face Mask	Protective Apron, Boots (avoid skin contact)

Work/Hygiene Practices

Always wash hands thoroughly after using chemical; never bring food, drink, or smoking materials into vicinity of chemicals.

GLYCEROL α-MONOCHLOROHYDRIN

$CH_2ClCH_2OCII_2Cl$ CAS: 96-24-2

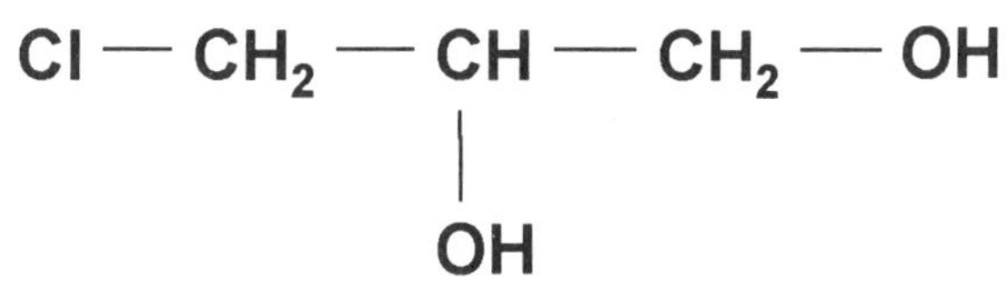

IDENTIFICATION AND TYPICAL USES

Glycerol α-monochlorohydrin is a clear, colorless, slightly viscous, unstable liquid with a faint, pleasant, chloroform-like odor and sweet taste. It is used in the synthesis of glycerol esters, amines, and other derivatives, as a general solvent, as an intermediate in organic synthesis, in the formulation of paints, varnishes, lacquers, as a water color binder, in photographic lacquers, to lower the freezing point of dynamite, and as a rodent chemosterilant.

RISK ASSESSMENT: HEALTH

General Assessment

Glycerol α-monochlorohydrin is a highly toxic chemical by *inhalation* and *ingestion.* It is suspected to cause carcinogenic effects in test animals and is a strong teratogen, causing serious adverse reproductive effects in experimental tests.

Inhalation causes irritation of the eyes, nose, throat, and respiratory system. Once absorbed through the respiration process, its toxic effects appear to focus directly on the reproductive system causing spermatogenesis, and possible injury to the sperm duct and testes. Skin contact causes mild to severe irritation with possible redness and rash. It is also a severe eye irritant. Ingestion can lead to gastrointestinal irritation and possibly systemic effects similar to that of inhalation exposure.

Acute Health Effects

The following acute (short-term) health effects may occur immediately or shortly after exposure to glycerol α-monochlorohydrin:

Skin: Local irritation at contact site.

Eye: Severe irritation and possible inflammation.

Lung: Irritation of the nose, throat, and respiratory tract.

Other: Ingestion causes headache, nausea, vomiting, and diarrhea. Short-term effects can lead to narcosis (drowsiness) and other symptoms of CNS depression.

Chronic Health Effects

The following chronic (long-term) health effects may occur at some time after exposure to glycerol monochlorohydrin and can last for months or even years:

Cancer Hazards: According to information presented in the references, glycerol α-monochlorohydrin has been shown to cause cancer in test animals. It appears to give rise to thyroid cancer in test animals. It is not known if it will have this effect on humans.

Reproductive Hazard: Glycerol α-monochlorohydrin has been shown to cause adverse reproductive effects in laboratory experiments. These include spermatogenesis-type effects to the testes, sperm duct, and Cowper's gland. It is not known if this effect will occur in humans exposed to glycerol α-monochlorohydrin.

Other Chronic Effects: Long-term exposure may cause weight loss and other chronic effects related to CNS depression.

Recommended Risk-Reduction Measures

Glycerol α-monochlorohydrin presents an extremely high health hazard risk. Personnel should avoid direct contact with glycerol α-monochlorohydrin. If a less toxic material or compound cannot be substituted for glycerol α-monochlorohydrin, then *engineering controls* are the most effective method of reducing exposures. The best protection is to enclose operations and/or provide local exhaust ventilation at the site of glycerol α-monochlorohydrin release. While not always operationally possible, isolating operations involving glycerol α-monochlorohydrin manufacturing or use can also reduce exposure. Personnel must follow proper industrial hygiene and good housekeeping work practices. Using respiratory protection is less effective than the controls mentioned above, but is still advisable whenever working with or around glycerol α-monochloro-hydrin. No exposure levels have been established for this chemical. However, this does not mean that exposure is without health hazard risks. At any exposure level to a suspect or probable carcinogen, an MSHA or NIOSH-approved powered supplied-air respirator or a self-contained breathing apparatus (SCBA) with full facepiece operated in pressure demand mode are recommended for the greatest possible respiratory protection. If a full facepiece is not available, then chemical goggles should be worn to protect the eyes. A face shield and protective apron should also be worn. Skin contact must not occur. To prevent hand and skin exposures, impervious gloves should be worn. Glove manufacturers should be contacted and permeation studies obtained *before* final glove selections have been made.

Administrative controls should also be in place to minimize the potential for human exposures. These may include written procedures or policies which specify the methods and techniques that will be practiced whenever personnel are to work with glycerol α-monochlorohydrin.

All personnel should receive *training* on the use, hazards, protective measures, emergency actions, and other precautions per 29 CFR 1910.1200 (Hazard Communication), prior to the first assignment in an area where glycerol α-monochlorohydrin is used or stored.

Before beginning initial assignments with this chemical and at regular intervals thereafter (e.g., annually), the following medical protocol are recommended:

☑ Lung function tests (establish initial baseline and compare thereafter).

☑ Thyroid function tests (establish initial baseline and compare thereafter).

If symptoms develop or overexposure is suspected, the following medical testing may be useful:

☑ Lung function tests (compare to initial and subsequent baseline tests).

☑ Thyroid function tests (compare to baseline).

Any evaluation should include a careful history of past and present symptoms with an examination. Medical tests that simply look for existing damage are not a substitute for controlling exposures.

Other methods to reduce exposure include:

☑ Where possible, enclose operations and use local exhaust ventilation at the site of chemical release. If local exhaust ventilation or enclosure is not used, respiratory protection should be mandatory.

☑ Always ensure that proper protective clothing is worn when using chemical substances and that personnel are trained in the proper use, care, and maintenance of their protective equipment.

☑ Wash thoroughly immediately after exposure to glycerol α-monochlorohydrin and at the end of the work shift or before eating, drinking, or smoking.

☑ Hazard warning information should be posted in the work area. In addition, as part of an on-going education and training program, all information on the health and safety hazards of glycerol α-monochlorohydrin should be communicated to all potentially exposed workers.

☑ Eye wash stations should be provided in the immediate work area for emergency use. If there is a possibility of skin exposure to glycerol α-monochlorohydrin, emergency shower facilities should also be provided in the work area.

☑ Workers whose clothing has been contaminated by glycerol α-monochlorohydrin should change into clean clothes *before* leaving work. Contaminated work clothing should never be worn home where family members can be contaminated. The clothing should be laundered only by individuals who have been informed of the hazards of exposure to glycerol α-monochlorohydrin.

RISK ASSESSMENT: ENVIRONMENT

General Assessment

The environment is at risk of exposure during transportation, storage, disposal, or destruction of glycerol α-monochlorohydrin. In almost every scenario, the threat of environmental exposure is contingent upon the proper handling of the chemical substance. Accidental spills, large or small, where glycerol α-monochlorohydrin contacts incompatible commodities can result in violent and even explosive reactions (depending upon conditions of spill), and possible contamination of the surrounding environmental mediums (water, soil, and air).

Glycerol α-monochlorohydrin is considered a combustible liquid. It can ignite when exposed to heat and support the burning of other materials in fire. It will react in contact with perchloric acid to form a sensitive explosive product. Therefore, extreme caution is required in handling, storage, transportation, and disposal of glycerol α-monochlorohydrin. Its physical hazards also require special consideration during any emergency situation involving a leak or spill of glycerol α-monochlorohydrin.

Glycerol α-monochlorohydrin may enter the environment through industrial effluents and from spills.

Acute Ecological Effects

Acute (short-term) toxic effects may include the death of animals, birds, or fish and death or low growth rate in plants. Acute effects are seen 2 to 4 days after animals or plants are exposed to glycerol α-monochlorohydrin.

Insufficient data are available on the short-term effects of glycerol α-monochlorohydrin to aquatic life, plants, land animals, and birds.

Chronic Ecological Effects

Chronic (long-term) toxic effects may include shortened life span, reproductive problems, lower fertility, and changes in appearance or behavior in exposed animals. These effects can be seen long after first exposure(s) to toxic chemicals.

Insufficient data are available to evaluate or predict the long-term effects of glycerol α-monochlorohydrin to aquatic life, plants, birds, and land animals.

Water Solubility

Glycerol α-monochlorohydrin is highly soluble in water. Concentrations of 1000 milligrams and even more will readily mix with a liter of water.

Persistence in the Environment

Glycerol α-monochlorohydrin is moderately persistent in the water, with a half-life between and 20 and 200 days. The half-life of a pollutant is the amount of time it takes for one-half of the chemical to be degraded.

Bioaccumulation in Aquatic Organisms

Some substances increase in concentration, or *bioaccumulate,* in living organisms as they breathe contaminated air, drink contaminated water, or eat contaminated food. These chemicals can become concentrated in the tissues and internal organs of animals as well as humans.

The concentration of glycerol α-monochlorohydrin found in fish tissues is expected to be about the same as the average concentration in the water from which the fish was taken.

Recommended Risk-Reduction Measures

Proper training of all transporters will reduce the likelihood of a mishap or accident resulting in a leak or spill to the environment. The correct use of DOT labeling (per the MSDS) on all containers, trucks, and rail cars will enable emergency responders to react properly and quickly to any disaster thereby reducing the potential risk to the environment and to personnel.

Storage of glycerol α-monochlorohydrin should be segregated from incompatible materials to minimize the risk of cross-contamination or contact. Storage facilities should be equipped with the proper fire protection equipment (sprinklers, alarms, extinguishers). To prevent static sparks, all equipment and containers used in shipping, receiving, or transferring should be electrically grounded and bonded. Workers must ensure that all necessary steps are taken to preclude the possibility of accidental spill or discharge into the environment.

If a spill or leak to the environment has occurred, fire department, emergency response and/or hazardous materials spill personnel should be notified immediately. Contaminated soils should be removed for incineration and replaced with clean soils. Cleanup should be attempted only by those trained in proper spill containment procedures. If glycerol α-monochlorohydrin should contact the water table, aquifer, or navigable waterway, reclamation procedures should be initiated as soon as practical. It is highly soluble in water and total remediation may not be possible. The local and/or state emergency response authorities must be notified. A comprehensive emergency response or disaster preparedness/recovery plan should be in place prior to any operations involving the use, transportation, storage, or disposal of glycerol α-monochlorohydrin. If glycerol α-monochlorohydrin is spilled or leaked, the following specific steps are recommended:

- ☑ Restrict persons not wearing protective clothing from area of spill or leak until cleanup is complete and area can be opened for normal work.
- ☑ Remove all ignition sources.
- ☑ Ventilate area of leak or spill.
- ☑ Absorb liquids in vermiculite or other compatible material and dispose of in sealed drums for disposal. Do NOT allow material or runoff to contact waterways or enter sewers.
- ☑ It may be necessary to dispose of glycerol α-monochlorohydrin as a hazardous waste. The responsible state agency or the regional office of the federal Environmental Protection Agency (EPA) should be contacted for specific recommendations.

RISK ASSESSMENT: BUSINESS

General Assessment

Accidents or mishaps involving glycerol α-monochlorohydrin can present a significant threat to business operations. The loss or damage of equipment or facilities can significantly affect fiscal viability. Lawsuits that may result from personnel injury/death, public exposures, and/or environmental contamination will also require a serious expenditure of resources. Media attention surrounding an injury, death, or environmental damage can also result in a loss of profits and loss of current as well as future business. Always remember that anytime the term "cancer," "carcinogen," or "reproductive hazard" are used, public emotion, hysteria, and ignorance can run equally high. This must be carefully considered when developing or implementing public relations policies.

Recommended Risk-Reduction Measures

Company attorneys, safety and health professionals, and environmental specialists should be involved in the development of any procedures for responding to chemical incidents. A company official should be predesignated as a public relations officer with specific training in dealing with the press. Corporate plans and policies should be developed, approved, and implemented long before any need for such arises.

REFERENCES

Ahlbom, A. 1993. *Biostatistics for Epidemiologists.* New York: CRC Press/Lewis Publishers.

American Conference of Governmental Industrial Hygienists. 1988. *Documentation of the Threshold Limit Values and Biological Exposure Indices,* 5th Edition (with updates). Cincinnati: ACGIH

Anthony, C. P., and N. J. Kolthoff. 1971. *Textbook of Anatomy and Physiology.* St. Louis: C. V. Mosby Company.

Asante-Duah, D. K. 1993. *Hazardous Waste Risk Assessment.* New York: CRC Press/Lewis Publishers.

Bloom, A. D. and F. J. DeSerres. 1995. *Ecotoxicity and Human Health.* New York: CRC Press/Lewis Publishers.

Brusick, D. J. 1994. *Methods for Genetic Risk Assessment.* New York: CRC Press/Lewis Publishers.

Bueche, J. R. 1972. *Principles of Physics,* 2nd Edition. New York: McGraw-Hill.

Burrell, R., D. K. Flaherty, and L. J. Sauers. 1992. *Toxicology of the Immune System.* New York: Van Nostrand Reinhold.

Calabrese, E. J. 1994. *Biological Effects of Low Level Exposures.* New York: CRC Press/Lewis Publishers.

Cockerman, L. G., B. S. Shane. 1994. *Basic Environmental Toxicology.* New York: CRC Press/Lewis Publishers.

Cothern, C. R. 1995. *Handbook for Environmental Risk Decisionmaking.* New York: CRC Press/Lewis Publishers.

Danse, I. R. 1991. *Common Sense Toxics in the Workplace.* New York: Van Nostrand Reinhold.

Environmental Protection Agency. 1983. *Chemical Hazard Information Profiles.* Washington, D.C.: U.S. EPA.

Fodden, C. R., and J. L. Weddell. 1992. *Hazardous Materials Emergency Action Data.* New York: CRC Press/Lewis Publishers.

Forseberg, K., and L. H. Keith. 1993. *Chemical; Protective Clothing Permeation and Degradation Database.* Chelsea, MI: Lewis Publishers.

Gardner, R. W. 1994. *Chemical Intolerance.* New York: CRC Press/Lewis Publishers.

Hodgson, E., R. B. Mailman, and Janice E. Chambers. 1988. *Dictionary of Toxicology.* London: The MacMillian Company.

Howard, P. H. 1995. *Handbook of Environmental Fate and Exposures Data for Organic Chemicals,* Vol. I-IV. New York: CRC Press/Lewis Publishers.

Hosty, J. W., and P. Foster. 1990. *A Practical Guide to Chemical Spill Response.* New York: Van Nostrand Reinhold.

Jones, F. E. 1994. *Toxic Organic Vapors in the Workplace.* New York: CRC Press/Lewis Publishers.

Kamrin, M. A. 1988. *Toxicology: A Primer on Principles and Applications.* Chelsea, Michigan: Lewis Publishers.

Lewis, .J., Sr. 1993. *Hawley's Condensed Chemical Dictionary,* 12th Edition. New York: Van Nostrand Reinhold.

Lewis, . J., Sr. 1992. *Sax's Dangerous Properties of Industrial Materials,* Volumes 1, 2, and 3, 8th Edition. New York: Van Nostrand Reinhold.

Mackay, D., W. Y. Shiu, and K. C. Ma. 1992. *Illustrated Handbook of Chemical-Physical Properties and Environmental Fate for Organic Chemicals.* New York: CRC Press/Lewis Publishers.

MATERIAL SAFETY DATA SHEET

CHEMICAL NAME	
GLYCIDOL	**CRC PRESS/LEWIS PUBLISHERS DISCLAIMER:** THE INFORMATION AND RECOMMENDATIONS PRESENTED HEREIN ARE BASED ON SOURCES BELIEVED TO BE RELIABLE. CRC MAKES NO REPRESENTATION ON ITS COMPLETENESS OR ACCURACY. IT IS THE RESPONSIBILITY OF THE USER TO DETERMINE THE CHEMICAL'S SUITABILITY FOR ITS INTENDED USE, THE CHEMICAL'S SAFE USE, AND THE CHEMICAL'S PROPER DISPOSAL. NO REPRESENTATIONS AND/OR WARRANTIES, EITHER EXPRESSED OR IMPLIED, OF THE MERCHANTABILITY OR FITNESS FOR A PARTICULAR PURPOSE, OR OF ANY OTHER NATURE, ARE MADE WITH RESPECT TO THE INFORMATION PROVIDED IN THIS MSDS OR TO THE CHEMICAL TO WHICH INFORMATION MAY REFER. CRC NEITHER ASSUMES NOR AUTHORIZES ANY OTHER PERSON TO ASSUME FOR IT, ANY OTHER ADDITIONAL RESPONSIBILITY OR LIABILITY FOR THE USE OF, OR RELIANCE UPON, THIS INFORMATION.

HAZARD WARNING INFORMATION

HEALTH	FIRE	REACTIVE	OTHER	DEGREE OF HAZARD	COLOR CODING	OTHER CODES
2	2	0		0 = Minimum Hazard 1 = Slight Hazard 2 = Moderate Hazard 3 = Serious Hazard 4 = Severe Hazard	HEALTH = BLUE FIRE = RED REACTIVITY = YELLOW OTHER = WHITE	OX = Oxidizer ACID = Acid ALK = Alkali COR = Corrosive W = Use No Water

SECTION I - GENERAL INFORMATION

Characterization	RCRA Number	EPA Class
Epoxy Compound	**None**	**Not Applicable**

DOT Proper Shipping Name **No Citation**	Chemical Abstract Service (CAS) Number **556-52-5**
DOT Hazard Class and Label Requirements **No Citation**	DOT Emergency Guide Code **No Citation**
DOT Identification Number **No Citation**	Chemical Formula $C_3H_6O_2$

Synonyms

2,3-Epoxy-1-propanol; epoxypropyl alcohol; glycide; hydroxymethyl ethylene oxide; 2-hydroxymethyl oxiran; 3-hydroxypropylene oxide; epihydrin alcohol.

SECTION II - HAZARDOUS INGREDIENTS AND IDENTITY INFORMATION

Hazardous Components (specific identity)	OSHA Exposure Criteria	NIOSH Exposure Criteria	Immediately Dangerous to Life and Health (IDLH)	ACGIH Exposure Criteria
Glycidol: (derivation: Treatment of monochlorohydrin with bases; reaction product of allyl alcohol and perbenzoic acid). 1 ppm = 3.08 mg/m^3	**PEL:** 50 ppm 150 mg/m^3 **STEL:** Not Established	**REL:** 25 ppm 75 mg/m^3 **STEL:** Not Established	150 ppm	**TLV:** 25 ppm 75 mg/m^3 **STEL:** Not Established

SECTION III - PHYSICAL AND CHEMICAL CHARACTERISTICS

Boiling Point **320°F (160°C) Decomposes**	Specific Gravity (H_2O = 1) **1.12**
Vapor Pressure (mm Hg) **0.9 at 77°F (25°C)**	Molecular Weight (atomic weight) **74.1**
Vapor Density (Air = 1) **2.6**	Melting Point **-49°F (-45°C)**

Solubility

Miscible in water, alcohol, and ether.

Appearance and Odor

Clear, colorless, odorless liquid.

SECTION IV - FIRE AND EXPLOSION HAZARD DATA

Flash Point (method used) **162°F (72°C) closed cup**	Explosive Limits in Air % by Volume **LEL: 2.7% UEL: Not Determined**
NFPA Classification **Class IIIA Combustible Liquid**	Autoignition Temperature **Not Determined**

Extinguishing Media

Use carbon dioxide, dry chemical, alcohol foam, water fog (water or foam may cause frothing).

Special Fire Fighting Procedures

Poisonous gases are emitted (including phosgene). Wear full protective clothing and self-contained breathing apparatus (SCBA). Remain clear of smoke, water fallout and water runoff. Move containers from fire area if it can be done without risk. Cool exposed containers. Fight fire from distance, if possible.

Unusual Fire and Explosion Hazards

Vapors form explosive mixtures in air. Vapors are heavier than air and may travel from some distance to an ignition source and flashback.

SECTION V - REACTIVITY DATA

Stability		Conditions to Avoid Glycidol is stable in closed containers under normal conditions of handling and storage. Keep away from heat and incompatible materials.
Stable X	Unstable	Incompatibility *(materials to avoid)* Reacts violently with strong oxidizers (such as chlorine, fluorine, bromine) and nitrates. Explodes when heated or in the presence of strong acids, bases, metals (copper, zinc), and metal salts.
Hazardous Polymerization		Conditions to Avoid Under normal conditions of temperature and pressure, hazardous polymerization of glycidol is not expected to occur.
May Occur	Will Not Occur X	Hazardous Decomposition or By-products At high temperatures, glycidol decomposes and may produce poisonous gases, including toxic oxides of carbon (monoxide and dioxide).

SECTION VI - HEALTH HAZARD DATA

Primary Route(s) of Entry:	Inhalation?	Absorption (skin/eye)?	Ingestion?
	X	X	X

Health Hazards

INHALATION: Irritation of the eyes, nose, and respiratory tract. Chemical pneumonitis possible. Other effects appear to focus on excitation of the central nervous system (CNS) followed closely by depression. It may cause narcosis (drowsiness), and other related symptoms. There is also evidence of reproductive and/or teratogenic and mutagenic effects.

ABSORPTION: A skin and eye irritant. Readily absorbed through the skin to cause CNS effects. Direct contact with the eye causes severe but reversible corneal damage (in rabbits).

INGESTION: Irritation of the gastrointestinal tract and CNS depression (see inhalation).

Carcinogenicity	NTP Listed?	IARC Cancer Review Group?	OSHA Regulated?	Target Organs?
No Evidence Human No Evidence Animal	No	No	29 CFR 1910.1000 Table Z-1	Resp. sys. skin, eyes, CNS, reproductive system.

Medical Conditions Generally Aggravated by Exposure

None Reported.

Emergency and First-aid Procedures

Eye contact: Do not allow victim to keep eyes tightly shut. Flush immediately with water for 15 minutes, seek medical attention. Skin contact: Remove all contaminated clothing. Immediately wash area with flooding amounts of water. Inhalation: Remove the person from exposure. Provide respiratory assistance and CPR. Transfer to medical facility. If swallowed: Contact poison control center. Unless advised otherwise and victim is conscious, give 1-2 glasses of water or milk as quickly as possible. Do NOT induce vomiting. Seek medical attention immediately. Never attempt to give anything by mouth to an unconscious or convulsing person and never try to make such a person vomit.

SECTION VII - PRECAUTIONS FOR SAFE HANDLING AND USE

Steps to be Taken in Case Material is Released or Spilled

Remove all sources of ignition. Ventilate area of spill. Restrict those not involved in cleanup from entering area. Notify appropriate authorities. Absorb in vermiculite or other suitable material and dispose of in sealed drums. Do NOT release to sewers, storm drains, surface waters, or soils.

Preferred Waste Disposal Method

Burn in a chemical incinerator equipped with an afterburner and scrubber.

Precautions to be Taken in Handling and Storage

Store in tightly closed containers in cool, well ventilated area away from heat and incompatibilities.

Other Precautions and Warnings

Prevent static sparks by grounding and bonding all containers. Treat "empty" containers as if full.

SECTION VIII - CONTROL MEASURES AND PERSONAL PROTECTIVE EQUIPMENT

Respiratory Protection *(specify type)*

For low exposure levels, use an organic vapor respirator. For higher exposures, a supplied-air respirator operated in positive pressure mode, or a SCBA with full facepiece operated in positive pressure mode.

Ventilation

Local exhaust (preferred) at site of chemical work or general ventilation.

Protective Gloves	Eye Protection	Other Protective Clothing
Impervious Rubber (test for adequacy)	Chemical Goggles or Face Mask	Protective Apron, Boots (avoid skin contact)

Work/Hygiene Practices

Always wash hands thoroughly after using chemical; never bring food, drink, or smoking materials into vicinity of chemicals.

GLYCIDOL

$C_3H_6O_2$ CAS: 556-52-5

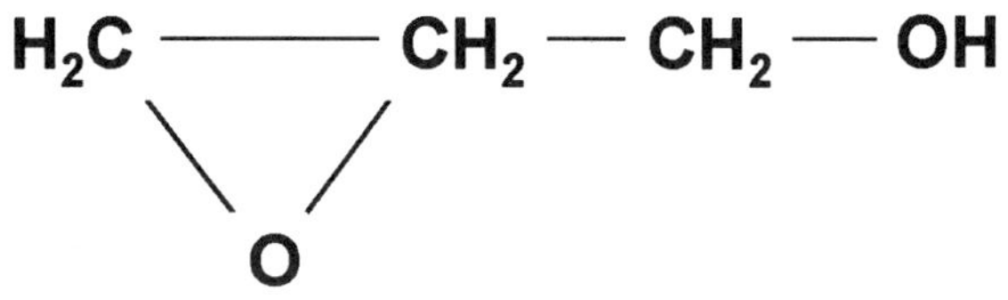

IDENTIFICATION AND TYPICAL USES

Glycidol is a clear, colorless liquid with no odor. It is used as a stabilizer for natural oils and vinyl polymers, as a de-emulsifier, and as a leveling agent for dyes.

RISK ASSESSMENT: HEALTH

General Assessment

Glycidol is toxic to humans by *inhalation* and by *ingestion*. It is also readily *absorbed* through intact skin to cause toxic effects. If absorption occurs in combination with other exposure routes, it can seriously compound the overall effects of exposure. The references regarding long-term toxicity are conflicting and inconclusive. Some indicate that no cumulative toxic effects are documented while others consider glycidol an experimental teratogen causing reproductive effects with human mutation data also reported. From the limited data available on this chemical, it would appear that the human health hazard exposure risk is primarily respiratory irritation, stimulation of the central nervous system followed by its depression.

Inhalation of glycidol vapors will cause irritation of the eyes, nose, and throat. Prolonged exposures to high concentrations will cause narcosis with symptoms of drowsiness and fatigue. The liquid is moderately irritating to the skin and a severe eye irritant. Ingesting glycidol will cause gastrointestinal irritation and may result in CNS depression.

Acute Health Effects

The following acute (short-term) health effects may occur immediately or shortly after exposure to glycidol:

Skin: Liquid can pass through unbroken skin. Contact can cause severe irritation and possible burns with redness and swelling.

Eye: Severe irritation and possible corneal damage.

Lung: Irritation of the nose and throat.

CNS: A central nervous system depressant. Symptoms include narcosis with drowsiness and fatigue.

Other: Ingestion can irritate the gastrointestinal tract and also lead to CNS depression.

Chronic Health Effects

The following chronic (long-term) health effects may occur at some time after exposure to glycidol and can last for months or even years:

Cancer Hazards: According to information presented in the references, glycidol has not been adequately tested for its ability to cause cancer in test animals. Human mutation data have been reported. Some scientists believe that long-term exposure to such chemicals may pose a cancer risk. More research is required before any determination can be made on the carcinogenic properties of glycidol.

Reproductive Hazard: Experimental studies have shown that exposure to glycidol may adversely affect reproduction and cause teratogenic effects. There are no human data on reproductive health risks.

Other Chronic Effects: There are no reports to conclude that any other chronic effects result from long-term exposure to glycidol.

Recommended Risk-Reduction Measures

The hazards to human health posed by exposure to glycidol are not clearly understood in the references. Personnel should avoid direct contact with glycidol. If a less toxic material or compound cannot be substituted for glycidol, then *engineering controls* are the most effective method of reducing exposures. The best protection is to enclose operations and/or provide local exhaust ventilation at the site of glycidol release. While not always operationally possible, isolating operations involving glycidol manufacturing or use can also reduce exposure. Personnel must follow proper industrial hygiene and good housekeeping work practices. Using respiratory protection is less effective than the controls mentioned above, but is still advisable whenever working with or around glycidol. At exposures above the PEL, an MSHA or NIOSH-approved organic vapor respirator may suffice. For higher exposure levels or when the level is not known (as in an emergency), a powered supplied-air respirator or a self-contained breathing apparatus (SCBA) with full facepiece and operated in pressure demand are recommended for the greatest possible respiratory protection. If a full facepiece is not available, then chemical goggles should be worn to protect the eyes. A face shield and protective apron should also be worn. Skin contact must not occur. To prevent hand and skin exposures, impervious gloves should be worn. Glove manufacturers should be contacted and permeation studies obtained *before* final glove selections are made.

Administrative controls should also be in place to minimize the potential for human exposures. These may include written procedures or policies which specify the methods and techniques that will be practiced whenever personnel are to work with glycidol.

All personnel should receive *training* on the use, hazards, protective measures, emergency actions, and other precautions per 29 CFR 1910.1200 (Hazard Communication), prior to the first assignment in an area where glycidol is used or stored.

Before beginning initial assignments with this chemical and at regular intervals thereafter (e.g., annually), the following medical protocol is recommended:

- ☑ Lung function tests (establish initial baseline and compare thereafter).

If symptoms develop or overexposure is suspected, the following medical testing may be useful:

- ☑ Lung function tests.
- ☑ Evaluation of central nervous system function.

Any evaluation should include a careful history of past and present symptoms with an examination. Medical tests that simply look for existing damage are not a substitute for controlling exposures. Also, since smoking can cause heart disease, emphysema, and other respiratory problems, personnel who smoke may show symptoms of exposure much quicker and more pronounced than those who do not. Prudent risk management requires careful consideration of *all* possible and potential exposure risk factors to ensure proper implementation of control measures and management of the exposure risk.

Other methods to reduce exposure include:

- ☑ Where possible, enclose operations and use local exhaust ventilation at the site of chemical release. If local exhaust ventilation or enclosure is not used, respiratory protection should be mandatory.
- ☑ Always ensure that proper protective clothing is worn when using chemical substances.
- ☑ Wash thoroughly immediately after exposure to glycidol and at the end of the work shift or before eating, drinking, or smoking.
- ☑ Hazard warning information should be posted in the work area. In addition, as part of an on-going education and training program, all information on the health and safety hazards of glycidol should be communicated to all potentially exposed workers.
- ☑ Eye wash stations should be provided in the immediate work area for emergency use. If there is a possibility of skin exposure to glycidol, emergency shower facilities should also be provided.
- ☑ Workers whose clothing has been contaminated by glycidol should change into clean clothes *before* leaving work. Contaminated work clothing should be laundered only by individuals who have been informed of the hazards of exposure to glycidol.

RISK ASSESSMENT: ENVIRONMENT

General Assessment

The environment is at risk of exposure during transportation, storage, disposal, or destruction of glycidol. In almost every scenario, the threat of environmental exposure is contingent upon the proper handling of the chemical substance. Accidental spills, large or small, where glycidol contacts incompatible commodities can result in violent and even explosive reactions (depending upon conditions of spill), and possible contamination of the surrounding environmental mediums (water, soil, and air).

Glycidol is considered a Class IIIA combustible liquid (per OSHA 29 CFR 1910.106). Its moderate flash point and boiling point present a fire risk in the presence of an ignition source, heat, or flame. Also, it can explode when heated in the presence of strong acids, bases, metals (copper, zinc), and metallic salts (aluminum chloride, iron (III) chloride, tin (IV) chloride). Therefore, extreme caution is required in handling, storage, transportation, and disposal of glycidol. These characteristics also require special consideration during any emergency situation involving a leak or spill of glycidol.

The proper disposal method is to burn glycidol in a chemical incinerator equipped with an afterburner and air scrubber.

Glycidol may enter the environment through industrial effluents and from spills.

Acute Ecological Effects

Acute (short-term) toxic effects may include the death of animals, birds, or fish and death or low growth rate in plants. Acute effects are seen 2 to 4 days after animals or plants are exposed to glycidol.

Insufficient data are available to evaluate or predict the short-term effects of glycidol to aquatic life, plants, land animals, and birds.

Chronic Ecological Effects

Chronic (long-term) toxic effects may include shortened life span, reproductive problems, lower fertility, and changes in appearance or behavior in exposed animals. These effects can be seen long after first exposure(s) to toxic chemicals.

Insufficient data are available to evaluate or predict the long-term effects of glycidol to aquatic life, plants, birds, and land animals.

Water Solubility

Glycidol is completely soluble in water. Concentrations of 1000 milligrams and more will mix with a liter of water.

Persistence in the Environment

Glycidol is moderately persistent in water, with a half-life between 2 and 20 days. The half-life of a pollutant is the amount of time it takes for one-half of the chemical to be degraded.

Bioaccumulation in Aquatic Organisms

Some substances increase in concentration, or *bioaccumulate,* in living organisms as they breathe contaminated air, drink contaminated water, or eat contaminated food. These chemicals can become concentrated in the tissues and internal organs of animals as well as humans exposed to chemical substances.

There is no information in the references on the bioaccumulation of glycidol in the edible tissues of fish.

Recommended Risk-Reduction Measures

Proper training of all transporters will reduce the likelihood of a mishap or accident resulting in a leak or spill to the environment. The correct use of labeling on all containers, trucks, and rail cars will enable emergency responders to react properly and quickly to any disaster thereby reducing the potential risk to the environment and to personnel.

Storage of glycidol should be segregated from incompatible materials to minimize the risk of cross-contamination or contact. Storage facilities should be equipped with the proper fire protection equipment (sprinklers, alarms, extinguishers). Proper ventilation and protective equipment should be used while handing glycidol. Non-sparking tools should be provided and used to ensure the prevention of sparks. Also, all equipment and containers used in shipping, receiving, or transferring should be electrically grounded and bonded.

If a spill or leak to the environment has occurred, fire department, emergency response and/or hazardous materials spill personnel should be notified immediately. Cleanup should be attempted only by those trained in proper spill containment procedures. If glycidol should contact the water table, aquifer, or navigable waterway, reclamation procedures should be initiated as soon as practical. It is highly soluble in water and total remediation may not be possible. The local and/or state emergency response authorities must be notified. A comprehensive emergency response or disaster preparedness/recovery plan should be in place prior to any operations involving the use, transportation, storage, or disposal of glycidol. If glycidol is spilled or leaked, the following specific steps are recommended:

- ☑ Restrict persons not wearing protective clothing from area of spill or leak until cleanup is complete and area can be opened for normal work.
- ☑ Remove all ignition sources and ventilate area.
- ☑ Absorb liquids in vermiculite or other compatible material and dispose of in sealed drums.
- ☑ It may be necessary to dispose of glycidol as a hazardous waste. The responsible state agency or the regional office of the federal Environmental Protection Agency (EPA) should be contacted for specific recommendations.

RISK ASSESSMENT: BUSINESS

General Assessment

Accidents or mishaps involving glycidol can present a significant threat to business operations. The loss or damage of equipment or facilities can significantly affect fiscal viability. Lawsuits that may result from personnel injury/death, public exposures, and/or environmental contamination will also require a serious expenditure of resources. Media attention surrounding an injury, death, or environmental damage can also result in a loss of profits and loss of current as well as future business. Always remember that anytime the term "reproductive hazard" is used, public emotion, hysteria, and ignorance can run equally high. This must be a consideration when developing public relations policies.

Recommended Risk-Reduction Measures

Company attorneys, safety and health professionals, and environmental specialists should be involved in the development of any procedures for responding to chemical incidents. A company official should be predesignated as a public relations officer with specific training in dealing with the press. Corporate plans and policies should be developed, approved, and implemented long before any need for such arises.

REFERENCES

Ahlbom, A. 1993. *Biostatistics for Epidemiologists.* New York: CRC Press/Lewis Publishers.

American Conference of Governmental Industrial Hygienists. 1988. *Documentation of the Threshold Limit Values and Biological Exposure Indices,* 5th Edition (with updates). Cincinnati: ACGIH

Anthony, C. P., and N. J. Kolthoff. 1971. *Textbook of Anatomy and Physiology.* St. Louis: C. V. Mosby Company.

Asante-Duah, D. K. 1993. *Hazardous Waste Risk Assessment.* New York: CRC Press/Lewis Publishers.

Bloom, A. D. and F. J. DeSerres. 1995. *Ecotoxicity and Human Health.* New York: CRC Press/Lewis Publishers.

Brusick, D. J. 1994. *Methods for Genetic Risk Assessment.* New York: CRC Press/Lewis Publishers.

Bueche, J. R. 1972. *Principles of Physics,* 2nd Edition. New York: McGraw-Hill.

Burrell, R., D. K. Flaherty, and L. J. Sauers. 1992. *Toxicology of the Immune System.* New York: Van Nostrand Reinhold.

Calabrese, E. J. 1994. *Biological Effects of Low Level Exposures.* New York: CRC Press/Lewis Publishers.

Cockerman, L. G., B. S. Shane. 1994. *Basic Environmental Toxicology.* New York: CRC Press/Lewis Publishers.

Cothern, C. R. 1995. *Handbook for Environmental Risk Decisionmaking.* New York: CRC Press/Lewis Publishers.

Danse, I. R. 1991. *Common Sense Toxics in the Workplace.* New York: Van Nostrand Reinhold.

Environmental Protection Agency. 1983. *Chemical Hazard Information Profiles.* Washington, D.C.: U.S. EPA.

Fodden, C. R., and J. L. Weddell. 1992. *Hazardous Materials Emergency Action Data.* New York: CRC Press/Lewis Publishers.

Forseberg, K., and L. H. Keith. 1993. *Chemical; Protective Clothing Permeation and Degradation Database.* Chelsea, MI: Lewis Publishers.

Gardner, R. W. 1994. *Chemical Intolerance.* New York: CRC Press/Lewis Publishers.

Gerarde, H. W. 1960. *Toxicology and Biochemistry of Aromatic Hydrocarbons.* New York: Elsevier.

Hallenbeck, W. H. 1993. *Quantitative Risk Assessment for Environmental and Occupational Health,* 2nd Edition. New York: CRC Press/Lewis Publishers.

Hodgson, E., R. B. Mailman, and Janice E. Chambers. 1988. *Dictionary of Toxicology.* London: The MacMillian Company.

Howard, P. H., et. al. 1991. *Handbook of Environmental Degradation Rates.* New York: CRC Press/Lewis Publishers.

Hosty, J. W., and P. Foster. 1990. *A Practical Guide to Chemical Spill Response.* New York: Van Nostrand Reinhold.

International Agency for Research on Cancer (IARC). 1987. *IARC Monographs on the Evaluation of Carcinogenic Risk to Humans, Overall Evaluations of Carcinogenicity: An updating of IARC Monographs Volumes 1-42.* Lyon: IARC.

Jones, F. E. 1994. *Toxic Organic Vapors in the Workplace.* New York: CRC Press/Lewis Publishers.

Kamrin, M. A. 1988. *Toxicology: A Primer on Principles and Applications.* Chelsea, Michigan: Lewis Publishers.

Keith, L. H. and D. B. Walters. (Editors) 1992. *National Toxicology Program's Chemical Database, Volumes I-VII.* Chelsea, MI: Lewis Publishers.

Landis, W. G., M. H. Yu. 1995. *Introduction to Environmental Toxicology.* New York: CRC Press/Lewis Publishers.

Langerman, N. 1994 *Precautionary Labels for Chemical Containers.* New York: CRC Press/Lewis Publishers.

Lauwerys, R. R., P. Hoet. 1993. *Industrial Chemical Exposure.* New York: CRC Press/Lewis Publishers.

Lewis, .J., Sr. 1993. *Hawley's Condensed Chemical Dictionary,* 12th Edition. New York: Van Nostrand Reinhold.

Lewis, . J., Sr. 1992. *Sax's Dangerous Properties of Industrial Materials,* Volumes 1, 2, and 3, 8th Edition. New York: Van Nostrand Reinhold.

U.S. Department of Health and Human Services, National Institute for Occupational Safety and Health. 1994. *NIOSH Pocket Guide to Chemical Hazards.* Washington, D.C.: U.S. Government Printing Office.

MATERIAL SAFETY DATA SHEET

CHEMICAL NAME	
GLYCOLONITRILE	**CRC PRESS/LEWIS PUBLISHERS DISCLAIMER:** THE INFORMATION AND RECOMMENDATIONS PRESENTED HEREIN ARE BASED ON SOURCES BELIEVED TO BE RELIABLE. CRC MAKES NO REPRESENTATION ON ITS COMPLETENESS OR ACCURACY. IT IS THE RESPONSIBILITY OF THE USER TO DETERMINE THE CHEMICAL'S SUITABILITY FOR ITS INTENDED USE, THE CHEMICAL'S SAFE USE, AND THE CHEMICAL'S PROPER DISPOSAL. NO REPRESENTATIONS AND/OR WARRANTIES, EITHER EXPRESSED OR IMPLIED, OF THE MERCHANTABILITY OR FITNESS FOR A PARTICULAR PURPOSE, OR OF ANY OTHER NATURE, ARE MADE WITH RESPECT TO THE INFORMATION PROVIDED IN THIS MSDS OR TO THE CHEMICAL TO WHICH INFORMATION MAY REFER. CRC NEITHER ASSUMES NOR AUTHORIZES ANY OTHER PERSON TO ASSUME FOR IT, ANY OTHER ADDITIONAL RESPONSIBILITY OR LIABILITY FOR THE USE OF, OR RELIANCE UPON, THIS INFORMATION.

HAZARD WARNING INFORMATION

HEALTH	FIRE	REACTIVE	OTHER	DEGREE OF HAZARD	COLOR CODING	OTHER CODES
4	1	2		0 = Minimum Hazard 1 = Slight Hazard 2 = Moderate Hazard 3 = Serious Hazard 4 = Severe Hazard	HEALTH = BLUE FIRE = RED REACTIVITY = YELLOW OTHER = WHITE	OX = Oxidizer ACID = Acid ALK = Alkali COR = Corrosive W = Use No Water

SECTION I - GENERAL INFORMATION

Characterization	**Cyanide, Organic (Nitrile)**	RCRA Number	**D003**
		EPA Class	**Characteristic (R) Waste**
DOT Proper Shipping Name	**No Citation**	Chemical Abstract Service (CAS) Number	**107-16-4**
DOT Hazard Class and Label Requirements	**No Citation**	DOT Emergency Guide Code	**No Citation**
DOT Identification Number	**No Citation**	Chemical Formula	**$HOCH_2CN$**

Synonyms

Cyanomethanol; formaldehyde cyanohydrin; glycolic nitrile; hydroxyacetonitrile; cyanomethanol; hydroxymethylinitrile.

SECTION II - HAZARDOUS INGREDIENTS AND IDENTITY INFORMATION

Hazardous Components (specific identity)	OSHA Exposure Criteria	NIOSH Exposure Criteria	Immediately Dangerous to Life and Health (IDLH)	ACGIH Exposure Criteria
Glycolonitrile (derivation: By reaction of formaldehyde and aqueous sodium cyanide in the presence of mineral acid). Commercial supply is a 70% aqueous solution **1 ppm = 2.37 mg/m³**	**PEL:** Not Established **STEL:** Not Established	**REL (ceiling):** 2 ppm, 5 mg/m³ (15 minutes) **STEL:** Not Applicable	**Not Determined**	**TLV:** Not Established **STEL:** Not Established

SECTION III - PHYSICAL AND CHEMICAL CHARACTERISTICS

Boiling Point	**361°F (183°C) Decomposes**	Specific Gravity (H_2O = 1)	**1.10**
Vapor Pressure (mm Hg)	**1 at 145°F (63°C)**	Molecular Weight	**57.1**
Vapor Density (Air = 1)	**Not Found**	Freezing Point	**<-80°F (<-72°C)**

Solubility

Soluble in water, ethanol and ether, insoluble in benzene and chloroform.

Appearance and Odor

Colorless to water white, mobile, odorless oil with a sweetish taste.

SECTION IV - FIRE AND EXPLOSION HAZARD DATA

Flash Point (method used)	**Not Determined**	Explosive Limits in Air % by Volume	**LEL: Not Determined UEL: Not Determined**
NFPA Classification	**Combustible Liquid**	Autoignition Temperature	**Not Determined**

Extinguishing Media

Dry chemical, carbon dioxide, water spray, fog, or foam.

Special Fire Fighting Procedures

Wear full protective clothing and self-contained breathing apparatus (SCBA). Use a water spray to keep fire-exposed containers cool. Poisonous gases are produced in fire. Do not release water runoff to sewers or waterways; dike and contain for collection and disposal.

Unusual Fire and Explosion Hazards

Containers may explode in fire due to excessive buildup of internal pressure. Evacuate immediately if rising sound is heard from venting device or fire is causing discoloration on sides of tank.

SECTION V - REACTIVITY DATA

Stability		Conditions to Avoid **Undergoes spontaneous, violent, and possibly explosive decomposition at room temperature and under normal or routine conditions of storage and handling. Do not store for prolonged periods of time.**
Stable	Unstable X	Incompatibility *(materials to avoid)* **Even small traces of alkali can cause violent polymerization.**
Hazardous Polymerization		Conditions to Avoid **Hazardous polymerization can occur. Signs of tarry formation or a dark red precipitate will usually precede explosion. Avoid polymerization by adding a stabilizer (phosphoric acid).**
May Occur X	Will Not Occur	Hazardous Decomposition or By-products **When heated to decomposition, glycolonitrile emits highly toxic fumes of nitrogen oxide and cyanide fumes.**

SECTION VI - HEALTH HAZARD DATA

Primary Route(s) of Entry:	Inhalation?	Absorption (skin)?	Ingestion?
	X	X	X

Health Hazards

INHALATION: At high concentrations, causes chemical asphyxia, nausea, vomiting, and tightness of chest. Hydrolysis causes release of cyanide. Causes pulse and breathing obstruction. CNS depression with giddiness, confusion, convulsions, dypsnea, weakness.

ABSORPTION: Skin contact can result in irritating rash. Can cause severe eye burns and/or irritation. Will pass through intact skin and can cause serious effects, up to and including death.

INGESTION: Abdominal pain, nausea, vomiting.

Carcinogenicity	NTP Listed?	IARC Cancer Review Group?	OSHA Regulated?	Target Organs?
Unknown Human Unknown Animal	No	No	No	Respiratory system, eyes, skin, CVS, CNS.

Medical Conditions Generally Aggravated by Exposure

None Reported.

Emergency and First-aid Procedures

Eye contact: Flush immediately with water for 15 minutes (minimum); seek medical attention. **Skin contact:** Remove all contaminated clothing. Immediately wash area with large amounts of soap and water. Seek medical assistance. For **inhalation:** Remove the person from exposure. Provide respiratory assistance and CPR. Transfer to medical facility. Glycolonitrile can cause fatal cyanide poisoning. If **swallowed:** Seek medical attention immediately. Do not give anything by mouth to an unconscious or convulsing person. DOCTOR: Diagnostic signs include bradycardia and absence of cyanosis.

SECTION VII - PRECAUTIONS FOR SAFE HANDLING AND USE

Steps to be Taken in Case Material is Released or Spilled

Contain spills using absorbent material (vermiculite, or other). Ventilate area of spill. Remove all ignition sources. Store materials in DOT-approved containers. Restrict those not involved in cleanup from entering area. Notify appropriate authorities, as required by SARA III, if applicable.

Preferred Waste Disposal Method

Burn in a chemical incinerator equipped with an afterburner and scrubber.

Precautions to be Taken in Handling and Storage

Store to avoid contact with alkali materials in tightly closed containers in cool, dry, well-ventilated area away from heat and sources of ignition. Prevent damage to storage containers.

Other Precautions and Warnings

Regularly monitor the level of polymerization inhibitor to prevent polymerization. Prior to working with glycolonitrile, personnel should be trained on its proper handling and storage.

SECTION VIII - CONTROL MEASURES AND PERSONAL PROTECTIVE EQUIPMENT

Respiratory Protection *(specify type)*

NIOSH ceiling limit is low and difficult to monitor. A cyanide gas respirator may suffice. For best protection, use an approved supplied-air respirator in continuous mode or use a self-contained breathing apparatus (SCBA) with full facepiece operated in pressure demand or other positive pressure mode.

Ventilation

Local exhaust or general mechanical systems recommended.

Protective Gloves	Eye Protection	Other Protective Clothing
Impervious Material	Chemical Goggles or Face Mask	Protective Apron or Suit

Work/Hygiene Practices

Always wash hands thoroughly after using chemical; never bring food, drink, or smoking materials into vicinity of chemicals.

GLYCOLONITRILE

$HOCH_2CN$ CAS: 107-16-4

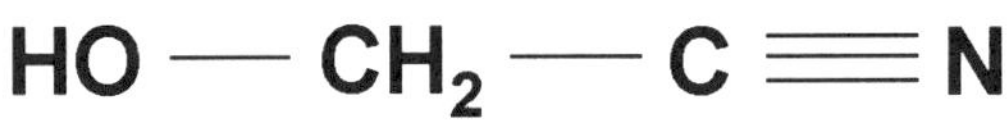

IDENTIFICATION AND TYPICAL USES

Glycolonitrile is a colorless to water-white, mobile, odorless oil with a sweet taste. It is used as a solvent, a chemical intermediate, and as a component of synthetic resins.

RISK ASSESSMENT: HEALTH

General Assessment

Glycolonitrile causes headache, dizziness, and nausea. At extremely high concentrations, it can cause convulsions, coma, and death. No data exist on the chemical's carcinogenicity in humans. The toxic effects of glycolonitrile are attributed to the metabolic release of cyanide, which in turn acts by inhibiting cytochrome oxidase and thus impairs cellular respiration.

Human exposure to glycolonitrile can occur through three routes of entry into the body: *inhalation, absorption,* and *ingestion.* At high concentrations, glycolonitrile can produce severe adverse effects. The target organs are the eyes, skin, lungs, central nervous system (CNS), and the cardiovascular system (CVS). In humans, inhalation of glycolonitrile vapors can result in asphyxia, nausea, vomiting, and tightness of the chest. These effects will most like manifest several hours after exposure. There can be weak and rapid pulse, irregular heartbeat, bright pink coloration of the skin, dilated pupils, dizziness, difficulty breathing, anxiety, and possible lockjaw. Symptoms may progress to convulsions, paralysis, coma, and cardiac arrhythmia. Death can occur within minutes or hours due to respiratory failure. It is excreted in the urine as a cyanate. The blood cyanide concentration does not always show any significant increase in cyanide at low exposure concentrations. Ingestion of glycolonitrile results in gastrointestinal pain, nausea, vomiting, stupor, convulsion, and weakness. Large doses can be fatal. Skin contact results in absorption and systemic effects. There can be redness, swelling, blistering, and inflammation. Severe eye irritation is likely.

Acute Health Effects

The following acute (short-term) health effects may occur immediately or shortly after exposure to glycolonitrile. (Note: Exposure can cause fatal cyanide poisoning. Symptoms of flushing of face, chest tightness, nausea, and vomiting, weakness, and shortness of breath may lead to death. These reactions may begin minutes to hours after an overexposure).

Skin: Irritation/rash on contact (contact dermatitis) with redness, swelling, and blistering.

Eye: Severe irritation, conjunctivitis, damage.

Lung: Irritation of the nose, throat, and lungs following exposure.

CNS: Inhalation of high concentrations of the vapor may cause sleepiness, dizziness, loss of consciousness, coma, respiratory failure, and death.

Chronic Health Effects

The following chronic (long-term) health effects may occur at some time after exposure to glycolonitrile and can last for months or even years:

Cancer Hazards: According to the references, glycolonitrile has not been tested for its ability to cause cancer in animals. Chronic effects of high exposures have not been documented in humans.

Reproductive Hazard: According to the references, glycolonitrile has not been significantly tested for its ability to adversely affect reproduction.

Other Chronic Effects: Repeated exposures may affect the thyroid gland. High exposures may damage the nervous system.

Recommended Risk-Reduction Measures

Personnel should avoid direct contact with glycolonitrile. Unless a less toxic chemical can be substituted for a hazardous substance, *engineering controls* are the most effective method of reducing exposures. The best protection is to enclose operations and/or provide local exhaust ventilation at the site of chemical release. While not always operationally feasible, isolating operations can also reduce exposure. Using respiratory protection is less effective than the controls mentioned above, but is still advisable whenever working with or around glycolonitrile. The NIOSH 15-minute ceiling is relatively low and difficult to monitor (2 ppm). For low or infrequent exposures, a cyanide gas respirator may provide suitable protection. However, for high concentrations, or when the exposure level is unknown (as in an emergency), a supplied-air respirator or a self-contained breathing apparatus (SCBA) with full facepiece operated in pressure demand are the recommended respiratory protection methods of choice. If a full facepiece is not available, then chemical goggles should be worn to protect the eyes. Whenever a chemical splash hazard exists, a face shield and a rubber apron should be worn. To prevent hand and skin exposures, impervious gloves should be used. Glove manufacturers should be contacted and permeation studies obtained *before* final glove selections are made.

Administrative controls should also be in place to minimize the potential for human exposures. These may include written procedures or policies which specify the methods and techniques that will be practiced whenever personnel are to work with glycolonitrile.

All personnel should receive *training* on the use, hazards, protective measures, emergency actions, and other precautions per 29 CFR 1910.1200 (Hazard Communication), prior to the first assignment in an area where glycolonitrile is used or stored.

Medical testing of personnel with frequent or potentially high exposure, should be offered before beginning work and at regular intervals thereafter. These tests should include:

☑ Lung function (pulmonary function) test.

If symptoms develop or overexposure is suspected, the following tests may be useful:

☑ Blood cyanide test and/or urine thiocyanate test (currently, a blood cyanide level over 0.1 mg/l or urine thiocyanate over 20 mg/l indicates overexposure).

☑ Maintain close medical surveillance. Slow release of cyanide from absorbed glycolonitrile may cause delayed symptoms.

☑ Liver and kidney function tests.

Other methods to reduce exposure include:

☑ Where possible, enclose operations and use local exhaust ventilation at the site of chemical release. If local exhaust ventilation or enclosure is not used, respiratory protection should be mandatory.

☑ Always ensure that proper protective clothing is worn when using chemical substances.

☑ Wash thoroughly immediately after exposure to glycolonitrile and at the end of the work shift or before eating, drinking, or smoking.

☑ Hazard warning information should be posted in the work area. In addition, as part of an on-going education and training program, all information on the health and safety hazards of glycolonitrile should be communicated to all exposed or potentially exposed workers.

RISK ASSESSMENT: ENVIRONMENT

General Assessment

The environment is at risk of exposure during transportation, storage, disposal, or destruction of glycolonitrile. In almost every scenario, the threat of environmental exposure is contingent upon the proper handling of the chemical substance. Accidental spills, large or small, can result in fire, explosion, and possible contamination of the surrounding environmental mediums (water, soil, and air).

Glycolonitrile is considered a combustible liquid and a poison. It reacts with trace levels of alkalis to polymerize violently and contact under closed conditions can cause explosion. These characteristics require special consideration during any emergency situation involving a leak or spill of glycolonitrile and caution is also required to ensure separation from incompatible materials during use, transportation, or storage, since the formation of highly toxic and/or highly explosive commodities is extremely possible.

The proper disposal/destruction method for glycolonitrile is to burn it in a chemical incinerator equipped with an afterburner and air scrubber.

Glycolonitrile can enter the environment through manufacturing effluents, municipal waste treatment discharges, and spills.

☠ *Acute Ecological Effects*

Acute (short-term) toxic effects may include the death of animals, birds, or fish and death or low growth rate in plants. Acute effects are seen 2 to 4 days after animals or plants are exposed to glycolonitrile.

This chemical has slight acute toxicity to aquatic life. Insufficient data are available to evaluate or predict the short-term effects of glycolonitrile to plants, birds, or terrestrial animals.

Chronic Ecological Effects

Chronic toxic effects may include shortened life span, reproductive problems, lower fertility, and changes in appearance or behavior in exposed animals. These effects can be seen long after first exposure(s) to toxic chemicals.

Glycolonitrile has slight chronic toxicity to aquatic life. Insufficient data are available to evaluate or predict the long-term effects of glycolonitrile to plants, birds, or terrestrial animals.

Water Solubility

Glycolonitrile is highly soluble in water. Concentrations of 1000 milligrams and more will mix with a liter of water.

Persistence in the Environment

Glycolonitrile is slightly persistent in water, with a half-life of between 2 to 20 days. The half-life of a pollutant is the amount of time it takes for one half of the chemical to be degraded.

Bioaccumulation in Aquatic Organisms

Some substances increase in concentration, or *bioaccumulate,* in living organisms as they breathe contaminated air, drink contaminated water, or eat contaminated food. These chemicals can become concentrated in the tissues and internal organs of animals as well as humans.

The concentration of glycolonitrile found in fish tissues is expected to be about the same as the average concentration of glycolonitrile in water from which the fish was taken.

Recommended Risk-Reduction Measures

Proper training of all transporters will reduce the likelihood of a mishap or accident resulting in a leak or spill to the environment. The correct use of DOT labeling (per the MSDS) on all containers, trucks, and rail cars will enable emergency responders to react properly and quickly to any disaster thereby reducing the potential risk to the environment and to personnel.

Storage of glycolonitrile should be segregated from incompatible chemicals to minimize the risk of cross-contamination or contact. Buildings designated for storage should be equipped with appropriate fire protection systems (alarms, sprinklers, emergency lighting, portable extinguishers). Glycolonitrile will polymerize violently but can be kept stable with an inhibitor (such as phosphoric acid). The inhibitor is usually expended quickly. It is therefore important to institute a consistent monitoring program of inhibitor levels to prevent polymerization.

If a spill or leak to the environment has occurred, fire department, emergency response, and/or hazardous materials spill personnel should be notified immediately. Cleanup should be attempted only by those trained in proper spill containment procedures. Contaminated soils should be removed for incineration and replaced with clean soil. If glycolonitrile should contact the water table, aquifer, or navigable waterway, time is of the essence. It is readily miscible in water and therefore, total containment and remediation may not be entirely possible. When such spills occur, the local and/or state emergency response authorities must be notified. A comprehensive emergency response or disaster preparedness/recovery plan should be in place prior to any operations involving the use, transportation, storage, or disposal of glycolonitrile. If glycolonitrile is spilled or leaked, the following specific steps are recommended:

- ☑ Restrict persons not wearing protective clothing from area of spill or leak until cleanup is complete and area can be opened for normal work.
- ☑ Remove all ignition sources.
- ☑ Ventilate area of spill or leak.
- ☑ Absorb liquids in vermiculite, dry sand, earth, or a similar material and deposit in sealed containers.
- ☑ It may be necessary to dispose of glycolonitrile as a hazardous waste. The responsible state agency or the regional office of the federal Environmental Protection Agency (EPA) should be contacted for specific recommendations.

RISK ASSESSMENT: BUSINESS

General Assessment

Accidents or mishaps involving glycolonitrile can present a significant threat to business operations. The loss or damage of equipment or facilities can significantly

affect fiscal viability. Lawsuits that may result from personnel injury/death, public exposures, and/or environmental contamination will also require a serious expenditure of resources. Media attention surrounding an injury, death, or environmental damage can also result in a loss of profits and loss of current as well as future business.

❶ *Recommended Risk-Reduction Measures*

Company attorneys, safety and health professionals, and environmental specialists should be involved in the development of any procedures dealing with safety or emergency response. A company official should be pre-designated as a public relations officer with specific training in dealing with the press. Corporate plans and policies should be developed, approved, and implemented long before any need for such arises.

REFERENCES

Ahlbom, A. 1993. *Biostatistics for Epidemiologists.* New York: CRC Press/Lewis Publishers.

American Conference of Governmental Industrial Hygienists. 1988. *Documentation of the Threshold Limit Values and Biological Exposure Indices,* 5th Edition (with updates). Cincinnati: ACGIH

Asante-Duah, D. K. 1993. *Hazardous Waste Risk Assessment.* New York: CRC Press/Lewis Publishers.

Brusick, D. J. 1994. *Methods for Genetic Risk Assessment.* New York: CRC Press/Lewis Publishers.

Bueche, J. R. 1972. *Principles of Physics,* 2nd Edition. New York: McGraw-Hill.

Burrell, R., D. K. Flaherty, and L. J. Sauers. 1992. *Toxicology of the Immune System.* New York: Van Nostrand Reinhold.

Calabrese, E. J., E. M. Kenyon. *Air Toxics and Risk Assessment.* New York: CRC Press/Lewis Publishers.

Calabrese, E. J. 1994. *Biological Effects of Low Level Exposures.* New York: CRC Press/Lewis Publishers.

Cockerman, L. G., B. S. Shane. 1994. *Basic Environmental Toxicology.* New York: CRC Press/Lewis Publishers.

Cothern, R. R. 1993. *Comparative Environmental Risk Assessment.* New York: CRC Press/Lewis Publishers.

Cothern, C. R. 1995. *Handbook for Environmental Risk Decisionmaking.* New York: CRC Press/Lewis Publishers.

Danse, I. R. 1991. *Common Sense Toxics in the Workplace.* New York: Van Nostrand Reinhold.

Environmental Protection Agency. 1983. *Chemical Hazard Information Profiles.* Washington, D.C.: U.S. EPA.

Fodden, C. R., and J. L. Weddell. 1992. *Hazardous Materials Emergency Action Data.* New York: CRC Press/Lewis Publishers.

Forseberg, K., and L. H. Keith. 1993. *Chemical; Protective Clothing Permeation and Degradation Database.* Chelsea, MI: Lewis Publishers.

Gardner, R. W. 1994. *Chemical Intolerance.* New York: CRC Press/Lewis Publishers.

Hallenbeck, W. H. 1993. *Quantitative Risk Assessment for Environmental and Occupational Health,* 2nd Edition. New York: CRC Press/Lewis Publishers.

Hosty, J. W., and P. Foster. 1990. *A Practical Guide to Chemical Spill Response.* New York: Van Nostrand Reinhold.

Howard, P. H. 1995. *Handbook of Environmental Fate and Exposures Data for Organic Chemicals,* Vol. I-IV. New York: CRC Press/Lewis Publishers.

Jones, F. E. 1994. *Toxic Organic Vapors in the Workplace.* New York: CRC Press/Lewis Publishers.

Kamrin, M. A. 1988. *Toxicology: A Primer on Principles and Applications.* Chelsea, Michigan: Lewis Publishers.

Keith, L. H. , M. T. Johnston, C. E. Hudak, and M. Conoley. 1992. *National Toxicology Program's Gloves+ Database.* Chelsea, MI: Lewis Publishers.

Langerman, N. 1994 *Precautionary Labels for Chemical Containers.* New York: CRC Press/Lewis Publishers.

Lewis, R. J., Sr. 1993. *Hawley's Condensed Chemical Dictionary,* 12th Edition. New York: Van Nostrand Reinhold.

Lewis, R. J., Sr. 1992. *Sax's Dangerous Properties of Industrial Materials,* Volumes 1, 2, and 3, 8th Edition. New York: Van Nostrand Reinhold.

Mackay, D., W. Y. Shiu, and K. C. Ma. 1992. *Illustrated Handbook of Chemical-Physical Properties and Environmental Fate for Organic Chemicals.* New York: CRC Press/Lewis Publishers.

Manahan, S. E. 1994. *Environmental Chemistry,* 6th Edition. New York: CRC Press/Lewis Publishers.

Mellström, G. A., J. E. Wahlberg, and H. I. Maibach. 1994. *Protective Gloves for Occupational Use.* New York: CRC Press/Lewis Publishers.

Meyer, E. 1990. *Chemistry of Hazardous Materials.* Englewood Cliffs, NJ: Prentice Hall.

Patnaik, P. 1992. *A Comprehensive Guide to the Hazardous Properties of Chemical Substances.* New York: Van Nostrand Reinhold.

Proctor, N. H., and James P. Hughes. 1991. *Chemical Hazards of the Workplace.* New York: Van Nostrand Reinhold.

Sherman, J. D. 1988. *Chemical Exposure and Disease.* New York: Van Nostrand Reinhold.

U.S. Department of Health and Human Services, National Institute for Occupational Safety and Health. 1978. *Criteria for a Recommended Standard - Occupational Exposure to Nitriles.* DHEW (NIOSH) Publication Number: 78-212. Washington, D.C.: U.S. Government Printing Office.

U.S. Department of Health and Human Services, National Institute for Occupational Safety and Health. 1994. *NIOSH Pocket Guide to Chemical Hazards.* Washington, D.C.: U.S. Government Printing Office.

MATERIAL SAFETY DATA SHEET

CHEMICAL NAME	
GLYOXAL	**CRC PRESS/LEWIS PUBLISHERS DISCLAIMER:** THE INFORMATION AND RECOMMENDATIONS PRESENTED HEREIN ARE BASED ON SOURCES BELIEVED TO BE RELIABLE. CRC MAKES NO REPRESENTATION ON ITS COMPLETENESS OR ACCURACY. IT IS THE RESPONSIBILITY OF THE USER TO DETERMINE THE CHEMICAL'S SUITABILITY FOR ITS INTENDED USE, THE CHEMICAL'S SAFE USE, AND THE CHEMICAL'S PROPER DISPOSAL. NO REPRESENTATIONS AND/OR WARRANTIES, EITHER EXPRESSED OR IMPLIED, OF THE MERCHANTABILITY OR FITNESS FOR A PARTICULAR PURPOSE, OR OF ANY OTHER NATURE, ARE MADE WITH RESPECT TO THE INFORMATION PROVIDED IN THIS MSDS OR TO THE CHEMICAL TO WHICH INFORMATION MAY REFER. CRC NEITHER ASSUMES NOR AUTHORIZES ANY OTHER PERSON TO ASSUME FOR IT, ANY OTHER ADDITIONAL RESPONSIBILITY OR LIABILITY FOR THE USE OF, OR RELIANCE UPON, THIS INFORMATION.

HAZARD WARNING INFORMATION

HEALTH	FIRE	REACTIVE	OTHER	DEGREE OF HAZARD	COLOR CODING	OTHER CODES
1	0	2	~~W~~	0 = Minimum Hazard 1 = Slight Hazard 2 = Moderate Hazard 3 = Serious Hazard 4 = Severe Hazard	HEALTH = BLUE FIRE = RED REACTIVITY = YELLOW OTHER = WHITE	OX = Oxidizer ACID = Acid ALK = Alkali COR = Corrosive ~~W~~ = Use No Water

SECTION I - GENERAL INFORMATION

Characterization	**Aldehyde**	RCRA Number	**None**
		EPA Class	**Not Applicable**
DOT Proper Shipping Name	**No Citation**	Chemical Abstract Service (CAS) Number	**107-22-2**
DOT Hazard Class and Label Requirements	**No Citation**	DOT Emergency Guide Code	**No Citation**
DOT Identification Number	**No Citation**	Chemical Formula	$C_2H_2O_2$

Synonyms
Ethanedial; 1,2-ethanedione; biformal; biformyl; oxal; oxaldehyde; glyoxylaldehyde; diformyl.

SECTION II - HAZARDOUS INGREDIENTS AND IDENTITY INFORMATION

Hazardous Components (specific identity)	OSHA Exposure Criteria	NIOSH Exposure Criteria	Immediately Dangerous to Life and Health (IDLH)	ACGIH Exposure Criteria
Glyoxal (derivation: By oxidation of aldehyde).	**PEL:** Not Established **STEL:** Not Established	**REL:** Not Established **STEL:** Not Established	**Not Determined**	**TLV:** Not Established **STEL:** Not Established

SECTION III - PHYSICAL AND CHEMICAL CHARACTERISTICS

Boiling Point	**124°F (51°C)**	Specific Gravity (H_2O = 1)	**1.14**
Vapor Pressure (mm Hg)	**Not Found**	Molecular Weight	**58.04**
Vapor Density (Air = 1)	**Not Found**	Freezing Point	**59°F (15°C)**

Solubility
Soluble in water (polymerizes), alcohol, and ether.

Appearance and Odor
Yellowish crystals (prisms) in solid form becoming white, or light yellow liquid with mild odor. Vapor has a green color and burns with a violet flame.

SECTION IV - FIRE AND EXPLOSION HAZARD DATA

Flash Point (method used)	**Not Determined**	Explosive Limits in Air % by Volume	**LEL: Not Determined UEL: Not Determined**
NFPA Classification	**Non-Combustible**	Autoignition Temperature	**Not Determined**

Extinguishing Media
Use dry chemical, alcohol foam, carbon dioxide. Use water spray only to cool closed containers.

Special Fire Fighting Procedures
Wear full protective clothing and self-contained breathing apparatus (SCBA). Heat will build pressure and may rupture closed storage containers. Keep fire-exposed containers cool with water spray. Move containers from fire area if it can be done without risk. Do NOT allow water to contact material.

Unusual Fire and Explosion Hazards
Corrosive to most metals. Heat may cause polymerization to a combustible viscous material. Reacts violently with water and air. Fight fire from distance if possible.

SECTION V - REACTIVITY DATA

Stability		Conditions to Avoid **Unstable at room temperature. May explode on standing in open air at ambient temperature. During storage, it may spontaneously polymerize and ignite.**
Stable	Unstable **X**	Incompatibility *(materials to avoid)* **Incompatible with strong oxidizers, strong acids, and bases. Reactions on contact can become violent. A vapor-air mixture is explosive.**
Hazardous Polymerization		Conditions to Avoid **Polymerizes on standing or when mixed with water. Polymerization is exothermic and can become violent if uncontrolled. Avoid contact with air or water. Keep in air-tight containers.**
May Occur **X**	Will Not Occur	Hazardous Decomposition or By-products **When heated to decomposition, glyoxal emits acrid, irritating smoke and fumes.**

SECTION VI - HEALTH HAZARD DATA

Primary Route(s) of Entry:	Inhalation?	Absorption (skin)?	Ingestion?
	X	X	X

Health Hazards

INHALATION: Eye, nose, and throat irritant. Slight smarting of the respiratory tract; the effect appears to be temporary.

EYES & SKIN: Skin contact can result in irritating rash or inflammation. Can cause severe eye burns or severe irritation. May stain the skin yellow.

INGESTION: Little data available. May cause somnolence and gastrointestinal irritation.

Carcinogenicity	NTP Listed?	IARC Cancer Review Group?	OSHA Regulated?	Target Organs?
Unknown Human **Unknown Animal**	**No**	**No**	**29 CFR 1910.1000 (Table Z-1)**	**Respiratory system; skin; eyes.**

Medical Conditions Generally Aggravated by Exposure

None Reported.

Emergency and First-aid Procedures

Eye contact: Flush immediately with water for 15 minutes (minimum); seek medical attention. Skin contact: Remove all contaminated clothing. Immediately wash area with large amounts of soap and water. Seek medical assistance. For inhalation: Remove the person from exposure. Provide respiratory assistance and CPR. Transfer to medical facility. If swallowed: Seek medical attention immediately. Unless advised otherwise, give victim milk or water. Never attempt to give an unconscious or convulsing person anything by mouth. Do not make a convulsing or unconscious person vomit.

SECTION VII - PRECAUTIONS FOR SAFE HANDLING AND USE

Steps to be Taken in Case Material is Released or Spilled

Contain spills using absorbent material (vermiculite, or other). Collect solid materials using HEPA vacuum. Do NOT dry sweep. Ventilate area of spill. Store materials in DOT-approved containers. Restrict those not involved in cleanup from entering area. Notify appropriate authorities, as required.

Preferred Waste Disposal Method

Mix with a combustible solvent and burn in a chemical incinerator with afterburner and scrubber.

Precautions to be Taken in Handling and Storage

Do not store with incompatible chemicals. Store in tightly closed (air-tight) containers in a cool, dry, well-ventilated area. Keep fire, flame, and heat away (including radiant heat). Keep away from water sources. Do NOT allow exposure to air.

Other Precautions and Warnings

Bulk storage of glyoxal is not recommended. Nitrogen or other inert gas should be used as an "inert blanket" over glyoxal in storage containers. Prevent damage to containers.

SECTION VIII - CONTROL MEASURES AND PERSONAL PROTECTIVE EQUIPMENT

Respiratory Protection *(specify type)*

No exposure levels established. Recommend using a supplied-air respirator or a self-contained breathing apparatus (SCBA) with full facepiece in pressure demand or other positive pressure mode.

Ventilation

Local exhaust or general mechanical systems recommended.

Protective Gloves	Eye Protection	Other Protective Clothing
Impervious Material	**Chemical Goggles or Face Mask**	**Protective Apron**

Work/Hygiene Practices

Always wash hands thoroughly after using chemical; never bring food, drink, or smoking materials into vicinity of chemicals.

GLYOXAL

$C_2H_2O_2$ CAS: 107-22-2

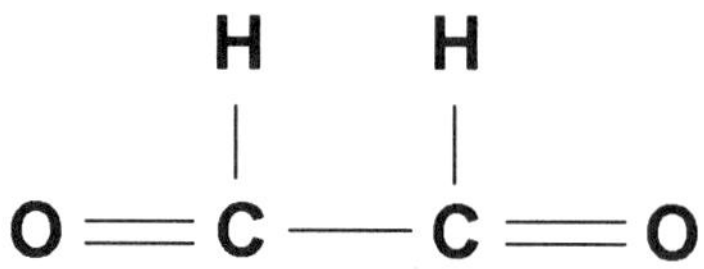

IDENTIFICATION AND TYPICAL USES

Glyoxal is a yellowish liquid or, in solid form, a pale yellow prism that becomes white. It has a mild odor. Its vapor has a green color and burns with a violet flame. It is used in the production of permanent-press fabrics and in the dimensional stabilization of rayon and other fabrics. It is also used as an insolubilizing agent for compounds containing polyhydroxyl groups (polyvinyl alcohol, starch, and cellulose materials), insolubilizing of proteins (casein, gelatin, and animal glue); in embalming fluids; in leather tanning; in paper coatings with hydroxycellulose; and as a reducing agent in dyeing textiles.

RISK ASSESSMENT: HEALTH

General Assessment

Glyoxal is moderately toxic by *ingestion* and skin contact (although absorption through the skin has not been reported as an entry route). *Inhalation* of its vapors can also pose a moderate health hazard risk. There are no data regarding its carcinogenic or teratogenic capabilities. Mutation data, however, have been recorded.

Glyoxal vapors are irritating to the skin and eyes and inhalation will cause irritation to the respiratory tract and may lead to symptoms of cough, sore throat, and congestion. It will severely irritate the eyes on contact. If not removed immediately, damage may occur to the cornea and surrounding tissue. Ingestion of this compound may cause somnolence and gastrointestinal pain with nausea and possible vomiting.

Acute Health Effects

The following acute (short-term) health effects may occur immediately or shortly after exposure to glyoxal:

Skin: Irritation and rash or burning sensation on contact.

Eye: Severe irritation and possible corneal damage.

Lung: Irritation causing coughing and/ congestion

Other: Ingestion of high concentrations may lead to somnolence and symptoms of gastrointestinal pain.

Chronic Health Effects

The following chronic (long-term) health effects may occur at some time after exposure to glyoxal and can last for months or even years:

Cancer Hazards: According to information presented in the references, glyoxal has not been adequately tested for its ability to cause cancer in test animals. Glyoxal has been shown to cause mutations (genetic changes). Such chemicals may have a cancer risk on long-term exposure. There is no evidence to indicate that glyoxal will cause cancer in humans or animals. More research is required.

Reproductive Hazard: According to information presented in the references, glyoxal has not been adequately tested for its ability to adversely affect reproduction in test animals.

Recommended Risk-Reduction Measures

Personnel should avoid direct contact with glyoxal. While its toxicity in humans is thought to be relatively low, the specific nature of its toxic effects is not clearly understood in the references. This fact alone warrants extreme caution when working with or around glyoxal. Unless a less toxic chemical can be substituted for a hazardous substance, *engineering controls* are the most effective method of reducing exposures. The best protection is to enclose operations and/or provide local exhaust ventilation at the site of chemical release. While not always operationally feasible, isolating operations can also reduce exposure.

Using respiratory protection is less effective than the controls mentioned above, but is still advisable whenever working with or around glyoxal. No exposure limits have been developed for glyoxal. This does not mean that exposure is without health hazard risk. For the best protection, use a supplied-air respirator or a self-contained breathing apparatus (SCBA) with full facepiece operated pressure demand or other positive pressure mode. If a full facepiece is not available, then chemical goggles should be worn to protect the eyes. Whenever a chemical splash hazard exists, a face shield and a rubber apron should be worn. To prevent hand and skin exposures, impervious gloves should be used. Glove manufacturers should be contacted and permeation studies obtained *before* final glove selections have been made.

Administrative controls should also be in place to minimize the potential for human exposures. These may include written procedures or policies which specify the methods and techniques that will be practiced whenever personnel are to work with glyoxal.

All personnel should receive *training* on the use, hazards, protective measures, emergency actions, and other precautions per 29 CFR 1910.1200 (Hazard Communication), prior to the first assignment in an area where glyoxal is used or stored.

Other methods to reduce exposure include:

- ☑ Where possible, enclose operations and use local exhaust ventilation at the site of chemical release. If local exhaust ventilation or enclosure is not used, respiratory protection should be mandatory.
- ☑ Always ensure that proper protective clothing is worn when using chemical substances and that personnel are trained in its use and care.
- ☑ Wash thoroughly immediately after exposure to glyoxal and at the end of the work shift or before eating, drinking, or smoking.
- ☑ Hazard warning information should be posted in the work area. In addition, as part of an on-going education and training program, all information on the health and safety hazards of glyoxal should be communicated to all potentially exposed workers.
- ☑ Never eat, drink, or smoke in areas were glyoxal is used, handled, or stored.

RISK ASSESSMENT: ENVIRONMENT

General Assessment

The environment is at risk of exposure during transportation, storage, disposal, or destruction of glyoxal. In almost every scenario, the threat of environmental exposure is contingent upon the proper handling of the chemical substance. Accidental spills, large or small, can result in fire, explosion, and possible contamination of the surrounding environmental mediums (water, soil, and air).

Glyoxal is considered a non-combustible liquid. However, it is highly unstable at room temperature and can polymerize violently on standing or when mixed with water. Polymerization is exothermic (liberates heat) and can become explosive if uncontrolled. Reactions with strong acids, bases, and oxidizers can also become extremely violent. It is a powerful reducing agent and may explode on contact with air. During storage for long periods, it may spontaneously polymerize and ignite. These characteristics require special consideration during any emergency situation involving a leak or spill of glyoxal. Should glyoxal ever come into contact with incompatible substances either during use, transportation, or storage, the formation of highly toxic and/or highly explosive commodities is extremely possible.

The proper disposal/destruction method for glyoxal is to mix it with a combustible solvent and then burn it in a chemical incinerator equipped with an afterburner and air scrubber.

Glyoxal can enter the environment through manufacturing, unchecked discharge into effluents, and through spills.

☠ *Acute Ecological Effects*

Acute (short-term) toxic effects may include the death of animals, birds, or fish, and death or low growth rate in plants. Acute effects are seen 2 to 4 days after animals or plants are exposed to glyoxal. Insufficient data are available to evaluate or predict the short-term effects of glyoxal to aquatic life, birds, plants, or terrestrial animals.

Chronic Ecological Effects

Chronic toxic effects may include shortened life span, reproductive problems, lower fertility, and changes in appearance or behavior in exposed animals. These effects can be seen long after first exposure(s) to toxic chemicals. Insufficient data are available to properly evaluate or predict the chronic (long-term) effects of glyoxal to aquatic life, plants, birds, or land animals.

Water Solubility

Glyoxal is highly reactive and decomposes in water.

⌛ *Persistence in the Environment*

Glyoxal decomposes violently in water and, therefore, is not expected to persist in the aquatic environment or settle in aquatic sediments. It reacts on standing and may explode on contact with air.

Bioaccumulation in Aquatic Organisms

Some substances increase in concentration, or *bioaccumulate,* in living organisms as they breathe contaminated air, drink contaminated water, or eat contaminated food. These chemicals can become concentrated in the tissues and internal organs of animals as well as humans.

Glyoxal is not expected to accumulate in the edible tissues of fish.

Recommended Risk-Reduction Measures

Proper training of all transporters will reduce the likelihood of a mishap or accident resulting in a leak or spill to the environment. The correct use of labeling on all containers, trucks, and rail cars will enable emergency responders to react properly and quickly to any disaster thereby reducing the potential risk to the environment and to personnel.

Storage of glyoxal should be segregated from incompatible chemicals to minimize the risk of cross-contamination or contact. Buildings designated for storage should be equipped with appropriate fire protection systems (alarms, sprinklers, emergency lighting, portable extinguishers). Glyoxal should be kept away from water and stored in air-tight containers. An inert nitrogen purge or "blanket" inside containers is recommended to prevent the condensation of moisture or contact with air.

If a spill or leak to the environment has occurred, fire department, emergency response, and/or hazardous materials spill personnel should be notified immediately. Cleanup should be attempted only by those trained in proper spill containment procedures. Contaminated soils should be removed for incineration and replaced with clean soil. If glyoxal should contact the water table, aquifer, or navigable waterway, time is of the essence. It will violently decompose in water and, therefore, total containment and remediation may not be entirely possible. When such spills occur, the local and/or state emergency response authorities must be notified. A comprehensive emergency response or disaster preparedness/recovery plan should be in place prior to any operations involving the use, transportation, storage, or disposal of glyoxal. If glyoxal is spilled or leaked, the following specific steps are recommended:

- ☑ Restrict persons not wearing protective clothing from area of spill or leak until cleanup is complete and area can be opened for normal work.
- ☑ Remove all ignition sources and ventilate area.
- ☑ Proceed with caution during cleanup. Glyoxal may cause violent explosions during spill conditions. Absorb liquids in vermiculite, dry sand, earth, or a similar material and deposit in sealed containers. Collect powdered materials using HEPA vacuum. Do NOT dry sweep.
- ☑ It may be necessary to dispose of glyoxal as a hazardous waste. The responsible state agency or the regional office of the federal Environmental Protection Agency (EPA) should be contacted for specific recommendations.

RISK ASSESSMENT: BUSINESS

General Assessment

Accidents or mishaps involving glyoxal can present a significant threat to business operations. The loss or damage of equipment or facilities can significantly affect fiscal viability. Lawsuits that may result from personnel injury/death, public exposures, and/or environmental contamination will also require a serious expenditure of resources. Media attention surrounding an injury, death, or environmental damage can also result in a loss of profits and loss of current as well as future business.

Recommended Risk-Reduction Measures

Company attorneys, safety and health professionals, and environmental specialists should be involved in the development of any procedures. A company official should be pre-designated as a public relations officer with specific training in dealing with the press. Corporate plans and policies should be developed, approved, and implemented long before any need for such arises.

REFERENCES

Ahlbom, A. 1993. *Biostatistics for Epidemiologists.* New York: CRC Press/Lewis Publishers.

American Conference of Governmental Industrial Hygienists. 1988. *Documentation of the Threshold Limit Values and Biological Exposure Indices,* 5th Edition (with updates). Cincinnati: ACGIH

Anthony, C. P., and N. J. Kolthoff. 1971. *Textbook of Anatomy and Physiology.* St. Louis: C. V. Mosby Company.

Asante-Duah, D. K. 1993. *Hazardous Waste Risk Assessment.* New York: CRC Press/Lewis Publishers.

Bloom, A. D. and F. J. DeSerres. 1995. *Ecotoxicity and Human Health.* New York: CRC Press/Lewis Publishers.

Brusick, D. J. 1994. *Methods for Genetic Risk Assessment.* New York: CRC Press/Lewis Publishers.

Bueche, J. R. 1972. *Principles of Physics,* 2nd Edition. New York: McGraw-Hill.

Burrell, R., D. K. Flaherty, and L. J. Sauers. 1992. *Toxicology of the Immune System.* New York: Van Nostrand Reinhold.

Calabrese, E. J., E. M. Kenyon. *Air Toxics and Risk Assessment.* New York: CRC Press/Lewis Publishers.

Calabrese, E. J. 1994. *Biological Effects of Low Level Exposures.* New York: CRC Press/Lewis Publishers.

Cockerman, L. G., B. S. Shane. 1994. *Basic Environmental Toxicology.* New York: CRC Press/Lewis Publishers.

Cothern, C. R., N. P. Ross. 1994. *Environmental Statistics, Assessment, and Forecasting.* New York: CRC Press/Lewis Publishers.

Cothern, R. R. 1993. *Comparative Environmental Risk Assessment.* New York: CRC Press/Lewis Publishers.

Cothern, C. R. 1995. *Handbook for Environmental Risk Decisionmaking.* New York: CRC Press/Lewis Publishers.

Danse, I. R. 1991. *Common Sense Toxics in the Workplace.* New York: Van Nostrand Reinhold.

Environmental Protection Agency. 1983. *Chemical Hazard Information Profiles.* Washington, D.C.: U.S. EPA.

Fodden, C. R., and J. L. Weddell. 1992. *Hazardous Materials Emergency Action Data.* New York: CRC Press/Lewis Publishers.

Forseberg, K., and L. H. Keith. 1993. *Chemical; Protective Clothing Permeation and Degradation Database.* Chelsea, MI: Lewis Publishers.

Gardner, R. W. 1994. *Chemical Intolerance.* New York: CRC Press/Lewis Publishers.

Hallenbeck, W. H. 1993. *Quantitative Risk Assessment for Environmental and Occupational Health,* 2nd Edition. New York: CRC Press/Lewis Publishers.

Hodgson, E., R. B. Mailman, and J. E. Chambers. 1988. *Dictionary of Toxicology.* London: The MacMillian Company.

Heck, H. 1985. Mechanisms of aldehyde toxicity: Structure activity Studies. *CIIT Activities,* 5(10):106.

Hosty, J. W., and P. Foster. 1990. *A Practical Guide to Chemical Spill Response.* New York: Van Nostrand Reinhold.

Howard, P. H. 1995. *Handbook of Environmental Fate and Exposures Data for Organic Chemicals,* Vol. I-IV. New York: CRC Press/Lewis Publishers.

International Agency for Research on Cancer (IARC). 1985. *IARC Monographs on the Evaluation of Carcinogenic Risk of Chemicals to Humans,* Vol. 36, *Allyl Compounds, Aldehydes, Epoxides, and Peroxides,* pp 101-136. Lyon: IARC.

International Agency for Research on Cancer (IARC). 1987. *IARC Monographs on the Evaluation of Carcinogenic Risk to Humans, Overall Evaluations of Carcinogenicity: An updating of IARC Monographs Volumes 1-42,* Supplement 7, pp 77-78. Lyon: IARC.

James, A., and M. Lord. 1992. *Index of Chemical and Physical Data.* New York: Van Nostrand Reinhold.

Jones, F. E. 1994. *Toxic Organic Vapors in the Workplace.* New York: CRC Press/Lewis Publishers.

Kamrin, M. 1 A. 1988. *Toxicology: A Primer on Principles and Applications.* Chelsea, Michigan: Lewis Publishers.

Keith, L. H. and M. M. Walker. 1995. *Handbook of Air Toxics.* New York: CRC Press/Lewis Publishers.

Landis, W. G., M. H. Yu. 1995. *Introduction to Environmental Toxicology.* New York: CRC Press/Lewis Publishers.

Lewis, .J., Sr. 1993. *Hawley's Condensed Chemical Dictionary,* 12th Edition. New York: Van Nostrand Reinhold.

Lewis, . J., Sr. 1992. *Sax's Dangerous Properties of Industrial Materials,* Volumes 1, 2, and 3, 8th Edition. New York: Van Nostrand Reinhold.

Lisella, F. S. 1994. *The VNR Dictionary of Environmental Health and Safety.* New York: Van Nostrand Reinhold.

Meyer, E. 1990. *Chemistry of Hazardous Materials.* Englewood Cliffs, NJ: Prentice Hall.

Mackay, D., W. Y. Shiu, and K. C. Ma. 1992. *Illustrated Handbook of Chemical-Physical Properties and Environmental Fate for Organic Chemicals.* New York: CRC Press/Lewis Publishers.

Manahan, S. E. 1994. *Environmental Chemistry,* 6th Edition. New York: CRC Press/Lewis Publishers.

Manahan, S. E. 1992. *Toxicological Chemistry,* 2nd Edition. New York: CRC Press/Lewis Publishers.

Mellström, G. A., J. E. Wahlberg, and H. I. Maibach. 1994. *Protective Gloves for Occupational Use.* New York: CRC Press/Lewis Publishers.

Patnaik, P. 1992. *A Comprehensive Guide to the Hazardous Properties of Chemical Substances.* New York: Van Nostrand Reinhold.

Proctor, N. H., and J. P. Hughes. 1991. *Chemical Hazards of the Workplace.* New York: Van Nostrand Reinhold.

Silverman, L., H. F. Schuttle, and M. W. First. 1946. Further studies on sensory response to certain industrial solvent vapors. *Journal of Industrial Hygiene Toxicology,* 28:262-266.

Sherman, J. D. 1988. *Chemical Exposure and Disease.* New York: Van Nostrand Reinhold.

Talbott, E., G. Craun. 1995. *An Introduction to Environmental Epidemiology.* New York: CRC Press/Lewis Publishers.

Torkelson, T. R. V. K. Rowe. 1981. *Patty's Industrial Hygiene and Toxicology,* Vol. 2B. New York: Wiley-Interscience.

U.S. Department of Health and Human Services, National Institute for Occupational Safety and Health. 1994. *NIOSH Pocket Guide to Chemical Hazards.* Washington, D.C.: U.S. Government Printing Office.

MATERIAL SAFETY DATA SHEET

CHEMICAL NAME	
GRAPHITE	**CRC PRESS/LEWIS PUBLISHERS DISCLAIMER:** THE INFORMATION AND RECOMMENDATIONS PRESENTED HEREIN ARE BASED ON SOURCES BELIEVED TO BE RELIABLE. CRC MAKES NO REPRESENTATION ON ITS COMPLETENESS OR ACCURACY. IT IS THE RESPONSIBILITY OF THE USER TO DETERMINE THE CHEMICAL'S SUITABILITY FOR ITS INTENDED USE, THE CHEMICAL'S SAFE USE, AND THE CHEMICAL'S PROPER DISPOSAL. NO REPRESENTATIONS AND/OR WARRANTIES, EITHER EXPRESSED OR IMPLIED, OF THE MERCHANTABILITY OR FITNESS FOR A PARTICULAR PURPOSE, OR OF ANY OTHER NATURE, ARE MADE WITH RESPECT TO THE INFORMATION PROVIDED IN THIS MSDS OR TO THE CHEMICAL TO WHICH INFORMATION MAY REFER. CRC NEITHER ASSUMES NOR AUTHORIZES ANY OTHER PERSON TO ASSUME FOR IT, ANY OTHER ADDITIONAL RESPONSIBILITY OR LIABILITY FOR THE USE OF, OR RELIANCE UPON, THIS INFORMATION.

HAZARD WARNING INFORMATION

HEALTH	FIRE	REACTIVE	OTHER	DEGREE OF HAZARD	COLOR CODING	OTHER CODES
1	0	0		0 = Minimum Hazard 1 = Slight Hazard 2 = Moderate Hazard 3 = Serious Hazard 4 = Severe Hazard	HEALTH = BLUE FIRE = RED REACTIVITY = YELLOW OTHER = WHITE	OX = Oxidizer ACID = Acid ALK = Alkali COR = Corrosive W = Use No Water

SECTION I - GENERAL INFORMATION

Characterization **Particulate**	RCRA Number **None**	EPA Class **Not Applicable**
DOT Proper Shipping Name **No Citation**	Chemical Abstract Service (CAS) Number **7782-42-5** (natural graphite); **7440-44-0** (synthetic graphite)	
DOT Hazard Class and Label Requirements **No Citation**	DOT Emergency Guide Code **32** (carbon, activated = synthetic graphite)	
DOT Identification Number **UN 1362** (synthetic graphite only)	Chemical Formula **C**	

Synonyms

Natural Graphite Synonyms: Black lead; mineral carbon; plumbago; silver graphite; stove black. Synthetic Graphite Synonyms: Activated carbon.

SECTION II - HAZARDOUS INGREDIENTS AND IDENTITY INFORMATION

Hazardous Components (specific identity)	OSHA Exposure Criteria	NIOSH Exposure Criteria	Immediately Dangerous to Life and Health (IDLH)	ACGIH Exposure Criteria
Graphite (derivation: A crystalline form of carbon.).	**PEL (synthetic):** 5 mg/m^3 (respirable) 15 mg/m^3 (total dust)	**REL (natural):** 2.5 mg/m^3 (respirable)	**Not Determined**	**TLV (synthetic):** 10 mg/m^3 (total dusts) **TLV (natural):** 2.5 mg/m^3 (respirable)

SECTION III - PHYSICAL AND CHEMICAL CHARACTERISTICS

Boiling Point **Sublimes**	Specific Gravity (H_2O = 1) **2.0 - 2.25** (natural) **1.5 - 1.8** (synthetic)
Vapor Pressure (mmHg) **0 (approx.) at 69°F (20°C)**	Molecular Weight **12.0**
Vapor Density (Air = 1) **Not Found**	Freezing Point **6602°F (3650°C) Sublimes**

Solubility

Insoluble in water.

Appearance and Odor

Steel gray to black, greasy feeling, odorless solid. May appear as a soft, greasy, black powder, scales, or sheets.

SECTION IV - FIRE AND EXPLOSION HAZARD DATA

Flash Point (method used) **Not Found**	Explosive Limits in Air % by Volume **LEL: Not Determined UEL: Not Determined**
NFPA Classification **Combustible Solid**	Autoignition Temperature **Not Determined**

Extinguishing Media

In solid form, graphite will not burn or support combustion. Use extinguishing media suitable to surrounding fire.

Special Fire Fighting Procedures

Poisonous gases are produced in fire. Wear self-contained breathing apparatus. Use water spray to keep containers cool. Graphite is a rather inert chemical, but breathing its dusts should be avoided.

Unusual Fire and Explosion Hazards

None Reported.

SECTION V - REACTIVITY DATA

Stability		Conditions to Avoid Graphite is normally stable in closed containers under normal conditions of temperature and pressure and routine conditions of handling and storage. Avoid contact with heat and incompatible materials.
Stable X	Unstable	Incompatibility *(materials to avoid)* Both the synthetic and natural forms of graphite are incompatible with very strong oxidizers such as chlorine trifluoride, fluorine, and potassium peroxide.
Hazardous Polymerization		Conditions to Avoid Under normal conditions of temperature and pressure, hazardous polymerization of graphite is not expected to occur.
May Occur	Will Not Occur X	Hazardous Decomposition or By-products When heated or involved in fire, graphite can emit toxic carbon monoxide.

SECTION VI - HEALTH HAZARD DATA

Primary Route(s) of Entry:	Inhalation? X	Absorption (skin)?	Ingestion?

Health Hazards

INHALATION: Synthetic graphite is a "nuisance dust" that may produce minor but reversible respiratory irritation. Natural graphite is not considered a nuisance dust since it has been shown to cause chronic lung damage. If either type contains more than 1% quartz, the exposure limits do not apply and more research is required. Its causes cough, dypsnea, and black sputum.

SKIN & EYES: From no significant effects to minor skin and eye irritation.

INGESTION: No documented effects.

Carcinogenicity	NTP Listed?	IARC Cancer Review Group?	OSHA Regulated?	Target Organs?
No Evidence Human No Evidence Animal	No	No	29 CFR 1910.1000 Table Z-1	Respiratory system

Medical Conditions Generally Aggravated by Exposure

Respiratory or pulmonary conditions may be aggravated by exposure to natural graphite.

Emergency and First-aid Procedures

Eye contact: Immediately flush with large amounts of water for 15 minutes (minimum), occasionally lifting eyelids; seek medical attention. Skin contact: Remove all contaminated clothing. Immediately wash area with large amounts of soap and water. For inhalation: Remove the person from exposure. Provide respiratory assistance and CPR. Transfer promptly to medical facility. If swallowed: Not likely to occur. Seek medical attention immediately. Never attempt to give an unconscious and convulsing person anything by mouth.

SECTION VII - PRECAUTIONS FOR SAFE HANDLING AND USE

Steps to be Taken in Case Material is Released or Spilled

Ventilate area. Cleanup crew should wear breathing protection. Collect powdered materials in safest manner possible. Do NOT dry sweep. Use HEPA vacuum instead. Damp mop residue and collect for disposal. Emergency personnel should avoid contact with airborne dusts (especially natural graphite).

Preferred Waste Disposal Method

No Citation.

Precautions to be Taken in Handling and Storage

Store in closed containers in cool, dry, well-ventilated area away from heat. Avoid handling activities that create dusty conditions. Respiratory protection should be worn by all personnel involved in handling and storage operations.

Other Precautions and Warnings

Keep away from incompatible materials.

SECTION VIII - CONTROL MEASURES AND PERSONAL PROTECTIVE EQUIPMENT

Respiratory Protection *(specify type)*

Use a half-mask respirator equipped with a high efficiency particulate air (HEPA) filter under normal working conditions. For particularly dusty conditions, a powered air-purifying respirator will provide greater protection.

Ventilation

Local exhaust or general mechanical systems recommended.

Protective Gloves	Eye Protection	Other Protective Clothing
Impervious Rubber	Chemical/Dust Goggles and/or Face Mask	Protective Apron

Work/Hygiene Practices

Always wash hands thoroughly after using chemical; never bring food, drink, or smoking materials into vicinity of chemicals.

GRAPHITE

C

CAS: 7440-44-0 (synthetic graphite)
CAS: 7782-42-5 (natural graphite)

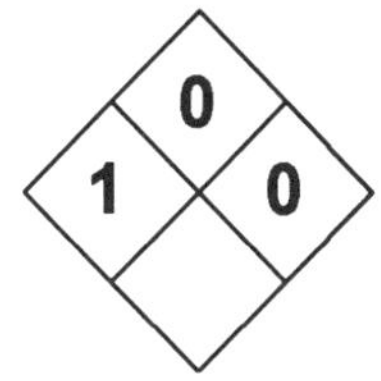

IDENTIFICATION AND TYPICAL USES

Graphite is a steel gray to black, greasy-feeling, odorless solid that can appear as a soft, greasy, black powder, in scales, or in sheets. It is the crystalline form of carbon and exists in two forms: natural graphite and synthetic graphic. Natural graphite is used to make lead pencils, stove polishes, matches, explosives, and refractory crucibles. It is also a lubricant in nuclear piles and is used to re-carbonize steel, in electroplating, in the manufacture of electrodes, and in other electrical equipment. Synthetic graphite is used as an adsorbent and in water purification.

RISK ASSESSMENT: HEALTH

General Assessment

Graphite is generally considered a "nuisance dust" or "inert" dust in its synthetic form because it is not expected to cause any chronic, irreversible, or long-term lung damage. However, natural graphite has been known to cause chronic lung damage and, therefore, cannot be classified in the same manner. Also, if either version of graphite happens to contain more than 1% quartz (crystalline silica), then the established exposure limits *do not* apply and more research must be performed on the health hazard risks associated with the material.

Primary exposure route is through *inhalation.* Ingestion is not likely to occur and skin or eye contact produces very few noticeable effects.

Inhaling synthetic graphite dusts may produce only minor but reversible respiratory irritation. Inhaling the dusts of natural graphite can cause cough, dypsnea, black sputum, decreased pulmonary function, and lung fibrosis (depending on silica content of dust). In high concentrations, it may cause pneumoconiosis and anthracosilicosis.

Acute Health Effects

The following acute (short-term) health effects may occur immediately or shortly after exposure to graphite:

Skin: No effect, to mild surface irritation.

Eye: Mild to moderate irritation.

Lung: Irritation of the mouth, nose, and throat causing coughing, dypsnea, black sputum, decreased respiratory function, and respiratory irritation.

Chronic Health Effects

The following chronic (long-term) health effects may occur at some time after exposure to graphite and can last for months or even years:

Cancer Hazards: According to information presented in the references, graphite has not been shown to cause cancer in test animals.

Reproductive Hazards: There are no reports to support any claims of reproductive hazards following exposure to graphite.

Other Chronic Effects: Long-term or repeated inhalation exposure to natural graphite dusts may lead to pneumoconiosis and anthracosilicosis.

Recommended Risk-Reduction Measures

Personnel should avoid direct contact with graphite dusts and powders. Although the toxicity of the synthetic material is considered relatively low, the natural version has been shown to cause moderate toxic systemic effects. Unless a less toxic chemical can be substituted for a hazardous substance, *engineering controls* are the most effective method of reducing exposures. The best protection is to enclose operations and/or provide local exhaust ventilation at the site of chemical release. While not always operationally feasible, isolating operations can also reduce exposure. Using respiratory protection is less effective than the controls mentioned above, but is still advisable whenever working with graphite. A half-mask respirator equipped with a high-efficiency particulate air (HEPA) filter and dust pre-filter should provide adequate protection at low-level exposures. Better protection is obtained using an MSHA/NIOSH-approved powered supplied-air respirator operated in continuous flow mode or a self-contained breathing apparatus (SCBA) operated in pressure demand mode. A full facepiece, face shield, or chemical/dust goggles would provide additional protection for the eyes. Whenever an airborne dust hazard exists, a protective apron should be worn. It is common to find graphite in a fine powder form which has a tendency to get under fingernails and into skin pores. To prevent hand and skin exposures, impervious protective gloves should be worn.

Administrative controls should also be in place to minimize the potential for human exposures. These may include written procedures or policies which specify the methods and techniques that will be practiced whenever personnel are to work with graphite.

All personnel should receive *training* on the use, hazards, protective measures, emergency actions, and other precautions per 29 CFR 1910.1200 (Hazard Communication), prior to the first assignment in an area where graphite is used or stored.

Before beginning employment and at regular intervals thereafter (e.g., annually), the following medical tests are recommended:

☑ Lung function tests.

If symptoms develop or overexposure is suspected, the following may be useful:

☑ Consider chest X-ray after acute overexposure.

Any evaluation should include a careful history of past and present symptoms with an examination. Medical tests that simply look for existing damage are not a substitute for controlling exposures. Also, since smoking can cause heart disease, lung cancer, emphysema, and other respiratory disorders, smokers exposed to graphite may experience symptoms more rapidly than non-smokers under the same conditions of exposure. Prudent risk management requires careful consideration of *all* possible risk factors.

Other methods to reduce exposure to chemicals include:

☑ Where possible, enclose operations and use local exhaust ventilation at the site of chemical release. If local exhaust ventilation or enclosure is not used, respiratory protection should be mandatory.

☑ Always ensure that proper protective clothing is worn when using chemical substances and that personnel are trained in its use and care.

☑ Wash thoroughly immediately after exposure to graphite and at the end of the work shift or before eating, drinking, or smoking. Work clothing contaminated with graphite should never be taken home for laundering. Only personnel trained in the hazards of exposure to chemical substances should launder contaminated clothing.

☑ Safety shower and eyewash stations should be readily available in work areas where graphite is used or stored.

☑ Hazard warning information should be posted in the work area. In addition, as part of an on-going education and training program, all information on the health and safety hazards of graphite should be communicated to all exposed and potentially exposed workers.

RISK ASSESSMENT: ENVIRONMENT

General Assessment

The environment is at risk of exposure during transportation, storage, disposal, or destruction of graphite. In almost every scenario, the threat of environmental exposure is contingent upon the proper handling of the chemical substance. Accidental releases, large or small, can result in fire, explosion, and possible contamination of the surrounding environmental mediums (air, water, and soil).

Graphite is considered to be a combustible solid, but it is difficult to ignite or sustain combustion once fire or ignition source is removed. Its powders or dusts in mixture with air may present a greater combustion hazard. Hazardous reactions occur when graphite comes into contact with very strong oxidizers, such as fluorine, chlorine trifluoride, and potassium peroxide. These characteristic require special consideration during any emergency involving a leak or release of graphite powder.

Graphite can be present naturally in the environment. High-grade natural graphite is mined in Canada, Ceylon, Germany, Madagascar, and Sri Lanka. It can also enter the environment through spills or leaks and unchecked discharges from industrial sources.

Acute Ecological Effects

Acute (short-term) toxic effects may include the death of animals, birds, or fish, and death or low growth rate in plants. Acute effects are seen 2 to 4 days after animals or plants are exposed to graphite.

Insufficient data are available to accurately evaluate or predict the short-term effects of graphite exposure to aquatic life, plants, birds, or land animals.

Chronic Ecological Effects

Chronic (long-term) toxic effects may include shortened life span, reproductive problems, lower fertility, and changes in appearance or behavior in exposed animals. These effects can be seen long after first exposure(s) to toxic chemicals.

Insufficient data are available to evaluate or predict the long-term (chronic) effects of graphite to aquatic life, plants, birds, or land animals.

Water Solubility

Graphite is insoluble in water. Concentrations of 1 milligram will not mix with a liter of water.

Persistence in the Environment

Graphite is highly persistent in water, with a half-life of greater than 200 days. The half-life of a pollutant is the amount of time it takes for one half of the chemical to be degraded. Virtually all of graphite will eventually end up in terrestrial soil and aquatic sediments.

Bioaccumulation in Aquatic Organisms

Some substances increase in concentration, or *bioaccumulate,* in living organisms as they breathe contaminated air, drink contaminated water, or eat contaminated food. These chemicals can become concentrated in the tissues and internal organs of animals as well as humans.

The concentration of graphite found in fish tissues is expected to be somewhat higher than the average concentration of graphite in the water from which the fish was taken.

Recommended Risk-Reduction Measures

Proper training of all transporters will reduce the likelihood of a mishap or accident resulting in a leak or release to the environment. The correct use of labeling on all containers, trucks, and rail cars will enable emergency responders to react properly and quickly to any disaster thereby reducing the potential risk to the environment and to personnel.

Storage of graphite should be segregated from sources or heat or ignition and kept away from incompatible materials. Containers should be kept closed and in a cool, dry, well-ventilated area. Buildings designated for storage should be equipped with appropriate fire protection systems (alarms, sprinklers, emergency lighting, portable extinguishers). Dusty conditions in storage areas should be avoided. All personnel should be briefed on the hazards of graphite powders.

If a spill or release to the environment has occurred, fire department, emergency response, and/or hazardous materials spill personnel should be notified immediately. Contaminated soils should be removed and replaced with clean soils. Cleanup should be attempted only by those trained in proper containment procedures. If graphite should contact the water table, aquifer, or navigable waterway, time is of the essence. It is insoluble in water and total containment and remediation may be possible. When such spills occur, the local and/or state emergency response authorities should be notified. A comprehensive emergency response or disaster preparedness/recovery plan should be in place prior to any operations involving the use, transportation, storage, or disposal of graphite.

If graphite is spilled or leaked, the following steps are recommended:

- ☑ Restrict persons not wearing protective clothing from area of spill or leak until cleanup is complete and area can be opened for normal work.
- ☑ Remove all ignition sources.
- ☑ Ventilate area of spill or leak.
- ☑ Avoid generating airborne dusts during cleanup operations.
- ☑ Collect powdered materials in safest manner possible. Hold for reclamation or arrange for disposal. Do NOT dry sweep (generates airborne dusts). Use a vacuum equipped with a high-efficiency particulate air (HEPA) filter instead. Damp mop residue and collect for disposal. Place in sealed containers.
- ☑ It may be necessary to dispose of graphite as a hazardous waste. The responsible state agency or the regional office of the federal Environmental Protection Agency (EPA) should be contacted for specific recommendations on the disposal requirements for graphite.

RISK ASSESSMENT: BUSINESS
General Assessment

Accidents or mishaps involving graphite can present a significant threat to business operations. The loss or damage of equipment or facilities can significantly affect fiscal viability. Lawsuits that may result from personnel injury, public exposures, and environmental contamination will also require a serious expenditure of resources. Media attention surrounding an injury, illness, or environmental damage can also result in a loss of profits and loss of current as well as future business.

Recommended Risk-Reduction Measures

Company attorneys, safety and health professionals, and environmental specialists should be involved in the development of any procedures dealing with safety or emergency response. A company official should be pre-designated as a public relations officer with specific training in dealing with the press. Corporate plans and policies should be developed, approved, and implemented long before any need for such arises.

REFERENCES

Ahlbom, A. 1993. *Biostatistics for Epidemiologists.* New York: CRC Press/Lewis Publishers.

American Conference of Governmental Industrial Hygienists. 1988. *Documentation of the Threshold Limit Values and Biological Exposure Indices,* 5th Edition (with updates). Cincinnati: ACGIH

Anthony, C. P., and N. J. Kolthoff. 1971. *Textbook of Anatomy and Physiology.* St. Louis: C. V. Mosby Company.

Asante-Duah, D. K. 1993. *Hazardous Waste Risk Assessment.* New York: CRC Press/Lewis Publishers.

Bloom, A. D. and F. J. DeSerres. 1995. *Ecotoxicity and Human Health.* New York: CRC Press/Lewis Publishers.

Brusick, D. J. 1994. *Methods for Genetic Risk Assessment.* New York: CRC Press/Lewis Publishers.

Bueche, J. R. 1972. *Principles of Physics,* 2nd Edition. New York: McGraw-Hill.

Burrell, R., D. K. Flaherty, and L. J. Sauers. 1992. *Toxicology of the Immune System.* New York: Van Nostrand Reinhold.

Calabrese, E. J. 1994. *Biological Effects of Low Level Exposures.* New York: CRC Press/Lewis Publishers.

Cockerman, L. G., B. S. Shane. 1994. *Basic Environmental Toxicology.* New York: CRC Press/Lewis Publishers.

Cothern, C. R. 1995. *Handbook for Environmental Risk Decisionmaking.* New York: CRC Press/Lewis Publishers.

Gardner, R. W. 1994. *Chemical Intolerance.* New York: CRC Press/Lewis Publishers.

Hallenbeck, W. H. 1993. *Quantitative Risk Assessment for Environmental and Occupational Health,* 2nd Edition. New York: CRC Press/Lewis Publishers.

Hodgson, E., R. B. Mailman, and J. E. Chambers. 1988. *Dictionary of Toxicology.* London: The MacMillian Company.

Hosty, J. W., and P. Foster. 1990. *A Practical Guide to Chemical Spill Response.* New York: Van Nostrand Reinhold.

Howard, P. H. 1995. *Handbook of Environmental Fate and Exposures Data for Organic Chemicals,* Vol. I-IV. New York: CRC Press/Lewis Publishers.

Howard, P. H., et. al. 1991. *Handbook of Environmental Degradation Rates.* New York: CRC Press/Lewis Publishers.

James, A., and M. Lord. 1992. *Index of Chemical and Physical Data.* New York: Van Nostrand Reinhold.

Jones, F. E. 1994. *Toxic Organic Vapors in the Workplace.* New York: CRC Press/Lewis Publishers.

Kamrin, M. A. 1988. *Toxicology: A Primer on Principles and Applications.* Chelsea, Michigan: Lewis Publishers.

Keith, L. H. and M. M. Walker. 1995. *Handbook of Air Toxics.* New York: CRC Press/Lewis Publishers.

Landis, W. G., M. H. Yu. 1995. *Introduction to Environmental Toxicology.* New York: CRC Press/Lewis Publishers.

Langerman, N. 1994 *Precautionary Labels for Chemical Containers.* New York: CRC Press/Lewis Publishers.

Lauwerys, R. R., P. Hoet. 1993. *Industrial Chemical Exposure.* New York: CRC Press/Lewis Publishers.

Lewis, R. J., Sr. 1993. *Hawley's Condensed Chemical Dictionary,* 12th Edition. New York: Van Nostrand Reinhold.

Lewis, R. J., Sr. 1992. *Sax's Dangerous Properties of Industrial Materials,* Volumes 1, 2, and 3, 8th Edition. New York: Van Nostrand Reinhold.

Lide, D. R. 1995. *Handbook of Chemistry and Physics,* 76th Edition. New York: CRC Press/Lewis Publishers.

Manahan, S. E. 1992. *Toxicological Chemistry,* 2nd Edition. New York: CRC Press/Lewis Publishers.

Meyer, E. 1990. *Chemistry of Hazardous Materials.* Englewood Cliffs, NJ: Prentice Hall.

Patnaik, P. 1992. *A Comprehensive Guide to the Hazardous Properties of Chemical Substances.* New York: Van Nostrand Reinhold.

Philp, R. B. 1995. *Environmental Hazards and Human Health.* New York: CRC Press/Lewis Publishers.

Proctor, N. H., and J. P. Hughes. 1991. *Chemical Hazards of the Workplace.* New York: Van Nostrand Reinhold.

Quigley, D. R. 1994. *Handbook of Emergency Chemical Management.* New York: CRC Press/Lewis Publishers.

Rea, W. J. 1995. *Chemical Sensitivity,* Vol. I-III. New York: CRC Press/Lewis Publishers.

Richardson, M. 1995. *Dictionary of Substance and Their Effects,* Vol. I-VI. New York: CRC Press/Lewis Publishers.

MATERIAL SAFETY DATA SHEET

CHEMICAL NAME	CRC PRESS/LEWIS PUBLISHERS DISCLAIMER
HAFNIUM	**CRC PRESS/LEWIS PUBLISHERS DISCLAIMER:** THE INFORMATION AND RECOMMENDATIONS PRESENTED HEREIN ARE BASED ON SOURCES BELIEVED TO BE RELIABLE. CRC MAKES NO REPRESENTATION ON ITS COMPLETENESS OR ACCURACY. IT IS THE RESPONSIBILITY OF THE USER TO DETERMINE THE CHEMICAL'S SUITABILITY FOR ITS INTENDED USE, THE CHEMICAL'S SAFE USE, AND THE CHEMICAL'S PROPER DISPOSAL. NO REPRESENTATIONS AND/OR WARRANTIES, EITHER EXPRESSED OR IMPLIED, OF THE MERCHANTABILITY OR FITNESS FOR A PARTICULAR PURPOSE, OR OF ANY OTHER NATURE, ARE MADE WITH RESPECT TO THE INFORMATION PROVIDED IN THIS MSDS OR TO THE CHEMICAL TO WHICH INFORMATION MAY REFER. CRC NEITHER ASSUMES NOR AUTHORIZES ANY OTHER PERSON TO ASSUME FOR IT, ANY OTHER ADDITIONAL RESPONSIBILITY OR LIABILITY FOR THE USE OF, OR RELIANCE UPON, THIS INFORMATION.

HAZARD WARNING INFORMATION

HEALTH	FIRE	REACTIVE	OTHER	DEGREE OF HAZARD	COLOR CODING	OTHER CODES
2	3	1		0 = Minimum Hazard 1 = Slight Hazard 2 = Moderate Hazard 3 = Serious Hazard 4 = Severe Hazard	HEALTH = BLUE FIRE = RED REACTIVITY = YELLOW OTHER = WHITE	OX = Oxidizer ACID = Acid ALK = Alkali COR = Corrosive W = Use No Water

SECTION I - GENERAL INFORMATION

Characterization	**Metal**	RCRA Number	**None**
		EPA Class	**Not Applicable**
DOT Proper Shipping Name	**Hafnium Metal** (dry); **Hafnium Metal** (wet)	Chemical Abstract Service (CAS) Number	**7440-58-6**
DOT Hazard Class and Label Requirements	**Flammable Solid; Spontaneously Combustible**	DOT Emergency Guide Code	**40 (dry) 32 (wet)**
DOT Identification Number	**UN 2545** (powder, dry); **UN 1326** (powder, wet)	Chemical Formula	**Hf**

Synonyms
Celtium; elemental hafnium; hafnium metal.

SECTION II - HAZARDOUS INGREDIENTS AND IDENTITY INFORMATION

Hazardous Components (specific identity)	OSHA Exposure Criteria	NIOSH Exposure Criteria	Immediately Dangerous to Life and Health (IDLH)	ACGIH Exposure Criteria
Hafnium (derivation: Extremely difficult process to separate from zirconium either by solvent extraction of the thiocyanates by hexone, or solvent extraction of the nitrates by tributyl phosphate; or fractional crystallization of the double fluorides).	**PEL:** 0.5 mg/m^3 **STEL:** Not Established	**REL:** 0.5 mg/m^3 **STEL:** Not Established	50 mg/m^3 as Hf	**TLV:** 0.5 mg/m^3 **STEL:** Not Established

SECTION III - PHYSICAL AND CHEMICAL CHARACTERISTICS

Boiling Point	**8316°F (4602°C)**	Specific Gravity (H_2O = 1)	**13.31**
Vapor Pressure (mm Hg)	**0 (approx.) at 69°F (20°C)**	Molecular Weight	**178.5**
Vapor Density (Air = 1)	**Not Found**	Melting Point	**4041°F (2227°C)**

Solubility
Insoluble in water.

Appearance and Odor
Silvery, highly lustrous, ductile, metallic solid.

SECTION IV - FIRE AND EXPLOSION HAZARD DATA

Flash Point (method used)	**Not Applicable**	Explosive Limits in Air % by Volume	**LEL: Not Applicable UEL: Not Applicable**
NFPA Classification	**No Citation**	Autoignition Temperature	**Not Determined**

Extinguishing Media
Wet Hafnium: Use carbon dioxide, foam, water spray, or fog. Dry hafnium: Use dry chemical, soda ash, lime, or sand.

Special Fire Fighting Procedures
Wear full protective clothing and self-contained breathing apparatus (SCBA). Use a water spray to keep containers cool. For large fires, if possible, withdraw and allow to burn. Remain clear of smoke.

Unusual Fire and Explosion Hazards
The powder ignites spontaneously with exposure to heat, flames, sparks, or exposure to air. The damp powder burns explosively. The powder may self-explode. The powder may explode when heated with nitrogen, phosphorous, oxygen, sulfur, non-metals, or halogens. May explode on contact with hot nitric acid and other oxidants.

SECTION V - REACTIVITY DATA

Stability		Conditions to Avoid Hafnium can ignite spontaneously on exposure to air, static electricity, or friction. Keep away from incompatible materials. Can be kept stable in closed, air-tight containers.
Stable	Unstable X	Incompatibility *(materials to avoid)* Incompatible with strong oxidizers (such as chlorine), nitrogen, phosphorous, oxygen, sulfur, non-metals, halogens, nitric acid, and other oxidants.
Hazardous Polymerization		Conditions to Avoid Under normal conditions of temperature and pressure, hazardous polymerization of hafnium is not known to occur.
May Occur	Will Not Occur X	Hazardous Decomposition or By-products Toxic fumes and gases are released when hafnium is involved in fire.

SECTION VI - HEALTH HAZARD DATA

Primary Route(s) of Entry:	Inhalation?	Absorption (skin)?	Ingestion?
	X		X

Health Hazards

INHALATION: Irritation of the respiratory system and mucous membranes. Exposure can cause allergic lung reaction with chest pain, coughing, wheezing, and shortness of breath. May also cause liver damage.

SKIN & EYES: An irritant to the eyes and skin. The powder may cause contact dermatitis.

INGESTION: Unspecified gastrointestinal effects.

Carcinogenicity	NTP Listed?	IARC Cancer Review Group?	OSHA Regulated?	Target Organs?
Unknown Human Unknown Animal	No	No	29 CFR 1910.1000 Table Z-1	Respiratory system, skin, eyes, mucosa, liver.

Medical Conditions Generally Aggravated by Exposure

None Reported.

Emergency and First-aid Procedures

Eye contact: Flush immediately with water for 15 minutes (minimum), seek medical attention. Skin contact: Remove all contaminated clothing. Immediately wash area with large amounts of soap and water. If rash develops, seek medical attention. For inhalation: Remove the person from exposure. Provide respiratory assistance and CPR. Transfer to medical facility. If swallowed: Provide breathing assistance if necessary. Never attempt to give a convulsing or unconscious person anything by mouth. Seek medical attention immediately. Contact poison control center for assistance.

SECTION VII - PRECAUTIONS FOR SAFE HANDLING AND USE

Steps to be Taken in Case Material is Released or Spilled

Collect powdered material in most convenient manner possible and deposit in sealed drum. Keep those not involved in cleanup from entering area. Do NOT dry sweep (creates airborne dusts and friction). Carefully use a vacuum equipped with a high efficiency particulate air (HEPA) filter. Exercise extreme caution; hafnium can self-ignite or explode in air or if heated. Use non-sparking tools.

Preferred Waste Disposal Method

No Citation.

Precautions to be Taken in Handling and Storage

Store in tightly closed containers in cool, well-ventilated area and protect from physical damage. Sources of heat or ignition are prohibited where hafnium is used, handled, or stored.

Other Precautions and Warnings

A regulated, marked area should be established where hafnium is handled, used, or stored.

SECTION VIII - CONTROL MEASURES AND PERSONAL PROTECTIVE EQUIPMENT

Respiratory Protection *(specify type)*

For exposure to low levels over the PEL, use a full facepiece respirator with a high efficiency particulate air (HEPA) filter. For higher or prolonged exposures, use a full facepiece supplied-air respirator or a self-contained breathing apparatus (SCBA) operated in positive pressure mode.

Ventilation

Local exhaust or general mechanical systems recommended.

Protective Gloves	Eye Protection	Other Protective Clothing
Impervious Material	Chemical/Dust Goggles or Face Mask	Protective Suit or Clothing

Work/Hygiene Practices

Always wash hands thoroughly after using chemical; never bring food, drink, or smoking materials into vicinity of chemicals.

HAFNIUM

Hf

CAS: 7440-58-6

IDENTIFICATION AND TYPICAL USES

Hafnium is a steel-gray, shining, lustrous, ductile solid that appears in the form of powder or dusts. It is metallic element of atomic number 72, group IVB of the periodic table, atomic weight 178.49, valances 2, 3, 4, with 6 stable isotopes. Used in control rods in water-cooled nuclear reactors, in light bulb filaments, in electrodes, special glasses, and as a "getter" in vacuum tubes.

RISK ASSESSMENT: HEALTH

General Assessment

Hafnium is a human poison by an unspecified route. The exposure routes are *inhalation* and *ingestion*. However, since it is not significantly soluble in water, it is not absorbed efficiently by ingestion. There are no data on its carcinogenic, mutagenic, and teratogenic capabilities.

Inhalation of hafnium dusts and powders will cause irritation of the nose, throat, respiratory tract, and associated mucous membranes. It has been known to cause liver damage in test animals. Skin contact results in surface irritation and may lead to dermatitis. The powders can irritate the eyes. Ingestion causes unspecified gastrointestinal effects.

Acute Health Effects

The following acute (short-term) health effects may occur immediately or shortly after exposure to hafnium:

Skin: Irritation, rash, and possible contact dermatitis.

Eye: Hafnium particles can irritate the eyes and may cause damage to vision if not quickly removed.

Lung: Inhalation of the dusts and fumes can irritate the nose, mouth, throat, lungs, and mucosa of the respiratory tract.

Chronic Health Effects

The following chronic (long-term) health effects may occur at some time after exposure to hafnium and can last for months or even years:

Cancer Hazards: According to information presented in the references, hafnium has not been shown to cause cancer in test animals. Liver damage has occurred in test animals, but no correlation to human exposure has been documented.

Reproductive Hazard: According to information available in the references, hafnium has not been tested for its ability to adversely affect reproduction in humans.

Other Chronic Effects: Possible liver damage and skin allergy on long-term exposure to hafnium.

Recommended Risk-Reduction Measures

Personnel should avoid direct contact with hafnium. The exact nature of its human toxicity is not clearly understood in the references. If the specific hafnium compound in question is unknown, adequate risk management dictates that it be handled and treated as though it posed a serious risk to human health. If a less toxic chemical cannot be substituted for a hazardous substance, then *engineering controls* are the most effective method of reducing exposures.

The best protection is to enclose operations and/or provide local exhaust ventilation at the site of chemical release. While not always operationally feasible, isolating operations can also reduce exposure. Using respiratory protection is less effective than the controls mentioned above, but is still advisable whenever working with or around hafnium. For exposure to hafnium at levels above the OSHA PEL (0.5 mg/m^3), an MSHA/NIOSH-approved respirator with a high efficiency particulate air (HEPA) filter is recommended. For higher exposures, or exposures to unknown levels (as in an emergency), use a supplied-air respirator with a full facepiece operated in positive pressure mode or an MSHA/NIOSH-approved self-contained breathing apparatus (SCBA) with full facepiece and pressure demand. If a full facepiece is not available, then chemical dust goggles should be worn to protect the eyes. Whenever an airborne chemical dust hazard exists, a face shield and a rubber apron should be worn. To prevent hand and skin exposures, impervious rubber gloves should be used.

Administrative controls should also be in place to minimize the potential for human exposures. These may include written procedures or policies which specify the methods and techniques that will be practiced whenever personnel are to work with hafnium.

All personnel should receive *training* on the use, hazards, protective measures, emergency actions, and other precautions per 29 CFR 1910.1200 (Hazard Communication), prior to the first assignment in an area where hafnium is used or stored.

Before beginning employment and at regular intervals thereafter (e.g., annually), for those with frequent or potentially high exposures (half the PEL or greater, or significant skin contact), the following medical tests are recommended:

☑ Lung function tests.

☑ Chest X-ray.

If symptoms develop or overexposure is suspected, the following may also be useful:

☑ Liver and lung function tests.

☑ Evaluation by a qualified allergist, including careful exposure history and special testing (may help diagnose skin allergy).

Any evaluation should include a careful history of past and present symptoms with a medical examination. Medical tests that look for damage already done are not a substitute for controlling exposure. Also, since smoking can cause heart disease, as well as lung cancer, emphysema, and other related respiratory problems, the effects of exposure to hafnium may be more pronounced in smokers than in non-smokers. Prudent risk management requires proper consideration of *all* possible factors that may be causing exposure symptoms in the workplace.

Other methods to reduce exposure include:

☑ Where possible, enclose operations and use local exhaust ventilation at the site of chemical release. If local exhaust ventilation or enclosure is not used, respiratory protection should be mandatory.

☑ Always ensure that proper protective clothing is worn when using chemical substances and that personnel are adequately trained on the use, care, and maintenance of their personnel protective equipment.

☑ Wash thoroughly immediately after exposure to hafnium and at the end of the work shift or before eating, drinking, or smoking.

☑ Hazard warning information should be posted in the work area. In addition, as part of an on-going education and training program, all information on the health and safety hazards of hafnium should be communicated to all potentially exposed workers.

☑ Where possible, automatically transfer hafnium from drums or other storage containers to process containers. Avoid contact with air and do not generate airborne dusts.

RISK ASSESSMENT: ENVIRONMENT

General Assessment

The environment is at risk of exposure during transportation, storage, disposal, or destruction of hafnium. In almost every scenario, the threat of environmental exposure is contingent upon the proper handling of the chemical substance. Accidental spills, large or small, can result in fire, explosion, and possible contamination of the surrounding environmental mediums (water, soil, and air).

Hafnium powder will explode spontaneously in air. It will ignite and is potentially explosive in contact with heat, friction, and sparks, or when heated in the presence of nitrogen, phosphorous, oxygen, sulfur, nonmetals, or halogens. It may explode on contact with hot nitric acid and other oxidants. It is also incompatible with strong oxidizers (such as chlorine) and violent reactions can occur on contact. These characteristics require special consideration during any emergency situation involving a leak or spill of hafnium.

Hafnium can occur in the environment as a natural element in certain ores in the earth's crust. It can also enter the environment through industrial and municipal discharges, or from spills.

☠ *Acute Ecological Effects*

Acute (short-term) toxic effects may include the death of animals, birds, or fish and death or low growth rate in plants. Acute effects are seen 2 to 4 days after animals or plants are exposed to hafnium.

Insufficient data are available to evaluate the short-term effects of hafnium exposure to aquatic life, plants, birds, or land animals.

Chronic Ecological Effects

Chronic (long-term) toxic effects may include shortened life span, reproductive problems, lower fertility, and changes in appearance or behavior in exposed animals. These effects can be seen long after first exposure(s) to toxic chemicals.

Insufficient data are available to evaluate the long-term effects of hafnium exposure to aquatic life, plants, birds, or land animals.

Water Solubility

Elemental hafnium and most of its compounds are insoluble in water. Concentrations of less than 1 milligram will not mix with one liter of water.

Persistence in the Environment

Hafnium and its compounds are highly persistent in the aquatic environment, with a half-life of greater than 200 days. The half-life of a pollutant is the amount of time it takes for one half of the chemical to be degraded.

Bioaccumulation in Aquatic Organisms

Some substances increase in concentration, or *bioaccumulate,* in living organisms as they breathe contaminated air, drink contaminated water, or eat contaminated food. These chemicals can become concentrated in the tissues and internal organs of animals as well as humans.

The concentration of hafnium found in fish is expected to be somewhat higher than the average concentration of hafnium in the water from which the fish was taken.

Recommended Risk-Reduction Measures

Proper training of all transporters will reduce the likelihood of a mishap or accidents resulting in a leak or spill to the environment. The correct use of DOT labeling (per the MSDS) on all containers, trucks, and rail cars will enable emergency responders to react properly and quickly to any disaster thereby reducing the potential risk to the environment and to personnel.

Storage of hafnium should be segregated from incompatible chemicals to minimize the risk of cross-contamination or contact. Bulk storage of hafnium should be avoided. Conditions which can generate dusty conditions must be avoided. Storage facilities should be designed with proper fire suppression and protection equipment. Containers must be protected from exposure to heat and from physical damage.

If a spill or leak to the environment has occurred, fire department, emergency response and/or hazardous materials spill personnel should be notified immediately. Use a vacuum method to reduce dust during cleanup. Do not dry sweep (generates airborne dusts and friction). *Carefully* and *slowly* use a vacuum equipped with a high efficiency particulate air (HEPA) filter, not a standard shop vacuum. Deposit collected materials in sealed drum for disposal. Cleanup should be attempted only by those trained in proper spill containment procedures using non-sparking tools. Contaminated soils should be removed and replaced with clean soil. If hafnium should contact the water table, aquifer, or navigable waterway, time is of the essence. Hafnium is not soluble in water and, therefore, total containment and remediation may be possible. When such spills occur, the local and/or state emergency response authorities must be notified. A comprehensive emergency response or disaster preparedness/recovery plan should be in place prior to any operations involving the use, transportation, storage, or disposal of hafnium. If hafnium is spilled or leaked, the following specific steps are recommended:

- ☑ Restrict persons not wearing protective clothing from area of spill or leak until cleanup is complete and area can be opened for normal work.
- ☑ Collect powdered materials in most convenient and safe manner possible and place in sealed drums for disposal. Do not dry sweep, use HEPA vacuum instead.
- ☑ It may be necessary to dispose of hafnium as a hazardous waste. The responsible state agency or the regional office of the federal Environmental Protection Agency (EPA) should be contacted for specific recommendations.

RISK ASSESSMENT: BUSINESS

General Assessment

Accidents or mishaps involving hafnium can present a significant threat to business operations. The loss or damage of equipment or facilities can significantly affect fiscal viability. Lawsuits that may result from personnel injury/death, public exposures, and/or environmental contamination will also require a serious expenditure of resources. Media attention surrounding an injury, death, or environmental damage can also result in a loss of profits and loss of current as well as future business.

Recommended Risk-Reduction Measures

Company attorneys, safety and health professionals, and environmental specialists should be involved in the development of any procedures dealing with safety or emergency response. A company official should be pre-designated as a public relations officer with specific training in dealing with the press. Corporate plans and policies should be developed, approved, and implemented long before any need for such arises.

REFERENCES

Ahlbom, A. 1993. *Biostatistics for Epidemiologists.* New York: CRC Press/Lewis Publishers.

American Conference of Governmental Industrial Hygienists. 1988. *Documentation of the Threshold Limit Values and Biological Exposure Indices,* 5th Edition (with updates). Cincinnati: ACGIH

Asante-Duah, D. K. 1993. *Hazardous Waste Risk Assessment.* New York: CRC Press/Lewis Publishers.

Bloom, A. D. and F. J. DeSerres. 1995. *Ecotoxicity and Human Health.* New York: CRC Press/Lewis Publishers.

Brusick, D. J. 1994. *Methods for Genetic Risk Assessment.* New York: CRC Press/Lewis Publishers.

Calabrese, E. J. 1994. *Biological Effects of Low Level Exposures.* New York: CRC Press/Lewis Publishers.

Cockerman, L. G., B. S. Shane. 1994. *Basic Environmental Toxicology.* New York: CRC Press/Lewis Publishers.

Cothern, C. R., N. P. Ross. 1994. *Environmental Statistics, Assessment, and Forecasting.* New York: CRC Press/Lewis Publishers.

Danse, I. R. 1991. *Common Sense Toxics in the Workplace.* New York: Van Nostrand Reinhold.

Environmental Protection Agency. 1983. *Chemical Hazard Information Profiles.* Washington, D.C.: U.S. EPA.

Fodden, C. R., and J. L. Weddell. 1992. *Hazardous Materials Emergency Action Data.* New York: CRC Press/Lewis Publishers.

Forseberg, K., and L. H. Keith. 1993. *Chemical; Protective Clothing Permeation and Degradation Database.* Chelsea, MI: Lewis Publishers.

Gardner, R. W. 1994. *Chemical Intolerance.* New York: CRC Press/Lewis Publishers.

Howard, P. H., et. al. 1991. *Handbook of Environmental Degradation Rates.* New York: CRC Press/Lewis Publishers.

Hosty, J. W., and P. Foster. 1990. *A Practical Guide to Chemical Spill Response.* New York: Van Nostrand Reinhold.

Landis, W. G., M. H. Yu. 1995. *Introduction to Environmental Toxicology.* New York: CRC Press/Lewis Publishers.

Langerman, N. 1994 *Precautionary Labels for Chemical Containers.* New York: CRC Press/Lewis Publishers.

Lauwerys, R. R., P. Hoet. 1993. *Industrial Chemical Exposure.* New York: CRC Press/Lewis Publishers.

Lewis, .J., Sr. 1993. *Hawley's Condensed Chemical Dictionary,* 12th Edition. New York: Van Nostrand Reinhold.

Lewis, . J., Sr. 1992. *Sax's Dangerous Properties of Industrial Materials,* Volumes 1, 2, and 3, 8th Edition. New York: Van Nostrand Reinhold.

Mackay, D., W. Y. Shiu, and K. C. Ma. 1992. *Illustrated Handbook of Chemical-Physical Properties and Environmental Fate for Organic Chemicals.* New York: CRC Press/Lewis Publishers.

Manahan, S. E. 1992. *Toxicological Chemistry,* 2nd Edition. New York: CRC Press/Lewis Publishers.

Mellström, G. A., J. E. Wahlberg, and H. I. Maibach. 1994. *Protective Gloves for Occupational Use.* New York: CRC Press/Lewis Publishers.

Meyer, E. 1990. *Chemistry of Hazardous Materials.* Englewood Cliffs, NJ: Prentice Hall.

Miller, B. F., M.D., and C. B. Keane, R.N. 1972. *Encyclopedia and Dictionary of Medicine and Nursing.* Philadelphia: W.B. Sanders Company.

Mycock, J. C., J. D. McKenna, and L. Theodore. 1995. *Handbook of Air Pollution Control Engineering and Technology.* New York: CRC Press/Lewis Publishers.

Ott, W. R. 1995. *Environmental Statistics and Data Analysis.* New York: CRC Press/Lewis Publishers.

Patnaik, P. 1992. *A Comprehensive Guide to the Hazardous Properties of Chemical Substances.* New York: Van Nostrand Reinhold.

Philp, R. B. 1995. *Environmental Hazards and Human Health.* New York: CRC Press/Lewis Publishers.

Quigley, D. R. 1994. *Handbook of Emergency Chemical Management.* New York: CRC Press/Lewis Publishers.

Rea, W. J. 1995. *Chemical Sensitivity,* Vol. I-III. New York: CRC Press/Lewis Publishers.

Reeves, A. L. 1979. *Handbook on the Toxicology of Metals.* New York: Elsevier North-Holland.

Royal Society of Chemistry. 1992. *Chemical Safety Data Sheets, Volumes 1-5.* Newcastle-upon-Tyne, UK: Athenaeum Press.

Rylander, R., R. R. Jacobs. 1994. *Organic Dusts.* New York: CRC Press/Lewis Publishers.

U.S. Department of Health and Human Services (DHHS), National Institute for Occupational Health (NIOHS). 1994. *NIOSH Pocket Guide to Hazardous Chemicals.* Publication Number 94-116. Washington, D.C.; U.S. Government Printing Office.

MATERIAL SAFETY DATA SHEET

CHEMICAL NAME	
HALOTHANE	**CRC PRESS/LEWIS PUBLISHERS DISCLAIMER:** THE INFORMATION AND RECOMMENDATIONS PRESENTED HEREIN ARE BASED ON SOURCES BELIEVED TO BE RELIABLE. CRC MAKES NO REPRESENTATION ON ITS COMPLETENESS OR ACCURACY. IT IS THE RESPONSIBILITY OF THE USER TO DETERMINE THE CHEMICAL'S SUITABILITY FOR ITS INTENDED USE, THE CHEMICAL'S SAFE USE, AND THE CHEMICAL'S PROPER DISPOSAL. NO REPRESENTATIONS AND/OR WARRANTIES, EITHER EXPRESSED OR IMPLIED, OF THE MERCHANTABILITY OR FITNESS FOR A PARTICULAR PURPOSE, OR OF ANY OTHER NATURE, ARE MADE WITH RESPECT TO THE INFORMATION PROVIDED IN THIS MSDS OR TO THE CHEMICAL TO WHICH INFORMATION MAY REFER. CRC NEITHER ASSUMES NOR AUTHORIZES ANY OTHER PERSON TO ASSUME FOR IT, ANY OTHER ADDITIONAL RESPONSIBILITY OR LIABILITY FOR THE USE OF, OR RELIANCE UPON, THIS INFORMATION.

HAZARD WARNING INFORMATION

HEALTH	FIRE	REACTIVE	OTHER	DEGREE OF HAZARD	COLOR CODING	OTHER CODES
2	0	0		0 = Minimum Hazard 1 = Slight Hazard 2 = Moderate Hazard 3 = Serious Hazard 4 = Severe Hazard	HEALTH = BLUE FIRE = RED REACTIVITY = YELLOW OTHER = WHITE	OX = Oxidizer ACID = Acid ALK = Alkali COR = Corrosive W = Use No Water

SECTION I - GENERAL INFORMATION

Characterization	RCRA Number	EPA Class
Halogenated Hydrocarbon	**None**	**Not Applicable**

DOT Proper Shipping Name	**No Citation**	Chemical Abstract Service (CAS) Number	**151-67-7**
DOT Hazard Class and Label Requirements	**No Citation**	DOT Emergency Guide Code	**No Citation**
DOT Identification Number	**No Citation**	Molecular Formula	**$CF_3CHBrCl$**

Synonyms

1-Bromo-1-chloro-2,2,2-trifluoroethane; 2-bromo-2-chloro-1,1,1-trifluoroethane; 1,1,1-trifluoro-2-bromo-2-chloroethane; 2,2,2-trifluoro-1-bromo-1-chloroethane.

SECTION II - HAZARDOUS INGREDIENTS AND IDENTITY INFORMATION

Hazardous Components (specific identity)	OSHA Exposure Criteria	NIOSH Exposure Criteria	Immediately Dangerous to Life and Health (IDLH)	ACGIH Exposure Criteria
Halothane **1 ppm = 8.20 mg/m^3**	**PEL:** Not Established **STEL:** Not Established	**REL (ceiling):** 2 ppm 16.2 mg/m^3 (60-minute) **REL for waste anesthetic gas**	**Not Determined**	**TLV:** 50 ppm 404 mg/m^3 **STEL:** 50 ppm 404 mg/m^3

SECTION III - PHYSICAL AND CHEMICAL CHARACTERISTICS

Boiling Point	**122°F (50°C)**	Specific Gravity (H_2O = 1)	**1.87 at 20°F (4°C)**
Vapor Pressure (mm Hg)	**243 at 69°F (20°C)**	Molecular Weight (atomic weight)	**197.4**
Vapor Density (Air = 1)	**Not Found**	Freezing Point	**-180°F (-118°C)**

Solubility

Nearly insoluble in water (0.3%). Soluble in many organic solvents.

Appearance and Odor

Clear, colorless, highly volatile liquid with a sweet, pleasant odor.

SECTION IV - FIRE AND EXPLOSION HAZARD DATA

Flash Point (method used)	**Not Determined**	Explosive Limits in Air % by Volume	**LEL: Not Determined UEL: Not Determined**
NFPA Classification	**Non-Combustible Liquid**	Autoignition Temperature	**Not Determined**

Extinguishing Media

Use extinguishing media suitable to surrounding fires. Halothane does not burn.

Special Fire Fighting Procedures

Poisonous gases are produced in fire. Wear self-contained breathing apparatus (SCBA) and full protective gear. Move containers from fire if it can be done without risk. Cool sides of fire-exposed containers with water until long after fire is out. Monitor smoke directions and stay clear. Be aware of runoff from fire control measures. Do NOT release runoff to sewers or waterways. Dike for collection and disposal.

Unusual Fire and Explosion Hazards

None Reported.

SECTION V - REACTIVITY DATA

Stability		Conditions to Avoid **Halothane is normally stable in closed containers (reacts to light, becomes opaque), under routine conditions of storage and handling. Light-sensitive halothane is normally stabilized with 0.01% thymol.**
Stable X	Unstable	Incompatibility *(materials to avoid)* **Halothane may attack certain types of rubber and some plastics. It is also sensitive to light.**
Hazardous Polymerization		Conditions to Avoid **Under normal conditions of temperature and pressure, hazardous polymerization of halothane cannot occur.**
May Occur	Will Not Occur X	Hazardous Decomposition or By-products **Poisonous gases and acrid fumes are produced in fire, including extremely toxic fumes of chloride, fluoride, and bromide.**

SECTION VI - HEALTH HAZARD DATA

	Inhalation?	Absorption (skin/eye)?	Ingestion?
Primary Route(s) of Entry:	X		X

Health Hazards

INHALATION: Inhalation causes nausea, hepatitis, and fever. An intoxicant and central nervous system (CNS) depressant. Causes confusion, drowsiness, dizziness, analgesia, anesthesia, cardiac arrhythmia, liver and kidney damage, decreased hearing and visual performance, cyanosis, and (in animals) reproductive effects.

SKIN & EYES: A severe eye and skin irritant.

INGESTION: CNS symptoms similar to that of inhalation.

Carcinogenicity	NTP Listed?	IARC Cancer Review Group?	OSHA Regulated?	Target Organs?
Questioned Human Questioned Animal	No	Reviewed Inadequate Evidence	No	Eyes, skin, resp. sys. liver, kidney, CVS, CNS, repro.

Medical Conditions Generally Aggravated by Exposure

None reported. However, liver, kidney, heart, and respiratory problems may be aggravated by exposure.

Emergency and First-aid Procedures

Eye contact: Do not allow victim to keep eyes tightly shut. Flush immediately with water for 15 minutes (minimum), seek medical attention. Skin contact: Remove all contaminated clothing, including shoes. Immediately wash area with flooding amounts of warm water (104°F). Do NOT rub affected area or use dry heat. Seek medical attention immediately. For inhalation: Remove the person from exposure. Provide respiratory assistance and CPR. Transfer to medical facility. If swallowed: Seek medical attention immediately. Give conscious and alert person 1-2 glasses of water and induce vomiting until vomitus is clear. Never give an unconscious or convulsing person anything by mouth.

SECTION VII - PRECAUTIONS FOR SAFE HANDLING AND USE

Steps to be Taken in Case Material is Released or Spilled

Wear SCBA respiratory protection. Restrict those not wearing protective equipment and who are not involved in cleanup from entering area. Stop leak if it can be done without risk. Spilled liquids are likely to evaporate quickly. Absorb with vermiculite and place in sealed container.

Preferred Waste Disposal Method

Mix with a combustible solid and burn in chemical incinerator equipped with afterburner and scrubber.

Precautions to be Taken in Handling and Storage

Store in tightly closed containers in cool, dark, well-ventilated location. Do NOT use plastic containers.

Other Precautions and Warnings

Ensure adequate ventilation in storage areas. Avoid vapor inhalation and use appropriate personal protective equipment if necessary.

SECTION VIII - CONTROL MEASURES AND PERSONAL PROTECTIVE EQUIPMENT

Respiratory Protection *(specify type)*

Use an MSHA/NIOSH-approved air purifying respirator (cartridge type) or use a supplied-air respirator or a self-contained breathing apparatus (SCBA) with full facepiece operated in positive pressure mode.

Ventilation

Local exhaust (preferred) at site of chemical work or general ventilation.

Protective Gloves	Eye Protection	Other Protective Clothing
Impervious Gloves	Chemical Goggles or Face Mask	Protective Apron if Splash is Likely

Work/Hygiene Practices

Always wash hands thoroughly after using chemical; never bring food, drink, or smoking materials into vicinity of chemicals.

HALOTHANE

$CF_3CHBrCl$

CAS: 151-67-7

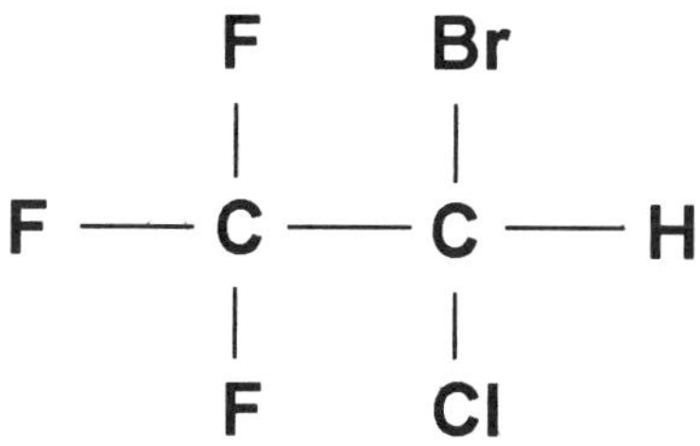

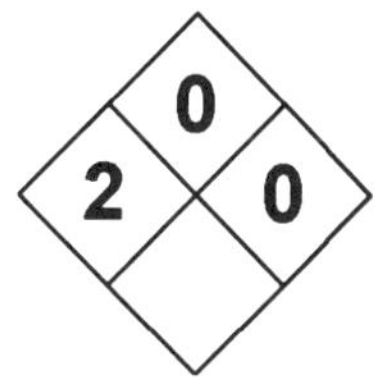

IDENTIFICATION AND TYPICAL USES

Halothane is a colorless, clear, highly volatile liquid with a sweet, pleasant odor. It is used as an inhalation anesthetic and may contain 0.01% thymol as a stabilizer.

RISK ASSESSMENT: HEALTH

General Assessment

Halothane is moderately toxic by *inhalation* and *ingestion.* If *injected* (intravenous administration) directly into the body, it can also produced serious and life-threatening effects. Although it has not been listed as a carcinogen, its carcinogenic properties have been questioned and the results are inconclusive. It is a known human mutagen (causes genetic changes) and has produced adverse reproductive and teratogenic effects in animal studies.

Inhalation causes nausea, vomiting, hepatitis, and fever. Other toxic effects by inhalation, ingestion, or injection include increased body temperature, cyanosis, central nervous system (CNS) depression with drowsiness, dizziness, excitement, confusion, amnesia, analgesia, anesthesia, and respiratory depression. There may also be kidney damage and liver dysfunction with either hepatitis or tissue necrosis. The cardiovascular system (CVS) effects may include potentially fatal heart disturbances and arrhythmia, depression, and dilated blood vessels (vasodilatation). There may also be adverse effects and decreased performance in the senses of hearing and sight. Although rare, halothane may be sensitized in some individuals. In its uses as an inhalation anesthetic in the health care industry, anesthetists appear to show a higher than normal frequency of hepatitis, allergies, intolerance to alcohol, and the use of sedative analgesics than the general population.

Acute Health Effects

The following acute (short-term) health effects may occur immediately or shortly after exposure to halothane:

Skin: Irritation with potential for redness.

Eye: Severe irritation, burning feeling, pain.

Lung: Nose, throat, and respiratory tract irritation.

CNS: Headaches, lightheadedness, dizziness, confusion, excitement, drowsiness, intoxication, anesthesia, amnesia, analgesia, respiratory depression.

CVS: Cardiac arrhythmia, depression of the cardiovascular system, vasodilatation, death.

Chronic Health Effects

The following chronic (long-term) health effects may occur at some time after exposure to halothane and can last for months or even years:

Cancer Hazards: Halothane has been tested by the IARC for its ability to cause cancer in test animals and the results were inconclusive. It has been shown to cause mutations (genetic changes in living cells) in humans. Some scientists believe that long-term exposure to mutagens may also pose a cancer risk. More research is required in this regard.

Reproductive Hazard: According to information presented in the references, halothane has shown to cause adverse reproductive and teratogenic effects in laboratory test animals. It is not known if human exposure will have this effect. There is some evidence to show that there is an increased incidence of spontaneous abortions and congenital abnormalities of offspring among operating room personnel.

Other Chronic Effects: Long-term exposure may lead to liver and kidney damage. Such damage has been demonstrated in animals; there has been a reported increase in hepatitis in operating room personnel. Also, this chemical may sensitize some individuals. Later re-exposure can cause a hypersensitive reaction leading to hepatitis, liver function failure, and death.

Ʋ *Recommended Risk-Reduction Measures*

Exposure to halothane presents a serious and even life-threatening risk to human health. Personnel should therefore avoid direct contact with halothane. Occupational poisoning occurs most commonly by inhalation. In most instances, a less toxic material or compound cannot be substituted for halothane in surgical and other health care applications. Therefore, *engineering controls* are the most effective method of reducing exposures. The best protection for workers is to ensure controlled administration of halothane and, wherever possible, enclose operations or provide adequate local exhaust ventilation at the site of halothane release. In surgical settings, the work area must be isolated from other areas for sterilization purposes. This will also reduce the likelihood of personnel exposure (only those in the room are at risk). Using respiratory protection is less effective than the controls mentioned above, but is still advisable whenever working with or around halothane. In the operating room, this is usually not feasible. But in manufacturing and production of halothane, respiratory protection is essential. Caution should be exercised when selecting respiratory protection since it is not known if exposure to halothane will cause cancer in humans. For exposure over the NIOSH ceiling (2 ppm), an MSHA/NIOSH-approved air-purifying respirator with organic vapor cartridge(s) should be used. For exposures to higher concentrations, or when the concentration is unknown, use a supplied-air respirator with full facepiece and mask operated in positive pressure mode or a self-contained breathing apparatus (SCBA) with full facepiece operated in pressure demand. If a full facepiece is not available, then chemical goggles should be worn to protect the eyes. A face shield should also be considered. To prevent hand and skin exposures impervious gloves should be used. Glove manufacturers should be contacted and permeation studies obtained *before* final glove selections are made. Halothane will attack some forms of rubber and plastic. This must also be considered when selecting personal protective equipment.

Administrative controls should also be in place to minimize the potential for human exposures. These may include written procedures or policies which specify the methods and techniques that will be practiced whenever personnel are to work with halothane.

All personnel should receive *training* on the use, hazards, protective measures, emergency actions, and other precautions per 29 CFR 1910.1200 (Hazard Communication), prior to the first assignment in an area where halothane is used or stored.

If symptoms develop or overexposure is suspected, the following recommended medical tests should be considered:

- ☑ Liver and kidney function tests.
- ☑ Special 24-hour electrocardiogram (EKG) using the Holter Monitor to evaluate for irregular heartbeat.
- ☑ Interview for brain effects, including recent changes in memory, mood (irritability, withdrawal), and concentration. Look for headaches, malaise, and altered sleeping patterns. Cerebellar, autonomic, and peripheral nervous system evaluation should be considered. Positive or borderline cases should be referred for further neuropsychological testing.

Any evaluation should include a careful history of past and present symptoms with an examination. Medical tests that simply look for existing damage are not a substitute for controlling exposures.

Other methods to reduce exposure include:

- ☑ Where possible, enclose operations and use local exhaust ventilation at the site of chemical release. If local exhaust ventilation or enclosure is not used, respiratory protection should be mandatory.
- ☑ Always ensure that proper protective clothing is worn when using chemical substances and that personnel are trained on its proper use, care, and maintenance.
- ☑ Wash thoroughly immediately after exposure to halothane and at the end of the work shift or before eating, drinking, or smoking.
- ☑ Hazard warning information should be posted in the work area. In addition, as part of an on-going education and training program, all information on the health and safety hazards of halothane should be communicated to all exposed workers.

- ☑ Eye wash stations should be provided in the immediate work area for emergency use. If there is a possibility of skin exposure to halothane, emergency shower facilities should also be provided in the immediate area.
- ☑ Workers whose clothing has been contaminated by halothane should change into clean clothes before leaving work. Contaminated work clothing should be laundered only by individuals who have been informed of the hazards of exposure to halothane.
- ☑ Where possible, automatically transfer liquids containing halothane from drums or other containers to process containers.

RISK ASSESSMENT: ENVIRONMENT

General Assessment

The environment is at risk of exposure during transportation, storage, disposal, or destruction of halothane. In almost every scenario, the threat of environmental exposure is contingent upon the proper handling of the chemical substance. Accidental spills, large or small, can result in contamination of the surrounding environmental mediums (water, soil, and air).

Halothane is non-combustible liquid. It has been known to attack some rubbers and plastics. If involved in fire, halothane can emit highly toxic fumes of chloride, fluoride, and bromide. Caution is always required in handling, storage, transportation, and disposal of halothane. Emergency responders should be made aware of the presence of halothane at any emergency response situation.

Halothane can enter the environment from industrial effluents and emissions, and from spills.

Acute Ecological Effects

Acute (short-term) toxic effects may include the death of animals, birds, or fish and death or low growth rate in plants. Acute effects are seen 2 to 4 days after animals or plants are exposed to halothane.

Insufficient data are available on the short-term effects of halothane on aquatic life, plants, birds, or land animals.

Chronic Ecological Effects

Chronic (long-term) toxic effects may include shortened life span, reproductive problems, lower fertility, and changes in appearance or behavior in exposed animals. These effects can be seen long after first exposure(s) to toxic chemicals.

Insufficient data are available to evaluate or predict the long-term effects of halothane to aquatic life, plants, birds, or terrestrial animals.

Water Solubility

Halothane is nearly insoluble in water. Concentrations between 1 and 100 milligrams may mix with a liter of water.

Persistence in the Environment

Halothane is non-persistent in water, with a half-life of less than 2 days. The half-life of a pollutant is the amount of time it takes for one half of the chemical to be degraded. About 99.8% of halothane will eventually end up in air; the remainder will end up in water.

Bioaccumulation in Aquatic Organisms

Some substances increase in concentration, or *bioaccumulate,* in living organisms as they breathe contaminated air, drink contaminated water, or eat contaminated food. These chemicals can become concentrated in the tissues and internal organs of animals as well as humans.

The concentration of halothane found in fish tissues is expected to be about the same as the average concentration of halothane in the water from which the fish was taken.

Recommended Risk-Reduction Measures

Proper training of all transporters will reduce the likelihood of a mishap or accident resulting in a leak or spill of halothane into the environment. Labels on all containers, trucks, and rail cars must meet DOT requirements and accurately reflect their contents to enable emergency responders to react properly and quickly to any disaster thereby reducing the potential risk to the environment and to personnel.

Storage of halothane should be segregated from other chemicals and materials (especially rubber and plastic) to minimize the risk of cross-contamination.

If a spill or leak to the environment has occurred, fire department, emergency response and/or hazardous materials spill personnel should be notified immediately. Cleanup should be attempted only by those trained in proper spill containment procedures. Contaminated soils should be removed for incineration and replaced with clean soil. If halothane should contact the water table, aquifer, or navigable waterway, time is of the essence. It is insoluble but highly volatile and total remediation may not be possible. The local and/or state emergency response authorities must be notified. A comprehensive emergency response or disaster preparedness/recovery plan should be in place prior to any operations involving the use, transportation, storage, or disposal of halothane.

For disposal, halothane may be mixed with a combustible solvent and burned in a chemical incinerator equipped with an afterburner and scrubber to remove the haloacids.

If halothane is spilled or leaked, the following specific steps are recommended:

- ☑ Restrict persons not wearing protective clothing from area of spill or leak until cleanup is complete and area can be opened for normal work.
- ☑ Absorb liquids in vermiculite, sand, earth, or similar material and deposit in sealed drum.
- ☑ Ventilate area of spill or leak.
- ☑ It may be necessary to dispose of halothane as a hazardous waste. The responsible state agency or the regional office of the federal Environmental Protection Agency (EPA) should be contacted for specific recommendations.

RISK ASSESSMENT: BUSINESS

General Assessment

Accidents or mishaps involving halothane can present a serious threat to business operations. The loss or damage of equipment or facilities can significantly affect fiscal viability. Lawsuits that may result from personnel illness, injury/death, public exposures, and/or environmental contamination will require a serious expenditure of resources. Media attention surrounding an injury, death, or environmental damage can also result in a loss of profits and loss of current as well as future business. Always remember that anytime the term "reproductive hazard" is used, public emotion, ignorance, and hysteria can run equally high. This must be carefully considered during the development of any public relations policies.

Recommended Risk-Reduction Measures

Company attorneys, safety and health professionals, and environmental specialists should be involved in the development of any procedures for responding to chemical incidents. A company official should be predesignated as a public relations officer with specific training in dealing with the press. Corporate plans and policies should be developed, approved, and implemented long before any need for such arises.

REFERENCES

American Conference of Governmental Industrial Hygienists. 1988. *Documentation of the Threshold Limit Values and Biological Exposure Indices,* 5th Edition (with updates). Cincinnati: ACGIH

Anthony, C. P., and N. J. Kolthoff. 1971. *Textbook of Anatomy and Physiology.* St. Louis: C. V. Mosby Company.

Bloom, A. D. and F. J. DeSerres. 1995. *Ecotoxicity and Human Health.* New York: CRC Press/Lewis Publishers.

Brusick, D. J. 1994. *Methods for Genetic Risk Assessment.* New York: CRC Press/Lewis Publishers.

Bueche, J. R. 1972. *Principles of Physics,* 2nd Edition. New York: McGraw-Hill.

Burrell, R., D. K. Flaherty, and L. J. Sauers. 1992. *Toxicology of the Immune System.* New York: Van Nostrand Reinhold.

Calabrese, E. J. 1994. *Biological Effects of Low Level Exposures.* New York: CRC Press/Lewis Publishers.

Cockerman, L. G., B. S. Shane. 1994. *Basic Environmental Toxicology.* New York: CRC Press/Lewis Publishers.

Cothern, C. R., N. P. Ross. 1994. *Environmental Statistics, Assessment, and Forecasting.* New York: CRC Press/Lewis Publishers.

Danse, I. R. 1991. *Common Sense Toxics in the Workplace.* New York: Van Nostrand Reinhold.

Environmental Protection Agency. 1983. *Chemical Hazard Information Profiles.* Washington, D.C.: U.S. EPA.

Gardner, R. W. 1994. *Chemical Intolerance.* New York: CRC Press/Lewis Publishers.

Hallenbeck, W. H. 1993. *Quantitative Risk Assessment for Environmental and Occupational Health,* 2nd Edition. New York: CRC Press/Lewis Publishers.

Lewis, R. J., Sr. 1993. *Hawley's Condensed Chemical Dictionary,* 12th Edition. New York: Van Nostrand Reinhold.

Lewis, R. J., Sr. 1992. *Sax's Dangerous Properties of Industrial Materials,* Volumes 1, 2, and 3, 8th Edition. New York: Van Nostrand Reinhold.

Meyer, E. 1990. *Chemistry of Hazardous Materials.* Englewood Cliffs, NJ: Prentice Hall.

Patnaik, P. 1992. *A Comprehensive Guide to the Hazardous Properties of Chemical Substances.* New York: Van Nostrand Reinhold.

Quigley, D. R. 1994. *Handbook of Emergency Chemical Management.* New York: CRC Press/Lewis Publishers.

U.S. Department of Health and Human Services, National Institute for Occupational Safety and Health. 1994. *NIOSH Pocket Guide to Chemical Hazards.* Washington, D.C.: U.S. Government Printing Office.

MATERIAL SAFETY DATA SHEET

CHEMICAL NAME	
HELIUM	**CRC PRESS/LEWIS PUBLISHERS DISCLAIMER:** THE INFORMATION AND RECOMMENDATIONS PRESENTED HEREIN ARE BASED ON SOURCES BELIEVED TO BE RELIABLE. CRC MAKES NO REPRESENTATION ON ITS COMPLETENESS OR ACCURACY. IT IS THE RESPONSIBILITY OF THE USER TO DETERMINE THE CHEMICAL'S SUITABILITY FOR ITS INTENDED USE, THE CHEMICAL'S SAFE USE, AND THE CHEMICAL'S PROPER DISPOSAL. NO REPRESENTATIONS AND/OR WARRANTIES, EITHER EXPRESSED OR IMPLIED, OF THE MERCHANTABILITY OR FITNESS FOR A PARTICULAR PURPOSE, OR OF ANY OTHER NATURE, ARE MADE WITH RESPECT TO THE INFORMATION PROVIDED IN THIS MSDS OR TO THE CHEMICAL TO WHICH INFORMATION MAY REFER. CRC NEITHER ASSUMES NOR AUTHORIZES ANY OTHER PERSON TO ASSUME FOR IT, ANY OTHER ADDITIONAL RESPONSIBILITY OR LIABILITY FOR THE USE OF, OR RELIANCE UPON, THIS INFORMATION.

HAZARD WARNING INFORMATION

HEALTH	FIRE	REACTIVE	OTHER	DEGREE OF HAZARD	COLOR CODING	OTHER CODES
2	0	0		0 = Minimum Hazard 1 = Slight Hazard 2 = Moderate Hazard 3 = Serious Hazard 4 = Severe Hazard	HEALTH = BLUE FIRE = RED REACTIVITY = YELLOW OTHER = WHITE	OX = Oxidizer ACID = Acid ALK = Alkali COR = Corrosive W = Use No Water

SECTION I - GENERAL INFORMATION

Characterization	RCRA Number	EPA Class
Gas	**None**	**Not Applicable**

DOT Proper Shipping Name **Helium (compressed); Helium (refrigerated liquid)**	Chemical Abstract Service (CAS) Number **7440-59-7**
DOT Hazard Class and Label Requirements **Non-Flammable Gas**	DOT Emergency Guide Code **15**
DOT Identification Number **UN1046 (compressed) UN1963 (refrigerated liquid)**	Atomic Formula **He**

Synonyms
None.

SECTION II - HAZARDOUS INGREDIENTS AND IDENTITY INFORMATION

Hazardous Components (specific identity)	OSHA Exposure Criteria	NIOSH Exposure Criteria	Immediately Dangerous to Life and Health (IDLH)	ACGIH Exposure Criteria
Helium (derivation: From natural gas, by liquefaction of all other components, followed by purification over activated charcoal).	**PEL:** Not Established **STEL:** Not Established	**REL:** Not Established **STEL:** Not Established	**Not Determined**	**TLV:** Not Established **STEL:** Not Established

SECTION III - PHYSICAL AND CHEMICAL CHARACTERISTICS

Boiling Point **-452°F (-269°C)**	Specific Gravity (H_2O = 1) **0.15 at -454°F (-270°C)**
Vapor Pressure (atmospheres) **2.61 (critical pressure)**	Atomic Weight **4.00**
Vapor Density (Air = 1) **0.138**	Freezing Point **-458°F (-272°C)**

Solubility
Very slightly soluble in water; insoluble in alcohol. Rate of diffusion through solids is three times that of air.

Appearance and Odor
Colorless, odorless, tasteless gas.

SECTION IV - FIRE AND EXPLOSION HAZARD DATA

Flash Point (method used) **Not Determined**	Explosive Limits in Air % by Volume **LEL: Not Determined UEL: Not Determined**
NFPA Classification **Non-Combustible Gas**	Autoignition Temperature **Not Determined**

Extinguishing Media
Use extinguishing media suitable to surrounding fires. Helium will not ignite.

Special Fire Fighting Procedures
Fire can produce toxic fumes: wear full protective clothing and self-contained breathing apparatus (SCBA). Contact with the liquid may cause frostbite. Heat will build pressure and may rupture closed storage containers. Keep fire-exposed containers cool with water spray. Containers should never reach temperatures of 125°F (52°C).

Unusual Fire and Explosion Hazards
Helium cylinders may explode in the heat of fire.

SECTION V - REACTIVITY DATA

Stability		Conditions to Avoid Helium is stable in closed cylinders under routine conditions of handling and storage. Keep away from heat and other gases. It can condense and solidify other gases. Use metal containers made for cryogenics.
Stable X	Unstable	Incompatibility *(materials to avoid)* Helium is a monatomic, chemically inactive, simple asphyxiant gas that is lighter than air. Some steels and other materials will become brittle in contact with the compressed gas or the refrigerated liquid.
Hazardous Polymerization		Conditions to Avoid Under normal conditions of temperature and pressure, hazardous polymerization of helium is not expected to occur.
May Occur	Will Not Occur X	Hazardous Decomposition or By-products None reported.

SECTION VI - HEALTH HAZARD DATA

Primary Route(s) of Entry:	Inhalation?	Absorption (skin)?	Ingestion?
	X		

Health Hazards

INHALATION: An asphyxiant gas. Initial symptoms are those of oxygen starvation. Feeling of "air hunger," diminished mental alertness, headache, dullness, shortness of breath, and an impairment in muscle coordination. Continued exposures lead to faulty judgment, depression of senses, rapid fatigue, unconsciousness, convulsions, coma, and death.

SKIN & EYES: Contact with the compressed gas or refrigerated liquid can cause frostbite and burns.

INGESTION: Not likely since helium is a gas at normal working (room) temperatures.

Carcinogenicity	NTP Listed?	IARC Cancer Review Group?	OSHA Regulated?	Target Organs?
No Evidence Human No Evidence Animal	No	No	No	Respiratory system; skin; eyes, CNS.

Medical Conditions Generally Aggravated by Exposure

None Reported.

Emergency and First-aid Procedures

Eye contact (liquid): Immediately flush with large amounts of tepid water for 15 minutes (minimum), occasionally lifting eyelids; seek medical attention. Skin contact (liquid): Remove all contaminated clothing. Immediately flush area with large amounts of tepid water. Do NOT rub! Seek medical assistance. For inhalation: Remove the person from exposure. Emergency responders must be protected from inert atmospheres. Provide respiratory assistance and CPR. Transfer promptly to medical facility. If swallowed: Treat frozen tissues and protect the airway. Seek medical attention immediately.

SECTION VII - PRECAUTIONS FOR SAFE HANDLING AND USE

Steps to be Taken in Case Material is Released or Spilled

Ventilate area. Stop flow of gas. If the source of leak is a cylinder and cannot be stopped, remove to safe place in open air; repair leak or let cylinder empty. Ensure response personnel are properly protected with self-contained breathing apparatus (SCBA).

Preferred Waste Disposal Method

No Citation.

Precautions to be Taken in Handling and Storage

Store in tightly closed containers in a cool, dry, well-ventilated area away from heat, direct sunlight, or other hot areas (water lines, vent pipes). Ensure proper storage and handling precautions are in place whenever working with compressed gas cylinders.

Other Precautions and Warnings

Containers may explode in fire or under conditions of extreme heat. Protect cylinders from physical damage. Use pressure relief on liquid containers. Periodically check vent valves for blockage (ice).

SECTION VIII - CONTROL MEASURES AND PERSONAL PROTECTIVE EQUIPMENT

Respiratory Protection *(specify type)*

No exposure limits have been established. Helium is a simple asphyxiant. Do NOT use cartridge-type respirators (ineffective for this type of gas). Use a powered supplied-air respirator or a self-contained breathing apparatus with full facepiece operated in pressure demand or other positive pressure mode.

Ventilation

Local exhaust or general mechanical systems recommended.

Protective Gloves	Eye Protection	Other Protective Clothing
Impervious, Thermal Insulated Gloves	Chemical Goggles and/or Face Mask	Protective Boots, Gauntlets, Apron

Work/Hygiene Practices

Always wash hands thoroughly after using chemical; never bring food, drink, or smoking materials into vicinity of chemicals.

HELIUM

He

CAS: 7440-59-7

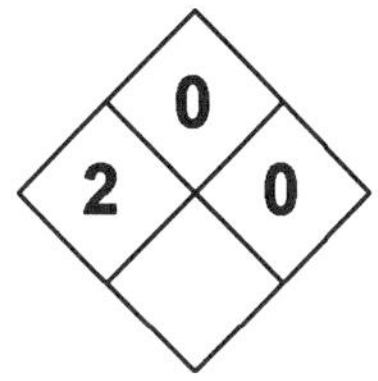

IDENTIFICATION AND TYPICAL USES

Helium is a colorless, odorless, and tasteless gas. It is used to pressurize rocket fuels, in welding, to create an inert atmosphere (in growing germanium and silicon crystals), inflating weather and research balloons, as a heat-transfer medium, in leak detection, in chromatography, in cryogenic research, in magnetohydrodynamics, in the manufacture of luminous signs, for geological dating, in aerodynamic research, in lasers (usually with neon), and in diving and space vehicle breathing equipment. It may also be used as a coolant for nuclear fusion power plants and in some electrical systems.

RISK ASSESSMENT: HEALTH

General Assessment

Helium is a simple asphyxiant gas. As such, its primary route of entry into the body is through *inhalation*. While it does not pose a significant physiological effect (i.e., no systemic poisoning or toxicity), it can displace oxygen levels to below that which is necessary to sustain life. Skin contact with the liquid can be destructive to tissues due to the extremely cold temperatures, but absorption is not a likely exposure route. Since helium usually exists as a gas at room temperatures, ingestion is not likely to occur either.

Inhalation of helium gas can cause headache, dullness, shortness of breath, and the symptoms of asphyxiation. These include rapid respiration and a feeling of "air hunger" or air starvation followed by a decrease in mental alertness, and impaired muscular coordination. If exposure continues, there can be nausea, vomiting, prostration, unconsciousness, coma, and death.

Skin or eye contact with the vapor poses no serious threat. The vapors have not been shown to cause any physiological response. However, if the liquid should contact either the skin or the eyes, there can be serious irritation, burns, tissue damage, and frostbite.

Acute Health Effects

The following acute (short-term) health effects may occur immediately or shortly after exposure to helium:

Skin: Liquid threat only. Redness, pain, swelling, blisters, and severe frostbite.

Eye: Liquid threat only. Causes frostbite, redness, irritation, severe pain, and blurred vision.

Lung: Vapor is non-irritating and therefore poses no threat of damage to the lungs. However, in high concentrations, helium can lead to symptoms of asphyxiation and oxygen deprivation up to and including death.

Chronic Health Effects

The following chronic (long-term) health effects may occur at some time after exposure to helium and can last for months or even years:

Cancer Hazards: According to information presented in the references, exposure to helium does not lead to carcinogenic effects (it is a simple asphyxiant).

Reproductive Hazards: There are no reports to support any claims of reproductive hazards.

Other Chronic Effects: None reported.

Recommended Risk-Reduction Measures

Personnel should avoid direct contact with helium. If another gas cannot be substituted for a helium, then *engineering controls* are the most effective method of reducing exposures. The best protection is to separate operations and provide local exhaust ventilation at the site of chemical release. While not always operationally feasible, isolating operations can also reduce exposure. Using respiratory protection is less effective than the controls mentioned above, but is still advisable whenever working with helium. Air-purifying respirators (e.g., the cartridge-type or gas masks) cannot be used with helium. These respirators only purify the air and can do nothing to protect in oxygen deficient atmospheres. A powered air-supplied respirator operated in pressure demand or other positive pressure mode, or a self-contained breathing apparatus (SCBA) with full facepiece operated in pressure demand mode are the recommended respiratory protection methods of choice. If a full facepiece is not available, then chemical goggles should be worn to protect the eyes. Whenever a chemical splash hazard exists, a face shield and a rubber or leather apron should be worn. To prevent hand and skin exposures with the liquid, thermal insulated gloves should be used.

Administrative controls should also be in place to minimize the potential for human exposures. These may include written procedures or policies which specify the methods and techniques that will be practiced whenever personnel are to work with helium.

All personnel should receive *training* on the use, hazards, protective measures, emergency actions, and other precautions per 29 CFR 1910.1200 (Hazard Communication), prior to the first assignment in an area where helium is used or stored.

Other methods to reduce exposure to chemicals include:

- ☑ Where possible, consider installing automatic sensors that warn personnel of dropping oxygen levels and use local exhaust ventilation at the site of chemical release. If local exhaust ventilation or enclosure is not used, respiratory protection should be mandatory.
- ☑ Always ensure that proper protective clothing is worn when using chemical substances.
- ☑ Safety shower and eyewash stations should be readily available in work areas where helium is used or stored.
- ☑ Hazard warning information should be posted in the work area. In addition, as part of an on-going education and training program, all information on the health and safety hazards of helium should be communicated to all potentially exposed workers.

RISK ASSESSMENT: ENVIRONMENT

General Assessment

The environment is at risk of exposure during transportation, storage, disposal, or destruction of helium. In almost every scenario, the threat of environmental exposure is contingent upon the proper handling of the chemical substance. Accidental releases, large or small, can result in the contamination of the surrounding environmental mediums (mostly air but, depending upon the spill, water and soil may be affected).

Helium is a non-flammable, non-combustible, non-explosive, and chemically inactive, inert gas. It presents no fire hazard. However, it can cause other gases to condense and solidify. It causes the condensation of oxygen from the air on non-insulated surfaces. Liquid oxygen is a fire hazard. Ice formation inside valves, vents, or critical components can be hazardous to equipment operation and to personnel. Also, because of its ability to displace oxygen, helium requires special consideration during any emergency involving a leak or release of helium gas or liquid especially in confined or enclosed spaces (such as sewers).

Helium can enter the environment through industrial discharges, unchecked venting to air, or through spills of the liquid. Since liquids will most likely vaporize rapidly, water and soil contamination are possible but improbable.

Acute Ecological Effects

Acute (short-term) toxic effects may include the death of animals, birds, or fish, and death or low growth rate in plants. Acute effects are seen 2 to 4 days after animals or plants are exposed to helium.

Insufficient data are available on the short-term effects of helium on aquatic life, plants, birds, or land animals. However, as a relatively non-toxic, simple asphyxiant, it might be assumed that any acute ecological toxicity will be relatively low.

Chronic Ecological Effects

Chronic (long-term) toxic effects may include shortened life span, reproductive problems, lower fertility, and changes in appearance or behavior in exposed animals. These effects can be seen long after first exposure(s) to toxic chemicals.

Insufficient data are available to evaluate the long-term effects of helium to aquatic life, plants, birds, or land animals. However, as a relatively non-toxic, simple asphyxiant, it might be assumed that any chronic ecological toxicity will be relatively low.

Water Solubility

Helium is only slightly soluble in water. Concentrations of 1 milligram or less may mix with a liter of water. It is normally a gas at room temperature and pressure and solubility may not be a concern.

⌛ *Persistence in the Environment*

Helium is non-persistent in water, with a half-life of less than 2 days. The half-life of a pollutant is the amount of time it takes for one half of the chemical to be degraded. Virtually all of helium will eventually end up in the air.

Bioaccumulation in Aquatic Organisms

Some substances increase in concentration, or *bioaccumulate,* in living organisms as they breathe contaminated air, drink contaminated water, or eat contaminated food. These chemicals can become concentrated in the tissues and internal organs of animals as well as humans.

Helium is not expected to accumulate in the edible tissues of fish.

Recommended Risk-Reduction Measures

Proper training of all transporters will reduce the likelihood of a mishap or accident resulting in a leak or release to the environment. The correct use of DOT labeling (per the MSDS) on all containers, trucks, and rail cars will enable emergency responders to react properly and quickly to any disaster thereby reducing the potential risk to the environment and to personnel.

Storage of helium should be segregated from other gases to prevent cross-contamination or contact. Buildings designated for storage should be equipped with monitors that provide ample warning (audible and visual) of oxygen deficient atmospheres.

If a spill or release to the environment has occurred, fire department, emergency response, and/or hazardous materials spill personnel should be notified immediately. Cleanup should be attempted only by those trained in proper containment procedures. If applicable, contaminated soils should be removed for incineration and replaced with clean soil. When spills or leaks occur, it may be necessary to contact the local and/or state emergency response authorities. A comprehensive emergency response or disaster preparedness/recovery plan should be in place prior to any operations involving the use, transportation, storage, or disposal of helium. If helium is spilled or leaked, the following specific steps are recommended:

- ☑ Restrict persons not wearing protective clothing from area of spill or leak until cleanup is complete and area can be opened for normal work.
- ☑ Ventilate area of spill or leak.
- ☑ If applicable, stop flow of leaking liquid or gas. If leak source is a cylinder and the leak cannot be stopped in place, remove leaking cylinder to a safe place in the open air, and repair or allow cylinder to empty.
- ☑ Keep helium out of a confined space, such as a sewer, because of the possibility of creating an inert atmosphere.
- ☑ It may be necessary to dispose of helium as a hazardous waste. The responsible state agency or the regional office of the federal Environmental Protection Agency (EPA) should be contacted for specific recommendations.

RISK ASSESSMENT: BUSINESS

General Assessment

Accidents or mishaps involving helium can present a significant threat to business operations. The loss or damage of equipment or facilities can significantly affect fiscal viability. Lawsuits that may result from personnel injury, public exposures, and environmental contamination will also require a serious expenditure of resources. Media attention surrounding an injury, illness, or environmental damage can also result in a loss of profits and loss of current as well as future business.

Recommended Risk-Reduction Measures

Company attorneys, safety and health professionals, and environmental specialists should be involved in the development of any procedures dealing with safety and emergency response. A company official should be pre-designated as a public relations officer with specific training in dealing with the press. Corporate plans and policies should be developed, approved, and implemented long before any need for such arises.

REFERENCES

Ahlbom, A. 1993. *Biostatistics for Epidemiologists.* New York: CRC Press/Lewis Publishers.

American Conference of Governmental Industrial Hygienists. 1988. *Documentation of the Threshold Limit Values and Biological Exposure Indices,* 5th Edition (with updates). Cincinnati: ACGIH

Anthony, C. P., and N. J. Kolthoff. 1971. *Textbook of Anatomy and Physiology.* St. Louis: C. V. Mosby Company.

Asante-Duah, D. K. 1993. *Hazardous Waste Risk Assessment.* New York: CRC Press/Lewis Publishers.

Bloom, A. D. and F. J. DeSerres. 1995. *Ecotoxicity and Human Health.* New York: CRC Press/Lewis Publishers.

Brusick, D. J. 1994. *Methods for Genetic Risk Assessment.* New York: CRC Press/Lewis Publishers.

Bueche, J. R. 1972. *Principles of Physics,* 2nd Edition. New York: McGraw-Hill.

Burrell, R., D. K. Flaherty, and L. J. Sauers. 1992. *Toxicology of the Immune System.* New York: Van Nostrand Reinhold.

Calabrese, E. J. 1994. *Biological Effects of Low Level Exposures.* New York: CRC Press/Lewis Publishers.

Cockerman, L. G., B. S. Shane. 1994. *Basic Environmental Toxicology.* New York: CRC Press/Lewis Publishers.

Cothern, C. R. 1995. *Handbook for Environmental Risk Decisionmaking.* New York: CRC Press/Lewis Publishers.

Danse, I. R. 1991. *Common Sense Toxics in the Workplace.* New York: Van Nostrand Reinhold.

Environmental Protection Agency. 1983. *Chemical Hazard Information Profiles.* Washington, D.C.: U.S. EPA.

Fodden, C. R., and J. L. Weddell. 1992. *Hazardous Materials Emergency Action Data.* New York: CRC Press/Lewis Publishers.

Forseberg, K., and L. H. Keith. 1993. *Chemical; Protective Clothing Permeation and Degradation Database.* Chelsea, MI: Lewis Publishers.

Gardner, R. W. 1994. *Chemical Intolerance.* New York: CRC Press/Lewis Publishers.

Hallenbeck, W. H. 1993. *Quantitative Risk Assessment for Environmental and Occupational Health,* 2nd Edition. New York: CRC Press/Lewis Publishers.

Hodgson, E., R. B. Mailman, and J. E. Chambers. 1988. *Dictionary of Toxicology.* London: The MacMillian Company.

Hosty, J. W., and P. Foster. 1990. *A Practical Guide to Chemical Spill Response.* New York: Van Nostrand Reinhold.

Howard, P. H. 1995. *Handbook of Environmental Fate and Exposures Data for Organic Chemicals,* Vol. I-IV. New York: CRC Press/Lewis Publishers.

James, A., and M. Lord. 1992. *Index of Chemical and Physical Data.* New York: Van Nostrand Reinhold.

Jones, F. E. 1994. *Toxic Organic Vapors in the Workplace.* New York: CRC Press/Lewis Publishers.

Kamrin, M. A. 1988. *Toxicology: A Primer on Principles and Applications.* Chelsea, Michigan: Lewis Publishers.

Keith, L. H. , M. T. Johnston, C. E. Hudak, and M. Conoley. 1992. *National Toxicology Program's Gloves+ Database.* Chelsea, MI: Lewis Publishers.

Keith, L. H. and M. M. Walker. 1995. *Handbook of Air Toxics.* New York: CRC Press/Lewis Publishers.

Keith, L. H. and D. B. Walters. (Editors) 1992. *National Toxicology Program's Chemical Database, Volumes I-VII.* Chelsea, MI: Lewis Publishers.

Landis, W. G., M. H. Yu. 1995. *Introduction to Environmental Toxicology.* New York: CRC Press/Lewis Publishers.

Langerman, N. 1994 *Precautionary Labels for Chemical Containers.* New York: CRC Press/Lewis Publishers.

Lauwerys, R. R., P. Hoet. 1993. *Industrial Chemical Exposure.* New York: CRC Press/Lewis Publishers.

Lewis, R. J., Sr. 1993. *Hawley's Condensed Chemical Dictionary,* 12th Edition. New York: Van Nostrand Reinhold.

Lewis, R. J., Sr. 1992. *Sax's Dangerous Properties of Industrial Materials,* Volumes 1, 2, and 3, 8th Edition. New York: Van Nostrand Reinhold.

Lide, D. R. 1995. *Handbook of Chemistry and Physics,* 76th Edition. New York: CRC Press/Lewis Publishers.

Manahan, S. E. 1994. *Environmental Chemistry,* 6th Edition. New York: CRC Press/Lewis Publishers.

Manahan, S. E. 1992. *Toxicological Chemistry,* 2nd Edition. New York: CRC Press/Lewis Publishers.

Mellström, G. A., J. E. Wahlberg, and H. I. Maibach. 1994. *Protective Gloves for Occupational Use.* New York: CRC Press/Lewis Publishers.

Meyer, E. 1990. *Chemistry of Hazardous Materials.* Englewood Cliffs, NJ: Prentice Hall.

Patnaik, P. 1992. *A Comprehensive Guide to the Hazardous Properties of Chemical Substances.* New York: Van Nostrand Reinhold.

Philp, R. B. 1995. *Environmental Hazards and Human Health.* New York: CRC Press/Lewis Publishers.

Proctor, N. H., and J. P. Hughes. 1991. *Chemical Hazards of the Workplace.* New York: Van Nostrand Reinhold.

Quigley, D. R. 1994. *Handbook of Emergency Chemical Management.* New York: CRC Press/Lewis Publishers.

Rea, W. J. 1995. *Chemical Sensitivity,* Vol. I-III. New York: CRC Press/Lewis Publishers.

Richardson, M. 1995. *Dictionary of Substance and Their Effects,* Vol. I-VI. New York: CRC Press/Lewis Publishers.

Sacarello, H. L. A. 1994. *The Comprehensive Handbook of Hazardous Materials.* New York: CRC Press/Lewis Publishers.

Sherman, J. D. 1988. *Chemical Exposure and Disease.* New York: Van Nostrand Reinhold.

Talbott, E., G. Craun. 1995. *An Introduction to Environmental Epidemiology.* New York: CRC Press/Lewis Publishers.

Torkelson, T. R. V. K. Rowe. 1981. *Patty's Industrial Hygiene and Toxicology,* Vol. 2B. New York: Wiley-Interscience.

U.S. Department of Health and Human Services, National Institute for Occupational Safety and Health. 1994. *NIOSH Pocket Guide to Chemical Hazards.* Washington, D.C.: U.S. Government Printing Office.

Witorsch, P. and S. V. Spagnolo. 1994. *Air Pollution and Lung Disease in Adults.* New York: CRC Press/Lewis Publishers.

MATERIAL SAFETY DATA SHEET

CHEMICAL NAME	CRC PRESS/LEWIS PUBLISHERS DISCLAIMER
HEPTACHLOR	

HAZARD WARNING INFORMATION

HEALTH	FIRE	REACTIVE	OTHER	DEGREE OF HAZARD	COLOR CODING	OTHER CODES
3	0	0		0 = Minimum Hazard 1 = Slight Hazard 2 = Moderate Hazard 3 = Serious Hazard 4 = Severe Hazard	HEALTH = BLUE FIRE = RED REACTIVITY = YELLOW OTHER = WHITE	OX = Oxidizer ACID = Acid ALK = Alkali COR = Corrosive W = Use No Water

SECTION I - GENERAL INFORMATION

Characterization	**Pesticide (polycyclic, chlorinated)**	RCRA Number	**P059**
		EPA Class	**Acute Hazardous Waste**
DOT Proper Shipping Name	**Organochlorine pesticides, solid, toxic, n.o.s.**	Chemical Abstract Service (CAS) Number	**76-44-8**
DOT Hazard Class and Label Requirements	**Poison**	DOT Emergency Guide Code	**55**
DOT Identification Number	**NA 2761**	Chemical Formula	**$C_{10}H_5Cl_7$**

Synonyms

1,4,5,6,7,8,8-Heptachloro-3a,4,7,7a-tetrahydo-4,7-methanoindene; Agroceres; basaklor; hepatachlorane; Hepta; hepatgranox; heptamul; heptsol; heptox; termide; solepax; 3-chlorochlordene.

SECTION II - HAZARDOUS INGREDIENTS AND IDENTITY INFORMATION

Hazardous Components (specific identity)	OSHA Exposure Criteria	NIOSH Exposure Criteria	Immediately Dangerous to Life and Health (IDLH)	ACGIH Exposure Criteria
Heptachlor (derivation: By chlorination of the adduct of hexachlorocyclopentadiene and cyclopentadiene using sulfonyl chloride in the presence of benzoyl peroxide or chlorine in the dark and in the presence of Fuller's earth).	**PEL:** 0.5 mg/m^3 (skin) **STEL:** Not Established	**REL:** 0.5 mg/m3 (skin) Possible Cancer Agent	35 mg/m^3	**TLV:** 0.05 mg/m^3 **STEL:** Not Established

SECTION III - PHYSICAL AND CHEMICAL CHARACTERISTICS

Boiling Point	**293°F (145°C)**	Specific Gravity (H_2O = 1)	**1.66**
Vapor Pressure (mm Hg)	**0.0003 (approx.) at 77°F (25°C)**	Molecular Weight (atomic weight)	**373.4**
Vapor Density (Air = 1)	**13.0**	Freezing Point	**203°F (95°C)**

Solubility

Insoluble in water (0.0006%). Soluble in hexane, petroleum distillates, xylene, alcohols, organic solvents.

Appearance and Odor

Pure heptachlor is a white, sand-like material with a camphor-like odor. Technical heptachlor is a white to light-tan crystalline, waxy, solid with a camphor-like odor.

SECTION IV - FIRE AND EXPLOSION HAZARD DATA

Flash Point (method used)	**Not Determined**	Explosive Limits in Air % by Volume	**LEL: Not Determined** **UEL: Not Determined**
NFPA Classification	**Non-Combustible Solid**	Autoignition Temperature	**Not Determined**

Extinguishing Media

Use extinguishing media suitable to surrounding fire.

Special Fire Fighting Procedures

Heptachlor is a non-combustible solid, but it is often dissolved in a flammable or combustible liquid. Wear full protective clothing and SCBA. Remain clear of smoke, water fallout and water runoff. Move containers from fire area if it can be done without risk. Cool exposed containers. Toxic hydrogen chloride and carbon monoxide are formed in fire.

Unusual Fire and Explosion Hazards

Containers may explode in fire.

SECTION V - REACTIVITY DATA

Stability		Conditions to Avoid
		Heptachlor is normally stable in closed containers under routine conditions of storage and handling. Avoid contact with incompatible materials and do not store near heat or heat source.
Stable	Unstable	Incompatibility *(materials to avoid)*
X		Incompatible with alkali metals. Corrosive to most metals. It can react with iron and its oxide to form hydrogen chloride gas. It is oxidized by chromic trioxide in acetic acid to produce heptachlor epoxide.
Hazardous Polymerization		Conditions to Avoid Under normal conditions of temperature and pressure, hazardous polymerization of heptachlor is not known to occur.
May Occur	Will Not Occur	Hazardous Decomposition or By-products
	X	At high temperatures, heptachlor decomposes and may produce poisonous gases, including toxic hydrogen chloride, carbonyl chloride, and carbon monoxide.

SECTION VI - HEALTH HAZARD DATA

	Inhalation?	Absorption (skin/eye)?	Ingestion?
Primary Route(s) of Entry:	X	X	X

Health Hazards

INHALATION: Breathing vapors can cause convulsions, unconsciousness, and death. Can irritate the nose, eyes, mouth, and throat. Long-term exposure can damage the liver.

ABSORPTION: Will pass through unbroken skin and enter the bloodstream. Contact may severely irritate the skin and eyes.

INGESTION: Poisonous to humans by ingestion. Systemic effects due to central nervous system action include tremors, convulsion, excitement, ataxia, and gastritis.

Carcinogenicity	NTP Listed?	IARC Cancer Review Group?	OSHA Regulated?	Target Organs?
Suspected Human Confirmed Animal	No	Group 3	29 CFR 1910.1000 Table Z-1	Central nervous system; liver.

Medical Conditions Generally Aggravated by Exposure

Kidney, liver, and lung diseases, and convulsive disorders may be aggravated by exposure to heptachlor.

Emergency and First-aid Procedures

Eye contact: Do not allow victim to keep eyes tightly shut. Flush immediately with water for 15 minutes; seek medical attention. Skin contact: Remove all contaminated clothing. Immediately wash area with flooding amounts of water, then triple-wash with soap and water. Inhalation: Remove the person from exposure. Provide respiratory assistance and CPR. Transfer to medical facility. If swallowed: If victim is conscious, do NOT induce vomiting (CNS depression may induce seizures). Seek medical attention immediately. Never give anything by mouth to an unconscious or convulsing person. DOCTOR: Suggest gastric lavage and use of activated charcoal and sodium sulfate. Do not give epinephrine since it may induce ventricular fibrillation. Enforce complete rest. Monitor CBC, liver, and kidney function (urinalysis).

SECTION VII - PRECAUTIONS FOR SAFE HANDLING AND USE

Steps to be Taken in Case Material is Released or Spilled

Remove all sources of heat. Ventilate area of spill. Restrict those not involved in cleanup from entering area. Notify appropriate authorities. Collect powder materials using HEPA vacuum. Do NOT dry sweep. Absorb any liquid solution spills in vermiculite and place in sealed drum for disposal.

Preferred Waste Disposal Method

No Citation.

Precautions to be Taken in Handling and Storage

Store in tightly closed containers in cool, dry, dark, well-ventilated area away from heat.

Other Precautions and Warnings

Avoid generating dusty conditions in storage areas. Protect containers from physical damage.

SECTION VIII - CONTROL MEASURES AND PERSONAL PROTECTIVE EQUIPMENT

Respiratory Protection *(specify type)*

For exposures above 0.5 mg/m^3, use a NIOSH-approved supplied-air respirator or a self-contained breathing apparatus (SCBA) with full facepiece operated in positive pressure mode.

Ventilation

Local exhaust (preferred) at site of chemical work or general ventilation.

Protective Gloves	Eye Protection	Other Protective Clothing
Rubber	Chemical Goggles or Face Mask	Protective Apron, Boots (avoid skin contact)

Work/Hygiene Practices

Always wash hands thoroughly after using chemical; never bring food, drink, or smoking materials into vicinity of chemicals.

HEPTACHLOR

$C_{10}H_5Cl_7$

CAS: 76-44-8

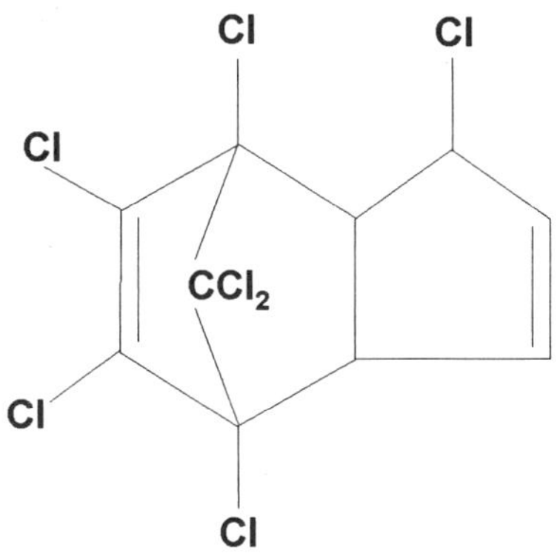

IDENTIFICATION AND TYPICAL USES

Pure heptachlor is a white, sand-like material with a camphor-like odor. Technical heptachlor is a white to light-tan crystalline, waxy, solid with a camphor-like odor. It was used for many years as an insecticide for seed treatment and in the control of certain insects that attack corn and other field-type crops. Since 1983, the only permitted use has been for the control of termites in subsurface soil treatments and for dipping non-food plants.

RISK ASSESSMENT: HEALTH

General Assessment

Heptachlor is highly toxic to humans by *ingestion*. It is moderately toxic by *inhalation* and skin *absorption*. It is a suspected human carcinogen with experimental carcinogenic data reported. Human mutation data have been reported. There is some evidence in the references that indicates heptachlor is a suspected human carcinogen with long-term effects primarily in the liver.

Heptachlor is a central nervous system stimulant. Symptoms of exposure by all routes include nausea, vomiting, abdominal pain, irritation, confusion, apprehension, excitability, paresthesia, dizziness, headache, disorientation, tremor, ataxia (loss of muscle coordination), stupor, coma, convulsions, respiratory collapse, and possibly death.

Skin contact will result in moderate to severe irritation with rash or burning sensation. Eye contact may also result in irritation. Caution is warranted when both inhalation and skin contact exposures are possible. Even though air levels may be within limits, personnel may be overexposed if heptachlor should contact the skin. The dangerous chronic dose in humans is unknown. May cause liver damage (fatty degeneration), characteristic of chronic alcoholism.

Acute Health Effects

The following acute (short-term) health effects may occur immediately or shortly after exposure to heptachlor:

Skin: Liquid can pass through unbroken skin. Contact may cause irritation and rash. Symptoms of exposure include headache, nausea, abdominal pain, and vomiting.

Eye: Severe irritation and possible damage on contact.

Lung: Irritation of the nose and throat. Breathing vapors can cause cough, convulsions, unconsciousness, and death.

CNS: A central nervous system stimulant. Symptoms include ataxia, excitement, tremors, convulsions, confusion, delirium, and irritability.

Chronic Health Effects

The following chronic (long-term) health effects may occur at some time after exposure to heptachlor and can last for months or even years:

Cancer Hazards: Heptachlor may be a carcinogen in humans since it has been shown to cause liver cancer in animals. There is no evidence in the references to confirm the human carcinogenicity of heptachlor.

Reproductive Hazard: There is no evidence that exposure to heptachlor can cause damage to the developing fetus in test animals. More research is required.

Other Chronic Effects: Long-term exposure to heptachlor may damage the liver and kidneys.

❶ *Recommended Risk-Reduction Measures*

Personnel should avoid direct contact with heptachlor. If a less toxic material or compound cannot be substituted for heptachlor, then *engineering controls* are the most effective method of reducing exposures. The best protection is to enclose operations and/or provide local exhaust ventilation at the site of heptachlor dust release. This may be practical in manufacturing operations, but is not feasible in application since it is used outdoors. While not always operationally possible, isolating operations involving heptachlor manufacturing can also reduce exposure. Using respiratory protection is less effective than the controls mentioned above, but is still advisable whenever working with or around heptachlor. At exposures above the PEL of 5 mg/m^3 (TWA), use an MSHA/NIOSH-approved supplied-air respirator or a self-contained breathing apparatus (SCBA) with full facepiece operated in pressure demand or other positive pressure mode for the greatest possible respiratory protection. If a full facepiece is not available, then chemical goggles should be worn to protect the eyes. A face shield and protective apron should also be worn. To prevent hand and skin exposures, impervious gloves should be worn. Glove manufacturers should be contacted and permeation studies obtained *before* final glove selections have been made.

Administrative controls should also be in place to minimize the potential for human exposures. These may include written procedures or policies which specify the methods and techniques that will be practiced whenever personnel are to work with heptachlor.

All personnel should receive *training* on the use, hazards, protective measures, emergency actions, and other precautions per 29 CFR 1910.1200 (Hazard Communication), prior to the first assignment in an area where heptachlor is used or stored.

If symptoms develop or overexposure is suspected, the following medical testing may be useful:

- ☑ Liver and kidney function tests.

Any evaluation should include a careful history of past and present symptoms with an examination. Medical tests that simply look for existing damage are not a substitute for controlling exposures.

Other methods to reduce exposure include:

- ☑ Where possible, enclose operations and use local exhaust ventilation at the site of chemical release. If local exhaust ventilation or enclosure is not used, respiratory protection should be mandatory.
- ☑ Always ensure that proper protective clothing is worn when using chemical substances.
- ☑ Wash thoroughly immediately after exposure to heptachlor and at the end of the work shift or before eating, drinking, or smoking.
- ☑ Hazard warning information should be posted in the work area. In addition, as part of an on-going education and training program, all information on the health and safety hazards of heptachlor should be communicated to all exposed and potentially exposed workers.
- ☑ Eye wash stations should be provided in the immediate work area for emergency use. If there is a possibility of skin exposure to heptachlor, emergency shower facilities should also be provided.
- ☑ Workers whose clothing has been contaminated by heptachlor should change into clean clothes before leaving work. Contaminated work clothing should be laundered only by individuals who have been informed of the hazards of exposure to heptachlor.
- ☑ Where possible, automatically pump liquid heptachlor from drums or other storage containers to process containers.
- ☑ Specific engineering controls are recommended by NIOSH. Refer to NIOSH Criteria Document: Occupational Exposure During the Manufacture and Formulation of Pesticides (Publication Number 78-174).

RISK ASSESSMENT: ENVIRONMENT

General Assessment

Because of its use as an insecticide, heptachlor may be present in the environment at any given time and at a variety of levels. The environment is also at risk of exposure during transportation, storage, disposal, or destruction of heptachlor. In almost every scenario, the threat of environmental exposure is contingent upon the proper handling of the chemical substance. Accidental spills, large or small, where heptachlor contacts incompatible commodities can result in violent and even explosive reactions (depending upon conditions of spill), and possible contamination of the surrounding environmental mediums (water, soil, and air).

Heptachlor is considered a non-combustible solid. It may also be present (dissolved) in some other flammable liquids such as benzene. Also, because it is incompatible with many common alkali metals, such as iron and its oxide, extreme caution is required in handling, storage, transportation, and disposal of heptachlor. These characteristics also require special consideration during any emergency situation involving a leak or spill of heptachlor powder or dust.

Heptachlor is an insecticide and, because of its once widespread, general usage in the past in areas where termites are treated, heptachlor enters the environment in agricultural and residential runoff.

Acute Ecological Effects

Acute (short-term) toxic effects may include the death of animals, birds, or fish and death or low growth rate in plants. Acute effects are seen 2 to 4 days after animals or plants are exposed to heptachlor.

Heptachlor has high acute toxicity to aquatic life. Heptachlor has caused injuries to agricultural crops and ornamental tree shrub species. Insufficient data are available to evaluate or predict on the short-term effects of heptachlor to birds and land animals.

Chronic Ecological Effects

Chronic (long-term) toxic effects may include shortened life span, reproductive problems, lower fertility, and changes in appearance or behavior in exposed animals. These effects can be seen long after first exposure(s) to toxic chemicals.

Heptachlor has high chronic toxicity to aquatic life. Insufficient data are available to evaluate or predict the long-term effects of heptachlor to plants, birds, or land animals.

Water Solubility

Heptachlor is very slightly soluble in water. Concentrations of 1 to 1000 milligrams will mix with a liter of water.

Persistence in the Environment

Heptachlor is highly persistent in the water, with a half-life of greater than 200 days. The half-life of a pollutant is the amount of time it takes for one-half of the chemical to be degraded. About 44.5% of heptachlor will eventually end up in terrestrial soil; approximately 41.5% will end up and aquatic sediments; the remainder will end up in water.

Bioaccumulation in Aquatic Organisms

Some substances increase in concentration, or *bioaccumulate,* in living organisms as they breathe contaminated air, drink contaminated water, or eat contaminated food. These chemicals can become concentrated in the tissues and internal organs of animals as well as humans.

The concentration of heptachlor found in fish tissues is expected to be considerably higher that the average concentration of heptachlor in the water from which the fish was taken.

Recommended Risk-Reduction Measures

Proper training of all transporters will reduce the likelihood of a mishap or accident resulting in a leak or spill of heptachlor into the environment. The correct use of DOT labeling (per the MSDS) on all containers, trucks, and rail cars will enable emergency responders to react properly and quickly to any disaster thereby reducing the potential risk to the environment and to personnel.

Storage of heptachlor should be segregated from incompatible materials to minimize the risk of cross-contamination or contact. Buildings or storage areas should be equipped with appropriate fire protection systems (alarms, sprinklers). Proper ventilation and protective equipment should be used while handing the powders or while preparing aqueous solutions.

If a spill or leak to the environment has occurred, fire department, emergency response and/or hazardous materials spill personnel should be notified immediately. Cleanup should be attempted only by those trained in proper spill containment procedures. If heptachlor should contact the water table, aquifer, or navigable waterway, reclamation procedures should be initiated as soon as practical. It is only slightly soluble in water and partial or total remediation may be possible. The local and/or state emergency response authorities must be notified. A comprehensive emergency response or disaster preparedness/recovery plan should be in place prior to any operations involving the use, transportation, storage, or disposal of heptachlor. If heptachlor is spilled or leaked, the following specific steps are recommended:

☑ Restrict persons not wearing protective clothing from area of spill or leak until cleanup is complete and area can be opened for normal work.

☑ Ventilate are of spill or leak.

☑ Collect powdered materials using HEPA vacuum. Do NOT dry sweep. Absorb solution spills in vermiculite or other material and place in sealed drum for disposal or reclamation.

☑ It may be necessary to dispose of heptachlor as a hazardous waste. The responsible state agency or the regional office of the federal Environmental Protection Agency (EPA) should be contacted for specific recommendations.

RISK ASSESSMENT: BUSINESS

General Assessment

Accidents or mishaps involving heptachlor can present a significant threat to business operations. The loss or damage of equipment or facilities can significantly affect fiscal viability. Lawsuits that may result from personnel injury/death, public exposures, and/or environmental contamination will also require a serious expenditure of resources. Media attention surrounding an injury, death, or environmental damage can also result in a loss of profits and loss of current as well as future business. Always remember that anytime to terms "cancer" or "carcinogen" are used, public emotion, hysteria, and ignorance can run equally high. This must be carefully considered whenever developing or implementing public relations policies.

Recommended Risk-Reduction Measures

Company attorneys, safety and health professionals, and environmental specialists should be involved in the development of any procedures for responding to chemical incidents. A company official should be predesignated as a public relations officer with specific training in dealing with the press. Corporate plans and policies should be developed, approved, and implemented long before any need for such arises.

REFERENCES

American Conference of Governmental Industrial Hygienists. 1988. *Documentation of the Threshold Limit Values and Biological Exposure Indices,* 5th Edition (with updates). Cincinnati: ACGIH

Brusick, D. J. 1994. *Methods for Genetic Risk Assessment.* New York: CRC Press/Lewis Publishers.

Bloom, A. D. and F. J. DeSerres. 1995. *Ecotoxicity and Human Health.* New York: CRC Press/Lewis Publishers.

Calabrese, E. J. 1994. *Biological Effects of Low Level Exposures.* New York: CRC Press/Lewis Publishers.

Cockerman, L. G., B. S. Shane. 1994. *Basic Environmental Toxicology.* New York: CRC Press/Lewis Publishers.

Cothern, C. R. 1995. *Handbook for Environmental Risk Decisionmaking.* New York: CRC Press/Lewis Publishers.

Gardner, R. W. 1994. *Chemical Intolerance.* New York: CRC Press/Lewis Publishers.

Howard, P. H., et. al. 1991. *Handbook of Environmental Degradation Rates.* New York: CRC Press/Lewis Publishers.

Keith, L. H. and M. M. Walker. 1995. *Handbook of Air Toxics.* New York: CRC Press/Lewis Publishers.

Landis, W. G., M. H. Yu. 1995. *Introduction to Environmental Toxicology.* New York: CRC Press/Lewis Publishers.

Lauwerys, R. R., P. Hoet. 1993. *Industrial Chemical Exposure.* New York: CRC Press/Lewis Publishers.

Lewis, R. J., Sr. 1992. *Sax's Dangerous Properties of Industrial Materials,* Volumes 1, 2, and 3, 8th Edition. New York: Van Nostrand Reinhold.

Lide, D. R. 1995. *Handbook of Chemistry and Physics,* 76th Edition. New York: CRC Press/Lewis Publishers.

Manahan, S. E. 1992. *Toxicological Chemistry,* 2nd Edition. New York: CRC Press/Lewis Publishers.

Mellström, G. A., J. E. Wahlberg, and H. I. Maibach. 1994. *Protective Gloves for Occupational Use.* New York: CRC Press/Lewis Publishers.

Meyer, E. 1990. *Chemistry of Hazardous Materials.* Englewood Cliffs, NJ: Prentice Hall.

Mycock, J. C., J. D. McKenna, and L. Theodore. 1995. *Handbook of Air Pollution Control Engineering and Technology.* New York: CRC Press/Lewis Publishers.

Nelson, G. O. 1992. *Gas Mixtures.* New York: CRC Press/Lewis Publishers.

Ott, W. R. 1995. *Environmental Statistics and Data Analysis.* New York: CRC Press/Lewis Publishers.

Patnaik, P. 1992. *A Comprehensive Guide to the Hazardous Properties of Chemical Substances.* New York: Van Nostrand Reinhold.

Quigley, D. R. 1994. *Handbook of Emergency Chemical Management.* New York: CRC Press/Lewis Publishers.

Richardson, M. 1995. *Dictionary of Substance and Their Effects,* Vol. I-VI. New York: CRC Press/Lewis Publishers.

Sacarello, H. L. A. 1994. *The Comprehensive Handbook of Hazardous Materials.* New York: CRC Press/Lewis Publishers.

Torkelson, T. R. V. K. Rowe. 1981. *Patty's Industrial Hygiene and Toxicology,* Vol. 2B. New York: Wiley-Interscience.

U.S. Department of Health and Human Services, National Institute for Occupational Health (NIOSH). 1978. *NIOSH Criteria Document: Occupational Exposure During the Manufacture and Formulation of Pesticides.* Publication Number 78-174. Washington, DC: U.S. Government Printing Office.

MATERIAL SAFETY DATA SHEET

CHEMICAL NAME	
HEPTANE	**CRC PRESS/LEWIS PUBLISHERS DISCLAIMER:** THE INFORMATION AND RECOMMENDATIONS PRESENTED HEREIN ARE BASED ON SOURCES BELIEVED TO BE RELIABLE. CRC MAKES NO REPRESENTATION ON ITS COMPLETENESS OR ACCURACY. IT IS THE RESPONSIBILITY OF THE USER TO DETERMINE THE CHEMICAL'S SUITABILITY FOR ITS INTENDED USE, THE CHEMICAL'S SAFE USE, AND THE CHEMICAL'S PROPER DISPOSAL. NO REPRESENTATIONS AND/OR WARRANTIES, EITHER EXPRESSED OR IMPLIED, OF THE MERCHANTABILITY OR FITNESS FOR A PARTICULAR PURPOSE, OR OF ANY OTHER NATURE, ARE MADE WITH RESPECT TO THE INFORMATION PROVIDED IN THIS MSDS OR TO THE CHEMICAL TO WHICH INFORMATION MAY REFER. CRC NEITHER ASSUMES NOR AUTHORIZES ANY OTHER PERSON TO ASSUME FOR IT, ANY OTHER ADDITIONAL RESPONSIBILITY OR LIABILITY FOR THE USE OF, OR RELIANCE UPON, THIS INFORMATION.

HAZARD WARNING INFORMATION

HEALTH	FIRE	REACTIVE	OTHER	DEGREE OF HAZARD	COLOR CODING	OTHER CODES
1	3	0		0 = Minimum Hazard 1 = Slight Hazard 2 = Moderate Hazard 3 = Serious Hazard 4 = Severe Hazard	HEALTH = BLUE FIRE = RED REACTIVITY = YELLOW OTHER = WHITE	OX = Oxidizer ACID = Acid ALK = Alkali COR = Corrosive W = Use No Water

SECTION I - GENERAL INFORMATION

Characterization	RCRA Number	EPA Class
Hydrocarbon, Aliphatic	**D001**	**Characteristic (I) Waste**

DOT Proper Shipping Name **Heptane**	Chemical Abstract Service (CAS) Number **142-82-5**
DOT Hazard Class and Label Requirements **Flammable Liquid**	DOT Emergency Guide Code **27**
DOT Identification Number **UN 1206**	Chemical Formula **$CH_3(CH_2)_5CH_3$**

Synonyms
***n*-Heptane; normal-heptane; dipropylmethane.**

SECTION II - HAZARDOUS INGREDIENTS AND IDENTITY INFORMATION

Hazardous Components (specific identity)	OSHA Exposure Criteria	NIOSH Exposure Criteria	Immediately Dangerous to Life and Health (IDLH)	ACGIH Exposure Criteria
Heptane (derivation: By fractional distillation of petroleum followed by purification by rectification). **1 ppm = 4.17 mg/m³**	**PEL:** **400 ppm** **1600 mg/m³** **STEL:** **500 ppm** **2000 mg/m³**	**REL:** **85 ppm** **350 mg/m³** **CEILING (15 min):** **440 ppm** **1800 mg/m³**	**750 ppm**	**TLV:** **400 ppm** **1600 mg/m³** **STEL:** **500 ppm** **2000 mg/m³**

SECTION III - PHYSICAL AND CHEMICAL CHARACTERISTICS

Boiling Point **209°F (98°C)**	Specific Gravity (H_2O = 1) **0.68**
Vapor Pressure (gas) **40 at 72°F (22°C)**	Molecular Weight **100.2**
Vapor Density (Air = 1) **3.45**	Melting Point **-131°F (-90°C)**

Solubility
Insoluble in water (0.0003%). Soluble in alcohol, ether, chloroform.

Appearance and Odor
A clear, colorless liquid with a mild, gasoline-like odor.

SECTION IV - FIRE AND EXPLOSION HAZARD DATA

Flash Point (method used) **25°F (-4°C) closed cup**	Explosive Limits in Air % by Volume **LEL: 1.05% UEL: 6.7%**
NFPA Classification **Class 1B Flammable Liquid**	Autoignition Temperature **433°F (223°C)**

Extinguishing Media
Use dry chemical, carbon dioxide, or alcohol foam. Water is ineffective and a stream may spread fire.

Special Fire Fighting Procedures
Heptane is a flammable liquid. Wear full protective clothing and self-contained breathing apparatus (SCBA). Poisonous gases are produced in fire. Use water spray to keep fire-exposed containers cool.

Unusual Fire and Explosion Hazards
Dangerous fire and explosion hazard when exposed to heat or flame. Containers may explode in fire. Chemical will react on contact with oxidizable materials. Moderate explosion hazard when in the form of vapor. Its vapors are heavier than air and may travel for some distance to flashback and cause fire.

SECTION V - REACTIVITY DATA

Stability		Conditions to Avoid: Keep away from heat or flame. Heptane is normally stable in closed containers under routine conditions of handling and storage.
Stable	Unstable	Incompatibility (materials to avoid)
X		Strong oxidizers, such as permanganates, peroxides, nitrates, chlorates, and perchlorates.
Hazardous Polymerization		Conditions to Avoid: Under normal conditions of temperature and pressure, hazardous polymerization of heptane is not expected to occur.
May Occur	Will Not Occur	Hazardous Decomposition or By-products
	X	When heated to decomposition, heptane can emit highly toxic/poisonous gases, including oxides of carbon.

SECTION VI - HEALTH HAZARD DATA

Primary Route(s) of Entry:	Inhalation?	Absorption (skin)?	Ingestion?
	X	X	X

Health Hazards

INHALATION: Low-order toxicity by inhalation. However, it does act as an irritant and a central nervous system depressant. Can cause nausea, dizziness, vomiting, and depression. Higher concentrations can lead to coma and death from respiratory failure.

ABSORPTION: Irritating to the skin, may cause drying, cracking, and chapping. Eye irritation likely.

INGESTION: Gastrointestinal irritation with possible systemic effects targeting the CNS.

Carcinogenicity	NTP Listed?	IARC Cancer Review Group?	OSHA Regulated?	Target Organs?
Unknown Human Unknown Animal	No	No	29 CFR 1910.1000 Table Z-1	Respiratory system, eyes, skin, CNS.

Medical Conditions Generally Aggravated by Exposure

None reported.

Emergency and First-aid Procedures

Eye contact: Flush immediately with water for 15 minutes (minimum); seek medical attention. Skin contact: Remove all contaminated clothing. Immediately wash area with large amounts of soap and water. Seek medical assistance. For inhalation: Remove the person from exposure. Provide respiratory assistance and CPR, if required. Transfer to medical facility. Keep victim warm and at rest. If swallowed: Seek medical attention immediately. Warning! Aspiration hazard, do NOT induce vomiting.

SECTION VII - PRECAUTIONS FOR SAFE HANDLING AND USE

Steps to be Taken in Case Material is Released or Spilled

For small spills, absorb in vermiculite and allow to evaporate in a safe place (fume hood). Absorb large spills in vermiculite, dry sand, earth, or similar material and deposit in sealed containers. Ventilate area of spill or leak. Restrict those not involved in cleanup from entering area. Remove all sources of ignition and use only non-sparking tools during cleanup. Notify proper authorities.

Preferred Waste Disposal Method

No Citation.

Precautions to be Taken in Handling and Storage

Store to avoid contact with incompatible materials in tightly closed containers in cool, well-ventilated area away from heat and flame. Use only non-sparking tools to open and close containers.

Other Precautions and Warnings

Sources of ignition are prohibited where heptane is used, handled, or stored. Heat may cause containers to build up pressure and explode. Containers must be grounded and bonded during transfer.

SECTION VIII - CONTROL MEASURES AND PERSONAL PROTECTIVE EQUIPMENT

Respiratory Protection *(specify type)*

For high exposures (above the PEL), use a MSHA/NIOSH-approved organic vapor respirator. Best protection is obtained using a supplied-air respirator with full facepiece in continuous flow mode, or a self-contained breathing apparatus (SCBA) operated in pressure demand or other positive pressure mode.

Ventilation

Local exhaust or general mechanical systems recommended.

Protective Gloves	Eye Protection	Other Protective Clothing
Neoprene or Rubber	Chemical Goggles or Face Mask	Rubber Apron

Work/Hygiene Practices

Always wash hands thoroughly after using chemical; never bring food, drink, or smoking materials into vicinity of chemicals.

HEPTANE

$CH_3(CH_2)_5CH_3$ CAS: 142-82-5

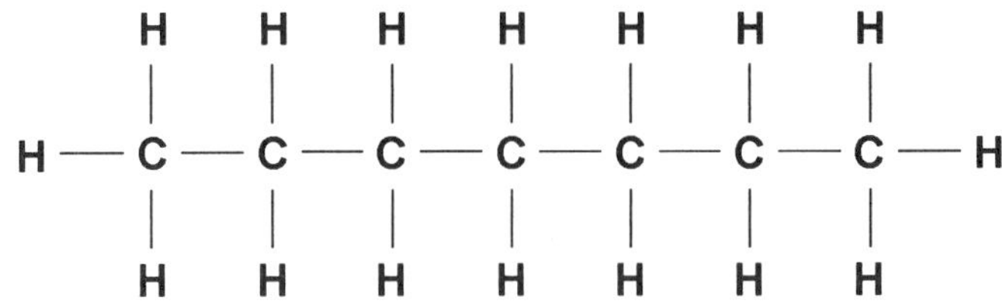

IDENTIFICATION AND TYPICAL USES

Heptane is a colorless liquid with a gasoline-like odor. It is used as a standard for octane rating determinations (pure, normal heptane has zero octane number). Also used as an anesthetic, a solvent, in organic synthesis, and in the preparation of laboratory reagents.

RISK ASSESSMENT: HEALTH

General Assessment

Heptane is an acute irritant of low-order toxicity by *inhalation* and *ingestion*. Skin contact can also produce some adverse topical effects and *absorption* is likely to occur. There is no evidence in the references that exposure to heptane will lead to carcinogenic, mutagenic, or teratogenic effects.

Significant inhalation may cause irritation of the respiratory tract and depress the central nervous system. Symptoms include nausea, dizziness, vomiting, lightheadedness, giddiness, stupor, vertigo, loss of coordination, loss of appetite, depression, hallucinations, and possibly chemical pneumonia (if the swallowed liquid is aspirated into the lungs). At high concentrations, it is a narcotic and can lead to loss of consciousness, coma, and death due to respiratory failure. It is of low chronic toxicity due to its efficient metabolism and excretion.

Skin and eye contact may produce mild to moderate irritation on contact. Prolonged or repeated contact with the skin can cause dryness, cracking, and dermatitis. Ingestion may cause gastrointestinal irritation.

Acute Health Effects

The following acute (short-term) health effects may occur immediately or shortly after exposure to heptane:

Skin: Mild to moderate irritation and absorption.

Eye: Mild to moderate irritation.

Lung: Direct irritation of the respiratory tract and associated mucosa. May cause dizziness, vomiting, and nausea. Primary result of inhalation of heptane is its action on the CNS.

CNS: A CNS depressant: headache, dizziness, depression, hallucinations, changes in mood, loss of consciousness, coma, and death (respiratory failure).

Chronic Health Effects

The following chronic (long-term) health effects may occur at some time after exposure to heptane and can last for months or even years:

Cancer Hazards: According to information presented in the references, heptane has not been adequately tested for its ability to cause cancer in animals.

Reproductive Hazard: According to information presented in the references, heptane has not been tested for its ability to adversely affect reproduction.

Other Chronic Effects: Prolonged skin contact may cause drying, cracking, and chapping of the affected area that may lead to dermatitis. Repeated or long-term exposures to heptane may affect the liver and kidneys.

Recommended Risk-Reduction Measures

Personnel should avoid direct contact with heptane. Even though the toxicity of this chemical is considered to be of relatively low order, it can still cause significant health effects if improperly used or if overexposures occur. If a less toxic chemical cannot be substituted for a hazardous substance, then *engineering controls* are the most effective method of reducing exposures. The best protection is to enclose operations and/or provide local exhaust ventilation at the site of chemical release. While not always operationally feasible, isolating operations can also reduce exposure. Using respiratory protection is less effective than the controls mentioned above, but is still advisable whenever working with or around heptane. For exposures above the PEL (400 ppm), an organic vapor respirator should be used (per OSHA 29 CFR 1910.134). Better protection is provided by a supplied-air respirator with full facepiece operated in continuous flow mode or a self-contained breathing apparatus (SCBA) with full facepiece in pressure demand. If a full facepiece is not available, then chemical goggles should be worn to protect the eyes. Whenever a chemical splash hazard exists, a face shield and a rubber apron should be worn. To prevent hand and skin exposures, impervious neoprene or rubber gloves should be used. Glove manufacturers should be contacted and permeation studies obtained *before* final glove selections are made.

Administrative controls should also be in place to minimize the potential for human exposures. These may include written procedures or policies which specify the methods and techniques that will be practiced whenever personnel are to work with heptane.

All personnel should receive *training* on the use, hazards, protective measures, emergency actions, and other precautions per 29 CFR 1910.1200 (Hazard Communication), prior to the first assignment in an area where heptane is used or stored.

If symptoms develop or overexposure is suspected, the following tests are recommended:

☑ Kidney and liver function tests.

☑ Evaluation of nervous system function.

☑ Evaluation by a qualified allergist with careful consideration of exposure history and special testing (may help diagnose skin allergy and/or dermatitis).

Any evaluation should include a careful history of past and present symptoms with an examination. Medical tests that simply look for existing damage are not a substitute for controlling exposures.

Other methods to reduce exposure include:

☑ Where possible, enclose operations and use local exhaust ventilation at the site of chemical release. If local exhaust ventilation or enclosure is not used, respiratory protection should be mandatory.

☑ Always ensure that proper protective clothing is worn when using chemical substances and personnel are trained on its use, care, and maintenance.

☑ Wash thoroughly immediately after exposure to heptane and at the end of the work shift or before eating, drinking, or smoking.

☑ Hazard warning information should be posted in the work area. In addition, as part of an on-going education and training program, all information on the health and safety hazards of heptane should be communicated to all potentially exposed workers.

☑ Eye wash stations should be provided in the immediate work area for emergency use. If there is a possibility of skin exposure to heptane, emergency shower facilities should also be provided.

☑ Personnel should never wear contaminated clothing home (family members can be exposed). Clothing should be laundered by personnel who have been trained on the hazards of heptane.

RISK ASSESSMENT: ENVIRONMENT

General Assessment

The environment is at risk of exposure during transportation, storage, disposal, or destruction of heptane. In almost every scenario, the threat of environmental exposure is contingent upon the proper handling of the chemical substance. Accidental spills, large or small, can result in fire, explosion, and possible contamination of the surrounding environmental mediums (water, soil, and air).

Heptane is considered a Class 1B flammable liquid (per OSHA 29 CFR 1910.106). Its extremely low flash point and relatively low boiling point make this chemical a serious fire and explosion hazard concern. These characteristics require extreme caution in handling, storage, transportation, and disposal. It is incompatible with common peroxides and other oxidizers and contact can cause fire or explosion. Therefore, special consideration is required during any emergency situation involving a leak or spill of heptane. Should heptane ever come into contact with these incompatible substances during use, transportation, storage, or disposal, the formation of highly toxic and/or highly explosive commodities is possible.

Heptane may enter the environment through industrial discharges or spills.

☠ *Acute Ecological Effects*

Acute (short-term) toxic effects may include the death of animals, birds, or fish and death or low growth rate

in plants. Acute effects are seen 2 to 4 days after animals or plants are exposed to heptane.

This chemical has high acute toxicity to aquatic life. It causes membrane damage in ornamental crop species. Insufficient data are available on the short-term effects of heptane exposure to birds or terrestrial animals.

Chronic Ecological Effects

Chronic (long-term) toxic effects may include shortened life span, reproductive problems, lower fertility, and changes in appearance or behavior in exposed animals. These effects can be seen long after first exposure(s) to toxic chemicals.

Heptane has moderate chronic toxicity in aquatic life. Insufficient data are available on the long-term effects of heptane to plants, birds, or land animals.

Water Solubility

Heptane is nearly insoluble in water. Concentrations of less than 10 milligrams may mix with a liter of water, depending on water pH and chemical concentration of heptane.

Persistence in the Environment

Heptane is non-persistent in water, with a half-life of less than 2 days. The half-life of a pollutant is the amount of time it takes for one half of the chemical to be degraded. Virtually 100% of heptane will eventually end up in air.

Bioaccumulation in Aquatic Organisms

Some substances increase in concentration, or *bioaccumulate,* in living organisms as they breathe contaminated air, drink contaminated water, or eat contaminated food. These chemicals can become concentrated in the tissues and internal organs of animals as well as humans.

The concentration of heptane found in fish tissue is expected to be somewhat higher than the average concentration of heptane in the water from which the fish was taken.

Recommended Risk-Reduction Measures

Proper training of all transporters will reduce the likelihood of a mishap or accident resulting in a leak or spill to the environment. The correct use of DOT labeling (per the MSDS) on all containers, trucks, and rail cars will enable emergency responders to react properly and quickly to any disaster thereby reducing the potential risk to the environment and to personnel.

Storage of heptane should be segregated from incompatible chemicals to minimize the risk of cross-contamination or contact. Buildings designated for storage should be equipped with appropriate fire protection systems (alarms, sprinklers, emergency lighting, portable extinguishers). All transfers of heptane from storage containers to process containers should be done automatically whenever possible. All containers and equipment must be grounded and bonded.

If a spill or leak to the environment has occurred, fire department, emergency response and/or hazardous materials spill personnel should be notified immediately. Cleanup should be attempted only by those trained in proper spill containment procedures using non-sparking tools. Contaminated soils should be removed for incineration and replaced with clean soil. If heptane should contact the water table, aquifer, or navigable waterway, the local and/or state emergency response authorities must be notified. A comprehensive emergency response or disaster preparedness/recovery plan should be in place prior to any operations involving the use, transportation, storage, or disposal of heptane. If heptane is spilled or leaked, the following specific steps are recommended:

- ☑ Restrict persons not wearing protective clothing from area of spill or leak until cleanup is complete and area can be opened for normal work.
- ☑ Remove all ignition sources, ventilate area of leak.
- ☑ Absorb small spills (laboratory spills) in vermiculite and allow evaporation in a safe place (a fume hood or open air). Dispose of residue in sealed drums. For larger spills, absorb liquids in vermiculite, dry sand, earth, or similar material and deposit in sealed containers. Collect materials using non-sparking tools.
- ☑ It may be necessary to dispose of heptane as a hazardous waste. The responsible state agency or the regional office of the federal Environmental Protection Agency (EPA) should be contacted for specific recommendations.

RISK ASSESSMENT: BUSINESS

General Assessment

Accidents or mishaps involving heptane can present a significant threat to business operations. The loss or

damage of equipment or facilities can significantly affect fiscal viability. Lawsuits that may result from personnel injury/death, public exposures, and/or environmental contamination will also require a serious expenditure of resources. Media attention surrounding an injury, death, or environmental damage can also result in a loss of profits and loss of current as well as future business.

❶ *Recommended Risk-Reduction Measures*

Company attorneys, safety and health professionals, and environmental specialists should be involved in the development of any procedures dealing with safety or emergency response. A company official should be pre-designated as a public relations officer with specific training in dealing with the press. Corporate plans and policies should be developed, approved, and implemented long before any need for such arises.

REFERENCES

American Conference of Governmental Industrial Hygienists. 1988. *Documentation of the Threshold Limit Values and Biological Exposure Indices,* 5th Edition (with updates). Cincinnati: ACGIH

Bueche, J. R. 1972. *Principles of Physics,* 2nd Edition. New York: McGraw-Hill.

Calabrese, E. J. 1994. *Biological Effects of Low Level Exposures.* New York: CRC Press/Lewis Publishers.

Cockerman, L. G., B. S. Shane. 1994. *Basic Environmental Toxicology.* New York: CRC Press/Lewis Publishers.

Cothern, C. R. 1995. *Handbook for Environmental Risk Decisionmaking.* New York: CRC Press/Lewis Publishers.

Danse, I. R. 1991. *Common Sense Toxics in the Workplace.* New York: Van Nostrand Reinhold.

Environmental Protection Agency. 1983. *Chemical Hazard Information Profiles.* Washington, D.C.: U.S. EPA.

Fodden, C. R., and J. L. Weddell. 1992. *Hazardous Materials Emergency Action Data.* New York: CRC Press/Lewis Publishers.

Forseberg, K., and L. H. Keith. 1993. *Chemical; Protective Clothing Permeation and Degradation Database.* Chelsea, MI: Lewis Publishers.

Gardner, R. W. 1994. *Chemical Intolerance.* New York: CRC Press/Lewis Publishers.

Hallenbeck, W. H. 1993. *Quantitative Risk Assessment for Environmental and Occupational Health,* 2nd Edition. New York: CRC Press/Lewis Publishers.

Hosty, J. W., and P. Foster. 1990. *A Practical Guide to Chemical Spill Response.* New York: Van Nostrand Reinhold.

Hodgson, E., R. B. Mailman, and Janice E. Chambers. 1988. *Dictionary of Toxicology.* London: The MacMillian Company.

Jones, F. E. 1994. *Toxic Organic Vapors in the Workplace.* New York: CRC Press/Lewis Publishers.

Kamrin, M. A. 1988. *Toxicology: A Primer on Principles and Applications.* Chelsea, Michigan: Lewis Publishers.

Keith, L. H. and M. M. Walker. 1995. *Handbook of Air Toxics.* New York: CRC Press/Lewis Publishers.

Landis, W. G., M. H. Yu. 1995. *Introduction to Environmental Toxicology.* New York: CRC Press/Lewis Publishers.

Langerman, N. 1994 *Precautionary Labels for Chemical Containers.* New York: CRC Press/Lewis Publishers.

Lauwerys, R. R., P. Hoet. 1993. *Industrial Chemical Exposure.* New York: CRC Press/Lewis Publishers.

Lewis, R. J., Sr. 1992. *Sax's Dangerous Properties of Industrial Materials,* Volumes 1, 2, and 3, 8th Edition. New York: Van Nostrand Reinhold.

Lide, D. R. 1995. *Handbook of Chemistry and Physics,* 76th Edition. New York: CRC Press/Lewis Publishers.

Mackay, D., W. Y. Shiu, and K. C. Ma. 1992. *Illustrated Handbook of Chemical-Physical Properties and Environmental Fate for Organic Chemicals.* New York: CRC Press/Lewis Publishers.

Manahan, S. E. 1994. *Environmental Chemistry,* 6th Edition. New York: CRC Press/Lewis Publishers.

Mellström, G. A., J. E. Wahlberg, and H. I. Maibach. 1994. *Protective Gloves for Occupational Use.* New York: CRC Press/Lewis Publishers.

Meyer, E. 1990. *Chemistry of Hazardous Materials.* Englewood Cliffs, NJ: Prentice Hall.

Mycock, J. C., J. D. McKenna, and L. Theodore. 1995. *Handbook of Air Pollution Control Engineering and Technology.* New York: CRC Press/Lewis Publishers.

Patnaik, P. 1992. *A Comprehensive Guide to the Hazardous Properties of Chemical Substances.* New York: Van Nostrand Reinhold.

Proctor, N. H., J. P. Hughes. 1991. *Chemical Hazards of the Workplace.* New York: Van Nostrand Reinhold.

Quigley, D. R. 1994. *Handbook of Emergency Chemical Management.* New York: CRC Press/Lewis Publishers.

Rea, W. J. 1995. *Chemical Sensitivity,* Vol. I-III. New York: CRC Press/Lewis Publishers.

Richardson, M. 1995. *Dictionary of Substance and Their Effects,* Vol. I-VI. New York: CRC Press/Lewis Publishers.

Sacarello, H. L. A. 1994. *The Comprehensive Handbook of Hazardous Materials.* New York: CRC Press/Lewis Publishers.

Sheftel, V. O. 1995. *Handbook of Toxic Properties of Monomers and Additives.* New York: CRC Press/Lewis Publishers.

Torkelson, T. R. V. K. Rowe. 1981. Halogenated aliphatic hydrocarbons. *Patty's Industrial Hygiene and Toxicology,* Vol. 2B, *Toxicology,* pp 3568-3572. New York: Wiley-Interscience.

U.S. Department of Health and Human Services, National Institute for Occupational Safety and Health. 1994. *NIOSH Pocket Guide to Chemical Hazards.* Washington, D.C.: U.S. Government Printing Office.

MATERIAL SAFETY DATA SHEET

CHEMICAL NAME

1-HEPTENE

CRC PRESS/LEWIS PUBLISHERS DISCLAIMER: THE INFORMATION AND RECOMMENDATIONS PRESENTED HEREIN ARE BASED ON SOURCES BELIEVED TO BE RELIABLE. CRC MAKES NO REPRESENTATION ON ITS COMPLETENESS OR ACCURACY. IT IS THE RESPONSIBILITY OF THE USER TO DETERMINE THE CHEMICAL'S SUITABILITY FOR ITS INTENDED USE, THE CHEMICAL'S SAFE USE, AND THE CHEMICAL'S PROPER DISPOSAL. NO REPRESENTATIONS AND/OR WARRANTIES, EITHER EXPRESSED OR IMPLIED, OF THE MERCHANTABILITY OR FITNESS FOR A PARTICULAR PURPOSE, OR OF ANY OTHER NATURE, ARE MADE WITH RESPECT TO THE INFORMATION PROVIDED IN THIS MSDS OR TO THE CHEMICAL TO WHICH INFORMATION MAY REFER. CRC NEITHER ASSUMES NOR AUTHORIZES ANY OTHER PERSON TO ASSUME FOR IT, ANY OTHER ADDITIONAL RESPONSIBILITY OR LIABILITY FOR THE USE OF, OR RELIANCE UPON, THIS INFORMATION.

HAZARD WARNING INFORMATION

HEALTH	FIRE	REACTIVE	OTHER	DEGREE OF HAZARD	COLOR CODING	OTHER CODES
0	3	1		0 = Minimum Hazard 1 = Slight Hazard 2 = Moderate Hazard 3 = Serious Hazard 4 = Severe Hazard	HEALTH = BLUE FIRE = RED REACTIVITY = YELLOW OTHER = WHITE	OX = Oxidizer ACID = Acid ALK = Alkali COR = Corrosive W = Use No Water

SECTION I - GENERAL INFORMATION

Characterization: **Hydrocarbon, Aliphatic**

RCRA Number: **None**

EPA Class: **Not Applicable**

DOT Proper Shipping Name: **1-Heptene**

Chemical Abstract Service (CAS) Number: **592-76-7**

DOT Hazard Class and Label Requirements: **Flammable Liquid**

DOT Emergency Guide Code: **27**

DOT Identification Number: **UN 2278**

Chemical Formula: **$CH_3(CH_2)_3CH{:}CHCH_3$**

Synonyms

1-Heptylene.

SECTION II - HAZARDOUS INGREDIENTS AND IDENTITY INFORMATION

Hazardous Components (specific identity)	OSHA Exposure Criteria	NIOSH Exposure Criteria	Immediately Dangerous to Life and Health (IDLH)	ACGIH Exposure Criteria
1-heptene	PEL: Not Established STEL: Not Established	REL: Not Established STEL: Not Established	Not Determined	TLV: Not Established STEL: Not Established

SECTION III - PHYSICAL AND CHEMICAL CHARACTERISTICS

Boiling Point: **199°F (93°C)**

Specific Gravity (H_2O = 1): **0.697**

Vapor Pressure (gas): **420 at 68°F (20°C)**

Molecular Weight: **98.19**

Vapor Density (Air = 1): **3.4**

Melting Point: **14°F (-10°C)**

Solubility

Insoluble in water. Soluble in alcohol, acetone, ether, petroleum, and coal-tar solvents.

Appearance and Odor

A clear, colorless liquid.

SECTION IV - FIRE AND EXPLOSION HAZARD DATA

Flash Point (method used): **32°F (0°C) closed cup**

Explosive Limits in Air % by Volume: **LEL: 1.0% UEL: Not Found**

NFPA Classification: **Class 1C Flammable Liquid**

Autoignition Temperature: **500°F (260°C)**

Extinguishing Media

Use dry chemical, carbon dioxide, foam, or water.

Special Fire Fighting Procedures

1-heptene is a flammable liquid. Wear full protective clothing and self-contained breathing apparatus (SCBA). Poisonous gases are produced in fire. Use water spray to keep fire-exposed containers cool.

Unusual Fire and Explosion Hazards

Dangerous fire and explosion hazard when exposed to heat or flame. Containers may explode in fire. Chemical will react on contact with oxidizable materials. Moderate explosion hazard when in the form of vapor. Its vapors are heavier than air and may travel for some distance to flashback and cause fire.

SECTION V - REACTIVITY DATA

Stability		Conditions to Avoid
		Keep away from heat or flame. 1-Heptene is normally stable in closed containers under routine conditions of handling and storage.
Stable	Unstable	Incompatibility *(materials to avoid)*
X		Strong oxidizers, such as permanganates, peroxides, nitrates, chlorates, and perchlorates.
Hazardous Polymerization		Conditions to Avoid Under normal conditions of temperature and pressure, hazardous polymerization of 1-heptene is not expected to occur.
May Occur	Will Not Occur	Hazardous Decomposition or By-products
	X	When heated to decomposition, 1-heptene can emit highly toxic/poisonous gases, including oxides of carbon.

SECTION VI - HEALTH HAZARD DATA

Primary Route(s) of Entry:	Inhalation?	Absorption (skin)?	Ingestion?
	X		X

Health Hazards

INHALATION: An asphyxiant gas. Initial symptoms are those of oxygen starvation. Feeling of "air hunger," diminished mental alertness, headache, dullness, shortness of breath, and an impairment in muscle coordination. Continued exposures lead to faulty judgment, depression of senses, rapid fatigue, unconsciousness, convulsions, coma, and death.

SKIN & EYES: Minimum hazard. On prolonged contact may cause smarting and reddening of the skin.

INGESTION: Possibly mildly toxic. Unspecified gastrointestinal effects.

Carcinogenicity	NTP Listed?	IARC Cancer Review Group?	OSHA Regulated?	Target Organs?
Unknown Human Unknown Animal	No	No	No	Respiratory system, CNS.

Medical Conditions Generally Aggravated by Exposure

None reported.

Emergency and First-aid Procedures

Eye contact: Flush immediately with water for 15 minutes (minimum), seek medical attention. Skin contact: Remove all contaminated clothing. Immediately wash area with large amounts of soap and water. Seek medical assistance. For inhalation: Remove the person from exposure. Provide respiratory assistance and CPR, if required. Transfer to medical facility. Keep victim warm and at rest. If swallowed: Seek medical attention immediately. Contact poison control center for assistance.

SECTION VII - PRECAUTIONS FOR SAFE HANDLING AND USE

Steps to be Taken in Case Material is Released or Spilled

For small spills, absorb in vermiculite and allow to evaporate in a safe place (fume hood). Absorb large spills in vermiculite, dry sand, earth, or similar material and deposit in sealed containers. Ventilate area of spill or leak. Restrict those not involved in cleanup from entering area. Remove all sources of ignition and use only non-sparking tools during cleanup. Notify proper authorities.

Preferred Waste Disposal Method

No Citation.

Precautions to be Taken in Handling and Storage

Store to avoid contact with incompatible materials in tightly closed containers in cool, well-ventilated area away from heat and flame. Use only non-sparking tools to open and close containers.

Other Precautions and Warnings

Sources of ignition are prohibited where 1-heptene is used, handled, or stored. Heat may cause containers to build up pressure and explode. Containers must be grounded and bonded during transfer.

SECTION VIII - CONTROL MEASURES AND PERSONAL PROTECTIVE EQUIPMENT

Respiratory Protection *(specify type)*

No exposure levels established and 1-heptene is an asphyxiant. Best protection is obtained using a supplied-air respirator with full facepiece in continuous flow mode, or a self-contained breathing apparatus (SCBA) operated in pressure demand or other positive pressure mode.

Ventilation

Local exhaust or general mechanical systems recommended.

Protective Gloves	Eye Protection	Other Protective Clothing
Impervious Material	Chemical Goggles or Face Mask	Protective Clothing

Work/Hygiene Practices

Always wash hands thoroughly after using chemical; never bring food, drink, or smoking materials into vicinity of chemicals.

1-HEPTENE

$CH_3(CH_2)_4CH{:}CH_2$ CAS: 592-76-7

IDENTIFICATION AND TYPICAL USES

1-Heptene is a colorless liquid used primarily in organic synthesis.

RISK ASSESSMENT: HEALTH

General Assessment

1-Heptene is a simple asphyxiant gas. As such, its primary route of entry into the body is through *inhalation.* While it does not pose a significant physiological effect (i.e., no systemic poisoning or toxicity), it can displace oxygen levels to below that which is necessary to sustain life. Skin contact with the liquid can cause mild irritation on prolonged contact, but absorption is not a likely exposure route. Ingestion of the liquid may also occur, but little data are available on the resulting effects. There are no data in the references on the carcinogenic, mutagenic, or teratogenic properties of 1-heptene in humans or animals.

Inhalation of 1-heptene vapors can cause mild irritation with headache, dullness, shortness of breath, and the symptoms of asphyxiation. These include rapid respiration and a feeling of "air hunger" or air starvation followed by a decrease in mental alertness, and impaired muscular coordination. If exposure continues, there can be nausea, vomiting, prostration, unconsciousness, coma, and death.

Skin or eye contact with the vapor poses no serious threat. The vapors have not been shown to cause any physiological response and the liquid has only been shown to cause mild to moderate irritation on prolonged contact. Soaked clothing should be removed immediately to avoid a reaction.

Acute Health Effects

The following acute health effects may occur immediately or shortly after exposure to 1-heptene:

Skin: Possible mild irritation.

Eye: Mild to moderate irritation.

Lung: Vapor is only slightly irritating and therefore poses little threat of damage to the lungs. However, in high concentrations, 1-heptene can lead to symptoms of asphyxiation and oxygen deprivation up to and including death.

CNS: Effects of asphyxiation cause adverse nervous system reactions.

Chronic Health Effects

The following chronic (long-term) health effects may occur at some time after exposure to 1-heptene and can last for months or even years:

Cancer Hazards: According to information presented in the references, exposure to 1-heptene does not lead to carcinogenic effects (it is a simple asphyxiant).

Reproductive Hazards: There are no reports to support any claims of reproductive hazards.

Other Chronic Effects: None reported.

Recommended Risk-Reduction Measures

Personnel should avoid direct contact with 1-heptene. Even though the toxicity of this chemical is considered to be of relatively low order, it can still cause significant health effects if improperly used or if overexposures occur. If a less toxic chemical cannot be substituted for a hazardous substance, then *engineering controls* are the most effective method of reducing exposures. The best protection is to enclose operations and/or provide local exhaust ventilation at the site of chemical release. However, because the vapors of 1-heptene can deplete oxygen levels in confined spaces, caution is warranted if enclosed operations is a chosen method of exposure control. While not always operationally feasible, isolating operations can also reduce exposure. Using respiratory protection is less effective than the controls mentioned above, but is still advisable whenever working with or around 1-heptene. Air-purifying respirators (e.g., the cartridge-type or gas masks) cannot be used with 1-heptene. These types of respirators only purify the air and can do nothing to protect in oxygen deficient atmospheres. A powered air-supplied respirator operated in pressure demand or other positive pressure mode, or a self-contained breathing apparatus (SCBA) with full facepiece operated in pressure demand mode are the recommended respiratory protection methods of choice. If a full facepiece is not available, then chemical goggles should be worn to protect the eyes. Whenever a chemical splash hazard exists, a face shield and a protective apron should be worn. To prevent hand and skin exposures with the liquid, impervious gloves should be used. Glove manufacturers should be contacted and permeation studies obtained *before* final glove selections have been made.

Administrative controls should also be in place to minimize the potential for human exposures. These may include written procedures or policies which specify the methods and techniques that will be practiced whenever personnel are to work with 1-heptene.

All personnel should receive *training* on the use, hazards, protective measures, emergency actions, and other precautions per 29 CFR 1910.1200 (Hazard Communication), prior to the first assignment in an area where 1-heptene is used or stored.

Other methods to reduce exposure include:

☑ Where possible, enclose operations and use local exhaust ventilation at the site of chemical release. Because exposure to 1-heptene can cause asphyxiation, the use of local exhaust ventilation is vital. If local exhaust ventilation or enclosure is not possible or feasible, respiratory protection should be mandatory.

☑ Always ensure that proper protective clothing is worn when using chemical substances and personnel are trained on its use, care, and maintenance.

☑ Wash thoroughly immediately after exposure to 1-heptene and at the end of the work shift or before eating, drinking, or smoking.

☑ Hazard warning information should be posted in the work area. In addition, as part of an on-going education and training program, all information on the health and safety hazards of 1-heptene should be communicated to all exposed and potentially exposed workers.

☑ Eye wash stations should be provided in the immediate work area for emergency use. If there is a possibility of skin exposure to 1-heptene, emergency shower facilities should also be provided.

☑ Personnel should never wear contaminated clothing home (family members can be exposed). Clothing should be laundered by personnel who have been trained on the hazards of 1-heptene.

RISK ASSESSMENT: ENVIRONMENT

General Assessment

The environment is at risk of exposure during transportation, storage, disposal, or destruction of 1-heptene. In almost every scenario, the threat of environmental exposure is contingent upon the proper handling of the chemical substance. Accidental spills, large or small, can result in fire, explosion, and possible contamination of the surrounding environmental mediums (water, soil, and air).

1-Heptene is considered a Class 1C flammable liquid (per OSHA 29 CFR 1910.106). Its extremely low flash point and relatively low boiling point make this chemical a serious fire and explosion hazard concern. These characteristics require extreme caution in handling, storage, transportation, and disposal. It is incompatible with common peroxides and other oxidizers and contact can cause fire or explosion. Therefore, special consideration is required during any emergency situation involving a leak or spill of 1-heptene. Should 1-heptene come into contact with these incompatible substances during use, transportation, storage, or disposal, the formation of highly toxic and/or highly explosive commodities is possible. 1-Heptene may enter the environment through industrial discharges or spills.

☠ *Acute Ecological Effects*

Acute (short-term) toxic effects may include the death of animals, birds, or fish and death or low growth rate in plants. Acute effects are seen 2 to 4 days after animals or plants are exposed to 1-heptene.

This chemical has high acute toxicity to aquatic life. It may cause membrane damage in ornamental crop species. Insufficient data are available on the short-term effects of 1-heptene exposure to birds or terrestrial animals.

Chronic Ecological Effects

Chronic (long-term) toxic effects may include shortened life span, reproductive problems, lower fertility, and changes in appearance or behavior in exposed animals. These effects can be seen long after first exposure(s) to toxic chemicals.

1-Heptene has moderate chronic toxicity in aquatic life. Insufficient data are available on the long-term effects of 1-heptene to plants, birds, or land animals.

Water Solubility

1-Heptene is insoluble in water. Concentrations of less than 1 milligram will not mix with a liter of water.

Persistence in the Environment

1-Heptene is non-persistent in water, with a half-life of less than 2 days. The half-life of a pollutant is the amount of time it takes for one half of the chemical to be degraded. There are no data on its soil sorption capabilities. It is assumed that virtually 100% of 1-heptene will eventually end up in air.

Bioaccumulation in Aquatic Organisms

Some substances increase in concentration, or *bioaccumulate,* in living organisms as they breathe contaminated air, drink contaminated water, or eat contaminated food. These chemicals can become concentrated in the tissues and internal organs of animals as well as humans.

There are no data available on the bioaccumulation of 1-heptene in aquatic organisms.

Recommended Risk-Reduction Measures

Proper training of all transporters will reduce the likelihood of a mishap or accident resulting in a leak or spill to the environment. The correct use of DOT labeling (per the MSDS) on all containers, trucks, and rail cars will enable emergency responders to react properly and quickly to any disaster thereby reducing the potential risk to the environment and to personnel.

Storage of 1-heptene should be segregated from incompatible chemicals to minimize the risk of cross-contamination or contact. Buildings designated for storage should be equipped with appropriate fire protection systems (alarms, sprinklers, emergency lighting, portable extinguishers). All transfers of 1-heptene from storage containers to process containers should be done automatically whenever possible. All containers and equipment must be grounded and bonded.

If a spill or leak to the environment has occurred, fire department, emergency response and/or hazardous materials spill personnel should be notified immediately. Cleanup should be attempted only by those trained in proper spill containment procedures using non-sparking tools. Contaminated soils should be removed for incineration and replaced with clean soil. If 1-heptene should contact the water table, aquifer, or navigable waterway, the local and/or state emergency response authorities must be notified. A comprehensive emergency response or disaster preparedness/recovery plan should be in place prior to any operations involving the use, transportation, storage, or disposal of 1-heptene. If 1-heptene is spilled or leaked, the following specific steps are recommended:

- ☑ Restrict persons not wearing protective clothing from area of spill or leak until cleanup is complete and area can be opened for normal work.
- ☑ Remove all ignition sources, ventilate area of leak.
- ☑ Absorb small spills (laboratory spills) in vermiculite and allow evaporation in a safe place (a fume hood or open air). Dispose of residue in sealed drums. For larger spills, absorb liquids in vermiculite, dry sand, earth, or similar material and deposit in sealed containers. Collect materials using non-sparking tools.
- ☑ It may be necessary to dispose of 1-heptene as a hazardous waste. The responsible state agency or the regional office of the federal Environmental Protection Agency (EPA) should be contacted for specific recommendations.

RISK ASSESSMENT: BUSINESS

General Assessment

Accidents or mishaps involving 1-heptene can present a significant threat to business operations. The loss or damage of equipment or facilities can significantly affect fiscal viability. Lawsuits that may result from personnel injury/death, public exposures, and/or environmental contamination will also require a serious expenditure of resources. Media attention surrounding an injury, death, or environmental damage can also result in a loss of profits and loss of current as well as future business.

℧ *Recommended Risk-Reduction Measures*

Company attorneys, safety and health professionals, and environmental specialists should be involved in the development of any procedures dealing with safety or emergency response. A company official should be pre-designated as a public relations officer with specific training in dealing with the press. Corporate plans and policies should be developed, approved, and implemented long before any need for such arises.

REFERENCES

American Conference of Governmental Industrial Hygienists. 1988. *Documentation of the Threshold Limit Values and Biological Exposure Indices,* 5th Edition (with updates). Cincinnati: ACGIH

Bueche, J. R. 1972. *Principles of Physics,* 2nd Edition. New York: McGraw-Hill.

Calabrese, E. J. 1994. *Biological Effects of Low Level Exposures.* New York: CRC Press/Lewis Publishers.

Cockerman, L. G., B. S. Shane. 1994. *Basic Environmental Toxicology.* New York: CRC Press/Lewis Publishers.

Cothern, C. R. 1995. *Handbook for Environmental Risk Decisionmaking.* New York: CRC Press/Lewis Publishers.

Danse, I. R. 1991. *Common Sense Toxics in the Workplace.* New York: Van Nostrand Reinhold.

Environmental Protection Agency. 1983. *Chemical Hazard Information Profiles.* Washington, D.C.: U.S. EPA.

Fodden, C. R., and J. L. Weddell. 1992. *Hazardous Materials Emergency Action Data.* New York: CRC Press/Lewis Publishers.

Forseberg, K., and L. H. Keith. 1993. *Chemical; Protective Clothing Permeation and Degradation Database.* Chelsea, MI: Lewis Publishers.

Gardner, R. W. 1994. *Chemical Intolerance.* New York: CRC Press/Lewis Publishers.

Hallenbeck, W. H. 1993. *Quantitative Risk Assessment for Environmental and Occupational Health,* 2nd Edition. New York: CRC Press/Lewis Publishers.

Hosty, J. W., and P. Foster. 1990. *A Practical Guide to Chemical Spill Response.* New York: Van Nostrand Reinhold.

Hodgson, E., R. B. Mailman, and Janice E. Chambers. 1988. *Dictionary of Toxicology.* London: The MacMillian Company.

Jones, F. E. 1994. *Toxic Organic Vapors in the Workplace.* New York: CRC Press/Lewis Publishers.

Kamrin, M. A. 1988. *Toxicology: A Primer on Principles and Applications.* Chelsea, Michigan: Lewis Publishers.

Keith, L. H. and M. M. Walker. 1995. *Handbook of Air Toxics.* New York: CRC Press/Lewis Publishers.

Landis, W. G., M. H. Yu. 1995. *Introduction to Environmental Toxicology.* New York: CRC Press/Lewis Publishers.

Langerman, N. 1994 *Precautionary Labels for Chemical Containers.* New York: CRC Press/Lewis Publishers.

Lauwerys, R. R., P. Hoet. 1993. *Industrial Chemical Exposure.* New York: CRC Press/Lewis Publishers.

Lewis, R. J., Sr. 1992. *Sax's Dangerous Properties of Industrial Materials,* Volumes 1, 2, and 3, 8th Edition. New York: Van Nostrand Reinhold.

Lide, D. R. 1995. *Handbook of Chemistry and Physics,* 76th Edition. New York: CRC Press/Lewis Publishers.

Mackay, D., W. Y. Shiu, and K. C. Ma. 1992. *Illustrated Handbook of Chemical-Physical Properties and Environmental Fate for Organic Chemicals.* New York: CRC Press/Lewis Publishers.

Manahan, S. E. 1994. *Environmental Chemistry,* 6th Edition. New York: CRC Press/Lewis Publishers.

Mellström, G. A., J. E. Wahlberg, and H. I. Maibach. 1994. *Protective Gloves for Occupational Use.* New York: CRC Press/Lewis Publishers.

Meyer, E. 1990. *Chemistry of Hazardous Materials.* Englewood Cliffs, NJ: Prentice Hall.

Mycock, J. C., J. D. McKenna, and L. Theodore. 1995. *Handbook of Air Pollution Control Engineering and Technology.* New York: CRC Press/Lewis Publishers.

Patnaik, P. 1992. *A Comprehensive Guide to the Hazardous Properties of Chemical Substances.* New York: Van Nostrand Reinhold.

Proctor, N. H., J. P. Hughes. 1991. *Chemical Hazards of the Workplace.* New York: Van Nostrand Reinhold.

Quigley, D. R. 1994. *Handbook of Emergency Chemical Management.* New York: CRC Press/Lewis Publishers.

Rea, W. J. 1995. *Chemical Sensitivity,* Vol. I-III. New York: CRC Press/Lewis Publishers.

Richardson, M. 1995. *Dictionary of Substance and Their Effects,* Vol. I-VI. New York: CRC Press/Lewis Publishers.

Sacarello, H. L. A. 1994. *The Comprehensive Handbook of Hazardous Materials.* New York: CRC Press/Lewis Publishers.

Sheftel, V. O. 1995. *Handbook of Toxic Properties of Monomers and Additives.* New York: CRC Press/Lewis Publishers.

Torkelson, T. R. V. K. Rowe. 1981. Halogenated aliphatic hydrocarbons. *Patty's Industrial Hygiene and Toxicology,* Vol. 2B, *Toxicology,* pp 3568-3572. New York: Wiley-Interscience.

U.S. Department of Health and Human Services, National Institute for Occupational Safety and Health. 1994. *NIOSH Pocket Guide to Chemical Hazards.* Washington, D.C.: U.S. Government Printing Office.

MATERIAL SAFETY DATA SHEET

CHEMICAL NAME

2-HEPTENE

HAZARD WARNING INFORMATION

HEALTH	FIRE	REACTIVE	OTHER	DEGREE OF HAZARD	COLOR CODING	OTHER CODES
0	3	1		0 = Minimum Hazard 1 = Slight Hazard 2 = Moderate Hazard 3 = Serious Hazard 4 = Severe Hazard	HEALTH = BLUE FIRE = RED REACTIVITY = YELLOW OTHER = WHITE	OX = Oxidizer ACID = Acid ALK = Alkali COR = Corrosive W = Use No Water

SECTION I - GENERAL INFORMATION

Characterization: **Hydrocarbon, Aliphatic**

RCRA Number: **None**

EPA Class: **Not Applicable**

DOT Proper Shipping Name: **No Citation**

Chemical Abstract Service (CAS) Number: **592-77-8**

DOT Hazard Class and Label Requirements: **No Citation**

DOT Emergency Guide Code: **No Citation**

DOT Identification Number: **No Citation**

Chemical Formula: **$CH_3(CH_2)_3CH{:}CHCH_3$**

Synonyms

2-Heptylene.

SECTION II - HAZARDOUS INGREDIENTS AND IDENTITY INFORMATION

Hazardous Components (specific identity)	OSHA Exposure Criteria	NIOSH Exposure Criteria	Immediately Dangerous to Life and Health (IDLH)	ACGIH Exposure Criteria
2-Heptene	**PEL:** Not Established **STEL:** Not Established	**REL:** Not Established **STEL:** Not Established	Not Determined	**TLV:** Not Established **STEL:** Not Established

SECTION III - PHYSICAL AND CHEMICAL CHARACTERISTICS

Boiling Point: **208°F (98°C)**

Specific Gravity (H_2O = 1): **0.697**

Vapor Pressure (gas): **420 at 68°F (20°C)**

Molecular Weight: **98.19**

Vapor Density (Air = 1): **3.4**

Melting Point: **14°F (-10°C)**

Solubility

Insoluble in water. Soluble in alcohol, acetone, ether, petroleum, and coal-tar solvents.

Appearance and Odor

A clear, colorless liquid.

SECTION IV - FIRE AND EXPLOSION HAZARD DATA

Flash Point (method used): **28°F (-2.2°C) closed cup**

Explosive Limits in Air % by Volume: **LEL: Not Found UEL: Not Found**

NFPA Classification: **Class 1C Flammable Liquid**

Autoignition Temperature: **707°F (375°C)**

Extinguishing Media

Use dry chemical, carbon dioxide, foam, or water.

Special Fire Fighting Procedures

2-heptene is a flammable liquid. Wear full protective clothing and self-contained breathing apparatus (SCBA). Poisonous gases are produced in fire. Use water spray to keep fire-exposed containers cool.

Unusual Fire and Explosion Hazards

Dangerous fire and explosion hazard when exposed to heat or flame. Containers may explode in fire. Chemical will react on contact with oxidizable materials. Moderate explosion hazard when in the form of vapor. Its vapors are heavier than air and may travel for some distance to flashback and cause fire.

SECTION V - REACTIVITY DATA

Stability		Conditions to Avoid Keep away from heat or flame. 2-heptene is normally stable in closed containers under routine conditions of handling and storage.
Stable X	Unstable	Incompatibility *(materials to avoid)* Strong oxidizers, such as permanganates, peroxides, nitrates, chlorates, and perchlorates.
Hazardous Polymerization		Conditions to Avoid Under normal conditions of temperature and pressure, hazardous polymerization of 2-heptene is not expected to occur.
May Occur	Will Not Occur X	Hazardous Decomposition or By-products When heated to decomposition, 2-heptene can emit highly toxic/poisonous gases, including oxides of carbon.

SECTION VI - HEALTH HAZARD DATA

Primary Route(s) of Entry:	Inhalation?	Absorption (skin)?	Ingestion?
	X		X

Health Hazards

INHALATION: An asphyxiant gas. Initial symptoms are those of oxygen starvation. Feeling of "air hunger," diminished mental alertness, headache, dullness, shortness of breath, and an impairment in muscle coordination. Continued exposures lead to faulty judgment, depression of senses, rapid fatigue, unconsciousness, convulsions, coma, and death.

SKIN & EYES: Minimum hazard. On prolonged contact may cause smarting and reddening of the skin.

INGESTION: Possibly mildly toxic. Unspecified gastrointestinal effects.

Carcinogenicity	NTP Listed?	IARC Cancer Review Group?	OSHA Regulated?	Target Organs?
Unknown Human Unknown Animal	No	No	No	Respiratory system, CNS.

Medical Conditions Generally Aggravated by Exposure

None reported.

Emergency and First-aid Procedures

Eye contact: Flush immediately with water for 15 minutes (minimum); seek medical attention. Skin contact: Remove all contaminated clothing. Immediately wash area with large amounts of soap and water. Seek medical assistance. For inhalation: Remove the person from exposure. Provide respiratory assistance and CPR, if required. Transfer to medical facility. Keep victim warm and at rest. If swallowed: Seek medical attention immediately. Contact poison control center for assistance.

SECTION VII - PRECAUTIONS FOR SAFE HANDLING AND USE

Steps to be Taken in Case Material is Released or Spilled

For small spills, absorb in vermiculite and allow to evaporate in a safe place (fume hood). Absorb large spills in vermiculite, dry sand, earth, or similar material and deposit in sealed containers. Ventilate area of spill or leak. Restrict those not involved in cleanup from entering area. Remove all sources of ignition and use only non-sparking tools during cleanup. Notify proper authorities.

Preferred Waste Disposal Method

No Citation.

Precautions to be Taken in Handling and Storage

Store to avoid contact with incompatible materials in tightly closed containers in cool, well-ventilated area away from heat and flame. Use only non-sparking tools to open and close containers.

Other Precautions and Warnings

Sources of ignition are prohibited where 2-heptene is used, handled, or stored. Heat may cause containers to build up pressure and explode. Containers must be grounded and bonded during transfer.

SECTION VIII - CONTROL MEASURES AND PERSONAL PROTECTIVE EQUIPMENT

Respiratory Protection *(specify type)*

No exposure levels established and 2-heptene is an asphyxiant. Best protection is obtained using a supplied-air respirator with full facepiece in continuous flow mode, or a self-contained breathing apparatus (SCBA) operated in pressure demand or other positive pressure mode.

Ventilation

Local exhaust or general mechanical systems recommended.

Protective Gloves	Eye Protection	Other Protective Clothing
Impervious Material	Chemical Goggles or Face Mask	Protective Clothing

Work/Hygiene Practices

Always wash hands thoroughly after using chemical; never bring food, drink, or smoking materials into vicinity of chemicals.

2-HEPTENE

$CH_3(CH_2)_3CH{:}CHCH_3$

CAS: 592-77-8

IDENTIFICATION AND TYPICAL USES

2-Heptene is a colorless liquid. Its suggested use is as a plant growth retardant.

RISK ASSESSMENT: HEALTH

General Assessment

2-Heptene is a simple asphyxiant gas. As such, its primary route of entry into the body is through *inhalation*. While it does not pose a significant physiological effect (i.e., no systemic poisoning or toxicity), it can displace oxygen levels to below that which is necessary to sustain life. Skin contact with the liquid can cause mild irritation on prolonged contact, but absorption is not a likely exposure route. Ingestion of the liquid may also occur, but little data are available on the resulting effects. There are no data in the references on the carcinogenic, mutagenic, or teratogenic properties of 2-heptene in humans or animals.

Inhalation of 2-heptene vapors can cause mild irritation with headache, dullness, shortness of breath, and the symptoms of asphyxiation. These include rapid respiration and a feeling of "air hunger" or air starvation followed by a decrease in mental alertness, and impaired muscular coordination. If exposure continues, there can be nausea, vomiting, prostration, unconsciousness, coma, and death.

Skin or eye contact with the vapor poses no serious threat. The vapors have not been shown to cause any physiological response and the liquid has only been shown to cause mild to moderate irritation on prolonged contact. Soaked clothing should be removed immediately to avoid a reaction.

Acute Health Effects

The following acute (short-term) health effects may occur immediately or shortly after exposure to chemicals such as 2-heptene:

Skin: Possible mild irritation.

Eye: Mild to moderate irritation.

Lung: Vapor is only slightly irritating and therefore poses little threat of damage to the lungs. However, in high concentrations, 2-heptene can lead to symptoms of asphyxiation and oxygen deprivation up to and including death.

CNS: Effects of asphyxiation cause adverse nervous system reactions.

Chronic Health Effects

The following chronic (long-term) health effects may occur at some time after exposure to 2-heptene and can last for months or even years:

Cancer Hazards: According to information presented in the references, exposure to 2-heptene does not lead to carcinogenic effects (it is a simple asphyxiant).

Reproductive Hazards: There are no reports to support any claims of reproductive hazards.

Other Chronic Effects: None reported.

Recommended Risk-Reduction Measures

Personnel should avoid direct contact with 2-heptene. Even though the toxicity of this chemical is considered to be of relatively low order, it can still cause significant health effects if improperly used or if overexposures occur. If a less toxic chemical cannot be substituted for a hazardous substance, then *engineering controls* are the most effective method of reducing exposures. The best protection is to enclose operations and/or provide local exhaust ventilation at the site of chemical release. However, because the vapors of 2-heptene can deplete oxygen levels in confined spaces, caution is warranted if enclosed operations is a chosen method of exposure control. While not always operationally feasible, isolating operations can also reduce exposure. Using respiratory protection is less effective than the controls mentioned above, but is still advisable whenever working with or around 2-heptene. Air-purifying respirators (e.g., the cartridge-type or gas masks) cannot be used with 2-heptene. These types of respirators only purify the air and can do nothing to protect in oxygen deficient atmospheres. A powered air-supplied respirator operated in pressure demand or other positive pressure mode, or a self-contained breathing apparatus (SCBA) with full facepiece operated in pressure demand mode are the recommended respiratory protection methods of choice. If a full facepiece is not available, then chemical goggles should be worn to protect the eyes. Whenever a chemical splash hazard exists, a face shield and a protective apron should be worn. To prevent hand and skin exposures with the liquid, impervious gloves should be used. Glove manufacturers should be contacted and permeation studies obtained *before* final glove selections have been made.

Administrative controls should also be in place to minimize the potential for human exposures. These may include written procedures or policies which specify the methods and techniques that will be practiced whenever personnel are to work with 2-heptene.

All personnel should receive *training* on the use, hazards, protective measures, emergency actions, and other precautions per 29 CFR 1910.1200 (Hazard Communication), prior to the first assignment in an area where 2-heptene is used or stored.

Other methods to reduce exposure include:

☑ Where possible, enclose operations and use local exhaust ventilation at the site of chemical release. Because exposure to 2-heptene can cause asphyxiation, the use of local exhaust ventilation is vital. If local exhaust ventilation or enclosure is not possible or feasible, respiratory protection should be mandatory.

☑ Always ensure that proper protective clothing is worn when using chemical substances and personnel are trained on its use, care, and maintenance.

☑ Wash thoroughly immediately after exposure to 2-heptene and at the end of the work shift or before eating, drinking, or smoking.

☑ Hazard warning information should be posted in the work area. In addition, as part of an on-going education and training program, all information on the health and safety hazards of 2-heptene should be communicated to all exposed and potentially exposed workers.

☑ Eye wash stations should be provided in the immediate work area for emergency use. If there is a possibility of skin exposure to 2-heptene, emergency shower facilities should also be provided.

☑ Personnel should never wear contaminated clothing home (family members can be exposed). Clothing should be laundered by personnel who have been trained on the hazards of 2-heptene.

RISK ASSESSMENT: ENVIRONMENT

General Assessment

The environment is at risk of exposure during transportation, storage, disposal, or destruction of 2-heptene. In almost every scenario, the threat of environmental exposure is contingent upon the proper handling of the chemical substance. Accidental spills, large or small, can result in fire, explosion, and possible contamination of the surrounding environmental mediums (water, soil, and air).

2-Heptene is considered a Class 1C flammable liquid (per OSHA 29 CFR 1910.106). Its extremely low flash point and relatively low boiling point make this chemical a serious fire and explosion hazard concern. These characteristics require extreme caution in handling, storage, transportation, and disposal. It is incompatible with a common peroxides and other oxidizers and contact can cause fire or explosion. Therefore, special consideration is required during any emergency situation involving a leak or spill of 2-heptene. Should 2-heptene come into contact with these incompatible substances during use, transportation, storage, or disposal, the formation of highly toxic and/or highly explosive commodities is possible. 2-heptene may enter the environment through industrial discharges or spills.

☠ *Acute Ecological Effects*

Acute (short-term) toxic effects may include the death of animals, birds, or fish and death or low growth rate in plants. Acute effects are seen 2 to 4 days after animals or plants are exposed to 2-heptene.

This chemical has high acute toxicity to aquatic life. It may cause membrane damage in ornamental crop species. Insufficient data are available on the short-term effects of 2-heptene exposure to birds or terrestrial animals.

Chronic Ecological Effects

Chronic (long-term) toxic effects may include shortened life span, reproductive problems, lower fertility, and changes in appearance or behavior in exposed animals. These effects can be seen long after first exposure(s) to toxic chemicals.

2-Heptene has moderate chronic toxicity in aquatic life. Insufficient data are available on the long-term effects of 2-heptene to plants, birds, or land animals.

Water Solubility

2-Heptene is insoluble in water. Concentrations of less than 1 milligram will not mix with a liter of water.

Persistence in the Environment

2-Heptene is non-persistent in water, with a half-life of less than 2 days. The half-life of a pollutant is the amount of time it takes for one half of the chemical to be degraded. There are no data on its soil sorption capabilities. It is assumed that virtually 100% of 2-heptene will eventually end up in air.

Bioaccumulation in Aquatic Organisms

Some substances increase in concentration, or *bioaccumulate,* in living organisms as they breathe contaminated air, drink contaminated water, or eat contaminated food. These chemicals can become concentrated in the tissues and internal organs of animals as well as humans.

There are no data available on the bioaccumulation of 2-heptene in aquatic organisms.

Recommended Risk-Reduction Measures

Proper training of all transporters will reduce the likelihood of a mishap or accident resulting in a leak or spill to the environment. The correct use of DOT labeling (per the MSDS) on all containers, trucks, and rail cars will enable emergency responders to react properly and quickly to any disaster thereby reducing the potential risk to the environment and to personnel.

Storage of 2-heptene should be segregated from incompatible chemicals to minimize the risk of cross-contamination or contact. Buildings designated for storage should be equipped with appropriate fire protection systems (alarms, sprinklers, emergency lighting, portable extinguishers). All transfers of 2-heptene from storage containers to process containers should be done automatically whenever possible. All containers and equipment must be grounded and bonded.

If a spill or leak to the environment has occurred, fire department, emergency response and/or hazardous materials spill personnel should be notified immediately. Cleanup should be attempted only by those trained in proper spill containment procedures using non-sparking tools. Contaminated soils should be removed for incineration and replaced with clean soil. If 2-heptene should contact the water table, aquifer, or navigable waterway, the local and/or state emergency response authorities must be notified. A comprehensive emergency response or disaster preparedness/recovery plan should be in place prior to any operations involving the use, transportation, storage, or disposal of 2-heptene. If 2-heptene is spilled or leaked, the following specific steps are recommended:

- ☑ Restrict persons not wearing protective clothing from area of spill or leak until cleanup is complete and area can be opened for normal work.
- ☑ Remove all ignition sources, ventilate area of leak.
- ☑ Absorb small spills (laboratory spills) in vermiculite and allow evaporation in a safe place (a fume hood or open air). Dispose of residue in sealed drums. For larger spills, absorb liquids in vermiculite, dry sand, earth, or similar material and deposit in sealed containers. Collect materials using non-sparking tools.
- ☑ It may be necessary to dispose of 2-heptene as a hazardous waste. The responsible state agency or the regional office of the federal Environmental Protection Agency (EPA) should be contacted for specific recommendations.

RISK ASSESSMENT: BUSINESS

General Assessment

Accidents or mishaps involving 2-heptene can present a significant threat to business operations. The loss or damage of equipment or facilities can significantly affect fiscal viability. Lawsuits that may result from personnel injury/death, public exposures, and/or environmental contamination will also require a serious expenditure of resources.

Media attention surrounding an injury, death, or environmental damage can also result in a loss of profits and loss of current as well as future business.

ⓤ Recommended Risk-Reduction Measures

Company attorneys, safety and health professionals, and environmental specialists should be involved in the development of any procedures dealing with safety or emergency response. A company official should be pre-designated as a public relations officer with specific training in dealing with the press. Corporate plans and policies should be developed, approved, and implemented long before any need for such arises.

REFERENCES

American Conference of Governmental Industrial Hygienists. 1988. *Documentation of the Threshold Limit Values and Biological Exposure Indices,* 5th Edition (with updates). Cincinnati: ACGIH

Bueche, J. R. 1972. *Principles of Physics,* 2nd Edition. New York: McGraw-Hill.

Calabrese, E. J. 1994. *Biological Effects of Low Level Exposures.* New York: CRC Press/Lewis Publishers.

Cockerman, L. G., B. S. Shane. 1994. *Basic Environmental Toxicology.* New York: CRC Press/Lewis Publishers.

Cothern, C. R. 1995. *Handbook for Environmental Risk Decisionmaking.* New York: CRC Press/Lewis Publishers.

Danse, I. R. 1991. *Common Sense Toxics in the Workplace.* New York: Van Nostrand Reinhold.

Environmental Protection Agency. 1983. *Chemical Hazard Information Profiles.* Washington, D.C.: U.S. EPA.

Fodden, C. R., and J. L. Weddell. 1992. *Hazardous Materials Emergency Action Data.* New York: CRC Press/Lewis Publishers.

Forseberg, K., and L. H. Keith. 1993. *Chemical; Protective Clothing Permeation and Degradation Database.* Chelsea, MI: Lewis Publishers.

Gardner, R. W. 1994. *Chemical Intolerance.* New York: CRC Press/Lewis Publishers.

Hallenbeck, W. H. 1993. *Quantitative Risk Assessment for Environmental and Occupational Health,* 2nd Edition. New York: CRC Press/Lewis Publishers.

Hosty, J. W., and P. Foster. 1990. *A Practical Guide to Chemical Spill Response.* New York: Van Nostrand Reinhold.

Hodgson, E., R. B. Mailman, and Janice E. Chambers. 1988. *Dictionary of Toxicology.* London: The MacMillian Company.

Jones, F. E. 1994. *Toxic Organic Vapors in the Workplace.* New York: CRC Press/Lewis Publishers.

Kamrin, M. A. 1988. *Toxicology: A Primer on Principles and Applications.* Chelsea, Michigan: Lewis Publishers.

Keith, L. H. and M. M. Walker. 1995. *Handbook of Air Toxics.* New York: CRC Press/Lewis Publishers.

Landis, W. G., M. H. Yu. 1995. *Introduction to Environmental Toxicology.* New York: CRC Press/Lewis Publishers.

Langerman, N. 1994 *Precautionary Labels for Chemical Containers.* New York: CRC Press/Lewis Publishers.

Lauwerys, R. R., P. Hoet. 1993. *Industrial Chemical Exposure.* New York: CRC Press/Lewis Publishers.

Lewis, R. J., Sr. 1992. *Sax's Dangerous Properties of Industrial Materials,* Volumes 1, 2, and 3, 8th Edition. New York: Van Nostrand Reinhold.

Lide, D. R. 1995. *Handbook of Chemistry and Physics,* 76th Edition. New York: CRC Press/Lewis Publishers.

Mackay, D., W. Y. Shiu, and K. C. Ma. 1992. *Illustrated Handbook of Chemical-Physical Properties and Environmental Fate for Organic Chemicals.* New York: CRC Press/Lewis Publishers.

Manahan, S. E. 1994. *Environmental Chemistry,* 6th Edition. New York: CRC Press/Lewis Publishers.

Mellström, G. A., J. E. Wahlberg, and H. I. Maibach. 1994. *Protective Gloves for Occupational Use.* New York: CRC Press/Lewis Publishers.

Meyer, E. 1990. *Chemistry of Hazardous Materials.* Englewood Cliffs, NJ: Prentice Hall.

Mycock, J. C., J. D. McKenna, and L. Theodore. 1995. *Handbook of Air Pollution Control Engineering and Technology.* New York: CRC Press/Lewis Publishers.

Patnaik, P. 1992. *A Comprehensive Guide to the Hazardous Properties of Chemical Substances.* New York: Van Nostrand Reinhold.

Proctor, N. H., J. P. Hughes. 1991. *Chemical Hazards of the Workplace.* New York: Van Nostrand Reinhold.

Quigley, D. R. 1994. *Handbook of Emergency Chemical Management.* New York: CRC Press/Lewis Publishers.

Rea, W. J. 1995. *Chemical Sensitivity,* Vol. I-III. New York: CRC Press/Lewis Publishers.

Richardson, M. 1995. *Dictionary of Substance and Their Effects,* Vol. I-VI. New York: CRC Press/Lewis Publishers.

Sacarello, H. L. A. 1994. *The Comprehensive Handbook of Hazardous Materials.* New York: CRC Press/Lewis Publishers.

Sheftel, V. O. 1995. *Handbook of Toxic Properties of Monomers and Additives.* New York: CRC Press/Lewis Publishers.

Torkelson, T. R. V. K. Rowe. 1981. Halogenated aliphatic hydrocarbons. *Patty's Industrial Hygiene and Toxicology,* Vol. 2B, *Toxicology,* pp 3568-3572. New York: Wiley-Interscience.

U.S. Department of Health and Human Services, National Institute for Occupational Safety and Health. 1994. *NIOSH Pocket Guide to Chemical Hazards.* Washington, D.C.: U.S. Government Printing Office.

MATERIAL SAFETY DATA SHEET

CHEMICAL NAME

3-HEPTENE
(mixed isomers)

HAZARD WARNING INFORMATION

HEALTH	FIRE	REACTIVE	OTHER	DEGREE OF HAZARD	COLOR CODING	OTHER CODES
0	3	1		0 = Minimum Hazard 1 = Slight Hazard 2 = Moderate Hazard 3 = Serious Hazard 4 = Severe Hazard	HEALTH = BLUE FIRE = RED REACTIVITY = YELLOW OTHER = WHITE	OX = Oxidizer ACID = Acid ALK = Alkali COR = Corrosive W = Use No Water

SECTION I - GENERAL INFORMATION

Characterization	Hydrocarbon, Aliphatic
RCRA Number	None
EPA Class	Not Applicable
DOT Proper Shipping Name	No Citation
Chemical Abstract Service (CAS) Number	25339-56-4
DOT Hazard Class and Label Requirements	No Citation
DOT Emergency Guide Code	No Citation
DOT Identification Number	No Citation
Chemical Formula	$C_3H_7CH{:}CHC_2H_5$

Synonyms
Heptylene.

SECTION II - HAZARDOUS INGREDIENTS AND IDENTITY INFORMATION

Hazardous Components (specific identity)	OSHA Exposure Criteria	NIOSH Exposure Criteria	Immediately Dangerous to Life and Health (IDLH)	ACGIH Exposure Criteria
3-Heptene	PEL: Not Established STEL: Not Established	REL: Not Established STEL: Not Established	Not Determined	TLV: Not Established STEL: Not Established

SECTION III - PHYSICAL AND CHEMICAL CHARACTERISTICS

Boiling Point	203°F (95°C)
Specific Gravity (H_2O = 1)	0.705
Vapor Pressure (gas)	Not Found
Molecular Weight	98.21
Vapor Density (Air = 1)	3.38
Melting Point	Not Found

Solubility
Insoluble in water. Soluble in alcohol, acetone, ether, petroleum, and coal-tar solvents.

Appearance and Odor
A clear, colorless liquid.

SECTION IV - FIRE AND EXPLOSION HAZARD DATA

Flash Point (method used)	21°F (-6.1°C) closed cup
Explosive Limits in Air % by Volume	LEL: Not Found UEL: Not Found
NFPA Classification	Class 1C Flammable Liquid
Autoignition Temperature	Not Found

Extinguishing Media
Use dry chemical, carbon dioxide, foam, or water.

Special Fire Fighting Procedures
3-Heptene is a flammable liquid. Wear full protective clothing and self-contained breathing apparatus (SCBA). Poisonous gases are produced in fire. Use water spray to keep fire-exposed containers cool.

Unusual Fire and Explosion Hazards
Dangerous fire and explosion hazard when exposed to heat or flame. Containers may explode in fire. Chemical will react on contact with oxidizable materials. Moderate explosion hazard when in the form of vapor. Its vapors are heavier than air and may travel for some distance to flashback and cause fire.

SECTION V - REACTIVITY DATA

Stability		Conditions to Avoid: Keep away from heat or flame. 3-heptene is normally stable in closed containers under routine conditions of handling and storage.
Stable	Unstable	Incompatibility *(materials to avoid)*
X		Strong oxidizers, such as permanganates, peroxides, nitrates, chlorates, and perchlorates.
Hazardous Polymerization		Conditions to Avoid: Under normal conditions of temperature and pressure, hazardous polymerization of 3-heptene is not expected to occur.
May Occur	Will Not Occur	Hazardous Decomposition or By-products
	X	When heated to decomposition, 3-heptene can emit highly toxic/poisonous gases, including oxides of carbon.

SECTION VI - HEALTH HAZARD DATA

Primary Route(s) of Entry:	Inhalation?	Absorption (skin)?	Ingestion?
	X		X

Health Hazards

INHALATION: An asphyxiant gas. Initial symptoms are those of oxygen starvation. Feeling of "air hunger," diminished mental alertness, headache, dullness, shortness of breath, and an impairment in muscle coordination. Continued exposures lead to faulty judgment, depression of senses, rapid fatigue, unconsciousness, convulsions, coma, and death.

SKIN & EYES: Minimum hazard. On prolonged contact may cause smarting and reddening of the skin.

INGESTION: Possibly mildly toxic. Unspecified gastrointestinal effects.

Carcinogenicity	NTP Listed?	IARC Cancer Review Group?	OSHA Regulated?	Target Organs?
Unknown Human Unknown Animal	No	No	No	Respiratory system, CNS.

Medical Conditions Generally Aggravated by Exposure

None reported.

Emergency and First-aid Procedures

Eye contact: Flush immediately with water for 15 minutes (minimum); seek medical attention. Skin contact: Remove all contaminated clothing. Immediately wash area with large amounts of soap and water. Seek medical assistance. For inhalation: Remove the person from exposure. Provide respiratory assistance and CPR, if required. Transfer to medical facility. Keep victim warm and at rest. If swallowed: Seek medical attention immediately. Contact poison control center for assistance.

SECTION VII - PRECAUTIONS FOR SAFE HANDLING AND USE

Steps to be Taken in Case Material is Released or Spilled

For small spills, absorb in vermiculite and allow to evaporate in a safe place (fume hood). Absorb large spills in vermiculite, dry sand, earth, or similar material and deposit in sealed containers. Ventilate area of spill or leak. Restrict those not involved in cleanup from entering area. Remove all sources of ignition and use only non-sparking tools during cleanup. Notify proper authorities.

Preferred Waste Disposal Method

No Citation.

Precautions to be Taken in Handling and Storage

Store to avoid contact with incompatible materials in tightly closed containers in cool, well-ventilated area away from heat and flame. Use only non-sparking tools to open and close containers.

Other Precautions and Warnings

Sources of ignition are prohibited where 3-heptene is used, handled, or stored. Heat may cause containers to buildup pressure and explode. Containers must be grounded and bonded during transfer.

SECTION VIII - CONTROL MEASURES AND PERSONAL PROTECTIVE EQUIPMENT

Respiratory Protection *(specify type)*

No exposure levels established and 3-heptene is an asphyxiant. Best protection is obtained using a supplied-air respirator with full facepiece in continuous flow mode, or a self-contained breathing apparatus (SCBA) operated in pressure demand or other positive pressure mode.

Ventilation

Local exhaust or general mechanical systems recommended.

Protective Gloves	Eye Protection	Other Protective Clothing
Impervious Material	Chemical Goggles or Face Mask	Protective Clothing

Work/Hygiene Practices

Always wash hands thoroughly after using chemical; never bring food, drink, or smoking materials into vicinity of chemicals.

3-HEPTENE (mixed isomers)

$C_3H_7CH{:}CHC_2H_5$ CAS: 25339-56-4

IDENTIFICATION AND TYPICAL USES

3-Heptene is a colorless liquid of mixed heptene monomers. Its suggested use is as a plant growth retardant.

RISK ASSESSMENT: HEALTH

General Assessment

3-Heptene is a simple asphyxiant gas. As such, its primary route of entry into the body is through *inhalation.* While it does not pose a significant physiological effect (i.e., no systemic poisoning or toxicity), it can displace oxygen levels to below that which is necessary to sustain life. Skin contact with the liquid can cause mild irritation on prolonged contact, but absorption is not a likely exposure route. Ingestion of the liquid may also occur, but little data are available on the resulting effects. There are no data in the references on the carcinogenic, mutagenic, or teratogenic properties of 3-heptene in humans or animals.

Inhalation of 3-heptene vapors is probably narcotic in high concentrations. It can cause mild irritation with headache, dullness, shortness of breath, and the symptoms of asphyxiation. These include rapid respiration and a feeling of "air hunger" or air starvation followed by a decrease in mental alertness, impaired muscular coordination with progression to nausea, vomiting, prostration, unconsciousness, coma, and death.

Skin or eye contact with the vapor poses no serious threat. The vapors have not been shown to cause any physiological response and the liquid has only been shown to cause mild to moderate irritation on prolonged contact. Soaked clothing should be removed immediately to avoid a reaction.

Acute Health Effects

The following acute (short-term) health effects may occur immediately or shortly after exposure to chemicals such as 3-heptene:

Skin: Possible mild irritation.

Eye: Mild to moderate irritation.

Lung: Vapor is only slightly irritating and therefore poses little threat of damage to the lungs. However, in high concentrations, 3-heptene can lead to symptoms of asphyxiation and oxygen deprivation up to and including death.

CNS: Effects of asphyxiation cause adverse nervous system reactions. Possible narcotic effects in high concentrations.

Chronic Health Effects

The following chronic (long-term) health effects may occur at some time after exposure to 3-heptene and can last for months or even years:

Cancer Hazards: According to information presented in the references, exposure to 3-heptene does not lead to carcinogenic effects (it is a simple asphyxiant).

Reproductive Hazards: There are no reports to support any claims of reproductive hazards.

Other Chronic Effects: None reported.

Recommended Risk-Reduction Measures

Personnel should avoid direct contact with 3-heptene. Even though the toxicity of this chemical is considered to be of relatively low order, it can still cause significant health effects if improperly used or if overexposures occur. If a less toxic chemical cannot be substituted for a hazardous substance, then *engineering controls* are the most effective method of reducing exposures. The best protection is to enclose operations and/or provide local exhaust ventilation at the site of chemical release. However, because the vapors of 3-heptene can deplete oxygen levels in confined spaces, caution is warranted if enclosed operations is a chosen method of exposure control. While not always operationally feasible, isolating operations can also reduce exposure. Using respiratory protection is less effective than the controls mentioned above, but is still advisable whenever working with or around 3-heptene. Air-purifying respirators (e.g., the cartridge-type or gas masks) cannot be used with 3-heptene. These types of respirators only purify the air and can do nothing to protect in oxygen deficient atmospheres. A powered air-supplied respirator operated in pressure demand or other positive pressure mode, or a self-contained breathing apparatus (SCBA) with full facepiece operated in pressure demand mode are the recommended respiratory protection methods of choice. If a full facepiece is not available, then chemical goggles should be worn to protect the eyes. Whenever a chemical splash hazard exists, a face shield and a protective apron should be worn. To prevent hand and skin exposures with the liquid, impervious gloves should be used. Glove manufacturers should be contacted and permeation studies obtained *before* final glove selections have been made.

Administrative controls should also be in place to minimize the potential for human exposures. These may include written procedures or policies which specify the methods and techniques that will be practiced whenever personnel are to work with 3-heptene.

All personnel should receive *training* on the use, hazards, protective measures, emergency actions, and other precautions per 29 CFR 1910.1200 (Hazard Communication), prior to the first assignment in an area where 3-heptene is used or stored.

Other methods to reduce exposure include:

- ☑ Where possible, enclose operations and use local exhaust ventilation at the site of chemical release. Because exposure to 3-heptene can cause asphyxiation, the use of local exhaust ventilation is vital. If local exhaust ventilation or enclosure is not possible or feasible, respiratory protection should be mandatory.
- ☑ Always ensure that proper protective clothing is worn when using chemical substances and personnel are trained on its use, care, and maintenance.
- ☑ Wash thoroughly immediately after exposure to 3-heptene and at the end of the work shift or before eating, drinking, or smoking.
- ☑ Hazard warning information should be posted in the work area. In addition, as part of an on-going education and training program, all information on the health and safety hazards of 3-heptene should be communicated to all exposed and potentially exposed workers.
- ☑ Eye wash stations should be provided in the immediate work area for emergency use. If there is a possibility of skin exposure to 3-heptene, emergency shower facilities should also be provided.
- ☑ Personnel should never wear contaminated clothing home (family members can be exposed). Clothing should be laundered by personnel who have been trained on the hazards of 3-heptene.

RISK ASSESSMENT: ENVIRONMENT

General Assessment

The environment is at risk of exposure during transportation, storage, disposal, or destruction of 3-heptene. In almost every scenario, the threat of environmental exposure is contingent upon the proper handling of the chemical substance. Accidental spills, large or small, can result in fire, explosion, and possible contamination of the surrounding environmental mediums (water, soil, and air).

3-Heptene is considered a Class 1C flammable liquid (per OSHA 29 CFR 1910.106). Its extremely low flash point and relatively low boiling point make this chemical a serious fire and explosion hazard concern. These characteristics require extreme caution in handling, storage, transportation, and disposal. It is incompatible with a common peroxides and other oxidizers and contact can cause fire or explosion. Therefore, special consideration is required during any emergency situation involving a leak or spill of 3-heptene. Should 3-heptene come into contact with these incompatible substances during use, transportation, storage, or disposal, the formation of highly toxic and/or highly explosive commodities is possible.

3-Heptene may enter the environment through industrial discharges or spills.

☠ *Acute Ecological Effects*

Acute (short-term) toxic effects may include the death of animals, birds, or fish and death or low growth rate in plants. Acute effects are seen 2 to 4 days after animals or plants are exposed to 3-heptene.

This chemical has high acute toxicity to aquatic life. It may cause membrane damage in ornamental crop species. Insufficient data are available on the short-term effects of 3-heptene exposure to birds or terrestrial animals.

Chronic Ecological Effects

Chronic (long-term) toxic effects may include shortened life span, reproductive problems, lower fertility, and changes in appearance or behavior in exposed animals. These effects can be seen long after first exposure(s) to toxic chemicals.

3-Heptene has moderate chronic toxicity in aquatic life. Insufficient data are available on the long-term effects of 3-heptene to plants, birds, or land animals.

Water Solubility

3-Heptene is insoluble in water. Concentrations of less than 1 milligram will not mix with a liter of water.

Persistence in the Environment

3-Heptene is non-persistent in water, with a half-life of less than 2 days. The half-life of a pollutant is the amount of time it takes for one half of the chemical to be degraded. There are no data on its soil sorption capabilities. It is assumed that virtually 100% of 3-heptene will eventually end up in air.

Bioaccumulation in Aquatic Organisms

Some substances increase in concentration, or *bioaccumulate,* in living organisms as they breathe contaminated air, drink contaminated water, or eat contaminated food. These chemicals can become concentrated in the tissues and internal organs of animals as well as humans.

There are no data available on the bioaccumulation of 3-heptene in aquatic organisms.

Recommended Risk-Reduction Measures

Proper training of all transporters will reduce the likelihood of a mishap or accident resulting in a leak or spill to the environment. The correct use of DOT labeling (per the MSDS) on all containers, trucks, and rail cars will enable emergency responders to react properly and quickly to any disaster thereby reducing the potential risk to the environment and to personnel.

Storage of 3-heptene should be segregated from incompatible chemicals to minimize the risk of cross-contamination or contact. Buildings designated for storage should be equipped with appropriate fire protection systems (alarms, sprinklers, emergency lighting, portable extinguishers). All transfers of 3-heptene from storage containers to process containers should be done automatically whenever possible. All containers and equipment must be grounded and bonded.

If a spill or leak to the environment has occurred, fire department, emergency response and/or hazardous materials spill personnel should be notified immediately. Cleanup should be attempted only by those trained in proper spill containment procedures using non-sparking tools. Contaminated soils should be removed for incineration and replaced with clean soil. If 3-heptene should contact the water table, aquifer, or navigable waterway, the local and/or state emergency response authorities must be notified. A comprehensive emergency response or disaster preparedness/recovery plan should be in place prior to any operations involving the use, transportation, storage, or disposal of 3-heptene. If 3-heptene is spilled or leaked, the following specific steps are recommended:

- ☑ Restrict persons not wearing protective clothing from area of spill or leak until cleanup is complete and area can be opened for normal work.
- ☑ Remove all ignition sources, ventilate area of leak.
- ☑ Absorb small spills (laboratory spills) in vermiculite and allow evaporation in a safe place (a fume hood or open air). Dispose of residue in sealed drums. For larger spills, absorb liquids in vermiculite, dry sand, earth, or similar material and deposit in sealed containers. Collect materials using non-sparking tools.
- ☑ It may be necessary to dispose of 3-heptene as a hazardous waste. The responsible state agency or the regional office of the federal Environmental Protection Agency (EPA) should be contacted for specific recommendations.

RISK ASSESSMENT: BUSINESS

General Assessment

Accidents or mishaps involving 3-heptene can present a significant threat to business operations. The loss or damage of equipment or facilities can significantly affect fiscal viability. Lawsuits that may result from personnel injury/death, public exposures, and/or environmental contamination will also require a serious expenditure of resources.

Media attention surrounding an injury, death, or environmental damage can also result in a loss of profits and loss of current as well as future business.

Recommended Risk-Reduction Measures

Company attorneys, safety and health professionals, and environmental specialists should be involved in the development of any procedures dealing with safety or emergency response. A company official should be pre-designated as a public relations officer with specific training in dealing with the press. Corporate plans and policies should be developed, approved, and implemented long before any need for such arises.

REFERENCES

American Conference of Governmental Industrial Hygienists. 1988. *Documentation of the Threshold Limit Values and Biological Exposure Indices,* 5th Edition (with updates). Cincinnati: ACGIH

Bueche, J. R. 1972. *Principles of Physics,* 2nd Edition. New York: McGraw-Hill.

Calabrese, E. J. 1994. *Biological Effects of Low Level Exposures.* New York: CRC Press/Lewis Publishers.

Cockerman, L. G., B. S. Shane. 1994. *Basic Environmental Toxicology.* New York: CRC Press/Lewis Publishers.

Cothern, C. R. 1995. *Handbook for Environmental Risk Decisionmaking.* New York: CRC Press/Lewis Publishers.

Danse, I. R. 1991. *Common Sense Toxics in the Workplace.* New York: Van Nostrand Reinhold.

Environmental Protection Agency. 1983. *Chemical Hazard Information Profiles.* Washington, D.C.: U.S. EPA.

Fodden, C. R., and J. L. Weddell. 1992. *Hazardous Materials Emergency Action Data.* New York: CRC Press/Lewis Publishers.

Forseberg, K., and L. H. Keith. 1993. *Chemical; Protective Clothing Permeation and Degradation Database.* Chelsea, MI: Lewis Publishers.

Gardner, R. W. 1994. *Chemical Intolerance.* New York: CRC Press/Lewis Publishers.

Hallenbeck, W. H. 1993. *Quantitative Risk Assessment for Environmental and Occupational Health,* 2nd Edition. New York: CRC Press/Lewis Publishers.

Hosty, J. W., and P. Foster. 1990. *A Practical Guide to Chemical Spill Response.* New York: Van Nostrand Reinhold.

Hodgson, E., R. B. Mailman, and Janice E. Chambers. 1988. *Dictionary of Toxicology.* London: The MacMillian Company.

Jones, F. E. 1994. *Toxic Organic Vapors in the Workplace.* New York: CRC Press/Lewis Publishers.

Kamrin, M. A. 1988. *Toxicology: A Primer on Principles and Applications.* Chelsea, Michigan: Lewis Publishers.

Keith, L. H. and M. M. Walker. 1995. *Handbook of Air Toxics.* New York: CRC Press/Lewis Publishers.

Landis, W. G., M. H. Yu. 1995. *Introduction to Environmental Toxicology.* New York: CRC Press/Lewis Publishers.

Langerman, N. 1994 *Precautionary Labels for Chemical Containers.* New York: CRC Press/Lewis Publishers.

Lauwerys, R. R., P. Hoet. 1993. *Industrial Chemical Exposure.* New York: CRC Press/Lewis Publishers.

Lewis, R. J., Sr. 1992. *Sax's Dangerous Properties of Industrial Materials,* Volumes 1, 2, and 3, 8th Edition. New York: Van Nostrand Reinhold.

Lide, D. R. 1995. *Handbook of Chemistry and Physics,* 76th Edition. New York: CRC Press/Lewis Publishers.

Mackay, D., W. Y. Shiu, and K. C. Ma. 1992. *Illustrated Handbook of Chemical-Physical Properties and Environmental Fate for Organic Chemicals.* New York: CRC Press/Lewis Publishers.

Manahan, S. E. 1994. *Environmental Chemistry,* 6th Edition. New York: CRC Press/Lewis Publishers.

Mellström, G. A., J. E. Wahlberg, and H. I. Maibach. 1994. *Protective Gloves for Occupational Use.* New York: CRC Press/Lewis Publishers.

Meyer, E. 1990. *Chemistry of Hazardous Materials.* Englewood Cliffs, NJ: Prentice Hall.

Mycock, J. C., J. D. McKenna, and L. Theodore. 1995. *Handbook of Air Pollution Control Engineering and Technology.* New York: CRC Press/Lewis Publishers.

Patnaik, P. 1992. *A Comprehensive Guide to the Hazardous Properties of Chemical Substances.* New York: Van Nostrand Reinhold.

Proctor, N. H., J. P. Hughes. 1991. *Chemical Hazards of the Workplace.* New York: Van Nostrand Reinhold.

Quigley, D. R. 1994. *Handbook of Emergency Chemical Management.* New York: CRC Press/Lewis Publishers.

Rea, W. J. 1995. *Chemical Sensitivity,* Vol. I-III. New York: CRC Press/Lewis Publishers.

Richardson, M. 1995. *Dictionary of Substance and Their Effects,* Vol. I-VI. New York: CRC Press/Lewis Publishers.

Sacarello, H. L. A. 1994. *The Comprehensive Handbook of Hazardous Materials.* New York: CRC Press/Lewis Publishers.

Sheftel, V. O. 1995. *Handbook of Toxic Properties of Monomers and Additives.* New York: CRC Press/Lewis Publishers.

Torkelson, T. R. V. K. Rowe. 1981. Halogenated aliphatic hydrocarbons. *Patty's Industrial Hygiene and Toxicology,* Vol. 2B, *Toxicology,* pp 3568-3572. New York: Wiley-Interscience.

U.S. Department of Health and Human Services, National Institute for Occupational Safety and Health. 1994. *NIOSH Pocket Guide to Chemical Hazards.* Washington, D.C.: U.S. Government Printing Office.

MATERIAL SAFETY DATA SHEET

CHEMICAL NAME

HEXACHLOROBENZENE

CRC PRESS/LEWIS PUBLISHERS DISCLAIMER: THE INFORMATION AND RECOMMENDATIONS PRESENTED HEREIN ARE BASED ON SOURCES BELIEVED TO BE RELIABLE. CRC MAKES NO REPRESENTATION ON ITS COMPLETENESS OR ACCURACY. IT IS THE RESPONSIBILITY OF THE USER TO DETERMINE THE CHEMICAL'S SUITABILITY FOR ITS INTENDED USE, THE CHEMICAL'S SAFE USE, AND THE CHEMICAL'S PROPER DISPOSAL. NO REPRESENTATIONS AND/OR WARRANTIES, EITHER EXPRESSED OR IMPLIED, OF THE MERCHANTABILITY OR FITNESS FOR A PARTICULAR PURPOSE, OR OF ANY OTHER NATURE, ARE MADE WITH RESPECT TO THE INFORMATION PROVIDED IN THIS MSDS OR TO THE CHEMICAL TO WHICH INFORMATION MAY REFER. CRC NEITHER ASSUMES NOR AUTHORIZES ANY OTHER PERSON TO ASSUME FOR IT, ANY OTHER ADDITIONAL RESPONSIBILITY OR LIABILITY FOR THE USE OF, OR RELIANCE UPON, THIS INFORMATION.

HAZARD WARNING INFORMATION

HEALTH	FIRE	REACTIVE	OTHER	DEGREE OF HAZARD	COLOR CODING	OTHER CODES
3	1	0		0 = Minimum Hazard 1 = Slight Hazard 2 = Moderate Hazard 3 = Serious Hazard 4 = Severe Hazard	HEALTH = BLUE FIRE = RED REACTIVITY = YELLOW OTHER = WHITE	OX = Oxidizer ACID = Acid ALK = Alkali COR = Corrosive W̶ = Use No Water

SECTION I - GENERAL INFORMATION

Characterization	Chlorinated Hydrocarbon, Aromatic
RCRA Number	U127
EPA Class	Toxic Waste
DOT Proper Shipping Name	Hexachlorobenzene
Chemical Abstract Service (CAS) Number	118-74-1
DOT Hazard Class and Label Requirements	Poison B; St. Andrews Cross
DOT Emergency Guide Code	53
DOT Identification Number	UN 2729
Chemical Formula	C_6Cl_6

Synonyms

Amatin; anticarie; bunt-care; bunt-no-more; pentachlorophenyl chloride; perchlorobenzene.

SECTION II - HAZARDOUS INGREDIENTS AND IDENTITY INFORMATION

Hazardous Components (specific identity)	OSHA Exposure Criteria	NIOSH Exposure Criteria	Immediately Dangerous to Life and Health (IDLH)	ACGIH Exposure Criteria
Hexachlorobenzene	PEL: Not Established STEL: Not Established	REL: Not Established STEL: Not Established	Not Determined	TLV: 0.025 mg/m^3 Suspected Human Carcinogen

SECTION III - PHYSICAL AND CHEMICAL CHARACTERISTICS

Boiling Point	613-619°F (323-326°C)
Specific Gravity (H_2O = 1)	2.44
Vapor Pressure (mm Hg)	1 at 237°F (114°C)
Molecular Weight	284.76
Vapor Density (Air = 1)	9.8
Melting Point	444°F (229°C)

Solubility

Insoluble in water. Soluble in benzene and boiling alcohol.

Appearance and Odor

White, monoclinic needles or prisms. Often dissolved in liquid solution.

SECTION IV - FIRE AND EXPLOSION HAZARD DATA

Flash Point (method used)	468°F (242°C) closed cup
Explosive Limits in Air % by Volume	LEL: Not Determined; UEL: Not Determined
NFPA Classification	Non-Combustible Solid
Autoignition Temperature	Not Determined

Extinguishing Media

Carbon dioxide, water spray, foam, or dry chemical.

Special Fire Fighting Procedures

Wear full protective clothing and self-contained breathing apparatus (SCBA). Approach fire from upwind position. Use a water spray to keep fire-exposed containers cool. Poisonous gases are produced in fire. Hexachlorobenzene is a suspected human carcinogen. Avoid smoke and runoff.

Unusual Fire and Explosion Hazards

Can become combustible when exposed to heat and flame. Move containers from fire if it can be done without risk. Violent and explosive reaction with dimethylformamide.

SECTION V - REACTIVITY DATA

Stability		Conditions to Avoid: Hexachlorobenzene is normally stable in closed containers under routine conditions of handling and storage. Avoid contact with heat and incompatible materials.
Stable	Unstable	Incompatibility (*materials to avoid*): Reacts violently and explosively with dimethylformamide. Becomes combustible when exposed to heat or flame.
X		
Hazardous Polymerization		Conditions to Avoid: Under normal conditions of temperature and pressure, hazardous polymerization of hexachlorobenzene is not known to occur.
May Occur	Will Not Occur	Hazardous Decomposition or By-products: When heated to decomposition, hexachlorobenzene can emit highly toxic/poisonous fumes including fumes of chloride.
	X	

SECTION VI - HEALTH HAZARD DATA

Primary Route(s) of Entry:	Inhalation?	Absorption (skin)?	Ingestion?
	X	X	X

Health Hazards

INHALATION: Irritation of the eyes, nose, throat, and respiratory tract. Possible damage to the liver, immune system, thyroid, kidneys, and the nervous system. Causes irritability, muscle weakness, tremors, pins and needles, and other nerve damage.

ABSORPTION: Passes through intact skin to cause toxic systemic effects. A condition called porphiria cutaneatarda can also occur (rash, changes in pigment, skin thickening, red or dark urine, and other body changes). Hexachlorobenzene accumulates in the body.

INGESTION: Toxic systemic effects (see inhalation).

Carcinogenicity	NTP Listed?	IARC Cancer Review Group?	OSHA Regulated?	Target Organs?
Suspected Human Confirmed Animal	5th Annual Report	Group 2B	No	Respiratory system, eyes, skin, CNS, liver, kidney.

Medical Conditions Generally Aggravated by Exposure

Exposure to sunlight shortly after exposure may increase the systemic effects of the chemical.

Emergency and First-aid Procedures

Eye contact: Flush immediately with water for 15 minutes (minimum); seek medical attention. Skin contact: Remove all contaminated clothing. Immediately wash area with large amounts of soap and water. Seek medical assistance. For inhalation: Remove the person from exposure. Provide respiratory assistance and CPR. Transfer to medical facility. If swallowed: Seek medical attention immediately. Gastric lavage followed by saline catharsis recommended.

SECTION VII - PRECAUTIONS FOR SAFE HANDLING AND USE

Steps to be Taken in Case Material is Released or Spilled

Absorb liquid solutions in vermiculite, dry sand, earth, or similar material and deposit in sealed containers. Collect solids using HEPA vacuum. Do NOT dry sweep. Damp-mop residues. Ventilate area of spill or leak. Remove all ignition sources. Restrict those not involved in cleanup from entering area.

Preferred Waste Disposal Method

Dissolve in combustible solvent and burn in a chemical incinerator (with an afterburner and scrubber).

Precautions to be Taken in Handling and Storage

Store in tightly closed containers in cool, well-ventilated area away from heat, sparks, or flames. Sources of ignition are prohibited where hexachlorobenzene is used, handled, or stored.

Other Precautions and Warnings

At temperatures above 149°F (65°C), hexachlorobenzene reacts violently with dimethylformamide.

SECTION VIII - CONTROL MEASURES AND PERSONAL PROTECTIVE EQUIPMENT

Respiratory Protection (*specify type*)

The ACGIH TLV is low and difficult to monitor. For any exposure to a suspected carcinogen, use an MSHA/NIOSH-approved supplied-air respirator with full facepiece, hood, or helmet in continuous flow mode or, use a self-contained breathing apparatus (SCBA) in continuous flow mode.

Ventilation

Local exhaust or general mechanical systems recommended.

Protective Gloves	Eye Protection	Other Protective Clothing
Impervious Material	Chemical Goggles or Face Mask	Protective Apron or Clothing

Work/Hygiene Practices

Always wash hands thoroughly after using chemical; never bring food, drink, or smoking materials into vicinity of chemicals.

HEXACHLOROBENZENE

C_6Cl_6

CAS: 118-74-1

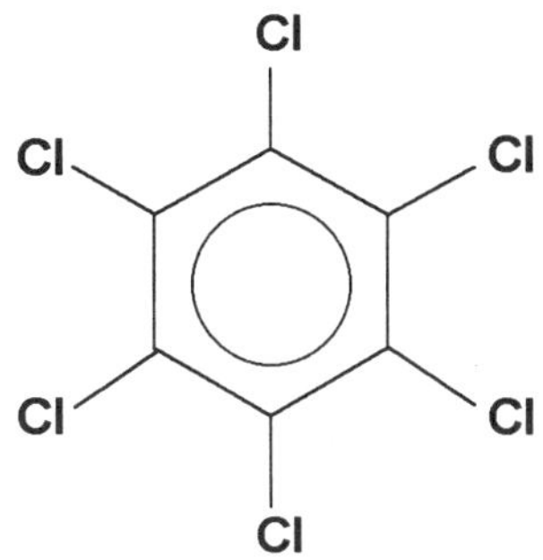

IDENTIFICATION AND TYPICAL USES

Hexachlorobenzene appears as a white crystalline solid in the form of needles or prisms. It is often mixed in solution. It is used as a wood preservative, as a fungicide for treating seeds, and as an intermediate in organic synthesis.

RISK ASSESSMENT: HEALTH

General Assessment

Hexachlorobenzene is an extremely toxic and poisonous substance. It can enter the body through *inhalation, absorption* (through intact skin), and by *ingestion.* It is a confirmed animal carcinogen and has caused teratogenic, reproductive, and mutagenic effects in animal experiments.

Inhalation of hexachlorobenzene vapors can cause irritation of the eyes, nose, throat, and mucosa of the respiratory tract. Exposure by all routes can lead to damage to the liver, immune system, thyroid, kidneys, and the central nervous system (CNS). There may be symptoms of irritability, muscle weakness, tremors, paresthesia (pins and needles), and other nerve damage.

Skin contact causes irritation. Exposure by absorption and possibly by other routes may lead to a condition known as porphiria cutaneatarda with rash, changes in skin pigment, skin thickening, red and dark urine, and other body changes. Hexachlorobenzene accumulates in the body over time.

☠ *Acute Health Effects*

The following acute (short-term) health effects may occur immediately or shortly after exposure to hexachlorobenzene:

Skin: Possible irritation with rash. There may be changes in skin pigment, thickening of the skin, and other changes. Exposure to sunlight with or soon after exposure will increase these effects. After this reaction, blistering of the skin and red or dark urine may be noticed.

Eye: Moderate to severe irritation.

Lung: Irritation of the nose, throat, and lungs following exposure.

CNS: A CNS depressant causing systemic effects including drowsiness, ataxia, weakness, lethargy, malaise, fatigue, lightheadedness, and loss of consciousness.

💣 *Chronic Health Effects*

The following chronic (long-term) health effects may occur at some time after exposure to hexachlorobenzene and can last for months or even years:

Cancer Hazards: Hexachlorobenzene is a known animal carcinogen and has been shown to cause mutations in living cells. Its human carcinogenicity is considered highly suspect.

Reproductive Hazard: According to the data presented in the references, hexachlorobenzene has been shown to cause teratogenic and adverse reproductive effects in test animals.

Other Chronic Effects: Long-term exposure to hexachlorobenzene may cause damage to the liver, immune system (increasing the risk of infection), thyroid, and kidneys. High or repeated exposure may damage the nervous system, and can cause irritability, difficulty with walking and coordination, muscle weakness, tremor, and/or a feeling of "pins and needles" (paresthesia) on the skin. Repeated skin exposures can lead to permanent skin changes, such as changes in pigment, tight, thickened skin, easy wrinkling, skin scarring, fragile skin, and increased hair growth (on the face and forearms).

Recommended Risk-Reduction Measures

Personnel should avoid direct contact with hexachlorobenzene. It is a suspected human carcinogen and has been shown to cause mutagenic, teratogenic, and reproductive effects in animals. If a less toxic chemical cannot be substituted for a hazardous substance, then *engineering controls* are the most effective method of reducing exposures.

The best protection is to enclose operations and/or provide local exhaust ventilation at the site of chemical release. While not always operationally feasible, isolating operations can also reduce exposure. Using respiratory protection is less effective than the controls mentioned above, but is still advisable whenever working with or around hexachlorobenzene. The ACGIH TLV for this chemical is extremely low (0.025 mg/m^3) and difficult to monitor reliably under most operating or emergency conditions. A supplied-air respirator with a full facepiece operated in positive pressure mode or a MSHA/NIOSH-approved self-contained breathing apparatus (SCBA) with full facepiece operated pressure demand or other positive pressure mode are the recommended respiratory protection methods of choice. In either case, if a full facepiece is not available, then chemical splash goggles should be worn to protect the eyes. Whenever a chemical splash hazard exists, a face shield and a protective apron should be worn. To prevent hand and skin exposures, impervious gloves should be used. Glove manufacturers should be contacted and permeation studies obtained *before* final glove selections are made.

Administrative controls should also be in place to minimize the potential for human exposures. These may include written procedures or policies which specify the methods and techniques that will be practiced whenever personnel are to work with hexachlorobenzene.

All personnel should receive *training* on the use, hazards, protective measures, emergency actions, and other precautions per 29 CFR 1910.1200 (Hazard Communication), prior to the first assignment in an area where hexachlorobenzene is used or stored.

If symptoms develop or overexposure is suspected, the following medical tests are recommended:

- ☑ Liver function tests.
- ☑ Interview for nervous system effects, including recent irritability, weakness, loss of coordination, paresthesia, and other such symptoms.
- ☑ Consider cerebellar and autonomic and peripheral nervous system evaluation. Positive and borderline individuals should be referred for neuropsychological testing.

Also, iron used as a dietary supplement could increase liver damage caused by hexachlorobenzene. Personnel who have potential exposure to hexachlorobenzene should be reminded to consult with their personal physicians before taking iron supplements.

Other methods to reduce exposure include:

- ☑ Where possible, enclose operations and use local exhaust ventilation at the site of chemical release. If local exhaust ventilation or enclosure is not used, respiratory protection should be mandatory.
- ☑ Hexachlorobenzene is often mixed in solution with a combustible solvent. Where possible, automatically transfer hexachlorobenzene from drums or other storage containers to process containers.
- ☑ Specific engineering controls are recommended by NIOSH in publication 78-174 (Occupational Exposure During the Manufacture and Formulation of Pesticides) and publication 76-147 (Working with Pesticides).
- ☑ Always ensure that proper protective clothing is worn when using chemical substances.
- ☑ Wash thoroughly immediately after exposure to hexachlorobenzene and at the end of the work shift or before eating, drinking, or smoking.
- ☑ Hazard warning information should be posted in the work area. In addition, as part of an on-going education and training program, all information on the health and safety hazards of hexachlorobenzene should be communicated to all exposed and potentially exposed workers.

RISK ASSESSMENT: ENVIRONMENT

General Assessment

The environment is at risk of exposure during transportation, storage, disposal, or destruction of hexachlorobenzene. In almost every scenario, the threat of environmental exposure is contingent upon the proper handling of the chemical substance. Accidental spills,

large or small, can result in fire, explosion, and possible contamination of the surrounding environmental mediums (water, soil, and air).

Hexachlorobenzene is considered a non-combustible solid but it is often mixed in solution with combustible or possibly flammable commodities. It can also react explosively when heated or when in contact with dimethylformamide. These characteristics require special consideration during any emergency situation involving a leak or spill of hexachlorobenzene.

Hexachlorobenzene has been used for many years as a fungicide and wood preservative. It has been detected in samples taken around the world and is generally recognized as a global pollutant. It may enter the environment from agricultural runoff or atmospheric deposition.

Acute Ecological Effects

Acute (short-term) toxic effects may include the death of animals, birds, or fish and death or low growth rate in plants. Acute effects are seen 2 to 4 days after animals or plants are exposed to hexachlorobenzene.

Hexachlorobenzene has high acute toxicity to aquatic life and moderate acute toxicity to land animals. Insufficient data are available to evaluate the short-term effects of hexachlorobenzene exposure to plants or birds.

Chronic Ecological Effects

Chronic (long-term) toxic effects may include shortened life span, reproductive problems, lower fertility, and changes in appearance or behavior in exposed animals. These effects can be seen long after first exposure(s) to toxic chemicals.

Hexachlorobenzene has high chronic toxicity in aquatic life. Insufficient data are available on the long-term effects of hexachlorobenzene to plants, birds, or terrestrial animals.

Water Solubility

Hexachlorobenzene is only very slightly soluble in the aquatic environment. Concentrations of less than 1 milligram will mix with a liter of water.

Persistence in the Environment

Hexachlorobenzene is highly persistent in the aquatic environment, with a half-life of greater than 200 days. The half-life of a pollutant is the amount of time it takes for one half of the chemical to be degraded.

Approximately 50% of hexachlorobenzene will eventually end up in terrestrial soils; about 46.75% will end up in aquatic sediments, about 3% will end up in air, and approximately 0.25% will end up in water.

Bioaccumulation in Aquatic Organisms

Some substances increase in concentration, or *bioaccumulate,* in living organisms as they breathe contaminated air, drink contaminated water, or eat contaminated food. These chemicals can become concentrated in the tissues and internal organs of animals as well as humans.

The concentration of hexachlorobenzene found in fish tissues is expected to be considerably higher than the average concentration of hexachlorobenzene in the water from which the fish was taken.

Recommended Risk-Reduction Measures

Proper training of all transporters will reduce the likelihood of a mishap or accident resulting in a leak or spill to the environment. The correct use of labeling on all containers, trucks, and rail cars will enable emergency responders to react properly and quickly to any disaster thereby reducing the potential risk to the environment and to personnel.

Storage of hexachlorobenzene should be segregated from incompatible chemicals to minimize the risk of cross-contamination or contact. Storage buildings or facilities should be equipped with proper fire protection equipment (sprinklers/extinguishers, alarms).

If a spill or leak to the environment has occurred, fire department, emergency response and/or hazardous materials spill personnel should be notified immediately. Cleanup should be attempted only by those trained in proper spill containment procedures. Contaminated soils should be removed for incineration and replaced with clean soil. If hexachlorobenzene should contact the water table, aquifer, or navigable waterway, time is of the essence. It is nearly insoluble in the aquatic environment and, therefore, total containment and remediation may be possible. When such spills occur, the local and/or state emergency response authorities must be notified. A comprehensive emergency response or disaster preparedness/recovery plan should be in place prior to any operations involving the use, transportation, storage, or disposal of hexachlorobenzene. The proper disposal method for hexachlorobenzene is to mix it in a combustible solvent and burn it in a chemical incinerator equipped with an afterburner and an air scrubber.

If hexachlorobenzene powders or liquid solutions are spilled or leaked, the following specific response steps are recommended:

☑ Restrict persons not wearing protective clothing from area of spill or leak until cleanup is complete and area can be opened for normal work.

☑ Remove all ignition sources.

☑ Ventilate area of spill or leak.

☑ If liquid is spilled, absorb in vermiculite, dry sand, earth, or similar material and deposit in sealed containers. Collect using non-sparking tools. For solid (powdered) spills, collect using a vacuum equipped with a high-efficiency particulate air (HEPA) filter. Do NOT dry sweep (generates dusts). Damp mop residues and collect for disposal.

☑ It may be necessary to dispose of hexachlorobenzene as a hazardous waste. The responsible state agency or the regional office of the federal Environmental Protection Agency (EPA) should be contacted for specific recommendations.

RISK ASSESSMENT: BUSINESS

General Assessment

Accidents or mishaps involving hexachlorobenzene can present a significant threat to business operations. The loss or damage of equipment or facilities can significantly affect fiscal viability. Lawsuits that may result from personnel injury/death, public exposures, and/or environmental contamination will also require a serious expenditure of resources. Media attention surrounding an injury, death, or environmental damage can also result in a loss of profits and loss of current as well as future business. Always remeber that anytime the terms "cancer," "carcinogen," or "reproductive hazard" are used, public emotion, hysteria, and ignorance can run equally high. This must be carefully considered whenever developing or implementing any public relations policies.

Recommended Risk-Reduction Measures

Company attorneys, safety and health professionals, and environmental specialists should be involved in the development of any procedures for responding to hazardous chemical incidents . A company official should be pre-designated as a public relations officer with specific training in dealing with the press. Corporate plans and policies should be developed, approved, and implemented long before any need for such arises.

REFERENCES

American Conference of Governmental Industrial Hygienists. 1988. *Documentation of the Threshold Limit Values and Biological Exposure Indices,* 5th Edition (with updates). Cincinnati: ACGIH

Bloom, A. D. and F. J. DeSerres. 1995. *Ecotoxicity and Human Health.* New York: CRC Press/Lewis Publishers.

Bueche, J. R. 1972. *Principles of Physics,* 2nd Edition. New York: McGraw-Hill.

Calabrese, E. J. 1994. *Biological Effects of Low Level Exposures.* New York: CRC Press/Lewis Publishers.

Cockerman, L. G., B. S. Shane. 1994. *Basic Environmental Toxicology.* New York: CRC Press/Lewis Publishers.

Cothern, R. R. 1993. *Comparative Environmental Risk Assessment.* New York: CRC Press/Lewis Publishers.

Gardner, R. W. 1994. *Chemical Intolerance.* New York: CRC Press/Lewis Publishers.

Hallenbeck, W. H. 1993. *Quantitative Risk Assessment for Environmental and Occupational Health,* 2nd Edition. New York: CRC Press/Lewis Publishers.

Howard, P. H., et. al. 1991. *Handbook of Environmental Degradation Rates.* New York: CRC Press/Lewis Publishers.

Jones, F. E. 1994. *Toxic Organic Vapors in the Workplace.* New York: CRC Press/Lewis Publishers.

Langerman, N. 1994 *Precautionary Labels for Chemical Containers.* New York: CRC Press/Lewis Publishers.

Lauwerys, R. R., P. Hoet. 1993. *Industrial Chemical Exposure.* New York: CRC Press/Lewis Publishers.

Lewis, R. J., Sr. 1992. *Sax's Dangerous Properties of Industrial Materials,* Volumes 1, 2, and 3, 8th Edition. New York: Van Nostrand Reinhold.

Manahan, S. E. 1992. *Toxicological Chemistry,* 2nd Edition. New York: CRC Press/Lewis Publishers.

Meyer, E. 1990. *Chemistry of Hazardous Materials.* Englewood Cliffs, NJ: Prentice Hall.

Patnaik, P. 1992. *A Comprehensive Guide to the Hazardous Properties of Chemical Substances.* New York: Van Nostrand Reinhold.

Rea, W. J. 1995. *Chemical Sensitivity,* Vol. I-III. New York: CRC Press/Lewis Publishers.

U.S. Department of Health and Human Services, National Institute for Occupational Safety and Health. 1978. Publication 78-174. *Occupational Exposure During the Manufacture and Formulation of Pesticides.* Washington, D.C.: U.S. Government Printing Office.

U.S. Department of Health and Human Services, National Institute for Occupational Safety and Health. 1976. Publication 76-147. *Working with Pesticides.* Washington, D.C.: U.S. Government Printing Office.

MATERIAL SAFETY DATA SHEET

CHEMICAL NAME	CRC PRESS/LEWIS PUBLISHERS DISCLAIMER
HEXACHLOROBUTADIENE	

HAZARD WARNING INFORMATION

HEALTH	FIRE	REACTIVE	OTHER	DEGREE OF HAZARD	COLOR CODING	OTHER CODES
2	1	1		0 = Minimum Hazard 1 = Slight Hazard 2 = Moderate Hazard 3 = Serious Hazard 4 = Severe Hazard	HEALTH = BLUE FIRE = RED REACTIVITY = YELLOW OTHER = WHITE	OX = Oxidizer ACID = Acid ALK = Alkali COR = Corrosive W = Use No Water

SECTION I - GENERAL INFORMATION

Characterization **Chlorinated Hydrocarbon, Aromatic**	RCRA Number **U128**	EPA Class **Toxic Waste**
DOT Proper Shipping Name **Hexachlorobutadiene**	Chemical Abstract Service (CAS) Number **87-68-3**	
DOT Hazard Class and Label Requirements **Poison B; St. Andrews Cross**	DOT Emergency Guide Code **55**	
DOT Identification Number **UN 2279**	Chemical Formula **Cl_2C:ClCCl:CCl_2**	

Synonyms
HCBD; hexachloro-1,3-butadiene; 1,3-hexachlorobutadiene; perchlorobutadiene.

SECTION II - HAZARDOUS INGREDIENTS AND IDENTITY INFORMATION

Hazardous Components (specific identity)	OSHA Exposure Criteria	NIOSH Exposure Criteria	Immediately Dangerous to Life and Health (IDLH)	ACGIH Exposure Criteria
Hexachlorobutadiene (derivation: A by-product of the chlorination of various hydrocarbons, such as tetrachloroethylene, trichloroethylene, and carbon trichloride). 1 ppm = 10.84 mg/m^3	PEL (skin): 0.02 ppm 0.24 mg/m^3 STEL: Not Established	REL (skin): 0.02 ppm 0.24 mg/m^3 Possible Human Carcinogen	Not Determined	TLV (skin): 0.02 ppm 0.021 mg/m^3 Suspected Human Carcinogen

SECTION III - PHYSICAL AND CHEMICAL CHARACTERISTICS

Boiling Point **419°F (215°C)**	Specific Gravity (H_2O = 1) **1.55**
Vapor Pressure (mm Hg) **0.2 at 68°F (20°C)**	Molecular Weight **260.74**
Vapor Density (Air = 1) **9.0**	Melting Point **-6°F (-21°C)**

Solubility
Insoluble in water. Soluble in alcohol and ether, compatible (miscible) with many resins.

Appearance and Odor
Clear, colorless, heavy liquid with a mild turpentine-like odor. Odor Threshold ≈ 1 ppm.

SECTION IV - FIRE AND EXPLOSION HAZARD DATA

Flash Point (method used) **Not Determined**	Explosive Limits in Air % by Volume **LEL: Not Determined UEL: Not Determined**
NFPA Classification **Combustible Liquid**	Autoignition Temperature **1130°F (610°C)**

Extinguishing Media
Carbon dioxide, water spray, foam, or dry chemical.

Special Fire Fighting Procedures
Wear full protective clothing and self-contained breathing apparatus (SCBA). Approach fire from upwind position. Use a water spray to keep fire-exposed containers cool. Poisonous gases are produced in fire. Hexachlorobutadiene is a suspected human carcinogen. Avoid smoke and runoff.

Unusual Fire and Explosion Hazards
Can become combustible when exposed to heat and flame. Move containers from fire if it can be done without risk. Containers may explode in fire.

SECTION V - REACTIVITY DATA

Stability		Conditions to Avoid **Hexachlorobutadiene is normally stable in closed containers under routine conditions of handling and storage. Avoid contact with heat and incompatible materials.**
Stable **X**	Unstable	Incompatibility *(materials to avoid)* **Reacts with bromine perchlorate to form an explosive product. Combustible when exposed to heat or flame. Can react vigorously with oxidizing materials.**
Hazardous Polymerization		Conditions to Avoid **Under normal conditions of temperature and pressure, hazardous polymerization of hexachlorobutadiene is not known to occur.**
May Occur	Will Not Occur **X**	Hazardous Decomposition or By-products **When heated to decomposition, hexachlorobutadiene can emit highly toxic/poisonous fumes including fumes of chloride and carbon dioxide.**

SECTION VI - HEALTH HAZARD DATA

Primary Route(s) of Entry:	Inhalation?	Absorption (skin)?	Ingestion?
	X	X	X

Health Hazards

INHALATION: Irritation of the eyes, nose, throat, and respiratory tract. Possible damage to the liver, kidneys, and the nervous system. Causes irritability, muscle weakness, tremors, pins and needles, and other nerve damage.

ABSORPTION: Passes through intact skin to cause toxic systemic effects. Contact will also cause surface irritation to the skin and may damage the eyes if not removed immediately.

INGESTION: Toxic systemic effects (see inhalation).

Carcinogenicity	NTP Listed?	IARC Cancer Review Group?	OSHA Regulated?	Target Organs?
Suspected Human **Confirmed Animal**	**No**	**Group 3**	**No**	**Respiratory system, eyes, skin, CNS, liver, kidney.**

Medical Conditions Generally Aggravated by Exposure

None Reported.

Emergency and First-aid Procedures

<u>Eye contact</u>: Flush immediately with water for 15 minutes (minimum); seek medical attention. <u>Skin contact</u>: Remove all contaminated clothing. Immediately wash area with large amounts of soap and water. For red or blistered skin, seek medical assistance. For <u>inhalation</u>: Remove the person from exposure. Provide respiratory assistance and CPR. Transfer to medical facility. If <u>swallowed</u>: Seek medical attention immediately. Contact poison control center. Unless advised otherwise, give 1 to 2 glasses of water and induce vomiting. Never make an unconscious or convulsing person vomit.

SECTION VII - PRECAUTIONS FOR SAFE HANDLING AND USE

Steps to be Taken in Case Material is Released or Spilled

Absorb liquid solutions in vermiculite, dry sand, earth, or similar material and deposit in sealed containers. Ventilate area of spill or leak. Remove all ignition sources. Restrict those not involved in cleanup from entering area. Hexachlorobutadiene is a hazardous waste; handle accordingly.

Preferred Waste Disposal Method

Burn in a chemical incinerator equipped with an afterburner and scrubber.

Precautions to be Taken in Handling and Storage

Store in tightly closed containers in cool, well-ventilated area away from heat, sparks, or flames. Sources of ignition are prohibited where hexachlorobutadiene is used, handled, or stored.

Other Precautions and Warnings

A regulated, marked controlled area should be established where hexachlorobutadiene is handled, used, or stored.

SECTION VIII - CONTROL MEASURES AND PERSONAL PROTECTIVE EQUIPMENT

Respiratory Protection *(specify type)*

The PEL is low and difficult to monitor. For any exposure to a suspected carcinogen, use an MSHA/NIOSH-approved supplied-air respirator with full facepiece, hood, or helmet in continuous flow mode, or use a self-contained breathing apparatus (SCBA) in continuous flow mode.

Ventilation

Local exhaust or general mechanical systems recommended.

Protective Gloves	Eye Protection	Other Protective Clothing
Impervious Material	**Chemical Goggles or Face Mask**	**Protective Apron or Clothing**

Work/Hygiene Practices

Always wash hands thoroughly after using chemical; never bring food, drink, or smoking materials into vicinity of chemicals.

HEXACHLOROBUTADIENE

$Cl_2C{:}CClCCl{:}CCl_2$

CAS: 87-68-3

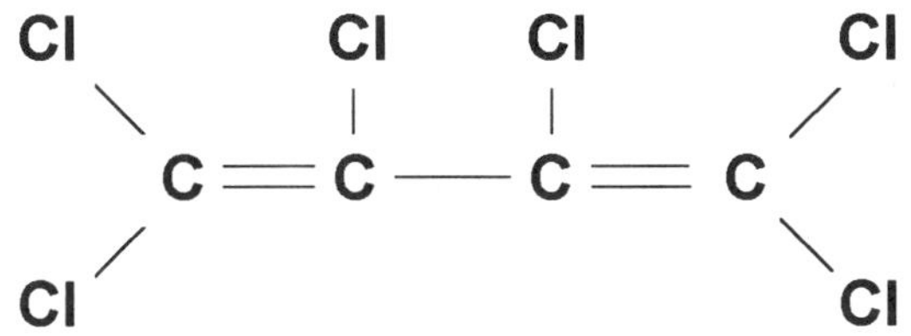

IDENTIFICATION AND TYPICAL USES

Hexachlorobutadiene appears as a clear, colorless, heavy liquid with a mild, turpentine-like odor. It is used as a solvent for elastomers, as a heat-transfer liquid, a transformer and hydraulic fluid, a wash liquor for removing C_4 and higher hydrocarbons.

RISK ASSESSMENT: HEALTH

General Assessment

Hexachlorobutadiene is an extremely toxic and poisonous substance. It can enter the body through *inhalation, absorption* (through intact skin), and by *ingestion.* It is a confirmed animal carcinogen and has caused teratogenic, reproductive, and mutagenic effects in animal experiments.

Inhalation of hexachlorobutadiene vapors can cause irritation of the eyes, nose, throat, and mucosa of the respiratory tract. It may lead to chronic bronchitis and possibly hepatitis. Exposure by all routes can lead to damage to the liver, kidneys, and the central nervous system (CNS). There may be symptoms of hypotension, cardiac disease, nervous system functional disturbances such as irritability, muscle weakness, tremors, paresthesia (pins and needles), and other nerve damage.

Skin contact causes localized irritation at site of contact. Simultaneous exposure by absorption and other routes can cause a compound effect on the body.

Acute Health Effects

The following acute (short-term) health effects may occur immediately or shortly after exposure to hexachlorobutadiene:

Skin: Possible irritation with rash.

Eye: Moderate to severe irritation.

Lung: Irritation of the nose, throat, and lungs following exposure.

CNS: A CNS depressant causing systemic effects including drowsiness, ataxia, weakness, lethargy, malaise, fatigue, lightheadedness, and loss of consciousness.

Chronic Health Effects

The following chronic (long-term) health effects may occur at some time after exposure to hexachlorobutadiene and can last for months or even years:

Cancer Hazards: Hexachlorobutadiene is a known animal carcinogen and has been shown to cause mutations in living cells. Its human carcinogenicity is considered highly suspect.

Reproductive Hazard: According to the data presented in the references, hexachlorobutadiene has been shown to cause teratogenic and adverse reproductive effects in test animals.

Other Chronic Effects: Long-term exposure to hexachlorobutadiene may cause damage to the liver and/or kidneys and may lead to cardiac disease and possibly hypotension.

Recommended Risk-Reduction Measures

Personnel should avoid direct contact with hexachlorobutadiene. It is a suspected human carcinogen and has been shown to cause mutagenic, teratogenic, and reproductive effects in animals. If a less toxic chemical cannot be substituted for a hazardous substance, then *engineering controls* are the most effective method of reducing exposures.

The best protection is to enclose operations and/or provide local exhaust ventilation at the site of chemical release. While not always operationally feasible, isolating operations can also reduce exposure. Using respiratory protection is less effective than the controls mentioned above, but is still advisable whenever working with or around hexachlorobutadiene. The PEL for this chemical is extremely low (0.02 ppm) and difficult to monitor reliably under most operating or emergency conditions. A supplied-air respirator with a full facepiece operated in positive pressure mode or a MSHA/NIOSH-approved self-contained breathing apparatus (SCBA) with full facepiece operated pressure demand or other positive pressure mode are the recommended respiratory protection methods of choice. In either case, if a full facepiece is not available, then chemical splash goggles should be worn to protect the eyes. Whenever a chemical splash hazard exists, a face shield and a protective apron should be worn. To prevent hand and skin exposures, impervious gloves should be used. Glove manufacturers should be contacted and permeation studies obtained *before* final glove selections are made.

Administrative controls should also be in place to minimize the potential for human exposures. These may include written procedures or policies which specify the methods and techniques that will be practiced whenever personnel are to work with hexachlorobutadiene.

All personnel should receive *training* on the use, hazards, protective measures, emergency actions, and other precautions per 29 CFR 1910.1200 (Hazard Communication), prior to the first assignment in an area where hexachlorobutadiene is used or stored.

If symptoms develop or overexposure is suspected, the following medical tests are recommended:

- ☑ Liver function tests.
- ☑ Kidney function tests.
- ☑ Interview for nervous system effects, including recent irritability, weakness, loss of coordination, paresthesia, and other such symptoms.
- ☑ Consider cerebellar and autonomic and peripheral nervous system evaluation. Positive and borderline individuals should be referred for neuropsychological testing.

Also, because more than light or casual alcohol consumption can lead to liver damage, drinking alcohol may increase the potential liver damage caused by hexachlorobutadiene. Prudent risk management requires proper consideration of *all* possible exposure factors when evaluating workplace health hazard risk.

Other methods to reduce exposure include:

- ☑ Where possible, enclose operations and use local exhaust ventilation at the site of chemical release. If local exhaust ventilation or enclosure is not used, respiratory protection should be mandatory.
- ☑ Where possible, automatically transfer hexachlorobutadiene from drums or other storage containers to process containers.
- ☑ Always ensure that proper protective clothing is worn when using chemical substances.
- ☑ Wash thoroughly immediately after exposure to hexachlorobutadiene and at the end of the work shift or before eating, drinking, or smoking.
- ☑ Hazard warning information should be posted in the work area. In addition, as part of an on-going education and training program, all information on the health and safety hazards of hexachlorobutadiene should be communicated to all exposed and potentially exposed workers.

RISK ASSESSMENT: ENVIRONMENT

General Assessment

The environment is at risk of exposure during transportation, storage, disposal, or destruction of hexachlorobutadiene. In almost every scenario, the threat of environmental exposure is contingent upon the proper handling of the chemical substance. Accidental spills, large or small, can result in fire, explosion, and possible contamination of the surrounding environmental mediums (water, soil, and air).

Hexachlorobutadiene is considered a combustible liquid. It can react in contact with bromine perchlorate to form an explosive product. It can react vigorously with oxidizing agents. These characteristics require special consideration during any emergency situation involving a leak or spill of hexachlorobutadiene.

Hexachlorobutadiene may enter the environment primarily during the disposal of wastes from the chlorinated hydrocarbon industry.

☠ *Acute Ecological Effects*

Acute (short-term) toxic effects may include the death of animals, birds, or fish and death or low growth rate in plants. Acute effects are seen 2 to 4 days after animals or plants are exposed to hexachlorobutadiene.

Hexachlorobutadiene has high acute toxicity to aquatic life. Insufficient data are available to evaluate the short-term effects of hexachlorobutadiene exposure to plants, birds, or land animals.

Chronic Ecological Effects

Chronic (long-term) toxic effects may include shortened life span, reproductive problems, lower fertility, and changes in appearance or behavior in exposed animals. These effects can be seen long after first exposure(s) to toxic chemicals.

Hexachlorobutadiene has high chronic toxicity in aquatic life. Insufficient data are available on the long-term effects of hexachlorobutadiene to plants, birds, or terrestrial animals.

Water Solubility

Hexachlorobutadiene is moderately soluble in the aquatic environment. Concentrations between 1 and 1000 milligrams will mix with a liter of water.

Persistence in the Environment

Hexachlorobutadiene is non-persistent in the aquatic environment, with a half-life of less than 2 days. The half-life of a pollutant is the amount of time it takes for one half of the chemical to be degraded.

Approximately 98.6% of hexachlorobutadiene will eventually end up in air; about 0.4% will end up in water; and the remaining 1% will be divided about evenly between terrestrial soils and aquatic sediments.

Bioaccumulation in Aquatic Organisms

Some substances increase in concentration, or *bioaccumulate,* in living organisms as they breathe contaminated air, drink contaminated water, or eat contaminated food. These chemicals can become concentrated in the tissues and internal organs of animals as well as humans.

The concentration of hexachlorobutadiene found in fish tissues is expected to be much higher than the average concentration of hexachlorobutadiene in the water from which the fish was taken.

Recommended Risk-Reduction Measures

Proper training of all transporters will reduce the likelihood of a mishap or accident resulting in a leak or spill to the environment. The correct use of labeling on all containers, trucks, and rail cars will enable emergency responders to react properly and quickly to any disaster thereby reducing the potential risk to the environment and to personnel.

Storage of hexachlorobutadiene should be segregated from incompatible chemicals to minimize the risk of cross-contamination or contact. Storage buildings or facilities should be equipped with proper fire protection equipment (sprinklers/extinguishers, alarms).

If a spill or leak to the environment has occurred, fire department, emergency response and/or hazardous materials spill personnel should be notified immediately. Cleanup should be attempted only by those trained in proper spill containment procedures. Contaminated soils should be removed for incineration and replaced with clean soil. If hexachlorobutadiene should contact the water table, aquifer, or navigable waterway, time is of the essence. It is moderately soluble in the aquatic environment and, therefore, total containment and remediation may not be possible. When such spills occur, the local and/or state emergency response authorities must be notified. A comprehensive emergency response or disaster preparedness/recovery plan should be in place prior to any operations involving the use, transportation, storage, or disposal of hexachlorobutadiene. The proper disposal method for hexachlorobutadiene is to burn it in a chemical incinerator equipped with an afterburner and an air scrubber.

If hexachlorobutadiene is spilled or leaked, the following specific response steps are recommended:

- ☑ Restrict persons not wearing protective clothing from area of spill or leak until cleanup is complete and area can be opened for normal work.
- ☑ Remove all ignition sources.
- ☑ Ventilate area of spill or leak.
- ☑ If liquid is spilled, absorb in vermiculite, dry sand, earth, or similar material and deposit in sealed containers.
- ☑ It may be necessary to dispose of hexachlorobutadiene as a hazardous waste. The responsible state agency or the regional office of the federal Environmental Protection Agency (EPA) should be contacted for specific recommendations.

RISK ASSESSMENT: BUSINESS

General Assessment

Accidents or mishaps involving hexachlorobutadiene can present a significant threat to business operations. The loss or damage of equipment or facilities can significantly affect fiscal viability. Lawsuits that may result from personnel injury/death, public exposures, and/or environmental contamination will also require a serious expenditure of resources. Media attention surrounding an injury, death, or environmental damage can also result in a loss of profits and loss of current as well as future business. Always remember that anytime the terms "cancer," "carcinogen," or "reproductive hazard" are used, public emotion, hysteria, and ignorance can run equally high. This must be carefully considered whenever developing or implementing any public relations policies.

Recommended Risk-Reduction Measures

Company attorneys, safety and health professionals, and environmental specialists should be involved in the development of any procedures for responding to hazardous chemical incidents . A company official should be pre-designated as a public relations officer with specific training in dealing with the press. Corporate plans and policies should be developed, approved, and implemented long before any need for such arises.

REFERENCES

American Conference of Governmental Industrial Hygienists. 1988. *Documentation of the Threshold Limit Values and Biological Exposure Indices,* 5th Edition (with updates). Cincinnati: ACGIH

Bloom, A. D. and F. J. DeSerres. 1995. *Ecotoxicity and Human Health.* New York: CRC Press/Lewis Publishers.

Bueche, J. R. 1972. *Principles of Physics,* 2nd Edition. New York: McGraw-Hill.

Calabrese, E. J. 1994. *Biological Effects of Low Level Exposures.* New York: CRC Press/Lewis Publishers.

Cockerman, L. G., B. S. Shane. 1994. *Basic Environmental Toxicology.* New York: CRC Press/Lewis Publishers.

Cothern, R. R. 1993. *Comparative Environmental Risk Assessment.* New York: CRC Press/Lewis Publishers.

Danse, I. R. 1991. *Common Sense Toxics in the Workplace.* New York: Van Nostrand Reinhold.

Environmental Protection Agency. 1983. *Chemical Hazard Information Profiles.* Washington, D.C.: U.S. EPA.

Fodden, C. R., and J. L. Weddell. 1992. *Hazardous Materials Emergency Action Data.* New York: CRC Press/Lewis Publishers.

Forseberg, K., and L. H. Keith. 1993. *Chemical; Protective Clothing Permeation and Degradation Database.* Chelsea, MI: Lewis Publishers.

Gardner, R. W. 1994. *Chemical Intolerance.* New York: CRC Press/Lewis Publishers.

Gerarde, H. W. 1960. *Toxicology and Biochemistry of Aromatic Hydrocarbons.* New York: Elsevier.

Hallenbeck, W. H. 1993. *Quantitative Risk Assessment for Environmental and Occupational Health,* 2nd Edition. New York: CRC Press/Lewis Publishers.

Howard, P. H., et. al. 1991. *Handbook of Environmental Degradation Rates.* New York: CRC Press/Lewis Publishers.

Hosty, J. W., and P. Foster. 1990. *A Practical Guide to Chemical Spill Response.* New York: Van Nostrand Reinhold.

International Agency for Research on Cancer (IARC). 1987. *IARC Monographs on the Evaluation of Carcinogenic Risk to Humans, Overall Evaluations of Carcinogenicity: An updating of IARC Monographs Volumes 1-42.* Lyon: IARC.

Jones, F. E. 1994. *Toxic Organic Vapors in the Workplace.* New York: CRC Press/Lewis Publishers.

Keith, L. H. and M. M. Walker. 1995. *Handbook of Air Toxics.* New York: CRC Press/Lewis Publishers.

Langerman, N. 1994 *Precautionary Labels for Chemical Containers.* New York: CRC Press/Lewis Publishers.

Lauwerys, R. R., P. Hoet. 1993. *Industrial Chemical Exposure.* New York: CRC Press/Lewis Publishers.

Lewis, R. J., Sr. 1992. *Sax's Dangerous Properties of Industrial Materials,* Volumes 1, 2, and 3, 8th Edition. New York: Van Nostrand Reinhold.

Mackay, D., W. Y. Shiu, and K. C. Ma. 1992. *Illustrated Handbook of Chemical-Physical Properties and Environmental Fate for Organic Chemicals.* New York: CRC Press/Lewis Publishers.

Manahan, S. E. 1992. *Toxicological Chemistry,* 2nd Edition. New York: CRC Press/Lewis Publishers.

Meyer, E. 1990. *Chemistry of Hazardous Materials.* Englewood Cliffs, NJ: Prentice Hall.

Miller, B. F., M.D., and C. B. Keane, R.N. 1972. *Encyclopedia and Dictionary of Medicine and Nursing.* Philadelphia: W.B. Sanders Company.

Mycock, J. C., J. D. McKenna, and L. Theodore. 1995. *Handbook of Air Pollution Control Engineering and Technology.* New York: CRC Press/Lewis Publishers.

Nelson, G. O. 1992. *Gas Mixtures.* New York: CRC Press/Lewis Publishers.

Ott, W. R. 1995. *Environmental Statistics and Data Analysis.* New York: CRC Press/Lewis Publishers.

Patnaik, P. 1992. *A Comprehensive Guide to the Hazardous Properties of Chemical Substances.* New York: Van Nostrand Reinhold.

Philp, R. B. 1995. *Environmental Hazards and Human Health.* New York: CRC Press/Lewis Publishers.

Rea, W. J. 1995. *Chemical Sensitivity,* Vol. I-III. New York: CRC Press/Lewis Publishers.

Sacarello, H. L. A. 1994. *The Comprehensive Handbook of Hazardous Materials.* New York: CRC Press/Lewis Publishers.

Sherman, J. D. 1988. *Chemical Exposure and Disease.* New York: Van Nostrand Reinhold.

U.S. Department of Health and Human Services, National Institute for Occupational Safety and Health. 1994. *NIOSH Pocket Guide to Chemical Hazards.* Washington, D.C.: U.S. Government Printing Office.

MATERIAL SAFETY DATA SHEET

CHEMICAL NAME	
HEXACHLORO-CYCLOPENTADIENE	**CRC PRESS/LEWIS PUBLISHERS DISCLAIMER:** THE INFORMATION AND RECOMMENDATIONS PRESENTED HEREIN ARE BASED ON SOURCES BELIEVED TO BE RELIABLE. CRC MAKES NO REPRESENTATION ON ITS COMPLETENESS OR ACCURACY. IT IS THE RESPONSIBILITY OF THE USER TO DETERMINE THE CHEMICAL'S SUITABILITY FOR ITS INTENDED USE, THE CHEMICAL'S SAFE USE, AND THE CHEMICAL'S PROPER DISPOSAL. NO REPRESENTATIONS AND/OR WARRANTIES, EITHER EXPRESSED OR IMPLIED, OF THE MERCHANTABILITY OR FITNESS FOR A PARTICULAR PURPOSE, OR OF ANY OTHER NATURE, ARE MADE WITH RESPECT TO THE INFORMATION PROVIDED IN THIS MSDS OR TO THE CHEMICAL TO WHICH INFORMATION MAY REFER. CRC NEITHER ASSUMES NOR AUTHORIZES ANY OTHER PERSON TO ASSUME FOR IT, ANY OTHER ADDITIONAL RESPONSIBILITY OR LIABILITY FOR THE USE OF, OR RELIANCE UPON, THIS INFORMATION.

HAZARD WARNING INFORMATION

HEALTH	FIRE	REACTIVE	OTHER	DEGREE OF HAZARD	COLOR CODING	OTHER CODES
3	0	1	~~W~~	0 = Minimum Hazard 1 = Slight Hazard 2 = Moderate Hazard 3 = Serious Hazard 4 = Severe Hazard	HEALTH = BLUE FIRE = RED REACTIVITY = YELLOW OTHER = WHITE	OX = Oxidizer ACID = Acid ALK = Alkali COR = Corrosive ~~W~~ = Use No Water

SECTION I - GENERAL INFORMATION

Characterization	RCRA Number	EPA Class
Chlorinated Aliphatic Hydrocarbon	**U130**	**Toxic Waste**

DOT Proper Shipping Name **Hexachlorocyclopentadiene**	Chemical Abstract Service (CAS) Number **77-47-4**
DOT Hazard Class and Label Requirements **Corrosive Material; Corrosive**	DOT Emergency Guide Code **55**
DOT Identification Number **UN 2646**	Chemical Formula **C_5Cl_6**

Synonyms

HCCPD; hexachloro-1,3-cyclopentadiene; 1,2,3,4,5,5-hexachloro-1,3-cyclopentadiene; Graphlox; perchlorocyclopentadiene;

SECTION II - HAZARDOUS INGREDIENTS AND IDENTITY INFORMATION

Hazardous Components (specific identity)	OSHA Exposure Criteria	NIOSH Exposure Criteria	Immediately Dangerous to Life and Health (IDLH)	ACGIH Exposure Criteria
Hexachlorocyclopentadiene (derivation: By chlorinating cyclopentadiene, followed by catalytic chlorination over phosphorus pentachloride or aresenic oxide and thermally dechlorinating the product). **1 ppm = 11.34 mg/m³**	**PEL:** **0.01 ppm** **0.1 mg/m³** **STEL:** **Not Established**	**REL:** **0.01 ppm** **0.1 mg/m³** **STEL:** **Not Established**	**Not Determined**	**TLV:** **0.01 ppm** **0.11 mg/m³** **STEL:** **Not Established**

SECTION III - PHYSICAL AND CHEMICAL CHARACTERISTICS

Boiling Point **462°F (239°C)**	Specific Gravity (H_2O = 1) **1.71**
Vapor Pressure (gas) **0.08 at 77°F (25°C)**	Molecular Weight **272.8**
Vapor Density (Air = 1) **9.42**	Melting Point **16°F (-9°C)**

Solubility

Practically insoluble in water (0.0002% at 77°F). Soluble in acetone, carbon tetrachloride, hexane, and methanol.

Appearance and Odor

Pale-yellow to amber-colored dense, oily liquid with a pungent, unpleasant odor (a solid below 16°F). Odor Threshold = 0.03 ppm.

SECTION IV - FIRE AND EXPLOSION HAZARD DATA

Flash Point (method used) **None Reported (opened cup)**	Explosive Limits in Air % by Volume **LEL: Not Determined** **UEL: Not Determined**
NFPA Classification **Non-Combustible Liquid**	Autoignition Temperature **Not Determined**

Extinguishing Media

Use dry chemical, foam, carbon dioxide. Do NOT use water spray unless no other means are available.

Special Fire Fighting Procedures

Wear full protective clothing and self-contained breathing apparatus (SCBA). Poisonous gases are produced in fire. Do not allow water or water spray to enter drums or storage containers.

Unusual Fire and Explosion Hazards

Will react with water to form explosive hydrogen gas. Container may explode violently in heat.

SECTION V - REACTIVITY DATA

Stability		Conditions to Avoid **Stable in dry, closed containers. Keep away from heat or flame and incompatible materials. Reacts in the presence of water or moisture. Forms explosive hydrogen gas on reaction with water.**
Stable **X**	Unstable	Incompatibility *(materials to avoid)* **May explode on contact with sodium. Forms hydrochloric acid in contact with water or moisture and can corrode iron and other metals. Also forms hydrogen gas, which is an explosion hazard.**
Hazardous Polymerization		Conditions to Avoid **Under normal conditions of temperature and pressure, hazardous polymerization of hexachlorocyclopentadiene is not expected to occur.**
May Occur	Will Not Occur **X**	Hazardous Decomposition or By-products **Hexachlorocyclopentadiene decomposes at high temperatures to produce toxic and acrid smoke and fumes, including hydrogen chloride, chloride, phosgene, and carbon monoxide fumes.**

SECTION VI - HEALTH HAZARD DATA

Primary Route(s) of Entry:	Inhalation?	Absorption (skin)?	Ingestion?
	X	X	X

Health Hazards

INHALATION: Very irritating to the eyes, nose, throat, and lungs. At low levels, there will be sneezing, lachrymation (tearing), headaches, and burning of the eyes and throat. Higher levels can lead to pulmonary edema (fluid in the lungs), which can be fatal.

ABSORPTION: Severe irritation and possible burns to the eyes and skin. Passes through intact skin to cause toxic systemic effects including dypsnea, salivation, nausea, vomiting.

INGESTION: Causes nausea, vomiting, diarrhea, depression, and gastrointestinal disturbances.

Carcinogenicity	NTP Listed?	IARC Cancer Review Group?	OSHA Regulated?	Target Organs?
Unknown Human **Unknown Animal**	**No**	**No**	**29 CFR 1910.1000** **Table Z-1-A**	**Respiratory system, eyes, skin, liver, kidneys.**

Medical Conditions Generally Aggravated by Exposure

None reported.

Emergency and First-aid Procedures

Eye contact: Flush immediately with water for 15 minutes (minimum); seek medical attention. Skin contact: Remove all contaminated clothing. Immediately wash area with large amounts of soap and water. Seek medical assistance, if necessary. For inhalation: Remove the person from exposure. Provide respiratory assistance and CPR, if required. Transfer to medical facility. Observe for 48 hours for lung effect (pulmonary edema). If swallowed: Never give an unconscious or convulsing person anything by mouth. Seek medical attention immediately. Give 1 to 2 glasses of water, then induce vomiting.

SECTION VII - PRECAUTIONS FOR SAFE HANDLING AND USE

Steps to be Taken in Case Material is Released or Spilled

Absorb spills in vermiculite, dry sand, earth, or similar material and deposit in sealed containers. Use water spray to knock down vapors, but do not direct water on liquid spills. Ventilate area of spill or leak. Restrict those not involved in cleanup from entering area.

Preferred Waste Disposal Method

No Citation.

Precautions to be Taken in Handling and Storage

Store in tightly closed containers in cool, dry, well-ventilated area away from heat and flame. Suggest using an inert gas (nitrogen) blanket inside storage containers to prevent moisture condensation.

Other Precautions and Warnings

Sources of ignition are prohibited where hexachlorocyclopentadiene is used, handled, or stored. Heat may cause containers to build up pressure and explode.

SECTION VIII - CONTROL MEASURES AND PERSONAL PROTECTIVE EQUIPMENT

Respiratory Protection *(specify type)*

For low exposures (to PEL), use an organic vapor respirator. For higher exposures, use a supplied-air respirator with full facepiece, hood, or helmet in continuous flow mode, or a self-contained breathing apparatus (SCBA) operated in pressure demand or other positive pressure mode.

Ventilation

Local exhaust or general mechanical systems recommended.

Protective Gloves	Eye Protection	Other Protective Clothing
Impervious Rubber	**Chemical Goggles or Face Mask**	**Rubber Apron**

Work/Hygiene Practices

Always wash hands thoroughly after using chemical; never bring food, drink, or smoking materials into vicinity of chemicals.

HEXACHLOROCYCLOPENTADIENE

C_5Cl_6

CAS: 77-47-4

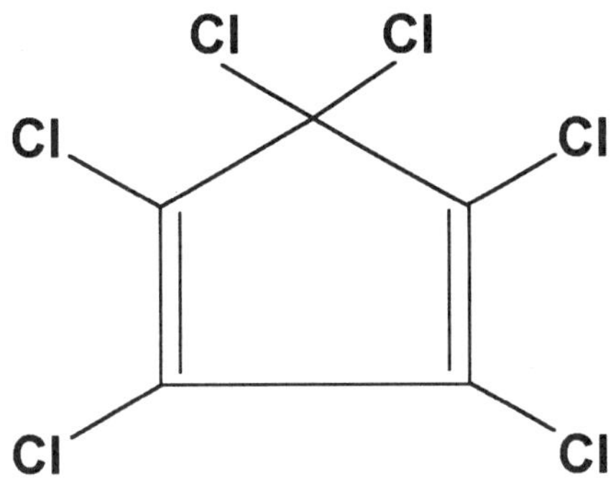

IDENTIFICATION AND TYPICAL USES

Hexachlorocyclopentadiene is a pale-yellow to amber-colored liquid with a pungent, unpleasant odor. It may appear as a solid below 16°F (-9°C). It is used as an intermediate in the synthesis of dyes, cyclodiene pesticides, (aldrin, dieldrin, endosulfan), fungicides, pharmaceuticals, and in the manufacture of chlorendic anhydride and chlorendic acid.

RISK ASSESSMENT: HEALTH

General Assessment

Hexachlorocyclopentadiene is considered a deadly poison by *inhalation* and is moderately toxic by *ingestion* and skin contact (*absorption*). There are no data in the references on its carcinogenic or mutagenic properties. However, there is some evidence that exposure causes teratogenic effects in test animals.

Inhalation of hexachlorocyclopentadiene vapors can be extremely irritating to the eyes, nose, throat, and lungs. Because of its distinguishable odor and irritating, intolerable properties, overexposure is not likely since personnel will normally remove themselves from exposure. However, such exposures are still possible, especially in emergency situations. Symptoms include dypsnea, cough, sneezing, chest pains and tightness, headache, bronchitis and bronchiolitis, and possible pulmonary edema (fluid in the lungs), which is a medical emergency and can be fatal. Development of pulmonary edema may be delayed up to 48 hours thereby creating a false sense of security with regard to health exposure risk. There may also be transient nervous system effects such as loss of memory and confusion. A condition known as proteinuria (excess protein in urine) may develop which could indicate latent kidney effects. Liver, nervous system, and heart damage are also suspect possibilities.

Skin contact will cause immediate irritation with possible burns and blistering. Absorption will occur and result in the toxic systemic effects noted for inhalation. Eye contact with the vapor or liquid will cause conjunctivitis, inflammation, and other possible damage to the surrounding tissues (depending upon length of contact). Ingestion will cause nausea, vomiting, diarrhea, and other unspecified gastrointestinal effects.

Acute Health Effects

The following acute (short-term) health effects may occur immediately or shortly after exposure to hexachlorocyclopentadiene:

<u>Skin</u>: Severe irritation and possible burns.

<u>Eye</u>: Moderate to severe irritation and possible burns. Conjunctivitis, inflammation, and possible tissue damage.

<u>Lung</u>: Irritation of the entire respiratory tract with a potential for the development of bronchitis (inflammation of the bronchial tubes), bronchiolitis (inflammation of the bronchioles), and pulmonary edema (fluid in the lungs).

<u>Other</u>: May cause kidney effects (proteinuria) and possible kidney and liver damage.

Chronic Health Effects

The following chronic (long-term) health effects may occur at some time after exposure to hexachlorocyclopentadiene and can last for months or even years:

<u>Cancer Hazards</u>: According to information presented in the references, hexachlorocyclopentadiene has not

been adequately tested for its ability to cause cancer in animals.

Reproductive Hazard: According to information presented in the references, hexachlorocyclopentadiene may cause teratogenic effects in test animals.

Other Chronic Effects: Repeated or long-term exposures to hexachlorocyclopentadiene may affect or damage the liver, kidneys, heart, and nervous system. Chronic effects may also include the development of chloro-acne.

Recommended Risk-Reduction Measures

Personnel should avoid direct contact with hexachlorocyclopentadiene. The long-term (chronic) toxicity of this chemical is not completely understood. Prudent risk management requires caution and a conservative approach when assessing the hazard of such chemicals.

If a less toxic chemical cannot be substituted for a hazardous substance, then *engineering controls* are the most effective method of reducing exposures. The best protection is to enclose operations and/or provide local exhaust ventilation at the site of chemical release. While not always operationally feasible, isolating operations can also reduce exposure. Using respiratory protection is less effective than the controls mentioned above, but is still advisable whenever working with or around hexachlorocyclopentadiene. The PEL is extremely low and difficult to monitor (0.01 ppm). For exposures above the PEL, a full facepiece respirator equipped with an organic vapor cartridge should be used. For high or prolonged exposures, or when exposure levels are unknown (as in an emergency), a supplied-air respirator with full facepiece operated in continuous flow mode, or a self-contained breathing apparatus (SCBA) with full facepiece and pressure demand are the recommended respiratory protection methods of choice. If a full facepiece is not available, then chemical goggles should be worn to protect the eyes. Whenever a chemical splash hazard exists, a face shield and a protective apron should be worn. To prevent hand and skin exposures, acid resistant gloves should be used. Glove manufacturers should be contacted and permeation studies obtained *before* final glove selections have been made.

Administrative controls should also be in place to minimize the potential for human exposures. These may include written procedures or policies which specify the methods and techniques that will be practiced whenever personnel are to work with or around chemicals such as hexachlorocyclopentadiene.

All personnel should receive *training* on the use, hazards, protective measures, emergency actions, and other precautions per 29 CFR 1910.1200 (Hazard Communication), prior to the first assignment in an area where hexachlorocyclopentadiene is used or stored. Before beginning employment and at regular intervals thereafter (e.g., annually), the following medical tests are recommended:

- ☑ Kidney and liver function tests.
- ☑ Lung function tests.

If symptoms develop or overexposure is suspected, the following tests are recommended:

- ☑ Kidney and liver function tests.
- ☑ Lung function tests (including Forced Vital Capacity, or "FVC").
- ☑ Examination of the nervous system.
- ☑ Consider chest X-ray following acute overexposure (may be negative if taken immediately after exposure due to delayed development of pulmonary edema).

Any evaluation should include a careful history of past and present symptoms with an examination. Medical tests that simply look for existing damage are not a substitute for controlling exposures. Also, because smoking can cause heart disease, emphysema, and other lung diseases, smokers may experience exposure symptoms more quickly than non-smokers under the same conditions of exposure. Risk managers must consider *all* possible exposure factors when evaluating workplace hazards.

Other methods to reduce exposure include:

- ☑ Where possible, enclose operations and use local exhaust ventilation at the site of chemical release. If local exhaust ventilation or enclosure is not used, respiratory protection should be mandatory. Also, if possible, liquids should be transferred automatically from storage containers to process containers.
- ☑ Always ensure that proper protective clothing is worn when using chemical substances.
- ☑ Wash thoroughly immediately after exposure to hexachlorocyclopentadiene and at the end of the work shift or before eating, drinking, or smoking.
- ☑ Hazard warning information should be posted in the work area. In addition, as part of an on-going education and training program, all information on

the health and safety hazards of hexachlorocyclopentadiene should be communicated to all potentially exposed workers.

☑ Eye wash stations should be provided in the immediate work area for emergency use. If there is a possibility of skin exposure to hexachlorocyclopentadiene, emergency shower facilities should also be provided.

☑ Personnel should never wear contaminated clothing home (family members can be exposed). Clothing should be laundered by personnel who have been trained on the hazards of hexachlorocyclopentadiene.

RISK ASSESSMENT: ENVIRONMENT

General Assessment

The environment is at risk of exposure during transportation, storage, disposal, or destruction of hexachlorocyclopentadiene. In almost every scenario, the threat of environmental exposure is contingent upon the proper handling of the chemical substance. Accidental spills, large or small, can result in fire, explosion, and possible contamination of the surrounding environmental mediums (water, soil, and air).

Hexachlorocyclopentadiene is considered a noncombustible liquid. But it can react in the presence of moisture or water to liberate corrosive hydrochloric acid and explosive hydrogen case. These characteristics require extreme caution in handling, storage, transportation, and disposal. It is a dangerous explosion hazard when exposed to heat and flame. Therefore, special consideration is required during any emergency situation involving a leak or spill of hexachlorocyclopentadiene.

Hexachlorocyclopentadiene may enter the environment through industrial discharges or spills.

Acute Ecological Effects

Acute (short-term) toxic effects may include the death of animals, birds, or fish and death or low growth rate in plants. Acute effects are seen 2 to 4 days after animals or plants are exposed to chemicals such as hexachlorocyclopentadiene.

Hexachlorocyclopentadiene has high acute toxicity to aquatic life. Insufficient data are available to evaluate or predict the acute effects of hexachlorocyclopentadiene exposure to plants, birds, or land animals.

Chronic Ecological Effects

Chronic (long-term) toxic effects may include shortened life span, reproductive problems, lower fertility, and changes in appearance or behavior in exposed animals. These effects can be seen long after first exposure(s) to toxic chemicals.

Hexachlorocyclopentadiene has high chronic toxicity to aquatic life. Insufficient data are available to evaluate pr predict the long-term effects of hexachlorocyclopentadiene to plants, birds, or land animals.

Water Solubility

Hexachlorocyclopentadiene is nearly insoluble in water. Concentrations of less than 100 milligrams may mix with a liter of water.

Persistence in the Environment

Hexachlorocyclopentadiene is virtually non-persistent in water, with a half-life of less than 2 days. The half-life of a pollutant is the amount of time it takes for one half of the chemical to be degraded. About 97.6% of this chemical will eventually end up in the air; about 0.4% will end up in water; and about 2% will be divided almost equally between terrestrial soil and aquatic sediments.

Bioaccumulation in Aquatic Organisms

Some substances increase in concentration, or *bioaccumulate,* in living organisms as they breathe contaminated air, drink contaminated water, or eat contaminated food. These chemicals can become concentrated in the tissues and internal organs of animals as well as humans.

The concentration of hexachlorocyclopentadiene found in fish tissue is expected to be much higher than the average concentration of hexachlorocyclopentadiene in the water from which the fish was taken.

Recommended Risk-Reduction Measures

Proper training of all transporters will reduce the likelihood of a mishap or accident resulting in a leak or spill to the environment. The correct use of DOT labeling (per the MSDS) on all containers, trucks, and rail cars will enable emergency responders to react properly and quickly to any disaster thereby reducing the potential risk to the environment and to personnel.

Storage of hexachlorocyclopentadiene should be segregated from incompatible materials (such as water) to minimize the risk of cross-contamination or contact. Buildings designated for storage should be equipped with appropriate fire protection systems (alarms, sprinklers, emergency lighting, portable extinguishers).

If a spill or leak to the environment has occurred, fire department, emergency response and/or hazardous materials spill personnel should be notified immediately. Cleanup should be attempted only by those trained in proper spill containment procedures. Contaminated soils should be removed for incineration and replaced with clean soil. If hexachlorocyclopentadiene should contact the water table, aquifer, or navigable waterway, the local and/or state emergency response authorities must be notified. A comprehensive emergency response or disaster preparedness/recovery plan should be in place prior to any operations involving the use, transportation, storage, or disposal of hexachlorocyclopentadiene. If hexachlorocyclopentadiene is spilled or leaked, the following specific steps are recommended:

- ☑ Restrict persons not wearing protective clothing from area of spill or leak until cleanup is complete and area can be opened for normal work.
- ☑ Remove all ignition sources, ventilate area of leak.
- ☑ Absorb liquids in vermiculite, dry sand, earth, or similar material and deposit in sealed containers.
- ☑ It may be necessary to dispose of hexachlorocyclopentadiene as a hazardous waste. The responsible state agency or the regional office of the federal Environmental Protection Agency (EPA) should be contacted for specific recommendations.

RISK ASSESSMENT: BUSINESS

General Assessment

Accidents or mishaps involving hexachlorocyclopentadiene can present a significant threat to business operations. The loss or damage of equipment or facilities can significantly affect fiscal viability. Lawsuits that may result from personnel injury/death, public exposures, and/or environmental contamination will also require a serious expenditure of resources. Media attention surrounding an injury, death, or environmental damage can also result in a loss of profits and loss of current as well as future business. Always remember that anytime the term "reproductive hazard' is used, public emotion, hysteria, and ignorance can run equally high. This must be properly considered when developing public relation policies and procedures.

Recommended Risk-Reduction Measures

Company attorneys, safety and health professionals, and environmental specialists should be involved in the development of any procedures dealing with safety or emergency response. A company official should be pre-designated as a public relations officer with specific training in dealing with the press. Corporate plans and policies should be developed, approved, and implemented long before any need for such arises.

REFERENCES

Asante-Duah, D. K. 1993. *Hazardous Waste Risk Assessment.* New York: CRC Press/Lewis Publishers.

Brusick, D. J. 1994. *Methods for Genetic Risk Assessment.* New York: CRC Press/Lewis Publishers.

Cockerman, L. G., B. S. Shane. 1994. *Basic Environmental Toxicology.* New York: CRC Press/Lewis Publishers.

Cothern, R. R. 1993. *Comparative Environmental Risk Assessment.* New York: CRC Press/Lewis Publishers.

Gardner, R. W. 1994. *Chemical Intolerance.* New York: CRC Press/Lewis Publishers.

Howard, P. H., et. al. 1991. *Handbook of Environmental Degradation Rates.* New York: CRC Press/Lewis Publishers.

Jones, F. E. 1994. *Toxic Organic Vapors in the Workplace.* New York: CRC Press/Lewis Publishers.

Keith, L. H. and M. M. Walker. 1995. *Handbook of Air Toxics.* New York: CRC Press/Lewis Publishers.

Lauwerys, R. R., P. Hoet. 1993. *Industrial Chemical Exposure.* New York: CRC Press/Lewis Publishers.

Lewis, R. J., Sr. 1992. *Sax's Dangerous Properties of Industrial Materials,* Volumes 1, 2, and 3, 8th Edition. New York: Van Nostrand Reinhold.

Lide, D. R. 1995. *Handbook of Chemistry and Physics,* 76th Edition. New York: CRC Press/Lewis Publishers.

Manahan, S. E. 1992. *Toxicological Chemistry,* 2nd Edition. New York: CRC Press/Lewis Publishers.

Meyer, E. 1990. *Chemistry of Hazardous Materials.* Englewood Cliffs, NJ: Prentice Hall.

Patnaik, P. 1992. *A Comprehensive Guide to the Hazardous Properties of Chemical Substances.* New York: Van Nostrand Reinhold.

Torkelson, T. R. V. K. Rowe. 1981. Halogenated aliphatic hydrocarbons. *Patty's Industrial Hygiene and Toxicology,* Vol. 2B, *Toxicology,* pp 3568-3572. New York: Wiley-Interscience.

U.S. Department of Health and Human Services, National Institute for Occupational Safety and Health. 1994. *NIOSH Pocket Guide to Chemical Hazards.* Washington, D.C.: U.S. Government Printing Office.

MATERIAL SAFETY DATA SHEET

CHEMICAL NAME	
HEXACHLOROETHANE	**CRC PRESS/LEWIS PUBLISHERS DISCLAIMER:** THE INFORMATION AND RECOMMENDATIONS PRESENTED HEREIN ARE BASED ON SOURCES BELIEVED TO BE RELIABLE. CRC MAKES NO REPRESENTATION ON ITS COMPLETENESS OR ACCURACY. IT IS THE RESPONSIBILITY OF THE USER TO DETERMINE THE CHEMICAL'S SUITABILITY FOR ITS INTENDED USE, THE CHEMICAL'S SAFE USE, AND THE CHEMICAL'S PROPER DISPOSAL. NO REPRESENTATIONS AND/OR WARRANTIES, EITHER EXPRESSED OR IMPLIED, OF THE MERCHANTABILITY OR FITNESS FOR A PARTICULAR PURPOSE, OR OF ANY OTHER NATURE, ARE MADE WITH RESPECT TO THE INFORMATION PROVIDED IN THIS MSDS OR TO THE CHEMICAL TO WHICH INFORMATION MAY REFER. CRC NEITHER ASSUMES NOR AUTHORIZES ANY OTHER PERSON TO ASSUME FOR IT, ANY OTHER ADDITIONAL RESPONSIBILITY OR LIABILITY FOR THE USE OF, OR RELIANCE UPON, THIS INFORMATION.

HAZARD WARNING INFORMATION

HEALTH	FIRE	REACTIVE	OTHER	DEGREE OF HAZARD	COLOR CODING	OTHER CODES
1	0	0		0 = Minimum Hazard 1 = Slight Hazard 2 = Moderate Hazard 3 = Serious Hazard 4 = Severe Hazard	HEALTH = BLUE FIRE = RED REACTIVITY = YELLOW OTHER = WHITE	OX = Oxidizer ACID = Acid ALK = Alkali COR = Corrosive W = Use No Water

SECTION I - GENERAL INFORMATION

Characterization	RCRA Number	EPA Class
Halogenated Hydrocarbon	**U131**	**Toxic Waste**

DOT Proper Shipping Name **Hexachloroethane**	Chemical Abstract Service (CAS) Number **67-72-1**
DOT Hazard Class and Label Requirements **ORM-A**	DOT Emergency Guide Code **53**
DOT Identification Number **NA 9037**	Molecular Formula **Cl_3CCCl_3**

Synonyms

Carbon hexachloride; ethane hexachloride; perchloroethane; Distokal; Distopan; hexachloroethylene; ethylene hexachloride; 1,1,1,2,2,2-hexachloroethane.

SECTION II - HAZARDOUS INGREDIENTS AND IDENTITY INFORMATION

Hazardous Components (specific identity)	OSHA Exposure Criteria	NIOSH Exposure Criteria	Immediately Dangerous to Life and Health (IDLH)	ACGIH Exposure Criteria
Hexachloroethane **1 ppm = 9.84 mg/m^3**	**PEL (skin):** **1 ppm** **10 mg/m^3** **STEL:** **Not Established**	**REL (skin):** **1 ppm** **10 mg/m^3** **Possible Human Carcinogen**	**300 ppm**	**TLV (skin):** **1 ppm** **10 mg/m^3** **Suspected Human Carcinogen**

SECTION III - PHYSICAL AND CHEMICAL CHARACTERISTICS

Boiling Point **365°F (185°C) sublimes**	Specific Gravity (H_2O = 1) **2.09 at 20°F (4°C)**
Vapor Pressure (mm Hg) **0.2 at 69°F (20°C)**	Molecular Weight (atomic weight) **236.7**
Vapor Density (Air = 1) **8.2**	Freezing Point **368°F (187°C)**

Solubility

Nearly insoluble in water (0.005%). Soluble in alcohol, benzene, chloroform, ether, and oils.

Appearance and Odor

Colorless rhombic, triclinic crystalline solid with a mild, camphor-like odor. Odor Threshold = 0.15 ppm.

SECTION IV - FIRE AND EXPLOSION HAZARD DATA

Flash Point (method used) **Not Determined**	Explosive Limits in Air % by Volume **LEL: Not Determined** **UEL: Not Determined**
NFPA Classification **Non-Combustible Liquid**	Autoignition Temperature **Not Determined**

Extinguishing Media

Use extinguishing media suitable to surrounding fires. Hexachloroethane does not burn.

Special Fire Fighting Procedures

Poisonous gases are produced in fire. Wear self-contained breathing apparatus (SCBA) and full protective gear. Move containers from fire if it can be done without risk. Cool sides of fire-exposed containers with water until long after fire is out. Monitor smoke directions and stay clear. Be aware of runoff from fire control measure. Do NOT release runoff to sewers or waterways. Dike for collection and disposal.

Unusual Fire and Explosion Hazards

Vapors are heavier than air and can travel some distance to ignition source to flash back.

SECTION V - REACTIVITY DATA

Stability		Conditions to Avoid **Hexachloroethane is normally stable in closed containers, under routine conditions of storage and handling. Avoid contact with heat and exposure to incompatible materials.**
Stable X	Unstable	Incompatibility *(materials to avoid)* **Hexachloroethane is incompatible with alkalis, metals (such as zinc, cadmium, aluminum, hot iron, and mercury). Also reacts with oxidizers.**
Hazardous Polymerization		Conditions to Avoid **Under normal conditions of temperature and pressure, hazardous polymerization of hexachloroethane cannot occur.**
May Occur	Will Not Occur X	Hazardous Decomposition or By-products **Poisonous gases and acrid fumes are produced in fire, including extremely toxic fumes and gases, such as oxides of carbon, hydrogen chloride gas, and phosgene.**

SECTION VI - HEALTH HAZARD DATA

Primary Route(s) of Entry:	Inhalation?	Absorption (skin/eye)?	Ingestion?
	X	X	X

Health Hazards

INHALATION: Irritation of the eyes, nose, throat, and lungs. An intoxicant and central nervous system (CNS) depressant. Causes confusion, drowsiness, dizziness, nausea, weakness, muscle twitching, staggering gait, and other adverse effects to the nervous system.

ABSORPTION: Severe eye and skin irritant. Irritation to the mucous membranes. Passes through intact skin to cause toxic systemic effects.

INGESTION: CNS symptoms similar to that of inhalation.

Carcinogenicity	NTP Listed?	IARC Cancer Review Group?	OSHA Regulated?	Target Organs?
Suspected Human Suspected Animal	**No**	**Group 3**	**29 CFR 1910.1000 Table Z-1**	**Eyes, skin, resp. sys. liver, kidney, CNS.**

Medical Conditions Generally Aggravated by Exposure

None reported.

Emergency and First-aid Procedures

Eye contact: Do not allow victim to keep eyes tightly shut. Flush immediately with water for 15 minutes (minimum), seek medical attention. Skin contact: Remove all contaminated clothing, including shoes. Immediately wash area with flooding amounts of warm water. Seek medical attention immediately. For inhalation: Remove the person from exposure. Provide respiratory assistance and CPR. Transfer to medical facility. If swallowed: Seek medical attention immediately. Give conscious and alert person 2-3 glasses of water to dilute. Never give an unconscious or convulsing person anything by mouth.

SECTION VII - PRECAUTIONS FOR SAFE HANDLING AND USE

Steps to be Taken in Case Material is Released or Spilled

Restrict those not wearing protective equipment and who are not involved in cleanup from entering area. Collect solids with HEPA vacuum and absorb liquids with vermiculite. Place in sealed container for disposal. Flush trace residues with copious amounts of water and collect for disposal.

Preferred Waste Disposal Method

Mix with a combustible solid and burn in chemical incinerator equipped with afterburner and scrubber.

Precautions to be Taken in Handling and Storage

Store in tightly closed containers in cool, dark, well-ventilated location. Protect containers from physical damage. Keep away from oxidizing materials and heat.

Other Precautions and Warnings

Ensure adequate ventilation in storage areas. Avoid vapor inhalation and use appropriate personal protective equipment if necessary.

SECTION VIII - CONTROL MEASURES AND PERSONAL PROTECTIVE EQUIPMENT

Respiratory Protection *(specify type)*

At any detectable level, use a supplied-air respirator or a self-contained breathing apparatus (SCBA) with full facepiece operated in positive pressure mode.

Ventilation

Local exhaust (preferred) at site of chemical work or general ventilation.

Protective Gloves	Eye Protection	Other Protective Clothing
Impervious Gloves	**Chemical Goggles or Face Mask**	**Protective Apron if Splash is Likely**

Work/Hygiene Practices

Always wash hands thoroughly after using chemical; never bring food, drink, or smoking materials into vicinity of chemicals.

HEXACHLOROETHANE

Cl_3CCCl_3

CAS: 67-72-1

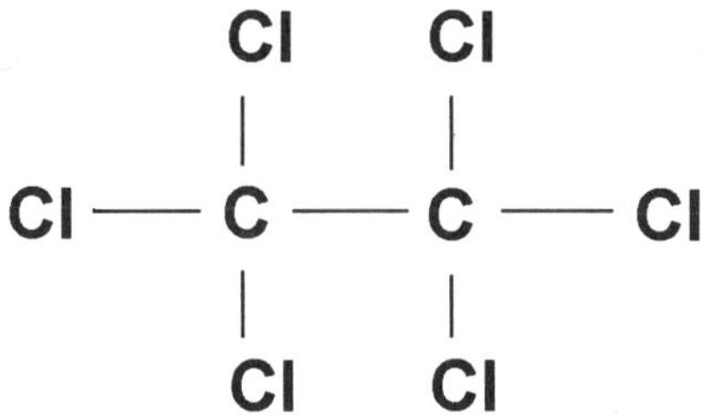

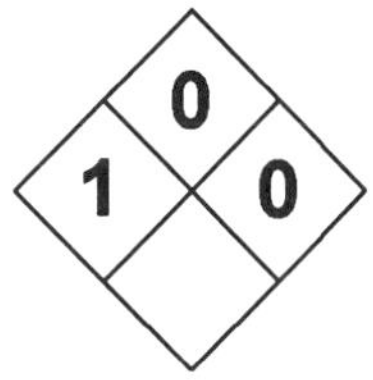

IDENTIFICATION AND TYPICAL USES

Hexachloroethane is a colorless to white crystalline (sand-like) solid or a dissolved liquid with a camphor-like odor. Used as a solvent, in the manufacture of fireworks and smoke devices, in explosives, in celluloid, as an insecticide, and as a rubber vulcanizing accelerator. It had once been used as an anthelmintic for livestock.

RISK ASSESSMENT: HEALTH

General Assessment

Hexachloroethane is toxic by *ingestion* and *absorption* through intact skin. It is also toxic through *inhalation,* but its specific toxicity is expected to be relatively low due to hexachloroethane's low vapor pressure. It is suspected to cause cancer in test animals and has been shown to adversely affect reproduction in laboratory tests.

Inhalation causes irritation of the eyes, nose, throat, and respiratory system. Exposure to high concentrations may cause narcosis with central nervous system (CNS) depression with symptoms of weakness, dizziness, lightheadedness, staggering gait, and twitching muscles. Loss of consciousness may also occur. Eye irritation from exposure to the vapors may lead to conjunctivitis, lachrymation, redness, and inflammation.

☠ *Acute Health Effects*

The following acute (short-term) health effects may occur immediately or shortly after exposure to hexachloroethane:

Skin: Irritation with possible rash and burning feeling. Absorption is likely to occur.

Eye: Severe irritation, conjunctivitis, burning feeling, pain. Corneal damage is not likely to occur.

Lung: Nose, throat, and respiratory tract irritation.

CNS: Headaches, lightheadedness, dizziness, confusion, excitement, drowsiness, intoxication, anesthesia, amnesia, analgesia, respiratory depression.

Chronic Health Effects

The following chronic (long-term) health effects may occur at some time after exposure to hexachloroethane and can last for months or even years:

Cancer Hazards: Hexachloroethane has been shown to cause liver cancer in test animals and it may have this effect in humans as well.

Reproductive Hazard: According to information presented in the references, hexachloroethane has been shown to cause adverse reproductive and teratogenic effects in laboratory test animals. It is not known if human exposure will have this effect.

Other Chronic Effects: Long-term exposure may lead to liver and kidney damage. This chemical has not been adequately tested to determine whether brain or other nerve damage could occur with repeated exposure. However, many solvents and other petroleum-based chemicals have been shown to cause such damage. Effects may include reduced memory and concentration, personality changes (withdrawal, irritability), fatigue, sleep disturbances, reduced coordination, possible effects on the nerves supplying internal organs (autonomic nerves), or the nerves to the arms and legs (weakness, "pins and needles").

Recommended Risk-Reduction Measures

Exposure to hexachloroethane presents a serious and even life-threatening risk to human health. Personnel should therefore avoid direct contact with hexachloroethane. Occupational poisoning occurs most commonly by inhalation and skin absorption. If a less toxic material or compound cannot be substituted for hexachloroethane, then *engineering controls* are the most effective method of reducing exposures. The best protection for workers is to ensure controlled use of hexachloroethane and, wherever possible, enclose operations or provide adequate local exhaust ventilation at the site of hexachloroethane release. Using respiratory protection is less effective than the controls mentioned above, but is still advisable whenever working with or around hexachloroethane. Caution should be exercised when selecting respiratory protection since it is not known if exposure to hexachloroethane will cause cancer in humans. The PEL is extremely low and difficult to accurately monitor (1 ppm). For exposures to any level, use a supplied-air respirator with full facepiece and mask operated in positive pressure mode or a self-contained breathing apparatus (SCBA) with full facepiece operated in pressure demand. If a full facepiece is not available, then chemical goggles should be worn to protect the eyes. A face shield should also be considered. To prevent hand and skin exposures impervious gloves should be used. Glove manufacturers should be contacted and permeation studies obtained *before* final glove selections are made.

Administrative controls should also be in place to minimize the potential for human exposures. These may include written procedures or policies which specify the methods and techniques that will be practiced whenever personnel are to work with hexachloroethane.

All personnel should receive *training* on the use, hazards, protective measures, emergency actions, and other precautions per 29 CFR 1910.1200 (Hazard Communication), prior to the first assignment in an area where hexachloroethane is used or stored.

For those personnel with frequent or potentially high exposures (half the PEL or greater), the following are recommended before beginning employment and at regular intervals thereafter (e.g., annually):

☑ Liver function tests.

If symptoms develop or overexposure is suspected, the following recommended medical tests should be considered:

☑ Liver and kidney function tests.

☑ Interview for brain effects, including recent changes in memory, mood (irritability, withdrawal), and concentration. Look for headaches, malaise, and altered sleeping patterns. Cerebellar, autonomic, and peripheral nervous system evaluation should be considered. Positive or borderline cases should be referred for further neuropsychological testing.

Any evaluation should include a careful history of past and present symptoms with an examination. Medical tests that simply look for existing damage are not a substitute for controlling exposures. Also, because more than light or casual alcohol consumption can lead to liver damage, drinking alcohol may increase the potential liver damage caused by hexachloroethane. Prudent risk management requires proper consideration of *all* possible exposure factors when evaluating workplace health hazard risk.

Other methods to reduce exposure include:

☑ Where possible, enclose operations and use local exhaust ventilation at the site of chemical release. If local exhaust ventilation or enclosure is not used, respiratory protection should be mandatory.

☑ Always ensure that proper protective clothing is worn when using chemical substances and that personnel are trained on its proper use and care.

☑ Wash thoroughly immediately after exposure to hexachloroethane and at the end of the work shift or before eating, drinking, or smoking.

☑ Hazard warning information should be posted in the work area. In addition, as part of an on-going education and training program, all information on the health and safety hazards of hexachloroethane should be communicated to all exposed workers.

☑ Eye wash stations should be provided in the immediate work area for emergency use. If there is a possibility of skin exposure to hexachloroethane, emergency shower facilities should also be provided in the immediate area.

☑ Workers whose clothing has been contaminated by hexachloroethane should change into clean clothes before leaving work. Contaminated work clothing should be laundered only by individuals who have been informed of the hazards of exposure to hexachloroethane.

☑ Where possible, automatically transfer liquids containing hexachloroethane from drums or other containers to process containers.

RISK ASSESSMENT: ENVIRONMENT

General Assessment

The environment is at risk of exposure during transportation, storage, disposal, or destruction of hexachloroethane. In almost every scenario, the threat of environmental exposure is contingent upon the proper handling of the chemical substance. Accidental spills, large or small, can result in contamination of the surrounding environmental mediums (water, soil, and air).

Hexachloroethane is a non-combustible solid. However, if it is involved in fire, hexachloroethane can emit highly toxic oxides of carbon, hydrogen chloride gas, and phosgene. It is incompatible with most oxidizers, alkalis, and many metals (such as zinc, cadmium, aluminum, hot iron, and mercury). Caution is always required in handling, storage, transportation, and disposal of hexachloroethane. Emergency responders should be made aware of the presence of hexachloroethane at any emergency response situation.

Hexachloroethane can enter the environment from chlorination of water at municipal treatment plants, industrial discharges, municipal waste treatment discharges, or spills.

Acute Ecological Effects

Acute (short-term) toxic effects may include the death of animals, birds, or fish and death or low growth rate in plants. Acute effects are seen 2 to 4 days after animals or plants are exposed to hexachloroethane.

Hexachloroethane has high acute toxicity to aquatic life. Insufficient data are available on the short-term effects of hexachloroethane to plants, birds, or land animals.

Chronic Ecological Effects

Chronic (long-term) toxic effects may include shortened life span, reproductive problems, lower fertility, and changes in appearance or behavior in exposed animals. These effects can be seen long after first exposure(s) to toxic chemicals.

Hexachloroethane has moderate chronic toxicity to aquatic life. Insufficient data are available to evaluate or predict the long-term effects of hexachloroethane to plants, birds, or terrestrial animals.

Water Solubility

Hexachloroethane is nearly insoluble in water. Concentrations between 1 and 100 milligrams may mix with a liter of water.

Persistence in the Environment

Hexachloroethane is moderately persistent in water, with a half-life between 20 and 200 days. The half-life of a pollutant is the amount of time it takes for one half of the chemical to be degraded. Virtually all (99.98%) of hexachloroethane will eventually end up in air.

Bioaccumulation in Aquatic Organisms

Some substances increase in concentration, or *bioaccumulate,* in living organisms as they breathe contaminated air, drink contaminated water, or eat contaminated food. These chemicals can become concentrated in the tissues and internal organs of animals as well as humans.

The concentration of hexachloroethane found in fish tissues is expected to be somewhat higher than the average concentration of hexachloroethane in the water from which the fish was taken.

Recommended Risk-Reduction Measures

Proper training of all transporters will reduce the likelihood of a mishap or accident resulting in a leak or spill of hexachloroethane into the environment. Labels on all containers, trucks, and rail cars must meet DOT requirements and accurately reflect their contents to enable emergency responders to react properly and quickly to any disaster thereby reducing the potential risk to the environment and to personnel.

Storage of hexachloroethane should be segregated from other chemicals and materials (especially oxidizers, alkalis, and metals) to minimize the risk of cross-contamination. Careful consideration is required when selecting containers for storage or use.

If a spill or leak to the environment has occurred, fire department, emergency response and/or hazardous materials spill personnel should be notified immediately. Cleanup should be attempted only by those trained in proper spill containment procedures. Contaminated soils should be removed for incineration and replaced with clean soil. If hexachloroethane should contact the water table, aquifer, or navigable waterway, time is of the essence. It is practically insoluble and total remediation may be possible. The local and/or state emergency response authorities must be notified. A comprehensive emergency response or disaster preparedness/recovery plan should be in place prior to any operations involving the use, transportation, storage, or disposal of hexachloroethane. For disposal, hexachloroethane may be mixed with a combustible solvent and burned in a chemical incinerator equipped with an afterburner and scrubber to remove the haloacids.

If hexachloroethane is spilled or leaked, the following specific steps are recommended:

☑ Restrict persons not wearing protective clothing from area of spill or leak until cleanup is complete and area can be opened for normal work.

☑ Collect solids using HEPA vacuum. Do NOT dry sweep (generates airborne dusts). Damp mop residue and collect for disposal. Absorb liquids in vermiculite, sand, earth, or similar material and deposit in sealed drum.

☑ Ventilate area of spill or leak.

☑ It may be necessary to dispose of hexachloroethane as a hazardous waste. The responsible state agency or the regional office of the federal Environmental Protection Agency (EPA) should be contacted for specific recommendations.

RISK ASSESSMENT: BUSINESS

General Assessment

Accidents or mishaps involving hexachloroethane can present a serious threat to business operations. The loss or damage of equipment or facilities can significantly affect fiscal viability. Lawsuits that may result from personnel illness, injury/death, public exposures, and/or environmental contamination will require a serious expenditure of resources. Media attention surrounding an injury, death, or environmental damage can also result in a loss of profits and loss of current as well as future business. Always remember that anytime the terms "cancer," "carcinogen," or "reproductive hazard" are used, public emotion, ignorance, and hysteria can run equally high. This must be carefully considered during the development of any public relations policies.

Recommended Risk-Reduction Measures

Company attorneys, safety and health professionals, and environmental specialists should be involved in the development of any procedures for responding to chemical incidents. A company official should be predesignated as a public relations officer with specific training in dealing with the press. Corporate plans and policies should be developed, approved, and implemented long before any need for such arises.

REFERENCES

American Conference of Governmental Industrial Hygienists. 1988. *Documentation of the Threshold Limit Values and Biological Exposure Indices,* 5th Edition (with updates). Cincinnati: ACGIH

Anthony, C. P., and N. J. Kolthoff. 1971. *Textbook of Anatomy and Physiology.* St. Louis: C. V. Mosby Company.

Bloom, A. D. and F. J. DeSerres. 1995. *Ecotoxicity and Human Health.* New York: CRC Press/Lewis Publishers.

Brusick, D. J. 1994. *Methods for Genetic Risk Assessment.* New York: CRC Press/Lewis Publishers.

Bueche, J. R. 1972. *Principles of Physics,* 2nd Edition. New York: McGraw-Hill.

Burrell, R., D. K. Flaherty, and L. J. Sauers. 1992. *Toxicology of the Immune System.* New York: Van Nostrand Reinhold.

Calabrese, E. J. 1994. *Biological Effects of Low Level Exposures.* New York: CRC Press/Lewis Publishers.

Cockerman, L. G., B. S. Shane. 1994. *Basic Environmental Toxicology.* New York: CRC Press/Lewis Publishers.

Cothern, C. R., N. P. Ross. 1994. *Environmental Statistics, Assessment, and Forecasting.* New York: CRC Press/Lewis Publishers.

Danse, I. R. 1991. *Common Sense Toxics in the Workplace.* New York: Van Nostrand Reinhold.

Environmental Protection Agency. 1983. *Chemical Hazard Information Profiles.* Washington, D.C.: U.S. EPA.

Gardner, R. W. 1994. *Chemical Intolerance.* New York: CRC Press/Lewis Publishers.

Hallenbeck, W. H. 1993. *Quantitative Risk Assessment for Environmental and Occupational Health,* 2nd Edition. New York: CRC Press/Lewis Publishers.

Lewis, R. J., Sr. 1993. *Hawley's Condensed Chemical Dictionary,* 12th Edition. New York: Van Nostrand Reinhold.

Lewis, R. J., Sr. 1992. *Sax's Dangerous Properties of Industrial Materials,* Volumes 1, 2, and 3, 8th Edition. New York: Van Nostrand Reinhold.

Meyer, E. 1990. *Chemistry of Hazardous Materials.* Englewood Cliffs, NJ: Prentice Hall.

Patnaik, P. 1992. *A Comprehensive Guide to the Hazardous Properties of Chemical Substances.* New York: Van Nostrand Reinhold.

Quigley, D. R. 1994. *Handbook of Emergency Chemical Management.* New York: CRC Press/Lewis Publishers.

U.S. Department of Health and Human Services, National Institute for Occupational Safety and Health. 1994. *NIOSH Pocket Guide to Chemical Hazards.* Washington, D.C.: U.S. Government Printing Office.

MATERIAL SAFETY DATA SHEET

CHEMICAL NAME	
HEXACHLORO-NAPHTHALENE	**CRC PRESS/LEWIS PUBLISHERS DISCLAIMER:** THE INFORMATION AND RECOMMENDATIONS PRESENTED HEREIN ARE BASED ON SOURCES BELIEVED TO BE RELIABLE. CRC MAKES NO REPRESENTATION ON ITS COMPLETENESS OR ACCURACY. IT IS THE RESPONSIBILITY OF THE USER TO DETERMINE THE CHEMICAL'S SUITABILITY FOR ITS INTENDED USE, THE CHEMICAL'S SAFE USE, AND THE CHEMICAL'S PROPER DISPOSAL. NO REPRESENTATIONS AND/OR WARRANTIES, EITHER EXPRESSED OR IMPLIED, OF THE MERCHANTABILITY OR FITNESS FOR A PARTICULAR PURPOSE, OR OF ANY OTHER NATURE, ARE MADE WITH RESPECT TO THE INFORMATION PROVIDED IN THIS MSDS OR TO THE CHEMICAL TO WHICH INFORMATION MAY REFER. CRC NEITHER ASSUMES NOR AUTHORIZES ANY OTHER PERSON TO ASSUME FOR IT, ANY OTHER ADDITIONAL RESPONSIBILITY OR LIABILITY FOR THE USE OF, OR RELIANCE UPON, THIS INFORMATION.

HAZARD WARNING INFORMATION

HEALTH	FIRE	REACTIVE	OTHER	DEGREE OF HAZARD	COLOR CODING	OTHER CODES
3	0	0		0 = Minimum Hazard 1 = Slight Hazard 2 = Moderate Hazard 3 = Serious Hazard 4 = Severe Hazard	HEALTH = BLUE FIRE = RED REACTIVITY = YELLOW OTHER = WHITE	OX = Oxidizer ACID = Acid ALK = Alkali COR = Corrosive W = Use No Water

SECTION I - GENERAL INFORMATION

Characterization	RCRA Number	EPA Class
Halogenated Hydrocarbon	**None**	**Not Applicable**

DOT Proper Shipping Name **No Citation**	Chemical Abstract Service (CAS) Number **1335-87-1**
DOT Hazard Class and Label Requirements **No Citation**	DOT Emergency Guide Code **No Citation**
DOT Identification Number **No Citation**	Molecular Formula $C_{10}H_2Cl_3$

Synonyms

Halowax ® 1014.

SECTION II - HAZARDOUS INGREDIENTS AND IDENTITY INFORMATION

Hazardous Components (specific identity)	OSHA Exposure Criteria	NIOSH Exposure Criteria	Immediately Dangerous to Life and Health (IDLH)	ACGIH Exposure Criteria
Hexachloronaphthalene (derivation: By chlorination of naphthalene in the presence of catalytic amounts of iron chloride or antimony chloride).	**PEL (skin):** $0.2\ mg/m^3$ **STEL:** Not Established	**REL (skin):** $0.2\ mg/m^3$ **STEL:** Not Established	$2\ mg/m^3$	**TLV (skin):** $0.2\ mg/m^3$ **STEL:** Not Established

SECTION III - PHYSICAL AND CHEMICAL CHARACTERISTICS

Boiling Point **650-730°F (343-387°C)**	Specific Gravity (H_2O = 1) **1.78**
Vapor Pressure (mm Hg) **<1 at 69°F (20°C)**	Molecular Weight (atomic weight) **334.9**
Vapor Density (Air = 1) **11.6**	Freezing Point **279°F (137°C)**

Solubility

Insoluble in water, soluble in organic solvents.

Appearance and Odor

White or light-yellow waxy solid with an aromatic odor.

SECTION IV - FIRE AND EXPLOSION HAZARD DATA

Flash Point (method used) **Not Determined**	Explosive Limits in Air % by Volume **LEL: Not Determined** **UEL: Not Determined**
NFPA Classification **Non-Combustible Liquid**	Autoignition Temperature **Not Determined**

Extinguishing Media

Use extinguishing media suitable to surrounding fires. Hexachloronaphthalene does not burn.

Special Fire Fighting Procedures

Poisonous gases are produced in fire. Wear self-contained breathing apparatus (SCBA) and full protective gear. Move containers from fire if it can be done without risk. Cool sides of fire-exposed containers with water until long after fire is out. Monitor smoke directions and stay clear. Be aware of runoff from fire control measure. Do NOT release runoff to sewers or waterways. Dike for collection and disposal.

Unusual Fire and Explosion Hazards

None Reported.

SECTION V - REACTIVITY DATA

Stability		Conditions to Avoid Hexachloronaphthalene is normally stable in closed containers, under routine conditions of storage and handling. Avoid contact with heat and exposure to incompatible materials.
Stable X	Unstable	Incompatibility *(materials to avoid)* Hexachloronaphthalene is incompatible with strong oxidizers (such as chlorine, bromine, and fluorine), strong acids (nitric, sulfuric, hydrochloric), and acid fumes.
Hazardous Polymerization		Conditions to Avoid Under normal conditions of temperature and pressure, hazardous polymerization of hexachloronaphthalene cannot occur.
May Occur	Will Not Occur X	Hazardous Decomposition or By-products Poisonous gases and acrid fumes are produced in fire, including extremely toxic fumes and gases, including carbon dioxide and toxic chloride fumes.

SECTION VI - HEALTH HAZARD DATA

Primary Route(s) of Entry:	Inhalation?	Absorption (skin/eye)?	Ingestion?
	X	X	X

Health Hazards

INHALATION: A severe irritant causing nausea, jaundice, indigestion, drowsiness, which can progress to central nervous system depression with confusion, fatigue, fever, and coma. A serious hepatoxic chemical. Severe liver damage and tissue necrosis can occur.

ABSORPTION: Severe eye and skin irritant. May cause chloro-acne (an acne-form dermatitis). Symptoms include dry skin with intense itching, blackheads, and cysts. Hepatitis can also occur.

INGESTION: Nausea, jaundice, liver effects and damage, coma, and possibly death.

Carcinogenicity	NTP Listed?	IARC Cancer Review Group?	OSHA Regulated?	Target Organs?
Unknown Human Unknown Animal	No	No	29 CFR 1910.1000 Table Z-1	Liver, skin, and CNS.

Medical Conditions Generally Aggravated by Exposure

Existing skin diseases or conditions (eczema, psoriasis) and liver dysfunction may be aggravated.

Emergency and First-aid Procedures

Eye contact: Do not allow victim to keep eyes tightly shut. Flush immediately with water for 15 minutes (minimum); seek medical attention. Skin contact: Remove all contaminated clothing, including shoes. Immediately wash area with flooding amounts of warm water. Seek medical attention immediately. For inhalation: Remove the person from exposure. Provide respiratory assistance and CPR. Transfer to medical facility. If swallowed: Seek medical attention immediately. Give conscious and alert person 2-3 glasses of water and induce vomiting with 40 to 45 mg (adult dose) of syrup of Ipecac. Never give an unconscious or convulsing person anything by mouth. Post-vomiting, give 1-2 mg/kg charcoal slurry.

SECTION VII - PRECAUTIONS FOR SAFE HANDLING AND USE

Steps to be Taken in Case Material is Released or Spilled

Restrict those not wearing protective equipment and who are not involved in cleanup from entering area. Collect solids with HEPA vacuum and damp mop residues. Do NOT dry sweep. Place in sealed container for disposal. Flush trace residues with copious amounts of water and collect for disposal.

Preferred Waste Disposal Method

Mix with a combustible solid and burn in chemical incinerator equipped with afterburner and scrubber.

Precautions to be Taken in Handling and Storage

Store in tightly closed containers in cool, dark, well-ventilated location. Protect containers from physical damage. Keep away from oxidizing materials and heat.

Other Precautions and Warnings

Ensure adequate ventilation in storage areas. Avoid vapor/dust inhalation and use appropriate personal protective equipment if necessary. Do not store in direct sunlight.

SECTION VIII - CONTROL MEASURES AND PERSONAL PROTECTIVE EQUIPMENT

Respiratory Protection *(specify type)*

For low exposures, use an air purifying respirator with a HEPA filter. Otherwise, use a supplied-air respirator or a self-contained breathing apparatus with full facepiece operated in positive pressure mode.

Ventilation

Local exhaust (preferred) at site of chemical work or general ventilation.

Protective Gloves	Eye Protection	Other Protective Clothing
Impervious Gloves	Chemical/Dust Goggles or Face Mask	Protective Apron if Splash is Likely

Work/Hygiene Practices

Always wash hands thoroughly after using chemical; never bring food, drink, or smoking materials into vicinity of chemicals.

HEXACHLORONAPHTHALENE

$C_{10}H_2Cl_6$

CAS: 1335-87-1

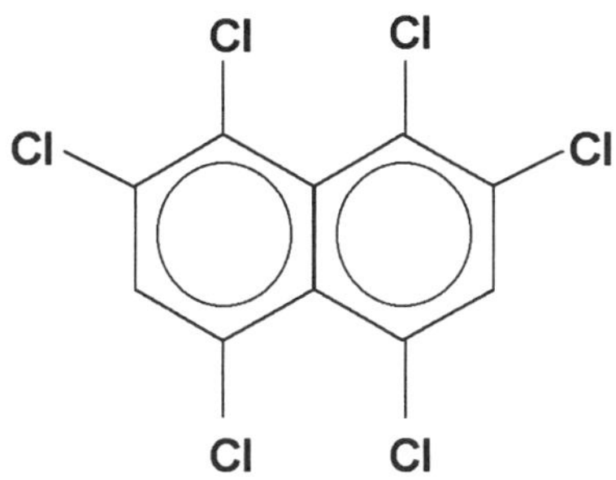

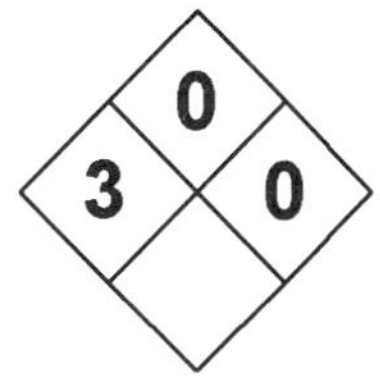

IDENTIFICATION AND TYPICAL USES

Hexachloronaphthalene is a white to light-yellow wax-like solid with a mild, aromatic, pleasant odor. It is used to insulate electrical equipment and wires, in flame proofing and water proofing, in paper coatings, in batteries, and as a lubricant additive.

RISK ASSESSMENT: HEALTH

General Assessment

Hexachloronaphthalene is toxic to humans by *ingestion, inhalation,* and *absorption* through intact skin. It is a severe irritant and a strong liver toxin (hepatoxin). There is no information in the references on its carcinogenic, teratogenic, or mutagenic capabilities.

Inhalation causes irritation of the eyes, nose, throat, and respiratory system. It can cause nausea, jaundice (yellowing of the skin and sclera), dark urine, indigestion, and drowsiness. There may be progression to central nervous system (CNS) depression with confusion, fever, and even coma. Possible liver damage, necrosis (tissue destruction), and/or atrophy (tissue death) may also occur.

Skin contact results in the absorption of the chemical into the body causing many of the same toxic systemic effects of inhalation, especially those resulting in liver and CNS damage. Skin contact can also cause a serious acne-like dermatitis known as chlor-acne. This condition is characterized by dry skin with intense itching and the appearance of blackheads and cysts containing keratin and sebaceous matter about the face, ears, neck, shoulders, arms, and stomach. Hepatitis is also a frequent result of skin contact. Ingestion will likely cause similar toxic systemic effects as inhalation and absorption with nausea and stomach pain as well.

Acute Health Effects

The following acute (short-term) health effects may occur immediately or shortly after exposure to hexachloronaphthalene:

Skin: Irritation with possible rash, intense itching, and acne-like rash with blackheads and cysts on the face, neck, ears, shoulders, and stomach.

Eye: Mild to moderate irritation (depending upon length of time contact occurs).

Lung: Nose, throat, and respiratory tract irritation.

CNS: Headaches, lightheadedness, dizziness, confusion, excitement, drowsiness, fever, coma, and possibly death.

Other: Exposure results in serious and possibly life-threatening effects to the liver. There can be jaundice, hepatitis, necrosis, and atrophy of the organ. Death can also occur, but is not likely on acute exposure (unless exposure levels are extremely high).

Chronic Health Effects

The following chronic (long-term) health effects may occur at some time after exposure to hexachloronaphthalene and can last for months or even years:

Cancer Hazards: According to information presented in the references, hexachloronaphthalene has not been adequately tested for its ability to cause cancer in laboratory animals.

Reproductive Hazard: According to information presented in the references, hexachloronaphthalene has not been adequately tested for its ability to adversely affect reproduction in laboratory animals.

Other Chronic Effects: Long-term exposure may lead to fatal damage to the central nervous system and/or the liver. Prolonged contact may cause chronic skin irritation and acne-like rash. The skin may become extremely sensitive to sunlight and any exposure may cause a persistent flare-up of acne symptoms.

Recommended Risk-Reduction Measures

Personnel should avoid direct contact with hexachloronaphthalene. Occupational poisoning occurs most commonly by inhalation and skin absorption. If a less toxic material or compound cannot be substituted for hexachloronaphthalene, then *engineering controls* are the most effective method of reducing exposures. The best protection for workers is to ensure controlled use of hexachloronaphthalene and, wherever possible, enclose operations or provide adequate local exhaust ventilation at the site of hexachloronaphthalene release. Using respiratory protection is less effective than the controls mentioned above, but is still advisable whenever working with or around hexachloronaphthalene. The PEL is extremely low and difficult to accurately monitor (0.2 mg/m^3). For exposures to low levels or infrequent exposures, use an air purifying respirator equipped with a dust/mist pre-filter or high-efficiency particulate air (HEPA) filter. For higher exposures, or when the level is unknown (as in an emergency), use a supplied-air respirator with full facepiece and mask operated in positive pressure mode or a self-contained breathing apparatus (SCBA) with full facepiece operated in pressure demand. If a full facepiece is not available, then chemical goggles should be worn to protect the eyes. A face shield should also be considered. To prevent hand and skin exposures impervious gloves should be used. Glove manufacturers should be contacted and permeation studies obtained *before* final glove selections are made.

Administrative controls should also be in place to minimize the potential for human exposures. These may include written procedures or policies which specify the methods and techniques that will be practiced whenever personnel are to work with hexachloronaphthalene.

All personnel should receive *training* on the use, hazards, protective measures, emergency actions, and other precautions per 29 CFR 1910.1200 (Hazard Communication), prior to the first assignment in an area where hexachloronaphthalene is used or stored.

For those personnel with frequent or potentially high exposures (half the PEL or greater), the following are recommended before beginning employment and at regular intervals thereafter (e.g., annually):

- ☑ Liver function tests.
- ☑ Evaluation of the skin by a dermatologist for indications or symptoms of chlor-acne.

If symptoms develop or overexposure is suspected, the following recommended medical tests should be considered:

- ☑ Liver and kidney function tests.
- ☑ Interview for nervous system effects and any changes in personality (irritability, moodiness) or normal function (weakness, fever, drowsiness).

Any evaluation should include a careful history of past and present symptoms with an examination. Medical tests that simply look for existing damage are not a substitute for controlling exposures. Also, because more than light or casual alcohol consumption can lead to liver damage, drinking alcohol may increase the potential liver damage caused by hexachloronaphthalene. Prudent risk management requires proper consideration of *all* possible exposure factors when evaluating workplace health hazard risk.

Other methods to reduce exposure include:

- ☑ Where possible, enclose operations and use local exhaust ventilation at the site of chemical release. If local exhaust ventilation or enclosure is not used, respiratory protection should be mandatory.
- ☑ Always ensure that proper protective clothing is worn when using chemical substances and that personnel are trained on its proper use and care.
- ☑ Wash thoroughly immediately after exposure to hexachloronaphthalene and at the end of the work shift or before eating, drinking, or smoking.
- ☑ Hazard warning information should be posted in the work area. In addition, as part of an on-going education and training program, all information on the health and safety hazards of hexachloronaphthalene should be communicated to all exposed workers.
- ☑ Eye wash stations should be provided in the immediate work area for emergency use. If there is a possibility of skin exposure to hexachloronaphthalene, emergency shower facilities should also be provided in the immediate area.
- ☑ Workers whose clothing has been contaminated by hexachloronaphthalene should change into clean

clothes before leaving work. Contaminated work clothing should be laundered only by individuals who have been informed of the hazards of exposure to hexachloronaphthalene.

RISK ASSESSMENT: ENVIRONMENT

General Assessment

The environment is at risk of exposure during transportation, storage, disposal, or destruction of hexachloronaphthalene. In almost every scenario, the threat of environmental exposure is contingent upon the proper handling of the chemical substance. Accidental spills, large or small, can result in contamination of the surrounding environmental mediums (water, soil, and air).

Hexachloronaphthalene is a non-combustible solid. However, if it is involved in fire, hexachloronaphthalene can emit highly toxic oxides of carbon and toxic fumes of chloride. It is incompatible with most oxidizers and must be kept away from permanganates, nitrates, peroxides, chlorates, and perchlorates. Caution is always required in handling, storage, transportation, and disposal of hexachloronaphthalene. Emergency responders should be made aware of the presence of hexachloronaphthalene at any emergency response situation.

Hexachloronaphthalene can enter the environment through industrial discharges or spills.

Acute Ecological Effects

Acute (short-term) toxic effects may include the death of animals, birds, or fish and death or low growth rate in plants. Acute effects are seen 2 to 4 days after animals or plants are exposed to hexachloronaphthalene.

Hexachloronaphthalene has high acute toxicity to aquatic life. Insufficient data are available on the short-term effects of hexachloronaphthalene to plants, birds, or land animals.

Chronic Ecological Effects

Chronic (long-term) toxic effects may include shortened life span, reproductive problems, lower fertility, and changes in appearance or behavior in exposed animals. These effects can be seen long after first exposure(s) to toxic chemicals.

Hexachloronaphthalene has high chronic toxicity to aquatic life. Insufficient data are available to evaluate or predict the long-term effects of hexachloronaphthalene to plants, birds, or terrestrial animals.

Water Solubility

Hexachloronaphthalene is nearly insoluble in water. Concentrations of 1 milligram will mix with a liter of water.

Persistence in the Environment

Hexachloronaphthalene is highly persistent in water, with a half-life greater than 200 days. The half-life of a pollutant is the amount of time it takes for one half of the chemical to be degraded. About 52% of hexachloronaphthalene will eventually end up in terrestrial soil; the remainder will end up in aquatic sediments.

Bioaccumulation in Aquatic Organisms

Some substances increase in concentration, or *bioaccumulate,* in living organisms as they breathe contaminated air, drink contaminated water, or eat contaminated food. These chemicals can become concentrated in the tissues and internal organs of animals as well as humans.

The concentration of hexachloronaphthalene found in fish tissues is expected to be considerably higher than the average concentration of hexachloronaphthalene in the water from which the fish was taken.

Recommended Risk-Reduction Measures

Proper training of all transporters will reduce the likelihood of a mishap or accident resulting in a leak or spill of hexachloronaphthalene into the environment. Labels on all containers, trucks, and rail cars should accurately reflect their contents to enable emergency responders to react properly and quickly to any disaster thereby reducing the potential risk to the environment and to personnel.

Storage of hexachloronaphthalene should be segregated from other chemicals and materials (especially oxidizers) to minimize the risk of cross-contamination.

If a spill or leak to the environment has occurred, fire department, emergency response and/or hazardous materials spill personnel should be notified immediately. Cleanup should be attempted only by those trained in proper spill containment procedures. Contaminated soils should be removed for incineration and replaced with clean soil. If hexachloronaphthalene should contact the water table, aquifer, or navigable waterway, time is of the essence. It is practically insoluble and total remediation may be possible. The local and/or state emergency response authorities must be notified. A comprehensive emergency response or disaster preparedness/recovery plan should be in place prior to any operations involving the use, transportation, storage, or disposal of hexachloronaphthalene. For disposal, hexachloronaphthalene may be dissolved in a combustible solvent and burned in a chemical incinerator equipped with an afterburner and scrubber to remove the haloacids.

If hexachloronaphthalene is spilled or leaked, the following specific steps are recommended:

- ☑ Restrict persons not wearing protective clothing from area of spill or leak until cleanup is complete and area can be opened for normal work.
- ☑ Collect solids using HEPA vacuum. Do NOT dry sweep (generates airborne dusts). Damp mop residue and collect for disposal. Absorb liquid solutions in vermiculite, sand, earth, or similar material and deposit in sealed drum.
- ☑ Ventilate area of spill or leak.
- ☑ It may be necessary to dispose of hexachloronaphthalene as a hazardous waste. The responsible state agency or the regional office of the federal Environmental Protection Agency (EPA) should be contacted for specific recommendations.

RISK ASSESSMENT: BUSINESS

General Assessment

Accidents or mishaps involving hexachloronaphthalene can present a serious threat to business operations. The loss or damage of equipment or facilities can significantly affect fiscal viability. Lawsuits that may result from personnel illness, injury/death, public exposures, and/or environmental contamination will require a serious expenditure of resources. Media attention surrounding an injury, death, or environmental damage can also result in a loss of profits and loss of current as well as future business.

Recommended Risk-Reduction Measures

Company attorneys, safety and health professionals, and environmental specialists should be involved in the development of any procedures for responding to chemical incidents. A company official should be predesignated as a public relations officer with specific training in dealing with the press. Corporate plans and policies should be developed, approved, and implemented long before any need for such arises.

REFERENCES

American Conference of Governmental Industrial Hygienists. 1988. *Documentation of the Threshold Limit Values and Biological Exposure Indices,* 5th Edition (with updates). Cincinnati: ACGIH

Anthony, C. P., and N. J. Kolthoff. 1971. *Textbook of Anatomy and Physiology.* St. Louis: C. V. Mosby Company.

Bloom, A. D. and F. J. DeSerres. 1995. *Ecotoxicity and Human Health.* New York: CRC Press/Lewis Publishers.

Brusick, D. J. 1994. *Methods for Genetic Risk Assessment.* New York: CRC Press/Lewis Publishers.

Bueche, J. R. 1972. *Principles of Physics,* 2nd Edition. New York: McGraw-Hill.

Burrell, R., D. K. Flaherty, and L. J. Sauers. 1992. *Toxicology of the Immune System.* New York: Van Nostrand Reinhold.

Calabrese, E. J. 1994. *Biological Effects of Low Level Exposures.* New York: CRC Press/Lewis Publishers.

Cockerman, L. G., B. S. Shane. 1994. *Basic Environmental Toxicology.* New York: CRC Press/Lewis Publishers.

Cothern, C. R., N. P. Ross. 1994. *Environmental Statistics, Assessment, and Forecasting.* New York: CRC Press/Lewis Publishers.

Danse, I. R. 1991. *Common Sense Toxics in the Workplace.* New York: Van Nostrand Reinhold.

Environmental Protection Agency. 1983. *Chemical Hazard Information Profiles.* Washington, D.C.: U.S. EPA.

Gardner, R. W. 1994. *Chemical Intolerance.* New York: CRC Press/Lewis Publishers.

Hallenbeck, W. H. 1993. *Quantitative Risk Assessment for Environmental and Occupational Health,* 2nd Edition. New York: CRC Press/Lewis Publishers.

Lewis, R. J., Sr. 1993. *Hawley's Condensed Chemical Dictionary,* 12th Edition. New York: Van Nostrand Reinhold.

Lewis, R. J., Sr. 1992. *Sax's Dangerous Properties of Industrial Materials,* Volumes 1, 2, and 3, 8th Edition. New York: Van Nostrand Reinhold.

Meyer, E. 1990. *Chemistry of Hazardous Materials.* Englewood Cliffs, NJ: Prentice Hall.

Patnaik, P. 1992. *A Comprehensive Guide to the Hazardous Properties of Chemical Substances.* New York: Van Nostrand Reinhold.

Quigley, D. R. 1994. *Handbook of Emergency Chemical Management.* New York: CRC Press/Lewis Publishers.

U.S. Department of Health and Human Services, National Institute for Occupational Safety and Health. 1994. *NIOSH Pocket Guide to Chemical Hazards.* Washington, D.C.: U.S. Government Printing Office.

MATERIAL SAFETY DATA SHEET

CHEMICAL NAME	
HEXAFLUOROACETONE	**CRC PRESS/LEWIS PUBLISHERS DISCLAIMER:** THE INFORMATION AND RECOMMENDATIONS PRESENTED HEREIN ARE BASED ON SOURCES BELIEVED TO BE RELIABLE. CRC MAKES NO REPRESENTATION ON ITS COMPLETENESS OR ACCURACY. IT IS THE RESPONSIBILITY OF THE USER TO DETERMINE THE CHEMICAL'S SUITABILITY FOR ITS INTENDED USE, THE CHEMICAL'S SAFE USE, AND THE CHEMICAL'S PROPER DISPOSAL. NO REPRESENTATIONS AND/OR WARRANTIES, EITHER EXPRESSED OR IMPLIED, OF THE MERCHANTABILITY OR FITNESS FOR A PARTICULAR PURPOSE, OR OF ANY OTHER NATURE, ARE MADE WITH RESPECT TO THE INFORMATION PROVIDED IN THIS MSDS OR TO THE CHEMICAL TO WHICH INFORMATION MAY REFER. CRC NEITHER ASSUMES NOR AUTHORIZES ANY OTHER PERSON TO ASSUME FOR IT, ANY OTHER ADDITIONAL RESPONSIBILITY OR LIABILITY FOR THE USE OF, OR RELIANCE UPON, THIS INFORMATION.

HAZARD WARNING INFORMATION

HEALTH	FIRE	REACTIVE	OTHER	DEGREE OF HAZARD	COLOR CODING	OTHER CODES
3	0	2	~~W~~	0 = Minimum Hazard 1 = Slight Hazard 2 = Moderate Hazard 3 = Serious Hazard 4 = Severe Hazard	HEALTH = BLUE FIRE = RED REACTIVITY = YELLOW OTHER = WHITE	OX = Oxidizer ACID = Acid ALK = Alkali COR = Corrosive ~~W~~ = Use No Water

SECTION I - GENERAL INFORMATION

Characterization	RCRA Number	EPA Class
Gas; Ketone	**None**	**Not Applicable**

DOT Proper Shipping Name **Hexafluoroacetone**	Chemical Abstract Service (CAS) Number **684-16-2**
DOT Hazard Class and Label Requirements **IMO Poison A; Poison Gas**	DOT Emergency Guide Code **15**
DOT Identification Number **UN 2420**	Molecular Formula **$(CF_3)_2CO$**

Synonyms
Hexafluoro-2-propanone; 1,1,1,3,3,3-hexafluoro-2-propanone; HFA; perfluoroacetone.

SECTION II - HAZARDOUS INGREDIENTS AND IDENTITY INFORMATION

Hazardous Components (specific identity)	OSHA Exposure Criteria	NIOSH Exposure Criteria	Immediately Dangerous to Life and Health (IDLH)	ACGIH Exposure Criteria
Hexafluoroacetone **1 ppm = 6.90 mg/m^3**	**PEL (skin):** **0.1 ppm** **0.7 mg/m^3** **STEL:** **Not Established**	**REL (skin):** **0.1 ppm** **0.7 mg/m^3** **STEL:** **Not Established**	**Not Determined**	**TLV:** **0.1 ppm** **0.7 mg/m^3** **STEL:** **Not Established**

SECTION III - PHYSICAL AND CHEMICAL CHARACTERISTICS

Boiling Point **-17°F (-27°C)**	Specific Gravity (H_2O = 1) **5.76 (Gas)**
Vapor Pressure (atmospheres) **5.8 at 69°F (20°C)**	Molecular Weight (atomic weight) **166.0**
Vapor Density (Air = 1) **Not Found**	Freezing Point **-188°F (-122°C)**

Solubility
Reacts violently in water and many other substances, releasing large amounts of heat in the process.

Appearance and Odor
Colorless gas with a musty odor, normally shipped as a liquefied compressed gas.

SECTION IV - FIRE AND EXPLOSION HAZARD DATA

Flash Point (method used) **Not Determined**	Explosive Limits in Air % by Volume **LEL: Not Determined UEL: Not Determined**
NFPA Classification **Non-Flammable Gas**	Autoignition Temperature **Not Determined**

Extinguishing Media
Use extinguishing media suitable to surrounding fires. Do NOT use water. Hexafluoroacetone does not burn but it is highly reactive with water and many other substances and will release heat on reaction.

Special Fire Fighting Procedures
Poisonous gases are produced in fire. Wear self-contained breathing apparatus (SCBA) and full protective gear. Move containers from fire if it can be done without risk. Cool sides of closed, fire-exposed containers with water. Monitor smoke directions and stay clear. Be aware of runoff from fire control measure. Do NOT release runoff to sewers or waterways. Dike for collection and disposal.

Unusual Fire and Explosion Hazards
Reactive gas can produce extreme heat in contact with incompatible commodities, including water.

SECTION V - REACTIVITY DATA

Stability		Conditions to Avoid
Stable	Unstable	Hexafluoroacetone is normally stable in closed containers, under routine conditions of storage and handling. Avoid contact with heat and exposure to incompatible materials. A very reactive gas.
X		**Incompatibility** *(materials to avoid)*: Hexafluoroacetone is incompatible with water and acids. Reaction will be exothermic (produces extreme heat). It is also hygroscopic (absorbs moisture from air) and will react to form highly acidic sesquilhydrate.
Hazardous Polymerization		**Conditions to Avoid**: Under normal conditions of temperature and pressure, hazardous polymerization of hexafluoroacetone cannot occur.
May Occur	Will Not Occur	**Hazardous Decomposition or By-products**: Poisonous gases and acrid fumes are produced in fire, including extremely toxic fumes and gases, including carbon monoxide and oxides of fluorine.
	X	

SECTION VI - HEALTH HAZARD DATA

Primary Route(s) of Entry:	Inhalation?	Absorption (skin/eye)?	Ingestion?
	X	X	

Health Hazards

INHALATION: Severe irritation of the eyes, nose, throat, and respiratory system. May react with the moisture-content on the nose, throat, and lungs to cause burns and destruction of tissue. It can also lead to a life-threatening buildup of fluid in the lungs (pulmonary edema).

ABSORPTION: Severe eye and skin irritant. Irritation to the mucous membranes. Passes through intact skin. Liquid will cause frostbite burns and tissue destruction.

INGESTION: Not likely since hexafluoroacetone is normally a gas at room temperature and pressure.

Carcinogenicity	NTP Listed?	IARC Cancer Review Group?	OSHA Regulated?	Target Organs?
Unknown Human Unknown Animal	No	No	No	Eyes, skin, resp. sys., kidney, reproductive system.

Medical Conditions Generally Aggravated by Exposure

None reported.

Emergency and First-aid Procedures

Eye contact: Do not allow victim to keep eyes tightly shut. Flush immediately with tepid water for 15 minutes (minimum); seek medical attention. **Skin contact:** Remove all contaminated clothing. Immediately wash area with flooding amounts of warm water. Do NOT rub affected areas or use forced air heat. Seek medical attention immediately. For **inhalation:** Remove the person from exposure. Provide respiratory assistance and CPR. Transfer to medical facility. Observe for 48 hours for lung effects. If **swallowed:** Not likely to occur.

SECTION VII - PRECAUTIONS FOR SAFE HANDLING AND USE

Steps to be Taken in Case Material is Released or Spilled

Stop flow of gas, if possible. Restrict those not wearing protective equipment and who are not involved in cleanup from entering area. Move leaking cylinders to a safe location (outdoors) and allow to vent until empty. Keep water away.

Preferred Waste Disposal Method

No Citation.

Precautions to be Taken in Handling and Storage

Store in tightly closed, air-tight, moisture-proof cylinders in a cool, dry, well-ventilated location away from water. Protect containers from physical damage.

Other Precautions and Warnings

Ensure adequate ventilation in storage areas. Avoid gas inhalation and use appropriate personal protective equipment if necessary. Always ensure stored cylinders are properly secured at all times.

SECTION VIII - CONTROL MEASURES AND PERSONAL PROTECTIVE EQUIPMENT

Respiratory Protection *(specify type)*

For low-level exposures, use a gas mask equipped for hexafluoroacetone. Otherwise, use a supplied-air respirator or a self-contained breathing apparatus with full facepiece in a positive pressure mode.

Ventilation

Local exhaust (preferred) at site of chemical work or general ventilation.

Protective Gloves	Eye Protection	Other Protective Clothing
Impervious Thermal Gloves	Chemical Goggles or Face Mask	Protective Apron

Work/Hygiene Practices

Always wash hands thoroughly after using chemical; never bring food, drink, or smoking materials into vicinity of chemicals.

HEXAFLUOROACETONE

$(CF_3)_2CO$ CAS: 684-16-2

IDENTIFICATION AND TYPICAL USES

Hexafluoroacetone is a colorless gas with a slight musty odor. It is normally shipped as a liquefied compressed gas. It is used almost exclusively as an intermediate in organic synthesis.

RISK ASSESSMENT: HEALTH

General Assessment

Hexafluoroacetone is normally a gas and is therefore toxic primarily by *inhalation*. However, it can also pass through intact skin (*absorption*) and, because of the super-cold temperature of the liquid, it presents a skin contact hazard as well. Ingestion is not a likely exposure route since hexafluoroacetone normally exists as a gas at room temperature and pressure. The compressed liquid can be expected to vaporize rather quickly when released to air. There are no reports of any carcinogenic effects resulting from exposure to this chemical. Animal studies have shown teratogenic effects in laboratory experiments.

Inhalation of the gas causes immediate and severe irritation to the moisture-rich mucosa of the eyes, nose, throat, and respiratory tract. This is due to reactive properties of hexafluoroacetone on contact with water. Because the reaction is exothermic (produces heat), there may also be burns and tissue destruction. The irritation may be severe enough to cause a dangerous (life-threatening) buildup of fluid in the lungs, a condition known as pulmonary edema. Symptoms of cough, tightness in chest, congestion, and dypsnea may be delayed up to 48 hours, thereby creating a false sense of security with regard to health exposure risk.

Skin or eye contact with the gas results in severe irritation. The liquid will cause burns due to frostbite. Tissue destruction, especially around the eyes, can be quite severe.

Acute Health Effects

The following acute (short-term) health effects may occur immediately or shortly after exposure to hexafluoroacetone:

Skin: Irritation, burn (frostbite), tissue destruction. Absorption is likely to occur.

Eye: Severe irritation and burn with possible temporary or permanent damage to eye tissues.

Lung: Nose, throat, and respiratory tract irritation and burns. May cause pulmonary edema.

Chronic Health Effects

The following chronic (long-term) health effects may occur at some time after exposure to hexafluoroacetone and can last for months or even years:

Cancer Hazards: According to information in the references, hexafluoroacetone has not been adequately tested for its ability to cause cancer in test animals.

Reproductive Hazard: According to information presented in the references, hexafluoroacetone has shown to cause teratogenic effects in test animals, including testicular atrophy. It is not known if exposure to hexafluoroacetone will have this effect on humans.

Other Chronic Effects: Repeated exposure has caused changes in the spleen, thymus, lymph nodes, and bone marrow of test animals. It is not known if human exposure will have this same effect.

Recommended Risk-Reduction Measures

Exposure to hexafluoroacetone presents a serious and even life-threatening risk to human health. Personnel should therefore avoid direct contact with hexafluoroacetone. Occupational poisoning occurs most commonly by inhalation and skin absorption. If a less toxic material or compound cannot be substituted for hexafluoroacetone, then *engineering controls* are the most effective method of reducing exposures. The best protection for workers is to ensure controlled use of hexafluoroacetone and, wherever possible, enclose operations or provide adequate local exhaust ventilation at the site of hexafluoroacetone release. Using respiratory protection is less effective than the controls mentioned above, but is still advisable whenever working with or around hexafluoroacetone. The PEL is extremely low and difficult to accurately monitor (0.1 ppm). For exposures to low levels, use a gas mask. Otherwise, use a supplied-air respirator with full facepiece and mask operated in positive pressure mode or a self-contained breathing apparatus (SCBA) with full facepiece operated in pressure demand. If a full facepiece is not available, then chemical goggles should be worn to protect the eyes. A face shield should also be considered. To prevent hand and skin exposures impervious, thermal protective gloves should be used. Glove manufacturers should be contacted and permeation studies obtained *before* final glove selections are made.

Administrative controls should also be in place to minimize the potential for human exposures. These may include written procedures or policies which specify the methods and techniques that will be practiced whenever personnel are to work with hexafluoroacetone.

All personnel should receive *training* on the use, hazards, protective measures, emergency actions, and other precautions per 29 CFR 1910.1200 (Hazard Communication), prior to the first assignment in an area where hexafluoroacetone is used or stored.

For those personnel with frequent or potentially high exposures (half the PEL or greater), the following are recommended before beginning employment and at regular intervals thereafter (e.g., annually):

☑ Lung function tests (establish baselines).

If symptoms develop or overexposure is suspected, the following recommended medical tests should be considered:

☑ Lung function tests.

☑ Consider chest X-ray after acute overexposure (may be negative if taken immediately following exposure due to delayed development of pulmonary edema).

Any evaluation should include a careful history of past and present symptoms with an examination. Medical tests that simply look for existing damage are not a substitute for controlling exposures. Also, because smoking can cause heart disease, emphysema, and other lung diseases, smokers exposed to hexafluoroacetone may experience symptoms more quickly than non-smokers under the same conditions of exposure. Prudent risk management requires proper consideration of *all* possible exposure factors when evaluating workplace health hazard risk.

Other methods to reduce exposure include:

☑ Where possible, enclose operations and use local exhaust ventilation at the site of chemical release. If local exhaust ventilation or enclosure is not used, respiratory protection should be mandatory.

☑ Always ensure that proper protective clothing is worn when using chemical substances and that personnel are trained on its proper use and care.

☑ Wash thoroughly immediately after exposure to hexafluoroacetone and at the end of the work shift or before eating, drinking, or smoking.

☑ Hazard warning information should be posted in the work area. In addition, as part of an on-going education and training program, all information on the health and safety hazards of hexafluoroacetone should be communicated to all exposed workers.

☑ Eye wash stations should be provided in the immediate work area for emergency use. If there is a possibility of skin exposure to hexafluoroacetone, emergency shower facilities should also be provided in the immediate area.

☑ Workers whose clothing has been contaminated by hexafluoroacetone should change into clean clothes before leaving work. Contaminated work clothing should be laundered only by individuals who have been informed of the hazards of exposure to hexafluoroacetone.

RISK ASSESSMENT: ENVIRONMENT

General Assessment

The environment is at risk of exposure during transportation, storage, disposal, or destruction of hexafluoroacetone. In almost every scenario, the threat of environmental exposure is contingent upon the proper handling of the chemical substance. Accidental spills, large or small, can result in contamination of the surrounding environmental mediums (primarily the air).

Hexafluoroacetone is a non-combustible gas. However, it is very reactive with water and many other materials and will produce excessive heat on contact.

Also, if it is involved in fire, hexafluoroacetone can emit highly toxic carbon monoxide and oxides of fluoride. It is incompatible with many common acids, such as nitric and sulfuric, and contact can liberate heat and acid fumes. It is hygroscopic (absorbs moisture from air) and will form a highly acidic sesquilhydrate. Caution is always required in handling, storage, transportation, and disposal of hexafluoroacetone. Emergency responders should be made aware of the presence of hexafluoroacetone at any emergency response situation.

Hexafluoroacetone can enter the environment from accidental discharges, leaking cylinders, and from spills.

Acute Ecological Effects

Acute (short-term) toxic effects may include the death of animals, birds, or fish and death or low growth rate in plants. Acute effects are seen 2 to 4 days after animals or plants are exposed to hexafluoroacetone.

Insufficient data are available on the short-term effects of hexafluoroacetone to aquatic life, plants, birds, or land animals.

Chronic Ecological Effects

Chronic (long-term) toxic effects may include shortened life span, reproductive problems, lower fertility, and changes in appearance or behavior in exposed animals. These effects can be seen long after first exposure(s) to toxic chemicals.

Insufficient data are available to evaluate or predict the long-term effects of hexafluoroacetone to aquatic life, plants, birds, or terrestrial animals.

Water Solubility

Hexafluoroacetone is highly reactive in water. It will liberate excessive heat and form highly corrosive compounds.

Persistence in the Environment

Hexafluoroacetone is non-persistent in water, with a half-life of less than 2 days. The half-life of a pollutant is the amount of time it takes for one half of the chemical to be degraded.

Bioaccumulation in Aquatic Organisms

Some substances increase in concentration, or *bioaccumulate,* in living organisms as they breathe contaminated air, drink contaminated water, or eat contaminated food. These chemicals can become concentrated in the tissues and internal organs of animals as well as humans.

Hexafluoroacetone is not expected to accumulate in the edible tissues of fish.

Recommended Risk-Reduction Measures

Proper training of all transporters will reduce the likelihood of a mishap or accident resulting in a leak or spill of hexafluoroacetone into the environment. Labels on all containers, trucks, and rail cars must meet DOT requirements and accurately reflect their contents to enable emergency responders to react properly and quickly to any disaster thereby reducing the potential risk to the environment and to personnel.

Storage of hexafluoroacetone should be segregated from other chemicals and materials (especially water and acids) to minimize the risk of cross-contamination.

If a spill or leak to the environment has occurred, fire department, emergency response and/or hazardous materials spill personnel should be notified immediately. Cleanup should be attempted only by those trained in proper spill containment procedures. Contaminated soils should be removed for incineration and replaced with clean soil. If hexafluoroacetone should contact the water table, aquifer, or navigable waterway, the local and/or state emergency response authorities must be notified. A comprehensive emergency response or disaster preparedness/recovery plan should be in place prior to any operations involving the use, transportation, storage, or disposal of hexafluoroacetone.

If hexafluoroacetone is spilled or leaked, the following specific steps are recommended:

- ☑ Restrict persons not wearing protective clothing from area of spill or leak until cleanup is complete and area can be opened for normal work.
- ☑ Stop leak of gas, if possible. Remove damage or leaking cylinders to open areas and allow to vent. Do NOT use water to knock down vapors and keep all water sources clear of the area.
- ☑ Ventilate area of spill or leak.

RISK ASSESSMENT: BUSINESS

General Assessment

Accidents or mishaps involving hexafluoroacetone can present a serious threat to business operations. The loss or damage of equipment or facilities can significantly affect fiscal viability. Lawsuits that may result from personnel illness, injury/death, public exposures, and/or environmental contamination will require a serious expenditure of resources. Media attention surrounding an injury, death, or environmental damage can also result in a loss of profits and loss of current as well as future business. Always remember that anytime the term "reproductive hazard" is used, public emotion, ignorance, and hysteria can run equally high. This must be carefully considered during the development of any public relations policies.

Recommended Risk-Reduction Measures

Company attorneys, safety and health professionals, and environmental specialists should be involved in the development of any procedures for responding to chemical incidents. A company official should be predesignated as a public relations officer with specific training in dealing with the press. Corporate plans and policies should be developed, approved, and implemented long before any need for such arises.

REFERENCES

American Conference of Governmental Industrial Hygienists. 1988. *Documentation of the Threshold Limit Values and Biological Exposure Indices,* 5th Edition (with updates). Cincinnati: ACGIH

Anthony, C. P., and N. J. Kolthoff. 1971. *Textbook of Anatomy and Physiology.* St. Louis: C. V. Mosby Company.

Bloom, A. D. and F. J. DeSerres. 1995. *Ecotoxicity and Human Health.* New York: CRC Press/Lewis Publishers.

Brusick, D. J. 1994. *Methods for Genetic Risk Assessment.* New York: CRC Press/Lewis Publishers.

Bueche, J. R. 1972. *Principles of Physics,* 2nd Edition. New York: McGraw-Hill.

Burrell, R., D. K. Flaherty, and L. J. Sauers. 1992. *Toxicology of the Immune System.* New York: Van Nostrand Reinhold.

Calabrese, E. J. 1994. *Biological Effects of Low Level Exposures.* New York: CRC Press/Lewis Publishers.

Cockerman, L. G., B. S. Shane. 1994. *Basic Environmental Toxicology.* New York: CRC Press/Lewis Publishers.

Cothern, C. R., N. P. Ross. 1994. *Environmental Statistics, Assessment, and Forecasting.* New York: CRC Press/Lewis Publishers.

Danse, I. R. 1991. *Common Sense Toxics in the Workplace.* New York: Van Nostrand Reinhold.

Environmental Protection Agency. 1983. *Chemical Hazard Information Profiles.* Washington, D.C.: U.S. EPA.

Gardner, R. W. 1994. *Chemical Intolerance.* New York: CRC Press/Lewis Publishers.

Hallenbeck, W. H. 1993. *Quantitative Risk Assessment for Environmental and Occupational Health,* 2nd Edition. New York: CRC Press/Lewis Publishers.

Hodgson, E., R. B. Mailman, and Janice E. Chambers. 1988. *Dictionary of Toxicology.* London: The MacMillian Company.

Jones, F. E. 1994. *Toxic Organic Vapors in the Workplace.* New York: CRC Press/Lewis Publishers.

Kamrin, M. A. 1988. *Toxicology: A Primer on Principles and Applications.* Chelsea, Michigan: Lewis Publishers.

Keith, L. H. and M. M. Walker. 1995. *Handbook of Air Toxics.* New York: CRC Press/Lewis Publishers.

Landis, W. G., M. H. Yu. 1995. *Introduction to Environmental Toxicology.* New York: CRC Press/Lewis Publishers.

Lewis, R. J., Sr. 1993. *Hawley's Condensed Chemical Dictionary,* 12th Edition. New York: Van Nostrand Reinhold.

Lewis, R. J., Sr. 1992. *Sax's Dangerous Properties of Industrial Materials,* Volumes 1, 2, and 3, 8th Edition. New York: Van Nostrand Reinhold.

Mackay, D., W. Y. Shiu, and K. C. Ma. 1992. *Illustrated Handbook of Chemical-Physical Properties and Environmental Fate for Organic Chemicals.* New York: CRC Press/Lewis Publishers.

Manahan, S. E. 1994. *Environmental Chemistry,* 6th Edition. New York: CRC Press/Lewis Publishers.

Meyer, E. 1990. *Chemistry of Hazardous Materials.* Englewood Cliffs, NJ: Prentice Hall.

Miller, B. F., M.D., and C. B. Keane, R.N. 1972. *Encyclopedia and Dictionary of Medicine and Nursing.* Philadelphia: W.B. Sanders Company.

Mycock, J. C., J. D. McKenna, and L. Theodore. 1995. *Handbook of Air Pollution Control Engineering and Technology.* New York: CRC Press/Lewis Publishers.

Nelson, G. O. 1992. *Gas Mixtures.* New York: CRC Press/Lewis Publishers.

Ott, W. R. 1995. *Environmental Statistics and Data Analysis.* New York: CRC Press/Lewis Publishers.

Patnaik, P. 1992. *A Comprehensive Guide to the Hazardous Properties of Chemical Substances.* New York: Van Nostrand Reinhold.

Philp, R. B. 1995. *Environmental Hazards and Human Health.* New York: CRC Press/Lewis Publishers.

Proctor, N. H., and J. P. Hughes. 1991. *Chemical Hazards of the Workplace.* New York: Van Nostrand Reinhold.

Quigley, D. R. 1994. *Handbook of Emergency Chemical Management.* New York: CRC Press/Lewis Publishers.

Rea, W. J. 1995. *Chemical Sensitivity,* Vol. I-III. New York: CRC Press/Lewis Publishers.

Richardson, M. 1995. *Dictionary of Substance and Their Effects,* Vol. I-VI. New York: CRC Press/Lewis Publishers.

U.S. Department of Health and Human Services, National Institute for Occupational Safety and Health. 1994. *NIOSH Pocket Guide to Chemical Hazards.* Washington, D.C.: U.S. Government Printing Office.

MATERIAL SAFETY DATA SHEET

CHEMICAL NAME

HEXAFLUOROETHANE

HAZARD WARNING INFORMATION

HEALTH	FIRE	REACTIVE	OTHER	DEGREE OF HAZARD	COLOR CODING	OTHER CODES
1	0	0		0 = Minimum Hazard 1 = Slight Hazard 2 = Moderate Hazard 3 = Serious Hazard 4 = Severe Hazard	HEALTH = BLUE FIRE = RED REACTIVITY = YELLOW OTHER = WHITE	OX = Oxidizer ACID = Acid ALK = Alkali COR = Corrosive W = Use No Water

SECTION I - GENERAL INFORMATION

Field	Value	Field	Value
Characterization	**Halogenated Hydrocarbon**	RCRA Number	**None**
		EPA Class	**Not Applicable**
DOT Proper Shipping Name	**Hexafluoroethane**	Chemical Abstract Service (CAS) Number	**76-16-4**
DOT Hazard Class and Label Requirements	**Non-Flammable Gas**	DOT Emergency Guide Code	**15**
DOT Identification Number	**UN 2193**	Molecular Formula	**CF_3CH_3**

Synonyms
Fluorocarbon 116; perfluoroethane; gaseous perfluorinated ethane.

SECTION II - HAZARDOUS INGREDIENTS AND IDENTITY INFORMATION

Hazardous Components (specific identity)	OSHA Exposure Criteria	NIOSH Exposure Criteria	Immediately Dangerous to Life and Health (IDLH)	ACGIH Exposure Criteria
Hexafluoroethane **99.6% Pure**	**PEL:** Not Established **STEL:** Not Established	**REL:** Not Established **STEL:** Not Established	**Not Determined**	**TLV:** Not Established **STEL:** Not Established

SECTION III - PHYSICAL AND CHEMICAL CHARACTERISTICS

Field	Value	Field	Value
Boiling Point	**-109°F (-78°C) at 1 atmosphere**	Specific Gravity (H_2O = 1)	**1.58 at -100°F (-73°C)**
Vapor Pressure (mm Hg)	**30.3 at 70°F (21°C)**	Molecular Weight (atomic weight)	**139.01**
Vapor Density (Air = 1)	**4.8**	Freezing Point	**-149°F (-100°C)**

Solubility
Insoluble in water, slightly soluble in alcohol.

Appearance and Odor
Exists as a gas at room temperature and pressure with a slight, mild, ethereal odor.

SECTION IV - FIRE AND EXPLOSION HAZARD DATA

Field	Value	Field	Value
Flash Point (method used)	**Non-Flammable**	Explosive Limits in Air % by Volume	**LEL: Not Determined** **UEL: Not Determined**
NFPA Classification	**Non-Flammable Gas**	Autoignition Temperature	**Non-Flammable**

Extinguishing Media
Use extinguishing media suitable to surrounding fires. Hexafluoroethane does not burn.

Special Fire Fighting Procedures
Poisonous gases are produced in fire. Wear self-contained breathing apparatus (SCBA) and full protective gear. Move containers from fire if it can be done without risk. Cool sides of fire-exposed containers with water until long after fire is out. Monitor smoke directions and stay clear. Be aware of runoff from fire control measure.

Unusual Fire and Explosion Hazards
Vapors are much heavier than air. They can travel low to the ground and accumulate in low-lying areas to levels that can present a serious toxic, corrosive, and suffocation hazard.

SECTION V - REACTIVITY DATA

Stability		Conditions to Avoid: Hexafluoroethane is normally stable in closed containers, under routine conditions of storage and handling. Avoid contact with heat and exposure to incompatible materials.
Stable	Unstable	Incompatibility *(materials to avoid)*: Hexafluoroethane is incompatible and will react violently with alkaline or alkali earth metals such as sodium or potassium. Finely powdered magnesium or aluminum may react at high temperatures.
X		
Hazardous Polymerization		Conditions to Avoid: Under normal conditions of temperature and pressure, hazardous polymerization of hexafluoroethane cannot occur.
May Occur	Will Not Occur	Hazardous Decomposition or By-products: Poisonous gases and acrid fumes are produced in fire, including toxic, corrosive, and dangerous materials such as fluorine, hydrogen fluoride, and carbonyl fluoride.
	X	

SECTION VI - HEALTH HAZARD DATA

Primary Route(s) of Entry:	Inhalation?	Absorption (skin/eye)?	Ingestion?
	X		

Health Hazards

INHALATION: An asphyxiant gas. Initial symptoms are those of oxygen starvation. Feeling of "air hunger," diminished mental alertness, headache, dullness, shortness of breath, and an impairment in muscle coordination. Continued exposures lead to faulty judgment, depression of senses, rapid fatigue, unconsciousness, convulsions, coma, and death.

SKIN & EYES: Contact with the compressed gas or refrigerated liquid can cause frostbite and burns.

INGESTION: Not likely since hexafluoroethane is a gas at normal working (room) temperatures.

Carcinogenicity	NTP Listed?	IARC Cancer Review Group?	OSHA Regulated?	Target Organs?
Unknown Human Unknown Animal	No	No	No	Skin, eyes (frostbite).

Medical Conditions Generally Aggravated by Exposure

None reported.

Emergency and First-aid Procedures

Eye contact (liquid): Immediately flush large amounts of tepid water for 15 minutes (minimum), occasionally lifting eyelids; seek medical attention. Skin contact (liquid): Remove all contaminated clothing. Immediately flush area with large amounts of tepid water. Do NOT rub! Seek medical assistance. For inhalation: Remove the person from exposure. Emergency responders must be protected from inert atmospheres. Provide respiratory assistance and CPR. Transfer promptly to medical facility. If swallowed: Treat frozen tissues and protect the airway. Seek medical attention immediately.

SECTION VII - PRECAUTIONS FOR SAFE HANDLING AND USE

Steps to be Taken in Case Material is Released or Spilled

Ventilate area. Stop flow of gas. If the source of leak is a cylinder and cannot be stopped, remove to safe place in open air; repair leak or let cylinder empty. Ensure response personnel are properly protected with self-contained breathing apparatus (SCBA).

Preferred Waste Disposal Method

No Citation.

Precautions to be Taken in Handling and Storage

Store in tightly closed steel pressurized containers in cool, dark, dry, well-ventilated location. Protect containers from physical damage. Exercise caution when working with asphyxiating gases.

Other Precautions and Warnings

Ensure adequate ventilation in storage areas. Avoid vapor inhalation and use appropriate personnel protective equipment if necessary. Hexafluoroethane is a high-density gas (nearly five times that of air) and can present an asphyxiation hazard, especially in confined areas. Keep out of direct sunlight.

SECTION VIII - CONTROL MEASURES AND PERSONAL PROTECTIVE EQUIPMENT

Respiratory Protection *(specify type)*

At any detectable level, use a supplied-air respirator or a self-contained breathing apparatus (SCBA) with full facepiece operated in positive pressure mode.

Ventilation

Local exhaust (preferred) at site of chemical work or general ventilation.

Protective Gloves	Eye Protection	Other Protective Clothing
Impervious Thermal Gloves	Chemical Goggles or Face Mask	Protective Apron if Splash is Likely

Work/Hygiene Practices

Always wash hands thoroughly after using chemical; never bring food, drink, or smoking materials into vicinity of chemicals.

HEXAFLUOROETHANE

CF_3CH_3

CAS: 76-16-4

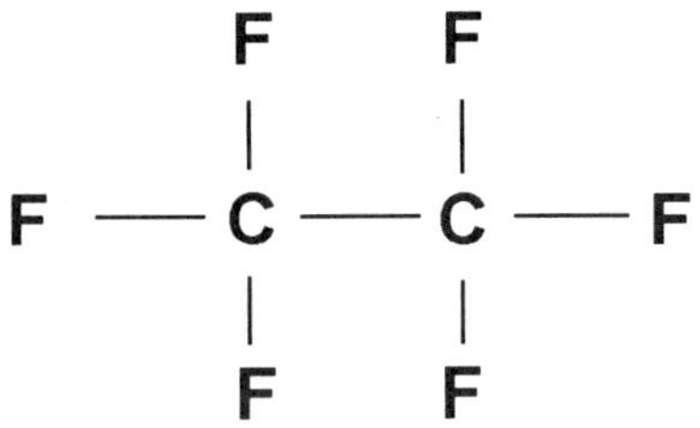

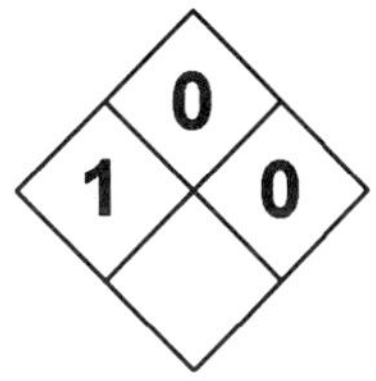

IDENTIFICATION AND TYPICAL USES

Hexafluoroethane is a gas at room temperature and pressure with a slight ethereal odor. It is used as a dielectric and as a coolant. It is also used as an aerosol propellant, and as a refrigerant.

RISK ASSESSMENT: HEALTH

General Assessment

Hexafluoroethane is an asphyxiating gas. As such, its primary route of entry into the body is through *inhalation.* While it does not pose a significant physiological effect (i.e., no systemic poisoning or toxicity), it can displace oxygen levels to below that which is necessary to sustain life. Skin contact with the liquid can be destructive to tissues due to the extremely cold temperatures but absorption is not a likely exposure route. Since hexafluoroethane usually exists as a gas at room temperature, ingestion is not likely to occur either.

Inhalation of hexafluoroethane gas can cause headache, dullness, shortness of breath, and the symptoms of asphyxiation. These include rapid respiration and a feeling of "air hunger" or air starvation followed by a decrease in mental alertness and impaired muscular coordination. If exposure continues, there can be nausea, vomiting, prostration, unconsciousness, coma, and death.

Skin or eye contact with the vapor poses no serious threat. The vapors have not been shown to cause any physiological response. However, if the liquid should contact either the skin or the eyes, there can be serious irritation, burns, tissue damage, and frostbite.

Acute Health Effects

The following acute (short-term) health effects may occur immediately or shortly after exposure to chemicals such as hexafluoroethane:

Skin: Liquid threat only. Redness, pain, swelling, blisters, and severe frostbite.

Eye: Liquid threat only. Causes frostbite, redness, irritation, severe pain, and blurred vision.

Lung: Vapor is non-irritating and therefore poses no threat of damage to the lungs. However, in high concentrations, hexafluoroethane can lead to symptoms of asphyxiation and oxygen depravation up to and including death.

Chronic Health Effects

The following chronic (long-term) health effects may occur at some time after exposure to hexafluoroethane and can last for months or even years:

Cancer Hazards: According to information presented in the references, hexafluoroethane has not been adequately tested for its ability to cause cancer in test animals. However, its primary effect appears to be asphyxiation and other such gases have not been shown to cause a cancer threat in humans or test animals.

Reproductive Hazard: According to information presented in the references, hexafluoroethane has not been adequately tested for its ability to cause adverse reproductive effects in humans. However, its primary effect appears to be asphyxiation and other such gases have not been shown to adversely affect reproduction in humans or test animals.

Other Chronic Effects: None reported.

Recommended Risk-Reduction Measures

Personnel should avoid direct contact with hexafluoroethane. If another gas cannot be substituted for a hexafluoroethane, then *engineering controls* are the most effective method of reducing exposures. The best protection is to separate operations and provide local exhaust ventilation at the site of chemical release. While not always operationally feasible, isolating operations can also reduce exposure. Using respiratory protection is less effective than the controls mentioned above, but is still advisable whenever working with hexafluoroethane. Air-purifying respirators (e.g., the cartridge-type or gas masks) *cannot* be used with hexafluoroethane. These types of respirators only purify the air and can do nothing to protect in oxygen deficient atmospheres. A powered air-supplied respirator operated in pressure demand or other positive pressure mode, or a self-contained breathing apparatus (SCBA) with full facepiece operated in pressure demand mode are the recommended respiratory protection methods of choice. If a full facepiece is not available, then chemical goggles should be worn to protect the eyes. Whenever a chemical splash hazard exists, a face shield and a rubber or leather apron should be worn. To prevent hand and skin exposures with the liquid, thermal insulated gloves should be used.

Administrative controls should also be in place to minimize the potential for human exposures. These may include written procedures or policies which specify the methods and techniques that will be practiced whenever personnel are to work with hexafluoroethane.

All personnel should receive *training* on the use, hazards, protective measures, emergency actions, and other precautions per 29 CFR 1910.1200 (Hazard Communication), prior to the first assignment in an area where hexafluoroethane is used or stored.

Other methods to reduce exposure to chemicals include:

☑ Where possible, consider installing automatic sensors that warn personnel of dropping oxygen levels and use local exhaust ventilation at the site of chemical release. If local exhaust ventilation or enclosure is not used, respiratory protection should be mandatory.

☑ Always ensure that proper protective clothing is worn when using chemical substances and that personnel are adequately trained on its proper use, care, and maintenance.

☑ Safety shower and eyewash stations should be readily available in work areas where hexafluoroethane is used or stored.

☑ Hazard warning information should be posted in the work area. In addition, as part of an on-going education and training program, all information on the health and safety hazards of hexafluoroethane should be communicated to all exposed or potentially exposed workers.

RISK ASSESSMENT: ENVIRONMENT

General Assessment

The environment is at risk of exposure during transportation, storage, disposal, or destruction of hexafluoroethane. In almost every scenario, the threat of environmental exposure is contingent upon the proper handling of the chemical substance. Accidental spills, large or small, can result in contamination of the surrounding environmental mediums (water, soil, and air).

Hexafluoroethane is non-combustible gas. However, if it is involved in fire, hexafluoroethane can emit highly toxic, corrosive, and dangerous materials such as fluorine, hydrogen fluoride, and carbonyl fluorides. It is incompatible with alkali and alkali earth metals such as sodium and potassium, and finely-ground or powdered magnesium or aluminum. Caution is always required in handling, storage, transportation, and disposal of hexafluoroethane. Emergency responders should be made aware of the presence of hexafluoroethane at any emergency response situation.

Hexafluoroethane can enter the environment through industrial discharges, unchecked venting to air, or through spills of the liquid. Since liquids will most likely vaporize rapidly, water and soil contamination are possible but improbable.

☠ *Acute Ecological Effects*

Acute (short-term) toxic effects may include the death of animals, birds, or fish, and death or low growth rate in plants. Acute effects are seen 2 to 4 days after animals or plants are exposed to hexafluoroethane.

Insufficient data are available on the short-term effects of hexafluoroethane on aquatic life, plants, birds, or land animals. However, as a relatively non-toxic, asphyxiant gas, it might be assumed that any acute ecological toxicity will be relatively low.

Chronic Ecological Effects

Chronic (long-term) toxic effects may include shortened life span, reproductive problems, lower fertility, and changes in appearance or behavior in exposed animals. These effects can be seen long after first exposure(s) to toxic chemicals.

Insufficient data are available to evaluate the long-term effects of hexafluoroethane to aquatic life, plants, birds, or land animals. However, as a relatively non-toxic, asphyxiant gas, it might be assumed that any chronic ecological toxicity will be relatively low.

Water Solubility

Hexafluoroethane is only slightly soluble in water. Concentrations of 1 milligram or less may mix with a liter of water. It is normally a gas at room temperature and pressure and solubility may not be a concern.

Persistence in the Environment

Hexafluoroethane is non-persistent in water, with a half-life of less than 2 days. The half-life of a pollutant is the amount of time it takes for one half of the chemical to be degraded. Virtually all of hexafluoroethane will eventually end up in the air.

Bioaccumulation in Aquatic Organisms

Some substances increase in concentration, or *bioaccumulate,* in living organisms as they breathe contaminated air, drink contaminated water, or eat contaminated food. These chemicals can become concentrated in the tissues and internal organs of animals as well as humans.

Hexafluoroethane is not expected to accumulate in the edible tissues of fish.

Recommended Risk-Reduction Measures

Proper training of all transporters will reduce the likelihood of a mishap or accident resulting in a leak or release to the environment. The correct use of DOT labeling (per the MSDS) on all containers, trucks, and rail cars will enable emergency responders to react properly and quickly to any disaster thereby reducing the potential risk to the environment and to personnel.

Storage of hexafluoroethane should be segregated from other gases to prevent cross-contamination or contact. Buildings designated for storage should be equipped with monitors that provide ample warning (audible and visual) of oxygen deficient atmospheres.

If a spill or release to the environment has occurred, fire department, emergency response, and/or hazardous materials spill personnel should be notified immediately. Cleanup should be attempted only by those trained in proper containment procedures. If applicable contaminated soils should be removed for incineration and replaced with clean soil. When spills or leaks occur, it may be necessary to contact the local and/or state emergency response authorities.

A comprehensive emergency response or disaster preparedness/recovery plan should be in place prior to any operations involving the use, transportation, storage, or disposal of hexafluoroethane. If hexafluoroethane is spilled or leaked, the following specific steps are recommended:

- ☑ Restrict persons not wearing protective clothing from area of spill or leak until cleanup is complete and area can be opened for normal work.
- ☑ Ventilate area of spill or leak.
- ☑ If applicable, stop flow of leaking liquid or gas. If the leak cannot be stopped in place, remove leaking container to a safe place in the open air, and repair or allow cylinder to empty.
- ☑ Keep hexafluoroethane out of a confined space, such as a sewer, because of the possibility of creating an inert atmosphere.

RISK ASSESSMENT: BUSINESS

General Assessment

Accidents or mishaps involving hexafluoroethane can present a significant threat to business operations. The loss or damage of equipment or facilities can significantly affect fiscal viability. Lawsuits that may result from personnel injury, public exposures, and environmental contamination will also require a serious expenditure of resources. Media attention surrounding an injury, illness, or environmental damage can also result in a loss of profits and loss of current as well as future business.

Recommended Risk-Reduction Measures

Company attorneys, safety and health professionals, and environmental specialists should be involved in the development of any procedures dealing with safety and emergency response. A company official should be pre-designated as a public relations officer with specific training in dealing with the press. Corporate plans and policies should be developed, approved, and implemented long before any need for such arises.

REFERENCES

American Conference of Governmental Industrial Hygienists. 1988. *Documentation of the Threshold Limit Values and Biological Exposure Indices,* 5th Edition (with updates). Cincinnati: ACGIH

Anthony, C. P., and N. J. Kolthoff. 1971. *Textbook of Anatomy and Physiology.* St. Louis: C. V. Mosby Company.

Bloom, A. D. and F. J. DeSerres. 1995. *Ecotoxicity and Human Health.* New York: CRC Press/Lewis Publishers.

Brusick, D. J. 1994. *Methods for Genetic Risk Assessment.* New York: CRC Press/Lewis Publishers.

Bueche, J. R. 1972. *Principles of Physics,* 2nd Edition. New York: McGraw-Hill.

Burrell, R., D. K. Flaherty, and L. J. Sauers. 1992. *Toxicology of the Immune System.* New York: Van Nostrand Reinhold.

Calabrese, E. J., E. M. Kenyon. *Air Toxics and Risk Assessment.* New York: CRC Press/Lewis Publishers.

Calabrese, E. J. 1994. *Biological Effects of Low Level Exposures.* New York: CRC Press/Lewis Publishers.

Cockerman, L. G., B. S. Shane. 1994. *Basic Environmental Toxicology.* New York: CRC Press/Lewis Publishers.

Cothern, C. R., N. P. Ross. 1994. *Environmental Statistics, Assessment, and Forecasting.* New York: CRC Press/Lewis Publishers.

Cothern, R. R. 1993. *Comparative Environmental Risk Assessment.* New York: CRC Press/Lewis Publishers.

Cothern, C. R. 1995. *Handbook for Environmental Risk Decisionmaking.* New York: CRC Press/Lewis Publishers.

Danse, I. R. 1991. *Common Sense Toxics in the Workplace.* New York: Van Nostrand Reinhold.

Environmental Protection Agency. 1983. *Chemical Hazard Information Profiles.* Washington, D.C.: U.S. EPA.

Fodden, C. R., and J. L. Weddell. 1992. *Hazardous Materials Emergency Action Data.* New York: CRC Press/Lewis Publishers.

Forseberg, K., and L. H. Keith. 1993. *Chemical; Protective Clothing Permeation and Degradation Database.* Chelsea, MI: Lewis Publishers.

Gardner, R. W. 1994. *Chemical Intolerance.* New York: CRC Press/Lewis Publishers.

Gerarde, H. W. 1960. *Toxicology and Biochemistry of Aromatic Hydrocarbons.* New York: Elsevier.

Hallenbeck, W. H. 1993. *Quantitative Risk Assessment for Environmental and Occupational Health,* 2nd Edition. New York: CRC Press/Lewis Publishers.

Hodgson, E., R. B. Mailman, and Janice E. Chambers. 1988. *Dictionary of Toxicology.* London: The MacMillian Company.

Howard, P. H. 1995. *Handbook of Environmental Fate and Exposures Data for Organic Chemicals,* Vol. I-IV. New York: CRC Press/Lewis Publishers.

Howard, P. H., et. al. 1991. *Handbook of Environmental Degradation Rates.* New York: CRC Press/Lewis Publishers.

Hosty, J. W., and P. Foster. 1990. *A Practical Guide to Chemical Spill Response.* New York: Van Nostrand Reinhold.

International Agency for Research on Cancer (IARC). 1987. *IARC Monographs on the Evaluation of Carcinogenic Risk to Humans, Overall Evaluations of Carcinogenicity: An updating of IARC Monographs Volumes 1-42.* Lyon: IARC.

James, A., and M. Lord. 1992. *Index of Chemical and Physical Data.* New York: Van Nostrand Reinhold.

Jones, F. E. 1994. *Toxic Organic Vapors in the Workplace.* New York: CRC Press/Lewis Publishers.

Kamrin, M. A. 1988. *Toxicology: A Primer on Principles and Applications.* Chelsea, Michigan: Lewis Publishers.

Keith, L. H. , M. T. Johnston, C. E. Hudak, and M. Conoley. 1992. *National Toxicology Program's Gloves+ Database.* Chelsea, MI: Lewis Publishers.

Keith, L. H. and M. M. Walker. 1995. *Handbook of Air Toxics.* New York: CRC Press/Lewis Publishers.

Keith, L. H. (Editor). 1992. *Integrated Risk Information System (IRIS) Chemical Information Database.* Prepared by the U.S. Environmental Protection Agency. Boca Raton, FL: CRC Press/Lewis Publishers.

Keith, L. H. and D. B. Walters. (Editors) 1992. *National Toxicology Program's Chemical Database, Volumes I-VII.* Chelsea, MI: Lewis Publishers.

Landis, W. G., M. H. Yu. 1995. *Introduction to Environmental Toxicology.* New York: CRC Press/Lewis Publishers.

Langerman, N. 1994 *Precautionary Labels for Chemical Containers.* New York: CRC Press/Lewis Publishers.

Lauwerys, R. R., P. Hoet. 1993. *Industrial Chemical Exposure.* New York: CRC Press/Lewis Publishers.

Lewis, .J., Sr. 1993. *Hawley's Condensed Chemical Dictionary,* 12th Edition. New York: Van Nostrand Reinhold.

Mackay, D., W. Y. Shiu, and K. C. Ma. 1992. *Illustrated Handbook of Chemical-Physical Properties and Environmental Fate for Organic Chemicals.* New York: CRC Press/Lewis Publishers.

Manahan, S. E. 1994. *Environmental Chemistry,* 6th Edition. New York: CRC Press/Lewis Publishers.

Manahan, S. E. 1992. *Toxicological Chemistry,* 2nd Edition. New York: CRC Press/Lewis Publishers.

Mellström, G. A., J. E. Wahlberg, and H. I. Maibach. 1994. *Protective Gloves for Occupational Use.* New York: CRC Press/Lewis Publishers.

Meyer, E. 1990. *Chemistry of Hazardous Materials.* Englewood Cliffs, NJ: Prentice Hall.

Miller, B. F., M.D., and C. B. Keane, R.N. 1972. *Encyclopedia and Dictionary of Medicine and Nursing.* Philadelphia: W.B. Sanders Company.

Mycock, J. C., J. D. McKenna, and L. Theodore. 1995. *Handbook of Air Pollution Control Engineering and Technology.* New York: CRC Press/Lewis Publishers.

Patnaik, P. 1992. *A Comprehensive Guide to the Hazardous Properties of Chemical Substances.* New York: Van Nostrand Reinhold.

Quigley, D. R. 1994. *Handbook of Emergency Chemical Management.* New York: CRC Press/Lewis Publishers.

U.S. Department of Health and Human Services, National Institute for Occupational Safety and Health. 1994. *NIOSH Pocket Guide to Chemical Hazards.* Washington, D.C.: U.S. Government Printing Office.

MATERIAL SAFETY DATA SHEET

CHEMICAL NAME	
HEXAMETHYLENE DIAMINE	**CRC PRESS/LEWIS PUBLISHERS DISCLAIMER:** THE INFORMATION AND RECOMMENDATIONS PRESENTED HEREIN ARE BASED ON SOURCES BELIEVED TO BE RELIABLE. CRC MAKES NO REPRESENTATION ON ITS COMPLETENESS OR ACCURACY. IT IS THE RESPONSIBILITY OF THE USER TO DETERMINE THE CHEMICAL'S SUITABILITY FOR ITS INTENDED USE, THE CHEMICAL'S SAFE USE, AND THE CHEMICAL'S PROPER DISPOSAL. NO REPRESENTATIONS AND/OR WARRANTIES, EITHER EXPRESSED OR IMPLIED, OF THE MERCHANTABILITY OR FITNESS FOR A PARTICULAR PURPOSE, OR OF ANY OTHER NATURE, ARE MADE WITH RESPECT TO THE INFORMATION PROVIDED IN THIS MSDS OR TO THE CHEMICAL TO WHICH INFORMATION MAY REFER. CRC NEITHER ASSUMES NOR AUTHORIZES ANY OTHER PERSON TO ASSUME FOR IT, ANY OTHER ADDITIONAL RESPONSIBILITY OR LIABILITY FOR THE USE OF, OR RELIANCE UPON, THIS INFORMATION.

HAZARD WARNING INFORMATION

HEALTH	FIRE	REACTIVE	OTHER	DEGREE OF HAZARD	COLOR CODING	OTHER CODES
2	2	0		0 = Minimum Hazard 1 = Slight Hazard 2 = Moderate Hazard 3 = Serious Hazard 4 = Severe Hazard	HEALTH = BLUE FIRE = RED REACTIVITY = YELLOW OTHER = WHITE	OX = Oxidizer ACID = Acid ALK = Alkali COR = Corrosive W = Use No Water

SECTION I - GENERAL INFORMATION

Field	Value	Field	Value
Characterization	**Aliphatic Amine**	RCRA Number	**None**
		EPA Class	**Not Applicable**
DOT Proper Shipping Name	**Hexamethylenediamine (solid or solution)**	Chemical Abstract Service (CAS) Number	**124-09-4**
DOT Hazard Class and Label Requirements	**Corrosive Material; Corrosive Poison**	DOT Emergency Guide Code	**68**
DOT Identification Number	**UN 2280 UN 1783**	Chemical Formula	$H_2N(CH_2)_6NH_2$

Synonyms
1,6-Diaminohexane; 1,6-hexanediamine; 1,6-diamino-*n*-hexane; 1,6-hexamethylenediamine; HMDA.

SECTION II - HAZARDOUS INGREDIENTS AND IDENTITY INFORMATION

Hazardous Components (specific identity)	OSHA Exposure Criteria	NIOSH Exposure Criteria	Immediately Dangerous to Life and Health (IDLH)	ACGIH Exposure Criteria
Hexamethylenediamine (derivation: Reaction of adipic acid and ammonia (catalytic vapor-phase) to yield adiponitrile, followed by liquid-phase catalytic hydrogenation; chlorination of butadiene followed by reaction with sodium cyanide (cuprous chloride catalyst) to 1,4-dicyanobutylene and hydrogenation).	**PEL:** Not Established **STEL:** Not Established	**REL:** Not Established **STEL:** Not Established	**Not Determined**	**TLV:** Not Established **STEL:** Not Established

SECTION III - PHYSICAL AND CHEMICAL CHARACTERISTICS

Field	Value	Field	Value
Boiling Point	**401°F (205°C)**	Specific Gravity (H_2O = 1)	**0.799 (solid) 0.933 (liquid)**
Vapor Pressure (mm Hg)	**Not Found**	Molecular Weight	**116.20**
Vapor Density (Air = 1)	**4.0**	Melting Point	**108°F (42°C)**

Solubility
Very soluble in water, slightly soluble in alcohol and benzene. Absorbs water and carbon dioxide from the air.

Appearance and Odor
Colorless, leaflet-shaped solid with a piperidine odor (weak, fishy odor). Also appears in solution form.

SECTION IV - FIRE AND EXPLOSION HAZARD DATA

Field	Value	Field	Value
Flash Point (method used)	**160°F (71°C) opened cup**	Explosive Limits in Air % by Volume	**LEL: 0.7% UEL: 6.3%**
NFPA Classification	**Combustible Solid**	Autoignition Temperature	**Not Determined**

Extinguishing Media
Use dry chemical, carbon dioxide, foam, or water.

Special Fire Fighting Procedures
Poisonous gases are produced in fire; wear full protective clothing and self-contained breathing apparatus (SCBA). Contact with hexamethylenediamine causes burns. Stay upwind and clear of smoke and runoff. Move containers from fire area if it can be done safely. Cool exposed container with water.

Unusual Fire and Explosion Hazards
Hexamethylenediamine is combustible when exposed to heat or flame and will react with oxidizing materials. Its low melting causes it to become liquid or molten in a fire. Exercise extreme caution when attempting to move or relocate containers that may have been exposed to high heat.

SECTION V - REACTIVITY DATA

Stability		Conditions to Avoid
Stability		Hexamethylenediamine is normally stable under routine conditions of handing and storage. Avoid contact with incompatible materials. Keep away from heat and flame.
Stable	Unstable	Incompatibility *(materials to avoid)*
X		Oxidizing agents (such as chlorine, bromine, and oxygen) and other oxidizable materials.
Hazardous Polymerization		Conditions to Avoid Under normal conditions of temperature and pressure, hazardous polymerization of hexamethylenediamine is not expected to occur.
May Occur	Will Not Occur	Hazardous Decomposition or By-products
	X	Toxic and irritating oxides of nitrogen are produced when hexamethylenediamine is heated to decomposition or involved in a fire.

SECTION VI - HEALTH HAZARD DATA

Primary Route(s) of Entry:	Inhalation?	Absorption (skin)?	Ingestion?
	X	X	X

Health Hazards

INHALATION: Irritation of the nose, throat, and lungs. Will cause coughing, nausea, headache, and may cause vomiting and fainting.

SKIN & EYES: A corrosive. Eye contact will cause irritation and inflammation. Skin contact causes severe irritation and possible dermatitis. First and second degree burns possible.

INGESTION: Poison by ingestion. Gastrointestinal irritation and irritation of the mouth and stomach.

Carcinogenicity	NTP Listed?	IARC Cancer Review Group?	OSHA Regulated?	Target Organs?
Unknown Human Unknown Animal	No	No	No	Eyes, skin, respiratory system mucosa.

Medical Conditions Generally Aggravated by Exposure

Chronic respiratory diseases and skin disorders may be aggravated by exposure.

Emergency and First-aid Procedures

Eye contact: Flush immediately with water for 15 minutes (minimum), seek medical attention. Skin contact: Remove all contaminated clothing. Immediately wash area with large amounts of soap and water. Seek medical assistance. For inhalation: Remove the person from exposure. Provide respiratory assistance and CPR. Transfer to medical facility. If swallowed: Call the poison control center and seek medical attention immediately. Never attempt to give an unconscious or convulsing person anything by mouth. Have conscious person drink 1-2 glasses of water and induce vomiting.

SECTION VII - PRECAUTIONS FOR SAFE HANDLING AND USE

Steps to be Taken in Case Material is Released or Spilled

Absorb solutions in vermiculite or other absorbent and deposit in DOT-approved drum. Dry spills should be collected in safest manner possible and deposited in sealed drums. Use HEPA vacuum, do NOT dry sweep. Ventilate area. Restrict those not involved in cleanup from entering area.

Preferred Waste Disposal Method

No citation.

Precautions to be Taken in Handling and Storage

Do not store hexamethylenediamine in the presence of incompatible chemicals or materials. Store in tightly closed containers in a cool, well-ventilated area. Protect containers from physical damage. Electrically ground and bond all storage containers and equipment.

Other Precautions and Warnings

A corrosive material.

SECTION VIII - CONTROL MEASURES AND PERSONAL PROTECTIVE EQUIPMENT

Respiratory Protection *(specify type)*

No exposure levels have been established. This does not mean exposures are without risk to heath. For low or infrequent exposures, use an organic vapor respirator with dust/mist pre-filter. Greater protection is provided with an MSHA/NIOSH-approved supplied-air respirator with full facepiece operated in continuous flow or other positive pressure mode (pressure demand), or a self-contained breathing apparatus (SCBA) with full facepiece operated in pressure demand or other positive pressure mode.

Ventilation

Local exhaust or general mechanical systems recommended.

Protective Gloves	Eye Protection	Other Protective Clothing
Impervious Gloves	Chemical/Dust Goggles or Face Mask	Protective Apron

Work/Hygiene Practices

Always wash hands thoroughly after using chemical; never bring food, drink, or smoking materials into vicinity of chemicals.

HEXAMETHYLENEDIAMINE

$H_2N(CH_2)_6NH_2$ CAS: 124-09-4

IDENTIFICATION AND TYPICAL USES

Hexamethylenediamine appears as colorless leaflets with an odor of piperidine (weak, fish-like). It may also appear mixed in liquid solution. It absorbs water and carbon dioxide from the air. It is used as a chemical intermediate for nylon-type polyamide resins and for the formation of high polymers such as nylon 66.

RISK ASSESSMENT: HEALTH

General Assessment

Hexamethylenediamine is a human poison and is toxic by *ingestion* and *inhalation.* It is also moderately toxic by skin contact (corrosive). It will irritate the nose and throat (mucous membranes) and is a severe eye and skin irritant. Experimental teratogenic data have also been reported.

Inhalation will cause irritation to the nose, throat, and the mucous membranes of the respiratory tract. It will cause cough, nausea, headache, and fainting. In addition to airway irritation, there is a possibility of abnormalities in white blood cells as well as liver and intestine irregularities.

Ingestion will result in digestive system irritation, and unspecified but adverse gastrointestinal effects with irritation of the mouth and stomach. There may be nausea, vomiting, cramps, and headache.

Skin contact results in severe irritation and may produce dermatitis with rash and itching. Eye contact can result in severe irritation, conjunctivitis, smarting pain, corneal edema, and burns. The severity of burns may range from first degree (brief exposures) and second degree (prolonged contact).

Acute Health Effects

The following acute (short-term) health effects may occur either immediately or shortly after (within hours or days) exposure to hexamethylenediamine:

Skin: Irritation, inflammation, redness, itching, and possible dermatitis. First or second degree burns are possible.

Eye: Severe irritation, inflammation, and possible damage to the cornea. Burns and conjunctivitis can also occur.

Lung: Irritation causing coughing and/or shortness of breath. There may be nausea, headache, fainting, and other adverse effects.

Other: In animal studies and limited human exposure data, there appears to be some propensity for liver and intestine irregularities and white blood cell abnormalities.

Chronic Health Effects

The following chronic (long-term) health effects may occur at some time after exposure to hexamethylenediamine and can last for months or even years:

Cancer Hazards: According to the information presented in the references, hexamethylenediamine has not been adequately tested for its ability to cause cancer in test animals.

Reproductive Hazard: According to the information presented in the references, hexamethylenediamine has been shown to cause teratogenic effects in test animals. It is not know if it will have this effect on humans.

Other Chronic Effects: Low, chronic exposures to skin may cause allergy or dermatitis. Once an allergy develops, even small future exposures will cause a recurrence of symptoms.

Recommended Risk-Reduction Measures

Unless a less toxic chemical can be substituted for a hazardous substance, *engineering controls* are the most effective method of reducing exposures. The best protection is to enclose operations and/or provide local exhaust ventilation at the site of chemical release. While not always operationally feasible, isolating operations can also reduce exposure.

Using respiratory protection is less effective than the controls mentioned above, but is still advisable whenever working with or around hexamethylenediamine. No exposure levels have been established for this chemical. This does not mean, however, that human exposures present no health risks. For relatively low, infrequent, or transient exposures, an organic vapor respirator with a dust/mist pre-filter may suffice. For higher or prolonged exposure, a self-contained breathing apparatus (SCBA) with full facepiece operated in pressure demand or a supplied-air respirator with full facepiece in continuous flow mode are the recommended respiratory protection methods of choice. If a full facepiece is not available, then chemical goggles should be worn to protect the eyes. Whenever a chemical splash hazard exists or an airborne dust condition is likely, a face shield and apron should be worn. To prevent hand and skin exposures, impervious gloves should be used. Glove manufacturers should be contacted and permeation studies obtained *before* final glove selections have been made.

Administrative controls should also be in place to minimize the potential for human exposures. These may include written procedures or policies which specify the methods and techniques that will be practiced whenever personnel are to work with hexamethylenediamine.

All personnel should receive *training* on the use, hazards, protective measures, emergency actions, and other precautions per 29 CFR 1910.1200 (Hazard Communication), prior to the first assignment in an area where hexamethylenediamine is used or stored.

Before beginning employment and at regular intervals thereafter (e.g. annually), the following medical tests are recommended:

☑ Lung function tests (establish a baseline).

If symptoms develop or overexposure is suspected, the following medical tests may be useful:

☑ Lung function tests.

☑ Consider chest X-ray after acute overexposure.

☑ Evaluation by a qualified allergist with careful evaluation of exposure history and administration of special tests (may help diagnose skin allergy).

Any evaluation should include a careful medical history of past and present symptoms with an examination. However, medical tests that evaluate existing damage are not a substitute for controlling exposure. Also, since smoking can cause heart disease, emphysema, and numerous other respiratory disorders, smokers may be affected more rapidly than non-smokers to the same exposure conditions. Prudent risk assessment and management requires careful consideration of *all* possible risk factors that may be causing the appearance of symptoms in exposed workers.

Other methods to reduce exposure include:

☑ Where possible, enclose operations and use local exhaust ventilation at the site of chemical release. If local exhaust ventilation or enclosure is not used, respiratory protection should be mandatory. Also, if possible, automatically transfer liquids from storage containers to process containers using non-sparking tools.

☑ Always ensure that proper protective clothing is worn when using chemical substances.

☑ Wash thoroughly immediately after exposure to hexamethylenediamine and at the end of the work shift or before eating, drinking, or smoking.

☑ Hazard warning information should be posted in the work area. In addition, as part of an on-going education and training program, all information on the health and safety hazards of hexamethylenediamine should be communicated to all potentially exposed workers.

☑ Eye wash stations should be provided in work areas. If the potential for whole body exposure exists, then safety showers should also be provided.

RISK ASSESSMENT: ENVIRONMENT

General Assessment

The environment is at risk of exposure during intended use, transportation, storage, disposal, or destruction of hexamethylenediamine. In almost every scenario, the threat of environmental exposure is contingent upon the proper handling of the chemical substance. Accidental spills, large or small, can result in fire, explosion, and possible contamination of the surrounding environmental mediums (water, soil, and air).

Hexamethylenediamine is considered a combustible solid. It can react with many common oxidizing materials to cause a fire or explosion. These characteristics require special consideration during any emergency situation involving a leak or spill of hexamethylenediamine. Should hexamethylenediamine ever come into contact with incompatible substances either during use, transportation, or storage, violent reactions can occur.

Hexamethylenediamine can enter the environment through industrial discharges and spills.

Acute Ecological Effects

Acute (short-term) toxic effects may include the death of animals, birds, or fish and death or low growth rate in plants. Acute effects are seen 2 to 4 days after animals or plants are exposed to hexamethylenediamine.

Insufficient data are available to evaluate or predict the short-term effects of hexamethylenediamine to aquatic life, plants, birds, or land animals.

Chronic Ecological Effects

Chronic toxic effects may include shortened life span, reproductive problems, lower fertility, and changes in appearance or behavior in exposed animals. These effects can be seen long after first exposure(s) to toxic chemicals.

Insufficient data are available to evaluate or predict the long-term effects of hexamethylenediamine to aquatic life, plants, birds, or land animals.

Water Solubility

Hexamethylenediamine is highly soluble in water. Concentrations of 1000 milligrams and more will mix with a liter of water.

Persistence in the Environment

Hexamethylenediamine is moderately persistent in water, with a half-life of between 20 to 200 days. The half-life of a pollutant is the amount of time it takes for one half of the chemical to be degraded.

Bioaccumulation in Aquatic Organisms

Some substances increase in concentration, or *bioaccumulate,* in living organisms as they breathe contaminated air, drink contaminated water, or eat contaminated food. These chemicals can become concentrated in the tissues and internal organs of animals as well as humans.

The concentration of hexamethylenediamine found in fish tissues is expected to be somewhat higher than the average concentration of hexamethylenediamine in the water from which the fish was taken.

Recommended Risk-Reduction Measures

Proper training of all transporters will reduce the likelihood of a mishap or accident resulting in a leak or spill to the environment. The correct use of DOT labeling (per the MSDS) on all containers, trucks, and rail cars will enable emergency responders to react properly and quickly to any disaster thereby reducing the potential risk to the environment and to personnel.

Storage of hexamethylenediamine should be segregated from incompatible chemicals to minimize the risk of cross-contamination or contact. Storage buildings should be equipped with the proper fire protection equipment (alarms, sprinklers, extinguishers, etc.).

If a spill or leak to the environment has occurred, fire department, emergency response and/or hazardous materials spill personnel should be notified immediately. Cleanup should be attempted only by those trained in proper spill containment procedures. This chemical will remain in soil for extended periods and may leach to underground aquifers. Contaminated soils should therefore be removed for incineration and replaced with clean soil. If hexamethylenediamine should contact the water table, aquifer, or navigable waterway, remediation activities should be prompt to ensure protection of aquatic sediments. When such spills occur, the local and/or state emergency response authorities must be notified. A comprehensive emergency response or disaster preparedness/recovery plan should be in place prior to any operations involving the use, transportation, storage, or disposal of hexamethylenediamine. If hexamethylenediamine is spilled or leaked, the following specific steps are recommended:

- ☑ Restrict persons not wearing protective clothing from area of spill or leak until cleanup is complete and area can be opened for normal work.
- ☑ Ventilate area and remove all ignition sources.
- ☑ Absorb liquid spills using vermiculite, dry earth, or sand to absorb and place in a sealed drum for disposal. Collect solids using a shovel or scoop. Do NOT dry sweep (creates airborne dusts). Recommend using a vacuum equipped with a high-efficiency particulate air (HEPA) filter.
- ☑ It may be necessary to dispose of hexamethylenediamine as a hazardous waste. The responsible state agency or the regional office of the federal Environmental Protection Agency (EPA) should be contacted for specific recommendations.

RISK ASSESSMENT: BUSINESS
General Assessment

Accidents or mishaps involving hexamethylenediamine can present a significant threat to business operations. The loss or damage of equipment or facilities can significantly affect fiscal viability. Lawsuits that may result from personnel injury/death, public exposures, and/or environmental contamination will also require a serious expenditure of resources. Media attention surrounding an injury, death, or environmental damage can also result in a loss of profits and loss of current as well as future business. Always remember that anytime the term "reproductive hazard" is used, public emotion, hysteria, and ignorance can run equally high. This must be carefully considered whenever developing public relations policies.

Recommended Risk-Reduction Measures

Company attorneys, safety and health professionals, and environmental specialists should be involved in the development of any procedures dealing with safety or emergency response. A company official should be pre-designated as a public relations officer with specific training in dealing with the press. Corporate plans and policies should be developed, approved, and implemented long before any need for such arises.

REFERENCES

Ahlbom, A. 1993. *Biostatistics for Epidemiologists.* New York: CRC Press/Lewis Publishers.

American Conference of Governmental Industrial Hygienists. 1988. *Documentation of the Threshold Limit Values and Biological Exposure Indices,* 5th Edition (with updates). Cincinnati: ACGIH

Anthony, C. P., and N. J. Kolthoff. 1971. *Textbook of Anatomy and Physiology.* St. Louis: C. V. Mosby Company.

Asante-Duah, D. K. 1993. *Hazardous Waste Risk Assessment.* New York: CRC Press/Lewis Publishers.

Bloom, A. D. and F. J. DeSerres. 1995. *Ecotoxicity and Human Health.* New York: CRC Press/Lewis Publishers.

Bueche, J. R. 1972. *Principles of Physics,* 2nd Edition. New York: McGraw-Hill.

Burrell, R., D. K. Flaherty, and L. J. Sauers. 1992. *Toxicology of the Immune System.* New York: Van Nostrand Reinhold.

Calabrese, E. J. 1994. *Biological Effects of Low Level Exposures.* New York: CRC Press/Lewis Publishers.

Cockerman, L. G., B. S. Shane. 1994. *Basic Environmental Toxicology.* New York: CRC Press/Lewis Publishers.

Cothern, R. R. 1993. *Comparative Environmental Risk Assessment.* New York: CRC Press/Lewis Publishers.

Gardner, R. W. 1994. *Chemical Intolerance.* New York: CRC Press/Lewis Publishers.

Hallenbeck, W. H. 1993. *Quantitative Risk Assessment for Environmental and Occupational Health,* 2nd Edition. New York: CRC Press/Lewis Publishers.

Hodgson, E., R. B. Mailman, and Janice E. Chambers. 1988. *Dictionary of Toxicology.* London: The MacMillian Company.

Howard, P. H., et. al. 1991. *Handbook of Environmental Degradation Rates.* New York: CRC Press/Lewis Publishers.

Jones, F. E. 1994. *Toxic Organic Vapors in the Workplace.* New York: CRC Press/Lewis Publishers.

Landis, W. G., M. H. Yu. 1995. *Introduction to Environmental Toxicology.* New York: CRC Press/Lewis Publishers.

Langerman, N. 1994 *Precautionary Labels for Chemical Containers.* New York: CRC Press/Lewis Publishers.

Lauwerys, R. R., P. Hoet. 1993. *Industrial Chemical Exposure.* New York: CRC Press/Lewis Publishers.

Lewis, R. J., Sr. 1993. *Hawley's Condensed Chemical Dictionary,* 12th Edition. New York: Van Nostrand Reinhold.

Lewis, R. J., Sr. 1992. *Sax's Dangerous Properties of Industrial Materials,* Volumes 1, 2, and 3, 8th Edition. New York: Van Nostrand Reinhold.

Manahan, S. E. 1992. *Toxicological Chemistry,* 2nd Edition. New York: CRC Press/Lewis Publishers.

Mellström, G. A., J. E. Wahlberg, and H. I. Maibach. 1994. *Protective Gloves for Occupational Use.* New York: CRC Press/Lewis Publishers.

Meyer, E. 1990. *Chemistry of Hazardous Materials.* Englewood Cliffs, NJ: Prentice Hall.

Nelson, G. O. 1992. *Gas Mixtures.* New York: CRC Press/Lewis Publishers.

Ott, W. R. 1995. *Environmental Statistics and Data Analysis.* New York: CRC Press/Lewis Publishers.

Philp, R. B. 1995. *Environmental Hazards and Human Health.* New York: CRC Press/Lewis Publishers.

Proctor, N. H., and James P. Hughes. 1991. *Chemical Hazards of the Workplace.* New York: Van Nostrand Reinhold.

Quigley, D. R. 1994. *Handbook of Emergency Chemical Management.* New York: CRC Press/Lewis Publishers.

Rea, W. J. 1995. *Chemical Sensitivity,* Vol. I-III. New York: CRC Press/Lewis Publishers.

Richardson, M. 1995. *Dictionary of Substance and Their Effects,* Vol. I-VI. New York: CRC Press/Lewis Publishers.

Rylander, R., R. R. Jacobs. 1994. *Organic Dusts.* New York: CRC Press/Lewis Publishers.

Sacarello, H. L. A. 1994. *The Comprehensive Handbook of Hazardous Materials.* New York: CRC Press/Lewis Publishers.

U.S. Department of Health and Human Services, National Institute for Occupational Safety and Health. 1994. *NIOSH Pocket Guide to Chemical Hazards.* Washington, D.C.: U.S. Government Printing Office.

MATERIAL SAFETY DATA SHEET

CHEMICAL NAME

HEXAMETHYLENE DIISOCYANATE

HAZARD WARNING INFORMATION

HEALTH	FIRE	REACTIVE	OTHER	DEGREE OF HAZARD	COLOR CODING	OTHER CODES
2	1	1		0 = Minimum Hazard 1 = Slight Hazard 2 = Moderate Hazard 3 = Serious Hazard 4 = Severe Hazard	HEALTH = BLUE FIRE = RED REACTIVITY = YELLOW OTHER = WHITE	OX = Oxidizer ACID = Acid ALK = Alkali COR = Corrosive ~~W~~ = Use No Water

SECTION I - GENERAL INFORMATION

Characterization **Isocyanate, Organic**	RCRA Number **None**	EPA Class **Not Applicable**
DOT Proper Shipping Name **Hexamethylenediisocyanate**	Chemical Abstract Service (CAS) Number **822-06-0**	
DOT Hazard Class and Label Requirements **Poison B**	DOT Emergency Guide Code **55**	
DOT Identification Number **UN 2281**	Chemical Formula **$OCN(CH_2)_6NCO$**	

Synonyms
1,6-Diisocyanatohexane; HDI; hexamethylene-1,6-dissocyanate; 1-6-hexamethylene diisocyanate; HMDI.

SECTION II - HAZARDOUS INGREDIENTS AND IDENTITY INFORMATION

Hazardous Components (specific identity)	OSHA Exposure Criteria	NIOSH Exposure Criteria	Immediately Dangerous to Life and Health (IDLH)	ACGIH Exposure Criteria
Hexamethylene diisocyanate **1 ppm = 6.99 mg/m^3**	**PEL:** Not Established **STEL:** Not Established	**REL:** 0.005 ppm, 0.034 mg/m^3 **CEILING: (10 minute)** 0.020 ppm, 0.14 mg/m^3	**Not Determined**	**TLV:** 0.005 ppm, 0.034 mg/m^3 **STEL:** Not Established

SECTION III - PHYSICAL AND CHEMICAL CHARACTERISTICS

Boiling Point **491°F (225°C)**	Specific Gravity (H_2O = 1) **1.04**
Vapor Pressure (mm Hg) **0.5 at 77°F (25°C)**	Molecular Weight **168.2**
Saturated Vapor Density (Air = 1.2 kg/m^3) **1.2002 kg/m^3**	Melting Point **-105°F (-67°C)**

Solubility
Decomposes in water; soluble in most organic solvents.

Appearance and Odor
Clear, colorless, to slightly yellow liquid with a sharp, pungent odor. Odor Threshold = 0.02 ppm.

SECTION IV - FIRE AND EXPLOSION HAZARD DATA

Flash Point (method used) **284°F (140°C) closed cup**	Explosive Limits in Air % by Volume **LEL: Not Determined UEL: Not Determined**
NFPA Classification **Class IIIB Combustible Liquid**	Autoignition Temperature **Not Determined**

Extinguishing Media
Dry chemical, water spray, regular foam, or fog.

Special Fire Fighting Procedures
Wear full protective clothing and self-contained breathing apparatus (SCBA). Approach fire from upwind position. Remove containers from fire if it can be done without risk. Use a water spray to keep fire-exposed containers cool. Poisonous gases are produced in fire. Avoid smoke and runoff.

Unusual Fire and Explosion Hazards
Combustible when exposed to heat and flame. Containers may explode in fire. Stay clear of ends of tanks. Fight fire from maximum distance, if possible.

SECTION V - REACTIVITY DATA

Stability		Conditions to Avoid **Hexamethylene diisocyanate is normally stable at room temperature and pressure. However, exposure to excessive heat will cause polymerization. Avoid contact with heat and incompatible materials.**
Stable **X**	Unstable	Incompatibility *(materials to avoid)* **Reacts with water (forms carbon dioxide), alcohols, strong bases, amines, carboxylic acids, metal compounds, and organotin catalysts.**
Hazardous Polymerization		Conditions to Avoid **Hexamethylene diisocyanate will polymerize at temperatures above 392°F (200°C). It has an average shelf-life of twelve months.**
May Occur **X**	Will Not Occur	Hazardous Decomposition or By-products **When heated to decomposition, hexamethylene diisocyanate can emit highly toxic/poisonous fumes including fumes of chloride and carbon dioxide.**

SECTION VI - HEALTH HAZARD DATA

Primary Route(s) of Entry:	Inhalation?	Absorption (skin)?	Ingestion?
	X	X (skin contact)	X

Health Hazards

INHALATION: Irritation of the eyes, nose, throat, and respiratory tract. Causes increased secretions, cough, painful respiration, pulmonary edema (fluid in the lungs), and possible death. A cholinesterase inhibitor that may lead to respiratory disease.

SKIN & EYES: Skin contact may cause redness, swelling, and painful blistering. Eye contact can lead to the development of conjunctivitis and inflammation of the eyelids.

INGESTION: Unspecified gastrointestinal effects.

Carcinogenicity	NTP Listed?	IARC Cancer Review Group?	OSHA Regulated?	Target Organs?
Unknown Human Unknown Animal	No	No	No	Respiratory system, eyes, skin.

Medical Conditions Generally Aggravated by Exposure

Respiratory system problems or diseases may be aggravated by exposure.

Emergency and First-aid Procedures

Eye contact: Flush immediately with water for 15 minutes (minimum), seek medical attention. Skin contact: Remove all contaminated clothing. Immediately wash area with large amounts of soap and water. For red or blistered skin, seek medical assistance. For inhalation: Remove the person from exposure. Provide respiratory assistance and CPR. Transfer to medical facility. Observe for 48 hours for lung effects. If swallowed: Contact poison control center. Unless advised otherwise, give conscious person 1 to 2 glasses of water to dilute. Do NOT induce vomiting (chemical is a severe irritant).

SECTION VII - PRECAUTIONS FOR SAFE HANDLING AND USE

Steps to be Taken in Case Material is Released or Spilled

Absorb liquid solutions in vermiculite, dry sand, earth, or similar material and deposit in sealed containers. Ventilate area of spill or leak. Remove all ignition sources. Restrict those not involved in cleanup from entering area.

Preferred Waste Disposal Method

Mix in combustible solvent, burn in a chemical incinerator equipped with an afterburner and scrubber.

Precautions to be Taken in Handling and Storage

Store in tightly closed containers in cool, dry, well-ventilated area away from heat, sparks, or flames. Sources of ignition are prohibited where hexamethylene diisocyanate is used, handled, or stored.

Other Precautions and Warnings

Keep away from water or moisture. Do NOT reseal a container if contamination is suspected. Package for disposal or reclamation. Protect containers from physical damage.

SECTION VIII - CONTROL MEASURES AND PERSONAL PROTECTIVE EQUIPMENT

Respiratory Protection *(specify type)*

The PEL is low and difficult to monitor. For any exposure level, use a MSHA/NIOSH-approved supplied-air respirator with full facepiece, hood, or helmet in continuous flow mode or, use a self-contained breathing apparatus (SCBA) in continuous flow mode.

Ventilation

Local exhaust or general mechanical systems recommended.

Protective Gloves	Eye Protection	Other Protective Clothing
Impervious Material	Chemical Goggles or Face Mask	Protective Apron or Clothing

Work/Hygiene Practices

Always wash hands thoroughly after using chemical; never bring food, drink, or smoking materials into vicinity of chemicals.

HEXAMETHYLENE DIISOCYANATE

$OCN(CH_2)_6NCO$ CAS: 822-06-0

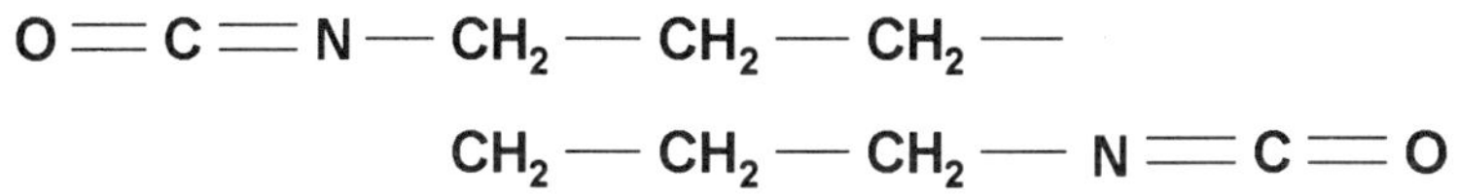

IDENTIFICATION AND TYPICAL USES

Hexamethylene diisocyanate appears as a clear, colorless liquid with a sharp, pungent odor. It is used primarily as a chemical intermediate and as a cross-linking agent in the production of polyurethane foam, dental materials, contact lenses, medical absorbents, and extremely high quality coatings.

RISK ASSESSMENT: HEALTH

General Assessment

Hexamethylene diisocyanate is an extremely toxic and poisonous substance. It can enter the body through *inhalation* and by *ingestion*. It will also cause adverse health effects on contact with the skin, although absorption has not been reported. There is no information on its carcinogenic, teratogenic, or mutagenic capabilities in humans or animals.

Inhalation of hexamethylene diisocyanate vapors can be highly irritating to the eyes, nose, throat, and mucosa of the respiratory tract. Symptoms of exposure include wheezing, cough, increased secretions, dyspnea, sweating, difficulty in breathing, insomnia, painful respiration, and possible life-threatening pulmonary edema (fluid in the lungs). Development of pulmonary edema may be delayed up to 48 hours thereby creating a false sense of security with regard to health exposure risk. In severe cases, there may be some restriction of air movement and eventual respiratory failure (due to edema). Prolonged contact can lead to chronic bronchitis, asthma-like symptoms, and allergic pulmonary sensitization. Hexamethylene diisocyanate is a cholinesterase inhibitor.

Skin contact causes irritation and painful burning with skin blisters on prolonged contact. If the liquid should contact the eye directly, there may be irritation, inflammation of the eyelids and surrounding tissues, conjunctivitis, and possible damage to the cornea. Ingestion causes unspecified gastrointestinal effects.

Acute Health Effects

The following acute (short-term) health effects may occur immediately or shortly after exposure to hexamethylene diisocyanate:

Skin: Possible irritation with painful burning and blistering.

Eye: Moderate to severe irritation. There can be inflammation, irritation of the eyelids, and conjunctivitis.

Lung: Irritation of the nose, throat, and lungs following exposure. There can be a delayed but dangerous buildup of fluid in the lungs (pulmonary edema), which is a medical emergency.

Chronic Health Effects

The following chronic (long-term) health effects may occur at some time after exposure to hexamethylene diisocyanate and can last for months or even years:

Cancer Hazards: There is insufficient information in the references on the carcinogenic or mutagenic properties of hexamethylene diisocyanate.

Reproductive Hazard: According to the data presented in the references, hexamethylene diisocyanate has not been adequately tested for its ability to adversely affect reproduction in test animals.

Other Chronic Effects: Long-term exposure to hexamethylene diisocyanate may lead to allergic sensitization of the respiratory system with initial symptoms of night coughs and difficulty breathing and possible progression to asthmatic bronchitis.

Recommended Risk-Reduction Measures

Personnel should avoid direct contact with hexamethylene diisocyanate. If a less toxic chemical cannot be substituted for a hazardous substance, then *engineering controls* are the most effective method of reducing exposures.

The best protection is to enclose operations and/or provide local exhaust ventilation at the site of chemical release. While not always operationally feasible, isolating operations can also reduce exposure. Using respiratory protection is less effective than the controls mentioned above, but is still advisable whenever working with or around hexamethylene diisocyanate. The ACGIH TLV for this chemical is extremely low (0.005 ppm) and difficult to monitor reliably under most operating or emergency conditions. A supplied-air respirator with a full facepiece operated in positive pressure mode or an MSHA/NIOSH-approved self-contained breathing apparatus (SCBA) with full facepiece operated in pressure demand or other positive pressure mode are the recommended respiratory protection methods of choice. In either case, if a full facepiece is not available, then chemical splash goggles should be worn to protect the eyes. Whenever a chemical splash hazard exists, a face shield and a protective apron should be worn. To prevent hand and skin exposures, impervious gloves should be used. Glove manufacturers should be contacted and permeation studies obtained *before* final glove selections are made.

Administrative controls should also be in place to minimize the potential for human exposures. These may include written procedures or policies which specify the methods and techniques that will be practiced whenever personnel are to work with hexamethylene diisocyanate.

All personnel should receive *training* on the use, hazards, protective measures, emergency actions, and other precautions per 29 CFR 1910.1200 (Hazard Communication), prior to the first assignment in an area where hexamethylene diisocyanate is used or stored. If symptoms develop or overexposure is suspected, the following medical tests are recommended:

☑ Lung function tests.

☑ Consider chest X-ray following acute over exposure (may be negative if taken immediately after exposure due to delayed development of pulmonary edema).

Also, since smoking can cause heart disease, emphysema, and other lung diseases, smokers exposed to hexamethylene diisocyanate may experience symptoms more quickly than non-smokers under the same conditions of exposure. Prudent risk management requires proper consideration of *all* possible exposure factors when evaluating workplace health hazard risk.

Other methods to reduce exposure include:

☑ Where possible, enclose operations and use local exhaust ventilation at the site of chemical release. If local exhaust ventilation or enclosure is not used, respiratory protection should be mandatory.

☑ Where possible, automatically transfer hexamethylene diisocyanate from drums or other storage containers to process containers.

☑ Always ensure that proper protective clothing is worn when using chemical substances and that personnel are trained on its use and care.

☑ Wash thoroughly immediately after exposure to hexamethylene diisocyanate and at the end of the work shift or before eating, drinking, or smoking.

☑ Hazard warning information should be posted in the work area. In addition, as part of an on-going education and training program, all information on the health and safety hazards of hexamethylene diisocyanate should be communicated to all exposed and potentially exposed workers.

RISK ASSESSMENT: ENVIRONMENT

General Assessment

The environment is at risk of exposure during transportation, storage, disposal, or destruction of hexamethylene diisocyanate. In almost every scenario, the threat of environmental exposure is contingent upon the proper handling of the chemical substance. Accidental spills, large or small, can result in fire, explosion, and possible contamination of the surrounding environmental mediums (water, soil, and air).

Hexamethylene diisocyanate is considered a Class IIIB combustible liquid (per OSHA 29 CFR 1910.106). It can react vigorously with oxidizing agents, water, alcohols, strong bases, amines, carboxylic acids, and organotin catalysts. In addition, pressure may build in closed containers. These characteristics require special consideration during any emergency situation involving a leak or spill of hexamethylene diisocyanate.

Hexamethylene diisocyanate may enter the environment primarily during the disposal of wastes from the chlorinated hydrocarbon industry.

☠ *Acute Ecological Effects*

Acute (short-term) toxic effects may include the death of animals, birds, or fish and death or low growth rate in plants. Acute effects are seen 2 to 4 days after ani-

mals or plants are exposed to hexamethylene diisocyanate.

Insufficient data are available to evaluate or predict the short-term effects of hexamethylene diisocyanate exposure to aquatic life, plants, birds, or land animals.

Chronic Ecological Effects

Chronic (long-term) toxic effects may include shortened life span, reproductive problems, lower fertility, and changes in appearance or behavior in exposed animals. These effects can be seen long after first exposure(s) to toxic chemicals.

Insufficient data are available to evaluate or predict the long-term effects of hexamethylene diisocyanate to aquatic life, plants, birds, or terrestrial animals.

Water Solubility

Hexamethylene diisocyanate reacts in water to form carbon dioxide. It is, therefore, only slightly soluble in the aquatic environment. Concentrations of less than 1 milligram will mix with a liter of water.

Persistence in the Environment

Hexamethylene diisocyanate is non-persistent in the aquatic environment, with a half-life of less than 2 days. The half-life of a pollutant is the amount of time it takes for one half of the chemical to be degraded.

Bioaccumulation in Aquatic Organisms

Some substances increase in concentration, or *bioaccumulate,* in living organisms as they breathe contaminated air, drink contaminated water, or eat contaminated food. These chemicals can become concentrated in the tissues and internal organs of animals as well as humans.

The concentration of hexamethylene diisocyanate found in fish tissues is expected to be much lower than the average concentration of hexamethylene diisocyanate in the water from which the fish was taken.

Recommended Risk-Reduction Measures

Proper training of all transporters will reduce the likelihood of a mishap or accident resulting in a leak or spill to the environment. The correct use of labeling on all containers, trucks, and rail cars will enable emergency responders to react properly and quickly to any disaster thereby reducing the potential risk to the environment and to personnel.

Storage of hexamethylene diisocyanate should be segregated from incompatible chemicals to minimize the risk of cross-contamination or contact. Storage buildings or facilities should be equipped with proper fire protection equipment (sprinklers/extinguishers, alarms).

If a spill or leak to the environment has occurred, fire department, emergency response and/or hazardous materials spill personnel should be notified immediately. Cleanup should be attempted only by those trained in proper spill containment procedures. Contaminated soils should be removed for incineration and replaced with clean soil. If hexamethylene diisocyanate should contact the water table, aquifer, or navigable waterway, time is of the essence. Even though it is only slightly soluble in the aquatic environment, it will slowly react in water and, therefore, total containment and remediation may not be possible. When such spills occur, the local and/or state emergency response authorities must be notified. A comprehensive emergency response or disaster preparedness/recovery plan should be in place prior to any operations involving the use, transportation, storage, or disposal of hexamethylene diisocyanate. The proper disposal method for hexamethylene diisocyanate is to burn it in a chemical incinerator equipped with an afterburner and an air scrubber.

If hexamethylene diisocyanate is spilled or leaked, the following specific response steps are recommended:

- ☑ Restrict persons not wearing protective clothing from area of spill or leak until cleanup is complete and area can be opened for normal work.
- ☑ Remove all ignition sources.
- ☑ Ventilate area of spill or leak.
- ☑ If liquid is spilled, absorb in vermiculite, dry sand, earth, or similar material and deposit in sealed containers. Use non-sparking tools.
- ☑ It may be necessary to dispose of hexamethylene diisocyanate as a hazardous waste. The responsible state agency or the regional office of the federal Environmental Protection Agency (EPA) should be contacted for specific recommendations.

RISK ASSESSMENT: BUSINESS

General Assessment

Accidents or mishaps involving hexamethylene diisocyanate can present a significant threat to business operations. The loss or damage of equipment or facilities can significantly affect fiscal viability. Lawsuits that may result from personnel injury/death, public exposures, and/or environmental contamination will also require a serious expenditure of resources. Media attention surrounding an injury, death, or environmental damage can also result in a loss of profits and loss of current as well as future business.

Recommended Risk-Reduction Measures

Company attorneys, safety and health professionals, and environmental specialists should be involved in the development of any procedures for responding to hazardous chemical incidents. A company official should be pre-designated as a public relations officer with specific training in dealing with the press. Corporate plans and policies should be developed, approved, and implemented long before any need for such arises.

REFERENCES

American Conference of Governmental Industrial Hygienists. 1988. *Documentation of the Threshold Limit Values and Biological Exposure Indices*, 5th Edition (with updates). Cincinnati: ACGIH

Bloom, A. D. and F. J. DeSerres. 1995. *Ecotoxicity and Human Health.* New York: CRC Press/Lewis Publishers.

Bueche, J. R. 1972. *Principles of Physics*, 2nd Edition. New York: McGraw-Hill.

Calabrese, E. J. 1994. *Biological Effects of Low Level Exposures.* New York: CRC Press/Lewis Publishers.

Cockerman, L. G., B. S. Shane. 1994. *Basic Environmental Toxicology.* New York: CRC Press/Lewis Publishers.

Cothern, R. R. 1993. *Comparative Environmental Risk Assessment.* New York: CRC Press/Lewis Publishers.

Danse, I. R. 1991. *Common Sense Toxics in the Workplace.* New York: Van Nostrand Reinhold.

Environmental Protection Agency. 1983. *Chemical Hazard Information Profiles.* Washington, D.C.: U.S. EPA.

Fodden, C. R., and J. L. Weddell. 1992. *Hazardous Materials Emergency Action Data.* New York: CRC Press/Lewis Publishers.

Forseberg, K., and L. H. Keith. 1993. *Chemical; Protective Clothing Permeation and Degradation Database.* Chelsea, MI: Lewis Publishers.

Gardner, R. W. 1994. *Chemical Intolerance.* New York: CRC Press/Lewis Publishers.

Gerarde, H. W. 1960. *Toxicology and Biochemistry of Aromatic Hydrocarbons.* New York: Elsevier.

Hallenbeck, W. H. 1993. *Quantitative Risk Assessment for Environmental and Occupational Health*, 2nd Edition. New York: CRC Press/Lewis Publishers.

Howard, P. H., et. al. 1991. *Handbook of Environmental Degradation Rates.* New York: CRC Press/Lewis Publishers.

Hosty, J. W., and P. Foster. 1990. *A Practical Guide to Chemical Spill Response.* New York: Van Nostrand Reinhold.

International Agency for Research on Cancer (IARC). 1987. *IARC Monographs on the Evaluation of Carcinogenic Risk to Humans, Overall Evaluations of Carcinogenicity: An updating of IARC Monographs Volumes 1-42.* Lyon: IARC.

Jones, F. E. 1994. *Toxic Organic Vapors in the Workplace.* New York: CRC Press/Lewis Publishers.

Keith, L. H. and M. M. Walker. 1995. *Handbook of Air Toxics.* New York: CRC Press/Lewis Publishers.

Langerman, N. 1994 *Precautionary Labels for Chemical Containers.* New York: CRC Press/Lewis Publishers.

Lauwerys, R. R., P. Hoet. 1993. *Industrial Chemical Exposure.* New York: CRC Press/Lewis Publishers.

Lewis, R. J., Sr. 1992. *Sax's Dangerous Properties of Industrial Materials*, Volumes 1, 2, and 3, 8th Edition. New York: Van Nostrand Reinhold.

Mackay, D., W. Y. Shiu, and K. C. Ma. 1992. *Illustrated Handbook of Chemical-Physical Properties and Environmental Fate for Organic Chemicals.* New York: CRC Press/Lewis Publishers.

Manahan, S. E. 1992. *Toxicological Chemistry*, 2nd Edition. New York: CRC Press/Lewis Publishers.

Meyer, E. 1990. *Chemistry of Hazardous Materials.* Englewood Cliffs, NJ: Prentice Hall.

Miller, B. F., M.D., and C. B. Keane, R.N. 1972. *Encyclopedia and Dictionary of Medicine and Nursing.* Philadelphia: W.B. Sanders Company.

Mycock, J. C., J. D. McKenna, and L. Theodore. 1995. *Handbook of Air Pollution Control Engineering and Technology.* New York: CRC Press/Lewis Publishers.

Nelson, G. O. 1992. *Gas Mixtures.* New York: CRC Press/Lewis Publishers.

Ott, W. R. 1995. *Environmental Statistics and Data Analysis.* New York: CRC Press/Lewis Publishers.

Patnaik, P. 1992. *A Comprehensive Guide to the Hazardous Properties of Chemical Substances.* New York: Van Nostrand Reinhold.

Philp, R. B. 1995. *Environmental Hazards and Human Health.* New York: CRC Press/Lewis Publishers.

Rea, W. J. 1995. *Chemical Sensitivity*, Vol. I-III. New York: CRC Press/Lewis Publishers.

Sacarello, H. L. A. 1994. *The Comprehensive Handbook of Hazardous Materials.* New York: CRC Press/Lewis Publishers.

Sherman, J. D. 1988. *Chemical Exposure and Disease.* New York: Van Nostrand Reinhold.

U.S. Department of Health and Human Services, National Institute for Occupational Safety and Health. 1994. *NIOSH Pocket Guide to Chemical Hazards.* Washington, D.C.: U.S. Government Printing Office.

MATERIAL SAFETY DATA SHEET

CHEMICAL NAME	
HEXAMETHYL PHOSPHORAMIDE	**CRC PRESS/LEWIS PUBLISHERS DISCLAIMER:** THE INFORMATION AND RECOMMENDATIONS PRESENTED HEREIN ARE BASED ON SOURCES BELIEVED TO BE RELIABLE. CRC MAKES NO REPRESENTATION ON ITS COMPLETENESS OR ACCURACY. IT IS THE RESPONSIBILITY OF THE USER TO DETERMINE THE CHEMICAL'S SUITABILITY FOR ITS INTENDED USE, THE CHEMICAL'S SAFE USE, AND THE CHEMICAL'S PROPER DISPOSAL. NO REPRESENTATIONS AND/OR WARRANTIES, EITHER EXPRESSED OR IMPLIED, OF THE MERCHANTABILITY OR FITNESS FOR A PARTICULAR PURPOSE, OR OF ANY OTHER NATURE, ARE MADE WITH RESPECT TO THE INFORMATION PROVIDED IN THIS MSDS OR TO THE CHEMICAL TO WHICH INFORMATION MAY REFER. CRC NEITHER ASSUMES NOR AUTHORIZES ANY OTHER PERSON TO ASSUME FOR IT, ANY OTHER ADDITIONAL RESPONSIBILITY OR LIABILITY FOR THE USE OF, OR RELIANCE UPON, THIS INFORMATION.

HAZARD WARNING INFORMATION

HEALTH	FIRE	REACTIVE	OTHER	DEGREE OF HAZARD	COLOR CODING	OTHER CODES
3	1	0		0 = Minimum Hazard 1 = Slight Hazard 2 = Moderate Hazard 3 = Serious Hazard 4 = Severe Hazard	HEALTH = BLUE FIRE = RED REACTIVITY = YELLOW OTHER = WHITE	OX = Oxidizer ACID = Acid ALK = Alkali COR = Corrosive W = Use No Water

SECTION I - GENERAL INFORMATION

Characterization: **Aromatic Amine**	RCRA Number: **None**	EPA Class: **Not Applicable**
DOT Proper Shipping Name: **Not Listed**	Chemical Abstract Service (CAS) Number: **680-31-9**	
DOT Hazard Class and Label Requirements: **Not Listed**	DOT Emergency Guide Code: **No Citation**	
DOT Identification Number: **Not Listed**	Molecular Formula: **$[(CH_3)_2N]_3PO$**	

Synonyms

HEMPA; hexamethylphosphoric-triamide; hexamethylphosphoro-triamide; tris(dimethylamino)phosphine oxide; hexametapol.

SECTION II - HAZARDOUS INGREDIENTS AND IDENTITY INFORMATION

Hazardous Components (specific identity)	OSHA Exposure Criteria	NIOSH Exposure Criteria	Immediately Dangerous to Life and Health (IDLH)	ACGIH Exposure Criteria
Hexamethylphosphoramide (derivation: By reaction of excess dimethylamine with phosphorus oxychloride). **1 ppm = 7.45 mg/m^3**	**PEL:** Not Established **STEL:** Not Established	**REL:** Possible Human Carcinogen **STEL:** Not Applicable	**Not Determined**	**TLV:** Suspected Human Carcinogen **STEL:** Not Applicable

SECTION III - PHYSICAL AND CHEMICAL CHARACTERISTICS

Boiling Point: **451°F (233°C)**	Specific Gravity (H_2O = 1): **1.03**
Vapor Pressure (mm Hg): **0.03 at 68°F (20°C)**	Molecular Weight (atomic weight): **179.2**
Vapor Density (Air = 1): **6.18**	Melting Point: **43°F (6°C)**

Solubility

Miscible in water, soluble in most organic liquid (polar and non-polar solvents).

Appearance and Odor

Clear, colorless mobile liquid with a mild, spicy, aromatic, amine-like odor. A solid below 43°F (6°C).

SECTION IV - FIRE AND EXPLOSION HAZARD DATA

Flash Point (method used): **222°F (105°C) closed cup**	Explosive Limits in Air % by Volume: **LEL: Not Determined UEL: Not Determined**
NFPA Classification: **Class IIIB Combustible Liquid**	Autoignition Temperature: **Not Determined**

Extinguishing Media

Use dry chemical, carbon dioxide, regular foam, water spray, or fog.

Special Fire Fighting Procedures

Poisonous gases are produced in fire. Structural fire-fighting clothing is permeable and may not provide adequate protection. Wear self-contained breathing apparatus (SCBA) operated in positive pressure mode. Move fire-exposed containers from fire if it can be done without risk. Cool with water spray. Be aware of runoff and stay clear of smoke. Do NOT release to sewers or waterways.

Unusual Fire and Explosion Hazards

None Reported.

SECTION V - REACTIVITY DATA

Stability		Conditions to Avoid **Hexamethylphosphoramide is stable at room temperature under normal conditions of handling and storage. Avoid heat and contact with incompatible materials.**
Stable **X**	Unstable	Incompatibility *(materials to avoid)* **Incompatible with oxidizers, strong acids, chemically-active metals (such as potassium, sodium, magnesium, and zinc).**
Hazardous Polymerization		Conditions to Avoid **Under normal conditions of temperature and pressure, hazardous polymerization of hexamethylphosphoramide is not expected to occur.**
May Occur	Will Not Occur **X**	Hazardous Decomposition or By-products **Poisonous gases, including oxides of carbon (carbon monoxide) are produced when hexamethylphosphoramide is involved in a fire.**

SECTION VI - HEALTH HAZARD DATA

Primary Route(s) of Entry:	Inhalation?	Absorption (skin/eye)?	Ingestion?
	X	X	X

Health Hazards

INHALATION: Non-specific symptoms of exposure. May cause nausea, abdominal pain, irritation and possible damage to the eyes, nose, throat, and lungs causing nasal discharges and lung changes. High exposures may lead to kidney damage. A possible carcinogen.

ABSORPTION: Irritation of the skin and eyes. Absorbs through the skin (see effects of inhalation).

INGESTION: Can cause similar systemic effects as noted for inhalation.

Carcinogenicity	NTP Listed?	IARC Cancer Review Group?	OSHA Regulated?	Target Organs?
Suspected Human Confirmed Animal	**5th Annual Report**	**Group 2B**	**No**	**Eyes, skin, respiratory system, CNS, GI tract.**

Medical Conditions Generally Aggravated by Exposure

None identified.

Emergency and First-aid Procedures

Eye contact: Do not allow victim to keep eyes tightly shut. Flush immediately with water for 15 minutes (minimum), seek medical attention. Skin contact: Remove all contaminated clothing, including shoes. Immediately wash area with flooding amounts of soap and water. Seek medical attention immediately. For inhalation: Remove the person from exposure. Provide respiratory assistance and CPR. Transfer to medical facility. If swallowed: Seek medical attention immediately. Have conscious person drink 1 - 2 glasses of water and induce vomiting. Never make an unconscious of convulsing person vomit.

SECTION VII - PRECAUTIONS FOR SAFE HANDLING AND USE

Steps to be Taken in Case Material is Released or Spilled

Remove all ignition sources. Ventilate area of spill. Wear SCBA respiratory protection. Restrict those not wearing protective equipment and who are not involved in cleanup from entering area. Avoid skin contact. Absorb liquids in vermiculite and deposit in sealed drums. Use non-sparking tools.

Preferred Waste Disposal Method

No Citation.

Precautions to be Taken in Handling and Storage

Store in tightly closed containers in cool, well-ventilated location away from heat (including direct sunlight). Protect containers from physical damage. Avoid skin contact and inhalation.

Other Precautions and Warnings

Sources of ignition (open flame or smoking), are prohibited where hexamethylphosphoramide is used or stored. Work and storage areas should be labeled as containing a suspected carcinogen.

SECTION VIII - CONTROL MEASURES AND PERSONAL PROTECTIVE EQUIPMENT

Respiratory Protection *(specify type)*

No exposure levels established and material is a suspected human carcinogen. For any exposure level, use a powered air purifying respirator or a self-contained breathing apparatus (SCBA) with full facepiece operated in positive pressure or other continuous flow mode.

Ventilation

Local exhaust (preferred) at site of chemical work or general ventilation.

Protective Gloves	Eye Protection	Other Protective Clothing
Impervious Gloves	**Chemical Goggles or Face Mask**	**Protective Clothing (prevent skin contact)**

Work/Hygiene Practices

Always wash hands thoroughly after using chemical; never bring food, drink, or smoking materials into vicinity of chemicals.

HEXAMETHYLPHOSPHORAMIDE

$[(CH_3)_2N]_2PO$ CAS: 680-31-9

IDENTIFICATION AND TYPICAL USES

Hexamethylphosphoramide (HEMPA) is a colorless, clear, water-white liquid with an aromatic, spicy, amine-like odor. It can appear as a solid at temperatures below 43°F (6°C).

RISK ASSESSMENT: HEALTH

General Assessment

HEMPA is a human poison by *inhalation* and *absorption* through intact skin. *Ingestion* is also a possible route of entry into the body. It is a known animal carcinogen and has also caused adverse reproductive effects in test animals. Human mutation data have also been reported for HEMPA.

Inhalation causes irritation of the eyes, nose, throat, and respiratory tract. There may be symptoms of dypsnea, cough, wheeze, abdominal pain, and possible damage to the eyes, nose, throat, and lungs causing nasal discharges and lung changes. Prolonged contact may lead to severe kidney damage.

Skin contact causes immediate irritation and possible burns with blisters. HEMPA passes through intact skin and will cause toxic systemic effects involving the central nervous system and possibly the kidneys. Eye contact will cause burning and intense pain. Damage to vision may occur if not removed immediately.

Acute Health Effects

The following acute (short-term) health effects may occur immediately or shortly after exposure to HEMPA:

Skin: Irritation with rash and localized inflammation. Absorption through intact skin is extremely likely on short exposures.

Eye: Severe irritation; pain and inflammation. Damage to vision is possible.

Lung: Nose, throat, and respiratory tract irritation causing coughing, sneezing, and likely absorption through the respiration process to cause toxic systemic effects.

Chronic Health Effects

The following chronic (long-term) health effects may occur at some time after exposure to HEMPA and can last for months or even years:

Cancer Hazards: According to the information presented in the references, there is limited evidence that exposure to HEMPA may cause cancer of the nose in test animals. The lungs may also be at risk. Human mutation data have also been reported. The specific human carcinogenicity of HEMPA is not clearly understood in the references.

Reproductive Hazard: According to information presented in the references, HEMPA has been shown to adversely affect reproduction in test animals causing developmental abnormalities. Its human teratogenicity is unknown and requires additional study.

Other Chronic Effects: Long-term exposure to HEMPA may lead to kidney damage.

Recommended Risk-Reduction Measures

Personnel should avoid direct contact with HEMPA. Occupational poisoning occurs most commonly by inhalation and skin absorption. If a less toxic material or compound cannot be substituted for HEMPA, then *engineering controls* are the most effective method of reducing exposures. The best protection is to enclose operations and/or provide local exhaust ventilation at the site of HEMPA release. While not always operationally feasible, isolating operations can also reduce exposure. Using respiratory protection is less effective than the controls mentioned above, but is still advisable whenever working with or around HEMPA. No exposure limits have been established for HEMPA. This does not mean that exposure is without risk. Its specific carcinogenic properties are not clearly understood. Therefore, an MSHA/NIOSH-approved self-contained breathing apparatus (SCBA) with full facepiece operated in pressure demand or a supplied-air respirator in continuous flow mode are appropriate levels of respiratory protection. If a full facepiece is not available, then chemical goggles should be worn to protect the eyes. A face shield should also be considered. To prevent hand and skin exposures, impervious gloves should be used. Glove manufacturers should be contracted and permeation studies obtained *before* final glove selections have been made.

Administrative controls should also be in place to minimize the potential for human exposures. These may include written procedures or policies which specify the methods and techniques that will be practiced whenever personnel are to work with HEMPA.

All personnel should receive *training* on the use, hazards, protective measures, emergency actions, and other precautions per 29 CFR 1910.1200 (Hazard Communication), prior to the first assignment in an area where HEMPA is used or stored.

Before beginning employment and at regular times thereafter, the following recommended medical tests should be considered:

☑ Examination of the nasal cavity.

☑ Lung function tests.

☑ Kidney function tests.

Any evaluation should include a careful history of past and present symptoms with an examination. Medical tests that simply look for existing damage are not a substitute for controlling exposures. Also, because smoking can cause heart disease, emphysema, and other respiratory disorders, smokers exposed to HEMPA may experience symptoms more quickly than non-smokers under the same conditions of exposure. Prudent risk management requires proper consideration of *all* factors which may be causing the appearance of exposure symptoms in the workplace.

Other methods to reduce exposure include:

☑ Where possible, enclose operations and use local exhaust ventilation at the site of chemical release. If local exhaust ventilation or enclosure is not used, respiratory protection should be mandatory.

☑ Always ensure that proper protective clothing is worn when using chemical substances.

☑ Wash thoroughly immediately after exposure to HEMPA and at the end of the work shift or before eating, drinking, or smoking.

☑ Hazard warning information should be posted in the work area. In addition, as part of an on-going education and training program, all information on the health and safety hazards of HEMPA should be communicated to all exposed and potentially exposed workers.

☑ Eye wash stations should be provided in the immediate work area for emergency use. If there is a possibility of skin exposure to HEMPA, emergency shower facilities should be provided.

☑ Workers whose clothing has been contaminated by HEMPA should change into clean clothes before leaving work. Contaminated work clothing should be laundered only by individuals who have been informed of the hazards of exposure to HEMPA.

RISK ASSESSMENT: ENVIRONMENT

General Assessment

The environment is at risk of exposure during transportation, storage, disposal, or destruction of HEMPA. In almost every scenario, the threat of environmental exposure is contingent upon the proper handling of the chemical substance. Accidental spills, large or small, can result in contamination of the surrounding environmental mediums (water, soil, and air).

HEMPA is a Class IIIB combustible solid per OSHA 29 CFR 1910.106. It is incompatible with many commonly encountered strong oxidizing materials, strong acids, chemically-active metals (such as potassium, sodium, magnesium, and zinc) and contact can result in violent and explosive reactions. Caution is always required in handling, storage, transportation, and disposal of HEMPA. Emergency responders should be made aware of the presence of HEMPA at any emergency response situation.

HEMPA can enter the environment from industrial effluents and from spills.

☠ *Acute Ecological Effects*

Acute (short-term) toxic effects may include the death of animals, birds, or fish and death or low growth rate in plants. Acute effects are seen 2 to 4 days after animals or plants are exposed to HEMPA. HEMPA has slight acute toxicity to aquatic life. It has also been shown to have high acute toxicity to redwing blackbirds and moderate-to-slight acute toxicity to coturnix quail. Insufficient data are available to evaluate the short-term effects of HEMPA on plants or terrestrial animals.

Chronic Ecological Effects

Chronic (long-term) toxic effects may include shortened life span, reproductive problems, lower fertility, and changes in appearance or behavior in exposed animals. These effects can be seen long after first exposure(s) to toxic chemicals. HEMPA has slight chronic toxicity to aquatic life. Insufficient data are available to evaluate or predict the long-term effects of HEMPA to plants, birds, or terrestrial animals.

Water Solubility

HEMPA is highly soluble in water. Concentrations of 1000 milligrams and more will mix with a liter of water.

Persistence in the Environment

HEMPA is slightly persistent in water, with a half-life between 2 to 20 days. The half-life of a pollutant is the amount of time it takes for one half of the chemical to be degraded. About 99.8% of HEMPA will eventually end up in water; the rest will end up in the air.

Bioaccumulation in Aquatic Organisms

Some substances increase in concentration, or *bioaccumulate,* in living organisms as they breathe contaminated air, drink contaminated water, or eat contaminated food. These chemicals can become concentrated in the tissues and internal organs of animals as well as humans.

The concentration of HEMPA found in fish tissues is expected to be about the same as the average concentration of HEMPA in the water from which the fish was taken.

Recommended Risk-Reduction Measures

Proper training of all transporters will reduce the likelihood of a mishap or accident resulting in a leak or spill of HEMPA into the environment. Labels on all containers, trucks, and rail cars must meet DOT requirements and accurately reflect their contents to enable emergency responders to react properly and quickly to any disaster thereby reducing the potential risk to the environment and to personnel.

Storage of HEMPA should be segregated from other chemicals and materials to minimize the risk of cross-contamination.

If a spill or leak to the environment has occurred, fire department, emergency response and/or hazardous materials spill personnel should be notified immediately. Cleanup should be attempted only by those trained in proper spill containment procedures using non-sparking tools. Contaminated soils should be removed for incineration and replaced with clean soil. If HEMPA should contact the water table, aquifer, or navigable waterway, time is of the essence. It is highly soluble in water and total containment and remediation may not be possible. The local and/or state emergency response authorities must be notified. A comprehensive emergency response or disaster preparedness/recovery plan should be in place prior to any operations involving the use, transportation, storage, or disposal of HEMPA.

If HEMPA is spilled or leaked, the following specific steps are recommended:

☑ Restrict persons not wearing protective clothing from area of spill or leak until cleanup is complete and area can be opened for normal work.

☑ Absorb liquids in vermiculite, dry earth, or other similar material, collect using non-sparking tools, and deposit in sealed drums for disposal.

☑ Ventilate area of spill or leak and remove all ignition sources and sources of heat.

☑ It may be necessary to dispose of HEMPA as a hazardous waste. The responsible state agency or the regional office of the federal Environmental Protection Agency (EPA) should be contacted for specific recommendations.

RISK ASSESSMENT: BUSINESS

General Assessment

Accidents or mishaps involving HEMPA can present a serious threat to business operations. The loss or damage of equipment or facilities can significantly affect fiscal viability. Lawsuits that may result from personnel illness (leukemia), injury/death, public exposures, and/or environmental contamination will require a serious expenditure of resources. Media attention surrounding an injury, death, or environmental damage can also result in a loss of profits and loss of current as well as future business. Always remember that anytime the terms “cancer,” “carcinogen,” or “reproductive hazard" are used, public emotion, hysteria, and ignorance can run equally high. This must be carefully considered whenever developing or implementing public relations policies.

Recommended Risk-Reduction Measures

Company attorneys, safety and health professionals, and environmental specialists should be involved in the development of any procedures for responding to chemical incidents. A company official should be predesignated as a public relations officer with specific training in dealing with the press. Corporate plans and policies should be developed, approved, and implemented long before any need for such arises.

REFERENCES

American Conference of Governmental Industrial Hygienists. 1988. *Documentation of the Threshold Limit Values and Biological Exposure Indices,* 5th Edition (with updates). Cincinnati: ACGIH

Bloom, A. D. and F. J. DeSerres. 1995. *Ecotoxicity and Human Health.* New York: CRC Press/Lewis Publishers.

Brusick, D. J. 1994. *Methods for Genetic Risk Assessment.* New York: CRC Press/Lewis Publishers.

Bueche, J. R. 1972. *Principles of Physics,* 2nd Edition. New York: McGraw-Hill.

Calabrese, E. J., E. M. Kenyon. *Air Toxics and Risk Assessment.* New York: CRC Press/Lewis Publishers.

Calabrese, E. J. 1994. *Biological Effects of Low Level Exposures.* New York: CRC Press/Lewis Publishers.

Cockerman, L. G., B. S. Shane. 1994. *Basic Environmental Toxicology.* New York: CRC Press/Lewis Publishers.

Cothern, C. R., N. P. Ross. 1994. *Environmental Statistics, Assessment, and Forecasting.* New York: CRC Press/Lewis Publishers.

Cothern, R. R. 1993. *Comparative Environmental Risk Assessment.* New York: CRC Press/Lewis Publishers.

Danse, I. R. 1991. *Common Sense Toxics in the Workplace.* New York: Van Nostrand Reinhold.

Gerarde, H. W. 1960. *Toxicology and Biochemistry of Aromatic Hydrocarbons.* New York: Elsevier.

Hallenbeck, W. H. 1993. *Quantitative Risk Assessment for Environmental and Occupational Health,* 2nd Edition. New York: CRC Press/Lewis Publishers.

Howard, P. H., et. al. 1991. *Handbook of Environmental Degradation Rates.* New York: CRC Press/Lewis Publishers.

International Agency for Research on Cancer (IARC). 1987. *IARC Monographs on the Evaluation of Carcinogenic Risk to Humans, Overall Evaluations of Carcinogenicity: An updating of IARC Monographs Volumes 1-42.* Lyon: IARC.

Jones, F. E. 1994. *Toxic Organic Vapors in the Workplace.* New York: CRC Press/Lewis Publishers.

Keith, L. H. and M. M. Walker. 1995. *Handbook of Air Toxics.* New York: CRC Press/Lewis Publishers.

Landis, W. G., M. H. Yu. 1995. *Introduction to Environmental Toxicology.* New York: CRC Press/Lewis Publishers.

Langerman, N. 1994 *Precautionary Labels for Chemical Containers.* New York: CRC Press/Lewis Publishers.

Lauwerys, R. R., P. Hoet. 1993. *Industrial Chemical Exposure.* New York: CRC Press/Lewis Publishers.

Lewis, R. J., Sr. 1993. *Hawley's Condensed Chemical Dictionary,* 12th Edition. New York: Van Nostrand Reinhold.

Lewis, R. J., Sr. 1992. *Sax's Dangerous Properties of Industrial Materials,* Volumes 1, 2, and 3, 8th Edition. New York: Van Nostrand Reinhold.

Manahan, S. E. 1994. *Environmental Chemistry,* 6th Edition. New York: CRC Press/Lewis Publishers.

Mellström, G. A., J. E. Wahlberg, and H. I. Maibach. 1994. *Protective Gloves for Occupational Use.* New York: CRC Press/Lewis Publishers.

Meyer, E. 1990. *Chemistry of Hazardous Materials.* Englewood Cliffs, NJ: Prentice Hall.

Ott, W. R. 1995. *Environmental Statistics and Data Analysis.* New York: CRC Press/Lewis Publishers.

Patnaik, P. 1992. *A Comprehensive Guide to the Hazardous Properties of Chemical Substances.* New York: Van Nostrand Reinhold.

Philp, R. B. 1995. *Environmental Hazards and Human Health.* New York: CRC Press/Lewis Publishers.

Proctor, N. H., J. P. Hughes. 1991. *Chemical Hazards of the Workplace.* New York: Van Nostrand Reinhold.

Quigley, D. R. 1994. *Handbook of Emergency Chemical Management.* New York: CRC Press/Lewis Publishers.

Sacarello, H. L. A. 1994. *The Comprehensive Handbook of Hazardous Materials.* New York: CRC Press/Lewis Publishers.

Torkelson, T. R. V. K. Rowe. 1981. *Patty's Industrial Hygiene and Toxicology,* Vol. 2B. New York: Wiley-Interscience.

U.S. Department of Health and Human Services, National Institute for Occupational Safety and Health. 1994. *NIOSH Pocket Guide to Chemical Hazards.* Washington, D.C.: U.S. Government Printing Office.

MATERIAL SAFETY DATA SHEET

CHEMICAL NAME	
HEXANE	**CRC PRESS/LEWIS PUBLISHERS DISCLAIMER:** THE INFORMATION AND RECOMMENDATIONS PRESENTED HEREIN ARE BASED ON SOURCES BELIEVED TO BE RELIABLE. CRC MAKES NO REPRESENTATION ON ITS COMPLETENESS OR ACCURACY. IT IS THE RESPONSIBILITY OF THE USER TO DETERMINE THE CHEMICAL'S SUITABILITY FOR ITS INTENDED USE, THE CHEMICAL'S SAFE USE, AND THE CHEMICAL'S PROPER DISPOSAL. NO REPRESENTATIONS AND/OR WARRANTIES, EITHER EXPRESSED OR IMPLIED, OF THE MERCHANTABILITY OR FITNESS FOR A PARTICULAR PURPOSE, OR OF ANY OTHER NATURE, ARE MADE WITH RESPECT TO THE INFORMATION PROVIDED IN THIS MSDS OR TO THE CHEMICAL TO WHICH INFORMATION MAY REFER. CRC NEITHER ASSUMES NOR AUTHORIZES ANY OTHER PERSON TO ASSUME FOR IT, ANY OTHER ADDITIONAL RESPONSIBILITY OR LIABILITY FOR THE USE OF, OR RELIANCE UPON, THIS INFORMATION.

HAZARD WARNING INFORMATION

HEALTH	FIRE	REACTIVE	OTHER	DEGREE OF HAZARD	COLOR CODING	OTHER CODES
1	3	0		0 = Minimum Hazard 1 = Slight Hazard 2 = Moderate Hazard 3 = Serious Hazard 4 = Severe Hazard	HEALTH = BLUE FIRE = RED REACTIVITY = YELLOW OTHER = WHITE	OX = Oxidizer ACID = Acid ALK = Alkali COR = Corrosive W = Use No Water

SECTION I - GENERAL INFORMATION

Characterization	RCRA Number	EPA Class
Hydrocarbon, Aliphatic	**D001**	**Characteristic (I) Waste**

DOT Proper Shipping Name **Hexane**	Chemical Abstract Service (CAS) Number **110-54-3**
DOT Hazard Class and Label Requirements **Flammable Liquid**	DOT Emergency Guide Code **27**
DOT Identification Number **UN 1208**	Chemical Formula **$CH_3(CH_2)_4CH_3$**

Synonyms

***n*-Hexane; hexyl hydride; normal-hexane.**

SECTION II - HAZARDOUS INGREDIENTS AND IDENTITY INFORMATION

Hazardous Components (specific identity)	OSHA Exposure Criteria	NIOSH Exposure Criteria	Immediately Dangerous to Life and Health (IDLH)	ACGIH Exposure Criteria
Hexane (derivation: By fractional distillation from petroleum using the molecular sieve process). **1 ppm = 3.58 mg/m³**	**PEL:** **500 ppm** **1800 mg/m³** **STEL:** **Not Established**	**REL:** **50 ppm** **180 mg/m³** **STEL:** **Not Established**	**1100 ppm**	**TLV:** **50 ppm** **180 mg/m³** **STEL:** **Not Established**

SECTION III - PHYSICAL AND CHEMICAL CHARACTERISTICS

Boiling Point **156°F (69°C)**	Specific Gravity (H_2O = 1) **0.66**
Vapor Pressure (gas) **124 at 72°F (22°C)**	Molecular Weight **86.2**
Vapor Density (Air = 1) **2.97**	Melting Point **-139°F (-96°C)**

Solubility

Insoluble in water (0.002%). Soluble in alcohol, ether, chloroform, acetone, and most non-polar solvents.

Appearance and Odor

A clear, colorless liquid with a mild, gasoline-like odor.

SECTION IV - FIRE AND EXPLOSION HAZARD DATA

Flash Point (method used) **-8°F (-22°C) closed cup**	Explosive Limits in Air % by Volume **LEL: 1.1% UEL: 7.5%**
NFPA Classification **Class 1B Flammable Liquid**	Autoignition Temperature **437°F (225°C)**

Extinguishing Media

Use dry chemical, carbon dioxide, water spray, or foam.

Special Fire Fighting Procedures

Hexane is a flammable liquid. Wear full protective clothing and self-contained breathing apparatus (SCBA). Poisonous gases are produced in fire. Use water spray to keep fire-exposed containers cool.

Unusual Fire and Explosion Hazards

Dangerous fire and explosion hazard when exposed to heat or flame. Containers may explode in fire. Chemical will react on contact with oxidizable materials. Moderate explosion hazard when in the form of vapor. Its vapors are heavier than air and may travel for some distance to flashback and cause fire.

SECTION V - REACTIVITY DATA

Stability		Conditions to Avoid Keep away from heat or flame. Hexane is normally stable in closed containers under routine conditions of handling and storage.
Stable X	Unstable	Incompatibility *(materials to avoid)* Strong oxidizers, such as permanganates, peroxides, nitrates, chlorates, and perchlorates.
Hazardous Polymerization		Conditions to Avoid Under normal conditions of temperature and pressure, hazardous polymerization of hexane is not expected to occur.
May Occur	Will Not Occur X	Hazardous Decomposition or By-products When heated to decomposition, hexane can emit highly toxic/poisonous gases, including oxides of carbon.

SECTION VI - HEALTH HAZARD DATA

Primary Route(s) of Entry:	Inhalation? X	Absorption (skin)? X	Ingestion? X

Health Hazards

INHALATION: An irritant to the eyes, nose, throat, and lungs, and a central nervous system depressant. Can cause nausea, dizziness, vomiting, and depression. Higher concentrations can lead to lightheadedness, peripheral neuropathy, numbing in the extremities, muscular weakness, giddiness, chemical pneumonia, coma, and death from respiratory failure.

ABSORPTION: Irritating to the skin, may cause drying, cracking, and chapping. Severe eye irritant.

INGESTION: Gastrointestinal irritation with possible systemic effects targeting the CNS.

Carcinogenicity	NTP Listed?	IARC Cancer Review Group?	OSHA Regulated?	Target Organs?
Unknown Human Unknown Animal	No	No	29 CFR 1910.1000 Table Z-1	Respiratory system, eyes, skin, CNS, PNS.

Medical Conditions Generally Aggravated by Exposure

Skin, CNS, PNS, and any existing respiratory diseases may be aggravated by exposure to hexane.

Emergency and First-aid Procedures

Eye contact: Flush immediately with water for 15 minutes (minimum); seek medical attention. Skin contact: Remove all contaminated clothing. Immediately wash area with large amounts of soap and water. Seek medical assistance. For inhalation: Remove the person from exposure. Provide respiratory assistance and CPR, if required. Transfer to medical facility. Keep victim warm and at rest. If swallowed: Seek medical attention immediately. Warning! Aspiration hazard; do NOT induce vomiting.

SECTION VII - PRECAUTIONS FOR SAFE HANDLING AND USE

Steps to be Taken in Case Material is Released or Spilled

Absorb in vermiculite, dry sand, earth, or similar material and deposit in sealed containers. Ventilate area of spill or leak. Restrict those not involved in cleanup from entering area. Remove all sources of ignition and use only non-sparking tools during cleanup. Notify proper authorities.

Preferred Waste Disposal Method

Burn in chemical incinerator equipped with an afterburner and scrubber.

Precautions to be Taken in Handling and Storage

Store to avoid contact with incompatible materials in tightly closed containers in cool, well-ventilated area away from heat and flame. Use only non-sparking tools to open and close containers.

Other Precautions and Warnings

Sources of ignition are prohibited where hexane is used, handled, or stored. Heat may cause containers to build up pressure and explode. Containers must be grounded and bonded during transfer.

SECTION VIII - CONTROL MEASURES AND PERSONAL PROTECTIVE EQUIPMENT

Respiratory Protection *(specify type)*

For high exposures (above the PEL), use a MSHA/NIOSH-approved organic vapor respirator. Best protection is obtained using a supplied-air respirator with full facepiece in continuous flow mode, or a self-contained breathing apparatus (SCBA) operated in pressure demand or other positive pressure mode.

Ventilation

Local exhaust or general mechanical systems recommended.

Protective Gloves	Eye Protection	Other Protective Clothing
Viton, PVC, or Chlorinated Polyethylene	Chemical Goggles or Face Mask	Protective Apron

Work/Hygiene Practices

Always wash hands thoroughly after using chemical; never bring food, drink, or smoking materials into vicinity of chemicals.

HEXANE

$CH_3(CH_2)_4CH_3$

CAS: 110-54-3

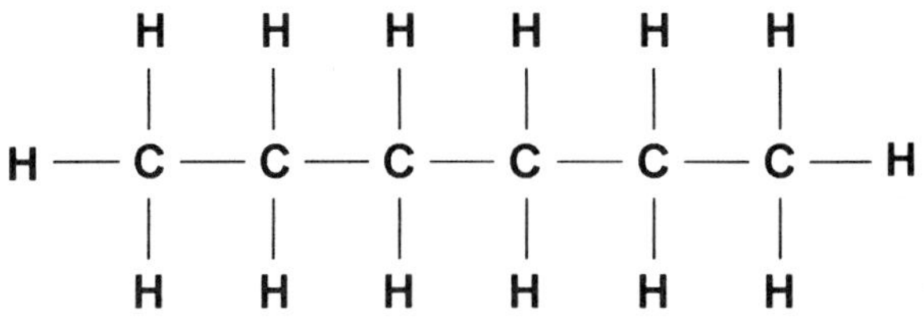

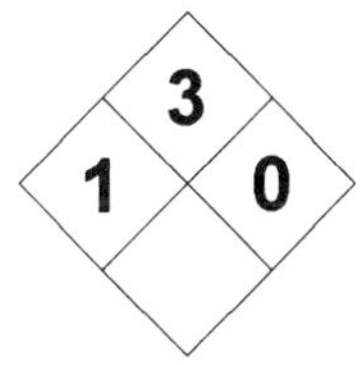

IDENTIFICATION AND TYPICAL USES

Hexane is a colorless liquid with a gasoline-like odor. It is a chief constituent of petroleum ether, gasoline, and rubber solvent. It is used as a solvent for adhesives, in low temperature thermometers, in calibration procedures, as a paint diluent, in vegetable oils, and in organic analysis, and for denaturing alcohol.

RISK ASSESSMENT: HEALTH

General Assessment

Hexane is a severe irritant by *inhalation* and *ingestion*. Skin contact can also produce some adverse topical effects and *absorption* is likely to occur. There is no clear evidence in the references to indicate that exposure to hexane will lead to carcinogenic, mutagenic, or teratogenic effects. However, there does appear to be an increased risk of contracting Leukemia associated with exposure to a number of chemical substances in the rubber industry, including hexane.

Significant inhalation may cause irritation of the respiratory tract and depress the central nervous system. Symptoms include nausea, dizziness, vomiting, lightheadedness, giddiness, stupor, vertigo, loss of coordination, loss of appetite, depression, hallucinations, and possibly chemical pneumonia (if the swallowed liquid is aspirated into the lungs). At high concentrations, it is a narcotic and can lead to muscle weakness, cold feeling in extremities, blurred vision, anorexia, headaches, polyneuropathy, loss of consciousness, coma, and death due to respiratory failure. Studies have shown that hexane will accumulate in fatty tissues as is evidenced by its affinity for the blood, liver, and brain where lipids are prevalent.

Skin and eye contact may produce immediate irritation with redness, painful burning, and possible blistering on contact. It is absorbed through the skin in both liquid and vapor form. Prolonged or repeated contact with the skin can cause dryness, cracking, and dermatitis. Ingestion will cause immediate gastrointestinal irritation and presents a serious aspiration hazard. If the liquid is aspirated into the lungs, oxygen can be displaced and the resultant asphyxiation can lead to brain damage or even cardiac arrest.

Acute Health Effects

The following acute (short-term) health effects may occur immediately or shortly after exposure to hexane:

Skin: Severe and painful irritation with possible burns and blistering. Absorption through intact skin is likely to occur on exposure to the liquid or the vapor.

Eye: Severe irritation and burning on contact.

Lung: Direct irritation of the respiratory tract and associated mucosa. May cause dizziness, vomiting, nausea, and can lead to chemical pneumonia. Primary result of inhalation of hexane is its action on the CNS.

CNS: A CNS depressant: headache, dizziness, depression, hallucinations, changes in mood, loss of consciousness, coma, and death (respiratory failure).

Chronic Health Effects

The following chronic (long-term) health effects may occur at some time after exposure to hexane and can last for months or even years:

Cancer Hazards: According to the references, hexane has not been adequately tested for its ability to cause cancer in animals. However, there is some evidence to indicate that exposure to hexane may be associated with an increased risk of contracting Leukemia.

Reproductive Hazard: According to information presented in the references, hexane has not been tested for its ability to adversely affect reproduction.

Other Chronic Effects: Prolonged exposure may lead to polyneuropathy with symptoms of muscle weakness, loss of motor functions, numbness and pain in the extremities. There may also be paresthesia, foot drop, decreased muscle tone and strength, visual disturbances including changes in color perception, retinal pigmentation, and blurred vision.

Recommended Risk-Reduction Measures

Personnel should avoid direct contact with hexane since it is known to cause significant adverse health effects, especially if improperly used or if overexposures occur. If a less toxic chemical cannot be substituted for a hazardous substance, then *engineering controls* are the most effective method of reducing exposures. The best protection is to enclose operations and/or provide local exhaust ventilation at the site of chemical release. While not always operationally feasible, isolating operations can also reduce exposure. Using respiratory protection is less effective than the controls mentioned above, but is still advisable whenever working with or around hexane. For exposures above the PEL (500 ppm), an organic vapor respirator should be used (per OSHA 29 CFR 1910.134). Better protection is provided by a supplied-air respirator with full facepiece operated in continuous flow mode or a self-contained breathing apparatus (SCBA) with full facepiece in pressure demand. If a full facepiece is not available, then chemical goggles should be worn to protect the eyes. Whenever a chemical splash hazard exists, a face shield and a rubber apron should be worn. To prevent hand and skin exposures, impervious gloves should be used. Nitrile, Viton, PVC, or chlorinated polyethylene gloves are recommended. Glove manufacturers should be contacted and permeation studies obtained *before* final glove selections are made.

Administrative controls should also be in place to minimize the potential for human exposures. These may include written procedures or policies which specify the methods and techniques that will be practiced whenever personnel are to work with hexane.

All personnel should receive *training* on the use, hazards, protective measures, emergency actions, and other precautions per 29 CFR 1910.1200 (Hazard Communication), prior to the first assignment in an area where hexane is used or stored.

If symptoms develop or overexposure is suspected, the following tests are recommended:

- ☑ Lung function tests.
- ☑ Evaluation of nervous system function.
- ☑ Evaluation by a qualified allergist with careful consideration of exposure history and special testing (may help diagnose skin allergy and/or dermatitis).

Any evaluation should include a careful history of past and present symptoms with an examination. Medical tests that simply look for existing damage are not a substitute for controlling exposures.

Other methods to reduce exposure include:

- ☑ Where possible, enclose operations and use local exhaust ventilation at the site of chemical release. If local exhaust ventilation or enclosure is not used, respiratory protection should be mandatory.
- ☑ Always ensure that proper protective clothing is worn when using chemical substances and personnel are trained on its use, care, and maintenance.
- ☑ Wash thoroughly immediately after exposure to hexane and at the end of the work shift or before eating, drinking, or smoking.
- ☑ Hazard warning information should be posted in the work area. In addition, as part of an on-going education and training program, all information on the health and safety hazards of hexane should be communicated to all potentially exposed workers.
- ☑ Eye wash stations should be provided in the immediate work area for emergency use. If there is a possibility of skin exposure to hexane, emergency shower facilities should also be provided.
- ☑ Personnel should never wear contaminated clothing home (family members can be exposed). Clothing should be laundered by personnel who have been trained on the hazards of hexane.

RISK ASSESSMENT: ENVIRONMENT

General Assessment

The environment is at risk of exposure during transportation, storage, disposal, or destruction of hexane. In almost every scenario, the threat of environmental exposure is contingent upon the proper handling of the chemical substance. Accidental spills, large or small, can result in fire, explosion, and possible contamination of the surrounding environmental mediums (water, soil, and air).

Hexane is considered a Class 1B flammable liquid (per OSHA 29 CFR 1910.106). Its extremely low flash point and relatively low boiling point make this chemical a serious fire and explosion hazard concern. These characteristics require extreme caution in handling, storage, transportation, and disposal. It is incompatible with common peroxides and other oxidizers and contact can cause fire or explosion. Therefore, special consideration is required during any emergency situation involving a leak or spill of hexane. Should hexane ever come into contact with these incompatible substances during use, transportation, storage, or disposal, the formation of highly toxic and/or highly explosive commodities is possible.

Hexane may enter the environment through industrial discharges or spills.

Acute Ecological Effects

Acute (short-term) toxic effects may include the death of animals, birds, or fish and death or low growth rate in plants. Acute effects are seen 2 to 4 days after animals or plants are exposed to hexane.

This chemical has high acute toxicity to aquatic life. It causes membrane damage in ornamental crop species. Insufficient data are available on the short-term effects of hexane exposure to birds or terrestrial animals.

Chronic Ecological Effects

Chronic (long-term) toxic effects may include shortened life span, reproductive problems, lower fertility, and changes in appearance or behavior in exposed animals. These effects can be seen long after first exposure(s) to toxic chemicals.

Hexane has moderate chronic toxicity in aquatic life. Insufficient data are available on the long-term effects of hexane to plants, birds, or land animals.

Water Solubility

Hexane is nearly insoluble in water. Concentrations of less than 10 milligrams may mix with a liter of water, depending on water pH and chemical concentration of hexane.

Persistence in the Environment

Hexane is non-persistent in water, with a half-life of less than 2 days. The half-life of a pollutant is the amount of time it takes for one half of the chemical to be degraded. Virtually 100% of hexane will eventually end up in air.

Bioaccumulation in Aquatic Organisms

Some substances increase in concentration, or *bioaccumulate,* in living organisms as they breathe contaminated air, drink contaminated water, or eat contaminated food. These chemicals can become concentrated in the tissues and internal organs of animals as well as humans.

The concentration of hexane found in fish tissue is expected to be about the same as the average concentration of hexane in the water from which the fish was taken.

Recommended Risk-Reduction Measures

Proper training of all transporters will reduce the likelihood of a mishap or accident resulting in a leak or spill to the environment. The correct use of DOT labeling (per the MSDS) on all containers, trucks, and rail cars will enable emergency responders to react properly and quickly to any disaster thereby reducing the potential risk to the environment and to personnel.

Storage of hexane should be segregated from incompatible chemicals to minimize the risk of cross-contamination or contact. Buildings designated for storage should be equipped with appropriate fire protection systems (alarms, sprinklers, emergency lighting, portable extinguishers). All transfers of hexane from storage containers to process containers should be done automatically whenever possible. All containers and equipment must be grounded and bonded.

If a spill or leak to the environment has occurred, fire department, emergency response and/or hazardous materials spill personnel should be notified immediately. Cleanup should be attempted only by those trained in proper spill containment procedures using non-sparking tools. Contaminated soils should be removed for incineration and replaced with clean soil. If hexane should contact the water table, aquifer, or navigable waterway, the local and/or state emergency response authorities must be notified. A comprehensive emergency response or disaster preparedness/recovery plan should be in place prior to any operations involving the use, transportation, storage, or disposal of hexane.

To dispose of hexane properly, it should be burned in a chemical incinerator equipped with an afterburner and scrubber to ensure appropriate emission control.

If hexane is spilled or leaked, the following specific steps are recommended:

☑ Restrict persons not wearing protective clothing from area of spill or leak until cleanup is complete and area can be opened for normal work.

☑ Remove all ignition sources, ventilate area of leak.

☑ Absorb liquids in vermiculite, dry sand, earth, or similar material and deposit in sealed containers. Collect materials using non-sparking tools.

☑ It may be necessary to dispose of hexane as a hazardous waste. The responsible state agency or the regional office of the federal Environmental Protection Agency (EPA) should be contacted for specific recommendations.

RISK ASSESSMENT: BUSINESS

General Assessment

Accidents or mishaps involving hexane can present a significant threat to business operations. The loss or damage of equipment or facilities can significantly affect fiscal viability. Lawsuits that may result from personnel injury/death, public exposures, and/or environmental contamination will also require a serious expenditure of resources. Media attention surrounding an injury, death, or environmental damage can also result in a loss of profits and loss of current as well as future business.

Recommended Risk-Reduction Measures

Company attorneys, safety and health professionals, and environmental specialists should be involved in the development of any procedures dealing with safety or emergency response. A company official should be pre-designated as a public relations officer with specific training in dealing with the press. Corporate plans and policies should be developed, approved, and implemented long before any need for such arises.

REFERENCES

American Conference of Governmental Industrial Hygienists. 1988. *Documentation of the Threshold Limit Values and Biological Exposure Indices*, 5th Edition (with updates). Cincinnati: ACGIH

Bueche, J. R. 1972. *Principles of Physics*, 2nd Edition. New York: McGraw-Hill.

Calabrese, E. J. 1994. *Biological Effects of Low Level Exposures.* New York: CRC Press/Lewis Publishers.

Cockerman, L. G., B. S. Shane. 1994. *Basic Environmental Toxicology.* New York: CRC Press/Lewis Publishers.

Cothern, C. R. 1995. *Handbook for Environmental Risk Decisionmaking.* New York: CRC Press/Lewis Publishers.

Environmental Protection Agency. 1983. *Chemical Hazard Information Profiles.* Washington, D.C.: U.S. EPA.

Gardner, R. W. 1994. *Chemical Intolerance.* New York: CRC Press/Lewis Publishers.

Hallenbeck, W. H. 1993. *Quantitative Risk Assessment for Environmental and Occupational Health*, 2nd Edition. New York: CRC Press/Lewis Publishers.

Hosty, J. W., and P. Foster. 1990. *A Practical Guide to Chemical Spill Response.* New York: Van Nostrand Reinhold.

Hodgson, E., R. B. Mailman, and Janice E. Chambers. 1988. *Dictionary of Toxicology.* London: The MacMillian Company.

Kamrin, M. A. 1988. *Toxicology: A Primer on Principles and Applications.* Chelsea, Michigan: Lewis Publishers.

Keith, L. H. and M. M. Walker. 1995. *Handbook of Air Toxics.* New York: CRC Press/Lewis Publishers.

Landis, W. G., M. H. Yu. 1995. *Introduction to Environmental Toxicology.* New York: CRC Press/Lewis Publishers.

Langerman, N. 1994 *Precautionary Labels for Chemical Containers.* New York: CRC Press/Lewis Publishers.

Lauwerys, R. R., P. Hoet. 1993. *Industrial Chemical Exposure.* New York: CRC Press/Lewis Publishers.

Lewis, R. J., Sr. 1992. *Sax's Dangerous Properties of Industrial Materials*, Volumes 1, 2, and 3, 8th Edition. New York: Van Nostrand Reinhold.

Manahan, S. E. 1994. *Environmental Chemistry*, 6th Edition. New York: CRC Press/Lewis Publishers.

Meyer, E. 1990. *Chemistry of Hazardous Materials.* Englewood Cliffs, NJ: Prentice Hall.

Mycock, J. C., J. D. McKenna, and L. Theodore. 1995. *Handbook of Air Pollution Control Engineering and Technology.* New York: CRC Press/Lewis Publishers.

Patnaik, P. 1992. *A Comprehensive Guide to the Hazardous Properties of Chemical Substances.* New York: Van Nostrand Reinhold.

Quigley, D. R. 1994. *Handbook of Emergency Chemical Management.* New York: CRC Press/Lewis Publishers.

Rea, W. J. 1995. *Chemical Sensitivity*, Vol. I-III. New York: CRC Press/Lewis Publishers.

Richardson, M. 1995. *Dictionary of Substance and Their Effects*, Vol. I-VI. New York: CRC Press/Lewis Publishers.

Sacarello, H. L. A. 1994. *The Comprehensive Handbook of Hazardous Materials.* New York: CRC Press/Lewis Publishers.

Torkelson, T. R. V. K. Rowe. 1981. Halogenated aliphatic hydrocarbons. *Patty's Industrial Hygiene and Toxicology*, Vol. 2B, *Toxicology*, pp 3568-3572. New York: Wiley-Interscience.

U.S. Department of Health and Human Services, National Institute for Occupational Safety and Health. 1994. *NIOSH Pocket Guide to Chemical Hazards.* Washington, D.C.: U.S. Government Printing Office.

MATERIAL SAFETY DATA SHEET

CHEMICAL NAME

HEXANE ISOMERS
(excluding *n*-hexane)

HAZARD WARNING INFORMATION

HEALTH	FIRE	REACTIVE	OTHER	DEGREE OF HAZARD	COLOR CODING	OTHER CODES
1	3	0		0 = Minimum Hazard 1 = Slight Hazard 2 = Moderate Hazard 3 = Serious Hazard 4 = Severe Hazard	HEALTH = BLUE FIRE = RED REACTIVITY = YELLOW OTHER = WHITE	OX = Oxidizer ACID = Acid ALK = Alkali COR = Corrosive W = Use No Water

SECTION I - GENERAL INFORMATION

Characterization
Hydrocarbon, Aliphatic

RCRA Number
D001

EPA Class
Characteristic (I) Waste

DOT Proper Shipping Name
Hexane

Chemical Abstract Service (CAS) Number
107-83-5

DOT Hazard Class and Label Requirements
Flammable Liquid

DOT Emergency Guide Code
27

DOT Identification Number
UN 1208

Chemical Formula
C_6H_{14}

Synonyms
Hexane isomers, other than *n*-hexane; diethylmethylmethane; diisopropyl; 2,2-dimethylbutane; isohexane; 2-methylpentane; 3-methylpentane.

SECTION II - HAZARDOUS INGREDIENTS AND IDENTITY INFORMATION

Hazardous Components (specific identity)	OSHA Exposure Criteria	NIOSH Exposure Criteria	Immediately Dangerous to Life and Health (IDLH)	ACGIH Exposure Criteria
Hexane isomers (derivation: By fractional distillation from petroleum using the molecular sieve process). **1 ppm = 3.58 mg/m³**	**PEL:** 500 ppm 1800 mg/m³ **STEL** (15 min): 1000 ppm 3600 mg/m³	**REL:** 100 ppm 350 mg/m³ **CEILING** (15 min): 510 ppm 1800 mg/m³	**Not Determined**	**TLV:** 500 ppm 1800 mg/m³ **STEL:** 1000 ppm 3600 mg/m³

SECTION III - PHYSICAL AND CHEMICAL CHARACTERISTICS

Boiling Point
122°F (50°C) to 145°F (63°C)

Specific Gravity (H_2O = 1)
0.66

Vapor Pressure (gas)
Not Found

Molecular Weight
86.2

Vapor Density (Air = 1)
3.0

Melting Point
-139°F (-96°C)

Solubility
Insoluble in water. Soluble in alcohol, ether, chloroform, acetone, and most non-polar solvents.

Appearance and Odor
A clear, colorless liquid with a mild, gasoline-like odor.

SECTION IV - FIRE AND EXPLOSION HAZARD DATA

Flash Point (method used)
-54°F (-48°C) to 19°F (-23°C) closed cup

Explosive Limits in Air % by Volume
LEL: 1.2% UEL: 8%

NFPA Classification
Class 1B Flammable Liquid

Autoignition Temperature
437°F (225°C)

Extinguishing Media
Use dry chemical, carbon dioxide, water spray, or foam.

Special Fire Fighting Procedures
Wear full protective clothing and self-contained breathing apparatus (SCBA). Poisonous gases are produced in fire. Use water spray to keep fire-exposed containers cool.

Unusual Fire and Explosion Hazards
Dangerous fire and explosion hazard when exposed to heat or flame. Containers may explode in fire. Chemical will react on contact with oxidizable materials. Moderate explosion hazard when in the form of vapor. Its vapors are heavier than air and may travel for some distance to flashback and cause fire.

SECTION V - REACTIVITY DATA

Stability		Conditions to Avoid: Keep away from heat or flame. Hexane isomers are normally stable in closed containers under routine conditions of handling and storage.
Stable: X	Unstable	Incompatibility *(materials to avoid)*: Strong oxidizers, such as permanganates, peroxides, nitrates, chlorates, and perchlorates.
Hazardous Polymerization		Conditions to Avoid: Under normal conditions of temperature and pressure, hazardous polymerization of hexane isomers is not expected to occur.
May Occur	Will Not Occur: X	Hazardous Decomposition or By-products: When heated to decomposition, hexane isomers can emit highly toxic/poisonous gases, including oxides of carbon.

SECTION VI - HEALTH HAZARD DATA

Primary Route(s) of Entry:	Inhalation?	Absorption (skin)?	Ingestion?
	X	X	X

Health Hazards

INHALATION: An irritant to the eyes, nose, throat, and lungs, and a central nervous system depressant. Can cause nausea, dizziness, vomiting, and depression. Higher concentrations can lead to lightheadedness, peripheral neuropathy, numbing in the extremities, muscular weakness, giddiness, chemical pneumonia, coma, and death from respiratory failure.

ABSORPTION: Irritating to the skin, may cause drying, cracking, and chapping. Severe eye irritant.

INGESTION: Gastrointestinal irritation with possible systemic effects targeting the CNS.

Carcinogenicity	NTP Listed?	IARC Cancer Review Group?	OSHA Regulated?	Target Organs?
Unknown Human Unknown Animal	No	No	29 CFR 1910.1000 Table Z-1	Respiratory system, eyes, skin, CNS, PNS.

Medical Conditions Generally Aggravated by Exposure

Skin, CNS, PNS, and existing respiratory diseases may be aggravated by exposure to hexane isomers.

Emergency and First-aid Procedures

Eye contact: Flush immediately with water for 15 minutes (minimum); seek medical attention. Skin contact: Remove all contaminated clothing. Immediately wash area with large amounts of soap and water. Seek medical assistance. For inhalation: Remove the person from exposure. Provide respiratory assistance and CPR, if required. Transfer to medical facility. Keep victim warm and at rest. If swallowed: Seek medical attention immediately. Warning! Aspiration hazard; do NOT induce vomiting.

SECTION VII - PRECAUTIONS FOR SAFE HANDLING AND USE

Steps to be Taken in Case Material is Released or Spilled

Absorb in vermiculite, dry sand, earth, or similar material and deposit in sealed containers. Ventilate area of spill or leak. Restrict those not involved in cleanup from entering area. Remove all sources of ignition and use only non-sparking tools during cleanup. Notify proper authorities.

Preferred Waste Disposal Method

Burn in chemical incinerator equipped with an afterburner and scrubber.

Precautions to be Taken in Handling and Storage

Store to avoid contact with incompatible materials in tightly closed containers in cool, well-ventilated area away from heat and flame. Use only non-sparking tools to open and close containers.

Other Precautions and Warnings

Sources of ignition are prohibited where hexane isomers are used, handled, or stored. Heat may cause containers to build up pressure and explode. Containers must be grounded and bonded during liquid transfer operations.

SECTION VIII - CONTROL MEASURES AND PERSONAL PROTECTIVE EQUIPMENT

Respiratory Protection *(specify type)*

For high exposures (above the PEL), use a MSHA/NIOSH-approved organic vapor respirator. Best protection is obtained using a supplied-air respirator with full facepiece in continuous flow mode, or a self-contained breathing apparatus (SCBA) operated in pressure demand or other positive pressure mode.

Ventilation

Local exhaust or general mechanical systems recommended.

Protective Gloves	Eye Protection	Other Protective Clothing
Viton, PVC, or Chlorinated Polyethylene	Chemical Goggles or Face Mask	Protective Apron

Work/Hygiene Practices

Always wash hands thoroughly after using chemical; never bring food, drink, or smoking materials into vicinity of chemicals.

HEXANE ISOMERS (excluding *n*-hexane)

C_6H_{14} CAS: 107-83-5

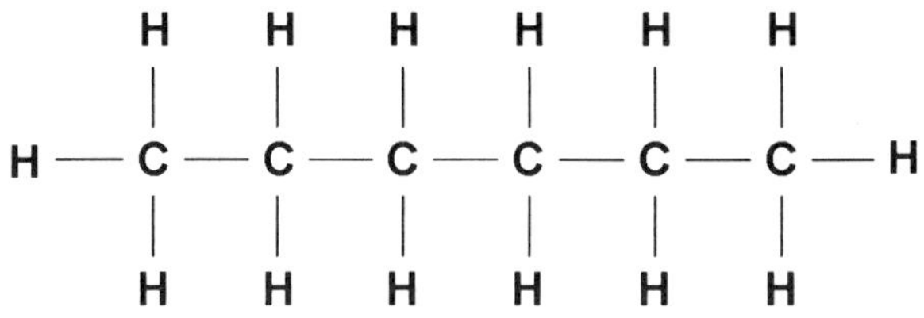

IDENTIFICATION AND TYPICAL USES

Hexane isomers appear as colorless liquids with a gasoline-like odor. They are used as commercial solvents for vegetable oils, glues, coatings, and paints. They are also found in gasoline and as intermediates for chemicals. Also constituents of rubber solvent and petroleum ether.

RISK ASSESSMENT: HEALTH

General Assessment

Hexane isomers are severe irritants by *inhalation* and *ingestion*. Skin contact can also produce some adverse topical effects and *absorption* is likely to occur. There is no clear evidence in the references to indicate that exposure to hexane isomers will lead to carcinogenic, mutagenic, or teratogenic effects.

Significant inhalation may cause irritation of the respiratory tract and depress the central nervous system. Symptoms include nausea, dizziness, vomiting, lightheadedness, giddiness, stupor, vertigo, loss of coordination, loss of appetite, depression, hallucinations, and possibly chemical pneumonia (if the swallowed liquid is aspirated into the lungs). At high concentrations, it is a narcotic and can lead to muscle weakness, cold feeling in extremities, blurred vision, anorexia, headaches, polyneuropathy, loss of consciousness, coma, and death due to respiratory failure. Studies have shown that hexane isomers will accumulate in fatty tissues as is evidenced by its affinity for the blood, liver, and brain where lipids are prevalent.

Skin and eye contact may produce immediate irritation with redness , painful burning, and possible blistering on contact. It is absorbed through the skin in both liquid and vapor form. Prolonged or repeated contact with the skin can cause dryness, cracking, and dermatitis. Ingestion will cause immediate gastrointestinal irritation and presents a serious aspiration hazard. If the liquid is aspirated into the lungs, oxygen can be displaced and the resultant asphyxiation can lead to brain damage or even cardiac arrest.

Acute Health Effects

The following acute (short-term) health effects may occur immediately or shortly after exposure to hexane isomers:

Skin: Severe and painful irritation with possible burns and blistering. Absorption through intact skin is likely to occur on exposure to the liquid or the vapor.

Eye: Severe irritation and burning on contact.

Lung: Direct irritation of the respiratory tract and associated mucosa. May cause dizziness, vomiting, nausea, and can lead to chemical pneumonia. Primary result of inhalation of hexane isomers is its action on the CNS.

CNS: A CNS depressant: headache, dizziness, depression, hallucinations, changes in mood, loss of consciousness, coma, and death (respiratory failure).

Chronic Health Effects

The following chronic (long-term) health effects may occur at some time after exposure to hexane isomers and can last for months or even years:

Cancer Hazards: According to information presented in the references, the hexane isomers have not been adequately tested for their ability to cause cancer in test animals.

Reproductive Hazard: According to information presented in the references, the hexane isomers have not been tested for their ability to affect reproduction.

Other Chronic Effects: Prolonged exposure may lead to polyneuropathy with symptoms of muscle weakness, loss of motor functions, numbness and pain in the extremities. There may also be contact dermatitis with de-fattening of the skin's protective oily surface layer and subsequent drying, cracking, chapping, and possible secondary infection.

Recommended Risk-Reduction Measures

Personnel should avoid direct contact with hexane isomers since they are known to cause significant adverse health effects, especially if improperly used or if overexposures occur. If a less toxic chemical cannot be substituted for a hazardous substance, then *engineering controls* are the most effective method of reducing exposures. The best protection is to enclose operations and/or provide local exhaust ventilation at the site of chemical release. While not always operationally feasible, isolating operations can also reduce exposure. Using respiratory protection is less effective than the controls mentioned above, but is still advisable whenever working with or around hexane isomers. For exposures above the PEL (500 ppm), an organic vapor respirator should be used (per OSHA 29 CFR 1910.134). Better protection is provided by a supplied-air respirator with full facepiece operated in continuous flow mode or a self-contained breathing apparatus (SCBA) with full facepiece in pressure demand. If a full facepiece is not available, then chemical goggles should be worn to protect the eyes. Whenever a chemical splash hazard exists, a face shield and a rubber apron should be worn. To prevent hand and skin exposures, impervious gloves should be used. Nitrile, Viton, PVC, or chlorinated polyethylene gloves are recommended. Glove manufacturers should be contacted and permeation studies obtained *before* final glove selections are made.

Administrative controls should also be in place to minimize the potential for human exposures. These may include written procedures or policies which specify the methods and techniques that will be practiced whenever personnel are to work with hexane isomers.

All personnel should receive *training* on the use, hazards, protective measures, emergency actions, and other precautions per 29 CFR 1910.1200 (Hazard Communication), prior to the first assignment in an area where hexane isomers are used or stored.

If symptoms develop or overexposure is suspected, the following tests are recommended:

- ☑ Lung function tests.
- ☑ Evaluation of nervous system function.
- ☑ Evaluation by a qualified allergist with careful consideration of exposure history and special testing (may help diagnose skin allergy and/or dermatitis).

Any evaluation should include a careful history of past and present symptoms with an examination. Medical tests that simply look for existing damage are not a substitute for controlling exposures.

Other methods to reduce exposure include:

- ☑ Where possible, enclose operations and use local exhaust ventilation at the site of chemical release. If local exhaust ventilation or enclosure is not used, respiratory protection should be mandatory.
- ☑ Always ensure that proper protective clothing is worn when using chemical substances and personnel are trained on its use, care, and maintenance.
- ☑ Wash thoroughly immediately after exposure to hexane isomers and at the end of the work shift or before eating, drinking, or smoking.
- ☑ Hazard warning information should be posted in the work area. In addition, as part of an on-going education and training program, all information on the health and safety hazards of hexane isomers should be communicated to all potentially exposed workers.
- ☑ Eye wash stations should be provided in the immediate work area for emergency use. If there is a possibility of skin exposure to hexane isomers, emergency shower facilities should also be provided.
- ☑ Personnel should never wear contaminated clothing home (family members can be exposed). Clothing should be laundered by personnel who have been trained on the hazards of hexane isomers.

RISK ASSESSMENT: ENVIRONMENT

General Assessment

The environment is at risk of exposure during transportation, storage, disposal, or destruction of hexane isomers. In almost every scenario, the threat of environmental exposure is contingent upon the proper handling of the chemical substance. Accidental spills, large or small, can result in fire, explosion, and possible contamination of the surrounding environmental mediums (water, soil, and air).

Hexane isomers are considered Class 1B flammable liquids (per OSHA 29 CFR 1910.106). Their extremely low flash point range and relatively low boiling point range make these chemicals a serious fire and explosion hazard concern. These characteristics require extreme caution in handling, storage, transportation, and disposal. Hexane isomers are incompatible with common peroxides and other oxidizers and contact can cause fire or explosion. Therefore, special consideration is required during any emergency situation involving a leak or spill of hexane isomers. Should hexane isomers ever come into contact with these incompatible substances during use, transportation, storage, or disposal, the formation of highly toxic and/or highly explosive commodities is possible.

Hexane isomers may enter the environment through industrial discharges or spills.

Acute Ecological Effects

Acute (short-term) toxic effects may include the death of animals, birds, or fish and death or low growth rate in plants. Acute effects are seen 2 to 4 days after animals or plants are exposed to hexane isomers.

These chemicals have high acute toxicity to aquatic life. Insufficient data are available on the short-term effects of hexane isomers exposure to plants, birds, or terrestrial animals.

Chronic Ecological Effects

Chronic (long-term) toxic effects may include shortened life span, reproductive problems, lower fertility, and changes in appearance or behavior in exposed animals. These effects can be seen long after first exposure(s) to toxic chemicals.

Insufficient data are available on the long-term effects of hexane isomers to aquatic life, plants, birds, or land animals.

Water Solubility

Hexane isomers are nearly insoluble in water. Concentrations of less than 1 milligram may mix with a liter of water, depending on water pH and chemical concentration of hexane isomers.

Persistence in the Environment

Hexane isomers are non-persistent in water, with a half-life of less than 2 days. The half-life of a pollutant is the amount of time it takes for one half of the chemical to be degraded. Virtually 100% of hexane isomers will eventually end up in air.

Bioaccumulation in Aquatic Organisms

Some substances increase in concentration, or *bioaccumulate,* in living organisms as they breathe contaminated air, drink contaminated water, or eat contaminated food. These chemicals can become concentrated in the tissues and internal organs of animals as well as humans.

The concentration of hexane isomers found in fish tissue is expected to be about the same as the average concentration of hexane isomers in the water from which the fish was taken.

Recommended Risk-Reduction Measures

Proper training of all transporters will reduce the likelihood of a mishap or accident resulting in a leak or spill to the environment. The correct use of DOT labeling (per the MSDS) on all containers, trucks, and rail cars will enable emergency responders to react properly and quickly to any disaster thereby reducing the potential risk to the environment and to personnel.

Storage of hexane isomers should be segregated from incompatible chemicals to minimize the risk of cross-contamination or contact. Buildings designated for storage should be equipped with appropriate fire protection systems (alarms, sprinklers, emergency lighting, portable extinguishers). All transfers of hexane isomers from storage containers to process containers should be done automatically whenever possible. All containers and equipment must be grounded and bonded.

If a spill or leak to the environment has occurred, fire department, emergency response and/or hazardous materials spill personnel should be notified immediately. Cleanup should be attempted only by those trained in proper spill containment procedures using non-sparking tools. Contaminated soils should be removed for incineration and replaced with clean soil. If hexane isomers should contact the water table, aquifer, or navigable waterway, the local and/or state emergency response authorities must be notified. A comprehensive emergency response or disaster preparedness/recovery plan should be in place prior to any operations involving the use, transportation, storage, or disposal of hexane isomers.

To dispose of hexane isomers properly, they should be burned in a chemical incinerator equipped with an afterburner and scrubber to ensure appropriate emission control.

If hexane isomers are spilled or leaked, the following specific steps are recommended:

- ☑ Restrict persons not wearing protective clothing from area of spill or leak until cleanup is complete and area can be opened for normal work.
- ☑ Remove all ignition sources, ventilate area of leak.
- ☑ Absorb liquids in vermiculite, dry sand, earth, or similar material and deposit in sealed containers. Collect materials using non-sparking tools.
- ☑ It may be necessary to dispose of hexane isomers as hazardous wastes. The responsible state agency or the regional office of the federal Environmental Protection Agency (EPA) should be contacted for specific recommendations.

RISK ASSESSMENT: BUSINESS

General Assessment

Accidents or mishaps involving hexane isomers can present a significant threat to business operations. The loss or damage of equipment or facilities can significantly affect fiscal viability. Lawsuits that may result from personnel injury/death, public exposures, and/or environmental contamination will also require a serious expenditure of resources. Media attention surrounding an injury, death, or environmental damage can also result in a loss of profits and loss of current as well as future business.

Recommended Risk-Reduction Measures

Company attorneys, safety and health professionals, and environmental specialists should be involved in the development of any procedures dealing with safety or emergency response. A company official should be pre-designated as a public relations officer with specific training in dealing with the press. Corporate plans and policies should be developed, approved, and implemented long before any need for such arises.

REFERENCES

American Conference of Governmental Industrial Hygienists. 1988. *Documentation of the Threshold Limit Values and Biological Exposure Indices,* 5th Edition (with updates). Cincinnati: ACGIH

Bueche, J. R. 1972. *Principles of Physics,* 2nd Edition. New York: McGraw-Hill.

Calabrese, E. J. 1994. *Biological Effects of Low Level Exposures.* New York: CRC Press/Lewis Publishers.

Cockerman, L. G., B. S. Shane. 1994. *Basic Environmental Toxicology.* New York: CRC Press/Lewis Publishers.

Cothern, C. R. 1995. *Handbook for Environmental Risk Decisionmaking.* New York: CRC Press/Lewis Publishers.

Environmental Protection Agency. 1983. *Chemical Hazard Information Profiles.* Washington, D.C.: U.S. EPA.

Gardner, R. W. 1994. *Chemical Intolerance.* New York: CRC Press/Lewis Publishers.

Hallenbeck, W. H. 1993. *Quantitative Risk Assessment for Environmental and Occupational Health,* 2nd Edition. New York: CRC Press/Lewis Publishers.

Hosty, J. W., and P. Foster. 1990. *A Practical Guide to Chemical Spill Response.* New York: Van Nostrand Reinhold.

Hodgson, E., R. B. Mailman, and Janice E. Chambers. 1988. *Dictionary of Toxicology.* London: The MacMillian Company.

Kamrin, M. A. 1988. *Toxicology: A Primer on Principles and Applications.* Chelsea, Michigan: Lewis Publishers.

Keith, L. H. and M. M. Walker. 1995. *Handbook of Air Toxics.* New York: CRC Press/Lewis Publishers.

Landis, W. G., M. H. Yu. 1995. *Introduction to Environmental Toxicology.* New York: CRC Press/Lewis Publishers.

Langerman, N. 1994 *Precautionary Labels for Chemical Containers.* New York: CRC Press/Lewis Publishers.

Lauwerys, R. R., P. Hoet. 1993. *Industrial Chemical Exposure.* New York: CRC Press/Lewis Publishers.

Lewis, R. J., Sr. 1992. *Sax's Dangerous Properties of Industrial Materials,* Volumes 1, 2, and 3, 8th Edition. New York: Van Nostrand Reinhold.

Manahan, S. E. 1994. *Environmental Chemistry,* 6th Edition. New York: CRC Press/Lewis Publishers.

Meyer, E. 1990. *Chemistry of Hazardous Materials.* Englewood Cliffs, NJ: Prentice Hall.

Patnaik, P. 1992. *A Comprehensive Guide to the Hazardous Properties of Chemical Substances.* New York: Van Nostrand Reinhold.

Quigley, D. R. 1994. *Handbook of Emergency Chemical Management.* New York: CRC Press/Lewis Publishers.

Rea, W. J. 1995. *Chemical Sensitivity,* Vol. I-III. New York: CRC Press/Lewis Publishers.

Richardson, M. 1995. *Dictionary of Substance and Their Effects,* Vol. I-VI. New York: CRC Press/Lewis Publishers.

Sacarello, H. L. A. 1994. *The Comprehensive Handbook of Hazardous Materials.* New York: CRC Press/Lewis Publishers.

Torkelson, T. R. V. K. Rowe. 1981. Halogenated aliphatic hydrocarbons. *Patty's Industrial Hygiene and Toxicology,* Vol. 2B, *Toxicology,* pp 3568-3572. New York: Wiley-Interscience.

U.S. Department of Health and Human Services, National Institute for Occupational Safety and Health. 1994. *NIOSH Pocket Guide to Chemical Hazards.* Washington, D.C.: U.S. Government Printing Office.

MATERIAL SAFETY DATA SHEET

CHEMICAL NAME	
HEXANOL	**CRC PRESS/LEWIS PUBLISHERS DISCLAIMER:** THE INFORMATION AND RECOMMENDATIONS PRESENTED HEREIN ARE BASED ON SOURCES BELIEVED TO BE RELIABLE. CRC MAKES NO REPRESENTATION ON ITS COMPLETENESS OR ACCURACY. IT IS THE RESPONSIBILITY OF THE USER TO DETERMINE THE CHEMICAL'S SUITABILITY FOR ITS INTENDED USE, THE CHEMICAL'S SAFE USE, AND THE CHEMICAL'S PROPER DISPOSAL. NO REPRESENTATIONS AND/OR WARRANTIES, EITHER EXPRESSED OR IMPLIED, OF THE MERCHANTABILITY OR FITNESS FOR A PARTICULAR PURPOSE, OR OF ANY OTHER NATURE, ARE MADE WITH RESPECT TO THE INFORMATION PROVIDED IN THIS MSDS OR TO THE CHEMICAL TO WHICH INFORMATION MAY REFER. CRC NEITHER ASSUMES NOR AUTHORIZES ANY OTHER PERSON TO ASSUME FOR IT, ANY OTHER ADDITIONAL RESPONSIBILITY OR LIABILITY FOR THE USE OF, OR RELIANCE UPON, THIS INFORMATION.

HAZARD WARNING INFORMATION

HEALTH	FIRE	REACTIVE	OTHER	DEGREE OF HAZARD	COLOR CODING	OTHER CODES
1	2	0		0 = Minimum Hazard 1 = Slight Hazard 2 = Moderate Hazard 3 = Serious Hazard 4 = Severe Hazard	HEALTH = BLUE FIRE = RED REACTIVITY = YELLOW OTHER = WHITE	OX = Oxidizer ACID = Acid ALK = Alkali COR = Corrosive W = Use No Water

SECTION I - GENERAL INFORMATION

Characterization	RCRA Number	EPA Class
Alcohol	**None**	**Not Applicable**

DOT Proper Shipping Name **Combustible Liquid, n.o.s.**	Chemical Abstract Service (CAS) Number **111-27-3**
DOT Hazard Class and Label Requirements **Combustible Liquid**	DOT Emergency Guide Code **26**
DOT Identification Number **NA 1993**	Chemical Formula **$C_6H_{14}O$**

Synonyms

Hexyl alcohol; pentylcarbinol; amylcarbinol; caproyl alcohol; 1-hydroxyhexane; *n*-hexanol; 1-hexanol.

SECTION II - HAZARDOUS INGREDIENTS AND IDENTITY INFORMATION

Hazardous Components (specific identity)	OSHA Exposure Criteria	NIOSH Exposure Criteria	Immediately Dangerous to Life and Health (IDLH)	ACGIH Exposure Criteria
Hexanol (derivation: By reduction of ethyl caproate; or from olefins by the Oxo process in which pentenes react with carbon monoxide and hydrogen using cobalt catalyst). 44% 1-hexanol 53% methy-1-pentanols 03% 2-ethyl-1-butanol	**PEL:** Not Established **STEL:** Not Established	**REL:** Not Established **STEL:** Not Established	**Not Reported**	**REL:** Not Established **STEL:** Not Established

SECTION III - PHYSICAL AND CHEMICAL CHARACTERISTICS

Boiling Point **315°F (157°C)**	Specific Gravity (H_2O = 1) **0.85**
Vapor Pressure (mm Hg) **1 at 76°F (24°C)**	Molecular Weight **102.17**
Vapor Density (Air = 1) **3.52**	Melting Point **-61°F (-52°C)**

Solubility

Slightly soluble in water. Soluble in alcohol, acetone, and chloroform; miscible with benzene and ether.

Appearance and Odor

Colorless liquid with a fatty, fruity odor and aromatic flavor. Odor Threshold = 0.01 ppm.

SECTION IV - FIRE AND EXPLOSION HAZARD DATA

Flash Point (method used) **145°F (63°C) closed cup**	Explosive Limits in Air % by Volume **LEL: 1.2%** **UEL: 7.7%**
NFPA Classification **Class IIIA Combustible Liquid**	Autoignition Temperature **559°F (292°C)**

Extinguishing Media

Dry chemical, carbon dioxide, or alcohol foam.

Special Fire Fighting Procedures

Poisonous gases are produced in fire. Wear full protective clothing and self-contained breathing apparatus (SCBA). Continue to cool containers with water after fire is extinguished. Move containers from fire area if it can be done without risk. For large fires, use unmanned hose apparatus, if possible. Evacuate non-essential personnel 1500 feet radius if fire. Consider down-wind conditions.

Unusual Fire and Explosion Hazards

Vapors are heavier than air and may travel a distances to ignition source to flashback. Vapors may explode if ignited in an enclosed area. Hexanol floats on water.

SECTION V - REACTIVITY DATA

Stability		Conditions to Avoid
		Hexanol is stable under normal conditions of storage and operation. Avoid contact with incompatible materials. Keep away from heat, flame, and other ignition sources.
Stable	Unstable	Incompatibility *(materials to avoid)*
X		Reacts violently in contact with oxidizers, such as fluorine, bromine, chlorine, and many peroxides and permanganates.
Hazardous Polymerization		Conditions to Avoid
		Under normal conditions of temperature and pressure, hexanol is not expected to undergo hazardous polymerization.
May Occur	Will Not Occur	Hazardous Decomposition or By-products
	X	When heated to decomposition, hexanol emits acrid and irritating fumes, including oxides of carbon (monoxide and dioxide).

SECTION VI - HEALTH HAZARD DATA

Primary Route(s) of Entry:	Inhalation?	Absorption (skin)?	Ingestion?
	X		X

Health Hazards

INHALATION: There may be slight irritation of the eyes, nose, and mucosa of the respiratory tract. Exposure can also cause CNS depressions with stupor, fatigue, sleepiness, nausea, vomiting, diarrhea, and hypotension. Pulmonary edema may also occur.

SKIN & EYES: As a solvent, causes dry skin, cracking, and possible dermatitis. Eye contact can result in burning, stinging, tearing, corneal injury, and hyperemia of the conjunctiva.

INGESTION: Nausea, vomiting, diarrhea, and gastrointestinal hemorrhage. Also causes headache, dizziness, giddiness, loss of muscular coordination, delirium, sedation, respiratory depression, and coma.

Carcinogenicity	NTP Listed?	IARC Cancer Review Group?	OSHA Regulated?	Target Organs?
Unknown Human Unknown Animal	No	No	No	Skin, eyes, mucosa, CNS, respiratory system.

Medical Conditions Generally Aggravated by Exposure

Skin and respiratory diseases may be aggravated by exposure.

Emergency and First-aid Procedures

Eye contact: Flush immediately with water for 15 minutes (minimum); seek medical attention. Skin contact: Remove all contaminated clothing. Immediately wash area with large amounts of soap and water. Seek medical assistance. For inhalation: Remove the person from exposure. Provide respiratory assistance and CPR. Transfer to medical facility. Observe for 48 hours for lung effects. If swallowed: Seek medical attention. Unless advised otherwise, induce vomiting using Syrup of Ipecac.

SECTION VII - PRECAUTIONS FOR SAFE HANDLING AND USE

Steps to be Taken in Case Material is Released or Spilled

Contain spills using absorbent material (vermiculite, or other). Ventilate area of spill. Have extinguishing agent available in case of fire. Store materials in DOT-approved containers.

Preferred Waste Disposal Method

Incineration in chemical incinerator equipped with scrubber and afterburner.

Precautions to be Taken in Handling and Storage

Store in tightly closed containers in a cool, dark, well-ventilated area. Protect containers from physical damage. Keep heat, fire, and flame away. Outside building or underground storage is recommended.

Other Precautions and Warnings

Ground and bond all metal containers. Ensure Class 1, Group D electrical installations.

SECTION VIII - CONTROL MEASURES AND PERSONAL PROTECTIVE EQUIPMENT

Respiratory Protection *(specify type)*

No exposure levels have been established. An organic vapor air-purifying respirator may suffice, but greater protection is obtained using a supplied-air respirator or a self-contained breathing apparatus (SCBA) with full facepiece operated in pressure demand or other positive pressure mode.

Ventilation

Local exhaust or general mechanical systems recommended.

Protective Gloves	Eye Protection	Other Protective Clothing
Impervious Gloves	Chemical Goggles or Face Mask	Impermeable Apron or Protective Suit

Work/Hygiene Practices

Always wash hands thoroughly after using chemical; never bring food, drink, or smoking materials into vicinity of chemicals.

HEXANOL

$C_6H_{13}OH$

CAS: 111-27-3

$$CH_3 - CH_2 - CH_2 - CH_2 - CH_2 - CH_2 - OH$$

IDENTIFICATION AND TYPICAL USES

Hexanol is a colorless liquid with a faint, fatty, fruity odor and aromatic flavor. It is used as a synthetic flavoring agent, a solvent, a plasticizer, an intermediate for textile and leather finishing, and as a chemical intermediate for hypnotic and antiseptic preparations.

RISK ASSESSMENT: HEALTH

General Assessment

Hexanol is primarily toxic to humans by *ingestion, inhalation,* and skin *absorption.* There are no reports of carcinogenic, teratogenic, or mutagenic effects following exposure to hexanol in animals or humans.

Inhalation can cause irritation of the eyes, nose, throat, upper respiratory tract, and associated mucosa. There may be adverse effects on the central nervous system (CNS) with symptoms of headache, nervousness, tremors, dizziness, tearing, hypotension, fatigue, loss of muscle coordination, somnolence, delirium, sedation, giddiness, and narcosis with stupor and loss of consciousness. Other symptoms of exposure can include respiratory depression, nausea, vomiting, and diarrhea. Lung tissue irritation may lead to a dangerous buildup of fluids in the lungs, a condition known as pulmonary edema. This is a medical emergency and can be fatal. Symptoms of cough, congestion, difficulty breathing, and tightness in chest may be delayed up to 48 hours after exposure thereby creating a false sense of security with regard to health exposure risk.

Liquid contact with the eyes causes immediate burning and stinging with lachrymation and reflex closure of the lids. There may be injury to the corneal epithelium and possible hyperemia (excessive blood) of the conjunctiva. Skin contact results in drying and cracking which can lead to secondary infections and dermatitis. Absorption may be quick and systemic effects target the CNS in the same way as inhalation or ingestion. If skin absorption occurs in conjunction with another exposure route, such as inhalation of hexanol vapors, then the resulting effects can be present a serious threat to health.

Ingestion of hexanol causes nausea, vomiting, diarrhea, and gastrointestinal hemorrhage. In addition, the ingested chemical can lead to CNS effects similar to that of inhalation. If the liquid is aspirated into the lungs (as is possible during vomiting episodes), hemorrhagic pneumonitis can occur.

Acute Health Effects

The following acute (short-term) health effects may occur immediately or shortly after exposure to hexanol:

Skin: Causes dryness and cracking leading to dermatitis and possible secondary infection.

Eye: Severe irritation with burning and stinging with possible damage to the cornea and conjunctiva.

Lung: Irritation of the eyes, nose, throat, and respiratory tract. Delayed pulmonary edema is possible which can be fatal.

CNS: High concentrations can cause depression of the CNS with symptoms of sleepiness and lack of concentration, giddiness, somnolence, loss of muscle coordination, delirium, depression, respiratory depression, and coma.

Chronic Health Effects

The following chronic (long-term) health effects may occur at some time after exposure to hexanol and can last for months or even years:

Cancer Hazards: According to information presented in the references, hexanol is not known to cause cancer in laboratory test animals.

Reproduction: According to the references, hexanol has not been adequately tested for its ability to cause cancer in test animals. Additional research is required in this area of study.

Other Chronic Effects: Very high or prolonged exposure may result in mucous membrane irritation, headache, and depression of the CNS with symptoms of somnolence and lack of concentration. Prolonged skin contact can cause dermatitis. In animal studies, hexanol is known to cause liver damage.

Recommended Risk-Reduction Measures

The best risk reduction measure is to use a less toxic chemical as a substitute for a hexanol. However, depending upon industrial applications and use requirements, substitution may not always be a practical alternative. Therefore, *engineering controls* are the most effective method of reducing exposures. The best protection is to enclose operations and/or provide local exhaust ventilation at the site of chemical release. While not always operationally feasible, isolating operations and restricting personnel access can also reduce exposure risk.

Using respiratory protection is less effective than the controls mentioned above, but is still advisable whenever working with or around hexanol. Even though no exposure levels have been established for this chemical, it does not mean that exposure to hexanol is without health risks. For low or infrequent exposures, an air-purifying respirator with an organic vapor cartridge may suffice. However, the best protection is obtained using a supplied-air respirator with full facepiece operated in positive pressure mode or a self-contained breathing apparatus (SCBA) with full facepiece operated in pressure demand mode. If a full facepiece is not available, then chemical goggles should be worn to protect the eyes. Whenever a chemical splash hazard exists, a face shield and a protective apron should be worn. To prevent hand and skin exposures, impervious gloves should be used. Glove manufacturers should be contacted and permeation studies obtained *before* final glove selections have been made.

Administrative controls should also be in place to minimize the potential for human exposures. These may include written procedures or policies which specify the methods and techniques that will be practiced whenever personnel are to work with hexanol.

All personnel should receive *training* on the use, hazards, protective measures, emergency actions, and other precautions per 29 CFR 1910.1200 (Hazard Communication), prior to the first assignment in an area where hexanol is used or stored.

If symptoms develop or overexposure is suspected, the following medical tests are recommended:

☑ Liver function tests.

☑ Lung function tests.

☑ Consider chest X-ray following acute overexposure (may be negative if taken immediately following exposure due to delayed development of pulmonary edema).

☑ Skin testing with dilute hexanol to help diagnose allergy (performed by a qualified allergist).

Any medical evaluation should include a careful history of past and present symptoms with an examination. Medical tests that look for existing damage are not a substitute for controlling exposures. Also, since smoking can cause lung cancer, heart disease, emphysema, and other respiratory disorders, smokers exposed to hexanol may experience symptoms of respiratory distress more quickly than non-smokers under the same conditions of exposure. Prudent risk management requires careful consideration of *all* possible factors which may be causing the appearance of exposure symptoms.

Other methods to reduce exposure include:

☑ Where possible, enclose operations and use local exhaust ventilation at the site of chemical release. If local exhaust ventilation or enclosure is not used, respiratory protection should be mandatory.

☑ Always ensure that proper protective clothing is worn when using chemical substances and that personnel have been trained on its use and care.

☑ Wash thoroughly immediately after exposure to hexanol and at the end of the work shift or before eating, drinking, or smoking.

☑ Hazard warning information should be posted in the work area. In addition, as part of an on-going education and training program, all information on the health and safety hazards of hexanol should be communicated to all potentially exposed workers.

RISK ASSESSMENT: ENVIRONMENT

General Assessment

The environment is at risk of exposure during transportation, storage, disposal, or destruction of hexanol. In almost every scenario, the threat of environmental exposure is contingent upon the proper handling of the chemical substance. Accidental spills, large or small, can result in fire, explosion, and possible contamination of the surrounding environmental mediums (water, soil, and air).

Hexanol is considered a Class IIIA combustible liquid (according to OSHA 29 CFR 1910.106). Its moderate flash point and boiling point present a fire and explosion hazard concern. Also, because it is incompatible with a number of common materials, especially strong oxidizers, contact can result in violent and explosive reactions. It can also form explosive mixtures in air and can ignite on contact with heat, fire, or sparks. Its vapors are heavier than air and can travel for some distance to an ignition source to flashback and cause fire or explosion. A boiling liquid expanding vapor explosion (BLEVE) is possible when hexanol is exposed to heat or fire in closed containers. Fire fighters should be notified of the presence of hexanol. These characteristics require special consideration during any emergency situation involving a leak or spill of hexanol.

Hexanol can enter the environment through unchecked industrial discharges into effluents and through spills.

Acute Ecological Effects

Acute (short-term) toxic effects may include the death of animals, birds, or fish, and death or low growth rate in plants. Acute effects are seen 2 to 4 days after animals or plants are exposed to hexanol.

This chemical has moderate acute toxicity to aquatic life. Insufficient data are available to evaluate or predict the short-term effects of hexanol to plants, birds, or terrestrial animals.

Chronic Ecological Effects

Chronic toxic effects may include shortened life span, reproductive problems, lower fertility, and changes in appearance or behavior in exposed animals. These effects can be seen long after first exposure(s) to toxic chemicals.

Hexanol has moderate chronic toxicity to aquatic life. Insufficient data are available to evaluate or predict the long-term effects of hexanol to plants, birds, or land animals.

Water Solubility

Hexanol is slightly soluble in water. Concentrations of 1 to 100 milligrams can be effectively mixed with a liter of water.

Persistence in the Environment

Hexanol is slightly persistent in water, with a half-life of between 2 to 20 days. The half-life of a pollutant is the amount of time it takes for one-half of the chemical to be degraded.

Bioaccumulation in Aquatic Organisms

Some substances increase in concentration, or *bioaccumulate,* in living organisms as they breathe contaminated air, drink contaminated water, or eat contaminated food. These chemicals can become concentrated in the tissues and internal organs of animals as well as humans.

The concentration of hexanol found in fish tissues is expected to be about the same as the average concentration of hexanol in water from which the fish was taken.

Recommended Risk-Reduction Measures

Proper training of all transporters will reduce the likelihood of a mishap or accident resulting in a leak or spill to the environment. The correct use of DOT labeling (per the MSDS) on all containers, trucks, and rail cars will enable emergency responders to react properly and quickly to any disaster thereby reducing the potential risk to the environment and to personnel.

Storage of hexanol should be segregated from incompatible chemicals to minimize the risk of cross-contamination or contact. Buildings designated for storage should be equipped with appropriate fire protection systems (alarms, sprinklers, emergency lighting, portable extinguishers). Equipment should be designed to meet explosion-proof standards.

If a spill or leak to the environment has occurred, fire department, emergency response, and/or hazardous materials spill personnel should be notified immediately. Cleanup should be attempted only by those trained in proper spill containment procedures using non-sparking tools. Contaminated soils should be removed for incineration and replaced with clean soil. If hexanol should contact the water table, aquifer, or navigable waterway, time is of the essence. It is only slightly soluble in water and, therefore, total containment and remediation may be possible. When such spills occur, the local and/or state emergency response authorities must be notified. A comprehensive emergency response or disaster preparedness/recovery plan should be in place prior to any operations involving the use, transportation, storage, or disposal of hexanol. If hexanol is spilled or leaked, the following specific steps are recommended:

☑ Restrict persons not wearing protective clothing from area of spill or leak until cleanup is complete and area can be opened for normal work.

☑ Ventilate area of spill or leak.

☑ Remove ignition sources.

☑ Absorb liquids in vermiculite, dry sand, earth, or a similar material and deposit in sealed containers. Use non-sparking tools.

☑ It may be necessary to dispose of hexanol as a hazardous waste. The responsible state agency or the regional office of the federal Environmental Protection Agency (EPA) should be contacted for specific recommendations.

RISK ASSESSMENT: BUSINESS

General Assessment

Accidents or mishaps involving hexanol can present a moderate threat to business operations. The loss or damage of equipment or facilities can significantly affect fiscal viability. Lawsuits that may result from personnel injury/death, public exposures, and/or environmental contamination will also require a serious expenditure of resources. Media attention surrounding an injury, death, or environmental damage can also result in a loss of profits and loss of current as well as future business.

Recommended Risk-Reduction Measures

Company attorneys, safety and health professionals, and environmental specialists should be involved in the development of any procedures or policies intended to manage the use of chemicals in the workplace. A company official should be pre-designated as a public relations officer with specific training in dealing with the press. Corporate plans and policies should be developed, approved, and implemented long before any need for such arises.

REFERENCES

American Conference of Governmental Industrial Hygienists. 1988. *Documentation of the Threshold Limit Values and Biological Exposure Indices,* 5th Edition (with updates). Cincinnati: ACGIH

Anthony, C. P., and N. J. Kolthoff. 1971. *Textbook of Anatomy and Physiology.* St. Louis: C. V. Mosby Company.

Bloom, A. D. and F. J. DeSerres. 1995. *Ecotoxicity and Human Health.* New York: CRC Press/Lewis Publishers.

Brusick, D. J. 1994. *Methods for Genetic Risk Assessment.* New York: CRC Press/Lewis Publishers.

Bueche, J. R. 1972. *Principles of Physics,* 2nd Edition. New York: McGraw-Hill.

Burrell, R., D. K. Flaherty, and L. J. Sauers. 1992. *Toxicology of the Immune System.* New York: Van Nostrand Reinhold.

Calabrese, E. J. 1994. *Biological Effects of Low Level Exposures.* New York: CRC Press/Lewis Publishers.

Cothern, C. R., N. P. Ross. 1994. *Environmental Statistics, Assessment, and Forecasting.* New York: CRC Press/Lewis Publishers.

Danse, I. R. 1991. *Common Sense Toxics in the Workplace.* New York: Van Nostrand Reinhold.

Environmental Protection Agency. 1983. *Chemical Hazard Information Profiles.* Washington, D.C.: U.S. EPA.

Fodden, C. R., and J. L. Weddell. 1992. *Hazardous Materials Emergency Action Data.* New York: CRC Press/Lewis Publishers.

Forseberg, K., and L. H. Keith. 1993. *Chemical; Protective Clothing Permeation and Degradation Database.* Chelsea, MI: Lewis Publishers.

Gardner, R. W. 1994. *Chemical Intolerance.* New York: CRC Press/Lewis Publishers.

Hodgson, E., R. B. Mailman, and J. E. Chambers. 1988. *Dictionary of Toxicology.* London: The MacMillian Company.

Hosty, J. W., and P. Foster. 1990. *A Practical Guide to Chemical Spill Response.* New York: Van Nostrand Reinhold.

Howard, P. H., et. al. 1991. *Handbook of Environmental Degradation Rates.* New York: CRC Press/Lewis Publishers.

International Agency for Research on Cancer (IARC). 1987. *IARC Monographs on the Evaluation of Carcinogenic Risk to Humans, Overall Evaluations of Carcinogenicity: An updating of IARC Monographs Volumes 1-42.* Lyon: IARC.

Keith, L. H. and D. B. Walters. (Editors) 1992. *National Toxicology Program's Chemical Database, Volumes I-VII.* Chelsea, MI: Lewis Publishers.

Landis, W. G., M. H. Yu. 1995. *Introduction to Environmental Toxicology.* New York: CRC Press/Lewis Publishers.

Lauwerys, R. R., P. Hoet. 1993. *Industrial Chemical Exposure.* New York: CRC Press/Lewis Publishers.

Lewis, Richard J., Sr. 1992. *Sax's Dangerous Properties of Industrial Materials,* Volumes 1, 2, and 3, 8th Edition. New York: Van Nostrand Reinhold.

Manahan, S. E. 1992. *Toxicological Chemistry,* 2nd Edition. New York: CRC Press/Lewis Publishers.

Meyer, Eugene. 1990. *Chemistry of Hazardous Materials.* Englewood Cliffs, NJ: Prentice Hall.

Patnaik, P. 1992. *A Comprehensive Guide to the Hazardous Properties of Chemical Substances.* New York: Van Nostrand Reinhold.

Philp, R. B. 1995. *Environmental Hazards and Human Health.* New York: CRC Press/Lewis Publishers.

MATERIAL SAFETY DATA SHEET

CHEMICAL NAME

2-HEXANONE

HAZARD WARNING INFORMATION

HEALTH	FIRE	REACTIVE	OTHER	DEGREE OF HAZARD	COLOR CODING	OTHER CODES
2	3	0		0 = Minimum Hazard 1 = Slight Hazard 2 = Moderate Hazard 3 = Serious Hazard 4 = Severe Hazard	HEALTH = BLUE FIRE = RED REACTIVITY = YELLOW OTHER = WHITE	OX = Oxidizer ACID = Acid ALK = Alkali COR = Corrosive W = Use No Water

SECTION I - GENERAL INFORMATION

Characterization: **Ketone**

RCRA Number: **D001**

EPA Class: **Characteristic (I) Waste**

DOT Proper Shipping Name: **Ketones, Liquid, N.O.S.**

Chemical Abstract Service (CAS) Number: **591-78-6**

DOT Hazard Class and Label Requirements: **Flammable Liquid**

DOT Emergency Guide Code: **26**

DOT Identification Number: **UN 1224**

Chemical Formula: **$CH_3CO[CH_2]_3CH_3$**

Synonyms

Methyl butyl ketone; methyl *n*-butyl ketone; butyl methyl ketone; MBK; MNBK; hexane-2-one; propylacetone; *n*-butyl methyl ketone; 2-oxohexane.

SECTION II - HAZARDOUS INGREDIENTS AND IDENTITY INFORMATION

Hazardous Components (specific identity)	OSHA Exposure Criteria	NIOSH Exposure Criteria	Immediately Dangerous to Life and Health (IDLH)	ACGIH Exposure Criteria
2 Hexanone (derivation: By reaction of acetyl chloride with butyl magnesium chloride or by catalyzed reaction of acetic acid and ethylene under pressure, followed by distillation.) 1 ppm = 4.17 mg/m^3	PEL (8 hour): 100 ppm 410 mg/m^3 STEL: Not Established	REL (10 hour): 1 ppm 4 mg/m^3 (skin) STEL: Not Established	1600 ppm	TLV: 5 ppm 20 mg/m^3 STEL: Not Established

SECTION III - PHYSICAL AND CHEMICAL CHARACTERISTICS

Boiling Point: **262°F (128°C)**

Specific Gravity (H_2O = 1): **0.81**

Vapor Pressure (mm Hg): **11 at 77°F (25°C)**

Molecular Weight: **100.2**

Vapor Density (Air = 1): **3.45**

Freezing Point: **-71°F (-57°C)**

Solubility

Slightly soluble (1.4%) in water. Soluble in alcohol, acetone, and ether.

Appearance and Odor

Clear, colorless liquid with an acetone-like odor. Odor Threshold = 0.076 ppm.

SECTION IV - FIRE AND EXPLOSION HAZARD DATA

Flash Point (method used): **77°F (25°C) closed cup**

Explosive Limits in Air % by Volume: **LEL: 1.22% UEL: 8%**

NFPA Classification: **Class IC Flammable Liquid**

Autoignition Temperature: **795°F (423°C)**

Extinguishing Media

Carbon dioxide, dry chemical, alcohol-resistant foam, or fog. Water spray may be ineffective on fire.

Special Fire Fighting Procedures

Wear full protective clothing and SCBA. Fight fire from a distance. Keep containers cool with water spray. A solid stream of water may spread fire. If possible, move fire-exposed containers from fire area.

Unusual Fire and Explosion Hazards

Containers may explode in fire. BLEVE (boiling liquid expanding vapor explosion) is possible. Immediately evacuate if rising sound is heard from venting device or a discoloration is observed on the sides of the tank or storage vessel.

SECTION V - REACTIVITY DATA

Stability		Conditions to Avoid
Stability		Normally stable. Do not allow 2-hexanone to come into contact with incompatible materials. Avoid contact with fire or high heat sources. Do not use in confined spaces.
Stable	Unstable	Incompatibility *(materials to avoid)*
X		Incompatible with strong oxidizers (bromine, chlorine, fluorine). Also note that ketones + nitric acid and hydrogen peroxide form highly explosive solids or oils.
Hazardous Polymerization		Conditions to Avoid: Under normal conditions of temperature and pressure, hazardous polymerization of 2-hexanone is not expected to occur.
May Occur	Will Not Occur	Hazardous Decomposition or By-products
	X	When heated to decomposition, 2-hexanone emits acrid, irritating smoke and fumes, including oxides of carbon.

SECTION VI - HEALTH HAZARD DATA

Primary Route(s) of Entry:	Inhalation?	Absorption (skin)?	Ingestion?
	X	X	X

Health Hazards

INHALATION: Irritation of the conjunctiva of the eyes and mucosa of the nose and throat. Symptoms of CNS depression include dizziness, headache, and unconsciousness. Neurotoxic effects to the peripheral nervous system may also occur.

ABSORPTION: Mild irritation on contact with the skin and eyes. Easily absorbed through intact skin.

INGESTION: Symptoms of gastrointestinal irritation with nausea and vomiting and CNS depression.

Carcinogenicity	NTP Listed?	IARC Cancer Review Group?	OSHA Regulated?	Target Organs?
Unknown Human Unknown Animal	No	No	29 CFR 1910.1000 Table Z-1	Respiratory system; skin; eyes; CNS, PNS.

Medical Conditions Generally Aggravated by Exposure

Skin conditions (dermatitis) or any nervous system disorders may be aggravated.

Emergency and First-aid Procedures

Eye contact: Immediately flush large amounts of water for 15 minutes (minimum), occasionally lifting eyelids. Skin contact: Remove all contaminated clothing. Immediately wash area with large amounts of soap and water. For inhalation: Remove the person from exposure. Provide respiratory assistance and CPR. Transfer promptly to medical facility. If swallowed: Seek medical attention immediately. Give victim several glasses of water. Do NOT induce vomiting (possibility of liquid aspiration to lungs). Never try to give an unconscious or convulsing person anything by mouth.

SECTION VII - PRECAUTIONS FOR SAFE HANDLING AND USE

Steps to be Taken in Case Material is Released or Spilled

Remove all ignition sources. Restrict persons not wearing protective equipment from entering area of spill or leak until cleanup is complete. Ventilate area. Absorb liquid in dry sand or other material and deposit in sealed containers. Use water spray to knock-down vapors.

Preferred Waste Disposal Method

Burn in chemical incinerator equipped with afterburner and scrubber.

Precautions to be Taken in Handling and Storage

Do not store with incompatible chemicals since violent reactions can occur. Store in tightly closed containers in a cool, dark, well-ventilated area away from fire, sparks, and flame. Use non-sparking tools on containers. Ground and bond metal containers during transfer operations.

Other Precautions and Warnings

Containers may explode in fire or under conditions of extreme heat (do not store outdoors or in direct sunlight). Protect containers from physical damage. Dike around outside storage areas.

SECTION VIII - CONTROL MEASURES AND PERSONAL PROTECTIVE EQUIPMENT

Respiratory Protection *(specify type)*

Exposures above 100 ppm: MSHA/NIOSH-approved full-facepiece respirator with organic vapor cartridge. Greater protection is obtained from a self-contained breathing apparatus (SCBA) with full facepiece and operated in pressure demand or other positive pressure mode.

Ventilation

Local exhaust or general mechanical systems recommended.

Protective Gloves	Eye Protection	Other Protective Clothing
Impervious Gloves	Chemical Goggles and/or Face Mask	Protective Apron

Work/Hygiene Practices

Always wash hands thoroughly after using chemical; never bring food, drink, or smoking materials into vicinity of chemicals.

2-HEXANONE

$CH_3CO[CH_2]_3CH_3$ CAS: 591-78-6

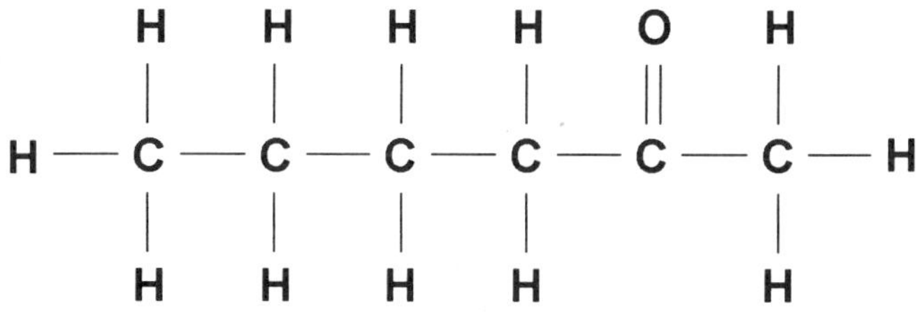

IDENTIFICATION AND TYPICAL USES

2-Hexanone is a colorless liquid with an acetone-like odor. It was formerly used as a solvent for nitrocellulose, resins, lacquers, oils, fats, and waxes. It is not currently produced in the United States. However, it may exist as a by-product in wood pulping, coal gassification, and oil shale processing and, therefore, exposure is still possible.

RISK ASSESSMENT: HEALTH

General Assessment

2-Hexanone is mildly toxic to humans by *inhalation,* skin contact (*absorption*), and *ingestion.* There are no reports in the references of carcinogenic or mutagenic effects. However, experimental teratogenic and reproductive effects have been reported.

Inhalation of 2-hexanone vapors causes mild to moderate irritation to the skin, eyes, and mucous membranes of the respiratory tract. Exposures to extremely high levels of vapor concentration can cause depression of the central nervous system (CNS) leading to symptoms of narcosis with headache, dizziness, nausea, loss of consciousness, and possible coma. Neurotoxic effects also appear to focus on the actions of the peripheral nervous system due to the metabolites of 2-hexanone.

Skin contact will result in mild irritation. Prolonged exposures can lead to dryness, cracking, contact dermatitis, and a possibility for secondary infection at the point of contact. 2-Hexanone is easily absorbed through intact skin. Eye contact causes irritation, even at concentrations as low as 1000 ppm for as little as fifteen minutes.

Ingestion can produce gastrointestinal irritation, nausea, and vomiting. It will also lead to central nervous system depression with symptoms similar to that of inhalation.

Acute Health Effects

The following acute (short-term) health effects may occur immediately or shortly after exposure to 2-hexanone:

Skin: Mild to moderate irritation and possible dermatitis. Absorption is likely.

Eye: Mild to moderate irritation.

Lung: Irritation of the mucous membranes of the respiratory tract.

CNS: A mild narcotic compound. Inhalation of high concentrations of the vapor may cause dizziness, lightheadedness, and possible loss of consciousness. Also appears to effect the peripheral nervous system (PNS) with symptoms of numbness and paresthesia in the extremities.

Chronic Health Effects

The following chronic (long-term) health effects may occur at some time after exposure to 2-hexanone and can last for months or even years:

Cancer Hazards: According to information presented in the references, exposure to 2-hexanone has not been shown to cause carcinogenic effects.

Reproduction: There is limited evidence that exposure to 2-hexanone causes teratogenic and adverse reproductive effects in test animals. It is not know if human exposure will have this effect.

Sensitivity: Repeated exposure to the skin may result in dermatitis. Repeated or prolonged exposures through inhalation or skin absorption may lead to damage to the peripheral nervous system.

ʊ *Recommended Risk-Reduction Measures*

Personnel should avoid direct contact with 2-hexanone. Unless a less toxic chemical can be substituted for a hazardous substance, *engineering controls* are the most effective method of reducing exposures. The best protection is to enclose operations and/or provide local exhaust ventilation at the site of chemical release. While not always operationally feasible, isolating operations can also reduce exposure. Using respiratory protection is less effective than the controls mentioned above, but is still advisable whenever working with 2-hexanone. For exposures to the PEL (100 ppm), use a NIOSH-approved, full facepiece organic vapor respirator. For higher or prolonged exposures, a supplied-air respirator with full facepiece operated in a positive pressure mode or a self-contained breathing apparatus (SCBA) with full facepiece and pressure demand are the recommended respiratory protection methods of choice. If a full facepiece is not available, then chemical goggles should be worn to protect the eyes. Whenever a chemical splash hazard exists, a face shield and a rubber apron should be worn. Natural rubber and polyvinylchloride (PVC) gloves will NOT protect against ketones. To prevent hand and skin exposures, impervious gloves should be worn. Glove manufacturers should be contacted and permeation studies obtained *before* final glove selections are made.

Administrative controls should also be in place to minimize the potential for human exposures. These may include written procedures or policies which specify the methods and techniques that will be practiced whenever personnel are to work with 2-hexanone.

All personnel should receive *training* on the use, hazards, protective measures, emergency actions, and other precautions per 29 CFR 1910.1200 (Hazard Communication), prior to the first assignment in an area where 2-hexanone is used or stored.

Other methods to reduce exposure include:

- ☑ Where possible, enclose operations and use local exhaust ventilation at the site of chemical release. If local exhaust ventilation or enclosure is not used, respiratory protection should be mandatory.
- ☑ Always ensure that proper protective clothing is worn when using chemical substances and that personnel are trained on its use, care, and maintenance.
- ☑ Wash thoroughly immediately after exposure to 2-hexanone and at the end of the work shift or before eating, drinking, or smoking.
- ☑ Hazard warning information should be posted in the work area. As part of an on-going education and training program, all information on the health and safety hazards of 2-hexanone should be communicated to all potentially exposed workers.

RISK ASSESSMENT: ENVIRONMENT

General Assessment

The environment is at risk of exposure during transportation, storage, disposal, or destruction of 2-hexanone. In almost every scenario, the threat of environmental exposure is contingent upon the proper handling of the chemical substance. Accidental spills, large or small, can result in fire, explosion, and possible contamination of the surrounding environmental mediums (water, soil, and air).

2-Hexanone is considered a Class IC flammable liquid (per OSHA 29 CFR 1910.106). It can react with strong oxidizers to cause a serious fire and explosion hazard. When mixed with nitric acid and hydrogen peroxide, it will form solids or oils that are highly explosive. When involved in a fire, it can release acrid and irritating oxides of carbon. A boiling liquid expanding vapor explosion (BLEVE) is also a possibility. These characteristics require special consideration during any emergency involving a leak or spill of 2-hexanone.

The proper disposal/destruction method for 2-hexanone is to burn it in a chemical incinerator equipped with an afterburner and air scrubber.

2-Hexanone can enter the environment through manufacturing, unchecked discharge into effluents from industrial or municipal waste treatment plants, and through spills.

☠ *Acute Ecological Effects*

Acute (short-term) toxic effects may include the death of animals, birds, or fish, and death or low growth rate in plants. Acute effects are seen 2 to 4 days after animals or plants are exposed to 2-hexanone. Insufficient data are available to evaluate or predict the short-term effects of 2-hexanone on aquatic life, plants, birds, or land animals.

Chronic Ecological Effects

Chronic (long-term) toxic effects may include shortened life span, reproductive problems, lower fertility, and changes in appearance or behavior in exposed animals.

These effects can be seen long after first exposure(s) to toxic chemicals. Insufficient data are available to evaluate or predict the long-term effects of 2-hexanone on aquatic life, plants, birds, or land animals.

♦ *Water Solubility*

2-Hexanone is slightly soluble in water. Concentrations of less than 100 milligrams will mix with a liter of water.

⌛ *Persistence in the Environment*

2-Hexanone is slightly persistent in water, with a half-life of between 2 to 20 days. The half-life of a pollutant is the amount of time it takes for one half of the chemical to be degraded. There is no information in the references on the specific persistence percentages for this chemical in the various environmental mediums (air, soil, water).

≈ *Bioaccumulation in Aquatic Organisms*

Some substances increase in concentration, or *bioaccumulate,* in living organisms as they breathe contaminated air, drink contaminated water, or eat contaminated food. These chemicals can become concentrated in the tissues and internal organs of animals as well as humans.

The concentration of 2-hexanone found in fish tissues is expected to be about the same as the average concentration of 2-hexanone in water from which the fish was taken.

Recommended Risk-Reduction Measures

Proper training of all transporters will reduce the likelihood of a mishap or accident resulting in a leak or spill to the environment. The correct use of DOT labeling (per the MSDS) on all containers, trucks, and rail cars will enable emergency responders to react properly and quickly to any disaster thereby reducing the potential risk to the environment and to personnel.

Storage of 2-hexanone should be segregated from incompatible chemicals to minimize the risk of cross-contamination or contact. Buildings designated for storage should be equipped with appropriate fire protection systems (alarms, sprinklers, emergency lighting, portable extinguishers).

If a spill or leak to the environment has occurred, fire department, emergency response, and/or hazardous materials spill personnel should be notified immediately. Cleanup should be attempted only by those trained in proper spill containment procedures using non-sparking tools. Contaminated soils should be removed for incineration and replaced with clean soil. If 2-hexanone should contact the water table, aquifer, or navigable waterway, time is of the essence. It is slightly soluble in water and, therefore, total containment and remediation may not be possible. When such spills occur, the local and/or state emergency response authorities must be notified. A comprehensive emergency response or disaster preparedness/recovery plan should be in place prior to any operations involving the use, transportation, storage, or disposal of 2-hexanone.

If 2-hexanone is spilled or leaked, the following specific steps are recommended:

- ☑ Restrict persons not wearing protective clothing from area of spill or leak until cleanup is complete and area can be opened for normal work.
- ☑ Remove all ignition sources.
- ☑ Ventilate area of spill or leak.
- ☑ Absorb liquids in vermiculite, dry sand, earth, or a similar material and deposit in sealed containers for disposal.
- ☑ For large spills, dike area far ahead of spill to ensure total containment. Do not allow spilled materials or runoff to reach waterways.
- ☑ It may be necessary to dispose of 2-hexanone as a hazardous waste. The responsible state agency or the regional office of the federal Environmental Protection Agency (EPA) should be contacted for specific recommendations.

RISK ASSESSMENT: BUSINESS

General Assessment

Accidents or mishaps involving 2-hexanone can present a moderate threat to business operations. The loss or damage of equipment or facilities can significantly affect fiscal viability. Lawsuits that may result from personnel injury, public exposures, and/or environmental contamination will also require a serious expenditure of resources. Media attention surrounding an injury, illness, or environmental damage can also result in a loss of profits and loss of current as well as future business. Always remember that anytime the term "reproductive hazard" is used, public emotion, ignorance, and hysteria can run equally high. This must be properly considered whenever developing or implementing public relations policies.

Ü *Recommended Risk-Reduction Measures*

Company attorneys, safety and health professionals, and environmental specialists should be involved in the development of any procedures dealing with safety, health, or emergency response. A company official should be pre-designated as a public relations officer with specific training in dealing with the press. Corporate plans and policies should be developed, approved, and implemented long before any need for such arises.

REFERENCES

Ahlbom, A. 1993. *Biostatistics for Epidemiologists.* New York: CRC Press/Lewis Publishers.

American Conference of Governmental Industrial Hygienists. 1988. *Documentation of the Threshold Limit Values and Biological Exposure Indices,* 5th Edition (with updates). Cincinnati: ACGIH

Anthony, C. P., and N. J. Kolthoff. 1971. *Textbook of Anatomy and Physiology.* St. Louis: C. V. Mosby Company.

Asante-Duah, D. K. 1993. *Hazardous Waste Risk Assessment.* New York: CRC Press/Lewis Publishers.

Bloom, A. D. and F. J. DeSerres. 1995. *Ecotoxicity and Human Health.* New York: CRC Press/Lewis Publishers.

Brusick, D. J. 1994. *Methods for Genetic Risk Assessment.* New York: CRC Press/Lewis Publishers.

Bueche, J. R. 1972. *Principles of Physics,* 2nd Edition. New York: McGraw-Hill.

Burrell, R., D. K. Flaherty, and L. J. Sauers. 1992. *Toxicology of the Immune System.* New York: Van Nostrand Reinhold.

Calabrese, E. J. 1994. *Biological Effects of Low Level Exposures.* New York: CRC Press/Lewis Publishers.

Cockerman, L. G., B. S. Shane. 1994. *Basic Environmental Toxicology.* New York: CRC Press/Lewis Publishers.

Cothern, C. R. 1995. *Handbook for Environmental Risk Decisionmaking.* New York: CRC Press/Lewis Publishers.

DiVincenzo, G. O., F. J. Yanno, and B. D. Astill. 1973. Exposure of man and dog to low concentrations of acetone vapor. *American Industrial Hygiene Journal.* 34:329-336.

Gardner, R. W. 1994. *Chemical Intolerance.* New York: CRC Press/Lewis Publishers.

Hallenbeck, W. H. 1993. *Quantitative Risk Assessment for Environmental and Occupational Health,* 2nd Edition. New York: CRC Press/Lewis Publishers.

Hodgson, E., R. B. Mailman, and J. E. Chambers. 1988. *Dictionary of Toxicology.* London: The MacMillian Company.

Hosty, J. W., and P. Foster. 1990. *A Practical Guide to Chemical Spill Response.* New York: Van Nostrand Reinhold.

Howard, P. H. 1995. *Handbook of Environmental Fate and Exposures Data for Organic Chemicals,* Vol. I-IV. New York: CRC Press/Lewis Publishers.

James, A., and M. Lord. 1992. *Index of Chemical and Physical Data.* New York: Van Nostrand Reinhold.

Jones, F. E. 1994. *Toxic Organic Vapors in the Workplace.* New York: CRC Press/Lewis Publishers.

Kamrin, M. A. 1988. *Toxicology: A Primer on Principles and Applications.* Chelsea, Michigan: Lewis Publishers.

Keith, L. H. and M. M. Walker. 1995. *Handbook of Air Toxics.* New York: CRC Press/Lewis Publishers.

Landis, W. G., M. H. Yu. 1995. *Introduction to Environmental Toxicology.* New York: CRC Press/Lewis Publishers.

Langerman, N. 1994 *Precautionary Labels for Chemical Containers.* New York: CRC Press/Lewis Publishers.

Lauwerys, R. R., P. Hoet. 1993. *Industrial Chemical Exposure.* New York: CRC Press/Lewis Publishers.

Lewis, R. J., Sr. 1993. *Hawley's Condensed Chemical Dictionary,* 12th Edition. New York: Van Nostrand Reinhold.

Lewis, R. J., Sr. 1992. *Sax's Dangerous Properties of Industrial Materials,* Volumes 1, 2, and 3, 8th Edition. New York: Van Nostrand Reinhold.

Manahan, S. E. 1992. *Toxicological Chemistry,* 2nd Edition. New York: CRC Press/Lewis Publishers.

Mellström, G. A., J. E. Wahlberg, and H. I. Maibach. 1994. *Protective Gloves for Occupational Use.* New York: CRC Press/Lewis Publishers.

Meyer, E. 1990. *Chemistry of Hazardous Materials.* Englewood Cliffs, NJ: Prentice Hall.

Patnaik, P. 1992. *A Comprehensive Guide to the Hazardous Properties of Chemical Substances.* New York: Van Nostrand Reinhold.

Philp, R. B. 1995. *Environmental Hazards and Human Health.* New York: CRC Press/Lewis Publishers.

Proctor, N. H., and J. P. Hughes. 1991. *Chemical Hazards of the Workplace.* New York: Van Nostrand Reinhold.

Quigley, D. R. 1994. *Handbook of Emergency Chemical Management.* New York: CRC Press/Lewis Publishers.

Raleigh, R. L. and W. A. McGee. 1972. Effects of short, high concentration exposures to acetone as determined by observation in the work area. *Journal of Occupational Medicine.* 14:607-610.

Rea, W. J. 1995. *Chemical Sensitivity,* Vol. I-III. New York: CRC Press/Lewis Publishers.

Richardson, M. 1995. *Dictionary of Substance and Their Effects,* Vol. I-VI. New York: CRC Press/Lewis Publishers.

U.S. Department of Health and Human Services, National Institute for Occupational Safety and Health. 1994. *NIOSH Pocket Guide to Chemical Hazards.* Washington, D.C.: U.S. Government Printing Office.

U.S. Department of Health and Human Services, National Institute for Occupational Safety and Health. 1978. *Criteria for a Recommended Standard - Occupational Exposure to Ketones.* DHEW Publication Number: 78-173. Washington, D.C.: U.S. Government Printing Office.

MATERIAL SAFETY DATA SHEET

CHEMICAL NAME	CRC PRESS/LEWIS PUBLISHERS DISCLAIMER
***sec*-HEXYL ACETATE**	

HAZARD WARNING INFORMATION

HEALTH	FIRE	REACTIVE	OTHER	DEGREE OF HAZARD	COLOR CODING	OTHER CODES
1	2	0		0 = Minimum Hazard 1 = Slight Hazard 2 = Moderate Hazard 3 = Serious Hazard 4 = Severe Hazard	HEALTH = BLUE FIRE = RED REACTIVITY = YELLOW OTHER = WHITE	OX = Oxidizer ACID = Acid ALK = Alkali COR = Corrosive W = Use No Water

SECTION I - GENERAL INFORMATION

Characterization	**Aliphatic Ester**	RCRA Number	**D001**
		EPA Class	**Characteristic (I) Waste**
DOT Proper Shipping Name	**Methylamyl Acetate**	Chemical Abstract Service (CAS) Number	**108-84-9**
DOT Hazard Class and Label Requirements	**Flammable Liquid**	DOT Emergency Guide Code	**26**
DOT Identification Number	**UN 1233**	Chemical Formula	$C_6H_{16}O_2$

Synonyms

1,3-Dimethylbutyl acetate; methylisoamyl acetate; methyl amyl acetate; methylisobutyl carbinol acetate; 4-methyl-2-pentanol acetate; acetic acid-1,3-dimethylbutyl ester.

SECTION II - HAZARDOUS INGREDIENTS AND IDENTITY INFORMATION

Hazardous Components (specific identity)	OSHA Exposure Criteria	NIOSH Exposure Criteria	Immediately Dangerous to Life and Health (IDLH)	ACGIH Exposure Criteria
***sec*-Hexyl Acetate** (derivation: By catalytic reaction of 4-methyl-2-pentanol and acetic acid. Occurs naturally in apples and in their leaves). **1 ppm = 5.99 mg/m^3**	**PEL:** **50 ppm** **300 mg/m^3** **STEL:** **Not Established**	**REL:** **50 ppm** **300 mg/m^3** **STEL:** **Not Established**	**500 ppm**	**TLV:** **50 ppm** **295 mg/m^3** **STEL:** **Not Established**

SECTION III - PHYSICAL AND CHEMICAL CHARACTERISTICS

Boiling Point	**297°F (146°C)**	Specific Gravity (H_2O = 1)	**0.86**
Vapor Pressure (mm Hg)	**3 at 68°F (20°C)**	Molecular Weight	**144.2**
Vapor Density (Air = 1)	**4.97**	Melting Point	**-83°F (-64°C)**

Solubility

Slightly soluble in water (0.08%). Soluble in alcohol and acetone (miscible).

Appearance and Odor

Clear, colorless liquid with a mild, pleasant, fruity, banana-like odor. Odor Threshold = 0.39 ppm.

SECTION IV - FIRE AND EXPLOSION HAZARD DATA

Flash Point (method used)	**113°F (45°C) closed cup**	Explosive Limits in Air % by Volume	**LEL: 0.9%** **UEL: 5.83%**
NFPA Classification	**Class II Combustible Liquid**	Autoignition Temperature	**510°F (265°C)**

Extinguishing Media

Dry chemical, alcohol foam, carbon dioxide, water fog. Use water to keep fire-exposed containers cool.

Special Fire Fighting Procedures

Poisonous gases are produced in fire. Wear full protective clothing and self-contained breathing apparatus (SCBA). Use blanketing effect with foam to smother large fires. Continue to cool containers with water after fire is extinguished. Water on fire itself may be ineffective and actually spread fire further.

Unusual Fire and Explosion Hazards

Immediately withdraw if rising sound from venting device is heard or if fire is causing discoloration to the tank. Evacuate 1500 feet radius if fire becomes uncontrollable or container is exposed to direct flame. Vapors are heavier than air and may travel a distance to ignition source to flashback.

SECTION V - REACTIVITY DATA

Stability		Conditions to Avoid *sec*-Hexyl acetate is stable under normal conditions of storage and operation. Avoid contact with incompatible materials. Keep away from heat, flame, and other sources of ignition.
Stable X	Unstable	Incompatibility *(materials to avoid)* Reacts violently in contact with nitrates, strong oxidizers, alkalis, and acids. Particularly reactive with chlorosulfonic acid, lithium aluminum hydride and 2-chloromethyl furan, oleum, and potassium t-butoxide.
Hazardous Polymerization		Conditions to Avoid Under normal conditions of temperature and pressure, hazardous polymerization of *sec*-hexyl acetate is not expected to occur.
May Occur	Will Not Occur X	Hazardous Decomposition or By-products When heated to decomposition, sec-hexyl acetate emits acrid and irritating fumes and oxides of carbon.

SECTION VI - HEALTH HAZARD DATA

Primary Route(s) of Entry:	Inhalation?	Absorption (skin)?	Ingestion?
	X		X

Health Hazards

INHALATION: Irritation of the eyes, nose, throat, and respiratory tract. Exposures above 100 ppm may cause narcosis and headache. Systemic effects include olfactory changes, conjunctiva irritation, and pulmonary changes. It is irritating to the mucosa, particularly the eyes, gums, and respiratory passages. Can also cause corneal clouding.

SKIN & EYES: Will irritate the skin. May cause contact dermatitis. Eye contact may cause corneal damage with cloudy swelling, and fatty degeneration of the viscera.

INGESTION: May cause headache, drowsiness, and unconsciousness.

Carcinogenicity	NTP Listed?	IARC Cancer Review Group?	OSHA Regulated?	Target Organs?
Unknown Human Unknown Animal	No	No	29 CFR 1910.1000 Table Z-1	Respiratory system; skin; eyes, CNS.

Medical Conditions Generally Aggravated by Exposure

None reported.

Emergency and First-aid Procedures

Eye contact: Flush immediately with water for 15 minutes (minimum); seek medical attention. Skin contact: Remove all contaminated clothing. Immediately wash area with large amounts of soap and water. Seek medical assistance. For inhalation: Remove the person from exposure. Provide respiratory assistance and CPR. Transfer to medical facility. If swallowed: Seek medical attention immediately. Call poison control center for advice. Give 1-2 glasses of milk or water and induce vomiting unless advised otherwise. Never give an unconscious or convulsing person anything by mouth.

SECTION VII - PRECAUTIONS FOR SAFE HANDLING AND USE

Steps to be Taken in Case Material is Released or Spilled

Contain spills using absorbent material (vermiculite, or other). Ventilate area of spill. Have water source available in case of fire. Store materials in DOT-approved containers. Restrict those not involved in cleanup from entering area. Notify appropriate authorities, as required.

Preferred Waste Disposal Method

Incineration in chemical incinerator equipped with scrubber and afterburner.

Precautions to be Taken in Handling and Storage

Store in tightly closed containers in a cool, dark, well-ventilated area. Protect containers from physical damage. Keep heat, fire, and flame away. If possible, automatically transfer liquids between containers.

Other Precautions and Warnings

Bulk storage of *sec*-hexyl acetate is not recommended. Ground and bond all metal containers.

SECTION VIII - CONTROL MEASURES AND PERSONAL PROTECTIVE EQUIPMENT

Respiratory Protection *(specify type)*

Where potential exists for exposure over 50 ppm, use an MSHA/NIOSH-approved full facepiece respirator with an organic vapor cartridge. For high exposures, use a self-contained breathing apparatus (SCBA) with full facepiece operated in pressure demand or other positive pressure mode.

Ventilation

Local exhaust or general mechanical systems recommended.

Protective Gloves	Eye Protection	Other Protective Clothing
Impervious Gloves	Chemical Goggles or Face Mask	Impermeable Apron

Work/Hygiene Practices

Always wash hands thoroughly after using chemical; never bring food, drink, or smoking materials into vicinity of chemicals.

sec-HEXYL ACETATE

$C_8H_{16}O_2$

CAS: 108-84-9

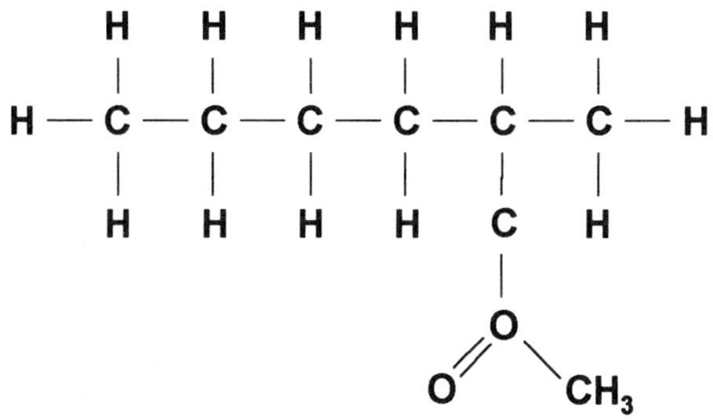

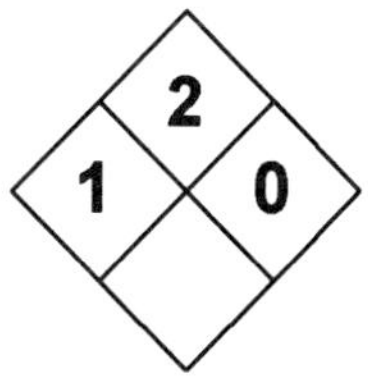

IDENTIFICATION AND TYPICAL USES

sec-Hexyl acetate is a clear, colorless liquid with a characteristic sweet, mild, pleasant, fruity, banana-like odor. It is used as a solvent for nitrocellulose and lacquers, as a fragrance in perfumes and cosmetics, as a component in insecticide formulations (particularly housefly insecticides), and as a component in attractants for the large spruce bark beetle.

RISK ASSESSMENT: HEALTH

General Assessment

sec-Hexyl acetate is mildly toxic by *inhalation* and *ingestion*. Skin contact can produce localized effects but absorption has not been reported. No specific carcinogenic, mutagenic, or teratogenic effects have been reported.

Inhalation of *sec*-hexyl acetate vapors will cause irritation of the eyes, nose, throat, and upper respiratory tract, including the mucosa (particularly the eyes, gums, and respiratory passages). Inhalation can cause changes in the sense of smell (olfactory changes), conjunctiva irritation, and pulmonary changes. In high concentrations, there can be depression of the central nervous system with mild narcotic effects with headache, dizziness, and nausea.

Skin contact results in the removal of the skin's natural protective oily layer on its surface with subsequent drying, cracking, and possible secondary infections and dermatitis. Eye contact can cause serious damage to the cornea with a possibility for permanent effects to vision.

Ingestion of *sec*-hexyl acetate may cause headache, drowsiness, unconsciousness, and unspecified gastrointestinal effects.

Acute Health Effects

The following acute (short-term) health effects may occur immediately or shortly after exposure to *sec*-hexyl acetate:

Skin: Mild to moderate irritation. Will remove the skin's protective oily layer to cause drying, cracking, and possible secondary infection.

Eye: Severe irritation which can lead to injury and permanent damage if allowed to remain in contact with the eyes.

Lung: Irritation of the eyes, nose, throat and lungs (especially the mucus membranes). May cause changes in the sense of smell and in pulmonary function.

CNS: High concentrations can depress the central nervous system leading to narcoses and possible loss of consciousness.

Chronic Health Effects

The following chronic (long-term) health effects may occur at some time after exposure to *sec*-hexyl acetate and can last for months or even years:

Cancer Hazards: According to the references, *sec*-hexyl acetate has not been adequately tested for its ability to cause cancer in test animals. Further research is required and caution is warranted in use, handling, and storage.

Reproduction: According to the references, *sec*-hexyl acetate has not been adequately tested for its ability to adversely affect reproduction in test animals.

Other Chronic Effects: There are reports that prolonged exposure in animal studies led to permanent damage to the CNS. It is not known if long-term human exposure will have this effect.

Recommended Risk-Reduction Measures

Personnel should avoid direct contact with *sec*-hexyl acetate. The best risk reduction measure is to use a less toxic chemical as a substitute for *sec*-hexyl acetate. If this is not possible or feasible, then *engineering controls* are the most effective method of reducing exposures. The best protection is to enclose operations and/or provide local exhaust ventilation at the site of chemical release. While not always operationally feasible, isolating operations can also reduce exposure risk.

Using respiratory protection is less effective than the controls mentioned above, but is still advisable whenever working with or around *sec*-hexyl acetate. For exposures over 50 ppm (PEL for 8 hours), an MSHA/NIOSH-approved chemical respirator with an organic vapor cartridge should be used. For higher exposures, a self-contained breathing apparatus (SCBA) with full facepiece operated in pressure demand mode or a supplied-air respirator with full facepiece operated in pressure demand are the recommended respiratory protection methods of choice. If a full facepiece is not available, then chemical goggles should be worn to protect the eyes. Whenever a chemical splash hazard exists, a face shield and a protective apron should be worn. To prevent hand and skin exposures, impervious gloves should be used. Caution is warranted in glove selection for *sec*-hexyl acetate. Glove manufacturers should be contacted to obtain permeation studies *before* gloves are selected.

Administrative controls should also be in place to minimize the potential for human exposures. These may include written procedures or policies which specify the methods and techniques that will be practiced whenever personnel are to work with *sec*-hexyl acetate.

All personnel should receive *training* on the use, hazards, protective measures, emergency actions, and other precautions per 29 CFR 1910.1200 (Hazard Communication), prior to the first assignment in an area where *sec*-hexyl acetate is used or stored.

If symptoms develop or overexposure is suspected, the following medical tests are recommended:

- ☑ Liver and kidney function tests.
- ☑ Evaluation of nervous system functions.
- ☑ Skin testing with dilute *sec*-hexyl acetate to help diagnose allergy (performed by a qualified allergist).

Any medical evaluation should include a careful history of past and present symptoms with an examination. Medical tests that look for existing damage are not a substitute for controlling exposures.

Other methods to reduce exposure include:

- ☑ Where possible, enclose operations and use local exhaust ventilation at the site of chemical release. If local exhaust ventilation or enclosure is not used, respiratory protection should be mandatory.
- ☑ Always ensure that proper protective clothing is worn when using chemical substances.
- ☑ Wash thoroughly immediately after exposure to *sec*-hexyl acetate and at the end of the work shift or before eating, drinking, or smoking.
- ☑ Hazard warning information should be posted in the work area. In addition, as part of an on-going education and training program, all information on the health and safety hazards of *sec*-hexyl acetate should be communicated to all potentially exposed workers.
- ☑ Never eat, drink, or smoke in areas were *sec*-hexyl acetate is used, handled or stored.

RISK ASSESSMENT: ENVIRONMENT

General Assessment

The environment is at risk of exposure during transportation, storage, disposal, or destruction of sec-hexyl acetate. In almost every scenario, the threat of environmental exposure is contingent upon the proper handling of the chemical substance.

Accidental spills, large or small, can result in fire, explosion, and possible contamination of the surrounding environmental mediums (water, soil, and air). *sec*-Hexyl acetate is considered a Class II combustible liquid (according to OSHA 29 CFR 1910.106). Its moderate flash point and boiling point present a fire and explosion hazard concern. Also, because it is incompatible with a number of common materials, especially strong oxidizers and acids, contact can result in violent and explosive reactions. It can also form explosive mixtures in air and can ignite on contact with heat, fire, or sparks. These characteristics require special consideration during any emergency situation involving a leak or spill of *sec*-hexyl acetate.

The proper disposal/destruction method for *sec*-hexyl acetate is to burn it in a chemical incinerator equipped with an afterburner and air scrubber. *sec*-Hexyl acetate can enter the environment through unchecked industrial discharges into effluents and through spills.

Acute Ecological Effects

Acute (short-term) toxic effects may include the death of animals, birds, or fish, and death or low growth rate in plants. Acute effects are seen 2 to 4 days after animals or plants are exposed to *sec*-hexyl acetate. Insufficient data are available to evaluate or predict the short-term effects of *sec*-hexyl acetate to aquatic life, plants, birds, or terrestrial animals.

Chronic Ecological Effects

Chronic toxic effects may include shortened life span, reproductive problems, lower fertility, and changes in appearance or behavior in exposed animals. These effects can be seen long after first exposure(s) to toxic chemicals. Insufficient data are available to evaluate or predict the long-term effects of *sec*-hexyl acetate to aquatic life, plants, birds, or land animals.

Water Solubility

sec-Hexyl acetate is only slightly soluble in water. Concentrations of 1 to 10 milligrams may mix with a liter of water.

Persistence in the Environment

sec-Hexyl acetate is slightly persistent in water, with a half-life of between 2 to 20 days. The half-life of a pollutant is the amount of time it takes for one-half of the chemical to be degraded.

Bioaccumulation in Aquatic Organisms

Some substances increase in concentration, or *bioaccumulate,* in living organisms as they breathe contaminated air, drink contaminated water, or eat contaminated food. These chemicals can become concentrated in the tissues and internal organs of animals as well as humans.

The concentration of *sec*-hexyl acetate found in fish tissues is expected to be somewhat higher then the average concentration of *sec*-hexyl acetate in water from which the fish was taken.

Recommended Risk-Reduction Measures

Proper training of all transporters will reduce the likelihood of a mishap or accident resulting in a leak or spill to the environment. The correct use of DOT labeling (per the MSDS) on all containers, trucks, and rail cars will enable emergency responders to react properly and quickly to any disaster thereby reducing the potential risk to the environment and to personnel.

Storage of *sec*-hexyl acetate should be segregated from incompatible chemicals to minimize the risk of cross-contamination or contact. Buildings designated for storage should be equipped with appropriate fire protection systems (alarms, sprinklers, emergency lighting, portable extinguishers).

If a spill or leak to the environment has occurred, fire department, emergency response, and/or hazardous materials spill personnel should be notified immediately. Cleanup should be attempted only by those trained in proper spill containment procedures using non-sparking tools. Contaminated soils should be removed for incineration and replaced with clean soil.

If *sec*-hexyl acetate should contact the water table, aquifer, or navigable waterway, time is of the essence. It is nearly insoluble in water and, therefore, total containment and remediation may be possible. When such spills occur, the local and/or state emergency response authorities must be notified. A comprehensive emergency response or disaster preparedness/recovery plan should be in place prior to any operations involving the use, transportation, storage, or disposal of *sec*-hexyl acetate. If *sec*-hexyl acetate is spilled or leaked, the following specific steps are recommended:

- ☑ Restrict persons not wearing protective clothing from area of spill or leak until cleanup is complete and area can be opened for normal work.
- ☑ Ventilate area of spill or leak.
- ☑ Remove all ignition sources from area.
- ☑ Absorb liquids in vermiculite, dry sand, earth, or a similar material and deposit in sealed containers. Use non-sparking tools.
- ☑ It may be necessary to dispose of sec-hexyl acetate as a hazardous waste. The responsible state agency or the regional office of the federal Environmental Protection Agency (EPA) should be contacted for specific recommendations.

RISK ASSESSMENT: BUSINESS

General Assessment

Accidents or mishaps involving *sec*-hexyl acetate can present a moderate threat to business operations. The loss or damage of equipment or facilities can significantly affect fiscal viability. Lawsuits that may result from personnel injury/death, public exposures, and/or environmental contamination will also require a serious expenditure of resources.

Media attention surrounding an injury, death, or environmental damage can also result in a loss of profits and loss of current as well as future business.

Recommended Risk-Reduction Measures

Company attorneys, safety and health professionals, and environmental specialists should be involved in the development of any procedures or policies intended to manage the use of chemicals in the workplace. A company official should be pre-designated as a public relations officer with specific training in dealing with the press. Corporate plans and policies should be developed, approved, and implemented long before any need for such arises.

REFERENCES

American Conference of Governmental Industrial Hygienists. 1988. *Documentation of the Threshold Limit Values and Biological Exposure Indices,* 5th Edition (with updates). Cincinnati: ACGIH

Anthony, C. P., and N. J. Kolthoff. 1971. *Textbook of Anatomy and Physiology.* St. Louis: C. V. Mosby Company.

Bloom, A. D. and F. J. DeSerres. 1995. *Ecotoxicity and Human Health.* New York: CRC Press/Lewis Publishers.

Brusick, D. J. 1994. *Methods for Genetic Risk Assessment.* New York: CRC Press/Lewis Publishers.

Bueche, J. R. 1972. *Principles of Physics,* 2nd Edition. New York: McGraw-Hill.

Burrell, R., D. K. Flaherty, and L. J. Sauers. 1992. *Toxicology of the Immune System.* New York: Van Nostrand Reinhold.

Calabrese, E. J. 1994. *Biological Effects of Low Level Exposures.* New York: CRC Press/Lewis Publishers.

Cockerman, L. G., B. S. Shane. 1994. *Basic Environmental Toxicology.* New York: CRC Press/Lewis Publishers.

Cothern, C. R., N. P. Ross. 1994. *Environmental Statistics, Assessment, and Forecasting.* New York: CRC Press/Lewis Publishers.

Danse, I. R. 1991. *Common Sense Toxics in the Workplace.* New York: Van Nostrand Reinhold.

Environmental Protection Agency. 1983. *Chemical Hazard Information Profile: Sec-hexyl acetate.* Washington, D.C.: U.S. EPA.

Gardner, R. W. 1994. *Chemical Intolerance.* New York: CRC Press/Lewis Publishers.

Hallenbeck, W. H. 1993. *Quantitative Risk Assessment for Environmental and Occupational Health,* 2nd Edition. New York: CRC Press/Lewis Publishers.

Hodgson, E., R. B. Mailman, and J. E. Chambers. 1988. *Dictionary of Toxicology.* London: The MacMillian Company.

Hosty, J. W., and P. Foster. 1990. *A Practical Guide to Chemical Spill Response.* New York: Van Nostrand Reinhold.

Howard, P. H., et. al. 1991. *Handbook of Environmental Degradation Rates.* New York: CRC Press/Lewis Publishers.

James, A., and M. Lord. 1992. *Index of Chemical and Physical Data.* New York: Van Nostrand Reinhold.

Jones, F. E. 1994. *Toxic Organic Vapors in the Workplace.* New York: CRC Press/Lewis Publishers.

Kamrin, M. 1 A. 1988. *Toxicology: A Primer on Principles and Applications.* Chelsea, Michigan: Lewis Publishers.

Keith, L. H. and M. M. Walker. 1995. *Handbook of Air Toxics.* New York: CRC Press/Lewis Publishers.

Landis, W. G., M. H. Yu. 1995. *Introduction to Environmental Toxicology.* New York: CRC Press/Lewis Publishers.

Langerman, N. 1994 *Precautionary Labels for Chemical Containers.* New York: CRC Press/Lewis Publishers.

Lauwerys, R. R., P. Hoet. 1993. *Industrial Chemical Exposure.* New York: CRC Press/Lewis Publishers.

Lide, D. R. 1995. *Handbook of Chemistry and Physics,* 76th Edition. New York: CRC Press/Lewis Publishers.

Lewis, Richard J., Sr. 1992. *Sax's Dangerous Properties of Industrial Materials,* Volumes 1, 2, and 3, 8th Edition. New York: Van Nostrand Reinhold.

Lisella, Frank S. 1994. *The VNR Dictionary of Environmental Health and Safety.* New York: Van Nostrand Reinhold.

Manahan, S. E. 1994. *Environmental Chemistry,* 6th Edition. New York: CRC Press/Lewis Publishers.

Manahan, S. E. 1992. *Toxicological Chemistry,* 2nd Edition. New York: CRC Press/Lewis Publishers.

Meyer, Eugene. 1990. *Chemistry of Hazardous Materials.* Englewood Cliffs, NJ: Prentice Hall.

Patnaik, P. 1992. *A Comprehensive Guide to the Hazardous Properties of Chemical Substances.* New York: Van Nostrand Reinhold.

Proctor, N. H., and J. P. Hughes. 1991. *Chemical Hazards of the Workplace.* New York: Van Nostrand Reinhold.

U.S. Department of Health and Human Services, National Institute for Occupational Safety and Health. 1994. *NIOSH Pocket Guide to Chemical Hazards.* Washington, D.C.: U.S. Government Printing Office.

MATERIAL SAFETY DATA SHEET

CHEMICAL NAME	
HEXYLENE GLYCOL	**CRC PRESS/LEWIS PUBLISHERS DISCLAIMER:** THE INFORMATION AND RECOMMENDATIONS PRESENTED HEREIN ARE BASED ON SOURCES BELIEVED TO BE RELIABLE. CRC MAKES NO REPRESENTATION ON ITS COMPLETENESS OR ACCURACY. IT IS THE RESPONSIBILITY OF THE USER TO DETERMINE THE CHEMICAL'S SUITABILITY FOR ITS INTENDED USE, THE CHEMICAL'S SAFE USE, AND THE CHEMICAL'S PROPER DISPOSAL. NO REPRESENTATIONS AND/OR WARRANTIES, EITHER EXPRESSED OR IMPLIED, OF THE MERCHANTABILITY OR FITNESS FOR A PARTICULAR PURPOSE, OR OF ANY OTHER NATURE, ARE MADE WITH RESPECT TO THE INFORMATION PROVIDED IN THIS MSDS OR TO THE CHEMICAL TO WHICH INFORMATION MAY REFER. CRC NEITHER ASSUMES NOR AUTHORIZES ANY OTHER PERSON TO ASSUME FOR IT, ANY OTHER ADDITIONAL RESPONSIBILITY OR LIABILITY FOR THE USE OF, OR RELIANCE UPON, THIS INFORMATION.

HAZARD WARNING INFORMATION

HEALTH	FIRE	REACTIVE	OTHER	DEGREE OF HAZARD	COLOR CODING	OTHER CODES
1	1	0		0 = Minimum Hazard 1 = Slight Hazard 2 = Moderate Hazard 3 = Serious Hazard 4 = Severe Hazard	HEALTH = BLUE FIRE = RED REACTIVITY = YELLOW OTHER = WHITE	OX = Oxidizer ACID = Acid ALK = Alkali COR = Corrosive W = Use No Water

SECTION I - GENERAL INFORMATION

Characterization	**Alcohol**	RCRA Number	**None**
		EPA Class	**Not Applicable**
DOT Proper Shipping Name	**No Citation**	Chemical Abstract Service (CAS) Number	**107-41-5**
DOT Hazard Class and Label Requirements	**No Citation**	DOT Emergency Guide Code	**No Citation**
DOT Identification Number	**No Citation**	Chemical Formula	**$(CH_3)_2COHCH_2CHOHCH_3$**

Synonyms

2,4-Dihydroxy-2-methylpentane; 2-methyl-2,4-pentanediol; 4-methyl-2,4-pentanediol; 2-methylpentane-2,4-diol; Pinakon; α,α,α'-trimethyltrimethylene glycol.

SECTION II - HAZARDOUS INGREDIENTS AND IDENTITY INFORMATION

Hazardous Components (specific identity)	OSHA Exposure Criteria	NIOSH Exposure Criteria	Immediately Dangerous to Life and Health (IDLH)	ACGIH Exposure Criteria
Hexylene glycol (derivation: By catalytic hydrogenation of diacetone alcohol). **1 ppm = 4.91 mg/m³**	**PEL (ceiling):** 25 ppm 125 mg/m³ **STEL:** Not Applicable	**REL (ceiling):** 25 ppm 125 mg/m³ **STEL:** Not Applicable	**Not Determined**	**TLV (ceiling):** 25 ppm 121 mg/m³ **STEL:** Not Applicable

SECTION III - PHYSICAL AND CHEMICAL CHARACTERISTICS

Boiling Point	**385°F (196°C)**	Specific Gravity (H_2O = 1)	**0.92**
Vapor Pressure (gas)	**0.05 at 69°F (20°C)**	Molecular Weight	**118.2**
Vapor Density (Air = 1)	**4**	Freezing Point	**-58°F (-50°C) sets to glass**

Solubility

Miscible in water. Soluble in ether, lower aliphatic hydrocarbons, and fatty acids.

Appearance and Odor

Clear, colorless, liquid with a mild, sweet odor. Hygroscopic (absorbs moisture from the air).

SECTION IV - FIRE AND EXPLOSION HAZARD DATA

Flash Point (method used)	**215°F (102°C) opened cup**	Explosive Limits in Air % by Volume	**LEL: 1.3% (calculated) UEL: 15.3% (estimated)**
NFPA Classification	**Class IIIB Combustible Liquid**	Autoignition Temperature	**Not Found**

Extinguishing Media

Use dry chemical, carbon dioxide, or alcohol foam (water may cause frothing).

Special Fire Fighting Procedures

Wear full protective clothing and self-contained breathing apparatus (SCBA). Poisonous gases are produced in fire. Use water spray to keep fire-exposed containers cool. Containers may explode in fire.

Unusual Fire and Explosion Hazards

Dangerous fire hazard when exposed to heat. Chemical will react on contact with oxidizable materials. Moderate explosion hazard when in the form of vapor. Vapors are heavier than air.

SECTION V - REACTIVITY DATA

Stability		Conditions to Avoid: Hexylene glycol is stable under normal conditions. Keep away from sources of heat or flame. Keep away from incompatible materials.
Stable	Unstable	Incompatibility *(materials to avoid)*: Strong oxidizers, such as potassium permanganate, chromium trioxide, sodium peroxide, nitrates, chlorates, and perchlorates. Hygroscopic (i.e., absorbs moisture from air).
X		
Hazardous Polymerization		Conditions to Avoid: Under normal conditions of temperature and pressure, hazardous polymerization of hexylene glycol is not expected to occur.
May Occur	Will Not Occur	Hazardous Decomposition or By-products: When heated to decomposition, hexylene glycol can emit highly toxic/poisonous and acrid fumes and gases, including carbon dioxide and carbon monoxide.
	X	

SECTION VI - HEALTH HAZARD DATA

Primary Route(s) of Entry:	Inhalation?	Absorption (skin)?	Ingestion?
	X		X

Health Hazards

INHALATION: Symptoms include lachrymation, irritation of the throat and upper respiratory tract, general anesthesia, headache, burning cough, respiratory stimulation, nausea or vomiting, liver, kidney, or pulmonary changes (edema), and coma (especially if vapors are heated).

SKIN & EYES: Contact with broken skin will facilitate absorption. Irritation to the eyes and skin.

INGESTION: Same symptoms as inhalation, only more severe. May also cause CNS depression with excitement, nausea, vomiting, drowsiness, coma, respiratory or kidney failure, death.

Carcinogenicity	NTP Listed?	IARC Cancer Review Group?	OSHA Regulated?	Target Organs?
No Evidence Human No Evidence Animal	No	No	29 CFR 1910.1000 Table Z-1-A	Resp. system, eyes, skin, liver, kidneys, CNS, CVS.

Medical Conditions Generally Aggravated by Exposure

Any liver, kidney, cardiovascular, or nervous system disorders may be affected.

Emergency and First-aid Procedures

Eye contact: Flush immediately with water for 15 minutes (minimum), seek medical attention. Skin contact: Remove all contaminated clothing. Immediately wash area with large amounts of soap and water. Seek medical assistance. For inhalation: Remove the person from exposure. Provide respiratory assistance and CPR, if required. Transfer to medical facility. Observe for 48 hours for lung effects (pulmonary edema). If swallowed: Seek medical attention immediately. Unless told otherwise, give a conscious victim 1-2 glasses of water. Do NOT induce vomiting (aspiration hazard).

SECTION VII - PRECAUTIONS FOR SAFE HANDLING AND USE

Steps to be Taken in Case Material is Released or Spilled

Absorb spills in vermiculite or similar material and deposit in sealed containers. Ventilate area of spill or leak. Restrict those not involved in cleanup from entering area. Remove all sources of ignition.

Preferred Waste Disposal Method

Burn in chemical incinerator equipped with an afterburner and scrubber.

Precautions to be Taken in Handling and Storage

Store to avoid contact with incompatible materials in tightly closed containers in cool, dry, well-ventilated area away from heat and flame. Use only non-sparking tools to open and close containers.

Other Precautions and Warnings

Sources of ignition are prohibited where hexylene glycol is used, handled, or stored. Heat may cause containers to build up pressure and explode. Containers must be grounded and bonded during transfer.

SECTION VIII - CONTROL MEASURES AND PERSONAL PROTECTIVE EQUIPMENT

Respiratory Protection *(specify type)*

For low exposures, uses a chemical cartridge respirator. Otherwise, use a MSHA/NIOSH-approved supplied-air respirator with full facepiece, hood, or helmet in continuous flow mode, or use a self-contained breathing apparatus (SCBA) operated in pressure demand or other positive pressure mode.

Ventilation

Local exhaust or general mechanical systems recommended.

Protective Gloves	Eye Protection	Other Protective Clothing
Impervious Material	Chemical Goggles or Face Mask	Protective Apron or Suit

Work/Hygiene Practices

Always wash hands thoroughly after using chemical; never bring food, drink, or smoking materials into vicinity of chemicals.

HEXYLENE GLYCOL

$(CH_3)_2COHCH_2CHOHCH_3$

CAS: 107-41-5

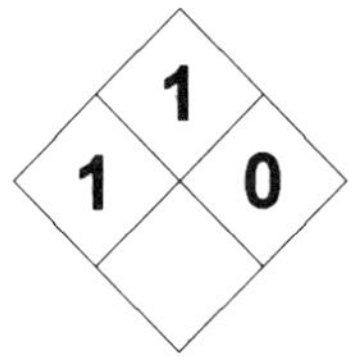

IDENTIFICATION AND TYPICAL USES

Hexylene glycol is a clear, colorless, hygroscopic, liquid with a sweet odor. It is used as a chemical intermediate, a selective solvent in petroleum refining, in hydraulic brake fluids, in printing inks, as a coupling agent and penetrant for textiles, a fuel and lubricant additive, and emulsifying agent, an inhibitor of ice formation in carburetors, and as a component in some cosmetics and pharmaceuticals.

RISK ASSESSMENT: HEALTH

General Assessment

Hexylene glycol is a dangerous human poison by *ingestion* and is moderately toxic by *inhalation*. It will also cause skin irritation and may easily pass through broken skin (*absorption*); however reports of toxic systemic effects by this route of exposure are lacking in the references. There are no reports in the references to indicate that exposure causes cancer in humans or animals. However, mutation data (genetic changes) have been reported and many scientists believe that such chemicals may pose a carcinogenic threat in the long-term. Data are conflicting and inconsistent in this regard.

Inhalation is rare because of this chemical's low vapor pressure. However, when it is heated or misted, it presents a significant heath hazard risk. There will be some initial irritation of the eyes, nose, and throat followed by lachrymation, feeling of intoxication or "drunkenness" (inebriation), burning cough, headache, nystagmus (rapid eye movements), and coma. Other symptoms include urine abnormalities, delayed onset of respiratory deterioration with pulmonary edema (fluid buildup in the lungs), heart enlargement, with onset of kidney failure. Symptoms of pulmonary edema may be delayed up to 48 hours, thereby creating a false sense of security with regard to health exposure risk.

Skin and eye contact may produce mild to moderate irritation on contact and may cause an allergic reaction and rash. It will pass through broken skin, but its ability to produce toxic systemic effects by this route has not been confirmed.

On ingestion, hexylene glycol poses a serious poison threat. Initial symptoms are stimulation or excitement, followed by depression of the central nervous system (CNS). Symptoms can progress to include cardiovascular and pulmonary system affects, mild hypertension, tachycardia (rapid heartbeat), tachypnea (rapid respiration), and systemic capillary damage. Finally, death may occur due to kidney complications leading to necrosis and renal failure (24-72 hours post-ingestion).

Acute Health Effects

The following acute (short-term) health effects may occur immediately or shortly after exposure to hexylene glycol:

Skin: Some irritation at site of contact. Allergic rash may develop. Absorption is possible when there is broken skin.

Eye: Mild to moderate irritation (liquid only).

Lung: Nose and throat irritation. Produces intoxicating effects and may cause lung effects, including pulmonary edema.

CNS: Stimulation and depression with symptoms of dizziness, headache, lightheadedness, nausea, vomiting, and coma.

Chronic Health Effects

The following chronic (long-term) health effects may occur at some time after exposure to hexylene glycol and can last for months or even years:

Cancer Hazards: According to information presented in the references, hexylene glycol has not been adequately tested for its ability to cause cancer in animals. Some studies indicate that it causes mutations in humans, while others are inconclusive in this regard.

Reproductive Hazard: According to information presented in the references, hexylene glycol has not been adequately tested for its ability to adversely affect reproduction.

Other Chronic Effects: Repeated or high exposures may cause kidney damage. Skin allergy can develop in the long-term. If this happens, even small future exposures can cause recurrence of symptoms (rash and itching). Cardiovascular effects include heart enlargement.

Recommended Risk-Reduction Measures

Personnel should avoid direct contact with hexylene glycol. This chemical is a dangerous human poison by two exposure routes (inhalation and ingestion). If a less toxic chemical cannot be substituted for a hazardous substance, then *engineering controls* are the most effective method of reducing exposures. The best protection is to enclose operations and/or provide local exhaust ventilation at the site of chemical release. While not always operationally feasible, isolating operations can also reduce exposure. Using respiratory protection is less effective than the controls mentioned above, but is still advisable whenever working with or around hexylene glycol. Chemical cartridge respirators may provide adequate protection up to the PEL. However, the best protection is provided by a supplied-air respirator with full facepiece operated in continuous flow mode, or a self-contained breathing apparatus (SCBA) with full facepiece operated in pressure demand or other positive pressure mode. If a full facepiece is not available, then chemical goggles should be worn to protect the eyes. Whenever a chemical splash hazard exists, a face shield and a rubber apron should be worn. To prevent hand and skin exposures, impervious gloves should be used. Glove manufacturers should be contacted and permeation studies obtained *before* final glove selections have been made.

Administrative controls should also be in place to minimize the potential for human exposures. These may include written procedures or policies which specify the methods and techniques that will be practiced whenever personnel are to work with hexylene glycol.

All personnel should receive *training* on the use, hazards, protective measures, emergency actions, and other precautions per 29 CFR 1910.1200 (Hazard Communication), prior to the first assignment in an area where hexylene glycol is used or stored.

If symptoms develop or overexposure is suspected, the following tests are recommended:

☑ Kidney function tests.

☑ Urine oxalate level. Urinalysis may show low specific gravity, proteinuria, pyuria, cylinduria, hematuria, calcium oxide, and hippuric crystals.

☑ Lung function tests.

☑ Consider chest X-ray following acute overexposure (may be negative if taken immediately after exposure due to delayed development of pulmonary edema).

Any evaluation should include a careful history of past and present symptoms with an examination. Medical tests that simply look for existing damage are not a substitute for controlling exposures. Medical management of acute exposures may include determination of anion gap and monitoring of heart rhythm, respiratory status, arterial blood gases, and both acute and delayed hepatic and renal function status.

Other methods to reduce exposure include:

☑ Where possible, enclose operations and use local exhaust ventilation at the site of chemical release. If local exhaust ventilation or enclosure is not used, respiratory protection should be mandatory.

☑ Where possible, automatically pump liquid hexylene glycol from drums to process containers.

☑ Always ensure that proper protective clothing is worn when using chemical substances and that personnel are trained in its use, care, and maintenance.

☑ Wash thoroughly immediately after exposure to hexylene glycol and at the end of the work shift or before eating, drinking, or smoking.

☑ Hazard warning information should be posted in the work area. In addition, as part of an on-going education and training program, all information on the health and safety hazards of hexylene glycol should be communicated to all exposed workers.

☑ Eye wash stations should be provided in the immediate work area for emergency use. If there is a possibility of skin exposure to hexylene glycol, emergency shower facilities should also be provided in the immediate work area.

☑ Personnel should never wear contaminated clothing home (family members can be exposed). Clothing should be laundered only by personnel who have been properly trained on the hazards of exposure to hexylene glycol.

RISK ASSESSMENT: ENVIRONMENT

General Assessment

The environment is at risk of exposure during transportation, storage, disposal, or destruction of hexylene glycol. In almost every scenario, the threat of environmental exposure is contingent upon the proper handling of the chemical substance. Accidental spills, large or small, can result in fire, explosion, and possible contamination of the surrounding environmental mediums (water, soil, and air).

Hexylene glycol is considered a Class IIIB combustible liquid (per OSHA 29 CFR 1910.106). Under normal operating conditions, it presents a mild fire hazard. However, when heated or misted, it becomes a serious fire and explosion hazard concern. These characteristics require extreme caution in handling, storage, transportation, and disposal. It is incompatible with many common peroxides, other oxidizers, and strong acids and contact can cause fire or explosion. Therefore, special consideration is required during any emergency situation involving a leak or spill of hexylene glycol. Should hexylene glycol ever come into contact with these incompatible substances during use, transportation, storage, or disposal, the formation of highly toxic and/or highly explosive commodities is extremely possible.

The proper disposal method for hexylene glycol is to burning it in a chemical incinerator equipped with an afterburner and scrubber.

Hexylene glycol may enter the environment through industrial discharges, municipal waste treatment plant discharges, agricultural runoff or disposal, or spills.

Acute Ecological Effects

Acute (short-term) toxic effects may include the death of animals, birds, or fish and death or low growth rate in plants. Acute effects are seen 2 to 4 days after animals or plants are exposed to hexylene glycol.

Hexylene glycol has moderate acute toxicity to aquatic life. Insufficient data are available on the short-term effects of hexylene glycol exposure to plants, birds or land animals.

Chronic Ecological Effects

Chronic (long-term) toxic effects may include shortened life span, reproductive problems, lower fertility, and changes in appearance or behavior in exposed animals. These effects can be seen long after first exposure(s) to toxic chemicals.

Hexylene glycol has moderate chronic toxicity to aquatic life. Insufficient data are available on the long-term effects of hexylene glycol exposure on birds, plants, or land animals.

Water Solubility

Hexylene glycol is highly soluble in water. Concentrations of 1000 milligrams and more will mix with a liter of water.

Persistence in the Environment

Hexylene glycol is slightly persistent in water, with a half-life between 2 and 20 days. The half-life of a pollutant is the amount of time it takes for one half of the chemical to be degraded. Virtually 100% of hexylene glycol will eventually end up in water.

Bioaccumulation in Aquatic Organisms

Some substances increase in concentration, or *bioaccumulate,* in living organisms as they breathe contaminated air, drink contaminated water, or eat contaminated food. These chemicals can become concentrated in the tissues and internal organs of animals as well as humans.

The concentration of hexylene glycol found in fish tissue is expected to be about the same as the average concentration of hexylene glycol in the water from which the fish was taken.

Recommended Risk-Reduction Measures

Proper training of all transporters will reduce the likelihood of a mishap or accident resulting in a leak or spill to the environment. The correct use of DOT labeling (per the MSDS) on all containers, trucks, and rail cars will enable emergency responders to react properly and quickly to any disaster thereby reducing the potential risk to the environment and to personnel.

Storage of hexylene glycol should be segregated from incompatible chemicals to minimize the risk of cross-contamination or contact. Buildings designated for storage should be equipped with appropriate fire protection systems (alarms, sprinklers, emergency lighting, portable extinguishers).

If a spill or leak to the environment has occurred, fire department, emergency response and/or hazardous materials spill personnel should be notified immediately. Cleanup should be attempted only by those trained in proper spill containment procedures using non-sparking tools. Contaminated soils should be removed for incineration and replaced with clean soil. If hexylene glycol should contact the water table, aquifer, or navigable waterway, cleanup should be started immediately. It is highly soluble in water and total remediation may not be possible. The local and/or state emergency response authorities must be notified. A comprehensive emergency response or disaster preparedness/recovery plan should be in place prior to any operations involving the use, transportation, storage, or disposal of hexylene glycol.

If hexylene glycol is spilled or leaked, the following specific steps are recommended:

☑ Restrict persons not wearing protective clothing from area of spill or leak until cleanup is complete and area can be opened for normal work.

☑ Remove all ignition sources.

☑ Ventilate area of spill or leak.

☑ Absorb liquids in vermiculite, dry sand, earth, or similar material and deposit in sealed containers using non-sparking tools.

☑ It may be necessary to dispose of hexylene glycol as a hazardous waste. The responsible state agency or the regional office of the federal Environmental Protection Agency (EPA) should be contacted for specific recommendations.

RISK ASSESSMENT: BUSINESS

General Assessment

Accidents or mishaps involving hexylene glycol can present a significant threat to business operations. The loss or damage of equipment or facilities can significantly affect fiscal viability. Lawsuits that may result from personnel injury/death, public exposures, and/or environmental contamination will also require a serious expenditure of resources. Media attention surrounding an injury, death, or environmental damage can also result in a loss of profits and loss of current as well as future business.

◗ *Recommended Risk-Reduction Measures*

Company attorneys, safety and health professionals, and environmental specialists should be involved in the development of any procedures dealing with safety or emergency response. A company official should be pre-designated as a public relations officer with specific training in dealing with the press. Corporate plans and policies should be developed, approved, and implemented long before any need for such arises.

REFERENCES

American Conference of Governmental Industrial Hygienists. 1988. *Documentation of the Threshold Limit Values and Biological Exposure Indices,* 5th Edition (with updates). Cincinnati: ACGIH

Bueche, J. R. 1972. *Principles of Physics,* 2nd Edition. New York: McGraw-Hill.

Calabrese, E. J. 1994. *Biological Effects of Low Level Exposures.* New York: CRC Press/Lewis Publishers.

Gardner, R. W. 1994. *Chemical Intolerance.* New York: CRC Press/Lewis Publishers.

Hallenbeck, W. H. 1993. *Quantitative Risk Assessment for Environmental and Occupational Health,* 2nd Edition. New York: CRC Press/Lewis Publishers.

Keith, L. H. and M. M. Walker. 1995. *Handbook of Air Toxics.* New York: CRC Press/Lewis Publishers.

Lauwerys, R. R., P. Hoet. 1993. *Industrial Chemical Exposure.* New York: CRC Press/Lewis Publishers.

Lewis, R. J., Sr. 1992. *Sax's Dangerous Properties of Industrial Materials,* Volumes 1, 2, and 3, 8th Edition. New York: Van Nostrand Reinhold.

Manahan, S. E. 1994. *Environmental Chemistry,* 6th Edition. New York: CRC Press/Lewis Publishers.

Meyer, E. 1990. *Chemistry of Hazardous Materials.* Englewood Cliffs, NJ: Prentice Hall.

Patnaik, P. 1992. *A Comprehensive Guide to the Hazardous Properties of Chemical Substances.* New York: Van Nostrand Reinhold.

Rea, W. J. 1995. *Chemical Sensitivity,* Vol. I-III. New York: CRC Press/Lewis Publishers.

Richardson, M. 1995. *Dictionary of Substance and Their Effects,* Vol. I-VI. New York: CRC Press/Lewis Publishers.

Sacarello, H. L. A. 1994. *The Comprehensive Handbook of Hazardous Materials.* New York: CRC Press/Lewis Publishers.

Torkelson, T. R. V. K. Rowe. 1981. Halogenated aliphatic hydrocarbons. *Patty's Industrial Hygiene and Toxicology,* Vol. 2B, *Toxicology,* pp 3568-3572. New York: Wiley-Interscience.

U.S. Department of Health and Human Services, National Institute for Occupational Safety and Health. 1994. *NIOSH Pocket Guide to Chemical Hazards.* Washington, D.C.: U.S. Government Printing Office.

MATERIAL SAFETY DATA SHEET

CHEMICAL NAME	
HYDRAZINE	**CRC PRESS/LEWIS PUBLISHERS DISCLAIMER:** THE INFORMATION AND RECOMMENDATIONS PRESENTED HEREIN ARE BASED ON SOURCES BELIEVED TO BE RELIABLE. CRC MAKES NO REPRESENTATION ON ITS COMPLETENESS OR ACCURACY. IT IS THE RESPONSIBILITY OF THE USER TO DETERMINE THE CHEMICAL'S SUITABILITY FOR ITS INTENDED USE, THE CHEMICAL'S SAFE USE, AND THE CHEMICAL'S PROPER DISPOSAL. NO REPRESENTATIONS AND/OR WARRANTIES, EITHER EXPRESSED OR IMPLIED, OF THE MERCHANTABILITY OR FITNESS FOR A PARTICULAR PURPOSE, OR OF ANY OTHER NATURE, ARE MADE WITH RESPECT TO THE INFORMATION PROVIDED IN THIS MSDS OR TO THE CHEMICAL TO WHICH INFORMATION MAY REFER. CRC NEITHER ASSUMES NOR AUTHORIZES ANY OTHER PERSON TO ASSUME FOR IT, ANY OTHER ADDITIONAL RESPONSIBILITY OR LIABILITY FOR THE USE OF, OR RELIANCE UPON, THIS INFORMATION.

HAZARD WARNING INFORMATION

HEALTH	FIRE	REACTIVE	OTHER	DEGREE OF HAZARD	COLOR CODING	OTHER CODES
3	3	2		0 = Minimum Hazard 1 = Slight Hazard 2 = Moderate Hazard 3 = Serious Hazard 4 = Severe Hazard	HEALTH = BLUE FIRE = RED REACTIVITY = YELLOW OTHER = WHITE	OX = Oxidizer ACID = Acid ALK = Alkali COR = Corrosive W̶ = Use No Water

SECTION I - GENERAL INFORMATION

Characterization	**Hypergolic Compound**	RCRA Number	**U133**
		EPA Class	**Toxic Waste**
DOT Proper Shipping Name	**Hydrazine (anhydrous) or Hydrazine (aqueous)**	Chemical Abstract Service (CAS) Number	**302-01-2**
DOT Hazard Class and Label Requirements	**Flammable Liquid and Poison; Corrosive**	DOT Emergency Guide Code	**28 (anhydrous) 59 (aqueous)**
DOT Identification Number	**UN 2029 (anhydrous) UN 2030 (aqueous hydrate)**	Chemical Formula	**N_2H_4**

Synonyms

Diamine; hydrazine, anhydrous; hydrazine base; diamide.

SECTION II - HAZARDOUS INGREDIENTS AND IDENTITY INFORMATION

Hazardous Components (specific identity)	OSHA Exposure Criteria	NIOSH Exposure Criteria	Immediately Dangerous to Life and Health (IDLH)	ACGIH Exposure Criteria
Hydrazine (derivation: Reaction of sodium hypochlorite and ammonia yielding chloramine and sodium hydroxide; reaction of chloramine, ammonia, and sodium hydroxide to yield hydrazine, sodium chloride, and water in the presence of colloidal materials (glue, starch) to prevent side reactions). **1 ppm = 1.33 mg/m^3**	**PEL (skin):** **1 ppm** **1.3 mg/m^3** **STEL:** **Not Applicable**	**REL (ceiling):** **0.06 ppm** **0.04 mg/m^3** **(2-hr ceiling)** **Possible Cancer Agent**	**50 ppm**	**TLV (skin):** **0.1 ppm** **0.13 mg/m^3** **Suspected Human Carcinogen**

SECTION III - PHYSICAL AND CHEMICAL CHARACTERISTICS

Boiling Point	**236°F (113°C)**	Specific Gravity (H_2O = 1)	**1.01**
Vapor Pressure (mm Hg)	**10 mm at 69°F (20°C)**	Molecular Weight	**32.1**
Vapor Density (Air = 1)	**1.1**	Melting Point	**36°F (2°C)**

Solubility

Miscible in water and alcohol. Insoluble in chloroform and ether.

Appearance and Odor

Colorless, fuming, oily liquid with an ammonia or fish-like odor. A solid below 36°F (2°C).

SECTION IV - FIRE AND EXPLOSION HAZARD DATA

Flash Point (method used)	**100°F (38°C) opened cup**	Explosive Limits in Air % by Volume	**LEL: 2.9% UEL: 98%**
NFPA Classification	**Class IC Flammable Liquid**	Autoignition Temperature	**Varies with contact surfaces**

Extinguishing Media

Use water spray in flooding amounts as fog, or use alcohol foam, carbon dioxide, or dry chemical.

Special Fire Fighting Procedures

Structural protective clothing is permeable. Remain clear of smoke, water fallout, and water runoff. Poisonous gases are produced in fire. Move containers if it can be done without risk. Cool fire-exposed containers with water spray and use water spray to disperse vapors. Evacuate non-essential personnel 2500 feet from the fire area (consider wind conditions).

Unusual Fire and Explosion Hazards

Contact with oxidizing agents can cause violent and explosive ignition. A volatile liquid that is flammable over a wide vapor-to-air range. Vapors are heavier than air and may travel a considerable distance to a source of ignition and flashback to cause fire or explosion. Burns with clear flame.

SECTION V - REACTIVITY DATA

Stability		Conditions to Avoid **Hydrazine is stable at room temperature in closed containers under normal conditions of storage and handling. Avoid contact with incompatible chemicals and materials and heat.**
Stable X	Unstable	Incompatibility *(materials to avoid)* **Highly reactive with oxidizers (chlorine, bromine, nitrogen dioxide), halogens, ferric oxide (iron rust), metallic mercury, fuming nitric acid, hydrogen peroxide (ignites spontaneously with oxidizers).**
Hazardous Polymerization		Conditions to Avoid **Under normal conditions of temperature and pressure, hazardous polymerization of hydrazine is not expected to occur.**
May Occur	Will Not Occur X	Hazardous Decomposition or By-products **When heated to decomposition, hydrazine will produce toxic and irritating gases, including ammonia and oxides of nitrogen.**

SECTION VI - HEALTH HAZARD DATA

Primary Route(s) of Entry:	Inhalation?	Absorption (skin)?	Ingestion?
	X	X	X

Health Hazards

INHALATION: Highly corrosive to the skin, eyes, and mucosa. Causes choking, chest pain, dyspnea, lethargy, nausea, convulsions, tremors, pulmonary edema, liver injury, and hypoxia.

ABSORPTION: Will pass through unbroken skin to cause toxic systemic effects as noted above. Extremely corrosive to skin tissue causing chemical burns (2nd and 3rd degree) on short contact. Very injurious to the eyes, causing corneal damage and permanent blindness.

INGESTION: A poison by ingestion. Delayed gastrointestinal irritation. Systemic effects can occur.

Carcinogenicity	NTP Listed?	IARC Cancer Review Group?	OSHA Regulated?	Target Organs?
Suspected Human Confirmed Animal	5th Annual Report	Group 2B	29 CFR 1910.1000 Table Z-1	CNS, skin, liver, eyes, GI tract, blood, resp. system.

Medical Conditions Generally Aggravated by Exposure

Existing deficiencies in CNS function, liver, kidneys, and the circulatory system may be aggravated.

Emergency and First-aid Procedures

Eye contact: Flush immediately with water for 15 minutes (minimum) while holding upper and lower eyelids open. Seek medical attention. Skin contact: Remove all contaminated clothing. Immediately wash area with large amounts of water (15 minutes minimum). For inhalation: Remove the person from exposure. Provide respiratory assistance and CPR. Transfer to medical facility. Observe for 24-48 hours for lung effects (pulmonary edema). If swallowed: Call the poison control center. Give 1-2 glasses of water or milk to a conscious and alert person and induce vomiting. Seek medical attention immediately. Never attempt to give anything by mouth to an unconscious or convulsing person.

SECTION VII - PRECAUTIONS FOR SAFE HANDLING AND USE

Steps to be Taken in Case Material is Released or Spilled

Evacuate area and deny entry to those not involved in cleanup activities. Protective clothing is mandatory. Remove all sources of ignition. Ventilate area of spill. Dilute with water to less than 40% hydrazine solution. Absorb liquids with toweling, vermiculite, or sand. Deposit in sealed drum for disposal.

Preferred Waste Disposal Method

Incineration in a chemical incinerator equipped with afterburner and scrubber.

Precautions to be Taken in Handling and Storage

Store in tightly closed containers in a cool, dark, dry, well-ventilated area. Personnel should be trained on the hazardous properties of hydrazine *prior* to working with the material.

Other Precautions and Warnings

Containers should be protected from physical damage. Use extreme caution when handling this compound. Use explosion-proof electrical equipment in areas where this chemical is used or stored.

SECTION VIII - CONTROL MEASURES AND PERSONAL PROTECTIVE EQUIPMENT

Respiratory Protection *(specify type)*

A carcinogenic agent. Exposure level is extremely low and difficult to monitor. For any exposures, use a supplied-air respirator set in positive pressure or continuous flow mode or a self-contained breathing apparatus (SCBA) with full facepiece operated in pressure demand or other positive pressure mode.

Ventilation

Local exhaust or general mechanical systems recommended.

Protective Gloves	Eye Protection	Other Protective Clothing
Impervious Gloves	**Chemical Goggles or Face Mask**	**Full Protective Clothing or Suit**

Work/Hygiene Practices

Always wash hands thoroughly after using chemical; never bring food, drink, or smoking materials into vicinity of chemicals.

HYDRAZINE

N_2H_4

CAS: 302-01-2

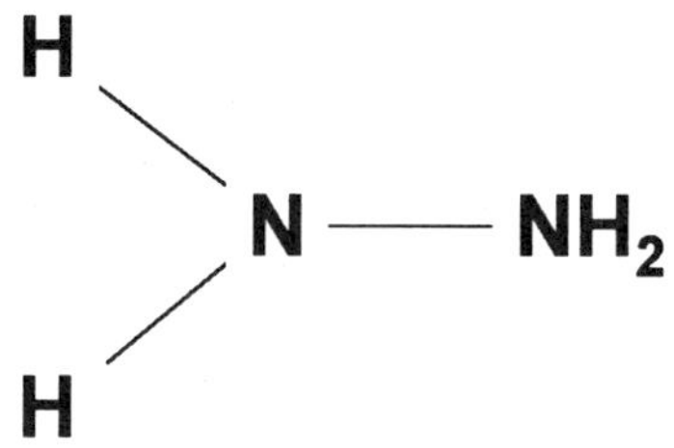

IDENTIFICATION AND TYPICAL USES

Hydrazine is a colorless, fuming, oily liquid with an ammonia or fish-like odor. It is used as a reducing agent for many transition metals and some nonmetals (arsenic, selenium, tellurium), as well as uranium and plutonium. It is a corrosion inhibitor in boiler feed water and reactor cooling water. It is used in wastewater treatment, in electroplating of metals on glass and plastics, in nuclear fuel processing, in redox reactions, as a polymerization catalyst, in fuel cells, as a blowing agent, a scavenger for gases, in the manufacture of certain drugs and agricultural chemicals (maleic hydrazide); as a photographic developer, as a component in high energy fuel cells, in dyes, in explosives, in the metallurgy industry, in spandex fibers, in diving equipment, in pesticides, in solder fluxes, and as a hypergolic rocket propellant.

RISK ASSESSMENT: HEALTH

General Assessment

Hydrazine is a suspected human and animal carcinogen. It is an experimental teratogen and human mutation data have also been reported. It is primarily toxic by *inhalation* and skin contact (*absorption*). It can also cause toxic and damaging effects by *ingestion*.

Inhalation will cause pulmonary irritation up to and including pulmonary edema (fluid buildup in the lungs), which is a medical emergency and can lead to death. Symptoms may be delayed up to 48 hours creating a false sense of security with regard to health exposure risk. Symptoms include choking, extreme chest pain, dyspnea (difficulty in breathing), lethargy, nausea, convulsions, tremors, liver injury, and hypoxia (reduced oxygen levels in the blood) leading to hemolytic anemia. Systemic intoxication, depression of the central nervous system (CNS), and delayed gastrointestinal irritation may also occur. Hydrazine may cause damage to the liver and destruction of red blood cells.

Aside from its ability to pass through intact skin (absorption), hydrazine is also extremely corrosive to skin tissue and can cause 2nd and 3rd degree burns on contact. Damage to the eyes and cornea can be severe causing partial or complete loss of vision. It is a powerful reducing agent which is corrosive to the eyes, skin, and mucous membranes. May also cause skin sensitization as well as systemic poisoning.

Ingestion is very unusual since the vapors of this chemical are so irritating, personnel are not likely to get close enough to ingest it. However, should it be swallowed, there can be burning of the mouth, throat, and esophagus, followed by gastrointestinal irritation.

Acute Health Effects

The following acute (short-term) health effects may occur immediately or shortly after exposure to hydrazine:

Skin: Severe irritation and burning. Extremely corrosive to tissues. Skin absorption will occur causing toxic systemic effects.

Eye: Severe irritation and tissue damage up to and including loss of vision.

Lung: Irritation of the upper respiratory tract causing choking, coughing, dypsnea, and pulmonary irritation leading to pulmonary edema (fluid in the lungs), which can be fatal.

CNS: Dizziness, convulsions, tremors, nausea, headaches, possible injury to the nervous system.

Chronic Health Effects

The following chronic (long-term) health effects may occur at some time after exposure to hydrazine and can last for months or even years:

Cancer Hazards: According to the information presented in the references, hydrazine has been shown to cause cancer in test animals. Its carcinogenicity in humans is expected to be at least equal to that of animals. Human mutation data have been reported for this compound.

Reproductive Hazard: According to the information presented in the references, this chemical has been shown to cause teratogenic effects as well as adversely affect reproduction in test animals.

Other Chronic Effects: Repeated exposures can lead to hemolytic anemia and convulsions. Chronic exposures may lead to cancer. There can also be chronic inflammation of the nasal, tracheal, and bronchial tissue resulting in chronic bronchitis. Repeated skin exposure can cause dermatitis with a characteristic rash.

Recommended Risk-Reduction Measures

Hydrazine is listed by the IARC as a confirmed animal carcinogen and the ACGIH and NTP as a suspected human carcinogen. Unless a less toxic chemical can be substituted for a hazardous substance, *engineering controls* are the most effective method of reducing exposures. The best protection is to enclose operations and/or provide local exhaust ventilation at the site of chemical release. While not always operationally feasible, isolating operations can also reduce exposure.

Using respiratory protection is less effective than the controls mentioned above, but is still advisable whenever working with or around hydrazine. The exposure level is extremely low (1 ppm) and difficult to detect and monitor. For any detectable exposures to a suspected carcinogenic agent, use a supplied-air respirator set in positive pressure or continuous flow mode, or use a self-contained breathing apparatus (SCBA) with full facepiece operated in pressure demand or other positive pressure mode. If a full facepiece is not available, then chemical goggles should be worn to protect the eyes. Whenever an eye or face contact hazard exists, a face shield and apron should be worn. To prevent hand and skin exposures, impervious gloves should be used. Glove manufacturers should be contacted and permeation studies obtained *before* final glove selection has been made.

Administrative controls should also be in place to minimize the potential for human exposures. These may include written procedures or policies which specify the methods and techniques as well as the exposure control measures that will be practiced whenever personnel are to work with or around hydrazine.

All personnel should receive *training* on the use, hazards, protective measures, emergency actions, and other precautions per 29 CFR 1910.1200 (Hazard Communication), prior to the first assignment in an area where hydrazine is used or stored.

Before beginning employment and at regular intervals thereafter (e.g. annually), the following medical tests are recommended:

- ☑ Lung function tests (establish a baseline).
- ☑ Complete blood count (CBC).
- ☑ Examination of the nervous system.

If symptoms develop or overexposure is suspected, the following medical tests may be useful:

- ☑ Lung function tests.
- ☑ Liver and kidney function tests.
- ☑ Consider chest X-ray following acute overexposure (may be negative if taken immediately after exposure due to delayed onset of pulmonary edema).
- ☑ Evaluation by a qualified allergist with careful consideration of exposure history and special tests (may help diagnose allergy).

Any evaluation should include a careful medical history of past and present symptoms with an examination. However, medical tests that evaluate existing damage are not a substitute for controlling exposure. Since smoking can cause heart disease, emphysema, and numerous other respiratory disorders, smokers may be affected more rapidly than non-smokers under the same exposure conditions. Also, because alcohol consumption can cause liver damage, drinking alcohol can increase the potential for liver damage caused by exposure to hydrazine. Prudent risk assessment and management requires careful consideration of *all* possible risk factors that may be causing the appearance of symptoms in exposed workers.

Other methods to reduce exposure include:

- ☑ Where possible, enclose operations and use local exhaust ventilation at the site of chemical release. If local exhaust ventilation or enclosure is not used, respiratory protection should be mandatory.
- ☑ Always ensure that proper protective clothing is worn when using chemical substances.
- ☑ Wash thoroughly immediately after exposure to hydrazine and at the end of the work shift or before eating, drinking, or smoking.

- ☑ Hazard warning information must be posted in the work area. It is recommended that a regulated, controlled work area should be established wherever this chemical is used, handled, or stored. In addition, as part of an on-going education and training program, all information on the health and safety hazards of hydrazine should be communicated to all potentially exposed workers.
- ☑ Eye wash stations should be provided in work areas. If the potential for whole body exposure exists, then safety showers should also be provided in the immediate area.
- ☑ Specific engineering controls are recommended for this chemical by NIOSH in Publication Number 78-172 (*Occupational Exposure to Hydrazine*).

RISK ASSESSMENT: ENVIRONMENT

General Assessment

The environment is at risk of exposure during intended use, transportation, storage, disposal, or destruction of hydrazine. In almost every scenario, the threat of environmental exposure is contingent upon the proper handling of the chemical substance. Accidental spills, large or small, can result in fire, explosion, and possible contamination of the surrounding environmental mediums (water, soil, and air).

Hydrazine should always be handled with extreme caution. It is a Class 1C flammable liquid. Its relatively low flash point and moderate boiling point make this chemical a fire and explosion hazard. Its vapors form explosive mixtures in air over a wide range of concentrations. It forms shock-sensitive explosive products when combined with metal perchlorates, alkali metals, or their dichromates. It is highly reactive with a number of substances, especially oxidizers and oxidizing agents. These characteristics require special consideration during any emergency situation involving a leak or spill of hydrazine.

Hydrazine can enter the environment through industrial discharges or through spills.

Acute Ecological Effects

Acute (short-term) toxic effects may include the death of animals, birds, or fish and death or low growth rate in plants. Acute effects are seen 2 to 4 days after animals or plants are exposed to hydrazine.

This chemical has high acute toxicity to aquatic life. Insufficient data are available to evaluate or predict the short-term effects of hydrazine to plant, birds, or land animals.

Chronic Ecological Effects

Chronic toxic effects may include shortened life span, reproductive problems, lower fertility, and changes in appearance or behavior in exposed animals. These effects can be seen long after first exposure(s) to toxic chemicals. Hydrazine has high chronic toxicity to aquatic life. Insufficient data are available to evaluate or predict the long-term effects of hydrazine to plants, birds, or land animals.

Water Solubility

Hydrazine is highly soluble in water. Concentrations of 1000 milligrams and more will mix with a liter of water.

Persistence in the Environment

Hydrazine is slightly persistent in water, with a half-life of between 2 to 20 days. The half-life of a pollutant is the amount of time it takes for one half of the chemical to be degraded. It is expected to absorb to sediment.

Bioaccumulation in Aquatic Organisms

Some substances increase in concentration, or *bioaccumulate,* in living organisms as they breathe contaminated air, drink contaminated water, or eat contaminated food. These chemicals can become concentrated in the tissues and internal organs of animals as well as humans.

Hydrazine is not expected to accumulate in the edible tissues of aquatic species consumed by humans.

Recommended Risk-Reduction Measures

Proper training of all transporters will reduce the likelihood of a mishap or accident resulting in a leak or spill to the environment. The correct use of labeling on all containers, trucks, and rail cars will enable emergency responders to react properly and quickly to any disaster thereby reducing the potential risk to the environment and to personnel.

Storage of hydrazine should be segregated from other chemicals to minimize the risk of cross-contamination or contact. Storage buildings should be equipped with proper fire protection and prevention equipment and explosion-proof electrical components and equipment. Bulk storage of hydrazine is not recommended. Hydrazine is hypergolic (ignites explosively) with incompatible materials such as oxidizers.

If a spill or leak to the environment has occurred, fire department, emergency response and/or hazardous materials spill personnel should be notified immediately. Cleanup should be attempted only by those trained in proper spill containment procedures and the health hazards associated with carcinogenic materials. Contaminated soils should be removed for incineration and replaced with clean soil.

If hydrazine should contact the water table, aquifer, or navigable waterway, remediation activities should be prompt to ensure protection of aquatic sediments. It is highly soluble in water and total containment and remediation may not be possible. When such spills occur, the local and/or state emergency response authorities must be notified. A comprehensive emergency response or disaster preparedness/recovery plan should be in place prior to any operations involving the use, transportation, storage, or disposal of hydrazine. Burning hydrazine in a chemical incinerator equipped with an afterburner and an air scrubber is an acceptable disposal method.

If hydrazine is spilled or leaked, the following specific steps are recommended:

- ☑ Restrict persons not wearing protective clothing from area of spill or leak until cleanup is complete and area can be opened for normal work.
- ☑ Ventilate area and remove all ignition sources.
- ☑ Absorb liquids in vermiculite or sand and place in a sealed drums for disposal.
- ☑ It may be necessary to dispose of hydrazine as a hazardous waste. The responsible state agency or the regional office of the federal Environmental Protection Agency (EPA) should be contacted for specific recommendations.

RISK ASSESSMENT: BUSINESS

General Assessment

Accidents or mishaps involving hydrazine can present a significant threat to business operations. The loss or damage of equipment or facilities can significantly affect fiscal viability. Lawsuits that may result from personnel injury/death, public exposures, and/or environmental contamination will also require a serious expenditure of resources. Media attention surrounding an injury, death, or environmental damage can also result in a loss of profits and loss of current as well as future business. Always remember that anytime the term "carcinogen," "cancer," or "reproductive hazard" are used, public emotion, hysteria, and ignorance can run equally high. This should be considered during the development of any public relations policies.

Recommended Risk-Reduction Measures

Company attorneys, safety and health professionals, and environmental specialists should be involved in the development of any procedures dealing with safety or emergency response. A company official should be pre-designated as a public relations officer with specific training in dealing with the press. Corporate plans and policies should be developed, approved, and implemented long before any need for such arises.

REFERENCES

American Conference of Governmental Industrial Hygienists. 1988. *Documentation of the Threshold Limit Values and Biological Exposure Indices,* 5th Edition (with updates). Cincinnati: ACGIH

Anthony, C. P., and N. J. Kolthoff. 1971. *Textbook of Anatomy and Physiology.* St. Louis: C. V. Mosby Company.

Brusick, D. J. 1994. *Methods for Genetic Risk Assessment.* New York: CRC Press/Lewis Publishers.

Bueche, J. R. 1972. *Principles of Physics,* 2nd Edition. New York: McGraw-Hill.

Burrell, R., D. K. Flaherty, and L. J. Sauers. 1992. *Toxicology of the Immune System.* New York: Van Nostrand Reinhold.

Calabrese, E. J. 1994. *Biological Effects of Low Level Exposures.* New York: CRC Press/Lewis Publishers.

Cothern, C. R. 1995. *Handbook for Environmental Risk Decisionmaking.* New York: CRC Press/Lewis Publishers.

Hodgson, E., R. B. Mailman, and Janice E. Chambers. 1988. *Dictionary of Toxicology.* London: The MacMillian Company.

Lewis, R. J., Sr. 1993. *Hawley's Condensed Chemical Dictionary,* 12th Edition. New York: Van Nostrand Reinhold.

Lewis, R. J., Sr. 1992. *Sax's Dangerous Properties of Industrial Materials,* Volumes 1, 2, and 3, 8th Edition. New York: Van Nostrand Reinhold.

Meyer, E. 1990. *Chemistry of Hazardous Materials.* Englewood Cliffs, NJ: Prentice Hall.

U.S. Department of Health and Human Services, National Institute for Occupational Safety and Health. 1994. *NIOSH Pocket Guide to Chemical Hazards.* Washington, D.C.: U.S. Government Printing Office.

U.S. Department of Health and Human Services, National Institute for Occupational Safety and Health. 1978. *Occupational Exposure Hydrazine.* Publication Number 78-172. Washington, D.C.: U.S. Government Printing Office.

MATERIAL SAFETY DATA SHEET

CHEMICAL NAME	CRC PRESS/LEWIS PUBLISHERS DISCLAIMER
HYDRAZINE SULFATE	

HAZARD WARNING INFORMATION

HEALTH	FIRE	REACTIVE	OTHER	DEGREE OF HAZARD	COLOR CODING	OTHER CODES
3	0	1		0 = Minimum Hazard 1 = Slight Hazard 2 = Moderate Hazard 3 = Serious Hazard 4 = Severe Hazard	HEALTH = BLUE FIRE = RED REACTIVITY = YELLOW OTHER = WHITE	OX = Oxidizer ACID = Acid ALK = Alkali COR = Corrosive W = Use No Water

SECTION I - GENERAL INFORMATION

Characterization	**Sulfate**	RCRA Number	**None**
		EPA Class	**Not Applicable**
DOT Proper Shipping Name	**No Citation**	Chemical Abstract Service (CAS) Number	**10034-93-2**
DOT Hazard Class and Label Requirements	**No Citation**	DOT Emergency Guide Code	**No Citation**
DOT Identification Number	**No Citation**	Chemical Formula	**$H_4N_2 \bullet H_2O_4S$**

Synonyms

Diamine sulfate; diamidogen sulfate; hydrazine hydrogen sulfate; hydrazine monosulfate.

SECTION II - HAZARDOUS INGREDIENTS AND IDENTITY INFORMATION

Hazardous Components (specific identity)	OSHA Exposure Criteria	NIOSH Exposure Criteria	Immediately Dangerous to Life and Health (IDLH)	ACGIH Exposure Criteria
Hydrazine sulfate **(NOTE: NIOSH Exposure Limits established for Hydrazines may be used as references. No specific limits established for hydrazine sulfate)**	**PEL:** Not Established **STEL:** Not Established	**REL (ceiling):** 0.06 ppm 0.04 mg/m^3 (2-hr ceiling) **Possible Cancer Agent**	**Not Determined**	**TLV:** Not Established **STEL:** Not Established

SECTION III - PHYSICAL AND CHEMICAL CHARACTERISTICS

Boiling Point	**Not Found**	Specific Gravity (H_2O = 1)	**1.37**
Vapor Pressure (mm Hg)	**Not Found**	Molecular Weight	**130.14**
Vapor Density (Air = 1)	**Not Found**	Melting Point	**185°F (85°C)**

Solubility

Soluble in water, very soluble in hot water. Insoluble in alcohol.

Appearance and Odor

White or colorless, crystalline (sand-like) powder.

SECTION IV - FIRE AND EXPLOSION HAZARD DATA

Flash Point (method used)	**Not Found**	Explosive Limits in Air % by Volume	**LEL: Not Found UEL: Not Found**
NFPA Classification	**Not Listed**	Autoignition Temperature	**Not Found**

Extinguishing Media

Use high expansion foam, low expansion foam, dry chemical, carbon dioxide, water, water fog, or water spray.

Special Fire Fighting Procedures

Structural protective clothing is permeable. Remain clear of smoke, water fallout, and water runoff. Poisonous gases are produced in fire. Move containers if it can be done without risk. Cool fire-exposed containers with water spray. Evacuate non-essential personnel 2500 feet from the fire area (consider wind conditions).

Unusual Fire and Explosion Hazards

Contact with oxidizing agents can cause violent and explosive ignition.

SECTION V - REACTIVITY DATA

Stability		Conditions to Avoid: Hydrazine sulfate is stable at room temperature in closed containers under normal conditions of storage and handling. Avoid contact with incompatible chemicals and materials and heat.
Stable	Unstable	Incompatibility *(materials to avoid)*: Highly reactive with oxidizers (chlorine, bromine, nitrogen dioxide, peroxides, permanganates, chlorates, and nitrates), and bases.
X		
Hazardous Polymerization		Conditions to Avoid: Under normal conditions of temperature and pressure, hazardous polymerization of hydrazine sulfate is not expected to occur.
May Occur	Will Not Occur	Hazardous Decomposition or By-products: When heated to decomposition, hydrazine sulfate will produce toxic and irritating gases, including oxides of nitrogen and sulfur.
	X	

SECTION VI - HEALTH HAZARD DATA

Primary Route(s) of Entry:	Inhalation?	Absorption (skin)?	Ingestion?
	X	X	X

Health Hazards

INHALATION: Highly irritating to the skin, eyes, and mucosa. Causes choking, chest pain, dyspnea, nausea, convulsions, tremors, blood cell injury, and possible methemoglobinemia. May affect the brain and the function of the central nervous system.

ABSORPTION: Will pass through unbroken skin to cause toxic systemic effects as noted above. Eye contact causes severe pain and possible damage to vision.

INGESTION: A poison by ingestion causing paresthesia, somnolence, nausea, and vomiting.

Carcinogenicity	NTP Listed?	IARC Cancer Review Group?	OSHA Regulated?	Target Organs?
Suspected Human Confirmed Animal	5th Annual Report	Animal Evidence	No	CNS, skin, liver, eyes, GI tract, blood, resp. system.

Medical Conditions Generally Aggravated by Exposure

Existing deficiencies in CNS function, liver, kidneys, and the circulatory system may be aggravated.

Emergency and First-aid Procedures

Eye contact: Flush immediately with water for 15 minutes (minimum) while holding upper and lower eyelids open. Seek medical attention. Skin contact: Remove all contaminated clothing. Immediately wash area with large amounts of water (15 minutes minimum). For inhalation: Remove the person from exposure. Provide respiratory assistance and CPR. Transfer to medical facility. If swallowed: Call the poison control center. Give 1-2 glasses of water or milk to a conscious and alert person to dilute. Do NOT induce vomiting. Seek medical attention immediately. Never attempt to give anything by mouth to an unconscious or convulsing person. Never make an unconscious or convulsing person vomit.

SECTION VII - PRECAUTIONS FOR SAFE HANDLING AND USE

Steps to be Taken in Case Material is Released or Spilled

Evacuate area and deny entry to those not involved in cleanup activities. Protective clothing is mandatory. Remove all sources of ignition. Ventilate area of spill. Collect powdered materials in safest manner possible. Do NOT dry sweep. Use HEPA vacuum instead. Deposit in sealed drum for disposal.

Preferred Waste Disposal Method

Incineration in a chemical incinerator equipped with afterburner and scrubber.

Precautions to be Taken in Handling and Storage

Store in tightly closed containers in a cool, dark, dry, well-ventilated area. Personnel should be trained on the hazardous properties of hydrazine sulfate *prior* to working with the material.

Other Precautions and Warnings

Containers should be protected from physical damage. Use extreme caution when handling this compound. Protect containers from physical damage and keep away from incompatible materials.

SECTION VIII - CONTROL MEASURES AND PERSONAL PROTECTIVE EQUIPMENT

Respiratory Protection *(specify type)*

A carcinogenic agent. No exposure levels have been established. For any exposures, use a supplied-air respirator set in positive pressure or continuous flow mode or a self-contained breathing apparatus (SCBA) with full facepiece operated in pressure demand or other positive pressure mode.

Ventilation

Local exhaust or general mechanical systems recommended.

Protective Gloves	Eye Protection	Other Protective Clothing
Impervious Gloves	Chemical/Dust Goggles or Face Mask	Full Protective Clothing or Suit

Work/Hygiene Practices

Always wash hands thoroughly after using chemical; never bring food, drink, or smoking materials into vicinity of chemicals.

HYDRAZINE SULFATE

$H_4N_2 \bullet H_2O_4S$ CAS: 10034-93-2

IDENTIFICATION AND TYPICAL USES

Hydrazine sulfate is a colorless or white crystalline, sand-like solid material. It is used the manufacture of chemicals, in condensation reactions, as a catalyst in making acetate fibers, in refining rare metals, as an antioxidant in soldering flux for light metals, in the determination of arsenic content in metals, for separating polonium from tellurium, as a reducing agent in the analysis of minerals and slags, in tests for blood, and as a fungicide and germicide.

RISK ASSESSMENT: HEALTH

General Assessment

Hydrazine sulfate is a suspected human and confirmed animal carcinogen. It is an experimental teratogen and human mutation data have also been reported. It is primarily toxic by *inhalation* of its airborne dusts and powders, by skin contact (*absorption*), and by direct *ingestion*.

Inhalation will cause irritation of the nose, throat, and upper respiratory tract. Symptoms include choking, extreme chest pain, dyspnea (difficulty in breathing), lethargy, nausea, and possible liver and kidney injury. It can also affect brain and nervous system function with symptoms of lightheadedness, dizziness, trembling, excitement, and even convulsions. Systemic intoxication, depression of the central nervous system (CNS), and delayed gastrointestinal irritation may also occur. Hydrazine sulfate may cause damage to the red blood cells and may result in a low blood count (anemia). It can also interfere with the ability of the blood to carry or transport oxygen throughout the body, a condition known as methemoglobinemia, causing weakness and shortness of breath.

Skin contact results in irritation and possible burns. Damage to the eyes and cornea can be severe causing partial or complete loss of vision. It is a powerful reducing agent which is corrosive to the eyes, skin, and mucous membranes. May also cause skin sensitization as well as systemic poisoning. Once absorbed through the skin, toxic systemic effects similar to that of inhalation can occur. If absorption occurs simultaneously with inhalation or ingestion, the resulting effects can be fatal.

Ingestion of hydrazine sulfate powders will cause paresthesia (pins and needles), somnolence, nausea, and vomiting.

Acute Health Effects

The following acute (short-term) health effects may occur immediately or shortly after exposure to hydrazine sulfate:

Skin: Severe irritation and burning. Skin absorption will occur causing toxic systemic effects.

Eye: Severe irritation and tissue damage up to and including loss of vision.

Lung: Irritation of the upper respiratory tract causing choking, coughing, dypsnea, and pulmonary irritation.

CNS: Dizziness, convulsions, tremors, nausea, headaches, possible injury to the brain and nervous system.

Chronic Health Effects

The following chronic (long-term) health effects may occur at some time after exposure to hydrazine sulfate and can last for months or even years:

Cancer Hazards: According to the information presented in the references, hydrazine sulfate has been shown to cause cancer in test animals. Its carcinogenicity in humans is expected to be at least equal to that of animals. Human mutation data have been reported for this compound.

Reproductive Hazard: According to the information presented in the references, this chemical has been shown to cause teratogenic effects as well as adversely affect reproduction in test animals.

Other Chronic Effects: Repeated exposures can lead to hemolytic anemia and convulsions. Chronic exposures may lead to cancer, liver or kidney damage, and methemoglobinemia. The latter can cause chronic fatigue, shortness of breath, bluish color in the face, lips, nose, finger tips, and toes (cyanosis) due to a lack of oxygen in the blood. Repeated skin exposure can cause dermatitis or skin allergy. Once an allergy develops, even the slightest future exposures can cause itching and rash.

Recommended Risk-Reduction Measures

Hydrazine sulfate is listed by the IARC as a confirmed animal carcinogen and the NTP as a suspected human carcinogen. Unless a less toxic chemical can be substituted for a hazardous substance, *engineering controls* are the most effective method of reducing exposures. The best protection is to enclose operations and/or provide local exhaust ventilation at the site of chemical release. While not always operationally feasible, isolating operations can also reduce exposure.

Using respiratory protection is less effective than the controls mentioned above, but is always required whenever working with or around hydrazine sulfate. No exposure levels have been established for hydrazine sulfate, although NIOSH has set a maximum ceiling (2-hours) for hydrazine at 0.04 mg/m^3 which may be used as a reference. For any possible or potential exposures to a suspected carcinogenic agent, use a supplied-air respirator set in positive pressure or continuous flow mode, or use a self-contained breathing apparatus (SCBA) with full facepiece operated in pressure demand or other positive pressure mode. If a full facepiece is not available, then chemical/dust goggles should be worn to protect the eyes. Whenever an eye or face contact hazard exists, a face shield and apron should be worn. To prevent hand and skin exposures, impervious gloves should be used. Glove manufacturers should be contacted and permeation studies obtained *before* final glove selection has been made.

Administrative controls should also be in place to minimize the potential for human exposures. These may include written procedures or policies which specify the methods and techniques as well as the exposure control measures that will be practiced whenever personnel are to work with or around hydrazine sulfate.

All personnel should receive *training* on the use, hazards, protective measures, emergency actions, and other precautions per 29 CFR 1910.1200 (Hazard Communication), prior to the first assignment in an area where hydrazine sulfate is used or stored.

Before beginning employment and at regular intervals thereafter (e.g. annually), the following medical tests are recommended:

☑ Liver and kidney function tests.

☑ Complete blood count (CBC).

If symptoms develop or overexposure is suspected, the following medical tests may be useful:

☑ Examination of nervous system function.

☑ Liver and kidney function tests.

☑ Blood methemoglobin level.

☑ Evaluation by a qualified allergist with careful consideration of exposure history and special tests (may help diagnose allergy).

Any evaluation should include a careful medical history of past and present symptoms with an examination. However, medical tests that evaluate existing damage are not a substitute for controlling exposure. Also, because more than light alcohol consumption can cause liver damage, drinking alcohol can increase the potential for liver damage caused by exposure to hydrazine sulfate. Prudent risk assessment and management requires careful consideration of *all* possible risk factors that may be causing the appearance of symptoms in exposed workers.

Other methods to reduce exposure include:

☑ Where possible, enclose operations and use local exhaust ventilation at the site of chemical release. If local exhaust ventilation or enclosure is not used, respiratory protection should be mandatory.

☑ Always ensure that proper protective clothing is worn when using chemical substances.

☑ Wash thoroughly immediately after exposure to hydrazine sulfate and at the end of the work shift or before eating, drinking, or smoking.

☑ Hazard warning information must be posted in the work area. It is recommended that a regulated, controlled work area should be established wherever this chemical is used, handled, or stored. In addition, as part of an on-going education and training program, all information on the health and safety hazards of hydrazine sulfate should be communicated to all potentially exposed workers.

☑ Eye wash stations should be provided in work areas. If the potential for whole body exposure exists, then safety showers should also be provided in the immediate area.

☑ Specific engineering controls are recommended for this chemical by NIOSH in Publication Number 78-172 (*Occupational Exposure to Hydrazine*).

RISK ASSESSMENT: ENVIRONMENT

General Assessment

The environment is at risk of exposure during intended use, transportation, storage, disposal, or destruction of hydrazine sulfate. In almost every scenario, the threat of environmental exposure is contingent upon the proper handling of the chemical substance. Accidental spills, large or small, can result in fire, explosion, and possible contamination of the surrounding environmental mediums (water, soil, and air).

Hydrazine sulfate should always be handled with extreme caution. It is highly reactive with a number of substances, especially oxidizers and oxidizing agents such as perchlorates, permanganates, chlorates, and nitrates. These characteristics require special consideration during any emergency situation involving a leak or spill of hydrazine sulfate.

Hydrazine sulfate can enter the environment through industrial and municipal discharges, spills, and through non-point sources due to its use as a fungicide.

Acute Ecological Effects

Acute (short-term) toxic effects may include the death of animals, birds, or fish and death or low growth rate in plants. Acute effects are seen 2 to 4 days after animals or plants are exposed to hydrazine sulfate.

This chemical has slight acute toxicity to aquatic life. Insufficient data are available to evaluate or predict the short-term effects of hydrazine sulfate to plant, birds, or land animals.

Chronic Ecological Effects

Chronic toxic effects may include shortened life span, reproductive problems, lower fertility, and changes in appearance or behavior in exposed animals. These effects can be seen long after first exposure(s) to toxic chemicals. Hydrazine sulfate has slight chronic toxicity to aquatic life. Insufficient data are available to evaluate or predict the long-term effects of hydrazine sulfate to plants, birds, or land animals.

Water Solubility

Hydrazine sulfate is highly soluble in water. Concentrations of 1000 milligrams and more will mix with a liter of water.

Persistence in the Environment

Hydrazine is slightly persistent in water, with a half-life of between 2 to 20 days. Sulfate is highly persistent in water with a half-life of greater than 200 days. The half-life of a pollutant is the amount of time it takes for one half of the chemical to be degraded.

Bioaccumulation in Aquatic Organisms

Some substances increase in concentration, or *bioaccumulate,* in living organisms as they breathe contaminated air, drink contaminated water, or eat contaminated food. These chemicals can become concentrated in the tissues and internal organs of animals as well as humans.

Neither hydrazine nor the sulfate are expected to accumulate in the edible tissues of aquatic species consumed by humans.

Recommended Risk-Reduction Measures

Proper training of all transporters will reduce the likelihood of a mishap or accident resulting in a leak or spill to the environment. The correct use of labeling on all containers, trucks, and rail cars will enable emergency responders to react properly and quickly to any disaster thereby reducing the potential risk to the environment and to personnel.

Storage of hydrazine sulfate should be segregated from other chemicals to minimize the risk of cross-contamination or contact. Storage buildings should be equipped with proper fire protection and prevention equipment and explosion-proof electrical components and equipment. Bulk storage of hydrazine sulfate is not recommended.

If a spill or leak to the environment has occurred, fire department, emergency response and/or hazardous materials spill personnel should be notified immediately. Cleanup should be attempted only by those trained in proper spill containment procedures and the health hazards associated with carcinogenic materials.

Contaminated soils should be removed for incineration and replaced with clean soil.

If hydrazine sulfate should contact the water table, aquifer, or navigable waterway, remediation activities should be prompt to ensure protection of aquatic sediments. It is highly soluble in water and total containment and remediation may not be possible. When such spills occur, the local and/or state emergency response authorities must be notified. A comprehensive emergency response or disaster preparedness/recovery plan should be in place prior to any operations involving the use, transportation, storage, or disposal of hydrazine sulfate. Burning hydrazine sulfate in a chemical incinerator equipped with an afterburner and an air scrubber is an acceptable disposal method.

If hydrazine sulfate is spilled or leaked, the following specific steps are recommended:

- ☑ Restrict persons not wearing protective clothing from area of spill or leak until cleanup is complete and area can be opened for normal work.
- ☑ Ventilate area and remove all ignition sources.
- ☑ Collect powders in safest manner possible. Do NOT dry sweep (generates air borne dusts). Use a vacuum equipped with a high-efficiency particulate air filter (HEPA) instead. Deposit in sealed drums for disposal.
- ☑ It may be necessary to dispose of hydrazine sulfate as a hazardous waste. The responsible state agency or the regional office of the federal Environmental Protection Agency (EPA) should be contacted for specific recommendations.

RISK ASSESSMENT: BUSINESS

General Assessment

Accidents or mishaps involving 4-dimethylaminoazobenzene can present a significant threat to business operations. The loss or damage of equipment or facilities can significantly affect fiscal viability. Lawsuits that may result from personnel injury/death, public exposures, and/or environmental contamination will also require a serious expenditure of resources. Media attention surrounding an injury, death, or environmental damage can also result in a loss of profits and loss of current as well as future business. Always remember that anytime the term "carcinogen," "cancer," or "reproductive hazard" are used, public emotion, hysteria, and ignorance can run equally high. This should be considered during the development of any public relations policies.

Recommended Risk-Reduction Measures

Company attorneys, safety and health professionals, and environmental specialists should be involved in the development of any procedures dealing with safety or emergency response. A company official should be pre-designated as a public relations officer with specific training in dealing with the press. Corporate plans and policies should be developed, approved, and implemented long before any need for such arises.

REFERENCES

American Conference of Governmental Industrial Hygienists. 1988. *Documentation of the Threshold Limit Values and Biological Exposure Indices,* 5th Edition (with updates). Cincinnati: ACGIH

Anthony, C. P., and N. J. Kolthoff. 1971. *Textbook of Anatomy and Physiology.* St. Louis: C. V. Mosby Company.

Brusick, D. J. 1994. *Methods for Genetic Risk Assessment.* New York: CRC Press/Lewis Publishers.

Bueche, J. R. 1972. *Principles of Physics,* 2nd Edition. New York: McGraw-Hill.

Burrell, R., D. K. Flaherty, and L. J. Sauers. 1992. *Toxicology of the Immune System.* New York: Van Nostrand Reinhold.

Calabrese, E. J. 1994. *Biological Effects of Low Level Exposures.* New York: CRC Press/Lewis Publishers.

Cothern, C. R. 1995. *Handbook for Environmental Risk Decisionmaking.* New York: CRC Press/Lewis Publishers.

Gardner, R. W. 1994. *Chemical Intolerance.* New York: CRC Press/Lewis Publishers.

Hodgson, E., R. B. Mailman, and Janice E. Chambers. 1988. *Dictionary of Toxicology.* London: The MacMillian Company.

Lewis, R. J., Sr. 1993. *Hawley's Condensed Chemical Dictionary,* 12th Edition. New York: Van Nostrand Reinhold.

Lewis, R. J., Sr. 1992. *Sax's Dangerous Properties of Industrial Materials,* Volumes 1, 2, and 3, 8th Edition. New York: Van Nostrand Reinhold.

Meyer, E. 1990. *Chemistry of Hazardous Materials.* Englewood Cliffs, NJ: Prentice Hall.

Rea, W. J. 1995. *Chemical Sensitivity,* Vol. I-III. New York: CRC Press/Lewis Publishers.

U.S. Department of Health and Human Services, National Institute for Occupational Safety and Health. 1994. *NIOSH Pocket Guide to Chemical Hazards.* Washington, D.C.: U.S. Government Printing Office.

U.S. Department of Health and Human Services, National Institute for Occupational Safety and Health. 1978. *Occupational Exposure Hydrazine.* Publication Number 78-172. Washington, D.C.: U.S. Government Printing Office.

MATERIAL SAFETY DATA SHEET

CHEMICAL NAME

HYDRIODIC ACID

CRC PRESS/LEWIS PUBLISHERS DISCLAIMER: THE INFORMATION AND RECOMMENDATIONS PRESENTED HEREIN ARE BASED ON SOURCES BELIEVED TO BE RELIABLE. CRC MAKES NO REPRESENTATION ON ITS COMPLETENESS OR ACCURACY. IT IS THE RESPONSIBILITY OF THE USER TO DETERMINE THE CHEMICAL'S SUITABILITY FOR ITS INTENDED USE, THE CHEMICAL'S SAFE USE, AND THE CHEMICAL'S PROPER DISPOSAL. NO REPRESENTATIONS AND/OR WARRANTIES, EITHER EXPRESSED OR IMPLIED, OF THE MERCHANTABILITY OR FITNESS FOR A PARTICULAR PURPOSE, OR OF ANY OTHER NATURE, ARE MADE WITH RESPECT TO THE INFORMATION PROVIDED IN THIS MSDS OR TO THE CHEMICAL TO WHICH INFORMATION MAY REFER. CRC NEITHER ASSUMES NOR AUTHORIZES ANY OTHER PERSON TO ASSUME FOR IT, ANY OTHER ADDITIONAL RESPONSIBILITY OR LIABILITY FOR THE USE OF, OR RELIANCE UPON, THIS INFORMATION.

HAZARD WARNING INFORMATION

HEALTH	FIRE	REACTIVE	OTHER	DEGREE OF HAZARD	COLOR CODING	OTHER CODES
3	0	0		0 = Minimum Hazard 1 = Slight Hazard 2 = Moderate Hazard 3 = Serious Hazard 4 = Severe Hazard	HEALTH = BLUE FIRE = RED REACTIVITY = YELLOW OTHER = WHITE	OX = Oxidizer ACID = Acid ALK = Alkali COR = Corrosive W̶ = Use No Water

SECTION I - GENERAL INFORMATION

Characterization **Iodide**	RCRA Number **D002**	EPA Class **Characteristic (C) Waste**
DOT Proper Shipping Name **Hydriodic Acid Solution**	Chemical Abstract Service (CAS) Number **10034-85-2**	
DOT Hazard Class and Label Requirements **Corrosive Material**	DOT Emergency Guide Code **60**	
DOT Identification Number **UN 1787**	Chemical Formula **HI**	

Synonyms

Hydrogen iodide; anhydrous hydriodic acid.

SECTION II - HAZARDOUS INGREDIENTS AND IDENTITY INFORMATION

Hazardous Components (specific identity)	OSHA Exposure Criteria	NIOSH Exposure Criteria	Immediately Dangerous to Life and Health (IDLH)	ACGIH Exposure Criteria
Hydriodic acid (derivation: By passing hydrogen with iodine vapor over warm platinum sponge and absorption in water; by the action of iodine on a solution of hydrogen sulfide). Typically 57% solution of hydrogen iodide.	**PEL:** Not Established **STEL:** Not Established	**REL:** Not Established **STEL:** Not Established	**Not Determined**	**TLV:** Not Established **STEL:** Not Established

SECTION III - PHYSICAL AND CHEMICAL CHARACTERISTICS

Boiling Point **-31°F (-35°C)**	Specific Gravity (H_2O = 1) **2.85 at 40°F (5°C)**
Vapor Pressure (mm Hg) **Not Found**	Molecular Weight **127.91**
Vapor Density (Air = 1) **Not Found**	Melting Point **-59°F (-51°C)**

Solubility

Miscible with water and alcohol.

Appearance and Odor

Colorless liquid solution (or gas) when freshly made, turning yellow-brown on exposure to air or light. May contain hypophosphorous acid to prevent discoloration.

SECTION IV - FIRE AND EXPLOSION HAZARD DATA

Flash Point (method used) **Not Determined**	Explosive Limits in Air % by Volume **LEL: Not Determined UEL: Not Determined**
NFPA Classification **Non-Combustible Liquid**	Autoignition Temperature **Not Determined**

Extinguishing Media

Carbon dioxide, dry chemical, fog, or regular foam. Reacts with water or steam to form toxic and corrosive fumes. Use water only as a last resort and take precautions to avoid inhalation or contact.

Special Fire Fighting Procedures

Wear full protective clothing and self-contained breathing apparatus (SCBA). Also, use a water spray to keep fire-exposed containers cool. Do not release runoff to sewers and waterways. Poisonous gases are produced in fire. Hydriodic acid will corrode natural rubber. Stay away from ends of tanks.

Unusual Fire and Explosion Hazards

Potentially explosive reaction with strong oxidizers, strong caustics, and concentrated sulfuric acid. Containers may explode in fire. Firefighters should avoid contact with vapors produced during fire.

SECTION V - REACTIVITY DATA

Stability		Conditions to Avoid **Hydriodic acid is mildly endothermic (absorbs heat). It must be kept away from light and air and storage temperatures should not exceed 30°F (-1°C). Keep away from natural rubber and natural rubber products.**
Stable	Unstable **X**	Incompatibility *(materials to avoid)* **Explodes with ethyl hydroperoxide; ignites on contact with magnesium, perchloric acid, when heated with potassium, potassium chlorate, and oxidizers such as fluorine, dinitrogen trioxide, and fuming nitric acid.**
Hazardous Polymerization		Conditions to Avoid **Under normal conditions of temperature and pressure, hazardous polymerization of hydriodic acid is not expected to occur.**
May Occur	Will Not Occur **X**	Hazardous Decomposition or By-products **When heated to decomposition, hydriodic acid can emit highly toxic/poisonous gases including carbon dioxide and toxic iodide fumes.**

SECTION VI - HEALTH HAZARD DATA

Primary Route(s) of Entry:	Inhalation?	Absorption (skin)?	Ingestion?
	X		X

Health Hazards

INHALATION: **Vapors cause irritation to the upper respiratory tract with coughing, dypsnea, headache, bronchitis, weakness, tearing, nasal discharge, nausea, cyanosis, salivation, giddiness, muscle weakness, nasal membrane ulceration, pulmonary edema, chemical pneumonia.**

SKIN & EYES: **Skin contact causes smarting, tingling, redness, burning, swelling, blistering, necrosis, and shock. Eye contact causes burning, watering, redness, corneal ulceration, blurred vision, and possible blindness.**

INGESTION: **Severe poisoning, kidney damage, salivation, bloody vomiting, burning in mouth, pain, and diarrhea. Severe cases may cause shock, dypsnea, circulatory collapse, and death.**

Carcinogenicity	NTP Listed?	IARC Cancer Review Group?	OSHA Regulated?	Target Organs?
Unknown Human **Unknown Animal**	**No**	**No**	**No**	**Respiratory system, eyes, skin, digestive system.**

Medical Conditions Generally Aggravated by Exposure

Respiratory impairments (bronchitis, asthma); Skin conditions (dermatitis).

Emergency and First-aid Procedures

<u>Eye contact</u>: Flush immediately with water for 15 minutes (minimum), seek medical attention. <u>Skin contact</u>: Remove all contaminated clothing. Immediately wash area with large amounts of water. Seek medical assistance. For <u>inhalation</u>: Remove the person from exposure. Provide respiratory assistance and CPR. Transfer to medical facility. Observe for 48 hours for lung effects. If <u>swallowed</u>: Seek medical attention; contact poison control center. Do NOT induce vomiting. Give conscious person 1-2 glasses of water to drink to dilute. Treat exposures as acid burn.

SECTION VII - PRECAUTIONS FOR SAFE HANDLING AND USE

Steps to be Taken in Case Material is Released or Spilled

Ventilate area (avoid vapors). Absorb liquids in vermiculite or other material, collect and deposit in sealed containers. Ventilate area of spill. Restrict those not involved in cleanup from entering area.

Preferred Waste Disposal Method

No Citation.

Precautions to be Taken in Handling and Storage

Hydriodic acid is a non-combustible liquid. Avoid contact with strong oxidizers such as chlorine since violent reactions occur. Store in tightly closed containers in cool, well-ventilated area away from heat.

Other Precautions and Warnings

Sources of ignition are prohibited where hydriodic acid is used, handled, or stored. Detached facility storage is preferred. Prevent damage to containers.

SECTION VIII - CONTROL MEASURES AND PERSONAL PROTECTIVE EQUIPMENT

Respiratory Protection *(specify type)*

No exposure levels established. Recommend using a supplied-air respirator or a self-contained breathing apparatus (SCBA) operated in pressure demand or other positive pressure mode.

Ventilation

Local exhaust or general mechanical systems recommended.

Protective Gloves	Eye Protection	Other Protective Clothing
Acid Resistant *Synthetic* Rubber	**Chemical Goggles or Face Mask**	**Synthetic Rubber Apron**

Work/Hygiene Practices

Always wash hands thoroughly after using chemical; never bring food, drink, or smoking materials into vicinity of chemicals.

HYDRIODIC ACID

HI

CAS: 10034-85-2

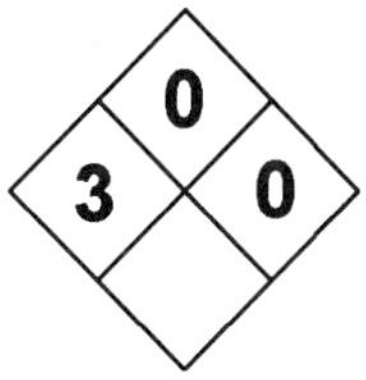

IDENTIFICATION AND TYPICAL USES

Hydriodic acid is a colorless liquid when freshly made but will rapidly turn to yellow-brown in the presence of light or air. It is used in the preparation of iodine salts, in organic preparations, as an analytical reagent, as a disinfectant, and in pharmaceuticals.

RISK ASSESSMENT: HEALTH

General Assessment

Exposure to hydriodic acid is possible through *inhalation* of its vapors and by skin contact (although absorption through intact skin has not been reported). It can also enter the body through *ingestion.* Although it is considered to be of moderate to severe toxicity, it is a highly caustic and, therefore, corrosive compound. There are no reports on its carcinogenic, mutagenic, or teratogenic properties.

Inhalation produces severe irritation of the eyes, nose, and mucosa of the respiratory tract with coughing, dypsnea (labored breathing), sneezing, lachrymation, nasal discharge, bronchitis, headache, cyanosis (bluish face and lips), salivation, giddiness, nausea, nasal membrane ulceration, muscle weakness and/or total body weakness. There can be progression to secondary chemical pneumonia which can lead to death. There may also be a life-threatening buildup of fluid in the lungs (pulmonary edema) which may be delayed up to 48 hours thereby creating a false sense of security with regard to health exposure risk.

Skin contact causes severe burning with possible vesiculation. Its extremely corrosive properties cause immediate stinging, burning, redness, swelling, membrane ulceration, painful skin blistering, and local necrosis (tissue destruction) upon contact. Shock may occur as a result of the intense pain. Eye contact causes immediate stinging and watering with tissue damage and the possibility of corneal ulceration. There will be redness and swelling of the lids, intense pain, and tissue destruction. There may be blurred vision resulting from corneal opaqueness, and there is a possibility of lost sight.

Ingestion is extremely dangerous, even life-threatening. It causes severe poisoning, burning of the lips, mouth, and throat. Other symptoms include painful swallowing, excessive salivation and drooling, nausea, ulceration of the mouth mucosa, intense thirst, painful stomach cramps, bloody vomiting (due to digestive hemorrhage), acute tissue damage, diarrhea with pain, shock, breathing difficulties, convulsions, coma, circulatory collapse, and death.

Acute Health Effects

The following acute (short-term) health effects may occur immediately or shortly after exposure to hydriodic acid:

Skin: Irritation/rash or burning feeling on contact with redness, blistering and vesiculation, and localized tissue destruction.

Eye: Severe irritation; may cause burns and damage with permanent loss of vision.

Lung: Severe irritation of the mucous membranes in the nose, throat, and lungs following exposure. May cause pulmonary edema and secondary chemical pneumonia.

Chronic Health Effects

The following chronic (long-term) health effects may occur at some time after exposure to hydriodic acid and can last for months or even years:

Cancer Hazards: According to information presented in the references, hydriodic acid has not been adequately tested for its ability to cause cancer in test animals.

Reproductive Hazard: According to information presented in the references, hydriodic acid has not been adequately tested for its ability to adversely affect reproduction in test animals.

Other Chronic Effects: Skin contact causes irritation and burns. There may be some lasting effects of chronic exposures such as sensitization. Once an individual becomes "sensitized," even the slightest future exposures can cause an allergic reaction.

Recommended Risk-Reduction Measures

Personnel should avoid direct contact with hydriodic acid. It is extremely corrosive and moderately toxic to humans. If a less toxic chemical cannot be substituted for a hazardous substance, then *engineering controls* are the most effective method of reducing exposures.

The best protection is to enclose operations and/or provide local exhaust ventilation at the site of chemical release. While not always operationally feasible, isolating operations can also reduce exposure. Using respiratory protection is less effective than the controls mentioned above, but is still advisable whenever working with or around hydriodic acid. No exposure levels have been established for this chemical. This does not mean that exposure to hydriodic acid is without risk. For best protection, a self-contained breathing apparatus (SCBA) or a supplied-air respirator with full facepiece operated in pressure demand mode are the recommended respiratory protection methods of choice. If a full facepiece is not available, then chemical goggles should be worn to protect the eyes. Whenever a chemical splash hazard exists, a face shield and a synthetic rubber apron should be worn. To prevent hand and skin exposures, acid resistant gloves should be used. Hydriodic acid will attack and corrode natural rubber and natural rubber products (such as gloves). Glove manufacturers should be contacted and permeation studies obtained *before* final glove selections are made.

Administrative controls should also be in place to minimize the potential for human exposures. These may include written procedures or policies which specify the methods and techniques that will be practiced whenever personnel are to work with hydriodic acid.

All personnel should receive *training* on the use, hazards, protective measures, emergency actions, and other precautions per 29 CFR 1910.1200 (Hazard Communication), prior to the first assignment in an area where hydriodic acid is used or stored.

Before beginning employment and at regular intervals thereafter (e.g., annually), the following tests are recommended:

- ☑ Lung function tests (look for chronic respiratory diseases).

If symptoms occur or overexposure is suspected, the following additional tests are suggested:

- ☑ Evaluation by a qualified allergist with careful consideration of exposure history and special testing (may help diagnose skin allergy).
- ☑ Lung function tests.
- ☑ Consider chest X-ray following acute overexposure (may be negative if taken immediately after exposure due to delayed onset of pulmonary edema).

Other methods to reduce exposure include:

- ☑ Where possible, enclose operations and use local exhaust ventilation at the site of chemical release. If local exhaust ventilation or enclosure is not used, respiratory protection should be mandatory.
- ☑ Always ensure that proper protective clothing is worn when using chemical substances.
- ☑ Wash thoroughly immediately after exposure to hydriodic acid and at the end of the work shift or before eating, drinking, or smoking.
- ☑ Hazard warning information should be posted in the work area. In addition, as part of an on-going education and training program, all information on the health and safety hazards of hydriodic acid should be communicated to all exposed workers.

RISK ASSESSMENT: ENVIRONMENT

General Assessment

The environment is at risk of exposure during transportation, storage, disposal, or destruction of hydriodic acid. In almost every scenario, the threat of environmental exposure is contingent upon the proper handling of the chemical substance. Accidental spills, large or small, can result in fire, explosion, and possible contamination of the surrounding environmental mediums (water, soil, air).

Hydriodic acid is considered a non-combustible liquid. However, it is extremely corrosive to many materials. It is incompatible with oxidizing agents and contact can cause fire or explosion. These characteristics require special consideration during any emergency situation involving a leak or spill of hydriodic acid Should hydriodic acid ever come into contact with incompatible substances either during use, transportation or storage, the formation of highly toxic and/or highly explosive commodities is extremely possible.

Hydriodic acid can enter the environment mainly through industrial discharges or spills.

Acute Ecological Effects

Acute (short-term) toxic effects may include the death of animals, birds, or fish and death or low growth rate in plants. Acute effects are seen 2 to 4 days after animals or plants are exposed to hydriodic acid.

Insufficient data are available to evaluate or predict the short-term effects of hydriodic acid exposure to aquatic life, plants, birds, or land animals.

Chronic Ecological Effects

Chronic (long-term) toxic effects may include shortened life span, reproductive problems, lower fertility, and changes in appearance or behavior in exposed animals. These effects can be seen long after first exposure(s) to toxic chemicals.

Insufficient data are available to evaluate or predict the long-term effects of hydriodic acid to aquatic life, plants, birds, or land animals.

Water Solubility

Hydriodic acid is highly soluble in water. Concentrations of over 1000 milligrams will mix with a liter of water.

Persistence in the Environment

Hydriodic acid is slightly persistent in the aquatic environment, with a half-life of between 2 to 20 days. The half-life of a pollutant is the amount of time it takes for one half of the chemical to be degraded.

Bioaccumulation in Aquatic Organisms

Some substances increase in concentration, or *bioaccumulate,* in living organisms as they breathe contaminated air, drink contaminated water, or eat contaminated food. These chemicals can become concentrated in the tissues and internal organs of animals as well as humans.

The concentration of hydriodic acid found in fish tissues is expected to be about the same as the average concentration of hydriodic acid in the water from which the fish was taken.

Recommended Risk-Reduction Measures

Proper training of all transporters will reduce the likelihood of a mishap or accident resulting in a leak or spill to the environment. The correct use of DOT labeling (per the MSDS) on all containers, trucks, and rail cars will enable emergency responders to react properly and quickly to any disaster thereby reducing the potential risk to the environment and to personnel.

Storage of hydriodic acid should be segregated from incompatible chemicals to minimize the risk of cross-contamination or contact. Buildings designated for storage should be equipped with appropriate fire protection systems (alarms, sprinklers, emergency lighting, portable extinguishers).

If a spill or leak to the environment has occurred, fire department, emergency response and/or hazardous materials spill personnel should be notified immediately. Cleanup should be attempted only by those trained in proper spill containment procedures. Contaminated soils should be removed for incineration and replaced with clean soil. If hydriodic acid should contact the water table, aquifer, or navigable waterway, time is of the essence. It is highly miscible in water and, therefore, total containment and remediation may not be possible. When such spills occur, the local and/or state emergency response authorities must be notified. A comprehensive emergency response or disaster preparedness/recovery plan should be in place prior to any operations involving the use, transportation, storage, or disposal of hydriodic acid. If hydriodic acid is spilled or leaked, the following specific steps are recommended:

- ☑ Restrict persons not wearing protective clothing from area of spill or leak until cleanup is complete and area can be opened for normal work.
- ☑ Remove all sources of heat.
- ☑ Ventilate area or spill or leak.
- ☑ Absorb liquids in vermiculite, dry sand, earth, or similar material and deposit in sealed containers. Neutralize remaining residue with soda ash or sodium bicarbonate. Do not flush to drains, sewers, or waterways.
- ☑ It may be necessary to dispose of hydriodic acid as a hazardous waste. The responsible state agency or the regional office of the federal Environmental Protection Agency (EPA) should be contacted for specific recommendations.

RISK ASSESSMENT: BUSINESS

General Assessment

Accidents or mishaps involving hydriodic acid can present a significant threat to business operations. The loss or damage of equipment or facilities can significantly affect fiscal viability. Lawsuits that may result from personnel injury/death, public exposures, and/or environmental contamination will also require a serious expenditure of resources. Media attention surrounding an injury, death, or environmental damage can also result in a loss of profits and loss of current as well as future business.

◐ *Recommended Risk-Reduction Measures*

Company attorneys, safety and health professionals, and environmental specialists should be involved in the development of any procedures dealing with safety or emergency response. A company official should be pre-designated as a public relations officer with specific training in dealing with the press. Corporate plans and policies should be developed, approved, and implemented long before any need for such arises.

REFERENCES

Ahlbom, A. 1993. *Biostatistics for Epidemiologists.* New York: CRC Press/Lewis Publishers.

American Conference of Governmental Industrial Hygienists. 1988. *Documentation of the Threshold Limit Values and Biological Exposure Indices,* 5th Edition (with updates). Cincinnati: ACGIH

Asante-Duah, D. K. 1993. *Hazardous Waste Risk Assessment.* New York: CRC Press/Lewis Publishers.

Bloom, A. D. and F. J. DeSerres. 1995. *Ecotoxicity and Human Health.* New York: CRC Press/Lewis Publishers.

Bueche, J. R. 1972. *Principles of Physics,* 2nd Edition. New York: McGraw-Hill.

Burrell, R., D. K. Flaherty, and Leonard J. Sauers. 1992. *Toxicology of the Immune System.* New York: Van Nostrand Reinhold.

Calabrese, E. J., E. M. Kenyon. *Air Toxics and Risk Assessment.* New York: CRC Press/Lewis Publishers.

Cockerman, L. G., B. S. Shane. 1994. *Basic Environmental Toxicology.* New York: CRC Press/Lewis Publishers.

Cothern, C. R. 1995. *Handbook for Environmental Risk Decisionmaking.* New York: CRC Press/Lewis Publishers.

Danse, I. R. 1991. *Common Sense Toxics in the Workplace.* New York: Van Nostrand Reinhold.

Gardner, R. W. 1994. *Chemical Intolerance.* New York: CRC Press/Lewis Publishers.

Hallenbeck, W. H. 1993. *Quantitative Risk Assessment for Environmental and Occupational Health,* 2nd Edition. New York: CRC Press/Lewis Publishers.

Hosty, J. W., P. Foster. 1990. *A Practical Guide to Chemical Spill Response.* New York: Van Nostrand Reinhold.

Howard, P. H., et. al. 1991. *Handbook of Environmental Degradation Rates.* New York: CRC Press/Lewis Publishers.

Keith, L. H. and M. M. Walker. 1995. *Handbook of Air Toxics.* New York: CRC Press/Lewis Publishers.

Landis, W. G., M. H. Yu. 1995. *Introduction to Environmental Toxicology.* New York: CRC Press/Lewis Publishers.

Lauwerys, R. R., P. Hoet. 1993. *Industrial Chemical Exposure.* New York: CRC Press/Lewis Publishers.

Lewis, R. J., Sr. 1993. *Hawley's Condensed Chemical Dictionary,* 12th Edition. New York: Van Nostrand Reinhold.

Lewis, R. J., Sr. 1992. *Sax's Dangerous Properties of Industrial Materials,* Volumes 1, 2, and 3, 8th Edition. New York: Van Nostrand Reinhold.

Manahan, S. E. 1992. *Toxicological Chemistry,* 2nd Edition. New York: CRC Press/Lewis Publishers.

Meyer, E. 1990. *Chemistry of Hazardous Materials.* Englewood Cliffs, NJ: Prentice Hall.

Mycock, J. C., J. D. McKenna, and L. Theodore. 1995. *Handbook of Air Pollution Control Engineering and Technology.* New York: CRC Press/Lewis Publishers.

Ott, W. R. 1995. *Environmental Statistics and Data Analysis.* New York: CRC Press/Lewis Publishers.

Patnaik, P. 1992. *A Comprehensive Guide to the Hazardous Properties of Chemical Substances.* New York: Van Nostrand Reinhold.

Philp, R. B. 1995. *Environmental Hazards and Human Health.* New York: CRC Press/Lewis Publishers.

Rea, W. J. 1995. *Chemical Sensitivity,* Vol. I-III. New York: CRC Press/Lewis Publishers.

Richardson, M. 1995. *Dictionary of Substance and Their Effects,* Vol. I-VI. New York: CRC Press/Lewis Publishers.

Rylander, R., R. R. Jacobs. 1994. *Organic Dusts.* New York: CRC Press/Lewis Publishers.

Sacarello, H. L. A. 1994. *The Comprehensive Handbook of Hazardous Materials.* New York: CRC Press/Lewis Publishers.

Sheftel, V. O. 1995. *Handbook of Toxic Properties of Monomers and Additives.* New York: CRC Press/Lewis Publishers.

Sherman, J. D. 1988. *Chemical Exposure and Disease.* New York: Van Nostrand Reinhold.

Talbott, E., G. Craun. 1995. *An Introduction to Environmental Epidemiology.* New York: CRC Press/Lewis Publishers.

U.S. Department of Health and Human Services, National Institute for Occupational Safety and Health. 1994. *NIOSH Pocket Guide to Chemical Hazards.* Washington, D.C.: U.S. Government Printing Office.

Witorsch, P. and S. V. Spagnolo. 1994. *Air Pollution and Lung Disease in Adults.* New York: CRC Press/Lewis Publishers.

MATERIAL SAFETY DATA SHEET

CHEMICAL NAME	
HYDROGEN BROMIDE	**CRC PRESS/LEWIS PUBLISHERS DISCLAIMER:** THE INFORMATION AND RECOMMENDATIONS PRESENTED HEREIN ARE BASED ON SOURCES BELIEVED TO BE RELIABLE. CRC MAKES NO REPRESENTATION ON ITS COMPLETENESS OR ACCURACY. IT IS THE RESPONSIBILITY OF THE USER TO DETERMINE THE CHEMICAL'S SUITABILITY FOR ITS INTENDED USE, THE CHEMICAL'S SAFE USE, AND THE CHEMICAL'S PROPER DISPOSAL. NO REPRESENTATIONS AND/OR WARRANTIES, EITHER EXPRESSED OR IMPLIED, OF THE MERCHANTABILITY OR FITNESS FOR A PARTICULAR PURPOSE, OR OF ANY OTHER NATURE, ARE MADE WITH RESPECT TO THE INFORMATION PROVIDED IN THIS MSDS OR TO THE CHEMICAL TO WHICH INFORMATION MAY REFER. CRC NEITHER ASSUMES NOR AUTHORIZES ANY OTHER PERSON TO ASSUME FOR IT, ANY OTHER ADDITIONAL RESPONSIBILITY OR LIABILITY FOR THE USE OF, OR RELIANCE UPON, THIS INFORMATION.

HAZARD WARNING INFORMATION

HEALTH	FIRE	REACTIVE	OTHER	DEGREE OF HAZARD	COLOR CODING	OTHER CODES
3	0	0	COR	0 = Minimum Hazard 1 = Slight Hazard 2 = Moderate Hazard 3 = Serious Hazard 4 = Severe Hazard	HEALTH = BLUE FIRE = RED REACTIVITY = YELLOW OTHER = WHITE	OX = Oxidizer ACID = Acid ALK = Alkali COR = Corrosive ~~W~~ = Use No Water

SECTION I - GENERAL INFORMATION

Characterization **Mineral Acid**	RCRA Number **None**	EPA Class **Not Applicable**
DOT Proper Shipping Name **Hydrogen Bromide**	Chemical Abstract Service (CAS) Number **10035-10-6**	
DOT Hazard Class and Label Requirements **Non-Flammable Gas or Corrosive Liquid; Poison Gas.**	DOT Emergency Guide Code **15 (anhydrous) 60 (solution)**	
DOT Identification Number **UN 1048 (anhydrous) UN 1788 (solution)**	Chemical Formula **BrH**	

Synonyms
Anhydrous hydrogen bromide; hydrobromic acid; anhydrous hydrobromic acid.

SECTION II - HAZARDOUS INGREDIENTS AND IDENTITY INFORMATION

Hazardous Components (specific identity)	OSHA Exposure Criteria	NIOSH Exposure Criteria	Immediately Dangerous to Life and Health (IDLH)	ACGIH Exposure Criteria
Hydrogen bromide (derivation: By passing hydrogen with bromine vapor over warm platinum sponge which acts as catalyst; as a by-product in the bromination of organic compounds). **1 ppm = 3.36 mg/m^3**	**PEL (ceiling):** **3 ppm** **10 mg/m^3** **STEL:** **Not Applicable**	**PEL (ceiling):** **3 ppm** **10 mg/m^3** **STEL:** **Not Applicable**	**30 ppm**	**TLV (ceiling):** **3 ppm** **10 mg/m^3** **STEL:** **Not Applicable**

SECTION III - PHYSICAL AND CHEMICAL CHARACTERISTICS

Boiling Point **-87°F (-66°C)**	Specific Gravity (H_2O = 1) **2.81**
Vapor Pressure (atmospheres) **20**	Molecular Weight **127.91**
Vapor Density (Air = 1) **2.71**	Melting Point **-124°F (-86°C)**

Solubility
Dissolves in water to form hydrobromic acid. Soluble in alcohol.

Appearance and Odor
Colorless gas with a sharp, irritating odor. Normally shipped as a liquefied compressed gas. Often used in aqueous solution. In solution, appears as a clear or slightly yellow fuming liquid.

SECTION IV - FIRE AND EXPLOSION HAZARD DATA

Flash Point (method used) **Not Determined**	Explosive Limits in Air % by Volume **LEL: Not Determined UEL: Not Determined**
NFPA Classification **Non-Flammable Gas**	Autoignition Temperature **Not Determined**

Extinguishing Media
Carbon dioxide, dry chemical, fog, or regular foam. Reacts with water or steam to form toxic and corrosive fumes. Use water only as a last resort and take precautions to avoid inhalation or contact.

Special Fire Fighting Procedures
Wear full protective clothing and self-contained breathing apparatus (SCBA). Also, use a water spray to keep fire-exposed containers cool. Poisonous gases are produced in fire.

Unusual Fire and Explosion Hazards
Potentially explosive reaction with strong oxidizers, strong caustics, and moisture. Containers may explode in fire. Firefighters should avoid contact with vapors produced during fire. Hydrogen bromide will react with metals and moisture to produce hydrogen.

SECTION V - REACTIVITY DATA

Stability		Conditions to Avoid **Hydrogen bromide is normally stable in closed, pressurized containers under routine conditions of handling and storage. Keep away from metals and other incompatible materials.**
Stable	Unstable	Incompatibility *(materials to avoid)*
X		**Incompatible with strong oxidizers, strong caustics, moisture, and most metals (especially copper, brass, and zinc). It will corrode most metals and form explosive hydrogen gas in the process.**
Hazardous Polymerization		Conditions to Avoid **Under normal conditions of temperature and pressure, hazardous polymerization of hydrogen bromide is not expected to occur.**
May Occur	Will Not Occur	Hazardous Decomposition or By-products
	X	**When heated to decomposition, hydrogen bromide can emit highly toxic/poisonous gases. The emission of explosive hydrogen gas during a fire creates a dangerous hazard to personnel and property.**

SECTION VI - HEALTH HAZARD DATA

Primary Route(s) of Entry:	Inhalation?	Absorption (skin)?	Ingestion?
	X		X

Health Hazards

INHALATION: Vapors cause irritation to the upper respiratory tract with coughing, dypsnea, headache, congestion, and possible burns to the nasal airway with a brownish discoloration of the affected tissues, including the mucous membranes of the upper respiratory.

SKIN & EYES: Skin contact causes smarting, tingling, redness, burning, swelling, blistering, necrosis, and shock. Eye contact causes burning, watering, redness, corneal ulceration, blurred vision, and possible blindness. The liquefied gas causes frostbite burns to tissues.

INGESTION: Not likely (hydrogen bromide is a gas at room temperature.

Carcinogenicity	NTP Listed?	IARC Cancer Review Group?	OSHA Regulated?	Target Organs?
Unknown Human Unknown Animal	No	No	29 CFR 1910.1000 Table Z-1	Upper respiratory tract, eyes, skin.

Medical Conditions Generally Aggravated by Exposure

Respiratory impairments (bronchitis, asthma); Skin conditions (dermatitis).

Emergency and First-aid Procedures

Eye contact: Flush immediately with water for 15 minutes (minimum), seek medical attention. Skin contact: Remove all contaminated clothing. Immediately wash area with large amounts of water. Seek medical assistance. For inhalation: Remove the person from exposure. Provide respiratory assistance and CPR. Transfer to medical facility. If swallowed: Not likely. If accidental ingestion of solutions occurs, seek medical attention or contact poison control center. Do NOT induce vomiting. Give conscious person 1-2 glasses of water or milk to drink to dilute.

SECTION VII - PRECAUTIONS FOR SAFE HANDLING AND USE

Steps to be Taken in Case Material is Released or Spilled

Ventilate area (avoid vapors). Shut off flow of gas if possible. Solution spills should be neutralized with sodium bicarbonate, soda ash, or slacked lime and deposited in sealed containers. Ventilate area of spill. Restrict those not involved in cleanup from entering area.

Preferred Waste Disposal Method

No Citation.

Precautions to be Taken in Handling and Storage

Hydrogen bromide is a non-flammable gas. Avoid contact with strong oxidizers such as chlorine since violent reactions occur. Store in tightly closed containers in cool, well-ventilated area away from heat.

Other Precautions and Warnings

Sources of ignition are prohibited where hydrogen bromide is used, handled, or stored. Detached facility storage is preferred. Prevent damage to containers.

SECTION VIII - CONTROL MEASURES AND PERSONAL PROTECTIVE EQUIPMENT

Respiratory Protection *(specify type)*

Exposure levels are low. Recommend using a supplied-air respirator or a self-contained breathing apparatus (SCBA) operated in pressure demand or other positive pressure mode.

Ventilation

Local exhaust or general mechanical systems recommended.

Protective Gloves	Eye Protection	Other Protective Clothing
Acid Resistant Rubber	**Chemical Goggles or Face Mask**	**Protective Apron**

Work/Hygiene Practices

Always wash hands thoroughly after using chemical; never bring food, drink, or smoking materials into vicinity of chemicals.

HYDROGEN BROMIDE

BrH

CAS: 10035-10-6

IDENTIFICATION AND TYPICAL USES

Hydrogen bromide is a colorless gas with a sharp, irritating odor. It is normally shipped as a liquefied compressed gas and is often used in aqueous solution. The solution (hydrobromic acid) is a clear or slightly yellow fuming liquid that will darken slowly on exposure to air or light. The fume from reaction with moist air has a sour taste. It is used in organic synthesis, to make bromides by direct reaction with alcohols, as a pharmaceutical intermediate, an alkylation and oxidation catalyst, and as a powerful reducing agent.

RISK ASSESSMENT: HEALTH

General Assessment

Exposure to hydrogen bromide is possible through *inhalation* of its vapors and by skin contact (although absorption through intact skin has not been reported). It can also enter the body through *ingestion*, but this exposure route is extremely unlikely since the chemical exists as a gas at room temperature and pressure. Although it is considered to be of moderate to severe toxicity, it is a highly caustic and, therefore, corrosive compound. There are no reports on its carcinogenic, mutagenic, or teratogenic properties.

Inhalation produces severe irritation of the eyes, nose, and mucosa of the upper respiratory tract with coughing, dypsnea (labored breathing), sneezing, lachrymation, and nausea. There may be a brownish discoloration of the affected tissues, including the mucous membranes that line the upper respiratory tract.

Skin contact with the vapor or the liquid solution causes severe burning due to hydrogen bromide's intense corrosive action on human tissue. The liquid may cause frostbite burns that will compound the corrosive effects. These extremely corrosive properties cause immediate stinging, burning, redness, swelling, membrane ulceration, painful skin blistering, and local necrosis (tissue destruction) upon contact. Shock may occur as a result of the intense pain. Eye contact causes immediate stinging and watering with tissue damage and the possibility of corneal ulceration. There will be redness and swelling of the lids, intense pain, and tissue destruction. Loss of vision is possible.

Ingestion is extremely unlikely under normal operating conditions. However, if liquid solutions are accidentally swallowed, there can be severe burns to the lips, mouth, throat, and stomach. If vomiting occurs and the liquid is aspirated into the lungs, the corrosive action on the lung tissues can be fatal.

Acute Health Effects

The following acute (short-term) health effects may occur immediately or shortly after exposure to hydrogen bromide:

Skin: Irritation/rash or burning feeling on contact with redness, blistering, and localized tissue destruction. The liquefied compressed gas can cause frostbite.

Eye: Severe irritation; may cause burns and damage with permanent loss of vision. Frostbite can also occur on contact with the compressed gas.

Lung: Severe irritation of the mucous membranes in the nose, throat, and lungs following exposure.

Chronic Health Effects

The following chronic (long-term) health effects may occur at some time after exposure to hydrogen bromide and can last for months or even years:

Cancer Hazards: According to information presented in the references, hydrogen bromide has not been adequately tested for its ability to cause cancer in test animals.

Reproductive Hazard: According to the references, hydrogen bromide has not been adequately tested for its ability to adversely affect reproduction in test animals.

Other Chronic Effects: None reported.

Recommended Risk-Reduction Measures

Personnel should avoid direct contact with hydrogen bromide. It is extremely corrosive and moderately toxic to humans. If a less toxic chemical cannot be substituted for a hazardous substance, then *engineering controls* are the most effective method of reducing exposures.

The best protection is to enclose operations and/or provide local exhaust ventilation at the site of chemical release. While not always operationally feasible, isolating operations can also reduce exposure. Using respiratory protection is less effective than the controls mentioned above, but is still advisable whenever working with or around hydrogen bromide. The exposure level is extremely low (3 ppm) and difficult to accurately monitor. For low or infrequent exposures, an acid vapor respirator or gas mask may suffice. However, better protection is obtained using a self-contained breathing apparatus (SCBA) or a supplied-air respirator with full facepiece operated in pressure demand mode. If a full facepiece is not available, then chemical goggles should be worn to protect the eyes. Whenever a chemical splash hazard exists, a face shield and a protective apron should be worn. To prevent hand and skin exposures, acid resistant gloves should be used. Glove manufacturers should be contacted and permeation studies obtained *before* final glove selections are made.

Administrative controls should also be in place to minimize the potential for human exposures. These may include written procedures or policies which specify the methods and techniques that will be practiced whenever personnel are to work with hydrogen bromide.

All personnel should receive *training* on the use, hazards, protective measures, emergency actions, and other precautions per 29 CFR 1910.1200 (Hazard Communication), prior to the first assignment in an area where hydrogen bromide is used or stored.

Before beginning employment and at regular intervals thereafter (e.g., annually), the following tests are recommended:

☑ Lung function tests (establish baseline).

If symptoms occur or overexposure is suspected, the following additional tests are suggested:

☑ Lung function tests (compare to baseline).

☑ Examination of the nasal airway for discoloration.

Other methods to reduce exposure include:

☑ Where possible, enclose operations and use local exhaust ventilation at the site of chemical release. If local exhaust ventilation or enclosure is not used, respiratory protection should be mandatory.

☑ Always ensure that proper protective clothing is worn when using chemical substances.

☑ Wash thoroughly immediately after exposure to hydrogen bromide and at the end of the work shift or before eating, drinking, or smoking.

☑ Hazard warning information should be posted in the work area. In addition, as part of an on-going education and training program, all information on the health and safety hazards of hydrogen bromide should be communicated to all exposed workers.

RISK ASSESSMENT: ENVIRONMENT

General Assessment

The environment is at risk of exposure during transportation, storage, disposal, or destruction of hydrogen bromide. In almost every scenario, the threat of environmental exposure is contingent upon the proper handling of the chemical substance. Accidental spills, large or small, can result in fire, explosion, and possible contamination of the surrounding environmental mediums (water, soil, and especially air).

Hydrogen bromide is considered a non-flammable gas. However, it is extremely corrosive to many materials. It is incompatible with oxidizing agents and contact can cause fire or explosion. In contact with many common metals (copper, brass, zinc) and moisture, it can liberate extremely explosive hydrogen gas. These characteristics require special consideration during any emergency situation involving a leak or spill of hydrogen bromide. Should hydrogen bromide ever come into contact with any incompatible substances either during use, transportation, or storage, the formation of highly toxic and/or highly explosive commodities is extremely possible.

Hydrogen bromide can enter the environment mainly industrial discharges or spills.

Acute Ecological Effects

Acute (short-term) toxic effects may include the death of animals, birds, or fish and death or low growth rate in plants. Acute effects are seen 2 to 4 days after animals or plants are exposed to hydrogen bromide.

Insufficient data are available to evaluate or predict the short-term effects of hydrogen bromide exposure to aquatic life, plants, birds, or land animals.

Chronic Ecological Effects

Chronic (long-term) toxic effects may include shortened life span, reproductive problems, lower fertility, and changes in appearance or behavior in exposed animals. These effects can be seen long after first exposure(s) to toxic chemicals.

Insufficient data are available to evaluate or predict the long-term effects of hydrogen bromide to aquatic life, plants, birds, or land animals.

Water Solubility

Hydrogen bromide is highly soluble in water. Concentrations of over 1000 milligrams will mix with a liter of water.

Persistence in the Environment

Hydrogen bromide is practically non-persistent in the aquatic environment, with a half-life of less than 2 days. The half-life of a pollutant is the amount of time it takes for one half of the chemical to be degraded. Virtually 100% of hydrogen bromide will eventually end up in the air.

Bioaccumulation in Aquatic Organisms

Some substances increase in concentration, or *bioaccumulate,* in living organisms as they breathe contaminated air, drink contaminated water, or eat contaminated food. These chemicals can become concentrated in the tissues and internal organs of animals as well as humans.

The concentration of hydrogen bromide found in fish tissues is expected to be much lower than the average concentration of hydrogen bromide in the water from which the fish was taken.

Recommended Risk-Reduction Measures

Proper training of all transporters will reduce the likelihood of a mishap or accident resulting in a leak or spill to the environment. The correct use of DOT labeling (per the MSDS) on all containers, trucks, and rail cars will enable emergency responders to react properly and quickly to any disaster thereby reducing the potential risk to the environment and to personnel.

Storage of hydrogen bromide should be segregated from incompatible chemicals to minimize the risk of cross-contamination or contact. Buildings designated for storage should be equipped with appropriate fire protection systems (alarms, sprinklers, emergency lighting, portable extinguishers).

If a spill or leak to the environment has occurred, fire department, emergency response and/or hazardous materials spill personnel should be notified immediately. Cleanup should be attempted only by those trained in proper spill containment procedures. Contaminated soils (solution spills) should be removed for incineration and replaced with clean soil. If hydrogen bromide should contact the water table, aquifer, or navigable waterway, time is of the essence. It is highly soluble in water and is extremely volatile. Therefore, total containment and remediation may not be possible. When such spills occur, the local and/or state emergency response authorities must be notified. A comprehensive emergency response or disaster preparedness/recovery plan should be in place prior to any operations involving the use, transportation, storage, or disposal of hydrogen bromide. If hydrogen bromide is spilled or leaked, the following specific steps are recommended:

- ☑ Restrict persons not wearing protective clothing from area of spill or leak until cleanup is complete and area can be opened for normal work.
- ☑ Remove all sources of heat and ignition.
- ☑ Ventilate area or spill or leak.
- ☑ Shut off gas leak, if possible. If leak is from a cylinder and it cannot be turned off, remove cylinder to safe area (outdoors) and allow to vent. Take care to avoid contact with the vapors or any escaping compressed liquids. Use water spray to protect personnel attempting to stop leak. Be aware of the possibility that hydrogen gas may be in the area (explosion hazard).
- ☑ Absorb solution spills in vermiculite, dry sand, earth, or similar material and deposit in sealed containers. *Carefully* neutralize remaining residue with soda ash, slacked lime, or sodium bicarbonate. Do not flush to drains, sewers, or waterways.
- ☑ It may be necessary to dispose of hydrogen bromide as a hazardous waste. The responsible state agency or the regional office of the federal Environmental Protection Agency (EPA) should be contacted for specific recommendations.

RISK ASSESSMENT: BUSINESS

General Assessment

Accidents or mishaps involving hydrogen bromide can present a significant threat to business operations. The loss or damage of equipment or facilities can significantly affect fiscal viability. Lawsuits that may result from personnel injury/death, public exposures, and/or environmental contamination will also require a serious expenditure of resources. Media attention surrounding an injury, death, or environmental damage can also result in a loss of profits and loss of current as well as future business.

Recommended Risk-Reduction Measures

Company attorneys, safety and health professionals, and environmental specialists should be involved in the development of any procedures dealing with safety or emergency response. A company official should be pre-designated as a public relations officer with specific training in dealing with the press. Corporate plans and policies should be developed, approved, and implemented long before any need for such arises.

REFERENCES

Ahlbom, A. 1993. *Biostatistics for Epidemiologists.* New York: CRC Press/Lewis Publishers.

American Conference of Governmental Industrial Hygienists. 1988. *Documentation of the Threshold Limit Values and Biological Exposure Indices,* 5th Edition (with updates). Cincinnati: ACGIH

Asante-Duah, D. K. 1993. *Hazardous Waste Risk Assessment.* New York: CRC Press/Lewis Publishers.

Bloom, A. D. and F. J. DeSerres. 1995. *Ecotoxicity and Human Health.* New York: CRC Press/Lewis Publishers.

Bueche, J. R. 1972. *Principles of Physics,* 2nd Edition. New York: McGraw-Hill.

Burrell, R., D. K. Flaherty, and Leonard J. Sauers. 1992. *Toxicology of the Immune System.* New York: Van Nostrand Reinhold.

Calabrese, E. J., E. M. Kenyon. *Air Toxics and Risk Assessment.* New York: CRC Press/Lewis Publishers.

Cockerman, L. G., B. S. Shane. 1994. *Basic Environmental Toxicology.* New York: CRC Press/Lewis Publishers.

Cothern, C. R. 1995. *Handbook for Environmental Risk Decisionmaking.* New York: CRC Press/Lewis Publishers.

Danse, I. R. 1991. *Common Sense Toxics in the Workplace.* New York: Van Nostrand Reinhold.

Gardner, R. W. 1994. *Chemical Intolerance.* New York: CRC Press/Lewis Publishers.

Hallenbeck, W. H. 1993. *Quantitative Risk Assessment for Environmental and Occupational Health,* 2nd Edition. New York: CRC Press/Lewis Publishers.

Hosty, J. W., P. Foster. 1990. *A Practical Guide to Chemical Spill Response.* New York: Van Nostrand Reinhold.

Howard, P. H., et. al. 1991. *Handbook of Environmental Degradation Rates.* New York: CRC Press/Lewis Publishers.

Keith, L. H. and M. M. Walker. 1995. *Handbook of Air Toxics.* New York: CRC Press/Lewis Publishers.

Landis, W. G., M. H. Yu. 1995. *Introduction to Environmental Toxicology.* New York: CRC Press/Lewis Publishers.

Lauwerys, R. R., P. Hoet. 1993. *Industrial Chemical Exposure.* New York: CRC Press/Lewis Publishers.

Lewis, R. J., Sr. 1993. *Hawley's Condensed Chemical Dictionary,* 12th Edition. New York: Van Nostrand Reinhold.

Lewis, R. J., Sr. 1992. *Sax's Dangerous Properties of Industrial Materials,* Volumes 1, 2, and 3, 8th Edition. New York: Van Nostrand Reinhold.

Manahan, S. E. 1992. *Toxicological Chemistry,* 2nd Edition. New York: CRC Press/Lewis Publishers.

Meyer, E. 1990. *Chemistry of Hazardous Materials.* Englewood Cliffs, NJ: Prentice Hall.

Mycock, J. C., J. D. McKenna, and L. Theodore. 1995. *Handbook of Air Pollution Control Engineering and Technology.* New York: CRC Press/Lewis Publishers.

Ott, W. R. 1995. *Environmental Statistics and Data Analysis.* New York: CRC Press/Lewis Publishers.

Patnaik, P. 1992. *A Comprehensive Guide to the Hazardous Properties of Chemical Substances.* New York: Van Nostrand Reinhold.

Philp, R. B. 1995. *Environmental Hazards and Human Health.* New York: CRC Press/Lewis Publishers.

Rea, W. J. 1995. *Chemical Sensitivity,* Vol. I-III. New York: CRC Press/Lewis Publishers.

Richardson, M. 1995. *Dictionary of Substance and Their Effects,* Vol. I-VI. New York: CRC Press/Lewis Publishers.

Sacarello, H. L. A. 1994. *The Comprehensive Handbook of Hazardous Materials.* New York: CRC Press/Lewis Publishers.

Sherman, J. D. 1988. *Chemical Exposure and Disease.* New York: Van Nostrand Reinhold.

Talbott, E., G. Craun. 1995. *An Introduction to Environmental Epidemiology.* New York: CRC Press/Lewis Publishers.

U.S. Department of Health and Human Services, National Institute for Occupational Safety and Health. 1994. *NIOSH Pocket Guide to Chemical Hazards.* Washington, D.C.: U.S. Government Printing Office.

Witorsch, P. and S. V. Spagnolo. 1994. *Air Pollution and Lung Disease in Adults.* New York: CRC Press/Lewis Publishers.

MATERIAL SAFETY DATA SHEET

CHEMICAL NAME	CRC PRESS/LEWIS PUBLISHERS DISCLAIMER
HYDROGEN CHLORIDE	

HAZARD WARNING INFORMATION

HEALTH	FIRE	REACTIVE	OTHER	DEGREE OF HAZARD	COLOR CODING	OTHER CODES
3	0	0	COR	0 = Minimum Hazard 1 = Slight Hazard 2 = Moderate Hazard 3 = Serious Hazard 4 = Severe Hazard	HEALTH = BLUE FIRE = RED REACTIVITY = YELLOW OTHER = WHITE	OX = Oxidizer ACID = Acid ALK = Alkali COR = Corrosive W = Use No Water

SECTION I - GENERAL INFORMATION

Item	Value	Item	Value
Characterization	**Mineral Acid**	RCRA Number	**D002**
		EPA Class	**Characteristic (C) Waste**
DOT Proper Shipping Name	**Hydrogen Chloride (anhydrous) ; Hydrochloric Acid**	Chemical Abstract Service (CAS) Number	**7647-01-0**
DOT Hazard Class and Label Requirements	**Non-Flammable Gas or Corrosive Liquid; Poison Gas.**	DOT Emergency Guide Code	**15 (anhydrous) 60 (solution)**
DOT Identification Number	**UN 1050 (anhydrous) UN 1789 (solution)**	Chemical Formula	**HCl**

Synonyms
Anhydrous hydrogen chloride; aqueous hydrogen chloride; hydrochloric acid, muriatic acid.

SECTION II - HAZARDOUS INGREDIENTS AND IDENTITY INFORMATION

Hazardous Components (specific identity)	OSHA Exposure Criteria	NIOSH Exposure Criteria	Immediately Dangerous to Life and Health (IDLH)	ACGIH Exposure Criteria
Hydrogen chloride (derivation: By-product of organic chlorination reactions; reaction of sodium chloride and sulfuric acid; burning hydrogen in an atmosphere of chlorine in the absence of air. The solution is derived by dissolving hydrogen chloride in water at various concentrations). **1 ppm = 1.52 mg/m^3**	**PEL (ceiling):** **5 ppm** **7 mg/m^3** **STEL:** **Not Applicable**	**REL (ceiling):** **5 ppm** **7 mg/m^3** **STEL:** **Not Applicable**	**50 ppm**	**TLV (ceiling):** **5 ppm** **7 mg/m^3** **STEL:** **Not Applicable**

SECTION III - PHYSICAL AND CHEMICAL CHARACTERISTICS

Item	Value	Item	Value
Boiling Point	**-121°F (-85°C)**	Specific Gravity (H_2O – 1)	**1.27**
Vapor Pressure (atmospheres)	**40.5**	Molecular Weight	**36.5**
Vapor Density (Air = 1)	**1.268**	Melting Point	**-174°F (-114°C)**

Solubility
Dissolves in water to form hydrochloric acid. Soluble in alcohol and ether.

Appearance and Odor
Colorless to slightly yellow gas with a pungent, irritating odor. Normally shipped as a liquefied compressed gas. Often used in aqueous solution. In solution, appears as a clear or yellow fuming liquid.

SECTION IV - FIRE AND EXPLOSION HAZARD DATA

Item	Value	Item	Value
Flash Point (method used)	**Not Determined**	Explosive Limits in Air % by Volume	**LEL: Not Determined UEL: Not Determined**
NFPA Classification	**Non-Flammable Gas**	Autoignition Temperature	**Not Determined**

Extinguishing Media
Use extinguishing agents suitable to surrounding fire.

Special Fire Fighting Procedures
Wear full protective clothing and self-contained breathing apparatus (SCBA). Also, use a water spray to keep fire-exposed containers cool. Poisonous gases are produced in fire.

Unusual Fire and Explosion Hazards
Potentially explosive reaction with hydroxides, amines, and alkalis. Corrosive to most metals. Containers may explode in fire. Firefighters should avoid contact with vapors produced during fire. In the presence of water or moisture, hydrogen chloride can react with metals such as iron to produce explosive hydrogen gas.

SECTION V - REACTIVITY DATA

Stability		Conditions to Avoid Hydrogen chloride is normally stable in closed, pressurized containers under routine conditions of handling and storage. Keep away from metals and other incompatible materials.
Stable	Unstable	Incompatibility (materials to avoid) Calcium carbide, cesium acetylene carbide, cesium carbide, lithium silicide, magnesium borate, mercuric sulfate, rubidium acetylene carbide, sodium, other hydroxides, alkalis, and many metals.
X		
Hazardous Polymerization		Conditions to Avoid Under normal conditions of temperature and pressure, hazardous polymerization of hydrogen chloride is not expected to occur.
May Occur	Will Not Occur	Hazardous Decomposition or By-products When heated to decomposition, hydrogen chloride can emit highly toxic/poisonous gases. The emission of explosive hydrogen gas during a fire creates a dangerous hazard to personnel and property.
	X	

SECTION VI - HEALTH HAZARD DATA

Primary Route(s) of Entry:	Inhalation?	Absorption (skin)?	Ingestion?
	X		X

Health Hazards

INHALATION: Vapors cause irritation to the upper respiratory tract with coughing, dypsnea, headache, congestion, ulcers in the nose, throat, and larynx, tissue dehydration, larynx spasm, and/or pulmonary edema. Chronic exposures may cause teeth erosion.

SKIN & EYES: Skin contact causes smarting, tingling, redness, burning, swelling, blistering, necrosis, and shock. Eye contact causes burning, watering, redness, corneal ulceration, blurred vision, and possible blindness. The liquefied gas causes frostbite burns to tissues.

INGESTION: Not likely (hydrogen chloride is a gas at room temperature).

Carcinogenicity	NTP Listed?	IARC Cancer Review Group?	OSHA Regulated?	Target Organs?
Unknown Human Unknown Animal	No	No	29 CFR 1910.1000 Table Z-1	Upper respiratory tract, eyes, skin.

Medical Conditions Generally Aggravated by Exposure

Respiratory impairments (bronchitis, asthma); Skin conditions (dermatitis).

Emergency and First-aid Procedures

Eye contact: Flush immediately with water for 15 minutes (minimum); seek medical attention. Skin contact: Remove all contaminated clothing. Immediately wash area with large amounts of water. Seek medical assistance. For inhalation: Remove the person from exposure. Provide respiratory assistance and CPR. Transfer to medical facility. Observe 48 hours for lung effects. If swallowed: Not likely. If accidental ingestion of solutions occurs, seek medical attention or contact poison control center. Do NOT induce vomiting. Give conscious person 1-2 glasses of water or milk to drink to dilute.

SECTION VII - PRECAUTIONS FOR SAFE HANDLING AND USE

Steps to be Taken in Case Material is Released or Spilled

Ventilate area (avoid vapors). Shut off flow of gas if possible. Solution spills should be neutralized with sodium bicarbonate, soda ash, or slacked lime and deposited in sealed containers. Restrict those not involved in cleanup from entering area.

Preferred Waste Disposal Method

No citation.

Precautions to be Taken in Handling and Storage

Hydrogen chloride is a non-flammable gas. Avoid contact with incompatible materials since violent reactions can occur. Store in tightly closed containers in cool, well-ventilated area away from heat.

Other Precautions and Warnings

Sources of ignition are prohibited where hydrogen chloride is used, handled, or stored. Detached facility storage is preferred. Prevent damage to containers.

SECTION VIII - CONTROL MEASURES AND PERSONAL PROTECTIVE EQUIPMENT

Respiratory Protection (specify type)

Exposure levels are relatively low. Recommend using a supplied-air respirator or a self-contained breathing apparatus (SCBA) operated in pressure demand or other positive pressure mode.

Ventilation

Local exhaust or general mechanical systems recommended.

Protective Gloves	Eye Protection	Other Protective Clothing
Acid Resistant Rubber	Chemical Goggles or Face Mask	Protective Apron

Work/Hygiene Practices

Always wash hands thoroughly after using chemical; never bring food, drink, or smoking materials into vicinity of chemicals.

HYDROGEN CHLORIDE

HCl CAS: 7647-01-0

IDENTIFICATION AND TYPICAL USES

Hydrogen chloride is a colorless gas with a sharp, characteristic, suffocating, pungent, and irritating odor. It is normally shipped as a liquefied compressed gas and is often used in aqueous solution. The solution (hydrochloric acid) is a clear or slightly yellow fuming liquid that will darken slowly on exposure to air or light. The fumes from reaction with moist air has a sour taste. It is used in the production of vinyl chlorides from acetylene and alkyl chlorides from olefins. It is also used in the hydrochlorination of certain rubber products, in polymerization, isomerization, alkylation, and nitration reactions. It is used in the manufacture of pharmaceutical hydrochlorides, and as a gaseous flux for babbiting operations.

RISK ASSESSMENT: HEALTH

General Assessment

Exposure to hydrogen chloride is possible through *inhalation* of its vapors and by skin contact (although absorption through intact skin has not been reported). It can also enter the body through *ingestion*, but this exposure route is extremely unlikely since the chemical exists as a gas at room temperature and pressure. Although it is considered to be of moderate to severe toxicity, it is a highly caustic and, therefore, corrosive compound. There are no reports on its carcinogenic, mutagenic, or teratogenic properties.

Inhalation severely dehydrates tissues and causes destructive corrosive damage to the nose, throat, and mucosa of the upper respiratory tract with ulceration of the nose, throat, and larynx. There can be coughing, dypsnea (labored breathing), lachrymation, and nausea. Severe exposures may cause spasm of the larynx and/or pulmonary edema (fluid in the lungs). The development of pulmonary edema can be delayed up to 48 hours after exposure, thereby creating a false sense of security with regard to health exposure risk.

Skin contact with the vapor or the liquid solution causes severe burning due to hydrogen chloride's intense corrosive action on human tissue. The liquid may cause frostbite burns that will compound the corrosive effects. These extremely corrosive properties cause immediate stinging, burning, redness, swelling, membrane ulceration, painful skin blistering, and local necrosis (tissue destruction) upon contact. Eye contact causes immediate stinging and watering with tissue damage and the possibility of corneal damage. There will be redness and swelling of the lids, intense pain, and tissue destruction. Loss of vision may occur.

Ingestion is extremely unlikely under normal operating conditions. However, if liquid solutions are accidentally swallowed, there can be severe burns to the lips, mouth, throat, and stomach. If vomiting occurs and the liquid is aspirated into the lungs, the corrosive action on the lung tissues can be fatal.

Acute Health Effects

The following acute (short-term) health effects may occur immediately or shortly after exposure to hydrogen chloride:

Skin: Irritation/rash or burning feeling on contact with redness, blistering, and localized tissue destruction. The liquefied compressed gas can cause frostbite.

Eye: Severe irritation; may cause burns and damage with permanent loss of vision. Frostbite burns can occur on contact with the compressed gas.

Lung: Severe irritation of the mucous membranes in the nose, throat, and lungs following exposure.

Chronic Health Effects

The following chronic (long-term) health effects may occur at some time after exposure to hydrogen chloride and can last for months or even years:

Cancer Hazards: According to information presented in the references, hydrogen chloride has not been adequately tested for its ability to cause cancer in laboratory test animals.

Reproductive Hazard: According to information presented in the references, hydrogen chloride has not been adequately tested for its ability to adversely affect reproduction in test animals.

Other Chronic Effects: Repeated or prolonged exposure may cause erosion of the teeth.

❶ *Recommended Risk-Reduction Measures*

Personnel should avoid direct contact with hydrogen chloride. It is extremely corrosive and moderately toxic to humans. If a less toxic chemical cannot be substituted for a hazardous substance, then *engineering controls* are the most effective method of reducing exposures.

The best protection is to enclose operations and/or provide local exhaust ventilation at the site of chemical release. While not always operationally feasible, isolating operations can also reduce exposure. Using respiratory protection is less effective than the controls mentioned above, but is still advisable whenever working with or around hydrogen chloride. The exposure level is extremely low (5 ppm) and difficult to accurately monitor. For low or infrequent exposures, an acid vapor respirator or gas mask may suffice. However, better protection is obtained using a self-contained breathing apparatus (SCBA) or a supplied-air respirator with full facepiece operated in pressure demand mode. If a full facepiece is not available, then chemical goggles should be worn to protect the eyes. Whenever a chemical splash hazard exists, a face shield and a protective apron should be worn. To prevent hand and skin exposures, acid resistant gloves should be used. Glove manufacturers should be contacted and permeation studies obtained *before* final glove selections are made.

Administrative controls should also be in place to minimize the potential for human exposures. These may include written procedures or policies which specify the methods and techniques that will be practiced whenever personnel are to work with hydrogen chloride.

All personnel should receive *training* on the use, hazards, protective measures, emergency actions, and other precautions per 29 CFR 1910.1200 (Hazard Communication), prior to the first assignment in an area where hydrogen chloride is used or stored.

Before beginning employment and at regular intervals thereafter (e.g., annually), the following tests are recommended:

☑ Lung function tests (establish baseline).

If symptoms occur or overexposure is suspected, the following additional tests are suggested:

☑ Lung function tests (compare to baseline).

☑ Consider chest X-ray following acute overexposure (may be negative it taken immediately after exposure due to delayed onset of pulmonary edema).

Other methods to reduce exposure include:

☑ Where possible, enclose operations and use local exhaust ventilation at the site of chemical release. If local exhaust ventilation or enclosure is not used, respiratory protection should be mandatory.

☑ Always ensure that proper protective clothing is worn when using chemical substances.

☑ Wash thoroughly immediately after exposure to hydrogen chloride and at the end of the work shift or before eating, drinking, or smoking.

☑ Hazard warning information should be posted in the work area. In addition, as part of an on-going education and training program, all information on the health and safety hazards of hydrogen chloride should be communicated to all exposed workers.

RISK ASSESSMENT: ENVIRONMENT

General Assessment

The environment is at risk of exposure during transportation, storage, disposal, or destruction of hydrogen chloride. In almost every scenario, the threat of environmental exposure is contingent upon the proper handling of the chemical substance. Accidental spills, large or small, can result in fire, explosion, and possible contamination of the surrounding environmental mediums (water, soil, and especially air).

Hydrogen chloride is considered a non-flammable gas. However, it is extremely corrosive to many materials. It is incompatible with hydroxides, amines, and alkalis and contact can cause fire or explosion. In contact with many common metals (copper, brass, zinc) and moisture, it can liberate extremely explosive hydrogen gas. These characteristics require special consideration during any emergency situation involving a leak or spill of hydrogen chloride. Should hydrogen chloride ever come into contact with any incompatible substances either during use, transportation, or storage, the formation of highly toxic and/or highly explosive commodities is extremely possible.

Hydrogen chloride can enter the environment mainly through industrial discharges or spills.

Acute Ecological Effects

Acute (short-term) toxic effects may include the death of animals, birds, or fish and death or low growth rate in plants. Acute effects are seen 2 to 4 days after animals or plants are exposed to hydrogen chloride.

Hydrogen chloride has slight acute toxicity to aquatic life. Insufficient data are available to evaluate or predict the short-term effects of hydrogen chloride exposure to plants, birds, or land animals.

Chronic Ecological Effects

Chronic (long-term) toxic effects may include shortened life span, reproductive problems, lower fertility, and changes in appearance or behavior in exposed animals. These effects can be seen long after first exposure(s) to toxic chemicals.

Hydrogen chloride has slight chronic toxicity to aquatic life. Insufficient data are available to evaluate or predict the long-term effects of hydrogen chloride to plants, birds, or land animals.

Water Solubility

Hydrogen chloride is highly soluble in water. Concentrations of over 1000 milligrams will mix with a liter of water.

Persistence in the Environment

Hydrogen chloride is slightly persistent in the aquatic environment, with a half-life of less than 2 and 20 days. The half-life of a pollutant is the amount of time it takes for one half of the chemical to be degraded. Virtually 100% of hydrogen chloride will eventually end up in the air.

Bioaccumulation in Aquatic Organisms

Some substances increase in concentration, or *bioaccumulate,* in living organisms as they breathe contaminated air, drink contaminated water, or eat contaminated food. These chemicals can become concentrated in the tissues and internal organs of animals as well as humans.

Small quantities of acids will be neutralized by the alkalinity in aquatic systems, but larger quantities can lower the pH for extended periods of time. Hydrogen chloride contains chloride whose concentrations in edible tissues of most aquatic species that are consumed by humans is expected to be about the same as the average concentration of hydrogen chloride in the water from which the fish was taken.

Recommended Risk-Reduction Measures

Proper training of all transporters will reduce the likelihood of a mishap or accident resulting in a leak or spill to the environment. The correct use of DOT labeling (per the MSDS) on all containers, trucks, and rail cars will enable emergency responders to react properly and quickly to any disaster thereby reducing the potential risk to the environment and to personnel.

Storage of hydrogen chloride should be segregated from incompatible chemicals to minimize the risk of cross-contamination or contact. Buildings designated for storage should be equipped with appropriate fire protection systems (alarms, sprinklers, emergency lighting, portable extinguishers). Containers should be protected against physical damage and stored away from direct sunlight or other heat sources.

If a spill or leak to the environment has occurred, fire department, emergency response and/or hazardous materials spill personnel should be notified immediately. Cleanup should be attempted only by those trained in proper spill containment procedures. Contaminated soils (solution spills) should be removed for incineration and replaced with clean soil. If hydrogen chloride should contact the water table, aquifer, or navigable waterway, time is of the essence. It is highly soluble in water and is extremely volatile. Therefore, total containment and remediation may not be possible. When such spills occur, the local and/or state emergency response authorities must be notified. A comprehensive emergency response or disaster preparedness/recovery plan should be in place prior to any operations involving the use, transportation, storage, or disposal of hydrogen chloride. If hydrogen chloride is spilled or leaked, the following specific steps are recommended:

- ☑ Restrict persons not wearing protective clothing from area of spill or leak until cleanup is complete and area can be opened for normal work.
- ☑ Remove all sources of heat and ignition.
- ☑ Ventilate area or spill or leak.

☑ Shut off gas leak, if possible. If leak is from a cylinder and it cannot be turned off, remove cylinder to safe area (outdoors) and allow to vent. Take care to avoid contact with the vapors or any escaping compressed liquids. Use water spray to protect personnel attempting to stop leak. Be aware of the possibility that hydrogen gas may be in the area (explosion hazard).

☑ Absorb solution spills in vermiculite, dry sand, earth, or similar material and deposit in sealed containers. *Carefully* neutralize remaining residue with soda ash, slacked lime, or sodium bicarbonate. Do not flush to drains, sewers, or waterways.

☑ It may be necessary to dispose of hydrogen chloride as a hazardous waste. The responsible state agency or the regional office of the federal Environmental Protection Agency (EPA) should be contacted for specific recommendations.

RISK ASSESSMENT: BUSINESS

General Assessment

Accidents or mishaps involving hydrogen chloride can present a significant threat to business operations. The loss or damage of equipment or facilities can significantly affect fiscal viability. Lawsuits that may result from personnel injury/death, public exposures, and/or environmental contamination will also require a serious expenditure of resources. Media attention surrounding an injury, death, or environmental damage can also result in a loss of profits and loss of current as well as future business.

Recommended Risk-Reduction Measures

Company attorneys, safety and health professionals, and environmental specialists should be involved in the development of any procedures dealing with safety or emergency response. A company official should be pre-designated as a public relations officer with specific training in dealing with the press. Corporate plans and policies should be developed, approved, and implemented long before any need for such arises.

REFERENCES

American Conference of Governmental Industrial Hygienists. 1988. *Documentation of the Threshold Limit Values and Biological Exposure Indices*, 5th Edition (with updates). Cincinnati: ACGIH

Asante-Duah, D. K. 1993. *Hazardous Waste Risk Assessment.* New York: CRC Press/Lewis Publishers.

Bloom, A. D. and F. J. DeSerres. 1995. *Ecotoxicity and Human Health.* New York: CRC Press/Lewis Publishers.

Bueche, J. R. 1972. *Principles of Physics,* 2nd Edition. New York: McGraw-Hill.

Burrell, R., D. K. Flaherty, and Leonard J. Sauers. 1992. *Toxicology of the Immune System.* New York: Van Nostrand Reinhold.

Calabrese, E. J., E. M. Kenyon. *Air Toxics and Risk Assessment.* New York: CRC Press/Lewis Publishers.

Cockerman, L. G., B. S. Shane. 1994. *Basic Environmental Toxicology.* New York: CRC Press/Lewis Publishers.

Danse, I. R. 1991. *Common Sense Toxics in the Workplace.* New York: Van Nostrand Reinhold.

Gardner, R. W. 1994. *Chemical Intolerance.* New York: CRC Press/Lewis Publishers.

Hosty, J. W., P. Foster. 1990. *A Practical Guide to Chemical Spill Response.* New York: Van Nostrand Reinhold.

Howard, P. H., et. al. 1991. *Handbook of Environmental Degradation Rates.* New York: CRC Press/Lewis Publishers.

Keith, L. H. and M. M. Walker. 1995. *Handbook of Air Toxics.* New York: CRC Press/Lewis Publishers.

Landis, W. G., M. H. Yu. 1995. *Introduction to Environmental Toxicology.* New York: CRC Press/Lewis Publishers.

Lauwerys, R. R., P. Hoet. 1993. *Industrial Chemical Exposure.* New York: CRC Press/Lewis Publishers.

Lewis, R. J., Sr. 1993. *Hawley's Condensed Chemical Dictionary,* 12th Edition. New York: Van Nostrand Reinhold.

Lewis, R. J., Sr. 1992. *Sax's Dangerous Properties of Industrial Materials,* Volumes 1, 2, and 3, 8th Edition. New York: Van Nostrand Reinhold.

Manahan, S. E. 1992. *Toxicological Chemistry,* 2nd Edition. New York: CRC Press/Lewis Publishers.

Meyer, E. 1990. *Chemistry of Hazardous Materials.* Englewood Cliffs, NJ: Prentice Hall.

Ott, W. R. 1995. *Environmental Statistics and Data Analysis.* New York: CRC Press/Lewis Publishers.

Patnaik, P. 1992. *A Comprehensive Guide to the Hazardous Properties of Chemical Substances.* New York: Van Nostrand Reinhold.

Philp, R. B. 1995. *Environmental Hazards and Human Health.* New York: CRC Press/Lewis Publishers.

Richardson, M. 1995. *Dictionary of Substance and Their Effects,* Vol. I-VI. New York: CRC Press/Lewis Publishers.

Sacarello, H. L. A. 1994. *The Comprehensive Handbook of Hazardous Materials.* New York: CRC Press/Lewis Publishers.

Sherman, J. D. 1988. *Chemical Exposure and Disease.* New York: Van Nostrand Reinhold.

Talbott, E., G. Craun. 1995. *An Introduction to Environmental Epidemiology.* New York: CRC Press/Lewis Publishers.

U.S. Department of Health and Human Services, National Institute for Occupational Safety and Health. 1994. *NIOSH Pocket Guide to Chemical Hazards.* Washington, D.C.: U.S. Government Printing Office.

MATERIAL SAFETY DATA SHEET

CHEMICAL NAME

HYDROGEN CYANIDE

HAZARD WARNING INFORMATION

HEALTH	FIRE	REACTIVE	OTHER	DEGREE OF HAZARD	COLOR CODING	OTHER CODES
4	4	2		0 = Minimum Hazard 1 = Slight Hazard 2 = Moderate Hazard 3 = Serious Hazard 4 = Severe Hazard	HEALTH = BLUE FIRE = RED REACTIVITY = YELLOW OTHER = WHITE	OX = Oxidizer ACID = Acid ALK = Alkali COR = Corrosive W = Use No Water

SECTION I - GENERAL INFORMATION

Characterization **Cyanides, Inorganic**	RCRA Number **P063**	EPA Class **Acute Hazardous Waste**
DOT Proper Shipping Name **Hydrocyanic Acid Solution**	Chemical Abstract Service (CAS) Number **74-90-8**	
DOT Hazard Class and Label Requirements **Poison A, Flammable Gas; Poison B, Flammable Liquid**	DOT Emergency Guide Code **13 (anhydrous) 55 (solution)**	
DOT Identification Number **UN 1051 (anhydrous) UN 1790 (solution)**	Chemical Formula **HCN**	

Synonyms
Hydrocyanic acid; formonitrile; prussic acid.

SECTION II - HAZARDOUS INGREDIENTS AND IDENTITY INFORMATION

Hazardous Components (specific identity)	OSHA Exposure Criteria	NIOSH Exposure Criteria	Immediately Dangerous to Life and Health (IDLH)	ACGIH Exposure Criteria
Hydrogen cyanide (derivation: By catalytic reaction of ammonia and air with methane or natural gas; by recovery from coke oven gases; from bituminous coal and ammonia at 2282°F (1250°C). Also occurs naturally in some plants such as almond and oleander). **1 ppm = 1.12 mg/m^3**	**PEL (skin):** **10 ppm** **11.2 mg/m^3** **STEL (skin):** **Not Established**	**REL (skin):** **4.7 ppm** **5 mg(CN)/m^3** **(10 min ceiling)** **STEL:** **Not Applicable**	**50 ppm**	**TLV (skin)):** **10 ppm** **10 mg/m^3** **(ceiling)** **STEL:** **Not Applicable**

SECTION III - PHYSICAL AND CHEMICAL CHARACTERISTICS

Boiling Point **79°F (29°C)**	Specific Gravity (H_2O = 1) **0.69**
Vapor Pressure (mm Hg) **630 at 68°F (20°C)**	Molecular Weight (atomic weight) **27.0**
Vapor Density (Air = 1) **0.941**	Melting Point **8°F (-13°C)**

Solubility
Miscible in water and alcohol, soluble in ether.

Appearance and Odor
White crystalline solid or white powder. Technical grade may appear as a gray-black powder.

SECTION IV - FIRE AND EXPLOSION HAZARD DATA

Flash Point (method used) **0°F (18°C) closed cup**	Explosive Limits in Air % by Volume **LEL: 5.6% UEL: 40%**
NFPA Classification **Class IA Flammable Liquid; Flammable Gas**	Autoignition Temperature **1000°F (538°C)**

Extinguishing Media
Carbon dioxide, dry chemical, or foams. Use water spray to flush the spill and disperse the vapors.

Special Fire Fighting Procedures
Wear full protective clothing and self-contained breathing apparatus (SCBA). Poisonous cyanide fumes and gases are produced in fire. Structural protective clothing is permeable and may not provide adequate protection. Avoid any contact with skin (fatal amounts can be readily absorbed through intact skin. Fight fire from distance or protected location, if possible.

Unusual Fire and Explosion Hazards
Hydrogen cyanide is an extremely flammable, toxic, and poisonous gas. Consider total evacuation of the incident area in large fires.

SECTION V - REACTIVITY DATA

Stability		Conditions to Avoid: Normally stable under routine conditions of handling and storage. Avoid contact with incompatible materials. Keep hydrogen cyanide away from open flame, heat, or other sources of ignition.
Stable	Unstable	Incompatibility *(materials to avoid)*: Reacts on contact with amines, oxidizers, acids, sodium hydroxide, calcium hydroxide, sodium carbonate, water, caustics, and ammonia.
X		
Hazardous Polymerization		Conditions to Avoid: Under normal conditions of temperature and pressure, hazardous polymerization of hydrogen cyanide will not occur. However, it can polymerize at 122-140°F (50-60°C).
May Occur	Will Not Occur	Hazardous Decomposition or By-products: When heated to decomposition, hydrogen cyanide releases toxic fumes of cyanide and nitric oxide. On contact with water or steam, toxic hydrogen cyanide will be released.
	X	

SECTION VI - HEALTH HAZARD DATA

Primary Route(s) of Entry:	Inhalation?	Absorption (skin/eye)?	Ingestion?
	X	X	X

Health Hazards

INHALATION: High concentrations will cause tachypnea (rapid breathing) which will further increase the amount of cyanide inhaled. This leads to dyspnea, paralysis, convulsions, and respiratory arrest with death occurring in minutes. Low concentrations may cause headache, vertigo, nausea, and vomiting. There could be asphyxiation, weakness, confusion, and gasping.

ABSORPTION: Will pass through unbroken skin and enter the bloodstream. Large exposures can be fatal.

INGESTION: A poison by ingestion. May cause toxic systemic effects and death.

Carcinogenicity	NTP Listed?	IARC Cancer Review Group?	OSHA Regulated?	Target Organs?
Unknown Human Unknown Animal	No	No	29 CFR 1910.1000 Table Z-1	CNS, CVS, thyroid, blood, all body cells.

Medical Conditions Generally Aggravated by Exposure

None reported.

Emergency and First-aid Procedures

Eye contact: Do not allow victim to keep eyes tightly shut. Flush immediately with water for 15 minutes (minimum); seek medical attention. Skin contact: Remove all contaminated clothing. Immediately wash area with flooding amounts of soap and water. Do not allow water to enter nose or mouth. For inhalation: Remove the person from exposure. Break amyl nitrate pearl in cloth and hold lightly under nose for 15 seconds; repeat 5 times at 15 second intervals. Give 100% oxygen. Do NOT administer CPR (caregiver can be poisoned). Transfer to medical facility. If swallowed: Break amyl nitrate pearl in cloth and hold lightly under nose for 15 seconds. Do NOT induce vomiting. Repeat inhalation of amyl nitrate 5 times at 15 second intervals. Seek medical attention immediately. Call poison control center for advice. Never give anything by mouth to an unconscious or convulsing person.

SECTION VII - PRECAUTIONS FOR SAFE HANDLING AND USE

Steps to be Taken in Case Material is Released or Spilled

Full protective clothing is required. Stop flow of gas. Evacuate area. Use water spray to knock down vapors. Restrict those not involved in cleanup from entering area. Notify emergency response personnel.

Preferred Waste Disposal Method

Destroy by burning in a combustion chamber with a fuel-air mix.

Precautions to be Taken in Handling and Storage

Store in air-tight, moisture-proof containers in cool, dry, well-ventilated area. Prevent physical damage.

Other Precautions and Warnings

Shield containers from exposure to heat, including direct sunlight. Limit access to storage areas.

SECTION VIII - CONTROL MEASURES AND PERSONAL PROTECTIVE EQUIPMENT

Respiratory Protection *(specify type)*

For any exposure level use an NIOSH/MSHA-approved self-contained breathing apparatus with full facepiece operated in positive pressure mode. Inhalation of dusts must be avoided.

Ventilation

Local exhaust (preferred) at site of chemical work or general ventilation.

Protective Gloves	Eye Protection	Other Protective Clothing
Impervious Material	Chemical/Dust Goggles or Face Mask	Protective Suit, Boots (avoid skin contact)

Work/Hygiene Practices

Always wash hands thoroughly after using chemical; never bring food, drink, or smoking materials into vicinity of chemicals.

HYDROGEN CYANIDE

HCN

CAS: 74-90-8

IDENTIFICATION AND TYPICAL USES

Hydrogen cyanide is a colorless gas or a pale-blue liquid (above 78°F) with a bitter, almond-like odor. It is used in the manufacture of acrylonitrile, acrylates, adiponitrile, cyanide salts, dyes, chelates, rodenticides, and pesticides. Used in the fumigation of ships, railroad cards, orchards, buildings, tobacco, and various food stocks. Can also be produced during petroleum refining, electroplating, metallurgy, and in photographic development.

RISK ASSESSMENT: HEALTH

General Assessment

Hydrogen cyanide is a highly poisonous compound. The toxic routes of entry are *inhalation, ingestion* and skin contact (*absorption*). Exposure by either route can lead to death in large enough doses. There are no reports of carcinogenic, mutagenic, or teratogenic effects resulting from exposure to hydrogen cyanide. It is a true "protoplasmic poison" which means it will combine in the tissues with the enzymes associated with cellular oxidation. This makes oxygen unavailable to the tissues and causes death through asphyxia. The effect on tissue oxidation lasts only as long as exposure occurs. If the person is removed from exposure before death has occurred, normal cellular oxidation will be restored.

Inhalation causes a burning sensation in the mouth and throat. Initial systemic response is rapid breathing which further increases the cyanide intake. Symptoms progress first to rapid pulse and high blood pressure followed by decreased pulse and lower blood pressure. There will be dypsnea, paralysis, convulsions, and respiratory arrest. Death can occur in minutes if the person is not removed from exposure. Lesser concentrations cause headache, vertigo, confusion, shock, convulsions, coma, nausea, vomiting, and gasping for breath. There may be thyroid and blood changes.

Poisonous reactions on exposure may cause cherry-red mucous membranes and blood due to the inability of the blood to remove its oxygen to the cells. High concentrations can be rapidly fatal.

Skin contact will cause burns and redness. Absorption occurs quickly causing toxic systemic effects similar to that of inhalation. It is also extremely toxic via ingestion.

Acute Health Effects

The following acute (short-term) health effects may occur immediately or shortly after exposure to hydrogen cyanide:

Skin: Severe irritation may occur. Absorption is rapid and the effects can be fatal.

Eye: Irritation and possible damage on contact.

Lung: High exposures can cause rapid and severe lung damage, with shortness of breath, chest pain, cough, loss of consciousness, and death.

Other: High exposure to hydrogen cyanide may be rapidly fatal. It appears to injure the central nervous system, liver, and kidneys as well as affect the blood and every cell in the body.

Chronic Health Effects

The following chronic (long-term) health effects may occur at some time after exposure to hydrogen cyanide and can last for months or even years:

Cancer Hazards: According to the information in the references, the data on the cancer-causing properties of hydrogen cyanide are inconclusive. However, this is probably due to the fact that it is highly poisonous and long-term exposures seldom occur (death is more likely).

Reproductive Hazard: It is not known if exposure to hydrogen cyanide will adversely affect the reproductive system of males or females.

Other Chronic Effects: Repeated exposures to low concentrations may result in enlargement of the thyroid gland, chronic fatigue, weakness, headache, changes in taste and smell, irritation of the throat, chronic dypsnea, nervous instability, abdominal colic, and other systemic chronic effects.

Recommended Risk-Reduction Measures

Personnel should avoid direct contact with hydrogen cyanide. If a less toxic material or compound cannot be substituted for hydrogen cyanide, then *engineering controls* are the most effective method of reducing exposures. The best protection is to enclose operations and/or provide local exhaust ventilation at the site of hydrogen cyanide dust release. While not always operationally feasible, isolating operations can also reduce exposure. Using respiratory protection is less effective than the controls mentioned above, but is still required whenever working with or around hydrogen cyanide. Although an air-purifying respirator may suffice for low concentrations, better protection is provided using a self-contained breathing apparatus (SCBA) with full facepiece operated in pressure demand mode or a supplied-air respirator operated in continuous flow or other positive pressure mode. If a full facepiece is not available, then chemical goggles should be worn to protect the eyes. A face shield and protective suit should also be used (skin contact must be prevented). To prevent hand and skin exposures, gloves should be worn. Glove manufacturers should be contacted and permeation studies obtained *before* final glove selections have been made.

Administrative controls should also be in place to minimize the potential for human exposures. These may include written procedures or policies which specify the methods and techniques that will be practiced whenever personnel are to work with hydrogen cyanide.

All personnel should receive *training* on the use, hazards, protective measures, emergency actions, and other precautions per 29 CFR 1910.1200 (Hazard Communication), prior to the first assignment in an area where hydrogen cyanide is used or stored.

If exposure occurs, break an amyl nitrate pearl into a cloth and hold it lightly under the victim's nose for 15 seconds. This should be repeated five times, at 15 second intervals.

For those with frequent or potentially high exposure (half the PEL or greater), or significant skin contact, the following medical tests are recommended before initial assignment and regularly thereafter:

- ☑ Urine thiocyanate test (most accurate if performed right after exposure).
- ☑ Lung function test and chest X-ray (for baseline).

If symptoms develop or overexposure is suspected, the following should also be considered:

- ☑ Blood cyanide level. (Note: Since cigarettes contain some cyanide, smokers may show an increase in cyanide level on testing).
- ☑ Evaluation of thyroid function.
- ☑ Lung function test and chest X-ray (compare to baseline).

Any evaluation should include a careful history of past and present symptoms with an examination. Medical tests that simply look for existing damage are not a substitute for controlling exposures. Also, personnel who smoke should avoid any unprotected contact with hydrogen cyanide powders.

Other methods to reduce exposure include:

- ☑ Where possible, enclose operations and use local exhaust ventilation at the site of chemical release. If local exhaust ventilation or enclosure is not used, respiratory protection should be mandatory.
- ☑ Always ensure that proper protective clothing is worn when using chemical substances and personnel must be trained on its correct use, care, and maintenance.
- ☑ Wash thoroughly immediately after exposure to hydrogen cyanide and at the end of the work shift or before eating, drinking, or smoking.
- ☑ Hazard warning information should be posted in the work area. In addition, as part of an on-going education and training program, all information on the health and safety hazards of hydrogen cyanide should be communicated to all exposed and potentially exposed workers.
- ☑ Eye wash stations should be provided in the immediate work area for emergency use. If there is a possibility of skin exposure to hydrogen cyanide, emergency shower facilities should also be provided in the immediate work area.
- ☑ Workers whose clothing has been contaminated by hydrogen cyanide should change into clean clothes before leaving work. Contaminated work clothing should be laundered only by individuals who have been informed of the hazards of exposure to hydrogen cyanide.

RISK ASSESSMENT: ENVIRONMENT

General Assessment

Industrial uses of hydrogen cyanide place the environment at risk of exposure during its transportation, storage, disposal, or destruction. In almost every scenario, the threat of environmental exposure is contingent upon the proper handling of the chemical substance. Accidental spills, large or small, where hydrogen cyanide contacts incompatible commodities can result in the release of hazardous fumes and gases (depending upon conditions of spill), and possible contamination of the surrounding environmental mediums (water, soil, and air).

Hydrogen cyanide is considered a Class IA flammable liquid (per OSHA 29 CFR 1910.106). Its extremely low flash point and low boiling point make hydrogen cyanide a dangerous fire and explosion hazard. Because it is incompatible with many commodities, including water, steam, acids, oxidizers, amines, sodium hydroxides, calcium hydroxide, and most caustics, extreme caution is required in handling, storage, transportation, and disposal of hydrogen cyanide. These characteristics also require special consideration during any emergency situation involving a leak or spill of hydrogen cyanide.

Should hydrogen cyanide ever come into contact with any incompatible substances either during use, transportation, storage, or disposal, release of poisonous fumes and possible violent reactions are possible.

Hydrogen cyanide can enter the environment from industrial and municipal waste treatment plant discharges and from spills. Also, as a fumigant and insecticide, it will also enter the environment as a result of its intended use.

Acute Ecological Effects

Acute (short-term) toxic effects may include the death of animals, birds, or fish and death or low growth rate in plants. Acute effects are seen 2 to 4 days after animals or plants are exposed to hydrogen cyanide. Hydrogen cyanide has high acute toxicity to aquatic life and terrestrial animals. Insufficient data are available to evaluate the short-term effects of hydrogen cyanide to plants and birds.

Chronic Ecological Effects

Chronic (long-term) toxic effects may include shortened life span, reproductive problems, lower fertility, and changes in appearance or behavior in exposed animals. These effects can be seen long after first exposure(s) to toxic chemicals.

Hydrogen cyanide has high chronic toxicity to aquatic life and terrestrial animals. Insufficient data are available to evaluate or predict the long-term effects of hydrogen cyanide to plants and birds.

Water Solubility

Hydrogen cyanide is highly soluble in water. Concentrations of 1000 milligrams will mix with a liter of water.

Persistence in the Environment

Hydrogen cyanide is highly persistent in water, with a half-life greater than 200 days. The half-life of a pollutant is the amount of time it takes for one half of the chemical to be degraded. No data are available on the estimated percentages of hydrogen cyanide persistence in the aquatic or terrestrial environments.

Bioaccumulation in Aquatic Organisms

Some substances increase in concentration, or *bioaccumulate,* in living organisms as they breathe contaminated air, drink contaminated water, or eat contaminated food. These chemicals can become concentrated in the tissues and internal organs of animals as well as humans.

The concentration of hydrogen cyanide found in fish tissues is expected to be much higher than the average concentration of hydrogen cyanide in the water from which the fish was taken.

Recommended Risk-Reduction Measures

Proper training of all transporters will reduce the likelihood of a mishap or accident resulting in a leak or spill of hydrogen cyanide dusts or powders into the environment. The correct use of DOT labeling (per the MSDS) on all containers, trucks, and rail cars will enable emergency responders to react properly and quickly to any disaster thereby reducing the potential risk to the environment and to personnel.

Storage of hydrogen cyanide should be segregated from incompatible chemicals to minimize the risk of cross-contamination or contact. It should be kept dry and in air-tight, moisture-proof containers. Physical damage to containers must be prevented. Proper ventilation and protective equipment should be used while handing the solid or while preparing aqueous solutions.

Hydrogen cyanide is destroyed by burning in a chemical incinerator equipped with an afterburner and scrubber.

If a spill or leak to the environment has occurred, fire department, emergency response and/or hazardous materials spill personnel should be notified immediately. A hydrogen cyanide spill control and countermeasure plan should be developed and practiced before the need for such arises. Cleanup should be attempted only by those trained in proper spill containment procedures using non-sparking tools. If hydrogen cyanide should contact the water table, aquifer, or navigable waterway, time is of the essence. It is highly soluble and total remediation may not be possible. The local and/or state emergency response authorities must be notified. A comprehensive emergency response or disaster preparedness/recovery plan should be in place prior to any operations involving the use, transportation, storage, or disposal of hydrogen cyanide. If hydrogen cyanide is spilled or leaked, the following specific steps are recommended:

- ☑ Restrict persons not wearing protective clothing from area of spill or leak until cleanup is complete and area can be opened for normal work.
- ☑ Remove all ignition sources.
- ☑ Gas leaks must be stopped. If this is not possible, the container should be removed to a safe area (outdoors) and allow to vent. Water spray should be used to knock down vapors and protect response personnel trying to stop leak.
- ☑ It may be necessary to dispose of hydrogen cyanide as a hazardous waste. The responsible state agency or the regional office of the federal Environmental Protection Agency (EPA) should be contacted for specific recommendations.

RISK ASSESSMENT: BUSINESS

General Assessment

Accidents or mishaps involving hydrogen cyanide can present a significant threat to business operations. The loss or damage of equipment or facilities can significantly affect fiscal viability. Any incident involving a chemical containing the word "cyanide" in its description can evoke an emotional reaction from the public, even when no exposures have occurred. Lawsuits that may result from personnel injury/death, public exposures, and/or environmental contamination will also require a serious expenditure of resources. Media attention surrounding an injury, death, or environmental damage can also result in a loss of profits and loss of current as well as future business.

Recommended Risk-Reduction Measures

Company attorneys, safety and health professionals, and environmental specialists should be involved in the development of any procedures for responding to chemical incidents. A company official should be predesignated as a public relations officer with specific training in dealing with the press. Corporate plans and policies should be developed, approved, and implemented long before any need for such arises.

REFERENCES

American Conference of Governmental Industrial Hygienists. 1988. *Documentation of the Threshold Limit Values and Biological Exposure Indices,* 5th Edition (with updates). Cincinnati: ACGIH

Brusick, D. J. 1994. *Methods for Genetic Risk Assessment.* New York: CRC Press/Lewis Publishers.

Bueche, J. R. 1972. *Principles of Physics,* 2nd Edition. New York: McGraw-Hill.

Calabrese, E. J., E. M. Kenyon. *Air Toxics and Risk Assessment.* New York: CRC Press/Lewis Publishers.

Cothern, C. R., N. P. Ross. 1994. *Environmental Statistics, Assessment, and Forecasting.* New York: CRC Press/Lewis Publishers.

Gardner, R. W. 1994. *Chemical Intolerance.* New York: CRC Press/Lewis Publishers.

Howard, P. H., et. al. 1991. *Handbook of Environmental Degradation Rates.* New York: CRC Press/Lewis Publishers.

Keith, L. H. and M. M. Walker. 1995. *Handbook of Air Toxics.* New York: CRC Press/Lewis Publishers.

Lauwerys, R. R., P. Hoet. 1993. *Industrial Chemical Exposure.* New York: CRC Press/Lewis Publishers.

Lewis, R. J., Sr. 1992. *Sax's Dangerous Properties of Industrial Materials,* Volumes 1, 2, and 3, 8th Edition. New York: Van Nostrand Reinhold.

Manahan, S. E. 1992. *Toxicological Chemistry,* 2nd Edition. New York: CRC Press/Lewis Publishers.

Meyer, E. 1990. *Chemistry of Hazardous Materials.* Englewood Cliffs, NJ: Prentice Hall.

Patnaik, P. 1992. *A Comprehensive Guide to the Hazardous Properties of Chemical Substances.* New York: Van Nostrand Reinhold.

Richardson, M. 1995. *Dictionary of Substance and Their Effects,* Vol. I-VI. New York: CRC Press/Lewis Publishers.

Sacarello, H. L. A. 1994. *The Comprehensive Handbook of Hazardous Materials.* New York: CRC Press/Lewis Publishers.

Torkelson, T. R. V. K. Rowe. 1981. *Patty's Industrial Hygiene and Toxicology,* Vol. 2B. New York: Wiley-Interscience.

MATERIAL SAFETY DATA SHEET

CHEMICAL NAME	CRC PRESS/LEWIS PUBLISHERS DISCLAIMER
HYDROGEN FLUORIDE	**CRC PRESS/LEWIS PUBLISHERS DISCLAIMER:** THE INFORMATION AND RECOMMENDATIONS PRESENTED HEREIN ARE BASED ON SOURCES BELIEVED TO BE RELIABLE. CRC MAKES NO REPRESENTATION ON ITS COMPLETENESS OR ACCURACY. IT IS THE RESPONSIBILITY OF THE USER TO DETERMINE THE CHEMICAL'S SUITABILITY FOR ITS INTENDED USE, THE CHEMICAL'S SAFE USE, AND THE CHEMICAL'S PROPER DISPOSAL. NO REPRESENTATIONS AND/OR WARRANTIES, EITHER EXPRESSED OR IMPLIED, OF THE MERCHANTABILITY OR FITNESS FOR A PARTICULAR PURPOSE, OR OF ANY OTHER NATURE, ARE MADE WITH RESPECT TO THE INFORMATION PROVIDED IN THIS MSDS OR TO THE CHEMICAL TO WHICH INFORMATION MAY REFER. CRC NEITHER ASSUMES NOR AUTHORIZES ANY OTHER PERSON TO ASSUME FOR IT, ANY OTHER ADDITIONAL RESPONSIBILITY OR LIABILITY FOR THE USE OF, OR RELIANCE UPON, THIS INFORMATION.

HAZARD WARNING INFORMATION

HEALTH	FIRE	REACTIVE	OTHER	DEGREE OF HAZARD	COLOR CODING	OTHER CODES
4	0	0		0 = Minimum Hazard 1 = Slight Hazard 2 = Moderate Hazard 3 = Serious Hazard 4 = Severe Hazard	HEALTH = BLUE FIRE = RED REACTIVITY = YELLOW OTHER = WHITE	OX = Oxidizer ACID = Acid ALK = Alkali COR = Corrosive W = Use No Water

SECTION I - GENERAL INFORMATION

Characterization **Fluoride**	RCRA Number **U134**	EPA Class **Toxic Waste**
DOT Proper Shipping Name **Hydrogen fluoride (anhydrous) ; Hydrochloric Acid**	Chemical Abstract Service (CAS) Number **7664-39-3**	
DOT Hazard Class and Label Requirements **Corrosive Material, Poison A; Poison Gas.**	DOT Emergency Guide Code **15 (anhydrous) 59 (solution)**	
DOT Identification Number **UN 1052 (anhydrous) UN 1780 (solution)**	Chemical Formula **HF**	

Synonyms
Anhydrous hydrogen fluoride; aqueous hydrogen fluoride; hydrofluoric acid, fluorohydric acid.

SECTION II - HAZARDOUS INGREDIENTS AND IDENTITY INFORMATION

Hazardous Components (specific identity)	OSHA Exposure Criteria	NIOSH Exposure Criteria	Immediately Dangerous to Life and Health (IDLH)	ACGIH Exposure Criteria
Hydrogen fluoride (derivation: By distillation from the reaction product of calcium fluoride and sulfuric acid, also from fluosilicic acid). **1 ppm = 0.83 mg/m^3**	**PEL:** **3 ppm** **2.5 mg/m^3** **CEILING:** **6 ppm (as F)** **5 mg/m^3**	**REL:** **3 ppm** **2.5 mg/m^3** **CEILING:** **6 ppm (as F)** **5 mg/m^3**	**30 ppm**	**TLV (ceiling):** **3 ppm** **2.5 mg/m^3** **STEL:** **Not Applicable**

SECTION III - PHYSICAL AND CHEMICAL CHARACTERISTICS

Boiling Point **67°F (19°C)**	Specific Gravity (H_2O = 1) **1.00 (liquid at 67°F/19°C)**
Vapor Pressure (mmHg) **783 at 68°F (20°C)**	Molecular Weight **20.0**
Vapor Density (Air = 1) **0.7 at 140°F (60°C)**	Melting Point **-118°F (-83°C)**

Solubility
Miscible in water, highly soluble in alcohol, and slightly soluble in ether.

Appearance and Odor
A colorless gas or a fuming liquid (below 67°F) with a strong irritating odor. Normally shipped as a compressed gas in cylinders. Odor Threshold = 5 ppm.

SECTION IV - FIRE AND EXPLOSION HAZARD DATA

Flash Point (method used) **Not Determined**	Explosive Limits in Air % by Volume **LEL: Not Determined UEL: Not Determined**
NFPA Classification **Non-Flammable Gas**	Autoignition Temperature **Not Determined**

Extinguishing Media
Use carbon dioxide, alcohol foam. Do NOT use water (generates hydrofluoric acid solution and heat).

Special Fire Fighting Procedures
Wear full protective clothing and self-contained breathing apparatus (SCBA). Also, use a water spray to keep fire-exposed containers cool. Poisonous gases are produced in fire.

Unusual Fire and Explosion Hazards
Potentially explosive reaction with metals such as yellow brass, lead, stainless steel, aluminum, and cast iron. Firefighters should avoid contact with vapors produced during fire. In the presence of water or moisture, hydrogen fluoride can produce explosive hydrogen gas.

SECTION V - REACTIVITY DATA

Stability		Conditions to Avoid
		Hydrogen fluoride is normally stable in closed, pressurized containers under routine conditions of handling and storage. Keep away from metals and other incompatible materials.
Stable	Unstable	Incompatibility *(materials to avoid)*
X		Hydrogen fluoride is hygroscopic and will absorb moisture from air to produce hydrofluoric acid and release heat. Reacts in contact with most metals, cyanides, and sulfides.
Hazardous Polymerization		Conditions to Avoid
		Under normal conditions of temperature and pressure, hazardous polymerization of hydrogen fluoride is not expected to occur.
May Occur	Will Not Occur	Hazardous Decomposition or By-products
	X	When heated to decomposition, hydrogen fluoride emits highly toxic/poisonous gases, including fluoride fumes. The emission of explosive hydrogen gas during a fire creates a hazard to personnel and property.

SECTION VI - HEALTH HAZARD DATA

Primary Route(s) of Entry:	Inhalation?	Absorption (skin)?	Ingestion?
	X	X (solution)	X

Health Hazards

INHALATION: Vapors cause irritation to the upper respiratory tract with coughing, dypsnea, headache, congestion, ulcers in the nose and throat, tissue dehydration, rhinitis, bronchitis, bone changes, congestion, fluorosis, and/or pulmonary edema.

ABSORPTION: Skin contact causes immediate swelling, blistering, necrosis, and burning that becomes more intense with prolonged contact. Burns can be bone-deep. Eye contact causes burning, watering, redness, corneal ulceration, blurred vision, and possible blindness.

INGESTION: Not likely (hydrogen fluoride is a gas at room temperature).

Carcinogenicity	NTP Listed?	IARC Cancer Review Group?	OSHA Regulated?	Target Organs?
Unknown Human Unknown Animal	No	No	29 CFR 1910.1000 Table Z-1	Respiratory system, eyes, skin; bone.

Medical Conditions Generally Aggravated by Exposure

Respiratory impairments (bronchitis, asthma); Skin conditions (dermatitis).

Emergency and First-aid Procedures

Eye contact: Flush immediately with water for 15 minutes (minimum), seek medical attention. Skin contact: Remove all contaminated clothing. Immediately wash area with large amounts of water for 2 to 4 hours or soak in iced Zephiran (0.13%), Epson salts or a 70% denatured ethyl alcohol solution for 1 to 4 hours. Seek medical assistance. For inhalation: Remove the person from exposure. Provide respiratory assistance and CPR. Transfer to medical facility. If swallowed: Not likely. If accidental ingestion of solutions occurs, seek medical attention or contact poison control center. Do NOT induce vomiting.

SECTION VII - PRECAUTIONS FOR SAFE HANDLING AND USE

Steps to be Taken in Case Material is Released or Spilled

Ventilate area (avoid vapors). Shut off flow of gas if possible. Solution spills should be covered with lime to form a slurry and deposited in sealed containers. Ventilate area of spill and remove all ignition sources (hydrogen gas may be present). Restrict those not involved in cleanup from entering area.

Preferred Waste Disposal Method

No Citation.

Precautions to be Taken in Handling and Storage

Avoid contact with incompatible materials since violent reactions can occur. Store in tightly closed containers in cool, dry, well-ventilated area away from heat and moisture.

Other Precautions and Warnings

Sources of ignition are prohibited where hydrogen fluoride is used, handled, or stored. Detached facility storage is preferred. Prevent damage to containers. Use non-sparking tools around containers.

SECTION VIII - CONTROL MEASURES AND PERSONAL PROTECTIVE EQUIPMENT

Respiratory Protection *(specify type)*

Exposure levels are relatively low. Recommend using a supplied-air respirator or a self-contained breathing apparatus (SCBA) operated in pressure demand or other positive pressure mode.

Ventilation

Local exhaust or general mechanical systems recommended.

Protective Gloves	Eye Protection	Other Protective Clothing
Acid Resistant Neoprene or PVC	Chemical Goggles or Face Mask	Protective Apron

Work/Hygiene Practices

Always wash hands thoroughly after using chemical; never bring food, drink, or smoking materials into vicinity of chemicals.

HYDROGEN FLUORIDE

HF

CAS: 7664-39-3

IDENTIFICATION AND TYPICAL USES

Hydrogen fluoride is a colorless gas or a fuming liquid (below 67°F) with a strong, irritating, pungent odor. It is used as a catalyst in alkylation, isomerization, condensation, dehydration, and polymerization reactions. It is also used as a fluorinating agent in organic and inorganic reactions, in the production of fluorine and aluminum fluoride, as an additive in liquid rocket propellants, and in the refining of uranium.

RISK ASSESSMENT: HEALTH

General Assessment

Hydrogen fluoride is a deadly human poison by *inhalation* and by skin contact (*absorption*). It can also enter the body through *ingestion*, but this exposure route is extremely unlikely since the chemical exists as a gas at room temperature and pressure. It is highly caustic and, therefore, an extremely corrosive and destructive compound. While there are no reports of any carcinogenic effects resulting from exposure to hydrogen fluoride, experimental teratogenic, reproductive, and mutagenic effects have been documented.

Inhalation severely dehydrates tissues and causes destructive corrosive damage to the nose, throat, and mucosa of the upper respiratory tract with possible ulceration of the nose and throat. There can be coughing, dypsnea (labored breathing), lachrymation, and nausea. Severe exposures may cause fluorosis and/or pulmonary edema (fluid in the lungs). The development of pulmonary edema can be delayed up to 48 hours after exposure, thereby creating a false sense of security with regard to health exposure risk.

Skin contact with the vapor or the liquid solution causes severe burning due to hydrogen fluoride's intense corrosive action on human tissue. The liquid may continue to burn and destroy skin tissues long after the exposure if it is not properly removed. Because of the fluoride ion's affinity for calcium, burns may be bone-deep in some instances. Extensive hydrogen fluoride burns can cause a calcium imbalance which can lead to cardiac irregularities. The extremely corrosive properties of hydrogen fluoride can cause immediate stinging, burning, redness, swelling, membrane ulceration, painful skin blistering, and local necrosis (tissue destruction) upon contact. Eye contact causes immediate stinging and watering with tissue damage and the possibility of corneal damage. There will be redness and swelling of the lids, intense pain, and tissue destruction. Loss of vision is likely.

Ingestion is extremely rare under normal operating conditions. However, if liquid solutions are accidentally swallowed, there can be severe burns to the lips, mouth, throat, and stomach. If vomiting occurs and the liquid is aspirated into the lungs, the corrosive action on the lung tissues can be fatal.

Acute Health Effects

The following acute (short-term) health effects may occur immediately or shortly after exposure to hydrogen fluoride:

Skin: Irritation/rash or burning feeling on contact with redness, blistering, localized tissue destruction, and deep-seated burns.

Eye: Severe irritation; may cause burns and damage with permanent loss of vision.

Lung: Severe irritation of the mucous membranes in the nose, throat, and lungs following exposure. Pulmonary edema is likely to occur within 48 hours of exposure.

Chronic Health Effects

The following chronic (long-term) health effects may occur at some time after exposure to hydrogen fluoride and can last for months or even years:

Cancer Hazards: According to information presented in the references, hydrogen fluoride has not been adequately tested for its ability to cause cancer in laboratory test animals. However, mutation data have been reported and many scientists believe that such chemicals may pose a cancer risk in the long-term. Additional research is required in this regard.

Reproductive Hazard: According to information presented in the references, hydrogen fluoride has been shown to cause teratogenic and other adverse reproductive effects in test animals.

Other Chronic Effects: Repeated or prolonged exposure may cause fluorosis with possible erosion of teeth enamel. Long-term exposure may damage the kidneys and liver.

Recommended Risk-Reduction Measures

Personnel should avoid direct contact with hydrogen fluoride. It is extremely corrosive and highly toxic to humans. If a less toxic chemical cannot be substituted for a hazardous substance, then *engineering controls* are the most effective method of reducing exposures.

The best protection is to enclose operations and/or provide local exhaust ventilation at the site of chemical release. While not always operationally feasible, isolating operations can also reduce exposure. Using respiratory protection is less effective than the controls mentioned above, but is still advisable whenever working with or around hydrogen fluoride. The exposure level is extremely low (3 ppm) and difficult to accurately monitor. It is not detectable by odor below this PEL and, therefore, by the time its presence is noticed by smell, an overexposure may have already occurred. While an acid vapor respirator or gas mask may suffice for low or infrequent exposures, the best protection is obtained using a self-contained breathing apparatus (SCBA) or a supplied-air respirator with full facepiece operated in pressure demand mode. If a full facepiece is not available, then chemical goggles should be worn to protect the eyes. Whenever a chemical splash hazard exists, a face shield and a protective apron should be worn. To prevent hand and skin exposures, acid resistant neoprene or polyvinyl chloride gloves should be used. Glove manufacturers should be contacted and permeation studies obtained *before* final glove selections are made.

Administrative controls should also be in place to minimize the potential for human exposures. These may include written procedures or policies which specify the methods and techniques that will be practiced whenever personnel are to work with hydrogen fluoride.

All personnel should receive *training* on the use, hazards, protective measures, emergency actions, and other precautions per 29 CFR 1910.1200 (Hazard Communication), prior to the first assignment in an area where hydrogen fluoride is used or stored.

Before beginning employment and at regular intervals thereafter (e.g., annually), the following tests are recommended:

- ☑ Lung function tests (establish baseline).

If symptoms occur or overexposure is suspected, the following additional tests are suggested:

- ☑ Lung function tests (compare to baseline).
- ☑ Consider chest X-ray following acute overexposure (may be negative it taken immediately after exposure due to delayed onset of pulmonary edema).
- ☑ Liver and kidney function tests.
- ☑ If skin exposure has occurred, affected area should be washed with flooding amounts of water for 2 to 4 hours. If possible, soak in iced Zephiran (0.13%), Epson salts, or 70% denatured ethyl alcohol for 1 to 4 hours.

Any evaluation should include a careful history of past and present symptoms with an examination. Medical tests that look for damage already done are not a substitute for controlling exposure. Also, because smoking can cause heart disease, emphysema, and other respiratory disorders, smokers exposed to hydrogen fluoride may experience symptoms more quickly than nonsmokers under the same conditions of exposure. Prudent risk management requires proper consideration of *all* possible exposure factors.

Other methods to reduce exposure include:

- ☑ Where possible, enclose operations and use local exhaust ventilation at the site of chemical release. If local exhaust ventilation or enclosure is not used, respiratory protection should be mandatory.
- ☑ Always ensure that proper protective clothing is worn when using chemical substances.
- ☑ Wash thoroughly immediately after exposure to hydrogen fluoride and at the end of the work shift or before eating, drinking, or smoking.
- ☑ Hazard warning information should be posted in the work area. In addition, as part of an on-going education and training program, all information on the health and safety hazards of hydrogen fluoride should be communicated to all exposed workers.

RISK ASSESSMENT: ENVIRONMENT

General Assessment

The environment is at risk of exposure during transportation, storage, disposal, or destruction of hydrogen fluoride. In almost every scenario, the threat of environmental exposure is contingent upon the proper handling of the chemical substance. Accidental spills, large or small, can result in fire, explosion, and possible contamination of the surrounding environmental mediums (water, soil, and especially air).

Hydrogen fluoride is considered a non-flammable gas. However, it is extremely corrosive to many materials. It is incompatible with water or steam and will liberate explosive hydrogen gas and heat on contact. It is corrosive to most metals, glass, and reacts with calcium silicate to form silicon tetrafluoride. Reaction with alkali metals can be explosive and violent. If it contacts porous materials such as wood or cement, the resultant continuation can remain indefinitely. These characteristics require special consideration during any emergency situation involving a leak or spill of hydrogen fluoride. Should hydrogen fluoride ever come into contact with any incompatible substances either during use, transportation, or storage, the formation of highly toxic and/or highly explosive commodities is extremely possible.

Hydrogen fluoride can enter the environment mainly industrial discharges or spills.

Acute Ecological Effects

Acute (short-term) toxic effects may include the death of animals, birds, or fish and death or low growth rate in plants. Acute effects are seen 2 to 4 days after animals or plants are exposed to hydrogen fluoride.

Insufficient data are available to evaluate or predict the short-term effects of hydrogen fluoride exposure to aquatic life, plants, birds, or land animals.

Chronic Ecological Effects

Chronic (long-term) toxic effects may include shortened life span, reproductive problems, lower fertility, and changes in appearance or behavior in exposed animals. These effects can be seen long after first exposure(s) to toxic chemicals.

Insufficient data are available to evaluate or predict the long-term effects of hydrogen fluoride to aquatic life, plants, birds, or land animals.

Water Solubility

Hydrogen fluoride is highly soluble in water. Concentrations of over 1000 milligrams will mix with a liter of water.

Persistence in the Environment

Hydrogen fluoride is slightly persistent in the aquatic environment, with a half-life between 2 and 20 days. The half-life of a pollutant is the amount of time it takes for one half of the chemical to be degraded.

Bioaccumulation in Aquatic Organisms

Some substances increase in concentration, or *bioaccumulate,* in living organisms as they breathe contaminated air, drink contaminated water, or eat contaminated food. These chemicals can become concentrated in the tissues and internal organs of animals as well as humans.

Small quantities of such acids as hydrogen fluoride will be neutralized by the alkalinity in aquatic systems, but larger quantities can lower the pH for extended periods of time. The concentration of fluoride found in the edible tissues of most aquatic species that are consumed by humans is expected to be about the same as the average concentration of hydrogen fluoride in the water from which the fish was taken.

Recommended Risk-Reduction Measures

Proper training of all transporters will reduce the likelihood of a mishap or accident resulting in a leak or spill to the environment. The correct use of DOT labeling (per the MSDS) on all containers, trucks, and rail cars will enable emergency responders to react properly and quickly to any disaster thereby reducing the potential risk to the environment and to personnel.

Storage of hydrogen fluoride should be segregated from incompatible chemicals to minimize the risk of cross-contamination or contact. Buildings designated for storage should be equipped with appropriate fire protection systems (alarms, sprinklers, emergency lighting, portable extinguishers). Containers should be protected against physical damage and stored away from direct sunlight or other heat sources. Because it will corrode and/or react with many materials, selection of storage containers requires proper consideration. Hydrogen fluoride will corrode glass and react with most metals. Among the substances it does not corrode are lead, platinum, wax, and polyethylene. It should therefore be stored in lead carboys or in waxed or polyethylene bottles.

If a spill or leak to the environment has occurred, fire department, emergency response and/or hazardous materials spill personnel should be notified immediately. Cleanup should be attempted only by those trained in proper spill containment procedures using non-sparking tools. Contaminated soils (solution spills) should be removed for incineration and replaced with clean soil. If hydrogen fluoride should contact the water table, aquifer, or navigable waterway, time is of the essence. It is highly soluble in water and, therefore, total containment and remediation may not be possible. When such spills occur, the local and/or state emergency response authorities must be notified. A comprehensive emergency response or disaster preparedness/recovery plan should be in place prior to any operations involving the use, transportation, storage, or disposal of hydrogen fluoride. If hydrogen fluoride is spilled or leaked, the following specific steps are recommended:

- ☑ Restrict persons not wearing protective clothing from area of spill or leak until cleanup is complete and area can be opened for normal work.
- ☑ Remove ignition sources and ventilate area.
- ☑ Shut off gas leak, if possible. If leak is from a cylinder and it cannot be turned off, remove cylinder to safe area (outdoors) and allow to vent. Take care to avoid contact with the vapors or any escaping compressed liquids. Be aware of the possibility that hydrogen gas may be in the area (explosion hazard).
- ☑ *Carefully* neutralize liquid spills with lime to form a slurry and collect for disposal. Do not flush to drains, sewers, or waterways.
- ☑ It may be necessary to dispose of hydrogen fluoride as a hazardous waste. The responsible state agency or the regional office of the federal Environmental Protection Agency (EPA) should be contacted for specific recommendations.

RISK ASSESSMENT: BUSINESS

General Assessment

Accidents or mishaps involving hydrogen fluoride can present a significant threat to business operations. The loss or damage of equipment or facilities can significantly affect fiscal viability. Lawsuits that may result from personnel injury/death, public exposures, and/or environmental contamination will also require a serious expenditure of resources. Media attention surrounding an injury, death, or environmental damage can also result in a loss of profits and loss of current as well as future business.

Recommended Risk-Reduction Measures

Company attorneys, safety and health professionals, and environmental specialists should be involved in the development of any procedures dealing with safety or emergency response. A company official should be pre-designated as a public relations officer with specific training in dealing with the press. Corporate plans and policies should be developed, approved, and implemented long before any need for such arises.

REFERENCES

American Conference of Governmental Industrial Hygienists. 1988. *Documentation of the Threshold Limit Values and Biological Exposure Indices*, 5th Edition (with updates). Cincinnati: ACGIH

Asante-Duah, D. K. 1993. *Hazardous Waste Risk Assessment.* New York: CRC Press/Lewis Publishers.

Bueche, J. R. 1972. *Principles of Physics*, 2nd Edition. New York: McGraw-Hill.

Burrell, R., D. K. Flaherty, and Leonard J. Sauers. 1992. *Toxicology of the Immune System.* New York: Van Nostrand Reinhold.

Calabrese, E. J., E. M. Kenyon. *Air Toxics and Risk Assessment.* New York: CRC Press/Lewis Publishers.

Gardner, R. W. 1994. *Chemical Intolerance.* New York: CRC Press/Lewis Publishers.

Howard, P. H., et. al. 1991. *Handbook of Environmental Degradation Rates.* New York: CRC Press/Lewis Publishers.

Keith, L. H. and M. M. Walker. 1995. *Handbook of Air Toxics.* New York: CRC Press/Lewis Publishers.

Lewis, R. J., Sr. 1993. *Hawley's Condensed Chemical Dictionary*, 12th Edition. New York: Van Nostrand Reinhold.

Lewis, R. J., Sr. 1992. *Sax's Dangerous Properties of Industrial Materials*, Volumes 1, 2, and 3, 8th Edition. New York: Van Nostrand Reinhold.

Meyer, E. 1990. *Chemistry of Hazardous Materials.* Englewood Cliffs, NJ: Prentice Hall.

Patnaik, P. 1992. *A Comprehensive Guide to the Hazardous Properties of Chemical Substances.* New York: Van Nostrand Reinhold.

U.S. Department of Health and Human Services, National Institute for Occupational Safety and Health. 1994. *NIOSH Pocket Guide to Chemical Hazards.* Washington, D.C.: U.S. Government Printing Office.

MATERIAL SAFETY DATA SHEET

CHEMICAL NAME

HYDROGEN PEROXIDE
(3-52%)

HAZARD WARNING INFORMATION

HEALTH	FIRE	REACTIVE	OTHER	DEGREE OF HAZARD	COLOR CODING	OTHER CODES
2	0	1	OX	0 = Minimum Hazard 1 = Slight Hazard 2 = Moderate Hazard 3 = Serious Hazard 4 = Severe Hazard	HEALTH = BLUE FIRE = RED REACTIVITY = YELLOW OTHER = WHITE	OX = Oxidizer ACID = Acid ALK = Alkali COR = Corrosive W = Use No Water

SECTION I - GENERAL INFORMATION

Characterization
Peroxide

RCRA Number
None

EPA Class
Not Applicable

DOT Proper Shipping Name
Hydrogen Peroxide (aqueous solution)

Chemical Abstract Service (CAS) Number
7722-84-1

DOT Hazard Class and Label Requirements
Oxidizer; Oxidizer and Corrosive

DOT Emergency Guide Code
45 (52-100% solution) 60 (8-20%)

DOT Identification Number
UN 2014 (20-52% solution) UN984 (8-20%)

Chemical Formula
H_2O_2

Synonyms
Hydrogen dioxide; hydrogen peroxide (aqueous); hydroperoxide; peroxide; Albone; carbamide peroxide; Interox; Inhibine; Kastone; Perhone; Perhydrol; Peroxaan; Superoxol, T-stuff; urea peroxide.

SECTION II - HAZARDOUS INGREDIENTS AND IDENTITY INFORMATION

Hazardous Components (specific identity)	OSHA Exposure Criteria	NIOSH Exposure Criteria	Immediately Dangerous to Life and Health (IDLH)	ACGIH Exposure Criteria
Hydrogen peroxide (derivation: By autoxidation of an alkyl anthrahydroquinone in which the quinone formed in the oxidation step is reduced to the starting material by hydrogen in the presence of a supported palladium catalyst; by an electrolytic process). **1 ppm = 1.41 mg/m^3**	**PEL:** 1 ppm 1.4 mg/m^3 **STEL:** Not Established	**REL:** 1 ppm 1.4 mg/m^3 **STEL:** Not Established	75 ppm	**TLV:** 1 ppm 1.4 mg/m^3 **STEL:** Not Established

SECTION III - PHYSICAL AND CHEMICAL CHARACTERISTICS

Boiling Point
226 to 237°F (108-114°C)

Specific Gravity (H_2O = 1)
1.29

Vapor Pressure (gas)
1 at 59°F (15°C)

Molecular Weight
34.0

Vapor Density (Air = 1)
1.2

Melting Point
-33 to -52°F (-27 to -61°C)

Solubility
Soluble in water, alcohol, and ether. Insoluble in petroleum ether. Decomposes in most solvents.

Appearance and Odor
Colorless liquid with a slight, sharp odor. Often used in aqueous solutions of varying concentrations.

SECTION IV - FIRE AND EXPLOSION HAZARD DATA

Flash Point (method used)
Not Applicable

Explosive Limits in Air % by Volume
LEL: Not Applicable UEL: Not Applicable

NFPA Classification
Non-Combustible Liquid

Autoignition Temperature
Not Applicable

Extinguishing Media
Use flooding amounts of water on surrounding fire. Do NOT use dry chemical.

Special Fire Fighting Procedures
Wear full protective clothing and self-contained breathing apparatus (SCBA). Poisonous gases are produced in fire. Move containers from fire if it can be done without risk. Remain clear of smoke. Concentrations of less than 50% hydrogen peroxide will not usually ignite combustible materials.

Unusual Fire and Explosion Hazards
A non-combustible liquid but a powerful oxidizer. Dangerous fire and explosion hazard when exposed to heat or flame. Containers may explode in fire. Flammable hydrogen gas may be formed in fire.

SECTION V - REACTIVITY DATA

Stability		Conditions to Avoid
		Hydrogen peroxide solution free of contaminants is normally stable but will still decompose to release oxygen when inhibited with acetanilide or sodium stannate. It must be stored in vented containers.
Stable	Unstable	Incompatibility *(materials to avoid)*
X		Reacts with alkalis, ammonia and its carbonates, alcohols, iodides, hypophosphites, lime water, sulfites, finely divided metals, permanganates, organic compounds (such as wood, oil), and other combustibles.
Hazardous Polymerization		Conditions to Avoid
		Under normal conditions of temperature and pressure, hazardous polymerization of hydrogen peroxide is not expected to occur.
May Occur	Will Not Occur	Hazardous Decomposition or By-products
	X	When heated to decomposition, explosion is possible due to the release of hydrogen gas. In fire, hydrogen peroxide produces water and ½ oxygen.

SECTION VI - HEALTH HAZARD DATA

Primary Route(s) of Entry:	Inhalation?	Absorption (skin)?	Ingestion?
	X		

Health Hazards

INHALATION: Irritation of the nose, mouth, and lungs. At low concentrations (3%), relatively non-toxic. Between 8-20%, very corrosive. The vapors can cause lachrymation, sneezing, headache, dypsnea, cyanosis, salivation, nausea and vomiting, giddiness, insomnia, tremors, coma, bleaching of hair, and pulmonary edema.

SKIN & EYES: A severe (caustic) skin irritant that can cause whitening of the surface, redness, blisters, and vesiculation. Eye contact can cause corneal ulceration and loss of sight.

INGESTION: Irritation and burning of the lips, mouth, and throat, painful swallowing, salivation, throat swelling, intense thirst, nausea and bloody vomiting, stomach distention and perforation, convulsions, coma, possible cerebral edema, and death.

Carcinogenicity	NTP Listed?	IARC Cancer Review Group?	OSHA Regulated?	Target Organs?
Unknown Human Questioned Animal	No	Group 3	29 CFR 1910.1000 Table Z-1	Respiratory system, eyes, skin, CNS.

Medical Conditions Generally Aggravated by Exposure

Chronic respiratory diseases or skin diseases may be aggravated by exposure.

Emergency and First-aid Procedures

Eye contact: Flush immediately with water for 15 minutes (minimum), lifting and holding eyelids open. Seek medical attention. Skin contact: Remove all contaminated clothing. Immediately wash area with large amounts of soap and water for 15 minutes (minimum). Seek medical assistance. For inhalation: Remove the person from exposure. Provide respiratory assistance and CPR. Transfer to medical facility. Observe for 48 hours for lung effects. If swallowed: Seek medical attention immediately. Give large amounts of water and induce vomiting (unless advised otherwise).

SECTION VII - PRECAUTIONS FOR SAFE HANDLING AND USE

Steps to be Taken in Case Material is Released or Spilled

Absorb liquids in vermiculite, or similar material and deposit in sealed containers. Ventilate area of spill or leak. Restrict those not involved in cleanup from entering area. Remove all sources of ignition.

Preferred Waste Disposal Method

No citation.

Precautions to be Taken in Handling and Storage

Store to avoid contact with incompatible materials. Store in tightly closed containers in cool, dry, well-ventilated area away from heat and flame. Prevent contamination to avoid violent decomposition.

Other Precautions and Warnings

Do NOT use near organic materials. Heat may cause containers to build up pressure and explode.

SECTION VIII - CONTROL MEASURES AND PERSONAL PROTECTIVE EQUIPMENT

Respiratory Protection *(specify type)*

Exposure level is low and difficult to monitor. Recommend use of a supplied-air respirator with full facepiece, in continuous flow mode, or a SCBA operated in pressure demand or positive pressure mode.

Ventilation

Local exhaust or general mechanical systems recommended.

Protective Gloves	Eye Protection	Other Protective Clothing
Butyl, Nitrile, Natural Rubber, Viton	Chemical Goggles or Face Mask	Protective Apron

Work/Hygiene Practices

Always wash hands thoroughly after using chemical; never bring food, drink, or smoking materials into vicinity of chemicals.

HYDROGEN PEROXIDE (3-52%)

H_2O_2

CAS: 7722-84-1

IDENTIFICATION AND TYPICAL USES

Hydrogen peroxide is a colorless liquid with a slight, sharp odor. It is often used in aqueous solutions of varying concentrations. It is used in bleaching and deodorizing of textiles, wood, pulp, hair, and fur. It is also a source of organic and inorganic peroxides. It is used extensively in the pulp and paper industries, as a plasticizer, in waste treatment, in the manufacture of foam rubber, in making glycerol, in dyeing, in electroplating, as an antiseptic (dilute solutions), as a laboratory reagent, in epoxidation, in hydroxylation, in oxidation, and reduction. It is used as a viscosity control for starch and cellulose derivatives, in refining and cleaning metals, as a bleaching and oxidizing agent in foods, as a neutralizing agent in wine distillation, as a seed disinfectant, and as a substitute for chlorine in water and sewage treatment.

RISK ASSESSMENT: HEALTH

General Assessment

Hydrogen peroxide at concentrations of 3% is relatively non-toxic. Concentrations over 8% is considered a severe irritant causing caustic destruction to tissues. Exposure can be by *inhalation* of its vapors, direct *ingestion* of the liquid, or by skin contact (although absorption through the skin has not been reported). At concentrations above 50% it has caused questionable carcinogenic effects in test animals but the extent of its human cancer-causing properties is unknown. Human mutation data have been reported.

Inhalation of concentrations above 8% causes mild to severe irritation and burns to the eyes, nose, and throat. There will be tearing (lachrymation), sneezing, cough with phlegm, headache, dypsnea, cyanosis, salivation, giddiness, nausea, vomiting, muscle weakness, ulceration of the nasal mucosa, insomnia, tremors, parasthesia, secondary chemical pneumonia, and pulmonary edema (fluid in the lungs). Symptoms of pulmonary edema may be delayed up to 48 hours thereby creating a false sense of security with regard to health exposure risk. Serious cases end in shock, convulsions, and death. The vapors may also bleach the hair white.

On skin contact, hydrogen peroxide will cause irritation and immediate erythema (redness). If allowed to remain in contact, there will be painful blisters, swelling, vesiculation, and corrosive tissue damage. Eye contact will also be irritating and painful and may cause corneal ulceration and possible damage to vision.

Ingestion of concentrations between 8 and 50% causes irritation and burning to the lips, mouth, throat, and gastrointestinal tract with painful swallowing, excessive salivation, thirst, throat swelling, nausea and bloody vomiting due to digestive hemorrhage, stomach distention and perforation, convulsions, possible cerebral edema (fluid on the brain), and death.

Acute Health Effects

The following acute (short-term) health effects may occur immediately or shortly after exposure to hydrogen peroxide:

Skin: Immediate irritation, rash, and burning feeling on contact. Corrosive destruction possible.

Eye: Irritation and possible damage if allowed to remain in contact. May cause burns.

Lung: Irritation of the respiratory tract, including the nose and mouth. There may be cough with phlegm, salivation, and shortness of breath. Pulmonary edema is possible at higher levels, which can be fatal.

CNS: Causes giddiness, parasthesia, muscle weakness, convulsions, cyanosis, tremors, coma, and death.

Other: Exposure to the vapors may bleach the hair and skin white.

Chronic Health Effects

The following chronic (long-term) health effects may occur at some time after exposure to hydrogen peroxide and can last for months or even years:

Cancer Hazards: According to information presented in the references, hydrogen peroxide concentrations greater than 8% may cause mutations (genetic changes) in living cells. Its carcinogenic properties are questioned in animals and unknown in humans.

Reproductive Hazard: According to information presented in the references, hydrogen peroxide has not been adequately tested for its ability to adversely affect reproduction.

Other Chronic Effects: None reported.

Recommended Risk-Reduction Measures

Personnel should avoid direct contact with hydrogen peroxide. The exact nature of its carcinogenic properties is unknown in humans. Solutions greater than 8% hydrogen peroxide are severely corrosive and destructive to human tissues. If a less toxic chemical cannot be substituted for a hazardous substance, then *engineering controls* are the most effective method of reducing exposures. The best protection is to enclose operations and/or provide local exhaust ventilation at the site of chemical release. While not always operationally feasible, isolating operations can also reduce exposure. Using respiratory protection is less effective than the controls mentioned above, but is still advisable whenever working with or around hydrogen peroxide, especially at concentrations above 8%. The exposure limit for hydrogen peroxide is relatively low (1 ppm) and difficult to accurately monitor. For any concentration, a self-contained breathing apparatus (SCBA) with full facepiece operated in pressure demand or a supplied-air respirator in continuous flow mode are the recommended respiratory protection methods of choice. If a full facepiece is not available, then chemical goggles should be worn to protect the eyes. Whenever a chemical splash hazard exists, a face shield and a rubber apron should be worn. To prevent hand and skin exposures, nitrile, Viton, butyl, or natural rubber gloves should be used. Glove manufacturers should be compensated and permeation studies obtained *before* glove selections have been made.

Administrative controls should also be in place to minimize the potential for human exposures. These may include written procedures or policies which specify the methods and techniques that will be practiced whenever personnel are to work with or around hydrogen peroxide.

All personnel should receive *training* on the use, hazards, protective measures, emergency actions, and other precautions per 29 CFR 1910.1200 (Hazard Communication), prior to the first assignment in an area where hydrogen peroxide is used or stored.

Before beginning employment and at regular intervals thereafter (e.g., annually), or for those having frequent or potentially high exposures, the following medical tests are recommended:

☑ Lung function tests.

If symptoms develop or overexposure is suspected, the following tests are recommended:

☑ Lung function tests.

☑ Consider chest X-ray after acute overexposure (may be negative if taken immediately after exposure due to delayed onset of pulmonary edema).

☑ Gastric lavage may be recommended if swallowed.

☑ Analysis of body fluids (gastric aspirates) using titanium chloride reaction will reveal peroxides if performed immediately.

Any evaluation should include a careful history of past and present symptoms with an examination. Medical tests that simply look for existing damage are not a substitute for controlling exposures. Also, because smoking can cause lung cancers, emphysema, heart disease, and a number of other respiratory problems, the effects of exposure to hydrogen peroxide in smokers may appear to be significantly higher than that which is seen in non-smokers. This must be taken into consideration for proper risk assessment and management.

Other methods to reduce exposure include:

☑ Where possible, enclose operations and use local exhaust ventilation at the site of chemical release. If local exhaust ventilation or enclosure is not used, respiratory protection should be mandatory.

☑ Always ensure that proper protective clothing is worn when using chemical substances.

☑ Wash thoroughly immediately after exposure to hydrogen peroxide and at the end of the work shift or before eating, drinking, or smoking.

☑ Hazard warning information should be posted in the work area. In addition, as part of an on-going education and training program, all information on the health and safety hazards of hydrogen peroxide should be communicated to all potentially exposed workers.

☑ Eye wash stations should be provided in the immediate work area for emergency use. If there is a possibility of skin exposure to hydrogen peroxide, emergency shower facilities should also be provided in the immediate area.

☑ Personnel should never wear contaminated clothing home (family members can be exposed). Clothing should be laundered by personnel who have been trained on the hazards of hydrogen peroxide.

RISK ASSESSMENT: ENVIRONMENT

General Assessment

The environment is at risk of exposure during transportation, storage, disposal, or destruction of hydrogen peroxide. In almost every scenario, the threat of environmental exposure is contingent upon the proper handling of the chemical substance. Accidental spills, large or small, can result in fire, explosion, and possible contamination of the surrounding environmental mediums (water, soil, and air).

Hydrogen peroxide is considered a non-combustible liquid but an extremely powerful oxidizer. Concentrated solutions can decompose violently if trace impurities are present. Contact with oxidizable materials may cause spontaneous combustion. Violent explosions may occur in contact with many common metals. These characteristics require extreme caution in handling, storage, transportation, and disposal. Also, contact with incompatible materials can cause violent fire or explosion and special consideration is required during any emergency situation involving a leak or spill of hydrogen peroxide. Should hydrogen peroxide ever come into contact with these incompatible substances during use, transportation, storage, or disposal, the formation of highly toxic and/or highly explosive commodities is also extremely possible.

Hydrogen peroxide may enter the environment through industrial discharges or spills.

Acute Ecological Effects

Acute (short-term) toxic effects may include the death of animals, birds, or fish and death or low growth rate in plants. Acute effects are seen 2 to 4 days after animals or plants are exposed to hydrogen peroxide.

Insufficient data are available on the short-term effects of hydrogen peroxide exposure to aquatic life, plants, birds, or land animals.

Chronic Ecological Effects

Chronic (long-term) toxic effects may include shortened life span, reproductive problems, lower fertility, and changes in appearance or behavior in exposed animals. These effects can be seen long after first exposure(s) to toxic chemicals.

Insufficient data are available on the long-term effects of hydrogen peroxide to aquatic life, plants, birds, or land animals.

Water Solubility

Hydrogen peroxide is highly soluble in water. Concentrations of 1000 milligrams or more will mix with a liter of water.

Persistence in the Environment

Insufficient data are available on the properties of hydrogen peroxide to estimate or predict is distribution in the environment. Limited data suggest that hydrogen peroxide will be moderately persistent in water, with a half-life of between 2 and 20 days. The half-life of a pollutant is the amount of time it takes for one half of the chemical to be degraded.

Bioaccumulation in Aquatic Organisms

Some substances increase in concentration, or *bioaccumulate,* in living organisms as they breathe contaminated air, drink contaminated water, or eat contaminated food. These chemicals can become concentrated in the tissues and internal organs of animals as well as humans.

Hydrogen peroxide is not expected to accumulate in the edible tissues of fish.

Recommended Risk-Reduction Measures

Proper training of all transporters will reduce the likelihood of a mishap or accidents resulting in a leak or spill to the environment. The correct use of DOT labeling (per the MSDS) on all containers, trucks, and rail cars will enable emergency responders to react properly and quickly to any disaster thereby reducing the potential risk to the environment and to personnel.

Storage of hydrogen peroxide should be segregated from incompatible chemicals to minimize the risk of

cross-contamination or contact. Buildings designated for storage should be equipped with appropriate fire protection systems (alarms, sprinklers, emergency lighting, portable extinguishers).

If a spill or leak to the environment has occurred, fire department, emergency response and/or hazardous materials spill personnel should be notified immediately. Cleanup should be attempted only by those trained in proper spill containment procedures. Contaminated soils should be removed for incineration and replaced with clean soil. If hydrogen peroxide should contact the water table, aquifer, or navigable waterway, the local and/or state emergency response authorities must be notified. A comprehensive emergency response or disaster preparedness/recovery plan should be in place prior to any operations involving the use, transportation, storage, or disposal of hydrogen peroxide. If hydrogen peroxide is spilled or leaked, the following specific steps are recommended:

- ☑ Restrict persons not wearing protective clothing from area of spill or leak until cleanup is complete and area can be opened for normal work.
- ☑ Remove all ignition sources and ventilate area.
- ☑ Absorb liquids in vermiculite, dry sand, earth, or similar material and deposit in sealed containers.
- ☑ It may be necessary to dispose of hydrogen peroxide as a hazardous waste. The responsible state agency or the regional office of the federal Environmental Protection Agency (EPA) should be contacted for specific recommendations.

RISK ASSESSMENT: BUSINESS

General Assessment

Accidents or mishaps involving hydrogen peroxide can present a significant threat to business operations. The loss or damage of equipment or facilities can significantly affect fiscal viability. Lawsuits that may result from personnel injury/death, public exposures, and/or environmental contamination will also require a serious expenditure of resources. Media attention surrounding an injury, death, or environmental damage can also result in a loss of profits and loss of current as well as future business.

Recommended Risk-Reduction Measures

Company attorneys, safety and health professionals, and environmental specialists should be involved in the development of any procedures dealing with safety or emergency response. A company official should be pre-designated as a public relations officer with specific training in dealing with the press. Corporate plans and policies should be developed, approved, and implemented long before any need for such arises.

REFERENCES

American Conference of Governmental Industrial Hygienists. 1988. *Documentation of the Threshold Limit Values and Biological Exposure Indices,* 5th Edition (with updates). Cincinnati: ACGIH

Bueche, J. R. 1972. *Principles of Physics,* 2nd Edition. New York: McGraw-Hill.

Calabrese, E. J. 1994. *Biological Effects of Low Level Exposures.* New York: CRC Press/Lewis Publishers.

Cockerman, L. G., B. S. Shane. 1994. *Basic Environmental Toxicology.* New York: CRC Press/Lewis Publishers.

Gardner, R. W. 1994. *Chemical Intolerance.* New York: CRC Press/Lewis Publishers.

Hallenbeck, W. H. 1993. *Quantitative Risk Assessment for Environmental and Occupational Health,* 2nd Edition. New York: CRC Press/Lewis Publishers.

Jones, F. E. 1994. *Toxic Organic Vapors in the Workplace.* New York: CRC Press/Lewis Publishers.

Keith, L. H. and M. M. Walker. 1995. *Handbook of Air Toxics.* New York: CRC Press/Lewis Publishers.

Lauwerys, R. R., P. Hoet. 1993. *Industrial Chemical Exposure.* New York: CRC Press/Lewis Publishers.

Mellström, G. A., J. E. Wahlberg, and H. I. Maibach. 1994. *Protective Gloves for Occupational Use.* New York: CRC Press/Lewis Publishers.

Meyer, E. 1990. *Chemistry of Hazardous Materials.* Englewood Cliffs, NJ: Prentice Hall.

Patnaik, P. 1992. *A Comprehensive Guide to the Hazardous Properties of Chemical Substances.* New York: Van Nostrand Reinhold.

Quigley, D. R. 1994. *Handbook of Emergency Chemical Management.* New York: CRC Press/Lewis Publishers.

Rea, W. J. 1995. *Chemical Sensitivity,* Vol. I-III. New York: CRC Press/Lewis Publishers.

Richardson, M. 1995. *Dictionary of Substance and Their Effects,* Vol. I-VI. New York: CRC Press/Lewis Publishers.

Sacarello, H. L. A. 1994. *The Comprehensive Handbook of Hazardous Materials.* New York: CRC Press/Lewis Publishers.

Torkelson, T. R. V. K. Rowe. 1981. Halogenated aliphatic hydrocarbons. *Patty's Industrial Hygiene and Toxicology,* Vol. 2B, *Toxicology,* pp 3568-3572. New York: Wiley-Interscience.

U.S. Department of Health and Human Services, National Institute for Occupational Safety and Health. 1994. *NIOSH Pocket Guide to Chemical Hazards.* Washington, D.C.: U.S. Government Printing Office.

MATERIAL SAFETY DATA SHEET

CHEMICAL NAME

HYDROGEN PEROXIDE (52-100%)

CRC PRESS/LEWIS PUBLISHERS DISCLAIMER: THE INFORMATION AND RECOMMENDATIONS PRESENTED HEREIN ARE BASED ON SOURCES BELIEVED TO BE RELIABLE. CRC MAKES NO REPRESENTATION ON ITS COMPLETENESS OR ACCURACY. IT IS THE RESPONSIBILITY OF THE USER TO DETERMINE THE CHEMICAL'S SUITABILITY FOR ITS INTENDED USE, THE CHEMICAL'S SAFE USE, AND THE CHEMICAL'S PROPER DISPOSAL. NO REPRESENTATIONS AND/OR WARRANTIES, EITHER EXPRESSED OR IMPLIED, OF THE MERCHANTABILITY OR FITNESS FOR A PARTICULAR PURPOSE, OR OF ANY OTHER NATURE, ARE MADE WITH RESPECT TO THE INFORMATION PROVIDED IN THIS MSDS OR TO THE CHEMICAL TO WHICH INFORMATION MAY REFER. CRC NEITHER ASSUMES NOR AUTHORIZES ANY OTHER PERSON TO ASSUME FOR IT, ANY OTHER ADDITIONAL RESPONSIBILITY OR LIABILITY FOR THE USE OF, OR RELIANCE UPON, THIS INFORMATION.

HAZARD WARNING INFORMATION

HEALTH	FIRE	REACTIVE	OTHER	DEGREE OF HAZARD	COLOR CODING	OTHER CODES
2	0	3	OX	0 = Minimum Hazard 1 = Slight Hazard 2 = Moderate Hazard 3 = Serious Hazard 4 = Severe Hazard	HEALTH = BLUE FIRE = RED REACTIVITY = YELLOW OTHER = WHITE	OX = Oxidizer ACID = Acid ALK = Alkali COR = Corrosive W = Use No Water

SECTION I - GENERAL INFORMATION

Characterization	RCRA Number	EPA Class
Peroxide	**D001**	**Characteristic (I) Waste**

DOT Proper Shipping Name **Hydrogen Peroxide (aqueous solution)**	Chemical Abstract Service (CAS) Number **7722-84-1**
DOT Hazard Class and Label Requirements **Oxidizer; Oxidizer and Corrosive**	DOT Emergency Guide Code **47 (52-100% solution)**
DOT Identification Number **UN 2015 (52-100% solution)**	Chemical Formula **H_2O_2**

Synonyms

High-strength hydrogen peroxide; hydrogen dioxide; hydrogen peroxide (aqueous); hydroperoxide; peroxide; Albone; carbamide peroxide; Interox; Inhibine; Kastone; Perhone; Perhydrol; Peroxaan.

SECTION II - HAZARDOUS INGREDIENTS AND IDENTITY INFORMATION

Hazardous Components (specific identity)	OSHA Exposure Criteria	NIOSH Exposure Criteria	Immediately Dangerous to Life and Health (IDLH)	ACGIH Exposure Criteria
Hydrogen peroxide (derivation: By autoxidation of an alkyl anthrahydroquinone in which the quinone formed in the oxidation step is reduced to the starting material by hydrogen in the presence of a supported palladium catalyst; by an electrolytic process). **1 ppm = 1.41 mg/m^3**	**PEL:** **1 ppm** **1.4 mg/m^3** **STEL:** **Not Established**	**REL:** **1 ppm** **1.4 mg/m^3** **STEL:** **Not Established**	**75 ppm**	**TLV:** **1 ppm** **1.4 mg/m^3** **STEL:** **Not Established**

SECTION III - PHYSICAL AND CHEMICAL CHARACTERISTICS

Boiling Point **306°F (152°C) pure H_2O_2**	Specific Gravity (H_2O = 1) **1.29**
Vapor Pressure (gas) **5 at 86°F (30°C)**	Molecular Weight **34.0**
Vapor Density (Air = 1) **1.2**	Melting Point **31°F (-0.4°C) pure H_2O_2**

Solubility

Soluble in water, alcohol, and ether. Insoluble in petroleum ether. Decomposes in most solvents.

Appearance and Odor

Colorless liquid with a slight, sharp odor. The pure compound is a crystalline solid below 31°F (-0.4°C).

SECTION IV - FIRE AND EXPLOSION HAZARD DATA

Flash Point (method used) **Not Applicable**	Explosive Limits in Air % by Volume **LEL: Not Applicable UEL: Not Applicable**
NFPA Classification **Non-Combustible Liquid**	Autoignition Temperature **Not Applicable**

Extinguishing Media

Use flooding amounts of water on surrounding fire. Do NOT use dry chemical.

Special Fire Fighting Procedures

Wear full protective clothing and self-contained breathing apparatus (SCBA). Poisonous gases are produced in fire. Move containers from fire if it can be done without risk. Remain clear of smoke. Fight fires from a distance or protected location, if possible. Use water to keep combustible materials wet.

Unusual Fire and Explosion Hazards

A non-combustible liquid but a powerful oxidizer. Dangerous fire and explosion hazard when exposed to heat or flame. Containers may explode in fire. Flammable hydrogen gas may be formed in fire.

SECTION V - REACTIVITY DATA

Stability		Conditions to Avoid **Hydrogen peroxide solution free of contaminants is normally stable but will still decompose to release oxygen when inhibited with acetanilide or sodium stannate. It must be stored in vented containers.**
Stable **X**	Unstable	Incompatibility *(materials to avoid)* **Oxidizable materials, ordinary organic materials (paper, wood), most metals (such as iron, copper, brass, bronze, chromium, zinc, lead, silver, manganese). Will ignite spontaneously in contact with organics.**
Hazardous Polymerization		Conditions to Avoid **Under normal conditions of temperature and pressure, hazardous polymerization of hydrogen peroxide is not expected to occur.**
May Occur	Will Not Occur **X**	Hazardous Decomposition or By-products **When heated to decomposition, explosion is possible due to the release of hydrogen gas. In fire, hydrogen peroxide produces water and ½ oxygen.**

SECTION VI - HEALTH HAZARD DATA

Primary Route(s) of Entry:	Inhalation?	Absorption (skin)?	Ingestion?
	X		

Health Hazards

INHALATION: Irritation of the nose, mouth, and lungs. Symptoms include nosebleeds, sore throat, hoarseness, cough with phlegm, increased saliva, and shortness of breath. The vapors may also cause lachrymation, sneezing, headache, dypsnea, cyanosis, salivation, nausea and vomiting, giddiness, insomnia, tremors, bleaching of hair, and pulmonary edema.

SKIN & EYES: A severe (caustic) skin irritant that can cause whitening of the surface, redness, blisters, and vesiculation. Eye contact can cause corneal ulceration and loss of sight.

INGESTION: Irritation and burning of the lips, mouth, and throat, painful swallowing, salivation, throat swelling, intense thirst, nausea and bloody vomiting, stomach distention and perforation, convulsions, coma, possible cerebral edema, and death.

Carcinogenicity	NTP Listed?	IARC Cancer Review Group?	OSHA Regulated?	Target Organs?
Unknown Human Questioned Animal	**No**	**Group 3**	**29 CFR 1910.1000 Table Z-1**	**Respiratory system, eyes, skin, CNS.**

Medical Conditions Generally Aggravated by Exposure

Chronic respiratory diseases or skin diseases may be aggravated by exposure.

Emergency and First-aid Procedures

Eye contact: Flush immediately with water for 15 minutes (minimum), lifting and holding eyelids open. Seek medical attention. Skin contact: Remove all contaminated clothing. Immediately wash area with large amounts of soap and water for 15 minutes (minimum). Seek medical assistance. For inhalation: Remove the person from exposure. Provide respiratory assistance and CPR. Transfer to medical facility. Observe for 48 hours for lung effects. If swallowed: Seek medical attention immediately. Give large amounts of water and induce vomiting (unless advised otherwise).

SECTION VII - PRECAUTIONS FOR SAFE HANDLING AND USE

Steps to be Taken in Case Material is Released or Spilled

Absorb liquids in vermiculite, or similar material and deposit in sealed containers. Ventilate area of spill or leak. Restrict those not involved in cleanup from entering area. Remove all sources of ignition.

Preferred Waste Disposal Method

No Citation.

Precautions to be Taken in Handling and Storage

Store to avoid contact with incompatible materials. Store in tightly closed containers in cool, dry, well-ventilated area away from heat and flame. Prevent contamination to avoid violent decomposition.

Other Precautions and Warnings

Do NOT use near organic materials. Heat may cause containers to build up pressure and explode.

SECTION VIII - CONTROL MEASURES AND PERSONAL PROTECTIVE EQUIPMENT

Respiratory Protection *(specify type)*

Exposure level is low and difficult to monitor. Recommend use of a supplied-air respirator with full facepiece, in continuous flow mode, or a SCBA operated in pressure demand or positive pressure mode.

Ventilation

Local exhaust or general mechanical systems recommended.

Protective Gloves	Eye Protection	Other Protective Clothing
Butyl, Nitrile, Natural Rubber, Viton	**Chemical Goggles or Face Mask**	**Protective Apron**

Work/Hygiene Practices

Always wash hands thoroughly after using chemical; never bring food, drink, or smoking materials into vicinity of chemicals.

HYDROGEN PEROXIDE (52-100%)

H_2O_2

CAS: 7722-84-1

IDENTIFICATION AND TYPICAL USES

Hydrogen peroxide is a colorless liquid with a slight, sharp odor. The pure compound appears as a crystalline solid at temperatures below 31°F (-0.4°C). It is often used in aqueous solutions of varying concentrations. It is used in bleaching and deodorizing of textiles, wood, pulp, hair, and fur. It is also a source of organic and inorganic peroxides. It is used extensively in the pulp and paper industries, as a plasticizer, in rocket fuels (90% solution), in the manufacture of foam rubber, in making glycerol, in dyeing, in electroplating, as an antiseptic (dilute solutions), as a laboratory reagent, in epoxidation, in hydroxylation, in oxidation, and reduction. It is used as a viscosity control for starch and cellulose derivatives, in refining and cleaning metals, as a bleaching and oxidizing agent in foods, as a neutralizing agent in wine distillation, as a seed disinfectant, and as a substitute for chlorine in water and sewage treatment.

RISK ASSESSMENT: HEALTH

General Assessment

Hydrogen peroxide is a severe irritant causing caustic destruction to tissues. Exposure can be by *inhalation* of its vapors, direct *ingestion* of the liquid, or by skin contact (although absorption through the skin has not been reported). It has caused questionable carcinogenic effects in test animals but the extent of its human cancer-causing properties is unknown. Human mutation data have been reported.

Inhalation causes severe irritation and even burns to the eyes, nose, and throat. There will be tearing (lachrymation), sneezing, cough with phlegm, headache, dypsnea, cyanosis, salivation, giddiness, nausea, vomiting, muscle weakness, ulceration of the nasal mucosa, insomnia, tremors, parasthesia, secondary chemical pneumonia, and possible pulmonary edema (fluid in the lungs). Symptoms of pulmonary edema may be delayed up to 48 hours thereby creating a false sense of security with regard to health exposure risk. Serious cases end in shock, convulsions, and death. The vapors may also bleach the hair white.

On skin contact, hydrogen peroxide will cause irritation and immediate erythema (redness). If allowed to remain in contact, there will be painful blisters, swelling, vesiculation, and corrosive tissue damage. Eye contact will also be irritating and painful. Contact may also cause corneal ulceration and permanent damage to vision.

Ingestion causes immediate irritation and burning to the lips, mouth, throat, and gastrointestinal tract with painful swallowing, excessive salivation, thirst, throat swelling, nausea and bloody vomiting due to digestive hemorrhage, stomach distention and perforation, convulsions, possible cerebral edema (fluid on the brain), and death.

Acute Health Effects

The following acute (short-term) health effects may occur immediately or shortly after exposure to hydrogen peroxide:

Skin: Immediate irritation, rash, and burning feeling on contact. Corrosive destruction possible.

Eye: Irritation and possible damage if allowed to remain in contact. May cause burns.

Lung: Irritation of the respiratory tract, including the nose and mouth. There may be cough with phlegm, salivation, and shortness of breath. Pulmonary edema possible at higher levels, which can be fatal.

CNS: Causes giddiness, parasthesia, muscle weakness, convulsions, and coma.

Other: Exposure to the vapors may bleach the hair and skin white.

Chronic Health Effects

The following chronic (long-term) health effects may occur at some time after exposure to hydrogen peroxide and can last for months or even years:

Cancer Hazards: According to information presented in the references, hydrogen peroxide may cause mutations (genetic changes) in living cells. Its carcinogenic properties are questioned in animals and unknown in humans and requires more testing.

Reproductive Hazard: According to information presented in the references, hydrogen peroxide has not been adequately tested for its ability to adversely affect reproduction.

Other Chronic Effects: None reported.

Recommended Risk-Reduction Measures

Personnel should avoid direct contact with hydrogen peroxide. The exact nature of its carcinogenic properties is unknown in humans. Solution greater than 52% hydrogen peroxide are severely corrosive and destructive to human tissues. If a less toxic chemical cannot be substituted for a hazardous substance, then *engineering controls* are the most effective method of reducing exposures. The best protection is to enclose operations and/or provide local exhaust ventilation at the site of chemical release. While not always operationally feasible, isolating operations can also reduce exposure. Using respiratory protection is less effective than the controls mentioned above, but is still advisable whenever working with or around hydrogen peroxide. The exposure limit for hydrogen peroxide is relatively low (1 ppm) and difficult to accurately monitor. For any concentration, a self-contained breathing apparatus (SCBA) with full facepiece operated in pressure demand or a supplied-air respirator in continuous flow mode are the recommended respiratory protection methods of choice. If a full facepiece is not available, then chemical goggles should be worn to protect the eyes. Whenever a chemical splash hazard exists, a face shield and a rubber apron should be worn. To prevent hand and skin exposures, nitrile, Viton, butyl, or natural rubber gloves should be used. Glove manufacturers should be contacted and permeation studies obtained *before* glove selections have been made.

Administrative controls should also be in place to minimize the potential for human exposures. These may include written procedures or policies which specify the methods and techniques that will be practiced whenever personnel are to work with hydrogen peroxide.

All personnel should receive *training* on the use, hazards, protective measures, emergency actions, and other precautions per 29 CFR 1910.1200 (Hazard Communication), prior to the first assignment in an area where hydrogen peroxide is used or stored. Before beginning employment and at regular intervals thereafter (e.g., annually), or for those having frequent or potentially high exposures, the following medical tests are recommended:

- ☑ Lung function tests.

If symptoms develop or overexposure is suspected, the following tests are recommended:

- ☑ Lung function tests.
- ☑ Consider chest X-ray after acute overexposure (may be negative if taken immediately after exposure due to delayed onset of pulmonary edema).
- ☑ Gastric lavage may be recommended if swallowed.
- ☑ Analysis of body fluids (gastric aspirates) using titanium chloride reaction will reveal peroxides if performed immediately.

Any evaluation should include a careful history of past and present symptoms with an examination. Medical tests that simply look for existing damage are not a substitute for controlling exposures. Also, because smoking can cause lung cancers, emphysema, heart disease, and a number of other respiratory problems, the effects of exposure to hydrogen peroxide in smokers may appear to be significantly higher than that which is seen in non-smokers. This must be taken into consideration for proper risk assessment and management.

Other methods to reduce exposure include:

- ☑ Where possible, enclose operations and use local exhaust ventilation at the site of chemical release. If local exhaust ventilation or enclosure is not used, respiratory protection should be mandatory.
- ☑ Always ensure that proper protective clothing is worn when using chemical substances.
- ☑ Wash thoroughly immediately after exposure to hydrogen peroxide and at the end of the work shift or before eating, drinking, or smoking.
- ☑ Hazard warning information should be posted in the work area. In addition, as part of an on-going education and training program, all information on the health and safety hazards of hydrogen perox-

ide should be communicated to all potentially exposed workers.

- ☑ Eye wash stations should be provided in the immediate work area for emergency use. If there is a possibility of skin exposure to hydrogen peroxide, emergency shower facilities should also be provided in the immediate area.
- ☑ Personnel should never wear contaminated clothing home (family members can be exposed). Clothing should be laundered by personnel who have been trained on the hazards of hydrogen peroxide.

RISK ASSESSMENT: ENVIRONMENT

General Assessment

The environment is at risk of exposure during transportation, storage, disposal, or destruction of hydrogen peroxide. In almost every scenario, the threat of environmental exposure is contingent upon the proper handling of the chemical substance. Accidental spills, large or small, can result in fire, explosion, and possible contamination of the surrounding environmental mediums (water, soil, and air).

Hydrogen peroxide is considered a non-combustible liquid but an extremely powerful oxidizer. Concentrated solutions can decompose violently if trace impurities are present. Contact with oxidizable materials may cause spontaneous combustion. Violent explosions may occur in contact with many common metals. These characteristics require extreme caution in handling, storage, transportation, and disposal. Also, contact with incompatible materials can cause violent fire or explosion and special consideration is required during any emergency situation involving a leak or spill of hydrogen peroxide. Should hydrogen peroxide ever come into contact with these incompatible substances during use, transportation, storage, or disposal, the formation of highly toxic and/or highly explosive commodities is also extremely possible.

Hydrogen peroxide may enter the environment through industrial discharges or spills.

Acute Ecological Effects

Acute (short-term) toxic effects may include the death of animals, birds, or fish and death or low growth rate in plants. Acute effects are seen 2 to 4 days after animals or plants are exposed to hydrogen peroxide.

Insufficient data are available on the short-term effects of hydrogen peroxide exposure to aquatic life, plants, birds, or land animals.

Chronic Ecological Effects

Chronic (long-term) toxic effects may include shortened life span, reproductive problems, lower fertility, and changes in appearance or behavior in exposed animals. These effects can be seen long after first exposure(s) to toxic chemicals.

Insufficient data are available on the long-term effects of hydrogen peroxide to aquatic life, plants, birds, or land animals.

Water Solubility

Hydrogen peroxide is highly soluble in water. Concentrations of 1000 milligrams or more will mix with a liter of water.

Persistence in the Environment

Insufficient data are available on the properties of hydrogen peroxide to estimate or predict is distribution in the environment. Limited data suggest that hydrogen peroxide will be moderately persistent in water, with a half-life of between 2 and 20 days. The half-life of a pollutant is the amount of time it takes for one half of the chemical to be degraded.

Bioaccumulation in Aquatic Organisms

Some substances increase in concentration, or *bioaccumulate,* in living organisms as they breathe contaminated air, drink contaminated water, or eat contaminated food. These chemicals can become concentrated in the tissues and internal organs of animals as well as humans.

Hydrogen peroxide is not expected to accumulate in the edible tissues of fish.

Recommended Risk-Reduction Measures

Proper training of all transporters will reduce the likelihood of a mishap or accidents resulting in a leak or spill to the environment. The correct use of DOT labeling (per the MSDS) on all containers, trucks, and rail cars will enable emergency responders to react properly and quickly to any disaster thereby reducing the potential risk to the environment and to personnel.

Storage of hydrogen peroxide should be segregated from incompatible chemicals to minimize the risk of

cross-contamination or contact. Buildings designated for storage should be equipped with appropriate fire protection systems (alarms, sprinklers, emergency lighting, portable extinguishers).

If a spill or leak to the environment has occurred, fire department, emergency response and/or hazardous materials spill personnel should be notified immediately. Cleanup should be attempted only by those trained in proper spill containment procedures. Contaminated soils should be removed for incineration and replaced with clean soil. If hydrogen peroxide should contact the water table, aquifer, or navigable waterway, the local and/or state emergency response authorities must be notified. A comprehensive emergency response or disaster preparedness/recovery plan should be in place prior to any operations involving the use, transportation, storage, or disposal of hydrogen peroxide. If hydrogen peroxide is spilled or leaked, the following specific steps are recommended:

☑ Restrict persons not wearing protective clothing from area of spill or leak until cleanup is complete and area can be opened for normal work.

☑ Remove all ignition sources and ventilate area.

☑ Absorb liquids in vermiculite, dry sand, earth, or similar material and deposit in sealed containers.

☑ It may be necessary to dispose of hydrogen peroxide as a hazardous waste. The responsible state agency or the regional office of the federal Environmental Protection Agency (EPA) should be contacted for specific recommendations.

RISK ASSESSMENT: BUSINESS

General Assessment

Accidents or mishaps involving hydrogen peroxide can present a significant threat to business operations. The loss or damage of equipment or facilities can significantly affect fiscal viability. Lawsuits that may result from personnel injury/death, public exposures, and/or environmental contamination will also require a serious expenditure of resources. Media attention surrounding an injury, death, or environmental damage can also result in a loss of profits and loss of current as well as future business.

Recommended Risk-Reduction Measures

Company attorneys, safety and health professionals, and environmental specialists should be involved in the development of any procedures dealing with safety or emergency response. A company official should be pre-designated as a public relations officer with specific training in dealing with the press. Corporate plans and policies should be developed, approved, and implemented long before any need for such arises.

REFERENCES

American Conference of Governmental Industrial Hygienists. 1988. *Documentation of the Threshold Limit Values and Biological Exposure Indices,* 5th Edition (with updates). Cincinnati: ACGIH

Bueche, J. R. 1972. *Principles of Physics,* 2nd Edition. New York: McGraw-Hill.

Calabrese, E. J. 1994. *Biological Effects of Low Level Exposures.* New York: CRC Press/Lewis Publishers.

Cockerman, L. G., B. S. Shane. 1994. *Basic Environmental Toxicology.* New York: CRC Press/Lewis Publishers.

Gardner, R. W. 1994. *Chemical Intolerance.* New York: CRC Press/Lewis Publishers.

Hallenbeck, W. H. 1993. *Quantitative Risk Assessment for Environmental and Occupational Health,* 2nd Edition. New York: CRC Press/Lewis Publishers.

Jones, F. E. 1994. *Toxic Organic Vapors in the Workplace.* New York: CRC Press/Lewis Publishers.

Keith, L. H. and M. M. Walker. 1995. *Handbook of Air Toxics.* New York: CRC Press/Lewis Publishers.

Lauwerys, R. R., P. Hoet. 1993. *Industrial Chemical Exposure.* New York: CRC Press/Lewis Publishers.

Mellström, G. A., J. E. Wahlberg, and H. I. Maibach. 1994. *Protective Gloves for Occupational Use.* New York: CRC Press/Lewis Publishers.

Meyer, E. 1990. *Chemistry of Hazardous Materials.* Englewood Cliffs, NJ: Prentice Hall.

Patnaik, P. 1992. *A Comprehensive Guide to the Hazardous Properties of Chemical Substances.* New York: Van Nostrand Reinhold.

Quigley, D. R. 1994. *Handbook of Emergency Chemical Management.* New York: CRC Press/Lewis Publishers.

Rea, W. J. 1995. *Chemical Sensitivity,* Vol. I-III. New York: CRC Press/Lewis Publishers.

Richardson, M. 1995. *Dictionary of Substance and Their Effects,* Vol. I-VI. New York: CRC Press/Lewis Publishers.

Sacarello, H. L. A. 1994. *The Comprehensive Handbook of Hazardous Materials.* New York: CRC Press/Lewis Publishers.

Torkelson, T. R. V. K. Rowe. 1981. Halogenated aliphatic hydrocarbons. *Patty's Industrial Hygiene and Toxicology,* Vol. 2B, *Toxicology,* pp 3568-3572. New York: Wiley-Interscience.

U.S. Department of Health and Human Services, National Institute for Occupational Safety and Health. 1994. *NIOSH Pocket Guide to Chemical Hazards.* Washington, D.C.: U.S. Government Printing Office.

MATERIAL SAFETY DATA SHEET

CHEMICAL NAME	
HYDROGEN SELENIDE	**CRC PRESS/LEWIS PUBLISHERS DISCLAIMER:** THE INFORMATION AND RECOMMENDATIONS PRESENTED HEREIN ARE BASED ON SOURCES BELIEVED TO BE RELIABLE. CRC MAKES NO REPRESENTATION ON ITS COMPLETENESS OR ACCURACY. IT IS THE RESPONSIBILITY OF THE USER TO DETERMINE THE CHEMICAL'S SUITABILITY FOR ITS INTENDED USE, THE CHEMICAL'S SAFE USE, AND THE CHEMICAL'S PROPER DISPOSAL. NO REPRESENTATIONS AND/OR WARRANTIES, EITHER EXPRESSED OR IMPLIED, OF THE MERCHANTABILITY OR FITNESS FOR A PARTICULAR PURPOSE, OR OF ANY OTHER NATURE, ARE MADE WITH RESPECT TO THE INFORMATION PROVIDED IN THIS MSDS OR TO THE CHEMICAL TO WHICH INFORMATION MAY REFER. CRC NEITHER ASSUMES NOR AUTHORIZES ANY OTHER PERSON TO ASSUME FOR IT, ANY OTHER ADDITIONAL RESPONSIBILITY OR LIABILITY FOR THE USE OF, OR RELIANCE UPON, THIS INFORMATION.

HAZARD WARNING INFORMATION

HEALTH	FIRE	REACTIVE	OTHER	DEGREE OF HAZARD	COLOR CODING	OTHER CODES
3	3	3		0 = Minimum Hazard 1 = Slight Hazard 2 = Moderate Hazard 3 = Serious Hazard 4 = Severe Hazard	HEALTH = BLUE FIRE = RED REACTIVITY = YELLOW OTHER = WHITE	OX = Oxidizer ACID = Acid ALK = Alkali COR = Corrosive W = Use No Water

SECTION I - GENERAL INFORMATION

Characterization	RCRA Number	EPA Class
Hydride; Selenium Compound	**None**	**Not Applicable**

DOT Proper Shipping Name	**Hydrogen Selenide**	Chemical Abstract Service (CAS) Number	**7783-07-5**
DOT Hazard Class and Label Requirements	**Flammable Gas**	DOT Emergency Guide Code	**13**
DOT Identification Number	**UN 2201**	Chemical Formula	**H_2Se**

Synonyms
Selenium dihydride; selenium hydride; Selenium hydride (anhydrous).

SECTION II - HAZARDOUS INGREDIENTS AND IDENTITY INFORMATION

Hazardous Components (specific identity)	OSHA Exposure Criteria	NIOSH Exposure Criteria	Immediately Dangerous to Life and Health (IDLH)	ACGIH Exposure Criteria
Hydrogen selenide (derivation: By action of acids on inorganic selenides; by reaction of selenium with organic matter; or by direct combination of the elements of hydrogen and selenium). **1 ppm = 3.37 mg/m^3**	**PEL (as Se):** **0.05 ppm** **0.2 mg/m^3** **STEL:** **Not Established**	**REL (as Se):** **0.05 ppm** **0.2 mg/m^3** **STEL:** **Not Established**	**2 ppm** **(6.7 mg/m^3)**	**TLV (as Se):** **0.05 ppm** **0.2 mg/m^3** **STEL:** **Not Established**

SECTION III - PHYSICAL AND CHEMICAL CHARACTERISTICS

Boiling Point	**-42°F (-41°C) Decomposes**	Specific Gravity (H_2O = 1)	**2.80**
Vapor Pressure (atmospheres)	**9.5 at 70°F (21°C)**	Molecular Weight	**81.0**
Vapor Density (Air = 1)	**2.79**	Melting Point	**-87°F (-66°C)**

Solubility
Slightly soluble water (0.9%). Soluble in carbon disulfide, carbonyl chloride, and phosgene.

Appearance and Odor
Colorless gas with an odor similar to that of decayed horseradish. Normally shipped as a liquefied compressed gas.

SECTION IV - FIRE AND EXPLOSION HAZARD DATA

Flash Point (method used)	**Not Determined (Gas)**	Explosive Limits in Air % by Volume	**LEL: Not Determined UEL: Not Determined**
NFPA Classification	**Flammable Gas**	Autoignition Temperature	**Not Determined**

Extinguishing Media
Stop flow of gas before attempting to extinguish fire! Use water spray to knock down vapors.

Special Fire Fighting Procedures
Extremely flammable and reactive gas. Poisonous gases are produced in fire; wear full protective clothing and self-contained breathing apparatus (SCBA). Move containers from fire area if it can be done without risk.

Unusual Fire and Explosion Hazards
Hydrogen selenide is denser than air and can travel low to the ground and collect in low-lying areas to present a serious explosion hazard. It can also travel to an ignition source and flashback to cause fire or explosion.

SECTION V - REACTIVITY DATA

Stability		Conditions to Avoid: Hydrogen selenide is normally stable under routine conditions of handling and storage. Avoid contact with sources of ignition and incompatible materials.
Stable	Unstable	Incompatibility *(materials to avoid)*: Reacts in contact with strong oxidizers, acids, water, and halogenated hydrocarbons. Also, as a gas, it will form explosive mixtures in air.
X		
Hazardous Polymerization		Conditions to Avoid: Under normal conditions of temperature and pressure, hazardous polymerization of hydrogen selenide is not expected to occur.
May Occur	Will Not Occur	Hazardous Decomposition or By-products: When heated to decomposition, hydrogen selenide emits toxic fumes and gases, including highly toxic selenium oxide.
	X	

SECTION VI - HEALTH HAZARD DATA

Primary Route(s) of Entry:	Inhalation?	Absorption (skin)?	Ingestion?
	X		

Health Hazards

INHALATION: Irritation of the eyes, nose, throat, and respiratory system. May cause nausea, vomiting, metallic taste, fatigue, runny nose, tearing, cough, sneezing, garlic breath, dizziness, lassitude, olfactory fatigue, diarrhea, pulmonary edema, lung or liver injury, and death either resulting from lung or liver damage or asphyxiation.

SKIN & EYES: The liquefied compressed gas can cause painful frostbite, burning, and blistering to the eyes and skin. Absorption of the vapor is not likely to occur.

INGESTION: Not likely (hydrogen selenide is a gas at room temperature and pressure)

Carcinogenicity	NTP Listed?	IARC Cancer Review Group?	OSHA Regulated?	Target Organs?
Unknown Human Unknown Animal	No	No	29 CFR 1910.1000 Table Z-1	Respiratory system, skin, eyes, CNS.

Medical Conditions Generally Aggravated by Exposure

None Reported.

Emergency and First-aid Procedures

<u>Eye contact</u>: Flush immediately with warm water for 15 minutes (minimum); seek medical attention. <u>Skin contact</u>: Remove all contaminated clothing. Wash area with large amounts of warm (tepid) water. Seek medical assistance. For <u>inhalation</u>: Remove the person from exposure. Provide respiratory assistance and CPR. Transfer to medical facility. Observe for 48 hours for lung effects. If <u>swallowed</u>: Not a likely exposure route.

SECTION VII - PRECAUTIONS FOR SAFE HANDLING AND USE

Steps to be Taken in Case Material is Released or Spilled

Stop flow of gas. Use water spray to knock down vapors and protect personnel. If leak is a cylinder and it cannot be stopped, remove it to a safe location (outdoors) and allow to vent until empty. Ventilate area and remove ignition sources. Restrict those not involved in cleanup from entering area.

Preferred Waste Disposal Method

No Citation.

Precautions to be Taken in Handling and Storage

Do not store hydrogen selenide in the presence of incompatible chemicals or materials. Store in tightly closed containers in a cool, dark, well-ventilated area. Personnel working with hydrogen selenide should be trained on its proper handing and storage *prior* to being assigned to such responsibilities.

Other Precautions and Warnings

Keep away from water and halogenated compounds. Bulk storage of hydrogen selenide is not recommended. Use non-sparking tools, especially when opening and closing containers.

SECTION VIII - CONTROL MEASURES AND PERSONAL PROTECTIVE EQUIPMENT

Respiratory Protection *(specify type)*

PEL is extremely low (0.05 ppm) and difficult to monitor. Recommend using a self-contained breathing apparatus (SCBA) with full facepiece operated in pressure demand or other positive pressure mode, or use a supplied-air respirator operated in continuous flow mode.

Ventilation

Local exhaust or general mechanical systems recommended.

Protective Gloves	Eye Protection	Other Protective Clothing
Impervious, Thermal Protective	Chemical Goggles or Face Mask	Protective Apron (avoid skin contact)

Work/Hygiene Practices

Always wash hands thoroughly after using chemical; never bring food, drink, or smoking materials into vicinity of chemicals.

HYDROGEN SELENIDE

H_2Se

CAS: 7783-07-5

IDENTIFICATION AND TYPICAL USES

Hydrogen selenide is a colorless gas with an extremely irritating odor resembling that of decaying horseradish. It is normally shipped as a liquefied compressed gas. It is used in the preparation of metallic selenides and organoselenium compounds. It is also used in doping as a mix for preparation of semiconductor materials containing controlled amounts of significant impurities.

RISK ASSESSMENT: HEALTH

General Assessment

Hydrogen selenide is a gas at room temperature and pressure and, therefore, presents a serious (life-threatening) hazard to human health through *inhalation.* The liquefied compressed gas can also cause serious health hazard concerns if it should contact the skin or eyes. Ingestion is not considered a probable route of entry. There are no reports in the references regarding the carcinogenic, mutagenic, or teratogenic effects of hydrogen selenide on humans or animals.

Inhalation of hydrogen selenide vapors causes irritation of the eyes, nose, throat, and mucous membranes of the respiratory tract. Other symptoms include a metallic taste in the mouth and garlic odor on breath, nausea, vomiting, diarrhea, olfactory fatigue, and other indications of central nervous system effects. These include dizziness, lassitude, headache, and fatigue. It is an allergen and may cause conjunctivitis. Higher exposures can lead to a dangerous buildup of fluid in the lungs in a condition known as pulmonary edema. This is a medical emergency and can be fatal. Symptoms may be delayed up to 48 hours thereby creating a false sense of security with regard to health exposure risk.

Skin and eye contact with the compressed liquefied gas can result in serious tissue destruction and burns from frostbite. There may be surface redness (erythema), blistering, and painful swelling. There may also be permanent damage or loss to vision.

Acute Health Effects

The following acute (short-term) health effects may occur immediately or shortly after exposure to hydrogen selenide:

Skin: The liquid causes painful frostbite and deep burns with tissue destruction.

Eye: The vapor can be extremely irritating and the liquid may cause tissue destruction as a result of frostbite burns. Loss of vision is possible.

Lung: Severe irritation of the nose and throat. May cause delayed pulmonary edema.

CNS: Dizziness, headaches, nausea, vomiting, fatigue, lassitude, and other related nervous system effects.

Chronic Health Effects

The following chronic (long-term) health effects may occur at some time after exposure to hydrogen selenide and can last for months or even years:

Cancer Hazards: According to the information presented in the references, hydrogen selenide has not been adequately tested for its ability to cause cancer in test animals.

Reproductive Hazard: According to the information presented in the references, hydrogen selenide has not been adequately tested for its ability to adversely affect reproduction.

Other Chronic Effects: Prolonged exposures may damage the liver and/or the lungs.

Recommended Risk-Reduction Measures

Personnel should avoid direct contact with hydrogen selenide. Unless a less toxic chemical can be substituted for a hazardous substance, *engineering controls* are the most effective method of reducing exposures. The best protection is to enclose operations and/or provide local exhaust ventilation at the site of chemical release. While not always operationally feasible, isolating operations can also reduce exposure. Care should be taken when isolating jobs to ensure an asphyxiation condition is not created.

Using respiratory protection is less effective than the controls mentioned above, but is still advisable whenever working with or around hydrogen selenide. This chemical has an extremely low and difficult to monitor permissible exposure limit (0.05 ppm). Therefore, prudent risk management requires a conservative approach to personal protection. A self-contained breathing apparatus (SCBA) with full facepiece operated in pressure demand or a supplied-air respirator in continuous flow mode are the recommended respiratory protection methods of choice. If a full facepiece is not available, then chemical/dust goggles should be worn to protect the eyes. Whenever compressed liquefied hydrogen selenide gas is being used, a chemical splash hazard exists and a face shield and protective apron should be worn. To prevent hand and skin exposures, impervious, thermal protective gloves should be worn. Glove manufacturers should be contacted and permeation studies obtained *before* final glove selections have been made.

Administrative controls should also be in place to minimize the potential for human exposures. These may include written procedures or policies which specify the methods and techniques that will be practiced whenever personnel are to work with hydrogen selenide.

All personnel should receive *training* on the use, hazards, protective measures, emergency actions, and other precautions per 29 CFR 1910.1200 (Hazard Communication), prior to the first assignment in an area where hydrogen selenide is used or stored.

If symptoms develop or overexposure is suspected, the following medical tests may be useful:

- ☑ Lung function tests
- ☑ Liver function tests.
- ☑ Consider chest X-ray following acute overexposure (may be negative if taken immediately after exposure due to delayed onset of pulmonary edema).

Any evaluation should include a careful medical history of past and present symptoms with an examination. However, medical tests that evaluate existing damage are not a substitute for controlling exposure. Also, since smoking can cause heart disease, lung cancer, emphysema, and other respiratory disorders, smokers may experience symptoms more quickly than nonsmokers under the same conditions of exposure. Prudent risk management requires careful consideration of *all* possible factors which might be causing the appearance of exposure symptoms in the workplace.

Other methods to reduce exposure include:

- ☑ Where possible, enclose operations and use local exhaust ventilation at the site of chemical release. If local exhaust ventilation or enclosure is not used, respiratory protection should be mandatory.
- ☑ Always ensure that proper protective clothing is worn when using chemical substances.
- ☑ Wash thoroughly immediately after exposure to hydrogen selenide and at the end of the work shift or before eating, drinking, or smoking.
- ☑ Hazard warning information should be posted in the work area. In addition, as part of an on-going education and training program, all information on the health and safety hazards of hydrogen selenide should be communicated to all exposed or potentially exposed workers.
- ☑ Eye wash stations should be provided in work areas. If the potential for whole body exposure exists, then safety showers should also be provided.

RISK ASSESSMENT: ENVIRONMENT

General Assessment

The environment is at risk of exposure during transportation, storage, disposal, or destruction of hydrogen selenide. In almost every scenario, the threat of environmental exposure is contingent upon the proper handling of the chemical substance. Accidental spills of the liquefied compressed gas or releases of the hydrogen selenide gas, large or small, can result in fire, explosion, and possible contamination of the surrounding environmental mediums (water, soil, and air).

Hydrogen selenide is considered a highly flammable and extremely reactive gas. It presents a serious fire and explosion hazard under normal conditions of temperature and pressure. As a gas, it can quickly and easily form explosive mixtures in air. In reaction to incompatible materials such as halogenated hydrocarbons, strong oxidizers, acids, and even water, it may ignite or explode. These characteristics require special consideration during any emergency situation involving a leak or spill of hydrogen selenide or hydrogen selenide mixtures. Should hydrogen selenide ever come into contact with any of these incompatible substances either during use, transportation, or storage, violent reactions can occur.

Hydrogen selenide can enter the environment through industrial discharges or releases, or from spills.

Acute Ecological Effects

Acute (short-term) toxic effects may include the death of animals, birds, or fish and death or low growth rate in plants. Acute effects are seen 2 to 4 days after animals or plants are exposed to hydrogen selenide.

Insufficient data are available to evaluate or predict the short-term effects of hydrogen selenide to aquatic life, plants, birds, or land animals.

Chronic Ecological Effects

Chronic toxic effects may include shortened life span, reproductive problems, lower fertility, and changes in appearance or behavior in exposed animals. These effects can be seen long after first exposure(s) to toxic chemicals.

Insufficient data are available to evaluate or predict the long-term effects of hydrogen selenide to aquatic life, plants, birds, or land animals.

Water Solubility

Hydrogen selenide is very slightly soluble in water. Concentrations of 1 milligram or less will mix with a liter of water.

Persistence in the Environment

Hydrogen selenide is non-persistent in water, with a half-life less than 2 days. The half-life of a pollutant is the amount of time it takes for one-half of the chemical to be degraded. Virtually 100% of hydrogen is expected to end up in the air. Selenium is moderately persistent in the environment with a half-life between 20 and 200 days. Selenium is expected to end up in aquatic sediments and terrestrial soils.

Bioaccumulation in Aquatic Organisms

Some substances increase in concentration, or *bioaccumulate,* in living organisms as they breathe contaminated air, drink contaminated water, or eat contaminated food. These chemicals can become concentrated in the tissues and internal organs of animals as well as humans.

The concentration of the selenide found in fish tissues is expected to be much higher than the average concentration of hydrogen selenide in the water from which the fish was taken.

Recommended Risk-Reduction Measures

Proper training of all transporters will reduce the likelihood of a mishap or accident resulting in a leak or spill to the environment. The correct use of DOT labeling (per the MSDS) on all containers, trucks, and rail cars will enable emergency responders to react properly and quickly to any disaster thereby reducing the potential risk to the environment and to personnel.

Storage of hydrogen selenide should be segregated from incompatible chemicals to minimize the risk of cross-contamination or contact.

If a spill or leak to the environment has occurred, fire department, emergency response and/or hazardous materials spill personnel should be notified immediately. Cleanup should be attempted only by those trained in proper spill containment procedures. Contaminated soils should be removed for incineration and replaced with clean soil using non-sparking tools. If hydrogen selenide should contact the water table, aquifer, or navigable waterway, remediation activities should be prompt to ensure protection of aquatic sediments. When such spills occur, the local and/or state emergency response authorities must be notified. A comprehensive emergency response or disaster preparedness/recovery plan should be in place prior to any operations involving the use, transportation, storage, or disposal of hydrogen selenide. If hydrogen selenide is spilled or leaked, the following specific steps are recommended:

- ☑ Restrict persons not wearing protective clothing from area of spill or leak until cleanup is complete and area can be opened for normal work.
- ☑ Stop the flow of gas before attempting cleanup procedures. If leak is from a cylinder and it cannot be stopped, remove cylinder to safe location (outdoors) and allow to vent until empty. Use water spray to knock down unignited vapors and to protect response personnel.

RISK ASSESSMENT: BUSINESS

General Assessment

Accidents or mishaps involving hydrogen selenide can present a significant threat to business operations. The loss or damage of equipment or facilities can significantly affect fiscal viability. Lawsuits that may result

from personnel injury/death, public exposures, and/or environmental contamination will also require a serious expenditure of resources. Media attention surrounding an injury, death, or environmental damage can also result in a loss of profits and loss of current as well as future business.

Recommended Risk-Reduction Measures

Company attorneys, safety and health professionals, and environmental specialists should be involved in the development of any procedures dealing with safety or emergency response. A company official should be pre-designated as a public relations officer with specific training in dealing with the press. Corporate plans and policies should be developed, approved, and implemented long before any need for such arises.

REFERENCES

Ahlbom, A. 1993. *Biostatistics for Epidemiologists.* New York: CRC Press/Lewis Publishers.

American Conference of Governmental Industrial Hygienists. 1988. *Documentation of the Threshold Limit Values and Biological Exposure Indices,* 5th Edition (with updates). Cincinnati: ACGIH

Anthony, C. P., and N. J. Kolthoff. 1971. *Textbook of Anatomy and Physiology.* St. Louis: C. V. Mosby Company.

Asante-Duah, D. K. 1993. *Hazardous Waste Risk Assessment.* New York: CRC Press/Lewis Publishers.

Bloom, A. D. and F. J. DeSerres. 1995. *Ecotoxicity and Human Health.* New York: CRC Press/Lewis Publishers.

Brusick, D. J. 1994. *Methods for Genetic Risk Assessment.* New York: CRC Press/Lewis Publishers.

Bueche, J. R. 1972. *Principles of Physics,* 2nd Edition. New York: McGraw-Hill.

Burrell, R., D. K. Flaherty, and L. J. Sauers. 1992. *Toxicology of the Immune System.* New York: Van Nostrand Reinhold.

Calabrese, E. J. 1994. *Biological Effects of Low Level Exposures.* New York: CRC Press/Lewis Publishers.

Cockerman, L. G., B. S. Shane. 1994. *Basic Environmental Toxicology.* New York: CRC Press/Lewis Publishers.

Cothern, C. R., N. P. Ross. 1994. *Environmental Statistics, Assessment, and Forecasting.* New York: CRC Press/Lewis Publishers.

Gardner, R. W. 1994. *Chemical Intolerance.* New York: CRC Press/Lewis Publishers.

Hallenbeck, W. H. 1993. *Quantitative Risk Assessment for Environmental and Occupational Health,* 2nd Edition. New York: CRC Press/Lewis Publishers.

Hodgson, E., R. B. Mailman, and Janice E. Chambers. 1988. *Dictionary of Toxicology.* London: The MacMillian Company.

Howard, P. H. 1995. *Handbook of Environmental Fate and Exposures Data for Organic Chemicals,* Vol. I-IV. New York: CRC Press/Lewis Publishers.

Kamrin, M. A. 1988. *Toxicology: A Primer on Principles and Applications.* Chelsea, Michigan: Lewis Publishers.

Landis, W. G., M. H. Yu. 1995. *Introduction to Environmental Toxicology.* New York: CRC Press/Lewis Publishers.

Langerman, N. 1994 *Precautionary Labels for Chemical Containers.* New York: CRC Press/Lewis Publishers.

Lauwerys, R. R., P. Hoet. 1993. *Industrial Chemical Exposure.* New York: CRC Press/Lewis Publishers.

Lide, D. R. 1995. *Handbook of Chemistry and Physics,* 76th Edition. New York: CRC Press/Lewis Publishers.

Lewis, R. J., Sr. 1993. *Hawley's Condensed Chemical Dictionary,* 12th Edition. New York: Van Nostrand Reinhold.

Lewis, R. J., Sr. 1992. *Sax's Dangerous Properties of Industrial Materials,* Volumes 1, 2, and 3, 8th Edition. New York: Van Nostrand Reinhold.

Manahan, S. E. 1992. *Toxicological Chemistry,* 2nd Edition. New York: CRC Press/Lewis Publishers.

Mellström, G. A., J. E. Wahlberg, and H. I. Maibach. 1994. *Protective Gloves for Occupational Use.* New York: CRC Press/Lewis Publishers.

Meyer, E. 1990. *Chemistry of Hazardous Materials.* Englewood Cliffs, NJ: Prentice Hall.

Nelson, G. O. 1992. *Gas Mixtures.* New York: CRC Press/Lewis Publishers.

Ott, W. R. 1995. *Environmental Statistics and Data Analysis.* New York: CRC Press/Lewis Publishers.

Patnaik, P. 1992. *A Comprehensive Guide to the Hazardous Properties of Chemical Substances.* New York: Van Nostrand Reinhold.

Philp, R. B. 1995. *Environmental Hazards and Human Health.* New York: CRC Press/Lewis Publishers.

Proctor, N. H., and James P. Hughes. 1991. *Chemical Hazards of the Workplace.* New York: Van Nostrand Reinhold.

Quigley, D. R. 1994. *Handbook of Emergency Chemical Management.* New York: CRC Press/Lewis Publishers.

Rea, W. J. 1995. *Chemical Sensitivity,* Vol. I-III. New York: CRC Press/Lewis Publishers.

Richardson, M. 1995. *Dictionary of Substance and Their Effects,* Vol. I-VI. New York: CRC Press/Lewis Publishers.

Sacarello, H. L. A. 1994. *The Comprehensive Handbook of Hazardous Materials.* New York: CRC Press/Lewis Publishers.

U.S. Department of Health and Human Services, National Institute for Occupational Safety and Health. 1994. *NIOSH Pocket Guide to Chemical Hazards.* Washington, D.C.: U.S. Government Printing Office.

MATERIAL SAFETY DATA SHEET

CHEMICAL NAME	
HYDROGEN SULFIDE	**CRC PRESS/LEWIS PUBLISHERS DISCLAIMER:** THE INFORMATION AND RECOMMENDATIONS PRESENTED HEREIN ARE BASED ON SOURCES BELIEVED TO BE RELIABLE. CRC MAKES NO REPRESENTATION ON ITS COMPLETENESS OR ACCURACY. IT IS THE RESPONSIBILITY OF THE USER TO DETERMINE THE CHEMICAL'S SUITABILITY FOR ITS INTENDED USE, THE CHEMICAL'S SAFE USE, AND THE CHEMICAL'S PROPER DISPOSAL. NO REPRESENTATIONS AND/OR WARRANTIES, EITHER EXPRESSED OR IMPLIED, OF THE MERCHANTABILITY OR FITNESS FOR A PARTICULAR PURPOSE, OR OF ANY OTHER NATURE, ARE MADE WITH RESPECT TO THE INFORMATION PROVIDED IN THIS MSDS OR TO THE CHEMICAL TO WHICH INFORMATION MAY REFER. CRC NEITHER ASSUMES NOR AUTHORIZES ANY OTHER PERSON TO ASSUME FOR IT, ANY OTHER ADDITIONAL RESPONSIBILITY OR LIABILITY FOR THE USE OF, OR RELIANCE UPON, THIS INFORMATION.

HAZARD WARNING INFORMATION

HEALTH	FIRE	REACTIVE	OTHER	DEGREE OF HAZARD	COLOR CODING	OTHER CODES
3	4	0		0 = Minimum Hazard 1 = Slight Hazard 2 = Moderate Hazard 3 = Serious Hazard 4 = Severe Hazard	HEALTH = BLUE FIRE = RED REACTIVITY = YELLOW OTHER = WHITE	OX = Oxidizer ACID = Acid ALK = Alkali COR = Corrosive W = Use No Water

SECTION I - GENERAL INFORMATION

Characterization	RCRA Number	EPA Class
Gas, Toxic and Flammable	**U135**	**Toxic Waste**

DOT Proper Shipping Name	Chemical Abstract Service (CAS) Number
Hydrogen Sulfide	**7783-06-4**
DOT Hazard Class and Label Requirements	DOT Emergency Guide Code
Flammable Gas; Poison Gas	**13**
DOT Identification Number	Chemical Formula
UN 1053	**H_2S**

Synonyms
Sulfuretted hydrogen; sulfur hydride; hydrogen sulfuric acid; stink damp.

SECTION II - HAZARDOUS INGREDIENTS AND IDENTITY INFORMATION

Hazardous Components (specific identity)	OSHA Exposure Criteria	NIOSH Exposure Criteria	Immediately Dangerous to Life and Health (IDLH)	ACGIH Exposure Criteria
Hydrogen sulfide (derivation: By action of dilute sulfuric acid on a sulfide, usually iron sulfide; by direct union of hydrogen and sulfur vapor at a definite temperature and pressure; as a byproduct of petroleum refining). **1 ppm = 1.42 mg/m^3**	**PEL (ceiling):** **20 ppm** **28 mg/m^3** **PEAK:** **50 ppm** **(10-minute)**	**REL (ceiling):** **10 ppm** **15 mg/m^3** **(10-minute)** **STEL:** **Not Applicable**	**100 ppm**	**TLV:** **10 ppm** **15 mg/m^3** **STEL:** **15 ppm** **21 mg/m^3**

SECTION III - PHYSICAL AND CHEMICAL CHARACTERISTICS

Boiling Point	Specific Gravity (H_2O = 1)
-76°F (-60°C)	**1.19**
Vapor Pressure (atmospheres)	Molecular Weight
18.5 at 68°F (20°C)	**34.1**
Vapor Density (Air = 1)	Melting Point
1.175	**-122°F (-86°C)**

Solubility
Slightly soluble water (0.4%). Soluble in ethyl alcohol, gasoline, kerosene, crude oil, ethylene glycol.

Appearance and Odor
Colorless gas with a characteristic odor similar to that of rotten eggs. Normally shipped as a liquefied compressed gas.

SECTION IV - FIRE AND EXPLOSION HAZARD DATA

Flash Point (method used)	Explosive Limits in Air % by Volume
Not Determined	**LEL: 4.0%** **UEL: 44.0%**
NFPA Classification	Autoignition Temperature
Flammable Gas	**500°F (260°C)**

Extinguishing Media
Stop flow of gas before attempting to extinguish fire! Use water spray to knock down vapors.

Special Fire Fighting Procedures
Extremely flammable gas. Burns with a blue flame and gives off sulfur dioxide and other poisonous gases. Wear full protective clothing and self-contained breathing apparatus (SCBA). Move containers from fire area if it can be done without risk. Use unmanned device if possible.

Unusual Fire and Explosion Hazards
Hydrogen sulfide is denser than air and can travel low to the ground and collect in low-lying areas to present a serious explosion hazard. It can also travel to an ignition source and flashback to cause fire or explosion. Cool containers with water spray long after fire is out. Contain runoff.

SECTION V - REACTIVITY DATA

Stability		Conditions to Avoid Hydrogen sulfide is normally stable under routine conditions of handling and storage. Avoid contact with heat and incompatible materials.
Stable X	Unstable	Incompatibility *(materials to avoid)* Reacts in contact with strong oxidizers, such as chlorine, fluorine, and bromine, strong nitric acid, and will attack most metals (forms sulfides).
Hazardous Polymerization		Conditions to Avoid Under normal conditions of temperature and pressure, hazardous polymerization of hydrogen sulfide is not expected to occur.
May Occur	Will Not Occur X	Hazardous Decomposition or By-products When heated to decomposition, hydrogen sulfide emits toxic fumes and gases, including highly toxic sulfur dioxide.

SECTION VI - HEALTH HAZARD DATA

Primary Route(s) of Entry:	Inhalation?	Absorption (skin)?	Ingestion?
	X		

Health Hazards

INHALATION: Irritation of the eyes, nose, throat, and respiratory system. May cause nausea, vomiting, cramps, dizziness, headache, diarrhea, sneezing, staggering, excitability, dry cough, pale complexion, muscular weakness, drowsiness, rhinitis, bronchitis, pharyngitis, and other lung effects such as pneumonia and pulmonary edema. Other effects include asphyxia, tremors, fatigue, weakness and numbing in the extremities, convulsions, coma, death.

SKIN & EYES: The liquefied compressed gas can cause painful frostbite, burning, and blistering to the eyes and skin. Absorption of the vapor is not likely to occur.

INGESTION: Not likely (hydrogen sulfide is a gas at room temperature and pressure)

Carcinogenicity	NTP Listed?	IARC Cancer Review Group?	OSHA Regulated?	Target Organs?
Unknown Human Unknown Animal	No	No	29 CFR 1910.1000 Table Z-1	Respiratory system, skin, eyes, CNS.

Medical Conditions Generally Aggravated by Exposure

Eye and nervous system disorders may be aggravated by exposure to hydrogen sulfide.

Emergency and First-aid Procedures

Eye contact: Flush immediately with warm water for 15 minutes (minimum); seek medical attention. Skin contact: Remove all contaminated clothing. Wash area with large amounts of warm (tepid) water. Seek medical assistance. For inhalation: Remove the person from exposure. Provide respiratory assistance and CPR. Transfer to medical facility. Observe 48 hours for lung effects. If swallowed: Not a likely exposure route. DOCTOR: Administration of 100% oxygen and hyperbaric oxygen if possible.

SECTION VII - PRECAUTIONS FOR SAFE HANDLING AND USE

Steps to be Taken in Case Material is Released or Spilled

Stop flow of gas. Use water spray to knock down vapors and protect personnel. If leak is a cylinder and it cannot be stopped, remove it to a safe location (outdoors) and allow to vent until empty. Ventilate area and remove ignition sources. Neutralize runoff with crushed limestone or slacked lime.

Preferred Waste Disposal Method

No Citation.

Precautions to be Taken in Handling and Storage

Do not store hydrogen sulfide in the presence of incompatible chemicals or materials. Store in tightly closed steel cylinders in a cool, dark, well-ventilated area. Personnel working with hydrogen sulfide should be trained on its proper handling and storage *prior* to being assigned to such responsibilities.

Other Precautions and Warnings

Prevent physical damage to containers. Use non-sparking tools, especially when opening and closing cylinders. Use class 1, group C electrical equipment in storage and use areas.

SECTION VIII - CONTROL MEASURES AND PERSONAL PROTECTIVE EQUIPMENT

Respiratory Protection *(specify type)*

Recommend using a self-contained breathing apparatus (SCBA) with full facepiece operated in pressure demand or other positive pressure mode, or use a supplied-air respirator operated in continuous flow mode. Air purifying respirators do NOT provide protection in oxygen deficient atmospheres.

Ventilation

Local exhaust or general mechanical systems recommended.

Protective Gloves	Eye Protection	Other Protective Clothing
Impervious, Thermal Protective	Chemical Goggles or Face Mask	Protective Apron (avoid skin contact)

Work/Hygiene Practices

Always wash hands thoroughly after using chemical; never bring food, drink, or smoking materials into vicinity of chemicals.

HYDROGEN SULFIDE

H_2S

CAS: 7783-06-4

IDENTIFICATION AND TYPICAL USES

Hydrogen sulfide is a colorless gas with an extremely irritating, characteristic odor resembling that of rotten eggs. It is normally shipped as a liquefied compressed gas. It is used in the purification of hydrochloric acid and sulfuric acid, in precipitating sulfides of metals, as an analytical reagent, in the manufacture of heavy water, and as a source of sulfur and hydrogen. It occurs in natural gas and in sewer gas. It is formed by the reaction of a metal sulfide with dilute mineral acid and in petroleum refining.

RISK ASSESSMENT: HEALTH

General Assessment

Hydrogen sulfide is a gas at room temperature and pressure and, therefore, presents a serious (life-threatening) hazard to human health through *inhalation*. The liquefied compressed gas can also cause serious health hazard risks if it should contact the skin or eyes. Ingestion is not considered a probable route of entry. There are no reports in the references regarding the carcinogenic, mutagenic, or teratogenic effects of hydrogen sulfide on humans or animals.

Inhalation of hydrogen sulfide vapors causes irritation of the eyes, nose, throat, and mucous membranes of the respiratory tract. Other symptoms include nausea, vomiting, diarrhea, olfactory fatigue, and other indications of central nervous system effects. These include dizziness, convulsions, staggering gait, excitability, pale complexion, muscular weakness, numbing in the extremities, headache, and fatigue. These effects can progress to unconsciousness and death due to respiratory paralysis. It is also an allergen and an asphyxiant. Higher exposures can lead to a dangerous buildup of fluid in the lungs in a condition known as pulmonary edema. This is a medical emergency and can be fatal. Symptoms may be delayed up to 48 hours thereby creating a false sense of security with regard to health exposure risk. Other respiratory system effects include asphyxia, rhinitis, bronchitis, pharyngitis, and pneumonia.

Skin and eye contact with the compressed liquefied gas can result in serious tissue destruction and burns from frostbite. There may be surface redness (erythema), blistering, vesiculation, and painful swelling. There may also be permanent damage or loss to vision. The eyes are susceptible to a condition known as gas eye (keratoconjunctivitis) with palpebral edema, bulbar conjunctivitis, and mucous-puss secretions.

Acute Health Effects

The following acute (short-term) health effects may occur immediately or shortly after exposure to hydrogen sulfide:

Skin: The liquid causes painful frostbite and deep burns with tissue destruction.

Eye: The vapor can be extremely irritating and the liquid may cause tissue destruction as a result of frostbite burns. Loss of vision is possible.

Lung: Severe irritation of the nose and throat. May cause delayed pulmonary edema, which can be fatal. Olfactory fatigue is also common.

CNS: Dizziness, headaches, nausea, vomiting, fatigue, and other related nervous system effects.

Chronic Health Effects

The following chronic (long-term) health effects may occur at some time after exposure to hydrogen sulfide and can last for months or even years:

Cancer Hazards: According to the information presented in the references, hydrogen sulfide has not been adequately tested for its ability to cause cancer in test animals.

Reproductive Hazard: According to the information presented in the references, hydrogen sulfide has not been adequately tested for its ability to adversely affect reproduction.

Other Chronic Effects: Prolonged exposures may lead to chronic headaches, fatigue, conjunctivitis, digestive disturbances, weight loss, dizziness, a grayish-green line on the gums of the mouth, and irritability.

Recommended Risk-Reduction Measures

Personnel should avoid direct contact with hydrogen sulfide. Unless a less toxic chemical can be substituted for a hazardous substance, *engineering controls* are the most effective method of reducing exposures. The best protection is to enclose operations and/or provide local exhaust ventilation at the site of chemical release. While not always operationally feasible, isolating operations can also reduce exposure. Care should be taken when isolating jobs to ensure an asphyxiation condition is not created.

Using respiratory protection is less effective than the controls mentioned above, but is still advisable whenever working with or around hydrogen sulfide. This chemical has a relatively low permissible exposure limit ceiling (20 ppm) and, because hydrogen sulfide will cause rapid olfactory fatigue, the sense of smell should NOT be relied upon to warn of the continuous presence of hydrogen sulfide gas. Therefore, prudent risk management requires a conservative approach to personal protection. A self-contained breathing apparatus (SCBA) with full facepiece operated in pressure demand or a supplied-air respirator in continuous flow mode are the recommended respiratory protection methods of choice. If a full facepiece is not available, then chemical/dust goggles should be worn to protect the eyes. Whenever compressed liquefied hydrogen sulfide gas is being used, a chemical splash hazard exists and a face shield and protective apron should be worn. To prevent hand and skin exposures, impervious, thermal protective gloves should be worn. Glove manufacturers should be contacted and permeation studies obtained *before* final glove selections have been made.

Administrative controls should also be in place to minimize the potential for human exposures. These may include written procedures or policies which specify the methods and techniques that will be practiced whenever personnel are to work with hydrogen sulfide.

All personnel should receive *training* on the use, hazards, protective measures, emergency actions, and other precautions per 29 CFR 1910.1200 (Hazard Communication), prior to the first assignment in an area where hydrogen sulfide is used or stored.

If symptoms develop or overexposure is suspected, the following medical tests may be useful:

- ☑ Lung function tests
- ☑ Consider chest X-ray following acute overexposure (may be negative if taken immediately after exposure due to delayed onset of pulmonary edema).

Any evaluation should include a careful medical history of past and present symptoms with an examination. However, medical tests that evaluate existing damage are not a substitute for controlling exposure. Also, since smoking can cause heart disease, lung cancer, emphysema, and other respiratory disorders, smokers may experience symptoms more quickly than non-smokers under the same conditions of exposure. Prudent risk management requires careful consideration of *all* possible factors which might be causing the appearance of exposure symptoms in the workplace.

Other methods to reduce exposure include:

- ☑ Where possible, enclose operations and use local exhaust ventilation at the site of chemical release. If local exhaust ventilation or enclosure is not used, respiratory protection should be mandatory.
- ☑ Always ensure that proper protective clothing is worn when using chemical substances.
- ☑ Wash thoroughly immediately after exposure to hydrogen sulfide and at the end of the work shift or before eating, drinking, or smoking.
- ☑ Hazard warning information should be posted in the work area. In addition, as part of an on-going education and training program, all information on the health and safety hazards of hydrogen sulfide should be communicated to all exposed workers.
- ☑ Eye wash stations should be provided in work areas. If the potential for whole body exposure exists, then safety showers should also be provided.

RISK ASSESSMENT: ENVIRONMENT

General Assessment

The environment is at risk of exposure during transportation, storage, disposal, or destruction of hydrogen sulfide. In almost every scenario, the threat of environmental exposure is contingent upon the proper handling of the chemical substance. Accidental spills of the liquefied compressed gas or releases of the hydrogen sulfide gas, large or small, can result in fire, explosion, and possible contamination of the surrounding environmental mediums (water, soil, and air).

Hydrogen sulfide is considered a highly flammable gas. It presents a serious fire and explosion hazard under normal conditions of temperature and pressure. As a gas, it can quickly and easily form explosive mixtures in air. In reaction to incompatible materials such as strong oxidizers, acids, and many metals, it may ignite or explode. These characteristics require special consideration during any emergency situation involving a leak or spill of hydrogen sulfide or hydrogen sulfide mixtures. Should hydrogen sulfide ever come into contact with any of these incompatible substances either during use, transportation, or storage, violent reactions can occur.

Hydrogen sulfide occurs naturally and can also enter the environment through industrial discharges or releases, or from spills.

Acute Ecological Effects

Acute (short-term) toxic effects may include the death of animals, birds, or fish and death or low growth rate in plants. Acute effects are seen 2 to 4 days after animals or plants are exposed to hydrogen sulfide.

Insufficient data are available to evaluate or predict the short-term effects of hydrogen sulfide to aquatic life, plants, birds, or land animals.

Chronic Ecological Effects

Chronic toxic effects may include shortened life span, reproductive problems, lower fertility, and changes in appearance or behavior in exposed animals. These effects can be seen long after first exposure(s) to toxic chemicals.

Insufficient data are available to evaluate or predict the long-term effects of hydrogen sulfide to aquatic life, plants, birds, or land animals.

Water Solubility

Hydrogen sulfide is very slightly soluble in water. Concentrations of 1 milligram or less will mix with a liter of water.

Persistence in the Environment

Hydrogen sulfide is non-persistent in water, with a half-life less than 2 days. The half-life of a pollutant is the amount of time it takes for one-half of the chemical to be degraded. Virtually 100% of hydrogen is expected to end up in the air. The sulfide is moderately persistent in the environment with a half-life between 20 and 200 days.

Bioaccumulation in Aquatic Organisms

Some substances increase in concentration, or *bioaccumulate,* in living organisms as they breathe contaminated air, drink contaminated water, or eat contaminated food. These chemicals can become concentrated in the tissues and internal organs of animals as well as humans.

The concentration of the sulfide found in fish tissues is expected to be much higher than the average concentration of hydrogen sulfide in the water from which the fish was taken.

Recommended Risk-Reduction Measures

Proper training of all transporters will reduce the likelihood of a mishap or accident resulting in a leak or spill to the environment. The correct use of DOT labeling (per the MSDS) on all containers, trucks, and rail cars will enable emergency responders to react properly and quickly to any disaster thereby reducing the potential risk to the environment and to personnel.

Storage of hydrogen sulfide should be segregated from incompatible chemicals to minimize the risk of cross-contamination or contact. Class 1, group C electrical equipment should be installed in storage areas.

If a spill or leak to the environment has occurred, fire department, emergency response and/or hazardous materials spill personnel should be notified immediately. Cleanup should be attempted only by those trained in proper spill containment procedures. Contaminated soils should be removed for incineration and replaced with clean soil. If hydrogen sulfide should contact the water table, aquifer, or navigable waterway, remediation activities should be prompt to ensure protection of aquatic sediments. When such spills occur, the local and/or state emergency response authorities must be notified. A comprehensive emergency response or disaster preparedness/recovery plan should be in place prior to any operations involving the use, transportation, storage, or disposal of hydrogen sulfide.

If hydrogen sulfide is spilled or leaked, the following specific steps are recommended:

- ☑ Restrict persons not wearing protective clothing from area of spill or leak until cleanup is complete and area can be opened for normal work.

☑ Remove all sources of ignition.

☑ Stop the flow of gas before attempting cleanup procedures. If leak is from a cylinder and it cannot be stopped, remove cylinder to safe location (outdoors) and allow to vent until empty. Use water spray to knock down unignited vapors and to protect response personnel.

RISK ASSESSMENT: BUSINESS

General Assessment

Accidents or mishaps involving hydrogen sulfide can present a significant threat to business operations. The loss or damage of equipment or facilities can significantly affect fiscal viability. Lawsuits that may result from personnel injury/death, public exposures, and/or environmental contamination will also require a serious expenditure of resources. Media attention surrounding an injury, death, or environmental damage can also result in a loss of profits and loss of current as well as future business.

Recommended Risk-Reduction Measures

Company attorneys, safety and health professionals, and environmental specialists should be involved in the development of any procedures dealing with safety or emergency response. A company official should be pre-designated as a public relations officer with specific training in dealing with the press. Corporate plans and policies should be developed, approved, and implemented long before any need for such arises.

REFERENCES

Ahlbom, A. 1993. *Biostatistics for Epidemiologists.* New York: CRC Press/Lewis Publishers.

American Conference of Governmental Industrial Hygienists. 1988. *Documentation of the Threshold Limit Values and Biological Exposure Indices,* 5th Edition (with updates). Cincinnati: ACGIH

Anthony, C. P., and N. J. Kolthoff. 1971. *Textbook of Anatomy and Physiology.* St. Louis: C. V. Mosby Company.

Bloom, A. D. and F. J. DeSerres. 1995. *Ecotoxicity and Human Health.* New York: CRC Press/Lewis Publishers.

Brusick, D. J. 1994. *Methods for Genetic Risk Assessment.* New York: CRC Press/Lewis Publishers.

Bueche, J. R. 1972. *Principles of Physics,* 2nd Edition. New York: McGraw-Hill.

Burrell, R., D. K. Flaherty, and L. J. Sauers. 1992. *Toxicology of the Immune System.* New York: Van Nostrand Reinhold.

Calabrese, E. J. 1994. *Biological Effects of Low Level Exposures.* New York: CRC Press/Lewis Publishers.

Cockerman, L. G., B. S. Shane. 1994. *Basic Environmental Toxicology.* New York: CRC Press/Lewis Publishers.

Cothern, C. R., N. P. Ross. 1994. *Environmental Statistics, Assessment, and Forecasting.* New York: CRC Press/Lewis Publishers.

Gardner, R. W. 1994. *Chemical Intolerance.* New York: CRC Press/Lewis Publishers.

Hallenbeck, W. H. 1993. *Quantitative Risk Assessment for Environmental and Occupational Health,* 2nd Edition. New York: CRC Press/Lewis Publishers.

Hodgson, E., R. B. Mailman, and Janice E. Chambers. 1988. *Dictionary of Toxicology.* London: The MacMillian Company.

Howard, P. H. 1995. *Handbook of Environmental Fate and Exposures Data for Organic Chemicals,* Vol. I-IV. New York: CRC Press/Lewis Publishers.

Kamrin, M. A. 1988. *Toxicology: A Primer on Principles and Applications.* Chelsea, Michigan: Lewis Publishers.

Landis, W. G., M. H. Yu. 1995. *Introduction to Environmental Toxicology.* New York: CRC Press/Lewis Publishers.

Langerman, N. 1994 *Precautionary Labels for Chemical Containers.* New York: CRC Press/Lewis Publishers.

Lauwerys, R. R., P. Hoet. 1993. *Industrial Chemical Exposure.* New York: CRC Press/Lewis Publishers.

Lewis, R. J., Sr. 1993. *Hawley's Condensed Chemical Dictionary,* 12th Edition. New York: Van Nostrand Reinhold.

Lewis, R. J., Sr. 1992. *Sax's Dangerous Properties of Industrial Materials,* Volumes 1, 2, and 3, 8th Edition. New York: Van Nostrand Reinhold.

Manahan, S. E. 1992. *Toxicological Chemistry,* 2nd Edition. New York: CRC Press/Lewis Publishers.

Meyer, E. 1990. *Chemistry of Hazardous Materials.* Englewood Cliffs, NJ: Prentice Hall.

Nelson, G. O. 1992. *Gas Mixtures.* New York: CRC Press/Lewis Publishers.

Ott, W. R. 1995. *Environmental Statistics and Data Analysis.* New York: CRC Press/Lewis Publishers.

Patnaik, P. 1992. *A Comprehensive Guide to the Hazardous Properties of Chemical Substances.* New York: Van Nostrand Reinhold.

Rea, W. J. 1995. *Chemical Sensitivity,* Vol. I-III. New York: CRC Press/Lewis Publishers.

Richardson, M. 1995. *Dictionary of Substance and Their Effects,* Vol. I-VI. New York: CRC Press/Lewis Publishers.

Sacarello, H. L. A. 1994. *The Comprehensive Handbook of Hazardous Materials.* New York: CRC Press/Lewis Publishers.

U.S. Department of Health and Human Services, National Institute for Occupational Safety and Health. 1994. *NIOSH Pocket Guide to Chemical Hazards.* Washington, D.C.: U.S. Government Printing Office.

MATERIAL SAFETY DATA SHEET

CHEMICAL NAME	
HYDROQUINONE	**CRC PRESS/LEWIS PUBLISHERS DISCLAIMER:** THE INFORMATION AND RECOMMENDATIONS PRESENTED HEREIN ARE BASED ON SOURCES BELIEVED TO BE RELIABLE. CRC MAKES NO REPRESENTATION ON ITS COMPLETENESS OR ACCURACY. IT IS THE RESPONSIBILITY OF THE USER TO DETERMINE THE CHEMICAL'S SUITABILITY FOR ITS INTENDED USE, THE CHEMICAL'S SAFE USE, AND THE CHEMICAL'S PROPER DISPOSAL. NO REPRESENTATIONS AND/OR WARRANTIES, EITHER EXPRESSED OR IMPLIED, OF THE MERCHANTABILITY OR FITNESS FOR A PARTICULAR PURPOSE, OR OF ANY OTHER NATURE, ARE MADE WITH RESPECT TO THE INFORMATION PROVIDED IN THIS MSDS OR TO THE CHEMICAL TO WHICH INFORMATION MAY REFER. CRC NEITHER ASSUMES NOR AUTHORIZES ANY OTHER PERSON TO ASSUME FOR IT, ANY OTHER ADDITIONAL RESPONSIBILITY OR LIABILITY FOR THE USE OF, OR RELIANCE UPON, THIS INFORMATION.

HAZARD WARNING INFORMATION

HEALTH	FIRE	REACTIVE	OTHER	DEGREE OF HAZARD	COLOR CODING	OTHER CODES
2	1	0		0 = Minimum Hazard 1 = Slight Hazard 2 = Moderate Hazard 3 = Serious Hazard 4 = Severe Hazard	HEALTH = BLUE FIRE = RED REACTIVITY = YELLOW OTHER = WHITE	OX = Oxidizer ACID = Acid ALK = Alkali COR = Corrosive ~~W~~ = Use No Water

SECTION I - GENERAL INFORMATION

Characterization	**Hydrocarbon**	RCRA Number	**None**
		EPA Class	**Not Applicable**
DOT Proper Shipping Name	**Hydroquinone, liquid and solid**	Chemical Abstract Service (CAS) Number	**123-31-9**
DOT Hazard Class and Label Requirements	**IMO Poison B; St. Andrews Cross**	DOT Emergency Guide Code	**53**
DOT Identification Number	**UN 2662**	Chemical Formula	$C_6H_4(OH)_2$

Synonyms
p-Benzenediol; 1,4-benzenediol; dihydroxybenzene; Quinol; hydroquinol; pyrogentisic acid.

SECTION II - HAZARDOUS INGREDIENTS AND IDENTITY INFORMATION

Hazardous Components (specific identity)	OSHA Exposure Criteria	NIOSH Exposure Criteria	Immediately Dangerous to Life and Health (IDLH)	ACGIH Exposure Criteria
Hydroquinone (derivation: By oxidation of aniline to quinone with manganese dioxide followed by reduction to hydroquinone with sulfurous acid).	**PEL (8-hour):** 2 mg/m^3 **STEL:** Not Established	**REL (ceiling):** 2 mg/m^3 (15 minute) **STEL:** Not Applicable	50 mg/m^3	**TLV:** 2 mg/m^3 **STEL:** Not Established

SECTION III - PHYSICAL AND CHEMICAL CHARACTERISTICS

Boiling Point	**546°F (286°C)**	Specific Gravity (H_2O = 1)	**1.33**
Vapor Pressure (gas)	**1 x 10^{-5} at 69°F (20°C)**	Molecular Weight	**98.2**
Vapor Density (Air = 1)	**3.81**	Melting Point	**338°F (170°C)**

Solubility
Soluble in water (7%), alcohol, acetone, carbon tetrachloride, and ether. Slightly soluble in benzene.

Appearance and Odor
Colorless, light gray, or light tan hexagonal crystals. Color change depends on degree of oxidation.

SECTION IV - FIRE AND EXPLOSION HAZARD DATA

Flash Point (method used)	**329°F (165°C) closed cup**	Explosive Limits in Air % by Volume	**LEL: Not Determined UEL: Not Determined**
NFPA Classification	**Combustible Solid**	Autoignition Temperature	**960°F (515°C)**

Extinguishing Media
Use dry chemical, carbon dioxide, water spray, or regular foam.

Special Fire Fighting Procedures
Wear full protective clothing and self-contained breathing apparatus (SCBA). Poisonous gases are produced in fire. Use water spray to keep fire-exposed containers cool. Containers may explode in fire.

Unusual Fire and Explosion Hazards
Combustible solid. Its dusts can form a cloud that may explode if ignited in a closed area. Move containers from fire area if it can be done without risk. Violent reactions can occur with oxidizers. Do not release runoff from fire methods to sewers or waterways.

SECTION V - REACTIVITY DATA

Stability		
Stability		Conditions to Avoid **Hydroquinone turns brown on exposure to light and air due to oxidation, which becomes more rapid in the presence of alkali. Avoid contact with incompatible materials. Do not generate dusty conditions.**
Stable	Unstable **X**	Incompatibility *(materials to avoid)* **Strong oxidizers (permanganates, peroxides, nitrates, chlorates, and perchlorates), and alkalis. Reacts violently with sodium hydroxide, forms *p*-benzoquinone with chlorine, and is explosive with heated oxygen.**
Hazardous Polymerization		Conditions to Avoid **Under normal conditions of temperature and pressure, hazardous polymerization of hydroquinone is not expected to occur.**
May Occur	Will Not Occur **X**	Hazardous Decomposition or By-products **When heated to decomposition, hydroquinone can emit highly toxic/poisonous and acrid fumes and gases including toxic quinone and carbon dioxide fumes.**

SECTION VI - HEALTH HAZARD DATA

Primary Route(s) of Entry:	Inhalation?	Absorption (skin)?	Ingestion?
	X		X

Health Hazards

INHALATION: **Symptoms include irritation of the nose, throat, and respirat[illegible]. There may also be central nervous system depression causing excitement, dizziness, and collapse.**

SKIN & EYES: **Irritating to the skin; may cause irritation and dermatitis. An eye irritant with possibility of corneal damage, conjunctivitis, and keratitis. May cause brownish staining of eye tissue.**

INGESTION: **Serious human poison causing ringing in the ears, nausea, dizziness, vomiting, pallor, a feeling of suffocation, increased respiration, muscular twitching, headache, dypsnea, cyanosis, delirium, collapse, urine color change (green or brownish green), convulsions, blurred speech, tremors, and severe gastrointestinal disturbances.**

Carcinogenicity	NTP Listed?	IARC Cancer Review Group?	OSHA Regulated?	Target Organs?
Unknown Human Questioned Animal	**No**	**Group 3**	**29 CFR 1910.1000 Table Z-1**	**Respiratory system, eyes, skin, CNS, bladder.**

Medical Conditions Generally Aggravated by Exposure

Skin diseases may be aggravated by exposure.

Emergency and First-aid Procedures

Eye contact: Flush immediately with water for 15 minutes (minimum); seek medical attention. Skin contact: Remove all contaminated clothing. Immediately wash area with large amounts of soap and water. Seek medical assistance. For inhalation: Remove the person from exposure. Provide respiratory assistance and CPR, if required. Transfer to medical facility. Keep victim warm and at rest. If swallowed: Seek medical attention immediately. Give conscious person 1 to 2 glasses of water to dilute. Do NOT induce vomiting. DOCTOR: Consider gastric lavage and a urine screen test.

SECTION VII - PRECAUTIONS FOR SAFE HANDLING AND USE

Steps to be Taken in Case Material is Released or Spilled

Collect powders using HEPA vacuum. Do NOT dry sweep. Absorb solution spills in vermiculite and place in sealed drum. Ventilate area of spill or leak. Remove all sources of ignition.

Preferred Waste Disposal Method

No Citation.

Precautions to be Taken in Handling and Storage

Store to avoid contact with incompatible materials in tightly closed containers in cool, dry, well-ventilated area away from heat, flame, and direct sunlight. Use only non-sparking tools.

Other Precautions and Warnings

Sources of ignition are prohibited where hydroquinone is used, handled, or stored.

SECTION VIII - CONTROL MEASURES AND PERSONAL PROTECTIVE EQUIPMENT

Respiratory Protection *(specify type)*

For exposures above the PEL, use a MSHA/NIOSH-approved supplied-air respirator with full facepiece, hood, or helmet in continuous flow mode, or use a self-contained breathing apparatus (SCBA) operated in pressure demand or other positive pressure mode.

Ventilation

Local exhaust or general mechanical systems recommended.

Protective Gloves	Eye Protection	Other Protective Clothing
PVC, Chlorinated Polyethylene	**Chemical/Dust Goggles or Face Mask**	**Protective Apron**

Work/Hygiene Practices

Always wash hands thoroughly after using chemical; never bring food, drink, or smoking materials into vicinity of chemicals.

HYDROQUINONE

$C_6H_4(OH)_2$

CAS: 123-31-9

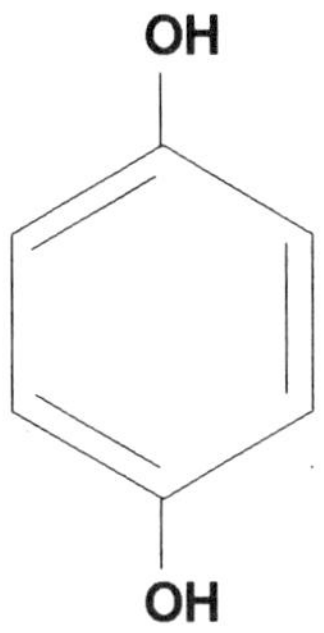

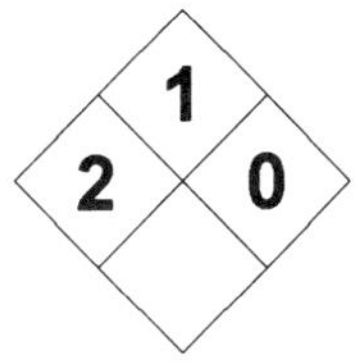

IDENTIFICATION AND TYPICAL USES

Hydroquinone is a colorless, light-tan, or light-gray, crystalline (hexagonal) solid in the shape of prisms. The degree of color change is dependent upon the level of oxidation. It is used as a reducing agent, in black and white photography development, as a dye intermediate, as an inhibitor, as a stabilizer in paints and varnishes, in motor fuels and oils, as an antioxidant for fats and oils, as an inhibitor of polymerization, and in skin hyper-pigmentation.

RISK ASSESSMENT: HEALTH

General Assessment

Hydroquinone is moderately toxic by *ingestion* and *inhalation*. It will cause moderate health hazard risk on skin or eye contact, but absorption through intact skin has not been reported. Its carcinogenicity in humans is not known but is questioned in some animal studies. It has been shown to cause mutations (genetic changes) in human cells. Experimental reproductive data have also been reported in animal tests. Hydroquinone is metabolized in the body to more toxic quinone which is responsible for the compound's systemic toxicity.

Inhalation of the dusts causes irritation of the nose, throat, and mucous membranes of the respiratory tract. It may also cause central nervous system disturbances similar to those caused by ingestion.

Skin and eye contact may produce mild to moderate irritation on contact. It can cause sensitization and eczematous dermatitis. Direct eye contact with hydroquinone particles causes immediate irritation and possible corneal ulceration.

Ingestion of hydroquinone causes serious toxic systemic effects. It is readily absorbed through the digestive process. There will be initial gastrointestinal irritation followed by central nervous system effects including blurred speech, ringing in the ears, tremors, a feeling of suffocation, headache, muscular twitching, vomiting, nausea, convulsions, dypsnea, cyanosis and methemoglobinemia, coma and unconsciousness, collapse, and possibly death due to respiratory failure. There may be changes in the color of the urine to green or brownish-green.

Acute Health Effects

The following acute (short-term) health effects may occur immediately or shortly after exposure to hydroquinone:

Skin: Will defat the skin and cause irritation and dermatitis (eczematous) and sensitization.

Eye: Severe irritation with a potential for corneal injury or damage.

Lung: Nose and throat irritation. Produces central nervous system effects and may cause lung effects as well.

CNS: Numerous effects by ingestion and inhalation. There may be headaches, muscle twitching, convulsions, dypsnea, blurred speech, ringing in the ears, tremors, coma, collapse, and death.

Chronic Health Effects

The following chronic (long-term) health effects may occur at some time after exposure to hydroquinone and can last for months or even years:

Cancer Hazards: According to information presented in the references, hydroquinone has been shown to cause cancer in some test animals. However, the data are conflicting and inconclusive in this regard. Some studies indicate that it causes mutations in humans. Some studies have listed its cancer causing properties as questioned animal studies.

Reproductive Hazard: According to information presented in the references, hydroquinone may have the ability to adversely affect reproduction in animals. The teratogenicity of hydroquinone in humans has not been adequately studied.

Other Chronic Effects: Prolonged contact may cause brownish staining of the conjunctiva followed by corneal changes and possible loss of visual acuity. Repeated skin contact can lead to chronic dermatitis and may cause de-pigmentation (in darker-skinned people) and hyper-pigmentation (in fair-skinned persons).

Recommended Risk-Reduction Measures

Personnel should avoid direct contact with hydroquinone. The exact nature of the toxicity of this chemical is not entirely understood in the references. Therefore, prudent risk management requires it be treated as though it posed a significant health risk in the event of overexposure. If a less toxic chemical cannot be substituted for a hazardous substance, then *engineering controls* are the most effective method of reducing exposures. The best protection is to enclose operations and/or provide local exhaust ventilation at the site of chemical release. While not always operationally feasible, isolating operations can also reduce exposure. Using respiratory protection is less effective than the controls mentioned above, but is still advisable whenever working with or around hydroquinone. The exposure ceiling is relatively low (2 mg/m^3) and difficult to accurately monitor. For all exposures, a supplied-air respirator with full facepiece operated in continuous flow mode, or a self-contained breathing apparatus (SCBA) with full facepiece and pressure demand are the recommended respiratory protection methods of choice. If a full facepiece is not available, then chemical goggles should be worn to protect the eyes. Whenever a chemical splash hazard exists, a face shield and a rubber apron should be worn. To prevent hand and skin exposures, impervious polyvinyl chloride or polyethylene gloves should be used. Glove manufacturers should be contacted to obtain permeation studies *before* glove selection has been made for hydroquinone.

Administrative controls should also be in place to minimize the potential for human exposures. These may include written procedures or policies which specify the methods and techniques that will be practiced whenever personnel are to work with hydroquinone.

All personnel should receive *training* on the use, hazards, protective measures, emergency actions, and other precautions per 29 CFR 1910.1200 (Hazard Communication), prior to the first assignment in an area where hydroquinone is used or stored.

If symptoms develop or overexposure is suspected, the following tests are recommended:

- ☑ Kidney function tests and urinalysis (check for color changes).
- ☑ Evaluation by a qualified allergist, with careful consideration of exposure history and special skin tests (may help diagnose allergy).
- ☑ Evaluation for nervous system disorders.

Any evaluation should include a careful history of past and present symptoms with an examination. Medical tests that simply look for existing damage are not a substitute for controlling exposures.

Other methods to reduce exposure include:

- ☑ Where possible, enclose operations and use local exhaust ventilation at the site of chemical release. If local exhaust ventilation or enclosure is not used, respiratory protection should be mandatory. Also, if possible, automatic transfer of hydroquinone from storage containers to process or work containers is recommended.
- ☑ Always ensure that proper protective clothing is worn when using chemical substances and that personnel are trained in its proper use and care.
- ☑ Wash thoroughly immediately after exposure to hydroquinone and at the end of the work shift or before eating, drinking, or smoking.
- ☑ Hazard warning information should be posted in the work area. In addition, as part of an on-going education and training program, all information on the health and safety hazards of hydroquinone should be communicated to all exposed and potentially exposed workers.
- ☑ Eye wash stations should be provided in the immediate work area for emergency use. If there is a possibility of skin exposure to hydroquinone, emergency shower facilities should also be provided in the immediate work area.
- ☑ Personnel should never wear contaminated clothing home (family members can be exposed). Clothing should be laundered by personnel who have been trained on the hazards of hydroquinone.

RISK ASSESSMENT: ENVIRONMENT

General Assessment

The environment is at risk of exposure during transportation, storage, disposal, or destruction of hydroquinone. In almost every scenario, the threat of environmental exposure is contingent upon the proper handling of the chemical substance. Accidental spills, large or small, can result in fire, explosion, and possible contamination of the surrounding environmental mediums (water, soil, and air).

Hydroquinone is considered a combustible solid. A cloud of hydroquinone dusts may explode if ignited in an enclosed area. It also presents a moderate fire and explosion hazard risk, especially in contact with incompatible materials. These characteristics require extreme caution in handling, storage, transportation, and disposal. It is incompatible with strong oxidizers and alkalis and contact can cause fire or explosion. It reacts violently with sodium hydroxide and forms *p*-benzoquinone with chlorine. Therefore, special consideration is required during any emergency situation involving a leak or spill of hydroquinone. Should hydroquinone ever come into contact with these incompatible substances during use, transportation, storage, or disposal, the formation of highly toxic and/or highly explosive commodities is extremely possible.

Hydroquinone may enter the environment through industrial discharges or spills.

Acute Ecological Effects

Acute (short-term) toxic effects may include the death of animals, birds, or fish and death or low growth rate in plants. Acute effects are seen 2 to 4 days after animals or plants are exposed to hydroquinone.

Hydroquinone has high acute toxicity to aquatic life and has caused germination decrease in agricultural crops. Insufficient data are available on the short-term effects of hydroquinone exposure to birds or terrestrial animals.

Chronic Ecological Effects

Chronic (long-term) toxic effects may include shortened life span, reproductive problems, lower fertility, and changes in appearance or behavior in exposed animals. These effects can be seen long after first exposure(s) to toxic chemicals.

Hydroquinone has high chronic toxicity to aquatic life. Insufficient data are available on the long-term effects of hydroquinone exposure to birds, plants, or terrestrial animals.

Water Solubility

Hydroquinone is highly soluble in water. Concentrations of 1000 milligrams and more will mix with a liter of water.

Persistence in the Environment

Hydroquinone is slightly persistent in water, with a half-life between 2 to 20 days. The half-life of a pollutant is the amount of time it takes for one half of the chemical to be degraded. About 99% of hydroquinone will eventually end up in water; the rest will end up in air.

Bioaccumulation in Aquatic Organisms

Some substances increase in concentration, or *bioaccumulate,* in living organisms as they breathe contaminated air, drink contaminated water, or eat contaminated food. These chemicals can become concentrated in the tissues and internal organs of animals as well as humans.

The concentration of hydroquinone found in fish tissue is expected to be about the same as the average concentration of hydroquinone in the water from which the fish was taken.

Recommended Risk-Reduction Measures

Proper training of all transporters will reduce the likelihood of a mishap or accident resulting in a leak or spill to the environment. The correct use of DOT labeling (per the MSDS) on all containers, trucks, and rail cars will enable emergency responders to react properly and quickly to any disaster thereby reducing the potential risk to the environment and to personnel.

Storage of hydroquinone should be segregated from incompatible chemicals to minimize the risk of cross-contamination or contact. Buildings designated for storage should be equipped with appropriate fire protection systems (alarms, sprinklers, emergency lighting, portable extinguishers).

If a spill or leak to the environment has occurred, fire department, emergency response and/or hazardous materials spill personnel should be notified immediately. Cleanup should be attempted only by those trained in proper spill containment procedures. Contaminated soils should be removed for incineration and

replaced with clean soil. If hydroquinone should contact the water table, aquifer, or navigable waterway, cleanup should be started immediately. It is highly soluble in water and total remediation may not be possible. The local and/or state emergency response authorities must be notified. A comprehensive emergency response or disaster preparedness/recovery plan should be in place prior to any operations involving the use, transportation, storage, or disposal of hydroquinone. If hydroquinone is spilled or leaked, the following specific steps are recommended:

- ☑ Restrict persons not wearing protective clothing from area of spill or leak until cleanup is complete and area can be opened for normal work.
- ☑ Remove all ignition sources.
- ☑ Ventilate area of leak.
- ☑ Collect powdered materials using a high-efficiency particulate air (HEPA) filter. Do NOT dry sweep (generates airborne dusts). Place in proper container for disposal. Absorb liquids containing hydroquinone in vermiculite, dry sand, earth, or similar material and deposit in sealed containers.
- ☑ It may be necessary to dispose of hydroquinone as a hazardous waste. The responsible state agency or the regional office of the federal Environmental Protection Agency (EPA) should be contacted for specific recommendations.

RISK ASSESSMENT: BUSINESS

General Assessment

Accidents or mishaps involving hydroquinone can present a significant threat to business operations. The mismanagement of chemical commodities can cause personnel injuries and/or property damage/loss. The loss or damage of equipment or facilities can significantly affect fiscal viability. Lawsuits that may result from personnel injury/death, public exposures, and/or environmental contamination will also require a serious expenditure of resources. Media attention surrounding an injury, death, or environmental damage can also result in a loss of profits and loss of current as well as future business prospects.

Always remember that anytime the terms "cancer," "carcinogen," or "reproductive hazard" are used, public emotion, hysteria, and ignorance can run equally high. This must be carefully considered whenever developing or implementing public relations policies.

Recommended Risk-Reduction Measures

Company attorneys, safety and health professionals, and environmental specialists should be involved in the development of any procedures dealing with safety or emergency response. A company official should be pre-designated as a public relations officer with specific training in dealing with the press. Corporate plans and policies should be developed, approved, and implemented long before any need for such arises.

REFERENCES

American Conference of Governmental Industrial Hygienists. 1988. *Documentation of the Threshold Limit Values and Biological Exposure Indices,* 5th Edition (with updates). Cincinnati: ACGIH

Bueche, J. R. 1972. *Principles of Physics,* 2nd Edition. New York: McGraw-Hill.

Calabrese, E. J. 1994. *Biological Effects of Low Level Exposures.* New York: CRC Press/Lewis Publishers.

Cockerman, L. G., B. S. Shane. 1994. *Basic Environmental Toxicology.* New York: CRC Press/Lewis Publishers.

Gardner, R. W. 1994. *Chemical Intolerance.* New York: CRC Press/Lewis Publishers.

Lauwerys, R. R., P. Hoet. 1993. *Industrial Chemical Exposure.* New York: CRC Press/Lewis Publishers.

Lewis, R. J., Sr. 1992. *Sax's Dangerous Properties of Industrial Materials,* Volumes 1, 2, and 3, 8th Edition. New York: Van Nostrand Reinhold.

Meyer, E. 1990. *Chemistry of Hazardous Materials.* Englewood Cliffs, NJ: Prentice Hall.

Patnaik, P. 1992. *A Comprehensive Guide to the Hazardous Properties of Chemical Substances.* New York: Van Nostrand Reinhold.

Proctor, N. H., J. P. Hughes. 1991. *Chemical Hazards of the Workplace.* New York: Van Nostrand Reinhold.

Quigley, D. R. 1994. *Handbook of Emergency Chemical Management.* New York: CRC Press/Lewis Publishers.

Rea, W. J. 1995. *Chemical Sensitivity,* Vol. I-III. New York: CRC Press/Lewis Publishers.

Richardson, M. 1995. *Dictionary of Substance and Their Effects,* Vol. I-VI. New York: CRC Press/Lewis Publishers.

Rylander, R., R. R. Jacobs. 1994. *Organic Dusts.* New York: CRC Press/Lewis Publishers.

Sacarello, H. L. A. 1994. *The Comprehensive Handbook of Hazardous Materials.* New York: CRC Press/Lewis Publishers.

Torkelson, T. R. V. K. Rowe. 1981. Halogenated aliphatic hydrocarbons. *Patty's Industrial Hygiene and Toxicology,* Vol. 2B, *Toxicology,* pp 3568-3572. New York: Wiley-Interscience.

U.S. Department of Health and Human Services, National Institute for Occupational Safety and Health. 1994. *NIOSH Pocket Guide to Chemical Hazards.* Washington, D.C.: U.S. Government Printing Office.

MATERIAL SAFETY DATA SHEET

CHEMICAL NAME	CRC PRESS/LEWIS PUBLISHERS DISCLAIMER:
HYDROXYLAMINE	THE INFORMATION AND RECOMMENDATIONS PRESENTED HEREIN ARE BASED ON SOURCES BELIEVED TO BE RELIABLE. CRC MAKES NO REPRESENTATION ON ITS COMPLETENESS OR ACCURACY. IT IS THE RESPONSIBILITY OF THE USER TO DETERMINE THE CHEMICAL'S SUITABILITY FOR ITS INTENDED USE, THE CHEMICAL'S SAFE USE, AND THE CHEMICAL'S PROPER DISPOSAL. NO REPRESENTATIONS AND/OR WARRANTIES, EITHER EXPRESSED OR IMPLIED, OF THE MERCHANTABILITY OR FITNESS FOR A PARTICULAR PURPOSE, OR OF ANY OTHER NATURE, ARE MADE WITH RESPECT TO THE INFORMATION PROVIDED IN THIS MSDS OR TO THE CHEMICAL TO WHICH INFORMATION MAY REFER. CRC NEITHER ASSUMES NOR AUTHORIZES ANY OTHER PERSON TO ASSUME FOR IT, ANY OTHER ADDITIONAL RESPONSIBILITY OR LIABILITY FOR THE USE OF, OR RELIANCE UPON, THIS INFORMATION.

HAZARD WARNING INFORMATION

HEALTH	FIRE	REACTIVE	OTHER	DEGREE OF HAZARD	COLOR CODING	OTHER CODES
2	0	3		0 = Minimum Hazard 1 = Slight Hazard 2 = Moderate Hazard 3 = Serious Hazard 4 = Severe Hazard	HEALTH = BLUE FIRE = RED REACTIVITY = YELLOW OTHER = WHITE	OX = Oxidizer ACID = Acid ALK = Alkali COR = Corrosive W = Use No Water

SECTION I - GENERAL INFORMATION

Characterization	**Amine**	RCRA Number	**D003**
		EPA Class	**Characteristic (R) Waste**
DOT Proper Shipping Name	**Not Listed**	Chemical Abstract Service (CAS) Number	**7803-49-8**
DOT Hazard Class and Label Requirements	**Not Listed**	DOT Emergency Guide Code	**No Citation**
DOT Identification Number	**Not Listed**	Molecular Formula	**H_3NO**

Synonyms
Oxammonium; hydroxylammonium.

SECTION II - HAZARDOUS INGREDIENTS AND IDENTITY INFORMATION

Hazardous Components (specific identity)	OSHA Exposure Criteria	NIOSH Exposure Criteria	Immediately Dangerous to Life and Health (IDLH)	ACGIH Exposure Criteria
Hydroxylamine (derivation: By decomposing hydroxylamine hydrofluoride or sulfate with a base and distilling in vacuum).	**PEL:** Not Established **STEL:** Not Established	**REL:** Not Established **STEL:** Not Established	Not Determined	**TLV:** Not Established **STEL:** Not Established

SECTION III - PHYSICAL AND CHEMICAL CHARACTERISTICS

Boiling Point	**110°F (56°C)**	Specific Gravity (H_2O = 1)	**1.23**
Vapor Pressure (mm Hg)	**10 at 117°F (47°C)**	Molecular Weight (atomic weight)	**33.03**
Vapor Density (Air = 1)	**1.1**	Melting Point	**93°F (34°C)**

Solubility
Soluble in water (forms alkaline solution); soluble in ammonia and methanol, slightly soluble in benzene, ether, chloroform, and carbon disulfide.

Appearance and Odor
Large, hygroscopic, white flakes or crystals. May also appear in liquid solution.

SECTION IV - FIRE AND EXPLOSION HAZARD DATA

Flash Point (method used)	**158°F (70°C) explodes**	Explosive Limits in Air % by Volume	**LEL: Not Determined UEL: Not Determined**
NFPA Classification	**Combustible Solid**	Autoignition Temperature	**Not Determined**

Extinguishing Media
Use dry chemical or alcohol foam. Water may cause frothing. Carbon dioxide initiates decomposition.

Special Fire Fighting Procedures
Poisonous gases are produced in fire. Structural fire-fighting clothing is permeable and may not provide adequate protection. Wear self-contained breathing apparatus (SCBA) operated in positive pressure mode. Move fire-exposed containers from fire if it can be done without risk. Cool with water spray.

Unusual Fire and Explosion Hazards
Highly explosive when heated. Containers may explode in fire. Fight fire from distance, if possible, using unmanned equipment. Do not direct a solid stream of water at bulk quantities of hydroxylamine.

SECTION V - REACTIVITY DATA

Stability		Conditions to Avoid
Stability		Hydroxylamine is highly unstable and will rapidly decompose on exposure to air at room temperature. This is especially true in the presence of moisture or carbon dioxide.
Stable	Unstable X	Incompatibility *(materials to avoid)*: Barium oxides, chlorine, copper (II) sulfate, lead dioxide, potassium dichromate, potassium permanganate, sodium, zinc, calcium, chlorates, bromates, and perchlorates in the presence of sulfuric acid.
Hazardous Polymerization		Conditions to Avoid: Under normal conditions of temperature and pressure, hazardous polymerization of hydroxylamine is not expected to occur.
May Occur	Will Not Occur X	Hazardous Decomposition or By-products: Poisonous gases, including oxides of nitrogen are produced when hydroxylamine is involved in a fire.

SECTION VI - HEALTH HAZARD DATA

Primary Route(s) of Entry:	Inhalation?	Absorption (skin/eye)?	Ingestion?
	X		X

Health Hazards

INHALATION: Irritation of the nose, throat, and respiratory tract. Symptoms include cough, sneezing, tachycardia, hypertension, cyanosis, headache, fatigue, anorexia, vomiting, emaciation, and bladder problems causing hematuria (blood in urine), proteinuria (excessive proteins in urine), and bladder injury. Can cause vertigo, tinnitus (ringing in the ears), and cyanosis due to methemoglobinemia (decreased oxygen-carrying capabilities of the blood).

SKIN & EYES: Irritation of the skin and eyes. A strong sensitizer causing eczema in some individuals.

INGESTION: Can cause similar systemic effects as noted for inhalation.

Carcinogenicity	NTP Listed?	IARC Cancer Review Group?	OSHA Regulated?	Target Organs?
Unknown Human Unknown Animal	No	No	No	Resp. sys. CNS, eyes, skin, blood, kidneys.

Medical Conditions Generally Aggravated by Exposure

Existing skin diseases may be aggravated by exposure to hydroxylamine.

Emergency and First-aid Procedures

Eye contact: Do not allow victim to keep eyes tightly shut. Flush immediately with water for 15 minutes (minimum); seek medical attention. Skin contact: Remove all contaminated clothing, including shoes. Immediately wash area with flooding amounts of soap and water. Seek medical attention immediately. For inhalation: Remove the person from exposure. Provide respiratory assistance and CPR. Transfer to medical facility. If swallowed: Seek medical attention immediately. Unless advised otherwise, give a conscious person 1 to 2 glasses of water to drink and induce vomiting.

SECTION VII - PRECAUTIONS FOR SAFE HANDLING AND USE

Steps to be Taken in Case Material is Released or Spilled

Remove all ignition sources. Wear SCBA respiratory protection. Restrict those not wearing protective equipment and who are not involved in cleanup from entering area. Avoid skin contact. Do not create dust cloud. Ventilate area of spill. Do NOT dry sweep. Use HEPA vacuum and deposit in sealed drums.

Preferred Waste Disposal Method

No Citation.

Precautions to be Taken in Handling and Storage

Store in tightly closed containers in cool, well-ventilated location away from heat (including direct sunlight). Protect containers from physical damage. Prevent dust buildup in storage and use areas.

Other Precautions and Warnings

Sources of ignition (open flame or smoking) are prohibited where hydroxylamine is used or stored. Metal containers should be used. Do NOT generate dust clouds or dusty conditions when working.

SECTION VIII - CONTROL MEASURES AND PERSONAL PROTECTIVE EQUIPMENT

Respiratory Protection *(specify type)*

Use a NIOSH-approved respirator equipped with HEPA filter, or a powered air purifying respirator, or use a self-contained breathing apparatus (SCBA) with full facepiece operated in positive pressure mode.

Ventilation

Local exhaust (preferred) at site of chemical work or general ventilation.

Protective Gloves	Eye Protection	Other Protective Clothing
Impervious Gloves	Chemical/Dust Goggles or Face Mask	Protective Clothing (prevent skin contact)

Work/Hygiene Practices

Always wash hands thoroughly after using chemical; never bring food, drink, or smoking materials into vicinity of chemicals.

HYDROXYLAMINE

H_3NO CAS: 7803-49-8

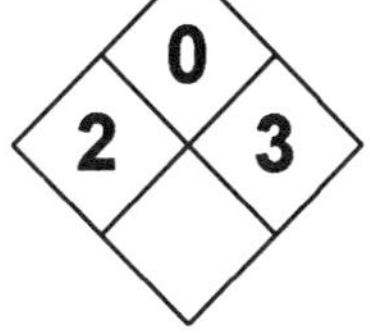

IDENTIFICATION AND TYPICAL USES

Hydroxylamine can appear as large, hygroscopic (absorbs moisture from the air), white flakes or crystals, or as a colorless liquid solution. It is used as a reducing agent, an analytical reagent, as a hair remover, and as an antioxidant for fatty acids and soaps. It is also used in the manufacture of synthetic rubber, in photographic developing solutions, in the preparation of paints and varnishes, in rust proofing, and in dilute solutions for the treatment of skin diseases.

RISK ASSESSMENT: HEALTH

General Assessment

Hydroxylamine is a human poison by *inhalation* and *ingestion*. It is a corrosive irritant to the eyes, skin, and mucosa. Its human carcinogenicity, mutagenicity, and teratogenicity are unknown.

Inhalation or ingestion of hydroxylamine dusts can cause toxic systemic effects with localized irritation of the skin, eyes, nose, throat, and the mucous membranes lining the respiratory tract. There can be eczema, tachycardia (rapid heartbeat) hypertension (increased blood pressure) coughing, sneezing, headache, fatigue, anorexia, diarrhea, loss of appetite with subsequent weight loss, low body temperature, cyanosis due to methemoglobinemia, jaundice, nausea and vomiting, emaciation, and death due to respiratory paralysis. There can also be serious bladder disorders, including hematuria (blood in urine) and proteinuria (excessive proteins in urine), as well as bladder damage. Central nervous system effects typically include vertigo, headache, tinnitus (ringing in the ears), and general fatigue.

Prolonged exposures can lead to contact dermatitis with erythema and squamous-type eczema. The trunk, forearms, and the back of the hands are usually the areas most affected by the dermatitis. Hydroxylamine is highly corrosive to the eye and surrounding tissues.

Acute Health Effects

The following acute (short-term) health effects may occur immediately or shortly after exposure to hydroxylamine:

Skin: Irritation with rash and localized inflammation. An allergen that may cause dermatitis in some individuals.

Eye: Severe irritation; pain and inflammation.

Lung: Nose, throat, and respiratory tract irritation causing coughing, sneezing, and likely absorption through the respiration process to cause toxic systemic effects.

CNS: Causes vertigo, tinnitus, headaches, fatigue, and other depression effects up to and including death from respiratory paralysis.

Chronic Health Effects

The following chronic (long-term) health effects may occur at some time after exposure to hydroxylamine and can last for months or even years:

Cancer Hazards: According to the information presented in the references, hydroxylamine has not been adequately tested for its ability to cause cancer in test animals.

Reproductive Hazard: According to information presented in the references, hydroxylamine has not been adequately tested for its ability to cause adverse reproductive effects in test animals.

Other Chronic Effects: Chronic exposure to skin can cause eczema.

Recommended Risk-Reduction Measures

Personnel should avoid direct contact with hydroxylamine. Occupational poisoning occurs most commonly by inhalation and ingestion. If a less toxic material or compound cannot be substituted for hydroxylamine, then *engineering controls* are the most effective method of reducing exposures. The best protection is to enclose operations and/or provide local exhaust ventilation at the site of hydroxylamine release. While not always operationally feasible, isolating operations can also reduce exposure. Using respiratory protection is less effective than the controls mentioned above, but is still advisable whenever working with or around hydroxylamine. No exposure levels have been established for this compound. This does not mean that exposure is without risk. As a minimum, personnel should wear an MSHA/NIOSH-approved respirator equipped with a high efficiency particulate air (HEPA) filter for low or infrequent exposures. An MSHA/NIOSH-approved self-contained breathing apparatus (SCBA) with full facepiece operated in pressure demand is a more appropriate level of respiratory protection when high exposures to hydroxylamine are anticipated. If a full facepiece is not available, then chemical/dust goggles should be worn to protect the eyes. A face shield should also be considered. To prevent hand and skin exposures, polyvinyl alcohol gloves should be worn.

Administrative controls should also be in place to minimize the potential for human exposures. These may include written procedures or policies which specify the methods and techniques that will be practiced whenever personnel are to work with hydroxylamine.

All personnel should receive *training* on the use, hazards, protective measures, emergency actions, and other precautions per 29 CFR 1910.1200 (Hazard Communication), prior to the first assignment in an area where hydroxylamine is used or stored.

Before beginning employment and at regular times thereafter, the following recommended medical tests should be considered:

- ☑ Complete blood count (CBC).
- ☑ Urinalysis to test for blood and protein levels in urine (establish baseline).

If symptoms should develop or overexposure is suspected, the following additional tests may be helpful:

- ☑ Complete blood count.
- ☑ Urinalysis (compare to baseline).
- ☑ Liver and kidney function tests.

Any evaluation should include a careful history of past and present symptoms with an examination. Medical tests that simply look for existing damage are not a substitute for controlling exposures.

Other methods to reduce exposure include:

- ☑ Where possible, enclose operations and use local exhaust ventilation at the site of chemical release. If local exhaust ventilation or enclosure is not used, respiratory protection should be mandatory.
- ☑ Always ensure that proper protective clothing is worn when using chemical substances.
- ☑ Wash thoroughly immediately after exposure to hydroxylamine and at the end of the work shift or before eating, drinking, or smoking.
- ☑ Hazard warning information should be posted in the work area. In addition, as part of an on-going education and training program, all information on the health and safety hazards of hydroxylamine should be communicated to all exposed and potentially exposed workers.
- ☑ Eye wash stations should be provided in the immediate work area for emergency use. If there is a possibility of skin exposure to hydroxylamine, emergency shower facilities should be provided.
- ☑ Workers whose clothing has been contaminated by hydroxylamine should change into clean clothes before leaving work. Contaminated work clothing should be laundered only by individuals who have been informed of the hazards of exposure to hydroxylamine.

RISK ASSESSMENT: ENVIRONMENT

General Assessment

The environment is at risk of exposure during transportation, storage, disposal, or destruction of hydroxylamine. In almost every scenario, the threat of environmental exposure is contingent upon the proper handling of the chemical substance. Accidental spills, large or small, can result in contamination of the surrounding environmental mediums (water, soil, and air).

Hydroxylamine is a combustible solid. It can present a dangerous explosion hazard if its dusts form a cloud and come into contact with an ignition source. It is also incompatible with oxidizing materials and contact can result in violent and explosive reactions. Caution is always required in handling, storage, transportation, and disposal of hydroxylamine. Emergency responders should be made aware of the presence of hydroxylamine at any emergency response situation.

Hydroxylamine can enter the environment from industrial effluents and from spills.

Acute Ecological Effects

Acute (short-term) toxic effects may include the death of animals, birds, or fish and death or low growth rate in plants. Acute effects are seen 2 to 4 days after animals or plants are exposed to hydroxylamine. Hydroxylamine has high acute toxicity to aquatic life. Insufficient data are available to evaluate the short-term effects of hydroxylamine on plants, birds, or terrestrial animals.

Chronic Ecological Effects

Chronic (long-term) toxic effects may include shortened life span, reproductive problems, lower fertility, and changes in appearance or behavior in exposed animals. These effects can be seen long after first exposure(s) to toxic chemicals. Hydroxylamine has high chronic toxicity to aquatic life. Insufficient data are available to evaluate or predict the long-term effects of hydroxylamine to plants, birds, or terrestrial animals.

Water Solubility

Hydroxylamine is highly soluble in water, especially cold water. Concentrations of 1000 milligrams and more will mix with a liter of water. It decomposes in water to form ammonium hydroxide, nitrous oxide, and nitrogen oxide.

Persistence in the Environment

Hydroxylamine is slightly persistent in water, with a half-life of between 2 and 20 days. The half-life of a pollutant is the amount of time it takes for one half of the chemical to be degraded. It is expected to decompose to form other compounds relatively quickly in the aquatic environment.

Bioaccumulation in Aquatic Organisms

Some substances increase in concentration, or *bioaccumulate,* in living organisms as they breathe contaminated air, drink contaminated water, or eat contaminated food. These chemicals can become concentrated in the tissues and internal organs of animals as well as humans.

Hydroxylamine is not expected to accumulate in the edible tissues of fish.

Recommended Risk-Reduction Measures

Proper training of all transporters will reduce the likelihood of a mishap or accident resulting in a leak or spill of hydroxylamine into the environment. Labels on all containers, trucks, and rail cars must meet DOT requirements and accurately reflect their contents to enable emergency responders to react properly and quickly to any disaster thereby reducing the potential risk to the environment and to personnel.

Storage of hydroxylamine should be segregated from other chemicals and materials to minimize the risk of cross-contamination. It should be stored to avoid contact with heat or ignition sources, including direct sunlight.

If a spill or leak to the environment has occurred, fire department, emergency response and/or hazardous materials spill personnel should be notified immediately. Cleanup should be attempted only by those trained in proper spill containment procedures. Contaminated soils should be removed for incineration and replaced with clean soil. If hydroxylamine should contact the water table, aquifer, or navigable waterway, time is of the essence. It decomposes in water and total containment and remediation may not be possible. The local and/or state emergency response authorities must be notified. A comprehensive emergency response or disaster preparedness/recovery plan should be in place prior to any operations involving the use, transportation, storage, or disposal of hydroxylamine.

If hydroxylamine is spilled or leaked, the following specific steps are recommended:

- ☑ Restrict persons not wearing protective clothing from area of spill or leak until cleanup is complete and area can be opened for normal work.
- ☑ Do NOT generate dusts. Dust clouds of the right proportions in air present a dangerous explosion hazard. Use a scoop or a vacuum equipped with a high efficiency particulate air filter (HEPA) to collect spills. Place materials in drum for recovery or disposal, as required. Absorb solution spills in vermiculite and place in a sealed container for disposal.
- ☑ Ventilate area of spill or leak.
- ☑ Remove all ignition sources and sources of heat from the hazard area.
- ☑ Containers should be removed to safer area if it can be done without increasing the risk.
- ☑ It may be necessary to dispose of hydroxylamine as a hazardous waste. The responsible state agency or the regional office of the federal Environmental Protection Agency (EPA) should be contacted for specific recommendations.

RISK ASSESSMENT: BUSINESS

General Assessment

Accidents or mishaps involving hydroxylamine can present a serious threat to business operations. The loss or damage of equipment or facilities can significantly affect fiscal viability. Lawsuits that may result from personnel illness, injury/death, public exposures, and/or environmental contamination will require a serious expenditure of resources. Media attention surrounding an injury, death, or environmental damage can also result in a loss of profits and loss of current as well as future business.

Recommended Risk-Reduction Measures

Company attorneys, safety and health professionals, and environmental specialists should be involved in the development of any procedures for responding to chemical incidents. A company official should be predesignated as a public relations officer with specific training in dealing with the press. Corporate plans and policies should be developed, approved, and implemented long before any need for such arises.

REFERENCES

American Conference of Governmental Industrial Hygienists. 1988. *Documentation of the Threshold Limit Values and Biological Exposure Indices,* 5th Edition (with updates). Cincinnati: ACGIH

Bloom, A. D. and F. J. DeSerres. 1995. *Ecotoxicity and Human Health.* New York: CRC Press/Lewis Publishers.

Brusick, D. J. 1994. *Methods for Genetic Risk Assessment.* New York: CRC Press/Lewis Publishers.

Bueche, J. R. 1972. *Principles of Physics,* 2nd Edition. New York: McGraw-Hill.

Calabrese, E. J., E. M. Kenyon. *Air Toxics and Risk Assessment.* New York: CRC Press/Lewis Publishers.

Calabrese, E. J. 1994. *Biological Effects of Low Level Exposures.* New York: CRC Press/Lewis Publishers.

Cockerman, L. G., B. S. Shane. 1994. *Basic Environmental Toxicology.* New York: CRC Press/Lewis Publishers.

Cothern, C. R., N. P. Ross. 1994. *Environmental Statistics, Assessment, and Forecasting.* New York: CRC Press/Lewis Publishers.

Cothern, R. R. 1993. *Comparative Environmental Risk Assessment.* New York: CRC Press/Lewis Publishers.

Danse, I. R. 1991. *Common Sense Toxics in the Workplace.* New York: Van Nostrand Reinhold.

Gerarde, H. W. 1960. *Toxicology and Biochemistry of Aromatic Hydrocarbons.* New York: Elsevier.

Hallenbeck, W. H. 1993. *Quantitative Risk Assessment for Environmental and Occupational Health,* 2nd Edition. New York: CRC Press/Lewis Publishers.

Howard, P. H., et. al. 1991. *Handbook of Environmental Degradation Rates.* New York: CRC Press/Lewis Publishers.

Jones, F. E. 1994. *Toxic Organic Vapors in the Workplace.* New York: CRC Press/Lewis Publishers.

Keith, L. H. and M. M. Walker. 1995. *Handbook of Air Toxics.* New York: CRC Press/Lewis Publishers.

Landis, W. G., M. H. Yu. 1995. *Introduction to Environmental Toxicology.* New York: CRC Press/Lewis Publishers.

Langerman, N. 1994 *Precautionary Labels for Chemical Containers.* New York: CRC Press/Lewis Publishers.

Lauwerys, R. R., P. Hoet. 1993. *Industrial Chemical Exposure.* New York: CRC Press/Lewis Publishers.

Lewis, R. J., Sr. 1993. *Hawley's Condensed Chemical Dictionary,* 12th Edition. New York: Van Nostrand Reinhold.

Lewis, R. J., Sr. 1992. *Sax's Dangerous Properties of Industrial Materials,* Volumes 1, 2, and 3, 8th Edition. New York: Van Nostrand Reinhold.

Manahan, S. E. 1994. *Environmental Chemistry,* 6th Edition. New York: CRC Press/Lewis Publishers.

Mellström, G. A., J. E. Wahlberg, and H. I. Maibach. 1994. *Protective Gloves for Occupational Use.* New York: CRC Press/Lewis Publishers.

Meyer, E. 1990. *Chemistry of Hazardous Materials.* Englewood Cliffs, NJ: Prentice Hall.

Ott, W. R. 1995. *Environmental Statistics and Data Analysis.* New York: CRC Press/Lewis Publishers.

Patnaik, P. 1992. *A Comprehensive Guide to the Hazardous Properties of Chemical Substances.* New York: Van Nostrand Reinhold.

Philp, R. B. 1995. *Environmental Hazards and Human Health.* New York: CRC Press/Lewis Publishers.

Proctor, N. H., J. P. Hughes. 1991. *Chemical Hazards of the Workplace.* New York: Van Nostrand Reinhold.

Quigley, D. R. 1994. *Handbook of Emergency Chemical Management.* New York: CRC Press/Lewis Publishers.

Sacarello, H. L. A. 1994. *The Comprehensive Handbook of Hazardous Materials.* New York: CRC Press/Lewis Publishers.

Torkelson, T. R. V. K. Rowe. 1981. *Patty's Industrial Hygiene and Toxicology,* Vol. 2B. New York: Wiley-Interscience.

U.S. Department of Health and Human Services, National Institute for Occupational Safety and Health. 1994. *NIOSH Pocket Guide to Chemical Hazards.* Washington, D.C.: U.S. Government Printing Office.

MATERIAL SAFETY DATA SHEET

CHEMICAL NAME

INDENE

HAZARD WARNING INFORMATION

HEALTH	FIRE	REACTIVE	OTHER	DEGREE OF HAZARD	COLOR CODING	OTHER CODES
1	2	1		0 = Minimum Hazard 1 = Slight Hazard 2 = Moderate Hazard 3 = Serious Hazard 4 = Severe Hazard	HEALTH = BLUE FIRE = RED REACTIVITY = YELLOW OTHER = WHITE	OX = Oxidizer ACID = Acid ALK = Alkali COR = Corrosive W = Use No Water

SECTION I - GENERAL INFORMATION

Characterization	RCRA Number	EPA Class
Hydrocarbon	**None**	**Not Applicable**

DOT Proper Shipping Name **Combustible Liquid, N.O.S.**	Chemical Abstract Service (CAS) Number **95-13-6**
DOT Hazard Class and Label Requirements **Combustible Liquid**	DOT Emergency Guide Code **No Citation**
DOT Identification Number **NA 1993**	Molecular Formula **C_9H_8**

Synonyms

Inden; indonaphthene.

SECTION II - HAZARDOUS INGREDIENTS AND IDENTITY INFORMATION

Hazardous Components (specific identity)	OSHA Exposure Criteria	NIOSH Exposure Criteria	Immediately Dangerous to Life and Health (IDLH)	ACGIH Exposure Criteria
Indene (derivation: Occurs naturally in tars from coal, lignite, and crude petroleum. Also prepared from acetylene over activated charcoal at 1157°F or by passing tetrahydronaphthalene over silicon dioxide and aluminum trioxide at 1238°F). 1 ppm = 4.83 mg/m^3	**PEL:** Not Established **STEL:** Not Established	**REL (10-hour):** 10 ppm 45 mg/m^3 **STEL:** Not Established	Not Determined	**TLV:** 10 ppm 45 mg/m^3 **STEL:** Not Established

SECTION III - PHYSICAL AND CHEMICAL CHARACTERISTICS

Boiling Point **359°F (182°C)**	Specific Gravity (H_2O = 1) **0.997**
Vapor Pressure (mm Hg) **Not Found**	Molecular Weight (atomic weight) **116.29**
Vapor Density (Air = 1) **Not Found**	Freezing Point **29°F (-2°C)**

Solubility

Insoluble in water. Soluble in alcohol, ether, carbon disulfide, benzene, pyrimidene, organic solvents.

Appearance and Odor

A colorless liquid with a pungent odor. A solid at temperatures below 29°F (-2°C).

SECTION IV - FIRE AND EXPLOSION HAZARD DATA

Flash Point (method used) **173°F (78°C) closed cup**	Explosive Limits in Air % by Volume **LEL: 0.9%** **UEL: Not Determined**
NFPA Classification **Class IIIA Combustible Liquid**	Autoignition Temperature **Not Determined**

Extinguishing Media

Use carbon dioxide, dry chemical, or foam. Indene floats on water making it an ineffective extinguisher.

Special Fire Fighting Procedures

Poisonous gases are produced in fire. Wear self-contained breathing apparatus (SCBA) and full protective gear. Move containers from fire if it can be done without risk. Cool sides of fire-exposed containers with water until long after fire is out. Monitor smoke directions and stay clear. Be aware of runoff from fire control measure. Do NOT release runoff to sewers or waterways. Dike for collection and disposal.

Unusual Fire and Explosion Hazards

None reported.

SECTION V - REACTIVITY DATA

Stability		Conditions to Avoid
Stable	Unstable	Indene is normally stable in closed containers, under routine conditions of storage and handling. Avoid contact with heat and exposure to incompatible materials. Keep away from strong oxidizers and acids.
X		Incompatibility *(materials to avoid)* Indene will explode during nitration with sulfuric acid + nitric acid. May react with strong oxidizers.
Hazardous Polymerization		Conditions to Avoid Hazardous polymerization and oxidation will occur on standing or when exposed to sunlight. Keep away from direct sunlight.
May Occur	Will Not Occur	Hazardous Decomposition or By-products
X		Poisonous gases and acrid fumes are produced in fire, including extremely toxic fumes and gases, including oxides of carbon.

SECTION VI - HEALTH HAZARD DATA

Primary Route(s) of Entry:	Inhalation?	Absorption (skin/eye)?	Ingestion?
	X		X

Health Hazards

INHALATION: Irritation of the eyes, nose, and respiratory tract. It may also cause chemical pneumonitis and possible delayed pulmonary edema (fluid in the lungs) which can be fatal. It can also cause damage to the liver, kidney, and spleen.

SKIN & EYES: Severe eye and skin irritant. May cause dermatitis and skin sensitization. Skin absorption is not likely to occur.

INGESTION: Not expected to be acutely toxic. Large amounts may cause adverse health effects.

Carcinogenicity	NTP Listed?	IARC Cancer Review Group?	OSHA Regulated?	Target Organs?
Unknown Human Unknown Animal	No	No	No	Respiratory system, skin, eyes, liver, kidneys, spleen.

Medical Conditions Generally Aggravated by Exposure

Existing skin diseases (eczema, psoriasis) and liver or kidney dysfunction may be aggravated.

Emergency and First-aid Procedures

Eye contact: Do not allow victim to keep eyes tightly shut. Flush immediately with water for 15 minutes (minimum); seek medical attention. Skin contact: Remove all contaminated clothing, including shoes. Immediately wash area with flooding amounts of warm water. Seek medical attention immediately. For inhalation: Remove the person from exposure. Provide respiratory assistance and CPR. Transfer to medical facility. Observe 48 hours for lung effects. If swallowed: Seek medical attention immediately. Give conscious and alert person 1 to 2 glasses of water. Do NOT induce vomiting unless advised otherwise (aspiration hazard). Never give an unconscious or convulsing person anything by mouth.

SECTION VII - PRECAUTIONS FOR SAFE HANDLING AND USE

Steps to be Taken in Case Material is Released or Spilled

Restrict those not wearing protective equipment and who are not involved in cleanup from entering area. Absorb liquids with vermiculite and place in sealed container for disposal. Flush trace residues with copious amounts of water and collect for disposal. Ventilate area and remove ignition sources.

Preferred Waste Disposal Method

No Citation.

Precautions to be Taken in Handling and Storage

Store in tightly closed containers in cool, dark, well-ventilated location. Protect containers from physical damage. Keep away from oxidizing materials, heat, and exposure to direct sunlight.

Other Precautions and Warnings

Ensure adequate ventilation in storage areas. Use non-sparking tools, especially when opening or closing containers. Personnel should be trained on the hazards of indene *before* working with or around it.

SECTION VIII - CONTROL MEASURES AND PERSONAL PROTECTIVE EQUIPMENT

Respiratory Protection *(specify type)*

For low exposures, use an air-purifying respirator. Otherwise, use a supplied-air respirator or a self-contained breathing apparatus (SCBA) with full facepiece operated in positive pressure mode.

Ventilation

Local exhaust (preferred) at site of chemical work or general ventilation.

Protective Gloves	Eye Protection	Other Protective Clothing
Impervious Gloves	Chemical Goggles or Face Mask	Protective Apron if Splash is Likely

Work/Hygiene Practices

Always wash hands thoroughly after using chemical; never bring food, drink, or smoking materials into vicinity of chemicals.

INDENE

C_9H_8

CAS: 95-13-6

IDENTIFICATION AND TYPICAL USES

Indene is a colorless liquid with a pungent odor. It can also exist as a solid at temperatures below 29°F (-2°C). It is used primarily in the preparation of coumarone-indene resins and as a chemical intermediate. It has also been used as an insecticide.

RISK ASSESSMENT: HEALTH

General Assessment

Indene is moderately toxic to humans by *ingestion,* and *inhalation.* Skin contact causes irritation but dermal absorption has not been reported as a route of entry. It is a severe irritant and may cause liver, kidney, or spleen damage. There is no information in the references on its carcinogenic, teratogenic, or mutagenic capabilities.

Inhalation causes irritation of the eyes, nose, throat, and respiratory system. Its severe irritating properties can lead to chemical pneumonitis, especially if the liquid is somehow aspirated into the lungs (as is possible when vomiting). Pulmonary edema with symptoms of coughing, wheezing, congestion, dypsnea, and tightness or pain in the chest, may develop and can be fatal. These symptoms may be delayed up to 48 hours thereby creating a false sense of security with regard to health exposure risk.

Skin contact results in surface irritation and may cause redness (erythema) and possible blistering. Eye contact may cause moderate to severe irritation. Ingestion of large amounts of indene will likely cause toxic systemic effects with nausea and stomach pain.

Acute Health Effects

The following acute (short-term) health effects may occur immediately or shortly after exposure to indene:

Skin: Irritation with possible redness and blistering.

Eye: Mild to moderate irritation (depending upon length of time contact occurs).

Lung: Nose, throat, and respiratory tract irritation. May cause chemical pneumonitis and even delayed pulmonary edema, which can be fatal.

Other: Acute exposures to high concentrations may lead to liver, kidney, or spleen injury.

Chronic Health Effects

The following chronic (long-term) health effects may occur at some time after exposure to indene and can last for months or even years:

Cancer Hazards: According to information presented in the references, indene has not been adequately tested for its ability to cause cancer in laboratory animals.

Reproductive Hazard: According to information presented in the references, indene has not been adequately tested for its ability to adversely affect reproduction in laboratory animals.

Other Chronic Effects: Long-term exposure may lead to permanent liver, kidney, or spleen dysfunction. Indene can also cause dermatitis in some individuals.

Recommended Risk-Reduction Measures

Personnel should avoid direct contact with indene. Occupational poisoning occurs most commonly by inhalation and ingestion. If a less toxic material or compound cannot be substituted for indene, then *engineering controls* are the most effective method of reducing exposures. The best protection for workers is to ensure controlled use of indene and, wherever possible, enclose operations or provide adequate local exhaust ventilation at the site of indene release. Using respiratory protection is less effective than the controls mentioned above, but is still advisable whenever working with or around indene. For exposures to low levels or infrequent exposures, use an air-purifying respirator equipped with an organic vapor cartridge. For higher exposures, or when the level is unknown (as in an emergency), use a supplied-air respirator with full facepiece and mask operated in positive pressure mode or a self-contained breathing apparatus (SCBA) with full facepiece operated in pressure demand. If a full facepiece is not available, then chemical goggles should be worn to protect the eyes. A face shield should also be considered. To prevent hand and skin exposures, impervious gloves should be used. Glove manufacturers should be contacted and permeation studies obtained *before* final glove selections are made.

Administrative controls should also be in place to minimize the potential for human exposures. These may include written procedures or policies which specify the methods and techniques that will be practiced whenever personnel are to work with indene.

All personnel should receive *training* on the use, hazards, protective measures, emergency actions, and other precautions per 29 CFR 1910.1200 (Hazard Communication), prior to the first assignment in an area where indene is used or stored.

For those personnel with frequent or potentially high exposures (half the PEL or greater), the following are recommended before beginning employment and at regular intervals thereafter (e.g., annually):

☑ Liver and kidney function tests.

☑ Lung function tests.

☑ Evaluation of the skin by a dermatologist for indications or symptoms of chlor-acne.

If symptoms develop or overexposure is suspected, the following recommended medical tests should be considered:

☑ Liver and kidney function tests.

☑ Lung function tests.

☑ Consider chest X-ray following acute overexposure (may be negative if taken immediately after exposure due to delayed onset of pulmonary edema).

Any evaluation should include a careful history of past and present symptoms with an examination. Medical tests that simply look for existing damage are not a substitute for controlling exposures. Because more than light or casual alcohol consumption can lead to liver damage, drinking alcohol may increase the potential liver damage caused by indene. Also, since smoking can cause heart disease, emphysema, and other respiratory disorders, smokers exposed to indene may experience symptoms more quickly than non-smokers under the same conditions of exposure. Prudent risk management requires proper consideration of *all* possible exposure factors when evaluating workplace health hazard risk.

Other methods to reduce exposure include:

☑ Where possible, enclose operations and use local exhaust ventilation at the site of chemical release. If local exhaust ventilation or enclosure is not used, respiratory protection should be mandatory.

☑ Always ensure that proper protective clothing is worn when using chemical substances and that personnel are trained on its proper use and care.

☑ Wash thoroughly immediately after exposure to indene and at the end of the work shift or before eating, drinking, or smoking.

☑ Hazard warning information should be posted in the work area. In addition, as part of an on-going education and training program, all information on the health and safety hazards of indene should be communicated to all exposed workers.

☑ Eye wash stations should be provided in the immediate work area for emergency use. If there is a possibility of skin exposure to indene, emergency shower facilities should also be provided in the immediate area.

☑ Workers whose clothing has been contaminated by indene should change into clean clothes before leaving work. Contaminated work clothing should be laundered only by individuals who have been informed of the hazards of exposure to indene.

RISK ASSESSMENT: ENVIRONMENT

General Assessment

The environment is at risk of exposure during transportation, storage, disposal, or destruction of indene. In almost every scenario, the threat of environmental exposure is contingent upon the proper handling of the chemical substance. Accidental spills, large or small, can result in contamination of the surrounding environmental mediums (water, soil, and air).

Indene is a Class IIIA combustible liquid (as per OSHA 29 CFR 1910.106). It can react in contact with heat, sunlight, and acids. It will explode in contact with sulfuric + nitric acid. If it is involved in fire, indene can emit highly toxic oxides of carbon. Because it is also incompatible with most oxidizers, it must be kept away from permanganates, nitrates, peroxides, chlorates, and perchlorates. Caution is always required in handling, storage, transportation, and disposal of indene. Emergency responders should be made aware of the presence of indene at any emergency response situation.

Indene can enter the environment through industrial discharges or spills.

Acute Ecological Effects

Acute (short-term) toxic effects may include the death of animals, birds, or fish and death or low growth rate in plants. Acute effects are seen 2 to 4 days after animals or plants are exposed to indene.

Insufficient data are available on the short-term effects of indene to aquatic life, plants, birds, or land animals.

Chronic Ecological Effects

Chronic (long-term) toxic effects may include shortened life span, reproductive problems, lower fertility, and changes in appearance or behavior in exposed animals. These effects can be seen long after first exposure(s) to toxic chemicals.

Insufficient data are available to evaluate or predict the long-term effects of indene to aquatic life, plants, birds, or terrestrial animals.

Water Solubility

Indene is insoluble in water. Concentrations of 1 milligram or less may not mix with a liter of water.

Persistence in the Environment

Indene is highly persistent in water, with a half-life greater than 200 days. The half-life of a pollutant is the amount of time it takes for one half of the chemical to be degraded.

Bioaccumulation in Aquatic Organisms

Some substances increase in concentration, or *bioaccumulate,* in living organisms as they breathe contaminated air, drink contaminated water, or eat contaminated food. These chemicals can become concentrated in the tissues and internal organs of animals as well as humans.

The concentration of indene found in fish tissues is expected to be considerably higher than the average concentration of indene in the water from which the fish was taken.

Recommended Risk-Reduction Measures

Proper training of all transporters will reduce the likelihood of a mishap or accident resulting in a leak or spill of indene into the environment. Labels on all containers, trucks, and rail cars should accurately reflect their contents to enable emergency responders to react properly and quickly to any disaster thereby reducing the potential risk to the environment and to personnel.

Storage of indene should be segregated from other chemicals and materials (especially oxidizers) to minimize the risk of cross-contamination. Caution must be taken to ensure indene is not stored in direct sunlight or near any source of heat or ignition.

If a spill or leak to the environment has occurred, fire department, emergency response and/or hazardous materials spill personnel should be notified immediately. Cleanup should be attempted only by those trained in proper spill containment procedures. Contaminated soils should be removed for incineration and replaced with clean soil. If indene should contact the water table, aquifer, or navigable waterway, time is of the essence. It is insoluble in water and total remediation may be possible. The local and/or state emergency response authorities must be notified. A comprehensive emergency response or disaster preparedness/recovery plan should be in place prior to any operations involving the use, transportation, storage, or disposal of indene.

If indene is spilled or leaked, the following specific steps are recommended:

- ☑ Restrict persons not wearing protective clothing from area of spill or leak until cleanup is complete and area can be opened for normal work.
- ☑ Absorb liquid solutions in vermiculite, sand, earth, or similar material and deposit in sealed drum. Flush residue with large amounts of water and collect for disposal. Do not release residue to waterways or sewers.
- ☑ Ventilate area of spill or leak and remove all sources of ignition.
- ☑ It may be necessary to dispose of indene as a hazardous waste. The responsible state agency or the regional office of the federal Environmental Protection Agency (EPA) should be contacted for specific recommendations.

RISK ASSESSMENT: BUSINESS

General Assessment

Accidents or mishaps involving indene can present a serious threat to business operations. The loss or damage of equipment or facilities can significantly affect fiscal viability. Lawsuits that may result from personnel illness, injury/death, public exposures, and/or environmental contamination will require a serious expenditure of resources. Media attention surrounding an injury, death, or environmental damage can also result in a loss of profits and loss of current as well as future business.

Recommended Risk-Reduction Measures

Company attorneys, safety and health professionals, and environmental specialists should be involved in the development of any procedures for responding to chemical incidents. A company official should be predesignated as a public relations officer with specific training in dealing with the press. Corporate plans and policies should be developed, approved, and implemented long before any need for such arises.

REFERENCES

American Conference of Governmental Industrial Hygienists. 1988. *Documentation of the Threshold Limit Values and Biological Exposure Indices,* 5th Edition (with updates). Cincinnati: ACGIH

Anthony, C. P., and N. J. Kolthoff. 1971. *Textbook of Anatomy and Physiology.* St. Louis: C. V. Mosby Company.

Bloom, A. D. and F. J. DeSerres. 1995. *Ecotoxicity and Human Health.* New York: CRC Press/Lewis Publishers.

Brusick, D. J. 1994. *Methods for Genetic Risk Assessment.* New York: CRC Press/Lewis Publishers.

Bueche, J. R. 1972. *Principles of Physics,* 2nd Edition. New York: McGraw-Hill.

Burrell, R., D. K. Flaherty, and L. J. Sauers. 1992. *Toxicology of the Immune System.* New York: Van Nostrand Reinhold.

Calabrese, E. J. 1994. *Biological Effects of Low Level Exposures.* New York: CRC Press/Lewis Publishers.

Cockerman, L. G., B. S. Shane. 1994. *Basic Environmental Toxicology.* New York: CRC Press/Lewis Publishers.

Cothern, C. R., N. P. Ross. 1994. *Environmental Statistics, Assessment, and Forecasting.* New York: CRC Press/Lewis Publishers.

Danse, I. R. 1991. *Common Sense Toxics in the Workplace.* New York: Van Nostrand Reinhold.

Environmental Protection Agency. 1983. *Chemical Hazard Information Profiles.* Washington, D.C.: U.S. EPA.

Fodden, C. R., and J. L. Weddell. 1992. *Hazardous Materials Emergency Action Data.* New York: CRC Press/Lewis Publishers.

Forseberg, K., and L. H. Keith. 1993. *Chemical; Protective Clothing Permeation and Degradation Database.* Chelsea, MI: Lewis Publishers.

Gardner, R. W. 1994. *Chemical Intolerance.* New York: CRC Press/Lewis Publishers.

Gerarde, H. W. 1960. *Toxicology and Biochemistry of Aromatic Hydrocarbons.* New York: Elsevier.

Hallenbeck, W. H. 1993. *Quantitative Risk Assessment for Environmental and Occupational Health,* 2nd Edition. New York: CRC Press/Lewis Publishers.

Jones, F. E. 1994. *Toxic Organic Vapors in the Workplace.* New York: CRC Press/Lewis Publishers.

Kamrin, M. A. 1988. *Toxicology: A Primer on Principles and Applications.* Chelsea, Michigan: Lewis Publishers.

Lauwerys, R. R., P. Hoet. 1993. *Industrial Chemical Exposure.* New York: CRC Press/Lewis Publishers.

Lewis, R. J., Sr. 1993. *Hawley's Condensed Chemical Dictionary,* 12th Edition. New York: Van Nostrand Reinhold.

Lewis, R. J., Sr. 1992. *Sax's Dangerous Properties of Industrial Materials,* Volumes 1, 2, and 3, 8th Edition. New York: Van Nostrand Reinhold.

Mackay, D., W. Y. Shiu, and K. C. Ma. 1992. *Illustrated Handbook of Chemical-Physical Properties and Environmental Fate for Organic Chemicals.* New York: CRC Press/Lewis Publishers.

Meyer, E. 1990. *Chemistry of Hazardous Materials.* Englewood Cliffs, NJ: Prentice Hall.

Miller, B. F., M.D., and C. B. Keane, R.N. 1972. *Encyclopedia and Dictionary of Medicine and Nursing.* Philadelphia: W.B. Sanders Company.

Mycock, J. C., J. D. McKenna, and L. Theodore. 1995. *Handbook of Air Pollution Control Engineering and Technology.* New York: CRC Press/Lewis Publishers.

Patnaik, P. 1992. *A Comprehensive Guide to the Hazardous Properties of Chemical Substances.* New York: Van Nostrand Reinhold.

Quigley, D. R. 1994. *Handbook of Emergency Chemical Management.* New York: CRC Press/Lewis Publishers.

Rea, W. J. 1995. *Chemical Sensitivity,* Vol. I-III. New York: CRC Press/Lewis Publishers.

Richardson, M. 1995. *Dictionary of Substance and Their Effects,* Vol. I-VI. New York: CRC Press/Lewis Publishers.

Royal Society of Chemistry. 1992. *Chemical Safety Data Sheets, Volumes 1-5.* Newcastle-upon-Tyne, UK: Athenaeum Press.

Sacarello, H. L. A. 1994. *The Comprehensive Handbook of Hazardous Materials.* New York: CRC Press/Lewis Publishers.

U.S. Department of Health and Human Services, National Institute for Occupational Safety and Health. 1994. *NIOSH Pocket Guide to Chemical Hazards.* Washington, D.C.: U.S. Government Printing Office.

MATERIAL SAFETY DATA SHEET

CHEMICAL NAME	
INDIUM	**CRC PRESS/LEWIS PUBLISHERS DISCLAIMER:** THE INFORMATION AND RECOMMENDATIONS PRESENTED HEREIN ARE BASED ON SOURCES BELIEVED TO BE RELIABLE. CRC MAKES NO REPRESENTATION ON ITS COMPLETENESS OR ACCURACY. IT IS THE RESPONSIBILITY OF THE USER TO DETERMINE THE CHEMICAL'S SUITABILITY FOR ITS INTENDED USE, THE CHEMICAL'S SAFE USE, AND THE CHEMICAL'S PROPER DISPOSAL. NO REPRESENTATIONS AND/OR WARRANTIES, EITHER EXPRESSED OR IMPLIED, OF THE MERCHANTABILITY OR FITNESS FOR A PARTICULAR PURPOSE, OR OF ANY OTHER NATURE, ARE MADE WITH RESPECT TO THE INFORMATION PROVIDED IN THIS MSDS OR TO THE CHEMICAL TO WHICH INFORMATION MAY REFER. CRC NEITHER ASSUMES NOR AUTHORIZES ANY OTHER PERSON TO ASSUME FOR IT, ANY OTHER ADDITIONAL RESPONSIBILITY OR LIABILITY FOR THE USE OF, OR RELIANCE UPON, THIS INFORMATION.

HAZARD WARNING INFORMATION

HEALTH	FIRE	REACTIVE	OTHER	DEGREE OF HAZARD	COLOR CODING	OTHER CODES
1	0	1		0 = Minimum Hazard 1 = Slight Hazard 2 = Moderate Hazard 3 = Serious Hazard 4 = Severe Hazard	HEALTH = BLUE FIRE = RED REACTIVITY = YELLOW OTHER = WHITE	OX = Oxidizer ACID = Acid ALK = Alkali COR = Corrosive W = Use No Water

SECTION I - GENERAL INFORMATION

Characterization	**Metal**	RCRA Number	**None**
		EPA Class	**Not Applicable**
DOT Proper Shipping Name	**No Citation**	Chemical Abstract Service (CAS) Number	**7440-74-6**
DOT Hazard Class and Label Requirements	**No Citation**	DOT Emergency Guide Code	**No Citation**
DOT Identification Number	**No Citation**	Chemical Formula	**In**

Synonyms
Indium metal.

SECTION II - HAZARDOUS INGREDIENTS AND IDENTITY INFORMATION

Hazardous Components (specific identity)	OSHA Exposure Criteria	NIOSH Exposure Criteria	Immediately Dangerous to Life and Health (IDLH)	ACGIH Exposure Criteria
Indium (derivation: Metallic element of atomic number 49, group IIIA of the periodic system. Not found native but in a variety of zinc and other ores).	**PEL:** Not Established **STEL:** Not Established	**REL:** 0.1 mg/m^3 **STEL:** Not Established	**Not Determined**	**TLV:** 0.1 mg/m^3 **STEL:** Not Established

SECTION III - PHYSICAL AND CHEMICAL CHARACTERISTICS

Boiling Point	**3767°F (2075°C)**	Specific Gravity (H_2O = 1)	**7.31**
Vapor Pressure (mm Hg)	**0 (approx.) at 69°F (20°C)**	Molecular Weight	**114.8**
Vapor Density (Air = 1)	**Not Found**	Melting Point	**314°F (157°C)**

Solubility
Insoluble in water and alkalis, soluble in acids.

Appearance and Odor
Silver-white, highly lustrous, ductile, metallic solid that is softer than lead.

SECTION IV - FIRE AND EXPLOSION HAZARD DATA

Flash Point (method used)	**Not Applicable**	Explosive Limits in Air % by Volume	**LEL: Not Applicable UEL: Not Applicable**
NFPA Classification	**Non-Combustible Solid**	Autoignition Temperature	**Not Determined**

Extinguishing Media
Use media suitable to surrounding fire. Bulk indium will not burn but powdered indium may ignite, especially when mixed with air.

Special Fire Fighting Procedures
Wear full protective clothing and self-contained breathing apparatus (SCBA). Use a water spray to keep containers cool. For large fires, if possible, withdraw and allow to burn. Remain clear of smoke.

Unusual Fire and Explosion Hazards
The powder ignites spontaneously with exposure to heat, flames, sparks, or exposure to air. Violent reactions with some oxides and metallic compounds. The powder may explode when heated with dinitrogen tetraoxide + acetonitrile. Mixtures with sulfur ignite when heated.

SECTION V - REACTIVITY DATA

Stability		Conditions to Avoid Indium is normally stable in closed containers under routine conditions of handling and storage but it will oxidize readily at higher temperatures. Keep away from heat and incompatible materials.
Stable X	Unstable	Incompatibility *(materials to avoid)* Incompatible with dinitrogen tetraoxide + acetonitrile, mercury (II) bromide (at high temperatures), and will ignite with sulfur when heated.
Hazardous Polymerization		Conditions to Avoid Under normal conditions of temperature and pressure, hazardous polymerization of indium is not known to occur.
May Occur	Will Not Occur X	Hazardous Decomposition or By-products Toxic fumes and gases are released when indium is involved in fire.

SECTION VI - HEALTH HAZARD DATA

Primary Route(s) of Entry:	Inhalation?	Absorption (skin)?	Ingestion?
	X		X

Health Hazards

INHALATION: Irritation of the respiratory system and mucous membranes. Exposure can cause allergic lung reaction with chest pain, coughing, wheezing, and shortness of breath. May also cause liver, kidney, heart, and blood effects. Pulmonary edema can occur.

SKIN & EYES: An irritant to the eyes and skin. The powder may cause contact dermatitis.

INGESTION: Unspecified gastrointestinal effects.

Carcinogenicity	NTP Listed?	IARC Cancer Review Group?	OSHA Regulated?	Target Organs?
Unknown Human Unknown Animal	No	No	No	Resp. sys., skin, eyes, liver, kidney, blood, heart.

Medical Conditions Generally Aggravated by Exposure

None Reported.

Emergency and First-aid Procedures

<u>Eye contact</u>: Flush immediately with water for 15 minutes (minimum); seek medical attention. <u>Skin contact</u>: Remove all contaminated clothing. Immediately wash area with large amounts of soap and water. If rash develops, seek medical attention. For <u>inhalation</u>: Remove the person from exposure. Provide respiratory assistance and CPR. Transfer to medical facility. Observe for 48 hours for lung effects. If <u>swallowed</u>: Provide breathing assistance if necessary. Never attempt to give a convulsing or unconscious person anything by mouth. Seek medical attention immediately.

SECTION VII - PRECAUTIONS FOR SAFE HANDLING AND USE

Steps to be Taken in Case Material is Released or Spilled

Collect powdered material in most convenient manner possible and deposit in sealed drum. Keep those not involved in cleanup from entering area. Do NOT dry sweep (creates airborne dusts and friction). Carefully use a vacuum equipped with a high efficiency particulate air (HEPA) filter. Exercise extreme caution, indium dust clouds can ignite or explode in air or if heated. Use non-sparking tools.

Preferred Waste Disposal Method

No citation.

Precautions to be Taken in Handling and Storage

Store in tightly closed containers in cool, well-ventilated area and protect from physical damage. Sources of heat or ignition are prohibited where indium is used, handled, or stored.

Other Precautions and Warnings

Avoid contact with incompatible materials.

SECTION VIII - CONTROL MEASURES AND PERSONAL PROTECTIVE EQUIPMENT

Respiratory Protection *(specify type)*

For exposure to low levels, use a full facepiece respirator with a high efficiency particulate air (HEPA) filter. For higher or prolonged exposures, use a full facepiece supplied-air respirator or a self-contained breathing apparatus (SCBA) operated in positive pressure mode.

Ventilation

Local exhaust or general mechanical systems recommended.

Protective Gloves	Eye Protection	Other Protective Clothing
Impervious Material	Chemical/Dust Goggles or Face Mask	Protective Suit or Clothing

Work/Hygiene Practices

Always wash hands thoroughly after using chemical; never bring food, drink, or smoking materials into vicinity of chemicals.

INDIUM

In

CAS: 7440-74-6

IDENTIFICATION AND TYPICAL USES

Indium is a silver-white, shining, lustrous, ductile, metallic solid that can also appear in the form of powder or dusts. It is softer than lead. It is metallic element of atomic number 49, group IIIA of the periodic table, atomic weight 114.82, valances 1, 3, with 2 stable isotopes. Used in automobile bearings, radiation detectors, electronic and semiconductor devices, low-melting brazing and soldering alloys, and in reactor control rods. Also used in electroplated coatings on silver-plated steel aircraft bearings which are tarnish-resistant.

RISK ASSESSMENT: HEALTH

General Assessment

Indium is a moderately toxic metal by *inhalation* and mildly toxic by *ingestion.* Since it is not significantly soluble in water, it is not absorbed efficiently by ingestion. Terratogenic effects have been reported but there are no data on its carcinogenicity or mutagenicity.

Inhalation of indium dusts and powders will cause irritation of the nose, throat, respiratory tract, and associated mucous membranes. It has been known to cause liver, kidney, heart, and blood effects in test animals. Lung effects can include the development of pulmonary edema (fluid in the lungs). Symptoms of cough, sneezing, congestion, difficulty in breathing, and tightness in the chest may be delayed up to 48 hours thereby creating a false sense of security with regard to health exposure risk.

Skin contact results in surface irritation and may lead to dermatitis. The powders can irritate the eyes and may cause redness and inflammation depending upon the duration of exposure. Ingestion causes unspecified gastrointestinal effects. There may be vomiting and stomach pain. Toxic systemic effects resulting from ingestion are possible but not likely to occur.

Acute Health Effects

The following acute (short-term) health effects may occur immediately or shortly after exposure to indium:

Skin: Irritation, rash, and possible contact dermatitis.

Eye: Indium particles can irritate the eyes and may cause redness if not quickly removed.

Lung: Irritation of the dusts and fumes can irritate the nose, mouth, throat, lungs, and mucosa of the respiratory tract. Delayed pulmonary edema can occur which is a medical emergency and can be fatal.

Chronic Health Effects

The following chronic (long-term) health effects may occur at some time after exposure to indium and can last for months or even years:

Cancer Hazards: According to information presented in the references, indium has not been shown to cause cancer in test animals. Liver damage has occurred in test animals but no correlation to human exposure has been documented.

Reproductive Hazard: According to information available in the references, indium has been shown to cause teratogenic effects in test animals.

Other Chronic Effects: Possible liver, kidney, heart, and blood effects.

Recommended Risk-Reduction Measures

Personnel should avoid direct contact with indium. The exact nature of its human toxicity is not clearly understood in the references. If the specific indium compound in question is unknown, adequate risk management dictates that it be handled and treated as though it posed a serious risk to human health. If a less toxic chemical cannot be substituted for a hazardous substance, then *engineering controls* are the most effective method of reducing exposures.

The best protection is to enclose operations and/or provide local exhaust ventilation at the site of chemical release. While not always operationally feasible, isolating operations can also reduce exposure. Using respiratory protection is less effective than the controls mentioned above, but is still advisable whenever working with or around indium. For exposure to indium at low or transient levels, an MSHA/NIOSH-approved respirator with a high efficiency particulate air (HEPA) filter is recommended. For higher exposures, or exposures to unknown levels (as in an emergency), use a supplied-air respirator with a full facepiece operated in positive pressure mode or an MSHA/NIOSH-approved self-contained breathing apparatus (SCBA) with full facepiece operated in pressure demand. If a full facepiece is not available, then chemical dust goggles should be worn to protect the eyes. Whenever an airborne chemical dust hazard exists, a face shield and a protective apron should be worn. To prevent hand and skin exposures, impervious rubber gloves should be used.

Administrative controls should also be in place to minimize the potential for human exposures. These may include written procedures or policies which specify the methods and techniques that will be practiced whenever personnel are to work with indium.

All personnel should receive *training* on the use, hazards, protective measures, emergency actions, and other precautions per 29 CFR 1910.1200 (Hazard Communication), prior to the first assignment in an area where indium is used or stored.

Before beginning employment and at regular intervals thereafter (e.g., annually), for those with frequent or potentially high exposures, the following medical tests are recommended:

☑ Lung function tests.

☑ Chest X-ray.

If symptoms develop or overexposure is suspected, the following may also be useful:

☑ Liver, kidney, and lung function tests.

☑ Consider chest X-ray following acute overexposure (may be negative if taken immediately after exposure due to delayed onset of pulmonary edema).

☑ Evaluation by a qualified allergist, including careful exposure history and special testing (may help diagnose skin allergy).

Any evaluation should include a careful history of past and present symptoms with a medical examination. Medical tests that look for damage already done are not a substitute for controlling exposure. Also, since smoking can cause heart disease, as well as lung cancer, emphysema, and other related respiratory problems, the effects of exposure to indium may be more pronounced in smokers than in non-smokers. Prudent risk management requires proper consideration of *all* possible factors that may be causing exposure symptoms in the workplace.

Other methods to reduce exposure include:

☑ Where possible, enclose operations and use local exhaust ventilation at the site of chemical release. If local exhaust ventilation or enclosure is not used, respiratory protection should be mandatory.

☑ Always ensure that proper protective clothing is worn when using chemical substances and that personnel are adequately trained on the use, care, and maintenance of their personnel protective equipment.

☑ Wash thoroughly immediately after exposure to indium and at the end of the work shift or before eating, drinking, or smoking.

☑ Hazard warning information should be posted in the work area. In addition, as part of an on-going education and training program, all information on the health and safety hazards of indium should be communicated to all potentially exposed workers.

☑ Where possible, automatically transfer indium from drums or other storage containers to process containers. Avoid contact with air and do not generate airborne dusts.

RISK ASSESSMENT: ENVIRONMENT

General Assessment

The environment is at risk of exposure during transportation, storage, disposal, or destruction of indium. In almost every scenario, the threat of environmental exposure is contingent upon the proper handling of the chemical substance. Accidental spills, large or small, can result in fire, explosion, and possible contamination of the surrounding environmental mediums (water, soil, and air).

Indium powder will explode in air when heated. It will also react explosively in the presence of dinitrogen tetraoxide + acetonitrile. It will react violently with mercury (II) bromide at high temperatures. Mixtures with sulfur ignite when heated. It oxidizes readily at high temperatures. These characteristics require special consideration during any emergency situation involving a leak or spill of indium.

Indium can occur in the environment as a natural element in certain ores in the earth's crust. Indium-bearing ores occur in western United States, Canada, Peru, Japan, Europe, and the former Soviet Union. It can also enter the environment through industrial and municipal discharges, or from spills.

Acute Ecological Effects

Acute (short-term) toxic effects may include the death of animals, birds, or fish and death or low growth rate in plants. Acute effects are seen 2 to 4 days after animals or plants are exposed to indium.

Insufficient data are available to evaluate the short-term effects of indium exposure to aquatic life, plants, birds, or land animals.

Chronic Ecological Effects

Chronic (long-term) toxic effects may include shortened life span, reproductive problems, lower fertility, and changes in appearance or behavior in exposed animals. These effects can be seen long after first exposure(s) to toxic chemicals.

Insufficient data are available to evaluate the long-term effects of indium exposure to aquatic life, plants, birds, or land animals.

Water Solubility

Elemental indium and most of its compounds are insoluble in water. Concentrations of less than 1 milligram will not mix with water.

Persistence in the Environment

Indium and its compounds are highly persistent in the aquatic environment, with a half-life of greater than 200 days. The half-life of a pollutant is the amount of time it takes for one half of the chemical to be degraded.

Bioaccumulation in Aquatic Organisms

Some substances increase in concentration, or *bioaccumulate,* in living organisms as they breathe contaminated air, drink contaminated water, or eat contaminated food. These chemicals can become concentrated in the tissues and internal organs of animals as well as humans.

The concentration of indium found in fish is expected to be somewhat higher than the average concentration of indium in the water from which the fish was taken.

Recommended Risk-Reduction Measures

Proper training of all transporters will reduce the likelihood of a mishap or accidents resulting in a leak or spill to the environment. The correct use of labeling on all containers, trucks, and rail cars will enable emergency responders to react properly and quickly to any disaster thereby reducing the potential risk to the environment and to personnel.

Storage of indium should be segregated from incompatible chemicals to minimize the risk of cross-contamination or contact. Bulk storage of indium should be avoided. Conditions which can generate dusty conditions must be avoided. Storage facilities should be designed with proper fire suppression and protection equipment. Containers must be protected from exposure to heat and from physical damage.

If a spill or leak to the environment has occurred, fire department, emergency response and/or hazardous materials spill personnel should be notified immediately. Use a vacuum method to reduce dust during cleanup. Do not dry sweep (generates airborne dusts and friction). *Carefully* and *slowly* use a vacuum equipped with a high efficiency particulate air (HEPA) filter, not a standard shop vacuum. Deposit collected materials in sealed drum for disposal. Cleanup should be attempted only by those trained in proper spill containment procedures using non-sparking tools. Contaminated soils should be removed and replaced with clean soil. If indium should contact the water table, aquifer, or navigable waterway, time is of the essence. Indium is not soluble in water and, therefore, total containment and remediation may be possible. When such spills occur, the local and/or state emergency response authorities must be notified. A comprehensive emergency response or disaster preparedness/recovery plan should be in place prior to any operations involving the use, transportation, storage, or disposal of indium.

If indium is spilled or leaked, the following specific steps are recommended:

☑ Restrict persons not wearing protective clothing from area of spill or leak until cleanup is complete and area can be opened for normal work.

☑ Collect powdered materials in most convenient and safe manner possible and place in sealed drums for disposal. Do not dry sweep, use HEPA vacuum instead.

☑ It may be necessary to dispose of indium as a hazardous waste. The responsible state agency or the regional office of the federal Environmental Protection Agency (EPA) should be contacted for specific recommendations.

RISK ASSESSMENT: BUSINESS

General Assessment

Accidents or mishaps involving indium can present a significant threat to business operations. The loss or damage of equipment or facilities can significantly affect fiscal viability. Lawsuits that may result from personnel injury/death, public exposures, and/or environmental contamination will also require a serious expenditure of resources. Media attention surrounding an injury, death, or environmental damage can also result in a loss of profits and loss of current as well as future business.

Recommended Risk-Reduction Measures

Company attorneys, safety and health professionals, and environmental specialists should be involved in the development of any procedures dealing with safety or emergency response. A company official should be pre-designated as a public relations officer with specific training in dealing with the press. Corporate plans and policies should be developed, approved, and implemented long before any need for such arises.

REFERENCES

Ahlbom, A. 1993. *Biostatistics for Epidemiologists.* New York: CRC Press/Lewis Publishers.

American Conference of Governmental Industrial Hygienists. 1988. *Documentation of the Threshold Limit Values and Biological Exposure Indices,* 5th Edition (with updates). Cincinnati: ACGIH

Asante-Duah, D. K. 1993. *Hazardous Waste Risk Assessment.* New York: CRC Press/Lewis Publishers.

Brusick, D. J. 1994. *Methods for Genetic Risk Assessment.* New York: CRC Press/Lewis Publishers.

Calabrese, E. J. 1994. *Biological Effects of Low Level Exposures.* New York: CRC Press/Lewis Publishers.

Cockerman, L. G., B. S. Shane. 1994. *Basic Environmental Toxicology.* New York: CRC Press/Lewis Publishers.

Cothern, C. R., N. P. Ross. 1994. *Environmental Statistics, Assessment, and Forecasting.* New York: CRC Press/Lewis Publishers.

Danse, I. R. 1991. *Common Sense Toxics in the Workplace.* New York: Van Nostrand Reinhold.

Environmental Protection Agency. 1983. *Chemical Hazard Information Profiles.* Washington, D.C.: U.S. EPA.

Fodden, C. R., and J. L. Weddell. 1992. *Hazardous Materials Emergency Action Data.* New York: CRC Press/Lewis Publishers.

Forseberg, K., and L. H. Keith. 1993. *Chemical; Protective Clothing Permeation and Degradation Database.* Chelsea, MI: Lewis Publishers.

Gardner, R. W. 1994. *Chemical Intolerance.* New York: CRC Press/Lewis Publishers.

Howard, P. H., et. al. 1991. *Handbook of Environmental Degradation Rates.* New York: CRC Press/Lewis Publishers.

Hosty, J. W., and P. Foster. 1990. *A Practical Guide to Chemical Spill Response.* New York: Van Nostrand Reinhold.

Landis, W. G., M. H. Yu. 1995. *Introduction to Environmental Toxicology.* New York: CRC Press/Lewis Publishers.

Langerman, N. 1994 *Precautionary Labels for Chemical Containers.* New York: CRC Press/Lewis Publishers.

Lauwerys, R. R., P. Hoet. 1993. *Industrial Chemical Exposure.* New York: CRC Press/Lewis Publishers.

Lewis, .J., Sr. 1993. *Hawley's Condensed Chemical Dictionary,* 12th Edition. New York: Van Nostrand Reinhold.

Lewis, . J., Sr. 1992. *Sax's Dangerous Properties of Industrial Materials,* Volumes 1, 2, and 3, 8th Edition. New York: Van Nostrand Reinhold.

Mackay, D., W. Y. Shiu, and K. C. Ma. 1992. *Illustrated Handbook of Chemical-Physical Properties and Environmental Fate for Organic Chemicals.* New York: CRC Press/Lewis Publishers.

Manahan, S. E. 1992. *Toxicological Chemistry,* 2nd Edition. New York: CRC Press/Lewis Publishers.

Meyer, E. 1990. *Chemistry of Hazardous Materials.* Englewood Cliffs, NJ: Prentice Hall.

Miller, B. F., M.D., and C. B. Keane, R.N. 1972. *Encyclopedia and Dictionary of Medicine and Nursing.* Philadelphia: W.B. Sanders Company.

Patnaik, P. 1992. *A Comprehensive Guide to the Hazardous Properties of Chemical Substances.* New York: Van Nostrand Reinhold.

Rea, W. J. 1995. *Chemical Sensitivity,* Vol. I-III. New York: CRC Press/Lewis Publishers.

Reeves, A. L. 1979. *Handbook on the Toxicology of Metals.* New York: Elsevier North-Holland.

Richardson, M. 1995. *Dictionary of Substance and Their Effects,* Vol. I-VI. New York: CRC Press/Lewis Publishers.

Rylander, R., R. R. Jacobs. 1994. *Organic Dusts.* New York: CRC Press/Lewis Publishers.

U.S. Department of Health and Human Services (DHHS), National Institute for Occupational Health (NIOHS). 1994. *NIOSH Pocket Guide to Hazardous Chemicals.* Publication Number 94-116. Washington, D.C.; U.S. Government Printing Office.

MATERIAL SAFETY DATA SHEET

CHEMICAL NAME	CRC PRESS/LEWIS PUBLISHERS DISCLAIMER:
IODINE	THE INFORMATION AND RECOMMENDATIONS PRESENTED HEREIN ARE BASED ON SOURCES BELIEVED TO BE RELIABLE. CRC MAKES NO REPRESENTATION ON ITS COMPLETENESS OR ACCURACY. IT IS THE RESPONSIBILITY OF THE USER TO DETERMINE THE CHEMICAL'S SUITABILITY FOR ITS INTENDED USE, THE CHEMICAL'S SAFE USE, AND THE CHEMICAL'S PROPER DISPOSAL. NO REPRESENTATIONS AND/OR WARRANTIES, EITHER EXPRESSED OR IMPLIED, OF THE MERCHANTABILITY OR FITNESS FOR A PARTICULAR PURPOSE, OR OF ANY OTHER NATURE, ARE MADE WITH RESPECT TO THE INFORMATION PROVIDED IN THIS MSDS OR TO THE CHEMICAL TO WHICH INFORMATION MAY REFER. CRC NEITHER ASSUMES NOR AUTHORIZES ANY OTHER PERSON TO ASSUME FOR IT, ANY OTHER ADDITIONAL RESPONSIBILITY OR LIABILITY FOR THE USE OF, OR RELIANCE UPON, THIS INFORMATION.

HAZARD WARNING INFORMATION

HEALTH	FIRE	REACTIVE	OTHER	DEGREE OF HAZARD	COLOR CODING	OTHER CODES
1	0	1	OX	0 = Minimum Hazard 1 = Slight Hazard 2 = Moderate Hazard 3 = Serious Hazard 4 = Severe Hazard	HEALTH = BLUE FIRE = RED REACTIVITY = YELLOW OTHER = WHITE	OX = Oxidizer ACID = Acid ALK = Alkali COR = Corrosive W = Use No Water

SECTION I - GENERAL INFORMATION

Characterization	**Halogen**	RCRA Number	**None**
		EPA Class	**Not Applicable**
DOT Proper Shipping Name	**Irritating Agent, N.O.S.**	Chemical Abstract Service (CAS) Number	**7553-56-2**
DOT Hazard Class and Label Requirements	**Irritating Material; Irritant**	DOT Emergency Guide Code	**No Citation**
DOT Identification Number	**NA 1693**	Chemical Formula	I_2

Synonyms
Iodine crystals; molecular iodine.

SECTION II - HAZARDOUS INGREDIENTS AND IDENTITY INFORMATION

Hazardous Components (specific identity)	OSHA Exposure Criteria	NIOSH Exposure Criteria	Immediately Dangerous to Life and Health (IDLH)	ACGIH Exposure Criteria
Iodine (derivation: From brine wells in Michigan, Oklahoma, Japan, Indonesia; from mother liquors of Chilean nitrate; can be extracted from kelp. Purified by sublimation). **1 ppm = 10.55 mg/m^3**	**PEL (ceiling): 0.1 ppm 1 mg/m^3** **STEL: Not Applicable**	**REL (ceiling): 0.1 ppm 1 mg/m^3** **STEL: Not Applicable**	**2 ppm**	**TLV (ceiling): 0.1 ppm 1 mg/m^3** **STEL: Not Applicable**

SECTION III - PHYSICAL AND CHEMICAL CHARACTERISTICS

Boiling Point	**363°F (184°C)**	Specific Gravity (H_2O = 1)	**4.93**
Vapor Pressure (mm Hg)	**0.3 at 77°F (25°C)**	Molecular Weight	**253.8**
Vapor Density (Air = 1)	**9.0**	Melting Point	**236°F (113°C)**

Solubility
Very slightly soluble in water (0.01%). Soluble in alcohol, carbon disulfide, chloroform, ether, carbon tetrachloride, glycerol, many other organic solvents, and alkaline iodide solutions.

Appearance and Odor
Violet solid with a sharp, irritating odor. Volatilizes at room temperature to a violet gas.

SECTION IV - FIRE AND EXPLOSION HAZARD DATA

Flash Point (method used)	**Not Applicable**	Explosive Limits in Air % by Volume	**LEL: Not Determined UEL: Not Determined**
NFPA Classification	**Non-Combustible Solid**	Autoignition Temperature	**Not Applicable**

Extinguishing Media
Use agent suitable to surrounding fire. Iodine itself will not burn.

Special Fire Fighting Procedures
Poisonous fumes and gases are emitted during fire. Wear full protective clothing and self-contained breathing apparatus. Heat will build pressure and may rupture closed storage containers. Keep fire-exposed containers cool with water spray. Continue to cool containers after fire is extinguished.

Unusual Fire and Explosion Hazards
Iodine is a strong oxidizing agent and will support combustion. Incompatible with a number of other chemicals. Violent and explosive reactions possible.

SECTION V - REACTIVITY DATA

Stability		Conditions to Avoid **Avoid contact with incompatible materials. Mixtures with some chemicals can produce violent explosive reactions and fires. Avoid contact with heat or flame.**
Stable **X**	Unstable	Incompatibility *(materials to avoid)* **Strong oxidizers, ammonia, acetylene, acetaldehyde, powdered aluminum, active metals (lithium, sodium, potassium), and liquid chlorine. Reacts with ammonium hydroxide to form shock-sensitive iodides.**
Hazardous Polymerization		Conditions to Avoid **Under normal conditions of temperature and pressure, hazardous polymerization of iodine is not expected to occur.**
May Occur	Will Not Occur **X**	Hazardous Decomposition or By-products **When heated to decomposition, iodine emits acrid, irritating smoke and fumes including iodine vapors and iodide.**

SECTION VI - HEALTH HAZARD DATA

Primary Route(s) of Entry:	Inhalation?	Absorption (skin)?	Ingestion?
	X	X	X

Health Hazards

INHALATION: Severe irritation of the respiratory system, mucous membranes, lachrymation, severe cough, headache, somnolence, and possibly pulmonary edema (fluid in the lungs). It may also affect cardiovascular function.

ABSORPTION: Corrosive. Will irritate and even burn the skin and eyes. May cause acne-form lesions that may resemble thermal burns. A skin sensitizer. Eye contact may cause staining to the corneal epithelium and the loss of the layer of tissue. May also stain the skin brown.

INGESTION: Nausea, abdominal pain, diarrhea, headache.

Carcinogenicity	NTP Listed?	IARC Cancer Review Group?	OSHA Regulated?	Target Organs?
Unknown Human Unknown Animal	No	No	29 CFR 1910.1000 Table Z-1	Respiratory system; skin; eyes; CNS; CVS.

Medical Conditions Generally Aggravated by Exposure

Diseases of the thyroid, lungs, and kidneys may be aggravated by exposure.

Emergency and First-aid Procedures

Eye contact: Flush immediately with water for 15 minutes (minimum); seek medical attention. Skin contact: Remove all contaminated clothing. Immediately wash area with large amounts of soap and water. Seek medical assistance. For inhalation: Remove the person from exposure. Provide respiratory assistance and CPR. Transfer to medical facility. Observe for 48 hours for lung effects. If swallowed: Seek medical attention immediately. Give conscious person milk to drink. Keep victim warm and at rest. Contact poison control center.

SECTION VII - PRECAUTIONS FOR SAFE HANDLING AND USE

Steps to be Taken in Case Material is Released or Spilled

Collect powdered materials using HEPA vacuum (do NOT dry sweep) and place in sealed containers. Damp mop residues. Avoid vapors. Restrict those not involved in cleanup from entering area.

Preferred Waste Disposal Method

No Citation

Precautions to be Taken in Handling and Storage

Do not store with incompatible chemicals. Store in tightly closed containers in a cool, dark, well-ventilated area. Keep fire and flame away. Keep separate from organic, combustible, and readily oxidizable substances. Keep out of direct sunlight.

Other Precautions and Warnings

Bulk storage of iodine is not recommended. Protect containers from physical damage. Avoid breathing vapors or dusts.

SECTION VIII - CONTROL MEASURES AND PERSONAL PROTECTIVE EQUIPMENT

Respiratory Protection *(specify type)*

Extremely corrosive and damaging to lung tissue. For any exposure, use a supplied-air respirator or a self-contained breathing apparatus (SCBA) with full facepiece operated in positive pressure mode.

Ventilation

Local exhaust or general mechanical systems recommended.

Protective Gloves	Eye Protection	Other Protective Clothing
Impervious Material	**Chemical/Dust Goggles or Face Mask**	**Impermeable Apron**

Work/Hygiene Practices

Always wash hands thoroughly after using chemical; never bring food, drink, or smoking materials into vicinity of chemicals.

IODINE

I_2

CAS: 7553-56-2

IDENTIFICATION AND TYPICAL USES

Iodine is a violet-colored solid with a sharp, characteristic odor. It volatilizes at room temperature to a violet gas. It is a nonmetallic halogen element of atomic number 53, group VIIA of the periodic table, atomic number 126.904, valances of 1,3,5,7, and no stable isotopes but many artificial isotopes. It is the least reactive of the halogens. It is used in dyes, as an alkylation and condensation catalyst, in iodides, iodates, antiseptics, and germicides. Also used as an X-ray contrast media, as a food and feed additive, in stabilizers, in photographic film, in water treatment, in pharmaceuticals, and in medicinal soaps.

RISK ASSESSMENT: HEALTH

General Assessment

Iodine is a human poison by *ingestion* and possibly by other routes of exposure. It is moderately toxic by *inhalation*. Skin contact causes serious effects because of the corrosive properties of iodine and dermal *absorption* may occur on prolonged contact. Human mutation data have been reported but there are no data on the carcinogenic or teratogenic capabilities of iodine.

Inhalation of iodine vapors will cause severe irritation of the eye, nose, throat, and mucosa of the respiratory tract with lachrymation (tearing), nasal secretion, tightness in the chest, sore throat, and headache. High concentrations can lead to a dangerous buildup of fluid in the lungs (pulmonary edema) which can be fatal. Symptoms can be delayed up to 48 hours thereby creating a false sense of security with regard to health hazard risk.

Skin contact causes serious corrosive destruction to tissues. There may be rash, burns, and skin sensitization. Absorption through intact skin has been reported leading to toxic systemic effects similar to those of inhalation and ingestion.

Ingestion causes diarrhea, abdominal pains, nausea, vomiting, purging, excessive thirst, circulatory failure, and possible hypersensitivity of the thyroid.

Acute Health Effects

The following acute (short-term) health effects may occur immediately or shortly after exposure to iodine:

Skin: Irritation, burns, possible ulcer-like wounds.

Eye: Irritation which can lead to permanent damage if allowed to remain in contact with the eyes.

Lung: Severe irritation of respiratory passages and lung tissue with coughing, nausea, and possible pulmonary edema (fluid in the lungs), which is a medical emergency and can be fatal.

Other: A human poison by ingestion causing diarrhea, stomach pain, nausea, vomiting, thirst, and a possibility of circulatory failure and death.

Chronic Health Effects

The following chronic (long-term) health effects may occur at some time after exposure to iodine and can last for months or even years:

Cancer Hazards: According to the references, iodine is not known to cause cancer in humans or test animals. Human mutation data have been reported and many scientists believe that such chemicals may pose a cancer risk in the long-term.

Reproduction: According to the references, iodine has not been shown to cause teratogenic and experimental reproductive effects in test animals.

Other Chronic Effects: It has been reported that long-term exposure to iodine may cause allergic sensitization in some individuals. Once sensitization has occurred, even the slightest exposures in the future can cause the reappearance of exposure symptoms.

Ⓤ *Recommended Risk-Reduction Measures*

Personnel should avoid direct contact with iodine under any circumstances. A less toxic chemical should be substituted for a hazardous substance. If this is not possible or feasible, then *engineering controls* are the most effective method of reducing exposures. The best protection is to enclose operations and/or provide local exhaust ventilation at the site of chemical release. While not always operationally feasible, isolating operations can also reduce exposure.

Using respiratory protection is less effective than the controls mentioned above, but is still advisable whenever working with or around iodine. At virtually any exposure level, a self-contained breathing apparatus (SCBA) with full facepiece operated in pressure demand is the recommended respiratory protection of choice. If a full facepiece is not available, then chemical goggles should be worn to protect the eyes. Whenever a chemical splash hazard exists, a face shield and a rubber apron should be worn. To prevent hand and skin exposures, protective gloves should be used. Glove manufacturers should be contacted and permeation studies obtained *before* final glove selections have been made.

Administrative controls should also be in place to minimize the potential for human exposures. These may include written procedures or policies which specify the methods and techniques that will be practiced whenever personnel are to work with iodine.

All personnel should receive *training* on the use, hazards, protective measures, emergency actions, and other precautions per 29 CFR 1910.1200 (Hazard Communication), prior to the first assignment in an area where iodine is used or stored. If overexposure is suspected, the following is recommended:

☑ Lung function test.

☑ Complete examination of nervous system function.

☑ Consider chest X-ray following acute overexposure (may be negative if taken immediately after exposure due to delayed onset of pulmonary edema).

☑ Evaluation by a qualified allergist with special testing and careful consideration of exposure history (may help diagnose skin allergy).

Other methods to reduce exposure include:

☑ Where possible, enclose operations and use local exhaust ventilation at the site of chemical release. If local exhaust ventilation or enclosure is not used, respiratory protection should be mandatory.

☑ Always ensure that proper protective clothing is worn when using chemical substances.

☑ Wash thoroughly immediately after exposure to iodine and at the end of the work shift or before eating, drinking, or smoking.

☑ Hazard warning information should be posted in the work area. In addition, as part of an on-going education and training program, all information on the health and safety hazards of iodine should be communicated to all potentially exposed workers.

☑ Never eat, drink, or smoke in areas were iodine is used, handled or stored.

RISK ASSESSMENT: ENVIRONMENT

General Assessment

The environment is at risk of exposure during transportation, storage, disposal, or destruction of iodine. In almost every scenario, the threat of environmental exposure is contingent upon the proper handling of the chemical substance. Accidental spills, large or small, can result in fire, explosion, and possible contamination of the surrounding environmental mediums (water, soil, and air).

Iodine is considered a non-combustible solid. However, as a powerful oxidizing agent, iodine will accelerate the burning of combustibles. These characteristics require special consideration during any emergency situation involving a leak or spill of iodine.

Should iodine ever come into contact with incompatible substances such as ammonia, acetylene, phosphorus, potassium, sodium, acetaldehyde, acrolein, acrylonitrile, or powdered aluminum either during use, transportation, or storage, violent and explosive reactions are extremely possible. Ammonium hydroxide will react with iodine to form shock-sensitive iodides that are explosive when dry.

Iodine can occur in the environment naturally as well as through manufacturing, unchecked discharge into effluents, and through spills.

☠ *Acute Ecological Effects*

Acute (short-term) toxic effects may include the death of animals, birds, or fish, and death or low growth rate in plants. Acute effects are seen 2 to 4 days after animals or plants are exposed to iodine.

There is insufficient information available to evaluate or predict the short-term effects of iodine on aquatic life, birds, plants or terrestrial animals.

Chronic Ecological Effects

Chronic toxic effects may include shortened life span, reproductive problems, lower fertility, and changes in appearance or behavior in exposed animals. These effects can be seen long after first exposure(s) to toxic chemicals.

There is insufficient information available to evaluate the long-term effects of iodine on aquatic life, birds, plants or terrestrial animals.

Water Solubility

Iodine is very slightly soluble in water. Concentrations of 1 to 100 milligrams may mix with a liter of water.

Persistence in the Environment

Iodine is highly persistent in water, with a half-life of between 20 to 200 days. The half-life of a pollutant is the amount of time it takes for one-half of the chemical to be degraded. No data are available on the percentages of iodine that will eventually end up in water.

Bioaccumulation in Aquatic Organisms

Some substances increase in concentration, or *bioaccumulate,* in living organisms as they breathe contaminated air, drink contaminated water, or eat contaminated food. These chemicals can become concentrated in the tissues and internal organs of animals as well as humans.

The concentration of iodine found in fish tissues is expected to be much higher than the average concentration of iodine in water from which the fish was taken.

Recommended Risk-Reduction Measures

Proper training of all transporters will reduce the likelihood of a mishap or accident resulting in a leak or spill to the environment. The correct use of DOT labeling (per the MSDS) on all containers, trucks, and rail cars will enable emergency responders to react properly and quickly to any disaster thereby reducing the potential risk to the environment and to personnel.

Storage of iodine should be segregated from incompatible chemicals to minimize the risk of cross-contamination or contact. Buildings designated for storage should be equipped with appropriate fire protection systems (alarms, sprinklers, emergency lighting, portable extinguishers).

If a spill or leak to the environment has occurred, fire department, emergency response, and/or hazardous materials spill personnel should be notified immediately. Cleanup should be attempted only by those trained in proper spill containment procedures. Contaminated soils should be removed for incineration and replaced with clean soil.

If iodine should contact the water table, aquifer, or navigable waterway, time is of the essence. It is nearly insoluble in water and, therefore, total containment and remediation may be possible. When such spills occur, the local and/or state emergency response authorities must be notified. A comprehensive emergency response or disaster preparedness/recovery plan should be in place prior to any operations involving the use, transportation, storage, or disposal of iodine. If iodine is spilled or leaked, the following specific steps are recommended:

- ☑ Restrict persons not wearing protective clothing from area of spill or leak until cleanup is complete and area can be opened for normal work.
- ☑ Remove all ignition sources.
- ☑ Ventilate area of spill or leak.
- ☑ Collect powdered materials in safest manner possible and place in sealed containers for reclamation or disposal. Do NOT dry sweep (generates airborne dusts). Use vacuum equipped with a high-efficiency particulate air (HEPA) filter instead.
- ☑ It may be necessary to dispose of iodine as a hazardous waste. The responsible state agency or the regional office of the federal Environmental Protection Agency (EPA) should be contacted for specific recommendations.

RISK ASSESSMENT: BUSINESS

General Assessment

Accidents or mishaps involving iodine can present a moderate threat to business operations. The loss or damage of equipment or facilities can significantly affect fiscal viability. Lawsuits that may result from personnel injury/death, public exposures, and/or environmental contamination will also require a serious expenditure of resources. Media attention surrounding an injury, death, or environmental damage can also result in a loss of profits and loss of current as well as future business. Always remember that anytime the term "reproductive hazard" is used, public emotion, hysteria, and ignorance can run equally high. This must be carefully considered when developing or implementing public relations policies.

Recommended Risk-Reduction Measures

Company attorneys, safety and health professionals, and environmental specialists should be involved in the development of any procedures or policies intended to manage the use of chemicals in the workplace. A company official should be pre-designated as a public relations officer with specific training in dealing with the press. Corporate plans and policies should be developed, approved, and implemented long before any need for such arises.

REFERENCES

Ahlbom, A. 1993. *Biostatistics for Epidemiologists.* New York: CRC Press/Lewis Publishers.

Anthony, C. P., and N. J. Kolthoff. 1971. *Textbook of Anatomy and Physiology.* St. Louis: C. V. Mosby Company.

Asante-Duah, D. K. 1993. *Hazardous Waste Risk Assessment.* New York: CRC Press/Lewis Publishers.

Bloom, A. D. and F. J. DeSerres. 1995. *Ecotoxicity and Human Health.* New York: CRC Press/Lewis Publishers.

Brusick, D. J. 1994. *Methods for Genetic Risk Assessment.* New York: CRC Press/Lewis Publishers.

Bueche, J. R. 1972. *Principles of Physics,* 2nd Edition. New York: McGraw-Hill.

Burrell, R., D. K. Flaherty, and L. J. Sauers. 1992. *Toxicology of the Immune System.* New York: Van Nostrand Reinhold.

Calabrese, E. J. 1994. *Biological Effects of Low Level Exposures.* New York: CRC Press/Lewis Publishers.

Cockerman, L. G., B. S. Shane. 1994. *Basic Environmental Toxicology.* New York: CRC Press/Lewis Publishers.

Cothern, R. R. 1993. *Comparative Environmental Risk Assessment.* New York: CRC Press/Lewis Publishers.

Danse, I. R. 1991. *Common Sense Toxics in the Workplace.* New York: Van Nostrand Reinhold.

Gardner, R. W. 1994. *Chemical Intolerance.* New York: CRC Press/Lewis Publishers.

Hallenbeck, W. H. 1993. *Quantitative Risk Assessment for Environmental and Occupational Health,* 2nd Edition. New York: CRC Press/Lewis Publishers.

Hodgson, E., R. B. Mailman, and J. E. Chambers. 1988. *Dictionary of Toxicology.* London: The MacMillian Company.

Hosty, J. W., and P. Foster. 1990. *A Practical Guide to Chemical Spill Response.* New York: Van Nostrand Reinhold.

Howard, P. H., et. al. 1991. *Handbook of Environmental Degradation Rates.* New York: CRC Press/Lewis Publishers.

James, A., and M. Lord. 1992. *Index of Chemical and Physical Data.* New York: Van Nostrand Reinhold.

Jones, F. E. 1994. *Toxic Organic Vapors in the Workplace.* New York: CRC Press/Lewis Publishers.

Kamrin, M. 1 A. 1988. *Toxicology: A Primer on Principles and Applications.* Chelsea, Michigan: Lewis Publishers.

Keith, L. H. and M. M. Walker. 1995. *Handbook of Air Toxics.* New York: CRC Press/Lewis Publishers.

Landis, W. G., M. H. Yu. 1995. *Introduction to Environmental Toxicology.* New York: CRC Press/Lewis Publishers.

Langerman, N. 1994 *Precautionary Labels for Chemical Containers.* New York: CRC Press/Lewis Publishers.

Lauwerys, R. R., P. Hoet. 1993. *Industrial Chemical Exposure.* New York: CRC Press/Lewis Publishers.

Lide, D. R. 1995. *Handbook of Chemistry and Physics,* 76th Edition. New York: CRC Press/Lewis Publishers.

Lewis, Richard J., Sr. 1993. *Hawley's Condensed Chemical Dictionary,* 12th Edition. New York: Van Nostrand Reinhold.

Lewis, Richard J., Sr. 1992. *Sax's Dangerous Properties of Industrial Materials,* Volumes 1, 2, and 3, 8th Edition. New York: Van Nostrand Reinhold.

Lisella, Frank S. 1994. *The VNR Dictionary of Environmental Health and Safety.* New York: Van Nostrand Reinhold.

Manahan, S. E. 1992. *Toxicological Chemistry,* 2nd Edition. New York: CRC Press/Lewis Publishers.

Mellström, G. A., J. E. Wahlberg, and H. I. Maibach. 1994. *Protective Gloves for Occupational Use.* New York: CRC Press/Lewis Publishers.

Meyer, Eugene. 1990. *Chemistry of Hazardous Materials.* Englewood Cliffs, NJ: Prentice Hall.

Mycock, J. C., J. D. McKenna, and L. Theodore. 1995. *Handbook of Air Pollution Control Engineering and Technology.* New York: CRC Press/Lewis Publishers.

Nelson, G. O. 1992. *Gas Mixtures.* New York: CRC Press/Lewis Publishers.

Ott, W. R. 1995. *Environmental Statistics and Data Analysis.* New York: CRC Press/Lewis Publishers.

Patnaik, P. 1992. *A Comprehensive Guide to the Hazardous Properties of Chemical Substances.* New York: Van Nostrand Reinhold.

Philp, R. B. 1995. *Environmental Hazards and Human Health.* New York: CRC Press/Lewis Publishers.

Proctor, N. H., and J. P. Hughes. 1991. *Chemical Hazards of the Workplace.* New York: Van Nostrand Reinhold.

Sherman, J. D. 1988. *Chemical Exposure and Disease.* New York: Van Nostrand Reinhold.

MATERIAL SAFETY DATA SHEET

CHEMICAL NAME

IODINE MONOCHLORIDE

HAZARD WARNING INFORMATION

HEALTH	FIRE	REACTIVE	OTHER	DEGREE OF HAZARD	COLOR CODING	OTHER CODES
1	0	1		0 = Minimum Hazard 1 = Slight Hazard 2 = Moderate Hazard 3 = Serious Hazard 4 = Severe Hazard	HEALTH = BLUE FIRE = RED REACTIVITY = YELLOW OTHER = WHITE	OX = Oxidizer ACID = Acid ALK = Alkali COR = Corrosive W̶ = Use No Water

SECTION I - GENERAL INFORMATION

Characterization	**Interhalogen Compound**	RCRA Number	**None**
		EPA Class	**Not Applicable**
DOT Proper Shipping Name	**Corrosive Material, N.O.S.**	Chemical Abstract Service (CAS) Number	**7790-99-0**
DOT Hazard Class and Label Requirements	**Corrosive Material; Corrosive**	DOT Emergency Guide Code	**No Citation**
DOT Identification Number	**UN 1792**	Chemical Formula	**ICl**

Synonyms

Iodine chloride; Wijs' chloride; crystalline iodine monochloride (α and β forms).

SECTION II - HAZARDOUS INGREDIENTS AND IDENTITY INFORMATION

Hazardous Components (specific identity)	OSHA Exposure Criteria	NIOSH Exposure Criteria	Immediately Dangerous to Life and Health (IDLH)	ACGIH Exposure Criteria
Iodine monochloride (derivation: By the action of dry chlorine in iodine).	**PEL:** Not Established **STEL:** Not Established	**REL:** Not Established **STEL:** Not Established	**Not Determined**	**TLV:** Not Established **STEL:** Not Established

SECTION III - PHYSICAL AND CHEMICAL CHARACTERISTICS

Boiling Point	**207°F (97°C)**	Specific Gravity (H_2O = 1)	**3.18 (α) 3.24 (β)**
Vapor Pressure (mm Hg)	**Not Found**	Molecular Weight	**162.38**
Vapor Density (Air = 1)	**Not Found**	Melting Point	**81°F/27°C (α form) 57°F/14°C (β form)**

Solubility

Soluble in water (causes decomposition), alcohol, ether, carbon disulfide, acetic acid, and dilute hydrochloric acid.

Appearance and Odor

Black crystals or needles (α and β forms), or a reddish-brown, oily liquid.

SECTION IV - FIRE AND EXPLOSION HAZARD DATA

Flash Point (method used)	**Not Applicable**	Explosive Limits in Air % by Volume	**LEL: Not Determined UEL: Not Determined**
NFPA Classification	**Non-Combustible Solid**	Autoignition Temperature	**Not Applicable**

Extinguishing Media

Use agent suitable to surrounding fire. Iodine monochloride itself will not burn.

Special Fire Fighting Procedures

Poisonous fumes and gases are emitted during fire. Wear full protective clothing and self-contained breathing apparatus. Heat will build pressure and may rupture closed storage containers. Keep fire-exposed containers cool with water spray. Continue to cool containers after fire is extinguished.

Unusual Fire and Explosion Hazards

Iodine monochloride is a moderately explosive when exposed to heat. Incompatible with a number of other chemicals and materials. Violent and explosive reactions possible.

SECTION V - REACTIVITY DATA

Stability		Conditions to Avoid **Avoid contact with incompatible materials. Mixtures with some chemicals can produce violent explosive reactions and fires. Avoid contact with heat or flame.**
Stable	Unstable	Incompatibility *(materials to avoid)*
X		**Reacts with water or steam to produce toxic and corrosive fumes. Dangerous reactions with metals such as sodium, potassium, aluminum, and other materials. Reacts violently with aluminum foil, rubber, cork.**
Hazardous Polymerization		Conditions to Avoid **Under normal conditions of temperature and pressure, hazardous polymerization of iodine monochloride is not expected to occur.**
May Occur	Will Not Occur	Hazardous Decomposition or By-products
	X	**When heated to decomposition, iodine monochloride emits acrid, irritating smoke and fumes including fumes of iodide and chloride, and may explode.**

SECTION VI - HEALTH HAZARD DATA

Primary Route(s) of Entry:	Inhalation?	Absorption (skin)?	Ingestion?
	X		X

Health Hazards

INHALATION: Severe irritation of the respiratory system, mucous membranes, lachrymation, severe cough, headache, somnolence, and possibly pulmonary edema (fluid in the lungs). It may also affect cardiovascular function.

SKIN & EYES: Corrosive. Will irritate and even burn the skin and eyes. May cause deep burns and dark patches similar to thermal burns. A skin sensitizer. Eye contact may cause staining to the corneal epithelium and the loss of the layer of tissue. May also stain the skin brown.

INGESTION: Poisonous by ingestion. Nausea, abdominal pain, diarrhea, headache.

Carcinogenicity	NTP Listed?	IARC Cancer Review Group?	OSHA Regulated?	Target Organs?
Unknown Human Unknown Animal	No	No	No	Respiratory system; skin; eyes; CNS; CVS.

Medical Conditions Generally Aggravated by Exposure

Diseases of the thyroid, lungs, and kidneys may be aggravated by exposure.

Emergency and First-aid Procedures

Eye contact: Flush immediately with water for 15 minutes (minimum); seek medical attention. Skin contact: Remove all contaminated clothing. Immediately wash area with large amounts of soap and water. Seek medical assistance. For inhalation: Remove the person from exposure. Provide respiratory assistance and CPR. Transfer to medical facility. Observe for 48 hours for lung effects. If swallowed: Seek medical attention immediately. Give conscious person milk to drink. Keep victim warm and at rest. Contact poison control center.

SECTION VII - PRECAUTIONS FOR SAFE HANDLING AND USE

Steps to be Taken in Case Material is Released or Spilled

Collect powdered materials using HEPA vacuum (do NOT dry sweep) and place in sealed containers. Damp mop residues. Avoid vapors. Absorb liquids in vermiculite and deposit in sealed drums.

Preferred Waste Disposal Method

No citation

Precautions to be Taken in Handling and Storage

Do not store with incompatible chemicals. Store in tightly closed containers in a cool, dark, well-ventilated area. Keep fire and flame away. Keep separate from organic, combustible, and readily oxidizable substances. Keep out of direct sunlight.

Other Precautions and Warnings

Bulk storage of iodine monochloride is not recommended. Protect containers from physical damage. Avoid breathing vapors or dusts.

SECTION VIII - CONTROL MEASURES AND PERSONAL PROTECTIVE EQUIPMENT

Respiratory Protection *(specify type)*

Extremely corrosive and damaging to lung tissue. For any exposure, use a supplied-air respirator or a self-contained breathing apparatus (SCBA) with full facepiece operated in positive pressure mode.

Ventilation

Local exhaust or general mechanical systems recommended.

Protective Gloves	Eye Protection	Other Protective Clothing
Impervious Material	Chemical/Dust Goggles or Face Mask	Impermeable Apron

Work/Hygiene Practices

Always wash hands thoroughly after using chemical; never bring food, drink, or smoking materials into vicinity of chemicals.

IODINE MONOCHLORIDE

ICl

CAS: 7790-99-0

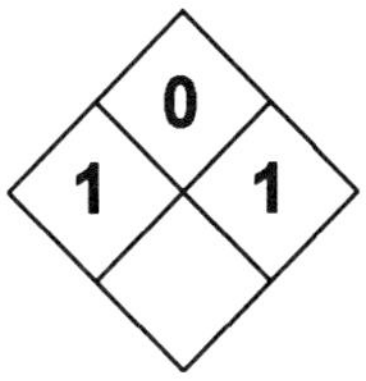

IDENTIFICATION AND TYPICAL USES

Iodine monochloride can appear as a reddish-brown oily liquid or as black crystals (the α form) or black needles (the β form). It is used in analytical chemistry and organic synthesis. Iodine monochloride is also used to estimate the iodine values of fats and oils and is a topical anti-infection agent.

RISK ASSESSMENT: HEALTH

General Assessment

Iodine monochloride is a human poison by *ingestion* and possibly by other routes of exposure. It is moderately toxic by *inhalation*. Skin contact causes serious effects because of the corrosive properties of iodine monochloride and dermal *absorption* may even be possible on prolonged contact (data inconclusive). There are no data on the carcinogenic, mutagenic, or teratogenic capabilities of iodine monochloride.

Inhalation of iodine monochloride vapors will cause severe irritation of the eye, nose, throat, and mucosa of the respiratory tract with lachrymation (tearing), nasal secretion, tightness in the chest, sore throat, and headache. High concentrations may cause significant damage to the mucous membranes and may lead to a dangerous buildup of fluid in the lungs (pulmonary edema) which can be fatal. Symptoms can be delayed up to 48 hours thereby creating a false sense of security with regard to health hazard risk.

Skin contact causes serious corrosive destruction to tissues. There may be rash, burns, the appearance of dark patches, and skin sensitization. Absorption of the liquid through intact skin may be a possible route of exposure. Eye contact may cause immediate pain with redness, swelling, and possible damage to vision.

Ingestion of the liquid or crystalline solid may cause diarrhea, abdominal pains, nausea, vomiting, and other unspecified gastrointestinal effects.

Acute Health Effects

The following acute (short-term) health effects may occur immediately or shortly after exposure to iodine monochloride:

Skin: Irritation, burns, dark patches on the surface of the skin.

Eye: Irritation which can lead to permanent damage if allowed to remain in contact with the eyes.

Lung: Severe irritation of respiratory passages and lung tissue with coughing, nausea, and possible pulmonary edema (fluid on the lungs), which is a medical emergency and can be fatal.

Other: A human poison by ingestion causing diarrhea, stomach pain, nausea, and vomiting.

Chronic Health Effects

The following chronic (long-term) health effects may occur at some time after exposure to iodine monochloride and can last for months or even years:

Cancer Hazards: According to the references, iodine monochloride has not been adequately tested for its ability to cause cancer in test animals.

Reproduction: According to the references, iodine monochloride has not been shown to cause teratogenic and experimental reproductive effects in test animals.

Other Chronic Effects: None reported.

Recommended Risk-Reduction Measures

Personnel should avoid direct contact with iodine monochloride under any circumstances. A less toxic chemical should be substituted for a hazardous substance. If this is not possible or feasible, then *engineering controls* are the most effective method of reducing exposures. The best protection is to enclose operations and/or provide local exhaust ventilation at the site of chemical release. While not always operationally feasible, isolating operations can also reduce exposure.

Using respiratory protection is less effective than the controls mentioned above, but is still advisable whenever working with or around iodine monochloride. No exposure levels have been established for this compound. However, this does not mean that exposure is without health risk. At virtually any exposure level, a powered supplied-air respirator in continuous flow mode or a self-contained breathing apparatus (SCBA) with full facepiece operated in pressure demand mode are the recommended respiratory protection methods of choice. If a full facepiece is not available, then chemical goggles should be worn to protect the eyes. Whenever a chemical splash or airborne dust hazard exists, a face shield and a protective apron should be worn. To prevent hand and skin exposures, protective gloves should be used. Iodine monochloride will react violently in contact with rubber and other organic materials. Glove manufacturers should therefore be contacted and permeation studies obtained *before* final glove selections have been made.

Administrative controls should also be in place to minimize the potential for human exposures. These may include written procedures or policies which specify the methods and techniques that will be practiced whenever personnel are to work with iodine monochloride.

All personnel should receive *training* on the use, hazards, protective measures, emergency actions, and other precautions per 29 CFR 1910.1200 (Hazard Communication), prior to the first assignment in an area where iodine monochloride is used or stored. If an overexposure is suspected, the following is recommended:

- ☑ Lung function test.
- ☑ Complete examination of nervous system function.
- ☑ Consider chest X-ray following acute overexposure (may be negative if taken immediately after exposure due to delayed onset of pulmonary edema).
- ☑ Evaluation by a qualified allergist with special testing and careful consideration of exposure history (may help diagnose skin allergy).

Other methods to reduce exposure include:

- ☑ Where possible, enclose operations and use local exhaust ventilation at the site of chemical release. If local exhaust ventilation or enclosure is not used, respiratory protection should be mandatory.
- ☑ Always ensure that proper protective clothing is worn when using chemical substances and that personnel are trained in the use, care, and maintenance of this equipment.
- ☑ Wash thoroughly immediately after exposure to iodine monochloride and at the end of the work shift or before eating, drinking, or smoking.
- ☑ Hazard warning information should be posted in the work area. In addition, as part of an on-going education and training program, all information on the health and safety hazards of iodine monochloride should be communicated to all exposed and potentially exposed workers.
- ☑ Never eat, drink, or smoke in areas were iodine monochloride is used, handled or stored.

RISK ASSESSMENT: ENVIRONMENT

General Assessment

The environment is at risk of exposure during transportation, storage, disposal, or destruction of iodine monochloride. In almost every scenario, the threat of environmental exposure is contingent upon the proper handling of the chemical substance. Accidental spills, large or small, can result in fire, explosion, and possible contamination of the surrounding environmental mediums (water, soil, and air).

Iodine monochloride is considered a non-combustible solid. However, as an oxidizing agent, iodine monochloride will accelerate the burning of combustibles. It will even react with water or steam to produce highly toxic and corrosive fumes. These characteristics require special consideration during any emergency situation involving a leak or spill of iodine monochloride. It will react violently with aluminum foil and organic matter such as rubber or cork. Should iodine monochloride ever come into contact with these or other incompatible substances such as metals like potassium (explodes on contact), sodium (mixture explodes on impact), or aluminum (ignition after a delay period), either during use, transportation, or storage, violent and explosive reactions are extremely possible.

Iodine monochloride can enter the environment through manufacturing, unchecked discharge into effluents, and through spills.

☠ *Acute Ecological Effects*

Acute (short-term) toxic effects may include the death of animals, birds, or fish, and death or low growth rate in plants. Acute effects are seen 2 to 4 days after animals or plants are exposed to iodine monochloride.

There is insufficient information available to evaluate or predict the short-term effects of iodine monochloride on aquatic life, birds, plants or terrestrial animals.

Chronic Ecological Effects

Chronic toxic effects may include shortened life span, reproductive problems, lower fertility, and changes in appearance or behavior in exposed animals. These effects can be seen long after first exposure(s) to toxic chemicals.

There is insufficient information available to evaluate the long-term effects of iodine monochloride on aquatic life, birds, plants or terrestrial animals.

Water Solubility

Iodine monochloride is moderately soluble in water (decomposes). Concentrations of 1 to 1000 milligrams will mix with a liter of water.

Persistence in the Environment

Iodine monochloride is highly persistent in water, with a half-life of between 20 to 200 days. The half-life of a pollutant is the amount of time it takes for one-half of the chemical to be degraded. No data are available on the percentages of iodine monochloride that will eventually end up in water or air.

Bioaccumulation in Aquatic Organisms

Some substances increase in concentration, or *bioaccumulate,* in living organisms as they breathe contaminated air, drink contaminated water, or eat contaminated food. These chemicals can become concentrated in the tissues and internal organs of animals as well as humans.

The concentration of iodine found in fish tissues is expected to be much higher than the average concentration of iodine in water from which the fish was taken. There is no information on the bioaccumulation of the monochloride in the edible tissues of fish.

Recommended Risk-Reduction Measures

Proper training of all transporters will reduce the likelihood of a mishap or accident resulting in a leak or spill to the environment. The correct use of DOT labeling (per the MSDS) on all containers, trucks, and rail cars will enable emergency responders to react properly and quickly to any disaster thereby reducing the potential risk to the environment and to personnel.

Storage of iodine monochloride should be segregated from incompatible chemicals to minimize the risk of cross-contamination or contact. Buildings designated for storage should be equipped with appropriate fire protection systems (alarms, sprinklers, emergency lighting, portable extinguishers).

If a spill or leak to the environment has occurred, fire department, emergency response, and/or hazardous materials spill personnel should be notified immediately. Cleanup should be attempted only by those trained in proper spill containment procedures. Contaminated soils should be removed for incineration and replaced with clean soil.

If iodine monochloride should contact the water table, aquifer, or navigable waterway, time is of the essence. It is moderately soluble in water and will decompose. Therefore, total containment and remediation may not be possible. When such spills occur, the local and/or state emergency response authorities must be notified. A comprehensive emergency response or disaster preparedness/recovery plan should be in place prior to any operations involving the use, transportation, storage, or disposal of iodine monochloride. If iodine monochloride is spilled or leaked, the following specific steps are recommended:

- ☑ Restrict persons not wearing protective clothing from area of spill or leak until cleanup is complete and area can be opened for normal work.
- ☑ Remove all ignition sources.
- ☑ Ventilate area of spill or leak.
- ☑ Collect powdered materials in safest manner possible and place in sealed containers for reclamation or disposal. Do NOT dry sweep (generates airborne dusts). Use vacuum equipped with a high-efficiency particulate air (HEPA) filter instead. Damp mop residue and collect for disposal, Do NOT allow runoff from cleanup methods to enter sewers or waterways.
- ☑ Absorb liquid spills with vermiculite, dry earth, sand, or other material and place in sealed drums for disposal.
- ☑ It may be necessary to dispose of iodine monochloride as a hazardous waste. The responsible state agency or the regional office of the federal Environmental Protection Agency (EPA) should be contacted for specific recommendations.

RISK ASSESSMENT: BUSINESS
General Assessment

Accidents or mishaps involving iodine monochloride can present a moderate threat to business operations. The loss or damage of equipment or facilities can significantly affect fiscal viability. Lawsuits that may result from personnel injury/death, public exposures, and/or environmental contamination will also require a serious expenditure of resources. Media attention surrounding an injury, death, or environmental damage can also result in a loss of profits and loss of current as well as future business.

Recommended Risk-Reduction Measures

Company attorneys, safety and health professionals, and environmental specialists should be involved in the development of any procedures or policies intended to manage the use of chemicals in the workplace. A company official should be pre-designated as a public relations officer with specific training in dealing with the press. Corporate plans and policies should be developed, approved, and implemented long before any need for such arises.

REFERENCES

Ahlbom, A. 1993. *Biostatistics for Epidemiologists.* New York: CRC Press/Lewis Publishers.

Anthony, C. P., and N. J. Kolthoff. 1971. *Textbook of Anatomy and Physiology.* St. Louis: C. V. Mosby Company.

Asante-Duah, D. K. 1993. *Hazardous Waste Risk Assessment.* New York: CRC Press/Lewis Publishers.

Bloom, A. D. and F. J. DeSerres. 1995. *Ecotoxicity and Human Health.* New York: CRC Press/Lewis Publishers.

Brusick, D. J. 1994. *Methods for Genetic Risk Assessment.* New York: CRC Press/Lewis Publishers.

Bueche, J. R. 1972. *Principles of Physics,* 2nd Edition. New York: McGraw-Hill.

Burrell, R., D. K. Flaherty, and L. J. Sauers. 1992. *Toxicology of the Immune System.* New York: Van Nostrand Reinhold.

Calabrese, E. J. 1994. *Biological Effects of Low Level Exposures.* New York: CRC Press/Lewis Publishers.

Cockerman, L. G., B. S. Shane. 1994. *Basic Environmental Toxicology.* New York: CRC Press/Lewis Publishers.

Cothern, R. R. 1993. *Comparative Environmental Risk Assessment.* New York: CRC Press/Lewis Publishers.

Danse, I. R. 1991. *Common Sense Toxics in the Workplace.* New York: Van Nostrand Reinhold.

Gardner, R. W. 1994. *Chemical Intolerance.* New York: CRC Press/Lewis Publishers.

Hallenbeck, W. H. 1993. *Quantitative Risk Assessment for Environmental and Occupational Health,* 2nd Edition. New York: CRC Press/Lewis Publishers.

Hodgson, E., R. B. Mailman, and J. E. Chambers. 1988. *Dictionary of Toxicology.* London: The MacMillian Company.

Hosty, J. W., and P. Foster. 1990. *A Practical Guide to Chemical Spill Response.* New York: Van Nostrand Reinhold.

Howard, P. H., et. al. 1991. *Handbook of Environmental Degradation Rates.* New York: CRC Press/Lewis Publishers.

James, A., and M. Lord. 1992. *Index of Chemical and Physical Data.* New York: Van Nostrand Reinhold.

Jones, F. E. 1994. *Toxic Organic Vapors in the Workplace.* New York: CRC Press/Lewis Publishers.

Kamrin, M. 1 A. 1988. *Toxicology: A Primer on Principles and Applications.* Chelsea, Michigan: Lewis Publishers.

Keith, L. H. and M. M. Walker. 1995. *Handbook of Air Toxics.* New York: CRC Press/Lewis Publishers.

Landis, W. G., M. H. Yu. 1995. *Introduction to Environmental Toxicology.* New York: CRC Press/Lewis Publishers.

Langerman, N. 1994 *Precautionary Labels for Chemical Containers.* New York: CRC Press/Lewis Publishers.

Lauwerys, R. R., P. Hoet. 1993. *Industrial Chemical Exposure.* New York: CRC Press/Lewis Publishers.

Lide, D. R. 1995. *Handbook of Chemistry and Physics,* 76th Edition. New York: CRC Press/Lewis Publishers.

Lewis, Richard J., Sr. 1993. *Hawley's Condensed Chemical Dictionary,* 12th Edition. New York: Van Nostrand Reinhold.

Lewis, Richard J., Sr. 1992. *Sax's Dangerous Properties of Industrial Materials,* Volumes 1, 2, and 3, 8th Edition. New York: Van Nostrand Reinhold.

Lisella, Frank S. 1994. *The VNR Dictionary of Environmental Health and Safety.* New York: Van Nostrand Reinhold.

Manahan, S. E. 1992. *Toxicological Chemistry,* 2nd Edition. New York: CRC Press/Lewis Publishers.

Mellström, G. A., J. E. Wahlberg, and H. I. Maibach. 1994. *Protective Gloves for Occupational Use.* New York: CRC Press/Lewis Publishers.

Meyer, Eugene. 1990. *Chemistry of Hazardous Materials.* Englewood Cliffs, NJ: Prentice Hall.

Mycock, J. C., J. D. McKenna, and L. Theodore. 1995. *Handbook of Air Pollution Control Engineering and Technology.* New York: CRC Press/Lewis Publishers.

Ott, W. R. 1995. *Environmental Statistics and Data Analysis.* New York: CRC Press/Lewis Publishers.

Patnaik, P. 1992. *A Comprehensive Guide to the Hazardous Properties of Chemical Substances.* New York: Van Nostrand Reinhold.

Sherman, J. D. 1988. *Chemical Exposure and Disease.* New York: Van Nostrand Reinhold.

MATERIAL SAFETY DATA SHEET

CHEMICAL NAME

IODOFORM

HAZARD WARNING INFORMATION

HEALTH	FIRE	REACTIVE	OTHER	DEGREE OF HAZARD	COLOR CODING	OTHER CODES
2	0	0		0 = Minimum Hazard 1 = Slight Hazard 2 = Moderate Hazard 3 = Serious Hazard 4 = Severe Hazard	HEALTH = BLUE FIRE = RED REACTIVITY = YELLOW OTHER = WHITE	OX = Oxidizer ACID = Acid ALK = Alkali COR = Corrosive W = Use No Water

SECTION I - GENERAL INFORMATION

Characterization: **Halogenated Hydrocarbon**	RCRA Number: **None**	EPA Class: **Not Applicable**
DOT Proper Shipping Name: **No Citation**	Chemical Abstract Service (CAS) Number: **75-47-8**	
DOT Hazard Class and Label Requirements: **No Citation**	DOT Emergency Guide Code: **No Citation**	
DOT Identification Number: **No Citation**	Molecular Formula: **CHI_3**	

Synonyms
Triiodomethane; carbon triiodide; methane, triiodo.

SECTION II - HAZARDOUS INGREDIENTS AND IDENTITY INFORMATION

Hazardous Components (specific identity)	OSHA Exposure Criteria	NIOSH Exposure Criteria	Immediately Dangerous to Life and Health (IDLH)	ACGIH Exposure Criteria
Iodoform (derivation: By heating acetone or methanol with iodine in the presence of an alkali or alkaline carbonates; electrolytically by passing a current through a solution containing potassium iodide, alcohol, and sodium carbonate). 1 ppm = 16.34 mg/m^3	**PEL (skin):** 0.6 ppm 10 mg/m^3 **STEL:** Not Established	**REL (skin):** 0.6 ppm 10 mg/m^3 **STEL:** Not Established	Not Determined	**TLV (skin):** 0.6 ppm 10 mg/m^3 **STEL:** Not Established

SECTION III - PHYSICAL AND CHEMICAL CHARACTERISTICS

Boiling Point: **410°F (210°C) sublimes**	Specific Gravity (H_2O = 1): **4.01**
Vapor Pressure (mm Hg): **Not Found**	Molecular Weight (atomic weight): **393.7**
Vapor Density (Air = 1): **Not Found**	Freezing Point: **246°F (119°C)**

Solubility
Slightly soluble (0.01%) in water, soluble in alcohol, acetone, acetic acid, benzene, chloroform, carbon disulfide, glycerol, ether, olive oil, and petroleum ether.

Appearance and Odor
Yellow to greenish-yellow powder or crystalline solid with a pungent, disagreeable odor.

SECTION IV - FIRE AND EXPLOSION HAZARD DATA

Flash Point (method used): **Not Determined**	Explosive Limits in Air % by Volume: **LEL: Not Determined** **UEL: Not Determined**
NFPA Classification: **Non-Combustible Solid**	Autoignition Temperature: **Not Determined**

Extinguishing Media
Use extinguishing media suitable to surrounding fires. Iodoform does not burn.

Special Fire Fighting Procedures
Poisonous gases are produced in fire. Wear self-contained breathing apparatus (SCBA) and full protective gear. Move containers from fire if it can be done without risk. Cool sides of fire-exposed containers with water until long after fire is out. Monitor smoke directions and stay clear. Be aware of runoff from fire control measure. Do NOT release runoff to sewers or waterways. Dike for collection and disposal.

Unusual Fire and Explosion Hazards
None Reported.

SECTION V - REACTIVITY DATA

Stability		Conditions to Avoid **Iodoform is normally stable in closed containers, under routine conditions of storage and handling. Avoid contact with heat and exposure to incompatible materials.**
Stable **X**	Unstable	Incompatibility *(materials to avoid)* **Iodoform is incompatible with strong oxidizers (such as chlorine, bromine, and fluorine), strong acids (nitric, sulfuric, hydrochloric), lithium, metallic salts (mercuric oxide, silver nitrate), strong bases, calomel, tannin.**
Hazardous Polymerization		Conditions to Avoid **Under normal and routine conditions of temperature and pressure, hazardous polymerization of iodoform will not occur.**
May Occur	Will Not Occur **X**	Hazardous Decomposition or By-products **Poisonous gases and acrid fumes are produced in fire, including iodine vapors.**

SECTION VI - HEALTH HAZARD DATA

Primary Route(s) of Entry:	Inhalation?	Absorption (skin/eye)?	Ingestion?
	X	X	X

Health Hazards

INHALATION: A severe irritant to the eyes, nose, throat, and mucosa of the respiratory tract. Data on the effects of inhalation are limited. It may cause central nervous system effects to include lightheadedness, dizziness, nausea, and incoordination. It may also cause dypsnea, and liver, kidney, and heart damage. Visual disturbances have also been reported.

ABSORPTION: Effects include vomiting, delirium, hallucinations, rapid heart rate, possible fever, coma, and death. High concentrations may cause liver, heart, and kidney damage.

INGESTION: May cause same effects as that noted for absorption.

Carcinogenicity	NTP Listed?	IARC Cancer Review Group?	OSHA Regulated?	Target Organs?
Unknown Human Unknown Animal	No	No	29 CFR 1910.1000 Table Z-1	Respiratory system, eyes, liver, kidneys, heart, CNS.

Medical Conditions Generally Aggravated by Exposure

None reported.

Emergency and First-aid Procedures

Eye contact: Do not allow victim to keep eyes tightly shut. Flush immediately with water for 15 minutes (minimum); seek medical attention. **Skin contact:** Remove all contaminated clothing, including shoes. Immediately wash area with flooding amounts of water. Seek medical attention immediately. For **inhalation:** Remove the person from exposure. Provide respiratory assistance and CPR. Transfer to medical facility. If **swallowed:** Seek medical attention immediately. Give conscious person 1-2 glasses of water and induce vomiting. Never give an unconscious or convulsing person anything by mouth.

SECTION VII - PRECAUTIONS FOR SAFE HANDLING AND USE

Steps to be Taken in Case Material is Released or Spilled

Restrict those not wearing protective equipment and who are not involved in cleanup from entering area. Collect solids with HEPA vacuum and damp mop residues. Do NOT dry sweep. Place in sealed container for disposal. Flush trace residues with copious amounts of water and collect for disposal.

Preferred Waste Disposal Method

Mix with a combustible solid and burn in chemical incinerator equipped with afterburner and scrubber.

Precautions to be Taken in Handling and Storage

Store in tightly closed containers in cool, dark, well-ventilated location. Protect containers from physical damage. Keep away from oxidizing materials and heat.

Other Precautions and Warnings

Ensure adequate ventilation in storage areas. Avoid vapor/dust inhalation and use appropriate personnel protective equipment if necessary. Do not store in direct sunlight.

SECTION VIII - CONTROL MEASURES AND PERSONAL PROTECTIVE EQUIPMENT

Respiratory Protection *(specify type)*

For low exposures, use an air-purifying respirator with a HEPA filter. Otherwise, use a supplied-air respirator or a self-contained breathing apparatus with full facepiece operated in positive pressure mode.

Ventilation

Local exhaust (preferred) at site of chemical work or general ventilation.

Protective Gloves	Eye Protection	Other Protective Clothing
Impervious Gloves	Chemical/Dust Goggles or Face Mask	Protective Apron if Splash is Likely

Work/Hygiene Practices

Always wash hands thoroughly after using chemical; never bring food, drink, or smoking materials into vicinity of chemicals.

IODOFORM

CHI_3

CAS: 75-47-8

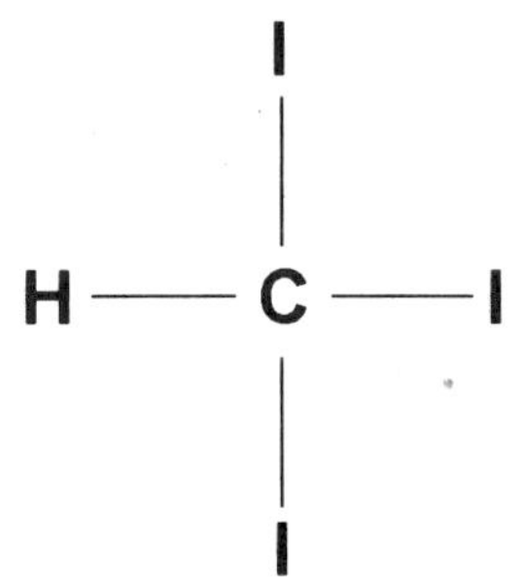

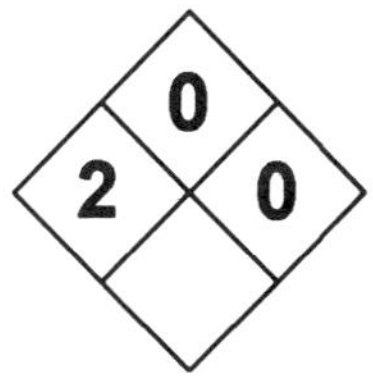

IDENTIFICATION AND TYPICAL USES

Iodoform is a yellow to greenish-yellow powder or crystalline solid with a pungent, disagreeable odor. It is used as a chemical intermediate, as a germicide and antiseptic for external use (in veterinary medicine), as a component in adhesives, and as a polymerization control agent.

RISK ASSESSMENT: HEALTH

General Assessment

Iodoform is toxic to humans by *ingestion, inhalation,* and *absorption* through intact skin. It is a severe irritant and a strong liver toxin (hepatoxin). There is no information in the references on its carcinogenic, teratogenic, or mutagenic capabilities. The data on human exposures are somewhat limited, especially with regard to inhalation exposures.

Inhalation causes irritation of the eyes, nose, throat, and respiratory system. It may cause dypsnea. There is also the possibility that inhalation will lead to central nervous system depression with lightheadedness, dizziness, nausea, incoordination, and other effects similar to that of ingestion.

Skin contact results in the absorption of the chemical into the body causing many of the same toxic systemic effects of inhalation and ingestion, especially those resulting in liver injury and CNS damage. Eye contact with iodoform may lead to visual impairment and possible transitory blindness. The optic nerve and retina may be affected.

Ingestion will likely cause similar toxic systemic effects as inhalation and absorption with nausea, vomiting, delirium, hallucinations, rapid heart rate, fever, coma, and death. High concentration absorbed through the skin can lead to injury of the liver, heart, and kidneys.

Acute Health Effects

The following acute (short-term) health effects may occur immediately or shortly after exposure to iodoform:

Skin: Irritation with possible rash and redness. Absorption leading to toxic systemic effects is likely to occur, especially on prolonged contact.

Eye: Mild to moderate irritation (depending upon length of time contact occurs). Visual impairment and even transient blindness are possible (effects seem to be temporary).

Lung: Nose, throat, and respiratory tract irritation. Absorption through the respiratory process may cause toxic systemic effects.

CNS: Headaches, lightheadedness, dizziness, confusion, excitement, drowsiness, fever, coma, and possibly death.

Other: Exposure to high concentrations may cause serious and possibly life-threatening effects to the liver, kidney, and heart.

Chronic Health Effects

The following chronic (long-term) health effects may occur at some time after exposure to iodoform and can last for months or even years:

Cancer Hazards: According to information presented in the references, iodoform has not been adequately tested for its ability to cause cancer in laboratory animals.

Reproductive Hazard: According to information presented in the references, iodoform has not been adequately tested for its ability to adversely affect reproduction in laboratory animals.

Other Chronic Effects: Repeated skin contact may cause dermatitis. Iodoform converts to carbon monoxide in the body. High exposures could lead to the production of carboxyhemoglobin at levels which may impair cardiovascular function.

Recommended Risk-Reduction Measures

Personnel should avoid direct contact with iodoform. Occupational poisoning occurs most commonly by inhalation and skin absorption. If a less toxic material or compound cannot be substituted for iodoform, then *engineering controls* are the most effective method of reducing exposures. The best protection for workers is to ensure controlled use of iodoform and, wherever possible, enclose operations or provide adequate local exhaust ventilation at the site of iodoform release. Using respiratory protection is less effective than the controls mentioned above, but is still advisable whenever working with or around iodoform. The PEL is extremely low and difficult to accurately monitor (0.6 ppm). For exposures to low levels or infrequent exposures, use an air-purifying respirator equipped with a dust/mist prefilter of high-efficiency particulate air (HEPA) filter. For higher exposures, or when the level is unknown (as in an emergency), use a supplied-air respirator with full facepiece and mask operated in positive pressure mode or a self-contained breathing apparatus (SCBA) with full facepiece operated in pressure demand. If a full facepiece is not available, then chemical goggles should be worn to protect the eyes. A face shield should also be considered. To prevent hand and skin exposures impervious gloves should be used. Glove manufacturers should be contacted and permeation studies obtained *before* final glove selections are made.

Administrative controls should also be in place to minimize the potential for human exposures. These may include written procedures or policies which specify the methods and techniques that will be practiced whenever personnel are to work with iodoform.

All personnel should receive *training* on the use, hazards, protective measures, emergency actions, and other precautions per 29 CFR 1910.1200 (Hazard Communication), prior to the first assignment in an area where iodoform is used or stored.

For those personnel with frequent or potentially high exposures (half the PEL or greater), the following are recommended before beginning employment and at regular intervals thereafter (e.g., annually):

- ☑ Liver function tests.
- ☑ Evaluation of the skin by a dermatologist for indications or symptoms of chlor-acne.

If symptoms develop or overexposure is suspected, the following recommended medical tests should be considered:

- ☑ Liver and kidney function tests.
- ☑ Interview for nervous system effects and any changes in personality (irritability, moodiness) or normal function (weakness, fever, drowsiness).

Any evaluation should include a careful history of past and present symptoms with an examination. Medical tests that simply look for existing damage are not a substitute for controlling exposures. Also, because more than light or casual alcohol consumption can lead to liver damage, drinking alcohol may increase the potential liver damage caused by iodoform. Prudent risk management requires proper consideration of *all* possible exposure factors when evaluating workplace health hazard risks.

Other methods to reduce exposure include:

- ☑ Where possible, enclose operations and use local exhaust ventilation at the site of chemical release. If local exhaust ventilation or enclosure is not used, respiratory protection should be mandatory.
- ☑ Always ensure that proper protective clothing is worn when using chemical substances and that personnel are trained on its proper use, care, and maintenance.
- ☑ Wash thoroughly immediately after exposure to iodoform and at the end of the work shift or before eating, drinking, or smoking.
- ☑ Hazard warning information should be posted in the work area. In addition, as part of an on-going education and training program, all information on the health and safety hazards of iodoform should be communicated to all exposed workers.
- ☑ Eye wash stations should be provided in the immediate work area for emergency use. If there is a possibility of skin exposure to iodoform, emergency shower facilities should be provided in the immediate area.

☑ Workers whose clothing has been contaminated by iodoform should change into clean clothes before leaving work. Contaminated work clothing should be laundered only by individuals who have been informed of the hazards of exposure to iodoform.

RISK ASSESSMENT: ENVIRONMENT

General Assessment

The environment is at risk of exposure during transportation, storage, disposal, or destruction of iodoform. In almost every scenario, the threat of environmental exposure is contingent upon the proper handling of the chemical substance. Accidental spills, large or small, can result in contamination of the surrounding environmental mediums (water, soil, and air).

Iodoform is a non-combustible solid. However, if it is involved in fire, iodoform can emit highly toxic oxides of carbon and toxic fumes of chloride. It is incompatible with most oxidizers and must be kept away from permanganates, nitrates, peroxides, chlorates, and perchlorates. Caution is always required in handling, storage, transportation, and disposal of iodoform. Emergency responders should be made aware of the presence of iodoform at any emergency response situation.

Iodoform can enter the environment through industrial discharges or spills.

Acute Ecological Effects

Acute (short-term) toxic effects may include the death of animals, birds, or fish and death or low growth rate in plants. Acute effects are seen 2 to 4 days after animals or plants are exposed to iodoform. Insufficient data are available on the short-term effects of iodoform to aquatic life, plants, birds, or land animals.

Chronic Ecological Effects

Chronic (long-term) toxic effects may include shortened life span, reproductive problems, lower fertility, and changes in appearance or behavior in exposed animals. These effects can be seen long after first exposure(s) to toxic chemicals. Insufficient data are available to evaluate or predict the long-term effects of iodoform to aquatic life, plants, birds, or terrestrial animals.

Water Solubility

Iodoform is nearly insoluble in water. Concentrations of 1 milligram will mix with a liter of water.

Persistence in the Environment

Iodoform is highly persistent in water, with a half-life greater than 200 days. The half-life of a pollutant is the amount of time it takes for one half of the chemical to be degraded. Iodoform is expected to be mobile in terrestrial soils.

Bioaccumulation in Aquatic Organisms

Some substances increase in concentration, or *bioaccumulate,* in living organisms as they breathe contaminated air, drink contaminated water, or eat contaminated food. These chemicals can become concentrated in the tissues and internal organs of animals as well as humans.

The concentration of iodoform found in fish tissues is expected to be about the same as the average concentration of iodoform in the water from which the fish was taken.

Recommended Risk-Reduction Measures

Proper training of all transporters will reduce the likelihood of a mishap or accident resulting in a leak or spill of iodoform into the environment. Labels on all containers, trucks, and rail cars should accurately reflect their contents to enable emergency responders to react properly and quickly to any disaster thereby reducing the potential risk to the environment and to personnel.

Storage of iodoform should be segregated from other chemicals and materials (especially oxidizers) to minimize the risk of cross-contamination.

If a spill or leak to the environment has occurred, fire department, emergency response and/or hazardous materials spill personnel should be notified immediately. Cleanup should be attempted only by those trained in proper spill containment procedures. Contaminated soils should be removed for incineration and replaced with clean soil. If iodoform should contact the water table, aquifer, or navigable waterway, time is of the essence. It is practically insoluble and total remediation may be possible. The local and/or state emergency response authorities must be notified. A comprehensive emergency response or disaster preparedness/recovery plan should be in place prior to any operations involving the use, transportation, storage, or disposal of iodoform. For disposal, iodoform may be dissolved in a combustible solvent and burned in an incinerator equipped with an afterburner and scrubber.

If iodoform is spilled or leaked, the following specific steps are recommended:

☑ Restrict persons not wearing protective clothing from area of spill or leak until cleanup is complete and area can be opened for normal work.

☑ Collect solids using HEPA vacuum. Do NOT dry sweep (generates airborne dusts). Damp mop residue and collect for disposal. Absorb liquid solutions in vermiculite, sand, earth, or similar material and deposit in sealed drum.

☑ Ventilate area of spill or leak.

☑ It may be necessary to dispose of iodoform as a hazardous waste. The responsible state agency or the regional office of the federal Environmental Protection Agency (EPA) should be contacted for specific recommendations.

RISK ASSESSMENT: BUSINESS

General Assessment

Accidents or mishaps involving iodoform can present a serious threat to business operations. The loss or damage of equipment or facilities can significantly affect fiscal viability. Lawsuits that may result from personnel illness, injury/death, public exposures, and/or environmental contamination will require a serious expenditure of resources. Media attention surrounding an injury, death, or environmental damage can also result in a loss of profits and loss of current as well as future business.

Recommended Risk-Reduction Measures

Company attorneys, safety and health professionals, and environmental specialists should be involved in the development of any procedures for responding to chemical incidents. A company official should be predesignated as a public relations officer with specific training in dealing with the press. Corporate plans and policies should be developed, approved, and implemented long before any need for such arises.

REFERENCES

American Conference of Governmental Industrial Hygienists. 1988. *Documentation of the Threshold Limit Values and Biological Exposure Indices*, 5th Edition (with updates). Cincinnati: ACGIH

Anthony, C. P., and N. J. Kolthoff. 1971. *Textbook of Anatomy and Physiology.* St. Louis: C. V. Mosby Company.

Bloom, A. D. and F. J. DeSerres. 1995. *Ecotoxicity and Human Health.* New York: CRC Press/Lewis Publishers.

Brusick, D. J. 1994. *Methods for Genetic Risk Assessment.* New York: CRC Press/Lewis Publishers.

Bueche, J. R. 1972. *Principles of Physics*, 2nd Edition. New York: McGraw-Hill.

Burrell, R., D. K. Flaherty, and L. J. Sauers. 1992. *Toxicology of the Immune System.* New York: Van Nostrand Reinhold.

Calabrese, E. J. 1994. *Biological Effects of Low Level Exposures.* New York: CRC Press/Lewis Publishers.

Cockerman, L. G., B. S. Shane. 1994. *Basic Environmental Toxicology.* New York: CRC Press/Lewis Publishers.

Cothern, C. R., N. P. Ross. 1994. *Environmental Statistics, Assessment, and Forecasting.* New York: CRC Press/Lewis Publishers.

Danse, I. R. 1991. *Common Sense Toxics in the Workplace.* New York: Van Nostrand Reinhold.

Environmental Protection Agency. 1983. *Chemical Hazard Information Profiles.* Washington, D.C.: U.S. EPA.

Fodden, C. R., and J. L. Weddell. 1992. *Hazardous Materials Emergency Action Data.* New York: CRC Press/Lewis Publishers.

Forseberg, K., and L. H. Keith. 1993. *Chemical; Protective Clothing Permeation and Degradation Database.* Chelsea, MI: Lewis Publishers.

Gardner, R. W. 1994. *Chemical Intolerance.* New York: CRC Press/Lewis Publishers.

Hodgson, E., R. B. Mailman, and Janice E. Chambers. 1988. *Dictionary of Toxicology.* London: The MacMillian Company.

Howard, P. H. 1995. *Handbook of Environmental Fate and Exposures Data for Organic Chemicals*, Vol. I-IV. New York: CRC Press/Lewis Publishers.

Hosty, J. W., and P. Foster. 1990. *A Practical Guide to Chemical Spill Response.* New York: Van Nostrand Reinhold.

Landis, W. G., M. H. Yu. 1995. *Introduction to Environmental Toxicology.* New York: CRC Press/Lewis Publishers.

Lewis, R. J., Sr. 1993. *Hawley's Condensed Chemical Dictionary*, 12th Edition. New York: Van Nostrand Reinhold.

Lewis, R. J., Sr. 1992. *Sax's Dangerous Properties of Industrial Materials*, Volumes 1, 2, and 3, 8th Edition. New York: Van Nostrand Reinhold.

Mackay, D., W. Y. Shiu, and K. C. Ma. 1992. *Illustrated Handbook of Chemical-Physical Properties and Environmental Fate for Organic Chemicals.* New York: CRC Press/Lewis Publishers.

Meyer, E. 1990. *Chemistry of Hazardous Materials.* Englewood Cliffs, NJ: Prentice Hall.

Mycock, J. C., J. D. McKenna, and L. Theodore. 1995. *Handbook of Air Pollution Control Engineering and Technology.* New York: CRC Press/Lewis Publishers.

Patnaik, P. 1992. *A Comprehensive Guide to the Hazardous Properties of Chemical Substances.* New York: Van Nostrand Reinhold.

Quigley, D. R. 1994. *Handbook of Emergency Chemical Management.* New York: CRC Press/Lewis Publishers.

U.S. Department of Health and Human Services, National Institute for Occupational Safety and Health. 1994. *NIOSH Pocket Guide to Chemical Hazards.* Washington, D.C.: U.S. Government Printing Office.

MATERIAL SAFETY DATA SHEET

CHEMICAL NAME	
IRON (powder and solid)	**CRC PRESS/LEWIS PUBLISHERS DISCLAIMER:** THE INFORMATION AND RECOMMENDATIONS PRESENTED HEREIN ARE BASED ON SOURCES BELIEVED TO BE RELIABLE. CRC MAKES NO REPRESENTATION ON ITS COMPLETENESS OR ACCURACY. IT IS THE RESPONSIBILITY OF THE USER TO DETERMINE THE CHEMICAL'S SUITABILITY FOR ITS INTENDED USE, THE CHEMICAL'S SAFE USE, AND THE CHEMICAL'S PROPER DISPOSAL. NO REPRESENTATIONS AND/OR WARRANTIES, EITHER EXPRESSED OR IMPLIED, OF THE MERCHANTABILITY OR FITNESS FOR A PARTICULAR PURPOSE, OR OF ANY OTHER NATURE, ARE MADE WITH RESPECT TO THE INFORMATION PROVIDED IN THIS MSDS OR TO THE CHEMICAL TO WHICH INFORMATION MAY REFER. CRC NEITHER ASSUMES NOR AUTHORIZES ANY OTHER PERSON TO ASSUME FOR IT, ANY OTHER ADDITIONAL RESPONSIBILITY OR LIABILITY FOR THE USE OF, OR RELIANCE UPON, THIS INFORMATION.

HAZARD WARNING INFORMATION

HEALTH	FIRE	REACTIVE	OTHER	DEGREE OF HAZARD	COLOR CODING	OTHER CODES
(powder) 2	(powder) 3	(powder) 3		0 = Minimum Hazard 1 = Slight Hazard 2 = Moderate Hazard 3 = Serious Hazard 4 = Severe Hazard	HEALTH = BLUE FIRE = RED REACTIVITY = YELLOW OTHER = WHITE	OX = Oxidizer ACID = Acid ALK = Alkali COR = Corrosive W = Use No Water

SECTION I - GENERAL INFORMATION

Characterization: **Metal**	RCRA Number: **None**	EPA Class: **Not Applicable**
DOT Proper Shipping Name: **Iron Mass or Sponge**	Chemical Abstract Service (CAS) Number: **7439-89-6**	
DOT Hazard Class and Label Requirements: **Flammable Solid**	DOT Emergency Guide Code: **No Citation**	
DOT Identification Number: **NA1383**	Atomic Formula: **Fe**	

Synonyms
Carbonyl metal; Armco Iron ®; Ancor 80/150 ®; iron, electrolytic.

SECTION II - HAZARDOUS INGREDIENTS AND IDENTITY INFORMATION

Hazardous Components (specific identity)	OSHA Exposure Criteria	NIOSH Exposure Criteria	Immediately Dangerous to Life and Health (IDLH)	ACGIH Exposure Criteria
Iron (derivation: Metallic element of atomic number 26, group VIII of the periodic system. Occurs naturally as the second most abundant metal in the earth's crust. Also from decomposition of iron pentacarbonyl. Found in Minnesota, Alabama, Labrador, Yukon, Europe, South Africa).	**PEL: 10 mg/m^3 as iron oxide fume** **STEL: Not Established**	**REL: 5 mg/m^3 as iron oxide fume** **STEL: Not Established**	**Not Determined**	**TLV: 5 mg/m^3 as iron oxide fume** **STEL: Not Established**

SECTION III - PHYSICAL AND CHEMICAL CHARACTERISTICS

Boiling Point: **4982°F (2750°C)**	Specific Gravity (H_2O = 1): **7.86**
Vapor Pressure (mm Hg): **1 at 3248°F (1787°C)**	Atomic Weight: **55.85**
Vapor Density (Air = 1): **Not Found**	Melting Point: **2795°F (1535°C)**

Solubility
Insoluble in water, dissolves in non-oxidizing acids (sulfuric and hydrochloric).

Appearance and Odor
Silver-white or gray, soft, ductile, malleable, slightly magnetic metal. The powder is black-gray.

SECTION IV - FIRE AND EXPLOSION HAZARD DATA

Flash Point (method used): **Not Determined**	Explosive Limits in Air % by Volume: **LEL: Not Determined UEL: Not Determined**
NFPA Classification: **Non-Combustible Solid (Pyrophoric Powder)**	Autoignition Temperature: **Not Determined**

Extinguishing Media
Use media suitable to surrounding fire. Bulk iron will not burn but powdered iron may ignite, especially when mixed with air.

Special Fire Fighting Procedures
Wear full protective clothing and self-contained breathing apparatus (SCBA). Use a water spray to keep containers cool until long after the fire has been extinguished. For large fires, if possible, withdraw and allow to burn. Remain clear of smoke.

Unusual Fire and Explosion Hazards
The powder ignites spontaneously with exposure to heat, flames, sparks, or exposure to air. Violent reactions with many other compounds. Fight fire from distance, if possible. Move containers from fire area if it can be done without risk.

SECTION V - REACTIVITY DATA

Stability		Conditions to Avoid **Iron is normally stable in dry air, but will readily oxidize in moist air to form rust. The finely divided powders are very unstable and can ignite spontaneously in air.**
Stable X	Unstable	Incompatibility *(materials to avoid)* **Iron is incompatible and will explode on contact with acetaldehyde, ammonium peroxodisulfate, chloroformamidinium, chloric acid, ammonium nitrate, halogens, steam, heat, and many other compounds.**
Hazardous Polymerization		Conditions to Avoid **Under normal conditions of temperature and pressure, hazardous polymerization of iron is not known to occur.**
May Occur	Will Not Occur X	Hazardous Decomposition or By-products **When heated to decomposition, iron can produce toxic iron oxide fumes.**

SECTION VI - HEALTH HAZARD DATA

Primary Route(s) of Entry:	Inhalation?	Absorption (skin)?	Ingestion?
	X		X

Health Hazards

INHALATION: Irritation of the respiratory system and mucous membranes. Exposures to iron dusts can lead to iron pneumoconiosis and possible siderosis (mottling or spotting of the lungs).

SKIN & EYES: An irritant to the eyes causing conjunctivitis with deposition of iron particles leaving a rust ring or brownish stain on the cornea. Skin contact can cause rash.

INGESTION: Bloody vomiting (hematemesis), lethargy, shock, acidosis, cyanosis, and fever. There can also be liver damage, gastrointestinal fibrosis, pancreatic fibrosis, obstruction of the respiratory tract, diabetes mellitus, liver cirrhosis, and possible cardiac toxicity.

Carcinogenicity	NTP Listed?	IARC Cancer Review Group?	OSHA Regulated?	Target Organs?
Unknown Human Unknown Animal	No	No	No	Respiratory system, skin, eyes, liver, pancreas.

Medical Conditions Generally Aggravated by Exposure

Existing chronic respiratory diseases may be aggravated by exposure.

Emergency and First-aid Procedures

Eye contact: Flush immediately with water for 15 minutes (minimum); seek medical attention. Skin contact: Remove all contaminated clothing. Immediately wash area with large amounts of soap and water. If rash develops, seek medical attention. For inhalation: Remove the person from exposure. Provide respiratory assistance and CPR. Transfer to medical facility. If swallowed: Provide breathing assistance if necessary. Give conscious person 1 to 2 glasses of water and induce vomiting. Never give a convulsing or unconscious person anything by mouth. Seek medical attention immediately.

SECTION VII - PRECAUTIONS FOR SAFE HANDLING AND USE

Steps to be Taken in Case Material is Released or Spilled

Collect powdered material in most convenient manner possible and deposit in sealed drum. Keep those not involved in cleanup from entering area. Do NOT dry sweep (creates airborne dusts). Use a HEPA vacuum. Iron dust clouds can ignite or explode in air or if heated. Use non-sparking tools.

Preferred Waste Disposal Method

No citation.

Precautions to be Taken in Handling and Storage

Store in tightly closed containers in cool, dry, well-ventilated area and protect from physical damage. Sources of heat or ignition are prohibited where iron is used, handled, or stored.

Other Precautions and Warnings

Avoid contact with incompatible materials. Avoid breathing iron dusts or powders.

SECTION VIII - CONTROL MEASURES AND PERSONAL PROTECTIVE EQUIPMENT

Respiratory Protection *(specify type)*

For exposure to low levels, use a full facepiece respirator with a high efficiency particulate air (HEPA) filter. For higher or prolonged exposures, use a full facepiece supplied-air respirator or a self-contained breathing apparatus (SCBA) operated in positive pressure mode.

Ventilation

Local exhaust or general mechanical systems recommended.

Protective Gloves	Eye Protection	Other Protective Clothing
Impervious Material	Chemical/Dust Goggles or Face Mask	Protective Suit or Clothing

Work/Hygiene Practices

Always wash hands thoroughly after using chemical; never bring food, drink, or smoking materials into vicinity of chemicals.

IRON

Fe

CAS: 7439-89-6

IDENTIFICATION AND TYPICAL USES

Iron is a silver-white or gray, shining, ductile, malleable metallic solid that can also appear in the form of a black-gray powder or dusts. It is the only metal that can be tempered. It is a metallic element of atomic number 26, group VIII of the periodic table, atomic weight 55.847, valances 2, 3, with 4 stable isotopes and 4 radioactive isotopes. The solid form is used to alloy with carbon, manganese, chromium, nickel, and other elements to form steel. Powdered iron is used in metallurgy products, in magnets, in high-frequency cores, in certain automobile parts, and as a catalyst in ammonia synthesis.

RISK ASSESSMENT: HEALTH

General Assessment

Iron is a mildly toxic metal by *inhalation* and moderately toxic by *ingestion*. It can also cause adverse health effects in contact with the eyes. It is potentially toxic by all routes of exposure in all forms (solid, powder/dust). The data on its specific carcinogenic properties are conflicting and inconclusive in the references.

Inhalation of iron dusts and powders will cause irritation of the nose, throat, respiratory tract, and associated mucous membranes. Inhalation of large amounts of iron dusts can cause iron pneumoconiosis and siderosis (mottling and spotting of the lungs).

Skin contact results in surface irritation and may lead to irritating rash. The powders can irritate the eyes and may cause redness and inflammation depending upon the duration of exposure. Eye contact may also cause conjunctivitis and the deposition of iron particles can leave a rust ring or brownish stain on the cornea.

Ingestion causes bloody vomiting (hematemesis), lethargy, shock, abdominal pain, bloody diarrhea, acidosis, cyanosis, fever, liver injury (possible cirrhosis), gastrointestinal fibrosis, pancreatic fibrosis, diabetes mellitus, and possible cardiac effects. There can also be obstruction of the digestive tract and large doses can be fatal.

Acute Health Effects

The following acute (short-term) health effects may occur immediately or shortly after exposure to iron:

Skin: Irritation, rash, and possible contact dermatitis.

Eye: Iron particles can irritate the eyes and may cause redness, conjunctivitis, and brown-staining of the cornea (rust rings).

Lung: Inhalation of the dusts and fumes can irritate the nose, mouth, throat, lungs, and mucosa of the respiratory tract. Iron pneumoconiosis (arc welder's lung) is possible.

Other: Ingestion can result in the systemic distribution of iron deposits throughout the body leading to conditions such as pancreatic scarring, diabetes mellitus, and cirrhosis of the liver.

Chronic Health Effects

The following chronic (long-term) health effects may occur at some time after exposure to iron and can last for months or even years:

Cancer Hazards: According to information presented in the references, iron's ability to cause cancer has not been clearly established. Liver damage has occurred in test animals as well as occupational exposures, but no correlation to cancer has been determined.

Reproductive Hazard: According to information available in the references, iron has been shown to cause teratogenic effects in test animals.

Other Chronic Effects: Possible liver, pancreas, lung, and digestive tract effects may occur in the long-term.

ʘ *Recommended Risk-Reduction Measures*

Personnel should avoid direct contact with iron, especially its dusts and powders. The exact nature of its human toxicity is not clearly understood in the references. If the specific iron compound in question is unknown, adequate risk management dictates that it be handled and treated as though it posed a serious risk to human health. If a less toxic chemical cannot be substituted for a hazardous substance, then *engineering controls* are the most effective method of reducing exposures. The best protection is to enclose operations and/or provide local exhaust ventilation at the site of chemical release. While not always operationally feasible, isolating operations can also reduce exposure. Using respiratory protection is less effective than the controls mentioned above, but is still advisable whenever working with or around iron compounds. For exposure to iron at low or transient levels, an MSHA/NIOSH-approved respirator with a high efficiency particulate air (HEPA) filter is recommended. For higher exposures, or exposures to unknown levels (as in an emergency), use a supplied-air respirator with a full facepiece operated in positive pressure mode or an MSHA/NIOSH-approved self-contained breathing apparatus (SCBA) with full facepiece operated in pressure demand. If a full facepiece is not available, then chemical dust goggles should be worn to protect the eyes. Whenever an airborne chemical dust hazard exists, a face shield and a protective apron should be worn. To prevent hand and skin exposures, impervious gloves should be used.

Administrative controls should also be in place to minimize the potential for human exposures. These may include written procedures or policies which specify the methods and techniques that will be practiced whenever personnel are to work with iron.

All personnel should receive *training* on the use, hazards, protective measures, emergency actions, and other precautions per 29 CFR 1910.1200 (Hazard Communication), prior to the first assignment in an area where iron is used or stored. Before beginning employment and at regular intervals thereafter (e.g., annually), for those with frequent or potentially high exposures, the following medical tests are recommended:

☑ Lung function tests.

☑ Chest X-ray.

If symptoms develop or overexposure is suspected, the following may also be useful:

☑ Liver, kidney, and lung function tests.

☑ Consider chest X-ray after acute overexposure.

☑ Complete blood count (CBC) and blood gases.

☑ Evaluation by a qualified allergist, including careful exposure history and special testing (may help diagnose skin allergy).

Any evaluation should include a careful history of past and present symptoms with a medical examination. Medical tests that look for damage already done are not a substitute for controlling exposure. Also, since smoking can cause heart disease, as well as lung cancer, emphysema, and other related respiratory problems, the effects of exposure to iron may be more pronounced in smokers than in non-smokers. Prudent risk management requires proper consideration of *all* possible factors that may be causing exposure symptoms in the workplace.

Other methods to reduce exposure include:

☑ Where possible, enclose operations and use local exhaust ventilation at the site of chemical release. If local exhaust ventilation or enclosure is not used, respiratory protection should be mandatory.

☑ Always ensure that proper protective clothing is worn when using chemical substances and that personnel are adequately trained on the use, care, and maintenance of their personnel protective equipment.

☑ Wash thoroughly immediately after exposure to iron and at the end of the work shift or before eating, drinking, or smoking.

☑ Hazard warning information should be posted in the work area. In addition, as part of an on-going education and training program, all information on the health and safety hazards of iron should be communicated to all potentially exposed workers.

☑ Where possible, automatically transfer iron from drums or other storage containers to process containers. Avoid contact with air and do not generate airborne dusts.

RISK ASSESSMENT: ENVIRONMENT

General Assessment

While iron obviously occurs naturally in the environment, the environment is at risk of exposure during transportation, storage, disposal, or destruction of iron. In almost every scenario, the threat of environmental exposure is contingent upon the proper handling of the chemical substance. Accidental spills, large or small, can result in fire, explosion, and possible contamination of the surrounding environmental mediums (water, soil, and air).

Iron powder will explode in air (pyrophoric). It will also react explosively in the presence of a number of compounds and materials, including ammonium nitrate + heat, chlorine, dinitrogen tetraoxide, liquid fluorine, hydrogen peroxide, and many other common compounds. Mixtures of iron dust with air and water may ignite on drying. It is a powerful reducing agent that oxidizes readily in moist air and reacts with steam to yield hydrogen and iron oxides. These characteristics require special consideration during any emergency situation involving a leak or spill of iron.

Iron can occur in the environment as a natural element. Iron occurs in Minnesota, Alabama, Labrador, Yukon, Europe, and South America. It can also enter the environment through industrial and municipal discharges, or from spills.

Acute Ecological Effects

Acute (short-term) toxic effects may include the death of animals, birds, or fish and death or low growth rate in plants. Acute effects are seen 2 to 4 days after animals or plants are exposed to iron.

Insufficient data are available to evaluate the short-term effects of iron exposure to aquatic life, plants, birds, or land animals.

Chronic Ecological Effects

Chronic (long-term) toxic effects may include shortened life span, reproductive problems, lower fertility, and changes in appearance or behavior in exposed animals. These effects can be seen long after first exposure(s) to toxic chemicals.

Insufficient data are available to evaluate the long-term effects of iron exposure to aquatic life, plants, birds, or land animals.

Water Solubility

Elemental iron and most of its compounds are insoluble in water. Concentrations of less than 1 milligram will not mix with water.

Persistence in the Environment

Iron and its compounds are highly persistent in the aquatic environment, with a half-life of greater than 200 days. The half-life of a pollutant is the amount of time it takes for one half of the chemical to be degraded. Virtually 100% of iron is expected to accumulate in either terrestrial soils or aquatic sediments.

Bioaccumulation in Aquatic Organisms

Some substances increase in concentration, or *bioaccumulate,* in living organisms as they breathe contaminated air, drink contaminated water, or eat contaminated food. These chemicals can become concentrated in the tissues and internal organs of animals as well as humans.

The concentration of iron found in fish is expected to be somewhat higher than the average concentration of iron in the water from which the fish was taken.

Recommended Risk-Reduction Measures

Proper training of all transporters will reduce the likelihood of a mishap or accidents resulting in a leak or spill to the environment. The correct use of labeling on all containers, trucks, and rail cars will enable emergency responders to react properly and quickly to any disaster thereby reducing the potential risk to the environment and to personnel.

Storage of iron should be segregated from incompatible chemicals to minimize the risk of cross-contamination or contact. Bulk storage of iron powders should be avoided. Conditions which can generate airborne dusts must also be avoided. Storage facilities should be designed with proper fire suppression and protection equipment. Containers must be protected from exposure to heat and from physical damage or abuse.

If a spill or leak to the environment has occurred, fire department, emergency response and/or hazardous materials spill personnel should be notified immediately. Use a vacuum method to reduce dust during cleanup. Do not dry sweep (generates airborne dusts and friction). *Carefully* and *slowly* use a vacuum equipped with a high efficiency particulate air (HEPA) filter, not a standard shop vacuum. Deposit collected materials in sealed drum for disposal. Cleanup should be attempted only by those trained in proper spill containment procedures using non-sparking tools. Contaminated soils should be removed and replaced with clean soil. If iron should contact the water table, aquifer, or navigable waterway, time is of the essence. Iron is not soluble in water and, therefore, total containment and remediation may be possible. When such spills occur, the local and/or state emergency response authorities must be notified. A comprehensive emergency response or disaster preparedness/recovery plan should be in place prior to any operations involving the use, transportation, storage, or disposal of iron.

If iron is spilled or leaked, the following specific steps are recommended:

☑ Restrict persons not wearing protective clothing from area of spill or leak until cleanup is complete and area can be opened for normal work.

☑ Collect powdered materials in safest manner possible and place in sealed drums for disposal. Do not dry sweep, use HEPA vacuum instead. Damp mop residue and collect for disposal.

☑ It may be necessary to dispose of iron as a hazardous waste. The responsible state agency or the regional office of the federal Environmental Protection Agency (EPA) should be contacted for specific recommendations.

RISK ASSESSMENT: BUSINESS

General Assessment

Accidents or mishaps involving iron can present a significant threat to business operations. The loss or damage of equipment or facilities can significantly affect fiscal viability. Lawsuits that may result from personnel injury/death, public exposures, and/or environmental contamination will also require a serious expenditure of resources. Media attention surrounding an injury, death, or environmental damage can also result in a loss of profits and loss of current as well as future business.

Recommended Risk-Reduction Measures

Company attorneys, safety and health professionals, and environmental specialists should be involved in the development of any procedures dealing with safety or emergency response. A company official should be pre-designated as a public relations officer with specific training in dealing with the press. Corporate plans and policies should be developed, approved, and implemented long before any need for such arises.

REFERENCES

Ahlbom, A. 1993. *Biostatistics for Epidemiologists.* New York: CRC Press/Lewis Publishers.

American Conference of Governmental Industrial Hygienists. 1988. *Documentation of the Threshold Limit Values and Biological Exposure Indices,* 5th Edition (with updates). Cincinnati: ACGIH

Brusick, D. J. 1994. *Methods for Genetic Risk Assessment.* New York: CRC Press/Lewis Publishers.

Calabrese, E. J. 1994. *Biological Effects of Low Level Exposures.* New York: CRC Press/Lewis Publishers.

Cockerman, L. G., B. S. Shane. 1994. *Basic Environmental Toxicology.* New York: CRC Press/Lewis Publishers.

Cothern, C. R., N. P. Ross. 1994. *Environmental Statistics, Assessment, and Forecasting.* New York: CRC Press/Lewis Publishers.

Danse, I. R. 1991. *Common Sense Toxics in the Workplace.* New York: Van Nostrand Reinhold.

Environmental Protection Agency. 1983. *Chemical Hazard Information Profiles.* Washington, D.C.: U.S. EPA.

Fodden, C. R., and J. L. Weddell. 1992. *Hazardous Materials Emergency Action Data.* New York: CRC Press/Lewis Publishers.

Forseberg, K., and L. H. Keith. 1993. *Chemical; Protective Clothing Permeation and Degradation Database.* Chelsea, MI: Lewis Publishers.

Gardner, R. W. 1994. *Chemical Intolerance.* New York: CRC Press/Lewis Publishers.

Howard, P. H., et. al. 1991. *Handbook of Environmental Degradation Rates.* New York: CRC Press/Lewis Publishers.

Hosty, J. W., and P. Foster. 1990. *A Practical Guide to Chemical Spill Response.* New York: Van Nostrand Reinhold.

Langerman, N. 1994 *Precautionary Labels for Chemical Containers.* New York: CRC Press/Lewis Publishers.

Lauwerys, R. R., P. Hoet. 1993. *Industrial Chemical Exposure.* New York: CRC Press/Lewis Publishers.

Lewis, .J., Sr. 1993. *Hawley's Condensed Chemical Dictionary,* 12th Edition. New York: Van Nostrand Reinhold.

Lewis, . J., Sr. 1992. *Sax's Dangerous Properties of Industrial Materials,* Volumes 1, 2, and 3, 8th Edition. New York: Van Nostrand Reinhold.

Manahan, S. E. 1992. *Toxicological Chemistry,* 2nd Edition. New York: CRC Press/Lewis Publishers.

Meyer, E. 1990. *Chemistry of Hazardous Materials.* Englewood Cliffs, NJ: Prentice Hall.

Miller, B. F., M.D., and C. B. Keane, R.N. 1972. *Encyclopedia and Dictionary of Medicine and Nursing.* Philadelphia: W.B. Sanders Company.

Patnaik, P. 1992. *A Comprehensive Guide to the Hazardous Properties of Chemical Substances.* New York: Van Nostrand Reinhold.

Reeves, A. L. 1979. *Handbook on the Toxicology of Metals.* New York: Elsevier North-Holland.

Richardson, M. 1995. *Dictionary of Substance and Their Effects,* Vol. I-VI. New York: CRC Press/Lewis Publishers.

U.S. Department of Health and Human Services (DHHS), National Institute for Occupational Health (NIOHS). 1994. *NIOSH Pocket Guide to Hazardous Chemicals.* Publication Number 94-116. Washington, D.C.; U.S. Government Printing Office.

MATERIAL SAFETY DATA SHEET

CHEMICAL NAME	
IRON (III) OXIDE	**CRC PRESS/LEWIS PUBLISHERS DISCLAIMER:** THE INFORMATION AND RECOMMENDATIONS PRESENTED HEREIN ARE BASED ON SOURCES BELIEVED TO BE RELIABLE. CRC MAKES NO REPRESENTATION ON ITS COMPLETENESS OR ACCURACY. IT IS THE RESPONSIBILITY OF THE USER TO DETERMINE THE CHEMICAL'S SUITABILITY FOR ITS INTENDED USE, THE CHEMICAL'S SAFE USE, AND THE CHEMICAL'S PROPER DISPOSAL. NO REPRESENTATIONS AND/OR WARRANTIES, EITHER EXPRESSED OR IMPLIED, OF THE MERCHANTABILITY OR FITNESS FOR A PARTICULAR PURPOSE, OR OF ANY OTHER NATURE, ARE MADE WITH RESPECT TO THE INFORMATION PROVIDED IN THIS MSDS OR TO THE CHEMICAL TO WHICH INFORMATION MAY REFER. CRC NEITHER ASSUMES NOR AUTHORIZES ANY OTHER PERSON TO ASSUME FOR IT, ANY OTHER ADDITIONAL RESPONSIBILITY OR LIABILITY FOR THE USE OF, OR RELIANCE UPON, THIS INFORMATION.

HAZARD WARNING INFORMATION

HEALTH	FIRE	REACTIVE	OTHER	DEGREE OF HAZARD	COLOR CODING	OTHER CODES
1	0	0		0 = Minimum Hazard 1 = Slight Hazard 2 = Moderate Hazard 3 = Serious Hazard 4 = Severe Hazard	HEALTH = BLUE FIRE = RED REACTIVITY = YELLOW OTHER = WHITE	OX = Oxidizer ACID = Acid ALK = Alkali COR = Corrosive W = Use No Water

SECTION I - GENERAL INFORMATION

Characterization	**Metallic Oxide**	RCRA Number	**None**
		EPA Class	**Not Applicable**
DOT Proper Shipping Name	**No Citation**	Chemical Abstract Service (CAS) Number	**1309-37-1**
DOT Hazard Class and Label Requirements	**No Citation**	DOT Emergency Guide Code	**37**
DOT Identification Number	**UN 1376**	Atomic Formula	**Fe_2O_3**

Synonyms
Ferric oxide; iron (III) oxide; iron sesquioxide; red iron oxide; anchred standard; anhydrous iron oxide.

SECTION II - HAZARDOUS INGREDIENTS AND IDENTITY INFORMATION

Hazardous Components (specific identity)	OSHA Exposure Criteria	NIOSH Exposure Criteria	Immediately Dangerous to Life and Health (IDLH)	ACGIH Exposure Criteria
Iron (III) oxide (derivation: By action of air, steam, or carbon dioxide on iron; especially pure grade is derived by precipitating hydrated ferric oxide from a solution of iron salts, dehydrating, and reducing with hydrogen; occurs in nature as the mineral magnetite.).	**PEL:** **10 mg(Fe)/m^3** **as iron oxide fume** **STEL:** **Not Established**	**REL:** **5 mg(Fe)/m^3** **as iron oxide fume** **STEL:** **Not Established**	**2500 mg/m^3**	**TLV:** **5 mg(Fe)/m^3** **as iron oxide fume** **STEL:** **Not Established**

SECTION III - PHYSICAL AND CHEMICAL CHARACTERISTICS

Boiling Point	**Not Found**	Specific Gravity (H_2O = 1)	**5.24**
Vapor Pressure (mm Hg)	**0 at 68°F (20°C)**	Atomic Weight	**159.7**
Vapor Density (Air = 1)	**Not Found**	Melting Point	**2664°F (1462°C)**

Solubility
Insoluble in water, alcohol, and ether. Soluble in acids.

Appearance and Odor
Reddish-brown solid, dust, or fume with a metallic taste.

SECTION IV - FIRE AND EXPLOSION HAZARD DATA

Flash Point (method used)	**Not Determined**	Explosive Limits in Air % by Volume	**LEL: Not Determined** **UEL: Not Determined**
NFPA Classification	**Non-Combustible Solid**	Autoignition Temperature	**Not Determined**

Extinguishing Media
Use media suitable to surrounding fire. Bulk iron (III) oxide will not burn.

Special Fire Fighting Procedures
Wear full protective clothing and self-contained breathing apparatus (SCBA). Use a water spray to keep containers cool until long after the fire has been extinguished. For large fires, if possible, withdraw and allow to burn. Remain clear of smoke.

Unusual Fire and Explosion Hazards
Explosive reaction when heated with guanidinium perchlorate. Reaction with carbon monoxide may form an explosive product. Potentially violent reaction with hydrogen peroxide. Fight fire from distance, if possible. Move containers from fire area if it can be done without risk.

SECTION V - REACTIVITY DATA

Stability		Conditions to Avoid Iron (III) oxide is normally stable in closed containers at room temperature under routine conditions of handling in storage. Avoid contact with heat and incompatible materials.
Stable X	Unstable	Incompatibility *(materials to avoid)* Incompatible with carbon monoxide, hydrogen peroxide, powdered aluminum, calcium hypochlorite, hydrazine, performic acid, bromine pentafluoride, cesium acetylide, and ethylene oxide.
Hazardous Polymerization		Conditions to Avoid Under normal conditions of temperature and pressure, hazardous polymerization of iron (III) oxide is not known to occur.
May Occur	Will Not Occur X	Hazardous Decomposition or By-products When heated to decomposition, iron (III) oxide can produce toxic and acrid fumes.

SECTION VI - HEALTH HAZARD DATA

Primary Route(s) of Entry:	Inhalation?	Absorption (skin)?	Ingestion?
	X		

Health Hazards

INHALATION: Irritation of the respiratory system and mucous membranes. Exposures may lead to the development of benign pneumoconiosis with X-ray shadows that are indistinguishable from fibrotic pneumoconiosis (siderosis). A nuisance dust at low exposures.

SKIN & EYES: An irritant to the eyes and skin that may cause redness and inflammation.

INGESTION: Not a probable exposure route.

Carcinogenicity	NTP Listed?	IARC Cancer Review Group?	OSHA Regulated?	Target Organs?
Unknown Human Unknown Animal	No	Group 3	29 CFR 1910.1000 Table Z-1	Respiratory system

Medical Conditions Generally Aggravated by Exposure

Existing chronic respiratory diseases may be aggravated by exposure.

Emergency and First-aid Procedures

Eye contact: Flush immediately with water for 15 minutes (minimum); seek medical attention. Skin contact: Remove all contaminated clothing. Immediately wash area with large amounts of soap and water. If rash develops, seek medical attention. For inhalation: Remove the person from exposure. Provide respiratory assistance and CPR. Transfer to medical facility. If swallowed: Not likely. Seek medical attention immediately. Unless advised otherwise, give large amounts of water and induce vomiting using an emetic. Never give a convulsing or unconscious person anything by mouth.

SECTION VII - PRECAUTIONS FOR SAFE HANDLING AND USE

Steps to be Taken in Case Material is Released or Spilled

Collect powdered material in most convenient manner possible and deposit in sealed drum. Keep those not involved in cleanup from entering area. Do NOT dry sweep (creates airborne dusts). Use a vacuum equipped with a high-efficiency particulate air (HEPA) filter. Damp mop residues and collect for disposal. Do NOT release to sewers or waterways. Notify appropriate authorities, if required.

Preferred Waste Disposal Method

No citation.

Precautions to be Taken in Handling and Storage

Store in tightly closed containers in cool, dry, well-ventilated area and protect from physical damage. Sources of heat or ignition are prohibited where iron (III) oxide is used, handled, or stored.

Other Precautions and Warnings

Avoid contact with incompatible materials. Do NOT create dusty conditions. Always use in areas with adequate ventilation.

SECTION VIII - CONTROL MEASURES AND PERSONAL PROTECTIVE EQUIPMENT

Respiratory Protection *(specify type)*

For exposure to low levels, use a full facepiece respirator with a high efficiency particulate air (HEPA) filter. For higher or prolonged exposures, use a full facepiece supplied-air respirator or a self-contained breathing apparatus (SCBA) operated in positive pressure mode.

Ventilation

Local exhaust or general mechanical systems recommended.

Protective Gloves	Eye Protection	Other Protective Clothing
Impervious Material	Chemical/Dust Goggles or Face Mask	Protective Suit or Clothing

Work/Hygiene Practices

Always wash hands thoroughly after using chemical; never bring food, drink, or smoking materials into vicinity of chemicals.

IRON (III) OXIDE

Fe_2O_3

CAS: 1309-37-1

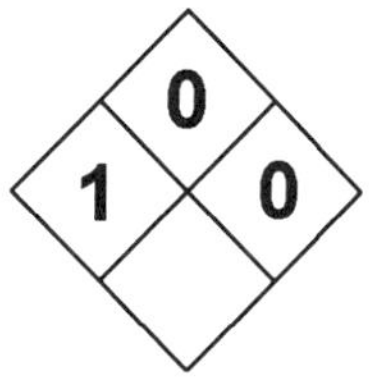

IDENTIFICATION AND TYPICAL USES

Iron (III) oxide is a reddish-brown solid, dust, or fume. It is used as a pigment, a polishing compound, in metallurgy, in magnetic inks, in ferrites for the electronic industry, in coatings for magnetic tape, and as a catalyst.

RISK ASSESSMENT: HEALTH

General Assessment

Iron (III) oxide is a mildly toxic compound by *inhalation*. Skin contact produced no serious health effects and ingestion is not considered a likely route of exposure. It is a questionable carcinogen with experimental tumorigenic data reported. It is considered a nuisance dust on low-level exposures. Higher concentrations may lead to pulmonary changes.

Inhalation of large quantities of iron (III) oxide dusts and powders over a prolonged period may cause the development a benign pneumoconiosis with X-ray shadows that may be indistinguishable from fibrotic pneumoconiosis (siderosis). There can be irritation of the eyes, nose, throat, and respiratory tract.

Acute Health Effects

The following acute (short-term) health effects may occur immediately or shortly after exposure to iron (III) oxide:

Skin: No serious effects reported. May cause irritation and rash in some individuals.

Eye: No serious effects reported. May cause irritation and redness in some individuals, depending upon the duration of exposure.

Lung: Inhalation of the dusts and fumes can irritate the nose, mouth, throat, lungs, and mucosa of the respiratory tract.

Chronic Health Effects

The following chronic (long-term) health effects may occur at some time after exposure to iron (III) oxide and can last for months or even years:

Cancer Hazards: According to information presented in the references, iron (III) oxide has not been adequately tested for its ability to cause cancer in test animals.

Reproductive Hazard: According to information available in the references, iron (III) oxide has been shown to cause teratogenic effects in test animals.

Other Chronic Effects: Prolonged exposure may cause the development of a benign pneumoconiosis that does not appear to be associated with pulmonary fibrosis or disability. It may cause the appearance of siderosis-type shadows on chest X-rays.

Recommended Risk-Reduction Measures

Personnel should avoid direct contact with iron (III) oxide, especially its dusts and powders. The exact nature of its human toxicity is not clearly understood in the references. Prudent risk management dictates that it be handled and treated as though it posed a serious risk to human health. If a less toxic chemical cannot be substituted for a hazardous substance, then *engineering controls* are the most effective method of reducing exposures.

The best protection is to enclose operations and/or provide local exhaust ventilation at the site of chemical release. While not always operationally feasible, isolating operations can also reduce exposure. Using respiratory protection is less effective than the controls mentioned above, but is still advisable whenever working with or around iron (III) oxide compounds. For exposure to iron (III) oxide at low or transient levels, an MSHA/NIOSH-approved respirator with a high efficiency particulate air (HEPA) filter is recommended.

For higher exposures, or exposures to unknown levels (as in an emergency), use a supplied-air respirator with a full facepiece operated in positive pressure mode or an MSHA/NIOSH-approved self-contained breathing apparatus (SCBA) with full facepiece operated in pressure demand. If a full facepiece is not available, then chemical dust goggles should be worn to protect the eyes. Whenever an airborne chemical dust hazard exists, a face shield and a protective apron should be worn. To prevent hand and skin exposures, impervious gloves should be used.

Administrative controls should also be in place to minimize the potential for human exposures. These may include written procedures or policies which specify the methods and techniques that will be practiced whenever personnel are to work with iron (III) oxide.

All personnel should receive *training* on the use, hazards, protective measures, emergency actions, and other precautions per 29 CFR 1910.1200 (Hazard Communication), prior to the first assignment in an area where iron (III) oxide is used or stored.

Before beginning employment and at regular intervals thereafter (e.g., annually), for those with frequent or potentially high exposures, the following medical tests are recommended:

☑ Lung function tests.

☑ Chest X-ray.

If symptoms develop or overexposure is suspected, the following may also be useful:

☑ Lung function tests.

☑ Consider chest X-ray after acute overexposure.

Any evaluation should include a careful history of past and present symptoms with a medical examination. Medical tests that look for damage already done are not a substitute for controlling exposure. Also, since smoking can cause heart disease, as well as lung cancer, emphysema, and other related respiratory problems, the effects of exposure to iron (III) oxide may be more pronounced in smokers than in non-smokers. Prudent risk management requires proper consideration of *all* possible factors that may be causing exposure symptoms in the workplace.

Other methods to reduce exposure include:

☑ Where possible, enclose operations and use local exhaust ventilation at the site of chemical release. If local exhaust ventilation or enclosure is not used, respiratory protection should be mandatory.

☑ Always ensure that proper protective clothing is worn when using chemical substances and that personnel are trained on its use and care.

☑ Wash thoroughly immediately after exposure to iron (III) oxide and at the end of the work shift or before eating, drinking, or smoking.

☑ Hazard warning information should be posted in the work area. In addition, as part of an on-going education and training program, all information on the health and safety hazards of iron (III) oxide should be communicated to all exposed workers.

☑ Where possible, automatically transfer iron (III) oxide from drums or other storage containers to process containers. Avoid contact with air and do not generate airborne dusts.

RISK ASSESSMENT: ENVIRON (III) OXIDEMENT

General Assessment

The environment is at risk of exposure during transportation, storage, disposal, or destruction of iron (III) oxide. In almost every scenario, the threat of environmental exposure is contingent upon the proper handling of the chemical substance. Accidental spills, large or small, can result in fire, explosion, and possible contamination of the surrounding environmental mediums (water, soil, and air).

Iron (III) oxide powder will explode in contact with many compounds and when heated. It will also react and possibly ignite in the presence of a number of compounds and materials, including guanidinium perchlorate. Reaction with carbon monoxide may form an explosive product. The wet oxide reacts explosively with molten aluminum-magnesium alloys. It reacts violently when heated with powdered aluminum, calcium disilicide, magnesium, metal acetylides, hydrazine, calcium hypochlorite, performic acid, and bromine pentafluoride. These characteristics require special consideration during any emergency situation involving a leak or spill of iron (III) oxide.

Iron can occur in the environment as a natural element. Iron (III) oxide can also enter the environment through industrial discharges, or from spills.

☠ *Acute Ecological Effects*

Acute (short-term) toxic effects may include the death of animals, birds, or fish and death or low growth rate

in plants. Acute effects are seen 2 to 4 days after animals or plants are exposed to iron (III) oxide.

Insufficient data are available to evaluate the short-term effects of iron (III) oxide exposure to aquatic life, plants, birds, or land animals.

Chronic Ecological Effects

Chronic (long-term) toxic effects may include shortened life span, reproductive problems, lower fertility, and changes in appearance or behavior in exposed animals. These effects can be seen long after first exposure(s) to toxic chemicals.

Insufficient data are available to evaluate the long-term effects of iron (III) oxide exposure to aquatic life, plants, birds, or land animals.

Water Solubility

Elemental iron and most of its compounds are insoluble in water. Concentrations of less than 1 milligram will not mix with water.

Persistence in the Environment

Iron and its compounds are highly persistent in the aquatic environment, with a half-life of greater than 200 days. The half-life of a pollutant is the amount of time it takes for one half of the chemical to be degraded.

Bioaccumulation in Aquatic Organisms

Some substances increase in concentration, or *bioaccumulate,* in living organisms as they breathe contaminated air, drink contaminated water, or eat contaminated food. These chemicals can become concentrated in the tissues and internal organs of animals as well as humans.

The concentration of iron found in fish is expected to be somewhat higher than the average concentration of iron in the water from which the fish was taken.

Recommended Risk-Reduction Measures

Proper training of all transporters will reduce the likelihood of a mishap or accidents resulting in a leak or spill to the environment. The correct use of labeling on all containers, trucks, and rail cars will enable emergency responders to react properly and quickly to any disaster thereby reducing the potential risk to the environment and to personnel.

Storage of iron (III) oxide should be segregated from incompatible chemicals to minimize the risk of cross-contamination or contact. Bulk storage of iron (III) oxide powders should be avoided. Storage facilities should be designed with proper fire suppression and protection equipment. Containers must be protected from exposure to heat and physical damage.

If a spill or leak to the environment has occurred, fire department, emergency response and/or hazardous materials spill personnel should be notified immediately. Use a vacuum method to reduce dust during cleanup. Do not dry sweep (generates airborne dusts and friction). Use a vacuum equipped with a high efficiency particulate air (HEPA) filter, not a standard shop vacuum. Deposit collected materials in sealed drum for disposal. Cleanup should be attempted only by those trained in proper spill containment procedures using non-sparking tools. Contaminated soils should be removed and replaced with clean soil. If iron (III) oxide should contact the water table, aquifer, or navigable waterway, time is of the essence. It is not soluble in water and total containment and remediation may be possible. When such spills occur, the local and/or state emergency response authorities must be notified. A comprehensive emergency response or disaster preparedness/recovery plan should be in place prior to any operations involving the use, transportation, storage, or disposal of iron (III) oxide.

If iron (III) oxide is spilled or leaked, the following specific steps are recommended:

☑ Restrict persons not wearing protective clothing from area of spill or leak until cleanup is complete and area can be opened for normal work.

☑ Collect powdered materials in and safest manner possible and place in sealed drums for disposal. Do not dry sweep; use HEPA vacuum instead. Damp mop residue and collect for disposal.

☑ It may be necessary to dispose of iron (III) oxide as a hazardous waste. The responsible state agency or the regional office of the federal Environmental Protection Agency (EPA) should be contacted for specific recommendations.

RISK ASSESSMENT: BUSINESS

General Assessment

Accidents or mishaps involving iron (III) oxide can present a significant threat to business operations. The loss or damage of equipment or facilities can significantly affect fiscal viability. Lawsuits that may result from personnel injury/death, public exposures, and/or environmental contamination will also require a serious expenditure of resources. Media attention surrounding an injury, death, or environmental damage can also result in a loss of profits and loss of current as well as future business.

Recommended Risk-Reduction Measures

Company attorneys, safety and health professionals, and environmental specialists should be involved in the development of any procedures dealing with safety or emergency response. A company official should be pre-designated as a public relations officer with specific training in dealing with the press. Corporate plans and policies should be developed, approved, and implemented long before any need for such arises.

REFERENCES

Ahlbom, A. 1993. *Biostatistics for Epidemiologists.* New York: CRC Press/Lewis Publishers.

American Conference of Governmental Industrial Hygienists. 1988. *Documentation of the Threshold Limit Values and Biological Exposure Indices,* 5th Edition (with updates). Cincinnati: ACGIH

Anthony, C. P., and N. J. Kolthoff. 1971. *Textbook of Anatomy and Physiology.* St. Louis: C. V. Mosby Company.

Asante-Duah, D. K. 1993. *Hazardous Waste Risk Assessment.* New York: CRC Press/Lewis Publishers.

Bloom, A. D. and F. J. DeSerres. 1995. *Ecotoxicity and Human Health.* New York: CRC Press/Lewis Publishers.

Brusick, D. J. 1994. *Methods for Genetic Risk Assessment.* New York: CRC Press/Lewis Publishers.

Burrell, R., D. K. Flaherty, and L. J. Sauers. 1992. *Toxicology of the Immune System.* New York: Van Nostrand Reinhold.

Calabrese, E. J. 1994. *Biological Effects of Low Level Exposures.* New York: CRC Press/Lewis Publishers.

Cockerman, L. G., B. S. Shane. 1994. *Basic Environmental Toxicology.* New York: CRC Press/Lewis Publishers.

Cothern, C. R., N. P. Ross. 1994. *Environmental Statistics, Assessment, and Forecasting.* New York: CRC Press/Lewis Publishers.

Danse, I. R. 1991. *Common Sense Toxics in the Workplace.* New York: Van Nostrand Reinhold.

Environmental Protection Agency. 1983. *Chemical Hazard Information Profiles.* Washington, D.C.: U.S. EPA.

Fodden, C. R., and J. L. Weddell. 1992. *Hazardous Materials Emergency Action Data.* New York: CRC Press/Lewis Publishers.

Forseberg, K., and L. H. Keith. 1993. *Chemical; Protective Clothing Permeation and Degradation Database.* Chelsea, MI: Lewis Publishers.

Gardner, R. W. 1994. *Chemical Intolerance.* New York: CRC Press/Lewis Publishers.

Hallenbeck, W. H. 1993. *Quantitative Risk Assessment for Environmental and Occupational Health,* 2nd Edition. New York: CRC Press/Lewis Publishers.

Hodgson, E., R. B. Mailman, and Janice E. Chambers. 1988. *Dictionary of Toxicology.* London: The MacMillian Company.

Howard, P. H., et. al. 1991. *Handbook of Environmental Degradation Rates.* New York: CRC Press/Lewis Publishers.

International Agency for Research on Cancer (IARC). 1987. *IARC Monographs on the Evaluation of Carcinogenic Risk to Humans, Overall Evaluations of Carcinogenicity: An updating of IARC Monographs Volumes 1-42.* Lyon: IARC.

Kamrin, M. A. 1988. *Toxicology: A Primer on Principles and Applications.* Chelsea, Michigan: Lewis Publishers.

Keith, L. H. , M. T. Johnston, C. E. Hudak, and M. Conoley. 1992. *National Toxicology Program's Gloves+ Database.* Chelsea, MI: Lewis Publishers.

Langerman, N. 1994 *Precautionary Labels for Chemical Containers.* New York: CRC Press/Lewis Publishers.

Lauwerys, R. R., P. Hoet. 1993. *Industrial Chemical Exposure.* New York: CRC Press/Lewis Publishers.

Lewis, .J., Sr. 1993. *Hawley's Condensed Chemical Dictionary,* 12th Edition. New York: Van Nostrand Reinhold.

Lewis, . J., Sr. 1992. *Sax's Dangerous Properties of Industrial Materials,* Volumes 1, 2, and 3, 8th Edition. New York: Van Nostrand Reinhold.

Mackay, D., W. Y. Shiu, and K. C. Ma. 1992. *Illustrated Handbook of Chemical-Physical Properties and Environmental Fate for Organic Chemicals.* New York: CRC Press/Lewis Publishers.

Manahan, S. E. 1992. *Toxicological Chemistry,* 2nd Edition. New York: CRC Press/Lewis Publishers.

Meyer, E. 1990. *Chemistry of Hazardous Materials.* Englewood Cliffs, NJ: Prentice Hall.

Miller, B. F., M.D., and C. B. Keane, R.N. 1972. *Encyclopedia and Dictionary of Medicine and Nursing.* Philadelphia: W.B. Sanders Company.

Mycock, J. C., J. D. McKenna, and L. Theodore. 1995. *Handbook of Air Pollution Control Engineering and Technology.* New York: CRC Press/Lewis Publishers.

Patnaik, P. 1992. *A Comprehensive Guide to the Hazardous Properties of Chemical Substances.* New York: Van Nostrand Reinhold.

Reeves, A. L. 1979. *Handbook on the Toxicology of Metals.* New York: Elsevier North-Holland.

Richardson, M. 1995. *Dictionary of Substance and Their Effects,* Vol. I-VI. New York: CRC Press/Lewis Publishers.

U.S. Department of Health and Human Services (DHHS), National Institute for Occupational Health (NIOHS). 1994. *NIOSH Pocket Guide to Hazardous Chemicals.* Publication Number 94-116. Washington, D.C.; U.S. Government Printing Office.

MATERIAL SAFETY DATA SHEET

CHEMICAL NAME	
IRON PENTACARBONYL	**CRC PRESS/LEWIS PUBLISHERS DISCLAIMER:** THE INFORMATION AND RECOMMENDATIONS PRESENTED HEREIN ARE BASED ON SOURCES BELIEVED TO BE RELIABLE. CRC MAKES NO REPRESENTATION ON ITS COMPLETENESS OR ACCURACY. IT IS THE RESPONSIBILITY OF THE USER TO DETERMINE THE CHEMICAL'S SUITABILITY FOR ITS INTENDED USE, THE CHEMICAL'S SAFE USE, AND THE CHEMICAL'S PROPER DISPOSAL. NO REPRESENTATIONS AND/OR WARRANTIES, EITHER EXPRESSED OR IMPLIED, OF THE MERCHANTABILITY OR FITNESS FOR A PARTICULAR PURPOSE, OR OF ANY OTHER NATURE, ARE MADE WITH RESPECT TO THE INFORMATION PROVIDED IN THIS MSDS OR TO THE CHEMICAL TO WHICH INFORMATION MAY REFER. CRC NEITHER ASSUMES NOR AUTHORIZES ANY OTHER PERSON TO ASSUME FOR IT, ANY OTHER ADDITIONAL RESPONSIBILITY OR LIABILITY FOR THE USE OF, OR RELIANCE UPON, THIS INFORMATION.

HAZARD WARNING INFORMATION

HEALTH	FIRE	REACTIVE	OTHER	DEGREE OF HAZARD	COLOR CODING	OTHER CODES
4	4	3	~~W~~	0 = Minimum Hazard 1 = Slight Hazard 2 = Moderate Hazard 3 = Serious Hazard 4 = Severe Hazard	HEALTH = BLUE FIRE = RED REACTIVITY = YELLOW OTHER = WHITE	OX = Oxidizer ACID = Acid ALK = Alkali COR = Corrosive ~~W~~ = Use No Water

SECTION I - GENERAL INFORMATION

Characterization	RCRA Number	EPA Class
Carbonyl	**None**	**Not Applicable**

DOT Proper Shipping Name	Chemical Abstract Service (CAS) Number
Iron Pentacarbonyl	**13463-40-6**
DOT Hazard Class and Label Requirements	DOT Emergency Guide Code
Poison B; Flammable Liquid and Poison	**57**
DOT Identification Number	Atomic Formula
UN 1994	**$Fe(CO)_5$**

Synonyms
Iron carbonyl; pentacarbonyl iron.

SECTION II - HAZARDOUS INGREDIENTS AND IDENTITY INFORMATION

Hazardous Components (specific identity)	OSHA Exposure Criteria	NIOSH Exposure Criteria	Immediately Dangerous to Life and Health (IDLH)	ACGIH Exposure Criteria
Iron pentacarbonyl (derivation: By treating finely divided iron with carbon monoxide in the presence of a catalyst such as ammonia). **1 ppm = 8.14 mg/m^3**	**PEL:** Not Established **STEL:** Not Established	**REL:** 0.1 ppm 0.8 mg/m^3 **STEL:** 0.2 ppm 1.6 mg/m^3	**Not Determined**	**TLV:** 0.1 ppm 0.8 mg/m^3 **STEL:** 0.2 ppm 1.6 mg/m^3

SECTION III - PHYSICAL AND CHEMICAL CHARACTERISTICS

Boiling Point	Specific Gravity (H_2O = 1)
217°F (103°C)	**1.46 - 1.52**
Vapor Pressure (mm Hg)	Atomic Weight
40 at 87°F (30.5°C)	**195.9**
Vapor Density (Air = 1)	Melting Point
Not Found	**-6°F (-21°C)**

Solubility
Insoluble in water. Soluble in nickel tetracarbonyl and most organic solvents. Soluble (decomposes) in acids and alkalis.

Appearance and Odor
Colorless to yellow to dark-red, mobile, oily liquid with no odor.

SECTION IV - FIRE AND EXPLOSION HAZARD DATA

Flash Point (method used)	Explosive Limits in Air % by Volume
5°F (-15°C) closed cup	**LEL: Not Determined UEL: Not Determined**
NFPA Classification	Autoignition Temperature
Class IB Flammable Liquid	**Not Determined**

Extinguishing Media
Use water, foam, dry chemical, or carbon dioxide.

Special Fire Fighting Procedures
Wear full protective clothing and self-contained breathing apparatus (SCBA). Use a water spray to keep containers cool. Fight fire from distance, if possible.

Unusual Fire and Explosion Hazards
Iron pentacarbonyl is pyrophoric in air. It will ignite and burn vigorously without any contact with an ignition source. It presents a serious fire and explosion hazard.

SECTION V - REACTIVITY DATA

Stability		Conditions to Avoid Iron pentacarbonyl will spontaneously ignite in air. It decomposes on exposure to sunlight or air, releasing carbon monoxide in the process. It is stable only in tightly closed, air-tight, moisture-proof containers.
Stable	Unstable X	Incompatibility *(materials to avoid)* Iron pentacarbonyl is incompatible with oxidizers, nitrogen oxide, zinc + metal halides (cobalt, rhodium, ruthenium) mixtures with acetic acid + water produce a pyrophoric powder, heat, and light.
Hazardous Polymerization		Conditions to Avoid Under normal conditions of temperature and pressure, hazardous polymerization of iron pentacarbonyl is improbable (both the liquid and the vapor are highly reactive).
May Occur	Will Not Occur X	Hazardous Decomposition or By-products When heated to decomposition, iron pentacarbonyl can produce high levels of toxic carbon monoxide gas.

SECTION VI - HEALTH HAZARD DATA

Primary Route(s) of Entry:	Inhalation?	Absorption (skin)?	Ingestion?
	X	X	X

Health Hazards

INHALATION: Symptoms of dizziness, headache, fever, cyanosis, nausea, vomiting, coughing, shortness of breath, difficulty breathing, degenerative changes in the central nervous system, unconsciousness, delayed chest pains, pneumonitis, circulatory collapse, and death with injury to the kidneys, liver, and brain.

ABSORPTION: An irritant to the eyes and skin. Absorption through intact skin causes toxic systemic effects similar to that of inhalation.

INGESTION: Toxic systemic effects that are similar to that of inhalation.

Carcinogenicity	NTP Listed?	IARC Cancer Review Group?	OSHA Regulated?	Target Organs?
Unknown Human Unknown Animal	No	No	No	Respiratory system, skin, eyes, liver, kidneys, CNS

Medical Conditions Generally Aggravated by Exposure

Liver diseases and nervous system disorders may be affected by exposure.

Emergency and First-aid Procedures

Eye contact: Flush immediately with water for 15 minutes (minimum); seek medical attention. Skin contact: Remove all contaminated clothing. Immediately wash area with large amounts of soap and water. If rash develops, seek medical attention. For inhalation: Remove the person from exposure. Provide respiratory assistance and CPR. Transfer to medical facility. If swallowed: Seek medical attention immediately. Never give a convulsing or unconscious person anything by mouth.

SECTION VII - PRECAUTIONS FOR SAFE HANDLING AND USE

Steps to be Taken in Case Material is Released or Spilled

Iron pentacarbonyl is pyrophoric and will ignite on exposure to air or light. Spill situations can present a serious danger to personnel and property. Evacuate area and keep those not involved in cleanup from entering area. Remove all ignition sources. Provide explosion-proof ventilation. Stop leak of iron pentacarbonyl at its source, if possible. Use non-sparking tools to collect the material.

Preferred Waste Disposal Method

No citation.

Precautions to be Taken in Handling and Storage

Store in air-tight, moisture-proof containers in cool, well-ventilated area. Protect from physical damage. Sources of heat or ignition are prohibited where iron pentacarbonyl is used, handled, or stored.

Other Precautions and Warnings

Avoid contact with incompatible materials. Do not store in direct sunlight.

SECTION VIII - CONTROL MEASURES AND PERSONAL PROTECTIVE EQUIPMENT

Respiratory Protection *(specify type)*

Exposure level is low and difficult to monitor. For any potential exposures recommend using a full facepiece supplied-air respirator or a self-contained breathing apparatus (SCBA) operated in positive pressure or continuous flow mode.

Ventilation

Local exhaust or general mechanical systems recommended.

Protective Gloves	Eye Protection	Other Protective Clothing
Impervious Material	Chemical Goggles or Face Mask	Protective Suit or Clothing

Work/Hygiene Practices

Always wash hands thoroughly after using chemical; never bring food, drink, or smoking materials into vicinity of chemicals.

IRON PENTACARBONYL

$Fe(CO)_5$ CAS: 13463-40-6

IDENTIFICATION AND TYPICAL USES

Iron pentacarbonyl is a colorless to yellow to dark-red, oily, mobile liquid with no detectable odor. It is used as a catalyst and a reagent in organic chemistry, as an anti-knock agent in motor fuels, and to produce finely divided iron (carbonyl iron) which, in turn, is used in the production of powdered iron cores for high-frequency coils in radios and televisions. It is also used as an ignitor for underground combustion petroleum recovery.

RISK ASSESSMENT: HEALTH

General Assessment

Iron pentacarbonyl is highly toxic by *inhalation* and skin *absorption*. It can also enter the body and cause adverse health effects through *ingestion*. The are no data on its carcinogenic, mutagenic, or teratogenic properties in the references.

Inhalation and dermal absorption of iron pentacarbonyl can cause immediate headache, dizziness, fever, cyanosis, coughing, shortness of breath, labored breathing, nausea, vomiting, delayed chest pains, degenerative changes in the central nervous system, pneumonitis, circulatory collapse, liver, kidney, and brain effects, and death.

Ingestion of this compound may cause similar toxic effects, although there are limited data available in the references on the specific health risks posed by this exposure route.

Acute Health Effects

The following acute (short-term) health effects may occur immediately or shortly after exposure to iron pentacarbonyl:

Skin: Irritation with a possibility of rash and other surface effects. Absorption through intact skin can occur quickly, causing serious (life-threatening) systemic effects.

Eye: No documented effects. May cause irritation, especially on prolonged contact.

Lung: Irritation to the nose, throat, and respiratory tract. Coughing, dypsnea, chest pain, and other lung-related conditions are possible.

Other: Systemic effects appear to target the central nervous system, the liver, the kidneys, and (possibly) the brain. Immediate headaches and dizziness, followed by fever, cyanosis, nausea, and vomiting. Death may occur due to circulatory system collapse.

Chronic Health Effects

The following chronic (long-term) health effects may occur at some time after exposure to iron pentacarbonyl and can last for months or even years:

Cancer Hazards: According to information presented in the references, the ability of iron pentacarbonyl to cause cancer has not been clearly established. Liver, kidney, and even brain injury has occurred in test animals, but no correlation to cancer has been determined.

Reproductive Hazard: According to information available in the references, iron pentacarbonyl has been shown to cause teratogenic effects in test animals.

Other Chronic Effects: There appears to be a possibility of cumulative effects in the central nervous system and the liver. Additional research is required.

Recommended Risk-Reduction Measures

Personnel should avoid direct contact with iron pentacarbonyl. If a less toxic chemical cannot be substituted for a hazardous substance, then *engineering controls* are the most effective method of reducing exposures in the workplace.

The best protection is to enclose operations and/or provide local exhaust ventilation at the site of chemical release. While not always operationally feasible, isolating operations can also reduce exposure. Using respiratory protection is less effective than the controls mentioned above, but is still advisable whenever working with or around iron pentacarbonyl compounds. The exposure level for this compound is extremely low (0.1 ppm) and difficult to accurately monitor. For any exposure level, a supplied-air respirator with a full facepiece operated in positive pressure mode or an MSHA/NIOSH-approved self-contained breathing apparatus (SCBA) with full facepiece operated in pressure demand mode are the recommended respiratory methods of choice. If a full facepiece is not available, then chemical goggles should be worn to protect the eyes. Whenever an airborne chemical hazard exists, a face shield and a protective apron should be worn. To prevent hand and skin exposures, impervious gloves should be used. Glove manufacturers should be contacted and permeation studies obtained *before* final glove selection have been made.

Administrative controls should also be in place to minimize the potential for human exposures. These may include written procedures or policies which specify the methods and techniques that will be practiced whenever personnel are to work with iron pentacarbonyl.

All personnel should receive *training* on the use, hazards, protective measures, emergency actions, and other precautions per 29 CFR 1910.1200 (Hazard Communication), prior to the first assignment in an area where iron pentacarbonyl is used or stored.

Before beginning employment and at regular intervals thereafter (e.g., annually), for those with frequent or potentially high exposures, the following medical tests are recommended:

☑ Lung function tests.

If symptoms develop or overexposure is suspected, the following may also be useful:

☑ Liver, kidney, and lung function tests.

☑ Consider chest X-ray after acute overexposure.

☑ Nervous system evaluation with careful consideration of exposure history and special testing of nervous system function by a qualified neurologist.

Any evaluation should include a careful history of past and present symptoms with a medical examination. Medical tests that look for damage already done are not a substitute for controlling exposure. Because more than light alcohol consumption can affect the function of the liver and lead to liver injury and/or disease, those persons who use alcohol regularly may experience symptoms more quickly than non-drinkers under the same conditions of exposure. Also, since smoking can cause heart disease, as well as lung cancer, emphysema, and other related respiratory problems, the effects of exposure to iron pentacarbonyl may be more pronounced in smokers than in non-smokers. Prudent risk management requires proper consideration of *all* possible factors that may be causing exposure symptoms in the workplace.

Other methods to reduce exposure include:

☑ Where possible, enclose operations and use local exhaust ventilation at the site of chemical release. If local exhaust ventilation or enclosure is not used, respiratory protection should be mandatory.

☑ Always ensure that proper protective clothing is worn when using chemical substances and that personnel are adequately trained on the use, care, and maintenance of their personnel protective equipment.

☑ Wash thoroughly immediately after exposure to iron pentacarbonyl and at the end of the work shift or before eating, drinking, or smoking.

☑ Hazard warning information should be posted in the work area. In addition, as part of an on-going education and training program, all information on the health and safety hazards of iron pentacarbonyl should be communicated to all workers.

☑ Iron pentacarbonyl is pyrophoric. Avoid contact with air. Where possible, automatically transfer iron pentacarbonyl from drums or other storage containers to process containers.

RISK ASSESSMENT: ENVIRONMENT

General Assessment

While iron pentacarbonyl obviously occurs naturally in the environment, the environment is at risk of exposure during transportation, storage, disposal, or destruction of iron pentacarbonyl. In almost every scenario, the threat of environmental exposure is contingent upon the proper handling of the chemical substance. Accidental spills, large or small, can result in fire, explosion, and possible contamination of the surrounding environmental mediums (water, soil, and air).

Iron pentacarbonyl will ignite and possibly explode on exposure to air or sunlight (pyrophoric). It will also react vigorously in the presence of a number of compounds and materials, including most oxidizable materials, nitrogen oxide, and zinc + metallic halides (such as calcium halides). Mixtures of iron pentacarbonyl with acetic acid + water produce a pyrophoric powder. It is also a powerful reducing agent. These characteristics require special consideration during any emergency situation involving a leak or spill of iron pentacarbonyl.

Iron pentacarbonyl can enter in the environment through industrial and municipal discharges, or from spills.

Acute Ecological Effects

Acute (short-term) toxic effects may include the death of animals, birds, or fish and death or low growth rate in plants. Acute effects are seen 2 to 4 days after animals or plants are exposed to iron pentacarbonyl.

Insufficient data are available to evaluate the short-term effects of iron pentacarbonyl exposure to aquatic life, plants, birds, or land animals.

Chronic Ecological Effects

Chronic (long-term) toxic effects may include shortened life span, reproductive problems, lower fertility, and changes in appearance or behavior in exposed animals. These effects can be seen long after first exposure(s) to toxic chemicals.

Insufficient data are available to evaluate the long-term effects of iron pentacarbonyl exposure to aquatic life, plants, birds, or land animals.

Water Solubility

Iron pentacarbonyl is insoluble in water. Concentrations of less than 1 milligram will not mix with water.

Persistence in the Environment

Iron pentacarbonyl is slightly persistent in the aquatic environment, with a half-life between 2 and 20 days. The half-life of a pollutant is the amount of time it takes for one half of the chemical to be degraded.

Bioaccumulation in Aquatic Organisms

Some substances increase in concentration, or *bioaccumulate,* in living organisms as they breathe contaminated air, drink contaminated water, or eat contaminated food. These chemicals can become concentrated in the tissues and internal organs of animals as well as humans.

The concentration of iron pentacarbonyl found in fish is expected to be somewhat higher than the average concentration of iron pentacarbonyl in the water from which the fish was taken.

Recommended Risk-Reduction Measures

Proper training of all transporters will reduce the likelihood of a mishap or accidents resulting in a leak or spill to the environment. The correct use of labeling on all containers, trucks, and rail cars will enable emergency responders to react properly and quickly to any disaster thereby reducing the potential risk to the environment and to personnel.

Storage of iron pentacarbonyl should be segregated from incompatible chemicals to minimize the risk of cross-contamination or contact. Bulk storage of iron pentacarbonyl powders should be avoided. Storage facilities should be designed with proper fire suppression and protection equipment. Containers must be protected from exposure to heat, air, and light, as well as physical damage or abuse.

If a spill or leak to the environment has occurred, fire department, emergency response and/or hazardous materials spill personnel should be notified immediately. Cleanup should be attempted only by those trained in proper spill containment procedures using non-sparking tools. Contaminated soils should be removed and replaced with clean soil. If iron pentacarbonyl should contact the water table, aquifer, or navigable waterway, time is of the essence. Iron pentacarbonyl is not soluble in water and, therefore, total containment and remediation may be possible. When such spills occur, the local and/or state emergency response authorities must be notified. A comprehensive emergency response or disaster preparedness/recovery plan should be in place prior to any operations involving the use, transportation, storage, or disposal of iron pentacarbonyl.

If iron pentacarbonyl is spilled or leaked, the following specific steps are recommended:

- ☑ Restrict persons not wearing protective clothing from area of spill or leak until cleanup is complete and area can be opened for normal work.
- ☑ The source of any leak should be located and stopped. Personnel must be aware of the pyrophoric action of iron pentacarbonyl and respond accordingly. A source of water should be avail-

able in case of fire. Personnel involved in cleanup activities should use non-sparking tools and collect the spilled materials in safest manner possible.

☑ It may be necessary to dispose of iron pentacarbonyl as a hazardous waste. The responsible state agency or the regional office of the federal Environmental Protection Agency (EPA) should be contacted for specific recommendations.

RISK ASSESSMENT: BUSINESS

General Assessment

Accidents or mishaps involving iron pentacarbonyl can present a significant threat to business operations. The loss or damage of equipment or facilities can significantly affect fiscal viability. Lawsuits that may result from personnel injury/death, public exposures, and/or environmental contamination will also require a serious expenditure of resources. Media attention surrounding an injury, death, or environmental damage can also result in a loss of profits and loss of current as well as future business.

Recommended Risk-Reduction Measures

Company attorneys, safety and health professionals, and environmental specialists should be involved in the development of any procedures dealing with safety or emergency response. A company official should be pre-designated as a public relations officer with specific training in dealing with the press. Corporate plans and policies should be developed, approved, and implemented long before any need for such arises.

REFERENCES

Ahlbom, A. 1993. *Biostatistics for Epidemiologists.* New York: CRC Press/Lewis Publishers.

American Conference of Governmental Industrial Hygienists. 1988. *Documentation of the Threshold Limit Values and Biological Exposure Indices,* 5th Edition (with updates). Cincinnati: ACGIH

Brusick, D. J. 1994. *Methods for Genetic Risk Assessment.* New York: CRC Press/Lewis Publishers.

Calabrese, E. J. 1994. *Biological Effects of Low Level Exposures.* New York: CRC Press/Lewis Publishers.

Cockerman, L. G., B. S. Shane. 1994. *Basic Environmental Toxicology.* New York: CRC Press/Lewis Publishers.

Cothern, C. R., N. P. Ross. 1994. *Environmental Statistics, Assessment, and Forecasting.* New York: CRC Press/Lewis Publishers.

Danse, I. R. 1991. *Common Sense Toxics in the Workplace.* New York: Van Nostrand Reinhold.

Environmental Protection Agency. 1983. *Chemical Hazard Information Profiles.* Washington, D.C.: U.S. EPA.

Fodden, C. R., and J. L. Weddell. 1992. *Hazardous Materials Emergency Action Data.* New York: CRC Press/Lewis Publishers.

Forseberg, K., and L. H. Keith. 1993. *Chemical; Protective Clothing Permeation and Degradation Database.* Chelsea, MI: Lewis Publishers.

Gardner, R. W. 1994. *Chemical Intolerance.* New York: CRC Press/Lewis Publishers.

Howard, P. H., et. al. 1991. *Handbook of Environmental Degradation Rates.* New York: CRC Press/Lewis Publishers.

Hosty, J. W., and P. Foster. 1990. *A Practical Guide to Chemical Spill Response.* New York: Van Nostrand Reinhold.

Langerman, N. 1994 *Precautionary Labels for Chemical Containers.* New York: CRC Press/Lewis Publishers.

Lauwerys, R. R., P. Hoet. 1993. *Industrial Chemical Exposure.* New York: CRC Press/Lewis Publishers.

Lewis, .J., Sr. 1993. *Hawley's Condensed Chemical Dictionary,* 12th Edition. New York: Van Nostrand Reinhold.

Lewis, . J., Sr. 1992. *Sax's Dangerous Properties of Industrial Materials,* Volumes 1, 2, and 3, 8th Edition. New York: Van Nostrand Reinhold.

Manahan, S. E. 1992. *Toxicological Chemistry,* 2nd Edition. New York: CRC Press/Lewis Publishers.

Meyer, E. 1990. *Chemistry of Hazardous Materials.* Englewood Cliffs, NJ: Prentice Hall.

Miller, B. F., M.D., and C. B. Keane, R.N. 1972. *Encyclopedia and Dictionary of Medicine and Nursing.* Philadelphia: W.B. Sanders Company.

Patnaik, P. 1992. *A Comprehensive Guide to the Hazardous Properties of Chemical Substances.* New York: Van Nostrand Reinhold.

Reeves, A. L. 1979. *Handbook on the Toxicology of Metals.* New York: Elsevier North-Holland.

Richardson, M. 1995. *Dictionary of Substance and Their Effects,* Vol. I-VI. New York: CRC Press/Lewis Publishers.

U.S. Department of Health and Human Services (DHHS), National Institute for Occupational Health (NIOHS). 1994. *NIOSH Pocket Guide to Hazardous Chemicals.* Publication Number 94-116. Washington, D.C.; U.S. Government Printing Office.

MATERIAL SAFETY DATA SHEET

CHEMICAL NAME	CRC PRESS/LEWIS PUBLISHERS DISCLAIMER
ISOAMYL ACETATE	

HAZARD WARNING INFORMATION

HEALTH	FIRE	REACTIVE	OTHER	DEGREE OF HAZARD	COLOR CODING	OTHER CODES
1	3	0		0 = Minimum Hazard 1 = Slight Hazard 2 = Moderate Hazard 3 = Serious Hazard 4 = Severe Hazard	HEALTH = BLUE FIRE = RED REACTIVITY = YELLOW OTHER = WHITE	OX = Oxidizer ACID = Acid ALK = Alkali COR = Corrosive W = Use No Water

SECTION I - GENERAL INFORMATION

Characterization	**Aliphatic Ester**	RCRA Number	**D001**
		EPA Class	**Characteristic (I) Waste**
DOT Proper Shipping Name	**Amyl Acetates**	Chemical Abstract Service (CAS) Number	**123-92-2**
DOT Hazard Class and Label Requirements	**Flammable Liquid**	DOT Emergency Guide Code	**26**
DOT Identification Number	**UN 1104**	Chemical Formula	**$CH_3COOCH_2CH_2CH(CH_3)_2$**

Synonyms

Banana oil; isopentyl acetate; 3-methyl-1-butanol acetate; 3-methylbutyl ester of acetic acid; 3-methylbutyl ethanoate; acetic acid, isoamyl ester; acetic acid, isopentyl ester.

SECTION II - HAZARDOUS INGREDIENTS AND IDENTITY INFORMATION

Hazardous Components (specific identity)	OSHA Exposure Criteria	NIOSH Exposure Criteria	Immediately Dangerous to Life and Health (IDLH)	ACGIH Exposure Criteria
Isoamyl acetate (derivation: By rectification of commercial amyl acetate). **1 ppm = 5.41 mg/m^3**	**PEL:** **100 ppm** **525 mg/m^3** **STEL:** **Not Established**	**REL:** **100 ppm** **525 mg/m^3** **STEL:** **Not Established**	**1000 ppm**	**TLV:** **100 ppm** **525 mg/m^3** **STEL:** **Not Established**

SECTION III - PHYSICAL AND CHEMICAL CHARACTERISTICS

Boiling Point	**288°F (142°C)**	Specific Gravity (H_2O = 1)	**0.87**
Vapor Pressure (mm Hg)	**4 at 68°F (20°C)**	Molecular Weight	**130.2**
Vapor Density (Air = 1)	**4.49**	Melting Point	**-109°F (-78°C)**

Solubility

Very slightly soluble in water. Soluble in ethyl acetate, amyl alcohol, acetone, ethanol, ether, fixed oils.

Appearance and Odor

Clear colorless liquid with a pear or banana-like odor and taste. Odor Threshold = <1 ppm.

SECTION IV - FIRE AND EXPLOSION HAZARD DATA

Flash Point (method used)	**77°F (25°C) closed cup**	Explosive Limits in Air % by Volume	**LEL: 1.0% at 212°F (100°C) UEL: 7.5%**
NFPA Classification	**Class IC Flammable Liquid**	Autoignition Temperature	**680°F (360°C)**

Extinguishing Media

Dry chemical, alcohol foam, carbon dioxide, water spray. Use water spray to cool exposed containers.

Special Fire Fighting Procedures

Poisonous gases are produced in fire. Wear full protective clothing with SCBA. Use blanketing effect with foam to smother large fires. Continue to cool containers with water after fire is extinguished.

Unusual Fire and Explosion Hazards

Immediately withdraw if rising sound from venting device is heard or if fire is causing discoloration to the tank. Evacuate 1500 feet radius if fire becomes uncontrollable or container is exposed to direct flame. Vapors are heavier than air and may travel a distances to ignition source to flashback.

SECTION V - REACTIVITY DATA

Stability		Conditions to Avoid
		Isoamyl acetate is stable under normal conditions of storage and operation. Avoid contact with incompatible materials. Keep away from heat, flame, and other sources of ignition.
Stable	Unstable	Incompatibility *(materials to avoid)*
X		Reacts violently in contact with nitrates, strong oxidizers, alkalis, and acids. Particularly reactive with chlorosulfonic acid and strong reducing agents.
Hazardous Polymerization		Conditions to Avoid Under normal conditions of temperature and pressure, hazardous polymerization of isoamyl acetate is not expected to occur. It will attack or break down many natural and artificial rubber products and coatings.
May Occur	Will Not Occur	Hazardous Decomposition or By-products
	X	When heated to decomposition, isoamyl acetate emits acrid and irritating fumes, including toxic oxides of carbon.

SECTION VI - HEALTH HAZARD DATA

Primary Route(s) of Entry:	Inhalation?	Absorption (skin)?	Ingestion?
	X		X

Health Hazards

INHALATION: Irritation of the eyes, nose, throat, and respiratory tract. Causes central nervous system depression with headache, weakness, dizziness, and loss of consciousness. It can also cause edema of the glottis, difficulty breathing, rapid pulse rate, and fatigue.

SKIN & EYES: Will irritate the skin. May cause contact dermatitis. Eye contact may cause irritation and possible visual disturbances that do not appear to be permanent.

INGESTION: Irritation of the mouth, esophagus, stomach, and digestive tract, nausea, and vomiting.

Carcinogenicity	NTP Listed?	IARC Cancer Review Group?	OSHA Regulated?	Target Organs?
Unknown Human Unknown Animal	No	No	29 CFR 1910.1000 Table Z-1	Respiratory system; skin; eyes, CNS.

Medical Conditions Generally Aggravated by Exposure

None reported.

Emergency and First-aid Procedures

Eye contact: Flush immediately with water for 15 minutes (minimum); seek medical attention. Skin contact: Remove all contaminated clothing. Immediately wash area with large amounts of soap and water. Seek medical assistance. For inhalation: Remove the person from exposure. Provide respiratory assistance and CPR. Transfer to medical facility. If swallowed: Seek medical attention immediately. Call poison control center for advice. Give 1-2 glasses of milk or water. Do NOT induce vomiting unless advised otherwise. Never give an unconscious or convulsing person anything by mouth.

SECTION VII - PRECAUTIONS FOR SAFE HANDLING AND USE

Steps to be Taken in Case Material is Released or Spilled

Contain spills using absorbent material (vermiculite or other). Ventilate area of spill, remove ignition sources. Have water source available in case of fire. Store materials in DOT-approved containers. Restrict those not involved in cleanup from entering area. Notify appropriate authorities, as required.

Preferred Waste Disposal Method

Incineration in chemical incinerator equipped with scrubber and afterburner.

Precautions to be Taken in Handling and Storage

Store in tightly closed containers in a cool, dark, well-ventilated area. Protect containers from physical damage. Keep heat, fire, and flame away. If possible, automatically transfer liquids between containers.

Other Precautions and Warnings

Bulk storage of isoamyl acetate is not recommended. Ground and bond all metal containers.

SECTION VIII - CONTROL MEASURES AND PERSONAL PROTECTIVE EQUIPMENT

Respiratory Protection *(specify type)*

Where potential exists for exposure over 100 ppm, use an MSHA/NIOSH-approved full facepiece respirator with an organic vapor cartridge. For higher exposures, use a self-contained breathing apparatus (SCBA) with full facepiece operated in pressure demand or other positive pressure mode.

Ventilation

Local exhaust or general mechanical systems recommended.

Protective Gloves	Eye Protection	Other Protective Clothing
Impervious Gloves (avoid rubber)	Chemical Goggles or Face Mask	Impermeable Apron

Work/Hygiene Practices

Always wash hands thoroughly after using chemical; never bring food, drink, or smoking materials into vicinity of chemicals.

ISOAMYL ACETATE

$CH_3COOCH_2CH_2CH(CH_3)_2$

CAS: 123-92-2

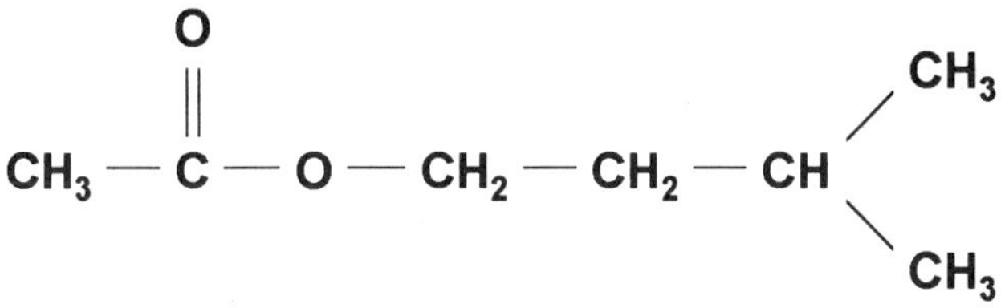

IDENTIFICATION AND TYPICAL USES

Isoamyl acetate is a colorless liquid with a characteristic pear or banana-like odor and taste. It is used in alcohol solution as a pear flavoring agent in mineral waters and syrups. It is also used as a solvent for old oil colors, tannins, nitrocellulose, lacquers, celluloid cements, waterproofing varnishes, bronzing liquids, and metallic paints. It is used in the production of smokeless powders, airplane dopes, and certain cleaning compounds.

RISK ASSESSMENT: HEALTH

General Assessment

Isoamyl acetate is mildly toxic by *inhalation* and *ingestion*. Skin contact can produce localized effects but absorption has not been reported. It is generally considered to be of low toxicity in comparison to other aliphatic esters. There are no reports of any carcinogenic, teratogenic, or mutagenic effects in humans or animals following exposure to isoamyl acetate.

Inhalation of isoamyl acetate vapors can cause irritation to the eyes, nose, throat, and respiratory tract. Depending upon the concentration and duration of the exposure, there can also be varying degrees of central nervous system depression. Symptoms may include headaches, dizziness, general weakness and fatigue, loss of consciousness, difficulty in breathing, and increased or rapid pulse rate.

Skin contact results in the removal of the skin's natural protective oily layer on its surface with subsequent drying, cracking, and possible secondary infections and dermatitis. Eye contact can cause irritation and the possibility for adverse effects to vision.

Ingestion of isoamyl acetate may cause irritation of the mouth, esophagus, stomach, and digestive tract with nausea and vomiting.

☠ *Acute Health Effects*

The following acute (short-term) health effects may occur immediately or shortly after exposure to isoamyl acetate:

Skin: Mild to moderate irritation. Will remove the skin's protective oily layer to cause drying, cracking, and possible secondary infection.

Eye: Irritation and possible adverse effects on visual acuity (temporary).

Lung: Irritation of the eyes, nose, throat, and lungs (especially the mucus membranes). May cause changes in the sense of smell and in pulmonary function.

CNS: High concentrations can depress the central nervous system leading to narcoses and possible loss of consciousness.

Chronic Health Effects

The following chronic (long-term) health effects may occur at some time after exposure to isoamyl acetate and can last for months or even years:

Cancer Hazards: According to the references, isoamyl acetate has not been adequately tested for its ability to cause cancer in test animals. This does not mean that isoamyl acetate is non-carcinogenic. It does mean that further research is required and caution is warranted in use, handling, and storage.

Reproduction: According to the references, isoamyl acetate has not been adequately tested for its ability to adversely affect reproduction in test animals.

Other Chronic Effects: Repeated or prolonged skin contact may cause the development of contact dermatitis with rash, itching, redness, and localized inflammation of the affected surface area.

☉ *Recommended Risk-Reduction Measures*

Personnel should avoid direct contact with isoamyl acetate. The best risk reduction measure is to use a less toxic chemical as a substitute for isoamyl acetate. If this is not possible or practical, then *engineering controls* are the most effective method of reducing exposures. The best protection is to enclose operations and/or provide local exhaust ventilation at the site of chemical release. While not always operationally feasible, isolating operations can also reduce exposure risks to other personnel.

Using respiratory protection is less effective than the controls mentioned above, but is still advisable whenever working with or around isoamyl acetate. For exposures over 100 ppm (8 hours), an MSHA/NIOSH-approved chemical respirator with an organic vapor cartridge should be used. For higher exposures, a self-contained breathing apparatus (SCBA) with full facepiece and operated in pressure demand mode or a supplied-air respirator with full facepiece operated in pressure demand are the recommended respiratory protection methods of choice. If a full facepiece is not available, then chemical goggles should be worn to protect the eyes. Whenever a chemical splash hazard exists, a face shield and a protective apron should be worn. To prevent hand and skin exposures, impervious gloves should be used. Caution is warranted in glove selection since isoamyl acetate will attack many rubber products. Butyl rubber appears to provide some protection for limited periods of time. Natural rubber, neoprene, nitrile, polyethylene, polyvinyl chloride, and Viton are NOT recommended. Glove manufacturers should be contacted to obtain permeation studies before gloves are selected.

Administrative controls should also be in place to minimize the potential for human exposures. These may include written procedures or policies which specify the methods and techniques that will be practiced whenever personnel are to work with isoamyl acetate.

All personnel should receive *training* on the use, hazards, protective measures, emergency actions, and other precautions per 29 CFR 1910.1200 (Hazard Communication), prior to the first assignment in an area where isoamyl acetate is used or stored.

If symptoms develop or overexposure is suspected, the following medical tests are recommended:

- ☑ Lung function tests.
- ☑ Skin testing with dilute isoamyl acetate to help diagnose allergy (performed by a qualified allergist).

Any medical evaluation should include a careful history of past and present symptoms with an examination. Medical tests that look for existing damage are not a substitute for controlling exposures.

Other methods to reduce exposure include:

- ☑ Where possible, enclose operations and use local exhaust ventilation at the site of chemical release. If local exhaust ventilation or enclosure is not used, respiratory protection should be mandatory.
- ☑ Always ensure that proper protective clothing is worn when using chemical substances.
- ☑ Wash thoroughly immediately after exposure to isoamyl acetate and at the end of the work shift or before eating, drinking, or smoking.
- ☑ Hazard warning information should be posted in the work area. In addition, as part of an on-going education and training program, all information on the health and safety hazards of isoamyl acetate should be communicated to all exposed and potentially exposed workers.
- ☑ Never eat, drink, or smoke in areas were isoamyl acetate is used, handled, or stored.

RISK ASSESSMENT: ENVIRONMENT

General Assessment

The environment is at risk of exposure during transportation, storage, disposal, or destruction of isoamyl acetate. In almost every scenario, the threat of environmental exposure is contingent upon the proper handling of the chemical substance.

Accidental spills, large or small, can result in fire, explosion, and possible contamination of the surrounding environmental mediums (water, soil, and air). Isoamyl acetate is considered a Class IC flammable liquid (according to OSHA 29 CFR 1910.106). Its moderately low flash point presents a serious fire and explosion hazard concern. Also, because it is incompatible with a number of common materials, especially strong oxidizers and acids, contact can result in violent and explosive reactions. It can also form explosive mixtures in air and can ignite on contact with heat, fire, or sparks. These characteristics require special consideration during any emergency situation involving a leak or spill of isoamyl acetate.

The proper disposal/destruction method for isoamyl acetate is to burn it in a chemical incinerator equipped with an afterburner and air scrubber. Isoamyl acetate can enter the environment through unchecked industrial discharges into effluents and through spills.

Acute Ecological Effects

Acute (short-term) toxic effects may include the death of animals, birds, or fish, and death or low growth rate in plants. Acute effects are seen 2 to 4 days after animals or plants are exposed to isoamyl acetate.

Insufficient data are available to evaluate or predict the short-term effects of isoamyl acetate to aquatic life, plants, birds, or terrestrial animals.

Chronic Ecological Effects

Chronic toxic effects may include shortened life span, reproductive problems, lower fertility, and changes in appearance or behavior in exposed animals. These effects can be seen long after first exposure(s) to toxic chemicals.

Insufficient data are available to evaluate or predict the long-term effects of isoamyl acetate to aquatic life, plants, birds, or land animals.

Water Solubility

Isoamyl acetate is only slightly soluble in water. Concentrations of 1 to 100 milligrams may mix with a liter of water.

Persistence in the Environment

Isoamyl acetate is slightly persistent in water, with a half-life of between 2 to 20 days. The half-life of a pollutant is the amount of time it takes for one-half of the chemical to be degraded.

Bioaccumulation in Aquatic Organisms

Some substances increase in concentration, or *bioaccumulate,* in living organisms as they breathe contaminated air, drink contaminated water, or eat contaminated food. These chemicals can become concentrated in the tissues and internal organs of animals as well as humans.

The concentration of isoamyl acetate found in fish tissues is expected to be somewhat higher then the average concentration of isoamyl acetate in water from which the fish was taken.

Recommended Risk-Reduction Measures

Proper training of all transporters will reduce the likelihood of a mishap or accident resulting in a leak or spill to the environment. The correct use of DOT labeling (per the MSDS) on all containers, trucks, and rail cars will enable emergency responders to react properly and quickly to any disaster thereby reducing the potential risk to the environment and to personnel.

Storage of isoamyl acetate should be segregated from incompatible chemicals to minimize the risk of cross-contamination or contact. Buildings designated for storage should be equipped with appropriate fire protection systems (alarms, sprinklers, emergency lighting, portable extinguishers).

If a spill or leak to the environment has occurred, fire department, emergency response, and/or hazardous materials spill personnel should be notified immediately. Cleanup should be attempted only by those trained in proper spill containment procedures using non-sparking tools. Contaminated soils should be removed for incineration and replaced with clean soil. If isoamyl acetate should contact the water table, aquifer, or navigable waterway, time is of the essence. It is only slightly soluble in water and, therefore, total containment and remediation may be possible. When such spills occur, the local and/or state emergency response authorities must be notified. A comprehensive emergency response or disaster preparedness/recovery plan should be in place prior to any operations involving the use, transportation, storage, or disposal of isoamyl acetate. If isoamyl acetate is spilled or leaked, the following specific steps are recommended:

- ☑ Restrict persons not wearing protective clothing from area of spill or leak until cleanup is complete and area can be opened for normal work.
- ☑ Ventilate area and remove all ignition sources.
- ☑ Absorb liquids in vermiculite, dry sand, earth, or a similar material and deposit in sealed containers. Use non-sparking tools.
- ☑ It may be necessary to dispose of isoamyl acetate as a hazardous waste. The responsible state agency or the regional office of the federal Environmental Protection Agency (EPA) should be contacted for specific recommendations.

RISK ASSESSMENT: BUSINESS

General Assessment

Accidents or mishaps involving isoamyl acetate can present a moderate threat to business operations. The loss or damage of equipment or facilities can significantly affect fiscal viability. Lawsuits that may result from personnel injury/death, public exposures, and/or environmental contamination will also require a serious expenditure of resources.

Media attention surrounding an injury, death, or environmental damage can also result in a loss of profits and loss of current as well as future business.

Recommended Risk-Reduction Measures

Company attorneys, safety and health professionals, and environmental specialists should be involved in the development of any procedures or policies intended to manage the use of chemicals in the workplace. A company official should be pre-designated as a public relations officer with specific training in dealing with the press. Corporate plans and policies should be developed, approved, and implemented long before any need for such arises.

REFERENCES

American Conference of Governmental Industrial Hygienists. 1988. *Documentation of the Threshold Limit Values and Biological Exposure Indices,* 5th Edition (with updates). Cincinnati: ACGIH

Anthony, C. P., and N. J. Kolthoff. 1971. *Textbook of Anatomy and Physiology.* St. Louis: C. V. Mosby Company.

Bloom, A. D. and F. J. DeSerres. 1995. *Ecotoxicity and Human Health.* New York: CRC Press/Lewis Publishers.

Brusick, D. J. 1994. *Methods for Genetic Risk Assessment.* New York: CRC Press/Lewis Publishers.

Bueche, J. R. 1972. *Principles of Physics,* 2nd Edition. New York: McGraw-Hill.

Burrell, R., D. K. Flaherty, and L. J. Sauers. 1992. *Toxicology of the Immune System.* New York: Van Nostrand Reinhold.

Calabrese, E. J. 1994. *Biological Effects of Low Level Exposures.* New York: CRC Press/Lewis Publishers.

Cockerman, L. G., B. S. Shane. 1994. *Basic Environmental Toxicology.* New York: CRC Press/Lewis Publishers.

Cothern, C. R., N. P. Ross. 1994. *Environmental Statistics, Assessment, and Forecasting.* New York: CRC Press/Lewis Publishers.

Danse, I. R. 1991. *Common Sense Toxics in the Workplace.* New York: Van Nostrand Reinhold.

Environmental Protection Agency. 1983. *Chemical Hazard Information Profile: Isoamyl acetate.* Washington, D.C.: U.S. EPA.

Gardner, R. W. 1994. *Chemical Intolerance.* New York: CRC Press/Lewis Publishers.

Hallenbeck, W. H. 1993. *Quantitative Risk Assessment for Environmental and Occupational Health,* 2nd Edition. New York: CRC Press/Lewis Publishers.

Hodgson, E., R. B. Mailman, and J. E. Chambers. 1988. *Dictionary of Toxicology.* London: The MacMillian Company.

Hosty, J. W., and P. Foster. 1990. *A Practical Guide to Chemical Spill Response.* New York: Van Nostrand Reinhold.

Howard, P. H., et. al. 1991. *Handbook of Environmental Degradation Rates.* New York: CRC Press/Lewis Publishers.

James, A., and M. Lord. 1992. *Index of Chemical and Physical Data.* New York: Van Nostrand Reinhold.

Jones, F. E. 1994. *Toxic Organic Vapors in the Workplace.* New York: CRC Press/Lewis Publishers.

Kamrin, M. 1 A. 1988. *Toxicology: A Primer on Principles and Applications.* Chelsea, Michigan: Lewis Publishers.

Keith, L. H. and M. M. Walker. 1995. *Handbook of Air Toxics.* New York: CRC Press/Lewis Publishers.

Landis, W. G., M. H. Yu. 1995. *Introduction to Environmental Toxicology.* New York: CRC Press/Lewis Publishers.

Langerman, N. 1994 *Precautionary Labels for Chemical Containers.* New York: CRC Press/Lewis Publishers.

Lauwerys, R. R., P. Hoet. 1993. *Industrial Chemical Exposure.* New York: CRC Press/Lewis Publishers.

Lide, D. R. 1995. *Handbook of Chemistry and Physics,* 76th Edition. New York: CRC Press/Lewis Publishers.

Lewis, Richard J., Sr. 1992. *Sax's Dangerous Properties of Industrial Materials,* Volumes 1, 2, and 3, 8th Edition. New York: Van Nostrand Reinhold.

Lisella, Frank S. 1994. *The VNR Dictionary of Environmental Health and Safety.* New York: Van Nostrand Reinhold.

Manahan, S. E. 1994. *Environmental Chemistry,* 6th Edition. New York: CRC Press/Lewis Publishers.

Manahan, S. E. 1992. *Toxicological Chemistry,* 2nd Edition. New York: CRC Press/Lewis Publishers.

Meyer, Eugene. 1990. *Chemistry of Hazardous Materials.* Englewood Cliffs, NJ: Prentice Hall.

Patnaik, P. 1992. *A Comprehensive Guide to the Hazardous Properties of Chemical Substances.* New York: Van Nostrand Reinhold.

Proctor, N. H., and J. P. Hughes. 1991. *Chemical Hazards of the Workplace.* New York: Van Nostrand Reinhold.

Sacarello, H. L. A. 1994. *The Comprehensive Handbook of Hazardous Materials.* New York: CRC Press/Lewis Publishers.

Torkelson, T. R. V. K. Rowe. 1981. *Patty's Industrial Hygiene and Toxicology,* Vol. 2B. New York: Wiley-Interscience.

U.S. Department of Health and Human Services, National Institute for Occupational Safety and Health. 1994. *NIOSH Pocket Guide to Chemical Hazards.* Washington, D.C.: U.S. Government Printing Office.

MATERIAL SAFETY DATA SHEET

CHEMICAL NAME

ISOAMYL ALCOHOL (primary)

CRC PRESS/LEWIS PUBLISHERS DISCLAIMER: THE INFORMATION AND RECOMMENDATIONS PRESENTED HEREIN ARE BASED ON SOURCES BELIEVED TO BE RELIABLE. CRC MAKES NO REPRESENTATION ON ITS COMPLETENESS OR ACCURACY. IT IS THE RESPONSIBILITY OF THE USER TO DETERMINE THE CHEMICAL'S SUITABILITY FOR ITS INTENDED USE, THE CHEMICAL'S SAFE USE, AND THE CHEMICAL'S PROPER DISPOSAL. NO REPRESENTATIONS AND/OR WARRANTIES, EITHER EXPRESSED OR IMPLIED, OF THE MERCHANTABILITY OR FITNESS FOR A PARTICULAR PURPOSE, OR OF ANY OTHER NATURE, ARE MADE WITH RESPECT TO THE INFORMATION PROVIDED IN THIS MSDS OR TO THE CHEMICAL TO WHICH INFORMATION MAY REFER. CRC NEITHER ASSUMES NOR AUTHORIZES ANY OTHER PERSON TO ASSUME FOR IT, ANY OTHER ADDITIONAL RESPONSIBILITY OR LIABILITY FOR THE USE OF, OR RELIANCE UPON, THIS INFORMATION.

HAZARD WARNING INFORMATION

HEALTH	FIRE	REACTIVE	OTHER	DEGREE OF HAZARD	COLOR CODING	OTHER CODES
1	2	0		0 = Minimum Hazard 1 = Slight Hazard 2 = Moderate Hazard 3 = Serious Hazard 4 = Severe Hazard	HEALTH = BLUE FIRE = RED REACTIVITY = YELLOW OTHER = WHITE	OX = Oxidizer ACID = Acid ALK = Alkali COR = Corrosive W = Use No Water

SECTION I - GENERAL INFORMATION

Characterization	**Alcohol**	RCRA Number / EPA Class	**None** / **Not Applicable**
DOT Proper Shipping Name	**Isoamyl Alcohol**	Chemical Abstract Service (CAS) Number	**123-51-3**
DOT Hazard Class and Label Requirements	**Flammable or Combustible Liquid**	DOT Emergency Guide Code	**26**
DOT Identification Number	**UN 1105**	Chemical Formula	**$(CH_3)_2CHCH_2CH_2OH$**

Synonyms

Fermentation amyl alcohol; fusel oil; isobutyl carbinol; isopentyl alcohol; 3-methyl-1-butanol.

SECTION II - HAZARDOUS INGREDIENTS AND IDENTITY INFORMATION

Hazardous Components (specific identity)	OSHA Exposure Criteria	NIOSH Exposure Criteria	Immediately Dangerous to Life and Health (IDLH)	ACGIH Exposure Criteria
Isoamyl alcohol (derivation: Distillation of fusel oil or the mixed alcohols resulting from the chlorination and hydrolysis of pentane). **1 ppm = 3.67 mg/m^3**	**PEL:** 100 ppm 360 mg/m^3 **STEL:** Not Established	**REL:** 100 ppm 360 mg/m^3 **STEL:** 125 ppm 450 mg/m^3	**500 ppm**	**TLV:** 100 ppm 360 mg/m^3 **STEL:** 125 ppm 450 mg/m^3

SECTION III - PHYSICAL AND CHEMICAL CHARACTERISTICS

Boiling Point	**270°F (132°C)**	Specific Gravity (H_2O = 1)	**0.81**
Vapor Pressure (mm Hg)	**28 at 68°F (20°C)**	Molecular Weight	**88.2**
Vapor Density (Air = 1)	**3.04**	Melting Point	**-179°F (-117°C)**

Solubility

Slightly soluble in water. Miscible with alcohol and ether.

Appearance and Odor

Colorless liquid with a disagreeable odor with an acrid, repulsive taste.

SECTION IV - FIRE AND EXPLOSION HAZARD DATA

Flash Point (method used)	**109°F (43°C) closed cup**	Explosive Limits in Air % by Volume	**LEL: 1.2% UEL: 9.0% at 212°F (100°C)**
NFPA Classification	**Class II Combustible Liquid**	Autoignition Temperature	**657°F (347°C)**

Extinguishing Media

Dry chemical, carbon dioxide, or alcohol foam, or water spray.

Special Fire Fighting Procedures

Poisonous gases are produced in fire. Wear full protective clothing and self-contained breathing apparatus (SCBA). Continue to cool containers with water after fire is extinguished. Move containers from fire area if it can be done without risk. For large fires, use unmanned hose apparatus, if possible. Evacuate non-essential personnel 1500 feet radius if fire. Consider down-wind conditions.

Unusual Fire and Explosion Hazards

Vapors are heavier than air and may travel a distances to ignition source to flashback. Vapors may explode if ignited in an enclosed area.

SECTION V - REACTIVITY DATA

<table>
<tr><td colspan="2" rowspan="1">Stability</td><td>Conditions to Avoid
Isoamyl alcohol is stable under normal conditions of storage and operation. Avoid contact with incompatible materials. Keep away from heat, flame, and other ignition sources.</td></tr>
<tr><td>Stable
X</td><td>Unstable</td><td>Incompatibility (materials to avoid)
Reacts violently in contact with oxidizers and many reducing materials. It will react explosively in contact with hydrogen trisulfide.</td></tr>
<tr><td colspan="2">Hazardous Polymerization</td><td>Conditions to Avoid
Under normal conditions of temperature and pressure, isoamyl alcohol is not expected to undergo hazardous polymerization.</td></tr>
<tr><td>May Occur</td><td>Will Not Occur
X</td><td>Hazardous Decomposition or By-products
When heated to decomposition, isoamyl alcohol emits acrid and irritating fumes.</td></tr>
</table>

SECTION VI - HEALTH HAZARD DATA

Primary Route(s) of Entry:	Inhalation?	Absorption (skin)?	Ingestion?
	X		X

Health Hazards

INHALATION: There may be slight irritation of the eyes, nose, and mucosa of the respiratory tract. Exposure can also cause stupor, fatigue, and sleepiness. Other effects include nausea, vomiting, flushing of skin, headaches, double vision, vertigo, weakness, sleepiness, olfactory changes, and death.

SKIN & EYES: As a solvent, causes dry skin, cracking, and possible dermatitis. Eye contact can result in burning, stinging, tearing, corneal injury, and hyperemia of the conjunctiva.

INGESTION: Sleep disorders, hallucinations, distorted perceptions, convulsions, ataxia and motor activity changes, coma, headache, and death.

Carcinogenicity	NTP Listed?	IARC Cancer Review Group?	OSHA Regulated?	Target Organs?
Unknown Human Questioned Animal	No	No	29 CFR 1910.1000 Table Z-1	Skin, eyes, respiratory. system, CNS.

Medical Conditions Generally Aggravated by Exposure

Chronic liver disease.

Emergency and First-aid Procedures

Eye contact: Flush immediately with water for 15 minutes (minimum); seek medical attention. Skin contact: Remove all contaminated clothing. Immediately wash area with large amounts of soap and water. Seek medical assistance. For inhalation: Remove the person from exposure. Provide respiratory assistance and CPR. Transfer to medical facility. If swallowed: Contact poison control center. Give 1-2 glasses of water and induce vomiting. Seek medical attention immediately.

SECTION VII - PRECAUTIONS FOR SAFE HANDLING AND USE

Steps to be Taken in Case Material is Released or Spilled

Contain spills using absorbent material (vermiculite or other). Ventilate area of spill. Have extinguishing agent available in case of fire. Store materials in DOT-approved containers.

Preferred Waste Disposal Method

Incineration in chemical incinerator equipped with scrubber and afterburner.

Precautions to be Taken in Handling and Storage

Store in tightly closed containers in a cool, dark, well-ventilated area. Protect containers from physical damage. Keep heat, fire, and flame away. Outside building or underground storage is recommended.

Other Precautions and Warnings

Ground and bond all metal containers. Avoid contact with skin and eyes.

SECTION VIII - CONTROL MEASURES AND PERSONAL PROTECTIVE EQUIPMENT

Respiratory Protection *(specify type)*

For vapor exposures, use an organic vapor air-purifying respirator. Greater protection is obtained using a supplied-air respirator or a self-contained breathing apparatus (SCBA) with full facepiece operated in pressure demand or other positive pressure mode.

Ventilation

Local exhaust or general mechanical systems recommended.

Protective Gloves	Eye Protection	Other Protective Clothing
Impervious Gloves	Chemical Goggles or Face Mask	Impermeable Apron or Protective Suit

Work/Hygiene Practices

Always wash hands thoroughly after using chemical; never bring food, drink, or smoking materials into vicinity of chemicals.

ISOAMYL ALCOHOL

$(CH_3)_2CHCH_2CH_2OH$ CAS: 123-51-3

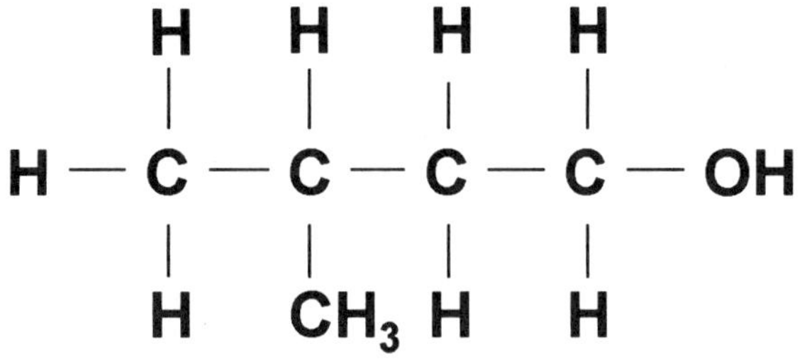

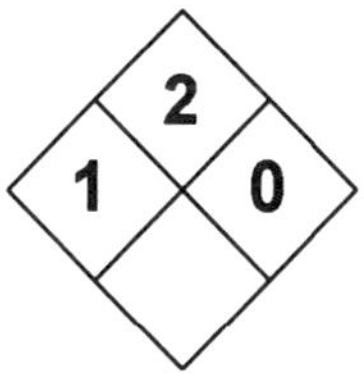

IDENTIFICATION AND TYPICAL USES

Isoamyl alcohol is a colorless or clear liquid with a disagreeable odor and pungent taste. It is used in the manufacture of isovaleric acid, isoamyl compounds, smokeless powders, mercury fulminate, pyroxin, lacquers, artificial silk, paints, varnishes, paint removers, shoe cements, perfumes, pharmaceuticals, rubber, plastics, fruit essences, and hydraulic fluid. It is also used for dehydrating celloidin solutions, in microscopy, for determining fat in milk, in photographic chemicals and products, and as a solvent for alkaloids, resins, and fats.

RISK ASSESSMENT: HEALTH

General Assessment

Isoamyl alcohol is primarily toxic to humans by *inhalation* and *ingestion*. Skin contact can cause adverse surface effects but dermal absorption has not been reported. Isoamyl alcohol is not currently listed as a carcinogen by OSHA, the NTP, or the IARC. However, there are reports in the references that this compound is under review and that its listing as a carcinogenic is imminent. Mutation data have also been reported.

Inhalation can cause irritation of the eyes, nose, throat, upper respiratory tract, and associated mucosa. There may also be depression of the central nervous system (CNS) with narcosis and associated symptoms, including headache, nervousness, flushing of the skin, double vision, vertigo, tremors, dizziness, fatigue and general weakness, nausea, somnolence, and narcosis with stupor, loss of consciousness, and death.

Liquid contact with the eyes causes immediate burning and stinging with lachrymation and reflex closure of the lids. There may be injury to the corneal epithelium and possible conjunctivitis, depending upon the duration of exposure and concentration strength. Skin contact results in drying and cracking which can lead to secondary infections and dermatitis.

Ingestion of isoamyl alcohol can lead to sleep disorders, hallucinations, distorted perceptions, ataxia, motor function changes, convulsions and tremors, coma, headaches, alteration of gastric secretions, glandular changes, nausea or vomiting, loss of body temperature, unconsciousness, and death.

Acute Health Effects

The following acute (short-term) health effects may occur immediately or shortly after exposure to isoamyl alcohol:

Skin: Causes dryness and cracking leading to dermatitis and possible infection.

Eye: Severe irritation with burning and stinging with possible damage to the cornea and conjunctiva.

Lung: Irritation of the eyes, nose, throat, and respiratory tract.

CNS: High concentrations can cause depression of the CNS with symptoms of sleepiness, vertigo, loss of consciousness, and death.

Chronic Health Effects

The following chronic (long-term) health effects may occur at some time after exposure to isoamyl alcohol and can last for months or even years:

Cancer Hazards: Isoamyl alcohol has been shown to cause cancer in experimental animals. It is not known whether it will have this effect in humans but its carcinogenicity is considered highly suspect. Human mutation data have been reported. Many scientists believe that such chemicals pose a cancer risk in the long-term.

Reproduction: According to the references, isoamyl alcohol has not been adequately tested for its ability to adversely affect reproduction in test animals.

Other Chronic Effects: Very high or prolonged exposure may result in mucous membrane irritation, headache, and depression of the CNS with symptoms of somnolence and lack of concentration. Prolonged skin contact can cause dermatitis.

Recommended Risk-Reduction Measures

Personnel should avoid direct contact with isoamyl alcohol. The exact nature of its toxicity is not clearly understood in the references and its carcinogenic ability is considered suspect. The best risk reduction measure is to use a less toxic chemical as a substitute for a isoamyl alcohol. However, if substitution is not possible or operationally feasible, then *engineering controls* are the most effective methods of reducing exposures. An alternative method of protection is to enclose operations and/or provide local exhaust ventilation at the site of chemical release. While not always operationally feasible, isolating operations can also reduce exposure risk.

Using respiratory protection is less effective than the controls mentioned above, but is still advisable whenever working with or around isoamyl alcohol. For concentrations over the PEL (100 ppm), an air-purifying respirator with an organic vapor cartridge will suffice. For higher exposures, a supplied-air respirator with full facepiece operated in positive pressure mode, or a self-contained breathing apparatus (SCBA) with full facepiece and operated in pressure demand mode are the recommended respiratory protection methods of choice. If a full facepiece is not available, then chemical goggles should be worn to protect the eyes. Whenever a chemical splash hazard exists, a face shield and a protective apron should be worn. To prevent hand and skin exposures, impervious gloves should be used.

Administrative controls should also be in place to minimize the potential for human exposures. These may include written procedures or policies which specify the methods and techniques that will be practiced whenever personnel are to work with isoamyl alcohol.

All personnel should receive *training* on the use, hazards, protective measures, emergency actions, and other precautions per 29 CFR 1910.1200 (Hazard Communication), prior to the first assignment in an area where isoamyl alcohol is used or stored.

If symptoms develop or overexposure is suspected, the following medical tests are recommended:

- ☑ Evaluation of nervous system function with careful consideration of exposure history and special testing by a qualified neurologist.
- ☑ Skin testing with dilute isoamyl alcohol to help diagnose allergy (performed by a qualified allergist).

Any medical evaluation should include a careful history of past and present symptoms with an examination. Medical tests that look for existing damage are not a substitute for controlling exposures.

Other methods to reduce exposure include:

- ☑ Where possible, enclose operations and use local exhaust ventilation at the site of chemical release. If local exhaust ventilation or enclosure is not used, respiratory protection should be mandatory.
- ☑ Always ensure that proper protective clothing is worn when using chemical substances and that personnel are trained in its use, care, and maintenance.
- ☑ Wash thoroughly immediately after exposure to isoamyl alcohol and at the end of the work shift or before eating, drinking, or smoking.
- ☑ Hazard warning information should be posted in the work area. In addition, as part of an on-going education and training program, all information on the health and safety hazards of isoamyl alcohol should be communicated to all exposed or potentially exposed workers.

RISK ASSESSMENT: ENVIRONMENT

General Assessment

The environment is at risk of exposure during transportation, storage, disposal, or destruction of isoamyl alcohol. In almost every scenario, the threat of environmental exposure is contingent upon the proper handling of the chemical substance. Accidental spills, large or small, can result in fire, explosion, and possible contamination of the surrounding environmental mediums (water, soil, and air).

Isoamyl alcohol is considered a Class II combustible liquid (according to OSHA 29 CFR 1910.106). Its relatively low flash point presents a moderate fire and explosion hazard concern. Also, because it is incompatible with a number of common materials, especially strong oxidizers and many reducing materials, contact can result in violent and explosive reactions. It can also react explosively with hydrogen trisulfide and can ignite on contact with heat, fire, or sparks. These characteristics require special consideration during any emergency situation involving a leak or spill of isoamyl alcohol.

Isoamyl alcohol can enter the environment through unchecked industrial discharges into effluents and through spills.

Acute Ecological Effects

Acute (short-term) toxic effects may include the death of animals, birds, or fish, and death or low growth rate in plants. Acute effects are seen 2 to 4 days after animals or plants are exposed to isoamyl alcohol.

Insufficient data are available to evaluate or predict the short-term effects of isoamyl alcohol to aquatic life, plants, birds, or terrestrial animals.

Chronic Ecological Effects

Chronic toxic effects may include shortened life span, reproductive problems, lower fertility, and changes in appearance or behavior in exposed animals. These effects can be seen long after first exposure(s) to toxic chemicals.

Insufficient data are available to evaluate or predict the long-term effects of isoamyl alcohol to aquatic life, plants, birds, or land animals.

Water Solubility

Isoamyl alcohol is slightly soluble in water. Concentrations of 1 to 100 milligrams can be expected to mix with a liter of water.

Persistence in the Environment

Isoamyl alcohol is slightly persistent in water, with a half-life of between 2 to 20 days. The half-life of a pollutant is the amount of time it takes for one-half of the chemical to be degraded. About 90% of isoamyl alcohol will eventually end up in the air; the remainder will end up in water.

Bioaccumulation in Aquatic Organisms

Some substances increase in concentration, or *bioaccumulate,* in living organisms as they breathe contaminated air, drink contaminated water, or eat contaminated food. These chemicals can become concentrated in the tissues and internal organs of animals as well as humans.

The concentration of isoamyl alcohol found in fish tissues is expected to be about the same as the average concentration of isoamyl alcohol in water from which the fish was taken.

Recommended Risk-Reduction Measures

Proper training of all transporters will reduce the likelihood of a mishap or accident resulting in a leak or spill to the environment. The correct use of DOT labeling (per the MSDS) on all containers, trucks, and rail cars will enable emergency responders to react properly and quickly to any disaster thereby reducing the potential risk to the environment and to personnel.

Storage of isoamyl alcohol should be segregated from incompatible chemicals to minimize the risk of cross-contamination or contact. Buildings designated for storage should be equipped with appropriate fire protection systems (alarms, sprinklers, emergency lighting, portable extinguishers).

If a spill or leak to the environment has occurred, fire department, emergency response, and/or hazardous materials spill personnel should be notified immediately. Cleanup should be attempted only by those trained in proper spill containment procedures using non-sparking tools. Contaminated soils should be removed for incineration and replaced with clean soil. If isoamyl alcohol should contact the water table, aquifer, or navigable waterway, time is of the essence. It is only slightly soluble in water and, therefore, total containment and remediation may be possible. When such spills occur, the local and/or state emergency response authorities must be notified. A comprehensive emergency response or disaster preparedness/recovery plan should be in place prior to any operations involving the use, transportation, storage, or disposal of isoamyl alcohol.

If isoamyl alcohol is spilled or leaked, the following specific steps are recommended to ensure the protection of personnel and the environment:

- ☑ Restrict persons not wearing protective clothing from area of spill or leak until cleanup is complete and area can be opened for normal work.
- ☑ Ventilate area and remove ignition sources.
- ☑ Absorb liquids in vermiculite, dry sand, earth, or a similar material and deposit in sealed containers for disposal. Use non-sparking tools.

☑ It may be necessary to dispose of isoamyl alcohol as a hazardous waste. The responsible state agency or the regional office of the federal Environmental Protection Agency (EPA) should be contacted for specific recommendations.

RISK ASSESSMENT: BUSINESS

General Assessment

Accidents or mishaps involving isoamyl alcohol can present a moderate threat to business operations. The loss or damage of equipment or facilities can significantly affect fiscal viability. Lawsuits that may result from personnel injury/death, public exposures, and/or environmental contamination will also require a serious expenditure of resources. Media attention surrounding an injury, death, or environmental damage can also result in a loss of profits and loss of current as well as future business. Always remember that anytime the terms "cancer," or "carcinogen" are used, public emotion, hysteria, and ignorance can run equally high. This must be properly considered during the development of any public relations policies.

Recommended Risk-Reduction Measures

Company attorneys, safety and health professionals, and environmental specialists should be involved in the development of any procedures or policies intended to manage the use of chemicals in the workplace. A company official should be pre-designated as a public relations officer with specific training in dealing with the press. Corporate plans and policies should be developed, approved, and implemented long before any need for such arises.

REFERENCES

American Conference of Governmental Industrial Hygienists. 1988. *Documentation of the Threshold Limit Values and Biological Exposure Indices,* 5th Edition (with updates). Cincinnati: ACGIH

Anthony, C. P., and N. J. Kolthoff. 1971. *Textbook of Anatomy and Physiology.* St. Louis: C. V. Mosby Company.

Bloom, A. D. and F. J. DeSerres. 1995. *Ecotoxicity and Human Health.* New York: CRC Press/Lewis Publishers.

Brusick, D. J. 1994. *Methods for Genetic Risk Assessment.* New York: CRC Press/Lewis Publishers.

Bueche, J. R. 1972. *Principles of Physics,* 2nd Edition. New York: McGraw-Hill.

Burrell, R., D. K. Flaherty, and L. J. Sauers. 1992. *Toxicology of the Immune System.* New York: Van Nostrand Reinhold.

Calabrese, E. J. 1994. *Biological Effects of Low Level Exposures.* New York: CRC Press/Lewis Publishers.

Cothern, C. R., N. P. Ross. 1994. *Environmental Statistics, Assessment, and Forecasting.* New York: CRC Press/Lewis Publishers.

Danse, I. R. 1991. *Common Sense Toxics in the Workplace.* New York: Van Nostrand Reinhold.

Environmental Protection Agency. 1983. *Chemical Hazard Information Profiles.* Washington, D.C.: U.S. EPA.

Fodden, C. R., and J. L. Weddell. 1992. *Hazardous Materials Emergency Action Data.* New York: CRC Press/Lewis Publishers.

Forseberg, K., and L. H. Keith. 1993. *Chemical; Protective Clothing Permeation and Degradation Database.* Chelsea, MI: Lewis Publishers.

Gardner, R. W. 1994. *Chemical Intolerance.* New York: CRC Press/Lewis Publishers.

Hodgson, E., R. B. Mailman, and J. E. Chambers. 1988. *Dictionary of Toxicology.* London: The MacMillian Company.

Hosty, J. W., and P. Foster. 1990. *A Practical Guide to Chemical Spill Response.* New York: Van Nostrand Reinhold.

Howard, P. H., et. al. 1991. *Handbook of Environmental Degradation Rates.* New York: CRC Press/Lewis Publishers.

International Agency for Research on Cancer (IARC). 1987. *IARC Monographs on the Evaluation of Carcinogenic Risk to Humans, Overall Evaluations of Carcinogenicity: An updating of IARC Monographs Volumes 1-42.* Lyon: IARC.

Keith, L. H. and D. B. Walters. (Editors) 1992. *National Toxicology Program's Chemical Database, Volumes I-VII.* Chelsea, MI: Lewis Publishers.

Landis, W. G., M. H. Yu. 1995. *Introduction to Environmental Toxicology.* New York: CRC Press/Lewis Publishers.

Lauwerys, R. R., P. Hoet. 1993. *Industrial Chemical Exposure.* New York: CRC Press/Lewis Publishers.

Lewis, Richard J., Sr. 1992. *Sax's Dangerous Properties of Industrial Materials,* Volumes 1, 2, and 3, 8th Edition. New York: Van Nostrand Reinhold.

Manahan, S. E. 1992. *Toxicological Chemistry,* 2nd Edition. New York: CRC Press/Lewis Publishers.

Meyer, Eugene. 1990. *Chemistry of Hazardous Materials.* Englewood Cliffs, NJ: Prentice Hall.

Patnaik, P. 1992. *A Comprehensive Guide to the Hazardous Properties of Chemical Substances.* New York: Van Nostrand Reinhold.

Philp, R. B. 1995. *Environmental Hazards and Human Health.* New York: CRC Press/Lewis Publishers.

Torkelson, T. R. V. K. Rowe. 1981. *Patty's Industrial Hygiene and Toxicology,* Vol. 2B. New York: Wiley-Interscience.

U.S. Department of Health and Human Services, National Institute for Occupational Safety and Health. 1994. *NIOSH Pocket Guide to Chemical Hazards.* Washington, D.C.: U.S. Government Printing Office.

MATERIAL SAFETY DATA SHEET

CHEMICAL NAME	CRC PRESS/LEWIS PUBLISHERS DISCLAIMER
ISOBUTANE	**CRC PRESS/LEWIS PUBLISHERS DISCLAIMER:** THE INFORMATION AND RECOMMENDATIONS PRESENTED HEREIN ARE BASED ON SOURCES BELIEVED TO BE RELIABLE. CRC MAKES NO REPRESENTATION ON ITS COMPLETENESS OR ACCURACY. IT IS THE RESPONSIBILITY OF THE USER TO DETERMINE THE CHEMICAL'S SUITABILITY FOR ITS INTENDED USE, THE CHEMICAL'S SAFE USE, AND THE CHEMICAL'S PROPER DISPOSAL. NO REPRESENTATIONS AND/OR WARRANTIES, EITHER EXPRESSED OR IMPLIED, OF THE MERCHANTABILITY OR FITNESS FOR A PARTICULAR PURPOSE, OR OF ANY OTHER NATURE, ARE MADE WITH RESPECT TO THE INFORMATION PROVIDED IN THIS MSDS OR TO THE CHEMICAL TO WHICH INFORMATION MAY REFER. CRC NEITHER ASSUMES NOR AUTHORIZES ANY OTHER PERSON TO ASSUME FOR IT, ANY OTHER ADDITIONAL RESPONSIBILITY OR LIABILITY FOR THE USE OF, OR RELIANCE UPON, THIS INFORMATION.

HAZARD WARNING INFORMATION

HEALTH	FIRE	REACTIVE	OTHER	DEGREE OF HAZARD	COLOR CODING	OTHER CODES
1	4	0		0 = Minimum Hazard 1 = Slight Hazard 2 = Moderate Hazard 3 = Serious Hazard 4 = Severe Hazard	HEALTH = BLUE FIRE = RED REACTIVITY = YELLOW OTHER = WHITE	OX = Oxidizer ACID = Acid ALK = Alkali COR = Corrosive W = Use No Water

SECTION I - GENERAL INFORMATION

Characterization	Gas (Hydrocarbon)	RCRA Number	D001
		EPA Class	Characteristic (I) Waste
DOT Proper Shipping Name	Liquefied Petroleum Gas	Chemical Abstract Service (CAS) Number	75-28-5
DOT Hazard Class and Label Requirements	Flammable Gas	DOT Emergency Guide Code	22
DOT Identification Number	UN 1075	Chemical Formula	$CH_3CH(CH_3)_2$

Synonyms
2-Methylpropane; 1,1-dimethylethane; liquefied petroleum gas; trimethylmethane.

SECTION II - HAZARDOUS INGREDIENTS AND IDENTITY INFORMATION

Hazardous Components (specific identity)	OSHA Exposure Criteria	NIOSH Exposure Criteria	Immediately Dangerous to Life and Health (IDLH)	ACGIH Exposure Criteria
Isobutane (derivation: An important component of natural gasoline, refinery gases, wet natural gas, also obtained by isomerization of butane). 1 ppm = 2.42 mg/m^3	PEL: Not Established STEL: Not Established	REL: 800 ppm 1900 mg/m^3 STEL: Not Established	Not Determined	TLV: Not Established STEL: Not Established

SECTION III - PHYSICAL AND CHEMICAL CHARACTERISTICS

Boiling Point	11°F (-12°C)	Specific Gravity (H_2O = 1)	2.06
Vapor Pressure (atmospheres)	3.1 at 70°F (21°C)	Molecular Weight	58.14
Vapor Density (Air = 1)	2.01	Freezing Point	-255°F (-160°C)

Solubility
Slightly soluble in water. Soluble in alcohol and ether, very soluble in organic solvents.

Appearance and Odor
Colorless gas with a gasoline-like or natural gas odor. Shipped as a liquefied compressed gas.

SECTION IV - FIRE AND EXPLOSION HAZARD DATA

Flash Point (method used)	-117°F (-83°C) closed cup	Explosive Limits in Air % by Volume	LEL: 1.6% UEL: 8.4%
NFPA Classification	Flammable Gas; Class IA Flammable Liquid	Autoignition Temperature	860°F (460°C)

Extinguishing Media
Stop flow of gas first (if possible). Use carbon dioxide, dry chemical, foam, or water spray on fire.

Special Fire Fighting Procedures
Isobutane is an extremely flammable, very dangerous fire and explosion hazard. Poisonous gases are evolved from fire. Wear a self-contained breathing apparatus (SCBA) with full facepiece. Ensure ALL sources of escaping gas are controlled. Monitor area long after fire is extinguished. Move containers from fire area if it can be done without risk. Use water spray to knock down vapors.

Unusual Fire and Explosion Hazards
Gas is explosive. Containers may explode in fire. Even if fire is extinguished, escaping gas can form explosive mixtures in air quickly and ignite without warning. Evacuate 2500 feet from fire area.

SECTION V - REACTIVITY DATA

Stability		Conditions to Avoid **Isobutane is normally stable in its closed containers, during routine operations (those that do not expose it to ignition sources). Avoid contact with heat, sparks, flame, lighted tobacco products, and steam lines.**
Stable X	Unstable	Incompatibility *(materials to avoid)* **Isobutane is reactive in contact with strong oxidizers (nitrates and perchlorates), chlorine or chlorine dioxide, fluorine, other oxidizers, nickel carbonyl + oxygen, heat, sparks, and open flame.**
Hazardous Polymerization		Conditions to Avoid **Under normal conditions of temperature and pressure, hazardous polymerization of isobutane is not expected to occur.**
May Occur	Will Not Occur X	Hazardous Decomposition or By-products **If involved in fire, isobutane can produce acrid smoke and irritating fumes.**

SECTION VI - HEALTH HAZARD DATA

Primary Route(s) of Entry:	Inhalation?	Absorption (skin)?	Ingestion?
	X		

Health Hazards

INHALATION: An asphyxiant gas. Initial symptoms are those of oxygen starvation. Feeling of "air hunger," diminished mental alertness, headache, dullness, shortness of breath, and an impairment in muscle coordination. Continued exposures lead to faulty judgment, depression of senses, rapid fatigue, unconsciousness, convulsions, coma, and death.

SKIN & EYES: Contact with the compressed gas or refrigerated liquid can cause frostbite and burns.

INGESTION: Not likely since isobutane is a gas at normal working (room) temperatures.

Carcinogenicity	NTP Listed?	IARC Cancer Review Group?	OSHA Regulated?	Target Organs?
Insufficient data	No	No	No	Central nervous system

Medical Conditions Generally Aggravated by Exposure

None Reported.

Emergency and First-aid Procedures

Eye contact: Immediately flush with large amounts of warm (tepid) water for 15 minutes (minimum), occasionally lifting eyelids; seek medical attention. Skin contact: Remove all contaminated clothing. Immediately wash area with large amounts of tepid water. If blisters occur, seek medical attention. For inhalation: Remove the person from exposure. Provide respiratory assistance and CPR. Transfer promptly to medical facility. If swallowed: Not a likely exposure route.

SECTION VII - PRECAUTIONS FOR SAFE HANDLING AND USE

Steps to be Taken in Case Material is Released or Spilled

Remove all ignition sources. Ventilate area. Stop flow of gas. If the source of leak is a cylinder and cannot be stopped, remove to safe place in open air; repair leak or let cylinder empty. If gas has not yet ignited, use a water spray to knock down vapors and direct them from ignition source.

Preferred Waste Disposal Method

None specified in the references (incineration of a gas may be the only alternative).

Precautions to be Taken in Handling and Storage

Do not store with incompatible materials or in the presence of heat since violent reactions or explosions can occur. Store in tightly closed pressurized containers in a cool, dark, well-ventilated area away from sunlight. Keep away from ignition sources such as fire, sparks, and flame.

Other Precautions and Warnings

Bulk storage of isobutane is not recommended. Containers may explode in fire or under conditions of extreme heat (do not store outdoors or in direct sunlight).

SECTION VIII - CONTROL MEASURES AND PERSONAL PROTECTIVE EQUIPMENT

Respiratory Protection *(specify type)*

Isobutane is a simple asphyxiant. Do NOT use cartridge-type respirators (ineffective for this type of gas). Use a powered supplied-air respirator or a self-contained breathing apparatus with full facepiece operated in pressure demand or other positive pressure mode.

Ventilation

Local exhaust or general mechanical systems recommended.

Protective Gloves	Eye Protection	Other Protective Clothing
Impervious, Thermal Protective	Goggles and/or Face Mask	Leather Apron

Work/Hygiene Practices

Always wash hands thoroughly after using chemical; never bring food, drink, or smoking materials into vicinity of chemicals.

ISOBUTANE

$CH_3CH(CH_3)_2$ CAS: 75-28-5

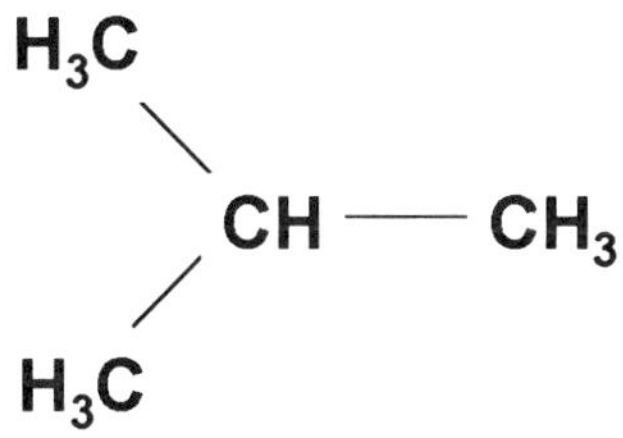

IDENTIFICATION AND TYPICAL USES

Isobutane is a colorless gas with a gasoline-like or natural gas odor. It is normally shipped as a liquefied compressed gas. It is a liquid below 11°F (-12°C). It is used in organic synthesis, as a refrigerant, in motor fuels, as a propellant for aerosols, in the manufacture of synthetic rubber, and as a calibration fluid.

RISK ASSESSMENT: HEALTH

General Assessment

Isobutane is a simple asphyxiant gas. As such, its primary route of entry into the body is through *inhalation.* It can displace oxygen levels to below that which is necessary to sustain life. At high concentrations, it is a narcotic and can cause anesthesia and unconsciousness. Skin contact with the liquid can be destructive to tissues due to the extremely cold temperatures but absorption is not a likely exposure route. Since isobutane usually exists as a gas at room temperature, ingestion is not likely to occur either.

Inhalation of isobutane gas can cause headache, dullness, shortness of breath, and the symptoms of asphyxiation. These include rapid respiration and a feeling of "air hunger" or air starvation followed by a decrease in mental alertness, and impaired muscular coordination. If exposure continues, there can be narcosis, anesthesia, faulty judgment, depression of the senses, emotional instability, nausea, vomiting, prostration, unconsciousness, convulsions, coma, and death.

Skin or eye contact with the vapor poses no serious threat. The vapors have not been shown to cause any physiological response. On exposure to air, the liquid evaporates rapidly and may not have time to cause damage. However, if the liquid should contact either the skin or the eyes, there can be serious irritation, burns, tissue damage, and frostbite.

Acute Health Effects

The following acute (short-term) health effects may occur immediately or shortly after exposure to isobutane:

Skin: Liquid threat only. Redness, pain, swelling, blisters, and severe frostbite.

Eye: Liquid threat only. Causes frostbite, redness, irritation, severe pain, and blurred vision.

Lung: Vapor is non-irritating and therefore poses no threat of damage to the lungs. However, in high concentrations, isobutane can lead to symptoms of asphyxiation and oxygen depravation up to and including irregular heartbeat, loss of consciousness, and death.

Chronic Health Effects

The following chronic (long-term) health effects may occur at some time after exposure to isobutane and can last for months or even years:

Cancer Hazards: There is insufficient evidence in the references to evaluate isobutane for its ability to cause cancer in animals. It is a simple asphyxiant.

Reproductive Hazards: There are no reports to support any claims of reproductive hazards.

Other Chronic Effects: Isobutane may be a cardiac sensitizer. More research is required in this regard.

Recommended Risk-Reduction Measures

Personnel should avoid direct contact with isobutane. If another gas cannot be substituted for isobutane, then *engineering controls* are the most effective method of reducing exposures. The best protection is to separate operations and provide local exhaust ventilation at the site of chemical release. While not always operationally feasible, isolating operations can also reduce exposure. Using equipment with explosion-proof designs and ensuring all metal containers and tools are electrically grounded and bonded can further reduce the potential for injury or damage. Using respiratory protection is less effective than the controls mentioned above, but is still advisable whenever working with isobutane. Air-purifying respirators (e.g., the cartridge-type or gas masks) cannot be used with isobutane. These types of respirators only purify the air and can do nothing to protect in oxygen deficient atmospheres. A powered air-supplied respirator operated in pressure demand or other positive pressure mode, or a self-contained breathing apparatus (SCBA) with full facepiece operated in pressure demand mode are the recommended respiratory protection methods of choice. If a full facepiece is not available, then chemical goggles should be worn to protect the eyes. Whenever a chemical splash hazard exists, a face shield and a rubber or leather apron should be worn. To prevent hand and skin exposures with the liquid, thermal gloves should be worn. Glove manufacturers should be contacted and permeation studies obtained *before* final glove selections have been made.

Administrative controls should also be in place to minimize the potential for human exposures. These may include written procedures or policies which specify the methods and techniques that will be practiced whenever personnel are to work with isobutane. Administrative controls as well as engineering controls must consider the effects of working with isobutane.

All personnel should receive *training* on the use, hazards, protective measures, emergency actions, and other precautions per 29 CFR 1910.1200 (Hazard Communication), prior to the first assignment in an area where isobutane is used or stored.

Other methods to reduce exposure to chemicals include:

- ☑ Where possible, consider installing automatic sensors that warn personnel of dropping oxygen levels and use local exhaust ventilation at the site of chemical release. If local exhaust ventilation or enclosure is not used, respiratory protection should be mandatory.
- ☑ Always ensure that proper protective clothing is worn when using chemical substances and that personnel are trained in its use and care.
- ☑ Wash thoroughly immediately after exposure to isobutane and at the end of the work shift or before eating, drinking, or smoking.
- ☑ Safety shower and eyewash stations should be readily available in work areas where isobutane is used or stored.
- ☑ Hazard warning information should be posted in the work area. In addition, as part of an on-going education and training program, all information on the health and safety hazards of isobutane should be communicated to all exposed and potentially exposed workers.

RISK ASSESSMENT: ENVIRONMENT

General Assessment

The environment is at risk of exposure during transportation, storage, disposal, or destruction of isobutane. In almost every scenario, the threat of environmental exposure is contingent upon the proper handling of the chemical substance. Accidental releases, large or small, can result in fire, explosion, and possible contamination of the surrounding environmental mediums (water, soil, and air). Isobutane is a highly flammable and explosive gas. Its extremely low flash point and boiling point present a serious fire hazard. It will float and boil on water to produce a flammable and visible vapor cloud. Also, since it can react violently in the presence of oxidizers, such as chlorine, isobutane requires special consideration during any emergency involving a leak or release of isobutane gas or liquid.

Isobutane can enter the environment through industrial discharges, unchecked venting to air, or through spills of the liquid. Since liquids will most likely vaporize rapidly, water and soil contamination are possible but improbable.

☠ *Acute Ecological Effects*

Acute (short-term) toxic effects may include the death of animals, birds, or fish, and death or low growth rate in plants. Acute effects are seen 2 to 4 days after animals or plants are exposed to isobutane.

Isobutane has slight acute toxicity to aquatic life. Insufficient data are available on the acute (short-term) effects of isobutane on plants, birds, or land animals.

Chronic Ecological Effects

Chronic (long-term) toxic effects may include shortened life span, reproductive problems, lower fertility, and changes in appearance or behavior in exposed animals. These effects can be seen long after first exposure(s) to toxic chemicals.

Insufficient data are available to evaluate or predict the long-term effects of isobutane to aquatic life, plants, birds, or land animals.

♦ *Water Solubility*

Isobutane is only slightly soluble in water. Concentrations of 1 to 100 milligrams will mix with a liter of water.

⌛ *Persistence in the Environment*

Isobutane is non-persistent in water, with a half-life of less than 2 days. The half-life of a pollutant is the amount of time it takes for one half of the chemical to be degraded. About 99.9% of isobutane will eventually end up in air; the rest will end up in the water.

Bioaccumulation in Aquatic Organisms

Some substances increase in concentration, or *bioaccumulate,* in living organisms as they breathe contaminated air, drink contaminated water, or eat contaminated food. These chemicals can become concentrated in the tissues and internal organs of animals as well as humans.

Isobutane is not expected to accumulate in the tissues of fish.

Recommended Risk-Reduction Measures

Proper training of all transporters will reduce the likelihood of a mishap or accident resulting in a leak or release to the environment. The correct use of DOT labeling (per the MSDS) on all containers, trucks, and rail cars will enable emergency responders to react properly and quickly to any disaster thereby reducing the potential risk to the environment and to personnel.

Storage of isobutane should be segregated from incompatible chemicals to minimize the risk of cross-contamination or contact. Buildings designated for storage should be equipped with appropriate fire protection systems (alarms, sprinklers, emergency lighting, portable extinguishers).

If a spill or release to the environment has occurred, fire department, emergency response, and/or hazardous materials spill personnel should be notified immediately. Cleanup should be attempted only by those trained in proper containment procedures. Contaminated soils should be removed for incineration and replaced with clean soil. When spills or leaks occur, it may be necessary to contact the local and/or state emergency response authorities. A comprehensive emergency response or disaster preparedness/recovery plan should be in place prior to any operations involving the use, transportation, storage, or disposal of isobutane. If isobutane is spilled or leaked, the following specific steps are recommended:

- ☑ Restrict persons not wearing protective clothing from area of spill or leak until cleanup is complete and area can be opened for normal work.
- ☑ Remove all ignition sources.
- ☑ Ventilate area of spill or leak.
- ☑ If applicable, stop flow of leaking liquid or gas. If leak source is a cylinder and the leak cannot be stopped in place, remove leaking cylinder to a safe place in the open air, and repair or allow cylinder to empty.
- ☑ Keep isobutane out of a confined space, such as a sewer, because of the possibility of explosion (unless the sewer is designed to prevent the buildup of explosive concentrations).
- ☑ It may be necessary to dispose of isobutane as a hazardous waste. The responsible state agency or the regional office of the federal Environmental Protection Agency (EPA) should be contacted for specific recommendations.

RISK ASSESSMENT: BUSINESS

General Assessment

Accidents or mishaps involving isobutane can present a significant threat to business operations. The loss or damage of equipment or facilities can significantly affect fiscal viability. Lawsuits that may result from personnel injury, public exposures, and environmental contamination will also require a serious expenditure of resources.

Media attention surrounding an injury, illness, or environmental damage can also result in a loss of profits and loss of current as well as future business.

Recommended Risk-Reduction Measures

Company attorneys, safety and health professionals, and environmental specialists should be involved in the development of any procedures dealing with safety or emergency response. A company official should be pre-designated as a public relations officer with specific training in dealing with the press. Corporate plans and policies should be developed, approved, and implemented long before any need for such arises.

REFERENCES

Ahlbom, A. 1993. *Biostatistics for Epidemiologists.* New York: CRC Press/Lewis Publishers.

American Conference of Governmental Industrial Hygienists. 1988. *Documentation of the Threshold Limit Values and Biological Exposure Indices,* 5th Edition (with updates). Cincinnati: ACGIH

Anthony, C. P., and N. J. Kolthoff. 1971. *Textbook of Anatomy and Physiology.* St. Louis: C. V. Mosby Company.

Asante-Duah, D. K. 1993. *Hazardous Waste Risk Assessment.* New York: CRC Press/Lewis Publishers.

Bloom, A. D. and F. J. DeSerres. 1995. *Ecotoxicity and Human Health.* New York: CRC Press/Lewis Publishers.

Brusick, D. J. 1994. *Methods for Genetic Risk Assessment.* New York: CRC Press/Lewis Publishers.

Bueche, J. R. 1972. *Principles of Physics,* 2nd Edition. New York: McGraw-Hill.

Burrell, R., D. K. Flaherty, and L. J. Sauers. 1992. *Toxicology of the Immune System.* New York: Van Nostrand Reinhold.

Calabrese, E. J. 1994. *Biological Effects of Low Level Exposures.* New York: CRC Press/Lewis Publishers.

Cockerman, L. G., B. S. Shane. 1994. *Basic Environmental Toxicology.* New York: CRC Press/Lewis Publishers.

Cothern, C. R. 1995. *Handbook for Environmental Risk Decisionmaking.* New York: CRC Press/Lewis Publishers.

Danse, I. R. 1991. *Common Sense Toxics in the Workplace.* New York: Van Nostrand Reinhold.

Environmental Protection Agency. 1983. *Chemical Hazard Information Profiles.* Washington, D.C.: U.S. EPA.

Fodden, C. R., and J. L. Weddell. 1992. *Hazardous Materials Emergency Action Data.* New York: CRC Press/Lewis Publishers.

Forseberg, K., and L. H. Keith. 1993. *Chemical; Protective Clothing Permeation and Degradation Database.* Chelsea, MI: Lewis Publishers.

Gardner, R. W. 1994. *Chemical Intolerance.* New York: CRC Press/Lewis Publishers.

Hallenbeck, W. H. 1993. *Quantitative Risk Assessment for Environmental and Occupational Health,* 2nd Edition. New York: CRC Press/Lewis Publishers.

Hodgson, E., R. B. Mailman, and J. E. Chambers. 1988. *Dictionary of Toxicology.* London: The MacMillian Company.

Howard, P. H. 1995. *Handbook of Environmental Fate and Exposures Data for Organic Chemicals,* Vol. I-IV. New York: CRC Press/Lewis Publishers.

James, A., and M. Lord. 1992. *Index of Chemical and Physical Data.* New York: Van Nostrand Reinhold.

Jones, F. E. 1994. *Toxic Organic Vapors in the Workplace.* New York: CRC Press/Lewis Publishers.

Keith, L. H. and M. M. Walker. 1995. *Handbook of Air Toxics.* New York: CRC Press/Lewis Publishers.

Keith, L. H. (Editor). 1992. *Integrated Risk Information System (IRIS) Chemical Information Database.* Prepared by the U.S. Environmental Protection Agency. Boca Raton, FL: CRC Press/Lewis Publishers.

Keith, L. H. and D. B. Walters. (Editors) 1992. *National Toxicology Program's Chemical Database, Volumes I-VII.* Chelsea, MI: Lewis Publishers.

Landis, W. G., M. H. Yu. 1995. *Introduction to Environmental Toxicology.* New York: CRC Press/Lewis Publishers.

Langerman, N. 1994 *Precautionary Labels for Chemical Containers.* New York: CRC Press/Lewis Publishers.

Lauwerys, R. R., P. Hoet. 1993. *Industrial Chemical Exposure.* New York: CRC Press/Lewis Publishers.

Lewis, R. J., Sr. 1993. *Hawley's Condensed Chemical Dictionary,* 12th Edition. New York: Van Nostrand Reinhold.

Lewis, R. J., Sr. 1992. *Sax's Dangerous Properties of Industrial Materials,* Volumes 1, 2, and 3, 8th Edition. New York: Van Nostrand Reinhold.

Manahan, S. E. 1992. *Toxicological Chemistry,* 2nd Edition. New York: CRC Press/Lewis Publishers.

Meyer, E. 1990. *Chemistry of Hazardous Materials.* Englewood Cliffs, NJ: Prentice Hall.

Patnaik, P. 1992. *A Comprehensive Guide to the Hazardous Properties of Chemical Substances.* New York: Van Nostrand Reinhold.

Philp, R. B. 1995. *Environmental Hazards and Human Health.* New York: CRC Press/Lewis Publishers.

Proctor, N. H., and J. P. Hughes. 1991. *Chemical Hazards of the Workplace.* New York: Van Nostrand Reinhold.

Quigley, D. R. 1994. *Handbook of Emergency Chemical Management.* New York: CRC Press/Lewis Publishers.

Rea, W. J. 1995. *Chemical Sensitivity,* Vol. I-III. New York: CRC Press/Lewis Publishers.

Richardson, M. 1995. *Dictionary of Substance and Their Effects,* Vol. I-VI. New York: CRC Press/Lewis Publishers.

Sacarello, H. L. A. 1994. *The Comprehensive Handbook of Hazardous Materials.* New York: CRC Press/Lewis Publishers.

Sherman, J. D. 1988. *Chemical Exposure and Disease.* New York: Van Nostrand Reinhold.

Torkelson, T. R. V. K. Rowe. 1981. *Patty's Industrial Hygiene and Toxicology,* Vol. 2B. New York: Wiley-Interscience.

U.S. Department of Health and Human Services, National Institute for Occupational Safety and Health. 1994. *NIOSH Pocket Guide to Chemical Hazards.* Washington, D.C.: U.S. Government Printing Office.

MATERIAL SAFETY DATA SHEET

CHEMICAL NAME	CRC PRESS/LEWIS PUBLISHERS DISCLAIMER
ISOBUTYL ACETATE	**CRC PRESS/LEWIS PUBLISHERS DISCLAIMER:** THE INFORMATION AND RECOMMENDATIONS PRESENTED HEREIN ARE BASED ON SOURCES BELIEVED TO BE RELIABLE. CRC MAKES NO REPRESENTATION ON ITS COMPLETENESS OR ACCURACY. IT IS THE RESPONSIBILITY OF THE USER TO DETERMINE THE CHEMICAL'S SUITABILITY FOR ITS INTENDED USE, THE CHEMICAL'S SAFE USE, AND THE CHEMICAL'S PROPER DISPOSAL. NO REPRESENTATIONS AND/OR WARRANTIES, EITHER EXPRESSED OR IMPLIED, OF THE MERCHANTABILITY OR FITNESS FOR A PARTICULAR PURPOSE, OR OF ANY OTHER NATURE, ARE MADE WITH RESPECT TO THE INFORMATION PROVIDED IN THIS MSDS OR TO THE CHEMICAL TO WHICH INFORMATION MAY REFER. CRC NEITHER ASSUMES NOR AUTHORIZES ANY OTHER PERSON TO ASSUME FOR IT, ANY OTHER ADDITIONAL RESPONSIBILITY OR LIABILITY FOR THE USE OF, OR RELIANCE UPON, THIS INFORMATION.

HAZARD WARNING INFORMATION

HEALTH	FIRE	REACTIVE	OTHER	DEGREE OF HAZARD	COLOR CODING	OTHER CODES
1	3	0		0 = Minimum Hazard 1 = Slight Hazard 2 = Moderate Hazard 3 = Serious Hazard 4 = Severe Hazard	HEALTH = BLUE FIRE = RED REACTIVITY = YELLOW OTHER = WHITE	OX = Oxidizer ACID = Acid ALK = Alkali COR = Corrosive W = Use No Water

SECTION I - GENERAL INFORMATION

Characterization	**Aliphatic Ester**	RCRA Number	**None**
		EPA Class	**Not Applicable**
DOT Proper Shipping Name	**Isobutyl Acetate**	Chemical Abstract Service (CAS) Number	**110-19-0**
DOT Hazard Class and Label Requirements	**Flammable Liquid**	DOT Emergency Guide Code	**26**
DOT Identification Number	**UN 1213**	Chemical Formula	**$CH_3COOCH_2CH(CH_3)_2$**

Synonyms

Isobutyl ester of acetic acid; 2-methylpropyl acetate; 2-methylpropyl ester of acetic acid; β-methylpropyl ethanoate; acetic acid isobutyl ester.

SECTION II - HAZARDOUS INGREDIENTS AND IDENTITY INFORMATION

Hazardous Components (specific identity)	OSHA Exposure Criteria	NIOSH Exposure Criteria	Immediately Dangerous to Life and Health (IDLH)	ACGIH Exposure Criteria
Isobutyl acetate (derivation: By treating isobutanol with acetic acid in the presence of catalysts). **1 ppm = 4.83 mg/m³**	**PEL:** **150 ppm** **700 mg/m³** **STEL:** **Not Established**	**REL:** **150 ppm** **700 mg/m³** **STEL:** **Not Established**	**1300 ppm**	**TLV:** **150 ppm** **700 mg/m³** **STEL:** **Not Established**

SECTION III - PHYSICAL AND CHEMICAL CHARACTERISTICS

Boiling Point	**243°F (117°C)**	Specific Gravity (H_2O = 1)	**0.87**
Vapor Pressure (mm Hg)	**13 at 68°F (20°C)**	Molecular Weight	**116.2**
Vapor Density (Air = 1)	**4.00**	Melting Point	**-145°F (-98°C)**

Solubility

Very slightly soluble in water (0.6%). Soluble in alcohol, ether, and hydrocarbons.

Appearance and Odor

Colorless, mobile liquid with a fruity, floral odor. Odor Threshold = 0.5 ppm.

SECTION IV - FIRE AND EXPLOSION HAZARD DATA

Flash Point (method used)	**64°F (18°C) closed cup**	Explosive Limits in Air % by Volume	**LEL: 1.3%** **UEL: 10.5%**
NFPA Classification	**Class IB Flammable Liquid**	Autoignition Temperature	**793°F (423°C)**

Extinguishing Media

Dry chemical, alcohol foam, carbon dioxide, water spray. Use water spray to cool exposed containers.

Special Fire Fighting Procedures

Poisonous gases are produced in fire. Wear full protective clothing with SCBA. Use blanketing effect with foam to smother large fires. Continue to cool containers with water after fire is extinguished.

Unusual Fire and Explosion Hazards

Immediately withdraw if rising sound from venting device is heard or if fire is causing discoloration to the tank. Evacuate 1500 feet radius if fire becomes uncontrollable or container is exposed to direct flame. Vapors are heavier than air and may travel a distances to ignition source to flashback.

SECTION V - REACTIVITY DATA

Stability		Conditions to Avoid **Isobutyl acetate is stable under normal conditions of storage and operation. Avoid contact with incompatible materials. Keep away from heat, flame, and other sources of ignition.**
Stable **X**	Unstable	Incompatibility *(materials to avoid)* **Reacts violently in contact with nitrates, strong oxidizers, alkalis, and acids. Particularly reactive with chlorosulfonic acid. It will dissolve many plastics.**
Hazardous Polymerization		Conditions to Avoid **Under normal conditions of temperature and pressure, hazardous polymerization of isobutyl acetate is not expected to occur.**
May Occur	Will Not Occur **X**	Hazardous Decomposition or By-products **When heated to decomposition, isobutyl acetate emits acrid and irritating fumes, including toxic oxides of carbon.**

SECTION VI - HEALTH HAZARD DATA

Primary Route(s) of Entry:	Inhalation?	Absorption (skin)?	Ingestion?
	X		X

Health Hazards

INHALATION: **Irritation of the eyes, nose, throat, and respiratory tract. Causes central nervous system depression with headache, weakness, dizziness, and loss of consciousness. It can also cause pulmonary edema, difficulty breathing, rapid pulse rate, and fatigue. There may also be liver damage, renal injury, and CNS effects (giddiness, confusion, delirium).**

SKIN & EYES: **Will irritate the skin. May cause contact dermatitis. Eye contact may cause irritation and possible visual disturbances that do not appear to be permanent.**

INGESTION: **Irritation of the mouth, esophagus, stomach, and digestive tract, nausea, and vomiting.**

Carcinogenicity	NTP Listed?	IARC Cancer Review Group?	OSHA Regulated?	Target Organs?
Unknown Human Unknown Animal	**No**	**No**	**29 CFR 1910.1000 Table Z-1**	**Respiratory system; skin; eyes, CNS.**

Medical Conditions Generally Aggravated by Exposure

None reported.

Emergency and First-aid Procedures

Eye contact: Flush immediately with water for 15 minutes (minimum); seek medical attention. Skin contact: Remove all contaminated clothing. Immediately wash area with large amounts of soap and water. Seek medical assistance. For inhalation: Remove the person from exposure. Provide respiratory assistance and CPR. Transfer to medical facility. If swallowed: Seek medical attention immediately. Call poison control center for advice. Give 1-2 glasses of water, induce vomiting and repeat until vomit is clear unless advised otherwise. Never give an unconscious or convulsing person anything by mouth.

SECTION VII - PRECAUTIONS FOR SAFE HANDLING AND USE

Steps to be Taken in Case Material is Released or Spilled

Contain spills using absorbent material (vermiculite, or other). Ventilate area of spill, remove ignition sources. Have water source available in case of fire. Store materials in DOT-approved containers. Restrict those not involved in cleanup from entering area. Notify appropriate authorities, as required.

Preferred Waste Disposal Method

Incineration in chemical incinerator equipped with scrubber and afterburner.

Precautions to be Taken in Handling and Storage

Store in tightly closed containers in a cool, dark, well-ventilated area. Protect containers from physical damage. Keep heat, fire, and flame away. If possible, automatically transfer liquids between containers.

Other Precautions and Warnings

Bulk storage of isobutyl acetate is not recommended. Ground and bond all metal containers.

SECTION VIII - CONTROL MEASURES AND PERSONAL PROTECTIVE EQUIPMENT

Respiratory Protection *(specify type)*

Where potential exists for exposure over 150 ppm (8 hours), use a MSHA/NIOSH-approved full facepiece respirator with an organic vapor cartridge. For higher exposures, use a self-contained breathing apparatus (SCBA) with full facepiece operated in pressure demand or other positive pressure mode.

Ventilation

Local exhaust or general mechanical systems recommended.

Protective Gloves	Eye Protection	Other Protective Clothing
Impervious Gloves	**Chemical Goggles or Face Mask**	**Impermeable Apron**

Work/Hygiene Practices

Always wash hands thoroughly after using chemical; never bring food, drink, or smoking materials into vicinity of chemicals.

ISOBUTYL ACETATE

$CH_3COOCH_2CH(CH_3)_2$ CAS: 110-19-0

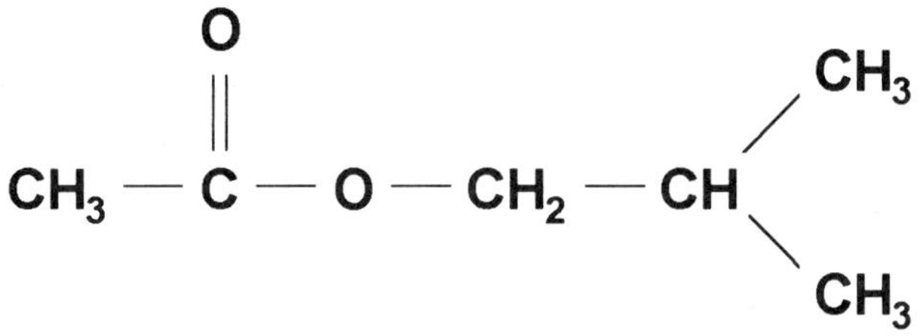

IDENTIFICATION AND TYPICAL USES

Isobutyl acetate is a colorless liquid with a characteristic fruity or floral odor. It is used as a solvent for nitrocellulose, in the manufacture of thinners, sealants, and topcoat lacquers, in perfumery, and as a flavoring agent.

RISK ASSESSMENT: HEALTH

General Assessment

Isobutyl acetate is mildly toxic by *inhalation* and *ingestion*. Skin contact can produce localized effects but absorption has not been reported. Upon absorption into the body (through respiration or digestion) it is hydrolyzed to acetic acid and isobutanol. There are no reports of any carcinogenic, teratogenic, or mutagenic effects in humans or animals following exposure to isobutyl acetate.

Inhalation of isobutyl acetate vapors can cause irritation to the eyes, nose, throat, and respiratory tract with nausea, vomiting, cough, shortness of breath, and unconsciousness. There may also be the development of pulmonary edema (fluid in the lungs) which can be fatal. Symptoms can be delayed up to 48 hours thereby creating a false sense of security with regard to health exposure risk. Depending upon the concentration and duration of the exposure, there can also be varying degrees of central nervous system depression. Symptoms may include headaches, dizziness, general weakness and fatigue, loss of consciousness, difficulty in breathing, increased or rapid pulse rate, unconsciousness, and death.

Skin contact results in the removal of the skin's natural protective oily layer on its surface with subsequent drying, cracking, and possible secondary infections and dermatitis. Eye contact can cause irritation and the possibility for adverse effects to vision.

Ingestion of isobutyl acetate may cause irritation of the mouth, esophagus, stomach, and digestive tract with nausea and vomiting. There can also be gastrointestinal hemorrhage, liver damage, renal damage with glycosuria (excessive sugar in urine), and cardiac failure.

Acute Health Effects

The following acute (short-term) health effects may occur immediately or shortly after exposure to isobutyl acetate:

Skin: Mild to moderate irritation. Will remove the skin's protective oily layer to cause drying, cracking, and possible secondary infection and dermatitis.

Eye: Irritation and possible inflammation.

Lung: Irritation of the eyes, nose, throat and lungs (especially the mucus membranes). May cause changes in the sense of smell and in pulmonary function. Delayed pulmonary edema is possible, which is a medical emergency.

CNS: High concentrations can depress the central nervous system leading to narcoses and possible loss of consciousness.

Chronic Health Effects

The following chronic (long-term) health effects may occur at some time after exposure to isobutyl acetate and can last for months or even years:

Cancer Hazards: According to the references, isobutyl acetate has not been adequately tested for its ability to cause cancer in test animals.

Reproduction: According to the references, isobutyl acetate has not been adequately tested for its ability to adversely affect reproduction in test animals.

Other Chronic Effects: Repeated or prolonged skin contact may cause the development of contact dermatitis with rash, itching, redness, and localized inflammation of the affected surface area.

Recommended Risk-Reduction Measures

Personnel should avoid direct contact with isobutyl acetate. The best risk reduction measure is to use a less toxic chemical as a substitute for isobutyl acetate. If this is not possible or practical, then *engineering controls* are the most effective method of reducing exposures. The best protection is to enclose operations and/or provide local exhaust ventilation at the site of chemical release. While not always operationally feasible, isolating operations can also reduce exposure risks to other personnel.

Using respiratory protection is less effective than the controls mentioned above, but is still advisable whenever working with or around isobutyl acetate. For exposures over 150 ppm (8-hours), an MSHA/NIOSH approved chemical respirator with an organic vapor cartridge should be used. For higher exposures, a self-contained breathing apparatus (SCBA) with full facepiece and operated in pressure demand mode or a supplied-air respirator with full facepiece operated in pressure demand are the recommended respiratory protection methods of choice. If a full facepiece is not available, then chemical goggles should be worn to protect the eyes. Whenever a chemical splash hazard exists, a face shield and a protective apron should be worn. To prevent hand and skin exposures, impervious gloves should be used. Caution is warranted in glove selection since isobutyl acetate will attack many plastic formulations. Glove manufacturers should be contacted to obtain permeation studies *before* gloves are selected.

Administrative controls should also be in place to minimize the potential for human exposures. These may include written procedures or policies which specify the methods and techniques that will be practiced whenever personnel are to work with isobutyl acetate.

All personnel should receive *training* on the use, hazards, protective measures, emergency actions, and other precautions per 29 CFR 1910.1200 (Hazard Communication), prior to the first assignment in an area where isobutyl acetate is used or stored.

If symptoms develop or overexposure is suspected, the following medical tests are recommended:

- ☑ Lung function tests.
- ☑ Liver and kidney function tests.
- ☑ Consider chest X-ray after acute overexposure (may be negative if taken immediately after exposure due to delayed onset of pulmonary edema).
- ☑ Skin testing with dilute isobutyl acetate to help diagnose allergy (performed by a qualified allergist).

Any medical evaluation should include a careful history of past and present symptoms with an examination. Medical tests that look for existing damage are not a substitute for controlling exposures. Also, since smoking can cause heart disease, emphysema, and other respiratory disorders, smokers exposed to isobutyl acetate may experience symptoms more quickly than non-smokers under the same conditions of exposure. Prudent risk management requires proper consideration of *all* factors which may be causing the appearance of exposure symptoms in the workplace.

Other methods to reduce exposure include:

- ☑ Where possible, enclose operations and use local exhaust ventilation at the site of chemical release. If local exhaust ventilation or enclosure is not used, respiratory protection should be mandatory.
- ☑ Always ensure that proper protective clothing is worn when using chemical substances.
- ☑ Wash thoroughly immediately after exposure to isobutyl acetate and at the end of the work shift or before eating, drinking, or smoking.
- ☑ Hazard warning information should be posted in the work area. In addition, as part of an on-going education and training program, all information on the health and safety hazards of isobutyl acetate should be communicated to all exposed and potentially exposed workers.
- ☑ Never eat, drink, or smoke in areas were isobutyl acetate is used, handled, or stored.

RISK ASSESSMENT: ENVIRONMENT

General Assessment

The environment is at risk of exposure during transportation, storage, disposal, or destruction of isobutyl acetate. In almost every scenario, the threat of environmental exposure is contingent upon the proper handling of the chemical substance.

Accidental spills, large or small, can result in fire, explosion, and possible contamination of the surrounding environmental mediums (water, soil, and air). Isobutyl acetate is considered a Class IB flammable liquid (according to OSHA 29 CFR 1910.106). Its moderately low flash point presents a serious fire and explosion hazard concern. Also, because it is incompatible with a number of common materials, especially strong oxidizers and acids, contact can result in violent and explosive reactions. It can also form explosive mixtures in air and can ignite on contact with heat, fire, or sparks. These characteristics require special consideration during any emergency situation involving a leak or spill of isobutyl acetate.

The proper disposal/destruction method for isobutyl acetate is to burn it in a chemical incinerator equipped with an afterburner and air scrubber. Isobutyl acetate can enter the environment through unchecked industrial discharges and through spills.

Acute Ecological Effects

Acute (short-term) toxic effects may include the death of animals, birds, or fish, and death or low growth rate in plants. Acute effects are seen 2 to 4 days after animals or plants are exposed to isobutyl acetate.

Insufficient data are available to evaluate or predict the short-term effects of isobutyl acetate to aquatic life, plants, birds, or terrestrial animals.

Chronic Ecological Effects

Chronic toxic effects may include shortened life span, reproductive problems, lower fertility, and changes in appearance or behavior in exposed animals. These effects can be seen long after first exposure(s) to toxic chemicals.

Insufficient data are available to evaluate or predict the long-term effects of isobutyl acetate to aquatic life, plants, birds, or land animals.

Water Solubility

Isobutyl acetate is only slightly soluble in water. Concentrations of 1 to 100 milligrams may mix with a liter of water.

Persistence in the Environment

Isobutyl acetate is slightly persistent in water, with a half-life of between 2 to 20 days. The half-life of a pollutant is the amount of time it takes for one-half of the chemical to be degraded. There are insufficient data to evaluate or predict the percentages of isobutyl acetate that will eventually end up in the various environmental mediums (air, water, and soil).

Bioaccumulation in Aquatic Organisms

Some substances increase in concentration, or *bioaccumulate,* in living organisms as they breathe contaminated air, drink contaminated water, or eat contaminated food. These chemicals can become concentrated in the tissues and internal organs of animals as well as humans.

The concentration of isobutyl acetate found in fish tissues is expected to be somewhat higher then the average concentration of isobutyl acetate in water from which the fish was taken.

Recommended Risk-Reduction Measures

Proper training of all transporters will reduce the likelihood of a mishap or accident resulting in a leak or spill to the environment. The correct use of DOT labeling (per the MSDS) on all containers, trucks, and rail cars will enable emergency responders to react properly and quickly to any disaster thereby reducing the potential risk to the environment and to personnel.

Storage of isobutyl acetate should be segregated from incompatible chemicals to minimize the risk of cross-contamination or contact. Buildings designated for storage should be equipped with appropriate fire protection systems (alarms, sprinklers, emergency lighting, portable extinguishers).

If a spill or leak to the environment has occurred, fire department, emergency response, and/or hazardous materials spill personnel should be notified immediately. Cleanup should be attempted only by those trained in proper spill containment procedures using non-sparking tools. Contaminated soils should be removed for incineration and replaced with clean soil. If isobutyl acetate should contact the water table, aquifer, or navigable waterway, time is of the essence. It is only slightly soluble in water and, therefore, total containment and remediation may be possible. When such spills occur, the local and/or state emergency response authorities must be notified. A comprehensive emergency response or disaster preparedness/recovery plan should be in place prior to any operations involving the use, transportation, storage, or disposal of isobutyl acetate. If isobutyl acetate is spilled or leaked, the following specific steps are recommended:

- ☑ Restrict persons not wearing protective clothing from area of spill or leak until cleanup is complete and area can be opened for normal work.
- ☑ Ventilate area and remove all ignition sources.
- ☑ Absorb liquids in vermiculite, dry sand, earth, or a similar material and deposit in sealed containers. Use non-sparking tools.
- ☑ It may be necessary to dispose of isobutyl acetate as a hazardous waste. The responsible state agency or the regional office of the federal Environmental Protection Agency (EPA) should be contacted for specific recommendations.

RISK ASSESSMENT: BUSINESS

General Assessment

Accidents or mishaps involving isobutyl acetate can present a moderate threat to business operations. The loss or damage of equipment or facilities can significantly affect fiscal viability. Lawsuits that may result from personnel injury/death, public exposures, and/or environmental contamination will also require a serious expenditure of resources. Media attention surrounding an injury, death, or environmental damage can also result in a loss of profits and loss of current as well as future business.

❶ *Recommended Risk-Reduction Measures*

Company attorneys, safety and health professionals, and environmental specialists should be involved in the development of any procedures or policies intended to manage the use of chemicals in the workplace. A company official should be pre-designated as a public relations officer with specific training in dealing with the press. Corporate plans and policies should be developed, approved, and implemented long before any need for such arises.

REFERENCES

American Conference of Governmental Industrial Hygienists. 1988. *Documentation of the Threshold Limit Values and Biological Exposure Indices,* 5th Edition (with updates). Cincinnati: ACGIH

Anthony, C. P., and N. J. Kolthoff. 1971. *Textbook of Anatomy and Physiology.* St. Louis: C. V. Mosby Company.

Bloom, A. D. and F. J. DeSerres. 1995. *Ecotoxicity and Human Health.* New York: CRC Press/Lewis Publishers.

Brusick, D. J. 1994. *Methods for Genetic Risk Assessment.* New York: CRC Press/Lewis Publishers.

Burrell, R., D. K. Flaherty, and L. J. Sauers. 1992. *Toxicology of the Immune System.* New York: Van Nostrand Reinhold.

Calabrese, E. J. 1994. *Biological Effects of Low Level Exposures.* New York: CRC Press/Lewis Publishers.

Cockerman, L. G., B. S. Shane. 1994. *Basic Environmental Toxicology.* New York: CRC Press/Lewis Publishers.

Cothern, C. R., N. P. Ross. 1994. *Environmental Statistics, Assessment, and Forecasting.* New York: CRC Press/Lewis Publishers.

Danse, I. R. 1991. *Common Sense Toxics in the Workplace.* New York: Van Nostrand Reinhold.

Environmental Protection Agency. 1983. *Chemical Hazard Information Profile: Isobutyl acetate.* Washington, D.C.: U.S. EPA.

Gardner, R. W. 1994. *Chemical Intolerance.* New York: CRC Press/Lewis Publishers.

Hallenbeck, W. H. 1993. *Quantitative Risk Assessment for Environmental and Occupational Health,* 2nd Edition. New York: CRC Press/Lewis Publishers.

Hodgson, E., R. B. Mailman, and J. E. Chambers. 1988. *Dictionary of Toxicology.* London: The MacMillian Company.

Howard, P. H., et. al. 1991. *Handbook of Environmental Degradation Rates.* New York: CRC Press/Lewis Publishers.

Jones, F. E. 1994. *Toxic Organic Vapors in the Workplace.* New York: CRC Press/Lewis Publishers.

Kamrin, M. 1 A. 1988. *Toxicology: A Primer on Principles and Applications.* Chelsea, Michigan: Lewis Publishers.

Keith, L. H. and M. M. Walker. 1995. *Handbook of Air Toxics.* New York: CRC Press/Lewis Publishers.

Langerman, N. 1994 *Precautionary Labels for Chemical Containers.* New York: CRC Press/Lewis Publishers.

Lauwerys, R. R., P. Hoet. 1993. *Industrial Chemical Exposure.* New York: CRC Press/Lewis Publishers.

Lewis, Richard J., Sr. 1992. *Sax's Dangerous Properties of Industrial Materials,* Volumes 1, 2, and 3, 8th Edition. New York: Van Nostrand Reinhold.

Manahan, S. E. 1994. *Environmental Chemistry,* 6th Edition. New York: CRC Press/Lewis Publishers.

Manahan, S. E. 1992. *Toxicological Chemistry,* 2nd Edition. New York: CRC Press/Lewis Publishers.

Patnaik, P. 1992. *A Comprehensive Guide to the Hazardous Properties of Chemical Substances.* New York: Van Nostrand Reinhold.

Sacarello, H. L. A. 1994. *The Comprehensive Handbook of Hazardous Materials.* New York: CRC Press/Lewis Publishers.

Torkelson, T. R. V. K. Rowe. 1981. *Patty's Industrial Hygiene and Toxicology,* Vol. 2B. New York: Wiley-Interscience.

U.S. Department of Health and Human Services, National Institute for Occupational Safety and Health. 1994. *NIOSH Pocket Guide to Chemical Hazards.* Washington, D.C.: U.S. Government Printing Office.

MATERIAL SAFETY DATA SHEET

CHEMICAL NAME

ISOBUTYL ALCOHOL

HAZARD WARNING INFORMATION

HEALTH	FIRE	REACTIVE	OTHER	DEGREE OF HAZARD	COLOR CODING	OTHER CODES
1	3	0		0 = Minimum Hazard 1 = Slight Hazard 2 = Moderate Hazard 3 = Serious Hazard 4 = Severe Hazard	HEALTH = BLUE FIRE = RED REACTIVITY = YELLOW OTHER = WHITE	OX = Oxidizer ACID = Acid ALK = Alkali COR = Corrosive W = Use No Water

SECTION I - GENERAL INFORMATION

Characterization: **Alcohol**

RCRA Number: **U140**

EPA Class: **Toxic Waste**

DOT Proper Shipping Name: **Isobutanol**

Chemical Abstract Service (CAS) Number: **78-83-1**

DOT Hazard Class and Label Requirements: **Flammable or Combustible Liquid**

DOT Emergency Guide Code: **26**

DOT Identification Number: **UN 1212**

Chemical Formula: **$(CH_3)_2CHCH_2OH$**

Synonyms

IBA; isobutanol; isopropylcarbinol; 2-methyl-1-propanol; fermentation butyl alcohol.

SECTION II - HAZARDOUS INGREDIENTS AND IDENTITY INFORMATION

Hazardous Components (specific identity)	OSHA Exposure Criteria	NIOSH Exposure Criteria	Immediately Dangerous to Life and Health (IDLH)	ACGIH Exposure Criteria
Isobutyl alcohol (derivation: By-product of synthetic methanol production, purified by rectification or by fermentation of carbohydrates). **1 ppm = 3.08 mg/m^3**	**PEL:** **100 ppm** **300 mg/m^3** **STEL:** **Not Established**	**REL:** **50 ppm** **150 mg/m^3** **STEL:** **Not Established**	**1600 ppm**	**TLV:** **50 ppm** **152 mg/m^3** **STEL:** **Not Established**

SECTION III - PHYSICAL AND CHEMICAL CHARACTERISTICS

Boiling Point: **270°F (132°C)**

Specific Gravity (H_2O = 1): **0.81**

Vapor Pressure (mm Hg): **28 at 68°F (20°C)**

Molecular Weight: **88.2**

Vapor Density (Air = 1): **3.04**

Melting Point: **-179°F (-117°C)**

Solubility

Soluble in water. Miscible in alcohol and ether.

Appearance and Odor

Colorless, oily liquid with a sweet, musty odor. Odor Threshold = 40ppm.

SECTION IV - FIRE AND EXPLOSION HAZARD DATA

Flash Point (method used): **82°F (28°C) closed cup**

Explosive Limits in Air % by Volume: **LEL: 1.7% at 123°F (51°C) UEL: 10.6% at 202°F (94°C)**

NFPA Classification: **Class IC Flammable Liquid**

Autoignition Temperature: **780°F (415°C)**

Extinguishing Media

Dry chemical, carbon dioxide, alcohol-resistant foam, or water spray.

Special Fire Fighting Procedures

Poisonous gases are produced in fire. Wear full protective clothing and self-contained breathing apparatus (SCBA). Continue to cool containers with water after fire is extinguished. Move containers from fire area if it can be done without risk. For large fires, use unmanned hose apparatus, if possible. Evacuate non-essential personnel one-half mile radius if fire. Consider down-wind conditions.

Unusual Fire and Explosion Hazards

Vapors are heavier than air and may travel a distances to ignition source to flashback. Vapors may explode if ignited in an enclosed area.

SECTION V - REACTIVITY DATA

Stability		Conditions to Avoid Isobutyl alcohol is stable under normal conditions of storage and operation. Avoid contact with incompatible materials. Keep away from heat, flame, and other ignition sources.
Stable X	Unstable	Incompatibility *(materials to avoid)* Reacts violently in contact with strong oxidizers, inorganic acids, aldehydes, and isocyanates. It ignites on contact with chromium trioxide and will form explosive hydrogen gas in contact with heated aluminum.
Hazardous Polymerization		Conditions to Avoid Under normal conditions of temperature and pressure, isobutyl alcohol is not expected to undergo hazardous polymerization.
May Occur	Will Not Occur X	Hazardous Decomposition or By-products When heated to decomposition, isobutyl alcohol emits acrid and irritating fumes including carbon dioxide and isobutylene.

SECTION VI - HEALTH HAZARD DATA

Primary Route(s) of Entry:	Inhalation?	Absorption (skin)?	Ingestion?
	X		X

Health Hazards

INHALATION: There may be slight irritation of the eyes, nose, and mucosa of the respiratory tract. Exposure can also cause stupor, fatigue, and sleepiness. Other effects include nausea, vomiting, flushing of skin, headaches, double vision, vertigo, weakness, sleepiness, olfactory changes, pulmonary edema, and death.

SKIN & EYES: As a solvent, causes skin dryness, cracking, and possible dermatitis. Eye contact can result in burning, stinging, tearing, corneal injury, and hyperemia of the conjunctiva.

INGESTION: Sleep disorders, hallucinations, distorted perceptions, convulsions, ataxia and motor activity changes, renal and liver damage, cardiac failure, coma, headache, and death.

Carcinogenicity	NTP Listed?	IARC Cancer Review Group?	OSHA Regulated?	Target Organs?
Unknown Human Questioned Animal	No	No	29 CFR 1910.1000 Table Z-1	Skin, eyes, respiratory. system, CNS.

Medical Conditions Generally Aggravated by Exposure

Chronic liver, kidney, and chronic respiratory diseases may be aggravated by exposure.

Emergency and First-aid Procedures

Eye contact: Flush immediately with water for 15 minutes (minimum); seek medical attention. Skin contact: Remove all contaminated clothing. Immediately wash area with large amounts of soap and water. Seek medical assistance. For inhalation: Remove the person from exposure. Provide respiratory assistance and CPR. Transfer to medical facility. Observe for 48 hours for lung effects. If swallowed: Contact poison control center. Give 1-2 glasses of water or milk. Do NOT induce vomiting. Seek medical attention immediately.

SECTION VII - PRECAUTIONS FOR SAFE HANDLING AND USE

Steps to be Taken in Case Material is Released or Spilled

Contain spills using absorbent material (vermiculite, or other). Ventilate area of spill. Have extinguishing agent available in case of fire. Store materials in DOT-approved containers.

Preferred Waste Disposal Method

Incineration in chemical incinerator equipped with scrubber and afterburner.

Precautions to be Taken in Handling and Storage

Store in tightly closed containers in a cool, dark, well-ventilated area. Protect containers from physical damage. Keep heat, fire, and flame away. Outside building or underground storage is recommended.

Other Precautions and Warnings

Ground and bond all metal containers. Avoid contact with skin and eyes.

SECTION VIII - CONTROL MEASURES AND PERSONAL PROTECTIVE EQUIPMENT

Respiratory Protection *(specify type)*

For vapor exposures, use an organic vapor air-purifying respirator. Greater protection is obtained using a supplied-air respirator or a self-contained breathing apparatus (SCBA) with full facepiece operated in pressure demand or other positive pressure mode.

Ventilation

Local exhaust or general mechanical systems recommended.

Protective Gloves	Eye Protection	Other Protective Clothing
Impervious Gloves	Chemical Goggles or Face Mask	Impermeable Apron or Protective Suit

Work/Hygiene Practices

Always wash hands thoroughly after using chemical; never bring food, drink, or smoking materials into vicinity of chemicals.

ISOBUTYL ALCOHOL

$(CH_3)_2CHCH_2OH$ CAS: 78-83-1

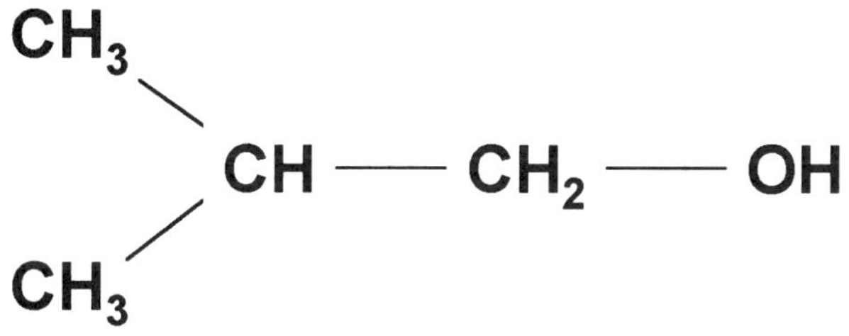

IDENTIFICATION AND TYPICAL USES

Isobutyl alcohol is a colorless or clear, oily liquid with a sweet, musty odor. It is used in organic synthesis, as a latent solvent in paints and lacquers, as an intermediate for amino coating resins, and as a substitute for n-butanol. It is also used in paint removers, fluorometric determination, in liquid chromatography, and in fruit flavor concentrates.

RISK ASSESSMENT: HEALTH

General Assessment

Isobutyl alcohol is primarily toxic to humans by *inhalation* and *ingestion*. Skin contact can cause adverse surface effects but dermal absorption has not been reported. Isobutyl alcohol is not currently listed as a carcinogen by OSHA, the NTP, or the IARC. However, there are reports in the references that this compound is under review because of its ability to cause tumors in rats and that its listing as a carcinogenic is imminent. Mutation data have also been reported.

Inhalation can cause irritation of the eyes, nose, throat, upper respiratory tract, and associated mucosa. There may also be depression of the central nervous system (CNS) with narcosis and associated symptoms, including headache, nervousness, flushing of the skin, double vision, vertigo, tremors, dizziness, fatigue and general weakness, nausea, somnolence, pulmonary edema (fluid on the lungs), and narcosis with stupor, loss of consciousness, and death. The symptoms of pulmonary edema may be delayed up to 48 hours which can create a false sense of security with regard to health exposure risk.

Liquid contact with the eyes causes immediate burning and stinging with lachrymation and reflex closure of the lids. There may be severe, irreversible damage, injury to the corneal epithelium and possible conjunctivitis, depending upon the duration of exposure and concentration strength. There may be a formation of vacuoles (small pockets of air or fluid) in cornea. Skin contact results in drying and cracking which can lead to secondary infections and dermatitis. There may be erythema (redness) and hypermia (and excessive blood in the skin).

Ingestion of isobutyl alcohol can lead to sleep disorders, hallucinations, ataxia, motor function changes, convulsions and tremors, coma, headaches, nausea or vomiting, loss of body temperature, renal and liver damage, gastrointestinal damage, cardiac failure, unconsciousness, and death.

☠ *Acute Health Effects*

The following acute (short-term) health effects may occur immediately or shortly after exposure to isobutyl alcohol:

Skin: Causes dryness and cracking leading to dermatitis and possible infection with erythema.

Eye: Severe irritation with burning and stinging with possible damage to the cornea and conjunctiva.

Lung: Irritation of the eyes, nose, throat, and respiratory tract. Delayed pulmonary edema is also a possible effect, which can be fatal.

CNS: High concentrations can cause depression of the CNS with symptoms of sleepiness, vertigo, loss of consciousness, and death.

Chronic Health Effects

The following chronic (long-term) health effects may occur at some time after exposure to isobutyl alcohol and can last for months or even years:

Cancer Hazards: Isobutyl alcohol has been shown to cause cancer in experimental animals. It is not known whether it will have this affect in humans but its carcinogenicity is considered questionable. Human mutation data have been reported. Many scientists believe that such chemicals pose a cancer risk in the long-term.

Reproduction: According to the references, isobutyl alcohol has not been adequately tested for its ability to adversely affect reproduction in test animals.

Other Chronic Effects: Very high or prolonged exposure may result in mucous membrane irritation, headache, and depression of the CNS with symptoms of somnolence and lack of concentration. Prolonged skin contact can cause dermatitis.

Recommended Risk-Reduction Measures

Personnel should avoid direct contact with isobutyl alcohol. The exact nature of its toxicity is not clearly understood in the references and its carcinogenic ability is considered suspect. The best risk reduction measure is to use a less toxic chemical as a substitute for isobutyl alcohol. However, if substitution is not possible or operationally feasible, then *engineering controls* are the most effective method of reducing exposures. An alternative method of protection is to enclose operations and/or provide local exhaust ventilation at the site of chemical release. While not always operationally feasible, isolating operations can also reduce exposure risk.

Using respiratory protection is less effective than the controls mentioned above, but is still advisable whenever working with or around isobutyl alcohol. For concentrations over the PEL (100 ppm), an air-purifying respirator with an organic vapor cartridge will suffice. For higher exposures, a supplied-air respirator with full facepiece operated in positive pressure mode, or a self-contained breathing apparatus (SCBA) with full facepiece and operated in pressure demand mode are the recommended respiratory protection methods of choice. If a full facepiece is not available, then chemical goggles should be worn to protect the eyes. Whenever a chemical splash hazard exists, a face shield and a protective apron should be worn. To prevent hand and skin exposures, impervious gloves should be used.

Administrative controls should also be in place to minimize the potential for human exposures. These may include written procedures or policies which specify the methods and techniques that will be practiced whenever personnel are to work with isobutyl alcohol. All personnel should receive *training* on the use, hazards, protective measures, emergency actions, and other precautions per 29 CFR 1910.1200 (Hazard Communication), prior to the first assignment in an area where isobutyl alcohol is used or stored.

If symptoms develop or overexposure is suspected, the following medical tests are recommended:

- ☑ Evaluation of nervous system function with careful consideration of exposure history and special testing by a qualified neurologist.
- ☑ Complete blood count (CBC) and determination of isobutyl alcohol levels (or its metabolites).
- ☑ Urinalysis to check for glycosuria (excessive sugar concentrations in urine).
- ☑ Electrocardiogram (EKG).
- ☑ Liver and lung function tests.
- ☑ Consider chest X-ray following acute overexposure (may be negative if taken immediately after exposure due to delayed onset of pulmonary edema).
- ☑ Skin testing with dilute isobutyl alcohol to help diagnose allergy (performed by a qualified allergist).

Any medical evaluation should include a careful history of past and present symptoms with an examination. Medical tests that look for existing damage are not a substitute for controlling exposures. Also, since smoking can cause heart disease, lung cancer, emphysema, and other respiratory diseases, smokers exposed to isobutyl alcohol may experience symptoms more rapidly than non-smokers under the same conditions of exposure.

Other methods to reduce exposure include:

- ☑ Where possible, enclose operations and use local exhaust ventilation at the site of chemical release. If local exhaust ventilation or enclosure is not used, respiratory protection should be mandatory.
- ☑ Always ensure that proper protective clothing is worn when using chemical substances and that personnel are trained in its use and care.
- ☑ Wash thoroughly immediately after exposure to isobutyl alcohol and at the end of the work shift or before eating, drinking, or smoking.
- ☑ Hazard warning information should be posted in the work area. In addition, as part of an on-going education and training program, all information on the health and safety hazards of isobutyl alcohol should be communicated to all exposed or potentially exposed workers.

RISK ASSESSMENT: ENVIRONMENT

General Assessment

The environment is at risk of exposure during transportation, storage, disposal, or destruction of isobutyl alcohol. In almost every scenario, the threat of environmental exposure is contingent upon the proper handling of the chemical substance. Accidental spills, large or small, can result in fire, explosion, and possible contamination of the surrounding environmental mediums (water, soil, and air).

Isobutyl alcohol is considered a Class IC flammable liquid (according to OSHA 29 CFR 1910.106). Its relatively low flash point presents a moderate fire and explosion hazard concern. Also, because it is incompatible with a number of common materials, especially strong oxidizers and many reducing materials, contact can result in violent and explosive reactions. It can also react explosively with hydrogen trisulfide and can ignite on contact with heat, fire, or sparks. These characteristics require special consideration during any emergency situation involving a leak or spill of isobutyl alcohol.

Isobutyl alcohol can enter the environment through unchecked industrial discharges into effluents and through spills.

Acute Ecological Effects

Acute (short-term) toxic effects may include the death of animals, birds, or fish, and death or low growth rate in plants. Acute effects are seen 2 to 4 days after animals or plants are exposed to isobutyl alcohol.

Insufficient data are available to evaluate or predict the short-term effects of isobutyl alcohol to aquatic life, plants, birds, or terrestrial animals.

Chronic Ecological Effects

Chronic toxic effects may include shortened life span, reproductive problems, lower fertility, and changes in appearance or behavior in exposed animals. These effects can be seen long after first exposure(s) to toxic chemicals.

Insufficient data are available to evaluate or predict the long-term effects of isobutyl alcohol to aquatic life, plants, birds, or land animals.

Water Solubility

Isobutyl alcohol is slightly soluble in water. Concentrations of 1 to 100 milligrams can be expected to mix with a liter of water.

Persistence in the Environment

Isobutyl alcohol is slightly persistent in water, with a half-life of between 2 to 20 days. The half-life of a pollutant is the amount of time it takes for one-half of the chemical to be degraded.

Bioaccumulation in Aquatic Organisms

Some substances increase in concentration, or *bioaccumulate,* in living organisms as they breathe contaminated air, drink contaminated water, or eat contaminated food. These chemicals can become concentrated in the tissues and internal organs of animals as well as humans.

The concentration of isobutyl alcohol found in fish tissues is expected to be about the same as the average concentration of isobutyl alcohol in water from which the fish was taken.

Recommended Risk-Reduction Measures

Proper training of all transporters will reduce the likelihood of a mishap or accident resulting in a leak or spill to the environment. The correct use of DOT labeling (per the MSDS) on all containers, trucks, and rail cars will enable emergency responders to react properly and quickly to any disaster thereby reducing the potential risk to the environment and to personnel.

Storage of isobutyl alcohol should be segregated from incompatible chemicals to minimize the risk of cross-contamination or contact. Buildings designated for storage should be equipped with appropriate fire protection systems (alarms, sprinklers, emergency lighting, portable extinguishers).

If a spill or leak to the environment has occurred, fire department, emergency response, and/or hazardous materials spill personnel should be notified immediately. Cleanup should be attempted only by those trained in proper spill containment procedures using non-sparking tools. Contaminated soils should be removed for incineration and replaced with clean soil.

If isobutyl alcohol should contact the water table, aquifer, or navigable waterway, time is of the essence. It is only slightly soluble in water and, therefore, total containment and remediation may be possible. When such spills occur, the local and/or state emergency response authorities must be notified. A comprehensive emergency response or disaster preparedness/recovery plan should be in place prior to any operations involving the use, transportation, storage, or disposal of isobutyl alcohol.

If isobutyl alcohol is spilled or leaked, the following specific steps are recommended to ensure the protection of personnel and the environment:

- ☑ Restrict persons not wearing protective clothing from area of spill or leak until cleanup is complete and area can be opened for normal work.
- ☑ Ventilate area of spill or leak.
- ☑ Remove all ignition sources.
- ☑ Absorb liquids in vermiculite, dry sand, earth, or a similar material and deposit in sealed containers for disposal. Use non-sparking (e.g., brass, plastic) tools.
- ☑ It may be necessary to dispose of isobutyl alcohol as a hazardous waste. The responsible state agency or the regional office of the federal Environmental Protection Agency (EPA) should be contacted for specific recommendations.

RISK ASSESSMENT: BUSINESS

General Assessment

Accidents or mishaps involving isobutyl alcohol can present a moderate threat to business operations. The loss or damage of equipment or facilities can significantly affect fiscal viability. Lawsuits that may result from personnel injury/death, public exposures, and/or environmental contamination will also require a serious expenditure of resources. Media attention surrounding an injury, death, or environmental damage can also result in a loss of profits and loss of current as well as future business. Always remember that any time the terms "cancer," or "carcinogen" are used, public emotion, hysteria, and ignorance can run equally high. This must be properly considered during the development of any public relations policies.

Recommended Risk-Reduction Measures

Company attorneys, safety and health professionals, and environmental specialists should be involved in the development of any procedures or policies intended to manage the use of chemicals in the workplace. A company official should be pre-designated as a public relations officer with specific training in dealing with the press. Corporate plans and policies should be developed, approved, and implemented long before any need for such arises.

REFERENCES

American Conference of Governmental Industrial Hygienists. 1988. *Documentation of the Threshold Limit Values and Biological Exposure Indices*, 5th Edition (with updates). Cincinnati: ACGIH

Anthony, C. P., and N. J. Kolthoff. 1971. *Textbook of Anatomy and Physiology.* St. Louis: C. V. Mosby Company.

Bloom, A. D. and F. J. DeSerres. 1995. *Ecotoxicity and Human Health.* New York: CRC Press/Lewis Publishers.

Brusick, D. J. 1994. *Methods for Genetic Risk Assessment.* New York: CRC Press/Lewis Publishers.

Bueche, J. R. 1972. *Principles of Physics*, 2nd Edition. New York: McGraw-Hill.

Calabrese, E. J. 1994. *Biological Effects of Low Level Exposures.* New York: CRC Press/Lewis Publishers.

Cothern, C. R., N. P. Ross. 1994. *Environmental Statistics, Assessment, and Forecasting.* New York: CRC Press/Lewis Publishers.

Environmental Protection Agency. 1983. *Chemical Hazard Information Profiles.* Washington, D.C.: U.S. EPA.

Forseberg, K., and L. H. Keith. 1993. *Chemical; Protective Clothing Permeation and Degradation Database.* Chelsea, MI: Lewis Publishers.

Gardner, R. W. 1994. *Chemical Intolerance.* New York: CRC Press/Lewis Publishers.

Hodgson, E., R. B. Mailman, and J. E. Chambers. 1988. *Dictionary of Toxicology.* London: The MacMillian Company.

Hosty, J. W., and P. Foster. 1990. *A Practical Guide to Chemical Spill Response.* New York: Van Nostrand Reinhold.

Howard, P. H., et. al. 1991. *Handbook of Environmental Degradation Rates.* New York: CRC Press/Lewis Publishers.

International Agency for Research on Cancer (IARC). 1987. *IARC Monographs on the Evaluation of Carcinogenic Risk to Humans, Overall Evaluations of Carcinogenicity: An updating of IARC Monographs Volumes 1-42.* Lyon: IARC.

Keith, L. H. and D. B. Walters. (Editors) 1992. *National Toxicology Program's Chemical Database, Volumes I-VII.* Chelsea, MI: Lewis Publishers.

Lauwerys, R. R., P. Hoet. 1993. *Industrial Chemical Exposure.* New York: CRC Press/Lewis Publishers.

Lewis, Richard J., Sr. 1992. *Sax's Dangerous Properties of Industrial Materials*, Volumes 1, 2, and 3, 8th Edition. New York: Van Nostrand Reinhold.

Meyer, Eugene. 1990. *Chemistry of Hazardous Materials.* Englewood Cliffs, NJ: Prentice Hall.

Patnaik, P. 1992. *A Comprehensive Guide to the Hazardous Properties of Chemical Substances.* New York: Van Nostrand Reinhold.

Philp, R. B. 1995. *Environmental Hazards and Human Health.* New York: CRC Press/Lewis Publishers.

Torkelson, T. R. V. K. Rowe. 1981. *Patty's Industrial Hygiene and Toxicology*, Vol. 2B. New York: Wiley-Interscience.

U.S. Department of Health and Human Services, National Institute for Occupational Safety and Health. 1994. *NIOSH Pocket Guide to Chemical Hazards.* Washington, D.C.: U.S. Government Printing Office.

MATERIAL SAFETY DATA SHEET

CHEMICAL NAME	CRC PRESS/LEWIS PUBLISHERS DISCLAIMER
ISOBUTYLAMINE	**CRC PRESS/LEWIS PUBLISHERS DISCLAIMER:** THE INFORMATION AND RECOMMENDATIONS PRESENTED HEREIN ARE BASED ON SOURCES BELIEVED TO BE RELIABLE. CRC MAKES NO REPRESENTATION ON ITS COMPLETENESS OR ACCURACY. IT IS THE RESPONSIBILITY OF THE USER TO DETERMINE THE CHEMICAL'S SUITABILITY FOR ITS INTENDED USE, THE CHEMICAL'S SAFE USE, AND THE CHEMICAL'S PROPER DISPOSAL. NO REPRESENTATIONS AND/OR WARRANTIES, EITHER EXPRESSED OR IMPLIED, OF THE MERCHANTABILITY OR FITNESS FOR A PARTICULAR PURPOSE, OR OF ANY OTHER NATURE, ARE MADE WITH RESPECT TO THE INFORMATION PROVIDED IN THIS MSDS OR TO THE CHEMICAL TO WHICH INFORMATION MAY REFER. CRC NEITHER ASSUMES NOR AUTHORIZES ANY OTHER PERSON TO ASSUME FOR IT, ANY OTHER ADDITIONAL RESPONSIBILITY OR LIABILITY FOR THE USE OF, OR RELIANCE UPON, THIS INFORMATION.

HAZARD WARNING INFORMATION

HEALTH	FIRE	REACTIVE	OTHER	DEGREE OF HAZARD	COLOR CODING	OTHER CODES
2	3	0		0 = Minimum Hazard 1 = Slight Hazard 2 = Moderate Hazard 3 = Serious Hazard 4 = Severe Hazard	HEALTH = BLUE FIRE = RED REACTIVITY = YELLOW OTHER = WHITE	OX = Oxidizer ACID = Acid ALK = Alkali COR = Corrosive W = Use No Water

SECTION I - GENERAL INFORMATION

Characterization	**Aliphatic Amine**	RCRA Number	**D001**
		EPA Class	**Characteristic (I) Waste**
DOT Proper Shipping Name	**Isobutylamine**	Chemical Abstract Service (CAS) Number	**78-81-9**
DOT Hazard Class and Label Requirements	**Flammable Liquid**	DOT Emergency Guide Code	**68**
DOT Identification Number	**UN 1214**	Chemical Formula	**$(CH_3)_2CHCH_2NH_2$**

Synonyms

1-Amino-2-methylpropane; monoisobutylamine; valamine; iso-butylamine; 2-methylpropylamine.

SECTION II - HAZARDOUS INGREDIENTS AND IDENTITY INFORMATION

Hazardous Components (specific identity)	OSHA Exposure Criteria	NIOSH Exposure Criteria	Immediately Dangerous to Life and Health (IDLH)	ACGIH Exposure Criteria
Isobutylamine (derivation: From isobutyl alcohol and ammonia. May also be isolated from fungi, and is a thermal decomposition product from valine and leucine).	**PEL:** Not Established **STEL:** Not Established	**REL:** Not Established **STEL:** Not Established	**Not Determined**	**TLV:** Not Established **STEL:** Not Established

SECTION III - PHYSICAL AND CHEMICAL CHARACTERISTICS

Boiling Point	**154°F (68°C)**	Specific Gravity (H_2O = 1)	**0.739**
Vapor Pressure (mm Hg)	**100 at 66°F (19°C)**	Molecular Weight	**73.16**
Vapor Density (Air = 1)	**2.5**	Melting Point	**-121°F (-85°C)**

Solubility

Completely soluble in water. Soluble in alcohol, ether, acetone, and benzene.

Appearance and Odor

Colorless or clear liquid with a strong, fish-like odor.

SECTION IV - FIRE AND EXPLOSION HAZARD DATA

Flash Point (method used)	**15°F (-9°C) closed cup**	Explosive Limits in Air % by Volume	**LEL: 3.4%** **UEL: 9%**
NFPA Classification	**Class IB Flammable Liquid**	Autoignition Temperature	**712°F (378°C)**

Extinguishing Media

Use dry chemical, carbon dioxide, foam, or water (water may not be effective by itself).

Special Fire Fighting Procedures

Poisonous gases are produced in fire; wear full protective clothing and self-contained breathing apparatus (SCBA). Vapors are heavier than air and may travel for some distance to ignition source and flashback. Move containers from fire area if it can be done safely. Cool exposed containers with water.

Unusual Fire and Explosion Hazards

Containers may explode in fire due to buildup of excessive internal pressures. Exercise extreme caution when attempting to move or relocate containers that may have been exposed to high heat. A boiling liquid expanding vapor explosion (BLEVE) can occur when isobutylamine is involved in fire. Evacuate area if rising sound is heard coming from venting device or tanks are discoloring due to fire.

SECTION V - REACTIVITY DATA

Stability		Conditions to Avoid
		Isobutylamine is normally stable under routine conditions of handling and storage. Avoid contact with incompatible materials. Keep away from heat and flame.
Stable	Unstable	Incompatibility *(materials to avoid)*
X		Oxidizing agents (such as chlorine, bromine, and oxygen) and other oxidizable materials.
Hazardous Polymerization		Conditions to Avoid
		Under normal conditions of temperature and pressure, hazardous polymerization of isobutylamine is not expected to occur.
May Occur	Will Not Occur	Hazardous Decomposition or By-products
	X	Toxic and irritating oxides of nitrogen are produced when isobutylamine is heated to decomposition or involved in a fire. These include oxides of carbon and nitrogen.

SECTION VI - HEALTH HAZARD DATA

Primary Route(s) of Entry:	Inhalation?	Absorption (skin)?	Ingestion?
	X		X

Health Hazards

INHALATION: Irritation of the nose, throat, and lungs. Will cause coughing, nausea, headache, and may cause vomiting. May cause fluid buildup in lungs (pulmonary edema) and death.

SKIN & EYES: A severe irritant. Eye contact will cause irritation and inflammation. Skin contact causes irritation, inflammation, redness, blistering, and possible dermatitis.

INGESTION: Poison by ingestion. Irritation of the mouth, throat, stomach, and digestive tract. Will also cause nausea and vomiting.

Carcinogenicity	NTP Listed?	IARC Cancer Review Group?	OSHA Regulated?	Target Organs?
Unknown Human Unknown Animal	No	No	No	Eyes; skin; respiratory tract; mucous membranes.

Medical Conditions Generally Aggravated by Exposure

Respiratory system disorders and any skin diseases may be aggravated by exposure.

Emergency and First-aid Procedures

Eye contact: Flush immediately with water for 15 minutes (minimum); seek medical attention. Skin contact: Remove all contaminated clothing. Immediately wash area with large amounts of soap and water. Seek medical assistance. For inhalation: Remove the person from exposure. Provide respiratory assistance and CPR. Transfer to medical facility. Observe for 24-48 hours for lung effects. If swallowed: Call the poison control center; seek medical attention immediately. Give 1-2 glasses of water to dilute. Do NOT induce vomiting. Never attempt to give an unconscious person anything by mouth.

SECTION VII - PRECAUTIONS FOR SAFE HANDLING AND USE

Steps to be Taken in Case Material is Released or Spilled

Absorb liquids in vermiculite or other absorbent and deposit in DOT-approved drum. Small spills may be absorbed in paper towels and evaporated in fume hood. Ventilate area and remove ignition sources. Restrict those not involved in cleanup from entering area. Notify appropriate authorities, if applicable.

Preferred Waste Disposal Method

Burn in a chemical incinerator equipped with a scrubber and afterburner.

Precautions to be Taken in Handling and Storage

Do not store isobutylamine in the presence of incompatible chemicals or materials. Store in tightly closed containers in a cool, well-ventilated area. Personnel working with isobutylamine should be trained on its proper handing and storage *prior* to being assigned to such responsibilities.

Other Precautions and Warnings

To prevent static sparks, electrically ground and bond all containers and equipment.

SECTION VIII - CONTROL MEASURES AND PERSONAL PROTECTIVE EQUIPMENT

Respiratory Protection *(specify type)*

No exposure levels have been established. This does not mean exposures are without risk to health. For low or infrequent exposures, use an organic vapor respirator. Greater protection at higher exposures is provided with an MSHA/NIOSH-approved supplied-air respirator with full facepiece operated in continuous flow or other positive pressure mode (pressure demand), or a self-contained breathing apparatus (SCBA) with full facepiece operated in pressure demand or other positive pressure mode.

Ventilation

Local exhaust or general mechanical systems recommended.

Protective Gloves	Eye Protection	Other Protective Clothing
Impervious Material	Chemical Goggles or Face Mask	Protective Apron

Work/Hygiene Practices

Always wash hands thoroughly after using chemical; never bring food, drink, or smoking materials into vicinity of chemicals.

ISOBUTYLAMINE

$(CH_3)_2CHCH_2NH_2$ CAS: 78-81-9

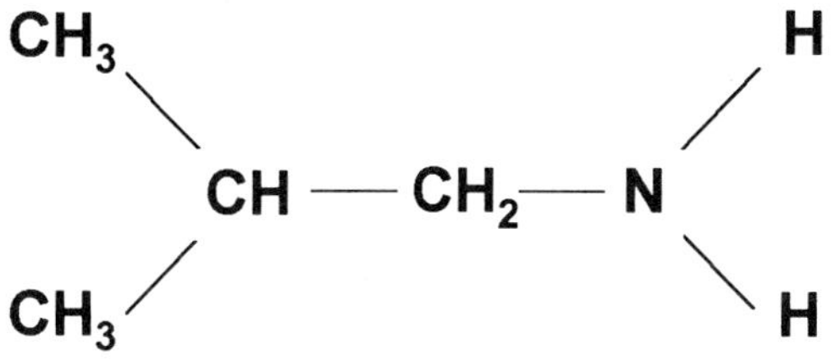

IDENTIFICATION AND TYPICAL USES

Isobutylamine is a colorless or clear liquid with a fish-like odor. It is used in organic synthesis, in insecticides, and in combination with crotonic acid to form interploymerization products suitable for wood sizing and sealing.

RISK ASSESSMENT: HEALTH

General Assessment

Isobutylamine is a human poison and is toxic by *ingestion* and *inhalation.* It is also moderately toxic by skin contact (irritant). It will irritate the nose and throat (mucous membranes) and is a severe eye and skin irritant. There are no data on its carcinogenic, mutagenic, or teratogenic capabilities.

Inhalation will cause irritation to the nose, throat, and especially the lungs. It will cause cough, nausea, and headache. Its respiratory toxicity is primarily due to pulmonary effects. On high exposures, it may cause a fluid buildup in the lungs (pulmonary edema) which is a medical emergency and can be fatal. Symptoms may be delayed up to 48 hours, thereby creating a false sense of security relative to exposure risk.

Ingestion will result in digestive system irritation, unspecified but adverse gastrointestinal effects with irritation of the mouth and stomach. There may be nausea, vomiting, cramps, and headache.

Skin contact results in severe irritation and may produce dermatitis with drying, redness, rash, and itching. Eye contact can result in severe irritation with corneal edema.

Acute Health Effects

The following acute (short-term) health effects may occur either immediately or shortly after (within hours or days) exposure to isobutylamine:

Skin: Irritation, inflammation, redness (erythema), itching, and possible dermatitis.

Eye: Severe irritation, inflammation, and possible damage to the cornea.

Lung: Irritation causing coughing and/or shortness of breath. Higher exposures can lead to respiratory problems, including pulmonary edema (fluid buildup in lungs) which is a medical emergency and can be fatal.

Chronic Health Effects

The following chronic (long-term) health effects may occur at some time after exposure to isobutylamine and can last for months or even years:

Cancer Hazards: According to the information presented in the references, isobutylamine has not been adequately tested for its ability to cause cancer in test animals.

Reproductive Hazard: According to the information presented in the references, isobutylamine has not been adequately tested for its ability to adversely affect reproduction.

Other Chronic Effects: Low, chronic exposures to skin may cause allergy or dermatitis.

Recommended Risk-Reduction Measures

Unless a less toxic chemical can be substituted for a hazardous substance, *engineering controls* are the most effective method of reducing exposures. The best protection is to enclose operations and/or provide local exhaust ventilation at the site of chemical release. While not always operationally feasible, isolating operations can also reduce exposure.

Using respiratory protection is less effective than the controls mentioned above, but is still advisable whenever working with or around isobutylamine. No exposure levels have been established for this chemical. This does not mean, however, that human exposures present no health risks. For relatively low, infrequent, or transient exposures, an organic vapor respirator may suffice. For higher or prolonged exposure, a self-contained breathing apparatus (SCBA) with full facepiece operated in pressure demand or a powered air-purifying respirator operated in continuous flow mode are the recommended respiratory protection methods of choice. If a full facepiece is not available, then chemical goggles should be worn to protect the eyes. Whenever a chemical splash hazard exists, a face shield and apron should be worn. To prevent hand and skin exposures, impervious gloves should be worn. Butyl rubber, natural rubber, neoprene, nitrile rubber, polyvinyl alcohol, polyvinyl chloride, and Viton are not suitable protection. Glove manufacturers should be contacted and permeation studies obtained *before* final glove selections have been made.

Administrative controls should also be in place to minimize the potential for human exposures. These may include written procedures or policies which specify the methods and techniques that will be practiced whenever personnel are to work with isobutylamine.

All personnel should receive *training* on the use, hazards, protective measures, emergency actions, and other precautions per 29 CFR 1910.1200 (Hazard Communication), prior to the first assignment in an area where isobutylamine is used or stored.

Before beginning employment and at regular intervals thereafter (e.g. annually), the following medical tests are recommended:

☑ Lung function tests (establish a baseline).

If symptoms develop or overexposure is suspected, the following medical tests may be useful:

☑ Lung function tests.

☑ Consider chest X-ray after acute overexposure (may be negative if performed immediately after exposure or if symptoms have not yet appeared).

☑ Evaluation by a qualified allergist with careful evaluation of exposure history and administration of special tests (may help diagnose skin allergy).

Any evaluation should include a careful medical history of past and present symptoms with an examination. However, medical tests that evaluate existing damage are not a substitute for controlling exposure. Also, since smoking can cause heart disease, emphysema, and numerous other respiratory disorders, smokers may be affected more rapidly than non-smokers to the same exposure conditions. Prudent risk assessment and management requires careful consideration of *all* possible risk factors that may be causing the appearance of symptoms in exposed workers.

Other methods to reduce exposure include:

☑ Where possible, enclose operations and use local exhaust ventilation at the site of chemical release. If local exhaust ventilation or enclosure is not used, respiratory protection should be mandatory. Also, if possible, automatically transfer liquids from storage containers to process containers using non-sparking tools.

☑ Always ensure that proper protective clothing is worn when using chemical substances and that personnel are trained on its use and care.

☑ Wash thoroughly immediately after exposure to isobutylamine and at the end of the work shift or before eating, drinking, or smoking.

☑ Hazard warning information should be posted in the work area. In addition, as part of an on-going education and training program, all information on the health and safety hazards of isobutylamine should be communicated to all exposed and potentially exposed workers.

☑ Eye wash stations should be provided in work areas. If the potential for whole body exposure exists, then safety showers should also be provided.

RISK ASSESSMENT: ENVIRONMENT

General Assessment

The environment is at risk of exposure during intended use, transportation, storage, disposal, or destruction of isobutylamine. In almost every scenario, the threat of environmental exposure is contingent upon the proper handling of the chemical substance. Accidental spills, large or small, can result in fire, explosion, and possible contamination of the surrounding environmental mediums (water, soil, and air).

Isobutylamine is considered a Class IB flammable liquid (per OSHA 29 CFR 1910.106). It can react with many common oxidizing materials to cause a fire or explosion. These characteristics require special consideration during any emergency situation involving a leak or spill of isobutylamine. Should isobutylamine ever come into contact with incompatible substances such as oxidizers and oxidizable materials, either during use, transportation, or storage, violent reactions can occur.

Isobutylamine can enter the environment through industrial discharges and spills.

Acute Ecological Effects

Acute (short-term) toxic effects may include the death of animals, birds, or fish and death or low growth rate in plants. Acute effects are seen 2 to 4 days after animals or plants are exposed to isobutylamine.

Isobutylamine has slight acute toxicity to aquatic life. Insufficient data are available to evaluate or predict the short-term effects of isobutylamine to plants, birds, or land animals.

Chronic Ecological Effects

Chronic toxic effects may include shortened life span, reproductive problems, lower fertility, and changes in appearance or behavior in exposed animals. These effects can be seen long after first exposure(s) to toxic chemicals.

Insufficient data are available to evaluate or predict the long-term effects of isobutylamine to aquatic life, plants, birds, or land animals.

Water Solubility

Isobutylamine is highly soluble in water. Concentrations of 1000 milligrams or more will mix with a liter of water.

Persistence in the Environment

Isobutylamine is moderately persistent in water, with a half-life of between 20 to 200 days. The half-life of a pollutant is the amount of time it takes for one half of the chemical to be degraded.

Bioaccumulation in Aquatic Organisms

Some substances increase in concentration, or *bioaccumulate,* in living organisms as they breathe contaminated air, drink contaminated water, or eat contaminated food. These chemicals can become concentrated in the tissues and internal organs of animals as well as humans.

Isobutylamine is not expected to accumulate in the edible tissues of fish.

Recommended Risk-Reduction Measures

Proper training of all transporters will reduce the likelihood of a mishap or accident resulting in a leak or spill to the environment. The correct use of DOT labeling (per the MSDS) on all containers, trucks, and rail cars will enable emergency responders to react properly and quickly to any disaster thereby reducing the potential risk to the environment and to personnel.

Storage of isobutylamine should be segregated from incompatible chemicals to minimize the risk of cross-contamination or contact. Storage buildings should be equipped with the proper fire protection equipment (alarms, sprinklers, extinguishers, etc.).

If a spill or leak to the environment has occurred, fire department, emergency response and/or hazardous materials spill personnel should be notified immediately. Cleanup should be attempted only by those trained in proper spill containment procedures. This chemical is not expected to adsorb significantly to terrestrial soil or aquatic sediments. It will remain in soil for extended periods and may leach to underground aquifers. Contaminated soils should therefore be removed for incineration and replaced with clean soil. If isobutylamine should contact the water table, aquifer, or navigable waterway, remediation activities should be prompt to ensure protection of aquatic sediments. When such spills occur, the local and/or state emergency response authorities must be notified. A comprehensive emergency response or disaster preparedness/recovery plan should be in place prior to any operations involving the use, transportation, storage, or disposal of isobutylamine. If isobutylamine is spilled or leaked, the following specific steps are recommended:

- ☑ Restrict persons not wearing protective clothing from area of spill or leak until cleanup is complete and area can be opened for normal work.
- ☑ Ventilate area and remove all ignition sources.
- ☑ Absorb liquid spills using vermiculite, dry earth, or sand and place in a sealed drum for disposal. Use non-sparking tools.
- ☑ It may be necessary to dispose of isobutylamine as a hazardous waste. The responsible state agency or the regional office of the federal Environmental Protection Agency (EPA) should be contacted for specific recommendations.

RISK ASSESSMENT: BUSINESS

General Assessment

Accidents or mishaps involving isobutylamine can present a significant threat to business operations. The loss or damage of equipment or facilities can significantly affect fiscal viability. Lawsuits that may result from personnel injury/death, public exposures, and/or environmental contamination will also require a serious expenditure of resources. Media attention surrounding an injury, death, or environmental damage can also result in a loss of profits and loss of current as well as future business.

Recommended Risk-Reduction Measures

Company attorneys, safety and health professionals, and environmental specialists should be involved in the development of any procedures dealing with safety or emergency response. A company official should be pre-designated as a public relations officer with specific training in dealing with the press. Corporate plans and policies should be developed, approved, and implemented long before any need for such arises.

REFERENCES

Ahlbom, A. 1993. *Biostatistics for Epidemiologists.* New York: CRC Press/Lewis Publishers.

American Conference of Governmental Industrial Hygienists. 1988. *Documentation of the Threshold Limit Values and Biological Exposure Indices,* 5th Edition (with updates). Cincinnati: ACGIH

Anthony, C. P., and N. J. Kolthoff. 1971. *Textbook of Anatomy and Physiology.* St. Louis: C. V. Mosby Company.

Asante-Duah, D. K. 1993. *Hazardous Waste Risk Assessment.* New York: CRC Press/Lewis Publishers.

Bloom, A. D. and F. J. DeSerres. 1995. *Ecotoxicity and Human Health.* New York: CRC Press/Lewis Publishers.

Bueche, J. R. 1972. *Principles of Physics,* 2nd Edition. New York: McGraw-Hill.

Burrell, R., D. K. Flaherty, and L. J. Sauers. 1992. *Toxicology of the Immune System.* New York: Van Nostrand Reinhold.

Calabrese, E. J., E. M. Kenyon. *Air Toxics and Risk Assessment.* New York: CRC Press/Lewis Publishers.

Calabrese, E. J. 1994. *Biological Effects of Low Level Exposures.* New York: CRC Press/Lewis Publishers.

Cockerman, L. G., B. S. Shane. 1994. *Basic Environmental Toxicology.* New York: CRC Press/Lewis Publishers.

Cothern, C. R., N. P. Ross. 1994. *Environmental Statistics, Assessment, and Forecasting.* New York: CRC Press/Lewis Publishers.

Cothern, R. R. 1993. *Comparative Environmental Risk Assessment.* New York: CRC Press/Lewis Publishers.

Cothern, C. R. 1995. *Handbook for Environmental Risk Decisionmaking.* New York: CRC Press/Lewis Publishers.

Gardner, R. W. 1994. *Chemical Intolerance.* New York: CRC Press/Lewis Publishers.

Hallenbeck, W. H. 1993. *Quantitative Risk Assessment for Environmental and Occupational Health,* 2nd Edition. New York: CRC Press/Lewis Publishers.

Hodgson, E., R. B. Mailman, and Janice E. Chambers. 1988. *Dictionary of Toxicology.* London: The MacMillian Company.

Howard, P. H. 1995. *Handbook of Environmental Fate and Exposures Data for Organic Chemicals,* Vol. I-IV. New York: CRC Press/Lewis Publishers.

Howard, P. H., et. al. 1991. *Handbook of Environmental Degradation Rates.* New York: CRC Press/Lewis Publishers.

Jones, F. E. 1994. *Toxic Organic Vapors in the Workplace.* New York: CRC Press/Lewis Publishers.

Landis, W. G., M. H. Yu. 1995. *Introduction to Environmental Toxicology.* New York: CRC Press/Lewis Publishers.

Langerman, N. 1994 *Precautionary Labels for Chemical Containers.* New York: CRC Press/Lewis Publishers.

Lauwerys, R. R., P. Hoet. 1993. *Industrial Chemical Exposure.* New York: CRC Press/Lewis Publishers.

Lewis, R. J., Sr. 1993. *Hawley's Condensed Chemical Dictionary,* 12th Edition. New York: Van Nostrand Reinhold.

Lewis, R. J., Sr. 1992. *Sax's Dangerous Properties of Industrial Materials,* Volumes 1, 2, and 3, 8th Edition. New York: Van Nostrand Reinhold.

Manahan, S. E. 1992. *Toxicological Chemistry,* 2nd Edition. New York: CRC Press/Lewis Publishers.

Mellström, G. A., J. E. Wahlberg, and H. I. Maibach. 1994. *Protective Gloves for Occupational Use.* New York: CRC Press/Lewis Publishers.

Meyer, E. 1990. *Chemistry of Hazardous Materials.* Englewood Cliffs, NJ: Prentice Hall.

Nelson, G. O. 1992. *Gas Mixtures.* New York: CRC Press/Lewis Publishers.

Ott, W. R. 1995. *Environmental Statistics and Data Analysis.* New York: CRC Press/Lewis Publishers.

Philp, R. B. 1995. *Environmental Hazards and Human Health.* New York: CRC Press/Lewis Publishers.

Proctor, N. H., and James P. Hughes. 1991. *Chemical Hazards of the Workplace.* New York: Van Nostrand Reinhold.

Quigley, D. R. 1994. *Handbook of Emergency Chemical Management.* New York: CRC Press/Lewis Publishers.

Rea, W. J. 1995. *Chemical Sensitivity,* Vol. I-III. New York: CRC Press/Lewis Publishers.

Richardson, M. 1995. *Dictionary of Substance and Their Effects,* Vol. I-VI. New York: CRC Press/Lewis Publishers.

Rylander, R., R. R. Jacobs. 1994. *Organic Dusts.* New York: CRC Press/Lewis Publishers.

Sacarello, H. L. A. 1994. *The Comprehensive Handbook of Hazardous Materials.* New York: CRC Press/Lewis Publishers.

U.S. Department of Health and Human Services, National Institute for Occupational Safety and Health. 1994. *NIOSH Pocket Guide to Chemical Hazards.* Washington, D.C.: U.S. Government Printing Office.

MATERIAL SAFETY DATA SHEET

CHEMICAL NAME	
ISOBUTYLENE	**CRC PRESS/LEWIS PUBLISHERS DISCLAIMER:** THE INFORMATION AND RECOMMENDATIONS PRESENTED HEREIN ARE BASED ON SOURCES BELIEVED TO BE RELIABLE. CRC MAKES NO REPRESENTATION ON ITS COMPLETENESS OR ACCURACY. IT IS THE RESPONSIBILITY OF THE USER TO DETERMINE THE CHEMICAL'S SUITABILITY FOR ITS INTENDED USE, THE CHEMICAL'S SAFE USE, AND THE CHEMICAL'S PROPER DISPOSAL. NO REPRESENTATIONS AND/OR WARRANTIES, EITHER EXPRESSED OR IMPLIED, OF THE MERCHANTABILITY OR FITNESS FOR A PARTICULAR PURPOSE, OR OF ANY OTHER NATURE, ARE MADE WITH RESPECT TO THE INFORMATION PROVIDED IN THIS MSDS OR TO THE CHEMICAL TO WHICH INFORMATION MAY REFER. CRC NEITHER ASSUMES NOR AUTHORIZES ANY OTHER PERSON TO ASSUME FOR IT, ANY OTHER ADDITIONAL RESPONSIBILITY OR LIABILITY FOR THE USE OF, OR RELIANCE UPON, THIS INFORMATION.

HAZARD WARNING INFORMATION

HEALTH	FIRE	REACTIVE	OTHER	DEGREE OF HAZARD	COLOR CODING	OTHER CODES
1	4	0		0 = Minimum Hazard 1 = Slight Hazard 2 = Moderate Hazard 3 = Serious Hazard 4 = Severe Hazard	HEALTH = BLUE FIRE = RED REACTIVITY = YELLOW OTHER = WHITE	OX = Oxidizer ACID = Acid ALK = Alkali COR = Corrosive W = Use No Water

SECTION I - GENERAL INFORMATION

Characterization: **Gas (Hydrocarbon)**	RCRA Number: **None**	EPA Class: **Not Applicable**
DOT Proper Shipping Name: **Isobutylene; Liquefied Petroleum Gas**	Chemical Abstract Service (CAS) Number: **115-11-7**	
DOT Hazard Class and Label Requirements: **Flammable Gas**	DOT Emergency Guide Code: **22**	
DOT Identification Number: **UN 1055**	Chemical Formula: **C_4H_8**	

Synonyms

2-Methylpropane; 1,1-dimethylethane; liquefied petroleum gas; trimethylmethane.

SECTION II - HAZARDOUS INGREDIENTS AND IDENTITY INFORMATION

Hazardous Components (specific identity)	OSHA Exposure Criteria	NIOSH Exposure Criteria	Immediately Dangerous to Life and Health (IDLH)	ACGIH Exposure Criteria
Isobutylene (derivation: By fractionation of refinery gases and steams).	**PEL:** Not Established **STEL:** Not Established	**REL:** Not Established **STEL:** Not Established	**Not Determined**	**TLV:** Not Established **STEL:** Not Established

SECTION III - PHYSICAL AND CHEMICAL CHARACTERISTICS

Boiling Point: **20°F (-7°C)**	Specific Gravity (H_2O = 1): **0.6**
Vapor Pressure (atmospheres): **Not Found**	Molecular Weight: **56.12**
Vapor Density (Air = 1): **1.9**	Freezing Point: **-220°F (-140°C)**

Solubility

Insoluble in water. Very soluble in alcohol, ether, and sulfuric acid.

Appearance and Odor

Colorless gas. Usually shipped as a liquefied compressed gas.

SECTION IV - FIRE AND EXPLOSION HAZARD DATA

Flash Point (method used): **-105°F (-76°C) closed cup**	Explosive Limits in Air % by Volume: **LEL: 1.8% UEL: 9.6%**
NFPA Classification: **Flammable Gas**	Autoignition Temperature: **869°F (465°C)**

Extinguishing Media

Stop flow of gas first (if possible). Use carbon dioxide, dry chemical, foam, or water spray on fire.

Special Fire Fighting Procedures

Isobutylene is an extremely flammable, very dangerous fire and explosion hazard. Poisonous gases are evolved from fire. Wear a self-contained breathing apparatus (SCBA) with full facepiece. Ensure ALL sources of escaping gas are controlled. Monitor area long after fire is extinguished. Move containers from fire area if it can be done without risk. Use water spray to knock down vapors.

Unusual Fire and Explosion Hazards

Gas is explosive. Containers may explode in fire. Even if fire is extinguished, escaping gas can form explosive mixtures in air quickly and ignite without warning. Evacuate 2500 feet from fire area.

SECTION V - REACTIVITY DATA

Stability		Conditions to Avoid
		Isobutylene is normally stable in its closed containers, during routine operations (those that do not expose it to ignition sources). Avoid contact with heat, sparks, flame, lighted tobacco products, and steam lines.
Stable	Unstable	Incompatibility *(materials to avoid)*
X		Isobutylene is reactive in contact with strong oxidizers (nitrates and perchlorates), chlorine or chlorine dioxide, fluorine, other oxidizers, nickel carbonyl + oxygen, heat, sparks, and open flame.
Hazardous Polymerization		Conditions to Avoid
		Under normal conditions of temperature and pressure, hazardous polymerization of isobutylene is not expected to occur.
May Occur	Will Not Occur	Hazardous Decomposition or By-products
	X	If involved in fire, isobutylene can produce acrid smoke and irritating fumes, such as carbon monoxide.

SECTION VI - HEALTH HAZARD DATA

Primary Route(s) of Entry:	Inhalation?	Absorption (skin)?	Ingestion?
	X		

Health Hazards

INHALATION: An asphyxiant gas. Initial symptoms are those of oxygen starvation. Feeling of "air hunger," diminished mental alertness, headache, dullness, shortness of breath, and an impairment in muscle coordination. Continued exposures lead to faulty judgment, depression of senses, rapid fatigue, unconsciousness, convulsions, coma, and death.

SKIN & EYES: Contact with the compressed gas or refrigerated liquid can cause frostbite and burns.

INGESTION: Not likely since isobutylene is a gas at normal working (room) temperatures.

Carcinogenicity	NTP Listed?	IARC Cancer Review Group?	OSHA Regulated?	Target Organs?
Insufficient data	No	No	No	Central nervous system

Medical Conditions Generally Aggravated by Exposure

None Reported.

Emergency and First-aid Procedures

Eye contact: Immediately flush with large amounts of warm (tepid) water for 15 minutes (minimum), occasionally lifting eyelids; seek medical attention. Skin contact: Remove all contaminated clothing. Immediately wash area with large amounts of tepid water. If blisters occur, seek medical attention. For inhalation: Remove the person from exposure. Provide respiratory assistance and CPR. Transfer promptly to medical facility. If swallowed: Not a likely occupational exposure route.

SECTION VII - PRECAUTIONS FOR SAFE HANDLING AND USE

Steps to be Taken in Case Material is Released or Spilled

Remove all ignition sources. Ventilate area. Stop flow of gas. If the source of leak is a cylinder and cannot be stopped, remove to safe place in open air; repair leak or let cylinder empty. If gas has not yet ignited, use a water spray to knock down vapors and direct them from ignition source.

Preferred Waste Disposal Method

None specified in the references (incineration of a gas may be the only alternative).

Precautions to be Taken in Handling and Storage

Do not store with incompatible materials or in the presence of heat since violent reactions or explosions can occur. Store in tightly closed pressurized containers in a cool, dark, well-ventilated area away from sunlight. Keep away from ignition sources such as fire, sparks, and flame.

Other Precautions and Warnings

Bulk storage of isobutylene is not recommended. Containers may explode in fire or under conditions of extreme heat (do not store outdoors or in direct sunlight).

SECTION VIII - CONTROL MEASURES AND PERSONAL PROTECTIVE EQUIPMENT

Respiratory Protection *(specify type)*

Isobutylene is a simple asphyxiant. Do NOT use cartridge-type respirators (ineffective for this type of gas). Use a powered supplied-air respirator or a self-contained breathing apparatus with full facepiece operated in pressure demand or other positive pressure mode.

Ventilation

Local exhaust or general mechanical systems recommended.

Protective Gloves	Eye Protection	Other Protective Clothing
Impervious, Thermal Protective	Goggles and/or Face Mask	Leather Apron

Work/Hygiene Practices

Always wash hands thoroughly after using chemical; never bring food, drink, or smoking materials into vicinity of chemicals.

ISOBUTYLENE

C_4H_8

CAS: 115-11-7

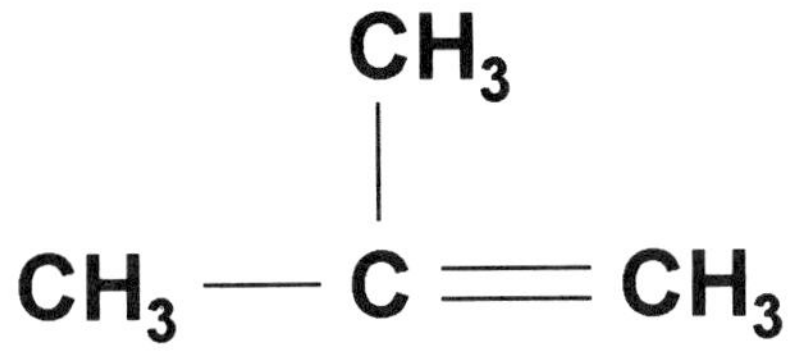

IDENTIFICATION AND TYPICAL USES

Isobutylene is an easily liquefied, colorless gas. It is usually shipped as a liquefied compressed gas. It is used in the production of other chemicals, including isooctane, high-octane aviation gasoline, butyl rubber, polyisobutene resins, tert-butyl chloride, tert-butanol methacrylates, copolymer resins with butadiene or acrylonitrile, and methyl-tert-butyl ether. Also used in the production of antioxidants for foods, plastics, and packaging food supplements.

RISK ASSESSMENT: HEALTH

General Assessment

Isobutylene is a simple asphyxiant gas. As such, its primary route of entry into the body is through *inhalation*. It can displace oxygen levels to below that which is necessary to sustain life. At high concentrations, it is a narcotic and can cause anesthesia and unconsciousness. Skin contact with the liquid can be destructive to tissues due to the extremely cold temperatures but absorption is not a likely exposure route. Since isobutylene usually exists as a gas at room temperatures, ingestion is not likely to occur either.

Inhalation of isobutylene gas can cause headache, dullness, shortness of breath, and the symptoms of asphyxiation. These include rapid respiration and a feeling of "air hunger" or air starvation followed by a decrease in mental alertness, and impaired muscular coordination. If exposure continues, there can be narcosis, anesthesia, faulty judgment, depression of the senses, emotional instability, nausea, vomiting, prostration, unconsciousness, convulsions, coma, and death.

Skin or eye contact with the vapor poses no serious threat. The vapors have not been shown to cause any physiological response. On exposure to air, the liquid evaporates rapidly and may not have time to cause damage. However, if the liquid should contact either the skin or the eyes, there can be serious irritation, burns, tissue damage, and frostbite.

Acute Health Effects

The following acute (short-term) health effects may occur immediately or shortly after exposure to isobutylene:

Skin: Liquid threat only. Redness, pain, swelling, blisters, and severe frostbite.

Eye: Liquid threat only. Causes frostbite, redness, irritation, severe pain, and blurred vision.

Lung: Vapor is non-irritating and therefore poses no threat of damage to the lungs. However, in high concentrations, isobutylene can lead to symptoms of asphyxiation and oxygen deprivation up to and including irregular heartbeat, loss of consciousness, and death.

Chronic Health Effects

The following chronic (long-term) health effects may occur at some time after exposure to isobutylene and can last for months or even years:

Cancer Hazards: There is insufficient evidence in the references to evaluate isobutylene for its ability to cause cancer in animals. It is a simple asphyxiant.

Reproductive Hazards: There are no reports to support any claims of reproductive hazards.

Other Chronic Effects: Isobutylene may be a cardiac sensitizer. More research is required in this regard.

Recommended Risk-Reduction Measures

Personnel should avoid direct contact with isobutylene. If another gas cannot be substituted for a isobutylene, then *engineering controls* are the most effective method of reducing exposures. The best protection is to separate operations and provide local exhaust ventilation at the site of chemical release. While not always operationally feasible, isolating operations can also reduce exposure. Using equipment with explosion-proof designs and ensuring all metal containers and tools are electrically grounded and bonded can further reduce the potential for injury or damage. Using respiratory protection is less effective than the controls mentioned above, but is still advisable whenever working with isobutylene. Air-purifying respirators (e.g., the cartridge-type or gas masks) cannot be used with isobutylene. These types of respirators only purify the air and can do nothing to protect in oxygen deficient atmospheres. A powered air-supplied respirator operated in pressure demand or other positive pressure mode, or a self-contained breathing apparatus (SCBA) with full facepiece operated in pressure demand mode are the recommended respiratory protection methods of choice. If a full facepiece is not available, then chemical goggles should be worn to protect the eyes. Whenever a chemical splash hazard exists, a face shield and a rubber or leather apron should be worn. To prevent hand and skin exposures with the liquid, thermal gloves should be worn. Glove manufacturers should be contacted and permeation studies obtained *before* final glove selections have been made.

Administrative controls should also be in place to minimize the potential for human exposures. These may include written procedures or policies which specify the methods and techniques that will be practiced whenever personnel are to work with isobutylene. Administrative controls as well as engineering controls must consider the effects of working with isobutylene.

All personnel should receive *training* on the use, hazards, protective measures, emergency actions, and other precautions per 29 CFR 1910.1200 (Hazard Communication), prior to the first assignment in an area where isobutylene is used or stored.

Other methods to reduce exposure to chemicals include:

- ☑ Where possible, consider installing automatic sensors that warn personnel of dropping oxygen levels and use local exhaust ventilation at the site of chemical release. If local exhaust ventilation or enclosure is not used, respiratory protection should be mandatory.
- ☑ Always ensure that proper protective clothing is worn when using chemical substances and that personnel are trained in its use and care.
- ☑ Wash thoroughly immediately after exposure to isobutylene and at the end of the work shift or before eating, drinking, or smoking. Clothing that has been contaminated with the vapor or compressed gas can present a serious fire hazard to personnel. Never wear contaminated clothing home where family members can be exposed.
- ☑ Safety shower and eyewash stations should be readily available in work areas where isobutylene is used or stored.
- ☑ Hazard warning information should be posted in the work area. In addition, as part of an on-going education and training program, all information on the health and safety hazards of isobutylene should be communicated to all exposed and potentially exposed workers.

RISK ASSESSMENT: ENVIRONMENT

General Assessment

The environment is at risk of exposure during transportation, storage, disposal, or destruction of isobutylene. In almost every scenario, the threat of environmental exposure is contingent upon the proper handling of the chemical substance. Accidental releases, large or small, can result in fire, explosion, and possible contamination of the surrounding environmental mediums (water, soil, and air). Isobutylene is a highly flammable and explosive gas. Its extremely low flash point and boiling point present a serious fire hazard. Also, since it can react violently in the presence of oxidizers, such as chlorine, isobutylene requires special consideration during any emergency involving a leak or release of isobutylene gas or liquid.

Isobutylene can enter the environment through industrial discharges, unchecked venting to air, or through spills of the liquid. Since liquids will most likely vaporize rapidly, water and soil contamination are possible but improbable.

☠ *Acute Ecological Effects*

Acute (short-term) toxic effects may include the death of animals, birds, or fish, and death or low growth rate in plants. Acute effects are seen 2 to 4 days after animals or plants are exposed to isobutylene.

Insufficient data are available on the acute (short-term) effects of isobutylene on aquatic life, plants, birds, or land animals.

Chronic Ecological Effects

Chronic (long-term) toxic effects may include shortened life span, reproductive problems, lower fertility, and changes in appearance or behavior in exposed animals. These effects can be seen long after first exposure(s) to toxic chemicals.

Insufficient data are available to evaluate or predict the long-term effects of isobutylene to aquatic life, plants, birds, or land animals.

Water Solubility

Isobutylene is only slightly soluble in water. Concentrations of 1 to 100 milligrams will mix with a liter of water.

Persistence in the Environment

Isobutylene is non-persistent in water, with a half-life of less than 2 days. The half-life of a pollutant is the amount of time it takes for one half of the chemical to be degraded. About 99.9% of isobutylene will eventually end up in air; the rest will end up in the water.

Bioaccumulation in Aquatic Organisms

Some substances increase in concentration, or *bioaccumulate,* in living organisms as they breathe contaminated air, drink contaminated water, or eat contaminated food. These chemicals can become concentrated in the tissues and internal organs of animals as well as humans.

Isobutylene is not expected to accumulate in the tissues of fish.

Recommended Risk-Reduction Measures

Proper training of all transporters will reduce the likelihood of a mishap or accident resulting in a leak or release to the environment. The correct use of DOT labeling (per the MSDS) on all containers, trucks, and rail cars will enable emergency responders to react properly and quickly to any disaster thereby reducing the potential risk to the environment and to personnel.

Storage of isobutylene should be segregated from incompatible chemicals to minimize the risk of cross-contamination or contact. Buildings designated for storage should be equipped with appropriate fire protection systems (alarms, sprinklers, emergency lighting, portable extinguishers).

If a spill or release to the environment has occurred, fire department, emergency response, and/or hazardous materials spill personnel should be notified immediately. Cleanup should be attempted only by those trained in proper containment procedures using non-sparking tools. Contaminated soils should be removed for incineration and replaced with clean soil. When spills or leaks occur, it may be necessary to contact the local and/or state emergency response authorities. A comprehensive emergency response or disaster preparedness/recovery plan should be in place prior to any operations involving the use, transportation, storage, or disposal of isobutylene.

If isobutylene is spilled or leaked, the following specific steps are recommended:

- ☑ Restrict persons not wearing protective clothing from area of spill or leak until cleanup is complete and area can be opened for normal work.
- ☑ Remove all ignition sources from the area immediately.
- ☑ Ventilate area of spill or leak.
- ☑ If applicable, stop flow of leaking liquid or gas. If leak source is a cylinder and the leak cannot be stopped in place, remove leaking cylinder to a safe place in the open air, and repair or allow cylinder to empty. Treat empty cylinders as containing some residue.
- ☑ Keep isobutylene out of a confined space, such as a sewer, because of the possibility of explosion (unless the sewer is designed to prevent the buildup of explosive concentrations).
- ☑ It may be necessary to dispose of isobutylene as a hazardous waste. The responsible state agency or the regional office of the federal Environmental Protection Agency (EPA) should be contacted for specific recommendations.

RISK ASSESSMENT: BUSINESS

General Assessment

Accidents or mishaps involving isobutylene can present a significant threat to business operations. The loss or damage of equipment or facilities can significantly affect fiscal viability. Lawsuits that may result from personnel injury, public exposures, and environmental contamination will also require a serious expenditure of resources.

Media attention surrounding an injury, illness, or environmental damage can also result in a loss of profits and loss of current as well as future business.

Recommended Risk-Reduction Measures

Company attorneys, safety and health professionals, and environmental specialists should be involved in the development of any procedures dealing with safety or emergency response. A company official should be pre-designated as a public relations officer with specific training in dealing with the press. Corporate plans and policies should be developed, approved, and implemented long before any need for such arises.

REFERENCES

Ahlbom, A. 1993. *Biostatistics for Epidemiologists.* New York: CRC Press/Lewis Publishers.

American Conference of Governmental Industrial Hygienists. 1988. *Documentation of the Threshold Limit Values and Biological Exposure Indices,* 5th Edition (with updates). Cincinnati: ACGIH

Anthony, C. P., and N. J. Kolthoff. 1971. *Textbook of Anatomy and Physiology.* St. Louis: C. V. Mosby Company.

Asante-Duah, D. K. 1993. *Hazardous Waste Risk Assessment.* New York: CRC Press/Lewis Publishers.

Bloom, A. D. and F. J. DeSerres. 1995. *Ecotoxicity and Human Health.* New York: CRC Press/Lewis Publishers.

Brusick, D. J. 1994. *Methods for Genetic Risk Assessment.* New York: CRC Press/Lewis Publishers.

Bueche, J. R. 1972. *Principles of Physics,* 2nd Edition. New York: McGraw-Hill.

Burrell, R., D. K. Flaherty, and L. J. Sauers. 1992. *Toxicology of the Immune System.* New York: Van Nostrand Reinhold.

Calabrese, E. J. 1994. *Biological Effects of Low Level Exposures.* New York: CRC Press/Lewis Publishers.

Cockerman, L. G., B. S. Shane. 1994. *Basic Environmental Toxicology.* New York: CRC Press/Lewis Publishers.

Cothern, C. R. 1995. *Handbook for Environmental Risk Decisionmaking.* New York: CRC Press/Lewis Publishers.

Danse, I. R. 1991. *Common Sense Toxics in the Workplace.* New York: Van Nostrand Reinhold.

Environmental Protection Agency. 1983. *Chemical Hazard Information Profiles.* Washington, D.C.: U.S. EPA.

Fodden, C. R., and J. L. Weddell. 1992. *Hazardous Materials Emergency Action Data.* New York: CRC Press/Lewis Publishers.

Forseberg, K., and L. H. Keith. 1993. *Chemical; Protective Clothing Permeation and Degradation Database.* Chelsea, MI: Lewis Publishers.

Gardner, R. W. 1994. *Chemical Intolerance.* New York: CRC Press/Lewis Publishers.

Hallenbeck, W. H. 1993. *Quantitative Risk Assessment for Environmental and Occupational Health,* 2nd Edition. New York: CRC Press/Lewis Publishers.

Hodgson, E., R. B. Mailman, and J. E. Chambers. 1988. *Dictionary of Toxicology.* London: The MacMillian Company.

Howard, P. H. 1995. *Handbook of Environmental Fate and Exposures Data for Organic Chemicals,* Vol. I-IV. New York: CRC Press/Lewis Publishers.

James, A., and M. Lord. 1992. *Index of Chemical and Physical Data.* New York: Van Nostrand Reinhold.

Jones, F. E. 1994. *Toxic Organic Vapors in the Workplace.* New York: CRC Press/Lewis Publishers.

Keith, L. H. and M. M. Walker. 1995. *Handbook of Air Toxics.* New York: CRC Press/Lewis Publishers.

Keith, L. H. (Editor). 1992. *Integrated Risk Information System (IRIS) Chemical Information Database.* Prepared by the U.S. Environmental Protection Agency. Boca Raton, FL: CRC Press/Lewis Publishers.

Keith, L. H. and D. B. Walters. (Editors) 1992. *National Toxicology Program's Chemical Database, Volumes I-VII.* Chelsea, MI: Lewis Publishers.

Landis, W. G., M. H. Yu. 1995. *Introduction to Environmental Toxicology.* New York: CRC Press/Lewis Publishers.

Langerman, N. 1994 *Precautionary Labels for Chemical Containers.* New York: CRC Press/Lewis Publishers.

Lauwerys, R. R., P. Hoet. 1993. *Industrial Chemical Exposure.* New York: CRC Press/Lewis Publishers.

Lewis, R. J., Sr. 1993. *Hawley's Condensed Chemical Dictionary,* 12th Edition. New York: Van Nostrand Reinhold.

Lewis, R. J., Sr. 1992. *Sax's Dangerous Properties of Industrial Materials,* Volumes 1, 2, and 3, 8th Edition. New York: Van Nostrand Reinhold.

Manahan, S. E. 1992. *Toxicological Chemistry,* 2nd Edition. New York: CRC Press/Lewis Publishers.

Meyer, E. 1990. *Chemistry of Hazardous Materials.* Englewood Cliffs, NJ: Prentice Hall.

Patnaik, P. 1992. *A Comprehensive Guide to the Hazardous Properties of Chemical Substances.* New York: Van Nostrand Reinhold.

Philp, R. B. 1995. *Environmental Hazards and Human Health.* New York: CRC Press/Lewis Publishers.

Proctor, N. H., and J. P. Hughes. 1991. *Chemical Hazards of the Workplace.* New York: Van Nostrand Reinhold.

Rea, W. J. 1995. *Chemical Sensitivity,* Vol. I-III. New York: CRC Press/Lewis Publishers.

Richardson, M. 1995. *Dictionary of Substance and Their Effects,* Vol. I-VI. New York: CRC Press/Lewis Publishers.

Sacarello, H. L. A. 1994. *The Comprehensive Handbook of Hazardous Materials.* New York: CRC Press/Lewis Publishers.

Sherman, J. D. 1988. *Chemical Exposure and Disease.* New York: Van Nostrand Reinhold.

Torkelson, T. R. V. K. Rowe. 1981. *Patty's Industrial Hygiene and Toxicology,* Vol. 2B. New York: Wiley-Interscience.

MATERIAL SAFETY DATA SHEET

CHEMICAL NAME

ISOBUTYRALDEHYDE

HAZARD WARNING INFORMATION

HEALTH	FIRE	REACTIVE	OTHER	DEGREE OF HAZARD	COLOR CODING	OTHER CODES
1	4	0		0 = Minimum Hazard 1 = Slight Hazard 2 = Moderate Hazard 3 = Serious Hazard 4 = Severe Hazard	HEALTH = BLUE FIRE = RED REACTIVITY = YELLOW OTHER = WHITE	OX = Oxidizer ACID = Acid ALK = Alkali COR = Corrosive ~~W~~ = Use No Water

SECTION I - GENERAL INFORMATION

Characterization: **Aldehyde**	RCRA Number: **None** / EPA Class: **Not Applicable**
DOT Proper Shipping Name: **Isobutyraldehyde**	Chemical Abstract Service (CAS) Number: **78-84-2**
DOT Hazard Class and Label Requirements: **Flammable Liquid**	DOT Emergency Guide Code: **26**
DOT Identification Number: **UN 2045**	Chemical Formula: **C_3H_7CHO**

Synonyms

2-Methylpropanol; 2-methyl-1-propanol; isobutylaldehyde; 2-methylpropionaldehyde; isobutyric aldehyde; isobutyl aldehyde; isobutyric aldehyde; valine aldehyde.

SECTION II - HAZARDOUS INGREDIENTS AND IDENTITY INFORMATION

Hazardous Components (specific identity)	OSHA Exposure Criteria	NIOSH Exposure Criteria	Immediately Dangerous to Life and Health (IDLH)	ACGIH Exposure Criteria
Isobutyraldehyde (derivation: By the oxo process reaction of propylene with carbon monoxide and hydrogen; dehydration of isobutanol).	**PEL:** Not Established **STEL:** Not Established	**REL:** Not Established **STEL:** Not Established	**Not Determined**	**TLV:** Not Established **STEL:** Not Established

SECTION III - PHYSICAL AND CHEMICAL CHARACTERISTICS

Boiling Point: **147°F (64°C)**	Specific Gravity (H_2O = 1): **0.783**
Vapor Pressure (mm Hg): **6 at 69°F (20°C)**	Molecular Weight: **72.1**
Vapor Density (Air = 1): **2.5**	Melting Point: **-87°F (-66°C)**

Solubility

Slightly soluble in water; soluble in alcohol, acetone, benzene, carbon disulfide, chloroform, and ether.

Appearance and Odor

Clear colorless mobile liquid with a pungent odor.

SECTION IV - FIRE AND EXPLOSION HAZARD DATA

Flash Point (method used): **-20°F (7°C) opened cup**	Explosive Limits in Air % by Volume: **LEL: Not Determined** **UEL: Not Determined**
NFPA Classification: **Class IB Flammable Liquid**	Autoignition Temperature: **Not Determined**

Extinguishing Media

Use water spray, dry chemical, alcohol foam, or carbon dioxide. Use water spray to disperse vapors.

Special Fire Fighting Procedures

Poisonous gases are produced in fire. Wear full protective clothing and self-contained breathing apparatus (SCBA). Heat will build pressure and may rupture closed storage containers. Continue to cool containers with water spray after fire is extinguished. Water on fire itself may be ineffective.

Unusual Fire and Explosion Hazards

Immediately withdraw if rising sound from venting device is heard or if fire is causing discoloration to the tank. Vapors may travel to a source of ignition and flash back.

SECTION V - REACTIVITY DATA

Stability		Conditions to Avoid
		Isobutyraldehyde is normally stable under routine conditions of storage and handling. Avoid contact with heat, flame, or other sources of ignition. In contact with air, it will slowly oxidize to isobutyric acid.
Stable	Unstable	Incompatibility *(materials to avoid)*
X		Reacts violently in contact with strong oxidizers (perchlorates, peroxides, permanganates, chlorates, and nitrates). Do not allow isobutyraldehyde to remain in contact with air for prolonged periods.
Hazardous Polymerization		Conditions to Avoid Under normal conditions of temperature and pressure, isobutyraldehyde is not known to undergo hazardous polymerization.
May Occur	Will Not Occur	Hazardous Decomposition or By-products
X		When heated to decomposition, isobutyraldehyde emits acrid and irritating smoke and fumes, including carbon monoxide.

SECTION VI - HEALTH HAZARD DATA

Primary Route(s) of Entry:	Inhalation?	Absorption (skin)?	Ingestion?
	X		X

Health Hazards

INHALATION: Vapors can irritate the nose, throat, and lungs. High exposures may cause a buildup of fluid in the lungs (pulmonary edema), a medical emergency. May also cause symptoms of dizziness and lightheadedness.

SKIN & EYES: Will cause severe skin and eye irritation and even burns. Eye contact can cause damage to the cornea and possibly affect visual acuity.

INGESTION: May cause vomiting and nausea. Large doses can cause adverse systemic effects.

Carcinogenicity	NTP Listed?	IARC Cancer Review Group?	OSHA Regulated?	Target Organs?
Unknown Human Unknown Animal	No	No	No	Respiratory system; skin; eyes.

Medical Conditions Generally Aggravated by Exposure

None reported. However, pre-existing lung ailments or skin disorders may be aggravated by exposure.

Emergency and First-aid Procedures

Eye contact: Flush immediately with water for 15 minutes (minimum); seek medical attention. Skin contact: Remove all contaminated clothing. Immediately wash area with large amounts of soap and water. Seek medical assistance. For inhalation: Remove the person from exposure. Provide respiratory assistance and CPR. Transfer to medical facility. Observe for 24-48 hours for lung effects (pulmonary edema). If swallowed: Seek medical attention immediately. Call poison control center for advice.

SECTION VII - PRECAUTIONS FOR SAFE HANDLING AND USE

Steps to be Taken in Case Material is Released or Spilled

Contain spills using absorbent material (vermiculite, or other). Ventilate area of spill. Remove all potential ignition sources. Store materials in DOT-approved containers. Restrict those not involved in cleanup from entering area. Notify appropriate authorities, as required.

Preferred Waste Disposal Method

Burn in a chemical incinerator equipped with an afterburner and scrubber.

Precautions to be Taken in Handling and Storage

Do not store with incompatible chemicals. Store in tightly closed containers in a cool, dark, well-ventilated area. Use explosion-proof equipment in areas where isobutyraldehyde is used or stored. Keep heat, fire, and flame away. If possible, automatically transfer liquids between containers.

Other Precautions and Warnings

Bulk storage of isobutyraldehyde is not recommended. Use non-sparking tools and equipment.

SECTION VIII - CONTROL MEASURES AND PERSONAL PROTECTIVE EQUIPMENT

Respiratory Protection *(specify type)*

No exposure limits have been established. This does not mean that unprotected exposure to isobutyraldehyde is permissible. For high exposures, use a self-contained breathing apparatus (SCBA) with full facepiece and pressure demand or other positive pressure mode.

Ventilation

Local exhaust or general mechanical systems recommended.

Protective Gloves	Eye Protection	Other Protective Clothing
Impervious Material	Chemical Goggles or Face Mask	Impermeable Apron

Work/Hygiene Practices

Always wash hands thoroughly after using chemical; never bring food, drink, or smoking materials into vicinity of chemicals.

ISOBUTYRALDEHYDE

C_3H_7CHO

CAS: 78-84-2

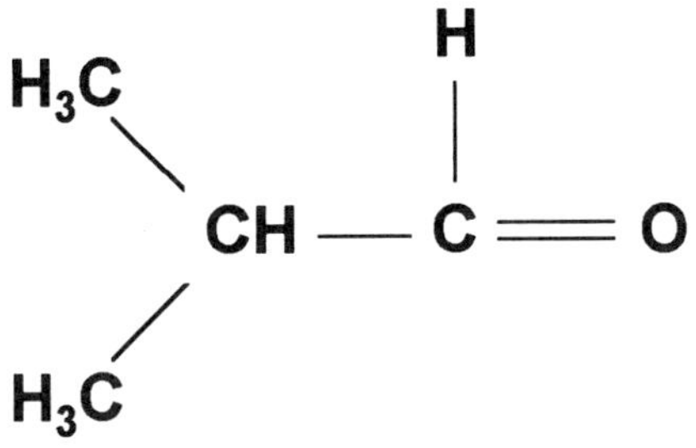

IDENTIFICATION AND TYPICAL USES

Isobutyraldehyde is a clear, colorless mobile liquid with a pungent odor. It is used as an intermediate for rubber antioxidants and accelerators, for neopentyl glycol, in the synthesis of amino acids, in cellulose esters, and it is a perfume, flavor, resin, and gasoline additive.

RISK ASSESSMENT: HEALTH

General Assessment

Isobutyraldehyde is moderately toxic *by ingestion, inhalation,* and by skin contact, although dermal absorption has not been reported. The effect of exposure to isobutyraldehyde is primarily narcotic. There are no reports on its carcinogenic, mutagenic, or teratogenic capabilities.

Inhalation may cause irritation of the nose and throat with possible nausea, vomiting, headaches, and loss of consciousness. High or prolonged exposures cause dizziness, delayed hypersensitivity, and lightheadedness. Such exposures may also cause fluid to buildup in the lungs (a condition known as pulmonary edema), which is a medical emergency and can be fatal. Symptoms of pulmonary edema may be delayed up to 48 hours, thereby creating a false sense of security with regard to exposure risk.

Isobutyraldehyde is a mild skin and eye irritant. However, prolonged contact may result in burns to the skin and eyes. Eye contact can also cause severe irritation and possible damage. If ingested, isobutyraldehyde can cause gastrointestinal irritation and may lead to toxic systemic effects. The data on the effects of human exposure are somewhat limited in the references and caution is therefore warranted whenever working with or around isobutyraldehyde.

Acute Health Effects

The following acute (short-term) health effects may occur immediately or shortly after exposure to isobutyraldehyde:

Skin: Irritation and possible burns.

Eye: Irritation and possible burns which can lead to injury and permanent damage if allowed to remain in contact with the eyes.

Lung: Irritation of the nose, throat and lungs. Can cause pulmonary edema (fluid buildup in the lungs) which can be fatal.

CNS: Insufficient evidence of nerve action on acute exposures. However, it is reported that exposures to high levels result in an anesthetic effect on test animals.

Chronic Health Effects

The following chronic (long-term) health effects may occur at some time after exposure to isobutyraldehyde and can last for months or even years:

Cancer Hazards: According to the references, isobutyraldehyde has not been adequately tested for its ability to cause cancer in test animals.

Reproduction: According to the references, isobutyraldehyde has not been adequately tested for its ability to adversely affect reproduction in test animals.

Other Chronic Effects: Isobutyraldehyde has not been tested for other chronic (long-term) health effects.

Recommended Risk-Reduction Measures

Personnel should avoid direct contact with isobutyraldehyde, especially skin and eye contact. The best risk reduction measure is to use a less toxic chemical as a substitute for a isobutyraldehyde. If this is not possible or feasible, then *engineering controls* are the most effective method of reducing exposures. The best protection is to enclose operations and/or provide local exhaust ventilation at the site of chemical release. While not always operationally feasible, isolating operations can also reduce exposure risk.

Using respiratory protection is less effective than the controls mentioned above, but is still advisable whenever working with or around isobutyraldehyde. No exposure levels have been established for this chemical. However, this does not mean that unprotected exposure is acceptable. For low level, infrequent exposures, an MSHA/NIOSH-approved chemical respirator with an organic vapor cartridge should be used. For higher exposures or when exposure levels are not known, a self-contained breathing apparatus (SCBA) with full facepiece operated in pressure demand or a powered air-purifying respirator operated in continuous flow mode are the recommended respiratory protection methods of choice. If a full facepiece is not available, then chemical goggles should be worn to protect the eyes. Whenever a chemical splash hazard exists, a face shield and a protective apron should be worn. To prevent hand and skin exposures, impervious gloves should be used. Glove manufacturers should be contacted and permeation studies obtained *before* final glove selections have been made.

Administrative controls should also be in place to minimize the potential for human exposures. These may include written procedures or policies which specify the methods and techniques that will be practiced whenever personnel are to work with isobutyraldehyde.

All personnel should receive *training* on the use, hazards, protective measures, emergency actions, and other precautions per 29 CFR 1910.1200 (Hazard Communication), prior to the first assignment in an area where isobutyraldehyde is used or stored.

If symptoms develop or overexposure is suspected, the following medical tests are recommended:

- ☑ Consider chest X-ray after acute exposure (may be negative if taken immediately following exposure due to delayed onset of pulmonary edema).

Any medical evaluation should include a careful history of past and present symptoms with an examination. Medical tests that look for existing damage are not a substitute for controlling exposures Also, since smoking can cause heart disease, lung cancer, emphysema, and other respiratory diseases, smokers exposed to isobutyraldehyde may experience symptoms more rapidly than non-smokers under the same conditions of exposure. Prudent risk management requires the proper consideration of *all* possible factors which may be causing the appearance of exposure symptoms in the workplace.

Other methods to reduce exposure include:

- ☑ Where possible, enclose operations and use local exhaust ventilation at the site of chemical release. If local exhaust ventilation or enclosure is not used, respiratory protection should be mandatory.
- ☑ Always ensure that proper protective clothing is worn when using chemical substances.
- ☑ Wash thoroughly immediately after exposure to isobutyraldehyde and at the end of the work shift or before eating, drinking, or smoking.
- ☑ Hazard warning information should be posted in the work area. In addition, as part of an on-going education and training program, all information on the health and safety hazards of isobutyraldehyde should be communicated to all exposed or potentially exposed workers.
- ☑ Never eat, drink, or smoke in areas were isobutyraldehyde is used, handled, or stored.
- ☑ Where possible, automatically pump liquid isobutyraldehyde from drums or other containers to process containers.
- ☑ Before entering a confined space where isobutyraldehyde may be present, verify that an explosive concentration does not exist.

RISK ASSESSMENT: ENVIRONMENT
General Assessment

The environment is at risk of exposure during transportation, storage, disposal, or destruction of isobutyraldehyde. In almost every scenario, the threat of environmental exposure is contingent upon the proper handling of the chemical substance.

Accidental spills, large or small, can result in fire, explosion, and possible contamination of the surrounding environmental mediums (water, soil, and air). Isobutyraldehyde is considered a Class IB flammable liquid (according to OSHA 29 CFR 1910.106) which presents a serious fire hazard concern. Also, because it is incompatible with a number of common materials, especially strong oxidizers (such as perchlorates, peroxides, permanganates, chlorates, and nitrates), contact can result in violent and explosive reactions. It can also form explosive mixtures in air. These characteristics require special consideration during any emergency situation involving a leak or spill of isobutyraldehyde.

Isobutyraldehyde can enter the environment as a result of unchecked industrial discharges into effluents and through spills.

Acute Ecological Effects

Acute (short-term) toxic effects may include the death of animals, birds, or fish, and death or low growth rate in plants. Acute effects are seen 2 to 4 days after animals or plants are exposed to isobutyraldehyde. This chemical has moderate acute toxicity to aquatic life. Insufficient data are available to evaluate or predict the short-term effects of isobutyraldehyde to plants, birds, or terrestrial animals.

Chronic Ecological Effects

Chronic toxic effects may include shortened life span, reproductive problems, lower fertility, and changes in appearance or behavior in exposed animals. These effects can be seen long after first exposure(s) to toxic chemicals. Isobutyraldehyde has moderate chronic toxicity to aquatic life. Insufficient data are available to evaluate or predict the long-term effects of isobutyraldehyde to plants, birds, or land animals.

Water Solubility

Isobutyraldehyde is highly soluble in water. Concentrations of 1000 milligrams and more will mix with a liter of water.

Persistence in the Environment

Isobutyraldehyde is slightly persistent in water, with a half-life of between 2 to 20 days. The half-life of a pollutant is the amount of time it takes for one-half of the chemical to be degraded. Approximately 88% of isobutyraldehyde that will eventually end up in air with the remainder ending up in water.

Bioaccumulation in Aquatic Organisms

Some substances increase in concentration, or *bioaccumulate,* in living organisms as they breathe contaminated air, drink contaminated water, or eat contaminated food. These chemicals can become concentrated in the tissues and internal organs of animals as well as humans.

The concentration of isobutyraldehyde found in fish tissues is expected to be about the same the average concentration of isobutyraldehyde in water from which the fish was taken.

Recommended Risk-Reduction Measures

Proper training of all transporters will reduce the likelihood of a mishap or accident resulting in a leak or spill to the environment. The correct use of DOT labeling (per the MSDS) on all containers, trucks, and rail cars will enable emergency responders to react properly and quickly to any disaster thereby reducing the potential risk to the environment and to personnel.

Storage of isobutyraldehyde should be segregated from incompatible chemicals to minimize the risk of cross-contamination or contact. Buildings designated for storage should be equipped with appropriate fire protection systems (alarms, sprinklers, emergency lighting, portable extinguishers). Explosion-proof equipment is recommended in areas where isobutyraldehyde is used, handled, or stored.

If a spill or leak to the environment has occurred, fire department, emergency response, and/or hazardous materials spill personnel should be notified immediately. Cleanup should be attempted only by those trained in proper spill containment procedures using non-sparking tools. Contaminated soils should be removed for incineration and replaced with clean soil.

If isobutyraldehyde should contact the water table, aquifer, or navigable waterway, time is of the essence. It is highly soluble in water and, therefore, total containment and remediation may not be possible. When such spills occur, the local and/or state emergency response authorities must be notified. A comprehensive emergency response or disaster preparedness/recovery plan should be in place prior to any operations involving the use, transportation, storage, or disposal of isobutyraldehyde. If isobutyraldehyde is spilled or leaked, the following specific steps are recommended:

- ☑ Restrict persons not wearing protective clothing from area of spill or leak until cleanup is complete and area can be opened for normal work.
- ☑ Remove all ignition sources.
- ☑ Ventilate area of spill or leak.
- ☑ Absorb liquids in vermiculite, dry sand, earth, or a similar material and deposit in sealed containers.

☑ It may be necessary to dispose of isobutyraldehyde as a hazardous waste. The responsible state agency or the regional office of the federal Environmental Protection Agency (EPA) should be contacted for specific recommendations.

RISK ASSESSMENT: BUSINESS

General Assessment

Accidents or mishaps involving isobutyraldehyde can present a serious threat to business operations. The loss or damage of equipment or facilities can significantly affect fiscal viability. Lawsuits that may result from personnel injury/death, public exposures, and/or environmental contamination will also require a serious expenditure of resources. Media attention surrounding an injury, death, or environmental damage can also result in a loss of profits and loss of current as well as future business.

Recommended Risk-Reduction Measures

Company attorneys, safety and health professionals, and environmental specialists should be involved in the development of any procedures or policies intended to manage the use of chemicals in the workplace. A company official should be pre-designated as a public relations officer with specific training in dealing with the press. Corporate plans and policies should be developed, approved, and implemented long before any need for such arises.

REFERENCES

American Conference of Governmental Industrial Hygienists. 1988. *Documentation of the Threshold Limit Values and Biological Exposure Indices,* 5th Edition (with updates). Cincinnati: ACGIH

Ahlbom, A. 1993. *Biostatistics for Epidemiologists.* New York: CRC Press/Lewis Publishers.

Anthony, C. P., and N. J. Kolthoff. 1971. *Textbook of Anatomy and Physiology.* St. Louis: C. V. Mosby Company.

Asante-Duah, D. K. 1993. *Hazardous Waste Risk Assessment.* New York: CRC Press/Lewis Publishers.

Bloom, A. D. and F. J. DeSerres. 1995. *Ecotoxicity and Human Health.* New York: CRC Press/Lewis Publishers.

Brusick, D. J. 1994. *Methods for Genetic Risk Assessment.* New York: CRC Press/Lewis Publishers.

Bueche, J. R. 1972. *Principles of Physics,* 2nd Edition. New York: McGraw-Hill.

Burrell, R., D. K. Flaherty, and L. J. Sauers. 1992. *Toxicology of the Immune System.* New York: Van Nostrand Reinhold.

Calabrese, E. J., E. M. Kenyon. *Air Toxics and Risk Assessment.* New York: CRC Press/Lewis Publishers.

Cockerman, L. G., B. S. Shane. 1994. *Basic Environmental Toxicology.* New York: CRC Press/Lewis Publishers.

Cothern, C. R. 1995. *Handbook for Environmental Risk Decisionmaking.* New York: CRC Press/Lewis Publishers.

Danse, I. R. 1991. *Common Sense Toxics in the Workplace.* New York: Van Nostrand Reinhold.

Gardner, R. W. 1994. *Chemical Intolerance.* New York: CRC Press/Lewis Publishers.

Hallenbeck, W. H. 1993. *Quantitative Risk Assessment for Environmental and Occupational Health,* 2nd Edition. New York: CRC Press/Lewis Publishers.

Hodgson, E., R. B. Mailman, and J. E. Chambers. 1988. *Dictionary of Toxicology.* London: The MacMillian Company.

Hosty, J. W., and P. Foster. 1990. *A Practical Guide to Chemical Spill Response.* New York: Van Nostrand Reinhold.

Howard, P. H. 1995. *Handbook of Environmental Fate and Exposures Data for Organic Chemicals,* Vol. I-IV. New York: CRC Press/Lewis Publishers.

Jones, F. E. 1994. *Toxic Organic Vapors in the Workplace.* New York: CRC Press/Lewis Publishers.

Kamrin, M. 1 A. 1988. *Toxicology: A Primer on Principles and Applications.* Chelsea, Michigan: Lewis Publishers.

Keith, L. H. and M. M. Walker. 1995. *Handbook of Air Toxics.* New York: CRC Press/Lewis Publishers.

Landis, W. G., M. H. Yu. 1995. *Introduction to Environmental Toxicology.* New York: CRC Press/Lewis Publishers.

Langerman, N. 1994 *Precautionary Labels for Chemical Containers.* New York: CRC Press/Lewis Publishers.

Lauwerys, R. R., P. Hoet. 1993. *Industrial Chemical Exposure.* New York: CRC Press/Lewis Publishers.

Lewis, Richard J., Sr. 1992. *Sax's Dangerous Properties of Industrial Materials,* Volumes 1, 2, and 3, 8th Edition. New York: Van Nostrand Reinhold.

Lisella, Frank S. 1994. *The VNR Dictionary of Environmental Health and Safety.* New York: Van Nostrand Reinhold.

Manahan, S. E. 1994. *Environmental Chemistry,* 6th Edition. New York: CRC Press/Lewis Publishers.

Manahan, S. E. 1992. *Toxicological Chemistry,* 2nd Edition. New York: CRC Press/Lewis Publishers.

Mellström, G. A., J. E. Wahlberg, and H. I. Maibach. 1994. *Protective Gloves for Occupational Use.* New York: CRC Press/Lewis Publishers.

Meyer, Eugene. 1990. *Chemistry of Hazardous Materials.* Englewood Cliffs, NJ: Prentice Hall.

Patnaik, P. 1992. *A Comprehensive Guide to the Hazardous Properties of Chemical Substances.* New York: Van Nostrand Reinhold.

Proctor, N. H., and J. P. Hughes. 1991. *Chemical Hazards of the Workplace.* New York: Van Nostrand Reinhold.

MATERIAL SAFETY DATA SHEET

CHEMICAL NAME

ISOBUTYRONITRILE

CRC PRESS/LEWIS PUBLISHERS DISCLAIMER: THE INFORMATION AND RECOMMENDATIONS PRESENTED HEREIN ARE BASED ON SOURCES BELIEVED TO BE RELIABLE. CRC MAKES NO REPRESENTATION ON ITS COMPLETENESS OR ACCURACY. IT IS THE RESPONSIBILITY OF THE USER TO DETERMINE THE CHEMICAL'S SUITABILITY FOR ITS INTENDED USE, THE CHEMICAL'S SAFE USE, AND THE CHEMICAL'S PROPER DISPOSAL. NO REPRESENTATIONS AND/OR WARRANTIES, EITHER EXPRESSED OR IMPLIED, OF THE MERCHANTABILITY OR FITNESS FOR A PARTICULAR PURPOSE, OR OF ANY OTHER NATURE, ARE MADE WITH RESPECT TO THE INFORMATION PROVIDED IN THIS MSDS OR TO THE CHEMICAL TO WHICH INFORMATION MAY REFER. CRC NEITHER ASSUMES NOR AUTHORIZES ANY OTHER PERSON TO ASSUME FOR IT, ANY OTHER ADDITIONAL RESPONSIBILITY OR LIABILITY FOR THE USE OF, OR RELIANCE UPON, THIS INFORMATION.

HAZARD WARNING INFORMATION

HEALTH	FIRE	REACTIVE	OTHER	DEGREE OF HAZARD	COLOR CODING	OTHER CODES
3	3	0		0 = Minimum Hazard 1 = Slight Hazard 2 = Moderate Hazard 3 = Serious Hazard 4 = Severe Hazard	HEALTH = BLUE FIRE = RED REACTIVITY = YELLOW OTHER = WHITE	OX = Oxidizer ACID = Acid ALK = Alkali COR = Corrosive ~~W~~ = Use No Water

SECTION I - GENERAL INFORMATION

Characterization: **Aliphatic Nitrile**

RCRA Number: **P030**

EPA Class: **Acute Hazardous Waste**

DOT Proper Shipping Name: **Isobutyronitrile**

Chemical Abstract Service (CAS) Number: **78-82-0**

DOT Hazard Class and Label Requirements: **Flammable Liquid; Poison**

DOT Emergency Guide Code: **28**

DOT Identification Number: **UN 2284**

Chemical Formula: **$(CH_3)_2CHCN$**

Synonyms

Isopropyl cyanide; 2-methylpropionitrile; 1-cyano-1-methylethane; 2-cyanopropane; dimethylacetonitrile; IBN; 2-methylpropanenitrile.

SECTION II - HAZARDOUS INGREDIENTS AND IDENTITY INFORMATION

Hazardous Components (specific identity)	OSHA Exposure Criteria	NIOSH Exposure Criteria	Immediately Dangerous to Life and Health (IDLH)	ACGIH Exposure Criteria
Isobutyronitrile (derivation: Obtained from isobutyraldehyde). **1 ppm = 2.87 mg/m^3**	**PEL:** Not Established **STEL:** Not Established	**REL:** 22 mg/m^3 (for nitriles) **STEL:** Not Established	**No Determined**	**TLV:** Not Established **STEL:** Not Established

SECTION III - PHYSICAL AND CHEMICAL CHARACTERISTICS

Boiling Point: **214°F (101°C)**

Specific Gravity (H_2O = 1): **0.773**

Vapor Pressure (mm Hg): **100 at 130°F (54°C)**

Molecular Weight: **69.12**

Vapor Density (Air = 1): **2.38**

Freezing Point: **-103°F (-75°C)**

Solubility

Slightly soluble in water. Soluble in alcohol, acetone, and ether.

Appearance and Odor

Colorless liquid with an odor of almonds.

SECTION IV - FIRE AND EXPLOSION HAZARD DATA

Flash Point (method used): **47°F (8°C) closed cup**

Explosive Limits in Air % by Volume: **LEL: Not Determined UEL: Not Determined**

NFPA Classification: **Class 1C Flammable Liquid**

Autoignition Temperature: **900°F (482°C)**

Extinguishing Media

Carbon dioxide, dry chemical, alcohol-resistant foam, or fog. Water by itself may be ineffective.

Special Fire Fighting Procedures

Poisonous gases are produced in fire. Wear full protective clothing and self-contained breathing apparatus (SCBA). Dangerous fire hazard. Use a water spray to keep fire-exposed containers cool. Move containers from fire area if it can be done without risk.

Unusual Fire and Explosion Hazards

Containers may explode in fire or when exposed to excessive amounts of heat. Vapors are heavier than air; flashback is likely.

SECTION V - REACTIVITY DATA

Stability		Conditions to Avoid **Isobutyronitrile is normally stable in closed containers under routine conditions of storage and handling. Avoid contact with incompatible materials or exposure to excessive heat, sparks, or flame.**
Stable X	Unstable	Incompatibility *(materials to avoid)* **Incompatible with strong oxidizers (such as chlorine, bromine, fluorine, perchlorates, nitrates, permanganates), and other oxidizable materials.**
Hazardous Polymerization		Conditions to Avoid **Under normal conditions of temperature and pressure, hazardous polymerization of isobutyronitrile is not expected to occur.**
May Occur	Will Not Occur X	Hazardous Decomposition or By-products **When heated to decomposition, isobutyronitrile emits highly toxic fumes of nitrogen oxide as well as cyanide gas.**

SECTION VI - HEALTH HAZARD DATA

Primary Route(s) of Entry:	Inhalation?	Absorption (skin)?	Ingestion?
	X	X	X

Health Hazards

INHALATION: Will cause irritation of the respiratory tract with symptoms of cyanide poisoning. These include seizures, dilated pupils, shock, reduced respiration, cyanosis, initial tachycardia (rapid heartbeat) and hypertension (high blood pressure) followed by hypotension (low blood pressure) and subsequent loss of consciousness and coma. There can also be symptoms of nausea, vomiting, liver injury, acidosis, and death (asphyxiation).

ABSORPTION: Skin contact can result in irritating rash. Can cause severe eye burns and/or irritation with conjunctivitis. Will pass through intact skin to cause systemic effects.

INGESTION: Irritation of the digestive tract and toxic systemic effects (reference inhalation above).

Carcinogenicity	NTP Listed?	IARC Cancer Review Group?	OSHA Regulated?	Target Organs?
Unknown Human Unknown Animal	No	No	No	Liver, CNS, respiratory tract, skin, eyes.

Medical Conditions Generally Aggravated by Exposure

Respiratory impairments, (asthma); skin conditions (dermatitis); liver dysfunction.

Emergency and First-aid Procedures

Eye contact: Flush immediately with water for 15 minutes (minimum); seek medical attention. **Skin contact:** Remove all contaminated clothing. Immediately wash area with large amounts of soap and water. Seek medical assistance. For **inhalation:** Remove the person from exposure. Provide respiratory assistance and CPR. Transfer to medical facility. Isobutyronitrile can cause fatal cyanide poisoning. If **swallowed:** Give conscious person 1-2 glasses of water to dilute. Do Not induce vomiting.

SECTION VII - PRECAUTIONS FOR SAFE HANDLING AND USE

Steps to be Taken in Case Material is Released or Spilled

Contain spills using absorbent material (vermiculite, or other). Ventilate area of spill and remove all ignition sources. Have water source available in case of fire. Store materials in DOT-approved containers. Restrict those not involved in cleanup from entering area.

Preferred Waste Disposal Method

Burn in chemical incineration equipped with an afterburner and scrubber.

Precautions to be Taken in Handling and Storage

Store in tightly closed containers in cool, well-ventilated area away from heat. Storage drums should have self-closing valves, pressure vacuum bungs, and flame arresters.

Other Precautions and Warnings

Bulk storage of isobutyronitrile is not recommended.

SECTION VIII - CONTROL MEASURES AND PERSONAL PROTECTIVE EQUIPMENT

Respiratory Protection *(specify type)*

No exposure levels have been established and the specific toxicity of this chemical is unclear. The best protection is obtained using a self-contained breathing apparatus (SCBA) or a supplied-air respirator with full facepiece operated in pressure demand or other positive pressure mode.

Ventilation

Local exhaust or general mechanical systems recommended.

Protective Gloves	Eye Protection	Other Protective Clothing
Impervious Material	Chemical Goggles or Face Mask	Protective Apron

Work/Hygiene Practices

Always wash hands thoroughly after using chemical; never bring food, drink, or smoking materials into vicinity of chemicals.

ISOBUTYRONITRILE

$(CH_3)_2CHCN$

CAS: 78-82-0

IDENTIFICATION AND TYPICAL USES

Isobutyronitrile is a colorless liquid with an almond-like odor. It is used as an intermediate for insecticides, in organic synthesis, in some plastics and synthetic fibers, and as a gasoline additive.

RISK ASSESSMENT: HEALTH

General Assessment

Human exposure to isobutyronitrile can occur through three routes of entry into the body: *inhalation, absorption,* and *ingestion.* The toxic effects of isobutyronitrile are attributed to the metabolic release of cyanide, which in turn acts by inhibiting cytochrome oxidase and thus impairs cellular respiration. The target organs are the liver, central nervous system (CNS), lungs, skin, and eyes. Isobutyronitrile causes headache, dizziness, and nausea. At extremely high concentrations, it can cause convulsions, coma, and death. There are no data in the references on the chemical's carcinogenicity, mutagenicity, or teratogenicity in humans.

In humans, inhalation of isobutyronitrile vapors can result in systemic symptoms as those for cyanide. These include seizures, dilated pupils, shock, decreased respiration, cyanosis (due to lack of oxygenated blood), initial tachycardia (rapid heartbeat) and hypertension followed by low blood pressure and coma. There can be asphyxia, liver toxicity, acidosis, nausea, vomiting, and tightness of the chest. Death can occur as a result of asphyxiation as a result of blockage of respiratory enzymes. It is excreted in the urine as a cyanate. The blood cyanide concentration does not always show any significant increase in cyanide at low exposure concentrations.

Ingestion of isobutyronitrile results in gastrointestinal pain, nausea, vomiting, stupor, convulsion, weakness, and other similar systemic effects. Large doses can be fatal. Skin contact results in absorption and systemic effects. Severe eye irritation is likely.

Acute Health Effects

The following acute (short-term) health effects may occur immediately or shortly after exposure to isobutyronitrile. (Note: Exposure can cause fatal cyanide poisoning. Symptoms of flushing of face, chest tightness, nausea, and vomiting, weakness, and shortness of breath may lead to death. These reactions may begin minutes or hours after an overexposure).

Skin: Irritation/rash on contact (contact dermatitis).

Eye: Severe irritation, conjunctivitis, damage.

Lung: Irritation of the nose, throat, and lungs following exposure.

CNS: High concentrations may cause severe depression of the central nervous system with sleepiness, dizziness, loss of consciousness, or death.

Chronic Health Effects

The following chronic (long-term) health effects may occur at some time after exposure to isobutyronitrile and can last for months or even years:

Cancer Hazards: According to the references, isobutyronitrile has not been tested for its ability to cause cancer in animals. Chronic effects of high exposures have not been documented in humans.

Reproductive Hazard: According to the references, isobutyronitrile has not been significantly tested for its ability to adversely affect reproduction.

Other Chronic Effects: High exposures may damage the nervous system, liver, and kidneys. Very irritating substances may affect the lungs. It is not known whether isobutyronitrile causes lung damage.

Recommended Risk-Reduction Measures

Personnel should avoid direct contact with isobutyronitrile. Unless a less toxic chemical can be substituted for a hazardous substance, *engineering controls* are the most effective method of reducing exposures. The best protection is to enclose operations and/or provide local exhaust ventilation at the site of chemical release. While not always operationally feasible, isolating operations can also reduce exposure. Using respiratory protection is less effective than the controls mentioned above, but is still advisable whenever working with or around isobutyronitrile. No exposure levels have been established for this compound. This does not mean that exposure is without risk. For any exposure, recommend using a self-contained breathing apparatus (SCBA) with full facepiece operated in pressure demand or a supplied-air respirator operated in continuous flow mode are the recommended respiratory protection methods of choice. If a full facepiece is not available, then chemical goggles should be worn to protect the eyes. Whenever a chemical splash hazard exists, a face shield and a protective apron should be worn. To prevent hand and skin exposures, impervious gloves should be used. Glove manufacturers should be contacted and permeation studies obtained *before* final glove selections have been made.

Administrative controls should also be in place to minimize the potential for human exposures. These may include written procedures or policies which specify the methods and techniques that will be practiced whenever personnel are to work with isobutyronitrile.

All personnel should receive *training* on the use, hazards, protective measures, emergency actions, and other precautions per 29 CFR 1910.1200 (Hazard Communication), prior to the first assignment in an area where isobutyronitrile is used or stored.

Medical testing of personnel with frequent or potentially high exposure (half the PEL or greater), should be offered before beginning work and at regular intervals thereafter. These tests should include:

- ☑ Lung function (pulmonary function) test.

If symptoms develop or overexposure is suspected, the following tests may be useful:

- ☑ Blood cyanide test and/or urine thiocyanate test (currently, a blood cyanide level over 0.1 mg/l or urine thiocyanate over 20 mg/l indicates overexposure).
- ☑ Maintain close medical surveillance. Slow release of cyanide from absorbed isobutyronitrile may cause delayed symptoms.
- ☑ Liver function tests.
- ☑ Kidney function tests.

Other methods to reduce exposure include:

- ☑ Where possible, enclose operations and use local exhaust ventilation at the site of chemical release. If local exhaust ventilation or enclosure is not used, respiratory protection should be mandatory.
- ☑ Always ensure that proper protective clothing is worn when using chemical substances.
- ☑ Wash thoroughly immediately after exposure to isobutyronitrile and at the end of the work shift or before eating, drinking, or smoking.
- ☑ Hazard warning information should be posted in the work area. In addition, as part of an on-going education and training program, all information on the health and safety hazards of isobutyronitrile should be communicated to all exposed or potentially exposed workers.

RISK ASSESSMENT: ENVIRONMENT

General Assessment

The environment is at risk of exposure during transportation, storage, disposal, or destruction of isobutyronitrile. In almost every scenario, the threat of environmental exposure is contingent upon the proper handling of the chemical substance. Accidental spills, large or small, can result in fire, explosion, and possible contamination of the surrounding environmental mediums (water, soil, and air).

Isobutyronitrile is considered a highly flammable liquid and a poison. Because of its low flash point and high boiling point, it is classified as a Class 1B flammable liquid (per OSHA 29 CFR 1910.106). Isobutyronitrile vapors form explosive mixtures in air. It reacts with strong oxidizers, liberating heat along with pressure increase. Thus, contact under closed conditions can cause explosion. These characteristics require special consideration during any emergency situation involving a leak or spill of isobutyronitrile. Should isobutyronitrile ever come into contact with incompatible substances either during use, transportation, or storage, the formation of highly toxic and/or highly explosive commodities is extremely possible.

The proper disposal/destruction method for isobutyronitrile is to burn it in a chemical incinerator equipped with an afterburner and air scrubber.

Isobutyronitrile can enter the environment through manufacturing effluents, municipal waste treatment discharges, and spills.

Acute Ecological Effects

Acute (short-term) toxic effects may include the death of animals, birds, or fish and death or low growth rate in plants. Acute effects are seen 2 to 4 days after animals or plants are exposed to isobutyronitrile.

This chemical has slight acute toxicity to aquatic life. No data are available on the short-term effects of isobutyronitrile to plants or land animals.

Chronic Ecological Effects

Chronic toxic effects may include shortened life span, reproductive problems, lower fertility, and changes in appearance or behavior in exposed animals. These effects can be seen long after first exposure(s) to toxic chemicals.

Isobutyronitrile has slight chronic toxicity in aquatic life. No data are available on the long-term effects of isobutyronitrile to plants, birds, or land animals.

Water Solubility

Isobutyronitrile is slightly soluble in water. Concentrations of 1 to 100 milligrams will mix with a liter of water.

Persistence in the Environment

Isobutyronitrile is slightly persistent in water, with a half-life of between 2 to 20 days. The half-life of a pollutant is the amount of time it takes for one half of the chemical to be degraded.

Bioaccumulation in Aquatic Organisms

Some substances increase in concentration, or *bioaccumulate,* in living organisms as they breathe contaminated air, drink contaminated water, or eat contaminated food. These chemicals can become concentrated in the tissues and internal organs of animals as well as humans.

The concentration of isobutyronitrile found in fish tissues is expected to be about the same as the average concentration of isobutyronitrile in water from which the fish was taken.

Recommended Risk-Reduction Measures

Proper training of all transporters will reduce the likelihood of a mishap or accident resulting in a leak or spill to the environment. The correct use of DOT labeling (per the MSDS) on all containers, trucks, and rail cars will enable emergency responders to react properly and quickly to any disaster thereby reducing the potential risk to the environment and to personnel.

Storage of isobutyronitrile should be segregated from incompatible chemicals to minimize the risk of cross-contamination or contact. Buildings designated for storage should be equipped with appropriate fire protection systems (alarms, sprinklers, emergency lighting, portable extinguishers).

If a spill or leak to the environment has occurred, fire department, emergency response, and/or hazardous materials spill personnel should be notified immediately. Cleanup should be attempted only by those trained in proper spill containment procedures using non-sparking tools. Contaminated soils should be removed for incineration and replaced with clean soil. If isobutyronitrile should contact the water table, aquifer, or navigable waterway, time is of the essence. It is readily miscible in water and, therefore, total containment and remediation may not be entirely possible. When such spills occur, the local and/or state emergency response authorities must be notified. A comprehensive emergency response or disaster preparedness/recovery plan should be in place prior to any operations involving the use, transportation, storage, or disposal of isobutyronitrile. If isobutyronitrile is spilled or leaked, the following specific steps are recommended:

- ☑ Restrict persons not wearing protective clothing from area of spill or leak until cleanup is complete and area can be opened for normal work.
- ☑ Remove ignition sources and ventilate area.
- ☑ Absorb liquids in vermiculite, dry sand, earth, or a similar material and deposit in sealed containers. Use non-sparking tools during cleanup.
- ☑ It may be necessary to dispose of isobutyronitrile as a hazardous waste. The responsible state agency or the regional office of the federal Environmental Protection Agency (EPA) should be contacted for specific recommendations.

RISK ASSESSMENT: BUSINESS

General Assessment

Accidents or mishaps involving isobutyronitrile can present a significant threat to business operations. The loss or damage of equipment or facilities can significantly affect fiscal viability. Lawsuits that may result from personnel injury/death, public exposures, and/or environmental contamination will also require a serious expenditure of resources. Media attention surrounding an injury, death, or environmental damage can also result in a loss of profits and loss of current as well as future business.

Recommended Risk-Reduction Measures

Company attorneys, safety and health professionals, and environmental specialists should be involved in the development of any procedures dealing with safety or emergency response. A company official should be pre-designated as a public relations officer with specific training in dealing with the press. Corporate plans and policies should be developed, approved, and implemented long before any need for such arises.

REFERENCES

Ahlbom, A. 1993. *Biostatistics for Epidemiologists.* New York: CRC Press/Lewis Publishers.

American Conference of Governmental Industrial Hygienists. 1988. *Documentation of the Threshold Limit Values and Biological Exposure Indices,* 5th Edition (with updates). Cincinnati: ACGIH

Amdur, M. L. 1959. Accidental group exposure to isobutyronitrile. *Journal of Occupational Medicine.* 1:627-633.

Asante-Duah, D. K. 1993. *Hazardous Waste Risk Assessment.* New York: CRC Press/Lewis Publishers.

Brusick, D. J. 1994. *Methods for Genetic Risk Assessment.* New York: CRC Press/Lewis Publishers.

Bueche, J. R. 1972. *Principles of Physics,* 2nd Edition. New York: McGraw-Hill.

Burrell, R., D. K. Flaherty, and L. J. Sauers. 1992. *Toxicology of the Immune System.* New York: Van Nostrand Reinhold.

Calabrese, E. J., E. M. Kenyon. *Air Toxics and Risk Assessment.* New York: CRC Press/Lewis Publishers.

Calabrese, E. J. 1994. *Biological Effects of Low Level Exposures.* New York: CRC Press/Lewis Publishers.

Cockerman, L. G., B. S. Shane. 1994. *Basic Environmental Toxicology.* New York: CRC Press/Lewis Publishers.

Cothern, R. R. 1993. *Comparative Environmental Risk Assessment.* New York: CRC Press/Lewis Publishers.

Cothern, C. R. 1995. *Handbook for Environmental Risk Decisionmaking.* New York: CRC Press/Lewis Publishers.

Danse, I. R. 1991. *Common Sense Toxics in the Workplace.* New York: Van Nostrand Reinhold.

Environmental Protection Agency. 1983. *Chemical Hazard Information Profiles.* Washington, D.C.: U.S. EPA.

Fodden, C. R., and J. L. Weddell. 1992. *Hazardous Materials Emergency Action Data.* New York: CRC Press/Lewis Publishers.

Forseberg, K., and L. H. Keith. 1993. *Chemical; Protective Clothing Permeation and Degradation Database.* Chelsea, MI: Lewis Publishers.

Gardner, R. W. 1994. *Chemical Intolerance.* New York: CRC Press/Lewis Publishers.

Hallenbeck, W. H. 1993. *Quantitative Risk Assessment for Environmental and Occupational Health,* 2nd Edition. New York: CRC Press/Lewis Publishers.

Hosty, J. W., and P. Foster. 1990. *A Practical Guide to Chemical Spill Response.* New York: Van Nostrand Reinhold.

Howard, P. H. 1995. *Handbook of Environmental Fate and Exposures Data for Organic Chemicals,* Vol. I-IV. New York: CRC Press/Lewis Publishers.

Jones, F. E. 1994. *Toxic Organic Vapors in the Workplace.* New York: CRC Press/Lewis Publishers.

Langerman, N. 1994 *Precautionary Labels for Chemical Containers.* New York: CRC Press/Lewis Publishers.

Lewis, R. J., Sr. 1993. *Hawley's Condensed Chemical Dictionary,* 12th Edition. New York: Van Nostrand Reinhold.

Lewis, R. J., Sr. 1992. *Sax's Dangerous Properties of Industrial Materials,* Volumes 1, 2, and 3, 8th Edition. New York: Van Nostrand Reinhold.

Mackay, D., W. Y. Shiu, and K. C. Ma. 1992. *Illustrated Handbook of Chemical-Physical Properties and Environmental Fate for Organic Chemicals.* New York: CRC Press/Lewis Publishers.

Manahan, S. E. 1994. *Environmental Chemistry,* 6th Edition. New York: CRC Press/Lewis Publishers.

Meyer, E. 1990. *Chemistry of Hazardous Materials.* Englewood Cliffs, NJ: Prentice Hall.

Miller, B. F., M.D., and C. B. Keane, R.N. 1972. *Encyclopedia and Dictionary of Medicine and Nursing.* Philadelphia: W.B. Sanders Company.

Mycock, J. C., J. D. McKenna, and L. Theodore. 1995. *Handbook of Air Pollution Control Engineering and Technology.* New York: CRC Press/Lewis Publishers.

Patnaik, P. 1992. *A Comprehensive Guide to the Hazardous Properties of Chemical Substances.* New York: Van Nostrand Reinhold.

Proctor, N. H., and James P. Hughes. 1991. *Chemical Hazards of the Workplace.* New York: Van Nostrand Reinhold.

Sherman, J. D. 1988. *Chemical Exposure and Disease.* New York: Van Nostrand Reinhold.

Union Carbide Corporation. 1965. *Toxicology Studies, Isobutyronitrile.* New York: Union Carbide Corporation.

U.S. Department of Health and Human Services, National Institute for Occupational Safety and Health. 1978. *Criteria for a Recommended Standard - Occupational Exposure to Nitriles.* DHEW (NIOSH) Publication Number: 78-212. Washington, D.C.: U.S. Government Printing Office.

U.S. Department of Health and Human Services, National Institute for Occupational Safety and Health. 1994. *NIOSH Pocket Guide to Chemical Hazards.* Washington, D.C.: U.S. Government Printing Office.

MATERIAL SAFETY DATA SHEET

CHEMICAL NAME	
ISOPENTANE	**CRC PRESS/LEWIS PUBLISHERS DISCLAIMER:** THE INFORMATION AND RECOMMENDATIONS PRESENTED HEREIN ARE BASED ON SOURCES BELIEVED TO BE RELIABLE. CRC MAKES NO REPRESENTATION ON ITS COMPLETENESS OR ACCURACY. IT IS THE RESPONSIBILITY OF THE USER TO DETERMINE THE CHEMICAL'S SUITABILITY FOR ITS INTENDED USE, THE CHEMICAL'S SAFE USE, AND THE CHEMICAL'S PROPER DISPOSAL. NO REPRESENTATIONS AND/OR WARRANTIES, EITHER EXPRESSED OR IMPLIED, OF THE MERCHANTABILITY OR FITNESS FOR A PARTICULAR PURPOSE, OR OF ANY OTHER NATURE, ARE MADE WITH RESPECT TO THE INFORMATION PROVIDED IN THIS MSDS OR TO THE CHEMICAL TO WHICH INFORMATION MAY REFER. CRC NEITHER ASSUMES NOR AUTHORIZES ANY OTHER PERSON TO ASSUME FOR IT, ANY OTHER ADDITIONAL RESPONSIBILITY OR LIABILITY FOR THE USE OF, OR RELIANCE UPON, THIS INFORMATION.

HAZARD WARNING INFORMATION

HEALTH	FIRE	REACTIVE	OTHER	DEGREE OF HAZARD	COLOR CODING	OTHER CODES
1	4	0		0 = Minimum Hazard 1 = Slight Hazard 2 = Moderate Hazard 3 = Serious Hazard 4 = Severe Hazard	HEALTH = BLUE FIRE = RED REACTIVITY = YELLOW OTHER = WHITE	OX = Oxidizer ACID = Acid ALK = Alkali COR = Corrosive W = Use No Water

SECTION I - GENERAL INFORMATION

Characterization	RCRA Number	EPA Class
Hydrocarbon, Aliphatic	**None**	**Not Applicable**

DOT Proper Shipping Name **Pentanes**	Chemical Abstract Service (CAS) Number **78-78-4**
DOT Hazard Class and Label Requirements **Flammable Liquid**	DOT Emergency Guide Code **26**
DOT Identification Number **UN 1265**	Chemical Formula **$(CH_3)_2CHCH_2CH_3$**

Synonyms

Ethyldimethylmethane; isoamyl hydride; 2-methylbutane; 1,1,2-trimethylethane.

SECTION II - HAZARDOUS INGREDIENTS AND IDENTITY INFORMATION

Hazardous Components (specific identity)	OSHA Exposure Criteria	NIOSH Exposure Criteria	Immediately Dangerous to Life and Health (IDLH)	ACGIH Exposure Criteria
Isopentane (derivation: Obtained from fractional distillation of crude petroleum, purified by rectification). **1 ppm = 2.95 mg/m³**	**PEL (as pentane):** **600 ppm** **1800 mg/m³** **STEL (15-min):** **750 ppm** **2250 mg/m³**	**REL (as alkanes):** **1.2 ppm** **350 mg/m³** **STEL:** **Not Established**	**Not Found**	**TLV:** **600 ppm** **1770 mg/m³** **STEL (15-min):** **750 ppm** **2210 mg/m³**

SECTION III - PHYSICAL AND CHEMICAL CHARACTERISTICS

Boiling Point **82°F (28°C)**	Specific Gravity (H_2O = 1) **0.62**
Vapor Pressure (gas) **595 at 70°F (21°C)**	Molecular Weight **72.15**
Vapor Density (Air = 1) **2.48**	Melting Point **-256°F (-160°C)**

Solubility

Insoluble in water. Soluble in hydrocarbons, oils, ether. Slightly soluble in alcohol.

Appearance and Odor

Colorless, mobile liquid with sweet, pleasant odor.

SECTION IV - FIRE AND EXPLOSION HAZARD DATA

Flash Point (method used) **<-60°F (-51°C) closed cup**	Explosive Limits in Air % by Volume **LEL: 1.4%** **UEL: 7.6%**
NFPA Classification **Class 1A Flammable Liquid**	Autoignition Temperature **788°F (420°C)**

Extinguishing Media

Use dry chemical, carbon dioxide, or foam. Use water spray to knock down vapors.

Special Fire Fighting Procedures

Isopentane is a flammable liquid. Wear full protective clothing and self-contained breathing apparatus (SCBA). Poisonous gases are produced in fire. Use water spray to keep fire-exposed containers cool.

Unusual Fire and Explosion Hazards

Dangerous fire and explosion hazard when exposed to heat or flame. Containers may explode in fire. Move containers from fire if it can be done without risk. Chemical will react on contact with oxidizable materials. Moderate explosion hazard when in the form of vapor.

SECTION V - REACTIVITY DATA

Stability		Conditions to Avoid **Isopentane is normally stable at room temperature in closed containers under routine conditions of storage and handling. Keep away from heat and incompatible materials.**
Stable	Unstable	Incompatibility *(materials to avoid)*
X		**Strong oxidizers, such as permanganates, peroxides, nitrates, chlorates, and perchlorates.**
Hazardous Polymerization		Conditions to Avoid **Under normal conditions of temperature and pressure, hazardous polymerization of isopentane is not expected to occur.**
May Occur	Will Not Occur	Hazardous Decomposition or By-products
	X	**When heated to decomposition, isopentane can emit highly toxic/poisonous gases.**

SECTION VI - HEALTH HAZARD DATA

Primary Route(s) of Entry:	Inhalation?	Absorption (skin)?	Ingestion?
	X		X

Health Hazards

INHALATION: **Low-order toxicity by inhalation. However, it does act as an irritant and a central nervous system depressant. Can cause nausea, dizziness, vomiting, and depression. Higher concentrations can lead to coma and death from respiratory failure or pulmonary edema.**

SKIN & EYES: **Irritating to the skin, may cause drying, cracking, and chapping. Eye irritation likely.**

INGESTION: **Irritation and narcotic effects with nausea, vomiting, abdominal swelling, headache, and a significant risk of aspiration and subsequent chemical pneumonitis.**

Carcinogenicity	NTP Listed?	IARC Cancer Review Group?	OSHA Regulated?	Target Organs?
Unknown Human Unknown Animal	**No**	**No**	**29 CFR 1910.1000 Table Z-1** (as pentane)	**Respiratory system, liver, skin, CNS.**

Medical Conditions Generally Aggravated by Exposure

None reported.

Emergency and First-aid Procedures

Eye contact: Flush immediately with water for 15 minutes (minimum); seek medical attention. Skin contact: Remove all contaminated clothing. Immediately wash area with large amounts of soap and water. Seek medical assistance. For inhalation: Remove the person from exposure. Provide respiratory assistance and CPR, if required. Transfer to medical facility. Observe 48 hours for lung effects. If swallowed: Seek medical attention immediately. Give conscious person water or milk; do NOT induce vomiting.

SECTION VII - PRECAUTIONS FOR SAFE HANDLING AND USE

Steps to be Taken in Case Material is Released or Spilled

For small spills, absorb in paper towels and evaporate in a safe place (fume hood). Afterwards, burn paper away from combustible materials. Absorb large spills in vermiculite, dry sand, earth, or similar material and deposit in sealed containers. Ventilate area of spill or leak. Restrict those not involved in cleanup from entering area. Remove all sources of ignition and ventilate area.

Preferred Waste Disposal Method

No citation.

Precautions to be Taken in Handling and Storage

Store to avoid contact with incompatible materials in tightly closed containers in cool, well-ventilated area away from heat and flame. Use only non-sparking tools to open and close containers.

Other Precautions and Warnings

Sources of ignition are prohibited where isopentane is used, handled, or stored. Heat may cause containers to build up pressure and explode. Containers must be grounded and bonded during transfer.

SECTION VIII - CONTROL MEASURES AND PERSONAL PROTECTIVE EQUIPMENT

Respiratory Protection *(specify type)*

For exposures above the PEL, use an MSHA/NIOSH-approved supplied-air respirator with full facepiece, hood, or helmet in continuous flow mode, or use a supplied-air respirator or a self-contained breathing apparatus (SCBA) operated in pressure demand or other positive pressure mode.

Ventilation

Local exhaust or general mechanical systems recommended.

Protective Gloves	Eye Protection	Other Protective Clothing
Impervious Material	**Chemical Goggles or Face Mask**	**Protective Apron**

Work/Hygiene Practices

Always wash hands thoroughly after using chemical; never bring food, drink, or smoking materials into vicinity of chemicals.

ISOPENTANE

$(CH_3)_2CHCH_2CH_3$ CAS: 78-78-4

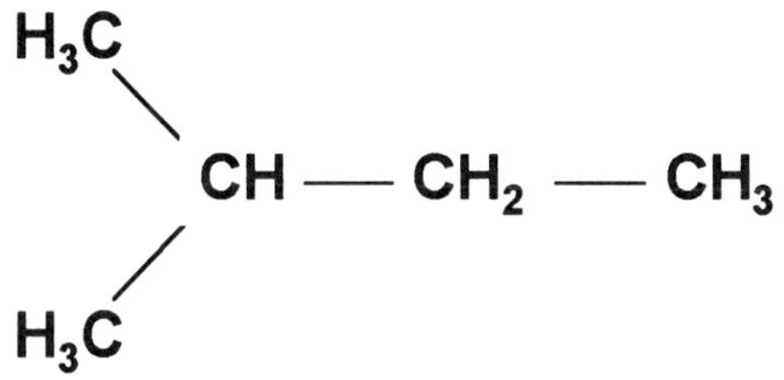

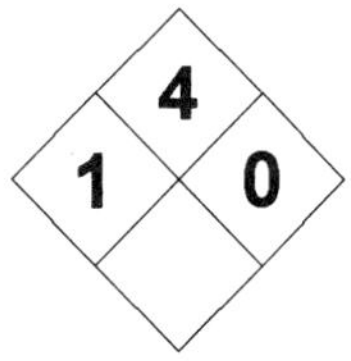

IDENTIFICATION AND TYPICAL USES

Isopentane is a colorless, mobile liquid with a pungent, sweet odor. It is used primarily as a solvent and in the manufacture of certain chlorinated derivatives. It is also a blowing agent for polystyrene.

RISK ASSESSMENT: HEALTH

General Assessment

Isopentane is an acute irritant of low-order toxicity by *inhalation* and *ingestion*. Skin contact can also produce some adverse topical effects. It is generally considered to be a safe solvent for industrial use when used according to manufacturer's instructions. However, in sufficient or high concentrations, there can be toxic systemic effects through both routes of exposure. There are no data on its carcinogenic, mutagenic, or teratogenic capabilities.

Significant inhalation may cause irritation of the respiratory tract and depress the central nervous system. Symptoms include cough, irregular heartbeat, and mild depression with nausea, dizziness, vomiting, and a possibility of pulmonary edema (fluid in the lungs). Development of pulmonary edema can be delayed up to 48 hours thereby creating a false sense of security with regard to health exposure risk. At high concentrations, it is a narcotic and can lead to coma and death due to respiratory failure. It has low chronic toxicity due to its efficient metabolism and excretion.

Skin and eye contact may produce mild to moderate irritation on contact. Repeated or prolonged contact with the skin can lead to drying, cracking, chapping, and secondary infection and/or dermatitis.

Ingestion can produce narcotic effects as those noted for inhalation above. Aspiration results in lung irritation, pulmonary edema, chemical pneumonitis, coughing, and central nervous system effects.

Acute Health Effects

The following acute (short-term) health effects may occur immediately or shortly after exposure to isopentane:

Skin: Mild irritation on contact. Can cause drying and cracking if not removed promptly.

Eye: Mild to moderate irritation.

Lung: Mild to moderate irritation of the nose, throat, mucosa, and respiratory tract causing cough and tightness in chest. May cause dizziness, vomiting, and nausea. Primary result of inhalation of isopentane is its action on the CNS.

CNS: A CNS depressant causing headache, dizziness, depression, and narcotic effects through inhalation or ingestion.

Chronic Health Effects

The following chronic (long-term) health effects may occur at some time after exposure to isopentane and can last for months or even years:

Cancer Hazards: According to information presented in the references, isopentane has not been adequately tested for its ability to cause cancer in animals. Limited studies indicate an increased incidence of Hodgkin's disease in workers exposed to organic solvents. It is not known, however, whether isopentane will have this effect.

Reproductive Hazard: According to information presented in the references, isopentane has not been tested for its ability to adversely affect reproduction.

Other Chronic Effects: Prolonged skin contact may cause drying, cracking, and chapping of the affected area creating the potential for secondary infection and/or dermatitis.

Recommended Risk-Reduction Measures

Personnel should avoid direct contact with isopentane. Even though the toxicity of this chemical is considered to be of relatively low order, it can still cause significant health effects if improperly used or if overexposures occur. If a less toxic chemical cannot be substituted for a hazardous substance, then *engineering controls* are the most effective method of reducing exposures. The best protection is to enclose operations and/or provide local exhaust ventilation at the site of chemical release. While not always operationally feasible, isolating operations can also reduce exposure. Using respiratory protection is less effective than the controls mentioned above, but is still advisable whenever working with or around isopentane. For exposures above the PEL (600 ppm), an air-purifying respirator with an organic vapor cartridge should be used. Better protection is obtained using a supplied-air respirator with full facepiece operated in continuous flow mode, or a self-contained breathing apparatus (SCBA) with full facepiece and pressure demand or other continuous flow mode. If a full facepiece is not available, then chemical goggles should be worn to protect the eyes. Whenever a chemical splash hazard exists, a face shield and a rubber apron should be worn. To prevent hand and skin exposures, impervious gloves should be used. Glove manufacturers should be contacted and permeation studies obtained *before* final glove selections have been made.

Administrative controls should also be in place to minimize the potential for human exposures. These may include written procedures or policies which specify the methods and techniques that will be practiced whenever personnel are to work with isopentane.

All personnel should receive *training* on the use, hazards, protective measures, emergency actions, and other precautions per 29 CFR 1910.1200 (Hazard Communication), prior to the first assignment in an area where isopentane is used or stored.

If symptoms develop or overexposure is suspected, the following tests are recommended:

- ☑ Lung function tests.
- ☑ Consider chest X-ray following acute overexposure (may be negative if taken immediately after exposure due to delayed onset of pulmonary edema).

Any evaluation should include a careful history of past and present symptoms with an examination. Medical tests that simply look for existing damage are not a substitute for controlling exposures. Also, since smoking can cause heart disease, lung cancer, emphysema, and other respiratory diseases, smokers exposed to isopentane may experience symptoms more rapidly than non-smokers under the same conditions of exposure. Prudent risk management requires proper consideration of *all* possible factors which may be causing the appearance of exposure symptoms in the workplace.

Other methods to reduce exposure include:

- ☑ Where possible, enclose operations and use local exhaust ventilation at the site of chemical release. If local exhaust ventilation or enclosure is not used, respiratory protection should be mandatory.
- ☑ Always ensure that proper protective clothing is worn when using chemical substances.
- ☑ Wash thoroughly immediately after exposure to isopentane and at the end of the work shift or before eating, drinking, or smoking.
- ☑ Hazard warning information should be posted in the work area. In addition, as part of an on-going education and training program, all information on the health and safety hazards of isopentane should be communicated to all exposed workers.
- ☑ Eye wash stations should be provided in the immediate work area for emergency use. If there is a possibility of skin exposure to isopentane, emergency shower facilities should also be provided.
- ☑ Personnel should never wear contaminated clothing home (family members can be exposed). Clothing should be laundered by personnel who have been trained on the hazards of isopentane.

RISK ASSESSMENT: ENVIRONMENT

General Assessment

The environment is at risk of exposure during transportation, storage, disposal, or destruction of isopentane. In almost every scenario, the threat of environmental exposure is contingent upon the proper handling of the chemical substance. Accidental spills, large or small, can result in fire, explosion, and possible contamination of the surrounding environmental mediums (water, soil, and air).

Isopentane is considered a Class 1A flammable liquid (per OSHA 29 CFR 1910.106). Its extremely low flash point and relatively low boiling point make this chemical a serious fire and explosion hazard concern. These characteristics require extreme caution in handling, storage, transportation, and disposal. It is incompatible with common peroxides and other oxidizers and contact can cause fire or explosion. Therefore, special consideration is required during any emergency situation involving a leak or spill of isopentane. Should isopentane ever come into contact with these incompatible substances during use, transportation, storage, or disposal, the formation of highly toxic and/or highly explosive commodities (including hydrogen chloride) is extremely possible.

Isopentane may enter the environment through industrial discharges or spills.

Acute Ecological Effects

Acute (short-term) toxic effects may include the death of animals, birds, or fish and death or low growth rate in plants. Acute effects are seen 2 to 4 days after animals or plants are exposed to isopentane.

This chemical has high acute toxicity to aquatic life. Insufficient data are available on the short-term effects of isopentane exposure to aquatic life, birds or land animals.

Chronic Ecological Effects

Chronic (long-term) toxic effects may include shortened life span, reproductive problems, lower fertility, and changes in appearance or behavior in exposed animals. These effects can be seen long after first exposure(s) to toxic chemicals.

Isopentane has moderate chronic toxicity in aquatic life. Insufficient data are available on the long-term effects of isopentane to plants, birds, or land animals.

Water Solubility

Isopentane is nearly insoluble in water. Concentrations of less than 100 milligrams may mix with a liter of water, depending on water pH and chemical concentration of isopentane.

Persistence in the Environment

Isopentane is non-persistent in water, with a half-life of less than 2 days. The half-life of a pollutant is the amount of time it takes for one half of the chemical to be degraded. Virtually 100% of isopentane will eventually end up in air.

Bioaccumulation in Aquatic Organisms

Some substances increase in concentration, or *bioaccumulate,* in living organisms as they breathe contaminated air, drink contaminated water, or eat contaminated food. These chemicals can become concentrated in the tissues and internal organs of animals as well as humans.

The concentration of isopentane found in fish tissue is expected to be somewhat lower than the average concentration of isopentane in the water from which the fish was taken.

Recommended Risk-Reduction Measures

Proper training of all transporters will reduce the likelihood of a mishap or accident resulting in a leak or spill to the environment. The correct use of DOT labeling (per the MSDS) on all containers, trucks, and rail cars will enable emergency responders to react properly and quickly to any disaster thereby reducing the potential risk to the environment and to personnel.

Storage of isopentane should be segregated from incompatible chemicals to minimize the risk of cross-contamination or contact. Buildings designated for storage should be equipped with appropriate fire protection systems (alarms, sprinklers, emergency lighting, portable extinguishers). Electrical equipment should be design to meet applicable explosion-proof requirements.

If a spill or leak to the environment has occurred, fire department, emergency response and/or hazardous materials spill personnel should be notified immediately. Cleanup should be attempted only by those trained in proper spill containment procedures using non-sparking tools. Contaminated soils should be removed for incineration and replaced with clean soil. If isopentane should contact the water table, aquifer, or navigable waterway, the local and/or state emergency response authorities must be notified. A comprehensive emergency response or disaster preparedness/recovery plan should be in place prior to any operations involving the use, transportation, storage, or disposal of isopentane. If isopentane is spilled or leaked, the following specific steps are recommended:

- ☑ Restrict persons not wearing protective clothing from area of spill or leak until cleanup is complete and area can be opened for normal work.
- ☑ Remove all ignition sources.
- ☑ Ventilate area of spill or leak using explosion-proof ventilation system, if possible.

- ☑ For small spills (laboratory spills), absorb liquid in paper towels. Allow evaporation in a safe place (a fume hood or open air). After sufficient time has passed for evaporation, burn paper towels away from other combustibles. For larger spills, absorb liquids in vermiculite, dry sand, earth, or similar material and deposit in sealed containers.
- ☑ It may be necessary to dispose of isopentane as a hazardous waste. The responsible state agency or the regional office of the federal Environmental Protection Agency (EPA) should be contacted for specific recommendations.

RISK ASSESSMENT: BUSINESS

General Assessment

Accidents or mishaps involving isopentane can present a significant threat to business operations. The loss or damage of equipment or facilities can significantly affect fiscal viability. Lawsuits that may result from personnel injury/death, public exposures, and/or environmental contamination will also require a serious expenditure of resources. Media attention surrounding an injury, death, or environmental damage can also result in a loss of profits and loss of current as well as future business.

Recommended Risk-Reduction Measures

Company attorneys, safety and health professionals, and environmental specialists should be involved in the development of any procedures dealing with safety or emergency response. A company official should be pre-designated as a public relations officer with specific training in dealing with the press. Corporate plans and policies should be developed, approved, and implemented long before any need for such arises.

REFERENCES

American Conference of Governmental Industrial Hygienists. 1988. *Documentation of the Threshold Limit Values and Biological Exposure Indices,* 5th Edition (with updates). Cincinnati: ACGIH

Bueche, J. R. 1972. *Principles of Physics,* 2nd Edition. New York: McGraw-Hill.

Calabrese, E. J. 1994. *Biological Effects of Low Level Exposures.* New York: CRC Press/Lewis Publishers.

Cockerman, L. G., B. S. Shane. 1994. *Basic Environmental Toxicology.* New York: CRC Press/Lewis Publishers.

Gardner, R. W. 1994. *Chemical Intolerance.* New York: CRC Press/Lewis Publishers.

Hallenbeck, W. H. 1993. *Quantitative Risk Assessment for Environmental and Occupational Health,* 2nd Edition. New York: CRC Press/Lewis Publishers.

Jones, F. E. 1994. *Toxic Organic Vapors in the Workplace.* New York: CRC Press/Lewis Publishers.

Keith, L. H. and M. M. Walker. 1995. *Handbook of Air Toxics.* New York: CRC Press/Lewis Publishers.

Landis, W. G., M. H. Yu. 1995. *Introduction to Environmental Toxicology.* New York: CRC Press/Lewis Publishers.

Langerman, N. 1994 *Precautionary Labels for Chemical Containers.* New York: CRC Press/Lewis Publishers.

Lauwerys, R. R., P. Hoet. 1993. *Industrial Chemical Exposure.* New York: CRC Press/Lewis Publishers.

Lewis, R. J., Sr. 1992. *Sax's Dangerous Properties of Industrial Materials,* Volumes 1, 2, and 3, 8th Edition. New York: Van Nostrand Reinhold.

Lide, D. R. 1995. *Handbook of Chemistry and Physics,* 76th Edition. New York: CRC Press/Lewis Publishers.

Manahan, S. E. 1994. *Environmental Chemistry,* 6th Edition. New York: CRC Press/Lewis Publishers.

Mellström, G. A., J. E. Wahlberg, and H. I. Maibach. 1994. *Protective Gloves for Occupational Use.* New York: CRC Press/Lewis Publishers.

Meyer, E. 1990. *Chemistry of Hazardous Materials.* Englewood Cliffs, NJ: Prentice Hall.

Patnaik, P. 1992. *A Comprehensive Guide to the Hazardous Properties of Chemical Substances.* New York: Van Nostrand Reinhold.

Proctor, N. H., J. P. Hughes. 1991. *Chemical Hazards of the Workplace.* New York: Van Nostrand Reinhold.

Quigley, D. R. 1994. *Handbook of Emergency Chemical Management.* New York: CRC Press/Lewis Publishers.

Rea, W. J. 1995. *Chemical Sensitivity,* Vol. I-III. New York: CRC Press/Lewis Publishers.

Richardson, M. 1995. *Dictionary of Substance and Their Effects,* Vol. I-VI. New York: CRC Press/Lewis Publishers.

Sacarello, H. L. A. 1994. *The Comprehensive Handbook of Hazardous Materials.* New York: CRC Press/Lewis Publishers.

Torkelson, T. R. V. K. Rowe. 1981. Halogenated aliphatic hydrocarbons. *Patty's Industrial Hygiene and Toxicology,* Vol. 2B, *Toxicology,* pp 3568-3572. New York: Wiley-Interscience.

U.S. Department of Health and Human Services, National Institute for Occupational Safety and Health. 1994. *NIOSH Pocket Guide to Chemical Hazards.* Washington, D.C.: U.S. Government Printing Office.

MATERIAL SAFETY DATA SHEET

CHEMICAL NAME	CRC PRESS/LEWIS PUBLISHERS DISCLAIMER
ISOPHORONE	THE INFORMATION AND RECOMMENDATIONS PRESENTED HEREIN ARE BASED ON SOURCES BELIEVED TO BE RELIABLE. CRC MAKES NO REPRESENTATION ON ITS COMPLETENESS OR ACCURACY. IT IS THE RESPONSIBILITY OF THE USER TO DETERMINE THE CHEMICAL'S SUITABILITY FOR ITS INTENDED USE, THE CHEMICAL'S SAFE USE, AND THE CHEMICAL'S PROPER DISPOSAL. NO REPRESENTATIONS AND/OR WARRANTIES, EITHER EXPRESSED OR IMPLIED, OF THE MERCHANTABILITY OR FITNESS FOR A PARTICULAR PURPOSE, OR OF ANY OTHER NATURE, ARE MADE WITH RESPECT TO THE INFORMATION PROVIDED IN THIS MSDS OR TO THE CHEMICAL TO WHICH INFORMATION MAY REFER. CRC NEITHER ASSUMES NOR AUTHORIZES ANY OTHER PERSON TO ASSUME FOR IT, ANY OTHER ADDITIONAL RESPONSIBILITY OR LIABILITY FOR THE USE OF, OR RELIANCE UPON, THIS INFORMATION.

HAZARD WARNING INFORMATION

HEALTH	FIRE	REACTIVE	OTHER	DEGREE OF HAZARD	COLOR CODING	OTHER CODES
2	2	0		0 = Minimum Hazard 1 = Slight Hazard 2 = Moderate Hazard 3 = Serious Hazard 4 = Severe Hazard	HEALTH = BLUE FIRE = RED REACTIVITY = YELLOW OTHER = WHITE	OX = Oxidizer ACID = Acid ALK = Alkali COR = Corrosive W = Use No Water

SECTION I - GENERAL INFORMATION

Characterization	**Ketone**	RCRA Number	**None**
		EPA Class	**Not Applicable**
DOT Proper Shipping Name	**Combustible Liquid, N.O.S.**	Chemical Abstract Service (CAS) Number	**78-59-1**
DOT Hazard Class and Label Requirements	**Combustible Liquid**	DOT Emergency Guide Code	**27**
DOT Identification Number	**NA 1993**	Chemical Formula	**$C_9H_{14}O$**

Synonyms

Isoacetophorone; 3,5,5-trimethyl-2-cyclohexenone; 3,5,5-trimethyl-2-cyclohexen-1-one; isoforone.

SECTION II - HAZARDOUS INGREDIENTS AND IDENTITY INFORMATION

Hazardous Components (specific identity)	OSHA Exposure Criteria	NIOSH Exposure Criteria	Immediately Dangerous to Life and Health (IDLH)	ACGIH Exposure Criteria
Isophorone **1 ppm = 5.74 mg/m³**	**PEL:** 25 ppm, 140 mg/m³ **STEL:** Not Established	**REL (ketones):** 4 ppm, 23 mg/m³ **STEL:** Not Established	**200 ppm**	**TLV (ceiling):** 5 ppm, 29 mg/m³ **STEL:** Not Applicable

SECTION III - PHYSICAL AND CHEMICAL CHARACTERISTICS

Boiling Point	**419°F (215°C)**	Specific Gravity (H_2O = 1)	**0.92**
Vapor Pressure (gas)	**0.3 at 69°F (20°C)**	Molecular Weight	**138.2**
Vapor Density (Air = 1)	**4.8**	Melting Point	**17°F (-8°C)**

Solubility

Slightly soluble in water (1%), cellulose esters, ether, and other solvents.

Appearance and Odor

Colorless to white liquid with an odor of acetone and peppermint. Odor Threshold = 0.54 ppm.

SECTION IV - FIRE AND EXPLOSION HAZARD DATA

Flash Point (method used)	**184°F (84°C) closed cup**	Explosive Limits in Air % by Volume	**LEL: 0.8% UEL: 3.8%**
NFPA Classification	**Class IIIA Combustible Liquid**	Autoignition Temperature	**860°F (460°C)**

Extinguishing Media

Use dry chemical, carbon dioxide, water spray, or alcohol foam (best for large fires).

Special Fire Fighting Procedures

Wear full protective clothing and self-contained breathing apparatus (SCBA). Poisonous gases are produced in fire. Use water spray to keep fire-exposed containers cool. Containers may explode in fire.

Unusual Fire and Explosion Hazards

Combustible liquid but a dangerous fire hazard when exposed to heat. At high temperatures, forms explosive mixtures with air. Violent reactions can occur with nitric acid. Chemical will react on contact with oxidizable materials. Moderate explosion hazard when in the form of vapor.

SECTION V - REACTIVITY DATA

Stability		Conditions to Avoid Normally stable. Keep away from sources of heat (including direct sunlight) or flame. Large quantities should be stored in metal tanks or drums. Keep away from incompatible materials.
Stable X	Unstable	Incompatibility *(materials to avoid)* Strong oxidizers (permanganates, peroxides, nitrates, chlorates, and perchlorates), acids (nitric and sulfuric), strong alkalis, and amines.
Hazardous Polymerization		Conditions to Avoid Under normal conditions of temperature and pressure, hazardous polymerization of isophorone is not expected to occur.
May Occur	Will Not Occur X	Hazardous Decomposition or By-products When heated to decomposition, isophorone can emit highly toxic/poisonous and acrid fumes and gases including oxides of carbon.

SECTION VI - HEALTH HAZARD DATA

Primary Route(s) of Entry:	Inhalation?	Absorption (skin)?	Ingestion?
	X		X

Health Hazards

INHALATION: Symptoms include irritation of the eyes, nose, throat, and respiratory system. May cause depression of the central nervous system with narcosis, inebriation, headache, fatigue, malaise, and dizziness. Liver and kidney damage are also possible. Other effects include olfactory changes, faintness, suffocation, tearing, and possible pulmonary edema.

SKIN & EYES: Irritating to the skin; may cause irritation and dermatitis. An eye irritant with possibility of corneal opacity and damage.

INGESTION: Irritation of the mouth and stomach, narcosis, anesthesia, and respiratory difficulties.

Carcinogenicity	NTP Listed?	IARC Cancer Review Group?	OSHA Regulated?	Target Organs?
Unknown Human Questioned Animal	No	No	29 CFR 1910.1000 Table Z-1	Respiratory system, eyes, skin, kidney, liver.

Medical Conditions Generally Aggravated by Exposure

None reported.

Emergency and First-aid Procedures

Eye contact: Flush immediately with water for 15 minutes (minimum); seek medical attention. Skin contact: Remove all contaminated clothing. Immediately wash area with large amounts of soap and water. Seek medical assistance. For inhalation: Remove the person from exposure. Provide respiratory assistance and CPR, if required. Transfer to medical facility. Keep victim warm and at rest. Observe for 24-48 hours for lungs effects (pulmonary edema). If swallowed: Seek medical attention immediately. Give 1-2 glasses of water to drink. Do NOT induce vomiting (aspiration hazard).

SECTION VII - PRECAUTIONS FOR SAFE HANDLING AND USE

Steps to be Taken in Case Material is Released or Spilled

Absorb in vermiculite, earth, or similar material and deposit in sealed containers. Ventilate area of spill or leak. Restrict those not involved in cleanup from entering area. Remove all sources of ignition.

Preferred Waste Disposal Method

Burn in chemical incinerator equipped with an afterburner and scrubber.

Precautions to be Taken in Handling and Storage

Store to avoid contact with incompatible materials in tightly closed containers in cool, well-ventilated area away from heat and flame. Use only non-sparking tools to open and close containers.

Other Precautions and Warnings

Sources of ignition are prohibited where isophorone is used, handled, or stored. Heat may cause containers to build up pressure and explode. Containers must be grounded and bonded during transfer.

SECTION VIII - CONTROL MEASURES AND PERSONAL PROTECTIVE EQUIPMENT

Respiratory Protection *(specify type)*

For exposures above the PEL, use an MSHA/NIOSH-approved supplied-air respirator with full facepiece, hood, or helmet in continuous flow mode, or use a self-contained breathing apparatus (SCBA) operated in pressure demand or other positive pressure mode.

Ventilation

Local exhaust or general mechanical systems recommended.

Protective Gloves	Eye Protection	Other Protective Clothing
Impervious Material	Chemical Goggles or Face Mask	Impervious Apron

Work/Hygiene Practices

Always wash hands thoroughly after using chemical; never bring food, drink, or smoking materials into vicinity of chemicals.

ISOPHORONE

$C_9H_{14}O$

CAS: 78-59-1

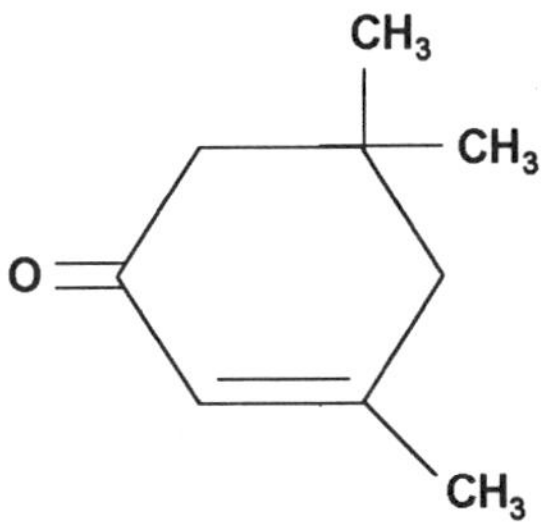

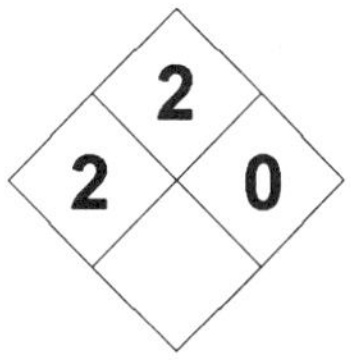

IDENTIFICATION AND TYPICAL USES

Isophorone is a clear or colorless liquid with an acetone, peppermint, or camphor-like odor. It is used in solvent mixtures for finishes, for polyvinyl and nitrocellulose resins, in pesticides, and in stoving lacquers.

RISK ASSESSMENT: HEALTH

General Assessment

Isophorone is moderately toxic by *ingestion* and skin contact (although absorption through intact skin has not been reported). It is also mildly toxic to humans by *inhalation.* Because of its low volatility, its vapors are not often present at dangerous levels in the workplace, unless heated or misted. Its carcinogenicity in humans is not known but is questioned in some animal studies. It is primarily a kidney poison. It has been shown to cause mutations (genetic changes) in living cells.

Inhalation can causes changes in the sense of smell, conjunctiva irritation, and possible respiratory changes. It will irritate the nose, throat, and eyes and fatigue the sense of smell. It may also cause a dangerous buildup of fluid in the lungs (pulmonary edema) which is a medical emergency and can be fatal. Symptoms can be delayed up to 48 hours, thereby creating a false sense of security with regard to health exposure risk. Mild narcotic properties have also been recorded, including pronounced headache, and it may cause depression of the central nervous system, although this effect is not completely confirmed in the references. Other symptoms of inhalation include CNS depression with nausea, dizziness, fatigue, malaise, narcosis, decreased breathing rate, and anesthesia. Kidney and liver damage with mild narcosis may also be possible.

Skin and eye contact may produce mild to moderate irritation on contact. Pure liquid in contact with the eye may result in corneal opacity and subsequent damage.

On ingestion, isophorone is expected to produce symptoms of irritation, nausea, vomiting, stomach pain, narcosis, anesthesia, and some respiratory difficulties.

☠ *Acute Health Effects*

The following acute (short-term) health effects may occur immediately or shortly after exposure to isophorone:

Skin: Will defat the skin and can cause dermatitis. Pure liquid will cause severe irritation.

Eye: Severe irritation with a potential for corneal injury or damage.

Lung: Nose and throat irritation. Produces intoxicating effects and may cause lung damage, including pulmonary edema.

CNS: There may be depression, anesthesia, hypothermia, narcosis, and other effects.

Chronic Health Effects

The following chronic (long-term) health effects may occur at some time after exposure to isophorone and can last for months or even years:

Cancer Hazards: According to information presented in the references, isophorone has been shown to cause cancer in test animals. Some studies indicate that it causes mutations in living cells, while others are inconclusive in this regard. Some studies have listed its cancer causing properties as questioned animal studies. Further studies are required in this regard.

Reproductive Hazard: According to information presented in the references, isophorone has not been adequately tested for its ability to adversely affect reproduction.

Other Chronic Effects: Prolonged skin contact may cause drying, cracking, and chapping of the affected area. There may be skin sensitization and/or dermatitis in some individuals. Repeated or long-term exposures to isophorone may affect the liver and kidneys.

Recommended Risk-Reduction Measures

Personnel should avoid direct contact with isophorone. The exact nature of the toxicity of this chemical is not entirely clear in the references. Therefore, prudent risk management requires it be treated as though it posed a significant health risk in the event of overexposure. If a less toxic chemical cannot be substituted for a hazardous substance, then *engineering controls* are the most effective method of reducing exposures. The best protection is to enclose operations and/or provide local exhaust ventilation at the site of chemical release. While not always operationally feasible, isolating operations can also reduce exposure. Using respiratory protection is less effective than the controls mentioned above, but is still advisable whenever working with or around isophorone. It is irritating at levels above 25 ppm. For exposures above the 25 ppm PEL, a supplied-air respirator with full facepiece operated in continuous flow mode, or a self-contained breathing apparatus (SCBA) with full facepiece and pressure demand are the recommended respiratory protection methods of choice. If a full facepiece is not available, then chemical goggles should be worn to protect the eyes. Whenever a chemical splash hazard exists, a face shield and a rubber apron should be worn. To prevent hand and skin exposures, impervious gloves should be used. Glove manufacturers should be contacted and permeation studies obtained *before* final glove selections have been made.

Administrative controls should also be in place to minimize the potential for human exposures. These may include written procedures or policies which specify the methods and techniques that will be practiced whenever personnel are to work with isophorone.

All personnel should receive *training* on the use, hazards, protective measures, emergency actions, and other precautions per 29 CFR 1910.1200 (Hazard Communication), prior to the first assignment in an area where isophorone is used or stored.

If symptoms develop or overexposure is suspected, the following tests are recommended:

- ☑ Kidney and liver function tests.
- ☑ Evaluation by a qualified allergist, with careful consideration of exposure history and special skin tests (may help diagnose allergy).
- ☑ Lung function tests, including forced vital capacity (FVC).
- ☑ Consider chest X-ray following acute overexposure (may be negative if taken immediately after exposure due to delayed onset of pulmonary edema).

Any evaluation should include a careful history of past and present symptoms with an examination. Medical tests that simply look for existing damage are not a substitute for controlling exposures. Also, since smoking can cause heart disease, emphysema, and other respiratory disorders, symptoms of exposure to isophorone may be more pronounced than those displayed by non-smokers. Proper risk assessment requires careful consideration of *all* possible factors that may be causing the appearance of symptoms in exposed workers.

Other methods to reduce exposure include:

- ☑ Where possible, enclose operations and use local exhaust ventilation at the site of chemical release. If local exhaust ventilation or enclosure is not used, respiratory protection should be mandatory. Also, if possible, automatic transfer of isophorone from storage containers to process or work containers is recommended.
- ☑ Always ensure that proper protective clothing is worn when using chemical substances.
- ☑ Wash thoroughly immediately after exposure to isophorone and at the end of the work shift or before eating, drinking, or smoking.
- ☑ Hazard warning information should be posted in the work area. In addition, as part of an on-going education and training program, all information on the health and safety hazards of isophorone should be communicated to all exposed and potentially exposed workers.
- ☑ Eye wash stations should be provided in the immediate work area for emergency use. If there is a possibility of skin exposure to isophorone, emergency shower facilities should also be provided in the immediate work area.

☑ Personnel should never wear contaminated clothing home (family members can be exposed). Clothing should be laundered by personnel who have been trained on the hazards of isophorone.

RISK ASSESSMENT: ENVIRONMENT

General Assessment

The environment is at risk of exposure and contamination during transportation, storage, disposal, or destruction of isophorone. In almost every scenario, the threat of environmental exposure is contingent upon the proper handling of the chemical substance. Accidental spills, large or small, can result in fire, explosion, and possible contamination of the surrounding environmental mediums (water, soil, and air).

Isophorone is considered a Class IIIA combustible liquid (per OSHA 29 CFR 1910.106). It presents a moderate fire and explosion hazard risk, especially in contact with incompatible materials. These characteristics require extreme caution in handling, storage, transportation, and disposal. It is incompatible with common peroxides and other oxidizers and contact can cause fire or explosion. It reacts violently when added to nitric or sulfuric acid. Therefore, special consideration is required during any emergency situation involving a leak or spill of isophorone. Should isophorone ever come into contact with these incompatible substances during use, transportation, storage, or disposal, the formation of highly toxic and/or highly explosive commodities is extremely possible.

The proper disposal method for isophorone is to burn it in a chemical incinerator equipped with an afterburner and scrubber.

Isophorone may enter the environment through industrial discharges or spills.

Acute Ecological Effects

Acute (short-term) toxic effects may include the death of animals, birds, or fish and death or low growth rate in plants. Acute effects are seen 2 to 4 days after animals or plants are exposed to isophorone.

Insufficient data are available on the short-term effects of isophorone exposure to aquatic life, birds, plants, or land animals.

Chronic Ecological Effects

Chronic (long-term) toxic effects may include shortened life span, reproductive problems, lower fertility, and changes in appearance or behavior in exposed animals. These effects can be seen long after first exposure(s) to toxic chemicals.

Insufficient data are available on the long-term effects of isophorone exposure to aquatic life, birds, plants, or land animals.

Water Solubility

Isophorone is nearly insoluble in water. Concentrations of less than 100 milligrams may mix with a liter of water, depending on water pH and chemical concentration of isophorone.

Persistence in the Environment

Isophorone is moderately persistent in water, with a half-life between 20 and 200 days. The half-life of a pollutant is the amount of time it takes for one half of the chemical to be degraded.

Bioaccumulation in Aquatic Organisms

Some substances increase in concentration, or *bioaccumulate,* in living organisms as they breathe contaminated air, drink contaminated water, or eat contaminated food. These chemicals can become concentrated in the tissues and internal organs of animals as well as humans.

The concentration of isophorone found in fish tissue is expected to be somewhat higher than the average concentration of isophorone in the water from which the fish was taken.

Recommended Risk-Reduction Measures

Proper training of all transporters will reduce the likelihood of a mishap or accident resulting in a leak or spill to the environment. The correct use of DOT labeling (per the MSDS) on all containers, trucks, and rail cars will enable emergency responders to react properly and quickly to any disaster thereby reducing the potential risk to the environment and to personnel.

Storage of isophorone should be segregated from incompatible chemicals to minimize the risk of cross-contamination or contact. Buildings designated for storage should be equipped with appropriate fire protection systems (alarms, sprinklers, emergency lighting, portable extinguishers).

If a spill or leak to the environment has occurred, fire department, emergency response and/or hazardous materials spill personnel should be notified immediately. Cleanup should be attempted only by those

trained in proper spill containment procedures using non-sparking tools. Contaminated soils should be removed for incineration and replaced with clean soil. If isophorone should contact the water table, aquifer, or navigable waterway, cleanup should be started immediately. It is only slightly soluble in water and total remediation may be possible. The local and/or state emergency response authorities must be notified. A comprehensive emergency response or disaster preparedness/recovery plan should be in place prior to any operations involving the use, transportation, storage, or disposal of isophorone. If isophorone is spilled or leaked, the following specific steps are recommended:

- ☑ Restrict persons not wearing protective clothing from area of spill or leak until cleanup is complete and area can be opened for normal work.
- ☑ Remove all ignition sources; ventilate area of leak.
- ☑ Absorb liquids in vermiculite, dry sand, earth, or similar material and deposit in sealed containers.
- ☑ It may be necessary to dispose of isophorone as a hazardous waste. The responsible state agency or the regional office of the federal Environmental Protection Agency (EPA) should be contacted for specific recommendations.

RISK ASSESSMENT: BUSINESS

General Assessment

Accidents or mishaps involving isophorone can present a significant threat to business operations. The loss or damage of equipment or facilities can significantly affect fiscal viability. Lawsuits that may result from personnel injury/death, public exposures, and/or environmental contamination will also require a serious expenditure of resources. Media attention surrounding an injury, death, or environmental damage can also result in a loss of profits and loss of current as well as future business.

Recommended Risk-Reduction Measures

Company attorneys, safety and health professionals, and environmental specialists should be involved in the development of any procedures dealing with safety or emergency response. A company official should be pre-designated as a public relations officer with specific training in dealing with the press. Corporate plans and policies should be developed, approved, and implemented long before any need for such arises.

REFERENCES

American Conference of Governmental Industrial Hygienists. 1988. *Documentation of the Threshold Limit Values and Biological Exposure Indices,* 5th Edition (with updates). Cincinnati: ACGIH

Bueche, J. R. 1972. *Principles of Physics,* 2nd Edition. New York: McGraw-Hill.

Calabrese, E. J. 1994. *Biological Effects of Low Level Exposures.* New York: CRC Press/Lewis Publishers.

Cockerman, L. G., B. S. Shane. 1994. *Basic Environmental Toxicology.* New York: CRC Press/Lewis Publishers.

Gardner, R. W. 1994. *Chemical Intolerance.* New York: CRC Press/Lewis Publishers.

Hallenbeck, W. H. 1993. *Quantitative Risk Assessment for Environmental and Occupational Health,* 2nd Edition. New York: CRC Press/Lewis Publishers.

Keith, L. H. and M. M. Walker. 1995. *Handbook of Air Toxics.* New York: CRC Press/Lewis Publishers.

Langerman, N. 1994 *Precautionary Labels for Chemical Containers.* New York: CRC Press/Lewis Publishers.

Lauwerys, R. R., P. Hoet. 1993. *Industrial Chemical Exposure.* New York: CRC Press/Lewis Publishers.

Lewis, R. J., Sr. 1992. *Sax's Dangerous Properties of Industrial Materials,* Volumes 1, 2, and 3, 8th Edition. New York: Van Nostrand Reinhold.

Meyer, E. 1990. *Chemistry of Hazardous Materials.* Englewood Cliffs, NJ: Prentice Hall.

Patnaik, P. 1992. *A Comprehensive Guide to the Hazardous Properties of Chemical Substances.* New York: Van Nostrand Reinhold.

Proctor, N. H., J. P. Hughes. 1991. *Chemical Hazards of the Workplace.* New York: Van Nostrand Reinhold.

Quigley, D. R. 1994. *Handbook of Emergency Chemical Management.* New York: CRC Press/Lewis Publishers.

Rea, W. J. 1995. *Chemical Sensitivity,* Vol. I-III. New York: CRC Press/Lewis Publishers.

Richardson, M. 1995. *Dictionary of Substance and Their Effects,* Vol. I-VI. New York: CRC Press/Lewis Publishers.

Sacarello, H. L. A. 1994. *The Comprehensive Handbook of Hazardous Materials.* New York: CRC Press/Lewis Publishers.

Sheftel, V. O. 1995. *Handbook of Toxic Properties of Monomers and Additives.* New York: CRC Press/Lewis Publishers.

Torkelson, T. R. V. K. Rowe. 1981. Halogenated aliphatic hydrocarbons. *Patty's Industrial Hygiene and Toxicology,* Vol. 2B, *Toxicology,* pp 3568-3572. New York: Wiley-Interscience.

U.S. Department of Health and Human Services, National Institute for Occupational Safety and Health. 1994. *NIOSH Pocket Guide to Chemical Hazards.* Washington, D.C.: U.S. Government Printing Office.

MATERIAL SAFETY DATA SHEET

CHEMICAL NAME

ISOPHORONE DIISOCYANATE

HAZARD WARNING INFORMATION

HEALTH	FIRE	REACTIVE	OTHER	DEGREE OF HAZARD	COLOR CODING	OTHER CODES
2	1	1		0 = Minimum Hazard 1 = Slight Hazard 2 = Moderate Hazard 3 = Serious Hazard 4 = Severe Hazard	HEALTH = BLUE FIRE = RED REACTIVITY = YELLOW OTHER = WHITE	OX = Oxidizer ACID = Acid ALK = Alkali COR = Corrosive W = Use No Water

SECTION I - GENERAL INFORMATION

Characterization **Isocyanate, Organic**	RCRA Number **None** / EPA Class **Not Applicable**
DOT Proper Shipping Name **Isophorone diisocyanate**	Chemical Abstract Service (CAS) Number **4098-71-9**
DOT Hazard Class and Label Requirements **Poison**	DOT Emergency Guide Code **55**
DOT Identification Number **UN 2290**	Chemical Formula $C_{12}H_{18}N_2O_2$

Synonyms
IPDI; 3-isocyanatomethyl-3,5,5-trimethylcyclohexylisocyanate; isophorone diamine diisocyanate.

SECTION II - HAZARDOUS INGREDIENTS AND IDENTITY INFORMATION

Hazardous Components (specific identity)	OSHA Exposure Criteria	NIOSH Exposure Criteria	Immediately Dangerous to Life and Health (IDLH)	ACGIH Exposure Criteria
Isophorone diisocyanate 1 ppm = 9.24 mg/m^3	PEL: Not Established STEL: Not Established	REL (skin): 0.005 ppm 0.045 mg/m^3 STEL: 0.02 ppm 0.18 mg/m^3	Not Determined	TLV (skin): 0.005 ppm 0.045 mg/m^3 STEL: Not Established

SECTION III - PHYSICAL AND CHEMICAL CHARACTERISTICS

Boiling Point **316°F (158°C)**	Specific Gravity (H_2O = 1) **1.06**
Vapor Pressure (mm Hg) **0.0003 at 68°F (20°C)**	Molecular Weight **222.3**
Saturated Vapor Density (Air = 1.2 kg/m^3) **Not Found**	Melting Point **-74°F (-60°C)**

Solubility
Decomposes in water; soluble in ethers, esters, ketones, and aliphatic and aromatic hydrocarbons.

Appearance and Odor
Clear, colorless, to slightly yellow liquid with a sharp, pungent odor.

SECTION IV - FIRE AND EXPLOSION HAZARD DATA

Flash Point (method used) **311°F (155°C) closed cup**	Explosive Limits in Air % by Volume **LEL: Not Determined** **UEL: Not Determined**
NFPA Classification **Class IIIB Combustible Liquid**	Autoignition Temperature **Not Determined**

Extinguishing Media
Dry chemical, water spray, regular foam, or fog.

Special Fire Fighting Procedures
Wear full protective clothing and self-contained breathing apparatus (SCBA). Approach fire from upwind position. Remove containers from fire if it can be done without risk. Use a water spray to keep fire-exposed containers cool. Poisonous gases are produced in fire. Avoid smoke and contain runoff. Do not allow water from control measures to enter sewers or waterways.

Unusual Fire and Explosion Hazards
Combustible when exposed to heat and flame. Containers may explode in fire. Stay clear of ends of tanks. Fight fire from maximum distance, if possible.

SECTION V - REACTIVITY DATA

Stability		Conditions to Avoid **Isophorone diisocyanate is normally stable at room temperature and pressure. Avoid contact with heat and incompatible materials.**
Stable X	Unstable	Incompatibility *(materials to avoid)* **Reacts with water (forms carbon dioxide), alcohols, phenols, amines, mercaptans, amides, urethane, ureas, and organo-metallic compounds (those containing aluminum, copper, or their alloys).**
Hazardous Polymerization		Conditions to Avoid **Under normal conditions of temperature and pressure, hazardous polymerization of isophorone diisocyanate is not expected to occur.**
May Occur	Will Not Occur X	Hazardous Decomposition or By-products **When heated to decomposition, isophorone diisocyanate can emit highly toxic/poisonous fumes including fumes of carbon dioxide, toxic nitrogen oxides, and cyanide compounds.**

SECTION VI - HEALTH HAZARD DATA

Primary Route(s) of Entry:	Inhalation?	Absorption (skin)?	Ingestion?
	X	X	X

Health Hazards

INHALATION: Irritation of the eyes, nose, throat, and respiratory tract with coughing, a choking sensation, deep chest pain or soreness, chest constriction and pain, severe bronchospasm, and pulmonary edema (fluid in the lungs). Asthmatic bronchitis can also occur.

ABSORPTION: Skin contact may cause redness, swelling, painful blistering, and possible dermatitis. If the liquid contacts the eyes, there can be the development of conjunctivitis with severe inflammation of the eyelids.

INGESTION: Unspecified gastrointestinal effects.

Carcinogenicity	NTP Listed?	IARC Cancer Review Group?	OSHA Regulated?	Target Organs?
Unknown Human Unknown Animal	No	No	No	Respiratory system, eyes, skin.

Medical Conditions Generally Aggravated by Exposure

Respiratory system problems or skin diseases may be aggravated by exposure.

Emergency and First-aid Procedures

Eye contact: Flush immediately with water for 15 minutes (minimum); seek medical attention. Skin contact: Remove all contaminated clothing. Immediately wash area with large amounts of soap and water. For red or blistered skin, seek medical assistance. For inhalation: Remove the person from exposure. Provide respiratory assistance and CPR. Transfer to medical facility. Observe for 48 hours for lung effects. If swallowed: Contact poison control center. Unless advised otherwise, give conscious person 1 to 2 glasses of water to dilute. Do NOT induce vomiting (chemical is a severe irritant).

SECTION VII - PRECAUTIONS FOR SAFE HANDLING AND USE

Steps to be Taken in Case Material is Released or Spilled

Absorb liquid solutions in vermiculite or similar material and place in sealed containers. Ventilate area of spill or leak. Remove all ignition sources. Restrict those not involved in cleanup from entering area.

Preferred Waste Disposal Method

Mix in combustible solvent, burn in a chemical incinerator equipped with an afterburner and scrubber.

Precautions to be Taken in Handling and Storage

Store in tightly closed containers in cool, dry, well-ventilated area away from heat, sparks, or flames. Sources of ignition are prohibited where isophorone diisocyanate is used, handled, or stored.

Other Precautions and Warnings

Keep away from water or moisture. Do NOT reseal a container if contamination is suspected. Package for disposal or reclamation. Do NOT store in aluminum or copper containers. Protect from damage.

SECTION VIII - CONTROL MEASURES AND PERSONAL PROTECTIVE EQUIPMENT

Respiratory Protection *(specify type)*

The NIOSH REL is low and difficult to monitor. For any exposure level, use an MSHA/NIOSH-approved supplied-air respirator with full facepiece, hood, or helmet in continuous flow mode, or use a self-contained breathing apparatus (SCBA) in continuous flow mode.

Ventilation

Local exhaust or general mechanical systems recommended.

Protective Gloves	Eye Protection	Other Protective Clothing
Rubber, Polyvinyl Chloride (PVC)	Chemical Goggles or Face Mask	Protective Apron or Clothing

Work/Hygiene Practices

Always wash hands thoroughly after using chemical; never bring food, drink, or smoking materials into vicinity of chemicals.

ISOPHORONE DIISOCYANATE

$C_{12}H_{18}N_2O_2$

CAS: 4098-71-9

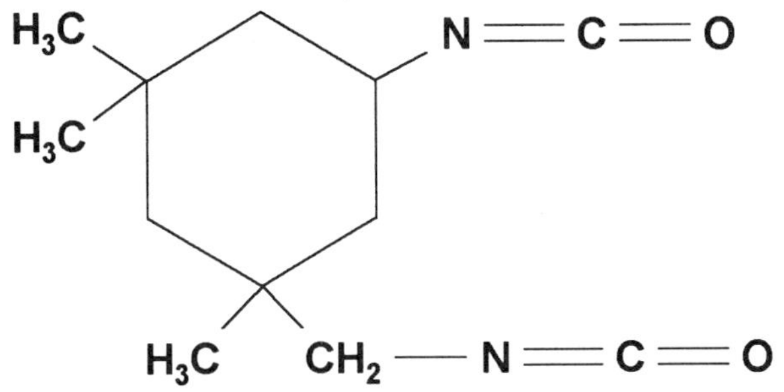

IDENTIFICATION AND TYPICAL USES

Isophorone diisocyanate appears as a clear, colorless to slightly yellow liquid with a sharp, pungent odor. It is used in the production of high-quality coatings, polyurethane paints, and varnishes that are resistant to light discoloration, abrasion, chemicals, chalking, and weathering. It is also used as an elastomer in casting compounds, mastics, and flexible tile coatings.

RISK ASSESSMENT: HEALTH

General Assessment

Isophorone diisocyanate is an extremely toxic and poisonous substance. It can enter the body through *inhalation, absorption,* and by *ingestion.* It will also cause adverse health effects on contact with the skin or eyes. There is no information on its carcinogenic, teratogenic, or mutagenic capabilities in humans or animals.

Inhalation of isophorone diisocyanate vapors can be highly irritating to the eyes, nose, throat, and mucosa of the respiratory tract. Symptoms of exposure include wheezing, cough, deep cheat pain, dyspnea, sweating, difficulty in breathing, severe bronchospasm, and possible life-threatening pulmonary edema (fluid in the lungs). Development of pulmonary edema may be delayed up to 48 hours thereby creating a false sense of security with regard to health exposure risk. In severe cases, there may be some restriction of air movement and eventual respiratory failure (due to edema). Prolonged contact can lead to chronic bronchitis, asthma-like symptoms, and allergic pulmonary sensitization. Isophorone diisocyanate is a cholinesterase inhibitor.

Skin contact causes irritation and painful burning with skin blisters and dermatitis on prolonged contact. If the liquid should contact the eye directly, there may be irritation, inflammation of the eyelids and surrounding tissues, conjunctivitis, and possible damage to the cornea and subsequent visual impairment.

Acute Health Effects

The following acute (short-term) health effects may occur immediately or shortly after exposure to isophorone diisocyanate:

Skin: Possible irritation with painful burning and blistering.

Eye: Moderate to severe irritation. There can be inflammation, irritation of the eyelids, and conjunctivitis.

Lung: Irritation of the nose, throat, and lungs following exposure. There can be a delayed but dangerous buildup of fluid in the lungs (pulmonary edema), which is a medical emergency.

Chronic Health Effects

The following chronic (long-term) health effects may occur at some time after exposure to isophorone diisocyanate and can last for months or even years:

Cancer Hazards: There are insufficient data in the references on the carcinogenic or mutagenic properties of isophorone diisocyanate.

Reproductive Hazard: According to the data presented in the references, isophorone diisocyanate has not been adequately tested for its ability to adversely affect reproduction in test animals.

Other Chronic Effects: Long-term exposure to isophorone diisocyanate may lead to allergic sensitization of the respiratory system with initial symptoms of night coughs and difficulty breathing and possible progression to asthmatic bronchitis. Dermal sensitization is also possible with rash and inflammation.

ⓘ *Recommended Risk-Reduction Measures*

Personnel should avoid direct contact with isophorone diisocyanate. If a less toxic chemical cannot be substituted for a hazardous substance, then *engineering controls* are the most effective method of reducing exposures. The best protection is to enclose operations and/or provide local exhaust ventilation at the site of chemical release. While not always operationally feasible, isolating operations can also reduce exposure. Using respiratory protection is less effective than the controls mentioned above, but is still advisable whenever working with or around isophorone diisocyanate. The NIOSH REL for this chemical is extremely low (0.005 ppm) and difficult to monitor reliably under most operating or emergency conditions. A supplied-air respirator with a full facepiece operated in positive pressure mode or an MSHA/NIOSH-approved self-contained breathing apparatus (SCBA) with full facepiece operated pressure demand or other positive pressure mode are the recommended respiratory protection methods of choice. In either case, if a full facepiece is not available, then chemical splash goggles should be worn to protect the eyes. Whenever a chemical splash hazard exists, a face shield and a protective apron should be worn. To prevent hand and skin exposures, impervious gloves should be used. Rubber and polyvinyl chloride materials should provide suitable protection. Glove manufacturers should be contacted and permeation studies obtained *before* final glove selections are made.

Administrative controls should also be in place to minimize the potential for human exposures. These may include written procedures or policies which specify the methods and techniques that will be practiced whenever personnel are to work with isophorone diisocyanate.

All personnel should receive *training* on the use, hazards, protective measures, emergency actions, and other precautions per 29 CFR 1910.1200 (Hazard Communication), prior to the first assignment in an area where isophorone diisocyanate is used or stored.

If symptoms develop or overexposure is suspected, the following medical tests are recommended:

- ☑ Lung function tests.
- ☑ Consider chest X-ray following acute overexposure (may be negative if taken immediately after exposure due to delayed onset of pulmonary edema).
- ☑ Monitor arterial blood gases.

Also, since smoking can cause heart disease, emphysema, and other lung diseases, smokers exposed to isophorone diisocyanate may experience symptoms more quickly than non-smokers under the same conditions of exposure. Prudent risk management requires proper consideration of *all* possible exposure factors when evaluating workplace health hazard risk.

Other methods to reduce exposure include:

- ☑ Where possible, enclose operations and use local exhaust ventilation at the site of chemical release. If local exhaust ventilation or enclosure is not used, respiratory protection should be mandatory.
- ☑ Where possible, automatically transfer isophorone diisocyanate from drums or other storage containers to process containers.
- ☑ Always ensure that proper protective clothing is worn when using chemical substances and that personnel are trained on its use and care.
- ☑ Wash thoroughly immediately after exposure to isophorone diisocyanate and at the end of the work shift or before eating, drinking, or smoking.
- ☑ Hazard warning information should be posted in the work area. In addition, as part of an on-going education and training program, all information on the health and safety hazards of isophorone diisocyanate should be communicated to all exposed and potentially exposed workers.

RISK ASSESSMENT: ENVIRONMENT

General Assessment

The environment is at risk of exposure during the transportation, storage, disposal, or destruction of isophorone diisocyanate. In almost every scenario, the threat of environmental exposure is contingent upon the proper handling of the chemical substance. Accidental spills, large or small, can result in fire, explosion, and possible contamination of the surrounding environmental mediums (water, soil, and air).

Isophorone diisocyanate is considered a Class IIIB combustible liquid (per OSHA 29 CFR 1910.106). It can react vigorously with oxidizing agents, water, alcohols, strong bases, amines, carboxylic acids, and organo-metallic catalysts. Pressure may build in closed containers. These characteristics require special consideration during any emergency situation involving a leak or spill of isophorone diisocyanate.

Isophorone diisocyanate may enter the environment primarily during the disposal of wastes from the chlorinated hydrocarbon industry.

Acute Ecological Effects

Acute (short-term) toxic effects may include the death of animals, birds, or fish and death or low growth rate in plants. Acute effects are seen 2 to 4 days after animals or plants are exposed to isophorone diisocyanate.

Insufficient data are available to evaluate or predict the short-term effects of isophorone diisocyanate exposure to aquatic life, plants, birds, or land animals.

Chronic Ecological Effects

Chronic (long-term) toxic effects may include shortened life span, reproductive problems, lower fertility, and changes in appearance or behavior in exposed animals. These effects can be seen long after first exposure(s) to toxic chemicals.

Insufficient data are available to evaluate or predict the long-term effects of isophorone diisocyanate to aquatic life, plants, birds, or terrestrial animals.

Water Solubility

Isophorone diisocyanate will sink and decompose in water to form its corrosive and toxic diamine and carbon dioxide. It is, therefore, only slightly soluble in the aquatic environment. Concentrations of less than 1 milligram will mix with a liter of water.

Persistence in the Environment

Isophorone diisocyanate is non-persistent in the aquatic environment, with a half-life of less than 2 days. The half-life of a pollutant is the amount of time it takes for one half of the chemical to be degraded.

Bioaccumulation in Aquatic Organisms

Some substances increase in concentration, or *bioaccumulate,* in living organisms as they breathe contaminated air, drink contaminated water, or eat contaminated food. These chemicals can become concentrated in the tissues and internal organs of animals as well as humans.

The concentration of isophorone diisocyanate found in fish tissues is expected to be much lower than the average concentration of isophorone diisocyanate in the water from which the fish was taken.

Recommended Risk-Reduction Measures

Proper training of all transporters will reduce the likelihood of a mishap or accident resulting in a leak or spill to the environment. The correct use of labeling on all containers, trucks, and rail cars will enable emergency responders to react properly and quickly to any disaster thereby reducing the potential risk to the environment and to personnel.

Storage of isophorone diisocyanate should be segregated from incompatible chemicals to minimize the risk of cross-contamination or contact. Storage buildings or facilities should be equipped with proper fire protection equipment (sprinklers/extinguishers, alarms).

If a spill or leak to the environment has occurred, fire department, emergency response and/or hazardous materials spill personnel should be notified immediately. Cleanup should be attempted only by those trained in proper spill containment procedures using non-sparking tools. Contaminated soils should be removed for incineration and replaced with clean soil. If isophorone diisocyanate should contact the water table, aquifer, or navigable waterway, time is of the essence. Even though it is only slightly soluble in the aquatic environment, it will slowly react in water and, therefore, total containment and remediation may not be possible. When such spills occur, the local and/or state emergency response authorities must be notified. A comprehensive emergency response or disaster preparedness/recovery plan should be in place prior to any operations involving the use, transportation, storage, or disposal of isophorone diisocyanate. The proper disposal method for isophorone diisocyanate is to burn it in a chemical incinerator equipped with an afterburner and an air scrubber.

If isophorone diisocyanate are spilled or leaked, the following specific response steps are recommended:

- ☑ Restrict persons not wearing protective clothing from area of spill or leak until cleanup is complete and area can be opened for normal work.
- ☑ Remove ignition sources and ventilate area.
- ☑ If liquid is spilled, absorb in vermiculite, dry sand, earth, or similar material and deposit in sealed containers. Use non-sparking tools.
- ☑ It may be necessary to dispose of isophorone diisocyanate as a hazardous waste. The responsible state agency or the regional office of the federal Environmental Protection Agency (EPA) should be contacted for specific recommendations.

RISK ASSESSMENT: BUSINESS

General Assessment

Accidents or mishaps involving isophorone diisocyanate can present a significant threat to business operations. The loss or damage of equipment or facilities can significantly affect fiscal viability. Lawsuits that may result from personnel injury/death, public exposures, and/or environmental contamination will also require a serious expenditure of resources. Media attention surrounding an injury, death, or environmental damage can also result in a loss of profits and loss of current as well as future business.

Recommended Risk-Reduction Measures

Company attorneys, safety and health professionals, and environmental specialists should be involved in the development of any procedures for responding to hazardous chemical incidents. A company official should be pre-designated as a public relations officer with specific training in dealing with the press. Corporate plans and policies should be developed, approved, and implemented long before any need for such arises.

REFERENCES

American Conference of Governmental Industrial Hygienists. 1988. *Documentation of the Threshold Limit Values and Biological Exposure Indices,* 5th Edition (with updates). Cincinnati: ACGIH

Bloom, A. D. and F. J. DeSerres. 1995. *Ecotoxicity and Human Health.* New York: CRC Press/Lewis Publishers.

Bueche, J. R. 1972. *Principles of Physics,* 2nd Edition. New York: McGraw-Hill.

Calabrese, E. J. 1994. *Biological Effects of Low Level Exposures.* New York: CRC Press/Lewis Publishers.

Cockerman, L. G., B. S. Shane. 1994. *Basic Environmental Toxicology.* New York: CRC Press/Lewis Publishers.

Cothern, R. R. 1993. *Comparative Environmental Risk Assessment.* New York: CRC Press/Lewis Publishers.

Danse, I. R. 1991. *Common Sense Toxics in the Workplace.* New York: Van Nostrand Reinhold.

Environmental Protection Agency. 1983. *Chemical Hazard Information Profiles.* Washington, D.C.: U.S. EPA.

Fodden, C. R., and J. L. Weddell. 1992. *Hazardous Materials Emergency Action Data.* New York: CRC Press/Lewis Publishers.

Forseberg, K., and L. H. Keith. 1993. *Chemical; Protective Clothing Permeation and Degradation Database.* Chelsea, MI: Lewis Publishers.

Gardner, R. W. 1994. *Chemical Intolerance.* New York: CRC Press/Lewis Publishers.

Hallenbeck, W. H. 1993. *Quantitative Risk Assessment for Environmental and Occupational Health,* 2nd Edition. New York: CRC Press/Lewis Publishers.

Howard, P. H., et. al. 1991. *Handbook of Environmental Degradation Rates.* New York: CRC Press/Lewis Publishers.

Hosty, J. W., and P. Foster. 1990. *A Practical Guide to Chemical Spill Response.* New York: Van Nostrand Reinhold.

International Agency for Research on Cancer (IARC). 1987. *IARC Monographs on the Evaluation of Carcinogenic Risk to Humans, Overall Evaluations of Carcinogenicity: An updating of IARC Monographs Volumes 1-42.* Lyon: IARC.

Jones, F. E. 1994. *Toxic Organic Vapors in the Workplace.* New York: CRC Press/Lewis Publishers.

Keith, L. H. and M. M. Walker. 1995. *Handbook of Air Toxics.* New York: CRC Press/Lewis Publishers.

Langerman, N. 1994 *Precautionary Labels for Chemical Containers.* New York: CRC Press/Lewis Publishers.

Lauwerys, R. R., P. Hoet. 1993. *Industrial Chemical Exposure.* New York: CRC Press/Lewis Publishers.

Lewis, R. J., Sr. 1992. *Sax's Dangerous Properties of Industrial Materials,* Volumes 1, 2, and 3, 8th Edition. New York: Van Nostrand Reinhold.

Mackay, D., W. Y. Shiu, and K. C. Ma. 1992. *Illustrated Handbook of Chemical-Physical Properties and Environmental Fate for Organic Chemicals.* New York: CRC Press/Lewis Publishers.

Manahan, S. E. 1992. *Toxicological Chemistry,* 2nd Edition. New York: CRC Press/Lewis Publishers.

Meyer, E. 1990. *Chemistry of Hazardous Materials.* Englewood Cliffs, NJ: Prentice Hall.

Miller, B. F., M.D., and C. B. Keane, R.N. 1972. *Encyclopedia and Dictionary of Medicine and Nursing.* Philadelphia: W.B. Sanders Company.

Mycock, J. C., J. D. McKenna, and L. Theodore. 1995. *Handbook of Air Pollution Control Engineering and Technology.* New York: CRC Press/Lewis Publishers.

Nelson, G. O. 1992. *Gas Mixtures.* New York: CRC Press/Lewis Publishers.

Ott, W. R. 1995. *Environmental Statistics and Data Analysis.* New York: CRC Press/Lewis Publishers.

Patnaik, P. 1992. *A Comprehensive Guide to the Hazardous Properties of Chemical Substances.* New York: Van Nostrand Reinhold.

Rea, W. J. 1995. *Chemical Sensitivity,* Vol. I-III. New York: CRC Press/Lewis Publishers.

Sacarello, H. L. A. 1994. *The Comprehensive Handbook of Hazardous Materials.* New York: CRC Press/Lewis Publishers.

Sherman, J. D. 1988. *Chemical Exposure and Disease.* New York: Van Nostrand Reinhold.

U.S. Department of Health and Human Services, National Institute for Occupational Safety and Health. 1994. *NIOSH Pocket Guide to Chemical Hazards.* Washington, D.C.: U.S. Government Printing Office.

MATERIAL SAFETY DATA SHEET

CHEMICAL NAME

2-ISOPROPOXYETHANOL

HAZARD WARNING INFORMATION

HEALTH	FIRE	REACTIVE	OTHER	DEGREE OF HAZARD	COLOR CODING	OTHER CODES
2	3	0		0 = Minimum Hazard 1 = Slight Hazard 2 = Moderate Hazard 3 = Serious Hazard 4 = Severe Hazard	HEALTH = BLUE FIRE = RED REACTIVITY = YELLOW OTHER = WHITE	OX = Oxidizer ACID = Acid ALK = Alkali COR = Corrosive W = Use No Water

SECTION I - GENERAL INFORMATION

Characterization	**Glycol Ether**	RCRA Number	**D001**
		EPA Class	**Characteristic (I) Waste**
DOT Proper Shipping Name	**No Citation**	Chemical Abstract Service (CAS) Number	**109-59-1**
DOT Hazard Class and Label Requirements	**No Citation**	DOT Emergency Guide Code	**No Citation**
DOT Identification Number	**No Citation**	Chemical Formula	$(CH_3)_2CHOCH_2CH_2OH$

Synonyms

Ethylene glycol isopropyl ether; β-hydroxyethylisopropyl ether; isopropyl Cellosolve ®; isopropyl glycol; isopropyl oxitol; *beta*-hydroxyethyl isopropyl ether; Dowanol Eipat ®; ethylene, 2-isopropoxy.

SECTION II - HAZARDOUS INGREDIENTS AND IDENTITY INFORMATION

Hazardous Components (specific identity)	OSHA Exposure Criteria	NIOSH Exposure Criteria	Immediately Dangerous to Life and Health (IDLH)	ACGIH Exposure Criteria
2-Isopropoxyethanol (derivation: By reacting ethylene oxide with ethanol). **1 ppm =4.33 mg/m³**	**PEL:** 25 ppm, 105 mg/m³ **STEL:** Not Established	**REL:** Not Established **STEL:** Not Established	**Not Determined**	**TLV:** 25 ppm, 106 mg/m³ **STEL:** Not Established

SECTION III - PHYSICAL AND CHEMICAL CHARACTERISTICS

Boiling Point	**283°F (140°C)**	Specific Gravity (H_2O = 1)	**0.90**
Vapor Pressure (gas)	**2.6 at 69°F (20°C)**	Molecular Weight	**104.2**
Vapor Density (Air = 1)	**3.6**	Freezing Point	**Not Found**

Solubility

Completely miscible in water.

Appearance and Odor

Clear, colorless, mobile liquid with a mild, ethereal odor.

SECTION IV - FIRE AND EXPLOSION HAZARD DATA

Flash Point (method used)	**92°F (33°C) opened cup**	Explosive Limits in Air % by Volume	**LEL: Not Determined UEL: Not Determined**
NFPA Classification	**Class IC Flammable Liquid**	Autoignition Temperature	**Not Determined**

Extinguishing Media

Use dry chemical, carbon dioxide, or alcohol-resistant foam (water spray may be ineffective).

Special Fire Fighting Procedures

Wear full protective clothing and self-contained breathing apparatus (SCBA). Poisonous gases are produced in fire. Use water spray to keep fire-exposed containers cool. Containers may explode in fire.

Unusual Fire and Explosion Hazards

Dangerous fire hazard when exposed to heat. Chemical will react on contact with oxidizable materials. Moderate explosion hazard when in the form of vapor. Vapors are heavier than air and can flashback.

SECTION V - REACTIVITY DATA

Stability		Conditions to Avoid
		2-Isopropoxyethanol is stable under normal conditions. Keep away from sources of heat or flame. Keep away from incompatible materials.
Stable	Unstable	Incompatibility *(materials to avoid)*
X		**Strong oxidizers, such as potassium permanganate, chromium trioxide, sodium peroxide, nitrates, chlorates, and perchlorates.**
Hazardous Polymerization		Conditions to Avoid **Under normal conditions of temperature and pressure, hazardous polymerization of 2-isopropoxyethanol is not expected to occur.**
May Occur	Will Not Occur	Hazardous Decomposition or By-products
	X	**When heated to decomposition, 2-isopropoxyethanol can emit highly toxic/poisonous and acrid fumes and gases, including toxic oxides of carbon.**

SECTION VI - HEALTH HAZARD DATA

Primary Route(s) of Entry:	Inhalation?	Absorption (skin)?	Ingestion?
	X	X	X

Health Hazards

INHALATION: Symptoms include lachrymation, irritation of the throat and upper respiratory tract, general anesthesia, headache, burning cough, respiratory stimulation, nausea or vomiting, liver, kidney, or pulmonary changes (edema), and coma (especially if vapors are heated).

ABSORPTION: Will pass through intact skin to cause toxic systemic effects as with inhalation.

INGESTION: Same symptoms as inhalation, only more severe. May also cause CNS depression with excitement, nausea, vomiting, drowsiness, coma, respiratory or kidney failure, and death.

Carcinogenicity	NTP Listed?	IARC Cancer Review Group?	OSHA Regulated?	Target Organs?
Unknown Human Unknown Animal	**No**	**No**	**No**	**Resp. system, eyes, skin, CNS, blood, kidneys.**

Medical Conditions Generally Aggravated by Exposure

Any liver, kidney, cardiovascular, or nervous system disorders may be affected.

Emergency and First-aid Procedures

Eye contact: Flush immediately with water for 15 minutes (minimum); seek medical attention. Skin contact: Remove all contaminated clothing. Immediately wash area with large amounts of soap and water. Seek medical assistance. For inhalation: Remove the person from exposure. Provide respiratory assistance and CPR, if required. Transfer to medical facility. Observe for 48 hours for lung effects (pulmonary edema). If swallowed: Seek medical attention immediately or contact poison control center. Unless told otherwise, give a conscious victim 1-2 glasses of water. Do NOT induce vomiting.

SECTION VII - PRECAUTIONS FOR SAFE HANDLING AND USE

Steps to be Taken in Case Material is Released or Spilled

Absorb large spills in vermiculite or similar material and deposit in sealed containers. Ventilate area of spill or leak. Restrict those not involved in cleanup from entering area. Remove all sources of ignition.

Preferred Waste Disposal Method

Burn in chemical incinerator equipped with an afterburner and scrubber.

Precautions to be Taken in Handling and Storage

Store to avoid contact with incompatible materials in tightly closed containers in cool, well-ventilated area away from heat and flame. Use only non-sparking tools to open and close containers.

Other Precautions and Warnings

Sources of ignition are prohibited where 2-isopropoxyethanol is used, handled, or stored. Heat may cause containers to build up pressure and explode. Containers and equipment must be grounded and bonded, especially during liquid transfer. Protect containers from physical damage.

SECTION VIII - CONTROL MEASURES AND PERSONAL PROTECTIVE EQUIPMENT

Respiratory Protection *(specify type)*

For low exposures, use a chemical cartridge respirator. Otherwise, use an MSHA/NIOSH-approved supplied-air respirator with full facepiece, hood, or helmet in continuous flow mode, or use a self-contained breathing apparatus (SCBA) operated in pressure demand or other positive pressure mode.

Ventilation

Local exhaust or general mechanical systems recommended.

Protective Gloves	Eye Protection	Other Protective Clothing
Impervious Material	**Chemical Goggles or Face Mask**	**Protective Apron**

Work/Hygiene Practices

Always wash hands thoroughly after using chemical; never bring food, drink, or smoking materials into vicinity of chemicals.

2-ISOPROPOXYETHANOL

$(CH_3)_2CHOCH_2CH_2OH$ CAS: 109-59-1

IDENTIFICATION AND TYPICAL USES

2-Isopropoxyethanol is a clear, colorless, mobile liquid with a mild, ethereal odor. It is used as a component of lacquers and other coatings and as a solvent for resins, dyes, and textiles.

RISK ASSESSMENT: HEALTH

General Assessment

2-Isopropoxyethanol is mildly toxic to humans by *ingestion* and *inhalation*. It will also pass through unbroken skin (*absorption*) to cause toxic systemic effects equal or greater than that caused by inhalation or ingestion. This is especially true if absorption occurs in tandem with some other exposure, such as inhalation. It is a severe central nervous system depressant. There are no reports in the references to indicate that exposure causes cancer, reproductive effects, or mutations in humans or animals.

Toxic systemic effects by all routes of exposure can include hemolytic anemia, digestive problems, dizziness, respiratory distress, and renal dysfunction. Symptoms of exposure can include abdominal pain, dehydration, nausea, vomiting, general weakness and muscle tenderness, acidosis and resultant hyperpnea (rapid, deep respiration), and pulmonary edema (fluid in the lungs). The development of pulmonary edema and associated symptoms may be delayed up to 48 hours, thereby creating a false sense of security with regard to health exposure risk. CNS depression may cause excitement or exhilaration, drunkenness and stupor, ataxia, dizziness, coma, and possible convulsions. Renal effects include lower back pain, excessive protein in urine (albuminuria), excessive blood in urine (hematuria), reduction in urine secretion (oliguria), and kidney failure leading to water retention in the extremities (peripheral edema). If death occurs, it is usually due to respiratory arrest or cardiovascular collapse.

Skin and eye contact may produce mild to moderate irritation on contact and may cause an allergic reaction and rash. It will pass through unbroken skin to cause many of the toxic systemic effects noted for inhalation.

Acute Health Effects

The following acute (short-term) health effects may occur immediately or shortly after exposure to 2-isopropoxyethanol:

Skin: Some irritation at site of contact. Allergic rash may develop. Absorption is likely.

Eye: Mild to moderate irritation.

Lung: Nose and throat irritation. Produces intoxicating effects and may cause lung effects, such as pulmonary edema.

CNS: Stimulation and depression with symptoms of dizziness, headache, lightheadedness, nausea, vomiting, and coma.

Chronic Health Effects

The following chronic (long-term) health effects may occur at some time after exposure to 2-isopropoxyethanol and can last for months or even years:

Cancer Hazards: According to information presented in the references, 2-isopropoxyethanol has not been adequately tested for its ability to cause cancer in animals.

Reproductive Hazard: According to information presented in the references, 2-isopropoxyethanol has not been adequately tested for its ability to adversely affect reproduction in test animals.

Other Chronic Effects: Repeated exposures may cause blood effects such as hemolytic anemia (destruction of red blood cells). Skin rash and/or dermatitis may develop following prolonged or repeated contact.

Recommended Risk-Reduction Measures

Personnel should avoid direct contact with 2-isopropoxyethanol. This chemical is toxic to humans by all exposure routes (inhalation, ingestion, and dermal absorption). If a less toxic chemical cannot be substituted for a hazardous substance, then *engineering controls* are the most effective method of reducing exposures. The best protection is to enclose operations and/or provide local exhaust ventilation at the site of chemical release. While not always operationally feasible, isolating operations can also reduce exposure. Using respiratory protection is less effective than the controls mentioned above, but is still advisable whenever working with or around 2-isopropoxyethanol. Chemical cartridge respirators may provide adequate protection up to the PEL. However, the best protection is obtained using a supplied-air respirator with full facepiece operated in continuous flow mode, or a self-contained breathing apparatus (SCBA) with full facepiece operated in pressure demand. If a full facepiece is not available, then chemical goggles should be worn to protect the eyes. Whenever a chemical splash hazard exists, a face shield and a protective apron should be worn. To prevent hand and skin exposures, impervious gloves should be used. Glove manufacturers should be contacted and permeation studies obtained *before* final glove selections have been made.

Administrative controls should also be in place to minimize the potential for human exposures. These may include written procedures or policies which specify the methods and techniques that will be practiced whenever personnel are to work with or around 2-isopropoxyethanol.

All personnel should receive *training* on the use, hazards, protective measures, emergency actions, and other precautions per 29 CFR 1910.1200 (Hazard Communication), prior to the first assignment in an area where 2-isopropoxyethanol is used or stored.

If symptoms develop or overexposure is suspected, the following tests are recommended:

- ☑ Kidney function tests.
- ☑ Urine oxalate level. Urinalysis may show low specific gravity, proteinuria, pyuria, cylinduria, hematuria, calcium oxide, and hippuric crystals.
- ☑ Lung function tests.
- ☑ Consider chest X-ray following acute overexposure (may be negative if taken immediately after exposure due to delayed onset of pulmonary edema).

Any evaluation should include a careful history of past and present symptoms with an examination. Medical tests that simply look for existing damage are not a substitute for controlling exposures. Sodium bicarbonate may be used intravenously to treat acidosis. Ethanol may be used in antedotal treatment. However, blood glucose levels must be monitored for hypoglycemia. Also, since smoking can cause heart disease, lung cancer, emphysema, and other respiratory disorders, smokers exposed to 2-isopropoxyethanol may experience symptoms more rapidly or more pronounced than non-smokers under the same conditions of exposure. Prudent risk management requires proper consideration of *all* possible factors which may be causing the appearance of exposure symptoms in the workplace.

Other methods to reduce exposure include:

- ☑ Where possible, enclose operations and use local exhaust ventilation at the site of chemical release. If local exhaust ventilation or enclosure is not used, respiratory protection should be mandatory.
- ☑ Where possible, automatically transfer liquid 2-isopropoxyethanol from storage drums to process containers. Electrically ground and bond all containers and equipment, especially during transfer operations.
- ☑ Always ensure that proper protective clothing is worn when using chemical substances and that personnel are trained in its use and care.
- ☑ Wash thoroughly immediately after exposure to 2-isopropoxyethanol and at the end of the work shift or before eating, drinking, or smoking.
- ☑ Hazard warning information should be posted in the work area. In addition, as part of an on-going education and training program, all information on the health and safety hazards associated with 2-isopropoxyethanol should be communicated to all potentially exposed workers.
- ☑ Eye wash stations should be provided in the immediate work area for emergency use. If there is a possibility of skin exposure to this chemical, then emergency shower facilities should also be provided in the immediate work area.
- ☑ Personnel should never wear contaminated clothing home (family members can be exposed). Clothing should be laundered only by personnel who have been properly trained on hazards associated with exposure to 2-isopropoxyethanol.

RISK ASSESSMENT: ENVIRONMENT

General Assessment

The environment is at risk of exposure during transportation, storage, disposal, or destruction of 2-isopropoxyethanol. In almost every scenario, the threat of environmental exposure is contingent upon the proper handling of the chemical substance. Accidental spills, large or small, can result in fire, explosion, and possible contamination of the surrounding environmental mediums (water, soil, and air).

2-Isopropoxyethanol is considered a Class IC flammable liquid (per OSHA 29 CFR 1910.106). Its relatively low flash point and moderate boiling point present a serious fire and explosion hazard, especially in confined spaces. Its vapors are heavier than air and can travel for great distances to an ignition source to flashback and cause fire or explosion. These characteristics require extreme caution in handling, storage, transportation, and disposal. It is incompatible with many common peroxides and other oxidizers and contact can cause fire or explosion. Therefore, special consideration is required during any emergency situation involving a leak or spill of 2-isopropoxyethanol. Should 2-isopropoxyethanol ever come into contact with any incompatible substances during use, transportation, storage, or disposal, the formation of highly toxic and/or highly explosive commodities is extremely possible. The proper method for disposing of waste 2-isopropoxyethanol is to burn it in a chemical incinerator equipped with an afterburner and scrubber.

2-Isopropoxyethanol may enter the environment through industrial or municipal waste treatment plant discharges, agricultural runoff or disposal, or spills.

Acute Ecological Effects

Acute (short-term) toxic effects may include the death of animals, birds, or fish and death or low growth rate in plants. Acute effects are seen 2 to 4 days after animals or plants are exposed to 2-isopropoxyethanol.

Insufficient data are available to evaluate or predict the short-term effects of 2-isopropoxyethanol exposure to aquatic life, plants, birds, or land animals.

Chronic Ecological Effects

Chronic (long-term) toxic effects may include shortened life span, reproductive problems, lower fertility, and changes in appearance or behavior in exposed animals. These effects can be seen long after (sometimes months or even years) first exposure(s) to toxic chemicals.

Insufficient data are available to evaluate or predict the long-term effects of 2-isopropoxyethanol exposure on aquatic life, birds, plants, or land animals.

Water Solubility

2-Isopropoxyethanol is highly soluble in water. Concentrations of 1000 milligrams and more will mix with a liter of water.

Persistence in the Environment

2-Isopropoxyethanol is slightly persistent in water, with a half-life between 2 and 20 days. The half-life of a pollutant is the amount of time it takes for one half of the chemical to be degraded.

Bioaccumulation in Aquatic Organisms

Some substances increase in concentration, or *bioaccumulate,* in living organisms as they breathe contaminated air, drink contaminated water, or eat contaminated food. These chemicals can become concentrated in the tissues and internal organs of animals as well as humans.

No data are available on the bioaccumulation properties of 2-isopropoxyethanol.

Recommended Risk-Reduction Measures

Proper training of all transporters will reduce the likelihood of a mishap or accident resulting in a leak or spill to the environment. The correct use of DOT labeling (per the MSDS) on all containers, trucks, and rail cars will enable emergency responders to react properly and quickly to any disaster thereby reducing the potential risk to the environment and to personnel.

Storage of 2-isopropoxyethanol should be segregated from incompatible chemicals to minimize the risk of cross-contamination or contact. Buildings designated for storage should be equipped with appropriate fire protection systems (alarms, sprinklers, emergency lighting, portable extinguishers).

If a spill or leak to the environment has occurred, fire department, emergency response and/or hazardous materials spill personnel should be notified immediately. Cleanup should be attempted only by those trained in proper spill containment procedures using non-sparking tools. Contaminated soils should be removed for incineration and replaced with clean soil. If 2-isopropoxyethanol should contact the water table, aquifer, or navigable waterway, cleanup should be started immediately. It is highly soluble in water and total remediation may not be possible. The local and/or state emergency response authorities must be notified. A comprehensive emergency response or disaster preparedness/recovery plan should be in place prior to any operations involving the use, transportation, storage, or disposal of 2-isopropoxyethanol. If this chemical is spilled or leaked, the following specific steps are recommended:

☑ Restrict persons not wearing protective clothing from area of spill or leak until cleanup is complete and area can be opened for normal work.

☑ Remove all ignition sources; ventilate area of leak.

☑ Absorb liquids in vermiculite, dry sand, earth, or similar material and deposit in sealed containers using non-sparking tools.

☑ It may be necessary to dispose of this compound as a hazardous waste. The responsible state agency or the regional office of the federal Environmental Protection Agency (EPA) should be contacted for specific recommendations.

RISK ASSESSMENT: BUSINESS

General Assessment

Accidents or mishaps involving 2-isopropoxyethanol can present a significant threat to business operations. The loss or damage of equipment or facilities can significantly affect fiscal viability. Lawsuits that may result from personnel injury/death, public exposures, and/or environmental contamination will also require a serious expenditure of resources.

Media attention surrounding an injury, death, or environmental damage can also result in a loss of profits and loss of current as well as future business.

Recommended Risk-Reduction Measures

Company attorneys, safety and health professionals, and environmental specialists should be involved in the development of any procedures dealing with safety or emergency response. A company official should be pre-designated as a public relations officer with specific training in dealing with the press. Corporate plans and policies should be developed, approved, and implemented long before any need for such arises.

REFERENCES

American Conference of Governmental Industrial Hygienists. 1988. *Documentation of the Threshold Limit Values and Biological Exposure Indices,* 5th Edition (with updates). Cincinnati: ACGIH

Bueche, J. R. 1972. *Principles of Physics,* 2nd Edition. New York: McGraw-Hill.

Calabrese, E. J. 1994. *Biological Effects of Low Level Exposures.* New York: CRC Press/Lewis Publishers.

Cockerman, L. G., B. S. Shane. 1994. *Basic Environmental Toxicology.* New York: CRC Press/Lewis Publishers.

Gardner, R. W. 1994. *Chemical Intolerance.* New York: CRC Press/Lewis Publishers.

Hallenbeck, W. H. 1993. *Quantitative Risk Assessment for Environmental and Occupational Health,* 2nd Edition. New York: CRC Press/Lewis Publishers.

Keith, L. H. and M. M. Walker. 1995. *Handbook of Air Toxics.* New York: CRC Press/Lewis Publishers.

Landis, W. G., M. H. Yu. 1995. *Introduction to Environmental Toxicology.* New York: CRC Press/Lewis Publishers.

Langerman, N. 1994 *Precautionary Labels for Chemical Containers.* New York: CRC Press/Lewis Publishers.

Lauwerys, R. R., P. Hoet. 1993. *Industrial Chemical Exposure.* New York: CRC Press/Lewis Publishers.

Lewis, R. J., Sr. 1992. *Sax's Dangerous Properties of Industrial Materials,* Volumes 1, 2, and 3, 8th Edition. New York: Van Nostrand Reinhold.

Manahan, S. E. 1994. *Environmental Chemistry,* 6th Edition. New York: CRC Press/Lewis Publishers.

Meyer, E. 1990. *Chemistry of Hazardous Materials.* Englewood Cliffs, NJ: Prentice Hall.

Patnaik, P. 1992. *A Comprehensive Guide to the Hazardous Properties of Chemical Substances.* New York: Van Nostrand Reinhold.

Quigley, D. R. 1994. *Handbook of Emergency Chemical Management.* New York: CRC Press/Lewis Publishers.

Rea, W. J. 1995. *Chemical Sensitivity,* Vol. I-III. New York: CRC Press/Lewis Publishers.

Richardson, M. 1995. *Dictionary of Substance and Their Effects,* Vol. I-VI. New York: CRC Press/Lewis Publishers.

Sacarello, H. L. A. 1994. *The Comprehensive Handbook of Hazardous Materials.* New York: CRC Press/Lewis Publishers.

Torkelson, T. R. V. K. Rowe. 1981. Halogenated aliphatic hydrocarbons. *Patty's Industrial Hygiene and Toxicology,* Vol. 2B, *Toxicology,* pp 3568-3572. New York: Wiley-Interscience.

U.S. Department of Health and Human Services, National Institute for Occupational Safety and Health. 1994. *NIOSH Pocket Guide to Chemical Hazards.* Washington, D.C.: U.S. Government Printing Office.

MATERIAL SAFETY DATA SHEET

CHEMICAL NAME

ISOPROPYL ACETATE

CRC PRESS/LEWIS PUBLISHERS DISCLAIMER: THE INFORMATION AND RECOMMENDATIONS PRESENTED HEREIN ARE BASED ON SOURCES BELIEVED TO BE RELIABLE. CRC MAKES NO REPRESENTATION ON ITS COMPLETENESS OR ACCURACY. IT IS THE RESPONSIBILITY OF THE USER TO DETERMINE THE CHEMICAL'S SUITABILITY FOR ITS INTENDED USE, THE CHEMICAL'S SAFE USE, AND THE CHEMICAL'S PROPER DISPOSAL. NO REPRESENTATIONS AND/OR WARRANTIES, EITHER EXPRESSED OR IMPLIED, OF THE MERCHANTABILITY OR FITNESS FOR A PARTICULAR PURPOSE, OR OF ANY OTHER NATURE, ARE MADE WITH RESPECT TO THE INFORMATION PROVIDED IN THIS MSDS OR TO THE CHEMICAL TO WHICH INFORMATION MAY REFER. CRC NEITHER ASSUMES NOR AUTHORIZES ANY OTHER PERSON TO ASSUME FOR IT, ANY OTHER ADDITIONAL RESPONSIBILITY OR LIABILITY FOR THE USE OF, OR RELIANCE UPON, THIS INFORMATION.

HAZARD WARNING INFORMATION

HEALTH	FIRE	REACTIVE	OTHER	DEGREE OF HAZARD	COLOR CODING	OTHER CODES
1	3	0		0 = Minimum Hazard 1 = Slight Hazard 2 = Moderate Hazard 3 = Serious Hazard 4 = Severe Hazard	HEALTH = BLUE FIRE = RED REACTIVITY = YELLOW OTHER = WHITE	OX = Oxidizer ACID = Acid ALK = Alkali COR = Corrosive W = Use No Water

SECTION I - GENERAL INFORMATION

Characterization	RCRA Number	EPA Class
Aliphatic Ester	**None**	**Not Applicable**

DOT Proper Shipping Name **Isopropyl acetate**	Chemical Abstract Service (CAS) Number **108-21-4**
DOT Hazard Class and Label Requirements **Flammable Liquid**	DOT Emergency Guide Code **26**
DOT Identification Number **UN 1220**	Chemical Formula **$CH_3COOCH(CH_3)_2$**

Synonyms

Isopropyl ester of acetic acid; 1-methylethyl ester of acetic acid; 2-propyl acetate; *sec*-propyl acetate.

SECTION II - HAZARDOUS INGREDIENTS AND IDENTITY INFORMATION

Hazardous Components (specific identity)	OSHA Exposure Criteria	NIOSH Exposure Criteria	Immediately Dangerous to Life and Health (IDLH)	ACGIH Exposure Criteria
Isopropyl acetate (derivation: By reacting isopropyl alcohol with acetic acid in the presence of catalysts). **1 ppm = 4.25 mg/m^3**	**PEL:** **250 ppm** **950 mg/m^3** **STEL:** **310 ppm** **1178 mg/m^3**	**REL:** **Not Established** **STEL:** **Not Established**	**1800 ppm**	**TLV:** **250 ppm** **950 mg/m^3** **STEL:** **310 ppm** **1178 mg/m^3**

SECTION III - PHYSICAL AND CHEMICAL CHARACTERISTICS

Boiling Point **194°F (90°C)**	Specific Gravity (H_2O = 1) **0.872**
Vapor Pressure (mm Hg) **42 at 68°F (20°C)**	Molecular Weight **102.2**
Vapor Density (Air = 1) **3.5**	Melting Point **-92°F (-69°C)**

Solubility

Slightly soluble in water (3%). Miscible in alcohol, ether, and fixed oils.

Appearance and Odor

Colorless, mobile liquid with an aromatic, fruity odor.

SECTION IV - FIRE AND EXPLOSION HAZARD DATA

Flash Point (method used) **36°F (2.2°C) closed cup**	Explosive Limits in Air % by Volume **LEL: 1.8% at 100°F (38°C)** **UEL: 8%**
NFPA Classification **Class IB Flammable Liquid**	Autoignition Temperature **860°F (460°C)**

Extinguishing Media

Dry chemical, alcohol foam, carbon dioxide, water spray. Use water spray to cool exposed containers.

Special Fire Fighting Procedures

Poisonous gases are produced in fire. Wear full protective clothing with SCBA. Use blanketing effect with foam to smother large fires. Continue to cool containers with water after fire is extinguished. Exercise caution in selecting equipment; isopropyl acetate will attack some forms of plastic and rubber.

Unusual Fire and Explosion Hazards

Immediately withdraw if rising sound from venting device is heard or if fire is causing discoloration to the tank. Evacuate 1500 feet radius if fire becomes uncontrollable or container is exposed to direct flame. Vapors are heavier than air and may travel a distances to ignition source to flashback.

SECTION V - REACTIVITY DATA

Stability		Conditions to Avoid Isopropyl acetate is stable under normal conditions of storage and operation. Avoid contact with incompatible materials. Keep away from heat, flame, and other sources of ignition.
Stable X	Unstable	Incompatibility *(materials to avoid)* Reacts violently in contact with nitrates, strong oxidizers, alkalis, and acids. It will dissolve many plastics, rubber, and coatings.
Hazardous Polymerization		Conditions to Avoid Under normal conditions of temperature and pressure, hazardous polymerization of isopropyl acetate is not expected to occur.
May Occur	Will Not Occur X	Hazardous Decomposition or By-products When heated to decomposition, isopropyl acetate emits acrid and irritating fumes, including toxic oxides of carbon.

SECTION VI - HEALTH HAZARD DATA

Primary Route(s) of Entry:	Inhalation?	Absorption (skin)?	Ingestion?
	X		X

Health Hazards

INHALATION: Irritation of the eyes, nose, throat, and respiratory tract. Causes central nervous system depression with headache, weakness, dizziness, and loss of consciousness. It causes difficulty breathing, rapid pulse rate, and fatigue. There may also be liver damage, renal injury, and CNS effects (giddiness, confusion, delirium).

SKIN & EYES: Will irritate the skin. May cause contact dermatitis. Eye contact may cause irritation and possible visual disturbances that do not appear to be permanent.

INGESTION: Irritation of the mouth, esophagus, stomach, and digestive tract, nausea, and vomiting.

Carcinogenicity	NTP Listed?	IARC Cancer Review Group?	OSHA Regulated?	Target Organs?
Unknown Human Unknown Animal	No	No	29 CFR 1910.1000 Table Z-1	Respiratory system, skin, eyes, CNS, liver.

Medical Conditions Generally Aggravated by Exposure

Respiratory system disorders, skin disease, and/or liver or kidney dysfunction may be aggravated.

Emergency and First-aid Procedures

Eye contact: Flush immediately with water for 15 minutes (minimum); seek medical attention. Skin contact: Remove all contaminated clothing. Immediately wash area with large amounts of soap and water. Seek medical assistance. For inhalation: Remove the person from exposure. Provide respiratory assistance and CPR. Transfer to medical facility. If swallowed: Seek medical attention immediately. Call poison control center for advice. Give 2-3 glasses of water or milk to drink. Do NOT induce vomiting unless advised otherwise. Never give an unconscious or convulsing person anything by mouth.

SECTION VII - PRECAUTIONS FOR SAFE HANDLING AND USE

Steps to be Taken in Case Material is Released or Spilled

Contain spills using absorbent material (vermiculite or other). Ventilate area of spill, remove ignition sources. Have water source available in case of fire. Store materials in DOT-approved containers. Restrict those not involved in cleanup from entering area. Notify appropriate authorities, as required.

Preferred Waste Disposal Method

Incineration in chemical incinerator equipped with scrubber and afterburner.

Precautions to be Taken in Handling and Storage

Store in tightly closed containers in a cool, dark, well-ventilated area. Protect containers from physical damage. Keep heat, fire, and flame away. If possible, automatically transfer liquids between containers.

Other Precautions and Warnings

Bulk storage of isopropyl acetate is not recommended. Ground and bond all metal containers.

SECTION VIII - CONTROL MEASURES AND PERSONAL PROTECTIVE EQUIPMENT

Respiratory Protection *(specify type)*

Where potential exists for exposure over 250 ppm (8 hours), use an MSHA/NIOSH-approved full facepiece respirator with an organic vapor cartridge. For higher exposures, use a self-contained breathing apparatus (SCBA) with full facepiece operated in pressure demand or other positive pressure mode.

Ventilation

Local exhaust or general mechanical systems recommended.

Protective Gloves	Eye Protection	Other Protective Clothing
Impervious Gloves	Chemical Goggles or Face Mask	Impermeable Apron

Work/Hygiene Practices

Always wash hands thoroughly after using chemical; never bring food, drink, or smoking materials into vicinity of chemicals.

ISOPROPYL ACETATE

$CH_3COOCH(CH_3)_2$

CAS: 108-21-4

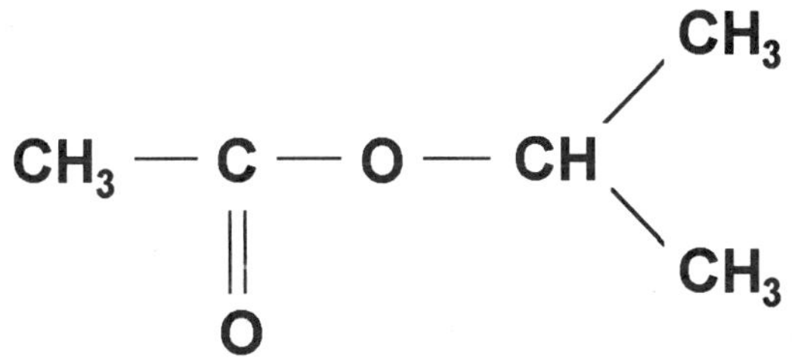

IDENTIFICATION AND TYPICAL USES

Isopropyl acetate is a colorless liquid with a characteristic fruity odor. It is used as a solvent for nitrocellulose, oils, fats, and resin gums; in the manufacture of thinners, sealants, paints, topcoat lacquers, and printing inks; in organic synthesis; in perfumery; and as a flavoring agent.

RISK ASSESSMENT: HEALTH

General Assessment

Isopropyl acetate is mildly toxic by *inhalation* and *ingestion*. Skin contact can produce localized effects but dermal absorption has not been reported. There are no reports of any carcinogenic, teratogenic, or mutagenic effects in humans or animals following exposure to isopropyl acetate.

Inhalation of isopropyl acetate vapors can cause irritation to the eyes, nose, throat, and respiratory tract with nausea, vomiting, cough, shortness of breath, and unconsciousness. Depending upon the concentration and duration of the exposure, there can also be varying degrees of central nervous system depression. Symptoms may include headaches, dizziness, general weakness and fatigue, loss of consciousness, difficulty in breathing, increased or rapid pulse rate, unconsciousness, and death. High or chronic exposure can lead to liver damage.

Skin contact results in the removal of the skin's natural protective oily layer on its surface with subsequent drying, cracking, and possible secondary infections and dermatitis. Eye contact can cause irritation and the possibility for adverse effects to vision.

Ingestion of isopropyl acetate may cause irritation of the mouth, esophagus, stomach, and digestive tract with nausea and vomiting. There can also be gastrointestinal hemorrhage, liver damage, and cardiac failure.

Acute Health Effects

The following acute (short-term) health effects may occur immediately or shortly after exposure to isopropyl acetate:

Skin: Mild to moderate irritation. Will remove the skin's protective oily layer causing drying, cracking, secondary infection, and dermatitis.

Eye: Irritation and possible inflammation.

Lung: Irritation of the eyes, nose, throat, and lungs (especially the mucus membranes). May cause olfactory changes and pulmonary effects.

CNS: High concentrations can depress the central nervous system leading to narcoses and possible loss of consciousness.

Chronic Health Effects

The following chronic (long-term) health effects may occur at some time after exposure to isopropyl acetate and can last for months or even years:

Cancer Hazards: According to the references, isopropyl acetate has not been adequately tested for its ability to cause cancer in test animals.

Reproduction: According to the references, isopropyl acetate has not been adequately tested for its ability to adversely affect reproduction in test animals.

Other Chronic Effects: Repeated or prolonged skin contact may cause the development of contact dermatitis with rash, redness, and inflammation of the affected surface area. Liver damage may also occur.

Recommended Risk-Reduction Measures

Personnel should avoid direct contact with isopropyl acetate. The best risk reduction measure is to use a less toxic chemical as a substitute for isopropyl acetate. If this is not possible or practical, then *engineering controls* are the most effective method of reducing exposures. The best protection is to enclose operations and/or provide local exhaust ventilation at the site of chemical release. While not always operationally feasible, isolating operations can also reduce exposure risks to other personnel. Using respiratory protection is less effective than the controls mentioned above, but is still advisable whenever working with or around isopropyl acetate. For exposures over 250 ppm (8 hours), a chemical respirator with an organic vapor cartridge should be used. For higher exposures, a self-contained breathing apparatus (SCBA) with full facepiece operated in pressure demand mode or a supplied-air respirator with full facepiece operated in positive pressure or other continuous flow mode are the recommended respiratory protection methods of choice. If a full facepiece is not available, then chemical goggles should be worn to protect the eyes. Whenever a chemical splash hazard exists, a face shield and a protective apron should be worn. To prevent hand and skin exposures, impervious gloves should be used. Caution is warranted in glove selection since isopropyl acetate will attack many plastic formulations and rubber materials. Glove manufacturers should always be contacted to obtain permeation studies *before* gloves are selected.

Administrative controls should also be in place to minimize the potential for human exposures. These may include written procedures or policies which specify the methods and techniques that will be practiced whenever personnel are to work with isopropyl acetate.

All personnel should receive *training* on the use, hazards, protective measures, emergency actions, and other precautions per 29 CFR 1910.1200 (Hazard Communication), prior to the first assignment in an area where isopropyl acetate is used or stored.

If symptoms develop or overexposure is suspected, the following medical tests are recommended:

- ☑ Lung function tests.
- ☑ Liver and kidney function tests.
- ☑ Skin testing with dilute isopropyl acetate to help diagnose allergy (performed by a qualified allergist under controlled conditions).

Any medical evaluation should include a careful history of past and present symptoms with an examination. Medical tests that look for existing damage are not a substitute for controlling exposures. Also, since smoking can cause heart disease, emphysema, and other respiratory disorders, smokers exposed to isopropyl acetate may experience symptoms more quickly than non-smokers under the same conditions of exposure. Prudent risk management requires proper consideration of *all* factors which may be causing the appearance of exposure symptoms in the workplace.

Other methods to reduce exposure include:

- ☑ Where possible, enclose operations and use local exhaust ventilation at the site of chemical release. If local exhaust ventilation or enclosure is not used, respiratory protection should be mandatory.
- ☑ Always ensure that proper protective clothing is worn when using chemical substances.
- ☑ Wash thoroughly immediately after exposure to isopropyl acetate and at the end of the work shift or before eating, drinking, or smoking.
- ☑ Hazard warning information should be posted in the work area. In addition, as part of an on-going education and training program, all information on the health and safety hazards of isopropyl acetate should be communicated to all exposed workers.
- ☑ Never eat, drink, or smoke in areas were isopropyl acetate is used, handled or stored.
- ☑ Never wear contaminated work clothing home where family members can be exposed. Clothing should be laundered only by personnel who have been briefed on the hazards of isopropyl acetate.

RISK ASSESSMENT: ENVIRONMENT

General Assessment

The environment is at risk of exposure during transportation, storage, disposal, or destruction of isopropyl acetate. In almost every scenario, the threat of environmental exposure is contingent upon the proper handling of the chemical substance.

Accidental spills, large or small, can result in fire, explosion, and possible contamination of the surrounding environmental mediums (water, soil, and air). Isopropyl acetate is considered a Class IB flammable liquid (according to OSHA 29 CFR 1910.106). Its moderately low flash point presents a serious fire and explosion hazard concern. Also, because it is incompatible with a number of common materials, especially strong oxidizers and acids, contact can result in violent and explosive reactions. It can also form explosive mixtures in air and can ignite on contact with heat, fire, or sparks. These characteristics require special consideration during any emergency situation involving a leak or spill of isopropyl acetate. The proper disposal method for isopropyl acetate is to burn it in a chemical incinerator equipped with an afterburner and air scrubber. Isopropyl acetate can enter the environment through unchecked industrial discharges and through accidental spills.

Acute Ecological Effects

Acute (short-term) toxic effects may include the death of animals, birds, or fish, and death or low growth rate in plants. Acute effects are seen 2 to 4 days after animals or plants are exposed to isopropyl acetate.

Insufficient data are available to evaluate or predict the short-term effects of isopropyl acetate to aquatic life, plants, birds, or terrestrial animals.

Chronic Ecological Effects

Chronic toxic effects may include shortened life span, reproductive problems, lower fertility, and changes in appearance or behavior in exposed animals. These effects can be seen long after first exposure(s) to toxic chemicals.

Insufficient data are available to evaluate or predict the long-term effects of isopropyl acetate to aquatic life, plants, birds, or land animals.

Water Solubility

Isopropyl acetate is only slightly soluble in water. Concentrations of 1 to 100 milligrams may mix with a liter of water.

Persistence in the Environment

Isopropyl acetate is highly persistent in water, with a half-life greater than 200 days. The half-life of a pollutant is the amount of time it takes for one-half of the chemical to be degraded.

There is insufficient information in the references to evaluate or predict the total percentages of isopropyl acetate that will eventually end up in the various environmental mediums (air, water, and soil).

Bioaccumulation in Aquatic Organisms

Some substances increase in concentration, or *bioaccumulate,* in living organisms as they breathe contaminated air, drink contaminated water, or eat contaminated food. These chemicals can become concentrated in the tissues and internal organs of animals as well as humans.

Insufficient data are available to evaluate or predict the accumulation rate of isopropyl acetate in the edible tissues of fish.

Recommended Risk-Reduction Measures

Proper training of all transporters will reduce the likelihood of a mishap or accident resulting in a leak or spill to the environment. The correct use of DOT labeling (per the MSDS) on all containers, trucks, and rail cars will enable emergency responders to react properly and quickly to any disaster thereby reducing the potential risk to the environment and to personnel.

Storage of isopropyl acetate should be segregated from incompatible chemicals to minimize the risk of cross-contamination or contact. Buildings designated for storage should be equipped with appropriate fire protection systems (alarms, sprinklers, emergency lighting, portable extinguishers).

If a spill or leak to the environment has occurred, fire department, emergency response, and/or hazardous materials spill personnel should be notified immediately. Cleanup should be attempted only by those trained in proper spill containment procedures using non-sparking tools. Contaminated soils should be removed for incineration and replaced with clean soil. If isopropyl acetate should contact the water table, aquifer, or navigable waterway, time is of the essence. It is only slightly soluble in water and, therefore, total containment and remediation may be possible. When such spills occur, the local and/or state emergency response authorities must be notified. A comprehensive emergency response or disaster preparedness/recovery plan should be in place prior to any operations involving the use, transportation, storage, or disposal of isopropyl acetate. If isopropyl acetate is spilled or leaked, the following specific steps are recommended:

- ☑ Restrict persons not wearing protective clothing from area of spill or leak until cleanup is complete and area can be opened for normal work.
- ☑ Ventilate area and remove all ignition sources.
- ☑ Absorb liquids in vermiculite, dry sand, earth, or a similar material and deposit in sealed containers. Use non-sparking tools!
- ☑ It may be necessary to dispose of isopropyl acetate as a hazardous waste. The responsible state

agency or the regional office of the federal Environmental Protection Agency (EPA) should be contacted for specific recommendations.

RISK ASSESSMENT: BUSINESS

General Assessment

Accidents or mishaps involving isopropyl acetate can present a moderate threat to business operations. The loss or damage of equipment or facilities can significantly affect fiscal viability. Lawsuits that may result from personnel injury/death, public exposures, and/or environmental contamination will also require a serious expenditure of resources. Media attention surrounding an injury, death, or environmental damage can also result in a loss of profits and loss of current as well as future business.

Recommended Risk-Reduction Measures

Company attorneys, safety and health professionals, and environmental specialists should be involved in the development of any procedures or policies intended to manage the use of chemicals in the workplace. A company official should be pre-designated as a public relations officer with specific training in dealing with the press. Corporate plans and policies should be developed, approved, and implemented long before any need for such arises.

REFERENCES

American Conference of Governmental Industrial Hygienists. 1988. *Documentation of the Threshold Limit Values and Biological Exposure Indices,* 5th Edition (with updates). Cincinnati: ACGIH

Anthony, C. P., and N. J. Kolthoff. 1971. *Textbook of Anatomy and Physiology.* St. Louis: C. V. Mosby Company.

Bloom, A. D. and F. J. DeSerres. 1995. *Ecotoxicity and Human Health.* New York: CRC Press/Lewis Publishers.

Brusick, D. J. 1994. *Methods for Genetic Risk Assessment.* New York: CRC Press/Lewis Publishers.

Burrell, R., D. K. Flaherty, and L. J. Sauers. 1992. *Toxicology of the Immune System.* New York: Van Nostrand Reinhold.

Calabrese, E. J. 1994. *Biological Effects of Low Level Exposures.* New York: CRC Press/Lewis Publishers.

Cockerman, L. G., B. S. Shane. 1994. *Basic Environmental Toxicology.* New York: CRC Press/Lewis Publishers.

Cothern, C. R., N. P. Ross. 1994. *Environmental Statistics, Assessment, and Forecasting.* New York: CRC Press/Lewis Publishers.

Danse, I. R. 1991. *Common Sense Toxics in the Workplace.* New York: Van Nostrand Reinhold.

Environmental Protection Agency. 1983. *Chemical Hazard Information Profile: Isopropyl acetate.* Washington, D.C.: U.S. EPA.

Fodden, C. R., and J. L. Weddell. 1992. *Hazardous Materials Emergency Action Data.* New York: CRC Press/Lewis Publishers.

Forseberg, K., and L. H. Keith. 1993. *Chemical; Protective Clothing Permeation and Degradation Database.* Chelsea, MI: Lewis Publishers.

Gardner, R. W. 1994. *Chemical Intolerance.* New York: CRC Press/Lewis Publishers.

Hallenbeck, W. H. 1993. *Quantitative Risk Assessment for Environmental and Occupational Health,* 2nd Edition. New York: CRC Press/Lewis Publishers.

Hodgson, E., R. B. Mailman, and J. E. Chambers. 1988. *Dictionary of Toxicology.* London: The MacMillian Company.

Howard, P. H., et. al. 1991. *Handbook of Environmental Degradation Rates.* New York: CRC Press/Lewis Publishers.

Hosty, J. W., and P. Foster. 1990. *A Practical Guide to Chemical Spill Response.* New York: Van Nostrand Reinhold.

Jones, F. E. 1994. *Toxic Organic Vapors in the Workplace.* New York: CRC Press/Lewis Publishers.

Kamrin, M. I A. 1988. *Toxicology: A Primer on Principles and Applications.* Chelsea, Michigan: Lewis Publishers.

Keith, L. H. and M. M. Walker. 1995. *Handbook of Air Toxics.* New York: CRC Press/Lewis Publishers.

Langerman, N. 1994 *Precautionary Labels for Chemical Containers.* New York: CRC Press/Lewis Publishers.

Lauwerys, R. R., P. Hoet. 1993. *Industrial Chemical Exposure.* New York: CRC Press/Lewis Publishers.

Lewis, Richard J., Sr. 1992. *Sax's Dangerous Properties of Industrial Materials,* Volumes 1, 2, and 3, 8th Edition. New York: Van Nostrand Reinhold.

Mackay, D., W. Y. Shiu, and K. C. Ma. 1992. *Illustrated Handbook of Chemical-Physical Properties and Environmental Fate for Organic Chemicals.* New York: CRC Press/Lewis Publishers.

Manahan, S. E. 1992. *Toxicological Chemistry,* 2nd Edition. New York: CRC Press/Lewis Publishers.

Meyer, E. 1990. *Chemistry of Hazardous Materials.* Englewood Cliffs, NJ: Prentice Hall.

Miller, B. F., M.D., and C. B. Keane, R.N. 1972. *Encyclopedia and Dictionary of Medicine and Nursing.* Philadelphia: W.B. Sanders Company.

Patnaik, P. 1992. *A Comprehensive Guide to the Hazardous Properties of Chemical Substances.* New York: Van Nostrand Reinhold.

Sacarello, H. L. A. 1994. *The Comprehensive Handbook of Hazardous Materials.* New York: CRC Press/Lewis Publishers.

Torkelson, T. R. V. K. Rowe. 1981. *Patty's Industrial Hygiene and Toxicology,* Vol. 2B. New York: Wiley-Interscience.

U.S. Department of Health and Human Services, National Institute for Occupational Safety and Health. 1994. *NIOSH Pocket Guide to Chemical Hazards.* Washington, D.C.: U.S. Government Printing Office.

MATERIAL SAFETY DATA SHEET

CHEMICAL NAME	
ISOPROPYL ALCOHOL	**CRC PRESS/LEWIS PUBLISHERS DISCLAIMER:** THE INFORMATION AND RECOMMENDATIONS PRESENTED HEREIN ARE BASED ON SOURCES BELIEVED TO BE RELIABLE. CRC MAKES NO REPRESENTATION ON ITS COMPLETENESS OR ACCURACY. IT IS THE RESPONSIBILITY OF THE USER TO DETERMINE THE CHEMICAL'S SUITABILITY FOR ITS INTENDED USE, THE CHEMICAL'S SAFE USE, AND THE CHEMICAL'S PROPER DISPOSAL. NO REPRESENTATIONS AND/OR WARRANTIES, EITHER EXPRESSED OR IMPLIED, OF THE MERCHANTABILITY OR FITNESS FOR A PARTICULAR PURPOSE, OR OF ANY OTHER NATURE, ARE MADE WITH RESPECT TO THE INFORMATION PROVIDED IN THIS MSDS OR TO THE CHEMICAL TO WHICH INFORMATION MAY REFER. CRC NEITHER ASSUMES NOR AUTHORIZES ANY OTHER PERSON TO ASSUME FOR IT, ANY OTHER ADDITIONAL RESPONSIBILITY OR LIABILITY FOR THE USE OF, OR RELIANCE UPON, THIS INFORMATION.

HAZARD WARNING INFORMATION

HEALTH	FIRE	REACTIVE	OTHER	DEGREE OF HAZARD	COLOR CODING	OTHER CODES
1	3	0		0 = Minimum Hazard 1 = Slight Hazard 2 = Moderate Hazard 3 = Serious Hazard 4 = Severe Hazard	HEALTH = BLUE FIRE = RED REACTIVITY = YELLOW OTHER = WHITE	OX = Oxidizer ACID = Acid ALK = Alkali COR = Corrosive W = Use No Water

SECTION I - GENERAL INFORMATION

Characterization	RCRA Number	EPA Class
Alcohol	**D001**	**Characteristic (I) Waste**

DOT Proper Shipping Name	**Isopropanol or Isopropyl Alcohol**	Chemical Abstract Service (CAS) Number	**67-63-0**
DOT Hazard Class and Label Requirements	**Flammable Liquid**	DOT Emergency Guide Code	**26**
DOT Identification Number	**UN 1219**	Chemical Formula	**$(CH_3)_2CHOH$**

Synonyms

IPA; isopropanol; dimethyl carbinol; 2-propanol; *sec*-propyl alcohol; rubbing alcohol.

SECTION II - HAZARDOUS INGREDIENTS AND IDENTITY INFORMATION

Hazardous Components (specific identity)	OSHA Exposure Criteria	NIOSH Exposure Criteria	Immediately Dangerous to Life and Health (IDLH)	ACGIH Exposure Criteria
Isopropyl alcohol (derivation: By treating propylene with sulfuric acid and then hydrolyzing, or by direct hydration of propylene using steam.). Commonly sold as Rubbing Alcohol at 70% **1 ppm = 2.5 mg/m^3**	**PEL (8-hour):** **400 ppm** **980 mg/m^3** **STEL:** **Not Established**	**REL (10-hour):** **400 ppm** **980 mg/m^3** **STEL:** **500 ppm** **1225 mg/m^3**	**12,000 ppm**	**TLV:** **400 ppm** **983 mg/m^3** **STEL:** **500 ppm** **1230 mg/m^3**

SECTION III - PHYSICAL AND CHEMICAL CHARACTERISTICS

Boiling Point	**181°F (83°C)**	Specific Gravity (H_2O = 1)	**0.79**
Vapor Pressure (mm Hg)	**33 at 68°F (20°C)**	Molecular Weight	**60.1**
Vapor Density (Air = 1)	**2.07**	Melting Point	**-127°F (-88.3°C)**

Solubility

Miscible in water. Soluble in alcohol, ether, chloroform, and benzene. Insoluble in salt solutions.

Appearance and Odor

Colorless liquid with a slight odor of rubbing alcohol and bitter taste. Odor Threshold = 22 ppm.

SECTION IV - FIRE AND EXPLOSION HAZARD DATA

Flash Point (method used)	**53°F (12°C) closed cup**	Explosive Limits in Air % by Volume	**LEL: 2.0% UEL: 12.7% at 200°F (93°C)**
NFPA Classification	**Class IB Flammable Liquid**	Autoignition Temperature	**750°F (399°C)**

Extinguishing Media

Dry chemical, carbon dioxide, or alcohol-resistant foam, fog, or water spray.

Special Fire Fighting Procedures

Poisonous gases are produced in fire. Wear full protective clothing and self-contained breathing apparatus (SCBA). Continue to cool containers with water after fire is extinguished. Move containers from fire area if it can be done without risk. For large fires, use unmanned hose apparatus, if possible. Evacuate non-essential personnel one-half mile radius if fire. Consider down-wind conditions.

Unusual Fire and Explosion Hazards

Vapors are heavier than air and may travel a distances to ignition source to flashback. Vapors may explode if ignited in an enclosed area. Evacuate area if rising sound is heard from venting device.

SECTION V - REACTIVITY DATA

Stability		Conditions to Avoid Isopropyl alcohol is stable under normal conditions of storage and operation. Avoid contact with incompatible materials. Keep away from heat, flame, and other ignition sources.
Stable X	Unstable	Incompatibility *(materials to avoid)* Reacts violently in contact with strong oxidizers, inorganic acids, aldehydes (especially acetaldehyde), and isocyanates. It ignites on contact with many metallic oxides.
Hazardous Polymerization		Conditions to Avoid Under normal conditions of temperature and pressure, isopropyl alcohol is not expected to undergo hazardous polymerization.
May Occur	Will Not Occur X	Hazardous Decomposition or By-products When heated to decomposition, isopropyl alcohol emits acrid and irritating fumes including toxic oxides of carbon.

SECTION VI - HEALTH HAZARD DATA

Primary Route(s) of Entry:	Inhalation?	Absorption (skin)?	Ingestion?
	X		X

Health Hazards

INHALATION: There may be slight irritation of the eyes, nose, and mucosa of the respiratory tract. Exposure can also cause stupor, fatigue, and sleepiness. Other effects include nausea, vomiting, flushing of skin, headaches, double vision, vertigo, weakness, sleepiness, olfactory changes (fatigue of sense of smell), pulmonary edema, and death.

SKIN & EYES: As a solvent, causes skin dryness, cracking, and possible dermatitis. Eye contact can result in burning, stinging, tearing, corneal injury, and possible corneal burns.

INGESTION: Sleep disorders, hallucinations, distorted perceptions, convulsions, ataxia and motor activity changes, renal and liver damage, cardiac failure, coma, headache, and death.

Carcinogenicity	NTP Listed?	IARC Cancer Review Group?	OSHA Regulated?	Target Organs?
Unknown Human Questioned Animal	No	Group 3	29 CFR 1910.1000 Table Z-1	Skin, eyes, respiratory system, CNS.

Medical Conditions Generally Aggravated by Exposure

Chronic liver, kidney, and chronic respiratory diseases may be aggravated by exposure.

Emergency and First-aid Procedures

Eye contact: Flush immediately with water for 15 minutes (minimum); seek medical attention. Skin contact: Remove all contaminated clothing. Immediately wash area with large amounts of soap and water. Seek medical assistance. For inhalation: Remove the person from exposure. Provide respiratory assistance and CPR. Transfer to medical facility. Observe for 48 hours for lung effects. If swallowed: Contact poison control center. Give 1-2 glasses of water or milk to dilute. Do NOT induce vomiting. Gastric lavage is preferred. Seek medical attention immediately.

SECTION VII - PRECAUTIONS FOR SAFE HANDLING AND USE

Steps to be Taken in Case Material is Released or Spilled

Contain spills using absorbent material (vermiculite). Ventilate area of spill and remove ignition sources. Have extinguishing agent available in case of fire. Store in DOT-approved containers.

Preferred Waste Disposal Method

Incineration in chemical incinerator equipped with scrubber and afterburner.

Precautions to be Taken in Handling and Storage

Store in tightly closed containers in a cool, dark, well-ventilated area. Protect containers from physical damage. Keep heat, fire, and flame away. Outside building or underground storage is recommended.

Other Precautions and Warnings

Ground and bond all metal containers. Avoid contact with skin and eyes.

SECTION VIII - CONTROL MEASURES AND PERSONAL PROTECTIVE EQUIPMENT

Respiratory Protection *(specify type)*

For vapor exposures, use an organic vapor air-purifying respirator. Greater protection is obtained using a supplied-air respirator or a self-contained breathing apparatus (SCBA) with full facepiece operated in pressure demand or other positive pressure mode.

Ventilation

Local exhaust or general mechanical systems recommended.

Protective Gloves	Eye Protection	Other Protective Clothing
Impervious Gloves	Chemical Goggles or Face Mask	Impermeable Apron or Protective Suit

Work/Hygiene Practices

Always wash hands thoroughly after using chemical; never bring food, drink, or smoking materials into vicinity of chemicals.

ISOPROPYL ALCOHOL

$(CH_3)_2CHOH$ CAS: 67-63-0

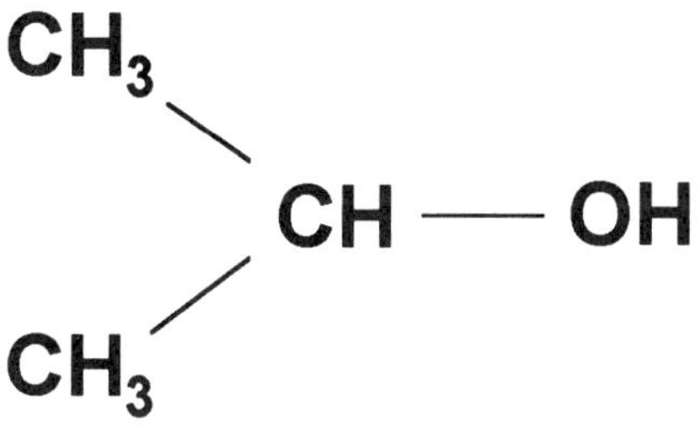

IDENTIFICATION AND TYPICAL USES

Isopropyl alcohol is a colorless or clear liquid with an odor of rubbing alcohol and a bitter taste. It is used in the manufacture of acetone and its derivatives and in the manufacture of glycerol and isopropyl acetate. It is also a strong and commonly used industrial solvent for cutting essential oils, other oils, alkaloids, gums, and resins. It is a latent solvent for cellulose derivatives and lacquers, a coatings solvent, a deicing agent for many liquid fuels, used in extraction processes, a dehydrating agent, a preservative, used in lotions, and as a denaturant.

RISK ASSESSMENT: HEALTH

General Assessment

Isopropyl alcohol is primarily toxic to humans by *inhalation* and *ingestion*. Skin contact can cause adverse surface effects but dermal absorption has not been reported. Isopropyl alcohol is a questioned animal carcinogen (the IARC reports insufficient evidence) and its human carcinogenic potential is unknown. Mutation data have also been reported as have experimental teratogenic and reproductive effects.

Inhalation can cause irritation of the eyes, nose, throat, upper respiratory tract, and associated mucosa. There may also be depression of the central nervous system (CNS) with narcosis and associated symptoms, including headache, nervousness, flushing of the skin, double vision, vertigo, tremors, dizziness, fatigue and general weakness, nausea, somnolence, and narcosis with stupor, loss of consciousness, and death. Pulmonary edema and/or chemical pneumonia can also occur, most often as a result of aspiration of the fluid into the lungs. The symptoms of pulmonary edema may be delayed up to 48 hours which can create a false sense of security with regard to health exposure risk.

Liquid contact with the eyes causes immediate burning and stinging with lachrymation and reflex closure of the lids. There may be severe, irreversible damage or injury to the cornea with possible conjunctivitis, depending upon the duration of exposure and concentration strength. Skin contact results in drying and cracking which can lead to secondary infections and dermatitis.

Ingestion of isopropyl alcohol can lead to symptoms of alcohol toxicity. These include nausea and vomiting, headache, flushing of the face, dizziness, decreased blood pressure, mental depression, hallucinations and distorted perceptions, dypsnea, respiratory depression, stupor, sleep disorders, ataxia, motor function changes, convulsions and tremors, coma, loss of body temperature, renal and liver damage, gastrointestinal damage, cardiac failure, unconsciousness, coma, and death. Kidney effects may include oliguria (reduction in urine secretion) or anuria (lack of urine secretion), nitrogen retention, and edema.

Acute Health Effects

The following acute (short-term) health effects may occur immediately or shortly after exposure to isopropyl alcohol:

Skin: Causes dryness and cracking leading to dermatitis and possible infection with erythema.

Eye: Severe irritation with burning, stinging, and possible damage to the cornea.

Lung: Irritation of the eyes, nose, throat, and respiratory tract. Delayed pulmonary edema is also a possible effect, especially on aspiration of the liquid, which can be fatal.

CNS: High concentrations either through inhalation or ingestion can cause depression of the CNS with symptoms of sleepiness, stupor, inebriation, vertigo, loss of consciousness, and death.

Chronic Health Effects

The following chronic (long-term) health effects may occur at some time after exposure to isopropyl alcohol and can last for months or even years:

Cancer Hazards: Isopropyl alcohol has been shown to cause cancer in limited experiments with test animals. However, more studies are required in this regard before any carcinogenic determinations can be made for animals or humans. Mutation data have been reported. Many scientists believe that such chemicals pose a cancer risk in the long-term.

Reproduction: According to the references, isopropyl alcohol has been shown to cause teratogenic and other adverse reproductive effects in animal experiments. It is not known whether or not it will have these effects in humans.

Other Chronic Effects: Very high or prolonged exposure may result in mucous membrane irritation, headache, and depression of the CNS with symptoms of somnolence and lack of concentration. Prolonged skin contact can cause drying, cracking, and dermatitis.

Recommended Risk-Reduction Measures

Personnel should avoid direct contact with isopropyl alcohol. The exact nature of its toxicity is not clearly understood in the references and its carcinogenic capability is considered suspect. The best risk reduction measure is to use a less toxic chemical as a substitute for a isopropyl alcohol. However, as one of the most commonly used industrial solvents, substitution may not be possible or operationally feasible. Therefore, *engineering controls* are the most effective methods of reducing exposures. An alternative method of protection is to enclose operations and/or provide local exhaust ventilation at the site of chemical release. While not always operationally feasible, isolating operations can also reduce exposure risk.

Using respiratory protection is less effective than the controls mentioned above, but is still advisable whenever working with or around isopropyl alcohol. For concentrations over the PEL (400 ppm), an air-purifying respirator with an organic vapor cartridge will suffice. For higher exposures, a supplied-air respirator with full facepiece operated in positive pressure mode, or a self-contained breathing apparatus (SCBA) with full facepiece and operated in pressure demand mode are the recommended respiratory protection methods of choice. If a full facepiece is not available, then chemical goggles should be worn to protect the eyes. Whenever a chemical splash hazard exists, a face shield and a protective apron should be worn. To prevent hand and skin exposures, impervious gloves should be used. Glove manufacturers should be contacted and permeation studies obtained *before* final glove selections have been made.

Administrative controls should also be in place to minimize the potential for human exposures. These may include written procedures or policies which specify the methods and techniques that will be practiced whenever personnel are to work with isopropyl alcohol.

All personnel should receive *training* on the use, hazards, protective measures, emergency actions, and other precautions per 29 CFR 1910.1200 (Hazard Communication), prior to the first assignment in an area where isopropyl alcohol is used or stored.

If symptoms develop or overexposure is suspected, the following medical tests are recommended:

- ☑ Evaluation of nervous system function with careful consideration of exposure history and special testing by a qualified neurologist.
- ☑ Complete blood count (CBC) and determination of isopropyl alcohol levels (or its metabolites).
- ☑ Urinalysis (check for acetone levels).
- ☑ Liver, kidney, and lung function tests.
- ☑ Consider chest X-ray following acute overexposure (may be negative if taken immediately after exposure due to delayed onset of pulmonary edema).
- ☑ Skin testing with dilute isopropyl alcohol to help diagnose allergy (performed by a qualified allergist).

Any medical evaluation should include a careful history of past and present symptoms with an examination. Medical tests that look for existing damage are not a substitute for controlling exposures. Also, since smoking can cause heart disease, lung cancer, emphysema, and other respiratory diseases, smokers exposed to isopropyl alcohol may experience symptoms more rapidly than non-smokers under the same conditions of exposure.

Other methods to reduce exposure include:

- ☑ Where possible, enclose operations and use local exhaust ventilation at the site of chemical release.

If local exhaust ventilation or enclosure is not used, respiratory protection should be mandatory.

☑ Always ensure that proper protective clothing is worn when using chemical substances.

☑ Wash thoroughly immediately after exposure to isopropyl alcohol and at the end of the work shift or before eating, drinking, or smoking.

☑ Hazard warning information should be posted in the work area. In addition, as part of an on-going education and training program, all information on the health and safety hazards of isopropyl alcohol should be communicated to all exposed workers.

RISK ASSESSMENT: ENVIRONMENT

General Assessment

The environment is at risk of exposure during transportation, storage, disposal, or destruction of isopropyl alcohol. In almost every scenario, the threat of environmental exposure is contingent upon the proper handling of the chemical substance. Accidental spills, large or small, can result in fire, explosion, and possible contamination of the surrounding environmental mediums (water, soil, and air).

Isopropyl alcohol is considered a Class IB flammable liquid (according to OSHA 29 CFR 1910.106). Its relatively low flash point presents a moderate fire and explosion hazard concern. Also, because it is incompatible with a number of common materials, especially strong oxidizers and many reducing materials, contact can result in violent and explosive reactions. It can also react explosively with many metallic oxides and can ignite on contact with heat, fire, or sparks. These characteristics require special consideration during any emergency situation involving a leak or spill of isopropyl alcohol. The proper destruction method is to burn it in an incinerator with an afterburner and scrubber.

Isopropyl alcohol can enter the environment through unchecked industrial discharges into effluents and through spills.

Acute Ecological Effects

Acute (short-term) toxic effects may include the death of animals, birds, or fish, and death or low growth rate in plants. Acute effects are seen 2 to 4 days after animals or plants are exposed to isopropyl alcohol.

Insufficient data are available to evaluate or predict the short-term effects of isopropyl alcohol to aquatic life, plants, birds, or terrestrial animals.

Chronic Ecological Effects

Chronic toxic effects may include shortened life span, reproductive problems, lower fertility, and changes in appearance or behavior in exposed animals. These effects can be seen long after first exposure(s) to toxic chemicals.

Insufficient data are available to evaluate or predict the long-term effects of isopropyl alcohol to aquatic life, plants, birds, or land animals.

Water Solubility

Isopropyl alcohol is highly soluble in water. Concentrations of 1000 milligrams and more can be expected to mix with a liter of water.

Persistence in the Environment

Isopropyl alcohol is slightly persistent in water, with a half-life of between 2 to 20 days. The half-life of a pollutant is the amount of time it takes for one-half of the chemical to be degraded.

Bioaccumulation in Aquatic Organisms

Some substances increase in concentration, or *bioaccumulate,* in living organisms as they breathe contaminated air, drink contaminated water, or eat contaminated food. These chemicals can become concentrated in the tissues and internal organs of animals as well as humans.

The concentration of isopropyl alcohol found in fish tissues is expected to be much lower than the average concentration of isopropyl alcohol in water from which the fish was taken.

Recommended Risk-Reduction Measures

Proper training of all transporters will reduce the likelihood of a mishap or accident resulting in a leak or spill to the environment. The correct use of DOT labeling (per the MSDS) on all containers, trucks, and rail cars will enable emergency responders to react properly and quickly to any disaster thereby reducing the potential risk to the environment and to personnel.

Storage of isopropyl alcohol should be segregated from incompatible chemicals to minimize the risk of cross-contamination or contact. Buildings designated for storage should be equipped with appropriate fire protection systems (alarms, sprinklers, emergency lighting, portable extinguishers) and designed to meet appropriate explosion-proof requirements.

If a spill or leak to the environment has occurred, fire department, emergency response, and/or hazardous materials spill personnel should be notified immediately. Cleanup should be attempted only by those trained in proper spill containment procedures using non-sparking tools. Contaminated soils should be removed for incineration and replaced with clean soil.

If isopropyl alcohol should contact the water table, aquifer, or navigable waterway, time is of the essence. It is highly soluble in water and, therefore, total containment and remediation may not be possible. When such spills occur, the local and/or state emergency response authorities must be notified. A comprehensive emergency response or disaster preparedness/recovery plan should be in place prior to any operations involving the use, transportation, storage, or disposal of isopropyl alcohol. If isopropyl alcohol is spilled or leaked, the following specific steps are recommended to ensure the protection of personnel and the environment:

☑ Restrict persons not wearing protective clothing from area of spill or leak until cleanup is complete and area can be opened for normal work.

☑ Ventilate area and remove ignition sources.

☑ Absorb liquids in vermiculite, dry sand, earth, or a similar material and deposit in sealed containers for disposal. Use non-sparking tools.

☑ It may be necessary to dispose of isopropyl alcohol as a hazardous waste. The responsible state agency or the regional office of the federal Environmental Protection Agency (EPA) should be contacted for specific recommendations.

RISK ASSESSMENT: BUSINESS

General Assessment

Accidents or mishaps involving isopropyl alcohol can present a moderate threat to business operations. The loss or damage of equipment or facilities can significantly affect fiscal viability. Lawsuits that may result from personnel injury/death, public exposures, and/or environmental contamination will also require a serious expenditure of resources. Media attention surrounding an injury, death, or environmental damage can also result in a loss of profits and loss of current as well as future business. Always remember that anytime the terms "cancer," "carcinogen," or reproductive hazard are used, public emotion, hysteria, and ignorance can run equally high. This must be properly considered during the development of any public relations policies.

Recommended Risk-Reduction Measures

Company attorneys, safety and health professionals, and environmental specialists should be involved in the development of any procedures or policies intended to manage the use of chemicals in the workplace. A company official should be pre-designated as a public relations officer with specific training in dealing with the press. Corporate plans and policies should be developed, approved, and implemented long before any need for such arises.

REFERENCES

American Conference of Governmental Industrial Hygienists. 1988. *Documentation of the Threshold Limit Values and Biological Exposure Indices,* 5th Edition (with updates). Cincinnati: ACGIH

Anthony, C. P., and N. J. Kolthoff. 1971. *Textbook of Anatomy and Physiology.* St. Louis: C. V. Mosby Company.

Bloom, A. D. and F. J. DeSerres. 1995. *Ecotoxicity and Human Health.* New York: CRC Press/Lewis Publishers.

Brusick, D. J. 1994. *Methods for Genetic Risk Assessment.* New York: CRC Press/Lewis Publishers.

Bueche, J. R. 1972. *Principles of Physics,* 2nd Edition. New York: McGraw-Hill.

Calabrese, E. J. 1994. *Biological Effects of Low Level Exposures.* New York: CRC Press/Lewis Publishers.

Cothern, C. R., N. P. Ross. 1994. *Environmental Statistics, Assessment, and Forecasting.* New York: CRC Press/Lewis Publishers.

Danse, I. R. 1991. *Common Sense Toxics in the Workplace.* New York: Van Nostrand Reinhold.

Environmental Protection Agency. 1983. *Chemical Hazard Information Profiles.* Washington, D.C.: U.S. EPA.

Gardner, R. W. 1994. *Chemical Intolerance.* New York: CRC Press/Lewis Publishers.

Hodgson, E., R. B. Mailman, and J. E. Chambers. 1988. *Dictionary of Toxicology.* London: The MacMillian Company.

Howard, P. H., et. al. 1991. *Handbook of Environmental Degradation Rates.* New York: CRC Press/Lewis Publishers.

International Agency for Research on Cancer (IARC). 1987. *IARC Monographs on the Evaluation of Carcinogenic Risk to Humans, Overall Evaluations of Carcinogenicity: An updating of IARC Monographs Volumes 1-42.* Lyon: IARC.

Lewis, Richard J., Sr. 1992. *Sax's Dangerous Properties of Industrial Materials,* Volumes 1, 2, and 3, 8th Edition. New York: Van Nostrand Reinhold.

Torkelson, T. R. V. K. Rowe. 1981. *Patty's Industrial Hygiene and Toxicology,* Vol. 2B. New York: Wiley-Interscience.

U.S. Department of Health and Human Services, National Institute for Occupational Safety and Health. 1994. *NIOSH Pocket Guide to Chemical Hazards.* Washington, D.C.: U.S. Government Printing Office.

MATERIAL SAFETY DATA SHEET

CHEMICAL NAME	
ISOPROPYLAMINE	**CRC PRESS/LEWIS PUBLISHERS DISCLAIMER:** THE INFORMATION AND RECOMMENDATIONS PRESENTED HEREIN ARE BASED ON SOURCES BELIEVED TO BE RELIABLE. CRC MAKES NO REPRESENTATION ON ITS COMPLETENESS OR ACCURACY. IT IS THE RESPONSIBILITY OF THE USER TO DETERMINE THE CHEMICAL'S SUITABILITY FOR ITS INTENDED USE, THE CHEMICAL'S SAFE USE, AND THE CHEMICAL'S PROPER DISPOSAL. NO REPRESENTATIONS AND/OR WARRANTIES, EITHER EXPRESSED OR IMPLIED, OF THE MERCHANTABILITY OR FITNESS FOR A PARTICULAR PURPOSE, OR OF ANY OTHER NATURE, ARE MADE WITH RESPECT TO THE INFORMATION PROVIDED IN THIS MSDS OR TO THE CHEMICAL TO WHICH INFORMATION MAY REFER. CRC NEITHER ASSUMES NOR AUTHORIZES ANY OTHER PERSON TO ASSUME FOR IT, ANY OTHER ADDITIONAL RESPONSIBILITY OR LIABILITY FOR THE USE OF, OR RELIANCE UPON, THIS INFORMATION.

HAZARD WARNING INFORMATION

HEALTH	FIRE	REACTIVE	OTHER	DEGREE OF HAZARD	COLOR CODING	OTHER CODES
3	4	0		0 = Minimum Hazard 1 = Slight Hazard 2 = Moderate Hazard 3 = Serious Hazard 4 = Severe Hazard	HEALTH = BLUE FIRE = RED REACTIVITY = YELLOW OTHER = WHITE	OX = Oxidizer ACID = Acid ALK = Alkali COR = Corrosive W = Use No Water

SECTION I - GENERAL INFORMATION

Characterization	RCRA Number	EPA Class
Aliphatic Amine	**D001**	**Characteristic (I) Waste**

DOT Proper Shipping Name **Isopropylamine**	Chemical Abstract Service (CAS) Number **75-31-0**
DOT Hazard Class and Label Requirements **Flammable Liquid**	DOT Emergency Guide Code **68**
DOT Identification Number **UN 1221**	Chemical Formula **$(CH_3)_2CHNH_2$**

Synonyms

2-Aminopropane; monoisopropylamine; 2-propylamine; *sec*-propylamine; 2-propanamine.

SECTION II - HAZARDOUS INGREDIENTS AND IDENTITY INFORMATION

Hazardous Components (specific identity)	OSHA Exposure Criteria	NIOSH Exposure Criteria	Immediately Dangerous to Life and Health (IDLH)	ACGIH Exposure Criteria
Isopropylamine (derivation: From the reaction of acetone and ammonia under pressure; by reacting isopropyl alcohol with ammonia in the presence of a dehydrating catalyst; or by reacting isopropyl alcohol and ammonia under pressure). **1 ppm = 2.46 mg/m^3**	**PEL:** **5 ppm** **12 mg/m^3** **STEL:** **Not Established**	**REL:** **Not Established** **STEL:** **Not Established**	**750 ppm**	**TLV:** **5 ppm** **12 mg/m^3** **STEL:** **10 ppm** **24 mg/m^3**

SECTION III - PHYSICAL AND CHEMICAL CHARACTERISTICS

Boiling Point **91°F (32°C)**	Specific Gravity (H_2O = 1) **0.69**
Vapor Pressure (mm Hg) **478 at 68°F (20°C)**	Molecular Weight **59.1**
Vapor Density (Air = 1) **2.0**	Melting Point **-150°F (-101°C)**

Solubility

Soluble in water, acetone, alcohol, benzene, chloroform, and ether.

Appearance and Odor

Colorless liquid with a strong ammonia-like odor. A gas above 91°F (32°C). Odor Threshold = 5 ppm.

SECTION IV - FIRE AND EXPLOSION HAZARD DATA

Flash Point (method used) **-35°F (-37°C) opened cup**	Explosive Limits in Air % by Volume **LEL: 2.3%** **UEL: 10.4%**
NFPA Classification **Class IA Flammable Liquid**	Autoignition Temperature **756°F (402°C)**

Extinguishing Media

Use dry chemical, carbon dioxide, regular foam, or water spray (water may not be effective by itself).

Special Fire Fighting Procedures

Poisonous gases are produced in fire; wear full protective clothing and self-contained breathing apparatus (SCBA). Vapors are heavier than air and may travel for some distance to ignition source and flashback. Move containers from fire area if it can be done safely. Cool exposed containers with water.

Unusual Fire and Explosion Hazards

Containers may explode in fire due to buildup of excessive internal pressures. Exercise extreme caution when attempting to move or relocate containers that may have been exposed to high heat. A boiling liquid expanding vapor explosion (BLEVE) can occur when isopropylamine is involved in fire. Evacuate area if rising sound is heard coming from venting device or tanks are discoloring due to fire.

SECTION V - REACTIVITY DATA

Stability		Conditions to Avoid **Isopropylamine is normally stable under routine conditions of handling and storage. Avoid contact with incompatible materials. Keep away from heat and flame.**
Stable X	Unstable	Incompatibility *(materials to avoid)* **Oxidizing agents (such as chlorine, bromine, and oxygen) and other oxidizable materials, strong acids, aldehydes, ketones, and epoxides.**
Hazardous Polymerization		Conditions to Avoid **Under normal conditions of temperature and pressure, hazardous polymerization of isopropylamine is not expected to occur.**
May Occur	Will Not Occur X	Hazardous Decomposition or By-products **Toxic and irritating oxides of nitrogen are produced when isopropylamine is heated to decomposition or involved in a fire. These include oxides of carbon and nitrogen.**

SECTION VI - HEALTH HAZARD DATA

Primary Route(s) of Entry:	Inhalation?	Absorption (skin)?	Ingestion?
	X	X	X

Health Hazards

INHALATION: Irritation of the nose, throat, and lungs. Will cause coughing, nausea, headache, and may cause vomiting. May cause fluid buildup in lungs (pulmonary edema) and death.

ABSORPTION: A severe irritant. Eye contact will cause irritation and inflammation. Skin contact causes irritation, inflammation, redness, blistering, and possible sensitization or dermatitis.

INGESTION: Poison by ingestion. Irritation of the mouth, throat, stomach, and digestive tract. Will also cause excessive salivation, nausea, and vomiting.

Carcinogenicity	NTP Listed?	IARC Cancer Review Group?	OSHA Regulated?	Target Organs?
Unknown Human Unknown Animal	No	No	29 CFR 1910.1000 Table Z-1	Eyes; skin; respiratory system.

Medical Conditions Generally Aggravated by Exposure

Respiratory system disorders and any skin diseases may be aggravated by exposure.

Emergency and First-aid Procedures

Eye contact: Flush immediately with water for 15 minutes (minimum); seek medical attention. Skin contact: Remove all contaminated clothing. Immediately wash area with large amounts of soap and water. Seek medical assistance. For inhalation: Remove the person from exposure. Provide respiratory assistance and CPR. Transfer to medical facility. Observe for 24-48 hours for lung effects. If swallowed: Call the poison control center; seek medical attention immediately. Give 1-2 glasses of water to dilute. Do NOT induce vomiting. Never attempt to give an unconscious person anything by mouth.

SECTION VII - PRECAUTIONS FOR SAFE HANDLING AND USE

Steps to be Taken in Case Material is Released or Spilled

Absorb liquids in vermiculite or other absorbent and deposit in DOT-approved drum. Small spills may be absorbed in paper towels and evaporate in fume hood. Ventilate area and remove ignition sources. Restrict those not involved in cleanup from entering area. Notify appropriate authorities, if applicable.

Preferred Waste Disposal Method

Burn in a chemical incinerator equipped with a scrubber and afterburner.

Precautions to be Taken in Handling and Storage

Do not store isopropylamine in the presence of incompatible chemicals or materials. Store in tightly closed containers in a cool, well-ventilated area. Personnel working with isopropylamine should be trained on its proper handling and storage *prior* to being assigned to such responsibilities.

Other Precautions and Warnings

To prevent static sparks, electrically ground and bond all containers and equipment.

SECTION VIII - CONTROL MEASURES AND PERSONAL PROTECTIVE EQUIPMENT

Respiratory Protection *(specify type)*

Exposure levels are low and difficult to accurately monitor. For low or infrequent exposures, use an organic vapor respirator. Greater protection at higher exposures is provided with an MSHA/NIOSH-approved supplied-air respirator with full facepiece operated in continuous flow or other positive pressure mode (pressure demand), or a self-contained breathing apparatus (SCBA) with full facepiece operated in pressure demand or other positive pressure mode.

Ventilation

Local exhaust or general mechanical systems recommended.

Protective Gloves	Eye Protection	Other Protective Clothing
Impervious Material	Chemical Goggles or Face Mask	Protective Apron

Work/Hygiene Practices

Always wash hands thoroughly after using chemical; never bring food, drink, or smoking materials into vicinity of chemicals.

ISOPROPYLAMINE

$(CH_3)_2CHNH_2$

CAS: 75-31-0

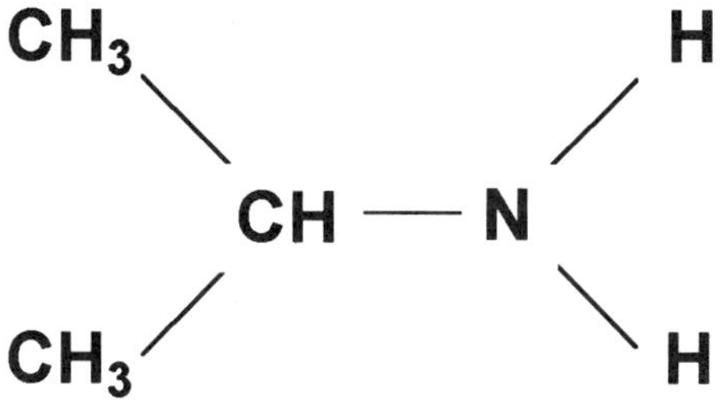

IDENTIFICATION AND TYPICAL USES

Isopropylamine is a colorless or clear liquid with a strong, ammonia-like odor. It can exist as a gas at temperatures above 91°F (32°C). It is used in the synthesis of rubber accelerators, in pharmaceuticals, in dyes, in insecticides, in certain textile specialties, in surface-active agents, and as a dehairing agent.

RISK ASSESSMENT: HEALTH

General Assessment

Isopropylamine is a human poison by skin contact (*absorption*), is moderately toxic by *ingestion* and mildly toxic by *inhalation*. It is also moderately toxic by skin contact (irritant). It will irritate the nose and throat (mucous membranes) and is a severe eye and skin irritant. There are no data on its carcinogenic, mutagenic, or teratogenic capabilities.

Inhalation will cause irritation to the nose, throat, and especially the lungs. It will cause cough, nausea, and headache. Its respiratory toxicity is primarily due to pulmonary effects. On high exposures, it may cause a fluid buildup in the lungs (pulmonary edema) which is a medical emergency and can be fatal. Symptoms may be delayed up to 48 hours, thereby creating a false sense of security relative to exposure risk.

Ingestion will result in digestive system irritation, unspecified but adverse gastrointestinal effects with irritation of the mouth, throat (esophagus), and stomach. There may be excessive salivation, nausea, vomiting, cramps, and headache.

Skin contact with the vapor or the liquid results in severe irritation and may produce dermatitis with drying, redness, rash, and itching. Some individuals may become sensitized. Eye contact with the vapor can result in severe irritation with corneal edema. The liquid may cause serious burns and permanent damage to eye tissues and possibly to vision.

☠ *Acute Health Effects*

The following acute (short-term) health effects may occur either immediately or shortly after (within hours or days) exposure to isopropylamine:

Skin: Irritation, inflammation, redness (erythema), itching, and possible sensitization and/or dermatitis.

Eye: Severe irritation, inflammation, and possible damage to the cornea and to vision.

Lung: Irritation causing coughing and/or shortness of breath. Higher exposures can lead to respiratory problems, including pulmonary edema (fluid buildup in lungs) which is a medical emergency and can be fatal.

Chronic Health Effects

The following chronic (long-term) health effects may occur at some time after exposure to isopropylamine and can last for months or even years:

Cancer Hazards: According to the information presented in the references, isopropylamine has not been adequately tested for its ability to cause cancer in test animals.

Reproductive Hazard: According to the information presented in the references, isopropylamine has not been adequately tested for its ability to adversely affect reproduction.

Other Chronic Effects: Low, chronic exposures to skin may cause allergy, sensitization, or dermatitis.

❶ *Recommended Risk-Reduction Measures*

Unless a less toxic chemical can be substituted for a hazardous substance, *engineering controls* are the most effective method of reducing exposures. The best protection is to enclose operations and/or provide local exhaust ventilation at the site of chemical release. While not always operationally feasible, isolating operations can also reduce exposure.

Using respiratory protection is less effective than the controls mentioned above, but is still advisable whenever working with or around isopropylamine. The exposure level for this compound is extremely low and difficult to accurately monitor. For relatively low, infrequent, or transient exposures, an organic vapor respirator may suffice. However, for higher or prolonged exposure, a self-contained breathing apparatus (SCBA) with full facepiece operated in pressure demand or a powered air-purifying respirator operated in continuous flow mode are the recommended respiratory protection methods of choice. If a full facepiece is not available, then chemical goggles should be worn to protect the eyes. Whenever a chemical splash hazard exists, a face shield and apron should be worn. To prevent hand and skin exposures, impervious gloves should be worn. Butyl rubber may be used with caution for short periods (less than 4 hours). Natural rubber, neoprene, nitrile rubber, polyvinyl alcohol, polyvinyl chloride, and Viton are *not* suitable protection. Glove manufacturers should be contacted and permeation studies obtained *before* final glove selections have been made.

Administrative controls should also be in place to minimize the potential for human exposures. These may include written procedures or policies which specify the methods and techniques that will be practiced whenever personnel are to work with isopropylamine.

All personnel should receive *training* on the use, hazards, protective measures, emergency actions, and other precautions per 29 CFR 1910.1200 (Hazard Communication), prior to the first assignment in an area where isopropylamine is used or stored.

Before beginning employment and at regular intervals thereafter (e.g. annually), the following medical tests are recommended:

- ☑ Lung function tests (establish a baseline).

If symptoms develop or overexposure is suspected, the following medical tests may be useful:

- ☑ Lung function tests (compare to baseline).
- ☑ Consider chest X-ray after acute overexposure (may be negative if performed immediately after exposure or if symptoms have not yet appeared).
- ☑ Evaluation by a qualified allergist with careful evaluation of exposure history and administration of special tests (may help diagnose skin allergy).
- ☑ Urinalysis (isopropylamine is metabolized as acetone and ammonia).
- ☑ Complete blood count (CBC) with sedimentation rate, serum electrolysis, calcium, phosphorus, albumin, total protein, and arterial blood gases.

Any evaluation should include a careful medical history of past and present symptoms with an examination. However, medical tests that evaluate existing damage are not a substitute for controlling exposure. Also, since smoking can cause heart disease, emphysema, and numerous other respiratory disorders, smokers may be affected more rapidly than non-smokers to the same exposure conditions. Prudent risk assessment and management requires careful consideration of *all* possible risk factors that may be causing the appearance of symptoms in exposed workers.

Other methods to reduce exposure include:

- ☑ Where possible, enclose operations and use local exhaust ventilation at the site of chemical release. If local exhaust ventilation or enclosure is not used, respiratory protection should be mandatory. Also, if possible, automatically transfer liquids from storage containers to process containers using non-sparking tools.
- ☑ Always ensure that proper protective clothing is worn when using chemical substances and that personnel are trained on its use and care.
- ☑ Wash thoroughly immediately after exposure to isopropylamine and at the end of the work shift or before eating, drinking, or smoking.
- ☑ Hazard warning information should be posted in the work area. In addition, as part of an on-going education and training program, all information on the health and safety hazards of isopropylamine should be communicated to all exposed and potentially exposed workers.
- ☑ Eye wash stations should be provided in work areas. If the potential for whole body exposure exists, then safety showers should also be provided.

RISK ASSESSMENT: ENVIRONMENT

General Assessment

The environment is at risk of exposure during intended use, transportation, storage, disposal, or destruction of isopropylamine. In almost every scenario, the threat of environmental exposure is contingent upon the proper handling of the chemical substance. Accidental spills, large or small, can result in fire, explosion, and possible contamination of the surrounding environmental mediums (water, soil, and air).

Isopropylamine is considered a Class IA flammable liquid (per OSHA 29 CFR 1910.106). Its extremely low flash point and relatively low boiling point present a dangerous fire and explosion hazard, especially in confined areas or under conditions of extreme heat. It can react with many common oxidizing materials to cause a fire or explosion. Its vapors are heavier than air and can travel for great distances to an ignition source to flashback and cause fire or explosion. These characteristics require special consideration during any emergency situation involving a leak or spill of isopropylamine. Should isopropylamine ever come into contact with incompatible substances such as oxidizers, acids, aldehydes, ketones, or epoxides, either during use, transportation, or storage, violent reactions can occur.

Isopropylamine can enter the environment through industrial discharges, municipal waste treatment plant discharges, and from spills.

Acute Ecological Effects

Acute (short-term) toxic effects may include the death of animals, birds, or fish and death or low growth rate in plants. Acute effects are seen 2 to 4 days after animals or plants are exposed to isopropylamine.

Isopropylamine has slight acute toxicity to aquatic life. Insufficient data are available to evaluate or predict the short-term effects of isopropylamine to plants, birds, or land animals.

Chronic Ecological Effects

Chronic toxic effects may include shortened life span, reproductive problems, lower fertility, and changes in appearance or behavior in exposed animals. These effects can be seen long after first exposure(s) to toxic chemicals.

Insufficient data are available to evaluate or predict the long-term effects of isopropylamine to aquatic life, plants, birds, or land animals.

Water Solubility

Isopropylamine is highly soluble in water. Concentrations of 1000 milligrams or more will mix with a liter of water.

Persistence in the Environment

Isopropylamine is moderately persistent in water, with a half-life of between 20 to 200 days. The half-life of a pollutant is the amount of time it takes for one half of the chemical to be degraded. It is expected to be highly mobile in terrestrial soils.

Bioaccumulation in Aquatic Organisms

Some substances increase in concentration, or *bioaccumulate,* in living organisms as they breathe contaminated air, drink contaminated water, or eat contaminated food. These chemicals can become concentrated in the tissues and internal organs of animals as well as humans.

The concentration of isopropylamine found in the edible tissues of fish is expected to be much lower than the average concentration in the water from which the fish was taken.

Recommended Risk-Reduction Measures

Proper training of all transporters will reduce the likelihood of a mishap or accident resulting in a leak or spill to the environment. The correct use of DOT labeling (per the MSDS) on all containers, trucks, and rail cars will enable emergency responders to react properly and quickly to any disaster thereby reducing the potential risk to the environment and to personnel.

Storage of isopropylamine should be segregated from incompatible chemicals to minimize the risk of cross-contamination or contact. Storage buildings should be equipped with the proper fire protection equipment (alarms, sprinklers, extinguishers, etc.).

If a spill or leak to the environment has occurred, fire department, emergency response and/or hazardous materials spill personnel should be notified immediately. Cleanup should be attempted only by those trained in proper spill containment procedures using non-sparking tools. Contaminated soils should be removed for incineration and replaced with clean soil. If isopropylamine should contact the water table, aquifer, or navigable waterway, remediation activities should be prompt to ensure protection of aquatic sediments. When such spills occur, the local and/or state emergency response authorities must be notified. A comprehensive emergency response or disaster preparedness/recovery plan should be in place prior to any operations involving the use, transportation, storage, or disposal of isopropylamine. If isopropylamine is spilled or leaked, the following specific steps are recommended:

- ☑ Restrict persons not wearing protective clothing from area of spill or leak until cleanup is complete and area can be opened for normal work.
- ☑ Ventilate area and remove all ignition sources.
- ☑ Absorb liquid spills using vermiculite, dry earth, or sand to absorb and place in a sealed drum for disposal. Use non-sparking tools.
- ☑ It may be necessary to dispose of isopropylamine as a hazardous waste. The responsible state agency or the regional office of the federal Environmental Protection Agency (EPA) should be contacted for specific recommendations.

RISK ASSESSMENT: BUSINESS

General Assessment

Accidents or mishaps involving isopropylamine can present a significant threat to business operations. The loss or damage of equipment or facilities can significantly affect fiscal viability. Lawsuits that may result from personnel injury/death, public exposures, and/or environmental contamination will also require a serious expenditure of resources. Media attention surrounding an injury, death, or environmental damage can also result in a loss of profits and loss of current as well as future business.

Recommended Risk-Reduction Measures

Company attorneys, safety and health professionals, and environmental specialists should be involved in the development of any procedures dealing with safety or emergency response. A company official should be pre-designated as a public relations officer with specific training in dealing with the press. Corporate plans and policies should be developed, approved, and implemented long before any need for such arises.

REFERENCES

American Conference of Governmental Industrial Hygienists. 1988. *Documentation of the Threshold Limit Values and Biological Exposure Indices*, 5th Edition (with updates). Cincinnati: ACGIH

Anthony, C. P., and N. J. Kolthoff. 1971. *Textbook of Anatomy and Physiology.* St. Louis: C. V. Mosby Company.

Bloom, A. D. and F. J. DeSerres. 1995. *Ecotoxicity and Human Health.* New York: CRC Press/Lewis Publishers.

Bueche, J. R. 1972. *Principles of Physics*, 2nd Edition. New York: McGraw-Hill.

Burrell, R., D. K. Flaherty, and L. J. Sauers. 1992. *Toxicology of the Immune System.* New York: Van Nostrand Reinhold.

Calabrese, E. J., E. M. Kenyon. *Air Toxics and Risk Assessment.* New York: CRC Press/Lewis Publishers.

Calabrese, E. J. 1994. *Biological Effects of Low Level Exposures.* New York: CRC Press/Lewis Publishers.

Cockerman, L. G., B. .S. Shane. 1994. *Basic Environmental Toxicology.* New York: CRC Press/Lewis Publishers.

Cothern, C. R., N. P. Ross. 1994. *Environmental Statistics, Assessment, and Forecasting.* New York: CRC Press/Lewis Publishers.

Gardner, R. W. 1994. *Chemical Intolerance.* New York: CRC Press/Lewis Publishers.

Hallenbeck, W. H. 1993. *Quantitative Risk Assessment for Environmental and Occupational Health*, 2nd Edition. New York: CRC Press/Lewis Publishers.

Hodgson, E., R. B. Mailman, and Janice E. Chambers. 1988. *Dictionary of Toxicology.* London: The MacMillian Company.

Howard, P. H., et. al. 1991. *Handbook of Environmental Degradation Rates.* New York: CRC Press/Lewis Publishers.

Jones, F. E. 1994. *Toxic Organic Vapors in the Workplace.* New York: CRC Press/Lewis Publishers.

Landis, W. G., M. H. Yu. 1995. *Introduction to Environmental Toxicology.* New York: CRC Press/Lewis Publishers.

Lewis, R. J., Sr. 1993. *Hawley's Condensed Chemical Dictionary*, 12th Edition. New York: Van Nostrand Reinhold.

Lewis, R. J., Sr. 1992. *Sax's Dangerous Properties of Industrial Materials*, Volumes 1, 2, and 3, 8th Edition. New York: Van Nostrand Reinhold.

Meyer, E. 1990. *Chemistry of Hazardous Materials.* Englewood Cliffs, NJ: Prentice Hall.

Ott, W. R. 1995. *Environmental Statistics and Data Analysis.* New York: CRC Press/Lewis Publishers.

Philp, R. B. 1995. *Environmental Hazards and Human Health.* New York: CRC Press/Lewis Publishers.

Rea, W. J. 1995. *Chemical Sensitivity*, Vol. I-III. New York: CRC Press/Lewis Publishers.

Richardson, M. 1995. *Dictionary of Substance and Their Effects*, Vol. I-VI. New York: CRC Press/Lewis Publishers.

Sacarello, H. L. A. 1994. *The Comprehensive Handbook of Hazardous Materials.* New York: CRC Press/Lewis Publishers.

U.S. Department of Health and Human Services, National Institute for Occupational Safety and Health. 1994. *NIOSH Pocket Guide to Chemical Hazards.* Washington, D.C.: U.S. Government Printing Office.

MATERIAL SAFETY DATA SHEET

CHEMICAL NAME	
ISOPROPYL ETHER	**CRC PRESS/LEWIS PUBLISHERS DISCLAIMER:** THE INFORMATION AND RECOMMENDATIONS PRESENTED HEREIN ARE BASED ON SOURCES BELIEVED TO BE RELIABLE. CRC MAKES NO REPRESENTATION ON ITS COMPLETENESS OR ACCURACY. IT IS THE RESPONSIBILITY OF THE USER TO DETERMINE THE CHEMICAL'S SUITABILITY FOR ITS INTENDED USE, THE CHEMICAL'S SAFE USE, AND THE CHEMICAL'S PROPER DISPOSAL. NO REPRESENTATIONS AND/OR WARRANTIES, EITHER EXPRESSED OR IMPLIED, OF THE MERCHANTABILITY OR FITNESS FOR A PARTICULAR PURPOSE, OR OF ANY OTHER NATURE, ARE MADE WITH RESPECT TO THE INFORMATION PROVIDED IN THIS MSDS OR TO THE CHEMICAL TO WHICH INFORMATION MAY REFER. CRC NEITHER ASSUMES NOR AUTHORIZES ANY OTHER PERSON TO ASSUME FOR IT, ANY OTHER ADDITIONAL RESPONSIBILITY OR LIABILITY FOR THE USE OF, OR RELIANCE UPON, THIS INFORMATION.

HAZARD WARNING INFORMATION

HEALTH	FIRE	REACTIVE	OTHER	DEGREE OF HAZARD	COLOR CODING	OTHER CODES
2	3	1		0 = Minimum Hazard 1 = Slight Hazard 2 = Moderate Hazard 3 = Serious Hazard 4 = Severe Hazard	HEALTH = BLUE FIRE = RED REACTIVITY = YELLOW OTHER = WHITE	OX = Oxidizer ACID = Acid ALK = Alkali COR = Corrosive W = Use No Water

SECTION I - GENERAL INFORMATION

Characterization	Ether	RCRA Number	None
		EPA Class	Not Applicable
DOT Proper Shipping Name	Diisopropyl Ether	Chemical Abstract Service (CAS) Number	108-20-3
DOT Hazard Class and Label Requirements	Flammable Liquid	DOT Emergency Guide Code	26
DOT Identification Number	UN 1159	Molecular Formula	$(CH_3)_2CHOCH(CH_3)_2$

Synonyms

Diisopropyl ether; diisopropyl oxide; 2-isopropoxy propane; isopropoxypropane; 2,2'-oxybis (propane).

SECTION II - HAZARDOUS INGREDIENTS AND IDENTITY INFORMATION

Hazardous Components (specific identity)	OSHA Exposure Criteria	NIOSH Exposure Criteria	Immediately Dangerous to Life and Health (IDLH)	ACGIH Exposure Criteria
Isopropyl ether: (derivation: By reacting isopropyl alcohol with sulfuric acid or as a by-product in the preparation of isopropyl alcohol from polypropylene). 1 ppm = 4.25 mg/m^3	PEL (8-hour): 500 ppm 2100 mg/m^3 STEL: Not Established	REL (10-hour): 500 ppm 2100 mg/m^3 STEL: Not Established	10,000 ppm	TLV: 250 ppm 1040 mg/m^3 STEL: 310 ppm 1300 mg/m^3

SECTION III - PHYSICAL AND CHEMICAL CHARACTERISTICS

Boiling Point	156°F (69°C)	Specific Gravity (H_2O = 1)	0.73
Vapor Pressure (mm Hg)	119 at 68°F (20°C)	Molecular Weight (atomic weight)	102.2
Vapor Density (Air = 1)	3.52	Melting Point	-76°F (-60°C)

Solubility

Slightly soluble in water (0.2%). Soluble in alcohol, acetone, ether, and dilute sulfuric acid.

Appearance and Odor

A clear, colorless, mobile, water white, extremely volatile liquid with a strong, sweet, aromatic, ether-like odor and burning sweet taste. Odor Threshold = <.1 ppm

SECTION IV - FIRE AND EXPLOSION HAZARD DATA

Flash Point (method used)	-18°F (-28°C) closed cup	Explosive Limits in Air % by Volume	LEL: 1.4% UEL: 7.9%
NFPA Classification	Class IB Flammable Liquid	Autoignition Temperature	830°F (443°C)

Extinguishing Media

Dry chemical, alcohol-resistant foam, or carbon dioxide. Use water spray to keep fire-exposed containers cool. Water itself may be ineffective on fire since isopropyl ether floats on water.

Special Fire Fighting Procedures

Wear full protective clothing and self-contained breathing apparatus (SCBA). Move container from fire area if it can be done without risk. Use water spray to flush spills from ignition sources and to knock down vapors that have not yet ignited. Treat as a gas fire; use blanketing effect to smother flames.

Unusual Fire and Explosion Hazards

Containers may explode in fire. Presents a vapor explosion hazard indoors, outdoors, and in sewers.

SECTION V - REACTIVITY DATA

Stability		Conditions to Avoid **Isopropyl ether is NOT stable under normal conditions of handing and storage. It can form unstable peroxides and hydroperoxides on exposure to air. Peroxide can explode if distilled or heated.**
Stable	Unstable **X**	Incompatibility *(materials to avoid)* **Incompatible with strong oxidizers, peroxides, permanganates, acids (chlorosulfonic acid, nitric acid), and propionyl chloride.**
Hazardous Polymerization		Conditions to Avoid **Under normal conditions of temperature and pressure, hazardous polymerization of isopropyl ether is not expected to occur. Do not store more than a six-month supply.**
May Occur	Will Not Occur **X**	Hazardous Decomposition or By-products **When heated to decomposition, isopropyl ether emits acrid smoke and fumes, including toxic carbon dioxide gas and fumes.**

SECTION VI - HEALTH HAZARD DATA

Primary Route(s) of Entry:	Inhalation?	Absorption (skin/eye)?	Ingestion?
	X		X

Health Hazards

INHALATION: Causes headache, nausea, vomiting, loss of consciousness and, at high concentrations, death. Depression of the central nervous system (CNS) causing anesthesia, exhaustion, loss of appetite, sleepiness, dizziness, excitation, anorexia, giddiness, narcotic effects.

SKIN & EYES: An eye and skin irritant. May lead to dermatitis on prolonged contact and it can also cause conjunctivitis. May cause skin burns and eye irritation (smarting).

INGESTION: Not likely, but possible. The liquid evaporates rapidly making ingestion unlikely.

Carcinogenicity	NTP Listed?	IARC Cancer Review Group?	OSHA Regulated?	Target Organs?
Unknown Human Unknown Animal	No	No	29 CFR 1910.1000 Table Z-1	Skin, eyes, respiratory system, CNS,

Medical Conditions Generally Aggravated by Exposure

Any nervous system disorders may be aggravated by exposure.

Emergency and First-aid Procedures

Eye contact: Do not allow victim to keep eyes tightly shut. Flush immediately with water for 15 minutes (minimum); seek medical attention. Skin contact: No serious hazard since evaporation is quick. Prolonged contact may cause drying, cracking, or burning Contaminated clothing should be removed (fire hazard). Wash area with large amounts of soap and water. For inhalation: Remove the person from exposure. Provide respiratory assistance and CPR. Transfer to medical facility. If swallowed: Not very likely. Seek medical attention immediately. Unless advised otherwise, give 1-2 glasses of water to dilute. Do NOT induce vomiting because of danger of aspiration (breathing the liquid into the lungs). Gastric lavage may be advised. Never attempt to give an unconscious or convulsing person anything by mouth.

SECTION VII - PRECAUTIONS FOR SAFE HANDLING AND USE

Steps to be Taken in Case Material is Released or Spilled

Wear self-contained breathing apparatus (SCBA). Restrict those not wearing protective equipment and who are not involved in cleanup from area. Avoid skin contact. Ventilate area of spill and remove ignition sources. Absorb liquids in vermiculite, collect with non-sparking tools, and deposit in sealed drum.

Preferred Waste Disposal Method

Mix with a higher-boiling solvent and burn in chemical incinerator with afterburner and scrubber.

Precautions to be Taken in Handling and Storage

Store in tightly closed (air-tight), metal containers in a cool, dry, well-ventilated area away from heat and incompatible materials. Control inventories. Protect containers from physical damage.

Other Precautions and Warnings

Before using bulk material, test small quantity to ascertain quality and peroxide content.

SECTION VIII - CONTROL MEASURES AND PERSONAL PROTECTIVE EQUIPMENT

Respiratory Protection *(specify type)*

Use a NIOSH-approved air-purifying respirator for low levels. Best protection is obtained using a supplied-air respirator or a SCBA with full facepiece operated in positive pressure mode.

Ventilation

Local exhaust (preferred) at site of chemical work or general ventilation.

Protective Gloves	Eye Protection	Other Protective Clothing
Impervious Material	Chemical Goggles and Face Mask	Protective Uniform or Apron

Work/Hygiene Practices

Always wash hands thoroughly after using chemical; never bring food, drink, or smoking materials into vicinity of chemicals.

ISOPROPYL ETHER

$(CH_3)_2CHOCH(CH_3)_2$

CAS: 108-20-3

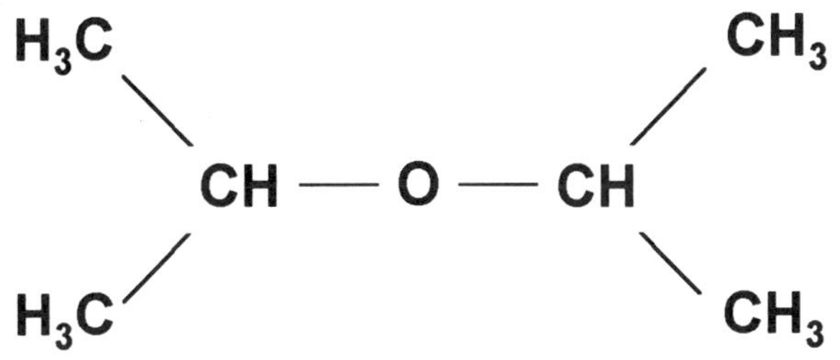

IDENTIFICATION AND TYPICAL USES

Isopropyl ether appears as a clear, colorless, volatile, mobile, water-white liquid with a characteristic, sweet, pungent, aromatic, ethereal odor and bittersweet taste. It is used as a solvent for animal, vegetable, and mineral oils, waxes, and resins. It is also used in the extraction of acetic acid from aqueous solutions. It is a solvent for dyes (in the presence of small quantities of alcohol). It is used in spotting compositions, and in the manufacture of rubber cements.

RISK ASSESSMENT: HEALTH

General Assessment

Isopropyl ether is a narcotic substance and is considered mildly to moderately toxic by *inhalation* and skin contact, although dermal absorption has not been reported as a route of entry into the body. *Ingestion* may also be an exposure route. However, the liquid is extremely volatile, evaporating quickly and making ingestion possible but improbable. There are no reports to indicate that this chemical poses carcinogenic, mutagenic, or teratogenic effects in humans.

Inhalation of isopropyl ether vapors causes anesthetic effects because of the chemical's ability to quickly and effectively depress the central nervous system. Symptoms of exposure include exhaustion, loss of appetite, sleepiness, dizziness, headache, giddiness, excitation, nausea, vomiting, anorexia, loss of consciousness, possible seizures and convulsions, psychic disturbances, and death due to respiratory arrest. Exposure can cause irritation of the eyes, nose, throat, and respiratory system with subsequent changes (numbing) of the sense of smell.

Skin contact with the liquid poses no serious threat because isopropyl ether will evaporate rather quickly. However, if allowed to remain in contact (in wet clothing, for example) for extended periods, there can be drying, cracking, even some burning of the skin's surface. Repeated skin contact may lead to dermatitis. The vapors will cause irritation and smarting of the eyes. Conjunctivitis is also possible, especially on exposure to heavy vapor concentrations.

Ingestion is not likely since the liquid will evaporate quickly, leaving very little time for ingestion. However, it is a remote possibility and the symptoms of inhalation can be expected to occur.

Acute Health Effects

The following acute (short-term) health effects may occur immediately or shortly after exposure to isopropyl ether:

Skin: Irritation with drying, cracking, secondary infection, and possible dermatitis.

Eye: Irritation, smarting of eye tissues and associated skin surfaces (lids), and conjunctivitis.

Lung: Irritation to respiratory tract causing cough and sore throat. May also cause olfactory changes (numbing of the sense of smell). Absorption through respiration causes anesthetic effects and narcosis (deep unconsciousness).

CNS: Rapid depression of the central nervous system causing a variety of toxic effects including loss of coordination and consciousness, and even death due to respiratory arrest.

Chronic Health Effects

The following chronic (long-term) health effects may occur at some time after exposure to isopropyl ether and can last for months or even years:

Cancer Hazards: There is no evidence in the references that exposure to isopropyl ether causes cancer in animals or humans. Additional research is required in this area of study before any conclusions can be made.

Reproductive Hazard: According to the references, isopropyl ether has not been adequately tested for its ability to adversely affect reproduction.

Other Chronic Effects: Repeated low-dose exposures can cause chronic dermatitis.

Recommended Risk-Reduction Measures

Personnel should avoid direct contact with isopropyl ether. Its specific toxicity is not clearly understood in the references. If a less toxic material or compound cannot be substituted for isopropyl ether, then *engineering controls* are the most effective method of reducing exposure risk. The best protection is to enclose operations and/or provide local exhaust ventilation at the site of isopropyl ether release. While not always operationally feasible, isolating operations can also reduce exposure. Using respiratory protection is less effective than the controls mentioned above, but is still required whenever working with or around isopropyl ether. For exposures to the PEL (500 ppm), an air-purifying respirator should provide adequate protection. For high exposures or when the concentration is unknown (as in an emergency), an MSHA/NIOSH-approved supplied-air respirator or a self-contained breathing apparatus (SCBA) with full facepiece and operated in pressure demand mode are the recommended methods of respiratory protection. If a full facepiece is not available, then chemical goggles should be worn to protect the eyes, especially when a splash hazard exists. A face shield should also be considered. To prevent hand and skin exposures, polyvinyl alcohol (PVA), polyvinyl chloride (PVC), nitrile rubber, or chlorinated polyethylene gloves should be used. Glove manufacturers should be contacted and permeation studies obtained *before* final glove selections have been made.

Administrative controls should also be in place to minimize the potential for human exposures. These may include written procedures or policies which specify the methods and techniques that will be practiced whenever personnel are to work with isopropyl ether.

All personnel should receive *training* on the use, hazards, protective measures, emergency actions, and other precautions per 29 CFR 1910.1200 (Hazard Communication) prior to the first assignment in an area where isopropyl ether is used or stored.

If symptoms develop or overexposure is suspected, the following may be useful:

- ☑ Liver, kidney, and lung function tests.
- ☑ Evaluation of the central nervous system.

It should be noted that medical tests that simply look for existing damage are not a substitute for controlling exposures. Medical histories are extremely important when assessing exposure risk.

Other methods to reduce exposure include:

- ☑ Always ensure that proper protective clothing is worn when using chemical substances and that personnel are trained in its use and care.
- ☑ Wash thoroughly immediately after exposure to isopropyl ether and at the end of the work shift or before eating, drinking, or smoking.
- ☑ Hazard warning information should be posted in the work area. In addition, as part of an on-going education and training program, all information on the health and safety hazards of isopropyl ether should be communicated to all exposed or potentially exposed workers.
- ☑ Eye wash stations should be provided in the immediate work area for emergency use. If there is a possibility of skin exposure to isopropyl ether, emergency shower facilities should also be provided in the work area.
- ☑ Workers whose clothing has been contaminated by isopropyl ether should change into clean clothes before leaving work. Contaminated work clothing presents a serious fire hazard and should be laundered only by individuals who have been informed of the hazards of exposure to isopropyl ether.

RISK ASSESSMENT: ENVIRONMENT

General Assessment

The environment is at risk of exposure during transportation, storage, disposal, or destruction of isopropyl ether. In almost every scenario, the threat of environmental exposure is contingent upon the proper handling of the chemical substance. Accidental spills, large or small, can result in contamination of the surrounding environmental mediums (primarily air, but water and soil contamination can also occur).

Isopropyl ether is Class IB flammable liquid (per OSHA 29 CFR 1910.106). This presents a dangerous fire and explosion hazard. Its vapors are heavier than air and can travel for some distance to an ignition source to flashback and cause fire or explosion. It can form explosive peroxides in contact with air or in the presence of a photosensitizer such as benzophene. Peroxides are shock-sensitive and can easily explode. It is incompatible with many oxidizers and acids and caution is always required in handling, storage, transportation, and disposal of isopropyl ether. When heated to decomposition it emits very acrid and irritating smoke and fumes. Emergency responders should be made aware of the presence of isopropyl ether at any emergency response situation. Isopropyl ether may enter the environment through industrial effluents and spills.

Acute Ecological Effects

Acute (short-term) toxic effects may include the death of animals, birds, or fish and death or low growth rate in plants. Acute effects are seen 2 to 4 days after animals or plants are exposed to isopropyl ether. Isopropyl ether has slight acute toxicity to aquatic life. Insufficient data are available to evaluate or predict the acute (short-term) effects of isopropyl ether on plants, birds, or land animals.

Chronic Ecological Effects

Chronic (long-term) toxic effects may include shortened life span, reproductive problems, lower fertility, and changes in appearance or behavior in exposed animals. These effects can be seen long after first exposure(s) to toxic chemicals. Insufficient data are available to evaluate the chronic (long-term) effects of isopropyl ether on aquatic life, plants, birds, or land animals.

Water Solubility

Isopropyl ether is slightly soluble in water. Concentrations of less than 100 milligrams will mix with a liter of water.

Persistence in the Environment

Isopropyl ether is non-persistent in water, with a half-life of less than 2 days. The half-life of a pollutant is the amount of time it takes for one half of the chemical to be degraded. Approximately 100% of isopropyl ether will eventually end up in air.

Bioaccumulation in Aquatic Organisms

Some substances increase in concentration, or *bioaccumulate,* in living organisms as they breathe contaminated air, drink contaminated water, or eat contaminated food. These chemicals can become concentrated in the tissues and internal organs of animals as well as humans.

The references do not address the bioaccumulation capabilities of isopropyl ether in aquatic organisms. However, considering the extremely volatile nature of this chemical, bioaccumulation may not be possible.

Recommended Risk-Reduction Measures

Proper training of all transporters will reduce the likelihood of a mishap or accident resulting in a leak or spill of isopropyl ether into the environment. Labels on all containers, trucks, and rail cars must meet DOT requirements and accurately reflect their contents to enable emergency responders to react properly and quickly to any disaster thereby reducing the potential risk to the environment and to personnel.

Storage of isopropyl ether should be segregated from other chemicals to minimize the risk of cross-contamination. Isopropyl ether must be stored to avoid contact with strong oxidizers, such as chlorine and bromine, since violent reactions can occur. Containers should be protected from physical damage and stored to avoid contact with heat, including sunlight. Isopropyl ether must be stored in air-tight containers to prevent the formation of explosive peroxides. Inventories should be controlled and no more than a six-month supply should be kept in storage.

If a spill or leak to the environment has occurred, fire department, emergency response and/or hazardous materials spill personnel should be notified immediately. Cleanup should be attempted only by those trained in proper spill containment procedures using non-sparking tools. Contaminated soils should be removed for incineration and replaced with clean soil.

If isopropyl ether should contact the water table, aquifer, or navigable waterway, time is of the essence. Its rapid evaporation may make total remediation impossible. The local and/or state emergency response authorities must be notified. A comprehensive emergency response or disaster preparedness/recovery plan should be in place prior to any operations involving the use, transportation, storage, or disposal of isopropyl ether.

Isopropyl ether may be disposed of by mixing with an excess of a higher-boiling solvent and then burning in a chemical incinerator equipped with an afterburner and scrubber. If isopropyl ether is spilled or leaked, the following specific steps are recommended:

- ☑ Restrict persons not wearing protective clothing from area of spill or leak until cleanup is complete. Avoid skin and eye contact.
- ☑ Absorb spilled liquids with vermiculite, dry sand, or a similar material and deposit in sealed drum for disposal. Use non-sparking tools!
- ☑ Ventilate area and remove all ignition sources.
- ☑ It may be necessary to dispose of isopropyl ether as a hazardous waste. The responsible state agency or the regional office of the federal Environmental Protection Agency (EPA) should be contacted for specific recommendations.

RISK ASSESSMENT: BUSINESS

General Assessment

Accidents or mishaps involving isopropyl ether can present a serious threat to business operations. The loss or damage of equipment or facilities can significantly affect fiscal viability. Lawsuits that may result from personnel illness, injury/death, public exposures, and/or environmental contamination will require a serious expenditure of resources. Media attention surrounding an injury, death, or environmental damage can also result in a loss of profits and loss of current as well as future business.

Recommended Risk-Reduction Measures

Company attorneys, safety and health professionals, and environmental specialists should be involved in the development of any procedures for responding to chemical incidents. A company official should be predesignated as a public relations officer with specific training in dealing with the press. Corporate plans and policies should be developed, approved, and implemented long before any need for such arises.

REFERENCES

American Conference of Governmental Industrial Hygienists. 1988. *Documentation of the Threshold Limit Values and Biological Exposure Indices,* 5th Edition (with updates). Cincinnati: ACGIH

Anthony, C. P., and N. J. Kolthoff. 1971. *Textbook of Anatomy and Physiology.* St. Louis: C. V. Mosby Company.

Bloom, A. D. and F. J. DeSerres. 1995. *Ecotoxicity and Human Health.* New York: CRC Press/Lewis Publishers.

Brusick, D. J. 1994. *Methods for Genetic Risk Assessment.* New York: CRC Press/Lewis Publishers.

Calabrese, E. J. 1994. *Biological Effects of Low Level Exposures.* New York: CRC Press/Lewis Publishers.

Cothern, C. R., N. P. Ross. 1994. *Environmental Statistics, Assessment, and Forecasting.* New York: CRC Press/Lewis Publishers.

Danse, I. R. 1991. *Common Sense Toxics in the Workplace.* New York: Van Nostrand Reinhold.

Gardner, R. W. 1994. *Chemical Intolerance.* New York: CRC Press/Lewis Publishers.

Hallenbeck, W. H. 1993. *Quantitative Risk Assessment for Environmental and Occupational Health,* 2nd Edition. New York: CRC Press/Lewis Publishers.

Hodgson, E., R. B. Mailman, and J. E. Chambers. 1988. *Dictionary of Toxicology.* London: The MacMillian Company.

Howard, P. H., et. al. 1991. *Handbook of Environmental Degradation Rates.* New York: CRC Press/Lewis Publishers.

Hosty, J. W., and P. Foster. 1990. *A Practical Guide to Chemical Spill Response.* New York: Van Nostrand Reinhold.

Jones, F. E. 1994. *Toxic Organic Vapors in the Workplace.* New York: CRC Press/Lewis Publishers.

Keith, L. H. and M. M. Walker. 1995. *Handbook of Air Toxics.* New York: CRC Press/Lewis Publishers.

Landis, W. G., M. H. Yu. 1995. *Introduction to Environmental Toxicology.* New York: CRC Press/Lewis Publishers.

Lauwerys, R. R., P. Hoet. 1993. *Industrial Chemical Exposure.* New York: CRC Press/Lewis Publishers.

Lewis, R. J., Sr. 1993. *Hawley's Condensed Chemical Dictionary,* 12th Edition. New York: Van Nostrand Reinhold.

Lewis, R. J., Sr. 1992. *Sax's Dangerous Properties of Industrial Materials,* Volumes 1, 2, and 3, 8th Edition. New York: Van Nostrand Reinhold.

Manahan, S. E. 1992. *Toxicological Chemistry,* 2nd Edition. New York: CRC Press/Lewis Publishers.

Mellström, G. A., J. E. Wahlberg, and H. I. Maibach. 1994. *Protective Gloves for Occupational Use.* New York: CRC Press/Lewis Publishers.

Meyer, E. 1990. *Chemistry of Hazardous Materials.* Englewood Cliffs, NJ: Prentice Hall.

Patnaik, P. 1992. *A Comprehensive Guide to the Hazardous Properties of Chemical Substances.* New York: Van Nostrand Reinhold.

Philp, R. B. 1995. *Environmental Hazards and Human Health.* New York: CRC Press/Lewis Publishers.

Quigley, D. R. 1994. *Handbook of Emergency Chemical Management.* New York: CRC Press/Lewis Publishers.

Richardson, M. 1995. *Dictionary of Substance and Their Effects,* Vol. I-VI. New York: CRC Press/Lewis Publishers.

Torkelson, T. R. V. K. Rowe. 1981. *Patty's Industrial Hygiene and Toxicology,* Vol. 2B. New York: Wiley-Interscience.

U.S. Department of Health and Human Services, National Institute for Occupational Safety and Health. 1994. *NIOSH Pocket Guide to Chemical Hazards.* Washington, D.C.: U.S. Government Printing Office.

MATERIAL SAFETY DATA SHEET

CHEMICAL NAME

ISOPROPYL GLYCIDYL ETHER

HAZARD WARNING INFORMATION

HEALTH	FIRE	REACTIVE	OTHER	DEGREE OF HAZARD	COLOR CODING	OTHER CODES
1	2	0		0 = Minimum Hazard 1 = Slight Hazard 2 = Moderate Hazard 3 = Serious Hazard 4 = Severe Hazard	HEALTH = BLUE FIRE = RED REACTIVITY = YELLOW OTHER = WHITE	OX = Oxidizer ACID = Acid ALK = Alkali COR = Corrosive W = Use No Water

SECTION I - GENERAL INFORMATION

Characterization	Glycol Ether
RCRA Number	None
EPA Class	Not Applicable
DOT Proper Shipping Name	No Citation
Chemical Abstract Service (CAS) Number	4016-14-2
DOT Hazard Class and Label Requirements	No Citation
DOT Emergency Guide Code	No Citation
DOT Identification Number	No Citation
Chemical Formula	$C_6H_{12}O_2$

Synonyms

1,2-Epoxy-3-isopropoxypropane; IGE; isopropoxymethyl oxirane; isopropyl epoxypropyl ether.

SECTION II - HAZARDOUS INGREDIENTS AND IDENTITY INFORMATION

Hazardous Components (specific identity)	OSHA Exposure Criteria	NIOSH Exposure Criteria	Immediately Dangerous to Life and Health (IDLH)	ACGIH Exposure Criteria
Isopropyl Glycidyl Ether: 1 ppm =4.83 mg/m³	PEL: 50 ppm 240 mg/m³ STEL: 75 ppm 362 mg/m³	REL (ceiling): 50 ppm 240 mg/m³ (15 minute) STEL: Not Applicable	400 ppm	TLV: 50 ppm 240 mg/m³ STEL: 75 ppm 362 mg/m³

SECTION III - PHYSICAL AND CHEMICAL CHARACTERISTICS

Boiling Point	279°F (137°C)
Specific Gravity (H_2O = 1)	0.92
Vapor Pressure (gas)	9 at 69°F (20°C)
Molecular Weight	116.2
Vapor Density (Air = 1)	4.0
Freezing Point	Not Found

Solubility

Soluble in water (19%) and most organic solvents.

Appearance and Odor

Clear, colorless, mobile liquid with a mild, ethereal odor.

SECTION IV - FIRE AND EXPLOSION HAZARD DATA

Flash Point (method used)	92°F (33°C) closed cup
Explosive Limits in Air % by Volume	LEL: Not Determined UEL: Not Determined
NFPA Classification	Class IC Flammable Liquid
Autoignition Temperature	525°F (274°C)

Extinguishing Media

Use dry chemical, carbon dioxide, alcohol foam, or water fog.

Special Fire Fighting Procedures

Wear full protective clothing and self-contained breathing apparatus (SCBA). Poisonous gases are produced in fire. Use water spray to keep fire-exposed containers cool and to knock down vapors. If it can be done without risk, move containers from fire area. Containers may explode in fire.

Unusual Fire and Explosion Hazards

Dangerous fire hazard when exposed to heat. Chemical will react on contact with oxidizable materials. Moderate explosion hazard when in the form of vapor. Vapors are heavier than air and can flashback.

SECTION V - REACTIVITY DATA

Stability		Conditions to Avoid **Isopropyl glycidyl ether is stable under normal conditions. Keep away from sources of heat or flame. Keep away from incompatible materials.**
Stable	Unstable	Incompatibility *(materials to avoid)*
X		**Strong oxidizers, such as potassium permanganate, chromium trioxide, sodium peroxide, nitrates, chlorates, and perchlorates, and strong caustics. May form explosive peroxides upon exposure to air or light.**
Hazardous Polymerization		Conditions to Avoid **Under normal conditions of temperature and pressure, hazardous polymerization of isopropyl glycidyl ether is not expected to occur.**
May Occur	Will Not Occur	Hazardous Decomposition or By-products
	X	**When heated to decomposition, isopropyl glycidyl ether can emit highly toxic/poisonous and acrid fumes and gases, including toxic oxides of carbon.**

SECTION VI - HEALTH HAZARD DATA

Primary Route(s) of Entry:	Inhalation?	Absorption (skin)?	Ingestion?
	X		X

Health Hazards

INHALATION: Symptoms include lachrymation, irritation of the throat and upper respiratory tract, general anesthesia, headache, burning cough, respiratory stimulation, nausea or vomiting, other central nervous system effects, liver damage or injury, and possible hematopoietic (blood-producing organ) changes.

SKIN & EYES: Eye and skin irritation. May cause dermatitis.

INGESTION: Unspecified toxic systemic effects. Some gastrointestinal irritation.

Carcinogenicity	NTP Listed?	IARC Cancer Review Group?	OSHA Regulated?	Target Organs?
Unknown Human Unknown Animal	No	No	29 CFR 1910.1000 Table Z-1	Resp. system, eye, liver, skin, CNS, blood, repro. sys.

Medical Conditions Generally Aggravated by Exposure

Any liver, kidney, cardiovascular, or nervous system disorders may be affected.

Emergency and First-aid Procedures

Eye contact: Flush immediately with water for 15 minutes (minimum); seek medical attention. Skin contact: Remove all contaminated clothing. Immediately wash area with large amounts of soap and water. Seek medical assistance. For inhalation: Remove the person from exposure. Provide respiratory assistance and CPR, if required. Transfer to medical facility. If swallowed: Seek medical attention immediately or contact poison control center. Unless told otherwise, give a conscious victim 1-2 glasses of water. Do NOT induce vomiting. Never give an unconscious or convulsing person anything by mouth.

SECTION VII - PRECAUTIONS FOR SAFE HANDLING AND USE

Steps to be Taken in Case Material is Released or Spilled

Absorb large spills in vermiculite or similar material and deposit in sealed containers. Ventilate area of spill or leak. Restrict those not involved in cleanup from entering area. Remove all sources of ignition.

Preferred Waste Disposal Method

Burn in chemical incinerator equipped with an afterburner and scrubber.

Precautions to be Taken in Handling and Storage

Store to avoid contact with incompatible materials in air-tight, closed containers in cool, dark, well-ventilated area away from heat and flame. Use only non-sparking tools to open and close containers.

Other Precautions and Warnings

Sources of ignition are prohibited where isopropyl glycidyl ether is used, handled, or stored. Heat may cause containers to build up pressure and explode. Containers and equipment must be grounded and bonded, especially during liquid transfer. Protect containers from physical damage.

SECTION VIII - CONTROL MEASURES AND PERSONAL PROTECTIVE EQUIPMENT

Respiratory Protection *(specify type)*

For low exposures, uses a chemical cartridge respirator. Otherwise, use an MSHA/NIOSH-approved supplied-air respirator with full facepiece, hood, or helmet in continuous flow mode, or use a self-contained breathing apparatus (SCBA) operated in pressure demand or other positive pressure mode.

Ventilation

Local exhaust or general mechanical systems recommended.

Protective Gloves	Eye Protection	Other Protective Clothing
Impervious Material	Chemical Goggles or Face Mask	Protective Apron

Work/Hygiene Practices

Always wash hands thoroughly after using chemical; never bring food, drink, or smoking materials into vicinity of chemicals.

ISOPROPYL GLYCIDYL ETHER

$C_6H_{12}O_2$

CAS: 4016-14-2

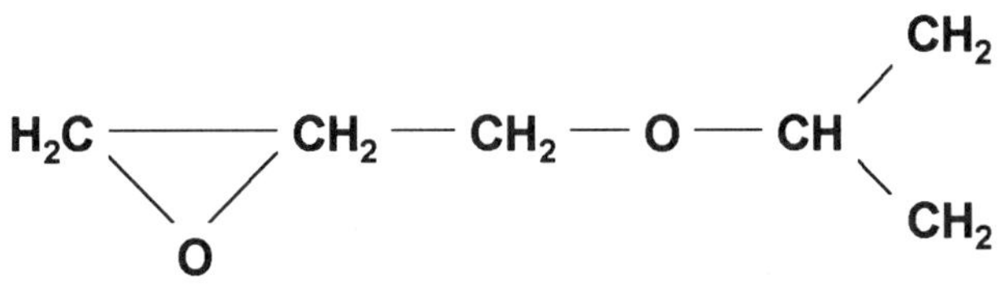

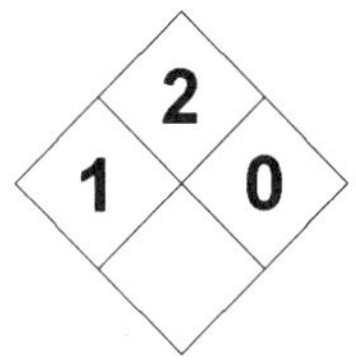

IDENTIFICATION AND TYPICAL USES

Isopropyl glycidyl ether is a clear, colorless, mobile liquid with a mild, ethereal odor. It is used as a stabilizer for chlorinated solvents and organic compounds, and a viscosity reducer for epoxy resins. It is also used as an intermediate for the synthesis of ether and esters.

RISK ASSESSMENT: HEALTH

General Assessment

Isopropyl glycidyl ether is mildly toxic to humans by *ingestion* and *inhalation*. It will also cause some health effects on contact with the skin, but dermal absorption has not been reported. It is a severe central nervous system depressant. There are no reports in the references to indicate that exposure causes cancer. However, mutation data and reproductive effects have been reported in test animals.

Toxic systemic effects by inhalation can include hemolytic anemia, digestive problems, dizziness, respiratory distress, and liver dysfunction. Symptoms of exposure can include abdominal pain, nausea, vomiting, general weakness and muscle tenderness, and possible changes in the hematopoietic system (blood-producing organs, such as bone marrow). CNS depression may cause excitement or exhilaration, drunkenness and stupor, ataxia, dizziness, coma, and possible convulsions.

Eye contact may produce mild to moderate irritation and skin contact may cause an allergic reaction, rash, and possible dermatitis. If ingested, this compound will cause unspecified gastrointestinal effects.

Acute Health Effects

The following acute (short-term) health effects may occur immediately or shortly after exposure to isopropyl glycidyl ether:

Skin: Some irritation at site of contact. Allergic rash with irritation can occur.

Eye: Mild to moderate irritation.

Lung: Nose and throat irritation. Produces intoxicating effects and may cause lung effects.

CNS: Stimulation and depression with symptoms of dizziness, headache, lightheadedness, nausea, vomiting, and coma.

Chronic Health Effects

The following chronic (long-term) health effects may occur at some time after exposure to isopropyl glycidyl ether and can last for months or even years:

Cancer Hazards: According to information presented in the references, isopropyl glycidyl ether has not been adequately tested for its ability to cause cancer in animals. Some mutation data have been reported in test animals. Some scientists believe that such chemicals may have the potential to cause cancer in the long-term. It is not known if exposure to isopropyl glycidyl ether will have this effect on humans.

Reproductive Hazard: According to information presented in the references, isopropyl glycidyl ether has been shown to adversely affect reproduction in test animals.

Other Chronic Effects: Repeated exposures may cause changes in blood-producing organs (bone marrow). These changes may include the development of hemolytic anemia (destruction of red blood cells). Skin rash and/or dermatitis may develop following prolonged or repeated contact.

Recommended Risk-Reduction Measures

Personnel should avoid direct contact with isopropyl glycidyl ether. This chemical is toxic to humans by inhalation and ingestion. If a less toxic chemical cannot be substituted for a hazardous substance, then *engineering controls* are the most effective method of reducing exposures. The best protection is to enclose operations and/or provide local exhaust ventilation at the site of chemical release. While not always operationally feasible, isolating operations can also reduce exposure. Using respiratory protection is less effective than the controls mentioned above, but is still advisable whenever working with or around isopropyl glycidyl ether. Chemical cartridge respirators may provide adequate protection over the PEL. However, the best protection is obtained using a supplied-air respirator with full facepiece operated in continuous flow mode, or a self-contained breathing apparatus (SCBA) with full facepiece operated in pressure demand. If a full facepiece is not available, then chemical goggles should be worn to protect the eyes. Whenever a chemical splash hazard exists, a face shield and a protective apron should be worn. To prevent hand and skin exposures, impervious gloves should be used. Glove manufacturers should be contacted and permeation studies obtained *before* final glove selections have been made.

Administrative controls should also be in place to minimize the potential for human exposures. These may include written procedures or policies which specify the methods and techniques that will be practiced whenever personnel are to work with or around isopropyl glycidyl ether.

All personnel should receive *training* on the use, hazards, protective measures, emergency actions, and other precautions per 29 CFR 1910.1200 (Hazard Communication), prior to the first assignment in an area where isopropyl glycidyl ether is used or stored.

If symptoms develop or overexposure is suspected, the following tests are recommended:

- ☑ Kidney function tests.
- ☑ Urinalysis.
- ☑ Evaluation by a qualified allergist with careful consideration of exposure history and special testing (may help diagnose skin allergy).

Any evaluation should include a careful history of past and present symptoms with an examination. Medical tests that simply look for existing damage are not a substitute for controlling exposures.

Other methods to reduce exposure include:

- ☑ Where possible, enclose operations and use local exhaust ventilation at the site of chemical release. If local exhaust ventilation or enclosure is not used, respiratory protection should be mandatory.
- ☑ Where possible, automatically transfer liquid isopropyl glycidyl ether from storage drums to process containers. Electrically ground and bond all containers and equipment, especially during transfer operations.
- ☑ Always ensure that proper protective clothing is worn when using chemical substances and that personnel are trained in its use and care.
- ☑ Wash thoroughly immediately after exposure to isopropyl glycidyl ether and at the end of the work shift or before eating, drinking, or smoking.
- ☑ Hazard warning information should be posted in the work area. In addition, as part of an on-going education and training program, all information on the health and safety hazards associated with isopropyl glycidyl ether should be communicated to all potentially exposed workers.
- ☑ Eye wash stations should be provided in the immediate work area for emergency use. If there is a possibility of skin exposure to this chemical, then emergency shower facilities should also be provided in the immediate work area.
- ☑ Personnel should never wear contaminated clothing home (family members can be exposed). Clothing should be laundered only by personnel who have been properly trained on hazards associated with exposure to isopropyl glycidyl ether.

RISK ASSESSMENT: ENVIRONMENT

General Assessment

The environment is at risk of exposure during transportation, storage, disposal, or destruction of isopropyl glycidyl ether. In almost every scenario, the threat of environmental exposure is contingent upon the proper handling of the chemical substance. Accidental spills, large or small, can result in fire, explosion, and possible contamination of the surrounding environmental mediums (water, soil, and air).

Isopropyl glycidyl ether is considered a Class IC flammable liquid (per OSHA 29 CFR 1910.106). Its relatively low flash point and moderate boiling point present a serious fire and explosion hazard, especially in confined spaces. Its vapors are heavier than air and can travel for great distances to an ignition source to flashback and cause fire or explosion. It can form explosive peroxides on exposure to air or light. These characteristics require extreme caution in handling, storage, transportation, and disposal. It is incompatible with many common peroxides and other oxidizers and contact can cause fire or explosion. Therefore, special consideration is required during any emergency situation involving a leak or spill of isopropyl glycidyl ether. Should isopropyl glycidyl ether ever come into contact with any incompatible substances during use, transportation, storage, or disposal, the formation of highly toxic and/or highly explosive commodities is extremely possible.

The proper method for disposing of waste isopropyl glycidyl ether is to burn it in a chemical incinerator equipped with an afterburner and scrubber.

Isopropyl glycidyl ether may enter the environment through industrial or municipal waste treatment plant discharges, agricultural runoff or disposal, or spills.

Acute Ecological Effects

Acute (short-term) toxic effects may include the death of animals, birds, or fish and death or low growth rate in plants. Acute effects are seen 2 to 4 days after animals or plants are exposed to isopropyl glycidyl ether.

Insufficient data are available to evaluate or predict the short-term effects of isopropyl glycidyl ether exposure to aquatic life, plants, birds, or land animals.

Chronic Ecological Effects

Chronic (long-term) toxic effects may include shortened life span, reproductive problems, lower fertility, and changes in appearance or behavior in exposed animals. These effects can be seen long after (sometimes months or even years) first exposure(s) to toxic chemicals.

Insufficient data are available to evaluate or predict the long-term effects of isopropyl glycidyl ether exposure on aquatic life, birds, plants, or land animals.

Water Solubility

Isopropyl glycidyl ether is moderately soluble in water. Concentrations up to 1000 milligrams will mix with a liter of water.

Persistence in the Environment

Isopropyl glycidyl ether is slightly persistent in water, with a half-life between 2 and 20 days. The half-life of a pollutant is the amount of time it takes for one half of the chemical to be degraded. There is insufficient data to evaluate the distribution of isopropyl glycidyl ether in the various environmental mediums (air, water, soil).

Bioaccumulation in Aquatic Organisms

Some substances increase in concentration, or *bioaccumulate,* in living organisms as they breathe contaminated air, drink contaminated water, or eat contaminated food. These chemicals can become concentrated in the tissues and internal organs of animals as well as humans.

No data are available on the bioaccumulation properties of isopropyl glycidyl ether.

Recommended Risk-Reduction Measures

Proper training of all transporters will reduce the likelihood of a mishap or accident resulting in a leak or spill to the environment. The correct use of DOT labeling (per the MSDS) on all containers, trucks, and rail cars will enable emergency responders to react properly and quickly to any disaster thereby reducing the potential risk to the environment and to personnel.

Storage of isopropyl glycidyl ether should be segregated from incompatible chemicals to minimize the risk of cross-contamination or contact. Buildings designated for storage should be equipped with appropriate fire protection systems (alarms, sprinklers, emergency lighting, portable extinguishers). Containers and equipment should be grounded and bonded, especially during transfer operations. Inventories should be monitored to ensure supplies are not kept in storage for extended periods (e.g., more than six months).

If a spill or leak to the environment has occurred, fire department, emergency response and/or hazardous materials spill personnel should be notified immediately. Cleanup should be attempted only by those trained in proper spill containment procedures using non-sparking tools. Contaminated soils should be removed for incineration and replaced with clean soil. If isopropyl glycidyl ether should contact the water table, aquifer, or navigable waterway, cleanup should be started immediately. It is highly soluble in water and total remediation may not be possible. The local and/or state emergency response authorities must be notified. A comprehensive emergency response or disaster preparedness/recovery plan should be in place prior to any operations involving the use, transportation, storage, or disposal of isopropyl glycidyl ether. If this chemical is spilled or leaked, the following specific steps are recommended:

- ☑ Restrict persons not wearing protective clothing from area of spill or leak until cleanup is complete and area can be opened for normal work.

- ☑ Remove all ignition sources.
- ☑ Ventilate area of spill or leak.
- ☑ Absorb liquids in vermiculite, dry sand, earth, or similar material and deposit in sealed containers using non-sparking tools.
- ☑ It may be necessary to dispose of this compound as a hazardous waste. The responsible state agency or the regional office of the federal Environmental Protection Agency (EPA) should be contacted for specific recommendations.

RISK ASSESSMENT: BUSINESS

General Assessment

Accidents or mishaps involving isopropyl glycidyl ether can present a significant threat to business operations. The loss or damage of equipment or facilities can significantly affect fiscal viability. Lawsuits that may result from personnel injury/death, public exposures, and/or environmental contamination will also require a serious expenditure of resources. Media attention surrounding an injury, death, or environmental damage can also result in a loss of profits and loss of current as well as future business. Always remember that anytime the term "reproductive hazard" is used, public hysteria, ignorance, and emotion can run equally high. This must be carefully considered whenever developing or implementing public relations policies.

Recommended Risk-Reduction Measures

Company attorneys, safety and health professionals, and environmental specialists should be involved in the development of any procedures dealing with safety or emergency response. A company official should be pre-designated as a public relations officer with specific training in dealing with the press. Corporate plans and policies should be developed, approved, and implemented long before any need for such arises.

REFERENCES

American Conference of Governmental Industrial Hygienists. 1988. *Documentation of the Threshold Limit Values and Biological Exposure Indices,* 5th Edition (with updates). Cincinnati: ACGIH

Bueche, J. R. 1972. *Principles of Physics,* 2nd Edition. New York: McGraw-Hill.

Calabrese, E. J. 1994. *Biological Effects of Low Level Exposures.* New York: CRC Press/Lewis Publishers.

Cockerman, L. G., B. S. Shane. 1994. *Basic Environmental Toxicology.* New York: CRC Press/Lewis Publishers.

Danse, I. R. 1991. *Common Sense Toxics in the Workplace.* New York: Van Nostrand Reinhold.

Environmental Protection Agency. 1983. *Chemical Hazard Information Profiles.* Washington, D.C.: U.S. EPA.

Fodden, C. R., and J. L. Weddell. 1992. *Hazardous Materials Emergency Action Data.* New York: CRC Press/Lewis Publishers.

Forseberg, K., and L. H. Keith. 1993. *Chemical; Protective Clothing Permeation and Degradation Database.* Chelsea, MI: Lewis Publishers.

Gardner, R. W. 1994. *Chemical Intolerance.* New York: CRC Press/Lewis Publishers.

Hallenbeck, W. H. 1993. *Quantitative Risk Assessment for Environmental and Occupational Health,* 2nd Edition. New York: CRC Press/Lewis Publishers.

Keith, L. H. and M. M. Walker. 1995. *Handbook of Air Toxics.* New York: CRC Press/Lewis Publishers.

Landis, W. G., M. H. Yu. 1995. *Introduction to Environmental Toxicology.* New York: CRC Press/Lewis Publishers.

Langerman, N. 1994 *Precautionary Labels for Chemical Containers.* New York: CRC Press/Lewis Publishers.

Lauwerys, R. R., P. Hoet. 1993. *Industrial Chemical Exposure.* New York: CRC Press/Lewis Publishers.

Lewis, .J., Sr. 1993. *Hawley's Condensed Chemical Dictionary,* 12th Edition. New York: Van Nostrand Reinhold.

Lewis, . J., Sr. 1992. *Sax's Dangerous Properties of Industrial Materials,* Volumes 1, 2, and 3, 8th Edition. New York: Van Nostrand Reinhold.

Manahan, S. E. 1994. *Environmental Chemistry,* 6th Edition. New York: CRC Press/Lewis Publishers.

Meyer, E. 1990. *Chemistry of Hazardous Materials.* Englewood Cliffs, NJ: Prentice Hall.

Patnaik, P. 1992. *A Comprehensive Guide to the Hazardous Properties of Chemical Substances.* New York: Van Nostrand Reinhold.

Rea, W. J. 1995. *Chemical Sensitivity,* Vol. I-III. New York: CRC Press/Lewis Publishers.

Richardson, M. 1995. *Dictionary of Substance and Their Effects,* Vol. I-VI. New York: CRC Press/Lewis Publishers.

Sacarello, H. L. A. 1994. *The Comprehensive Handbook of Hazardous Materials.* New York: CRC Press/Lewis Publishers.

Torkelson, T. R. V. K. Rowe. 1981. Halogenated aliphatic hydrocarbons. *Patty's Industrial Hygiene and Toxicology,* Vol. 2B, *Toxicology,* pp 3568-3572. New York: Wiley-Interscience.

U.S. Department of Health and Human Services, National Institute for Occupational Safety and Health. 1994. *NIOSH Pocket Guide to Chemical Hazards.* Washington, D.C.: U.S. Government Printing Office.

MATERIAL SAFETY DATA SHEET

CHEMICAL NAME	
JET FUEL **Jet A and Jet A-1**	**CRC PRESS/LEWIS PUBLISHERS DISCLAIMER:** THE INFORMATION AND RECOMMENDATIONS PRESENTED HEREIN ARE BASED ON SOURCES BELIEVED TO BE RELIABLE. CRC MAKES NO REPRESENTATION ON ITS COMPLETENESS OR ACCURACY. IT IS THE RESPONSIBILITY OF THE USER TO DETERMINE THE CHEMICAL'S SUITABILITY FOR ITS INTENDED USE, THE CHEMICAL'S SAFE USE, AND THE CHEMICAL'S PROPER DISPOSAL. NO REPRESENTATIONS AND/OR WARRANTIES, EITHER EXPRESSED OR IMPLIED, OF THE MERCHANTABILITY OR FITNESS FOR A PARTICULAR PURPOSE, OR OF ANY OTHER NATURE, ARE MADE WITH RESPECT TO THE INFORMATION PROVIDED IN THIS MSDS OR TO THE CHEMICAL TO WHICH INFORMATION MAY REFER. CRC NEITHER ASSUMES NOR AUTHORIZES ANY OTHER PERSON TO ASSUME FOR IT, ANY OTHER ADDITIONAL RESPONSIBILITY OR LIABILITY FOR THE USE OF, OR RELIANCE UPON, THIS INFORMATION.

HAZARD WARNING INFORMATION

HEALTH	FIRE	REACTIVE	OTHER	DEGREE OF HAZARD	COLOR CODING	OTHER CODES
0	2	0		0 = Minimum Hazard 1 = Slight Hazard 2 = Moderate Hazard 3 = Serious Hazard 4 = Severe Hazard	HEALTH = BLUE FIRE = RED REACTIVITY = YELLOW OTHER = WHITE	OX = Oxidizer ACID = Acid ALK = Alkali COR = Corrosive ~~W~~ = Use No Water

SECTION I - GENERAL INFORMATION

Characterization	**Hydrocarbon (fuel, solvent)**	RCRA Number	**None**
		EPA Class	**Not Applicable**
DOT Proper Shipping Name	**No Citation**	Chemical Abstract Service (CAS) Number	**8008-20-6**
DOT Hazard Class and Label Requirements	**Combustible Liquid**	DOT Emergency Guide Code	**27 (as kerosene)**
DOT Identification Number	**UN 1223 (as kerosene)**	Molecular Formula	**Varies with grade.**

Synonyms
Kerosine; kerosene fuel.

SECTION II - HAZARDOUS INGREDIENTS AND IDENTITY INFORMATION

Hazardous Components (specific identity)	OSHA Exposure Criteria	NIOSH Exposure Criteria	Immediately Dangerous to Life and Health (IDLH)	ACGIH Exposure Criteria
Jet fuel (derivation: A petroleum distillate similar to kerosene composed of 5 to 16 carbons, aliphatics, monocycloparaffins, aromatics, and olefins). 1	**PEL:** Not Established **STEL:** Not Established	**REL (as kerosene):** 14 ppm 100 mg/m^3 **STEL:** Not Established	**Not Determined**	**TLV:** Not Established **STEL:** Not Established

SECTION III - PHYSICAL AND CHEMICAL CHARACTERISTICS

Boiling Point	**300 to 550°F (149 to 228°C)**	Specific Gravity (H_2O = 1)	**0.80**
Vapor Pressure (mm Hg)	**0.1 at 68°F (20°C)**	Molecular Weight (atomic weight)	**170 (approx.)**
Vapor Density (Air = 1)	**1**	Freezing Point	**-50°F (-45.5°C) (as kerosene)**

Solubility
Negligible solubility in water. Miscible with other petroleum solvents.

Appearance and Odor
Colorless or clear, oily liquid with a strong, characteristic, hydrocarbon odor.

SECTION IV - FIRE AND EXPLOSION HAZARD DATA

Flash Point (method used)	**100 to 150°F (43 to 66°C) opened cup**	Explosive Limits in Air % by Volume	**LEL: 0.6% UEL: 3.7%**
NFPA Classification	**Class II Combustible Liquid**	Autoignition Temperature	**446°F (230°C)**

Extinguishing Media
Use regular foam, dry chemical, carbon dioxide, or water spray on small fires. For larger fires, use water spray, fog, or regular foam. Use a blanketing or smothering technique with foam for best results.

Special Fire Fighting Procedures
Isolate fire area. Wear self-contained breathing apparatus (SCBA). Poisonous gases are produced in fire. Use water spray to cool fire-exposed containers until long after fire is out. Use non-sparking tools. Move containers from fire area if it can be done without risk. Fight fire from distance, if possible.

Unusual Fire and Explosion Hazards
Dangerous fire and explosion hazard when exposed to heat or flame. Containers may explode in fire. The vapor is heavier than air and may travel for some distance to cause fire in flashback to source.

SECTION V - REACTIVITY DATA

Stability		Conditions to Avoid **Jet fuel is normally stable under routine conditions of handling and storage. Avoid contact with heat, flame, other sources of ignition, and contact with incompatible materials and chemicals.**
Stable X	Unstable	Incompatibility *(materials to avoid)* **Incompatible with strong oxidizers (such as peroxide), oxidizing materials, halogens, strong acids, and alkalines. Violent reaction with fluorine.**
Hazardous Polymerization		Conditions to Avoid **Hazardous polymerization has not been reported to occur under normal conditions of temperature and pressure.**
May Occur	Will Not Occur X	Hazardous Decomposition or By-products **When heated to decomposition, jet fuel emits toxic and acrid fumes, such as oxides of carbon, partially oxidized hydrocarbons, and trace quantities of sulfur dioxide.**

SECTION VI - HEALTH HAZARD DATA

Primary Route(s) of Entry:	Inhalation?	Absorption (skin/eye)?	Ingestion?
	X		X

Health Hazards

INHALATION: **Symptoms include irritation of the nose, throat, and mucosa of the respiratory tract with headache and drowsiness. High concentrations can cause feeling of suffocation, coma, pulmonary edema, and death (respiratory arrest). Chronic exposures can lead to serious central nervous system effects and neurological impairment.**

SKIN & EYES: **Irritation with drying, cracking, redness, blistering (prolonged contact). May lead to contact dermatitis and secondary infection.**

INGESTION: **May cause nausea, vomiting, headache and gastrointestinal irritation. Aspiration of vomit into the lungs can cause serious pneumonitis and even pulmonary hemorrhage.**

Carcinogenicity	NTP Listed?	IARC Cancer Review Group?	OSHA Regulated?	Target Organs?
Unknown Human **Unknown Animal**	**No**	**No**	**No**	**Eyes, skin, respiratory sys., liver, CNS, kidneys.**

Medical Conditions Generally Aggravated by Exposure

Chronic pulmonary disease may be aggravated by exposure.

Emergency and First-aid Procedures

Eye contact: Do not allow victim to keep eyes tightly shut. Flush immediately with water for 15 minutes (minimum); seek medical attention. Skin contact: Remove all contaminated clothing, including shoes. Immediately wash area with flooding amounts of soap and water. Seek medical attention if blisters develop. For inhalation: Remove the person from exposure. Provide respiratory assistance and CPR. Transfer to medical facility. Observe for 48 hours for lung effects. If swallowed: Seek medical attention immediately. Give conscious person milk or water to drink. Do NOT induce vomiting.

SECTION VII - PRECAUTIONS FOR SAFE HANDLING AND USE

Steps to be Taken in Case Material is Released or Spilled

Remove all ignition sources. Wear SCBA respiratory protection. Avoid skin contact. Stop leak if it can be done without risk. Provide maximum explosion-proof ventilation. Absorb with vermiculite or other similar material and deposit in sealed containers. Use non-sparking tools during cleanup.

Preferred Waste Disposal Method

No citation.

Precautions to be Taken in Handling and Storage

Store in tightly closed containers in cool, well-ventilated location. Keep away from sources of heat.

Other Precautions and Warnings

Sources of ignition such (open flame or smoking), are prohibited where jet fuel is used or stored. Metal containers should be bonded and grounded during transfer operations. Bulk storage not recommended.

SECTION VIII - CONTROL MEASURES AND PERSONAL PROTECTIVE EQUIPMENT

Respiratory Protection *(specify type)*

Use an MSHA/NIOSH-approved air-purifying organic vapor respirator, a powered air-purifying respirator, or a self-contained breathing apparatus (SCBA) with full facepiece operated in positive pressure mode.

Ventilation

Local exhaust (preferred) at site of chemical work or general ventilation.

Protective Gloves	Eye Protection	Other Protective Clothing
Impervious Material	**Chemical Goggles or Face Mask**	**Protective Apron if Splash is Likely**

Work/Hygiene Practices

Always wash hands thoroughly after using chemical; never bring food, drink, or smoking materials into vicinity of chemicals. Barrier creams may also be useful in protecting exposed skin surfaces.

JET FUEL

Formula varies with grade (may be C_5 through C_{16}) CAS: 8008-20-6

IDENTIFICATION AND TYPICAL USES

Jet fuel is a clear or colorless, oily liquid with a strong, characteristic, petroleum odor. It is used primarily as a fuel for rocket engines and high performance jet aircraft.

RISK ASSESSMENT: HEALTH

General Assessment

Jet fuel is considered moderately toxic by *inhalation* of its vapors and by *ingestion* of the liquid. Skin contact will also cause localized damage at site of contact, but dermal absorption through intact skin has not been reported. Jet fuel is not reported to be a carcinogenic, mutagenic, or teratogenic compound. However, kerosene, a close derivative, is suspected to cause cancer in animals, with mutation data also being reported.

Inhalation of the jet fuel vapors can cause mild to severe irritation of the eyes, nose, throat, and respiratory tract mucosa. Systemic effects include headache, drowsiness, burning sensation in the chest, nausea, weakness, restlessness, incoordination, and other effects of central nervous system depression such as staggering gait, slurred speech, and mental confusion. High exposures can lead to a feeling of suffocation, coma, and death due to respiratory arrest. Pulmonary edema (fluid in the lungs) can develop approximately 24 to 48 hours after exposure. This is a medical emergency and can also be fatal. There may be coughing, bloody sputum, chest pain, and congestion.

Skin contact may cause irritation producing erythema (redness), burning, and, in severe cases, edema and even blistering. Contact dermatitis is possible, as is secondary infection. Eye contact can cause irritation of the conjunctiva with redness, swelling, tearing, and localized inflammation of eye tissues.

Ingestion may cause nausea, vomiting, headaches, dizziness, gastrointestinal irritation, somnolence, hallucinations and distorted perceptions, coughing, drunkenness, fever, confusion, and diarrhea. If vomiting occurs, aspiration of the liquid into the lungs can cause choking, difficulty breathing or shortness of breath (dypsnea), tachypnea (rapid respiration), tachycardia (rapid heartbeat), fever, pneumonitis, and even pulmonary hemorrhage (bleeding in the lungs).

Acute Health Effects

The following acute (short-term) health effects may occur immediately or shortly after exposure to jet fuel:

Skin: Irritation with potential for redness, blistering, and burning.

Eye: Severe irritation, conjunctivitis, redness, possible pain, and visual disturbances.

Lung: Nose, throat, and respiratory tract irritation causing difficulty in breathing, and possible pulmonary effects if liquid enters the lungs.

CNS: Incbriation, dizziness, stupor, hallucinations, slurred speech, mental confusion, loss of consciousness, and possible convulsions.

Other: Ingestion can cause serious nervous system effects and possible injury to the lungs should any vomitus be aspirated into the lungs. Other organs that may be damaged by exposure include the liver, kidney, pancreas, and spleen.

Chronic Health Effects

The following chronic (long-term) health effects may occur at some time after exposure to jet fuel and can last for months or even years:

Cancer Hazards: According to information presented in the references, jet fuel has not been shown to cause cancer in test animal studies. Kerosene, a close derivative, has been shown to cause cancer in test animals. Its ability to cause cancer in humans is considered to be at least equal to that of animals. It is not known if jet fuel will have this effect on humans.

Reproductive Hazard: There is no evidence in the references to indicate that exposure to jet fuel causes any adverse affects to reproduction.

Other Chronic Effects: Chronic inhalation of hydrocarbon vapors can lead to neurological impairment. Repeated skin contact causes blistering, drying, cracking, and possible secondary infections and contact dermatitis.

Recommended Risk-Reduction Measures

Personnel should avoid direct contact with or exposure to jet fuel. Occupational poisoning occurs most commonly by inhalation. Substitution may not always be a viable option with jet fuel since its use is often as a substitute for even more toxic or flammable compounds. However, a less toxic material or compound should be used in place of jet fuel whenever possible. If substitution is not possible, *engineering controls* are the most effective method of reducing exposures. The best protection is to enclose operations and/or provide local exhaust ventilation at the site of jet fuel release. While not always operationally feasible, isolating operations can also reduce exposure. Using respiratory protection is less effective than the controls mentioned above, but is still advisable whenever working with or around jet fuel. No exposure levels have been established for jet fuel. However, this does not mean that exposure is without risk. An MSHA/NIOSH-approved air-purifying respirator with an organic vapor cartridge, a supplied-air respirator operated in continuous flow mode, or a self-contained breathing apparatus (SCBA) with full facepiece operated in pressure demand are the appropriate methods of respiratory protection when exposure to jet fuel is anticipated. If a full facepiece is not available, then chemical goggles should be worn to protect the eyes. A face shield should also be considered. To prevent hand and skin exposures, protective gloves should be used. Glove manufacturers should be contacted and permeation studies obtained *before* final glove selections are made.

Administrative controls should also be in place to minimize the potential for human exposures. These may include written procedures or policies which specify the methods and techniques that will be practiced whenever personnel are to work with jet fuel.

All personnel should receive *training* on the use, hazards, protective measures, emergency actions, and other precautions per 29 CFR 1910.1200 (Hazard Communication), prior to the first assignment in an area where jet fuel is used or stored.

Before beginning employment and at regular times thereafter, the following recommended medical tests should be considered:

☑ Lung function tests (establish baseline).

If symptoms occur or overexposure is suspected, the following additional tests are suggested:

☑ Lung function tests (compare to baseline).

☑ Consider chest X-ray following acute overexposure (may be negative if taken immediately after exposure due to delayed onset of pulmonary edema).

☑ Liver and kidney function tests.

Any evaluation should include a careful history of past and present symptoms with an examination. Medical tests that simply look for existing damage are not a substitute for controlling exposures. Also, since smoking can cause heart disease, emphysema, lung cancer, and other respiratory disorders, smokers exposed to jet fuel may experience symptoms more rapidly and more pronounced than non-smokers under the same conditions of exposure. Prudent risk management requires careful consideration of *all* possible factors that may be causing the appearance of exposure symptoms in the workplace.

Other methods to reduce exposure include:

☑ Where possible, enclose operations and use local exhaust ventilation at the site of chemical release. If local exhaust ventilation or enclosure is not used, respiratory protection should be mandatory.

☑ Always ensure that proper protective clothing is worn when using chemical substances and that personnel are trained in its use, care, and maintenance.

☑ Wash thoroughly immediately after exposure to jet fuel and at the end of the work shift or before eating, drinking, or smoking.

☑ Hazard warning information should be posted in the work area. In addition, as part of an on-going education and training program, all information on the health and safety hazards of jet fuel should be communicated to all exposed workers.

☑ Eye wash stations should be provided in the immediate work area for emergency use. If there is a possibility of skin exposure to jet fuel, emergency shower facilities should also be provided.

☑ Workers whose clothing has been contaminated by jet fuel should change into clean clothes before leaving work. Contaminated work clothing should be laundered only by individuals who have been informed of the hazards of exposure to jet fuel.

☑ Where possible, automatically transfer liquids containing jet fuel from drums or other containers to process containers. Always ground and bond all containers and equipment. Use non-sparking tools.

RISK ASSESSMENT: ENVIRONMENT

General Assessment

The environment is at risk of exposure during transportation, storage, disposal, or destruction of jet fuel. In almost every scenario, the threat of environmental exposure is contingent upon the proper handling of the chemical substance. Accidental spills, large or small, can result in contamination of the surrounding environmental mediums (water, soil, and air).

Jet fuel is a Class II combustible liquid (per OSHA 29 CFR 1910.106). It can present a dangerous fire hazard and moderate explosion hazard when exposed to heat or flame. It is incompatible with commonly encountered oxidizers and caution is always required in handling, storage, transportation, and disposal of jet fuel. Emergency responders should be made aware of the presence of jet fuel at any emergency response situation.

Jet fuel can enter the environment from industrial effluents and from spills.

Acute Ecological Effects

Acute (short-term) toxic effects may include the death of animals, birds, or fish and death or low growth rate in plants. Acute effects are seen 2 to 4 days after animals or plants are exposed to jet fuel. Jet fuel has high acute toxicity to aquatic life. It can cause death in plants and roots and membrane damage in leaves of various agricultural crops. Insufficient data are available to evaluate the short-term effects of jet fuel on birds or land animals.

Chronic Ecological Effects

Chronic (long-term) toxic effects may include shortened life span, reproductive problems, lower fertility, and changes in appearance or behavior in exposed animals. These effects can be seen long after first exposure(s) to toxic chemicals. Jet fuel has high chronic toxicity to aquatic life. Insufficient data are available to evaluate or predict the long-term effects of jet fuel to plants, birds, or terrestrial animals.

Water Solubility

Jet fuel is not soluble in water. Concentrations of 1 milligram will not mix with a liter of water.

Persistence in the Environment

Jet fuel is slightly persistent in water, with a half-life of between 2 to 20 days. The half-life of a pollutant is the amount of time it takes for one half of the chemical to be degraded. About 99.5% of jet fuel will eventually end up in air; the rest will end up in water.

Bioaccumulation in Aquatic Organisms

Some substances increase in concentration, or *bioaccumulate,* in living organisms as they breathe contaminated air, drink contaminated water, or eat contaminated food. These chemicals can become concentrated in the tissues and internal organs of animals as well as humans.

The concentration of jet fuel found in fish tissues is expected to be somewhat higher than the average concentration of jet fuel in the water from which the fish was taken.

Recommended Risk-Reduction Measures

Proper training of all transporters will reduce the likelihood of a mishap or accident resulting in a leak or spill of jet fuel into the environment. Labels on all containers, trucks, and rail cars must meet DOT requirements and accurately reflect their contents to enable emergency responders to react properly and quickly to any disaster thereby reducing the potential risk to the environment and to personnel.

Storage of jet fuel should be segregated from other chemicals and materials to minimize the risk of cross-contamination. Facilities should be designed to meet explosion-proof requirements.

If a spill or leak to the environment has occurred, fire department, emergency response and/or hazardous materials spill personnel should be notified immediately. Cleanup should be attempted only by those trained in proper spill containment procedures using non-sparking tools. Contaminated soils should be removed for incineration and replaced with clean soil. If jet fuel should contact the water table, aquifer, or navigable waterway, time is of the essence. It is not soluble and total remediation may be possible. In general, jet fuel will float on the surface and may be captured if response actions are prompt. The local and/or state emergency response authorities must be notified. A comprehensive emergency response or disaster preparedness/recovery plan should be in place prior to any operations involving the use, transportation, storage, or disposal of jet fuel.

If jet fuel is spilled or leaked, the following specific steps are recommended:

- ☑ Restrict persons not wearing protective clothing from area of spill or leak until cleanup is complete and area can be opened for normal work.
- ☑ Absorb liquids in vermiculite, sand, earth, or similar material and deposit in sealed drum. Use non-sparking tools during cleanup procedures.
- ☑ Ventilate area of spill or leak. Container should be removed to safer area if it can be done without increasing the risk.
- ☑ Use water spray to knock down vapors.
- ☑ It may be necessary to dispose of jet fuel as a hazardous waste. The responsible state agency or the regional office of the federal Environmental Protection Agency (EPA) should be contacted for specific recommendations.

RISK ASSESSMENT: BUSINESS

General Assessment

Accidents or mishaps involving jet fuel can present a serious threat to business operations. The loss or damage of equipment or facilities can significantly affect fiscal viability. Lawsuits that may result from personnel illness, injury/death, public exposures, and/or environmental contamination will require a serious expenditure of resources. Media attention surrounding an injury, death, or environmental damage can also result in a loss of profits and loss of current as well as future business.

Recommended Risk-Reduction Measures

Company attorneys, safety and health professionals, and environmental specialists should be involved in the development of any procedures for responding to chemical incidents. A company official should be predesignated as a public relations officer with specific training in dealing with the press. Corporate plans and policies should be developed, approved, and implemented long before any need for such arises.

REFERENCES

American Conference of Governmental Industrial Hygienists. 1988. *Documentation of the Threshold Limit Values and Biological Exposure Indices,* 5th Edition (with updates). Cincinnati: ACGIH

Anthony, C. P., and N. J. Kolthoff. 1971. *Textbook of Anatomy and Physiology.* St. Louis: C. V. Mosby Company.

Bueche, J. R. 1972. *Principles of Physics,* 2nd Edition. New York: McGraw-Hill.

Burrell, R., D. K. Flaherty, and L. J. Sauers. 1992. *Toxicology of the Immune System.* New York: Van Nostrand Reinhold.

Calabrese, E. J., E. M. Kenyon. *Air Toxics and Risk Assessment.* New York: CRC Press/Lewis Publishers.

Cockerman, L. G., B. S. Shane. 1994. *Basic Environmental Toxicology.* New York: CRC Press/Lewis Publishers.

Cothern, C. R., N. P. Ross. 1994. *Environmental Statistics, Assessment, and Forecasting.* New York: CRC Press/Lewis Publishers.

Danse, I. R. 1991. *Common Sense Toxics in the Workplace.* New York: Van Nostrand Reinhold.

Environmental Protection Agency. 1983. *Chemical Hazard Information Profiles.* Washington, D.C.: U.S. EPA.

Fodden, C. R., and J. L. Weddell. 1992. *Hazardous Materials Emergency Action Data.* New York: CRC Press/Lewis Publishers.

Forseberg, K., and L. H. Keith. 1993. *Chemical; Protective Clothing Permeation and Degradation Database.* Chelsea, MI: Lewis Publishers.

Howard, P. H., et. al. 1991. *Handbook of Environmental Degradation Rates.* New York: CRC Press/Lewis Publishers.

International Agency for Research on Cancer (IARC). 1987. *IARC Monographs on the Evaluation of Carcinogenic Risk to Humans, Overall Evaluations of Carcinogenicity: An updating of IARC Monographs Volumes 1-42.* Lyon: IARC.

Kamrin, M. A. 1988. *Toxicology: A Primer on Principles and Applications.* Chelsea, Michigan: Lewis Publishers.

Keith, L. H. , M. T. Johnston, C. E. Hudak, and M. Conoley. 1992. *National Toxicology Program's Gloves+ Database.* Chelsea, MI: Lewis Publishers.

Lauwerys, R. R., P. Hoet. 1993. *Industrial Chemical Exposure.* New York: CRC Press/Lewis Publishers.

Lewis, R. J., Sr. 1993. *Hawley's Condensed Chemical Dictionary,* 12th Edition. New York: Van Nostrand Reinhold.

Lewis, R. J., Sr. 1992. *Sax's Dangerous Properties of Industrial Materials,* Volumes 1, 2, and 3, 8th Edition. New York: Van Nostrand Reinhold.

Meyer, E. 1990. *Chemistry of Hazardous Materials.* Englewood Cliffs, NJ: Prentice Hall.

Quigley, D. R. 1994. *Handbook of Emergency Chemical Management.* New York: CRC Press/Lewis Publishers.

Sacarello, H. L. A. 1994. *The Comprehensive Handbook of Hazardous Materials.* New York: CRC Press/Lewis Publishers.

Torkelson, T. R. V. K. Rowe. 1981. *Patty's Industrial Hygiene and Toxicology,* Vol. 2B. New York: Wiley-Interscience.

MATERIAL SAFETY DATA SHEET

CHEMICAL NAME	CRC PRESS/LEWIS PUBLISHERS DISCLAIMER
KEPONE	**CRC PRESS/LEWIS PUBLISHERS DISCLAIMER:** THE INFORMATION AND RECOMMENDATIONS PRESENTED HEREIN ARE BASED ON SOURCES BELIEVED TO BE RELIABLE. CRC MAKES NO REPRESENTATION ON ITS COMPLETENESS OR ACCURACY. IT IS THE RESPONSIBILITY OF THE USER TO DETERMINE THE CHEMICAL'S SUITABILITY FOR ITS INTENDED USE, THE CHEMICAL'S SAFE USE, AND THE CHEMICAL'S PROPER DISPOSAL. NO REPRESENTATIONS AND/OR WARRANTIES, EITHER EXPRESSED OR IMPLIED, OF THE MERCHANTABILITY OR FITNESS FOR A PARTICULAR PURPOSE, OR OF ANY OTHER NATURE, ARE MADE WITH RESPECT TO THE INFORMATION PROVIDED IN THIS MSDS OR TO THE CHEMICAL TO WHICH INFORMATION MAY REFER. CRC NEITHER ASSUMES NOR AUTHORIZES ANY OTHER PERSON TO ASSUME FOR IT, ANY OTHER ADDITIONAL RESPONSIBILITY OR LIABILITY FOR THE USE OF, OR RELIANCE UPON, THIS INFORMATION.

HAZARD WARNING INFORMATION

HEALTH	FIRE	REACTIVE	OTHER	DEGREE OF HAZARD	COLOR CODING	OTHER CODES
N/A	N/A	N/A		0 = Minimum Hazard 1 = Slight Hazard 2 = Moderate Hazard 3 = Serious Hazard 4 = Severe Hazard	HEALTH = BLUE FIRE = RED REACTIVITY = YELLOW OTHER = WHITE	OX = Oxidizer ACID = Acid ALK = Alkali COR = Corrosive W = Use No Water

SECTION I - GENERAL INFORMATION

Characterization	RCRA Number	EPA Class
Fungicide; Insecticide	**U142**	**Toxic Waste**

DOT Proper Shipping Name	**No Citation**	Chemical Abstract Service (CAS) Number	**143-50-0**
DOT Hazard Class and Label Requirements	**No Citation**	DOT Emergency Guide Code	**No Citation**
DOT Identification Number	**No Citation**	Chemical Formula	**$C_{10}Cl_{10}O$**

Synonyms

Chlordecone; GC-1189; ENT 16391; Merex; decachlorooctahydro-kepone-2-one; decachlorooctahydro-1,3,4-metheno-2H-cyclobuta(cd)-pentalen-2-one.

SECTION II - HAZARDOUS INGREDIENTS AND IDENTITY INFORMATION

Hazardous Components (specific identity)	OSHA Exposure Criteria	NIOSH Exposure Criteria	Immediately Dangerous to Life and Health (IDLH)	ACGIH Exposure Criteria
Kepone	**PEL:** Not Established **STEL:** Not Established	**REL (ceiling):** 0.001 mg/m^3 (15 minute) **STEL:** Not Applicable	**Not Determined**	**TLV:** Not Established **STEL:** Not Established

SECTION III - PHYSICAL AND CHEMICAL CHARACTERISTICS

Boiling Point	**Sublimes**	Specific Gravity (H_2O = 1)	**Not Found**
Vapor Pressure (mm Hg)	**<3 x 10^{-7} at 77°F (25°C)**	Molecular Weight	**490.6**
Vapor Density (Air = 1)	**Not Found**	Melting Point	**662°F (350°C)**

Solubility

Slightly soluble in water (0.5% at 212°F). Soluble in acetic acid, alcohols, ketones, and acetone.

Appearance and Odor

Tan to white crystalline, odorless solid.

SECTION IV - FIRE AND EXPLOSION HAZARD DATA

Flash Point (method used)	**Not Determined**	Explosive Limits in Air % by Volume	**LEL: Not Determined UEL: Not Determined**
NFPA Classification	**Non-Combustible Solid**	Autoignition Temperature	**Not Determined**

Extinguishing Media

Use agent suitable to surrounding area. Kepone does not burn.

Special Fire Fighting Procedures

Poisonous gases may be produced in fire; wear full protective clothing and self-contained breathing apparatus (SCBA). Move containers from fire area if it can be done safely. Use water spray to cool fire-exposed containers. Do not release runoff from fire control measures to sewers or waterways; dike far ahead and contain for proper disposal.

Unusual Fire and Explosion Hazards

None reported.

SECTION V - REACTIVITY DATA

Stability		Conditions to Avoid
		Normally stable in closed containers under routine conditions of storage and handling. It readily hydrates on exposure to room temperature and humidity. Avoid contact with incompatible materials.
Stable	Unstable	Incompatibility *(materials to avoid)*
X		Incompatible with acids (nitric, sulfuric, hydrofluoric, chlorosulfonic) and acid fumes. Contact with either will produce highly toxic chloride fumes.
Hazardous Polymerization		Conditions to Avoid Under normal conditions of temperature and pressure, hazardous polymerization of kepone is not expected to occur.
May Occur	Will Not Occur	Hazardous Decomposition or By-products
	X	Toxic and irritating gases, smoke, and fumes are produced when kepone is involved in fire.

SECTION VI - HEALTH HAZARD DATA

Primary Route(s) of Entry:	Inhalation?	Absorption (skin)?	Ingestion?
	X	X	X

Health Hazards

INHALATION: Causes central nervous system, liver, and kidney damage with symptoms of headache, tremors, ataxia, skin changes, hyperexcitability, hyperactivity, muscle spasms, visual disturbances, chest pain, testicular atrophy, low sperm count, estrogenic effects, breast enlargements, sterility, liver lesions, and cancer.

ABSORPTION: Headache, dizziness, nausea, vomiting, tremor, ataxia, convulsions, CNS depression, respiratory failure, and may cause cardiac irregularities. May cause skin erythema.

INGESTION: May cause, nausea, vomiting, muscle twitching, coma, and similar effects of inhalation including liver and CNS effects and liver damage.

Carcinogenicity	NTP Listed?	IARC Cancer Review Group?	OSHA Regulated?	Target Organs?
Suspected Human Confirmed Animal	5th Annual Report	Group 2B	No	Eyes, skin, resp. sys., CNS, liver, kidney. repro.

Medical Conditions Generally Aggravated by Exposure

None Reported.

Emergency and First-aid Procedures

<u>Eye contact</u>: Flush immediately with water for 15 minutes (minimum); seek medical attention. <u>Skin contact</u>: Remove all contaminated clothing. Immediately wash area with large amounts of soap and water. Seek medical assistance. For <u>inhalation</u>: Remove the person from exposure. Provide respiratory assistance and CPR. Transfer to medical facility. If <u>swallowed</u>: If victim is conscious, give large amounts of milk or water and induce vomiting. Never attempt to give an unconscious or convulsing person anything by mouth. DOCTOR: Suggest gastric lavage and saline cathartic.

SECTION VII - PRECAUTIONS FOR SAFE HANDLING AND USE

Steps to be Taken in Case Material is Released or Spilled

Collect solid materials using a vacuum equipped with a HEPA filter. Damp mop residue. Do NOT dry sweep (generates airborne dusts). Ventilate area. Restrict those not involved in cleanup from entering area. Notify appropriate authorities, if applicable.

Preferred Waste Disposal Method

Mix with a combustible solvent and burn in a chemical incinerator with an afterburner and scrubber.

Precautions to be Taken in Handling and Storage

Do not store kepone in the presence of incompatible chemicals or materials. Store in tightly closed containers in a cool, dark, well-ventilated area. Personnel working with kepone should be trained on its proper handing and storage *prior* to being assigned to such responsibilities.

Other Precautions and Warnings

Manufacture and use are prohibited in the United States. Bulk storage is not recommended.

SECTION VIII - CONTROL MEASURES AND PERSONAL PROTECTIVE EQUIPMENT

Respiratory Protection *(specify type)*

Exposure levels are low and compound is carcinogenic. At any exposure level use a supplied-air respirator with full facepiece operated in pressure demand mode or a self-contained breathing apparatus (SCBA) with full facepiece operated in pressure demand or other positive pressure mode.

Ventilation

Local exhaust or general mechanical systems recommended.

Protective Gloves	Eye Protection	Other Protective Clothing
Protective Material	Chemical/Dust Goggles or Face Mask	Protective Apron

Work/Hygiene Practices

Always wash hands thoroughly after using chemical; never bring food, drink, or smoking materials into vicinity of chemicals.

KEPONE

$C_{10}Cl_{10}O$

CAS: 143-50-0

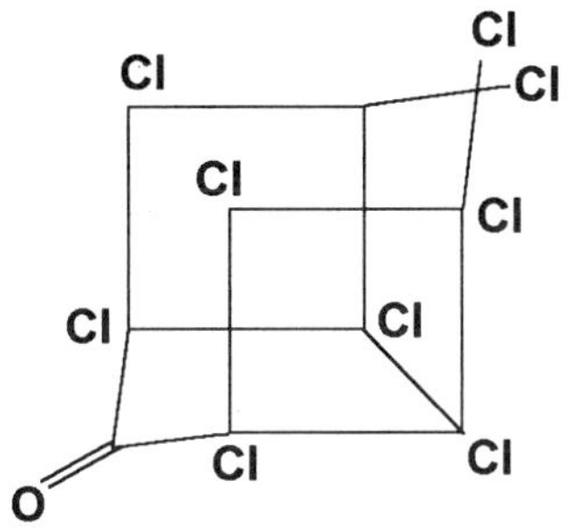

IDENTIFICATION AND TYPICAL USES

Kepone is a tan to white, crystalline, odorless solid. It was used as an insecticide and fungicide. However, its manufacture and use is now prohibited and its registration by the U.S. Environmental Protection Agency has been suspended.

RISK ASSESSMENT: HEALTH

General Assessment

Kepone is a deadly human poison by all routes of exposure (*ingestion, inhalation,* and dermal *absorption*). It is a known carcinogen in animals and is suspected to have the same effect on humans. Experimental teratogenic data have been reported along with adverse effects on the reproductive system. Mutagenic data have also been confirmed. Its primary systemic effect is damage or injury to the central nervous system and/or the liver and kidneys.

Exposure by all routes can cause depression of the central nervous system (CNS) with symptoms of headache, vertigo, blurred vision, lachrymation, salivation, sweating, muscular weakness and ataxia with spasms, dypsnea, diarrhea, abdominal cramps, vomiting, tremors, skin changes, hyperexcitability, hyperactivity, anxiety, irritability, memory disturbances, stuttering and slurred speech, abnormal gait, testicular atrophy, low sperm count, estrogenic effects, sterility, breast enlargement, fetal toxicity, anemia, liver lesions with hepatocellular carcinomas and hyperplasia, brain effects, and cancer.

Skin or eye contact results in severe irritation and discomfort. Kepone may be absorbed through unbroken skin to cause toxic systemic effects up to and including liver and CNS damage.

Acute Health Effects

The following acute (short-term) health effects may occur immediately or shortly after exposure to kepone:

Skin: Irritation and possible allergy. Most effect from skin contact is internal due to absorption of kepone. Symptoms include vomiting, nausea, malaise, and muscular twitching.

Eye: Possible irritation and damage if not removed immediately.

Lung: Irritation causing coughing and/or shortness of breath. Higher exposures can lead to respiratory problems.

CNS: Dizziness, headaches, nausea, vomiting, muscle jerks, convulsions, tremors, ataxia, memory disturbances, irritability, excitement, anxiety, and speech problems.

Other: Sterility, fetal toxicity, estrogenic effects, testicular atrophy, and other reproductive effects.

Chronic Health Effects

The following chronic (long-term) health effects may occur at some time after exposure to kepone and can last for months or even years:

Cancer Hazards: According to the information presented in the references, kepone has been shown to cause cancer in test animals. Its ability to cause cancer in humans is considered to be at least suspect. It has also been shown to cause mutations (genetic changes).

Reproductive Hazard: Kepone is an experimental teratogen in animals causing fetal changes as well as a reproductive hazard causing testicular atrophy, estrogenic changes, breast enlargement, and sterility.

Other Chronic Effects: Repeated or long-term exposures can lead to liver, kidney, brain, and central nervous system damage.

Ʊ *Recommended Risk-Reduction Measures*

Kepone is extremely toxic to humans and animals such that its use and manufacture is no longer permitted. However, some stock may exist especially in the agricultural sector and exposures are still possible. Using a less toxic compound in its place is the best way to prevent exposures. But when a less toxic chemical is not available or identified for substitution, *engineering controls* become the most effective method of reducing exposures. The best protection is to enclose operations and/or provide local exhaust ventilation at the site of chemical release. This may not be feasible during use as an insecticide outdoors, but it may be possible during handling and storage of kepone. While not always operationally feasible, isolating operations can also reduce exposure.

Using respiratory protection is less effective than the controls mentioned above, but is still advisable whenever working with or around kepone. The NIOSH REL is extremely low (0.001 mg/m^3) and difficult to accurately monitor. For any exposure to a known carcinogen, a supplied-air respirator or a self-contained breathing apparatus (SCBA) with full facepiece operated in pressure demand are the recommended respiratory protection methods of choice. If a full facepiece is not available, then chemical goggles should be worn to protect the eyes. Whenever a chemical splash hazard exists a face shield and protective apron should be worn. To prevent hand and skin exposures, rubber gloves should be used.

Administrative controls should also be in place to minimize the potential for human exposures. These may include written procedures or policies which specify the methods and techniques that will be practiced whenever personnel are to work with kepone.

All personnel should receive *training* on the use, hazards, protective measures, emergency actions, and other precautions per 29 CFR 1910.1200 (Hazard Communication), prior to the first assignment in an area where kepone is used or stored.

If symptoms develop or overexposure is suspected, the following medical tests may be useful:

- ☑ Liver and kidney function tests.
- ☑ Evaluation by a qualified neurologist with careful evaluation of exposure history and administration of special tests.

Any evaluation should include a careful medical history of past and present symptoms with an examination. However, medical tests that evaluate existing damage are not a substitute for controlling exposure.

Other methods to reduce exposure include:

- ☑ Where possible, enclose operations and use local exhaust ventilation at the site of chemical release. If local exhaust ventilation or enclosure is not used, respiratory protection should be mandatory.
- ☑ Always ensure that proper protective clothing is worn when using chemical substances.
- ☑ Wash thoroughly immediately after exposure to kepone and at the end of the work shift or before eating, drinking, or smoking.
- ☑ Hazard warning information should be posted in the work area. In addition, as part of an on-going education and training program, all information on the health and safety hazards of kepone should be communicated to all exposed or potentially exposed workers.
- ☑ Eye wash stations should be provided in work areas. If the potential for whole body exposure exists, then safety showers should also be provided.

RISK ASSESSMENT: ENVIRONMENT

General Assessment

The environment is at risk of exposure during intended use, transportation, storage, disposal, or destruction of kepone. In almost every scenario, the threat of environmental exposure is contingent upon the proper handling of the chemical substance. Accidental spills, large or small, can result in fire, explosion, and possible contamination of the surrounding environmental mediums (water, soil, and air).

Kepone is considered a non-combustible solid. However, it is reactive in contact with acid or acid fumes and will produce toxic chloride fumes. It may also form poisonous phosgene gas on exposure to heat. These characteristics require special consideration during any emergency situation involving a leak or spill of kepone or kepone mixtures. Should kepone ever come into contact with incompatible substances either during use, transportation, or storage, violent reaction can occur.

Kepone can enter the environment through its normal, intended use as an insecticide and fungicide, and through industrial discharges, urban runoff, and spills.

Acute Ecological Effects

Acute (short-term) toxic effects may include the death of animals, birds, or fish and death or low growth rate in plants. Acute effects are seen 2 to 4 days after animals or plants are exposed to kepone.

Kepone has high acute toxicity to aquatic life, birds, and land animals. Insufficient data are available to evaluate or predict the short-term effects of kepone to plant life.

Chronic Ecological Effects

Chronic toxic effects may include shortened life span, reproductive problems, lower fertility, and changes in appearance or behavior in exposed animals. These effects can be seen long after first exposure(s) to toxic chemicals.

Kepone has high chronic toxicity to aquatic life. Insufficient data are available to evaluate or predict the long-term effects of kepone to plants, birds, or land animals.

Water Solubility

Kepone is very slightly soluble in hot water. Concentrations up to 100 milligrams will mix with a liter of hot water.

Persistence in the Environment

Kepone is highly persistent in water, with a half-greater than 200 days. The half-life of a pollutant is the amount of time it takes for one-half of the chemical to be degraded.

Bioaccumulation in Aquatic Organisms

Some substances increase in concentration, or *bioaccumulate,* in living organisms as they breathe contaminated air, drink contaminated water, or eat contaminated food. These chemicals can become concentrated in the tissues and internal organs of animals as well as humans.

The concentration of kepone found in fish tissues is expected to be much higher than the average concentration of kepone in the water from which the fish was taken.

Recommended Risk-Reduction Measures

Proper training of all transporters will reduce the likelihood of a mishap or accident resulting in a leak or spill to the environment. The correct use of labeling on all containers, trucks, and rail cars will enable emergency responders to react properly and quickly to any disaster thereby reducing the potential risk to the environment and to personnel.

Storage of kepone should be segregated from incompatible chemicals to minimize the risk of cross-contamination or contact.

If a spill or leak to the environment has occurred, fire department, emergency response and/or hazardous materials spill personnel should be notified immediately. Cleanup should be attempted only by those trained in proper spill containment procedures. Contaminated soils should be removed for incineration and replaced with clean soil. If kepone should contact the water table, aquifer, or navigable waterway, remediation activities should be prompt to ensure protection of aquatic sediments. When such spills occur, the local and/or state emergency response authorities must be notified. A comprehensive emergency response or disaster preparedness/recovery plan should be in place prior to any operations involving the use, transportation, storage, or disposal of kepone. The correct disposal method is to mix kepone with a combustible solvent and burn it in a chemical incinerator equipped with an afterburner and scrubber.

If kepone is spilled or leaked, the following specific steps are recommended:

- ☑ Restrict persons not wearing protective clothing from area of spill or leak until cleanup is complete and area can be opened for normal work.
- ☑ Collect solid materials in safest manner possible and deposit in sealed drums. Do NOT dry sweep (creates airborne dusts). Use a vacuum equipped with a high-efficiency particulate air (HEPA) filter instead. Damp mop residue and collect for disposal. Avoid unprotected contact with kepone.
- ☑ It may be necessary to dispose of kepone as a hazardous waste. The responsible state agency or the regional office of the federal Environmental Protection Agency (EPA) should be contacted for specific recommendations.

RISK ASSESSMENT: BUSINESS

General Assessment

Accidents or mishaps involving kepone can present a significant threat to business operations. The loss or damage of equipment or facilities can significantly affect fiscal viability. Lawsuits that may result from personnel injury/death, public exposures, and/or environmental contamination will also require a serious expenditure of resources. Media attention surrounding an injury, death, or environmental damage can also result in a loss of profits and loss of current as well as future business. Always remember that anytime the terms "cancer," carcinogen," or "reproductive hazard" are used, public hysteria, emotion, and ignorance can run equally high. This must be carefully considered whenever developing or implementing public relation policy.

Recommended Risk-Reduction Measures

Company attorneys, safety and health professionals, and environmental specialists should be involved in the development of any procedures dealing with safety or emergency response. A company official should be pre-designated as a public relations officer with specific training in dealing with the press. Corporate plans and policies should be developed, approved, and implemented long before any need for such arises.

REFERENCES

Ahlbom, A. 1993. *Biostatistics for Epidemiologists.* New York: CRC Press/Lewis Publishers.

American Conference of Governmental Industrial Hygienists. 1988. *Documentation of the Threshold Limit Values and Biological Exposure Indices,* 5th Edition (with updates). Cincinnati: ACGIH

Anthony, C. P., and N. J. Kolthoff. 1971. *Textbook of Anatomy and Physiology.* St. Louis: C. V. Mosby Company.

Asante-Duah, D. K. 1993. *Hazardous Waste Risk Assessment.* New York: CRC Press/Lewis Publishers.

Bloom, A. D. and F. J. DeSerres. 1995. *Ecotoxicity and Human Health.* New York: CRC Press/Lewis Publishers.

Brusick, D. J. 1994. *Methods for Genetic Risk Assessment.* New York: CRC Press/Lewis Publishers.

Bueche, J. R. 1972. *Principles of Physics,* 2nd Edition. New York: McGraw-Hill.

Burrell, R., D. K. Flaherty, and L. J. Sauers. 1992. *Toxicology of the Immune System.* New York: Van Nostrand Reinhold.

Calabrese, E. J. 1994. *Biological Effects of Low Level Exposures.* New York: CRC Press/Lewis Publishers.

Cockerman, L. G., B. S. Shane. 1994. *Basic Environmental Toxicology.* New York: CRC Press/Lewis Publishers.

Cothern, C. R. 1995. *Handbook for Environmental Risk Decisionmaking.* New York: CRC Press/Lewis Publishers.

Gardner, R. W. 1994. *Chemical Intolerance.* New York: CRC Press/Lewis Publishers.

Hallenbeck, W. H. 1993. *Quantitative Risk Assessment for Environmental and Occupational Health,* 2nd Edition. New York: CRC Press/Lewis Publishers.

Hodgson, E., R. B. Mailman, and Janice E. Chambers. 1988. *Dictionary of Toxicology.* London: The MacMillian Company.

Howard, P. H., et. al. 1991. *Handbook of Environmental Degradation Rates.* New York: CRC Press/Lewis Publishers.

Jones, F. E. 1994. *Toxic Organic Vapors in the Workplace.* New York: CRC Press/Lewis Publishers.

Kamrin, M. A. 1988. *Toxicology: A Primer on Principles and Applications.* Chelsea, Michigan: Lewis Publishers.

Keith, L. H. and M. M. Walker. 1995. *Handbook of Air Toxics.* New York: CRC Press/Lewis Publishers.

Landis, W. G., M. H. Yu. 1995. *Introduction to Environmental Toxicology.* New York: CRC Press/Lewis Publishers.

Langerman, N. 1994 *Precautionary Labels for Chemical Containers.* New York: CRC Press/Lewis Publishers.

Lauwerys, R. R., P. Hoet. 1993. *Industrial Chemical Exposure.* New York: CRC Press/Lewis Publishers.

Lewis, R. J., Sr. 1993. *Hawley's Condensed Chemical Dictionary,* 12th Edition. New York: Van Nostrand Reinhold.

Lewis, R. J., Sr. 1992. *Sax's Dangerous Properties of Industrial Materials,* Volumes 1, 2, and 3, 8th Edition. New York: Van Nostrand Reinhold.

Manahan, S. E. 1992. *Toxicological Chemistry,* 2nd Edition. New York: CRC Press/Lewis Publishers.

Meyer, E. 1990. *Chemistry of Hazardous Materials.* Englewood Cliffs, NJ: Prentice Hall.

Ott, W. R. 1995. *Environmental Statistics and Data Analysis.* New York: CRC Press/Lewis Publishers.

Patnaik, P. 1992. *A Comprehensive Guide to the Hazardous Properties of Chemical Substances.* New York: Van Nostrand Reinhold.

Philp, R. B. 1995. *Environmental Hazards and Human Health.* New York: CRC Press/Lewis Publishers.

Proctor, N. H., and James P. Hughes. 1991. *Chemical Hazards of the Workplace.* New York: Van Nostrand Reinhold.

Quigley, D. R. 1994. *Handbook of Emergency Chemical Management.* New York: CRC Press/Lewis Publishers.

Rea, W. J. 1995. *Chemical Sensitivity,* Vol. I-III. New York: CRC Press/Lewis Publishers.

Richardson, M. 1995. *Dictionary of Substance and Their Effects,* Vol. I-VI. New York: CRC Press/Lewis Publishers.

Rylander, R., R. R. Jacobs. 1994. *Organic Dusts.* New York: CRC Press/Lewis Publishers.

Sacarello, H. L. A. 1994. *The Comprehensive Handbook of Hazardous Materials.* New York: CRC Press/Lewis Publishers.

U.S. Department of Health and Human Services, National Institute for Occupational Safety and Health. 1994. *NIOSH Pocket Guide to Chemical Hazards.* Washington, D.C.: U.S. Government Printing Office.

MATERIAL SAFETY DATA SHEET

CHEMICAL NAME	CRC PRESS/LEWIS PUBLISHERS DISCLAIMER
KEROSENE	THE INFORMATION AND RECOMMENDATIONS PRESENTED HEREIN ARE BASED ON SOURCES BELIEVED TO BE RELIABLE. CRC MAKES NO REPRESENTATION ON ITS COMPLETENESS OR ACCURACY. IT IS THE RESPONSIBILITY OF THE USER TO DETERMINE THE CHEMICAL'S SUITABILITY FOR ITS INTENDED USE, THE CHEMICAL'S SAFE USE, AND THE CHEMICAL'S PROPER DISPOSAL. NO REPRESENTATIONS AND/OR WARRANTIES, EITHER EXPRESSED OR IMPLIED, OF THE MERCHANTABILITY OR FITNESS FOR A PARTICULAR PURPOSE, OR OF ANY OTHER NATURE, ARE MADE WITH RESPECT TO THE INFORMATION PROVIDED IN THIS MSDS OR TO THE CHEMICAL TO WHICH INFORMATION MAY REFER. CRC NEITHER ASSUMES NOR AUTHORIZES ANY OTHER PERSON TO ASSUME FOR IT, ANY OTHER ADDITIONAL RESPONSIBILITY OR LIABILITY FOR THE USE OF, OR RELIANCE UPON, THIS INFORMATION.

HAZARD WARNING INFORMATION

HEALTH	FIRE	REACTIVE	OTHER	DEGREE OF HAZARD	COLOR CODING	OTHER CODES
0	2	0		0 = Minimum Hazard 1 = Slight Hazard 2 = Moderate Hazard 3 = Serious Hazard 4 = Severe Hazard	HEALTH = BLUE FIRE = RED REACTIVITY = YELLOW OTHER = WHITE	OX = Oxidizer ACID = Acid ALK = Alkali COR = Corrosive W = Use No Water

SECTION I - GENERAL INFORMATION

Characterization	RCRA Number	EPA Class
Hydrocarbon (fuel, solvent)	**D001**	**Characteristic (I) Waste**

DOT Proper Shipping Name **Kerosene**	Chemical Abstract Service (CAS) Number **8008-20-6**
DOT Hazard Class and Label Requirements **Combustible Liquid**	DOT Emergency Guide Code **27**
DOT Identification Number **UN 1223**	Molecular Formula **Varies with grade.**

Synonyms
Fuel oil number 1; range oil; JP-1; kerosine; illuminating oil; coal oil; deobase; straight-run kerosene.

SECTION II - HAZARDOUS INGREDIENTS AND IDENTITY INFORMATION

Hazardous Components (specific identity)	OSHA Exposure Criteria	NIOSH Exposure Criteria	Immediately Dangerous to Life and Health (IDLH)	ACGIH Exposure Criteria
Kerosene (derivation: A mixture of various distilled petroleum hydrocarbons, mainly of the methane series with 10 to 16 carbon atoms per molecule). **1 ppm = 7.07 mg/m^3 (approx.)**	**PEL:** Not Established **STEL:** Not Established	**REL:** 14 ppm 100 mg/m^3 **STEL:** Not Established	**Not Determined**	**TLV:** Not Established **STEL:** Not Established

SECTION III - PHYSICAL AND CHEMICAL CHARACTERISTICS

Boiling Point **347 to 617°F (175 to 325°C)**	Specific Gravity (H_2O = 1) **0.81**
Vapor Pressure (mm Hg) **5 at 100°F (38°C)**	Molecular Weight (atomic weight) **170 (approx.)**
Vapor Density (Air = 1) **4.5**	Freezing Point **-50°F (-45.5°C)**

Solubility
Insoluble in water. Miscible with other petroleum solvents.

Appearance and Odor
Colorless to yellowish, oily liquid with a strong, characteristic odor.

SECTION IV - FIRE AND EXPLOSION HAZARD DATA

Flash Point (method used) **100 to 162°F (43 to 72°C) closed cup**	Explosive Limits in Air % by Volume **LEL: 0.7% UEL: 5.0%**
NFPA Classification **Class II Combustible Liquid**	Autoignition Temperature **536 to 853°F (280 to 456°C)**

Extinguishing Media
Use regular foam, dry chemical, carbon dioxide, or water spray on small fires. For larger fires, use water spray, fog, or regular foam. Use a blanketing or smothering technique with foam for best results.

Special Fire Fighting Procedures
Isolate fire area. Wear self-contained breathing apparatus (SCBA). Poisonous gases are produced in fire. Use water spray to cool fire-exposed containers until long after fire is out. Use non-sparking tools. Move containers from fire area if it can be done without risk. Fight fire from distance, if possible.

Unusual Fire and Explosion Hazards
Dangerous fire and explosion hazard when exposed to heat or flame. Containers may explode in fire. The vapor is heavier than air and may travel for some distance to cause fire in flashback to source.

SECTION V - REACTIVITY DATA

Stability		Conditions to Avoid Kerosene is normally stable under routine conditions of handling and storage. Avoid contact with heat, flame, other sources of ignition, and contact with incompatible materials and chemicals.
Stable X	Unstable	Incompatibility (*materials to avoid*) Incompatible with strong oxidizers (such as peroxides) and oxidizing materials.
Hazardous Polymerization		Conditions to Avoid Hazardous polymerization has not been reported to occur under normal conditions of temperature and pressure.
May Occur	Will Not Occur X	Hazardous Decomposition or By-products When heated to decomposition, kerosene emits toxic and acrid fumes, such as oxides of carbon, partially oxidized hydrocarbons, and trace quantities of sulfur dioxide.

SECTION VI - HEALTH HAZARD DATA

Primary Route(s) of Entry:	Inhalation?	Absorption (skin/eye)?	Ingestion?
	X		X

Health Hazards

INHALATION: Symptoms include irritation of the nose, throat, and mucosa of the respiratory tract with headache and drowsiness. High concentrations can cause feeling of suffocation, coma, and death (respiratory arrest). Chronic exposures can lead to neurological impairment.

SKIN & EYES: Irritation with drying, cracking, redness, blistering (prolonged contact). May lead to contact dermatitis and secondary infection.

INGESTION: May cause nausea, vomiting, headache and gastrointestinal irritation. Aspiration of vomit into the lungs can cause serious pneumonitis and even pulmonary hemorrhage.

Carcinogenicity	NTP Listed?	IARC Cancer Review Group?	OSHA Regulated?	Target Organs?
Suspected Human Suspected Animal	No	Group 2A	No	Eyes, skin, respiratory sys., blood, CNS, kidneys.

Medical Conditions Generally Aggravated by Exposure

None identified.

Emergency and First-aid Procedures

Eye contact: Do not allow victim to keep eyes tightly shut. Flush immediately with water for 15 minutes (minimum); seek medical attention. Skin contact: Remove all contaminated clothing, including shoes. Immediately wash area with flooding amounts of soap and water. Seek medical attention if blisters develop. For inhalation: Remove the person from exposure. Provide respiratory assistance and CPR. Transfer to medical facility. If swallowed: Seek medical attention immediately. Give conscious person milk or water to drink. Do NOT induce vomiting (possible aspiration of vomitus/chemical into the lungs).

SECTION VII - PRECAUTIONS FOR SAFE HANDLING AND USE

Steps to be Taken in Case Material is Released or Spilled

Remove all ignition sources. Wear SCBA respiratory protection. Restrict those not wearing protective equipment and who are not involved in cleanup from entering area. Avoid skin contact. Stop leak if it can be done without risk. Provide maximum explosion-proof ventilation. Absorb with vermiculite or other similar material and deposit in sealed containers. Use non-sparking tools during cleanup.

Preferred Waste Disposal Method

No Citation.

Precautions to be Taken in Handling and Storage

Store in tightly closed containers in cool, well-ventilated location. Keep away from sources of heat.

Other Precautions and Warnings

Sources of ignition such (open flame or smoking), are prohibited where kerosene is used or stored. Metal containers should be bonded and grounded during transfer operations. Bulk storage not recommended.

SECTION VIII - CONTROL MEASURES AND PERSONAL PROTECTIVE EQUIPMENT

Respiratory Protection (*specify type*)

Use an MSHA/NIOSH-approved air-purifying organic vapor respirator, a powered air purifying respirator, or a self-contained breathing apparatus (SCBA) with full facepiece operated in positive pressure mode.

Ventilation

Local exhaust (preferred) at site of chemical work or general ventilation.

Protective Gloves	Eye Protection	Other Protective Clothing
Impervious Material	Chemical Goggles or Face Mask	Protective Apron if Splash is Likely

Work/Hygiene Practices

Always wash hands thoroughly after using chemical; never bring food, drink, or smoking materials into vicinity of chemicals. Barrier creams may also be useful in protecting exposed skin surfaces.

KEROSENE

Formula varies with grade (may be C_{10} through C_{16}) CAS: 8008-20-6

IDENTIFICATION AND TYPICAL USES

Kerosene is a clear or colorless to slightly yellow, oily liquid with a strong, characteristic, petroleum odor. It is used primarily as a fuel for rocket engines, kerosene lamps, flares, portable cooking stoves, diesel engines, and tractors. It is also used as an industrial solvent and degreasing agent, a cleaner, a mold-releasing agent, a solvent for asphalt coatings, enamels, paints, polishes, thinners, and varnishes. It is sometimes used by veterinarians for decontamination.

RISK ASSESSMENT: HEALTH

General Assessment

Kerosene is considered moderately toxic by *inhalation* of its vapors and by *ingestion* of the liquid. Skin contact will also cause localized damage at site of contact, but dermal absorption through intact skin has not been reported. It is suspected to cause cancer in animals with mutation data also being reported. Its teratogenic properties have not been documented in the references.

Inhalation of the kerosene vapors can cause mild to severe irritation of the eyes, nose, throat, and respiratory tract mucosa. Systemic effects include headache, drowsiness, burning sensation in the chest, nausea, weakness, restlessness, incoordination, and other effects of central nervous system depression. High exposures can lead to a feeling of suffocation, coma, and death due to respiratory arrest.

Skin contact may cause irritation producing erythema (redness), burning, and, in severe cases, edema and even blistering. Contact dermatitis is possible, as is secondary infection. Eye contact can cause irritation of the conjunctiva with redness, swelling, tearing, and localized inflammation of eye tissues.

Ingestion may cause nausea, vomiting, headaches, dizziness, gastrointestinal irritation, somnolence, hallucinations and distorted perceptions, coughing, fever, drunkenness, confusion, and diarrhea. If vomiting occurs, aspiration of the liquid into the lungs can cause choking, difficulty breathing or shortness of breath (dypsnea), tachypnea (rapid respiration), tachycardia (rapid heartbeat), fever, pneumonitis, and even pulmonary hemorrhage (bleeding in the lungs).

Acute Health Effects

The following acute (short-term) health effects may occur immediately or shortly after exposure to kerosene:

Skin: Irritation with potential for redness, blistering, and burning.

Eye: Severe irritation, conjunctivitis, redness, possible pain, and visual disturbances.

Lung: Nose, throat, and respiratory tract irritation causing difficulty in breathing, possible pulmonary effects if liquid enters the lungs.

CNS: Inebriation, dizziness, stupor, hallucinations, slurred speech, mental confusion, loss of consciousness, and possible convulsions.

Other: Ingestion can cause serious nervous system effects and possible injury to the lungs should any vomitus be aspirated into the lungs.

Chronic Health Effects

The following chronic (long-term) health effects may occur at some time after exposure to kerosene and can last for months or even years:

Cancer Hazards: Kerosene is a suspected carcinogen in animal studies. Its ability to cause cancer in humans is considered to be at least equal to that of animals.

Reproductive Hazard: There is no evidence in the references to indicate that exposure to kerosene causes any adverse effects to reproduction.

Other Chronic Effects: Chronic inhalation of hydrocarbon vapors can lead to neurological impairment.

Repeated skin contact causes blistering, drying, cracking, and possible secondary infections and contact dermatitis.

❶ *Recommended Risk-Reduction Measures*

Personnel should avoid direct contact with or exposure to kerosene. Occupational poisoning occurs most commonly by inhalation. Substitution may not always be a viable option with kerosene since its use is often as a substitute for even more toxic or flammable compounds. However, a less toxic material or compound should be used in place of kerosene whenever possible. If substitution is not possible, *engineering controls* are the most effective method of reducing exposures. The best protection is to enclose operations and/or provide local exhaust ventilation at the site of kerosene release. While not always operationally feasible, isolating operations can also reduce exposure. Using respiratory protection is less effective than the controls mentioned above, but is still advisable whenever working with or around kerosene. An MSHA/NIOSH-approved air-purifying respirator with an organic vapor cartridge, a supplied-air respirator operated in continuous flow mode, or a self-contained breathing apparatus (SCBA) with full facepiece operated in pressure demand are the appropriate methods of respiratory protection when exposure to kerosene is anticipated. If a full facepiece is not available, then chemical goggles should be worn to protect the eyes. A face shield should also be considered. To prevent hand and skin exposures, protective gloves should be used. Glove manufacturers should be contacted and permeation studies obtained *before* final glove selections are made.

Administrative controls should also be in place to minimize the potential for human exposures. These may include written procedures or policies which specify the methods and techniques that will be practiced whenever personnel are to work with kerosene.

All personnel should receive *training* on the use, hazards, protective measures, emergency actions, and other precautions per 29 CFR 1910.1200 (Hazard Communication), prior to the first assignment in an area where kerosene is used or stored.

Before beginning employment and at regular times thereafter, the following recommended medical tests should be considered:

☑ Lung function tests (establish baseline).

If symptoms occur or overexposure is suspected, the following additional tests are suggested:

☑ Lung function tests (compare to baseline).

☑ Liver and kidney function tests.

Any evaluation should include a careful history of past and present symptoms with an examination. Medical tests that simply look for existing damage are not a substitute for controlling exposures.

Other methods to reduce exposure include:

☑ Where possible, enclose operations and use local exhaust ventilation at the site of chemical release. If local exhaust ventilation or enclosure is not used, respiratory protection should be mandatory.

☑ Always ensure that proper protective clothing is worn when using chemical substances.

☑ Wash thoroughly immediately after exposure to kerosene and at the end of the work shift or before eating, drinking, or smoking.

☑ Hazard warning information should be posted in the work area. In addition, as part of an on-going education and training program, all information on the health and safety hazards of kerosene should be communicated to all exposed workers.

☑ Eye wash stations should be provided in the immediate work area for emergency use. If there is a possibility of skin exposure to kerosene, emergency shower facilities should also be provided.

☑ Workers whose clothing has been contaminated by kerosene should change into clean clothes before leaving work. Contaminated work clothing should be laundered only by individuals who have been informed of the hazards of exposure to kerosene.

☑ Where possible, automatically transfer liquids containing kerosene from drums or other containers to process containers. Always ground and bond all containers and equipment. Use non-sparking tools.

RISK ASSESSMENT: ENVIRONMENT

General Assessment

The environment is at risk of exposure during transportation, storage, disposal, or destruction of kerosene. In almost every scenario, the threat of environmental exposure is contingent upon the proper handling of the chemical substance. Accidental spills, large or small, can result in contamination of the surrounding environmental mediums (water, soil, and air).

Kerosene is a Class II combustible liquid (per OSHA 29 CFR 1910.106). It can present a dangerous fire hazard and moderate explosion hazard when exposed to heat or flame. It is incompatible with commonly encountered oxidizers and caution is always required in handling, storage, transportation, and disposal of kerosene. Emergency responders should be made aware of the presence of kerosene at any emergency response situation.

Kerosene can enter the environment from industrial effluents and from spills.

Acute Ecological Effects

Acute (short-term) toxic effects may include the death of animals, birds, or fish and death or low growth rate in plants. Acute effects are seen 2 to 4 days after animals or plants are exposed to kerosene. Kerosene has high acute toxicity to aquatic life. It can cause death in plants and roots and membrane damage in leaves of various agricultural crops. Insufficient data are available to evaluate the short-term effects of kerosene on birds or land animals.

Chronic Ecological Effects

Chronic (long-term) toxic effects may include shortened life span, reproductive problems, lower fertility, and changes in appearance or behavior in exposed animals. These effects can be seen long after first exposure(s) to toxic chemicals. Kerosene has high chronic toxicity to aquatic life. Insufficient data are available to evaluate or predict the long-term effects of kerosene to plants, birds, or terrestrial animals.

Water Solubility

Kerosene is not soluble in water. Concentrations of 1 milligram will not mix with a liter of water.

Persistence in the Environment

Kerosene is slightly persistent in water, with a half-life of between 2 to 20 days. The half-life of a pollutant is the amount of time it takes for one half of the chemical to be degraded. About 99.5% of kerosene will eventually end up in air; the rest will end up in water.

Bioaccumulation in Aquatic Organisms

Some substances increase in concentration, or *bioaccumulate,* in living organisms as they breathe contaminated air, drink contaminated water, or eat contaminated food. These chemicals can become concentrated in the tissues and internal organs of animals as well as humans.

The concentration of kerosene found in fish tissues is expected to be somewhat higher than the average concentration of kerosene in the water from which the fish was taken.

Recommended Risk-Reduction Measures

Proper training of all transporters will reduce the likelihood of a mishap or accident resulting in a leak or spill of kerosene into the environment. Labels on all containers, trucks, and rail cars must meet DOT requirements and accurately reflect their contents to enable emergency responders to react properly and quickly to any disaster thereby reducing the potential risk to the environment and to personnel.

Storage of kerosene should be segregated from other chemicals and materials to minimize the risk of cross-contamination. Facilities should be designed to meet explosion-proof requirements.

If a spill or leak to the environment has occurred, fire department, emergency response and/or hazardous materials spill personnel should be notified immediately. Cleanup should be attempted only by those trained in proper spill containment procedures using non-sparking tools. Contaminated soils should be removed for incineration and replaced with clean soil. If kerosene should contact the water table, aquifer, or navigable waterway, time is of the essence. It is not soluble and total remediation may be possible. In general, kerosene will float on the surface and may be captured if response actions are prompt. The local and/or state emergency response authorities must be notified. A comprehensive emergency response or disaster preparedness/recovery plan should be in place prior to any operations involving the use, transportation, storage, or disposal of kerosene.

If kerosene is spilled or leaked, the following specific steps are recommended:

- ☑ Restrict persons not wearing protective clothing from area of spill or leak until cleanup is complete and area can be opened for normal work.
- ☑ Absorb liquids in vermiculite, sand, earth, or similar material and deposit in sealed drum. Use non-sparking tools during cleanup procedures.
- ☑ Ventilate area of spill or leak. Container should be removed to safer area if it can be done without increasing the risk.

- ☑ Use water spray to knock down vapors.
- ☑ It may be necessary to dispose of kerosene as a hazardous waste. The responsible state agency or the regional office of the federal Environmental Protection Agency (EPA) should be contacted for specific recommendations.

RISK ASSESSMENT: BUSINESS

General Assessment

Accidents or mishaps involving kerosene can present a serious threat to business operations. The loss or damage of equipment or facilities can significantly affect fiscal viability. Lawsuits that may result from personnel illness, injury/death, public exposures, and/or environmental contamination will require a serious expenditure of resources.

Media attention surrounding an injury, death, or environmental damage can also result in a loss of profits and loss of current as well as future business. Always remember that anytime the terms "cancer," or "carcinogen" are used, public emotion, hysteria, and ignorance can run equally high. This must be carefully considered when developing or implementing any public relations policies.

Recommended Risk-Reduction Measures

Company attorneys, safety and health professionals, and environmental specialists should be involved in the development of any procedures for responding to chemical incidents. A company official should be predesignated as a public relations officer with specific training in dealing with the press. Corporate plans and policies should be developed, approved, and implemented long before any need for such arises.

REFERENCES

Ahlbom, A. 1993. *Biostatistics for Epidemiologists.* New York: CRC Press/Lewis Publishers.

American Conference of Governmental Industrial Hygienists. 1988. *Documentation of the Threshold Limit Values and Biological Exposure Indices,* 5th Edition (with updates). Cincinnati: ACGIH

Anthony, C. P., and N. J. Kolthoff. 1971. *Textbook of Anatomy and Physiology.* St. Louis: C. V. Mosby Company.

Asante-Duah, D. K. 1993. *Hazardous Waste Risk Assessment.* New York: CRC Press/Lewis Publishers.

Bloom, A. D. and F. J. DeSerres. 1995. *Ecotoxicity and Human Health.* New York: CRC Press/Lewis Publishers.

Bueche, J. R. 1972. *Principles of Physics,* 2nd Edition. New York: McGraw-Hill.

Burrell, R., D. K. Flaherty, and L. J. Sauers. 1992. *Toxicology of the Immune System.* New York: Van Nostrand Reinhold.

Calabrese, E. J., E. M. Kenyon. *Air Toxics and Risk Assessment.* New York: CRC Press/Lewis Publishers.

Cockerman, L. G., B. S. Shane. 1994. *Basic Environmental Toxicology.* New York: CRC Press/Lewis Publishers.

Cothern, C. R., N. P. Ross. 1994. *Environmental Statistics, Assessment, and Forecasting.* New York: CRC Press/Lewis Publishers.

Danse, I. R. 1991. *Common Sense Toxics in the Workplace.* New York: Van Nostrand Reinhold.

Environmental Protection Agency. 1983. *Chemical Hazard Information Profiles.* Washington, D.C.: U.S. EPA.

Fodden, C. R., and J. L. Weddell. 1992. *Hazardous Materials Emergency Action Data.* New York: CRC Press/Lewis Publishers.

Forseberg, K., and L. H. Keith. 1993. *Chemical; Protective Clothing Permeation and Degradation Database.* Chelsea, MI: Lewis Publishers.

Howard, P. H., et. al. 1991. *Handbook of Environmental Degradation Rates.* New York: CRC Press/Lewis Publishers.

International Agency for Research on Cancer (IARC). 1987. *IARC Monographs on the Evaluation of Carcinogenic Risk to Humans, Overall Evaluations of Carcinogenicity: An updating of IARC Monographs Volumes 1-42.* Lyon: IARC.

Jones, F. E. 1994. *Toxic Organic Vapors in the Workplace.* New York: CRC Press/Lewis Publishers.

Kamrin, M. A. 1988. *Toxicology: A Primer on Principles and Applications.* Chelsea, Michigan: Lewis Publishers.

Keith, L. H. , M. T. Johnston, C. E. Hudak, and M. Conoley. 1992. *National Toxicology Program's Gloves+ Database.* Chelsea, MI: Lewis Publishers.

Lauwerys, R. R., P. Hoet. 1993. *Industrial Chemical Exposure.* New York: CRC Press/Lewis Publishers.

Lewis, R. J., Sr. 1993. *Hawley's Condensed Chemical Dictionary,* 12th Edition. New York: Van Nostrand Reinhold.

Lewis, R. J., Sr. 1992. *Sax's Dangerous Properties of Industrial Materials,* Volumes 1, 2, and 3, 8th Edition. New York: Van Nostrand Reinhold.

Meyer, E. 1990. *Chemistry of Hazardous Materials.* Englewood Cliffs, NJ: Prentice Hall.

Patnaik, P. 1992. *A Comprehensive Guide to the Hazardous Properties of Chemical Substances.* New York: Van Nostrand Reinhold.

Quigley, D. R. 1994. *Handbook of Emergency Chemical Management.* New York: CRC Press/Lewis Publishers.

Sacarello, H. L. A. 1994. *The Comprehensive Handbook of Hazardous Materials.* New York: CRC Press/Lewis Publishers.

Torkelson, T. R. V. K. Rowe. 1981. *Patty's Industrial Hygiene and Toxicology,* Vol. 2B. New York: Wiley-Interscience.

U.S. Department of Health and Human Services, National Institute for Occupational Safety and Health. 1994. *NIOSH Pocket Guide to Chemical Hazards.* Washington, D.C.: U.S. Government Printing Office.

MATERIAL SAFETY DATA SHEET

CHEMICAL NAME

KETENE

HAZARD WARNING INFORMATION

HEALTH	FIRE	REACTIVE	OTHER	DEGREE OF HAZARD	COLOR CODING	OTHER CODES
4	2	3		0 = Minimum Hazard 1 = Slight Hazard 2 = Moderate Hazard 3 = Serious Hazard 4 = Severe Hazard	HEALTH = BLUE FIRE = RED REACTIVITY = YELLOW OTHER = WHITE	OX = Oxidizer ACID = Acid ALK = Alkali COR = Corrosive ~~W~~ = Use No Water

SECTION I - GENERAL INFORMATION

Characterization **Ketone**	RCRA Number **None**	EPA Class **Not Applicable**
DOT Proper Shipping Name **Poison Gas, N.O.S. (Ketene)**	Chemical Abstract Service (CAS) Number **463-51-4**	
DOT Hazard Class and Label Requirements **Poison A**	DOT Emergency Guide Code **26**	
DOT Identification Number **NA 1955**	Chemical Formula **C_2H_2O**	

Synonyms
Carbomethane; ethenone; keto-ethylene.

SECTION II - HAZARDOUS INGREDIENTS AND IDENTITY INFORMATION

Hazardous Components (specific identity)	OSHA Exposure Criteria	NIOSH Exposure Criteria	Immediately Dangerous to Life and Health (IDLH)	ACGIH Exposure Criteria
Ketene (derivation: By pyrolysis of acetone or acetic acid by passing its vapor through a tube at 500 to 600°C). **1 ppm = 1.75 mg/m^3**	**PEL (8-hour):** **0.5 ppm** **0.9 mg/m^3** **STEL:** **Not Established**	**REL (10 hour):** **0.5 ppm** **0.9 mg/m^3** **STEL:** **1.5 ppm** **3 mg/m^3**	**5 ppm**	**TLV:** **0.5 ppm** **0.9 mg/m^3** **STEL:** **1.5 ppm** **3 mg/m^3**

SECTION III - PHYSICAL AND CHEMICAL CHARACTERISTICS

Boiling Point **-69°F (-56°C)**	Specific Gravity (H_2O = 1) **1.45**
Vapor Pressure (atmospheres) **>1 atmospheres**	Molecular Weight **42.0**
Vapor Density (Air = 1) **1.45**	Freezing Point **-238°F (-150°C)**

Solubility
Reacts (decomposes) in water and alcohol; soluble in water and ether.

Appearance and Odor
Colorless gas with a sharp, disagreeable, penetrating odor.

SECTION IV - FIRE AND EXPLOSION HAZARD DATA

Flash Point (method used) **Not Determined (Gas)**	Explosive Limits In Air % by Volume **LEL: Not Determined UEL: Not Determined**
NFPA Classification **Flammable Gas**	Autoignition Temperature **Not Determined**

Extinguishing Media
Stop flow of gas before attempting to extinguish fire. Use water spray to knock down vapors.

Special Fire Fighting Procedures
Extremely poisonous gas. Wear full protective clothing and self-contained breathing apparatus (SCBA). Heat will build pressure and may rupture closed storage containers. Fight fire from an explosion-proof distance if possible. Use unmanned equipment to control fires. Fire cannot be stopped until gas release is stopped first. Keep fire-exposed containers cool with water spray.

Unusual Fire and Explosion Hazards
Ketene will react and decompose to acetic acid in contact with water. Avoid using water as an extinguishing agent until gas flow can be stopped.

SECTION V - REACTIVITY DATA

Stability		Conditions to Avoid
		Ketene is normally stable in closed containers under routine conditions of handling or storage. Avoid contact with incompatible materials, heat, or ignition sources.
Stable	Unstable	Incompatibility *(materials to avoid)*
X		Reacts with water to form acetic acid, decomposes in alcohols, ammonia, and reacts with hydrogen peroxide to forms explosive diacetyl peroxide.
Hazardous Polymerization		Conditions to Avoid
		Ketene readily polymerizes under normal conditions of temperature and pressure.
May Occur	Will Not Occur	Hazardous Decomposition or By-products
X		When heated to decomposition, ketene emits acrid, irritating smoke and fumes, including toxic gases such as carbon monoxide.

SECTION VI - HEALTH HAZARD DATA

Primary Route(s) of Entry:	Inhalation?	Absorption (skin)?	Ingestion?
	X		

Health Hazards

INHALATION: Extremely irritating to the eyes, nose, throat, and especially the upper respiratory tract. Can cause difficulty in breathing, cough, and pulmonary edema (fluid in the lungs).

SKIN & EYES: Skin contact can result in irritating rash or burning feeling. Can cause severe eye burns leading to permanent damage. Serious damage occurs as a result of frostbite.

INGESTION: Not likely, ketene is a gas.

Carcinogenicity	NTP Listed?	IARC Cancer Review Group?	OSHA Regulated?	Target Organs?
Unknown Human Unknown Animal	No	No	29 CFR 1910.1000 (Table Z-1)	Respiratory system; skin; eyes.

Medical Conditions Generally Aggravated by Exposure

Respiratory function impairments (asthma, etc.) may be aggravated by exposure.

Emergency and First-aid Procedures

Eye contact: Immediately flush with large amounts of warm, tepid water for 15 minutes (minimum); occasionally lifting eyelids; seek medical attention immediately. Skin contact: Remove all contaminated clothing. Immediately wash area with large amounts of warm, tepid water. Treat for frostbite. Do NOT rub affected area and do NOT attempt to warm skin using hot, blown air. For inhalation: Remove the person from exposure. Provide respiratory assistance and CPR. Transfer promptly to medical facility. Observe 48 for hours for lung effects. If swallowed: Unlikely to occur (ketene is a gas).

SECTION VII - PRECAUTIONS FOR SAFE HANDLING AND USE

Steps to be Taken in Case Material is Released or Spilled

Remove all ignition sources. Restrict persons not wearing protective equipment from entering area of gas leak. Ventilate area. Stop flow of leak. Use water spray to knock down vapors and protect personnel who are attempting to stop leak (use caution since ketene will react with water to form acetic acid).

Preferred Waste Disposal Method

No citation.

Precautions to be Taken in Handling and Storage

Ketene is a highly flammable and poisonous gas. Do not store with incompatible chemicals since violent reactions can occur. Store in tightly closed containers in a cool, dark, well-ventilated area away from fire, sparks, and flame. Use non-sparking tools on containers. Ground and bond metal containers during transfer operations. Use external or detached facility storage if possible.

Other Precautions and Warnings

Bulk storage of ketene is not recommended. Containers may explode in fire or under conditions of extreme heat (do not store outdoors or in direct sunlight).

SECTION VIII - CONTROL MEASURES AND PERSONAL PROTECTIVE EQUIPMENT

Respiratory Protection *(specify type)*

Exposure levels are extremely low (0.5 ppm) and ketene is extremely irritating. Use a supplied-air respiratory operated in positive pressure mode or a self-contained breathing apparatus (SCBA) with full facepiece and pressure demand or other positive pressure mode.

Ventilation

Local exhaust or general mechanical systems recommended.

Protective Gloves	Eye Protection	Other Protective Clothing
Impervious, Thermal Gloves	Chemical Goggles and/or Face Mask	Protective Apron

Work/Hygiene Practices

Always wash hands thoroughly after using chemical; never bring food, drink, or smoking materials into vicinity of chemicals.

KETENE

C_2H_2O

CAS: 463-51-4

IDENTIFICATION AND TYPICAL USES

Ketene is a colorless gas with a penetrating, disagreeable, sharp odor. It is used as an acetylating agent in the production of cellulose acetate, aspirin, acetic anhydride, and in various organic syntheses. It generally reacts with ammonia to produce acetamide. It is also a starting point for making various commercial products such as acetic anhydride and acetate esters.

RISK ASSESSMENT: HEALTH

General Assessment

Ketene is a highly toxic and extremely irritating gas. As a gas, it can enter the body primarily through *inhalation*. Skin contact with the compressed gas can cause serious, frostbite-type burns and damage, but dermal absorption has not been reported. There are no reports of carcinogenic, mutagenic, or teratogenic effects occurring as a result of exposure to ketene.

Inhalation will cause immediate and possibly severe irritation of the eyes, nose, throat, and upper respiratory tract. Concentrations of only 10 to 15 ppm for several minutes can injure the respiratory tract and lungs leading to a dangerous buildup of fluid in the lungs (pulmonary edema) which can be fatal. Symptoms of headache, cough, dizziness, congestion, and tightness in chest may be delayed up to 48 hours, thereby causing a false sense of security with regard to health exposure risk.

Skin or eye contact will cause irritation, rash, burns, and possible tissue destruction (frostbite) as a result of the cryogenic properties of compressed ketene.

Acute Health Effects

The following acute (short-term) health effects may occur immediately or shortly after exposure to ketene:

Skin: Irritation/rash or burning sensation, frostbite, and cryogenic burns on contact.

Eye: Burns which can lead to permanent damage.

Lung: Irritation causing coughing and/or shortness of breath leading to pulmonary edema and death.

Chronic Health Effects

The following chronic (long-term) health effects may occur at some time after exposure to ketene and can last for months or even years:

Cancer Hazards: According to information presented in the references, exposure to ketene does not currently lead to carcinogenic or mutagenic effects.

Reproductive Hazards: There are no reports in the references to support any claims of reproductive hazards resulting from exposure to ketene gas.

Other Chronic Effects: Human data on chronic exposures are limited in the references. However, animal studies have shown an increase in emphysema and pulmonary fibrosis resulting from prolonged or repeated exposure to low levels of ketene gas.

Recommended Risk-Reduction Measures

Personnel should avoid direct contact with ketene. It is extremely toxic and can damage the respiratory system. Unless a less toxic chemical can be substituted for a hazardous substance, *engineering controls* are the most effective method of reducing exposures. The best protection is to enclose operations and/or provide local exhaust ventilation at the site of chemical release. While not always operationally feasible, isolating operations can also reduce exposure. Using respiratory protection is less effective than the controls mentioned above, but is still advisable whenever working with ketene. The exposure level is extremely low (0.5 ppm) and difficult to accurately monitor. For the best protection, use a NIOSH/MSHA-approved, full facepiece supplied-air respirator operated in continuous flow mode or a self-contained breathing apparatus (SCBA) with full facepiece operated in pressure demand or other positive pressure mode. If a full facepiece is not available, then chemical goggles should be worn to protect the eyes. Whenever a compressed gas hazard exists, a face shield and a protective apron should be worn. To prevent hand and skin exposures, impervious, thermal protective gloves should be used. Glove manufacturers should be contacted and permeation studies obtained *before* final glove selections are made.

Administrative controls should also be in place to minimize the potential for human exposures. These may include written procedures or policies which specify the methods and techniques that will be practiced whenever personnel are to work with ketene.

All personnel should receive *training* on the use, hazards, protective measures, emergency actions, and other precautions per 29 CFR 1910.1200 (Hazard Communication), prior to the first assignment in an area where ketene is used or stored.

If symptoms develop or overexposure is suspected, the following medical tests are recommended:

☑ Lung function tests.

☑ Consider chest X-ray following acute overexposure (may be negative if taken immediately after exposure due to delayed onset of pulmonary edema).

Also, since smoking can cause heart disease, emphysema, lung cancer, and other respiratory disorders, smokers exposed to ketene may experience symptoms more rapidly and more pronounced than non-smokers under the same conditions of exposure. Prudent risk management requires careful consideration of *all* possible factors that may be causing the appearance of exposure symptoms in the workplace.

Other methods to reduce exposure include:

☑ Where possible, enclose operations and use local exhaust ventilation at the site of chemical release. If local exhaust ventilation or enclosure is not used, respiratory protection should be mandatory.

☑ Always ensure that proper protective clothing is worn when using chemical substances.

☑ Wash thoroughly immediately after exposure to ketene and at the end of the work shift or before eating, drinking, or smoking. Never wear clothes contaminated with ketene home.

☑ Hazard warning information should be posted in the work area. As part of an on-going education and training program, all information on the health and safety hazards of ketene should be communicated to all potentially exposed workers.

RISK ASSESSMENT: ENVIRONMENT

General Assessment

The environment is at risk of exposure during transportation, storage, disposal, or destruction of ketene. In almost every scenario, the threat of environmental exposure is contingent upon the proper handling of the chemical substance. Accidental spills can result in fire, explosion, and possible contamination of the various environmental mediums (water, soil, and primarily air).

Ketene is considered a highly flammable gas and presents a dangerous fire and explosion hazard. It is also incompatible with a number of commodities, including water, alcohols, and ammonia. It will react with hydrogen peroxide to form explosive diacetyl peroxide. These characteristics require special consideration during any emergency involving a leak of ketene.

Ketene can enter the environment through unchecked discharge and emissions from industrial or municipal waste treatment plants, and through spills.

Acute Ecological Effects

Acute (short-term) toxic effects may include the death of animals, birds, or fish, and death or low growth rate in plants. Acute effects are seen 2 to 4 days after animals or plants are exposed to ketene.

Insufficient data are available to evaluate or predict the short-term effects of ketene on aquatic life, birds, plants, or land animals.

Chronic Ecological Effects

Chronic (long-term) toxic effects may include shortened life span, reproductive problems, lower fertility, and changes in appearance or behavior in exposed animals. These effects can be seen long after first exposure(s) to toxic chemicals.

Insufficient data are available to evaluate or predict the long-term effects of ketene on aquatic life, birds, plants, or land animals.

Water Solubility

Ketene is insoluble in water. It will react and decompose to form acetic acid which is highly soluble in water. Acetic acid concentrations up to 1000 milligrams and possibly more will readily mix with a liter of water.

Persistence in the Environment

Ketene is non-persistent in water, with a half-life of less than 2 days. However, acetic acid, its decomposition product, is highly persistent in water with a half life between 20 and 200 days. The half-life of a pollutant is the amount of time it takes for one half of the chemical to be degraded.

Bioaccumulation in Aquatic Organisms

Some substances increase in concentration, or *bioaccumulate,* in living organisms as they breathe contaminated air, drink contaminated water, or eat contaminated food. These chemicals can become concentrated in the tissues and internal organs of animals as well as humans.

The concentration of ketene found in fish tissues is expected to be much lower than the average concentration of ketene in water from which the fish was taken. The concentration of acetic acid found in fish tissues is expected to be equal to or greater than the average concentration of acetic acid in the water from which the fish was taken.

Recommended Risk-Reduction Measures

Proper training of all transporters will reduce the likelihood of a mishap or accident resulting in a leak or spill to the environment. The correct use of DOT labeling (per the MSDS) on all containers, trucks, and rail cars will enable emergency responders to react properly and quickly to any disaster thereby reducing the potential risk to the environment and to personnel.

Storage of ketene should be segregated from incompatible chemicals to minimize the risk of cross-contamination or contact. Buildings designated for storage should be equipped with appropriate fire protection systems (alarms, sprinklers, emergency lighting, portable extinguishers).

If a spill or leak to the environment has occurred, fire department, emergency response, and/or hazardous materials spill personnel should be notified immediately. Cleanup should be attempted only by those trained in proper spill containment procedures using non-sparking tools. Contaminated soils should be removed for incineration and replaced with clean soil. If ketene should contact the water table, aquifer, or navigable waterway, time is of the essence. It is readily miscible in water and, therefore, total containment and remediation may not be entirely possible. When such spills occur, the local and/or state emergency response authorities must be notified. A comprehensive emergency response or disaster preparedness/recovery plan should be in place prior to any operations involving the use, transportation, storage, or disposal of ketene. If ketene is spilled or leaked, the following specific steps are recommended:

- ☑ Restrict persons not wearing protective clothing from area of spill or leak until cleanup is complete and area can be opened for normal work.
- ☑ Remove all ignition sources.
- ☑ Provide maximum explosion-proof ventilation in area of spill or leak.
- ☑ Stop flow of gas. If leak is from a cylinder and cannot be stopped, move cylinder to a safe location (outdoors) if it can be done without risk. Allow to vent until empty. Use water spray to knock down vapors and to protect personnel working to stop leak.
- ☑ Keep ketene out of a confined space, such as a sewer, because of the possibility of explosion (unless the sewer is designed to prevent the buildup of explosive concentrations).
- ☑ It may be necessary to dispose of ketene as a hazardous waste. The responsible state agency or the regional office of the federal Environmental Protection Agency (EPA) should be contacted for specific recommendations.

RISK ASSESSMENT: BUSINESS

General Assessment

Accidents or mishaps involving ketene can present a significant threat to business operations. The loss or damage of equipment or facilities can significantly affect fiscal viability. Lawsuits that may result from personnel injury, public exposures, and/or environmental contamination will also require a serious expenditure of resources.

Media attention surrounding an injury, illness, or environmental damage can also result in a loss of profits and loss of current as well as future business.

Recommended Risk-Reduction Measures

Company attorneys, safety and health professionals, and environmental specialists should be involved in the development of any procedures that deal with or address safety, emergency response, or employee health issues. A company official should be pre-designated as a public relations officer with specific training in dealing with the press. Corporate plans and policies should be developed, approved, and implemented long before any need for such arises.

REFERENCES

Ahlbom, A. 1993. *Biostatistics for Epidemiologists.* New York: CRC Press/Lewis Publishers.

American Conference of Governmental Industrial Hygienists. 1988. *Documentation of the Threshold Limit Values and Biological Exposure Indices,* 5th Edition (with updates). Cincinnati: ACGIH

Anthony, C. P., and N. J. Kolthoff. 1971. *Textbook of Anatomy and Physiology.* St. Louis: C. V. Mosby Company.

Asante-Duah, D. K. 1993. *Hazardous Waste Risk Assessment.* New York: CRC Press/Lewis Publishers.

Bloom, A. D. and F. J. DeSerres. 1995. *Ecotoxicity and Human Health.* New York: CRC Press/Lewis Publishers.

Brusick, D. J. 1994. *Methods for Generic Risk Assessment.* New York: CRC Press/Lewis Publishers.

Bueche, J. R. 1972. *Principles of Physics,* 2nd Edition. New York: McGraw-Hill.

Burrell, R., D. K. Flaherty, and L. J. Sauers. 1992. *Toxicology of the Immune System.* New York: Van Nostrand Reinhold.

Calabrese, E. J., E. M. Kenyon. *Air Toxics and Risk Assessment.* New York: CRC Press/Lewis Publishers.

Calabrese, E. J. 1994. *Biological Effects of Low Level Exposures.* New York: CRC Press/Lewis Publishers.

Cockerman, L. G., B. S. Shane. 1994. *Basic Environmental Toxicology.* New York: CRC Press/Lewis Publishers.

Cothern, R. R. 1993. *Comparative Environmental Risk Assessment.* New York: CRC Press/Lewis Publishers.

Cothern, C. R. 1995. *Handbook for Environmental Risk Decisionmaking.* New York: CRC Press/Lewis Publishers.

DiVincenzo, G. O., F. J. Yanno, and B. D. Astill. 1973. Exposure of man and dog to low concentrations of ketene vapor. *American Industrial Hygiene Journal.* 34:329-336.

Gardner, R. W. 1994. *Chemical Intolerance.* New York: CRC Press/Lewis Publishers.

Hallenbeck, W. H. 1993. *Quantitative Risk Assessment for Environmental and Occupational Health,* 2nd Edition. New York: CRC Press/Lewis Publishers.

Hodgson, E., R. B. Mailman, and J. E. Chambers. 1988. *Dictionary of Toxicology.* London: The MacMillian Company.

Hosty, J. W., and P. Foster. 1990. *A Practical Guide to Chemical Spill Response.* New York: Van Nostrand Reinhold.

Howard, P. H. 1995. *Handbook of Environmental Fate and Exposures Data for Organic Chemicals,* Vol. I-IV. New York: CRC Press/Lewis Publishers.

Howard, P. H., et. al. 1991. *Handbook of Environmental Degradation Rates.* New York: CRC Press/Lewis Publishers.

James, A., and M. Lord. 1992. *Index of Chemical and Physical Data.* New York: Van Nostrand Reinhold.

Jones, F. E. 1994. *Toxic Organic Vapors in the Workplace.* New York: CRC Press/Lewis Publishers.

Kamrin, M. A. 1988. *Toxicology: A Primer on Principles and Applications.* Chelsea, Michigan: Lewis Publishers.

Keith, L. H. and M. M. Walker. 1995. *Handbook of Air Toxics.* New York: CRC Press/Lewis Publishers.

Landis, W. G., M. H. Yu. 1995. *Introduction to Environmental Toxicology.* New York: CRC Press/Lewis Publishers.

Langerman, N. 1994 *Precautionary Labels for Chemical Containers.* New York: CRC Press/Lewis Publishers.

Lauwerys, R. R., P. Hoet. 1993. *Industrial Chemical Exposure.* New York: CRC Press/Lewis Publishers.

Lewis, R. J., Sr. 1993. *Hawley's Condensed Chemical Dictionary,* 12th Edition. New York: Van Nostrand Reinhold.

Lewis, R. J., Sr. 1992. *Sax's Dangerous Properties of Industrial Materials,* Volumes 1, 2, and 3, 8th Edition. New York: Van Nostrand Reinhold.

Lide, D. R. 1995. *Handbook of Chemistry and Physics,* 76th Edition. New York: CRC Press/Lewis Publishers.

Manahan, S. E. 1994. *Environmental Chemistry,* 6th Edition. New York: CRC Press/Lewis Publishers.

Manahan, S. E. 1992. *Toxicological Chemistry,* 2nd Edition. New York: CRC Press/Lewis Publishers.

Mellström, G. A., J. E. Wahlberg, and H. I. Maibach. 1994. *Protective Gloves for Occupational Use.* New York: CRC Press/Lewis Publishers.

Meyer, E. 1990. *Chemistry of Hazardous Materials.* Englewood Cliffs, NJ: Prentice Hall.

Patnaik, P. 1992. *A Comprehensive Guide to the Hazardous Properties of Chemical Substances.* New York: Van Nostrand Reinhold.

Philp, R. B. 1995. *Environmental Hazards and Human Health.* New York: CRC Press/Lewis Publishers.

Proctor, N. H., and J. P. Hughes. 1991. *Chemical Hazards of the Workplace.* New York: Van Nostrand Reinhold.

Quigley, D. R. 1994. *Handbook of Emergency Chemical Management.* New York: CRC Press/Lewis Publishers.

Raleigh, R. L. and W. A. McGee. 1972. Effects of short, high concentration exposures to ketene as determined by observation in the work area. *Journal of Occupational Medicine.* 14:607-610.

Rea, W. J. 1995. *Chemical Sensitivity,* Vol. I-III. New York: CRC Press/Lewis Publishers.

Richardson, M. 1995. *Dictionary of Substance and Their Effects,* Vol. I-VI. New York: CRC Press/Lewis Publishers.

U.S. Department of Health and Human Services, National Institute for Occupational Safety and Health. 1994. *NIOSH Pocket Guide to Chemical Hazards.* Washington, D.C.: U.S. Government Printing Office.

U.S. Department of Health and Human Services, National Institute for Occupational Safety and Health. 1978. *Criteria for a Recommended Standard - Occupational Exposure to Ketones.* DHEW Publication Number: 78-173. Washington, D.C.: U.S. Government Printing Office.

MATERIAL SAFETY DATA SHEET

CHEMICAL NAME	CRC PRESS/LEWIS PUBLISHERS DISCLAIMER
LACTIC ACID	

HAZARD WARNING INFORMATION

HEALTH	FIRE	REACTIVE	OTHER	DEGREE OF HAZARD	COLOR CODING	OTHER CODES
2	0	0		0 = Minimum Hazard 1 = Slight Hazard 2 = Moderate Hazard 3 = Serious Hazard 4 = Severe Hazard	HEALTH = BLUE FIRE = RED REACTIVITY = YELLOW OTHER = WHITE	OX = Oxidizer ACID = Acid ALK = Alkali COR = Corrosive ~~W~~ = Use No Water

SECTION I - GENERAL INFORMATION

Characterization	**Acid**	RCRA Number	**None**
		EPA Class	**Not Applicable**
DOT Proper Shipping Name	**Corrosive Liquid, N.O.S.**	Chemical Abstract Service (CAS) Number	**50-21-5**
DOT Hazard Class and Label Requirements	**Corrosive**	DOT Emergency Guide Code	**60**
DOT Identification Number	**UN 1760**	Chemical Formula	**CH_3CH_2OCOOH**

Synonyms

Milk acid; α-hydroxypropionic; acetonic acid; ethylidenelactic acid; 1-hydroxyethanecarboxylic acid.

SECTION II - HAZARDOUS INGREDIENTS AND IDENTITY INFORMATION

Hazardous Components (specific identity)	OSHA Exposure Criteria	NIOSH Exposure Criteria	Immediately Dangerous to Life and Health (IDLH)	ACGIH Exposure Criteria
Lactic acid (derivation: By fermenting starch, milk whey, molasses, potatoes, etc., and neutralizing the acid as soon as formed with calcium or zinc carbonate. The solution of lactates is concentrated and decomposed with sulfuric acid; synthetically by hydrolysis of lactonitrile).	**PEL:** Not Established **STEL:** Not Established	**REL:** Not Established **STEL:** Not Established	**Not Established**	**TLV:** Not Established **STEL:** Not Established

SECTION III - PHYSICAL AND CHEMICAL CHARACTERISTICS

Boiling Point	**252°F (122°C)**	Specific Gravity (H_2O = 1)	**1.249**
Vapor Pressure (mm Hg)	**Not Found**	Molecular Weight	**90.09**
Vapor Density (Air = 1)	**Not Found**	Melting Point	**62°F (17°C)**

Solubility

Soluble in water, furfurol, alcohol, glycerol, ester. Nearly insoluble in chloroform, petroleum ether, and carbon disulfide.

Appearance and Odor

Yellow or colorless crystalline solid or a thick, syrupy, hygroscopic liquid with a slightly acrid odor.

SECTION IV - FIRE AND EXPLOSION HAZARD DATA

Flash Point (method used)	**Not Determined**	Explosive Limits in Air % by Volume	**LEL: Not Determined UEL: Not Determined**
NFPA Classification	**Non-Combustible Solid**	Autoignition Temperature	**Not Determined**

Extinguishing Media

Use water, foam, dry chemical, or agent suitable to surrounding media. Lactic acid itself does not burn.

Special Fire Fighting Procedures

Poisonous gases are produced in fire. Wear full protective clothing and self-contained breathing apparatus (SCBA). Use water spray to flush and dilute a spill and to disperse the vapors. Also, use a water spray to keep fire-exposed containers cool. Do not release runoff from fire control measure to sewers or waterways. Firefighters should avoid contact with vapors produced during fire.

Unusual Fire and Explosion Hazards

None reported.

SECTION V - REACTIVITY DATA

Stability		Conditions to Avoid Lactic acid is normally stable under routine conditions of handling and storage. Avoid contact with heat and ignition sources and incompatible materials.
Stable X	Unstable	Incompatibility (materials to avoid) Mixtures with nitric acid + hydrofluoric acid reacts vigorously. Corrodes some metals (steel, aluminum, cast iron, carbon steel).
Hazardous Polymerization		Conditions to Avoid Under normal conditions of temperature and pressure, hazardous polymerization of lactic acid is not expected to occur.
May Occur	Will Not Occur X	Hazardous Decomposition or By-products When heated to decomposition, lactic acid can emit highly toxic and acrid smoke and irritating fumes, including toxic oxides of carbon.

SECTION VI - HEALTH HAZARD DATA

Primary Route(s) of Entry:	Inhalation?	Absorption (skin)?	Ingestion?
	X		X

Health Hazards

INHALATION: Extremely irritating to the eyes, nose, throat, and respiratory system. Inhaled dusts or mists may cause coughing, bronchial irritation, and inflammation of associated mucosa.

EYES & SKIN: Skin and eye contact causes irritation. Eye contact produces burns, conjunctivitis, and possible corneal abrasions. Skin contact causes burns and may cause eczema.

INGESTION: Causes burns to the mouth, throat, esophagus, and stomach with nausea and vomiting, sweating, dypsnea, cyanosis, and vascular effects.

Carcinogenicity	NTP Listed?	IARC Cancer Review Group?	OSHA Regulated?	Target Organs?
Unknown Human Unknown Animal	No	No	No	Respiratory system, eyes, skin.

Medical Conditions Generally Aggravated by Exposure

None reported.

Emergency and First-aid Procedures

Eye contact: Flush immediately with water for 15 minutes (minimum); seek medical attention. Skin contact: Remove all contaminated clothing. Immediately flush area with large amounts of water. For red or blistered skin, wash with soap and water. Seek medical assistance. For inhalation: Remove the person from exposure. Provide respiratory assistance and CPR. Transfer to medical facility. If swallowed: Contact poison control center. Unless advised otherwise, give 1 to 2 glasses of water or milk. Do NOT induce vomiting. Never attempt to give anything by mouth to an unconscious or convulsing person.

SECTION VII - PRECAUTIONS FOR SAFE HANDLING AND USE

Steps to be Taken in Case Material is Released or Spilled

Collect powdered materials in safest manner possible. Do NOT dry sweep (generates dusts). Use a HEPA vacuum instead. Absorb solution spills in vermiculite or similar materials and deposit in sealed containers. Ventilate area of spill or leak. Restrict those not involved in cleanup from entering area.

Preferred Waste Disposal Method

Mix with a combustible liquid and burn in a chemical incinerator with afterburner and scrubber.

Precautions to be Taken in Handling and Storage

Avoid contact with incompatible materials, especially nitric and hydrofluoric acid. Store in tightly closed containers in cool, well-ventilated area away from heat. Protect container from damage.

Other Precautions and Warnings

Avoid generating dusty conditions in storage areas. Do not store in aluminum, cast iron, or carbon steel containers. Keep containers tightly closed.

SECTION VIII - CONTROL MEASURES AND PERSONAL PROTECTIVE EQUIPMENT

Respiratory Protection *(specify type)*

No exposure levels have been established for lactic acid. An air purifying respirator with acid cartridge and a HEPA filter may suffice. However, better protection is provided by a supplied-air respirator with full facepiece operated in pressure demand or other positive pressure mode.

Ventilation

Local exhaust or general mechanical systems recommended.

Protective Gloves	Eye Protection	Other Protective Clothing
Polyethylene or Teflon	Chemical/Dust Goggles or Face Mask	Protective Apron

Work/Hygiene Practices

Always wash hands thoroughly after using chemical; never bring food, drink, or smoking materials into vicinity of chemicals.

LACTIC ACID

CH_3CH_2OCOOH CAS: 50-21-5

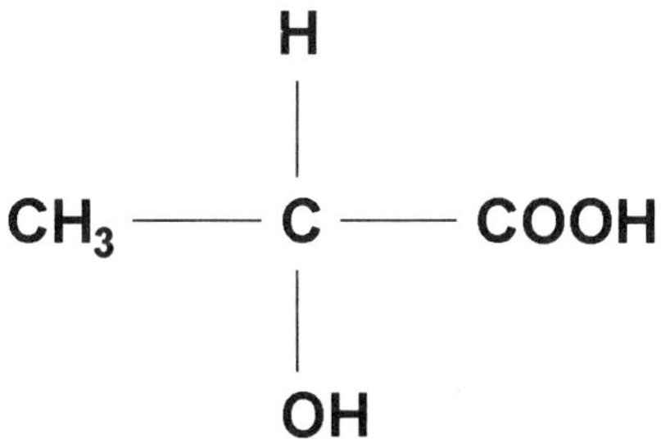

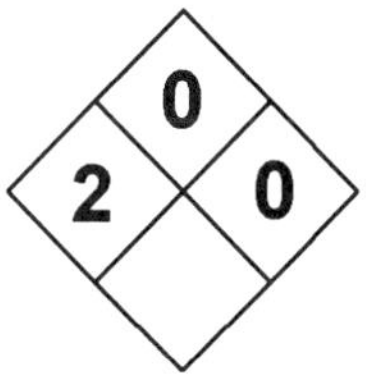

IDENTIFICATION AND TYPICAL USES

Lactic acid appears as a colorless or slightly yellow crystalline solid or a thick, syrupy, hygroscopic liquid with a slightly acrid odor. It is used in cultured dairy products, as a mordant in dyeing baths, adhesives, plastics, and textiles, as an acidulant, as a general purpose food additive, a solvent for cellulose formate, as a flux for soft solder, in the manufacture of cheese and confectionery, in dehairing, plumbing, and decalcifying hides, in the manufacture of lactates for food products, medicine, and as a solvent, a plasticizer, and as a catalyst in casting of phenolaldehyde resins.

RISK ASSESSMENT: HEALTH

General Assessment

Exposure to lactic acid is possible through *inhalation* of its dusts or mists and by *ingestion.* The data in the references are limited regarding its specific toxicity. However, there are no reports of animal or human carcinogenicity or teratogenicity. Mutation data have been reported.

Inhalation of lactic acid dusts or mists will cause irritation of the eyes, nose, throat, and mucous membranes of the upper respiratory tract. Bronchial tube irritation and coughing may also occur. Lactic acid can cause dental erosion.

Skin and eye contact with concentrated solutions can result in mild to moderate irritation. On the skin, lactic acid can cause burns, irritation, and possible eczema. If allowed to remain in contact with the eye for prolonged periods, corneal edema, burns, and conjunctivitis may occur.

Ingestion of diluted solutions can cause burns to the mouth, throat, esophagus, and stomach with abdominal cramps, nausea, vomiting, sweating, dypsnea, cyanosis, and vascular collapse.

Acute Health Effects

The following acute (short-term) health effects may occur immediately or shortly after exposure to lactic acid:

Skin: Irritation/rash or burning feeling on contact. With possible development of eczema.

Eye: Severe irritation; may cause burns, conjunctivitis, and corneal damage.

Lung: Irritation of the nose, throat, and mucosa of the lungs and respiratory tract.

Other: Ingestion causes burns to the digestive tract with vomiting, nausea, cramps, shortness of breath, cyanosis, and vascular effects. Dental erosion is also a possibility.

Chronic Health Effects

The following chronic (long-term) health effects may occur at some time after exposure to lactic acid and can last for months or even years:

Cancer Hazards: According to information presented in the references, lactic acid has not been adequately tested for its ability to cause cancer in animals. Mutation data have been reported and some scientists believe that long-term exposure to mutagens may pose a cancer risk. There is no indication that exposure to lactic acid will have this effect. Additional research is required.

Reproductive Hazard: According to information presented in the references, lactic acid has not been adequately tested for its ability to adversely affect reproduction in laboratory animals.

☛ *Recommended Risk-Reduction Measures*

Personnel should avoid direct contact with lactic acid. The exact nature of its toxicity is not clearly understood in the references. If a less toxic chemical cannot be substituted for a hazardous substance, then *engineering controls* are the most effective method of reducing exposures.

The best protection is to enclose operations and/or provide local exhaust ventilation at the site of chemical release. While not always operationally feasible, isolating operations can also reduce exposure. Using respiratory protection is less effective than the controls mentioned above, but is still advisable whenever working with or around lactic acid. No exposure levels have been established for lactic acid. This does not mean that exposure is without health risk. Prudent risk management may require a conservative approach to respiratory protection until further data are known about lactic acid. For transient exposures, an air-purifying respirator equipped with an acid cartridge, a HEPA filter, and a dust/mist pre-filter may suffice. Greater protection is obtained using a supplied-air respirator or a self-contained breathing apparatus (SCBA) with full facepiece operated in pressure demand or other positive pressure mode. If a full facepiece is not available, then chemical goggles should be worn to protect the eyes. Whenever a chemical splash/dust hazard exists, a face shield and a rubber apron should be worn. To prevent hand and skin exposures, acid resistant gloves should be worn. Glove manufacturers should be contacted and permeation studies obtained *before* final glove selections have been made.

Administrative controls should also be in place to minimize the potential for human exposures. These may include written procedures or policies which specify the methods and techniques that will be practiced whenever personnel are to work with lactic acid.

All personnel should receive *training* on the use, hazards, protective measures, emergency actions, and other precautions per 29 CFR 1910.1200 (Hazard Communication), prior to the first assignment in an area where lactic acid is used or stored. Other methods to reduce exposure include:

☑ Where possible, enclose operations and use local exhaust ventilation at the site of chemical release. If local exhaust ventilation or enclosure is not used, respiratory protection should be mandatory.

☑ Always ensure that proper protective clothing is worn when using chemical substances.

☑ Wash thoroughly immediately after exposure to lactic acid and at the end of the work shift or before eating, drinking, or smoking.

☑ Hazard warning information should be posted in the work area. In addition, as part of an on-going education and training program, all information on the health and safety hazards of lactic acid should be communicated to all exposed or potentially exposed workers.

RISK ASSESSMENT: ENVIRONMENT

General Assessment

The environment is at risk of exposure during transportation, storage, disposal, or destruction of lactic acid. In almost every scenario, the threat of environmental exposure is contingent upon the proper handling of the chemical substance. Accidental spills, large or small, can result in fire, explosion, and possible contamination of the surrounding environmental mediums (water, soil, and air).

Lactic acid is considered a non-combustible solid. It may be corrosive, especially when mixed in solution. It is incompatible with nitric and hydrofluoric acid and contact can cause violent reaction. It will also corrode many common metals, including aluminum, cast iron, steel (stainless steel appears to resist corrosion). These characteristics require special consideration during any emergency situation involving a leak or spill of lactic acid.

Lactic acid can enter the environment mainly industrial discharges or spills.

☠ *Acute Ecological Effects*

Acute (short-term) toxic effects may include the death of animals, birds, or fish and death or low growth rate in plants. Acute effects are seen 2 to 4 days after animals or plants are exposed to lactic acid.

Lactic acid has moderate acute toxicity to aquatic life. Insufficient data are available to evaluate or predict the short-term effects of lactic acid exposure to birds, plants, or land animals.

Chronic Ecological Effects

Chronic (long-term) toxic effects may include shortened life span, reproductive problems, lower fertility, and changes in appearance or behavior in exposed animals. These effects can be seen long after first exposure(s) to toxic chemicals. Insufficient data are available to

evaluate or predict the long-term effects of lactic acid exposure to aquatic life, birds, plants, or land animals.

Water Solubility

Lactic acid is soluble in water. Concentrations of up to 1000 milligrams and more will mix with a liter of water.

Persistence in the Environment

Lactic acid is slightly persistent in the aquatic environment, with a half-life of between 2 to 20 days. The half-life of a pollutant is the amount of time it takes for one half of the chemical to be degraded.

Bioaccumulation in Aquatic Organisms

Some substances increase in concentration, or *bioaccumulate,* in living organisms as they breathe contaminated air, drink contaminated water, or eat contaminated food. These chemicals can become concentrated in the tissues and internal organs of animals as well as humans.

Lactic acid can be expected to biodegrade quickly and should not accumulate in the edible tissues of fish.

Recommended Risk-Reduction Measures

Proper training of all transporters will reduce the likelihood of a mishap or accident resulting in a leak or spill to the environment. The correct use of DOT labeling on all containers, trucks, and rail cars will enable emergency responders to react properly and quickly to any disaster thereby reducing the potential risk to the environment and to personnel.

Storage of lactic acid should be segregated from incompatible chemicals to minimize the risk of cross-contamination or contact. Buildings designated for storage should be equipped with appropriate fire protection systems (alarms, sprinklers, emergency lighting, portable extinguishers).

If a spill or leak to the environment has occurred, fire department, emergency response and/or hazardous materials spill personnel should be notified immediately. Cleanup should be attempted only by those trained in proper spill containment procedures. Contaminated soils should be removed for incineration and replaced with clean soil. If lactic acid should contact the water table, aquifer, or navigable waterway, time is of the essence. It is miscible in water and, therefore, total containment and remediation may not be possible. When such spills occur, the local and/or state emergency response authorities must be notified. A comprehensive emergency response or disaster preparedness/recovery plan should be in place prior to any operations involving the use, transportation, storage, or disposal of lactic acid. If lactic acid is spilled or leaked, the following specific steps are recommended:

- ☑ Restrict persons not wearing protective clothing from area of spill or leak until cleanup is complete and area can be opened for normal work.
- ☑ Remove ignition sources and ventilate area.
- ☑ Collect all powdered materials in safest and most convenient manner possible and deposit in sealed drums for disposal. Do NOT dry sweep (generates airborne dusts). Use a vacuum equipped with a high efficiency particulate air (HEPA) filter instead. Damp mop residue and collect for disposal. Absorb any solution spills in vermiculite, dry sand, earth, or similar material and deposit in sealed containers.
- ☑ It may be necessary to dispose of lactic acid as a hazardous waste. The responsible state agency or the regional office of the federal Environmental Protection Agency (EPA) should be contacted for specific recommendations.

RISK ASSESSMENT: BUSINESS

General Assessment

Accidents or mishaps involving lactic acid can present a significant threat to business operations. The loss or damage of equipment or facilities can significantly affect fiscal viability. Lawsuits that may result from personnel injury/death, public exposures, and/or environmental contamination will also require a serious expenditure of resources. Media attention surrounding an injury, death, or environmental damage can also result in a loss of profits and loss of current as well as future business.

Recommended Risk-Reduction Measures

Company attorneys, safety and health professionals, and environmental specialists should be involved in the development of any procedures dealing with safety or emergency response. A company official should be pre-designated as a public relations officer with specific training in dealing with the press. Corporate plans and policies should be developed, approved, and implemented long before any need for such arises.

REFERENCES

Ahlbom, A. 1993. *Biostatistics for Epidemiologists.* New York: CRC Press/Lewis Publishers.

American Conference of Governmental Industrial Hygienists. 1988. *Documentation of the Threshold Limit Values and Biological Exposure Indices,* 5th Edition (with updates). Cincinnati: ACGIH

Asante-Duah, D. K. 1993. *Hazardous Waste Risk Assessment.* New York: CRC Press/Lewis Publishers.

Bloom, A. D. and F. J. DeSerres. 1995. *Ecotoxicity and Human Health.* New York: CRC Press/Lewis Publishers.

Brusick, D. J. 1994. *Methods for Genetic Risk Assessment.* New York: CRC Press/Lewis Publishers.

Bueche, J. R. 1972. *Principles of Physics,* 2nd Edition. New York: McGraw-Hill.

Burrell, R., D. K. Flaherty, and Leonard J. Sauers. 1992. *Toxicology of the Immune System.* New York: Van Nostrand Reinhold.

Calabrese, E. J. 1994. *Biological Effects of Low Level Exposures.* New York: CRC Press/Lewis Publishers.

Cockerman, L. G., B. S. Shane. 1994. *Basic Environmental Toxicology.* New York: CRC Press/Lewis Publishers.

Cothern, C. R. 1995. *Handbook for Environmental Risk Decisionmaking.* New York: CRC Press/Lewis Publishers.

Danse, I. R. 1991. *Common Sense Toxics in the Workplace.* New York: Van Nostrand Reinhold.

Gardner, R. W. 1994. *Chemical Intolerance.* New York: CRC Press/Lewis Publishers.

Hallenbeck, W. H. 1993. *Quantitative Risk Assessment for Environmental and Occupational Health,* 2nd Edition. New York: CRC Press/Lewis Publishers.

Hosty, J. W., P. Foster. 1990. *A Practical Guide to Chemical Spill Response.* New York: Van Nostrand Reinhold.

Howard, P. H., et. al. 1991. *Handbook of Environmental Degradation Rates.* New York: CRC Press/Lewis Publishers.

James, A., M. Lord. 1992. *Index of Chemical and Physical Data.* New York: Van Nostrand Reinhold.

Jones, F. E. 1994. *Toxic Organic Vapors in the Workplace.* New York: CRC Press/Lewis Publishers.

Keith, L. H. and M. M. Walker. 1995. *Handbook of Air Toxics.* New York: CRC Press/Lewis Publishers.

Landis, W. G., M. H. Yu. 1995. *Introduction to Environmental Toxicology.* New York: CRC Press/Lewis Publishers.

Langerman, N. 1994 *Precautionary Labels for Chemical Containers.* New York: CRC Press/Lewis Publishers.

Lauwerys, R. R., P. Hoet. 1993. *Industrial Chemical Exposure.* New York: CRC Press/Lewis Publishers.

Lewis, R. J., Sr. 1993. *Hawley's Condensed Chemical Dictionary,* 12th Edition. New York: Van Nostrand Reinhold.

Lewis, R. J., Sr. 1992. *Sax's Dangerous Properties of Industrial Materials,* Volumes 1, 2, and 3, 8th Edition. New York: Van Nostrand Reinhold.

Lide, D. R. 1995. *Handbook of Chemistry and Physics,* 76th Edition. New York: CRC Press/Lewis Publishers.

Manahan, S. E. 1994. *Environmental Chemistry,* 6th Edition. New York: CRC Press/Lewis Publishers.

Manahan, S. E. 1992. *Toxicological Chemistry,* 2nd Edition. New York: CRC Press/Lewis Publishers.

Mellström, G. A., J. E. Wahlberg, and H. I. Maibach. 1994. *Protective Gloves for Occupational Use.* New York: CRC Press/Lewis Publishers.

Meyer, E. 1990. *Chemistry of Hazardous Materials.* Englewood Cliffs, NJ: Prentice Hall.

Mycock, J. C., J. D. McKenna, and L. Theodore. 1995. *Handbook of Air Pollution Control Engineering and Technology.* New York: CRC Press/Lewis Publishers.

Nelson, G. O. 1992. *Gas Mixtures.* New York: CRC Press/Lewis Publishers.

Ott, W. R. 1995. *Environmental Statistics and Data Analysis.* New York: CRC Press/Lewis Publishers.

Patnaik, P. 1992. *A Comprehensive Guide to the Hazardous Properties of Chemical Substances.* New York: Van Nostrand Reinhold.

Philp, R. B. 1995. *Environmental Hazards and Human Health.* New York: CRC Press/Lewis Publishers.

Proctor, N. H., J. P. Hughes. 1991. *Chemical Hazards of the Workplace.* New York: Van Nostrand Reinhold.

Quigley, D. R. 1994. *Handbook of Emergency Chemical Management.* New York: CRC Press/Lewis Publishers.

Rea, W. J. 1995. *Chemical Sensitivity,* Vol. I-III. New York: CRC Press/Lewis Publishers.

Richardson, M. 1995. *Dictionary of Substance and Their Effects,* Vol. I-VI. New York: CRC Press/Lewis Publishers.

Rylander, R., R. R. Jacobs. 1994. *Organic Dusts.* New York: CRC Press/Lewis Publishers.

Sacarello, H. L. A. 1994. *The Comprehensive Handbook of Hazardous Materials.* New York: CRC Press/Lewis Publishers.

Sheftel, V. O. 1995. *Handbook of Toxic Properties of Monomers and Additives.* New York: CRC Press/Lewis Publishers.

Sherman, J. D. 1988. *Chemical Exposure and Disease.* New York: Van Nostrand Reinhold.

Talbott, E., G. Craun. 1995. *An Introduction to Environmental Epidemiology.* New York: CRC Press/Lewis Publishers.

U.S. Department of Health and Human Services, National Institute for Occupational Safety and Health. 1980. *Lactic acid. Registry of Toxic Effects of Chemical Substances,* Vol. 1. Washington, D.C.: U.S. Government Printing Office.

U.S. Department of Health and Human Services, National Institute for Occupational Safety and Health. 1994. *NIOSH Pocket Guide to Chemical Hazards.* Washington, D.C.: U.S. Government Printing Office.

Witorsch, P. and S. V. Spagnolo. 1994. *Air Pollution and Lung Disease in Adults.* New York: CRC Press/Lewis Publishers.

MATERIAL SAFETY DATA SHEET

CHEMICAL NAME	
LACTONITRILE	**CRC PRESS/LEWIS PUBLISHERS DISCLAIMER:** THE INFORMATION AND RECOMMENDATIONS PRESENTED HEREIN ARE BASED ON SOURCES BELIEVED TO BE RELIABLE. CRC MAKES NO REPRESENTATION ON ITS COMPLETENESS OR ACCURACY. IT IS THE RESPONSIBILITY OF THE USER TO DETERMINE THE CHEMICAL'S SUITABILITY FOR ITS INTENDED USE, THE CHEMICAL'S SAFE USE, AND THE CHEMICAL'S PROPER DISPOSAL. NO REPRESENTATIONS AND/OR WARRANTIES, EITHER EXPRESSED OR IMPLIED, OF THE MERCHANTABILITY OR FITNESS FOR A PARTICULAR PURPOSE, OR OF ANY OTHER NATURE, ARE MADE WITH RESPECT TO THE INFORMATION PROVIDED IN THIS MSDS OR TO THE CHEMICAL TO WHICH INFORMATION MAY REFER. CRC NEITHER ASSUMES NOR AUTHORIZES ANY OTHER PERSON TO ASSUME FOR IT, ANY OTHER ADDITIONAL RESPONSIBILITY OR LIABILITY FOR THE USE OF, OR RELIANCE UPON, THIS INFORMATION.

HAZARD WARNING INFORMATION

HEALTH	FIRE	REACTIVE	OTHER	DEGREE OF HAZARD	COLOR CODING	OTHER CODES
4	2	1		0 = Minimum Hazard 1 = Slight Hazard 2 = Moderate Hazard 3 = Serious Hazard 4 = Severe Hazard	HEALTH = BLUE FIRE = RED REACTIVITY = YELLOW OTHER = WHITE	OX = Oxidizer ACID = Acid ALK = Alkali COR = Corrosive W = Use No Water

SECTION I - GENERAL INFORMATION

Characterization **Nitrile**	RCRA Number **D003**	EPA Class **Characteristic (R) Waste**
DOT Proper Shipping Name **Poisonous Liquids, Flammable, N.O.S.**	Chemical Abstract Service (CAS) Number **78-97-7**	
DOT Hazard Class and Label Requirements **Poison; Flammable Liquid**	DOT Emergency Guide Code **28**	
DOT Identification Number **UN 2929**	Chemical Formula **$C_2H_4(OH)CN$**	

Synonyms
Acetaldehyde cyanohydrin; acetocyanohydrin; ethylidene cyanohydrin; 2-hydroxypropanenitrile; alpha-hydroxypropionitrile; α-hydroxypropionitrile.

SECTION II - HAZARDOUS INGREDIENTS AND IDENTITY INFORMATION

Hazardous Components (specific identity)	OSHA Exposure Criteria	NIOSH Exposure Criteria	Immediately Dangerous to Life and Health (IDLH)	ACGIH Exposure Criteria
Lactonitrile (derivation: From acetaldehyde and hydrocyanic acid).	**PEL:** Not Established **STEL:** Not Established	**REL:** Not Established **STEL:** Not Established	**Not Determined**	**TLV:** Not Established **STEL:** Not Established

SECTION III - PHYSICAL AND CHEMICAL CHARACTERISTICS

Boiling Point **361°F (183°C) with decomposition**	Specific Gravity (H_2O = 1) **0.98**
Vapor Pressure (mm Hg) **0.28 at 77°F (25°C)**	Molecular Weight **71.09**
Vapor Density (Air = 1) **2.45**	Freezing Point **-40°F (-40°C)**

Solubility
Soluble in water, alcohol, acetone, and ether. Insoluble in petroleum ether and carbon disulfide.

Appearance and Odor
Colorless to straw-colored liquid.

SECTION IV - FIRE AND EXPLOSION HAZARD DATA

Flash Point (method used) **171°F (77°C) opened cup**	Explosive Limits in Air % by Volume **LEL: Not Determined UEL: Not Determined**
NFPA Classification **Class IIIA Combustible Liquid**	Autoignition Temperature **Not Determined**

Extinguishing Media
Carbon dioxide, dry chemical, or regular foam.

Special Fire Fighting Procedures
Wear full protective clothing and self-contained breathing apparatus (SCBA). Use a water spray to keep fire-exposed containers cool. Poisonous gases are produced in fire, including cyanide and nitrogen oxide. Vapors are heavier than air; flashback is likely.

Unusual Fire and Explosion Hazards
Potentially explosive reaction with when exposed to oxidizers and oxidizing materials.

SECTION V - REACTIVITY DATA

Stability		Conditions to Avoid
		Lactonitrile in normally stable under routine conditions of handling and storage. Avoid contact with heat, ignition sources, and incompatible materials.
Stable	Unstable	Incompatibility *(materials to avoid)*
X		Strong oxidizers, acids, chlorates, and nitrates. It will also react with alkalis to evolve toxic hydrogen cyanide gas.
Hazardous Polymerization		Conditions to Avoid
		Under normal conditions of temperature and pressure, hazardous polymerization of lactonitrile will not occur.
May Occur	Will Not Occur	Hazardous Decomposition or By-products
	X	When heated to decomposition, lactonitrile emits highly toxic fumes of cyanide and nitrogen oxide.

SECTION VI - HEALTH HAZARD DATA

Primary Route(s) of Entry:	Inhalation?	Absorption (skin)?	Ingestion?
	X	X	X

Health Hazards

INHALATION: Effects are similar to that of cyanide poisoning. Symptoms include weakness, nausea, vomiting, headaches, rapid respiration, and symptoms of asphyxiation with a feeling of air hunger, faulty judgment, perception difficulties, blurred vision, unconsciousness, gasping breath, and death.

ABSORPTION: Skin and eye contact can result in irritation. Absorption can produce similar systemic effects as listed for inhalation.

INGESTION: Causes toxic systemic effects similar to that of inhalation.

Carcinogenicity	NTP Listed?	IARC Cancer Review Group?	OSHA Regulated?	Target Organs?
Human: not tested Animal: possible	No	No	No	Skin, eyes, respiratory system, CVS, and CNS.

Medical Conditions Generally Aggravated by Exposure

None reported.

Emergency and First-aid Procedures

Eye contact: Flush immediately with water for 15 minutes (minimum); seek medical attention. Skin contact: Remove all contaminated clothing. Immediately wash area with large amounts of soap and water. Seek medical assistance. For inhalation: Remove the person from exposure. Provide respiratory assistance and CPR. Transfer to medical facility. Lactonitrile can cause fatal cyanide poisoning. If swallowed: Contact poison control center. Unless advised otherwise, give 1 to 2 glasses of water to dilute. Do NOT induce vomiting. Use Lilly cyanide antidote kit. Seek medical attention immediately.

SECTION VII - PRECAUTIONS FOR SAFE HANDLING AND USE

Steps to be Taken in Case Material is Released or Spilled

Contain spills using absorbent material (vermiculite, or other). Ventilate area of spill. Have water source available in case of fire. Store materials in DOT-approved containers. Restrict those not involved in cleanup from entering area. Notify appropriate authorities, as required by SARA III, if applicable.

Preferred Waste Disposal Method

No citation.

Precautions to be Taken in Handling and Storage

Lactonitrile is highly volatile. Store to avoid contact with strong oxidizers. Store in tightly closed containers in cool, well-ventilated area away from heat. Use non-sparking tools.

Other Precautions and Warnings

Prior to working with lactonitrile, personnel should be trained on its proper handling and storage.

SECTION VIII - CONTROL MEASURES AND PERSONAL PROTECTIVE EQUIPMENT

Respiratory Protection *(specify type)*

For low exposures, NIOSH-approved full facepiece respirator with organic vapor cartridge. For higher exposures, use a supplied-air respirator in continuous flow mode or a self-contained breathing apparatus (SCBA) with full facepiece and pressure demand or other positive pressure mode.

Ventilation

Local exhaust or general mechanical systems recommended.

Protective Gloves	Eye Protection	Other Protective Clothing
Impervious Material	Chemical Goggles or Face Mask	Protective Apron

Work/Hygiene Practices

Always wash hands thoroughly after using chemical; never bring food, drink, or smoking materials into vicinity of chemicals.

LACTONITRILE

$C_2H_4(OH)CN$ CAS: 78-97-7

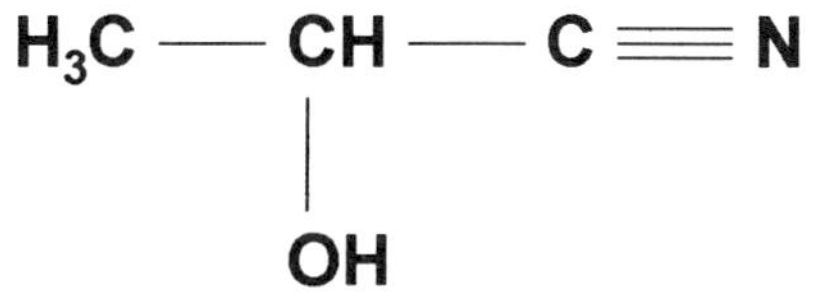

IDENTIFICATION AND TYPICAL USES

Lactonitrile is a colorless to straw-colored liquid. It is used as a solvent and as an intermediate in the production of ethyl lactate and lactic acid.

RISK ASSESSMENT: HEALTH

General Assessment

Lactonitrile is toxic by all common routes of exposure *(ingestion, inhalation,* and dermal *absorption*). The toxic effects of lactonitrile are attributed to the metabolic release of cyanide, which in turn acts by inhibiting cytochrome oxidase and thus impairs cellular respiration. There are no reports to indicate that exposure causes carcinogenic, mutagenic, or teratogenic effects.

Human exposure to lactonitrile can produce severe adverse effects. The target organs are the central nervous system (CNS), lungs, cardiovascular system (CVS), skin, and eyes. In humans, exposure to lactonitrile vapors can cause respiratory stimulation, dypsnea, ataxia, convulsion, decreased body temperature, and spastic paralysis. There can be nausea, vomiting, and tightness of the chest. It is excreted in the urine as a cyanate. The blood cyanide concentration does not always show any significant increase in cyanide at low exposure concentrations. Large doses can be fatal. Skin contact results in absorption and systemic effects. Severe eye irritation is likely.

Acute Health Effects

The following acute (short-term) health effects may occur immediately or shortly after exposure to lactonitrile. (Note: Exposure by any route can cause fatal cyanide poisoning. Symptoms of flushed face, chest tightness, nausea, vomiting, weakness, and shortness of breath may lead to death. These reactions may begin hours after an overexposure.)

Skin: Irritation/rash on contact (contact dermatitis) and absorption.

Eye: Severe irritation, conjunctivitis, and damage.

Lung: Irritation of the nose, throat, and lungs following exposure.

CNS: Inhalation of high concentrations of the vapor may cause sleepiness, dizziness, loss of consciousness, or death.

Chronic Health Effects

The following chronic (long-term) health effects may occur at some time after exposure to lactonitrile and can last for months or even years:

Cancer Hazards: According to the references, lactonitrile has not been tested for its ability to cause cancer in animals.

Reproductive Hazard: According to the references, lactonitrile has not been significantly tested for its ability to adversely affect reproduction.

Other Chronic Effects: High exposures may damage the nervous system, liver, and kidneys. Very irritating substances may affect the lungs. It is not known whether lactonitrile causes permanent lung damage. Prolonged or repeated skin contact may cause itching, rash, and possible dermatitis.

Recommended Risk-Reduction Measures

Personnel should avoid direct contact with lactonitrile. Unless a less toxic chemical can be substituted for a hazardous substance, *engineering controls* are the most effective method of reducing exposures. The best protection is to enclose operations and/or provide local exhaust ventilation at the site of chemical release. While not always operationally feasible, isolating operations can also reduce exposure. Using respiratory protection is less effective than the controls mentioned above, but is still advisable whenever working with or around lactonitrile. No exposure levels have been established for lactonitrile. However, this does not mean that exposure is without risk. For the best protection, a supplied-air respirator operated in continuous flow mode or a self-contained breathing apparatus (SCBA) with full facepiece operated in pressure demand are the recommended respiratory protection methods of choice. If a full facepiece is not available, then chemical goggles should be worn to protect the eyes. Whenever a chemical splash hazard exists, a face shield and a protective apron should be worn. To prevent hand and skin exposures, impervious gloves should be used. Glove manufacturers should be contacted and permeation studies obtained *before* final glove selections have been made.

Administrative controls should also be in place to minimize the potential for human exposures. These may include written procedures or policies which specify the methods and techniques that will be practiced whenever personnel are to work with lactonitrile.

All personnel should receive *training* on the use, hazards, protective measures, emergency actions, and other precautions per 29 CFR 1910.1200 (Hazard Communication), prior to the first assignment in an area where lactonitrile is used or stored.

Medical testing of personnel with frequent or potentially high exposure, should be provided before beginning work and at regular intervals thereafter (e.g., annually). These tests should include:

☑ Lung function (pulmonary function) test.

If symptoms develop or overexposure is suspected, the following tests may be useful:

☑ Blood cyanide test and/or urine thiocyanate test (currently, a blood cyanide level over 0.1 mg/l or urine thiocyanate over 20 mg/l indicates overexposure).

☑ Maintain close medical surveillance. Slow release of cyanide from absorbed lactonitrile may cause delayed symptoms.

☑ Liver and kidney function tests.

Other methods to reduce exposure include:

☑ Where possible, enclose operations and use local exhaust ventilation at the site of chemical release. If local exhaust ventilation or enclosure is not used, respiratory protection should be mandatory.

☑ Always ensure that proper protective clothing is worn when using chemical substances.

☑ Wash thoroughly immediately after exposure to lactonitrile and at the end of the work shift or before eating, drinking, or smoking.

☑ Hazard warning information should be posted in the work area. In addition, as part of an on-going education and training program, all information on the health and safety hazards of lactonitrile should be communicated to all exposed workers.

RISK ASSESSMENT: ENVIRONMENT

General Assessment

The environment is at risk of exposure during transportation, storage, disposal, or destruction of lactonitrile. In almost every scenario, the threat of environmental exposure is contingent upon the proper handling of the chemical substance. Accidental spills, large or small, can result in fire, explosion, and possible contamination of the surrounding environmental mediums (water, soil, and air).

Lactonitrile is considered a highly flammable liquid and a poison. Because of its low flash point and high boiling point, it is classified as a Class IIIA combustible liquid (per OSHA 29 CFR 1910.106). Lactonitrile vapors can form explosive mixtures in air. It reacts with strong oxidizers and acids, liberating heat along with pressure increase. It will also react with alkalis to evolve hydrogen cyanide. Thus, contact with these incompatible materials, especially under closed conditions, can cause dangerous reactions. These characteristics require special consideration during any emergency situation involving a leak or spill of lactonitrile. Should lactonitrile come into contact with incompatible materials either during use, transportation, or storage, the formation of highly toxic and/or highly explosive commodities is extremely possible.

The proper disposal/destruction method for lactonitrile is to mix it with a combustible solvent and burn the mixture in a chemical incinerator equipped with an afterburner and air scrubber.

Lactonitrile can enter the environment through manufacturing effluents, municipal waste treatment discharges, and spills.

☠ *Acute Ecological Effects*

Acute (short-term) toxic effects may include the death of animals, birds, or fish and death or low growth rate in plants. Acute effects are seen 2 to 4 days after animals or plants are exposed to lactonitrile. This chemical has slight acute toxicity to aquatic life. No data are available on the short-term effects of lactonitrile to plants or land animals.

Chronic Ecological Effects

Chronic toxic effects may include shortened life span, reproductive problems, lower fertility, and changes in appearance or behavior in exposed animals. These effects can be seen long after first exposure(s) to toxic chemicals. Lactonitrile has slight chronic toxicity in aquatic life. No data are available on the long-term effects of lactonitrile to plants, birds, or land animals.

Water Solubility

Lactonitrile is highly soluble in water. Concentrations of 1000 milligrams and more will mix with a liter of water.

Persistence in the Environment

Lactonitrile is slightly persistent in water, with a half-life of between 2 to 20 days. The half-life of a pollutant is the amount of time it takes for one half of the chemical to be degraded. In alkaline waters, it is expected to degrade to form hydrogen cyanide.

Bioaccumulation in Aquatic Organisms

Some substances increase in concentration, or *bioaccumulate,* in living organisms as they breathe contaminated air, drink contaminated water, or eat contaminated food. These chemicals can become concentrated in the tissues and internal organs of animals as well as humans.

The concentration of lactonitrile found in fish tissues is expected to be about the same as the average concentration of lactonitrile in water from which the fish was taken.

Recommended Risk-Reduction Measures

Proper training of all transporters will reduce the likelihood of a mishap or accident resulting in a leak or spill to the environment. The correct use of DOT labeling (per the MSDS) on all containers, trucks, and rail cars will enable emergency responders to react properly and quickly to any disaster thereby reducing the potential risk to the environment and to personnel.

Storage of lactonitrile should be segregated from incompatible chemicals to minimize the risk of cross-contamination or contact. Buildings designated for storage should be equipped with appropriate fire protection systems (alarms, sprinklers, emergency lighting, portable extinguishers).

If a spill or leak to the environment has occurred, fire department, emergency response, and/or hazardous materials spill personnel should be notified immediately. Cleanup should be attempted only by those trained in proper spill containment procedures using non-sparking tools. Contaminated soils should be removed for incineration and replaced with clean soil. If lactonitrile should contact the water table, aquifer, or navigable waterway, time is of the essence. It is readily miscible in water and, therefore, total containment and remediation may not be possible. When such spills occur, the local and/or state emergency response authorities must be notified. A comprehensive emergency response or disaster preparedness/recovery plan should be in place prior to any operations involving the use, transportation, storage, or disposal of lactonitrile. If lactonitrile is spilled or leaked, the following specific steps are recommended:

- ☑ Restrict persons not wearing protective clothing from area of spill or leak until cleanup is complete and area can be opened to normal work.
- ☑ Remove all ignition sources.
- ☑ Ventilate area of spill or leak.
- ☑ Absorb liquids in vermiculite, dry sand, earth, or a similar material and deposit in sealed containers.
- ☑ Keep lactonitrile out of a confined space, such as a sewer, because of the possibility of explosion (unless the sewer is designed to prevent the buildup of explosive concentrations).
- ☑ It may be necessary to dispose of lactonitrile as a hazardous waste. The responsible state agency or the regional office of the federal Environmental Protection Agency (EPA) should be contacted for specific recommendations.

RISK ASSESSMENT: BUSINESS

General Assessment

Accidents or mishaps involving lactonitrile can present a significant threat to business operations. The loss or damage of equipment or facilities can significantly affect fiscal viability. Lawsuits that may result from personnel injury/death, public exposures, and/or environmental contamination will also require a serious expenditure of resources. Media attention surrounding an

injury, death, or environmental damage can also result in a loss of profits and loss of current as well as future business.

Recommended Risk-Reduction Measures

Company attorneys, safety and health professionals, and environmental specialists should be involved in the development of any procedures dealing with safety or emergency response. A company official should be pre-designated as a public relations officer with specific training in dealing with the press. Corporate plans and policies should be developed, approved, and implemented long before any need for such arises.

REFERENCES

Ahlbom, A. 1993. *Biostatistics for Epidemiologists.* New York: CRC Press/Lewis Publishers.

American Conference of Governmental Industrial Hygienists. 1988. *Documentation of the Threshold Limit Values and Biological Exposure Indices,* 5th Edition (with updates). Cincinnati: ACGIH

Asante-Duah, D. K. 1993. *Hazardous Waste Risk Assessment.* New York: CRC Press/Lewis Publishers.

Brusick, D. J. 1994. *Methods for Genetic Risk Assessment.* New York: CRC Press/Lewis Publishers.

Bueche, J. R. 1972. *Principles of Physics,* 2nd Edition. New York: McGraw-Hill.

Burrell, R., D. K. Flaherty, and L. J. Sauers. 1992. *Toxicology of the Immune System.* New York: Van Nostrand Reinhold.

Calabrese, E. J., E. M. Kenyon. *Air Toxics and Risk Assessment.* New York: CRC Press/Lewis Publishers.

Calabrese, E. J. 1994. *Biological Effects of Low Level Exposures.* New York: CRC Press/Lewis Publishers.

Cockerman, L. G., B. S. Shane. 1994. *Basic Environmental Toxicology.* New York: CRC Press/Lewis Publishers.

Cothern, R. R. 1993. *Comparative Environmental Risk Assessment.* New York: CRC Press/Lewis Publishers.

Cothern, C. R. 1995. *Handbook for Environmental Risk Decisionmaking.* New York: CRC Press/Lewis Publishers.

Danse, I. R. 1991. *Common Sense Toxics in the Workplace.* New York: Van Nostrand Reinhold.

Environmental Protection Agency. 1983. *Chemical Hazard Information Profiles.* Washington, D.C.: U.S. EPA.

Fodden, C. R., and J. L. Weddell. 1992. *Hazardous Materials Emergency Action Data.* New York: CRC Press/Lewis Publishers.

Forseberg, K., and L. H. Keith. 1993. *Chemical; Protective Clothing Permeation and Degradation Database.* Chelsea, MI: Lewis Publishers.

Gardner, R. W. 1994. *Chemical Intolerance.* New York: CRC Press/Lewis Publishers.

Hallenbeck, W. H. 1993. *Quantitative Risk Assessment for Environmental and Occupational Health,* 2nd Edition. New York: CRC Press/Lewis Publishers.

Hosty, J. W., and P. Foster. 1990. *A Practical Guide to Chemical Spill Response.* New York: Van Nostrand Reinhold.

Howard, P. H. 1995. *Handbook of Environmental Fate and Exposures Data for Organic Chemicals,* Vol. I-IV. New York: CRC Press/Lewis Publishers.

Hosty, J. W., and P. Foster. 1990. *A Practical Guide to Chemical Spill Response.* New York: Van Nostrand Reinhold.

Jones, F. E. 1994. *Toxic Organic Vapors in the Workplace.* New York: CRC Press/Lewis Publishers.

Kamrin, M. A. 1988. *Toxicology: A Primer on Principles and Applications.* Chelsea, Michigan: Lewis Publishers.

Keith, L. H. (Editor). 1992. *Integrated Risk Information System (IRIS) Chemical Information Database.* Prepared by the U.S. Environmental Protection Agency. Boca Raton, FL: CRC Press/Lewis Publishers.

Keith, L. H. and D. B. Walters. (Editors) 1992. *National Toxicology Program's Chemical Database, Volumes I-VII.* Chelsea, MI: Lewis Publishers.

Landis, W. G., M. H. Yu. 1995. *Introduction to Environmental Toxicology.* New York: CRC Press/Lewis Publishers.

Langerman, N. 1994 *Precautionary Labels for Chemical Containers.* New York: CRC Press/Lewis Publishers.

Lewis, R. J., Sr. 1993. *Hawley's Condensed Chemical Dictionary,* 12th Edition. New York: Van Nostrand Reinhold.

Lewis, R. J., Sr. 1992. *Sax's Dangerous Properties of Industrial Materials,* Volumes 1, 2, and 3, 8th Edition. New York: Van Nostrand Reinhold.

Mackay, D., W. Y. Shiu, and K. C. Ma. 1992. *Illustrated Handbook of Chemical-Physical Properties and Environmental Fate for Organic Chemicals.* New York: CRC Press/Lewis Publishers.

Manahan, S. E. 1994. *Environmental Chemistry,* 6th Edition. New York: CRC Press/Lewis Publishers.

Meyer, E. 1990. *Chemistry of Hazardous Materials.* Englewood Cliffs, NJ: Prentice Hall.

Patnaik, P. 1992. *A Comprehensive Guide to the Hazardous Properties of Chemical Substances.* New York: Van Nostrand Reinhold.

Proctor, N. H., and James P. Hughes. 1991. *Chemical Hazards of the Workplace.* New York: Van Nostrand Reinhold.

Sherman, J. D. 1988. *Chemical Exposure and Disease.* New York: Van Nostrand Reinhold.

Torkelson, T. R. V. K. Rowe. 1981. *Patty's Industrial Hygiene and Toxicology,* Vol. 2B. New York: Wiley-Interscience.

U.S. Department of Health and Human Services, National Institute for Occupational Safety and Health. 1978. *Criteria for a Recommended Standard - Occupational Exposure to Nitriles.* DHEW (NIOSH) Publication Number: 78-212. Washington, D.C.: U.S. Government Printing Office.

MATERIAL SAFETY DATA SHEET

CHEMICAL NAME	
LAURYL ALCOHOL	**CRC PRESS/LEWIS PUBLISHERS DISCLAIMER:** THE INFORMATION AND RECOMMENDATIONS PRESENTED HEREIN ARE BASED ON SOURCES BELIEVED TO BE RELIABLE. CRC MAKES NO REPRESENTATION ON ITS COMPLETENESS OR ACCURACY. IT IS THE RESPONSIBILITY OF THE USER TO DETERMINE THE CHEMICAL'S SUITABILITY FOR ITS INTENDED USE, THE CHEMICAL'S SAFE USE, AND THE CHEMICAL'S PROPER DISPOSAL. NO REPRESENTATIONS AND/OR WARRANTIES, EITHER EXPRESSED OR IMPLIED, OF THE MERCHANTABILITY OR FITNESS FOR A PARTICULAR PURPOSE, OR OF ANY OTHER NATURE, ARE MADE WITH RESPECT TO THE INFORMATION PROVIDED IN THIS MSDS OR TO THE CHEMICAL TO WHICH INFORMATION MAY REFER. CRC NEITHER ASSUMES NOR AUTHORIZES ANY OTHER PERSON TO ASSUME FOR IT, ANY OTHER ADDITIONAL RESPONSIBILITY OR LIABILITY FOR THE USE OF, OR RELIANCE UPON, THIS INFORMATION.

HAZARD WARNING INFORMATION

HEALTH	FIRE	REACTIVE	OTHER	DEGREE OF HAZARD	COLOR CODING	OTHER CODES
0	1	0		0 = Minimum Hazard 1 = Slight Hazard 2 = Moderate Hazard 3 = Serious Hazard 4 = Severe Hazard	HEALTH = BLUE FIRE = RED REACTIVITY = YELLOW OTHER = WHITE	OX = Oxidizer ACID = Acid ALK = Alkali COR = Corrosive W = Use No Water

SECTION I - GENERAL INFORMATION

Characterization	**Alcohol**	RCRA Number	**None**
		EPA Class	**Not Applicable**
DOT Proper Shipping Name	**No Citation**	Chemical Abstract Service (CAS) Number	**112-53-8**
DOT Hazard Class and Label Requirements	**No Citation**	DOT Emergency Guide Code	**No Citation**
DOT Identification Number	**No Citation**	Chemical Formula	$CH_3(CH_2)_{10}CH_2OH$

Synonyms

Alcohol 12; Cachalot L-50; *n*-dodecanol; 1-dodecanol; duodecyl alcohol; Dytol J-68; Epal 12; laurinic alcohol; Lorol; Sipol L 12; Alfol 12; *n*-lauryl alcohol, primary.

SECTION II - HAZARDOUS INGREDIENTS AND IDENTITY INFORMATION

Hazardous Components (specific identity)	OSHA Exposure Criteria	NIOSH Exposure Criteria	Immediately Dangerous to Life and Health (IDLH)	ACGIH Exposure Criteria
Lauryl alcohol (derivation: By reduction of lauric acid esters with sodium and absolute alcohol or catalytic hydrogenation of coconut oil fatty acids or their esters under high pressure).	**PEL:** Not Established **STEL:** Not Established	**REL:** Not Established **STEL:** Not Established	**Not Established**	**TLV:** Not Established **STEL:** Not Established

SECTION III - PHYSICAL AND CHEMICAL CHARACTERISTICS

Boiling Point	**491°F (255°C)**	Specific Gravity (H_2O = 1)	**0.83**
Vapor Pressure (mm Hg)	**1 at 196°F (91°C)**	Molecular Weight	**186.33**
Vapor Density (Air = 1)	**Not Found**	Melting Point	**75°F (24°C)**

Solubility

Insoluble in water and glycerin. Soluble in ether and in 2 parts 70% alcohol, fixed oils, propylene glycol.

Appearance and Odor

Colorless solid, floral odor, and fatty, waxy taste. Liquid above 75°F (24°C). Odor Threshold = 7.1 ppm.

SECTION IV - FIRE AND EXPLOSION HAZARD DATA

Flash Point (method used)	**260°F (127°C) closed cup**	Explosive Limits in Air % by Volume	**LEL: Not Determined UEL: Not Determined**
NFPA Classification	**Combustible Liquid**	Autoignition Temperature	**527°F (275°C)**

Extinguishing Media

Dry chemical or carbon dioxide. Water or foam may cause frothing; use in flooding quantities.

Special Fire Fighting Procedures

Poisonous gases are produced in fire. Wear full protective clothing and self-contained breathing apparatus (SCBA). Continue to cool containers with water after fire is extinguished. Move containers from fire area if it can be done without risk. For large fires, use unmanned hose apparatus, if possible. Evacuate non-essential personnel 1500 feet radius if fire. Consider down-wind conditions.

Unusual Fire and Explosion Hazards

None reported.

SECTION V - REACTIVITY DATA

Stability		Conditions to Avoid
		Lauryl alcohol is stable under normal conditions of storage and operation. Avoid contact with incompatible materials. Keep away from heat, flame, and other ignition sources.
Stable	Unstable	Incompatibility *(materials to avoid)*
X		Reacts violently in contact with oxidizers and oxidizing materials.
Hazardous Polymerization		Conditions to Avoid
		Under normal conditions of temperature and pressure, lauryl alcohol is not expected to undergo hazardous polymerization.
May Occur	Will Not Occur	Hazardous Decomposition or By-products
	X	When heated to decomposition, lauryl alcohol emits acrid and irritating fumes.

SECTION VI - HEALTH HAZARD DATA

Primary Route(s) of Entry:	Inhalation?	Absorption (skin)?	Ingestion?
	X		X

Health Hazards

INHALATION: Low toxicity (no reported effects) by inhalation, unless heated. Heated vapors can lead to irritation of the respiratory tract (including the nose and throat). Systemic effects have not been studied in humans. Experimental carcinogenic effects are reported.

SKIN & EYES: As a solvent, cause skin dryness, cracking, and possible dermatitis. Eye contact can result in burning, stinging, tearing, corneal injury, and severe damage.

INGESTION: Believed to be of mild toxicity by ingestion. Experimental and inconclusive carcinogenic data reported.

Carcinogenicity	NTP Listed?	IARC Cancer Review Group?	OSHA Regulated?	Target Organs?
Unknown Human Questioned Animal	No	No	No	Skin, eyes, respiratory system.

Medical Conditions Generally Aggravated by Exposure

None reported.

Emergency and First-aid Procedures

Eye contact: Flush immediately with water for 15 minutes (minimum); seek medical attention. Skin contact: Remove all contaminated clothing. Immediately wash area with large amounts of soap and water. Seek medical assistance. For inhalation: Remove the person from exposure. Provide respiratory assistance and CPR. Transfer to medical facility. If swallowed: Seek medical attention immediately. Give 1-2 glasses of water to dilute. Low toxicity makes vomiting unnecessary.

SECTION VII - PRECAUTIONS FOR SAFE HANDLING AND USE

Steps to be Taken in Case Material is Released or Spilled

Contain spills using absorbent material (vermiculite, or other). Ventilate area of spill. Have extinguishing agent available in case of fire. Store materials in DOT-approved containers.

Preferred Waste Disposal Method

Incineration in chemical incinerator equipped with scrubber and afterburner.

Precautions to be Taken in Handling and Storage

Store in tightly closed containers in a cool, dark, well-ventilated area. Protect containers from physical damage. Keep heat, fire, and flame away.

Other Precautions and Warnings

Ground and bond all metal containers, especially during transfer operations.

SECTION VIII - CONTROL MEASURES AND PERSONAL PROTECTIVE EQUIPMENT

Respiratory Protection *(specify type)*

For vapor exposures, use an organic vapor air-purifying respirator. Greater protection is obtained using a supplied-air respirator or a self-contained breathing apparatus (SCBA) with full facepiece operated in pressure demand or other positive pressure mode.

Ventilation

Local exhaust or general mechanical systems recommended.

Protective Gloves	Eye Protection	Other Protective Clothing
Impervious Butyl Rubber Gloves	Chemical Goggles or Face Mask	Impermeable Apron or Protective Suit

Work/Hygiene Practices

Always wash hands thoroughly after using chemical; never bring food, drink, or smoking materials into vicinity of chemicals.

LAURYL ALCOHOL

$CH_3(CH_2)_{10}CH_2OH$ CAS: 112-53-8

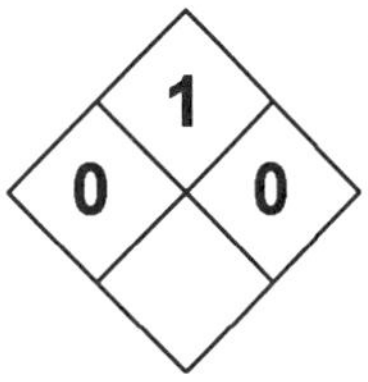

IDENTIFICATION AND TYPICAL USES

Lauryl alcohol is a colorless liquid with a floral odor and fatty, waxy taste. It can also be a colorless solid below 75°F (24°C). It is used in synthetic detergents, in lubrication additives, in pharmaceuticals, in rubber and textiles, in the manufacture of perfumes, and as a flavoring agent.

RISK ASSESSMENT: HEALTH

General Assessment

The toxicity of lauryl alcohol is much lower in comparison to most other alcohols. Lauryl alcohol is primarily toxic to humans by *inhalation* of its heated vapors. *Ingestion* can also produced mild toxicity. Skin contact can cause topical damage but dermal absorption is not likely. There are conflicting reports in the references to indicate that lauryl alcohol may cause carcinogenic effects in test animals.

Inhalation can cause irritation of the eyes, nose, throat, upper respiratory tract, and associated mucosa. Vapor exposure can also increase the toxic effects of other chemicals being inhaled.

Liquid contact with the eyes causes immediate burning and stinging with lachrymation and reflex closure of the lids. There may be injury to the corneal epithelium and possible hyperemia (excessive blood) of the conjunctiva. Skin contact results in drying and cracking which can lead to secondary infections and dermatitis.

Acute Health Effects

The following acute (short-term) health effects may occur immediately or shortly after exposure to lauryl alcohol:

Skin: Causes dryness and cracking leading to dermatitis and possible infection.

Eye: Severe irritation with burning and stinging with possible damage to the cornea and conjunctiva.

Lung: Irritation of the eyes, nose, throat, and respiratory tract.

Chronic Health Effects

The following chronic (long-term) health effects may occur at some time after exposure to lauryl alcohol and can last for months or even years:

Cancer Hazards: Lauryl alcohol has been shown to cause cancer in limited test studies on laboratory animals. However, data are conflicting and inconclusive and no determinations have been made on its carcinogenic potential in humans.

Reproduction: According to the references, lauryl alcohol has not been adequately tested for its ability to adversely affect reproduction in test animals.

Other Chronic Effects: Prolonged skin contact can cause dermatitis.

Recommended Risk-Reduction Measures

Even though lauryl alcohol is considered to be of relatively low toxicity, personnel should still avoid direct contact. The exact nature of its long-term effects on humans is not clearly understood on the references. The data are conflicting and inconclusive in this regard.

The best risk reduction measure is to use a less toxic chemical as a substitute for a lauryl alcohol. However, based upon the fact that the toxicity of lauryl alcohol is considered to be much lower compared to other alcohols, substitution may not be an alternative. Therefore, *engineering controls* are the most effective methods of reducing exposures. The best protection is to enclose operations and/or provide local exhaust ventilation at the site of chemical release. While not always operationally feasible, isolating operations can also reduce exposure risk.

Using respiratory protection is less effective than the controls mentioned above, but is still advisable whenever working with or around lauryl alcohol. No exposure levels have been established for lauryl alcohol. However, this does not mean that exposure is without risk. For low or infrequent exposures, an air-purifying respirator with an organic vapor cartridge will suffice. For higher exposures, a supplied-air respirator with full facepiece operated in positive pressure mode, or a self-contained breathing apparatus (SCBA) with full facepiece and operated in pressure demand mode are the recommended respiratory protection methods of choice. If a full facepiece is not available, then chemical goggles should be worn to protect the eyes. Whenever a chemical splash hazard exists, a face shield and a protective apron should be worn. To prevent hand and skin exposures, impervious butyl rubber gloves should be used. Glove manufacturers should be contacted and permeation studies obtained *before* final glove selections have been made.

Administrative controls should also be in place to minimize the potential for human exposures. These may include written procedures or policies which specify the methods and techniques that will be practiced whenever personnel are to work with lauryl alcohol.

All personnel should receive *training* on the use, hazards, protective measures, emergency actions, and other precautions per 29 CFR 1910.1200 (Hazard Communication), prior to the first assignment in an area where lauryl alcohol is used or stored.

If symptoms develop or overexposure is suspected, the following medical tests are recommended:

☑ Skin testing with dilute lauryl alcohol to help diagnose allergy (performed by a qualified allergist).

Any medical evaluation should include a careful history of past and present symptoms with an examination. Medical tests that look for existing damage are not a substitute for controlling exposures.

Other methods to reduce exposure include:

☑ Where possible, enclose operations and use local exhaust ventilation at the site of chemical release. If local exhaust ventilation or enclosure is not used, respiratory protection should be mandatory.

☑ Always ensure that proper protective clothing is worn when using chemical substances.

☑ Wash thoroughly immediately after exposure to lauryl alcohol and at the end of the work shift or before eating, drinking, or smoking.

☑ Hazard warning information should be posted in the work area. In addition, as part of an on-going education and training program, all information on the health and safety hazards of lauryl alcohol should be communicated to all exposed workers.

RISK ASSESSMENT: ENVIRONMENT

General Assessment

The environment is at risk of exposure during transportation, storage, disposal, or destruction of lauryl alcohol. In almost every scenario, the threat of environmental exposure is contingent upon the proper handling of the chemical substance. Accidental spills, large or small, can result in fire, explosion, and possible contamination of the surrounding environmental mediums (water, soil, and air).

Lauryl alcohol is considered a combustible liquid (according to OSHA 29 CFR 1910.106). Its high flash point and boiling point present a mild fire and explosion hazard concern. Also, because it is incompatible with a number of common materials, especially strong oxidizers, contact can result in violent and explosive reactions. It can also form explosive mixtures with air in enclosed or confined areas and may ignite on contact with heat, fire, or sparks. These characteristics require special consideration during any emergency situation involving a leak or spill of lauryl alcohol.

Lauryl alcohol can enter the environment through unchecked industrial discharges into effluents and through spills.

Acute Ecological Effects

Acute (short-term) toxic effects may include the death of animals, birds, or fish, and death or low growth rate in plants. Acute effects are seen 2 to 4 days after animals or plants are exposed to lauryl alcohol. Lauryl alcohol has slight acute toxicity to aquatic life. Insufficient data are available to evaluate the short-term effects of lauryl alcohol plants, birds, or land animals.

Chronic Ecological Effects

Chronic toxic effects may include shortened life span, reproductive problems, lower fertility, and changes in appearance or behavior in exposed animals. These effects can be seen long after first exposure(s) to toxic chemicals. Insufficient data are available to evaluate or predict the long-term effects of lauryl alcohol to aquatic life, plants, birds, or land animals.

♦ *Water Solubility*

Lauryl alcohol is highly soluble in water. Concentrations of 1000 milligrams and more can be expected to mix with a liter of water.

⌛ *Persistence in the Environment*

Lauryl alcohol is slightly persistent in water, with a half-life of between 2 to 20 days. The half-life of a pollutant is the amount of time it takes for one-half of the chemical to be degraded. About 90% of lauryl alcohol will eventually end up in the air; the remainder will end up in water.

Bioaccumulation in Aquatic Organisms

Some substances increase in concentration, or *bioaccumulate,* in living organisms as they breathe contaminated air, drink contaminated water, or eat contaminated food. These chemicals can become concentrated in the tissues and internal organs of animals as well as humans.

The concentration of lauryl alcohol found in fish tissues is expected to be about the same as the average concentration of lauryl alcohol in water from which the fish was taken.

Recommended Risk-Reduction Measures

Proper training of all transporters will reduce the likelihood of a mishap or accident resulting in a leak or spill to the environment. The correct use of labeling on all containers, trucks, and rail cars will enable emergency responders to react properly and quickly to any disaster thereby reducing the potential risk to the environment and to personnel.

Storage of lauryl alcohol should be segregated from incompatible chemicals to minimize the risk of cross-contamination or contact. Buildings designated for storage should be equipped with appropriate fire protection systems (alarms, sprinklers, emergency lighting, portable extinguishers).

If a spill or leak to the environment has occurred, fire department, emergency response, and/or hazardous materials spill personnel should be notified immediately. Cleanup should be attempted only by those trained in proper spill containment procedures. Contaminated soils should be removed for incineration and replaced with clean soil. If lauryl alcohol should contact the water table, aquifer, or navigable waterway, time is of the essence. It is highly soluble in water and, therefore, containment and remediation may not be entirely possible. When such spills occur, the local and/or state emergency response authorities must be notified. A comprehensive emergency response or disaster preparedness/recovery plan should be in place prior to any operations involving the use, transportation, storage, or disposal of lauryl alcohol. If lauryl alcohol is spilled or leaked, the following specific steps are recommended:

- ☑ Restrict persons not wearing protective clothing from area of spill or leak until cleanup is complete and area can be opened for normal work.
- ☑ Ventilate area of spill or leak.
- ☑ Remove all ignition sources.
- ☑ Absorb liquids in vermiculite, dry sand, earth, or a similar material and deposit in sealed containers. Use non-sparking tools.
- ☑ It may be necessary to dispose of lauryl alcohol as a hazardous waste. The responsible state agency or the regional office of the federal Environmental Protection Agency (EPA) should be contacted for specific recommendations.

RISK ASSESSMENT: BUSINESS

General Assessment

Accidents or mishaps involving lauryl alcohol can present a moderate threat to business operations. The loss or damage of equipment or facilities can significantly affect fiscal viability. Lawsuits that may result from personnel injury/death, public exposures, and/or environmental contamination will also require a serious expenditure of resources. Media attention surrounding an injury, death, or environmental damage can also result in a loss of profits and loss of current as well as future business. Always remember that anytime the terms "cancer" or "carcinogen" are used, public emotion, hysteria, and ignorance can run equally high. This should be carefully considered whenever developing or implementing public relations policies.

Recommended Risk-Reduction Measures

Company attorneys, safety and health professionals, and environmental specialists should be involved in the development of any procedures or policies intended to manage the use of chemicals in the workplace. A company official should be pre-designated as a public relations officer with specific training in dealing with the press. Corporate plans and policies should be developed, approved, and implemented long before any need for such arises.

REFERENCES

Ahlbom, A. 1993. *Biostatistics for Epidemiologists.* New York: CRC Press/Lewis Publishers.

American Conference of Governmental Industrial Hygienists. 1988. *Documentation of the Threshold Limit Values and Biological Exposure Indices,* 5th Edition (with updates). Cincinnati: ACGIH

Anthony, C. P., and N. J. Kolthoff. 1971. *Textbook of Anatomy and Physiology.* St. Louis: C. V. Mosby Company.

Asante-Duah, D. K. 1993. *Hazardous Waste Risk Assessment.* New York: CRC Press/Lewis Publishers.

Bloom, A. D. and F. J. DeSerres. 1995. *Ecotoxicity and Human Health.* New York: CRC Press/Lewis Publishers.

Brusick, D. J. 1994. *Methods for Genetic Risk Assessment.* New York: CRC Press/Lewis Publishers.

Bueche, J. R. 1972. *Principles of Physics,* 2nd Edition. New York: McGraw-Hill.

Burrell, R., D. K. Flaherty, and L. J. Sauers. 1992. *Toxicology of the Immune System.* New York: Van Nostrand Reinhold.

Calabrese, E. J. 1994. *Biological Effects of Low Level Exposures.* New York: CRC Press/Lewis Publishers.

Cockerman, L. G., B. S. Shane. 1994. *Basic Environmental Toxicology.* New York: CRC Press/Lewis Publishers.

Cothern, C. R., N. P. Ross. 1994. *Environmental Statistics, Assessment, and Forecasting.* New York: CRC Press/Lewis Publishers.

Danse, I. R. 1991. *Common Sense Toxics in the Workplace.* New York: Van Nostrand Reinhold.

Environmental Protection Agency. 1983. *Chemical Hazard Information Profiles.* Washington, D.C.: U.S. EPA.

Fodden, C. R., and J. L. Weddell. 1992. *Hazardous Materials Emergency Action Data.* New York: CRC Press/Lewis Publishers.

Forseberg, K., and L. H. Keith. 1993. *Chemical; Protective Clothing Permeation and Degradation Database.* Chelsea, MI: Lewis Publishers.

Gardner, R. W. 1994. *Chemical Intolerance.* New York: CRC Press/Lewis Publishers.

Hodgson, E., R. B. Mailman, and J. E. Chambers. 1988. *Dictionary of Toxicology.* London: The MacMillian Company.

Hosty, J. W., and P. Foster. 1990. *A Practical Guide to Chemical Spill Response.* New York: Van Nostrand Reinhold.

Howard, P. H., et. al. 1991. *Handbook of Environmental Degradation Rates.* New York: CRC Press/Lewis Publishers.

International Agency for Research on Cancer (IARC). 1987. *IARC Monographs on the Evaluation of Carcinogenic Risk to Humans, Overall Evaluations of Carcinogenicity: An updating of IARC Monographs Volumes 1-42.* Lyon: IARC.

Jones, F. E. 1994. *Toxic Organic Vapors in the Workplace.* New York: CRC Press/Lewis Publishers.

Kamrin, M. A. 1988. *Toxicology: A Primer on Principles and Applications.* Chelsea, Michigan: Lewis Publishers.

Keith, L. H. and D. B. Walters. (Editors) 1992. *National Toxicology Program's Chemical Database, Volumes I-VII.* Chelsea, MI: Lewis Publishers.

Landis, W. G., M. H. Yu. 1995. *Introduction to Environmental Toxicology.* New York: CRC Press/Lewis Publishers.

Langerman, N. 1994 *Precautionary Labels for Chemical Containers.* New York: CRC Press/Lewis Publishers.

Lauwerys, R. R., P. Hoet. 1993. *Industrial Chemical Exposure.* New York: CRC Press/Lewis Publishers.

Lewis, .J., Sr. 1993. *Hawley's Condensed Chemical Dictionary,* 12th Edition. New York: Van Nostrand Reinhold.

Lewis, . J., Sr. 1992. *Sax's Dangerous Properties of Industrial Materials,* Volumes 1, 2, and 3, 8th Edition. New York: Van Nostrand Reinhold.

Lide, D. R. 1995. *Handbook of Chemistry and Physics,* 76th Edition. New York: CRC Press/Lewis Publishers.

Mackay, D., W. Y. Shiu, and K. C. Ma. 1992. *Illustrated Handbook of Chemical-Physical Properties and Environmental Fate for Organic Chemicals.* New York: CRC Press/Lewis Publishers.

Manahan, S. E. 1994. *Environmental Chemistry,* 6th Edition. New York: CRC Press/Lewis Publishers.

Manahan, S. E. 1992. *Toxicological Chemistry,* 2nd Edition. New York: CRC Press/Lewis Publishers.

Mellström, G. A., J. E. Wahlberg, and H. I. Maibach. 1994. *Protective Gloves for Occupational Use.* New York: CRC Press/Lewis Publishers.

Meyer, E. 1990. *Chemistry of Hazardous Materials.* Englewood Cliffs. NJ: Prentice Hall.

Miller, B. F., M.D., and C. B. Keane, R.N. 1972. *Encyclopedia and Dictionary of Medicine and Nursing.* Philadelphia: W.B. Sanders Company.

Mycock, J. C., J. D. McKenna, and L. Theodore. 1995. *Handbook of Air Pollution Control Engineering and Technology.* New York: CRC Press/Lewis Publishers.

Patnaik, P. 1992. *A Comprehensive Guide to the Hazardous Properties of Chemical Substances.* New York: Van Nostrand Reinhold.

Philp, R. B. 1995. *Environmental Hazards and Human Health.* New York: CRC Press/Lewis Publishers.

Quigley, D. R. 1994. *Handbook of Emergency Chemical Management.* New York: CRC Press/Lewis Publishers.

Rea, W. J. 1995. *Chemical Sensitivity,* Vol. I-III. New York: CRC Press/Lewis Publishers.

Richardson, M. 1995. *Dictionary of Substance and Their Effects,* Vol. I-VI. New York: CRC Press/Lewis Publishers.

Royal Society of Chemistry. 1992. *Chemical Safety Data Sheets, Volumes 1-5.* Newcastle-upon-Tyne, UK: Athenaeum Press.

MATERIAL SAFETY DATA SHEET

CHEMICAL NAME	
LEAD	**CRC PRESS/LEWIS PUBLISHERS DISCLAIMER:** THE INFORMATION AND RECOMMENDATIONS PRESENTED HEREIN ARE BASED ON SOURCES BELIEVED TO BE RELIABLE. CRC MAKES NO REPRESENTATION ON ITS COMPLETENESS OR ACCURACY. IT IS THE RESPONSIBILITY OF THE USER TO DETERMINE THE CHEMICAL'S SUITABILITY FOR ITS INTENDED USE, THE CHEMICAL'S SAFE USE, AND THE CHEMICAL'S PROPER DISPOSAL. NO REPRESENTATIONS AND/OR WARRANTIES, EITHER EXPRESSED OR IMPLIED, OF THE MERCHANTABILITY OR FITNESS FOR A PARTICULAR PURPOSE, OR OF ANY OTHER NATURE, ARE MADE WITH RESPECT TO THE INFORMATION PROVIDED IN THIS MSDS OR TO THE CHEMICAL TO WHICH INFORMATION MAY REFER. CRC NEITHER ASSUMES NOR AUTHORIZES ANY OTHER PERSON TO ASSUME FOR IT, ANY OTHER ADDITIONAL RESPONSIBILITY OR LIABILITY FOR THE USE OF, OR RELIANCE UPON, THIS INFORMATION.

HAZARD WARNING INFORMATION

HEALTH	FIRE	REACTIVE	OTHER	DEGREE OF HAZARD	COLOR CODING	OTHER CODES
3	1	0		0 = Minimum Hazard 1 = Slight Hazard 2 = Moderate Hazard 3 = Serious Hazard 4 = Severe Hazard	HEALTH = BLUE FIRE = RED REACTIVITY = YELLOW OTHER = WHITE	OX = Oxidizer ACID = Acid ALK = Alkali COR = Corrosive W = Use No Water

SECTION I - GENERAL INFORMATION

Characterization	RCRA Number	EPA Class
Metal	**D008**	**Characteristic (T) Waste**

DOT Proper Shipping Name	Chemical Abstract Service (CAS) Number
Lead Compounds, N.O.S.	**7439-92-1**
DOT Hazard Class and Label Requirements	DOT Emergency Guide Code
St. Andrews Cross (X)	**No Citation**
DOT Identification Number	Chemical Symbol
UN 2291	**Pb**

Synonyms

Lead metal; plumbum; glover; lead flake.

SECTION II - HAZARDOUS INGREDIENTS AND IDENTITY INFORMATION

Hazardous Components (specific identity)	OSHA Exposure Criteria	NIOSH Exposure Criteria	Immediately Dangerous to Life and Health (IDLH)	ACGIH Exposure Criteria
Lead [derivation: Occurs naturally in a number or ores. Also obtained by roasting and reduction of galena (lead sulfide), anglesite (lead sulfate), and cerussite (lead carbonate). Also obtained from scrap].	**PEL (as Pb):** **0.050 mg/m^3** **STEL:** **Not Established**	**PEL (as Pb):** **0.100 mg/m^3** **STEL:** **Not Established**	**100 mg/m^3 (as Pb)**	**TLV (as Pb):** **0.15 mg/m^3** **BEI:** **50 μg(lead) per liter blood**

SECTION III - PHYSICAL AND CHEMICAL CHARACTERISTICS

Boiling Point	Specific Gravity (H_2O = 1)
3164°F (1740°C)	**11.34**
Vapor Pressure (mm Hg)	Molecular Weight
1.77 at 1783°F (973°C)	**207.20**
Vapor Density (Air = 1)	Melting Point
Not Applicable	**621°F (327°C)**

Solubility

Insoluble in hot or cold water, but dissolves slowly in water containing weak acid. Soluble in dilute nitric acid.

Appearance and Odor

Bluish-white or gray, heavy, soft, ductile metallic solid.

SECTION IV - FIRE AND EXPLOSION HAZARD DATA

Flash Point (method used)	Explosive Limits in Air % by Volume
Not Applicable	**LEL: N/A UEL: N/A**
NFPA Classification	Autoignition Temperature
Non-Combustible Solid	**Not Applicable**

Extinguishing Media

Use dry chemical, carbon dioxide, sand, or regular foam on surrounding fires.

Special Fire Fighting Procedures

Powdered lead or lead dusts are potentially flammable and moderately explosive. Great caution should be exercised when fighting fires involving lead powders/dusts.

Unusual Fire and Explosion Hazards

Powdered lead undergoes explosive reactions in the presence of numerous chemical substances. It also forms sensitive explosive mixtures with some oxidants. Dust is moderately flammable/explosive by heat, flame, or chemical reaction with powerful oxidizers. May ignite in the presence of nitric acid.

SECTION V - REACTIVITY DATA

Stability		Conditions to Avoid Lead is normally stable under routine conditions of handling and storage. It will tarnish over time on exposure to air. Avoid use of gloves containing impregnated lead (may ignite in the presence of nitric acid).
Stable X	Unstable	Incompatibility *(materials to avoid)* Strong oxidizers (chlorine, fluorine, bromine) and acids (especially nitric); hydrogen peroxide, chlorine trifluoride, sodium acetylide, ammonium nitrate, and pure water in the presence of oxygen.
Hazardous Polymerization		Conditions to Avoid Under normal conditions of temperature and pressure, hazardous polymerization of lead will not occur. Avoid exposure to incompatible materials.
May Occur	Will Not Occur X	Hazardous Decomposition or By-products Thermal decomposition of lead will produce toxic and irritating smoke and fumes, including highly toxic fumes of lead.

SECTION VI - HEALTH HAZARD DATA

Primary Route(s) of Entry:	Inhalation?	Absorption (skin)?	Ingestion?
	X		X

Health Hazards

INHALATION: Acute encephalopathy with seizures, coma, and death. The may be ataxia, headache, repeated vomiting, stupor, hallucinations, tremors, convulsions, and coma. Chronic exposures lead to central nervous system and cardiovascular system effects and may cause kidney damage. Inhaling dusts can cause gastritis and liver changes. There can be loss of appetite, malaise, insomnia, facial pallor, irritability, and anemia.

SKIN & EYES: Skin and eye contact with powdered fibers can result in scratches and minor irritation.

INGESTION: Ingested lead is poorly absorbed. If it does enter the bloodstream, it can have many of effects noted for inhalation. It is a systemic poison that accumulates in body organs.

Carcinogenicity	NTP Listed?	IARC Cancer Review Group?	OSHA Regulated?	Target Organs?
Probable Human Questioned Animal	No	Group 2B	29 CFR 1910.1025	Eyes, GI tract, blood, CNS, CVS, gingival tissues.

Medical Conditions Generally Aggravated by Exposure

Nervous system disorders (epilepsy), kidney diseases, hypertension, infertility, and anemia.

Emergency and First-aid Procedures

Eye contact: Flush immediately with water for 15 minutes (minimum); seek medical attention. Skin contact: Remove all contaminated clothing. Wash contaminated skin with soap and water. For inhalation: Remove the person from exposure. Provide respiratory assistance and CPR. Transfer to medical facility. If swallowed: Never attempt to give anything by mouth to an unconscious or convulsing person. If large quantities are ingested, induce vomiting with Syrup of Ipecac and call physician.

SECTION VII - PRECAUTIONS FOR SAFE HANDLING AND USE

Steps to be Taken in Case Material is Released or Spilled

Remove all sources of ignition. Restrict those not involved in cleanup from entering area. Collect powdered material in the most convenient way possible and deposit in sealed containers. Do NOT dry sweep (creates airborne dust). Use HEPA vacuum instead. Damp mop and collect residue.

Preferred Waste Disposal Method

Lead is a naturally occurring metal and does not pose a significant threat to the environment.

Precautions to be Taken in Handling and Storage

Containers should be protected against physical damage. Store in tightly closed containers in a cool, dry, well-ventilated area away from acids, caustics, heat, flame, and direct sunlight. Keep lead dry.

Other Precautions and Warnings

Whenever bulk or finely powdered lead is used, handled, manufactured, or stored, use explosion-proof electrical equipment and fittings.

SECTION VIII - CONTROL MEASURES AND PERSONAL PROTECTIVE EQUIPMENT

Respiratory Protection *(specify type)*

Exposure level is low and difficult to monitor. Use a respirator with particulate (dust/fume/mist) filter. Otherwise use a supplied-air respirator or self-contained breathing apparatus in continuous flow mode.

Ventilation

Local exhaust or general mechanical systems recommended.

Protective Gloves	Eye Protection	Other Protective Clothing
Suitable to protect from skin contact	Chemical/Dust Goggles	N/A

Work/Hygiene Practices

Always wash hands thoroughly after working with lead powders and dusts; never bring food, drink, or smoking materials into vicinity of lead powders.

LEAD

Pb

CAS: 7439-92-1

IDENTIFICATION AND TYPICAL USES

Lead is a bluish-white or gray, heavy, soft, ductile metallic solid. It is metallic element of atomic number 82, group IVA of the periodic table, atomic weight 207.2, valance of 2,4, and four stable isotopes. It is used in storage batteries, tertaethylead (a former gasoline additive), in chemical reaction equipment (piping, tank linings), in solder, in fusible alloys, in type metal, for vibration damping in heavy construction, in foil, as a protective barrier against certain radioactive isotopes as well as against non-ionizing radiation (X-rays), and in petroleum refining.

RISK ASSESSMENT: HEALTH

General Assessment

Lead is a systemic human poison through *inhalation* of finely divided powder and, to a lesser degree, *ingestion* either of foods or liquids containing lead or direct ingestion of lead itself (not a typical occupational exposure). Skin contact with lead dusts may result in scratching, itching, and possible scarring. The same holds true for eye contact. Ingested lead is poorly absorbed, with most of it being eliminated in the feces. The majority of the lead that is absorbed is collected in the liver and excreted in the bile. Therefore, large doses over a long period are required to significantly affect human health through ingestion. The IARC lists lead as a probable human carcinogen with insufficient evidence (Group 2B). Human mutation data have been reported and it is an experimental teratogen.

Inhalation causes acute encephalopathy with ataxia, repeated vomiting, headaches, stupor, hallucinations, tremors, convulsions or seizures, coma, and death. Other central nervous system effects include weight loss, malnutrition, restlessness, irritability, tingling in the wrists and ankles, and possible permanent brain damage. There can be kidney degradation with glomular atrophy, interstitial fibrosis, and sclerosis of vessels. There can be gastritis, changes in the liver, and accumulation in the bones and teeth. Heavy intoxication can often be detected by the formation of a dark line on the gum margins (the "lead line"). Lead causes the red blood cells to become brittle and more fragile. This increased fragility allows them to be destroyed more rapidly than normal, causing anemia. Lead has an experimental half-life in the human body of 1460 days.

Acute Health Effects

The following acute (short-term) health effects may occur immediately or shortly after exposure to lead:

Skin: Irritation/scratching on contact with particles.

Eye: Irritation/scratching on contact with particles.

Lung: Possible irritation of the nose, throat, and lungs following exposure to fibers and dusts.

CNS: A variety of effects including stupor, hallucinations, delirium, parasthesia, headaches, convulsions and seizures, coma, and death.

Chronic Health Effects

The following chronic (long-term) health effects may occur at some time after exposure to lead and can last for months or even years:

Cancer Hazards: According to information presented in the references, lead has been tested for its ability to cause cancer in test animals and the results are conflicting and inconclusive. IARC lists it as a probable human carcinogen with insufficient evidence (Group 2B). Human mutation data have been reported and many scientists believe that exposure to mutagenic compounds may pose a cancer risk in the long-term.

Reproductive Hazard: According to information presented in the references, lead has been shown to adversely affect reproduction in test animals. There may be a decreased sex drive, impotence, sterility in males, decreased fertility and abnormal menstrual cycles in females and possible spontaneous fetal abortions in females. Unborn children may suffer neurological damage or developmental problems due to excessive lead exposures by pregnant women.

Other Chronic Effects: Long-term exposure can lead to appetite loss, nausea, metallic taste in the mouth, a dark "lead line" on the gingival tissues in the mouth, constipation, anxiety, anemia, facial pallor, tiredness, weakness, insomnia, chronic headache, nervousness, irritability, tremors, tingling and numbness in the extremities, muscle and joint pain, colic, abdominal pain, kidney disease, and liver damage.

ⓤ *Recommended Risk-Reduction Measures*

Personnel should avoid direct contact with lead. *Engineering controls* are always the most effective method of reducing exposures. The best protection, especially when working with fine powders and dusts of lead, is to enclose operations and/or provide explosion-proof, local exhaust ventilation at the site of powder/dust release. While not always operationally feasible, isolating operations can also reduce exposure. Using respiratory protection is less effective than the controls mentioned above, but is still advisable whenever working with or around lead. The exposure level is extremely low (0.050 mg/m^3) and difficult to accurately monitor. A respirator equipped with a dust/mist/fume filter may suffice for low or infrequent (transient) exposures. However, a supplied-air respirator operated in continuous flow mode or a self-contained breathing apparatus (SCBA) with full facepiece in pressure demand mode will provide the best respiratory protection. If a full facepiece respirator is not available, then dust goggles or safety glasses with side shields attached should be worn to protect the eyes. To prevent hand and skin exposures, suitable gloves should be used.

Administrative controls should also be in place to minimize the potential for human exposures. These may include written procedures or policies which specify the methods and techniques that will be practiced whenever personnel are to work with lead.

All personnel should receive *training* on the use, hazards, protective measures, emergency actions, and other precautions per 29 CFR 1910.1200 (Hazard Communication), prior to the first assignment in an area where lead is used or stored.

Before beginning employment with lead, and at regular intervals thereafter (e.g., annually), the following medical tests are recommended:

- ☑ Blood lead levels and blood pressure (baseline).
- ☑ Complete blood count (establish baseline).

If symptoms develop or overexposure is suspected, the following additional tests may be useful:

- ☑ Blood lead levels and blood pressure (compare to baseline).
- ☑ Complete blood count for microcytic anemia and basophilic stippling, blood urea nitrogen (BUN) of creatinine, and zinc protoporphyrin (ZPP).
- ☑ Examination of the central and peripheral nervous systems.
- ☑ Visual examination of the gingival gum line for the appearance of a dark "lead line" on the tissues.

Other methods to reduce exposure include:

- ☑ Where possible, enclose operations and use local exhaust ventilation at the site where lead powder or dust is worked. If local exhaust ventilation or enclosure is not used, respiratory protection should be mandatory.
- ☑ Always ensure that proper protective clothing is worn when working with lead compounds.
- ☑ Wash thoroughly immediately after exposure to lead and at the end of the work shift or before eating, drinking, or smoking. Clothing contaminated with lead dusts or powders should not be taken home.
- ☑ Hazard warning information should be posted in the work area. In addition, as part of an on-going education and training program, all information on the health and safety hazards of lead should be communicated to all potentially exposed workers.
- ☑ Additional exposure control measures and engineering controls are provided by OSHA in 29 CFR 1910.1025 (Lead).

RISK ASSESSMENT: ENVIRONMENT

General Assessment

Lead occurs naturally in the environment. However, the environment is still at risk of exposure during transportation, storage, disposal, or destruction of lead. In almost every scenario, the threat of environmental exposure is contingent upon the proper handling of lead materials, especially the dusts and powders. Accidental contact with a variety of incompatible materials can result in fire, explosion, and possible contamination of the surrounding environmental mediums (water, soil, and, to a lesser degree, air).

Lead is considered a non-combustible solid in bulk form. If its dusts and powders become airborne in sufficient quantities or in enclosed areas, there may be an explosion hazard. It is incompatible with oxidizing agents and contact can cause fire or explosion. These characteristics require special consideration during any emergency situation involving a leak or spill of lead. Should lead ever come into contact with incompatible substances such as oxidizers, acids, or hydrogen peroxide, during transportation or storage, the formation of highly explosive commodities is extremely possible.

Lead is one of the most abundant metals in the earth's crust. It is used extensively across a broad spectrum of industries for a large variety of purposes. Because of its frequent use and common occurrence, lead enters the environment from point and non-point sources.

☠ *Acute Ecological Effects*

Acute (short-term) toxic effects may include the death of animals, birds, or fish and death or low growth rate in plants. Acute effects are seen 2 to 4 days after animals or plants are exposed to lead.

This metal has high acute toxicity to aquatic life and high toxicity to birds. Insufficient data are available on the short-term effects of lead exposure to plants or land animals.

Chronic Ecological Effects

Chronic (long-term) toxic effects may include shortened life span, reproductive problems, lower fertility, and changes in appearance or behavior in exposed animals. These effects can be seen long after first exposure(s) to toxic chemicals.

Lead has high chronic toxicity in aquatic life. Lead causes nerve and behavioral effects in humans and could cause similar long-term effects in birds and land animals. Insufficient data are available on the long-term effects of lead to plants.

Water Solubility

Lead and its compounds range in their respective water solubilities from highly soluble to practically insoluble.

⌛ *Persistence in the Environment*

Lead and its compounds are highly persistent in water, with a half-life greater than 200 days. The half-life of a pollutant is the amount of time it takes for one half of the chemical to be degraded.

Bioaccumulation in Aquatic Organisms

Some substances increase in concentration, or *bioaccumulate,* in living organisms as they breathe contaminated air, drink contaminated water, or eat contaminated food. These chemicals can become concentrated in the tissues and internal organs of animals as well as humans.

The concentration of lead found in the edible tissues of fish is expected to be much higher than the average concentration of lead found in the water from which the fish was taken.

Recommended Risk-Reduction Measures

Proper training of all transporters will reduce the likelihood of a mishap or accident resulting in the contamination of the environment. The correct use of DOT labeling (per the MSDS) on all containers, trucks, and rail cars will enable emergency responders to react properly and quickly to any disaster thereby reducing the potential risk to the environment and to personnel.

Storage of lead should be segregated from incompatible chemicals to minimize the risk of cross-contamination or contact. Buildings designated for storage should be equipped with appropriate fire protection systems (alarms, sprinklers, emergency lighting, portable extinguishers).

If a spill or leak to the environment has occurred, fire department, emergency response and/or hazardous materials spill personnel should be notified immediately. Cleanup should be attempted only by those trained in proper spill containment procedures. Contaminated soils should be removed and replaced with clean soil. When environmental contamination occurs, the local and/or state emergency response authorities may require notification. A comprehensive emergency response or disaster preparedness/recovery plan should be in place prior to any operations involving the use, transportation, storage, or disposal of lead. If lead is spilled or leaked, the following specific steps are recommended:

- ☑ Restrict persons not wearing protective clothing from area of spill or leak until cleanup is complete and area can be opened for normal work.
- ☑ Remove all ignition sources.
- ☑ Collect powdered materials in the safest manner possible and deposit in sealed containers. Do NOT dry sweep (generates airborne dust). Use a vacuum equipped with a high-efficiency particulate air (HEPA) filter. Damp mop residues.

☑ It may be necessary to dispose of lead as a hazardous waste. The responsible state agency or the regional office of the federal Environmental Protection Agency (EPA) should be contacted for specific recommendations.

RISK ASSESSMENT: BUSINESS

General Assessment

Accidents or mishaps involving lead can present a significant threat to business operations. The loss or damage of equipment or facilities due to explosion or fire can significantly affect fiscal viability. Lawsuits that may result from personnel injury/death, public exposures, and/or environmental contamination will also require a serious expenditure of resources. Media attention surrounding an injury, death, or environmental damage can also result in a loss of profits and loss of current as well as future business. Always remember that anytime the terms "cancer," "carcinogen," or "reproductive hazard" are used, public emotion, hysteria, and ignorance can run equally high. This is also true whenever the words "lead" and "exposure" are used together. This must be carefully considered whenever developing or implementing a public relations policy.

Recommended Risk-Reduction Measures

Company attorneys, safety and health professionals, and environmental specialists should be involved in the development of any procedures for dealing with safety or emergency response. A company official should be pre-designated as a public relations officer with specific training in dealing with the press. Corporate plans and policies should be developed, approved, and implemented long before any need for such arises.

REFERENCES

Ahlbom, A. 1993. *Biostatistics for Epidemiologists.* New York: CRC Press/Lewis Publishers.

American Conference of Governmental Industrial Hygienists. 1988. *Documentation of the Threshold Limit Values and Biological Exposure Indices,* 5th Edition (with updates). Cincinnati: ACGIH

Asante-Duah, D. K. 1993. *Hazardous Waste Risk Assessment.* New York: CRC Press/Lewis Publishers.

Bloom, A. D. and F. J. DeSerres. 1995. *Ecotoxicity and Human Health.* New York: CRC Press/Lewis Publishers.

Brusick, D. J. 1994. *Methods for Genetic Risk Assessment.* New York: CRC Press/Lewis Publishers.

Burrell, R., D. K. Flaherty, and Leonard J. Sauers. 1992. *Toxicology of the Immune System.* New York: Van Nostrand Reinhold.

Calabrese, E. J. 1994. *Biological Effects of Low Level Exposures.* New York: CRC Press/Lewis Publishers.

Cockerman, L. G., B. S. Shane. 1994. *Basic Environmental Toxicology.* New York: CRC Press/Lewis Publishers.

Cothern, C. R. 1995. *Handbook for Environmental Risk Decisionmaking.* New York: CRC Press/Lewis Publishers.

Danse, I. R. 1991. *Common Sense Toxics in the Workplace.* New York: Van Nostrand Reinhold.

Gardner, R. W. 1994. *Chemical Intolerance.* New York: CRC Press/Lewis Publishers.

Hallenbeck, W. H. 1993. *Quantitative Risk Assessment for Environmental and Occupational Health,* 2nd Edition. New York: CRC Press/Lewis Publishers.

Howard, P. H., et. al. 1991. *Handbook of Environmental Degradation Rates.* New York: CRC Press/Lewis Publishers.

James, A., M. Lord. 1992. *Index of Chemical and Physical Data.* New York: Van Nostrand Reinhold.

Keith, L. H. and M. M. Walker. 1995. *Handbook of Air Toxics.* New York: CRC Press/Lewis Publishers.

Landis, W. G., M. H. Yu. 1995. *Introduction to Environmental Toxicology.* New York: CRC Press/Lewis Publishers.

Langerman, N. 1994 *Precautionary Labels for Chemical Containers.* New York: CRC Press/Lewis Publishers.

Lauwerys, R. R., P. Hoet. 1993. *Industrial Chemical Exposure.* New York: CRC Press/Lewis Publishers.

Lewis, R. J., Sr. 1993. *Hawley's Condensed Chemical Dictionary,* 12th Edition. New York: Van Nostrand Reinhold.

Lewis, R. J., Sr. 1992. *Sax's Dangerous Properties of Industrial Materials,* Volumes 1, 2, and 3, 8th Edition. New York: Van Nostrand Reinhold.

Manahan, S. E. 1992. *Toxicological Chemistry,* 2nd Edition. New York: CRC Press/Lewis Publishers.

Meyer, E. 1990. *Chemistry of Hazardous Materials.* Englewood Cliffs, NJ: Prentice Hall.

Patnaik, P. 1992. *A Comprehensive Guide to the Hazardous Properties of Chemical Substances.* New York: Van Nostrand Reinhold.

Rea, W. J. 1995. *Chemical Sensitivity,* Vol. I-III. New York: CRC Press/Lewis Publishers.

Richardson, M. 1995. *Dictionary of Substance and Their Effects,* Vol. I-VI. New York: CRC Press/Lewis Publishers.

Sacarello, H. L. A. 1994. *The Comprehensive Handbook of Hazardous Materials.* New York: CRC Press/Lewis Publishers.

Sherman, J. D. 1988. *Chemical Exposure and Disease.* New York: Van Nostrand Reinhold.

U.S. Department of Health and Human Services, National Institute for Occupational Safety and Health. 1994. *NIOSH Pocket Guide to Chemical Hazards.* Washington, D.C.: U.S. Government Printing Office.

MATERIAL SAFETY DATA SHEET

CHEMICAL NAME	
LEAD ACETATE	**CRC PRESS/LEWIS PUBLISHERS DISCLAIMER:** THE INFORMATION AND RECOMMENDATIONS PRESENTED HEREIN ARE BASED ON SOURCES BELIEVED TO BE RELIABLE. CRC MAKES NO REPRESENTATION ON ITS COMPLETENESS OR ACCURACY. IT IS THE RESPONSIBILITY OF THE USER TO DETERMINE THE CHEMICAL'S SUITABILITY FOR ITS INTENDED USE, THE CHEMICAL'S SAFE USE, AND THE CHEMICAL'S PROPER DISPOSAL. NO REPRESENTATIONS AND/OR WARRANTIES, EITHER EXPRESSED OR IMPLIED, OF THE MERCHANTABILITY OR FITNESS FOR A PARTICULAR PURPOSE, OR OF ANY OTHER NATURE, ARE MADE WITH RESPECT TO THE INFORMATION PROVIDED IN THIS MSDS OR TO THE CHEMICAL TO WHICH INFORMATION MAY REFER. CRC NEITHER ASSUMES NOR AUTHORIZES ANY OTHER PERSON TO ASSUME FOR IT, ANY OTHER ADDITIONAL RESPONSIBILITY OR LIABILITY FOR THE USE OF, OR RELIANCE UPON, THIS INFORMATION.

HAZARD WARNING INFORMATION

HEALTH	FIRE	REACTIVE	OTHER	DEGREE OF HAZARD	COLOR CODING	OTHER CODES
1	0	0		0 = Minimum Hazard 1 = Slight Hazard 2 = Moderate Hazard 3 = Serious Hazard 4 = Severe Hazard	HEALTH = BLUE FIRE = RED REACTIVITY = YELLOW OTHER = WHITE	OX = Oxidizer ACID = Acid ALK = Alkali COR = Corrosive W = Use No Water

SECTION I - GENERAL INFORMATION

Characterization	RCRA Number	EPA Class
Metal	**U144**	**Toxic Waste**

DOT Proper Shipping Name	Chemical Abstract Service (CAS) Number
Lead Acetate	**301-04-2**
DOT Hazard Class and Label Requirements	DOT Emergency Guide Code
St. Andrew's Cross (X)	**No Citation**
DOT Identification Number	Chemical Symbol
UN 1616	$Pb(C_2H_3O_2)_2$

Synonyms

Sugar of lead; acetic acid, lead salt; lead acetate trihydrate; neutral lead acetate; normal lead acetate; plumbous acetate; salt of saturn.

SECTION II - HAZARDOUS INGREDIENTS AND IDENTITY INFORMATION

Hazardous Components (specific identity)	OSHA Exposure Criteria	NIOSH Exposure Criteria	Immediately Dangerous to Life and Health (IDLH)	ACGIH Exposure Criteria
Lead acetate (derivation: Produced by reacting acetic acid with litharge or thin lead plates, lead monoxide, or lead carbonate; the trihydrate is formed by dissolving lead oxide in hot, dilute acetic acid solution.)	**PEL (as Pb):** **0.050 mg/m^3** **STEL:** **Not Established**	**PEL (as Pb):** **0.100 mg/m^3** **STEL:** **Not Established**	**100 mg/m^3 (as Pb)**	**TLV (as Pb):** **0.15 mg/m^3** **BEI:** **50 μg(lead) per liter blood**

SECTION III - PHYSICAL AND CHEMICAL CHARACTERISTICS

Boiling Point	Specific Gravity (H_2O = 1)
536°F (280°C)	**2.5**
Vapor Pressure (mm Hg)	Molecular Weight
Not Found	**1510.39**
Vapor Density (Air = 1)	Melting Point
13.0	**167°F (75°C)**

Solubility

Soluble in water, slightly soluble in alcohol, freely soluble in glycerol.

Appearance and Odor

White crystals or flakes with a slight vinegar-like odor. The commercial grades may appear as brown or gray lumps.

SECTION IV - FIRE AND EXPLOSION HAZARD DATA

Flash Point (method used)	Explosive Limits in Air % by Volume
Not Applicable	**LEL: N/A UEL: N/A**
NFPA Classification	Autoignition Temperature
Non-Combustible Solid	**Not Applicable**

Extinguishing Media

Use water spray, dry chemical, carbon dioxide, fog, or foam.

Special Fire Fighting Procedures

Poisonous gases are produced in fire. Wear self-contained breathing apparatus (SCBA) operated in positive pressure mode. Do NOT release runoff from fire fighting measures to sewers or waterways. Dike for containment and disposal.

Unusual Fire and Explosion Hazards

None reported.

SECTION V - REACTIVITY DATA

Stability		Conditions to Avoid **Lead acetate is normally stable under routine conditions of handling and storage. Avoid contact with heat and incompatible materials.**
Stable **X**	Unstable	Incompatibility *(materials to avoid)* **Strong oxidizers, acids, hydrogen peroxide, reactive metals (sodium, potassium), resorcinol, salicylic acid, phenol, chloral hydrate, sulfites, vegetable infusions, alkalis, tannin, phosphates, and soluble sulfates.**
Hazardous Polymerization		Conditions to Avoid **Under normal conditions of temperature and pressure, hazardous polymerization of lead acetate will not occur.**
May Occur	Will Not Occur **X**	Hazardous Decomposition or By-products **Thermal decomposition of lead acetate will produce toxic and irritating smoke and fumes, including highly toxic fumes of lead.**

SECTION VI - HEALTH HAZARD DATA

Primary Route(s) of Entry:	Inhalation?	Absorption (skin)?	Ingestion?
	X	X	

Health Hazards

INHALATION: Symptoms include headache, insomnia, persistent vomiting, blurred vision and other visual disturbances, irritability, facial pallor, pale lips, anemia, hypertension, anxiety, restlessness, delirium, hallucinations, convulsions, seizures, coma, and the appearance of a dark line along the gums in the mouth ("lead line").

ABSORPTION: Skin and eye contact can cause irritation. Dermal absorption can occur.

INGESTION: Unlikely. If it occurs, can cause metallic taste, dry throat, abdominal pain, nausea and vomiting, diarrhea, constipation, peripheral circulatory system collapse, muscle weakness, headache, insomnia, kidney damage, coma, and death.

Carcinogenicity	NTP Listed?	IARC Cancer Review Group?	OSHA Regulated?	Target Organs?
Probable Human Questioned Animal	NTP-2	Group 2B	29 CFR 1910.1025 (Lead)	GI tract, blood, CNS, kidneys, gingival tissues.

Medical Conditions Generally Aggravated by Exposure

Nervous system disorders (epilepsy), kidney diseases, hypertension, infertility, and anemia.

Emergency and First-aid Procedures

Eye contact: Flush immediately with water for 15 minutes (minimum); seek medical attention. Skin contact: Remove all contaminated clothing. Wash contaminated skin with soap and water. For inhalation: Remove the person from exposure. Provide respiratory assistance and CPR. Transfer to medical facility. If swallowed: Never attempt to give anything by mouth to an unconscious or convulsing person. If large quantities are ingested, induce vomiting with Syrup of Ipecac and call physician.

SECTION VII - PRECAUTIONS FOR SAFE HANDLING AND USE

Steps to be Taken in Case Material is Released or Spilled

Remove all sources of ignition. Restrict those not involved in cleanup from entering area. Collect powdered material in the most convenient way possible and deposit in sealed containers. Do NOT dry sweep (creates airborne dust). Use HEPA vacuum instead. Damp mop and collect residue.

Preferred Waste Disposal Method

Convert to nitrates by adding nitric acid. Evaporate to a paste in a fume hood. Add water and hydrogen sulfide. Filter, wash, and dry the precipitate (lead sulfide), package, and ship to supplier.

Precautions to be Taken in Handling and Storage

Containers should be protected against physical damage. Store in tightly closed containers in a cool, dry, well-ventilated area away from acids, caustics, heat, flame, and direct sunlight.

Other Precautions and Warnings

Wear appropriate personal protective equipment.

SECTION VIII - CONTROL MEASURES AND PERSONAL PROTECTIVE EQUIPMENT

Respiratory Protection *(specify type)*

Exposure level is low and difficult to monitor. Use a respirator with particulate (dust/fume/mist) filter. Otherwise use a supplied-air respirator or self-contained breathing apparatus in continuous flow mode.

Ventilation

Local exhaust or general mechanical systems recommended.

Protective Gloves	Eye Protection	Other Protective Clothing
Suitable to protect from skin contact	Chemical/Dust Goggles	N/A

Work/Hygiene Practices

Always wash hands thoroughly after working with lead acetate powders and dusts; never bring food, drink, or smoking materials into vicinity of lead acetate powders.

LEAD ACETATE

$Pb(C_2H_3O_2)_2$

CAS: 301-04-2

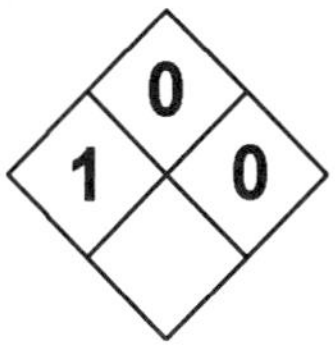

IDENTIFICATION AND TYPICAL USES

Lead acetate appears as white crystals or flakes with a slight, vinegar-like odor. The commercial grades may be brown or gray lumps. It is used in the dyeing of textiles, in waterproofing, in the making of certain varnishes, in chrome pigments, in the gold cyanidation process, in insecticides, in anti-fouling paints, as an analytical reagent, and in hair dyes.

RISK ASSESSMENT: HEALTH

General Assessment

Lead acetate is toxic through *inhalation*, *ingestion,* and dermal *absorption*. It should be noted that ingestion poses no real toxic effect in small amounts, but repeated (chronic) ingestion can cause serious problems subsequent to lead poisoning. The IARC lists lead acetate as a probable human carcinogen with insufficient evidence (Group 2B). Mutation data have been reported. While no specific teratogenic data are documented for the acetate, lead itself is considered a teratogen with reproductive effects also being reported.

Acute poisoning is rare by any route of exposure. Inhalation of lead and possibly the acetate causes acute encephalopathy with ataxia, repeated vomiting, headaches, stupor, hallucinations, tremors, convulsions or seizures, coma, and death. Other central nervous system effects include weight loss, malnutrition, restlessness, irritability, tingling in the wrists and ankles, and possible permanent brain damage. There can be kidney degradation with glomular atrophy, interstitial fibrosis, and sclerosis of vessels. There may also be gastritis, changes in the liver, and accumulation of lead in the bones and teeth. Heavy intoxication can often be detected by the formation of a dark line on the gum margins (the "lead line"). Lead causes the red blood cells to become brittle and more fragile. This increased fragility allows them to be destroyed more rapidly than normal causing anemia. Lead accumulates in bone on prolonged or repeated exposures. It has an experimental half-life in the body of 1460 days.

Acute Health Effects

The following acute (short-term) health effects may occur immediately or shortly after exposure to lead:

Skin: Irritation/scratching on contact with particles.

Eye: Irritation/scratching on contact with particles.

Lung: Possible irritation of the nose, throat, and lungs following exposure to powders and dusts.

CNS: A variety of effects including stupor, hallucinations, delirium, parasthesia, headaches, convulsions and seizures, coma, and death.

Chronic Health Effects

The following chronic (long-term) health effects may occur at some time after exposure to lead and can last for months or even years:

Cancer Hazards: According to information presented in the references, lead has been tested for its ability to cause cancer in test animals and the results are conflicting and inconclusive. IARC lists it as a probable human carcinogen with insufficient evidence (Group 2B). Human mutation data have been reported and many scientists believe that exposure to mutagenic compounds may pose a cancer risk in the long-term.

Reproductive Hazard: According to information presented in the references, lead has been shown to adversely affect reproduction in test animals. There may be a decreased sex drive, impotence, sterility in males, decreased fertility and abnormal menstrual cycles in females and possible spontaneous fetal abortions in females. Unborn children may suffer neurological damage or developmental problems due to excessive lead exposures by pregnant women.

Other Chronic Effects: Long-term exposure can lead to appetite loss, nausea, metallic taste in the mouth, a dark "lead acetate line" on the gingival tissues in the mouth, constipation, anxiety, anemia, facial pallor, tiredness, weakness, insomnia, chronic headache, nervousness, irritability, tremors, tingling and numbness in the extremities, muscle and joint pain, colic, abdominal pain, kidney disease, and liver damage.

❶ *Recommended Risk-Reduction Measures*

Personnel should avoid direct contact with lead acetate. *Engineering controls* are always the most effective method of reducing exposures. The best protection, especially when working with fine powders and dusts of lead acetate, is to enclose operations and/or provide explosion-proof, local exhaust ventilation at the site of powder/dust release. While not always operationally feasible, isolating operations can also reduce exposure. Using respiratory protection is less effective than the controls mentioned above, but is still advisable whenever working with or around lead acetate. The exposure level (for lead) is extremely low (0.050 mg/m^3) and difficult to accurately monitor. A respirator equipped with a dust/mist/fume filter may suffice for low or infrequent (transient) exposures. However, a supplied-air respirator operated in continuous flow mode or a self-contained breathing apparatus (SCBA) with full facepiece in pressure demand mode will provide the best respiratory protection. If a full facepiece respirator is not available, then dust goggles or safety glasses with side shields attached should be worn to protect the eyes. To prevent hand and skin exposures, suitable gloves should be used.

Administrative controls should also be in place to minimize the potential for human exposures. These may include written procedures or policies which specify the methods and techniques that will be practiced whenever personnel are to work with lead acetate.

All personnel should receive *training* on the use, hazards, protective measures, emergency actions, and other precautions per 29 CFR 1910.1200 (Hazard Communication), prior to the first assignment in an area where lead acetate is used or stored.

Before beginning employment with lead acetate, and at regular intervals thereafter (e.g., annually), the following medical tests are recommended:

- ☑ Blood lead levels and blood pressure (baseline).
- ☑ Complete blood count (establish baseline).

If symptoms develop or overexposure is suspected, the following additional tests may be useful:

- ☑ Blood lead levels and blood pressure (compare to baseline).
- ☑ Complete blood count for microcytic anemia and basophilic stippling, blood urea nitrogen (BUN) of creatinine, and zinc protoporphyrin (ZPP).
- ☑ Examination of the central and peripheral nervous systems.
- ☑ Visual examination of the gingival gum line for the appearance of a dark "lead line" on the tissues.

Other methods to reduce exposure include:

- ☑ Where possible, enclose operations and use local exhaust ventilation at the site where lead acetate powder or dust is worked. If local exhaust ventilation or enclosure is not used, respiratory protection should be mandatory.
- ☑ Always ensure that proper protective clothing is worn when working with lead compounds.
- ☑ Wash thoroughly immediately after exposure to lead acetate and at the end of the work shift or before eating, drinking, or smoking. Clothing contaminated with lead acetate dusts or powders should not be taken home.
- ☑ Hazard warning information should be posted in the work area. In addition, as part of an on-going education and training program, all information on the health and safety hazards of lead acetate should be communicated to all potentially exposed workers.
- ☑ Additional exposure control measures and engineering controls are provided by OSHA in 29 CFR 1910.1025 (Lead).

RISK ASSESSMENT: ENVIRONMENT

General Assessment

Lead occurs naturally in the environment. However, the environment is still at risk of exposure during transportation, storage, disposal, or destruction of lead acetate. In almost every scenario, the threat of environmental exposure is contingent upon the proper handling of lead acetate materials, especially the dusts and powders. Accidental contact with a variety of incompatible materials can result in fire, explosion, and possible contamination of the surrounding environmental mediums (water, soil, and, to a lesser degree, air).

Lead acetate is considered a non-combustible solid in bulk form. If its dusts and powders become airborne in sufficient quantities or in enclosed areas, there may be an explosion hazard. It is incompatible with oxidizing agents, acids, hydrogen peroxide, reactive metals (sodium, potassium), resorcinol, salicylic acid, phenol, chloral hydrate, sulfites, vegetable infusions, alkalis, tannin, phosphates, soluble sulfates, citrates, chlorides, carbonates, and tartrates, and contact can cause fire or explosion. These characteristics require special consideration during any emergency situation involving a leak or spill of lead acetate. Should lead acetate ever come into contact with these incompatible substances either during use, transportation, or storage, the formation of highly explosive commodities is extremely possible.

Lead is one of the most abundant metals in the earth's crust. It is used extensively across a broad spectrum of industries for a large variety of purpose. Because of its frequent use and common occurrence, lead enters the environment from point and non-point sources.

Acute Ecological Effects

Acute (short-term) toxic effects may include the death of animals, birds, or fish and death or low growth rate in plants. Acute effects are seen 2 to 4 days after animals or plants are exposed to lead.

This metal has high acute toxicity to aquatic life and high toxicity to birds. Insufficient data are available on the short-term effects of lead exposure to plants or land animals.

Chronic Ecological Effects

Chronic (long-term) toxic effects may include shortened life span, reproductive problems, lower fertility, and changes in appearance or behavior in exposed animals. These effects can be seen long after first exposure(s) to toxic chemicals.

Lead has high chronic toxicity in aquatic life. Lead causes nerve and behavioral effects in humans and could cause similar long-term effects in birds and land animals. Insufficient data are available on the long-term effects of lead to plants.

Water Solubility

Lead and its compounds range in their respective water solubilities from highly soluble to practically insoluble.

Persistence in the Environment

Lead is highly persistent in water, with a half-life greater than 200 days. The half-life of a pollutant is the amount of time it takes for one half of the chemical to be degraded. Soluble lead compounds, such as the acetate, may decompose to insoluble sulfate or phosphate derivatives in soil where they will settle and become immobile.

Bioaccumulation in Aquatic Organisms

Some substances increase in concentration, or *bioaccumulate,* in living organisms as they breathe contaminated air, drink contaminated water, or eat contaminated food. These chemicals can become concentrated in the tissues and internal organs of animals as well as humans.

The concentration of lead found in the edible tissues of fish is expected to be much higher than the average concentration of lead found in the water from which the fish was taken.

Recommended Risk-Reduction Measures

Proper training of all transporters will reduce the likelihood of a mishap or accident resulting in the contamination of the environment. The correct use of DOT labeling (per the MSDS) on all containers, trucks, and rail cars will enable emergency responders to react properly and quickly to any disaster thereby reducing the potential risk to the environment and to personnel.

Storage of lead acetate should be segregated from incompatible chemicals to minimize the risk of cross-contamination or contact. Buildings designated for storage should be equipped with appropriate fire protection systems (alarms, sprinklers, emergency lighting, portable extinguishers).

If a spill or leak to the environment has occurred, fire department, emergency response and/or hazardous materials spill personnel should be notified immediately. Cleanup should be attempted only by those trained in proper spill containment procedures. Contaminated soils should be removed and replaced with clean soil. When environmental contamination occurs, the local and/or state emergency response authorities may require notification. A comprehensive emergency response or disaster preparedness/recovery plan should be in place prior to any operations involving the use, transportation, storage, or disposal of lead acetate.

For disposal, lead acetate is a poor candidate for incineration. It is suggested that it be converted to nitrates by adding a small amount of concentrated nitric acid. The mixture should then be evaporated in a fume hood to form a thin paste. Next, water (500 ml) should be added and the paste saturated with hydrogen sulfide. The resulting precipitate is lead sulfite which can be filtered, washed, and dried. Finally, the material can be packaged and sent back to a supplier for reclamation.

If lead acetate is spilled or leaked, the following specific steps are recommended:

- ☑ Restrict persons not wearing protective clothing from area of spill or leak until cleanup is complete and area can be opened for normal work.
- ☑ Remove all ignition sources.
- ☑ Collect powdered materials in the safest manner possible and deposit in sealed containers. Do NOT dry sweep (generates airborne dust). Use a vacuum equipped with a high-efficiency particulate air (HEPA) filter. Damp mop residues.
- ☑ It may be necessary to dispose of lead acetate as a hazardous waste. The responsible state agency or the regional office of the federal Environmental Protection Agency (EPA) should be contacted for specific recommendations.

RISK ASSESSMENT: BUSINESS

General Assessment

Accidents or mishaps involving lead acetate can present a significant threat to business operations. The loss or damage of equipment or facilities due to explosion or fire can significantly affect fiscal viability. Lawsuits that may result from personnel injury/death, public exposures, and/or environmental contamination will also require a serious expenditure of resources. Media attention surrounding an injury, death, or environmental damage can also result in a loss of profits and loss of current as well as future business. Always remember that anytime the terms "cancer," "carcinogen," or "reproductive hazard" are used, public emotion, hysteria, and ignorance can run equally high. This is also true whenever the words "lead" and "exposure" are used together. This must be carefully considered when developing or implementing public relations policy.

Recommended Risk-Reduction Measures

Company attorneys, safety and health professionals, and environmental specialists should be involved in the development of any procedures for dealing with safety or emergency response. A company official should be pre-designated as a public relations officer with specific training in dealing with the press. Corporate plans and policies should be developed, approved, and implemented long before any need for such arises.

REFERENCES

Ahlbom, A. 1993. *Biostatistics for Epidemiologists.* New York: CRC Press/Lewis Publishers.

American Conference of Governmental Industrial Hygienists. 1988. *Documentation of the Threshold Limit Values and Biological Exposure Indices,* 5th Edition (with updates). Cincinnati: ACGIH

Brusick, D. J. 1994. *Methods for Genetic Risk Assessment.* New York: CRC Press/Lewis Publishers.

Burrell, R., D. K. Flaherty, and Leonard J. Sauers. 1992. *Toxicology of the Immune System.* New York: Van Nostrand Reinhold.

Calabrese, E. J. 1994. *Biological Effects of Low Level Exposures.* New York: CRC Press/Lewis Publishers.

Cockerman, L. G., B. S. Shane. 1994. *Basic Environmental Toxicology.* New York: CRC Press/Lewis Publishers.

Cothern, C. R. 1995. *Handbook for Environmental Risk Decisionmaking.* New York: CRC Press/Lewis Publishers.

Gardner, R. W. 1994. *Chemical Intolerance.* New York: CRC Press/Lewis Publishers.

Howard, P. H., et. al. 1991. *Handbook of Environmental Degradation Rates.* New York: CRC Press/Lewis Publishers.

Langerman, N. 1994 *Precautionary Labels for Chemical Containers.* New York: CRC Press/Lewis Publishers.

Lauwerys, R. R., P. Hoet. 1993. *Industrial Chemical Exposure.* New York: CRC Press/Lewis Publishers.

Lewis, R. J., Sr. 1993. *Hawley's Condensed Chemical Dictionary,* 12th Edition. New York: Van Nostrand Reinhold.

Lewis, R. J., Sr. 1992. *Sax's Dangerous Properties of Industrial Materials,* Volumes 1, 2, and 3, 8th Edition. New York: Van Nostrand Reinhold.

Manahan, S. E. 1992. *Toxicological Chemistry,* 2nd Edition. New York: CRC Press/Lewis Publishers.

Meyer, E. 1990. *Chemistry of Hazardous Materials.* Englewood Cliffs, NJ: Prentice Hall.

Ott, W. R. 1995. *Environmental Statistics and Data Analysis.* New York: CRC Press/Lewis Publishers.

Patnaik, P. 1992. *A Comprehensive Guide to the Hazardous Properties of Chemical Substances.* New York: Van Nostrand Reinhold.

Philp, R. B. 1995. *Environmental Hazards and Human Health.* New York: CRC Press/Lewis Publishers.

Quigley, D. R. 1994. *Handbook of Emergency Chemical Management.* New York: CRC Press/Lewis Publishers.

Rea, W. J. 1995. *Chemical Sensitivity,* Vol. I-III. New York: CRC Press/Lewis Publishers.

Richardson, M. 1995. *Dictionary of Substance and Their Effects,* Vol. I-VI. New York: CRC Press/Lewis Publishers.

Sacarello, H. L. A. 1994. *The Comprehensive Handbook of Hazardous Materials.* New York: CRC Press/Lewis Publishers.

Sherman, J. D. 1988. *Chemical Exposure and Disease.* New York: Van Nostrand Reinhold.

U.S. Department of Health and Human Services, National Institute for Occupational Safety and Health. 1994. *NIOSH Pocket Guide to Chemical Hazards.* Washington, D.C.: U.S. Government Printing Office.

MATERIAL SAFETY DATA SHEET

CHEMICAL NAME	
LEAD CHLORIDE	**CRC PRESS/LEWIS PUBLISHERS DISCLAIMER:** THE INFORMATION AND RECOMMENDATIONS PRESENTED HEREIN ARE BASED ON SOURCES BELIEVED TO BE RELIABLE. CRC MAKES NO REPRESENTATION ON ITS COMPLETENESS OR ACCURACY. IT IS THE RESPONSIBILITY OF THE USER TO DETERMINE THE CHEMICAL'S SUITABILITY FOR ITS INTENDED USE, THE CHEMICAL'S SAFE USE, AND THE CHEMICAL'S PROPER DISPOSAL. NO REPRESENTATIONS AND/OR WARRANTIES, EITHER EXPRESSED OR IMPLIED, OF THE MERCHANTABILITY OR FITNESS FOR A PARTICULAR PURPOSE, OR OF ANY OTHER NATURE, ARE MADE WITH RESPECT TO THE INFORMATION PROVIDED IN THIS MSDS OR TO THE CHEMICAL TO WHICH INFORMATION MAY REFER. CRC NEITHER ASSUMES NOR AUTHORIZES ANY OTHER PERSON TO ASSUME FOR IT, ANY OTHER ADDITIONAL RESPONSIBILITY OR LIABILITY FOR THE USE OF, OR RELIANCE UPON, THIS INFORMATION.

HAZARD WARNING INFORMATION

HEALTH	FIRE	REACTIVE	OTHER	DEGREE OF HAZARD	COLOR CODING	OTHER CODES
4	**0**	**0**		0 = Minimum Hazard 1 = Slight Hazard 2 = Moderate Hazard 3 = Serious Hazard 4 = Severe Hazard	HEALTH = BLUE FIRE = RED REACTIVITY = YELLOW OTHER = WHITE	OX = Oxidizer ACID = Acid ALK = Alkali COR = Corrosive ~~W~~ = Use No Water

SECTION I - GENERAL INFORMATION

Characterization	RCRA Number	EPA Class
Metallic Chloride	**None**	**Not Applicable**

DOT Proper Shipping Name	Chemical Abstract Service (CAS) Number
Lead Chloride	**7758-95-4**
DOT Hazard Class and Label Requirements	DOT Emergency Guide Code
ORM-B	**No Citation**
DOT Identification Number	Chemical Symbol
UN 2291	**$PbCl_2$**

Synonyms
Lead dichloride; plumbous chloride; lead (II) chloride.

SECTION II - HAZARDOUS INGREDIENTS AND IDENTITY INFORMATION

Hazardous Components (specific identity)	OSHA Exposure Criteria	NIOSH Exposure Criteria	Immediately Dangerous to Life and Health (IDLH)	ACGIH Exposure Criteria
Lead chloride (derivation: Produced by the addition of hydrochloric acid or sodium chloride to a solution of a lead salt with subsequent crystallization.)	**PEL (as Pb): 0.050 mg/m^3** **STEL: Not Established**	**PEL (as Pb): 0.100 mg/m^3** **STEL: Not Established**	**700 mg/m^3 (as Pb)**	**TLV (as Pb): 0.15 mg/m^3** **BEI: 50 μg(lead) per liter blood**

SECTION III - PHYSICAL AND CHEMICAL CHARACTERISTICS

Boiling Point	Specific Gravity (H_2O = 1)
1742°F (950°C)	**5.85**
Vapor Pressure (mm Hg)	Molecular Weight
1 at 1017°F (547°C)	**278.09**
Vapor Density (Air = 1)	Melting Point
9.6	**934°F (501°C)**

Solubility
Soluble in cold water, more soluble in hot water. Very soluble in ammonium chloride, ammonium nitrate, and alkali hydroxides. Slightly soluble in glycerol and dilute hydrochloric acid. Insoluble in alcohol.

Appearance and Odor
White crystalline powder.

SECTION IV - FIRE AND EXPLOSION HAZARD DATA

Flash Point (method used)	Explosive Limits in Air % by Volume
Not Applicable	**LEL: N/A UEL: N/A**
NFPA Classification	Autoignition Temperature
Non-Combustible Solid	**Not Applicable**

Extinguishing Media
Use extinguishing agent suitable to surrounding fire. Lead chloride does not burn.

Special Fire Fighting Procedures
Poisonous gases are produced in fire. Wear self-contained breathing apparatus (SCBA) operated in positive pressure mode. If possible and without risk, move containers from fire area. Do NOT release runoff from fire fighting measures to sewers or waterways. Dike for containment and disposal.

Unusual Fire and Explosion Hazards
None reported.

SECTION V - REACTIVITY DATA

Stability		Conditions to Avoid Lead chloride is normally stable under routine conditions of handling and storage. Avoid contact with heat and incompatible materials.
Stable X	Unstable	Incompatibility *(materials to avoid)* Lead chloride can react explosively in contact with heated calcium.
Hazardous Polymerization		Conditions to Avoid Under normal conditions of temperature and pressure, hazardous polymerization of lead chloride will not occur.
May Occur	Will Not Occur X	Hazardous Decomposition or By-products Thermal decomposition of lead chloride will produce toxic and irritating smoke and fumes, including highly toxic fumes of lead and chloride.

SECTION VI - HEALTH HAZARD DATA

Primary Route(s) of Entry:	Inhalation?	Absorption (skin)?	Ingestion?
	X		X

Health Hazards

INHALATION: Symptoms of lead poisoning such as headache, insomnia, persistent vomiting, blurred vision and other visual disturbances, irritability, facial pallor, pale lips, anemia, anxiety, hypertension, restlessness, delirium, hallucinations, convulsions, seizures, coma, and the appearance of a dark line along the gums in the mouth ("lead line").

SKIN & EYES: Skin and eye contact can cause irritation.

INGESTION: Unlikely. If it occurs, can cause metallic taste, dry throat, abdominal pain, nausea and vomiting, diarrhea, constipation, peripheral circulatory system collapse, muscle weakness, headache, insomnia, kidney damage, coma, and death.

Carcinogenicity	NTP Listed?	IARC Cancer Review Group?	OSHA Regulated?	Target Organs?
Probable Human Questioned Animal	No	Group 2B	29 CFR 1910.1025 (Lead)	GI tract, blood, CNS, kidneys, gingival tissues.

Medical Conditions Generally Aggravated by Exposure

Nervous system disorders (epilepsy), kidney diseases, hypertension, infertility, and anemia.

Emergency and First-aid Procedures

Eye contact: Flush immediately with water for 15 minutes (minimum); seek medical attention. Skin contact: Remove all contaminated clothing. Wash contaminated skin with soap and water. For inhalation: Remove the person from exposure. Provide respiratory assistance and CPR. Transfer to medical facility. If swallowed: Never attempt to give anything by mouth to an unconscious or convulsing person. If large quantities are ingested, induce vomiting with Syrup of Ipecac and call physician.

SECTION VII - PRECAUTIONS FOR SAFE HANDLING AND USE

Steps to be Taken in Case Material is Released or Spilled

Remove all sources of ignition. Restrict those not involved in cleanup from entering area. Collect powdered material in the most convenient way possible and deposit in sealed containers. Do NOT dry sweep (creates airborne dust). Use vacuum equipped with HEPA filter instead. Damp mop residue and collect for disposal.

Preferred Waste Disposal Method

No citation.

Precautions to be Taken in Handling and Storage

Containers should be protected against physical damage. Store in tightly closed containers in a cool, dry, well-ventilated area away from acids, caustics, heat, flame, and direct sunlight.

Other Precautions and Warnings

Wear appropriate personal protective equipment.

SECTION VIII - CONTROL MEASURES AND PERSONAL PROTECTIVE EQUIPMENT

Respiratory Protection *(specify type)*

Exposure level is low and difficult to monitor. Use a respirator with particulate (dust/fume/mist) filter. Otherwise use a supplied-air respirator or self-contained breathing apparatus in continuous flow mode.

Ventilation

Local exhaust or general mechanical systems recommended.

Protective Gloves	Eye Protection	Other Protective Clothing
Suitable to protect from skin contact	Chemical/Dust Goggles	N/A

Work/Hygiene Practices

Always wash hands thoroughly after working with lead chloride powders and dusts; never bring food, drink, or smoking materials into vicinity of lead chloride powders.

LEAD CHLORIDE

$PbCl_2$

CAS: 7758-95-4

IDENTIFICATION AND TYPICAL USES

Lead chloride appears as a white crystalline powder. It is used in the preparation of lead salts, in pigments, paints, ceramics, in metallurgy, as a solder, flux, and as an analytical reagent.

RISK ASSESSMENT: HEALTH

General Assessment

The toxicity of lead chloride is primarily due to lead itself. It can enter the body through *inhalation* and *ingestion*. It should be noted that ingestion poses no real toxic effect in small amounts, but repeated (chronic) ingestion can cause serious problems subsequent to lead poisoning. The IARC lists lead as a probable human carcinogen with insufficient evidence (Group 2B). Mutation data have been reported. While no specific teratogenic data are documented for the chloride, lead itself is considered a teratogen with reproductive effects also being reported.

Acute poisoning is rare, but possible by any route of exposure. Inhalation of lead and possibly its soluble salts (such as the chloride) causes acute encephalopathy with ataxia, repeated vomiting, headaches, stupor, hallucinations, tremors, convulsions or seizures, coma, and death. Other central nervous system effects include weight loss, malnutrition, restlessness, irritability, tingling in the wrists and ankles, and possible permanent brain damage. There can be kidney degradation with glomular atrophy, interstitial fibrosis, and sclerosis of vessels. There may also be gastritis, changes in the liver, and accumulation of lead in the bones and teeth. Heavy lead intoxication can often be detected by the formation of a dark line on the gum margins (the "lead line").

Lead causes the red blood cells to become brittle and more fragile. This increased fragility of the cells allows them to be destroyed more rapidly than normal, causing anemia. Lead accumulates in bone and teeth on prolonged or repeated exposures. It has an experimental half-life in the body of 1460 days.

Acute Health Effects

The following acute (short-term) health effects may occur immediately or shortly after exposure to lead:

<u>Skin</u>: Irritation/scratching on contact with particles.

<u>Eye</u>: Irritation/scratching on contact with particles.

<u>Lung</u>: Possible irritation of the nose, throat, and lungs following exposure to powders and dusts.

<u>CNS</u>: A variety of effects including stupor, hallucinations, delirium, parasthesia, headaches, convulsions and seizures, coma, and death.

Chronic Health Effects

The following chronic (long-term) health effects may occur at some time after exposure to lead and can last for months or even years:

<u>Cancer Hazards</u>: According to information presented in the references, lead has been tested for its ability to cause cancer in test animals and the results are conflicting and inconclusive. IARC lists it as a probable human carcinogen with insufficient evidence (Group 2B). Human mutation data have been reported and many scientists believe that exposure to mutagenic compounds may pose a cancer risk in the long-term.

<u>Reproductive Hazard</u>: According to information presented in the references, lead has been shown to adversely affect reproduction in test animals. There may be a decreased sex drive, impotence, sterility in males, decreased fertility and abnormal menstrual cycles in females, and possible spontaneous fetal abortions in females. Unborn children may suffer neurological damage or developmental problems due to excessive lead exposures by pregnant women.

<u>Other Chronic Effects</u>: Long-term exposure can lead to appetite loss, nausea, metallic taste in the mouth, a dark "lead chloride line" on the gingival tissues in the mouth, constipation, anxiety, anemia, facial pallor, tiredness, weakness, insomnia, chronic headache, nervousness, irritability, tremors, tingling and numbness in the ex-

tremities, muscle and joint pain, colic, abdominal pain, kidney disease, and liver damage.

ʊ *Recommended Risk-Reduction Measures*

Personnel should avoid direct contact with lead chloride. *Engineering controls* are always the most effective method of reducing exposures. The best protection, especially when working with fine powders and dusts of lead chloride, is to enclose operations and/or provide explosion-proof, local exhaust ventilation at the site of powder/dust release. While not always operationally feasible, isolating operations can also reduce exposure. Using respiratory protection is less effective than the controls mentioned above, but is still advisable whenever working with or around lead chloride. The exposure level (for lead) is extremely low (0.050 mg/m^3) and difficult to accurately monitor. A respirator equipped with a dust/mist/fume filter may suffice for low or infrequent (transient) exposures. However, a supplied-air respirator operated in continuous flow mode or a self-contained breathing apparatus (SCBA) with full facepiece in pressure demand mode will provide the best respiratory protection. If a full facepiece respirator is not available, then dust goggles or safety glasses with side shields attached should be worn to protect the eyes. To prevent hand and skin exposures, suitable gloves should be used.

Administrative controls should also be in place to minimize the potential for human exposures. These may include written procedures or policies which specify the methods and techniques that will be practiced whenever personnel are to work with lead chloride.

All personnel should receive *training* on the use, hazards, protective measures, emergency actions, and other precautions per 29 CFR 1910.1200 (Hazard Communication), prior to the first assignment in an area where lead chloride is used or stored.

Before beginning employment with lead chloride, and at regular intervals thereafter (e.g., annually), the following medical tests are recommended:

- ☑ Blood lead levels and blood pressure (baseline).
- ☑ Complete blood count (establish baseline).

If symptoms develop or overexposure is suspected, the following additional tests may be useful:

- ☑ Blood lead levels and blood pressure (compare to baseline).
- ☑ Complete blood count for microcytic anemia and basophilic stippling, blood urea nitrogen (BUN) of creatinine, and zinc protoporphyrin (ZPP).
- ☑ Examination of the central and peripheral nervous systems.
- ☑ Visual examination of the gingival gum line for the appearance of a dark "lead line" on the tissues.

Other methods to reduce exposure include:

- ☑ Where possible, enclose operations and use local exhaust ventilation at the site where lead chloride powder or dust is worked. If local exhaust ventilation or enclosure is not used, respiratory protection should be mandatory.
- ☑ Always ensure that proper protective clothing is worn when working with lead compounds.
- ☑ Wash thoroughly immediately after exposure to lead chloride and at the end of the work shift or before eating, drinking, or smoking. Clothing contaminated with lead chloride dusts or powders should not be taken home.
- ☑ Hazard warning information should be posted in the work area. In addition, as part of an on-going education and training program, all information on the health and safety hazards of lead chloride should be communicated to all potentially exposed workers.
- ☑ Additional exposure control measures and engineering controls are provided by OSHA in 29 CFR 1910.1025 (Lead).

RISK ASSESSMENT: ENVIRONMENT

General Assessment

Lead occurs naturally in the environment. However, the environment is still at risk of exposure during transportation, storage, disposal, or destruction of lead chloride. In almost every scenario, the threat of environmental exposure is contingent upon the proper handling of lead chloride materials, especially the dusts and powders. Accidental contact with a variety of incompatible materials can result in fire, explosion, and possible contamination of the surrounding environmental mediums (water, soil, and, to a lesser degree, air).

Lead chloride is considered a non-combustible solid in bulk form. If its dusts and powders become airborne in sufficient quantities or in enclosed areas, there may be an explosion hazard. Lead chloride reacts explosively with calcium when heated slightly. These characteristics require special consideration during any emergency situation involving a leak or spill of lead chloride. Contact with incompatible materials can occur either during use, transportation, or storage, or disposal of lead chloride.

Lead is one of the most abundant metals in the earth's crust. It is used extensively across a broad spectrum of industries for a large variety of purpose. Because of its frequent use and common occurrence, lead enters the environment from point and non-point sources.

Acute Ecological Effects

Acute (short-term) toxic effects may include the death of animals, birds, or fish and death or low growth rate in plants. Acute effects are seen 2 to 4 days after animals or plants are exposed to lead.

This metal has high acute toxicity to aquatic life and high toxicity to birds. Insufficient data are available on the short-term effects of lead exposure to plants or land animals.

Chronic Ecological Effects

Chronic (long-term) toxic effects may include shortened life span, reproductive problems, lower fertility, and changes in appearance or behavior in exposed animals. These effects can be seen long after first exposure(s) to toxic chemicals.

Lead has high chronic toxicity in aquatic life. Lead causes nerve and behavioral effects in humans and could cause similar long-term effects in birds and land animals. Insufficient data are available on the long-term effects of lead to plants.

Water Solubility

Lead and its compounds range in their respective water solubilities from highly soluble to practically insoluble.

Persistence in the Environment

Lead is highly persistent in water, with a half-life greater than 200 days. The half-life of a pollutant is the amount of time it takes for one half of the chemical to be degraded. Soluble lead compounds, such as the chloride, may decompose to insoluble sulfate or phosphate derivatives in soil where they will settle and become immobile. Lead is expected to accumulate in aquatic and terrestrial sediments.

Bioaccumulation in Aquatic Organisms

Some substances increase in concentration, or *bioaccumulate,* in living organisms as they breathe contaminated air, drink contaminated water, or eat contaminated food. These chemicals can become concentrated in the tissues and internal organs of animals as well as humans.

The concentration of lead found in the edible tissues of fish is expected to be much higher than the average concentration of lead found in the water from which the fish was taken.

Recommended Risk-Reduction Measures

Proper training of all transporters will reduce the likelihood of a mishap or accident resulting in the contamination of the environment. The correct use of DOT labeling (per the MSDS) on all containers, trucks, and rail cars will enable emergency responders to react properly and quickly to any disaster thereby reducing the potential risk to the environment and to personnel.

Storage of lead chloride should be segregated from incompatible chemicals to minimize the risk of cross-contamination or contact. Buildings designated for storage should be equipped with appropriate fire protection systems (alarms, sprinklers, emergency lighting, portable extinguishers).

If a spill or leak to the environment has occurred, fire department, emergency response and/or hazardous materials spill personnel should be notified immediately. Cleanup should be attempted only by those trained in proper spill containment procedures. Contaminated soils should be removed and replaced with clean soil. When environmental contamination occurs, the local and/or state emergency response authorities may require notification. A comprehensive emergency response or disaster preparedness/recovery plan should be in place prior to any operations involving the use, transportation, storage, or disposal of lead chloride.

If lead chloride is spilled or leaked, the following specific steps are recommended:

- ☑ Restrict persons not wearing protective clothing from area of spill or leak until cleanup is complete and area can be opened for normal work.
- ☑ Remove all ignition sources and ventilate area.
- ☑ Collect powdered materials in the safest manner possible and deposit in sealed containers. Do NOT dry sweep (generates airborne dust). Use a vacuum equipped with a high-efficiency particulate air (HEPA) filter. Damp mop residues.
- ☑ It may be necessary to dispose of lead chloride as a hazardous waste. The responsible state agency or the regional office of the federal Environmental

Protection Agency (EPA) should be contacted for specific recommendations.

RISK ASSESSMENT: BUSINESS

General Assessment

Accidents or mishaps involving lead can present a significant threat to business operations. The loss or damage of equipment or facilities due to explosion or fire can significantly affect fiscal viability. Lawsuits that may result from personnel injury/death, public exposures, and/or environmental contamination will also require a serious expenditure of resources. Media attention surrounding an injury, death, or environmental damage can also result in a loss of profits and loss of current as well as future business. Always remember that anytime the terms "cancer," "carcinogen," or "reproductive hazard" are used, public emotion, hysteria, and ignorance can run equally high. This is also true whenever the words "lead" and "exposure" are used together. This must be carefully considered when developing or implementing public relations policy.

Recommended Risk-Reduction Measures

Company attorneys, safety and health professionals, and environmental specialists should be involved in the development of any procedures for dealing with safety or emergency response. A company official should be pre-designated as a public relations officer with specific training in dealing with the press. Corporate plans and policies should be developed, approved, and implemented long before any need for such arises.

REFERENCES

Ahlbom, A. 1993. *Biostatistics for Epidemiologists.* New York: CRC Press/Lewis Publishers.

American Conference of Governmental Industrial Hygienists. 1988. *Documentation of the Threshold Limit Values and Biological Exposure Indices,* 5th Edition (with updates). Cincinnati: ACGIH

Brusick, D. J. 1994. *Methods for Genetic Risk Assessment.* New York: CRC Press/Lewis Publishers.

Burrell, R., D. K. Flaherty, and Leonard J. Sauers. 1992. *Toxicology of the Immune System.* New York: Van Nostrand Reinhold.

Calabrese, E. J. 1994. *Biological Effects of Low Level Exposures.* New York: CRC Press/Lewis Publishers.

Cockerman, L. G., B. S. Shane. 1994. *Basic Environmental Toxicology.* New York: CRC Press/Lewis Publishers.

Cothern, C. R. 1995. *Handbook for Environmental Risk Decisionmaking.* New York: CRC Press/Lewis Publishers.

Danse, I. R. 1991. *Common Sense Toxics in the Workplace.* New York: Van Nostrand Reinhold.

Environmental Protection Agency. 1983. *Chemical Hazard Information Profiles.* Washington, D.C.: U.S. EPA.

Fodden, C. R., and J. L. Weddell. 1992. *Hazardous Materials Emergency Action Data.* New York: CRC Press/Lewis Publishers.

Forseberg, K., and L. H. Keith. 1993. *Chemical; Protective Clothing Permeation and Degradation Database.* Chelsea, MI: Lewis Publishers.

Gardner, R. W. 1994. *Chemical Intolerance.* New York: CRC Press/Lewis Publishers.

Howard, P. H., et. al. 1991. *Handbook of Environmental Degradation Rates.* New York: CRC Press/Lewis Publishers.

International Agency for Research on Cancer (IARC). 1987. *IARC Monographs on the Evaluation of Carcinogenic Risk to Humans, Overall Evaluations of Carcinogenicity: An updating of IARC Monographs Volumes 1-42.* Lyon: IARC.

Jones, F. E. 1994. *Toxic Organic Vapors in the Workplace.* New York: CRC Press/Lewis Publishers.

Kamrin, M. A. 1988. *Toxicology: A Primer on Principles and Applications.* Chelsea, Michigan: Lewis Publishers.

Keith, L. H. , M. T. Johnston, C. E. Hudak, and M. Conoley. 1992. *National Toxicology Program's Gloves+ Database.* Chelsea, MI: Lewis Publishers.

Langerman, N. 1994 *Precautionary Labels for Chemical Containers.* New York: CRC Press/Lewis Publishers.

Lauwerys, R. R., P. Hoet. 1993. *Industrial Chemical Exposure.* New York: CRC Press/Lewis Publishers.

Lewis, R. J., Sr. 1993. *Hawley's Condensed Chemical Dictionary,* 12th Edition. New York: Van Nostrand Reinhold.

Lewis, R. J., Sr. 1992. *Sax's Dangerous Properties of Industrial Materials,* Volumes 1, 2, and 3, 8th Edition. New York: Van Nostrand Reinhold.

Manahan, S. E. 1992. *Toxicological Chemistry,* 2nd Edition. New York: CRC Press/Lewis Publishers.

Meyer, E. 1990. *Chemistry of Hazardous Materials.* Englewood Cliffs, NJ: Prentice Hall.

Patnaik, P. 1992. *A Comprehensive Guide to the Hazardous Properties of Chemical Substances.* New York: Van Nostrand Reinhold.

Philp, R. B. 1995. *Environmental Hazards and Human Health.* New York: CRC Press/Lewis Publishers.

Rea, W. J. 1995. *Chemical Sensitivity,* Vol. I-III. New York: CRC Press/Lewis Publishers.

Richardson, M. 1995. *Dictionary of Substance and Their Effects,* Vol. I-VI. New York: CRC Press/Lewis Publishers.

Sacarello, H. L. A. 1994. *The Comprehensive Handbook of Hazardous Materials.* New York: CRC Press/Lewis Publishers.

Sherman, J. D. 1988. *Chemical Exposure and Disease.* New York: Van Nostrand Reinhold.

U.S. Department of Health and Human Services, National Institute for Occupational Safety and Health. 1994. *NIOSH Pocket Guide to Chemical Hazards.* Washington, D.C.: U.S. Government Printing Office.

MATERIAL SAFETY DATA SHEET

CHEMICAL NAME

LEAD IODIDE

CRC PRESS/LEWIS PUBLISHERS DISCLAIMER: THE INFORMATION AND RECOMMENDATIONS PRESENTED HEREIN ARE BASED ON SOURCES BELIEVED TO BE RELIABLE. CRC MAKES NO REPRESENTATION ON ITS COMPLETENESS OR ACCURACY. IT IS THE RESPONSIBILITY OF THE USER TO DETERMINE THE CHEMICAL'S SUITABILITY FOR ITS INTENDED USE, THE CHEMICAL'S SAFE USE, AND THE CHEMICAL'S PROPER DISPOSAL. NO REPRESENTATIONS AND/OR WARRANTIES, EITHER EXPRESSED OR IMPLIED, OF THE MERCHANTABILITY OR FITNESS FOR A PARTICULAR PURPOSE, OR OF ANY OTHER NATURE, ARE MADE WITH RESPECT TO THE INFORMATION PROVIDED IN THIS MSDS OR TO THE CHEMICAL TO WHICH INFORMATION MAY REFER. CRC NEITHER ASSUMES NOR AUTHORIZES ANY OTHER PERSON TO ASSUME FOR IT, ANY OTHER ADDITIONAL RESPONSIBILITY OR LIABILITY FOR THE USE OF, OR RELIANCE UPON, THIS INFORMATION.

HAZARD WARNING INFORMATION

HEALTH	FIRE	REACTIVE	OTHER	DEGREE OF HAZARD	COLOR CODING	OTHER CODES
4	0	0		0 = Minimum Hazard 1 = Slight Hazard 2 = Moderate Hazard 3 = Serious Hazard 4 = Severe Hazard	HEALTH = BLUE FIRE = RED REACTIVITY = YELLOW OTHER = WHITE	OX = Oxidizer ACID = Acid ALK = Alkali COR = Corrosive W = Use No Water

SECTION I - GENERAL INFORMATION

Characterization **Iodide**	RCRA Number **None** / EPA Class **Not Applicable**
DOT Proper Shipping Name **Lead Compounds, N.O.S.**	Chemical Abstract Service (CAS) Number **10101-63-0**
DOT Hazard Class and Label Requirements **St. Andrew's Cross (X)**	DOT Emergency Guide Code **No Citation**
DOT Identification Number **UN 2291**	Chemical Symbol **PbI_2**

Synonyms
Lead diiodide; plumbous iodide; lead (II) iodide.

SECTION II - HAZARDOUS INGREDIENTS AND IDENTITY INFORMATION

Hazardous Components (specific identity)	OSHA Exposure Criteria	NIOSH Exposure Criteria	Immediately Dangerous to Life and Health (IDLH)	ACGIH Exposure Criteria
Lead iodide (derivation: By interaction of lead acetate and potassium iodide; or by treating a water-soluble lead compound with hydroiodic acid or a soluble metal iodide followed by recrystallization in water)	**PEL (as Pb): 0.050 mg/m^3** **STEL: Not Established**	**PEL (as Pb): 0.100 mg/m^3** **STEL: Not Established**	**700 mg/m^3 (as Pb)**	**TLV (as Pb): 0.15 mg/m^3** **BEI: 50 μg(lead) per liter blood**

SECTION III - PHYSICAL AND CHEMICAL CHARACTERISTICS

Boiling Point **1749°F (954°C)**	Specific Gravity (H_2O = 1) **6.16**
Vapor Pressure (mm Hg) **1 at 894°F (479°C)**	Molecular Weight **461.01**
Vapor Density (Air = 1) **8.4**	Melting Point **756°F (402°C)**

Solubility
Slightly soluble in cold water, more soluble in boiling water. Soluble in potassium iodide and concentrated sodium acetate solutions.

Appearance and Odor
Heavy, hexagonal, odorless, golden yellow crystals or powder.

SECTION IV - FIRE AND EXPLOSION HAZARD DATA

Flash Point (method used) **Not Applicable**	Explosive Limits in Air % by Volume **LEL: Not Applicable** **UEL: Not Applicable**
NFPA Classification **Non-Combustible Solid**	Autoignition Temperature **Not Applicable**

Extinguishing Media
Use extinguishing agent suitable to surrounding fire (dry chemical, carbon dioxide, water spray, or regular foam). Lead iodide does not burn.

Special Fire Fighting Procedures
Poisonous gases are produced in fire. Wear self-contained breathing apparatus (SCBA) operated in positive pressure mode. If possible and without risk, move containers from fire area. Do NOT release runoff from fire fighting measures to sewers or waterways. Dike for containment and disposal.

Unusual Fire and Explosion Hazards
None reported.

SECTION V - REACTIVITY DATA

Stability		Conditions to Avoid Lead iodide is normally stable under routine conditions of handling and storage. Avoid contact with heat and incompatible materials.
Stable X	Unstable	Incompatibility (*materials to avoid*) Lead iodide can react with hydrogen peroxide, strong oxidizers, sodium and potassium, fluorine (decomposes).
Hazardous Polymerization		Conditions to Avoid Under normal conditions of temperature and pressure, hazardous polymerization of lead iodide will not occur. Lead iodide is light sensitive (store in dark area and away from sunlight).
May Occur	Will Not Occur X	Hazardous Decomposition or By-products Thermal decomposition of lead iodide will produce toxic and irritating smoke and fumes, including highly toxic fumes of lead and iodide.

SECTION VI - HEALTH HAZARD DATA

Primary Route(s) of Entry:	Inhalation?	Absorption (skin)?	Ingestion?
	X		X

Health Hazards

INHALATION: Symptoms of lead poisoning such as headache, insomnia, persistent vomiting, blurred vision and other visual disturbances, irritability, facial pallor, pale lips, anemia, anxiety, hypertension, restlessness, delirium, hallucinations, convulsions, seizures, coma, an the appearance of a dark line along the gums in the mouth ("lead line").

SKIN & EYES: Skin and eye contact can cause irritation.

INGESTION: Unlikely. If it occurs, can cause metallic taste, dry throat, abdominal pain, nausea and vomiting, diarrhea, constipation, peripheral circulatory system collapse, muscle weakness, headache, insomnia, kidney damage, coma, and death.

Carcinogenicity	NTP Listed?	IARC Cancer Review Group?	OSHA Regulated?	Target Organs?
Probable Human (Pb) Questioned Animal	No	Group 2B	29 CFR 1910.1025 (Lead)	GI tract, blood, CNS, kidneys, gingival tissues.

Medical Conditions Generally Aggravated by Exposure

Nervous system disorders (epilepsy), kidney diseases, hypertension, infertility, and anemia.

Emergency and First-aid Procedures

Eye contact: Flush immediately with water for 15 minutes (minimum); seek medical attention. Skin contact: Remove all contaminated clothing. Wash contaminated skin with soap and water. For inhalation: Remove the person from exposure. Provide respiratory assistance and CPR. Transfer to medical facility. If swallowed: Never attempt to give anything by mouth to an unconscious or convulsing person. If large quantities are ingested, induce vomiting with Syrup of Ipecac and call physician.

SECTION VII - PRECAUTIONS FOR SAFE HANDLING AND USE

Steps to be Taken in Case Material is Released or Spilled

Isolate area and stay upwind. Restrict those not involved in cleanup from entering area. Collect powdered material in the most convenient way possible and deposit in sealed containers. Do NOT dry sweep (creates airborne dust). Use vacuum equipped with HEPA filter instead. Damp mop residue and collect. Neutralize solution spills with agricultural lime, crushed lime, or sodium bicarbonate.

Preferred Waste Disposal Method

No citation.

Precautions to be Taken in Handling and Storage

Containers should be protected against physical damage. Store in tightly closed containers in a cool, dry, *dark,* well-ventilated area away from acids, caustics, heat, flame, and direct sunlight.

Other Precautions and Warnings

Wear appropriate personal protective equipment.

SECTION VIII - CONTROL MEASURES AND PERSONAL PROTECTIVE EQUIPMENT

Respiratory Protection (*specify type*)

Exposure level is low and difficult to monitor. Use a respirator with particulate (dust/fume/mist) filter. Otherwise use a supplied-air respirator or self-contained breathing apparatus in continuous flow mode.

Ventilation

Local exhaust or general mechanical systems recommended.

Protective Gloves	Eye Protection	Other Protective Clothing
Suitable to protect from skin contact	Chemical/Dust Goggles	N/A

Work/Hygiene Practices

Always wash hands thoroughly after working with lead iodide powders and dusts; never bring food, drink, or smoking materials into vicinity of lead iodide powders.

LEAD IODIDE

PbI_2

CAS: 10101-63-0

IDENTIFICATION AND TYPICAL USES

Lead iodide appears as a heavy, hexagonal, odorless, yellow crystals or powder. It is used in bronzing, printing, photography, cloud seeding, gold pencils, mosaic gold, in the manufacture of lead iodide film, in asbestos brake linings, in lubricating greases, in thermal batteries, and as a reagent in microchemical testing for physostigmine.

RISK ASSESSMENT: HEALTH

General Assessment

Once absorbed into the body, lead iodide dissociates to form lead and iodide. The specific toxicity of lead iodide is primarily due to lead itself. It can enter the body through *inhalation* and *ingestion*. It should be noted that ingestion poses no real toxic effect in small amounts, but repeated (chronic) ingestion can cause serious problems subsequent to lead poisoning. While the compound lead iodide is not currently reported as carcinogenic, the IARC does list lead as a probable human carcinogen with insufficient evidence (Group 2B). Mutation data have been reported. Also, while no specific teratogenic data are documented for the iodide, lead itself is considered a teratogen with reproductive effects also being reported.

Acute poisoning is rare, but possible by both routes of exposure. Inhalation of lead and possibly its soluble salts (such as the iodide) causes acute encephalopathy with ataxia, repeated vomiting, headaches, stupor, hallucinations, tremors, convulsions or seizures, coma, and death. Other central nervous system effects include weight loss, malnutrition, restlessness, irritability, tingling in the wrists and ankles, and possible permanent brain damage. There can be kidney degradation with glomular atrophy, interstitial fibrosis, and sclerosis of vessels. There may also be astringency, intense thirst, constipation, diarrhea, gastritis, changes in the liver, and accumulation of lead in the bones and teeth. Heavy lead intoxication can often be detected by the formation of a dark line on the gum margins (the "lead line").

Blood complications can be severe, including anemia and hemoglobinuria (blood in urine). Lead causes the red blood cells to become brittle and more fragile. This increased fragility of the cells allows them to be destroyed more rapidly than normal, causing anemia. Lead accumulates in bone and teeth on prolonged or repeated exposures. It has an experimental half-life in the body of 1460 days.

Acute Health Effects

The following acute (short-term) health effects may occur immediately or shortly after exposure to lead:

Skin: Irritation/scratching on contact.

Eye: Irritation/scratching on contact.

Lung: Possible irritation of the nose, throat, and lungs following exposure to powders and dusts.

CNS: A variety of effects including stupor, hallucinations, delirium, parasthesia, headaches, convulsions and seizures, coma, and death.

Other: Acute lead poisoning is rare but possible. There can be severe blood and renal system effects as well as brain effects.

Chronic Health Effects

The following chronic (long-term) health effects may occur at some time after exposure to lead and can last for months or even years:

Cancer Hazards: According to information presented in the references, lead has been tested for its ability to cause cancer in test animals and the results are conflicting and inconclusive. IARC lists it as a probable human carcinogen with insufficient evidence (Group 2B). Human mutation data have been reported and many scientists believe that exposure to mutagenic compounds may pose a cancer risk in the long-term.

Reproductive Hazard: According to information presented in the references, lead has been shown to adversely affect reproduction in test animals. There may be a decreased sex drive, impotence, sterility in males, decreased fertility and abnormal menstrual cycles in females, and possible spontaneous fetal abortions in females. Unborn children may suffer neurological damage or developmental problems due to excessive lead exposures by pregnant women.

Other Chronic Effects: Long-term exposure can lead to appetite loss, nausea, metallic taste in the mouth, a dark "lead iodide line" on the gingival tissues in the mouth, constipation, anxiety, anemia, facial pallor, tiredness, weakness, insomnia, chronic headache, nervousness, irritability, tremors, tingling and numbness in the extremities, muscle and joint pain, colic, abdominal pain, kidney disease, and liver damage.

Recommended Risk-Reduction Measures

Personnel should avoid direct contact with lead iodide. *Engineering controls* are always the most effective method of reducing exposures. The best protection, especially when working with fine powders and dusts of lead iodide, is to enclose operations and/or provide explosion-proof, local exhaust ventilation at the site of powder/dust release. While not always operationally feasible, isolating operations can also reduce exposure. Using respiratory protection is less effective than the controls mentioned above, but is still advisable whenever working with or around lead iodide. The exposure level (for lead) is extremely low (0.050 mg/m^3) and difficult to accurately monitor. A respirator equipped with a dust/mist/fume filter may suffice for low or infrequent (transient) exposures. However, a supplied-air respirator operated in continuous flow mode or a self-contained breathing apparatus (SCBA) with full facepiece in pressure demand mode will provide the best respiratory protection. If a full facepiece respirator is not available, then dust goggles or safety glasses with side shields attached should be worn to protect the eyes. To prevent hand and skin exposures, suitable gloves should be used.

Administrative controls should also be in place to minimize the potential for human exposures. These may include written procedures or policies which specify the methods and techniques that will be practiced whenever personnel are to work with lead iodide.

All personnel should receive *training* on the use, hazards, protective measures, emergency actions, and other precautions per 29 CFR 1910.1200 (Hazard Communication), prior to the first assignment in an area where lead iodide is used or stored.

Before beginning employment with lead iodide, and at regular intervals thereafter (e.g., annually), the following medical tests are recommended:

- ☑ Blood lead levels and blood pressure (baseline).
- ☑ Complete blood count (establish baseline).

If symptoms develop or overexposure is suspected, the following additional tests may be useful:

- ☑ Blood lead levels and blood pressure.
- ☑ Complete blood count for microcytic anemia and basophilic stippling, blood urea nitrogen (BUN) of creatinine, and zinc protoporphyrin (ZPP).
- ☑ Examination of the central and peripheral nervous systems.
- ☑ Visual examination of the gingival gum line for the appearance of a dark "lead line" on the tissues.

Other methods to reduce exposure include:

- ☑ Where possible, enclose operations and use local exhaust ventilation at the site where lead iodide powder or dust is worked. If local exhaust ventilation or enclosure is not used, respiratory protection should be mandatory.
- ☑ Always ensure that proper protective clothing is worn when working with lead compounds.
- ☑ Wash thoroughly immediately after exposure to lead iodide and at the end of the work shift or before eating, drinking, or smoking. Clothing contaminated with lead iodide dusts or powders should not be taken home.
- ☑ Hazard warning information should be posted in the work area. In addition, as part of an on-going education and training program, all information on the health and safety hazards of lead iodide should be communicated to all exposed workers.
- ☑ Additional exposure control measures and engineering controls are provided by OSHA in 29 CFR 1910.1025 (Lead).

RISK ASSESSMENT: ENVIRONMENT

General Assessment

Lead occurs naturally in the environment. However, the environment is still at risk of exposure during transportation, storage, disposal, or destruction of lead iodide. In almost every scenario, the threat of environmental exposure is contingent upon the proper handling of lead iodide materials, especially the dusts and powders. Accidental contact with a variety of incompatible materials can result in fire, explosion, and possible contamination of the surrounding environmental mediums (water, soil, and, to a lesser degree, air).

Lead iodide is considered a non-combustible solid in bulk form. If its dusts and powders become airborne in sufficient quantities or in enclosed areas, there may be an explosion hazard. Lead iodide reacts with hydrogen peroxide, strong oxidizers (especially fluorine), sodium, and potassium. These characteristics require special consideration during any emergency situation involving a leak or spill of lead iodide. Contact with incompatible materials can occur either during use, transportation, or storage, or disposal of lead iodide.

Lead is one of the most abundant metals in the earth's crust. It is used extensively across a broad spectrum of industries for a large variety of purpose. Because of its frequent use and common occurrence, lead enters the environment from point and non-point sources.

Acute Ecological Effects

Acute (short-term) toxic effects may include the death of animals, birds, or fish and death or low growth rate in plants. Acute effects are seen 2 to 4 days after animals or plants are exposed to lead.

This metal has high acute toxicity to aquatic life and high toxicity to birds. It is not expected to be taken up by plant life. Insufficient data are available on the short-term effects of lead exposure to terrestrial animals.

Chronic Ecological Effects

Chronic (long-term) toxic effects may include shortened life span, reproductive problems, lower fertility, and changes in appearance or behavior in exposed animals. These effects can be seen long after first exposure(s) to toxic chemicals.

Lead has high chronic toxicity in aquatic life. Lead causes nerve and behavioral effects in humans and could cause similar long-term effects in birds and land animals. Insufficient data are available on the long-term effects of lead to plants.

Water Solubility

Lead and its compounds range in their respective water solubilities from highly soluble to practically insoluble.

Persistence in the Environment

Lead is highly persistent in water, with a half-life greater than 200 days. The half-life of a pollutant is the amount of time it takes for one half of the chemical to be degraded. Soluble lead compounds, such as the iodide, may decompose to lead and iodide in soil where they will settle and become immobile. Lead is expected to accumulate in aquatic and terrestrial sediments. Because it absorbs in soil so well, it will not easily leach into aquatic systems nor be taken up by plant life.

Bioaccumulation in Aquatic Organisms

Some substances increase in concentration, or *bioaccumulate,* in living organisms as they breathe contaminated air, drink contaminated water, or eat contaminated food. These chemicals can become concentrated in the tissues and internal organs of animals as well as humans.

The concentration of lead found in the edible tissues of fish is expected to be much higher than the average concentration of lead found in the water from which the fish was taken.

Recommended Risk-Reduction Measures

Proper training of all transporters will reduce the likelihood of a mishap or accident resulting in the contamination of the environment. The correct use of DOT labeling (per the MSDS) on all containers, trucks, and rail cars will enable emergency responders to react properly and quickly to any disaster thereby reducing the potential risk to the environment and to personnel.

Storage of lead iodide should be segregated from incompatible chemicals to minimize the risk of cross-contamination or contact. Buildings designated for storage should be equipped with appropriate fire protection systems (alarms, sprinklers, emergency lighting, portable extinguishers).

If a spill or leak to the environment has occurred, fire department, emergency response and/or hazardous materials spill personnel should be notified immediately. Cleanup should be attempted only by those trained in proper spill containment procedures. Contaminated soils should be removed and replaced with clean soil. When environmental contamination occurs, the local and/or state emergency response authorities may require notification. A comprehensive emergency response or disaster preparedness/recovery plan should be in place prior to any operations involving the use, transportation, storage, or disposal of lead iodide.

If lead iodide is spilled or leaked, the following specific steps are recommended:

☑ Restrict persons not wearing protective clothing from area of spill or leak until cleanup is complete and area can be opened for normal work.

☑ Remove all ignition sources and ventilate area.

☑ Collect powdered materials in the safest manner possible and deposit in sealed containers. Do NOT dry sweep (generates airborne dust). Use a vacuum equipped with a high-efficiency particulate air (HEPA) filter. Damp mop residues. Solution spills should be neutralized using agricultural lime, crushed lime, or sodium bicarbonate.

☑ It may be necessary to dispose of lead iodide as a hazardous waste. The responsible state agency or the regional office of the federal Environmental Protection Agency (EPA) should be contacted for specific recommendations.

RISK ASSESSMENT: BUSINESS

General Assessment

Accidents or mishaps involving lead iodide can present a significant threat to business operations. The loss or damage of equipment or facilities due to explosion or fire can significantly affect fiscal viability. Lawsuits that may result from personnel injury/death, public exposures, and/or environmental contamination will also require a serious expenditure of resources. Media attention surrounding an injury, death, or environmental damage can also result in a loss of profits and loss of current as well as future business. Always remember that anytime the terms "cancer," "carcinogen," or "reproductive hazard" are used, public emotion, hysteria, and ignorance can run equally high. This is also true whenever the words "lead" and "exposure" are used together. This must be carefully considered when developing or implementing public relations policy.

Recommended Risk-Reduction Measures

Company attorneys, safety and health professionals, and environmental specialists should be involved in the development of any procedures for dealing with safety or emergency response. A company official should be pre-designated as a public relations officer with specific training in dealing with the press. Corporate plans and policies should be developed, approved, and implemented long before any need for such arises.

REFERENCES

American Conference of Governmental Industrial Hygienists. 1988. *Documentation of the Threshold Limit Values and Biological Exposure Indices,* 5th Edition (with updates). Cincinnati: ACGIH

Brusick, D. J. 1994. *Methods for Genetic Risk Assessment.* New York: CRC Press/Lewis Publishers.

Burrell, R., D. K. Flaherty, and Leonard J. Sauers. 1992. *Toxicology of the Immune System.* New York: Van Nostrand Reinhold.

Calabrese, E. J. 1994. *Biological Effects of Low Level Exposures.* New York: CRC Press/Lewis Publishers.

Cockerman, L. G., B. S. Shane. 1994. *Basic Environmental Toxicology.* New York: CRC Press/Lewis Publishers.

Environmental Protection Agency. 1983. *Chemical Hazard Information Profiles.* Washington, D.C.: U.S. EPA.

Fodden, C. R., and J. L. Weddell. 1992. *Hazardous Materials Emergency Action Data.* New York: CRC Press/Lewis Publishers.

Forseberg, K., and L. H. Keith. 1993. *Chemical; Protective Clothing Permeation and Degradation Database.* Chelsea, MI: Lewis Publishers.

Gardner, R. W. 1994. *Chemical Intolerance.* New York: CRC Press/Lewis Publishers.

Howard, P. H., et. al. 1991. *Handbook of Environmental Degradation Rates.* New York: CRC Press/Lewis Publishers.

International Agency for Research on Cancer (IARC). 1987. *IARC Monographs on the Evaluation of Carcinogenic Risk to Humans, Overall Evaluations of Carcinogenicity: An updating of IARC Monographs Volumes 1-42.* Lyon: IARC.

Kamrin, M. A. 1988. *Toxicology: A Primer on Principles and Applications.* Chelsea, Michigan: Lewis Publishers.

Langerman, N. 1994 *Precautionary Labels for Chemical Containers.* New York: CRC Press/Lewis Publishers.

Lauwerys, R. R., P. Hoet. 1993. *Industrial Chemical Exposure.* New York: CRC Press/Lewis Publishers.

Lewis, R. J., Sr. 1993. *Hawley's Condensed Chemical Dictionary,* 12th Edition. New York: Van Nostrand Reinhold.

Lewis, R. J., Sr. 1992. *Sax's Dangerous Properties of Industrial Materials,* Volumes 1, 2, and 3, 8th Edition. New York: Van Nostrand Reinhold.

Manahan, S. E. 1992. *Toxicological Chemistry,* 2nd Edition. New York: CRC Press/Lewis Publishers.

Meyer, E. 1990. *Chemistry of Hazardous Materials.* Englewood Cliffs, NJ: Prentice Hall.

Ott, W. R. 1995. *Environmental Statistics and Data Analysis.* New York: CRC Press/Lewis Publishers.

Patnaik, P. 1992. *A Comprehensive Guide to the Hazardous Properties of Chemical Substances.* New York: Van Nostrand Reinhold.

Rea, W. J. 1995. *Chemical Sensitivity,* Vol. I-III. New York: CRC Press/Lewis Publishers.

Richardson, M. 1995. *Dictionary of Substance and Their Effects,* Vol. I-VI. New York: CRC Press/Lewis Publishers.

MATERIAL SAFETY DATA SHEET

CHEMICAL NAME	
LEAD NITRATE	**CRC PRESS/LEWIS PUBLISHERS DISCLAIMER:** THE INFORMATION AND RECOMMENDATIONS PRESENTED HEREIN ARE BASED ON SOURCES BELIEVED TO BE RELIABLE. CRC MAKES NO REPRESENTATION ON ITS COMPLETENESS OR ACCURACY. IT IS THE RESPONSIBILITY OF THE USER TO DETERMINE THE CHEMICAL'S SUITABILITY FOR ITS INTENDED USE, THE CHEMICAL'S SAFE USE, AND THE CHEMICAL'S PROPER DISPOSAL. NO REPRESENTATIONS AND/OR WARRANTIES, EITHER EXPRESSED OR IMPLIED, OF THE MERCHANTABILITY OR FITNESS FOR A PARTICULAR PURPOSE, OR OF ANY OTHER NATURE, ARE MADE WITH RESPECT TO THE INFORMATION PROVIDED IN THIS MSDS OR TO THE CHEMICAL TO WHICH INFORMATION MAY REFER. CRC NEITHER ASSUMES NOR AUTHORIZES ANY OTHER PERSON TO ASSUME FOR IT, ANY OTHER ADDITIONAL RESPONSIBILITY OR LIABILITY FOR THE USE OF, OR RELIANCE UPON, THIS INFORMATION.

HAZARD WARNING INFORMATION

HEALTH	FIRE	REACTIVE	OTHER	DEGREE OF HAZARD	COLOR CODING	OTHER CODES
1	**0**	**0**	**OX**	0 = Minimum Hazard 1 = Slight Hazard 2 = Moderate Hazard 3 = Serious Hazard 4 = Severe Hazard	HEALTH = BLUE FIRE = RED REACTIVITY = YELLOW OTHER = WHITE	OX = Oxidizer ACID = Acid ALK = Alkali COR = Corrosive W = Use No Water

SECTION I - GENERAL INFORMATION

Characterization	RCRA Number	EPA Class
Nitrate	**D001**	**Characteristic (I) Waste (oxidizer)**

DOT Proper Shipping Name **Lead Nitrate**	Chemical Abstract Service (CAS) Number **10099-74-8**
DOT Hazard Class and Label Requirements **Oxidizer; Poison**	DOT Emergency Guide Code **No Citation**
DOT Identification Number **UN 1469**	Chemical Symbol $Pb(NO_3)_2$

Synonyms

Lead dinitrate; lead (2+) nitrate; lead (II) nitrate; nitric acid, lead (2+) salt.

SECTION II - HAZARDOUS INGREDIENTS AND IDENTITY INFORMATION

Hazardous Components (specific identity)	OSHA Exposure Criteria	NIOSH Exposure Criteria	Immediately Dangerous to Life and Health (IDLH)	ACGIH Exposure Criteria
Lead nitrate (derivation: By dissolving lead metal, lead oxide, or lead carbonate in nitric acid followed by cooling and evaporation). Lead nitrate: 98% Trace impurities: copper, iron, and chloride	**PEL (as Pb):** **0.050 mg/m^3** **STEL:** **Not Established**	**PEL (as Pb):** **0.100 mg/m^3** **STEL:** **Not Established**	**700 mg/m^3 (as Pb)**	**TLV (as Pb):** **0.15 mg/m^3** **BEI:** **50 μg(lead) per liter blood**

SECTION III - PHYSICAL AND CHEMICAL CHARACTERISTICS

Boiling Point **Decomposes**	Specific Gravity (H_2O = 1) **4.53**
Vapor Pressure (mm Hg) **Not Found**	Molecular Weight **331.21**
Vapor Density (Air = 1) **11.0**	Melting Point **878°F (470°C)**

Solubility

Slightly soluble in cold water, more soluble in boiling water. Soluble in alkalis and ammonia. Insoluble in concentrated nitric acid. Slightly soluble in absolute alcohol and absolute methanol.

Appearance and Odor

White, translucent crystalline solid with no odor.

SECTION IV - FIRE AND EXPLOSION HAZARD DATA

Flash Point (method used) **Not Applicable**	Explosive Limits in Air % by Volume **LEL: Not Applicable UEL: Not Applicable**
NFPA Classification **Non-Combustible Solid**	Autoignition Temperature **Not Applicable**

Extinguishing Media

Use water spray (not a solid stream). Lead nitrate is non-combustible but is a powerful oxidizer. Do Not use dry chemical or carbon dioxide.

Special Fire Fighting Procedures

Poisonous gases are produced in fire. Wear self-contained breathing apparatus (SCBA) operated in positive pressure mode. If possible and without risk, move containers from fire area. Do NOT release runoff from fire fighting measures to sewers or waterways. Dike for containment and disposal.

Unusual Fire and Explosion Hazards

Because it is a powerful oxidizer, lead nitrate can intensify any fire by accelerating the burning of combustible materials. Stay away from ends of tanks. Lead nitrate may explode under pressure and heat.

SECTION V - REACTIVITY DATA

Stability		Conditions to Avoid **Lead nitrate is normally stable under routine conditions of handling and storage. Avoid contact with heat and incompatible materials.**
Stable **X**	Unstable	Incompatibility *(materials to avoid)* **Incompatible with ammonium thiocyanate, heated carbon, calcium-silicon alloy, cyclopentadienyl sodium, lead hypophosphite, potassium acetate + heat, hydrogen peroxide, sodium, potassium, organic materials.**
Hazardous Polymerization		Conditions to Avoid **Under normal conditions of temperature and pressure, hazardous polymerization of lead nitrate will not occur.**
May Occur	Will Not Occur **X**	Hazardous Decomposition or By-products **Thermal decomposition of lead nitrate will produce toxic and irritating smoke and fumes, including highly toxic fumes of lead and nitrogen.**

SECTION VI - HEALTH HAZARD DATA

Primary Route(s) of Entry:	Inhalation?	Absorption (skin)?	Ingestion?
	X		X

Health Hazards

INHALATION: **Symptoms of lead poisoning such as headache, insomnia, persistent vomiting, blurred vision and other visual disturbances, irritability, facial pallor, pale lips, anemia, anxiety, hypertension, restlessness, delirium, hallucinations, convulsions, seizures, coma, and the appearance of a dark line along the gums in the mouth ("lead line").**

SKIN & EYES: **Skin and eye contact can cause irritation.**

INGESTION: **Unlikely. If it occurs, can cause metallic taste, dry throat, abdominal pain, nausea and vomiting, diarrhea, constipation, peripheral circulatory system collapse, muscle weakness, headache, insomnia, kidney damage, coma, and death.**

Carcinogenicity	NTP Listed?	IARC Cancer Review Group?	OSHA Regulated?	Target Organs?
Probable Human (Pb) Questioned Animal	**No**	**Group 2B**	**29 CFR 1910.1025 (Lead)**	**GI tract, blood, CNS, kidneys, gingival tissues.**

Medical Conditions Generally Aggravated by Exposure

Nervous system disorders (epilepsy), kidney diseases, hypertension, infertility, and anemia.

Emergency and First-aid Procedures

Eye contact: Flush immediately with water for 15 minutes (minimum); seek medical attention. Skin contact: Remove all contaminated clothing. Wash contaminated skin with soap and water. For inhalation: Remove the person from exposure. Provide respiratory assistance and CPR. Transfer to medical facility. If swallowed: Never attempt to give anything by mouth to an unconscious or convulsing person. If large quantities are ingested, induce vomiting with Syrup of Ipecac and call physician.

SECTION VII - PRECAUTIONS FOR SAFE HANDLING AND USE

Steps to be Taken in Case Material is Released or Spilled

Isolate area and stay upwind. Restrict those not involved in cleanup from entering area. Collect powdered material in the most convenient way possible and deposit in sealed containers. Do NOT dry sweep (creates airborne dust). Use vacuum equipped with HEPA filter instead. Damp mop residue and collect. Neutralize solution spills with agricultural lime, crushed lime, or sodium bicarbonate.

Preferred Waste Disposal Method

Precipitate with carbonate or sulfide, filter residue, wash, dry, package, and ship to supplier.

Precautions to be Taken in Handling and Storage

Containers should be protected against physical damage. Store in tightly closed containers in a cool, dry, dark, well-ventilated area away from combustible materials (do not store on wooden shelves).

Other Precautions and Warnings

Wear appropriate personal protective equipment. Keep away from heat and ignition sources.

SECTION VIII - CONTROL MEASURES AND PERSONAL PROTECTIVE EQUIPMENT

Respiratory Protection *(specify type)*

Exposure level is low and difficult to monitor. Use a respirator with particulate (dust/fume/mist) filter. Otherwise use a supplied-air respirator or self-contained breathing apparatus in continuous flow mode.

Ventilation

Local exhaust or general mechanical systems recommended.

Protective Gloves	Eye Protection	Other Protective Clothing
Suitable to protect from skin contact	**Chemical/Dust Goggles**	**N/A**

Work/Hygiene Practices

Always wash hands thoroughly after working with lead nitrate powders and dusts; never bring food, drink, or smoking materials into vicinity of lead nitrate powders.

LEAD NITRATE

$Pb(NO_3)_2$ CAS: 10099-74-8

IDENTIFICATION AND TYPICAL USES

Lead nitrate is a white, translucent, odorless, crystalline solid. It is used in lead salts, as a mordant in dyeing and printing calico and for staining mother of pearl, in matches, as an oxidizer in the dye industry, a sensitizer in photography, in explosives, tanning, process engraving, and in lithography.

RISK ASSESSMENT: HEALTH

General Assessment

The specific toxicity of lead nitrate is primarily due to lead itself. It can enter the body through *inhalation* and *ingestion*. It should be noted that ingestion poses no real toxic effect in small amounts, but repeated (chronic) ingestion or extremely large single (acute) doses can cause serious problems subsequent to lead poisoning. While the compound lead nitrate is not currently reported as carcinogenic, the IARC does list lead as a probable human carcinogen with insufficient evidence (Group 2B). Mutation data have been reported. Also, while no specific teratogenic data are documented for the nitrate, lead itself is considered a teratogen with reproductive effects also being reported.

Acute poisoning is rare, but possible by both routes of exposure. Inhalation of lead and possibly its soluble salts (such as the nitrate) causes acute encephalopathy with ataxia, repeated vomiting, headaches, stupor, hallucinations, tremors, convulsions or seizures, coma, and death. Other central nervous system effects include weight loss, malnutrition, restlessness, irritability, tingling in the wrists and ankles, and possible permanent brain damage. There can be kidney degradation with glomular atrophy, interstitial fibrosis, and sclerosis of vessels. There may also be astringency, intense thirst, constipation, diarrhea, gastritis, changes in the liver, and accumulation of lead in the bones and teeth. Heavy lead intoxication can often be detected by the formation of a dark line on the gum margins (the "lead line").

Blood complications can be severe, including anemia and hemoglobinuria (blood in urine). Lead causes the red blood cells to become brittle and more fragile. This increased fragility of the cells allows them to be destroyed more rapidly than normal, causing anemia. Lead accumulates in bone and teeth on prolonged or repeated exposures. It has an experimental half-life in the body of 1460 days.

Acute Health Effects

The following acute (short-term) health effects may occur immediately or shortly after exposure to lead:

Skin: Irritation/scratching on contact.

Eye: Irritation/scratching on contact.

Lung: Possible irritation of the nose, throat, and lungs following exposure to powders and dusts.

CNS: A variety of effects including stupor, hallucinations, delirium, parasthesia, headaches, convulsions and seizures, coma, and death.

Other: Acute lead poisoning is rare but possible. There can be severe blood and renal system effects as well as brain effects.

Chronic Health Effects

The following chronic (long-term) health effects may occur at some time after exposure to lead and can last for months or even years:

Cancer Hazards: According to information presented in the references, lead has been tested for its ability to cause cancer in test animals and the results are conflicting and inconclusive. IARC lists it as a probable human carcinogen with insufficient evidence (Group 2B). Human mutation data have been reported and many scientists believe that exposure to mutagenic compounds may pose a cancer risk in the long-term.

Reproductive Hazard: According to information presented in the references, lead has been shown to adversely affect reproduction in test animals. There may be a decreased sex drive, impotence, low sperm count and sterility in males, decreased fertility and abnormal menstrual cycles in females, and possible spontaneous fetal abortions in females. Unborn children may suffer neurological damage or developmental problems due to excessive lead exposures by pregnant women.

Other Chronic Effects: Long-term exposure can lead to appetite loss, nausea, metallic taste in the mouth, a dark "lead nitrate line" on the gingival tissues in the mouth, constipation, anxiety, anemia, facial pallor, tiredness, weakness, insomnia, chronic headache, nervousness, irritability, tremors, tingling and numbness in the extremities, muscle and joint pain, colic, abdominal pain, kidney disease, and liver damage.

ⓘ *Recommended Risk-Reduction Measures*

Personnel should avoid direct contact with lead nitrate. *Engineering controls* are always the most effective method of reducing exposures. The best protection, especially when working with fine powders and dusts of lead nitrate, is to enclose operations and/or provide explosion-proof, local exhaust ventilation at the site of powder/dust release. While not always operationally feasible, isolating operations can also reduce exposure. Using respiratory protection is less effective than the controls mentioned above, but is still advisable whenever working with or around lead nitrate. The exposure level (for lead) is extremely low (0.050 mg/m^3) and difficult to accurately monitor. A respirator equipped with a dust/mist/fume filter may suffice for low or infrequent (transient) exposures. However, a supplied-air respirator operated in continuous flow mode or a self-contained breathing apparatus (SCBA) with full facepiece in pressure demand mode will provide the best respiratory protection. If a full facepiece respirator is not available, then dust goggles or safety glasses with side shields attached should be worn to protect the eyes. To prevent hand and skin exposures, suitable gloves should be used.

Administrative controls should also be in place to minimize the potential for human exposures. These may include written procedures or policies which specify the methods and techniques that will be practiced whenever personnel are to work with lead nitrate.

All personnel should receive *training* on the use, hazards, protective measures, emergency actions, and other precautions per 29 CFR 1910.1200 (Hazard Communication), prior to the first assignment in an area where lead nitrate is used or stored.

Before beginning employment with lead nitrate, and at regular intervals thereafter (e.g., annually), the following medical tests are recommended:

- ☑ Blood lead levels and blood pressure (baseline).
- ☑ Complete blood count (establish baseline).

If symptoms develop or overexposure is suspected, the following additional tests may be useful:

- ☑ Blood lead levels and blood pressure.
- ☑ Complete blood count for microcytic anemia and basophilic stippling, blood urea nitrogen (BUN) of creatinine, and zinc protoporphyrin (ZPP).
- ☑ Examination of the central nervous system.
- ☑ Visual examination of the gingival gum line for the appearance of a dark "lead line" on the tissues.

Other methods to reduce exposure include:

- ☑ Where possible, enclose operations and use local exhaust ventilation at the site where lead nitrate powder or dust is worked. If local exhaust ventilation or enclosure is not used, respiratory protection should be mandatory.
- ☑ Always ensure that proper protective clothing is worn when working with lead compounds.
- ☑ Wash thoroughly immediately after exposure to lead nitrate and at the end of the work shift or before eating, drinking, or smoking. Clothing contaminated with lead nitrate dusts or powders should not be taken home.
- ☑ Hazard warning information should be posted in the work area. In addition, as part of an on-going education and training program, all information on the health and safety hazards of lead nitrate should be communicated to all exposed workers.
- ☑ Additional exposure control measures and engineering controls are provided by OSHA in 29 CFR 1910.1025 (Lead).

RISK ASSESSMENT: ENVIRONMENT

General Assessment

Lead occurs naturally in the environment. However, the environment is still at risk of exposure during transportation, storage, disposal, or destruction of lead nitrate. In almost every scenario, the threat of environmental exposure is contingent upon the proper handling of lead nitrate materials, especially the dusts and powders. Accidental contact with a variety of incompatible materials can result in fire, explosion, and possible contamination of the surrounding environmental mediums (water, soil, and, to a lesser degree, air).

Lead nitrate is considered a non-combustible solid in bulk form. If its dusts and powders become airborne in sufficient quantities or in enclosed areas, there may be an explosion hazard. Lead nitrate is a powerful oxidizers and can greatly accelerate the combustion of combustible materials. It can react with hydrogen peroxide, ammonium thiocyanate, calcium-silicon alloy, sodium, and potassium. It can explode in contact with heated carbon. These characteristics require special consideration during any emergency situation involving a leak or spill of lead nitrate. Contact with incompatible materials can occur either during use, transportation, or storage, or disposal of lead nitrate.

Lead is one of the most abundant metals in the earth's crust. It is used extensively across a broad spectrum of industries for a large variety of purpose. Because of its frequent use and common occurrence, lead enters the environment from point and non-point sources.

Acute Ecological Effects

Acute (short-term) toxic effects may include the death of animals, birds, or fish and death or low growth rate in plants. Acute effects are seen 2 to 4 days after animals or plants are exposed to lead. This metal has high acute toxicity to aquatic life and high toxicity to birds. It is not expected to be taken up by plant life. Insufficient data are available on the short-term effects of lead exposure to land animals.

Chronic Ecological Effects

Chronic (long-term) toxic effects may include shortened life span, reproductive problems, lower fertility, and changes in appearance or behavior in exposed animals. These effects can be seen long after first exposure(s) to toxic chemicals. Lead has high chronic toxicity in aquatic life. Lead causes nerve and behavioral effects in humans and could cause similar long-term effects in birds and land animals. Insufficient data are available on the long-term effects of lead to plants.

Water Solubility

Lead and its compounds range in their respective water solubilities from highly soluble to practically insoluble.

Persistence in the Environment

Lead is highly persistent in water, with a half-life greater than 200 days. The half-life of a pollutant is the amount of time it takes for one half of the chemical to be degraded. Soluble lead compounds, such as the nitrate, may decompose to lead and nitrate in soil where they will settle and become immobile. Lead is expected to accumulate in aquatic and terrestrial sediments. Because it absorbs in soil so well, it will not easily leach into aquatic systems nor be taken up by plant life.

Bioaccumulation in Aquatic Organisms

Some substances increase in concentration, or *bioaccumulate,* in living organisms as they breathe contaminated air, drink contaminated water, or eat contaminated food. These chemicals can become concentrated in the tissues and internal organs of animals as well as humans. The concentration of lead found in the edible tissues of fish is expected to be much higher than the average concentration of lead found in the water from which the fish was taken.

Recommended Risk-Reduction Measures

Proper training of all transporters will reduce the likelihood of a mishap or accident resulting in the contamination of the environment. The correct use of DOT labeling (per the MSDS) on all containers, trucks, and rail cars will enable emergency responders to react properly and quickly to any disaster thereby reducing the potential risk to the environment and to personnel.

Storage of lead nitrate should be segregated from incompatible chemicals to minimize the risk of cross-contamination or contact. Buildings designated for storage should be equipped with appropriate fire protection systems (alarms, sprinklers, emergency lighting, portable extinguishers).

If a spill or leak to the environment has occurred, fire department, emergency response and/or hazardous materials spill personnel should be notified immediately. Cleanup should be attempted only by those trained in proper spill containment procedures. Contaminated soils should be removed and replaced with clean soil. When environmental contamination occurs, the local and/or state emergency response authorities may require notification. A comprehensive emergency response or disaster preparedness/recovery plan should be in place prior to any operations involving the use, transportation, storage, or disposal of lead nitrate.

If lead nitrate is spilled or leaked, the following specific steps are recommended:

☑ Restrict persons not wearing protective clothing from area of spill or leak until cleanup is complete and area can be opened for normal work.

☑ Remove all ignition sources and ventilate area.

☑ Collect powdered materials in the safest manner possible and deposit in sealed containers. Do NOT dry sweep (generates airborne dust). Use a vacuum equipped with a high-efficiency particulate air (HEPA) filter. Damp mop residues. Solution spills should be neutralized with soda ash and taken up with sand, vermiculite, or other absorbent material.

☑ It may be necessary to dispose of lead nitrate as a hazardous waste. The responsible state agency or the regional office of the federal Environmental Protection Agency (EPA) should be contacted for specific recommendations.

RISK ASSESSMENT: BUSINESS

General Assessment

Accidents or mishaps involving lead can present a significant threat to business operations. The loss or damage of equipment or facilities due to explosion or fire can significantly affect fiscal viability. Lawsuits that may result from personnel injury/death, public exposures, and/or environmental contamination will also require a serious expenditure of resources. Media attention surrounding an injury, death, or environmental damage can also result in a loss of profits and loss of current as well as future business. Always remember that anytime the terms "cancer," "carcinogen," or "reproductive hazard" are used, public emotion, hysteria, and ignorance can run equally high. This is also true whenever the words "lead" and "exposure" are used together. This must be carefully considered when developing or implementing public relations policy.

Recommended Risk-Reduction Measures

Company attorneys, safety and health professionals, and environmental specialists should be involved in the development of any procedures for dealing with safety or emergency response. A company official should be pre-designated as a public relations officer with specific training in dealing with the press. Corporate plans and policies should be developed, approved, and implemented long before any need for such arises.

REFERENCES

American Conference of Governmental Industrial Hygienists. 1988. *Documentation of the Threshold Limit Values and Biological Exposure Indices,* 5th Edition (with updates). Cincinnati: ACGIH

Brusick, D. J. 1994. *Methods for Genetic Risk Assessment.* New York: CRC Press/Lewis Publishers.

Burrell, R., D. K. Flaherty, and Leonard J. Sauers. 1992. *Toxicology of the Immune System.* New York: Van Nostrand Reinhold.

Calabrese, E. J. 1994. *Biological Effects of Low Level Exposures.* New York: CRC Press/Lewis Publishers.

Cockerman, L. G., B. S. Shane. 1994. *Basic Environmental Toxicology.* New York: CRC Press/Lewis Publishers.

Environmental Protection Agency. 1983. *Chemical Hazard Information Profiles.* Washington, D.C.: U.S. EPA.

Fodden, C. R., and J. L. Weddell. 1992. *Hazardous Materials Emergency Action Data.* New York: CRC Press/Lewis Publishers.

Forseberg, K., and L. H. Keith. 1993. *Chemical; Protective Clothing Permeation and Degradation Database.* Chelsea, MI: Lewis Publishers.

Gardner, R. W. 1994. *Chemical Intolerance.* New York: CRC Press/Lewis Publishers.

Howard, P. H., et. al. 1991. *Handbook of Environmental Degradation Rates.* New York: CRC Press/Lewis Publishers.

International Agency for Research on Cancer (IARC). 1987. *IARC Monographs on the Evaluation of Carcinogenic Risk to Humans, Overall Evaluations of Carcinogenicity: An updating of IARC Monographs Volumes 1-42.* Lyon: IARC.

Kamrin, M. A. 1988. *Toxicology: A Primer on Principles and Applications.* Chelsea, Michigan: Lewis Publishers.

Langerman, N. 1994 *Precautionary Labels for Chemical Containers.* New York: CRC Press/Lewis Publishers.

Lauwerys, R. R., P. Hoet. 1993. *Industrial Chemical Exposure.* New York: CRC Press/Lewis Publishers.

Lewis, R. J., Sr. 1993. *Hawley's Condensed Chemical Dictionary,* 12th Edition. New York: Van Nostrand Reinhold.

Lewis, R. J., Sr. 1992. *Sax's Dangerous Properties of Industrial Materials,* Volumes 1, 2, and 3, 8th Edition. New York: Van Nostrand Reinhold.

Manahan, S. E. 1992. *Toxicological Chemistry,* 2nd Edition. New York: CRC Press/Lewis Publishers.

Meyer, E. 1990. *Chemistry of Hazardous Materials.* Englewood Cliffs, NJ: Prentice Hall.

Ott, W. R. 1995. *Environmental Statistics and Data Analysis.* New York: CRC Press/Lewis Publishers.

Patnaik, P. 1992. *A Comprehensive Guide to the Hazardous Properties of Chemical Substances.* New York: Van Nostrand Reinhold.

Rea, W. J. 1995. *Chemical Sensitivity,* Vol. I-III. New York: CRC Press/Lewis Publishers.

Richardson, M. 1995. *Dictionary of Substance and Their Effects,* Vol. I-VI. New York: CRC Press/Lewis Publishers.

Sacarello, H. L. A. 1994. *The Comprehensive Handbook of Hazardous Materials.* New York: CRC Press/Lewis Publishers.

Sherman, J. D. 1988. *Chemical Exposure and Disease.* New York: Van Nostrand Reinhold.

MATERIAL SAFETY DATA SHEET

CHEMICAL NAME

LEAD SULFATE

HAZARD WARNING INFORMATION

HEALTH	FIRE	REACTIVE	OTHER	DEGREE OF HAZARD	COLOR CODING	OTHER CODES
3	0	0		0 = Minimum Hazard 1 = Slight Hazard 2 = Moderate Hazard 3 = Serious Hazard 4 = Severe Hazard	HEALTH = BLUE FIRE = RED REACTIVITY = YELLOW OTHER = WHITE	OX = Oxidizer ACID = Acid ALK = Alkali COR = Corrosive ~~W~~ = Use No Water

SECTION I - GENERAL INFORMATION

Characterization	**Sulfate**	RCRA Number / EPA Class	**None** / **Not Applicable**
DOT Proper Shipping Name	**Lead Sulfate, Solid** (>3% free acid)	Chemical Abstract Service (CAS) Number	**7446-14-2**
DOT Hazard Class and Label Requirements	**ORM-E; Corrosive Material; Corrosive**	DOT Emergency Guide Code	**No Citation**
DOT Identification Number	**UN 1794**	Chemical Symbol	**$PbSO_4$**

Synonyms

Anglisite; Fast White®; CI Pigment White 3®; Freemans White Lead®; lead bottoms; Milk White®; Mulhouse White®; sulfuric acid, lead (2+) salt (1:1); lead dross.

SECTION II - HAZARDOUS INGREDIENTS AND IDENTITY INFORMATION

Hazardous Components (specific identity)	OSHA Exposure Criteria	NIOSH Exposure Criteria	Immediately Dangerous to Life and Health (IDLH)	ACGIH Exposure Criteria
Lead sulfate (derivation: By interaction of sodium sulfate and led nitrate solutions).	**PEL (as Pb): 0.050 mg/m^3** **STEL: Not Established**	**PEL (as Pb): 0.100 mg/m^3** **STEL: Not Established**	**700 mg/m^3 (as Pb)**	**TLV (as Pb): 0.15 mg/m^3** **BEI: 50 μg(lead) per liter blood**

SECTION III - PHYSICAL AND CHEMICAL CHARACTERISTICS

Boiling Point	**Decomposes**	Specific Gravity (H_2O = 1)	**4.53**
Vapor Pressure (mm Hg)	**Not Found**	Molecular Weight	**331.21**
Vapor Density (Air = 1)	**11.0**	Melting Point	**878°F (470°C)**

Solubility

Slightly soluble in hot water. Insoluble in alcohol. Soluble in sodium hydroxide solution and concentrated hydriodic acid.

Appearance and Odor

White, rhombic, heavy crystals or powder.

SECTION IV - FIRE AND EXPLOSION HAZARD DATA

Flash Point (method used)	**Not Applicable**	Explosive Limits in Air % by Volume	**LEL: Not Applicable** **UEL: Not Applicable**
NFPA Classification	**Non-Combustible Solid**	Autoignition Temperature	**Not Applicable**

Extinguishing Media

Use extinguishing media suitable to surrounding fire. Lead sulfate does not burn.

Special Fire Fighting Procedures

Poisonous gases are produced in fire. Wear self-contained breathing apparatus (SCBA) operated in positive pressure mode. If possible and without risk, move containers from fire area. Do NOT release runoff from fire fighting measures to sewers or waterways. Dike for containment and disposal.

Unusual Fire and Explosion Hazards

Use water spray to cool containers that cannot be moved from fire area. Stay away from ends of tanks. Lead sulfate produces toxic fumes of lead and sulfur oxides when involved in fire.

SECTION V - REACTIVITY DATA

Stability		Conditions to Avoid **Lead sulfate is normally stable under routine conditions of handling and storage. Avoid contact with heat and incompatible materials.**
Stable	Unstable	Incompatibility *(materials to avoid)*
X		**Violent or explosive reactions with potassium.**
Hazardous Polymerization		Conditions to Avoid **Under normal conditions of temperature and pressure, hazardous polymerization of lead sulfate will not occur.**
May Occur	Will Not Occur	Hazardous Decomposition or By-products
	X	**Thermal decomposition of lead sulfate will produce toxic and irritating smoke and fumes, including highly toxic fumes of lead and sulfur oxides.**

SECTION VI - HEALTH HAZARD DATA

Primary Route(s) of Entry:	Inhalation?	Absorption (skin)?	Ingestion?
	X		X

Health Hazards

INHALATION: **Symptoms of lead poisoning such as headache, insomnia, persistent vomiting, blurred vision and other visual disturbances, irritability, facial pallor, pale lips, anemia, anxiety, hypertension, restlessness, delirium, hallucinations, convulsions, seizures, coma, and the appearance of a dark line along the gums in the mouth ("lead line").**

SKIN & EYES: **Corrosive irritant to the skin and eyes.**

INGESTION: **Unlikely. If it occurs, can cause metallic taste, dry throat, abdominal pain, nausea and vomiting, diarrhea, constipation, peripheral circulatory system collapse, muscle weakness, headache, insomnia, kidney damage, coma, and death.**

Carcinogenicity	NTP Listed?	IARC Cancer Review Group?	OSHA Regulated?	Target Organs?
Probable Human (Pb) Questioned Animal	**No**	**Group 2B**	**29 CFR 1910.1025 (Lead)**	**GI tract, blood, CNS, kidneys, gingival tissues.**

Medical Conditions Generally Aggravated by Exposure

Nervous system disorders (epilepsy), kidney diseases, hypertension, infertility, and anemia.

Emergency and First-aid Procedures

Eye contact: Flush immediately with water for 15 minutes (minimum); seek medical attention. Skin contact: Remove all contaminated clothing. Wash contaminated skin with soap and water. For inhalation: Remove the person from exposure. Provide respiratory assistance and CPR. Transfer to medical facility. If swallowed: Never attempt to give anything by mouth to an unconscious or convulsing person. If large quantities are ingested, induce vomiting with Syrup of Ipecac and call physician.

SECTION VII - PRECAUTIONS FOR SAFE HANDLING AND USE

Steps to be Taken in Case Material is Released or Spilled

Isolate area and stay upwind. Restrict those not involved in cleanup from entering area. Collect powdered material in the most convenient way possible and deposit in sealed containers. Do NOT dry sweep (creates airborne dust). Use vacuum equipped with HEPA filter instead. Damp mop residue and collect for disposal.

Preferred Waste Disposal Method

No citation.

Precautions to be Taken in Handling and Storage

Containers should be protected against physical damage. Store in tightly closed containers in a cool, dry, dark, well-ventilated area. Do not store in direct sunlight.

Other Precautions and Warnings

Wear appropriate personal protective equipment. Keep away from heat and ignition sources.

SECTION VIII - CONTROL MEASURES AND PERSONAL PROTECTIVE EQUIPMENT

Respiratory Protection *(specify type)*

Exposure level is low and difficult to monitor. Use a respirator with particulate (dust/fume/mist) filter. Otherwise use a supplied-air respirator or self-contained breathing apparatus in continuous flow mode.

Ventilation

Local exhaust or general mechanical systems recommended.

Protective Gloves	Eye Protection	Other Protective Clothing
Suitable to protect from skin contact	**Chemical/Dust Goggles**	**N/A**

Work/Hygiene Practices

Always wash hands thoroughly after working with lead sulfate powders and dusts; never bring food, drink, or smoking materials into vicinity of lead sulfate powders.

LEAD SULFATE

$PbSO_4$ CAS: 7446-14-2

IDENTIFICATION AND TYPICAL USES

Lead sulfate is a white, heavy, rhombic, crystalline solid. It is used in storage batteries, paint pigments (whites), lithography, in the manufacture of minium, in weighting fabrics, and in the preparation of rapidly drying oil-based varnishes.

RISK ASSESSMENT: HEALTH

General Assessment

The specific toxicity of lead sulfate is primarily due to lead itself. It can enter the body through *inhalation* and *ingestion*. It is also extremely corrosive to skin and eye tissues, but dermal absorption has not been reported. It should be noted that ingestion poses no real toxic effect in small amounts, but repeated (chronic) ingestion or extremely large single (acute) doses can cause serious problems subsequent to lead poisoning. While the compound lead sulfate is not currently reported as carcinogenic, the IARC does list lead as a probable human carcinogen with insufficient evidence (Group 2B). Mutation data have been reported. Also, while no specific teratogenic data are documented for the sulfate, lead itself is considered a teratogen with reproductive effects also being reported.

Acute poisoning is rare, but possible by both routes of exposure. Inhalation of lead and possibly its soluble salts (such as the sulfate) causes acute encephalopathy with ataxia, repeated vomiting, headaches, stupor, hallucinations, tremors, convulsions or seizures, coma, and death. Other central nervous system effects include weight loss, malnutrition, restlessness, irritability, tingling in the wrists and ankles, and possible permanent brain damage. There can be kidney degradation with glomular atrophy, interstitial fibrosis, and sclerosis of vessels. There may also be astringency, intense thirst, constipation, diarrhea, gastritis, changes in the liver, and accumulation of lead in the bones and teeth. Heavy lead intoxication can often be detected by the formation of a dark line on the gum margins (the "lead line").

Blood complications can be severe, including anemia and hemoglobinuria (blood in urine). Lead causes the red blood cells to become brittle and more fragile. This increased fragility of the cells allows them to be destroyed more rapidly than normal, causing anemia. Once it enters the body, lead circulates to various organs and then settles and accumulates in bone and teeth on prolonged or repeated exposures. It has an experimental half-life in the body of 1460 days.

Acute Health Effects

The following acute (short-term) health effects may occur immediately or shortly after exposure to lead:

Skin: Irritation and possible corrosive destruction.

Eye: Corrosive to eye tissues. It may cause damage to vision on prolonged contact.

Lung: Possible irritation of the nose, throat, and lungs following exposure to powders and dusts.

CNS: A variety of effects including stupor, hallucinations, delirium, parasthesia, headaches, convulsions and seizures, coma, and death.

Other: Acute lead poisoning is rare but possible. There can be severe blood and renal system effects as well as brain effects. Lead-induced anemia and its subsequent effects on blood pressure can serious aggravate existing cardiovascular diseases.

Chronic Health Effects

The following chronic (long-term) health effects may occur at some time after exposure to lead and can last for months or even years:

Cancer Hazards: According to information presented in the references, lead has been tested for its ability to cause cancer in test animals and the results are conflicting and inconclusive. IARC lists it as a probable human carcinogen with insufficient evidence (Group 2B). Human mutation data have been reported and many scientists believe that exposure to mutagenic compounds may pose a cancer risk in the long-term.

Reproductive Hazard: According to information presented in the references, lead has been shown to adversely affect reproduction in test animals. There may be a decreased sex drive, impotence, low sperm count and sterility in males, decreased fertility and abnormal menstrual cycles in females, and possible spontaneous fetal abortions in females. Unborn children may suffer neurological damage or developmental problems due to excessive lead exposures by pregnant women.

Other Chronic Effects: Long-term exposure can lead to appetite loss, nausea, metallic taste in the mouth, a dark "lead sulfate line" on the gingival tissues in the mouth, constipation, anxiety, anemia, facial pallor, tiredness, weakness, insomnia, chronic headache, nervousness, irritability, tremors, tingling and numbness in the extremities, muscle and joint pain, colic, abdominal pain, kidney disease, and liver damage.

Recommended Risk-Reduction Measures

Personnel should avoid direct contact with lead sulfate. *Engineering controls* are always the most effective method of reducing exposures. The best protection, especially when working with fine powders and dusts of lead sulfate, is to enclose operations and/or provide explosion-proof, local exhaust ventilation at the site of powder/dust release. While not always operationally feasible, isolating operations can also reduce exposure. Using respiratory protection is less effective than the controls mentioned above, but is still advisable whenever working with or around lead sulfate. The exposure level (for lead) is extremely low (0.050 mg/m^3) and difficult to accurately monitor. A respirator equipped with a dust/mist/fume filter may suffice for low or infrequent (transient) exposures. However, a supplied-air respirator operated in continuous flow mode or a self-contained breathing apparatus (SCBA) with full facepiece in pressure demand mode will provide the best respiratory protection. If a full facepiece respirator is not available, then dust goggles or safety glasses with side shields attached should be worn to protect the eyes. To prevent hand and skin exposures, suitable gloves should be used. Glove manufacturers should be contacted for glove permeation studies.

Administrative controls should also be in place to minimize the potential for human exposures. These may include written procedures or policies which specify the methods and techniques that will be practiced whenever personnel are to work with lead sulfate.

All personnel should receive *training* on the use, hazards, protective measures, emergency actions, and other precautions per 29 CFR 1910.1200 (Hazard Communication), prior to the first assignment in an area where lead sulfate is used or stored.

Before beginning employment with lead sulfate, and at regular intervals thereafter (e.g., annually), the following medical tests are recommended:

- ☑ Blood lead levels and blood pressure (baseline).
- ☑ Complete blood count (establish baseline).

If symptoms develop or overexposure is suspected, the following additional tests may be useful:

- ☑ Blood lead levels and blood pressure.
- ☑ Complete blood count for microcytic anemia and basophilic stippling, blood urea nitrogen (BUN) of creatinine, and zinc protoporphyrin (ZPP).
- ☑ Examination of the central nervous system.
- ☑ Visual examination of the gingival gum line for the appearance of a dark "lead line" on the tissues.

Other methods to reduce exposure include:

- ☑ Where possible, enclose operations and use local exhaust ventilation at the site where lead sulfate powder or dust is worked. If local exhaust ventilation or enclosure is not used, respiratory protection should be mandatory.
- ☑ Always ensure that proper protective clothing is worn when working with lead compounds.
- ☑ Wash thoroughly immediately after exposure to lead sulfate and at the end of the work shift or before eating, drinking, or smoking. Clothing contaminated with lead sulfate dusts or powders should not be taken home.
- ☑ Hazard warning information should be posted in the work area. In addition, as part of an on-going education and training program, all information on the health and safety hazards of lead sulfate should be communicated to all exposed workers.
- ☑ Additional exposure control measures and engineering controls are provided by OSHA in 29 CFR 1910.1025 (Lead).

RISK ASSESSMENT: ENVIRONMENT

General Assessment

Lead occurs naturally in the environment. However, the environment is still at risk of exposure during transportation, storage, disposal, or destruction of lead sulfate. In almost every scenario, the threat of environmental exposure is contingent upon the proper handling of lead sulfate materials, especially the dusts and powders. Accidental contact with a variety of incompatible materials can result in fire, explosion, and possible contamination of the surrounding environmental mediums (water, soil, and, to a lesser degree, air).

Lead sulfate is considered a non-combustible solid in bulk form. If its dusts and powders become airborne in sufficient quantities or in enclosed areas, there may be an explosion hazard. Lead sulfate can explode in contact with potassium. These characteristics require special consideration during any emergency situation involving a leak or spill of lead sulfate. Contact with incompatible materials can occur either during use, transportation, or storage, or disposal of lead sulfate.

Lead is one of the most abundant metals in the earth's crust. It is used extensively across a broad spectrum of industries for a large variety of purpose. Because of its frequent use and common occurrence, lead enters the environment from point and non-point sources.

Acute Ecological Effects

Acute (short-term) toxic effects may include the death of animals, birds, or fish and death or low growth rate in plants. Acute effects are seen 2 to 4 days after animals or plants are exposed to lead. This metal has high acute toxicity to aquatic life and high toxicity to birds. It is not expected to be taken up by plant life. Insufficient data are available on the short-term effects of lead exposure to land animals.

Chronic Ecological Effects

Chronic (long-term) toxic effects may include shortened life span, reproductive problems, lower fertility, and changes in appearance or behavior in exposed animals. These effects can be seen long after first exposure(s) to toxic chemicals. Lead has high chronic toxicity in aquatic life. Lead causes nerve and behavioral effects in humans and could cause similar long-term effects in birds and land animals. Insufficient data are available on the long-term effects of lead to plants.

Water Solubility

Lead sulfate in slightly soluble in water, depending upon water temperature. Concentrations between 1 and 100 milligrams will mix with a liter of water.

Persistence in the Environment

Lead is highly persistent in water, with a half-life greater than 200 days. The half-life of a pollutant is the amount of time it takes for one half of the chemical to be degraded. Soluble lead compounds, such as the sulfate, may decompose to lead and sulfate in soil where they will settle and become immobile. Lead is expected to accumulate in aquatic and terrestrial sediments. Because it absorbs in soil and forms complexes with resident organic matter, it will not easily leach into aquatic systems nor be taken up by plant life.

Bioaccumulation in Aquatic Organisms

Some substances increase in concentration, or *bioaccumulate,* in living organisms as they breathe contaminated air, drink contaminated water, or eat contaminated food. These chemicals can become concentrated in the tissues and internal organs of animals as well as humans. The concentration of lead found in the edible tissues of fish is expected to be much higher than the average concentration of lead found in the water from which the fish was taken.

Recommended Risk-Reduction Measures

Proper training of all transporters will reduce the likelihood of a mishap or accident resulting in the contamination of the environment. The correct use of DOT labeling (per the MSDS) on all containers, trucks, and rail cars will enable emergency responders to react properly and quickly to any disaster thereby reducing the potential risk to the environment and to personnel.

Storage of lead sulfate should be segregated from incompatible chemicals to minimize the risk of cross-contamination or contact. Buildings designated for storage should be equipped with appropriate fire protection systems (alarms, sprinklers, emergency lighting, portable extinguishers).

If a spill or leak to the environment has occurred, fire department, emergency response and/or hazardous materials spill personnel should be notified immediately. Cleanup should be attempted only by those trained in proper spill containment procedures. Contaminated soils should be removed and replaced with clean soil. When environmental contamination occurs, the local and/or state emergency response authorities may require notification. A comprehensive emergency response or disaster preparedness/recovery plan should be in place prior to any operations involving the use, transportation, storage, or disposal of lead sulfate.

If lead sulfate is spilled or leaked, the following specific steps are recommended:

☑ Restrict persons not wearing protective clothing from area of spill or leak until cleanup is complete and area can be opened for normal work.

☑ Remove all ignition sources and ventilate area.

☑ Collect powdered materials in the safest manner possible and deposit in sealed containers. Do NOT dry sweep (generates airborne dust). Use a vacuum equipped with a high-efficiency particulate air (HEPA) filter. Damp mop residues.

☑ It may be necessary to dispose of lead sulfate as a hazardous waste. The responsible state agency or the regional office of the federal Environmental Protection Agency (EPA) should be contacted for specific recommendations.

RISK ASSESSMENT: BUSINESS

General Assessment

Accidents or mishaps involving lead sulfate can present a significant threat to business operations. The loss or damage of equipment or facilities due to explosion or fire can significantly affect fiscal viability. Lawsuits that may result from personnel injury/death, public exposures, and/or environmental contamination will also require a serious expenditure of resources. Media attention surrounding an injury, death, or environmental damage can also result in a loss of profits and loss of current as well as future business. Always remember that anytime the terms "cancer," "carcinogen," or "reproductive hazard" are used, public emotion, hysteria, and ignorance can run equally high. This is also true whenever the words "lead" and "exposure" are used together. This must be carefully considered when developing or implementing public relations policy.

Recommended Risk-Reduction Measures

Company attorneys, safety and health professionals, and environmental specialists should be involved in the development of any procedures for dealing with safety or emergency response. A company official should be pre-designated as a public relations officer with specific training in dealing with the press. Corporate plans and policies should be developed, approved, and implemented long before any need for such arises.

REFERENCES

American Conference of Governmental Industrial Hygienists. 1988. *Documentation of the Threshold Limit Values and Biological Exposure Indices*, 5th Edition (with updates). Cincinnati: ACGIH

Brusick, D. J. 1994. *Methods for Genetic Risk Assessment.* New York: CRC Press/Lewis Publishers.

Burrell, R., D. K. Flaherty, and Leonard J. Sauers. 1992. *Toxicology of the Immune System.* New York: Van Nostrand Reinhold.

Calabrese, E. J. 1994. *Biological Effects of Low Level Exposures.* New York: CRC Press/Lewis Publishers.

Cockerman, L. G., B. S. Shane. 1994. *Basic Environmental Toxicology.* New York: CRC Press/Lewis Publishers.

Environmental Protection Agency. 1983. *Chemical Hazard Information Profiles.* Washington, D.C.: U.S. EPA.

Fodden, C. R., and J. L. Weddell. 1992. *Hazardous Materials Emergency Action Data.* New York: CRC Press/Lewis Publishers.

Forseberg, K., and L. H. Keith. 1993. *Chemical; Protective Clothing Permeation and Degradation Database.* Chelsea, MI: Lewis Publishers.

Gardner, R. W. 1994. *Chemical Intolerance.* New York: CRC Press/Lewis Publishers.

Howard, P. H., et. al. 1991. *Handbook of Environmental Degradation Rates.* New York: CRC Press/Lewis Publishers.

International Agency for Research on Cancer (IARC). 1987. *IARC Monographs on the Evaluation of Carcinogenic Risk to Humans, Overall Evaluations of Carcinogenicity: An updating of IARC Monographs Volumes 1-42.* Lyon: IARC.

Kamrin, M. A. 1988. *Toxicology: A Primer on Principles and Applications.* Chelsea, Michigan: Lewis Publishers.

Langerman, N. 1994 *Precautionary Labels for Chemical Containers.* New York: CRC Press/Lewis Publishers.

Lauwerys, R. R., P. Hoet. 1993. *Industrial Chemical Exposure.* New York: CRC Press/Lewis Publishers.

Lewis, R. J., Sr. 1993. *Hawley's Condensed Chemical Dictionary*, 12th Edition. New York: Van Nostrand Reinhold.

Lewis, R. J., Sr. 1992. *Sax's Dangerous Properties of Industrial Materials*, Volumes 1, 2, and 3, 8th Edition. New York: Van Nostrand Reinhold.

Meyer, E. 1990. *Chemistry of Hazardous Materials.* Englewood Cliffs, NJ: Prentice Hall.

Patnaik, P. 1992. *A Comprehensive Guide to the Hazardous Properties of Chemical Substances.* New York: Van Nostrand Reinhold.

Rea, W. J. 1995. *Chemical Sensitivity*, Vol. I-III. New York: CRC Press/Lewis Publishers.

Richardson, M. 1995. *Dictionary of Substance and Their Effects*, Vol. I-VI. New York: CRC Press/Lewis Publishers.

Sacarello, H. L. A. 1994. *The Comprehensive Handbook of Hazardous Materials.* New York: CRC Press/Lewis Publishers.

Sherman, J. D. 1988. *Chemical Exposure and Disease.* New York: Van Nostrand Reinhold.

MATERIAL SAFETY DATA SHEET

CHEMICAL NAME	
LINDANE	**CRC PRESS/LEWIS PUBLISHERS DISCLAIMER:** THE INFORMATION AND RECOMMENDATIONS PRESENTED HEREIN ARE BASED ON SOURCES BELIEVED TO BE RELIABLE. CRC MAKES NO REPRESENTATION ON ITS COMPLETENESS OR ACCURACY. IT IS THE RESPONSIBILITY OF THE USER TO DETERMINE THE CHEMICAL'S SUITABILITY FOR ITS INTENDED USE, THE CHEMICAL'S SAFE USE, AND THE CHEMICAL'S PROPER DISPOSAL. NO REPRESENTATIONS AND/OR WARRANTIES, EITHER EXPRESSED OR IMPLIED, OF THE MERCHANTABILITY OR FITNESS FOR A PARTICULAR PURPOSE, OR OF ANY OTHER NATURE, ARE MADE WITH RESPECT TO THE INFORMATION PROVIDED IN THIS MSDS OR TO THE CHEMICAL TO WHICH INFORMATION MAY REFER. CRC NEITHER ASSUMES NOR AUTHORIZES ANY OTHER PERSON TO ASSUME FOR IT, ANY OTHER ADDITIONAL RESPONSIBILITY OR LIABILITY FOR THE USE OF, OR RELIANCE UPON, THIS INFORMATION.

HAZARD WARNING INFORMATION

HEALTH	FIRE	REACTIVE	OTHER	DEGREE OF HAZARD	COLOR CODING	OTHER CODES
3	0	0		0 = Minimum Hazard 1 = Slight Hazard 2 = Moderate Hazard 3 = Serious Hazard 4 = Severe Hazard	HEALTH = BLUE FIRE = RED REACTIVITY = YELLOW OTHER = WHITE	OX = Oxidizer ACID = Acid ALK = Alkali COR = Corrosive W = Use No Water

SECTION I - GENERAL INFORMATION

Characterization	**Pesticide**	RCRA Number	**U129**
		EPA Class	**Toxic Waste**
DOT Proper Shipping Name	**Organochlorine Pesticides (solid, toxic, n.o.s.)**	Chemical Abstract Service (CAS) Number	**58-89-9**
DOT Hazard Class and Label Requirements	**ORM-A; Keep Away From Food**	DOT Emergency Guide Code	**55**
DOT Identification Number	**UN 2761**	Chemical Formula	**$C_6H_6Cl_6$**

Synonyms

γ-Benzene hexachloride; Agrocide; Devoran; *gamma*-benzene hexachloride; *gamma*-BHC; *gamma*-HCH; 1,2,3,4,5,6-hexachlorocyclhexane, gamma isomer; Kwell; Novigam; Lentox; Streunex.

SECTION II - HAZARDOUS INGREDIENTS AND IDENTITY INFORMATION

Hazardous Components (specific identity)	OSHA Exposure Criteria	NIOSH Exposure Criteria	Immediately Dangerous to Life and Health (IDLH)	ACGIH Exposure Criteria
Lindane: (derivation: The gamma isomer of hexachlorocyclohexane. Produced by reaction of chlorine with excess benzene in the presence of sunlight followed by treatment with methanol or acetic acid and then fractional crystallization or chromatographic adsorption).	**PEL (8-hour):** 0.5 mg/m^3 (skin) **STEL:** Not Established	**REL (10-hour):** 0.5 mg/m^3 (skin) **STEL:** Not Established	**50 mg/m^3**	**TLV:** 0.5 mg/m^3 (skin) **STEL:** Not Established

SECTION III - PHYSICAL AND CHEMICAL CHARACTERISTICS

Boiling Point	**614°F (323°C)**	Specific Gravity (H_2O = 1)	**1.85**
Vapor Pressure (mm Hg)	**9.4×10^{-6} at 68°F (20°C)**	Molecular Weight (atomic weight)	**290.8**
Vapor Density (Air = 1)	**Not Found**	Melting Point	**235°F (113°C)**

Solubility

Nearly insoluble (0.001%) in water. Soluble in xylene, toluene, petroleum ether, ethyl acetate, carbon tetrachloride, cyclohexanone, acetic acid, dioxane, acetone, benzene, chloroform, ether, and ethanol.

Appearance and Odor

White to yellow crystalline powder with a slight, musty odor.

SECTION IV - FIRE AND EXPLOSION HAZARD DATA

Flash Point (method used)	**Not Applicable**	Explosive Limits in Air % by Volume	**LEL: Not Applicable UEL: Not Applicable**
NFPA Classification	**Non Combustible Solid**	Autoignition Temperature	**Not Determined**

Extinguishing Media

Use agent suitable to surrounding fire (lindane itself does not burn but it may be dissolved in a flammable solvent).

Special Fire Fighting Procedures

Wear full protective clothing and self-contained breathing apparatus (SCBA). Structural fire-fighting equipment is permeable and may not provide sufficient protection. Remain clear of smoke, water fallout and water runoff. Move containers from fire area if it can be done without risk. Cool exposed containers.

Unusual Fire and Explosion Hazards

Containers may explode in fire. Stay away from ends of tanks.

SECTION V - REACTIVITY DATA

Stability		Conditions to Avoid Lindane can decompose slowly at ordinary pH of 7 and more rapidly under alkaline conditions. Avoid exposure to high temperatures or incompatible materials.
Stable X	Unstable	Incompatibility *(materials to avoid)* Lindane is decomposed by alkalis to form trichlorobenzenes and reacts with powdered aluminum, iron, and zinc. It is corrosive to metals.
Hazardous Polymerization		Conditions to Avoid Under normal conditions of temperature and pressure, hazardous polymerization of lindane is not known to occur.
May Occur	Will Not Occur X	Hazardous Decomposition or By-products At high temperatures, lindane decomposes and may produce poisonous gases, including toxic carbon oxides, chlorine, hydrogen chloride, and phosgene gas.

SECTION VI - HEALTH HAZARD DATA

Primary Route(s) of Entry:	Inhalation?	Absorption (skin/eye)?	Ingestion?
	X	X	X

Health Hazards

INHALATION: Breathing dusts can cause irritation of the nose and throat. Repeated contact will lead to toxic systemic effects. These include headache, dizziness, nausea, vomiting, malaise, profuse sweating, fatigue, clonic or tonic convulsions, myoclonic limb jerks, coma, and respiratory failure. May cause tremors, fainting, muscle spasms, weight loss, and damage to the liver and kidneys.

ABSORPTION: Irritation and possible sensitization. Solution can pass through intact skin to cause toxic effects similar to inhalation. Absorption coupled with other exposures can be fatal.

INGESTION: Nausea, vomiting, diarrhea, and systemic effects (see inhalation).

Carcinogenicity	NTP Listed?	IARC Cancer Review Group?	OSHA Regulated?	Target Organs?
Questioned Human Questioned Animal	5th Annual Report	Group 2B	29 CFR 1910.1000 Table Z-1	CNS, liver, kidneys, eyes, skin, resp. system, blood.

Medical Conditions Generally Aggravated by Exposure

Central nervous system disorders (seizures) and kidney or liver diseases may be aggravated.

Emergency and First-aid Procedures

Eye contact: Do not allow victim to keep eyes tightly shut. Flush immediately with water for 15 minutes; seek medical attention. Skin contact: Remove all contaminated clothing. Immediately wash area with flooding amounts of water. If red skin or blisters appear, seek medical assistance. Inhalation: Remove the person from exposure. Provide respiratory assistance and CPR. Transfer to medical facility. If swallowed: Contact poison control center. Unless advised otherwise, give 1-2 glasses of water. The decision to induce vomiting should be made on a case-by-case bases because of the possibility of seizures. Never give anything by mouth to an unconscious or convulsing person.

SECTION VII - PRECAUTIONS FOR SAFE HANDLING AND USE

Steps to be Taken in Case Material is Released or Spilled

Remove all sources of ignition. Ventilate area of spill. Restrict those not involved in cleanup from entering area. Notify appropriate authorities. Carefully scoop or collect powdered materials with a vacuum equipped with HEPA filter (do NOT generate dusts). Dispose of in sealed drums.

Preferred Waste Disposal Method

Rotary kiln incineration is recommended.

Precautions to be Taken in Handling and Storage

Store in tightly closed containers in cool, well-ventilated area away from heat and incompatibilities.

Other Precautions and Warnings

Avoid generating dusts. Do not store near powdered aluminum, iron, or zinc.

SECTION VIII - CONTROL MEASURES AND PERSONAL PROTECTIVE EQUIPMENT

Respiratory Protection *(specify type)*

For exposures above 0.5 mg/m^3, use a NISOH-approved chemical respirator. Better protection is obtained using a supplied-air respirator or a SCBA with full facepiece operated in positive pressure mode.

Ventilation

Local exhaust (preferred) at site of chemical work or general ventilation.

Protective Gloves	Eye Protection	Other Protective Clothing
Polycarbonate or Butyl Rubber	Chemical/Dust Goggles or Face Mask	Protective Apron, Boots (avoid skin contact)

Work/Hygiene Practices

Always wash hands thoroughly after using chemical; never bring food, drink, or smoking materials into vicinity of chemicals.

LINDANE

$C_6H_6Cl_6$

CAS: 58-89-9

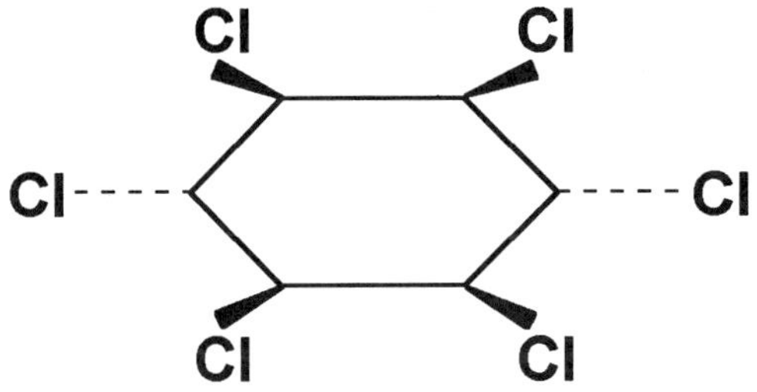

IDENTIFICATION AND TYPICAL USES

Lindane is a white to yellow, crystalline powder with a slight musty odor. It is used as an insecticide and pesticide for controlling insects in wood, cotton, and other foliar plants. It is also used in the treatment of vegetable and fruit seeds and for controlling termites. Weak solution (1%) may be used to kill lice and mites on the human body. Its use may be restricted in some areas.

RISK ASSESSMENT: HEALTH

General Assessment

Lindane is highly toxic to humans by all routes of exposure (*inhalation, ingestion,* and skin *absorption*). It is a questionable carcinogen in animals and in humans, with experimental teratogenic and mutagenic data also reported. Because of the chemical's long persistence in the environment, unknown exposures can occur through contact with contaminated soils, water, and vegetation, at hazardous waste sites, or contaminated manufacturing sites. Should human exposure ever occur, lindane can cause serious toxic and poisonous systemic effects up to and including death.

Lindane is a central nervous system stimulant. Symptoms of exposure by both inhalation and skin contact (absorption) include respiratory depression with convulsions, hypersalivation, hyperirritability, profuse sweating, myoclonic limb jerks, clonic/tonic convulsions, and coma. There may also be headache, dizziness, fatigue, blurred vision, tremors, and respiratory failure. Inhalation also causes irritation of the eyes, nose, and throat. Skin absorption may produce greater toxicity than inhalation. Combined exposures (e.g., inhalation *and* absorption) are extremely dangerous and can lead to death in some cases. Caution is therefore warranted when inhalation and skin contact exposures are possible. Even though air levels may be within limits, personnel may be overexposed if lindane should contact the skin.

In addition to absorption and subsequent systemic effects, skin contact will result in moderate to severe irritation. Eye contact may also result in irritation. The chemical may even be absorbed through direct eye contact. Ingestion causes nausea, vomiting, diarrhea, and systemic effects of inhalation and skin absorption.

Acute Health Effects

The following acute (short-term) health effects may occur immediately or shortly after exposure to lindane:

Skin: Liquid can pass through unbroken skin. Contact may cause irritation and rash. Symptoms of exposure include headache, nausea, abdominal pain, vomiting, and CNS stimulation.

Eye: Mild to severe irritation. Absorption through direct eye contact is possible.

Lung: Irritation of the nose and throat. Breathing dusts or solution vapors can cause cough, convulsions, and adverse effects on the liver.

CNS: A central nervous system stimulant. Symptoms include, irritability, blurred vision, tremors, convulsions, and mental confusion.

Chronic Health Effects

The following chronic (long-term) health effects may occur at some time after exposure to lindane and can last for months or even years:

Cancer Hazards: Lindane may be a carcinogen in humans since it has been shown to cause cancer in animals. Mutation data have also been reported and many scientists believe that exposure to such chemicals may pose a cancer risk in the long-term. Further research is required in this regard.

Reproductive Hazard: There is some evidence that exposure to lindane causes damage to the developing fetus in test animals. Lindane may decrease fertility in males and females.

Other Chronic Effects: Repeated exposures may lead to chronic fainting, muscle spasms, tremors, and weight loss. Lindane is stored primarily in the fatty tissues and remains relatively unchanged.

ʊ *Recommended Risk-Reduction Measures*

Personnel should avoid direct contact with lindane. If a less toxic material or compound cannot be substituted for lindane, then *engineering controls* are the most effective method of reducing exposures. The best protection is to enclose operations and/or provide local exhaust ventilation at the site of lindane dust release. This may be practical in manufacturing operations, but is not feasible in application since it is primarily used outdoors. While not always operationally possible, isolating operations involving lindane manufacturing can also reduce exposure. Using respiratory protection is less effective than the controls mentioned above, but is still advisable whenever working with or around lindane. The PEL is extremely low (0.5 mg/m^3) and difficult to accurately monitor. For low or infrequent exposures, a chemical cartridge respirator equipped with a dust/mist pre-filter may suffice. For higher or prolonged exposures, or exposures to unknown levels (as in an emergency), an MSHA or NIOSH-approved supplied-air respirator operated in continuous flow or a self-contained breathing apparatus (SCBA) with full facepiece operated in positive pressure mode are recommended for the greatest possible respiratory protection. If a full facepiece is not available, then chemical goggles should be worn to protect the eyes. A face shield and protective apron should also be worn. To prevent hand and skin exposures, impervious gloves should be used. Polycarbonate or butyl rubber may provide the best protection. Glove manufacturers should be contacted and permeation studies obtained *before* final glove selections have been made.

Administrative controls should also be in place to minimize the potential for human exposures. These may include written procedures or policies which specify the methods and techniques that will be practiced whenever personnel are to work with lindane.

All personnel should receive *training* on the use, hazards, protective measures, emergency actions, and other precautions per 29 CFR 1910.1200 (Hazard Communication), prior to the first assignment in an area where lindane is used or stored.

If symptoms develop or overexposure is suspected, the following medical testing may be useful:

- ☑ Liver and kidney function tests.
- ☑ Urinalysis for halogenated compounds.
- ☑ Neurological evaluation of central nervous system functions. Precautions against seizures should be taken in the event of acute overexposure.
- ☑ Complete blood count to include white blood count, red blood count and differential.

Any evaluation should include a careful history of past and present symptoms with an examination. Medical tests that simply look for existing damage are not a substitute for controlling exposures.

Other methods to reduce exposure include:

- ☑ Where possible, enclose operations and use local exhaust ventilation at the site of chemical release. If local exhaust ventilation or enclosure is not used, respiratory protection should be mandatory.
- ☑ Always ensure that proper protective clothing is worn when using chemical substances and that personnel are trained in its use and care.
- ☑ Wash thoroughly immediately after exposure to lindane and at the end of the work shift or before eating, drinking, or smoking.
- ☑ Hazard warning information should be posted in the work area. In addition, as part of an on-going education and training program, all information on the health and safety hazards of lindane should be communicated to all potentially exposed workers.
- ☑ Eye wash stations should be provided in the immediate work area for emergency use. If there is a possibility of skin exposure to lindane, emergency shower facilities should also be provided.
- ☑ Workers whose clothing has been contaminated by lindane should change into clean clothes before leaving work. Contaminated work clothing should be laundered only by individuals who have been informed of the hazards of exposure to lindane.
- ☑ Specific engineering controls are recommended by NIOSH. Refer to NIOSH Criteria Document: Occupational Exposure During the Manufacture and Formulation of Pesticides (Publication Number 78-174).

RISK ASSESSMENT: ENVIRONMENT

General Assessment

Because of its use as an insecticide and pesticide, lindane may be present in the environment. The environment is also at risk of exposure during transportation, storage, disposal, or destruction of lindane. In almost every scenario, the threat of environmental exposure is contingent upon the proper handling of the chemical substance. Accidental spills, large or small, where lindane contacts incompatible commodities can result in violent and even explosive reactions (depending upon conditions of spill), and possible contamination of the surrounding environmental mediums (water, soil, and air).

Lindane is considered a non-combustible solid. However, it is common practice to mix it with some flammable liquids such as benzene or xylene. Also, because it is incompatible with many alkalis, as well as some powdered metals, extreme caution is required in handling, storage, transportation, and disposal of lindane. These characteristics also require special consideration during any emergency situation involving a leak or spill of lindane powder or solutions.

Lindane may enter the environment in agricultural run off or through spills.

Acute Ecological Effects

Acute (short-term) toxic effects may include the death of animals, birds, or fish and death or low growth rate in plants. Acute effects are seen 2 to 4 days after animals or plants are exposed to lindane. Lindane has high acute toxicity to aquatic life and to birds. Insufficient data are available on the short-term effects of lindane to plants or land animals.

Chronic Ecological Effects

Chronic (long-term) toxic effects may include shortened life span, reproductive problems, lower fertility, and changes in appearance or behavior in exposed animals. These effects can be seen long after first exposure(s) to toxic chemicals. Lindane has high chronic toxicity to aquatic life. Insufficient data are available to evaluate or predict the long-term effects of lindane to birds, plants, or land animals.

Water Solubility

Lindane is moderately soluble in water. Concentrations of 1 to 1000 milligrams will mix with a liter of water.

Persistence in the Environment

Lindane is moderately persistent in the water, with a half-life between 20 and 200 days. The half-life of a pollutant is the amount of time it takes for one-half of the chemical to be degraded. Lindane is more persistent in dry, sandy soils than in moist soils. It can leach to ground water and absorb in terrestrial and aquatic sediments. About 52.3% of lindane will eventually end up in water; approximately 22% will end up in aquatic sediments; and about 2.3% will end up in air.

Bioaccumulation in Aquatic Organisms

Some substances increase in concentration, or *bioaccumulate,* in living organisms as they breathe contaminated air, drink contaminated water, or eat contaminated food. These chemicals can become concentrated in the tissues and internal organs of animals as well as humans.

The concentration of lindane found in fish tissues is expected to be somewhat higher that the average concentration of lindane in the water from which the fish was taken.

Recommended Risk-Reduction Measures

Environmental contamination involving lindane can pose a serious, long-lasting impact on affected ecosystems. This is based upon its persistence in the environment and the fact that it can be present in contaminated soils, at hazardous waste sites, and in some stockpiles awaiting destruction.

Storage of lindane should be segregated from incompatible materials to minimize the risk of cross-contamination or contact. Buildings or storage areas should be equipped with appropriate fire protection systems (alarms, sprinklers). Proper ventilation and protective equipment should be used while handling the powders or while preparing aqueous solutions.

If a spill or leak to the environment has occurred, fire department, emergency response and/or hazardous materials spill personnel should be notified immediately. Cleanup should be attempted only by those trained in proper spill containment procedures. If lindane should contact the water table, aquifer, or navigable waterway, reclamation procedures should be initiated as soon as practical. It is nearly insoluble in water and total remediation may be possible. The local and/or state emergency response authorities must be notified. A comprehensive emergency response or disaster preparedness/recovery plan should be in place prior to any operations involving the use, transportation, storage, or disposal of lindane. If lindane is spilled or leaked, the following specific steps are recommended:

☑ Restrict persons not wearing protective clothing from area of spill or leak until cleanup is complete and area can be opened for normal work.

☑ Remove all ignition sources and ventilate area.

☑ Collect solids using a HEPA vacuum to avoid the generation of dusts. Do NOT dry sweep. Damp mop any residues and collect for disposal.

☑ It may be necessary to dispose of lindane as a hazardous waste. The responsible state agency or the regional office of the federal Environmental Protection Agency (EPA) should be contacted for specific recommendations.

RISK ASSESSMENT: BUSINESS

General Assessment

Accidents or mishaps involving lindane can present a significant threat to business operations. The loss or damage of equipment or facilities can significantly affect fiscal viability. Lawsuits that may result from personnel injury/death, public exposures, and/or environmental contamination will also require a serious expenditure of resources. Media attention surrounding an injury, death, or environmental damage can also result in a loss of profits and loss of current as well as future business. Always remember that anytime the terms "cancer," "carcinogen," or "reproductive hazard" are used, public emotion, hysteria, and ignorance can run equally high. This must be a consideration when developing public relations policies.

Recommended Risk-Reduction Measures

Company attorneys, safety and health professionals, and environmental specialists should be involved in the development of any procedures for responding to chemical incidents. A company official should be predesignated as a public relations officer with specific training in dealing with the press. Corporate plans and policies should be developed, approved, and implemented long before any need for such arises.

REFERENCES

American Conference of Governmental Industrial Hygienists. 1988. *Documentation of the Threshold Limit Values and Biological Exposure Indices,* 5th Edition (with updates). Cincinnati: ACGIH

Brusick, D. J. 1994. *Methods for Genetic Risk Assessment.* New York: CRC Press/Lewis Publishers.

Calabrese, E. J. 1994. *Biological Effects of Low Level Exposures.* New York: CRC Press/Lewis Publishers.

Cockerman, L. G., B. S. Shane. 1994. *Basic Environmental Toxicology.* New York: CRC Press/Lewis Publishers.

Cothern, C. R. 1995. *Handbook for Environmental Risk Decisionmaking.* New York: CRC Press/Lewis Publishers.

Danse, I. R. 1991. *Common Sense Toxics in the Workplace.* New York: Van Nostrand Reinhold.

Environmental Protection Agency. 1983. *Chemical Hazard Information Profiles.* Washington, D.C.: U.S. EPA.

Fodden, C. R., and J. L. Weddell. 1992. *Hazardous Materials Emergency Action Data.* New York: CRC Press/Lewis Publishers.

Forseberg, K., and L. H. Keith. 1993. *Chemical; Protective Clothing Permeation and Degradation Database.* Chelsea, MI: Lewis Publishers.

Gardner, R. W. 1994. *Chemical Intolerance.* New York: CRC Press/Lewis Publishers.

Howard, P. H., et. al. 1991. *Handbook of Environmental Degradation Rates.* New York: CRC Press/Lewis Publishers.

Lauwerys, R. R., P. Hoet. 1993. *Industrial Chemical Exposure.* New York: CRC Press/Lewis Publishers.

Lewis, R. J., Sr. 1992. *Sax's Dangerous Properties of Industrial Materials,* Volumes 1, 2, and 3, 8th Edition. New York: Van Nostrand Reinhold.

Manahan, S. E. 1992. *Toxicological Chemistry,* 2nd Edition. New York: CRC Press/Lewis Publishers.

Meyer, E. 1990. *Chemistry of Hazardous Materials.* Englewood Cliffs, NJ: Prentice Hall.

Ott, W. R. 1995. *Environmental Statistics and Data Analysis.* New York: CRC Press/Lewis Publishers.

Patnaik, P. 1992. *A Comprehensive Guide to the Hazardous Properties of Chemical Substances.* New York: Van Nostrand Reinhold.

Richardson, M. 1995. *Dictionary of Substance and Their Effects,* Vol. I-VI. New York: CRC Press/Lewis Publishers.

U.S. Department of Health and Human Services, National Institute for Occupational Health (NIOSH). 1978. *NIOS Criteria Document: Occupational Exposure During the Manufacture and Formulation of Pesticides.* Publication Number 78-174. Washington, DC: U.S. Government Printing Office.

U.S. Department of Health and Human Services, National Institute for Occupational Safety and Health. 1994. *NIOSH Pocket Guide to Chemical Hazards.* Washington, D.C.: U.S. Government Printing Office.

U.S. Congress: Office of Technology Assessment Task Force. 1988. *Identifying and Regulating Carcinogens.* Chelsea, MI: Lewis Publishers.

Witorsch, P. and S. V. Spagnolo. 1994. *Air Pollution and Lung Disease in Adults.* New York: CRC Press/Lewis Publishers.

MATERIAL SAFETY DATA SHEET

CHEMICAL NAME	
LITHIUM	**CRC PRESS/LEWIS PUBLISHERS DISCLAIMER:** THE INFORMATION AND RECOMMENDATIONS PRESENTED HEREIN ARE BASED ON SOURCES BELIEVED TO BE RELIABLE. CRC MAKES NO REPRESENTATION ON ITS COMPLETENESS OR ACCURACY. IT IS THE RESPONSIBILITY OF THE USER TO DETERMINE THE CHEMICAL'S SUITABILITY FOR ITS INTENDED USE, THE CHEMICAL'S SAFE USE, AND THE CHEMICAL'S PROPER DISPOSAL. NO REPRESENTATIONS AND/OR WARRANTIES, EITHER EXPRESSED OR IMPLIED, OF THE MERCHANTABILITY OR FITNESS FOR A PARTICULAR PURPOSE, OR OF ANY OTHER NATURE, ARE MADE WITH RESPECT TO THE INFORMATION PROVIDED IN THIS MSDS OR TO THE CHEMICAL TO WHICH INFORMATION MAY REFER. CRC NEITHER ASSUMES NOR AUTHORIZES ANY OTHER PERSON TO ASSUME FOR IT, ANY OTHER ADDITIONAL RESPONSIBILITY OR LIABILITY FOR THE USE OF, OR RELIANCE UPON, THIS INFORMATION.

HAZARD WARNING INFORMATION

HEALTH	FIRE	REACTIVE	OTHER	DEGREE OF HAZARD	COLOR CODING	OTHER CODES
3	2	2	~~W~~	0 = Minimum Hazard 1 = Slight Hazard 2 = Moderate Hazard 3 = Serious Hazard 4 = Severe Hazard	HEALTH = BLUE FIRE = RED REACTIVITY = YELLOW OTHER = WHITE	OX = Oxidizer ACID = Acid ALK = Alkali COR = Corrosive ~~W~~ = Use No Water

SECTION I - GENERAL INFORMATION

Characterization	**Metal, Reactive**	RCRA Number	**D003**
		EPA Class	**Characteristic (R) Waste**
DOT Proper Shipping Name	**Lithium, metal**	Chemical Abstract Service (CAS) Number	**7439-93-2**
DOT Hazard Class and Label Requirements	**Flammable Solid**	DOT Emergency Guide Code	**40**
DOT Identification Number	**UN 1415**	Chemical Symbol	**Li**

Synonyms

Lithium metal.

SECTION II - HAZARDOUS INGREDIENTS AND IDENTITY INFORMATION

Hazardous Components (specific identity)	OSHA Exposure Criteria	NIOSH Exposure Criteria	Immediately Dangerous to Life and Health (IDLH)	ACGIH Exposure Criteria
Lithium (derivation: By electrolysis of a mixture of lithium chloride and potassium chloride, high-temperature extraction from spodumene by sodium carbonate, solar evaporation of lake brines).	**PEL:** Not Established **STEL:** Not Established	**REL:** Not Established **STEL:** Not Established	**Not Established**	**TLV:** Not Established **STEL:** Not Established

SECTION III - PHYSICAL AND CHEMICAL CHARACTERISTICS

Boiling Point	**2403°F (1317°C)**	Specific Gravity (H_2O = 1)	**0.534**
Vapor Pressure (mm Hg)	**1 at 1333°F (723°C)**	Atomic Weight	**6.94**
Vapor Density (Air = 1)	**Not Applicable**	Melting Point	**357°F (181°C)**

Solubility

Decomposes in water (forms the hydroxide and free hydrogen). Soluble in liquid ammonia and acids, decomposes in alcohol.

Appearance and Odor

Soft, silver-colored, light-weight, alkali metal turning yellow on exposure to air.

SECTION IV - FIRE AND EXPLOSION HAZARD DATA

Flash Point (method used)	**Not Determined (Flammable)**	Explosive Limits in Air % by Volume	**LEL: Not Determined UEL: Not Determined**
NFPA Classification	**Flammable Metal**	Autoignition Temperature	**354°F (180°C)**

Extinguishing Media

Use mixture of dry chemical, soda ash, and graphite, dry clay, or a commercial mixture ("Lith-X"). Do NOT use carbon dioxide, water, fog, foam, or halogens (there can be an explosion!).

Special Fire Fighting Procedures

Poisonous gases are released in fire. Wear self-contained breathing apparatus.

Unusual Fire and Explosion Hazards

Dangerous fire hazard when exposed to heat or flame. The powder may ignite spontaneously in air. It will burn in oxygen, nitrogen, and carbon dioxide. It will continue to burn in sand and sodium carbonate. Molten lithium is extremely reactive and will attack inert materials (sand, concrete, and ceramics).

SECTION V - REACTIVITY DATA

Stability		Conditions to Avoid Lithium metal is normally stable under routine conditions of storage and handling. Avoid contact with incompatible materials and keep away from heat and flame.
Stable X	Unstable	Incompatibility *(materials to avoid)* Attacked by acids (dilute hydrochloric and sulfuric). Reacts vigorously with water or steam (releases heat and explosive hydrogen gas). Attacks inert materials (sand, concrete).
Hazardous Polymerization		Conditions to Avoid Under normal conditions of temperature and pressure, hazardous polymerization of lithium metal is not expected to occur. Avoid contact with atmospheric gases (oxygen, nitrogen, hydrogen, carbon dioxide).
May Occur	Will Not Occur X	Hazardous Decomposition or By-products When heated to decomposition, lithium can produce toxic and poisonous gases and fumes, including lithium dioxide and lithium hydroxide.

SECTION VI - HEALTH HAZARD DATA

Primary Route(s) of Entry:	Inhalation?	Absorption (skin)?	Ingestion?
	X		

Health Hazards

INHALATION: Causes nausea, vomiting, abdominal pain, bloody diarrhea, weakness, fatigue, thirst and dehydration with dry mouth. Exposure can lead to CNS disturbances and kidney effects. There may be tremor, incoordination, giddiness, tinnitus, drowsiness, slurred speech, blurred vision, light sensitivity, and possible temporary loss of vision.

SKIN & EYES: Causes chemical burns, especially in contact with moist skin and the eyes. Skin rash may also develop after contact with lithium.

INGESTION: Not a likely exposure route. Ingestion causes kidney injury, especially when sodium intake is limited.

Carcinogenicity	NTP Listed?	IARC Cancer Review Group?	OSHA Regulated?	Target Organs?
Unknown Human Unknown Animal	No	No	No	Eyes, skin, kidney, central nervous system (CNS).

Medical Conditions Generally Aggravated by Exposure

None specifically cited. However, existing kidney diseases may be aggravated by exposure.

Emergency and First-aid Procedures

Eye contact: Flush immediately with water for 15 minutes (minimum); seek medical attention. Skin contact: Remove all contaminated clothing. Wash contaminated skin with soap and water. For inhalation: Remove the person from exposure. Provide respiratory assistance and CPR. Transfer to medical facility. Never attempt to give anything by mouth to an unconscious or convulsing person.

SECTION VII - PRECAUTIONS FOR SAFE HANDLING AND USE

Steps to be Taken in Case Material is Released or Spilled

Collect powdered material in the most convenient way possible and deposit in sealed containers. Do NOT dry sweep; use HEPA vacuum instead. Use non-sparking tools. Remove all sources of ignition. Restrict those not involved in cleanup from entering area. Notify appropriate authorities, as required.

Preferred Waste Disposal Method

No citation.

Precautions to be Taken in Handling and Storage

Containers should be protected against physical damage. Store in tightly closed containers in a cool, dry, well-ventilated area away from incompatible materials and heat. Storage under mineral oil, kerosene of other water-free liquid will prevent reactions (with oxygen or water).

Other Precautions and Warnings

Whenever bulk or finely powdered lithium is used, handled, manufactured, or stored, use explosion-proof electrical equipment and fittings. Electrically ground and bond containers and equipment.

SECTION VIII - CONTROL MEASURES AND PERSONAL PROTECTIVE EQUIPMENT

Respiratory Protection *(specify type)*

No exposures levels established. For best protection use a NIOSH-approved respirator with particulate (dust/fume/mist) filters. Otherwise, use a self-contained breathing apparatus in continuous flow mode.

Ventilation

Local exhaust or general mechanical systems recommended.

Protective Gloves	Eye Protection	Other Protective Clothing
Suitable to protect from skin contact	Chemical/Dust Goggles	N/A

Work/Hygiene Practices

Always wash hands thoroughly after working with lithium powders and dusts; never bring food, drink, or smoking materials into vicinity of lithium powders.

LITHIUM

Li

CAS: 7439-93-2

IDENTIFICATION AND TYPICAL USES

Lithium is a soft, silvery-white alkali metal that turns yellow on exposure to air. It can exist as ingots, wire, ribbon, pellets, or rods. Lithium is metallic element of atomic number 3, group IA of the periodic table, atomic weight 6.941, valance of 1, with two stable isotopes. It is the least reactive of the alkali metals and the lightest solid element. It is used in the production of tritium, in reducing and hydrogenating agents, in alloy hardeners, and certain pharmaceuticals (psychoactive drugs for treatment of manic depression). It is a scavenger for stainless steel and mild steels in the molten state. It is used as a deodorizer in copper and copper alloys, as a catalyst, in storage batteries, rocket propellants, silver solders, and a nuclear reactor coolant.

RISK ASSESSMENT: HEALTH

General Assessment

Lithium is primarily toxic through *inhalation* and skin contact (although dermal absorption has not been reported). *Ingestion* is also possible but is not considered a likely exposure route. There are no reports regarding its carcinogenic or mutagenic effects. Teratogenic effects have been documented.

Inhalation can cause nausea, vomiting, severe abdominal pain, diarrhea (possibly bloody), weakness and fatigue, dehydration, intense thirst, and dry mouth. Primary effects of exposure appear to be its toxic action on the central nervous system. These include tremors of the hands, sleepiness, vertigo, increased urine volume, apathy, anorexia, fatigue, incoordination, giddiness, tinnitus, slurred speech, blurred vision, light-sensitivity, possible temporary blindness, lethargy, muscular weakness, hyperactive reflexes, muscular weakness and twitching, stupor, and changes in electrocardiogram (EKG) readings. Long-term exposures can lead to hypothyroidism, leukocytosis, edema, weight gain, polydipsia (increased water intake) leading to polyuria (increase urinary output), memory impairment, epileptic seizures, kidney damage, shock, hypotension, cardiac arrhythmia, coma, and death.

Lithium can react with the moisture on the skin or in the eyes to produce corrosive hydroxide causing serious burns and tissue damage. While ingestion is not considered a likely exposure route, it can cause serious kidney injury and possible CNS effects.

Acute Health Effects

The following acute (short-term) health effects may occur immediately or shortly after exposure to lithium:

Skin: Irritation, burning, and corrosive destruction of tissues on contact. Extent of damage is contingent on moisture content on skin's surface.

Eye: Irritation and burning on contact. Water content in eyes will produce corrosive hydroxide which can seriously damage eye tissues and may cause temporary blindness.

Lung: Possible irritation of the nose, throat, and lungs following exposure to lithium dusts.

CNS: Broad range of serious and possibly life-threatening effects resulting from central nervous system depression, including death.

Chronic Health Effects

The following chronic (long-term) health effects may occur at some time after exposure to lithium and can last for months or even years:

Cancer Hazards: According to information presented in the references, lithium has not been shown to cause cancer in test animals. It is not known if long-term exposure will lead to carcinogenic effects in humans.

Reproductive Hazard: According to information presented in the references, lithium may have some adverse effects on the developing fetus. It has been suggested that pregnant women and those breast feeding their infants avoid contact with lithium.

Other Chronic Effects: Long-term exposure to lithium can lead to chronic nervous system disorders. It can also cause kidney dysfunction and injury. Goiter has occurred in patients treated with lithium.

Recommended Risk-Reduction Measures

Personnel should avoid direct contact with lithium. *Engineering controls* are always the most effective method of reducing exposures. The best protection, especially when working with fine powders and dusts of lithium, is to enclose operations and/or provide explosion-proof, local exhaust ventilation at the site of powder/dust release. While not always operationally feasible, isolating operations can also reduce exposure. Using respiratory protection is less effective than the controls mentioned above, but is still advisable whenever working with or around lithium. No exposure levels have been established for lithium. However, this does not mean that exposure is without risk. For low or transient exposures, an MSHA/NIOSH-approved respirator equipped with a dust/mist pre-filter may provide adequate protection. Better protection is obtained using a self-contained breathing apparatus (SCBA) with full facepiece operated in pressure demand or a supplied-air respirator. If a full facepiece respirator is not available, then dust goggles or safety glasses with side shields attached should be worn to protect the eyes. To prevent hand and skin exposures, suitable gloves should be used.

Administrative controls should also be in place to minimize the potential for human exposures. These may include written procedures or policies which specify the methods and techniques that will be practiced whenever personnel are to work with lithium.

All personnel should receive *training* on the use, hazards, protective measures, emergency actions, and other precautions per 29 CFR 1910.1200 (Hazard Communication), prior to the first assignment in an area where lithium is used or stored.

If symptoms develop or overexposure is suspected, the following tests might be considered:

- ☑ Complete blood count.
- ☑ Kidney function tests, including urinalysis.
- ☑ Electrocardiogram (EKG) for abnormal arrhythmia and other irregularities.
- ☑ Nervous system evaluation.

Other methods to reduce exposure include:

- ☑ Where possible, enclose operations and use local exhaust ventilation at the site where lithium powder or dust is worked. If local exhaust ventilation or enclosure is not used, respiratory protection should be mandatory.
- ☑ Always ensure that proper protective clothing is worn when working with lithium compounds.
- ☑ Wash thoroughly immediately after exposure to lithium and at the end of the work shift or before eating, drinking, or smoking. Clothing contaminated with lithium dusts or powders should not be taken home.
- ☑ Hazard warning information should be posted in the work area. In addition, as part of an on-going education and training program, all information on the health and safety hazards of lithium should be communicated to all potentially exposed workers.

RISK ASSESSMENT: ENVIRONMENT

General Assessment

The environment is at risk of exposure during transportation, storage, disposal, or destruction of lithium. In almost every scenario, the threat of environmental exposure is contingent upon the proper handling of lithium materials, especially the dusts and powders. Accidental contact with a variety of incompatible materials can result in fire, explosion, and possible contamination of the surrounding environmental mediums (water, soil, and, to a lesser degree, air).

Lithium is considered a flammable solid. It is incompatible with a number of commonly encountered materials, including water and air. When heated to the molten state, it is highly reactive with ordinary inert materials such as sand, cement, and ceramics. These characteristics require special consideration during any emergency situation involving a leak or spill of lithium. Should lithium ever come into contact with incompatible substances such as acids, halogenated compounds, or other reactive metals either during use, transportation, or storage, the formation of highly explosive commodities is extremely possible.

Lithium can enter the environment through spills and industrial waste discharges.

Acute Ecological Effects

Acute (short-term) toxic effects may include the death of animals, birds, or fish and death or low growth rate in plants. Acute effects are seen 2 to 4 days after animals or plants are exposed to lithium.

Insufficient data are available on the short-term effects of lithium exposure to aquatic life, plants, birds, or land animals.

Chronic Ecological Effects

Chronic (long-term) toxic effects may include shortened life span, reproductive problems, lower fertility, and changes in appearance or behavior in exposed animals. These effects can be seen long after first exposure(s) to toxic chemicals.

Insufficient data are available on the long-term effects of lithium to aquatic life, plants, birds, or land animals.

Water Solubility

Lithium is not soluble in water. It reacts and decomposes to form lithium hydroxide and free hydrogen gas, which is explosive.

Persistence in the Environment

Lithium is non-persistent in water, with a half-life of less than 2 days. The half-life of a pollutant is the amount of time it takes for one half of the chemical to be degraded.

Bioaccumulation in Aquatic Organisms

Some substances increase in concentration, or *bioaccumulate,* in living organisms as they breathe contaminated air, drink contaminated water, or eat contaminated food. These chemicals can become concentrated in the tissues and internal organs of animals as well as humans.

Lithium will not accumulate in fish tissues.

Recommended Risk-Reduction Measures

Proper training of all transporters will reduce the likelihood of a mishap or accident resulting in the contamination of the environment. The correct use of DOT labeling (per the MSDS) on all containers, trucks, and rail cars will enable emergency responders to react properly and quickly to any disaster thereby reducing the potential risk to the environment and to personnel.

Storage of lithium should be segregated from incompatible chemicals to minimize the risk of cross-contamination or contact. It should be stored under mineral oil, kerosene or other liquid free of water and air. Buildings designated for storage should be properly equipped with appropriate fire protection systems (alarms, sprinklers, emergency lighting, portable extinguishers).

If a spill or leak to the environment has occurred, fire department, emergency response and/or hazardous materials spill personnel should be notified immediately. Cleanup should be attempted only by those trained in proper spill containment procedures using non-sparking tools. Contaminated soils should be removed and replaced with clean soil. When environmental contamination occurs, the local and/or state emergency response authorities may require notification. A comprehensive emergency response or disaster preparedness/recovery plan should be in place prior to any operations involving the use, transportation, storage, or disposal of lithium. If lithium is spilled or leaked, the following specific steps are recommended:

- ☑ Restrict persons not wearing protective clothing from area of spill or leak until cleanup is complete and area can be opened for normal work.
- ☑ Remove all ignition sources.
- ☑ Collect powdered materials in the safest manner possible and deposit in sealed containers. Do NOT dry sweep (generates airborne dusts). Use a vacuum equipped with a high-efficiency particulate air (HEPA) filter. Use non-sparking tools and keep water away.
- ☑ It may be necessary to dispose of lithium as a hazardous waste. The responsible state agency or the regional office of the federal Environmental Protection Agency (EPA) should be contacted for specific recommendations.

RISK ASSESSMENT: BUSINESS

General Assessment

Accidents or mishaps involving lithium can present a significant threat to business operations. The loss or damage of equipment or facilities due to explosion or fire can significantly affect fiscal viability. Lawsuits that may result from personnel injury/death, public exposures, and/or environmental contamination will also require a serious expenditure of resources. Media attention surrounding an injury, death, or environmental damage can also result in a loss of profits and loss of current as well as future business.

Recommended Risk-Reduction Measures

Company attorneys, safety and health professionals, and environmental specialists should be involved in the development of any procedures dealing with safety or emergency response. A company official should be pre-designated as a public relations officer with specific

training in dealing with the press. Corporate plans and policies should be developed, approved, and implemented long before any need for such arises.

REFERENCES

American Conference of Governmental Industrial Hygienists. 1988. *Documentation of the Threshold Limit Values and Biological Exposure Indices,* 5th Edition (with updates). Cincinnati: ACGIH

Asante-Duah, D. K. 1993. *Hazardous Waste Risk Assessment.* New York: CRC Press/Lewis Publishers.

Bloom, A. D. and F. J. DeSerres. 1995. *Ecotoxicity and Human Health.* New York: CRC Press/Lewis Publishers.

Brusick, D. J. 1994. *Methods for Genetic Risk Assessment.* New York: CRC Press/Lewis Publishers.

Bueche, J. R. 1972. *Principles of Physics,* 2nd Edition. New York: McGraw-Hill.

Burrell, R., D. K. Flaherty, and Leonard J. Sauers. 1992. *Toxicology of the Immune System.* New York: Van Nostrand Reinhold.

Calabrese, E. J., E. M. Kenyon. *Air Toxics and Risk Assessment.* New York: CRC Press/Lewis Publishers.

Cockerman, L. G., B. S. Shane. 1994. *Basic Environmental Toxicology.* New York: CRC Press/Lewis Publishers.

Cothern, C. R., N. P. Ross. 1994. *Environmental Statistics, Assessment, and Forecasting.* New York: CRC Press/Lewis Publishers.

Danse, I. R. 1991. *Common Sense Toxics in the Workplace.* New York: Van Nostrand Reinhold.

Gardner, R. W. 1994. *Chemical Intolerance.* New York: CRC Press/Lewis Publishers.

Hallenbeck, W. H. 1993. *Quantitative Risk Assessment for Environmental and Occupational Health,* 2nd Edition. New York: CRC Press/Lewis Publishers.

Hosty, J. W., P. Foster. 1990. *A Practical Guide to Chemical Spill Response.* New York: Van Nostrand Reinhold.

Howard, P. H., et. al. 1991. *Handbook of Environmental Degradation Rates.* New York: CRC Press/Lewis Publishers.

James, A., M. Lord. 1992. *Index of Chemical and Physical Data.* New York: Van Nostrand Reinhold.

Jones, F. E. 1994. *Toxic Organic Vapors in the Workplace.* New York: CRC Press/Lewis Publishers.

Keith, L. H. and M. M. Walker. 1995. *Handbook of Air Toxics.* New York: CRC Press/Lewis Publishers.

Landis, W. G., M. H. Yu. 1995. *Introduction to Environmental Toxicology.* New York: CRC Press/Lewis Publishers.

Langerman, N. 1994 *Precautionary Labels for Chemical Containers.* New York: CRC Press/Lewis Publishers.

Lauwerys, R. R., P. Hoet. 1993. *Industrial Chemical Exposure.* New York: CRC Press/Lewis Publishers.

Lewis, R. J., Sr. 1993. *Hawley's Condensed Chemical Dictionary,* 12th Edition. New York: Van Nostrand Reinhold.

Lewis, R. J., Sr. 1992. *Sax's Dangerous Properties of Industrial Materials,* Volumes 1, 2, and 3, 8th Edition. New York: Van Nostrand Reinhold.

Lide, D. R. 1995. *Handbook of Chemistry and Physics,* 76th Edition. New York: CRC Press/Lewis Publishers.

Manahan, S. E. 1994. *Environmental Chemistry,* 6th Edition. New York: CRC Press/Lewis Publishers.

Manahan, S. E. 1992. *Toxicological Chemistry,* 2nd Edition. New York: CRC Press/Lewis Publishers.

Mellström, G. A., J. E. Wahlberg, and H. I. Maibach. 1994. *Protective Gloves for Occupational Use.* New York: CRC Press/Lewis Publishers.

Meyer, E. 1990. *Chemistry of Hazardous Materials.* Englewood Cliffs, NJ: Prentice Hall.

Mycock, J. C., J. D. McKenna, and L. Theodore. 1995. *Handbook of Air Pollution Control Engineering and Technology.* New York: CRC Press/Lewis Publishers.

Ott, W. R. 1995. *Environmental Statistics and Data Analysis.* New York: CRC Press/Lewis Publishers.

Patnaik, P. 1992. *A Comprehensive Guide to the Hazardous Properties of Chemical Substances.* New York: Van Nostrand Reinhold.

Philp, R. B. 1995. *Environmental Hazards and Human Health.* New York: CRC Press/Lewis Publishers.

Proctor, N. H., J. P. Hughes. 1991. *Chemical Hazards of the Workplace.* New York: Van Nostrand Reinhold.

Quigley, D. R. 1994. *Handbook of Emergency Chemical Management.* New York: CRC Press/Lewis Publishers.

Rea, W. J. 1995. *Chemical Sensitivity,* Vol. I-III. New York: CRC Press/Lewis Publishers.

Richardson, M. 1995. *Dictionary of Substance and Their Effects,* Vol. I-VI. New York: CRC Press/Lewis Publishers.

Rylander, R., R. R. Jacobs. 1994. *Organic Dusts.* New York: CRC Press/Lewis Publishers.

Sacarello, H. L. A. 1994. *The Comprehensive Handbook of Hazardous Materials.* New York: CRC Press/Lewis Publishers.

Sheftel, V. O. 1995. *Handbook of Toxic Properties of Monomers and Additives.* New York: CRC Press/Lewis Publishers.

Sherman, J. D. 1988. *Chemical Exposure and Disease.* New York: Van Nostrand Reinhold.

Talbott, E., G. Craun. 1995. *An Introduction to Environmental Epidemiology.* New York: CRC Press/Lewis Publishers.

U.S. Department of Health and Human Services, National Institute for Occupational Safety and Health. 1994. *NIOSH Pocket Guide to Chemical Hazards.* Washington, D.C.: U.S. Government Printing Office.

Witorsch, P. and S. V. Spagnolo. 1994. *Air Pollution and Lung Disease in Adults.* New York: CRC Press/Lewis Publishers.

MATERIAL SAFETY DATA SHEET

CHEMICAL NAME	
LITHIUM BROMIDE	**CRC PRESS/LEWIS PUBLISHERS DISCLAIMER:** THE INFORMATION AND RECOMMENDATIONS PRESENTED HEREIN ARE BASED ON SOURCES BELIEVED TO BE RELIABLE. CRC MAKES NO REPRESENTATION ON ITS COMPLETENESS OR ACCURACY. IT IS THE RESPONSIBILITY OF THE USER TO DETERMINE THE CHEMICAL'S SUITABILITY FOR ITS INTENDED USE, THE CHEMICAL'S SAFE USE, AND THE CHEMICAL'S PROPER DISPOSAL. NO REPRESENTATIONS AND/OR WARRANTIES, EITHER EXPRESSED OR IMPLIED, OF THE MERCHANTABILITY OR FITNESS FOR A PARTICULAR PURPOSE, OR OF ANY OTHER NATURE, ARE MADE WITH RESPECT TO THE INFORMATION PROVIDED IN THIS MSDS OR TO THE CHEMICAL TO WHICH INFORMATION MAY REFER. CRC NEITHER ASSUMES NOR AUTHORIZES ANY OTHER PERSON TO ASSUME FOR IT, ANY OTHER ADDITIONAL RESPONSIBILITY OR LIABILITY FOR THE USE OF, OR RELIANCE UPON, THIS INFORMATION.

HAZARD WARNING INFORMATION

HEALTH	FIRE	REACTIVE	OTHER	DEGREE OF HAZARD	COLOR CODING	OTHER CODES
2	**0**	**0**		0 = Minimum Hazard 1 = Slight Hazard 2 = Moderate Hazard 3 = Serious Hazard 4 = Severe Hazard	HEALTH = BLUE FIRE = RED REACTIVITY = YELLOW OTHER = WHITE	OX = Oxidizer ACID = Acid ALK = Alkali COR = Corrosive ~~W~~ = Use No Water

SECTION I - GENERAL INFORMATION

Characterization: **Bromide**	RCRA Number: **None**	EPA Class: **Not Applicable**
DOT Proper Shipping Name: **No Citation**	Chemical Abstract Service (CAS) Number: **7550-35-8**	
DOT Hazard Class and Label Requirements: **No Citation**	DOT Emergency Guide Code: **No Citation**	
DOT Identification Number: **No Citation**	Chemical Symbol: **LiBr**	

Synonyms

Not Applicable

SECTION II - HAZARDOUS INGREDIENTS AND IDENTITY INFORMATION

Hazardous Components (specific identity)	OSHA Exposure Criteria	NIOSH Exposure Criteria	Immediately Dangerous to Life and Health (IDLH)	ACGIH Exposure Criteria
Lithium bromide (derivation: By reaction of hydrobromic acid with lithium carbonate).	**PEL:** Not Established **STEL:** Not Established	**REL:** Not Established **STEL:** Not Established	**Not Established**	**TLV:** Not Established **STEL:** Not Established

SECTION III - PHYSICAL AND CHEMICAL CHARACTERISTICS

Boiling Point: **1017°F (547°C)**	Specific Gravity (H_2O = 1): **3.464**
Vapor Pressure (mm Hg): **1 at 1378°F (748°C)**	Molecular Weight: **86.85**
Vapor Density (Air = 1): **Not Applicable**	Melting Point: **2309°F (1265°C)**

Solubility

Very soluble in water, alcohol, amyl alcohol, and ether. Soluble in methanol, acetone, and glycol. Slightly soluble in pyridine.

Appearance and Odor

White, cubic, deliquescent (absorbs moisture from air) crystals or white to pink granular powder with no odor and a sharp, bitter taste.

SECTION IV - FIRE AND EXPLOSION HAZARD DATA

Flash Point (method used): **Not Determined**	Explosive Limits in Air % by Volume: **LEL: Not Determined** **UEL: Not Determined**
NFPA Classification: **Non-Combustible Solid**	Autoignition Temperature: **Not Determined**

Extinguishing Media

Use agent suitable to surrounding fire. Lithium bromide itself does not burn.

Special Fire Fighting Procedures

Poisonous gases are released in fire. Wear self-contained breathing apparatus. Move containers from fire area if it can be done without risk. Otherwise, use water spray to cool containers. Be aware of disposition of fire control measures; do not release runoff to waterways or sewers.

Unusual Fire and Explosion Hazards

None reported.

SECTION V - REACTIVITY DATA

Stability		Conditions to Avoid: Lithium bromide is normally stable under routine conditions of storage and handling. Containers should be kept air-tight to prevent moisture absorption. Avoid contact with incompatible materials and heat.
Stable: X	Unstable:	Incompatibility *(materials to avoid)*: Incompatible with ammonia, amines, and other metallic salts (copper bromide, mercuric bromide, mercuric iodide, and mercuric cyanide). When heated in solution, it will react with cellulose.
Hazardous Polymerization		Conditions to Avoid: Under normal conditions of temperature and pressure, hazardous polymerization of lithium bromide is not expected to occur.
May Occur:	Will Not Occur: X	Hazardous Decomposition or By-products: When heated to decomposition, lithium bromide can produce toxic and poisonous gases and fumes, including toxic fumes of bromine.

SECTION VI - HEALTH HAZARD DATA

Primary Route(s) of Entry:	Inhalation?	Absorption (skin)?	Ingestion?
	X	X	X

Health Hazards

INHALATION: Causes severe irritation of the nasal cavity and may lead to perforation of the nasal septum with sneezing, coughing, dypsnea, and respiratory effects (bronchitis or tracheitis), nausea, vomiting, diarrhea, abdominal pain, weight loss, tremors, muscular weakness, loss of coordination, and slurred speech. There may also be other CNS effects.

ABSORPTION: Bromide causes depression of the central nervous system. There may be skin eruptions and a possibility of an electrolyte imbalance in the blood.

INGESTION: Causes CNS effects similar to that of inhalation.

Carcinogenicity	NTP Listed?	IARC Cancer Review Group?	OSHA Regulated?	Target Organs?
Unknown Human Unknown Animal	No	No	No	Eyes, skin, kidney, CNS, GI tract, resp. sys., blood.

Medical Conditions Generally Aggravated by Exposure

None cited. However, existing kidney diseases or nervous system disorders may be aggravated.

Emergency and First-aid Procedures

Eye contact: Flush immediately with water for 15 minutes (minimum); seek medical attention. Skin contact: Remove all contaminated clothing. Wash contaminated skin with soap and water. For inhalation: Remove the person from exposure. Provide respiratory assistance and CPR. Transfer to medical facility. If swallowed: Give 1 to 2 glasses of water and induce vomiting. Never attempt to give anything by mouth to an unconscious or convulsing person. Seek medical attention immediately.

SECTION VII - PRECAUTIONS FOR SAFE HANDLING AND USE

Steps to be Taken in Case Material is Released or Spilled

Collect using a HEPA vacuum; do NOT dry sweep (generates dusts). Damp mop residues. Absorb solutions in vermiculite and place in sealed containers. Remove all sources of ignition and ventilate area. Restrict those not involved in cleanup from entering area. Notify appropriate authorities, as required.

Preferred Waste Disposal Method

No citation.

Precautions to be Taken in Handling and Storage

Containers should be protected against physical damage. Store in air-tight containers in a cool, dry, well-ventilated area away from incompatible materials, moisture, and heat.

Other Precautions and Warnings

Personal protective equipment should be used by all persons involved in the storage and handling of lithium bromide.

SECTION VIII - CONTROL MEASURES AND PERSONAL PROTECTIVE EQUIPMENT

Respiratory Protection *(specify type)*

No exposures levels established. For best protection use a NIOSH-approved respirator with particulate (dust/fume/mist) filters. Otherwise, use a self-contained breathing apparatus in continuous flow mode.

Ventilation

Local exhaust or general mechanical systems recommended.

Protective Gloves	Eye Protection	Other Protective Clothing
Suitable to protect from skin contact	Chemical/Dust Goggles	N/A

Work/Hygiene Practices

Always wash hands thoroughly after working with lithium bromide powders and dusts; never bring food, drink, or smoking materials into vicinity of lithium bromide powders or solutions.

LITHIUM BROMIDE

LiBr

CAS: 7550-35-8

IDENTIFICATION AND TYPICAL USES

Lithium bromide appears as white, cubic, deliquescent crystals or as a white to pink granular powder with no odor and a bitter taste. It is used in pharmaceuticals, in air conditioning, as a humectant, a drying agent, in batteries, as a low-temperature heat-exchange medium, and as a sedative.

RISK ASSESSMENT: HEALTH

General Assessment

Lithium bromide is toxic through *inhalation,* skin contact (*absorption*), and *ingestion*. The primary systemic effects by all routes of exposure is depression of the central nervous system. There are no reports regarding its carcinogenic, mutagenic, or teratogenic properties in humans or animals.

Inhalation, ingestion, and dermal absorption can cause nausea, vomiting, severe abdominal pain, diarrhea, general weakness, and fatigue. Inhalation can cause respiratory system injury, including bronchitis (inflammation of the bronchial tubes) and/or tracheitis (inflammation of the trachea). Irritation of the nasal cavity and perforation of the nasal septum have also been documented with sneezing, coughing, and shortness of breath (dypsnea). Symptoms of central nervous system depression can include tremors, weight loss, muscular weakness and loss of coordination, slurred speech, and possible seizures and coma.

Lithium bromide can readily pass through intact skin to the bloodstream and cause many of the effects noted above. If dermal absorption should occur in combination with another exposure route (e.g., inhalation), the effects can be serious, even life-threatening.

Acute Health Effects

The following acute (short-term) health effects may occur immediately or shortly after exposure to lithium bromide:

Skin: Irritation and inflammation with possible rash. Absorption is likely to occur.

Eye: Irritation and burning on contact. Injury or damage to vision may occur if not removed immediately.

Lung: Irritation of the nose, throat, and lungs following exposure to lithium bromide dusts. Perforation of the nasal septum, bronchitis, or tracheitis are all possible, especially on high exposure.

CNS: Broad range of serious and possibly life-threatening effects resulting from central nervous system depression up to and including death.

Chronic Health Effects

The following chronic (long-term) health effects may occur at some time after exposure to lithium bromide and can last for months or even years:

Cancer Hazards: According to information presented in the references, lithium has not been shown to cause cancer in test animals. It is not known if long-term exposure to lithium bromide will lead to carcinogenic effects in humans.

Reproductive Hazard: According to information presented in the references, lithium bromide has not been adequately tested for its ability to adversely affect reproduction in test animals.

Other Chronic Effects: Long-term skin exposure to lithium bromide may cause skin eruptions and chronic CNS disturbances (resulting from the bromide) and may cause a disturbances in blood electrolyte balance.

Recommended Risk-Reduction Measures

Personnel should avoid direct contact with lithium bromide. *Engineering controls* are always the most effective method of reducing exposures. The best protection, especially when working with fine powders and dusts of lithium bromide, is to enclose operations and/or provide explosion-proof, local exhaust ventilation at the site of powder/dust release. While not always operationally feasible, isolating operations can also reduce exposure. Using respiratory protection is less effective than the controls mentioned above, but is still advisable whenever working with or around lithium bromide. No exposure levels have been established for lithium bromide. However, this does not mean that exposure is without risk. For low or transient exposures, an MSHA/NIOSH-approved respirator equipped with a dust/mist pre-filter may provide adequate protection. Better protection is obtained using a self-contained breathing apparatus (SCBA) with full facepiece operated in pressure demand or a supplied-air respirator. If a full facepiece respirator is not available, then dust goggles or safety glasses with side shields attached should be worn to protect the eyes. To prevent hand and skin exposures, suitable gloves should be used.

Administrative controls should also be in place to minimize the potential for human exposures. These may include written procedures or policies which specify the methods and techniques that will be practiced whenever personnel are to work with lithium bromide.

All personnel should receive *training* on the use, hazards, protective measures, emergency actions, and other precautions per 29 CFR 1910.1200 (Hazard Communication), prior to the first assignment in an area where lithium bromide is used or stored.

If symptoms develop or overexposure is suspected, the following tests might be considered:

☑ Complete blood count.

☑ Kidney function tests, including urinalysis.

☑ Nervous system evaluation.

Other methods to reduce exposure include:

☑ Where possible, enclose operations and use local exhaust ventilation at the site where lithium bromide powder or dust is worked. If local exhaust ventilation or enclosure is not used, respiratory protection should be mandatory.

☑ Always ensure that proper protective clothing is worn when working with lithium bromide and that personnel are trained in its use and care.

☑ Wash thoroughly immediately after exposure to lithium bromide and at the end of the work shift or before eating, drinking, or smoking. Clothing contaminated with lithium bromide dusts or powders should not be taken home where family members can be exposed.

☑ Hazard warning information should be posted in the work area. In addition, as part of an on-going education and training program, all information on the health and safety hazards of lithium bromide should be communicated to all exposed and potentially exposed workers.

RISK ASSESSMENT: ENVIRONMENT

General Assessment

The environment is at risk of exposure during transportation, storage, disposal, or destruction of lithium bromide. In almost every scenario, the threat of environmental exposure is contingent upon the proper handling of lithium bromide, especially the dusts and powders. Accidental contact with a variety of incompatible materials can result in fire, explosion, and possible contamination of the surrounding environmental mediums (water, soil, and, to a lesser degree, air).

Lithium bromide is considered a non-combustible solid. However, it is incompatible with a number of commonly encountered materials. It will absorb moisture when exposure to air. It will react with ammonia and amines, and will form double salts in contact with other metallic salts (such as copper bromide, mercuric bromide, mercuric iodide, mercuric cyanide). When heated, lithium bromide will dissolve cellulose. These characteristics require special consideration during any emergency situation involving a leak or spill of lithium bromide.

Lithium bromide can enter the environment through spills and industrial waste discharges.

☠ *Acute Ecological Effects*

Acute (short-term) toxic effects may include the death of animals, birds, or fish and death or low growth rate in plants. Acute effects are seen 2 to 4 days after animals or plants are exposed to lithium bromide.

Insufficient data are available on the short-term effects of lithium bromide exposure to aquatic life, plants, birds, or land animals.

Chronic Ecological Effects

Chronic (long-term) toxic effects may include shortened life span, reproductive problems, lower fertility, and changes in appearance or behavior in exposed animals.

These effects can be seen long after first exposure(s) to toxic chemicals.

Insufficient data are available on the long-term effects of lithium bromide to aquatic life, plants, birds, or land animals.

♦ *Water Solubility*

Lithium bromide is highly soluble in water. Concentrations of 1000 milligrams and more may mix with a liter of water.

⌛ *Persistence in the Environment*

No information is available in the references on the persistence of lithium bromide in the various environmental mediums.

≈ *Bioaccumulation in Aquatic Organisms*

Some substances increase in concentration, or *bioaccumulate,* in living organisms as they breathe contaminated air, drink contaminated water, or eat contaminated food. These chemicals can become concentrated in the tissues and internal organs of animals as well as humans.

Lithium bromide will not accumulate in fish tissues.

Recommended Risk-Reduction Measures

Proper training of all transporters will reduce the likelihood of a mishap or accident resulting in the contamination of the environment. The correct use of DOT labeling (per the MSDS) on all containers, trucks, and rail cars will enable emergency responders to react properly and quickly to any disaster thereby reducing the potential risk to the environment and to personnel.

Storage of lithium bromide should be segregated from incompatible chemicals to minimize the risk of cross-contamination or contact. It should be kept in air-tight containers to avoid moisture absorption. Buildings designated for storage should be properly equipped with appropriate fire protection systems (alarms, signals, sprinklers, emergency lighting, portable extinguishers).

If a spill or leak to the environment has occurred, fire department, emergency response and/or hazardous materials spill personnel should be notified immediately. Cleanup should be attempted only by those trained in proper spill containment procedures. Contaminated soils should be removed and replaced with clean soil. When environmental contamination occurs, the local and/or state emergency response authorities may require notification. A comprehensive emergency response or disaster preparedness/recovery plan should be in place prior to any operations involving the use, transportation, storage, or disposal of lithium bromide. If lithium bromide is spilled or leaked, the following specific steps are recommended:

- ☑ Restrict persons not wearing protective clothing from area of spill or leak until cleanup is complete and area can be opened for normal work.
- ☑ Remove all ignition sources.
- ☑ Ventilate area of spill or leak.
- ☑ Collect powdered materials in the safest manner possible and deposit in sealed containers. Do NOT dry sweep (generates airborne dusts). Use a vacuum equipped with a high-efficiency particulate air (HEPA) filter. Absorb solution spills in vermiculite, sand, dry earth, or other absorbent and place in sealed drums for disposal.
- ☑ It may be necessary to dispose of lithium bromide as a hazardous waste. The responsible state agency or the regional office of the federal Environmental Protection Agency (EPA) should be contacted for specific recommendations.

RISK ASSESSMENT: BUSINESS

General Assessment

Accidents or mishaps involving lithium bromide can present a significant threat to business operations. The loss or damage of equipment or facilities due to explosion or fire can significantly affect fiscal viability. Lawsuits that may result from personnel injury/death, public exposures, and/or environmental contamination will also require a serious expenditure of resources.

Media attention surrounding an injury, death, or environmental damage can also result in a loss of profits and loss of current as well as future business.

Recommended Risk-Reduction Measures

Company attorneys, safety and health professionals, and environmental specialists should be involved in the development of any procedures dealing with safety or emergency response. A company official should be pre-designated as a public relations officer with specific training in dealing with the press. Corporate plans and policies should be developed, approved, and implemented long before any need for such arises.

REFERENCES

American Conference of Governmental Industrial Hygienists. 1988. *Documentation of the Threshold Limit Values and Biological Exposure Indices,* 5th Edition (with updates). Cincinnati: ACGIH

Asante-Duah, D. K. 1993. *Hazardous Waste Risk Assessment.* New York: CRC Press/Lewis Publishers.

Bloom, A. D. and F. J. DeSerres. 1995. *Ecotoxicity and Human Health.* New York: CRC Press/Lewis Publishers.

Brusick, D. J. 1994. *Methods for Genetic Risk Assessment.* New York: CRC Press/Lewis Publishers.

Bueche, J. R. 1972. *Principles of Physics,* 2nd Edition. New York: McGraw-Hill.

Burrell, R., D. K. Flaherty, and Leonard J. Sauers. 1992. *Toxicology of the Immune System.* New York: Van Nostrand Reinhold.

Calabrese, E. J., E. M. Kenyon. *Air Toxics and Risk Assessment.* New York: CRC Press/Lewis Publishers.

Cockerman, L. G., B. S. Shane. 1994. *Basic Environmental Toxicology.* New York: CRC Press/Lewis Publishers.

Cothern, C. R., N. P. Ross. 1994. *Environmental Statistics, Assessment, and Forecasting.* New York: CRC Press/Lewis Publishers.

Danse, I. R. 1991. *Common Sense Toxics in the Workplace.* New York: Van Nostrand Reinhold.

Gardner, R. W. 1994. *Chemical Intolerance.* New York: CRC Press/Lewis Publishers.

Hallenbeck, W. H. 1993. *Quantitative Risk Assessment for Environmental and Occupational Health,* 2nd Edition. New York: CRC Press/Lewis Publishers.

Hosty, J. W., P. Foster. 1990. *A Practical Guide to Chemical Spill Response.* New York: Van Nostrand Reinhold.

Howard, P. H., et. al. 1991. *Handbook of Environmental Degradation Rates.* New York: CRC Press/Lewis Publishers.

James, A., M. Lord. 1992. *Index of Chemical and Physical Data.* New York: Van Nostrand Reinhold.

Jones, F. E. 1994. *Toxic Organic Vapors in the Workplace.* New York: CRC Press/Lewis Publishers.

Keith, L. H. and M. M. Walker. 1995. *Handbook of Air Toxics.* New York: CRC Press/Lewis Publishers.

Landis, W. G., M. H. Yu. 1995. *Introduction to Environmental Toxicology.* New York: CRC Press/Lewis Publishers.

Langerman, N. 1994 *Precautionary Labels for Chemical Containers.* New York: CRC Press/Lewis Publishers.

Lauwerys, R. R., P. Hoet. 1993. *Industrial Chemical Exposure.* New York: CRC Press/Lewis Publishers.

Lewis, R. J., Sr. 1993. *Hawley's Condensed Chemical Dictionary,* 12th Edition. New York: Van Nostrand Reinhold.

Lewis, R. J., Sr. 1992. *Sax's Dangerous Properties of Industrial Materials,* Volumes 1, 2, and 3, 8th Edition. New York: Van Nostrand Reinhold.

Lide, D. R. 1995. *Handbook of Chemistry and Physics,* 76th Edition. New York: CRC Press/Lewis Publishers.

Manahan, S. E. 1994. *Environmental Chemistry,* 6th Edition. New York: CRC Press/Lewis Publishers.

Manahan, S. E. 1992. *Toxicological Chemistry,* 2nd Edition. New York: CRC Press/Lewis Publishers.

Mellström, G. A., J. E. Wahlberg, and H. I. Maibach. 1994. *Protective Gloves for Occupational Use.* New York: CRC Press/Lewis Publishers.

Meyer, E. 1990. *Chemistry of Hazardous Materials.* Englewood Cliffs, NJ: Prentice Hall.

Mycock, J. C., J. D. McKenna, and L. Theodore. 1995. *Handbook of Air Pollution Control Engineering and Technology.* New York: CRC Press/Lewis Publishers.

Ott, W. R. 1995. *Environmental Statistics and Data Analysis.* New York: CRC Press/Lewis Publishers.

Patnaik, P. 1992. *A Comprehensive Guide to the Hazardous Properties of Chemical Substances.* New York: Van Nostrand Reinhold.

Philp, R. B. 1995. *Environmental Hazards and Human Health.* New York: CRC Press/Lewis Publishers.

Proctor, N. H., J. P. Hughes. 1991. *Chemical Hazards of the Workplace.* New York: Van Nostrand Reinhold.

Quigley, D. R. 1994. *Handbook of Emergency Chemical Management.* New York: CRC Press/Lewis Publishers.

Rea, W. J. 1995. *Chemical Sensitivity,* Vol. I-III. New York: CRC Press/Lewis Publishers.

Richardson, M. 1995. *Dictionary of Substance and Their Effects,* Vol. I-VI. New York: CRC Press/Lewis Publishers.

Rylander, R., R. R. Jacobs. 1994. *Organic Dusts.* New York: CRC Press/Lewis Publishers.

Sacarello, H. L. A. 1994. *The Comprehensive Handbook of Hazardous Materials.* New York: CRC Press/Lewis Publishers.

Sheftel, V. O. 1995. *Handbook of Toxic Properties of Monomers and Additives.* New York: CRC Press/Lewis Publishers.

Sherman, J. D. 1988. *Chemical Exposure and Disease.* New York: Van Nostrand Reinhold.

Talbott, E., G. Craun. 1995. *An Introduction to Environmental Epidemiology.* New York: CRC Press/Lewis Publishers.

U.S. Department of Health and Human Services, National Institute for Occupational Safety and Health. 1994. *NIOSH Pocket Guide to Chemical Hazards.* Washington, D.C.: U.S. Government Printing Office.

Witorsch, P. and S. V. Spagnolo. 1994. *Air Pollution and Lung Disease in Adults.* New York: CRC Press/Lewis Publishers.

MATERIAL SAFETY DATA SHEET

CHEMICAL NAME

LITHIUM FLUORIDE

HAZARD WARNING INFORMATION

HEALTH	FIRE	REACTIVE	OTHER	DEGREE OF HAZARD	COLOR CODING	OTHER CODES
1	0	0		0 = Minimum Hazard 1 = Slight Hazard 2 = Moderate Hazard 3 = Serious Hazard 4 = Severe Hazard	HEALTH = BLUE FIRE = RED REACTIVITY = YELLOW OTHER = WHITE	OX = Oxidizer ACID = Acid ALK = Alkali COR = Corrosive W = Use No Water

SECTION I - GENERAL INFORMATION

Characterization	RCRA Number	EPA Class
Fluoride	**None**	**Not Applicable**

DOT Proper Shipping Name **No Citation**	Chemical Abstract Service (CAS) Number **7789-24-4**
DOT Hazard Class and Label Requirements **No Citation**	DOT Emergency Guide Code **No Citation**
DOT Identification Number **No Citation**	Chemical Symbol **LiF**

Synonyms
Not Applicable

SECTION II - HAZARDOUS INGREDIENTS AND IDENTITY INFORMATION

Hazardous Components (specific identity)	OSHA Exposure Criteria	NIOSH Exposure Criteria	Immediately Dangerous to Life and Health (IDLH)	ACGIH Exposure Criteria
Lithium fluoride (derivation: By reaction of hydrogen fluoride with lithium carbonate).	**PEL (8-hour): 2.5 mg/m^3 (as Fluorine)** **STEL: Not Established**	**REL (10-hour): 2.5 mg/m^3 (as Fluorine)** **STEL: Not Established**	**500 mg/m^3 (as Fluorine)**	**REL (10-hour): 2.5 mg/m^3 (as Fluorine)** **STEL: Not Established**

SECTION III - PHYSICAL AND CHEMICAL CHARACTERISTICS

Boiling Point **3040°F (1670°C)**	Specific Gravity (H_2O = 1) **2.64**
Vapor Pressure (mm Hg) **1 at 1917°F (1047°C)**	Molecular Weight **25.94**
Vapor Density (Air = 1) **Not Applicable**	Melting Point **1553°F (845°C)**

Solubility
Slightly soluble in water, soluble in acids, insoluble in alcohol.

Appearance and Odor
Fine white, fluffy powder or cubic crystals.

SECTION IV - FIRE AND EXPLOSION HAZARD DATA

Flash Point (method used) **Not Determined**	Explosive Limits in Air % by Volume **LEL: Not Determined UEL: Not Determined**
NFPA Classification **Non-Combustible Solid**	Autoignition Temperature **Not Determined**

Extinguishing Media
Use agent suitable to surrounding fire (carbon dioxide, dry chemical, foam, water spray). Lithium fluoride itself does not burn. Use water spray to knock down dust clouds or fumes.

Special Fire Fighting Procedures
Poisonous gases are released in fire, including corrosive hydrogen fluoride (hydrofluoric acid). Wear self-contained breathing apparatus. Move containers from fire area if it can be done without risk. Otherwise, use water spray to cool containers. Be aware of disposition of fire control measures; do not release runoff to waterways or sewers.

Unusual Fire and Explosion Hazards
None reported.

SECTION V - REACTIVITY DATA

Stability		Conditions to Avoid **Lithium fluoride is normally stable under routine conditions of storage and handling. Avoid contact with heat and incompatible materials. Do not generate dusty conditions in work or storage areas.**
Stable **X**	Unstable	Incompatibility *(materials to avoid)* **Incompatible with strong acids (nitric, sulfuric).**
Hazardous Polymerization		Conditions to Avoid **Under normal conditions of temperature and pressure, hazardous polymerization of lithium fluoride is not expected to occur.**
May Occur	Will Not Occur **X**	Hazardous Decomposition or By-products **When heated to decomposition, lithium fluoride can produce toxic and poisonous gases and fumes, including toxic and corrosive hydrogen fluoride.**

SECTION VI - HEALTH HAZARD DATA

Primary Route(s) of Entry:	Inhalation?	Absorption (skin)?	Ingestion?
	X		

Health Hazards

INHALATION: Causes severe irritation of the nasal cavity, throat, and respiratory tract. It may cause the mucous membranes to dry and lead to nosebleeds. Excessive exposures can have a cumulative effect and may damage the kidneys. There can also be abdominal pain with diarrhea, nausea, salivation, thirst, sweating, stiff spine and joints, and calcification of the rib and pelvic ligaments.

SKIN & EYES: Causes irritation and burns to the eyes and skin similar to that of hydrofluoric acid. The burn can be serious and deep due to the fluoride anion. Dermatitis can occur.

INGESTION: Not likely, but possible. Can cause serious gastrointestinal inflammation.

Carcinogenicity	NTP Listed?	IARC Cancer Review Group?	OSHA Regulated?	Target Organs?
Unknown Human **Unknown Animal**	**No**	**No**	**No**	**Eyes, skin, kidney, CNS, resp. sys., skeletal sys.**

Medical Conditions Generally Aggravated by Exposure

Existing kidney dysfunction may be aggravated by exposure to lithium fluoride.

Emergency and First-aid Procedures

Eye contact: Flush immediately with water for 15 minutes (minimum); seek medical attention. Skin contact: Remove all contaminated clothing. Immediately wash area with large amounts of water for 2 to 4 hours or soak in iced Zephiran (0.13%), Epson salts or a 70% denatured ethyl alcohol solution for 1 to 4 hours. Seek medical assistance. For inhalation: Remove the person from exposure. Provide respiratory assistance and CPR. Transfer to medical facility. If swallowed: Not likely. If accidental ingestion of solutions occurs, seek medical attention or contact poison control center. Do NOT induce vomiting.

SECTION VII - PRECAUTIONS FOR SAFE HANDLING AND USE

Steps to be Taken in Case Material is Released or Spilled

Collect using a HEPA vacuum; do NOT dry sweep (generates dusts). Damp mop residues and place in sealed containers. Ventilate area. Restrict those not involved in cleanup from entering area.

Preferred Waste Disposal Method

No citation.

Precautions to be Taken in Handling and Storage

Containers should be protected against physical damage. Store in tightly containers in a cool, dry, well-ventilated area away from incompatible materials.

Other Precautions and Warnings

Personal protective equipment should be used by all persons involved in the storage and handling of lithium fluoride.

SECTION VIII - CONTROL MEASURES AND PERSONAL PROTECTIVE EQUIPMENT

Respiratory Protection *(specify type)*

For best protection use a NIOSH-approved respirator with particulate (dust/fume/mist) filters. Otherwise, use a supplied-air respirator or a self-contained breathing apparatus in continuous flow mode.

Ventilation

Local exhaust or general mechanical systems recommended.

Protective Gloves	Eye Protection	Other Protective Clothing
Suitable to protect from skin contact	**Chemical/Dust Goggles**	**Protective Apron or Suit**

Work/Hygiene Practices

Always wash hands thoroughly after working with lithium fluoride powders and dusts; never bring food, drink, or smoking materials into vicinity of lithium fluoride powders.

LITHIUM FLUORIDE

LiF

CAS: 7789-24-4

IDENTIFICATION AND TYPICAL USES

Lithium fluoride appears as a white, fluffy powder of cubic crystals. It is used in welding and soldering flux, in ceramics, as a heat-exchange media, as synthetic crystals in infrared and ultraviolet instruments, as a rocket fuel component, and in radiation dosimetry. It is also a component of fuels used for molten salt reactors and in X-ray diffraction.

RISK ASSESSMENT: HEALTH

General Assessment

Lithium fluoride is primarily toxic through *inhalation,* and skin contact (although dermal absorption leading to toxic systemic effects has not been reported). Ingestion is unlikely but possible. The primary systemic effects appear to result from the fluoride content. There are no reports regarding its carcinogenic, mutagenic, or teratogenic properties in humans or animals.

Inhalation of low concentrations does not appear to cause any acute toxic effects. This is thought to be due to its low water solubility. However, it can be a cumulative poison. This is especially true for individuals with low sodium intake. Excessive lithium exposure can damage the kidneys and lead to central nervous system depression. Exposure to fluorides cause irritation to the eyes, nose, throat, and mucosa of the respiratory tract. There can be nasal cavity dryness and nosebleeds. Systemic effects include nausea, severe abdominal pain with vomiting and diarrhea, excessive salivation and extreme thirst, sweating, stiffening of the spine and joints with eventual calcification of the rib and pelvic ligaments.

Skin contact results in immediate irritation and burning rash. Contact with diluted solutions as low as 20% lithium fluoride may not burn immediately. However, if not completely removed, the fluoride anion's affinity for calcium can cause deep and painful burns. Dermatitis may also develop. Eye contact can cause irritation, burns, and possible damage to vision.

Acute Health Effects

The following acute (short-term) health effects may occur immediately or shortly after exposure to lithium fluoride:

Skin: Irritation and inflammation with possible deep burns and rash.

Eye: Irritation and burning on contact. Injury or damage to vision may occur if not removed immediately.

Lung: Irritation of the nose, throat, and lungs following exposure to lithium fluoride dusts. Dryness and nosebleeds have occurred.

CNS: Broad range of possible effects resulting from the lithium content.

Chronic Health Effects

The following chronic (long-term) health effects may occur at some time after exposure to lithium fluoride and can last for months or even years:

Cancer Hazards: According to information presented in the references, lithium has not been shown to cause cancer in test animals. It is not known if long-term exposure to lithium fluoride will lead to carcinogenic effects in humans.

Reproductive Hazard: According to information presented in the references, lithium fluoride has not been adequately tested for its ability to adversely affect reproduction in test animals.

Other Chronic Effects: Prolonged exposures to fluorides can lead to fluorosis which causes changes in the bones and ligaments, and possible kidney damage.

Recommended Risk-Reduction Measures

Personnel should avoid direct contact with lithium fluoride. *Engineering controls* are always the most effective method of reducing exposures. The best protection, especially when working with fine powders and dusts of lithium fluoride, is to enclose operations and/or provide explosion-proof, local exhaust ventilation at the site of powder/dust release. While not always operationally feasible, isolating operations can also reduce exposure. Using respiratory protection is less effective than the controls mentioned above, but is still advisable whenever working with or around lithium fluoride. For exposures over the PEL (2.5 mg/m^3 for fluorides), a chemical cartridge respirator equipped with a dust/mist pre-filter may provide adequate protection. For better protection, use of a self-contained breathing apparatus (SCBA) with full facepiece operated in pressure demand or a supplied-air respirator is recommended. If a full facepiece respirator is not available, then dust goggles or safety glasses with side shields attached should be worn to protect the eyes. To prevent hand and skin exposures, suitable gloves should be used. Glove manufacturers should be contacted an permeation studies obtained *before* final glove selections are made.

Administrative controls should also be in place to minimize the potential for human exposures. These may include written procedures or policies which specify the methods and techniques that will be practiced whenever personnel are to work with lithium fluoride.

All personnel should receive *training* on the use, hazards, protective measures, emergency actions, and other precautions per 29 CFR 1910.1200 (Hazard Communication), prior to the first assignment in an area where lithium fluoride is used or stored.

Before beginning employment and at regular intervals thereafter (e.g., semi-annual), the following medical testing is recommended:

☑ Urinalysis.

☑ Musculoskeletal examination.

☑ Lung function tests.

☑ Kidney function tests.

If symptoms develop or overexposure is suspected, the following tests might be considered:

☑ Nervous system evaluation.

Other methods to reduce exposure include:

☑ Where possible, enclose operations and use local exhaust ventilation at the site where lithium fluoride powder or dust is worked. If local exhaust ventilation or enclosure is not used, respiratory protection should be mandatory.

☑ Always ensure that proper protective clothing is worn when working with lithium fluoride and that personnel are trained in its use and care.

☑ Wash thoroughly immediately after exposure to lithium fluoride and at the end of the work shift or before eating, drinking, or smoking. Clothing contaminated with lithium fluoride dusts or powders should not be taken home where family members can be exposed.

☑ Hazard warning information should be posted in the work area. In addition, as part of an on-going education and training program, all information on the health and safety hazards of lithium fluoride should be communicated to all exposed and potentially exposed workers.

RISK ASSESSMENT: ENVIRONMENT

General Assessment

The environment is at risk of exposure during transportation, storage, disposal, or destruction of lithium fluoride. In almost every scenario, the threat of environmental exposure is contingent upon the proper handling of lithium fluoride, especially the dusts and powders. Accidental contact with a variety of incompatible materials can result in fire, explosion, and possible contamination of the surrounding environmental mediums (water, soil, and, to a lesser degree, air).

Lithium fluoride is considered a non-combustible solid. However, it is incompatible with a number of commonly encountered acids. Contact with acids, especially when heated, can produce highly toxic and corrosive hydrogen fluoride gas. Improper handling during use, transportation, storage, or disposal can result in injury to personnel and/or damage to the environment. These characteristics require special consideration during any emergency situation involving a leak or spill of lithium fluoride. Lithium fluoride can enter the environment through spills and industrial discharges.

☠ *Acute Ecological Effects*

Acute (short-term) toxic effects may include the death of animals, birds, or fish and death or low growth rate in plants. Acute effects are seen 2 to 4 days after animals or plants are exposed to lithium fluoride.

Insufficient data are available on the short-term effects of lithium fluoride exposure to aquatic life, plants, birds, or land animals.

Chronic Ecological Effects

Chronic (long-term) toxic effects may include shortened life span, reproductive problems, lower fertility, and changes in appearance or behavior in exposed animals. These effects can be seen long after first exposure(s) to toxic chemicals.

Insufficient data are available on the long-term effects of lithium fluoride to aquatic life, plants, birds, or land animals.

Water Solubility

Lithium fluoride is slightly soluble in water. Concentrations of 1 to 100 milligrams will mix with a liter of water.

Persistence in the Environment

No information is available in the references on the persistence of lithium fluoride in the various environmental mediums. However, hydrogen fluoride is slightly persistent in the aquatic environment, with a half-life of less than 2 and 20 days. The half-life of a pollutant is the amount of time it takes for one half of the chemical to be degraded.

Bioaccumulation in Aquatic Organisms

Some substances increase in concentration, or *bioaccumulate,* in living organisms as they breathe contaminated air, drink contaminated water, or eat contaminated food. These chemicals can become concentrated in the tissues and internal organs of animals as well as humans.

Lithium fluoride will not accumulate in fish tissues.

Recommended Risk-Reduction Measures

Proper training of all transporters will reduce the likelihood of a mishap or accident resulting in the contamination of the environment. The correct use of labeling on all containers, trucks, and rail cars will enable emergency responders to react properly and quickly to any disaster thereby reducing the potential risk to the environment and to personnel.

Storage of lithium fluoride should be segregated from incompatible chemicals to minimize the risk of cross-contamination or contact. It should be kept in tightly-closed in a cool, dry location. Buildings designated for storage should be properly equipped with appropriate fire protection systems (alarms, signals, sprinklers, emergency lighting, portable extinguishers).

If a spill or leak to the environment has occurred, fire department, emergency response and/or hazardous materials spill personnel should be notified immediately. Cleanup should be attempted only by those trained in proper spill containment procedures. Contaminated soils should be removed and replaced with clean soil. When environmental contamination occurs, the local and/or state emergency response authorities may require notification. A comprehensive emergency response or disaster preparedness/recovery plan should be in place prior to any operations involving the use, transportation, storage, or disposal of lithium fluoride. If lithium fluoride is spilled or leaked, the following specific steps are recommended:

- ☑ Restrict persons not wearing protective clothing from area of spill or leak until cleanup is complete and area can be opened for normal work.
- ☑ Ventilate area of spill or leak.
- ☑ Collect powdered materials in the safest manner possible and deposit in sealed containers. Do NOT dry sweep (generates airborne dusts). Use a vacuum equipped with a high-efficiency particulate air (HEPA) filter. Damp mop residues and collect for disposal. A water spray can be effective in controlling dust clouds and protecting response workers. Be aware of runoff. Collect for disposal.
- ☑ Absorb solution spills in vermiculite, sand, dry earth, or other absorbent and place is sealed drums for disposal. Avoid skin contact during cleanup operations.
- ☑ It may be necessary to dispose of lithium fluoride as a hazardous waste. The responsible state agency or the regional office of the federal Environmental Protection Agency (EPA) should be contacted for specific recommendations.

RISK ASSESSMENT: BUSINESS

General Assessment

Accidents or mishaps involving lithium fluoride can present a significant threat to business operations. The loss or damage of equipment or facilities due to explosion or fire can significantly affect fiscal viability. Lawsuits that may result from personnel injury/death, public exposures, and/or environmental contamination will also require a serious expenditure of resources. Media attention surrounding an injury, death, or environmental damage can also result in a loss of profits and loss of current as well as future business.

Recommended Risk-Reduction Measures

Company attorneys, safety and health professionals, and environmental specialists should be involved in the development of any procedures dealing with safety or emergency response. A company official should be pre-designated as a public relations officer with specific training in dealing with the press. Corporate plans and policies should be developed, approved, and implemented long before any need for such arises.

REFERENCES

American Conference of Governmental Industrial Hygienists. 1988. *Documentation of the Threshold Limit Values and Biological Exposure Indices*, 5th Edition (with updates). Cincinnati: ACGIH

Asante-Duah, D. K. 1993. *Hazardous Waste Risk Assessment.* New York: CRC Press/Lewis Publishers.

Brusick, D. J. 1994. *Methods for Genetic Risk Assessment.* New York: CRC Press/Lewis Publishers.

Bueche, J. R. 1972. *Principles of Physics*, 2nd Edition. New York: McGraw-Hill.

Burrell, R., D. K. Flaherty, and Leonard J. Sauers. 1992. *Toxicology of the Immune System.* New York: Van Nostrand Reinhold.

Calabrese, E. J., E. M. Kenyon. *Air Toxics and Risk Assessment.* New York: CRC Press/Lewis Publishers.

Cockerman, L. G., B. S. Shane. 1994. *Basic Environmental Toxicology.* New York: CRC Press/Lewis Publishers.

Cothern, C. R., N. P. Ross. 1994. *Environmental Statistics, Assessment, and Forecasting.* New York: CRC Press/Lewis Publishers.

Gardner, R. W. 1994. *Chemical Intolerance.* New York: CRC Press/Lewis Publishers.

Hallenbeck, W. H. 1993. *Quantitative Risk Assessment for Environmental and Occupational Health*, 2nd Edition. New York: CRC Press/Lewis Publishers.

Hosty, J. W., P. Foster. 1990. *A Practical Guide to Chemical Spill Response.* New York: Van Nostrand Reinhold.

Howard, P. H., et. al. 1991. *Handbook of Environmental Degradation Rates.* New York: CRC Press/Lewis Publishers.

James, A., M. Lord. 1992. *Index of Chemical and Physical Data.* New York: Van Nostrand Reinhold.

Jones, F. E. 1994. *Toxic Organic Vapors in the Workplace.* New York: CRC Press/Lewis Publishers.

Keith, L. H. and M. M. Walker. 1995. *Handbook of Air Toxics.* New York: CRC Press/Lewis Publishers.

Landis, W. G., M. H. Yu. 1995. *Introduction to Environmental Toxicology.* New York: CRC Press/Lewis Publishers.

Langerman, N. 1994 *Precautionary Labels for Chemical Containers.* New York: CRC Press/Lewis Publishers.

Lauwerys, R. R., P. Hoet. 1993. *Industrial Chemical Exposure.* New York: CRC Press/Lewis Publishers.

Lewis, R. J., Sr. 1993. *Hawley's Condensed Chemical Dictionary*, 12th Edition. New York: Van Nostrand Reinhold.

Lewis, R. J., Sr. 1992. *Sax's Dangerous Properties of Industrial Materials*, Volumes 1, 2, and 3, 8th Edition. New York: Van Nostrand Reinhold.

Lide, D. R. 1995. *Handbook of Chemistry and Physics*, 76th Edition. New York: CRC Press/Lewis Publishers.

Manahan, S. E. 1994. *Environmental Chemistry*, 6th Edition. New York: CRC Press/Lewis Publishers.

Mellström, G. A., J. E. Wahlberg, and H. I. Maibach. 1994. *Protective Gloves for Occupational Use.* New York: CRC Press/Lewis Publishers.

Meyer, E. 1990. *Chemistry of Hazardous Materials.* Englewood Cliffs, NJ: Prentice Hall.

Mycock, J. C., J. D. McKenna, and L. Theodore. 1995. *Handbook of Air Pollution Control Engineering and Technology.* New York: CRC Press/Lewis Publishers.

Ott, W. R. 1995. *Environmental Statistics and Data Analysis.* New York: CRC Press/Lewis Publishers.

Patnaik, P. 1992. *A Comprehensive Guide to the Hazardous Properties of Chemical Substances.* New York: Van Nostrand Reinhold.

Philp, R. B. 1995. *Environmental Hazards and Human Health.* New York: CRC Press/Lewis Publishers.

Proctor, N. H., J. P. Hughes. 1991. *Chemical Hazards of the Workplace.* New York: Van Nostrand Reinhold.

Quigley, D. R. 1994. *Handbook of Emergency Chemical Management.* New York: CRC Press/Lewis Publishers.

Rea, W. J. 1995. *Chemical Sensitivity*, Vol. I-III. New York: CRC Press/Lewis Publishers.

Richardson, M. 1995. *Dictionary of Substance and Their Effects*, Vol. I-VI. New York: CRC Press/Lewis Publishers.

Rylander, R., R. R. Jacobs. 1994. *Organic Dusts.* New York: CRC Press/Lewis Publishers.

Sacarello, H. L. A. 1994. *The Comprehensive Handbook of Hazardous Materials.* New York: CRC Press/Lewis Publishers.

Sheftel, V. O. 1995. *Handbook of Toxic Properties of Monomers and Additives.* New York: CRC Press/Lewis Publishers.

Sherman, J. D. 1988. *Chemical Exposure and Disease.* New York: Van Nostrand Reinhold.

Talbott, E., G. Craun. 1995. *An Introduction to Environmental Epidemiology.* New York: CRC Press/Lewis Publishers.

U.S. Department of Health and Human Services, National Institute for Occupational Safety and Health. 1994. *NIOSH Pocket Guide to Chemical Hazards.* Washington, D.C.: U.S. Government Printing Office.

Witorsch, P. and S. V. Spagnolo. 1994. *Air Pollution and Lung Disease in Adults.* New York: CRC Press/Lewis Publishers.

MATERIAL SAFETY DATA SHEET

CHEMICAL NAME	
LITHIUM HYDRIDE	**CRC PRESS/LEWIS PUBLISHERS DISCLAIMER:** THE INFORMATION AND RECOMMENDATIONS PRESENTED HEREIN ARE BASED ON SOURCES BELIEVED TO BE RELIABLE. CRC MAKES NO REPRESENTATION ON ITS COMPLETENESS OR ACCURACY. IT IS THE RESPONSIBILITY OF THE USER TO DETERMINE THE CHEMICAL'S SUITABILITY FOR ITS INTENDED USE, THE CHEMICAL'S SAFE USE, AND THE CHEMICAL'S PROPER DISPOSAL. NO REPRESENTATIONS AND/OR WARRANTIES, EITHER EXPRESSED OR IMPLIED, OF THE MERCHANTABILITY OR FITNESS FOR A PARTICULAR PURPOSE, OR OF ANY OTHER NATURE, ARE MADE WITH RESPECT TO THE INFORMATION PROVIDED IN THIS MSDS OR TO THE CHEMICAL TO WHICH INFORMATION MAY REFER. CRC NEITHER ASSUMES NOR AUTHORIZES ANY OTHER PERSON TO ASSUME FOR IT, ANY OTHER ADDITIONAL RESPONSIBILITY OR LIABILITY FOR THE USE OF, OR RELIANCE UPON, THIS INFORMATION.

HAZARD WARNING INFORMATION

HEALTH	FIRE	REACTIVE	OTHER	DEGREE OF HAZARD	COLOR CODING	OTHER CODES
3	2	2	~~W~~	0 = Minimum Hazard 1 = Slight Hazard 2 = Moderate Hazard 3 = Serious Hazard 4 = Severe Hazard	HEALTH = BLUE FIRE = RED REACTIVITY = YELLOW OTHER = WHITE	OX = Oxidizer ACID = Acid ALK = Alkali COR = Corrosive ~~W~~ = Use No Water

SECTION I - GENERAL INFORMATION

Characterization	RCRA Number	EPA Class
Hydride	**D003**	**Characteristic (R) Waste**

DOT Proper Shipping Name **Lithium hydride**	Chemical Abstract Service (CAS) Number **7580-67-8**
DOT Hazard Class and Label Requirements **Dangerous When Wet**	DOT Emergency Guide Code **40**
DOT Identification Number **UN 1414**	Chemical Formula **LiH**

Synonyms
Lithium monohydride.

SECTION II - HAZARDOUS INGREDIENTS AND IDENTITY INFORMATION

Hazardous Components (specific identity)	OSHA Exposure Criteria	NIOSH Exposure Criteria	Immediately Dangerous to Life and Health (IDLH)	ACGIH Exposure Criteria
Lithium hydride (derivation: By reaction of molten lithium with hydrogen).	**PEL (8-hour):** **0.025 mg/m³** **STEL:** **Not Established**	**REL (10-hour):** **0.025 mg/m³** **STEL:** **Not Established**	**0.5 mg/m³**	**TLV:** **0.025 mg/m³** **STEL:** **Not Established**

SECTION III - PHYSICAL AND CHEMICAL CHARACTERISTICS

Boiling Point **932°F (500°C) Decomposes**	Specific Gravity (H_2O = 1) **0.78**
Vapor Pressure (mm Hg) **0 at 69°F (20°C)**	Molecular Weight **7.95**
Vapor Density (Air = 1) **Not Found**	Melting Point **1256°F (680°C)**

Solubility
Decomposes in water rapidly to form lithium hydroxide and hydrogen. Soluble in ether and slightly soluble in acid. Insoluble in benzene and toluene.

Appearance and Odor
Odorless, off-white or light blue-gray, translucent, crystalline solid or white powder.

SECTION IV - FIRE AND EXPLOSION HAZARD DATA

Flash Point (method used) **Not Determined**	Explosive Limits in Air % by Volume **LEL: Not Determined** **UEL: Not Determined**
NFPA Classification **Combustible Solid**	Autoignition Temperature **392°F (200°C)**

Extinguishing Media
Use special mixture of dry chemicals (Lith-X® powder, G-1 graphite powder) or smother with sand.

Special Fire Fighting Procedures
Poisonous gases are produced in fire; wear full protective clothing and self-contained breathing apparatus (SCBA). Isolate area and deny entry to those not involved in response procedures. May give off toxic fumes of unburned material. Move containers from fire area if it can be done without risk. Do NOT use water, carbon dioxide, or halogenated agents to fight fire.

Unusual Fire and Explosion Hazards
Reacts rapidly with water to form flammable hydrogen gas. Lithium hydride ignites spontaneously in air and may re-ignite after fire has been extinguished.

SECTION V - REACTIVITY DATA

Stability		Conditions to Avoid
Stability		Thermally unstable, darkens on exposure to light, and can ignite spontaneously on exposure to air or heat. Avoid contact with incompatible materials.
Stable	Unstable	Incompatibility *(materials to avoid)*
	X	Strong oxidizers (chlorine), water, halogenated compounds, nitrous oxide, lower alcohols, carboxylic acids, ammonia, dinitrogen oxide, oxygen, moisture, and wet combustible materials.
Hazardous Polymerization		Conditions to Avoid: Under normal conditions of temperature and pressure, hazardous polymerization of lithium hydride is not expected to occur.
May Occur	Will Not Occur	Hazardous Decomposition or By-products
	X	When heated to decomposition, lithium hydride emits toxic and irritating alkali fumes and explosive hydrogen gas.

SECTION VI - HEALTH HAZARD DATA

Primary Route(s) of Entry:	Inhalation?	Absorption (skin)?	Ingestion?
	X		X

Health Hazards

INHALATION: Causes irritation of the eyes, nose, throat, and respiratory system with coughing, sneezing, dypsnea, mucous membrane irritation in the nasal cavity leading to ulceration of the septum, chemical bronchitis, or delayed pulmonary edema. Nervous system effects also possible, including mental confusion, nausea, muscular twitching, and blurred vision.

SKIN & EYES: Skin contact causes serious caustic burns, itchy rash, and surface inflammation. Eye contact may cause burns, irritation, corneal scarring, and perhaps permanent damage.

INGESTION: Burning of the mouth, esophagus, and digestive tract. Pain in the chest, stomach, and painful swallowing.

Carcinogenicity	NTP Listed?	IARC Cancer Review Group?	OSHA Regulated?	Target Organs?
Unknown Human Unknown Animal	No	No	29 CFR 1910.1000 Table Z-1 (skin)	CNS, skin, eyes, respiratory system.

Medical Conditions Generally Aggravated by Exposure

Existing chronic respiratory diseases may be aggravated by exposure.

Emergency and First-aid Procedures

Eye contact: Flush immediately with water for 15 minutes (minimum); seek medical attention. Skin contact: Remove all contaminated clothing. Wash area with large amounts of soap and water. Seek medical assistance. For inhalation: Remove the person from exposure. Provide respiratory assistance and CPR. Transfer to medical facility. Observe for 48 hours for lung effects. If swallowed: Contact poison control center. Unless told otherwise, give 1-2 glasses of water to dilute. Do NOT induce vomiting. Never attempt to give an unconscious or convulsing person anything by mouth.

SECTION VII - PRECAUTIONS FOR SAFE HANDLING AND USE

Steps to be Taken in Case Material is Released or Spilled

Collect powdered material using a vacuum equipped with HEPA filter and deposit in sealed drums for disposal. Avoid skin contact. Do NOT dry sweep (generates dusts). Keep water away. Ventilate area. Restrict those not involved in cleanup from entering area.

Preferred Waste Disposal Method

No citation.

Precautions to be Taken in Handling and Storage

Do not store lithium hydride in the presence of incompatible chemicals or materials. Store in tightly closed containers in a cool, dark, well-ventilated area. Personnel working with lithium hydride should be trained on its proper handling and storage *prior* to being assigned to such responsibilities.

Other Precautions and Warnings

Keep away from water and halogenated compounds. Bulk storage is of lithium hydride is not recommended. Keep stored under inert atmosphere, if possible.

SECTION VIII - CONTROL MEASURES AND PERSONAL PROTECTIVE EQUIPMENT

Respiratory Protection *(specify type)*

PEL is extremely low (0.025 ppm). Use a self-contained breathing apparatus (SCBA) with full facepiece or a supplied-air respirator operated in pressure demand or other positive pressure mode.

Ventilation

Local exhaust or general mechanical systems recommended.

Protective Gloves	Eye Protection	Other Protective Clothing
Impervious Material	Chemical/Dust Goggles or Face Mask	Protective Apron

Work/Hygiene Practices

Always wash hands thoroughly after using chemical; never bring food, drink, or smoking materials into vicinity of chemicals.

LITHIUM HYDRIDE

LiH

CAS: 7580-67-8

IDENTIFICATION AND TYPICAL USES

Lithium hydride appears as an odorless, off-white or light blue-gray, translucent, crystalline solid or white powder. It is used as a desiccant, as a source of hydrogen, as a condensing agent in organic synthesis, in the preparation of lithium amide and double hydrides, in nuclear shielding material, and as a reducing agent.

RISK ASSESSMENT: HEALTH

General Assessment

Lithium hydride is a human poison primarily by *inhalation,* causing toxic systemic effects targeting the respiratory system. This is due to the formation of highly caustic lithium hydroxide when lithium hydride contacts moisture. Therefore, the eyes, mucous membranes, and lungs are particularly at risk. Its carcinogenicity, teratogenicity, and mutagenicity are currently unknown in humans.

Inhalation of lithium hydride dusts can be irritating to the nose, throat, mucosa, and the upper respiratory tract. Symptoms can include coughing, sneezing, dypsnea, ulceration of the nasal septa, chemical bronchitis, and the possibility of pulmonary edema (a dangerous buildup of fluid in the lungs). Pulmonary edema can be delayed up to 48 hours or more, thereby creating a false sense of security with regard to health exposure risk. There can also be effects subsequent to lithium's ability to depress the central nervous system. These effects can include headache, nausea, dizziness, drowsiness, lightheadedness, muscular twitching, tremors, mental confusion, blurred vision, coma, and, in severe cases, death.

Ingestion of lithium hydride can cause burning of the mouth, esophagus, and gastrointestinal tract with deep, burning pain in the chest and stomach and pain on swallowing.

Skin contact causes painful caustic burns, especially if the contact area is moist (sweat). Even at relatively low concentrations (0.2 to 0.5 mg/m^3) the skin may become itchy and inflamed. Eye contact can result in serious burns, corneal scarring, and possible damage to loss or even loss of vision.

Acute Health Effects

The following acute (short-term) health effects may occur immediately or shortly after exposure to lithium hydride:

Skin: Irritation and possible rash. Moisture on skin's surface forms lithium hydroxide which will cause serious caustic burns.

Eye: Severe burns, corneal scarring, and possible damage to vision.

Lung: Severe irritation of the nose, throat, and esophagus. Possible pulmonary edema.

CNS: Dizziness, headaches, nausea, vomiting, muscle spasms, tremor, fatigue, convulsions, dypsnea, nervousness, drowsiness, loss of coordination, coma, and even death.

Chronic Health Effects

The following chronic (long-term) health effects may occur at some time after exposure to lithium hydride and can last for months or even years:

Cancer Hazards: According to the information presented in the references, lithium hydride has not been adequately tested for its ability to cause cancer in test animals.

Reproductive Hazard: According to the information presented in the references, lithium hydride has not been adequately tested for its ability to adversely affect reproduction. However, lithium has been shown to cross the placenta barrier.

Other Chronic Effects: Prolonged contact with low-level concentration of lithium hydride can lead to the development of emphysema.

Recommended Risk-Reduction Measures

Personnel should avoid direct contact with lithium hydride. It is a toxic, human poison on inhalation. If a less toxic chemical cannot be substituted for a hazardous substance, *engineering controls* are the most effective method of reducing exposures. The best protection is to enclose operations and/or provide local exhaust ventilation at the site of chemical release. While not always operationally feasible, isolating operations can also reduce exposure.

Using respiratory protection is less effective than the controls mentioned above, but is still advisable whenever working with or around lithium hydride. This chemical has an extremely low permissible exposure limit (0.025 ppm) that is difficult to accurately monitor. Therefore, prudent risk management requires a conservative approach to personal protection. While a chemical respirator with a dust/mist pre-filter may suffice, the best protection is obtained using a supplied-air respirator operated in continuous flow mode or a self-contained breathing apparatus (SCBA) with full facepiece operated in pressure demand or positive pressure mode. If a full facepiece is not available, then chemical/dust goggles should be worn to protect the eyes. Whenever lithium hydride is being mixed with liquids, a chemical splash hazard exists and a face shield and rubber apron should be worn. To prevent hand and skin exposures, protective gloves should be used. Glove manufacturers should be contacted and permeation studies obtained *before* final glove selections have been made.

Administrative controls should also be in place to minimize the potential for human exposures. These may include written procedures or policies which specify the methods and techniques that will be practiced whenever personnel are to work with lithium hydride.

All personnel should receive *training* on the use, hazards, protective measures, emergency actions, and other precautions per 29 CFR 1910.1200 (Hazard Communication), prior to the first assignment in an area where lithium hydride is used or stored.

If symptoms develop or overexposure is suspected, the following medical tests may be useful:

- ☑ Lung function tests.
- ☑ Kidney function tests.
- ☑ Consider chest X-ray after an acute overexposure to lithium hydride (may be negative if taken immediately following exposure due to delayed development of pulmonary edema).

Any evaluation should include a careful medical history of past and present symptoms with an examination. However, medical tests that evaluate existing damage are not a substitute for controlling exposure. Also, since smoking can cause heart disease, emphysema, lung cancer, and other respiratory diseases, smokers may experience symptoms more rapidly than nonsmokers under the same conditions of exposure. Prudent risk management requires proper consideration of *all* possible factors that may be causing the appearance of exposure symptoms in the workplace.

Other methods to reduce exposure include:

- ☑ Where possible, enclose operations and use local exhaust ventilation at the site of chemical release. If local exhaust ventilation or enclosure is not used, respiratory protection should be mandatory.
- ☑ Always ensure that proper protective clothing is worn when using chemical substances.
- ☑ Wash thoroughly immediately after exposure to lithium hydride and at the end of the work shift or before eating, drinking, or smoking.
- ☑ Hazard warning information should be posted in the work area. In addition, as part of an on-going education and training program, all information on the health and safety hazards of lithium hydride should be communicated to all exposed or potentially exposed workers.
- ☑ Eye wash stations should be provided in work areas. If the potential for whole body exposure exists, then safety showers should also be provided.

RISK ASSESSMENT: ENVIRONMENT

General Assessment

The environment is at risk of exposure during transportation, storage, disposal, or destruction of lithium hydride. In almost every scenario, the threat of environmental exposure is contingent upon the proper handling of the chemical substance. Accidental spills, large or small, can result in fire, explosion, and possible contamination of the surrounding environmental mediums (water, soil, and air).

Lithium hydride is considered a combustible solid. It can ignite spontaneously in air or on exposure to heat. In reaction to halogenated compounds, it yields products that may explode on impact (hydrogen gas). If it should contact water or moisture, it will produce extremely caustic lithium hydroxide as well as flammable and explosive hydrogen gas. These characteristics require special consideration during any emergency situation involving a leak or spill of lithium hydride or lithium hydride mixtures. Should lithium hydride ever come into contact with other incompatible substances, such as strong oxidizers or acids, either during use, transportation, or storage, violent reactions can occur.

Lithium hydride can enter the environment through industrial discharges and from spills.

Acute Ecological Effects

Acute (short-term) toxic effects may include the death of animals, birds, or fish and death or low growth rate in plants. Acute effects are seen 2 to 4 days after animals or plants are exposed to lithium hydride.

Insufficient data are available to evaluate or predict the short-term effects of lithium hydride to aquatic life, plants, birds, or land animals.

Chronic Ecological Effects

Chronic toxic effects may include shortened life span, reproductive problems, lower fertility, and changes in appearance or behavior in exposed animals. These effects can be seen long after first exposure(s) to toxic chemicals.

Insufficient data are available to evaluate or predict the long-term effects of lithium hydride to aquatic life, plants, birds, or land animals.

Water Solubility

Lithium hydride is not soluble in water. It reacts and decomposes to form lithium hydroxide and free hydrogen gas, which is explosive.

Persistence in the Environment

Lithium hydride is non-persistent in water, with a half-life of less than 2 days. The half-life of a pollutant is the amount of time it takes for one half of the chemical to be degraded.

Bioaccumulation in Aquatic Organisms

Some substances increase in concentration, or *bioaccumulate,* in living organisms as they breathe contaminated air, drink contaminated water, or eat contaminated food. These chemicals can become concentrated in the tissues and internal organs of animals as well as humans.

Lithium is not expected to accumulate in the edible tissues of fish normally consumed by humans.

Recommended Risk-Reduction Measures

Proper training of all transporters will reduce the likelihood of a mishap or accident resulting in a leak or spill to the environment. The correct use of DOT labeling (per the MSDS) on all containers, trucks, and rail cars will enable emergency responders to react properly and quickly to any disaster thereby reducing the potential risk to the environment and to personnel.

Storage of lithium hydride should be segregated from incompatible chemicals to minimize the risk of cross-contamination or contact. A nitrogen purge is recommended inside all storage vessels and bulk storage containers, if possible. Precautions must be taken to ensure all contact with water or moisture is prevented in use or storage.

If a spill or leak to the environment has occurred, fire department, emergency response and/or hazardous materials spill personnel should be notified immediately. Cleanup should be attempted only by those trained in proper spill containment procedures. Contaminated soils should be removed for incineration and replaced with clean soil. If lithium hydride should contact the water table, aquifer, or navigable waterway, remediation activities should be prompt. When such spills occur, the local and/or state emergency response authorities must be notified. A comprehensive emergency response or disaster preparedness/recovery plan should be in place prior to any operations involving the use, transportation, storage, or disposal of lithium hydride. If lithium hydride is spilled or leaked, the following specific steps are recommended:

- ☑ Restrict persons not wearing protective clothing from area of spill or leak until cleanup is complete and area can be opened to normal work.
- ☑ Collect powdered material in the most convenient (safe) manner possible and deposit in sealed containers. Do NOT dry sweep (generates airborne dusts). Use a vacuum equipped with a HEPA filter instead. If liquid solutions are spilled, use vermiculite to absorb and place in a sealed drum.
- ☑ It may be necessary to dispose of lithium hydride as a hazardous waste. The responsible state agency or the regional office of the federal Environmental Protection Agency (EPA) should be contacted for specific recommendations.

RISK ASSESSMENT: BUSINESS

General Assessment

Accidents or mishaps involving lithium hydride can present a significant threat to business operations. The loss or damage of equipment or facilities can significantly affect fiscal viability. Lawsuits that may result from personnel injury/death, public exposures, and/or environmental contamination will also require a serious expenditure of resources. Media attention surrounding an injury, death, or environmental damage can also result in a loss of profits and loss of current as well as future business.

ʘ *Recommended Risk-Reduction Measures*

Company attorneys, safety and health professionals, and environmental specialists should be involved in the development of any procedures dealing with safety or emergency response. A company official should be pre-designated as a public relations officer with specific training in dealing with the press. Corporate plans and policies should be developed, approved, and implemented long before any need for such arises.

REFERENCES

American Conference of Governmental Industrial Hygienists. 1988. *Documentation of the Threshold Limit Values and Biological Exposure Indices,* 5th Edition (with updates). Cincinnati: ACGIH

Anthony, C. P., and N. J. Kolthoff. 1971. *Textbook of Anatomy and Physiology.* St. Louis: C. V. Mosby Company.

Asante-Duah, D. K. 1993. *Hazardous Waste Risk Assessment.* New York: CRC Press/Lewis Publishers.

Bloom, A. D. and F. J. DeSerres. 1995. *Ecotoxicity and Human Health.* New York: CRC Press/Lewis Publishers.

Brusick, D. J. 1994. *Methods for Genetic Risk Assessment.* New York: CRC Press/Lewis Publishers.

Bueche, J. R. 1972. *Principles of Physics,* 2nd Edition. New York: McGraw-Hill.

Burrell, R., D. K. Flaherty, and L. J. Sauers. 1992. *Toxicology of the Immune System.* New York: Van Nostrand Reinhold.

Calabrese, E. J. 1994. *Biological Effects of Low Level Exposures.* New York: CRC Press/Lewis Publishers.

Cockerman, L. G., B. S. Shane. 1994. *Basic Environmental Toxicology.* New York: CRC Press/Lewis Publishers.

Cothern, C. R., N. P. Ross. 1994. *Environmental Statistics, Assessment, and Forecasting.* New York: CRC Press/Lewis Publishers.

Gardner, R. W. 1994. *Chemical Intolerance.* New York: CRC Press/Lewis Publishers.

Hallenbeck, W. H. 1993. *Quantitative Risk Assessment for Environmental and Occupational Health,* 2nd Edition. New York: CRC Press/Lewis Publishers.

Hodgson, E., R. B. Mailman, and Janice E. Chambers. 1988. *Dictionary of Toxicology.* London: The MacMillian Company.

Howard, P. H. 1995. *Handbook of Environmental Fate and Exposures Data for Organic Chemicals,* Vol. I-IV. New York: CRC Press/Lewis Publishers.

Kamrin, M. A. 1988. *Toxicology: A Primer on Principles and Applications.* Chelsea, Michigan: Lewis Publishers.

Landis, W. G., M. H. Yu. 1995. *Introduction to Environmental Toxicology.* New York: CRC Press/Lewis Publishers.

Langerman, N. 1994 *Precautionary Labels for Chemical Containers.* New York: CRC Press/Lewis Publishers.

Lauwerys, R. R., P. Hoet. 1993. *Industrial Chemical Exposure.* New York: CRC Press/Lewis Publishers.

Lide, D. R. 1995. *Handbook of Chemistry and Physics,* 76th Edition. New York: CRC Press/Lewis Publishers.

Lewis, R. J., Sr. 1993. *Hawley's Condensed Chemical Dictionary,* 12th Edition. New York: Van Nostrand Reinhold.

Lewis, R. J., Sr. 1992. *Sax's Dangerous Properties of Industrial Materials,* Volumes 1, 2, and 3, 8th Edition. New York: Van Nostrand Reinhold.

Manahan, S. E. 1992. *Toxicological Chemistry,* 2nd Edition. New York: CRC Press/Lewis Publishers.

Mellström, G. A., J. E. Wahlberg, and H. I. Maibach. 1994. *Protective Gloves for Occupational Use.* New York: CRC Press/Lewis Publishers.

Meyer, E. 1990. *Chemistry of Hazardous Materials.* Englewood Cliffs, NJ: Prentice Hall.

Ott, W. R. 1995. *Environmental Statistics and Data Analysis.* New York: CRC Press/Lewis Publishers.

Patnaik, P. 1992. *A Comprehensive Guide to the Hazardous Properties of Chemical Substances.* New York: Van Nostrand Reinhold.

Philp, R. B. 1995. *Environmental Hazards and Human Health.* New York: CRC Press/Lewis Publishers.

Proctor, N. H., and James P. Hughes. 1991. *Chemical Hazards of the Workplace.* New York: Van Nostrand Reinhold.

Quigley, D. R. 1994. *Handbook of Emergency Chemical Management.* New York: CRC Press/Lewis Publishers.

Rea, W. J. 1995. *Chemical Sensitivity,* Vol. I-III. New York: CRC Press/Lewis Publishers.

Richardson, M. 1995. *Dictionary of Substance and Their Effects,* Vol. I-VI. New York: CRC Press/Lewis Publishers.

Rylander, R., R. R. Jacobs. 1994. *Organic Dusts.* New York: CRC Press/Lewis Publishers.

Sacarello, H. L. A. 1994. *The Comprehensive Handbook of Hazardous Materials.* New York: CRC Press/Lewis Publishers.

U.S. Department of Health and Human Services, National Institute for Occupational Safety and Health. 1994. *NIOSH Pocket Guide to Chemical Hazards.* Washington, D.C.: U.S. Government Printing Office.

MATERIAL SAFETY DATA SHEET

CHEMICAL NAME	
MAGNESITE	**CRC PRESS/LEWIS PUBLISHERS DISCLAIMER:** THE INFORMATION AND RECOMMENDATIONS PRESENTED HEREIN ARE BASED ON SOURCES BELIEVED TO BE RELIABLE. CRC MAKES NO REPRESENTATION ON ITS COMPLETENESS OR ACCURACY. IT IS THE RESPONSIBILITY OF THE USER TO DETERMINE THE CHEMICAL'S SUITABILITY FOR ITS INTENDED USE, THE CHEMICAL'S SAFE USE, AND THE CHEMICAL'S PROPER DISPOSAL. NO REPRESENTATIONS AND/OR WARRANTIES, EITHER EXPRESSED OR IMPLIED, OF THE MERCHANTABILITY OR FITNESS FOR A PARTICULAR PURPOSE, OR OF ANY OTHER NATURE, ARE MADE WITH RESPECT TO THE INFORMATION PROVIDED IN THIS MSDS OR TO THE CHEMICAL TO WHICH INFORMATION MAY REFER. CRC NEITHER ASSUMES NOR AUTHORIZES ANY OTHER PERSON TO ASSUME FOR IT, ANY OTHER ADDITIONAL RESPONSIBILITY OR LIABILITY FOR THE USE OF, OR RELIANCE UPON, THIS INFORMATION.

HAZARD WARNING INFORMATION

HEALTH	FIRE	REACTIVE	OTHER	DEGREE OF HAZARD	COLOR CODING	OTHER CODES
1	0	0		0 = Minimum Hazard 1 = Slight Hazard 2 = Moderate Hazard 3 = Serious Hazard 4 = Severe Hazard	HEALTH = BLUE FIRE = RED REACTIVITY = YELLOW OTHER = WHITE	OX = Oxidizer ACID = Acid ALK = Alkali COR = Corrosive W = Use No Water

SECTION I - GENERAL INFORMATION

Characterization	RCRA Number	EPA Class
Metallic Carbonate	**None**	**Not Applicable**

DOT Proper Shipping Name **No Citation**	Chemical Abstract Service (CAS) Number **546-93-0**
DOT Hazard Class and Label Requirements **No Citation**	DOT Emergency Guide Code **No Citation**
DOT Identification Number **No Citation**	Atomic Symbol **$MgCO_3$**

Synonyms

Carbonate magnesium; hydromagnesite; magnesium carbonate; natural magnesium carbonate; carbonic acid, magnesium salt; magnesia alba; magnesium carbonate-precipitated.

SECTION II - HAZARDOUS INGREDIENTS AND IDENTITY INFORMATION

Hazardous Components (specific identity)	OSHA Exposure Criteria	NIOSH Exposure Criteria	Immediately Dangerous to Life and Health (IDLH)	ACGIH Exposure Criteria
Magnesite (derivation: Mined as a natural material; by carbonation of magnesium oxide or magnesium hydroxide with carbon dioxide; or reaction of a soluble magnesium salt solution with sodium carbonate or sodium bicarbonate).	**PEL (8-hour): 15 mg(Mg)/m^3 (for oxide, fume, or total particulate)**	**REL (10-hour): 10 mg(Mg)/m^3 (for oxide, fume, or total particulate)**	**Not Determined**	**TLV: 10 mg(Mg)/m^3 (for oxide, fume, or total particulate)**
42% Magnesium Oxide 58% Carbon Dioxide	**5 mg(Mg)/m^3 (respirable fraction)**	**5 mg(Mg)/m^3 (respirable fraction)**		**5 mg(Mg)/m^3 (respirable fraction)**

SECTION III - PHYSICAL AND CHEMICAL CHARACTERISTICS

Boiling Point **1652°F (90°C) Decomposes**	Specific Gravity (H_2O = 1) **2.96**
Vapor Pressure (mm Hg) **0 (approx.) at 68°F (20°C)**	Molecular Weight (atomic weight) **84.32**
Vapor Density (Air = 1) **Not Found**	Melting Point **662°F (350°C) Decomposes**

Solubility

Very slightly soluble in water (0.01%). Soluble in acids. Insoluble in alcohols.

Appearance and Odor

Can appear as white, friable masses or a crystalline, bulky powder with no odor.

SECTION IV - FIRE AND EXPLOSION HAZARD DATA

Flash Point (method used) **Not Found**	Explosive Limits in Air % by Volume **LEL: Not Found UEL: Not Found**
NFPA Classification **Non-Combustible Solid**	Autoignition Temperature **Not Found**

Extinguishing Media

Use extinguishing media suitable to surrounding fire. Use water spray to wet down material.

Special Fire Fighting Procedures

Wear full protective clothing and self-contained breathing apparatus (SCBA). Poisonous gases are produced in fire. Use water mist to keep fire-exposed containers cool. Remove containers from fire if it can be done without risk. Fight fire from distance, if possible.

Unusual Fire and Explosion Hazards

None reported.

SECTION V - REACTIVITY DATA

<table>
<tr><td colspan="2">Stability</td><td>Conditions to Avoid
Magnesite is normally stable under routine conditions of handling and storage. Avoid contact with incompatible materials.</td></tr>
<tr><td>Stable
X</td><td>Unstable</td><td>Incompatibility (materials to avoid)
Incompatible with most acids and formaldehyde (reacts to liberate carbon dioxide).</td></tr>
<tr><td colspan="2">Hazardous Polymerization</td><td>Conditions to Avoid
Hazardous polymerization of magnesite will not occur.</td></tr>
<tr><td>May Occur</td><td>Will Not Occur
X</td><td>Hazardous Decomposition or By-products
Poisonous gases are emitted during fire, including toxic and irritating smoke and fumes.</td></tr>
</table>

SECTION VI - HEALTH HAZARD DATA

Primary Route(s) of Entry:	Inhalation?	Absorption (skin/eye)?	Ingestion?
	X		X

Health Hazards

INHALATION: Toxicity is considered to be low, but excessive dust inhalation can cause irritation to the nose and mucous membranes of the eyes, nose, throat, and respiratory tract.

EYES & SKIN: Exposure to the eyes and skin may cause mild irritation.

INGESTION: Excessive doses may cause diarrhea and possible depression of the CNS with loss of reflexes, and progression to respiratory paralysis or cardiac arrest and death.

Carcinogenicity	NTP Listed?	IARC Cancer Review Group?	OSHA Regulated?	Target Organs?
Unknown Human Unknown Animal	No	No	No	Respiratory system, eyes, skin, CNS, kidneys.

Medical Conditions Generally Aggravated by Exposure

Existing kidney dysfunction may be aggravated by ingestion of magnesium salts.

Emergency and First-aid Procedures

Eye contact: Do not allow victim to keep eyes tightly shut. Flush immediately with water for 15 minutes (minimum); seek medical attention. Skin contact: Remove all contaminated clothing. Immediately wash area with flooding amounts of water. For reddened or blistered skin, seek medical assistance. For inhalation: Remove the person from exposure. Provide respiratory assistance and CPR. Transfer to medical facility. If swallowed: Seek medical attention. Give 1 to 2 glasses of water and induce vomiting. Never give anything by mouth to an unconscious or convulsing person.

SECTION VII - PRECAUTIONS FOR SAFE HANDLING AND USE

Steps to be Taken in Case Material is Released or Spilled

Collect powdered materials in most convenient and safe manner and deposit in sealed containers. Dry sweep not recommended. Use HEPA vacuum. Ventilate area of spill. Restrict those not involved in cleanup from entering area. Notify appropriate authorities.

Preferred Waste Disposal Method

No citation.

Precautions to be Taken in Handling and Storage

Store to avoid contact with incompatible materials. Prevent physical damage to containers. Keep containers tightly closed.

Other Precautions and Warnings

Avoid excessive dust generation in storage and handling areas. Personnel should be trained on the proper precautions for working with magnesium compounds *before* initial assignment.

SECTION VIII - CONTROL MEASURES AND PERSONAL PROTECTIVE EQUIPMENT

Respiratory Protection (specify type)

For dust exposures over PEL, use a respirator with a high efficiency particulate air (HEPA) filter; otherwise, use a self-contained breathing apparatus with full facepiece operated in positive pressure mode.

Ventilation

Local exhaust (preferred) at site of chemical work or general ventilation.

Protective Gloves	Eye Protection	Other Protective Clothing
Impervious Material	Chemical/Dust Goggles or Face Mask	Protective Apron, Boots

Work/Hygiene Practices

Always wash hands thoroughly after using chemical; never bring food, drink, or smoking materials into vicinity of chemicals.

MAGNESITE

$MgCO_3$

CAS: 546-93-0

IDENTIFICATION AND TYPICAL USES

Magnesite appears as a white, friable mass or a crystalline, bulky powder with no odor. It is mined as a natural material. The synthetic variety is known as magnesium carbonate. It occurs in North America in California, Washington, and Nevada, as well as in Austria and Greece. It is used to make various grades of magnesium oxide and in the production of carbon dioxide. The synthetic variety is used in foods as a drying, color retention, and anti-caking agent. It is also used in the manufacture of magnesium citrate, in certain inks, in the making of glass, pharmaceuticals, dentifrice, cosmetics, and in heat insulation.

RISK ASSESSMENT: HEALTH

General Assessment

Magnesite exhibits relatively low toxicity in humans by *inhalation* and *ingestion*. It is generally considered a nuisance particulate. However, exposure to sufficient quantities can pose a serious health risk. There is no information on its carcinogenic, mutagenic, or teratogenic capabilities.

Inhalation causes irritation to the eyes, nose, throat, and mucosa of the upper respiratory tract with cough and congestion. If ingested, magnesium salts will be slowly absorbed. However, large amounts can cause systemic poisoning with symptoms of diarrhea, kidney effects, and depression of the central nervous system (CNS). This can lead to failed reflexes with progression to respiratory failure or cardiac arrest and death.

Acute Health Effects

The following acute (short-term) health effects may occur immediately or shortly after human exposure to magnesite:

Skin: May cause mild irritation, depending on the conditions and length of exposure.

Eye: Mild to moderate irritation.

Lung: Irritation of the mucosa of the eyes, nose, throat, and respiratory tract.

Other: Ingestion of excessive quantities of magnesite can cause gastrointestinal effects, kidney problems, and depression of the central nervous system that can lead to death.

Chronic Health Effects

The following chronic (long-term) health effects may occur at some time after exposure to magnesite and can last for months or even years:

Cancer Hazards: According to the references, magnesite has not been shown to cause cancer in animals.

Reproductive Hazard: According to the references, there are no reports of any teratogenic or reproductive effects resulting from exposure to magnesite.

Other Chronic Effects: None reported.

Recommended Risk-Reduction Measures

Even though the toxicity of magnesite is considered to be relatively low, personnel should avoid direct contact with magnesite. If a less toxic material or compound cannot be substituted for magnesite, then *engineering controls* are the most effective method of reducing exposures. The best protection is to enclose operations and/or provide local exhaust ventilation at the site of magnesite dust release. While not always operationally feasible, isolating operations can also reduce exposure. Using respiratory protection is less effective than the controls mentioned above, but is still advisable whenever working with or around magnesite. A supplied-air respirator operated in continuous flow mode or a self-contained breathing apparatus (SCBA) with full facepiece operated in pressure demand is recommended for the greatest possible respiratory protection. However, for low or infrequent exposure, an MSHA/NIOSH-approved full facepiece respirator with a high efficiency particulate air (HEPA) filter may be used. If a full facepiece is not available, then chemical/dust goggles should be worn to protect the eyes. A face shield and protective apron should also be worn. To prevent hand and skin exposures, gloves should be used.

Administrative controls should also be in place to minimize the potential for human exposures. These may include written procedures or policies which specify the methods and techniques that will be practiced whenever personnel are to work with magnesite.

All personnel should receive *training* on the use, hazards, protective measures, emergency actions, and other precautions per 29 CFR 1910.1200 (Hazard Communication), prior to the first assignment in an area where magnesite is used or stored.

Before beginning employment working with magnesite, and at regular intervals thereafter (e.g., annually), those personnel with frequent or potentially high exposures should be provided the following recommended medical tests:

☑ Kidney function tests (establish baseline).

If symptoms develop or overexposure is suspected, consider the following additional tests:

☑ Kidney function tests (compare to baseline).

☑ Neurological examination.

☑ Lung function tests.

Any evaluation should include a careful history of past and present symptoms with an examination. Medical tests that simply look for existing damage are not a substitute for controlling exposures. Also, since smoking can cause heart disease as well as lung cancer, emphysema, and other known respiratory problems, smokers exposure to magnesite may experience symptoms more rapidly than non-smokers under the same conditions of exposure. Prudent risk management requires careful consideration of *all* factors which may be causing exposure symptoms in the workplace.

Other methods to reduce exposure include:

☑ Where possible, enclose operations and use local exhaust ventilation at the site of chemical release. If local exhaust ventilation or enclosure is not used, respiratory protection should be mandatory.

☑ Always ensure that proper protective clothing is worn when using chemical substances. Proper maintenance and care of protective equipment is essential to its proper function.

☑ Wash thoroughly immediately after exposure to magnesite and at the end of the work shift or before eating, drinking, or smoking.

☑ Hazard warning information should be posted in the work area. In addition, as part of an on-going education and training program, all information on the health and safety hazards of magnesite should be communicated to all exposed and potentially exposed workers.

☑ Eye wash stations should be provided in the immediate work area for emergency use. If there is a possibility of skin exposure to magnesite, emergency shower facilities should be provided.

☑ Workers whose clothing has been contaminated by magnesite should change into clean clothes before leaving work. Contaminated work clothing should be laundered only by individuals who have been informed of the hazards of exposure to magnesite.

RISK ASSESSMENT: ENVIRONMENT

General Assessment

As a naturally occurring mineral, magnesite poses very little threat to the environment. However, accidental spills or releases to the environment can occur during transportation, storage, disposal, or destruction of magnesite. In almost every scenario, the threat of environmental exposure is contingent upon the proper handling of the chemical substance. Accidental spills, large or small, where magnesite contacts incompatible commodities can result in release of toxic products (depending upon conditions of spill), and possible contamination of the surrounding environmental mediums (water, soil, and air).

Magnesite is considered a non-combustible solid. However, finely divided dusts and powders may create an explosive mixture in air in closed spaces. It is incompatible with many acids and formaldehyde, and extreme caution is required in handling, storage, transportation, and disposal of magnesite. These characteristics also require special consideration during any emergency situation involving a leak or spill of magnesite powder or dust. Should magnesite ever come into contact with incompatible substances either during use, transportation, storage, or disposal, violent and even explosive reactions are possible.

Magnesite may enter the environment as a result of industrial and municipal waste treatment plant discharges and from spills.

Acute Ecological Effects

Acute (short-term) toxic effects may include the death of animals, birds, or fish and death or low growth rate in plants. Acute effects are seen 2 to 4 days after animals or plants are exposed to magnesite. Magnesium and its salts have slight acute toxicity to aquatic life. Insufficient data are available to evaluate the short-term effects of magnesite to plants, birds, or land animals.

Chronic Ecological Effects

Chronic (long-term) toxic effects may include shortened life span, reproductive problems, lower fertility, and changes in appearance or behavior in exposed animals. These effects can be seen long after first exposure(s) to toxic chemicals. Magnesium and its salts have slight chronic toxicity to aquatic life. Insufficient data are available to evaluate or predict the long-term effects of magnesite to plants, birds, or land animals.

Water Solubility

Magnesite is nearly insoluble (0.01%) in water. Concentrations of less than 1 milligram may mix with a liter of water.

Persistence in the Environment

Magnesite is highly persistent in water, with a half-life greater than 200 days. The half-life of a pollutant is the amount of time it takes for one half of the chemical to be degraded. Magnesite is also naturally present in soils, with its three hydrates (barrington, nesquehonite, and landsforite).

Bioaccumulation in Aquatic Organisms

Some substances increase in concentration, or *bioaccumulate,* in living organisms as they breathe contaminated air, drink contaminated water, or eat contaminated food. These chemicals can become concentrated in the tissues and internal organs of animals as well as humans.

The concentration of magnesite found in fish tissues is expected to be about the same as the average concentration of magnesite in the water from which the fish was taken.

Recommended Risk-Reduction Measures

Proper training of all transporters will reduce the likelihood of a mishap or accident resulting in a leak or spill of magnesite dusts or powders into the environment. The correct use of labeling on all containers, trucks, and rail cars will enable emergency responders to react properly and quickly to any disaster thereby reducing the potential risk to the environment and to personnel.

Storage of magnesite should be segregated from incompatible chemicals to minimize the risk of cross-contamination or contact. Care should be taken to prevent physical damage to storage containers. Also, dusty conditions must be avoided in storage and use areas. When magnesite comes into contact with acids, it will readily liberate excessive amounts of carbon dioxide which is not only toxic, it can also cause a dangerous pressure buildup inside closed containers and may lead to an explosion.

If a spill or leak to the environment has occurred, fire department, emergency response and/or hazardous materials spill personnel should be notified immediately. Cleanup should be attempted only by those trained in proper spill containment procedures. Contaminated soils should be removed for decontamination and replaced with clean soil. If magnesite should contact the water table, aquifer, or navigable waterway, time is of the essence. It is slightly soluble and total remediation may not be possible. The local and/or state emergency response authorities may require notification. A comprehensive emergency response or disaster preparedness/recovery plan should be in place prior to any operations involving the use, transportation, storage, or disposal of magnesite.

If magnesite is spilled or leaked, the following specific steps are recommended:

- ☑ Restrict persons not wearing protective clothing from area of spill or leak until cleanup is complete and area can be opened for normal work.
- ☑ Remove all ignition sources from area.
- ☑ Ventilate area of spill or leak.
- ☑ Collect powdered materials in the safest manner possible and deposit in sealed containers. Recommend using a vacuum equipped with a high-efficiency particulate air (HEPA) filter. Do not dry sweep (generates airborne dusts). Damp mop residues and collect for disposal. Water spray can be used to knock down vapors. Runoff should also be collected for disposal.
- ☑ It may be necessary to dispose of magnesite as a hazardous waste. The responsible state agency or the regional office of the federal Environmental Protection Agency (EPA) should be contacted for specific recommendations.

RISK ASSESSMENT: BUSINESS

General Assessment

Accidents or mishaps involving magnesite can present a significant threat to business operations. The loss or damage of equipment or facilities can significantly affect fiscal viability. Lawsuits that may result from personnel injury/death, public exposures, and/or environmental contamination will also require a serious expenditure of resources. Media attention surrounding an injury, death, or environmental damage can also result in a loss of profits and loss of current as well as future business.

Recommended Risk-Reduction Measures

Company attorneys, safety and health professionals, and environmental specialists should be involved in the development of any procedures for responding to chemical incidents. A company official should be predesignated as a public relations officer with specific training in dealing with the press. Corporate plans and policies should be developed, approved, and implemented long before any need for such arises.

REFERENCES

Ahlbom, A. 1993. *Biostatistics for Epidemiologists.* New York: CRC Press/Lewis Publishers.

American Conference of Governmental Industrial Hygienists. 1988. *Documentation of the Threshold Limit Values and Biological Exposure Indices,* 5th Edition (with updates). Cincinnati: ACGIH

Anthony, C. P., and N. J. Kolthoff. 1971. *Textbook of Anatomy and Physiology.* St. Louis: C. V. Mosby Company.

Asante-Duah, D. K. 1993. *Hazardous Waste Risk Assessment.* New York: CRC Press/Lewis Publishers.

Bloom, A. D. and F. J. DeSerres. 1995. *Ecotoxicity and Human Health.* New York: CRC Press/Lewis Publishers.

Brusick, D. J. 1994. *Methods for Genetic Risk Assessment.* New York: CRC Press/Lewis Publishers.

Bueche, J. R. 1972. *Principles of Physics,* 2nd Edition. New York: McGraw-Hill.

Calabrese, E. J. 1994. *Biological Effects of Low Level Exposures.* New York: CRC Press/Lewis Publishers.

Cockerman, L. G., B. S. Shane. 1994. *Basic Environmental Toxicology.* New York: CRC Press/Lewis Publishers.

Cothern, C. R., N. P. Ross. 1994. *Environmental Statistics, Assessment, and Forecasting.* New York: CRC Press/Lewis Publishers.

Cothern, R. R. 1993. *Comparative Environmental Risk Assessment.* New York: CRC Press/Lewis Publishers.

Danse, I. R. 1991. *Common Sense Toxics in the Workplace.* New York: Van Nostrand Reinhold.

Environmental Protection Agency. 1983. *Chemical Hazard Information Profiles.* Washington, D.C.: U.S. EPA.

Fodden, C. R., and J. L. Weddell. 1992. *Hazardous Materials Emergency Action Data.* New York: CRC Press/Lewis Publishers.

Forseberg, K., and L. H. Keith. 1993. *Chemical; Protective Clothing Permeation and Degradation Database.* Chelsea, MI: Lewis Publishers.

Gardner, R. W. 1994. *Chemical Intolerance.* New York: CRC Press/Lewis Publishers.

Howard, P. H. 1995. *Handbook of Environmental Fate and Exposures Data for Organic Chemicals,* Vol. I-IV. New York: CRC Press/Lewis Publishers.

Langerman, N. 1994 *Precautionary Labels for Chemical Containers.* New York: CRC Press/Lewis Publishers.

Lauwerys, R. R., P. Hoet. 1993. *Industrial Chemical Exposure.* New York: CRC Press/Lewis Publishers.

Lewis, R. J., Sr. 1993. *Hawley's Condensed Chemical Dictionary,* 12th Edition. New York: Van Nostrand Reinhold.

Lewis, R. J., Sr. 1992. *Sax's Dangerous Properties of Industrial Materials,* Volumes 1, 2, and 3, 8th Edition. New York: Van Nostrand Reinhold.

Mackay, D., W. Y. Shiu, and K. C. Ma. 1992. *Illustrated Handbook of Chemical-Physical Properties and Environmental Fate for Organic Chemicals.* New York: CRC Press/Lewis Publishers.

Manahan, S. E. 1994. *Environmental Chemistry,* 6th Edition. New York: CRC Press/Lewis Publishers.

Meyer, E. 1990. *Chemistry of Hazardous Materials.* Englewood Cliffs, NJ: Prentice Hall.

Miller, B. F., M.D., and C. B. Keane, R.N. 1972. *Encyclopedia and Dictionary of Medicine and Nursing.* Philadelphia: W.B. Sanders Company.

Mycock, J. C., J. D. McKenna, and L. Theodore. 1995. *Handbook of Air Pollution Control Engineering and Technology.* New York: CRC Press/Lewis Publishers.

Quigley, D. R. 1994. *Handbook of Emergency Chemical Management.* New York: CRC Press/Lewis Publishers.

Richardson, M. 1995. *Dictionary of Substance and Their Effects,* Vol. I-VI. New York: CRC Press/Lewis Publishers.

Sacarello, H. L. A. 1994. *The Comprehensive Handbook of Hazardous Materials.* New York: CRC Press/Lewis Publishers.

Talbott, E., G. Craun. 1995. *An Introduction to Environmental Epidemiology.* New York: CRC Press/Lewis Publishers.

Torkelson, T. R. V. K. Rowe. 1981. *Patty's Industrial Hygiene and Toxicology,* Vol. 2B. New York: Wiley-Interscience.

U.S. Department of Health and Human Services, National Institute for Occupational Safety and Health. 1994. *NIOSH Pocket Guide to Chemical Hazards.* Washington, D.C.: U.S. Government Printing Office.

MATERIAL SAFETY DATA SHEET

CHEMICAL NAME

MAGNESIUM

CRC PRESS/LEWIS PUBLISHERS DISCLAIMER: THE INFORMATION AND RECOMMENDATIONS PRESENTED HEREIN ARE BASED ON SOURCES BELIEVED TO BE RELIABLE. CRC MAKES NO REPRESENTATION ON ITS COMPLETENESS OR ACCURACY. IT IS THE RESPONSIBILITY OF THE USER TO DETERMINE THE CHEMICAL'S SUITABILITY FOR ITS INTENDED USE, THE CHEMICAL'S SAFE USE, AND THE CHEMICAL'S PROPER DISPOSAL. NO REPRESENTATIONS AND/OR WARRANTIES, EITHER EXPRESSED OR IMPLIED, OF THE MERCHANTABILITY OR FITNESS FOR A PARTICULAR PURPOSE, OR OF ANY OTHER NATURE, ARE MADE WITH RESPECT TO THE INFORMATION PROVIDED IN THIS MSDS OR TO THE CHEMICAL TO WHICH INFORMATION MAY REFER. CRC NEITHER ASSUMES NOR AUTHORIZES ANY OTHER PERSON TO ASSUME FOR IT, ANY OTHER ADDITIONAL RESPONSIBILITY OR LIABILITY FOR THE USE OF, OR RELIANCE UPON, THIS INFORMATION.

HAZARD WARNING INFORMATION

HEALTH	FIRE	REACTIVE	OTHER	DEGREE OF HAZARD	COLOR CODING	OTHER CODES
0	1	2	~~W~~	0 = Minimum Hazard 1 = Slight Hazard 2 = Moderate Hazard 3 = Serious Hazard 4 = Severe Hazard	HEALTH = BLUE FIRE = RED REACTIVITY = YELLOW OTHER = WHITE	OX = Oxidizer ACID = Acid ALK = Alkali COR = Corrosive ~~W~~ = Use No Water

SECTION I - GENERAL INFORMATION

Characterization: **Metal (mineral)**	RCRA Number: **D001**	EPA Class: **Characteristic (I) Waste**
DOT Proper Shipping Name: **Magnesium Metal**	Chemical Abstract Service (CAS) Number: **7439-95-4**	
DOT Hazard Class and Label Requirements: **Flammable Solid**	DOT Emergency Guide Code: **No Citation**	
DOT Identification Number: **UN 1869**	Atomic Symbol: **Mg**	

Synonyms

Magnesium borings; magnesium clippings; magnesium pellets; magnesium powder; magnesium ribbons; magnesium scalpings; magnesium scrap; magnesium shavings; magnesium sheets.

SECTION II - HAZARDOUS INGREDIENTS AND IDENTITY INFORMATION

Hazardous Components (specific identity)	OSHA Exposure Criteria	NIOSH Exposure Criteria	Immediately Dangerous to Life and Health (IDLH)	ACGIH Exposure Criteria
Magnesium (derivation: By electrolysis of fused magnesium chloride; by reduction of magnesium oxide with ferrosilicon).	**PEL:** Not Established **STEL:** Not Established	**REL:** Not Established **STEL:** Not Established	**Not Determined**	**TLV:** Not Established **STEL:** Not Established

SECTION III - PHYSICAL AND CHEMICAL CHARACTERISTICS

Boiling Point: **2025°F (1107°C)**	Specific Gravity (H_2O = 1): **1.74**
Vapor Pressure (mm Hg): **1 at 1150°F (621°C)**	Atomic Weight: **24.31**
Vapor Density (Air = 1): **Not Found**	Melting Point: **1300°F (649°C)**

Solubility

Insoluble in water; soluble in acids.

Appearance and Odor

Silvery white, moderately hard, alkaline-earth metal. Odorless.

SECTION IV - FIRE AND EXPLOSION HAZARD DATA

Flash Point (method used): **Not Found**	Explosive Limits in Air % by Volume: **LEL: Not Found UEL: Not Found**
NFPA Classification: **Combustible Solid**	Autoignition Temperature: **Not Found**

Extinguishing Media

Use G-1® graphite powder, dry graphite, powdered sodium chloride, soda ash, or powdered talc. Do NOT use water, carbon dioxide, or other common extinguishing agents.

Special Fire Fighting Procedures

Wear full protective clothing and self-contained breathing apparatus (SCBA). Poisonous gases are produced in fire. Do NOT look directly into magnesium fires. Fight fire from distance, if possible.

Unusual Fire and Explosion Hazards

Extremely dangerous when heated. May react with moisture or acids to evolve explosive hydrogen gas. Finely divided flakes or particles in air can ignite readily by spark or flame and burn rapidly.

SECTION V - REACTIVITY DATA

Stability		Conditions to Avoid
Stable	Unstable	Magnesium is normally stable under routine conditions of handling and storage. Avoid contact with incompatible materials, heat, flame, sparks, or other ignition sources, and keep moisture away.
X		Incompatibility *(materials to avoid)*: Incompatible with most acids, chlorinated solvents, methanol, hydrogen peroxide, oxidizing agents, sulfur compounds, animal and vegetable oils, water, metal oxides and their salts, metal cyanides, oxygen.
Hazardous Polymerization		Conditions to Avoid: Hazardous polymerization of magnesium will not occur. May ignite spontaneously in air or moisture.
May Occur	Will Not Occur	Hazardous Decomposition or By-products
	X	Poisonous gases are emitted during fire, including toxic and irritating smoke and fumes of magnesium oxide.

SECTION VI - HEALTH HAZARD DATA

Primary Route(s) of Entry:	Inhalation?	Absorption (skin/eye)?	Ingestion?
	X		X

Health Hazards

INHALATION: The dusts can produce irritation of the eyes and mucous membranes. It may react with the water content in the bronchial passages to form caustic magnesium hydroxide which can damage the lungs. Inhalation of the fumes can cause metal fume fever with symptoms of cough, a feeling of chest pressure, leukocytosis, and other febrile (feverish) conditions.

EYES & SKIN: Causes skin sores that resist healing and accumulate gas under the skin (small blisters). The powder metal ignites easily on moist skin and eyes causing serious burns.

INGESTION: Large doses can cause gastrointestinal irritation and other unspecified effects.

Carcinogenicity	NTP Listed?	IARC Cancer Review Group?	OSHA Regulated?	Target Organs?
Unknown Human Unknown Animal	No	No	No	Respiratory system, eyes, skin, CNS, kidneys.

Medical Conditions Generally Aggravated by Exposure

Existing kidney dysfunction may be aggravated by ingestion of magnesium salts.

Emergency and First-aid Procedures

Eye contact: Do not allow victim to keep eyes tightly shut. Flush immediately with water for 15 minutes (minimum); seek medical attention. Skin contact: Remove all contaminated clothing. Immediately wash area with flooding amounts of water. For reddened or blistered skin, seek medical assistance. For inhalation: Remove the person from exposure. Provide respiratory assistance and CPR. Transfer to medical facility. If swallowed: Seek medical attention. Rinse mouth with water. Give 1 to 2 glasses of water. Do NOT induce vomiting. Never give anything by mouth to an unconscious or convulsing person.

SECTION VII - PRECAUTIONS FOR SAFE HANDLING AND USE

Steps to be Taken in Case Material is Released or Spilled

Collect powdered materials in most convenient and safe manner and deposit in clean, dry, sealed, metal containers. Dry sweep not recommended. Use HEPA vacuum. Ventilate area of spill. Restrict those not involved in cleanup from entering area. Do NOT use water.

Preferred Waste Disposal Method

Reclaim scrap.

Precautions to be Taken in Handling and Storage

Store to avoid contact with incompatible materials. Prevent physical damage to containers. Keep containers tightly closed. Keep moisture and sources of heat away. Use non-sparking tools.

Other Precautions and Warnings

Avoid excessive dust generation in storage and handling areas. Personnel should be trained on the proper precautions for working with magnesium compounds *before* initial assignment.

SECTION VIII - CONTROL MEASURES AND PERSONAL PROTECTIVE EQUIPMENT

Respiratory Protection *(specify type)*

For dust exposures, use a respirator with a HEPA filter; otherwise, use a supplied-air respirator or a self-contained breathing apparatus with full facepiece operated in positive pressure mode.

Ventilation

Local exhaust (preferred) at site of chemical work or general ventilation.

Protective Gloves	Eye Protection	Other Protective Clothing
Impervious Material	Chemical/Dust Goggles or Face Mask	Protective Apron, Boots

Work/Hygiene Practices

Always wash hands thoroughly after using chemical; never bring food, drink, or smoking materials into vicinity of chemicals.

MAGNESIUM

Mg

CAS: 7439-95-4

IDENTIFICATION AND TYPICAL USES

Magnesium appears as a silvery-white, moderately hard, alkaline earth metal. It is metallic element of atomic number 12, group IIA of the periodic table, atomic weight 24.305, and valance of 2 with 3 isotopes. It is used in aluminum alloys for structural parts, in diecast auto parts, in missiles, space vehicles, in powder for pyrotechnics and flares, and in flash photography. It is also used in the production of iron, nickel, zinc, titanium, zirconium, in antiknock gasoline additives, in making magnesium compounds and Gringard syntheses, in cathodic protection, as a reducing agent, in desulfurizing iron in steel manufacturing, in precision instruments, in optical mirrors, and in dry and wet batteries.

RISK ASSESSMENT: HEALTH

General Assessment

The toxicity of magnesium metal in humans is not easily characterized. It can enter the body and cause varying degrees of adverse health effects through *inhalation* and *ingestion*. There is no information on its carcinogenic, mutagenic, or teratogenic capabilities.

Inhalation of the dusts and particles of magnesium can irritate the eyes and nasal passages. Magnesium may react with the water content in the bronchial passages to form caustic magnesium hydroxide which can cause severe burns and tissue damage. Magnesium fumes in the oxide form can also irritate the eyes, nose, throat, and respiratory tract and may cause metal fume fever with cough, tightness and pressure in the chest, nausea, vomiting, and other febrile (feverish) symptoms. Leukocytosis (increase of leukocytes in the blood) may also develop.

Magnesium particles that become imbedded in the skin can cause the development of slow healing sores and gaseous blebs (small blisters). If the skin is moist, burns may be serious and destructive.

Acute Health Effects

The following acute (short-term) health effects may occur immediately or shortly after human exposure to magnesium:

Skin: May cause serious irritation and even burns, especially if the skin is moist. Small blisters can develop that will heal slowly and may become infected.

Eye: Severe irritation and possible burns due to the eyes' moisture content. Damage can be serious and permanent.

Lung: Irritation of the mucosa of the eyes, nose, throat, and respiratory tract.

CNS: Breathing heated fumes can cause metal fume fever with specific action on the central nervous system.

Chronic Health Effects

The following chronic (long-term) health effects may occur at some time after exposure to magnesium and can last for months or even years:

Cancer Hazards: According to the references, magnesium has not been shown to cause cancer in animals.

Reproductive Hazard: According to the references, there are no reports of any teratogenic or adverse reproductive effects resulting from exposure to magnesium.

Other Chronic Effects: None reported.

Recommended Risk-Reduction Measures

The specific human toxicity of magnesium is not easily characterized and, therefore, personnel should avoid direct contact with magnesium. If a less toxic material or compound cannot be substituted for magnesium, then *engineering controls* are the most effective method of reducing exposures. The best protection is to enclose operations and/or provide local exhaust ventilation at the site of magnesium dust release. While not always operationally feasible, isolating operations can also reduce exposure. Using respiratory protection is less effective than the controls mentioned above, but is still advisable whenever working with or around magnesium. No exposure limits have been established specifically for magnesium metal. This does not mean that exposure is without a health risk. Other magnesium compounds have levels set relatively low (10 mg/m^3 for magnesium oxide, for example). For low or transient exposures, a chemical cartridge respirator equipped with a high-efficiency particulate air (HEPA) filter and a dust/mist pre-filter may provide adequate respiratory protection. Better protection is obtained using a supplied-air respirator operated in continuous flow mode or a self-contained breathing apparatus (SCBA) with full facepiece operated in pressure demand. If a full facepiece is not available, then chemical/dust goggles should be worn to protect the eyes. A face shield and protective apron should also be worn. To prevent hand and skin exposures, gloves should be used. Glove manufacturers should be contacted *before* selections are made.

Administrative controls should also be in place to minimize the potential for human exposures. These may include written procedures or policies which specify the methods and techniques that will be practiced whenever personnel are to work with magnesium.

All personnel should receive *training* on the use, hazards, protective measures, emergency actions, and other precautions per 29 CFR 1910.1200 (Hazard Communication), prior to the first assignment in an area where magnesium is used or stored.

Before beginning employment working with magnesium, and at regular intervals thereafter (e.g., annually), those personnel with frequent or potentially high exposures should be provided the following recommended medical tests:

☑ Kidney function tests (establish baseline).

If symptoms develop or overexposure is suspected, consider the following additional tests:

☑ Kidney function tests (compare to baseline).

☑ Neurological examination.

☑ Lung function tests.

Any evaluation should include a careful history of past and present symptoms with an examination. Medical tests that simply look for existing damage are not a substitute for controlling exposures. Also, since smoking can cause heart disease as well as lung cancer, emphysema, and other known respiratory problems, smokers exposure to magnesium may experience symptoms more rapidly than non-smokers under the same conditions of exposure. Prudent risk management requires careful consideration of *all* factors which may be causing exposure symptoms in the workplace.

Other methods to reduce exposure include:

☑ Where possible, enclose operations and use local exhaust ventilation at the site of chemical release. If local exhaust ventilation or enclosure is not used, respiratory protection should be mandatory.

☑ Always ensure that proper protective clothing is worn when using chemical substances. Proper maintenance and care of protective equipment is essential to its proper function.

☑ Wash thoroughly immediately after exposure to magnesium and at the end of the work shift or before eating, drinking, or smoking.

☑ Hazard warning information should be posted in the work area. In addition, as part of an on-going education and training program, all information on the health and safety hazards of magnesium should be communicated to all exposed and potentially exposed workers.

☑ Eye wash stations should be provided in the immediate work area for emergency use. If there is a possibility of skin exposure to magnesium, emergency shower facilities should be provided.

☑ Workers whose clothing has been contaminated by magnesium should change into clean clothes before leaving work. Contaminated work clothing should be laundered only by individuals who have been informed of the hazards of exposure to magnesium.

RISK ASSESSMENT: ENVIRONMENT

General Assessment

As a naturally occurring mineral, magnesium poses very little threat to the environment. However, accidental spills or releases to the environment can occur during transportation, storage, disposal, or destruction of magnesium. In almost every scenario, the threat of environmental exposure is contingent upon the proper handling of the chemical substance. Accidental spills, large or small, where magnesium contacts incompatible commodities can result in release of toxic products (depending upon conditions of spill), and possible contamination of the surrounding environmental mediums (water, soil, and air).

Magnesium is considered a combustible solid. Finely divided dusts and powders may create an explosive mixture in air, in contact with moisture, oxidizing materials, or in contact with an ignition source. It reacts with water, moisture, and acids to evolve explosive hydrogen gas. It is also incompatible with many other materials, including chlorinated solvents, methanol, hydrogen peroxide, sulfur compounds, animal and vegetable oils, the oxides, salts, and cyanides of other metals, and even oxygen. Extreme caution is required in handling, storage, transportation, and disposal of magnesium. These characteristics also require special consideration during any emergency situation involving a leak or spill of magnesium powder or dust. Should magnesium ever come into contact with these various incompatible substances either during use, transportation, storage, or disposal, violent and even explosive reactions are possible.

Magnesium occurs naturally in the environment (terrestrial soils and aquatic sediments) and may enter the environment as a result of industrial and municipal waste treatment plant discharges and from spills.

Acute Ecological Effects

Acute (short-term) toxic effects may include the death of animals, birds, or fish and death or low growth rate in plants. Acute effects are seen 2 to 4 days after animals or plants are exposed to magnesium.

Magnesium and its salts have slight acute toxicity to aquatic life. Insufficient data are available to evaluate the short-term effects of magnesium to plants, birds, or land animals. However, it should be noted that the U.S. Environmental Protection Agency does NOT list magnesium as a hazardous substance.

Chronic Ecological Effects

Chronic (long-term) toxic effects may include shortened life span, reproductive problems, lower fertility, and changes in appearance or behavior in exposed animals. These effects can be seen long after first exposure(s) to toxic chemicals.

Magnesium and its salts have slight chronic toxicity to aquatic life. Insufficient data are available to evaluate or predict the long-term effects of magnesium to plants, birds, or land animals. However, it should be noted that the U.S. Environmental Protection Agency does NOT list magnesium as a hazardous substance.

Water Solubility

Magnesium is nearly insoluble (0.01%) in water. Concentrations of less than 1 milligram may mix with a liter of water.

Persistence in the Environment

Magnesium is highly persistent in water, with a half-life greater than 200 days. The half-life of a pollutant is the amount of time it takes for one half of the chemical to be degraded. Magnesium is also naturally present in terrestrial soils and aquatic sediments.

Bioaccumulation in Aquatic Organisms

Some substances increase in concentration, or *bioaccumulate,* in living organisms as they breathe contaminated air, drink contaminated water, or eat contaminated food. These chemicals can become concentrated in the tissues and internal organs of animals as well as humans.

The concentration of magnesium found in fish tissues is expected to be about the same as the average concentration of magnesium in the water from which the fish was taken.

Recommended Risk-Reduction Measures

Proper training of all transporters will reduce the likelihood of a mishap or accident resulting in a leak or spill of magnesium dusts or powders into the environment. The correct use of labeling on all containers, trucks, and rail cars will enable emergency responders to react properly and quickly to any disaster thereby reducing the potential risk to the environment and to personnel.

Storage of magnesium should be segregated from incompatible chemicals to minimize the risk of cross-contamination or contact. Care should be taken to prevent physical damage to storage containers. Also, dusty conditions must be avoided in storage and use areas. When magnesium comes into contact with water or acids, it will readily liberate excessive amounts of highly flammable and explosive hydrogen gas.

If a spill or leak to the environment has occurred, fire department, emergency response and/or hazardous materials spill personnel should be notified immediately. Cleanup should be attempted only by those trained in proper spill containment procedures using non-sparking tools. Contaminated soils should be removed for decontamination and replaced with clean soil. If magnesium should contact the water table, aquifer, or navigable waterway, time is of the essence. It is only slightly soluble and total remediation may be possible. The local and/or state emergency response authorities may require notification. A comprehensive emergency response or disaster preparedness/recovery plan should be in place prior to any operations involving the use, transportation, storage, or disposal of magnesium.

If magnesium is spilled or leaked, the following specific steps are recommended:

- ☑ Restrict persons not wearing protective clothing from area of spill or leak until cleanup is complete and area can be opened for normal work.
- ☑ Remove all ignition sources from area.
- ☑ Ventilate area of spill or leak.
- ☑ Collect powdered materials in the safest manner possible and deposit in sealed containers. Recommend using a vacuum equipped with a high-efficiency particulate air (HEPA) filter. Do not dry sweep (generates airborne dusts). Keep water away (do NOT damp mop residues).
- ☑ It may be necessary to dispose of magnesium as a hazardous waste. The responsible state agency or the regional office of the federal Environmental Protection Agency (EPA) should be contacted for specific recommendations.

RISK ASSESSMENT: BUSINESS

General Assessment

Accidents or mishaps involving magnesium can present a significant threat to business operations. The loss or damage of equipment or facilities can significantly affect fiscal viability. Lawsuits that may result from personnel injury/death, public exposures, and/or environmental contamination will also require a serious expenditure of resources. Media attention surrounding an injury, death, or environmental damage can also result in a loss of profits and loss of current as well as future business.

Recommended Risk-Reduction Measures

Company attorneys, safety and health professionals, and environmental specialists should be involved in the development of any procedures for responding to chemical incidents. A company official should be predesignated as a public relations officer with specific training in dealing with the press. Corporate plans and policies should be developed, approved, and implemented long before any need for such arises.

REFERENCES

American Conference of Governmental Industrial Hygienists. 1988. *Documentation of the Threshold Limit Values and Biological Exposure Indices,* 5th Edition (with updates). Cincinnati: ACGIH

Anthony, C. P., and N. J. Kolthoff. 1971. *Textbook of Anatomy and Physiology.* St. Louis: C. V. Mosby Company.

Bloom, A. D. and F. J. DeSerres. 1995. *Ecotoxicity and Human Health.* New York: CRC Press/Lewis Publishers.

Bueche, J. R. 1972. *Principles of Physics,* 2nd Edition. New York: McGraw-Hill.

Calabrese, E. J. 1994. *Biological Effects of Low Level Exposures.* New York: CRC Press/Lewis Publishers.

Cockerman, L. G., B. S. Shane. 1994. *Basic Environmental Toxicology.* New York: CRC Press/Lewis Publishers.

Cothern, R. R. 1993. *Comparative Environmental Risk Assessment.* New York: CRC Press/Lewis Publishers.

Danse, I. R. 1991. *Common Sense Toxics in the Workplace.* New York: Van Nostrand Reinhold.

Environmental Protection Agency. 1983. *Chemical Hazard Information Profiles.* Washington, D.C.: U.S. EPA.

Gardner, R. W. 1994. *Chemical Intolerance.* New York: CRC Press/Lewis Publishers.

Langerman, N. 1994 *Precautionary Labels for Chemical Containers.* New York: CRC Press/Lewis Publishers.

Lauwerys, R. R., P. Hoet. 1993. *Industrial Chemical Exposure.* New York: CRC Press/Lewis Publishers.

Lewis, R. J., Sr. 1993. *Hawley's Condensed Chemical Dictionary,* 12th Edition. New York: Van Nostrand Reinhold.

Lewis, R. J., Sr. 1992. *Sax's Dangerous Properties of Industrial Materials,* Volumes 1, 2, and 3, 8th Edition. New York: Van Nostrand Reinhold.

Mackay, D., W. Y. Shiu, and K. C. Ma. 1992. *Illustrated Handbook of Chemical-Physical Properties and Environmental Fate for Organic Chemicals.* New York: CRC Press/Lewis Publishers.

Manahan, S. E. 1994. *Environmental Chemistry,* 6th Edition. New York: CRC Press/Lewis Publishers.

Meyer, E. 1990. *Chemistry of Hazardous Materials.* Englewood Cliffs, NJ: Prentice Hall.

Richardson, M. 1995. *Dictionary of Substance and Their Effects,* Vol. I-VI. New York: CRC Press/Lewis Publishers.

Sacarello, H. L. A. 1994. *The Comprehensive Handbook of Hazardous Materials.* New York: CRC Press/Lewis Publishers.

Talbott, E., G. Craun. 1995. *An Introduction to Environmental Epidemiology.* New York: CRC Press/Lewis Publishers.

Torkelson, T. R. V. K. Rowe. 1981. *Patty's Industrial Hygiene and Toxicology,* Vol. 2B. New York: Wiley-Interscience.

MATERIAL SAFETY DATA SHEET

CHEMICAL NAME

MAGNESIUM CHLORIDE

HAZARD WARNING INFORMATION

HEALTH	FIRE	REACTIVE	OTHER	DEGREE OF HAZARD	COLOR CODING	OTHER CODES
1	0	0		0 = Minimum Hazard 1 = Slight Hazard 2 = Moderate Hazard 3 = Serious Hazard 4 = Severe Hazard	HEALTH = BLUE FIRE = RED REACTIVITY = YELLOW OTHER = WHITE	OX = Oxidizer ACID = Acid ALK = Alkali COR = Corrosive W = Use No Water

SECTION I - GENERAL INFORMATION

Characterization: **Magnesium Compound**	RCRA Number: **None** — EPA Class: **Not Applicable**
DOT Proper Shipping Name: **Not Listed**	Chemical Abstract Service (CAS) Number: **7786-30-3**
DOT Hazard Class and Label Requirements: **No Citation**	DOT Emergency Guide Code: **No Citation**
DOT Identification Number: **No Citation**	Atomic Formula: **$MgCl_2$**

Synonyms
Magnogene; dus-top.

SECTION II - HAZARDOUS INGREDIENTS AND IDENTITY INFORMATION

Hazardous Components (specific identity)	OSHA Exposure Criteria	NIOSH Exposure Criteria	Immediately Dangerous to Life and Health (IDLH)	ACGIH Exposure Criteria
Magnesium chloride (derivation: By action of hydrochloric acid on magnesium oxide or hydroxide, especially the latter when precipitated from sea water).	**PEL:** Not Established **STEL:** Not Established	**REL:** Not Established **STEL:** Not Established	**Not Determined**	**TLV:** Not Established **STEL:** Not Established

SECTION III - PHYSICAL AND CHEMICAL CHARACTERISTICS

Boiling Point: **2574°F (1412°C) Decomposes**	Specific Gravity (H_2O = 1): **2.4**
Vapor Pressure (mm Hg): **Not Found**	Molecular Weight: **95.21**
Vapor Density (Air = 1): **Not Found**	Melting Point: **1314°F (712°C)**

Solubility
Soluble in water (evolves excessive heat) and alcohol.

Appearance and Odor
Soft, white, deliquescent, crystalline solid with no detectable odor.

SECTION IV - FIRE AND EXPLOSION HAZARD DATA

Flash Point (method used): **Not Determined**	Explosive Limits in Air % by Volume: **LEL: Not Applicable UEL: Not Applicable**
NFPA Classification: **Non-Flammable Solid**	Autoignition Temperature: **Not Determined**

Extinguishing Media
Use agent suitable to surrounding fire. Do NOT use water (reaction is exothermic). Dry chemical or foam may suffice for small fires. For large fires, try fog or regular foam.

Special Fire Fighting Procedures
Poisonous gases are produced in fire. Wear full protective clothing and self-contained breathing apparatus (SCBA). Only use a water spray to keep fire-exposed containers cool. Do NOT get water inside containers. Do NOT release runoff from fire-fighting measures to sewers or waterways.

Unusual Fire and Explosion Hazards
Can release extreme heat when wet or upon contact with incompatible materials. When combined with some incompatible materials, explosion and fire can be severe.

SECTION V - REACTIVITY DATA

Stability		Conditions to Avoid **Magnesium chloride is stable under normal conditions of temperature and pressure. Because it will readily absorb moisture, containers must be kept tightly closed. Keep away from incompatible materials.**
Stable **X**	Unstable	Incompatibility *(materials to avoid)* **Incompatible with 2-furan percarboxylic acid. Also, in humid environments, magnesium chloride can cause steel to rust rapidly.**
Hazardous Polymerization		Conditions to Avoid **Under normal conditions of temperature and pressure, hazardous polymerization of magnesium chloride is not known to occur.**
May Occur	Will Not Occur **X**	Hazardous Decomposition or By-products **When heated to decomposition, magnesium chloride can emit toxic/poisonous fumes and gases. These include toxic chloride fumes, magnesium fumes, and oxides of chlorine.**

SECTION VI - HEALTH HAZARD DATA

Primary Route(s) of Entry:	Inhalation? X	Absorption (skin)?	Ingestion? X

Health Hazards

INHALATION: The dusts can produce irritation of the eyes and mucous membranes. There may be cough, congestion, and other symptoms of respiratory system irritation. Some of the irritant effects resemble that of simple mechanical irritation (a nuisance dust).

EYES & SKIN: Causes skin sores that resist healing and accumulate gas under the skin (small blisters).

INGESTION: Large doses can cause gastrointestinal irritation and other unspecified effects.

Carcinogenicity	NTP Listed?	IARC Cancer Review Group?	OSHA Regulated?	Target Organs?
Unknown Human Unknown Animal	No	No	No	Skin, eyes, and respiratory system (mucosa).

Medical Conditions Generally Aggravated by Exposure

None reported.

Emergency and First-aid Procedures

Eye contact: Flush immediately with water for 15 minutes (minimum); seek medical attention. **Skin contact:** Remove all contaminated clothing. Immediately wash area with large amounts of soap and water. For **inhalation:** Remove the person from exposure. Provide respiratory assistance and CPR. Transfer to medical facility. If **swallowed:** Provide breathing assistance if necessary. Never attempt to give a convulsing or unconscious person anything by mouth. If victim is conscious, give 1-2 glasses of water to dilute and induce vomiting with Syrup of Ipecac. Seek medical attention immediately.

SECTION VII - PRECAUTIONS FOR SAFE HANDLING AND USE

Steps to be Taken in Case Material is Released or Spilled

Collect powdered material in most convenient manner possible and deposit in sealed drum. Keep those not involved in cleanup from entering area. Do NOT dry sweep (creates airborne dusts). Use a vacuum equipped with a high-efficiency particulate air (HEPA) filter. Do NOT use water.

Preferred Waste Disposal Method

No citation.

Precautions to be Taken in Handling and Storage

Store in tightly closed containers in cool, well-ventilated area. Sources of ignition are prohibited where magnesium chloride is stored. Keep away from water. Storage with desiccant is recommended.

Other Precautions and Warnings

Outside or detached storage facility is recommended. Protect containers from physical damage.

SECTION VIII - CONTROL MEASURES AND PERSONAL PROTECTIVE EQUIPMENT

Respiratory Protection *(specify type)*

No exposure levels established. For best protection, use a supplied-air respirator with a full facepiece, hood, or helmet or a self-contained breathing apparatus (SCBA) operated in positive pressure mode.

Ventilation

Local exhaust or general mechanical systems recommended.

Protective Gloves	Eye Protection	Other Protective Clothing
Impervious Material	**Chemical/Dust Goggles or Face Mask**	**Protective Suit or Clothing**

Work/Hygiene Practices

Always wash hands thoroughly after using chemical; never bring food, drink, or smoking materials into vicinity of chemicals.

MAGNESIUM CHLORIDE

$MgCl_2$

CAS: 7786-30-3

IDENTIFICATION AND TYPICAL USES

Magnesium chloride appears as a soft, white, deliquescent, crystalline solid with no detectable odor. Used as a source of magnesium metal, in disinfectants, in fire extinguishers, for fire-proofing wood, in the manufacture of magnesium oxychloride cement, in refrigerating brines, in the manufacture of ceramics, for cooling drilling tools, in the textile industry, in the manufacture of parchment paper, in floor sweeping compounds, as a flocculating agent, and as a catalyst.

RISK ASSESSMENT: HEALTH

General Assessment

Magnesium chloride is moderately toxic by *ingestion* and *inhalation*. Skin contact with a solution of magnesium chloride is also dangerous due to the corrosive and destructive nature of this compound (absorption has not been reported as an entry route). Its carcinogenic potential is not clearly understood in the references. Some mutation data have been reported.

Inhalation of solution vapors causes sneezing, cough, and irritation of the nose and throat. There may be lachrymation (tearing), and a burning sensation in the nasal cavity often accompanied by nosebleeds Inhalation of the dusts and particles of magnesium can irritate the eyes and nasal passages. Magnesium may react with the water content in the bronchial passages to form caustic magnesium hydroxide which can cause severe burns and tissue damage.

Magnesium chloride particles that become imbedded in the skin can cause the development of slow healing sores and gaseous blebs (small blisters). Skin contact with a concentrated solution causes irritation, inflammation, and the appearance of small ulcerations on the skin's surface. There may be erythema (redness) and possible peeling of the facial skin. Eye contact will cause tearing and inflammation of the eyelids.

Acute Health Effects

The following acute (short-term) health effects may occur immediately or shortly after exposure to magnesium chloride:

Skin: Irritation, rash, and possible redness and the appearance of small blisters.

Eye: Irritation, inflammation, watering, and swelling.

Lung: Inhalation of magnesium chloride dusts may cause irritation of the mucous membranes in the respiratory tract. There may be burning in the nasal cavity with nosebleeds.

Chronic Health Effects

The following chronic (long-term) health effects may occur at some time after exposure to magnesium chloride and can last for months or even years:

Cancer Hazards: According to the information presented in the references, magnesium chloride has not been adequately tested for its ability to cause cancer in test animals. The data are conflicting and inconclusive in this regard. Mutation (causes genetic changes) data have been documented and many scientists believe that such chemicals may pose a cancer risk in the long-term. Additional study and research is required.

Reproductive Hazard: According to information available in the references, magnesium chloride has not been adequately tested for its ability to adversely affect reproduction in laboratory animals.

Recommended Risk-Reduction Measures

Personnel should avoid direct contact with magnesium chloride. If a less toxic chemical cannot be substituted for a hazardous material, then *engineering controls* are the most effective method of reducing exposures.

The best protection is to enclose operations and/or provide local exhaust ventilation at the site of chemical release. While not always operationally feasible, isolating operations can also reduce exposure. Using respiratory protection is less effective than the controls mentioned above, but is still advisable whenever working with or around magnesium chloride. No exposure levels have been established for magnesium chloride. This does not mean that exposure is without health risk. A full facepiece respirator equipped with a high efficiency particulate air (HEPA) filter equipped with a dust/mist/fume pre-filter may suffice. However, for the best protection use a supplied-air respirator with a full facepiece operated in positive pressure mode or an MSHA/NIOSH-approved self-contained breathing apparatus (SCBA) with full facepiece operated in pressure demand or other continuous flow mode. If a full facepiece is not available, then chemical/dust goggles should be worn to protect the eyes. Whenever a chemical dust hazard exists, a face shield and a protective apron should be worn. To prevent hand and skin exposures, impervious rubber gloves should be used. Glove manufacturers should be contacted and permeation studies obtained *before* final glove selections are made.

Administrative controls should also be in place to minimize the potential for human exposures. These may include written procedures or policies which specify the methods and techniques that will be practiced whenever personnel are to work with magnesium chloride dusts or solutions.

All personnel should receive *training* on the use, hazards, protective measures, emergency actions, and other precautions per 29 CFR 1910.1200 (Hazard Communication), prior to the first assignment in an area where magnesium chloride is used or stored.

If symptoms develop or overexposure is suspected, the following may also be useful:

☑ Urine and serum magnesium levels.

☑ Lung function tests.

☑ Consider chest X-ray after acute overexposure.

Any evaluation should include a careful history of past and present symptoms with a medical examination. Medical tests that look for damage already done are not a substitute for controlling exposure. Also, since smoking can cause heart disease, emphysema, and a number of other respiratory disorders, smokers exposed to lung irritants such as magnesium chloride dusts and fumes may experience symptoms much more quickly than non-smokers under the same conditions of exposure. Prudent risk management requires proper consideration of *all* factors which may be causing the appearance of exposures symptoms in the work place.

Other methods to reduce exposure include:

☑ Where possible, enclose operations and use local exhaust ventilation at the site of chemical release. If local exhaust ventilation or enclosure is not used, respiratory protection should be mandatory.

☑ Always ensure that proper protective clothing is worn when using chemical substances and that personnel are trained in its use and care.

☑ Wash thoroughly immediately after exposure to magnesium chloride and at the end of the work shift or before eating, drinking, or smoking.

☑ Hazard warning information should be posted in the work area. In addition, as part of an on-going education and training program, all information on the health and safety hazards of magnesium chloride should be communicated to all exposed and potentially exposed workers.

☑ Eye wash stations and emergency showers should be located in the immediate work area.

☑ Work clothing that has been contaminated with magnesium chloride should never be worn home where family members can be exposed as well. Contaminated clothing should be laundered at work or by an authorized service by personnel who have been trained in the hazards associated with exposure to magnesium chloride.

RISK ASSESSMENT: ENVIRONMENT

General Assessment

The environment is at risk of exposure during transportation, storage, disposal, or destruction of magnesium chloride. In almost every scenario, the threat of environmental exposure is contingent upon the proper handling of the chemical substance. Accidental spills, large or small, can result in fire, explosion, and possible contamination of the surrounding environmental mediums (water, soil, and air).

Magnesium chloride may react violently in contact with many common acids. Reactions with water and moisture will evolve extreme heat. In a fire, magnesium chloride will emit toxic and poisonous gases. These characteristics require special consideration during any emergency situation involving a leak or spill of magnesium chloride.

Magnesium chloride can enter the environment through industrial discharge, spills, or leaks.

Acute Ecological Effects

Acute (short-term) toxic effects may include the death of animals, birds, or fish and death or low growth rate in plants. Acute effects are seen 2 to 4 days after animals or plants are exposed to magnesium chloride. Insufficient data are available to evaluate the short-term effects of magnesium chloride exposure to aquatic life, plants, birds or land animals.

Chronic Ecological Effects

Chronic (long-term) toxic effects may include shortened life span, reproductive problems, lower fertility, and changes in appearance or behavior in exposed animals. These effects can be seen long after first exposure(s) to toxic chemicals. Insufficient data are available to evaluate the long-term effects of magnesium chloride exposure to aquatic life, plants, birds or land animals.

Water Solubility

Magnesium chloride is soluble in water, decomposing and producing heat. Concentrations of 1000 milligrams or more will mix with a liter of water.

Persistence in the Environment

Magnesium chloride is only slightly persistent in the aquatic environment, with a half-life of less than 2 days. The half-life of a pollutant is the amount of time it takes for one half of the chemical to be degraded.

Bioaccumulation in Aquatic Organisms

Some substances increase in concentration, or *bioaccumulate,* in living organisms as they breathe contaminated air, drink contaminated water, or eat contaminated food. These chemicals can become concentrated in the tissues and internal organs of animals as well as humans.

The concentration of magnesium chloride found in fish is expected to be much lower than the average concentration of magnesium chloride in the water from which the fish was taken.

Recommended Risk-Reduction Measures

Proper training of all transporters will reduce the likelihood of a mishap or accidents resulting in a leak or spill to the environment. Proper labeling on all containers, trucks, and rail cars will enable emergency responders to react properly and quickly to any disaster thereby reducing the potential risk to the environment and to personnel.

If a spill or leak to the environment has occurred, fire department, emergency response and/or hazardous materials spill personnel should be notified immediately. Do not dry sweep. When vacuuming, a high efficiency particulate air (HEPA) filter should be used, not a standard shop vacuum. Do NOT use water to rinse residue. Deposit collected materials in sealed, moisture-free drums for disposal. Cleanup should be attempted only by those trained in proper spill containment procedures. Contaminated soils may need to be removed and replaced with clean soil. If magnesium chloride should contact the water table, aquifer, or navigable waterway, time is of the essence. Magnesium chloride is soluble in water and will decompose. Therefore, total containment and remediation may not be possible. When such spills occur, the local and/or state emergency response authorities should be notified. A comprehensive emergency response or disaster preparedness/recovery plan should be in place prior to any operations involving the use, transportation, storage, or disposal of magnesium chloride. If magnesium chloride is spilled or leaked, the following specific steps are recommended:

- ☑ Restrict persons not wearing protective clothing from area of spill or leak until cleanup is complete and area can be opened for normal work.
- ☑ Ventilate area of spill or leak and remove all ignition sources.
- ☑ Collect powdered materials in most convenient and safe manner possible and place in sealed drums for disposal. Do NOT dry sweep; use HEPA vacuum instead. Do NOT use water to clean residue or damp mop.
- ☑ It may be necessary to dispose of magnesium chloride as a hazardous waste. The responsible state agency or the regional office of the federal Environmental Protection Agency (EPA) should be contacted for specific recommendations.

RISK ASSESSMENT: BUSINESS

General Assessment

Accidents or mishaps involving magnesium chloride can present a significant threat to business operations. The loss or damage of equipment or facilities can significantly affect fiscal viability. Lawsuits that may result from personnel injury/death, public exposures, and/or environmental contamination will also require a serious expenditure of resources. Media attention surrounding an injury, death, or environmental damage can also result in a loss of profits and loss of current as well as future business.

ʊ *Recommended Risk-Reduction Measures*

Company attorneys, safety and health professionals, and environmental specialists should be involved in the development of such procedures dealing with safety or emergency response. A company official should be pre-designated as a public relations officer with specific training in dealing with the press. Corporate plans and policies should be developed, approved, and implemented long before any need for such arises.

REFERENCES

Ahlbom, A. 1993. *Biostatistics for Epidemiologists.* New York: CRC Press/Lewis Publishers.

American Conference of Governmental Industrial Hygienists. 1988. *Documentation of the Threshold Limit Values and Biological Exposure Indices,* 5th Edition (with updates). Cincinnati: ACGIH

Asante-Duah, D. K. 1993. *Hazardous Waste Risk Assessment.* New York: CRC Press/Lewis Publishers.

Bloom, A. D. and F. J. DeSerres. 1995. *Ecotoxicity and Human Health.* New York: CRC Press/Lewis Publishers.

Brusick, D. J. 1994. *Methods for Genetic Risk Assessment.* New York: CRC Press/Lewis Publishers.

Calabrese, E. J., E. M. Kenyon. *Air Toxics and Risk Assessment.* New York: CRC Press/Lewis Publishers.

Calabrese, E. J. 1994. *Biological Effects of Low Level Exposures.* New York: CRC Press/Lewis Publishers.

Cockerman, L. G., B. S. Shane. 1994. *Basic Environmental Toxicology.* New York: CRC Press/Lewis Publishers.

Cothern, C. R., N. P. Ross. 1994. *Environmental Statistics, Assessment, and Forecasting.* New York: CRC Press/Lewis Publishers.

Cothern, R. R. 1993. *Comparative Environmental Risk Assessment.* New York: CRC Press/Lewis Publishers.

Cothern, C. R. 1995. *Handbook for Environmental Risk Decisionmaking.* New York: CRC Press/Lewis Publishers.

Danse, I. R. 1991. *Common Sense Toxics in the Workplace.* New York: Van Nostrand Reinhold.

Environmental Protection Agency. 1983. *Chemical Hazard Information Profiles.* Washington, D.C.: U.S. EPA.

Fodden, C. R., and J. L. Weddell. 1992. *Hazardous Materials Emergency Action Data.* New York: CRC Press/Lewis Publishers.

Forseberg, K., and L. H. Keith. 1993. *Chemical; Protective Clothing Permeation and Degradation Database.* Chelsea, MI: Lewis Publishers.

Gardner, R. W. 1994. *Chemical Intolerance.* New York: CRC Press/Lewis Publishers.

Hallenbeck, W. H. 1993. *Quantitative Risk Assessment for Environmental and Occupational Health,* 2nd Edition. New York: CRC Press/Lewis Publishers.

Howard, P. H., et. al. 1991. *Handbook of Environmental Degradation Rates.* New York: CRC Press/Lewis Publishers.

Howard, P. H. 1995. *Handbook of Environmental Fate and Exposures Data for Organic Chemicals,* Vol. I-IV. New York: CRC Press/Lewis Publishers.

Jones, F. E. 1994. *Toxic Organic Vapors in the Workplace.* New York: CRC Press/Lewis Publishers.

Keith, L. H. and M. M. Walker. 1995. *Handbook of Air Toxics.* New York: CRC Press/Lewis Publishers.

Landis, W. G., M. H. Yu. 1995. *Introduction to Environmental Toxicology.* New York: CRC Press/Lewis Publishers.

Langerman, N. 1994 *Precautionary Labels for Chemical Containers.* New York: CRC Press/Lewis Publishers.

Lauwerys, R. R., P. Hoet. 1993. *Industrial Chemical Exposure.* New York: CRC Press/Lewis Publishers.

Lewis, R. J., Sr. 1992. *Sax's Dangerous Properties of Industrial Materials,* Volumes 1, 2, and 3, 8th Edition. New York: Van Nostrand Reinhold.

Manahan, S. E. 1994. *Environmental Chemistry,* 6th Edition. New York: CRC Press/Lewis Publishers.

Manahan, S. E. 1992. *Toxicological Chemistry,* 2nd Edition. New York: CRC Press/Lewis Publishers.

Meyer, E. 1990. *Chemistry of Hazardous Materials.* Englewood Cliffs, NJ: Prentice Hall.

Patnaik, P. 1992. *A Comprehensive Guide to the Hazardous Properties of Chemical Substances.* New York: Van Nostrand Reinhold.

Quigley, D. R. 1994. *Handbook of Emergency Chemical Management.* New York: CRC Press/Lewis Publishers.

Rea, W. J. 1995. *Chemical Sensitivity,* Vol. I-III. New York: CRC Press/Lewis Publishers.

Sacarello, H. L. A. 1994. *The Comprehensive Handbook of Hazardous Materials.* New York: CRC Press/Lewis Publishers.

Sherman, J. D. 1988. *Chemical Exposure and Disease.* New York: Van Nostrand Reinhold.

Talbott, E., G. Craun. 1995. *An Introduction to Environmental Epidemiology.* New York: CRC Press/Lewis Publishers.

MATERIAL SAFETY DATA SHEET

CHEMICAL NAME	
MAGNESIUM HYDROXIDE	**CRC PRESS/LEWIS PUBLISHERS DISCLAIMER:** THE INFORMATION AND RECOMMENDATIONS PRESENTED HEREIN ARE BASED ON SOURCES BELIEVED TO BE RELIABLE. CRC MAKES NO REPRESENTATION ON ITS COMPLETENESS OR ACCURACY. IT IS THE RESPONSIBILITY OF THE USER TO DETERMINE THE CHEMICAL'S SUITABILITY FOR ITS INTENDED USE, THE CHEMICAL'S SAFE USE, AND THE CHEMICAL'S PROPER DISPOSAL. NO REPRESENTATIONS AND/OR WARRANTIES, EITHER EXPRESSED OR IMPLIED, OF THE MERCHANTABILITY OR FITNESS FOR A PARTICULAR PURPOSE, OR OF ANY OTHER NATURE, ARE MADE WITH RESPECT TO THE INFORMATION PROVIDED IN THIS MSDS OR TO THE CHEMICAL TO WHICH INFORMATION MAY REFER. CRC NEITHER ASSUMES NOR AUTHORIZES ANY OTHER PERSON TO ASSUME FOR IT, ANY OTHER ADDITIONAL RESPONSIBILITY OR LIABILITY FOR THE USE OF, OR RELIANCE UPON, THIS INFORMATION.

HAZARD WARNING INFORMATION

HEALTH	FIRE	REACTIVE	OTHER	DEGREE OF HAZARD	COLOR CODING	OTHER CODES
1	0	0		0 = Minimum Hazard 1 = Slight Hazard 2 = Moderate Hazard 3 = Serious Hazard 4 = Severe Hazard	HEALTH = BLUE FIRE = RED REACTIVITY = YELLOW OTHER = WHITE	OX = Oxidizer ACID = Acid ALK = Alkali COR = Corrosive ~~W~~ = Use No Water

SECTION I - GENERAL INFORMATION

Characterization	**Hydroxide**	RCRA Number	**None**
		EPA Class	**Not Applicable**
DOT Proper Shipping Name	**No Citation**	Chemical Abstract Service (CAS) Number	**1309-42-8**
DOT Hazard Class and Label Requirements	**No Citation**	DOT Emergency Guide Code	**No Citation**
DOT Identification Number	**No Citation**	Atomic Symbol	**H_2MgO_2**

Synonyms

Magnesia magma; magnesium hydrate; Milk of Magnesia ®.

SECTION II - HAZARDOUS INGREDIENTS AND IDENTITY INFORMATION

Hazardous Components (specific identity)	OSHA Exposure Criteria	NIOSH Exposure Criteria	Immediately Dangerous to Life and Health (IDLH)	ACGIH Exposure Criteria
Magnesium hydroxide (derivation: By precipitation from a solution of a magnesium salt by sodium hydroxide; or by precipitation from sea water with lime. It occurs naturally as brucite).	**PEL:** Not Established **STEL:** Not Established	**REL:** Not Established **STEL:** Not Established	**Not Determined**	**TLV:** Not Established **STEL:** Not Established

SECTION III - PHYSICAL AND CHEMICAL CHARACTERISTICS

Boiling Point	**Not Found**	Specific Gravity (H_2O = 1)	**2.36**
Vapor Pressure (mm Hg)	**Not Found**	Molecular Weight (atomic weight)	**58.33**
Vapor Density (Air = 1)	**Not Found**	Melting Point	**662°F (350°C) Decomposes**

Solubility

Slightly soluble in water (0.01%) and alcohol. Soluble in solutions of ammonium salts and dilute acids.

Appearance and Odor

White, odorless powder.

SECTION IV - FIRE AND EXPLOSION HAZARD DATA

Flash Point (method used)	**Not Found**	Explosive Limits in Air % by Volume	**LEL: Not Found UEL: Not Found**
NFPA Classification	**Non-Combustible Solid**	Autoignition Temperature	**Not Found**

Extinguishing Media

Use extinguishing media suitable to surrounding fire. Use water spray to wet down material.

Special Fire Fighting Procedures

Wear full protective clothing and self-contained breathing apparatus (SCBA). Poisonous gases are produced in fire. Use water mist to keep fire-exposed containers cool. Remove containers from fire if it can be done without risk. Fight fire from distance, if possible.

Unusual Fire and Explosion Hazards

None reported.

SECTION V - REACTIVITY DATA

Stability		Conditions to Avoid: Magnesium hydroxide is normally stable under routine conditions of handling and storage. Avoid contact with incompatible materials.
Stable	Unstable	Incompatibility *(materials to avoid)*: Incompatible with maleic anhydride, sodium hydroxide, and phosphorus.
X		
Hazardous Polymerization		Conditions to Avoid: Hazardous polymerization of magnesium hydroxide will not occur.
May Occur	Will Not Occur	Hazardous Decomposition or By-products: None reported.
	X	

SECTION VI - HEALTH HAZARD DATA

Primary Route(s) of Entry:	Inhalation?	Absorption (skin/eye)?	Ingestion?
	X		X

Health Hazards

INHALATION: Toxicity is considered to be low, but excessive dust inhalation can cause irritation to the nose and mucous membranes of the eyes, nose, throat, and respiratory tract.

EYES & SKIN: Exposure to the eyes and skin may cause mild irritation.

INGESTION: Excessive doses may cause diarrhea, constipation, abdominal distention, obstruction of the intestine, and possible depression of the CNS with loss of reflexes, and progression to respiratory paralysis or cardiac arrest and death. Elevated serum magnesium levels may cause a drop in blood pressure, pulse, respiration, and mental status.

Carcinogenicity	NTP Listed?	IARC Cancer Review Group?	OSHA Regulated?	Target Organs?
Unknown Human Unknown Animal	No	No	No	Respiratory system, eyes, skin, CNS, kidneys.

Medical Conditions Generally Aggravated by Exposure

Existing kidney dysfunction may be aggravated by ingestion of magnesium salts.

Emergency and First-aid Procedures

Eye contact: Do not allow victim to keep eyes tightly shut. Flush immediately with water for 15 minutes (minimum); seek medical attention. Skin contact: Remove all contaminated clothing. Immediately wash area with flooding amounts of water. For reddened or blistered skin, seek medical assistance. For inhalation: Remove the person from exposure. Provide respiratory assistance and CPR. Transfer to medical facility. If swallowed: Seek medical attention. Give 1 to 2 glasses of water and induce vomiting. Never give anything by mouth to an unconscious or convulsing person.

SECTION VII - PRECAUTIONS FOR SAFE HANDLING AND USE

Steps to be Taken in Case Material is Released or Spilled

Collect powdered materials in most convenient and safe manner and deposit in sealed containers. Dry sweep not recommended. Use HEPA vacuum. Ventilate area of spill. Restrict those not involved in cleanup from entering area. Notify appropriate authorities.

Preferred Waste Disposal Method

No citation.

Precautions to be Taken in Handling and Storage

Store to avoid contact with incompatible materials. Prevent physical damage to containers. Keep containers tightly closed.

Other Precautions and Warnings

Avoid excessive dust generation in storage and handling areas. Personnel should be trained on the proper precautions for working with magnesium compounds *before* initial assignment.

SECTION VIII - CONTROL MEASURES AND PERSONAL PROTECTIVE EQUIPMENT

Respiratory Protection *(specify type)*

No exposure levels established: Use a respirator with a high-efficiency particulate air (HEPA) filter; otherwise, use a self-contained breathing apparatus with full facepiece operated in positive pressure mode.

Ventilation

Local exhaust (preferred) at site of chemical work or general ventilation.

Protective Gloves	Eye Protection	Other Protective Clothing
Impervious Material	Chemical/Dust Goggles or Face Mask	Protective Apron, Boots

Work/Hygiene Practices

Always wash hands thoroughly after using chemical; never bring food, drink, or smoking materials into vicinity of chemicals.

MAGNESIUM HYDROXIDE

H_2MgO_2

CAS: 1309-42-8

IDENTIFICATION AND TYPICAL USES

Magnesium hydroxide appears as a white odorless powder. It is an intermediate for obtaining magnesium metal, in the refining of sugar, in medicine (antacids, laxatives), as a residual fuel oil additive, in sulfite pulp, in uranium processing, in the making of dentifrice, in foods (as a drying agent and for color retention) and in frozen desserts.

RISK ASSESSMENT: HEALTH

General Assessment

Magnesium hydroxide exhibits relatively low toxicity in humans by *inhalation* and *ingestion*. It is generally considered a nuisance particulate. However, exposure to sufficient quantities, especially by ingestion, can pose a serious health risk. There is no information on its carcinogenic, mutagenic, or teratogenic capabilities.

Inhalation causes irritation to the eyes, nose, throat, and mucosa of the upper respiratory tract with cough and congestion. If ingested, magnesium salts will be slowly absorbed. However, large amounts can cause systemic poisoning. Elevated blood magnesium levels may cause a drop in blood pressure, pulse, and respiration. Other symptoms include diarrhea, constipation, abdominal distention, intestinal blockage, possible kidney effects, and depression of the central nervous system (CNS). This can lead to failed reflexes with progression to respiratory failure or cardiac arrest and death.

Acute Health Effects

The following acute (short-term) health effects may occur immediately or shortly after human exposure to magnesium hydroxide:

Skin: May cause mild irritation, depending on the conditions and length of exposure.

Eye: Mild to moderate irritation.

Lung: Irritation of the mucosa of the eyes, nose, throat, and respiratory tract.

Other: Ingestion of excessive quantities of magnesium hydroxide can cause gastrointestinal effects, kidney problems, and depression of the central nervous system that can lead to death.

Chronic Health Effects

The following chronic (long-term) health effects may occur at some time after exposure to magnesium hydroxide and can last for months or even years:

Cancer Hazards: According to information presented in the references, magnesium hydroxide has not been shown to cause cancer in animals.

Reproductive Hazard: According to information presented in the references, there are no reports of any teratogenic or reproductive effects resulting from exposure to magnesium hydroxide.

Other Chronic Effects: Repeated or prolonged contact may cause chronic gastrointestinal problems and may lead to kidney dysfunction.

Recommended Risk-Reduction Measures

Even though the toxicity of magnesium hydroxide is considered to be relatively low, personnel should avoid direct contact with magnesium hydroxide. If a less toxic material or compound cannot be substituted for magnesium hydroxide, then *engineering controls* are the most effective method of reducing exposures. The best protection is to enclose operations and/or provide local exhaust ventilation at the site of magnesium hydroxide dust release. While not always operationally feasible, isolating operations can also reduce exposure. Using respiratory protection is less effective than the controls mentioned above, but is still advisable whenever working with or around magnesium hydroxide. A supplied-air respirator operated in continuous flow mode or a self-contained breathing apparatus (SCBA) with full facepiece operated in pressure demand are recommended for the greatest possible respiratory protection. However, for low or infrequent exposure, an MSHA/NIOSH-approved full facepiece respirator with a high efficiency particulate air (HEPA) filter may be used. If a full facepiece is not available, then chemical/dust goggles should be worn to protect the eyes. A face shield and protective apron should also be worn. To prevent hand and skin exposures, gloves should be used.

Administrative controls should also be in place to minimize the potential for human exposures. These may include written procedures or policies which specify the methods and techniques that will be practiced whenever personnel are to work with magnesium hydroxide.

All personnel should receive *training* on the use, hazards, protective measures, emergency actions, and other precautions per 29 CFR 1910.1200 (Hazard Communication), prior to the first assignment in an area where magnesium hydroxide is used or stored.

Before beginning employment working with magnesium hydroxide, and at regular intervals thereafter (e.g., annually), those personnel with frequent or potentially high exposures should be provided the following recommended medical tests:

☑ Kidney function tests (establish baseline)

If symptoms develop or overexposure is suspected, consider the following additional tests:

☑ Kidney function tests (compare to baseline).

☑ Magnesium serum levels.

☑ Lung function tests.

Any evaluation should include a careful history of past and present symptoms with an examination. Medical tests that simply look for existing damage are not a substitute for controlling exposures. Also, since smoking can cause heart disease as well as lung cancer, emphysema, and other known respiratory problems, smokers exposure to magnesium hydroxide may experience symptoms more rapidly than non-smokers under the same conditions of exposure. Prudent risk management requires careful consideration of *all* factors which may be causing exposure symptoms in the workplace.

Other methods to reduce exposure include:

☑ Where possible, enclose operations and use local exhaust ventilation at the site of chemical release. If local exhaust ventilation or enclosure is not used, respiratory protection should be mandatory.

☑ Always ensure that proper protective clothing is worn when using chemical substances. Proper maintenance and care of protective equipment is essential to its proper function.

☑ Wash thoroughly immediately after exposure to magnesium hydroxide and at the end of the work shift or before eating, drinking, or smoking.

☑ Hazard warning information should be posted in the work area. In addition, as part of an on-going education and training program, all information on the health and safety hazards of magnesium hydroxide should be communicated to all exposed and potentially exposed workers.

☑ Eye wash stations should be provided in the immediate work area for emergency use. If there is a possibility of skin exposure to magnesium hydroxide, emergency shower facilities should be provided in the immediate area.

☑ Workers whose clothing has been contaminated by magnesium hydroxide should change into clean clothes before leaving work. Contaminated work clothing should be laundered only by individuals who have been informed of the hazards of exposure to magnesium hydroxide.

RISK ASSESSMENT: ENVIRONMENT

General Assessment

As a naturally occurring mineral, magnesium poses very little threat to the environment. However, accidental spills or releases of magnesium hydroxide to the environment can occur during transportation, storage, disposal, or destruction of magnesium hydroxide. In almost every scenario, the threat of environmental exposure is contingent upon the proper handling of the chemical substance. Accidental spills, large or small, where magnesium hydroxide contacts incompatible commodities can result in release of toxic products (depending upon conditions of spill), and possible contamination of the surrounding environmental mediums (water, soil, and air).

Magnesium hydroxide is considered a noncombustible solid. It is incompatible with maleic anhydride, sodium hydroxide, and phosphorus, and extreme caution is required in handling, storage, transportation, and disposal of magnesium hydroxide. These characteristics also require special consideration during any emergency situation involving a leak or spill of magnesium hydroxide powder or dust. Should magnesium hydroxide ever come into contact with incompatible substances either during use, transportation, storage, or disposal, violent and even explosive reactions are possible.

Magnesium hydroxide may enter the environment as a result of industrial and municipal waste treatment plant discharges and from spills.

☠ *Acute Ecological Effects*

Acute (short-term) toxic effects may include the death of animals, birds, or fish and death or low growth rate in plants. Acute effects are seen 2 to 4 days after animals or plants are exposed to magnesium hydroxide. Magnesium and its salts have slight acute toxicity to aquatic life. Insufficient data are available to evaluate the short-term effects of magnesium hydroxide to plants, birds, or land animals.

Chronic Ecological Effects

Chronic (long-term) toxic effects may include shortened life span, reproductive problems, lower fertility, and changes in appearance or behavior in exposed animals. These effects can be seen long after first exposure(s) to toxic chemicals. Magnesium and its salts have slight chronic toxicity to aquatic life. Insufficient data are available to evaluate or predict the long-term effects of magnesium hydroxide to plants, birds, or land animals.

Water Solubility

Magnesium hydroxide is nearly insoluble (0.01%) in water. Concentrations of less than 1 milligram may mix with a liter of water.

⌛ *Persistence in the Environment*

Magnesium is highly persistent in water, with a half-life greater than 200 days. The half-life of a pollutant is the amount of time it takes for one half of the chemical to be degraded. Magnesium is also naturally present in soils, with its three hydrates (barrington, nesquehonite, and landsforite).

Bioaccumulation in Aquatic Organisms

Some substances increase in concentration, or *bioaccumulate,* in living organisms as they breathe contaminated air, drink contaminated water, or eat contaminated food. These chemicals can become concentrated in the tissues and internal organs of animals as well as humans.

The concentration of magnesium hydroxide found in fish tissues is expected to be about the same as the average concentration of magnesium hydroxide in the water from which the fish was taken.

Recommended Risk-Reduction Measures

Proper training of all transporters will reduce the likelihood of a mishap or accident resulting in a leak or spill of magnesium hydroxide dusts or powders into the environment. The correct use of labeling on all containers, trucks, and rail cars will enable emergency responders to react properly and quickly to any disaster thereby reducing the potential risk to the environment and to personnel.

Storage of magnesium hydroxide should be segregated from incompatible chemicals to minimize the risk of cross-contamination or contact. Care should be taken to prevent physical damage to storage containers. Also, dusty conditions must be avoided in storage and use areas.

If a spill or leak to the environment has occurred, fire department, emergency response and/or hazardous materials spill personnel should be notified immediately. Cleanup should be attempted only by those trained in proper spill containment procedures. Contaminated soils should be removed for decontamination and replaced with clean soil. If magnesium hydroxide should contact the water table, aquifer, or navigable waterway, time is of the essence. It is slightly soluble and total remediation may not be possible. The local and/or state emergency response authorities may require notification. A comprehensive emergency response or disaster preparedness/recovery plan should be in place prior to any operations involving the use, transportation, storage, or disposal of magnesium hydroxide.

If magnesium hydroxide is spilled or leaked, the following specific steps are recommended:

- ☑ Restrict persons not wearing protective clothing from area of spill or leak until cleanup is complete and area can be opened for normal work.
- ☑ Remove all ignition sources from area.
- ☑ Ventilate area of spill or leak.
- ☑ Collect powdered materials in the safest manner possible and deposit in sealed containers. Recommend using a vacuum equipped with a high-

efficiency particulate air (HEPA) filter. Do not dry sweep (generates airborne dusts). Damp mop residues and collect for disposal. Water spray can be used to knock down vapors. Runoff should also be collected for disposal.

☑ It may be necessary to dispose of magnesium hydroxide as a hazardous waste. The responsible state agency or the regional office of the federal Environmental Protection Agency (EPA) should be contacted for specific recommendations.

RISK ASSESSMENT: BUSINESS

General Assessment

Accidents or mishaps involving magnesium hydroxide can present a significant threat to business operations. The loss or damage of equipment or facilities can significantly affect fiscal viability. Lawsuits that may result from personnel injury/death, public exposures, and/or environmental contamination will also require a serious expenditure of resources. Media attention surrounding an injury, death, or environmental damage can also result in a loss of profits and loss of current as well as future business.

Recommended Risk-Reduction Measures

Company attorneys, safety and health professionals, and environmental specialists should be involved in the development of any procedures for responding to chemical incidents. A company official should be predesignated as a public relations officer with specific training in dealing with the press. Corporate plans and policies should be developed, approved, and implemented long before any need for such arises.

REFERENCES

Ahlbom, A. 1993. *Biostatistics for Epidemiologists.* New York: CRC Press/Lewis Publishers.

American Conference of Governmental Industrial Hygienists. 1988. *Documentation of the Threshold Limit Values and Biological Exposure Indices,* 5th Edition (with updates). Cincinnati: ACGIH

Anthony, C. P., and N. J. Kolthoff. 1971. *Textbook of Anatomy and Physiology.* St. Louis: C. V. Mosby Company.

Asante-Duah, D. K. 1993. *Hazardous Waste Risk Assessment.* New York: CRC Press/Lewis Publishers.

Bloom, A. D. and F. J. DeSerres. 1995. *Ecotoxicity and Human Health.* New York: CRC Press/Lewis Publishers.

Brusick, D. J. 1994. *Methods for Genetic Risk Assessment.* New York: CRC Press/Lewis Publishers.

Bueche, J. R. 1972. *Principles of Physics,* 2nd Edition. New York: McGraw-Hill.

Calabrese, E. J. 1994. *Biological Effects of Low Level Exposures.* New York: CRC Press/Lewis Publishers.

Cockerman, L. G., B. S. Shane. 1994. *Basic Environmental Toxicology.* New York: CRC Press/Lewis Publishers.

Cothern, R. R. 1993. *Comparative Environmental Risk Assessment.* New York: CRC Press/Lewis Publishers.

Danse, I. R. 1991. *Common Sense Toxics in the Workplace.* New York: Van Nostrand Reinhold.

Environmental Protection Agency. 1983. *Chemical Hazard Information Profiles.* Washington, D.C.: U.S. EPA.

Forseberg, K., and L. H. Keith. 1993. *Chemical; Protective Clothing Permeation and Degradation Database.* Chelsea, MI: Lewis Publishers.

Gardner, R. W. 1994. *Chemical Intolerance.* New York: CRC Press/Lewis Publishers.

Howard, P. H. 1995. *Handbook of Environmental Fate and Exposures Data for Organic Chemicals,* Vol. I-IV. New York: CRC Press/Lewis Publishers.

Lauwerys, R. R., P. Hoet. 1993. *Industrial Chemical Exposure.* New York: CRC Press/Lewis Publishers.

Lewis, R. J., Sr. 1993. *Hawley's Condensed Chemical Dictionary,* 12th Edition. New York: Van Nostrand Reinhold.

Lewis, R. J., Sr. 1992. *Sax's Dangerous Properties of Industrial Materials,* Volumes 1, 2, and 3, 8th Edition. New York: Van Nostrand Reinhold.

Mackay, D., W. Y. Shiu, and K. C. Ma. 1992. *Illustrated Handbook of Chemical-Physical Properties and Environmental Fate for Organic Chemicals.* New York: CRC Press/Lewis Publishers.

Manahan, S. E. 1994. *Environmental Chemistry,* 6th Edition. New York: CRC Press/Lewis Publishers.

Meyer, E. 1990. *Chemistry of Hazardous Materials.* Englewood Cliffs, NJ: Prentice Hall.

Miller, B. F., M.D., and C. B. Keane, R.N. 1972. *Encyclopedia and Dictionary of Medicine and Nursing.* Philadelphia: W.B. Sanders Company.

Mycock, J. C., J. D. McKenna, and L. Theodore. 1995. *Handbook of Air Pollution Control Engineering and Technology.* New York: CRC Press/Lewis Publishers.

Quigley, D. R. 1994. *Handbook of Emergency Chemical Management.* New York: CRC Press/Lewis Publishers.

Richardson, M. 1995. *Dictionary of Substance and Their Effects,* Vol. I-VI. New York: CRC Press/Lewis Publishers.

Sacarello, H. L. A. 1994. *The Comprehensive Handbook of Hazardous Materials.* New York: CRC Press/Lewis Publishers.

Talbott, E., G. Craun. 1995. *An Introduction to Environmental Epidemiology.* New York: CRC Press/Lewis Publishers.

Torkelson, T. R. V. K. Rowe. 1981. *Patty's Industrial Hygiene and Toxicology,* Vol. 2B. New York: Wiley-Interscience.

MATERIAL SAFETY DATA SHEET

CHEMICAL NAME

MAGNESIUM OXIDE

CRC PRESS/LEWIS PUBLISHERS DISCLAIMER: THE INFORMATION AND RECOMMENDATIONS PRESENTED HEREIN ARE BASED ON SOURCES BELIEVED TO BE RELIABLE. CRC MAKES NO REPRESENTATION ON ITS COMPLETENESS OR ACCURACY. IT IS THE RESPONSIBILITY OF THE USER TO DETERMINE THE CHEMICAL'S SUITABILITY FOR ITS INTENDED USE, THE CHEMICAL'S SAFE USE, AND THE CHEMICAL'S PROPER DISPOSAL. NO REPRESENTATIONS AND/OR WARRANTIES, EITHER EXPRESSED OR IMPLIED, OF THE MERCHANTABILITY OR FITNESS FOR A PARTICULAR PURPOSE, OR OF ANY OTHER NATURE, ARE MADE WITH RESPECT TO THE INFORMATION PROVIDED IN THIS MSDS OR TO THE CHEMICAL TO WHICH INFORMATION MAY REFER. CRC NEITHER ASSUMES NOR AUTHORIZES ANY OTHER PERSON TO ASSUME FOR IT, ANY OTHER ADDITIONAL RESPONSIBILITY OR LIABILITY FOR THE USE OF, OR RELIANCE UPON, THIS INFORMATION.

HAZARD WARNING INFORMATION

HEALTH	FIRE	REACTIVE	OTHER	DEGREE OF HAZARD	COLOR CODING	OTHER CODES
2	0	0		0 = Minimum Hazard 1 = Slight Hazard 2 = Moderate Hazard 3 = Serious Hazard 4 = Severe Hazard	HEALTH = BLUE FIRE = RED REACTIVITY = YELLOW OTHER = WHITE	OX = Oxidizer ACID = Acid ALK = Alkali COR = Corrosive W = Use No Water

SECTION I - GENERAL INFORMATION

Characterization **Metallic Carbonate**	RCRA Number **None**	EPA Class **Not Applicable**
DOT Proper Shipping Name **No Citation**	Chemical Abstract Service (CAS) Number **1309-48-4**	
DOT Hazard Class and Label Requirements **No Citation**	DOT Emergency Guide Code **No Citation**	
DOT Identification Number **No Citation**	Atomic Symbol **MgO**	

Synonyms

Calcined brucite; calcined magnesium; magnesia; calcined magnesia; calcined magnesite; sea water magnesia.

SECTION II - HAZARDOUS INGREDIENTS AND IDENTITY INFORMATION

Hazardous Components (specific identity)	OSHA Exposure Criteria	NIOSH Exposure Criteria	Immediately Dangerous to Life and Health (IDLH)	ACGIH Exposure Criteria
Magnesium oxide (derivation: By calcining magnesium carbonate or magnesium hydroxide; by treating magnesium chloride with lime and heating or by heating it in air; or from sea water via the hydroxide. It can also occur in nature as the mineral periclase).	**PEL (8-hour): 15 mg/m^3 (for oxide, fume, or total particulate)** **5 mg/m^3 (respirable fraction)**	**REL (10-hour): 10 mg/m^3 (for oxide, fume, or total particulate)** **5 mg(Mg)/m^3 (respirable fraction)**	**750 mg/m^3**	**TLV: 10 mg/m^3 (for oxide, fume, or total particulate)**

SECTION III - PHYSICAL AND CHEMICAL CHARACTERISTICS

Boiling Point **6512°F (3600°C)**	Specific Gravity (H_2O = 1) **3.58**
Vapor Pressure (mm Hg) **0 (approx.) at 68°F (20°C)**	Molecular Weight (atomic weight) **40.3**
Vapor Density (Air = 1) **Not Found**	Melting Point **5072°F (2800°C)**

Solubility

Very slightly soluble in water (0.009% at 86°F). Soluble in dilute acids. Insoluble in alcohols.

Appearance and Odor

Finely divided, white particulate powder with no odor. Appears as a fume in air when heated.

SECTION IV - FIRE AND EXPLOSION HAZARD DATA

Flash Point (method used) **Not Found**	Explosive Limits in Air % by Volume **LEL: Not Found** **UEL: Not Found**
NFPA Classification **Non-Combustible Solid**	Autoignition Temperature **Not Found**

Extinguishing Media

Use extinguishing media suitable to surrounding fire.

Special Fire Fighting Procedures

Wear full protective clothing and self-contained breathing apparatus (SCBA). Poisonous gases are produced in fire. Use water mist to keep fire-exposed containers cool. Remove containers from fire if it can be done without risk. Fight fire from distance, if possible.

Unusual Fire and Explosion Hazards

Avoid use of water or carbon dioxide where exposed magnesium oxide is present. It will combine with water to form magnesium hydroxide.

SECTION V - REACTIVITY DATA

Stability		Conditions to Avoid
		Magnesium oxide is normally stable under routine conditions of handling and storage. Avoid contact with incompatible materials. It can absorb carbon dioxide and water from the air.
Stable	Unstable	Incompatibility *(materials to avoid)*
X		Incompatible and will ignite on contact with interhalogens such as chlorine trifluoride and bromine pentafluoride. It will react incandescently with phosphorus pentachloride.
Hazardous Polymerization		Conditions to Avoid
		Under normal conditions of temperature and pressure, hazardous polymerization of magnesium oxide cannot occur.
May Occur	Will Not Occur	Hazardous Decomposition or By-products
	X	Poisonous gases are emitted during fire, including toxic and irritating smoke and fumes.

SECTION VI - HEALTH HAZARD DATA

Primary Route(s) of Entry:	Inhalation?	Absorption (skin/eye)?	Ingestion?
	X		

Health Hazards

INHALATION: The dusts can produce irritation of the eyes and mucous membranes. It may react with the water content in the bronchial passages to form caustic magnesium hydroxide which can damage the lungs. Inhalation of the fumes can cause metal fume fever with symptoms of cough, pressure in chest, leukocytosis, chills, and other febrile (feverish) conditions.

EYES & SKIN: Causes skin sores that resist healing and accumulate gas under the skin (small blisters). The powder metal ignites easily on moist skin and eyes causing serious burns.

INGESTION: Large doses can cause gastrointestinal irritation and other unspecified effects.

Carcinogenicity	NTP Listed?	IARC Cancer Review Group?	OSHA Regulated?	Target Organs?
Unknown Human Unknown Animal	No	No	No	Respiratory system, eyes, skin, CNS, kidneys.

Medical Conditions Generally Aggravated by Exposure

Existing kidney dysfunction may be aggravated by ingestion of magnesium salts.

Emergency and First-aid Procedures

Eye contact: Do not allow victim to keep eyes tightly shut. Flush immediately with water for 15 minutes (minimum); seek medical attention. Skin contact: Remove all contaminated clothing. Immediately wash area with flooding amounts of water. For reddened or blistered skin, seek medical assistance. For inhalation: Remove the person from exposure. Provide respiratory assistance and CPR. Transfer to medical facility. If swallowed: Seek medical attention. Give 1 to 2 glasses of water to dilute. Never give anything by mouth to an unconscious or convulsing person.

SECTION VII - PRECAUTIONS FOR SAFE HANDLING AND USE

Steps to be Taken in Case Material is Released or Spilled

Collect powdered materials in most convenient and safe manner and deposit in sealed containers. Dry sweep not recommended. Use HEPA vacuum. Ventilate area of spill. Flush residue with flooding amounts of water and collect. Restrict those not involved in cleanup from entering area.

Preferred Waste Disposal Method

No citation.

Precautions to be Taken in Handling and Storage

Store to avoid contact with incompatible materials. Prevent physical damage to containers. Keep containers tightly closed (air-tight) to prevent contact with moisture.

Other Precautions and Warnings

Avoid excessive dust generation in storage and handling areas. Personnel should be trained on the proper precautions for working with magnesium compounds *before* initial assignment.

SECTION VIII - CONTROL MEASURES AND PERSONAL PROTECTIVE EQUIPMENT

Respiratory Protection *(specify type)*

For dust exposures over PEL, use a respirator with a high-efficiency particulate air (HEPA) filter; otherwise, use a self-contained breathing apparatus with full facepiece operated in positive pressure mode.

Ventilation

Local exhaust (preferred) at site of chemical work or general ventilation.

Protective Gloves	Eye Protection	Other Protective Clothing
Impervious Material	Chemical/Dust Goggles or Face Mask	Protective Apron, Boots

Work/Hygiene Practices

Always wash hands thoroughly after using chemical; never bring food, drink, or smoking materials into vicinity of chemicals.

MAGNESIUM OXIDE

MgO

CAS: 1309-48-4

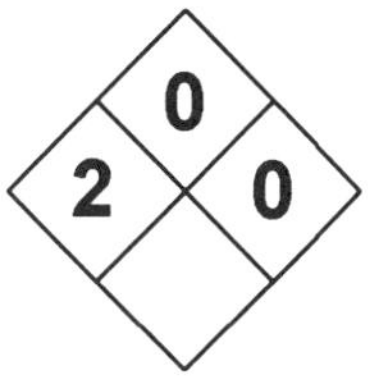

IDENTIFICATION AND TYPICAL USES

Magnesium oxide appears as a finely divided white powder with no odor that can be dispersed in air as a fume when heated. It is used in the manufacture of steel linings, in the production of polycrystalline ceramic for aircraft windshields, in electrical insulation, in pharmaceuticals and cosmetics, as an accelerator in the making of inorganic rubber materials, in oxychloride and oxysulfate cements, in paper, in fertilizers, for removing sulfur dioxide from stack gases, in semiconductors, and as a food and feed additive.

RISK ASSESSMENT: HEALTH

General Assessment

Magnesium oxide can enter the body and cause varying degrees of adverse health effects through *inhalation* and *ingestion*. There is no information on its teratogenic or mutagenic capabilities. While no agencies have listed magnesium oxide as a carcinogen, there is limited evidence that it has caused some development of tumors in test animals.

Inhalation of the dusts and particles of magnesium oxide can irritate the eyes and nasal passages. Magnesium oxide may react with the water content in the bronchial passages to form caustic magnesium hydroxide which can cause severe burns and tissue damage. Magnesium fumes in the oxide form can also irritate the eyes, nose, throat, and respiratory tract and may cause metal fume fever with cough, muscle pain, tightness and pressure in the chest, nausea, vomiting, chills, and other febrile (feverish) symptoms. These symptoms can begin 4 to 6 hours after exposure and persist for 24 hours. Leukocytosis (increase of leukocytes in the blood) may also develop.

Magnesium particles can become imbedded in the skin causing the development of slow healing sores and gaseous blebs (small blisters). If the skin is moist, burns may be serious and destructive.

Acute Health Effects

The following acute (short-term) health effects may occur immediately or shortly after human exposure to magnesium oxide:

Skin: May cause serious irritation and even burns, especially if the skin is moist. Small blisters can develop that will heal slowly and may become infected.

Eye: Severe irritation and possible burns due to the eyes' moisture content. Damage can be serious and permanent.

Lung: Irritation of the mucosa of the eyes, nose, throat, and respiratory tract.

CNS: Breathing heated fumes can cause metal fume fever with specific action on the central nervous system.

Chronic Health Effects

The following chronic (long-term) health effects may occur at some time after exposure to magnesium oxide and can last for months or even years:

Cancer Hazards: According to the references, magnesium oxide has been shown to cause tumors in experimental test animals. Its human carcinogenicity is not known.

Reproductive Hazard: According to the references, there are no reports of any teratogenic or reproductive effects resulting from exposure to magnesium oxide.

Other Chronic Effects: None reported.

Recommended Risk-Reduction Measures

Even though the toxicity of magnesium oxide is considered to be relatively low, personnel should avoid direct contact with magnesium oxide. If a less toxic material or compound cannot be substituted for magnesium oxide, then *engineering controls* are the most effective method of reducing exposures. The best protection is to enclose operations and/or provide local exhaust ventilation at the site of magnesium oxide dust release. While not always operationally feasible, isolating operations can also reduce exposure. Using respiratory protection is less effective than the controls mentioned above, but is still advisable whenever working with or around magnesium oxide. A supplied-air respirator operated in continuous flow mode or a self-contained breathing apparatus (SCBA) with full facepiece operated in pressure demand are recommended for the best possible respiratory protection. However, for low or infrequent exposure, an MSHA/NIOSH-approved full facepiece respirator with a high-efficiency particulate air (HEPA) filter may be used. If a full facepiece is not available, then chemical/dust goggles should be worn to protect the eyes. A face shield and protective apron should also be used. To prevent hand and skin exposures, gloves should be used.

Administrative controls should also be in place to minimize the potential for human exposures. These may include written procedures or policies which specify the methods and techniques that will be practiced whenever personnel are to work with magnesium oxide.

All personnel should receive *training* on the use, hazards, protective measures, emergency actions, and other precautions per 29 CFR 1910.1200 (Hazard Communication), prior to the first assignment in an area where magnesium oxide is used or stored.

Before beginning employment working with magnesium oxide, and at regular intervals thereafter (e.g., annually), those personnel with frequent or potentially high exposures should be provided the following recommended medical tests:

- ☑ Kidney function tests (establish baseline).

If symptoms develop or overexposure is suspected, consider the following additional tests:

- ☑ Kidney function tests (compare to baseline).
- ☑ Magnesium serum levels.
- ☑ Lung function tests.

Any evaluation should include a careful history of past and present symptoms with an examination. Medical tests that simply look for existing damage are not a substitute for controlling exposures. Also, since smoking can cause heart disease as well as lung cancer, emphysema, and other known respiratory problems, smokers exposure to magnesium oxide may experience symptoms more rapidly than non-smokers under the same conditions of exposure. Prudent risk management requires careful consideration of *all* factors which may be causing exposure symptoms in the workplace.

Other methods to reduce exposure include:

- ☑ Where possible, enclose operations and use local exhaust ventilation at the site of chemical release. If local exhaust ventilation or enclosure is not used, respiratory protection should be mandatory.
- ☑ Always ensure that proper protective clothing is worn when using chemical substances. Proper maintenance and care of protective equipment is essential to its proper function.
- ☑ Wash thoroughly immediately after exposure to magnesium oxide and at the end of the work shift or before eating, drinking, or smoking.
- ☑ Hazard warning information should be posted in the work area. In addition, as part of an on-going education and training program, all information on the health and safety hazards of magnesium oxide should be communicated to all exposed and potentially exposed workers.
- ☑ Eye wash stations should be provided in the immediate work area for emergency use. If there is a possibility of skin exposure to magnesium oxide, emergency shower facilities should be provided.
- ☑ Workers whose clothing has been contaminated by magnesium oxide should change into clean clothes before leaving work. Contaminated work clothing should be laundered only by individuals who have been informed of the hazards of exposure to magnesium oxide.

RISK ASSESSMENT: ENVIRONMENT

General Assessment

As the naturally occurring mineral periclase, magnesium oxide poses very little threat to the environment. However, accidental spills or releases to the environment can occur during transportation, storage, disposal, or destruction of magnesium oxide. In almost every scenario, the threat of environmental exposure is contingent upon the proper handling of the chemical substance. Accidental spills, large or small, where magnesium oxide contacts incompatible commodities can result in release of toxic products (depending upon conditions of spill), and possible contamination of the surrounding environmental mediums (water, soil, and air).

Magnesium oxide is considered a non-combustible solid. However, finely divided dusts and powders may create an explosive mixture in air in closed spaces. It is incompatible and reacts violently with many interhalogen compounds, including chlorine trifluoride and bromine trifluoride, and extreme caution is required in handling, storage, transportation, and disposal of magnesium oxide. These characteristics also require special consideration during any emergency situation involving a leak or spill of magnesium oxide powder or dust. Should magnesium oxide ever come into contact with incompatible substances either during use, transportation, storage, or disposal, violent and even explosive reactions are possible.

Magnesium oxide may enter the environment as a result of industrial and municipal waste treatment plant discharges and from spills.

Acute Ecological Effects

Acute (short-term) toxic effects may include the death of animals, birds, or fish and death or low growth rate in plants. Acute effects are seen 2 to 4 days after animals or plants are exposed to magnesium oxide. Magnesium and its salts have slight acute toxicity to aquatic life. Insufficient data are available to evaluate the short-term effects of magnesium oxide to plants, birds, or land animals.

Chronic Ecological Effects

Chronic (long-term) toxic effects may include shortened life span, reproductive problems, lower fertility, and changes in appearance or behavior in exposed animals. These effects can be seen long after first exposure(s) to toxic chemicals. Magnesium and its salts have slight chronic toxicity to aquatic life. Insufficient data are available to evaluate or predict the long-term effects of magnesium oxide to plants, birds, or land animals.

Water Solubility

Magnesium oxide is nearly insoluble (0.009% at 86°F) in water. Concentrations of less than 1 milligram may mix with a liter of water. It will react in water to evolve magnesium hydroxide, a caustic.

Persistence in the Environment

Magnesium oxide is highly persistent in water, with a half-life greater than 200 days. The half-life of a pollutant is the amount of time it takes for one half of the chemical to be degraded. Magnesium oxide is also naturally present in terrestrial soils and aquatic sediments as the mineral periclase.

Bioaccumulation in Aquatic Organisms

Some substances increase in concentration, or *bioaccumulate,* in living organisms as they breathe contaminated air, drink contaminated water, or eat contaminated food. These chemicals can become concentrated in the tissues and internal organs of animals as well as humans.

The concentration of magnesium found in fish tissues is expected to be about the same as the average concentration of magnesium in the water from which the fish was taken.

Recommended Risk-Reduction Measures

Proper training of all transporters will reduce the likelihood of a mishap or accident resulting in a leak or spill of magnesium oxide dusts or powders into the environment. The correct use of labeling on all containers, trucks, and rail cars will enable emergency responders to react properly and quickly to any disaster thereby reducing the potential risk to the environment and to personnel.

Storage of magnesium oxide should be segregated from incompatible chemicals to minimize the risk of cross-contamination or contact. Care should be taken to prevent physical damage to storage containers. Also, dusty conditions must be avoided in storage and use areas. Water and moisture should be kept away.

If a spill or leak to the environment has occurred, fire department, emergency response and/or hazardous materials spill personnel should be notified immediately. Cleanup should be attempted only by those trained in proper spill containment procedures. Contaminated soils should be removed for decontamination and replaced with clean soil. If magnesium oxide should contact the water table, aquifer, or navigable waterway, time is of the essence. It is slightly soluble and total remediation may not be possible. The local and/or state emergency response authorities should be notified. A comprehensive emergency response or disaster preparedness/recovery plan should be in place prior to any operations involving the use, transportation, storage, or disposal of magnesium compounds.

If magnesium oxide is spilled or leaked, the following specific steps are recommended:

- ☑ Restrict persons not wearing protective clothing from area of spill or leak until cleanup is complete and area can be opened for normal work.
- ☑ Remove all ignition sources and ventilate area of spill or leak.

☑ Ventilate area of spill or leak.

☑ Collect powdered materials in the safest manner possible and deposit in sealed containers. Recommend using a vacuum equipped with a high-efficiency particulate air (HEPA) filter. Do not dry sweep (generates airborne dusts). Using flooding amounts of water to wash away residues. Runoff should also be collected for disposal.

☑ It may be necessary to dispose of magnesium oxide as a hazardous waste. The responsible state agency or the regional office of the federal Environmental Protection Agency (EPA) should be contacted for specific recommendations.

RISK ASSESSMENT: BUSINESS

General Assessment

Accidents or mishaps involving magnesium oxide can present a significant threat to business operations. The loss or damage of equipment or facilities can significantly affect fiscal viability. Lawsuits that may result from personnel injury/death, public exposures, and/or environmental contamination will also require a serious expenditure of resources. Media attention surrounding an injury, death, or environmental damage can also result in a loss of profits and loss of current as well as future business.

Ⓘ Recommended Risk-Reduction Measures

Company attorneys, safety and health professionals, and environmental specialists should be involved in the development of any procedures for responding to chemical incidents. A company official should be predesignated as a public relations officer with specific training in dealing with the press. Corporate plans and policies should be developed, approved, and implemented long before any need for such arises.

REFERENCES

Ahlbom, A. 1993. *Biostatistics for Epidemiologists.* New York: CRC Press/Lewis Publishers.

American Conference of Governmental Industrial Hygienists. 1988. *Documentation of the Threshold Limit Values and Biological Exposure Indices,* 5th Edition (with updates). Cincinnati: ACGIH

Anthony, C. P., and N. J. Kolthoff. 1971. *Textbook of Anatomy and Physiology.* St. Louis: C. V. Mosby Company.

Asante-Duah, D. K. 1993. *Hazardous Waste Risk Assessment.* New York: CRC Press/Lewis Publishers.

Bloom, A. D. and F. J. DeSerres. 1995. *Ecotoxicity and Human Health.* New York: CRC Press/Lewis Publishers.

Brusick, D. J. 1994. *Methods for Genetic Risk Assessment.* New York: CRC Press/Lewis Publishers.

Bueche, J. R. 1972. *Principles of Physics,* 2nd Edition. New York: McGraw-Hill.

Calabrese, E. J. 1994. *Biological Effects of Low Level Exposures.* New York: CRC Press/Lewis Publishers.

Cockerman, L. G., B. S. Shane. 1994. *Basic Environmental Toxicology.* New York: CRC Press/Lewis Publishers.

Cothern, R. R. 1993. *Comparative Environmental Risk Assessment.* New York: CRC Press/Lewis Publishers.

Danse, I. R. 1991. *Common Sense Toxics in the Workplace.* New York: Van Nostrand Reinhold.

Environmental Protection Agency. 1983. *Chemical Hazard Information Profiles.* Washington, D.C.: U.S. EPA.

Fodden, C. R., and J. L. Weddell. 1992. *Hazardous Materials Emergency Action Data.* New York: CRC Press/Lewis Publishers.

Forseberg, K., and L. H. Keith. 1993. *Chemical; Protective Clothing Permeation and Degradation Database.* Chelsea, MI: Lewis Publishers.

Gardner, R. W. 1994. *Chemical Intolerance.* New York: CRC Press/Lewis Publishers.

Howard, P. H. 1995. *Handbook of Environmental Fate and Exposures Data for Organic Chemicals,* Vol. I-IV. New York: CRC Press/Lewis Publishers.

Lauwerys, R. R., P. Hoet. 1993. *Industrial Chemical Exposure.* New York: CRC Press/Lewis Publishers.

Lewis, R. J., Sr. 1993. *Hawley's Condensed Chemical Dictionary,* 12th Edition. New York: Van Nostrand Reinhold.

Lewis, R. J., Sr. 1992. *Sax's Dangerous Properties of Industrial Materials,* Volumes 1, 2, and 3, 8th Edition. New York: Van Nostrand Reinhold.

Manahan, S. E. 1994. *Environmental Chemistry,* 6th Edition. New York: CRC Press/Lewis Publishers.

Meyer, E. 1990. *Chemistry of Hazardous Materials.* Englewood Cliffs, NJ: Prentice Hall.

Miller, B. F., M.D., and C. B. Keane, R.N. 1972. *Encyclopedia and Dictionary of Medicine and Nursing.* Philadelphia: W.B. Sanders Company.

Mycock, J. C., J. D. McKenna, and L. Theodore. 1995. *Handbook of Air Pollution Control Engineering and Technology.* New York: CRC Press/Lewis Publishers.

Richardson, M. 1995. *Dictionary of Substance and Their Effects,* Vol. I-VI. New York: CRC Press/Lewis Publishers.

Sacarello, H. L. A. 1994. *The Comprehensive Handbook of Hazardous Materials.* New York: CRC Press/Lewis Publishers.

Torkelson, T. R. V. K. Rowe. 1981. *Patty's Industrial Hygiene and Toxicology,* Vol. 2B. New York: Wiley-Interscience.

U.S. Department of Health and Human Services, National Institute for Occupational Safety and Health. 1994. *NIOSH Pocket Guide to Chemical Hazards.* Washington, D.C.: U.S. Government Printing Office.

MATERIAL SAFETY DATA SHEET

CHEMICAL NAME	
MAGNESIUM SULFATE	**CRC PRESS/LEWIS PUBLISHERS DISCLAIMER:** THE INFORMATION AND RECOMMENDATIONS PRESENTED HEREIN ARE BASED ON SOURCES BELIEVED TO BE RELIABLE. CRC MAKES NO REPRESENTATION ON ITS COMPLETENESS OR ACCURACY. IT IS THE RESPONSIBILITY OF THE USER TO DETERMINE THE CHEMICAL'S SUITABILITY FOR ITS INTENDED USE, THE CHEMICAL'S SAFE USE, AND THE CHEMICAL'S PROPER DISPOSAL. NO REPRESENTATIONS AND/OR WARRANTIES, EITHER EXPRESSED OR IMPLIED, OF THE MERCHANTABILITY OR FITNESS FOR A PARTICULAR PURPOSE, OR OF ANY OTHER NATURE, ARE MADE WITH RESPECT TO THE INFORMATION PROVIDED IN THIS MSDS OR TO THE CHEMICAL TO WHICH INFORMATION MAY REFER. CRC NEITHER ASSUMES NOR AUTHORIZES ANY OTHER PERSON TO ASSUME FOR IT, ANY OTHER ADDITIONAL RESPONSIBILITY OR LIABILITY FOR THE USE OF, OR RELIANCE UPON, THIS INFORMATION.

HAZARD WARNING INFORMATION

HEALTH	FIRE	REACTIVE	OTHER	DEGREE OF HAZARD	COLOR CODING	OTHER CODES
1	0	0		0 = Minimum Hazard 1 = Slight Hazard 2 = Moderate Hazard 3 = Serious Hazard 4 = Severe Hazard	HEALTH = BLUE FIRE = RED REACTIVITY = YELLOW OTHER = WHITE	OX = Oxidizer ACID = Acid ALK = Alkali COR = Corrosive W = Use No Water

SECTION I - GENERAL INFORMATION

Characterization	RCRA Number	EPA Class
Sulfate	**None**	**Not Applicable**

DOT Proper Shipping Name	Chemical Abstract Service (CAS) Number
No Citation	**7487-88-9**
DOT Hazard Class and Label Requirements	**DOT Emergency Guide Code**
No Citation	**No Citation**
DOT Identification Number	**Atomic Symbol**
No Citation	**$MgSO_4$**

Synonyms

Epsom salts; magnesium sulphate (anhydrous); bitter salts; sulfuric acid magnesium salt (1:1) heptahydrate; magnesium (II) sulfate heptahydrate.

SECTION II - HAZARDOUS INGREDIENTS AND IDENTITY INFORMATION

Hazardous Components (specific identity)	OSHA Exposure Criteria	NIOSH Exposure Criteria	Immediately Dangerous to Life and Health (IDLH)	ACGIH Exposure Criteria
Magnesium sulfate (derivation: By reacting sulfuric acid with magnesium oxide, magnesium hydroxide, or magnesium carbonate. The monohydrate occurs naturally as the mineral kieserite and the heptahydrate occurs naturally as the mineral epsomite).	**PEL:** Not Established **STEL:** Not Established	**REL:** Not Established **STEL:** Not Established	**Not Determined**	**TLV:** Not Established **STEL:** Not Established

SECTION III - PHYSICAL AND CHEMICAL CHARACTERISTICS

Boiling Point	Specific Gravity (H_2O = 1)
Not Found	**2.65**
Vapor Pressure (mm Hg)	Molecular Weight (atomic weight)
Not Found	**120.37**
Vapor Density (Air = 1)	Melting Point
Not Found	**2055°F (1124°C) Decomposes**

Solubility

Very soluble in water, soluble in glycerol, sparingly soluble in alcohol.

Appearance and Odor

Colorless or opaque needles, crystals, or powder with a cooling, saline, bitter taste, and no odor.

SECTION IV - FIRE AND EXPLOSION HAZARD DATA

Flash Point (method used)	Explosive Limits in Air % by Volume
Not Found	**LEL: Not Found UEL: Not Found**
NFPA Classification	Autoignition Temperature
Non-Combustible Solid	**Not Found**

Extinguishing Media

Use extinguishing media suitable to surrounding fire. Use water spray to wet down material.

Special Fire Fighting Procedures

Wear full protective clothing and self-contained breathing apparatus (SCBA). Poisonous gases are produced in fire. Use water mist to keep fire-exposed containers cool. Remove containers from fire if it can be done without risk. Fight fire from distance, if possible.

Unusual Fire and Explosion Hazards

None reported.

SECTION V - REACTIVITY DATA

Stability		Conditions to Avoid **Magnesium sulfate is normally stable under routine conditions of handling and storage. Avoid contact with incompatible materials. The hyptahydrate will lose water in dry air at ordinary room temperatures.**
Stable **X**	Unstable	Incompatibility *(materials to avoid)* **Incompatible with ethoxyethynl alcohols (reacts explosively). Also reacts with arsenates, phosphates, and tartrates to form their magnesium salts. Reacts with lead, barium, strontium, calcium to form their sulfates.**
Hazardous Polymerization		Conditions to Avoid **Hazardous polymerization of magnesium sulfate will not occur.**
May Occur	Will Not Occur **X**	Hazardous Decomposition or By-products **When heated to decomposition, magnesium sulfate can emit toxic and acrid smoke and fumes, including oxides of magnesium and sulfur.**

SECTION VI - HEALTH HAZARD DATA

Primary Route(s) of Entry:	Inhalation? X	Absorption (skin/eye)?	Ingestion? X

Health Hazards

INHALATION: **Toxicity is considered to be low, but excessive dust inhalation can cause irritation to the nose and mucous membranes of the eyes, nose, throat, and respiratory tract.**

EYES & SKIN: **Exposure to the eyes and skin may cause mild irritation.**

INGESTION: **Excessive doses may cause diarrhea, constipation, abdominal distention, obstruction of the intestine, and possible depression of the CNS with loss of reflexes, and progression to respiratory paralysis or cardiac arrest and death. Elevated serum magnesium levels may cause a drop in blood pressure, pulse, respiration, and mental status.**

Carcinogenicity	NTP Listed?	IARC Cancer Review Group?	OSHA Regulated?	Target Organs?
Unknown Human **Unknown Animal**	**No**	**No**	**No**	**Respiratory system, eyes, skin, GI tract, CNS.**

Medical Conditions Generally Aggravated by Exposure

None reported. However, existing kidney disease is aggravated by ingestion of other magnesium salts.

Emergency and First-aid Procedures

Eye contact: Do not allow victim to keep eyes tightly shut. Flush immediately with water for 15 minutes (minimum); seek medical attention. Skin contact: Remove all contaminated clothing. Immediately wash area with flooding amounts of water. For reddened or blistered skin, seek medical assistance. For inhalation: Remove the person from exposure. Provide respiratory assistance and CPR. Transfer to medical facility. If swallowed: Seek medical attention. Give 1 to 2 glasses of water to dilute. Vomiting may be spontaneous (without inducement). Never give anything by mouth to an unconscious or convulsing person.

SECTION VII - PRECAUTIONS FOR SAFE HANDLING AND USE

Steps to be Taken in Case Material is Released or Spilled

Collect powdered materials in most convenient and safe manner and deposit in sealed containers. Dry sweep NOT recommended. Use HEPA vacuum. Damp mop residue and collect for disposal. Ventilate area of spill. Restrict those not involved in cleanup from entering area. Notify appropriate authorities.

Preferred Waste Disposal Method

No citation.

Precautions to be Taken in Handling and Storage

Store to avoid contact with incompatible materials. Prevent physical damage to containers. Keep containers tightly closed.

Other Precautions and Warnings

Avoid excessive dust generation in storage and handling areas. Personnel should be trained on the proper precautions for working with magnesium compounds *before* initial assignment.

SECTION VIII - CONTROL MEASURES AND PERSONAL PROTECTIVE EQUIPMENT

Respiratory Protection *(specify type)*

No exposure levels established: Use a respirator with a high-efficiency particulate air (HEPA) filter; otherwise, use a self-contained breathing apparatus with full facepiece operated in positive pressure mode.

Ventilation

Local exhaust (preferred) at site of chemical work or general ventilation.

Protective Gloves	Eye Protection	Other Protective Clothing
Impervious Material	**Chemical/Dust Goggles or Face Mask**	**Protective Apron, Boots**

Work/Hygiene Practices

Always wash hands thoroughly after using chemical; never bring food, drink, or smoking materials into vicinity of chemicals.

MAGNESIUM SULFATE

$MgSO_4$ CAS: 7487-88-9

IDENTIFICATION AND TYPICAL USES

Magnesium sulfate appears as colorless or opaque needles, crystals, or powder with a cooling, saline, bitter taste, and no odor. It is used in the textile industry for warp-sizing and loading of cotton goods, weighting silk, dyeing, and calico printing. It is also used for fireproofing other materials, in mineral waters, as a catalyst carrier, in ceramics, in fertilizers, for paper sizing, in certain cosmetic lotions, and as a dietary supplement.

RISK ASSESSMENT: HEALTH

General Assessment

Because of its poor absorption, magnesium sulfate exhibits relatively low toxicity in humans by *inhalation* and *ingestion*. It is generally considered a nuisance particulate. However, exposure to sufficient quantities, especially by ingestion, can pose a serious health risk. There is no information on its carcinogenic, mutagenic, or teratogenic capabilities.

Inhalation causes mechanical (as opposed to chemical) irritation to the eyes, nose, throat, and mucosa of the upper respiratory tract. If ingested and sufficient quantities are absorbed, magnesium sulfate can cause systemic poisoning. Elevated blood magnesium levels may cause a drop in blood pressure, pulse, and respiration. Other symptoms include hypocalcemia (a deficiency in blood calcium levels), diarrhea, constipation, nausea, vomiting, intestinal blockage due to blocked bowel or loss of muscle contraction, and depression of the central nervous system (CNS). This can lead to flushing of the face, sweating, failed reflexes with progression to respiratory failure or cardiac arrest and death.

Acute Health Effects

The following acute (short-term) health effects may occur immediately or shortly after human exposure to magnesium sulfate:

Skin: May cause mild irritation, depending on the conditions and length of exposure.

Eye: Mild to moderate irritation.

Lung: Mechanical irritation of the mucosa of the eyes, nose, throat, and respiratory tract.

Other: Ingestion of excessive quantities of magnesium sulfate can cause gastrointestinal effects, kidney injury, and depression of the central nervous system that can lead to death.

Chronic Health Effects

The following chronic (long-term) health effects may occur at some time after exposure to magnesium sulfate and can last for months or even years:

Cancer Hazards: According to information presented in the references, magnesium sulfate has not been shown to cause cancer in animals.

Reproductive Hazard: According to information presented in the references, there are no reports of any teratogenic or reproductive effects resulting from exposure to magnesium sulfate.

Other Chronic Effects: Since magnesium sulfate is excreted by the kidneys, persons with kidney function problems may experience systemic toxicity.

Recommended Risk-Reduction Measures

Even though the toxicity of magnesium sulfate is considered to be relatively low, personnel should avoid direct contact with magnesium sulfate. If a less toxic material or compound cannot be substituted for magnesium sulfate, then *engineering controls* are the most effective method of reducing exposures. The best protection is to enclose operations and/or provide local exhaust ventilation at the site of magnesium sulfate release. While not always operationally feasible, isolating operations can also reduce exposure. Using respiratory protection is less effective than the controls mentioned above, but is still advisable whenever working with or around magnesium sulfate. No exposure levels have been established for this compound. This does not mean that exposure is without risk. For low or infrequent exposure, an MSHA/NIOSH-approved full facepiece respirator with a high-efficiency particulate air (HEPA) filter may provide adequate protection. A supplied-air respirator operated in continuous flow mode or a self-contained breathing apparatus (SCBA) with full facepiece operated in pressure demand are recommended for the greatest possible respiratory protection. If a full facepiece is not available, then chemical/dust goggles should be worn to protect the eyes. A face shield and protective apron should also be worn. To prevent hand and skin exposures, protective gloves should be used.

Administrative controls should also be in place to minimize the potential for human exposures. These may include written procedures or policies which specify the methods and techniques that will be practiced whenever personnel are to work with magnesium sulfate.

All personnel should receive *training* on the use, hazards, protective measures, emergency actions, and other precautions per 29 CFR 1910.1200 (Hazard Communication), prior to the first assignment in an area where magnesium sulfate is used or stored.

Before beginning employment working with magnesium sulfate, and at regular intervals thereafter (e.g., annually), those personnel with frequent or potentially high exposures should be provided the following recommended medical tests:

☑ Kidney function tests (establish baseline).

If symptoms develop or overexposure is suspected, consider the following additional tests:

☑ Kidney function tests (compare to baseline).

☑ Magnesium serum levels.

☑ Lung function tests.

Any evaluation should include a careful history of past and present symptoms with an examination. Medical tests that simply look for existing damage are not a substitute for controlling exposures. Also, since smoking can cause heart disease as well as lung cancer, emphysema, and other known respiratory problems, smokers exposed to magnesium sulfate may experience symptoms more rapidly than non-smokers under the same conditions of exposure. Prudent risk management requires careful consideration of *all* factors when exposure symptoms appear in the workplace.

Other methods to reduce exposure include:

☑ Where possible, enclose operations and use local exhaust ventilation at the site of chemical release. If local exhaust ventilation or enclosure is not used, respiratory protection should be mandatory.

☑ Always ensure that proper protective clothing is worn when using chemical substances. Proper maintenance and care of protective equipment is essential to its proper function.

☑ Wash thoroughly immediately after exposure to magnesium sulfate and at the end of the work shift or before eating, drinking, or smoking.

☑ Hazard warning information should be posted in the work area. In addition, as part of an on-going education and training program, all information on the health and safety hazards of magnesium sulfate should be communicated to all exposed and potentially exposed workers.

☑ Eye wash stations should be provided in the immediate work area for emergency use. If there is a possibility of skin exposure to magnesium sulfate, emergency shower facilities should be provided in the immediate area.

☑ Workers whose clothing has been contaminated by magnesium sulfate should change into clean clothes before leaving work. Contaminated work clothing should be laundered only by individuals who have been informed of the hazards of exposure to magnesium sulfate.

RISK ASSESSMENT: ENVIRONMENT

General Assessment

As a naturally occurring mineral, magnesium poses very little threat to the environment. However, accidental spills or releases of magnesium sulfate to the environment can occur during transportation, storage, disposal, or destruction of magnesium sulfate. In almost every scenario, the threat of environmental exposure is contingent upon the proper handling of the chemical substance. Accidental spills, large or small, where magnesium sulfate contacts incompatible commodities can result in release of toxic products (depending upon conditions of spill), and possible contamination of the surrounding environmental mediums (water, soil, and air).

Magnesium sulfate is considered a non-combustible solid. However, potentially explosive reactions can occur when it is heated in the presence of ethoxyethynl alcohols. It is also incompatible with arsenates, phosphates, and tartrates. It can react with many other metals such as lead, barium, and calcium to precipitate the formation of their respective sulfates. Therefore, extreme caution is required in handling, storage, transportation, and disposal of magnesium sulfate. These characteristics also require special consideration during any emergency situation involving a leak or spill of magnesium sulfate powder or dust. Should magnesium sulfate ever come into contact with these incompatible substances either during use, transportation, storage, or disposal, violent and even explosive reactions are extremely possible.

Magnesium sulfate may enter the environment as a result of industrial and municipal waste treatment plant discharges and from spills.

Acute Ecological Effects

Acute (short-term) toxic effects may include the death of animals, birds, or fish and death or low growth rate in plants. Acute effects are seen 2 to 4 days after animals or plants are exposed to magnesium sulfate. Magnesium and its salts have slight acute toxicity to aquatic life. Insufficient data are available to evaluate the short-term effects of magnesium sulfate to plants, birds, or land animals.

Chronic Ecological Effects

Chronic (long-term) toxic effects may include shortened life span, reproductive problems, lower fertility, and changes in appearance or behavior in exposed animals. These effects can be seen long after first exposure(s) to toxic chemicals. Magnesium and its salts have slight chronic toxicity to aquatic life. Insufficient data are available to evaluate or predict the long-term effects of magnesium sulfate to plants, birds, or land animals.

Water Solubility

Magnesium sulfate is highly soluble in water. Concentrations of 1000 milligrams and more will mix with a liter of water.

Persistence in the Environment

Magnesium is highly persistent in water, with a half-life greater than 200 days. The half-life of a pollutant is the amount of time it takes for one half of the chemical to be degraded. Magnesium is also naturally present in terrestrial soils and aquatic sediments.

Bioaccumulation in Aquatic Organisms

Some substances increase in concentration, or *bioaccumulate,* in living organisms as they breathe contaminated air, drink contaminated water, or eat contaminated food. These chemicals can become concentrated in the tissues and internal organs of animals as well as humans.

The concentration of magnesium sulfate found in fish tissues is expected to be about the same as the average concentration of magnesium sulfate in the water from which the fish was taken.

Recommended Risk-Reduction Measures

Proper training of all transporters will reduce the likelihood of a mishap or accident resulting in a leak or spill of magnesium sulfate dusts or powders into the environment. The correct use of labeling on all containers, trucks, and rail cars will enable emergency responders to react properly and quickly to any disaster thereby reducing the potential risk to the environment and to personnel.

Storage of magnesium sulfate should be segregated from incompatible chemicals to minimize the risk of cross-contamination or contact. Care should be taken to prevent physical damage to storage containers. Also, dusty conditions must be avoided in storage and use areas.

If a spill or leak to the environment has occurred, fire department, emergency response and/or hazardous materials spill personnel should be notified immediately. Cleanup should be attempted only by those trained in proper spill containment procedures. Contaminated soils should be removed for decontamination and replaced with clean soil. If magnesium sulfate should contact the water table, aquifer, or navigable waterway, time is of the essence. It is highly soluble and total remediation may not be possible. The local and/or state emergency response authorities may require notification. A comprehensive emergency response or disaster preparedness/recovery plan should be in place prior to any operations involving the use, transportation, storage, or disposal of magnesium compounds.

If magnesium sulfate is spilled or leaked, the following specific steps are recommended:

☑ Restrict persons not wearing protective clothing from area of spill or leak until cleanup is complete and area can be opened for normal work.

☑ Remove all ignition sources from area.

☑ Ventilate area of spill or leak.

☑ Collect powdered materials in the safest manner possible and deposit in sealed containers. Recommend using a vacuum equipped with a high-efficiency particulate air (HEPA) filter. Do not dry sweep (generates airborne dusts). Damp mop residues and collect for disposal. Water spray can be used to knock down vapors. Runoff should also be collected for disposal.

☑ It may be necessary to dispose of magnesium sulfate as a hazardous waste. The responsible state agency or the regional office of the federal Environmental Protection Agency (EPA) should be contacted for specific recommendations.

RISK ASSESSMENT: BUSINESS

General Assessment

Accidents or mishaps involving magnesium sulfate can present a significant threat to business operations. The loss or damage of equipment or facilities can significantly affect fiscal viability. Lawsuits that may result from personnel injury/death, public exposures, and/or environmental contamination will also require a serious expenditure of resources. Media attention surrounding an injury, death, or environmental damage can also result in a loss of profits and loss of current as well as future business.

Recommended Risk-Reduction Measures

Company attorneys, safety and health professionals, and environmental specialists should be involved in the development of any procedures for responding to chemical incidents. A company official should be predesignated as a public relations officer with specific training in dealing with the press. Corporate plans and policies should be developed, approved, and implemented long before any need for such arises.

REFERENCES

Ahlbom, A. 1993. *Biostatistics for Epidemiologists.* New York: CRC Press/Lewis Publishers.

American Conference of Governmental Industrial Hygienists. 1988. *Documentation of the Threshold Limit Values and Biological Exposure Indices,* 5th Edition (with updates). Cincinnati: ACGIH

Anthony, C. P., and N. J. Kolthoff. 1971. *Textbook of Anatomy and Physiology.* St. Louis: C. V. Mosby Company.

Bloom, A. D. and F. J. DeSerres. 1995. *Ecotoxicity and Human Health.* New York: CRC Press/Lewis Publishers.

Brusick, D. J. 1994. *Methods for Genetic Risk Assessment.* New York: CRC Press/Lewis Publishers.

Bueche, J. R. 1972. *Principles of Physics,* 2nd Edition. New York: McGraw-Hill.

Calabrese, E. J. 1994. *Biological Effects of Low Level Exposures.* New York: CRC Press/Lewis Publishers.

Cockerman, L. G., B. S. Shane. 1994. *Basic Environmental Toxicology.* New York: CRC Press/Lewis Publishers.

Cothern, R. R. 1993. *Comparative Environmental Risk Assessment.* New York: CRC Press/Lewis Publishers.

Environmental Protection Agency. 1983. *Chemical Hazard Information Profiles.* Washington, D.C.: U.S. EPA.

Forseberg, K., and L. H. Keith. 1993. *Chemical; Protective Clothing Permeation and Degradation Database.* Chelsea, MI: Lewis Publishers.

Gardner, R. W. 1994. *Chemical Intolerance.* New York: CRC Press/Lewis Publishers.

Howard, P. H. 1995. *Handbook of Environmental Fate and Exposures Data for Organic Chemicals,* Vol. I-IV. New York: CRC Press/Lewis Publishers.

Lauwerys, R. R., P. Hoet. 1993. *Industrial Chemical Exposure.* New York: CRC Press/Lewis Publishers.

Lewis, R. J., Sr. 1993. *Hawley's Condensed Chemical Dictionary,* 12th Edition. New York: Van Nostrand Reinhold.

Lewis, R. J., Sr. 1992. *Sax's Dangerous Properties of Industrial Materials,* Volumes 1, 2, and 3, 8th Edition. New York: Van Nostrand Reinhold.

Mackay, D., W. Y. Shiu, and K. C. Ma. 1992. *Illustrated Handbook of Chemical-Physical Properties and Environmental Fate for Organic Chemicals.* New York: CRC Press/Lewis Publishers.

Manahan, S. E. 1994. *Environmental Chemistry,* 6th Edition. New York: CRC Press/Lewis Publishers.

Meyer, E. 1990. *Chemistry of Hazardous Materials.* Englewood Cliffs, NJ: Prentice Hall.

Miller, B. F., M.D., and C. B. Keane, R.N. 1972. *Encyclopedia and Dictionary of Medicine and Nursing.* Philadelphia: W.B. Sanders Company.

Mycock, J. C., J. D. McKenna, and L. Theodore. 1995. *Handbook of Air Pollution Control Engineering and Technology.* New York: CRC Press/Lewis Publishers.

Quigley, D. R. 1994. *Handbook of Emergency Chemical Management.* New York: CRC Press/Lewis Publishers.

Richardson, M. 1995. *Dictionary of Substance and Their Effects,* Vol. I-VI. New York: CRC Press/Lewis Publishers.

Sacarello, H. L. A. 1994. *The Comprehensive Handbook of Hazardous Materials.* New York: CRC Press/Lewis Publishers.

Talbott, E., G. Craun. 1995. *An Introduction to Environmental Epidemiology.* New York: CRC Press/Lewis Publishers.

Torkelson, T. R. V. K. Rowe. 1981. *Patty's Industrial Hygiene and Toxicology,* Vol. 2B. New York: Wiley-Interscience.

MATERIAL SAFETY DATA SHEET

CHEMICAL NAME	
MALATHION®	**CRC PRESS/LEWIS PUBLISHERS DISCLAIMER:** THE INFORMATION AND RECOMMENDATIONS PRESENTED HEREIN ARE BASED ON SOURCES BELIEVED TO BE RELIABLE. CRC MAKES NO REPRESENTATION ON ITS COMPLETENESS OR ACCURACY. IT IS THE RESPONSIBILITY OF THE USER TO DETERMINE THE CHEMICAL'S SUITABILITY FOR ITS INTENDED USE, THE CHEMICAL'S SAFE USE, AND THE CHEMICAL'S PROPER DISPOSAL. NO REPRESENTATIONS AND/OR WARRANTIES, EITHER EXPRESSED OR IMPLIED, OF THE MERCHANTABILITY OR FITNESS FOR A PARTICULAR PURPOSE, OR OF ANY OTHER NATURE, ARE MADE WITH RESPECT TO THE INFORMATION PROVIDED IN THIS MSDS OR TO THE CHEMICAL TO WHICH INFORMATION MAY REFER. CRC NEITHER ASSUMES NOR AUTHORIZES ANY OTHER PERSON TO ASSUME FOR IT, ANY OTHER ADDITIONAL RESPONSIBILITY OR LIABILITY FOR THE USE OF, OR RELIANCE UPON, THIS INFORMATION.

HAZARD WARNING INFORMATION

HEALTH	FIRE	REACTIVE	OTHER	DEGREE OF HAZARD	COLOR CODING	OTHER CODES
2	1	1		0 = Minimum Hazard 1 = Slight Hazard 2 = Moderate Hazard 3 = Serious Hazard 4 = Severe Hazard	HEALTH = BLUE FIRE = RED REACTIVITY = YELLOW OTHER = WHITE	OX = Oxidizer ACID = Acid ALK = Alkali COR = Corrosive W = Use No Water

SECTION I - GENERAL INFORMATION

Characterization	RCRA Number	EPA Class
Pesticide	**None**	**Not Applicable**

DOT Proper Shipping Name **Organophosphorous Pesticides, Liquid, Toxic, N.O.S.**	Chemical Abstract Service (CAS) Number **121-75-5**
DOT Hazard Class and Label Requirements **ORM-A; Poison**	DOT Emergency Guide Code **55**
DOT Identification Number **NA 2783**	Chemical Formula $(CH_3O)_2P(S)SCH(COOC_2H_5)CH_2COOC_2H_5$

Synonyms

Cythion; carbophos; O,O-dimethyl dithiophosphate of diethyl mercaptosuccinate; O,O-dimethyl-S-(1-2-dicarbethoxyethyl) phosphorodithioate; mercaptothion; Maldison; carbethoxy malathion.

SECTION II - HAZARDOUS INGREDIENTS AND IDENTITY INFORMATION

Hazardous Components (specific identity)	OSHA Exposure Criteria	NIOSH Exposure Criteria	Immediately Dangerous to Life and Health (IDLH)	ACGIH Exposure Criteria
Malathion: (derivation: By condensing iso-butyramidine with ethyl-3-oxobutryrate and treating the resulting pyrimidine with o,o-diethylphosphorochloriodothioate). **1 ppm = 13.73 mg/m³**	**PEL (8-hour):** 15 mg/m³ (skin) **STEL:** Not Established	**REL (10-hour):** 10 mg/m³ (skin) **STEL:** Not Established	**250 mg/m³**	**TLV:** 10 mg/m³ (skin) **STEL:** Not Established

SECTION III - PHYSICAL AND CHEMICAL CHARACTERISTICS

Boiling Point **140°F (60°C) Decomposes**	Specific Gravity (H_2O = 1) **1.21**
Vapor Pressure (mm Hg) **1.25×10^{-4} at 69°F (20°C)**	Molecular Weight **330.38**
Vapor Density (Air = 1) **11**	Melting Point **37°F (3°C)**

Solubility

Slightly soluble in water (0.02%). Soluble in alcohol, ketones, ethers, vegetable oils, most organic solvents, aromatic and alkylated aromatic hydrocarbons. Sparingly soluble in paraffin hydrocarbons.

Appearance and Odor

Colorless (pure) to dark brown to yellow liquid with a garlic-like odor. Odor Threshold = 13.5 mg/m³.

SECTION IV - FIRE AND EXPLOSION HAZARD DATA

Flash Point (method used) **>325°F (>163°C) opened cup**	Explosive Limits in Air % by Volume **LEL: Not Determined UEL: Not Determined**
NFPA Classification **Class IIIB Combustible Liquid**	Autoignition Temperature **Not Determined**

Extinguishing Media

Use dry chemical, carbon dioxide, water spray, or foam.

Special Fire Fighting Procedures

Poisonous gases are produced in fire; wear full protective clothing and self-contained breathing apparatus (SCBA). If possible, fight fire from a distance and avoid smoke (consider down-wind conditions). Move containers from fire area if it can be done safely. Use chlorine bleach solution to decontaminate.

Unusual Fire and Explosion Hazards

Containers may explode when exposed to heat of a fire. Use water spray to cool fire-exposed containers that cannot be moved from fire area. Evacuate non-essential personnel 2000 feet from fire area.

SECTION V - REACTIVITY DATA

Stability		Conditions to Avoid Malathion is kept stable in aqueous solution that has been buffered to pH of 5.25. Avoid contact with incompatible materials. Keep away from heat and flame.
Stable X	Unstable	Incompatibility *(materials to avoid* Incompatible with strong oxidizers, magnesium, alkaline pesticides, and is corrosive to many metals including steel, tin plate, lead, and copper. It will attack some forms of plastic, rubber, and coatings.
Hazardous Polymerization		Conditions to Avoid Under normal conditions of temperature and pressure, hazardous polymerization of malathion will not occur.
May Occur	Will Not Occur X	Hazardous Decomposition or By-products Acrid and irritating gases and fumes are produced when malathion is involved in fire, including sulfur dioxide, phosphoric acid, carbon monoxide, and phosphorus oxides. Decomposition may be explosive.

SECTION VI - HEALTH HAZARD DATA

Primary Route(s) of Entry:	Inhalation?	Absorption (skin)?	Ingestion?
	X	X	X

Health Hazards

INHALATION: Tightness in chest, wheezing, coughing, bluish discoloration of the skin, small pupils, an aching in and behind the eyes, blurred vision, tearing, runny nose, headache, an increase in pulmonary secretions, and watering in the mouth. CNS effects may include weakness, generalized twitching and paralysis, dizziness, confusion, slurred speech, staggering gait, sweating, irregular heartbeat, convulsions, stopped breathing, coma. Also inhibits blood cholinesterase.

ABSORPTION: Irritation to the skin and eyes. May absorb to cause toxic systemic effects, including sweating and twitching. Mild sensitization and dermatitis may occur at high exposures.

INGESTION: Loss of appetite, nausea, vomiting, abdominal cramps, and diarrhea within 2 hours.

Carcinogenicity	NTP Listed?	IARC Cancer Review Group?	OSHA Regulated?	Target Organs?
Unknown Human Questioned Animal	No	Group 3	No	Resp. sys., liver, eyes, skin, CNS, CVS, blood, GI.

Medical Conditions Generally Aggravated by Exposure

Existing central nervous system disorders may be aggravated by exposure.

Emergency and First-aid Procedures

Eye contact: Flush immediately with water for 15 minutes (minimum); seek medical attention. Skin contact: Remove all contaminated clothing. Immediately wash area with large amounts of soap and water. Seek medical assistance. For inhalation: Remove the person from exposure. Provide respiratory assistance and CPR. Transfer to medical facility. If swallowed: Contact poison control center. Unless advised otherwise, do NOT induce vomiting. Consider gastric lavage *after* control of seizures.

SECTION VII - PRECAUTIONS FOR SAFE HANDLING AND USE

Steps to be Taken in Case Material is Released or Spilled

Absorb liquids in vermiculite or other absorbent and deposit in DOT-approved drum. Ventilate area and remove ignition sources. Restrict those not involved in cleanup from entering area. Do not use plastic or rubber tools during cleanup since malathion can attack such materials.

Preferred Waste Disposal Method

Mix in flammable solvent and burn in chemical incinerator equipped with afterburner and scrubber.

Precautions to be Taken in Handling and Storage

Store in tightly closed containers in a cool, well-ventilated, moisture-free area. Personnel working with malathion should be trained on its hazards *prior* to being assigned to such responsibilities.

Other Precautions and Warnings

Do not store in steel, iron, tin plate, lead, or copper containers. Also keep away from rubber and plastic.

SECTION VIII - CONTROL MEASURES AND PERSONAL PROTECTIVE EQUIPMENT

Respiratory Protection *(specify type)*

For relatively low or infrequent exposure to powders, use a full facepiece respirator equipped with a high-efficiency particulate air (HEPA) filter and dust/mist pre-filter. For greater protection, use a supplied-air respirator with full facepiece operated in pressure demand mode or a self-contained breathing apparatus (SCBA) with full facepiece operated in pressure demand or other positive pressure mode.

Ventilation

Local exhaust or general mechanical systems recommended.

Protective Gloves	Eye Protection	Other Protective Clothing
Impervious Material	Chemical/Dust Goggles or Face Mask	Whole Body Protection

Work/Hygiene Practices

Always wash hands thoroughly after using chemical; never bring food, drink, or smoking materials into vicinity of chemicals.

MALATHION

$(CH_3O)_2P(S)SCH(COOC_2H_5)CH_2COOC_2H_5$ CAS: 121-75-5

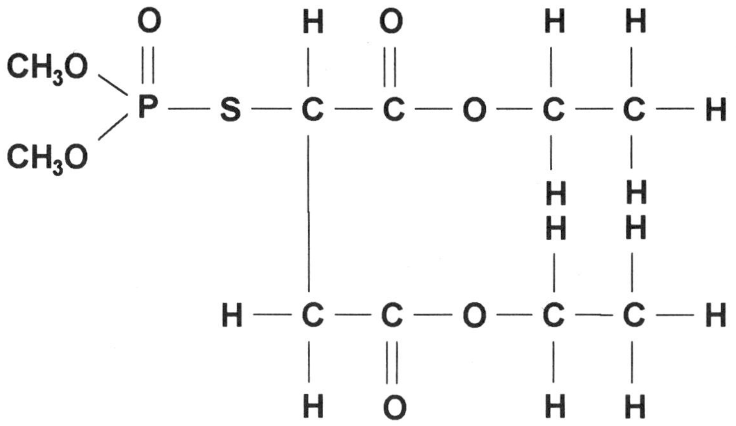

IDENTIFICATION AND TYPICAL USES

Malathion is a colorless (in pure form) to dark brown to yellow liquid with a garlic or skunk-like odor. It is an insecticide used for control of mosquitoes, the Mediterranean fruit fly, spiders, animal ectoparasites, and human head and body lice or mites.

RISK ASSESSMENT: HEALTH

General Assessment

Malathion is primarily toxic to humans by *ingestion*. However, its vapors can also cause adverse systemic effects by *inhalation,* and skin contact may result in *absorption* which will also cause adverse systemic effects. Contact with the eyes or skin will result in irritation. It is a questionable carcinogen with human mutation data also reported. It is also an experimental teratogen with positive evidence of reproductive effects in animals.

Inhalation of malathion vapors will cause tightness in the chest, wheezing, increased pulmonary secretions, bluish color in the skin, contricted pupils, an aching feeling in and behind the eyes, blurred vision, lachrymation (tearing), runny nose, headache, and watering mouth. It will cause systemic effects with action on the central nervous system. Action on the central nervous system can include weakness, general twitching and paralysis, dizziness, confusion, staggering gait, slurred speech, sweating, irregular heartbeat, convulsions, stopped breathing, and coma. This compound is also an inhibitor of cholinesterase, an enzyme that is essential to the proper functioning of the muscles.

Acute Health Effects

The following acute (short-term) health effects may occur immediately or shortly after exposure to malathion:

Skin: Irritation and possible inflammation. May absorb through unbroken skin to cause toxic systemic effects, including sweating and twitching at site of absorption.

Eye: Irritation and possible inflammation.

Lung: Irritation of the mucosa of the nose, throat, and upper respiratory tract. Effects are more systemic than localized, with action on the blood, central nervous system, and (possibly) the cardiovascular system.

CNS: A depressant, causing headache, blurred vision, giddiness, nervousness, and other systemic effects.

Other: Ingestion of malathion will result in loss of appetite, nausea, vomiting, abdominal cramps, and diarrhea with 2 hours of exposure.

Chronic Health Effects

The following chronic (long-term) health effects may occur at some time after exposure to malathion and can last for months or even years:

Cancer Hazards: According to the information presented in the references, malathion has been shown to cause cancer in test animals. It has been shown to cause mutations (genetic changes) in humans and may pose a cancer risk as well.

Reproductive Hazard: According to the information presented in the references, malathion may damage the developing fetus in humans since it has been shown to be fetotoxic in animals.

Other Chronic Effects: Chronic exposure to malathion has not been sufficiently studied. However, there is some indication that long-term exposure or frequent high exposures may inhibit the essential enzyme cholinesterase. It may also cause dermal sensitization as well as dermatitis in some people.

Recommended Risk-Reduction Measures

Malathion is less toxic to humans than most other anticholinesterase agents. However, less toxic substances should always be sought as a first step towards risk reduction. If a less toxic chemical cannot be substituted for a hazardous substance, *engineering controls* are the most effective method of reducing exposures. The best protection is to enclose operations and/or provide local exhaust ventilation at the site of chemical release. While not always operationally feasible, isolating operations can also reduce exposure. These controls certainly may not be practical or applicable when using malathion as an insecticide, especially outdoors. However, during manufacture, transportation, or storage, these engineering controls may be effective in reducing exposure risk.

Using respiratory protection is less effective than the controls mentioned above, but is still advisable whenever working with or around malathion. The exposure level for this compound is relatively low and may be difficult to measure and even more difficult to protect against. At low or infrequent exposures, a full facepiece respirator equipped with a high-efficiency particulate air (HEPA) filter and a dust/mist pre-filter may suffice. However, for maximum exposure risk reduction or for exposures to high concentrations, a self-contained breathing apparatus (SCBA) with full facepiece operated in pressure demand is the recommended respiratory protection of choice. If a full facepiece is not available, then chemical goggles should be worn to protect the eyes. Whenever a chemical splash hazard exists, a face shield and protective apron should be worn. To prevent hand and skin exposures, impervious gloves should be worn. Caution is warranted in glove selection since malathion will attack rubbers, some plastics, and many coatings. Glove manufacturers should be contacted and permeation studies obtained *before* final glove selections are made.

Administrative controls should also be in place to minimize the potential for human exposures. These may include written procedures or policies which specify the methods and techniques that will be practiced whenever personnel are to work with malathion.

All personnel should receive *training* on the use, hazards, protective measures, emergency actions, and other precautions per 29 CFR 1910.1200 (Hazard Communication), prior to the first assignment in an area where malathion is used or stored.

If symptoms develop or overexposure is suspected, the following medical tests may be useful:

☑ Complete blood count (CBC).

Any evaluation should include a careful medical history of past and present symptoms with an examination. However, medical tests that evaluate existing damage are not a substitute for controlling exposure.

Other methods to reduce exposure include:

☑ Where possible, enclose operations and use local exhaust ventilation at the site of chemical release. If local exhaust ventilation or enclosure is not used, respiratory protection should be mandatory.

☑ Always ensure that proper protective clothing is worn when using chemical substances.

☑ Wash thoroughly immediately after exposure to malathion and at the end of the work shift or before eating, drinking, or smoking.

☑ Hazard warning information should be posted in the work area. In addition, as part of an on-going education and training program, all information on the health and safety hazards of malathion should be communicated to all exposed workers.

☑ Eye wash stations should be provided in work areas. If the potential for whole body exposure exists, then safety showers should also be provided.

RISK ASSESSMENT: ENVIRONMENT

General Assessment

The environment is at risk of exposure during the intended use as an insecticide, as well as during transportation, storage, disposal, or destruction of malathion. In almost every scenario, the threat of environmental exposure is contingent upon the proper handling of the chemical substance. Accidental spills, large or small, can result in fire, explosion, and possible contamination of the surrounding environmental mediums (water, soil, and, to a lesser extent, air).

Malathion is considered a Class IIIB combustible liquid (per OSHA 29 CFR 1910.106). These characteristics require special consideration during any emergency situation involving a leak or spill of malathion. Should malathion ever come into contact with incompatible substances such as strong oxidizers, alkaline pesticides, or metals such as magnesium, iron, steel, tin plate, lead, or copper, either during use, transportation, or storage, violent reactions can occur. Malathion will slowly hydrolyze in water to dilute maleic acid.

Malathion can enter the environment through its intended use as an insecticide, through industrial discharges, and from spills.

☠ *Acute Ecological Effects*

Acute (short-term) toxic effects may include the death of animals, birds, or fish and death or low growth rate in plants. Acute effects are seen 2 to 4 days after animals or plants are exposed to malathion.

This chemical has high acute toxicity to aquatic life and moderate acute toxicity to birds. Insufficient data are available to evaluate or predict the short-term effects of malathion to plants or land animals.

Chronic Ecological Effects

Chronic toxic effects may include shortened life span, reproductive problems, lower fertility, and changes in appearance or behavior in exposed animals. These effects can be seen long after first exposure(s) to toxic chemicals.

Malathion has high chronic toxicity to aquatic life. Insufficient data are available to evaluate or predict the long-term effects of malathion to plants, birds, or land animals.

Water Solubility

Malathion is very slightly soluble in water. Concentrations of 1 milligram or less will mix with a liter of water.

⌛ *Persistence in the Environment*

Malathion is slightly persistent in water, with a half-life of between 2 to 20 days. The half-life of a pollutant is the amount of time it takes for one half of the chemical to be degraded.

Bioaccumulation in Aquatic Organisms

Some substances increase in concentration, or *bioaccumulate,* in living organisms as they breathe contaminated air, drink contaminated water, or eat contaminated food. These chemicals can become concentrated in the tissues and internal organs of animals as well as humans.

Malathion is not expected to accumulate in the edible tissues of fish normally consumed by humans.

Recommended Risk-Reduction Measures

Proper training of all transporters will reduce the likelihood of a mishap or accident resulting in a leak or spill to the environment. The correct use of DOT labeling (per the MSDS) on all containers, trucks, and rail cars will enable emergency responders to react properly and quickly to any disaster thereby reducing the potential risk to the environment and to personnel.

Storage of malathion should be segregated from incompatible chemicals to minimize the risk of cross-contamination or contact. It should be kept in dry areas away from water and moisture.

If a spill or leak to the environment has occurred, fire department, emergency response and/or hazardous materials spill personnel should be notified immediately. Cleanup should be attempted only by those trained in proper spill containment procedures. Contaminated soils should be removed for incineration and replaced with clean soil. If malathion should contact the water table, aquifer, or navigable waterway, remediation activities should be prompt to ensure protection of aquatic sediments. When such spills occur, the local and/or state emergency response authorities must be notified. A comprehensive emergency response or disaster preparedness/recovery plan should be in place prior to any operations involving the use, transportation, storage, or disposal of malathion. If malathion is spilled or leaked, the following specific steps are recommended:

- ☑ Restrict persons not wearing protective clothing from area of spill or leak until cleanup is complete and area can be opened for normal work.
- ☑ Ventilate area and remove ignition sources.
- ☑ Absorb liquid spills using vermiculite, dry earth, or sand and place in a sealed drum for disposal.

Avoid using plastic or rubber tools during cleanup since malathion can attack such materials.

☑ It may be necessary to dispose of malathion as a hazardous waste. The responsible state agency or the regional office of the federal Environmental Protection Agency (EPA) should be contacted for specific recommendations.

RISK ASSESSMENT: BUSINESS

General Assessment

Accidents or mishaps involving malathion can present a significant threat to business operations. The loss or damage of equipment or facilities can significantly affect fiscal viability. Lawsuits that may result from personnel injury/death, public exposures, and/or environmental contamination will also require a serious expenditure of resources. Media attention surrounding an injury, death, or environmental damage can also result in a loss of profits and loss of current as well as future business. Always remember that anytime the terms "cancer," "carcinogen," or "reproductive hazard" are used, public emotion, hysteria, and ignorance can run equally high. This should be carefully considered whenever developing or implementing public relations polices.

Recommended Risk-Reduction Measures

Company attorneys, safety and health professionals, and environmental specialists should be involved in the development of any procedures dealing with safety or emergency response. A company official should be pre-designated as a public relations officer with specific training in dealing with the press. Corporate plans and policies should be developed, approved, and implemented long before any need for such arises.

REFERENCES

Ahlbom, A. 1993. *Biostatistics for Epidemiologists.* New York: CRC Press/Lewis Publishers.

American Conference of Governmental Industrial Hygienists. 1988. *Documentation of the Threshold Limit Values and Biological Exposure Indices,* 5th Edition (with updates). Cincinnati: ACGIH

Asante-Duah, D. K. 1993. *Hazardous Waste Risk Assessment.* New York: CRC Press/Lewis Publishers.

Bloom, A. D. and F. J. DeSerres. 1995. *Ecotoxicity and Human Health.* New York: CRC Press/Lewis Publishers.

Bueche, J. R. 1972. *Principles of Physics,* 2nd Edition. New York: McGraw-Hill.

Burrell, R., D. K. Flaherty, and L. J. Sauers. 1992. *Toxicology of the Immune System.* New York: Van Nostrand Reinhold.

Calabrese, E. J., E. M. Kenyon. *Air Toxics and Risk Assessment.* New York: CRC Press/Lewis Publishers.

Cockerman, L. G., B. S. Shane. 1994. *Basic Environmental Toxicology.* New York: CRC Press/Lewis Publishers.

Cothern, C. R. 1995. *Handbook for Environmental Risk Decisionmaking.* New York: CRC Press/Lewis Publishers.

Gardner, R. W. 1994. *Chemical Intolerance.* New York: CRC Press/Lewis Publishers.

Hallenbeck, W. H. 1993. *Quantitative Risk Assessment for Environmental and Occupational Health,* 2nd Edition. New York: CRC Press/Lewis Publishers.

Howard, P. H. 1995. *Handbook of Environmental Fate and Exposures Data for Organic Chemicals,* Vol. I-IV. New York: CRC Press/Lewis Publishers.

Keith, L. H. and M. M. Walker. 1995. *Handbook of Air Toxics.* New York: CRC Press/Lewis Publishers.

Landis, W. G., M. H. Yu. 1995. *Introduction to Environmental Toxicology.* New York: CRC Press/Lewis Publishers.

Lauwerys, R. R., P. Hoet. 1993. *Industrial Chemical Exposure.* New York: CRC Press/Lewis Publishers.

Lewis, R. J., Sr. 1993. *Hawley's Condensed Chemical Dictionary,* 12th Edition. New York: Van Nostrand Reinhold.

Lewis, R. J., Sr. 1992. *Sax's Dangerous Properties of Industrial Materials,* Volumes 1, 2, and 3, 8th Edition. New York: Van Nostrand Reinhold.

Manahan, S. E. 1992. *Toxicological Chemistry,* 2nd Edition. New York: CRC Press/Lewis Publishers.

Meyer, E. 1990. *Chemistry of Hazardous Materials.* Englewood Cliffs, NJ: Prentice Hall.

Mycock, J. C., J. D. McKenna, and L. Theodore. 1995. *Handbook of Air Pollution Control Engineering and Technology.* New York: CRC Press/Lewis Publishers.

Ott, W. R. 1995. *Environmental Statistics and Data Analysis.* New York: CRC Press/Lewis Publishers.

Philp, R. B. 1995. *Environmental Hazards and Human Health.* New York: CRC Press/Lewis Publishers.

Quigley, D. R. 1994. *Handbook of Emergency Chemical Management.* New York: CRC Press/Lewis Publishers.

Rea, W. J. 1995. *Chemical Sensitivity,* Vol. I-III. New York: CRC Press/Lewis Publishers.

Richardson, M. 1995. *Dictionary of Substance and Their Effects,* Vol. I-VI. New York: CRC Press/Lewis Publishers.

Rylander, R., R. R. Jacobs. 1994. *Organic Dusts.* New York: CRC Press/Lewis Publishers.

Sacarello, H. L. A. 1994. *The Comprehensive Handbook of Hazardous Materials.* New York: CRC Press/Lewis Publishers.

U.S. Department of Health and Human Services, National Institute for Occupational Safety and Health. 1994. *NIOSH Pocket Guide to Chemical Hazards.* Washington, D.C.: U.S. Government Printing Office.

MATERIAL SAFETY DATA SHEET

CHEMICAL NAME	CRC PRESS/LEWIS PUBLISHERS DISCLAIMER
MALEIC ACID	**CRC PRESS/LEWIS PUBLISHERS DISCLAIMER:** THE INFORMATION AND RECOMMENDATIONS PRESENTED HEREIN ARE BASED ON SOURCES BELIEVED TO BE RELIABLE. CRC MAKES NO REPRESENTATION ON ITS COMPLETENESS OR ACCURACY. IT IS THE RESPONSIBILITY OF THE USER TO DETERMINE THE CHEMICAL'S SUITABILITY FOR ITS INTENDED USE, THE CHEMICAL'S SAFE USE, AND THE CHEMICAL'S PROPER DISPOSAL. NO REPRESENTATIONS AND/OR WARRANTIES, EITHER EXPRESSED OR IMPLIED, OF THE MERCHANTABILITY OR FITNESS FOR A PARTICULAR PURPOSE, OR OF ANY OTHER NATURE, ARE MADE WITH RESPECT TO THE INFORMATION PROVIDED IN THIS MSDS OR TO THE CHEMICAL TO WHICH INFORMATION MAY REFER. CRC NEITHER ASSUMES NOR AUTHORIZES ANY OTHER PERSON TO ASSUME FOR IT, ANY OTHER ADDITIONAL RESPONSIBILITY OR LIABILITY FOR THE USE OF, OR RELIANCE UPON, THIS INFORMATION.

HAZARD WARNING INFORMATION

HEALTH	FIRE	REACTIVE	OTHER	DEGREE OF HAZARD	COLOR CODING	OTHER CODES
2	1	0		0 = Minimum Hazard 1 = Slight Hazard 2 = Moderate Hazard 3 = Serious Hazard 4 = Severe Hazard	HEALTH = BLUE FIRE = RED REACTIVITY = YELLOW OTHER = WHITE	OX = Oxidizer ACID = Acid ALK = Alkali COR = Corrosive W = Use No Water

SECTION I - GENERAL INFORMATION

Characterization	**Acid**	RCRA Number	**None**
		EPA Class	**Not Applicable**
DOT Proper Shipping Name	**Maleic Acid**	Chemical Abstract Service (CAS) Number	**110-16-7**
DOT Hazard Class and Label Requirements	**ORM-A**	DOT Emergency Guide Code	**60**
DOT Identification Number	**NA 2215**	Chemical Formula	**$C_4H_4O_4$**

Synonyms

***cis*-Butenedioic acid; hydroxysuccinic acid; *cis*-1,2-ethylenedicarboxylic acid; maleinic acid.**

SECTION II - HAZARDOUS INGREDIENTS AND IDENTITY INFORMATION

Hazardous Components (specific identity)	OSHA Exposure Criteria	NIOSH Exposure Criteria	Immediately Dangerous to Life and Health (IDLH)	ACGIH Exposure Criteria
Maleic acid (derivation: By catalytic oxidation of benzene over vanadium pentoxide).	**PEL:** Not Established **STEL:** Not Established	**REL:** Not Established **STEL:** Not Established	**Not Established**	**TLV:** Not Established **STEL:** Not Established

SECTION III - PHYSICAL AND CHEMICAL CHARACTERISTICS

Boiling Point	**275°F (135°C) Decomposes**	Specific Gravity (H_2O = 1)	**1.59**
Vapor Pressure (mm Hg)	**Not Found**	Molecular Weight	**116.08**
Vapor Density (Air = 1)	**4.0**	Melting Point	**266°F (130°C)**

Solubility

Soluble in water, alcohol, glacial acetic acid, and acetone. Very slightly soluble in benzene.

Appearance and Odor

Colorless or white crystalline solid with a faint, acidulous odor and a characteristic, astringent, repulsive taste.

SECTION IV - FIRE AND EXPLOSION HAZARD DATA

Flash Point (method used)	**Not Determined**	Explosive Limits in Air % by Volume	**LEL: Not Determined UEL: Not Determined**
NFPA Classification	**Non-Combustible Solid**	Autoignition Temperature	**Not Determined**

Extinguishing Media

Use agent suitable to surrounding media. Maleic acid itself presents only a slight fire hazard.

Special Fire Fighting Procedures

Poisonous gases are produced in fire. Wear full protective clothing and self-contained breathing apparatus (SCBA). Use water spray to flush and dilute a spill and to disperse the vapors. Also, use a water spray to keep fire-exposed containers cool. Do not release runoff from fire control measure to sewers or waterways. Firefighters should avoid contact with vapors produced during fire.

Unusual Fire and Explosion Hazards

None reported.

SECTION V - REACTIVITY DATA

Stability		Conditions to Avoid **Maleic acid is normally stable under routine conditions of handling and storage. Avoid contact with heat and ignition sources and incompatible materials.**
Stable X	Unstable	Incompatibility *(materials to avoid)* **Maleic acid can react violently in contact with strong oxidizers (chlorine, bromine, fluorine) and other oxidizing materials. It may corrode metals when wet.**
Hazardous Polymerization		Conditions to Avoid **Under normal conditions of temperature and pressure, hazardous polymerization of maleic acid is not expected to occur.**
May Occur	Will Not Occur X	Hazardous Decomposition or By-products **When heated to decomposition, maleic acid can emit highly toxic and acrid smoke and irritating fumes, including toxic oxides of carbon.**

SECTION VI - HEALTH HAZARD DATA

Primary Route(s) of Entry:	Inhalation?	Absorption (skin)?	Ingestion?
	X		X

Health Hazards

INHALATION: Extremely irritating to the eyes, nose, throat, and respiratory system. Inhaled dusts or mists may cause coughing, bronchial irritation, and inflammation of associated mucosa.

EYES & SKIN: Skin and eye contact causes irritation. Eye contact produces burns, conjunctivitis, and possible corneal abrasions. Skin contact causes acid burns and dermatitis.

INGESTION: Ingestion is unlikely but possible. Can cause burns to the mouth, throat, esophagus, and stomach with nausea and vomiting.

Carcinogenicity	NTP Listed?	IARC Cancer Review Group?	OSHA Regulated?	Target Organs?
Unknown Human Unknown Animal	No	No	No	Respiratory system, eyes, skin.

Medical Conditions Generally Aggravated by Exposure

None reported.

Emergency and First-aid Procedures

Eye contact: Flush immediately with water for 15 minutes (minimum); seek medical attention. **Skin contact:** Remove all contaminated clothing. Immediately flush area with large amounts of water. For red or blistered skin, wash with soap and water. Seek medical assistance. For **inhalation:** Remove the person from exposure. Provide respiratory assistance and CPR. Transfer to medical facility. If **swallowed:** Contact poison control center. Unless advised otherwise, give 1 to 2 glasses of water or milk. Do NOT induce vomiting. Never attempt to give anything by mouth to an unconscious or convulsing person.

SECTION VII - PRECAUTIONS FOR SAFE HANDLING AND USE

Steps to be Taken in Case Material is Released or Spilled

Collect powdered materials in safest manner possible. Do NOT dry sweep (generates dusts). Use a HEPA vacuum instead. Absorb solution spills in vermiculite or similar materials and deposit in sealed containers. Ventilate area of spill or leak. Restrict those not involved in cleanup from entering area.

Preferred Waste Disposal Method

Mix with a combustible liquid and burn in a chemical incinerator with afterburner and scrubber.

Precautions to be Taken in Handling and Storage

Avoid contact with incompatible materials, especially strong oxidizers. Store in tightly closed containers in cool, well-ventilated area away from heat and sources of ignition. Protect containers from damage. Do not allow persons who have not been trained to work with or around hazardous materials.

Other Precautions and Warnings

Avoid generating dusty conditions in storage areas. Keep containers tightly closed.

SECTION VIII - CONTROL MEASURES AND PERSONAL PROTECTIVE EQUIPMENT

Respiratory Protection *(specify type)*

No exposure levels have been established for maleic acid. An air purifying respirator with acid cartridge and a HEPA filter may suffice. However, better protection is provided by a supplied-air respirator with full facepiece operated in pressure demand or other positive pressure mode.

Ventilation

Local exhaust or general mechanical systems recommended.

Protective Gloves	Eye Protection	Other Protective Clothing
Impervious Material	Chemical/Dust Goggles or Face Mask	Protective Apron

Work/Hygiene Practices

Always wash hands thoroughly after using chemical; never bring food, drink, or smoking materials into vicinity of chemicals.

MALEIC ACID

$C_4H_4O_4$

CAS: 110-16-7

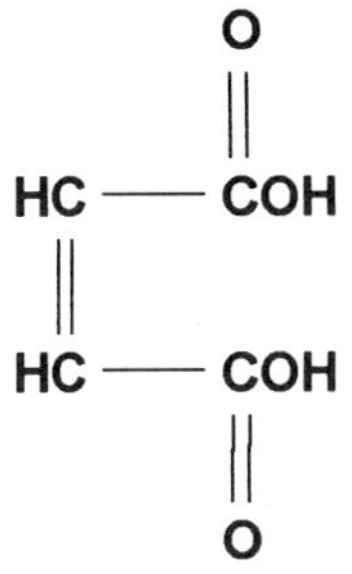

IDENTIFICATION AND TYPICAL USES

Maleic acid appears as a white crystalline solid with a faint acidulous odor and a repulsive, astringent, characteristic taste. It is used in organic synthesis of severe acids (malic, succinic, aspartic, tartaric, propionic, lactic, malonic, and acrylic). It is also used in the textile industry for dyeing and finishing of cotton, wool, and silk. It is a preservative for oils and fats.

RISK ASSESSMENT: HEALTH

General Assessment

Exposure to maleic acid is possible through *inhalation* of its dusts or mists and possibly by *ingestion,* although the latter is not considered a likely exposure route. The data in the references are limited regarding its specific toxicity. However, there are no reports of animal or human carcinogenicity, mutagenicity, or teratogenicity.

Inhalation of maleic acid dusts or mists will cause severe irritation of the eyes, nose, throat, and mucous membranes of the upper respiratory tract. Because of its acidic nature, bronchial tube irritation with coughing and phlegm may also occur.

Skin and eye contact with concentrated solutions can result in mild to moderate irritation. On the skin, maleic acid can cause burns and irritation, especially if the skin is moist (sweat). Repeated contact or high exposures may cause dermatitis. If allowed to remain in contact with the eye for prolonged periods, corneal edema, burns, and conjunctivitis may occur.

Acute Health Effects

The following acute (short-term) health effects may occur immediately or shortly after exposure to maleic acid:

Skin: Irritation/rash or burning feeling on contact, with possible development of dermatitis.

Eye: Severe irritation; may cause burns, conjunctivitis, and corneal damage.

Lung: Irritation of the nose, throat, and mucosa of the lungs and respiratory tract. Because of the acidic nature, bronchial tube irritation with cough and phlegm may occur.

Chronic Health Effects

The following chronic (long-term) health effects may occur at some time after exposure to maleic acid and can last for months or even years:

Cancer Hazards: According to information presented in the references, maleic acid has not been adequately tested for its ability to cause cancer in animals. Additional research is required.

Reproductive Hazard: According to information presented in the references, maleic acid has not been adequately tested for its ability to adversely affect reproduction in laboratory animals.

Recommended Risk-Reduction Measures

Personnel should avoid direct contact with maleic acid. The exact nature of its toxicity is not clearly understood in the references. If a less toxic chemical cannot be substituted for a hazardous substance, then *engineering controls* are the most effective method of reducing exposures.

The best protection is to enclose operations and/or provide local exhaust ventilation at the site of chemical release. While not always operationally feasible, isolating operations can also reduce exposure. Using respiratory protection is less effective than the controls

mentioned above, but is still advisable whenever working with or around maleic acid. No exposure levels have been established for maleic acid. This does not mean that exposure is without health risk. Prudent risk management may require a conservative approach to respiratory protection until further data are known about maleic acid. For transient exposures, an air-purifying respirator equipped with an acid cartridge, a high-efficiency particulate air (HEPA) filter, and a dust/mist pre-filter may suffice. Greater protection is obtained using a supplied-air respirator or a self-contained breathing apparatus (SCBA) with full facepiece operated in pressure demand or other positive pressure mode. If a full facepiece is not available, then chemical goggles should be worn to protect the eyes. Whenever a chemical splash/dust hazard exists, a face shield and a rubber apron should be worn. To prevent hand and skin exposures, acid resistant gloves should be worn. Glove manufacturers should be contacted and permeation studies obtained *before* final glove selections have been made.

Administrative controls should also be in place to minimize the potential for human exposures. These may include written procedures or policies which specify the methods and techniques that will be practiced whenever personnel are to work with maleic acid.

All personnel should receive *training* on the use, hazards, protective measures, emergency actions, and other precautions per 29 CFR 1910.1200 (Hazard Communication), prior to the first assignment in an area where maleic acid is used or stored. Other methods to reduce exposure include:

☑ Where possible, enclose operations and use local exhaust ventilation at the site of chemical release. If local exhaust ventilation or enclosure is not used, then the use of respiratory protection should be mandatory.

☑ Always ensure that proper protective clothing is worn when using chemical substances and that personnel are trained in its use and care.

☑ Wash thoroughly immediately after exposure to maleic acid and at the end of the work shift or before eating, drinking, or smoking.

☑ Hazard warning information should be posted in the work area. In addition, as part of an on-going education and training program, all information on the health and safety hazards of maleic acid should be communicated to all exposed or potentially exposed workers.

RISK ASSESSMENT: ENVIRONMENT

General Assessment

The environment is at risk of exposure during transportation, storage, disposal, or destruction of maleic acid. In almost every scenario, the threat of environmental exposure is contingent upon the proper handling of the chemical substance. Accidental spills, large or small, can result in fire, explosion, and possible contamination of the surrounding environmental mediums (water, soil, and air).

Maleic acid is considered a non-combustible solid. It may be corrosive, especially when mixed in solution. It is incompatible with oxidizers such as chlorine or fluorine and contact can cause a violent reaction. It may corrode metals when wet. These characteristics require special consideration during any emergency situation involving a leak or spill of maleic acid.

Maleic acid can enter the environment mainly industrial discharges or spills.

Acute Ecological Effects

Acute (short-term) toxic effects may include the death of animals, birds, or fish and death or low growth rate in plants. Acute effects are seen 2 to 4 days after animals or plants are exposed to maleic acid.

Maleic acid has moderate acute toxicity to aquatic life. Insufficient data are available to evaluate or predict the short-term effects of maleic acid exposure to birds, plants, or land animals.

Chronic Ecological Effects

Chronic (long-term) toxic effects may include shortened life span, reproductive problems, lower fertility, and changes in appearance or behavior in exposed animals. These effects can be seen long after first exposure(s) to toxic chemicals. Insufficient data are available to evaluate or predict the long-term effects of maleic acid exposure to aquatic life, birds, plants, or land animals.

Water Solubility

Maleic acid is slightly soluble in water. Concentrations of 1 milligram will mix with 1 liter of water.

Persistence in the Environment

Maleic acid is moderately persistent in the aquatic environment, with a half-life of between 2 to 20 days. The half-life of a pollutant is the amount of time it takes for one half of the chemical to be degraded.

🌊 *Bioaccumulation in Aquatic Organisms*

Some substances increase in concentration, or *bioaccumulate,* in living organisms as they breathe contaminated air, drink contaminated water, or eat contaminated food. These chemicals can become concentrated in the tissues and internal organs of animals as well as humans.

Maleic acid can be expected to biodegrade quickly and should not accumulate in the edible tissues of fish.

Recommended Risk-Reduction Measures

Proper training of all transporters will reduce the likelihood of a mishap or accident resulting in a leak or spill to the environment. The correct use of DOT labeling on all containers, trucks, and rail cars will enable emergency responders to react properly and quickly to any disaster thereby reducing the potential risk to the environment and to personnel.

Storage of maleic acid should be segregated from incompatible chemicals to minimize the risk of cross-contamination or contact. Buildings designated for storage should be equipped with appropriate fire protection systems (alarms, sprinklers, emergency lighting, portable extinguishers).

If a spill or leak to the environment has occurred, fire department, emergency response and/or hazardous materials spill personnel should be notified immediately. Cleanup should be attempted only by those trained in proper spill containment procedures. Contaminated soils should be removed for incineration and replaced with clean soil. If maleic acid should contact the water table, aquifer, or navigable waterway, time is of the essence. It is miscible in water and, therefore, total containment and remediation may not be possible. When such spills occur, the local and/or state emergency response authorities must be notified. A comprehensive emergency response or disaster preparedness/recovery plan should be in place prior to any operations involving the use, transportation, storage, or disposal of maleic acid. If maleic acid is spilled or leaked, the following specific steps are recommended:

- ☑ Restrict persons not wearing protective clothing from area of spill or leak until cleanup is complete and area can be opened for normal work.
- ☑ Remove ignition sources and ventilate area.
- ☑ Collect all powdered materials in safest and most convenient manner possible and deposit in sealed drums for disposal. Do NOT dry sweep (generates airborne dusts). Use a vacuum equipped with a high efficiency particulate air (HEPA) filter instead. Damp mop residue and collect for disposal. Absorb any solution spills in vermiculite, dry sand, earth, or similar material and deposit in sealed containers.
- ☑ It may be necessary to dispose of maleic acid as a hazardous waste. The responsible state agency or the regional office of the federal Environmental Protection Agency (EPA) should be contacted for specific recommendations.

RISK ASSESSMENT: BUSINESS

General Assessment

Accidents or mishaps involving maleic acid can present a significant threat to business operations. The loss or damage of equipment or facilities can significantly affect fiscal viability. Lawsuits that may result from personnel injury/death, public exposures, and/or environmental contamination will also require a serious expenditure of resources. Media attention surrounding an injury, death, or environmental damage can also result in a loss of profits and loss of current as well as future business.

Recommended Risk-Reduction Measures

Company attorneys, safety and health professionals, and environmental specialists should be involved in the development of any procedures dealing with safety or emergency response. A company official should be pre-designated as a public relations officer with specific training in dealing with the press. Corporate plans and policies should be developed, approved, and implemented long before any need for such arises.

REFERENCES

Ahlbom, A. 1993. *Biostatistics for Epidemiologists.* New York: CRC Press/Lewis Publishers.

American Conference of Governmental Industrial Hygienists. 1988. *Documentation of the Threshold Limit Values and Biological Exposure Indices,* 5th Edition (with updates). Cincinnati: ACGIH

Asante-Duah, D. K. 1993. *Hazardous Waste Risk Assessment.* New York: CRC Press/Lewis Publishers.

Bloom, A. D. and F. J. DeSerres. 1995. *Ecotoxicity and Human Health.* New York: CRC Press/Lewis Publishers.

Brusick, D. J. 1994. *Methods for Genetic Risk Assessment.* New York: CRC Press/Lewis Publishers.

Bueche, J. R. 1972. *Principles of Physics,* 2nd Edition. New York: McGraw-Hill.

Burrell, R., D. K. Flaherty, and Leonard J. Sauers. 1992. *Toxicology of the Immune System.* New York: Van Nostrand Reinhold.

Calabrese, E. J. 1994. *Biological Effects of Low Level Exposures.* New York: CRC Press/Lewis Publishers.

Cockerman, L. G., B. S. Shane. 1994. *Basic Environmental Toxicology.* New York: CRC Press/Lewis Publishers.

Cothern, C. R. 1995. *Handbook for Environmental Risk Decisionmaking.* New York: CRC Press/Lewis Publishers.

Danse, I. R. 1991. *Common Sense Toxics in the Workplace.* New York: Van Nostrand Reinhold.

Gardner, R. W. 1994. *Chemical Intolerance.* New York: CRC Press/Lewis Publishers.

Hallenbeck, W. H. 1993. *Quantitative Risk Assessment for Environmental and Occupational Health,* 2nd Edition. New York: CRC Press/Lewis Publishers.

Hosty, J. W., P. Foster. 1990. *A Practical Guide to Chemical Spill Response.* New York: Van Nostrand Reinhold.

Howard, P. H., et. al. 1991. *Handbook of Environmental Degradation Rates.* New York: CRC Press/Lewis Publishers.

James, A., M. Lord. 1992. *Index of Chemical and Physical Data.* New York: Van Nostrand Reinhold.

Jones, F. E. 1994. *Toxic Organic Vapors in the Workplace.* New York: CRC Press/Lewis Publishers.

Keith, L. H. and M. M. Walker. 1995. *Handbook of Air Toxics.* New York: CRC Press/Lewis Publishers.

Landis, W. G., M. H. Yu. 1995. *Introduction to Environmental Toxicology.* New York: CRC Press/Lewis Publishers.

Langerman, N. 1994 *Precautionary Labels for Chemical Containers.* New York: CRC Press/Lewis Publishers.

Lauwerys, R. R., P. Hoet. 1993. *Industrial Chemical Exposure.* New York: CRC Press/Lewis Publishers.

Lewis, R. J., Sr. 1993. *Hawley's Condensed Chemical Dictionary,* 12th Edition. New York: Van Nostrand Reinhold.

Lewis, R. J., Sr. 1992. *Sax's Dangerous Properties of Industrial Materials,* Volumes 1, 2, and 3, 8th Edition. New York: Van Nostrand Reinhold.

Lide, D. R. 1995. *Handbook of Chemistry and Physics,* 76th Edition. New York: CRC Press/Lewis Publishers.

Manahan, S. E. 1994. *Environmental Chemistry,* 6th Edition. New York: CRC Press/Lewis Publishers.

Manahan, S. E. 1992. *Toxicological Chemistry,* 2nd Edition. New York: CRC Press/Lewis Publishers.

Mellström, G. A., J. E. Wahlberg, and H. I. Maibach. 1994. *Protective Gloves for Occupational Use.* New York: CRC Press/Lewis Publishers.

Meyer, E. 1990. *Chemistry of Hazardous Materials.* Englewood Cliffs, NJ: Prentice Hall.

Mycock, J. C., J. D. McKenna, and L. Theodore. 1995. *Handbook of Air Pollution Control Engineering and Technology.* New York: CRC Press/Lewis Publishers.

Nelson, G. O. 1992. *Gas Mixtures.* New York: CRC Press/Lewis Publishers.

Ott, W. R. 1995. *Environmental Statistics and Data Analysis.* New York: CRC Press/Lewis Publishers.

Patnaik, P. 1992. *A Comprehensive Guide to the Hazardous Properties of Chemical Substances.* New York: Van Nostrand Reinhold.

Philp, R. B. 1995. *Environmental Hazards and Human Health.* New York: CRC Press/Lewis Publishers.

Proctor, N. H., J. P. Hughes. 1991. *Chemical Hazards of the Workplace.* New York: Van Nostrand Reinhold.

Quigley, D. R. 1994. *Handbook of Emergency Chemical Management.* New York: CRC Press/Lewis Publishers.

Rea, W. J. 1995. *Chemical Sensitivity,* Vol. I-III. New York: CRC Press/Lewis Publishers.

Richardson, M. 1995. *Dictionary of Substance and Their Effects,* Vol. I-VI. New York: CRC Press/Lewis Publishers.

Rylander, R., R. R. Jacobs. 1994. *Organic Dusts.* New York: CRC Press/Lewis Publishers.

Sacarello, H. L. A. 1994. *The Comprehensive Handbook of Hazardous Materials.* New York: CRC Press/Lewis Publishers.

Sheftel, V. O. 1995. *Handbook of Toxic Properties of Monomers and Additives.* New York: CRC Press/Lewis Publishers.

Sherman, J. D. 1988. *Chemical Exposure and Disease.* New York: Van Nostrand Reinhold.

Talbott, E., G. Craun. 1995. *An Introduction to Environmental Epidemiology.* New York: CRC Press/Lewis Publishers.

U.S. Department of Health and Human Services, National Institute for Occupational Safety and Health. 1980. *Maleic acid. Registry of Toxic Effects of Chemical Substances,* Vol. 1. Washington, D.C.: U.S. Government Printing Office.

U.S. Department of Health and Human Services, National Institute for Occupational Safety and Health. 1994. *NIOSH Pocket Guide to Chemical Hazards.* Washington, D.C.: U.S. Government Printing Office.

Witorsch, P. and S. V. Spagnolo. 1994. *Air Pollution and Lung Disease in Adults.* New York: CRC Press/Lewis Publishers.

MATERIAL SAFETY DATA SHEET

CHEMICAL NAME	
MALEIC ANHYDRIDE	**CRC PRESS/LEWIS PUBLISHERS DISCLAIMER:** THE INFORMATION AND RECOMMENDATIONS PRESENTED HEREIN ARE BASED ON SOURCES BELIEVED TO BE RELIABLE. CRC MAKES NO REPRESENTATION ON ITS COMPLETENESS OR ACCURACY. IT IS THE RESPONSIBILITY OF THE USER TO DETERMINE THE CHEMICAL'S SUITABILITY FOR ITS INTENDED USE, THE CHEMICAL'S SAFE USE, AND THE CHEMICAL'S PROPER DISPOSAL. NO REPRESENTATIONS AND/OR WARRANTIES, EITHER EXPRESSED OR IMPLIED, OF THE MERCHANTABILITY OR FITNESS FOR A PARTICULAR PURPOSE, OR OF ANY OTHER NATURE, ARE MADE WITH RESPECT TO THE INFORMATION PROVIDED IN THIS MSDS OR TO THE CHEMICAL TO WHICH INFORMATION MAY REFER. CRC NEITHER ASSUMES NOR AUTHORIZES ANY OTHER PERSON TO ASSUME FOR IT, ANY OTHER ADDITIONAL RESPONSIBILITY OR LIABILITY FOR THE USE OF, OR RELIANCE UPON, THIS INFORMATION.

HAZARD WARNING INFORMATION

HEALTH	FIRE	REACTIVE	OTHER	DEGREE OF HAZARD	COLOR CODING	OTHER CODES
3	1	1		0 = Minimum Hazard 1 = Slight Hazard 2 = Moderate Hazard 3 = Serious Hazard 4 = Severe Hazard	HEALTH = BLUE FIRE = RED REACTIVITY = YELLOW OTHER = WHITE	OX = Oxidizer ACID = Acid ALK = Alkali COR = Corrosive W = Use No Water

SECTION I - GENERAL INFORMATION

Characterization	RCRA Number	EPA Class
Anhydride	**U147**	**Toxic Waste**

DOT Proper Shipping Name **Maleic anhydride**	Chemical Abstract Service (CAS) Number **108-31-6**
DOT Hazard Class and Label Requirements **ORM-A**	DOT Emergency Guide Code **60**
DOT Identification Number **UN 2215**	Chemical Formula $C_4H_2O_3$

Synonyms
***cis*-Butenedioic anhydride; 2,5-dihydrofuran-2,5-dione; 2,5-furandione; toxilic anhydride.**

SECTION II - HAZARDOUS INGREDIENTS AND IDENTITY INFORMATION

Hazardous Components (specific identity)	OSHA Exposure Criteria	NIOSH Exposure Criteria	Immediately Dangerous to Life and Health (IDLH)	ACGIH Exposure Criteria
Maleic anhydride (derivation: By catalytic vapor-phase oxidation of benzene or other hydrocarbons). **1 ppm = 4.08 mg/m^3**	**PEL (8-hour):** **0.25 ppm** **1.0 mg/m^3** **STEL:** **Not Established**	**REL (10-hour):** **0.25ppm** **1.0 mg/m^3** **STEL:** **Not Established**	**10 mg/m^3**	**TLV:** **0.25ppm** **1.0 mg/m^3** **STEL:** **Not Established**

SECTION III - PHYSICAL AND CHEMICAL CHARACTERISTICS

Boiling Point **396°F (202°C)**	Specific Gravity (H_2O = 1) **1.48**
Vapor Pressure (mm Hg) **0.16 at 69°F (20°C)**	Molecular Weight **98.1**
Vapor Density (Air = 1) **3.4**	Melting Point **127°F (53°C)**

Solubility
Soluble in water (forms maleic acid). Soluble in dioxane, very slightly soluble in alcohol.

Appearance and Odor
Odorless, colorless, needles or crystals, white lumps, or pellets with an irritating, choking odor. Odor Threshold = .4 ppm. May be shipped in molten form.

SECTION IV - FIRE AND EXPLOSION HAZARD DATA

Flash Point (method used) **218°F (103°C)**	Explosive Limits in Air % by Volume **LEL: 1.4% UEL: 7.1%**
NFPA Classification **Combustible Solid**	Autoignition Temperature **891°F (477°C)**

Extinguishing Media
Use alcohol foam or carbon dioxide. Do NOT use dry chemical (exothermic reaction).

Special Fire Fighting Procedures
Poisonous gases are produced in fire; wear full protective clothing and self-contained breathing apparatus (SCBA). Isolate area and deny entry to those not involved in response procedures. Move containers from fire area if it can be done without risk. If this is not possible, use water spray to keep fire-exposed containers cool. Do NOT release runoff from fire control measures to sewers or waterways.

Unusual Fire and Explosion Hazards
Reacts with water to form corrosive maleic acid. Maleic anhydride in the form of vapor, cloud, or mist may ignite spontaneously in air and may re-ignite after fire has been extinguished.

SECTION V - REACTIVITY DATA

Stability		Conditions to Avoid Extremely volatile solid but kept stable if containers are properly vented. Avoid contact with water and moisture, and keep away from heat and ignition sources.
Stable X	Unstable	Incompatibility *(materials to avoid)* Strong oxidizers (chlorine), water. Alkali, alkaline earth, metal ions, ammonium ions, caustics, amines (at temperatures above 150°F).
Hazardous Polymerization		Conditions to Avoid Maleic anhydride does not self-polymerize. However, it can undergo uncontrolled co-polymerization when mixed with olefins and catalysts. Explosive decomposition can occur in the presence of alkalis.
May Occur X	Will Not Occur	Hazardous Decomposition or By-products When heated to decomposition, maleic anhydride emits toxic and irritating fumes and vapors and carbon monoxide.

SECTION VI - HEALTH HAZARD DATA

Primary Route(s) of Entry:	Inhalation?	Absorption (skin)?	Ingestion?
	X		X

Health Hazards

INHALATION: Causes irritation of the eyes, nose, throat, and respiratory system with coughing, sneezing, dypsnea, mucous membrane irritation in the nasal cavity leading to ulceration of the septum, chemical bronchitis, or delayed pulmonary edema. Nervous system effects also possible, including mental confusion, nausea, muscular twitching, and blurred vision.

SKIN & EYES: Skin contact causes serious caustic burns, itchy rash, and dermatitis. Eye contact may cause burns, irritation, corneal scarring, and perhaps permanent damage.

INGESTION: Burning of the mouth, esophagus, and digestive tract. Pain in the chest, stomach, and painful swallowing.

Carcinogenicity	NTP Listed?	IARC Cancer Review Group?	OSHA Regulated?	Target Organs?
Unknown Human Unknown Animal	No	No	29 CFR 1910.1000 Table Z-1	Skin, eyes, respiratory system.

Medical Conditions Generally Aggravated by Exposure

Existing chronic respiratory diseases may be aggravated by exposure.

Emergency and First-aid Procedures

Eye contact: Flush immediately with water for 15 minutes (minimum); seek medical attention. Skin contact: Remove all contaminated clothing. Wash area with large amounts of soap and water. Seek medical assistance. For inhalation: Remove the person from exposure. Provide respiratory assistance and CPR. Transfer to medical facility. Observe 48 hours for lung effects. If swallowed: Contact poison control center. Unless told otherwise, give 1-2 glasses of water to dilute then induce vomiting. Save some of the vomitus for analysis. After the vomiting, have the person drink 2 tablespoons of activated charcoal in 8 oz. of water. Never give an unconscious or convulsing person anything by mouth.

SECTION VII - PRECAUTIONS FOR SAFE HANDLING AND USE

Steps to be Taken in Case Material is Released or Spilled

Collect powdered material using a vacuum equipped with HEPA filter; deposit in sealed drums for disposal. Avoid skin contact. Do NOT dry sweep (generates dusts). Allow molten materials to solidify, then treat as a solid. Absorb liquids in vermiculite and collect for disposal. Keep water away. Ventilate area.

Preferred Waste Disposal Method

No citation.

Precautions to be Taken in Handling and Storage

Do not store maleic anhydride in the presence of incompatible chemicals or materials. Store in tightly closed containers in a cool, dark, well-ventilated area. Personnel working with maleic anhydride should be trained on its proper handling and storage *prior* to being assigned to such responsibilities.

Other Precautions and Warnings

Keep away from water and incompatible materials.

SECTION VIII - CONTROL MEASURES AND PERSONAL PROTECTIVE EQUIPMENT

Respiratory Protection *(specify type)*

PEL is extremely low (0.25 ppm). Use a self-contained breathing apparatus (SCBA) with full facepiece or a supplied-air respirator operated in pressure demand or other positive pressure mode.

Ventilation

Local exhaust or general mechanical systems recommended.

Protective Gloves	Eye Protection	Other Protective Clothing
Impervious Material	Chemical/Dust Goggles or Face Mask	Protective Apron

Work/Hygiene Practices

Always wash hands thoroughly after using chemical; never bring food, drink, or smoking materials into vicinity of chemicals.

MALEIC ANHYDRIDE

$C_4H_2O_3$

CAS: 108-31-6

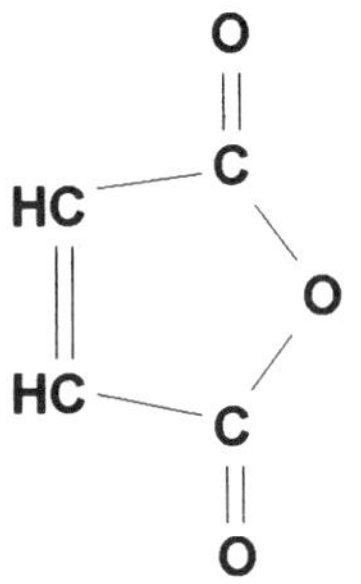

IDENTIFICATION AND TYPICAL USES

Maleic anhydride appears as colorless needles or crystals, white lumps, pellets, briquettes, rods, or flakes with an acrid, irritating, choking odor. It is often shipped in molten form. It is used primarily in the manufacture of polyester resins, in alkyd coating resins, in fumaric and tartaric acid, in pesticides, as a preservative in oils and fats, and in permanent-press (textiles).

RISK ASSESSMENT: HEALTH

General Assessment

Maleic anhydride is a human poison primarily by *inhalation,* causing toxic systemic effects targeting the respiratory system. It can also cause toxic effects by *ingestion*. It is a questionable carcinogen with human mutation data reported. There are no reports of its teratogenic effects in humans or animals.

Inhalation of maleic anhydride dusts can be irritating to the nose, throat, mucosa, and the upper respiratory tract. Asthmatic-like symptoms include coughing, sneezing, dypsnea, ulceration of the nasal septa, chemical bronchitis, and the possibility of pulmonary edema (a dangerous buildup of fluid in the lungs). Pulmonary edema can be delayed up to 48 hours or more, thereby creating a false sense of security with regard to health exposure risk. There can be severe headache, nose bleeds, nausea, and temporary vision impairment (photophobia, double vision, blurred vision). There can also be effects subsequent to depression of the central nervous system. These effects can include headache, nausea, dizziness, drowsiness, lightheadedness, muscular twitching, tremors, mental confusion, coma, and, in severe cases, death.

Ingestion of maleic anhydride can cause burning of the mouth, esophagus, and gastrointestinal tract with deep, burning pain in the chest and stomach and pain on swallowing.

Skin contact can cause blistering, dermatitis, inflammation, and thermal burns. Sensitization can also occur in some individuals.

Acute Health Effects

The following acute (short-term) health effects may occur immediately or shortly after exposure to maleic anhydride:

Skin: Irritation and possible rash. Moisture on skin's surface will enhance the reaction.

Eye: Severe burns, corneal scarring, and possible damage to vision.

Lung: Severe irritation of the nose, throat, and esophagus. Nasal cavity irritation can lead to ulceration of the septa. Delayed pulmonary edema is possible.

CNS: Dizziness, headaches, nausea, vomiting, muscle spasms, tremor, fatigue, convulsions, dypsnea, nervousness, drowsiness, loss of coordination, coma, and even death.

Chronic Health Effects

The following chronic (long-term) health effects may occur at some time after exposure to maleic anhydride and can last for months or even years:

Cancer Hazards: According to the information presented in the references, maleic anhydride has been shown to cause cancer in test animals. Human mutation data have also been reported.

Reproductive Hazard: According to the information presented in the references, maleic anhydride has not been adequately tested for its ability to adversely affect reproduction.

Other Chronic Effects: Prolonged contact with maleic anhydride can lead to the development of bronchitis and asthma-like symptoms. Dermal and respiratory sensitization are also possible.

ʊ *Recommended Risk-Reduction Measures*

Personnel should avoid direct contact with maleic anhydride. It is a toxic, human poison on inhalation. Its human carcinogenic properties are not clearly understood in the references. If a less toxic chemical cannot be substituted for a hazardous substance, *engineering controls* are the most effective method of reducing exposures. The best protection is to enclose operations and/or provide local exhaust ventilation at the site of chemical release. While not always operationally feasible, isolating operations can also reduce exposure.

Using respiratory protection is less effective than the controls mentioned above, but is still advisable whenever working with or around maleic anhydride. This chemical has an extremely low permissible exposure limit (0.25 ppm) that is difficult to accurately monitor. Therefore, prudent risk management requires a conservative approach to personal protection. While a chemical respirator with a dust/mist pre-filter may suffice, the best protection is obtained using a supplied-air respirator operated in continuous flow mode or a self-contained breathing apparatus (SCBA) with full facepiece operated in pressure demand or positive pressure mode. If a full facepiece is not available, then chemical/dust goggles should be worn to protect the eyes. Whenever maleic anhydride is being mixed with liquids, a chemical splash hazard exists and a face shield and rubber apron should be worn. To prevent hand and skin exposures, protective gloves should be used. Glove manufacturers should be contacted and permeation studies obtained *before* final glove selections have been made.

Administrative controls should also be in place to minimize the potential for human exposures. These may include written procedures or policies which specify the methods and techniques that will be practiced whenever personnel are to work with maleic anhydride.

All personnel should receive *training* on the use, hazards, protective measures, emergency actions, and other precautions per 29 CFR 1910.1200 (Hazard Communication), prior to the first assignment in an area where maleic anhydride is used or stored.

If symptoms develop or overexposure is suspected, the following medical tests may be useful:

☑ Lung function tests.

☑ Consider chest X-ray after an acute overexposure to maleic anhydride (may be negative if taken immediately following exposure due to delayed development of pulmonary edema).

☑ Evaluation by a qualified allergist with careful consideration of exposure history and special testing (may help diagnose skin allergy).

Any evaluation should include a careful medical history of past and present symptoms with an examination. However, medical tests that evaluate existing damage are not a substitute for controlling exposure. Also, since smoking can cause heart disease, emphysema, lung cancer, and other respiratory diseases, smokers may experience symptoms more rapidly than non-smokers under the same conditions of exposure. Prudent risk management requires proper consideration of *all* possible factors that may be causing the appearance of exposure symptoms in the workplace.

Other methods to reduce exposure include:

☑ Where possible, enclose operations and use local exhaust ventilation at the site of chemical release. If local exhaust ventilation or enclosure is not used, then respiratory protection should be mandatory and its use enforced.

☑ Always ensure that proper protective clothing is worn when using chemical substances and that personnel are trained in its use, care, and maintenance.

☑ Wash thoroughly immediately after exposure to maleic anhydride and at the end of the work shift or before eating, drinking, or smoking.

☑ Hazard warning information should be posted in the work area. In addition, as part of an on-going education and training program, all information on the health and safety hazards of maleic anhydride should be communicated to all exposed or potentially exposed workers.

☑ Eye wash stations should be provided in work areas. If the potential for whole body exposure exists, then safety showers should also be provided in the immediate work area.

RISK ASSESSMENT: ENVIRONMENT

General Assessment

The environment is at risk of exposure during transportation, storage, disposal, or destruction of maleic anhydride. In almost every scenario, the threat of environmental exposure is contingent upon the proper handling of the chemical substance. Accidental spills, large or small, can result in fire, explosion, and possible contamination of the surrounding environmental mediums (water, soil, and air).

Maleic anhydride is considered a combustible solid. It can react exothermically with water or moisture to produce maleic acid which is corrosive to most metals. Maleic anhydride mist, dust, or vapor can explode in mixtures with air in contact with an ignition source. These characteristics require special consideration during any emergency situation involving a leak or spill of maleic anhydride or maleic anhydride mixtures. Should maleic anhydride ever come into contact with other incompatible substances, such as strong oxidizers, alkalis, caustics, or heated amines, either during use, transportation, or storage, violent and even explosive reactions can occur. Maleic anhydride can enter the environment from industrial discharges and spills.

Acute Ecological Effects

Acute (short-term) toxic effects may include the death of animals, birds, or fish and death or low growth rate in plants. Acute effects are seen 2 to 4 days after animals or plants are exposed to maleic anhydride.

Maleic anhydride has slight acute toxicity to aquatic life. It has caused bud-kill in tung trees (an agricultural crop). Insufficient data are available to evaluate or predict the short-term effects of maleic anhydride to birds or land animals.

Chronic Ecological Effects

Chronic toxic effects may include shortened life span, reproductive problems, lower fertility, and changes in appearance or behavior in exposed animals. These effects can be seen long after first exposure(s) to toxic chemicals. Insufficient data are available to evaluate or predict the long-term effects of maleic anhydride to aquatic life, plants, birds, or land animals.

Water Solubility

Maleic anhydride is only slightly soluble in cold (86°F/30°C) water. It reacts to form maleic acid which is corrosive.

Persistence in the Environment

Maleic anhydride is non-persistent in water, with a half-life of less than 2 days. The half-life of a pollutant is the amount of time it takes for one half of the chemical to be degraded.

Bioaccumulation in Aquatic Organisms

Some substances increase in concentration, or *bioaccumulate,* in living organisms as they breathe contaminated air, drink contaminated water, or eat contaminated food. These chemicals can become concentrated in the tissues and internal organs of animals as well as humans.

Maleic anhydride is not expected to accumulate in the edible tissues of fish normally consumed by humans.

Recommended Risk-Reduction Measures

Proper training of all transporters will reduce the likelihood of a mishap or accident resulting in a leak or spill to the environment. The correct use of DOT labeling (per the MSDS) on all containers, trucks, and rail cars will enable emergency responders to react properly and quickly to any disaster thereby reducing the potential risk to the environment and to personnel.

Storage of maleic anhydride should be segregated from incompatible chemicals to minimize the risk of cross-contamination or contact. Precautions must be taken to ensure all contact with water or moisture is prevented in use or storage.

If a spill or leak to the environment has occurred, fire department, emergency response and/or hazardous materials spill personnel should be notified immediately. Cleanup should be attempted only by those trained in proper spill containment procedures. Contaminated soils should be removed for incineration and replaced with clean soil. If maleic anhydride should contact the water table, aquifer, or navigable waterway, remediation activities should be prompt. When such spills occur, the local and/or state emergency response authorities must be notified. A comprehensive emergency response or disaster preparedness/recovery plan should be in place prior to any operations involving the use, transportation, storage, or disposal of maleic anhydride. If maleic anhydride is spilled or leaked, the following specific steps are recommended:

- ☑ Restrict persons not wearing protective clothing from area of spill or leak until cleanup is complete and area can be opened to normal work.

☑ Collect powdered material in the most convenient (safe) manner possible and deposit in sealed containers. Do NOT dry sweep (generates airborne dusts). Use a vacuum equipped with a HEPA filter instead. If liquid solutions are spilled, use vermiculite to absorb and place in a sealed drum.

☑ It may be necessary to dispose of maleic anhydride as a hazardous waste. The responsible state agency or the regional office of the federal Environmental Protection Agency (EPA) should be contacted for specific recommendations.

RISK ASSESSMENT: BUSINESS

General Assessment

Accidents or mishaps involving maleic anhydride can present a significant threat to business operations. The loss or damage of equipment or facilities can significantly affect fiscal viability. Lawsuits that may result from personnel injury/death, public exposures, and/or environmental contamination will also require a serious expenditure of resources. Media attention surrounding an injury, death, or environmental damage can also result in a loss of profits and loss of current as well as future business. Always remember that anytime the terms "cancer" or "carcinogen" are used, public hysteria, emotion, and ignorance can run equally high. This should be carefully considered when developing or implementing any public relations policies.

Recommended Risk-Reduction Measures

Company attorneys, safety and health professionals, and environmental specialists should be involved in the development of any procedures dealing with safety or emergency response. A company official should be pre-designated as a public relations officer with specific training in dealing with the press. Corporate plans and policies should be developed, approved, and implemented long before any need for such arises.

REFERENCES

American Conference of Governmental Industrial Hygienists. 1988. *Documentation of the Threshold Limit Values and Biological Exposure Indices*, 5th Edition (with updates). Cincinnati: ACGIH

Anthony, C. P., and N. J. Kolthoff. 1971. *Textbook of Anatomy and Physiology.* St. Louis: C. V. Mosby Company.

Asante-Duah, D. K. 1993. *Hazardous Waste Risk Assessment.* New York: CRC Press/Lewis Publishers.

Bloom, A. D. and F. J. DeSerres. 1995. *Ecotoxicity and Human Health.* New York: CRC Press/Lewis Publishers.

Brusick, D. J. 1994. *Methods for Genetic Risk Assessment.* New York: CRC Press/Lewis Publishers.

Bueche, J. R. 1972. *Principles of Physics,* 2nd Edition. New York: McGraw-Hill.

Burrell, R., D. K. Flaherty, and L. J. Sauers. 1992. *Toxicology of the Immune System.* New York: Van Nostrand Reinhold.

Calabrese, E. J. 1994. *Biological Effects of Low Level Exposures.* New York: CRC Press/Lewis Publishers.

Cockerman, L. G., B. S. Shane. 1994. *Basic Environmental Toxicology.* New York: CRC Press/Lewis Publishers.

Cothern, C. R., N. P. Ross. 1994. *Environmental Statistics, Assessment, and Forecasting.* New York: CRC Press/Lewis Publishers.

Gardner, R. W. 1994. *Chemical Intolerance.* New York: CRC Press/Lewis Publishers.

Hodgson, E., R. B. Mailman, and Janice E. Chambers. 1988. *Dictionary of Toxicology.* London: The MacMillian Company.

Howard, P. H. 1995. *Handbook of Environmental Fate and Exposures Data for Organic Chemicals,* Vol. I-IV. New York: CRC Press/Lewis Publishers.

Langerman, N. 1994 *Precautionary Labels for Chemical Containers.* New York: CRC Press/Lewis Publishers.

Lauwerys, R. R., P. Hoet. 1993. *Industrial Chemical Exposure.* New York: CRC Press/Lewis Publishers.

Lewis, R. J., Sr. 1993. *Hawley's Condensed Chemical Dictionary,* 12th Edition. New York: Van Nostrand Reinhold.

Lewis, R. J., Sr. 1992. *Sax's Dangerous Properties of Industrial Materials,* Volumes 1, 2, and 3, 8th Edition. New York: Van Nostrand Reinhold.

Manahan, S. E. 1992. *Toxicological Chemistry,* 2nd Edition. New York: CRC Press/Lewis Publishers.

Meyer, E. 1990. *Chemistry of Hazardous Materials.* Englewood Cliffs, NJ: Prentice Hall.

Patnaik, P. 1992. *A Comprehensive Guide to the Hazardous Properties of Chemical Substances.* New York: Van Nostrand Reinhold.

Proctor, N. H., and James P. Hughes. 1991. *Chemical Hazards of the Workplace.* New York: Van Nostrand Reinhold.

Quigley, D. R. 1994. *Handbook of Emergency Chemical Management.* New York: CRC Press/Lewis Publishers.

Rea, W. J. 1995. *Chemical Sensitivity,* Vol. I-III. New York: CRC Press/Lewis Publishers.

Richardson, M. 1995. *Dictionary of Substance and Their Effects,* Vol. I-VI. New York: CRC Press/Lewis Publishers.

Rylander, R., R. R. Jacobs. 1994. *Organic Dusts.* New York: CRC Press/Lewis Publishers.

Sacarello, H. L. A. 1994. *The Comprehensive Handbook of Hazardous Materials.* New York: CRC Press/Lewis Publishers.

U.S. Department of Health and Human Services, National Institute for Occupational Safety and Health. 1994. *NIOSH Pocket Guide to Chemical Hazards.* Washington, D.C.: U.S. Government Printing Office.

MATERIAL SAFETY DATA SHEET

CHEMICAL NAME	
MANGANESE	**CRC PRESS/LEWIS PUBLISHERS DISCLAIMER:** THE INFORMATION AND RECOMMENDATIONS PRESENTED HEREIN ARE BASED ON SOURCES BELIEVED TO BE RELIABLE. CRC MAKES NO REPRESENTATION ON ITS COMPLETENESS OR ACCURACY. IT IS THE RESPONSIBILITY OF THE USER TO DETERMINE THE CHEMICAL'S SUITABILITY FOR ITS INTENDED USE, THE CHEMICAL'S SAFE USE, AND THE CHEMICAL'S PROPER DISPOSAL. NO REPRESENTATIONS AND/OR WARRANTIES, EITHER EXPRESSED OR IMPLIED, OF THE MERCHANTABILITY OR FITNESS FOR A PARTICULAR PURPOSE, OR OF ANY OTHER NATURE, ARE MADE WITH RESPECT TO THE INFORMATION PROVIDED IN THIS MSDS OR TO THE CHEMICAL TO WHICH INFORMATION MAY REFER. CRC NEITHER ASSUMES NOR AUTHORIZES ANY OTHER PERSON TO ASSUME FOR IT, ANY OTHER ADDITIONAL RESPONSIBILITY OR LIABILITY FOR THE USE OF, OR RELIANCE UPON, THIS INFORMATION.

HAZARD WARNING INFORMATION

HEALTH	FIRE	REACTIVE	OTHER	DEGREE OF HAZARD	COLOR CODING	OTHER CODES
3	2	1	~~W~~	0 = Minimum Hazard 1 = Slight Hazard 2 = Moderate Hazard 3 = Serious Hazard 4 = Severe Hazard	HEALTH = BLUE FIRE = RED REACTIVITY = YELLOW OTHER = WHITE	OX = Oxidizer ACID = Acid ALK = Alkali COR = Corrosive ~~W~~ = Use No Water

SECTION I - GENERAL INFORMATION

Characterization	RCRA Number	EPA Class
Metal (mineral)	**None**	**Not Applicable**

DOT Proper Shipping Name **No Citation**	Chemical Abstract Service (CAS) Number **7439-96-5**
DOT Hazard Class and Label Requirements **No Citation**	DOT Emergency Guide Code **No Citation**
DOT Identification Number **No Citation**	Atomic Symbol **Mn**

Synonyms
Manganese metal powder; colloidal manganese; manganese-55; magnacat.

SECTION II - HAZARDOUS INGREDIENTS AND IDENTITY INFORMATION

Hazardous Components (specific identity)	OSHA Exposure Criteria	NIOSH Exposure Criteria	Immediately Dangerous to Life and Health (IDLH)	ACGIH Exposure Criteria
Manganese (derivation: By reduction of the oxide with aluminum or carbon. Pure manganese is obtained electrolytically from sulfate or chloride solution. It is also found natural in iron ores).	**PEL (ceiling):** **5 mg(Mn)/m^3** **(dust/compounds)** **STEL:** **Not Applicable**	**REL (ceiling):** **5 mg(Mn)/m^3** **(dust/compounds)** **STEL:** **Not Applicable**	**500 mg(Mn)/m^3**	**TLV:** **5 mg(Mn)/m^3** **(dust/compounds)** **1 mg(Mn)/m^3** **(fumes)** **STEL:** **3 mg(Mn)/m^3** **(fumes)**

SECTION III - PHYSICAL AND CHEMICAL CHARACTERISTICS

Boiling Point **3564°F (1962°C)**	Specific Gravity (H_2O = 1) **7.20**
Vapor Pressure (mm Hg) **1 at 2358°F (1292°C)**	Atomic Weight **54.94**
Vapor Density (Air = 1) **Not Found**	Melting Point **2271°F (1244°C)**

Solubility
Insoluble in water (decomposes); readily dissolves in dilute mineral acids.

Appearance and Odor
Silvery-gray or reddish-gray powder or lustrous metal with no odor.

SECTION IV - FIRE AND EXPLOSION HAZARD DATA

Flash Point (method used) **Not Found**	Explosive Limits in Air % by Volume **LEL: Not Found** **UEL: Not Found**
NFPA Classification **Combustible Solid**	Autoignition Temperature **Not Found**

Extinguishing Media
Use a dry chemical extinguishing agent suitable for metal fires (Class D).

Special Fire Fighting Procedures
Wear full protective clothing and self-contained breathing apparatus (SCBA). Poisonous gases are produced in fire. Do NOT look directly into manganese fires. Fight fire from distance, if possible.

Unusual Fire and Explosion Hazards
Manganese dusts or powders are flammable and can present a dust explosion hazard in mixtures with air or when heated in carbon dioxide. Mixtures with aluminum dusts can explode in air. It can react slowly with water to form explosive hydrogen gas.

SECTION V - REACTIVITY DATA

Stability		Conditions to Avoid **Manganese is normally stable under routine conditions of handling and storage. Avoid contact with incompatible materials, heat, flame, sparks, or other ignition sources, and keep moisture away.**
Stable **X**	Unstable	Incompatibility *(materials to avoid)* **Incompatible with hydrogen peroxide, bromine pentafluoride, fluorine, chlorine with heat, and sulfur dioxide with heat. It is incompatible with oxidants, nitrogen dioxide, nitric acid, phosphorus, and nitryl fluoride.**
Hazardous Polymerization		Conditions to Avoid **Hazardous polymerization of manganese will not occur. May ignite spontaneously in air or moisture (forms explosive hydrogen gas on contact with water).**
May Occur	Will Not Occur **X**	Hazardous Decomposition or By-products **Poisonous gases are emitted during fire, including toxic and irritating smoke and fumes of manganese oxide.**

SECTION VI - HEALTH HAZARD DATA

Primary Route(s) of Entry:	Inhalation? X	Absorption (skin/eye)?	Ingestion? X

Health Hazards

INHALATION: **The dusts can produce irritation of the eyes and mucous membranes. The fumes can lead to metal fume fever with symptoms of fever, chills, nausea, weakness, headache, blurred vision, back stiffness and pain, muscle pain, shallow respiration, dryness in mouth and throat, intense thirst, dry cough, metallic taste, chest tightness, and Mn pneumonitis. It can also lead to degenerative brain effects with symptoms similar to Parkinson's disease.**

EYES & SKIN: **Causes skin and eye irritation. There can be dermatitis, conjunctivitis, corneal damage.**

INGESTION: **Large doses can cause gastrointestinal irritation and other unspecified effects.**

Carcinogenicity	NTP Listed?	IARC Cancer Review Group?	OSHA Regulated?	Target Organs?
Unknown Human Questioned Animal	**No**	**No**	**No**	**Respiratory system, eyes, skin, CNS, kidneys, blood.**

Medical Conditions Generally Aggravated by Exposure

Existing kidney dysfunction, brain, or muscle problems may be aggravated by exposure.

Emergency and First-aid Procedures

Eye contact: Do not allow victim to keep eyes tightly shut. Flush immediately with water for 15 minutes (minimum); seek medical attention. Skin contact: Remove all contaminated clothing. Immediately wash area with flooding amounts of water. For reddened or blistered skin, seek medical assistance. For inhalation: Remove the person from exposure. Provide respiratory assistance and CPR. Transfer to medical facility. If swallowed: Seek medical attention. Rinse mouth with water. Give 1 to 2 glasses of water and induce vomiting until vomit is clear. Never give anything by mouth to a convulsing person.

SECTION VII - PRECAUTIONS FOR SAFE HANDLING AND USE

Steps to be Taken in Case Material is Released or Spilled

Collect powdered materials in most convenient and safe manner and deposit in clean, dry, sealed, metal containers. Dry sweep not recommended. Use HEPA vacuum. Ventilate area of spill. Restrict those not involved in cleanup from entering area.

Preferred Waste Disposal Method

Reclaim scrap.

Precautions to be Taken in Handling and Storage

Store to avoid contact with incompatible materials. Prevent physical damage to containers. Keep containers tightly closed. Keep moisture and sources of heat away. Use non-sparking tools.

Other Precautions and Warnings

Avoid excessive dust generation in storage and handling areas. Personnel should be trained on the proper precautions for working with manganese compounds *before* initial assignment.

SECTION VIII - CONTROL MEASURES AND PERSONAL PROTECTIVE EQUIPMENT

Respiratory Protection *(specify type)*

For dust exposures, use a respirator with a HEPA filter; otherwise, use a supplied-air respirator or a self-contained breathing apparatus with full facepiece operated in positive pressure mode.

Ventilation

Local exhaust (preferred) at site of chemical work or general ventilation.

Protective Gloves	Eye Protection	Other Protective Clothing
Impervious Material	**Chemical/Dust Goggles or Face Mask**	**Protective Apron, Boots**

Work/Hygiene Practices

Always wash hands thoroughly after using chemical; never bring food, drink, or smoking materials into vicinity of chemicals.

MANGANESE

Mn

CAS: 7439-96-5

IDENTIFICATION AND TYPICAL USES

Manganese appears as a silvery-white or reddish gray, brittle, solid or powder with no odor. It is metallic element of atomic number 25, group VIIB of the periodic table, atomic weight 64.9380, and valance of 2, 3, 4, 6, 7 and no stable isotopes. Pure manganese cannot be fabricated. It is used in the manufacture of steel (ferroalloys) as well as for improving corrosion resistance and hardness. It is a high-purity salt for various chemical uses, is used as a purifying and scavenging agent in metal production, and in the manufacture of aluminum by the Toth process.

RISK ASSESSMENT: HEALTH

General Assessment

Even though manganese is an essential mineral in human nutrition, it can also be toxic by *inhalation* of its fumes and/or by *ingestion* of the powders and dusts. While it has been shown to cause tumors in experimental test animals with mutation data also reported, the exact nature of its human carcinogenic properties is unknown.

Inhalation of manganese fumes can cause acute or chronic poisoning with primary effects on the immune system characterized by symptoms of metal fume fever. These can include fever, chills, general weakness, body aches, frontal headache, blurred vision, back pain and stiffness, muscular cramps, shallow and labored respiration, dryness in the mouth and throat with intense thirst, dry cough, metallic taste, and tightness in the chest occurring several hours after exposure. Lung irritation can become severe with inflammation causing manganese pneumonitis. Other effects of central nervous system depression include headache, insomnia, changes in personality, emotions, and attitude, irritability, inappropriate laughter or crying, hallucinations, effects on vision (double vision, blurred vision), uncontrolled impulse behavior, euphoria, excessive salivation, trembling, staggering gait, and signs and symptoms normally associated with Parkinson's disease.

Acute Health Effects

The following acute (short-term) health effects may occur immediately or shortly after human exposure to manganese:

Skin: A skin irritant that can cause dermatitis in some individuals.

Eye: Irritation to the eye and associated tissues. It can cause conjunctivitis and corneal damage.

Lung: Irritation of the mucosa of the eyes, nose, throat, and respiratory tract.

CNS: Breathing heated fumes can cause metal fume fever with specific action on the central nervous system.

Chronic Health Effects

The following chronic (long-term) health effects may occur at some time after exposure to manganese and can last for months or even years:

Cancer Hazards: According to the references, manganese has been shown to cause tumors in test animals. Its human carcinogenic potential is unknown. Mutation data have been reported.

Reproductive Hazard: According to the references, there is some limited evidence that chronic overexposure to manganese may cause impotence in human males.

Other Chronic Effects: Prolonged or repeated exposure can damage the central nervous system and cause impaired mental function and degenerative brain changes with personality disorders and physical changes in motor control, vision, and other body functions. Chronic exposure may cause dermatitis.

ⓘ *Recommended Risk-Reduction Measures*

While manganese is an essential mineral in human nutrition, it can be extremely toxic under certain exposure conditions and personnel should, therefore, avoid direct occupational contact with manganese. If a less toxic material or compound cannot be substituted for manganese, then *engineering controls* are the most effective method of reducing exposures. The best protection is to enclose operations and/or provide local exhaust ventilation at the site of manganese dust release. While not always operationally feasible, isolating operations can also reduce exposure. Using respiratory protection is less effective than the controls mentioned above, but is still advisable whenever working with or around manganese. The exposure ceiling for manganese is relatively low (5mg/m^3) and difficult to accurately monitor. For low or transient exposures, a chemical cartridge respirator equipped with a high-efficiency particulate air (HEPA) filter and a dust/mist pre-filter may provide adequate respiratory protection. Better protection is obtained using a supplied-air respirator operated in continuous flow mode or a self-contained breathing apparatus (SCBA) with full facepiece operated in pressure demand. If a full facepiece is not available, then chemical/dust goggles should be worn to protect the eyes. A face shield and protective apron should also be worn. To prevent hand and skin exposures, gloves should be used. Glove manufacturers should be contacted *before* selections are made.

Administrative controls should also be in place to minimize the potential for human exposures. These may include written procedures or policies which specify the methods and techniques that will be practiced whenever personnel are to work with manganese.

All personnel should receive *training* on the use, hazards, protective measures, emergency actions, and other precautions per 29 CFR 1910.1200 (Hazard Communication), prior to the first assignment in an area where manganese is used or stored.

Before beginning employment working with manganese, and at regular intervals thereafter (e.g., annually), those personnel with frequent or potentially high exposures should be provided the following recommended medical tests:

☑ Kidney function tests (establish baseline).

☑ Complete blood count (establish baseline).

If symptoms develop or overexposure is suspected, consider the following additional tests:

☑ Kidney function tests (compare to baseline).

☑ Complete blood count (compare to baseline).

☑ Neurological examination.

☑ Lung function tests.

Any evaluation should include a careful history of past and present symptoms with an examination. Medical tests that simply look for existing damage are not a substitute for controlling exposures.

Other methods to reduce exposure include:

☑ Where possible, enclose operations and use local exhaust ventilation at the site of chemical release. If local exhaust ventilation or enclosure is not used, respiratory protection should be mandatory.

☑ Always ensure that proper protective clothing is worn when using chemical substances. Proper maintenance and care of protective equipment is essential to its proper function.

☑ Wash thoroughly immediately after exposure to manganese and at the end of the work shift or before eating, drinking, or smoking.

☑ Hazard warning information should be posted in the work area. In addition, as part of an on-going education and training program, all information on the health and safety hazards of manganese should be communicated to all exposed and potentially exposed workers.

☑ Eye wash stations should be provided in the immediate work area for emergency use. If there is a possibility of skin exposure to manganese, emergency shower facilities should be provided.

☑ Workers whose clothing has been contaminated by manganese should change into clean clothes before leaving work. Never wear contaminated clothing home where family members can be exposed. Contaminated work clothing should be laundered only by individuals who have been informed of the hazards of exposure to manganese.

RISK ASSESSMENT: ENVIRONMENT

General Assessment

As a naturally occurring mineral, manganese poses very little threat to the environment. However, accidental spills or releases to the environment can occur during transportation, storage, disposal, or destruction of manganese. In almost every scenario, the threat of environmental exposure is contingent upon the proper handling of the chemical substance. Accidental spills, large or small, where manganese contacts incompatible commodities can result in release of toxic products (depending upon conditions of spill), and possible contamination of the surrounding environmental mediums (water, soil, and air).

Manganese is considered a combustible solid. Finely divided dusts and powders may create an explosive mixture in air, in contact with moisture, oxidizing materials, or in contact with an ignition source. Mixtures of manganese dust and aluminum dusts (not an improbable combination) can explode in contact with air. It reacts with water and moisture to evolve explosive hydrogen gas. It is also incompatible with many other materials, including nitrogen dioxide, nitric acid, phosphorus, and nitryl fluoride. Extreme caution is always required in handling, storage, transportation, and disposal of manganese. These characteristics also require special consideration during any emergency situation involving a leak or spill of manganese powder or dust. Should manganese ever come into contact with these various incompatible substances either during use, transportation, storage, or disposal, violent and even explosive reactions are possible.

Manganese occurs naturally in the environment (terrestrial soils and aquatic sediments) and may enter the environment as a result of industrial and municipal waste treatment plant discharges and from spills.

☠ *Acute Ecological Effects*

Acute (short-term) toxic effects may include the death of animals, birds, or fish and death or low growth rate in plants. Acute effects are seen 2 to 4 days after animals or plants are exposed to manganese.

Manganese and its compounds have moderate acute toxicity to aquatic life. Manganese is considered essential for plant and animal life. However, insufficient data are available on the acute effects of excessive manganese contamination on the other ecosystems (plants, birds, or land animals).

Chronic Ecological Effects

Chronic (long-term) toxic effects may include shortened life span, reproductive problems, lower fertility, and changes in appearance or behavior in exposed animals. These effects can be seen long after first exposure(s) to toxic chemicals.

Manganese and its compounds have moderate chronic toxicity to aquatic life. Manganese is considered essential for plant and animal life. However, insufficient data are available on the chronic effects of excessive manganese contamination on the other ecosystems (plants, birds, or land animals).

Water Solubility

Manganese and its compounds vary in their solubility in water from very soluble to nearly insoluble. In water, it reacts and decomposes to form explosive hydrogen gas.

⌛ *Persistence in the Environment*

Manganese and its compounds are highly persistent in water, with a half-life greater than 200 days. The half-life of a pollutant is the amount of time it takes for one half of the chemical to be degraded. Manganese is also naturally present in terrestrial soils and aquatic sediments.

Bioaccumulation in Aquatic Organisms

Some substances increase in concentration, or *bioaccumulate,* in living organisms as they breathe contaminated air, drink contaminated water, or eat contaminated food. These chemicals can become concentrated in the tissues and internal organs of animals as well as humans.

The concentration of manganese found in fish tissues is expected to be about the same as the average concentration of manganese in the water from which the fish was taken.

Recommended Risk-Reduction Measures

Proper training of all transporters will reduce the likelihood of a mishap or accident resulting in a leak or spill of manganese dusts or powders into the environment. The correct use of labeling on all containers, trucks, and rail cars will enable emergency responders to react properly and quickly to any disaster thereby reducing the potential risk to the environment and to personnel.

Storage of manganese should be segregated from incompatible chemicals to minimize the risk of cross-contamination or contact. Care should be taken to prevent physical damage to storage containers. Also, dusty conditions must be avoided in storage and use areas. When manganese comes into contact with water, it will readily liberate excessive amounts of highly flammable and explosive hydrogen gas.

If a spill or leak to the environment has occurred, fire department, emergency response and/or hazardous materials spill personnel should be notified immediately. Cleanup should be attempted only by those

trained in proper spill containment procedures using non-sparking tools. Contaminated soils should be removed for decontamination and replaced with clean soil. If manganese should contact the water table, aquifer, or navigable waterway, time is of the essence. It will slowly decompose in water and total remediation may not be possible. The local and/or state emergency response authorities may require notification. A comprehensive emergency response or disaster preparedness/recovery plan should be in place prior to any operations involving the use, transportation, storage, or disposal of manganese.

If manganese is spilled or leaked, the following specific steps are recommended:

☑ Restrict persons not wearing protective clothing from area of spill or leak until cleanup is complete and area can be opened for normal work.

☑ Remove all ignition sources from area.

☑ Ventilate area of spill or leak.

☑ Collect powdered materials in the safest manner possible and deposit in sealed containers. Recommend using a vacuum equipped with a high-efficiency particulate air (HEPA) filter. Do not dry sweep (generates airborne dusts). Keep water away (do NOT damp mop residues).

☑ It may be necessary to dispose of manganese as a hazardous waste. The responsible state agency or the regional office of the federal Environmental Protection Agency (EPA) should be contacted for specific recommendations.

RISK ASSESSMENT: BUSINESS

General Assessment

Accidents or mishaps involving manganese can present a significant threat to business operations. The loss or damage of equipment or facilities can significantly affect fiscal viability. Lawsuits that may result from personnel injury/death, public exposures, and/or environmental contamination will also require a serious expenditure of resources. Media attention surrounding an injury, death, or environmental damage can also result in a loss of profits and loss of current as well as future business. Always remember that anytime the terms "cancer," "carcinogen," or "reproductive hazard" are used, public hysteria, emotion, and ignorance can run equally high. This should be carefully considered during the development of public relations policies.

Recommended Risk-Reduction Measures

Company attorneys, safety and health professionals, and environmental specialists should be involved in the development of any procedures for responding to chemical incidents. A company official should be predesignated as a public relations officer with specific training in dealing with the press. Corporate plans and policies should be developed, approved, and implemented long before any need for such arises.

REFERENCES

American Conference of Governmental Industrial Hygienists. 1988. *Documentation of the Threshold Limit Values and Biological Exposure Indices*, 5th Edition (with updates). Cincinnati: ACGIH

Anthony, C. P., and N. J. Kolthoff. 1971. *Textbook of Anatomy and Physiology*. St. Louis: C. V. Mosby Company.

Bloom, A. D. and F. J. DeSerres. 1995. *Ecotoxicity and Human Health*. New York: CRC Press/Lewis Publishers.

Bueche, J. R. 1972. *Principles of Physics*, 2nd Edition. New York: McGraw-Hill.

Calabrese, E. J. 1994. *Biological Effects of Low Level Exposures*. New York: CRC Press/Lewis Publishers.

Cockerman, L. G., B. S. Shane. 1994. *Basic Environmental Toxicology*. New York: CRC Press/Lewis Publishers.

Cothern, R. R. 1993. *Comparative Environmental Risk Assessment*. New York: CRC Press/Lewis Publishers.

Environmental Protection Agency. 1983. *Chemical Hazard Information Profiles*. Washington, D.C.: U.S. EPA.

Gardner, R. W. 1994. *Chemical Intolerance*. New York: CRC Press/Lewis Publishers.

Lauwerys, R. R., P. Hoet. 1993. *Industrial Chemical Exposure*. New York: CRC Press/Lewis Publishers.

Lewis, R. J., Sr. 1993. *Hawley's Condensed Chemical Dictionary*, 12th Edition. New York: Van Nostrand Reinhold.

Lewis, R. J., Sr. 1992. *Sax's Dangerous Properties of Industrial Materials*, Volumes 1, 2, and 3, 8th Edition. New York: Van Nostrand Reinhold.

Mackay, D., W. Y. Shiu, and K. C. Ma. 1992. *Illustrated Handbook of Chemical-Physical Properties and Environmental Fate for Organic Chemicals*. New York: CRC Press/Lewis Publishers.

Manahan, S. E. 1994. *Environmental Chemistry*, 6th Edition. New York: CRC Press/Lewis Publishers.

Meyer, E. 1990. *Chemistry of Hazardous Materials*. Englewood Cliffs, NJ: Prentice Hall.

Richardson, M. 1995. *Dictionary of Substance and Their Effects*, Vol. I-VI. New York: CRC Press/Lewis Publishers.

Torkelson, T. R. V. K. Rowe. 1981. *Patty's Industrial Hygiene and Toxicology*, Vol. 2B. New York: Wiley-Interscience.

MATERIAL SAFETY DATA SHEET

CHEMICAL NAME	
MANGANESE CARBONATE	**CRC PRESS/LEWIS PUBLISHERS DISCLAIMER:** THE INFORMATION AND RECOMMENDATIONS PRESENTED HEREIN ARE BASED ON SOURCES BELIEVED TO BE RELIABLE. CRC MAKES NO REPRESENTATION ON ITS COMPLETENESS OR ACCURACY. IT IS THE RESPONSIBILITY OF THE USER TO DETERMINE THE CHEMICAL'S SUITABILITY FOR ITS INTENDED USE, THE CHEMICAL'S SAFE USE, AND THE CHEMICAL'S PROPER DISPOSAL. NO REPRESENTATIONS AND/OR WARRANTIES, EITHER EXPRESSED OR IMPLIED, OF THE MERCHANTABILITY OR FITNESS FOR A PARTICULAR PURPOSE, OR OF ANY OTHER NATURE, ARE MADE WITH RESPECT TO THE INFORMATION PROVIDED IN THIS MSDS OR TO THE CHEMICAL TO WHICH INFORMATION MAY REFER. CRC NEITHER ASSUMES NOR AUTHORIZES ANY OTHER PERSON TO ASSUME FOR IT, ANY OTHER ADDITIONAL RESPONSIBILITY OR LIABILITY FOR THE USE OF, OR RELIANCE UPON, THIS INFORMATION.

HAZARD WARNING INFORMATION

HEALTH	FIRE	REACTIVE	OTHER	DEGREE OF HAZARD	COLOR CODING	OTHER CODES
2	0	0		0 = Minimum Hazard 1 = Slight Hazard 2 = Moderate Hazard 3 = Serious Hazard 4 = Severe Hazard	HEALTH = BLUE FIRE = RED REACTIVITY = YELLOW OTHER = WHITE	OX = Oxidizer ACID = Acid ALK = Alkali COR = Corrosive W = Use No Water

SECTION I - GENERAL INFORMATION

Characterization	RCRA Number	EPA Class
Metallic Carbonate	**None**	**Not Applicable**

DOT Proper Shipping Name	Chemical Abstract Service (CAS) Number
No Citation	**598-62-9**
DOT Hazard Class and Label Requirements	DOT Emergency Guide Code
No Citation	**No Citation**
DOT Identification Number	Atomic Symbol
No Citation	**$MnCO_3$**

Synonyms

Carbonic acid, magnesium (2+) salt; manganous carbonate; rhodochrosite.

SECTION II - HAZARDOUS INGREDIENTS AND IDENTITY INFORMATION

Hazardous Components (specific identity)	OSHA Exposure Criteria	NIOSH Exposure Criteria	Immediately Dangerous to Life and Health (IDLH)	ACGIH Exposure Criteria
Manganese carbonate (derivation: A precipitate from the addition of sodium carbonate to a solution of a manganese salt; by hydrometallurgical treatment of manganiferrous iron ore).	**PEL (ceiling):** 5 mg(Mn)/m^3 (dust/compounds) **STEL:** Not Applicable	**REL (ceiling):** 5 mg(Mn)/m^3 (dust/compounds) **STEL:** 3 mg(Mn)/m^3 (fumes)	**Not Determined**	**TLV:** 5 mg(Mn)/m^3 (dust/compounds) 1 mg(Mn)/m^3 (fumes) **STEL:** 3 mg(Mn)/m^3 (fumes)

SECTION III - PHYSICAL AND CHEMICAL CHARACTERISTICS

Boiling Point	Specific Gravity (H_2O = 1)
Not Found	**3.125**
Vapor Pressure (mm Hg)	Atomic Weight
Not Found	**114.9**
Vapor Density (Air = 1)	Melting Point
Not Found	**Decomposes**

Solubility

Practically insoluble in water; Soluble in dilute inorganic acids; insoluble in common organic acids, alcohol, and ammonia.

Appearance and Odor

Rose-colored crystals, (when precipitated, white crystals). Gradually turns light brown on exposure to air.

SECTION IV - FIRE AND EXPLOSION HAZARD DATA

Flash Point (method used)	Explosive Limits in Air % by Volume
Not Found	**LEL: Not Found** **UEL: Not Found**
NFPA Classification	Autoignition Temperature
Combustible Solid	**Not Found**

Extinguishing Media

Use a dry chemical extinguishing agent suitable for metal fires (Class D).

Special Fire Fighting Procedures

Wear full protective clothing and self-contained breathing apparatus (SCBA). Poisonous gases are produced in fire. Cool fire-exposed containers that cannot be safely moved (use water spray). Do NOT release runoff from fire-fighting procedures to sewers or waterways. Dike and collect for disposal.

Unusual Fire and Explosion Hazards

None reported.

SECTION V - REACTIVITY DATA

Stability		Conditions to Avoid
		Manganese carbonate is normally stable under routine conditions of handling and storage. Avoid contact with incompatible materials and heat. It will slowly decompose in air to manganese oxide and carbon dioxide.
Stable	Unstable	Incompatibility *(materials to avoid)*
X		Incompatible with sulfur dioxide (causes oxidation to the more toxic sulfur trioxide). Causes violent decomposition of hydrogen peroxide. Also incompatible with acids (can generate carbon dioxide).
Hazardous Polymerization		Conditions to Avoid
		Hazardous polymerization of manganese carbonate will not occur.
May Occur	Will Not Occur	Hazardous Decomposition or By-products
	X	Poisonous gases are emitted during fire, including toxic and irritating smoke and fumes of manganese oxide and carbon dioxide gas.

SECTION VI - HEALTH HAZARD DATA

Primary Route(s) of Entry:	Inhalation?	Absorption (skin/eye)?	Ingestion?
	X		X

Health Hazards

INHALATION: The dusts can produce irritation of the eyes and mucous membranes. The fumes can lead to metal fume fever with symptoms of fever, chills, nausea, weakness, headache, blurred vision, back stiffness and pain, muscle pain, shallow respiration, dryness in mouth and throat, intense thirst, dry cough, metallic taste, chest tightness, and Mn pneumonitis. It can also lead to degenerative brain effects with symptoms similar to Parkinson's disease.

EYES & SKIN: Causes skin and eye irritation. There can be dermatitis, conjunctivitis, corneal damage.

INGESTION: Large doses can cause gastrointestinal irritation and other effects similar to inhalation.

Carcinogenicity	NTP Listed?	IARC Cancer Review Group?	OSHA Regulated?	Target Organs?
Unknown Human Unknown Animal	No	No	No	Respiratory system, eyes, skin, CNS, kidneys, blood.

Medical Conditions Generally Aggravated by Exposure

Existing kidney dysfunction, brain, or muscle problems may be aggravated by exposure.

Emergency and First-aid Procedures

Eye contact: Do not allow victim to keep eyes tightly shut. Flush immediately with water for 15 minutes (minimum); seek medical attention. Skin contact: Remove all contaminated clothing. Immediately wash area with flooding amounts of water. For reddened or blistered skin, seek medical assistance. For inhalation: Remove the person from exposure. Provide respiratory assistance and CPR. Transfer to medical facility. If swallowed: Seek medical attention. Rinse mouth with water. Give 1 to 2 glasses of water and induce vomiting until vomit is clear. Never give anything by mouth to an convulsing person.

SECTION VII - PRECAUTIONS FOR SAFE HANDLING AND USE

Steps to be Taken in Case Material is Released or Spilled

Collect powdered materials in most convenient and safe manner and deposit in clean, dry, sealed, metal containers. Dry sweep not recommended. Use vacuum equipped with a high-efficiency particulate air (HEPA) filter. Ventilate area of spill. Restrict those not involved in cleanup from entering area.

Preferred Waste Disposal Method

Reclaim for scrap.

Precautions to be Taken in Handling and Storage

Store to avoid contact with incompatible materials. Prevent physical damage to containers. Keep containers tightly closed. Keep sources of heat away.

Other Precautions and Warnings

Avoid excessive dust generation in storage and handling areas. Personnel should be trained on the proper precautions for working with manganese carbonate compounds *before* initial assignment.

SECTION VIII - CONTROL MEASURES AND PERSONAL PROTECTIVE EQUIPMENT

Respiratory Protection *(specify type)*

For dust exposures, use a respirator with a HEPA filter, otherwise use a supplied-air respirator or a self-contained breathing apparatus with full facepiece operated in positive pressure mode.

Ventilation

Local exhaust (preferred) at site of chemical work or general ventilation.

Protective Gloves	Eye Protection	Other Protective Clothing
Impervious Material	Chemical/Dust Goggles or Face Mask	Protective Apron, Boots

Work/Hygiene Practices

Always wash hands thoroughly after using chemical; never bring food, drink, or smoking materials into vicinity of chemicals.

MANGANESE CARBONATE

$MnCO_3$

CAS: 598-62-9

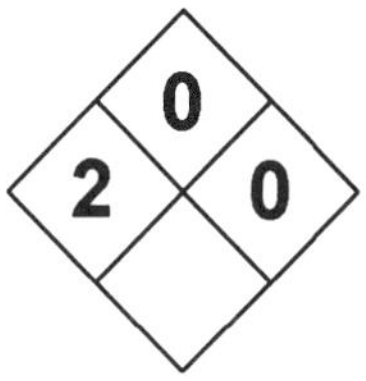

IDENTIFICATION AND TYPICAL USES

Manganese carbonate appears as rose-colored, rhombohedral crystals. The precipitated form is a white crystalline solid. Manganese carbonate gradually turns light brown on exposure to air as it decomposes to form manganese oxide and carbon dioxide. It is used as a pigment in paints, as a drier for varnishes, as a source of acid-soluble manganese, as a nutrient for plants and crops, in feeds, in the making of ceramics, in the production of ferrite, in the production of manganese salts and pharmaceuticals (medicines to stimulate blood formation).

RISK ASSESSMENT: HEALTH

General Assessment

Manganese carbonate does not absorb well through the digestive or respiratory processes. It therefore presents very low acute toxicity. However, in large or chronic doses (1 to 2 years of exposure) by *inhalation* or *ingestion*, manganese poisoning can be quite severe. There is little information in the references regarding its carcinogenic, mutagenic, or teratogenic capabilities in animals or humans.

Inhalation of manganese carbonate over prolonged periods of time can lead to the development of manganism. The early symptoms are difficult to recognize because they are generally nonspecific in nature and may occur gradually over a long period of time. It usually begins with the nonspecific neurologic symptoms such as headache, general apathy, weakness of the legs, progressing to psychosis, with final symptoms similar to that of Parkinson's disease. There can be pain and stiffness in the back and joints, anorexia, weakness, sleepiness, muscle spasms, psychosis, emotional swings (laughter or crying for no apparent reason), euphoria, mental confusion, hallucinations, speech disturbances, muscle rigidity, salivation, trembling, staggering gait, and profuse sweating.

Acute Health Effects

Acute poisoning with manganese carbonate are rare. The following acute (short-term) health effects may occur immediately or shortly after human exposure to manganese carbonate:

Skin: Possible irritation and dermatitis (in some individuals).

Eye: Irritation to the eye and associated tissues. It can cause conjunctivitis and corneal damage.

Lung: Irritation of the mucosa of the eyes, nose, throat, and respiratory tract.

CNS: There may be no noticeable acute effects. However, large doses may result in generally adverse nervous system effects.

Chronic Health Effects

The following chronic (long-term) health effects may occur at some time after exposure to manganese carbonate and can last for months or even years:

Cancer Hazards: According to the references, manganese carbonate has not been shown to cause cancer in test animals. Its human carcinogenic potential is unknown.

Reproductive Hazard: According to the references, manganese carbonate has not been adequately tested for its ability to cause adverse reproductive effects in test animals.

Other Chronic Effects: Prolonged or repeated exposure can damage the central nervous system and cause impaired mental function and degenerative brain changes with personality disorders and physical changes in motor control, vision, and other body functions.

Recommended Risk-Reduction Measures

While manganese is an essential mineral in human nutrition, the base element and the carbonate as well can be extremely toxic under certain exposure conditions and personnel should, therefore, avoid direct occupational contact with manganese carbonate. If a less toxic material or compound cannot be substituted for manganese carbonate, then *engineering controls* are the most effective method of reducing exposures. The best protection is to enclose operations and/or provide local exhaust ventilation at the site of manganese carbonate dust release. While not always operationally feasible, isolating operations can also reduce exposure. Using respiratory protection is less effective than the controls mentioned above, but is still advisable whenever working with or around manganese carbonate. The exposure ceiling for manganese is relatively low ($5mg/m^3$) and difficult to accurately monitor. For low or transient exposures, a chemical cartridge respirator equipped with a high-efficiency particulate air (HEPA) filter and a dust/mist pre-filter may provide adequate respiratory protection. Better protection is obtained using a supplied-air respirator operated in continuous flow mode or a self-contained breathing apparatus (SCBA) with full facepiece operated in pressure demand. If a full facepiece is not available, then chemical/dust goggles should be worn to protect the eyes. A face shield and protective apron should also be worn. To prevent hand and skin exposures, gloves should be used. Glove manufacturers should be contacted *before* selections are made.

Administrative controls should also be in place to minimize the potential for human exposures. These may include written procedures or policies which specify the methods and techniques that will be practiced whenever personnel are to work with manganese carbonate.

All personnel should receive *training* on the use, hazards, protective measures, emergency actions, and other precautions per 29 CFR 1910.1200 (Hazard Communication), prior to the first assignment in an area where manganese carbonate is used or stored.

Before beginning employment working with manganese carbonate, and at regular intervals thereafter (e.g., annually), those personnel with frequent or potentially high exposures should be provided the following recommended medical tests:

- ☑ Complete blood count (establish baseline).

If symptoms develop or overexposure is suspected, consider the following additional tests:

- ☑ Complete blood count (compare to baseline).
- ☑ Neurological examination.
- ☑ Lung function tests.

Any evaluation should include a careful history of past and present symptoms with an examination. Medical tests that simply look for existing damage are not a substitute for controlling exposures.

Other methods to reduce exposure include:

- ☑ Where possible, enclose operations and use local exhaust ventilation at the site of chemical release. If local exhaust ventilation or enclosure is not used, respiratory protection should be mandatory.
- ☑ Always ensure that proper protective clothing is worn when using chemical substances. Proper maintenance and care of protective equipment is essential to its proper function.
- ☑ Wash thoroughly immediately after exposure to manganese carbonate and at the end of the work shift or before eating, drinking, or smoking.
- ☑ Hazard warning information should be posted in the work area. In addition, as part of an on-going education and training program, all information on the health and safety hazards of manganese carbonate should be communicated to all exposed and potentially exposed workers.
- ☑ Eye wash stations should be provided in the immediate work area for emergency use. If there is a possibility of skin exposure to manganese carbonate, emergency shower facilities should be provided in the immediate area.
- ☑ Workers whose clothing has been contaminated by manganese carbonate should change into clean clothes before leaving work. Never wear contaminated clothing home where family members can be exposed. Contaminated work clothing should be laundered only by individuals who have been informed of the hazards of exposure to manganese carbonate.

RISK ASSESSMENT: ENVIRONMENT

General Assessment

As a naturally occurring mineral, manganese itself poses very little threat to the environment. However, accidental spills or releases to the environment can occur during transportation, storage, disposal, or destruction of manganese carbonate. In almost every scenario, the threat of environmental exposure is contingent upon the proper handling of the chemical substance. Accidental spills, large or small, where manganese carbonate contacts incompatible commodities can result in release of toxic products (depending upon conditions of spill), and possible contamination of the surrounding environmental mediums (water, soil, and air).

Manganese carbonate is considered a noncombustible solid. It reacts with acid to generate toxic carbon dioxide. It is also incompatible with many other materials, including sulfur dioxide which oxidizes to the even more toxic sulfur trioxide. It will cause the violent decomposition of hydrogen peroxide (>52% solutions). Extreme caution is always required in handling, storage, transportation, and disposal of manganese carbonate. These characteristics also require special consideration during any emergency situation involving a leak or spill of manganese carbonate powder or dust. Should manganese carbonate ever come into contact with these various incompatible substances either during use, transportation, storage, or disposal, violent and even explosive reactions are possible.

Manganese occurs naturally in the environment (terrestrial soils and aquatic sediments) and may enter the environment as a result of industrial and municipal waste treatment plant discharges and from accidental spills.

Acute Ecological Effects

Acute (short-term) toxic effects may include the death of animals, birds, or fish and death or low growth rate in plants. Acute effects are seen 2 to 4 days after animals or plants are exposed to manganese.

Manganese and its compounds have moderate acute toxicity to aquatic life. Manganese is considered essential for plant and animal life. However, insufficient data are available on the acute effects of excessive manganese carbonate contamination on plants, birds, or land animals.

Chronic Ecological Effects

Chronic (long-term) toxic effects may include shortened life span, reproductive problems, lower fertility, and changes in appearance or behavior in exposed animals. These effects can be seen long after first exposure(s) to toxic chemicals.

Manganese and its compounds have moderate chronic toxicity to aquatic life. Manganese is considered essential for plant and animal life. However, insufficient data are available on the chronic effects of excessive manganese contamination on plants, birds, or land animals.

Water Solubility

Manganese and its compounds vary in their solubility in water from very soluble to nearly insoluble. In water, manganese reacts and decomposes to form explosive hydrogen gas. Manganese carbonate easily dissolves in acidic water and becomes highly mobile.

Persistence in the Environment

Manganese and its compounds are highly persistent in water, with a half-life greater than 200 days. The half-life of a pollutant is the amount of time it takes for one half of the chemical to be degraded. Manganese is also naturally present in terrestrial soils and aquatic sediments.

Bioaccumulation in Aquatic Organisms

Some substances increase in concentration, or *bioaccumulate,* in living organisms as they breathe contaminated air, drink contaminated water, or eat contaminated food. These chemicals can become concentrated in the tissues and internal organs of animals as well as humans.

The concentration of manganese found in fish tissues is expected to be about the same as the average concentration of manganese found in the water from which the fish was taken.

Recommended Risk-Reduction Measures

Proper training of all transporters will reduce the likelihood of a mishap or accident resulting in a leak or spill of manganese carbonate dusts or powders into the environment. The correct use of labeling on all containers, trucks, and rail cars will enable emergency responders to react properly and quickly to any disaster thereby reducing the potential risk to the environment and to personnel.

Storage of manganese carbonate should be segregated from incompatible chemicals to minimize the risk of cross-contamination or contact. Care should be taken to prevent physical damage to storage containers. Also, dusty conditions must be avoided in storage and use areas.

If a spill or leak to the environment has occurred, fire department, emergency response and/or hazardous materials spill personnel should be notified immediately. Cleanup should be attempted only by those trained in proper spill containment procedures using non-sparking tools. Contaminated soils should be removed for decontamination and replaced with clean soil. If manganese carbonate should contact the water table, aquifer, or navigable waterway, time is of the essence. It will slowly dissolve in water and total remediation may not be possible. The local and/or state emergency response authorities may require notification. A comprehensive emergency response or disaster preparedness/recovery plan should be in place prior to any operations involving the use, transportation, storage, or disposal of manganese carbonate.

If manganese carbonate is spilled or leaked, the following specific steps are recommended:

☑ Restrict persons not wearing protective clothing from area of spill or leak until cleanup is complete and area can be opened for normal work.

☑ Ventilate area of spill or leak.

☑ Collect powdered materials in the safest manner possible and deposit in sealed containers. Recommend using a vacuum equipped with a high-efficiency particulate air (HEPA) filter. Do not dry sweep (generates airborne dusts).

☑ It may be necessary to dispose of manganese carbonate as a hazardous waste. The responsible state agency or the regional office of the federal Environmental Protection Agency (EPA) should be contacted for specific recommendations.

RISK ASSESSMENT: BUSINESS

General Assessment

Accidents or mishaps involving manganese carbonate can present a significant threat to business operations. The loss or damage of equipment or facilities can significantly affect fiscal viability. Lawsuits that may result from personnel injury/death, public exposures, and/or environmental contamination will also require a serious expenditure of resources. Media attention surrounding an injury, death, or environmental damage can also result in a loss of profits and loss of current as well as future business.

Recommended Risk-Reduction Measures

Company attorneys, safety and health professionals, and environmental specialists should be involved in the development of any procedures for responding to chemical incidents. A company official should be predesignated as a public relations officer with specific training in dealing with the press. Corporate plans and policies should be developed, approved, and implemented long before any need for such arises.

REFERENCES

Ahlbom, A. 1993. *Biostatistics for Epidemiologists.* New York: CRC Press/Lewis Publishers.

American Conference of Governmental Industrial Hygienists. 1988. *Documentation of the Threshold Limit Values and Biological Exposure Indices,* 5th Edition (with updates). Cincinnati: ACGIH

Anthony, C. P., and N. J. Kolthoff. 1971. *Textbook of Anatomy and Physiology.* St. Louis: C. V. Mosby Company.

Asante-Duah, D. K. 1993. *Hazardous Waste Risk Assessment.* New York: CRC Press/Lewis Publishers.

Bloom, A. D. and F. J. DeSerres. 1995. *Ecotoxicity and Human Health.* New York: CRC Press/Lewis Publishers.

Bueche, J. R. 1972. *Principles of Physics,* 2nd Edition. New York: McGraw-Hill.

Calabrese, E. J. 1994. *Biological Effects of Low Level Exposures.* New York: CRC Press/Lewis Publishers.

Cockerman, L. G., B. S. Shane. 1994. *Basic Environmental Toxicology.* New York: CRC Press/Lewis Publishers.

Cothern, R. R. 1993. *Comparative Environmental Risk Assessment.* New York: CRC Press/Lewis Publishers.

Environmental Protection Agency. 1983. *Chemical Hazard Information Profiles.* Washington, D.C.: U.S. EPA.

Gardner, R. W. 1994. *Chemical Intolerance.* New York: CRC Press/Lewis Publishers.

Lauwerys, R. R., P. Hoet. 1993. *Industrial Chemical Exposure.* New York: CRC Press/Lewis Publishers.

Lewis, R. J., Sr. 1993. *Hawley's Condensed Chemical Dictionary,* 12th Edition. New York: Van Nostrand Reinhold.

Lewis, R. J., Sr. 1992. *Sax's Dangerous Properties of Industrial Materials,* Volumes 1, 2, and 3, 8th Edition. New York: Van Nostrand Reinhold.

Mackay, D., W. Y. Shiu, and K. C. Ma. 1992. *Illustrated Handbook of Chemical-Physical Properties and Environmental Fate for Organic Chemicals.* New York: CRC Press/Lewis Publishers.

Manahan, S. E. 1994. *Environmental Chemistry,* 6th Edition. New York: CRC Press/Lewis Publishers.

Meyer, E. 1990. *Chemistry of Hazardous Materials.* Englewood Cliffs, NJ: Prentice Hall.

Richardson, M. 1995. *Dictionary of Substance and Their Effects,* Vol. I-VI. New York: CRC Press/Lewis Publishers.

Torkelson, T. R. V. K. Rowe. 1981. *Patty's Industrial Hygiene and Toxicology,* Vol. 2B. New York: Wiley-Interscience.

MATERIAL SAFETY DATA SHEET

CHEMICAL NAME	
MANGANESE DIOXIDE	**CRC PRESS/LEWIS PUBLISHERS DISCLAIMER:** THE INFORMATION AND RECOMMENDATIONS PRESENTED HEREIN ARE BASED ON SOURCES BELIEVED TO BE RELIABLE. CRC MAKES NO REPRESENTATION ON ITS COMPLETENESS OR ACCURACY. IT IS THE RESPONSIBILITY OF THE USER TO DETERMINE THE CHEMICAL'S SUITABILITY FOR ITS INTENDED USE, THE CHEMICAL'S SAFE USE, AND THE CHEMICAL'S PROPER DISPOSAL. NO REPRESENTATIONS AND/OR WARRANTIES, EITHER EXPRESSED OR IMPLIED, OF THE MERCHANTABILITY OR FITNESS FOR A PARTICULAR PURPOSE, OR OF ANY OTHER NATURE, ARE MADE WITH RESPECT TO THE INFORMATION PROVIDED IN THIS MSDS OR TO THE CHEMICAL TO WHICH INFORMATION MAY REFER. CRC NEITHER ASSUMES NOR AUTHORIZES ANY OTHER PERSON TO ASSUME FOR IT, ANY OTHER ADDITIONAL RESPONSIBILITY OR LIABILITY FOR THE USE OF, OR RELIANCE UPON, THIS INFORMATION.

HAZARD WARNING INFORMATION

HEALTH	FIRE	REACTIVE	OTHER	DEGREE OF HAZARD	COLOR CODING	OTHER CODES
2	0	1	OX	0 = Minimum Hazard 1 = Slight Hazard 2 = Moderate Hazard 3 = Serious Hazard 4 = Severe Hazard	HEALTH = BLUE FIRE = RED REACTIVITY = YELLOW OTHER = WHITE	OX = Oxidizer ACID = Acid ALK = Alkali COR = Corrosive W = Use No Water

SECTION I - GENERAL INFORMATION

Characterization	**Manganese Compound**	RCRA Number	**None**
		EPA Class	**Not Applicable**
DOT Proper Shipping Name	**No Citation**	Chemical Abstract Service (CAS) Number	**1313-13-9**
DOT Hazard Class and Label Requirements	**No Citation**	DOT Emergency Guide Code	**35**
DOT Identification Number	**NA 1479**	Atomic Symbol	**MnO_2**

Synonyms
Battery manganese; manganese binoxide; manganese black; manganese peroxide.

SECTION II - HAZARDOUS INGREDIENTS AND IDENTITY INFORMATION

Hazardous Components (specific identity)	OSHA Exposure Criteria	NIOSH Exposure Criteria	Immediately Dangerous to Life and Health (IDLH)	ACGIH Exposure Criteria
Manganese dioxide (derivation: Occurs naturally and pyrolusite and as a special African ore of different atomic structure used exclusively for the battery grade; by electrolysis; by heating manganese oxide in the presence of oxygen; or by decomposition of manganese nitrate).	**PEL (ceiling):** 5 mg(Mn)/m^3 (dust/compounds) **STEL:** Not Applicable	**REL (ceiling):** 5 mg(Mn)/m^3 (dust/compounds) **STEL:** Not Applicable	**500 mg(Mn)/m^3**	**TLV:** 5 mg(Mn)/m^3 (dust/compounds) 1 mg(Mn)/m^3 (fumes) **STEL:** 3 mg(Mn)/m^3 (fumes)

SECTION III - PHYSICAL AND CHEMICAL CHARACTERISTICS

Boiling Point	**Not Found**	Specific Gravity (H_2O = 1)	**5.026**
Vapor Pressure (mm Hg)	**Not Found**	Molecular Weight	**86.94**
Vapor Density (Air = 1)	**Not Found**	Melting Point	**995°F (535°C) Decomposes**

Solubility
Insoluble in water, nitric, or cold sulfuric acid. Soluble in hydrochloric acid.

Appearance and Odor
Fine black or brownish black powder or crystals. Lumps appear as steel-gray. There is no odor.

SECTION IV - FIRE AND EXPLOSION HAZARD DATA

Flash Point (method used)	**Not Found**	Explosive Limits in Air % by Volume	**LEL: Not Found UEL: Not Found**
NFPA Classification	**Non-Combustible Solid**	Autoignition Temperature	**Not Found**

Extinguishing Media
Use dry chemical or carbon dioxide.

Special Fire Fighting Procedures
Wear full protective clothing and self-contained breathing apparatus (SCBA). Poisonous gases are produced in fire. Be aware of runoff; do NOT release to sewers or waterways.

Unusual Fire and Explosion Hazards
Manganese dioxide is non-combustible but is a powerful oxidizer. It can increase the burning intensity and rate of ordinary combustible items. It can also cause the re-ignition of such items after the fire has been extinguished. Oxygen released during decomposition will also facilitate combustion of other items.

SECTION V - REACTIVITY DATA

Stability		Conditions to Avoid: Manganese dioxide is normally stable under routine conditions of handling and storage. Avoid contact with incompatible materials, heat, flame, sparks, or other ignition sources, and keep organic materials away.
Stable	Unstable	Incompatibility *(materials to avoid)*: Incompatible with hydrogen sulfide, heated aluminum, hydrogen peroxide, chlorates + heat, peroxomonosulfuric acid, anilinium perchlorate, oxidizers, diboron tetrafluoride, warm potassium azide, chlorine trifluoride.
X		
Hazardous Polymerization		Conditions to Avoid: Under normal conditions of temperature and pressure, hazardous polymerization of manganese dioxide will not occur.
May Occur	Will Not Occur	Hazardous Decomposition or By-products: Poisonous gases are emitted during fire, including toxic and irritating smoke and fumes. Oxygen is also release during decomposition which can greatly accelerate combustion of other items.
	X	

SECTION VI - HEALTH HAZARD DATA

Primary Route(s) of Entry:	Inhalation?	Absorption (skin/eye)?	Ingestion?
	X		X

Health Hazards

INHALATION: The dusts can irritate the respiratory system with progression to lung damage, pleuritis, bronchitis, and possible respiratory infection. The fumes can lead to metal fume fever with symptoms of fever, chills, nausea, weakness, headache, blurred vision, back stiffness and pain, muscular pain, shallow respiration, dryness in mouth and throat, intense thirst, dry cough, metallic taste, chest tightness, and Mn pneumonitis. It can also lead to degenerative brain effects with symptoms similar to Parkinson's disease.

EYES & SKIN: Causes eye irritation. Skin contact causes very little noticeable effect.

INGESTION: Large doses can cause gastrointestinal irritation and other unspecified effects.

Carcinogenicity	NTP Listed?	IARC Cancer Review Group?	OSHA Regulated?	Target Organs?
Unknown Human Unknown Animal	No	No	No	Respiratory system, eyes, CNS.

Medical Conditions Generally Aggravated by Exposure

Central nervous system disorders, psychiatric problems, and respiratory diseases may be aggravated.

Emergency and First-aid Procedures

Eye contact: Do not allow victim to keep eyes tightly shut. Flush immediately with water for 15 minutes (minimum); seek medical attention. Skin contact: Remove all contaminated clothing. Immediately wash area with flooding amounts of water. For reddened or blistered skin, seek medical assistance. For inhalation: Remove the person from exposure. Provide respiratory assistance and CPR. Transfer to medical facility. If swallowed: Seek medical attention. Rinse mouth with water. Give 2 to 3 glasses of water and induce vomiting. Never give anything by mouth to an unconscious or convulsing person.

SECTION VII - PRECAUTIONS FOR SAFE HANDLING AND USE

Steps to be Taken in Case Material is Released or Spilled

Cover with clean, moist sand, mix, collect in most convenient and safe manner and deposit in clean, sealed, metal containers. Dry sweep not recommended. Use scoop for bulk materials and a HEPA vacuum on residue. Ventilate area of spill. Restrict those not involved in cleanup from entering area.

Preferred Waste Disposal Method

No Citation.

Precautions to be Taken in Handling and Storage

Store to avoid contact with incompatible materials. Prevent physical damage to containers. Keep containers tightly closed. Keep organic materials, other oxidizing materials, and sources of heat away.

Other Precautions and Warnings

Avoid excessive dust generation in storage and handling areas.

SECTION VIII - CONTROL MEASURES AND PERSONAL PROTECTIVE EQUIPMENT

Respiratory Protection *(specify type)*

For dust exposures, use a respirator with a HEPA filter; otherwise, use a supplied-air respirator or a self-contained breathing apparatus with full facepiece operated in positive pressure mode.

Ventilation

Local exhaust (preferred) at site of chemical work or general ventilation.

Protective Gloves	Eye Protection	Other Protective Clothing
Impervious Material	Chemical/Dust Goggles or Face Mask	Protective Apron, Boots

Work/Hygiene Practices

Always wash hands thoroughly after using chemical; never bring food, drink, or smoking materials into vicinity of chemicals.

MANGANESE DIOXIDE

MnO_3

CAS: 1313-13-9

IDENTIFICATION AND TYPICAL USES

Manganese dioxide appears as a black or brownish-black fine powder or crystals, or, it can appear as steel-gray lumps. There is no odor. It is used as an oxidizing agent, in some dry cell batteries (African and synthetic types), in the manufacture of pyrotechnics, in matches, as a catalysts, a laboratory reagent, in textile dyeing, and as a source of metallic manganese.

RISK ASSESSMENT: HEALTH

General Assessment

Even though manganese is an essential mineral in human nutrition, it can also be toxic by *inhalation* of its fumes and/or by *ingestion* of the powders and dusts. While manganese has been shown to cause tumors in experimental test animals with mutation data also reported, the exact nature of manganese dioxide's human carcinogenic properties is unknown.

Inhalation of manganese dioxide dusts can irritate the respiratory tract with progression to lung damage, pleuritis, bronchitis, and possible respiratory infections. The fumes can cause acute or chronic poisoning with primary effects on the immune system characterized by symptoms of metal fume fever. These can include fever, chills, general weakness, body aches, frontal headache, blurred vision, back pain and stiffness, muscular cramps, shallow and labored respiration, dryness in the mouth and throat with intense thirst, dry cough, metallic taste, and tightness in the chest occurring several hours after exposure. Lung irritation can become severe with inflammation causing manganese dioxide pneumonitis. Other effects of central nervous system depression include headache, insomnia, changes in personality, emotions, and attitude, irritability, inappropriate laughter or crying, hallucinations, effects on vision (double vision, blurred vision), uncontrolled impulse behavior, euphoria, excessive salivation, trembling, staggering gait, clumsiness, and signs and symptoms normally associated with Parkinson's disease.

Acute Health Effects

The following acute (short-term) health effects may occur immediately or shortly after human exposure to manganese dioxide:

Skin: Manganese dioxide appears to be completely innocuous to intact (unbroken) skin.

Eye: Irritation to the eye and associated tissues. Effects appear to be minor and temporary.

Lung: Irritation of the mucosa of the eyes, nose, throat, and respiratory tract.

CNS: Breathing heated fumes can cause metal fume fever with specific action on the central nervous system.

Chronic Health Effects

The following chronic (long-term) health effects may occur at some time after exposure to manganese dioxide and can last for months or even years:

Cancer Hazards: According to the references, manganese has been shown to cause tumors in test animals. There are no reports of the specific carcinogenic properties of manganese dioxide.

Reproductive Hazard: According to the references, there is some limited evidence that chronic overexposure to manganese dioxide may cause adverse reproductive effects in test animals.

Other Chronic Effects: Prolonged or repeated exposure can damage the central nervous system and cause impaired mental function and degenerative brain changes with personality disorders, emotional instability, awkward gestures, and physical changes in motor control, vision, and other body functions.

Recommended Risk-Reduction Measures

While manganese is an essential mineral in human nutrition, the base mineral as well as many of its compounds (including in the dioxide) can be extremely toxic under certain exposure conditions. Personnel should, therefore, avoid direct occupational contact with manganese dioxide. If a less toxic material or compound cannot be substituted for manganese dioxide, then *engineering controls* are the most effective method of reducing exposures. The best protection is to enclose operations and/or provide local exhaust ventilation at the site of manganese dioxide dust release. While not always operationally feasible, isolating operations can also reduce exposure. Using respiratory protection is less effective than the controls mentioned above, but is still advisable whenever working with or around manganese dioxide. The exposure ceiling for manganese is relatively low (5mg/m^3) and difficult to accurately monitor. For low or transient exposures, a chemical cartridge respirator equipped with a high-efficiency particulate air (HEPA) filter and a dust/mist pre-filter may provide adequate respiratory protection. Better protection is obtained using a supplied-air respirator operated in continuous flow mode or a self-contained breathing apparatus (SCBA) with full facepiece operated in pressure demand. If a full facepiece is not available, then chemical/dust goggles should be worn to protect the eyes. A face shield and protective apron should also be worn. To prevent hand and skin exposures, gloves should be used. Glove manufacturers should be contacted *before* selections are made.

Administrative controls should also be in place to minimize the potential for human exposures. These may include written procedures or policies which specify the methods and techniques that will be practiced whenever personnel are to work with manganese dioxide.

All personnel should receive *training* on the use, hazards, protective measures, emergency actions, and other precautions per 29 CFR 1910.1200 (Hazard Communication), prior to the first assignment in an area where manganese dioxide is used or stored.

Before beginning employment working with manganese dioxide, and at regular intervals thereafter (e.g., annually), those personnel with frequent or potentially high exposures should be provided the following recommended medical tests:

- ☑ Complete blood count (establish baseline).
- ☑ Complete examination of the nervous system.
- ☑ Lung function tests (may be normal if the person is not having an attack at the time of the test)
- ☑ Kidney function tests.

If symptoms develop or overexposure is suspected, consider the following additional tests:

- ☑ Complete blood count (compare to baseline).
- ☑ Consider chest X-ray after acute overexposure.
- ☑ Lung function tests.

Any evaluation should include a careful history of past and present symptoms with an examination. Medical tests that simply look for existing damage are not a substitute for controlling exposures. Also, since smoking causes heart disease, lung cancer, emphysema, and other respiratory diseases, smokers exposed to manganese dioxide may experience symptoms more rapidly than non-smokers under the same conditions of exposure. Prudent risk management requires careful consideration of *all* factors which may be causing exposure symptoms in the workplace.

Other methods to reduce exposure include:

- ☑ Where possible, enclose operations and use local exhaust ventilation at the site of chemical release. If local exhaust ventilation or enclosure is not used, respiratory protection should be mandatory.
- ☑ Always ensure that proper protective clothing is worn when using chemical substances. Proper maintenance and care of protective equipment is essential to its proper function.
- ☑ Wash thoroughly immediately after exposure to manganese dioxide and at the end of the work shift or before eating, drinking, or smoking.
- ☑ Hazard warning information should be posted in the work area. In addition, as part of an on-going education and training program, all information on the health and safety hazards of manganese dioxide should be communicated to all exposed and potentially exposed workers.
- ☑ Eye wash stations should be provided in the immediate work area for emergency use. If there is a possibility of skin exposure to manganese dioxide, emergency shower facilities should be provided.
- ☑ Workers whose clothing has been contaminated by manganese dioxide should change into clean clothes before leaving work. Never wear contaminated clothing home where family members can be exposed. Contaminated work clothing

should be laundered only by individuals who have been informed of the hazards of exposure to manganese dioxide.

RISK ASSESSMENT: ENVIRONMENT

General Assessment

As a naturally occurring mineral, manganese poses very little threat to the environment. However, accidental spills or releases to the environment can occur during transportation, storage, disposal, or destruction of manganese dioxide. In almost every scenario, the threat of environmental exposure is contingent upon the proper handling of the chemical substance. Accidental spills, large or small, where manganese dioxide contacts incompatible commodities can result in release of toxic products (depending upon conditions of spill), and possible contamination of the surrounding environmental mediums (water, soil, and air).

Manganese dioxide is considered a noncombustible solid. However, it is a powerful oxidizer and it can greatly accelerate the burning of ordinary combustible materials. It is incompatible with sulfur, sulfides, phosphides, hypophosphites, chlorates, and peroxides. It can also react violently in contact with other oxidizers. Extreme caution is always required in handling, storage, transportation, and disposal of manganese dioxide. These characteristics also require special consideration during any emergency situation involving a leak or spill of manganese dioxide powder or dust. Should manganese dioxide ever come into contact with these various incompatible substances either during use, transportation, storage, or disposal, violent and even explosive reactions are possible.

Manganese occurs naturally in many minerals that are widely distributed in the earth's crust and, in trace amounts, is an essential element for both plants and animals. The wide variety of possible manganese compounds may enter the environment from natural and industrial sources.

Acute Ecological Effects

Acute (short-term) toxic effects may include the death of animals, birds, or fish and death or low growth rate in plants. Acute effects are seen 2 to 4 days after animals or plants are exposed to manganese dioxide.

Manganese and its compounds have moderate acute toxicity to aquatic life. While manganese dioxide is considered essential for plant and animal life, insufficient data are available on the acute effects of excessive manganese dioxide contamination on the other ecosystems (plants, birds, or land animals).

Chronic Ecological Effects

Chronic (long-term) toxic effects may include shortened life span, reproductive problems, lower fertility, and changes in appearance or behavior in exposed animals. These effects can be seen long after first exposure(s) to toxic chemicals.

Manganese and its compounds have moderate chronic toxicity to aquatic life. Manganese dioxide is considered essential for plant and animal life. However, insufficient data are available on the chronic effects of excessive manganese dioxide contamination on the other ecosystems (plants, birds, or land animals).

Water Solubility

Manganese and its compounds vary in their solubility in water from very soluble to nearly insoluble. In water, it reacts and decomposes to form explosive hydrogen gas.

Persistence in the Environment

Manganese and its compounds are highly persistent in water, with a half-life greater than 200 days. The half-life of a pollutant is the amount of time it takes for one half of the chemical to be degraded. Manganese is also naturally present in terrestrial soils and aquatic sediments.

Bioaccumulation in Aquatic Organisms

Some substances increase in concentration, or *bioaccumulate,* in living organisms as they breathe contaminated air, drink contaminated water, or eat contaminated food. These chemicals can become concentrated in the tissues and internal organs of animals as well as humans.

The concentration of manganese dioxide found in fish tissues is expected to be about the same as the average concentration of manganese dioxide in the water from which the fish was taken.

Recommended Risk-Reduction Measures

Proper training of all transporters will reduce the likelihood of a mishap or accident resulting in a leak or spill of manganese dioxide dusts or powders into the environment. The correct use of labeling on all containers, trucks, and rail cars will enable emergency responders to react properly and quickly to any disaster thereby reducing the potential risk to the environment and to personnel.

Storage of manganese dioxide should be segregated from incompatible chemicals to minimize the risk of cross-contamination or contact. Care should be taken to prevent physical damage to storage containers. Also, dusty conditions must be avoided in storage and use areas.

If a spill or leak to the environment has occurred, fire department, emergency response and/or hazardous materials spill personnel should be notified immediately. Cleanup should be attempted only by those trained in proper spill containment procedures. Contaminated soils should be removed for decontamination and replaced with clean soil. If manganese dioxide should contact the water table, aquifer, or navigable waterway, response actions should be prompt. The local and/or state emergency response authorities may require notification. A comprehensive emergency response or disaster preparedness/recovery plan should be in place prior to any operations involving the use, transportation, storage, or disposal of manganese dioxide.

If manganese dioxide is spilled or leaked, the following specific steps are recommended:

☑ Restrict persons not wearing protective clothing from area of spill or leak until cleanup is complete and area can be opened for normal work.

☑ Remove ignition sources and ventilate area.

☑ Cover spilled materials with clean, moist sand. Mix and collect powdered materials in the safest manner possible and deposit in sealed containers. Recommend using a vacuum equipped with a high-efficiency particulate air (HEPA) filter on residues. Do NOT dry sweep (generates airborne dusts).

☑ It may be necessary to dispose of manganese dioxide as a hazardous waste. The responsible state agency or the regional office of the federal Environmental Protection Agency (EPA) should be contacted for specific recommendations.

RISK ASSESSMENT: BUSINESS

General Assessment

Accidents or mishaps involving manganese dioxide can present a significant threat to business operations. The loss or damage of equipment or facilities can significantly affect fiscal viability. Lawsuits that may result from personnel injury/death, public exposures, and/or environmental contamination will also require a serious expenditure of resources. Media attention surrounding an injury, death, or environmental damage can also result in a loss of profits and loss of current as well as future business. Always remember that anytime the terms "cancer," "carcinogen," or "reproductive hazard" are used, public hysteria, emotion, and ignorance can run equally high. This should be carefully considered during the development of public relations policies.

Recommended Risk-Reduction Measures

Company attorneys, safety and health professionals, and environmental specialists should be involved in the development of any procedures for responding to chemical incidents. A company official should be predesignated as a public relations officer with specific training in dealing with the press. Corporate plans and policies should be developed, approved, and implemented long before any need for such arises.

REFERENCES

American Conference of Governmental Industrial Hygienists. 1988. *Documentation of the Threshold Limit Values and Biological Exposure Indices*, 5th Edition (with updates). Cincinnati: ACGIH

Anthony, C. P., and N. J. Kolthoff. 1971. *Textbook of Anatomy and Physiology.* St. Louis: C. V. Mosby Company.

Bloom, A. D. and F. J. DeSerres. 1995. *Ecotoxicity and Human Health.* New York: CRC Press/Lewis Publishers.

Calabrese, E. J. 1994. *Biological Effects of Low Level Exposures.* New York: CRC Press/Lewis Publishers.

Cothern, R. R. 1993. *Comparative Environmental Risk Assessment.* New York: CRC Press/Lewis Publishers.

Environmental Protection Agency. 1983. *Chemical Hazard Information Profiles.* Washington, D.C.: U.S. EPA.

Lauwerys, R. R., P. Hoet. 1993. *Industrial Chemical Exposure.* New York: CRC Press/Lewis Publishers.

Lewis, R. J., Sr. 1993. *Hawley's Condensed Chemical Dictionary*, 12th Edition. New York: Van Nostrand Reinhold.

Lewis, R. J., Sr. 1992. *Sax's Dangerous Properties of Industrial Materials*, Volumes 1, 2, and 3, 8th Edition. New York: Van Nostrand Reinhold.

Meyer, E. 1990. *Chemistry of Hazardous Materials.* Englewood Cliffs, NJ: Prentice Hall.

Richardson, M. 1995. *Dictionary of Substance and Their Effects*, Vol. I-VI. New York: CRC Press/Lewis Publishers.

Torkelson, T. R. V. K. Rowe. 1981. *Patty's Industrial Hygiene and Toxicology*, Vol. 2B. New York: Wiley-Interscience.

MATERIAL SAFETY DATA SHEET

CHEMICAL NAME	
MANGANESE NITRATE	**CRC PRESS/LEWIS PUBLISHERS DISCLAIMER:** THE INFORMATION AND RECOMMENDATIONS PRESENTED HEREIN ARE BASED ON SOURCES BELIEVED TO BE RELIABLE. CRC MAKES NO REPRESENTATION ON ITS COMPLETENESS OR ACCURACY. IT IS THE RESPONSIBILITY OF THE USER TO DETERMINE THE CHEMICAL'S SUITABILITY FOR ITS INTENDED USE, THE CHEMICAL'S SAFE USE, AND THE CHEMICAL'S PROPER DISPOSAL. NO REPRESENTATIONS AND/OR WARRANTIES, EITHER EXPRESSED OR IMPLIED, OF THE MERCHANTABILITY OR FITNESS FOR A PARTICULAR PURPOSE, OR OF ANY OTHER NATURE, ARE MADE WITH RESPECT TO THE INFORMATION PROVIDED IN THIS MSDS OR TO THE CHEMICAL TO WHICH INFORMATION MAY REFER. CRC NEITHER ASSUMES NOR AUTHORIZES ANY OTHER PERSON TO ASSUME FOR IT, ANY OTHER ADDITIONAL RESPONSIBILITY OR LIABILITY FOR THE USE OF, OR RELIANCE UPON, THIS INFORMATION.

HAZARD WARNING INFORMATION

HEALTH	FIRE	REACTIVE	OTHER	DEGREE OF HAZARD	COLOR CODING	OTHER CODES
2	0	0		0 = Minimum Hazard 1 = Slight Hazard 2 = Moderate Hazard 3 = Serious Hazard 4 = Severe Hazard	HEALTH = BLUE FIRE = RED REACTIVITY = YELLOW OTHER = WHITE	OX = Oxidizer ACID = Acid ALK = Alkali COR = Corrosive W = Use No Water

SECTION I - GENERAL INFORMATION

Characterization	RCRA Number	EPA Class
Metallic Nitrate	**None**	**Not Applicable**

DOT Proper Shipping Name	Chemical Abstract Service (CAS) Number
Manganese Nitrate	**10377-66-9**
DOT Hazard Class and Label Requirements	DOT Emergency Guide Code
Non-Combustible Solid	**35**
DOT Identification Number	Atomic Symbol
UN2724	**$Mn(NO_3)_2 \cdot 6H_2O$**

Synonyms
Manganous nitrate; manganese (II) nitrate; nitric acid, manganese(2+) salt.

SECTION II - HAZARDOUS INGREDIENTS AND IDENTITY INFORMATION

Hazardous Components (specific identity)	OSHA Exposure Criteria	NIOSH Exposure Criteria	Immediately Dangerous to Life and Health (IDLH)	ACGIH Exposure Criteria
Manganese nitrate (derivation: By the action of nitric acid on manganous hydroxide or carbonate).	**PEL (ceiling): 5 mg(Mn)/m^3 (dust/compounds)** **STEL: Not Applicable**	**REL (ceiling): 5 mg(Mn)/m^3 (dust/compounds)** **STEL: 3 mg(Mn)/m^3 (fumes)**	**500 mg(Mn)/m^3**	**TLV: 5 mg(Mn)/m^3 (dust/compounds) 1 mg(Mn)/m^3 (fumes)** **STEL: 3 mg(Mn)/m^3 (fumes)**

SECTION III - PHYSICAL AND CHEMICAL CHARACTERISTICS

Boiling Point	Specific Gravity (H_2O = 1)
264°F (129°C)	**1.82**
Vapor Pressure (mm Hg)	Molecular Weight
Not Found	**214.88**
Vapor Density (Air = 1)	Melting Point
Not Found	**79°F (26°C)**

Solubility
Very soluble in water, insoluble in alcohol.

Appearance and Odor
Colorless, pink, or rose-colored deliquescent crystals with no odor.

SECTION IV - FIRE AND EXPLOSION HAZARD DATA

Flash Point (method used)	Explosive Limits in Air % by Volume
Not Found	**LEL: Not Found UEL: Not Found**
NFPA Classification	Autoignition Temperature
Non-Combustible Solid	**Not Found**

Extinguishing Media
Manganese nitrate may burn, but does not readily ignite. Use agent suitable to surrounding fire (dry chemical, carbon dioxide, water spray, or foam).

Special Fire Fighting Procedures
Wear full protective clothing and self-contained breathing apparatus (SCBA). Poisonous gases are produced in fire. Cool fire-exposed containers that cannot be safely moved (use water spray). Do NOT release runoff from fire-fighting procedures to sewers or waterways. Dike and collect for disposal.

Unusual Fire and Explosion Hazards
At high temperatures, releases toxic fumes of nitric oxides. Use water spray to knock-down vapors.

SECTION V - REACTIVITY DATA

Stability		Conditions to Avoid **Manganese nitrate is normally stable under routine conditions of handling and storage. Avoid contact with incompatible materials and heat.**
Stable	Unstable	Incompatibility *(materials to avoid)*
X		**Manganese nitrate is incompatible with organic materials (wood, paper, oil) and may react violently or explosively on contact.**
Hazardous Polymerization		Conditions to Avoid **Hazardous polymerization of manganese nitrate will not occur.**
May Occur	Will Not Occur	Hazardous Decomposition or By-products
	X	**Poisonous gases are emitted during fire, including toxic and irritating smoke and fumes of carbon dioxide and toxic fumes of nitric oxides.**

SECTION VI - HEALTH HAZARD DATA

Primary Route(s) of Entry:	Inhalation?	Absorption (skin/eye)?	Ingestion?
	X		X

Health Hazards

INHALATION: **The dusts can produce irritation of the nose and respiratory system causing nose bleeds, dryness of the throat, bronchial irritation, cough with phlegm, and possibly pulmonary edema. The fumes can cause metal fume fever with, chills, nausea, weakness, headache, fever, blurred vision, back stiffness and pain, muscle pain, shallow respiration, dryness in mouth and throat, intense thirst, metallic taste, chest tightness, and Mn pneumonitis. It can also lead to degenerative brain effects with symptoms similar to Parkinson's disease. It may also cause methemoglobinemia and subsequent cyanosis.**

EYES & SKIN: **Causes skin and eye irritation and burns. There can be conjunctivitis, corneal damage.**

INGESTION: **Large doses can cause gastrointestinal irritation and other effects similar to inhalation.**

Carcinogenicity	NTP Listed?	IARC Cancer Review Group?	OSHA Regulated?	Target Organs?
Unknown Human Questioned Animal	**No**	**No**	**No**	**Respiratory system, eyes, skin, CNS, kidneys, blood.**

Medical Conditions Generally Aggravated by Exposure

Neurologic, psychological, or pulmonary disorders, and anemia may be aggravated by exposure.

Emergency and First-aid Procedures

Eye contact: Do not allow victim to keep eyes tightly shut. Flush immediately with water for 15 minutes (minimum); seek medical attention. Skin contact: Remove all contaminated clothing. Immediately wash area with flooding amounts of water. For reddened or blistered skin, seek medical assistance. For inhalation: Remove the person from exposure. Provide respiratory assistance and CPR. Transfer to medical facility. Observe for 48 hours for lung effects. If swallowed: Seek medical attention.

SECTION VII - PRECAUTIONS FOR SAFE HANDLING AND USE

Steps to be Taken in Case Material is Released or Spilled

Collect powdered materials in most convenient and safe manner and deposit in clean, dry, sealed, metal containers. Dry sweep not recommended. Use HEPA vacuum. For solutions, absorb in vermiculite and place in sealed drum. Ventilate area of spill. Restrict those not involved in cleanup from entering area.

Preferred Waste Disposal Method

No Citation.

Precautions to be Taken in Handling and Storage

Store to avoid contact with incompatible materials. Prevent physical damage to containers. Keep containers tightly closed. Keep sources of heat away.

Other Precautions and Warnings

Avoid excessive dust generation in storage and handling areas.

SECTION VIII - CONTROL MEASURES AND PERSONAL PROTECTIVE EQUIPMENT

Respiratory Protection *(specify type)*

For dust exposures, use a respirator with a HEPA filter; otherwise, use a supplied-air respirator or a self-contained breathing apparatus with full facepiece operated in positive pressure mode.

Ventilation

Local exhaust (preferred) at site of chemical work or general ventilation.

Protective Gloves	Eye Protection	Other Protective Clothing
Impervious Material	**Chemical/Dust Goggles or Face Mask**	**Protective Apron, Boots**

Work/Hygiene Practices

Always wash hands thoroughly after using chemical; never bring food, drink, or smoking materials into vicinity of chemicals.

MANGANESE NITRATE

$Mn(NO_3)_2 \cdot 6H_2O$ CAS: 10377-66-9

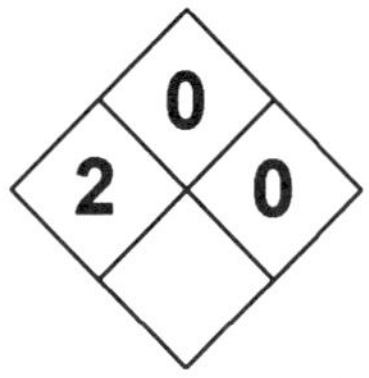

IDENTIFICATION AND TYPICAL USES

Manganese nitrate appears as colorless, pink, or rose-colored deliquescent crystalline solid with no detectable odor. It is used in ceramics, as an intermediate, as a catalyst, and as a source of magnesium dioxide.

RISK ASSESSMENT: HEALTH

General Assessment

Manganese is highly toxic to humans by *inhalation* or *ingestion*. High exposures can produce quite severe poisoning with action on the respiratory and central nervous system, depending upon the conditions of exposure. There are no reports of any carcinogenic, mutagenic, or teratogenic effects in humans or animals.

Inhalation of manganese nitrate dusts can cause irritation to the eyes, nose, throat, and mucosa of the respiratory tract. There can be dryness of the nose (with nosebleeds), throat, and bronchial passages producing cough and phlegm. High exposures can lead to a dangerous buildup of fluid in the lungs (pulmonary edema) which can be fatal. Symptoms may not develop for up to 48 hours or more, thereby creating a false sense of security with regard to health exposure risk. Exposure over prolonged periods of time can lead to the development of manganism. The early symptoms are difficult to recognize because they are generally nonspecific in nature and may occur gradually over a long period of time. It often begins with nonspecific neurologic symptoms such as headache, general apathy, weakness of the legs, progressing to psychosis, with final symptoms similar to that of Parkinson's disease. There can be pain and stiffness in the back and joints, anorexia, weakness, sleepiness, muscle spasms, psychosis, anorexia, emotional swings (laughter or crying for no apparent reason), euphoria, mental confusion, hallucinations, speech disturbances, muscle rigidity, excessive salivation, trembling and tremors, staggering gait and balance disturbances, and profuse sweating.

Acute Health Effects

Acute poisoning with manganese nitrate is rare. The following acute (short-term) health effects may occur immediately or shortly after human exposure to manganese nitrate:

Skin: Irritation, burns, and blisters.

Eye: Irritation to the eye and associated tissues. It can cause burns (especially the concentrated solutions), conjunctivitis, and corneal damage.

Lung: Irritation of the mucosa of the eyes, nose, throat, and respiratory tract. Can cause dryness, nosebleeds, cough with phlegm, and potentially fatal pulmonary edema.

CNS: There may be no noticeable acute effects. However, large doses may result in generally adverse nervous system effects.

Chronic Health Effects

The following chronic (long-term) health effects may occur at some time after exposure to manganese nitrate and can last for months or even years:

Cancer Hazards: According to the references, manganese nitrate has not been adequately tested for its ability to cause cancer in test animals.

Reproductive Hazard: According to the references, manganese nitrate has not been adequately tested for its ability to cause adverse reproductive effects in test animals.

Other Chronic Effects: Prolonged or repeated exposure can damage the central nervous system and cause impaired mental function and degenerative brain changes with personality disorders and physical changes in motor control, vision, and other body functions. High or repeated exposures may damage the lungs, cause changes in the blood (methemoglobinemia), and damage the kidneys or liver.

ʊ *Recommended Risk-Reduction Measures*

While manganese is an essential mineral in human nutrition, the base element and the nitrate can be extremely toxic under certain exposure conditions and personnel should, therefore, avoid direct occupational contact with manganese nitrate. If a less toxic material or compound cannot be substituted for manganese nitrate, then *engineering controls* are the most effective method of reducing exposures. The best protection is to enclose operations and/or provide local exhaust ventilation at the site of manganese nitrate dust release. While not always operationally feasible, isolating operations can also reduce exposure. Using respiratory protection is less effective than the controls mentioned above, but is still advisable whenever working with or around manganese nitrate. The exposure ceiling for manganese compounds is relatively low (5mg/m^3) and difficult to accurately monitor. For low or transient exposures, a chemical cartridge respirator equipped with a high-efficiency particulate air (HEPA) filter and a dust/mist pre-filter may provide adequate respiratory protection. Better protection is obtained using a supplied-air respirator operated in continuous flow mode or a self-contained breathing apparatus (SCBA) with full facepiece operated in pressure demand. If a full facepiece is not available, then chemical/dust goggles should be worn to protect the eyes. A face shield and protective apron should also be worn. To prevent hand and skin exposures, gloves should be used. Glove manufacturers should be contacted *before* selections are made.

Administrative controls should also be in place to minimize the potential for human exposures. These may include written procedures or policies which specify the methods and techniques that will be practiced whenever personnel are to work with manganese nitrate.

All personnel should receive *training* on the use, hazards, protective measures, emergency actions, and other precautions per 29 CFR 1910.1200 (Hazard Communication), prior to the first assignment in an area where manganese nitrate is used or stored.

Before beginning employment working with manganese nitrate, and at regular intervals thereafter (e.g., annually), those personnel with frequent or potentially high exposures should be provided the following recommended medical tests:

- ☑ Complete blood count (establish baseline).
- ☑ Complete examination of the nervous system.
- ☑ Lung function tests (may be normal if the person is not having an attack at the time of the test).
- ☑ Kidney function tests.

If symptoms develop or overexposure is suspected, consider the following additional tests:

- ☑ Complete blood count (compare to baseline).
- ☑ Consider chest X-ray after acute overexposure (may be negative if taken immediately after exposure due to delayed onset of pulmonary edema).
- ☑ Lung function tests.

Any evaluation should include a careful history of past and present symptoms with an examination. Medical tests that simply look for existing damage are not a substitute for controlling exposures. Also, since smoking causes heart disease, lung cancer, emphysema, and other respiratory diseases, smokers exposed to manganese nitrate may experience symptoms more rapidly than non-smokers under the same conditions of exposure. Prudent risk management requires careful consideration of *all* factors which may be causing exposure symptoms in the workplace.

Other methods to reduce exposure include:

- ☑ Where possible, enclose operations and use local exhaust ventilation at the site of chemical release. If local exhaust ventilation or enclosure is not used, respiratory protection should be mandatory.
- ☑ Always ensure that proper protective clothing is worn when using chemical substances. Proper maintenance and care of protective equipment is essential to its function. Therefore, ensure that personnel have received the necessary training on personal protective equipment use and care.
- ☑ Wash thoroughly immediately after exposure to manganese nitrate and at the end of the work shift or before eating, drinking, or smoking.
- ☑ Hazard warning information should be posted in the work area. In addition, as part of an on-going education and training program, all information on the health and safety hazards of manganese nitrate should be communicated to all exposed and potentially exposed workers.
- ☑ Eye wash stations should be provided in the immediate work area for emergency use. If there is a possibility of skin exposure to manganese nitrate, emergency shower facilities should be provided in the immediate area.

☑ Workers whose clothing has been contaminated by manganese nitrate should change into clean clothes before leaving work. Never wear contaminated clothing home where family members can be exposed. Contaminated work clothing should be laundered only by individuals who have been informed of the hazards of exposure to manganese nitrate.

RISK ASSESSMENT: ENVIRONMENT

General Assessment

As a naturally occurring mineral, manganese itself poses very little threat to the environment. However, accidental spills or releases to the environment can occur during transportation, storage, disposal, or destruction of manganese nitrate. In almost every scenario, the threat of environmental exposure is contingent upon the proper handling of the chemical substance. Accidental spills, large or small, where manganese nitrate contacts incompatible commodities can result in release of toxic products (depending upon conditions of spill), and possible contamination of the surrounding environmental mediums (water, soil, and air).

Manganese nitrate is considered a noncombustible solid. But it will react violently in contact with ordinary combustible materials. Extreme caution is always required in handling, storage, transportation, and disposal of manganese nitrate. These characteristics also require special consideration during any emergency situation involving a leak or spill of manganese nitrate powder or dust. Should manganese nitrate ever come into contact with organic materials either during use, transportation, storage, or disposal, violent and even explosive reactions are possible.

Manganese occurs naturally in many minerals that are widely distributed in the earth's crust and, in trace amounts, is an essential element for both plants and animals. The wide variety of possible manganese compounds may enter the environment from natural and industrial sources.

Acute Ecological Effects

Acute (short-term) toxic effects may include the death of animals, birds, or fish and death or low growth rate in plants. Acute effects are seen 2 to 4 days after animals or plants are exposed to manganese.

Manganese and its compounds have moderate acute toxicity to aquatic life. While manganese is considered essential for plant and animal life, insufficient data are available on the acute effects of excessive manganese nitrate contamination on plants, birds, or land animals.

Chronic Ecological Effects

Chronic (long-term) toxic effects may include shortened life span, reproductive problems, lower fertility, and changes in appearance or behavior in exposed animals. These effects can be seen long after first exposure(s) to toxic chemicals.

Manganese and its compounds have moderate chronic toxicity to aquatic life. Manganese is considered essential for plant and animal life. However, insufficient data are available on the chronic effects of excessive manganese contamination on plants, birds, or land animals.

Water Solubility

Manganese and its compounds vary in their solubility in water from very soluble to nearly insoluble. In water, manganese reacts and decomposes to form explosive hydrogen gas. Manganese nitrate easily dissolves in water and becomes highly mobile.

Persistence in the Environment

Manganese and its compounds are highly persistent in water, with a half-life greater than 200 days. The half-life of a pollutant is the amount of time it takes for one half of the chemical to be degraded. Manganese is also naturally present in terrestrial soils and aquatic sediments.

Bioaccumulation in Aquatic Organisms

Some substances increase in concentration, or *bioaccumulate,* in living organisms as they breathe contaminated air, drink contaminated water, or eat contaminated food. These chemicals can become concentrated in the tissues and internal organs of animals as well as humans.

The concentration of manganese found in fish tissues is expected to be about the same as the average concentration of manganese found in the water from which the fish was taken.

Recommended Risk-Reduction Measures

Proper training of all transporters will reduce the likelihood of a mishap or accident resulting in a leak or spill of manganese nitrate dusts or powders into the environment. The correct use of labeling on all containers, trucks, and rail cars will enable emergency responders to react properly and quickly to any disaster thereby reducing the potential risk to the environment and to personnel.

Storage of manganese nitrate should be segregated from incompatible chemicals to minimize the risk of cross-contamination or contact. Care should be taken to prevent physical damage to storage containers. Also, dusty conditions must be avoided in storage and use areas.

If a spill or leak to the environment has occurred, fire department, emergency response and/or hazardous materials spill personnel should be notified immediately. Cleanup should be attempted only by those trained in proper spill containment procedures. Contaminated soils should be removed for decontamination and replaced with clean soil. If manganese nitrate should contact the water table, aquifer, or navigable waterway, response activities should be prompt. The local and/or state emergency response authorities may require notification. A comprehensive emergency response or disaster preparedness/recovery plan should be in place prior to any operations involving the use, transportation, storage, or disposal of manganese nitrate.

If manganese nitrate is spilled or leaked, the following specific steps are recommended:

- ☑ Restrict persons not wearing protective clothing from area of spill or leak until cleanup is complete and area can be opened for normal work.
- ☑ Ventilate area of spill or leak.
- ☑ Collect powdered materials in the safest manner possible and deposit in sealed containers. Recommend using a vacuum equipped with a high-efficiency particulate air (HEPA) filter. Do not dry sweep (generates airborne dusts). Absorb solution spills with vermiculite or other material, collect, and place in sealed drum for disposal.
- ☑ It may be necessary to dispose of manganese nitrate as a hazardous waste. The responsible state agency or the regional office of the federal Environmental Protection Agency (EPA) should be contacted for specific recommendations.

RISK ASSESSMENT: BUSINESS

General Assessment

Accidents or mishaps involving manganese nitrate can present a significant threat to business operations. The loss or damage of equipment or facilities can significantly affect fiscal viability. Lawsuits that may result from personnel injury/death, public exposures, and/or environmental contamination will also require a serious expenditure of resources. Media attention surrounding an injury, death, or environmental damage can also result in a loss of profits and loss of current as well as future business.

Recommended Risk-Reduction Measures

Company attorneys, safety and health professionals, and environmental specialists should be involved in the development of any procedures for responding to chemical incidents. A company official should be predesignated as a public relations officer with specific training in dealing with the press. Corporate plans and policies should be developed, approved, and implemented long before any need for such arises.

REFERENCES

American Conference of Governmental Industrial Hygienists. 1988. *Documentation of the Threshold Limit Values and Biological Exposure Indices,* 5th Edition (with updates). Cincinnati: ACGIH

Asante-Duah, D. K. 1993. *Hazardous Waste Risk Assessment.* New York: CRC Press/Lewis Publishers.

Bloom, A. D. and F. J. DeSerres. 1995. *Ecotoxicity and Human Health.* New York: CRC Press/Lewis Publishers.

Calabrese, E. J. 1994. *Biological Effects of Low Level Exposures.* New York: CRC Press/Lewis Publishers.

Cockerman, L. G., B. S. Shane. 1994. *Basic Environmental Toxicology.* New York: CRC Press/Lewis Publishers.

Environmental Protection Agency. 1983. *Chemical Hazard Information Profiles.* Washington, D.C.: U.S. EPA.

Lauwerys, R. R., P. Hoet. 1993. *Industrial Chemical Exposure.* New York: CRC Press/Lewis Publishers.

Lewis, R. J., Sr. 1993. *Hawley's Condensed Chemical Dictionary,* 12th Edition. New York: Van Nostrand Reinhold.

Lewis, R. J., Sr. 1992. *Sax's Dangerous Properties of Industrial Materials,* Volumes 1, 2, and 3, 8th Edition. New York: Van Nostrand Reinhold.

Mackay, D., W. Y. Shiu, and K. C. Ma. 1992. *Illustrated Handbook of Chemical-Physical Properties and Environmental Fate for Organic Chemicals.* New York: CRC Press/Lewis Publishers.

Meyer, E. 1990. *Chemistry of Hazardous Materials.* Englewood Cliffs, NJ: Prentice Hall.

Richardson, M. 1995. *Dictionary of Substance and Their Effects,* Vol. I-VI. New York: CRC Press/Lewis Publishers.

Torkelson, T. R. V. K. Rowe. 1981. *Patty's Industrial Hygiene and Toxicology,* Vol. 2B. New York: Wiley-Interscience.

MATERIAL SAFETY DATA SHEET

CHEMICAL NAME	
MANGANESE SULFATE	**CRC PRESS/LEWIS PUBLISHERS DISCLAIMER:** THE INFORMATION AND RECOMMENDATIONS PRESENTED HEREIN ARE BASED ON SOURCES BELIEVED TO BE RELIABLE. CRC MAKES NO REPRESENTATION ON ITS COMPLETENESS OR ACCURACY. IT IS THE RESPONSIBILITY OF THE USER TO DETERMINE THE CHEMICAL'S SUITABILITY FOR ITS INTENDED USE, THE CHEMICAL'S SAFE USE, AND THE CHEMICAL'S PROPER DISPOSAL. NO REPRESENTATIONS AND/OR WARRANTIES, EITHER EXPRESSED OR IMPLIED, OF THE MERCHANTABILITY OR FITNESS FOR A PARTICULAR PURPOSE, OR OF ANY OTHER NATURE, ARE MADE WITH RESPECT TO THE INFORMATION PROVIDED IN THIS MSDS OR TO THE CHEMICAL TO WHICH INFORMATION MAY REFER. CRC NEITHER ASSUMES NOR AUTHORIZES ANY OTHER PERSON TO ASSUME FOR IT, ANY OTHER ADDITIONAL RESPONSIBILITY OR LIABILITY FOR THE USE OF, OR RELIANCE UPON, THIS INFORMATION.

HAZARD WARNING INFORMATION

HEALTH	FIRE	REACTIVE	OTHER	DEGREE OF HAZARD	COLOR CODING	OTHER CODES
2	0	0		0 = Minimum Hazard 1 = Slight Hazard 2 = Moderate Hazard 3 = Serious Hazard 4 = Severe Hazard	HEALTH = BLUE FIRE = RED REACTIVITY = YELLOW OTHER = WHITE	OX = Oxidizer ACID = Acid ALK = Alkali COR = Corrosive W = Use No Water

SECTION I - GENERAL INFORMATION

Characterization: **Metallic Sulfate**	RCRA Number: **None**	EPA Class: **Not Applicable**
DOT Proper Shipping Name: **No Citation**	Chemical Abstract Service (CAS) Number: **7785-87-7**	
DOT Hazard Class and Label Requirements: **No Citation**	DOT Emergency Guide Code: **No Citation**	
DOT Identification Number: **No Citation**	Atomic Symbol: **$MnSO_4$**	

Synonyms

Manganous sulfate; Man-Gro ®; manganese (II) sulfate; Sorba-Spray manganese ®; sulfuric acid, manganese(2+) salt.

SECTION II - HAZARDOUS INGREDIENTS AND IDENTITY INFORMATION

Hazardous Components (specific identity)	OSHA Exposure Criteria	NIOSH Exposure Criteria	Immediately Dangerous to Life and Health (IDLH)	ACGIH Exposure Criteria
Manganese sulfate (derivation: A by-product of production of hydroquinone; or by the action of sulfuric acid on manganous hydroxide or carbonate).	**PEL (ceiling):** 5 mg(Mn)/m^3 (dust/compounds) **STEL:** Not Applicable	**REL (ceiling):** 5 mg(Mn)/m^3 (dust/compounds) **STEL:** 3 mg(Mn)/m^3 (fumes)	500 mg(Mn)/m^3	**TLV:** 5 mg(Mn)/m^3 (dust/compounds) 1 mg(Mn)/m^3 (fumes) **STEL:** 3 mg(Mn)/m^3 (fumes)

SECTION III - PHYSICAL AND CHEMICAL CHARACTERISTICS

Boiling Point: **1562°F (850°C) Decomposes**	Specific Gravity (H_2O = 1): **3.25**
Vapor Pressure (mm Hg): **Not Found**	Molecular Weight: **151**
Vapor Density (Air = 1): **Not Found**	Melting Point: **1292°F (700°C)**

Solubility

Very soluble in water (especially boiling water) to form several hydrates, insoluble in alcohol.

Appearance and Odor

Pink, red, or rose-colored translucent powder or crystals with no odor.

SECTION IV - FIRE AND EXPLOSION HAZARD DATA

Flash Point (method used): **Not Found**	Explosive Limits in Air % by Volume: **LEL: Not Found** **UEL: Not Found**
NFPA Classification: **Non-Combustible Solid**	Autoignition Temperature: **Not Found**

Extinguishing Media

Use agent suitable to surrounding fire (dry chemical, carbon dioxide, water spray, or foam).

Special Fire Fighting Procedures

Wear full protective clothing and self-contained breathing apparatus (SCBA). Poisonous gases are produced in fire. Cool fire-exposed containers that cannot be safely moved (use water spray). Do NOT release runoff from fire-fighting procedures to sewers or waterways. Dike and collect for disposal.

Unusual Fire and Explosion Hazards

At high temperatures, releases toxic fumes of sulfur oxides. Use water spray to knock down vapors.

SECTION V - REACTIVITY DATA

Stability		Conditions to Avoid: Manganese sulfate is normally stable under routine conditions of handling and storage. Avoid contact with incompatible materials and heat. Avoid vibration.
Stable	Unstable	Incompatibility *(materials to avoid)*
X		Contact with carbon monoxide, silicon dioxide, fluorine, copper, lead, X-rays, vibration will increase the toxic effects of manganese sulfate. Contact will cause violent decomposition of >52% hydrogen peroxide.
Hazardous Polymerization		Conditions to Avoid: Hazardous polymerization of manganese sulfate will not occur.
May Occur	Will Not Occur	Hazardous Decomposition or By-products
	X	Poisonous gases are emitted during fire, including toxic and irritating smoke and fumes of carbon dioxide and toxic fumes of sulfur oxides.

SECTION VI - HEALTH HAZARD DATA

Primary Route(s) of Entry:	Inhalation?	Absorption (skin/eye)?	Ingestion?
	X		X

Health Hazards

INHALATION: The dusts can produce irritation of the eyes and mucous membranes. The fumes can lead to metal fume fever with symptoms of fever, chills, nausea, weakness, headache, blurred vision, back stiffness and pain, muscle pain, shallow respiration, dryness in mouth and throat, intense thirst, dry cough, metallic taste, chest tightness, and Mn pneumonitis. It can also lead to degenerative brain effects with symptoms similar to Parkinson's disease.

EYES & SKIN: Causes skin and eye irritation. There can be dermatitis, conjunctivitis, corneal damage.

INGESTION: Large doses can cause gastrointestinal irritation and other effects similar to inhalation.

Carcinogenicity	NTP Listed?	IARC Cancer Review Group?	OSHA Regulated?	Target Organs?
Unknown Human Questioned Animal	No	No	No	Respiratory system, eyes, skin, CNS, kidneys, blood.

Medical Conditions Generally Aggravated by Exposure

Neurologic, psychological, or pulmonary disorders, and anemia may be aggravated by exposure.

Emergency and First-aid Procedures

Eye contact: Do not allow victim to keep eyes tightly shut. Flush immediately with water for 15 minutes (minimum); seek medical attention. Skin contact: Remove all contaminated clothing. Immediately wash area with flooding amounts of water. For reddened or blistered skin, seek medical assistance. For inhalation: Remove the person from exposure. Provide respiratory assistance and CPR. Transfer to medical facility. If swallowed: Seek medical attention. Rinse mouth with water. Give 1 to 2 glasses of water and induce vomiting until vomit is clear. Never give anything by mouth to a convulsing person.

SECTION VII - PRECAUTIONS FOR SAFE HANDLING AND USE

Steps to be Taken in Case Material is Released or Spilled

Collect powdered materials in most convenient and safe manner and deposit in clean, dry, sealed, metal containers. Dry sweep not recommended. Use HEPA vacuum. For solutions, absorb in vermiculite and place in sealed drum. Ventilate area of spill. Restrict those not involved in cleanup from entering area.

Preferred Waste Disposal Method

No citation.

Precautions to be Taken in Handling and Storage

Store to avoid contact with incompatible materials. Prevent physical damage to containers. Keep containers tightly closed. Keep sources of heat away.

Other Precautions and Warnings

Avoid excessive dust generation in storage and handling areas. Personnel should be trained on the proper precautions for working with manganese sulfate compounds *before* initial assignment.

SECTION VIII - CONTROL MEASURES AND PERSONAL PROTECTIVE EQUIPMENT

Respiratory Protection *(specify type)*

For dust exposures, use a respirator with a HEPA filter; otherwise, use a supplied-air respirator or a self-contained breathing apparatus with full facepiece operated in positive pressure mode.

Ventilation

Local exhaust (preferred) at site of chemical work or general ventilation.

Protective Gloves	Eye Protection	Other Protective Clothing
Impervious Material	Chemical/Dust Goggles or Face Mask	Protective Apron, Boots

Work/Hygiene Practices

Always wash hands thoroughly after using chemical; never bring food, drink, or smoking materials into vicinity of chemicals.

MANGANESE SULFATE

$MnSO_4$

CAS: 7785-87-7

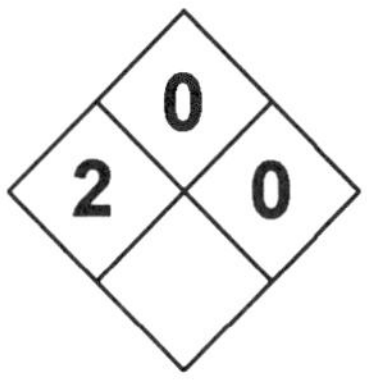

IDENTIFICATION AND TYPICAL USES

Manganese sulfate appears as pink, red, or rose-colored, translucent efflorescent prisms or crystals with no detectable odor. It is used in fertilizers and fungicides, as a feed additive, in paints and varnishes, in ceramics, in textile dyes, medicines, for ore flotation, as a catalyst in viscose process, and in the manufacture of synthetic manganese dioxide.

RISK ASSESSMENT: HEALTH

General Assessment

Manganese sulfate does not absorb well through the digestive or respiratory processes. It therefore presents very low acute toxicity. However, in large or chronic doses (1 to 2 years of exposure) by *inhalation* or *ingestion*, manganese poisoning can be quite severe. It is considered a questionable carcinogen by at least one source (Lewis 1992) with experimental neoplastic and mutagenic data reported. It has also been shown to cause experimental teratogenic effects. The data are inconclusive and conflicting with regard to its human carcinogenic potential.

Inhalation of manganese sulfate over prolonged periods of time can lead to the development of manganism. The early symptoms are difficult to recognize because they are generally nonspecific in nature and may occur gradually over a long period of time. It often begins with nonspecific neurologic symptoms such as headache, general apathy, weakness of the legs, progressing to psychosis, with final symptoms similar to that of Parkinson's disease. There can be pain and stiffness in the back and joints, anorexia, weakness, sleepiness, muscle spasms, psychosis, anorexia, emotional swings (laughter or crying for no apparent reason), euphoria, mental confusion, hallucinations, speech disturbances, muscle rigidity, excessive salivation, trembling and tremors, staggering gait and balance disturbances, and profuse sweating.

Acute Health Effects

Acute poisoning with manganese sulfate is rare. The following acute (short-term) health effects may occur immediately or shortly after human exposure to manganese sulfate:

Skin: Possible irritation and dermatitis (in some individuals).

Eye: Irritation to the eye and associated tissues. It can cause conjunctivitis and corneal damage.

Lung: Irritation of the mucosa of the eyes, nose, throat, and respiratory tract.

CNS: There may be no noticeable acute effects. However, large doses may result in generally adverse nervous system effects.

Chronic Health Effects

The following chronic (long-term) health effects may occur at some time after exposure to manganese sulfate and can last for months or even years:

Cancer Hazards: According to the references, manganese sulfate may cause cancer in test animals. Its human carcinogenic potential is unknown.

Reproductive Hazard: According to the references, manganese sulfate has been shown to cause adverse reproductive effects in test animals.

Other Chronic Effects: Prolonged or repeated exposure can damage the central nervous system and cause impaired mental function and degenerative brain changes with personality disorders and physical changes in motor control, vision, and other body functions.

Recommended Risk-Reduction Measures

While manganese is an essential mineral in human nutrition, the base element and the sulfate can be extremely toxic under certain exposure conditions and personnel should, therefore, avoid direct occupational contact with manganese sulfate. If a less toxic material or compound cannot be substituted for manganese sulfate, then *engineering controls* are the most effective method of reducing exposures. The best protection is to enclose operations and/or provide local exhaust ventilation at the site of manganese sulfate dust release. While not always operationally feasible, isolating operations can also reduce exposure. Using respiratory protection is less effective than the controls mentioned above, but is still advisable whenever working with or around manganese sulfate. The exposure ceiling for manganese compounds is relatively low ($5mg/m^3$) and difficult to accurately monitor. For low or transient exposures, a chemical cartridge respirator equipped with a high-efficiency particulate air (HEPA) filter and a dust/mist pre-filter may provide adequate respiratory protection. Better protection is obtained using a supplied-air respirator operated in continuous flow mode or a self-contained breathing apparatus (SCBA) with full facepiece operated in pressure demand. If a full facepiece is not available, then chemical/dust goggles should be worn to protect the eyes. A face shield and protective apron should also be worn. To prevent hand and skin exposures, gloves should be used. Glove manufacturers should be contacted *before* selections are made.

Administrative controls should also be in place to minimize the potential for human exposures. These may include written procedures or policies which specify the methods and techniques that will be practiced whenever personnel are to work with manganese sulfate.

All personnel should receive *training* on the use, hazards, protective measures, emergency actions, and other precautions per 29 CFR 1910.1200 (Hazard Communication), prior to the first assignment in an area where manganese sulfate is used or stored.

Before beginning employment working with manganese sulfate, and at regular intervals thereafter (e.g., annually), those personnel with frequent or potentially high exposures should be provided the following recommended medical tests:

- ☑ Complete blood count (establish baseline).
- ☑ Complete examination of the nervous system.
- ☑ Lung function tests (may be normal if the person is not having an attack at the time of the test)
- ☑ Kidney function tests.

If symptoms develop or overexposure is suspected, consider the following additional tests:

- ☑ Complete blood count (compare to baseline).
- ☑ Consider chest X-ray after acute overexposure.
- ☑ Lung function tests.

Any evaluation should include a careful history of past and present symptoms with an examination. Medical tests that simply look for existing damage are not a substitute for controlling exposures. Also, since smoking causes heart disease, lung cancer, emphysema, and other respiratory diseases, smokers exposed to manganese sulfate may experience symptoms more rapidly than non-smokers under the same conditions of exposure. Prudent risk management requires careful consideration of *all* factors which may be causing exposure symptoms in the workplace.

Other methods to reduce exposure include:

- ☑ Where possible, enclose operations and use local exhaust ventilation at the site of chemical release. If local exhaust ventilation or enclosure is not used, respiratory protection should be mandatory.
- ☑ Always ensure that proper protective clothing is worn when using chemical substances. Proper maintenance and care of protective equipment is essential to its function. Therefore, ensure that personnel have received the necessary training on personal protective equipment use and care.
- ☑ Wash thoroughly immediately after exposure to manganese sulfate and at the end of the work shift or before eating, drinking, or smoking.
- ☑ Hazard warning information should be posted in the work area. In addition, as part of an on-going education and training program, all information on the health and safety hazards of manganese sulfate should be communicated to all exposed and potentially exposed workers.
- ☑ Eye wash stations should be provided in the immediate work area for emergency use. If there is a possibility of skin exposure to manganese sulfate, emergency shower facilities should be provided in the immediate area.
- ☑ Workers whose clothing has been contaminated by manganese sulfate should change into clean clothes before leaving work. Never wear contaminated clothing home where family members can be exposed. Contaminated work clothing should be laundered only by individuals who have been informed of the hazards of exposure to manganese sulfate.

RISK ASSESSMENT: ENVIRONMENT

General Assessment

As a naturally occurring mineral, manganese itself poses very little threat to the environment. However, accidental spills or releases to the environment can occur during transportation, storage, disposal, or destruction of manganese sulfate. In almost every scenario, the threat of environmental exposure is contingent upon the proper handling of the chemical substance. Accidental spills, large or small, where manganese sulfate contacts incompatible commodities can result in release of toxic products (depending upon conditions of spill), and possible contamination of the surrounding environmental mediums (water, soil, and air).

Manganese sulfate is considered a noncombustible solid. It will react and violently and cause the violent decomposition of hydrogen peroxide (>52% solutions). Contact with vibrations, X-rays, carbon monoxide, silicon dioxide, fluorine, copper, and lead, will greatly increase the toxicity of manganese sulfate. Extreme caution is always required in handling, storage, transportation, and disposal of manganese sulfate. These characteristics also require special consideration during any emergency situation involving a leak or spill of manganese sulfate powder or dust. Should manganese sulfate ever come into contact with these various incompatible substances either during use, transportation, storage, or disposal, violent and even explosive reactions are possible. Manganese occurs naturally in many minerals that are widely distributed in the earth's crust and, in trace amounts, is an essential element for both plants and animals. The wide variety of possible manganese compounds may enter the environment from natural and industrial sources.

Acute Ecological Effects

Acute (short-term) toxic effects may include the death of animals, birds, or fish and death or low growth rate in plants. Acute effects are seen 2 to 4 days after animals or plants are exposed to manganese.

Manganese and its compounds have moderate acute toxicity to aquatic life. While manganese is considered essential for plant and animal life, insufficient data are available on the acute effects of excessive manganese sulfate contamination on plants, birds, or land animals.

Chronic Ecological Effects

Chronic (long-term) toxic effects may include shortened life span, reproductive problems, lower fertility, and changes in appearance or behavior in exposed animals. These effects can be seen long after first exposure(s) to toxic chemicals.

Manganese and its compounds have moderate chronic toxicity to aquatic life. Manganese is considered essential for plant and animal life. However, insufficient data are available on the chronic effects of excessive manganese contamination on plants, birds, or land animals.

Water Solubility

Manganese and its compounds vary in their solubility in water from very soluble to nearly insoluble. In water, manganese reacts and decomposes to form explosive hydrogen gas. Manganese sulfate easily dissolves in hot (boiling) water and becomes highly mobile.

Persistence in the Environment

Manganese and its compounds are highly persistent in water, with a half-life greater than 200 days. The half-life of a pollutant is the amount of time it takes for one half of the chemical to be degraded. Manganese is also naturally present in terrestrial soils and aquatic sediments.

Bioaccumulation in Aquatic Organisms

Some substances increase in concentration, or *bioaccumulate,* in living organisms as they breathe contaminated air, drink contaminated water, or eat contaminated food. These chemicals can become concentrated in the tissues and internal organs of animals as well as humans.

The concentration of manganese found in fish tissues is expected to be about the same as the average concentration of manganese found in the water from which the fish was taken.

Recommended Risk-Reduction Measures

Proper training of all transporters will reduce the likelihood of a mishap or accident resulting in a leak or spill of manganese sulfate dusts or powders into the environment. The correct use of labeling on all containers, trucks, and rail cars will enable emergency responders to react properly and quickly to any disaster thereby reducing the potential risk to the environment and to personnel.

Storage of manganese sulfate should be segregated from incompatible chemicals to minimize the risk of cross-contamination or contact. Care should be taken to prevent physical damage to storage containers. Also, dusty conditions must be avoided in storage and use areas.

If a spill or leak to the environment has occurred, fire department, emergency response and/or hazardous materials spill personnel should be notified immediately. Cleanup should be attempted only by those trained in proper spill containment procedures. Contaminated soils should be removed for decontamination and replaced with clean soil. If manganese sulfate should contact the water table, aquifer, or navigable waterway, response activities should be prompt. The local and/or state emergency response authorities may require notification. A comprehensive emergency response or disaster preparedness/recovery plan should be in place prior to any operations involving the use, transportation, storage, or disposal of manganese sulfate.

If manganese sulfate is spilled or leaked, the following specific steps are recommended:

☑ Restrict persons not wearing protective clothing from area of spill or leak until cleanup is complete and area can be opened for normal work.

☑ Ventilate area of spill or leak.

☑ Collect powdered materials in the safest manner possible and deposit in sealed containers. Recommend using a vacuum equipped with a high-efficiency particulate air (HEPA) filter. Do not dry sweep (generates airborne dusts). Absorb solution spills with vermiculite or other material, collect, and place in sealed drum for disposal.

☑ It may be necessary to dispose of manganese sulfate as a hazardous waste. The responsible state agency or the regional office of the federal Environmental Protection Agency (EPA) should be contacted for specific recommendations.

RISK ASSESSMENT: BUSINESS

General Assessment

Accidents or mishaps involving manganese sulfate can present a significant threat to business operations. The loss or damage of equipment or facilities can significantly affect fiscal viability. Lawsuits that may result from personnel injury/death, public exposures, and/or environmental contamination will also require a serious expenditure of resources. Media attention surrounding an injury, death, or environmental damage can also result in a loss of profits and loss of current as well as future business. Always remember that anytime the terms "cancer," "carcinogen," or "reproductive hazard" are used, public emotion, hysteria, and ignorance can run equally high. This must be carefully considered whenever developing public relations policies.

Recommended Risk-Reduction Measures

Company attorneys, safety and health professionals, and environmental specialists should be involved in the development of any procedures for responding to chemical incidents. A company official should be predesignated as a public relations officer with specific training in dealing with the press. Corporate plans and policies should be developed, approved, and implemented long before any need for such arises.

REFERENCES

American Conference of Governmental Industrial Hygienists. 1988. *Documentation of the Threshold Limit Values and Biological Exposure Indices,* 5th Edition (with updates). Cincinnati: ACGIH

Asante-Duah, D. K. 1993. *Hazardous Waste Risk Assessment.* New York: CRC Press/Lewis Publishers.

Bloom, A. D. and F. J. DeSerres. 1995. *Ecotoxicity and Human Health.* New York: CRC Press/Lewis Publishers.

Calabrese, E. J. 1994. *Biological Effects of Low Level Exposures.* New York: CRC Press/Lewis Publishers.

Cockerman, L. G., B. S. Shane. 1994. *Basic Environmental Toxicology.* New York: CRC Press/Lewis Publishers.

Environmental Protection Agency. 1983. *Chemical Hazard Information Profiles.* Washington, D.C.: U.S. EPA.

Gardner, R. W. 1994. *Chemical Intolerance.* New York: CRC Press/Lewis Publishers.

Lauwerys, R. R., P. Hoet. 1993. *Industrial Chemical Exposure.* New York: CRC Press/Lewis Publishers.

Lewis, R. J., Sr. 1993. *Hawley's Condensed Chemical Dictionary,* 12th Edition. New York: Van Nostrand Reinhold.

Lewis, R. J., Sr. 1992. *Sax's Dangerous Properties of Industrial Materials,* Volumes 1, 2, and 3, 8th Edition. New York: Van Nostrand Reinhold.

Mackay, D., W. Y. Shiu, and K. C. Ma. 1992. *Illustrated Handbook of Chemical-Physical Properties and Environmental Fate for Organic Chemicals.* New York: CRC Press/Lewis Publishers.

Meyer, E. 1990. *Chemistry of Hazardous Materials.* Englewood Cliffs, NJ: Prentice Hall.

Richardson, M. 1995. *Dictionary of Substance and Their Effects,* Vol. I-VI. New York: CRC Press/Lewis Publishers.

Torkelson, T. R. V. K. Rowe. 1981. *Patty's Industrial Hygiene and Toxicology,* Vol. 2B. New York: Wiley-Interscience.

MATERIAL SAFETY DATA SHEET

CHEMICAL NAME	
MANGANESE TETROXIDE	**CRC PRESS/LEWIS PUBLISHERS DISCLAIMER:** THE INFORMATION AND RECOMMENDATIONS PRESENTED HEREIN ARE BASED ON SOURCES BELIEVED TO BE RELIABLE. CRC MAKES NO REPRESENTATION ON ITS COMPLETENESS OR ACCURACY. IT IS THE RESPONSIBILITY OF THE USER TO DETERMINE THE CHEMICAL'S SUITABILITY FOR ITS INTENDED USE, THE CHEMICAL'S SAFE USE, AND THE CHEMICAL'S PROPER DISPOSAL. NO REPRESENTATIONS AND/OR WARRANTIES, EITHER EXPRESSED OR IMPLIED, OF THE MERCHANTABILITY OR FITNESS FOR A PARTICULAR PURPOSE, OR OF ANY OTHER NATURE, ARE MADE WITH RESPECT TO THE INFORMATION PROVIDED IN THIS MSDS OR TO THE CHEMICAL TO WHICH INFORMATION MAY REFER. CRC NEITHER ASSUMES NOR AUTHORIZES ANY OTHER PERSON TO ASSUME FOR IT, ANY OTHER ADDITIONAL RESPONSIBILITY OR LIABILITY FOR THE USE OF, OR RELIANCE UPON, THIS INFORMATION.

HAZARD WARNING INFORMATION

HEALTH	FIRE	REACTIVE	OTHER	DEGREE OF HAZARD	COLOR CODING	OTHER CODES
1	0	0		0 = Minimum Hazard 1 = Slight Hazard 2 = Moderate Hazard 3 = Serious Hazard 4 = Severe Hazard	HEALTH = BLUE FIRE = RED REACTIVITY = YELLOW OTHER = WHITE	OX = Oxidizer ACID = Acid ALK = Alkali COR = Corrosive W = Use No Water

SECTION I - GENERAL INFORMATION

Characterization	**Manganese Compound**	RCRA Number	**None**
		EPA Class	**Not Applicable**
DOT Proper Shipping Name	**No Citation**	Chemical Abstract Service (CAS) Number	**1317-35-7**
DOT Hazard Class and Label Requirements	**No Citation**	DOT Emergency Guide Code	**35**
DOT Identification Number	**No Citation**	Atomic Symbol	**Mn_3O_4**

Synonyms

Manganese oxide; manganomanganic oxide; trimanganese tetroxide; trimanganese tetroxide.

SECTION II - HAZARDOUS INGREDIENTS AND IDENTITY INFORMATION

Hazardous Components (specific identity)	OSHA Exposure Criteria	NIOSH Exposure Criteria	Immediately Dangerous to Life and Health (IDLH)	ACGIH Exposure Criteria
Manganese tetroxide (derivation: From the pouring and casting of molten ferromanganese).	**PEL (ceiling):** 5 mg(Mn)/m^3 (dust/compounds) **STEL:** Not Applicable	**REL (ceiling):** Not Established **STEL:** Not Applicable	**Not Determined**	**TLV:** 5 mg(Mn)/m^3 (dust/compounds) 1 mg(Mn)/m^3 (fumes) **STEL:** 3 mg(Mn)/m^3 (fumes)

SECTION III - PHYSICAL AND CHEMICAL CHARACTERISTICS

Boiling Point	**Not Found**	Specific Gravity (H_2O = 1)	**4.88**
Vapor Pressure (mm Hg)	**0 (approx.) at 68°F (20°C)**	Molecular Weight	**228.8**
Vapor Density (Air = 1)	**Not Found**	Melting Point	**2847°F (1564°C)**

Solubility

Insoluble in water, soluble in hydrochloric acid with evolution of chlorine.

Appearance and Odor

Fine black or brownish black powder. There is no odor.

SECTION IV - FIRE AND EXPLOSION HAZARD DATA

Flash Point (method used)	**Not Found**	Explosive Limits in Air % by Volume	**LEL: Not Found** **UEL: Not Found**
NFPA Classification	**Non-Combustible Solid**	Autoignition Temperature	**Not Found**

Extinguishing Media

Use extinguishing agent suitable for surrounding fires (dry chemical, carbon dioxide, water spray, foam). Manganese tetroxide is not likely to burn.

Special Fire Fighting Procedures

Wear full protective clothing and self-contained breathing apparatus (SCBA) operated in positive pressure or continuous flow mode. Poisonous gases and fumes are produced in fire. Be aware of runoff, do NOT release to sewers or waterways.

Unusual Fire and Explosion Hazards

None reported.

SECTION V - REACTIVITY DATA

Stability		Conditions to Avoid: Manganese tetroxide is normally stable under routine conditions of handling and storage. Avoid contact with incompatible materials, heat, flame, sparks, or other ignition sources.
Stable	Unstable	Incompatibility *(materials to avoid)*
X		Incompatible with strong reducing agents, acids, and bases.
Hazardous Polymerization		Conditions to Avoid: Under normal conditions of temperature and pressure, hazardous polymerization of manganese tetroxide will not occur. Be cautious of welding, cutting, casting, pouring, or finishing operations.
May Occur	Will Not Occur	Hazardous Decomposition or By-products
	X	Poisonous gases are emitted during fire, including toxic and irritating smoke and fumes. Oxygen is also release during decomposition which can greatly accelerate combustion of other items.

SECTION VI - HEALTH HAZARD DATA

Primary Route(s) of Entry:	Inhalation?	Absorption (skin/eye)?	Ingestion?
	X		X

Health Hazards

INHALATION: The dusts can irritate the respiratory system with progression to lung damage, pleuritis, bronchitis, and possible respiratory infection. The fumes can lead to metal fume fever with symptoms of fever, chills, nausea, weakness, headache, blurred vision, back stiffness and pain, muscular pain, shallow respiration, dryness in mouth and throat, intense thirst, dry cough, metallic taste, chest tightness, and Mn pneumonitis. It can also lead to degenerative brain effects with symptoms similar to Parkinson's disease.

EYES & SKIN: Causes eye irritation. Particles embedded in the skin cause inflammation and redness.

INGESTION: Large doses are required for toxic effects (digestive tract irritation and other effects).

Carcinogenicity	NTP Listed?	IARC Cancer Review Group?	OSHA Regulated?	Target Organs?
Unknown Human Unknown Animal	No	No	No	Respiratory system, eyes, CNS.

Medical Conditions Generally Aggravated by Exposure

Central nervous system disorders, psychiatric problems, and respiratory diseases may be aggravated.

Emergency and First-aid Procedures

Eye contact: Do not allow victim to keep eyes tightly shut. Flush immediately with water for 15 minutes (minimum); seek medical attention. Skin contact: Remove all contaminated clothing. Immediately wash area with flooding amounts of water. For reddened or blistered skin, seek medical assistance. For inhalation: Remove the person from exposure. Provide respiratory assistance and CPR. Transfer to medical facility. If swallowed: Seek medical attention. Rinse mouth with water. Give 2 to 3 glasses of water and induce vomiting. Never give anything by mouth to an unconscious or convulsing person.

SECTION VII - PRECAUTIONS FOR SAFE HANDLING AND USE

Steps to be Taken in Case Material is Released or Spilled

Collect in most convenient and safe manner and deposit in clean, sealed, metal containers. Dry sweep not recommended. Use scoop for bulk materials and a HEPA vacuum on residue. Ventilate area of spill. Restrict those not involved in cleanup from entering area.

Preferred Waste Disposal Method

No citation.

Precautions to be Taken in Handling and Storage

Store to avoid contact with incompatible materials. Prevent physical damage to containers. Keep containers tightly closed. Personal protective equipment is required when working with this compound.

Other Precautions and Warnings

Avoid excessive dust generation in storage and handling areas.

SECTION VIII - CONTROL MEASURES AND PERSONAL PROTECTIVE EQUIPMENT

Respiratory Protection *(specify type)*

For dust exposures, use a respirator with a HEPA filter; otherwise, use a supplied-air respirator or a self-contained breathing apparatus with full facepiece operated in positive pressure mode.

Ventilation

Local exhaust (preferred) at site of chemical work or general ventilation.

Protective Gloves	Eye Protection	Other Protective Clothing
Impervious Material	Chemical/Dust Goggles or Face Mask	Protective Apron, Boots

Work/Hygiene Practices

Always wash hands thoroughly after using chemical; never bring food, drink, or smoking materials into vicinity of chemicals.

MANGANESE TETROXIDE

Mn_3O_4

CAS: 1317-35-7

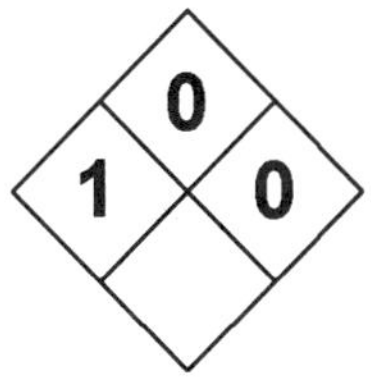

IDENTIFICATION AND TYPICAL USES

Manganese tetroxide appears as a black or brownish-black fine powder with no detectable odor. Manganese tetroxide is created whenever manganese oxides are heated to high temperatures, as in welding, cutting, brazing, or other hot work, especially involving the release of ferromanganese fumes. There is no commercial use for manganese tetroxide. It is simply a hazardous by-product of other processes.

RISK ASSESSMENT: HEALTH

General Assessment

Even though manganese is an essential mineral in human nutrition, it can also be toxic by *inhalation* of its fumes and/or by *ingestion* of the powders and dusts. While manganese has been shown to cause tumors in experimental test animals with mutation data also reported, the exact nature of manganese tetroxide's human carcinogenic properties is unknown. Experimental reproductive effects have been documented.

Inhalation of manganese tetroxide dusts can irritate the respiratory tract with progression to lung damage, pleuritis, bronchitis, and possible respiratory infections and fatal pneumonia. The fumes can cause acute or chronic poisoning with primary effects on the immune system characterized by symptoms of metal fume fever. These can include fever, chills, general weakness, body aches, frontal headache, blurred vision, back pain and stiffness, muscular cramps, shallow and labored respiration, dryness in the mouth and throat with intense thirst, dry cough, metallic taste, and tightness in the chest occurring several hours after exposure. Lung irritation can become severe with inflammation causing manganese tetroxide pneumonitis. Other effects of central nervous system depression include headache, insomnia, changes in personality, emotions, and attitude, irritability, inappropriate laughter or crying, hallucinations, effects on vision (double vision, blurred vision), uncontrolled impulse behavior, euphoria, salivation, trembling, staggering gait, and symptoms normally associated with Parkinson's disease.

Acute Health Effects

The following acute (short-term) health effects may occur immediately or shortly after human exposure to manganese tetroxide:

Skin: Manganese tetroxide particles can become embedded under the skin causing localized inflammation and redness.

Eye: Irritation to the eye and associated tissues. Effects appear to be minor and temporary.

Lung: Irritation of the mucosa of the eyes, nose, throat, and respiratory tract.

CNS: Breathing heated fumes can cause metal fume fever with specific action on the central nervous system.

Chronic Health Effects

The following chronic (long-term) health effects may occur at some time after exposure to manganese tetroxide and can last for months or even years:

Cancer Hazards: According to the references, manganese has been shown to cause tumors in test animals. There are no reports of the specific carcinogenic properties of manganese tetroxide.

Reproductive Hazard: According to the references, there is some limited evidence that chronic overexposure to manganese tetroxide may cause adverse reproductive effects in test animals.

Other Chronic Effects: Prolonged or repeated exposure can damage the central nervous system and cause impaired mental function and degenerative brain changes with emotional instability, and changes in motor control, vision, and other body functions.

☉ Recommended Risk-Reduction Measures

While manganese is an essential mineral in human nutrition, the base mineral as well as many of its compounds (including in the tetroxide) can be extremely toxic under certain exposure conditions. Personnel should, therefore, avoid direct occupational contact with manganese tetroxide. If a less toxic material or compound cannot be substituted for manganese tetroxide, then *engineering controls* are the most effective method of reducing exposures. The best protection is to enclose operations and/or provide local exhaust ventilation at the site of manganese tetroxide dust release. While not always operationally feasible, isolating operations can also reduce exposure. Using respiratory protection is less effective than the controls mentioned above, but is still advisable whenever working with or around manganese tetroxide. Manganese tetroxide is generated whenever manganese oxides are heated to high temperatures. Respiratory protection should be provided whenever welding, cutting, brazing, or other hot work occurs. The exposure ceiling for manganese is relatively low (5mg/m^3) and difficult to accurately monitor. For low or transient exposures, a chemical cartridge respirator equipped with a high-efficiency particulate air (HEPA) filter and a dust/mist pre-filter may provide adequate respiratory protection. Better protection is obtained using a supplied-air respirator operated in continuous flow mode or a self-contained breathing apparatus (SCBA) with full facepiece operated in pressure demand. If a full facepiece is not available, then chemical/dust goggles should be worn to protect the eyes. A face shield and protective apron should also be worn. To prevent hand and skin exposures, gloves should be used. Glove manufacturers should be contacted *before* selections are made.

Administrative controls should also be in place to minimize the potential for human exposures. These may include written procedures or policies which specify the methods and techniques that will be practiced whenever personnel are to work with manganese tetroxide.

All personnel should receive *training* on the use, hazards, protective measures, emergency actions, and other precautions per 29 CFR 1910.1200 (Hazard Communication), prior to the first assignment in an area where manganese tetroxide is used or stored.

Before beginning employment working with manganese tetroxide, and at regular intervals thereafter (e.g., annually), those personnel with frequent or potentially high exposures should be provided the following recommended medical tests:

- ☑ Complete blood count (establish baseline).
- ☑ Complete examination of the nervous system.
- ☑ Lung function tests (may be normal if the person is not having an attack at the time of the test)
- ☑ Kidney function tests.

If symptoms develop or overexposure is suspected, consider the following additional tests:

- ☑ Complete blood count (compare to baseline).
- ☑ Consider chest X-ray after acute overexposure.
- ☑ Lung function tests.

Any evaluation should include a careful history of past and present symptoms with an examination. Medical tests that simply look for existing damage are not a substitute for controlling exposures. Also, since smoking causes heart disease, lung cancer, emphysema, and other respiratory diseases, smokers exposed to manganese tetroxide may experience symptoms more rapidly than non-smokers under the same conditions of exposure. Prudent risk management requires careful consideration of *all* factors which may be causing exposure symptoms in the workplace.

Other methods to reduce exposure include:

- ☑ Where possible, enclose operations and use local exhaust ventilation at the site of chemical release. If local exhaust ventilation or enclosure is not used, respiratory protection should be mandatory.
- ☑ Always ensure that proper protective clothing is worn when using chemical substances. Proper maintenance and care of protective equipment is essential to its proper function.
- ☑ Wash thoroughly immediately after exposure to manganese tetroxide and at the end of the work shift or before eating, drinking, or smoking.
- ☑ Hazard warning information should be posted in the work area. In addition, as part of an on-going education and training program, all information on the health and safety hazards of manganese tetroxide should be communicated to all exposed and potentially exposed workers.
- ☑ Eye wash stations should be provided in the immediate work area for emergency use. If there is a possibility of skin exposure to manganese tetrox-

ide, emergency shower facilities should be provided.

☑ Workers whose clothing has been contaminated by manganese tetroxide should change into clean clothes before leaving work. Never wear contaminated clothing home where family members can be exposed. Contaminated work clothing should be laundered only by individuals who have been informed of the hazards of exposure to manganese tetroxide.

RISK ASSESSMENT: ENVIRONMENT

General Assessment

As a naturally occurring mineral, manganese poses very little threat to the environment. However, accidental spills or releases to the environment can occur during transportation, storage, disposal, or destruction of manganese tetroxide. In almost every scenario, the threat of environmental exposure is contingent upon the proper handling of the chemical substance. Accidental spills, large or small, where manganese tetroxide contacts incompatible commodities can result in release of toxic products (depending upon conditions of spill), and possible contamination of the surrounding environmental mediums (water, soil, and air).

Manganese tetroxide is considered a noncombustible solid. It is incompatible with strong reducing agents, acids, and bases. Extreme caution is always required in handling, storage, transportation, and disposal of manganese tetroxide. Its reactive characteristics also require special consideration during any emergency situation involving a leak or spill of manganese tetroxide powder or dust. Should manganese tetroxide ever come into contact with incompatible substances either during use, transportation, storage, or disposal, violent and even explosive reactions are possible.

Manganese occurs naturally in many minerals that are widely distributed in the earth's crust and, in trace amounts, is an essential element for both plants and animals. The wide variety of possible manganese compounds may enter the environment from natural and industrial sources.

Acute Ecological Effects

Acute (short-term) toxic effects may include the death of animals, birds, or fish and death or low growth rate in plants. Acute effects are seen 2 to 4 days after animals or plants are exposed to manganese tetroxide.

Manganese and its compounds have moderate acute toxicity to aquatic life. While manganese tetroxide is considered essential for plant and animal life, insufficient data are available on the acute effects of excessive manganese tetroxide contamination on the other ecosystems (plants, birds, or land animals).

Chronic Ecological Effects

Chronic (long-term) toxic effects may include shortened life span, reproductive problems, lower fertility, and changes in appearance or behavior in exposed animals. These effects can be seen long after first exposure(s) to toxic chemicals.

Manganese and its compounds have moderate chronic toxicity to aquatic life. Manganese tetroxide is considered essential for plant and animal life. However, insufficient data are available on the chronic effects of excessive manganese tetroxide contamination on the other ecosystems (plants, birds, or land animals).

Water Solubility

Manganese and its compounds vary in their solubility in water from very soluble to nearly insoluble.

Persistence in the Environment

Manganese and its compounds are highly persistent in water, with a half-life greater than 200 days. The half-life of a pollutant is the amount of time it takes for one half of the chemical to be degraded.

Bioaccumulation in Aquatic Organisms

Some substances increase in concentration, or *bioaccumulate,* in living organisms as they breathe contaminated air, drink contaminated water, or eat contaminated food. These chemicals can become concentrated in the tissues and internal organs of animals as well as humans.

The concentration of manganese tetroxide found in fish tissues is expected to be about the same as the average concentration of manganese tetroxide in the water from which the fish was taken.

Recommended Risk-Reduction Measures

Proper training of all transporters will reduce the likelihood of a mishap or accident resulting in a leak or spill of manganese tetroxide dusts or powders into the environment. The correct use of labeling on all containers, trucks, and rail cars will enable emergency responders to react properly and quickly to any disaster thereby reducing the potential risk to the environment and to personnel.

Storage of manganese tetroxide should be segregated from incompatible chemicals to minimize the risk of cross-contamination or contact. Care should be taken to prevent physical damage to storage containers. Also, dusty conditions must be avoided in storage and use areas.

If a spill or leak to the environment has occurred, fire department, emergency response and/or hazardous materials spill personnel should be notified immediately. Cleanup should be attempted only by those trained in proper spill containment procedures. Contaminated soils should be removed for decontamination and replaced with clean soil. If manganese tetroxide should contact the water table, aquifer, or navigable waterway, response actions should be prompt. The local and/or state emergency response authorities may require notification. A comprehensive emergency response or disaster preparedness/recovery plan should be in place prior to any operations involving the use, transportation, storage, or disposal of manganese tetroxide.

If manganese tetroxide is spilled or leaked, the following specific steps are recommended:

- ☑ Restrict persons not wearing protective clothing from area of spill or leak until cleanup is complete and area can be opened for normal work.
- ☑ Remove ignition sources and ventilate area.
- ☑ Cover spilled materials with clean, moist sand. Mix and collect powdered materials in the safest manner possible and deposit in sealed containers. Recommend using a vacuum equipped with a high-efficiency particulate air (HEPA) filter on residues. Do NOT dry sweep (generates dusts).
- ☑ It may be necessary to dispose of manganese tetroxide as a hazardous waste. The responsible state agency or the regional office of the federal Environmental Protection Agency (EPA) should be contacted for specific recommendations.

RISK ASSESSMENT: BUSINESS

General Assessment

Accidents or mishaps involving manganese tetroxide can present a significant threat to business operations. The loss or damage of equipment or facilities can significantly affect fiscal viability. Lawsuits that may result from personnel injury/death, public exposures, and/or environmental contamination will also require a serious expenditure of resources. Media attention surrounding an injury, death, or environmental damage can also result in a loss of profits and loss of current as well as future business. Always remember that anytime the term "reproductive hazard" is used, public hysteria, emotion, and ignorance can run equally high. This should be carefully considered during the development of public relations policies.

Recommended Risk-Reduction Measures

Company attorneys, safety and health professionals, and environmental specialists should be involved in the development of any procedures for responding to chemical incidents. A company official should be predesignated as a public relations officer with specific training in dealing with the press. Corporate plans and policies should be developed, approved, and implemented long before any need for such arises.

REFERENCES

American Conference of Governmental Industrial Hygienists. 1988. *Documentation of the Threshold Limit Values and Biological Exposure Indices,* 5th Edition (with updates). Cincinnati: ACGIH

Anthony, C. P., and N. J. Kolthoff. 1971. *Textbook of Anatomy and Physiology.* St. Louis: C. V. Mosby Company.

Bloom, A. D. and F. J. DeSerres. 1995. *Ecotoxicity and Human Health.* New York: CRC Press/Lewis Publishers.

Calabrese, E. J. 1994. *Biological Effects of Low Level Exposures.* New York: CRC Press/Lewis Publishers.

Cothern, R. R. 1993. *Comparative Environmental Risk Assessment.* New York: CRC Press/Lewis Publishers.

Environmental Protection Agency. 1983. *Chemical Hazard Information Profiles.* Washington, D.C.: U.S. EPA.

Lauwerys, R. R., P. Hoet. 1993. *Industrial Chemical Exposure.* New York: CRC Press/Lewis Publishers.

Lewis, R. J., Sr. 1993. *Hawley's Condensed Chemical Dictionary,* 12th Edition. New York: Van Nostrand Reinhold.

Lewis, R. J., Sr. 1992. *Sax's Dangerous Properties of Industrial Materials,* Volumes 1, 2, and 3, 8th Edition. New York: Van Nostrand Reinhold.

Meyer, E. 1990. *Chemistry of Hazardous Materials.* Englewood Cliffs, NJ: Prentice Hall.

Richardson, M. 1995. *Dictionary of Substance and Their Effects,* Vol. I-VI. New York: CRC Press/Lewis Publishers.

Torkelson, T. R. V. K. Rowe. 1981. *Patty's Industrial Hygiene and Toxicology,* Vol. 2B. New York: Wiley-Interscience.

MATERIAL SAFETY DATA SHEET

CHEMICAL NAME	
MERCURIC ACETATE	**CRC PRESS/LEWIS PUBLISHERS DISCLAIMER:** THE INFORMATION AND RECOMMENDATIONS PRESENTED HEREIN ARE BASED ON SOURCES BELIEVED TO BE RELIABLE. CRC MAKES NO REPRESENTATION ON ITS COMPLETENESS OR ACCURACY. IT IS THE RESPONSIBILITY OF THE USER TO DETERMINE THE CHEMICAL'S SUITABILITY FOR ITS INTENDED USE, THE CHEMICAL'S SAFE USE, AND THE CHEMICAL'S PROPER DISPOSAL. NO REPRESENTATIONS AND/OR WARRANTIES, EITHER EXPRESSED OR IMPLIED, OF THE MERCHANTABILITY OR FITNESS FOR A PARTICULAR PURPOSE, OR OF ANY OTHER NATURE, ARE MADE WITH RESPECT TO THE INFORMATION PROVIDED IN THIS MSDS OR TO THE CHEMICAL TO WHICH INFORMATION MAY REFER. CRC NEITHER ASSUMES NOR AUTHORIZES ANY OTHER PERSON TO ASSUME FOR IT, ANY OTHER ADDITIONAL RESPONSIBILITY OR LIABILITY FOR THE USE OF, OR RELIANCE UPON, THIS INFORMATION.

HAZARD WARNING INFORMATION

HEALTH	FIRE	REACTIVE	OTHER	DEGREE OF HAZARD	COLOR CODING	OTHER CODES
3	0	1		0 = Minimum Hazard 1 = Slight Hazard 2 = Moderate Hazard 3 = Serious Hazard 4 = Severe Hazard	HEALTH = BLUE FIRE = RED REACTIVITY = YELLOW OTHER = WHITE	OX = Oxidizer ACID = Acid ALK = Alkali COR = Corrosive W = Use No Water

SECTION I - GENERAL INFORMATION

Characterization **Mercury Compound**	RCRA Number **None**	EPA Class **Not Applicable**
DOT Proper Shipping Name **Mercuric Acetate**	Chemical Abstract Service (CAS) Number **1600-27-7**	
DOT Hazard Class and Label Requirements **Poison B; Poison**	DOT Emergency Guide Code **60 (for Hg compounds)**	
DOT Identification Number **UN1629**	Molecular Formula $Hg(C_2H_3O_2)_2$	

Synonyms

Mercury acetate; acetic acid, mercury(2+) salt; *bis*(acetyloxy)mercury; diacetoxymercury; mercuriacetate; mercuric diacetate; mercury(2+)acetate; mercuryl acetate.

SECTION II - HAZARDOUS INGREDIENTS AND IDENTITY INFORMATION

Hazardous Components (specific identity)	OSHA Exposure Criteria	NIOSH Exposure Criteria	Immediately Dangerous to Life and Health (IDLH)	ACGIH Exposure Criteria
Mercuric acetate (derivation: Not found).	**PEL:** 0.01 mg/m^3 (as Hg vapor) **CEILING (skin):** 0.05 mg/m^3 (Hg compounds)	**REL (skin):** 0.05 mg/m^3 (as Hg vapor) **STEL:** Not Established	10 mg/m^3 (as Hg)	**TLV (skin):** 0.01 mg/m^3 (Hg compounds) **STEL:** 0.03 mg/m^3 (Hg compounds)

SECTION III - PHYSICAL AND CHEMICAL CHARACTERISTICS

Boiling Point **Decomposes**	Specific Gravity (H_2O = 1) **3.280**
Vapor Pressure (mm Hg) **Not Found**	Molecular Weight **318.69**
Vapor Density (Air = 1) **11.0**	Melting Point **354°F (179°C)**

Solubility

Soluble in alcohol and water.

Appearance and Odor

Colorless or white crystalline powder (sugar or sand-like) with no odor.

SECTION IV - FIRE AND EXPLOSION HAZARD DATA

Flash Point (method used) **Not Determined**	Explosive Limits in Air % by Volume **LEL: Not Applicable UEL: Not Applicable**
NFPA Classification **Non-Combustible Solid**	Autoignition Temperature **Not Determined**

Extinguishing Media

Use agent suitable to surrounding fire. Dry chemical or foam may suffice for small fires. For large fires, try fog or regular foam. Exercise caution if water is used (it may increase contamination of area).

Special Fire Fighting Procedures

Poisonous gases are produced in fire. Wear full protective clothing and self-contained breathing apparatus (SCBA). Move containers from fire area if it can be done without risk. Otherwise, use a water spray to keep fire-exposed containers cool. Do NOT release runoff from fire-fighting measures to sewers or waterways.

Unusual Fire and Explosion Hazards

Mercuric acetate can react explosively in contact with oxidizers, light, heat, and acids.

SECTION V - REACTIVITY DATA

Stability		Conditions to Avoid
		Mercuric acetate is stable under normal conditions of temperature and pressure. However, it will gradually decompose to metallic mercury when heated or exposed to sunlight. It is volatile at 354°F.
Stable	Unstable	Incompatibility *(materials to avoid)*
X		Incompatible with strong oxidizers (such as perchlorates, peroxides, permanganates, chlorates, and nitrates), light, heat, and acids.
Hazardous Polymerization		Conditions to Avoid Under normal conditions of temperature and pressure, hazardous polymerization of mercuric acetate is not known to occur.
May Occur	Will Not Occur	Hazardous Decomposition or By-products
	X	When heated to decomposition, mercuric acetate can emit toxic/poisonous fumes and gases. These include toxic fumes of mercury.

SECTION VI - HEALTH HAZARD DATA

Primary Route(s) of Entry:	Inhalation?	Absorption (skin)?	Ingestion?
	X	X	X

Health Hazards

INHALATION: Causes headaches, shortness of breath, cough, chills, fever, congestion, tightness in chest, possible pneumonitis and pulmonary edema. Prolonged contact can cause CNS effects with vertigo, depression, ataxia, emotional instability, and other effects.

ABSORPTION: Passes through intact skin to cause effects similar to that of inhalation and ingestion. There can also be visual disturbances, eye burns, and dermatitis.

INGESTION: Abdominal pain, diarrhea (possibly bloody), digestive tract irritation, inflammation of the mouth and gums, liver damage, loose teeth, jaw necrosis, reduced urine excretion.

Carcinogenicity	NTP Listed?	IARC Cancer Review Group?	OSHA Regulated?	Target Organs?
Unknown Human Unknown Animal	No	No	29 CFR 1910.1000 Table Z-1 (as Hg)	Kidney, liver, gastrointestinal tract, CNS.

Medical Conditions Generally Aggravated by Exposure

Mercury-sensitive people, existing dental and oral problems, kidney disease, respiratory problems, and central nervous system disorders may be aggravated by exposure to mercuric acetate.

Emergency and First-aid Procedures

Eye contact: Flush immediately with water for 15 minutes (minimum); seek medical attention. Skin contact: Remove all contaminated clothing. Immediately wash area with large amounts of soap and water. For inhalation: Remove the person from exposure. Provide respiratory assistance and CPR. Transfer to medical facility. Observe for 48 hours for lung effects. If swallowed: Provide breathing assistance if necessary. Never attempt to give a convulsing or unconscious person anything by mouth. If victim is conscious, give 1-2 glasses of water to dilute and induce vomiting.

SECTION VII - PRECAUTIONS FOR SAFE HANDLING AND USE

Steps to be Taken in Case Material is Released or Spilled

Collect powdered material and deposit in sealed drum. Keep those not involved in cleanup from entering area. Do NOT dry sweep (creates airborne dusts). Use a vacuum equipped with a (HEPA) filter. Absorb solution spills in vermiculite or other materials.

Preferred Waste Disposal Method

No Citation.

Precautions to be Taken in Handling and Storage

Store in tightly closed containers in cool, well-ventilated area. Sources of ignition are prohibited where mercuric acetate is stored. Keep away from light.

Other Precautions and Warnings

Outside or detached storage facility is recommended. Protect containers from physical damage.

SECTION VIII - CONTROL MEASURES AND PERSONAL PROTECTIVE EQUIPMENT

Respiratory Protection *(specify type)*

Exposure levels are low and difficult to monitor. For best protection, use a supplied-air respirator with a full facepiece or a self-contained breathing apparatus (SCBA) operated in positive pressure mode.

Ventilation

Local exhaust or general mechanical systems recommended.

Protective Gloves	Eye Protection	Other Protective Clothing
Impervious Material	Chemical/Dust Goggles or Face Mask	Protective Suit or Clothing

Work/Hygiene Practices

Always wash hands thoroughly after using chemical; never bring food, drink, or smoking materials into vicinity of chemicals. Skin contact must be avoided.

MERCURIC ACETATE

$Hg(C_2H_3O_2)_2$

CAS: 1600-27-7

IDENTIFICATION AND TYPICAL USES

Mercuric acetate appears as a white crystalline, sugar or sand-like solid with no detectable odor. It is used in the manufacture of other chemicals and pharmaceuticals.

RISK ASSESSMENT: HEALTH

General Assessment

Mercuric acetate is a toxic human poison by three routes of exposure: *inhalation, ingestion,* and dermal *absorption*. Mercury accumulates in the body with each exposure. While it is not currently listed as a carcinogenic compound, mutation data have been reported for many mercury compounds. Organic mercury compounds (those that contain carbon), such as mercuric acetate have caused adverse reproductive effects (miscarriages and/or fetal damage) in test animals.

Symptoms of exposure by inhalation of mercury acetate vapors are headache, dypsnea, cough, chills, fever, pneumonitis, and delayed pulmonary edema (fluid in the lungs), which can be fatal. Symptoms of pulmonary edema (cough, chest pain, congestion, nausea, dizziness) may be delayed up to 48 hours thereby creating a false sense of security with regard to health exposure risk.

Ingestion of mercury or its compounds causes abdominal pain, bloody diarrhea, gastrointestinal irritation, vomiting, and cardiac arrhythmia.

Repeated exposure from inhalation, absorption, and/or ingestion can cause central nervous system (CNS) effects including vertigo, anxiety, depression, ataxia, emotional instability, kidney injury, brain damage, tremors (the "shakes"), memory loss, metallic taste, discoloration of the skin, sore gums, personality changes, liver damage, loose teeth, necrosis of the jaw, decreased urine excretion, visual disturbances, and possible development of dermatitis.

Acute Health Effects

The following acute (short-term) health effects may occur immediately or shortly after exposure to mercuric acetate:

Skin: Irritation, rash, dermatitis, and possible redness and the appearance of small blisters. Dermal absorption can be quick and will cause toxic systemic effects.

Eye: Irritation, inflammation, watering, and swelling, burns, and tissue damage.

Lung: Inhalation of mercuric acetate dusts may cause irritation of the mucous membranes in the respiratory tract causing cough, phlegm, and lung tissue irritation. It can also lead to pneumonitis as well as pulmonary edema, which is a medical emergency and can be fatal.

Chronic Health Effects

The following chronic (long-term) health effects may occur at some time after exposure to mercuric acetate and can last for months or even years:

Cancer Hazards: According to the information presented in the references, mercuric acetate has not been adequately tested for its ability to cause cancer in test animals. However, mutation (causes genetic changes) data have been documented and many scientists believe that such chemicals may pose a cancer risk in the long-term. More research is required.

Reproductive Hazards: According to information available in the references, organic mercury compounds (such as mercuric acetate) have been shown to adversely affect reproduction in humans and animals. They can cause the early termination of pregnancies (miscarriages) and may injure the fetus. Mercury may also reduce fertility in both males and females. While inorganic mercury substances (those without carbon), have not specifically been shown to be human teratogens, they should still be handled with caution as they may cause reproductive problems in males and females.

Other Chronic Hazards: High or repeated exposure can cause kidney damage or mercury poisoning with symptoms of sore gums, "shakes" (especially noticeable in handwriting), irritability, increased saliva production, memory loss, extreme shyness, weakness, loss of teeth, poor appetite, and metallic taste. Serious personality changes and brain damage can occur, especially if exposure continues. Repeated skin contact may cause gray discoloration and a skin allergy may develop. If this happens, even small future exposures can cause rash. Repeated exposures may also cause reduced peripheral vision and brown staining of the eye. Mercury can accumulate in the body on prolonged exposure and it may take months or years for the body to dispose of excess mercury.

Recommended Risk-Reduction Measures

Personnel should avoid direct contact with mercuric acetate. Mercury is an extremely toxic, cumulative poison. If a less toxic chemical cannot be substituted for a hazardous material, then *engineering controls* are the most effective method of reducing exposures.

The best protection is to enclose operations and/or provide local exhaust ventilation at the site of chemical release. While not always operationally feasible, isolating operations can also reduce exposure. Using respiratory protection is less effective than the controls mentioned above, but is still advisable whenever working with or around mercuric acetate. Exposure levels for mercury are extremely low (0.05 mg/m^3) and difficult to monitor accurately. A full facepiece respirator with a high-efficiency particulate air (HEPA) filter equipped with a dust/mist/fume pre-filter may suffice. However, for the best protection use a supplied-air respirator with a full facepiece operated in positive pressure mode or an MSHA/NIOSH-approved self-contained breathing apparatus (SCBA) with full facepiece operated in pressure demand or other continuous flow mode. If a full facepiece is not available, then chemical/dust goggles should be worn to protect the eyes. Whenever a chemical dust hazard exists, a face shield and a protective apron should be worn. To prevent hand and skin exposures, impervious rubber gloves should be used. Glove manufacturers should be contacted and permeation studies obtained *before* final glove selections are made. Skin contact must be prevented. Protective suits and clothing should be used.

Administrative controls should also be in place to minimize the potential for human exposures. These may include written procedures or policies which specify the methods and techniques that will be practiced whenever personnel are to work with mercuric acetate dusts or solutions. All personnel should receive *training* on the use, hazards, protective measures, emergency actions, and other precautions per 29 CFR 1910.1200 (Hazard Communication), prior to the first assignment in an area where mercuric acetate is used or stored. Before first exposure and every six to twelve months thereafter, a complete medical history and examination is recommended, including the following:

- ☑ Examination of the nervous system (taking note of any problems with person's handwriting).
- ☑ Routine urinalysis and special urine test for mercury (should be less than 0.02 mg/liter).
- ☑ Eye examination.
- ☑ Lung function tests for frequent exposures.

If symptoms develop or overexposure is suspected, the above listed tests should be repeated, along with the following recommended tests:

- ☑ Consider nerve conduction tests.
- ☑ Urinary enzyme and neurological behavior tests.
- ☑ Serum mercury tests.
- ☑ Consider chest X-ray following acute overexposure (may be negative if taken immediately after exposure due to delayed onset of pulmonary edema).

Any evaluation should include a careful history of past and present symptoms with a medical examination. Medical tests that look for damage already done are not a substitute for controlling exposure. Also, since smoking can cause heart disease, emphysema, and a number of other respiratory disorders, smokers exposed to lung irritants such as mercuric acetate dusts and fumes may experience symptoms much more quickly than non-smokers under the same conditions of exposure. Prudent risk management requires proper consideration of *all* factors which may be causing the appearance of exposures symptoms in the work place.

Other methods to reduce exposure include:

- ☑ Where possible, enclose operations and use local exhaust ventilation at the site of chemical release. If local exhaust ventilation or enclosure is not used, respiratory protection should be provided and its use mandatory.

- ☑ Always ensure that proper protective clothing is worn when using chemical substances and that personnel are trained in its use and care.
- ☑ Wash thoroughly immediately after exposure to mercuric acetate and at the end of the work shift or before eating, drinking, or smoking.
- ☑ Hazard warning information should be posted in the work area. In addition, as part of an on-going education and training program, all information on the health and safety hazards of mercuric acetate should be communicated to all exposed and potentially exposed workers.
- ☑ Eye wash stations and emergency showers should be located in the immediate work area.
- ☑ Work clothing that has been contaminated with mercuric acetate should never be worn home where family members can be exposed as well. Contaminated clothing should be laundered at work or by an authorized service by personnel who have been trained in the hazards associated with exposure to mercuric acetate.
- ☑ The drug n-acetyl penicillamine (NAP) has been used to treat mercury poisoning, with mixed success.
- ☑ Specific engineering controls are recommended for mercury by NIOSH. Refer to NIOSH Criteria Document, "Inorganic Mercury" (Publication Number 73-11024).

RISK ASSESSMENT: ENVIRONMENT

General Assessment

The environment is at risk of exposure during transportation, storage, disposal, or destruction of mercuric acetate. In almost every scenario, the threat of environmental exposure is contingent upon the proper handling of the chemical substance. Accidental spills, large or small, can result in fire, explosion, and possible contamination of the surrounding environmental mediums (water, soil, and air).

Mercuric acetate is incompatible with many common oxidizers and acids. Reactions can be vigorous and violent. It is also sensitive to heat and light. In a fire, mercuric acetate will emit toxic and poisonous mercury fumes. These characteristics require special consideration during any emergency situation involving a leak or spill of mercuric acetate.

Elemental mercury is a heavy and relatively inert liquid which is oxidized to inorganic mercury (II) under natural conditions. Mercury (II) may combine with an organic fraction to form methyl mercury. Both mercury (II) and methyl mercury can cause environmental contamination. Mercury (II) may enter the environment in industrial or municipal waste treatment discharges, from previously contaminated sediments, and from the weathering of natural rocks. Bacteria may then convert it to methyl mercury. The concentration of mercury (II) in water may be elevated with acid rain due to the scouring of mercury from the air and increased partitioning from the sediment into the water.

☠ *Acute Ecological Effects*

Acute (short-term) toxic effects may include the death of animals, birds, or fish and death or low growth rate in plants. Acute effects are seen 2 to 4 days after animals or plants are exposed to mercuric acetate. Mercury (II) and methyl mercury have high acute toxicity to aquatic life. Insufficient data are available to evaluate the short-term effects of mercury (II) and methyl mercury exposure to plants, birds or terrestrial animals.

💣 *Chronic Ecological Effects*

Chronic (long-term) toxic effects may include shortened life span, reproductive problems, lower fertility, and changes in appearance or behavior in exposed animals. These effects can be seen long after first exposure(s) to toxic chemicals. Mercury (II) and methyl mercury have high chronic toxicity to aquatic life. Eating fish contaminated with mercury residues has caused secondary poisoning in humans. Birds or land animals similarly exposed to mercury and its compounds could also be subject to such effects. Insufficient data are available to evaluate the long-term effects of mercuric acetate exposure to plants.

💧 *Water Solubility*

Mercuric acetate is soluble in water. Concentrations of 1000 milligrams or more will mix with a liter of water. Mercury is insoluble in the aquatic environment.

⌛ *Persistence in the Environment*

Mercury is highly persistent in water, with a half-life greater than 200 days. The half-life of a pollutant is the amount of time it takes for one half of the chemical to be degraded. When released into water, microorganisms convert mercuric salts into methyl mercury which is rapidly taken up by algae and enters the food chain. It concentrates in the edible tissues of fish and, eventually, can cause human poisoning as well.

Bioaccumulation in Aquatic Organisms

Some substances increase in concentration, or *bioaccumulate,* in living organisms as they breathe contaminated air, drink contaminated water, or eat contaminated food. These chemicals can become concentrated in the tissues and internal organs of animals as well as humans. The concentration of mercury and methyl mercury found in fish tissues is expected to be considerably higher than the average concentration in the water from which the fish was taken.

Recommended Risk-Reduction Measures

Proper training of all transporters will reduce the likelihood of a mishap or accidents resulting in a leak or spill to the environment. Proper labeling on all containers, trucks, and rail cars will enable emergency responders to react properly and quickly to any disaster thereby reducing the potential risk to the environment and to personnel.

If a spill or leak to the environment has occurred, fire department, emergency response and/or hazardous materials spill personnel should be notified immediately. Do not dry sweep. When vacuuming, a high efficiency particulate air (HEPA) filter should be used, not a standard shop vacuum. Deposit collected materials in sealed drums for disposal. Cleanup should be attempted only by those trained in proper spill containment procedures. Contaminated soils may need to be removed and replaced with clean soil. If mercuric acetate should contact the water table, aquifer, or navigable waterway, time is of the essence. Mercuric acetate is soluble in water and total containment and remediation may not be possible. When such spills occur, the local and/or state emergency response authorities should be notified. A comprehensive emergency response or disaster preparedness/recovery plan should be in place prior to any operations involving the use, transportation, storage, or disposal of mercuric acetate.

If mercuric acetate is spilled or leaked, the following specific steps are recommended:

- ☑ Restrict persons not wearing protective clothing from area of spill or leak until cleanup is complete and area can be opened for normal work.
- ☑ Ventilate area of spill or leak and remove all ignition sources.
- ☑ Collect powdered materials in safe manner and place in sealed drums for disposal. Do NOT dry sweep; use HEPA vacuum instead.
- ☑ It may be necessary to dispose of mercuric acetate as a hazardous waste. The responsible state agency or the regional office of the federal Environmental Protection Agency (EPA) should be contacted for specific recommendations.

RISK ASSESSMENT: BUSINESS

General Assessment

Accidents or mishaps involving mercuric acetate can present a significant threat to business operations. The loss or damage of equipment or facilities can significantly affect fiscal viability. Lawsuits that may result from personnel injury/death, public exposures, and/or environmental contamination will also require a serious expenditure of resources. Media attention surrounding an injury, death, or environmental damage can also result in a loss of profits and loss of current as well as future business. Always remember that anytime the term "reproductive hazard" is used, public emotion, hysteria, and ignorance can run equally high. This must be properly considered whenever developing public relations policies.

Recommended Risk-Reduction Measures

Company attorneys, safety and health professionals, and environmental specialists should be involved in the development of such procedures dealing with safety or emergency response. A company official should be pre-designated as a public relations officer with specific training in dealing with the press. Corporate plans and policies should be developed, approved, and implemented long before any need for such arises.

REFERENCES

American Conference of Governmental Industrial Hygienists. 1988. *Documentation of the Threshold Limit Values and Biological Exposure Indices,* 5th Edition (with updates). Cincinnati: ACGIH

Environmental Protection Agency. 1983. *Chemical Hazard Information Profiles.* Washington, D.C.: U.S. EPA.

U.S. Department of Health and Human Services, National Institute for Occupational Safety and Health. 1973. NIOSH Criteria Document. *Inorganic Mercury.* Publication Number 73-11024. Washington, D.C.: U.S. Government Printing Office.

MATERIAL SAFETY DATA SHEET

CHEMICAL NAME

MERCURIC CHLORIDE

HAZARD WARNING INFORMATION

HEALTH	FIRE	REACTIVE	OTHER	DEGREE OF HAZARD	COLOR CODING	OTHER CODES
3	0	1		0 = Minimum Hazard 1 = Slight Hazard 2 = Moderate Hazard 3 = Serious Hazard 4 = Severe Hazard	HEALTH = BLUE FIRE = RED REACTIVITY = YELLOW OTHER = WHITE	OX = Oxidizer ACID = Acid ALK = Alkali COR = Corrosive W = Use No Water

SECTION I - GENERAL INFORMATION

Characterization: **Mercury Compound**

RCRA Number: **None**

EPA Class: **Not Applicable**

DOT Proper Shipping Name: **Mercuric Chloride**

Chemical Abstract Service (CAS) Number: **7487-94-7**

DOT Hazard Class and Label Requirements: **Poison B; Poison**

DOT Emergency Guide Code: **60 (for Hg compounds)**

DOT Identification Number: **UN1624**

Molecular Formula: **$HgCl_2$**

Synonyms

Mercury bichloride; mercury chloride; corrosive sublimate; corrosive mercury chloride; mercury perchloride; sublimate; Sulem ®, Calcochlor ®.

SECTION II - HAZARDOUS INGREDIENTS AND IDENTITY INFORMATION

Hazardous Components (specific identity)	OSHA Exposure Criteria	NIOSH Exposure Criteria	Immediately Dangerous to Life and Health (IDLH)	ACGIH Exposure Criteria
Mercuric chloride (derivation: By direct combination of chlorine with mercury heated to volatilization point; by subliming mercuric sulfate with common salt).	**PEL (skin):** 0.05 mg/m^3 (as Hg vapor) **CEILING:** 0.1 mg/m^3 (Hg compounds)	**REL (skin):** 0.05 mg/m^3 (as Hg vapor) **STEL:** Not Established	28 mg/m^3 (as Hg)	**TLV (skin):** 0.1 mg/m^3 (Hg compounds) **STEL:** Not Established

SECTION III - PHYSICAL AND CHEMICAL CHARACTERISTICS

Boiling Point: **576°F (302°C) Decomposes**

Specific Gravity (H_2O = 1): **5.44**

Vapor Pressure (mm Hg): **1 at 277°F (136°C)**

Molecular Weight: **271.52**

Vapor Density (Air = 1): **8.7**

Melting Point: **529°F (276°C)**

Solubility

Soluble in water, alcohol, ether, pyridine, glycerol, and acetic acid.

Appearance and Odor

Colorless or white crystals or powder with no odor.

SECTION IV - FIRE AND EXPLOSION HAZARD DATA

Flash Point (method used): **Not Determined**

Explosive Limits in Air % by Volume: **LEL: Not Applicable UEL: Not Applicable**

NFPA Classification: **Non-Flammable Solid**

Autoignition Temperature: **Not Determined**

Extinguishing Media

Use agent suitable to surrounding fire. Dry chemical or foam may suffice for small fires. For large fires, try fog or regular foam. Exercise caution if water is used (it may increase contamination of area).

Special Fire Fighting Procedures

Poisonous gases are produced in fire. Wear full protective clothing and self-contained breathing apparatus (SCBA). Only use a water spray to keep fire-exposed containers cool. Do NOT release runoff from fire-fighting measures to sewers or waterways.

Unusual Fire and Explosion Hazards

Mercuric chloride can explode in conditions of friction, heat, or impact with phosphorus, antimony, arsenic, silver salts, sulfides, acetylene, and ammonia.

SECTION V - REACTIVITY DATA

Stability		Conditions to Avoid **Mercuric chloride is stable under normal conditions of temperature and pressure. However, it will gradually decompose to metallic mercury in the presence of organic matter and sunlight. It is volatile at 572°F.**
Stable X	Unstable	Incompatibility *(materials to avoid)* **Incompatible with formates, sulfites, phosphates, albumin, ammonia, gelatin, carbonates, hypophosphites, sulfides, alkalis, alkaloid salts, lime water, arsenic, antimony, bromides, and many other compounds.**
Hazardous Polymerization		Conditions to Avoid **Under normal conditions of temperature and pressure, hazardous polymerization of mercuric chloride is not known to occur.**
May Occur	Will Not Occur X	Hazardous Decomposition or By-products **When heated to decomposition, mercuric chloride can emit toxic/poisonous fumes and gases. These include toxic fumes of mercury.**

SECTION VI - HEALTH HAZARD DATA

Primary Route(s) of Entry:	Inhalation?	Absorption (skin)?	Ingestion?
	X	X	X

Health Hazards

INHALATION: Causes headaches, shortness of breath, cough, chills, fever, congestion, tightness in chest, possible pneumonitis and pulmonary edema. Prolonged contact can cause CNS effects with vertigo, depression, ataxia, emotional instability, and other effects.

ABSORPTION: Passes through intact skin to cause effects similar to that of inhalation and ingestion. There can also be visual disturbances and dermatitis.

INGESTION: Abdominal pain, diarrhea (possibly bloody), digestive tract irritation, inflammation of the mouth and gums, liver damage, loose teeth, jaw necrosis, reduced urine excretion.

Carcinogenicity	NTP Listed?	IARC Cancer Review Group?	OSHA Regulated?	Target Organs?
Unknown Human Unknown Animal	No	No	29 CFR 1910.1000 Table Z-1 (as Hg)	Kidney, liver, gastrointestinal tract, CNS.

Medical Conditions Generally Aggravated by Exposure

Mercury-sensitive people, existing dental and oral problems, kidney disease, respiratory problems, and central nervous system disorders may be aggravated by exposure to mercuric chloride.

Emergency and First-aid Procedures

Eye contact: Flush immediately with water for 15 minutes (minimum); seek medical attention. Skin contact: Remove all contaminated clothing. Immediately wash area with large amounts of soap and water. For inhalation: Remove the person from exposure. Provide respiratory assistance and CPR. Transfer to medical facility. Observe for 48 hours for lung effects. If swallowed: Provide breathing assistance if necessary. Never attempt to give a convulsing or unconscious person anything by mouth. If victim is conscious, give 1-2 glasses of water to dilute and induce vomiting.

SECTION VII - PRECAUTIONS FOR SAFE HANDLING AND USE

Steps to be Taken in Case Material is Released or Spilled

Collect powdered material and deposit in sealed drum. Keep those not involved in cleanup from entering area. Do NOT dry sweep (creates airborne dusts). Use a vacuum equipped with a (HEPA) filter. Absorb solution spills in vermiculite or other materials. Wash area with dilute calcium sulfide solutions.

Preferred Waste Disposal Method

No citation.

Precautions to be Taken in Handling and Storage

Store in tightly closed containers in cool, well-ventilated area. Sources of ignition are prohibited where mercuric chloride is stored.

Other Precautions and Warnings

Outside or detached storage facility is recommended. Protect containers from physical damage.

SECTION VIII - CONTROL MEASURES AND PERSONAL PROTECTIVE EQUIPMENT

Respiratory Protection *(specify type)*

Exposure levels are low and difficult to monitor. For best protection, use a supplied-air respirator with a full facepiece or a self-contained breathing apparatus (SCBA) operated in positive pressure mode.

Ventilation

Local exhaust or general mechanical systems recommended.

Protective Gloves	Eye Protection	Other Protective Clothing
Impervious Material	Chemical/Dust Goggles or Face Mask	Protective Suit or Clothing

Work/Hygiene Practices

Always wash hands thoroughly after using chemical; never bring food, drink, or smoking materials into vicinity of chemicals. Skin contact must be avoided.

MERCURIC CHLORIDE

$HgCl_2$

CAS: 7487-94-7

IDENTIFICATION AND TYPICAL USES

Mercuric chloride appears as a white crystalline solid with no detectable odor. It is used in the manufacture of calomel and other mercury compounds, as a disinfectant, in organic synthesis, as an analytical reagent, in metallurgy, in tanning, as a catalyst for vinyl chloride, in sterilizing seed potatoes, as a fungicide, insecticide, as a wood preservative, in embalming fluids, in textile printing, dry batteries, photography, process engraving, and lithography.

RISK ASSESSMENT: HEALTH

General Assessment

Mercuric chloride is a toxic human poison by three routes of exposure: *inhalation, ingestion,* and dermal *absorption.* Mercury accumulates in the body with each exposure. While it is not currently listed as a carcinogenic compound, human mutation data have been reported for inorganic mercury compounds. Mercuric chloride has caused adverse human reproductive effects (miscarriages and/or fetal damage).

Symptoms of exposure by inhalation of mercury chloride vapors include headache, dypsnea, cough, chills, fever, pneumonitis, and delayed pulmonary edema (fluid in the lungs), which can be fatal. Symptoms of pulmonary edema may be delayed up to 48 hours thereby creating a false sense of security with regard to health exposure risk.

Ingestion causes abdominal pain, bloody diarrhea, gastrointestinal irritation, vomiting, and cardiac arrhythmia. Repeated exposure from inhalation, absorption, and/or ingestion can cause CNS effects including vertigo, anxiety, depression, ataxia, emotional instability, kidney injury, brain damage, tremors (the "shakes"), sore gums, personality changes, liver damage, loose teeth, necrosis of the jaw, decreased urine excretion, visual disturbances, and dermatitis.

Acute Health Effects

The following acute (short-term) health effects may occur immediately or shortly after exposure to mercuric chloride:

Skin: Irritation, rash, dermatitis, and possible redness and the appearance of small blisters. Dermal absorption can be quick and will cause toxic systemic effects.

Eye: Irritation, inflammation, watering, and swelling.

Lung: Inhalation of mercuric chloride dusts may cause irritation of the mucous membranes in the respiratory tract causing cough, phlegm, and lung tissue irritation. It can also lead to pneumonitis as well as pulmonary edema, which can be fatal.

Chronic Health Effects

The following chronic (long-term) health effects may occur at some time after exposure to mercuric chloride and can last for months or even years:

Cancer Hazards: According to the information presented in the references, mercuric chloride has not been adequately tested for its ability to cause cancer in test animals. However, human mutation (causes genetic changes) data have been documented and many scientists believe that such chemicals may pose a cancer risk in the long-term.

Reproductive Hazards: According to information available in the references, mercuric chloride has been shown to adversely affect reproduction in humans and animals. It causes the early termination of pregnancies (miscarriages) and may injure the fetus. It may also reduce fertility in both males and females. While inorganic mercury substances (those without carbon), such as mercury chloride, have not specifically been shown to be human teratogens, they should still be handled with caution as they may cause reproductive problems in males and females.

Other Chronic Hazards: High or repeated exposure can cause kidney damage or mercury poisoning with symptoms of sore gums, "shakes" (especially noticeable in handwriting), irritability, increased saliva production, memory loss, extreme shyness, weakness, loss of teeth, poor appetite, and metallic taste. Serious personality changes and brain damage can occur, especially if exposure continues. Repeated skin contact may cause gray discoloration and a skin allergy may develop. If this happens, even small future exposures can cause rash. Repeated exposures may also cause reduced peripheral vision and brown staining of the eye. Mercury can accumulate in the body on prolonged exposure and it may take months or years for the body to dispose of excess mercury.

Recommended Risk-Reduction Measures

Personnel should avoid direct contact with mercuric chloride. It is an extremely toxic, cumulative poison. If a less toxic chemical cannot be substituted for a hazardous material, then *engineering controls* are the most effective method of reducing exposures.

The best protection is to enclose operations and/or provide local exhaust ventilation at the site of chemical release. While not always operationally feasible, isolating operations can also reduce exposure. Using respiratory protection is less effective than the controls mentioned above, but is still advisable whenever working with or around mercuric chloride. Exposure levels for mercury are extremely low (0.05 mg/m^3) and difficult to monitor accurately. A full facepiece respirator with a high-efficiency particulate air (HEPA) filter equipped with a dust/mist/fume pre-filter may suffice. However, for the best protection use a supplied-air respirator with a full facepiece operated in positive pressure mode or an MSHA/NIOSH-approved self-contained breathing apparatus (SCBA) with full facepiece operated in pressure demand or other continuous flow mode. If a full facepiece is not available, then chemical/dust goggles should be worn to protect the eyes. Whenever a chemical dust hazard exists, a face shield and a protective apron should be worn. To prevent hand and skin exposures, impervious rubber gloves should be used. Glove manufacturers should be contacted and permeation studies obtained *before* final glove selections are made. Skin contact must be prevented. Protective suits and clothing should be used.

Administrative controls should also be in place to minimize the potential for human exposures. These may include written procedures or policies which specify the methods and techniques that will be practiced whenever personnel are to work with mercuric chloride dusts or solutions.

All personnel should receive *training* on the use, hazards, protective measures, emergency actions, and other precautions per 29 CFR 1910.1200 (Hazard Communication), prior to the first assignment in an area where mercuric chloride is used or stored. Before first exposure and every six to twelve months thereafter, a complete medical history and examination is recommended, including the following:

- ☑ Examination of the nervous system (taking note of any problems with person's handwriting).
- ☑ Routine urinalysis and special urine test for mercury (should be less than 0.02 mg/liter).
- ☑ Eye examination.
- ☑ Consider lung function tests for persons with frequent exposure.

If symptoms develop or overexposure is suspected, the above listed tests should be repeated, along with the following recommended tests:

- ☑ Consider nerve conduction tests.
- ☑ Urinary enzyme and neurological behavior tests.
- ☑ Consider chest X-ray following acute overexposure (may be negative if taken immediately after exposure due to delayed onset of pulmonary edema).

Any evaluation should include a careful history of past and present symptoms with a medical examination. Medical tests that look for damage already done are not a substitute for controlling exposure. Also, since smoking can cause heart disease, emphysema, and a number of other respiratory disorders, smokers exposed to lung irritants such as mercuric chloride dusts and fumes may experience symptoms much more quickly than non-smokers under the same conditions of exposure. Prudent risk management requires proper consideration of *all* factors which may be causing the appearance of exposures symptoms in the work place.

Other methods to reduce exposure include:

- ☑ Where possible, enclose operations and use local exhaust ventilation at the site of chemical release. If local exhaust ventilation or enclosure is not used, respiratory protection should be provided and its use mandatory.

- ☑ Always ensure that proper protective clothing is worn when using chemical substances and that personnel are trained in its use and care.
- ☑ Wash thoroughly immediately after exposure to mercuric chloride and at the end of the work shift or before eating, drinking, or smoking.
- ☑ Hazard warning information should be posted in the work area. In addition, as part of an on-going education and training program, all information on the health and safety hazards of mercuric chloride should be communicated to all exposed and potentially exposed workers.
- ☑ Eye wash stations and emergency showers should be located in the immediate work area.
- ☑ Work clothing that has been contaminated with mercuric chloride should never be worn home where family members can be exposed as well. Contaminated clothing should be laundered at work or by an authorized service by personnel who have been trained in the hazards associated with exposure to mercuric chloride.
- ☑ Specific engineering controls are recommended for mercury by NIOSH. Refer to NIOSH Criteria Document, "Inorganic Mercury" (Publication Number 73-11024).

RISK ASSESSMENT: ENVIRONMENT

General Assessment

The environment is at risk of exposure during transportation, storage, disposal, or destruction of mercuric chloride. In almost every scenario, the threat of environmental exposure is contingent upon the proper handling of the chemical substance. Accidental spills, large or small, can result in fire, explosion, and possible contamination of the surrounding environmental mediums (water, soil, and air).

Mercuric chloride is incompatible with many compounds, including formates, sulfites, phosphates, albumin, ammonia, other metals, even vegetable astringents. Reactions can be vigorous and violent. In a fire, mercuric chloride will emit toxic and poisonous mercury fumes. These characteristics require special consideration during any emergency situation involving a leak or spill of mercuric chloride.

Elemental mercury is a heavy and relatively inert liquid which is oxidized to inorganic mercury (II) under natural conditions. Mercury (II) may combine with an organic fraction to form methyl mercury. Both mercury (II) and methyl mercury can cause environmental contamination. Mercury (II) may enter the environment in industrial or municipal waste treatment discharges, from previously contaminated sediments, and from the weathering of natural rocks. Bacteria may then convert it to methyl mercury. The concentration of mercury (II) in water may be elevated with acid rain due to the scouring of mercury from the air and increased partitioning from the sediment into the water.

Acute Ecological Effects

Acute (short-term) toxic effects may include the death of animals, birds, or fish and death or low growth rate in plants. Acute effects are seen 2 to 4 days after animals or plants are exposed to mercuric chloride.

Mercury (II) and methyl mercury have high acute toxicity to aquatic life. Insufficient data are available to evaluate the short-term effects of mercury (II) and methyl mercury exposure to plants, birds or terrestrial animals.

Chronic Ecological Effects

Chronic (long-term) toxic effects may include shortened life span, reproductive problems, lower fertility, and changes in appearance or behavior in exposed animals. These effects can be seen long after first exposure(s) to toxic chemicals.

Mercury (II) and methyl mercury have high chronic toxicity to aquatic life. Eating fish contaminated with mercury residues has caused secondary poisoning in humans. Birds or land animals similarly exposed to mercury and its compounds could also be subject to such effects. Insufficient data are available to evaluate the long-term effects of mercuric chloride exposure to plants.

Water Solubility

Mercuric chloride is soluble in water. Concentrations of 1000 milligrams or more will mix with a liter of water. Mercury itself is insoluble in the water.

Persistence in the Environment

Mercury is highly persistent in water, with a half-life greater than 200 days. The half-life of a pollutant is the amount of time it takes for one half of the chemical to be degraded. When released into water, microorganisms convert mercuric salts into methyl mercury which is rapidly taken up by algae and enters the food chain. It concentrates in the edible tissues of fish and, eventually, can cause human poisoning as well.

Bioaccumulation in Aquatic Organisms

Some substances increase in concentration, or *bioaccumulate,* in living organisms as they breathe contaminated air, drink contaminated water, or eat contaminated food. These chemicals can become concentrated in the tissues and internal organs of animals as well as humans. The concentration of mercury and methyl mercury found in fish tissues is expected to be considerably higher than the average concentration in the water from which the fish was taken.

Recommended Risk-Reduction Measures

Proper training of all transporters will reduce the likelihood of a mishap or accidents resulting in a leak or spill to the environment. Proper labeling on all containers, trucks, and rail cars will enable emergency responders to react properly and quickly to any disaster thereby reducing the potential risk to the environment and to personnel.

If a spill or leak to the environment has occurred, fire department, emergency response and/or hazardous materials spill personnel should be notified immediately. Do not dry sweep. When vacuuming, a high efficiency particulate air (HEPA) filter should be used, not a standard shop vacuum. Deposit collected materials in sealed drums for disposal. Cleanup should be attempted only by those trained in proper spill containment procedures. Contaminated soils may need to be removed and replaced with clean soil. If mercuric chloride should contact the water table, aquifer, or navigable waterway, time is of the essence. Mercuric chloride is soluble in water and total containment and remediation may not be possible. When such spills occur, the local and/or state emergency response authorities should be notified. A comprehensive emergency response or disaster preparedness/recovery plan should be in place prior to any operations involving the use, transportation, storage, or disposal of mercuric chloride. If mercuric chloride is spilled or leaked, the following specific steps are recommended:

- ☑ Restrict persons not wearing protective clothing from area of spill or leak until cleanup is complete and area can be opened for normal work.
- ☑ Ventilate area of spill or leak and remove all ignition sources.
- ☑ Collect powdered materials in most convenient and safe manner possible and place in sealed drums for disposal. Do NOT dry sweep; use HEPA vacuum instead.
- ☑ It may be necessary to dispose of mercuric chloride as a hazardous waste. The responsible state agency or the regional office of the federal Environmental Protection Agency (EPA) should be contacted for specific recommendations.

RISK ASSESSMENT: BUSINESS

General Assessment

Accidents or mishaps involving mercuric chloride can present a significant threat to business operations. The loss or damage of equipment or facilities can significantly affect fiscal viability. Lawsuits that may result from personnel injury/death, public exposures, and/or environmental contamination will also require a serious expenditure of resources. Media attention surrounding an injury, death, or environmental damage can also result in a loss of profits and loss of current as well as future business. Always remember that anytime the term "reproductive hazard" is used, public emotion, hysteria, and ignorance can run equally high. This must be properly considered whenever developing public relations policies.

Recommended Risk-Reduction Measures

Company attorneys, safety and health professionals, and environmental specialists should be involved in the development of such procedures dealing with safety or emergency response. A company official should be pre-designated as a public relations officer with specific training in dealing with the press. Corporate plans and policies should be developed, approved, and implemented long before any need for such arises.

REFERENCES

American Conference of Governmental Industrial Hygienists. 1988. *Documentation of the Threshold Limit Values and Biological Exposure Indices,* 5th Edition (with updates). Cincinnati: ACGIH

Environmental Protection Agency. 1983. *Chemical Hazard Information Profiles.* Washington, D.C.: U.S. EPA.

U.S. Department of Health and Human Services, National Institute for Occupational Safety and Health. 1973. NIOSH Criteria Document. *Inorganic Mercury.* Publication Number 73-11024. Washington, D.C.: U.S. Government Printing Office.

MATERIAL SAFETY DATA SHEET

CHEMICAL NAME	
MERCURIC CYANIDE	**CRC PRESS/LEWIS PUBLISHERS DISCLAIMER:** THE INFORMATION AND RECOMMENDATIONS PRESENTED HEREIN ARE BASED ON SOURCES BELIEVED TO BE RELIABLE. CRC MAKES NO REPRESENTATION ON ITS COMPLETENESS OR ACCURACY. IT IS THE RESPONSIBILITY OF THE USER TO DETERMINE THE CHEMICAL'S SUITABILITY FOR ITS INTENDED USE, THE CHEMICAL'S SAFE USE, AND THE CHEMICAL'S PROPER DISPOSAL. NO REPRESENTATIONS AND/OR WARRANTIES, EITHER EXPRESSED OR IMPLIED, OF THE MERCHANTABILITY OR FITNESS FOR A PARTICULAR PURPOSE, OR OF ANY OTHER NATURE, ARE MADE WITH RESPECT TO THE INFORMATION PROVIDED IN THIS MSDS OR TO THE CHEMICAL TO WHICH INFORMATION MAY REFER. CRC NEITHER ASSUMES NOR AUTHORIZES ANY OTHER PERSON TO ASSUME FOR IT, ANY OTHER ADDITIONAL RESPONSIBILITY OR LIABILITY FOR THE USE OF, OR RELIANCE UPON, THIS INFORMATION.

HAZARD WARNING INFORMATION

HEALTH	FIRE	REACTIVE	OTHER	DEGREE OF HAZARD	COLOR CODING	OTHER CODES
3	0	0		0 = Minimum Hazard 1 = Slight Hazard 2 = Moderate Hazard 3 = Serious Hazard 4 = Severe Hazard	HEALTH = BLUE FIRE = RED REACTIVITY = YELLOW OTHER = WHITE	OX = Oxidizer ACID = Acid ALK = Alkali COR = Corrosive ~~W~~ = Use No Water

SECTION I - GENERAL INFORMATION

Characterization	**Mercury Compound**	RCRA Number	**None**
		EPA Class	**Not Applicable**
DOT Proper Shipping Name	**Mercuric Cyanide**	Chemical Abstract Service (CAS) Number	**592-04-1**
DOT Hazard Class and Label Requirements	**Poison B; Poison**	DOT Emergency Guide Code	**60 (for Hg compounds)**
DOT Identification Number	**UN1636**	Molecular Formula	**$Hg(CN)_2$**

Synonyms

Mercury cyanide; cianurina; mercury (II) cyanide.

SECTION II - HAZARDOUS INGREDIENTS AND IDENTITY INFORMATION

Hazardous Components (specific identity)	OSHA Exposure Criteria	NIOSH Exposure Criteria	Immediately Dangerous to Life and Health (IDLH)	ACGIH Exposure Criteria
Mercuric cyanide (derivation: By interaction of mercuric oxide and an aqueous solution of hydrogen cyanide).	**PEL:** 0.1 mg/m^3 (as Hg vapor) **CEILING (skin):** 0.1 mg/m^3 (Hg compounds)	**REL (skin):** 0.05 mg/m^3 (as Hg vapor) **STEL:** Not Established	**Not Determined**	**TLV (skin):** 0.1 mg/m^3 (Hg compounds) 0.05 mg/m^3 (Hg vapors) **STEL:** None

SECTION III - PHYSICAL AND CHEMICAL CHARACTERISTICS

Boiling Point	**Decomposes**	Specific Gravity (H_2O = 1)	**3.996**
Vapor Pressure (mm Hg)	**Not Found**	Molecular Weight	**252.63**
Vapor Density (Air = 1)	**8.7**	Melting Point	**608°F (320°C) Decomposes**

Solubility

Soluble in alcohol and water.

Appearance and Odor

Colorless or white crystalline solid (prisms) that darkens on exposure to light.

SECTION IV - FIRE AND EXPLOSION HAZARD DATA

Flash Point (method used)	**Not Determined**	Explosive Limits in Air % by Volume	**LEL: Not Applicable UEL: Not Applicable**
NFPA Classification	**Non-Flammable Solid**	Autoignition Temperature	**Not Determined**

Extinguishing Media

Use agent suitable to surrounding fire. Dry chemical or foam may suffice for small fires. For large fires, try fog or regular foam. Exercise caution if water is used (it may increase contamination of area).

Special Fire Fighting Procedures

Poisonous gases are produced in fire. Wear full protective clothing and self-contained breathing apparatus (SCBA). Move containers from fire area if it can be done without risk. Otherwise, use a water spray to keep fire-exposed containers cool. Do NOT release runoff from fire-fighting measures to sewers or waterways.

Unusual Fire and Explosion Hazards

Explosive reactions occur in contact with incompatible materials. In contact with acid or heat, it can release flammable and toxic hydrogen cyanide gas.

SECTION V - REACTIVITY DATA

Stability		Conditions to Avoid Mercuric cyanide is stable under normal conditions of temperature and pressure. However, it will gradually decompose to metallic mercury when heated or exposed to light.
Stable X	Unstable	Incompatibility *(materials to avoid)* Incompatible with strong oxidizers (such as fluorine, perchlorates, peroxides, permanganates, chlorates, and nitrates), magnesium, sodium nitrate, light, heat, and acids.
Hazardous Polymerization		Conditions to Avoid Under normal conditions of temperature and pressure, hazardous polymerization of mercuric cyanide is not known to occur.
May Occur	Will Not Occur X	Hazardous Decomposition or By-products When heated to decomposition, mercuric cyanide can emit toxic/poisonous fumes and gases. These include toxic fumes of mercury and cyanide.

SECTION VI - HEALTH HAZARD DATA

Primary Route(s) of Entry:	Inhalation?	Absorption (skin)?	Ingestion?
	X	X	X

Health Hazards

INHALATION: Causes headaches, shortness of breath, cough, chills, fever, congestion, tightness in chest. Prolonged contact can cause CNS effects with vertigo, depression, ataxia, emotional instability, and other effects. Over exposure to cyanide can cause sudden death.

ABSORPTION: Passes through intact skin to cause effects similar to that of inhalation and ingestion. There can also be visual disturbances, eye burns, and dermatitis.

INGESTION: Abdominal pain, diarrhea (possibly bloody), digestive tract irritation, inflammation of the mouth and gums, liver damage, loose teeth, jaw necrosis, reduced urine excretion.

Carcinogenicity	NTP Listed?	IARC Cancer Review Group?	OSHA Regulated?	Target Organs?
Unknown Human Unknown Animal	No	No	29 CFR 1910.1000 Table Z-1 (as Hg)	Kidney, liver, gastrointestinal tract, CNS.

Medical Conditions Generally Aggravated by Exposure

Mercury-sensitive people, existing dental and oral problems, kidney disease, respiratory problems, and central nervous system disorders may be aggravated by exposure to mercuric cyanide.

Emergency and First-aid Procedures

Eye contact: Flush immediately with water for 15 minutes (minimum); seek medical attention. Skin contact: Remove all contaminated clothing. Immediately wash area with large amounts of soap and water. For inhalation: Remove the person from exposure. Provide respiratory assistance and CPR. Transfer to medical facility. Observe for 48 hours for lung effects. If swallowed: Provide breathing assistance if necessary. Never attempt to give a convulsing or unconscious person anything by mouth. If victim is conscious, give 1-2 glasses of water to dilute and induce vomiting.

SECTION VII - PRECAUTIONS FOR SAFE HANDLING AND USE

Steps to be Taken in Case Material is Released or Spilled

Collect powdered material and deposit in sealed drum. Keep those not involved in cleanup from entering area. Do NOT dry sweep (creates airborne dusts). Use a vacuum equipped with a (HEPA) filter. Absorb solution spills in vermiculite or other materials.

Preferred Waste Disposal Method

No citation.

Precautions to be Taken in Handling and Storage

Store in tightly closed containers in cool, well-ventilated area. Sources of ignition are prohibited where mercuric cyanide is stored. Keep away from light. Keep water and moisture away.

Other Precautions and Warnings

Outside or detached storage facility is recommended. Protect containers from physical damage.

SECTION VIII - CONTROL MEASURES AND PERSONAL PROTECTIVE EQUIPMENT

Respiratory Protection *(specify type)*

Exposure levels are low and difficult to monitor. For best protection, use a supplied-air respirator with a full facepiece or a self-contained breathing apparatus (SCBA) operated in positive pressure mode.

Ventilation

Local exhaust or general mechanical systems recommended.

Protective Gloves	Eye Protection	Other Protective Clothing
Impervious Material	Chemical/Dust Goggles or Face Mask	Protective Suit or Clothing

Work/Hygiene Practices

Always wash hands thoroughly after using chemical; never bring food, drink, or smoking materials into vicinity of chemicals. Skin contact must be avoided.

MERCURIC CYANIDE

$Hg(CN)_2$

CAS: 592-04-1

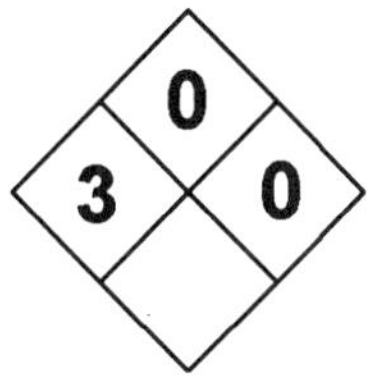

IDENTIFICATION AND TYPICAL USES

Mercuric cyanide appears as a colorless or white crystalline, transparent solid (prisms) with no detectable odor. It is used in the manufacture of medicines (an antiseptic), in germicidal soaps, in manufacturing of cyanogen gas, and in photography.

RISK ASSESSMENT: HEALTH

General Assessment

Mercuric cyanide is a toxic human poison by three routes of exposure: *inhalation, ingestion,* and dermal *absorption.* Mercury accumulates in the body with each exposure and overexposures to cyanide compounds can cause sudden death. While mercury is not currently listed as a carcinogenic compound, mutation data have been reported for many mercury compounds. Organic mercury compounds (those that contain carbon), such as mercuric cyanide, have caused adverse reproductive effects (miscarriages and/or fetal damage) in test animals.

Symptoms of exposure by all inhalation of mercury cyanide vapors can include headache, dypsnea, cough, chills, fever, and pneumonitis. Cyanide causes bronchial irritation with cough and phlegm, and/or lung tissue irritation.

Ingestion of mercury or its compounds causes abdominal pain, bloody diarrhea, gastrointestinal irritation, vomiting, and cardiac arrhythmia.

Repeated exposure from inhalation, absorption, and/or ingestion of mercuric cyanide can cause central nervous system (CNS) effects due to both the cyanide and the mercury base. These include vertigo, anxiety, depression, ataxia, emotional instability, kidney injury, brain damage, tremors (the "shakes"), memory loss, metallic taste, discoloration of the skin, sore gums, personality changes, liver damage, loose teeth, decreased urine excretion, visual disturbances, and dermatitis.

Acute Health Effects

The following acute (short-term) health effects may occur immediately or shortly after exposure to mercuric cyanide:

Skin: Irritation, rash, dermatitis, and possible redness and the appearance of small blisters. Dermal absorption can be quick and will cause toxic systemic effects.

Eye: Irritation, inflammation, watering, and swelling, burns, and tissue damage.

Lung: Inhalation of mercuric cyanide causes irritation of the mucosa in the respiratory tract causing cough, phlegm, and lung tissue irritation.

Chronic Health Effects

The following chronic (long-term) health effects may occur at some time after exposure to mercuric cyanide and can last for months or even years:

Cancer Hazards: According to the information presented in the references, mercuric cyanide has not been adequately tested for its ability to cause cancer in test animals. However, mutation (causes genetic changes) data have been documented and many scientists believe that such chemicals may pose a cancer risk in the long-term. More research is required.

Reproductive Hazards: According to information available in the references, organic mercury compounds (such as mercuric cyanide) have been shown to adversely affect reproduction in humans and animals. They can cause the early termination of pregnancies (miscarriages) and may injure the fetus. Mercury may also reduce fertility in both males and females. While inorganic mercury substances (those without carbon), have not specifically been shown to be human teratogens, they should still be handled with caution as they may cause reproductive problems in males and females.

Other Chronic Hazards: High or repeated exposure can cause kidney damage or mercury poisoning with symptoms of sore gums, "shakes" (especially noticeable in handwriting), irritability, increased saliva production, memory loss, extreme shyness, weakness, loss of teeth, poor appetite, and metallic taste. Serious personality changes and brain damage can occur, especially if exposure continues. Repeated skin contact may cause gray discoloration and a skin allergy may develop. If this happens, even small future exposures can cause rash. Repeated exposures may also cause reduced peripheral vision and brown staining of the eye. Mercury can accumulate in the body on prolonged exposure and it may take months or years for the body to dispose of excess mercury.

Recommended Risk-Reduction Measures

Personnel should avoid direct contact with mercuric cyanide. Mercury is an extremely toxic, cumulative poison. If a less toxic chemical cannot be substituted for a hazardous material, then *engineering controls* are the most effective method of reducing exposures.

The best protection is to enclose operations and/or provide local exhaust ventilation at the site of chemical release. While not always operationally feasible, isolating operations can also reduce exposure. Using respiratory protection is less effective than the controls mentioned above, but is still advisable whenever working with or around mercuric cyanide. Exposure levels for mercury are extremely low (0.05 mg/m^3) and difficult to monitor accurately. A full facepiece respirator with a high efficiency particulate air (HEPA) filter equipped with a dust/mist/fume pre-filter may suffice. However, for the best protection use a supplied-air respirator with a full facepiece operated in positive pressure mode or an MSHA/NIOSH-approved self-contained breathing apparatus (SCBA) with full facepiece operated in pressure demand or other continuous flow mode. If a full facepiece is not available, then chemical/dust goggles should be worn to protect the eyes. Whenever a chemical dust hazard exists, a face shield and a protective apron should be worn. To prevent hand and skin exposures, impervious rubber gloves should be used. Glove manufacturers should be contacted and permeation studies obtained *before* final glove selections are made. Skin contact must be prevented. Protective suits and clothing should be used.

Administrative controls should also be in place to minimize the potential for human exposures. These may include written procedures or policies which specify the methods and techniques that will be practiced whenever personnel are to work with mercuric cyanide dusts or solutions. All personnel should receive *training* on the use, hazards, protective measures, emergency actions, and other precautions per 29 CFR 1910.1200 (Hazard Communication), prior to the first assignment in an area where mercuric cyanide is used or stored. Before first exposure and every six to twelve months thereafter, a complete medical history and examination is recommended, including the following:

- ☑ Examination of the nervous system (taking note of any problems with person's handwriting).
- ☑ Routine urinalysis and special urine test for mercury (should be less than 0.02 mg/liter).
- ☑ Eye examination.
- ☑ Lung function tests for frequent exposures.

If symptoms develop or overexposure is suspected, the above listed tests should be repeated, along with the following recommended tests:

- ☑ Consider nerve conduction tests.
- ☑ Urinary enzyme and neurological behavior tests.
- ☑ Serum mercury tests.

Any evaluation should include a careful history of past and present symptoms with a medical examination. Medical tests that look for damage already done are not a substitute for controlling exposure. Also, since smoking can cause heart disease, emphysema, and a number of other respiratory disorders, smokers exposed to lung irritants such as mercuric cyanide dusts and fumes may experience symptoms much more quickly than non-smokers under the same conditions of exposure. Other mixed exposures occur when personnel use creams to whiten or bleach the skin that contain mercury. If so, their use increases exposure risk. Persons who are allergic to mercury may react to Mercurochrome or Merthiolate, which also contain mercury. Prudent risk management requires proper consideration of *all* factors which may be causing the appearance of exposures symptoms in the work place.

Other methods to reduce exposure include:

- ☑ Where possible, enclose operations and use local exhaust ventilation at the site of chemical release. If local exhaust ventilation or enclosure is not used, respiratory protection should be mandatory.
- ☑ Always ensure that proper protective clothing is worn when using chemical substances.

☑ Wash thoroughly immediately after exposure to mercuric cyanide and at the end of the work shift or before eating, drinking, or smoking.

☑ Hazard warning information should be posted in the work area. In addition, as part of an on-going education and training program, all information on the health and safety hazards of mercuric cyanide should be communicated to all exposed and potentially exposed workers.

☑ Eye wash stations and emergency showers should be located in the immediate work area.

☑ Work clothing that has been contaminated with mercuric cyanide should never be worn home where family members can be exposed as well. Contaminated clothing should be laundered at work or by an authorized service by personnel who have been trained in the hazards associated with exposure to mercuric cyanide.

☑ The drug n-acetyl penicillamine (NAP) has been used to treat mercury poisoning, with mixed success.

☑ Specific engineering controls are recommended for mercury and cyanide by NIOSH. Refer to NIOSH Criteria Document, "Inorganic Mercury" (Publication Number 73-11024) and NIOSH Criteria Document, "Occupational Exposure to Hydrogen Cyanide and Cyanide Salts (Publication Number 77-108).

RISK ASSESSMENT: ENVIRONMENT

General Assessment

The environment is at risk of exposure during transportation, storage, disposal, or destruction of mercuric cyanide. In almost every scenario, the threat of environmental exposure is contingent upon the proper handling of the chemical substance. Accidental spills, large or small, can result in fire, explosion, and possible contamination of the surrounding environmental mediums (water, soil, and air).

Mercuric cyanide is incompatible with many common oxidizers (fluorine), other metals and their salts (magnesium and sodium nitrate) and acids. It is also friction- and impact-sensitive and can easily explode under such conditions. It is also sensitive to heat and light. It reacts with moisture and steam to release explosive hydrogen cyanide gas. In a fire, mercuric cyanide will emit toxic and poisonous cyanide and mercury fumes. These characteristics require special consideration during any emergency situation involving a leak or spill of mercuric cyanide.

Elemental mercury is a heavy and relatively inert liquid which is oxidized to inorganic mercury (II) under natural conditions. Mercury (II) may combine with an organic fraction to form methyl mercury. Both mercury (II) and methyl mercury can cause environmental contamination. Mercury (II) may enter the environment in industrial or municipal waste treatment discharges, from previously contaminated sediments, and from the weathering of natural rocks. Bacteria may then convert it to methyl mercury. The concentration of mercury (II) in water may be elevated with acid rain due to the scouring of mercury from the air and increased partitioning from the sediment into the water.

Acute Ecological Effects

Acute (short-term) toxic effects may include the death of animals, birds, or fish and death or low growth rate in plants. Acute effects are seen 2 to 4 days after animals or plants are exposed to mercuric cyanide. Mercury (II) and methyl mercury have high acute toxicity to aquatic life. Insufficient data are available to evaluate the short-term effects of mercury (II) and methyl mercury exposure to plants, birds or terrestrial animals.

Chronic Ecological Effects

Chronic (long-term) toxic effects may include shortened life span, reproductive problems, lower fertility, and changes in appearance or behavior in exposed animals. These effects can be seen long after first exposure(s) to toxic chemicals. Mercury (II) and methyl mercury have high chronic toxicity to aquatic life. Eating fish contaminated with mercury residues has caused secondary poisoning in humans. Birds or land animals similarly exposed to mercury and its compounds could also be subject to such effects. Insufficient data are available to evaluate the long-term effects of mercuric cyanide exposure to plants.

Water Solubility

Mercuric cyanide is soluble in water. Concentrations of 1000 milligrams or more will mix with a liter of water. Mercury itself is insoluble in water

Persistence in the Environment

Mercury is highly persistent in water, with a half-life greater than 200 days. The half-life of a pollutant is the amount of time it takes for one half of the chemical to be degraded. When released into water, microorganisms convert mercuric salts into methyl mercury which is rapidly taken up by algae and enters the food chain. It concentrates in the edible tissues of fish and, eventually, can cause human poisoning as well.

Bioaccumulation in Aquatic Organisms

Some substances increase in concentration, or *bioaccumulate,* in living organisms as they breathe contaminated air, drink contaminated water, or eat contaminated food. These chemicals can become concentrated in the tissues and internal organs of animals as well as humans. The concentration of mercury and methyl mercury found in fish tissues is expected to be considerably higher than the average concentration in the water from which the fish was taken.

Recommended Risk-Reduction Measures

Proper training of all transporters will reduce the likelihood of a mishap or accidents resulting in a leak or spill to the environment. Proper labeling on all containers, trucks, and rail cars will enable emergency responders to react properly and quickly to any disaster thereby reducing the potential risk to the environment and to personnel.

If a spill or leak to the environment has occurred, fire department, emergency response and/or hazardous materials spill personnel should be notified immediately. Do not dry sweep. When vacuuming, a high efficiency particulate air (HEPA) filter should be used, not a standard shop vacuum. Deposit collected materials in sealed drums for disposal. Cleanup should be attempted only by those trained in proper spill containment procedures. Contaminated soils may need to be removed and replaced with clean soil. If mercuric cyanide should contact the water table, aquifer, or navigable waterway, time is of the essence. Mercuric cyanide is soluble in water and total containment and remediation may not be possible. When such spills occur, the local and/or state emergency response authorities should be notified. A comprehensive emergency response or disaster preparedness/recovery plan should be in place prior to any operations involving the use, transportation, storage, or disposal of mercuric cyanide. If mercuric cyanide is spilled or leaked, the following specific steps are recommended:

- ☑ Restrict persons not wearing protective clothing from area of spill or leak until cleanup is complete and area can be opened for normal work.
- ☑ Ventilate area and remove ignition sources.
- ☑ It may be necessary to dispose of mercuric cyanide as a hazardous waste. The responsible state agency or the regional office of the federal Environmental Protection Agency (EPA) should be contacted for specific recommendations.

RISK ASSESSMENT: BUSINESS

General Assessment

Accidents or mishaps involving mercuric cyanide can present a significant threat to business operations. The loss or damage of equipment or facilities can significantly affect fiscal viability. Lawsuits that may result from personnel injury/death, public exposures, and/or environmental contamination will also require a serious expenditure of resources. Media attention surrounding an injury, death, or environmental damage can also result in a loss of profits and loss of current as well as future business. Always remember that anytime the term "reproductive hazard" is used, public emotion, hysteria, and ignorance can run equally high. This must be properly considered whenever developing public relations policies.

Recommended Risk-Reduction Measures

Company attorneys, safety and health professionals, and environmental specialists should be involved in the development of such procedures dealing with safety or emergency response. A company official should be pre-designated as a public relations officer with specific training in dealing with the press. Corporate plans and policies should be developed, approved, and implemented long before any need for such arises.

REFERENCES

Environmental Protection Agency. 1983. *Chemical Hazard Information Profiles.* Washington, D.C.: U.S. EPA.

U.S. Department of Health and Human Services, National Institute for Occupational Safety and Health. 1973. NIOSH Criteria Document. *Inorganic Mercury.* Publication Number 73-11024. Washington, D.C.: U.S. Government Printing Office.

U.S. Department of Health and Human Services, National Institute for Occupational Safety and Health. 1977. NIOSH Criteria Document. *Hydrogen Cyanide and Cyanide Salts.* Publication Number 77-108. Washington, D.C.: U.S. Government Printing Office.

MATERIAL SAFETY DATA SHEET

CHEMICAL NAME	
MERCURIC NITRATE	**CRC PRESS/LEWIS PUBLISHERS DISCLAIMER:** THE INFORMATION AND RECOMMENDATIONS PRESENTED HEREIN ARE BASED ON SOURCES BELIEVED TO BE RELIABLE. CRC MAKES NO REPRESENTATION ON ITS COMPLETENESS OR ACCURACY. IT IS THE RESPONSIBILITY OF THE USER TO DETERMINE THE CHEMICAL'S SUITABILITY FOR ITS INTENDED USE, THE CHEMICAL'S SAFE USE, AND THE CHEMICAL'S PROPER DISPOSAL. NO REPRESENTATIONS AND/OR WARRANTIES, EITHER EXPRESSED OR IMPLIED, OF THE MERCHANTABILITY OR FITNESS FOR A PARTICULAR PURPOSE, OR OF ANY OTHER NATURE, ARE MADE WITH RESPECT TO THE INFORMATION PROVIDED IN THIS MSDS OR TO THE CHEMICAL TO WHICH INFORMATION MAY REFER. CRC NEITHER ASSUMES NOR AUTHORIZES ANY OTHER PERSON TO ASSUME FOR IT, ANY OTHER ADDITIONAL RESPONSIBILITY OR LIABILITY FOR THE USE OF, OR RELIANCE UPON, THIS INFORMATION.

HAZARD WARNING INFORMATION

HEALTH	FIRE	REACTIVE	OTHER	DEGREE OF HAZARD	COLOR CODING	OTHER CODES
3	0	0	OX	0 = Minimum Hazard 1 = Slight Hazard 2 = Moderate Hazard 3 = Serious Hazard 4 = Severe Hazard	HEALTH = BLUE FIRE = RED REACTIVITY = YELLOW OTHER = WHITE	OX = Oxidizer ACID = Acid ALK = Alkali COR = Corrosive W = Use No Water

SECTION I - GENERAL INFORMATION

Characterization	**Mercury Compound**	RCRA Number	**None**
		EPA Class	**Characteristic (T) Waste**
DOT Proper Shipping Name	**Mercuric Nitrate**	Chemical Abstract Service (CAS) Number	**10045-94-0**
DOT Hazard Class and Label Requirements	**Poison B; Poison**	DOT Emergency Guide Code	**42**
DOT Identification Number	**UN1625**	Molecular Formula	**$Hg(NO_3)_2$**

Synonyms
Mercury nitrate; mercury (II) nitrate; mercury pernitrate; nitric acid, mercury (II) salt.

SECTION II - HAZARDOUS INGREDIENTS AND IDENTITY INFORMATION

Hazardous Components (specific identity)	OSHA Exposure Criteria	NIOSH Exposure Criteria	Immediately Dangerous to Life and Health (IDLH)	ACGIH Exposure Criteria
Mercuric nitrate (derivation: By action of hot nitric acid on mercury).	**PEL (ceiling):** 0.1 mg/m^3 (skin) (as Hg compounds) **STEL:** Not Applicable	**REL (ceiling):** 0.1 mg/m^3 (skin) (as Hg compounds) **STEL:** Not Applicable	**28 mg/m^3 (as Hg)**	**TLV (ceiling):** 0.1 mg/m^3 (skin) (as Hg compounds) **STEL:** Not Applicable

SECTION III - PHYSICAL AND CHEMICAL CHARACTERISTICS

Boiling Point	**Decomposes**	Specific Gravity (H_2O = 1)	**4.3**
Vapor Pressure (mm Hg)	**Not Found**	Molecular Weight	**324.66**
Vapor Density (Air = 1)	**11.0**	Melting Point	**174°F (79°C)**

Solubility
Soluble in water (especially cold water), nitric acid, acetone, and ammonia.

Appearance and Odor
White or yellow deliquescent, crystalline powder (sugar or sand-like) with no odor.

SECTION IV - FIRE AND EXPLOSION HAZARD DATA

Flash Point (method used)	**Not Determined**	Explosive Limits in Air % by Volume	**LEL: Not Applicable UEL: Not Applicable**
NFPA Classification	**Non-Combustible Solid**	Autoignition Temperature	**Not Determined**

Extinguishing Media
Use agent suitable to surrounding fire. Dry chemical or foam may suffice for small fires. For large fires, try fog or regular foam. Exercise caution if water is used (it may increase contamination of area).

Special Fire Fighting Procedures
Poisonous gases are produced in fire. Wear full protective clothing and self-contained breathing apparatus (SCBA). Move containers from fire area if it can be done without risk. Otherwise, use a water spray to keep fire-exposed containers cool until long after fire has been extinguished. Do NOT release runoff from fire-fighting measures to sewers or waterways.

Unusual Fire and Explosion Hazards
Mercuric nitrate is a powerful oxidizer and can greatly accelerate the combustion of ordinary combustible materials.

SECTION V - REACTIVITY DATA

Stability		Conditions to Avoid
		Mercuric nitrate is stable under normal conditions of temperature and pressure. However, it will gradually decompose to metallic mercury when heated. Avoid contact with incompatible materials.
Stable	Unstable	Incompatibility *(materials to avoid)*
X		**Reacts with acetylene (forms explsive mercury acetylide), ethanol (forms explosive mercury fulminate), phosphine, cyanide, ferrocene, sulfur, isobutene, phosphinic acid, hypophosphoric acid, aromatics.**
Hazardous Polymerization		Conditions to Avoid
		Under normal conditions of temperature and pressure, hazardous polymerization of mercuric nitrate is not known to occur.
May Occur	Will Not Occur	Hazardous Decomposition or By-products
	X	**When heated to decomposition, mercuric nitrate can emit toxic/poisonous fumes and gases. These include toxic fumes of mercury and nitrogen oxides.**

SECTION VI - HEALTH HAZARD DATA

Primary Route(s) of Entry:	Inhalation?	Absorption (skin)?	Ingestion?
	X	X	X

Health Hazards

INHALATION: Causes headaches, shortness of breath, cough, chills, fever, congestion, tightness in chest, possible pneumonitis and pulmonary edema. Prolonged contact can cause CNS effects with vertigo, depression, ataxia, emotional instability, and other effects.

ABSORPTION: Passes through intact skin to cause effects similar to that of inhalation and ingestion. There can also be visual disturbances, eye burns, and dermatitis.

INGESTION: Abdominal pain, diarrhea (possibly bloody), digestive tract irritation, inflammation of the mouth and gums, liver and kidney damage, loose teeth, reduced urine excretion.

Carcinogenicity	NTP Listed?	IARC Cancer Review Group?	OSHA Regulated?	Target Organs?
Unknown Human Unknown Animal	No	No	29 CFR 1910.1000 Table Z-1 (as Hg)	Kidney, liver, gastrointestinal tract, CNS.

Medical Conditions Generally Aggravated by Exposure

Mercury-sensitive people, existing dental and oral problems, kidney disease, respiratory problems, and central nervous system disorders may be aggravated by exposure to mercuric nitrate.

Emergency and First-aid Procedures

Eye contact: Flush immediately with water for 15 minutes (minimum); seek medical attention. Skin contact: Remove all contaminated clothing. Immediately wash area with large amounts of soap and water. For inhalation: Remove the person from exposure. Provide respiratory assistance and CPR. Transfer to medical facility. Observe for 48 hours for lung effects. If swallowed: Provide breathing assistance if necessary. Never attempt to give a convulsing or unconscious person anything by mouth. If victim is conscious, give 1-2 glasses of water to dilute and induce vomiting.

SECTION VII - PRECAUTIONS FOR SAFE HANDLING AND USE

Steps to be Taken in Case Material is Released or Spilled

Collect powdered material and deposit in sealed drum. Keep those not involved in cleanup from entering area. Do NOT dry sweep (creates airborne dusts). Use a vacuum equipped with a (HEPA) filter. Absorb solution spills in vermiculite or other materials.

Preferred Waste Disposal Method

No Citation.

Precautions to be Taken in Handling and Storage

Store in tightly closed containers in cool, well-ventilated area. Sources of ignition are prohibited where mercuric nitrate is stored. Keep away from light. Do NOT store on wooden floors.

Other Precautions and Warnings

Outside or detached storage facility is recommended. Protect containers from physical damage.

SECTION VIII - CONTROL MEASURES AND PERSONAL PROTECTIVE EQUIPMENT

Respiratory Protection *(specify type)*

Exposure levels are low and difficult to monitor. For best protection, use a supplied-air respirator with a full facepiece or a self-contained breathing apparatus (SCBA) operated in positive pressure mode.

Ventilation

Local exhaust or general mechanical systems recommended.

Protective Gloves	Eye Protection	Other Protective Clothing
Impervious Material	Chemical/Dust Goggles or Face Mask	Protective Suit or Clothing

Work/Hygiene Practices

Always wash hands thoroughly after using chemical; never bring food, drink, or smoking materials into vicinity of chemicals. Skin contact must be avoided.

MERCURIC NITRATE

$Hg(NO_3)_2$

CAS: 10045-94-0

IDENTIFICATION AND TYPICAL USES

Mercuric nitrate appears as a white or yellow deliquescent crystalline, sugar, sand-like, or powder-like solid with no detectable odor. It is used in the manufacture of felt, as an analytical reagent, in bronzing, in the destruction of phylloxera, and in the manufacture of mercury fulminate.

RISK ASSESSMENT: HEALTH

General Assessment

Mercuric nitrate is a toxic human poison by three routes of exposure: *inhalation, ingestion,* and dermal *absorption.* Mercury accumulates in the body with each exposure. While it is not currently listed as a carcinogenic compound, mutation data have been reported for many mercury compounds. Organic mercury compounds (those that contain carbon) have caused adverse reproductive effects (miscarriages and/or fetal damage) in test animals.

Symptoms of exposure by inhalation of mercury nitrate vapors can include headache, dypsnea, cough, chills, fever, pneumonitis, and delayed pulmonary edema (fluid in the lungs), which can be fatal. Symptoms of pulmonary edema (cough, chest pain, congestion, nausea, dizziness) may be delayed up to 48 hours thereby creating a false sense of security with regard to health exposure risk. Ingestion of mercury or its compounds causes abdominal pain, bloody diarrhea, gastrointestinal irritation, vomiting, cardiac arrhythmia, shock, renal failure, and circulatory collapse.

Repeated exposure from inhalation, absorption, and/or ingestion can cause central nervous system (CNS) effects including vertigo, anxiety, depression, ataxia, emotional instability, kidney injury, brain damage, tremors (the "shakes"), memory loss, metallic taste, discoloration of the skin, sore gums, personality changes, liver damage, loose teeth, decreased urine excretion, visual disturbances, and possible dermatitis.

Acute Health Effects

The following acute (short-term) health effects may occur immediately or shortly after exposure to mercuric nitrate:

<u>Skin</u>: Irritation, rash, dermatitis, redness and the appearance of small blisters. Dermal absorption can be quick, causing toxic systemic effects.

<u>Eye</u>: Irritation, inflammation, watering, and swelling, burns, and tissue damage. Ulceration of the conjunctiva and cornea, discoloration of the cornea and lens, tremor of the eyelids, and disturbances in vision.

<u>Lung</u>: Inhalation of mercuric nitrate dusts may cause irritation of the mucous membranes in the respiratory tract causing cough, phlegm, and lung tissue irritation. It can also lead to pneumonitis as well as pulmonary edema, which is a medical emergency and can be fatal.

Chronic Health Effects

The following chronic (long-term) health effects may occur at some time after exposure to mercuric nitrate and can last for months or even years:

<u>Cancer Hazards</u>: According to the information presented in the references, mercuric nitrate has not been adequately tested for its ability to cause cancer in test animals. However, mutation (causes genetic changes) data have been documented and many scientists believe that such chemicals may pose a cancer risk in the long-term. More research is required.

<u>Reproductive Hazards</u>: According to information available in the references, organic mercury compounds have been shown to adversely affect reproduction in humans and animals. They can cause the early termination of pregnancies (miscarriages) and may injure the fetus. Mercury may also reduce fertility in both males and females. While inorganic mercury substances

(those without carbon), such as mercuric nitrate, have not specifically been shown to be human teratogens, they should still be handled with caution as they may cause reproductive problems in males and females.

Other Chronic Hazards: High or repeated exposure can cause kidney damage or mercury poisoning with symptoms of sore gums, "shakes" (especially noticeable in handwriting), irritability, increased saliva production, memory loss, extreme shyness, weakness, loss of teeth, poor appetite, and metallic taste. Serious personality changes and brain damage can occur, especially if exposure continues. Repeated skin contact may cause gray discoloration and a skin allergy may develop. If this happens, even small future exposures can cause rash. Repeated exposures may also cause reduced peripheral vision and brown staining of the eye. Mercury can accumulate in the body on prolonged exposure and it may take years for the body to dispose of mercury.

Recommended Risk-Reduction Measures

Personnel should avoid direct contact with mercuric nitrate. Mercury is an extremely toxic, cumulative poison. If a less toxic chemical cannot be substituted for a hazardous material, then *engineering controls* are the most effective method of reducing exposures.

The best protection is to enclose operations and/or provide local exhaust ventilation at the site of chemical release. While not always operationally feasible, isolating operations can also reduce exposure. Using respiratory protection is less effective than the controls mentioned above, but is still advisable whenever working with or around mercuric nitrate. Exposure levels for mercury are extremely low (0.1 mg/m^3) and difficult to monitor accurately. A full facepiece respirator with a high-efficiency particulate air (HEPA) filter equipped with a dust/mist/fume pre-filter may suffice. However, for the best protection use a supplied-air respirator with a full facepiece operated in positive pressure mode or an MSHA/NIOSH-approved self-contained breathing apparatus (SCBA) with full facepiece operated in pressure demand or other continuous flow mode. If a full facepiece is not available, then chemical/dust goggles should be worn to protect the eyes. Whenever a chemical dust hazard exists, a face shield and a protective apron should be worn. To prevent hand and skin exposures, impervious rubber gloves should be used. Glove manufacturers should be contacted and permeation studies obtained *before* final glove selections are made. Skin contact must be prevented. Protective suits and clothing should be used.

Administrative controls should also be in place to minimize the potential for human exposures. These may include written procedures or policies which specify the methods and techniques that will be practiced whenever personnel are to work with mercuric nitrate dusts or solutions. All personnel should receive *training* on the use, hazards, protective measures, emergency actions, and other precautions per 29 CFR 1910.1200 (Hazard Communication), prior to the first assignment in an area where mercuric nitrate is used or stored. Before first exposure and every six to twelve months thereafter, a complete medical history and examination is recommended, including the following:

- ☑ Examination of the nervous system (taking note of any problems with person's handwriting).
- ☑ Routine urinalysis and special urine test for mercury (should be less than 0.02 mg/liter).
- ☑ Eye examination.
- ☑ Lung function tests for frequent exposures.

If symptoms develop or overexposure is suspected, the following additional tests are recommended:

- ☑ Consider nerve conduction tests.
- ☑ Urinary enzyme and neurological behavior tests.
- ☑ Serum mercury tests.
- ☑ Consider chest X-ray following acute overexposure (may be negative if taken immediately after exposure due to delayed onset of pulmonary edema).

Any evaluation should include a careful history of past and present symptoms with a medical examination. Medical tests that look for damage already done are not a substitute for controlling exposure. Also, since smoking can cause heart disease, emphysema, and a number of other respiratory disorders, smokers exposed to lung irritants such as mercuric nitrate dusts and fumes may experience symptoms much more quickly than non-smokers under the same conditions of exposure. Other mixed exposures occur when personnel use creams to whiten or bleach the skin that contain mercury. If so, their use increases exposure risk. Persons who are allergic to mercury may react to Mercurochrome or Merthiolate, which also contain mercury. Prudent risk management requires proper consideration of *all* exposure factors.

Other methods to reduce exposure include:

- ☑ Where possible, enclose operations and use local exhaust ventilation at the site of chemical release.

If local exhaust ventilation or enclosure is not used, respiratory protection should be provided and its use mandatory.

☑ Always ensure that proper protective clothing is worn when using chemical substances.

☑ Wash thoroughly immediately after exposure to mercuric nitrate and at the end of the work shift or before eating, drinking, or smoking.

☑ Hazard warning information should be posted in the work area. In addition, as part of an on-going education and training program, all information on the health and safety hazards of mercuric nitrate should be communicated to all exposed and potentially exposed workers.

☑ Eye wash stations and emergency showers should be located in the immediate work area.

☑ Work clothing that has been contaminated with mercuric nitrate should never be worn home where family members can be exposed as well. Contaminated clothing should be laundered at work or by an authorized service by personnel who have been trained in the hazards associated with exposure to mercuric nitrate.

☑ The drug n-acetyl penicillamine (NAP) has been used in the past to treat mercury poisoning, with mixed success.

☑ Specific engineering controls are recommended for mercury by NIOSH. Refer to NIOSH Criteria Document, "Inorganic Mercury" (Publication Number 73-11024).

RISK ASSESSMENT: ENVIRONMENT

General Assessment

The environment is at risk of exposure during transportation, storage, disposal, or destruction of mercuric nitrate. In almost every scenario, the threat of environmental exposure is contingent upon the proper handling of the chemical substance. Accidental spills, large or small, can result in fire, explosion, and possible contamination of the surrounding environmental mediums (water, soil, and air).

Mercuric nitrate is incompatible with many common materials. It can react with acetylene, ethanol, phosphine, potassium cyanide, ferrocene, and sulfur. Reactions can be vigorous and violent. It is also sensitive to heat and light. In a fire, mercuric nitrate will emit toxic and poisonous fumes of mercury and nitrogen. These characteristics require special consideration during any emergency situation involving a leak or spill of mercuric nitrate.

Elemental mercury is a heavy and relatively inert liquid which is oxidized to inorganic mercury (II) under natural conditions. Mercury (II) may combine with an organic fraction to form methyl mercury. Both mercury (II) and methyl mercury can cause environmental contamination. Mercury (II) may enter the environment in industrial or municipal waste treatment discharges, from previously contaminated sediments, and from the weathering of natural rocks. Bacteria may then convert it to methyl mercury. The concentration of mercury (II) in water may be elevated with acid rain due to the scouring of mercury from the air and increased partitioning from the sediment into the water.

Acute Ecological Effects

Acute (short-term) toxic effects may include the death of animals, birds, or fish and death or low growth rate in plants. Acute effects are seen 2 to 4 days after animals or plants are exposed to mercuric nitrate. Mercury (II) and methyl mercury have high acute toxicity to aquatic life. Insufficient data are available to evaluate the short-term effects of mercury (II) and methyl mercury exposure to plants, birds or terrestrial animals.

Chronic Ecological Effects

Chronic (long-term) toxic effects may include shortened life span, reproductive problems, lower fertility, and changes in appearance or behavior in exposed animals. These effects can be seen long after first exposure(s) to toxic chemicals. Mercury (II) and methyl mercury have high chronic toxicity to aquatic life. Eating fish contaminated with mercury residues has caused secondary poisoning in humans. Birds or land animals similarly exposed to mercury and its compounds could also be subject to such effects. Insufficient data are available to evaluate the long-term effects of mercuric nitrate exposure to plants.

Water Solubility

Mercuric nitrate is soluble in water. Concentrations of 1000 milligrams or more will mix with a liter of water.

Persistence in the Environment

Mercury is highly persistent in water, with a half-life greater than 200 days. The half-life of a pollutant is the amount of time it takes for one half of the chemical to be degraded. When released into water, microorganisms convert mercuric salts into methyl mercury which is rapidly taken up by algae and enters the food chain. It concentrates in the edible tissues of fish and, eventually, can cause human poisoning as well.

≈ *Bioaccumulation in Aquatic Organisms*

Some substances increase in concentration, or *bioaccumulate,* in living organisms as they breathe contaminated air, drink contaminated water, or eat contaminated food. These chemicals can become concentrated in the tissues and internal organs of animals as well as humans. The concentration of mercury and methyl mercury found in fish tissues is expected to be considerably higher than the average concentration in the water from which the fish was taken.

Recommended Risk-Reduction Measures

Proper training of all transporters will reduce the likelihood of a mishap or accidents resulting in a leak or spill to the environment. Proper labeling on all containers, trucks, and rail cars will enable emergency responders to react properly and quickly to any disaster thereby reducing the potential risk to the environment and to personnel.

If a spill or leak to the environment has occurred, fire department, emergency response and/or hazardous materials spill personnel should be notified immediately. Do not dry sweep. When vacuuming, a high-efficiency particulate air (HEPA) filter should be used, not a standard shop vacuum. Deposit collected materials in sealed drums for disposal. Cleanup should be attempted only by those trained in proper spill containment procedures. Contaminated soils may need to be removed and replaced with clean soil. If mercuric nitrate should contact the water table, aquifer, or navigable waterway, time is of the essence. Mercuric nitrate is soluble in water and total containment and remediation may not be possible. When such spills occur, the local and/or state emergency response authorities should be notified. A comprehensive emergency response or disaster preparedness/recovery plan should be in place prior to any operations involving the use, transportation, storage, or disposal of mercuric nitrate. If mercuric nitrate is spilled or leaked, the following specific steps are recommended:

- ☑ Restrict persons not wearing protective clothing from area of spill or leak until cleanup is complete and area can be opened for normal work.
- ☑ Ventilate area and remove all ignition sources.
- ☑ It may be necessary to dispose of mercuric nitrate as a hazardous waste. The responsible state agency or the regional office of the federal Environmental Protection Agency (EPA) should be contacted for specific recommendations.

RISK ASSESSMENT: BUSINESS

General Assessment

Accidents or mishaps involving mercuric nitrate can present a significant threat to business operations. The loss or damage of equipment or facilities can significantly affect fiscal viability. Lawsuits that may result from personnel injury/death, public exposures, and/or environmental contamination will also require a serious expenditure of resources. Media attention surrounding an injury, death, or environmental damage can also result in a loss of profits and loss of current as well as future business. Always remember that anytime the term "reproductive hazard" is used, public emotion, hysteria, and ignorance can run equally high. This must be properly considered whenever developing public relations policies.

Recommended Risk-Reduction Measures

Company attorneys, safety and health professionals, and environmental specialists should be involved in the development of such procedures dealing with safety or emergency response. A company official should be pre-designated as a public relations officer with specific training in dealing with the press. Corporate plans and policies should be developed, approved, and implemented long before any need for such arises.

REFERENCES

American Conference of Governmental Industrial Hygienists. 1988. *Documentation of the Threshold Limit Values and Biological Exposure Indices,* 5th Edition (with updates). Cincinnati: ACGIH

Environmental Protection Agency. 1983. *Chemical Hazard Information Profiles.* Washington, D.C.: U.S. EPA.

Patnaik, P. 1992. *A Comprehensive Guide to the Hazardous Properties of Chemical Substances.* New York: Van Nostrand Reinhold.

Reeves, A. L. 1979. *Handbook on the Toxicology of Metals.* New York: Elsevier North-Holland.

U.S. Department of Health and Human Services, National Institute for Occupational Safety and Health. 1973. NIOSH Criteria Document. *Inorganic Mercury.* Publication Number 73-11024. Washington, D.C.: U.S. Government Printing Office.

MATERIAL SAFETY DATA SHEET

CHEMICAL NAME

MERCURIC OXIDE

HAZARD WARNING INFORMATION

HEALTH	FIRE	REACTIVE	OTHER	DEGREE OF HAZARD	COLOR CODING	OTHER CODES
3	0	0	OX	0 = Minimum Hazard 1 = Slight Hazard 2 = Moderate Hazard 3 = Serious Hazard 4 = Severe Hazard	HEALTH = BLUE FIRE = RED REACTIVITY = YELLOW OTHER = WHITE	OX = Oxidizer ACID = Acid ALK = Alkali COR = Corrosive W = Use No Water

SECTION I - GENERAL INFORMATION

Characterization **Mercury Compound**	RCRA Number **None**	EPA Class **Not Applicable**
DOT Proper Shipping Name **Mercurous Oxide, Black, Solid**	Chemical Abstract Service (CAS) Number **15829-53-5**	
DOT Hazard Class and Label Requirements **Poison B; Poison**	DOT Emergency Guide Code **42**	
DOT Identification Number **UN1641**	Molecular Formula **Hg_2O**	

Synonyms
Mercury oxide, black; mercury (I) oxide; mercurous oxide.

SECTION II - HAZARDOUS INGREDIENTS AND IDENTITY INFORMATION

Hazardous Components (specific identity)	OSHA Exposure Criteria	NIOSH Exposure Criteria	Immediately Dangerous to Life and Health (IDLH)	ACGIH Exposure Criteria
Mercuric oxide (derivation: By action of sodium hydroxide on mercurous nitrate).	**PEL (ceiling):** 0.1 mg/m^3 (skin) (as Hg compounds) **STEL:** Not Applicable	**REL (ceiling):** 0.1 mg/m^3 (skin) (as Hg compounds) **STEL:** Not Applicable	**28 mg/m^3 (as Hg)**	**TLV (ceiling):** 0.1 mg/m^3 (skin) (as Hg compounds) **STEL:** Not Applicable

SECTION III - PHYSICAL AND CHEMICAL CHARACTERISTICS

Boiling Point **Decomposes**	Specific Gravity (H_2O = 1) **11.0**
Vapor Pressure (mm Hg) **Not Found**	Molecular Weight **417.22**
Vapor Density (Air = 1) **7.5**	Melting Point **932°F (500°C) Decomposes**

Solubility
Insoluble in water. Soluble in dilute hydrochloric and nitric acids. Insoluble in alcohol, acetone, ammonia, and ether.

Appearance and Odor
Black to grayish-black powder. No odor.

SECTION IV - FIRE AND EXPLOSION HAZARD DATA

Flash Point (method used) **Not Determined**	Explosive Limits in Air % by Volume **LEL: Not Applicable UEL: Not Applicable**
NFPA Classification **Non-Combustible Solid**	Autoignition Temperature **Not Determined**

Extinguishing Media
Use agent suitable to surrounding fire. Dry chemical, carbon dioxide, alcohol foam, or water.

Special Fire Fighting Procedures
Poisonous gases are produced in fire. Wear full protective clothing and self-contained breathing apparatus (SCBA). Move containers from fire area if it can be done without risk. Otherwise, use a water spray to keep fire-exposed containers cool until long after fire has been extinguished. Do NOT release runoff from fire-fighting measures to sewers or waterways.

Unusual Fire and Explosion Hazards
Mercuric oxide is a powerful oxidizer and can greatly accelerate the combustion of ordinary combustible materials. Evacuate nonessential personnel a 1500-foot radius away from fire danger area.

SECTION V - REACTIVITY DATA

Stability		Conditions to Avoid
Stable	Unstable	Mercuric oxide is stable under normal conditions of temperature and pressure. However, it will gradually decompose to metallic mercury when heated or exposed to light. Keep away from incompatible materials.
X		Incompatibility *(materials to avoid)*: Hydrogen peroxide, chlorine + ethylene, molten potassium, molten sodium, sulfur, hydrogen sulfide + barium oxide + air. Forms explosive mixtures with non-metals, alkali metals, and reducing materials.

Hazardous Polymerization		Conditions to Avoid
May Occur	Will Not Occur	Under normal conditions of temperature and pressure, hazardous polymerization of mercuric oxide is not known to occur.
	X	Hazardous Decomposition or By-products: When heated to decomposition, mercuric oxide can emit toxic/poisonous fumes and gases. These include toxic fumes of mercury.

SECTION VI - HEALTH HAZARD DATA

Primary Route(s) of Entry:	Inhalation?	Absorption (skin)?	Ingestion?
	X	X	X

Health Hazards

INHALATION: Causes headaches, shortness of breath, cough, chills, fever, congestion, tightness in chest, possible pneumonitis and pulmonary edema. Prolonged contact can cause CNS effects with vertigo, depression, ataxia, emotional instability, and other effects.

ABSORPTION: Passes through intact skin to cause effects similar to that of inhalation and ingestion. There can also be visual disturbances, eye burns, and dermatitis.

INGESTION: Abdominal pain, diarrhea (possibly bloody), digestive tract irritation, inflammation of the mouth and gums, vomiting with blood, shallow breathing, paleness, collapse, death.

Carcinogenicity	NTP Listed?	IARC Cancer Review Group?	OSHA Regulated?	Target Organs?
Unknown Human Unknown Animal	No	No	29 CFR 1910.1000 Table Z-1 (as Hg)	Kidney, liver, gastrointestinal tract, CNS, skin.

Medical Conditions Generally Aggravated by Exposure

Mercury-sensitive people, existing dental and oral problems, kidney disease, respiratory problems, and central nervous system disorders may be aggravated by exposure to mercuric oxide.

Emergency and First-aid Procedures

Eye contact: Flush immediately with water for 15 minutes (minimum); seek medical attention. Skin contact: Remove all contaminated clothing. Immediately wash area with large amounts of soap and water. For inhalation: Remove the person from exposure. Provide respiratory assistance and CPR. Transfer to medical facility. Observe for 48 hours for lung effects. If swallowed: Provide breathing assistance if necessary. Never attempt to give a convulsing or unconscious person anything by mouth. If victim is conscious, give 1-2 glasses of water to dilute and induce vomiting.

SECTION VII - PRECAUTIONS FOR SAFE HANDLING AND USE

Steps to be Taken in Case Material is Released or Spilled

Collect powdered material and deposit in sealed drum. Keep those not involved in cleanup from entering area. Do NOT dry sweep (creates airborne dusts). Use a vacuum equipped with a (HEPA) filter. Damp mop residue. Absorb solution spills in vermiculite or other materials.

Preferred Waste Disposal Method

Add chlorine to oxidize; absorb with activated carbon/sulfur. Return spent resin to the supplier.

Precautions to be Taken in Handling and Storage

Store in tightly closed containers in cool, well-ventilated area. Sources of ignition are prohibited where mercuric oxide is stored. Keep away from light. Do NOT store on wooden floors.

Other Precautions and Warnings

Outside or detached storage facility is recommended. Protect containers from physical damage.

SECTION VIII - CONTROL MEASURES AND PERSONAL PROTECTIVE EQUIPMENT

Respiratory Protection *(specify type)*

Exposure levels are low and difficult to monitor. For best protection, use a supplied-air respirator with a full facepiece or a self-contained breathing apparatus (SCBA) operated in positive pressure mode.

Ventilation

Local exhaust or general mechanical systems recommended.

Protective Gloves	Eye Protection	Other Protective Clothing
Impervious Material	Chemical/Dust Goggles or Face Mask	Protective Suit or Clothing

Work/Hygiene Practices

Always wash hands thoroughly after using chemical; never bring food, drink, or smoking materials into vicinity of chemicals. Skin contact must be avoided.

MERCURIC OXIDE

Hg_2O CAS: 15829-53-5

IDENTIFICATION AND TYPICAL USES

Mercuric oxide appears as a black crystalline, sugar or sand-like powder with no detectable odor. It is used in paint and ceramic pigment, as an analytical reagent, in batteries, as a fixative in selenium determination, as a chemical intermediate, a preservative in cosmetics, and an antiseptic in pharmaceuticals.

RISK ASSESSMENT: HEALTH

General Assessment

Mercuric oxide is a toxic human poison by three routes of exposure: *inhalation, ingestion,* and dermal *absorption*. Mercury accumulates in the body with each exposure. While it is not currently listed as a carcinogenic compound, mutation data have been reported for many mercury compounds. Organic mercury compounds (those that contain carbon) have caused adverse reproductive effects (miscarriages and/or fetal damage) in test animals.

Inhalation of mercury oxide vapors can cause immediate irritation to the eyes, nose, and mucosa of the respiratory tract. It may cause neurologic disturbances, weight loss, and possible low sperm count. Other symptoms may include headache, dypsnea, cough, chills, and fever. Ingestion of mercury or its compounds causes toxicity that generally begins with burning mouth pain, feeling of tightness or constriction, and an ashen-gray coloration of the mucosa. Symptoms progress to abdominal pain with bloody vomiting (due to digestive hemorrhage), intense thirst, bloody diarrhea, metallic taste the mouth, excessive salivation, rapid and weak pulse, difficulty breathing, pale complexion, collapse, and death.

Repeated exposure from inhalation, absorption, and/or ingestion can cause central nervous system (CNS) effects including vertigo, anxiety, depression, ataxia, emotional instability, kidney injury, brain damage, tremors (the "shakes"), memory loss, discoloration of the skin, sore gums and ulcerative gingivitis, personality changes, liver damage, loose teeth, decreased urine excretion, visual disturbances, and possible dermatitis.

Acute Health Effects

The following acute (short-term) health effects may occur immediately or shortly after exposure to mercuric oxide:

Skin: Irritation, rash, dermatitis, redness and the appearance of small blisters. Dermal absorption can be quick, causing toxic systemic effects.

Eye: Irritation, inflammation, watering, and swelling, burns, and tissue damage. Ulceration of the conjunctiva and cornea, discoloration of the cornea and lens, tremor of the eyelids, and disturbances in vision.

Lung: Inhalation of mercuric oxide dusts may cause irritation of the mucous membranes in the respiratory tract causing cough, phlegm, and lung tissue irritation.

Chronic Health Effects

The following chronic (long-term) health effects may occur at some time after exposure to mercuric oxide and can last for months or even years:

Cancer Hazards: According to the information presented in the references, mercuric oxide has not been adequately tested for its ability to cause cancer in test animals. However, mutation (causes genetic changes) data have been documented for many mercury compounds and many scientists believe that such chemicals may pose a cancer risk in the long-term. More research is required.

Reproductive Hazards: According to information available in the references, organic mercury compounds have been shown to adversely affect reproduction in humans and animals. They can cause the early termination of pregnancies (miscarriages) and may injure the fetus. Mercury may also reduce fertility in both males and females. While inorganic mercury substances (those without carbon), such as mercuric oxide, have not specifically been shown to be human teratogens, they should still be handled with caution as they may cause reproductive problems in males and females.

Other Chronic Hazards: High or repeated exposure can cause kidney damage or mercury poisoning with symptoms of sore gums, "shakes" (especially noticeable in handwriting), irritability, increased saliva production, memory loss, extreme shyness, weakness, loss of teeth, poor appetite, and metallic taste. Serious personality changes and brain damage can occur, especially if exposure continues. Repeated skin contact may cause gray discoloration and a skin allergy may develop. If this happens, even small future exposures can cause rash. Repeated exposures may also cause reduced peripheral vision and brown staining of the eye. Mercury can accumulate in the body on prolonged exposure and it may take years for the body to dispose of mercury.

Recommended Risk-Reduction Measures

Personnel should avoid direct contact with mercuric oxide. Mercury is an extremely toxic, cumulative poison. If a less toxic chemical cannot be substituted for a hazardous material, then *engineering controls* are the most effective method of reducing exposures.

The best protection is to enclose operations and/or provide local exhaust ventilation at the site of chemical release. While not always operationally feasible, isolating operations can also reduce exposure. Using respiratory protection is less effective than the controls mentioned above, but is still advisable whenever working with or around mercuric oxide. Exposure levels for mercury are extremely low (0.1 mg/m^3) and difficult to monitor accurately. A full facepiece respirator with a high-efficiency particulate air (HEPA) filter equipped with a dust/mist/fume pre-filter may suffice. However, for the best protection use a supplied-air respirator with a full facepiece operated in positive pressure mode or an MSHA/NIOSH-approved self-contained breathing apparatus (SCBA) with full facepiece operated in pressure demand or other continuous flow mode. If a full facepiece is not available, then chemical/dust goggles should be worn to protect the eyes. Whenever a chemical dust hazard exists, a face shield and a protective apron should be worn. To prevent hand and skin exposures, impervious rubber gloves should be used. Glove manufacturers should be contacted and permeation studies obtained *before* final glove selections are made. Skin contact must be prevented. Protective suits and clothing should be used.

Administrative controls should also be in place to minimize the potential for human exposures. These may include written procedures or policies which specify the methods and techniques that will be practiced whenever personnel are to work with mercuric oxide dusts or solutions. All personnel should receive *training* on the use, hazards, protective measures, emergency actions, and other precautions per 29 CFR 1910.1200 (Hazard Communication), prior to the first assignment in an area where mercuric oxide is used or stored. Before first exposure and every six to twelve months thereafter, a complete medical history and examination is recommended, including the following:

☑ Examination of the nervous system (taking note of any problems with person's handwriting).

☑ Routine urinalysis and special urine test for mercury (should be less than 0.02 mg/liter).

☑ Eye examination.

☑ Lung function tests for frequent exposures.

If symptoms develop or overexposure is suspected, the following additional tests are recommended:

☑ Consider nerve conduction tests.

☑ Urinary enzyme and neurological behavior tests.

☑ Serum mercury tests.

Any evaluation should include a careful history of past and present symptoms with a medical examination. Medical tests that look for damage already done are not a substitute for controlling exposure. Also, mixed exposures occur when personnel use creams to whiten or bleach the skin that contain mercury. If so, their use increases exposure risk. Persons who are allergic to mercury may react to Mercurochrome or Merthiolate, which also contain mercury. Prudent risk management requires proper consideration of *all* exposure factors.

Other methods to reduce exposure include:

☑ Where possible, enclose operations and use local exhaust ventilation at the site of chemical release. If local exhaust ventilation or enclosure is not used, respiratory protection should be provided and its use mandatory.

☑ Always ensure that proper protective clothing is worn when using chemical substances and that personnel are trained in its use and care.

- ☑ Wash thoroughly immediately after exposure to mercuric oxide and at the end of the work shift or before eating, drinking, or smoking.
- ☑ Hazard warning information should be posted in the work area. In addition, as part of an on-going education and training program, all information on the health and safety hazards of mercuric oxide should be communicated to all exposed and potentially exposed workers.
- ☑ Eye wash stations and emergency showers should be located in the immediate work area.
- ☑ Work clothing that has been contaminated with mercuric oxide should never be worn home where family members can be exposed as well. Contaminated clothing should be laundered at work or by an authorized service by personnel who have been trained in the hazards associated with exposure to mercuric oxide.
- ☑ The drug n-acetyl penicillamine (NAP) has been used in the past to treat mercury poisoning, with mixed success.
- ☑ Specific engineering controls are recommended for mercury by NIOSH. Refer to NIOSH Criteria Document, "Inorganic Mercury" (Publication Number 73-11024).

RISK ASSESSMENT: ENVIRONMENT

General Assessment

The environment is at risk of exposure during transportation, storage, disposal, or destruction of mercuric oxide. In almost every scenario, the threat of environmental exposure is contingent upon the proper handling of the chemical substance. Accidental spills, large or small, can result in fire, explosion, and possible contamination of the surrounding environmental mediums (water, soil, and air).

Mercuric oxide is incompatible with many common materials. It can react explosively with hydrogen peroxide, chlorine plus ethylene, molten potassium, molten sodium, and sulfur. Reactions with non-metals and alkaline metals can also be vigorous and violent. It is also sensitive to heat and light and will decompose to mercury and oxygen when exposed to either. In a fire, mercuric oxide will emit toxic and poisonous fumes of mercury. These characteristics require special consideration during any emergency situation involving a leak or spill of mercuric oxide.

Elemental mercury is a heavy and relatively inert liquid which is oxidized to inorganic mercury (II) under natural conditions. Mercury (II) may combine with an organic fraction to form methyl mercury. Both mercury (II) and methyl mercury can cause environmental contamination. Mercury (II) may enter the environment in industrial or municipal waste treatment discharges, from previously contaminated sediments, and from the weathering of natural rocks. Bacteria may then convert it to methyl mercury. The concentration of mercury (II) in water may be elevated with acid rain due to the scouring of mercury from the air and increased partitioning from the sediment into the water.

☠ *Acute Ecological Effects*

Acute (short-term) toxic effects may include the death of animals, birds, or fish and death or low growth rate in plants. Acute effects are seen 2 to 4 days after animals or plants are exposed to mercuric oxide.

Mercury (II) and methyl mercury have high acute toxicity to aquatic life. Insufficient data are available to evaluate the short-term effects of mercury (II) and methyl mercury exposure to plants, birds or terrestrial animals.

💣 *Chronic Ecological Effects*

Chronic (long-term) toxic effects may include shortened life span, reproductive problems, lower fertility, and changes in appearance or behavior in exposed animals. These effects can be seen long after first exposure(s) to toxic chemicals.

Mercury (II) and methyl mercury have high chronic toxicity to aquatic life. Eating fish contaminated with mercury residues has caused secondary poisoning in humans. Birds or land animals similarly exposed to mercury and its compounds could also be subject to such effects. Insufficient data are available to evaluate the long-term effects of mercuric oxide exposure to plants.

💧 *Water Solubility*

Mercuric oxide is insoluble in water. Concentrations of less than 1 milligram will not mix with a liter of water.

⌛ *Persistence in the Environment*

Mercury is highly persistent in water, with a half-life greater than 200 days. The half-life of a pollutant is the amount of time it takes for one half of the chemical to be degraded. When released into water, microorganisms convert mercuric salts into methyl mercury which is rapidly taken up by algae and enters the food chain. It concentrates in the edible tissues of fish and, eventually, can cause human poisoning as well.

Bioaccumulation in Aquatic Organisms

Some substances increase in concentration, or *bioaccumulate,* in living organisms as they breathe contaminated air, drink contaminated water, or eat contaminated food. These chemicals can become concentrated in the tissues and internal organs of animals as well as humans.

The concentration of mercury and methyl mercury found in fish tissues is expected to be considerably higher than the average concentration in the water from which the fish was taken.

Recommended Risk-Reduction Measures

Proper training of all transporters will reduce the likelihood of a mishap or accidents resulting in a leak or spill to the environment. Proper labeling on all containers, trucks, and rail cars will enable emergency responders to react properly and quickly to any disaster thereby reducing the potential risk to the environment and to personnel.

If a spill or leak to the environment has occurred, fire department, emergency response and/or hazardous materials spill personnel should be notified immediately. Do not dry sweep. When vacuuming, a high-efficiency particulate air (HEPA) filter should be used, not a standard shop vacuum. Alternatively, commercially available mercury spill cleanup kits can be used. Deposit collected materials in sealed drums for disposal. Cleanup should be attempted only by those trained in proper spill containment procedures. Contaminated soils may need to be removed and replaced with clean soil. If mercuric oxide should contact the water table, aquifer, or navigable waterway, time is of the essence. Mercuric oxide is insoluble in water and total containment and remediation may be possible. When such spills occur, the local and/or state emergency response authorities should be notified. A comprehensive emergency response or disaster preparedness/recovery plan should be in place prior to any operations involving the use, transportation, storage, or disposal of mercuric oxide.

If mercuric oxide is spilled or leaked, the following specific steps are recommended:

- ☑ Restrict persons not wearing protective clothing from area of spill or leak until cleanup is complete and area can be opened for normal work.
- ☑ Ventilate area and remove all ignition sources.
- ☑ It may be necessary to dispose of mercuric oxide as a hazardous waste. The responsible state agency or the regional office of the federal Environmental Protection Agency (EPA) should be contacted for specific recommendations.

RISK ASSESSMENT: BUSINESS

General Assessment

Accidents or mishaps involving mercuric oxide can present a significant threat to business operations. The loss or damage of equipment or facilities can significantly affect fiscal viability. Lawsuits that may result from personnel injury/death, public exposures, and/or environmental contamination will also require a serious expenditure of resources. Media attention surrounding an injury, death, or environmental damage can also result in a loss of profits and loss of current as well as future business. Always remember that anytime the term "reproductive hazard" is used, public emotion, hysteria, and ignorance can run equally high. This must be properly considered whenever developing public relations policies.

Recommended Risk-Reduction Measures

Company attorneys, safety and health professionals, and environmental specialists should be involved in the development of such procedures dealing with safety or emergency response. A company official should be pre-designated as a public relations officer with specific training in dealing with the press. Corporate plans and policies should be developed, approved, and implemented long before any need for such arises.

REFERENCES

American Conference of Governmental Industrial Hygienists. 1988. *Documentation of the Threshold Limit Values and Biological Exposure Indices,* 5th Edition (with updates). Cincinnati: ACGIH

Environmental Protection Agency. 1983. *Chemical Hazard Information Profiles.* Washington, D.C.: U.S. EPA.

U.S. Department of Health and Human Services, National Institute for Occupational Safety and Health. 1973. NIOSH Criteria Document. *Inorganic Mercury.* Publication Number 73-11024. Washington, D.C.: U.S. Government Printing Office.

MATERIAL SAFETY DATA SHEET

CHEMICAL NAME	
MERCURIC SULFATE	**CRC PRESS/LEWIS PUBLISHERS DISCLAIMER:** THE INFORMATION AND RECOMMENDATIONS PRESENTED HEREIN ARE BASED ON SOURCES BELIEVED TO BE RELIABLE. CRC MAKES NO REPRESENTATION ON ITS COMPLETENESS OR ACCURACY. IT IS THE RESPONSIBILITY OF THE USER TO DETERMINE THE CHEMICAL'S SUITABILITY FOR ITS INTENDED USE, THE CHEMICAL'S SAFE USE, AND THE CHEMICAL'S PROPER DISPOSAL. NO REPRESENTATIONS AND/OR WARRANTIES, EITHER EXPRESSED OR IMPLIED, OF THE MERCHANTABILITY OR FITNESS FOR A PARTICULAR PURPOSE, OR OF ANY OTHER NATURE, ARE MADE WITH RESPECT TO THE INFORMATION PROVIDED IN THIS MSDS OR TO THE CHEMICAL TO WHICH INFORMATION MAY REFER. CRC NEITHER ASSUMES NOR AUTHORIZES ANY OTHER PERSON TO ASSUME FOR IT, ANY OTHER ADDITIONAL RESPONSIBILITY OR LIABILITY FOR THE USE OF, OR RELIANCE UPON, THIS INFORMATION.

HAZARD WARNING INFORMATION

HEALTH	FIRE	REACTIVE	OTHER	DEGREE OF HAZARD	COLOR CODING	OTHER CODES
2	0	1		0 = Minimum Hazard 1 = Slight Hazard 2 = Moderate Hazard 3 = Serious Hazard 4 = Severe Hazard	HEALTH = BLUE FIRE = RED REACTIVITY = YELLOW OTHER = WHITE	OX = Oxidizer ACID = Acid ALK = Alkali COR = Corrosive W = Use No Water

SECTION I - GENERAL INFORMATION

Characterization	**Mercury Compound**	RCRA Number	**None**
		EPA Class	**Not Applicable**
DOT Proper Shipping Name	**Mercuric Sulfate, Solid**	Chemical Abstract Service (CAS) Number	**7783-35-9**
DOT Hazard Class and Label Requirements	**Poison B; Poison**	DOT Emergency Guide Code	**60 (for Hg compounds)**
DOT Identification Number	**UN1645**	Molecular Formula	**$HgSO_4$**

Synonyms

Mercury persulfate; mercury sulfate; mercury bisulfate; sulfuric acid, mercury (+2) salt.

SECTION II - HAZARDOUS INGREDIENTS AND IDENTITY INFORMATION

Hazardous Components (specific identity)	OSHA Exposure Criteria	NIOSH Exposure Criteria	Immediately Dangerous to Life and Health (IDLH)	ACGIH Exposure Criteria
Mercuric sulfate (derivation: By reaction of mercuric acid on mercury with subsequent crystallization).	**PEL:** 0.1 mg/m^3 (as Hg vapor) **CEILING (skin):** 0.1 mg/m^3 (Hg compounds)	**REL (skin):** 0.05 mg/m^3 (as Hg vapor) **STEL:** Not Established	28 mg/m^3 (as Hg)	**TLV (skin):** 0.1 mg/m^3 (Hg compounds) **STEL:** Not Established

SECTION III - PHYSICAL AND CHEMICAL CHARACTERISTICS

Boiling Point	**Decomposes at red heat**	Specific Gravity (H_2O = 1)	**6.47**
Vapor Pressure (mm Hg)	**Not Found**	Molecular Weight	**296.68**
Vapor Density (Air = 1)	**10.0**	Melting Point	**Decomposes**

Solubility

Decomposes in water. Soluble in acids, insoluble in alcohol.

Appearance and Odor

Colorless or white granules or crystalline powder (sugar or sand-like) with no odor.

SECTION IV - FIRE AND EXPLOSION HAZARD DATA

Flash Point (method used)	**Not Determined**	Explosive Limits in Air % by Volume	**LEL: Not Applicable UEL: Not Applicable**
NFPA Classification	**Combustible Solid**	Autoignition Temperature	**Not Determined**

Extinguishing Media

Mercuric sulfate burns, but does not easily ignite. Use dry chemical carbon dioxide, water spray or regular foam for small fires. For larger fires, try water spray, fog, or regular foam.

Special Fire Fighting Procedures

Poisonous gases are produced in fire. Wear full protective clothing and self-contained breathing apparatus (SCBA). Move containers from fire area if it can be done without risk. Otherwise, use a water spray to keep fire-exposed containers cool. Do NOT release runoff from fire-fighting measures to sewers or waterways.

Unusual Fire and Explosion Hazards

None reported.

SECTION V - REACTIVITY DATA

Stability		Conditions to Avoid: Mercuric sulfate is stable in dark, dry, closed containers under normal conditions of temperature and pressure. Avoid contact with incompatible materials, including water. Keep away from heat and flame.
Stable	Unstable	Incompatibility *(materials to avoid)*: Decomposes in water, violently absorbs gaseous hydrogen chloride at high temperatures, aluminum foil in the presence of moisture, sodium carbonate; is corrosive to iron, zinc, copper, lead, and other metals.
X		
Hazardous Polymerization		Conditions to Avoid: Under normal conditions of temperature and pressure, hazardous polymerization of mercuric sulfate is not known to occur.
May Occur	Will Not Occur	Hazardous Decomposition or By-products: When heated to decomposition, mercuric sulfate can emit toxic/poisonous fumes and gases. These include toxic fumes of mercury and sulfur oxide.
	X	

SECTION VI - HEALTH HAZARD DATA

Primary Route(s) of Entry:	Inhalation?	Absorption (skin)?	Ingestion?
	X	X	X

Health Hazards

INHALATION: Causes immediate irritation to the nose, throat, and mucosa of the respiratory system. Once absorbed into the bloodstream, kidney damage can occur. Prolonged contact can cause CNS effects with vertigo, depression, ataxia, and emotional instability.

ABSORPTION: Passes through intact skin to cause effects similar to that of inhalation and ingestion. There can also be visual disturbances, eye burns, and dermatitis.

INGESTION: Abdominal pain, diarrhea (possibly bloody), digestive tract irritation, inflammation of the mouth and gums, esophageal burns, liver damage, loose teeth, reduced urine excretion.

Carcinogenicity	NTP Listed?	IARC Cancer Review Group?	OSHA Regulated?	Target Organs?
Unknown Human Unknown Animal	No	No	29 CFR 1910.1000 Table Z-1 (as Hg)	Kidney, liver, gastrointestinal tract, CNS.

Medical Conditions Generally Aggravated by Exposure

Mercury-sensitive people, existing dental and oral problems, kidney disease, respiratory problems, and central nervous system disorders may be aggravated by exposure to mercuric sulfate.

Emergency and First-aid Procedures

Eye contact: Flush immediately with water for 15 minutes (minimum); seek medical attention. Skin contact: Remove all contaminated clothing. Immediately wash area with large amounts of soap and water. For inhalation: Remove the person from exposure. Provide respiratory assistance and CPR. Transfer to medical facility. Observe 48 for hours for lung effects. If swallowed: Provide breathing assistance if necessary. Never attempt to give a convulsing or unconscious person anything by mouth. If victim is conscious, give 1-2 glasses of water to dilute and induce vomiting.

SECTION VII - PRECAUTIONS FOR SAFE HANDLING AND USE

Steps to be Taken in Case Material is Released or Spilled

Collect powdered material and deposit in sealed drum. Keep those not involved in cleanup from entering area. Do NOT dry sweep (creates airborne dusts). Use a vacuum equipped with a (HEPA) filter or mercury spill cleanup kit. Wash area with dilute calcium sulfide and collect for disposal.

Preferred Waste Disposal Method

No citation.

Precautions to be Taken in Handling and Storage

Store in tightly closed containers in cool, dark, dry, well-ventilated area. Sources of ignition are prohibited where mercuric sulfate is stored. Do NOT store on wood or porous floors.

Other Precautions and Warnings

Outside or detached storage facility is recommended. Protect containers from physical damage.

SECTION VIII - CONTROL MEASURES AND PERSONAL PROTECTIVE EQUIPMENT

Respiratory Protection *(specify type)*

Exposure levels are low and difficult to monitor. For best protection, use a supplied-air respirator with a full facepiece or a self-contained breathing apparatus (SCBA) operated in positive pressure mode.

Ventilation

Local exhaust or general mechanical systems recommended.

Protective Gloves	Eye Protection	Other Protective Clothing
Impervious Material	Chemical/Dust Goggles or Face Mask	Protective Suit or Clothing

Work/Hygiene Practices

Always wash hands thoroughly after using chemical; never bring food, drink, or smoking materials into vicinity of chemicals. Skin contact must be avoided.

MERCURIC SULFATE

$HgSO_4$ CAS: 7783-35-9

IDENTIFICATION AND TYPICAL USES

Mercuric sulfate appears as a white crystalline, sugar or sand-like solid with no detectable odor. It is used as a calomel and corrosive sublimate, as a catalyst in the conversion of acetylene to acetaldehyde, for extracting gold and silver from roasted pyrites, and as a battery electrolyte.

RISK ASSESSMENT: HEALTH

General Assessment

Mercuric sulfate is a toxic human poison by three routes of exposure: *inhalation, ingestion,* and dermal *absorption*. Mercury accumulates in the body with each exposure. While it is not currently listed as a carcinogenic compound, mutation data have been reported for many mercury compounds. Organic mercury compounds (those that contain carbon) have caused adverse reproductive effects (miscarriages and/or fetal damage) in test animals.

Symptoms of exposure by inhalation of mercury sulfate vapors can include immediate irritation of the mouth and throat, headache, dypsnea, cough, chills, fever, and bronchial irritation.

Ingestion of mercury or its compounds causes abdominal pain, bloody diarrhea, gastrointestinal irritation, vomiting, and cardiac arrhythmia. There will be almost immediate necrosis in the mouth, throat, esophagus, and stomach. Up to 72 hours after ingestion, there can still be inflammation of the mouth tissues and kidney inflammation or disease. Death can occur from total renal failure.

Repeated exposure from inhalation, absorption, and/or ingestion can cause central nervous system (CNS) effects including vertigo, anxiety, depression, ataxia, emotional instability, kidney injury, brain damage, tremors (the "shakes"), memory loss, metallic taste, discoloration of the skin, sore gums, personality changes, liver damage, loose teeth, necrosis of the jaw, decreased urine excretion, visual disturbances, and possible development of dermatitis.

Acute Health Effects

The following acute (short-term) health effects may occur immediately or shortly after exposure to mercuric sulfate:

Skin: Corrosion to the surface, irritation, rash, dermatitis, and possible redness and the appearance of small blisters. Dermal absorption can be quick and will cause toxic systemic effects.

Eye: Irritation, inflammation, swelling, conjunctiva and corneal ulceration, burns, and tissue damage.

Lung: Inhalation of mercuric sulfate dusts may cause irritation of the mucous membranes in the respiratory tract causing cough, phlegm, and lung tissue irritation.

Chronic Health Effects

The following chronic (long-term) health effects may occur at some time after exposure to mercuric sulfate and can last for months or even years:

Cancer Hazards: According to the information presented in the references, mercuric sulfate has not been adequately tested for its ability to cause cancer in test animals. However, mutation (causes genetic changes) data have been documented and many scientists believe that such chemicals may pose a cancer risk in the long-term. More research is required.

Reproductive Hazards: According to information available in the references, mercuric sulfate has not been specifically tested for its ability to cause cancer in test animals. However, some organic mercury compounds have been shown to adversely affect reproduction in humans and animals. They can cause the early termination of pregnancies (miscarriages) and may injure the fetus. Mercury may also reduce fertility in both males and females. While inorganic mercury substances (those without carbon), have not specifically been shown to be human teratogens, they should still be handled with caution as they may cause reproductive problems in males and females.

Other Chronic Hazards: High or repeated exposure can cause kidney damage or mercury poisoning with symptoms of sore gums, "shakes" (especially noticeable in handwriting), irritability, increased saliva production, memory loss, extreme shyness, weakness, loss of teeth, poor appetite, and metallic taste. Serious personality changes and brain damage can occur, especially if exposure continues. Repeated skin contact may cause gray discoloration and a skin allergy may develop. If this happens, even small future exposures can cause rash. Repeated exposures may also cause reduced peripheral vision and brown staining of the eye. Mercury can accumulate in the body on prolonged exposure and it may take months or years for the body to dispose of excess mercury.

Recommended Risk-Reduction Measures

Personnel should avoid direct contact with mercuric sulfate. Mercury is an extremely toxic, cumulative poison. If a less toxic chemical cannot be substituted for a hazardous material, then *engineering controls* are the most effective method of reducing exposures.

The best protection is to enclose operations and/or provide local exhaust ventilation at the site of chemical release. While not always operationally feasible, isolating operations can also reduce exposure. Using respiratory protection is less effective than the controls mentioned above, but is still advisable whenever working with or around mercuric sulfate. Exposure levels for mercury are extremely low (0.1 mg/m^3) and difficult to monitor accurately. A full facepiece respirator with a high-efficiency particulate air (HEPA) filter equipped with a dust/mist/fume pre-filter may suffice. However, for the best protection use a supplied-air respirator with a full facepiece operated in positive pressure mode or an MSHA/NIOSH-approved self-contained breathing apparatus (SCBA) with full facepiece operated in pressure demand or other continuous flow mode. If a full facepiece is not available, then chemical/dust goggles should be worn to protect the eyes. Whenever a chemical dust hazard exists, a face shield and a protective apron should be worn. To prevent hand and skin exposures, impervious rubber gloves should be used. Glove manufacturers should be contacted and permeation studies obtained *before* final glove selections are made. Mercury salts such as mercuric sulfate are among mercury's most irritating and acutely toxic compounds. They are corrosive to skin tissues. Skin contact must be prevented.

Administrative controls should also be in place to minimize the potential for human exposures. These may include written procedures or policies which specify the methods and techniques that will be practiced whenever personnel are to work with mercuric sulfate dusts or solutions. All personnel should receive *training* on the use, hazards, protective measures, emergency actions, and other precautions per 29 CFR 1910.1200 (Hazard Communication), prior to the first assignment in an area where mercuric sulfate is used or stored. Before first exposure and every six to twelve months thereafter, a complete medical history and examination is recommended, including the following:

- ☑ Examination of the nervous system (taking note of any problems with person's handwriting).
- ☑ Routine urinalysis and special urine test for mercury (should be less than 0.02 mg/liter).
- ☑ Eye examination.
- ☑ Lung function tests for frequent exposures.

If symptoms develop or overexposure is suspected, the above listed tests should be repeated, along with the following recommended tests:

- ☑ Consider nerve conduction tests.
- ☑ Urinary enzyme and neurological behavior tests.
- ☑ Serum mercury tests.

Any evaluation should include a careful history of past and present symptoms with a medical examination. Medical tests that look for damage already done are not a substitute for controlling exposure. Also, mixed exposures occur when personnel use creams to whiten or bleach the skin that contain mercury. If so, their use increases exposure risk. Persons who are allergic to mercury may react to Mercurochrome or Merthiolate, which also contain mercury. Prudent risk management requires proper consideration of *all* exposure factors.

Other methods to reduce exposure include:

- ☑ Where possible, enclose operations and use local exhaust ventilation at the site of chemical release. If local exhaust ventilation or enclosure is not used, respiratory protection should be provided and its use mandatory.

☑ Always ensure that proper protective clothing is worn when using chemical substances and that personnel are trained in its use and care.

☑ Wash thoroughly immediately after exposure to mercuric sulfate and at the end of the work shift or before eating, drinking, or smoking.

☑ Hazard warning information should be posted in the work area. In addition, as part of an on-going education and training program, all information on the health and safety hazards of mercuric sulfate should be communicated to all exposed and potentially exposed workers.

☑ Eye wash stations and emergency showers should be located in the immediate work area.

☑ Work clothing that has been contaminated with mercuric sulfate should never be worn home where family members can be exposed as well. Contaminated clothing should be laundered at work or by an authorized service by personnel who have been trained in the hazards associated with exposure to mercuric sulfate.

☑ The drug n-acetyl penicillamine (NAP) has been used to treat mercury poisoning, with mixed success.

☑ Specific engineering controls are recommended for mercury by NIOSH. Refer to NIOSH Criteria Document, "Inorganic Mercury" (Publication Number 73-11024).

RISK ASSESSMENT: ENVIRONMENT

General Assessment

The environment is at risk of exposure during transportation, storage, disposal, or destruction of mercuric sulfate. In almost every scenario, the threat of environmental exposure is contingent upon the proper handling of the chemical substance. Accidental spills, large or small, can result in fire, explosion, and possible contamination of the surrounding environmental mediums (water, soil, and air).

Mercuric sulfate is incompatible with many common materials. Reactions can be vigorous and violent. It decomposes in water into a yellow, insoluble, basic sulfate and free sulfuric acid. It will violently absorb gaseous hydrogen chlorides at high temperatures. It is corrosive to many metals, including aluminum, zinc, copper, and lead. In a fire, mercuric sulfate will emit toxic and poisonous fumes of mercury and sulfur oxide. These characteristics require special consideration during any emergency situation involving a leak or spill of mercuric sulfate.

Elemental mercury is a heavy and relatively inert liquid which is oxidized to inorganic mercury (II) under natural conditions. Mercury (II) may combine with an organic fraction to form methyl mercury. Both mercury (II) and methyl mercury can cause environmental contamination. Mercury (II) may enter the environment in industrial or municipal waste treatment discharges, from previously contaminated sediments, and from the weathering of natural rocks. Bacteria may then convert it to methyl mercury. The concentration of mercury (II) in water may be elevated with acid rain due to the scouring of mercury from the air and increased partitioning from the sediment into the water.

☠ *Acute Ecological Effects*

Acute (short-term) toxic effects may include the death of animals, birds, or fish and death or low growth rate in plants. Acute effects are seen 2 to 4 days after animals or plants are exposed to mercuric sulfate.

Mercury (II) and methyl mercury have high acute toxicity to aquatic life. Insufficient data are available to evaluate the short-term effects of mercury (II) and methyl mercury exposure to plants, birds or terrestrial animals.

Chronic Ecological Effects

Chronic (long-term) toxic effects may include shortened life span, reproductive problems, lower fertility, and changes in appearance or behavior in exposed animals. These effects can be seen long after first exposure(s) to toxic chemicals.

Mercury (II) and methyl mercury have high chronic toxicity to aquatic life. Eating fish contaminated with mercury residues has caused secondary poisoning in humans. Birds or land animals exposed to mercury and its compounds could also be subject to such effects. Insufficient data are available to evaluate the long-term effects of mercuric sulfate exposure to plants.

Water Solubility

Mercuric sulfate is soluble in water. Concentrations of 1000 milligrams or more will mix with a liter of water. Mercury is insoluble in the aquatic environment.

⌛ Persistence in the Environment

Mercury is highly persistent in water, with a half-life greater than 200 days. The half-life of a pollutant is the amount of time it takes for one half of the chemical to be degraded. When released into water, microorganisms convert mercuric salts into methyl mercury which is rapidly taken up by algae and enters the food chain. It concentrates in the edible tissues of fish and, eventually, can cause human poisoning as well.

Bioaccumulation in Aquatic Organisms

Some substances increase in concentration, or *bioaccumulate,* in living organisms as they breathe contaminated air, drink contaminated water, or eat contaminated food. These chemicals can become concentrated in the tissues and internal organs of animals as well as humans. The concentration of mercury and methyl mercury found in fish tissues is expected to be considerably higher than the average concentration in the water from which the fish was taken.

Recommended Risk-Reduction Measures

Proper training of all transporters will reduce the likelihood of a mishap or accidents resulting in a leak or spill to the environment. Proper labeling on all containers, trucks, and rail cars will enable emergency responders to react properly and quickly to any disaster thereby reducing the potential risk to the environment and to personnel.

If a spill or leak to the environment has occurred, fire department, emergency response and/or hazardous materials spill personnel should be notified immediately. Do not dry sweep. When vacuuming, a high efficiency particulate air (HEPA) filter should be used, not a standard shop vacuum. Otherwise, commercially available mercury spill cleanup kits should be considered. Deposit collected materials in sealed drums for disposal. Cleanup should be attempted only by those trained in proper spill containment procedures. Contaminated soils may need to be removed and replaced with clean soil. If mercuric sulfate should contact the water table, aquifer, or navigable waterway, response activities should be prompt. When such spills occur, the local and/or state emergency response authorities should be notified. A comprehensive emergency response or disaster preparedness/recovery plan should be in place prior to any operations involving the use, transportation, storage, or disposal of mercuric sulfate.

If mercuric sulfate is spilled or leaked, the following additional steps are recommended:

- ☑ Ventilate area and remove all ignition sources.
- ☑ It may be necessary to dispose of mercuric sulfate as a hazardous waste. The responsible state agency or the regional office of the federal Environmental Protection Agency (EPA) should be contacted for specific recommendations.

RISK ASSESSMENT: BUSINESS

General Assessment

Accidents or mishaps involving mercuric sulfate can present a significant threat to business operations. The loss or damage of equipment or facilities can significantly affect fiscal viability. Lawsuits that may result from personnel injury/death, public exposures, and/or environmental contamination will also require a serious expenditure of resources. Media attention surrounding an injury, death, or environmental damage can also result in a loss of profits and loss of current as well as future business. Always remember that anytime the term "reproductive hazard" is used, public emotion, hysteria, and ignorance can run equally high. This must be properly considered whenever developing public relations policies.

Recommended Risk-Reduction Measures

Company attorneys, safety and health professionals, and environmental specialists should be involved in the development of such procedures dealing with safety or emergency response. A company official should be pre-designated as a public relations officer with specific training in dealing with the press. Corporate plans and policies should be developed, approved, and implemented long before any need for such arises.

REFERENCES

American Conference of Governmental Industrial Hygienists. 1988. *Documentation of the Threshold Limit Values and Biological Exposure Indices,* 5th Edition (with updates). Cincinnati: ACGIH

Environmental Protection Agency. 1983. *Chemical Hazard Information Profiles.* Washington, D.C.: U.S. EPA.

U.S. Department of Health and Human Services, National Institute for Occupational Safety and Health. 1973. NIOSH Criteria Document. *Inorganic Mercury.* Publication Number 73-11024. Washington, D.C.: U.S. Government Printing Office.

MATERIAL SAFETY DATA SHEET

CHEMICAL NAME	
MERCURY	**CRC PRESS/LEWIS PUBLISHERS DISCLAIMER:** THE INFORMATION AND RECOMMENDATIONS PRESENTED HEREIN ARE BASED ON SOURCES BELIEVED TO BE RELIABLE. CRC MAKES NO REPRESENTATION ON ITS COMPLETENESS OR ACCURACY. IT IS THE RESPONSIBILITY OF THE USER TO DETERMINE THE CHEMICAL'S SUITABILITY FOR ITS INTENDED USE, THE CHEMICAL'S SAFE USE, AND THE CHEMICAL'S PROPER DISPOSAL. NO REPRESENTATIONS AND/OR WARRANTIES, EITHER EXPRESSED OR IMPLIED, OF THE MERCHANTABILITY OR FITNESS FOR A PARTICULAR PURPOSE, OR OF ANY OTHER NATURE, ARE MADE WITH RESPECT TO THE INFORMATION PROVIDED IN THIS MSDS OR TO THE CHEMICAL TO WHICH INFORMATION MAY REFER. CRC NEITHER ASSUMES NOR AUTHORIZES ANY OTHER PERSON TO ASSUME FOR IT, ANY OTHER ADDITIONAL RESPONSIBILITY OR LIABILITY FOR THE USE OF, OR RELIANCE UPON, THIS INFORMATION.

HAZARD WARNING INFORMATION

HEALTH	FIRE	REACTIVE	OTHER	DEGREE OF HAZARD	COLOR CODING	OTHER CODES
4	0	0		0 = Minimum Hazard 1 = Slight Hazard 2 = Moderate Hazard 3 = Serious Hazard 4 = Severe Hazard	HEALTH = BLUE FIRE = RED REACTIVITY = YELLOW OTHER = WHITE	OX = Oxidizer ACID = Acid ALK = Alkali COR = Corrosive W = Use No Water

SECTION I - GENERAL INFORMATION

Characterization	RCRA Number	EPA Class
Mercury Compound	**U151**	**Toxic Waste**

DOT Proper Shipping Name	Chemical Abstract Service (CAS) Number
Mercury, Metallic	**7439-97-6**
DOT Hazard Class and Label Requirements	**DOT Emergency Guide Code**
Corrosive Material	**60**
DOT Identification Number	**Atomic Formula**
UN2809	**Hg**

Synonyms

Mercury metal; collodial mercury; elemental mercury; metallic mercury; quicksilver; hydrargyrum.

SECTION II - HAZARDOUS INGREDIENTS AND IDENTITY INFORMATION

Hazardous Components (specific identity)	OSHA Exposure Criteria	NIOSH Exposure Criteria	Immediately Dangerous to Life and Health (IDLH)	ACGIH Exposure Criteria
Mercury (derivation: By heating cinnabar in air or with lime and condensing the liquid. Also occurs naturally in Spain, Yugoslavia, Mexico, Canada, and Algeria.).	**PEL (ceiling):** 0.1 mg/m^3 (skin) (as vapor) **CEILING (skin):** Not Applicable	**REL (skin):** 0.05 mg/m^3 (as vapor) 0.1 mg/m^3 (as compounds) **STEL:** Not Established	**10 mg/m^3 (as Hg)**	**TLV (skin):** 0.025 mg/m^3 (compounds) **STEL:** Not Established

SECTION III - PHYSICAL AND CHEMICAL CHARACTERISTICS

Boiling Point	Specific Gravity (H_2O = 1)
674°F (357°C)	**13.534**
Vapor Pressure (mm Hg)	Atomic Weight
0.002 at 77°F (25°C)	**200.59**
Vapor Density (Air = 1)	Melting Point
7.0	**-38°F (-39°C)**

Solubility

Insoluble in water, alcohol, hydrochloric acid, and ether. Soluble in lipids, in sulfuric acid (on boiling), readily soluble in nitric acid.

Appearance and Odor

Silver-colored, extremely heavy, odorless liquid metal.

SECTION IV - FIRE AND EXPLOSION HAZARD DATA

Flash Point (method used)	Explosive Limits in Air % by Volume
Not Determined	**LEL: Not Applicable UEL: Not Applicable**
NFPA Classification	Autoignition Temperature
Combustible Solid	**Not Determined**

Extinguishing Media

Use agent suitable to surrounding fires.

Special Fire Fighting Procedures

Poisonous gases are produced in fire. Wear full protective clothing and self-contained breathing apparatus (SCBA). Move containers from fire area if it can be done without risk. Otherwise, use a water spray to keep fire-exposed containers cool. Do NOT release runoff from fire-fighting measures to sewers or waterways.

Unusual Fire and Explosion Hazards

None reported.

SECTION V - REACTIVITY DATA

Stability		Conditions to Avoid **Mercury is normally stable at room temperature and pressure. When heated to near its boiling point, it will slowly oxidize to mercuric oxide.**
Stable **X**	Unstable	Incompatibility *(materials to avoid)* **Forms amalgamate alloys with most metals, except iron. Also incompatible with oxidizers (bromine, 3-bromopropyne, methylsilane + oxygen, chlorine, chlorine dioxide), nitric acid, other acids.**
Hazardous Polymerization		Conditions to Avoid **Under normal conditions of temperature and pressure, hazardous polymerization of mercury is not known to occur.**
May Occur	Will Not Occur **X**	Hazardous Decomposition or By-products **When heated to decomposition, mercury can emit toxic/poisonous fumes and gases. These include toxic fumes of mercuric oxide.**

SECTION VI - HEALTH HAZARD DATA

Primary Route(s) of Entry:	Inhalation?	Absorption (skin)?	Ingestion?
	X	X	

Health Hazards

INHALATION: Causes immediate and severe respiratory damage. Also affects the central nervous with symptoms of sleeplessness, muscle weakness, anorexia, headache, tinnitus, headache, diarrhea, liver changes, fever, gum and mouth inflammation, chest pain, dypsnea, cough and phlegm, salivation, bronchitis, pulmonary edema, and pneumonitis.

ABSORPTION: Passes through intact skin to cause effects similar to that of inhalation and ingestion. There can also be visual disturbances, eye burns, and dermatitis. Skin irritation can be severe if mercury remains in contact for long periods.

INGESTION: Generally innocuous on ingestion (unless large amounts are ingested).

Carcinogenicity	NTP Listed?	IARC Cancer Review Group?	OSHA Regulated?	Target Organs?
Unknown Human Unknown Animal	No	No	29 CFR 1910.1000 Table Z-1 (as Hg)	Skin, eyes, kidney, liver, GI tract, CNS, resp. sys.

Medical Conditions Generally Aggravated by Exposure

Central nervous system disorders, and existing mercury allergies may be aggravated by exposure.

Emergency and First-aid Procedures

Eye contact: Flush immediately with water for 15 minutes (minimum); seek medical attention. Skin contact: Remove all contaminated clothing. Immediately wash area with large amounts of soap and water. For reddened or blistered skin, seek medical assistance. For inhalation: Remove the person from exposure. Provide respiratory assistance and CPR. Transfer to medical facility. Observe for 48 hours for lung effects. If swallowed: Contact poison control center for advice. Mercury will generally pass through the digestive system without incident. However, large doses may cause some toxic effects.

SECTION VII - PRECAUTIONS FOR SAFE HANDLING AND USE

Steps to be Taken in Case Material is Released or Spilled

Collect using commercially available mercury spill kit. Use kit's aspirator vacuum trap with a copper or copper-plated brush to sweep up excess mercury towards vacuum trap device. Wash spill area with dilute calcium sulfide or nitric acid solution.

Preferred Waste Disposal Method

Contact manufacturer or supplier for proper disposal methods.

Precautions to be Taken in Handling and Storage

Store in tightly closed containers in cool, dark, dry, well-ventilated area. Sources of ignition are prohibited where mercury is stored. Do NOT store on wood, porous floors, or unpainted metal shelves.

Other Precautions and Warnings

Outside or detached storage facility is recommended. Protect containers from physical damage.

SECTION VIII - CONTROL MEASURES AND PERSONAL PROTECTIVE EQUIPMENT

Respiratory Protection *(specify type)*

Exposure levels are low. Use a respirator equipped with a fume filter. For best protection, use a supplied-air respirator with a full facepiece or an SCBA operated in positive pressure mode.

Ventilation

Local exhaust or general mechanical systems recommended.

Protective Gloves	Eye Protection	Other Protective Clothing
Impervious Material	Chemical Goggles or Face Mask	Protective Suit or Clothing

Work/Hygiene Practices

Always wash hands thoroughly after using chemical; never bring food, drink, or smoking materials into vicinity of chemicals. Skin contact must be avoided.

MERCURY

Hg

CAS: 7439-97-6

IDENTIFICATION AND TYPICAL USES

Mercury appears as a silver-colored, extremely heavy, odorless liquid metal. It is used in amalgams, as a catalyst, in electrical equipment, in cathodes for production of chlorine and caustic soda, in many different types of instruments (thermometers, barometers, thermostats), in mercury vapor lamps, in extractive metallurgy, in mirror coatings, arc lamps, boilers, and as a coolant and neutron absorber in nuclear power plants.

RISK ASSESSMENT: HEALTH

General Assessment

Mercury is a toxic human poison by *inhalation* and dermal *absorption*. Mercury accumulates in the body with each exposure. While it is not currently listed as a carcinogen, human mutation data have been reported for mercury and its ability to cause cancer has been questioned. It is also an experimental teratogen and may cause adverse reproductive effects.

Symptoms of exposure by inhalation of mercury vapors can include sleeplessness, muscle weakness, anorexia, headaches, tinnitus, hypermotility, diarrhea, liver changes, dermatitis, and fever. Exposure to high levels can irritate the lungs, causing cough, tightness in chest, with possible progression to bronchitis, pneumonitis, and/or pulmonary edema (fluid in the lungs). Symptoms of pulmonary edema can be delayed up to 48 hours or more, thereby creating a false sense of security with regard to health exposure risk.

Skin contact can cause severe irritation if allowed to remain on the skin's surface. Skin absorption will occur and result in adverse nervous system effects.

Repeated exposure from inhalation, absorption, or both can cause central nervous system (CNS) effects including vertigo, anxiety, depression, ataxia, emotional instability, kidney injury, brain damage, tremors (the "shakes"), memory loss, metallic taste, discoloration of the skin, sore gums, personality changes, liver damage, loose teeth, necrosis of the jaw, decreased urine excretion, visual disturbances and changes in eye lens color, and possible development of dermatitis.

Acute Health Effects

The following acute (short-term) health effects may occur immediately or shortly after exposure to mercury:

Skin: Irritation, rash, dermatitis, and possible redness and the appearance of small blisters. Dermal absorption will cause toxic systemic effects.

Eye: Irritation, inflammation, watering, swelling, conjunctivitis.

Lung: Inhalation of mercury vapors may cause irritation of the mucous membranes in the respiratory tract causing cough, phlegm, and lung tissue irritation leading to bronchitis, pneumonitis, and even delayed pulmonary edema.

Chronic Health Effects

The following chronic (long-term) health effects may occur at some time after exposure to mercury and can last for months or even years:

Cancer Hazards: According to the information presented in the references, mercury has not been adequately tested for its ability to cause cancer in test animals. However, human mutation data (causes genetic changes) have been documented and many scientists believe that such chemicals may pose a cancer risk in the long-term. More research is required.

Reproductive Hazards: According to information available in the references, mercury has been shown to cause cancer in test animals. There is limited evidence that mercury may cause an increase in spontaneous abortions in exposed women. Some organic mercury compounds have been identified as human teratogens and can cause the early termination of pregnancies (miscarriages) and may injure the fetus. Mercury may also reduce fertility in both males and females. While inorganic mercury substances (those without carbon), have not specifically been shown to be human teratogens, they should still be handled with caution as they may cause reproductive problems in males and females.

Other Chronic Hazards: High or repeated exposure can cause kidney damage or mercury poisoning with symptoms of sore gums, "shakes" (especially noticeable in handwriting), irritability, increased saliva production, memory loss, extreme shyness, weakness, loss of teeth, poor appetite, and metallic taste. Serious personality changes and brain damage can occur, especially if exposure continues. Repeated skin contact may cause gray discoloration and a skin allergy may develop. If this happens, even small future exposures can cause rash. Repeated exposures may also cause reduced peripheral vision and brown staining of the eye. Mercury can accumulate in the body on prolonged exposure and it may take months or years for the body to dispose of excess mercury.

Recommended Risk-Reduction Measures

Personnel should avoid direct contact with mercury. Mercury is an extremely toxic, cumulative poison. If a less toxic chemical cannot be substituted for a hazardous material, then *engineering controls* are the most effective method of reducing exposures.

The best protection is to enclose operations and/or provide local exhaust ventilation at the site of chemical release. While not always operationally feasible, isolating operations can also reduce exposure. Using respiratory protection is less effective than the controls mentioned above, but is still advisable whenever working with or around mercury. Exposure levels for mercury are extremely low (0.1 mg/m^3) and difficult to monitor accurately. A full facepiece respirator with a high-efficiency particulate air (HEPA) filter equipped with a dust/mist/fume pre-filter may suffice. However, for the best protection use a supplied-air respirator with a full facepiece operated in positive pressure mode or an MSHA/NIOSH-approved self-contained breathing apparatus (SCBA) with full facepiece operated in pressure demand or other continuous flow mode. If a full facepiece is not available, then chemical/dust goggles should be worn to protect the eyes. Whenever a chemical dust hazard exists, a face shield and a protective apron should be worn. To prevent hand and skin exposures, impervious rubber gloves should be used. Glove manufacturers should be contacted and permeation studies obtained *before* final glove selections are made. Mercury salts such as mercury are among mercury's most irritating and acutely toxic compounds. They are corrosive to skin tissues. Skin contact must be prevented.

Administrative controls should also be in place to minimize the potential for human exposures. These may include written procedures or policies which specify the methods and techniques that will be practiced whenever personnel are to work with mercury dusts or solutions. All personnel should receive *training* on the use, hazards, protective measures, emergency actions, and other precautions per 29 CFR 1910.1200 (Hazard Communication), prior to the first assignment in an area where mercury is used or stored. Before first exposure and every six to twelve months thereafter, a complete medical history and examination is recommended, including the following:

☑ Examination of the nervous system (taking note of any problems with person's handwriting).

☑ Routine urinalysis and special urine test for mercury (should be less than 0.02 mg/liter).

☑ Eye examination.

☑ Lung function tests for frequent exposures.

If symptoms develop or overexposure is suspected, the above listed tests should be repeated, along with the following recommended tests:

☑ Consider nerve conduction tests.

☑ Urinary enzyme and neurological behavior tests.

☑ Serum mercury tests.

☑ Lung function tests.

☑ Consider chest X-ray following acute overexposure (may be negative if taken immediately after exposure due to delayed onset of pulmonary edema).

Any evaluation should include a careful history of past and present symptoms with a medical examination. Medical tests that look for damage already done are not a substitute for controlling exposure. Also, since smoking can cause heart disease, lung cancer, emphysema, and other respiratory disorders, smokers exposed to mercury may experience symptoms more rapidly than non-smokers under the same conditions of exposure. Other mixed exposures occur when personnel use creams to whiten or bleach the skin that contain mercury. If so, their use increases exposure risk. Persons who are allergic to mercury may react to Mercurochrome or Merthiolate, which also contain mercury. Prudent risk management requires proper consideration of *all* exposure factors.

Other methods to reduce exposure include:

- ☑ Where possible, enclose operations and use local exhaust ventilation at the site of chemical release. If local exhaust ventilation or enclosure is not used, respiratory protection should be provided.
- ☑ Always ensure that proper protective clothing is worn when using chemical substances.
- ☑ Wash thoroughly immediately after exposure to mercury and at the end of the work shift or before eating, drinking, or smoking.
- ☑ Hazard warning information should be posted in the work area. In addition, as part of an on-going education and training program, all information on the health and safety hazards of mercury should be communicated to all exposed workers.
- ☑ Eye wash stations and emergency showers should be located in the immediate work area.
- ☑ Work clothing that has been contaminated with mercury should never be worn home where family members can be exposed as well. Contaminated clothing should be laundered at work or by an authorized service by personnel who have been trained in the hazards associated with mercury.
- ☑ The drug n-acetyl penicillamine (NAP) has been used to treat mercury poisoning, with mixed success.
- ☑ Specific engineering controls are recommended for mercury by NIOSH. Refer to NIOSH Criteria Document, "Inorganic Mercury" (Publication Number 73-11024).

RISK ASSESSMENT: ENVIRONMENT

General Assessment

The environment is at risk of exposure during transportation, storage, disposal, or destruction of mercury. In almost every scenario, the threat of environmental exposure is contingent upon the proper handling of the chemical substance. Accidental spills of mercury, large or small, can result in fire, explosion, and possible contamination of the surrounding environmental mediums (water, soil, and air).

Mercury is incompatible with many common materials, including most metals (except iron). Reactions with oxidizers and acids can be vigorous and violent. These characteristics require special consideration during any emergency situation involving a leak or spill of mercury.

Elemental mercury is a heavy and relatively inert liquid which is oxidized to inorganic mercury (II) under natural conditions. Mercury (II) may combine with an organic fraction to form methyl mercury. Both mercury (II) and methyl mercury can cause environmental contamination. Mercury (II) may enter the environment in industrial or municipal waste treatment discharges, from previously contaminated sediments, and from the weathering of natural rocks. Bacteria may then convert it to methyl mercury. The concentration of mercury (II) in water may be elevated with acid rain due to the scouring of mercury from the air and increased partitioning from the sediment into the water.

Acute Ecological Effects

Acute (short-term) toxic effects may include the death of animals, birds, or fish and death or low growth rate in plants. Acute effects are seen 2 to 4 days after animals or plants are exposed to mercury.

Mercury (II) and methyl mercury have high acute toxicity to aquatic life. Insufficient data are available to evaluate the short-term effects of mercury (II) and methyl mercury exposure to plants, birds, or terrestrial animals.

Chronic Ecological Effects

Chronic (long-term) toxic effects may include shortened life span, reproductive problems, lower fertility, and changes in appearance or behavior in exposed animals. These effects can be seen long after first exposure(s) to toxic chemicals.

Mercury (II) and methyl mercury have high chronic toxicity to aquatic life. Eating fish contaminated with mercury has caused secondary poisoning in humans. Birds or land animals similarly exposed to mercury and its compounds could also be subject to such effects. Insufficient data are available to evaluate the long-term effects of mercury exposure to plants.

Water Solubility

Mercury is insoluble in water. Concentrations of 1 milligram will not mix with a liter of water.

⌛ *Persistence in the Environment*

Mercury is highly persistent in water, with a half-life greater than 200 days. The half-life of a pollutant is the amount of time it takes for one half of the chemical to be degraded. Mercury will volatilize rapidly when deposited in soil. When released into water, microorganisms convert mercuric salts into methyl mercury which is rapidly taken up by algae and enters the food chain. It concentrates in the edible tissues of fish and, eventually, can cause human poisoning as well.

Bioaccumulation in Aquatic Organisms

Some substances increase in concentration, or *bioaccumulate,* in living organisms as they breathe contaminated air, drink contaminated water, or eat contaminated food. These chemicals can become concentrated in the tissues and internal organs of animals as well as humans. The concentration of mercury and methyl mercury found in fish tissues is expected to be considerably higher than the average concentration in the water from which the fish was taken.

Recommended Risk-Reduction Measures

Proper training of all transporters will reduce the likelihood of a mishap or accidents resulting in a leak or spill to the environment. Proper labeling on all containers, trucks, and rail cars will enable emergency responders to react properly and quickly to any disaster thereby reducing the potential risk to the environment and to personnel.

If a spill or leak to the environment has occurred, fire department, emergency response and/or hazardous materials spill personnel should be notified immediately. Do not dry sweep. Commercially available mercury spill cleanup kits should be considered. Otherwise, when vacuuming, a high-efficiency particulate air (HEPA) filter should be used, not a standard shop vacuum. Deposit collected materials in sealed drums for disposal. Cleanup should be attempted only by those trained in proper mercury spill containment procedures. Contaminated soils may need to be removed and replaced with clean soil. If mercury should contact the water table, aquifer, or navigable waterway, response activities should be prompt. When such spills occur, the local and/or state emergency response authorities should be notified. A comprehensive emergency response or disaster preparedness/recovery plan should be in place prior to any operations involving the use, transportation, storage, or disposal of mercury. If mercury is spilled or leaked, the following additional steps are recommended:

- ☑ Ventilate area and remove all ignition sources.
- ☑ It may be necessary to dispose of mercury as a hazardous waste. The responsible state agency or the regional office of the federal Environmental Protection Agency (EPA) should be contacted for specific recommendations.

RISK ASSESSMENT: BUSINESS

General Assessment

Accidents or mishaps involving mercury can present a significant threat to business operations. The loss or damage of equipment or facilities can significantly affect fiscal viability. Lawsuits that may result from personnel injury/death, public exposures, and/or environmental contamination will also require a serious expenditure of resources. Media attention surrounding an injury, death, or environmental damage can also result in a loss of profits and loss of current as well as future business. Always remember that anytime the term "reproductive hazard" is used, public emotion, hysteria, and ignorance can run equally high. This must be properly considered whenever developing public relations policies.

Recommended Risk-Reduction Measures

Company attorneys, safety and health professionals, and environmental specialists should be involved in the development of such procedures dealing with safety or emergency response. A company official should be pre-designated as a public relations officer with specific training in dealing with the press. Corporate plans and policies should be developed, approved, and implemented long before any need for such arises.

REFERENCES

American Conference of Governmental Industrial Hygienists. 1988. *Documentation of the Threshold Limit Values and Biological Exposure Indices,* 5th Edition (with updates). Cincinnati: ACGIH

Environmental Protection Agency. 1983. *Chemical Hazard Information Profiles.* Washington, D.C.: U.S. EPA.

U.S. Department of Health and Human Services, National Institute for Occupational Safety and Health. 1973. NIOSH Criteria Document. *Inorganic Mercury.* Publication Number 73-11024. Washington, D.C.: U.S. Government Printing Office.

MATERIAL SAFETY DATA SHEET

CHEMICAL NAME	
MESITYL OXIDE	**CRC PRESS/LEWIS PUBLISHERS DISCLAIMER:** THE INFORMATION AND RECOMMENDATIONS PRESENTED HEREIN ARE BASED ON SOURCES BELIEVED TO BE RELIABLE. CRC MAKES NO REPRESENTATION ON ITS COMPLETENESS OR ACCURACY. IT IS THE RESPONSIBILITY OF THE USER TO DETERMINE THE CHEMICAL'S SUITABILITY FOR ITS INTENDED USE, THE CHEMICAL'S SAFE USE, AND THE CHEMICAL'S PROPER DISPOSAL. NO REPRESENTATIONS AND/OR WARRANTIES, EITHER EXPRESSED OR IMPLIED, OF THE MERCHANTABILITY OR FITNESS FOR A PARTICULAR PURPOSE, OR OF ANY OTHER NATURE, ARE MADE WITH RESPECT TO THE INFORMATION PROVIDED IN THIS MSDS OR TO THE CHEMICAL TO WHICH INFORMATION MAY REFER. CRC NEITHER ASSUMES NOR AUTHORIZES ANY OTHER PERSON TO ASSUME FOR IT, ANY OTHER ADDITIONAL RESPONSIBILITY OR LIABILITY FOR THE USE OF, OR RELIANCE UPON, THIS INFORMATION.

HAZARD WARNING INFORMATION

HEALTH	FIRE	REACTIVE	OTHER	DEGREE OF HAZARD	COLOR CODING	OTHER CODES
3	3	0		0 = Minimum Hazard 1 = Slight Hazard 2 = Moderate Hazard 3 = Serious Hazard 4 = Severe Hazard	HEALTH = BLUE FIRE = RED REACTIVITY = YELLOW OTHER = WHITE	OX = Oxidizer ACID = Acid ALK = Alkali COR = Corrosive W = Use No Water

SECTION I - GENERAL INFORMATION

Characterization **Ketone**	RCRA Number **None**	EPA Class **Not Applicable**
DOT Proper Shipping Name **Mesityl Oxide**	Chemical Abstract Service (CAS) Number **141-79-7**	
DOT Hazard Class and Label Requirements **Flammable Liquid**	DOT Emergency Guide Code **26**	
DOT Identification Number **UN1229**	Chemical Formula $(CH_3)_2$:$CHCOCH_3$	

Synonyms
Methyl isobutenyl ketone; isobutenyl methyl ketone; 4-methyl-3-pentene-2-one; isopropyledene acetone.

SECTION II - HAZARDOUS INGREDIENTS AND IDENTITY INFORMATION

Hazardous Components (specific identity)	OSHA Exposure Criteria	NIOSH Exposure Criteria	Immediately Dangerous to Life and Health (IDLH)	ACGIH Exposure Criteria
Mesityl oxide (derivation: By dehydration of acetone or diacetone alcohol) 1 ppm = 4.08 mg/m³	PEL: 25 ppm 100 mg/m³ STEL: Not Established	REL: 10 ppm 40 mg/m³ STEL: Not Established	1400 ppm	TLV: 15 ppm 60 mg/m³ STEL: 25 ppm 100 mg/m³

SECTION III - PHYSICAL AND CHEMICAL CHARACTERISTICS

Boiling Point **266°F (130°C)**	Specific Gravity (H_2O = 1) **0.86**
Vapor Pressure (gas) **8 at 69°F (20°C)**	Molecular Weight **98.2**
Vapor Density (Air = 1) **3.4**	Melting Point **-43°F (-41°C)**

Solubility
Slightly soluble in water (3%). Soluble in alcohols, ethers, and with most organic liquids.

Appearance and Odor
Oily, colorless to light yellow liquid with a peppermint or honey-like odor. Odor Threshold = 12 ppm.

SECTION IV - FIRE AND EXPLOSION HAZARD DATA

Flash Point (method used) **87°F (31°C) closed cup**	Explosive Limits in Air % by Volume **LEL: 1.4% UEL: 7.2%**
NFPA Classification **Class IC Flammable Liquid**	Autoignition Temperature **652°F (344°C)**

Extinguishing Media
Use alcohol foam, carbon dioxide, or dry chemical. Water spray or stream may spread fire.

Special Fire Fighting Procedures
Wear full protective clothing and self-contained breathing apparatus (SCBA). Poisonous gases are produced in fire. Use water spray to keep fire-exposed containers cool. Containers may explode in fire.

Unusual Fire and Explosion Hazards
Flammable liquid and a dangerous fire hazard when exposed to heat. At high temperatures, forms explosive mixtures with air. Violent reactions can occur with nitric acid. Chemical will react on contact with oxidizable materials. Moderate explosion hazard when in the form of vapor. Vapors are heavier than air and can travel for great distances to an ignition source and flashback to cause fire or explosion.

SECTION V - REACTIVITY DATA

Stability		Conditions to Avoid **Normally stable. Keep away from sources of heat or flame. Large quantities should be stored in metal tanks or drums. Keep away from incompatible materials.**
Stable **X**	Unstable	Incompatibility *(materials to avoid)* **Strong oxidizers (permanganates, peroxides, nitrates, chlorates, and perchlorates), acids (nitric and sulfuric), strong alkalis, and amines.**
Hazardous Polymerization		Conditions to Avoid **Under normal conditions of temperature and pressure, hazardous polymerization of mesityl oxide is not expected to occur.**
May Occur	Will Not Occur **X**	Hazardous Decomposition or By-products **When heated to decomposition, mesityl oxide can emit highly toxic/poisonous and acrid fumes and gases including oxides of carbon.**

SECTION VI - HEALTH HAZARD DATA

Primary Route(s) of Entry:	Inhalation?	Absorption (skin)?	Ingestion?
	X		X

Health Hazards

INHALATION: Symptoms include irritation of the eyes, nose, throat, and respiratory system. May cause depression of the central nervous system with narcosis, inebriation, headache, fatigue, malaise, and dizziness. Liver and kidney damage are also possible. Other effects include olfactory changes, faintness, dypsnea, vertigo, possible pulmonary edema, coma.

SKIN & EYES: Irritating to the skin, may cause irritation and dermatitis. An eye irritant with possibility of corneal opacity and damage.

INGESTION: Not a likely exposure route. Irritation of the mouth and stomach, narcosis, anesthesia.

Carcinogenicity	NTP Listed?	IARC Cancer Review Group?	OSHA Regulated?	Target Organs?
Unknown Human Unknown Animal	No	No	29 CFR 1910.1000 Table Z-1	Respiratory system, eyes, skin, kidney, liver, CNS.

Medical Conditions Generally Aggravated by Exposure

None reported.

Emergency and First-aid Procedures

Eye contact: Flush immediately with water for 15 minutes (minimum); seek medical attention. Skin contact: Remove all contaminated clothing. Immediately wash area with large amounts of soap and water. Seek medical assistance. For inhalation: Remove the person from exposure. Provide respiratory assistance and CPR, if required. Transfer to medical facility. Observe for 24-48 hours for lungs effects. If swallowed: Seek medical attention immediately. Give large amounts of water or milk to dilute. Contact poison control center. Unless advised otherwise, do NOT induce vomiting (aspiration hazard).

SECTION VII - PRECAUTIONS FOR SAFE HANDLING AND USE

Steps to be Taken in Case Material is Released or Spilled

Absorb in vermiculite or similar material and deposit in sealed containers. Ventilate area of spill. Restrict those not involved in cleanup from area. Remove all sources of ignition. Use non-sparking tools.

Preferred Waste Disposal Method

Burn in chemical incinerator equipped with an afterburner and scrubber.

Precautions to be Taken in Handling and Storage

Store to avoid contact with incompatible materials in tightly closed containers in cool, well-ventilated area away from heat and flame. Use only non-sparking tools to open and close containers.

Other Precautions and Warnings

Sources of ignition are prohibited where mesityl oxide is used, handled, or stored. Heat may cause containers to build up pressure and explode. Containers must be grounded and bonded during transfer.

SECTION VIII - CONTROL MEASURES AND PERSONAL PROTECTIVE EQUIPMENT

Respiratory Protection *(specify type)*

For exposures above the PEL, use an MSHA/NIOSH-approved organic vapor respirator, a supplied-air respirator with full facepiece, hood, or helmet in continuous flow mode, or use a self-contained breathing apparatus (SCBA) operated in pressure demand or other positive pressure mode.

Ventilation

Local exhaust or general mechanical systems recommended.

Protective Gloves	Eye Protection	Other Protective Clothing
Impervious Material	Chemical Goggles or Face Mask	Impervious Apron

Work/Hygiene Practices

Always wash hands thoroughly after using chemical; never bring food, drink, or smoking materials into vicinity of chemicals.

MESITYL OXIDE

$(CH_3)_2$:$CHCOCH_3$

CAS: 141-79-7

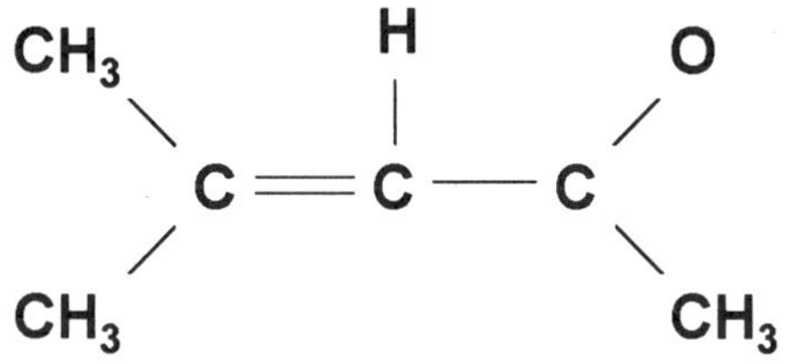

IDENTIFICATION AND TYPICAL USES

Mesityl oxide is an oily, colorless to light yellow liquid with a peppermint-like or honey-like odor. It is used as a solvent for cellulose ethers, vinyl resins, lacquers, roll-coating inks, and stains. It is also used in ore flotation, paint and varnish removers, insect repellent, and in the manufacture of methyl isobutyl ketone.

RISK ASSESSMENT: HEALTH

General Assessment

Mesityl oxide is moderately toxic by *inhalation* and skin contact (dermal *absorption*). Ingestion is, of course, possible, but few reports exist documenting ingestion of mesityl oxide as a likely exposure route. Of the ketone compounds, mesityl oxide is considered to be among the most toxic in humans. Its carcinogenicity, mutagenicity, and teratogenicity in humans have not been reported.

Inhalation can causes changes in the sense of smell, conjunctiva irritation, and possible respiratory changes. It will irritate the nose, throat, and eyes and fatigue the sense of smell. It may also cause a dangerous buildup of fluid in the lungs (pulmonary edema) which is a medical emergency and can be fatal. Symptoms can be delayed up to 48 hours, thereby creating a false sense of security with regard to health exposure risk. Mild to severe narcotic properties have also been recorded, including pronounced headache and depression of the central nervous system. Other symptoms on inhalation and CNS depression include nausea, dizziness, fatigue, malaise, narcosis, vertigo, dypsnea, decreased breathing rate, and anesthesia. Kidney and liver damage with mild narcosis may also be possible.

Skin and eye contact may produce mild to moderate irritation on contact. Pure liquid in contact with the eye may result in corneal opacity and subsequent damage. Skin contact can cause defatting, dryness, cracking, and dermatitis. First and even second degree burns are possible on prolonged contact.

On ingestion, mesityl oxide is expected to produce symptoms of irritation of the mouth and stomach, nausea, vomiting, stomach pain, narcosis, and anesthesia.

☠ *Acute Health Effects*

The following acute (short-term) health effects may occur immediately or shortly after exposure to mesityl oxide:

Skin: Will defat the skin and can cause dermatitis. Pure liquid will cause severe irritation and even first or second degree burns.

Eye: Severe irritation with a potential for corneal injury or damage.

Lung: Nose and throat irritation. Produces intoxicating effects and may cause lung damage, including pulmonary edema.

CNS: There may be depression, anesthesia, hypothermia, narcosis, and other effects.

Chronic Health Effects

The following chronic (long-term) health effects may occur at some time after exposure to mesityl oxide and can last for months or even years:

Cancer Hazards: According to information presented in the references, mesityl oxide has not been adequately tested for its ability to cause cancer in test animals.

Reproductive Hazard: According to information presented in the references, mesityl oxide has not been adequately tested for its ability to adversely affect reproduction.

Other Chronic Effects: Prolonged skin contact may cause drying, cracking, and chapping of the affected area. There may be skin sensitization and/or dermatitis in some individuals. Repeated or long-term exposures to mesityl oxide may affect the liver and kidneys.

⓿ *Recommended Risk-Reduction Measures*

Personnel should avoid direct contact with mesityl oxide. Of the ketones, mesityl oxide is among the most toxic. If a less toxic chemical cannot be substituted for a hazardous substance, then *engineering controls* are the most effective method of reducing exposures. The best protection is to enclose operations and/or provide local exhaust ventilation at the site of chemical release. While not always operationally feasible, isolating operations can also reduce exposure. Using respiratory protection is less effective than the controls mentioned above, but is still advisable whenever working with or around mesityl oxide. It is irritating at levels above 15 ppm and an organic vapor respirator should always be used to ensure exposure control. For exposures above the 25-ppm PEL, a supplied-air respirator with full facepiece operated in continuous flow mode, or a self-contained breathing apparatus (SCBA) with full facepiece operated in pressure demand are the recommended respiratory protection methods of choice. If a full facepiece is not available, then chemical goggles should be worn to protect the eyes. Whenever a chemical splash hazard exists, a face shield and a rubber apron should be worn. To prevent hand and skin exposures, impervious gloves should be used. Glove manufacturers should be contacted and permeation studies obtained *before* final glove selections have been made.

Administrative controls should also be in place to minimize the potential for human exposures. These may include written procedures or policies which specify the methods and techniques that will be practiced whenever personnel are to work with mesityl oxide.

All personnel should receive *training* on the use, hazards, protective measures, emergency actions, and other precautions per 29 CFR 1910.1200 (Hazard Communication), prior to the first assignment in an area where mesityl oxide is used or stored.

If symptoms develop or overexposure is suspected, the following tests are recommended:

☑ Kidney and liver function tests.

☑ Evaluation by a qualified allergist, with careful consideration of exposure history and special skin tests (may help diagnose allergy).

☑ Lung function tests.

☑ Consider chest X-ray following acute overexposure (may be negative if taken immediately after exposure due to delayed onset of pulmonary edema).

Any evaluation should include a careful history of past and present symptoms with an examination. Medical tests that simply look for existing damage are not a substitute for controlling exposures. Also, since smoking can cause heart disease, emphysema, and other respiratory disorders, symptoms of exposure to mesityl oxide may be more pronounced than those displayed by non-smokers. Proper risk assessment requires careful consideration of *all* possible factors that may be causing the appearance of symptoms in exposed workers.

Other methods to reduce exposure include:

☑ Where possible, enclose operations and use local exhaust ventilation at the site of chemical release. If local exhaust ventilation or enclosure is not used, respiratory protection should be mandatory. Also, if possible, automatic transfer of mesityl oxide from storage containers to process or work containers is recommended.

☑ Always ensure that proper protective clothing is worn when using chemical substances.

☑ Wash thoroughly immediately after exposure to mesityl oxide and at the end of the work shift or before eating, drinking, or smoking.

☑ Hazard warning information should be posted in the work area. In addition, as part of an on-going education and training program, all information on the health and safety hazards of mesityl oxide should be communicated to all exposed and potentially exposed workers.

☑ Eye wash stations should be provided in the immediate work area for emergency use. If there is a possibility of skin exposure to mesityl oxide, emergency shower facilities should also be provided in the immediate work area.

☑ Personnel should never wear contaminated clothing home (family members can be exposed). Clothing should be laundered by personnel who have been trained on the hazards of mesityl oxide.

RISK ASSESSMENT: ENVIRONMENT
General Assessment

The environment is at risk of exposure and contamination during transportation, storage, disposal, or destruction of mesityl oxide. In almost every scenario, the threat of environmental exposure is contingent upon the proper handling of the chemical substance. Accidental spills, large or small, can result in fire, explosion, and possible contamination of the surrounding environmental mediums (water, soil, and air).

Mesityl oxide is considered a Class IC flammable liquid (per OSHA 29 CFR 1910.106). It presents a moderate fire and explosion hazard risk, especially in contact with incompatible materials. These characteristics require extreme caution in handling, storage, transportation, and disposal. It is incompatible with common peroxides and other oxidizers and contact can cause fire or explosion. It reacts violently when added to nitric or sulfuric acid. Therefore, special consideration is required during any emergency situation involving a leak or spill of mesityl oxide. Should mesityl oxide ever come into contact with these or other incompatible substances during use, transportation, storage, or disposal, the formation of highly toxic and/or highly explosive commodities is extremely possible.

The proper disposal method for mesityl oxide is to burn it in a chemical incinerator equipped with an afterburner and scrubber.

Mesityl oxide may enter the environment through industrial discharges or spills.

Acute Ecological Effects

Acute (short-term) toxic effects may include the death of animals, birds, or fish and death or low growth rate in plants. Acute effects are seen 2 to 4 days after animals or plants are exposed to mesityl oxide.

Insufficient data are available on the short-term effects of mesityl oxide exposure to aquatic life, birds, plants, or land animals.

Chronic Ecological Effects

Chronic (long-term) toxic effects may include shortened life span, reproductive problems, lower fertility, and changes in appearance or behavior in exposed animals. These effects can be seen long after first exposure(s) to toxic chemicals.

Insufficient data are available on the long-term effects of mesityl oxide exposure to aquatic life, birds, plants, or land animals.

Water Solubility

Mesityl oxide is nearly insoluble in water. Concentrations of less than 100 milligrams may mix with a liter of water, depending on water pH and chemical concentration of mesityl oxide.

Persistence in the Environment

Mesityl oxide is moderately persistent in water, with a half-life between 20 and 200 days. The half-life of a pollutant is the amount of time it takes for one half of the chemical to be degraded.

Bioaccumulation in Aquatic Organisms

Some substances increase in concentration, or *bioaccumulate,* in living organisms as they breathe contaminated air, drink contaminated water, or eat contaminated food. These chemicals can become concentrated in the tissues and internal organs of animals as well as humans.

The concentration of mesityl oxide found in fish tissue is expected to be somewhat higher than the average concentration of mesityl oxide in the water from which the fish was taken.

Recommended Risk-Reduction Measures

Proper training of all transporters will reduce the likelihood of a mishap or accident resulting in a leak or spill to the environment. The correct use of DOT labeling (per the MSDS) on all containers, trucks, and rail cars will enable emergency responders to react properly and quickly to any disaster thereby reducing the potential risk to the environment and to personnel.

Storage of mesityl oxide should be segregated from incompatible chemicals to minimize the risk of cross-contamination or contact. Buildings designated for storage should be equipped with appropriate fire protection systems (alarms, sprinklers, emergency lighting, portable extinguishers).

If a spill or leak to the environment has occurred, fire department, emergency response and/or hazardous materials spill personnel should be notified immediately. Cleanup should be attempted only by those trained in proper spill containment procedures using non-sparking tools. Contaminated soils should be removed for incineration and replaced with clean soil. If mesityl oxide should contact the water table, aquifer, or navigable waterway, cleanup should be started immediately. It is only slightly soluble in water and total remediation may be possible. The local and/or state emergency response authorities must be notified. A comprehensive emergency response or disaster preparedness/recovery plan should be in place prior to any operations involving the use, transportation, storage, or disposal of mesityl oxide. If mesityl oxide is spilled or leaked, the following specific steps are recommended:

- ☑ Restrict persons not wearing protective clothing from area of spill or leak until cleanup is complete and area can be opened for normal work.
- ☑ Remove all ignition sources, ventilate area of leak.
- ☑ Absorb liquids in vermiculite, dry sand, earth, or similar material and deposit in sealed containers using non-sparking tools.
- ☑ It may be necessary to dispose of mesityl oxide as a hazardous waste. The responsible state agency or the regional office of the federal Environmental Protection Agency (EPA) should be contacted for specific recommendations.

RISK ASSESSMENT: BUSINESS

General Assessment

Accidents or mishaps involving mesityl oxide can present a significant threat to business operations. The loss or damage of equipment or facilities can significantly affect fiscal viability. Lawsuits that may result from personnel injury/death, public exposures, and/or environmental contamination will also require a serious expenditure of resources. Media attention surrounding an injury, death, or environmental damage can also result in a loss of profits and loss of current as well as future business.

Recommended Risk-Reduction Measures

Company attorneys, safety and health professionals, and environmental specialists should be involved in the development of any procedures dealing with safety or emergency response. A company official should be pre-designated as a public relations officer with specific training in dealing with the press. Corporate plans and policies should be developed, approved, and implemented long before any need for such arises.

REFERENCES

Ahlbom, A. 1993. *Biostatistics for Epidemiologists.* New York: CRC Press/Lewis Publishers.

American Conference of Governmental Industrial Hygienists. 1988. *Documentation of the Threshold Limit Values and Biological Exposure Indices,* 5th Edition (with updates). Cincinnati: ACGIH

Bueche, J. R. 1972. *Principles of Physics,* 2nd Edition. New York: McGraw-Hill.

Calabrese, E. J. 1994. *Biological Effects of Low Level Exposures.* New York: CRC Press/Lewis Publishers.

Cockerman, L. G., B. S. Shane. 1994. *Basic Environmental Toxicology.* New York: CRC Press/Lewis Publishers.

Cothern, C. R. 1995. *Handbook for Environmental Risk Decisionmaking.* New York: CRC Press/Lewis Publishers.

Danse, I. R. 1991. *Common Sense Toxics in the Workplace.* New York: Van Nostrand Reinhold.

Environmental Protection Agency. 1983. *Chemical Hazard Information Profiles.* Washington, D.C.: U.S. EPA.

Fodden, C. R., and J. L. Weddell. 1992. *Hazardous Materials Emergency Action Data.* New York: CRC Press/Lewis Publishers.

Forseberg, K., and L. H. Keith. 1993. *Chemical; Protective Clothing Permeation and Degradation Database.* Chelsea, MI: Lewis Publishers.

Gardner, R. W. 1994. *Chemical Intolerance.* New York: CRC Press/Lewis Publishers.

Hallenbeck, W. H. 1993. *Quantitative Risk Assessment for Environmental and Occupational Health,* 2nd Edition. New York: CRC Press/Lewis Publishers.

Keith, L. H. and M. M. Walker. 1995. *Handbook of Air Toxics.* New York: CRC Press/Lewis Publishers.

Langerman, N. 1994 *Precautionary Labels for Chemical Containers.* New York: CRC Press/Lewis Publishers.

Lauwerys, R. R., P. Hoet. 1993. *Industrial Chemical Exposure.* New York: CRC Press/Lewis Publishers.

Lewis, R. J., Sr. 1992. *Sax's Dangerous Properties of Industrial Materials,* Volumes 1, 2, and 3, 8th Edition. New York: Van Nostrand Reinhold.

Meyer, E. 1990. *Chemistry of Hazardous Materials.* Englewood Cliffs, NJ: Prentice Hall.

Patnaik, P. 1992. *A Comprehensive Guide to the Hazardous Properties of Chemical Substances.* New York: Van Nostrand Reinhold.

Proctor, N. H., J. P. Hughes. 1991. *Chemical Hazards of the Workplace.* New York: Van Nostrand Reinhold.

Quigley, D. R. 1994. *Handbook of Emergency Chemical Management.* New York: CRC Press/Lewis Publishers.

Rea, W. J. 1995. *Chemical Sensitivity,* Vol. I-III. New York: CRC Press/Lewis Publishers.

Richardson, M. 1995. *Dictionary of Substance and Their Effects,* Vol. I-VI. New York: CRC Press/Lewis Publishers.

Sacarello, H. L. A. 1994. *The Comprehensive Handbook of Hazardous Materials.* New York: CRC Press/Lewis Publishers.

Sheftel, V. O. 1995. *Handbook of Toxic Properties of Monomers and Additives.* New York: CRC Press/Lewis Publishers.

Torkelson, T. R. V. K. Rowe. 1981. Halogenated aliphatic hydrocarbons. *Patty's Industrial Hygiene and Toxicology,* Vol. 2B, *Toxicology,* pp 3568-3572. New York: Wiley-Interscience.

U.S. Department of Health and Human Services, National Institute for Occupational Safety and Health. 1994. *NIOSH Pocket Guide to Chemical Hazards.* Washington, D.C.: U.S. Government Printing Office.

MATERIAL SAFETY DATA SHEET

CHEMICAL NAME	
METHACRYLIC ACID	**CRC PRESS/LEWIS PUBLISHERS DISCLAIMER:** THE INFORMATION AND RECOMMENDATIONS PRESENTED HEREIN ARE BASED ON SOURCES BELIEVED TO BE RELIABLE. CRC MAKES NO REPRESENTATION ON ITS COMPLETENESS OR ACCURACY. IT IS THE RESPONSIBILITY OF THE USER TO DETERMINE THE CHEMICAL'S SUITABILITY FOR ITS INTENDED USE, THE CHEMICAL'S SAFE USE, AND THE CHEMICAL'S PROPER DISPOSAL. NO REPRESENTATIONS AND/OR WARRANTIES, EITHER EXPRESSED OR IMPLIED, OF THE MERCHANTABILITY OR FITNESS FOR A PARTICULAR PURPOSE, OR OF ANY OTHER NATURE, ARE MADE WITH RESPECT TO THE INFORMATION PROVIDED IN THIS MSDS OR TO THE CHEMICAL TO WHICH INFORMATION MAY REFER. CRC NEITHER ASSUMES NOR AUTHORIZES ANY OTHER PERSON TO ASSUME FOR IT, ANY OTHER ADDITIONAL RESPONSIBILITY OR LIABILITY FOR THE USE OF, OR RELIANCE UPON, THIS INFORMATION.

HAZARD WARNING INFORMATION

HEALTH	FIRE	REACTIVE	OTHER	DEGREE OF HAZARD	COLOR CODING	OTHER CODES
3	2	2		0 = Minimum Hazard 1 = Slight Hazard 2 = Moderate Hazard 3 = Serious Hazard 4 = Severe Hazard	HEALTH = BLUE FIRE = RED REACTIVITY = YELLOW OTHER = WHITE	OX = Oxidizer ACID = Acid ALK = Alkali COR = Corrosive W = Use No Water

SECTION I - GENERAL INFORMATION

Characterization	RCRA Number	EPA Class
Acid, Carboxylic	**None**	**Not Applicable**

DOT Proper Shipping Name **Methacrylic Acid, Inhibited**	Chemical Abstract Service (CAS) Number **79-41-4**
DOT Hazard Class and Label Requirements **Corrosive Material, Corrosive**	DOT Emergency Guide Code **60 (inhibited)**
DOT Identification Number **UN 2531**	Chemical Formula **$H_2C{:}C(CH_3)COOH$**

Synonyms

Methacrylic acid (glacial); methacrylic acid (inhibited); 2-methylacrylic acid; α-methacrylic acid; 2-methylpropenoic acid; monomer; 2-methyl-2-propenoic acid.

SECTION II - HAZARDOUS INGREDIENTS AND IDENTITY INFORMATION

Hazardous Components (specific identity)	OSHA Exposure Criteria	NIOSH Exposure Criteria	Immediately Dangerous to Life and Health (IDLH)	ACGIH Exposure Criteria
Methacrylic acid (derivation: By reaction of acetone cyanohydrin and dilute sulfuric acid; by oxidation of isobutylene). **1 ppm = 3.58 mg/m³**	**PEL (skin):** **20 ppm** **70 mg/m³** **STEL:** **Not Established**	**REL (skin):** **20 ppm** **70 mg/m³** **STEL:** **Not Established**	**Not Established**	**TLV (skin):** **20 ppm** **70 mg/m³** **STEL:** **Not Established**

SECTION III - PHYSICAL AND CHEMICAL CHARACTERISTICS

Boiling Point **325°F (163°C)**	Specific Gravity (H_2O = 1) **1.02**
Vapor Pressure (mm Hg) **1 at 69°F (20°C)**	Molecular Weight **86.1**
Vapor Density (Air = 1) **Not Found**	Melting Point **61°F (16°C)**

Solubility

Soluble water (9% at 77°F), alcohol, ether, and most organic solvents.

Appearance and Odor

Colorless liquid or solid (below 61°F) with an acrid, repulsive odor.

SECTION IV - FIRE AND EXPLOSION HAZARD DATA

Flash Point (method used) **171°F (77°C) opened cup**	Explosive Limits in Air % by Volume **LEL: 1.6% UEL: 8.8%**
NFPA Classification **Class IIIA Combustible Liquid**	Autoignition Temperature **154°F (68°C)**

Extinguishing Media

Carbon dioxide, dry chemical, water spray, or foam.

Special Fire Fighting Procedures

Wear full protective clothing and self-contained breathing apparatus (SCBA). Use water spray to flush and dilute a spill and to disperse the vapors. Also, use a water spray to keep fire-exposed containers cool. Poisonous gases are produced in fire. Evacuate 1500 feet radius if fire becomes uncontrollable.

Unusual Fire and Explosion Hazards

Potentially explosive reaction with strong oxidizing compounds, especially at elevated temperatures. Containers may explode in fire. Firefighters should avoid contact with vapors produced during fire.

SECTION V - REACTIVITY DATA

Stability		Conditions to Avoid
		Unstable at room temperatures. Can be stored safely (with or without an inhibitor) in temperatures below 59°F (15°C). Keep away from incompatible materials and heat.
Stable	Unstable	Incompatibility *(materials to avoid)*
	X	Strong oxidizers (especially peroxides), high temperatures, strong acids (such as hydrochloric acid), alkalis, and sunlight.
Hazardous Polymerization		Conditions to Avoid
		Hazardous polymerization can occur under normal conditions of temperature and pressure. Elevated temperatures, oxidizers, peroxides, hydrochloric acid, or sunlight can accelerate polymerization.
May Occur	Will Not Occur	Hazardous Decomposition or By-products
X		When heated to decomposition, methacrylic acid can emit highly toxic/poisonous gases, including carbon monoxide and acrid fumes.

SECTION VI - HEALTH HAZARD DATA

Primary Route(s) of Entry:	Inhalation?	Absorption (skin)?	Ingestion?
	X	X	X

Health Hazards

INHALATION: Causes irritation to the eyes, nose, throat, and respiratory system. Other effects include headache, nausea, vomiting, confusion, dizziness, and loss of consciousness. Chronic exposure can lead to bronchospasm and wheezing.

ABSORPTION: Skin contact results in irritation and burns. Causes severe eye burns and/or irritation. Absorbed methacrylic acid is rapidly metabolized as a short chain fatty acid. Possible skin sensitizer. Corrosive to skin. Chronic exposure can lead to dermatitis.

INGESTION: Poisonous by ingestion. Gastrointestinal irritation with unspecified systemic effects.

Carcinogenicity	NTP Listed?	IARC Cancer Review Group?	OSHA Regulated?	Target Organs?
Unknown Human Unknown Animal	No	No	29 CFR 1910.1000 Table Z-1	Respiratory system, eyes, skin.

Medical Conditions Generally Aggravated by Exposure

Respiratory impairments (bronchitis, asthma) or skin conditions (dermatitis) may be aggravated.

Emergency and First-aid Procedures

Eye contact: Flush immediately with water for 15 minutes (minimum); seek medical attention. Skin contact: Remove all contaminated clothing. Immediately wash area with large amounts of soap and water. Seek medical assistance. For inhalation: Remove the person from exposure. Provide respiratory assistance and CPR. Transfer to medical facility. If swallowed: Seek medical attention. Give 1 to 2 glasses of water to dilute. Never give an unconscious or convulsing person anything by mouth.

SECTION VII - PRECAUTIONS FOR SAFE HANDLING AND USE

Steps to be Taken in Case Material is Released or Spilled

Absorb liquids in vermiculite or similar material and deposit in sealed containers. Ventilate area of spill or leak. Restrict those not involved in cleanup from entering area. Spills can be neutralized with crushed limestone, soda ash, or lime. Non-sparking tools should be used during cleanup.

Preferred Waste Disposal Method

Incineration (with afterburner and scrubber).

Precautions to be Taken in Handling and Storage

To keep stable, store in cool (below 59°F), dry, well-ventilated area away from heat. Protect containers from physical damage. Storage in separate facility is recommended. Keep out of direct sunlight.

Other Precautions and Warnings

Sources of ignition are prohibited where methacrylic acid is used, handled, or stored.

SECTION VIII - CONTROL MEASURES AND PERSONAL PROTECTIVE EQUIPMENT

Respiratory Protection *(specify type)*

When potential for high exposures (above 20 ppm) exist, use an MSHA/NIOSH-approved supplied-air respirator with full facepiece, hood, or helmet in continuous flow mode, or use a full facepiece operated in pressure demand or other positive pressure mode. Use half-mask respirator with acid cartridge for low or transient exposures (cartridge should be designed to protect against acid vapor or mist).

Ventilation

Local exhaust or general mechanical systems recommended.

Protective Gloves	Eye Protection	Other Protective Clothing
Acid resistant synthetic rubber	Chemical Goggles or Face Mask	Protective Rubber Apron

Work/Hygiene Practices

Always wash hands thoroughly after using chemical; never bring food, drink, or smoking materials into vicinity of chemicals.

METHACRYLIC ACID

$H_2C{:}C(CH_3)COOH$ CAS: 79-41-4

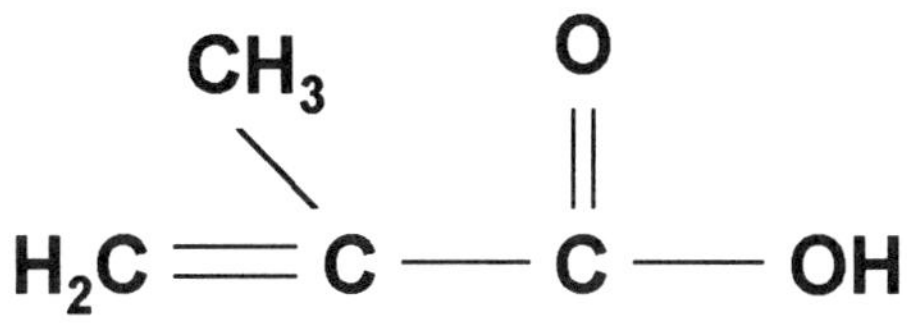

IDENTIFICATION AND TYPICAL USES

Methacrylic acid is a colorless liquid or solid (below 61°F) with an acrid, repulsive odor. It is used as a monomer for large-volume resins and polymers, as a co-monomer in emulsion polymers for adhesives, paints, paper manufacture, polishes, and in the textile industries. Many of the polymers are based on esters of the acid, as the methyl, butyl, or isobutyl esters. It is also used in organic synthesis.

RISK ASSESSMENT: HEALTH

General Assessment

Exposure to methacrylic acid is possible through *inhalation* of its vapors, dermal *absorption* of the liquid, or by direct *ingestion*. There are no data regarding is carcinogenic or teratogenic properties. However, mutation data have been reported for this compound.

Inhalation of its vapors can irritate the nose and throat as well as the lungs. High concentrations can cause headaches, nausea, vomiting, confusion, dizziness, loss of consciousness, and coma. Methacrylic acid may also cause bronchospasm and wheezing. It is a serious eye hazard. Splashing into the eyes can cause damage and loss of vision. Skin contact can cause mild to moderate irritation. Depending upon the conditions of exposure (duration, concentration), there may also be burns.

Acute Health Effects

The following acute (short-term) health effects may occur immediately or shortly after exposure to methacrylic acid:

Skin: Irritation/rash or burning feeling on contact. May cause sensitization and dermatitis.

Eye: Severe irritation; may cause burns and tissue damage, depending upon duration of contact.

Lung: Irritation of the nose, throat, and lungs following exposure.

CNS: Inconclusive (possible lethargy, weight loss and other symptoms of nervous system depression).

Chronic Health Effects

The following chronic (long-term) health effects may occur at some time after exposure to methacrylic acid and can last for months or even years:

Cancer Hazards: According to information presented in the references, methacrylic acid has not been adequately tested for its ability to cause cancer in test animals. Mutation data have been reported and many scientists believe that exposure to such chemicals may pose a cancer risk in the long term.

Reproductive Hazard: There is limited evidence to show that exposure to methacrylic acid causes teratogenic in laboratory animals.

Other Chronic Hazard: High, excessive, or long-term exposures to other carboxylic acids similar to methacrylic acid may damage the kidneys and lungs. Exposure to the skin may cause an allergic reaction in some people (sensitization) and dermatitis. Once sensitized, even very small future exposures to such compounds on the skin surface can result in a rash.

Recommended Risk-Reduction Measures

Personnel should avoid direct contact with methacrylic acid. If a less toxic chemical cannot be substituted for a hazardous substance, then *engineering controls* are the most effective method of reducing exposures.

The best protection is to enclose operations and/or provide local exhaust ventilation at the site of chemical release. While not always operationally feasible, isolating operations can also reduce exposure. Using respiratory protection is less effective than the controls

mentioned above, but is still advisable whenever working with or around methacrylic acid. A powered, air-supplied respirator or a self-contained breathing apparatus (SCBA) with full facepiece operated in pressure demand or other positive pressure mode is the recommended respiratory protection of choice. If a full facepiece is not available, then chemical goggles should be worn to protect the eyes. For low concentrations (below 20 ppm), an acid gas cartridge respirator (equipped with a mist/vapor filter) is acceptable protection. Whenever a chemical splash hazard exists, a face shield and a rubber apron should be worn. To prevent hand and skin exposures, acid resistant gloves should be use. Glove manufacturers should be contacted and permeation studies obtained *before* final glove selections have been made.

Administrative controls should also be in place to minimize the potential for human exposures. These may include written procedures or policies which specify the methods and techniques that will be practiced whenever personnel are to work with methacrylic acid.

All personnel should receive *training* on the use, hazards, protective measures, emergency actions, and other precautions per 29 CFR 1910.1200 (Hazard Communication), prior to the first assignment in an area where methacrylic acid is used or stored. Other methods to reduce exposure include:

☑ Where possible, enclose operations and use local exhaust ventilation at the site of chemical release. If local exhaust ventilation or enclosure is not used, respiratory protection should be mandatory.

☑ Always ensure that proper protective clothing is worn when using chemical substances.

☑ Wash thoroughly immediately after exposure to methacrylic acid and at the end of the work shift or before eating, drinking, or smoking.

☑ Hazard warning information should be posted in the work area. In addition, as part of an on-going education and training program, all information on the health and safety hazards of methacrylic acid should be communicated to all exposed or potentially exposed workers.

RISK ASSESSMENT: ENVIRONMENT

General Assessment

The environment is at risk of exposure during transportation, storage, disposal, or destruction of methacrylic acid. In almost every scenario, the threat of environmental exposure is contingent upon the proper handling of the chemical substance. Accidental spills, large or small, can result in fire, explosion, and possible contamination of the surrounding environmental mediums (water, soil, and, to a lesser degree, air).

Methacrylic acid is considered a class IIIA combustible liquid (per OSHA 29 CFR 1910.106). However, its vapors can form explosives mixtures especially in confined or enclosed spaces. It is also incompatible with oxidizing agents and strong acid. Contact with either can cause violent reactions, including fire or explosion. These characteristics require special consideration during any emergency situation involving a leak or spill of methacrylic acid.

Should methacrylic acid ever come into contact with incompatible substances such as oxidizers, acid, or even elevated temperatures either during use, transportation, or storage, the formation of highly toxic and/or highly explosive commodities is extremely possible.

The proper disposal/destruction method for methacrylic acid waste is to burn it in a chemical incinerator equipped with an afterburner and air scrubber.

Methacrylic acid can enter the environment mainly industrial discharges or spills.

☠ *Acute Ecological Effects*

Acute (short-term) toxic effects may include the death of animals, birds, or fish and death or low growth rate in plants. Acute effects are seen 2 to 4 days after animals or plants are exposed to methacrylic acid.

Insufficient data are available to evaluate or predict the short-term effects of methacrylic acid exposure to aquatic life, plants, birds, or land animals.

Chronic Ecological Effects

Chronic (long-term) toxic effects may include shortened life span, reproductive problems, lower fertility, and changes in appearance or behavior in exposed animals. These effects can be seen long after first exposure(s) to toxic chemicals.

Insufficient data are available to evaluate or predict the long-term effects of methacrylic acid to aquatic life, plants, birds, or land animals.

Water Solubility

Methacrylic acid is highly soluble in water. Concentrations of 1000 milligrams and more will mix with a liter of water.

⌛ *Persistence in the Environment*

Methacrylic acid is slightly persistent in the aquatic environment, with a half-life of between 2 to 20 days. The half-life of a pollutant is the amount of time it takes for one half of the chemical to be degraded.

Approximately 90% of methacrylic acid will eventually end up in water; the remainder will end up in air.

Bioaccumulation in Aquatic Organisms

Some substances increase in concentration, or *bioaccumulate,* in living organisms as they breathe contaminated air, drink contaminated water, or eat contaminated food. These chemicals can become concentrated in the tissues and internal organs of animals as well as humans.

The concentration of methacrylic acid found in fish tissues is expected to be about the same as the average concentration of methacrylic acid in the water from which the fish was taken.

Recommended Risk-Reduction Measures

Proper training of all transporters will reduce the likelihood of a mishap or accident resulting in a leak or spill to the environment. The correct use of DOT labeling (per the MSDS) on all containers, trucks, and rail cars will enable emergency responders to react properly and quickly to any disaster thereby reducing the potential risk to the environment and to personnel.

Storage of methacrylic acid should be segregated from incompatible chemicals to minimize the risk of cross-contamination or contact. Buildings designated for storage should be equipped with appropriate fire protection systems (alarms, sprinklers, emergency lighting, portable extinguishers).

If a spill or leak to the environment has occurred, fire department, emergency response and/or hazardous materials spill personnel should be notified immediately. Cleanup should be attempted only by those trained in proper spill containment procedures using non-sparking tools. Contaminated soils should be removed for incineration and replaced with clean soil. If methacrylic acid should contact the water table, aquifer, or navigable waterway, time is of the essence. It is highly miscible in water and, therefore, total containment and remediation may not be possible. When such spills occur, the local and/or state emergency response authorities must be notified. A comprehensive emergency response or disaster preparedness/recovery plan should be in place prior to any operations involving the use, transportation, storage, or disposal of methacrylic acid. If methacrylic acid is spilled or leaked, the following specific steps are recommended:

- ☑ Restrict persons not wearing protective clothing from area of spill or leak until cleanup is complete and area can be opened for normal work.
- ☑ Remove all ignition sources, ventilate area.
- ☑ Absorb liquids in vermiculite, dry sand, earth, or similar material and deposit in sealed containers. Spilled materials can be neutralized using crushed limestone, soda ash, or lime.
- ☑ It may be necessary to dispose of methacrylic acid as a hazardous waste. The responsible state agency or the regional office of the federal Environmental Protection Agency (EPA) should be contacted for specific recommendations.

RISK ASSESSMENT: BUSINESS

General Assessment

Accidents or mishaps involving methacrylic acid can present a significant threat to business operations. The loss or damage of equipment or facilities can significantly affect fiscal viability. Lawsuits that may result from personnel injury/death, public exposures, and/or environmental contamination will also require a serious expenditure of resources. Media attention surrounding an injury, death, or environmental damage can also result in a loss of profits and loss of current as well as future business.

Recommended Risk-Reduction Measures

Company attorneys, safety and health professionals, and environmental specialists should be involved in the development of any procedures dealing with safety or emergency response. A company official should be pre-designated as a public relations officer with specific training in dealing with the press. Corporate plans and policies should be developed, approved, and implemented long before any need for such arises.

REFERENCES

Ahlbom, A. 1993. *Biostatistics for Epidemiologists.* New York: CRC Press/Lewis Publishers.

American Conference of Governmental Industrial Hygienists. 1988. *Documentation of the Threshold Limit Values and Biological Exposure Indices,* 5th Edition (with updates). Cincinnati: ACGIH

Asante-Duah, D. K. 1993. *Hazardous Waste Risk Assessment.* New York: CRC Press/Lewis Publishers.

Bloom, A. D. and F. J. DeSerres. 1995. *Ecotoxicity and Human Health.* New York: CRC Press/Lewis Publishers.

Brusick, D. J. 1994. *Methods for Genetic Risk Assessment.* New York: CRC Press/Lewis Publishers.

Bueche, J. R. 1972. *Principles of Physics,* 2nd Edition. New York: McGraw-Hill.

Burrell, R., D. K. Flaherty, and Leonard J. Sauers. 1992. *Toxicology of the Immune System.* New York: Van Nostrand Reinhold.

Calabrese, E. J., E. M. Kenyon. *Air Toxics and Risk Assessment.* New York: CRC Press/Lewis Publishers.

Calabrese, E. J. 1994. *Biological Effects of Low Level Exposures.* New York: CRC Press/Lewis Publishers.

Cockerman, L. G., B. S. Shane. 1994. *Basic Environmental Toxicology.* New York: CRC Press/Lewis Publishers.

Cothern, C. R., N. P. Ross. 1994. *Environmental Statistics, Assessment, and Forecasting.* New York: CRC Press/Lewis Publishers.

Cothern, R. R. 1993. *Comparative Environmental Risk Assessment.* New York: CRC Press/Lewis Publishers.

Cothern, C. R. 1995. *Handbook for Environmental Risk Decisionmaking.* New York: CRC Press/Lewis Publishers.

Danse, I. R. 1991. *Common Sense Toxics in the Workplace.* New York: Van Nostrand Reinhold.

Gardner, R. W. 1994. *Chemical Intolerance.* New York: CRC Press/Lewis Publishers.

Hallenbeck, W. H. 1993. *Quantitative Risk Assessment for Environmental and Occupational Health,* 2nd Edition. New York: CRC Press/Lewis Publishers.

Hosty, J. W., P. Foster. 1990. *A Practical Guide to Chemical Spill Response.* New York: Van Nostrand Reinhold.

Howard, P. H. 1995. *Handbook of Environmental Fate and Exposures Data for Organic Chemicals,* Vol. I-IV. New York: CRC Press/Lewis Publishers.

Howard, P. H., et. al. 1991. *Handbook of Environmental Degradation Rates.* New York: CRC Press/Lewis Publishers.

James, A., M. Lord. 1992. *Index of Chemical and Physical Data.* New York: Van Nostrand Reinhold.

Jones, F. E. 1994. *Toxic Organic Vapors in the Workplace.* New York: CRC Press/Lewis Publishers.

Keith, L. H. and M. M. Walker. 1995. *Handbook of Air Toxics.* New York: CRC Press/Lewis Publishers.

Landis, W. G., M. H. Yu. 1995. *Introduction to Environmental Toxicology.* New York: CRC Press/Lewis Publishers.

Langerman, N. 1994 *Precautionary Labels for Chemical Containers.* New York: CRC Press/Lewis Publishers.

Lauwerys, R. R., P. Hoet. 1993. *Industrial Chemical Exposure.* New York: CRC Press/Lewis Publishers.

Lewis, R. J., Sr. 1993. *Hawley's Condensed Chemical Dictionary,* 12th Edition. New York: Van Nostrand Reinhold.

Lewis, R. J., Sr. 1992. *Sax's Dangerous Properties of Industrial Materials,* Volumes 1, 2, and 3, 8th Edition. New York: Van Nostrand Reinhold.

Lide, D. R. 1995. *Handbook of Chemistry and Physics,* 76th Edition. New York: CRC Press/Lewis Publishers.

Manahan, S. E. 1994. *Environmental Chemistry,* 6th Edition. New York: CRC Press/Lewis Publishers.

Manahan, S. E. 1992. *Toxicological Chemistry,* 2nd Edition. New York: CRC Press/Lewis Publishers.

Mellström, G. A., J. E. Wahlberg, and H. I. Maibach. 1994. *Protective Gloves for Occupational Use.* New York: CRC Press/Lewis Publishers.

Meyer, E. 1990. *Chemistry of Hazardous Materials.* Englewood Cliffs, NJ: Prentice Hall.

Mycock, J. C., J. D. McKenna, and L. Theodore. 1995. *Handbook of Air Pollution Control Engineering and Technology.* New York: CRC Press/Lewis Publishers.

Nelson, G. O. 1992. *Gas Mixtures.* New York: CRC Press/Lewis Publishers.

Ott, W. R. 1995. *Environmental Statistics and Data Analysis.* New York: CRC Press/Lewis Publishers.

Patnaik, P. 1992. *A Comprehensive Guide to the Hazardous Properties of Chemical Substances.* New York: Van Nostrand Reinhold.

Philp, R. B. 1995. *Environmental Hazards and Human Health.* New York: CRC Press/Lewis Publishers.

Proctor, N. H., J. P. Hughes. 1991. *Chemical Hazards of the Workplace.* New York: Van Nostrand Reinhold.

Quigley, D. R. 1994. *Handbook of Emergency Chemical Management.* New York: CRC Press/Lewis Publishers.

Rea, W. J. 1995. *Chemical Sensitivity,* Vol. I-III. New York: CRC Press/Lewis Publishers.

Richardson, M. 1995. *Dictionary of Substance and Their Effects,* Vol. I-VI. New York: CRC Press/Lewis Publishers.

Rylander, R., R. R. Jacobs. 1994. *Organic Dusts.* New York: CRC Press/Lewis Publishers.

Sacarello, H. L. A. 1994. *The Comprehensive Handbook of Hazardous Materials.* New York: CRC Press/Lewis Publishers.

Sheftel, V. O. 1995. *Handbook of Toxic Properties of Monomers and Additives.* New York: CRC Press/Lewis Publishers.

Sherman, J. D. 1988. *Chemical Exposure and Disease.* New York: Van Nostrand Reinhold.

Talbott, E., G. Craun. 1995. *An Introduction to Environmental Epidemiology.* New York: CRC Press/Lewis Publishers.

The Merck Index, 10th edition. 1983. Rahway, NJ: Merck & Company.

U.S. Department of Health and Human Services, National Institute for Occupational Safety and Health. 1980. *Methacrylic acid. Registry of Toxic Effects of Chemical Substances,* Vol. 1. Washington, D.C.: U.S. Government Printing Office.

U.S. Department of Health and Human Services, National Institute for Occupational Safety and Health. 1994. *NIOSH Pocket Guide to Chemical Hazards.* Washington, D.C.: U.S. Government Printing Office.

MATERIAL SAFETY DATA SHEET

CHEMICAL NAME	
METHANE	**CRC PRESS/LEWIS PUBLISHERS DISCLAIMER:** THE INFORMATION AND RECOMMENDATIONS PRESENTED HEREIN ARE BASED ON SOURCES BELIEVED TO BE RELIABLE. CRC MAKES NO REPRESENTATION ON ITS COMPLETENESS OR ACCURACY. IT IS THE RESPONSIBILITY OF THE USER TO DETERMINE THE CHEMICAL'S SUITABILITY FOR ITS INTENDED USE, THE CHEMICAL'S SAFE USE, AND THE CHEMICAL'S PROPER DISPOSAL. NO REPRESENTATIONS AND/OR WARRANTIES, EITHER EXPRESSED OR IMPLIED, OF THE MERCHANTABILITY OR FITNESS FOR A PARTICULAR PURPOSE, OR OF ANY OTHER NATURE, ARE MADE WITH RESPECT TO THE INFORMATION PROVIDED IN THIS MSDS OR TO THE CHEMICAL TO WHICH INFORMATION MAY REFER. CRC NEITHER ASSUMES NOR AUTHORIZES ANY OTHER PERSON TO ASSUME FOR IT, ANY OTHER ADDITIONAL RESPONSIBILITY OR LIABILITY FOR THE USE OF, OR RELIANCE UPON, THIS INFORMATION.

HAZARD WARNING INFORMATION

HEALTH	FIRE	REACTIVE	OTHER	DEGREE OF HAZARD	COLOR CODING	OTHER CODES
1	4	0		0 = Minimum Hazard 1 = Slight Hazard 2 = Moderate Hazard 3 = Serious Hazard 4 = Severe Hazard	HEALTH = BLUE FIRE = RED REACTIVITY = YELLOW OTHER = WHITE	OX = Oxidizer ACID = Acid ALK = Alkali COR = Corrosive W = Use No Water

SECTION I - GENERAL INFORMATION

Characterization	RCRA Number	EPA Class
Gas, Hydrocarbon	**None**	**Not Applicable**

DOT Proper Shipping Name **Methane, Compressed or Methane, Refrigerated Liquid**	Chemical Abstract Service (CAS) Number **74-82-8**
DOT Hazard Class and Label Requirements **Flammable Gas (both forms)**	DOT Emergency Guide Code **15**
DOT Identification Number **UN 1971 (compressed); UN 1972 (liquid)**	Chemical Formula **CH_4**

Synonyms
Fire damp; marsh gas; methyl hydride.

SECTION II - HAZARDOUS INGREDIENTS AND IDENTITY INFORMATION

Hazardous Components (specific identity)	OSHA Exposure Criteria	NIOSH Exposure Criteria	Immediately Dangerous to Life and Health (IDLH)	ACGIH Exposure Criteria
Methane (derivation: From natural gas by absorption or adsorption; from coal mines for use as fuel gas; from a mixture of carbon monoxide and hydrogen (synthesis gas) obtained by reaction of hot coal with steam, the mixed gas is passed over a nickel-based catalyst at high temperature).	**PEL:** Not Established **STEL:** Not Established	**REL:** Not Established **STEL:** Not Established	**Not Determined**	**TLV:** Not Established **STEL:** Not Established

SECTION III - PHYSICAL AND CHEMICAL CHARACTERISTICS

Boiling Point **-259°F (-167°C)**	Specific Gravity (H_2O = 1) **.926**
Vapor Pressure (atmospheres) **Not Found**	Molecular Weight **16.05**
Vapor Density (Air = 1) **0.6 at 32°F (0°C)**	Freezing Point **-296°F (-182°C)**

Solubility
Only slightly soluble in water. Soluble in alcohol and ether.

Appearance and Odor
Colorless, odorless, tasteless gas.

SECTION IV - FIRE AND EXPLOSION HAZARD DATA

Flash Point (method used) **-213°F (-136°C) closed cup**	Explosive Limits in Air % by Volume **LEL: 5% UEL: 15%**
NFPA Classification **Flammable Gas**	Autoignition Temperature **999°F (537°C)**

Extinguishing Media
Stop flow of gas to stop fire. If this is not possible, letting fire burn itself out may be only alternative.

Special Fire Fighting Procedures
Methane is an extremely flammable, very dangerous fire and explosion hazard. Poisonous gases are evolved from fire. Wear a self-contained breathing apparatus (SCBA) with full facepiece. Ensure ALL sources of escaping gas are controlled. Monitor area long after fire is extinguished. Move containers from fire area if it can be done without risk. Use water spray to cool fire-exposed containers.

Unusual Fire and Explosion Hazards
Gas is explosive. Containers may explode in flame. Even if fire is extinguished, escaping gas can form explosive mixtures in air quickly and ignite without warning.

SECTION V - REACTIVITY DATA

Stability		Conditions to Avoid Methane is normally stable in its closed containers, during routine operations (those that do not expose it to ignition sources). Avoid contact with heat, sparks, flame, lighted tobacco products, and steam lines.
Stable X	Unstable	Incompatibility *(materials to avoid)* Methane is reactive in contact with chlorine or chlorine dioxide, bromine pentafluoride, nitrogen trifluoride, liquid oxygen, oxygen difluoride, other oxidizers, heat, sparks, and open flame.
Hazardous Polymerization		Conditions to Avoid Under normal conditions of temperature and pressure, hazardous polymerization of methane is not expected to occur.
May Occur	Will Not Occur X	Hazardous Decomposition or By-products If involved in fire, methane can produce toxic oxides of carbon (carbon monoxide and carbon dioxide).

SECTION VI - HEALTH HAZARD DATA

Primary Route(s) of Entry:	Inhalation? X	Absorption (skin)?	Ingestion?

Health Hazards

INHALATION: An asphyxiant gas. Initial symptoms are those of oxygen starvation. Feeling of "air hunger," diminished mental alertness, headache, dullness, shortness of breath, and an impairment in muscle coordination. Continued exposures lead to faulty judgment, depression of senses, rapid fatigue, unconsciousness, convulsions, coma, and death.

SKIN & EYES: Contact with the compressed gas or refrigerated liquid can cause frostbite and burns.

INGESTION: Not likely since methane is a gas at normal working (room) temperatures.

Carcinogenicity	NTP Listed?	IARC Cancer Review Group?	OSHA Regulated?	Target Organs?
Unknown Human Unknown Animal	No	No	No	None Reported

Medical Conditions Generally Aggravated by Exposure

None Reported.

Emergency and First-aid Procedures

Eye contact: Immediately flush with large amounts of warm (tepid) water for 15 minutes (minimum), occasionally lifting eyelids; seek medical attention. Skin contact: Remove all contaminated clothing. Immediately wash area with large amounts of warm (tepid) water. Do NOT rub or use air-blown heat on affected area. For inhalation: Remove the person from exposure. Provide respiratory assistance and CPR. Transfer promptly to medical facility. If swallowed: Not likely.

SECTION VII - PRECAUTIONS FOR SAFE HANDLING AND USE

Steps to be Taken in Case Material is Released or Spilled

Remove all ignition sources. Ventilate area. Stop flow of gas. If the source of leak is a cylinder and cannot be stopped, remove to safe place in open air; repair leak or let cylinder empty. If gas has not yet ignited, use a water spray to knock down vapors and direct them from ignition source.

Preferred Waste Disposal Method

None specified in the references (incineration of a gas may be the only alternative).

Precautions to be Taken in Handling and Storage

Do not store with incompatible materials or in the presence of heat since violent reactions or explosions can occur. Store in tightly closed pressurized containers in a cool, dark, well-ventilated area away from sunlight. Keep away from ignition sources such as fire, sparks, and flame.

Other Precautions and Warnings

Bulk storage of methane is not recommended. Containers may explode in fire or under conditions of extreme heat (do not store outdoors or in direct sunlight).

SECTION VIII - CONTROL MEASURES AND PERSONAL PROTECTIVE EQUIPMENT

Respiratory Protection *(specify type)*

No exposure limits have been established. Methane is a simple asphyxiant. Do NOT use cartridge-type respirators (ineffective for this type of gas). Use a powered supplied-air respirator or a self-contained breathing apparatus with full facepiece operated in pressure demand or other positive pressure mode.

Ventilation

Local exhaust or general mechanical systems recommended.

Protective Gloves	Eye Protection	Other Protective Clothing
Insulated Rubber	Goggles and/or Face Mask	Leather Apron

Work/Hygiene Practices

Always wash hands thoroughly after using chemical; never bring food, drink, or smoking materials into vicinity of chemicals.

METHANE

CH_4

CAS: 74-82-8

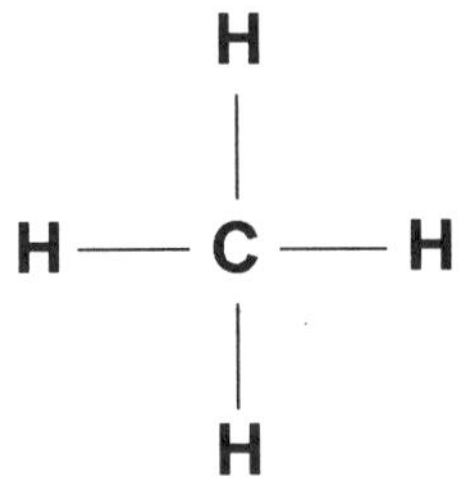

IDENTIFICATION AND TYPICAL USES

Methane is a colorless, tasteless, and odorless gas. It is used as a source of petrochemicals by conversion to hydrogen and carbon monoxide by steam cracking or partial oxidation. Products derived from methane include methanol, acetylene, hydrogen cyanide, and ammonia. As a natural gas, methane is used as a fuel, as a source of carbon black, and as the starting material for the manufacture of synthetic proteins.

RISK ASSESSMENT: HEALTH

General Assessment

Methane is a simple asphyxiant gas. As such, its primary route of entry into the body is through *inhalation.* While it does not pose a significant physiological effect (i.e., no systemic poisoning or toxicity), it can displace oxygen levels to below that which is necessary to sustain life. Skin contact with the liquid can be destructive to tissues due to the extremely cold temperatures but absorption is not a likely exposure route. Since methane usually exists as a gas at room temperatures, ingestion is not likely to occur either. It is not known to cause cancer, mutations, or reproductive effects in humans or test animals. It is an asphyxiant.

Inhalation of methane gas can cause headache, dullness, shortness of breath, and the symptoms of asphyxiation. These include rapid respiration and a feeling of "air hunger" or air starvation followed by a decrease in mental alertness, and impaired muscular coordination. If exposure continues, there can be nausea, vomiting, prostration, unconsciousness, coma, and death.

Skin or eye contact with the vapor poses no serious threat. The vapors have not been shown to cause any physiological response. However, if the liquid should contact either the skin or the eyes, there can be serious irritation, burns, tissue damage, and frostbite.

Acute Health Effects

The following acute (short-term) health effects may occur immediately or shortly after exposure to methane:

Skin: Liquid threat only. Redness, pain, swelling, blisters, and severe frostbite.

Eye: Liquid threat only. Causes frostbite, redness, irritation, severe pain, and blurred vision.

Lung: Vapor is non-irritating and therefore poses no threat of damage to the lungs. However, in high concentrations, methane can lead to symptoms of asphyxiation and oxygen deprivation up to and including death.

Chronic Health Effects

The following chronic (long-term) health effects may occur at some time after exposure to methane and can last for months or even years:

Cancer Hazards: According to information presented in the references, exposure to methane does not lead to carcinogenic effects (it is a simple asphyxiant).

Reproductive Hazards: There are no reports to support any claims of reproductive hazards.

Other Chronic Effects: None reported.

Recommended Risk-Reduction Measures

Personnel should avoid direct contact with methane. If another gas cannot be substituted for a methane, then *engineering controls* are the most effective method of reducing exposures. The best protection is to separate operations and provide local, explosion-proof exhaust ventilation at the site of chemical release. While not always operationally feasible, isolating operations can also reduce exposure. Using equipment with explosion-proof designs and ensuring all metal containers and tools are electrically grounded and bonded can further reduce the potential for injury or damage. Using respiratory protection is less effective than the controls mentioned above, but is still advisable whenever working with methane. Air-purifying respirators (e.g., the cartridge-type or gas masks) cannot be used with methane. These type respirators only purify the air and can do nothing to protect in oxygen deficient atmospheres. A powered air-supplied respirator operated in pressure demand or other positive pressure mode, or a self-contained breathing apparatus (SCBA) with full facepiece operated in pressure demand mode are the recommended respiratory protection methods of choice. If a full facepiece is not available, then chemical goggles should be worn to protect the eyes. Whenever a chemical splash hazard exists, a face shield and a rubber or leather apron should be worn. To prevent hand and skin exposures with the liquid, thermal protective gloves should be used.

Administrative controls should also be in place to minimize the potential for human exposures. These may include written procedures or policies which specify the methods and techniques that will be practiced whenever personnel are to work with methane.

All personnel should receive *training* on the use, hazards, protective measures, emergency actions, and other precautions per 29 CFR 1910.1200 (Hazard Communication), prior to the first assignment in an area where methane is used or stored.

Other methods to reduce exposure to chemicals include:

☑ Where possible, consider installing automatic sensors that warn personnel of dropping oxygen levels and use local exhaust ventilation at the site of chemical release. If local exhaust ventilation or enclosure is not used, respiratory protection should be mandatory.

☑ Always ensure that proper protective clothing is worn when using chemical substances and that personnel have been trained in its use, care, and maintenance.

☑ Wash thoroughly immediately after exposure to methane and at the end of the work shift or before eating, drinking, or smoking.

☑ Safety shower and eyewash stations should be readily available in work areas where methane is used or stored.

☑ Hazard warning information should be posted in the work area. In addition, as part of an on-going education and training program, all information on the health and safety hazards of methane should be communicated to all exposed or potentially exposed workers.

RISK ASSESSMENT: ENVIRONMENT

General Assessment

The environment is at risk of exposure during transportation, storage, disposal, or destruction of methane. In almost every scenario, the threat of environmental exposure is contingent upon the proper handling of the chemical substance. Accidental releases, large or small, can result in fire, explosion, and possible contamination of the surrounding environmental mediums (water, soil, and air). Methane is an extremely flammable and highly explosive gas. Its extremely low flash point and boiling point present a serious fire hazard. It has a relatively wide flammability range in air. Also, since it can react violently in the presence of oxidizers, such as chlorine, methane requires special consideration during any emergency involving a leak or release of methane gas or liquid.

Methane can enter the environment through industrial discharges, unchecked venting to air, or through spills of the liquid. Since liquids will most likely vaporize rapidly, water and soil contamination are possible but not probable.

☠ *Acute Ecological Effects*

Acute (short-term) toxic effects may include the death of animals, birds, or fish, and death or low growth rate in plants. Acute effects are seen 2 to 4 days after animals or plants are exposed to methane.

Insufficient data are available on the short-term effects of methane on aquatic life, plants, birds, or land animals. However, as a relatively non-toxic, simple asphyxiant, it might be assumed that ecological toxicity will be relatively low.

💣 *Chronic Ecological Effects*

Chronic (long-term) toxic effects may include shortened life span, reproductive problems, lower fertility, and changes in appearance or behavior in exposed animals. These effects can be seen long after first exposure(s) to toxic chemicals.

Insufficient data are available to evaluate the long-term effects of methane to aquatic life, plants, birds, or land animals. However, as a relatively non-toxic, simple asphyxiant, it might be assumed that any chronic ecological toxicity will be relatively low.

💧 *Water Solubility*

Methane is only slightly soluble in water. Concentrations of 1 milligram will mix with a liter of water.

⌛ *Persistence in the Environment*

Methane is non-persistent in water, with a half-life of less than 2 days. The half-life of a pollutant is the amount of time it takes for one half of the chemical to be degraded. Virtually 100% of methane will eventually end up in the air.

≈ *Bioaccumulation in Aquatic Organisms*

Some substances increase in concentration, or *bioaccumulate,* in living organisms as they breathe contaminated air, drink contaminated water, or eat contaminated food. These chemicals can become concentrated in the tissues and internal organs of animals as well as humans.

Methane is not expected to accumulate in the tissues of fish.

Recommended Risk-Reduction Measures

Proper training of all transporters will reduce the likelihood of a mishap or accident resulting in a leak or release to the environment. The correct use of DOT labeling (per the MSDS) on all containers, trucks, and rail cars will enable emergency responders to react properly and quickly to any disaster thereby reducing the potential risk to the environment and to personnel.

Storage of methane should be segregated from incompatible chemicals to minimize the risk of cross-contamination or contact. Buildings designated for storage should be equipped with appropriate fire protection systems (alarms, sprinklers, emergency lighting, portable extinguishers).

If a spill or release to the environment has occurred, fire department, emergency response, and/or hazardous materials spill personnel should be notified immediately. Cleanup should be attempted only by those trained in proper containment procedures using non-sparking tools. Contaminated soils should be removed for incineration and replaced with clean soil. When spills or leaks occur, it may be necessary to contact the local and/or state emergency response authorities. A comprehensive emergency response or disaster preparedness/recovery plan should be in place prior to any operations involving the use, transportation, storage, or disposal of methane. If methane is spilled or leaked, the following specific steps are recommended:

- ☑ Restrict persons not wearing protective clothing from area of spill or leak until cleanup is complete and area can be opened for normal work.
- ☑ Remove ignition sources, and ventilate area
- ☑ If applicable, stop flow of leaking liquid or gas. If leak source is a cylinder and the leak cannot be stopped in place, remove leaking cylinder to a safe place in the open air, and allow cylinder to empty.
- ☑ Keep methane out of a confined space, such as a sewer, because of the possibility of explosion (unless the sewer is designed to prevent the buildup of explosive concentrations).
- ☑ It may be necessary to dispose of methane as a hazardous waste. The responsible state agency or the regional office of the federal Environmental Protection Agency (EPA) should be contacted for specific recommendations.

RISK ASSESSMENT: BUSINESS

General Assessment

Accidents or mishaps involving methane can present a significant threat to business operations. The loss or damage of equipment or facilities can significantly affect fiscal viability. Lawsuits that may result from personnel injury, public exposures, and environmental contamination will also require a serious expenditure of resources. Media attention surrounding an injury, illness, or environmental damage can also result in a loss of profits and loss of current as well as future business.

Recommended Risk-Reduction Measures

Company attorneys, safety and health professionals, and environmental specialists should be involved in the development of any procedures dealing with safety and emergency response. A company official should be pre-designated as a public relations officer with specific training in dealing with the press. Corporate plans and policies should be developed, approved, and implemented long before any need for such arises.

REFERENCES

Ahlbom, A. 1993. *Biostatistics for Epidemiologists.* New York: CRC Press/Lewis Publishers.

American Conference of Governmental Industrial Hygienists. 1988. *Documentation of the Threshold Limit Values and Biological Exposure Indices,* 5th Edition (with updates). Cincinnati: ACGIH

Anthony, C. P., and N. J. Kolthoff. 1971. *Textbook of Anatomy and Physiology.* St. Louis: C. V. Mosby Company.

Asante-Duah, D. K. 1993. *Hazardous Waste Risk Assessment.* New York: CRC Press/Lewis Publishers.

Bloom, A. D. and F. J. DeSerres. 1995. *Ecotoxicity and Human Health.* New York: CRC Press/Lewis Publishers.

Brusick, D. J. 1994. *Methods for Genetic Risk Assessment.* New York: CRC Press/Lewis Publishers.

Bueche, J. R. 1972. *Principles of Physics,* 2nd Edition. New York: McGraw-Hill.

Burrell, R., D. K. Flaherty, and L. J. Sauers. 1992. *Toxicology of the Immune System.* New York: Van Nostrand Reinhold.

Calabrese, E. J., E. M. Kenyon. *Air Toxics and Risk Assessment.* New York: CRC Press/Lewis Publishers.

Calabrese, E. J. 1994. *Biological Effects of Low Level Exposures.* New York: CRC Press/Lewis Publishers.

Cockerman, L. G., B. S. Shane. 1994. *Basic Environmental Toxicology.* New York: CRC Press/Lewis Publishers.

Cothern, R. R. 1993. *Comparative Environmental Risk Assessment.* New York: CRC Press/Lewis Publishers.

Cothern, C. R. 1995. *Handbook for Environmental Risk Decisionmaking.* New York: CRC Press/Lewis Publishers.

Danse, I. R. 1991. *Common Sense Toxics in the Workplace.* New York: Van Nostrand Reinhold.

Environmental Protection Agency. 1983. *Chemical Hazard Information Profiles.* Washington, D.C.: U.S. EPA.

Fodden, C. R., and J. L. Weddell. 1992. *Hazardous Materials Emergency Action Data.* New York: CRC Press/Lewis Publishers.

Forseberg, K., and L. H. Keith. 1993. *Chemical; Protective Clothing Permeation and Degradation Database.* Chelsea, MI: Lewis Publishers.

Gardner, R. W. 1994. *Chemical Intolerance.* New York: CRC Press/Lewis Publishers.

Hallenbeck, W. H. 1993. *Quantitative Risk Assessment for Environmental and Occupational Health,* 2nd Edition. New York: CRC Press/Lewis Publishers.

Hodgson, E., R. B. Mailman, and J. E. Chambers. 1988. *Dictionary of Toxicology.* London: The MacMillian Company.

Hosty, J. W., and P. Foster. 1990. *A Practical Guide to Chemical Spill Response.* New York: Van Nostrand Reinhold.

Howard, P. H. 1995. *Handbook of Environmental Fate and Exposures Data for Organic Chemicals,* Vol. I-IV. New York: CRC Press/Lewis Publishers.

Howard, P. H., et. al. 1991. *Handbook of Environmental Degradation Rates.* New York: CRC Press/Lewis Publishers.

James, A., and M. Lord. 1992. *Index of Chemical and Physical Data.* New York: Van Nostrand Reinhold.

Jones, F. E. 1994. *Toxic Organic Vapors in the Workplace.* New York: CRC Press/Lewis Publishers.

Kamrin, M. A. 1988. *Toxicology: A Primer on Principles and Applications.* Chelsea, Michigan: Lewis Publishers.

Keith, L. H. and M. M. Walker. 1995. *Handbook of Air Toxics.* New York: CRC Press/Lewis Publishers.

Keith, L. H. (Editor). 1992. *Integrated Risk Information System (IRIS) Chemical Information Database.* Prepared by the U.S. Environmental Protection Agency. Boca Raton, FL: CRC Press/Lewis Publishers.

Keith, L. H. and D. B. Walters. (Editors) 1992. *National Toxicology Program's Chemical Database, Volumes I-VII.* Chelsea, MI: Lewis Publishers.

Landis, W. G., M. H. Yu. 1995. *Introduction to Environmental Toxicology.* New York: CRC Press/Lewis Publishers.

Langerman, N. 1994 *Precautionary Labels for Chemical Containers.* New York: CRC Press/Lewis Publishers.

Lauwerys, R. R., P. Hoet. 1993. *Industrial Chemical Exposure.* New York: CRC Press/Lewis Publishers.

Lewis, R. J., Sr. 1993. *Hawley's Condensed Chemical Dictionary,* 12th Edition. New York: Van Nostrand Reinhold.

Lewis, R. J., Sr. 1992. *Sax's Dangerous Properties of Industrial Materials,* Volumes 1, 2, and 3, 8th Edition. New York: Van Nostrand Reinhold.

Manahan, S. E. 1992. *Toxicological Chemistry,* 2nd Edition. New York: CRC Press/Lewis Publishers.

Meyer, E. 1990. *Chemistry of Hazardous Materials.* Englewood Cliffs, NJ: Prentice Hall.

Patnaik, P. 1992. *A Comprehensive Guide to the Hazardous Properties of Chemical Substances.* New York: Van Nostrand Reinhold.

Philp, R. B. 1995. *Environmental Hazards and Human Health.* New York: CRC Press/Lewis Publishers.

Proctor, N. H., and J. P. Hughes. 1991. *Chemical Hazards of the Workplace.* New York: Van Nostrand Reinhold.

Rea, W. J. 1995. *Chemical Sensitivity,* Vol. I-III. New York: CRC Press/Lewis Publishers.

Richardson, M. 1995. *Dictionary of Substance and Their Effects,* Vol. I-VI. New York: CRC Press/Lewis Publishers.

Sacarello, H. L. A. 1994. *The Comprehensive Handbook of Hazardous Materials.* New York: CRC Press/Lewis Publishers.

Sherman, J. D. 1988. *Chemical Exposure and Disease.* New York: Van Nostrand Reinhold.

Talbott, E., G. Craun. 1995. *An Introduction to Environmental Epidemiology.* New York: CRC Press/Lewis Publishers.

Torkelson, T. R. V. K. Rowe. 1981. *Patty's Industrial Hygiene and Toxicology,* Vol. 2B. New York: Wiley-Interscience.

Witorsch, P. and S. V. Spagnolo. 1994. *Air Pollution and Lung Disease in Adults.* New York: CRC Press/Lewis Publishers.

MATERIAL SAFETY DATA SHEET

CHEMICAL NAME

METHOXYCHLOR

HAZARD WARNING INFORMATION

HEALTH	FIRE	REACTIVE	OTHER	DEGREE OF HAZARD	COLOR CODING	OTHER CODES
1	0	0		0 = Minimum Hazard 1 = Slight Hazard 2 = Moderate Hazard 3 = Serious Hazard 4 = Severe Hazard	HEALTH = BLUE FIRE = RED REACTIVITY = YELLOW OTHER = WHITE	OX = Oxidizer ACID = Acid ALK = Alkali COR = Corrosive W = Use No Water

SECTION I - GENERAL INFORMATION

Characterization	RCRA Number	EPA Class
Pesticide (polycyclic, chlorinated)	**U247**	**Toxic Waste**

DOT Proper Shipping Name **Organochlorine pesticides, solid, toxic, n.o.s.**	Chemical Abstract Service (CAS) Number **72-43-5**
DOT Hazard Class and Label Requirements **Poison**	DOT Emergency Guide Code **55**
DOT Identification Number **NA 2761**	Chemical Formula **$(C_6H_4OCH_3)_2CHCCl_3$**

Synonyms

p,p'-Dimethoxydiphenyltrichloroethane; DMDT; Marlate 50 methoxy-DDT; 2,2-*bis*(p-methoxyphenyl)-1,1,1-trichloroethane; 1,1,1-trichloro-2,2-*bis*-(p-methoxyphenyl)ethane; Metox; methoxo.

SECTION II - HAZARDOUS INGREDIENTS AND IDENTITY INFORMATION

Hazardous Components (specific identity)	OSHA Exposure Criteria	NIOSH Exposure Criteria	Immediately Dangerous to Life and Health (IDLH)	ACGIH Exposure Criteria
Methoxychlor (derivation: By reaction of methyl phenyl ether and chloral hydrate; by condensation of anisole with chloral in t he presence of sulfuric acid).	**PEL:** 15 mg/m^3 (total dust) **STEL:** Not Established	**REL:** Lowest Feasible Limit Possible Cancer Agent	5000 mg/m^3	**TLV:** 10 mg/m^3 **STEL:** Not Established

SECTION III - PHYSICAL AND CHEMICAL CHARACTERISTICS

Boiling Point **Decomposes**	Specific Gravity (H_2O = 1) **1.41**
Vapor Pressure (mm Hg) **1.4×10^{-6} a 77°F (25°C)**	Molecular Weight (atomic weight) **345.7**
Vapor Density (Air = 1) **12.0**	Freezing Point **170°F (77°C)**

Solubility

Nearly insoluble in water. Soluble in alcohol, chloroform, methanol, xylene, most organic solvents.

Appearance and Odor

Pure methoxychlor is a white, crystalline, sand-like material with a slight, fruity odor. Technical methoxychlor is a pale buff flaky powder with a slight, fruity odor.

SECTION IV - FIRE AND EXPLOSION HAZARD DATA

Flash Point (method used) **Non-Combustible**	Explosive Limits in Air % by Volume **LEL: Not Determined UEL: Not Determined**
NFPA Classification **Non-Combustible Solid**	Autoignition Temperature **Not Determined**

Extinguishing Media

Use extinguishing media suitable to surrounding fire.

Special Fire Fighting Procedures

Methoxychlor is a non-combustible solid but it is often dissolved in a flammable or combustible liquid. Wear full protective clothing and SCBA. Remain clear of smoke, water fall out and water runoff. Move containers from fire area if it can be done without risk. Cool exposed containers. Toxic hydrogen chloride and carbon monoxide are formed in fire.

Unusual Fire and Explosion Hazards

None reported.

SECTION V - REACTIVITY DATA

<table>
<tr><td colspan="2">Stability</td><td>Conditions to Avoid
Methoxychlor is normally stable in closed containers under routine conditions of storage and handling. It will turn tan or pink on exposure to light. Avoid contact with heat and incompatible materials.</td></tr>
<tr><td>Stable
X</td><td>Unstable</td><td>Incompatibility (materials to avoid)
Incompatible with alkaline materials (decomposes). Also incompatible with strong oxidizers. Corrosive to aluminum and iron. Attacks some forms of rubber, plastic, and coatings.</td></tr>
<tr><td colspan="2">Hazardous Polymerization</td><td>Conditions to Avoid
Under normal conditions of temperature and pressure, hazardous polymerization of methoxychlor is not known to occur.</td></tr>
<tr><td>May Occur</td><td>Will Not Occur
X</td><td>Hazardous Decomposition or By-products
At high temperatures, methoxychlor decomposes and may produce poisonous gases, including toxic hydrogen chloride, carbonyl chloride, and carbon monoxide.</td></tr>
</table>

SECTION VI - HEALTH HAZARD DATA

Primary Route(s) of Entry:	Inhalation?	Absorption (skin/eye)?	Ingestion?
	X	X	X

Health Hazards

INHALATION: Human data inconclusive. May cause anxiety, dizziness, headache, disorientation, weakness, parasthesia ("pins and needles") muscle twitching, and tremor. High levels can lead to seizures, loss of consciousness, and death. Kidney and liver damage probable.

ABSORPTION: Will pass through unbroken skin and enter the bloodstream. Contact may severely irritate the skin and eyes. Systemic effects dominated by somnolence and depression.

INGESTION: Poisonous to humans by ingestion. Systemic effects due to central nervous system action include tremors, convulsion, excitement, ataxia, and gastritis.

Carcinogenicity	NTP Listed?	IARC Cancer Review Group?	OSHA Regulated?	Target Organs?
Questioned Human Suspected Animal	No	Group 3	29 CFR 1910.1000 Table Z-1	Skin, CNS, liver, kidney, reproductive system.

Medical Conditions Generally Aggravated by Exposure

Kidney and liver diseases, and convulsive disorders may be aggravated by exposure.

Emergency and First-aid Procedures

Eye contact: Do not allow victim to keep eyes tightly shut. Flush immediately with water for 15 minutes; seek medical attention. Skin contact: Remove all contaminated clothing. Immediately wash area with flooding amounts of water, then triple-wash with soap and water. Inhalation: Remove the person from exposure. Provide respiratory assistance and CPR. Transfer to medical facility. If swallowed: Contact poison control center. Unless advised otherwise and victim is conscious, give 1-2 glasses of water. The decision to induce vomiting must be carefully considered since CNS depression may induce seizures. Seek medical attention. Never give anything by mouth to an unconscious or convulsing person.

SECTION VII - PRECAUTIONS FOR SAFE HANDLING AND USE

Steps to be Taken in Case Material is Released or Spilled

Remove all sources of heat. Ventilate area of spill. Restrict those not involved in cleanup from entering area. Notify appropriate authorities. Collect powder materials using HEPA vacuum. Do NOT dry sweep. Absorb any liquid solution spills in vermiculite and place in sealed drum for disposal.

Preferred Waste Disposal Method

No citation.

Precautions to be Taken in Handling and Storage

Store in tightly closed containers in cool, dry, dark, well-ventilated area away from heat.

Other Precautions and Warnings

Avoid generating dusty conditions in storage areas. Protect containers from physical damage.

SECTION VIII - CONTROL MEASURES AND PERSONAL PROTECTIVE EQUIPMENT

Respiratory Protection *(specify type)*

For exposures above 15 mg/m^3 for extended periods, use a NISOH-approved supplied-air respirator or a self-contained breathing apparatus (SCBA) with full facepiece operated in positive pressure mode.

Ventilation

Local exhaust (preferred) at site of chemical work or general ventilation.

Protective Gloves	Eye Protection	Other Protective Clothing
Polycarbonate or Butyl Rubber	Chemical/Dust Goggles or Face Mask	Protective Apron, Boots (avoid skin contact)

Work/Hygiene Practices

Always wash hands thoroughly after using chemical; never bring food, drink, or smoking materials into vicinity of chemicals.

METHOXYCHLOR

$(C_6H_4OCH_3)_2CHCCl_3$

CAS: 72-43-5

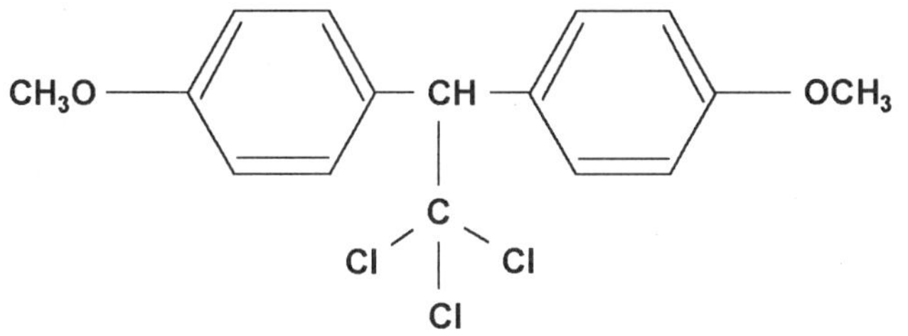

IDENTIFICATION AND TYPICAL USES

Pure methoxychlor is a white, sand-like material with a slight, fruit-like odor. Technical methoxychlor is a pale buff or gray flaky powder with a fruit-like odor. It is used as an insecticide and is effective against mosquito larvae and house flies. It is also used in dairy barns to control pests to livestock and to control insects for fruits, vegetables, and forage crops. Because it is less persistent in the environment and generally less toxic than DDT, but structurally similar, it has replaced DDT in many applications.

RISK ASSESSMENT: HEALTH

General Assessment

Methoxychlor is moderately toxic to humans by *ingestion, inhalation,* and skin *absorption.* It is a questionable human carcinogen with experimental carcinogenic data reported. Human data are limited and inconclusive with regard to overall toxicity. Animal studies have shown some mutation effects. There is some evidence in the references that indicates methoxychlor is a suspected animal carcinogen with long-term effects primarily in the liver.

Methoxychlor has been shown to be a central nervous system stimulant in animal studies. Symptoms of exposure by all routes include nausea, vomiting, abdominal pain, irritation, confusion, apprehension, excitability, paresthesia, dizziness, headache, disorientation, tremor, ataxia (loss of muscle coordination), stupor, coma, convulsions, respiratory collapse, and possibly death.

Skin contact will result in moderate to severe irritation with rash or burning sensation. Eye contact may also result in irritation. Caution is warranted when both inhalation and skin contact exposures are possible. Even though air levels may be within limits, personnel may be overexposed if methoxychlor should contact the skin. The dangerous chronic dose in humans is unknown. May cause liver damage (fatty degeneration), characteristic of chronic alcoholism.

Acute Health Effects

The following acute (short-term) health effects may occur immediately or shortly after exposure to methoxychlor:

Skin: Liquid can pass through unbroken skin. Contact may cause irritation and rash. Symptoms of exposure include headache, nausea, abdominal pain, and vomiting.

Eye: Severe irritation and possible damage on contact.

Lung: Irritation of the nose and throat. Breathing vapors can cause cough, convulsions, unconsciousness, and death.

CNS: A central nervous system stimulant. Symptoms include ataxia, excitement, tremors, convulsions, confusion, delirium, and irritability.

Chronic Health Effects

The following chronic (long-term) health effects may occur at some time after exposure to methoxychlor and can last for months or even years:

Cancer Hazards: Methoxychlor may be a carcinogen in humans since it has been shown to cause cancer in animals. There is no evidence in the references to confirm the human carcinogenicity of methoxychlor. More research is required in this area.

Reproductive Hazard: According to information presented in the references, exposure to methoxychlor may adversely affect reproduction.

Other Chronic Effects: Long-term exposure to methoxychlor may damage the liver, kidneys, reproductive system, lungs, prostate, ovaries, and may cause tumors. These effects have been documented in animals. There is no evidence of such effects in humans.

Ü Recommended Risk-Reduction Measures

While methoxychlor is considered to be of relatively low toxicity in comparison to the structurally similar pesticide DDT, personnel should still avoid direct contact with this compound. If a less toxic material or compound cannot be substituted for methoxychlor, then *engineering controls* are the most effective method of reducing exposures. The best protection is to enclose operations and/or provide local exhaust ventilation at the site of methoxychlor dust release. This may be practical in manufacturing operations, but is not feasible in application since it is primarily used outdoors. While not always operationally possible, isolating operations involving methoxychlor manufacturing can also reduce exposure. Using respiratory protection is less effective than the controls mentioned above, but is still advisable whenever working with or around methoxychlor. At exposures that exceed the PEL of 15 mg/m^3 (TWA), use an MSHA/NIOSH-approved chemical cartridge respirator equipped with an organic vapor filter and a dust/mist pre-filter. While a high-efficiency particulate air (HEPA) filter may not be entirely necessary for this compound, its use would certainly provide near optimum protection. The best protection is provided using a supplied-air respirator or a self-contained breathing apparatus (SCBA) with full facepiece operated in pressure demand or other positive pressure mode. If a full facepiece is not available, then chemical goggles should be worn to protect the eyes. A face shield and protective apron should also be worn. To prevent hand and skin exposures, impervious gloves should be used. Polycarbonate or butyl rubber appear to provide the best protection. However, since it will attack some rubbers and plastic materials, glove manufacturers should be contacted and permeation studies obtained *before* final glove selections have been made.

Administrative controls should also be in place to minimize the potential for human exposures. These may include written procedures or policies which specify the methods and techniques that will be practiced whenever personnel are to work with methoxychlor.

All personnel should receive *training* on the use, hazards, protective measures, emergency actions, and other precautions per 29 CFR 1910.1200 (Hazard Communication), prior to the first assignment in an area where methoxychlor is used or stored.

There are no specific medical testing protocols established or recommended for this compound. However, if symptoms develop or overexposure is suspected, medical attention is highly recommended. Any evaluation should include a careful history of past and present symptoms with an examination. Medical tests that simply look for existing damage are not a substitute for controlling exposures.

Other methods to reduce exposure include:

- ☑ Where possible, enclose operations and use local exhaust ventilation at the site of chemical release. If local exhaust ventilation or enclosure is not used, respiratory protection should be mandatory.
- ☑ Always ensure that proper protective clothing is worn when using chemical substances.
- ☑ Wash thoroughly immediately after exposure to methoxychlor and at the end of the work shift or before eating, drinking, or smoking.
- ☑ Hazard warning information should be posted in the work area. In addition, as part of an on-going education and training program, all information on the health and safety hazards of methoxychlor should be communicated to all exposed workers.
- ☑ Eye wash stations should be provided in the immediate work area for emergency use. If there is a possibility of skin exposure, then emergency shower facilities should also be provided.
- ☑ Workers whose clothing has been contaminated by methoxychlor should change into clean clothes before leaving work. Contaminated work clothing should be laundered only by individuals who have been informed of the hazards of methoxychlor.
- ☑ Specific engineering controls are recommended by NIOSH. Refer to NIOSH Criteria Document: Occupational Exposure During the Manufacture and Formulation of Pesticides (Publication Number 78-174)

RISK ASSESSMENT: ENVIRONMENT

General Assessment

Because of its use as an insecticide, methoxychlor may be present in the environment at any given time and at a variety of levels. The environment is also at risk of exposure during transportation, storage, disposal, or destruction of methoxychlor. In almost every scenario, the threat of environmental exposure is contingent upon the proper handling of the chemical substance. Accidental spills, large or small, where methoxychlor contacts incompatible commodities can result in violent and even explosive reactions (depending upon conditions of spill), and possible contamination of the surrounding environmental mediums (water, soil, and air).

Methoxychlor is considered a non-combustible solid. However, it may be present (dissolved) in some flammable liquids such as benzene. Also, because it is incompatible with many strong oxidizers, common alkaline materials, and some metals, such as iron and aluminum, extreme caution is required in handling, storage, transportation, and disposal of methoxychlor. It will also attack some forms of rubber and plastic. These characteristics also require special consideration during any emergency situation involving a leak or spill of methoxychlor powder or dust.

Methoxychlor is a chlorinated hydrocarbon insecticide used to control a variety of pests. It has served as a replacement for DDT in most cases. Methoxychlor enters the environment in agricultural runoff, industrial discharges, and spills.

Acute Ecological Effects

Acute (short-term) toxic effects may include the death of animals, birds, or fish and death or low growth rate in plants. Acute effects are seen 2 to 4 days after animals or plants are exposed to methoxychlor.

Methoxychlor has high acute toxicity to aquatic life. Methoxychlor has caused injuries to agricultural crops and tree species. Insufficient data are available to evaluate or predict on the short-term effects of methoxychlor to birds and land animals.

Chronic Ecological Effects

Chronic (long-term) toxic effects may include shortened life span, reproductive problems, lower fertility, and changes in appearance or behavior in exposed animals. These effects can be seen long after first exposure(s) to toxic chemicals.

Methoxychlor has high chronic toxicity to aquatic life. Insufficient data are available to evaluate or predict the long-term effects of methoxychlor to plants, birds, or land animals.

Water Solubility

Methoxychlor is only very slightly soluble in water. Concentrations of less than 1 milligram will mix with a liter of water.

Persistence in the Environment

Methoxychlor is moderately persistent in the water, with a half-life of 20 to 200 days. The half-life of a pollutant is the amount of time it takes for one-half of the chemical to be degraded. About 50% of methoxychlor will eventually end up in terrestrial soil; approximately 46.8% will end up and aquatic sediments; about 3% in water; and the remainder will end up in the suspended aquatic solids.

Bioaccumulation in Aquatic Organisms

Some substances increase in concentration, or *bioaccumulate,* in living organisms as they breathe contaminated air, drink contaminated water, or eat contaminated food. These chemicals can become concentrated in the tissues and internal organs of animals as well as humans.

The concentration of methoxychlor found in fish tissues is expected to be much higher that the average concentration of methoxychlor in the water from which the fish was taken.

Recommended Risk-Reduction Measures

Proper training of all transporters will reduce the likelihood of a mishap or accident resulting in a leak or spill of methoxychlor into the environment. The correct use of DOT labeling (per the MSDS) on all containers, trucks, and rail cars will enable emergency responders to react properly and quickly to any disaster thereby reducing the potential risk to the environment and to personnel.

Storage of methoxychlor should be segregated from incompatible materials to minimize the risk of cross-contamination or contact. Buildings or storage areas should be equipped with appropriate fire protection systems (alarms, sprinklers). Proper ventilation and protective equipment should be used while handling the powders or while preparing aqueous solutions.

If a spill or leak to the environment has occurred, fire department, emergency response and/or hazardous materials spill personnel should be notified immediately. Cleanup should be attempted only by those trained in proper spill containment procedures. If methoxychlor should contact the water table, aquifer, or navigable waterway, reclamation procedures should be initiated as soon as practical. It is only very slightly soluble in water and partial or even total remediation may be possible. The local and/or state emergency response authorities must be notified. A comprehensive emergency response or disaster preparedness/recovery plan should be in place prior to any operations involving the use, transportation, storage, or disposal of methoxychlor. If methoxychlor is spilled or leaked, the following specific steps are recommended:

☑ Restrict persons not wearing protective clothing from area of spill or leak until cleanup is complete and area can be opened for normal work.

☑ Ventilate are of spill or leak.

☑ Collect powdered materials using a vacuum equipped with a HEPA filter. Do NOT dry sweep. Damp mop residues. Absorb solution spills in vermiculite or other material and place in sealed drum for disposal or reclamation.

☑ It may be necessary to dispose of methoxychlor as a hazardous waste. The responsible state agency or the regional office of the federal Environmental Protection Agency (EPA) should be contacted for specific recommendations.

RISK ASSESSMENT: BUSINESS

General Assessment

Accidents or mishaps involving methoxychlor can present a significant threat to business operations. The loss or damage of equipment or facilities can significantly affect fiscal viability. Lawsuits that may result from personnel injury/death, public exposures, and/or environmental contamination will also require a serious expenditure of resources. Media attention surrounding an injury, death, or environmental damage can also result in a loss of profits and loss of current as well as future business. Always remember that anytime to terms "cancer" or "carcinogen" are used, public emotion, hysteria, and ignorance can run equally high. This must be carefully considered whenever developing or implementing public relations policies.

Recommended Risk-Reduction Measures

Company attorneys, safety and health professionals, and environmental specialists should be involved in the development of any procedures for responding to chemical incidents. A company official should be predesignated as a public relations officer with specific training in dealing with the press. Corporate plans and policies should be developed, approved, and implemented long before any need for such arises.

REFERENCES

American Conference of Governmental Industrial Hygienists. 1988. *Documentation of the Threshold Limit Values and Biological Exposure Indices,* 5th Edition (with updates). Cincinnati: ACGIH

Brusick, D. J. 1994. *Methods for Genetic Risk Assessment.* New York: CRC Press/Lewis Publishers.

Bloom, A. D. and F. J. DeSerres. 1995. *Ecotoxicity and Human Health.* New York: CRC Press/Lewis Publishers.

Calabrese, E. J. 1994. *Biological Effects of Low Level Exposures.* New York: CRC Press/Lewis Publishers.

Cockerman, L. G., B. S. Shane. 1994. *Basic Environmental Toxicology.* New York: CRC Press/Lewis Publishers.

Cothern, C. R. 1995. *Handbook for Environmental Risk Decisionmaking.* New York: CRC Press/Lewis Publishers.

Gardner, R. W. 1994. *Chemical Intolerance.* New York: CRC Press/Lewis Publishers.

Howard, P. H., et. al. 1991. *Handbook of Environmental Degradation Rates.* New York: CRC Press/Lewis Publishers.

Keith, L. H. and M. M. Walker. 1995. *Handbook of Air Toxics.* New York: CRC Press/Lewis Publishers.

Landis, W. G., M. H. Yu. 1995. *Introduction to Environmental Toxicology.* New York: CRC Press/Lewis Publishers.

Lauwerys, R. R., P. Hoet. 1993. *Industrial Chemical Exposure.* New York: CRC Press/Lewis Publishers.

Lewis, R. J., Sr. 1992. *Sax's Dangerous Properties of Industrial Materials,* Volumes 1, 2, and 3, 8th Edition. New York: Van Nostrand Reinhold.

Lide, D. R. 1995. *Handbook of Chemistry and Physics,* 76th Edition. New York: CRC Press/Lewis Publishers.

Manahan, S. E. 1992. *Toxicological Chemistry,* 2nd Edition. New York: CRC Press/Lewis Publishers.

Meyer, E. 1990. *Chemistry of Hazardous Materials.* Englewood Cliffs, NJ: Prentice Hall.

Patnaik, P. 1992. *A Comprehensive Guide to the Hazardous Properties of Chemical Substances.* New York: Van Nostrand Reinhold.

Rea, W. J. 1995. *Chemical Sensitivity,* Vol. I-III. New York: CRC Press/Lewis Publishers.

Richardson, M. 1995. *Dictionary of Substance and Their Effects,* Vol. I-VI. New York: CRC Press/Lewis Publishers.

Sacarello, H. L. A. 1994. *The Comprehensive Handbook of Hazardous Materials.* New York: CRC Press/Lewis Publishers.

Torkelson, T. R. V. K. Rowe. 1981. *Patty's Industrial Hygiene and Toxicology,* Vol. 2B. New York: Wiley-Interscience.

U.S. Department of Health and Human Services, National Institute for Occupational Health (NIOSH). 1978. *NIOSH Criteria Document: Occupational Exposure During the Manufacture and Formulation of Pesticides.* Publication Number 78-174. Washington, DC: U.S. Government Printing Office.

U.S. Department of Health and Human Services, National Institute for Occupational Safety and Health. 1994. *NIOSH Pocket Guide to Chemical Hazards.* Washington, D.C.: U.S. Government Printing Office.

MATERIAL SAFETY DATA SHEET

CHEMICAL NAME	CRC PRESS/LEWIS PUBLISHERS DISCLAIMER
METHYL ACETATE	THE INFORMATION AND RECOMMENDATIONS PRESENTED HEREIN ARE BASED ON SOURCES BELIEVED TO BE RELIABLE. CRC MAKES NO REPRESENTATION ON ITS COMPLETENESS OR ACCURACY. IT IS THE RESPONSIBILITY OF THE USER TO DETERMINE THE CHEMICAL'S SUITABILITY FOR ITS INTENDED USE, THE CHEMICAL'S SAFE USE, AND THE CHEMICAL'S PROPER DISPOSAL. NO REPRESENTATIONS AND/OR WARRANTIES, EITHER EXPRESSED OR IMPLIED, OF THE MERCHANTABILITY OR FITNESS FOR A PARTICULAR PURPOSE, OR OF ANY OTHER NATURE, ARE MADE WITH RESPECT TO THE INFORMATION PROVIDED IN THIS MSDS OR TO THE CHEMICAL TO WHICH INFORMATION MAY REFER. CRC NEITHER ASSUMES NOR AUTHORIZES ANY OTHER PERSON TO ASSUME FOR IT, ANY OTHER ADDITIONAL RESPONSIBILITY OR LIABILITY FOR THE USE OF, OR RELIANCE UPON, THIS INFORMATION.

HAZARD WARNING INFORMATION

HEALTH	FIRE	REACTIVE	OTHER	DEGREE OF HAZARD	COLOR CODING	OTHER CODES
1	3	0		0 = Minimum Hazard 1 = Slight Hazard 2 = Moderate Hazard 3 = Serious Hazard 4 = Severe Hazard	HEALTH = BLUE FIRE = RED REACTIVITY = YELLOW OTHER = WHITE	OX = Oxidizer ACID = Acid ALK = Alkali COR = Corrosive W = Use No Water

SECTION I - GENERAL INFORMATION

Characterization	RCRA Number	EPA Class
Aliphatic Ester	**D001**	**Characteristic (I) Waste**

DOT Proper Shipping Name: **Methyl Acetate**	Chemical Abstract Service (CAS) Number: **79-20-9**
DOT Hazard Class and Label Requirements: **Flammable Liquid**	DOT Emergency Guide Code: **26**
DOT Identification Number: **UN 1231**	Chemical Formula: **CH_3COOCH_3**

Synonyms
Methyl ester of acetic acid; methyl ethanoate; acetic acid methyl ester; Deveton; Tereton.

SECTION II - HAZARDOUS INGREDIENTS AND IDENTITY INFORMATION

Hazardous Components (specific identity)	OSHA Exposure Criteria	NIOSH Exposure Criteria	Immediately Dangerous to Life and Health (IDLH)	ACGIH Exposure Criteria
Methyl acetate (derivation: By heating methanol and acetic acid in presence of sulfuric acid and distilling). **1 ppm = 3.08 mg/m^3**	**PEL (8-hour):** **200 ppm** **610 mg/m^3** **STEL (15-min):** **250 ppm** **760 mg/m^3**	**REL (10-hour):** **200 ppm** **610 mg/m^3** **STEL:** **250 ppm** **760 mg/m^3**	**3100 ppm**	**TLV:** **200 ppm** **606 mg/m^3** **STEL:** **250 ppm** **760 mg/m^3**

SECTION III - PHYSICAL AND CHEMICAL CHARACTERISTICS

Boiling Point: **135°F (57°C)**	Specific Gravity (H_2O = 1): **0.93**
Vapor Pressure (mm Hg): **173 at 68°F (20°C)**	Molecular Weight: **74.1**
Vapor Density (Air = 1): **2.55**	Freezing Point: **-144°F (-98°C)**

Solubility
Soluble in water (30% at 20°C). Soluble in chloroform, alcohol, benzene, acetone, and ether.

Appearance and Odor
Clear colorless liquid with an strong ether-like, fragrant, or fruity odor. Odor Threshold = 180 ppm.

SECTION IV - FIRE AND EXPLOSION HAZARD DATA

Flash Point (method used): **14°F (-10°C) closed cup**	Explosive Limits in Air % by Volume: **LEL: 3.1%** **UEL: 16%**
NFPA Classification: **Class IB Flammable Liquid**	Autoignition Temperature: **935°F (501°C)**

Extinguishing Media
Dry chemical, alcohol-resistant foam, carbon dioxide, water spray, or fog.

Special Fire Fighting Procedures
Poisonous gases are produced in fire. Wear full protective clothing and self-contained breathing apparatus (SCBA). Use blanketing effect with foam to smother large fires. Continue to cool containers with water after fire is extinguished. Water on fire itself may be ineffective and actually spread fire further.

Unusual Fire and Explosion Hazards
Immediately withdraw if rising sound from venting device is heard or if fire is causing discoloration to the tank. Evacuate 1500 feet radius if fire becomes uncontrollable or container is exposed to direct flame. Vapors are heavier than air and may travel a distances to ignition source to flashback.

SECTION V - REACTIVITY DATA

Stability		Conditions to Avoid: Methyl acetate is stable under normal conditions of storage and operation. Avoid contact with incompatible materials. Keep away from heat, flame, and other sources of ignition.
Stable: X	Unstable:	Incompatibility (*materials to avoid*): Reacts violently in contact with nitrates, strong oxidizers, alkalis, and acids. Reacts slowly with water to form acetic acid and methanol.
Hazardous Polymerization		Conditions to Avoid: Under normal conditions of temperature and pressure, hazardous polymerization of methyl acetate is not expected to occur.
May Occur:	Will Not Occur: X	Hazardous Decomposition or By-products: When heated to decomposition, methyl acetate emits toxic oxides of carbon (dioxide and monoxide) and acrid and irritating fumes.

SECTION VI - HEALTH HAZARD DATA

Primary Route(s) of Entry:	Inhalation?	Absorption (skin)?	Ingestion?
	X		X

Health Hazards

INHALATION: Irritation of the eyes, nose, throat, and respiratory tract with a possibility for narcosis and headache. Systemic effects include olfactory changes, conjunctiva inflammation, chest tightness, nervous irritation, dizziness, drowsiness, dypsnea, palpitation, and corneal edema.

SKIN & EYES: Will irritate the skin. May cause contact dermatitis. Eye contact may cause corneal damage with cloudy swelling, and fatty degeneration of the viscera.

INGESTION: May cause headache, drowsiness, and unconsciousness.

Carcinogenicity	NTP Listed?	IARC Cancer Review Group?	OSHA Regulated?	Target Organs?
Unknown Human Unknown Animal	No	No	29 CFR 1910.1000 Table Z-1	Respiratory system, skin, eyes, CNS.

Medical Conditions Generally Aggravated by Exposure

None reported.

Emergency and First-aid Procedures

Eye contact: Flush immediately with water for 15 minutes (minimum); seek medical attention. Skin contact: Remove all contaminated clothing. Immediately wash area with large amounts of soap and water. Seek medical assistance. For inhalation: Remove the person from exposure. Provide respiratory assistance and CPR. Transfer to medical facility. If swallowed: Seek medical attention immediately. Call poison control center for advice. Give 1-2 glasses of milk or water to dilute. Do NOT induce vomiting unless advised otherwise. Never give an unconscious or convulsing person anything by mouth.

SECTION VII - PRECAUTIONS FOR SAFE HANDLING AND USE

Steps to be Taken in Case Material is Released or Spilled

Contain spills using absorbent material (vermiculite or other). Ventilate area of spill. Remove all ignition sources. Have water source available in case of fire. Store materials in DOT-approved containers. Restrict those not involved in cleanup from entering area. Notify appropriate authorities, as required.

Preferred Waste Disposal Method

Incineration in chemical incinerator equipped with scrubber and afterburner.

Precautions to be Taken in Handling and Storage

Store in tightly closed containers in a cool, dark, well-ventilated area. Protect containers from physical damage. Keep heat, fire, and flame away. If possible, automatically transfer liquids between containers.

Other Precautions and Warnings

Bulk storage of methyl acetate is not recommended. Ground and bond all metal containers.

SECTION VIII - CONTROL MEASURES AND PERSONAL PROTECTIVE EQUIPMENT

Respiratory Protection (*specify type*)

Where potential exists for exposure over 400 ppm, use an MSHA/NIOSH-approved full facepiece respirator with an organic vapor cartridge. For high exposures, use a self-contained breathing apparatus (SCBA) with full facepiece operated in pressure demand or other positive pressure mode.

Ventilation

Local exhaust or general mechanical systems recommended.

Protective Gloves	Eye Protection	Other Protective Clothing
Butyl Rubber or Neoprene	Chemical Goggles or Face Mask	Impermeable Apron

Work/Hygiene Practices

Always wash hands thoroughly after using chemical; never bring food, drink, or smoking materials into vicinity of chemicals.

METHYL ACETATE

CH_3COOCH_3

CAS: 79-20-9

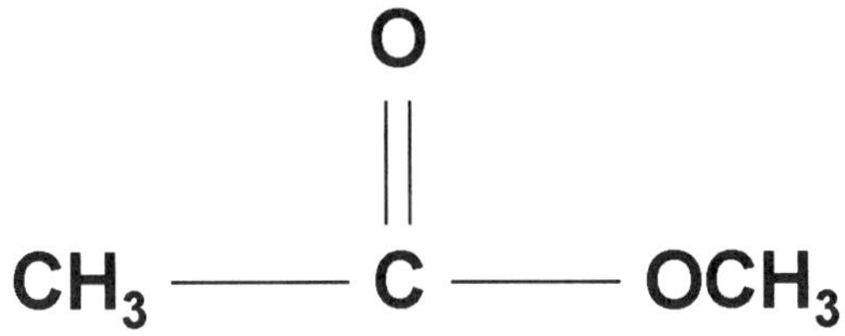

IDENTIFICATION AND TYPICAL USES

Methyl acetate is a colorless liquid with a characteristic, fruity, ether-like odor. It is used as a solvent for lacquers, resins, oils, and nitrocellulose. It is also used in paint removers, as a flavoring agent, in agriculture, and in the manufacture of artificial leather.

RISK ASSESSMENT: HEALTH

General Assessment

Methyl acetate is a human systemic irritant by *inhalation* and may cause toxic systemic effects. It is mildly toxic by *ingestion*. Skin contact can produce localized effects but dermal absorption has not been reported. While its specific carcinogenic and teratogenic properties are unknown, mutation data have been reported.

Inhalation of methyl acetate vapors is not likely to occur under normal operating conditions. The extremely irritating properties of the vapor will cause personnel to vacate the area before toxic exposure levels occur. However, should an exposure occur by inhalation, it will likely result in irritation of the eyes, nose, throat, and upper respiratory tract, including the mucosa (particularly the eyes, gums, and respiratory passages). Inhalation can cause changes in the sense of smell (olfactory changes), conjunctiva irritation, and pulmonary changes. In high concentrations, there can be additional pulmonary effects and/or depression of the central nervous system with mild narcotic effects including dizziness, drowsiness, tightness in chest, palpitation, dypsnea, conjunctivitis, and corneal edema.

Skin contact results in the mild irritation and, if allowed to remain in contact with the liquid for prolonged periods, removal of the skin's natural protective oily layer on its surface with subsequent drying, cracking, and possible secondary infections and dermatitis. Eye contact can cause serious damage to the cornea with a possibility for permanent effects to vision.

Ingestion of methyl acetate may cause headache, drowsiness, unconsciousness, and unspecified gastrointestinal effects.

Acute Health Effects

The following acute (short-term) health effects may occur immediately or shortly after exposure to methyl acetate:

Skin: Mild to moderate irritation. Will remove the skin's protective oily layer to cause drying, cracking, and possible secondary infection.

Eye: Severe irritation which can lead to injury and permanent damage if allowed to remain in contact with the eyes.

Lung: Irritation of the eyes, nose, throat and lungs (especially the mucus membranes). May cause changes in the sense of smell and in pulmonary function.

CNS: High concentrations can depress the central nervous system leading to narcoses and possible loss of consciousness.

Chronic Health Effects

The following chronic (long-term) health effects may occur at some time after exposure to methyl acetate and can last for months or even years:

Cancer Hazards: According to the references, methyl acetate has not been adequately tested for its ability to cause cancer in test animals. However, mutation data have been reported. This does not mean that methyl acetate is carcinogenic. It does mean that further research is required and caution is warranted.

Reproduction: According to the references, methyl acetate has not been adequately tested for its ability to adversely affect reproduction in test animals.

Other Chronic Effects: Very high or prolonged exposure may lead to chronic skin irritation and possible dermatitis.

Recommended Risk-Reduction Measures

Personnel should avoid direct contact with methyl acetate. The best risk reduction measure is to use a less toxic chemical as a substitute for methyl acetate. If this is not possible or feasible, then *engineering controls* are the most effective method of reducing exposures. The best protection is to enclose operations and/or provide local exhaust ventilation at the site of chemical release. While not always operationally feasible, isolating operations can also reduce exposure risk.

Using respiratory protection is less effective than the controls mentioned above, but is still advisable whenever working with or around methyl acetate. For exposures over 200 ppm (8 hours), an MSHA/NIOSH-approved chemical respirator with an organic vapor cartridge should be used. For higher exposures, or when the exposure level is unknown (as in an emergency), a self-contained breathing apparatus (SCBA) with full facepiece and operated in pressure demand mode or a supplied-air respirator with full facepiece operated in pressure demand are the recommended respiratory protection methods of choice. If a full facepiece is not available, then chemical goggles should be worn to protect the eyes. Whenever a chemical splash hazard exists, a face shield and a protective apron should be worn. To prevent hand and skin exposures, impervious gloves should be used. Neoprene, butyl rubber, or chlorinated polyethylene products should provide adequate protection. However, glove manufacturers should be contacted to obtain permeation studies *before* gloves are selected.

Administrative controls should also be in place to minimize the potential for human exposures. These may include written procedures or policies which specify the methods and techniques that will be practiced whenever personnel are to work with methyl acetate.

All personnel should receive *training* on the use, hazards, protective measures, emergency actions, and other precautions per 29 CFR 1910.1200 (Hazard Communication), prior to the first assignment in an area where methyl acetate is used or stored.

If symptoms develop or overexposure is suspected, the following medical tests are recommended:

- ☑ Liver and kidney function tests.
- ☑ Lung function tests.
- ☑ Skin testing with dilute methyl acetate to help diagnose allergy (performed by a qualified allergist).
- ☑ Methyl acetate is hydrolyzed to methanol in the body. Monitoring serum methanol and urine methanol levels may be useful in determining exposure.

Any medical evaluation should include a careful history of past and present symptoms with an examination. Medical tests that look for existing damage are not a substitute for controlling exposures.

Other methods to reduce exposure include:

- ☑ Where possible, enclose operations and use local exhaust ventilation at the site of chemical release. If local exhaust ventilation or enclosure is not used, respiratory protection should be mandatory.
- ☑ Always ensure that proper protective clothing is worn when using chemical substances and that personnel are trained in its use and care.
- ☑ Wash thoroughly immediately after exposure to methyl acetate and at the end of the work shift or before eating, drinking, or smoking.
- ☑ Hazard warning information should be posted in the work area. In addition, as part of an on-going education and training program, all information on the health and safety hazards of methyl acetate should be communicated to all exposed and potentially exposed workers.
- ☑ Never eat, drink, or smoke in areas were methyl acetate is used, handled, or stored.

RISK ASSESSMENT: ENVIRONMENT

General Assessment

The environment is at risk of exposure during transportation, storage, disposal, or destruction of methyl acetate. In almost every scenario, the threat of environmental exposure is contingent upon the proper handling of the chemical substance.

Accidental spills, large or small, can result in fire, explosion, and possible contamination of the surrounding environmental mediums (water, soil, and air). Methyl acetate is considered a Class IB flammable liquid (according to OSHA 29 CFR 1910.106). Its low flash point and relatively low boiling point present a serious fire and explosion hazard concern. Also, because it is incompatible with a number of common materials, especially strong oxidizers and acids, contact can result in violent and explosive reactions. It can also form explosive mixtures in air and can ignite on contact with heat, fire, or sparks. These characteristics require special consideration during any emergency situation involving a leak or spill of methyl acetate.

The proper disposal/destruction method for methyl acetate is to burn it in a chemical incinerator equipped with an afterburner and air scrubber. Methyl acetate can enter the environment through unchecked industrial discharges into effluents and through spills.

☠ *Acute Ecological Effects*

Acute (short-term) toxic effects may include the death of animals, birds, or fish, and death or low growth rate in plants. Acute effects are seen 2 to 4 days after animals or plants are exposed to methyl acetate. This chemical has moderate acute toxicity to aquatic life. Insufficient data are available to evaluate or predict the short-term effects of methyl acetate to plants, birds, or terrestrial animals.

Chronic Ecological Effects

Chronic toxic effects may include shortened life span, reproductive problems, lower fertility, and changes in appearance or behavior in exposed animals. These effects can be seen long after first exposure(s) to toxic chemicals. Methyl acetate has moderate chronic toxicity to aquatic life. Insufficient data are available to evaluate or predict the long-term effects of methyl acetate to plants, birds, or land animals.

Water Solubility

Methyl acetate is moderately soluble in water. Concentrations of 1 to 1000 milligrams will mix with a liter of water.

⌛ *Persistence in the Environment*

Methyl acetate is slightly persistent in water, with a half-life of between 2 to 20 days. The half-life of a pollutant is the amount of time it takes for one-half of the chemical to be degraded.

Bioaccumulation in Aquatic Organisms

Some substances increase in concentration, or *bioaccumulate,* in living organisms as they breathe contaminated air, drink contaminated water, or eat contaminated food. These chemicals can become concentrated in the tissues and internal organs of animals as well as humans.

The concentration of methyl acetate found in fish tissues is expected to be somewhat higher then the average concentration of methyl acetate in water from which the fish was taken.

Recommended Risk-Reduction Measures

Proper training of all transporters will reduce the likelihood of a mishap or accident resulting in a leak or spill to the environment. The correct use of DOT labeling (per the MSDS) on all containers, trucks, and rail cars will enable emergency responders to react properly and quickly to any disaster thereby reducing the potential risk to the environment and to personnel.

Storage of methyl acetate should be segregated from incompatible chemicals to minimize the risk of cross-contamination or contact. Buildings designated for storage should be equipped with appropriate fire protection systems (alarms, sprinklers, emergency lighting, portable extinguishers).

If a spill or leak to the environment has occurred, fire department, emergency response, and/or hazardous materials spill personnel should be notified immediately. Cleanup should be attempted only by those trained in proper spill containment procedures using non-sparking tools. Contaminated soils should be removed for incineration and replaced with clean soil.

If methyl acetate should contact the water table, aquifer, or navigable waterway, time is of the essence. It is moderately soluble in water and, therefore, total containment and remediation may not be possible. When such spills occur, the local and/or state emergency response authorities must be notified. A comprehensive emergency response or disaster preparedness/recovery plan should be in place prior to any operations involving the use, transportation, storage, or disposal of methyl acetate. If methyl acetate is spilled or leaked, the following specific steps are recommended:

- ☑ Restrict persons not wearing protective clothing from area of spill or leak until cleanup is complete and area can be opened for normal work.
- ☑ Ventilate area of spill or leak.
- ☑ Remove all ignition sources from area immediately.
- ☑ Absorb liquids in vermiculite, dry sand, earth, or a similar material and deposit in sealed containers.

Use non-sparking tools. Keep water away if possible since methyl acetate will slowly react to form acetic acid and methanol in contact with water.

☑ It may be necessary to dispose of methyl acetate as a hazardous waste. The responsible state agency or the regional office of the federal Environmental Protection Agency (EPA) should be contacted for specific recommendations.

RISK ASSESSMENT: BUSINESS

General Assessment

Accidents or mishaps involving methyl acetate can present a moderate threat to business operations. The loss or damage of equipment or facilities can significantly affect fiscal viability. Lawsuits that may result from personnel injury/death, public exposures, and/or environmental contamination will also require a serious expenditure of resources.

Media attention surrounding an injury, death, or environmental damage can also result in a loss of profits and loss of current as well as future business.

Recommended Risk-Reduction Measures

Company attorneys, safety and health professionals, and environmental specialists should be involved in the development of any procedures or policies intended to manage the use of chemicals in the workplace. A company official should be pre-designated as a public relations officer with specific training in dealing with the press. Corporate plans and policies should be developed, approved, and implemented long before any need for such arises.

REFERENCES

American Conference of Governmental Industrial Hygienists. 1988. *Documentation of the Threshold Limit Values and Biological Exposure Indices,* 5th Edition (with updates). Cincinnati: ACGIH

Anthony, C. P., and N. J. Kolthoff. 1971. *Textbook of Anatomy and Physiology.* St. Louis: C. V. Mosby Company.

Bloom, A. D. and F. J. DeSerres. 1995. *Ecotoxicity and Human Health.* New York: CRC Press/Lewis Publishers.

Brusick, D. J. 1994. *Methods for Genetic Risk Assessment.* New York: CRC Press/Lewis Publishers.

Bueche, J. R. 1972. *Principles of Physics,* 2nd Edition. New York: McGraw-Hill.

Burrell, R., D. K. Flaherty, and L. J. Sauers. 1992. *Toxicology of the Immune System.* New York: Van Nostrand Reinhold.

Calabrese, E. J. 1994. *Biological Effects of Low Level Exposures.* New York: CRC Press/Lewis Publishers.

Cockerman, L. G., B. S. Shane. 1994. *Basic Environmental Toxicology.* New York: CRC Press/Lewis Publishers.

Cothern, C. R., N. P. Ross. 1994. *Environmental Statistics, Assessment, and Forecasting.* New York: CRC Press/Lewis Publishers.

Danse, I. R. 1991. *Common Sense Toxics in the Workplace.* New York: Van Nostrand Reinhold.

Environmental Protection Agency. 1983. *Chemical Hazard Information Profile: Methyl acetate.* Washington, D.C.: U.S. EPA.

Gardner, R. W. 1994. *Chemical Intolerance.* New York: CRC Press/Lewis Publishers.

Hallenbeck, W. H. 1993. *Quantitative Risk Assessment for Environmental and Occupational Health,* 2nd Edition. New York: CRC Press/Lewis Publishers.

Hodgson, E., R. B. Mailman, and J. E. Chambers. 1988. *Dictionary of Toxicology.* London: The MacMillian Company.

Hosty, J. W., and P. Foster. 1990. *A Practical Guide to Chemical Spill Response.* New York: Van Nostrand Reinhold.

Howard, P. H., et. al. 1991. *Handbook of Environmental Degradation Rates.* New York: CRC Press/Lewis Publishers.

James, A., and M. Lord. 1992. *Index of Chemical and Physical Data.* New York: Van Nostrand Reinhold.

Jones, F. E. 1994. *Toxic Organic Vapors in the Workplace.* New York: CRC Press/Lewis Publishers.

Kamrin, M. l A. 1988. *Toxicology: A Primer on Principles and Applications.* Chelsea, Michigan: Lewis Publishers.

Keith, L. H. and M. M. Walker. 1995. *Handbook of Air Toxics.* New York: CRC Press/Lewis Publishers.

Landis, W. G., M. H. Yu. 1995. *Introduction to Environmental Toxicology.* New York: CRC Press/Lewis Publishers.

Langerman, N. 1994 *Precautionary Labels for Chemical Containers.* New York: CRC Press/Lewis Publishers.

Lauwerys, R. R., P. Hoet. 1993. *Industrial Chemical Exposure.* New York: CRC Press/Lewis Publishers.

Lide, D. R. 1995. *Handbook of Chemistry and Physics,* 76th Edition. New York: CRC Press/Lewis Publishers.

Lewis, Richard J., Sr. 1992. *Sax's Dangerous Properties of Industrial Materials,* Volumes 1, 2, and 3, 8th Edition. New York: Van Nostrand Reinhold.

Lisella, Frank S. 1994. *The VNR Dictionary of Environmental Health and Safety.* New York: Van Nostrand Reinhold.

Manahan, S. E. 1994. *Environmental Chemistry,* 6th Edition. New York: CRC Press/Lewis Publishers.

Manahan, S. E. 1992. *Toxicological Chemistry,* 2nd Edition. New York: CRC Press/Lewis Publishers.

Meyer, Eugene. 1990. *Chemistry of Hazardous Materials.* Englewood Cliffs, NJ: Prentice Hall.

Patnaik, P. 1992. *A Comprehensive Guide to the Hazardous Properties of Chemical Substances.* New York: Van Nostrand Reinhold.

MATERIAL SAFETY DATA SHEET

CHEMICAL NAME	
METHYL ACRYLATE	**CRC PRESS/LEWIS PUBLISHERS DISCLAIMER:** THE INFORMATION AND RECOMMENDATIONS PRESENTED HEREIN ARE BASED ON SOURCES BELIEVED TO BE RELIABLE. CRC MAKES NO REPRESENTATION ON ITS COMPLETENESS OR ACCURACY. IT IS THE RESPONSIBILITY OF THE USER TO DETERMINE THE CHEMICAL'S SUITABILITY FOR ITS INTENDED USE, THE CHEMICAL'S SAFE USE, AND THE CHEMICAL'S PROPER DISPOSAL. NO REPRESENTATIONS AND/OR WARRANTIES, EITHER EXPRESSED OR IMPLIED, OF THE MERCHANTABILITY OR FITNESS FOR A PARTICULAR PURPOSE, OR OF ANY OTHER NATURE, ARE MADE WITH RESPECT TO THE INFORMATION PROVIDED IN THIS MSDS OR TO THE CHEMICAL TO WHICH INFORMATION MAY REFER. CRC NEITHER ASSUMES NOR AUTHORIZES ANY OTHER PERSON TO ASSUME FOR IT, ANY OTHER ADDITIONAL RESPONSIBILITY OR LIABILITY FOR THE USE OF, OR RELIANCE UPON, THIS INFORMATION.

HAZARD WARNING INFORMATION

HEALTH	FIRE	REACTIVE	OTHER	DEGREE OF HAZARD	COLOR CODING	OTHER CODES
2	3	2		0 = Minimum Hazard 1 = Slight Hazard 2 = Moderate Hazard 3 = Serious Hazard 4 = Severe Hazard	HEALTH = BLUE FIRE = RED REACTIVITY = YELLOW OTHER = WHITE	OX = Oxidizer ACID = Acid ALK = Alkali COR = Corrosive W = Use No Water

SECTION I - GENERAL INFORMATION

Characterization	RCRA Number	EPA Class
Aliphatic Ester	**D001, D003**	**Characteristic Waste (I) (R)**

DOT Proper Shipping Name **Methyl Acrylate, Inhibited**	Chemical Abstract Service (CAS) Number **96-33-3**
DOT Hazard Class and Label Requirements **Flammable Liquid**	DOT Emergency Guide Code **26 (Inhibited)**
DOT Identification Number **UN 1919**	Chemical Formula **CH_2:$CHCOOCH_3$**

Synonyms

Methyl acrylate (inhibited); methoxycarbonylethylene; methyl ester of acrylic acid; methyl propenoate.

SECTION II - HAZARDOUS INGREDIENTS AND IDENTITY INFORMATION

Hazardous Components (specific identity)	OSHA Exposure Criteria	NIOSH Exposure Criteria	Immediately Dangerous to Life and Health (IDLH)	ACGIH Exposure Criteria
Methyl acrylate (derivation: From ethylene cyanohydrin, methanol, and dilute sulfuric acid; Oxo reaction of acetylene, carbon monoxide, and methanol in the presence of nickel or cobalt catalyst; from β-propiolacetone). **1 ppm = 3.58 mg/m^3**	**PEL (skin):** **10 ppm** **35 mg/m^3** **STEL:** **Not Determined**	**REL (skin):** **10 ppm** **35 mg/m^3** **STEL:** **Not Determined**	**250 ppm**	**TLV (skin):** **10 ppm** **35 mg/m^3** **STEL:** **Not Determined**

SECTION III - PHYSICAL AND CHEMICAL CHARACTERISTICS

Boiling Point **177°F (80.5°C)**	Specific Gravity (H_2O = 1) **.96**
Vapor Pressure (mm Hg) **65 at 68°F (20°C)**	Molecular Weight **86.1**
Vapor Density (Air = 1) **2.97**	Melting Point **-106°F (-76°C)**

Solubility

Soluble in water (6% at 20°C). Soluble in alcohol, acetone, and ether.

Appearance and Odor

Clear, colorless liquid with a sharp, acrid odor. Odor Threshold = 20 ppm.

SECTION IV - FIRE AND EXPLOSION HAZARD DATA

Flash Point (method used) **27°F (-3°C) opened cup**	Explosive Limits in Air % by Volume **LEL: 2.8%** **UEL: 25%**
NFPA Classification **Class IB Flammable Liquid**	Autoignition Temperature **875°F (468°C)**

Extinguishing Media

Dry chemical, alcohol foam, or carbon dioxide. Water streams may be ineffective and may spread fire.

Special Fire Fighting Procedures

Poisonous gases are produced in fire. Wear full protective clothing and self-contained breathing apparatus (SCBA). Use blanketing effect with foam to smother large fires. Continue to cool containers with water after fire is extinguished. Move containers from fire area if it can be done without risk.

Unusual Fire and Explosion Hazards

Evacuate 2500 feet radius if fire becomes uncontrollable or container is exposed to direct flame. Vapors are heavier than air and may travel a distance to ignition source to flashback. Heat and/or an insufficient concentration of inhibitor can cause violent polymerization with the rupture of container.

SECTION V - REACTIVITY DATA

Stability		Conditions to Avoid **Methyl acrylate is kept stable under normal conditions of storage and operation with an inhibitor. Avoid contact with incompatible materials. Keep away from heat, flame, and other sources of ignition.**
Stable **X**	Unstable	Incompatibility *(materials to avoid)* **Reacts violently in contact with oxidizers (peroxides, nitrates, perchlorates, chlorates, permanganates) polymerizers, strong alkalis (sodium hydroxide, potassium hydroxide), moisture, chlorosulfonic acid.**
Hazardous Polymerization		Conditions to Avoid **Explosive polymerization can occur if inhibitor levels become too low or when exposed to heat, light, oxidizers, and peroxides.**
May Occur **X**	Will Not Occur	Hazardous Decomposition or By-products **When heated to decomposition, methyl acrylate emits acrid and irritating fumes, including oxides of carbon (monoxide and dioxide).**

SECTION VI - HEALTH HAZARD DATA

Primary Route(s) of Entry:	Inhalation?	Absorption (skin)?	Ingestion?
	X	X	X

Health Hazards

INHALATION: Severe irritation to the eyes, nose, throat, and respiratory system. Exposure can cause drowsiness, headache, vomiting, and nausea. Extreme exposures can cause the development of pulmonary edema (fluid in the lungs), which can be fatal. There may be dypsnea, cyanosis, convulsions, lachrymation, coughing, fatigue, wheezing, and laryngitis.

ABSORPTION: Severe irritation of the skin and eyes. Will pass through unbroken skin. May cause skin sensitization. Systemic effects include dypsnea, cyanosis, and convulsive movements. Eye contact can cause inflammation, tearing, chemical burns, and impaired vision.

INGESTION: Severe local irritation of the gastrointestinal tract with irritation of the lips, mouth, throat, painful swallowing, abdominal pain and cramps, nausea, vomiting, diarrhea, shock.

Carcinogenicity	NTP Listed?	IARC Cancer Review Group?	OSHA Regulated?	Target Organs?
Unknown Human Questioned Animal	**No**	**Group 3**	**29 CFR 1910.1000 Table Z-1**	**Respiratory system; skin; eyes.**

Medical Conditions Generally Aggravated by Exposure

Existing chronic skin, respiratory, liver, and/or kidney disorders may be aggravated by exposure.

Emergency and First-aid Procedures

Eye contact: Flush immediately with water for 15 minutes (minimum); seek medical attention. Skin contact: Remove all contaminated clothing. Immediately wash area with large amounts of soap and water. Seek medical assistance. For inhalation: Remove the person from exposure. Provide respiratory assistance and CPR. Transfer to medical facility. Observe for 48 hours for lung effects. If swallowed: Contact poison control center. Give 1 to 2 glasses of water. Seek medical attention immediately.

SECTION VII - PRECAUTIONS FOR SAFE HANDLING AND USE

Steps to be Taken in Case Material is Released or Spilled

Contain spills using absorbent material (vermiculite or other). Ventilate area of spill and remove ignition sources. Have extinguishing agent available. Use non-sparking tools. Store materials in DOT-approved containers. Restrict those not involved in cleanup from entering area.

Preferred Waste Disposal Method

Incineration in chemical incinerator equipped with scrubber and afterburner.

Precautions to be Taken in Handling and Storage

Store in tightly closed containers in a cool, dark, well-ventilated area. Protect containers from physical damage. Keep heat, fire, and flame away. Outside building storage is recommended.

Other Precautions and Warnings

Bulk storage of methyl acrylate is not recommended. Ground and bond all metal containers.

SECTION VIII - CONTROL MEASURES AND PERSONAL PROTECTIVE EQUIPMENT

Respiratory Protection *(specify type)*

Carcinogenic properties unknown. Use a supplied-air respirator or a self-contained breathing apparatus (SCBA) with full facepiece operated in positive pressure mode for the best protection.

Ventilation

Local exhaust or general mechanical systems recommended.

Protective Gloves	Eye Protection	Other Protective Clothing
Impervious Gloves	**Chemical Goggles or Face Mask**	**Impermeable Apron or Protective Suit**

Work/Hygiene Practices

Always wash hands thoroughly after using chemical; never bring food, drink, or smoking materials into vicinity of chemicals.

METHYL ACRYLATE

$CH_2{:}CHCOOCH_3$ CAS: 96-33-3

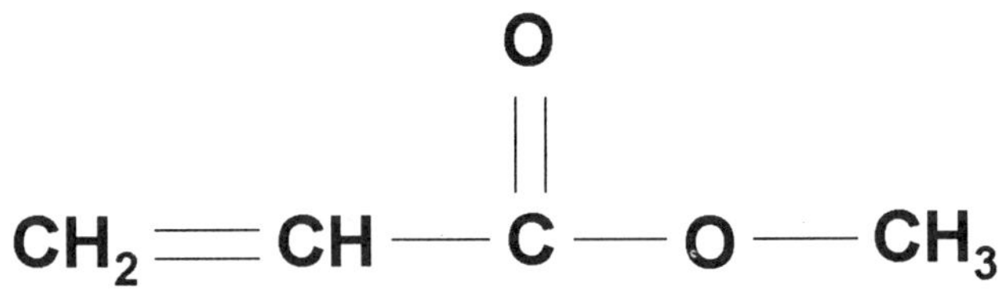

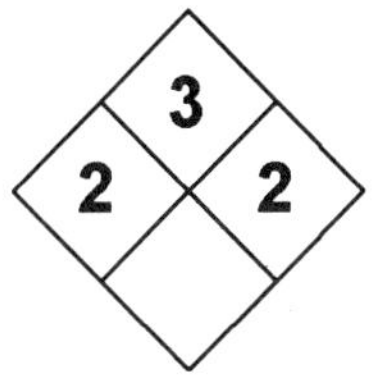

IDENTIFICATION AND TYPICAL USES

Methyl acrylate is a colorless liquid with an acrid odor. It is used as a monomer in the manufacture of leather finish resins, textile and paper coatings, and plastic films. It is also a resin in the purification and decolorization of industrial effluents or an aid in timed release and disintegration of pesticides, for the polymerization of radioactive wastes, and in the preparation of acrylic polymers.

RISK ASSESSMENT: HEALTH

General Assessment

Methyl acrylate is a toxic poison to humans by *inhalation, ingestion,* and by skin contact (*absorption*). It is a questioned animal carcinogen, but there is insufficient data regarding its carcinogenicity in humans. Prolonged exposures may lead to congestion of the lungs and changes in the heart, kidney, and liver.

Inhalation results in irritation of the mucus membranes of the eyes, nose, throat, and respiratory system. It can cause changes in the sense of smell and in pulmonary function. In addition, there may be symptoms such as lethargy, coughing, wheezing, fatigue, cyanosis, and convulsive movements. High concentrations can cause dizziness with dypsnea, nausea, headache, vomiting, and can lead to death as a result of chemical pneumonitis or pulmonary edema (fluid in the lungs). Symptoms of pulmonary edema can be delayed up to 48 hours following exposure which may create a false sense of security with regard to health exposure risk.

Skin contact can cause serious localized irritation at the site of contact with marked erythema, blisters and edema, itching, and inflammation. There may be first degree burns on short exposures and second degree burns on prolonged exposures. Repeated exposures may cause sensitivity and dermatitis.

Ingestion of methyl acrylate causes irritation of the gastrointestinal tract with irritation of the mouth, lips, throat, painful swallowing, abdominal cramps, nausea, vomiting, diarrhea, shock, and possible loss of consciousness. Subsequent absorption through the digestive process can lead to toxic systemic effects, especially changes to liver, lung, and kidney tissues.

Acute Health Effects

The following acute (short-term) health effects may occur immediately or shortly after exposure to methyl acrylate:

Skin: Serious irritation and possible first or second degree burns. There can be marked localized irritation, erythema, edema, and sensitization.

Eye: Severe irritation which can lead to injury, burns, painful irritation, and permanent damage if allowed to remain in contact.

Lung: Irritation of the eyes, nose, throat and lungs (especially the mucus membranes). May cause changes in the sense of smell and in pulmonary function. Can lead to chemical pneumonitis and/or pulmonary edema (fluid in the lungs).

CNS: High concentrations may cause dizziness, headache, nausea, vomiting, dypsnea, lethargy, and convulsive movements.

Chronic Health Effects

The following chronic (long-term) health effects may occur at some time after exposure to methyl acrylate and can last for months or even years:

Cancer Hazards: According to the references, methyl acrylate has been investigated for its ability to cause cancer in test animals and there is insufficient evidence to determine its carcinogenic effects. Its carcinogenicity is considered questioned. Mutation data have been reported and many scientists believe that such chemicals may pose a cancer risk

Reproduction: According to the references, methyl acrylate has not been adequately tested for its ability to adversely affect reproduction in test animals.

Other Chronic Effects: Very high or prolonged exposure may lead to congestion of the lungs and possible changes in the liver or kidneys.

Recommended Risk-Reduction Measures

Because of its unknown but questioned carcinogenicity, as well as its known ability to cause serious topical and systemic effects, personnel should avoid direct contact with methyl acrylate. The best risk reduction measure is to use a less toxic chemical as a substitute for methyl acrylate. If this is not possible or feasible, then *engineering controls* are the most effective method of reducing exposures. The best protection is to enclose operations and/or provide local exhaust ventilation at the site of chemical release. While not always operationally feasible, isolating operations can also reduce exposure risk.

Using respiratory protection is less effective than the controls mentioned above, but is still advisable whenever working with or around methyl acrylate. For exposures exceeding the 10 ppm PEL (8-hour TWA), a supplied-air respirator with full facepiece operated in positive pressure mode, or a self-contained breathing apparatus (SCBA) with full facepiece and operated in pressure demand mode are the recommended respiratory protection methods of choice. If a full facepiece is not available, then chemical goggles should be worn to protect the eyes. Whenever a chemical splash hazard exists, a face shield and a protective apron should be worn. To prevent hand and skin exposures, impervious gloves should be used. Glove manufacturers should be contacted to obtain permeation studies before gloves are selected.

Administrative controls should also be in place to minimize the potential for human exposures. These may include written procedures or policies which specify the methods and techniques that will be practiced whenever personnel are to work with methyl acrylate.

All personnel should receive *training* on the use, hazards, protective measures, emergency actions, and other precautions per 29 CFR 1910.1200 (Hazard Communication), prior to the first assignment in an area where methyl acrylate is used or stored.

Before beginning employment and at regular intervals thereafter (e.g., annually), those persons having frequent or potentially high exposures should be given the following medical tests:

- ☑ Lung function tests.

If symptoms develop or overexposure is suspected, the following medical tests are recommended:

- ☑ Liver and kidney function tests.
- ☑ Lung function tests.
- ☑ Skin testing with dilute methyl acrylate to help diagnose allergy (performed by a qualified allergist).
- ☑ Consider chest X-ray following acute overexposure (may be negative if taken immediately after exposure due to delayed onset of pulmonary edema).

Any medical evaluation should include a careful history of past and present symptoms with an examination. Medical tests that look for existing damage are not a substitute for controlling exposures. Also, since smoking can cause heart disease, emphysema, and other respiratory disorders, smokers exposed to methyl acrylate may develop symptoms much quicker and with greater intensity than non-smokers under identical exposure conditions. Prudent risk management requires careful consideration of *all* possible factors which may be causing the appearance of exposure symptoms.

Other methods to reduce exposure include:

- ☑ Where possible, enclose operations and use local exhaust ventilation at the site of chemical release. If local exhaust ventilation or enclosure is not used, respiratory protection should be mandatory.
- ☑ Always ensure that proper protective clothing is worn when using chemical substances.
- ☑ Wash thoroughly immediately after exposure to methyl acrylate and at the end of the work shift or before eating, drinking, or smoking.
- ☑ Hazard warning information should be posted in the work area. In addition, as part of an on-going education and training program, all information on the health and safety hazards of methyl acrylate should be communicated to all exposed workers.
- ☑ Never eat, drink, or smoke in areas were methyl acrylate is used, handled or stored.
- ☑ Work clothing contaminated with methyl acrylate should never be worn home where family members can be exposed. It should be laundered only by

those who have been trained on the health risks associated with methyl acrylate.

RISK ASSESSMENT: ENVIRONMENT

General Assessment

The environment is at risk of exposure during transportation, storage, disposal, or destruction of methyl acrylate. In almost every scenario, the threat of environmental exposure is contingent upon the proper handling of the chemical substance. Accidental spills, large or small, can result in fire, explosion, and possible contamination of the surrounding environmental mediums (water, soil, and air).

Methyl acrylate is considered a Class IB flammable liquid (according to OSHA 29 CFR 1910.106). Its low flash point and relatively low boiling point present a serious fire and explosion hazard concern. Also, because it is incompatible with a number of common materials, especially strong oxidizers, contact can result in violent and explosive reactions. It can also form explosive mixtures in air and can ignite on contact with heat, fire, or sparks. Heat can cause methyl acrylate to react by itself. If this happens in a closed container, an explosion can occur. Methyl acrylate usually contains an inhibitor such as hydroquinone or its methyl ether to prevent a self-reaction. If the inhibitor is absent or present in too low a quantity, the explosive reaction may occur without the application of heat. These characteristics require special consideration during any emergency situation involving a leak or spill of methyl acrylate.

Methyl acrylate can enter the environment through unchecked industrial discharges and through spills.

Acute Ecological Effects

Acute (short-term) toxic effects may include the death of animals, birds, or fish, and death or low growth rate in plants. Acute effects are seen 2 to 4 days after animals or plants are exposed to methyl acrylate. This chemical has moderate acute toxicity to aquatic life. Insufficient data are available to evaluate or predict the short-term effects of methyl acrylate to plants, birds, or terrestrial animals.

Chronic Ecological Effects

Chronic toxic effects may include shortened life span, reproductive problems, lower fertility, and changes in appearance or behavior in exposed animals. These effects can be seen long after first exposure(s) to toxic chemicals. Methyl acrylate has moderate chronic toxicity to aquatic life. Insufficient data are available to evaluate or predict the long-term effects of methyl acrylate to plants, birds, or land animals.

Water Solubility

Methyl acrylate is moderately soluble in water. Concentrations of 1 to 1000 milligrams will mix with a liter of water.

Persistence in the Environment

Methyl acrylate is slightly persistent in water, with a half-life of between 2 to 20 days. The half-life of a pollutant is the amount of time it takes for one-half of the chemical to be degraded. About 85% of methyl acrylate will eventually end up in the air; the remainder will end up in water.

Bioaccumulation in Aquatic Organisms

Some substances increase in concentration, or *bioaccumulate,* in living organisms as they breathe contaminated air, drink contaminated water, or eat contaminated food. These chemicals can become concentrated in the tissues and internal organs of animals as well as humans.

The concentration of methyl acrylate found in fish tissues is expected to be about the same as the average concentration of methyl acrylate in water from which the fish was taken.

Recommended Risk-Reduction Measures

Proper training of all transporters will reduce the likelihood of a mishap or accident resulting in a leak or spill to the environment. The correct use of DOT labeling (per the MSDS) on all containers, trucks, and rail cars will enable emergency responders to react properly and quickly to any disaster thereby reducing the potential risk to the environment and to personnel.

Storage of methyl acrylate should be segregated from incompatible chemicals to minimize the risk of cross-contamination or contact. Buildings designated for storage should be equipped with appropriate fire protection systems (alarms, sprinklers, emergency lighting, portable extinguishers). Equipment should be designed to meet explosion-proof standards.

If a spill or leak to the environment has occurred, fire department, emergency response, and/or hazardous materials spill personnel should be notified immediately. Cleanup should be attempted only by those trained in proper spill containment procedures using non-sparking tools. Contaminated soils should be removed for incineration and replaced with clean soil. If methyl acrylate should contact the water table, aquifer, or navigable waterway, time is of the essence. It is moderately soluble in water and, therefore, containment and remediation may not be entirely possible. When such spills occur, the local and/or state emergency response authorities must be notified. A comprehensive emergency response or disaster preparedness/recovery plan should be in place prior to any operations involving the use, transportation, storage, or disposal of methyl acrylate.

If methyl acrylate is spilled or leaked, the following specific steps are recommended:

- ☑ Restrict persons not wearing protective clothing from area of spill or leak until cleanup is complete and area can be opened for normal work.
- ☑ Ventilate area of spill or leak.
- ☑ Remove all ignition sources.
- ☑ Absorb liquids in vermiculite, dry sand, earth, or a similar material and deposit in sealed containers. Use non-sparking tools.
- ☑ It may be necessary to dispose of methyl acrylate as a hazardous waste. The responsible state agency or the regional office of the federal Environmental Protection Agency (EPA) should be contacted for specific recommendations.

RISK ASSESSMENT: BUSINESS

General Assessment

Accidents or mishaps involving methyl acrylate can present a moderate threat to business operations. The loss or damage of equipment or facilities can significantly affect fiscal viability. Lawsuits that may result from personnel injury/death, public exposures, and/or environmental contamination will also require a serious expenditure of resources.

Media attention surrounding an injury, death, or environmental damage can also result in a loss of profits and loss of current as well as future business. Always remember that anytime the terms "cancer" or "carcinogen" are used, public emotion, hysteria, and ignorance can run equally high. This should be carefully considered when developing or implementing a public relations policy.

Recommended Risk-Reduction Measures

Company attorneys, safety and health professionals, and environmental specialists should be involved in the development of any procedures or policies intended to manage the use of chemicals in the workplace. A company official should be pre-designated as a public relations officer with specific training in dealing with the press. Corporate plans and policies should be developed, approved, and implemented long before any need for such arises.

REFERENCES

American Conference of Governmental Industrial Hygienists. 1988. *Documentation of the Threshold Limit Values and Biological Exposure Indices,* 5th Edition (with updates). Cincinnati: ACGIH

Anthony, C. P., and N. J. Kolthoff. 1971. *Textbook of Anatomy and Physiology.* St. Louis: C. V. Mosby Company.

Brusick, D. J. 1994. *Methods for Genetic Risk Assessment.* New York: CRC Press/Lewis Publishers.

Bueche, J. R. 1972. *Principles of Physics,* 2nd Edition. New York: McGraw-Hill.

Burrell, R., D. K. Flaherty, and L. J. Sauers. 1992. *Toxicology of the Immune System.* New York: Van Nostrand Reinhold.

Calabrese, E. J. 1994. *Biological Effects of Low Level Exposures.* New York: CRC Press/Lewis Publishers.

Cothern, C. R., N. P. Ross. 1994. *Environmental Statistics, Assessment, and Forecasting.* New York: CRC Press/Lewis Publishers.

Gardner, R. W. 1994. *Chemical Intolerance.* New York: CRC Press/Lewis Publishers.

Hodgson, E., R. B. Mailman, and J. E. Chambers. 1988. *Dictionary of Toxicology.* London: The MacMillian Company.

Hosty, J. W., and P. Foster. 1990. *A Practical Guide to Chemical Spill Response.* New York: Van Nostrand Reinhold.

Howard, P. H., et. al. 1991. *Handbook of Environmental Degradation Rates.* New York: CRC Press/Lewis Publishers.

Keith, L. H. and D. B. Walters. (Editors) 1992. *National Toxicology Program's Chemical Database, Volumes I-VII.* Chelsea, MI: Lewis Publishers.

Lauwerys, R. R., P. Hoet. 1993. *Industrial Chemical Exposure.* New York: CRC Press/Lewis Publishers.

Lewis, Richard J., Sr. 1992. *Sax's Dangerous Properties of Industrial Materials,* Volumes 1, 2, and 3, 8th Edition. New York: Van Nostrand Reinhold.

Manahan, S. E. 1992. *Toxicological Chemistry,* 2nd Edition. New York: CRC Press/Lewis Publishers.

Meyer, Eugene. 1990. *Chemistry of Hazardous Materials.* Englewood Cliffs, NJ: Prentice Hall.

Patnaik, P. 1992. *A Comprehensive Guide to the Hazardous Properties of Chemical Substances.* New York: Van Nostrand Reinhold.

MATERIAL SAFETY DATA SHEET

CHEMICAL NAME	CRC PRESS/LEWIS PUBLISHERS DISCLAIMER
METHYLACRYLONITRILE	

HAZARD WARNING INFORMATION

HEALTH	FIRE	REACTIVE	OTHER	DEGREE OF HAZARD	COLOR CODING	OTHER CODES
2	3	2		0 = Minimum Hazard 1 = Slight Hazard 2 = Moderate Hazard 3 = Serious Hazard 4 = Severe Hazard	HEALTH = BLUE FIRE = RED REACTIVITY = YELLOW OTHER = WHITE	OX = Oxidizer ACID = Acid ALK = Alkali COR = Corrosive W = Use No Water

SECTION I - GENERAL INFORMATION

Characterization	**Organic Cyanides (Nitriles)**	RCRA Number / EPA Class	**U152** / **Toxic Waste**
DOT Proper Shipping Name	**Flammable Liquid, Poisonous, N.O.S.**	Chemical Abstract Service (CAS) Number	**126-98-7**
DOT Hazard Class and Label Requirements	**Flammable Liquid and Poison**	DOT Emergency Guide Code	**28 (Inhibited)**
DOT Identification Number	**UN 1992**	Chemical Formula	**CH_2:C(CH_3)CN**

Synonyms

2-Cyanopropene-1,isoprene cyanide; methyacrylonitrile; α-methylacrylonitrile; 2-methylpropenenitrile.

SECTION II - HAZARDOUS INGREDIENTS AND IDENTITY INFORMATION

Hazardous Components (specific identity)	OSHA Exposure Criteria	NIOSH Exposure Criteria	Immediately Dangerous to Life and Health (IDLH)	ACGIH Exposure Criteria
Methylacrylonitrile (derived: By vapor-phase catalytic oxidation of methylamine). **1 ppm = 2.79 mg/m³**	**PEL:** Not Established **STEL:** Not Established	**REL (skin):** 1 ppm, 3 mg/m³ **STEL:** Not Established	**Not Determined**	**TLV (skin):** 1 ppm, 3 mg/m³ **STEL:** Not Established

SECTION III - PHYSICAL AND CHEMICAL CHARACTERISTICS

Boiling Point	**194°F (90°C)**	Specific Gravity (H_2O = 1)	**0.80**
Vapor Pressure (mm Hg)	**71 at 77°F (25°C)**	Molecular Weight	**67.1**
Vapor Density (Air = 1)	**Not Found**	Freezing Point	**-32°F (-36°C)**

Solubility

Partially soluble (37%) in water. Soluble in acetone, thermoplastic, resistant to acids and alkalis.

Appearance and Odor

Clear, colorless liquid with an odor of bitter almonds.

SECTION IV - FIRE AND EXPLOSION HAZARD DATA

Flash Point (method used)	**34°F (1°C) closed cup**	Explosive Limits in Air % by Volume	**LEL: 2%** **UEL: 6.8%**
NFPA Classification	**Class 1B Flammable Liquid**	Autoignition Temperature	**Not Found**

Extinguishing Media

Carbon dioxide, dry chemical, alcohol foam. Water may be ineffective.

Special Fire Fighting Procedures

Poisonous gases are produced in fire. Wear full protective clothing and self-contained breathing apparatus (SCBA). Keep fire-exposed containers cool with water spray. Do not release runoff from fire fighting efforts to sewers or waterways. Contain and collect for disposal.

Unusual Fire and Explosion Hazards

Dilute solutions are also flammable. Vapors form explosive mixtures in air. Heat will build pressure and may rupture closed storage containers. Vapors are heavier than air; they may travel a distance to cause fire or explosion due to flashback phenomenon. Heat of fire may cause violent polymerization to occur.

SECTION V - REACTIVITY DATA

Stability		Conditions to Avoid
		Normally stable under routine conditions. Dangerous fire hazard when exposed to heat and flame. Avoid contact with incompatible materials.
Stable	Unstable	Incompatibility *(materials to avoid)*
	X	Incompatible with strong oxidizers (chlorine, fluorine, bromine), strong bases, strong acids, and strong reducing agents.
Hazardous Polymerization		Conditions to Avoid Under normal conditions of temperature and pressure, hazardous polymerization of methylacrylonitrile MAY occur. Violent polymerization occurs in the presence of caustic alkalis and at elevated temperatures.
May Occur	Will Not Occur	Hazardous Decomposition or By-products
X		Polymerizes violently in presence of concentrated alkali or at elevated temperature. It undergoes photochemical polymerization, liberating heat. Violent rupture of glass containers can occur.

SECTION VI - HEALTH HAZARD DATA

Primary Route(s) of Entry:	Inhalation?	Absorption (skin)?	Ingestion?
	X	X	X

Health Hazards

INHALATION: Asphyxia, headache, chest pains, dypsnea, lightheadedness, coughing, weakness, and fatigue. High concentrations can cause lachrymation, somnolence (sleepiness), jaundice, diarrhea, nausea, vomiting, and death. Extremely irritating to the eyes, nose, throat, and mucosa of the respiratory tract.

ABSORPTION: Headache, lightheadedness, sneezing, weakness, nausea, and vomiting. Severe intactaction can result in loss of consciousness, convulsions, respiratory arrest, and death. Methylacrylonitrile is a skin irritant. It may cause blistering after a few hours.

INGESTION: Gastrointestinal irritation and inflammation of the digestive tract.

Carcinogenicity	NTP Listed?	IARC Cancer Review Group?	OSHA Regulated?	Target Organs?
Unknown Human Unknown Animal	No	No	No	Respiratory system; skin; eyes; CNS.

Medical Conditions Generally Aggravated by Exposure

Respiratory impairments, (asthma; skin conditions (dermatitis); and cardiovascular disorders.

Emergency and First-aid Procedures

Eye contact: Flush immediately with water for 15 minutes (minimum); seek medical attention. **Skin contact**: Remove all contaminated clothing. Immediately wash area with large amounts of soap and water. Seek medical assistance. For **inhalation**: Remove the person from exposure. Provide respiratory assistance and CPR. Transfer to medical facility. If **swallowed**: Seek medical attention immediately. Give 1 or 2 glasses of water or milk. Do NOT induce vomiting, unless told to do so by a doctor.

SECTION VII - PRECAUTIONS FOR SAFE HANDLING AND USE

Steps to be Taken in Case Material is Released or Spilled

Contain spills using absorbent material (vermiculite, or other). Ventilate area of spill. Have water source available in case of fire. Place materials in DOT-approved containers. Restrict those not involved in cleanup from entering area. Remove all ignition sources.

Preferred Waste Disposal Method

Chemical incinerator equipped with scrubber and afterburner.

Precautions to be Taken in Handling and Storage

Methylacrylonitrile is highly volatile. Store in tightly closed containers in a cool, dark, well-ventilated area. Keep fire and flame away. Use non-sparking tools to open or close containers.

Other Precautions and Warnings

If exposed to heat or light, uninhibited methylacrylonitrile can react violently by itself, leading to an explosion of closed containers. Ground and bond all metal containers during transfer operations.

SECTION VIII - CONTROL MEASURES AND PERSONAL PROTECTIVE EQUIPMENT

Respiratory Protection *(specify type)*

Exposure level is low and difficult to monitor. At any exposure level, recommend using a self-contained breathing apparatus (SCBA) with full facepiece in pressure demand or other positive pressure mode.

Ventilation

Local exhaust or general mechanical systems recommended.

Protective Gloves	Eye Protection	Other Protective Clothing
Impervious Material	Chemical Goggles or Face Mask	Protective Apron

Work/Hygiene Practices

Always wash hands thoroughly after using chemical; never bring food, drink, or smoking materials into vicinity of chemicals.

METHYLACRYLONITRILE

CHI_2:C(CHI_3)CN

CAS: 126-98-7

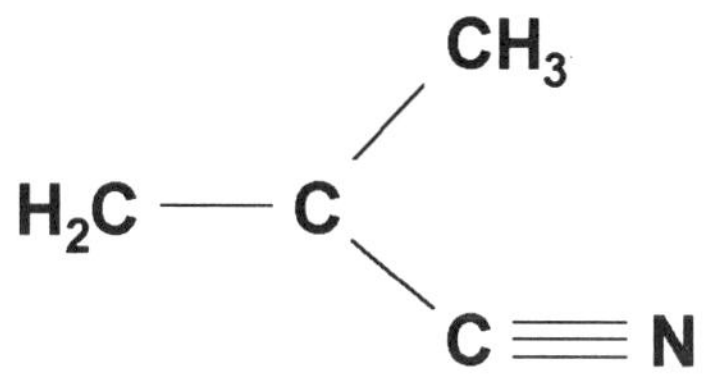

IDENTIFICATION AND TYPICAL USES

Methylacrylonitrile is a clear, colorless liquid with an odor similar to bitter almonds. It is used in the manufacture of coatings and elastomer and as an intermediate in the preparation of acids, amines, amides, and esters. It is also used in making dyes, acrylic fibers, fertilizers, and in organic synthesis.

RISK ASSESSMENT: HEALTH

General Assessment

Human exposure to methylacrylonitrile can occur through all routes of entry into the body: *inhalation, ingestion,* and dermal *absorption.* Each present a moderate level of toxicity. Methylacrylonitrile is readily absorbed through the respiratory tract, the gastrointestinal tract, and through unbroken skin. Absorption through the skin can result in toxic symptoms such as headache, lightheadedness, sneezing, weakness, nausea, and vomiting. In humans, many symptoms are nonspecific but appear to be related to the central nervous system (CNS), the respiratory tract, gastrointestinal tract, and the skin.

Inhalation of this compound can cause asphyxia, headache, mild to severe chest pains, weakness, lachrymation (tearing), dypsnea (labored breath), lightheadedness, and coughing. Severe intoxication can cause jaundice, diarrhea, loss of consciousness, respiratory arrest, convulsions, and death. There is also evidence to show that methylacrylonitrile can cause congestion in all types of organs and damage to the liver, blood, or the central nervous system (CNS).

Acute Health Effects

The following acute (short-term) health effects may occur immediately or shortly after exposure to methylacrylonitrile:

Skin: Severe irritation and blistering a few hours after exposure.

Eye: Burns leading to permanent damage.

Lung: Irritation causing coughing, congestion, and shortness of breath.

CNS: Inhalation of high concentrations of the vapor may cause sleepiness, dizziness, loss of consciousness, or death. Lower exposures can cause weakness, headache, confusion, nausea, vomiting, and can also lead to death.

Chronic Health Effects

The following chronic (long-term) health effects may occur at some time after exposure to methylacrylonitrile and can last for months or even years:

Cancer Hazards: According to information presented in the references, methylacrylonitrile has not been adequately tested for its ability to cause cancer in test animals.

Reproductive Hazard: There is no evidence that methylacrylonitrile adversely affects reproduction.

Other Chronic Effects: None reported.

Recommended Risk-Reduction Measures

Personnel should avoid direct contact with methylacrylonitrile. Unless a less toxic chemical can be substituted for a hazardous substance, *engineering controls* are the most effective method of reducing exposures. The best protection is to enclose operations and/or provide local exhaust ventilation at the site of chemical release. While not always operationally feasible, isolating operations can also reduce exposure.

Using respiratory protection is less effective than the controls mentioned above, but is still advisable whenever working with or around methylacrylonitrile. The exposure level is extremely low (1 ppm) and difficult to accurately monitor. A self-contained breathing apparatus (SCBA) with full facepiece operated in pressure demand or a supplied-air respirator operated in continuous flow mode are the recommended respiratory protection of choice. If a full facepiece is not available, then chemical goggles should be worn to protect the eyes. Whenever a chemical splash hazard exists, a face shield and a rubber apron should be worn. To prevent hand and skin exposures, gloves should be used. Glove manufacturers should be contacted and permeation studies obtained *before* final glove selections are made.

Administrative controls should also be in place to minimize the potential for human exposures. These may include written procedures or policies which specify the methods and techniques that will be practiced whenever personnel are to work with methylacrylonitrile.

All personnel should receive *training* on the use, hazards, protective measures, emergency actions, and other precautions per 29 CFR 1910.1200 (Hazard Communication), prior to the first assignment in an area where methylacrylonitrile is used or stored.

Medical testing of personnel should be performed before the first exposure and at least annually thereafter. As per OSHA requirements, these tests must be provided by the employer and include:

☑ Complete physical examination.

☑ Stool test for blood (employees over 40).

In addition, for those with frequent or potentially high exposure (half the TLV or greater, or significant skin contact), the following are recommended before beginning work and at regular intervals thereafter:

☑ Urine thiocyanate levels.

☑ Lung function (pulmonary function) tests.

If symptoms develop or overexposure is suspected, the following additional tests may be useful:

☑ Blood cyanide level.

☑ Thyroid function test.

☑ Consider chest X-ray after acute overexposure.

Any evaluation should include a careful medical history of past and present symptoms with an examination. However, medical tests that evaluate existing damage are not a substitute for controlling exposure. Antidotes following exposure to cyanides include:

☑ Sistine and *N*-acetyl Sistine are somewhat effective in combating cyanide poisoning.

Other methods to reduce exposure include:

☑ Use amyl nitrate capsules as antidote if symptoms develop.

☑ Employees should be trained regularly in emergency measures for cyanide poisoning and in CPR.

☑ A cyanide antidote kit must be rapidly available and ingredients replaced every 1-2 years to ensure freshness.

☑ Medical personnel trained in treatment of cyanide poisoning should be quickly available

☑ Where possible, enclose operations and use local exhaust ventilation at the site of chemical release. If local exhaust ventilation or enclosure is not used, respiratory protection should be mandatory.

☑ Always ensure that proper protective clothing is worn when using chemical substances and that personnel are trained in its use, care, and maintenance.

☑ Wash thoroughly immediately after exposure to methylacrylonitrile and at the end of the work shift or before eating, drinking, or smoking.

☑ Hazard warning information should be posted in the work area. In addition, as part of an on-going education and training program, all information on the health and safety hazards of methylacrylonitrile should be communicated to all exposed and potentially exposed workers.

RISK ASSESSMENT: ENVIRONMENT

General Assessment

The environment is at risk of exposure during transportation, storage, disposal, or destruction of methylacrylonitrile. In almost every scenario, the threat of environmental exposure is contingent upon the proper handling of the chemical substance. Accidental spills, large or small, can result in fire, explosion, and possible contamination of the surrounding environmental mediums (water, soil, and air).

Methylacrylonitrile is considered a highly flammable liquid. Because of its low flash point and high boiling point, it is classified as a Class 1B flammable liquid (per OSHA 29 CFR 1910.106). It is also a dangerous fire and explosion hazard, highly reactive, and it polymerizes easily. These characteristics require special consideration during any emergency situation involving a leak or spill of methylacrylonitrile. Should methylacrylonitrile ever come into contact with incompatible substances such as oxidizers, acids, bases, bromine, and amines either during use, transportation, or storage, the formation of highly toxic and/or highly explosive commodities is extremely possible.

The proper disposal/destruction method for methylacrylonitrile is to burn it in a chemical incinerator equipped with an afterburner and air scrubber.

Methylacrylonitrile can enter the environment through industrial effluents and through spills.

Acute Ecological Effects

Acute (short-term) toxic effects may include the death of animals, birds, or fish and death or low growth rate in plants. Acute effects are seen 2 to 4 days after animals or plants are exposed to methylacrylonitrile.

This chemical has high acute toxicity to aquatic life. Insufficient data are available to evaluate or predict the short-term effects of methylacrylonitrile to plants, animals, or birds.

Chronic Ecological Effects

Chronic toxic effects may include shortened life span, reproductive problems, lower fertility, and changes in appearance or behavior in exposed animals. These effects can be seen long after first exposure(s) to toxic chemicals.

Methylacrylonitrile has high chronic toxicity in aquatic life. Insufficient data are available on the long-term effects of methylacrylonitrile to plants, birds, or land animals.

Water Solubility

Methylacrylonitrile is moderately soluble in water. Concentrations of 1 to 1000 milligrams will mix with a liter of water.

Persistence in the Environment

Methylacrylonitrile is slightly persistent in water, with a half-life of between 2 to 20 days. The half-life of a pollutant is the amount of time it takes for one half of the chemical to be degraded.

Bioaccumulation in Aquatic Organisms

Some substances increase in concentration, or *bioaccumulate,* in living organisms as they breathe contaminated air, drink contaminated water, or eat contaminated food. These chemicals can become concentrated in the tissues and internal organs of animals as well as humans.

The concentration of methylacrylonitrile found in fish tissues is expected to be somewhat higher than the average concentration of methylacrylonitrile in the water from which the fish was taken.

Recommended Risk-Reduction Measures

Proper training of all transporters will reduce the likelihood of a mishap or accident resulting in a leak or spill to the environment. The correct use of DOT labeling (per the MSDS) on all containers, trucks, and rail cars will enable emergency responders to react properly and quickly to any disaster thereby reducing the potential risk to the environment and to personnel.

Storage of methylacrylonitrile should be segregated from incompatible chemicals to minimize the risk of cross-contamination or contact. Buildings designated for storage should be equipped with appropriate fire protection systems (alarms, sprinklers, emergency lighting, portable extinguishers).

If a spill or leak to the environment has occurred, fire department, emergency response and/or hazardous materials spill personnel should be notified immediately. Cleanup should be attempted only by those trained in proper spill containment procedures using non-sparking tools. Contaminated soils should be removed for incineration and replaced with clean soil. If methylacrylonitrile should contact the water table, aquifer, or navigable waterway, time is of the essence. It is moderately soluble in water and, therefore, total containment and remediation may not be possible. When such spills occur, the local and/or state emergency response authorities must be notified. A comprehensive emergency response or disaster preparedness/recovery plan should be in place prior to any operations involving the use, transportation, storage, or disposal of methylacrylonitrile. If methylacrylonitrile is spilled or leaked, the following specific steps are recommended:

- ☑ Restrict persons not wearing protective clothing from area of spill or leak until cleanup is complete and area can be opened for normal work.
- ☑ Remove all ignition sources.
- ☑ Ventilate area of spill or leak.
- ☑ Absorb liquids in vermiculite, dry sand, earth, or a similar material and deposit in sealed containers.
- ☑ Keep methylacrylonitrile out of a confined space, such as a sewer, because of the possibility of ex-

plosion (unless the sewer is designed to prevent the buildup of explosive concentrations).

☑ It may be necessary to dispose of methylacrylonitrile as a hazardous waste. The responsible state agency or the regional office of the federal Environmental Protection Agency (EPA) should be contacted for specific recommendations.

RISK ASSESSMENT: BUSINESS

General Assessment

Accidents or mishaps involving methylacrylonitrile can present a significant threat to business operations. The loss or damage of equipment or facilities can significantly affect fiscal viability. Lawsuits that may result from personnel injury/death, public exposures, and/or environmental contamination will also require a serious expenditure of resources. Media attention surrounding an injury, death, or environmental damage can also result in a loss of profits and loss of current as well as future business.

Recommended Risk-Reduction Measures

Company attorneys, safety and health professionals, and environmental specialists should be involved in the development of any procedures dealing with safety or emergency response. A company official should be pre-designated as a public relations officer with specific training in dealing with the press. Corporate plans and policies should be developed, approved, and implemented long before any need for such arises.

REFERENCES

American Conference of Governmental Industrial Hygienists. 1988. *Documentation of the Threshold Limit Values and Biological Exposure Indices,* 5th Edition (with updates). Cincinnati: ACGIH

Anthony, C. P., and N. J. Kolthoff. 1971. *Textbook of Anatomy and Physiology.* St. Louis: C. V. Mosby Company.

Bloom, A. D. and F. J. DeSerres. 1995. *Ecotoxicity and Human Health.* New York: CRC Press/Lewis Publishers.

Brusick, D. J. 1994. *Methods for Genetic Risk Assessment.* New York: CRC Press/Lewis Publishers.

Bueche, J. R. 1972. *Principles of Physics,* 2nd Edition. New York: McGraw-Hill.

Burrell, R., D. K. Flaherty, and L. J. Sauers. 1992. *Toxicology of the Immune System.* New York: Van Nostrand Reinhold.

Calabrese, E. J. 1994. *Biological Effects of Low Level Exposures.* New York: CRC Press/Lewis Publishers.

Cockerman, L. G., B. S. Shane. 1994. *Basic Environmental Toxicology.* New York: CRC Press/Lewis Publishers.

Cothern, C. R., N. P. Ross. 1994. *Environmental Statistics, Assessment, and Forecasting.* New York: CRC Press/Lewis Publishers.

Cothern, R. R. 1993. *Comparative Environmental Risk Assessment.* New York: CRC Press/Lewis Publishers.

Cothern, C. R. 1995. *Handbook for Environmental Risk Decisionmaking.* New York: CRC Press/Lewis Publishers.

Environmental Protection Agency. October 1983. *Health Assessment for Methylacrylonitrile.* Final Report: EPA 600/8-82-007F. NTIS PB84-149152. Springfield, VA: National Technical Information Service.

Gardner, R. W. 1994. *Chemical Intolerance.* New York: CRC Press/Lewis Publishers.

Hodgson, E., R. B. Mailman, and Janice E. Chambers. 1988. *Dictionary of Toxicology.* London: The MacMillian Company.

Jones, F. E. 1994. *Toxic Organic Vapors in the Workplace.* New York: CRC Press/Lewis Publishers.

Lewis, R. J., Sr. 1993. *Hawley's Condensed Chemical Dictionary,* 12th Edition. New York: Van Nostrand Reinhold.

Lewis, R. J., Sr. 1992. *Sax's Dangerous Properties of Industrial Materials,* Volumes 1, 2, and 3, 8th Edition. New York: Van Nostrand Reinhold.

Meyer, E. 1990. *Chemistry of Hazardous Materials.* Englewood Cliffs, NJ: Prentice Hall.

Patnaik, P. 1992. *A Comprehensive Guide to the Hazardous Properties of Chemical Substances.* New York: Van Nostrand Reinhold.

Proctor, N. H., and James P. Hughes. 1991. *Chemical Hazards of the Workplace.* New York: Van Nostrand Reinhold.

Quigley, D. R. 1994. *Handbook of Emergency Chemical Management.* New York: CRC Press/Lewis Publishers.

U.S. Department of Health and Human Services, National Institute for Occupational Safety and Health. 1994. *NIOSH Pocket Guide to Chemical Hazards.* Washington, D.C.: U.S. Government Printing Office.

Willhite, C. C. Toxicological updates. Methylacrylonitrile. *Journal of Applied Toxicology.* 2:54-56, 1982.

MATERIAL SAFETY DATA SHEET

CHEMICAL NAME	
METHYL ALCOHOL	**CRC PRESS/LEWIS PUBLISHERS DISCLAIMER:** THE INFORMATION AND RECOMMENDATIONS PRESENTED HEREIN ARE BASED ON SOURCES BELIEVED TO BE RELIABLE. CRC MAKES NO REPRESENTATION ON ITS COMPLETENESS OR ACCURACY. IT IS THE RESPONSIBILITY OF THE USER TO DETERMINE THE CHEMICAL'S SUITABILITY FOR ITS INTENDED USE, THE CHEMICAL'S SAFE USE, AND THE CHEMICAL'S PROPER DISPOSAL. NO REPRESENTATIONS AND/OR WARRANTIES, EITHER EXPRESSED OR IMPLIED, OF THE MERCHANTABILITY OR FITNESS FOR A PARTICULAR PURPOSE, OR OF ANY OTHER NATURE, ARE MADE WITH RESPECT TO THE INFORMATION PROVIDED IN THIS MSDS OR TO THE CHEMICAL TO WHICH INFORMATION MAY REFER. CRC NEITHER ASSUMES NOR AUTHORIZES ANY OTHER PERSON TO ASSUME FOR IT, ANY OTHER ADDITIONAL RESPONSIBILITY OR LIABILITY FOR THE USE OF, OR RELIANCE UPON, THIS INFORMATION.

HAZARD WARNING INFORMATION

HEALTH	FIRE	REACTIVE	OTHER	DEGREE OF HAZARD	COLOR CODING	OTHER CODES
1	3	0		0 = Minimum Hazard 1 = Slight Hazard 2 = Moderate Hazard 3 = Serious Hazard 4 = Severe Hazard	HEALTH = BLUE FIRE = RED REACTIVITY = YELLOW OTHER = WHITE	OX = Oxidizer ACID = Acid ALK = Alkali COR = Corrosive W = Use No Water

SECTION I - GENERAL INFORMATION

Characterization	**Alcohol**	RCRA Number	**U154**
		EPA Class	**Toxic Waste**
DOT Proper Shipping Name	**Methyl alcohol**	Chemical Abstract Service (CAS) Number	**67-56-1**
DOT Hazard Class and Label Requirements	**Flammable Liquid; Poison**	DOT Emergency Guide Code	**28**
DOT Identification Number	**UN 1230**	Chemical Formula	**CH_3OH**

Synonyms
Carbinol; Columbian spirits; methanol; pyroligneous spirit; wood alcohol; wood naphtha; wood spirit.

SECTION II - HAZARDOUS INGREDIENTS AND IDENTITY INFORMATION

Hazardous Components (specific identity)	OSHA Exposure Criteria	NIOSH Exposure Criteria	Immediately Dangerous to Life and Health (IDLH)	ACGIH Exposure Criteria
Methyl alcohol (derivation: By high-pressure catalytic synthesis from carbon monoxide and hydrogen; partial oxidation of natural gas hydrocarbons; by gasification of wood, peat, and lignite; from methane with molybdenum catalyst). **1 ppm = 1.33 mg/m^3**	**PEL (skin):** **200 ppm** **260 mg/m^3** **STEL (15-min):** **250 ppm** **310 mg/m^3**	**REL (skin):** **200 ppm** **260 mg/m^3** **STEL:** **250 ppm** **325 mg/m^3**	**6000 ppm**	**TLV:** **200 ppm** **262 mg/m^3** **STEL:** **250 ppm** **328 mg/m^3**

SECTION III - PHYSICAL AND CHEMICAL CHARACTERISTICS

Boiling Point	**148°F (64°C)**	Specific Gravity (H_2O = 1)	**0.79**
Vapor Pressure (mm Hg)	**29 at 68°F (20°C)**	Molecular Weight	**46.1**
Vapor Density (Air = 1)	**1.11**	Melting Point	**-144°F (-98°C)**

Solubility
Soluble in water, ethanol, alcohols, ether, benzene, ketones, and most organic solvents.

Appearance and Odor
Clear, colorless liquid with a characteristic, pungent odor. Odor Threshold = 10 ppm.

SECTION IV - FIRE AND EXPLOSION HAZARD DATA

Flash Point (method used)	**54°F (12°C) closed cup**	Explosive Limits in Air % by Volume	**LEL: 6.0%** **UEL: 36%**
NFPA Classification	**Class IB Flammable Liquid**	Autoignition Temperature	**878°F (470°C)**

Extinguishing Media
Dry chemical, carbon dioxide, alcohol resistant foam, water spray (use as fog in flooding quantities).

Special Fire Fighting Procedures
Poisonous gases are produced in fire. Wear full protective clothing and self-contained breathing apparatus (SCBA). Continue to cool containers with water after fire is extinguished. Move containers from fire area if it can be done without risk. For large fires, use unmanned hose apparatus, if possible. Evacuate non-essential personnel 1500 feet radius if fire. Consider down-wind conditions.

Unusual Fire and Explosion Hazards
Vapors are heavier than air and may travel a distances to ignition source to flashback. Vapors may explode if ignited in an enclosed area.

SECTION V - REACTIVITY DATA

Stability		Conditions to Avoid: Methyl alcohol is stable under normal conditions of storage and operation. Avoid contact with incompatible materials. Keep away from heat, flame, and other ignition sources.
Stable: X	Unstable:	Incompatibility (*materials to avoid*): Reacts violently in contact with oxidizers (bromine, chlorine, fluorine), beryllium dihydride, metals (potassium, magnesium), potassium tertbutoxide, carbon tetrachloride + metals, chloroform + heat.
Hazardous Polymerization		Conditions to Avoid: Under normal conditions of temperature and pressure, methyl alcohol is not expected to undergo hazardous polymerization.
May Occur:	Will Not Occur: X	Hazardous Decomposition or By-products: When heated to decomposition, methyl alcohol emits acrid and irritating fumes, including oxides of carbon (monoxide and dioxide) and formaldehyde.

SECTION VI - HEALTH HAZARD DATA

Primary Route(s) of Entry:	Inhalation?	Absorption (skin)?	Ingestion?
	X	X	X

Health Hazards

INHALATION: Irritation of the eyes, nose, throat, and respiratory tract. CNS effects include headache, nausea, visual disturbances and impairment (optic nerve damage can progress up to and including permanent blindness), acidosis, convulsions, circulatory collapse, respiratory failure, and death.

ABSORPTION: As a solvent, cause skin dryness, cracking, and possible dermatitis. Eye contact can result in burning, stinging, tearing, corneal injury, and painful photophobia. Absorption leads to systemic effects similar to inhalation.

INGESTION: Gastrointestinal irritation, kidney and brain effects, and similar effects to inhalation.

Carcinogenicity	NTP Listed?	IARC Cancer Review Group?	OSHA Regulated?	Target Organs?
Unknown Human Unknown Animal	No	No	29 CFR 1910.1000 Table Z-1	Skin, eyes, CNS, respiratory system, GI tract.

Medical Conditions Generally Aggravated by Exposure

Existing visual disorders or disturbances may be quickly aggravated by exposure to methyl alcohol.

Emergency and First-aid Procedures

Eye contact: Flush immediately with water for 15 minutes (minimum); seek medical attention. Skin contact: Remove all contaminated clothing. Immediately wash area with large amounts of soap and water. Seek medical assistance. For inhalation: Remove the person from exposure. Provide respiratory assistance and CPR. Transfer to medical facility. If swallowed: Seek medical attention immediately. Give 1-2 glasses of water and induce vomiting. Seek medical attention immediately.

SECTION VII - PRECAUTIONS FOR SAFE HANDLING AND USE

Steps to be Taken in Case Material is Released or Spilled

Full protective clothing required with SCBA. Contain spills using absorbent material (vermiculite or other). Ventilate area of spill and remove all ignition sources. Have extinguishing agent available in case of fire. Store materials in DOT-approved containers.

Preferred Waste Disposal Method

Incineration in chemical incinerator equipped with scrubber and afterburner.

Precautions to be Taken in Handling and Storage

Store in tightly closed containers in a cool, dark, well-ventilated area. Protect containers from physical damage. Keep heat, fire, and flame away. Outside building or underground storage is recommended.

Other Precautions and Warnings

Ground and bond all metal containers. Ensure explosion-proof electrical installations.

SECTION VIII - CONTROL MEASURES AND PERSONAL PROTECTIVE EQUIPMENT

Respiratory Protection (*specify type*)

For vapor exposures, use an organic vapor air-purifying respirator. Greater protection is obtained using a supplied-air respirator or a self-contained breathing apparatus (SCBA) with full facepiece operated in pressure demand or other positive pressure mode.

Ventilation

Local exhaust or general mechanical systems recommended.

Protective Gloves	Eye Protection	Other Protective Clothing
Impervious Rubber Gloves	Chemical Goggles or Face Mask	Impermeable Apron or Protective Suit

Work/Hygiene Practices

Always wash hands thoroughly after using chemical; never bring food, drink, or smoking materials into vicinity of chemicals.

METHYL ALCOHOL

CH_3OH

CAS: 67-56-1

CH_3 — OH

IDENTIFICATION AND TYPICAL USES

Methyl alcohol is a colorless or clear liquid with a characteristic pungent odor. It is used in the manufacture of formaldehyde, acetic acid, and dimethyl terephthalate. Also used in chemical synthesis of methyl amines, methyl chloride, and methylacrylate. Used in antifreeze, as an octane booster for gasoline, as a solvent for nitrocellulose, ethylcellulose, polyvinyl butyral, shellac, rosin, manila resin, and dyes. Also a denaturant for ethanol, a dehydrator for natural gas, a fuel for utility plants (methyl fuel), in feedstock for manufacture of synthetic proteins by continuous fermentation, as a source of hydrogen for fuel cells, and a home heating oil extender.

RISK ASSESSMENT: HEALTH

General Assessment

The toxicity of methyl alcohol is much higher in comparison to ethanol. Methyl alcohol is toxic to humans by all routes of exposure (*ingestion, inhalation,* and dermal *absorption*). Once in the body, methyl alcohol oxidizes to formaldehyde and formic acid. The toxicity of methyl alcohol is attributed primarily to the metabolites of these two toxic compounds. It is a narcotic. There are no report regarding its carcinogenic potential. However, it has caused mutations (genetic changes) and is an experimental teratogen and reproductive toxin.

Inhalation can cause irritation of the eyes, nose, throat, upper respiratory tract, and associated mucosa. There may be significant action on the nervous system, especially the optic nerve with damage progressing up to and including blindness. Other central nervous system effects include headache, nervousness, tremors, acidosis, convulsions, dizziness, tearing, fatigue, nausea, somnolence, narcosis with stupor and loss of consciousness, circulatory system collapse, respiratory failure, and death.

Liquid contact with the eyes causes immediate burning and stinging with lachrymation and reflex closure, and inflammation of the lids with painful photophobia. Skin contact results in drying and cracking which can lead to secondary infections and dermatitis. Dermal absorption causes many of the symptoms of inhalation. A combined absorption exposure with inhalation or ingestion can present a life-threatening exposure risk.

Ingestion of methyl alcohol affects the brain, lungs, kidney, gastrointestinal tract, eyes, and respiratory system and can cause coma, blindness, and death.

Acute Health Effects

The following acute (short-term) health effects may occur immediately or shortly after exposure to methyl alcohol:

Skin: Causes dryness and cracking leading to dermatitis and possible infection. Absorption can occur quickly causing toxic systemic effects.

Eye: Severe irritation with burning and stinging with possible damage to the cornea and conjunctiva.

Lung: Irritation of the eyes, nose, throat, and respiratory tract.

CNS: High concentrations can cause depression of the CNS with symptoms of sleepiness and lack of concentration.

Chronic Health Effects

The following chronic (long-term) health effects may occur at some time after exposure to methyl alcohol and can last for months or even years:

Cancer Hazards: According to information presented in the references, methyl alcohol has not been shown to cause cancer in test animals. However, human mutation data have been reported and many scientists believe that exposure to such chemicals may pose a cancer risk in the long-term.

Reproduction: According to the references, methyl alcohol may affect reproduction as well as adversely affect a developing fetus.

Other Chronic Effects: Very high or prolonged exposure may result in mucous membrane irritation, headache, and depression of the CNS with symptoms of somnolence and lack of concentration. Chronic inhalation or skin absorption may lead to visual impairment or even complete blindness.

Recommended Risk-Reduction Measures

Industrial exposures to methyl alcohol by all routes must be avoided. The best risk reduction measure is to use a less toxic chemical as a substitute for a methyl alcohol. However, substitution may not always be an alternative. Therefore, *engineering controls* are the most effective methods of reducing exposures. The best protection is to enclose operations and/or provide local, explosion-proof exhaust ventilation at the site of chemical release. While not always operationally feasible, isolating operations can also reduce exposure risk.

Using respiratory protection is less effective than the controls mentioned above, but is still advisable whenever working with or around methyl alcohol. For concentrations over the PEL (200 ppm), an air-purifying respirator with an organic vapor cartridge will suffice. For higher concentrations, exposures in confined spaces, or when the exposure level is unknown (as in an emergency), a supplied-air respirator with full facepiece operated in positive pressure mode, or a self-contained breathing apparatus (SCBA) with full facepiece and operated in pressure demand mode are the recommended respiratory protection methods of choice. If a full facepiece is not available, then chemical goggles should be worn to protect the eyes. Whenever a chemical splash hazard exists, a face shield and a protective apron should be worn. To prevent hand and skin exposures, impervious gloves should be used. Glove manufacturers should be contacted and permeation studies obtained *before* final glove selections have been made.

Administrative controls should also be in place to minimize the potential for human exposures. These may include written procedures or policies which specify the methods and techniques that will be practiced whenever personnel are to work with methyl alcohol.

All personnel should receive *training* on the use, hazards, protective measures, emergency actions, and other precautions per 29 CFR 1910.1200 (Hazard Communication), prior to the first assignment in an area where methyl alcohol is used or stored.

If symptoms develop or overexposure is suspected, the following medical tests are recommended:

- ☑ Urinalysis, including pH (lowered when poisoning has occurred) and formic acid excretion.
- ☑ Blood pH and plasma bicarbonate.
- ☑ Hemodialysis may be helpful treatment for severe poisoning. Ethyl alcohol may be partially effective as an antidote (it appears to compete with methyl alcohol to inhibit its oxidation in the body).
- ☑ Treat severe acidosis by oral administration of sodium bicarbonate and magnesium sulfate, or by intravenous administration of sodium bicarbonate.

Any medical evaluation should include a careful history of past and present symptoms with an examination. Medical tests that look for existing damage are not a substitute for controlling exposures.

Other methods to reduce exposure include:

- ☑ Where possible, enclose operations and use local exhaust ventilation at the site of chemical release. If local exhaust ventilation or enclosure is not used, respiratory protection should be mandatory.
- ☑ Always ensure that proper protective clothing is worn when using chemical substances.
- ☑ Wash thoroughly immediately after exposure to methyl alcohol and at the end of the work shift or before eating, drinking, or smoking.
- ☑ Hazard warning information should be posted in the work area. In addition, as part of an on-going education and training program, all information on the health and safety hazards of methyl alcohol should be communicated to all exposed or potentially exposed workers.

RISK ASSESSMENT: ENVIRONMENT

General Assessment

The environment is at risk of exposure during transportation, storage, disposal, or destruction of methyl alcohol. In almost every scenario, the threat of environmental exposure is contingent upon the proper handling of the chemical substance. Accidental spills,

large or small, can result in fire, explosion, and possible contamination of the surrounding environmental mediums (water, soil, and air).

Methyl alcohol is considered a Class IB flammable liquid (according to OSHA 29 CFR 1910.106). Its low flash point and relatively low boiling point present a serious fire and explosion hazard concern. Also, because it is incompatible with a number of common materials, especially strong oxidizers, and contact can result in violent and explosive reactions. It can form explosive mixtures in air and can ignite on contact with heat, fire, or sparks. It will react and then explode in contact with chloroform + heat. It also reacts violently with alkyl aluminum salts, acetyl bromide, nitric acid, and many other compounds. These characteristics require special consideration during any emergency situation involving a leak or spill of methyl alcohol.

Methyl alcohol can enter the environment through unchecked industrial discharges into effluents and through spills.

Acute Ecological Effects

Acute (short-term) toxic effects may include the death of animals, birds, or fish, and death or low growth rate in plants. Acute effects are seen 2 to 4 days after animals or plants are exposed to methyl alcohol.

This chemical has slight acute toxicity to aquatic life. It has caused germination and size decrease, and other injury to agricultural and ornamental crops. Insufficient data are available to evaluate or predict the short-term effects of methyl alcohol to birds or terrestrial animals.

Chronic Ecological Effects

Chronic toxic effects may include shortened life span, reproductive problems, lower fertility, and changes in appearance or behavior in exposed animals. These effects can be seen long after first exposure(s) to toxic chemicals.

Methyl alcohol has slight chronic toxicity to aquatic life. Insufficient data are available to evaluate or predict the long-term effects of methyl alcohol to plants, birds, or land animals.

Water Solubility

Methyl alcohol is highly soluble in water. Concentrations of 1000 milligrams and more can be expected to mix with a liter of water.

Persistence in the Environment

Methyl alcohol is slightly persistent in water, with a half-life of between 2 to 20 days. The half-life of a pollutant is the amount of time it takes for one-half of the chemical to be degraded. About 86.5% of methyl alcohol will eventually end up in the water; the remainder will end up in air.

Bioaccumulation in Aquatic Organisms

Some substances increase in concentration, or *bioaccumulate,* in living organisms as they breathe contaminated air, drink contaminated water, or eat contaminated food. These chemicals can become concentrated in the tissues and internal organs of animals as well as humans. The concentration of methyl alcohol found in fish tissues is expected to be about the same as the average concentration of methyl alcohol in water from which the fish was taken.

Recommended Risk-Reduction Measures

Proper training of all transporters will reduce the likelihood of a mishap or accident resulting in a leak or spill to the environment. The correct use of DOT labeling (per the MSDS) on all containers, trucks, and rail cars will enable emergency responders to react properly and quickly to any disaster thereby reducing the potential risk to the environment and to personnel.

Storage of methyl alcohol should be segregated from incompatible chemicals to minimize the risk of cross-contamination or contact. Buildings designated for storage should be equipped with appropriate fire protection systems (alarms, sprinklers, emergency lighting, portable extinguishers). Equipment should be designed to meet explosion-proof standards.

If a spill or leak to the environment has occurred, fire department, emergency response, and/or hazardous materials spill personnel should be notified immediately. Cleanup should be attempted only by those trained in proper spill containment procedures using non-sparking tools. Contaminated soils should be removed for incineration and replaced with clean soil. If methyl alcohol should contact the water table, aquifer, or navigable waterway, time is of the essence. It is highly soluble in water and, therefore, total containment and remediation may not be possible. When such spills occur, the local and/or state emergency response authorities should be notified. A comprehensive emergency response or disaster preparedness/recovery plan should be in place prior to any operations involving the use, transportation, storage, or disposal of methyl alcohol. If methyl alcohol is spilled or leaked, the following specific steps are recommended:

- ☑ Restrict persons not wearing protective clothing from area of spill or leak until cleanup is complete and area can be opened for normal work.
- ☑ Ventilate area and remove ignition sources.
- ☑ Absorb liquids in vermiculite, dry sand, earth, or a similar material and deposit in sealed containers. Use non-sparking tools.
- ☑ It may be necessary to dispose of methyl alcohol as a hazardous waste. The responsible state agency or the regional office of the federal Environmental Protection Agency (EPA) should be contacted for specific recommendations.

RISK ASSESSMENT: BUSINESS

General Assessment

Accidents or mishaps involving methyl alcohol can present a moderate threat to business operations. The loss or damage of equipment or facilities can significantly affect fiscal viability. Lawsuits that may result from personnel injury/death, public exposures, and/or environmental contamination will also require a serious expenditure of resources. Media attention surrounding an injury, death, or environmental damage can also result in a loss of profits and loss of current as well as future business.

Recommended Risk-Reduction Measures

Company attorneys, safety and health professionals, and environmental specialists should be involved in the development of any procedures or policies intended to manage the use of chemicals in the workplace. A company official should be pre-designated as a public relations officer with specific training in dealing with the press. Corporate plans and policies should be developed, approved, and implemented long before any need for such arises.

REFERENCES

American Conference of Governmental Industrial Hygienists. 1988. *Documentation of the Threshold Limit Values and Biological Exposure Indices,* 5th Edition (with updates). Cincinnati: ACGIH

Anthony, C. P., and N. J. Kolthoff. 1971. *Textbook of Anatomy and Physiology.* St. Louis: C. V. Mosby Company.

Bloom, A. D. and F. J. DeSerres. 1995. *Ecotoxicity and Human Health.* New York: CRC Press/Lewis Publishers.

Brusick, D. J. 1994. *Methods for Genetic Risk Assessment.* New York: CRC Press/Lewis Publishers.

Bueche, J. R. 1972. *Principles of Physics,* 2nd Edition. New York: McGraw-Hill.

Burrell, R., D. K. Flaherty, and L. J. Sauers. 1992. *Toxicology of the Immune System.* New York: Van Nostrand Reinhold.

Calabrese, E. J. 1994. *Biological Effects of Low Level Exposures.* New York: CRC Press/Lewis Publishers.

Cothern, C. R., N. P. Ross. 1994. *Environmental Statistics, Assessment, and Forecasting.* New York: CRC Press/Lewis Publishers.

Danse, I. R. 1991. *Common Sense Toxics in the Workplace.* New York: Van Nostrand Reinhold.

Environmental Protection Agency. 1983. *Chemical Hazard Information Profiles.* Washington, D.C.: U.S. EPA.

Fodden, C. R., and J. L. Weddell. 1992. *Hazardous Materials Emergency Action Data.* New York: CRC Press/Lewis Publishers.

Forseberg, K., and L. H. Keith. 1993. *Chemical; Protective Clothing Permeation and Degradation Database.* Chelsea, MI: Lewis Publishers.

Gardner, R. W. 1994. *Chemical Intolerance.* New York: CRC Press/Lewis Publishers.

Hodgson, E., R. B. Mailman, and J. E. Chambers. 1988. *Dictionary of Toxicology.* London: The MacMillian Company.

Hosty, J. W., and P. Foster. 1990. *A Practical Guide to Chemical Spill Response.* New York: Van Nostrand Reinhold.

Howard, P. H., et. al. 1991. *Handbook of Environmental Degradation Rates.* New York: CRC Press/Lewis Publishers.

International Agency for Research on Cancer (IARC). 1987. *IARC Monographs on the Evaluation of Carcinogenic Risk to Humans, Overall Evaluations of Carcinogenicity: An updating of IARC Monographs Volumes 1-42.* Lyon: IARC.

Keith, L. H. and D. B. Walters. (Editors) 1992. *National Toxicology Program's Chemical Database, Volumes I-VII.* Chelsea, MI: Lewis Publishers.

Landis, W. G., M. H. Yu. 1995. *Introduction to Environmental Toxicology.* New York: CRC Press/Lewis Publishers.

Lauwerys, R. R., P. Hoet. 1993. *Industrial Chemical Exposure.* New York: CRC Press/Lewis Publishers.

Lewis, Richard J., Sr. 1992. *Sax's Dangerous Properties of Industrial Materials,* Volumes 1, 2, and 3, 8th Edition. New York: Van Nostrand Reinhold.

Manahan, S. E. 1992. *Toxicological Chemistry,* 2nd Edition. New York: CRC Press/Lewis Publishers.

Meyer, Eugene. 1990. *Chemistry of Hazardous Materials.* Englewood Cliffs, NJ: Prentice Hall.

Patnaik, P. 1992. *A Comprehensive Guide to the Hazardous Properties of Chemical Substances.* New York: Van Nostrand Reinhold.

Philp, R. B. 1995. *Environmental Hazards and Human Health.* New York: CRC Press/Lewis Publishers.

MATERIAL SAFETY DATA SHEET

CHEMICAL NAME

METHYLAMINE

HAZARD WARNING INFORMATION

HEALTH	FIRE	REACTIVE	OTHER	DEGREE OF HAZARD	COLOR CODING	OTHER CODES
3	4	0		0 = Minimum Hazard 1 = Slight Hazard 2 = Moderate Hazard 3 = Serious Hazard 4 = Severe Hazard	HEALTH = BLUE FIRE = RED REACTIVITY = YELLOW OTHER = WHITE	OX = Oxidizer ACID = Acid ALK = Alkali COR = Corrosive W = Use No Water

SECTION I - GENERAL INFORMATION

Characterization: **Aliphatic Amine**

RCRA Number: **None**

EPA Class: **Not Applicable**

DOT Proper Shipping Name: **Methylamine (anhydrous or aqueous solution)**

Chemical Abstract Service (CAS) Number: **74-89-5**

DOT Hazard Class and Label Requirements: **Flammable Gas (or Flammable Liquid)**

DOT Emergency Guide Code: **19 (anhydrous) 65 (aqueous)**

DOT Identification Number: **UN 1061 (anhydrous); UN 1235 (aqueous)**

Chemical Formula: **CH_3NH_2**

Synonyms: **Aminomethane; anhydrous methylamine; aqueous methylamine; monomethylamine; carbinamine.**

SECTION II - HAZARDOUS INGREDIENTS AND IDENTITY INFORMATION

Hazardous Components (specific identity)	OSHA Exposure Criteria	NIOSH Exposure Criteria	Immediately Dangerous to Life and Health (IDLH)	ACGIH Exposure Criteria
Methylamine (derivation: By interaction of methanol and ammonia over a catalyst at high temperature). **1 ppm = 1.29 mg/m^3**	**PEL:** 10 ppm 12 mg/m^3 **STEL:** Not Established	**REL:** 10 ppm 12 mg/m^3 **STEL:** Not Established	**100 ppm**	**TLV:** 10 ppm 12 mg/m^3 **STEL:** Not Established

SECTION III - PHYSICAL AND CHEMICAL CHARACTERISTICS

Boiling Point: **19°F (-7°C)**

Specific Gravity (H_2O = 1): **0.70 (as liquid) at 13°F (-10°C)**

Vapor Pressure (atmospheres): **3.0 at 68°F (20°C)**

Molecular Weight: **31.07**

Vapor Density (Air = 1): **1.07**

Freezing Point: **-136°F (-93°C)**

Solubility: **Completely soluble in water. Soluble in alcohol and ether.**

Appearance and Odor: **Colorless gas or a volatile, water-white liquid with a fish or ammonia-like odor.**

SECTION IV - FIRE AND EXPLOSION HAZARD DATA

Flash Point (method used): **34°F (1.1°C) - liquid; 14°F (-10°C) - gas (CC)**

Explosive Limits in Air % by Volume: **LEL: 4.9% UEL: 20.7%**

NFPA Classification: **Flammable Gas or Class 1A Flammable Liquid**

Autoignition Temperature: **806°F (430°C)**

Extinguishing Media: **Stop flow of gas first (if possible). If liquid, use foam, carbon dioxide, or dry chemical.**

Special Fire Fighting Procedures: **Methylamine is an extremely flammable, very dangerous fire and explosion hazard. Poisonous gases are evolved from fire. Wear a self-contained breathing apparatus with full facepiece. Ensure ALL sources of escaping gas are controlled. Move containers from fire area if it can be done without risk. Water may be ineffective on fire. Cool containers with water until long after fire is out.**

Unusual Fire and Explosion Hazards: **Gas is explosive. Containers may explode in flame. Vapors are heavier than air and will travel for a great distance to an ignition source and flashback to cause fire or explosion.**

SECTION V - REACTIVITY DATA

<table>
<tr><td colspan="2">Stability</td><td>Conditions to Avoid
Methylamine is normally stable in its closed containers, during routine operations (those that do not expose it to ignition sources). Avoid contact with heat, sparks, flame, and other sources of ignition.</td></tr>
<tr><td>Stable
X</td><td>Unstable</td><td>Incompatibility (materials to avoid)
Incompatible with strong oxidizing materials (bromine, chlorine, fluorine, peroxides, permanganates), and mercury. Corrosive to copper and zinc alloys, aluminum, and galvanized surfaces.</td></tr>
<tr><td colspan="2">Hazardous Polymerization</td><td>Conditions to Avoid
Under normal conditions of temperature and pressure, hazardous polymerization of methylamine is not expected to occur.</td></tr>
<tr><td>May Occur</td><td>Will Not Occur
X</td><td>Hazardous Decomposition or By-products
If involved in fire, methylamine can produce toxic oxides of carbon and nitrogen, and ammonia derivatives.</td></tr>
</table>

SECTION VI - HEALTH HAZARD DATA

Primary Route(s) of Entry:	Inhalation? X	Absorption (skin)? X (liquid)	Ingestion? X (liquid)

Health Hazards

INHALATION: Vapors are irritating to the eyes, skin, and respiratory tract. Causes pulmonary edema (fluid in the lungs), which is a medical emergency and can lead to death. There may also be blurred or hazy vision (temporary) resulting from corneal epithelium.

SKIN & EYES: Contact with the compressed gas or liquid can cause frostbite and first degree burns.

INGESTION: Not likely since methylamine is a gas at normal room temperatures. If it occurs, there can be severe burns of the mouth and stomach that can be fatal.

Carcinogenicity	NTP Listed?	IARC Cancer Review Group?	OSHA Regulated?	Target Organs?
Unknown Human Unknown Animal	No	No	29 CFR 1910.1000 Table Z-1	Eyes, skin, respiratory system

Medical Conditions Generally Aggravated by Exposure

None Reported.

Emergency and First-aid Procedures

Eye contact: For frostbite, rinse with warm (104°F/40°C) water. Skin contact: Remove all contaminated clothing. Immediately wash area with large amounts of warm (104°F/40°C) water. For inhalation: Remove the person from exposure. Provide respiratory assistance and CPR. Transfer promptly to medical facility. Observe for 24-48 hours for lung effects (pulmonary edema). If swallowed: Contact poison control center. Unless advised otherwise, give 1-2 glasses of water to dilute. Do NOT induce vomiting.

SECTION VII - PRECAUTIONS FOR SAFE HANDLING AND USE

Steps to be Taken in Case Material is Released or Spilled

Remove all ignition sources. Ventilate area. Stop flow of gas. If the source of leak is a cylinder and cannot be stopped, remove to safe place in open air; repair leak or let cylinder empty. If gas has not yet ignited, use a water spray to knock down vapors and direct them from ignition source.

Preferred Waste Disposal Method

None specified in the references (incineration of a gas may be the only alternative).

Precautions to be Taken in Handling and Storage

Do not store with incompatible materials or in the presence of heat since violent reactions or explosions can occur. Store in tightly closed pressurized containers in a cool, dark, well-ventilated area away from sunlight. Keep away from ignition sources such as fire, sparks, and flame.

Other Precautions and Warnings

Bulk storage of methylamine is not recommended. Containers may explode in fire or under conditions of extreme heat (do not store outdoors or in direct sunlight).

SECTION VIII - CONTROL MEASURES AND PERSONAL PROTECTIVE EQUIPMENT

Respiratory Protection *(specify type)*

For exposures to the PEL, use a cartridge-type respirator For higher exposures, use a powered supplied-air respirator or a self-contained breathing apparatus with full facepiece operated in pressure demand or other positive pressure mode.

Ventilation

Local exhaust or general mechanical systems recommended.

Protective Gloves	Eye Protection	Other Protective Clothing
Insulated Impervious Material	Goggles and/or Face Mask	Leather or Impervious Apron

Work/Hygiene Practices

Always wash hands thoroughly after using chemical; never bring food, drink, or smoking materials into vicinity of chemicals.

METHYLAMINE

CH_3NH_2

CAS: 74-89-5

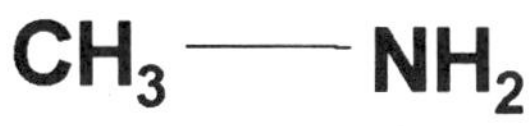

IDENTIFICATION AND TYPICAL USES

Methylamine is a colorless gas with a fish or ammonia-like odor. A liquid below 21°F (-6°C). It is normally shipped as a liquefied compressed gas. It is used as an intermediate for accelerators, dyes, pharmaceuticals, insecticides, fungicides, surface active agents, in tanning, for the dyeing of acetate textiles, as a fuel additive, and as a polymerization inhibitor. It is also used as a component in paint removers, solvents, as a photographic developer, and as a rocket propellant.

RISK ASSESSMENT: HEALTH

General Assessment

Methylamine normally appears as a gas. As such, its primary route of entry into the body is through *inhalation.* Skin contact with the liquid can be destructive to tissues due to the extremely cold temperatures and irritating properties of methylamine, but absorption is not a likely exposure route. Since methylamine usually exists as a gas at room temperatures, ingestion is not likely under normal conditions. It is not known if exposure to methylamine will lead to carcinogenic or teratogenic effects in humans or animals. However, mutation data have been reported and caution is warranted in use and handling.

Inhalation of methylamine vapors causes irritation of the eyes, nose, throat, and respiratory tract. Exposure can lead to pulmonary edema (fluid buildup in the lungs) which is a medical emergency and can be fatal. Symptoms of congestion, cough phlegm, tightness and possibly pain in chest, and difficulty breathing may be delayed up to 48 hours thereby creating a false sense of security with regard to health hazard/exposure risk. There is also a possibility that exposure can cause blurred or hazy vision, apparently as a result of corneal epithelium disturbances.

Skin or eye contact with the vapor causes irritation. Liquid contact with either the skin or the eyes causes serious irritation and even first degree burns with subsequent tissue damage, and frostbite.

Although ingestion is unlikely to occur under normal conditions (methylamine is a gas at room temperature), swallowing it will cause severe burns to the mouth and stomach and can lead to death.

Acute Health Effects

The following acute (short-term) health effects may occur immediately or shortly after exposure to methylamine:

Skin: Redness, smarting, pain, possible dermatitis, first degree burns, and severe frostbite.

Eye: Causes frostbite, redness, irritation, severe pain, and blurred vision.

Lung: Vapor is seriously irritating and therefore poses a significant threat of damage to the lungs. Irritation can lead to pulmonary edema (fluid buildup in the lungs).

Other: Methylamine solutions are strongly alkaline and can cause deep tissue destruction.

Chronic Health Effects

The following chronic (long-term) health effects may occur at some time after exposure to methylamine and can last for months or even years:

Cancer Hazards: According to information presented in the references, it is not known if exposure to methylamine will lead to carcinogenic or mutagenic effects. Mutation data have been reported and many scientists believe that exposure to such chemicals may pose a cancer risk in the long-term.

Reproductive Hazards: According to information presented in the references, methylamine has not been adequately tested for its ability to adversely affect reproduction in animals or humans. There are no reports to support any claims of reproductive hazards.

Ʋ *Recommended Risk-Reduction Measures*

Personnel should avoid direct contact with methylamine. If a less toxic chemical cannot be substituted for a hazardous substance, then *engineering controls* are the most effective method of reducing exposures. The best protection is to separate operations and provide local exhaust ventilation at the site of chemical release. While not always operationally feasible, isolating operations can also reduce exposure. Using equipment with explosion-proof designs and ensuring all metal containers and tools are electrically grounded and bonded can further reduce the potential for injury or damage. Using respiratory protection is less effective than the controls mentioned above, but is still advisable whenever working with methylamine. For exposures above the PEL (10-ppm), use an MSHA/NIOSH-approved chemical cartridge respirator with a full facepiece and cartridges suitable for protection against amines. For higher or more frequent exposures, use an powered supplied-air respirator operated in pressure demand or other positive pressure mode, or a self-contained breathing apparatus (SCBA) with full facepiece operated in pressure demand mode. If a full facepiece is not available, then chemical goggles should be worn to protect the eyes. Whenever a chemical splash hazard exists, a face shield and a rubber or leather apron should be worn. To prevent hand and skin exposures with the liquid, thermal gloves should be used. Glove manufacturers should be contacted and consulted *before* final glove selections have been made.

Administrative controls should also be in place to minimize the potential for human exposures. These may include written procedures or policies which specify the methods and techniques that will be practiced whenever personnel are to work with methylamine.

All personnel should receive *training* on the use, hazards, protective measures, emergency actions, and other precautions per 29 CFR 1910.1200 (Hazard Communication), prior to the first assignment in an area where methylamine is used or stored. Before beginning employment and at regular intervals thereafter (e.g., annually), the following medical tests are recommended:

☑ Lung function tests (establish baseline).

If symptoms develop or overexposure is suspected, the following additional tests may be helpful:

☑ Lung function tests (compare to baseline).

☑ Consider chest X-ray after acute overexposure (may be negative if taken immediately following exposure due to delayed onset of pulmonary edema).

Other methods to reduce exposure to chemicals include:

☑ Where possible, use local exhaust ventilation at the site of chemical release. If local exhaust ventilation or enclosure is not used, respiratory protection should be mandatory.

☑ Prevent static sparks by electrically grounding and bonding all equipment used with or around methylamine.

☑ Always ensure that proper protective clothing is worn when using chemical substances.

☑ Wash thoroughly immediately after exposure to methylamine and at the end of the work shift or before eating, drinking, or smoking.

☑ Safety shower and eyewash stations should be readily available in work areas where methylamine is used or stored.

☑ Hazard warning information should be posted in the work area. In addition, as part of an on-going education and training program, all information on the health and safety hazards of methylamine should be communicated to all exposed and potentially exposed workers.

RISK ASSESSMENT: ENVIRONMENT

General Assessment

The environment is at risk of exposure during transportation, storage, disposal, or destruction of methylamine. In almost every scenario, the threat of environmental exposure is contingent upon the proper handling of the chemical substance. Accidental releases, large or small, can result in fire, explosion, and possible contamination of the surrounding environmental mediums (water, soil, and air).

Methylamine is an extremely flammable and highly explosive gas. As a liquid, it is considered a Class 1A flammable liquid (per OSHA 29 CFR 1910.106). Its extremely low flash point and boiling point present a serious fire hazard. Also, since it can react violently in the presence of oxidizers, such as chlorine, methylamine requires special consideration during any emergency involving a leak or release of methylamine gas or liquid.

Methylamine can enter the environment through industrial discharges, unchecked venting to air, or through spills of the liquid.

Acute Ecological Effects

Acute (short-term) toxic effects may include the death of animals, birds, or fish, and death or low growth rate in plants. Acute effects are seen 2 to 4 days after animals or plants are exposed to methylamine. Insufficient data are available on the short-term effects of methylamine on aquatic life, plants, birds, or land animals.

Chronic Ecological Effects

Chronic (long-term) toxic effects may include shortened life span, reproductive problems, lower fertility, and changes in appearance or behavior in exposed animals. These effects can be seen long after first exposure(s) to toxic chemicals. Insufficient data are available to evaluate the long-term effects of methylamine to aquatic life, plants, birds, or land animals.

Water Solubility

Methylamine is highly soluble in water. Concentrations of 1000 milligrams and more will mix with a liter of water.

Persistence in the Environment

Methylamine is slightly persistent in the aquatic environment with a half-life of 2 to 20 days. The half-life of a pollutant is the amount of time it takes for one half of the chemical to be degraded.

Bioaccumulation in Aquatic Organisms

Some substances increase in concentration, or *bioaccumulate,* in living organisms as they breathe contaminated air, drink contaminated water, or eat contaminated food. These chemicals can become concentrated in the tissues and internal organs of animals as well as humans.

Methylamine is not expected to accumulate in the tissues of fish.

Recommended Risk-Reduction Measures

Proper training of all transporters will reduce the likelihood of a mishap or accident resulting in a leak or release to the environment. The correct use of DOT labeling (per the MSDS) on all containers, trucks, and rail cars will enable emergency responders to react properly and quickly to any disaster thereby reducing the potential risk to the environment and to personnel.

Storage of methylamine should be segregated from incompatible chemicals to minimize the risk of cross-contamination or contact. Buildings designated for storage should be equipped with appropriate fire protection systems (alarms, sprinklers, emergency lighting, portable extinguishers).

If a spill or release to the environment has occurred, fire department, emergency response, and/or hazardous materials spill personnel should be notified immediately. Cleanup should be attempted only by those trained in proper containment procedures. Leaking cylinders must be repaired or removed to a safe location and allowed to vent until empty. Contaminated soils should be removed for incineration and replaced with clean soil. If methylamine should contact the water table, aquifer, or navigable waterway, time is of the essence. It is readily miscible in water and total containment and remediation may not be possible. When such spills occur, it may be necessary to contact the local and/or state emergency response authorities. A comprehensive emergency response or disaster preparedness/recovery plan should be in place prior to any operations involving the use, transportation, storage, or disposal of methylamine. If methylamine is spilled or leaked, the following specific steps are recommended:

☑ Restrict persons not wearing protective clothing from area of spill or leak until cleanup is complete and area can be opened for normal work.

☑ Remove all ignition sources.

☑ Ventilate area of spill or leak.

☑ If applicable, stop flow of leaking liquid or gas. If leak source is a cylinder and the leak cannot be stopped in place, remove leaking cylinder to a safe place in the open air, and repair or allow cylinder to empty. Use water spray to knock down vapors and to protect emergency response personnel attempting to shut off leak.

☑ Keep methylamine out of a confined space, such as a sewer, because of the possibility of explosion (unless the sewer is designed to prevent the buildup of explosive concentrations).

RISK ASSESSMENT: BUSINESS

General Assessment

Accidents or mishaps involving methylamine can present a significant threat to business operations. The loss or damage of equipment or facilities can significantly affect fiscal viability. Lawsuits that may result from personnel injury, public exposures, and environmental contamination will also require a serious expenditure of resources. Media attention surrounding an injury, illness, or environmental damage can also result in a loss of profits and loss of current as well as future business.

Recommended Risk-Reduction Measures

Company attorneys, safety and health professionals, and environmental specialists should be involved in the development of any procedures dealing with safety or emergency response. A company official should be pre-designated as a public relations officer with specific training in dealing with the press. Corporate plans and policies should be developed, approved, and implemented long before any need for such arises.

REFERENCES

American Conference of Governmental Industrial Hygienists. 1988. *Documentation of the Threshold Limit Values and Biological Exposure Indices,* 5th Edition (with updates). Cincinnati: ACGIH

Anthony, C. P., and N. J. Kolthoff. 1971. *Textbook of Anatomy and Physiology.* St. Louis: C. V. Mosby Company.

Bloom, A. D. and F. J. DeSerres. 1995. *Ecotoxicity and Human Health.* New York: CRC Press/Lewis Publishers.

Brusick, D. J. 1994. *Methods for Genetic Risk Assessment.* New York: CRC Press/Lewis Publishers.

Bueche, J. R. 1972. *Principles of Physics,* 2nd Edition. New York: McGraw-Hill.

Burrell, R., D. K. Flaherty, and L. J. Sauers. 1992. *Toxicology of the Immune System.* New York: Van Nostrand Reinhold.

Calabrese, E. J. 1994. *Biological Effects of Low Level Exposures.* New York: CRC Press/Lewis Publishers.

Cockerman, L. G., B. S. Shane. 1994. *Basic Environmental Toxicology.* New York: CRC Press/Lewis Publishers.

Cothern, R. R. 1993. *Comparative Environmental Risk Assessment.* New York: CRC Press/Lewis Publishers.

Danse, I. R. 1991. *Common Sense Toxics in the Workplace.* New York: Van Nostrand Reinhold.

Environmental Protection Agency. 1983. *Chemical Hazard Information Profiles.* Washington, D.C.: U.S. EPA.

Fodden, C. R., and J. L. Weddell. 1992. *Hazardous Materials Emergency Action Data.* New York: CRC Press/Lewis Publishers.

Forseberg, K., and L. H. Keith. 1993. *Chemical; Protective Clothing Permeation and Degradation Database.* Chelsea, MI: Lewis Publishers.

Gardner, R. W. 1994. *Chemical Intolerance.* New York: CRC Press/Lewis Publishers.

Hallenbeck, W. H. 1993. *Quantitative Risk Assessment for Environmental and Occupational Health,* 2nd Edition. New York: CRC Press/Lewis Publishers.

Hodgson, E., R. B. Mailman, and J. E. Chambers. 1988. *Dictionary of Toxicology.* London: The MacMillian Company.

Hosty, J. W., and P. Foster. 1990. *A Practical Guide to Chemical Spill Response.* New York: Van Nostrand Reinhold.

Howard, P. H., et. al. 1991. *Handbook of Environmental Degradation Rates.* New York: CRC Press/Lewis Publishers.

James, A., and M. Lord. 1992. *Index of Chemical and Physical Data.* New York: Van Nostrand Reinhold.

Jones, F. E. 1994. *Toxic Organic Vapors in the Workplace.* New York: CRC Press/Lewis Publishers.

Kamrin, M. A. 1988. *Toxicology: A Primer on Principles and Applications.* Chelsea, Michigan: Lewis Publishers.

Keith, L. H. (Editor). 1992. *Integrated Risk Information System (IRIS) Chemical Information Database.* Prepared by the U.S. Environmental Protection Agency. Boca Raton, FL: CRC Press/Lewis Publishers.

Landis, W. G., M. H. Yu. 1995. *Introduction to Environmental Toxicology.* New York: CRC Press/Lewis Publishers.

Langerman, N. 1994 *Precautionary Labels for Chemical Containers.* New York: CRC Press/Lewis Publishers.

Lauwerys, R. R., P. Hoet. 1993. *Industrial Chemical Exposure.* New York: CRC Press/Lewis Publishers.

Lewis, R. J., Sr. 1993. *Hawley's Condensed Chemical Dictionary,* 12th Edition. New York: Van Nostrand Reinhold.

Lewis, R. J., Sr. 1992. *Sax's Dangerous Properties of Industrial Materials,* Volumes 1, 2, and 3, 8th Edition. New York: Van Nostrand Reinhold.

Manahan, S. E. 1992. *Toxicological Chemistry,* 2nd Edition. New York: CRC Press/Lewis Publishers.

Meyer, E. 1990. *Chemistry of Hazardous Materials.* Englewood Cliffs, NJ: Prentice Hall.

Patnaik, P. 1992. *A Comprehensive Guide to the Hazardous Properties of Chemical Substances.* New York: Van Nostrand Reinhold.

Philp, R. B. 1995. *Environmental Hazards and Human Health.* New York: CRC Press/Lewis Publishers.

Rea, W. J. 1995. *Chemical Sensitivity,* Vol. I-III. New York: CRC Press/Lewis Publishers.

Richardson, M. 1995. *Dictionary of Substance and Their Effects,* Vol. I-VI. New York: CRC Press/Lewis Publishers.

Sacarello, H. L. A. 1994. *The Comprehensive Handbook of Hazardous Materials.* New York: CRC Press/Lewis Publishers.

Sherman, J. D. 1988. *Chemical Exposure and Disease.* New York: Van Nostrand Reinhold.

Talbott, E., G. Craun. 1995. *An Introduction to Environmental Epidemiology.* New York: CRC Press/Lewis Publishers.

Torkelson, T. R. V. K. Rowe. 1981. *Patty's Industrial Hygiene and Toxicology,* Vol. 2B. New York: Wiley-Interscience.

U.S. Department of Health and Human Services, National Institute for Occupational Safety and Health. 1994. *NIOSH Pocket Guide to Chemical Hazards.* Washington, D.C.: U.S. Government Printing Office.

MATERIAL SAFETY DATA SHEET

CHEMICAL NAME	
METHYL *N*-AMYL KETONE	**CRC PRESS/LEWIS PUBLISHERS DISCLAIMER:** THE INFORMATION AND RECOMMENDATIONS PRESENTED HEREIN ARE BASED ON SOURCES BELIEVED TO BE RELIABLE. CRC MAKES NO REPRESENTATION ON ITS COMPLETENESS OR ACCURACY. IT IS THE RESPONSIBILITY OF THE USER TO DETERMINE THE CHEMICAL'S SUITABILITY FOR ITS INTENDED USE, THE CHEMICAL'S SAFE USE, AND THE CHEMICAL'S PROPER DISPOSAL. NO REPRESENTATIONS AND/OR WARRANTIES, EITHER EXPRESSED OR IMPLIED, OF THE MERCHANTABILITY OR FITNESS FOR A PARTICULAR PURPOSE, OR OF ANY OTHER NATURE, ARE MADE WITH RESPECT TO THE INFORMATION PROVIDED IN THIS MSDS OR TO THE CHEMICAL TO WHICH INFORMATION MAY REFER. CRC NEITHER ASSUMES NOR AUTHORIZES ANY OTHER PERSON TO ASSUME FOR IT, ANY OTHER ADDITIONAL RESPONSIBILITY OR LIABILITY FOR THE USE OF, OR RELIANCE UPON, THIS INFORMATION.

HAZARD WARNING INFORMATION

HEALTH	FIRE	REACTIVE	OTHER	DEGREE OF HAZARD	COLOR CODING	OTHER CODES
1	2	0		0 = Minimum Hazard 1 = Slight Hazard 2 = Moderate Hazard 3 = Serious Hazard 4 = Severe Hazard	HEALTH = BLUE FIRE = RED REACTIVITY = YELLOW OTHER = WHITE	OX = Oxidizer ACID = Acid ALK = Alkali COR = Corrosive ~~W~~ = Use No Water

SECTION I - GENERAL INFORMATION

Characterization **Ketone**	RCRA Number **D001**	EPA Class **Characteristic (I) Waste**
DOT Proper Shipping Name **Methyl Amyl Ketone**	Chemical Abstract Service (CAS) Number **110-43-0**	
DOT Hazard Class and Label Requirements **Combustible Liquid**	DOT Emergency Guide Code **26**	
DOT Identification Number **UN 1110**	Chemical Formula **$CH_3CO[CH_2]_4CH_3$**	

Synonyms
Amyl methyl ketone; *n*-amyl methyl ketone; 2-heptanone; methyl pentyl ketone.

SECTION II - HAZARDOUS INGREDIENTS AND IDENTITY INFORMATION

Hazardous Components (specific identity)	OSHA Exposure Criteria	NIOSH Exposure Criteria	Immediately Dangerous to Life and Health (IDLH)	ACGIH Exposure Criteria
Methyl n-amyl ketone (derivation: By catalytic dehydration of 2-heptanol or by decomposition of ethyl butylacetoacetate. Occurs naturally in oil of clove and cinnamon bark oil and also causes the peppery odor of Roquefort cheese). **1 ppm = 4.75 mg/m^3**	**PEL:** **100 ppm** **465 mg/m^3** **STEL:** **Not Established**	**REL:** **100 ppm** **465 mg/m^3** **STEL:** **Not Established**	**800 ppm**	**TLV:** **50 ppm** **233 mg/m^3** **STEL:** **Not Established**

SECTION III - PHYSICAL AND CHEMICAL CHARACTERISTICS

Boiling Point **305°F (152°C)**	Specific Gravity (H_2O = 1) **0.81**
Vapor Pressure (mm Hg) **3 at 69°F (20°C)**	Molecular Weight **114**
Vapor Density (Air = 1) **4.97**	Freezing Point **-31°F (-35°C)**

Solubility
Slightly soluble (0.4%) in water. Soluble in benzene, ethanol, diethyl ether, and most organic solvents.

Appearance and Odor
Colorless or water-white liquid with a spicy, banana-like odor and pear-like taste.

SECTION IV - FIRE AND EXPLOSION HAZARD DATA

Flash Point (method used) **102°F (39°C) closed cup**	Explosive Limits in Air % by Volume **LEL: 1.1% at 151°F UEL: 7.9% at 250°F**
NFPA Classification **Class II Combustible Liquid**	Autoignition Temperature **991°F (532°C)**

Extinguishing Media
Large fires: Alcohol foam or water spray. Small fires: Dry chemical, carbon dioxide, water spray, foam.

Special Fire Fighting Procedures
Poisonous gases are produced in fire. Wear full protective clothing and self-contained breathing apparatus. Fight fire from a distance if possible. Keep fire-exposed containers cool with water spray. A solid stream of water may spread fire. If it can be done safely, move fire-exposed containers from area.

Unusual Fire and Explosion Hazards
Containers may explode in fire due to the buildup of internal pressures. BLEVE (boiling liquid expanding vapor explosion) is possible. Immediately evacuate if rising sound is heard from venting device or a noticeable discoloration is observed on the sides of the tank or storage vessel.

SECTION V - REACTIVITY DATA

Stability		Conditions to Avoid **Normally stable. Do not allow methyl n-amyl ketone to come into contact with incompatible materials. Avoid contact with fire or high heat sources. Do not use in confined spaces.**
Stable **X**	Unstable	Incompatibility *(materials to avoid)* **Incompatible with strong oxidizers (bromine, chlorine, fluorine), alkalis, and strong acids. It will also attack some forms of plastic.**
Hazardous Polymerization		Conditions to Avoid **Under normal conditions of temperature and pressure, hazardous polymerization of methyl n-amyl ketone is not expected to occur.**
May Occur	Will Not Occur **X**	Hazardous Decomposition or By-products **When heated to decomposition, methyl n-amyl ketone emits acrid, irritating smoke and fumes, including oxides of carbon.**

SECTION VI - HEALTH HAZARD DATA

Primary Route(s) of Entry:	Inhalation?	Absorption (skin)?	Ingestion?
	X		X

Health Hazards

INHALATION: Irritation of the mucosa of the respiratory tract causing respiratory depression with gasping, choking, and dypsnea. Higher concentrations causes CNS depression with headache, dizziness, fainting, tremor, chemical pneumonitis, coma, and possibly death.

SKIN & EYES: Skin contact with liquid or vapor can result in irritation and may cause dermatitis and paresthesia (tingling). Eye contact causes pain, irritation, corneal damage, and tearing.

INGESTION: Unlikely. Symptoms of irritation, sore throat, nausea, coughing, salivation, vomiting.

Carcinogenicity	NTP Listed?	IARC Cancer Review Group?	OSHA Regulated?	Target Organs?
Unknown Human Unknown Animal	No	No	29 CFR 1910.1000 Table Z-1	Respiratory system; skin; eyes; CNS, PNS.

Medical Conditions Generally Aggravated by Exposure

Skin conditions (dermatitis) may be aggravated.

Emergency and First-aid Procedures

Eye contact: Immediately flush with large amounts of water for 15 minutes (minimum), occasionally lifting eyelids. Skin contact: Remove all contaminated clothing. Immediately wash area with large amounts of soap and water. For inhalation: Remove the person from exposure. Provide respiratory assistance and CPR. Transfer promptly to medical facility. If swallowed: Seek medical attention immediately. Give a conscious and alert victim several glasses of milk or water. The decision to induce vomiting must be carefully considered (it may cause aspiration of the liquid leading to pneumonitis).

SECTION VII - PRECAUTIONS FOR SAFE HANDLING AND USE

Steps to be Taken in Case Material is Released or Spilled

Remove all ignition sources. Restrict persons not wearing protective equipment from entering area of spill or leak until cleanup is complete. Ventilate area. Absorb liquid in dry sand or other material and deposit in sealed containers. Damp mop residue. Alert appropriate state or federal response agencies.

Preferred Waste Disposal Method

Burn in a chemical incinerator equipped with an afterburner and scrubber.

Precautions to be Taken in Handling and Storage

Do not store with incompatible chemicals since violent reactions can occur. Store in tightly closed containers in a cool, dark, well-ventilated area away from fire, sparks, and flame. Use non-sparking tools on containers. Ground and bond metal containers during transfer operations.

Other Precautions and Warnings

Containers may explode in fire or under conditions of extreme heat (do not store outdoors or in direct sunlight). Protect containers from physical damage.

SECTION VIII - CONTROL MEASURES AND PERSONAL PROTECTIVE EQUIPMENT

Respiratory Protection *(specify type)*

Exposures above 100 ppm: MSHA/NIOSH-approved full-facepiece respirator with organic vapor cartridge. Greater protection is obtained from a self-contained breathing apparatus (SCBA) with full facepiece and pressure demand or other positive pressure mode.

Ventilation

Local exhaust or general mechanical systems recommended.

Protective Gloves	Eye Protection	Other Protective Clothing
Impervious Materials	Chemical Goggles and/or Face Mask	Protective Apron

Work/Hygiene Practices

Always wash hands thoroughly after using chemical; never bring food, drink, or smoking materials into vicinity of chemicals.

METHYL *N*-AMYL KETONE

$CH_3CO[CH_2]_4CH_3$ CAS: 110-43-0

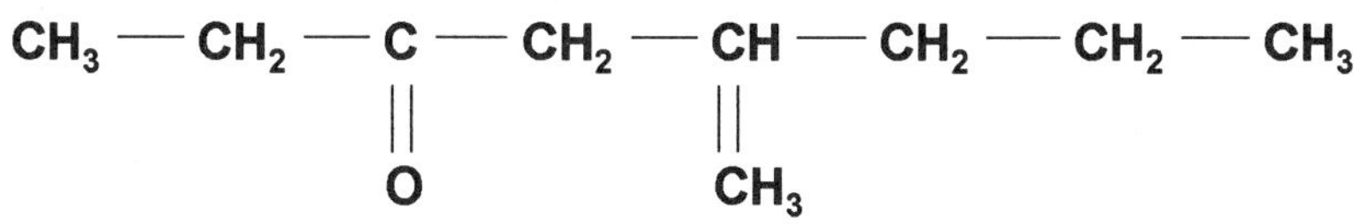

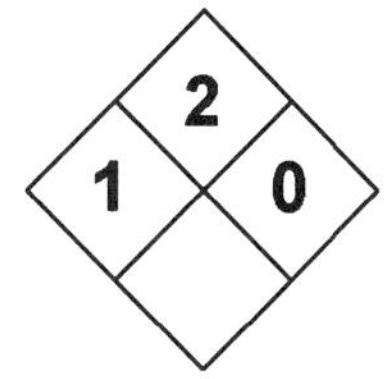

IDENTIFICATION AND TYPICAL USES

Methyl *n*-amyl ketone is a colorless or watery-white liquid with a spicy, banana-like odor and a pear-like taste. It is used as a solvent for nitrocellulose lacquers and synthetic resins, and as a flavoring agent in food and drink. It is also sometimes used as scenting agent in perfumes.

RISK ASSESSMENT: HEALTH

General Assessment

Methyl *n*-amyl ketone is moderately toxic to humans by *inhalation* and *ingestion*. There are no reports in the references that indicate dermal absorption is a risk factor. However, skin or eye contact can cause surface tissue damage and inflammation. The references contain no information on the carcinogenic, mutagenic, or teratogenic properties of methyl *n*-amyl ketone. Overexposure to this compound is not common. Its irritating properties and penetrating odor are usually present at levels low enough to serve as a warning to personnel.

Inhalation of methyl *n*-amyl ketone causes irritation of the mucosa of the respiratory tract leading to respiratory depression with symptoms of gasping, choking, and dypsnea. Exposure to higher concentrations can cause depression of the central nervous system (CNS) with symptoms of headache, lachrymation, somnolence, decreased breathing, dizziness, fainting, tremor, coma, and possibly death (rare but possible). If the liquid should be aspirated, it can lead to chemical pneumonitis.

Skin contact with the liquid or vapor can lead to chemical dermatitis and localized paresthesia with symptoms of prickling or tingling with no apparent stimulus and associated sensory nerve damage. Eye contact causes irritation with lachrymation and, in prolonged contact, may lead to ocular pain, corneal damage, and possible loss of vision.

Although ingestion is unlikely in the occupational setting, it can cause irritation of the mouth, sore throat, coughing, salivation, nausea, and vomiting.

Acute Health Effects

The following acute (short-term) health effects may occur immediately or shortly after exposure to methyl *n*-amyl ketone:

Skin: Irritation, dermatitis, and possible localized paresthesia.

Eye: Irritation with pain and inflammation that can lead to corneal damage and possible loss of vision.

Lung: Irritation causing coughing and/or shortness of breath. Higher exposures can cause dizziness, headache, and, possibly, coma and death.

CNS: A mild narcotic compound. It is not, however, considered neurotoxic. Inhalation of high concentrations of the vapor may cause dizziness, lightheadedness, and possible loss of consciousness.

Chronic Health Effects

The following chronic (long-term) health effects may occur at some time after exposure to methyl *n*-amyl ketone and can last for months or even years:

Cancer Hazards: According to information presented in the references, methyl *n*-amyl ketone has not been adequately tested for its ability to cause cancer in laboratory test animals.

Reproduction: There are no reports to support any claims of reproductive hazards in humans or animals.

Sensitivity: Studies with methyl *n*-amyl ketone have shown that long-term skin exposure may produce dermatitis. However, unlike other ketones, methyl *n*-amyl ketone is not considered a sensitizer.

⓿ *Recommended Risk-Reduction Measures*

Personnel should avoid direct contact with methyl *n*-amyl ketone. Unless a less toxic chemical can be substituted for a hazardous substance, *engineering controls* are the most effective method of reducing exposures. The best protection is to enclose operations and/or provide local exhaust ventilation at the site of chemical release. While not always operationally feasible, isolating operations can also reduce exposure. Using respiratory protection is less effective than the controls mentioned above, but is still advisable whenever working with methyl *n*-amyl ketone. For exposures to the PEL (100 ppm), use a NIOSH-approved, full facepiece organic vapor respirator. For higher or prolonged exposures, or when exposure levels are unknown (as in an emergency) a self-contained breathing apparatus (SCBA) with full facepiece operated in pressure demand mode, or a supplied-air respirator operated in continuous flow mode are the recommended respiratory protection methods of choice. If a full facepiece is not available, then chemical goggles should be worn to protect the eyes. Whenever a chemical splash hazard exists, a face shield and a rubber apron should be worn. To prevent hand and skin exposures, butyl rubber gloves should be used. Glove manufacturers should be contacted and permeation studies obtained *before* final glove selections are made.

Administrative controls should also be in place to minimize the potential for human exposures. These may include written procedures or policies which specify the methods and techniques that will be practiced whenever personnel are to work with methyl *n*-amyl ketone.

All personnel should receive *training* on the use, hazards, protective measures, emergency actions, and other precautions per 29 CFR 1910.1200 (Hazard Communication), prior to the first assignment in an area where methyl *n*-amyl ketone is used or stored.

If symptoms develop or overexposure is suspected, the following medical tests may be helpful:

- ☑ Skin testing by a qualified allergist with special testing and careful evaluation of exposure history (may help diagnose allergy).
- ☑ Neurological evaluation of the central nervous system.

Other methods to reduce exposure include:

- ☑ Where possible, enclose operations and use local exhaust ventilation at the site of chemical release. Explosion-proof design is recommended. If local exhaust ventilation or enclosure is not used, respiratory protection should be mandatory.
- ☑ Always ensure that proper protective clothing is worn when using chemical substances.
- ☑ Wash thoroughly immediately after exposure to methyl *n*-amyl ketone and at the end of the work shift or before eating, drinking, or smoking. Never wear clothes contaminated with methyl *n*-amyl ketone home. Family members can be exposed.
- ☑ Hazard warning information should be posted in the work area. As part of an *on*-going education and training program, all information on the health and safety hazards of methyl *n*-amyl ketone should be communicated to all exposed workers.

RISK ASSESSMENT: ENVIRONMENT

General Assessment

The environment is at risk of exposure during transportation, storage, disposal, or destruction of methyl *n*-amyl ketone. In almost every scenario, the threat of environmental exposure is contingent upon the proper handling of the chemical substance. Accidental spills, large or small, can result in fire, explosion, and possible contamination of the surrounding environmental mediums (water, soil, and air).

Methyl *n*-amyl ketone is considered a Class II combustible liquid (per OSHA 29 CFR 1910.106). It can react with strong oxidizers and acids to cause a serious fire and explosion hazard. When involved in a fire, it can release acrid and irritating oxides of carbon. A boiling liquid expanding vapor explosion (BLEVE) is also a possibility. These characteristics require special consideration during any emergency involving a leak or spill of methyl *n*-amyl ketone.

The proper disposal/destruction method for methyl *n*-amyl ketone is to burn it in a chemical incinerator (with molten salt treatment) equipped with an afterburner and air scrubber.

Methyl *n*-amyl ketone can enter the environment through manufacturing, unchecked discharge into effluents from industrial or municipal waste treatment plants, and through spills.

☠ *Acute Ecological Effects*

Acute (short-term) toxic effects may include the death of animals, birds, or fish, and death or low growth rate in plants. Acute effects are seen 2 to 4 days after animals or plants are exposed to methyl *n*-amyl ketone. This chemical has a slight acute toxicity to aquatic life. No data are available on the short-term effects of methyl *n*-amyl ketone on plants, birds, or land animals.

Chronic Ecological Effects

Chronic (long-term) toxic effects may include shortened life span, reproductive problems, lower fertility, and changes in appearance or behavior in exposed animals. These effects can be seen long after first exposure(s) to toxic chemicals. Methyl *n*-amyl ketone has slight chronic toxicity in aquatic life. No data are available to evaluate the long-term effects of methyl *n*-amyl ketone to plants, birds, or land animals.

Water Solubility

Methyl *n*-amyl ketone is nearly insoluble in water. Concentrations of less than 1 milligram may mix with a liter of water.

Persistence in the Environment

Methyl *n*-amyl ketone is moderately persistent in water, with a half-life of between 20 to 200 days. The half-life of a pollutant is the amount of time it takes for one half of the chemical to be degraded.

Bioaccumulation in Aquatic Organisms

Some substances increase in concentration, or *bioaccumulate,* in living organisms as they breathe contaminated air, drink contaminated water, or eat contaminated food. These chemicals can become concentrated in the tissues and internal organs of animals as well as humans.

The concentration of methyl *n*-amyl ketone found in fish tissues is expected to be about the same as the average concentration of methyl *n*-amyl ketone in water from which the fish was taken.

Recommended Risk-Reduction Measures

Proper training of all transporters will reduce the likelihood of a mishap or accident resulting in a leak or spill to the environment. The correct use of DOT labeling (per the MSDS) on all containers, trucks, and rail cars will enable emergency responders to react properly and quickly to any disaster thereby reducing the potential risk to the environment and to personnel.

Storage of methyl *n*-amyl ketone should be segregated from incompatible chemicals to minimize the risk of cross-contamination or contact. Buildings designated for storage should be equipped with appropriate fire protection systems (alarms, sprinklers, emergency lighting, portable extinguishers).

If a spill or leak to the environment has occurred, fire department, emergency response, and/or hazardous materials spill personnel should be notified immediately. Cleanup should be attempted only by those trained in proper spill containment procedures. Contaminated soils should be removed for incineration and replaced with clean soil. If methyl *n*-amyl ketone should contact the water table, aquifer, or navigable waterway, time is of the essence. It is practically insoluble in water and, therefore, near total containment and remediation may be possible. When such spills occur, the local and/or state emergency response authorities may require notification. A comprehensive emergency response or disaster preparedness/recovery plan should be in place prior to any operations involving the use, transportation, storage, or disposal of methyl *n*-amyl ketone.

If methyl *n*-amyl ketone is spilled or leaked, the following specific steps are recommended:

- ☑ Restrict persons not wearing protective clothing from area of spill or leak until cleanup is complete and area can be opened for normal work.
- ☑ Remove all ignition sources.
- ☑ Ventilate area of spill or leak.
- ☑ Absorb liquids in vermiculite, dry sand, earth, or a similar material and deposit in sealed containers for disposal.
- ☑ Keep methyl *n*-amyl ketone out of a confined space, such as a sewer, because of the possibility of explosion (unless the sewer is designed to prevent the buildup of explosive concentrations).
- ☑ It may be necessary to dispose of methyl *n*-amyl ketone as a hazardous waste. The responsible state agency or the regional office of the federal Environmental Protection Agency (EPA) should be contacted for specific recommendations.

RISK ASSESSMENT: BUSINESS

General Assessment

Accidents or mishaps involving methyl *n*-amyl ketone can present a significant threat to business operations. The loss or damage of equipment or facilities can significantly affect fiscal viability. Lawsuits that may result from personnel injury, public exposures, and/or environmental contamination will also require a serious expenditure of resources. Media attention surrounding an injury, illness, or environmental damage can also result in a loss of profits and loss of current as well as future business prospects.

ʋ *Recommended Risk-Reduction Measures*

Company attorneys, safety and health professionals, and environmental specialists should be involved in the development of any procedures dealing with safety or emergency response. A company official should be pre-designated as a public relations officer with specific training in dealing with the press. Corporate plans and policies should be developed, approved, and implemented long before any need for such arises.

REFERENCES

Ahlbom, A. 1993. *Biostatistics for Epidemiologists.* New York: CRC Press/Lewis Publishers.

American Conference of Governmental Industrial Hygienists. 1988. *Documentation of the Threshold Limit Values and Biological Exposure Indices,* 5th Edition (with updates). Cincinnati: ACGIH

Anthony, C. P., and N. J. Kolthoff. 1971. *Textbook of Anatomy and Physiology.* St. Louis: C. V. Mosby Company.

Asante-Duah, D. K. 1993. *Hazardous Waste Risk Assessment.* New York: CRC Press/Lewis Publishers.

Bloom, A. D. and F. J. DeSerres. 1995. *Ecotoxicity and Human Health.* New York: CRC Press/Lewis Publishers.

Brusick, D. J. 1994. *Methods for Genetic Risk Assessment.* New York: CRC Press/Lewis Publishers.

Bueche, J. R. 1972. *Principles of Physics,* 2nd Edition. New York: McGraw-Hill.

Burrell, R., D. K. Flaherty, and L. J. Sauers. 1992. *Toxicology of the Immune System.* New York: Van Nostrand Reinhold.

Calabrese, E. J., E. M. Kenyon. *Air Toxics and Risk Assessment.* New York: CRC Press/Lewis Publishers.

Cockerman, L. G., B. S. Shane. 1994. *Basic Environmental Toxicology.* New York: CRC Press/Lewis Publishers.

Cothern, C. R. 1995. *Handbook for Environmental Risk Decisionmaking.* New York: CRC Press/Lewis Publishers.

Gardner, R. W. 1994. *Chemical Intolerance.* New York: CRC Press/Lewis Publishers.

Hodgson, E., R. B. Mailman, and J. E. Chambers. 1988. *Dictionary of Toxicology.* London: The MacMillian Company.

Hosty, J. W., and P. Foster. 1990. *A Practical Guide to Chemical Spill Response.* New York: Van Nostrand Reinhold.

Howard, P. H. 1995. *Handbook of Environmental Fate and Exposures Data for Organic Chemicals,* Vol. I-IV. New York: CRC Press/Lewis Publishers.

Howard, P. H., et. al. 1991. *Handbook of Environmental Degradation Rates.* New York: CRC Press/Lewis Publishers.

Jones, F. E. 1994. *Toxic Organic Vapors in the Workplace.* New York: CRC Press/Lewis Publishers.

Kamrin, M. A. 1988. *Toxicology: A Primer on Principles and Applications.* Chelsea, Michigan: Lewis Publishers.

Keith, L. H. and M. M. Walker. 1995. *Handbook of Air Toxics.* New York: CRC Press/Lewis Publishers.

Landis, W. G., M. H. Yu. 1995. *Introduction to Environmental Toxicology.* New York: CRC Press/Lewis Publishers.

Langerman, N. 1994 *Precautionary Labels for Chemical Containers.* New York: CRC Press/Lewis Publishers.

Lauwerys, R. R., P. Hoet. 1993. *Industrial Chemical Exposure.* New York: CRC Press/Lewis Publishers.

Lewis, R. J., Sr. 1993. *Hawley's Condensed Chemical Dictionary,* 12th Edition. New York: Van Nostrand Reinhold.

Lewis, R. J., Sr. 1992. *Sax's Dangerous Properties of Industrial Materials,* Volumes 1, 2, and 3, 8th Edition. New York: Van Nostrand Reinhold.

Lide, D. R. 1995. *Handbook of Chemistry and Physics,* 76th Edition. New York: CRC Press/Lewis Publishers.

Manahan, S. E. 1992. *Toxicological Chemistry,* 2nd Edition. New York: CRC Press/Lewis Publishers.

Mellström, G. A., J. E. Wahlberg, and H. I. Maibach. 1994. *Protective Gloves for Occupational Use.* New York: CRC Press/Lewis Publishers.

Meyer, E. 1990. *Chemistry of Hazardous Materials.* Englewood Cliffs, NJ: Prentice Hall.

Patnaik, P. 1992. *A Comprehensive Guide to the Hazardous Properties of Chemical Substances.* New York: Van Nostrand Reinhold.

Philp, R. B. 1995. *Environmental Hazards and Human Health.* New York: CRC Press/Lewis Publishers.

Quigley, D. R. 1994. *Handbook of Emergency Chemical Management.* New York: CRC Press/Lewis Publishers.

Raleigh, R. L. and W. A. McGee. 1972. Effects of short, high concentration exposures to acetone as determined by observation in the work area. *Journal of Occupational Medicine.* 14:607-610.

Rea, W. J. 1995. *Chemical Sensitivity,* Vol. I-III. New York: CRC Press/Lewis Publishers.

Richardson, M. 1995. *Dictionary of Substance and Their Effects,* Vol. I-VI. New York: CRC Press/Lewis Publishers.

U.S. Department of Health and Human Services, National Institute for Occupational Safety and Health. 1994. *NIOSH Pocket Guide to Chemical Hazards.* Washington, D.C.: U.S. Government Printing Office.

U.S. Department of Health and Human Services, National Institute for Occupational Safety and Health. 1978. *Criteria for a Recommended Standard - Occupational Exposure to Ketones.* DHEW Publication Number: 78-173. Washington, D.C.: U.S. Government Printing Office.

MATERIAL SAFETY DATA SHEET

CHEMICAL NAME	
METHYL BROMIDE	**CRC PRESS/LEWIS PUBLISHERS DISCLAIMER:** THE INFORMATION AND RECOMMENDATIONS PRESENTED HEREIN ARE BASED ON SOURCES BELIEVED TO BE RELIABLE. CRC MAKES NO REPRESENTATION ON ITS COMPLETENESS OR ACCURACY. IT IS THE RESPONSIBILITY OF THE USER TO DETERMINE THE CHEMICAL'S SUITABILITY FOR ITS INTENDED USE, THE CHEMICAL'S SAFE USE, AND THE CHEMICAL'S PROPER DISPOSAL. NO REPRESENTATIONS AND/OR WARRANTIES, EITHER EXPRESSED OR IMPLIED, OF THE MERCHANTABILITY OR FITNESS FOR A PARTICULAR PURPOSE, OR OF ANY OTHER NATURE, ARE MADE WITH RESPECT TO THE INFORMATION PROVIDED IN THIS MSDS OR TO THE CHEMICAL TO WHICH INFORMATION MAY REFER. CRC NEITHER ASSUMES NOR AUTHORIZES ANY OTHER PERSON TO ASSUME FOR IT, ANY OTHER ADDITIONAL RESPONSIBILITY OR LIABILITY FOR THE USE OF, OR RELIANCE UPON, THIS INFORMATION.

HAZARD WARNING INFORMATION

HEALTH	FIRE	REACTIVE	OTHER	DEGREE OF HAZARD	COLOR CODING	OTHER CODES
3	1	0		0 = Minimum Hazard 1 = Slight Hazard 2 = Moderate Hazard 3 = Serious Hazard 4 = Severe Hazard	HEALTH = BLUE FIRE = RED REACTIVITY = YELLOW OTHER = WHITE	OX = Oxidizer ACID = Acid ALK = Alkali COR = Corrosive W = Use No Water

SECTION I - GENERAL INFORMATION

Characterization	**Halogenated Hydrocarbon**	RCRA Number	**U029**
		EPA Class	**Toxic Waste**
DOT Proper Shipping Name	**Methyl Bromide, Liquid**	Chemical Abstract Service (CAS) Number	**74-83-9**
DOT Hazard Class and Label Requirements	**Poison A; Poison Gas**	DOT Emergency Guide Code	**55**
DOT Identification Number	**UN 1062**	Molecular Formula	**CH_3Br**

Synonyms

Bromomethane; monobromomethane; embafume; M-B-C fumigant; Dawson 100; Halon 1001; isobrome.

SECTION II - HAZARDOUS INGREDIENTS AND IDENTITY INFORMATION

Hazardous Components (specific identity)	OSHA Exposure Criteria	NIOSH Exposure Criteria	Immediately Dangerous to Life and Health (IDLH)	ACGIH Exposure Criteria
Methyl bromide (derivation: By action of bromine on methanol in the presence of phosphorus with subsequent distillation). **1 ppm = 3.95 mg/m^3**	**PEL (skin):** **20 ppm** **80 mg/m^3** **STEL:** **Not Established**	**REL:** **Reduce to Lowest Level** **Possible Human Carcinogen**	**250 ppm**	**TLV (skin):** **5 ppm** **20 mg/m^3** **STEL:** **Not Established**

SECTION III - PHYSICAL AND CHEMICAL CHARACTERISTICS

Boiling Point	**38°F (3.6°C)**	Specific Gravity (H_2O = 1)	**1.73 (liquid at 38°F)**
Vapor Pressure (mm Hg)	**1420 at 69°F (20°C)**	Molecular Weight (atomic weight)	**95.0**
Vapor Density (Air = 1)	**3.27**	Melting Point	**-137°F (-94°C)**

Solubility

Slightly soluble in water (2%). Miscible with most organic solvents.

Appearance and Odor

Colorless gas with a chloroform-like odor at high concentrations. A liquid below 38°F (3.6°C) normally shipped as a liquefied compressed gas.

SECTION IV - FIRE AND EXPLOSION HAZARD DATA

Flash Point (method used)	**Not Determined (Flammable Gas)**	Explosive Limits in Air % by Volume	**LEL: 10% UEL: 16%**
NFPA Classification	**Flammable Gas**	Autoignition Temperature	**998°F (537°C)**

Extinguishing Media

Stop flow of gas. Use agents suitable to surrounding fire. Use water spray to knock down vapors.

Special Fire Fighting Procedures

Poisonous gases are produced in fire. Wear self-contained breathing apparatus (SCBA) and full protective gear. Move containers from fire if it can be done without risk. Containers may explode in fire. Cool sides of fire-exposed containers with water until long after fire is out.

Unusual Fire and Explosion Hazards

Although a flammable gas, methyl bromide will not ignite under normal operating conditions. But, in the presence of a high energy ignition source, it can ignite and burn rapidly.

SECTION V - REACTIVITY DATA

Stability		Conditions to Avoid: Under normal conditions of handling and storage, methyl bromide is considered stable. Avoid heat, flame, other sources of ignition, and contact with incompatible materials and chemicals.
Stable: X	Unstable:	Incompatibility (*materials to avoid*): Strong oxidizers (chlorine, bromine, fluorine), chemically active metals (potassium, aluminum, zinc, and magnesium). Attacks aluminum to form aluminum trimethyl which is spontaneously flammable.
Hazardous Polymerization		Conditions to Avoid: Hazardous polymerization of methyl bromide has not been reported to occur under normal conditions of temperature and pressure. The liquid attacks some plastics, rubber, and coatings.
May Occur:	Will Not Occur: X	Hazardous Decomposition or By-products: Poisonous gases and acrid fumes are produced in fire, including carbon dioxide, carbon monoxide, hydrogen bromide, and toxic oxides of bromine.

SECTION VI - HEALTH HAZARD DATA

Primary Route(s) of Entry:	Inhalation?	Absorption (skin/eye)?	Ingestion?
	X	X	

Health Hazards

INHALATION: Poisonous. Symptoms of headache, nausea, dizziness, vomiting, blurred vision, slurred speech, convulsions, lung irritation with congestion, coughing, chest pain, shortness of breath, loss of consciousness, and death. Lung effects may be delayed. Can cause kidney damage. A neurotoxin causing muscle weakness and pain, behavioral changes.

ABSORPTION: The liquid or gas can pass through intact skin to cause CNS effects. Liquid causes burns and blisters due to intense cold (frostbite) with subsequent destruction of tissues.

INGESTION: Unspecified gastrointestinal effects. An unlikely exposure route, but still possible.

Carcinogenicity	NTP Listed?	IARC Cancer Review Group?	OSHA Regulated?	Target Organs?
Suspected Human Suspected Animal	No	Group 3	29 CFR 1910.1000 Table Z-1	Eyes, skin, respiratory sys., Liver, kidney, CNS, PNS.

Medical Conditions Generally Aggravated by Exposure

Liver, kidney, heart, and respiratory problems may be aggravated by exposure.

Emergency and First-aid Procedures

Eye contact: Do not allow victim to keep eyes tightly shut. Flush immediately with warm (tepid) water for 15 minutes (minimum); seek medical attention. Skin contact: Remove all contaminated clothing, including shoes. Immediately wash area with flooding warm (tepid) water. Do NOT rub area or use air-blown heat in an attempt to warm skin. Seek medical attention immediately. For inhalation: Remove the person from exposure. Provide respiratory assistance and CPR. Transfer to medical facility. Observe for 24-48 hours for lung effects. If swallowed: Unlikely exposure route.

SECTION VII - PRECAUTIONS FOR SAFE HANDLING AND USE

Steps to be Taken in Case Material is Released or Spilled

Remove all ignition sources, wear SCBA respiratory protection. Restrict those not wearing protective equipment and who are not involved in cleanup from entering area. Stop leak if it can be done without risk. Provide explosion-proof ventilation. Absorb with-vermiculite and deposit in sealed containers.

Preferred Waste Disposal Method

Mix with combustible solvent; burn in chemical incinerator equipped with an afterburner and scrubber.

Precautions to be Taken in Handling and Storage

Store in tightly closed containers in cool, well-ventilated location. Keep away from sources of heat.

Other Precautions and Warnings

Sources of ignition such as open flame or smoking, are prohibited where methyl bromide is used or stored. Metal containers should be bonded and grounded. Empty containers contain residue; handle with care.

SECTION VIII - CONTROL MEASURES AND PERSONAL PROTECTIVE EQUIPMENT

Respiratory Protection (*specify type*)

Use an MSHA/NIOSH-approved air purifying respirator (cartridge type) or use a supplied-air respirator or a self-contained breathing apparatus (SCBA) with full facepiece operated in positive pressure mode.

Ventilation

Local exhaust (preferred) at site of chemical work or general ventilation.

Protective Gloves	Eye Protection	Other Protective Clothing
Impervious Gloves	Chemical Goggles or Face Mask	Protective Apron if Splash is Likely

Work/Hygiene Practices

Always wash hands thoroughly after using chemical; never bring food, drink, or smoking materials into vicinity of chemicals.

METHYL BROMIDE

CH_3Br

CAS: 74-83-9

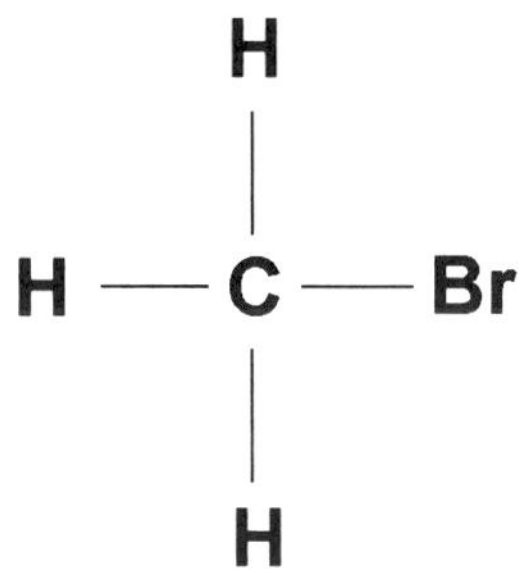

IDENTIFICATION AND TYPICAL USES

Methyl bromide is a colorless gas with a sweet, chloroform-like odor at high concentrations. It can also exist as a liquid below 38°F (3.6°C) and is usually shipped as a liquefied compressed gas. It is used as a fumigant for pest control, for degreasing wool, and as a methylating agent. It has also been used as a refrigerant and in fire extinguishers, but these uses have been restrained or curtailed due to risks to human health and the environment posed by methyl bromide.

RISK ASSESSMENT: HEALTH

General Assessment

Methyl bromide is a human poison by *inhalation*. Both the gas and the liquid have been known to pass through intact skin (dermal *absorption*) to cause toxic system effects with particular action on the central nervous system (CNS). Skin contact can also cause localized damage to tissues due to the extreme cold temperatures of the compressed or liquefied gas. Methyl bromide is a suspected animal carcinogen. Its human carcinogenic properties should also be considered suspect until further studies are performed. Human mutation data have been reported. Physiologically, it is a neurotoxin and is also thought to pose risk of injury or damage to the lungs, liver, and kidney. The primary human response to methyl bromide is depression of the central nervous system (CNS).

Inhalation causes irritation of the lungs, marked congestion and symptoms CNS depression. These can include drowsiness, slurred speech, lack of coordination, and possible loss of consciousness. There may also be headache, nausea, vomiting, anorexia, ataxia, numbness in the legs and arms, convulsions, and death. Irritation to the eyes and respiratory system is also know to occur on inhalation. In fact, pulmonary edema (fluid buildup in the lungs) has been reported. This is a medical emergency that can be fatal. Symptoms of congestion, difficulty breathing, cough with phlegm, and chest pain may be delayed for up to 48 hours following exposure. This may create a false sense of security with regard to health risk.

Skin contact may cause localized burning and frostbite at site of contact with irritation and possible tissue destruction. Eye contact can result in severe irritation and/or damage to the cornea, with redness and swelling. Damage can be serious if not immediately flushed from eye surface. Oral administration of the liquid to test animals caused gastrointestinal tumors.

Acute Health Effects

The following acute (short-term) health effects may occur immediately or shortly after exposure to methyl bromide:

Skin: Irritation and burning (frostbite) with potential for redness, pain, and inflammation.

Eye: Severe irritation, burning, pain, and permanent damage possible.

Lung: Nose, throat, and respiratory tract irritation causing difficulty in breathing, congestion, chest pains, possible pulmonary edema (fluid in lungs) on high exposure.

CNS: Narcosis, headaches, lightheadedness, dizziness, vertigo, coma, and death.

Chronic Health Effects

The following chronic (long-term) health effects may occur at some time after exposure to methyl bromide and can last for months or even years:

Cancer Hazards: According to information presented in the references, exposure to methyl bromide may cause cancer in test animals. Human data are lacking but human carcinogenicity may be suspect. More research is required since the data in the references are inconclusive and conflicting in this regard.

Reproductive Hazard: According to information presented in the references, methyl bromide has not been adequately tested for its ability to cause adverse reproductive effects in test animals.

Other Chronic Effects: Long-term exposure may lead to CNS damage with speech difficulties, ataxia, mental confusion, and hallucinations. Skin damage may consist of chronic acneform eruptions. Liver and kidney damage may also occur. Extremely irritating substances have been known to damage lung tissues. It is not known whether exposure to methyl bromide will have this effect on humans.

Recommended Risk-Reduction Measures

Personnel should avoid direct contact with methyl bromide. Occupational poisoning occurs most commonly by inhalation or skin absorption. If a less toxic material or compound cannot be substituted for methyl bromide, then *engineering controls* are the most effective method of reducing exposures. The best protection is to enclose operations and/or provide local exhaust ventilation at the site of methyl bromide release. While not always operationally feasible, isolating operations can also reduce exposure. Using respiratory protection is less effective than the controls mentioned above, but is still advisable whenever working with or around methyl bromide. Caution should be exercised when selecting respiratory protection since it is not known if exposure to methyl bromide will cause cancer in humans. For exposure over the PEL (20 ppm), an MSHA/NIOSH-approved air-purifying respiratory with organic vapor cartridge(s) will suffice. For exposures to higher concentrations, or when the concentration is unknown (as in an emergency), use a supplied-air respirator with full facepiece and mask operated in positive pressure mode or a self-contained breathing apparatus (SCBA) with full facepiece operated in pressure demand. If a full facepiece is not available, then chemical goggles should be worn to protect the eyes. A face shield should also be considered. To prevent hand and skin exposures, impervious thermal protective gloves should be worn. Glove manufacturers should be contacted and permeation studies obtained *before* final glove selections have been made.

Administrative controls should also be in place to minimize the potential for human exposures. These may include written procedures or policies which specify the methods and techniques that will be practiced whenever personnel are to work with methyl bromide.

All personnel should receive *training* on the use, hazards, protective measures, emergency actions, and other precautions per 29 CFR 1910.1200 (Hazard Communication), prior to the first assignment in an area where methyl bromide is used or stored.

If symptoms develop or overexposure is suspected, the following recommended medical tests should be considered:

- ☑ Liver and kidney function tests.
- ☑ Lung function tests.
- ☑ Consider chest X-ray following acute overexposure (may be negative if taken immediately after exposure due to delayed onset of pulmonary edema).

Any evaluation should include a careful history of past and present symptoms with an examination. Medical tests that simply look for existing damage are not a substitute for controlling exposures. Also, since smoking causes heart disease, emphysema, and other respiratory problems, smokers exposed to methyl bromide may experience symptoms quicker and more pronounced than non-smokers under the same conditions of exposure. Prudent risk management requires careful consideration of *all* possible factors that may be causing the appearance of exposures symptoms in the work place.

Other methods to reduce exposure include:

- ☑ Where possible, enclose operations and use local exhaust ventilation at the site of chemical release. If local exhaust ventilation or enclosure is not used, respiratory protection should be mandatory.
- ☑ Always ensure that proper protective clothing is worn when using chemical substances.
- ☑ Wash thoroughly immediately after exposure to methyl bromide and at the end of the work shift or before eating, drinking, or smoking.
- ☑ Hazard warning information should be posted in the work area. In addition, as part of an on-going education and training program, all information on the health and safety hazards of methyl bromide should be communicated to all exposed workers.
- ☑ Eye wash stations should be provided in the immediate work area for emergency use. If there is a

possibility of skin exposure to methyl bromide, emergency shower facilities should be provided.

☑ Workers whose clothing has been contaminated by methyl bromide should change into clean clothes before leaving work. Contaminated work clothing should be laundered only by individuals who have been informed of the hazards of methyl bromide.

RISK ASSESSMENT: ENVIRONMENT

General Assessment

The environment is at risk of exposure during transportation, storage, disposal, or destruction of methyl bromide. In almost every scenario, the threat of environmental exposure is contingent upon the proper handling of the chemical substance. Accidental spills, large or small, can result in contamination of the surrounding environmental mediums (water, soil, and air).

Methyl bromide is a moderately flammable gas. It is a dangerous fire hazard and moderate explosion hazard when exposed to heat or flame. Although it is considered a flammable gas, it does not easily ignite. It is incompatible with strong oxidizers, such as chlorine, fluorine, and bromine, and many chemically-active metals, such as sodium, potassium, calcium, powdered aluminum, zinc, and magnesium. It attacks aluminum to form aluminum triethyl which is spontaneously flammable. Caution is always required in handling, storage, transportation, and disposal of methyl bromide. Emergency responders should be made aware of the presence of methyl bromide at any emergency response situation.

Methyl bromide can enter the environment from industrial effluents and from spills.

Acute Ecological Effects

Acute (short-term) toxic effects may include the death of animals, birds, or fish and death or low growth rate in plants. Acute effects are seen 2 to 4 days after animals or plants are exposed to methyl bromide. Methyl bromide has high acute toxicity to aquatic life. It has caused injury to various agricultural crops. Insufficient data are available to evaluate the short-term effects of methyl bromide on birds or land animals.

Chronic Ecological Effects

Chronic (long-term) toxic effects may include shortened life span, reproductive problems, lower fertility, and changes in appearance or behavior in exposed animals. These effects can be seen long after first exposure(s) to toxic chemicals. Methyl bromide has high chronic toxicity to aquatic life. Insufficient data are available to evaluate or predict the long-term effects of methyl bromide to plants, birds, or terrestrial animals.

Water Solubility

Methyl bromide is only slightly soluble in water. Concentrations between 1 and 100 milligrams will mix with a liter of water.

Persistence in the Environment

Methyl bromide is slightly persistent in water, with a half-life between 2 and 20 days. The half-life of a pollutant is the amount of time it takes for one half of the chemical to be degraded.

Bioaccumulation in Aquatic Organisms

Some substances increase in concentration, or *bioaccumulate,* in living organisms as they breathe contaminated air, drink contaminated water, or eat contaminated food. These chemicals can become concentrated in the tissues and internal organs of animals as well as humans.

The concentration of methyl bromide found in fish tissues is expected to be about the same as the average concentration of methyl bromide in the water from which the fish was taken.

Recommended Risk-Reduction Measures

Proper training of all transporters will reduce the likelihood of a mishap or accident resulting in a leak or spill of methyl bromide into the environment. Labels on all containers, trucks, and rail cars must meet DOT requirements and accurately reflect their contents to enable emergency responders to react properly and quickly to any disaster thereby reducing the potential risk to the environment and to personnel.

Storage of methyl bromide should be segregated from other chemicals and materials to minimize the risk of cross-contamination.

If a spill or leak to the environment has occurred, fire department, emergency response and/or hazardous materials spill personnel should be notified immediately. Cleanup should be attempted only by those trained in proper spill containment procedures. Contaminated soils should be removed for incineration and replaced with clean soil. If methyl bromide should contact the water table, aquifer, or navigable waterway, time is of the essence. It is slightly soluble and total remediation may not be possible. The local and/or state emergency response authorities must be notified. A comprehensive emergency response or disaster preparedness/recovery plan should be in place prior to any operations involving the use, transportation, storage, or disposal of methyl bromide.

For disposal, methyl bromide may be mixed with a combustible solvent and burned in a chemical incinerator equipped with an afterburner and scrubber.

If methyl bromide is spilled or leaked, the following specific steps are recommended:

☑ Restrict persons not wearing protective clothing from area of spill or leak until cleanup is complete. Use non-sparking tools.

☑ Stop flow of gas. Use water to knock down vapors and to protect response crews attempting to shut off gas leak. If leak is from a cylinder and it cannot be stopped, remove to a safe area outdoors, if possible, and allow to vent until empty. Absorb any liquids in vermiculite, sand, earth, or similar material and deposit in sealed drum.

☑ Ventilate area of spill or leak and remove all ignition sources.

☑ It may be necessary to dispose of methyl bromide as a hazardous waste. The responsible state agency or the regional office of the federal Environmental Protection Agency (EPA) should be contacted for specific recommendations.

RISK ASSESSMENT: BUSINESS

General Assessment

Accidents or mishaps involving methyl bromide can present a serious threat to business operations. The loss or damage of equipment or facilities can significantly affect fiscal viability. Lawsuits that may result from personnel illness, injury/death, public exposures, and/or environmental contamination will require a serious expenditure of resources.

Media attention surrounding an injury, death, or environmental damage can also result in a loss of profits and loss of current as well as future business. Always remember that anytime terms such as "cancer" or "carcinogen" are used, public emotion, hysteria, and ignorance can run equally high. This must be carefully considered whenever developing or implementing a public relations policy.

Recommended Risk-Reduction Measures

Company attorneys, safety and health professionals, and environmental specialists should be involved in the development of any procedures for responding to chemical incidents. A company official should be predesignated as a public relations officer with specific training in dealing with the press. Corporate plans and policies should be developed, approved, and implemented long before any need for such arises.

REFERENCES

American Conference of Governmental Industrial Hygienists. 1988. *Documentation of the Threshold Limit Values and Biological Exposure Indices,* 5th Edition (with updates). Cincinnati: ACGIH

Anthony, C. P., and N. J. Kolthoff. 1971. *Textbook of Anatomy and Physiology.* St. Louis: C. V. Mosby Company.

Bloom, A. D. and F. J. DeSerres. 1995. *Ecotoxicity and Human Health.* New York: CRC Press/Lewis Publishers.

Calabrese, E. J. 1994. *Biological Effects of Low Level Exposures.* New York: CRC Press/Lewis Publishers.

Cockerman, L. G., B. S. Shane. 1994. *Basic Environmental Toxicology.* New York: CRC Press/Lewis Publishers.

Cothern, C. R., N. P. Ross. 1994. *Environmental Statistics, Assessment, and Forecasting.* New York: CRC Press/Lewis Publishers.

Hallenbeck, W. H. 1993. *Quantitative Risk Assessment for Environmental and Occupational Health,* 2nd Edition. New York: CRC Press/Lewis Publishers.

Howard, P. H., et. al. 1991. *Handbook of Environmental Degradation Rates.* New York: CRC Press/Lewis Publishers.

Lewis, R. J., Sr. 1993. *Hawley's Condensed Chemical Dictionary,* 12th Edition. New York: Van Nostrand Reinhold.

Lewis, R. J., Sr. 1992. *Sax's Dangerous Properties of Industrial Materials,* Volumes 1, 2, and 3, 8th Edition. New York: Van Nostrand Reinhold.

Manahan, S. E. 1994. *Environmental Chemistry,* 6th Edition. New York: CRC Press/Lewis Publishers.

Meyer, E. 1990. *Chemistry of Hazardous Materials.* Englewood Cliffs, NJ: Prentice Hall.

Patnaik, P. 1992. *A Comprehensive Guide to the Hazardous Properties of Chemical Substances.* New York: Van Nostrand Reinhold.

Quigley, D. R. 1994. *Handbook of Emergency Chemical Management.* New York: CRC Press/Lewis Publishers.

Sacarello, H. L. A. 1994. *The Comprehensive Handbook of Hazardous Materials.* New York: CRC Press/Lewis Publishers.

Torkelson, T. R. V. K. Rowe. 1981. *Patty's Industrial Hygiene and Toxicology,* Vol. 2B. New York: Wiley-Interscience.

U.S. Department of Health and Human Services, National Institute for Occupational Safety and Health. 1994. *NIOSH Pocket Guide to Chemical Hazards.* Washington, D.C.: U.S. Government Printing Office.

MATERIAL SAFETY DATA SHEET

CHEMICAL NAME	
METHYL CHLORIDE	**CRC PRESS/LEWIS PUBLISHERS DISCLAIMER:** THE INFORMATION AND RECOMMENDATIONS PRESENTED HEREIN ARE BASED ON SOURCES BELIEVED TO BE RELIABLE. CRC MAKES NO REPRESENTATION ON ITS COMPLETENESS OR ACCURACY. IT IS THE RESPONSIBILITY OF THE USER TO DETERMINE THE CHEMICAL'S SUITABILITY FOR ITS INTENDED USE, THE CHEMICAL'S SAFE USE, AND THE CHEMICAL'S PROPER DISPOSAL. NO REPRESENTATIONS AND/OR WARRANTIES, EITHER EXPRESSED OR IMPLIED, OF THE MERCHANTABILITY OR FITNESS FOR A PARTICULAR PURPOSE, OR OF ANY OTHER NATURE, ARE MADE WITH RESPECT TO THE INFORMATION PROVIDED IN THIS MSDS OR TO THE CHEMICAL TO WHICH INFORMATION MAY REFER. CRC NEITHER ASSUMES NOR AUTHORIZES ANY OTHER PERSON TO ASSUME FOR IT, ANY OTHER ADDITIONAL RESPONSIBILITY OR LIABILITY FOR THE USE OF, OR RELIANCE UPON, THIS INFORMATION.

HAZARD WARNING INFORMATION

HEALTH	FIRE	REACTIVE	OTHER	DEGREE OF HAZARD	COLOR CODING	OTHER CODES
2	4	0		0 = Minimum Hazard 1 = Slight Hazard 2 = Moderate Hazard 3 = Serious Hazard 4 = Severe Hazard	HEALTH = BLUE FIRE = RED REACTIVITY = YELLOW OTHER = WHITE	OX = Oxidizer ACID = Acid ALK = Alkali COR = Corrosive ~~W~~ = Use No Water

SECTION I - GENERAL INFORMATION

Characterization	RCRA Number	EPA Class
Halogenated Hydrocarbon	**U045**	**Toxic Waste**

DOT Proper Shipping Name **Methyl chloride**	Chemical Abstract Service (CAS) Number **74-87-3**
DOT Hazard Class and Label Requirements **Flammable Gas**	DOT Emergency Guide Code **18**
DOT Identification Number **UN 1063**	Molecular Formula CH_3Cl

Synonyms
Chloromethane; monochloromethane; methylene dichloride.

SECTION II - HAZARDOUS INGREDIENTS AND IDENTITY INFORMATION

Hazardous Components (specific identity)	OSHA Exposure Criteria	NIOSH Exposure Criteria	Immediately Dangerous to Life and Health (IDLH)	ACGIH Exposure Criteria
Methyl chloride (derivation: By chlorination of methyl chloride and subsequent distillation). **1 ppm = 2.10 mg/m³**	**PEL:** **100 ppm** **210 mg/m³** **STEL:** **200 ppm** **420 mg/m³**	**REL:** **Reduce to Lowest Level** **Possible Cancer Agent**	**2000 ppm**	**TLV:** **50 ppm** **105 mg/m³** **STEL:** **100 ppm** **210 mg/m³**

SECTION III - PHYSICAL AND CHEMICAL CHARACTERISTICS

Boiling Point **-12°F (-24°C)**	Specific Gravity (H_2O = 1) **0.997**
Vapor Pressure (atmospheres) **5.0 at 69°F (20°C)**	Molecular Weight (atomic weight) **50.5**
Vapor Density (Air = 1) **1.78**	Freezing Point **-144°F (-97°C)**

Solubility
Slightly soluble in water (0.5%). Soluble in alcohol, ether, chloroform, and glacial acetic acid.

Appearance and Odor
Colorless gas, with a faint, sweet ether-like odor that is not noticeable at dangerous levels. Normally shipped as a liquefied compressed gas. Odor Threshold = 10 ppm.

SECTION IV - FIRE AND EXPLOSION HAZARD DATA

Flash Point (method used) **<32°F (<0°C) closed cup**	Explosive Limits in Air % by Volume **LEL: 8.1%** **UEL: 17%**
NFPA Classification **Flammable Gas**	Autoignition Temperature **1170°F (632°C)**

Extinguishing Media
Stop flow of gas first, if possible. Use dry chemical, carbon dioxide, water spray, or foam.

Special Fire Fighting Procedures
Poisonous gases are produced in fire. Wear self-contained breathing apparatus (SCBA) and full protective gear. Move containers from fire if it can be done without risk. Cool sides of fire-exposed containers with water until long after fire is out. Reacts with water to form hydrochloric acid.

Unusual Fire and Explosion Hazards
Dangerous fire and explosion hazard. Containers may explode in fire. Its vapors are heavier than air and can travel for great distances to an ignition source to flashback and cause fire or explosion.

SECTION V - REACTIVITY DATA

Stability		Conditions to Avoid Methyl chloride is normally stable. Avoid heat, flame, other sources of ignition, and contact with incompatible materials and chemicals.
Stable X	Unstable	Incompatibility *(materials to avoid)* Strong oxidizers (chlorine, bromine, fluorine), chemically active metals (such as sodium, potassium, calcium, powdered aluminum, zinc, and magnesium). Water or steam (reacts to form hydrochloric acid).
Hazardous Polymerization		Conditions to Avoid Under normal conditions of temperature and pressure, hazardous polymerization of methyl chloride cannot occur.
May Occur	Will Not Occur X	Hazardous Decomposition or By-products Poisonous gases and acrid fumes are produced in fire, including extremely toxic fumes of chloride.

SECTION VI - HEALTH HAZARD DATA

Primary Route(s) of Entry:	Inhalation?	Absorption (skin/eye)?	Ingestion?
	X		

Health Hazards

INHALATION: Causes depression of the central nervous system (CNS) with stupor, eye irritation, incoordination, abdominal cramps, anesthetic effects, unconsciousness, and cardiac arrest. May damage liver and kidneys. Vapors may cause irritation to the eyes, nose, and throat.

EYES & SKIN: Irritation, redness and pain from frostbite to the skin. Eye contact may result in severe irritation and possible permanent damage due to frostbite.

INGESTION: Unspecified gastrointestinal effects. An unlikely exposure route, but still possible.

Carcinogenicity	NTP Listed?	IARC Cancer Review Group?	OSHA Regulated?	Target Organs?
Unknown Human Questioned Animal	No	Group 3	29 CFR 1910.1000 Table Z-1	Eyes, skin, respiratory sys., liver, kidney, CVS, CNS.

Medical Conditions Generally Aggravated by Exposure

None reported. However, liver, kidney, heart, and respiratory problems may be aggravated by exposure.

Emergency and First-aid Procedures

Eye contact: Do not allow victim to keep eyes tightly shut. Flush immediately with tepid water (104°F), for 15 minutes (minimum); seek medical attention. Skin contact: Remove all contaminated clothing, including shoes. Immediately wash area with flooding amounts of warm water (104°F). Do NOT rub affected area or use dry heat. Seek medical attention immediately. For inhalation: Remove the person from exposure. Provide respiratory assistance and CPR. Transfer to medical facility. If swallowed: Seek medical attention immediately. Give conscious and alert person 1-2 glasses of water and induce vomiting. Never give an unconscious or convulsing person anything by mouth.

SECTION VII - PRECAUTIONS FOR SAFE HANDLING AND USE

Steps to be Taken in Case Material is Released or Spilled

Remove all ignition sources; wear SCBA respiratory protection. Restrict those not wearing protective equipment and who are not involved in cleanup from entering area. Stop leak if it can be done without risk. Spilled liquids are likely to evaporate quickly. Absorb with vermiculite and place in sealed container.

Preferred Waste Disposal Method

Burn in chemical incinerator equipped with an afterburner and scrubber.

Precautions to be Taken in Handling and Storage

Store in tightly closed containers in cool, well-ventilated location. Keep away from sources of heat.

Other Precautions and Warnings

Sources of ignition such as open flame or smoking, are prohibited where methyl chloride is used. Metal containers should be bonded and grounded. Empty cylinders contain residue; handle with care.

SECTION VIII - CONTROL MEASURES AND PERSONAL PROTECTIVE EQUIPMENT

Respiratory Protection *(specify type)*

Use an MSHA/NIOSH-approved air purifying respirator (cartridge type) or use a supplied-air respirator or a self-contained breathing apparatus (SCBA) with full facepiece operated in positive pressure mode.

Ventilation

Local exhaust (preferred) at site of chemical work or general ventilation.

Protective Gloves	Eye Protection	Other Protective Clothing
Viton, Nitrile, or Butyl Rubber Gloves	Chemical Goggles or Face Mask	Protective Apron if Splash is Likely

Work/Hygiene Practices

Always wash hands thoroughly after using chemical; never bring food, drink, or smoking materials into vicinity of chemicals.

METHYL CHLORIDE

CH_3Cl CAS: 74-87-3

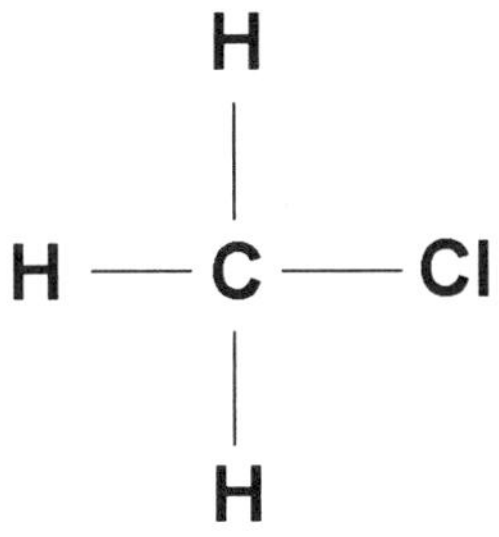

IDENTIFICATION AND TYPICAL USES

Methyl chloride is a colorless gas with a faint, sweet, ether-like odor. It is used as a catalyst in low-temperature polymerization (butyl rubber), in the manufacture of tetramethyl lead, in silicones, as a refrigerant, as a fluid for thermometric and thermostatic equipment, as a methylating agent in organic synthesis (such as that of methylcellulose), as an extractant and low-temperature solvent, an herbicide, and a topical anesthetic.

RISK ASSESSMENT: HEALTH

General Assessment

Methyl chloride is mildly toxic by *inhalation*. Although it is normally a gas at room temperature, it is normally shipped as a compressed liquid at temperatures below -144°F (-97°C) and, therefore, methyl chloride can cause serious damage on skin contact (localized tissue damage due to the extremely cold temperatures of the compressed or liquefied gas).

Inhalation is irritating to the eyes, nose, and mucosa of the respiratory tract. High concentrations of methyl chloride causes depression of the central nervous system and cardiac arrhythmia. Symptoms include dizziness, lightheadedness, headache, stupor, drunkenness and inebriation, loss of coordination, abdominal cramps, blurred vision, delayed pulmonary edema (fluid buildup in the lungs), unconsciousness, and possible cardiac arrest, liver, and kidney damage. Symptoms of pulmonary edema can be delayed up to 48 hours, thereby creating a false sense of security with regard to health exposure risk. It should also be noted that such symptoms as inebriation and drunkenness can also lead to accidents. Personnel may injure themselves or others, or damage equipment and property, as a result of the intoxicating effects of exposure.

Skin contact can produce allergic eczema and severe tissue damage due to frostbite. Eye contact will also cause serious injury to tissues as a result of the extremely cold temperatures of both the liquid and the compressed gas.

Acute Health Effects

The following acute (short-term) health effects may occur immediately or shortly after exposure to methyl chloride:

Skin: Irritation with potential for redness, pain, frostbite, and tissue damage. Possible allergic eczema can develop.

Eye: Severe irritation, burning (frostbite), pain, and smarting of the conjunctiva. Permanent damage to visual acuity is possible.

Lung: Nose, throat, and respiratory tract irritation. Absorption through the lungs is quick.

CNS: Headaches, lightheadedness, dizziness, inebriation, loss of coordination, stupor, and drunkenness. Death can occur due to cardiac arrhythmia and possible cardiac arrest.

Chronic Health Effects

The following chronic (long-term) health effects may occur at some time after exposure to methyl chloride and can last for months or even years:

Cancer Hazards: The National Institute for Occupational Safety and Health (NIOSH) considers methyl chloride to be a carcinogen and the International Agency for Research on Cancer (IARC) has conducted studies yielding insufficient evidence that methyl chloride is a tumorigenic agent. However, no carcinogenic designations have been assigned to this chemical. Further study and confirmation is required. It is recommended that users proceed with due caution until more is known about the carcinogenic properties of methyl chloride.

Reproductive Hazard: According to information presented in the references, methyl chloride may cause damage to the testes and other reproductive or possible teratogenic effects.

Other Chronic Effects: Repeated exposure to low concentrations may cause damage to central nervous system and, less frequently, to the liver, kidneys, bone marrow, and cardiovascular system. Hemorrhages into the lungs, intestinal tract, and dura have been reported.

Recommended Risk-Reduction Measures

Personnel should avoid direct contact with methyl chloride. Occupational poisoning occurs most commonly by inhalation. If a less toxic material or compound cannot be substituted for methyl chloride, then *engineering controls* are the most effective method of reducing exposures. The best protection is to enclose operations and/or provide local exhaust ventilation at the site of methyl chloride release. While not always operationally feasible, isolating operations can also reduce exposure. Using respiratory protection is less effective than the controls mentioned above, but is still required whenever working with or around methyl chloride. Caution should be exercised when selecting respiratory protection since it is not known if exposure to methyl chloride will cause cancer in humans. For exposure over the PEL (100 ppm), an MSHA/NIOSH-approved air-purifying respiratory with organic vapor cartridge(s) may suffice. For exposures to higher concentrations, or when the concentration is unknown (as in an emergency), use a supplied-air respirator with full facepiece and mask operated in positive pressure mode or a self-contained breathing apparatus (SCBA) with full facepiece operated in pressure demand. If a full facepiece is not available, then chemical goggles should be worn to protect the eyes. A face shield should also be considered. To prevent hand and skin exposures, nitrile, Viton, or butyl rubber gloves should be used. Glove manufacturers should be contacted and permeation studies obtained *before* selecting gloves.

Administrative controls should also be in place to minimize the potential for human exposures. These may include written procedures or policies which specify the methods and techniques that will be practiced whenever personnel are to work with methyl chloride.

All personnel should receive *training* on the use, hazards, protective measures, emergency actions, and other precautions per 29 CFR 1910.1200 (Hazard Communication), prior to the first assignment in an area where methyl chloride is used or stored.

If symptoms develop or overexposure is suspected, the following recommended medical tests should be considered:

- ☑ Lung function tests.
- ☑ Liver and kidney function tests.
- ☑ Consider chest X-ray following acute overexposure (may be negative if taken immediately after exposure due to delayed onset of pulmonary edema).
- ☑ Interview for nervous system and brain effects, including recent changes in memory, mood (irritability, withdrawal), and concentration. Look for headaches, malaise, and altered sleeping patterns. Cerebellar, autonomic, and peripheral nervous system evaluation should be considered. Positive or borderline cases should be referred for further neuropsychological testing.
- ☑ Complete blood count (CBC).

Any evaluation should include a careful history of past and present symptoms with an examination. Medical tests that simply look for existing damage are not a substitute for controlling exposures. Also, since smoking causes lung cancer, emphysema, heart disease, and other adverse health effects, smokers exposed to methyl chloride may experience symptoms more rapidly than non-smokers under the same conditions of exposure. Prudent risk management requires proper consideration of *all* factors which may be responsible for the appearance of exposure symptoms in the workplace.

Other methods to reduce exposure include:

- ☑ Where possible, enclose operations and use local exhaust ventilation at the site of chemical release. If local exhaust ventilation or enclosure is not used, respiratory protection should be mandatory.
- ☑ Always ensure that proper protective clothing is worn when using chemical substances.
- ☑ Wash thoroughly immediately after exposure to methyl chloride and at the end of the work shift or before eating, drinking, or smoking.
- ☑ Hazard warning information should be posted in the work area. In addition, as part of an on-going

education and training program, all information on the health and safety hazards of methyl chloride should be communicated to all exposed and potentially exposed workers.

☑ Eye wash stations should be provided in the immediate work area for emergency use. If there is a possibility of skin exposure to methyl chloride, emergency shower facilities should also be provided in the immediate area.

☑ Workers whose clothing has been contaminated by methyl chloride should change into clean clothes before leaving work. Contaminated work clothing should be laundered only by individuals who have been informed of the hazards of exposure to methyl chloride.

☑ Where possible, automatically transfer liquids containing methyl chloride from drums or other containers to process containers.

RISK ASSESSMENT: ENVIRONMENT

General Assessment

Although methyl chloride normally exists as a gas, it is usually shipped as a liquefied compressed gas. The environment is at risk of exposure during transportation, storage, disposal, or destruction of methyl chloride. In almost every scenario, the threat of environmental exposure is contingent upon the proper handling of the chemical substance. Accidental spills, large or small, can result in contamination of the surrounding environmental mediums (water, soil, and air).

Methyl chloride is very dangerous flammable gas. It is a serious fire hazard and moderate explosion hazard when exposed to heat or flame. It is incompatible with strong oxidizers, such as chlorine, fluorine, and bromine, and many chemically-active metals, such as sodium, potassium, calcium, powdered aluminum, zinc, and magnesium. Caution is always required in handling, storage, transportation, and disposal of methyl chloride. Emergency responders should be made aware of the presence of methyl chloride at any emergency response situation.

Methyl chloride can enter the environment from industrial discharges and from spills.

Acute Ecological Effects

Acute (short-term) toxic effects may include the death of animals, birds, or fish and death or low growth rate in plants. Acute effects are seen 2 to 4 days after animals or plants are exposed to methyl chloride.

Methyl chloride has slight acute toxicity to aquatic life. Insufficient data are available to evaluate or predict the short-term effects of methyl chloride on plants, birds, or land animals.

Chronic Ecological Effects

Chronic (long-term) toxic effects may include shortened life span, reproductive problems, lower fertility, and changes in appearance or behavior in exposed animals. These effects can be seen long after first exposure(s) to toxic chemicals.

Methyl chloride has slight chronic toxicity to aquatic life. Insufficient data are available to evaluate or predict the long-term effects of methyl chloride to plants, birds, or terrestrial animals.

Water Solubility

Methyl chloride is moderately soluble in water. Concentrations between 1 and 1000 milligrams will mix with a liter of water.

Persistence in the Environment

Methyl chloride is non-persistent in water, with a half-life of less than 2 days. The half-life of a pollutant is the amount of time it takes for one half of the chemical to be degraded. About 99.8% of methyl chloride will eventually end up in air; the remainder will end up in water.

Bioaccumulation in Aquatic Organisms

Some substances increase in concentration, or *bioaccumulate,* in living organisms as they breathe contaminated air, drink contaminated water, or eat contaminated food. These chemicals can become concentrated in the tissues and internal organs of animals as well as humans.

The concentration of methyl chloride found in fish tissues is expected to be about the same as the average concentration of methyl chloride in the water from which the fish was taken.

Recommended Risk-Reduction Measures

Proper training of all transporters will reduce the likelihood of a mishap or accident resulting in a leak or spill of methyl chloride into the environment. Labels on all containers, trucks, and rail cars must meet DOT requirements and accurately reflect their contents to enable emergency responders to react properly and quickly to any disaster thereby reducing the potential risk to the environment and to personnel.

Storage of methyl chloride should be segregated from other chemicals and materials to minimize the risk of cross-contamination.

If a spill or leak to the environment has occurred, fire department, emergency response and/or hazardous materials spill personnel should be notified immediately. Cleanup should be attempted only by those trained in proper spill containment procedures using non-sparking tools. Contaminated soils should be removed for incineration and replaced with clean soil. If methyl chloride should contact the water table, aquifer, or navigable waterway, time is of the essence. It is highly soluble and total remediation may not be possible. The local and/or state emergency response authorities must be notified. A comprehensive emergency response or disaster preparedness/recovery plan should be in place prior to any operations involving the use, transportation, storage, or disposal of methyl chloride.

For disposal, methyl chloride may be burned in a chemical incinerator equipped with an afterburner and scrubber. It should be noted that complete destruction by incineration is difficult unless a direct gas feed to the incinerator is used. These are normally available in permit-approved facilities only. Complete combustion is necessary to prevent the formation of phosgene gas.

If methyl chloride is spilled or leaked, the following specific steps are recommended:

☑ Restrict persons not wearing protective clothing from area of spill or leak until cleanup is complete. Use only non-sparking tools.

☑ If source of leak is a cylinder, attempt to shut off leak if it can be done without risk. If not, try to move cylinder to a safe place (outdoors) and allow to vent. If fire occurs, use water spray to knock down vapors and direct away from personnel. Absorb liquids in vermiculite, sand, earth, or similar material and deposit in sealed drum.

☑ Ventilate area of spill or leak and remove all ignition sources.

☑ It may be necessary to dispose of methyl chloride as a hazardous waste. The responsible state agency or the regional office of the federal Environmental Protection Agency (EPA) should be contacted for specific recommendations.

RISK ASSESSMENT: BUSINESS

General Assessment

Accidents or mishaps involving methyl chloride can present a serious threat to business operations. The loss or damage of equipment or facilities can significantly affect fiscal viability. Lawsuits that may result from personnel illness, injury/death, public exposures, and/or environmental contamination will require a serious expenditure of resources. Media attention surrounding an injury, death, or environmental damage can also result in a loss of profits and loss of current as well as future business.

Recommended Risk-Reduction Measures

Company attorneys, safety and health professionals, and environmental specialists should be involved in the development of any procedures for responding to chemical incidents. A company official should be predesignated as a public relations officer with specific training in dealing with the press. Corporate plans and policies should be developed, approved, and implemented long before any need for such arises.

REFERENCES

American Conference of Governmental Industrial Hygienists. 1988. *Documentation of the Threshold Limit Values and Biological Exposure Indices,* 5th Edition (with updates). Cincinnati: ACGIH

Bueche, J. R. 1972. *Principles of Physics,* 2nd Edition. New York: McGraw-Hill.

Calabrese, E. J. 1994. *Biological Effects of Low Level Exposures.* New York: CRC Press/Lewis Publishers.

Cockerman, L. G., B. S. Shane. 1994. *Basic Environmental Toxicology.* New York: CRC Press/Lewis Publishers.

Cothern, C. R., N. P. Ross. 1994. *Environmental Statistics, Assessment, and Forecasting.* New York: CRC Press/Lewis Publishers.

Hallenbeck, W. H. 1993. *Quantitative Risk Assessment for Environmental and Occupational Health,* 2nd Edition. New York: CRC Press/Lewis Publishers.

Lewis, R. J., Sr. 1992. *Sax's Dangerous Properties of Industrial Materials,* Volumes 1, 2, and 3, 8th Edition. New York: Van Nostrand Reinhold.

Meyer, E. 1990. *Chemistry of Hazardous Materials.* Englewood Cliffs, NJ: Prentice Hall.

Patnaik, P. 1992. *A Comprehensive Guide to the Hazardous Properties of Chemical Substances.* New York: Van Nostrand Reinhold.

U.S. Department of Health and Human Services, National Institute for Occupational Safety and Health. 1994. *NIOSH Pocket Guide to Chemical Hazards.* Washington, D.C.: U.S. Government Printing Office.

MATERIAL SAFETY DATA SHEET

CHEMICAL NAME

METHYL CHLOROFORM

HAZARD WARNING INFORMATION

HEALTH	FIRE	REACTIVE	OTHER	DEGREE OF HAZARD	COLOR CODING	OTHER CODES
2	1	1		0 = Minimum Hazard 1 = Slight Hazard 2 = Moderate Hazard 3 = Serious Hazard 4 = Severe Hazard	HEALTH = BLUE FIRE = RED REACTIVITY = YELLOW OTHER = WHITE	OX = Oxidizer ACID = Acid ALK = Alkali COR = Corrosive W = Use No Water

SECTION I - GENERAL INFORMATION

Characterization: **Halogenated Hydrocarbon**	RCRA Number: **U226**	EPA Class: **Toxic Waste**
DOT Proper Shipping Name: **1,1,1-Trichloroethane**	Chemical Abstract Service (CAS) Number: **71-55-6**	
DOT Hazard Class and Label Requirements: **ORM-A**	DOT Emergency Guide Code: **74**	
DOT Identification Number: **UN 2831**	Molecular Formula: **CH_3CCl_3**	

Synonyms

Chloroethane; 1,1,1-trichloroethane; 1.1.1-trichloroethane (stabilized); 1,1,1-TCE; TCE; Strobane.

SECTION II - HAZARDOUS INGREDIENTS AND IDENTITY INFORMATION

Hazardous Components (specific identity)	OSHA Exposure Criteria	NIOSH Exposure Criteria	Immediately Dangerous to Life and Health (IDLH)	ACGIH Exposure Criteria
Methyl chloroform (derivation: By catalytic addition of hydrogen chloride to 1,1-dichloroethylene; by re-fluxing chlorine monoxide with carbon tetrachloride and chloroethane). **1 ppm = 5.55 mg/m^3**	**PEL:** **350 ppm** **1900 mg/m^3** **STEL (15-min):** **450 ppm** **2450 mg/m^3**	**REL (ceiling):** **350 ppm** **1900 mg/m^3** **STEL:** **Not Applicable**	**700 ppm**	**TLV:** **350 ppm** **1910 mg/m^3** **STEL:** **450 ppm** **2460 mg/m^3**

SECTION III - PHYSICAL AND CHEMICAL CHARACTERISTICS

Boiling Point: **165°F (75°C)**	Specific Gravity (H_2O = 1): **1.34**
Vapor Pressure (atmospheres): **100 at 69°F (20°C)**	Molecular Weight (atomic weight): **133.4**
Vapor Density (Air = 1): **4.55**	Freezing Point: **-23°F (-30°C)**

Solubility

Slightly soluble in water (0.4%). Soluble in alcohol, ether, acetone, benzene, carbon tetrachloride, and carbon disulfide.

Appearance and Odor

Colorless liquid with a sweet, chloroform-like odor. Odor Threshold = 44 ppm.

SECTION IV - FIRE AND EXPLOSION HAZARD DATA

Flash Point (method used): **Not Determined**	Explosive Limits in Air % by Volume: **LEL: 7.7% UEL: 12.5%**
NFPA Classification: **Non-Combustible Liquid**	Autoignition Temperature: **932°F (500°C)**

Extinguishing Media

Use dry chemical or carbon dioxide on small fires. For large fires, use fog or regular foam.

Special Fire Fighting Procedures

Poisonous gases are produced in fire. Wear self-contained breathing apparatus (SCBA) and full protective gear. Move containers from fire if it can be done without risk. Cool sides of fire-exposed containers with water until long after fire is out. Reacts with water to form hydrochloric acid.

Unusual Fire and Explosion Hazards

Containers may explode in fire. Its vapors are heavier than air and can travel for great distances to an ignition source to flashback and cause fire or explosion. Keep away from ends of tanks.

SECTION V - REACTIVITY DATA

Stability		Conditions to Avoid: Methyl chloroform is normally stable. Avoid heat, flame, other sources of ignition, and contact with incompatible materials and chemicals.
Stable	Unstable	Incompatibility *(materials to avoid)*: Strong oxidizers (chlorine, fluorine), chemically active metals (sodium, potassium, calcium, powdered aluminum, zinc, and magnesium), and strong caustics. Water or steam (reacts to form hydrochloric acid).
X		
Hazardous Polymerization		Conditions to Avoid: Hazardous polymerization of methyl chloroform may occur in contact with aluminum trichloride. Also, keep away from strong ignition sources and arc-welding areas. Forms shock sensitive mixtures with potassium.
May Occur	Will Not Occur	Hazardous Decomposition or By-products: When heated to decomposition, methyl chloroform can produce carbon dioxide, toxic dichloroacetylene, hydrogen chloride, and deadly phosgene gas.
X		

SECTION VI - HEALTH HAZARD DATA

Primary Route(s) of Entry:	Inhalation?	Absorption (skin/eye)?	Ingestion?
	X		X

Health Hazards

INHALATION: Causes depression of the central nervous system (CNS) with stupor, eye irritation, incoordination, abdominal cramps, anesthetic effects, unconsciousness, and cardiac arrest. May damage liver and kidneys. Vapors may cause irritation to the eyes, nose, and throat.

SKIN & EYES: Causes de-fatting of skin's surface leading to irritation, redness, dryness, and scaling. Eye contact causes irritation and mild conjunctivitis.

INGESTION: An unlikely exposure route. May cause nausea, vomiting, diarrhea, and throat burns.

Carcinogenicity	NTP Listed?	IARC Cancer Review Group?	OSHA Regulated?	Target Organs?
Unknown Human Questioned Animal	No	Group 3	29 CFR 1910.1000 Table Z-1	Eyes, skin, liver, kidney, CVS, CNS.

Medical Conditions Generally Aggravated by Exposure

None reported; however, liver, kidney, heart, and skin problems may be aggravated by exposure.

Emergency and First-aid Procedures

Eye contact: Do not allow victim to keep eyes tightly shut. Flush immediately with water for 15 minutes (minimum). Seek medical attention. Skin contact: Remove all contaminated clothing, including shoes. Immediately wash area with flooding amounts of water, then wash area with soap and water. Seek medical attention, if necessary. For inhalation: Remove the person from exposure. Provide respiratory assistance and CPR. Transfer to medical facility. If swallowed: Seek medical attention immediately. Give conscious and alert person 1-2 glasses of water to dilute. Never give an unconscious or convulsing person anything by mouth. Contact poison control center or seek medical attention for further instructions.

SECTION VII - PRECAUTIONS FOR SAFE HANDLING AND USE

Steps to be Taken in Case Material is Released or Spilled

Remove all ignition sources; wear SCBA respiratory protection. Restrict those not wearing protective equipment and who are not involved in cleanup from entering area. Spilled liquids are likely to evaporate quickly. Absorb with vermiculite and place in sealed container.

Preferred Waste Disposal Method

Mix with combustible solvent; burn in chemical incinerator equipped with an afterburner and scrubber.

Precautions to be Taken in Handling and Storage

Store in tightly closed containers in cool, well-ventilated location. Keep away from sources of heat.

Other Precautions and Warnings

Liquid is non-combustible but vapor can easily ignite. Sources of ignition such as open flame or smoking, are prohibited where methyl chloroform is used. Metal containers should be bonded and grounded.

SECTION VIII - CONTROL MEASURES AND PERSONAL PROTECTIVE EQUIPMENT

Respiratory Protection *(specify type)*

Use an MSHA/NIOSH-approved air purifying respirator (cartridge type) or use a supplied-air respirator or a self-contained breathing apparatus (SCBA) with full facepiece operated in positive pressure mode.

Ventilation

Local exhaust (preferred) at site of chemical work or general ventilation.

Protective Gloves	Eye Protection	Other Protective Clothing
Viton or Butyl Rubber Gloves	Chemical Goggles or Face Mask	Protective Apron if Splash is Likely

Work/Hygiene Practices

Always wash hands thoroughly after using chemical; never bring food, drink, or smoking materials into vicinity of chemicals.

METHYL CHLOROFORM

CH_3CCl_3

CAS: 71-55-6

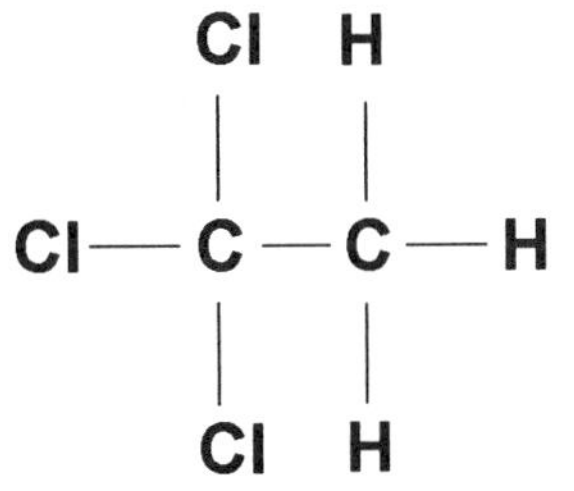

IDENTIFICATION AND TYPICAL USES

Methyl chloroform is a colorless liquid with a sweet, chloroform-like odor. It is primarily an industrial cleaning solvent for metals and plastic molds. It is also used in pesticides, as a component in drain cleaners and inks, and in textile processing. It has also been used extensively as a substitute for carbon tetrachloride.

RISK ASSESSMENT: HEALTH

General Assessment

Methyl chloroform is moderately toxic by *inhalation.* Although an unlikely exposure route, *ingestion* can also cause adverse systemic effects. While skin contact certainly causes localized, topical effects, the data on its ability to pass through intact skin (absorption) are conflicting and inconclusive. Methyl chloroform is an experimental teratogen and a known mutagen. There are insufficient data regarding is carcinogenic potential.

Inhalation is irritating to the eyes, nose, and mucosa of the respiratory tract. High concentrations of methyl chloroform vapors causes inflammation of the conjunctiva, depression of the central nervous system, and cardiac arrhythmia. Symptoms include dizziness, lightheadedness, headache, stupor, hallucinations, loss of coordination, abdominal cramps, irritability, equilibrium disturbances, aggressive behavior, hypermotility, unconsciousness, possible cardiac arrest, and liver and kidney damage. Ingestion can also cause these symptoms as well as nausea, vomiting, diarrhea, and gastrointestinal irritation. It is narcotic at high concentrations. It causes proarrhythmic activity which sensitizes the heart to epinephrine-induced arrhythmia. This can lead to cardiac arrest.

Skin contact result in the removal of the skin's natural, protective oily barrier (de-fatting) which can then lead to irritation, dryness, redness, cracking, and possible secondary infections. Eye contact will also cause serious irritation and inflammation of surrounding tissue and possible burning and stinging of the eye.

Acute Health Effects

The following acute (short-term) health effects may occur immediately or shortly after exposure to methyl chloroform:

Skin: Irritation with potential for redness, pain, dryness, scaling, cracking, and infection. Dermatitis may occur in some individuals.

Eye: Severe irritation, burning, pain, and smarting of the conjunctiva. Permanent damage to visual acuity is possible.

Lung: Nose, throat, and respiratory tract irritation. High concentrations can cause severe irritation.

CNS: Headaches, lightheadedness, dizziness, hallucinations, loss of coordination, aggressive behavior, motor changes. Death can occur due to cardiac arrhythmia and possible cardiac arrest.

Chronic Health Effects

The following chronic (long-term) health effects may occur at some time after exposure to methyl chloroform and can last for months or even years:

Cancer Hazards: According to information presented in the references, there is limited but insufficient evidence to conclude that exposure to methyl chloroform causes cancer in test animals. It is known to cause mutations (genetic changes) and many scientists believe that such chemicals may pose a cancer risk in the long-term. More research is required before any determinations can be made regarding carcinogenicity.

Reproductive Hazard: According to information presented in the references, methyl chloroform may cause reproductive damage or possible teratogenic effects in test animals. Human teratogenicity is unknown.

Other Chronic Effects: High or repeated exposures may damage the liver and kidneys. Prolonged skin contact can cause thickening and cracking of the skin.

Recommended Risk-Reduction Measures

Personnel should avoid direct contact with methyl chloroform. The exact nature of toxic effects on human exposure is not clearly understood in the references. Occupational poisoning occurs most commonly by inhalation. If a less toxic material or compound cannot be substituted for methyl chloroform, then *engineering controls* are the most effective method of reducing exposures. The best protection is to enclose operations and/or provide local exhaust ventilation at the site of methyl chloroform release. While not always operationally feasible, isolating operations can also reduce exposure. Using respiratory protection is less effective than the controls mentioned above, but is still required whenever working with or around methyl chloroform. Caution should be exercised when selecting respiratory protection since it is not known if exposure to methyl chloroform will cause cancer in humans. For exposure over the PEL (350 ppm), an MSHA/NIOSH-approved air-purifying respiratory with organic vapor cartridge(s) may suffice. For exposures to higher concentrations, or when the concentration is unknown (as in an emergency), use a supplied-air respirator with full facepiece and mask operated in positive pressure mode or a self-contained breathing apparatus (SCBA) with full facepiece operated in pressure demand. If a full facepiece is not available, then chemical goggles should be worn to protect the eyes. A face shield should also be considered. To prevent hand and skin exposures, nitrile, Viton, or butyl rubber gloves should be used. Glove manufacturers should be contacted and permeation studies obtained *before* selecting gloves.

Administrative controls should also be in place to minimize the potential for human exposures. These may include written procedures or policies which specify the methods and techniques that will be practiced whenever personnel are to work with methyl chloroform.

All personnel should receive *training* on the use, hazards, protective measures, emergency actions, and other precautions per 29 CFR 1910.1200 (Hazard Communication), prior to the first assignment in an area where methyl chloroform is used or stored.

If symptoms develop or overexposure is suspected, the following recommended medical tests should be considered:

- ☑ Liver and kidney function tests.
- ☑ Interview for nervous system depression and indications of recent mood changes or changes in behavior.

Any evaluation should include a careful history of past and present symptoms with an examination. Medical tests that simply look for existing damage are not a substitute for controlling exposures. Also, persons with heart disease may be at increased risk of irregular heartbeat from very high exposures to methyl chloroform. Proper risk management requires careful consideration of exposure potential during personnel selection and placement.

Other methods to reduce exposure include:

- ☑ Where possible, enclose operations and use local exhaust ventilation at the site of chemical release. If local exhaust ventilation or enclosure is not used, respiratory protection should be mandatory.
- ☑ Always ensure that proper protective clothing is worn when using chemical substances and that personnel are trained on its use and care.
- ☑ Wash thoroughly immediately after exposure to methyl chloroform and at the end of the work shift or before eating, drinking, or smoking.
- ☑ Hazard warning information should be posted in the work area. In addition, as part of an on-going education and training program, all information on the health and safety hazards of methyl chloroform should be communicated to all exposed and potentially exposed workers.
- ☑ Eye wash stations should be provided in the immediate work area for emergency use. If there is a possibility of skin exposure to methyl chloroform, emergency shower facilities should also be provided in the immediate area.
- ☑ Workers whose clothing has been contaminated by methyl chloroform should change into clean clothes before leaving work. Contaminated work clothing should be laundered only by individuals who have been informed of the hazards of exposure to methyl chloroform.

☑ Where possible, automatically transfer liquids containing methyl chloroform from drums or other containers to process containers.

RISK ASSESSMENT: ENVIRONMENT

General Assessment

The environment is at risk of exposure during transportation, storage, disposal, or destruction of methyl chloroform. In almost every scenario, the threat of environmental exposure is contingent upon the proper handling of the chemical substance. Accidental spills, large or small, can result in contamination of the surrounding environmental mediums (water, soil, and air).

Methyl chloroform is considered a non-combustible liquid, although it may ignite with some degree of difficulty. Its vapors, however, can form flammable mixtures in air that can easily ignite, especially in confined areas. It is incompatible with strong oxidizers, such as chlorine, fluorine, and bromine, and many chemically-active metals, such as sodium, potassium, calcium, powdered aluminum, zinc, and magnesium. Caution is always required in handling, storage, transportation, and disposal of methyl chloroform. Emergency responders should be made aware of the presence of methyl chloroform at any emergency response situation.

Methyl chloroform can enter the environment from industrial and municipal waste treatment plant discharges and from spills.

Acute Ecological Effects

Acute (short-term) toxic effects may include the death of animals, birds, or fish and death or low growth rate in plants. Acute effects are seen 2 to 4 days after animals or plants are exposed to methyl chloroform.

Methyl chloroform has moderate acute toxicity to aquatic life. Insufficient data are available to evaluate or predict the short-term effects of methyl chloroform on plants, birds, or land animals.

Chronic Ecological Effects

Chronic (long-term) toxic effects may include shortened life span, reproductive problems, lower fertility, and changes in appearance or behavior in exposed animals. These effects can be seen long after first exposure(s) to toxic chemicals.

Methyl chloroform has moderate chronic toxicity to aquatic life. Insufficient data are available to evaluate or predict the long-term effects of methyl chloroform to plants, birds, or terrestrial animals.

Water Solubility

Methyl chloroform is moderately soluble in the aquatic environment. Concentrations between 1 and 1000 milligrams will mix with a liter of water.

Persistence in the Environment

Methyl chloroform is non-persistent in water, with a half-life of less than 2 days. The half-life of a pollutant is the amount of time it takes for one half of the chemical to be degraded. About 99.9% of methyl chloroform will eventually end up in air; the remainder will end up in water.

Bioaccumulation in Aquatic Organisms

Some substances increase in concentration, or *bioaccumulate,* in living organisms as they breathe contaminated air, drink contaminated water, or eat contaminated food. These chemicals can become concentrated in the tissues and internal organs of animals as well as humans.

The concentration of methyl chloroform found in fish tissues is expected to be somewhat higher than the average concentration of methyl chloroform in the water from which the fish was taken.

Recommended Risk-Reduction Measures

Proper training of all transporters will reduce the likelihood of a mishap or accident resulting in a leak or spill of methyl chloroform into the environment. Labels on all containers, trucks, and rail cars must meet DOT requirements and accurately reflect their contents to enable emergency responders to react properly and quickly to any disaster thereby reducing the potential risk to the environment and to personnel.

Storage of methyl chloroform should be segregated from other chemicals and materials to minimize the risk of cross-contamination.

If a spill or leak to the environment has occurred, fire department, emergency response and/or hazardous materials spill personnel should be notified immediately. Cleanup should be attempted only by those trained in proper spill containment procedures using non-sparking tools. Contaminated soils should be removed for incineration and replaced with clean soil. Should methyl chloroform contact the water table, aquifer, or navigable waterway, time is of the essence. It is moderately soluble and total remediation may not be possible. The local and/or state emergency response authorities must be notified. A comprehensive emergency response or disaster preparedness/recovery plan should be in place prior to any operations involving the use, transportation, storage, or disposal of methyl chloroform.

For disposal, methyl chloroform may be mixed with a combustible solvent and burned in a chemical incinerator equipped with an afterburner and scrubber system. Complete combustion is necessary to prevent the formation of phosgene gas.

If methyl chloroform is spilled or leaked, the following specific steps are recommended:

- ☑ Restrict persons not wearing protective clothing from area of spill or leak until cleanup is complete. Use only non-sparking tools.
- ☑ If fire occurs, use water spray to knock down vapors and direct away from personnel. Absorb liquids in vermiculite, sand, earth, or similar material and deposit in sealed drum. Collect using non-sparking tools.
- ☑ Ventilate area of spill or leak and remove all ignition sources.
- ☑ It may be necessary to dispose of methyl chloroform as a hazardous waste. The responsible state agency or the regional office of the federal Environmental Protection Agency (EPA) should be contacted for specific recommendations.

RISK ASSESSMENT: BUSINESS

General Assessment

Accidents or mishaps involving methyl chloroform can present a serious threat to business operations. The loss or damage of equipment or facilities can significantly affect fiscal viability. Lawsuits that may result from personnel illness, injury/death, public exposures, and/or environmental contamination will require a serious expenditure of resources. Media attention surrounding an injury, death, or environmental damage can also result in a loss of profits and loss of current as well as future business. Always remember that anytime the terms "cancer" or "reproductive hazard" are used, public emotion, hysteria, and ignorance can run equally high. This must be carefully considered whenever developing or implementing public relations policies.

❶ *Recommended Risk-Reduction Measures*

Company attorneys, safety and health professionals, and environmental specialists should be involved in the development of any procedures for responding to chemical incidents. A company official should be predesignated as a public relations officer with specific training in dealing with the press. Corporate plans and policies should be developed, approved, and implemented long before any need for such arises.

REFERENCES

American Conference of Governmental Industrial Hygienists. 1988. *Documentation of the Threshold Limit Values and Biological Exposure Indices,* 5th Edition (with updates). Cincinnati: ACGIH

Asante-Duah, D. K. 1993. *Hazardous Waste Risk Assessment.* New York: CRC Press/Lewis Publishers.

Bloom, A. D. and F. J. DeSerres. 1995. *Ecotoxicity and Human Health.* New York: CRC Press/Lewis Publishers.

Bueche, J. R. 1972. *Principles of Physics,* 2nd Edition. New York: McGraw-Hill.

Burrell, R., D. K. Flaherty, and L. J. Sauers. 1992. *Toxicology of the Immune System.* New York: Van Nostrand Reinhold.

Calabrese, E. J., E. M. Kenyon. *Air Toxics and Risk Assessment.* New York: CRC Press/Lewis Publishers.

Cockerman, L. G., B. S. Shane. 1994. *Basic Environmental Toxicology.* New York: CRC Press/Lewis Publishers.

Cothern, C. R., N. P. Ross. 1994. *Environmental Statistics, Assessment, and Forecasting.* New York: CRC Press/Lewis Publishers.

Danse, I. R. 1991. *Common Sense Toxics in the Workplace.* New York: Van Nostrand Reinhold.

Environmental Protection Agency. 1983. *Chemical Hazard Information Profiles.* Washington, D.C.: U.S. EPA.

Fodden, C. R., and J. L. Weddell. 1992. *Hazardous Materials Emergency Action Data.* New York: CRC Press/Lewis Publishers.

Forseberg, K., and L. H. Keith. 1993. *Chemical; Protective Clothing Permeation and Degradation Database.* Chelsea, MI: Lewis Publishers.

Gardner, R. W. 1994. *Chemical Intolerance.* New York: CRC Press/Lewis Publishers.

Hallenbeck, W. H. 1993. *Quantitative Risk Assessment for Environmental and Occupational Health,* 2nd Edition. New York: CRC Press/Lewis Publishers.

Lewis, R. J., Sr. 1992. *Sax's Dangerous Properties of Industrial Materials,* Volumes 1, 2, and 3, 8th Edition. New York: Van Nostrand Reinhold.

Meyer, E. 1990. *Chemistry of Hazardous Materials.* Englewood Cliffs, NJ: Prentice Hall.

Patnaik, P. 1992. *A Comprehensive Guide to the Hazardous Properties of Chemical Substances.* New York: Van Nostrand Reinhold.

U.S. Department of Health and Human Services, National Institute for Occupational Safety and Health. 1994. *NIOSH Pocket Guide to Chemical Hazards.* Washington, D.C.: U.S. Government Printing Office.

MATERIAL SAFETY DATA SHEET

CHEMICAL NAME

METHYLCYCLOHEXANE

CRC PRESS/LEWIS PUBLISHERS DISCLAIMER: THE INFORMATION AND RECOMMENDATIONS PRESENTED HEREIN ARE BASED ON SOURCES BELIEVED TO BE RELIABLE. CRC MAKES NO REPRESENTATION ON ITS COMPLETENESS OR ACCURACY. IT IS THE RESPONSIBILITY OF THE USER TO DETERMINE THE CHEMICAL'S SUITABILITY FOR ITS INTENDED USE, THE CHEMICAL'S SAFE USE, AND THE CHEMICAL'S PROPER DISPOSAL. NO REPRESENTATIONS AND/OR WARRANTIES, EITHER EXPRESSED OR IMPLIED, OF THE MERCHANTABILITY OR FITNESS FOR A PARTICULAR PURPOSE, OR OF ANY OTHER NATURE, ARE MADE WITH RESPECT TO THE INFORMATION PROVIDED IN THIS MSDS OR TO THE CHEMICAL TO WHICH INFORMATION MAY REFER. CRC NEITHER ASSUMES NOR AUTHORIZES ANY OTHER PERSON TO ASSUME FOR IT, ANY OTHER ADDITIONAL RESPONSIBILITY OR LIABILITY FOR THE USE OF, OR RELIANCE UPON, THIS INFORMATION.

HAZARD WARNING INFORMATION

HEALTH	FIRE	REACTIVE	OTHER	DEGREE OF HAZARD	COLOR CODING	OTHER CODES
2	3	0		0 = Minimum Hazard 1 = Slight Hazard 2 = Moderate Hazard 3 = Serious Hazard 4 = Severe Hazard	HEALTH = BLUE FIRE = RED REACTIVITY = YELLOW OTHER = WHITE	OX = Oxidizer ACID = Acid ALK = Alkali COR = Corrosive W = Use No Water

SECTION I - GENERAL INFORMATION

Characterization	Hydrocarbon, Aliphatic	RCRA Number	D001
		EPA Class	Characteristic (I) Waste
DOT Proper Shipping Name	Methylcyclohexane	Chemical Abstract Service (CAS) Number	108-87-2
DOT Hazard Class and Label Requirements	Flammable Liquid	DOT Emergency Guide Code	27
DOT Identification Number	UN 2296	Chemical Formula	$CH_3C_6H_{11}$

Synonyms
Cyclohexylmethane; hexahydrotoluene; Sextone B; toluene hexahydride.

SECTION II - HAZARDOUS INGREDIENTS AND IDENTITY INFORMATION

Hazardous Components (specific identity)	OSHA Exposure Criteria	NIOSH Exposure Criteria	Immediately Dangerous to Life and Health (IDLH)	ACGIH Exposure Criteria
Methylcyclohexane (derivation: By hydrogenation of toluene, acidic hydrocracking of polycyclic aromatics, or reaction between benzene and methane at high temperatures. Also occurs in certain crude petroleum oils and is separated by distillation). **1 ppm = 4.08 mg/m³**	**PEL:** **500 ppm** **2000 mg/m³** **STEL:** **Not Established**	**REL:** **400 ppm** **1600 mg/m³** **STEL:** **Not Established**	**1200 ppm**	**TLV:** **400 ppm** **1610 mg/m³** **STEL:** **Not Established**

SECTION III - PHYSICAL AND CHEMICAL CHARACTERISTICS

Boiling Point	214°F (101°C)	Specific Gravity (H_2O = 1)	0.77
Vapor Pressure (gas)	43 at 77°F (25°C)	Molecular Weight	98.2
Vapor Density (Air = 1)	3.39	Freezing Point	-196°F (-126°C)

Solubility
Insoluble in water (floats); soluble in ethyl alcohol and ethyl ether; miscible in petroleum ether, acetone, benzene, and carbon tetrachloride.

Appearance and Odor
Colorless liquid with a faint, benzene-like odor. Odor threshold = 500 ppm.

SECTION IV - FIRE AND EXPLOSION HAZARD DATA

Flash Point (method used)	25°F (-4°C) closed cup	Explosive Limits in Air % by Volume	LEL: 1.2% UEL: 6.7%
NFPA Classification	Class 1B Flammable Liquid	Autoignition Temperature	482°F (250°C)

Extinguishing Media
Use dry chemical, carbon dioxide, or regular foam. Water may be ineffective or actually spread fires.

Special Fire Fighting Procedures
Wear full protective clothing and self-contained breathing apparatus (SCBA). Poisonous gases are produced in fire. Use water spray to keep fire-exposed containers cool.

Unusual Fire and Explosion Hazards
Dangerous fire and explosion hazard when exposed to heat or flame. Containers may explode in fire. Moderate explosion hazard when in the form of vapor. Boiling liquid expanding vapor explosion (BLEVE) is possible. Immediately evacuate if rising sound is heard from venting device.

SECTION V - REACTIVITY DATA

Stability		Conditions to Avoid Normally stable in closed containers under routine conditions of handing and storage. Keep away from heat, sources of ignition, and incompatible materials.
Stable	Unstable	Incompatibility *(materials to avoid)*
X		Strong oxidizers, such as permanganates, peroxides, nitrates, chlorates, and perchlorates.
Hazardous Polymerization		Conditions to Avoid Under normal conditions of temperature and pressure, hazardous polymerization of methylcyclohexane cannot occur.
May Occur	Will Not Occur	Hazardous Decomposition or By-products
	X	When heated to decomposition, methylcyclohexane can emit highly toxic/ carbon dioxide and acrid smoke and fumes.

SECTION VI - HEALTH HAZARD DATA

Primary Route(s) of Entry:	Inhalation?	Absorption (skin)?	Ingestion?
	X		X

Health Hazards

INHALATION: Minor irritation to the eyes and nose. A central nervous system depressant causing mild narcosis. It can lead to nausea, dizziness, vomiting, and depression. Higher concentrations can lead to unconsciousness and may cause kidney lesions.

SKIN & EYES: Irritating to the skin, may cause drying, cracking, and chapping. Eye irritation likely.

INGESTION: Unspecified gastrointestinal effects. Possible systemic effects similar to inhalation

Carcinogenicity	NTP Listed?	IARC Cancer Review Group?	OSHA Regulated?	Target Organs?
Unknown Human Unknown Animal	No	No	29 CFR 1910.1000 Table Z-1	Respiratory system, eyes, skin, CNS.

Medical Conditions Generally Aggravated by Exposure

None reported.

Emergency and First-aid Procedures

Eye contact: Flush immediately with water for 15 minutes (minimum); seek medical attention. Skin contact: Remove all contaminated clothing. Immediately wash area with large amounts of soap and water. Seek medical assistance. For inhalation: Remove the person from exposure. Provide respiratory assistance and CPR, if required. Transfer to medical facility. Keep victim warm and at rest. If swallowed: Seek medical attention immediately or contact poison control center for instructions.

SECTION VII - PRECAUTIONS FOR SAFE HANDLING AND USE

Steps to be Taken in Case Material is Released or Spilled

Absorb spills in vermiculite, dry sand, earth, or similar material and deposit in sealed containers. Ventilate area of spill or leak. Restrict those not involved in cleanup from entering area. Remove all sources of ignition and stay up wind. Water spray may be used to knock down or reduce airborne vapor concentrations and to protect emergency response crews.

Preferred Waste Disposal Method

No Citation.

Precautions to be Taken in Handling and Storage

Store to avoid contact with incompatible materials in tightly closed containers in cool, well-ventilated area away from heat and flame. Use only non-sparking tools to open and close containers.

Other Precautions and Warnings

Sources of ignition are prohibited where methylcyclohexane is used, handled, or stored. Heat may cause containers to build up pressure and explode. Bond and ground all containers during transfer.

SECTION VIII - CONTROL MEASURES AND PERSONAL PROTECTIVE EQUIPMENT

Respiratory Protection *(specify type)*

For exposures above the PEL, use an MSHA/NIOSH-approved organic vapor respirator or a supplied-air respirator with full facepiece, hood, or helmet in continuous flow mode, or use a self-contained breathing apparatus (SCBA) operated in pressure demand or other positive pressure mode.

Ventilation

Local exhaust or general mechanical systems recommended.

Protective Gloves	Eye Protection	Other Protective Clothing
Impervious Material	Chemical Goggles or Face Mask	Protective Apron

Work/Hygiene Practices

Always wash hands thoroughly after using chemical; never bring food, drink, or smoking materials into vicinity of chemicals.

METHYLCYCLOHEXANE

$C_3C_6H_{11}$

CAS: 108-87-2

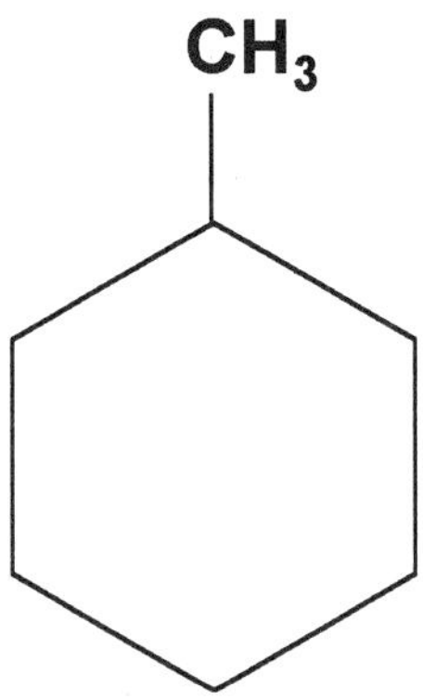

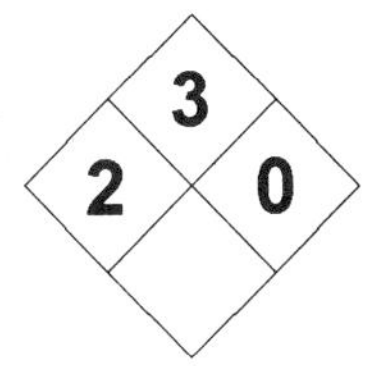

IDENTIFICATION AND TYPICAL USES

Methylcyclohexane is a colorless liquid with a faint, benzene-like odor. It is used as a solvent for cellulose ethers and as an intermediate in organic synthesis.

RISK ASSESSMENT: HEALTH

General Assessment

Methylcyclohexane is moderately toxic by *ingestion* and mildly toxic by *inhalation*. There are no reports of any carcinogenic, mutagenic, or teratogenic effects in humans or test animals.

Significant inhalation may cause irritation of the respiratory tract and depress the central nervous system. Symptoms include nausea, lightheadedness, drowsiness, dizziness, vomiting, and depression. At high concentrations, it is a narcotic and can lead to unconsciousness. It is of low chronic toxicity due to its efficient metabolism and excretion. It does *not* produce toxic changes to nerve cells and peripheral neuropathy normally associated with exposure to *n*-hexane. Vascular degenerative kidney lesions were reported in test animals.

Skin and eye contact may produce mild to moderate irritation on contact. There may be dermal thickening and ulceration of the skin surface. Eye contact may cause irritation and possible inflammation.

Acute Health Effects

The following acute (short-term) health effects may occur immediately or shortly after exposure to methylcyclohexane:

Skin: Mild to moderate irritation.

Eye: Mild to moderate irritation.

Lung: High concentrations may irritate the lungs and nasal passages. Primary result of inhalation of methylcyclohexane is its action on the CNS.

CNS: A CNS depressant causing headache, dizziness, depression, excitement, dizziness, drowsiness, lightheadedness, changes in mood, and other narcotic effects up to and including loss of consciousness (at high concentrations).

Chronic Health Effects

The following chronic (long-term) health effects may occur at some time after exposure to methylcyclohexane and can last for months or even years:

Cancer Hazards: According to information presented in the references, methylcyclohexane has not been adequately tested for its ability to cause cancer in animals.

Reproductive Hazard: According to information presented in the references, methylcyclohexane has not been tested for its ability to adversely affect reproduction.

Other Chronic Effects: Prolonged skin contact may cause drying, cracking, and chapping of the affected area. Repeated or long-term exposures to methylcyclohexane may affect the liver and kidneys.

Recommended Risk-Reduction Measures

Personnel should avoid direct contact with methylcyclohexane. The specific toxicity of this chemical is clearly understood in the references but it is believed to have the capability of causing significant health effects if improperly used or if overexposures occur. If a less toxic chemical cannot be substituted for a hazardous substance, then *engineering controls* are the most effective method of reducing exposures. The best protection is to enclose operations and/or provide local exhaust ventilation at the site of chemical release. While not always operationally feasible, isolating operations can also reduce exposure. Using respiratory protection is less effective than the controls mentioned above, but is still advisable whenever working with or around methylcyclohexane. For exposures above the PEL (500 ppm), an organic vapor respirator, a supplied-air respirator with full facepiece operated in continuous flow mode, or a self-contained breathing apparatus (SCBA) with full facepiece and pressure demand are the recommended respiratory protection methods of choice. If a full facepiece is not available, then chemical goggles should be worn to protect the eyes. Whenever a chemical splash hazard exists, a face shield and a rubber apron should be worn. To prevent hand and skin exposures, impervious gloves should be used. Glove manufacturers should be contacted and permeation studies obtained *before* final glove selections have been made.

Administrative controls should also be in place to minimize the potential for human exposures. These may include written procedures or policies which specify the methods and techniques that will be practiced whenever personnel are to work with methylcyclohexane.

All personnel should receive *training* on the use, hazards, protective measures, emergency actions, and other precautions per 29 CFR 1910.1200 (Hazard Communication), prior to the first assignment in an area where methylcyclohexane is used or stored.

If symptoms develop or overexposure is suspected, the following tests are recommended:

☑ Kidney and liver function tests.

Any evaluation should include a careful history of past and present symptoms with an examination. Medical tests that simply look for existing damage are not a substitute for controlling exposures.

Other methods to reduce exposure include:

☑ Where possible, enclose operations and use local exhaust ventilation at the site of chemical release. If local exhaust ventilation or enclosure is not used, respiratory protection should be mandatory.

☑ Always ensure that proper protective clothing is worn when using chemical substances and that personnel are trained on its proper use and care.

☑ Wash thoroughly immediately after exposure to methylcyclohexane and at the end of the work shift or before eating, drinking, or smoking.

☑ Hazard warning information should be posted in the work area. In addition, as part of an on-going education and training program, all information on the health and safety hazards of methylcyclohexane should be communicated to all exposed and potentially exposed workers.

☑ Eye wash stations should be provided in the immediate work area for emergency use. If there is a possibility of skin exposure to methylcyclohexane, emergency shower facilities should also be provided in the work area.

☑ Personnel should never wear contaminated clothing home (family members can be exposed). Clothing should be laundered by personnel who have been trained on the hazards of methylcyclohexane.

RISK ASSESSMENT: ENVIRONMENT

General Assessment

The environment is at risk of exposure during transportation, storage, disposal, or destruction of methylcyclohexane. In almost every scenario, the threat of environmental exposure is contingent upon the proper handling of the chemical substance. Accidental spills, large or small, can result in fire, explosion, and possible contamination of the surrounding environmental mediums (water, soil, and air).

Methylcyclohexane is considered a Class 1B flammable liquid (per OSHA 29 CFR 1910.106). Its extremely low flash point and relatively low boiling point make this chemical a serious fire and explosion hazard concern. During tank or rail car fires, a boiling liquid expanding vapor explosion (BLEVE) is possible. The area should be evacuated when a rising sound is heard from a venting device or a discoloration of the tank ends becomes noticeable. These characteristics require extreme caution in handling, storage, transportation, and disposal. It is incompatible with common oxidizers and contact can cause fire or explosion. Therefore, special consideration is required during any emergency situation involving a leak or spill of methylcyclohexane. Should methylcyclohexane ever come into contact with these incompatible substances during use, transportation, storage, or disposal, the formation of highly toxic and/or highly explosive commodities is extremely possible.

Methylcyclohexane may enter the environment through industrial discharges or spills.

Acute Ecological Effects

Acute (short-term) toxic effects may include the death of animals, birds, or fish and death or low growth rate in plants. Acute effects are seen 2 to 4 days after animals or plants are exposed to methylcyclohexane.

Insufficient data are available to evaluate or predict the short-term effects of methylcyclohexane exposure to aquatic life, plants, birds, or land animals.

Chronic Ecological Effects

Chronic (long-term) toxic effects may include shortened life span, reproductive problems, lower fertility, and changes in appearance or behavior in exposed animals. These effects can be seen long after first exposure(s) to toxic chemicals.

Insufficient data are available to evaluate or predict the long-term effects of methylcyclohexane to aquatic life, plants, birds, or land animals.

Water Solubility

Methylcyclohexane is insoluble in water. Concentrations of less than 1 milligram will not mix with a liter of water.

Persistence in the Environment

Methylcyclohexane is non-persistent in water, with a half-life of less than 2 days. The half-life of a pollutant is the amount of time it takes for one half of the chemical to be degraded. Virtually 100% of methylcyclohexane will eventually end up in air.

Bioaccumulation in Aquatic Organisms

Some substances increase in concentration, or *bioaccumulate,* in living organisms as they breathe contaminated air, drink contaminated water, or eat contaminated food. These chemicals can become concentrated in the tissues and internal organs of animals as well as humans.

The concentration of methylcyclohexane found in fish tissue is expected to be about the same as the average concentration of methylcyclohexane in the water from which the fish was taken.

Recommended Risk-Reduction Measures

Proper training of all transporters will reduce the likelihood of a mishap or accident resulting in a leak or spill to the environment. The correct use of DOT labeling (per the MSDS) on all containers, trucks, and rail cars will enable emergency responders to react properly and quickly to any disaster thereby reducing the potential risk to the environment and to personnel.

Storage of methylcyclohexane should be segregated from incompatible chemicals to minimize the risk of cross-contamination or contact. Buildings designated for storage should be equipped with appropriate fire protection systems (alarms, sprinklers, emergency lighting, portable extinguishers).

If a spill or leak to the environment has occurred, fire department, emergency response and/or hazardous materials spill personnel should be notified immediately. Cleanup should be attempted only by those trained in proper spill containment procedures using non-sparking tools. Contaminated soils should be removed for incineration and replaced with clean soil. If methylcyclohexane should contact the water table, aquifer, or navigable waterway, the local and/or state emergency response authorities must be notified. A comprehensive emergency response or disaster preparedness/recovery plan should be in place prior to any operations involving the use, transportation, storage, or disposal of methylcyclohexane. If methylcyclohexane is spilled or leaked, the following specific steps are recommended:

- ☑ Restrict persons not wearing protective clothing from area of spill or leak until cleanup is complete and area can be opened for normal work.
- ☑ Remove all ignition sources
- ☑ Ventilate area of leak or spill. Depending upon amount and location of spill, explosion-proof ventilation system may be necessary.
- ☑ Absorb liquids in vermiculite, dry sand, earth, or similar material and deposit in sealed containers.
- ☑ It may be necessary to dispose of methylcyclohexane as a hazardous waste. The responsible state agency or the regional office of the federal Environmental Protection Agency (EPA) should be contacted for specific recommendations.

RISK ASSESSMENT: BUSINESS

General Assessment

Accidents or mishaps involving methylcyclohexane can present a significant threat to business operations. The loss or damage of equipment or facilities can significantly affect fiscal viability. Lawsuits that may result from personnel injury/death, public exposures, and/or environmental contamination will also require a serious expenditure of resources. Media attention surrounding an injury, death, or environmental damage can also result in a loss of profits and loss of current as well as future business.

Recommended Risk-Reduction Measures

Company attorneys, safety and health professionals, and environmental specialists should be involved in the development of any procedures dealing with safety or emergency response. A company official should be pre-designated as a public relations officer with specific training in dealing with the press. Corporate plans and policies should be developed, approved, and implemented long before any need for such arises.

REFERENCES

American Conference of Governmental Industrial Hygienists. 1988. *Documentation of the Threshold Limit Values and Biological Exposure Indices,* 5th Edition (with updates). Cincinnati: ACGIH

Bueche, J. R. 1972. *Principles of Physics,* 2nd Edition. New York: McGraw-Hill.

Calabrese, E. J. 1994. *Biological Effects of Low Level Exposures.* New York: CRC Press/Lewis Publishers.

Cockerman, L. G., B. S. Shane. 1994. *Basic Environmental Toxicology.* New York: CRC Press/Lewis Publishers.

Danse, I. R. 1991. *Common Sense Toxics in the Workplace.* New York: Van Nostrand Reinhold.

Environmental Protection Agency. 1983. *Chemical Hazard Information Profiles.* Washington, D.C.: U.S. EPA.

Fodden, C. R., and J. L. Weddell. 1992. *Hazardous Materials Emergency Action Data.* New York: CRC Press/Lewis Publishers.

Forseberg, K., and L. H. Keith. 1993. *Chemical; Protective Clothing Permeation and Degradation Database.* Chelsea, MI: Lewis Publishers.

Gardner, R. W. 1994. *Chemical Intolerance.* New York: CRC Press/Lewis Publishers.

Hallenbeck, W. H. 1993. *Quantitative Risk Assessment for Environmental and Occupational Health,* 2nd Edition. New York: CRC Press/Lewis Publishers.

Jones, F. E. 1994. *Toxic Organic Vapors in the Workplace.* New York: CRC Press/Lewis Publishers.

Keith, L. H. and M. M. Walker. 1995. *Handbook of Air Toxics.* New York: CRC Press/Lewis Publishers.

Landis, W. G., M. H. Yu. 1995. *Introduction to Environmental Toxicology.* New York: CRC Press/Lewis Publishers.

Langerman, N. 1994 *Precautionary Labels for Chemical Containers.* New York: CRC Press/Lewis Publishers.

Lauwerys, R. R., P. Hoet. 1993. *Industrial Chemical Exposure.* New York: CRC Press/Lewis Publishers.

Lewis, R. J., Sr. 1992. *Sax's Dangerous Properties of Industrial Materials,* Volumes 1, 2, and 3, 8th Edition. New York: Van Nostrand Reinhold.

Lide, D. R. 1995. *Handbook of Chemistry and Physics,* 76th Edition. New York: CRC Press/Lewis Publishers.

Manahan, S. E. 1994. *Environmental Chemistry,* 6th Edition. New York: CRC Press/Lewis Publishers.

Mellström, G. A., J. E. Wahlberg, and H. I. Maibach. 1994. *Protective Gloves for Occupational Use.* New York: CRC Press/Lewis Publishers.

Meyer, E. 1990. *Chemistry of Hazardous Materials.* Englewood Cliffs, NJ: Prentice Hall.

Patnaik, P. 1992. *A Comprehensive Guide to the Hazardous Properties of Chemical Substances.* New York: Van Nostrand Reinhold.

Proctor, N. H., J. P. Hughes. 1991. *Chemical Hazards of the Workplace.* New York: Van Nostrand Reinhold.

Quigley, D. R. 1994. *Handbook of Emergency Chemical Management.* New York: CRC Press/Lewis Publishers.

Rea, W. J. 1995. *Chemical Sensitivity,* Vol. I-III. New York: CRC Press/Lewis Publishers.

Richardson, M. 1995. *Dictionary of Substance and Their Effects,* Vol. I-VI. New York: CRC Press/Lewis Publishers.

Sacarello, H. L. A. 1994. *The Comprehensive Handbook of Hazardous Materials.* New York: CRC Press/Lewis Publishers.

Sheftel, V. O. 1995. *Handbook of Toxic Properties of Monomers and Additives.* New York: CRC Press/Lewis Publishers.

Torkelson, T. R. V. K. Rowe. 1981. Halogenated aliphatic hydrocarbons. *Patty's Industrial Hygiene and Toxicology,* Vol. 2B, *Toxicology,* pp 3568-3572. New York: Wiley-Interscience.

U.S. Department of Health, Education, and Welfare, National Institute for Occupational Safety and Health. 1976. *Criteria for a Recommended Standard: Occupational Exposure to Cyclohexane.* DEHW (NIOSH) Publication Number: 77-210. Washington, D.C.: U.S. Government Printing Office.

U.S. Department of Health and Human Services, National Institute for Occupational Safety and Health. 1994. *NIOSH Pocket Guide to Chemical Hazards.* Washington, D.C.: U.S. Government Printing Office.

MATERIAL SAFETY DATA SHEET

CHEMICAL NAME	
METHYLCYCLOHEXANOL	**CRC PRESS/LEWIS PUBLISHERS DISCLAIMER:** THE INFORMATION AND RECOMMENDATIONS PRESENTED HEREIN ARE BASED ON SOURCES BELIEVED TO BE RELIABLE. CRC MAKES NO REPRESENTATION ON ITS COMPLETENESS OR ACCURACY. IT IS THE RESPONSIBILITY OF THE USER TO DETERMINE THE CHEMICAL'S SUITABILITY FOR ITS INTENDED USE, THE CHEMICAL'S SAFE USE, AND THE CHEMICAL'S PROPER DISPOSAL. NO REPRESENTATIONS AND/OR WARRANTIES, EITHER EXPRESSED OR IMPLIED, OF THE MERCHANTABILITY OR FITNESS FOR A PARTICULAR PURPOSE, OR OF ANY OTHER NATURE, ARE MADE WITH RESPECT TO THE INFORMATION PROVIDED IN THIS MSDS OR TO THE CHEMICAL TO WHICH INFORMATION MAY REFER. CRC NEITHER ASSUMES NOR AUTHORIZES ANY OTHER PERSON TO ASSUME FOR IT, ANY OTHER ADDITIONAL RESPONSIBILITY OR LIABILITY FOR THE USE OF, OR RELIANCE UPON, THIS INFORMATION.

HAZARD WARNING INFORMATION

HEALTH	FIRE	REACTIVE	OTHER	DEGREE OF HAZARD	COLOR CODING	OTHER CODES
2	2	0		0 = Minimum Hazard 1 = Slight Hazard 2 = Moderate Hazard 3 = Serious Hazard 4 = Severe Hazard	HEALTH = BLUE FIRE = RED REACTIVITY = YELLOW OTHER = WHITE	OX = Oxidizer ACID = Acid ALK = Alkali COR = Corrosive W = Use No Water

SECTION I - GENERAL INFORMATION

Characterization	RCRA Number	EPA Class
Alcohol	**None**	**Not Applicable**

DOT Proper Shipping Name **Methylcyclohexanol**	Chemical Abstract Service (CAS) Number **25639-42-3**
DOT Hazard Class and Label Requirements **Flammable or Combustible Liquid**	DOT Emergency Guide Code **26**
DOT Identification Number **UN 2617**	Chemical Formula **$CH_3C_6H_{10}OH$**

Synonyms
Hexahydrocresol; hexahydromethyl phenol; methyl hexalin.

SECTION II - HAZARDOUS INGREDIENTS AND IDENTITY INFORMATION

Hazardous Components (specific identity)	OSHA Exposure Criteria	NIOSH Exposure Criteria	Immediately Dangerous to Life and Health (IDLH)	ACGIH Exposure Criteria
Methylcyclohexanol (derivation: A mixture of three isomeric cyclic and secondary alcohols made by the hydrogenation of cresol; by catalytic oxidation of methylcyclohexane). **1 ppm = 4.75 mg/m³**	**PEL:** **100 ppm** **470 mg/m³** **STEL:** **Not Established**	**REL:** **50 ppm** **235 mg/m³** **STEL:** **Not Established**	**500 ppm**	**TLV:** **50 ppm** **235 mg/m³** **STEL:** **Not Established**

SECTION III - PHYSICAL AND CHEMICAL CHARACTERISTICS

Boiling Point **311-356°F (155-180°C)**	Specific Gravity (H_2O = 1) **0.92**
Vapor Pressure (gas) **2 at 86°F (8°C)**	Molecular Weight **114.2**
Vapor Density (Air = 1) **3.9**	Melting Point **-13°F (-25°C)**

Solubility
Slightly soluble in water (3-4%). Soluble in alcohol and ether.

Appearance and Odor
Straw-colored liquid with a weak odor like that of coconut oil.

SECTION IV - FIRE AND EXPLOSION HAZARD DATA

Flash Point (method used) **154°F (68°C) closed cup**	Explosive Limits in Air % by Volume **LEL: 2.4%** **UEL: Not Found**
NFPA Classification **Class IIIA Combustible Liquid**	**Autoignition Temperature** **565°F (296°C)**

Extinguishing Media
Use dry chemical, carbon dioxide, water fog, or alcohol foam.

Special Fire Fighting Procedures
Do not use solid stream of water on fire since it may actually spread the fire rather than extinguish it. Wear full protective clothing and self-contained breathing apparatus (SCBA). Poisonous gases are produced in fire. Use water spray to keep fire-exposed containers cool. Containers may explode in fire.

Unusual Fire and Explosion Hazards
Dangerous fire hazard when exposed to heat. Chemical will react on contact with oxidizable materials. Moderate explosion hazard when in the form of vapor.

SECTION V - REACTIVITY DATA

Stability		Conditions to Avoid Normally stable. Keep away from sources of heat (including direct sunlight) or flame. Large quantities should be stored in metal tanks or drums. Keep away from incompatible materials.
Stable X	Unstable	Incompatibility *(materials to avoid)* Strong oxidizers and oxidizing agents, such as chlorine, bromine, fluorine, permanganates, peroxides, nitrates, chlorates, and perchlorates.
Hazardous Polymerization		Conditions to Avoid Under normal conditions of temperature and pressure, hazardous polymerization of methylcyclohexanol is not expected to occur.
May Occur	Will Not Occur X	Hazardous Decomposition or By-products When heated to decomposition, methylcyclohexanol can emit highly toxic/poisonous and acrid fumes and gases, including oxides of carbon.

SECTION VI - HEALTH HAZARD DATA

Primary Route(s) of Entry:	Inhalation?	Absorption (skin)?	Ingestion?
	X	X	X

Health Hazards

INHALATION: Can produce human systemic effects, including conjunctivitis and changes in the olfactory and respiratory systems. Can damage the kidneys and liver. At high concentrations, it has narcotic-like action (headaches, dizziness). Also, lachrymation, convulsions loss of coordination, and possible effects on target organs.

ABSORPTION: Irritating to the skin, absorbed in toxic amounts. Eye contact may injure cornea.

INGESTION: Irritation and narcosis, including drowsiness, salivation, and liver damage.

Carcinogenicity	NTP Listed?	IARC Cancer Review Group?	OSHA Regulated?	Target Organs?
Unknown Human Unknown Animal	No	No	29 CFR 1910.1000 Table Z-1	Respiratory system, eyes, skin, liver, kidney, CNS.

Medical Conditions Generally Aggravated by Exposure

None reported. Skin conditions (dermatitis) or respiratory problems (asthma) may be aggravated.

Emergency and First-aid Procedures

Eye contact: Flush immediately with water for 15 minutes (minimum); seek medical attention. Skin contact: Remove all contaminated clothing. Immediately wash area with large amounts of soap and water. Seek medical assistance. For inhalation: Remove the person from exposure. Provide respiratory assistance and CPR, if required. Transfer to medical facility. Keep victim warm and at rest. If swallowed: Seek medical attention immediately. Do not induce vomiting unless directed to do so.

SECTION VII - PRECAUTIONS FOR SAFE HANDLING AND USE

Steps to be Taken in Case Material is Released or Spilled

Absorb large spills in vermiculite, dry sand, earth, or similar material and deposit in sealed containers. Ventilate area of spill or leak. Restrict those not involved in cleanup from entering area. Remove all sources of ignition.

Preferred Waste Disposal Method

Burn in chemical incinerator equipped with an afterburner and scrubber.

Precautions to be Taken in Handling and Storage

Store to avoid contact with incompatible materials in tightly closed containers in cool, well-ventilated area away from heat and flame. Use only non-sparking tools to open and close containers.

Other Precautions and Warnings

Sources of ignition are prohibited where methylcyclohexanol is used or stored. Heat may cause containers to build up pressure and explode. Containers must be grounded and bonded during transfer.

SECTION VIII - CONTROL MEASURES AND PERSONAL PROTECTIVE EQUIPMENT

Respiratory Protection *(specify type)*

For exposures above the PEL, use an MSHA/NIOSH-approved organic vapor respirator or a supplied-air respirator with full facepiece, hood, or helmet in continuous flow mode, or use a self-contained breathing apparatus (SCBA) operated in pressure demand or other positive pressure mode.

Ventilation

Local exhaust or general mechanical systems recommended.

Protective Gloves	Eye Protection	Other Protective Clothing
Impervious Material	Chemical Goggles or Face Mask	Protective Apron

Work/Hygiene Practices

Always wash hands thoroughly after using chemical; never bring food, drink, or smoking materials into vicinity of chemicals.

METHYLCYCLOHEXANOL

$CH_3C_6H_{10}OH$ CAS: 25639-42-3

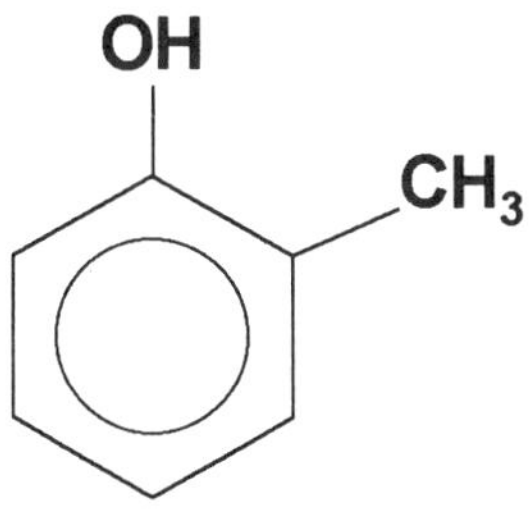

o-isomer

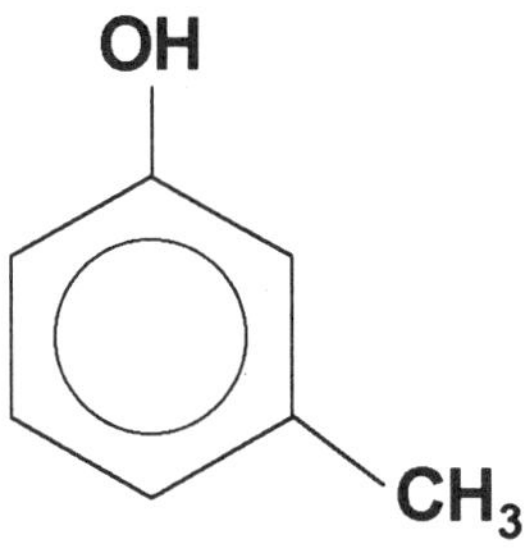

m-isomer

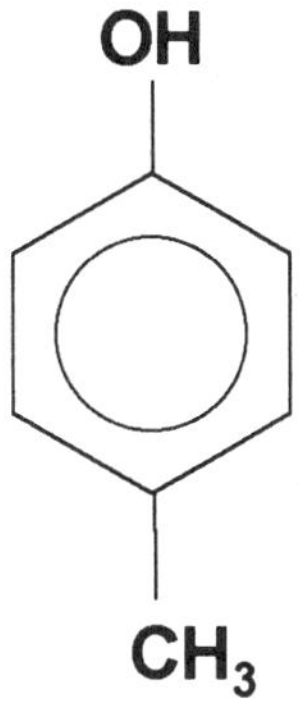

p-isomer

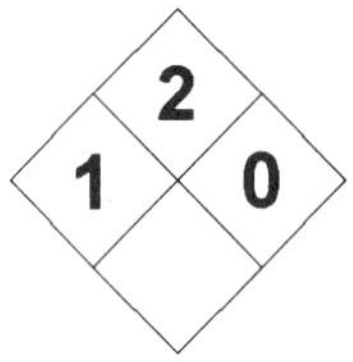

IDENTIFICATION AND TYPICAL USES

Methylcyclohexanol is a straw-colored liquid with a weak odor like that of coconut oil. It is used as a blending agent in textile soaps, as an antioxidant in lubricants, and as a solvent for lacquers.

RISK ASSESSMENT: HEALTH

General Assessment

Methylcyclohexanol is toxic by *ingestion* and *inhalation.* It will also pass through unbroken skin (*absorption*) to cause toxic systemic effects. Its carcinogenicity in humans or animals is not currently known. There are no reports of any carcinogenic or mutagenic effects resulting from exposure to methylcyclohexanol.

Inhalation results in lachrymation, irritation of the throat and upper respiratory tract, headaches, and burning cough. It will irritate the nose, throat, and eyes. Liver and kidney damage can also occur following high or prolonged exposures. Narcotic effects include pronounced dizziness and possible depression of the central nervous system.

Skin and eye contact may produce mild to moderate irritation on contact. It will pass through unbroken skin to cause toxic systemic effects similar to that of inhalation. Absorption through the skin occurring simultaneously with inhalation of methylcyclohexanol vapors may cause serious toxic reactions in the brain, liver, and kidneys.

If ingested, methylcyclohexanol is expected to produce symptoms of irritation and narcosis with salivation, lachrymation, and liver damage.

☠ *Acute Health Effects*

The following acute (short-term) health effects may occur immediately or shortly after exposure to methylcyclohexanol:

Skin: Moderate irritation with redness and possibly painful rash. Will pass through unbroken skin.

Eye: Severe irritation and possible corneal injury which may be irreversible depending upon length of contact and concentration of solution.

Lung: Nose and throat irritation. Produces intoxicating effects and may damage the lungs.

CNS: Unconfirmed effects by all routes of exposure. There may be tremor, dizziness, headache, and other unspecified effects.

Chronic Health Effects

The following chronic (long-term) health effects may occur at some time after exposure to methylcyclohexanol and can last for months or even years:

Cancer Hazards: According to information presented in the references, methylcyclohexanol has not been adequately tested for its ability to cause cancer in animals.

Reproductive Hazard: According to information presented in the references, methylcyclohexanol has not been adequately tested for its ability to cause adverse reproductive effects in test animals.

Other Chronic Effects: Prolonged skin contact may cause drying, cracking, and chapping of the affected area. Repeated or long-term exposures to methylcyclohexanol may affect the liver and kidneys.

⓿ *Recommended Risk-Reduction Measures*

Personnel should avoid direct contact with methylcyclohexanol. Although the acute toxicity of this chemical is believed to be relatively low, the exact nature of its chronic toxicity is not entirely understood in the references. Therefore, prudent risk management requires it be treated as though it posed a significant health risk in the event of overexposure. If a less toxic chemical cannot be substituted for a hazardous substance, then *engineering controls* are the most effective method of reducing exposures. The best protection is to enclose operations and/or provide local exhaust ventilation at the site of chemical release. While not always operationally feasible, isolating operations can also reduce exposure. Using respiratory protection is less effective than the controls mentioned above, but is still advisable whenever working with or around methylcyclohexanol. For exposures above the PEL, use an organic vapor respirator, a supplied-air respirator with full facepiece operated in continuous flow mode, or a self-contained breathing apparatus (SCBA) with full facepiece and pressure demand as the recommended respiratory protection methods of choice. If a full facepiece is not available, then chemical goggles should be worn to protect the eyes. Whenever a chemical splash hazard exists, a face shield and a protective apron should be worn. To prevent hand and skin exposures, protective gloves should be used. Glove manufacturers should be contacted and permeation studies obtained *before* final glove selections have been made.

Administrative controls should also be in place to minimize the potential for human exposures. These may include written procedures or policies which specify the methods and techniques that will be practiced whenever personnel are to work with methylcyclohexanol.

All personnel should receive *training* on the use, hazards, protective measures, emergency actions, and other precautions per 29 CFR 1910.1200 (Hazard Communication), prior to the first assignment in an area where methylcyclohexanol is used or stored.

If symptoms develop or overexposure is suspected, the following tests are recommended:

☑ Kidney and liver function tests.

☑ Evaluation for nervous system effects.

Any evaluation should include a careful history of past and present symptoms with an examination. Medical tests that simply look for existing damage are not a substitute for controlling exposures.

Other methods to reduce exposure include:

☑ Where possible, enclose operations and use local exhaust ventilation at the site of chemical release. If local exhaust ventilation or enclosure is not used, respiratory protection should be mandatory.

☑ Always ensure that proper protective clothing is worn when using chemical substances.

☑ Wash thoroughly immediately after exposure to methylcyclohexanol and at the end of the work shift or before eating, drinking, or smoking.

☑ Hazard warning information should be posted in the work area. In addition, as part of an on-going education and training program, all information on the health and safety hazards of methylcyclohexanol should be communicated to all potentially exposed workers.

☑ Eye wash stations should be provided in the immediate work area for emergency use. If there is a possibility of skin exposure to methylcyclohexanol, emergency shower facilities should also be provided in the immediate area.

☑ Personnel should never wear contaminated clothing home (family members can be exposed). Clothing should be laundered by personnel who have been trained on the hazards of methylcyclohexanol.

RISK ASSESSMENT: ENVIRONMENT

General Assessment

The environment is at risk of exposure during transportation, storage, disposal, or destruction of methylcyclohexanol. In almost every scenario, the threat of environmental exposure is contingent upon the proper handling of the chemical substance. Accidental spills, large or small, can result in fire, explosion, and possible contamination of the surrounding environmental mediums (water, soil, and air).

Methylcyclohexanol is considered a Class IIIA combustible liquid (per OSHA 29 CFR 1910.106). It presents a moderate fire hazard in contact with incompatible materials or in vapor form. These characteristics require extreme caution in handling, storage, transportation, and disposal. It is incompatible with common peroxides and other oxidizers and contact can cause fire or explosion. Therefore, special consideration is required during any emergency situation involving a leak or spill of methylcyclohexanol. Should methylcyclohexanol ever come into contact with these incompatible substances during use, transportation, storage, or disposal, the formation of highly toxic and/or highly explosive commodities (including hydrogen chloride) is extremely possible.

The proper disposal method for methylcyclohexanol is to burn it in a chemical incinerator equipped with an afterburner and scrubber.

Methylcyclohexanol may enter the environment through industrial discharges or spills.

Acute Ecological Effects

Acute (short-term) toxic effects may include the death of animals, birds, or fish and death or low growth rate in plants. Acute effects are seen 2 to 4 days after animals or plants are exposed to methylcyclohexanol.

Insufficient data are available on the short-term effects of methylcyclohexanol exposure to aquatic life, birds, plants, or land animals.

Chronic Ecological Effects

Chronic (long-term) toxic effects may include shortened life span, reproductive problems, lower fertility, and changes in appearance or behavior in exposed animals. These effects can be seen long after first exposure(s) to toxic chemicals.

Insufficient data are available on the long-term effects of methylcyclohexanol exposure to aquatic life, birds, plants, or land animals.

Water Solubility

Methylcyclohexanol is nearly insoluble in water. Concentrations of less than 100 milligrams may mix with a liter of water, depending on water pH and chemical concentration of methylcyclohexanol.

Persistence in the Environment

Methylcyclohexanol is non-persistent in water, with a half-life of less than 2 days. The half-life of a pollutant is the amount of time it takes for one half of the chemical to be degraded. Virtually 100% of methylcyclohexanol will eventually end up in air.

Bioaccumulation in Aquatic Organisms

Some substances increase in concentration, or *bioaccumulate,* in living organisms as they breathe contaminated air, drink contaminated water, or eat contaminated food. These chemicals can become concentrated in the tissues and internal organs of animals as well as humans.

The concentration of methylcyclohexanol found in fish tissue is expected to be about the same as the average concentration of methylcyclohexanol in the water from which the fish was taken.

Recommended Risk-Reduction Measures

Proper training of all transporters will reduce the likelihood of a mishap or accident resulting in a leak or spill to the environment. The correct use of DOT labeling (per the MSDS) on all containers, trucks, and rail cars will enable emergency responders to react properly and quickly to any disaster thereby reducing the potential risk to the environment and to personnel.

Storage of methylcyclohexanol should be segregated from incompatible chemicals to minimize the risk of cross-contamination or contact. Buildings designated for storage should be equipped with appropriate fire protection systems (alarms, sprinklers, emergency lighting, portable extinguishers).

If a spill or leak to the environment has occurred, fire department, emergency response and/or hazardous materials spill personnel should be notified immediately. Cleanup should be attempted only by those trained in proper spill containment procedures using non-sparking tools. Contaminated soils should be removed for incineration and replaced with clean soil. If methylcyclohexanol should contact the water table, aquifer, or navigable waterway, cleanup should be started immediately. It is only slightly soluble in water and total remediation may be possible. The local and/or state emergency response authorities must be notified. A comprehensive emergency response or disaster preparedness/recovery plan should be in place prior to any operations involving the use, transportation, storage, or disposal of methylcyclohexanol. If methylcyclohexanol is spilled or leaked, the following specific steps are recommended.

- ☑ Restrict persons not wearing protective clothing from area of spill or leak until cleanup is complete and area can be opened for normal work.
- ☑ Remove all ignition sources.
- ☑ Ventilate area of spill or leak. If in confined area, maximum explosion-proof ventilation system may be required.
- ☑ Absorb liquids in vermiculite, dry sand, earth, or similar material and deposit in sealed containers. Use non-sparking tools.

☑ It may be necessary to dispose of methylcyclohexanol as a hazardous waste. The responsible state agency or the regional office of the federal Environmental Protection Agency (EPA) should be contacted for specific recommendations.

RISK ASSESSMENT: BUSINESS

General Assessment

Accidents or mishaps involving methylcyclohexanol can present a significant threat to business operations. The loss or damage of equipment or facilities can significantly affect fiscal viability. Lawsuits that may result from personnel injury/death, public exposures, and/or environmental contamination will also require a serious expenditure of resources. Media attention surrounding an injury, death, or environmental damage can also result in a loss of profits and loss of current as well as future business.

Recommended Risk-Reduction Measures

Company attorneys, safety and health professionals, and environmental specialists should be involved in the development of any procedures dealing with safety or emergency response. A company official should be pre-designated as a public relations officer with specific training in dealing with the press. Corporate plans and policies should be developed, approved, and implemented long before any need for such arises.

REFERENCES

American Conference of Governmental Industrial Hygienists. 1988. *Documentation of the Threshold Limit Values and Biological Exposure Indices,* 5th Edition (with updates). Cincinnati: ACGIH

Bueche, J. R. 1972. *Principles of Physics,* 2nd Edition. New York: McGraw-Hill.

Calabrese, E. J. 1994. *Biological Effects of Low Level Exposures.* New York: CRC Press/Lewis Publishers.

Cockerman, L. G., B. S. Shane. 1994. *Basic Environmental Toxicology.* New York: CRC Press/Lewis Publishers.

Gardner, R. W. 1994. *Chemical Intolerance.* New York: CRC Press/Lewis Publishers.

Hallenbeck, W. H. 1993. *Quantitative Risk Assessment for Environmental and Occupational Health,* 2nd Edition. New York: CRC Press/Lewis Publishers.

Jones, F. E. 1994. *Toxic Organic Vapors in the Workplace.* New York: CRC Press/Lewis Publishers.

Keith, L. H. and M. M. Walker. 1995. *Handbook of Air Toxics.* New York: CRC Press/Lewis Publishers.

Landis, W. G., M. H. Yu. 1995. *Introduction to Environmental Toxicology.* New York: CRC Press/Lewis Publishers.

Langerman, N. 1994 *Precautionary Labels for Chemical Containers.* New York: CRC Press/Lewis Publishers.

Lauwerys, R. R., P. Hoet. 1993. *Industrial Chemical Exposure.* New York: CRC Press/Lewis Publishers.

Lewis, R. J., Sr. 1992. *Sax's Dangerous Properties of Industrial Materials,* Volumes 1, 2, and 3, 8th Edition. New York: Van Nostrand Reinhold.

Lide, D. R. 1995. *Handbook of Chemistry and Physics,* 76th Edition. New York: CRC Press/Lewis Publishers.

Manahan, S. E. 1994. *Environmental Chemistry,* 6th Edition. New York: CRC Press/Lewis Publishers.

Mellström, G. A., J. E. Wahlberg, and H. I. Maibach. 1994. *Protective Gloves for Occupational Use.* New York: CRC Press/Lewis Publishers.

Meyer, E. 1990. *Chemistry of Hazardous Materials.* Englewood Cliffs, NJ: Prentice Hall.

Patnaik, P. 1992. *A Comprehensive Guide to the Hazardous Properties of Chemical Substances.* New York: Van Nostrand Reinhold.

Proctor, N. H., J. P. Hughes. 1991. *Chemical Hazards of the Workplace.* New York: Van Nostrand Reinhold.

Quigley, D. R. 1994. *Handbook of Emergency Chemical Management.* New York: CRC Press/Lewis Publishers.

Rea, W. J. 1995. *Chemical Sensitivity,* Vol. I-III. New York: CRC Press/Lewis Publishers.

Richardson, M. 1995. *Dictionary of Substance and Their Effects,* Vol. I-VI. New York: CRC Press/Lewis Publishers.

Sacarello, H. L. A. 1994. *The Comprehensive Handbook of Hazardous Materials.* New York: CRC Press/Lewis Publishers.

Sheftel, V. O. 1995. *Handbook of Toxic Properties of Monomers and Additives.* New York: CRC Press/Lewis Publishers.

Torkelson, T. R. V. K. Rowe. 1981. Halogenated aliphatic hydrocarbons. *Patty's Industrial Hygiene and Toxicology,* Vol. 2B, *Toxicology,* pp 3568-3572. New York: Wiley-Interscience.

U.S. Department of Health and Human Services, National Institute for Occupational Safety and Health. 1994. *NIOSH Pocket Guide to Chemical Hazards.* Washington, D.C.: U.S. Government Printing Office.

MATERIAL SAFETY DATA SHEET

CHEMICAL NAME

o-METHYLCYCLOHEXANONE

HAZARD WARNING INFORMATION

HEALTH	FIRE	REACTIVE	OTHER	DEGREE OF HAZARD	COLOR CODING	OTHER CODES
1	2	0		0 = Minimum Hazard 1 = Slight Hazard 2 = Moderate Hazard 3 = Serious Hazard 4 = Severe Hazard	HEALTH = BLUE FIRE = RED REACTIVITY = YELLOW OTHER = WHITE	OX = Oxidizer ACID = Acid ALK = Alkali COR = Corrosive W = Use No Water

SECTION I - GENERAL INFORMATION

Characterization	**Ketone**	RCRA Number	**D001**
		EPA Class	**Characteristic (I) Waste**
DOT Proper Shipping Name	**Methyl cyclohexanone**	Chemical Abstract Service (CAS) Number	**583-60-8**
DOT Hazard Class and Label Requirements	**Flammable Liquid**	DOT Emergency Guide Code	**26**
DOT Identification Number	**UN 2297**	Chemical Formula	**$CH_3C_6H_9O$**

Synonyms

2-Methylcyclohexanone; α-methylcyclohexanone; 2-methyl-1-cyclohexanone; methylanon; Sexton-B.

SECTION II - HAZARDOUS INGREDIENTS AND IDENTITY INFORMATION

Hazardous Components (specific identity)	OSHA Exposure Criteria	NIOSH Exposure Criteria	Immediately Dangerous to Life and Health (IDLH)	ACGIH Exposure Criteria
***o*-Methylcyclohexanone** (derivation: By high-temperature, catalytic hydrogenation of cresols or by dehydrogenation of meythylcyclohexanol). **1 ppm = 4.66 mg/m³**	**PEL (skin):** **100 ppm** **460 mg/m³** **STEL:** **Not Established**	**REL (skin):** **50 ppm** **230 mg/m³** **STEL:** **75 ppm** **3450 mg/m³**	**600ppm**	**TLV (skin):** **50 ppm** **229 mg/m³** **STEL:** **75 ppm** **344 mg/m³**

SECTION III - PHYSICAL AND CHEMICAL CHARACTERISTICS

Boiling Point	**325°F (163°C)**	Specific Gravity (H_2O = 1)	**0.93**
Vapor Pressure (gas)	**1 at 69°F (20°C)**	Molecular Weight	**112.2**
Vapor Density (Air = 1)	**Not Found**	Melting Point	**7°F (-14°C)**

Solubility

Insoluble in water. Soluble in alcohol, ether, and diethyl ether.

Appearance and Odor

Colorless, water-white, or pale yellow oily liquid with an odor of peppermint.

SECTION IV - FIRE AND EXPLOSION HAZARD DATA

Flash Point (method used)	**118°F (48°C) closed cup**	Explosive Limits in Air % by Volume	**LEL: Not Determined UEL: Not Determined**
NFPA Classification	**Class II Combustible Liquid**	Autoignition Temperature	**Not Found**

Extinguishing Media

Use dry chemical, carbon dioxide, water spray, or alcohol foam (best for large fires).

Special Fire Fighting Procedures

Wear full protective clothing and self-contained breathing apparatus (SCBA). Poisonous gases are produced in fire. Use water spray to keep fire-exposed containers cool. Containers may explode in fire.

Unusual Fire and Explosion Hazards

Combustible liquid but a dangerous fire hazard when exposed to heat. At high temperatures, forms explosive mixtures with air. Move containers from fire if it can be done without risk. Chemical will react on contact with oxidizable materials. Moderate explosion hazard when in the form of vapor.

SECTION V - REACTIVITY DATA

Stability		Conditions to Avoid **Normally stable. Keep away from sources of heat (including direct sunlight) or flame. Large quantities should be stored in metal tanks or drums. Keep away from incompatible materials.**
Stable **X**	Unstable	Incompatibility *(materials to avoid)* **Strong oxidizers (permanganates, peroxides, nitrates, chlorates, and perchlorates).**
Hazardous Polymerization		Conditions to Avoid **Under normal conditions of temperature and pressure, hazardous polymerization of *o*-methylcyclohexanone is not expected to occur.**
May Occur	Will Not Occur **X**	Hazardous Decomposition or By-products **When heated to decomposition, *o*-methylcyclohexanone can emit highly toxic/poisonous and acrid fumes and gases, including oxides of carbon.**

SECTION VI - HEALTH HAZARD DATA

Primary Route(s) of Entry:	Inhalation? X	Absorption (skin)? X	Ingestion? X

Health Hazards

INHALATION: Symptoms include salivation, hypothermia, decreased breathing rate, and anesthesia. May also cause central nervous system depression with lung congestion. There can be slight kidney and liver damage with mild narcosis.

ABSORPTION: Irritating to the skin; may cause irritation and dermatitis. An eye irritant with possibility of corneal damage.

INGESTION: Irritation and narcosis, anesthesia, and respiratory difficulties.

Carcinogenicity	NTP Listed?	IARC Cancer Review Group?	OSHA Regulated?	Target Organs?
Unknown Human Unknown Animal	No	No	29 CFR 1910.1000 Table Z-1	Respiratory system, eyes, skin, liver, kidney, CNS.

Medical Conditions Generally Aggravated by Exposure

None reported. Skin conditions (dermatitis) or respiratory problems (asthma) may be aggravated.

Emergency and First-aid Procedures

Eye contact: Flush immediately with water for 15 minutes (minimum); seek medical attention. Skin contact: Remove all contaminated clothing. Immediately wash area with large amounts of soap and water. Seek medical assistance. For inhalation: Remove the person from exposure. Provide respiratory assistance and CPR, if required. Transfer to medical facility. Keep victim warm and at rest. If swallowed: Seek medical attention immediately. Give 1 to 2 glasses of water, then induce vomiting. Never give anything by mouth to an unconscious or convulsing person.

SECTION VII - PRECAUTIONS FOR SAFE HANDLING AND USE

Steps to be Taken in Case Material is Released or Spilled

Absorb in vermiculite, earth, or similar material and deposit in sealed containers. Ventilate area of spill or leak. Restrict those not involved in cleanup from entering area. Remove all sources of ignition.

Preferred Waste Disposal Method

Burn in chemical incinerator equipped with an afterburner and scrubber.

Precautions to be Taken in Handling and Storage

Store to avoid contact with incompatible materials in tightly closed containers in cool, well-ventilated area away from heat and flame. Use only non-sparking tools to open and close containers.

Other Precautions and Warnings

Sources of ignition are prohibited where *o*-methylcyclohexanone is used, handled, or stored. Heat may cause containers to build up pressure and explode. Ground and bond containers during transfer.

SECTION VIII - CONTROL MEASURES AND PERSONAL PROTECTIVE EQUIPMENT

Respiratory Protection *(specify type)*

For exposures above the PEL, use an MSHA/NIOSH-approved organic vapor full facepiece respirator. Better protection is obtained from a supplied-air respirator with full facepiece, hood, or helmet in continuous flow mode, or use a self-contained breathing apparatus (SCBA) operated in pressure demand or other positive pressure mode.

Ventilation

Local exhaust or general mechanical systems recommended.

Protective Gloves	Eye Protection	Other Protective Clothing
Impervious Material	Chemical Goggles or Face Mask	Protective Apron

Work/Hygiene Practices

Always wash hands thoroughly after using chemical; never bring food, drink, or smoking materials into vicinity of chemicals.

o-METHYLCYCLOHEXANONE

$CH_3C_6H_9O$

CAS: 583-60-8

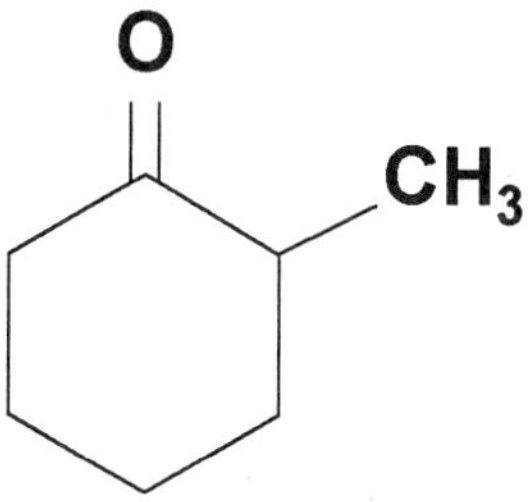

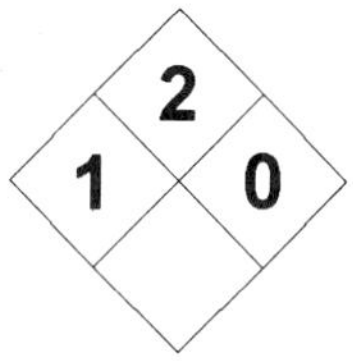

IDENTIFICATION AND TYPICAL USES

o-Methylcyclohexanone is a colorless, water-white, to pale yellow, oily liquid with a peppermint-like odor. It is used as a solvent for varnishes, lacquers, and plastics. It is also used in the leather industry and as a rust remover.

RISK ASSESSMENT: HEALTH

General Assessment

o-Methylcyclohexanone is moderately toxic by *ingestion* and *inhalation*. It will also pass through unbroken skin (*absorption*) to cause toxic systemic effects. Its carcinogenicity in humans and animals is not known. There are no reports in the references on any mutagenic or teratogenic effects in humans or animals resulting from exposure to *o*-methylcyclohexanone

Inhalation will irritate the nose, throat, and eyes. Mild narcotic properties have also been recorded, including pronounced headache and may cause depression of the central nervous system with symptoms such as headache, dizziness, tiredness, tremor, and loss of consciousness. Kidney and liver damage with mild narcosis may also be possible.

Skin and eye contact may produce mild to moderate irritation. There are also reports of skin absorption occurring on exposure to *o*-methylcyclohexanone. Pure liquid in contact with the eye may result in corneal damage. If ingested, *o*-methylcyclohexanone will produce unspecified gastrointestinal effects.

Acute Health Effects

The following acute (short-term) health effects may occur immediately or shortly after exposure to *o*-methylcyclohexanone:

Skin: Will defat the skin and cause irritation and dermatitis. Absorption is likely.

Eye: Severe irritation with a potential for corneal injury or damage.

Lung: Nose and throat irritation. Produces intoxicating effects and may cause lung effects.

CNS: Unconfirmed effects by all routes of exposure. There may be depression, anesthesia, dizziness, headache, and other unspecified effects.

Chronic Health Effects

The following chronic (long-term) health effects may occur at some time after exposure to *o*-methylcyclohexanone and can last for months or even years:

Cancer Hazards: According to information presented in the references, *o*-methylcyclohexanone has not been adequately tested for its ability to cause cancer in animals.

Reproductive Hazard: According to information presented in the references, *o*-methylcyclohexanone has not been adequately tested for its ability to adversely affect reproduction in test animals.

Other Chronic Effects: Prolonged skin contact may cause drying, cracking, and chapping of the affected area. There may be skin sensitization and/or dermatitis in some individuals. Repeated or long-term exposures to *o*-methylcyclohexanone may affect the liver and kidneys.

Recommended Risk-Reduction Measures

Personnel should avoid direct contact with *o*-methylcyclohexanone. The exact nature of the toxicity of this chemical is not entirely understood in the references. Therefore, prudent risk management requires it be treated as though it posed a significant health risk in the event of overexposure. If a less toxic chemical cannot be substituted for a hazardous substance, then *engineering controls* are the most effective method of reducing exposures. The best protection is to enclose operations and/or provide local exhaust ventilation at the site of chemical release. While not always operationally feasible, isolating operations can also reduce exposure. Using respiratory protection is less effective than the controls mentioned above, but is still advisable whenever working with or around chemicals such as *o*-methylcyclohexanone. For exposures above the PEL, a chemical cartridge respirator with a full facepiece, or a supplied-air respirator with full facepiece operated in continuous flow mode, or a self-contained breathing apparatus (SCBA) with full facepiece and pressure demand are the recommended respiratory protection methods of choice. If a full facepiece is not available, then chemical goggles should be worn to protect the eyes. Whenever a chemical splash hazard exists, a face shield and a rubber apron should be worn. To prevent hand and skin exposures, impervious gloves should be worn. Glove manufacturers should be contacted to obtain permeation studies *before* glove selection has been made.

Administrative controls should also be in place to minimize the potential for human exposures. These may include written procedures or policies which specify the methods and techniques that will be practiced whenever personnel are to work with or around *o*-methylcyclohexanone.

All personnel should receive *training* on the use, hazards, protective measures, emergency actions, and other precautions per 29 CFR 1910.1200 (Hazard Communication), prior to the first assignment in an area where *o*-methylcyclohexanone is used or stored.

If symptoms develop or overexposure is suspected, the following tests are recommended:

- ☑ Kidney and liver function tests.
- ☑ Evaluation by a qualified allergist, with careful consideration of exposure history and special skin tests (may help diagnose allergy).
- ☑ Lung function tests, including forced vital capacity (FVC).

Any evaluation should include a careful history of past and present symptoms with an examination. Medical tests that simply look for existing damage are not a substitute for controlling exposures. Also, since smoking can cause heart disease, emphysema, and other respiratory disorders, symptoms of exposure to *o*-methylcyclohexanone may be more pronounced than those displayed by non-smokers. Proper risk assessment requires careful consideration of *all* possible factors that may be causing the appearance of symptoms in exposed workers.

Other methods to reduce exposure include:

- ☑ Where possible, enclose operations and use local exhaust ventilation at the site of chemical release. If local exhaust ventilation or enclosure is not used, respiratory protection should be mandatory. Also, if possible, automatic transfer of *o*-methylcyclohexanone from storage containers to process or work containers is recommended.
- ☑ Always ensure that proper protective clothing is worn when using chemical substances.
- ☑ Wash thoroughly immediately after exposure to *o*-methylcyclohexanone and at the end of the work shift or before eating, drinking, or smoking.
- ☑ Hazard warning information should be posted in the work area. In addition, as part of an on-going education and training program, all information on the health and safety hazards of *o*-methylcyclohexanone should be communicated to all potentially exposed workers.
- ☑ Eye wash stations should be provided in the immediate work area for emergency use. If there is a possibility of skin exposure to *o*-methylcyclohexanone, emergency shower facilities should also be provided in the immediate area.
- ☑ Personnel should never wear contaminated clothing home (family members can be exposed). Clothing should be laundered by personnel who have been trained on the hazards of *o*-methylcyclohexanone.

RISK ASSESSMENT: ENVIRONMENT

General Assessment

The environment is at risk of exposure during transportation, storage, disposal, or destruction of *o*-methylcyclohexanone. In almost every scenario, the threat of environmental exposure is contingent upon the proper handling of the chemical substance. Accidental spills, large or small, can result in fire, explosion, and possible contamination of the surrounding environmental mediums (water, soil, and air).

o-Methylcyclohexanone is considered a Class II combustible liquid (per OSHA 29 CFR 1910.106). It presents a moderate fire and explosion hazard risk, especially in contact with incompatible materials or when its vapors mix with air. These characteristics require extreme caution in handling, storage, transportation, and disposal. It is incompatible with common peroxides and other oxidizers and contact can cause fire or explosion. Therefore, special consideration is required during any emergency situation involving a leak or spill of *o*-methylcyclohexanone. Should this chemical ever come into contact with these incompatible substances during use, transportation, storage, or disposal, the formation of highly toxic and/or highly explosive commodities (including hydrogen chloride) is also extremely possible.

The proper disposal method for *o*-methylcyclohexanone is to burn it in a chemical incinerator equipped with an afterburner and scrubber.

o-Methylcyclohexanone may enter the environment through industrial discharges or spills.

Acute Ecological Effects

Acute (short-term) toxic effects may include the death of animals, birds, or fish and death or low growth rate in plants. Acute effects are seen 2 to 4 days after animals or plants are exposed to *o*-methylcyclohexanone.

Insufficient data are available on the short-term effects of *o*-methylcyclohexanone exposure to aquatic life, birds, plants, or land animals.

Chronic Ecological Effects

Chronic (long-term) toxic effects may include shortened life span, reproductive problems, lower fertility, and changes in appearance or behavior in exposed animals. These effects can be seen long after first exposure(s) to toxic chemicals.

Insufficient data are available on the long-term effects of *o*-methylcyclohexanone exposure to aquatic life, birds, plants, or land animals.

Water Solubility

o-Methylcyclohexanone is insoluble in water. Concentrations of less than 1 milligram will not mix with a liter of water.

Persistence in the Environment

No information is available in the references on the persistence of *o*-methylcyclohexanone in the environment.

Bioaccumulation in Aquatic Organisms

Some substances increase in concentration, or *bioaccumulate,* in living organisms as they breathe contaminated air, drink contaminated water, or eat contaminated food. These chemicals can become concentrated in the tissues and internal organs of animals as well as humans.

No information is available in the references on the bioaccumulation of *o*-methylcyclohexanone in aquatic organisms.

Recommended Risk-Reduction Measures

Proper training of all transporters will reduce the likelihood of a mishap or accident resulting in a leak or spill to the environment. The correct use of DOT labeling (per the MSDS) on all containers, trucks, and rail cars will enable emergency responders to react properly and quickly to any disaster thereby reducing the potential risk to the environment and to personnel.

Storage of *o*-methylcyclohexanone should be segregated from incompatible chemicals to minimize the risk of cross-contamination or contact. Buildings designated for storage should be equipped with appropriate fire protection systems (alarms, sprinklers, emergency lighting, portable extinguishers).

If a spill or leak to the environment has occurred, fire department, emergency response and/or hazardous materials spill personnel should be notified immediately. Cleanup should be attempted only by those trained in proper spill containment procedures using non-sparking tools. Contaminated soils should be removed for incineration and replaced with clean soil. If *o*-methylcyclohexanone should contact the water table, aquifer, or navigable waterway, cleanup should be started immediately. It is insoluble in water (but highly volatile) and partial remediation may be possible. The local and/or state emergency response authorities must be notified. A comprehensive emergency response or disaster preparedness/recovery plan should be in place prior to any operations involving the use, transportation, storage, or disposal of *o*-methylcyclohexanone. If *o*-methylcyclohexanone is spilled or leaked, the following specific steps are recommended:

- ☑ Restrict persons not wearing protective clothing from area of spill or leak until cleanup is complete and area can be opened for normal work.
- ☑ Remove all ignition sources, ventilate area of leak.
- ☑ Absorb liquids in vermiculite, dry sand, earth, or similar material and deposit in sealed containers.

☑ It may be necessary to dispose of *o*-methylcyclohexanone as a hazardous waste. The responsible state agency or the regional office of the federal Environmental Protection Agency (EPA) should be contacted for specific recommendations.

RISK ASSESSMENT: BUSINESS

General Assessment

Accidents or mishaps involving *o*-methylcyclohexanone can present a significant threat to business operations. The loss or damage of equipment or facilities can significantly affect fiscal viability. Lawsuits that may result from personnel injury/death, public exposures, and/or environmental contamination will also require a serious expenditure of resources. Media attention surrounding an injury, death, or environmental damage can also result in a loss of profits and loss of current as well as future business.

Recommended Risk-Reduction Measures

Company attorneys, safety and health professionals, and environmental specialists should be involved in the development of any procedures dealing with safety or emergency response. A company official should be pre-designated as a public relations officer with specific training in dealing with the press. Corporate plans and policies should be developed, approved, and implemented long before any need for such arises.

REFERENCES

American Conference of Governmental Industrial Hygienists. 1988. *Documentation of the Threshold Limit Values and Biological Exposure Indices,* 5th Edition (with updates). Cincinnati: ACGIH

Bueche, J. R. 1972. *Principles of Physics,* 2nd Edition. New York: McGraw-Hill.

Calabrese, E. J. 1994. *Biological Effects of Low Level Exposures.* New York: CRC Press/Lewis Publishers.

Cockerman, L. G., B. S. Shane. 1994. *Basic Environmental Toxicology.* New York: CRC Press/Lewis Publishers.

Cothern, C. R. 1995. *Handbook for Environmental Risk Decisionmaking.* New York: CRC Press/Lewis Publishers.

Danse, I. R. 1991. *Common Sense Toxics in the Workplace.* New York: Van Nostrand Reinhold.

Environmental Protection Agency. 1983. *Chemical Hazard Information Profiles.* Washington, D.C.: U.S. EPA.

Fodden, C. R., and J. L. Weddell. 1992. *Hazardous Materials Emergency Action Data.* New York: CRC Press/Lewis Publishers.

Forseberg, K., and L. H. Keith. 1993. *Chemical; Protective Clothing Permeation and Degradation Database.* Chelsea, MI: Lewis Publishers.

Gardner, R. W. 1994. *Chemical Intolerance.* New York: CRC Press/Lewis Publishers.

Hallenbeck, W. H. 1993. *Quantitative Risk Assessment for Environmental and Occupational Health,* 2nd Edition. New York: CRC Press/Lewis Publishers.

Keith, L. H. and M. M. Walker. 1995. *Handbook of Air Toxics.* New York: CRC Press/Lewis Publishers.

Langerman, N. 1994 *Precautionary Labels for Chemical Containers.* New York: CRC Press/Lewis Publishers.

Lauwerys, R. R., P. Hoet. 1993. *Industrial Chemical Exposure.* New York: CRC Press/Lewis Publishers.

Lewis, R. J., Sr. 1992. *Sax's Dangerous Properties of Industrial Materials,* Volumes 1, 2, and 3, 8th Edition. New York: Van Nostrand Reinhold.

Meyer, E. 1990. *Chemistry of Hazardous Materials.* Englewood Cliffs, NJ: Prentice Hall.

Patnaik, P. 1992. *A Comprehensive Guide to the Hazardous Properties of Chemical Substances.* New York: Van Nostrand Reinhold.

Proctor, N. H., J. P. Hughes. 1991. *Chemical Hazards of the Workplace.* New York: Van Nostrand Reinhold.

Quigley, D. R. 1994. *Handbook of Emergency Chemical Management.* New York: CRC Press/Lewis Publishers.

Rea, W. J. 1995. *Chemical Sensitivity,* Vol. I-III. New York: CRC Press/Lewis Publishers.

Richardson, M. 1995. *Dictionary of Substance and Their Effects,* Vol. I-VI. New York: CRC Press/Lewis Publishers.

Sacarello, H. L. A. 1994. *The Comprehensive Handbook of Hazardous Materials.* New York: CRC Press/Lewis Publishers.

Sheftel, V. O. 1995. *Handbook of Toxic Properties of Monomers and Additives.* New York: CRC Press/Lewis Publishers.

Torkelson, T. R. V. K. Rowe. 1981. Halogenated aliphatic hydrocarbons. *Patty's Industrial Hygiene and Toxicology,* Vol. 2B, *Toxicology,* pp 3568-3572. New York: Wiley-Interscience.

U.S. Department of Health and Human Services, National Institute for Occupational Safety and Health. 1994. *NIOSH Pocket Guide to Chemical Hazards.* Washington, D.C.: U.S. Government Printing Office.

MATERIAL SAFETY DATA SHEET

CHEMICAL NAME	
METHYLENE BROMIDE	**CRC PRESS/LEWIS PUBLISHERS DISCLAIMER:** THE INFORMATION AND RECOMMENDATIONS PRESENTED HEREIN ARE BASED ON SOURCES BELIEVED TO BE RELIABLE. CRC MAKES NO REPRESENTATION ON ITS COMPLETENESS OR ACCURACY. IT IS THE RESPONSIBILITY OF THE USER TO DETERMINE THE CHEMICAL'S SUITABILITY FOR ITS INTENDED USE, THE CHEMICAL'S SAFE USE, AND THE CHEMICAL'S PROPER DISPOSAL. NO REPRESENTATIONS AND/OR WARRANTIES, EITHER EXPRESSED OR IMPLIED, OF THE MERCHANTABILITY OR FITNESS FOR A PARTICULAR PURPOSE, OR OF ANY OTHER NATURE, ARE MADE WITH RESPECT TO THE INFORMATION PROVIDED IN THIS MSDS OR TO THE CHEMICAL TO WHICH INFORMATION MAY REFER. CRC NEITHER ASSUMES NOR AUTHORIZES ANY OTHER PERSON TO ASSUME FOR IT, ANY OTHER ADDITIONAL RESPONSIBILITY OR LIABILITY FOR THE USE OF, OR RELIANCE UPON, THIS INFORMATION.

HAZARD WARNING INFORMATION

HEALTH	FIRE	REACTIVE	OTHER	DEGREE OF HAZARD	COLOR CODING	OTHER CODES
3	0	0		0 = Minimum Hazard 1 = Slight Hazard 2 = Moderate Hazard 3 = Serious Hazard 4 = Severe Hazard	HEALTH = BLUE FIRE = RED REACTIVITY = YELLOW OTHER = WHITE	OX = Oxidizer ACID = Acid ALK = Alkali COR = Corrosive W = Use No Water

SECTION I - GENERAL INFORMATION

Characterization **Halogenated Hydrocarbon**	RCRA Number **U068**	EPA Class **Toxic Waste**
DOT Proper Shipping Name **Methylene bromide**	Chemical Abstract Service (CAS) Number **74-95-3**	
DOT Hazard Class and Label Requirements **Poison B: St. Andrew's Cross**	DOT Emergency Guide Code **55**	
DOT Identification Number **UN 2664**	Molecular Formula CH_2Br_2	

Synonyms
Dibromomethane; methylene dibromide; methylene bromide.

SECTION II - HAZARDOUS INGREDIENTS AND IDENTITY INFORMATION

Hazardous Components (specific identity)	OSHA Exposure Criteria	NIOSH Exposure Criteria	Immediately Dangerous to Life and Health (IDLH)	ACGIH Exposure Criteria
Methylene bromide (derivation: By action of bromine on methyl bromide and subsequent distillation).	**PEL:** Not Established **STEL** Not Established	**REL:** Not Established **STEL** Not Established	**Not Determined**	**TLV:** Not Established **STEL** Not Established

SECTION III - PHYSICAL AND CHEMICAL CHARACTERISTICS

Boiling Point **203°F (95°C)**	Specific Gravity (H_2O = 1) **2.485 at 77°F (25°C)**
Vapor Pressure (mm Hg) **48 at 72°F (22°C)**	Molecular Weight (atomic weight) **173.85**
Vapor Density (Air = 1) **6.05**	Freezing Point **122°F (<50°C)**

Solubility
Slightly soluble in water (0.4%). Miscible with alcohol, ether, chloroform, acetone, and most other organic solvents.

Appearance and Odor
Colorless, heavy, viscous liquid (or solid below 50°C) with a sweet, pleasant, and agreeable chloroform-like odor.

SECTION IV - FIRE AND EXPLOSION HAZARD DATA

Flash Point (method used) **Not Applicable**	Explosive Limits in Air % by Volume **LEL: Not Applicable UEL: Not Applicable**
NFPA Classification **Not Applicable**	Autoignition Temperature **Not Applicable**

Extinguishing Media
Methylene bromide may burn, but not readily. Use extinguishing agent suitable to surrounding fire.

Special Fire Fighting Procedures
Poisonous gases are produced in fire. Wear self-contained breathing apparatus (SCBA) and full protective gear. Move containers from fire if it can be done without risk. Cool sides of fire-exposed containers with water until long after fire is out. Reacts with potassium.

Unusual Fire and Explosion Hazards
None reported. Methylene bromide is a non-flammable and non-combustible liquid.

SECTION V - REACTIVITY DATA

Stability		Conditions to Avoid **Methylene bromide is stable under normal conditions of handling and storage. Avoid exposing methylene bromide to heat or light in the presence of moisture (it may slowly hydrolyze to produce hydrogen bromide).**
Stable **X**	Unstable	Incompatibility *(materials to avoid)* **High heat sources and potassium (may form mixtures with potassium that can explode on light impact or the introduction of friction).**
Hazardous Polymerization		Conditions to Avoid **Under normal conditions of temperature and pressure, hazardous polymerization of methylene bromide cannot occur.**
May Occur	Will Not Occur **X**	Hazardous Decomposition or By-products **Poisonous gases and acrid fumes are produced in fire, including extremely toxic fumes of bromine and bromide.**

SECTION VI - HEALTH HAZARD DATA

Primary Route(s) of Entry:	Inhalation?	Absorption (skin/eye)?	Ingestion?
	X		X

Health Hazards

INHALATION: **Depression of the central nervous system (CNS) with general weakness, headache, loss of appetite and subsequent weight loss, depression, vomiting, diarrhea, chest pain, cough, shortness of breath, abdominal pain, unconsciousness, and death. May also cause the heart to beat irregularly or even stop.**

SKIN & EYES: **Irritation and blistering of skin. Eye contact may result in severe irritation.**

INGESTION: **Not a likely exposure route but possible. May cause nausea, vomiting, abdominal pain, and other unspecified gastrointestinal effects**

Carcinogenicity	NTP Listed?	IARC Cancer Review Group?	OSHA Regulated?	Target Organs?
Unknown Human **Unknown Animal**	**No**	**No**	**No**	**Eyes, skin, resp. sys., liver, kidney, CNS, CVS.**

Medical Conditions Generally Aggravated by Exposure

None reported.

Emergency and First-aid Procedures

Eye contact: Do not allow victim to keep eyes tightly shut. Flush immediately with water, for 15 minutes (minimum); seek medical attention. Skin contact: Remove all contaminated clothing, including shoes. Immediately wash area with flooding amounts of soap and water. Seek medical attention immediately. For inhalation: Remove the person from exposure. Provide respiratory assistance and CPR. Transfer to medical facility. If swallowed: Seek medical attention immediately. Give conscious and alert person 1-2 glasses of water to dilute. Contact poison control center for advice before inducing vomiting.

SECTION VII - PRECAUTIONS FOR SAFE HANDLING AND USE

Steps to be Taken in Case Material is Released or Spilled

Wear SCBA respiratory protection. Restrict those not wearing protective equipment and who are not involved in cleanup from entering area. Absorb with vermiculite and place in sealed container. Vapors are likely to collect in low lying areas in high concentrations. Exercise caution.

Preferred Waste Disposal Method

Mix with a combustible solvent, burn in chemical incinerator equipped with an afterburner and scrubber.

Precautions to be Taken in Handling and Storage

Store in tightly closed containers in cool, well-ventilated location. Keep away from sources of heat.

Other Precautions and Warnings

Workers must be trained on the hazards of working with methylene bromide *before* being assigned to areas where exposure is possible.

SECTION VIII - CONTROL MEASURES AND PERSONAL PROTECTIVE EQUIPMENT

Respiratory Protection *(specify type)*

An MSHA/NIOSH-approved air purifying respirator (cartridge type), is permissible. However, since no exposure levels have been established, best protection is provided using a supplied-air respirator or a self-contained breathing apparatus (SCBA) with full facepiece operated in positive pressure mode.

Ventilation

Local exhaust (preferred) at site of chemical work or general ventilation.

Protective Gloves	Eye Protection	Other Protective Clothing
Impervious Rubber Gloves	**Chemical Goggles or Face Mask**	**Protective Apron if Splash is Likely**

Work/Hygiene Practices

Always wash hands thoroughly after using chemical; never bring food, drink, or smoking materials into vicinity of chemicals.

METHYLENE BROMIDE

CH_2Br_2

CAS: 74-95-3

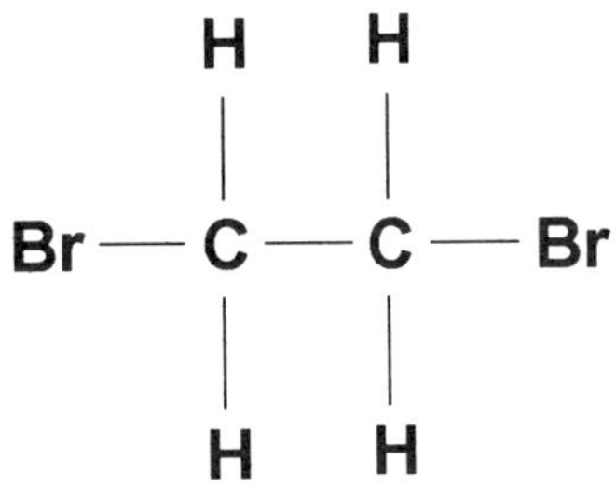

IDENTIFICATION AND TYPICAL USES

Methylene bromide is a colorless, heavy, viscous liquid with a sweet, pleasant, chloroform-like odor. It is used in organic synthesis and as a solvent.

RISK ASSESSMENT: HEALTH

General Assessment

Methylene bromide is moderately toxic by *inhalation.* While it is somewhat irritating to the skin's surface, dermal absorption has not been reported Although unlikely under normal operating conditions, *ingestion* is also a possible exposure route. There are no reports of any carcinogenic or teratogenic symptoms resulting from exposure to methylene bromide.

Inhalation is irritating to the eyes, nose, and mucosa of the respiratory tract. High concentrations of methylene bromide causes depression of the central nervous system. Subsequent symptoms include dizziness, lightheadedness, headache, loss of appetite and subsequent weight loss, abdominal pain with progression to major organ system damage (primarily the liver and kidneys), unconsciousness, and death due to liver and kidney damage (necrosis). Other symptoms include general weakness and fatigue, drowsiness, nausea and vomiting, diarrhea, chest pains, cough, and shortness of breath.

Skin contact can produce severe localized irritation and possible dermatitis. Eye contact will also cause irritation to the conjunctiva and surrounding tissues.

Acute Health Effects

The following acute (short-term) health effects may occur immediately or shortly after exposure to methylene bromide:

Skin: Irritation with potential for redness, pain, smarting, and dermatitis.

Eye: Irritation and smarting of the conjunctiva. No evidence of permanent damage.

Lung: Irritation of the mucous membranes of the nose, throat, and respiratory tract causing coughing, chest pains, and difficulty breathing.

CNS: Depression of the central nervous system with symptoms of headaches, lightheadedness, dizziness, nausea, vomiting, diarrhea, loss of appetite and weight.

Other: High exposures can lead to rapid progressive damage to the kidneys, liver, heart, and lungs which can lead to death. The heart may beat irregularly or even stop, causing death.

Chronic Health Effects

The following chronic (long-term) health effects may occur at some time after exposure to methylene bromide and can last for months or even years:

Cancer Hazards: According to the references, methylene bromide has not been adequately tested for its ability to cause cancer in test animals. However, mutation data have been reported and some scientists believe that such chemicals may pose a cancer risk in the long-term. More research is required in this regard.

Reproductive Hazard: According to information presented in the references, methylene bromide has not been adequately tested for its ability to cause reproductive effects in test animals.

Other Chronic Effects: Repeated skin contact can cause dryness, itching, and irritation. Long-term exposure may lead to liver and kidney damage. Repeated high exposures can cause both bromine and carbon monoxide to accumulate in body tissues.

Recommended Risk-Reduction Measures

Personnel should avoid direct contact with methylene bromide. Occupational poisoning occurs most commonly by inhalation. If a less toxic material or compound cannot be substituted for methylene bromide, then *engineering controls* are the most effective method of reducing exposures. The best protection is to enclose operations and/or provide local exhaust ventilation at the site of methylene bromide release. While not always operationally feasible, isolating operations can also reduce exposure. Using respiratory protection is less effective than the controls mentioned above, but is still advisable whenever working with or around methylene bromide. Caution should be exercised when selecting respiratory protection since the specific toxic nature of this chemical is not well understood or documented in the references. No exposure limits have been established. However, this does not mean that exposure to methylene bromide is without risk. For best protection, use an MSHA/NIOSH-approved supplied-air respirator with full facepiece and mask operated in positive pressure mode or a self-contained breathing apparatus (SCBA) with full facepiece operated in pressure demand. If a full facepiece is not available, then chemical goggles should be worn to protect the eyes. A face shield should also be considered. To prevent hand and skin exposures, impervious rubber gloves should be used. Glove manufacturers should be contacted and permeation studies obtained *before* final glove selections have been made.

Administrative controls should also be in place to minimize the potential for human exposures. These may include written procedures or policies which specify the methods and techniques that will be practiced whenever personnel are to work with methylene bromide.

All personnel should receive *training* on the use, hazards, protective measures, emergency actions, and other precautions per 29 CFR 1910.1200 (Hazard Communication), prior to the first assignment in an area where methylene bromide is used or stored.

Before beginning employment with methylene bromide and at regular intervals thereafter (e.g., annually), the following medical tests are recommended:

☑ Liver and kidney function tests (establish baseline).

If symptoms develop or overexposure is suspected, the following tests should be considered:

☑ Liver and kidney function tests.

☑ Carboxyhemoglobin level tests.

☑ Suggest use of Holter monitor (a special, 24-hour EKG monitor that can detect irregular heartbeat).

Any evaluation should include a careful history of past and present symptoms with an examination. Medical tests that simply look for existing damage are not a substitute for controlling exposures. Also, since smoking can cause heart disease, lung cancer, emphysema, and increase carboxyhemoglobin levels, smokers may be affected more rapidly and with more pronounced symptoms than non-smokers. Prudent risk management requires careful consideration of *all* risk factors that might be causing the appearance of symptoms in the work force.

Other methods to reduce exposure include:

☑ Where possible, enclose operations and use local exhaust ventilation at the site of chemical release. If local exhaust ventilation or enclosure is not used, respiratory protection should be mandatory.

☑ Always ensure that proper protective clothing is worn when using chemical substances.

☑ Wash thoroughly immediately after exposure to methylene bromide and at the end of the work shift or before eating, drinking, or smoking.

☑ Hazard warning information should be posted in the work area. In addition, as part of an on-going education and training program, all information on the hazards of methylene bromide should be communicated to all exposed workers.

☑ Eye wash stations should be provided in the immediate work area for emergency use. If there is a possibility of skin exposure to methylene bromide, emergency shower facilities should also be provided in the immediate area.

☑ Workers whose clothing has been contaminated by methylene bromide should change into clean clothes before leaving work. Contaminated work clothing should be laundered only by individuals who have been informed of the hazards of exposure to methylene bromide.

RISK ASSESSMENT: ENVIRONMENT

General Assessment

The environment is at risk of exposure during transportation, storage, disposal, or destruction of methylene bromide. In almost every scenario, the threat of environmental exposure is contingent upon the proper handling of the chemical substance. Accidental spills, large or small, can result in contamination of the surrounding environmental mediums (water, soil, and air).

Methylene bromide is a non-combustible liquid (it may burn, but does not easily ignite). However, it can react in the presence of chemically-active metals such potassium to form potentially explosive mixtures. In the presence of heat, it can produce toxic hydrogen bromide. Caution is always required in handling, storage, transportation, and disposal of methylene bromide. Emergency responders should be made aware of the presence of methylene bromide at any emergency response situation.

Methylene bromide can enter the environment from industrial effluents and from spills.

☠ *Acute Ecological Effects*

Acute (short-term) toxic effects may include the death of animals, birds, or fish and death or low growth rate in plants. Acute effects are seen 2 to 4 days after animals or plants are exposed to methylene bromide.

Methylene bromide has slight acute toxicity to aquatic life. Insufficient data are available to evaluate the short-term effects of methylene bromide on plants, birds, or land animals.

Chronic Ecological Effects

Chronic (long-term) toxic effects may include shortened life span, reproductive problems, lower fertility, and changes in appearance or behavior in exposed animals. These effects can be seen long after first exposure(s) to toxic chemicals.

Methylene bromide has slight chronic toxicity to aquatic life. Insufficient data are available to evaluate or predict the long-term effects of methylene bromide to plants, birds, or terrestrial animals.

Water Solubility

Methylene bromide is slight soluble in water. Concentrations of 1 to 100 milligrams will mix with a liter of water.

⌛ *Persistence in the Environment*

Methylene bromide is slightly persistent in water, with a half-life between 2 and 20 days. The half-life of a pollutant is the amount of time it takes for one half of the chemical to be degraded.

Bioaccumulation in Aquatic Organisms

Some substances increase in concentration, or *bioaccumulate,* in living organisms as they breathe contaminated air, drink contaminated water, or eat contaminated food. These chemicals can become concentrated in the tissues and internal organs of animals as well as humans.

The concentration of methylene bromide found in fish tissues is expected to be about the same as the average concentration of methylene bromide in the water from which the fish was taken.

Recommended Risk-Reduction Measures

Proper training of all transporters will reduce the likelihood of a mishap or accident resulting in a leak or spill of methylene bromide into the environment. Labels on all containers, trucks, and rail cars must meet DOT requirements and accurately reflect their contents to enable emergency responders to react properly and quickly to any disaster thereby reducing the potential risk to the environment and to personnel.

Storage of methylene bromide should be segregated from other chemicals and materials to minimize the risk of cross-contamination.

If a spill or leak to the environment has occurred, fire department, emergency response and/or hazardous materials spill personnel should be notified immediately. Cleanup should be attempted only by those trained in proper spill containment procedures. Contaminated soils should be removed for incineration and replaced with clean soil. If methylene bromide should contact the water table, aquifer, or navigable waterway, the local and/or state emergency response authorities must be notified. A comprehensive emergency response or disaster preparedness/recovery plan should be in place prior to any operations involving the use, transportation, storage, or disposal of methylene bromide.

For disposal, methylene bromide may be burned in a chemical incinerator equipped with an afterburner and scrubber.

If methylene bromide is spilled or leaked, the following specific steps are recommended:

- ☑ Restrict persons not wearing protective clothing from area of spill until cleanup is complete and area can be opened for normal work.
- ☑ Absorb liquids in vermiculite, sand, earth, or similar material and deposit in sealed drum. Avoid breathing vapors of methylene bromide.
- ☑ Ventilate area of spill or leak.

- ☑ Remove all sources of heat and ignition.
- ☑ It may be necessary to dispose of methylene bromide as a hazardous waste. The responsible state agency or the regional office of the federal Environmental Protection Agency (EPA) should be contacted for specific recommendations.

RISK ASSESSMENT: BUSINESS

General Assessment

Accidents or mishaps involving methylene bromide can present a serious threat to business operations. The loss or damage of equipment or facilities can significantly affect fiscal viability. Lawsuits that may result from personnel illness, injury/death, public exposures, and/or environmental contamination will require a serious expenditure of resources. Media attention surrounding an injury, death, or environmental damage can also result in a loss of profits and loss of current as well as future business.

❶ *Recommended Risk-Reduction Measures*

Company attorneys, safety and health professionals, and environmental specialists should be involved in the development of any procedures for responding to chemical incidents. A company official should be predesignated as a public relations officer with specific training in dealing with the press. Corporate plans and policies should be developed, approved, and implemented long before any need for such arises.

REFERENCES

American Conference of Governmental Industrial Hygienists. 1988. *Documentation of the Threshold Limit Values and Biological Exposure Indices*, 5th Edition (with updates). Cincinnati: ACGIH

Anthony, C. P., and N. J. Kolthoff. 1971. *Textbook of Anatomy and Physiology.* St. Louis: C. V. Mosby Company.

Bloom, A. D. and F. J. DeSerres. 1995. *Ecotoxicity and Human Health.* New York: CRC Press/Lewis Publishers.

Brusick, D. J. 1994. *Methods for Genetic Risk Assessment.* New York: CRC Press/Lewis Publishers.

Bueche, J. R. 1972. *Principles of Physics*, 2nd Edition. New York: McGraw-Hill.

Burrell, R., D. K. Flaherty, and L. J. Sauers. 1992. *Toxicology of the Immune System.* New York: Van Nostrand Reinhold.

Calabrese, E. J. 1994. *Biological Effects of Low Level Exposures.* New York: CRC Press/Lewis Publishers.

Cockerman, L. G., B. S. Shane. 1994. *Basic Environmental Toxicology.* New York: CRC Press/Lewis Publishers.

Cothern, C. R. 1995. *Handbook for Environmental Risk Decisionmaking.* New York: CRC Press/Lewis Publishers.

Danse, I. R. 1991. *Common Sense Toxics in the Workplace.* New York: Van Nostrand Reinhold.

Environmental Protection Agency. 1983. *Chemical Hazard Information Profiles.* Washington, D.C.: U.S. EPA.

Fodden, C. R., and J. L. Weddell. 1992. *Hazardous Materials Emergency Action Data.* New York: CRC Press/Lewis Publishers.

Forseberg, K., and L. H. Keith. 1993. *Chemical; Protective Clothing Permeation and Degradation Database.* Chelsea, MI: Lewis Publishers.

Gardner, R. W. 1994. *Chemical Intolerance.* New York: CRC Press/Lewis Publishers.

Hallenbeck, W. H. 1993. *Quantitative Risk Assessment for Environmental and Occupational Health*, 2nd Edition. New York: CRC Press/Lewis Publishers.

Hodgson, E., R. B. Mailman, and Janice E. Chambers. 1988. *Dictionary of Toxicology.* London: The MacMillian Company.

Hosty, J. W., and P. Foster. 1990. *A Practical Guide to Chemical Spill Response.* New York: Van Nostrand Reinhold.

Howard, P. H. 1995. *Handbook of Environmental Fate and Exposures Data for Organic Chemicals*, Vol. I-IV. New York: CRC Press/Lewis Publishers.

Jones, F. E. 1994. *Toxic Organic Vapors in the Workplace.* New York: CRC Press/Lewis Publishers.

Kamrin, M. A. 1988. *Toxicology: A Primer on Principles and Applications.* Chelsea, Michigan: Lewis Publishers.

Landis, W. G., M. H. Yu. 1995. *Introduction to Environmental Toxicology.* New York: CRC Press/Lewis Publishers.

Langerman, N. 1994 *Precautionary Labels for Chemical Containers.* New York: CRC Press/Lewis Publishers.

Lauwerys, R. R., P. Hoet. 1993. *Industrial Chemical Exposure.* New York: CRC Press/Lewis Publishers.

Lewis, R. J., Sr. 1993. *Hawley's Condensed Chemical Dictionary*, 12th Edition. New York: Van Nostrand Reinhold.

Lewis, R. J., Sr. 1992. *Sax's Dangerous Properties of Industrial Materials*, Volumes 1, 2, and 3, 8th Edition. New York: Van Nostrand Reinhold.

Meyer, E. 1990. *Chemistry of Hazardous Materials.* Englewood Cliffs, NJ: Prentice Hall.

Patnaik, P. 1992. *A Comprehensive Guide to the Hazardous Properties of Chemical Substances.* New York: Van Nostrand Reinhold.

Quigley, D. R. 1994. *Handbook of Emergency Chemical Management.* New York: CRC Press/Lewis Publishers.

U.S. Department of Health and Human Services, National Institute for Occupational Safety and Health. 1994. *NIOSH Pocket Guide to Chemical Hazards.* Washington, D.C.: U.S. Government Printing Office.

MATERIAL SAFETY DATA SHEET

CHEMICAL NAME

METHYLENE CHLORIDE

HAZARD WARNING INFORMATION

HEALTH	FIRE	REACTIVE	OTHER	DEGREE OF HAZARD	COLOR CODING	OTHER CODES
2	1	0		0 = Minimum Hazard 1 = Slight Hazard 2 = Moderate Hazard 3 = Serious Hazard 4 = Severe Hazard	HEALTH = BLUE FIRE = RED REACTIVITY = YELLOW OTHER = WHITE	OX = Oxidizer ACID = Acid ALK = Alkali COR = Corrosive W = Use No Water

SECTION I - GENERAL INFORMATION

Characterization	RCRA Number	EPA Class
Halogenated Hydrocarbon	**U080**	**Toxic Waste**

DOT Proper Shipping Name **Methylene chloride**	Chemical Abstract Service (CAS) Number **75-09-2**
DOT Hazard Class and Label Requirements **Poison B: St. Andrew's Cross**	DOT Emergency Guide Code **74**
DOT Identification Number **UN 1593**	Molecular Formula CH_2Cl_2

Synonyms

Dichloromethane; methylene dichloride; methylene bichloride; Solmethine.

SECTION II - HAZARDOUS INGREDIENTS AND IDENTITY INFORMATION

Hazardous Components (specific identity)	OSHA Exposure Criteria	NIOSH Exposure Criteria	Immediately Dangerous to Life and Health (IDLH)	ACGIH Exposure Criteria
Methylene chloride (derivation: By chlorination of methyl chloride and subsequent distillation). **1 ppm = 3.53 mg/m^3**	**PEL:** **500 ppm** **1765 mg/m^3** **CEILING:** **1000 ppm** **PEAK (5min/3hr):** **2000 ppm**	**REL:** **Reduce to Lowest Level** **Possible Human Carcinogen**	**2300 ppm**	**TLV:** **50 ppm** **174 mg/m^3** **Suspected Human Carcinogen**

SECTION III - PHYSICAL AND CHEMICAL CHARACTERISTICS

Boiling Point **104°F (40°C)**	Specific Gravity (H_2O = 1) **1.33**
Vapor Pressure (mm Hg) **350 at 69°F (20°C)**	Molecular Weight (atomic weight) **84.9**
Vapor Density (Air = 1) **2.93**	Freezing Point **-139°F (-95°C)**

Solubility

Slightly soluble in water (2%). Soluble in alcohol, acetone, chloroform, carbon tetrachloride, ether, and dimethylformamide.

Appearance and Odor

Colorless or clear liquid with a sweet, pleasant odor. Odor Threshold = 205 ppm.

SECTION IV - FIRE AND EXPLOSION HAZARD DATA

Flash Point (method used) **Not Determined**	Explosive Limits in Air % by Volume **LEL: 13%** **UEL: 23%**
NFPA Classification **Combustible Liquid**	Autoignition Temperature **775°F (413°C)**

Extinguishing Media

Dry chemical, carbon dioxide, alcohol foam, water spray/fog, or dry sand. Direct water may be ineffective.

Special Fire Fighting Procedures

Poisonous gases are produced in fire. Wear self-contained breathing apparatus (SCBA) and full protective gear. Move containers from fire if it can be done without risk. Cool sides of fire-exposed containers with water until long after fire is out.

Unusual Fire and Explosion Hazards

Containers may explode in fire. Stay away from ends of tanks. Evacuate 1500 feet if rising sound is heard coming from venting device or sides of tank become discolored due to heat of fire.

SECTION V - REACTIVITY DATA

Stability		Conditions to Avoid
		Methylene chloride is normally stable. Avoid heat, flame, other sources of ignition, and contact with incompatible materials and chemicals.
Stable	Unstable	Incompatibility *(materials to avoid)*
X		Strong oxidizers (chlorine, bromine, fluorine), chemically active metals (such as sodium, potassium, calcium, powdered aluminum, zinc, and magnesium), caustic agents, and concentrated nitric acid.
Hazardous Polymerization		Conditions to Avoid Under normal conditions of temperature and pressure, hazardous polymerization of methylene chloride cannot occur.
May Occur	Will Not Occur	Hazardous Decomposition or By-products
	X	Poisonous gases and acrid fumes are produced in fire, including extremely toxic phosgene gas and carbon monoxide fumes.

SECTION VI - HEALTH HAZARD DATA

Primary Route(s) of Entry:	Inhalation?	Absorption (skin/eye)?	Ingestion?
	X	X	X

Health Hazards

INHALATION: Respiratory irritation, intoxication, narcotic and anesthetic effects, vomiting, dizziness, depression, diarrhea, CNS depression, eye irritation, digestive disturbances, leukocytosis, fatigue, lack of motivation, psychosomatic symptoms, increase in carboxyhemoglobin content of the blood, numbness, pulmonary edema, unconsciousness, and death.

ABSORPTION: Irritation, de-fatting, and burning of the skin surface. Eye contact may result in severe irritation and possible damage. Will pass through intact skin and cause toxic effects.

INGESTION: Gastrointestinal irritation along with symptoms similar to that of inhalation.

Carcinogenicity	NTP Listed?	IARC Cancer Review Group?	OSHA Regulated?	Target Organs?
Suspected Human Confirmed Animal	Class 2	Group 2B	29 CFR 1910.1000 Table Z-1	Eyes, skin, resp. sys., liver, kidney, CVS, CNS.

Medical Conditions Generally Aggravated by Exposure

Skin and cardiovascular disorders may be aggravated by exposure.

Emergency and First-aid Procedures

Eye contact: Do not allow victim to keep eyes shut. Flush immediately with water for 15 minutes minimum; seek medical attention. Skin contact: Remove all contaminated clothing. Immediately wash area with flooding amounts of soap and water. Seek medical attention immediately. For inhalation: Remove the person from exposure. Provide respiratory assistance and CPR. Transfer to medical facility. Observe for 48 hours for lung effects. If swallowed: Seek medical attention immediately. Induce vomiting in conscious and alert person. Never give an unconscious or convulsing person anything by mouth.

SECTION VII - PRECAUTIONS FOR SAFE HANDLING AND USE

Steps to be Taken in Case Material is Released or Spilled

Remove all ignition sources, wear SCBA respiratory protection. Restrict those not wearing protective equipment who are not involved in cleanup from entering area. Absorb with vermiculite and place in sealed container. Flush residues with water (keep away from sewers).

Preferred Waste Disposal Method

Burn in chemical incinerator equipped with an afterburner and scrubber.

Precautions to be Taken in Handling and Storage

Store in tightly closed containers in cool, well-ventilated location. Keep away from sources of heat.

Other Precautions and Warnings

Sources of ignition are prohibited where methylene chloride is used or stored. Metal containers should be bonded and grounded. Empty cylinders contain residue; handle with care.

SECTION VIII - CONTROL MEASURES AND PERSONAL PROTECTIVE EQUIPMENT

Respiratory Protection *(specify type)*

Use an MSHA/NIOSH-approved air purifying respirator (cartridge type) or use a supplied-air respirator or a self-contained breathing apparatus (SCBA) with full facepiece operated in positive pressure mode.

Ventilation

Local exhaust (preferred) at site of chemical work or general ventilation.

Protective Gloves	Eye Protection	Other Protective Clothing
Impervious Rubber Gloves	Chemical Goggles or Face Mask	Protective Apron if Splash is Likely

Work/Hygiene Practices

Always wash hands thoroughly after using chemical; never bring food, drink, or smoking materials into vicinity of chemicals.

METHYLENE CHLORIDE

CH_2Cl_2

CAS: 75-09-2

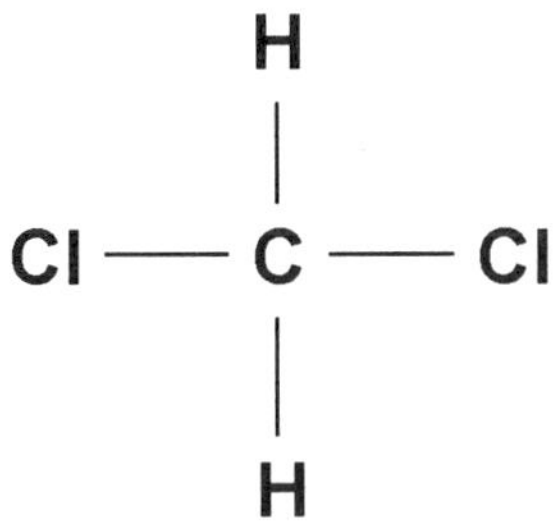

IDENTIFICATION AND TYPICAL USES

Methylene chloride is a colorless liquid with a sweet odor. It is widely used as a solvent, a degreaser, and a cleaning reagent. It is also used in paint removers, and in extractions of organic compounds from water for analyses. It is a blowing agent for foams, a solvent for cellulose acetate, and is used as an aerosol propellant.

RISK ASSESSMENT: HEALTH

General Assessment

Methylene chloride is mildly toxic by *inhalation*. The liquid is also known the pass through the skin (*absorption*) to cause toxic systemic effects, and it is considered moderately toxic by *ingestion*. This chemical is known to cause cancer in test animals (liver and lung) and this effect is suspected to occur in humans as well. Human mutation data have been reported. In addition, methylene chloride has been shown to adversely affect reproduction in test animals. The systemic effects of overexposure can be seen in the liver, kidneys, digestive tract, blood, lungs, and in the central nervous system.

Inhalation is irritating to the eyes, nose, and mucosa of the respiratory tract and may produce ulceration of the mucous membrane. High concentrations of methylene chloride causes depression of the central nervous system (CNS). Symptoms include dizziness, intoxication, narcotic and anesthetic effects, lightheadedness, headache, somnolence, cyanosis, unconsciousness, coma, and death.

This chemical poses a significant hepatoxic threat (injurious to the liver). Irritation to lung tissues can lead to a dangerous buildup of fluids in the lungs (pulmonary edema), which is a medical emergency and can be fatal. Symptoms may be delayed up to 48 hours creating a false sense of security with regard to health risk. Exposure can also cause the heart to beat irregularly or even stop which, of course, can be fatal.

Skin contact can produce severe localized irritation, burns, de-fatting of the surface leading to possible cracking, drying, and secondary infections. Absorption will occur rapidly causing an increase in toxicity. If coupled with inhalation, the systemic effects can be quite severe. Eye contact will also cause irritation and, if not removed immediately, conjunctivitis.

Ingestion can result in gastrointestinal irritation and systemic symptoms similar to that of inhalation.

Acute Health Effects

The following acute (short-term) health effects may occur immediately or shortly after exposure to methylene chloride:

<u>Skin</u>: Irritation with potential for redness, pain, burns, and tissue damage.

<u>Eye</u>: Irritation and smarting of the conjunctiva. Can cause tearing, redness, and pain.

<u>Lung</u>: Irritation of the mucous membranes of the nose, throat, and respiratory tract causing possible ulceration, coughing, chest pains, difficulty breathing, and pulmonary edema.

<u>CNS</u>: Depression with headaches, lightheadedness, dizziness, nausea, vomiting, diarrhea, narcotic and anesthetic effects.

Chronic Health Effects

The following chronic (long-term) health effects may occur at some time after exposure to methylene chloride and can last for months or even years:

Cancer Hazards: According to the references, methylene chloride in a confirmed animal carcinogen and several agencies consider methylene chloride to be a potential or suspect human carcinogen. These include the IARC (International Association for Research on Cancer), the NTP (National Toxicology Program), and the ACGIH (American Conference of Governmental Industrial Hygienists). It has caused cancer of the lung and liver in animals. Human mutation data have also been reported.

Reproductive Hazard: According to information presented in the references, methylene chloride is a possible teratogen in animals and has been shown to adversely affect reproduction in test animals. It is not known if exposure to methylene chloride will have this effect on humans.

Other Chronic Effects: Methylene chloride may irritate the lungs. This is more likely to occur if the chemical is used near a combustion source (furnace, welding) since phosgene (a lung-damaging gas) is released. Repeated exposures may cause bronchitis with cough, phlegm, and/or shortness of breath. Higher exposures may damage the liver. Long-term exposure may damage the brain, causing memory loss, poor coordination, and reduced thinking ability. Repeated skin exposure can cause thickening and cracking of the skin.

Recommended Risk-Reduction Measures

Personnel should avoid direct contact with methylene chloride. It is a highly toxic, known cancer-causing agent. There may be no safe exposure level to methylene chloride. Occupational poisoning occurs most commonly by inhalation and skin absorption. If a less toxic material or compound cannot be substituted for methylene chloride, then *engineering controls* are the most effective method of reducing exposures. The best protection is to enclose operations and/or provide local exhaust ventilation at the site of methylene chloride release. While not always operationally feasible, isolating operations can also reduce exposure. Using respiratory protection is less effective than the controls mentioned above, but is still advisable whenever working with or around methylene chloride. Caution should be exercised when selecting respiratory protection since it is not known if exposure to methylene chloride will cause cancer in humans. Many reference sources also indicate that the established PEL of 500 ppm is outdated and insufficient to protect against harmful exposures. Risk managers should consider the known evidence regarding methylene chloride's toxicity and act accordingly, regardless of established exposure levels. A conservative approach to a carcinogenic material may be prudent. For the best protection against a *possible carcinogen,* especially when the concentration is unknown, use a supplied-air respirator with full facepiece and mask operated in positive pressure mode or a self-contained breathing apparatus (SCBA) with full facepiece operated in pressure demand. If a full facepiece is not available, then chemical goggles should be worn to protect the eyes. A face shield should also be considered. To prevent hand and skin exposures, impervious rubber gloves should be used. Glove manufacturers should be contacted and permeation studies obtained *before* final glove selections have been made.

Administrative controls should also be in place to minimize the potential for human exposures. These may include written procedures or policies which specify the methods and techniques that will be practiced whenever personnel are to work with methylene chloride.

All personnel should receive *training* on the use, hazards, protective measures, emergency actions, and other precautions per 29 CFR 1910.1200 (Hazard Communication), prior to the first assignment in an area where methylene chloride is used or stored.

Before beginning employment with methylene chloride and at regular intervals thereafter (e.g., annually), the following medical tests are recommended:

☑ Liver and kidney function tests (establish baseline).

☑ Lung function tests (establish baseline).

If symptoms develop or overexposure is suspected, the following tests should be considered:

☑ Liver and kidney function tests.

☑ Suggest Holter monitor (special 24-hour EKG) to detect irregular heartbeat.

☑ Blood carboxyhemoglobin within a few hours of exposure.

☑ Interview for brain effects, including recent memory or mood changes (irritability, withdrawal), concentration problems, headaches, malaise, and altered sleep patterns. Consider cerebellar, autonomic, and peripheral nervous system evaluation. Positive and borderline individuals should be referred neuropsychological testing.

☑ Consider chest X-ray after acute overexposure (may be negative if taken immediately after exposure due to delayed onset of pulmonary edema).

☑ Lung function tests (compare to baseline).

Any evaluation should include a careful history of past and present symptoms with an examination. Medical tests that simply look for existing damage are not a substitute for controlling exposures. Also, since smoking can cause heart disease, lung cancer, increase carboxyhemoglobin levels, emphysema, and other respiratory problems, smokers may be affected more rapidly and with more pronounced symptoms than non-smokers. Prudent risk management requires careful consideration of *all* risk factors that might be causing the appearance of symptoms in the work force.

Other methods to reduce exposure include:

- ☑ Where possible, enclose operations and use local exhaust ventilation at the site of chemical release. If local exhaust ventilation or enclosure is not feasible, then respiratory protection should be provided and its use mandatory.
- ☑ Always ensure that proper protective clothing is worn when using chemical substances.
- ☑ Wash thoroughly immediately after exposure to methylene chloride and at the end of the work shift or before eating, drinking, or smoking.
- ☑ Hazard warning information should be posted in the work area. In addition, as part of an on-going education and training program, all information on the health and safety hazards of methylene chloride should be communicated to all potentially exposed workers.
- ☑ Eye wash stations should be provided in the immediate work area for emergency use. If there is a possibility of skin exposure to methylene chloride, emergency shower facilities should also be provided in the immediate work area.
- ☑ Workers whose clothing has been contaminated by methylene chloride should change into clean clothes before leaving work. Contaminated work clothing should be laundered only by individuals who have been informed of the hazards of exposure to methylene chloride.

RISK ASSESSMENT: ENVIRONMENT

General Assessment

The environment is at risk of exposure during transportation, storage, disposal, or destruction of methylene chloride. In almost every scenario, the threat of environmental exposure is contingent upon the proper handling of the chemical substance. Accidental spills, large or small, can result in contamination of the surrounding environmental mediums (water, soil, and air).

Methylene chloride is a mildly combustible liquid but it can form explosive mixtures in air. It is a moderate explosion hazard in closed containers when exposed to heat or flame. It can also react violently in the presence of chemically-active metals such as sodium, potassium, calcium, and strong oxidizing agents. Caution is always required in handling, storage, transportation, and disposal of methylene chloride. A phenomenon known as a boiling liquid expanding vapor explosion (BLEVE) can occur when tanks, containers, or rail cars containing methylene chloride are involved in a fire. The pressure that builds inside the container can explode violently and without much warning. Emergency responders should, therefore, be made aware of the presence of methylene chloride at any emergency response situation. Methylene chloride can enter the environment from industrial effluents and from spills.

Acute Ecological Effects

Acute (short-term) toxic effects may include the death of animals, birds, or fish and death or low growth rate in plants. Acute effects are seen 2 to 4 days after animals or plants are exposed to methylene chloride. Methylene chloride has moderate acute toxicity to aquatic life. It has caused both increases and decreases in seed germination in various agricultural crops. Insufficient data are available to evaluate the short-term effects of methylene chloride on birds or land animals.

Chronic Ecological Effects

Chronic (long-term) toxic effects may include shortened life span, reproductive problems, lower fertility, and changes in appearance or behavior in exposed animals. These effects can be seen long after first exposure(s) to toxic chemicals. Methylene chloride has moderate chronic toxicity to aquatic life. Insufficient data are available to evaluate the long-term effects of methylene chloride to plants, birds, or terrestrial animals.

Water Solubility

Methylene chloride is slightly soluble in water. Concentrations of 1 to 100 milligrams will mix with a liter of water.

Persistence in the Environment

Methylene chloride is slightly persistent in water, with a half-life between 2 and 200 days. The half-life of a pollutant is the amount of time it takes for one half of the chemical to be degraded. About 99% of methylene chloride will eventually end up in air, the rest will end up in the water.

Bioaccumulation in Aquatic Organisms

Some substances increase in concentration, or *bioaccumulate,* in living organisms as they breathe contaminated air, drink contaminated water, or eat contaminated food. These chemicals can become concentrated in the tissues and internal organs of animals as well as humans. The concentration of methylene chloride found in fish tissues is expected to be about the same as the average concentration of methylene chloride in the water from which the fish was taken.

Recommended Risk-Reduction Measures

Proper training of all transporters will reduce the likelihood of a mishap or accident resulting in a leak or spill of methylene chloride into the environment. Labels on all containers, trucks, and rail cars must meet DOT requirements and accurately reflect their contents to enable emergency responders to react properly and quickly to any disaster thereby reducing the potential risk to the environment and to personnel.

Storage of methylene chloride should be segregated from other chemicals and materials to minimize the risk of cross-contamination.

If a spill or leak to the environment has occurred, fire department, emergency response and/or hazardous materials spill personnel should be notified immediately. Cleanup should be attempted only by those trained in proper spill containment procedures. Contaminated soils should be removed for incineration and replaced with clean soil. If methylene chloride should contact the water table, aquifer, or navigable waterway, time is of the essence. It is only slightly soluble and total remediation may be possible. The local and/or state emergency response authorities must be notified. A comprehensive emergency response or disaster preparedness/recovery plan should be in place prior to any operations involving the use, transportation, storage, or disposal of methylene chloride.

For disposal, methylene chloride may be burned in a chemical incinerator equipped with an afterburner and scrubber. If methylene chloride is spilled or leaked, the following specific steps are recommended:

- ☑ Restrict persons not wearing protective clothing from area of spill until cleanup is complete.
- ☑ Absorb liquids in vermiculite, sand, earth, or similar material and deposit in sealed drum.
- ☑ Remove ignition sources and ventilate area of spill or leak.
- ☑ It may be necessary to dispose of methylene chloride as a hazardous waste. The responsible state agency or the regional office of the federal Environmental Protection Agency (EPA) should be contacted for specific recommendations.

RISK ASSESSMENT: BUSINESS

General Assessment

Accidents or mishaps involving methylene chloride can present a serious threat to business operations. The loss or damage of equipment or facilities can significantly affect fiscal viability. Lawsuits that may result from personnel illness, injury/death, public exposures, and/or environmental contamination will require a serious expenditure of resources. Media attention surrounding an injury, death, or environmental damage can also result in a loss of profits and loss of current as well as future business. Always remember that anytime the terms "cancer," "carcinogen," or "reproductive hazard" are used, public emotion, hysteria, and ignorance can run equally high. This must be carefully considered when developing any public relation policies.

Recommended Risk-Reduction Measures

Company attorneys, safety and health professionals, and environmental specialists should be involved in the development of any procedures for responding to chemical incidents. A company official should be predesignated as a public relations officer with specific training in dealing with the press. Corporate plans and policies should be developed, approved, and implemented long before any need for such arises.

REFERENCES

American Conference of Governmental Industrial Hygienists. 1988. *Documentation of the Threshold Limit Values and Biological Exposure Indices,* 5th Edition (with updates). Cincinnati: ACGIH

Lewis, R. J., Sr. 1992. *Sax's Dangerous Properties of Industrial Materials,* Volumes 1, 2, and 3, 8th Edition. New York: Van Nostrand Reinhold.

U.S. Department of Health and Human Services, National Institute for Occupational Safety and Health. 1994. *NIOSH Pocket Guide to Chemical Hazards.* Washington, D.C.: U.S. Government Printing Office.

MATERIAL SAFETY DATA SHEET

CHEMICAL NAME	
4,4-METHYLENEDIANILINE	**CRC PRESS/LEWIS PUBLISHERS DISCLAIMER:** THE INFORMATION AND RECOMMENDATIONS PRESENTED HEREIN ARE BASED ON SOURCES BELIEVED TO BE RELIABLE. CRC MAKES NO REPRESENTATION ON ITS COMPLETENESS OR ACCURACY. IT IS THE RESPONSIBILITY OF THE USER TO DETERMINE THE CHEMICAL'S SUITABILITY FOR ITS INTENDED USE, THE CHEMICAL'S SAFE USE, AND THE CHEMICAL'S PROPER DISPOSAL. NO REPRESENTATIONS AND/OR WARRANTIES, EITHER EXPRESSED OR IMPLIED, OF THE MERCHANTABILITY OR FITNESS FOR A PARTICULAR PURPOSE, OR OF ANY OTHER NATURE, ARE MADE WITH RESPECT TO THE INFORMATION PROVIDED IN THIS MSDS OR TO THE CHEMICAL TO WHICH INFORMATION MAY REFER. CRC NEITHER ASSUMES NOR AUTHORIZES ANY OTHER PERSON TO ASSUME FOR IT, ANY OTHER ADDITIONAL RESPONSIBILITY OR LIABILITY FOR THE USE OF, OR RELIANCE UPON, THIS INFORMATION.

HAZARD WARNING INFORMATION

HEALTH	FIRE	REACTIVE	OTHER	DEGREE OF HAZARD	COLOR CODING	OTHER CODES
3	1	0		0 = Minimum Hazard 1 = Slight Hazard 2 = Moderate Hazard 3 = Serious Hazard 4 = Severe Hazard	HEALTH = BLUE FIRE = RED REACTIVITY = YELLOW OTHER = WHITE	OX = Oxidizer ACID = Acid ALK = Alkali COR = Corrosive W = Use No Water

SECTION I - GENERAL INFORMATION

Characterization	RCRA Number	EPA Class
Aromatic Amine	**None**	**Not Applicable**

DOT Proper Shipping Name	Chemical Abstract Service (CAS) Number
Methylenedianiline	**101-77-9**
DOT Hazard Class and Label Requirements	**DOT Emergency Guide Code**
Poison B: St. Andrew's Cross	**No Citation**
DOT Identification Number	**Chemical Formula**
UN 2651	**$CH_2(C_6H_4NH_2)_2$**

Synonyms

4,4'-Diaminodiphenylmethane; p,p'-diaminodiphenylmethane: para, para'-diaminodiphenylmethane; dianilinomethane; 4,4'-diphenylmethanediamine; MDA.

SECTION II - HAZARDOUS INGREDIENTS AND IDENTITY INFORMATION

Hazardous Components (specific identity)	OSHA Exposure Criteria	NIOSH Exposure Criteria	Immediately Dangerous to Life and Health (IDLH)	ACGIH Exposure Criteria
4,4-Methylenedianiline (derivation: From aniline and formaldehyde). **1 ppm = 8 mg/m³**	**PEL (skin):** **0.010 ppm** **0.080 mg/m³** **STEL:** **0.1 ppm** **0.8 mg/m³**	**REL (skin):** **Reduce to Lowest Level** **Possible Human Cancer Agent**	**Not Determined**	**TLV (skin):** **0.1 ppm** **0.8 mg/m³** **STEL:** **Not Established**

SECTION III - PHYSICAL AND CHEMICAL CHARACTERISTICS

Boiling Point	Specific Gravity (H_2O = 1)
748°F (398°C)	**1.06**
Vapor Pressure (mm Hg)	Molecular Weight
10×2^{-7} at 77°F (25°C)	**198.3**
Vapor Density (Air = 1)	Freezing Point
Not Found	**198°F (92°C)**

Solubility

Slightly soluble in cold water (0.1%). Very soluble in alcohol, benzene, and ether.

Appearance and Odor

Light brown, tan, or amber crystalline solid with a faint, characteristic amine odor.

SECTION IV - FIRE AND EXPLOSION HAZARD DATA

Flash Point (method used)	Explosive Limits in Air % by Volume
428°F (220°C) closed cup	**LEL: Not Determined** **UEL: Not Determined**
NFPA Classification	Autoignition Temperature
Combustible Solid	**Not Determined**

Extinguishing Media

Use water spray, carbon dioxide, alcohol foam, or dry chemical.

Special Fire Fighting Procedures

Wear full protective clothing and self-contained breathing apparatus (SCBA). Use a water spray to keep fire-exposed containers cool. Poisonous gases are produced in fire. Can become combustible when exposed to heat or flame.

Unusual Fire and Explosion Hazards

None reported.

SECTION V - REACTIVITY DATA

Stability		Conditions to Avoid Under normal conditions of handling and storage, 4,4-methylenedianiline is stable at room temperature and pressure. Avoid exposure to incompatible chemicals, heat, and ignition sources.
Stable X	Unstable	Incompatibility *(materials to avoid)* Strong oxidizers (such as perchlorates, peroxides, permanganates, chlorates and nitrates); strong acids (such as hydrochloric, sulfuric, and nitric).
Hazardous Polymerization		Conditions to Avoid Under normal conditions of temperature and pressure, hazardous polymerization of 4,4-methylenedianiline is not expected to occur.
May Occur	Will Not Occur X	Hazardous Decomposition or By-products When heated to decomposition, 4,4-methylenedianiline can emit highly toxic/poisonous gases, including carbon monoxide and toxic aniline vapors.

SECTION VI - HEALTH HAZARD DATA

Primary Route(s) of Entry:	Inhalation?	Absorption (skin)?	Ingestion?
	X	X	

Health Hazards

INHALATION: Causes abrupt onset of upper abdominal pain accompanied by fever, jaundice, rigidity, and liver changes. There can also be visual disturbances (loss of visual acuity, color discrimination, and dark adaptation). A dangerous liver toxin. Also causes an increase in the carboxyhemoglobin levels of the blood.

ABSORPTION: Absorbs freely through intact skin. Also causes headache and symptoms similar to that of inhalation. Skin discoloration is also likely (to yellow or faint brown).

INGESTION: Unspecified effects resulting from ingestion. Liver damage is suspect.

Carcinogenicity	NTP Listed?	IARC Cancer Review Group?	OSHA Regulated?	Target Organs?
Suspected Human Confirmed Animal	5th Annual Report	Group 2B	29 CFR 1910.1050	Eyes, liver, CVS, spleen, skin.

Medical Conditions Generally Aggravated by Exposure

Existing liver diseases may be aggravated by exposure.

Emergency and First-aid Procedures

Eye contact: Flush immediately with water for 15 minutes (minimum); seek medical attention. Skin contact: Remove all contaminated clothing. Immediately wash area with large amounts of soap and water. Seek medical assistance. For inhalation: Remove the person from exposure. Provide respiratory assistance and CPR. Transfer to medical facility. If swallowed: Seek medical attention immediately. Unless advised otherwise, give 2-3 glasses of water to a conscious person and induce vomiting.

SECTION VII - PRECAUTIONS FOR SAFE HANDLING AND USE

Steps to be Taken in Case Material is Released or Spilled

Shovel, vacuum, or damp mop spilled methylenedianiline and place in sealed drums for disposal. Avoid generating airborne dusts. Ventilate area of spill or leak. Restrict those not involved in cleanup from entering area.

Preferred Waste Disposal Method

No citation.

Precautions to be Taken in Handling and Storage

Store in tightly closed containers in cool, well-ventilated area away from heat. Protect containers from physical damage and protect from exposure to direct sunlight. This chemical is suspected to cause cancer in humans. Avoid skin contact and inhalation of airborne dusts or fumes.

Other Precautions and Warnings

Sources of ignition are prohibited where 4,4-methylenedianiline is used, handled, or stored. Bulk storage of 4,4-methylenedianiline should be kept in isolated or detached buildings or facilities.

SECTION VIII - CONTROL MEASURES AND PERSONAL PROTECTIVE EQUIPMENT

Respiratory Protection *(specify type)*

Exposure level is extremely low and difficult to monitor. Suggest a self-contained breathing apparatus with full facepiece operated in positive pressure mode for protection against suspected carcinogen.

Ventilation

Local exhaust at site of chemical work.

Protective Gloves	Eye Protection	Other Protective Clothing
Impervious Rubber	Chemical Goggles or Face Mask	Rubber Apron

Work/Hygiene Practices

Always wash hands thoroughly after using chemical; never bring food, drink, or smoking materials into vicinity of chemicals.

4,4-METHYLENEDIANILINE

$CH_2(C_6H_4NH_2)_2$

CAS: 101-77-9

IDENTIFICATION AND TYPICAL USES

4,4-Methylenedianiline is a light brown, tan, or amber crystalline solid with a faint, characteristic amine odor. It is used in the determination of tungsten and sulfates, as a polymer and dye intermediate, as a corrosion inhibitor, as an epoxy-resin hardening agent, in isocyanate resins, and in polymides.

RISK ASSESSMENT: HEALTH

General Assessment

4,4-Methylenedianiline is a human poison by *inhalation* and skin contact (*absorption*). It is known to cause cancer in test animals and is considered a suspected human carcinogen. Mutation data have also been reported. There is no information in the references regarding its teratogenic effects on animals or humans.

Inhalation and dermal absorption is likely to cause systemic poisoning characterized by symptoms that reflect liver involvement. These include jaundice, upper abdominal pain, and hepatitis. Other symptoms of exposure include rigidity, visual disturbances (loss of central visual acuity, color discrimination, and dark adaptation), and a marked increase in the carboxyhemoglobin content of the blood. The latter reduces the blood ability to transport oxygen to the cells of the body and can lead to serious illness.

Eye and skin contact will result in irritation. More importantly, however, is the fact that this compound is absorbed through intact skin on prolonged exposure causing the same toxic systemic effects as that noted for inhalation. Combined exposures (inhalation and absorption) can therefore pose a serious health risk.

Acute Health Effects

The following acute (short-term) health effects may occur immediately or shortly after exposure to 4,4-methylenedianiline:

Skin: Irritation and absorption through unbroken skin. Skin discoloration (to light brown or yellow is common.

Eye: Severe irritation with possible corneal damage.

Lung: Irritation of the nose, throat, and upper respiratory tract. Easily absorbed and distributed through the respiration process to cause toxic systemic effects primarily involving the liver.

CVS: 4,4-Methylenedianiline affects the blood's ability to transport oxygen (increase in carboxyhemoglobin content of blood).

Chronic Health Effects

The following chronic (long-term) health effects may occur at some time after exposure to 4,4-methylenedianiline and can last for months or even years:

Cancer Hazards: 4,4-Methylenedianiline has been shown to cause cancer in test animals and its carcinogenic potential in humans is considered highly suspect. Mutation data have also been confirmed.

Reproductive Hazard: According to information presented in the references, 4,4-methylenedianiline has not been adequately tested for its ability to adversely affect reproduction.

Other Chronic Effects: Continuous exposure to small doses of 4,4-methylenedianiline may produce anemia, loss of energy, digestive disturbance, and headache. Repeated exposures may cause anemia and an increase in carboxyhemoglobin levels to occur gradually over time rather than all at once. Skin discoloration to light brown or yellow is common on prolonged exposures. There may also be the appearance of vertical brown streaks under the finger nails. This indicates repeated overexposures. Liver involvement may also be occurring at this stage.

ⓘ Recommended Risk-Reduction Measures

Personnel should avoid direct contact with 4,4-methylenedianiline. It is a suspected human carcinogen and there may be no safe exposure level. If a less toxic chemical cannot be substituted for a hazardous substance, then *engineering controls* are the most effective method of reducing exposures. The best protection is to enclose operations and/or provide local exhaust ventilation at the site of chemical release. While not always operationally feasible, isolating operations can also reduce exposure. Using respiratory protection is less effective than the controls mentioned above, but is still advisable whenever working with or around 4,4-methylenedianiline. The exposure level is extremely low (0.010 ppm) and difficult to accurately monitor, especially in emergency situations. For the best protection against suspected carcinogens, recommend using a self-contained breathing apparatus (SCBA) with full facepiece operated in pressure demand or a supplied-air respirator operated in positive pressure mode. If a full facepiece is not available, then chemical goggles should be worn to protect the eyes. 4,4'-Methylenedianiline is often mixed in a thick liquid with other chemical agents and, therefore, whenever a chemical splash hazard exists, a face shield and a rubber apron should be worn. Skin contact must be prevented. To prevent hand and skin exposures, impervious rubber gloves should be used. This chemical is known to have peculiar permeation characteristics. Glove manufacturers should therefore be contacted and permeation studies reviewed *before* final glove selections are made. Providing the wrong glove creates a false sense of security and can present a more serious health hazard risk than no protection at all.

Administrative controls should also be in place to minimize the potential for human exposures. These may include written procedures or policies which specify the methods and techniques that will be practiced whenever personnel are to work with or around 4,4-methylenedianiline.

All personnel should receive *training* on the use, hazards, protective measures, emergency actions, and other precautions per 29 CFR 1910.1200 (Hazard Communication) as well as OSHA 1910.1050 (4.4'-Methylenedianiline), prior to their first assignment in an area where 4,4-methylenedianiline is used or stored.

If symptoms develop or overexposure is suspected, the following medical tests may be useful:

☑ Carboxyhemoglobin level tests.

☑ Liver function tests

Any evaluation should include a careful history of past and present symptoms with an examination. Medical tests that simply look for existing damage are not a substitute for controlling exposures. Also, since smoking can increase carboxyhemoglobin levels as well as cause other serious health effects, smokers exposed to 4,4-methylinedianiline may experience symptoms more rapidly than non-smokers under the same conditions of exposure. Prudent risk management requires careful consideration of *all* possible risk factors.

Other methods to reduce exposure include:

☑ Where possible, enclose operations and use local exhaust ventilation at the site of chemical release. If local exhaust ventilation or enclosure is not used, respiratory protection should be mandatory.

☑ Always ensure that proper protective clothing is worn when using chemical substances.

☑ Safety showers should be provided in areas where work is performed around 4,4-methylenedianiline.

☑ Wash thoroughly immediately after exposure to 4,4-methylenedianiline and at the end of the work shift or before eating, drinking, or smoking.

☑ Hazard warning information should be posted in the work area. In addition, as part of an on-going education and training program, all information on the health and safety hazards associated with 4,4-methylenedianiline should be communicated to all potentially exposed workers.

☑ Eye wash stations should be provided in the immediate work area for emergency use. If there is a possibility of skin exposure of any degree to 4,4-methylenedianiline, emergency shower facilities should also be provided.

RISK ASSESSMENT: ENVIRONMENT

General Assessment

The environment is at risk of exposure during transportation, storage, disposal, or destruction of 4,4-methylenedianiline. In almost every scenario, the threat of environmental exposure is contingent upon the proper handling of the chemical substance. Accidental spills, large or small, can result in fire, explosion, and possible contamination of the surrounding environmental mediums (water, soil, and air).

4,4-Methylenedianiline is considered a combustible solid. However, because it is incompatible with so very many other common commodities (such as oxidizers and acids), extreme caution is required in handling, storage, transportation, and disposal of 4,4-methylenedianiline. These characteristics also require special consideration during any emergency situation involving a leak or spill of 4,4-methylenedianiline. Should 4,4-methylenedianiline ever come into contact with incompatible substances either during use, transportation, storage, or disposal, the formation of highly toxic and/or highly explosive commodities is possible.

4,4-Methylenedianiline can enter the environment through industrial and municipal waste treatment plant discharges, or from spills.

Acute Ecological Effects

Acute (short-term) toxic effects may include the death of animals, birds, or fish and death or low growth rate in plants. Acute effects are seen 2 to 4 days after animals or plants are exposed to 4,4-methylenedianiline. Insufficient data are available to evaluate or predict the short-term effects of 4,4-methylenedianiline to aquatic life, plants, birds, or land animals.

Chronic Ecological Effects

Chronic (long-term) toxic effects may include shortened life span, reproductive problems, lower fertility, and changes in appearance or behavior in exposed animals. These effects can be seen long after first exposure(s) to toxic chemicals. Insufficient data are available on the long-term effects of 4,4-methylenedianiline to aquatic life, plants, birds, or land animals.

Water Solubility

4,4-methylenedianiline is slightly soluble in water. Concentrations of 1 to 100 milligrams will mix with a liter of water.

Persistence in the Environment

4,4-Methylenedianiline is highly persistent in water, with a half-life between 20 and 200 days. The half-life of a pollutant is the amount of time it takes for one half of the chemical to be degraded.

Bioaccumulation in Aquatic Organisms

Some substances increase in concentration, or *bioaccumulate,* in living organisms as they breathe contaminated air, drink contaminated water, or eat contaminated food. These chemicals can become concentrated in the tissues and internal organs of animals as well as humans.

The concentration of 4,4-methylenedianiline found in fish tissues is expected to be about the same as the average concentration of 4,4-methylenedianiline in the water from which the fish was taken.

Recommended Risk-Reduction Measures

Proper training of all transporters will reduce the likelihood of a mishap or accident resulting in a leak or spill to the environment. The correct use of DOT labeling (per the MSDS) on all containers, trucks, and rail cars will enable emergency responders to react properly and quickly to any disaster thereby reducing the potential risk to the environment and to personnel.

Storage of 4,4-methylenedianiline should be segregated from incompatible chemicals to minimize the risk of cross-contamination or contact.

If a spill or leak to the environment has occurred, fire department, emergency response and/or hazardous materials spill personnel should be notified immediately. Cleanup should be attempted only by those trained in proper spill containment procedures. Contaminated soils should be removed for incineration and replaced with clean soil. If 4,4-methylenedianiline should contact the water table, aquifer, or navigable waterway, the local and/or state emergency response authorities must be notified. A comprehensive emergency response or disaster preparedness/recovery plan should be in place prior to any operations involving the use, transportation, storage, or disposal of 4,4-methylenedianiline. If 4,4-methylenedianiline is spilled or leaked, the following specific steps are recommended:

- ☑ Restrict persons not wearing protective clothing from area of spill or leak until cleanup is complete and area can be opened for normal.
- ☑ Remove all ignition sources.
- ☑ Shovel, vacuum, or damp mop spilled materials and place in sealed drums for reclamation or disposal. Avoid generating excessive amounts of airborne dusts during cleanup operations.
- ☑ It may be necessary to dispose of 4,4-methylenedianiline as a hazardous waste. The responsible state agency or the regional office of the federal Environmental Protection Agency (EPA) should be contacted for specific recommendations.

RISK ASSESSMENT: BUSINESS

General Assessment

Accidents or mishaps involving 4,4-methylenedianiline can present a significant threat to business operations. The loss or damage of equipment or facilities can significantly affect fiscal viability. Lawsuits that may result from personnel injury/death, public exposures, and/or environmental contamination will also require a serious expenditure of resources. Media attention surrounding an injury, death, or environmental damage can also result in a loss of profits and loss of current as well as future business. Always remember that anytime the terms "cancer" or "carcinogen" are used, public hysteria, emotion, and ignorance can run equally high. This should be carefully considered whenever developing or implementing public relation policies.

Recommended Risk-Reduction Measures

Company attorneys, safety and health professionals, and environmental specialists should be involved in the development of any procedures dealing with safety or emergency response. A company official should be pre-designated as a public relations officer with specific training in dealing with the press. Corporate plans and policies should be developed, approved, and implemented long before any need for such arises.

REFERENCES

Ahlbom, A. 1993. *Biostatistics for Epidemiologists.* New York: CRC Press/Lewis Publishers.

American Conference of Governmental Industrial Hygienists. 1988. *Documentation of the Threshold Limit Values and Biological Exposure Indices,* 5th Edition (with updates). Cincinnati: ACGIH

Asante-Duah, D. K. 1993. *Hazardous Waste Risk Assessment.* New York: CRC Press/Lewis Publishers.

Bloom, A. D. and F. J. DeSerres. 1995. *Ecotoxicity and Human Health.* New York: CRC Press/Lewis Publishers.

Brusick, D. J. 1994. *Methods for Genetic Risk Assessment.* New York: CRC Press/Lewis Publishers.

Bueche, J. R. 1972. *Principles of Physics,* 2nd Edition. New York: McGraw-Hill.

Calabrese, E. J. 1994. *Biological Effects of Low Level Exposures.* New York: CRC Press/Lewis Publishers.

Cockerman, L. G., B. S. Shane. 1994. *Basic Environmental Toxicology.* New York: CRC Press/Lewis Publishers.

Cothern, R. R. 1993. *Comparative Environmental Risk Assessment.* New York: CRC Press/Lewis Publishers.

Gardner, R. W. 1994. *Chemical Intolerance.* New York: CRC Press/Lewis Publishers.

Hallenbeck, W. H. 1993. *Quantitative Risk Assessment for Environmental and Occupational Health,* 2nd Edition. New York: CRC Press/Lewis Publishers.

Howard, P. H. 1995. *Handbook of Environmental Fate and Exposures Data for Organic Chemicals,* Vol. I-IV. New York: CRC Press/Lewis Publishers.

Jones, F. E. 1994. *Toxic Organic Vapors in the Workplace.* New York: CRC Press/Lewis Publishers.

Keith, L. H. and M. M. Walker. 1995. *Handbook of Air Toxics.* New York: CRC Press/Lewis Publishers.

Landis, W. G., M. H. Yu. 1995. *Introduction to Environmental Toxicology.* New York: CRC Press/Lewis Publishers.

Langerman, N. 1994 *Precautionary Labels for Chemical Containers.* New York: CRC Press/Lewis Publishers.

Lauwerys, R. R., P. Hoet. 1993. *Industrial Chemical Exposure.* New York: CRC Press/Lewis Publishers.

Lewis, R. J., Sr. 1993. *Hawley's Condensed Chemical Dictionary,* 12th Edition. New York: Van Nostrand Reinhold.

Lewis, R. J., Sr. 1992. *Sax's Dangerous Properties of Industrial Materials,* Volumes 1, 2, and 3, 8th Edition. New York: Van Nostrand Reinhold.

Manahan, S. E. 1992. *Toxicological Chemistry,* 2nd Edition. New York: CRC Press/Lewis Publishers.

Mellström, G. A., J. E. Wahlberg, and H. I. Maibach. 1994. *Protective Gloves for Occupational Use.* New York: CRC Press/Lewis Publishers.

Meyer, E. 1990. *Chemistry of Hazardous Materials.* Englewood Cliffs, NJ: Prentice Hall.

National Cancer Institute. 1978. *Carcinogenesis Technical Report Series. Bioassay of Aniline for Possible Carcinogenicity.* DEHW (NIH) Publication Number: 78-1323, p 53. Washington, DC: U.S. Government Printing Office.

Ott, W. R. 1995. *Environmental Statistics and Data Analysis.* New York: CRC Press/Lewis Publishers.

Philp, R. B. 1995. *Environmental Hazards and Human Health.* New York: CRC Press/Lewis Publishers.

Quigley, D. R. 1994. *Handbook of Emergency Chemical Management.* New York: CRC Press/Lewis Publishers.

Rea, W. J. 1995. *Chemical Sensitivity,* Vol. I-III. New York: CRC Press/Lewis Publishers.

Richardson, M. 1995. *Dictionary of Substance and Their Effects,* Vol. I-VI. New York: CRC Press/Lewis Publishers.

Sacarello, H. L. A. 1994. *The Comprehensive Handbook of Hazardous Materials.* New York: CRC Press/Lewis Publishers.

Sheftel, V. O. 1995. *Handbook of Toxic Properties of Monomers and Additives.* New York: CRC Press/Lewis Publishers.

Talbott, E., G. Craun. 1995. *An Introduction to Environmental Epidemiology.* New York: CRC Press/Lewis Publishers.

U.S. Department of Health and Human Services, National Institute for Occupational Safety and Health. 1994. *NIOSH Pocket Guide to Chemical Hazards.* Washington, D.C.: U.S. Government Printing Office.

MATERIAL SAFETY DATA SHEET

CHEMICAL NAME	
METHYL ETHYL KETONE	**CRC PRESS/LEWIS PUBLISHERS DISCLAIMER:** THE INFORMATION AND RECOMMENDATIONS PRESENTED HEREIN ARE BASED ON SOURCES BELIEVED TO BE RELIABLE. CRC MAKES NO REPRESENTATION ON ITS COMPLETENESS OR ACCURACY. IT IS THE RESPONSIBILITY OF THE USER TO DETERMINE THE CHEMICAL'S SUITABILITY FOR ITS INTENDED USE, THE CHEMICAL'S SAFE USE, AND THE CHEMICAL'S PROPER DISPOSAL. NO REPRESENTATIONS AND/OR WARRANTIES, EITHER EXPRESSED OR IMPLIED, OF THE MERCHANTABILITY OR FITNESS FOR A PARTICULAR PURPOSE, OR OF ANY OTHER NATURE, ARE MADE WITH RESPECT TO THE INFORMATION PROVIDED IN THIS MSDS OR TO THE CHEMICAL TO WHICH INFORMATION MAY REFER. CRC NEITHER ASSUMES NOR AUTHORIZES ANY OTHER PERSON TO ASSUME FOR IT, ANY OTHER ADDITIONAL RESPONSIBILITY OR LIABILITY FOR THE USE OF, OR RELIANCE UPON, THIS INFORMATION.

HAZARD WARNING INFORMATION

HEALTH	FIRE	REACTIVE	OTHER	DEGREE OF HAZARD	COLOR CODING	OTHER CODES
1	3	0		0 = Minimum Hazard 1 = Slight Hazard 2 = Moderate Hazard 3 = Serious Hazard 4 = Severe Hazard	HEALTH = BLUE FIRE = RED REACTIVITY = YELLOW OTHER = WHITE	OX = Oxidizer ACID = Acid ALK = Alkali COR = Corrosive W = Use No Water

SECTION I - GENERAL INFORMATION

Characterization	RCRA Number	EPA Class
Ketone	**U159**	**Toxic Waste**

DOT Proper Shipping Name **Methyl Ethyl Ketone**	Chemical Abstract Service (CAS) Number **78-93-3**
DOT Hazard Class and Label Requirements **Flammable Liquid**	DOT Emergency Guide Code **26**
DOT Identification Number **UN 1193**	Chemical Formula **C_4H_8O**

Synonyms

2-Butanone; butan-2-one; ethyl methyl ketone; methylacetone; 2-oxobutane; MEK.

SECTION II - HAZARDOUS INGREDIENTS AND IDENTITY INFORMATION

Hazardous Components (specific identity)	OSHA Exposure Criteria	NIOSH Exposure Criteria	Immediately Dangerous to Life and Health (IDLH)	ACGIH Exposure Criteria
Methyl ethyl ketone (derivation: From mixed n-butylenes and sulfuric acid to cause hydrolysis followed by distillation to separate sec-butyl alcohol which is dehydrogenated; by controlled oxidation of butane; by fermentation). **1 ppm = 2.95 mg/m^3**	**PEL:** 200 ppm 590 mg/m^3 **STEL:** 300 ppm 885 mg/m^3	**REL:** 200 ppm 590 mg/m^3 **STEL:** 300 ppm 885 mg/m^3	**3000 ppm**	**TLV:** 200 ppm 590 mg/m^3 **STEL:** 300 ppm 885 mg/m^3

SECTION III - PHYSICAL AND CHEMICAL CHARACTERISTICS

Boiling Point **176°F (80°C)**	Specific Gravity (H_2O = 1) **0.80**
Vapor Pressure (mm Hg) **71 at 69°F (20°C)**	Molecular Weight **72.12**
Vapor Density (Air = 1) **2.42**	Freezing Point **-123°F (-86°C)**

Solubility

Soluble in water (28%). Soluble in alcohol, benzene, and ether. Miscible in fixed oils.

Appearance and Odor

Colorless liquid with a sweet mint or acetone-like odor. Odor Threshold = 25 ppm.

SECTION IV - FIRE AND EXPLOSION HAZARD DATA

Flash Point (method used) **16°F (-9°C) closed cup**	Explosive Limits in Air % by Volume **LEL: 1.4% at 200 UEL: 11.4% at 200°F**
NFPA Classification **Class 1BFlammable Liquid**	Autoignition Temperature **759°F (404°C)**

Extinguishing Media

Dry chemical, carbon dioxide, water spray, fog, or alcohol-resistant foam.

Special Fire Fighting Procedures

Poisonous gases are produced in fire. Wear full protective clothing and self-contained breathing apparatus. Fight fire from a distance if possible. Keep fire-exposed containers cool with water spray. A solid stream of water may spread fire. If it can be done safely, move fire-exposed containers from area.

Unusual Fire and Explosion Hazards

Containers may explode in fire due to the buildup of internal pressures. BLEVE (boiling liquid expanding vapor explosion) is possible. Immediately evacuate if rising sound is heard from venting device or a noticeable discoloration is observed on the sides of the tank or storage vessel.

SECTION V - REACTIVITY DATA

Stability		Conditions to Avoid Normally stable. Do not allow methyl ethyl ketone to come into contact with incompatible materials. Avoid contact with fire or high heat sources. Do not use in confined spaces.
Stable X	Unstable	Incompatibility *(materials to avoid)* Incompatible with chlorosulfonic acid, fuming sulfuric acid, potassium-*t*-butoxide, hydrogen peroxide + nitric acid, 2-propanol, chloroform + alkali, amines, ammonia, inorganic acids, caustics, copper, oxidizers.
Hazardous Polymerization		Conditions to Avoid Under normal conditions of temperature and pressure, hazardous polymerization of methyl ethyl ketone is not expected to occur. Methyl ethyl ketone will dissolve some plastics.
May Occur	Will Not Occur X	Hazardous Decomposition or By-products When heated to decomposition, methyl ethyl ketone emits acrid, irritating smoke and fumes, including oxides of carbon.

SECTION VI - HEALTH HAZARD DATA

Primary Route(s) of Entry:	Inhalation?	Absorption (skin)?	Ingestion?
	X	X	

Health Hazards

INHALATION: Irritation of the mucosa of the respiratory tract causing respiratory depression with gasping, choking, and dypsnea. Higher concentrations causes CNS depression with headache, dizziness, fainting, tremor, chemical pneumonitis, coma, and possibly death.

ABSORPTION: Skin contact with liquid or vapor can result in irritation and may cause dermatitis and paresthesia (tingling). Eye contact causes pain, irritation, and corneal damage, tearing.

INGESTION: Unlikely. Symptoms of irritation, sore throat, nausea, coughing, salivation, vomiting.

Carcinogenicity	NTP Listed?	IARC Cancer Review Group?	OSHA Regulated?	Target Organs?
Unknown Human Unknown Animal	No	No	29 CFR 1910.1000 Table Z-1	Respiratory system; skin; eyes; CNS.

Medical Conditions Generally Aggravated by Exposure

Skin conditions (dermatitis) may be aggravated.

Emergency and First-aid Procedures

Eye contact: Immediately flush large amounts of water for 15 minutes (minimum); occasionally lifting eyelids. Skin contact: Remove all contaminated clothing. Immediately wash area with large amounts of soap and water. For inhalation: Remove the person from exposure. Provide respiratory assistance and CPR. Transfer promptly to medical facility. If swallowed: Seek medical attention immediately. Give a conscious and alert victim several glasses of milk or water. The decision to induce vomiting must be carefully considered (it may cause direct aspiration of the liquid leading to chemical pneumonitis).

SECTION VII - PRECAUTIONS FOR SAFE HANDLING AND USE

Steps to be Taken in Case Material is Released or Spilled

Remove all ignition sources. Restrict persons not wearing protective equipment from entering area of spill or leak until cleanup is complete. Ventilate area. Absorb liquid in dry sand or other material and deposit in sealed containers. Damp mop residue. Alert appropriate state or federal response agencies.

Preferred Waste Disposal Method

Burn in a chemical incinerator equipped with an afterburner and scrubber.

Precautions to be Taken in Handling and Storage

Do not store with incompatible chemicals since violent reactions can occur. Store in tightly closed containers in a cool, dark, well-ventilated area away from fire, sparks, and flame. Use non-sparking tools on containers. Ground and bond metal containers during transfer operations.

Other Precautions and Warnings

Containers may explode in fire or under conditions of extreme heat (do not store outdoors or in direct sunlight). Protect containers from physical damage.

SECTION VIII - CONTROL MEASURES AND PERSONAL PROTECTIVE EQUIPMENT

Respiratory Protection *(specify type)*

Exposures above 200 ppm: MSHA/NIOSH-approved full-facepiece respirator with organic vapor cartridge. Greater protection is obtained from a self-contained breathing apparatus (SCBA) with full facepiece and pressure demand or other positive pressure mode.

Ventilation

Local exhaust or general mechanical systems recommended.

Protective Gloves	Eye Protection	Other Protective Clothing
Impervious Materials	Chemical Goggles and/or Face Mask	Protective Apron

Work/Hygiene Practices

Always wash hands thoroughly after using chemical; never bring food, drink, or smoking materials into vicinity of chemicals.

METHYL ETHYL KETONE

C_4H_8O CAS: 78-93-3

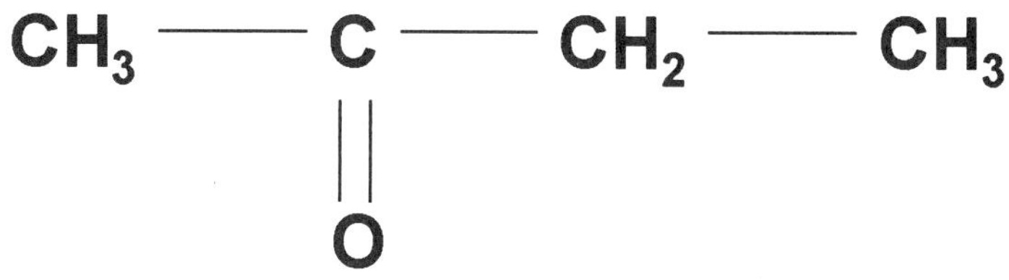

IDENTIFICATION AND TYPICAL USES

Methyl ethyl ketone is a colorless liquid with a sweet mint or acetone-like odor. It is used as an industrial solvent, in the manufacture of smokeless powder and colorless synthetic resins, and as a surface coating. It is also used in the production of plastics, textiles, cements and adhesives, artificial leather, cosmetics, lubricating oils, pharmaceuticals, cleaning fluids, paint removers, and paints.

RISK ASSESSMENT: HEALTH

General Assessment

Methyl ethyl ketone is moderately toxic to humans by *inhalation* and *absorption*. Skin or eye contact can also cause surface tissue damage and inflammation. There are no reports in the references to indicate that ingestion is a likely exposure route. The references contain no information on the carcinogenic or mutagenic properties of methyl ethyl ketone. However, it is suggested that it be treated as a suspected teratogen. Overexposure to this compound is not common. Its irritating properties and penetrating odor are usually present at levels low enough to serve as a warning to personnel.

Inhalation of methyl ethyl ketone causes irritation of the mucosa of the respiratory tract leading to respiratory depression with symptoms of gasping, choking, and dypsnea. Exposure to higher concentrations can cause depression of the central nervous system (CNS) with symptoms of headache, lachrymation, somnolence, decreased breathing, dizziness, fainting, blurred vision, tremor, parasthesia (numbness or tingling in the extremities), coma, and possibly death (rare but possible). If the liquid should be aspirated (during vomiting, for example), it can lead to chemical pneumonitis.

Skin contact with the liquid or vapor can lead to chemical dermatitis and localized paresthesia with symptoms of prickling or tingling with no apparent stimulus and associated sensory nerve damage. Methyl ethyl ketone can pass relatively quickly through intact skin to cause toxic systemic effects. When combined with other exposure routes, such as inhalation, the compound effects can be dangerous. Eye contact causes irritation with lachrymation and, in prolonged contact, may lead to ocular pain, corneal damage, and possible loss of vision.

Although ingestion is unlikely in the occupational setting, it can cause irritation of the mouth, sore throat, coughing, salivation, nausea, and vomiting.

☠ *Acute Health Effects*

The following acute (short-term) health effects may occur immediately or shortly after exposure to methyl ethyl ketone:

<u>Skin</u>: Irritation, dermatitis, and possible localized paresthesia.

<u>Eye</u>: Irritation with pain and inflammation that can lead to corneal damage and possible loss of vision.

<u>Lung</u>: Irritation causing coughing and/or shortness of breath. Higher exposures can cause dizziness, headache, and, possibly, coma and death.

<u>CNS</u>: A mild to moderate narcotic compound producing neurotoxicity as it is partially metabolized to methanol.

Chronic Health Effects

The following chronic (long-term) health effects may occur at some time after exposure to methyl ethyl ketone and can last for months or even years:

Cancer Hazards: According to information presented in the references, methyl ethyl ketone has not been adequately tested for its ability to cause cancer in laboratory test animals.

Reproduction: It has been suggested that this chemical be treated as a potential teratogen since it has been shown to adversely affect fetal development in limited experimental animal studies.

Other Chronic Effects: Studies with methyl ethyl ketone have shown that long-term skin exposure may produce dermatitis. However, unlike other ketones, methyl ethyl ketone is not considered a sensitizer. Also, methyl ethyl ketone has not been adequately tested to determine whether brain or other nerve damage could occur with repeated exposure. However, many solvents and other petroleum-based chemicals have been shown to cause such damage. Symptoms may include reduced memory and concentration, personality changes (withdrawal, irritability) fatigue, sleep disturbances, reduced concentration, and/or effects on nerves supplying internal organs (autonomic nerves) or the nerves to the arms and legs (weakness, tingling).

Recommended Risk-Reduction Measures

Personnel should avoid direct contact with methyl ethyl ketone. Unless a less toxic chemical can be substituted for a hazardous substance, *engineering controls* are the most effective method of reducing exposures. The best protection is to enclose operations and/or provide local exhaust ventilation at the site of chemical release. Because of its low flash point, maximum explosion-proof ventilation systems should be considered. While not always operationally feasible, isolating operations can also reduce exposure. Using respiratory protection is less effective than the controls mentioned above, but is still advisable whenever working with methyl ethyl ketone. For exposures to the PEL (200 ppm), use a NIOSH-approved, full facepiece organic vapor respirator. For higher or prolonged exposures, or when exposure levels are unknown (as in an emergency) a self-contained breathing apparatus (SCBA) with full facepiece operated in pressure demand mode, or a supplied-air respirator operated in continuous flow mode are the recommended respiratory protection methods of choice. If a full facepiece is not available, then chemical goggles should be worn to protect the eyes. Whenever a chemical splash hazard exists, a face shield and a rubber apron should be worn. To prevent hand and skin exposures, butyl rubber gloves should be used. Glove manufacturers should be contacted and permeation studies obtained *before* final glove selections are made.

Administrative controls should also be in place to minimize the potential for human exposures. These may include written procedures or policies which specify the methods and techniques that will be practiced whenever personnel are to work with or around methyl ethyl ketone.

All personnel should receive *training* on the use, hazards, protective measures, emergency actions, and other precautions per 29 CFR 1910.1200 (Hazard Communication), prior to the first assignment in an area where methyl ethyl ketone is used or stored.

If symptoms develop or overexposure is suspected, the following medical tests may be helpful:

- ☑ Skin testing by a qualified allergist with special testing and careful evaluation of exposure history (may help diagnose allergy).
- ☑ Neurological evaluation of the central nervous system.

Other methods to reduce exposure include:

- ☑ Where possible, enclose operations and use local exhaust ventilation at the site of chemical release. Explosion-proof design is recommended. If local exhaust ventilation or enclosure is not used, respiratory protection should be mandatory.
- ☑ Always ensure that proper protective clothing is worn when using chemical substances.
- ☑ Wash thoroughly immediately after exposure to methyl ethyl ketone and at the end of the work shift or before eating, drinking, or smoking. Never wear clothes contaminated with methyl ethyl ketone home.
- ☑ Hazard warning information should be posted in the work area. As part of an on-going education and training program, all information on the health and safety hazards of methyl ethyl ketone should be communicated to all exposed workers.

RISK ASSESSMENT: ENVIRONMENT

General Assessment

The environment is at risk of exposure during transportation, storage, disposal, or destruction of methyl ethyl ketone. In almost every scenario, the threat of environmental exposure is contingent upon the proper handling of the chemical substance. Accidental spills,

large or small, can result in fire, explosion, and possible contamination of the surrounding environmental mediums (water, soil, and air).

Methyl ethyl ketone is considered a Class 1B flammable liquid (per OSHA 29 CFR 1910.106). Its low flash point and relatively low boiling point make this chemical an extremely dangerous fire and explosion hazard. It can react with strong oxidizers and acids to cause a serious fire and explosion hazard. When involved in a fire, it can release acrid and irritating oxides of carbon. A boiling liquid expanding vapor explosion (BLEVE) is also a possibility. These characteristics require special consideration during any emergency involving a leak or spill of methyl ethyl ketone.

The proper disposal/destruction method for methyl ethyl ketone is to burn it in a chemical incinerator equipped with an afterburner and air scrubber.

Methyl ethyl ketone can enter the environment through manufacturing, unchecked discharge into effluents from industrial or municipal waste treatment plants, and through spills.

Acute Ecological Effects

Acute (short-term) toxic effects may include the death of animals, birds, or fish, and death or low growth rate in plants. Acute effects are seen 2 to 4 days after animals or plants are exposed to methyl ethyl ketone. This chemical has a slight acute toxicity to aquatic life. No data are available on the short-term effects of methyl ethyl ketone on plants, birds, or land animals.

Chronic Ecological Effects

Chronic (long-term) toxic effects may include shortened life span, reproductive problems, lower fertility, and changes in appearance or behavior in exposed animals. These effects can be seen long after first exposure(s) to toxic chemicals. Methyl ethyl ketone has slight chronic toxicity in aquatic life. No data are available to evaluate the long-term effects of methyl ethyl ketone to plants, birds, or land animals.

Water Solubility

Methyl ethyl ketone is highly soluble in water. Concentrations of 1000 milligrams and more will mix with a liter of water.

Persistence in the Environment

Methyl ethyl ketone is slightly persistent in water, with a half-life of between 2 to 20 days. The half-life of a pollutant is the amount of time it takes for one half of the chemical to be degraded. Approximately 59% of methyl ethyl ketone will eventually end up in the air; the rest will end up in the water.

Bioaccumulation in Aquatic Organisms

Some substances increase in concentration, or *bioaccumulate,* in living organisms as they breathe contaminated air, drink contaminated water, or eat contaminated food. These chemicals can become concentrated in the tissues and internal organs of animals as well as humans.

The concentration of methyl ethyl ketone found in fish tissues is expected to be about the same as the average concentration of methyl ethyl ketone in water from which the fish was taken.

Recommended Risk-Reduction Measures

Proper training of all transporters will reduce the likelihood of a mishap or accident resulting in a leak or spill to the environment. The correct use of DOT labeling (per the MSDS) on all containers, trucks, and rail cars will enable emergency responders to react properly and quickly to any disaster thereby reducing the potential risk to the environment and to personnel.

Storage of methyl ethyl ketone should be segregated from incompatible chemicals to minimize the risk of cross-contamination or contact. Buildings designated for storage should be equipped with appropriate fire protection systems (alarms, sprinklers, emergency lighting, portable extinguishers). Maximum explosion-proof design of ventilation systems is recommended.

If a spill or leak to the environment has occurred, fire department, emergency response, and/or hazardous materials spill personnel should be notified immediately. Cleanup should be attempted only by those trained in proper spill containment procedures. Contaminated soils should be removed for incineration and replaced with clean soil. If methyl ethyl ketone should contact the water table, aquifer, or navigable waterway, time is of the essence. It is highly soluble in water and, therefore, total containment and remediation may not be possible. When such spills occur, the local and/or state emergency response authorities may require notification. A comprehensive emergency response or disaster preparedness/recovery plan should be in place prior to any operations involving the use, transportation, storage, or disposal of methyl ethyl ketone.

If methyl ethyl ketone is spilled or leaked, the following specific steps are recommended:

☑ Restrict persons not wearing protective clothing from area of spill or leak until cleanup is complete and area can be opened for normal work.

☑ Remove all ignition sources.

☑ Ventilate area of spill or leak.

☑ Absorb liquids in vermiculite, dry sand, earth, or a similar material and deposit in sealed containers for disposal. Use non-sparking tools!

☑ Keep methyl ethyl ketone out of a confined space, such as a sewer, because of the possibility of explosion (unless the sewer is designed to prevent the buildup of explosive concentrations).

☑ It may be necessary to dispose of methyl ethyl ketone as a hazardous waste. The responsible state agency or the regional office of the federal Environmental Protection Agency (EPA) should be contacted for specific recommendations.

RISK ASSESSMENT: BUSINESS

General Assessment

Accidents or mishaps involving methyl ethyl ketone can present a significant threat to business operations. The loss or damage of equipment or facilities can significantly affect fiscal viability. Lawsuits that may result from personnel injury, public exposures, and/or environmental contamination will also require a serious expenditure of resources. Media attention surrounding an injury, illness, or environmental damage can also result in a loss of profits and loss of current as well as future business prospects.

Recommended Risk-Reduction Measures

Company attorneys, safety and health professionals, and environmental specialists should be involved in the development of any procedures dealing with safety or emergency response. A company official should be pre-designated as a public relations officer with specific training in dealing with the press. Corporate plans and policies should be developed, approved, and implemented long before any need for such arises.

REFERENCES

Ahlbom, A. 1993. *Biostatistics for Epidemiologists.* New York: CRC Press/Lewis Publishers.

American Conference of Governmental Industrial Hygienists. 1988. *Documentation of the Threshold Limit Values and Biological Exposure Indices,* 5th Edition (with updates). Cincinnati: ACGIH.

Bloom, A. D. and F. J. DeSerres. 1995. *Ecotoxicity and Human Health.* New York: CRC Press/Lewis Publishers.

Burrell, R., D. K. Flaherty, and L. J. Sauers. 1992. *Toxicology of the Immune System.* New York: Van Nostrand Reinhold.

Calabrese, E. J., E. M. Kenyon. *Air Toxics and Risk Assessment.* New York: CRC Press/Lewis Publishers.

Cockerman, L. G., B. S. Shane. 1994. *Basic Environmental Toxicology.* New York: CRC Press/Lewis Publishers.

Gardner, R. W. 1994. *Chemical Intolerance.* New York: CRC Press/Lewis Publishers.

Hodgson, E., R. B. Mailman, and J. E. Chambers. 1988. *Dictionary of Toxicology.* London: The MacMillian Company.

Hosty, J. W., and P. Foster. 1990. *A Practical Guide to Chemical Spill Response.* New York: Van Nostrand Reinhold.

Howard, P. H., et. al. 1991. *Handbook of Environmental Degradation Rates.* New York: CRC Press/Lewis Publishers.

Jones, F. E. 1994. *Toxic Organic Vapors in the Workplace.* New York: CRC Press/Lewis Publishers.

Keith, L. H. and M. M. Walker. 1995. *Handbook of Air Toxics.* New York: CRC Press/Lewis Publishers.

Lauwerys, R. R., P. Hoet. 1993. *Industrial Chemical Exposure.* New York: CRC Press/Lewis Publishers.

Lewis, R. J., Sr. 1993. *Hawley's Condensed Chemical Dictionary,* 12th Edition. New York: Van Nostrand Reinhold.

Lewis, R. J., Sr. 1992. *Sax's Dangerous Properties of Industrial Materials,* Volumes 1, 2, and 3, 8th Edition. New York: Van Nostrand Reinhold.

Meyer, E. 1990. *Chemistry of Hazardous Materials.* Englewood Cliffs, NJ: Prentice Hall.

Patnaik, P. 1992. *A Comprehensive Guide to the Hazardous Properties of Chemical Substances.* New York: Van Nostrand Reinhold.

Philp, R. B. 1995. *Environmental Hazards and Human Health.* New York: CRC Press/Lewis Publishers.

Quigley, D. R. 1994. *Handbook of Emergency Chemical Management.* New York: CRC Press/Lewis Publishers.

Raleigh, R. L. and W. A. McGee. 1972. Effects of short, high concentration exposures to acetone as determined by observation in the work area. *Journal of Occupational Medicine.* 14:607-610.

Rea, W. J. 1995. *Chemical Sensitivity,* Vol. I-III. New York: CRC Press/Lewis Publishers.

Richardson, M. 1995. *Dictionary of Substance and Their Effects,* Vol. I-VI. New York: CRC Press/Lewis Publishers.

U.S. Department of Health and Human Services, National Institute for Occupational Safety and Health. 1994. *NIOSH Pocket Guide to Chemical Hazards.* Washington, D.C.: U.S. Government Printing Office.

U.S. Department of Health and Human Services, National Institute for Occupational Safety and Health. 1978. *Criteria for a Recommended Standard - Occupational Exposure to Ketones.* DHEW Publication Number: 78-173. Washington, D.C.: U.S. Government Printing Office.

MATERIAL SAFETY DATA SHEET

CHEMICAL NAME

METHYL HYDRAZINE

CRC PRESS/LEWIS PUBLISHERS DISCLAIMER: THE INFORMATION AND RECOMMENDATIONS PRESENTED HEREIN ARE BASED ON SOURCES BELIEVED TO BE RELIABLE. CRC MAKES NO REPRESENTATION ON ITS COMPLETENESS OR ACCURACY. IT IS THE RESPONSIBILITY OF THE USER TO DETERMINE THE CHEMICAL'S SUITABILITY FOR ITS INTENDED USE, THE CHEMICAL'S SAFE USE, AND THE CHEMICAL'S PROPER DISPOSAL. NO REPRESENTATIONS AND/OR WARRANTIES, EITHER EXPRESSED OR IMPLIED, OF THE MERCHANTABILITY OR FITNESS FOR A PARTICULAR PURPOSE, OR OF ANY OTHER NATURE, ARE MADE WITH RESPECT TO THE INFORMATION PROVIDED IN THIS MSDS OR TO THE CHEMICAL TO WHICH INFORMATION MAY REFER. CRC NEITHER ASSUMES NOR AUTHORIZES ANY OTHER PERSON TO ASSUME FOR IT, ANY OTHER ADDITIONAL RESPONSIBILITY OR LIABILITY FOR THE USE OF, OR RELIANCE UPON, THIS INFORMATION.

HAZARD WARNING INFORMATION

HEALTH	FIRE	REACTIVE	OTHER	DEGREE OF HAZARD	COLOR CODING	OTHER CODES
3	3	2		0 = Minimum Hazard 1 = Slight Hazard 2 = Moderate Hazard 3 = Serious Hazard 4 = Severe Hazard	HEALTH = BLUE FIRE = RED REACTIVITY = YELLOW OTHER = WHITE	OX = Oxidizer ACID = Acid ALK = Alkali COR = Corrosive ~~W~~ = Use No Water

SECTION I - GENERAL INFORMATION

Characterization
Hypergolic Compound

RCRA Number
P068

EPA Class
Acute Toxic Waste

DOT Proper Shipping Name
Methyl hydrazine

Chemical Abstract Service (CAS) Number
60-34-4

DOT Hazard Class and Label Requirements
Flammable Liquid and Poison; Corrosive

DOT Emergency Guide Code
57

DOT Identification Number
UN 1244

Chemical Formula
CH_3NHNH_2

Synonyms
Monomethylhydrazine; MMH; hydrazomethane; 1-methyl hydrazine.

SECTION II - HAZARDOUS INGREDIENTS AND IDENTITY INFORMATION

Hazardous Components (specific identity)	OSHA Exposure Criteria	NIOSH Exposure Criteria	Immediately Dangerous to Life and Health (IDLH)	ACGIH Exposure Criteria
Methyl hydrazine (derivation: By reaction of monochloramine and monoethylamine at elevated temperatures and pressures; from primary amines through substituted ureas by treatment with nitrous acid followed by reduction and hydrolysis with HCl). **1 ppm = 1.92 mg/m^3**	**PEL (skin):** **0.2 ppm** **0.35 mg/m^3** **CEILING** **STEL:** **Not Applicable**	**REL (ceiling):** **0.04 ppm** **0.08 mg/m^3** **(2-hr ceiling)** **Possible Cancer Agent**	**20 ppm**	**TLV (skin):** **0.2 ppm** **0.38 mg/m^3** **Suspected Human Carcinogen**

SECTION III - PHYSICAL AND CHEMICAL CHARACTERISTICS

Boiling Point
190°F (88°C)

Specific Gravity (H_2O = 1)
0.87

Vapor Pressure (mm Hg)
38 mm at 69°F (20°C)

Molecular Weight
46.1

Vapor Density (Air = 1)
1.6

Melting Point
-626°F (-52°C)

Solubility
Miscible in water. Hydrazines, ether, and alcohol.

Appearance and Odor
Colorless, fuming, oily liquid with an ammonia or fish-like odor. Odor Threshold = 1.7 ppm.

SECTION IV - FIRE AND EXPLOSION HAZARD DATA

Flash Point (method used)
17°F (-8.3°C) closed cup

Explosive Limits in Air % by Volume
LEL: 2.5% UEL: 92%

NFPA Classification
Class IB Flammable Liquid

Autoignition Temperature
385°F (196°C)

Extinguishing Media
Use water spray in flooding amounts as fog, or use alcohol foam, carbon dioxide, or dry chemical.

Special Fire Fighting Procedures
Structural protective clothing is permeable. Remain clear of smoke, water fallout, and water runoff. Poisonous gases are produced in fire. Move containers if it can be done without risk. Cool fire-exposed containers with water spray and use water spray to disperse vapors. Evacuate non-essential personnel 2500 feet from the fire area (consider wind conditions).

Unusual Fire and Explosion Hazards
Contact with oxidizing agents can cause violent and explosive ignition. A volatile liquid that is flammable over a wide vapor-to-air range. Vapors are heavier than air and may travel a considerable distance to a source of ignition and flashback to cause fire or explosion. Burns with clear flame.

SECTION V - REACTIVITY DATA

Stability		Conditions to Avoid **Methyl hydrazine is stable at room temperature in closed containers under normal conditions of storage and handling. Avoid contact with incompatible chemicals and materials and heat.**
Stable **X**	Unstable	Incompatibility *(materials to avoid)* **Highly reactive with oxidizers (chlorine, bromine, nitrogen dioxide), halogens, ferric oxide (iron rust), metallic mercury, fuming nitric acid, hydrogen peroxide (ignites spontaneously with oxidizers).**
Hazardous Polymerization		Conditions to Avoid **Under normal conditions of temperature and pressure, hazardous polymerization of methyl hydrazine is not expected to occur. Do NOT expose to air (may ignite without warning).**
May Occur	Will Not Occur **X**	Hazardous Decomposition or By-products **When heated to decomposition, methyl hydrazine will produce toxic and irritating gases, including oxides of carbon and toxic fumes of nitrogen oxides.**

SECTION VI - HEALTH HAZARD DATA

Primary Route(s) of Entry:	Inhalation?	Absorption (skin)?	Ingestion?
	X	X	X

Health Hazards

INHALATION: Highly corrosive to the skin, eyes, and mucosa. Causes choking, chest pain, dyspnea, lethargy, nausea, convulsions, tremors, pulmonary edema, liver injury, hypoxia, blood disorders, and death.

ABSORPTION: Will pass through unbroken skin to cause toxic systemic effects as noted above. Extremely corrosive to skin tissue causing chemical burns (2nd and 3rd degree) on short contact. Very injurious to the eyes, causing corneal damage and permanent blindness.

INGESTION: A poison by ingestion. Delayed gastrointestinal irritation. Systemic effects can occur.

Carcinogenicity	NTP Listed?	IARC Cancer Review Group?	OSHA Regulated?	Target Organs?
Suspected Human Confirmed Animal	**No**	**No**	**29 CFR 1910.1000 Table Z-1**	**CNS, skin, liver, eyes, CVS, blood, resp. system.**

Medical Conditions Generally Aggravated by Exposure

Existing deficiencies in CNS function, liver, kidneys, and the circulatory system may be aggravated.

Emergency and First-aid Procedures

Eye contact: Flush immediately with water for 15 minutes (minimum) while holding upper and lower eyelids open. Seek medical attention. Skin contact: Remove all contaminated clothing. Immediately wash area with large amounts of water (15 minutes minimum). For inhalation: Remove the person from exposure. Provide respiratory assistance and CPR. Transfer to medical facility. Observe for 24-48 hours for lung effects (pulmonary edema). If swallowed: Call the poison control center. Give 1-2 glasses of water or milk to a conscious and alert person. Seek medical attention immediately. Never attempt to give anything by mouth to an unconscious or convulsing person.

SECTION VII - PRECAUTIONS FOR SAFE HANDLING AND USE

Steps to be Taken in Case Material is Released or Spilled

Evacuate area and deny entry to those not involved in cleanup activities. Protective clothing is mandatory. Remove all sources of ignition. Ventilate area of spill. Dilute with water to less than 40% methyl hydrazine solution. Absorb liquids with vermiculite or sand. Deposit in sealed drum for disposal.

Preferred Waste Disposal Method

Incineration in a chemical incinerator equipped with afterburner and scrubber.

Precautions to be Taken in Handling and Storage

Store in tightly closed containers in a cool, dark, dry, well-ventilated area. Personnel should be trained on the hazardous properties of methyl hydrazine *prior* to working with the material.

Other Precautions and Warnings

Containers should be protected from physical damage. Use extreme caution when handling this compound. Use explosion-proof electrical equipment in areas where this chemical is used or stored.

SECTION VIII - CONTROL MEASURES AND PERSONAL PROTECTIVE EQUIPMENT

Respiratory Protection *(specify type)*

A carcinogenic agent. Exposure level is extremely low and difficult to monitor. For any exposures, use a supplied-air respirator set in positive pressure or continuous flow mode or a self-contained breathing apparatus (SCBA) with full facepiece operated in pressure demand or other positive pressure mode.

Ventilation

Local exhaust or general mechanical systems recommended.

Protective Gloves	Eye Protection	Other Protective Clothing
Impervious Gloves	**Chemical Goggles or Face Mask**	**Full Protective Clothing or Suit**

Work/Hygiene Practices

Always wash hands thoroughly after using chemical; never bring food, drink, or smoking materials into vicinity of chemicals.

METHYL HYDRAZINE

CH_3NHNH_2

CAS: 60-34-4

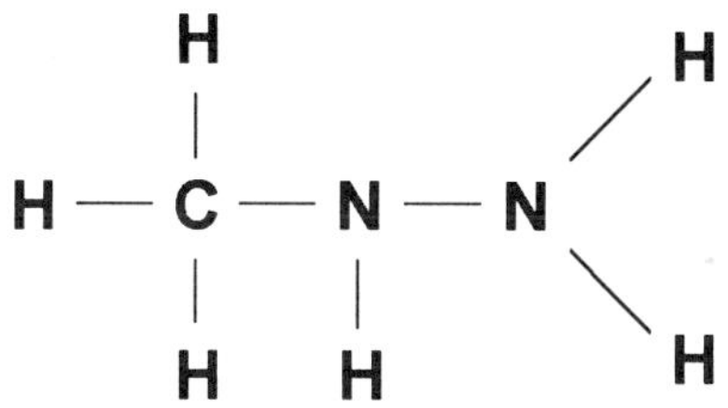

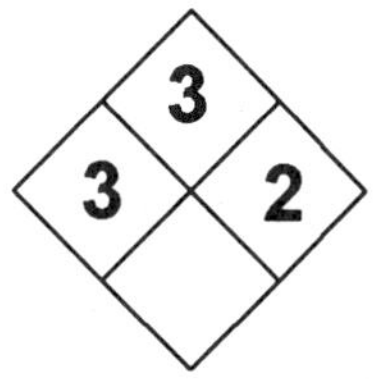

IDENTIFICATION AND TYPICAL USES

Methyl hydrazine is a colorless, fuming, oily liquid with an ammonia or fish-like odor. It is used primarily as a propellant in rocket fuels. It is a powerful reducing agent, a solvent, and an intermediate in chemical synthesis.

RISK ASSESSMENT: HEALTH

General Assessment

Methyl hydrazine is a suspected human and animal carcinogen. It is an experimental teratogen and human mutation data have also been reported. It is primarily toxic by *inhalation* and skin contact (*absorption*). Although an unlikely exposure route, it can also cause toxic and damaging effects by *ingestion*.

Inhalation will cause pulmonary irritation up to and including pulmonary edema (fluid buildup in the lungs), which is a medical emergency and can lead to death. Symptoms may be delayed up to 48 hours creating a false sense of security with regard to health exposure risk. Symptoms include choking, extreme chest pain, dyspnea (difficulty in breathing), lethargy, nausea, convulsions, tremors, liver injury, and hypoxia (reduced oxygen levels in the blood) leading to hemolytic anemia. Systemic intoxication, depression of the central nervous system (CNS), and delayed gastrointestinal irritation may also occur. Methyl hydrazine may cause damage to the liver and destruction of red blood cells.

Aside from its ability to pass through intact skin (absorption), methyl hydrazine is also extremely corrosive to skin tissue and can cause second and third degree burns on contact. Damage to the eyes and cornea can be severe causing partial or complete loss of vision. It is a powerful reducing agent which is corrosive to the eyes, skin, and mucous membranes. May also cause skin sensitization as well as systemic poisoning.

Ingestion is very unusual since the vapors of this chemical are so irritating, personnel are not likely to get close enough to ingest it. However, should it be swallowed, there can be burning of the mouth, throat, and esophagus, followed by gastrointestinal irritation.

Acute Health Effects

The following acute (short-term) health effects may occur immediately or shortly after exposure to methyl hydrazine:

Skin: Severe irritation and burning. Extremely corrosive to tissues. Skin absorption will occur causing toxic systemic effects.

Eye: Severe irritation and tissue damage up to and including loss of vision.

Lung: Irritation of the upper respiratory tract causing choking, coughing, dypsnea, and pulmonary irritation leading to pulmonary edema (fluid in the lungs), which can be fatal.

CNS: Dizziness, convulsions, tremors, nausea, headaches, possible injury to the nervous system.

CVS: It can interfere with the proper function of the circulatory system and lead to cardiovascular collapse and death.

Chronic Health Effects

The following chronic (long-term) health effects may occur at some time after exposure to methyl hydrazine and can last for months or even years:

Cancer Hazards: According to the information presented in the references, methyl hydrazine has been shown to cause cancer in test animals. Its carcinogenicity in humans is expected to be at least equal to that of animals. Human mutation data have been reported for this compound.

Reproductive Hazard: According to the information presented in the references, this chemical has been shown to cause teratogenic effects as well as adversely affect reproduction in test animals.

Other Chronic Effects: Repeated exposures can lead to hemolytic anemia and convulsions. Chronic exposures may lead to cancer and/or liver and kidney damage. There can also be chronic inflammation of the nasal, tracheal, and bronchial tissue resulting in chronic bronchitis. Repeated skin exposure can cause dermatitis with a characteristic rash.

Recommended Risk-Reduction Measures

Methyl hydrazine is listed by the ACGIH as a confirmed animal carcinogen and a suspected human carcinogen. Unless a less toxic chemical can be substituted for a hazardous substance, *engineering controls* are the most effective method of reducing exposures. The best protection is to enclose operations and/or provide local explosion-proof exhaust ventilation at the site of chemical release. While not always operationally feasible, isolating operations can also reduce exposure.

Using respiratory protection is less effective than the controls mentioned above, but is still advisable whenever working with or around methyl hydrazine. The exposure level is extremely low (0.2 ppm) and difficult to detect and monitor. For any detectable exposures to a suspected carcinogenic agent, use a supplied-air respirator set in positive pressure or continuous flow mode, or use a self-contained breathing apparatus (SCBA) with full facepiece operated in pressure demand or other positive pressure mode. If a full facepiece is not available, then chemical goggles should be worn to protect the eyes. Whenever an eye or face contact hazard exists, a face shield and apron should be worn. To prevent hand and skin exposures, impervious gloves should be used. Glove manufacturers should be contacted and permeation studies obtained *before* final glove selection has been made.

Administrative controls should also be in place to minimize the potential for human exposures. These may include written procedures or policies which specify the methods and techniques as well as the exposure control measures that will be practiced whenever personnel are to work with or around methyl hydrazine.

All personnel should receive *training* on the use, hazards, protective measures, emergency actions, and other precautions per 29 CFR 1910.1200 (Hazard Communication), prior to the first assignment in an area where methyl hydrazine is used or stored. Before beginning employment and at regular intervals thereafter (e.g. annually), the following medical tests are recommended:

- ☑ Lung function tests (establish a baseline).
- ☑ Complete blood count (CBC).
- ☑ Examination of the nervous system.

If symptoms develop or overexposure is suspected, the following medical tests may be useful:

- ☑ Lung function tests.
- ☑ Liver and kidney function tests.
- ☑ Consider chest X-ray following acute overexposure (may be negative if taken immediately after exposure due to delayed onset of pulmonary edema).
- ☑ Evaluation by a qualified allergist with careful consideration of exposure history and special tests (may help diagnose allergy).

Any evaluation should include a careful medical history of past and present symptoms with an examination. However, medical tests that evaluate existing damage are not a substitute for controlling exposure. Since smoking can cause heart disease, emphysema, and numerous other respiratory disorders, smokers may be affected more rapidly than non-smokers under the same exposure conditions. Also, because alcohol consumption can cause liver damage, drinking alcohol can increase the potential for liver damage caused by exposure to methyl hydrazine. Prudent risk assessment and management requires careful consideration of *all* possible risk factors that may be causing the appearance of symptoms in exposed workers.

Other methods to reduce exposure include:

- ☑ Where possible, enclose operations and use local exhaust ventilation at the site of chemical release. If local exhaust ventilation or enclosure is not used, respiratory protection should be mandatory.
- ☑ Always ensure that proper protective clothing is worn when using chemical substances.
- ☑ Wash thoroughly immediately after exposure to methyl hydrazine and at the end of the work shift or before eating, drinking, or smoking.
- ☑ Hazard warning information must be posted in the work area. It is recommended that a regulated, controlled work area should be established wherever this chemical is used, handled, or stored. In addition, as part of an on-going education and

training program, all information on the health and safety hazards of methyl hydrazine should be communicated to all potentially exposed workers.

☑ Eye wash stations should be provided in work areas. If the potential for whole body exposure exists, then safety showers should also be provided in the immediately area.

☑ Specific engineering controls are recommended for a similar chemical by NIOSH in Publication Number 78-172 (*Occupational Exposure to Hydrazine*).

RISK ASSESSMENT: ENVIRONMENT

General Assessment

The environment is at risk of exposure during intended use, transportation, storage, disposal, or destruction of methyl hydrazine. In almost every scenario, the threat of environmental exposure is contingent upon the proper handling of the chemical substance. Accidental spills, large or small, can result in fire, explosion, and possible contamination of the surrounding environmental mediums (water, soil, and air).

Methyl hydrazine should always be handled with extreme caution. It is a Class1B flammable liquid (per OSHA 29 CFR 1910.106). Its relatively low flash point and boiling point make this chemical a dangerous fire and explosion hazard. Its vapors form explosive mixtures in air over a wide range of concentrations in air and can easily ignite on exposure to air. It is highly reactive with a number of substances, especially oxidizers and oxidizing agents. It can also be quickly absorbed into porous materials such as wood and cloth. Its viscosity will make it difficult to remove from such materials. It attacks proteins such as leather and can remain in products such as belts and shoes for a long period of time. These characteristics require special consideration during any emergency situation involving a leak or spill of methyl hydrazine.

Methyl hydrazine can enter the environment through industrial discharges or through spills.

Acute Ecological Effects

Acute (short-term) toxic effects may include the death of animals, birds, or fish and death or low growth rate in plants. Acute effects are seen 2 to 4 days after animals or plants are exposed to methyl hydrazine.

This chemical has high acute toxicity to aquatic life. Insufficient data are available to evaluate or predict the short-term effects of methyl hydrazine to plant, birds, or land animals.

Chronic Ecological Effects

Chronic toxic effects may include shortened life span, reproductive problems, lower fertility, and changes in appearance or behavior in exposed animals. These effects can be seen long after first exposure(s) to toxic chemicals. Methyl hydrazine has high chronic toxicity to aquatic life. Insufficient data are available to evaluate or predict the long-term effects of methyl hydrazine to plants, birds, or land animals.

Water Solubility

Methyl hydrazine is highly soluble in water. Concentrations of 1000 milligrams and more will mix with a liter of water.

Persistence in the Environment

Methyl hydrazine is slightly persistent in water, with a half-life of less than 2 days. The half-life of a pollutant is the amount of time it takes for one half of the chemical to be degraded. Approximately 96.4% of methyl hydrazine will eventually end up in water; the rest will end up in the air.

Bioaccumulation in Aquatic Organisms

Some substances increase in concentration, or *bioaccumulate,* in living organisms as they breathe contaminated air, drink contaminated water, or eat contaminated food. These chemicals can become concentrated in the tissues and internal organs of animals as well as humans.

The concentration of methyl hydrazine found in fish tissues is expected to be about the same as the average concentration of methyl hydrazine found in the water from which the fish was taken.

Recommended Risk-Reduction Measures

Proper training of all transporters will reduce the likelihood of a mishap or accident resulting in a leak or spill to the environment. The correct use of DOT labeling (per the MSDS) on all containers, trucks, and rail cars will enable emergency responders to react properly and quickly to any disaster thereby reducing the potential risk to the environment and to personnel.

Storage of methyl hydrazine should be segregated from other chemicals, especially oxidizers, to minimize the risk of cross-contamination or contact. Storage buildings should be equipped with proper fire protection and prevention equipment and explosion-proof electrical components and equipment. Bulk storage of methyl hydrazine is not recommended. Methyl hydrazine is hypergolic (ignites explosively) with incompatible materials such as oxidizers.

If a spill or leak to the environment has occurred, fire department, emergency response and/or hazardous materials spill personnel should be notified immediately. Cleanup should be attempted only by those trained in proper spill containment procedures and the health hazards associated with carcinogenic materials. Contaminated soils should be removed for incineration and replaced with clean soil. Use non-sparking tools. If methyl hydrazine should contact the water table, aquifer, or navigable waterway, remediation activities should be prompt to ensure protection of aquatic sediments. It is highly soluble in water and total containment and remediation may not be possible. When such spills occur, the local and/or state emergency response authorities must be notified. A comprehensive emergency response or disaster preparedness/recovery plan should be in place prior to any operations involving the use, transportation, storage, or disposal of methyl hydrazine. Burning methyl hydrazine in a chemical incinerator equipped with an afterburner and an air scrubber is an acceptable disposal method.

If methyl hydrazine is spilled or leaked, the following specific steps are recommended:

- ☑ Restrict persons not wearing protective clothing from area of spill or leak until cleanup is complete and area can be opened for normal work.
- ☑ Ventilate area and remove all ignition sources.
- ☑ Absorb liquids in vermiculite or sand and place in a sealed drums for disposal.
- ☑ It may be necessary to dispose of methyl hydrazine as a hazardous waste. The responsible state agency or the regional office of the federal Environmental Protection Agency (EPA) should be contacted for specific recommendations.

RISK ASSESSMENT: BUSINESS

General Assessment

Accidents or mishaps involving methyl hydrazine can present a significant threat to business operations. The loss or damage of equipment or facilities can significantly affect fiscal viability. Lawsuits that may result from personnel injury/death, public exposures, and/or environmental contamination will also require a serious expenditure of resources. Media attention surrounding an injury, death, or environmental damage can also result in a loss of profits and loss of current as well as future business.

Always remember that anytime the term "carcinogen," "cancer," or "reproductive hazard" are used, public emotion, hysteria, and ignorance can run equally high. This should be considered during the development of any public relations policies.

Recommended Risk-Reduction Measures

Company attorneys, safety and health professionals, and environmental specialists should be involved in the development of any procedures dealing with safety or emergency response. A company official should be pre-designated as a public relations officer with specific training in dealing with the press. Corporate plans and policies should be developed, approved, and implemented long before any need for such arises.

REFERENCES

American Conference of Governmental Industrial Hygienists. 1988. *Documentation of the Threshold Limit Values and Biological Exposure Indices*, 5th Edition (with updates). Cincinnati: ACGIH

Brusick, D. J. 1994. *Methods for Genetic Risk Assessment.* New York: CRC Press/Lewis Publishers.

Burrell, R., D. K. Flaherty, and L. J. Sauers. 1992. *Toxicology of the Immune System.* New York: Van Nostrand Reinhold.

Calabrese, E. J. 1994. *Biological Effects of Low Level Exposures.* New York: CRC Press/Lewis Publishers.

Cothern, C. R. 1995. *Handbook for Environmental Risk Decisionmaking.* New York: CRC Press/Lewis Publishers.

Hodgson, E., R. B. Mailman, and Janice E. Chambers. 1988. *Dictionary of Toxicology.* London: The MacMillian Company.

Lewis, R. J., Sr. 1993. *Hawley's Condensed Chemical Dictionary,* 12th Edition. New York: Van Nostrand Reinhold.

Lewis, R. J., Sr. 1992. *Sax's Dangerous Properties of Industrial Materials,* Volumes 1, 2, and 3, 8th Edition. New York: Van Nostrand Reinhold.

Meyer, E. 1990. *Chemistry of Hazardous Materials.* Englewood Cliffs, NJ: Prentice Hall.

U.S. Department of Health and Human Services, National Institute for Occupational Safety and Health. 1994. *NIOSH Pocket Guide to Chemical Hazards.* Washington, D.C.: U.S. Government Printing Office.

U.S. Department of Health and Human Services, National Institute for Occupational Safety and Health. 1978. *Occupational Exposure Hydrazine.* Publication Number 78-172. Washington, D.C.: U.S. Government Printing Office.

MATERIAL SAFETY DATA SHEET

CHEMICAL NAME	
METHYL IODIDE	**CRC PRESS/LEWIS PUBLISHERS DISCLAIMER:** THE INFORMATION AND RECOMMENDATIONS PRESENTED HEREIN ARE BASED ON SOURCES BELIEVED TO BE RELIABLE. CRC MAKES NO REPRESENTATION ON ITS COMPLETENESS OR ACCURACY. IT IS THE RESPONSIBILITY OF THE USER TO DETERMINE THE CHEMICAL'S SUITABILITY FOR ITS INTENDED USE, THE CHEMICAL'S SAFE USE, AND THE CHEMICAL'S PROPER DISPOSAL. NO REPRESENTATIONS AND/OR WARRANTIES, EITHER EXPRESSED OR IMPLIED, OF THE MERCHANTABILITY OR FITNESS FOR A PARTICULAR PURPOSE, OR OF ANY OTHER NATURE, ARE MADE WITH RESPECT TO THE INFORMATION PROVIDED IN THIS MSDS OR TO THE CHEMICAL TO WHICH INFORMATION MAY REFER. CRC NEITHER ASSUMES NOR AUTHORIZES ANY OTHER PERSON TO ASSUME FOR IT, ANY OTHER ADDITIONAL RESPONSIBILITY OR LIABILITY FOR THE USE OF, OR RELIANCE UPON, THIS INFORMATION.

HAZARD WARNING INFORMATION

HEALTH	FIRE	REACTIVE	OTHER	DEGREE OF HAZARD	COLOR CODING	OTHER CODES
4	0	1		0 = Minimum Hazard 1 = Slight Hazard 2 = Moderate Hazard 3 = Serious Hazard 4 = Severe Hazard	HEALTH = BLUE FIRE = RED REACTIVITY = YELLOW OTHER = WHITE	OX = Oxidizer ACID = Acid ALK = Alkali COR = Corrosive W = Use No Water

SECTION I - GENERAL INFORMATION

Characterization	RCRA Number	EPA Class
Halogenated Hydrocarbon	**U138**	**Toxic Waste**

DOT Proper Shipping Name **Methyl iodide**	Chemical Abstract Service (CAS) Number **74-88-4**
DOT Hazard Class and Label Requirements **Poison**	DOT Emergency Guide Code **56**
DOT Identification Number **UN 2644**	Molecular Formula **CH_3I**

Synonyms
Idomethane; monoiodomethane.

SECTION II - HAZARDOUS INGREDIENTS AND IDENTITY INFORMATION

Hazardous Components (specific identity)	OSHA Exposure Criteria	NIOSH Exposure Criteria	Immediately Dangerous to Life and Health (IDLH)	ACGIH Exposure Criteria
Methyl iodide (derivation: By interaction of methanol, sodium iodide, and sulfuric acid with subsequent distillation). **1 ppm = 5.90 mg/m^3**	**PEL (skin):** **5 ppm** **28 mg/m^3** **STEL:** **Not Established**	**REL (skin):** **2 ppm** **10 mg/m^3** **Possible Human Carcinogen**	**100 ppm**	**TLV:** **2 ppm** **10 mg/m^3** **Suspected Human Carcinogen**

SECTION III - PHYSICAL AND CHEMICAL CHARACTERISTICS

Boiling Point **108°F (42°C)**	Specific Gravity (H_2O = 1) **2.28**
Vapor Pressure (mm Hg) **400 at 77°F (25°C)**	Molecular Weight (atomic weight) **141.9**
Vapor Density (Air = 1) **4.89**	Freezing Point **-87°F (-66°C)**

Solubility
Slightly soluble in water (1.8%). Soluble in alcohol and ether.

Appearance and Odor
Colorless or clear liquid, turning yellow, red, or brown upon exposure to light, heat, or moisture. It has a sweet, pungent, ether-like odor.

SECTION IV - FIRE AND EXPLOSION HAZARD DATA

Flash Point (method used) **Not Determined**	Explosive Limits in Air % by Volume **LEL: Not Determined** **UEL: Not Determined**
NFPA Classification **Non-Combustible Liquid**	Autoignition Temperature **665°F (352°C)**

Extinguishing Media
Use extinguishing media suitable to surrounding area. Methylene iodide itself does not burn.

Special Fire Fighting Procedures
Poisonous gases are produced in fire. Wear self-contained breathing apparatus (SCBA) and full protective gear. Move containers from fire if it can be done without risk. Cool sides of fire-exposed containers with water until long after fire is out.

Unusual Fire and Explosion Hazards
Containers may explode in fire. Stay away from ends of tanks. Evacuate 1500 feet if rising sound is heard coming from venting device or sides of tank become discolored due to heat of fire.

SECTION V - REACTIVITY DATA

Stability		Conditions to Avoid
		Methyl iodide is normally stable. Avoid heat, flame, other sources of ignition, and contact with incompatible materials and chemicals. Turns yellow, red, or brown on exposure to light.
Stable	Unstable	Incompatibility *(materials to avoid)*
X		**Strong oxidizers (chlorine, bromine, fluorine), silver chlorite, and zinc.**
Hazardous Polymerization		Conditions to Avoid
		Under normal conditions of temperature and pressure, hazardous polymerization of methyl iodide cannot occur.
May Occur	Will Not Occur	Hazardous Decomposition or By-products
	X	**Poisonous gases and acrid fumes are produced in fire, including extremely toxic gases of iodide and hydrogen iodide.**

SECTION VI - HEALTH HAZARD DATA

Primary Route(s) of Entry:	Inhalation?	Absorption (skin/eye)?	Ingestion?
	X	X	X

Health Hazards

INHALATION: **Irritating to the eyes, skin, and respiratory system. It also causes central nervous system effects including convulsions, dizziness, drowsiness, mental confusion, nausea, vomiting, diarrhea, coma, and death.**

ABSORPTION: **Eye contact causes burns and distortion of vision. Skin contact causes inflammation, burning, stinging, and slight reddening (erythema). Can be absorbed in toxic amounts.**

INGESTION: **Unspecified gastrointestinal effects.**

Carcinogenicity	NTP Listed?	IARC Cancer Review Group?	OSHA Regulated?	Target Organs?
Suspected Human Confirmed Animal	**5th Annual Report**	**Group 3**	**29 CFR 1910.1000 Table Z-1**	**Eyes, skin, respiratory system, CNS.**

Medical Conditions Generally Aggravated by Exposure

Skin disorders may be aggravated by exposure.

Emergency and First-aid Procedures

Eye contact: Do not allow victim to keep eyes shut. Flush immediately with water for 15 minutes minimum; seek medical attention. Skin contact: Remove all contaminated clothing. Immediately wash area with flooding amounts of soap and water. Seek medical attention immediately. For inhalation: Remove the person from exposure. Provide respiratory assistance and CPR. Observe for 24 - 48 hours for lung effects (pulmonary edema). Transfer to medical facility. If swallowed: Seek medical attention immediately. Never give an unconscious or convulsing person anything by mouth.

SECTION VII - PRECAUTIONS FOR SAFE HANDLING AND USE

Steps to be Taken in Case Material is Released or Spilled

Remove all ignition sources, wear SCBA respiratory protection. Restrict those not wearing protective equipment who are not involved in cleanup from entering area. Absorb with vermiculite and place in sealed container. Damp mop residue. Flush residues with water (keep away from sewers).

Preferred Waste Disposal Method

Burn in chemical incinerator equipped with an afterburner and scrubber.

Precautions to be Taken in Handling and Storage

Store in tightly closed containers in cool, well-ventilated location. Keep away from sources of heat.

Other Precautions and Warnings

Personnel must be trained on the hazards of working with methyl iodide *before* being assigned to areas where this chemical is used or stored. Empty cylinders contain residue; handle with care.

SECTION VIII - CONTROL MEASURES AND PERSONAL PROTECTIVE EQUIPMENT

Respiratory Protection *(specify type)*

Possible human carcinogen. For best protection, use a supplied-air respirator with full facepiece or a self-contained breathing apparatus (SCBA) with full facepiece operated in positive pressure mode.

Ventilation

Local exhaust (preferred) at site of chemical work or general ventilation.

Protective Gloves	Eye Protection	Other Protective Clothing
Impervious Rubber Gloves	**Chemical Goggles or Face Mask**	**Protective Apron if Splash is Likely**

Work/Hygiene Practices

Always wash hands thoroughly after using chemical; never bring food, drink, or smoking materials into vicinity of chemicals.

METHYL IODIDE

CH_3I

CAS: 74-88-4

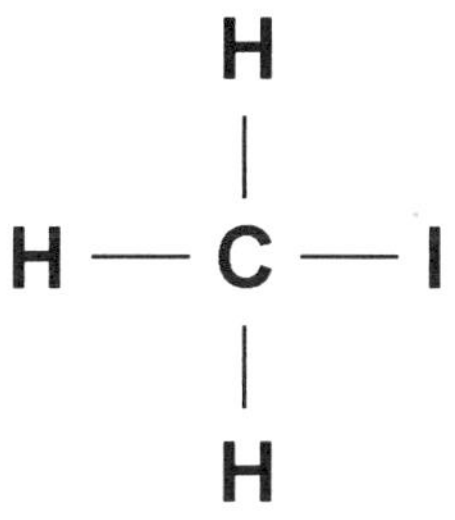

IDENTIFICATION AND TYPICAL USES

Methyl iodide is a colorless liquid that can turn yellow, brown, or red upon exposure to light, heat, or moisture. It has a sweet, pungent, ether-like odor. It is used primarily in organic synthesis, in microscopy, and in testing for pyridine.

RISK ASSESSMENT: HEALTH

General Assessment

Methyl iodide is moderately toxic by *inhalation.* The liquid is also known the pass through the skin (*absorption*) to cause toxic systemic effects, and it is considered a human poison by *ingestion.* Methyl iodide is known to cause lung cancer in test animals and this effect is suspected to occur in humans as well. Human mutation data have been reported. It is a strong narcotic and anesthetic. The systemic effects of overexposure can be seen in the central nervous system. It is more toxic than methyl bromide and is considered to be at least ten times as toxic as carbon tetrachloride.

Inhalation is irritating to the eyes, nose, and mucosa of the respiratory tract. Symptoms include nausea, vomiting, diarrhea, ataxia, drowsiness, slurred speech, visual disturbances, and tremor. Irritation to lung tissues can lead to a dangerous buildup of fluids in the lungs (pulmonary edema), which is a medical emergency and can be fatal. Symptoms may be delayed up to 48 hours creating a false sense of security with regard to health risk. There can be coma and death following exposures to high concentrations.

Skin contact can produce severe localized irritation, burns, de-fatting of the surface leading to possible cracking, drying, and secondary infections. Absorption will occur rapidly causing an increase in toxicity. If coupled with inhalation, the systemic effects can be quite severe. Eye contact will also cause irritation and, if not removed immediately, possible damage to vision.

Ingestion can result in gastrointestinal irritation and may lead to systemic symptoms similar to that of inhalation.

Acute Health Effects

The following acute (short-term) health effects may occur immediately or shortly after exposure to methyl iodide:

Skin: Irritation with potential for redness, pain, burns, and tissue damage.

Eye: Irritation and smarting of the conjunctiva. Can cause tearing, redness, pain, burns, and distortion to vision.

Lung: Irritation of the mucous membranes of the nose, throat, and respiratory tract causing possible ulceration, coughing, chest pains, difficulty breathing, and pulmonary edema.

CNS: Depression with headaches, lightheadedness, dizziness, nausea, vomiting, diarrhea, narcotic and anesthetic effects.

Chronic Health Effects

The following chronic (long-term) health effects may occur at some time after exposure to methyl iodide and can last for months or even years:

Cancer Hazards: According to the references, methyl iodide is a confirmed animal carcinogen and several agencies consider methyl iodide to be a potential or suspect human carcinogen. These include the IARC (International Association for Research on Cancer), the

NTP (National Toxicology Program), and the ACGIH (American Conference of Governmental Industrial Hygienists). It has caused cancer of the lung in animals. Human mutation data have also been reported.

Reproductive Hazard: According to information presented in the references, methyl iodide has not been adequately tested for its ability to adversely affect reproduction.

Other Chronic Effects: Long-term exposure may damage the central nervous system, leading to disorientation and psychotic behavior. Repeated skin exposure can cause thickening and cracking of the skin.

☉ *Recommended Risk-Reduction Measures*

Personnel should avoid direct contact with methyl iodide. It is a highly toxic, known cancer-causing agent. There may be no safe exposure level to methyl iodide. Occupational poisoning occurs most commonly by inhalation and skin absorption. If a less toxic material or compound cannot be substituted for methyl iodide, then *engineering controls* are the most effective method of reducing exposures. The best protection is to enclose operations and/or provide local exhaust ventilation at the site of methyl iodide release. While not always operationally feasible, isolating operations can also reduce exposure. Using respiratory protection is less effective than the controls mentioned above, but is still advisable whenever working with or around methyl iodide. Caution should be exercised when selecting respiratory protection since it is not known if exposure to methyl iodide will cause cancer in humans. A conservative approach to protection against exposure to a carcinogenic material may be prudent. For the best protection against a *possible carcinogen,* especially when the concentration is unknown, use a supplied-air respirator with full facepiece and mask operated in positive pressure mode or a self-contained breathing apparatus (SCBA) with full facepiece operated in pressure demand. If a full facepiece is not available, then chemical goggles should be worn to protect the eyes. A face shield should also be considered. To prevent hand and skin exposures, impervious rubber gloves should be used. Glove manufacturers should be contacted and permeation studies obtained *before* final glove selections have been made.

Administrative controls should also be in place to minimize the potential for human exposures. These may include written procedures or policies which specify the methods and techniques that will be practiced whenever personnel are to work with methyl iodide.

All personnel should receive *training* on the use, hazards, protective measures, emergency actions, and other precautions per 29 CFR 1910.1200 (Hazard Communication), prior to the first assignment in an area where methyl iodide is used or stored.

Before beginning employment with methyl iodide and at regular intervals thereafter (e.g., annually), the following medical tests are recommended:

☑ Lung function tests (establish baseline).

If symptoms develop or overexposure is suspected, the following tests should be considered:

☑ Consider urine test for iodine.

☑ Consider chest X-ray after acute overexposure (may be negative if taken immediately after exposure due to delayed onset of pulmonary edema).

☑ Lung function tests (compare to baseline).

Any evaluation should include a careful history of past and present symptoms with an examination. Medical tests that simply look for existing damage are not a substitute for controlling exposures. Also, since smoking can cause heart disease, lung cancer, emphysema, and other respiratory problems, smokers may be affected more rapidly and with more pronounced symptoms than non-smokers. Prudent risk management requires careful consideration of *all* risk factors that might be causing the appearance of symptoms among workers.

Other methods to reduce exposure include:

☑ Where possible, enclose operations and use local exhaust ventilation at the site of chemical release. If local exhaust ventilation or enclosure is not feasible, then respiratory protection should be provided and its use mandatory.

☑ Always ensure that proper protective clothing is worn when using chemical substances.

☑ Wash thoroughly immediately after exposure to methyl iodide and at the end of the work shift or before eating, drinking, or smoking.

☑ Hazard warning information should be posted in the work area. In addition, as part of an on-going education and training program, all information on the health and safety hazards of methyl iodide should be communicated to all potentially exposed workers.

☑ Eye wash stations should be provided in the immediate work area for emergency use. If there is a

possibility of skin exposure to methyl iodide, emergency shower facilities should also be provided in the immediate work area.

☑ Workers whose clothing has been contaminated by methyl iodide should change into clean clothes before leaving work. Contaminated work clothing should be laundered only by individuals who have been informed of the hazards of exposure to methyl iodide.

RISK ASSESSMENT: ENVIRONMENT

General Assessment

The environment is at risk of exposure during transportation, storage, disposal, or destruction of methyl iodide. In almost every scenario, the threat of environmental exposure is contingent upon the proper handling of the chemical substance. Accidental spills, large or small, can result in contamination of the surrounding environmental mediums (water, soil, and air).

Methyl iodide is a non-combustible liquid. However, it can also react violently in the presence of strong oxidizers such as chlorine, chlorine dioxide, bromine, and nitrates, and permanganates. It is also likely to react on contact with silver chloride and sodium. Caution is always required in handling, storage, transportation, and disposal of methyl iodide. A phenomenon known as a boiling liquid expanding vapor explosion (BLEVE) can occur when tanks, containers, or rail cars containing methyl iodide are involved in a fire. The pressure that builds inside the container can explode violently and without much warning. Emergency responders should, therefore, be made aware of the presence of methyl iodide at any emergency response situation.

Methyl iodide can enter the environment from industrial effluents and from spills.

☠ *Acute Ecological Effects*

Acute (short-term) toxic effects may include the death of animals, birds, or fish and death or low growth rate in plants. Acute effects are seen 2 to 4 days after animals or plants are exposed to methyl iodide.

Methyl iodide has slight acute toxicity to aquatic life. Insufficient data are available to evaluate the short-term effects of methyl iodide on aquatic life, birds, or land animals.

Chronic Ecological Effects

Chronic (long-term) toxic effects may include shortened life span, reproductive problems, lower fertility, and changes in appearance or behavior in exposed animals. These effects can be seen long after first exposure(s) to toxic chemicals.

Methyl iodide has slight chronic toxicity to aquatic life. Insufficient data are available to evaluate the long-term effects of methyl iodide to plants, birds, or terrestrial animals.

Water Solubility

Methyl iodide is highly soluble in water. Concentrations of 1000 milligrams and more will mix with a liter of water.

⌛ *Persistence in the Environment*

Methyl iodide is non-persistent in water, with a half-life of less than 2 days. The half-life of a pollutant is the amount of time it takes for one half of the chemical to be degraded. About 99.5% of methyl iodide will eventually end up in air; the rest will end up in the water.

Bioaccumulation in Aquatic Organisms

Some substances increase in concentration, or *bioaccumulate,* in living organisms as they breathe contaminated air, drink contaminated water, or eat contaminated food. These chemicals can become concentrated in the tissues and internal organs of animals as well as humans.

The concentration of methyl iodide found in fish tissues is expected to be about the same as the average concentration of methyl iodide in the water from which the fish was taken.

Recommended Risk-Reduction Measures

Proper training of all transporters will reduce the likelihood of a mishap or accident resulting in a leak or spill of methyl iodide into the environment. Labels on all containers, trucks, and rail cars must meet DOT requirements and accurately reflect their contents to enable emergency responders to react properly and quickly to any disaster thereby reducing the potential risk to the environment and to personnel.

Storage of methyl iodide should be segregated from other chemicals and materials to minimize the risk of cross-contamination.

If a spill or leak to the environment has occurred, fire department, emergency response and/or hazardous materials spill personnel should be notified immediately. Cleanup should be attempted only by those trained in proper spill containment procedures. Contaminated soils should be removed for incineration and replaced with clean soil. If methyl iodide should contact the water table, aquifer, or navigable waterway, time is of the essence. It is only highly and volatile soluble and, therefore, total remediation may not be possible. The local and/or state emergency response authorities must be notified. A comprehensive emergency response or disaster preparedness/recovery plan should be in place prior to any operations involving the use, transportation, storage, or disposal of methyl iodide.

For disposal, methyl iodide may be burned in a chemical incinerator equipped with an afterburner and scrubber. If methyl iodide is spilled or leaked, the following specific steps are recommended:

- ☑ Restrict persons not wearing protective clothing from area of spill until cleanup is complete and area can be opened for normal work
- ☑ Absorb liquids in vermiculite, sand, earth, or similar material and deposit in sealed drum. Damp mop residue and collect for disposal. Direct cleanup activities away from sewers or waterways.
- ☑ Remove ignition sources and ventilate area of spill or leak.
- ☑ It may be necessary to dispose of methyl iodide as a hazardous waste. The responsible state agency or the regional office of the federal Environmental Protection Agency (EPA) should be contacted for specific recommendations.

RISK ASSESSMENT: BUSINESS

General Assessment

Accidents or mishaps involving methyl iodide can present a serious threat to business operations. The loss or damage of equipment or facilities can significantly affect fiscal viability. Lawsuits that may result from personnel illness, injury/death, public exposures, and/or environmental contamination will require a serious expenditure of resources. Media attention surrounding an injury, death, or environmental damage can also result in a loss of profits and loss of current as well as future business. Always remember that anytime the terms "cancer," "carcinogen," or "reproductive hazard" are used, public emotion, hysteria, and ignorance can run equally high. This must be carefully considered when developing any public relation policies.

Recommended Risk-Reduction Measures

Company attorneys, safety and health professionals, and environmental specialists should be involved in the development of any procedures for responding to chemical incidents. A company official should be predesignated as a public relations officer with specific training in dealing with the press. Corporate plans and policies should be developed, approved, and implemented long before any need for such arises.

REFERENCES

American Conference of Governmental Industrial Hygienists. 1988. *Documentation of the Threshold Limit Values and Biological Exposure Indices,* 5th Edition (with updates). Cincinnati: ACGIH

Anthony, C. P., and N. J. Kolthoff. 1971. *Textbook of Anatomy and Physiology.* St. Louis: C. V. Mosby Company.

Asante-Duah, D. K. 1993. *Hazardous Waste Risk Assessment.* New York: CRC Press/Lewis Publishers.

Bloom, A. D. and F. J. DeSerres. 1995. *Ecotoxicity and Human Health.* New York: CRC Press/Lewis Publishers.

Brusick, D. J. 1994. *Methods for Genetic Risk Assessment.* New York: CRC Press/Lewis Publishers.

Bueche, J. R. 1972. *Principles of Physics,* 2nd Edition. New York: McGraw-Hill.

Burrell, R., D. K. Flaherty, and L. J. Sauers. 1992. *Toxicology of the Immune System.* New York: Van Nostrand Reinhold.

Calabrese, E. J., E. M. Kenyon. *Air Toxics and Risk Assessment.* New York: CRC Press/Lewis Publishers.

Cothern, C. R. 1995. *Handbook for Environmental Risk Decisionmaking.* New York: CRC Press/Lewis Publishers.

Danse, I. R. 1991. *Common Sense Toxics in the Workplace.* New York: Van Nostrand Reinhold.

Environmental Protection Agency. 1983. *Chemical Hazard Information Profiles.* Washington, D.C.: U.S. EPA.

Fodden, C. R., and J. L. Weddell. 1992. *Hazardous Materials Emergency Action Data.* New York: CRC Press/Lewis Publishers.

Forseberg, K., and L. H. Keith. 1993. *Chemical; Protective Clothing Permeation and Degradation Database.* Chelsea, MI: Lewis Publishers.

Lewis, R. J., Sr. 1992. *Sax's Dangerous Properties of Industrial Materials,* Volumes 1, 2, and 3, 8th Edition. New York: Van Nostrand Reinhold.

U.S. Department of Health and Human Services, National Institute for Occupational Safety and Health. 1994. *NIOSH Pocket Guide to Chemical Hazards.* Washington, D.C.: U.S. Government Printing Office.

MATERIAL SAFETY DATA SHEET

CHEMICAL NAME	
METHYL ISOBUTYL KETONE	CRC PRESS/LEWIS PUBLISHERS DISCLAIMER: THE INFORMATION AND RECOMMENDATIONS PRESENTED HEREIN ARE BASED ON SOURCES BELIEVED TO BE RELIABLE. CRC MAKES NO REPRESENTATION ON ITS COMPLETENESS OR ACCURACY. IT IS THE RESPONSIBILITY OF THE USER TO DETERMINE THE CHEMICAL'S SUITABILITY FOR ITS INTENDED USE, THE CHEMICAL'S SAFE USE, AND THE CHEMICAL'S PROPER DISPOSAL. NO REPRESENTATIONS AND/OR WARRANTIES, EITHER EXPRESSED OR IMPLIED, OF THE MERCHANTABILITY OR FITNESS FOR A PARTICULAR PURPOSE, OR OF ANY OTHER NATURE, ARE MADE WITH RESPECT TO THE INFORMATION PROVIDED IN THIS MSDS OR TO THE CHEMICAL TO WHICH INFORMATION MAY REFER. CRC NEITHER ASSUMES NOR AUTHORIZES ANY OTHER PERSON TO ASSUME FOR IT, ANY OTHER ADDITIONAL RESPONSIBILITY OR LIABILITY FOR THE USE OF, OR RELIANCE UPON, THIS INFORMATION.

HAZARD WARNING INFORMATION

HEALTH	FIRE	REACTIVE	OTHER	DEGREE OF HAZARD	COLOR CODING	OTHER CODES
2	3	0		0 = Minimum Hazard 1 = Slight Hazard 2 = Moderate Hazard 3 = Serious Hazard 4 = Severe Hazard	HEALTH = BLUE FIRE = RED REACTIVITY = YELLOW OTHER = WHITE	OX = Oxidizer ACID = Acid ALK = Alkali COR = Corrosive W = Use No Water

SECTION I - GENERAL INFORMATION

Characterization	**Ketone**	RCRA Number	**U161**
		EPA Class	**Toxic Waste**
DOT Proper Shipping Name	**Methyl isobutyl ketone**	Chemical Abstract Service (CAS) Number	**108-10-1**
DOT Hazard Class and Label Requirements	**Flammable Liquid**	DOT Emergency Guide Code	**26**
DOT Identification Number	**UN 1245**	Chemical Formula	$(CH_3)_2CHCH_2COCH_3$

Synonyms

Hexone; 4-methy-2-pentanone; isopropylacetone; MIBK; 4-methyl-2-oxopentane.

SECTION II - HAZARDOUS INGREDIENTS AND IDENTITY INFORMATION

Hazardous Components (specific identity)	OSHA Exposure Criteria	NIOSH Exposure Criteria	Immediately Dangerous to Life and Health (IDLH)	ACGIH Exposure Criteria
Methyl isobutyl ketone: (derivation: by mild hydrogenation of mesityl oxide over nickel at high temperature; by oxidation of methyl isobutyl carbinol). 1 ppm = 4.1 mg/m^3	PEL: 50 ppm 205 mg/m^3 STEL (15-min): 75 ppm 300 mg/m^3	REL: 50 ppm 205 mg/m^3 STEL: 75 ppm 307 mg/m^3	3000 ppm	TLV: 50 ppm 205 mg/m^3 STEL: 75 ppm 307 mg/m^3

SECTION III - PHYSICAL AND CHEMICAL CHARACTERISTICS

Boiling Point	**242°F (117°C)**	Specific Gravity (H_2O = 1)	**0.80**
Vapor Pressure (mm Hg)	**15 at 69°F (20°C)**	Molecular Weight	**100.16**
Vapor Density (Air = 1)	**1.26**	Freezing Point	**-120°F (-85°C)**

Solubility

Slightly soluble in water (floats and slowly mixes). Miscible in alcohol, acetone, benzene, chloroform.

Appearance and Odor

Colorless liquid with a sweet, camphor-like odor. Odor Threshold = 0.68 ppm.

SECTION IV - FIRE AND EXPLOSION HAZARD DATA

Flash Point (method used)	**64°F (18°C) closed cup**	Explosive Limits in Air % by Volume	**LEL: 1.2% at 200 UEL: 8% at 200°F**
NFPA Classification	**Class 1BFlammable Liquid**	Autoignition Temperature	**840°F (448°C)**

Extinguishing Media

Dry chemical, carbon dioxide, water spray, fog, or alcohol-resistant foam.

Special Fire Fighting Procedures

Poisonous gases are produced in fire. Wear full protective clothing and self-contained breathing apparatus. Fight fire from a distance if possible. Keep fire-exposed containers cool with water spray. A solid stream of water may spread fire. If it can be done safely, move fire-exposed containers from area.

Unusual Fire and Explosion Hazards

Containers may explode in fire due to the buildup of internal pressures. BLEVE (boiling liquid expanding vapor explosion) is possible. Immediately evacuate if rising sound is heard from venting device or a noticeable discoloration is observed on the sides of the tank or storage vessel.

SECTION V - REACTIVITY DATA

Stability		Conditions to Avoid
Stability		Normally stable. Do not allow methyl isobutyl ketone to come into contact with incompatible materials. Avoid contact with fire or high heat sources. Do not use in confined spaces.
Stable	Unstable	Incompatibility *(materials to avoid)*
X		Incompatible with strong oxidizers (chlorine, bromine, fluorine), potassium *tert*-butoxide, and can react vigorously with reducing materials. It forms explosive mixtures in air.
Hazardous Polymerization		Conditions to Avoid Under normal conditions of temperature and pressure, hazardous polymerization of methyl isobutyl ketone is not expected to occur.
May Occur	Will Not Occur	Hazardous Decomposition or By-products
	X	When heated to decomposition, methyl isobutyl ketone emits acrid, irritating smoke and fumes, including oxides of carbon.

SECTION VI - HEALTH HAZARD DATA

Primary Route(s) of Entry:	Inhalation?	Absorption (skin)?	Ingestion?
	X	X	

Health Hazards

INHALATION: Causes irritation to the eyes and mucosa. Symptoms include excitability, headache, loss of appetite, sore throat, visual disturbances, pale complexion, tiredness, nausea, and vomiting. High concentrations cause central nervous system depression with symptoms of intoxication, stupor, staggering, mental confusion, loss of consciousness, coma, and death. Animal studies suggest liver, brain, lung, and spleen injuries.

ABSORPTION: Skin contact with liquid or vapor can result in irritation and may cause dermatitis and redness (erythema). Eye contact causes pain, irritation, and corneal damage, tearing.

INGESTION: Unlikely. Symptoms of irritation, sore throat, nausea, coughing, salivation, vomiting.

Carcinogenicity	NTP Listed?	IARC Cancer Review Group?	OSHA Regulated?	Target Organs?
Unknown Human Unknown Animal	No	No	29 CFR 1910.1000 Table Z-1	Respiratory system; skin; eyes; CNS.

Medical Conditions Generally Aggravated by Exposure

Skin conditions (dermatitis) may be aggravated.

Emergency and First-aid Procedures

Eye contact: Immediately flush with large amounts of water for 15 minutes (minimum), occasionally lifting eyelids. Skin contact: Remove all contaminated clothing. Immediately wash area with large amounts of soap and water. For inhalation: Remove the person from exposure. Provide respiratory assistance and CPR. Transfer promptly to medical facility. If swallowed: Seek medical attention. Give a conscious and alert victim several glasses of water and then induce vomiting with Ipecac syrup.

SECTION VII - PRECAUTIONS FOR SAFE HANDLING AND USE

Steps to be Taken in Case Material is Released or Spilled

Remove all ignition sources. Restrict persons not wearing protective equipment from entering area of spill or leak until cleanup is complete. Ventilate area. Absorb liquid in dry sand or other material and deposit in sealed containers. Damp mop residue. Alert appropriate state or federal response agencies.

Preferred Waste Disposal Method

Burn in a chemical incinerator equipped with an afterburner and scrubber.

Precautions to be Taken in Handling and Storage

Do not store with incompatible chemicals since violent reactions can occur. Store in tightly closed containers in a cool, dark, well-ventilated area away from fire, sparks, and flame. Use non-sparking tools on containers. Ground and bond metal containers during transfer operations.

Other Precautions and Warnings

Containers may explode in fire or under conditions of extreme heat.

SECTION VIII - CONTROL MEASURES AND PERSONAL PROTECTIVE EQUIPMENT

Respiratory Protection *(specify type)*

Exposures above 50 ppm: MSHA/NIOSH-approved full-facepiece respirator with organic vapor cartridge. Greater protection is obtained from a self-contained breathing apparatus (SCBA) with full facepiece and pressure demand or other positive pressure mode.

Ventilation

Local exhaust or general mechanical systems recommended.

Protective Gloves	Eye Protection	Other Protective Clothing
Impervious Materials	Chemical Goggles and/or Face Mask	Protective Apron

Work/Hygiene Practices

Always wash hands thoroughly after using chemical; never bring food, drink, or smoking materials into vicinity of chemicals.

METHYL ISOBUTYL KETONE

$(CH_3)_2CHCH_2COCH_3$ CAS: 108-10-1

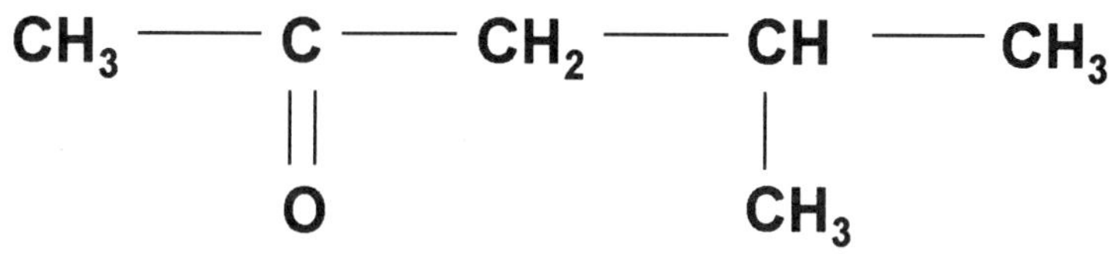

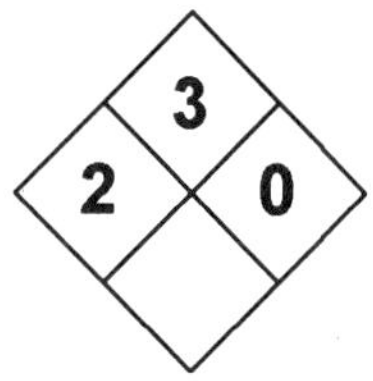

IDENTIFICATION AND TYPICAL USES

Methyl isobutyl ketone (MIBK) is a colorless liquid with a sweet, camphor-like odor. It is used as a solvent for gums, resins, nitrocellulose and cellulose ethers, paints, varnishes, lacquers, protective coatings, and various fats, oils, and waxes. It is also used in the manufacture of methyl amyl alcohol and antibiotics, as a synthetic flavoring agent, and in the plastic and petroleum industries.

RISK ASSESSMENT: HEALTH

General Assessment

MIBK is mildly toxic to humans by *inhalation* and *absorption*. Skin or eye contact can also cause surface tissue damage and inflammation. It is a human poison by *ingestion*. The references contain no information on the carcinogenic or mutagenic properties of MIBK. However, it is suggested that it be treated as a suspected teratogen. Overexposure to this compound is not common. Its irritating properties and penetrating odor are usually present at levels low enough to serve as a warning to personnel.

Inhalation of MIBK causes irritation of the mucosa of the respiratory tract leading to respiratory depression with symptoms of gasping, choking, and dypsnea. Exposure to higher concentrations can cause depression of the central nervous system (CNS) with symptoms of excitability, headache, appetite loss, sore throat, visual disturbances, pale complexion, fatigue, nausea, and vomiting. Higher concentrations can lead to dizziness, intoxication, stupor, staggering gait, loss of coordination, mental confusion, coma, and death. There may also be liver, spleen, brain, and lung injuries.

Skin contact with the liquid can lead to chemical dermatitis, inflammation, redness, and rash. MIBK can pass relatively quickly through intact skin to cause toxic systemic effects. When combined with other exposure routes, such as inhalation, the compound effects can be dangerous. Eye contact causes irritation with lachrymation and, in prolonged contact, may lead to ocular pain, corneal damage, and possible visual disturbances (blurred vision). Although ingestion is unlikely in the occupational setting, it can cause irritation of the mouth and throat with nausea and vomiting.

☠ *Acute Health Effects*

The following acute (short-term) health effects may occur immediately or shortly after exposure to MIBK:

Skin: Irritation and dermatitis with possible localized rash and redness. Absorption is likely.

Eye: Irritation with pain and inflammation that can lead to corneal damage and possible visual disturbances.

Lung: Irritation causing coughing and/or shortness of breath. Higher exposures can cause dizziness, headache, and, possibly, coma and death.

CNS: A mild to moderate narcotic compound producing neurotoxicity as it is partially metabolized to methanol.

Chronic Health Effects

The following chronic (long-term) health effects may occur at some time after exposure to MIBK and can last for months or even years:

Cancer Hazards: According to information presented in the references, MIBK has not been adequately tested for its ability to cause cancer in laboratory test animals.

Reproduction: It has been suggested that this chemical be treated as a potential teratogen since it has been shown to adversely affect fetal development in experimental animal studies.

Other Chronic Effects: Studies with MIBK have shown that long-term skin exposure may produce dermatitis. Long-term exposure has caused fatty-degeneration of the liver, brain, lungs, and spleen in test animals. It is not known if human exposure to MIBK will have this effect. Also, MIBK has not been adequately tested to determine whether brain or other nerve damage could occur with repeated exposure. However, many solvents and other petroleum-based chemicals have been shown to cause such damage. Symptoms may include reduced memory and concentration, personality changes (withdrawal, irritability) fatigue, sleep disturbances, reduced concentration, and/or effects on nerves supplying internal organs (autonomic nerves) or the nerves to the arms and legs (weakness, tingling).

❍ *Recommended Risk-Reduction Measures*

Personnel should avoid direct contact with MIBK. Unless a less toxic chemical can be substituted for a hazardous substance, *engineering controls* are the most effective method of reducing exposures. The best protection is to enclose operations and/or provide local exhaust ventilation at the site of chemical release. Because of its low flash point, maximum explosion-proof ventilation systems should be considered. While not always operationally feasible, isolating operations can also reduce exposure. Using respiratory protection is less effective than the controls mentioned above, but is still advisable whenever working with MIBK. For exposures to the PEL (50 ppm), use a NIOSH-approved, full facepiece organic vapor respirator. For higher or prolonged exposures, or when exposure levels are unknown (as in an emergency) a self-contained breathing apparatus (SCBA) with full facepiece operated in pressure demand mode, or a supplied-air respirator operated in continuous flow mode are the recommended respiratory protection methods of choice. If a full facepiece is not available, then chemical goggles should be worn to protect the eyes. Whenever a chemical splash hazard exists, a face shield and a rubber apron should be worn. To prevent hand and skin exposures, butyl rubber gloves should be used. Polyvinyl alcohol gloves have proven effective. However, glove manufacturers should be contacted and permeation studies obtained *before* final glove selections are made.

Administrative controls should also be in place to minimize the potential for human exposures. These may include written procedures or policies which specify the methods and techniques that will be practiced whenever personnel are to work with or around MIBK.

All personnel should receive *training* on the use, hazards, protective measures, emergency actions, and other precautions per 29 CFR 1910.1200 (Hazard Communication), prior to the first assignment in an area where MIBK is used or stored.

If symptoms develop or overexposure is suspected, the following medical tests may be helpful:

- ☑ Skin testing by a qualified allergist with special testing and careful evaluation of exposure history (may help diagnose allergy).
- ☑ Neurological evaluation of the central nervous system.
- ☑ Liver and kidney function tests.

Other methods to reduce exposure include:

- ☑ Where possible, enclose operations and use local exhaust ventilation at the site of chemical release. Explosion-proof design is recommended. If local exhaust ventilation or enclosure is not used, respiratory protection should be mandatory.
- ☑ Always ensure that proper protective clothing is worn when using chemical substances.
- ☑ Wash thoroughly immediately after exposure to MIBK and at the end of the work shift or before eating, drinking, or smoking. Never wear clothes contaminated with MIBK home. Family members can be exposed.
- ☑ Hazard warning information should be posted in the work area. As part of an on-going education and training program, all information on the health and safety hazards of MIBK should be communicated to all exposed workers.

RISK ASSESSMENT: ENVIRONMENT

General Assessment

The environment is at risk of exposure during transportation, storage, disposal, or destruction of MIBK. In almost every scenario, the threat of environmental exposure is contingent upon the proper handling of the chemical substance. Accidental spills, large or small, can result in fire, explosion, and possible contamination of the surrounding environmental mediums (water, soil, and air).

MIBK is considered a Class 1B flammable liquid (per OSHA 29 CFR 1910.106). Its low flash point and relatively low boiling point make this chemical an extremely dangerous fire and explosion hazard. It can react with strong oxidizers and acids to cause a serious fire and explosion hazard. When involved in a fire, it can release acrid and irritating oxides of carbon. A boiling liquid expanding vapor explosion (BLEVE) is also a possibility. These characteristics require special consideration during any emergency involving a leak or spill of MIBK.

The proper disposal/destruction method for MIBK is to burn it in a chemical incinerator equipped with an afterburner and air scrubber.

MIBK can enter the environment through manufacturing, unchecked discharge into effluents from industrial or municipal waste treatment plants, and through spills.

Acute Ecological Effects

Acute (short-term) toxic effects may include the death of animals, birds, or fish, and death or low growth rate in plants. Acute effects are seen 2 to 4 days after animals or plants are exposed to MIBK.

This chemical has a slight acute toxicity to aquatic life and high acute toxicity to birds. No data are available on the short-term effects of MIBK on plants or land animals.

Chronic Ecological Effects

Chronic (long-term) toxic effects may include shortened life span, reproductive problems, lower fertility, and changes in appearance or behavior in exposed animals. These effects can be seen long after first exposure(s) to toxic chemicals.

MIBK has slight chronic toxicity in aquatic life. No data are available to evaluate the long-term effects of MIBK to plants, birds, or land animals.

Water Solubility

MIBK is highly soluble in water. Concentrations of 1000 milligrams and more will mix with a liter of water.

Persistence in the Environment

MIBK is slightly persistent in water, with a half-life of between 2 to 20 days. The half-life of a pollutant is the amount of time it takes for one half of the chemical to be degraded. Approximately 61.5% of MIBK will eventually end up in the air; the rest will end up in the water. Accumulation of MIBK in aquatic or terrestrial sediments has not been demonstrated.

Bioaccumulation in Aquatic Organisms

Some substances increase in concentration, or *bioaccumulate,* in living organisms as they breathe contaminated air, drink contaminated water, or eat contaminated food. These chemicals can become concentrated in the tissues and internal organs of animals as well as humans.

The concentration of MIBK found in fish tissues is expected to be about the same as the average concentration of MIBK in water from which the fish was taken.

Recommended Risk-Reduction Measures

Proper training of all transporters will reduce the likelihood of a mishap or accident resulting in a leak or spill to the environment. The correct use of DOT labeling (per the MSDS) on all containers, trucks, and rail cars will enable emergency responders to react properly and quickly to any disaster thereby reducing the potential risk to the environment and to personnel.

Storage of MIBK should be segregated from incompatible chemicals to minimize the risk of cross-contamination or contact. Buildings designated for storage should be equipped with appropriate fire protection systems (alarms, sprinklers, emergency lighting, portable extinguishers). Maximum explosion-proof design of ventilation systems is recommended.

If a spill or leak to the environment has occurred, fire department, emergency response, and/or hazardous materials spill personnel should be notified immediately. Cleanup should be attempted only by those trained in proper spill containment procedures using non-sparking tools. Contaminated soils should be removed for incineration and replaced with clean soil. If MIBK should contact the water table, aquifer, or navigable waterway, time is of the essence. It is highly soluble in water and, therefore, total containment and remediation may not be possible. When such spills occur, the local and/or state emergency response authorities may require notification. A comprehensive emergency response or disaster preparedness/recovery plan should be in place prior to any operations involving the use, transportation, storage, or disposal of MIBK.

If MIBK is spilled or leaked, the following specific steps are recommended:

☑ Restrict persons not wearing protective clothing from area of spill or leak until cleanup is complete and area can be opened for normal work.

- ☑ Remove all ignition sources.
- ☑ Ventilate area of spill or leak.
- ☑ Absorb liquids in vermiculite, dry sand, earth, or a similar material and deposit in sealed containers for disposal. Use non-sparking tools!
- ☑ Keep MIBK out of a confined space, such as a sewer, because of the possibility of explosion (unless the sewer is designed to prevent the buildup of explosive concentrations).
- ☑ It may be necessary to dispose of MIBK as a hazardous waste. The responsible state agency or the regional office of the federal Environmental Protection Agency (EPA) should be contacted for specific recommendations.

RISK ASSESSMENT: BUSINESS

General Assessment

Accidents or mishaps involving MIBK can present a significant threat to business operations. The loss or damage of equipment or facilities can significantly affect fiscal viability. Lawsuits that may result from personnel injury, public exposures, and/or environmental contamination will also require a serious expenditure of resources. Media attention surrounding an injury, illness, or environmental damage can also result in a loss of profits and loss of current as well as future business prospects. Always remember that anytime the term "reproductive hazard" is used, public hysteria, emotion, and ignorance can run equally high. This must be carefully considered when adopting any public relations policies.

Recommended Risk-Reduction Measures

Company attorneys, safety and health professionals, and environmental specialists should be involved in the development of any procedures dealing with safety or emergency response. A company official should be pre-designated as a public relations officer with specific training in dealing with the press. Corporate plans and policies should be developed, approved, and implemented long before any need for such arises.

REFERENCES

American Conference of Governmental Industrial Hygienists. 1988. *Documentation of the Threshold Limit Values and Biological Exposure Indices,* 5th Edition (with updates). Cincinnati: ACGIH.

Bloom, A. D. and F. J. DeSerres. 1995. *Ecotoxicity and Human Health.* New York: CRC Press/Lewis Publishers.

Burrell, R., D. K. Flaherty, and L. J. Sauers. 1992. *Toxicology of the Immune System.* New York: Van Nostrand Reinhold.

Calabrese, E. J., E. M. Kenyon. *Air Toxics and Risk Assessment.* New York: CRC Press/Lewis Publishers.

Cockerman, L. G., B. S. Shane. 1994. *Basic Environmental Toxicology.* New York: CRC Press/Lewis Publishers.

Gardner, R. W. 1994. *Chemical Intolerance.* New York: CRC Press/Lewis Publishers.

Hodgson, E., R. B. Mailman, and J. E. Chambers. 1988. *Dictionary of Toxicology.* London: The MacMillian Company.

Hosty, J. W., and P. Foster. 1990. *A Practical Guide to Chemical Spill Response.* New York: Van Nostrand Reinhold.

Howard, P. H., et. al. 1991. *Handbook of Environmental Degradation Rates.* New York: CRC Press/Lewis Publishers.

Jones, F. E. 1994. *Toxic Organic Vapors in the Workplace.* New York: CRC Press/Lewis Publishers.

Keith, L. H. and M. M. Walker. 1995. *Handbook of Air Toxics.* New York: CRC Press/Lewis Publishers.

Lauwerys, R. R., P. Hoet. 1993. *Industrial Chemical Exposure.* New York: CRC Press/Lewis Publishers.

Lewis, R. J., Sr. 1993. *Hawley's Condensed Chemical Dictionary,* 12th Edition. New York: Van Nostrand Reinhold.

Lewis, R. J., Sr. 1992. *Sax's Dangerous Properties of Industrial Materials,* Volumes 1, 2, and 3, 8th Edition. New York: Van Nostrand Reinhold.

Meyer, E. 1990. *Chemistry of Hazardous Materials.* Englewood Cliffs, NJ: Prentice Hall.

Patnaik, P. 1992. *A Comprehensive Guide to the Hazardous Properties of Chemical Substances.* New York: Van Nostrand Reinhold.

Philp, R. B. 1995. *Environmental Hazards and Human Health.* New York: CRC Press/Lewis Publishers.

Quigley, D. R. 1994. *Handbook of Emergency Chemical Management.* New York: CRC Press/Lewis Publishers.

Raleigh, R. L. and W. A. McGee. 1972. Effects of short, high concentration exposures to acetone as determined by observation in the work area. *Journal of Occupational Medicine.* 14:607-610.

Rea, W. J. 1995. *Chemical Sensitivity,* Vol. I-III. New York: CRC Press/Lewis Publishers.

Richardson, M. 1995. *Dictionary of Substance and Their Effects,* Vol. I-VI. New York: CRC Press/Lewis Publishers.

U.S. Department of Health and Human Services, National Institute for Occupational Safety and Health. 1994. *NIOSH Pocket Guide to Chemical Hazards.* Washington, D.C.: U.S. Government Printing Office.

U.S. Department of Health and Human Services, National Institute for Occupational Safety and Health. 1978. *Criteria for a Recommended Standard - Occupational Exposure to Ketones.* DHEW Publication Number: 78-173. Washington, D.C.: U.S. Government Printing Office.

MATERIAL SAFETY DATA SHEET

CHEMICAL NAME	
METHYL ISOCYANATE	**CRC PRESS/LEWIS PUBLISHERS DISCLAIMER:** THE INFORMATION AND RECOMMENDATIONS PRESENTED HEREIN ARE BASED ON SOURCES BELIEVED TO BE RELIABLE. CRC MAKES NO REPRESENTATION ON ITS COMPLETENESS OR ACCURACY. IT IS THE RESPONSIBILITY OF THE USER TO DETERMINE THE CHEMICAL'S SUITABILITY FOR ITS INTENDED USE, THE CHEMICAL'S SAFE USE, AND THE CHEMICAL'S PROPER DISPOSAL. NO REPRESENTATIONS AND/OR WARRANTIES, EITHER EXPRESSED OR IMPLIED, OF THE MERCHANTABILITY OR FITNESS FOR A PARTICULAR PURPOSE, OR OF ANY OTHER NATURE, ARE MADE WITH RESPECT TO THE INFORMATION PROVIDED IN THIS MSDS OR TO THE CHEMICAL TO WHICH INFORMATION MAY REFER. CRC NEITHER ASSUMES NOR AUTHORIZES ANY OTHER PERSON TO ASSUME FOR IT, ANY OTHER ADDITIONAL RESPONSIBILITY OR LIABILITY FOR THE USE OF, OR RELIANCE UPON, THIS INFORMATION.

HAZARD WARNING INFORMATION

HEALTH	FIRE	REACTIVE	OTHER	DEGREE OF HAZARD	COLOR CODING	OTHER CODES
4	3	2		0 = Minimum Hazard 1 = Slight Hazard 2 = Moderate Hazard 3 = Serious Hazard 4 = Severe Hazard	HEALTH = BLUE FIRE = RED REACTIVITY = YELLOW OTHER = WHITE	OX = Oxidizer ACID = Acid ALK = Alkali COR = Corrosive W = Use No Water

SECTION I - GENERAL INFORMATION

Characterization **Isocyanate, Organic**	RCRA Number **P064**	EPA Class **Acute Toxic Waste**
DOT Proper Shipping Name **Methyl isocyanate**	Chemical Abstract Service (CAS) Number **624-83-9**	
DOT Hazard Class and Label Requirements **Flammable Liquid**	DOT Emergency Guide Code **30**	
DOT Identification Number **UN 2480**	Chemical Formula **CH_3NCO**	

Synonyms
Methyl ester of isocyanic acid; MIC; isocyanotomethane; methyl carbylamine.

SECTION II - HAZARDOUS INGREDIENTS AND IDENTITY INFORMATION

Hazardous Components (specific identity)	OSHA Exposure Criteria	NIOSH Exposure Criteria	Immediately Dangerous to Life and Health (IDLH)	ACGIH Exposure Criteria
Methyl isocyanate (derivation: By heating N,N-diphenyl-N-methyl urea or by phosgenation of bis (trimethyl silyl) methylamine). **1 ppm = 2.37 mg/m^3**	**PEL (skin):** **0.02 ppm** **0.05 mg/m^3** **STEL:** **Not Established**	**REL (skin):** **0.02 ppm** **0.05 mg/m^3** **STEL** **Not Established**	**3 ppm**	**TLV (skin):** **0.02 ppm** **0.05 mg/m^3** **STEL:** **Not Established**

SECTION III - PHYSICAL AND CHEMICAL CHARACTERISTICS

Boiling Point **109°F (39°C)**	Specific Gravity (H_2O = 1) **0.96**
Vapor Pressure (mm Hg) **348 at 68°F (20°C)**	Molecular Weight **57.1**
Saturated Vapor Density (Air = 1.2 kg/m^3) **1.73 kg/m^3**	Melting Point **-112°F (-80°C)**

Solubility
Decomposes in water; soluble in hexane, isooctane, chloroform, and most organic solvents.

Appearance and Odor
Clear, colorless liquid with a sharp, pungent odor. Odor Threshold = 0. 2 ppm.

SECTION IV - FIRE AND EXPLOSION HAZARD DATA

Flash Point (method used) **20°F (-7°C) closed cup**	Explosive Limits in Air % by Volume **LEL: 5.3% UEL: 26%**
NFPA Classification **Class IB Combustible Liquid**	Autoignition Temperature **994°F (534°C)**

Extinguishing Media
Dry chemical, carbon dioxide, water spray, regular foam, or fog.

Special Fire Fighting Procedures
Wear full protective clothing and self-contained breathing apparatus (SCBA). Approach fire from upwind position. Remove containers from fire if it can be done without risk. Use a water spray to keep fire-exposed containers cool. Poisonous gases are produced in fire. Avoid smoke and runoff.

Unusual Fire and Explosion Hazards
Combustible when exposed to heat and flame. Containers may explode in fire. Stay clear of ends of tanks. Withdraw immediately 1500 feet if rising sound is heard from venting device or ends of tank become discolored in heat of fire. Fight fire from maximum distance, if possible.

SECTION V - REACTIVITY DATA

Stability		Conditions to Avoid **Methyl isocyanate is reactive and unstable. It decomposes with water and reacts vigorously with acids, bases, amines, metals and their salts. It forms explosive mixtures in air. Avoid contact with heat.**
Stable	Unstable **X**	Incompatibility *(materials to avoid)* **Incompatible with water (exothermic reaction), oxidizers, acids, alkalis, amines, iron, tin, copper, and will attack most elastomers. Reacts with alcohols and glycols to form urethanes.**
Hazardous Polymerization		Conditions to Avoid **Methyl isocyanate will polymerize readily under normal conditions of temperature and pressure. Phosgene appears to be an effective inhibitor of polymerization.**
May Occur **X**	Will Not Occur	Hazardous Decomposition or By-products **When heated to decomposition, methyl isocyanate can emit highly toxic/poisonous fumes including fumes of nitrogen oxide and hydrogen cyanide gas.**

SECTION VI - HEALTH HAZARD DATA

Primary Route(s) of Entry:	Inhalation?	Absorption (skin)?	Ingestion?
	X	X	X

Health Hazards

INHALATION: Irritation of the eyes, nose, throat, and respiratory tract. High levels can irritate and burn the lungs. Causes cough, running nose, chest pain, incoordination, dypsnea, nausea, vomiting, colicky abdominal cramps, convulsions, depression, and possible damage to the lungs as a result of pulmonary edema (fluid in the lungs), coma, and death. It may lead to respiratory sensitization (asthma), decreased liver function, and tubule necrosis of the kidney. Spontaneous abortion and fetal abnormalities have been reported.

ABSORPTION: Skin contact causes redness, burning, and painful blistering. Eye contact can lead to the development of conjunctivitis and inflammation of the eyelids. Absorption occurs.

INGESTION: Unspecified gastrointestinal effects.

Carcinogenicity	NTP Listed?	IARC Cancer Review Group?	OSHA Regulated?	Target Organs?
Unknown Human Unknown Animal	**No**	**No**	**29 CFR 1910.1000 Table Z-1**	**Respiratory system, eyes, skin, liver, kidney.**

Medical Conditions Generally Aggravated by Exposure

Respiratory system problems or diseases may be aggravated by exposure.

Emergency and First-aid Procedures

Eye contact: Flush immediately with water for 15 minutes (minimum); seek medical attention. Skin contact: Remove all contaminated clothing. Immediately wash area with large amounts of soap and water. For red or blistered skin, seek medical assistance. For inhalation: Remove the person from exposure. Provide respiratory assistance and CPR. Transfer to medical facility. Observe for 48 hours for lung effects. If swallowed: Contact poison control center. Unless advised otherwise, give conscious person 1 to 2 glasses of water to dilute then induce vomiting.

SECTION VII - PRECAUTIONS FOR SAFE HANDLING AND USE

Steps to be Taken in Case Material is Released or Spilled

Absorb liquid in vermiculite or similar material and deposit in sealed containers. Ventilate area of spill or leak. Remove all ignition sources. Restrict those not involved in cleanup from entering area.

Preferred Waste Disposal Method

Burn in a chemical incinerator equipped with an afterburner and scrubber.

Precautions to be Taken in Handling and Storage

Store in tightly closed containers in cool, dry, well-ventilated area away from heat, sparks, or flames.

Other Precautions and Warnings

Keep away from water or moisture. Do NOT reseal a container if contamination is suspected. Package for disposal or reclamation. Protect containers from physical damage.

SECTION VIII - CONTROL MEASURES AND PERSONAL PROTECTIVE EQUIPMENT

Respiratory Protection *(specify type)*

For any exposure level, use an MSHA/NIOSH-approved supplied air respirator with full facepiece in continuous flow mode or, use a self-contained breathing apparatus (SCBA) in continuous flow mode.

Ventilation

Local exhaust or general mechanical systems recommended.

Protective Gloves	Eye Protection	Other Protective Clothing
Impervious Material	**Chemical Goggles or Face Mask**	**Protective Apron or Clothing**

Work/Hygiene Practices

Always wash hands thoroughly after using chemical; never bring food, drink, or smoking materials into vicinity of chemicals.

METHYL ISOCYANATE

CH_3NCO CAS: 624-83-9

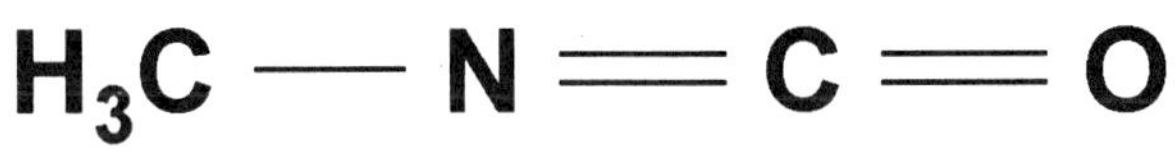

IDENTIFICATION AND TYPICAL USES

Methyl isocyanate appears as a clear, colorless liquid with a sharp, pungent odor. It is used primarily as an intermediate in the manufacture of pesticides and insecticides (including carbaryl), in organic synthesis, and in the production of plastics and polyurethane foams and insulation products.

RISK ASSESSMENT: HEALTH

General Assessment

Methyl isocyanate is an extremely toxic and poisonous substance. It can enter the body through *inhalation* skin contact (*absorption*), and by *ingestion*. It will also cause adverse topical health effects on contact with the skin. There is no information on its carcinogenic capabilities in humans or animals. However, incidents of spontaneous abortions and fetal abnormalities have been reported following acute overexposures. Mutation data have also been reported.

Inhalation exposures to high concentrations of the vapor can be extremely irritating to the eyes, nose, throat, and mucosa of the respiratory tract. Blindness has been reported. Symptoms of exposure include wheezing, coughing, increased secretions (runny nose), dyspnea, sweating, difficulty in breathing, insomnia, painful respiration, and possible life-threatening pulmonary edema (fluid in the lungs). Development of pulmonary edema may be delayed up to 48 hours thereby creating a false sense of security with regard to health exposure risk. In severe cases, there may be some restriction of air movement and eventual respiratory failure (due to edema). Prolonged contact can lead to chronic bronchitis, asthma-like symptoms, permanent fibrosis, and allergic pulmonary sensitization.

Skin contact causes irritation and painful burning with skin blisters on prolonged contact. If the liquid should contact the eye directly, there may be irritation, inflammation of the eyelids and surrounding tissues, conjunctivitis, and possible damage to the cornea. Ingestion causes unspecified gastrointestinal effects. Human data on exposure to methyl isocyanate are well documented as a result of an accidental release of the substance in Bhopal, India. This chemical presents a serious and life-threatening risk to human health.

Acute Health Effects

The following acute (short-term) health effects may occur immediately or shortly after exposure to methyl isocyanate:

Skin: Possible irritation with painful burning and blistering. Sensitization is possible.

Eye: Moderate to severe irritation. There can be inflammation, irritation of the eyelids, photophobia, tearing, conjunctivitis, and blindness.

Lung: Irritation of the nose, throat, and lungs following exposure. There can be a delayed but dangerous buildup of fluid in the lungs (pulmonary edema), which is a medical emergency.

Chronic Health Effects

The following chronic (long-term) health effects may occur at some time after exposure to methyl isocyanate and can last for months or even years:

Cancer Hazards: There is insufficient information in the references on the carcinogenic properties of methyl isocyanate. However, mutation data have been reported and many scientists believe that such chemicals may pose a cancer risk in the long-term.

Reproductive Hazard: According to the data presented in the references, methyl isocyanate may cause spontaneous abortions or lead to the development of fetal abnormalities.

Other Chronic Effects: Long-term exposure to methyl isocyanate may lead to allergic sensitization of the respiratory system with initial symptoms of night coughs and difficulty breathing and possible progression to asthmatic bronchitis. Repeated exposure may lead to permanent lung damage. Skin sensitization may occur on prolonged contact.

Recommended Risk-Reduction Measures

Personnel should avoid direct contact with methyl isocyanate. It is an extremely toxic and poisonous substance that must be handled with extreme care. If a less toxic chemical cannot be substituted for a hazardous substance, then *engineering controls* are the most effective method of reducing exposures.

The best protection is to enclose operations and/or provide local exhaust ventilation at the site of chemical release. While not always operationally feasible, isolating operations can also reduce exposure. Using respiratory protection is less effective than the controls mentioned above, but is still advisable whenever working with or around methyl isocyanate. The PEL for this chemical is extremely low (0.02 ppm) and difficult to monitor reliably under most operating or emergency conditions. A supplied-air respirator with a full facepiece operated in positive pressure mode or an MSHA/NIOSH-approved self-contained breathing apparatus (SCBA) with full facepiece operated in pressure demand or other positive pressure mode are the recommended respiratory protection methods of choice. In either case, if a full facepiece is not available, then chemical splash goggles should be worn to protect the eyes. Whenever a chemical splash hazard exists, a face shield and a protective apron should be worn. To prevent hand and skin exposures, impervious gloves should be used. Glove manufacturers should be contacted and permeation studies obtained *before* final glove selections are made.

Administrative controls should also be in place to minimize the potential for human exposures. These may include written procedures or policies which specify the methods and techniques that will be practiced whenever personnel are to work with methyl isocyanate.

All personnel should receive *training* on the use, hazards, protective measures, emergency actions, and other precautions per 29 CFR 1910.1200 (Hazard Communication), prior to the first assignment in an area where methyl isocyanate is used or stored.

If symptoms develop or overexposure is suspected, the following medical tests are recommended:

- ☑ Lung function tests (establish baseline)
- ☑ Consider chest X-ray after acute over exposure (may be negative if taken immediately after exposure due to delayed onset of pulmonary edema).
- ☑ Evaluation by a qualified allergist with special testing and careful consideration of exposure history (may help diagnose allergy).

Also, since smoking can cause heart disease, emphysema, and other lung diseases, smokers exposed to methyl isocyanate may experience symptoms more quickly than non-smokers under the same conditions of exposure. Prudent risk management requires proper consideration of *all* possible exposure factors when evaluating workplace health hazard risk.

Other methods to reduce exposure include:

- ☑ Where possible, enclose operations and use local exhaust ventilation at the site of chemical release. If local exhaust ventilation or enclosure is not used, respiratory protection should be mandatory.
- ☑ Where possible, automatically transfer methyl isocyanate from drums or other storage containers to process containers.
- ☑ Always ensure that proper protective clothing is worn when using chemical substances and that personnel are trained on its use and care.
- ☑ Wash thoroughly immediately after exposure to methyl isocyanate and at the end of the work shift or before eating, drinking, or smoking.
- ☑ Hazard warning information should be posted in the work area. In addition, as part of an on-going education and training program, all information on the health and safety hazards of methyl isocyanate should be communicated to all exposed and potentially exposed workers.

RISK ASSESSMENT: ENVIRONMENT

General Assessment

The environment is at risk of exposure during transportation, storage, disposal, or destruction of methyl isocyanate. In almost every scenario, the threat of environmental exposure is contingent upon the proper handling of the chemical substance. Accidental spills, large or small, can result in fire, explosion, and possible contamination of the surrounding environmental mediums (water, soil, and air).

Methyl isocyanate is considered a Class IB combustible liquid (per OSHA 29 CFR 1910.106). It can react vigorously with oxidizing agents, water, alcohols, strong bases, amines, alkalis, iron, tin, copper, elastomers, alcohols, and glycols. In addition, pressure may build in closed containers. These characteristics require special consideration during any emergency situation involving a leak or spill of methyl isocyanate.

Methyl isocyanate may enter the environment primarily during the disposal of wastes form the chlorinated hydrocarbon industry.

Acute Ecological Effects

Acute (short-term) toxic effects may include the death of animals, birds, or fish and death or low growth rate in plants. Acute effects are seen 2 to 4 days after animals or plants are exposed to methyl isocyanate.

Insufficient data are available to evaluate or predict the short-term effects of methyl isocyanate exposure to aquatic life, plants, birds, or land animals.

Chronic Ecological Effects

Chronic (long-term) toxic effects may include shortened life span, reproductive problems, lower fertility, and changes in appearance or behavior in exposed animals. These effects can be seen long after first exposure(s) to toxic chemicals.

Insufficient data are available to evaluate or predict the long-term effects of methyl isocyanate to aquatic life, plants, birds, or terrestrial animals.

Water Solubility

Methyl isocyanate decomposes exothermically in water. It is, therefore, only slightly soluble in the aquatic environment. Concentrations of less than 1 to 100 milligrams will mix with a liter of water.

Persistence in the Environment

Methyl isocyanate is non-persistent in the aquatic environment, with a half-life of less than 2 days. The half-life of a pollutant is the amount of time it takes for one half of the chemical to be degraded. Because methyl isocyanate will react with water, it is not expected to accumulate in the environment.

Bioaccumulation in Aquatic Organisms

Some substances increase in concentration, or *bioaccumulate,* in living organisms as they breathe contaminated air, drink contaminated water, or eat contaminated food. These chemicals can become concentrated in the body tissues and, especially, in the internal organs of animals as well as humans.

Because methyl isocyanate will react with water, it is not expected to accumulate in aquatic organisms.

Recommended Risk-Reduction Measures

Proper training of all transporters will reduce the likelihood of a mishap or accident resulting in a leak or spill to the environment. The correct use of labeling on all containers, trucks, and rail cars will enable emergency responders to react properly and quickly to any disaster thereby reducing the potential risk to the environment and to personnel.

Storage of methyl isocyanate should be segregated from incompatible chemicals to minimize the risk of cross-contamination or contact. Storage buildings or facilities should be equipped with proper fire protection equipment (sprinklers/extinguishers, alarms).

If a spill or leak to the environment has occurred, fire department, emergency response and/or hazardous materials spill personnel should be notified immediately. Cleanup should be attempted only by those trained in proper spill containment procedures. Contaminated soils should be removed for incineration and replaced with clean soil. If methyl isocyanate should contact the water table, aquifer, or navigable waterway, time is of the essence. Even though it is only slightly soluble in the aquatic environment, it will quickly decompose in water and, therefore, total containment and remediation will not be possible. When such spills occur, the local and/or state emergency response authorities must be notified. A comprehensive emergency response or disaster preparedness/recovery plan should be in place prior to any operations involving the use, transportation, storage, or disposal of methyl isocyanate. The proper disposal method for methyl isocyanate is to burn it in a chemical incinerator equipped with an afterburner and an air scrubber.

If methyl isocyanate are spilled or leaked, the following specific response steps are recommended:

- ☑ Restrict persons not wearing protective clothing from area of spill or leak until cleanup is complete and area can be opened for normal work.
- ☑ Remove all ignition sources.
- ☑ Ventilate area of spill or leak.
- ☑ If liquid is spilled, absorb in vermiculite, dry sand, earth, or similar material and deposit in sealed containers. Use non-sparking tools.

☑ It may be necessary to dispose of methyl isocyanate as a hazardous waste. The responsible state agency or the regional office of the federal Environmental Protection Agency (EPA) should be contacted for specific recommendations.

RISK ASSESSMENT: BUSINESS

General Assessment

Accidents or mishaps involving methyl isocyanate can present a significant threat to business operations. The loss or damage of equipment or facilities can significantly affect fiscal viability. Lawsuits that may result from personnel injury/death, public exposures, and/or environmental contamination will also require a serious expenditure of resources. Media attention surrounding an injury, death, or environmental damage can also result in a loss of profits and loss of current as well as future business. Always remember that anytime the term "reproductive hazard" is used, public hysteria, emotion, and ignorance can run equally high. This is also particularly true with the mere mention of methyl isocyanate as people recall the tragedy of Bhopal, India. These facts must be carefully considered whenever developing or implementing public relations policies or procedures.

Recommended Risk-Reduction Measures

Company attorneys, safety and health professionals, and environmental specialists should be involved in the development of any procedures for responding to hazardous chemical incidents. A company official should be pre-designated as a public relations officer with specific training in dealing with the press. Corporate plans and policies should be developed, approved, and implemented long before any need for such arises.

REFERENCES

American Conference of Governmental Industrial Hygienists. 1988. *Documentation of the Threshold Limit Values and Biological Exposure Indices,* 5th Edition (with updates). Cincinnati: ACGIH

Bloom, A. D. and F. J. DeSerres. 1995. *Ecotoxicity and Human Health.* New York: CRC Press/Lewis Publishers.

Bueche, J. R. 1972. *Principles of Physics,* 2nd Edition. New York: McGraw-Hill.

Calabrese, E. J. 1994. *Biological Effects of Low Level Exposures.* New York: CRC Press/Lewis Publishers.

Cockerman, L. G., B. S. Shane. 1994. *Basic Environmental Toxicology.* New York: CRC Press/Lewis Publishers.

Cothern, R. R. 1993. *Comparative Environmental Risk Assessment.* New York: CRC Press/Lewis Publishers.

Danse, I. R. 1991. *Common Sense Toxics in the Workplace.* New York: Van Nostrand Reinhold.

Environmental Protection Agency. 1983. *Chemical Hazard Information Profiles.* Washington, D.C.: U.S. EPA.

Fodden, C. R., and J. L. Weddell. 1992. *Hazardous Materials Emergency Action Data.* New York: CRC Press/Lewis Publishers.

Forseberg, K., and L. H. Keith. 1993. *Chemical; Protective Clothing Permeation and Degradation Database.* Chelsea, MI: Lewis Publishers.

Gardner, R. W. 1994. *Chemical Intolerance.* New York: CRC Press/Lewis Publishers.

Hallenbeck, W. H. 1993. *Quantitative Risk Assessment for Environmental and Occupational Health,* 2nd Edition. New York: CRC Press/Lewis Publishers.

Howard, P. H., et. al. 1991. *Handbook of Environmental Degradation Rates.* New York: CRC Press/Lewis Publishers.

Hosty, J. W., and P. Foster. 1990. *A Practical Guide to Chemical Spill Response.* New York: Van Nostrand Reinhold.

International Agency for Research on Cancer (IARC). 1987. *IARC Monographs on the Evaluation of Carcinogenic Risk to Humans, Overall Evaluations of Carcinogenicity: An updating of IARC Monographs Volumes 1-42.* Lyon: IARC.

Keith, L. H. and M. M. Walker. 1995. *Handbook of Air Toxics.* New York: CRC Press/Lewis Publishers.

Lauwerys, R. R., P. Hoet. 1993. *Industrial Chemical Exposure.* New York: CRC Press/Lewis Publishers.

Lewis, R. J., Sr. 1992. *Sax's Dangerous Properties of Industrial Materials,* Volumes 1, 2, and 3, 8th Edition. New York: Van Nostrand Reinhold.

Meyer, E. 1990. *Chemistry of Hazardous Materials.* Englewood Cliffs, NJ: Prentice Hall.

Miller, B. F., M.D., and C. B. Keane, R.N. 1972. *Encyclopedia and Dictionary of Medicine and Nursing.* Philadelphia: W.B. Sanders Company.

Mycock, J. C., J. D. McKenna, and L. Theodore. 1995. *Handbook of Air Pollution Control Engineering and Technology.* New York: CRC Press/Lewis Publishers.

Ott, W. R. 1995. *Environmental Statistics and Data Analysis.* New York: CRC Press/Lewis Publishers.

Patnaik, P. 1992. *A Comprehensive Guide to the Hazardous Properties of Chemical Substances.* New York: Van Nostrand Reinhold.

Rea, W. J. 1995. *Chemical Sensitivity,* Vol. I-III. New York: CRC Press/Lewis Publishers.

Sacarello, H. L. A. 1994. *The Comprehensive Handbook of Hazardous Materials.* New York: CRC Press/Lewis Publishers.

Sherman, J. D. 1988. *Chemical Exposure and Disease.* New York: Van Nostrand Reinhold.

U.S. Department of Health and Human Services, National Institute for Occupational Safety and Health. 1994. *NIOSH Pocket Guide to Chemical Hazards.* Washington, D.C.: U.S. Government Printing Office.

MATERIAL SAFETY DATA SHEET

CHEMICAL NAME	
METHYL METHACRYLATE	**CRC PRESS/LEWIS PUBLISHERS DISCLAIMER:** THE INFORMATION AND RECOMMENDATIONS PRESENTED HEREIN ARE BASED ON SOURCES BELIEVED TO BE RELIABLE. CRC MAKES NO REPRESENTATION ON ITS COMPLETENESS OR ACCURACY. IT IS THE RESPONSIBILITY OF THE USER TO DETERMINE THE CHEMICAL'S SUITABILITY FOR ITS INTENDED USE, THE CHEMICAL'S SAFE USE, AND THE CHEMICAL'S PROPER DISPOSAL. NO REPRESENTATIONS AND/OR WARRANTIES, EITHER EXPRESSED OR IMPLIED, OF THE MERCHANTABILITY OR FITNESS FOR A PARTICULAR PURPOSE, OR OF ANY OTHER NATURE, ARE MADE WITH RESPECT TO THE INFORMATION PROVIDED IN THIS MSDS OR TO THE CHEMICAL TO WHICH INFORMATION MAY REFER. CRC NEITHER ASSUMES NOR AUTHORIZES ANY OTHER PERSON TO ASSUME FOR IT, ANY OTHER ADDITIONAL RESPONSIBILITY OR LIABILITY FOR THE USE OF, OR RELIANCE UPON, THIS INFORMATION.

HAZARD WARNING INFORMATION

HEALTH	FIRE	REACTIVE	OTHER	DEGREE OF HAZARD	COLOR CODING	OTHER CODES
2	**3**	**2**		0 = Minimum Hazard 1 = Slight Hazard 2 = Moderate Hazard 3 = Serious Hazard 4 = Severe Hazard	HEALTH = BLUE FIRE = RED REACTIVITY = YELLOW OTHER = WHITE	OX = Oxidizer ACID = Acid ALK = Alkali COR = Corrosive W = Use No Water

SECTION I - GENERAL INFORMATION

Characterization	RCRA Number	EPA Class
Aliphatic Ester	**U162**	**Toxic Waste**

DOT Proper Shipping Name **Methyl methacrylate monomer, Inhibited**	Chemical Abstract Service (CAS) Number **80-62-6**
DOT Hazard Class and Label Requirements **Flammable Liquid**	DOT Emergency Guide Code **26 (Inhibited)**
DOT Identification Number **UN 1247**	Chemical Formula **$CH_2{=}C(CH_3)COOCH_3$**

Synonyms

Methyl methacrylate (inhibited); methacrylate monomer; methyl methacrylate monomer; methyl ester of methacrylic acid; methyl-2-methyl-2-2propenoate.

SECTION II - HAZARDOUS INGREDIENTS AND IDENTITY INFORMATION

Hazardous Components (specific identity)	OSHA Exposure Criteria	NIOSH Exposure Criteria	Immediately Dangerous to Life and Health (IDLH)	ACGIH Exposure Criteria
Methyl methacrylate (derivation: From acetone cyanohydrin, methanol, and dilute sulfuric acid; by oxidation of tertbutyl alcohol to methacrolein and then to methacrylic acid, followed by reaction with methanol). **1 ppm = 4.16 mg/m^3**	**PEL:** **100 ppm** **410 mg/m^3** **STEL:** **Not Determined**	**REL:** **100 ppm** **4105 mg/m^3** **STEL:** **Not Determined**	**1000 ppm**	**TLV:** **100 ppm** **410 mg/m^3** **STEL:** **Not Determined**

SECTION III - PHYSICAL AND CHEMICAL CHARACTERISTICS

Boiling Point **214°F (101°C)**	Specific Gravity (H_2O = 1) **0.94**
Vapor Pressure (mm Hg) **29 at 68°F (20°C)**	Molecular Weight **100.1**
Vapor Density (Air = 1) **3.45**	Melting Point **-54°F (-48°C)**

Solubility

Slightly soluble in water (1.5%). Soluble in most organic solvents.

Appearance and Odor

Clear, colorless liquid with a sharp, acrid, fruity odor. Odor Threshold = 0.083 ppm.

SECTION IV - FIRE AND EXPLOSION HAZARD DATA

Flash Point (method used) **50°F (-3°C) opened cup**	Explosive Limits in Air % by Volume **LEL: 1.78%** **UEL: 8.2%**
NFPA Classification **Class IB Flammable Liquid**	Autoignition Temperature **790°F (421°C)**

Extinguishing Media

Dry chemical, alcohol foam, or carbon dioxide. Use blanketing effect with foam to smother large fires.

Special Fire Fighting Procedures

Poisonous gases are produced in fire. Wear full protective clothing and (SCBA). Continue to cool containers with water after fire is out. Move containers from fire area if it can be done without risk.

Unusual Fire and Explosion Hazards

Evacuate 2500 feet radius if fire becomes uncontrollable or container is exposed to direct flame. Vapors are heavier than air and may travel a distance to ignition source to flashback. Heat and/or an insufficient concentration of inhibitor can cause violent polymerization with the rupture of container.

SECTION V - REACTIVITY DATA

Stability		Conditions to Avoid Methyl methacrylate is kept stable under normal conditions of storage and operation with an inhibitor. Avoid contact with incompatible materials. Keep away from heat, flame, and other sources of ignition.
Stable X	Unstable	Incompatibility *(materials to avoid)* Reacts violently in contact with oxidizers (peroxides, nitrates, perchlorates, chlorates, permanganates) polymerizers, strong alkalis (sodium hydroxide, potassium hydroxide), moisture, chlorosulfonic acid.
Hazardous Polymerization		Conditions to Avoid Explosive polymerization can occur if inhibitor levels become too low or when exposed to heat, light, oxidizers, and peroxides.
May Occur X	Will Not Occur	Hazardous Decomposition or By-products When heated to decomposition, methyl methacrylate emits acrid and irritating fumes, including oxides of carbon (monoxide and dioxide).

SECTION VI - HEALTH HAZARD DATA

Primary Route(s) of Entry:	Inhalation?	Absorption (skin)?	Ingestion?
	X		X

Health Hazards

INHALATION: Severe irritation to the eyes, nose, throat, and respiratory system. Exposure can cause drowsiness, headache, vomiting, and nausea. Severe exposures may cause depression of the central nervous system with anorexia, irritability, and narcosis. There also may be dypsnea, cyanosis, convulsions, lachrymation, coughing, fatigue, and wheezing.

SKIN & EYES: Skin and eye irritation. Eye tissue damage may occur on contact. Parasthesia (tingling sensation on the skin) may occur on prolonged or repeated contact.

INGESTION: Severe local irritation of the gastrointestinal tract with irritation of the lips, mouth, throat, painful swallowing, abdominal pain and cramps, nausea, vomiting, and diarrhea.

Carcinogenicity	NTP Listed?	IARC Cancer Review Group?	OSHA Regulated?	Target Organs?
Unknown Human Questioned Animal	No	Group 3	29 CFR 1910.1000 Table Z-1	Respiratory system; skin; eyes.

Medical Conditions Generally Aggravated by Exposure

None reported.

Emergency and First-aid Procedures

Eye contact: Flush immediately with water for 15 minutes (minimum); seek medical attention. Skin contact: Remove all contaminated clothing. Immediately wash area with large amounts of soap and water. Seek medical assistance. For inhalation: Remove the person from exposure. Provide respiratory assistance and CPR. Transfer to medical facility. If swallowed: Contact poison control center. Give 1 to 2 glasses of water; induce vomiting until vomitus is clear. Seek medical attention immediately.

SECTION VII - PRECAUTIONS FOR SAFE HANDLING AND USE

Steps to be Taken in Case Material is Released or Spilled

Contain spills using absorbent material (vermiculite or other). Ventilate area of spill and remove ignition sources. Have extinguishing agent available. Use non-sparking tools. Store materials in DOT-approved containers. Restrict those not involved in cleanup from entering area.

Preferred Waste Disposal Method

Incineration in chemical incinerator equipped with scrubber and afterburner.

Precautions to be Taken in Handling and Storage

Store in tightly closed containers in a cool, dark, well-ventilated area. Protect containers from physical damage. Keep heat, fire, and flame away. Outside building storage is recommended.

Other Precautions and Warnings

Bulk storage of methyl methacrylate is not recommended. Ground and bond all metal containers.

SECTION VIII - CONTROL MEASURES AND PERSONAL PROTECTIVE EQUIPMENT

Respiratory Protection *(specify type)*

Carcinogenic properties unknown. Use a supplied-air respirator or a self-contained breathing apparatus (SCBA) with full facepiece operated in positive pressure mode for the best protection.

Ventilation

Local exhaust or general mechanical systems recommended.

Protective Gloves	Eye Protection	Other Protective Clothing
Impervious Gloves	Chemical Goggles or Face Mask	Impermeable Apron or Protective Suit

Work/Hygiene Practices

Always wash hands thoroughly after using chemical; never bring food, drink, or smoking materials into vicinity of chemicals.

METHYL METHACRYLATE

$CH_2{=}C(CH_3)COOCH_3$ CAS: 80-62-6

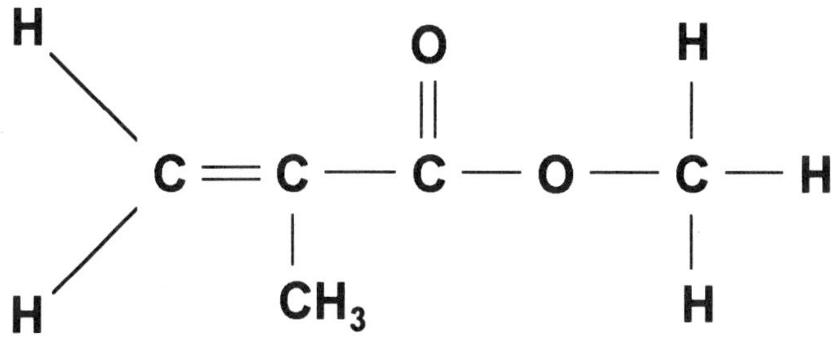

IDENTIFICATION AND TYPICAL USES

Methyl methacrylate is a clear, colorless liquid with an acrid, sweet, sharp, fruity odor. It is used as a monomer in the manufacture of polymethacrylate resins. It is also used in the impregnation of concrete, as a bone cement, as a dental ceramic filler or cement, as a medicinal spray adhesive or nonirritant bandage solvent, and also as a coating for contact lenses.

RISK ASSESSMENT: HEALTH

General Assessment

Methyl methacrylate is a moderately toxic to humans by *inhalation* and mildly toxic by *ingestion* and skin contact (although dermal absorption has not been reported). It is a questioned animal carcinogen, but there are insufficient data regarding its carcinogenicity in humans. It is known to cause tumorigenic, teratogenic, and reproductive effects. Mutation data are also reported.

Inhalation results in irritation of the mucus membranes of the eyes, nose, throat, and respiratory system. There may be symptoms such as lethargy, coughing, wheezing, fatigue, cyanosis, and convulsive movements. High concentrations can cause dizziness with dypsnea, nausea, headache, vomiting, and can lead to sleep effects, excitement, anorexia, blood pressure decrease, and other symptoms of central nervous system depression.

Skin contact can cause serious localized irritation at the site of contact with inflammation, rash, and possible sensitization. Repeated exposures may cause parasthesia (tingling or pricking sensation on the skin). Eye contact is dangerous and there may be damage on contact with the liquid or vapor.

Ingestion of methyl methacrylate is not likely but is possible. It causes irritation of the mouth, lips, throat and gastrointestinal tract with painful swallowing, abdominal cramps, nausea, vomiting, and diarrhea.

☠ *Acute Health Effects*

The following acute (short-term) health effects may occur immediately or shortly after exposure to methyl methacrylate:

Skin: Serious localized irritation, erythema, rash, and sensitization.

Eye: Severe irritation which can lead to injury, burns, painful irritation, and permanent damage if allowed to remain in contact.

Lung: Irritation of the eyes, nose, throat and lungs (especially the mucus membranes). Will cause coughing, wheezing, congestion, and possible lung damage.

CNS: High concentrations may cause dizziness, headache, nausea, vomiting, dypsnea, lethargy, and convulsive movements.

Chronic Health Effects

The following chronic (long-term) health effects may occur at some time after exposure to methyl methacrylate and can last for months or even years:

Cancer Hazards: According to the references, methyl methacrylate has been investigated for its ability to cause cancer in test animals and there is insufficient evidence to determine its carcinogenic effects. Its carcinogenicity is considered questioned in animals. Mutation data have been reported and many scientists believe that such chemicals may pose a cancer risk

Reproduction: According to the references, methyl methacrylate has shown to cause teratogenic effects in test animals as well as an ability to adversely affect the reproductive process.

Other Chronic Effects: Very high or prolonged exposure may lead to skin sensitization. Damage to the central nervous system can also occur in the long-term.

Recommended Risk-Reduction Measures

Because of its unknown but questioned carcinogenicity, as well as its known ability to cause teratogenic, mutagenic, and other systemic effects, personnel should avoid direct contact with methyl methacrylate. The best risk reduction measure is to use a less toxic chemical as a substitute for methyl methacrylate. If this is not possible or feasible, then *engineering controls* are the most effective method of reducing exposures. The best protection is to enclose operations and/or provide local exhaust ventilation at the site of chemical release. While not always operationally feasible, isolating operations can also reduce exposure risk.

Using respiratory protection is less effective than the controls mentioned above, but is still advisable whenever working with or around methyl methacrylate. For exposures exceeding the 100 ppm PEL (8-hour TWA), an organic vapor respirator may suffice. For higher concentrations, a supplied-air respirator with full facepiece operated in positive pressure mode, or a self-contained breathing apparatus (SCBA) with full facepiece and operated in pressure demand mode are the recommended respiratory protection methods of choice. If a full facepiece is not available, then chemical goggles should be worn to protect the eyes. Whenever a chemical splash hazard exists, a face shield and a protective apron should be worn. To prevent hand and skin exposures, impervious gloves should be used. Glove manufacturers should be contacted to obtain permeation studies before gloves are selected.

Administrative controls should also be in place to minimize the potential for human exposures. These may include written procedures or policies which specify the methods and techniques that will be practiced whenever personnel are to work with methyl methacrylate.

All personnel should receive *training* on the use, hazards, protective measures, emergency actions, and other precautions per 29 CFR 1910.1200 (Hazard Communication), prior to the first assignment in an area where methyl methacrylate is used or stored. Before beginning employment and at regular intervals thereafter (e.g., annually), those persons having frequent or potentially high exposures should be given the following medical tests:

- ☑ Lung function tests.
- ☑ Interview for brain effects, including recent memory problems, mood changes (irritability, withdrawal), difficulty with concentration, headaches, malaise, and altered sleep patterns. Cerebellar, autonomic, and peripheral nervous system evaluation is recommended. Positive or borderline cases should be referred for neuropsychological testing.

If symptoms develop or overexposure is suspected, the following medical tests are recommended:

- ☑ Examination of the nervous system.
- ☑ Lung function tests.
- ☑ Evaluation by a qualified allergist, including careful exposure history and special testing (may help diagnose skin allergy).

Any medical evaluation should include a careful history of past and present symptoms with an examination. Medical tests that look for existing damage are not a substitute for controlling exposures.

Other methods to reduce exposure include:

- ☑ Where possible, enclose operations and use local exhaust ventilation at the site of chemical release. If local exhaust ventilation or enclosure is not used, respiratory protection should be mandatory.
- ☑ Always ensure that proper protective clothing is worn when using chemical substances and that personnel are trained on its use and care.
- ☑ Wash thoroughly immediately after exposure to methyl methacrylate and at the end of the work shift or before eating, drinking, or smoking.
- ☑ Hazard warning information should be posted in the work area. In addition, as part of an on-going education and training program, all information on the health and safety hazards of methyl methacrylate should be communicated to all exposed workers.
- ☑ Never eat, drink, or smoke in areas were methyl methacrylate is used, handled or stored.
- ☑ Work clothing contaminated with methyl methacrylate should never be worn home where family members can be exposed. It should be laundered only by those who have been trained on the health risks associated with methyl methacrylate.

RISK ASSESSMENT: ENVIRONMENT

General Assessment

The environment is at risk of exposure during transportation, storage, disposal, or destruction of methyl methacrylate. In almost every scenario, the threat of environmental exposure is contingent upon the proper handling of the chemical substance. Accidental spills, large or small, can result in fire, explosion, and possible contamination of the surrounding environmental mediums (water, soil, and air).

Methyl methacrylate is considered a Class IB flammable liquid (according to OSHA 29 CFR 1910.106). Its low flash point and relatively low boiling point present a serious fire and explosion hazard concern. Also, because it is incompatible with a number of common materials, especially strong oxidizers, contact can result in violent and explosive reactions. It can also form explosive mixtures in air and can ignite on contact with heat, fire, or sparks. Heat can cause methyl methacrylate to react by itself. If this happens in a closed container, an explosion can occur. Methyl methacrylate usually contains an inhibitor such as hydroquinone or its methyl ether to prevent a self reaction. If the inhibitor is absent or present in too low a quantity, the explosive reaction may occur without the application of heat. These characteristics require special consideration during any emergency situation involving a leak or spill of methyl methacrylate.

Methyl methacrylate can enter the environment through unchecked industrial discharges and through spills.

Acute Ecological Effects

Acute (short-term) toxic effects may include the death of animals, birds, or fish, and death or low growth rate in plants. Acute effects are seen 2 to 4 days after animals or plants are exposed to methyl methacrylate. This chemical has slight acute toxicity to aquatic life. Insufficient data are available to evaluate or predict the short-term effects of methyl methacrylate to plants, birds, or terrestrial animals.

Chronic Ecological Effects

Chronic toxic effects may include shortened life span, reproductive problems, lower fertility, and changes in appearance or behavior in exposed animals. These effects can be seen long after first exposure(s) to toxic chemicals. Methyl methacrylate has slight chronic toxicity to aquatic life. Insufficient data are available to evaluate or predict the long-term effects of methyl methacrylate to plants, birds, or land animals.

Water Solubility

Methyl methacrylate is moderately soluble in water. Concentrations of 1 to 1000 milligrams will mix with a liter of water.

Persistence in the Environment

Methyl methacrylate is slightly persistent in water, with a half-life of between 2 to 20 days. The half-life of a pollutant is the amount of time it takes for one-half of the chemical to be degraded. About 86% of methyl methacrylate will eventually end up in the air; the remainder will end up in the water.

Bioaccumulation in Aquatic Organisms

Some substances increase in concentration, or *bioaccumulate,* in living organisms as they breathe contaminated air, drink contaminated water, or eat contaminated food. These chemicals can become concentrated in the tissues and internal organs of animals as well as humans.

The concentration of methyl methacrylate found in fish tissues is expected to be about the same as the average concentration of methyl methacrylate in water from which the fish was taken.

Recommended Risk-Reduction Measures

Proper training of all transporters will reduce the likelihood of a mishap or accident resulting in a leak or spill to the environment. The correct use of DOT labeling (per the MSDS) on all containers, trucks, and rail cars will enable emergency responders to react properly and quickly to any disaster thereby reducing the potential risk to the environment and to personnel.

Storage of methyl methacrylate should be segregated from incompatible chemicals to minimize the risk of cross-contamination or contact. Buildings designated for storage should be equipped with appropriate fire protection systems (alarms, sprinklers, emergency lighting, portable extinguishers). Equipment should be designed to meet explosion-proof standards.

If a spill or leak to the environment has occurred, fire department, emergency response, and/or hazardous materials spill personnel should be notified immediately. Cleanup should be attempted only by those trained in proper spill containment procedures using non-sparking tools. Contaminated soils should be removed for incineration and replaced with clean soil. If methyl methacrylate should contact the water table, aquifer, or navigable waterway, time is of the essence. It is moderately soluble in water and, therefore, containment and remediation may not be entirely possible. When such spills occur, the local and/or state emergency response authorities must be notified. A comprehensive emergency response or disaster preparedness/recovery plan should be in place prior to any operations involving the use, transportation, storage, or disposal of methyl methacrylate.

If methyl methacrylate is spilled or leaked, the following specific steps are recommended:

- ☑ Restrict persons not wearing protective clothing from area of spill or leak until cleanup is complete and area can be opened for normal work.
- ☑ Ventilate area of spill or leak.
- ☑ Remove all ignition sources.
- ☑ Absorb liquids in vermiculite, dry sand, earth, or a similar material and deposit in sealed containers for proper disposal. Use non-sparking tools during cleanup procedures.
- ☑ It may be necessary to dispose of methyl methacrylate as a hazardous waste. The responsible state agency or the regional office of the federal Environmental Protection Agency (EPA) should be contacted for specific recommendations.

RISK ASSESSMENT: BUSINESS

General Assessment

Accidents or mishaps involving methyl methacrylate can present a moderate threat to business operations. The loss or damage of equipment or facilities can significantly affect fiscal viability. Lawsuits that may result from personnel injury/death, public exposures, and/or environmental contamination will also require a serious expenditure of resources.

Media attention surrounding an injury, death, or environmental damage can also result in a loss of profits and loss of current as well as future business. Always remember that anytime the terms "cancer," "carcinogen," or "reproductive hazard" are used, public emotion, hysteria, and ignorance can run equally high. This should be carefully considered when developing or implementing a public relations policy.

Recommended Risk-Reduction Measures

Company attorneys, safety and health professionals, and environmental specialists should be involved in the development of any procedures or policies intended to manage the use of chemicals in the workplace. A company official should be pre-designated as a public relations officer with specific training in dealing with the press. Corporate plans and policies should be developed, approved, and implemented long before any need for such arises.

REFERENCES

American Conference of Governmental Industrial Hygienists. 1988. *Documentation of the Threshold Limit Values and Biological Exposure Indices,* 5th Edition (with updates). Cincinnati: ACGIH

Anthony, C. P., and N. J. Kolthoff. 1971. *Textbook of Anatomy and Physiology.* St. Louis: C. V. Mosby Company.

Brusick, D. J. 1994. *Methods for Genetic Risk Assessment.* New York: CRC Press/Lewis Publishers.

Bueche, J. R. 1972. *Principles of Physics,* 2nd Edition. New York: McGraw-Hill.

Burrell, R., D. K. Flaherty, and L. J. Sauers. 1992. *Toxicology of the Immune System.* New York: Van Nostrand Reinhold.

Calabrese, E. J. 1994. *Biological Effects of Low Level Exposures.* New York: CRC Press/Lewis Publishers.

Cothern, C. R., N. P. Ross. 1994. *Environmental Statistics, Assessment, and Forecasting.* New York: CRC Press/Lewis Publishers.

Gardner, R. W. 1994. *Chemical Intolerance.* New York: CRC Press/Lewis Publishers.

Hodgson, E., R. B. Mailman, and J. E. Chambers. 1988. *Dictionary of Toxicology.* London: The MacMillian Company.

Hosty, J. W., and P. Foster. 1990. *A Practical Guide to Chemical Spill Response.* New York: Van Nostrand Reinhold.

Howard, P. H., et. al. 1991. *Handbook of Environmental Degradation Rates.* New York: CRC Press/Lewis Publishers.

Keith, L. H. and D. B. Walters. (Editors) 1992. *National Toxicology Program's Chemical Database, Volumes I-VII.* Chelsea, MI: Lewis Publishers.

Lauwerys, R. R., P. Hoet. 1993. *Industrial Chemical Exposure.* New York: CRC Press/Lewis Publishers.

Lewis, Richard J., Sr. 1992. *Sax's Dangerous Properties of Industrial Materials,* Volumes 1, 2, and 3, 8th Edition. New York: Van Nostrand Reinhold.

Manahan, S. E. 1992. *Toxicological Chemistry,* 2nd Edition. New York: CRC Press/Lewis Publishers.

Meyer, Eugene. 1990. *Chemistry of Hazardous Materials.* Englewood Cliffs, NJ: Prentice Hall.

Patnaik, P. 1992. *A Comprehensive Guide to the Hazardous Properties of Chemical Substances.* New York: Van Nostrand Reinhold.

MATERIAL SAFETY DATA SHEET

CHEMICAL NAME	
NAPHTHALENE	**CRC PRESS/LEWIS PUBLISHERS DISCLAIMER:** THE INFORMATION AND RECOMMENDATIONS PRESENTED HEREIN ARE BASED ON SOURCES BELIEVED TO BE RELIABLE. CRC MAKES NO REPRESENTATION ON ITS COMPLETENESS OR ACCURACY. IT IS THE RESPONSIBILITY OF THE USER TO DETERMINE THE CHEMICAL'S SUITABILITY FOR ITS INTENDED USE, THE CHEMICAL'S SAFE USE, AND THE CHEMICAL'S PROPER DISPOSAL. NO REPRESENTATIONS AND/OR WARRANTIES, EITHER EXPRESSED OR IMPLIED, OF THE MERCHANTABILITY OR FITNESS FOR A PARTICULAR PURPOSE, OR OF ANY OTHER NATURE, ARE MADE WITH RESPECT TO THE INFORMATION PROVIDED IN THIS MSDS OR TO THE CHEMICAL TO WHICH INFORMATION MAY REFER. CRC NEITHER ASSUMES NOR AUTHORIZES ANY OTHER PERSON TO ASSUME FOR IT, ANY OTHER ADDITIONAL RESPONSIBILITY OR LIABILITY FOR THE USE OF, OR RELIANCE UPON, THIS INFORMATION.

NATIONAL FIRE PROTECTION ASSOCIATION (NFPA) LABELING INFORMATION

HEALTH	FIRE	REACTIVE	OTHER	DEGREE OF HAZARD	COLOR CODING	OTHER CODES
2	**2**	**0**		0 = Minimum Hazard 1 = Slight Hazard 2 = Moderate Hazard 3 = Serious Hazard 4 = Severe Hazard	HEALTH = BLUE FIRE = RED REACTIVITY = YELLOW OTHER = WHITE	OX = Oxidizer ACID = Acid ALK = Alkali COR = Corrosive W = Use No Water

SECTION I - GENERAL INFORMATION

Characterization	**Polynuclear Aromatic**	RCRA Number	**U165**
		EPA Class	**Toxic Waste**
DOT Proper Shipping Name	**Naphthalene**	Chemical Abstract Service (CAS) Number	**91-20-3**
DOT Hazard Class and Label Requirements	**ORM-A; Flammable Solid**	DOT Emergency Guide Code	**32**
DOT Identification Number	**UN 1334 (crude) UN2304 (molten)**	Chemical Formula	$C_{10}H_8$

Synonyms
Napthalin; tar camphor; white tar; mothballs.

SECTION II - HAZARDOUS INGREDIENTS AND IDENTITY INFORMATION

Hazardous Components (specific identity)	OSHA Exposure Criteria	NIOSH Exposure Criteria	Immediately Dangerous to Life and Health (IDLH)	ACGIH Exposure Criteria
Naphthalene (derivation: From commercial anthracene and coal tar). **1 ppm = 5.33 mg/m³**	**PEL:** 10 ppm 50 mg/m³ **STEL:** 15 ppm 75 mg/m³	**REL:** 10 ppm 50 mg/m³ **STEL:** 15 ppm 75 mg/m³	**250 ppm**	**TLV:** 10 ppm 50 mg/m³ **STEL:** 15 ppm 75 mg/m³

SECTION III - PHYSICAL AND CHEMICAL CHARACTERISTICS

Boiling Point	**424°F (218°C)**	Specific Gravity (H_2O = 1)	**1.15**
Vapor Pressure (mm Hg)	**0.08 at 69°F (20°C)**	Molecular Weight	**128.2**
Vapor Density (Air = 1)	**4.4**	Melting Point	**176°F (80°C)**

Solubility
Slightly soluble in water (0.003%); dissolves in most organic solvents.

Appearance and Odor
White or colorless to brownish-orange crystalline solid flakes with an odor of mothballs. It is often shipped as a molten solid. Odor Threshold = 0.084 ppm.

SECTION IV - FIRE AND EXPLOSION HAZARD DATA

Flash Point (method used)	**174°F (79°C) opened cup**	Explosive Limits in Air % by Volume	**LEL: 0.9% UEL: 5.9%**
NFPA Classification	**Combustible Solid**	Autoignition Temperature	**979°F (526°C)**

Extinguishing Media
Carbon dioxide, dry chemical, water, or alcohol foam. Foam and water may cause extensive foaming.

Special Fire Fighting Procedures
Wear full protective clothing and self-contained breathing apparatus (SCBA). Poisonous gases are produced in fire.

Unusual Fire and Explosion Hazards
Naphthalene is moderately explosive when exposed to flame. The vapors are much heavier than air and can collect in low areas or travel great distances to an ignition source and flashback to cause fire or explosion.

SECTION V - REACTIVITY DATA

Stability		Conditions to Avoid **Naphthalene is normally stable under routine conditions of storage and handling. Avoid contact with heat and incompatible materials.**
Stable **X**	Unstable	Incompatibility *(materials to avoid)* **Incompatible with strong oxidizers (fluorine, bromine, chlorine), chromic anhydride, and mixtures of aluminum trichloride, and benzoyl chloride.**
Hazardous Polymerization		Conditions to Avoid **Under normal conditions of temperature and pressure, hazardous polymerization of naphthalene is not expected to occur.**
May Occur	Will Not Occur **X**	Hazardous Decomposition or By-products **When heated to decomposition, naphthalene can emit carbon monoxide and acrid, irritating smoke.**

SECTION VI - HEALTH HAZARD DATA

Primary Route(s) of Entry:	Inhalation?	Absorption (skin/eye)?	Ingestion?
	X	X	X

Health Hazards

INHALATION: Irritation of the respiratory tract, headache, nausea, vomiting, loss of appetite, slowed reactions, excitement, confusion, and other central nervous system effects. There is also risk of renal failure, blood disorders, jaundice, and eye damage.

ABSORPTION: Coal tar derivatives may absorb through unbroken skin. Irritation with burning, itching, and edema (fluid buildup). Irritates the conjunctiva of the eye.

INGESTION: Gastrointestinal tract irritation.

Carcinogenicity	NTP Listed?	IARC Cancer Review Group?	OSHA Regulated?	Target Organs?
Unknown Human Questioned Animal	No	No	29 CFR 1910.1000 Table Z-1	Eyes, skin, blood, liver, kidneys, CNS

Medical Conditions Generally Aggravated by Exposure

Diseases of the blood, liver, and kidneys may be aggravated upon exposure to naphthalene.

Emergency and First-aid Procedures

Eye contact: Do not allow victim to keep eyes tightly shut. Flush immediately with water for 15 minutes (minimum); seek medical attention. Skin contact: Remove all contaminated clothing. Immediately wash area with large amounts of soap and water. For red or blistered skin, contact physician. For inhalation: Remove the person from exposure. Provide respiratory assistance and CPR. Transfer to medical facility. If swallowed: Seek medical attention immediately. Never give anything by mouth to an unconscious or convulsing person.

SECTION VII - PRECAUTIONS FOR SAFE HANDLING AND USE

Steps to be Taken in Case Material is Released or Spilled

Collect powdered materials in most convenient and safe manner and deposit in sealed containers. Ventilate area of spill or leak. Remove all ignition sources. Restrict those not involved in cleanup from entering area. Notify appropriate authorities, as required by applicable environmental regulations.

Preferred Waste Disposal Method

Incinerator equipped with scrubber and afterburner.

Precautions to be Taken in Handling and Storage

Naphthalene is a combustible solid. Store to avoid contact with strong oxidizers, since violent reactions can occur. Store in tightly closed containers in cool, well-ventilated area away from heat.

Other Precautions and Warnings

Sources of ignition (heat, flame, cigarettes, incompatibles) are prohibited where naphthalene is used, handled, or stored. Bulk storage of naphthalene is not recommended.

SECTION VIII - CONTROL MEASURES AND PERSONAL PROTECTIVE EQUIPMENT

Respiratory Protection *(specify type)*

Use an MSHA/NIOSH-approved supplied-air respirator with full facepiece in continuous flow mode, or use a self-contained breathing apparatus (SCBA) with full facepiece operated in positive pressure mode.

Ventilation

Local exhaust (preferred) at site of chemical work or general ventilation.

Protective Gloves	Eye Protection	Other Protective Clothing
Impervious Gloves	Chemical/Dust Goggles or Face Mask	Boots; Aprons; Gauntlets

Work/Hygiene Practices

Always wash hands thoroughly after using chemical; never bring food, drink, or smoking materials into vicinity of chemicals.

NAPHTHALENE

$C_{10}H_8$

CAS: 91-20-3

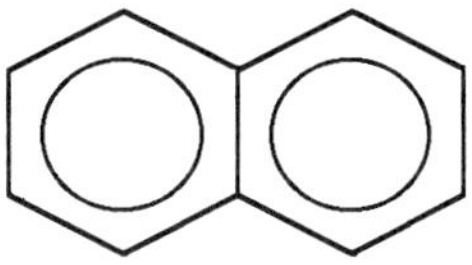

IDENTIFICATION AND TYPICAL USES

Naphthalene can appear as colorless or white to brownish-orange crystalline solid flakes with an odor of mothballs. It is normally shipped as a molten solid. It is used as an intermediate in the production of chlorinated naphthalenes, naphthyl, and naphthol derivatives, and dyes. It is also used as a moth repellent, a fungicide, a smokeless powder, a cutting fluid, a lubricant, in synthetic resins, for synthetic tanning, as a preservative, in textile chemicals, in emulsion breakers, in scintillation counters, and as an antiseptic.

RISK ASSESSMENT: HEALTH

General Assessment

Naphthalene is a human poison by *ingestion*. It is also moderately toxic by *inhalation*. It is irritating to the skin and, on prolonged contact, can be *absorbed* through intact skin to cause toxic systemic effects. It is an experimental teratogen and has also caused adverse reproductive effects. Mutation data have been reported and naphthalene is also a questionable carcinogen in experimental tests.

Inhalation of naphthalene vapors causes central nervous system effects, including excitement, confusion, headache, nausea, and loss of appetite. Exposure by all routes of exposure may lead to renal shutdown (kidney failure), hemolytic effects (breakdown of the red blood cells), hematuria (blood in the urine), oliguria (decreased urine volume), jaundice, eye damage, and possible cataracts, vomiting, convulsions, and coma.

Acute Health Effects

The following acute (short-term) health effects may occur immediately or shortly after exposure to naphthalene (within hours or days):

Skin: Irritation with burning, itching, and edema (buildup of fluid), redness and swelling.

Eye: Severe irritation of the conjunctiva with burning, itching, and watering. Direct contact with heated "fumes" can irritate the eyes.

Lung: Irritation of the bronchial tubes and other members of the respiratory tract. May cause headache, loss of appetite, nausea, vomiting, slowed reactions, and a loss of strength.

CNS: Depression of the nervous system causing excitement, convulsions, and possibly coma.

Chronic Health Effects

The following chronic (long-term) health effects may occur at some time after exposure to naphthalene and can last for months or even years:

Cancer Hazards: Naphthalene may cause cancer in animals and mutations (genetic changes) in living cells. The data on its human carcinogenicity are inconclusive.

Reproductive Hazard: According to the references, naphthalene has been shown to cause teratogenic effects and adversely affect reproduction in test animals.

Other Chronic Effects: Continuous exposures to naphthalene may lead to an increased incidence of cataracts (opacities of the lens) and optical neuritis. There may be gastrointestinal effects, abdominal pain, irritated bladder, profuse sweating, and dermatitis.

Recommended Risk-Reduction Measures

Personnel should avoid direct contact with naphthalene. It is a questionable teratogen, mutagen, and carcinogen. If a less toxic chemical cannot be substituted for a hazardous substance, then *engineering controls* are the most effective method of reducing exposures. The best protection is to enclose operations and/or provide local exhaust ventilation at the site of chemical release. While not always operationally feasible, isolating operations can also reduce exposure. Using respiratory protection is less effective than the controls mentioned above, but is still advisable whenever working with or around naphthalene. Air-purifying respirators will provide protection for exposures above the PEL. However, a self-contained breathing apparatus (SCBA) with full facepiece operated in pressure demand or a supplied-air respirator operated in continuous flow mode are the recommended respiratory protection methods of choice when working with a questioned carcinogen. If a full facepiece is not available, then chemical/dust goggles should be worn to protect the eyes. A face shield and a protective apron should also be worn. To prevent hand and skin exposures, impervious gloves should be used. Glove manufacturers should be contacted and permeation studies obtained *before* final glove selections have been made.

Administrative controls should also be in place to minimize the potential for human exposures. These may include written procedures or policies which specify the methods and techniques that will be practiced whenever personnel are to work with naphthalene.

All personnel should receive *training* on the use, hazards, protective measures, emergency actions, and other precautions per 29 CFR 1910.1200 (Hazard Communication), prior to the first assignment in an area where naphthalene is used or stored.

Before beginning employment, and at regular intervals thereafter (e.g., annually), the following medical tests are recommended for personnel working with naphthalene:

☑ Anatomical examination of the eye and vision testing (establish a baseline).

If symptoms develop or overexposure is suspected, the following may be useful:

☑ Liver and kidney function tests.

☑ Complete blood count (CBC).

☑ Evaluation by a qualified allergist, including careful evaluation of exposure history and special testing (may help diagnose skin allergy).

Any evaluation should include a careful history of past and present symptoms with an examination. Medical tests that simply look for existing damage are not a substitute for controlling exposures.

Other methods to reduce exposure include:

☑ Where possible, enclose operations and use local exhaust ventilation at the site of chemical release. If local exhaust ventilation or enclosure is not used, respiratory protection should be mandatory.

☑ Always ensure that proper protective clothing is worn when using chemical substances.

☑ Wash thoroughly immediately after exposure to naphthalene and at the end of the work shift or before eating, drinking, or smoking. Avoid contact with direct sunlight.

☑ Hazard warning information should be posted in the work area. In addition, as part of an on-going education and training program, all information on the health and safety hazards of naphthalene should be communicated to all exposed workers.

☑ Eye wash stations should be provided in the immediate work area for emergency use. If there is a possibility of skin exposure to naphthalene, emergency shower facilities should also be provided.

RISK ASSESSMENT: ENVIRONMENT

General Assessment

The environment is at risk of exposure during transportation, storage, disposal, or destruction of naphthalene. In almost every scenario, the threat of environmental exposure is contingent upon the proper handling of the chemical substance. Accidental spills, large or small, can result in fire, explosion (depending upon conditions of spill), and possible contamination of the surrounding environmental mediums (water, soil, and air).

Naphthalene is considered a combustible solid. Alone, its high flash point and autoignition temperature do not present a serious fire or explosion hazard. However, because it is incompatible with many other commodities, extreme caution is required in handling, storage, transportation, and disposal of naphthalene. These characteristics also require special consideration during any emergency situation involving a leak or spill of naphthalene crystals. Even as a solid, it has a considerably high vapor pressure. Vapors are heavier than air and can collect in low lying areas. Should naphthalene ever come into contact with incompatible substances (such as oxidizers) either during use, transportation, storage, or disposal, the formation of highly toxic and/or highly explosive commodities is extremely possible.

The proper disposal/destruction method for naphthalene waste is to burn it in a chemical incinerator equipped with an afterburner and air scrubber.

Naphthalene can enter the environment from industrial and municipal plant discharges and from spills.

Acute Ecological Effects

Acutc (short-term) toxic effects may include the death of animals, birds, or fish and death or low growth rate in plants. Acute effects are seen 2 to 4 days after animals or plants are exposed to naphthalene.

This chemical has moderate acute toxicity to aquatic life and has caused injury or death to various types of agricultural and ornamental crops. Insufficient data are available to evaluate or predict the short-term effects of naphthalene to birds or land animals.

Chronic Ecological Effects

Chronic (long-term) toxic effects may include shortened life span, reproductive problems, lower fertility, and changes in appearance or behavior in exposed animals. These effects can be seen long after first exposure(s) to toxic chemicals.

Naphthalene has moderate chromic toxicity to aquatic life. Insufficient data are available to evaluate or predict the long-term effects of naphthalene to plants, birds, or land animals.

Water Solubility

Naphthalene is very slightly soluble in water. Concentrations of 1 milligram or less will mix with a liter of water.

Persistence in the Environment

Naphthalene is slightly persistent in water, with a half-life between 2 and 20 days. The half-life of a pollutant is the amount of time it takes for one half of the chemical to be degraded. Approximately 96% of naphthalene will eventually end up in air; about 3% will end up in water; and about 0.5% will end up in both aquatic sediments and terrestrial soils, respectively.

Bioaccumulation in Aquatic Organisms

Some substances increase in concentration, or *bioaccumulate,* in living organisms as they breathe contaminated air, drink contaminated water, or eat contaminated food. These chemicals can become concentrated in the tissues and internal organs of animals as well as humans.

The concentration of naphthalene found in fish tissues is expected to be somewhat higher than the average concentration of naphthalene in the water from which the fish was taken.

Recommended Risk-Reduction Measures

Proper training of all transporters will reduce the likelihood of a mishap or accident resulting in a leak or spill of naphthalene into the environment. The correct use of DOT labeling (per the MSDS) on all containers, trucks, and rail cars will enable emergency responders to react properly and quickly to any disaster thereby reducing the potential risk to the environment and to personnel.

Storage of naphthalene should be segregated from incompatible chemicals to minimize the risk of cross-contamination or contact. Personnel working in storage areas must be trained on the hazards of naphthalene.

If a spill or leak to the environment has occurred, fire department, emergency response and/or hazardous materials spill personnel should be notified immediately. Cleanup should be attempted only by those trained in proper spill containment procedures. Contaminated soils should be removed for incineration and replaced with clean soil. If naphthalene should contact the water table, aquifer, or navigable waterway, the local and/or state emergency response authorities must be notified. A comprehensive emergency response or disaster preparedness/recovery plan should be in place prior to any operations involving the use, transportation, storage, or disposal of naphthalene. If naphthalene is spilled or leaked, the following specific steps are recommended:

- ☑ Restrict persons not wearing protective clothing from area of spill or leak until cleanup is complete and area can be opened for normal work.
- ☑ Remove all ignition sources.
- ☑ Collect powdered materials in the safest manner possible and deposit in sealed containers. Do NOT dry sweep. Use vacuum equipped with a HEPA filter instead. Damp mop residue.
- ☑ It may be necessary to dispose of naphthalene as a hazardous waste. The responsible state agency or the regional office of the federal Environmental Protection Agency (EPA) should be contacted for specific recommendations.

RISK ASSESSMENT: BUSINESS

General Assessment

Accidents or mishaps involving naphthalene can present a significant threat to business operations. The loss or damage of equipment or facilities can significantly affect fiscal viability. Lawsuits that may result from personnel injury/death, public exposures, and/or environmental contamination will also require a serious expenditure of resources.

Media attention surrounding an injury, death, or environmental damage can also result in a loss of profits and loss of current as well as future business. Always remember that anytime the terms "cancer," "carcinogen," or "reproductive hazard" are used, public emotion, hysteria, and ignorance can run equally high. This should be carefully considered when drafting or implementing public relations policies.

Recommended Risk-Reduction Measures

Company attorneys, safety and health professionals, and environmental specialists should be involved in the development of any safety, health or environmental policies or procedures. A company official should be pre-designated as a public relations officer with specific training in dealing with the press. Corporate plans and policies should be developed, approved, and implemented long before any need for such arises.

REFERENCES

Ahlbom, A. 1993. *Biostatistics for Epidemiologists.* New York: CRC Press/Lewis Publishers.

American Conference of Governmental Industrial Hygienists. 1988. *Documentation of the Threshold Limit Values and Biological Exposure Indices,* 5th Edition (with updates). Cincinnati: ACGIH

Asante-Duah, D. K. 1993. *Hazardous Waste Risk Assessment.* New York: CRC Press/Lewis Publishers.

Bloom, A. D. and F. J. DeSerres. 1995. *Ecotoxicity and Human Health.* New York: CRC Press/Lewis Publishers.

Brusick, D. J. 1994. *Methods for Genetic Risk Assessment.* New York: CRC Press/Lewis Publishers.

Bueche, J. R. 1972. *Principles of Physics,* 2nd Edition. New York: McGraw-Hill.

Calabrese, E. J. 1994. *Biological Effects of Low Level Exposures.* New York: CRC Press/Lewis Publishers.

Cockerman, L. G., B. S. Shane. 1994. *Basic Environmental Toxicology.* New York: CRC Press/Lewis Publishers.

Cothern, R. R. 1993. *Comparative Environmental Risk Assessment.* New York: CRC Press/Lewis Publishers.

Gardner, R. W. 1994. *Chemical Intolerance.* New York: CRC Press/Lewis Publishers.

Hallenbeck, W. H. 1993. *Quantitative Risk Assessment for Environmental and Occupational Health,* 2nd Edition. New York: CRC Press/Lewis Publishers.

Howard, P. H. 1995. *Handbook of Environmental Fate and Exposures Data for Organic Chemicals,* Vol. I-IV. New York: CRC Press/Lewis Publishers.

Jones, F. E. 1994. *Toxic Organic Vapors in the Workplace.* New York: CRC Press/Lewis Publishers.

Keith, L. H. and M. M. Walker. 1995. *Handbook of Air Toxics.* New York: CRC Press/Lewis Publishers.

Langerman, N. 1994 *Precautionary Labels for Chemical Containers.* New York: CRC Press/Lewis Publishers.

Lewis, R. J., Sr. 1993. *Hawley's Condensed Chemical Dictionary,* 12th Edition. New York: Van Nostrand Reinhold.

Lewis, R. J., Sr. 1992. *Sax's Dangerous Properties of Industrial Materials,* Volumes 1, 2, and 3, 8th Edition. New York: Van Nostrand Reinhold.

Lide, D. R. 1995. *Handbook of Chemistry and Physics,* 76th Edition. New York: CRC Press/Lewis Publishers.

Manahan, S. E. 1992. *Toxicological Chemistry,* 2nd Edition. New York: CRC Press/Lewis Publishers.

Mellström, G. A., J. E. Wahlberg, and H. I. Maibach. 1994. *Protective Gloves for Occupational Use.* New York: CRC Press/Lewis Publishers.

Meyer, E. 1990. *Chemistry of Hazardous Materials.* Englewood Cliffs, NJ: Prentice Hall.

Mycock, J. C., J. D. McKenna, and L. Theodore. 1995. *Handbook of Air Pollution Control Engineering and Technology.* New York: CRC Press/Lewis Publishers.

National Cancer Institute. 1978. *Carcinogenesis Technical Report Series. Bioassay of Naphthalene for Possible Carcinogenicity.* DEHW (NIH) Publication Number: 78-1323, p 53. Washington, DC: U.S. Government Printing Office.

Ott, W. R. 1995. *Environmental Statistics and Data Analysis.* New York: CRC Press/Lewis Publishers.

Patnaik, P. 1992. *A Comprehensive Guide to the Hazardous Properties of Chemical Substances.* New York: Van Nostrand Reinhold.

Philp, R. B. 1995. *Environmental Hazards and Human Health.* New York: CRC Press/Lewis Publishers.

Proctor, N. H., J. P. Hughes. 1991. *Chemical Hazards of the Workplace.* New York: Van Nostrand Reinhold.

Quigley, D. R. 1994. *Handbook of Emergency Chemical Management.* New York: CRC Press/Lewis Publishers.

Rea, W. J. 1995. *Chemical Sensitivity,* Vol. I-III. New York: CRC Press/Lewis Publishers.

Richardson, M. 1995. *Dictionary of Substance and Their Effects,* Vol. I-VI. New York: CRC Press/Lewis Publishers.

Sacarello, H. L. A. 1994. *The Comprehensive Handbook of Hazardous Materials.* New York: CRC Press/Lewis Publishers.

Talbott, E., G. Craun. 1995. *An Introduction to Environmental Epidemiology.* New York: CRC Press/Lewis Publishers.

Torkelson, T. R. V. K. Rowe. 1981. *Patty's Industrial Hygiene and Toxicology,* Vol. 2B. New York: Wiley-Interscience.

MATERIAL SAFETY DATA SHEET

CHEMICAL NAME	CRC PRESS/LEWIS PUBLISHERS DISCLAIMER:
1-NAPHTHYLAMINE	THE INFORMATION AND RECOMMENDATIONS PRESENTED HEREIN ARE BASED ON SOURCES BELIEVED TO BE RELIABLE. CRC MAKES NO REPRESENTATION ON ITS COMPLETENESS OR ACCURACY. IT IS THE RESPONSIBILITY OF THE USER TO DETERMINE THE CHEMICAL'S SUITABILITY FOR ITS INTENDED USE, THE CHEMICAL'S SAFE USE, AND THE CHEMICAL'S PROPER DISPOSAL. NO REPRESENTATIONS AND/OR WARRANTIES, EITHER EXPRESSED OR IMPLIED, OF THE MERCHANTABILITY OR FITNESS FOR A PARTICULAR PURPOSE, OR OF ANY OTHER NATURE, ARE MADE WITH RESPECT TO THE INFORMATION PROVIDED IN THIS MSDS OR TO THE CHEMICAL TO WHICH INFORMATION MAY REFER. CRC NEITHER ASSUMES NOR AUTHORIZES ANY OTHER PERSON TO ASSUME FOR IT, ANY OTHER ADDITIONAL RESPONSIBILITY OR LIABILITY FOR THE USE OF, OR RELIANCE UPON, THIS INFORMATION.

HAZARD WARNING INFORMATION

HEALTH	FIRE	REACTIVE	OTHER	DEGREE OF HAZARD	COLOR CODING	OTHER CODES
2	1	0		0 = Minimum Hazard 1 = Slight Hazard 2 = Moderate Hazard 3 = Serious Hazard 4 = Severe Hazard	HEALTH = BLUE FIRE = RED REACTIVITY = YELLOW OTHER = WHITE	OX = Oxidizer ACID = Acid ALK = Alkali COR = Corrosive W = Use No Water

SECTION I - GENERAL INFORMATION

Characterization	**Aromatic Amine**	RCRA Number	**U167**
		EPA Class	**Toxic Waste**
DOT Proper Shipping Name	**Naphthylamine (alpha)**	Chemical Abstract Service (CAS) Number	**134-32-7**
DOT Hazard Class and Label Requirements	**Poison B; St. Andrews Cross**	DOT Emergency Guide Code	**55**
DOT Identification Number	**UN 2077**	Molecular Formula	$C_{10}H_7NH_2$

Synonyms

1-Aminonaphthalene; α-naphthylamine; *alpha*-naphthylamine; naphthalidine.

SECTION II - HAZARDOUS INGREDIENTS AND IDENTITY INFORMATION

Hazardous Components (specific identity)	OSHA Exposure Criteria	NIOSH Exposure Criteria	Immediately Dangerous to Life and Health (IDLH)	ACGIH Exposure Criteria
1-Naphthylamine (derivation: By reduction of α-nitro-naphthalene with iron and hydrochloric acid. The mass is then mixed with milk of lime and distilled).	**PEL:** Suspected Cancer Agent **STEL:** Not Applicable	**REL:** Possible Cancer Agent Reduce to Lowest Level	**Not Applicable**	**TLV:** Not Listed **STEL:** Not Listed

SECTION III - PHYSICAL AND CHEMICAL CHARACTERISTICS

Boiling Point	**574°F (301°C)**	Specific Gravity (H_2O = 1)	**1.12**
Vapor Pressure (mm Hg)	**1 at 219°F (104°C)**	Molecular Weight (atomic weight)	**143.2**
Vapor Density (Air = 1)	**4.93**	Melting Point	**122°F (50°C)**

Solubility

Nearly insoluble in water (0.002%). Soluble in alcohol and ether.

Appearance and Odor

Colorless crystals with an unpleasant, ammonia-like odor. Darkens to reddish-purple on exposure to air.

SECTION IV - FIRE AND EXPLOSION HAZARD DATA

Flash Point (method used)	**315°F (157°C) closed cup**	Explosive Limits in Air % by Volume	**LEL: Not Determined UEL: Not Determined**
NFPA Classification	**Combustible Solid**	Autoignition Temperature	**Not Determined**

Extinguishing Media

Use dry chemical, carbon dioxide, alcohol foam, or water (spray, mist or fog).

Special Fire Fighting Procedures

Poisonous gases are produced in fire. Structural fire-fighting clothing is permeable and may not provide adequate protection. Wear self-contained breathing apparatus (SCBA) operated in positive pressure mode. Move fire-exposed containers from fire if it can be done without risk. Cool with water spray.

Unusual Fire and Explosion Hazards

Dusts can form extremely explosive mixtures if dispersed through the air in proper proportions and exposed to an ignition source.

SECTION V - REACTIVITY DATA

Stability		Conditions to Avoid: 1-Naphthylamine is stable at room temperature under normal conditions of handling and storage. Avoid the formation of dust clouds of 1-naphthylamine and prevent contact with ignition sources or oxidizers.
Stable	Unstable	Incompatibility *(materials to avoid)*: Violent reactions in contact with strong oxidizers, acids, and bases.
X		
Hazardous Polymerization		Conditions to Avoid: Under normal conditions of temperature and pressure, hazardous polymerization of 1-naphthylamine is not expected to occur.
May Occur	Will Not Occur	Hazardous Decomposition or By-products: Poisonous gases, including oxides of carbon (carbon monoxide) and oxides of nitrogen are produced when 1-naphthylamine is involved in a fire.
	X	

SECTION VI - HEALTH HAZARD DATA

Primary Route(s) of Entry:	Inhalation?	Absorption (skin/eye)?	Ingestion?
	X	X	X

Health Hazards

INHALATION: Irritation of the nose, throat, and respiratory tract. Symptoms include cough, sneezing, tachycardia, hypertension, cyanosis, headache, fatigue, anorexia, vomiting, emaciation, and bladder problems causing hematuria (blood in urine), proteinuria (excessive proteins in urine), and bladder injury and/or cancer.

ABSORPTION: Irritation of the skin and eyes. Absorbs through the skin to cause effects of inhalation and may lead to acute hemorrhagic cystitis, dypsnea, ataxia, dysuria, and hematuria.

INGESTION: Gastrointestinal irritation and symptoms of absorption have been reported.

Carcinogenicity	NTP Listed?	IARC Cancer Review Group?	OSHA Regulated?	Target Organs?
Suspected Human Confirmed Animal	No	Group 3	29 CFR 1910.1004	Bladder and skin

Medical Conditions Generally Aggravated by Exposure

May be especially harmful to persons with pre-existing bladder damage.

Emergency and First-aid Procedures

<u>Eye contact</u>: Do not allow victim to keep eyes tightly shut. Flush immediately with water for 15 minutes (minimum); seek medical attention. <u>Skin contact</u>: Remove all contaminated clothing, including shoes. Immediately wash area with flooding amounts of soap and water. Seek medical attention immediately. For <u>inhalation</u>: Remove the person from exposure. Provide respiratory assistance and CPR. Transfer to medical facility. If <u>swallowed</u>: Call poison control center. Unless advised otherwise, give conscious person one or two glasses of milk or water to drink and induce vomiting.

SECTION VII - PRECAUTIONS FOR SAFE HANDLING AND USE

Steps to be Taken in Case Material is Released or Spilled

Remove all ignition sources. Wear SCBA respiratory protection. Restrict those not wearing protective equipment and who are not involved in cleanup from entering area. Avoid skin contact. Do not create dust cloud. Ventilate area of spill. Do NOT dry sweep. Use HEPA vacuum and deposit in sealed drums.

Preferred Waste Disposal Method

Consider reclamation or recycling, if possible.

Precautions to be Taken in Handling and Storage

Store in tightly closed containers in cool, well-ventilated location away from heat. Protect containers from physical damage. A regulated controlled area must be established per OSHA 29 CFR 1910.1004.

Other Precautions and Warnings

Sources of ignition (open flame or smoking), are prohibited where 1-naphthylamine is used or stored. Metal containers should be used. Do NOT generate dust clouds or dusty conditions when working.

SECTION VIII - CONTROL MEASURES AND PERSONAL PROTECTIVE EQUIPMENT

Respiratory Protection *(specify type)*

Use a NIOSH-approved powered air-purifying respirator or a self-contained breathing apparatus (SCBA) with full facepiece operated in positive pressure mode.

Ventilation

Local exhaust (preferred) at site of chemical work or general ventilation.

Protective Gloves	Eye Protection	Other Protective Clothing
Impervious Gloves	Chemical/Dust Goggles or Face Mask	Protective Clothing (prevent skin contact)

Work/Hygiene Practices

Always wash hands thoroughly after using chemical; never bring food, drink, or smoking materials into vicinity of chemicals.

1-NAPHTHYLAMINE

$C_{10}H_7NH_2$

CAS: 134-32-7

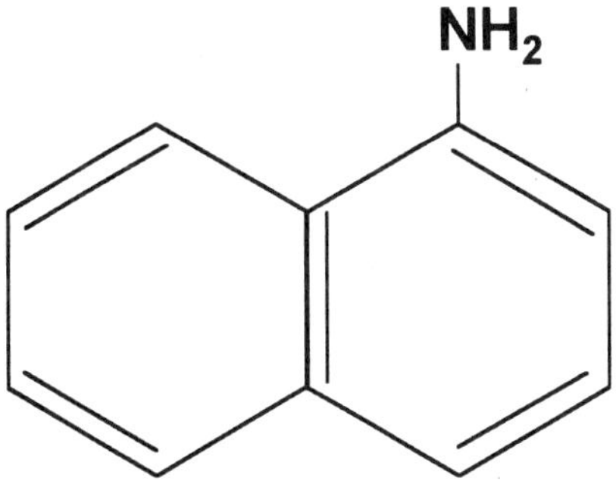

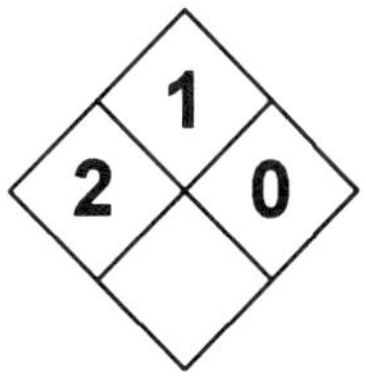

IDENTIFICATION AND TYPICAL USES

1-Naphthylamine is a colorless crystalline solid with an unpleasant, ammonia-like odor. It will darken to reddish-purple when exposed to air. It is used in the manufacture of dyes, intermediates, and tonic prints. It is also used in agricultural chemicals.

RISK ASSESSMENT: HEALTH

General Assessment

1-Naphthylamine is a human poison by *inhalation, ingestion,* and *absorption* through intact skin. It has been suggested that exposure to this chemical may lead to bladder cancer. Mutation data have also been reported. Its human teratogenicity and reproductive health risks are unknown. 1-Naphthylamine is soluble in fats and lipids.

Inhalation of the dusts, as well as skin absorption, causes irritation of the skin, eyes, nose, throat, and the mucous membranes lining the respiratory tract. There can be eczema, tachycardia (rapid heartbeat) hypertension (increased blood pressure) cough, sneezing, headache, fatigue, anorexia, painful diarrhea, loss of appetite with subsequent weight loss, low body temperature, cyanosis (blue discoloration of the skin, especially on face, hands, and feet due to oxygen deficiency), vomiting, and emaciation. There can also be serious bladder disorders, including hematuria (blood in urine) and proteinuria (excessive proteins in urine), as well as bladder damage (including cancer).

Ingestion can lead to symptoms of inhalation and absorption, as well as acute hemorrhagic cystitis, dypsnea, ataxia, and dysuria (infrequent urination).

Acute Health Effects

The following acute (short-term) health effects may occur immediately or shortly after exposure to 1-naphthylamine:

Skin: Irritation with rash and localized inflammation. Absorption through intact skin is likely on short exposures.

Eye: Severe irritation, pain, and inflammation.

Lung: Nose, throat, and respiratory tract irritation causing coughing, sneezing, and toxic systemic effects.

Other: Ingestion may cause hemorrhagic cystitis.

Chronic Health Effects

The following chronic (long-term) health effects may occur at some time after exposure to 1-naphthylamine and can last for months or even years:

Cancer Hazards: According to the information presented in the references, 1-naphthylamine is suspected to cause cancer of the bladder. Mutation data have also been recorded and many scientists believe that such chemicals pose a cancer risk in the long-term.

Reproductive Hazard: According to information presented in the references, 1-naphthylamine has not been adequately tested for its ability to cause cancer in test animals.

Recommended Risk-Reduction Measures

Personnel should avoid direct contact with 1-naphthylamine. It is suspected to cause cancer of the bladder. Occupational poisoning occurs most commonly by all routes of exposure. Substitution with a non-cancer agent is recommended. However, if a less toxic material or compound cannot be substituted for 1-naphthylamine, then *engineering controls* are the most effective method of reducing exposures. The best protection is to enclose operations and/or provide local exhaust ventilation at the site of 1-naphthylamine release. While not always operationally feasible, isolating operations can also reduce exposure. Using respiratory protection is less effective than the controls mentioned above, but is still advisable whenever working with or around 1-naphthylamine. For the best protection, use an MSHA/NIOSH-approved self-contained breathing apparatus (SCBA) with full facepiece operated in pressure demand or a supplied-air respirator operated in continuous flow, especially when high exposures to 1-naphthylamine are anticipated. If a full facepiece is not available, then chemical/dust goggles should be worn to protect the eyes. A face shield should also be considered. To prevent hand and skin exposures, impervious gloves should be used. Glove manufacturers should be contacted and permeation studies obtained *before* glove selections are made.

Administrative controls should also be in place to minimize the potential for human exposures. These may include written procedures or policies which specify the methods and techniques that will be practiced whenever personnel are to work with 1-naphthylamine.

All personnel should receive *training* on the use, hazards, protective measures, emergency actions, and other precautions per 29 CFR 1910.1200 (Hazard Communication), prior to the first assignment in an area where 1-naphthylamine is used or stored.

Before beginning employment and at regular times thereafter, the following recommended medical tests should be considered:

☑ Complete blood count (CBC).

☑ Urinalysis to test for blood and protein levels in urine (establish baseline).

If symptoms should develop or overexposure is suspected, the following additional tests may be helpful:

☑ Complete blood count.

☑ Urinalysis (compare to baseline).

Any evaluation should include a careful history of past and present symptoms with an examination. Medical tests that simply look for existing damage are not a substitute for controlling exposures.

Other methods to reduce exposure include:

☑ Where possible, enclose operations and use local exhaust ventilation at the site of chemical release. If local exhaust ventilation or enclosure is not used, respiratory protection should be mandatory.

☑ Always ensure that proper protective clothing is worn when using chemical substances and that personnel are trained on its use and care.

☑ Wash thoroughly immediately after exposure to 1-naphthylamine and at the end of the work shift or before eating, drinking, or smoking.

☑ A regulated, marked controlled area must be established wherever 1-naphthylamine is used, handled or stored. Other requirements are established by OSHA (29 CFR 1910.1004).

☑ Hazard warning information should be posted in the work area. In addition, as part of an on-going education and training program, all information on the health and safety hazards of 1-naphthylamine should be communicated to all potentially exposed workers.

☑ Eye wash stations should be provided in the immediate work area for emergency use. If there is a possibility of skin exposure to 1-naphthylamine, emergency shower facilities should be provided.

☑ Workers whose clothing has been contaminated by 1-naphthylamine should change into clean clothes before leaving work. Contaminated work clothing should be laundered only by individuals who have been informed of the hazards of exposure to 1-naphthylamine.

RISK ASSESSMENT: ENVIRONMENT

General Assessment

The environment is at risk of exposure during transportation, storage, disposal, or destruction of 1-naphthylamine. In almost every scenario, the threat of environmental exposure is contingent upon the proper handling of the chemical substance. Accidental spills, large or small, can result in contamination of the surrounding environmental mediums (water, soil, and air).

1-Naphthylamine is a combustible solid. It can present a dangerous explosion hazard if its dusts form a cloud and come into contact with an ignition source. It is also incompatible with oxidizing materials and contact can result in violent and explosive reactions. Caution is always required in handling, storage, transportation, and disposal of 1-naphthylamine. Emergency responders should be made aware of the presence of 1-naphthylamine at any emergency response situation.

1-Naphthylamine can enter the environment from industrial effluents and from spills.

Acute Ecological Effects

Acute (short-term) toxic effects may include the death of animals, birds, or fish and death or low growth rate in plants. Acute effects are seen 2 to 4 days after animals or plants are exposed to 1-naphthylamine.

Insufficient data are available to evaluate or predict the short-term effects of 1-naphthylamine to aquatic life, plants, birds, or terrestrial animals.

Chronic Ecological Effects

Chronic (long-term) toxic effects may include shortened life span, reproductive problems, lower fertility, and changes in appearance or behavior in exposed animals. These effects can be seen long after first exposure(s) to toxic chemicals.

Insufficient data are available to evaluate or predict the long-term effects of 1-naphthylamine to aquatic life, plants, birds, or terrestrial animals.

Water Solubility

1-Naphthylamine is nearly insoluble in water. Concentrations of less than 1 milligram will mix with a liter of water.

Persistence in the Environment

1-Naphthylamine is highly persistent in water, with a half-life of greater than 200 days. The half-life of a pollutant is the amount of time it takes for one half of the chemical to be degraded.

Bioaccumulation in Aquatic Organisms

Some substances increase in concentration, or *bioaccumulate,* in living organisms as they breathe contaminated air, drink contaminated water, or eat contaminated food. These chemicals can become concentrated in the tissues and internal organs of animals as well as humans.

The concentration of 1-naphthylamine found in fish tissues is expected to be somewhat higher than the average concentration of 1-naphthylamine in the water from which the fish was taken.

Recommended Risk-Reduction Measures

Proper training of all transporters will reduce the likelihood of a mishap or accident resulting in a leak or spill of 1-naphthylamine into the environment. Labels on all containers, trucks, and rail cars must meet DOT requirements and accurately reflect their contents to enable emergency responders to react properly and quickly to any disaster thereby reducing the potential risk to the environment and to personnel.

Storage of 1-naphthylamine should be segregated from other chemicals and materials to minimize the risk of cross-contamination.

If a spill or leak to the environment has occurred, fire department, emergency response and/or hazardous materials spill personnel should be notified immediately. Cleanup should be attempted only by those trained in proper spill containment procedures. Contaminated soils should be removed for incineration and replaced with clean soil. If 1-naphthylamine should contact the water table, aquifer, or navigable waterway, time is of the essence. It is practically insoluble and total containment and remediation may be possible. The local and/or state emergency response authorities must be notified. A comprehensive emergency response or disaster preparedness/recovery plan should be in place prior to any operations involving the use, transportation, storage, or disposal of 1-naphthylamine.

If 1-naphthylamine is spilled or leaked, the following specific steps are recommended:

☑ Restrict persons not wearing protective clothing from area of spill or leak until cleanup is complete and area can be opened for normal work.

☑ Do NOT generate dusts. Dust clouds of the right proportions in air present a dangerous explosion hazard. Use a scoop or a vacuum equipped with a high efficiency particulate air filter (HEPA) to collect spills. Place materials in drum for recovery or disposal, as required.

☑ Ventilate area and remove ignition sources.

☑ Containers should be removed to safer area if it can be done without increasing the risk.

☑ It may be necessary to dispose of 1-naphthylamine as a hazardous waste. The responsible state agency or the regional office of the federal Environmental Protection Agency (EPA) should be contacted for specific recommendations.

RISK ASSESSMENT: BUSINESS
General Assessment

Accidents or mishaps involving 1-naphthylamine can present a serious threat to business operations. The loss or damage of equipment or facilities can significantly affect fiscal viability. Lawsuits that may result from personnel illness, injury/death, public exposures, and/or environmental contamination will require a serious expenditure of resources. Media attention surrounding an injury, death, or environmental damage can also result in a loss of profits and loss of current as well as future business. Always remember that anytime the terms "cancer" or "carcinogen" are used, public emotion, hysteria, and ignorance can run equally high. This must be carefully considered when planning or implementing public relations policies.

Recommended Risk-Reduction Measures

Company attorneys, safety and health professionals, and environmental specialists should be involved in the development of any procedures for responding to chemical incidents. A company official should be predesignated as a public relations officer with specific training in dealing with the press. Corporate plans and policies should be developed, approved, and implemented long before any need for such arises.

REFERENCES

American Conference of Governmental Industrial Hygienists. 1988. *Documentation of the Threshold Limit Values and Biological Exposure Indices,* 5th Edition (with updates). Cincinnati: ACGIH

Bloom, A. D. and F. J. DeSerres. 1995. *Ecotoxicity and Human Health.* New York: CRC Press/Lewis Publishers.

Brusick, D. J. 1994. *Methods for Genetic Risk Assessment.* New York: CRC Press/Lewis Publishers.

Bueche, J. R. 1972. *Principles of Physics,* 2nd Edition. New York: McGraw-Hill.

Calabrese, E. J., E. M. Kenyon. *Air Toxics and Risk Assessment.* New York: CRC Press/Lewis Publishers.

Calabrese, E. J. 1994. *Biological Effects of Low Level Exposures.* New York: CRC Press/Lewis Publishers.

Cockerman, L. G., B. S. Shane. 1994. *Basic Environmental Toxicology.* New York: CRC Press/Lewis Publishers.

Cothern, C. R., N. P. Ross. 1994. *Environmental Statistics, Assessment, and Forecasting.* New York: CRC Press/Lewis Publishers.

Cothern, R. R. 1993. *Comparative Environmental Risk Assessment.* New York: CRC Press/Lewis Publishers.

Danse, I. R. 1991. *Common Sense Toxics in the Workplace.* New York: Van Nostrand Reinhold.

Gerarde, H. W. 1960. *Toxicology and Biochemistry of Aromatic Hydrocarbons.* New York: Elsevier.

Hallenbeck, W. H. 1993. *Quantitative Risk Assessment for Environmental and Occupational Health,* 2nd Edition. New York: CRC Press/Lewis Publishers.

Howard, P. H., et. al. 1991. *Handbook of Environmental Degradation Rates.* New York: CRC Press/Lewis Publishers.

Jones, F. E. 1994. *Toxic Organic Vapors in the Workplace.* New York: CRC Press/Lewis Publishers.

Keith, L. H. and M. M. Walker. 1995. *Handbook of Air Toxics.* New York: CRC Press/Lewis Publishers.

Landis, W. G., M. H. Yu. 1995. *Introduction to Environmental Toxicology.* New York: CRC Press/Lewis Publishers.

Langerman, N. 1994 *Precautionary Labels for Chemical Containers.* New York: CRC Press/Lewis Publishers.

Lauwerys, R. R., P. Hoet. 1993. *Industrial Chemical Exposure.* New York: CRC Press/Lewis Publishers.

Lewis, R. J., Sr. 1993. *Hawley's Condensed Chemical Dictionary,* 12th Edition. New York: Van Nostrand Reinhold.

Lewis, R. J., Sr. 1992. *Sax's Dangerous Properties of Industrial Materials,* Volumes 1, 2, and 3, 8th Edition. New York: Van Nostrand Reinhold.

Manahan, S. E. 1994. *Environmental Chemistry,* 6th Edition. New York: CRC Press/Lewis Publishers.

Mellström, G. A., J. E. Wahlberg, and H. I. Maibach. 1994. *Protective Gloves for Occupational Use.* New York: CRC Press/Lewis Publishers.

Meyer, E. 1990. *Chemistry of Hazardous Materials.* Englewood Cliffs, NJ: Prentice Hall.

Ott, W. R. 1995. *Environmental Statistics and Data Analysis.* New York: CRC Press/Lewis Publishers.

Patnaik, P. 1992. *A Comprehensive Guide to the Hazardous Properties of Chemical Substances.* New York: Van Nostrand Reinhold.

Proctor, N. H., J. P. Hughes. 1991. *Chemical Hazards of the Workplace.* New York: Van Nostrand Reinhold.

Sacarello, H. L. A. 1994. *The Comprehensive Handbook of Hazardous Materials.* New York: CRC Press/Lewis Publishers.

Torkelson, T. R. V. K. Rowe. 1981. *Patty's Industrial Hygiene and Toxicology,* Vol. 2B. New York: Wiley-Interscience.

U.S. Department of Health and Human Services, National Institute for Occupational Safety and Health. 1994. *NIOSH Pocket Guide to Chemical Hazards.* Washington, D.C.: U.S. Government Printing Office.

MATERIAL SAFETY DATA SHEET

CHEMICAL NAME

2-NAPHTHYLAMINE

HAZARD WARNING INFORMATION

HEALTH	FIRE	REACTIVE	OTHER	DEGREE OF HAZARD	COLOR CODING	OTHER CODES
2	1	0		0 = Minimum Hazard 1 = Slight Hazard 2 = Moderate Hazard 3 = Serious Hazard 4 = Severe Hazard	HEALTH = BLUE FIRE = RED REACTIVITY = YELLOW OTHER = WHITE	OX = Oxidizer ACID = Acid ALK = Alkali COR = Corrosive W̶ = Use No Water

SECTION I - GENERAL INFORMATION

Characterization	**Aromatic Amine**	RCRA Number	**U168**
		EPA Class	**Toxic Waste**
DOT Proper Shipping Name	**Naphthylamine (beta)**	Chemical Abstract Service (CAS) Number	**91-59-8**
DOT Hazard Class and Label Requirements	**Poison B; Poison**	DOT Emergency Guide Code	**55**
DOT Identification Number	**UN 1650**	Molecular Formula	**$C_{10}H_7NH_2$**

Synonyms

2-Aminonaphthalene; β-naphthylamine; *beta*-naphthylamine.

SECTION II - HAZARDOUS INGREDIENTS AND IDENTITY INFORMATION

Hazardous Components (specific identity)	OSHA Exposure Criteria	NIOSH Exposure Criteria	Immediately Dangerous to Life and Health (IDLH)	ACGIH Exposure Criteria
2-Naphthylamine (derivation: From β-naphthol by heating in an autoclave with ammonium sulfite and ammonia).	**PEL:** Suspected Cancer Agent **STEL:** Not Applicable	**REL:** Possible Cancer Agent Reduce to Lowest Level	**Not Applicable**	**TLV:** Confirmed Human Carcinogen **STEL:** Not Applicable

SECTION III - PHYSICAL AND CHEMICAL CHARACTERISTICS

Boiling Point	**583°F (306°C)**	Specific Gravity (H_2O = 1)	**1.06**
Vapor Pressure (mm Hg)	**1 at 226°F (108°C)**	Molecular Weight (atomic weight)	**143.2**
Vapor Density (Air = 1)	**Not Found**	Melting Point	**232°F (111°C)**

Solubility

Miscible in water. Soluble in alcohol and ether.

Appearance and Odor

Odorless, white to red crystals with a faint, aromatic odor. Darkens to reddish purple color in air.

SECTION IV - FIRE AND EXPLOSION HAZARD DATA

Flash Point (method used)	**315°F (157°C) closed cup**	Explosive Limits in Air % by Volume	**LEL: Not Determined UEL: Not Determined**
NFPA Classification	**Combustible Solid**	Autoignition Temperature	**Not Determined**

Extinguishing Media

Use dry chemical, carbon dioxide, alcohol foam, or water (spray, mist or fog).

Special Fire Fighting Procedures

Poisonous gases are produced in fire. Structural fire-fighting clothing is permeable and may not provide adequate protection. Wear self-contained breathing apparatus (SCBA) operated in positive pressure mode. Move fire-exposed containers from fire if it can be done without risk. Cool with water spray.

Unusual Fire and Explosion Hazards

Dusts can form extremely explosive mixtures if dispersed through the air in proper proportions and exposed to an ignition source.

SECTION V - REACTIVITY DATA

Stability		Conditions to Avoid 2-Naphthylamine is stable at room temperature under normal conditions of handling and storage. Avoid the formation of dust clouds of 2-naphthylamine and prevent contact with ignition sources or oxidizers.
Stable X	Unstable	Incompatibility *(materials to avoid)* Incompatible with nitrous acid.
Hazardous Polymerization		Conditions to Avoid Under normal conditions of temperature and pressure, hazardous polymerization of 2-naphthylamine is not expected to occur.
May Occur	Will Not Occur X	Hazardous Decomposition or By-products Poisonous gases, including oxides of carbon (carbon monoxide) and oxides of nitrogen are produced when 2-naphthylamine is involved in a fire.

SECTION VI - HEALTH HAZARD DATA

Primary Route(s) of Entry:	Inhalation?	Absorption (skin/eye)?	Ingestion?
	X	X	X

Health Hazards

INHALATION: Irritation of the nose, throat, and respiratory tract. Symptoms include cough, sneezing, tachycardia, hypertension, cyanosis, headache, fatigue, anorexia, vomiting, emaciation, and bladder problems causing hematuria (blood in urine), proteinuria (excessive proteins in urine), and bladder injury and/or cancer.

ABSORPTION: Irritation of the skin and eyes. Absorbs through the skin to cause effects of inhalation and may lead to acute hemorrhagic cystitis, dypsnea, ataxia, dysuria, and hematuria.

INGESTION: Gastrointestinal irritation and symptoms of absorption have been reported.

Carcinogenicity	NTP Listed?	IARC Cancer Review Group?	OSHA Regulated?	Target Organs?
Confirmed Human Confirmed Animal	5th Annual Report	Group 1	29 CFR 1910.1009	Bladder and skin

Medical Conditions Generally Aggravated by Exposure

May be especially harmful to persons with pre-existing bladder damage.

Emergency and First-aid Procedures

Eye contact: Do not allow victim to keep eyes tightly shut. Flush immediately with water for 15 minutes (minimum); seek medical attention. Skin contact: Remove all contaminated clothing, including shoes. Immediately wash area with flooding amounts of soap and water. Seek medical attention immediately. For inhalation: Remove the person from exposure. Provide respiratory assistance and CPR. Transfer to medical facility. If swallowed: Call poison control center. Unless advised otherwise, give conscious person one or two glasses of milk or water to drink and induce vomiting.

SECTION VII - PRECAUTIONS FOR SAFE HANDLING AND USE

Steps to be Taken in Case Material is Released or Spilled

Remove all ignition sources. Wear SCBA respiratory protection. Restrict those not wearing protective equipment and who are not involved in cleanup from entering area. Avoid skin contact. Do not create dust cloud. Ventilate area of spill. Do NOT dry sweep. Use HEPA vacuum and deposit in sealed drums.

Preferred Waste Disposal Method

Consider reclamation or recycling, if possible.

Precautions to be Taken in Handling and Storage

Store in tightly closed containers in cool, well-ventilated location away from heat. Protect containers from physical damage. A regulated controlled area must be established per OSHA 29 CFR 1910.1009.

Other Precautions and Warnings

Sources of ignition (open flame or smoking) are prohibited where 2-naphthylamine is used or stored. Metal containers should be used. Do NOT generate dust clouds or dusty conditions when working.

SECTION VIII - CONTROL MEASURES AND PERSONAL PROTECTIVE EQUIPMENT

Respiratory Protection *(specify type)*

This is a confirmed human carcinogen. Use a NIOSH-approved powered air-purifying respirator or a self-contained breathing apparatus (SCBA) with full facepiece operated in positive pressure mode.

Ventilation

Local exhaust (preferred) at site of chemical work or general ventilation.

Protective Gloves	Eye Protection	Other Protective Clothing
Impervious Gloves	Chemical/Dust Goggles or Face Mask	Protective Clothing (prevent skin contact)

Work/Hygiene Practices

Always wash hands thoroughly after using chemical; never bring food, drink, or smoking materials into vicinity of chemicals.

2-NAPHTHYLAMINE

$C_{10}H_7NH_2$

CAS: 91-59-8

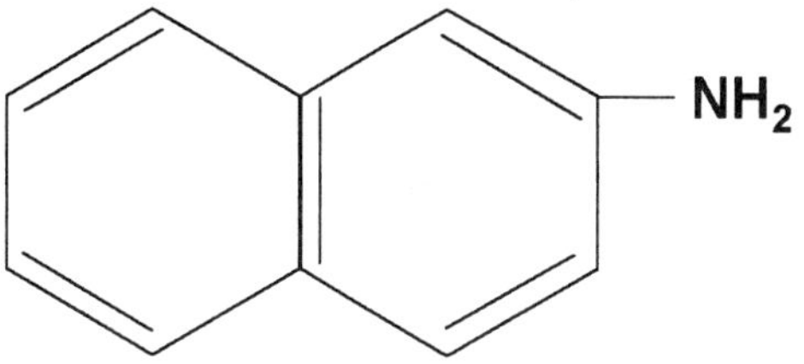

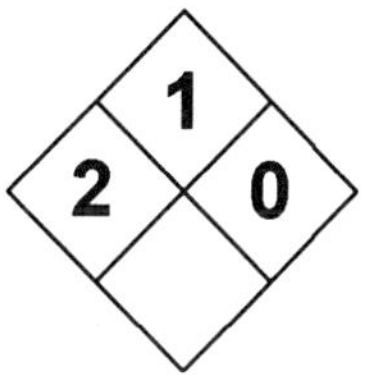

IDENTIFICATION AND TYPICAL USES

2-Naphthylamine is a white or colorless crystalline solid with a faint, unpleasant, aromatic odor. It will darken to a reddish-purple when exposed to air. It is widely used in the manufacture of dyes and in rubber. Its use is currently curtailed because of its hazardous (carcinogenic) properties.

RISK ASSESSMENT: HEALTH

General Assessment

2-Naphthylamine is a human poison by *inhalation, ingestion,* and *absorption* through intact skin. It is a severe health hazard due to its known carcinogenic properties. There is sufficient evidence that exposure leads to bladder cancer in humans. It also causes tumors in the kidney, liver, lungs, skin, and blood tissues in test animals. Its human teratogenicity and reproductive health risk are unknown.

Inhalation of the dusts, as well as skin absorption, causes irritation of the skin, eyes, nose, throat, and the mucous membranes lining the respiratory tract. There can be eczema, tachycardia (rapid heartbeat) hypertension (increased blood pressure) cough, sneezing, headache, fatigue, anorexia, painful diarrhea, loss of appetite with subsequent weight loss, low body temperature, cyanosis (blue discoloration of the skin, especially on face, hands, and feet due to oxygen deficiency), vomiting, and emaciation. There can also be serious bladder disorders, including hematuria (blood in urine) and proteinuria (excessive proteins in urine), as well as bladder damage (including cancer).

Ingestion can lead to symptoms of inhalation and absorption, as well as acute hemorrhagic cystitis, methemoglobinemia (causing hypoxia or inadequate supply of oxygen to tissues), and respiratory distress, dypsnea, ataxia, and dysuria (infrequent urination).

Acute Health Effects

The following acute (short-term) health effects may occur immediately or shortly after exposure to 2-naphthylamine:

Skin: Irritation with rash and localized inflammation. Absorption through intact skin is likely on short exposures.

Eye: Severe irritation, pain, and inflammation.

Lung: Nose, throat, and respiratory tract irritation causing coughing, sneezing, and toxic systemic effects.

Other: Ingestion may cause hemorrhagic cystitis.

Chronic Health Effects

The following chronic (long-term) health effects may occur at some time after exposure to 2-naphthylamine and can last for months or even years:

Cancer Hazards: According to the information presented in the references, 2-naphthylamine is suspected to cause cancer of the bladder. It may also cause tumors of the kidney, liver, lungs, skin, and blood tissues. Mutation data have also been recorded

Reproductive Hazard: According to information presented in the references, 2-naphthylamine has not been adequately tested for its ability to adversely affect reproduction

Recommended Risk-Reduction Measures

Personnel should avoid direct contact with 2-naphthylamine. It is suspected to cause cancer of the bladder. Occupational poisoning occurs most commonly by all routes of exposure. Substitution with a non-cancer agent is recommended. However, if a less toxic material or compound cannot be substituted for 2-naphthylamine, then *engineering controls* are the most effective method of reducing exposures. The best protection is to enclose operations and/or provide local exhaust ventilation at the site of 2-naphthylamine release. While not always operationally feasible, isolating operations can also reduce exposure. Using respiratory protection is less effective than the controls mentioned above, but is still advisable whenever working with or around 2-naphthylamine. For the best protection, use an MSHA/NIOSH-approved self-contained breathing apparatus (SCBA) with full facepiece operated in pressure demand or a supplied-air respirator operated in continuous flow, especially when high exposures to 2-naphthylamine are anticipated. If a full facepiece is not available, then chemical/dust goggles should be worn to protect the eyes. A face shield should also be considered. To prevent hand and skin exposures, impervious gloves should be used. Glove manufacturers should be contacted and permeation studies obtained *before* glove selections are made.

Administrative controls should also be in place to minimize the potential for human exposures. These may include written procedures or policies which specify the methods and techniques that will be practiced whenever personnel are to work with 2-naphthylamine.

All personnel should receive *training* on the use, hazards, protective measures, emergency actions, and other precautions per 29 CFR 1910.1200 (Hazard Communication), prior to the first assignment in an area where 2-naphthylamine is used or stored.

Before beginning employment and at regular times thereafter, the following recommended medical tests should be considered:

- ☑ Complete blood count (CBC).
- ☑ Urinalysis to test for blood and protein levels in urine (establish baseline).

If symptoms should develop or overexposure is suspected, the following additional tests may be helpful:

- ☑ Complete blood count.
- ☑ Urinalysis (compare to baseline).

Any evaluation should include a careful history of past and present symptoms with an examination. Medical tests that simply look for existing damage are not a substitute for controlling exposures.

Other methods to reduce exposure include:

- ☑ Where possible, enclose operations and use local exhaust ventilation at the site of chemical release. If local exhaust ventilation or enclosure is not used, respiratory protection should be mandatory.
- ☑ Always ensure that proper protective clothing is worn when using chemical substances and that personnel are trained on its use and care.
- ☑ Wash thoroughly immediately after exposure to 2-naphthylamine and at the end of the work shift or before eating, drinking, or smoking.
- ☑ A regulated, marked controlled area must be established wherever 2-naphthylamine is used, handled or stored. Other requirements are established by OSHA (29 CFR 1910.1009).
- ☑ Hazard warning information should be posted in the work area. In addition, as part of an on-going education and training program, all information on the health and safety hazards of 2-naphthylamine should be communicated to all potentially exposed workers.
- ☑ Eye wash stations should be provided in the immediate work area for emergency use. If there is a possibility of skin exposure to 2-naphthylamine, emergency shower facilities should be provided.
- ☑ Workers whose clothing has been contaminated by 2-naphthylamine should change into clean clothes before leaving work. Contaminated work clothing should be laundered only by individuals who have been informed of the hazards of exposure to 2-naphthylamine.

RISK ASSESSMENT: ENVIRONMENT

General Assessment

The environment is at risk of exposure during transportation, storage, disposal, or destruction of 2-naphthylamine. In almost every scenario, the threat of environmental exposure is contingent upon the proper handling of the chemical substance. Accidental spills, large or small, can result in contamination of the surrounding environmental mediums (water, soil, and air).

2-Naphthylamine is a combustible solid. It can present a dangerous explosion hazard if its dusts form a cloud and come into contact with an ignition source. It is also incompatible with oxidizing materials and contact can result in violent and explosive reactions. Caution is always required in handling, storage, transportation, and disposal of 2-naphthylamine. Emergency responders should be made aware of the presence of 2-naphthylamine at any emergency response situation.

2-Naphthylamine can enter the environment from industrial effluents and from spills.

Acute Ecological Effects

Acute (short-term) toxic effects may include the death of animals, birds, or fish and death or low growth rate in plants. Acute effects are seen 2 to 4 days after animals or plants are exposed to 2-naphthylamine.

Insufficient data are available to evaluate or predict the short-term effects of 2-naphthylamine to aquatic life, plants, birds, or terrestrial animals.

Chronic Ecological Effects

Chronic (long-term) toxic effects may include shortened life span, reproductive problems, lower fertility, and changes in appearance or behavior in exposed animals. These effects can be seen long after first exposure(s) to toxic chemicals.

Insufficient data are available to evaluate or predict the long-term effects of 2-naphthylamine to aquatic life, plants, birds, or terrestrial animals.

Water Solubility

2-Naphthylamine is nearly insoluble in water. Concentrations of less than 1 milligram will mix with a liter of water.

Persistence in the Environment

2-Naphthylamine is highly persistent in water, with a half-life of greater than 200 days. The half-life of a pollutant is the amount of time it takes for one half of the chemical to be degraded.

Bioaccumulation in Aquatic Organisms

Some substances increase in concentration, or *bioaccumulate,* in living organisms as they breathe contaminated air, drink contaminated water, or eat contaminated food. These chemicals can become concentrated in the tissues and internal organs of animals as well as humans.

The concentration of 2-naphthylamine found in fish tissues is expected to be somewhat higher than the average concentration of 2-naphthylamine in the water from which the fish was taken.

Recommended Risk-Reduction Measures

Proper training of all transporters will reduce the likelihood of a mishap or accident resulting in a leak or spill of 2-naphthylamine into the environment. Labels on all containers, trucks, and rail cars must meet DOT requirements and accurately reflect their contents to enable emergency responders to react properly and quickly to any disaster thereby reducing the potential risk to the environment and to personnel.

Storage of 2-naphthylamine should be segregated from other chemicals and materials to minimize the risk of cross-contamination.

If a spill or leak to the environment has occurred, fire department, emergency response and/or hazardous materials spill personnel should be notified immediately. Cleanup should be attempted only by those trained in proper spill containment procedures. Contaminated soils should be removed for incineration and replaced with clean soil. If 2-naphthylamine should contact the water table, aquifer, or navigable waterway, time is of the essence. It is practically insoluble and total containment and remediation may be possible. The local and/or state emergency response authorities must be notified. A comprehensive emergency response or disaster preparedness/recovery plan should be in place prior to any operations involving the use, transportation, storage, or disposal of 2-naphthylamine.

If 2-naphthylamine is spilled or leaked, the following specific steps are recommended:

- ☑ Restrict persons not wearing protective clothing from area of spill or leak until cleanup is complete and area can be opened for normal work.
- ☑ Do NOT generate dusts. Dust clouds of the right proportions in air present a dangerous explosion hazard. Use a scoop or a vacuum equipped with a high efficiency particulate air filter (HEPA) to collect spills. Place materials in drum for recovery or disposal, as required.
- ☑ Ventilate area and remove ignition sources.
- ☑ Containers should be removed to safer area if it can be done without increasing the risk.
- ☑ It may be necessary to dispose of 2-naphthylamine as a hazardous waste. The responsible state agency or the regional office of the federal Environmental Protection Agency (EPA) should be contacted for specific recommendations.

RISK ASSESSMENT: BUSINESS

General Assessment

Accidents or mishaps involving 2-naphthylamine can present a serious threat to business operations. The loss or damage of equipment or facilities can significantly affect fiscal viability. Lawsuits that may result from personnel illness, injury/death, public exposures, and/or environmental contamination will require a serious expenditure of resources. Media attention surrounding an injury, death, or environmental damage can also result in a loss of profits and loss of current as well as future business. Always remember that anytime the terms "cancer" or "carcinogen" are used, public emotion, hysteria, and ignorance can run equally high. This must be carefully considered when planning or implementing public relations policies.

Recommended Risk-Reduction Measures

Company attorneys, safety and health professionals, and environmental specialists should be involved in the development of any procedures for responding to chemical incidents. A company official should be predesignated as a public relations officer with specific training in dealing with the press. Corporate plans and policies should be developed, approved, and implemented long before any need for such arises.

REFERENCES

American Conference of Governmental Industrial Hygienists. 1988. *Documentation of the Threshold Limit Values and Biological Exposure Indices,* 5th Edition (with updates). Cincinnati: ACGIH

Bloom, A. D. and F. J. DeSerres. 1995. *Ecotoxicity and Human Health.* New York: CRC Press/Lewis Publishers.

Brusick, D. J. 1994. *Methods for Genetic Risk Assessment.* New York: CRC Press/Lewis Publishers.

Bueche, J. R. 1972. *Principles of Physics,* 2nd Edition. New York: McGraw-Hill.

Calabrese, E. J., E. M. Kenyon. *Air Toxics and Risk Assessment.* New York: CRC Press/Lewis Publishers.

Calabrese, E. J. 1994. *Biological Effects of Low Level Exposures.* New York: CRC Press/Lewis Publishers.

Cockerman, L. G., B. S. Shane. 1994. *Basic Environmental Toxicology.* New York: CRC Press/Lewis Publishers.

Cothern, C. R., N. P. Ross. 1994. *Environmental Statistics, Assessment, and Forecasting.* New York: CRC Press/Lewis Publishers.

Cothern, R. R. 1993. *Comparative Environmental Risk Assessment.* New York: CRC Press/Lewis Publishers.

Danse, I. R. 1991. *Common Sense Toxics in the Workplace.* New York: Van Nostrand Reinhold.

Gerarde, H. W. 1960. *Toxicology and Biochemistry of Aromatic Hydrocarbons.* New York: Elsevier.

Hallenbeck, W. H. 1993. *Quantitative Risk Assessment for Environmental and Occupational Health,* 2nd Edition. New York: CRC Press/Lewis Publishers.

Howard, P. H., et. al. 1991. *Handbook of Environmental Degradation Rates.* New York: CRC Press/Lewis Publishers.

Jones, F. E. 1994. *Toxic Organic Vapors in the Workplace.* New York: CRC Press/Lewis Publishers.

Keith, L. H. and M. M. Walker. 1995. *Handbook of Air Toxics.* New York: CRC Press/Lewis Publishers.

Landis, W. G., M. H. Yu. 1995. *Introduction to Environmental Toxicology.* New York: CRC Press/Lewis Publishers.

Langerman, N. 1994 *Precautionary Labels for Chemical Containers.* New York: CRC Press/Lewis Publishers.

Lauwerys, R. R., P. Hoet. 1993. *Industrial Chemical Exposure.* New York: CRC Press/Lewis Publishers.

Lewis, R. J., Sr. 1993. *Hawley's Condensed Chemical Dictionary,* 12th Edition. New York: Van Nostrand Reinhold.

Lewis, R. J., Sr. 1992. *Sax's Dangerous Properties of Industrial Materials,* Volumes 1, 2, and 3, 8th Edition. New York: Van Nostrand Reinhold.

Manahan, S. E. 1994. *Environmental Chemistry,* 6th Edition. New York: CRC Press/Lewis Publishers.

Mellström, G. A., J. E. Wahlberg, and H. I. Maibach. 1994. *Protective Gloves for Occupational Use.* New York: CRC Press/Lewis Publishers.

Meyer, E. 1990. *Chemistry of Hazardous Materials.* Englewood Cliffs, NJ: Prentice Hall.

Ott, W. R. 1995. *Environmental Statistics and Data Analysis.* New York: CRC Press/Lewis Publishers.

Patnaik, P. 1992. *A Comprehensive Guide to the Hazardous Properties of Chemical Substances.* New York: Van Nostrand Reinhold.

Philp, R. B. 1995. *Environmental Hazards and Human Health.* New York: CRC Press/Lewis Publishers.

Proctor, N. H., J. P. Hughes. 1991. *Chemical Hazards of the Workplace.* New York: Van Nostrand Reinhold.

Quigley, D. R. 1994. *Handbook of Emergency Chemical Management.* New York: CRC Press/Lewis Publishers.

Sacarello, H. L. A. 1994. *The Comprehensive Handbook of Hazardous Materials.* New York: CRC Press/Lewis Publishers.

Torkelson, T. R. V. K. Rowe. 1981. *Patty's Industrial Hygiene and Toxicology,* Vol. 2B. New York: Wiley-Interscience.

U.S. Department of Health and Human Services, National Institute for Occupational Safety and Health. 1994. *NIOSH Pocket Guide to Chemical Hazards.* Washington, D.C.: U.S. Government Printing Office.

MATERIAL SAFETY DATA SHEET

CHEMICAL NAME	
NAPTHA	**CRC PRESS/LEWIS PUBLISHERS DISCLAIMER:** THE INFORMATION AND RECOMMENDATIONS PRESENTED HEREIN ARE BASED ON SOURCES BELIEVED TO BE RELIABLE. CRC MAKES NO REPRESENTATION ON ITS COMPLETENESS OR ACCURACY. IT IS THE RESPONSIBILITY OF THE USER TO DETERMINE THE CHEMICAL'S SUITABILITY FOR ITS INTENDED USE, THE CHEMICAL'S SAFE USE, AND THE CHEMICAL'S PROPER DISPOSAL. NO REPRESENTATIONS AND/OR WARRANTIES, EITHER EXPRESSED OR IMPLIED, OF THE MERCHANTABILITY OR FITNESS FOR A PARTICULAR PURPOSE, OR OF ANY OTHER NATURE, ARE MADE WITH RESPECT TO THE INFORMATION PROVIDED IN THIS MSDS OR TO THE CHEMICAL TO WHICH INFORMATION MAY REFER. CRC NEITHER ASSUMES NOR AUTHORIZES ANY OTHER PERSON TO ASSUME FOR IT, ANY OTHER ADDITIONAL RESPONSIBILITY OR LIABILITY FOR THE USE OF, OR RELIANCE UPON, THIS INFORMATION.

HAZARD WARNING INFORMATION

HEALTH	FIRE	REACTIVE	OTHER	DEGREE OF HAZARD	COLOR CODING	OTHER CODES
3	3	0		0 = Minimum Hazard 1 = Slight Hazard 2 = Moderate Hazard 3 = Serious Hazard 4 = Severe Hazard	HEALTH = BLUE FIRE = RED REACTIVITY = YELLOW OTHER = WHITE	OX = Oxidizer ACID = Acid ALK = Alkali COR = Corrosive W = Use No Water

SECTION I - GENERAL INFORMATION

Characterization	RCRA Number	EPA Class
Aliphatic Hydrocarbon	**Not Listed**	**Not Applicable**

DOT Proper Shipping Name **Naptha or Petroleum Distillate**	Chemical Abstract Service (CAS) Number **8030-30-6**
DOT Hazard Class and Label Requirements **Flammable Liquid**	DOT Emergency Guide Code **27**
DOT Identification Number **UN 1268**	Molecular Formula **Varies with grade.**

Synonyms
Crude solvent coal tar naptha; high solvent naptha; lacquer diluent.

SECTION II - HAZARDOUS INGREDIENTS AND IDENTITY INFORMATION

Hazardous Components (specific identity)	OSHA Exposure Criteria	NIOSH Exposure Criteria	Immediately Dangerous to Life and Health (IDLH)	ACGIH Exposure Criteria
Naptha (derivation: Petroleum distillate consisting of various mixtures of aliphatic hydrocarbons). **1 ppm = 4.57 mg/m^3 (approx.)**	**PEL:** **100 ppm** **400 mg/m^3** **STEL:** **Not Established**	**REL:** **100 ppm** **400 mg/m^3** **STEL:** **Not Established**	**1000 ppm**	**TLV:** **300 ppm** **1370 mg/m^3** **STEL:** **Not Established**

SECTION III - PHYSICAL AND CHEMICAL CHARACTERISTICS

Boiling Point **100-280°F (38-138°C)**	Specific Gravity (H_2O = 1) **0.89 to 0.97 at 69°F (20°C)**
Vapor Pressure (mm Hg) **5 at 69°F (20°C)**	Molecular Weight (atomic weight) **110 (approx.)**
Vapor Density (Air = 1) **>3.2**	Melting Point **Not Found**

Solubility
Insoluble in water. Freely soluble in benzene, toluene, and xylene.

Appearance and Odor
A clear light colored or colorless liquid, or reddish-brown mobile liquid with an aromatic odor.

SECTION IV - FIRE AND EXPLOSION HAZARD DATA

Flash Point (method used) **-40°F (-40°C) closed cup (can vary with grade)**	Explosive Limits in Air % by Volume **LEL: 1%** **UEL: 7.6%**
NFPA Classification **Class 1B Flammable Liquid**	Autoignition Temperature **450°F (232°C)**

Extinguishing Media
Use alcohol foam, dry chemical, or carbon dioxide. Water may be ineffective and may actually spread fire. A water spray may be used to knock down vapors.

Special Fire Fighting Procedures
Isolate fire area. Wear self-contained breathing apparatus (SCBA). Poisonous gases are produced in fire. Use water spray to cool fire-exposed containers long after fire is out. Use non-sparking tools.

Unusual Fire and Explosion Hazards
Dangerous fire and explosion hazard when exposed to heat or flame. Containers may explode in fire and there is a possibility of a boiling liquid expanding vapor explosion (BLEVE). The vapor is heavier than air and may travel for some distance to cause fire to flashback to source.

SECTION V - REACTIVITY DATA

Stability		Conditions to Avoid
		Naptha is normally stable under routine conditions of handling and storage. Avoid contact with heat, flame, other sources of ignition, and contact with incompatible materials and chemicals.
Stable	Unstable	Incompatibility *(materials to avoid)*
X		**Strong oxidizers (such as chlorine, bromine, fluorine, and peroxides).**
Hazardous Polymerization		Conditions to Avoid
		Hazardous polymerization has not been reported to occur under normal conditions of temperature and pressure.
May Occur	Will Not Occur	Hazardous Decomposition or By-products
	X	**When heated to decomposition, naptha emits toxic and acrid fumes, such as oxides of carbon (monoxide and dioxide).**

SECTION VI - HEALTH HAZARD DATA

	Inhalation?	Absorption (skin/eye)?	Ingestion?
Primary Route(s) of Entry:	X	X	X

Health Hazards

INHALATION: Irritation of the eyes, nose, throat, and mucous membranes of the respiratory tract. Also stimulates and then depresses the central nervous system with dizziness, nausea, headache, anesthesia, euphoria, cardiac rhythm disruptions, respiratory arrest, and death.

SKIN & EYES: Irritation with drying, cracking, redness, blistering (prolonged contact). May lead to contact dermatitis and secondary infection.

INGESTION: May cause nausea, vomiting, headache, dizziness, gastrointestinal irritation; blurring of vision, gaseous tumors, and other effects of central nervous system depression.

Carcinogenicity	NTP Listed?	IARC Cancer Review Group?	OSHA Regulated?	Target Organs?
Unknown Human **Unknown Animal**	**No**	**No**	**29 CFR 1910.1000**	**Eyes, skin, respiratory sys., liver, CNS, kidneys.**

Medical Conditions Generally Aggravated by Exposure

None identified.

Emergency and First-aid Procedures

Eye contact: Do not allow victim to keep eyes tightly shut. Flush immediately with water for 15 minutes (minimum); seek medical attention. Skin contact: Remove all contaminated clothing, including shoes. Immediately wash area with flooding amounts of soap and water. Seek medical attention immediately. For inhalation: Remove the person from exposure. Provide respiratory assistance and CPR. Transfer to medical facility. If swallowed: Seek medical attention immediately. Do not induce vomiting due to possible aspiration of vomitus/chemical into the lungs which can cause chemical pneumonitis.

SECTION VII - PRECAUTIONS FOR SAFE HANDLING AND USE

Steps to be Taken in Case Material is Released or Spilled

Remove all ignition sources. Wear SCBA respiratory protection. Restrict those not wearing protective equipment and who are not involved in cleanup from entering area. Avoid skin contact. Stop leak if it can be done without risk. Provide maximum explosion-proof ventilation. Absorb with vermiculite or other similar material and deposit in sealed containers. Use non-sparking tools during cleanup.

Preferred Waste Disposal Method

No citation.

Precautions to be Taken in Handling and Storage

Store in tightly closed containers in cool, well-ventilated location. Keep away from sources of heat.

Other Precautions and Warnings

Sources of ignition (open flame or smoking) are prohibited where naptha is used or stored. Metal containers should be bonded and grounded during transfer operations. Bulk storage not recommended.

SECTION VIII - CONTROL MEASURES AND PERSONAL PROTECTIVE EQUIPMENT

Respiratory Protection *(specify type)*

Use an MSHA/NIOSH-approved organic vapor respirator, a powered air purifying respirator, or use a self-contained breathing apparatus (SCBA) with full facepiece operated in positive pressure mode.

Ventilation

Local exhaust (preferred) at site of chemical work or general ventilation.

Protective Gloves	Eye Protection	Other Protective Clothing
Nitrile or Polyvinyl Alcohol Gloves	**Chemical Goggles or Face Mask**	**Protective Apron if Splash is Likely**

Work/Hygiene Practices

Always wash hands thoroughly after using chemical; never bring food, drink, or smoking materials into vicinity of chemicals.

NAPTHA

Formula varies with grade CAS: 8030-30-6

IDENTIFICATION AND TYPICAL USES

Naptha is a clear, light colored to colorless or reddish-brown, mobile, aromatic, volatile liquid with a characteristic odor. Naptha is a general term applied to refined, partly refined, or unrefined petroleum products and liquid products of natural gas, not less than 10% of which distill below 347°F (175°C) and not less than 95% of which distill below 464°F (240°C) when subjected to distillation in accordance with the Standard Method of Test for Distillation of Gasoline, Naptha, Kerosene, and Similar Petroleum Products (ASTM D86). It is used as a source (by various cracking processes) of gasoline, special napthas, petroleum chemicals (especially ethylene).

RISK ASSESSMENT: HEALTH

General Assessment

Naptha is considered mildly to moderately toxic by *inhalation* of its vapors or by *ingestion* of the liquid. Skin contact will also cause localized damage at site of contact. There are no data regarding its carcinogenic, teratogenic, or mutagenic properties in animals or humans.

Inhalation of the naptha vapors can cause irritation of the eyes, nose, throat, and respiratory tract mucosa. Systemic effects include cough, dypsnea, nausea, headache, respiratory arrest, and death. Naptha is also a central nervous system depressant causing euphoria, anesthesia, cardiac rhythm disruptions, dizziness, intoxication, stupor, drunkenness, cyanosis, and possible loss of consciousness. Chronic lung dysfunction may occur if naptha is aspirated into the lungs.

Skin contact may cause dermatitis resulting from the defatting of the skin's natural protective oily barrier. Eye contact can lead to permanent damage depending upon the length of exposure. There may be irritation of the conjunctiva with redness, swelling, tearing, and localized inflammation of eye tissues on short exposures.

Ingestion may cause nausea, vomiting, diarrhea, and abdominal pain. If aspirated into the lungs, there may also be hypoxia, infection, chemical pneumonitis, and the possible formation of gaseous tumors. Other possible effects include headaches, dizziness, gastrointestinal irritation, drunkenness, cyanosis, and symptoms of CNS depression similar to that caused by inhalation.

Acute Health Effects

The following acute (short-term) health effects may occur immediately or shortly after exposure to naptha:

Skin: Irritation with potential for redness, blistering, burning, and dermatitis.

Eye: Severe irritation, conjunctivitis, redness, possible pain, and visual disturbances.

Lung: Nose, throat, and respiratory tract irritation causing difficulty in breathing, coughing, congestion, and possible respiratory arrest.

CNS: Inebriation, dizziness, stupor, headaches, mental confusion, loss of consciousness, and possible cardiac arrhythmia abnormalities.

Other: Ingestion followed by vomiting may cause some of the chemical to be aspirated into the lungs which can lead to chemical pneumonitis and other serious effects.

Chronic Health Effects

The following chronic (long-term) health effects may occur at some time after exposure to naptha and can last for months or even years:

Cancer Hazards: According to information presented in the references, naptha has not been adequately tested for its ability to cause cancer in test animals.

Reproductive Hazard: There is no evidence in the references to indicate that exposure to naptha causes any adverse effects to reproduction.

Other Chronic Effects: There is no evidence of systemic poisoning resulting from long-term exposures. There may be chronic headache, loss of appetite, dizziness, sleepiness, indigestion, and nausea. Chronic lung dysfunction may occur if naptha is aspirated into the lungs. Repeated skin contact causes blistering, drying, and lesions.

Recommended Risk-Reduction Measures

Personnel should avoid direct contact with or exposure to naptha. Occupational exposures occur most commonly through inhalation. A less toxic material or compound should be used in place of naptha whenever possible. If substitution cannot be done, *engineering controls* are the most effective method of reducing exposures. The best protection is to enclose operations and/or provide local explosion-proof exhaust ventilation at the site of naptha release. While not always operationally feasible, isolating operations can also reduce exposure. Using respiratory protection is less effective than the controls mentioned above, but is still advisable whenever working with or around naptha. An MSHA/NIOSH-approved air-purifying respirator with an organic vapor cartridge, a supplied-air respirator operated in continuous flow mode, or a self-contained breathing apparatus (SCBA) with full facepiece operated in pressure demand are the appropriate methods of respiratory protection when exposure to naptha is anticipated. If a full facepiece is not available, then chemical goggles should be worn to protect the eyes. A face shield should also be considered. To prevent hand and skin exposures, nitrile or polyvinyl alcohol gloves should be worn. Glove manufacturers should be contacted and permeation studies obtained *before* final glove selections are made.

Administrative controls should also be in place to minimize the potential for human exposures. These may include written procedures or policies which specify the methods and techniques that will be practiced whenever personnel are to work with naptha.

All personnel should receive *training* on the use, hazards, protective measures, emergency actions, and other precautions per 29 CFR 1910.1200 (Hazard Communication), prior to the first assignment in an area where naptha is used or stored.

Before beginning employment and at regular times thereafter, the following recommended medical tests should be considered:

☑ Lung function tests (establish baseline).

If symptoms occur or overexposure is suspected, the following additional tests are suggested:

☑ Lung function tests (compare to baseline).

☑ Evaluation for nervous system disorders.

☑ Liver and kidney function tests.

Any evaluation should include a careful history of past and present symptoms with an examination. Medical tests that simply look for existing damage are not a substitute for controlling exposures. Also, since smoking can cause heart disease, lung cancer, emphysema, and other respiratory disorders, smokers exposed to naptha may experience symptoms more rapidly than non-smokers under the same conditions of exposure. Prudent risk management requires careful consideration of *all* possible risk factors.

Other methods to reduce exposure include:

☑ Where possible, enclose operations and use local exhaust ventilation at the site of chemical release. If local exhaust ventilation or enclosure is not used, respiratory protection should be mandatory.

☑ Always ensure that proper protective clothing is worn when using chemical substances and that personnel have been trained on its use, care, and maintenance.

☑ Wash thoroughly immediately after exposure to naptha and at the end of the work shift or before eating, drinking, or smoking.

☑ Hazard warning information should be posted in the work area. In addition, as part of an on-going education and training program, all information on the health and safety hazards of naptha should be communicated to all potentially exposed workers.

☑ Eye wash stations should be provided in the immediate work area for emergency use. If there is a possibility of skin exposure to naptha, emergency shower facilities should also be provided.

☑ Workers whose clothing has been contaminated by naptha should change into clean clothes before leaving work. Contaminated work clothing should be laundered only by individuals who have been informed of the hazards of exposure to naptha.

☑ Where possible, automatically transfer liquids containing naptha from drums or other containers to process containers. Always practice grounding and bonding techniques. Use non-sparking tools.

RISK ASSESSMENT: ENVIRONMENT

General Assessment

The environment is at risk of exposure during transportation, storage, disposal, or destruction of naptha. In almost every scenario, the threat of environmental exposure is contingent upon the proper handling of the chemical substance. Accidental spills, large or small, can result in contamination of the surrounding environmental mediums (water, soil, and air).

Naptha is a moderately flammable liquid. Because of its extremely low flash point and high boiling point, it is classified as a Class 1B flammable liquid (per OSHA 29 CFR 1910.106). Containers can explode when exposed to excessive heat. A boiling liquid expanding vapor explosion (BLEVE) may occur when tankers or rail cars are involved in or are parked near a fire. It is a dangerous fire hazard and moderate explosion hazard when exposed to heat or flame. It is incompatible with commonly encountered oxidizers. Caution is always required in handling, storage, transportation, and disposal of naptha. Emergency responders should be made aware of the presence of naptha at any emergency response situation.

Naptha can enter the environment from industrial effluents and from spills.

Acute Ecological Effects

Acute (short-term) toxic effects may include the death of animals, birds, or fish and death or low growth rate in plants. Acute effects are seen 2 to 4 days after animals or plants are exposed to naptha.

Insufficient data are available to evaluate the short-term effects of naptha to aquatic life, plants, birds, or land animals.

Chronic Ecological Effects

Chronic (long-term) toxic effects may include shortened life span, reproductive problems, lower fertility, and changes in appearance or behavior in exposed animals. These effects can be seen long after first exposure(s) to toxic chemicals.

Insufficient data are available to evaluate or predict the long-term effects of naptha to aquatic life, plants, birds, or terrestrial animals.

Water Solubility

Naptha is not soluble in water. Concentrations of 1 milligram will not mix with a liter of water.

Persistence in the Environment

Naptha is slightly persistent in water, with a half-life of between 2 to 20 days. The half-life of a pollutant is the amount of time it takes for one half of the chemical to be degraded.

Bioaccumulation in Aquatic Organisms

Some substances increase in concentration, or *bioaccumulate,* in living organisms as they breathe contaminated air, drink contaminated water, or eat contaminated food. These chemicals can become concentrated in the tissues and internal organs of animals as well as humans.

The concentration of naptha found in fish tissues is expected to be about the same as the average concentration of naptha in the water from which the fish was taken.

Recommended Risk-Reduction Measures

Proper training of all transporters will reduce the likelihood of a mishap or accident resulting in a leak or spill of naptha into the environment. Labels on all containers, trucks, and rail cars must meet DOT requirements and accurately reflect their contents to enable emergency responders to react properly and quickly to any disaster thereby reducing the potential risk to the environment and to personnel.

Storage of naptha should be segregated from other chemicals and materials to minimize the risk of cross-contamination. Facilities should be designed to meet maximum explosion-proof requirements.

If a spill or leak to the environment has occurred, fire department, emergency response and/or hazardous materials spill personnel should be notified immediately. Cleanup should be attempted only by those trained in proper spill containment procedures using non-sparking tools. Contaminated soils should be removed for incineration and replaced with clean soil. If naptha should contact the water table, aquifer, or navigable waterway, time is of the essence. It is not soluble and total remediation may be possible. The local and/or state emergency response authorities must be notified. A comprehensive emergency response or disaster preparedness/recovery plan should be in place prior to any operations involving the use, transportation, storage, or disposal of naptha.

If naptha is spilled or leaked, the following specific steps are recommended:

- ☑ Restrict persons not wearing protective clothing from area of spill or leak until cleanup is complete and area can be opened for normal work.
- ☑ Absorb liquids in vermiculite, sand, earth, or similar material and deposit in sealed drum. Use non-sparking tools during cleanup procedures.
- ☑ Ventilate area of spill or leak. Container should be removed to safer area if it can be done without increasing the risk.
- ☑ Use water spray to knock down vapors.
- ☑ It may be necessary to dispose of naptha as a hazardous waste. The responsible state agency or the regional office of the federal Environmental Protection Agency (EPA) should be contacted for specific recommendations.

RISK ASSESSMENT: BUSINESS

General Assessment

Accidents or mishaps involving naptha can present a serious threat to business operations. The loss or damage of equipment or facilities can significantly affect fiscal viability. Lawsuits that may result from personnel illness, injury/death, public exposures, and/or environmental contamination will require a serious expenditure of resources. Media attention surrounding an injury, death, or environmental damage can also result in a loss of profits and loss of current as well as future business.

Recommended Risk-Reduction Measures

Company attorneys, safety and health professionals, and environmental specialists should be involved in the development of any procedures for responding to chemical incidents. A company official should be predesignated as a public relations officer with specific training in dealing with the press. Corporate plans and policies should be developed, approved, and implemented long before any need for such arises.

REFERENCES

American Conference of Governmental Industrial Hygienists. 1988. *Documentation of the Threshold Limit Values and Biological Exposure Indices,* 5th Edition (with updates). Cincinnati: ACGIH

Bloom, A. D. and F. J. DeSerres. 1995. *Ecotoxicity and Human Health.* New York: CRC Press/Lewis Publishers.

Bueche, J. R. 1972. *Principles of Physics,* 2nd Edition. New York: McGraw-Hill.

Cockerman, L. G., B. S. Shane. 1994. *Basic Environmental Toxicology.* New York: CRC Press/Lewis Publishers.

Howard, P. H., et. al. 1991. *Handbook of Environmental Degradation Rates.* New York: CRC Press/Lewis Publishers.

Jones, F. E. 1994. *Toxic Organic Vapors in the Workplace.* New York: CRC Press/Lewis Publishers.

Lauwerys, R. R., P. Hoet. 1993. *Industrial Chemical Exposure.* New York: CRC Press/Lewis Publishers.

Lewis, R. J., Sr. 1993. *Hawley's Condensed Chemical Dictionary,* 12th Edition. New York: Van Nostrand Reinhold.

Lewis, R. J., Sr. 1992. *Sax's Dangerous Properties of Industrial Materials,* Volumes 1, 2, and 3, 8th Edition. New York: Van Nostrand Reinhold.

Meyer, E. 1990. *Chemistry of Hazardous Materials.* Englewood Cliffs, NJ: Prentice Hall.

Patnaik, P. 1992. *A Comprehensive Guide to the Hazardous Properties of Chemical Substances.* New York: Van Nostrand Reinhold.

Proctor, N. H., J. P. Hughes. 1991. *Chemical Hazards of the Workplace.* New York: Van Nostrand Reinhold.

Quigley, D. R. 1994. *Handbook of Emergency Chemical Management.* New York: CRC Press/Lewis Publishers.

Sacarello, H. L. A. 1994. *The Comprehensive Handbook of Hazardous Materials.* New York: CRC Press/Lewis Publishers.

Torkelson, T. R. V. K. Rowe. 1981. *Patty's Industrial Hygiene and Toxicology,* Vol. 2B. New York: Wiley-Interscience.

U.S. Department of Health and Human Services, National Institute for Occupational Safety and Health. 1994. *NIOSH Pocket Guide to Chemical Hazards.* Washington, D.C.: U.S. Government Printing Office.

MATERIAL SAFETY DATA SHEET

CHEMICAL NAME

NEOCHROMIUM

HAZARD WARNING INFORMATION

HEALTH	FIRE	REACTIVE	OTHER	DEGREE OF HAZARD	COLOR CODING	OTHER CODES
For Chromium **3**	For Chromium **0**	For Chromium **0**		0 = Minimum Hazard 1 = Slight Hazard 2 = Moderate Hazard 3 = Serious Hazard 4 = Severe Hazard	HEALTH = BLUE FIRE = RED REACTIVITY = YELLOW OTHER = WHITE	OX = Oxidizer ACID = Acid ALK = Alkali COR = Corrosive ~~W~~ = Use No Water

SECTION I - GENERAL INFORMATION

Characterization	RCRA Number	EPA Class
Metal	**None**	**Not Applicable**

DOT Proper Shipping Name	**Neochromium**	Chemical Abstract Service (CAS) Number	**64093-79-4**
DOT Hazard Class and Label Requirements	**No Citation**	DOT Emergency Guide Code	**No Citation**
DOT Identification Number	**No Citation**	Chemical Formula	**$CrHO_5S$**

Synonyms

Basic chromic sulfate; basic chromium sulfate; chromium hydroxide sulfate; chromium sulfate; chromic sulfate; monobasic chromium sulfate.

SECTION II - HAZARDOUS INGREDIENTS AND IDENTITY INFORMATION

Hazardous Components (specific identity)	OSHA Exposure Criteria	NIOSH Exposure Criteria	Immediately Dangerous to Life and Health (IDLH)	ACGIH Exposure Criteria
Neochromium (derivation: By reacting sulfuric acid with chromium hydroxide and subsequent crystallization. Typically hydrated with between 15 and 18 water molecules).	**PEL (8 hour):** **0.5 mg(Cr)/m^3** **STEL:** **Not Established**	**REL (10 hour):** **0.5 mg(Cr)/m^3** **STEL:** **Not Established**	**250 mg/m^3 as Cr**	**TLV (8 hour):** **0.5 mg(Cr)/m^3** **STEL:** **Not Established**

SECTION III - PHYSICAL AND CHEMICAL CHARACTERISTICS

Boiling Point	**212°F (100°C) loses water**	Specific Gravity (H_2O = 1)	**3.01 (anhydrous)**
Vapor Pressure (mm Hg)	**0 at 69°F (20°C) for Chromium**	Molecular Weight	**165.07**
Vapor Density (Air = 1)	**Not Found**	Melting Point	**153°F (67°C) Decomposes**

Solubility

Anhydrous form is insoluble in water and most acids. Soluble in alcohol.

Appearance and Odor

Hard, brittle, semi-gray, lustrous metal. It often appears as a powder.

SECTION IV - FIRE AND EXPLOSION HAZARD DATA

Flash Point (method used)	**Not Applicable**	Explosive Limits in Air % by Volume	**LEL: Not Determined UEL: Not Determined**
NFPA Classification	**No Citation**	Autoignition Temperature	**Not Determined**

Extinguishing Media

Neochromium may burn, but does not readily ignite. Use agents suitable to surrounding fires.

Special Fire Fighting Procedures

Poisonous gases are produced in fire. Wear full protective clothing and self-contained breathing apparatus (SCBA). Use a water spray to keep fire-exposed containers cool. Water may be ineffective if used directly on fires involving neochromium. Remain clear of smoke, water fallout and water runoff. Do not release runoff to sewers or waterways; contain for proper disposal.

Unusual Fire and Explosion Hazards

None reported.

SECTION V - REACTIVITY DATA

Stability		Conditions to Avoid Neochromium is stable in closed containers under routine conditions of handling and storage. Keep away from incompatible materials since violent and or explosive reactions can occur.
Stable X	Unstable	Incompatibility (*materials to avoid*) Incompatible with strong oxidizers (such as chlorine, bromine, hydrogen peroxide, and fluorine), and most alkali materials. It is readily oxidized to chromic acid by hypochlorites, peroxides, or pressurized oxygen.
Hazardous Polymerization		Conditions to Avoid Under normal conditions of temperature and pressure, hazardous polymerization of neochromium is not known to occur.
May Occur	Will Not Occur X	Hazardous Decomposition or By-products When heated to decomposition, neochromium can emit highly toxic/poisonous fumes and gases including oxides of sulfur and chromium.

SECTION VI - HEALTH HAZARD DATA

Primary Route(s) of Entry:	Inhalation?	Absorption (skin)?	Ingestion?
	X		X

Health Hazards

INHALATION: Irritation of the respiratory system. Exposure to some chromium compounds can cause sores or "chrome holes" to appear in the nasal cavity. Respiratory irritation is primarily due to mechanical rather than chemical action. Overexposure to high concentrations of neochromium may cause delayed pulmonary edema (fluid in lungs), which can be fatal.

SKIN & EYES: A minor irritant to the eyes and skin. May cause sensitization leading to itchy rash.

INGESTION: Irritation of the gastrointestinal tract.

Carcinogenicity	NTP Listed?	IARC Cancer Review Group?	OSHA Regulated?	Target Organs?
Unknown Human Questioned Animal	No	Group 3	29 CFR 1910.1000 Table Z-1 (Chromium)	Respiratory system, skin, eyes.

Medical Conditions Generally Aggravated by Exposure

None reported. However, existing skin conditions or respiratory problems may be aggravated.

Emergency and First-aid Procedures

Eye contact: Flush immediately with water for 15 minutes (minimum); seek medical attention. Skin contact: Remove all contaminated clothing. Immediately wash area with large amounts of soap and water. For inhalation: Remove the person from exposure. Provide respiratory assistance and CPR. Transfer to medical facility. Observe for 48 hours for lung effects. If swallowed: Provide breathing assistance if necessary. Never attempt to give a convulsing or unconscious person anything by mouth.

SECTION VII - PRECAUTIONS FOR SAFE HANDLING AND USE

Steps to be Taken in Case Material is Released or Spilled

Collect powdered material in most convenient manner possible and deposit in sealed drum. Keep those not involved in cleanup from entering area. Do NOT dry sweep (creates airborne dusts). Use a vacuum equipped with a high efficiency particulate air (HEPA) filter.

Preferred Waste Disposal Method

No citation.

Precautions to be Taken in Handling and Storage

Store in tightly closed containers in cool, well-ventilated area and protect from physical damage. Sources of ignition, such as smoking and open flames, are prohibited where neochromium is used, handled, or stored.

Other Precautions and Warnings

Personnel must be trained on the hazards of chromium compounds *before* being assigned to work in areas where neochromium is used, handled, or stored.

SECTION VIII - CONTROL MEASURES AND PERSONAL PROTECTIVE EQUIPMENT

Respiratory Protection (*specify type*)

At any exposure level, recommend using an MSHA/NIOSH-approved supplied-air respirator with a full facepiece, hood, or helmet operated in the continuous flow mode or a self-contained breathing apparatus (SCBA) operated in positive pressure mode.

Ventilation

Local exhaust or general mechanical systems recommended.

Protective Gloves	Eye Protection	Other Protective Clothing
Impervious Material	Chemical/Dust Goggles or Face Mask	Protective Suit or Clothing

Work/Hygiene Practices

Always wash hands thoroughly after using chemical; never bring food, drink, or smoking materials into vicinity of chemicals.

NEOCHROMIUM

$CrHO_5S$

CAS: 64093-79-4

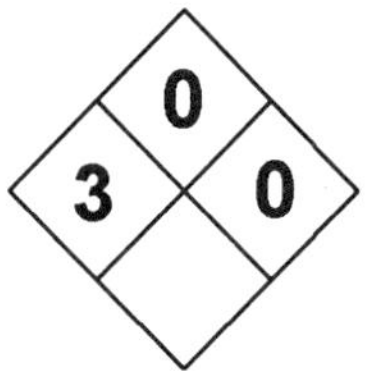

IDENTIFICATION AND TYPICAL USES

Neochromium appears as a peach-red to violet powder in anhydrous form, or as dark green, amorphous scales or crystals. It may also take the appearance of dark cubes. It is used in the textile industry for dyeing, printing, moth-proofing, and water-proofing. It is also used in metal treating and polishing, in the tanning of leather, in the manufacture of other chromium compounds, in the production of green varnishes, paints, inks, and glazes (for porcelain), and as a sensitizer in photographic processes.

RISK ASSESSMENT: HEALTH

General Assessment

Neochromium is a questioned animal carcinogen. It is harmful to humans by *ingestion* with gastrointestinal effects. It can also be toxic by *inhalation*. The toxicity of chromium alloys and compounds varies significantly. Neochromium has not been shown to cause significant toxicity. However, human mutation data have been reported and, because the nature of its human carcinogenic properties is unknown, it should be handled with extreme caution. Exposure to the dusts of neochromium may cause dangerous lung irritation and possibly lead to disease, including pulmonary fibrosis and pneumoconiosis.

Occupational exposure to certain chromium(III) compounds can produce skin ulceration, dermatitis, perforation of the nasal septa, and lung damage. Overexposure to high concentrations of neochromium dusts can cause a dangerous buildup of fluids in the lungs (pulmonary edema), which is a medical emergency and can be fatal. Symptoms of cough, congestion, tightness in the chest, and difficulty breathing may be delayed up to 48 hours. This can create a false sense of security with regard to health exposure risk.

Inhalation of chromium fumes can cause "metal fume fever," which is a flu-like illness lasting up to 24 hours with characteristic symptoms of chills, aches, cough, and fever. It is not known if exposure to heated neochromium fumes will have this same effect.

Acute Health Effects

The following acute (short-term) health effects may occur immediately or shortly after exposure to neochromium:

Skin: Irritation, rash, and possible contact dermatitis.

Eye: Neochromium particles can irritate the eyes.

Lung: Inhalation of neochromium will irritate the lungs and may cause a condition known as pulmonary edema (fluid in the lungs).

Chronic Health Effects

The following chronic (long-term) health effects may occur at some time after exposure to neochromium and can last for months or even years:

Cancer Hazards: Many chromium compounds are confirmed human carcinogens. They have been shown to cause cancer of the lung and throat in humans and test animals. It is not known if neochromium will cause cancer in humans. However, human mutation data have been reported and many scientists believe that exposure to such chemicals may pose a cancer risk in the long term.

Reproductive Hazard: According to information available in the references, there is no evidence to indicate that neochromium adversely affects reproduction.

Other Chronic Effects: Long-term exposures may case skin sensitization characterized by a red, itchy rash resembling eczema. Once this develops, even small future exposure can cause symptoms to reappear. Very irritating substances may affect the lungs. It is not known whether neochromium causes lung damage.

Ʋ *Recommended Risk-Reduction Measures*

Personnel should avoid direct contact with neochromium. Because many related chromium compounds are known human carcinogens, prudent risk management dictates that neochromium be handled and treated with extreme caution. If a less toxic chemical cannot be substituted for a hazardous substance, then *engineering controls* are the most effective method of reducing exposures.

The best protection is to enclose operations and/or provide local exhaust ventilation at the site of chemical release. While not always operationally feasible, isolating operations can also reduce exposure. Using respiratory protection is less effective than the controls mentioned above, but is still advisable whenever working with or around neochromium. The permissible exposure level (PEL) for chromium compounds is extremely low (0.5 mg/m^3) and difficult to accurately monitor. For any exposure level, recommend using a supplied-air respirator with a full facepiece operated in positive pressure mode or an MSHA/NIOSH-approved self-contained breathing apparatus (SCBA) with full facepiece in pressure demand mode as the respiratory protection methods of choice. If a full facepiece is not available, then chemical splash goggles should be worn to protect the eyes. Whenever a chemical dust hazard exists, a face shield and a rubber apron should be worn. To prevent hand and skin exposures, impervious rubber gloves should be used.

Administrative controls should also be in place to minimize the potential for human exposures. These may include written procedures or policies which specify the methods and techniques that will be practiced whenever personnel are to work with neochromium.

All personnel should receive *training* on the use, hazards, protective measures, emergency actions, and other precautions per 29 CFR 1910.1200 (Hazard Communication), prior to the first assignment in an area where neochromium is used or stored.

Before beginning employment and at regular intervals thereafter (e.g., annually), for those with frequent or potentially high exposures (half the PEL or greater, or significant skin contact), the following medical tests are recommended:

- ☑ Lung function tests (these may be normal if person is not displaying symptoms at the time of the test).

If symptoms develop or overexposure is suspected, the following may also be useful:

- ☑ Consider urine chromium levels. Persons without workplace exposure usually have levels below 15 micrograms per liter of urine.
- ☑ Lung function tests (these may be normal if person is not displaying symptoms at the time of the test).
- ☑ Consider chest X-ray after acute overexposure (may be negative if taken immediately after exposure due to delayed onset of pulmonary edema).

Any evaluation should include a careful history of past and present symptoms with a medical examination. Medical tests that look for damage already done are not a substitute for controlling exposure. Also, since smoking can cause heart disease, as well as lung cancer, emphysema, and other related respiratory problems, the effects of exposure to neochromium may be more pronounced in smokers than in non-smokers. Prudent risk management requires proper consideration of *all* possible factors which may be responsible for the appearance of exposure symptoms in the workplace.

Other methods to reduce exposure include:

- ☑ Where possible, enclose operations and use local exhaust ventilation at the site of chemical release. If local exhaust ventilation or enclosure is not used, respiratory protection should be mandatory.
- ☑ Always ensure that proper protective clothing is worn when using chemical substances.
- ☑ Wash thoroughly immediately after exposure to neochromium and at the end of the work shift or before eating, drinking, or smoking.
- ☑ Hazard warning information should be posted in the work area. In addition, as part of an on-going education and training program, all information on the health and safety hazards of neochromium should be communicated to all exposed and potentially exposed workers.
- ☑ Where possible, automatically transfer neochromium from drums or other storage containers to process containers.
- ☑ Specific engineering controls are recommended by NIOSH for chromium and some information may be useful in controlling the risks of exposure to neochromium. Refer to NIOSH Criteria Documents: Occupational Exposure to Chromium. Publication Number 76-129.

RISK ASSESSMENT: ENVIRONMENT

General Assessment

The environment is at risk of exposure during transportation, storage, disposal, or destruction of neochromium. In almost every scenario, the threat of environmental exposure is contingent upon the proper handling of the chemical substance. Accidental spills, large or small, can result in fire, explosion, and possible contamination of the surrounding environmental mediums (water, soil, and air).

Neochromium will not readily ignite but can burn if involved in a fire. Finely divided powders may form explosive mixtures with air, especially in confined areas. It can be readily oxidized on exposure to hypochlorites, peroxides, and oxygen under high pressure and temperature. In a fire, neochromium will emit toxic and poisonous gases. It is also incompatible with strong oxidizers (such as chlorine, fluorine, bromine, and hydrogen peroxide), and many alkali metals (zinc, magnesium, and aluminum in acid solution). These characteristics require special consideration during any emergency situation involving a leak or spill of neochromium.

Neochromium can enter the environment through effluents from the metal plating industries, from municipal waste treatment plant discharges, or from spills.

Acute Ecological Effects

Acute (short-term) toxic effects may include the death of animals, birds, or fish and death or low growth rate in plants. Acute effects are seen 2 to 4 days after animals or plants are exposed to neochromium.

Chromium is more toxic in soft water than in hard water. Neochromium has moderate acute toxicity to aquatic life. Insufficient data are available to evaluate the short-term effects of neochromium exposure to plants, birds or land animals.

Chronic Ecological Effects

Chronic (long-term) toxic effects may include shortened life span, reproductive problems, lower fertility, and changes in appearance or behavior in exposed animals. These effects can be seen long after first exposure(s) to toxic chemicals.

Neochromium has high chronic toxicity to aquatic life. Insufficient data are available to evaluate the long-term effects of neochromium exposure to plants, birds or land animals.

Water Solubility

Water solubility of chromium and its salts ranges from low to high.

Persistence in the Environment

Neochromium is highly persistent in the aquatic environment, with a half-life of greater than 200 days. The half-life of a pollutant is the amount of time it takes for one half of the chemical to be degraded.

Bioaccumulation in Aquatic Organisms

Some substances increase in concentration, or *bioaccumulate,* in living organisms as they breathe contaminated air, drink contaminated water, or eat contaminated food. These chemicals can become concentrated in the tissues and internal organs of animals as well as humans.

The concentration of neochromium found in fish is expected to be somewhat higher than the average concentration of neochromium in the water from which the fish was taken.

Recommended Risk-Reduction Measures

Proper training of all transporters will reduce the likelihood of a mishap or accidents resulting in a leak or spill to the environment. The correct use of DOT labeling (per the MSDS) on all containers, trucks, and rail cars will enable emergency responders to react properly and quickly to any disaster thereby reducing the potential risk to the environment and to personnel.

Storage of neochromium should be segregated from incompatible chemicals to minimize the risk of cross-contamination or contact. Bulk storage of neochromium should be avoided.

If a spill or leak to the environment has occurred, fire department, emergency response and/or hazardous materials spill personnel should be notified immediately. Use a vacuum or wet mop method to reduce dust during cleanup. Do not dry sweep. When vacuuming, a high efficiency particulate air (HEPA) filter should be used, not a standard shop vacuum. Deposit collected materials in sealed drum for disposal. Cleanup should be attempted only by those trained in proper spill containment procedures. Contaminated soils should be removed and replaced with clean soil. If neochromium should contact the water table, aquifer, or navigable waterway, time is of the essence. Chromium compounds may be moderately to highly soluble in water and, therefore, total containment and remediation may not be entirely possible. When such spills occur, the local and/or state emergency response authorities must be notified. A comprehensive emergency response or disaster preparedness/recovery plan should be in place prior to any operations involving the use, transportation, storage, or disposal of neochromium.

If neochromium is spilled or leaked, the following specific steps are recommended:

☑ Restrict persons not wearing protective clothing from area of spill or leak until cleanup is complete and area can be opened for normal work.

☑ Collect powdered materials in most convenient and safe manner possible and place in sealed drums for disposal. Do not dry sweep, use HEPA vacuum instead.

☑ It may be necessary to dispose of neochromium as a hazardous waste. The responsible state agency or the regional office of the federal Environmental Protection Agency (EPA) should be contacted for specific recommendations.

RISK ASSESSMENT: BUSINESS

General Assessment

Accidents or mishaps involving neochromium can present a significant threat to business operations. The loss or damage of equipment or facilities can significantly affect fiscal viability. Lawsuits that may result from personnel injury/death, public exposures, and/or environmental contamination will also require a serious expenditure of resources. Media attention surrounding an injury, death, or environmental damage can also result in a loss of profits and loss of current as well as future business. Always remember that any time the terms "cancer" or "carcinogen" are used, public emotion, hysteria, and ignorance can run equally high. This must be carefully considered when drafting or implementing public relations policies.

Recommended Risk-Reduction Measures

Company attorneys, safety and health professionals, and environmental specialists should be involved in the development of any procedures dealing with safety and emergency response. A company official should be pre-designated as a public relations officer with specific training in dealing with the press. Corporate plans and policies should be developed, approved, and implemented long before any need for such arises.

REFERENCES

American Conference of Governmental Industrial Hygienists. 1988. *Documentation of the Threshold Limit Values and Biological Exposure Indices,* 5th Edition (with updates). Cincinnati: ACGIH

Asante-Duah, D. K. 1993. *Hazardous Waste Risk Assessment.* New York: CRC Press/Lewis Publishers.

Bloom, A. D. and F. J. DeSerres. 1995. *Ecotoxicity and Human Health.* New York: CRC Press/Lewis Publishers.

Brusick, D. J. 1994. *Methods for Genetic Risk Assessment.* New York: CRC Press/Lewis Publishers.

Calabrese, E. J., E. M. Kenyon. *Air Toxics and Risk Assessment.* New York: CRC Press/Lewis Publishers.

Calabrese, E. J. 1994. *Biological Effects of Low Level Exposures.* New York: CRC Press/Lewis Publishers.

Cockerman, L. G., B. S. Shane. 1994. *Basic Environmental Toxicology.* New York: CRC Press/Lewis Publishers.

Cothern, C. R., N. P. Ross. 1994. *Environmental Statistics, Assessment, and Forecasting.* New York: CRC Press/Lewis Publishers.

Gardner, R. W. 1994. *Chemical Intolerance.* New York: CRC Press/Lewis Publishers.

Hallenbeck, W. H. 1993. *Quantitative Risk Assessment for Environmental and Occupational Health,* 2nd Edition. New York: CRC Press/Lewis Publishers.

Howard, P. H., et. al. 1991. *Handbook of Environmental Degradation Rates.* New York: CRC Press/Lewis Publishers.

Landis, W. G., M. H. Yu. 1995. *Introduction to Environmental Toxicology.* New York: CRC Press/Lewis Publishers.

Lauwerys, R. R., P. Hoet. 1993. *Industrial Chemical Exposure.* New York: CRC Press/Lewis Publishers.

Lewis, R. J., Sr. 1992. *Sax's Dangerous Properties of Industrial Materials,* Volumes 1, 2, and 3, 8th Edition. New York: Van Nostrand Reinhold.

Manahan, S. E. 1992. *Toxicological Chemistry,* 2nd Edition. New York: CRC Press/Lewis Publishers.

Meyer, E. 1990. *Chemistry of Hazardous Materials.* Englewood Cliffs, NJ: Prentice Hall.

Sacarello, H. L. A. 1994. *The Comprehensive Handbook of Hazardous Materials.* New York: CRC Press/Lewis Publishers.

Sherman, J. D. 1988. *Chemical Exposure and Disease.* New York: Van Nostrand Reinhold.

U.S. Department of Health, Education, and Welfare (DHEW), National Institute for Occupational Health (NIOHS). 1976. *NIOSH Criteria Document: Occupational Exposure to Chromium (VI).* Publication Number 76-129. Washington, D.C.; U.S. Government Printing Office.

MATERIAL SAFETY DATA SHEET

CHEMICAL NAME	
NEON	**CRC PRESS/LEWIS PUBLISHERS DISCLAIMER:** THE INFORMATION AND RECOMMENDATIONS PRESENTED HEREIN ARE BASED ON SOURCES BELIEVED TO BE RELIABLE. CRC MAKES NO REPRESENTATION ON ITS COMPLETENESS OR ACCURACY. IT IS THE RESPONSIBILITY OF THE USER TO DETERMINE THE CHEMICAL'S SUITABILITY FOR ITS INTENDED USE, THE CHEMICAL'S SAFE USE, AND THE CHEMICAL'S PROPER DISPOSAL. NO REPRESENTATIONS AND/OR WARRANTIES, EITHER EXPRESSED OR IMPLIED, OF THE MERCHANTABILITY OR FITNESS FOR A PARTICULAR PURPOSE, OR OF ANY OTHER NATURE, ARE MADE WITH RESPECT TO THE INFORMATION PROVIDED IN THIS MSDS OR TO THE CHEMICAL TO WHICH INFORMATION MAY REFER. CRC NEITHER ASSUMES NOR AUTHORIZES ANY OTHER PERSON TO ASSUME FOR IT, ANY OTHER ADDITIONAL RESPONSIBILITY OR LIABILITY FOR THE USE OF, OR RELIANCE UPON, THIS INFORMATION.

HAZARD WARNING INFORMATION

HEALTH	FIRE	REACTIVE	OTHER	DEGREE OF HAZARD	COLOR CODING	OTHER CODES
1	0	0		0 = Minimum Hazard 1 = Slight Hazard 2 = Moderate Hazard 3 = Serious Hazard 4 = Severe Hazard	HEALTH = BLUE FIRE = RED REACTIVITY = YELLOW OTHER = WHITE	OX = Oxidizer ACID = Acid ALK = Alkali COR = Corrosive W = Use No Water

SECTION I - GENERAL INFORMATION

Characterization	**Gas**	RCRA Number	**None**
		EPA Class	**Not Applicable**
DOT Proper Shipping Name	**Neon (compressed); Neon (refrigerated liquid)**	Chemical Abstract Service (CAS) Number	**7440-01-9**
DOT Hazard Class and Label Requirements	**Non Flammable Gas**	DOT Emergency Guide Code	**15**
DOT Identification Number	**UN1065 (compressed) UN1913 (refrigerated liquid)**	Atomic Formula	**Ne**
Synonyms	**None.**		

SECTION II - HAZARDOUS INGREDIENTS AND IDENTITY INFORMATION

Hazardous Components (specific identity)	OSHA Exposure Criteria	NIOSH Exposure Criteria	Immediately Dangerous to Life and Health (IDLH)	ACGIH Exposure Criteria
Neon (derivation: By fractional distillation of liquid air. It constitutes 0.0012% of normal air).	**PEL:** Not Established **STEL:** Not Established	**REL:** Not Established **STEL:** Not Established	**Not Determined**	**TLV:** Not Established **STEL:** Not Established

SECTION III - PHYSICAL AND CHEMICAL CHARACTERISTICS

Boiling Point	**-411°F (-246°C)**	Specific Gravity (H_2O = 1)	**1.204 at -246°F (-154°C)**
Vapor Pressure (atmospheres)	**Not Listed**	Atomic Weight	**20.18**
Vapor Density (Air = 1)	**0.90 at 32°F (0°C)**	Freezing Point	**-416°F (-249°C)**
Solubility	**Very slightly soluble in water.**		
Appearance and Odor	**Colorless, odorless, tasteless gas.**		

SECTION IV - FIRE AND EXPLOSION HAZARD DATA

Flash Point (method used)	**Not Determined**	Explosive Limits in Air % by Volume	**LEL: Not Determined UEL: Not Determined**
NFPA Classification	**Non-Combustible Gas**	Autoignition Temperature	**Not Determined**

Extinguishing Media

Use extinguishing media suitable to surrounding fires. Neon itself will not ignite or support combustion.

Special Fire Fighting Procedures

Neon is inert and will not support combustion. However, it is recommended that self-contained breathing apparatus (SCBA) be worn. Contact with the liquid may cause frostbite. Heat will build pressure and may rupture closed storage containers. Keep fire-exposed containers cool with water spray.

Unusual Fire and Explosion Hazards

Neon cylinders may explode in the heat of fire.

SECTION V - REACTIVITY DATA

Stability		Conditions to Avoid Neon is stable in closed cylinders under routine conditions of handling and storage. Keep away from heat and other gases. It can condense and solidify other gases. Use metal containers made for cryogenics.
Stable X	Unstable	Incompatibility *(materials to avoid)* Neon is a monatomic, chemically inactive, simple asphyxiant gas that is lighter than air. Some steels and other materials will become brittle in contact with the compressed gas or the refrigerated liquid.
Hazardous Polymerization		Conditions to Avoid Under normal conditions of temperature and pressure, hazardous polymerization of neon is not expected to occur. Prevent any accidental or uncontrollably rapid release of neon gas.
May Occur	Will Not Occur X	Hazardous Decomposition or By-products None reported.

SECTION VI - HEALTH HAZARD DATA

Primary Route(s) of Entry:	Inhalation?	Absorption (skin)?	Ingestion?
	X		

Health Hazards

INHALATION: An asphyxiant gas. Initial symptoms are those of oxygen starvation. Feeling of "air hunger," diminished mental alertness, headache, dullness, shortness of breath, and an impairment in muscle coordination. Continued exposures lead to faulty judgment, depression of senses, rapid fatigue, unconsciousness, convulsions, coma, and death.

SKIN & EYES: Contact with the compressed gas or refrigerated liquid can cause frostbite and burns.

INGESTION: Not likely since neon is a gas at normal working (room) temperatures.

Carcinogenicity	NTP Listed?	IARC Cancer Review Group?	OSHA Regulated?	Target Organs?
No Evidence Human No Evidence Animal	No	No	No	Respiratory system, skin, eyes, CNS.

Medical Conditions Generally Aggravated by Exposure

None Reported.

Emergency and First-aid Procedures

Eye contact (liquid): Immediately flush with large amounts of tepid water for 15 minutes (minimum), occasionally lifting eyelids; seek medical attention. Skin contact (liquid): Remove all contaminated clothing. Immediately flush area with large amounts of tepid water. Do NOT rub! Seek medical assistance. For inhalation: Remove the person from exposure. Emergency responders must be protected from inert atmospheres. Provide respiratory assistance and CPR. Transfer promptly to medical facility. If swallowed: Treat frozen tissues and protect the airway. Seek medical attention immediately.

SECTION VII - PRECAUTIONS FOR SAFE HANDLING AND USE

Steps to be Taken in Case Material is Released or Spilled

Ventilate area. Stop flow of gas. If the source of leak is a cylinder and cannot be stopped, remove to safe place in open air; repair leak or let cylinder empty. Ensure response personnel are properly protected with self-contained breathing apparatus (SCBA).

Preferred Waste Disposal Method

No citation.

Precautions to be Taken in Handling and Storage

Store in tightly closed containers in a cool, dry, well-ventilated area away from heat, direct sunlight, or other hot areas (water lines, vent pipes). Ensure proper storage and handling precautions are in place whenever working with compressed gas cylinders.

Other Precautions and Warnings

Containers may explode in fire or under conditions of extreme heat. Protect cylinders from physical damage. Use pressure relief on liquid containers. Periodically check vent valves for blockage (ice).

SECTION VIII - CONTROL MEASURES AND PERSONAL PROTECTIVE EQUIPMENT

Respiratory Protection *(specify type)*

No exposure limits have been established. Neon is a simple asphyxiant. Do NOT use cartridge-type respirators (ineffective for this type of gas). Use a powered supplied-air respirator or a self-contained breathing apparatus with full facepiece operated in pressure demand or other positive pressure mode.

Ventilation

Local exhaust or general mechanical systems recommended.

Protective Gloves	Eye Protection	Other Protective Clothing
Impervious, Thermal Insulated Gloves	Chemical Goggles and/or Face Mask	Protective Boots, Gauntlets, Apron

Work/Hygiene Practices

Always wash hands thoroughly after using chemical; never bring food, drink, or smoking materials into vicinity of chemicals.

NEON

Ne

CAS: 7440-01-9

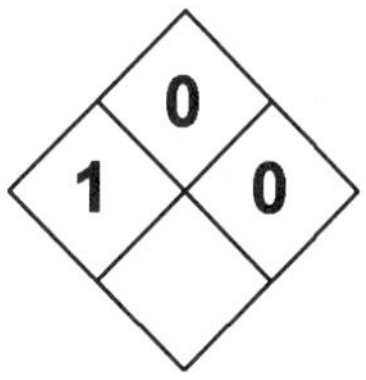

IDENTIFICATION AND TYPICAL USES

Neon is a colorless, odorless, and tasteless gas. It is an inert element of atomic number 10 of the noble gas group of the periodic system with three stable isotopes. It is used in luminescent electric tubes and photoelectric bulbs, in the electronic industry, in high-voltage indicators for high-tension power lines, in certain lasers, as an ingredient of gaseous fillers for anti-fog devices, in warning signals, in electrical current detectors, in lightening arresters and, as a liquid, in cryogenic research.

RISK ASSESSMENT: HEALTH

General Assessment

Neon is a simple asphyxiant gas. As such, its primary route of entry into the body is through *inhalation*. While it does not pose a significant physiological effect (i.e., no systemic poisoning or toxicity), it can displace oxygen levels to below that which is necessary to sustain life. Skin contact with the liquid can be destructive to tissues due to the extremely cold temperatures but absorption is not a likely exposure route. Since neon usually exists as a gas at room temperatures, ingestion is not likely to occur either.

Inhalation of neon gas can cause headache, dullness, shortness of breath, and the symptoms of asphyxiation. These include rapid respiration and a feeling of "air hunger" or air starvation followed by a decrease in mental alertness, and impaired muscular coordination. If exposure continues, there can be nausea, vomiting, prostration, unconsciousness, coma, and death.

Skin or eye contact with the vapor poses no serious threat. The vapors have not been shown to cause any physiological response. However, if the liquid should contact either the skin or the eyes, there can be serious irritation, burns, tissue damage, and frostbite.

Acute Health Effects

The following acute (short-term) health effects may occur immediately or shortly after exposure to neon:

Skin: Liquid threat only. Redness, pain, swelling, blisters, and severe frostbite.

Eye: Liquid threat only. Causes frostbite, redness, irritation, severe pain, and blurred vision.

Lung: Vapor is non-irritating and therefore poses no threat of damage to the lungs. However, in high concentrations, neon can lead to symptoms of asphyxiation and oxygen deprivation up to and including death.

Chronic Health Effects

The following chronic (long-term) health effects may occur at some time after exposure to neon and can last for months or even years:

Cancer Hazards: According to information presented in the references, exposure to neon does not lead to carcinogenic effects (it is a simple asphyxiant).

Reproductive Hazards: There are no reports to support any claims of reproductive hazards.

Other Chronic Effects: None reported.

Recommended Risk-Reduction Measures

Personnel should avoid direct contact with neon. If another gas cannot be substituted for neon, then *engineering controls* are the most effective method of reducing exposures. The best protection is to separate operations and provide local exhaust ventilation at the site of chemical release. While not always operationally feasible, isolating operations can also reduce exposure. Using respiratory protection is less effective than the controls mentioned above, but is still advisable whenever working with neon. Air-purifying respirators (e.g., the cartridge-type or gas masks) cannot be used with neon. These types of respirators only purify the air and can do nothing to protect in oxygen deficient atmospheres. A powered air-supplied respirator operated in pressure demand or other positive pressure mode, or a self-contained breathing apparatus (SCBA) with full facepiece operated in pressure demand mode are the recommended respiratory protection methods of choice. If a full facepiece is not available, then chemical goggles should be worn to protect the eyes. Whenever a chemical splash hazard exists, a face shield and a rubber or leather apron should be worn. To prevent hand and skin exposures with the liquid, thermal insulated gloves should be used.

Administrative controls should also be in place to minimize the potential for human exposures. These may include written procedures or policies which specify the methods and techniques that will be practiced whenever personnel are to work with neon.

All personnel should receive *training* on the use, hazards, protective measures, emergency actions, and other precautions per 29 CFR 1910.1200 (Hazard Communication), prior to the first assignment in an area where neon is used or stored.

Other methods to reduce exposure to chemicals include:

☑ Where possible, consider installing automatic sensors that warn personnel of dropping oxygen levels and use local exhaust ventilation at the site of chemical release. If local exhaust ventilation or enclosure is not used, respiratory protection should be mandatory.

☑ Always ensure that proper protective clothing is worn when using chemical substances.

☑ Safety shower and eyewash stations should be readily available in work areas where neon is used or stored.

☑ Hazard warning information should be posted in the work area. In addition, as part of an on-going education and training program, all information on the health and safety hazards of neon should be communicated to all potentially exposed workers.

RISK ASSESSMENT: ENVIRONMENT

General Assessment

Even though neon is a naturally occurring element that is present in the environment, the environment is at risk of exposure during transportation, storage, disposal, or destruction of neon. In almost every scenario, the threat of environmental exposure is contingent upon the proper handling of the chemical substance. Accidental releases, large or small, may cause possible contamination of the surrounding environmental mediums (mostly air but, depending upon the nature of the spill, water and soil may be affected).

Neon is a non-flammable, non-combustible, non-explosive, and chemically inactive, inert gas. It presents no fire hazard. However, it can cause other gases to condense and solidify. It causes the condensation of oxygen from the air on non-insulated surfaces. Liquid oxygen is a fire hazard. Ice formation inside valves, vents, or critical components can be hazardous to equipment operation and to personnel. Also, because of its ability to displace oxygen, neon requires special consideration during any emergency involving a leak or release of neon gas or liquid especially in confined or enclosed spaces (such as sewers). Neon can enter the environment through industrial discharges, unchecked venting to air, or through spills of the liquid. Since liquids will most likely vaporize rapidly, water and soil contamination are possible but improbable.

Acute Ecological Effects

Acute (short-term) toxic effects may include the death of animals, birds, or fish, and death or low growth rate in plants. Acute effects are seen 2 to 4 days after animals or plants are exposed to neon. Insufficient data are available on the short-term effects of neon on aquatic life, plants, birds, or land animals. However, as a relatively non-toxic, simple asphyxiant, it might be assumed that any acute ecological toxicity will be relatively low.

Chronic Ecological Effects

Chronic (long-term) toxic effects may include shortened life span, reproductive problems, lower fertility, and changes in appearance or behavior in exposed animals. These effects can be seen long after first exposure(s) to toxic chemicals. Insufficient data are available to evaluate the long-term effects of neon to aquatic life, plants, birds, or land animals. However, as a relatively non-toxic, simple asphyxiant, it might be assumed that any chronic ecological toxicity will be relatively low.

Water Solubility

Neon is only slightly soluble in water. Concentrations of 1 milligram or less may mix with a liter of water. It is normally a gas at room temperature and pressure and solubility may not be a concern.

⌛ *Persistence in the Environment*

Neon is non-persistent in water, with a half-life of less than 2 days. The half-life of a pollutant is the amount of time it takes for one half of the chemical to be degraded. Virtually all of neon will eventually end up in the air.

Bioaccumulation in Aquatic Organisms

Some substances increase in concentration, or *bioaccumulate,* in living organisms as they breathe contaminated air, drink contaminated water, or eat contaminated food. These chemicals can become concentrated in the tissues and internal organs of animals as well as humans.

Neon is not expected to accumulate in the edible tissues of fish.

Recommended Risk-Reduction Measures

Proper training of all transporters will reduce the likelihood of a mishap or accident resulting in a leak or release to the environment. The correct use of DOT labeling (per the MSDS) on all containers, trucks, and rail cars will enable emergency responders to react properly and quickly to any disaster thereby reducing the potential risk to the environment and to personnel.

Storage of neon should be segregated from other gases to prevent cross-contamination or contact. Buildings designated for storage should be equipped with monitors that provide ample warning (audible and visual) of oxygen deficient atmospheres.

If a spill or release to the environment has occurred, fire department, emergency response, and/or hazardous materials spill personnel should be notified immediately. Cleanup should be attempted only by those trained in proper containment procedures. If applicable contaminated soils should be removed for incineration and replaced with clean soil. When spills or leaks occur, it may be necessary to contact the local and/or state emergency response authorities. A comprehensive emergency response or disaster preparedness/recovery plan should be in place prior to any operations involving the use, transportation, storage, or disposal of neon.

If neon is spilled or leaked, the following specific steps are recommended:

- ☑ Restrict persons not wearing protective clothing from area of spill or leak until cleanup is complete and area can be opened for normal work.
- ☑ Ventilate area of spill or leak.
- ☑ If applicable, stop flow of leaking liquid or gas. If leak source is a cylinder and the leak cannot be stopped in place, remove leaking cylinder to a safe place in the open air, and repair or allow cylinder to empty.
- ☑ Keep neon out of a confined space, such as a sewer, because of the possibility of creating an inert atmosphere.

RISK ASSESSMENT: BUSINESS

General Assessment

Accidents or mishaps involving neon can present a significant threat to business operations. The loss or damage of equipment or facilities can significantly affect fiscal viability. Lawsuits that may result from personnel injury, public exposures, and environmental contamination will also require a serious expenditure of resources. Media attention surrounding an injury, illness, or environmental damage can also result in a loss of profits and loss of current as well as future business.

Recommended Risk-Reduction Measures

Company attorneys, safety and health professionals, and environmental specialists should be involved in the development of any procedures dealing with safety and emergency response. A company official should be pre-designated as a public relations officer with specific training in dealing with the press. Corporate plans and policies should be developed, approved, and implemented long before any need for such arises.

REFERENCES

Ahlbom, A. 1993. *Biostatistics for Epidemiologists.* New York: CRC Press/Lewis Publishers.

American Conference of Governmental Industrial Hygienists. 1988. *Documentation of the Threshold Limit Values and Biological Exposure Indices,* 5th Edition (with updates). Cincinnati: ACGIH

Anthony, C. P., and N. J. Kolthoff. 1971. *Textbook of Anatomy and Physiology.* St. Louis: C. V. Mosby Company.

Asante-Duah, D. K. 1993. *Hazardous Waste Risk Assessment.* New York: CRC Press/Lewis Publishers.

Bloom, A. D. and F. J. DeSerres. 1995. *Ecotoxicity and Human Health.* New York: CRC Press/Lewis Publishers.

Brusick, D. J. 1994. *Methods for Genetic Risk Assessment.* New York: CRC Press/Lewis Publishers.

Bueche, J. R. 1972. *Principles of Physics,* 2nd Edition. New York: McGraw-Hill.

Burrell, R., D. K. Flaherty, and L. J. Sauers. 1992. *Toxicology of the Immune System.* New York: Van Nostrand Reinhold.

Calabrese, E. J. 1994. *Biological Effects of Low Level Exposures.* New York: CRC Press/Lewis Publishers.

Cockerman, L. G., B. S. Shane. 1994. *Basic Environmental Toxicology.* New York: CRC Press/Lewis Publishers.

Cothern, C. R. 1995. *Handbook for Environmental Risk Decisionmaking.* New York: CRC Press/Lewis Publishers.

Danse, I. R. 1991. *Common Sense Toxics in the Workplace.* New York: Van Nostrand Reinhold.

Environmental Protection Agency. 1983. *Chemical Hazard Information Profiles.* Washington, D.C.: U.S. EPA.

Fodden, C. R., and J. L. Weddell. 1992. *Hazardous Materials Emergency Action Data.* New York: CRC Press/Lewis Publishers.

Forseberg, K., and L. H. Keith. 1993. *Chemical; Protective Clothing Permeation and Degradation Database.* Chelsea, MI: Lewis Publishers.

Gardner, R. W. 1994. *Chemical Intolerance.* New York: CRC Press/Lewis Publishers.

Hallenbeck, W. H. 1993. *Quantitative Risk Assessment for Environmental and Occupational Health,* 2nd Edition. New York: CRC Press/Lewis Publishers.

Hodgson, E., R. B. Mailman, and J. E. Chambers. 1988. *Dictionary of Toxicology.* London: The MacMillian Company.

Hosty, J. W., and P. Foster. 1990. *A Practical Guide to Chemical Spill Response.* New York: Van Nostrand Reinhold.

Howard, P. H. 1995. *Handbook of Environmental Fate and Exposures Data for Organic Chemicals,* Vol. I-IV. New York: CRC Press/Lewis Publishers.

James, A., and M. Lord. 1992. *Index of Chemical and Physical Data.* New York: Van Nostrand Reinhold.

Jones, F. E. 1994. *Toxic Organic Vapors in the Workplace.* New York: CRC Press/Lewis Publishers.

Kamrin, M. A. 1988. *Toxicology: A Primer on Principles and Applications.* Chelsea, Michigan: Lewis Publishers.

Keith, L. H. , M. T. Johnston, C. E. Hudak, and M. Conoley. 1992. *National Toxicology Program's Gloves+ Database.* Chelsea, MI: Lewis Publishers.

Keith, L. H. and M. M. Walker. 1995. *Handbook of Air Toxics.* New York: CRC Press/Lewis Publishers.

Keith, L. H. and D. B. Walters. (Editors) 1992. *National Toxicology Program's Chemical Database, Volumes I-VII.* Chelsea, MI: Lewis Publishers.

Landis, W. G., M. H. Yu. 1995. *Introduction to Environmental Toxicology.* New York: CRC Press/Lewis Publishers.

Langerman, N. 1994 *Precautionary Labels for Chemical Containers.* New York: CRC Press/Lewis Publishers.

Lauwerys, R. R., P. Hoet. 1993. *Industrial Chemical Exposure.* New York: CRC Press/Lewis Publishers.

Lewis, R. J., Sr. 1993. *Hawley's Condensed Chemical Dictionary,* 12th Edition. New York: Van Nostrand Reinhold.

Lewis, R. J., Sr. 1992. *Sax's Dangerous Properties of Industrial Materials,* Volumes 1, 2, and 3, 8th Edition. New York: Van Nostrand Reinhold.

Lide, D. R. 1995. *Handbook of Chemistry and Physics,* 76th Edition. New York: CRC Press/Lewis Publishers.

Manahan, S. E. 1994. *Environmental Chemistry,* 6th Edition. New York: CRC Press/Lewis Publishers.

Manahan, S. E. 1992. *Toxicological Chemistry,* 2nd Edition. New York: CRC Press/Lewis Publishers.

Mellström, G. A., J. E. Wahlberg, and H. I. Maibach. 1994. *Protective Gloves for Occupational Use.* New York: CRC Press/Lewis Publishers.

Meyer, E. 1990. *Chemistry of Hazardous Materials.* Englewood Cliffs, NJ: Prentice Hall.

Patnaik, P. 1992. *A Comprehensive Guide to the Hazardous Properties of Chemical Substances.* New York: Van Nostrand Reinhold.

Philp, R. B. 1995. *Environmental Hazards and Human Health.* New York: CRC Press/Lewis Publishers.

Proctor, N. H., and J. P. Hughes. 1991. *Chemical Hazards of the Workplace.* New York: Van Nostrand Reinhold.

Quigley, D. R. 1994. *Handbook of Emergency Chemical Management.* New York: CRC Press/Lewis Publishers.

Rea, W. J. 1995. *Chemical Sensitivity,* Vol. I-III. New York: CRC Press/Lewis Publishers.

Richardson, M. 1995. *Dictionary of Substance and Their Effects,* Vol. I-VI. New York: CRC Press/Lewis Publishers.

Sacarello, H. L. A. 1994. *The Comprehensive Handbook of Hazardous Materials.* New York: CRC Press/Lewis Publishers.

Sherman, J. D. 1988. *Chemical Exposure and Disease.* New York: Van Nostrand Reinhold.

Talbott, E., G. Craun. 1995. *An Introduction to Environmental Epidemiology.* New York: CRC Press/Lewis Publishers.

Torkelson, T. R. V. K. Rowe. 1981. *Patty's Industrial Hygiene and Toxicology,* Vol. 2B. New York: Wiley-Interscience.

U.S. Department of Health and Human Services, National Institute for Occupational Safety and Health. 1994. *NIOSH Pocket Guide to Chemical Hazards.* Washington, D.C.: U.S. Government Printing Office.

Witorsch, P. and S. V. Spagnolo. 1994. *Air Pollution and Lung Disease in Adults.* New York: CRC Press/Lewis Publishers.

MATERIAL SAFETY DATA SHEET

CHEMICAL NAME	CRC PRESS/LEWIS PUBLISHERS DISCLAIMER:
NICKEL	THE INFORMATION AND RECOMMENDATIONS PRESENTED HEREIN ARE BASED ON SOURCES BELIEVED TO BE RELIABLE. CRC MAKES NO REPRESENTATION ON ITS COMPLETENESS OR ACCURACY. IT IS THE RESPONSIBILITY OF THE USER TO DETERMINE THE CHEMICAL'S SUITABILITY FOR ITS INTENDED USE, THE CHEMICAL'S SAFE USE, AND THE CHEMICAL'S PROPER DISPOSAL. NO REPRESENTATIONS AND/OR WARRANTIES, EITHER EXPRESSED OR IMPLIED, OF THE MERCHANTABILITY OR FITNESS FOR A PARTICULAR PURPOSE, OR OF ANY OTHER NATURE, ARE MADE WITH RESPECT TO THE INFORMATION PROVIDED IN THIS MSDS OR TO THE CHEMICAL TO WHICH INFORMATION MAY REFER. CRC NEITHER ASSUMES NOR AUTHORIZES ANY OTHER PERSON TO ASSUME FOR IT, ANY OTHER ADDITIONAL RESPONSIBILITY OR LIABILITY FOR THE USE OF, OR RELIANCE UPON, THIS INFORMATION.

HAZARD WARNING INFORMATION

HEALTH	FIRE	REACTIVE	OTHER	DEGREE OF HAZARD	COLOR CODING	OTHER CODES
2	4	0		0 = Minimum Hazard 1 = Slight Hazard 2 = Moderate Hazard 3 = Serious Hazard 4 = Severe Hazard	HEALTH = BLUE FIRE = RED REACTIVITY = YELLOW OTHER = WHITE	OX = Oxidizer ACID = Acid ALK = Alkali COR = Corrosive W = Use No Water

SECTION I - GENERAL INFORMATION

Characterization	RCRA Number	EPA Class
Metal	**None**	**Not Applicable**

DOT Proper Shipping Name **Nickel Catalyst, Dry**	Chemical Abstract Service (CAS) Number **7440-02-0**
DOT Hazard Class and Label Requirements **Flammable Solid (dust)**	DOT Emergency Guide Code **53**
DOT Identification Number **UN 2881**	Chemical Symbol **Ni**

Synonyms
Raney alloy; Raney nickel; nickel dust; nickel particles; nickel sponge.

SECTION II - HAZARDOUS INGREDIENTS AND IDENTITY INFORMATION

Hazardous Components (specific identity)	OSHA Exposure Criteria	NIOSH Exposure Criteria	Immediately Dangerous to Life and Health (IDLH)	ACGIH Exposure Criteria
Nickel (derivation: Nickel ores are of two types, sulfur and oxide. Sulfur nickel accounts for two-thirds of the world's consumption. Sulfide ores are refined by flotation and roasting to sintered nickel oxide).	**PEL:** **0.1 mg/m^3** **(soluble nickel)** **1 mg/m^3** **(insoluble nickel)** **STEL:** **Not Established**	**REL:** **0.015 mg/m^3** **(inorganic nickel)** **STEL:** **Not Established**	**Not Established**	**TLV:** **1 mg/m^3** **Suspected Human Carcinogen**

SECTION III - PHYSICAL AND CHEMICAL CHARACTERISTICS

Boiling Point **4046°F (2730°C)**	Specific Gravity (H_2O = 1) **8.90 at 77°F (25°C)**
Vapor Pressure (mm Hg) **1 at 3290°F (1810°C)**	Molecular Weight **58.71**
Vapor Density (Air = 1) **Not Found**	Melting Point **2651°F (1455°C)**

Solubility
Insoluble in water.

Appearance and Odor
Malleable, silvery-white, hard, ductile metal.

SECTION IV - FIRE AND EXPLOSION HAZARD DATA

Flash Point (method used) **Not Applicable**	Explosive Limits in Air % by Volume **LEL: Not Determined UEL: Not Determined**
NFPA Classification **Combustible Solid**	Autoignition Temperature **Not Applicable**

Extinguishing Media
Smother fires using suitable dry powder or, if no other choice, use large amounts of water to flood fire.

Special Fire Fighting Procedures
Isolate area and deny entry to non-essential personnel. Nickel dusts can be toxic. Proper self-contained breathing apparatus must be worn. Do NOT release runoff from fire control measures to sewer or waterway. Contain for proper disposal.

Unusual Fire and Explosion Hazards
Powdered nickel or nickel dusts can be highly combustible in proper mixtures with air and may ignite spontaneously. Dust is moderately flammable/explosive by heat, flame, or chemical reaction with powerful oxidizers.

SECTION V - REACTIVITY DATA

Stability		Conditions to Avoid Nickel is normally stable in closed containers under routine conditions of handling and storage. Keep away from incompatible materials, especially strong oxidizers. Avoid generating dusty conditions.
Stable X	Unstable	Incompatibility *(materials to avoid)* Strong oxidizers (fluorine) and oxidants, ammonium nitrate, hydrogen + dioxane, performic acid, selenium, sulfur, ammonia, hydrazine, phosphorus, and titanium + potassium chlorate.
Hazardous Polymerization		Conditions to Avoid Under normal conditions of temperature and pressure, hazardous polymerization of nickel cannot occur.
May Occur	Will Not Occur X	Hazardous Decomposition or By-products Thermal oxidative decomposition of nickel can yield highly toxic nickel carbonyl.

SECTION VI - HEALTH HAZARD DATA

Primary Route(s) of Entry:	Inhalation?	Absorption (skin)?	Ingestion?
	X		X

Health Hazards

INHALATION: Severe lung irritation that may lead to cancer. Other symptoms include irritation of the upper respiratory tract, asthma, cough, rapid breathing, dypsnea, tightness in chest, metal fume fever, pneumonia, conjunctivitis, nausea, and abdominal pain.

SKIN & EYES: Skin and eye contact with powdered dusts can result in contact dermatitis and irritation.

INGESTION: Large doses can cause nausea, vomiting, and diarrhea.

Carcinogenicity	NTP Listed?	IARC Cancer Review Group?	OSHA Regulated?	Target Organs?
Suspected Human Confirmed Animal	5th Annual Report	Group 1	29 CFR 1910.1000 Table Z-1	Respiratory system (nasal cavity), eyes, skin.

Medical Conditions Generally Aggravated by Exposure

Respiratory impairments (bronchitis, asthma); Skin conditions (dermatitis).

Emergency and First-aid Procedures

Eye contact: Flush immediately with water for 15 minutes (minimum); seek medical attention. Skin contact: Remove all contaminated clothing. Wash contaminated skin with soap and water. For inhalation: Remove the person from exposure. Provide respiratory assistance and CPR. Transfer to medical facility. If swallowed: Oral toxicity is relatively low and inducing vomiting is seldom necessary. If intense vomiting develops, treat for fluid replacement and seek medical attention. Never attempt to give anything by mouth to an unconscious or convulsing person.

SECTION VII - PRECAUTIONS FOR SAFE HANDLING AND USE

Steps to be Taken in Case Material is Released or Spilled

Collect powdered material in the most convenient way possible and deposit in sealed containers. Remove all sources of ignition. Avoid generating dusts. Wet mop residue or use HEPA vacuum. Restrict those not involved in cleanup from entering area. Notify appropriate authorities, as required.

Preferred Waste Disposal Method

No citation.

Precautions to be Taken in Handling and Storage

Containers should be protected against physical damage. Store in tightly closed containers in a cool, dry, well-ventilated area away incompatible materials. Storage under an inert gas is recommended, away from heat or direct sunlight.

Other Precautions and Warnings

Personnel must be trained on the hazards of nickel and nickel compounds *before* being assigned to areas where nickel is used, handled, or stored.

SECTION VIII - CONTROL MEASURES AND PERSONAL PROTECTIVE EQUIPMENT

Respiratory Protection *(specify type)*

For exposures near or over 1 mg/m^3: NIOSH-approved respirator with particulate (dust/fume/mist) filters. Otherwise, use a self-contained breathing apparatus in continuous flow mode.

Ventilation

Local exhaust or general mechanical systems recommended.

Protective Gloves	Eye Protection	Other Protective Clothing
Suitable to protect from skin contact	Chemical/Dust Goggles	N/A

Work/Hygiene Practices

Always wash hands thoroughly after working with nickel powders and dusts; never bring food, drink, or smoking materials into vicinity of nickel powders.

NICKEL

Ni

CAS: 7440-02-0

IDENTIFICATION AND TYPICAL USES

Nickel is a malleable, silvery-white, hard, ductile metal. It is metallic element of atomic number 28, group VIII of the periodic table, atomic weight 58.70, with valance 2,4 and five stable isotopes. It is used in alloys (low-alloy steels, stainless steel, copper and brass, permanent magnets, and electrical resistance alloys). Also used in electroplated protective coatings, electroformed coatings, alkaline storage batteries, fuel cell electrodes, as a catalyst in the methanation of fuel gases and the hydrogenation of vegetable oils.

RISK ASSESSMENT: HEALTH

General Assessment

Nickel is a severe human poison by *ingestion* and *inhalation* of its finely divided powders and dusts. Skin contact with the dusts of nickel may result in scratching ("nickel scratch") and scarring. Nickel is a confirmed carcinogen in animals causing lung and nasal tumors. Its carcinogenic potential in humans is considered highly suspect. Nickel has also been shown to cause teratogenic effects in test animals as well as mutations (genetic changes) under experimental conditions.

Inhalation of nickel dusts or fumes can cause severe inflammation and irritation of the upper respiratory tract with nonproductive cough, rapid breathing, dypsnea, and tightness in the chest. The fumes may cause metal fume fever with chills, high fever, and flu-like symptoms that can last up to 24 hours after exposure. There may also be asthma-like symptoms, inflammation of the lungs (noninfectious), conjunctivitis, nausea, vomiting, and abdominal pain.

Ingested nickel is poorly absorbed, and there appears to be no retention of nickel from nutritional sources in individuals with normal kidneys. Dusts of metallic nickel and nickel oxide are not significantly absorbed systemically, although fumes from welding nickel are absorbed through the lungs, producing an increase in nickel levels in plasma and in urine. Symptoms of ingestion (large doses) may include nausea, vomiting, and diarrhea.

Acute Health Effects

The following acute (short-term) health effects may occur immediately or shortly after exposure to nickel:

Skin: Irritation/scratching on contact with dusts. May cause "nickel itch" or, nickel dermatitis.

Eye: Irritation and scratching with possible progression to conjunctivitis.

Lung: Irritation of the nose, throat, and lungs on exposure to powders and dusts. The fumes can cause flu-like conditions ("metal fume fever").

Chronic Health Effects

The following chronic (long-term) health effects may occur at some time after exposure to nickel and can last for months or even years:

Cancer Hazards: According to information presented in the references, nickel is known to cause nasal sinus and lung cancer in animals. It is thought to be able to cause at least the same damage in humans. Mutation data have also been reported and many scientists believe that such chemicals cause a cancer risk in the long-term.

Reproductive Hazard: According to information presented in the references, exposure to nickel and its salts may damage a developing fetus.

Other Chronic Effects: Skin contact can cause allergy. Symptoms include burning, itching, redness and bumps, or other rash-like developments. The rash may spread to other areas and remain for weeks after the exposure has ended. Repeated inhalation exposure can cause chronic pulmonary irritation, chronic thickening of the mucous membranes of the nose, nasal sinusitis, anosmia (loss of the sense of smell), and perforation (holes) in the nasal septum. Lung allergy (asthma) occasionally develops with wheezing and/or tightness in the chest. Long-term exposure to the dusts or fumes may cause carcinoma of the paranasal sinuses, larynx, and lungs. Single high or repeated lower exposures may cause damage to the heart muscle, liver, and/or kidney.

℧ *Recommended Risk-Reduction Measures*

Personnel should avoid direct contact with nickel. It is a confirmed carcinogen and human exposure must be avoided or controlled. *Engineering controls* are always the most effective method of reducing exposures. The best protection, especially when working with fine powders and dusts of nickel, is to enclose operations and/or provide explosion-proof, local exhaust ventilation at the site of powder/dust release. While not always operationally feasible, isolating operations can also reduce exposure. Using respiratory protection is less effective than the controls mentioned above, but is still advisable whenever working with or around nickel. A self-contained breathing apparatus (SCBA) with full facepiece operated in pressure demand is the best respiratory protection. However, for exposures near or over the 1 mg/m^3 PEL for dusts, a respirator with a dust/mist/fume filter may suffice. If a full facepiece respirator is not available, then dust goggles or safety glasses with side shields attached should be worn to protect the eyes. To prevent hand and skin exposures, suitable gloves should be used.

Administrative controls should also be in place to minimize the potential for human exposures. These may include written procedures or policies which specify the methods and techniques that will be practiced whenever personnel are to work with nickel.

All personnel should receive *training* on the use, hazards, protective measures, emergency actions, and other precautions per 29 CFR 1910.1200 (Hazard Communication), prior to the first assignment in an area where nickel is used or stored. Before beginning work with nickel, and at regular intervals thereafter (e.g., annually), the following medical tests are recommended:

- ☑ Lung function tests (may be normal if person is not experiencing symptoms at the time of the test).
- ☑ Urinalysis or plasma (blood) tests for nickel. Normal nickel content in unexposed population should be less than 10 micrograms per liter.

If symptoms develop or overexposure is suspected, the following additional tests may be useful:

- ☑ Daily urine nickel monitoring for several days following last exposure. Persons with urine nickel over 100 micrograms per liter require medical attention.
- ☑ Consider chest X-ray after acute overexposure.
- ☑ Evaluation by a qualified allergist with special testing and careful consideration of exposure history (may help diagnose allergy).
- ☑ Liver and kidney function tests.

Also, because smoking can cause heart disease, emphysema, lung cancer, and other respiratory disorders, smokers who are exposed to nickel may experience symptoms more quickly and more pronounced than non-smokers under the same conditions of exposure. Persons who are allergic to nickel may also react to nickel-coated jewelry, watchbands, and, in some cases, prolonged contact with keys and coins.

Other methods to reduce exposure include:

- ☑ Where possible, enclose operations and use local exhaust ventilation at the site where nickel powder or dust is worked. If local exhaust ventilation or enclosure is not used, respiratory protection should be mandatory.
- ☑ Always ensure that proper protective clothing is worn when working with nickel compounds.
- ☑ Wash thoroughly immediately after exposure to nickel and at the end of the work shift or before eating, drinking, or smoking. Clothing contaminated with nickel dusts or powders should not be taken home.
- ☑ Hazard warning information should be posted in the work area. In addition, as part of an on-going education and training program, all information on the health and safety hazards of nickel should be communicated to all exposed workers.
- ☑ Specific engineering controls are recommended for nickel by NIOSH in Criteria Document: Inorganic Nickel (Publication Number 77-164).

RISK ASSESSMENT: ENVIRONMENT

General Assessment

The environment is at risk of exposure during transportation, storage, disposal, or destruction of nickel. In almost every scenario, the threat of environmental exposure is contingent upon the proper handling of nickel materials, especially the dusts and powders. Accidental contact with a variety of incompatible materials can result in fire, explosion, and possible contamination of the surrounding environmental mediums (water, soil, and, to a lesser degree, air).

Nickel is considered a combustible solid. It is incompatible with oxidizing agents and contact can cause fire or explosion. These characteristics require special consideration during any emergency situation involving a leak or spill of nickel. Should nickel ever come into contact with incompatible substances such as oxidizers, ammonium nitrate, acids, or sulfur compounds either during use, transportation, or storage, the formation of highly explosive commodities is extremely possible.

Elemental nickel is one of the most common metals found in surface waters, occurring naturally due to the weathering of rocks. Other sources of nickel in the environment include the burning of coal and other fossil fuels and discharges from such industries as electroplating and smelting.

Acute Ecological Effects

Acute (short-term) toxic effects may include the death of animals, birds, or fish and death or low growth rate in plants. Acute effects are seen 2 to 4 days after animals or plants are exposed to nickel.

Nickel and its compounds have high acute toxicity to aquatic life. The softer the water, the higher the toxicity. Insufficient data are available on the short-term effects of nickel exposure to plants, birds, or land animals.

Chronic Ecological Effects

Chronic (long-term) toxic effects may include shortened life span, reproductive problems, lower fertility, and changes in appearance or behavior in exposed animals. These effects can be seen long after first exposure(s) to toxic chemicals.

Nickel and its compounds have high chronic toxicity in aquatic life. The softer the water, the greater the toxicity. Insufficient data are available on the long-term effects of nickel to plants, birds, or land animals.

Water Solubility

Nickel and its compounds have water solubilities ranging from extremely low (negligible) to relatively high.

Persistence in the Environment

Nickel is highly persistent in water, with a half-life greater than 200 days. The half-life of a pollutant is the amount of time it takes for one half of the chemical to be degraded.

Bioaccumulation in Aquatic Organisms

Some substances increase in concentration, or *bioaccumulate,* in living organisms as they breathe contaminated air, drink contaminated water, or eat contaminated food. These chemicals can become concentrated in the tissues and internal organs of animals as well as humans.

The concentration of nickel and its compounds found in fish tissues is expected to be somewhat higher than the average concentration of nickel in the water from which the fish was taken.

Recommended Risk-Reduction Measures

Proper training of all transporters will reduce the likelihood of a mishap or accident resulting in the contamination of the environment. The correct use of DOT labeling (per the MSDS) on all containers, trucks, and rail cars will enable emergency responders to react properly and quickly to any disaster thereby reducing the potential risk to the environment and to personnel.

Storage of nickel should be segregated from incompatible chemicals to minimize the risk of cross-contamination or contact. Buildings designated for storage should be equipped with appropriate fire protection systems (alarms, sprinklers, emergency lighting, portable extinguishers).

If a spill or leak to the environment has occurred, fire department, emergency response and/or hazardous materials spill personnel should be notified immediately. Cleanup should be attempted only by those trained in proper spill containment procedures. Contaminated soils should be removed and replaced with clean soil. When environmental contamination occurs, the local and/or state emergency response authorities may require notification. A comprehensive emergency response or disaster preparedness/recovery plan should be in place prior to any operations involving the use, transportation, storage, or disposal of nickel.

If nickel is spilled or leaked, the following specific steps are recommended:

- ☑ Restrict persons not wearing protective clothing from area of spill or leak until cleanup is complete and area can be opened for normal work.
- ☑ Remove all ignition sources and ventilate area of nickel powder spill or leak.
- ☑ Collect powdered materials in the safest manner possible and deposit in sealed containers. Do NOT dry sweep residues (generates dusts). Damp mop or use HEPA vacuum instead.
- ☑ It may be necessary to dispose of nickel as a hazardous waste. The responsible state agency or the regional office of the federal Environmental Protection Agency (EPA) should be contacted for specific recommendations.

RISK ASSESSMENT: BUSINESS

General Assessment

Accidents or mishaps involving nickel can present a significant threat to business operations. The loss or damage of equipment or facilities due to explosion or fire can significantly affect fiscal viability. Lawsuits that may result from personnel injury/death, public exposures, and/or environmental contamination will also require a serious expenditure of resources.

Media attention surrounding an injury, death, or environmental damage can also result in a loss of profits and loss of current as well as future business. Always remember that any time the terms "cancer," "carcinogen," or "reproductive hazard" are used, public emotion, hysteria, and ignorance can run equally high. This must be carefully considered whenever developing or implementing public relations policies.

Recommended Risk-Reduction Measures

Company attorneys, safety and health professionals, and environmental specialists should be involved in the development of any procedures dealing with safety or emergency response. A company official should be pre-designated as a public relations officer with specific training in dealing with the press. Corporate plans and policies should be developed, approved, and implemented long before any need for such arises.

REFERENCES

Ahlbom, A. 1993. *Biostatistics for Epidemiologists.* New York: CRC Press/Lewis Publishers.

American Conference of Governmental Industrial Hygienists. 1988. *Documentation of the Threshold Limit Values and Biological Exposure Indices,* 5th Edition (with updates). Cincinnati: ACGIH

Bloom, A. D. and F. J. DeSerres. 1995. *Ecotoxicity and Human Health.* New York: CRC Press/Lewis Publishers.

Brusick, D. J. 1994. *Methods for Genetic Risk Assessment.* New York: CRC Press/Lewis Publishers.

Burrell, R., D. K. Flaherty, and Leonard J. Sauers. 1992. *Toxicology of the Immune System.* New York: Van Nostrand Reinhold.

Calabrese, E. J. 1994. *Biological Effects of Low Level Exposures.* New York: CRC Press/Lewis Publishers.

Cockerman, L. G., B. S. Shane. 1994. *Basic Environmental Toxicology.* New York: CRC Press/Lewis Publishers.

Cothern, C. R. 1995. *Handbook for Environmental Risk Decisionmaking.* New York: CRC Press/Lewis Publishers.

Danse, I. R. 1991. *Common Sense Toxics in the Workplace.* New York: Van Nostrand Reinhold.

Gardner, R. W. 1994. *Chemical Intolerance.* New York: CRC Press/Lewis Publishers.

Hallenbeck, W. H. 1993. *Quantitative Risk Assessment for Environmental and Occupational Health,* 2nd Edition. New York: CRC Press/Lewis Publishers.

Hosty, J. W., P. Foster. 1990. *A Practical Guide to Chemical Spill Response.* New York: Van Nostrand Reinhold.

Howard, P. H., et. al. 1991. *Handbook of Environmental Degradation Rates.* New York: CRC Press/Lewis Publishers.

Jones, F. E. 1994. *Toxic Organic Vapors in the Workplace.* New York: CRC Press/Lewis Publishers.

Keith, L. H. and M. M. Walker. 1995. *Handbook of Air Toxics.* New York: CRC Press/Lewis Publishers.

Landis, W. G., M. H. Yu. 1995. *Introduction to Environmental Toxicology.* New York: CRC Press/Lewis Publishers.

Langerman, N. 1994 *Precautionary Labels for Chemical Containers.* New York: CRC Press/Lewis Publishers.

Lauwerys, R. R., P. Hoet. 1993. *Industrial Chemical Exposure.* New York: CRC Press/Lewis Publishers.

Lewis, R. J., Sr. 1993. *Hawley's Condensed Chemical Dictionary,* 12th Edition. New York: Van Nostrand Reinhold.

Lewis, R. J., Sr. 1992. *Sax's Dangerous Properties of Industrial Materials,* Volumes 1, 2, and 3, 8th Edition. New York: Van Nostrand Reinhold.

Manahan, S. E. 1992. *Toxicological Chemistry,* 2nd Edition. New York: CRC Press/Lewis Publishers.

Meyer, E. 1990. *Chemistry of Hazardous Materials.* Englewood Cliffs, NJ: Prentice Hall.

Mycock, J. C., J. D. McKenna, and L. Theodore. 1995. *Handbook of Air Pollution Control Engineering and Technology.* New York: CRC Press/Lewis Publishers.

U.S. Department of Health and Human Services (DHHS), National Institute for Occupational Health (NIOSH). 1977. *Criteria Document: Inorganic Nickel.* Publication Number 77-164. Washington, D.C.; U.S. Government Printing Office.

MATERIAL SAFETY DATA SHEET

CHEMICAL NAME

NICKEL AMMONIUM SULFATE

HAZARD WARNING INFORMATION

HEALTH	FIRE	REACTIVE	OTHER	DEGREE OF HAZARD	COLOR CODING	OTHER CODES
Not Found	Not Found	Not Found		0 = Minimum Hazard 1 = Slight Hazard 2 = Moderate Hazard 3 = Serious Hazard 4 = Severe Hazard	HEALTH = BLUE FIRE = RED REACTIVITY = YELLOW OTHER = WHITE	OX = Oxidizer ACID = Acid ALK = Alkali COR = Corrosive W = Use No Water

SECTION I - GENERAL INFORMATION

Characterization: **Metallic Salt**

RCRA Number: **None**

EPA Class: **Not Applicable**

DOT Proper Shipping Name: **Nickel ammonium sulfate**

Chemical Abstract Service (CAS) Number: **15699-18-0**

DOT Hazard Class and Label Requirements: **Flammable Solid (dust)**

DOT Emergency Guide Code: **53**

DOT Identification Number: **NA 9138**

Chemical Symbol: **$NiSO_4 \cdot (NH_4)_2SO_4 \cdot 6H_2O$**

Synonyms
Nickel salts; double ammonium nickel sulfate; ammonium disulfatonickelate(II).

SECTION II - HAZARDOUS INGREDIENTS AND IDENTITY INFORMATION

Hazardous Components (specific identity)	OSHA Exposure Criteria	NIOSH Exposure Criteria	Immediately Dangerous to Life and Health (IDLH)	ACGIH Exposure Criteria
Nickel ammonium sulfate (derivation: An aqueous solution of nickel sulfate is acidified with sulfuric acid, then an aqueous solution of ammonium sulfate is added. When concentrated, the crystals of nickel ammonium sulfate separate out).	**PEL:** 0.1 mg/m^3 (soluble nickel) 1 mg/m^3 (insoluble nickel) **STEL:** Not Established	**REL:** 0.015 mg/m^3 (inorganic nickel) **STEL:** Not Established	Not Established	**TLV:** 1 mg/m^3 Suspected Human Carcinogen

SECTION III - PHYSICAL AND CHEMICAL CHARACTERISTICS

Boiling Point: **Decomposes**

Specific Gravity (H_2O = 1): **1.929 at 77°F (25°C)**

Vapor Pressure (mm Hg): **Not Found**

Molecular Weight: **286.93**

Vapor Density (Air = 1): **Not Found**

Melting Point: **Not Found**

Solubility
Soluble in water, slightly soluble in ammonium sulfate solution. Insoluble in alcohol.

Appearance and Odor
Green, odorless, crystalline powder.

SECTION IV - FIRE AND EXPLOSION HAZARD DATA

Flash Point (method used): **Not Applicable**

Explosive Limits in Air % by Volume: **LEL: Not Determined UEL: Not Determined**

NFPA Classification: **Combustible Solid**

Autoignition Temperature: **Not Applicable**

Extinguishing Media
Smother fires using suitable dry powder or, if no other choice, use large amounts of water to flood fire.

Special Fire Fighting Procedures
Isolate area and deny entry to non-essential personnel. Nickel ammonium sulfate dusts can be toxic. Proper self-contained breathing apparatus must be worn. Do NOT release runoff from fire control measures to sewer or waterway. Contain for proper disposal.

Unusual Fire and Explosion Hazards
Powdered nickel ammonium sulfate or nickel ammonium sulfate dusts can be highly combustible in proper mixtures with air and may ignite spontaneously. Dust is moderately flammable/explosive by heat, flame, or chemical reaction with powerful oxidizers.

SECTION V - REACTIVITY DATA

Stability		Conditions to Avoid Nickel ammonium sulfate is normally stable in closed containers under routine conditions of handling and storage. Keep away from incompatible materials. Avoid generating dusty conditions.
Stable X	Unstable	Incompatibility *(materials to avoid)* Strong oxidizers (fluorine) and oxidants, ammonium nitrate, hydrogen + dioxane, performic acid, selenium, sulfur, ammonia, hydrazine, phosphorus, and titanium + potassium chlorate.
Hazardous Polymerization		Conditions to Avoid Under normal conditions of temperature and pressure, hazardous polymerization of nickel ammonium sulfate cannot occur.
May Occur	Will Not Occur X	Hazardous Decomposition or By-products Thermal oxidative decomposition of nickel ammonium sulfate can yield highly toxic oxides of sulfur, nitrogen, and nickel.

SECTION VI - HEALTH HAZARD DATA

Primary Route(s) of Entry:	Inhalation?	Absorption (skin)?	Ingestion?
	X		X

Health Hazards

INHALATION: Severe lung irritation that may lead to cancer. Other symptoms include irritation of the upper respiratory tract, asthma, cough, rapid breathing, dypsnea, tightness in chest, metal fume fever, pneumonia, pulmonary edema, conjunctivitis, and abdominal pain.

SKIN & EYES: Skin and eye contact with powdered dusts can result contact dermatitis and irritation.

INGESTION: Large doses can cause nausea, vomiting, and diarrhea.

Carcinogenicity	NTP Listed?	IARC Cancer Review Group?	OSHA Regulated?	Target Organs?
Suspected Human Confirmed Animal	5th Annual Report	Group 1	29 CFR 1910.1000 Table Z-1 (nickel)	Respiratory system (nasal cavity), eyes, skin.

Medical Conditions Generally Aggravated by Exposure

Respiratory impairments (bronchitis, asthma); Skin conditions (dermatitis).

Emergency and First-aid Procedures

Eye contact: Flush immediately with water for 15 minutes (minimum); seek medical attention. Skin contact: Remove all contaminated clothing. Wash contaminated skin with soap and water. For inhalation: Remove the person from exposure. Provide respiratory assistance and CPR. Transfer to medical facility. Observe 48 hours for lung effects. If swallowed: Oral toxicity is relatively low and inducing vomiting is seldom necessary. If intense vomiting develops, treat for fluid replacement and seek medical attention. Never attempt to give anything by mouth to an unconscious or convulsing person.

SECTION VII - PRECAUTIONS FOR SAFE HANDLING AND USE

Steps to be Taken in Case Material is Released or Spilled

Collect powdered material in the most convenient way possible and deposit in sealed containers. Remove all sources of ignition. Avoid generating dusts. Wet mop residue or use HEPA vacuum. Restrict those not involved in cleanup from entering area. Notify appropriate authorities, as required.

Preferred Waste Disposal Method

No citation.

Precautions to be Taken in Handling and Storage

Containers should be protected against physical damage. Store in tightly closed containers in a cool, dry, well-ventilated area away incompatible materials. Storage under an inert gas is recommended, away from heat or direct sunlight.

Other Precautions and Warnings

Personnel must be trained on the hazards of nickel ammonium sulfate and nickel ammonium sulfate compounds *before* being assigned to areas where nickel ammonium sulfate is used, handled, or stored.

SECTION VIII - CONTROL MEASURES AND PERSONAL PROTECTIVE EQUIPMENT

Respiratory Protection *(specify type)*

For exposures near or over 1 mg/m^3: NIOSH-approved respirator with particulate (dust/fume/mist) filters. Otherwise, use a self-contained breathing apparatus in continuous flow mode.

Ventilation

Local exhaust or general mechanical systems recommended.

Protective Gloves	Eye Protection	Other Protective Clothing
Suitable to protect from skin contact	Chemical/Dust Goggles	N/A

Work/Hygiene Practices

Always wash hands thoroughly after working with nickel ammonium sulfate powders and dusts; never bring food, drink, or smoking materials into vicinity of nickel ammonium sulfate powders.

NICKEL AMMONIUM SULFATE

$NiSO_4 \cdot (NH_4)_2SO_4 \cdot 6H_2O$ CAS: 15699-18-0

IDENTIFICATION AND TYPICAL USES

Nickel ammonium sulfate is a green, odorless, crystalline powder. It is used primarily as a nickel electrolyte for electroplating.

RISK ASSESSMENT: HEALTH

General Assessment

Nickel ammonium sulfate is a severe human poison by *ingestion* and *inhalation* of its finely divided powders and dusts. Skin contact with the dusts of nickel salts may result in scratching ("nickel ammonium sulfate scratch") and scarring. Nickel ammonium sulfate is a confirmed carcinogen in animals causing lung and nasal tumors. Its carcinogenic potential in humans is considered highly suspect. Nickel compounds have also been shown to cause teratogenic effects in test animals as well as mutations (genetic changes) under experimental conditions. It is not known if the ammonium sulfate will have this effect, but caution is certainly warranted.

Inhalation of nickel ammonium sulfate dusts or fumes can cause severe inflammation and irritation of the upper respiratory tract with nonproductive cough, rapid breathing, dypsnea, and tightness in the chest. The fumes of nickel and its compounds may cause metal fume fever with chills, high fever, and flu-like symptoms that can last up to 24 hours after exposure. There may also be asthma-like symptoms, inflammation of the lungs (noninfectious), conjunctivitis, nausea, vomiting, and abdominal pain. Symptoms of pulmonary edema, such as cough and shortness of breath, may be delayed up to 48 hours thereby creating a false sense of security with regard to health exposure risk.

Ingested nickel is poorly absorbed, and there appears to be no retention of the metal from nutritional sources in individuals with normal kidneys. Dusts of metallic nickel and nickel oxide are not significantly absorbed systemically, although fumes from welding nickel is absorbed through the lungs, producing an increase in nickel levels in plasma and in urine. Symptoms of ingestion (large doses) may include nausea, vomiting, and diarrhea.

Acute Health Effects

The following acute (short-term) health effects may occur immediately or shortly after exposure to nickel ammonium sulfate:

Skin: Irritation/scratching on contact with dusts. May cause "nickel itch" or nickel dermatitis.

Eye: Irritation and scratching with possible progression to conjunctivitis.

Lung: Irritation of the nose, throat, and lungs with the possible development of pulmonary edema. The fumes can cause flu-like conditions ("metal fume fever").

Chronic Health Effects

The following chronic (long-term) health effects may occur at some time after exposure to nickel ammonium sulfate and can last for months or even years:

Cancer Hazards: According to information presented in the references, nickel is known to cause nasal sinus and lung cancer in animals. It is thought to be able to cause at least the same damage in humans. Mutation data have also been reported and many scientists believe that such chemicals cause a cancer risk in the long-term. It is not known if the ammonium sulfate specifically will cause cancer. However, caution is warranted in use, handling, and storage.

Reproductive Hazard: According to information presented in the references, exposure to nickel and its salts may damage a developing fetus and reduce fertility in males and females.

Other Chronic Effects: Skin contact can cause allergy. Symptoms include burning, itching, redness and bumps, or other rash-like developments. The rash may spread to other areas and remain for weeks after the exposure has ended. Repeated inhalation exposure can cause chronic pulmonary irritation, chronic thickening of the mucous membranes of the nose, nasal sinusitis, anosmia (loss of the sense of smell), and perforation (holes) in the nasal septum. Lung allergy (asthma) occasionally develops with wheezing and/or tightness in the chest. Long-term exposure to the dusts or fumes may cause carcinoma of the paranasal sinuses, larynx, and lungs. Single high or repeated lower exposures may cause damage to the heart muscle, liver, and/or kidney.

Recommended Risk-Reduction Measures

Personnel should avoid direct contact with nickel ammonium sulfate. It is a confirmed animal carcinogen and human exposure must be avoided or controlled. *Engineering controls* are always the most effective method of reducing exposures. The best protection, especially when working with fine powders and dusts of nickel ammonium sulfate, is to enclose operations and/or provide explosion-proof, local exhaust ventilation at the site of powder/dust release. While not always operationally feasible, isolating operations can also reduce exposure. Using respiratory protection is less effective than the controls mentioned above, but is still advisable whenever working with or around nickel ammonium sulfate. A self-contained breathing apparatus (SCBA) with full facepiece operated in pressure demand is the best respiratory protection. However, for exposures near or over the 1 mg/m^3 PEL for dusts, a respirator with a dust/mist/fume filter may suffice. If a full facepiece respirator is not available, then dust goggles or safety glasses with side shields attached should be worn to protect the eyes. To prevent hand and skin exposures, suitable gloves should be used.

Administrative controls should also be in place to minimize the potential for human exposures. These may include written procedures or policies which specify the methods and techniques that will be practiced whenever personnel are to work with nickel ammonium sulfate.

All personnel should receive *training* on the use, hazards, protective measures, emergency actions, and other precautions per 29 CFR 1910.1200 (Hazard Communication), prior to the first assignment in an area where nickel ammonium sulfate is used or stored.

Before personnel beginning working with nickel ammonium sulfate, and at regular intervals thereafter (e.g., annually), the following medical tests are recommended to assist in exposure management:

- ☑ Lung function tests (may be normal if person is not experiencing symptoms at the time of the test).
- ☑ Urinalysis or plasma (blood) tests for nickel. Normal nickel content in unexposed population should be less than 10 micrograms per liter.

If symptoms develop or overexposure is suspected, the following additional tests may be useful:

- ☑ Daily urine nickel monitoring for several days following last exposure. Persons with urine nickel over 100 micrograms per liter require medical attention.
- ☑ Consider chest X-ray after acute overexposure (may be negative if taken immediately after exposure due to delayed onset of pulmonary edema).
- ☑ Evaluation by a qualified allergist with special testing and careful consideration of exposure history (may help diagnose allergy).
- ☑ Liver and kidney function tests.

Also, because smoking can cause heart disease, emphysema, lung cancer, and other respiratory disorders, smokers who are exposed to nickel ammonium sulfate may experience symptoms more quickly and more pronounced than non-smokers under the same conditions of exposure. Persons who are allergic to nickel ammonium sulfate may also react to nickel ammonium sulfate-coated jewelry, watchbands, and, in some cases, prolonged contact with keys and coins.

Other methods to reduce exposure include:

- ☑ Where possible, enclose operations and use local exhaust ventilation at the site where nickel ammonium sulfate powder or dust is worked. If local exhaust ventilation or enclosure is not used, respiratory protection should be mandatory.
- ☑ Always ensure that proper protective clothing is worn when working with nickel ammonium sulfate compounds.
- ☑ Wash thoroughly immediately after exposure to nickel ammonium sulfate and at the end of the work shift or before eating, drinking, or smoking. Clothing contaminated with nickel ammonium sulfate dusts or powders should not be taken home.
- ☑ Hazard warning information should be posted in the work area. In addition, as part of an on-going education and training program, all information on the health and safety hazards of nickel ammonium

sulfate should be communicated to all exposed and potentially exposed workers.

☑ Specific engineering controls are recommended for this chemical by NIOSH. Refer to NIOSH Criteria Document: Inorganic Nickel (Publication Number 77-164).

RISK ASSESSMENT: ENVIRONMENT

General Assessment

The environment is at risk of exposure during transportation, storage, disposal, or destruction of nickel ammonium sulfate. In almost every scenario, the threat of environmental exposure is contingent upon the proper handling of nickel ammonium sulfate materials, especially the dusts and powders. Accidental contact with a variety of incompatible materials can result in fire, explosion, and possible contamination of the surrounding environmental mediums (water, soil, and, to a lesser degree, air).

Nickel ammonium sulfate is considered a combustible solid. It is incompatible with oxidizing agents and acids and contact can cause fire or explosion. These characteristics require special consideration during any emergency situation involving a leak or spill of nickel ammonium sulfate. Should nickel ammonium sulfate ever come into contact with incompatible substances such as oxidizers, ammonium nitrate, acids, or sulfur compounds either during use, transportation, or storage, the formation of highly explosive commodities is extremely possible.

Elemental nickel is one of the most common metals found in surface waters, occurring naturally due to the weathering of rocks. Other sources of nickel in the environment include the burning of coal and other fossil fuels and discharges from such industries as electroplating and smelting.

Acute Ecological Effects

Acute (short-term) toxic effects may include the death of animals, birds, or fish and death or low growth rate in plants. Acute effects are seen 2 to 4 days after animals or plants are exposed to nickel ammonium sulfate.

Nickel and its compounds have high acute toxicity to aquatic life. The softer the water, the higher the toxicity. Insufficient data are available on the short-term effects of nickel exposure to plants, birds, or land animals.

Chronic Ecological Effects

Chronic (long-term) toxic effects may include shortened life span, reproductive problems, lower fertility, and changes in appearance or behavior in exposed animals. These effects can be seen long after first exposure(s) to toxic chemicals.

Nickel and its compounds have high chronic toxicity in aquatic life. The softer the water, the greater the toxicity. Insufficient data are available on the long-term effects of nickel ammonium sulfate to plants, birds, or land animals.

Water Solubility

Nickel and its compounds have water solubilities ranging from extremely low (negligible) to relatively high (almost miscible).

Persistence in the Environment

Nickel is highly persistent in water, with a half-life greater than 200 days. The half-life of a pollutant is the amount of time it takes for one half of the chemical to be degraded.

Bioaccumulation in Aquatic Organisms

Some substances increase in concentration, or *bioaccumulate,* in living organisms as they breathe contaminated air, drink contaminated water, or eat contaminated food. These chemicals can become concentrated in the tissues and internal organs of animals as well as humans.

The concentration of nickel and its compounds found in fish tissues is expected to be somewhat higher than the average concentration of nickel in the water from which the fish was taken.

Recommended Risk-Reduction Measures

Proper training of all transporters will reduce the likelihood of a mishap or accident resulting in the contamination of the environment. The correct use of DOT labeling (per the MSDS) on all containers, trucks, and rail cars will enable emergency responders to react properly and quickly to any disaster thereby reducing the potential risk to the environment and to personnel.

Storage of nickel ammonium sulfate should be segregated from incompatible chemicals to minimize the risk of cross-contamination or contact. Buildings designated for storage should be equipped with appropriate fire protection systems (alarms, sprinklers, emergency lighting, portable extinguishers).

If a spill or leak to the environment has occurred, fire department, emergency response and/or hazardous materials spill personnel should be notified immediately. Cleanup should be attempted only by those trained in proper spill containment procedures. Contaminated soils should be removed and replaced with clean soil. When environmental contamination occurs, the local and/or state emergency response authorities may require notification. A comprehensive emergency response or disaster preparedness/recovery plan should be in place prior to any operations involving the use, transportation, storage, or disposal of nickel ammonium sulfate. If nickel ammonium sulfate is spilled or leaked, the following specific steps are recommended:

☑ Restrict persons not wearing protective clothing from area of spill or leak until cleanup is complete and area can be opened for normal work.

☑ Remove all ignition sources and ventilate area of nickel ammonium sulfate powder spill or leak.

☑ Collect powdered materials in the safest manner possible and deposit in sealed containers. Do NOT dry sweep residues (generates dusts). Damp mop or use HEPA vacuum instead.

☑ It may be necessary to dispose of nickel ammonium sulfate as a hazardous waste. The responsible state agency or the regional office of the federal Environmental Protection Agency (EPA) should be contacted for specific recommendations.

RISK ASSESSMENT: BUSINESS

General Assessment

Accidents or mishaps involving nickel ammonium sulfate can present a significant threat to business operations. The loss or damage of equipment or facilities due to explosion or fire can significantly affect fiscal viability. Lawsuits that may result from personnel injury/death, public exposures, and/or environmental contamination will also require a serious expenditure of resources. Media attention surrounding an injury, death, or environmental damage can also result in a loss of profits and loss of current as well as future business. Always remember that any time the terms "cancer," "carcinogen," or "reproductive hazard" are used, public emotion, hysteria, and ignorance can run equally high. This must be carefully considered whenever developing or implementing public relations policies.

Recommended Risk-Reduction Measures

Company attorneys, safety and health professionals, and environmental specialists should be involved in the development of any procedures dealing with safety or emergency response. A company official should be pre-designated as a public relations officer with specific training in dealing with the press. Corporate plans and policies should be developed, approved, and implemented long before any need for such arises.

REFERENCES

American Conference of Governmental Industrial Hygienists. 1988. *Documentation of the Threshold Limit Values and Biological Exposure Indices,* 5th Edition (with updates). Cincinnati: ACGIH

Burrell, R., D. K. Flaherty, and Leonard J. Sauers. 1992. *Toxicology of the Immune System.* New York: Van Nostrand Reinhold.

Calabrese, E. J. 1994. *Biological Effects of Low Level Exposures.* New York: CRC Press/Lewis Publishers.

Cothern, C. R. 1995. *Handbook for Environmental Risk Decisionmaking.* New York: CRC Press/Lewis Publishers.

Gardner, R. W. 1994. *Chemical Intolerance.* New York: CRC Press/Lewis Publishers.

Hallenbeck, W. H. 1993. *Quantitative Risk Assessment for Environmental and Occupational Health,* 2nd Edition. New York: CRC Press/Lewis Publishers.

Hosty, J. W., P. Foster. 1990. *A Practical Guide to Chemical Spill Response.* New York: Van Nostrand Reinhold.

Howard, P. H., et. al. 1991. *Handbook of Environmental Degradation Rates.* New York: CRC Press/Lewis Publishers.

Landis, W. G., M. H. Yu. 1995. *Introduction to Environmental Toxicology.* New York: CRC Press/Lewis Publishers.

Lauwerys, R. R., P. Hoet. 1993. *Industrial Chemical Exposure.* New York: CRC Press/Lewis Publishers.

Lewis, R. J., Sr. 1992. *Sax's Dangerous Properties of Industrial Materials,* Volumes 1, 2, and 3, 8th Edition. New York: Van Nostrand Reinhold.

Manahan, S. E. 1992. *Toxicological Chemistry,* 2nd Edition. New York: CRC Press/Lewis Publishers.

Meyer, E. 1990. *Chemistry of Hazardous Materials.* Englewood Cliffs, NJ: Prentice Hall.

Sherman, J. D. 1988. *Chemical Exposure and Disease.* New York: Van Nostrand Reinhold.

Torkelson, T. R. V. K. Rowe. 1981. *Patty's Industrial Hygiene and Toxicology,* Vol. 2B. New York: Wiley-Interscience.

U.S. Department of Health and Human Services, National Institute for Occupational Safety and Health. 1977. *NIOSH Criteria Document: Inorganic Nickel.* Publication Number 77-164. Washington, D.C.: U.S. Government Printing Office.

MATERIAL SAFETY DATA SHEET

CHEMICAL NAME	
NICKEL CARBONYL	**CRC PRESS/LEWIS PUBLISHERS DISCLAIMER:** THE INFORMATION AND RECOMMENDATIONS PRESENTED HEREIN ARE BASED ON SOURCES BELIEVED TO BE RELIABLE. CRC MAKES NO REPRESENTATION ON ITS COMPLETENESS OR ACCURACY. IT IS THE RESPONSIBILITY OF THE USER TO DETERMINE THE CHEMICAL'S SUITABILITY FOR ITS INTENDED USE, THE CHEMICAL'S SAFE USE, AND THE CHEMICAL'S PROPER DISPOSAL. NO REPRESENTATIONS AND/OR WARRANTIES, EITHER EXPRESSED OR IMPLIED, OF THE MERCHANTABILITY OR FITNESS FOR A PARTICULAR PURPOSE, OR OF ANY OTHER NATURE, ARE MADE WITH RESPECT TO THE INFORMATION PROVIDED IN THIS MSDS OR TO THE CHEMICAL TO WHICH INFORMATION MAY REFER. CRC NEITHER ASSUMES NOR AUTHORIZES ANY OTHER PERSON TO ASSUME FOR IT, ANY OTHER ADDITIONAL RESPONSIBILITY OR LIABILITY FOR THE USE OF, OR RELIANCE UPON, THIS INFORMATION.

HAZARD WARNING INFORMATION

HEALTH	FIRE	REACTIVE	OTHER	DEGREE OF HAZARD	COLOR CODING	OTHER CODES
4	**3**	**3**		0 = Minimum Hazard 1 = Slight Hazard 2 = Moderate Hazard 3 = Serious Hazard 4 = Severe Hazard	HEALTH = BLUE FIRE = RED REACTIVITY = YELLOW OTHER = WHITE	OX = Oxidizer ACID = Acid ALK = Alkali COR = Corrosive ~~W~~ = Use No Water

SECTION I - GENERAL INFORMATION

Characterization **Metal Carbonyl**	RCRA Number **P073**	EPA Class **Acute Hazardous Waste**
DOT Proper Shipping Name **Nickel Carbonyl**	Chemical Abstract Service (CAS) Number **13463-39-3**	
DOT Hazard Class and Label Requirements **Flammable Gas: Poison B, Flammable Liquid**	DOT Emergency Guide Code **57**	
DOT Identification Number **UN 1259**	Chemical Formula **$Ni(CO)_4$**	

Synonyms
Nickel tetracarbonyl; tetracarbonylnickel.

SECTION II - HAZARDOUS INGREDIENTS AND IDENTITY INFORMATION

Hazardous Components (specific identity)	OSHA Exposure Criteria	NIOSH Exposure Criteria	Immediately Dangerous to Life and Health (IDLH)	ACGIH Exposure Criteria
Nickel carbonyl (derivation: By passing carbon monoxide gas over finely divided nickel). **1 ppm = 7.10 mg/m^3**	**PEL (8-hour):** **0.001 ppm** **0.007 mg/m^3** **(as nickel)** **STEL:** **Not Established**	**REL (8-hour):** **0.001 ppm** **0.007 mg/m^3** **(as nickel)** **Possible Human Carcinogen**	**2 ppm**	**TLV:** **0.05 ppm** **0.35 mg/m^3** **(as nickel)** **Suspected Human Carcinogen**

SECTION III - PHYSICAL AND CHEMICAL CHARACTERISTICS

Boiling Point **110°F (43°C)**	Specific Gravity (H_2O = 1) **1.32**
Vapor Pressure (mm Hg) **315 at 68°F (20°C)**	Molecular Weight **170.7**
Vapor Density (Air = 1) **5.45**	Melting Point **-2.7°F (-19°C)**

Solubility
Very slightly soluble (0.05%). Soluble in concentrated nitric acid, alcohol, and most organic solvents.

Appearance and Odor
Colorless to yellow liquid with a musty odor. A gas above 110°F (43°C). Odor Threshold = 30 ppb.

SECTION IV - FIRE AND EXPLOSION HAZARD DATA

Flash Point (method used) **-4°F (-20°C)**	Explosive Limits in Air % by Volume **LEL: 2%** **UEL: Not Determined**
NFPA Classification **Class IB Flammable Liquid**	Autoignition Temperature **Not Determined**

Extinguishing Media
Use water, dry chemical, alcohol foam, are carbon dioxide. Exercise extreme caution.

Special Fire Fighting Procedures
Poisonous gases are produced in fire. Wear full protective clothing and self-contained breathing apparatus (SCBA). Isolate area and deny entry. Use a water spray to keep fire-exposed containers cool long after the fire has been extinguished. Remain clear of smoke, water fallout and water runoff.

Unusual Fire and Explosion Hazards
Nickel carbonyl is an extremely dangerous explosive and flammable liquid (or gas) that rapidly evaporates at ambient temperatures. The vapors are more than five times heavier than air.

SECTION V - REACTIVITY DATA

Stability		Conditions to Avoid: Nickel carbonyl is normally stable in closed, pressurized containers under routine conditions of handling and storage. Keep away from heat and sources of ignition.
Stable	Unstable	Incompatibility *(materials to avoid)*: Incompatible with oxidizers (especially bromine, chlorine, and fluorine), nitric acid, air, oxygen, and flammable materials. In air, it will form an explosive peroxide that will decompose and ignite.
X		
Hazardous Polymerization		Conditions to Avoid: Under normal conditions of temperature and pressure, hazardous polymerization of nickel carbonyl is not known to occur.
May Occur	Will Not Occur	Hazardous Decomposition or By-products: When heated to decomposition, nickel carbonyl will emit toxic and acrid fumes and gases, including carbon monoxide and oxides of nickel and nitrogen.
	X	

SECTION VI - HEALTH HAZARD DATA

Primary Route(s) of Entry:	Inhalation?	Absorption (skin)?	Ingestion?
	X	X	X

Health Hazards

INHALATION: Delayed pneumonitis with chest pain, extremely difficult breathing, rapid respiration, intense exhaustion, cyanosis, loss of consciousness, and death. Other symptoms include headache, vertigo, sweating, weakness, coughing, dypsnea, nausea, stomach pain, and vomiting. It may lead to cancer of the lungs and sinuses.

SKIN & EYES: An irritant to the eyes and skin. Exposure may cause contact dermatitis. Absorption is possible, with symptoms similar to that of inhalation. Eye contact causes irritation.

INGESTION: A poison with CNS and gastrointestinal effects with nausea and vomiting.

Carcinogenicity	NTP Listed?	IARC Cancer Review Group?	OSHA Regulated?	Target Organs?
Suspected Human Confirmed Animal	5th Annual Report	Group 1	29 CFR 1910.1000 Table Z-1	Lungs, sinus, respiratory system, CNS, skin.

Medical Conditions Generally Aggravated by Exposure

Existing problems with the upper respiratory tract may be aggravated by exposure.

Emergency and First-aid Procedures

Eye contact: Flush immediately with water for 15 minutes (minimum); seek medical attention. Skin contact: Remove all contaminated clothing. Immediately wash area with large amounts of soap and water. For inhalation: Remove the person from exposure. Provide respiratory assistance and CPR. Transfer to medical facility. If swallowed: Unlikely, but possible. Contact poison control center for assistance. Never attempt to give a convulsing or unconscious person anything by mouth.

SECTION VII - PRECAUTIONS FOR SAFE HANDLING AND USE

Steps to be Taken in Case Material is Released or Spilled

Restrict those not involved in cleanup activities from area. Remove all ignition sources and provide explosion-proof ventilation if possible. Absorb spills using vermiculite or other similar material and place in sealed drums from disposal. Use non-sparking tools during cleanup operations.

Preferred Waste Disposal Method

No citation.

Precautions to be Taken in Handling and Storage

Store in tightly closed containers in cool, well-ventilated area and protect from physical damage. Use external or detached storage facility. Keep fire, heat, and ignition sources away (including direct sunlight). Store cylinders upright and secure them properly (never secure around the neck of the cylinder).

Other Precautions and Warnings

Bulk storage of nickel carbonyl is not recommended.

SECTION VIII - CONTROL MEASURES AND PERSONAL PROTECTIVE EQUIPMENT

Respiratory Protection *(specify type)*

Exposure levels are low and difficult to accurately monitor. For the best protection, use a full facepiece supplied-air respirator or a self-contained breathing apparatus (SCBA) operated in positive pressure or other continuous flow mode.

Ventilation

Local exhaust or general mechanical systems recommended.

Protective Gloves	Eye Protection	Other Protective Clothing
Impervious Material	Chemical Goggles or Face Mask	Protective Suit or Clothing

Work/Hygiene Practices

Always wash hands thoroughly after using chemical; never bring food, drink, or smoking materials into vicinity of chemicals.

NICKEL CARBONYL

$NiCO_4$

CAS: 13463-39-3

IDENTIFICATION AND TYPICAL USES

Nickel carbonyl is a colorless to yellow liquid with a characteristic musty odor. It can also appear as a gas above 110°F (43°C). It is used in the production of high-purity nickel powder by the Mond process. Also used in continuous nickel coatings on steel and other metals.

RISK ASSESSMENT: HEALTH

General Assessment

Nickel carbonyl is a human poison by *inhalation* and by dermal *absorption.* While *ingestion* is an unlikely exposure route because nickel carbonyl will vaporize rapidly at ambient temperatures, it is still possible. Because it is lipid soluble, it can easily cross biological membranes (such as the lung alveolus, blood-brain barrier, and placenta barrier). It is a confirmed carcinogen in animals and is likely to cause the same effects in humans. Tumors of the lung and nasal cavity have been well documented in experimental animal studies. Human mutation data have also been reported and nickel carbonyl is an experimental teratogen causing damage to the developing fetus in test animals.

Inhalation of nickel carbonyl vapors causes depression of the central nervous system (CNS) with symptoms of somnolence, fever, headache, dizziness (vertigo), sweating, weakness, and nausea. Pulmonary effects include delayed pneumonitis, coughing, dypsnea, pain in or behind sternum, irritation, congestion and hemorrhage of the lungs, delayed pulmonary edema (fluid in the lungs), tachycardia (rapid pulse), cyanosis, increased body temperature, hyperpnea, stomach pain, and vomiting. Symptoms of pulmonary edema can be delayed up to 48 hours, thereby creating a false sense of security with regard to health exposure risk. There may also be liver or brain affects on exposure to nickel carbonyl vapors.

Skin contact is likely to cause localized irritation and redness with dermal absorption possible. If absorption occurs, the symptoms noted for inhalation are likely to follow. Mixed exposures (e.g., inhalation and absorption) can therefore be extremely dangerous.

Acute Health Effects

The following acute (short-term) health effects may occur immediately or shortly after exposure to nickel carbonyl:

Skin: Irritation, rash, and possible contact dermatitis. Dermal absorption is possible.

Eye: Nickel carbonyl particles can irritate the eyes and may cause disturbances to visual acuity

Lung: Irritation of the vapors can irritate the nose, mouth, and throat. High levels can irritate the lungs, causing a buildup of fluid (pulmonary edema), which is a medical emergency and can be fatal.

CNS: Symptoms include vertigo, headache, somnolence, nausea, and vomiting.

Chronic Health Effects

The following chronic (long-term) health effects may occur at some time after exposure to nickel carbonyl and can last for months or even years:

Cancer Hazards: Nickel carbonyl is a probable human carcinogen. There is limited evidence in the references that exposure can cause lung and nasal sinus cancer in humans and it has been shown to cause lung cancer in test animals. Human mutation data have also been reported and many scientists believe that such chemicals pose a cancer risk in the long-term.

Reproductive Hazard: According to information available in the references, nickel carbonyl has been shown to cause teratogenic effects in animal studies.

Other Chronic Effects: Permanent lung damage may occur after a single high exposure or repeated exposures to low concentration of nickel carbonyl. Chronic exposure may lead to a skin allergy in some individuals. The allergy may begin as an itch followed by a rash or redness with small bumps. Once such an allergy develops, even the slightest future exposures can cause a reappearance. The rash may persist for weeks after the exposure has stopped with gradual improvement over time (3-4 weeks). High or repeated exposures may cause damage to the heart muscle, liver and/or kidney.

❶ *Recommended Risk-Reduction Measures*

Personnel should avoid direct contact with nickel carbonyl since it is considered a probable human carcinogen by more than one recognized authority (the ACGIH, NIOSH, the NTP, and the IARC). Prudent risk management dictates that it be handled and treated as though it were a confirmed human carcinogen. If a less toxic chemical cannot be substituted for a hazardous substance, then *engineering controls* are the most effective method of reducing exposures.

The best protection is to enclose operations and/or provide local exhaust ventilation at the site of chemical release. While not always operationally feasible, isolating operations can also reduce exposure. Using respiratory protection is less effective than the controls mentioned above, but is still advisable whenever working with or around nickel carbonyl. The permissible exposure limit (PEL) is extremely low and difficult to accurately monitor (0.001 ppm). For the best protection, use a supplied-air respirator equipped with a full facepiece and operated in positive pressure mode or an MSHA/NIOSH-approved self-contained breathing apparatus (SCBA) with full facepiece operated in pressure demand mode. If a full facepiece is not available, then chemical splash goggles should be worn to protect the eyes. Whenever a chemical dust hazard exists, a face shield and a rubber apron should be worn. To prevent hand and skin exposures, impervious rubber gloves should be used. Glove manufacturers should be contacted and permeation studies obtained *before* final glove selections are made.

Administrative controls should also be in place to minimize the potential for human exposures. These may include written procedures or policies which specify the methods and techniques that will be practiced whenever personnel are to work with nickel carbonyl. A regulated, controlled work area should be established where nickel compounds are used, handled, or stored.

All personnel should receive *training* on the use, hazards, protective measures, emergency actions, and other precautions per 29 CFR 1910.1200 (Hazard Communication), prior to the first assignment in an area where nickel carbonyl is used or stored.

Before beginning employment and at regular intervals thereafter (e.g., annually), for those with frequent or potentially high exposures (half the PEL or greater, or significant skin contact), the following medical tests are recommended:

- ☑ Lung function tests (may be normal if person is not experiencing symptoms at the time of the test).
- ☑ Urinalysis or plasma (blood) tests for nickel. Normal nickel content in unexposed population should be less than 10 micrograms per liter.

If symptoms develop or overexposure is suspected, the following additional tests may be useful:

- ☑ Daily urine nickel monitoring for several days following last exposure. Persons with urine nickel ammonium sulfate over 100 micrograms per liter require medical attention.
- ☑ Consider chest X-ray after acute overexposure (may be negative if taken immediately after exposure due to delayed onset of pulmonary edema).
- ☑ Evaluation by a qualified allergist with special testing and careful consideration of exposure history (may help diagnose allergy).
- ☑ Liver and kidney function tests.

Also, because smoking can cause heart disease, emphysema, lung cancer, and other respiratory disorders, smokers who are exposed to nickel ammonium sulfate may experience symptoms more quickly and more pronounced than non-smokers under the same conditions of exposure. Persons who are allergic to nickel ammonium sulfate may also react to nickel ammonium sulfate-coated jewelry, watchbands, and, in some cases, prolonged contact with keys and coins.

Other methods to reduce exposure include:

- ☑ Where possible, enclose operations and use local exhaust ventilation at the site of chemical release. If local exhaust ventilation or enclosure is not used, respiratory protection should be mandatory.
- ☑ Always ensure that proper protective clothing is worn when using chemical substances and that personnel have been trained on its use, care, and maintenance.

- ☑ Wash thoroughly immediately after exposure to nickel carbonyl and at the end of the work shift or before eating, drinking, or smoking.
- ☑ Hazard warning information should be posted in the work area. In addition, as part of an on-going education and training program, all information on the health and safety hazards of nickel carbonyl should be communicated to all exposed workers.
- ☑ Where possible, automatically transfer nickel carbonyl from drums or other storage containers to process containers.
- ☑ Specific engineering controls are recommended for nickel by NIOSH. Refer to NIOSH Criteria Document: Inorganic Nickel (Publication Number 77-164).

RISK ASSESSMENT: ENVIRONMENT

General Assessment

The environment is at risk of exposure during transportation, storage, disposal, or destruction of nickel carbonyl. In almost every scenario, the threat of environmental exposure is contingent upon the proper handling of the chemical substance. Accidental spills, large or small, can result in fire, explosion, and possible contamination of the surrounding environmental mediums (water, soil, and air).

Nickel carbonyl is considered an extremely flammable liquid. It is considered a Class IB flammable liquid per OSHA 29 CFR 1910.106. Its extremely low flash point and relatively low boiling point present an extremely dangerous fire and explosion hazard. It is also incompatible with acids and oxidizers and contact can cause fire or explosion. Its vapors are more than five times heavier than air. They can travel for long distances, settle in low-lying areas, and, if contact with an ignition source is made, there can be flashback to the chemical's source with resulting fire or explosion. The liquid can also explode when heated in closed containers. These characteristics require special consideration during any emergency situation involving a leak or spill of nickel carbonyl.

Nickel is one of the most common metals found in surface waters, due to the weathering of rocks. Other sources of nickel in the environment include the burning of coal and other fossil fuels and discharges from such industries as electroplating and smelting.

Acute Ecological Effects

Acute (short-term) toxic effects may include the death of animals, birds, or fish and death or low growth rate in plants. Acute effects are seen 2 to 4 days after animals or plants are exposed to nickel ammonium sulfate.

Nickel and its compounds have high acute toxicity to aquatic life. The softer the water, the higher the toxicity. Insufficient data are available on the short-term effects of nickel exposure to plants, birds, or land animals.

Chronic Ecological Effects

Chronic (long-term) toxic effects may include shortened life span, reproductive problems, lower fertility, and changes in appearance or behavior in exposed animals. These effects can be seen long after first exposure(s) to toxic chemicals.

Nickel and its compounds have high chronic toxicity in aquatic life. The softer the water, the greater the toxicity. Insufficient data are available on the long-term effects of nickel ammonium sulfate to plants, birds, or land animals.

Water Solubility

Nickel and its compounds have water solubilities ranging from extremely low (negligible) to relatively high (almost miscible).

Persistence in the Environment

Nickel is highly persistent in water, with a half-life greater than 200 days. The half-life of a pollutant is the amount of time it takes for one half of the chemical to be degraded.

Bioaccumulation in Aquatic Organisms

Some substances increase in concentration, or *bioaccumulate,* in living organisms as they breathe contaminated air, drink contaminated water, or eat contaminated food. These chemicals can become concentrated in the tissues and internal organs of animals as well as humans.

The concentration of nickel and its compounds found in fish tissues is expected to be somewhat higher than the average concentration of nickel in the water from which the fish was taken.

Recommended Risk-Reduction Measures

Proper training of all transporters will reduce the likelihood of a mishap or accidents resulting in a leak or spill to the environment. The correct use of DOT labeling (per the MSDS) on all containers, trucks, and rail cars will enable emergency responders to react properly and quickly to any disaster thereby reducing the potential risk to the environment and to personnel.

Storage of nickel compounds should be segregated from incompatible chemicals to minimize the risk of cross-contamination or contact.

If a spill or leak to the environment has occurred, fire department, emergency response and/or hazardous materials spill personnel should be notified immediately. Cleanup should be attempted only by those trained in proper spill containment procedures using non-sparking tools. Contaminated soils should be removed and replaced with clean soil. If nickel carbonyl should contact the water table, aquifer, or navigable waterway, time is of the essence. It is nearly insoluble in water and, therefore, total containment and remediation may be possible. When such spills occur, the local and/or state emergency response authorities must be notified. A comprehensive emergency response or disaster preparedness/recovery plan should be in place prior to any operations involving the use, transportation, storage, or disposal of nickel carbonyl. If nickel carbonyl is spilled or leaked, the following specific steps are recommended:

- ☑ Restrict persons not wearing protective clothing from area of spill or leak until cleanup is complete and area can be opened for normal work.
- ☑ Absorb liquids using vermiculite or other similar material and deposit in sealed drums for disposal. Use non-sparking tools.
- ☑ It may be necessary to dispose of nickel carbonyl as a hazardous waste. The responsible state agency or the regional office of the federal Environmental Protection Agency (EPA) should be contacted for specific recommendations.

RISK ASSESSMENT: BUSINESS

General Assessment

Accidents or mishaps involving nickel carbonyl can present a significant threat to business operations. The loss or damage of equipment or facilities can significantly affect fiscal viability. Lawsuits that may result from personnel injury/death, public exposures, and/or environmental contamination will also require a serious expenditure of resources. Media attention surrounding an injury, death, or environmental damage can also result in a loss of profits and loss of current as well as future business. Always remember that any time the terms “cancer,” “carcinogen,” or "reproductive hazard" are used, public emotion, hysteria, ignorance, and anxiety can run equally high. This must be considered when dealing with the public following an incident involving nickel carbonyl.

Recommended Risk-Reduction Measures

Company attorneys, safety and health professionals, and environmental specialists should be involved in the development of any procedures dealing with safety or emergency response. A company official should be pre-designated as a public relations officer with specific training in dealing with the press. Corporate plans and policies should be developed, approved, and implemented long before any need for such arises.

REFERENCES

American Conference of Governmental Industrial Hygienists. 1988. *Documentation of the Threshold Limit Values and Biological Exposure Indices*, 5th Edition (with updates). Cincinnati: ACGIH

Brusick, D. J. 1994. *Methods for Genetic Risk Assessment.* New York: CRC Press/Lewis Publishers.

Cockerman, L. G., B. S. Shane. 1994. *Basic Environmental Toxicology.* New York: CRC Press/Lewis Publishers.

Gardner, R. W. 1994. *Chemical Intolerance.* New York: CRC Press/Lewis Publishers.

Howard, P. H., et. al. 1991. *Handbook of Environmental Degradation Rates.* New York: CRC Press/Lewis Publishers.

Keith, L. H. and M. M. Walker. 1995. *Handbook of Air Toxics.* New York: CRC Press/Lewis Publishers.

Lewis, R. J., Sr. 1992. *Sax's Dangerous Properties of Industrial Materials,* Volumes 1, 2, and 3, 8th Edition. New York: Van Nostrand Reinhold.

Meyer, E. 1990. *Chemistry of Hazardous Materials.* Englewood Cliffs, NJ: Prentice Hall.

U.S. Department of Health and Human Services (DHHS), National Institute for Occupational Health (NIOHS). 1994. *NIOSH Pocket Guide to Hazardous Chemicals.* Publication Number 94-116. Washington, D.C.; U.S. Government Printing Office.

U.S. Department of Health and Human Services (DHHS), National Institute for Occupational Health (NIOSH). 1977. *Criteria Document: Inorganic Nickel.* Publication Number 77-164. Washington, D.C.; U.S. Government Printing Office.

MATERIAL SAFETY DATA SHEET

CHEMICAL NAME	
NICKEL SULFATE	**CRC PRESS/LEWIS PUBLISHERS DISCLAIMER:** THE INFORMATION AND RECOMMENDATIONS PRESENTED HEREIN ARE BASED ON SOURCES BELIEVED TO BE RELIABLE. CRC MAKES NO REPRESENTATION ON ITS COMPLETENESS OR ACCURACY. IT IS THE RESPONSIBILITY OF THE USER TO DETERMINE THE CHEMICAL'S SUITABILITY FOR ITS INTENDED USE, THE CHEMICAL'S SAFE USE, AND THE CHEMICAL'S PROPER DISPOSAL. NO REPRESENTATIONS AND/OR WARRANTIES, EITHER EXPRESSED OR IMPLIED, OF THE MERCHANTABILITY OR FITNESS FOR A PARTICULAR PURPOSE, OR OF ANY OTHER NATURE, ARE MADE WITH RESPECT TO THE INFORMATION PROVIDED IN THIS MSDS OR TO THE CHEMICAL TO WHICH INFORMATION MAY REFER. CRC NEITHER ASSUMES NOR AUTHORIZES ANY OTHER PERSON TO ASSUME FOR IT, ANY OTHER ADDITIONAL RESPONSIBILITY OR LIABILITY FOR THE USE OF, OR RELIANCE UPON, THIS INFORMATION.

HAZARD WARNING INFORMATION

HEALTH	FIRE	REACTIVE	OTHER	DEGREE OF HAZARD	COLOR CODING	OTHER CODES
3	0	0		0 = Minimum Hazard 1 = Slight Hazard 2 = Moderate Hazard 3 = Serious Hazard 4 = Severe Hazard	HEALTH = BLUE FIRE = RED REACTIVITY = YELLOW OTHER = WHITE	OX = Oxidizer ACID = Acid ALK = Alkali COR = Corrosive W = Use No Water

SECTION I - GENERAL INFORMATION

Characterization: **Metallic Compound**	RCRA Number: **None**	EPA Class: **Not Applicable**
DOT Proper Shipping Name: **Nickel Sulfate**	Chemical Abstract Service (CAS) Number: **7786-81-4**	
DOT Hazard Class and Label Requirements: **ORM-E; No Label**	DOT Emergency Guide Code: **No Citation**	
DOT Identification Number: **UN 3077**	Chemical Formula: **$NiSO_4$ (anhydrous)**	

Synonyms
Nickelous sulfate; sulfuric acid, nickel (II) salt; nickel salts, single; blue salt.

SECTION II - HAZARDOUS INGREDIENTS AND IDENTITY INFORMATION

Hazardous Components (specific identity)	OSHA Exposure Criteria	NIOSH Exposure Criteria	Immediately Dangerous to Life and Health (IDLH)	ACGIH Exposure Criteria
Nickel sulfate (derivation: By dissolving nickel oxide in sulfuric acid and concentrating the solution to precipitate the hexahydrate which is then heated to the anhydrous form, adding nickel powder to dilute sulfuric acid, or the reaction of nickel carbonyl, sulfuric acid, and oxygen in the gas phase at 212°F/100°C).	**PEL:** 0.1 mg/m^3 (soluble nickel) 1 mg/m^3 (insoluble nickel) **STEL:** Not Established	**REL:** 0.015 mg/m^3 (inorganic nickel) **STEL:** Not Established	Not Established	**TLV:** 0.1 mg/m^3 Suspected Human Carcinogen

SECTION III - PHYSICAL AND CHEMICAL CHARACTERISTICS

Boiling Point: **1544°F (840°C) Loses SO_3**	Specific Gravity (H_2O = 1): **3.68**
Vapor Pressure (mm Hg): **Not Available**	Molecular Weight: **154.77**
Vapor Density (Air = 1): **Not Available**	Melting Point: **Decomposes**

Solubility
Soluble in water. The anhydrous form is not soluble in alcohol or ether. Aqueous forms are soluble in alcohol.

Appearance and Odor
Can appear as either yellow-green crystals (anhydrous form), or blue or emerald-green crystals (aqueous form).

SECTION IV - FIRE AND EXPLOSION HAZARD DATA

Flash Point (method used): **Not Applicable**	Explosive Limits in Air % by Volume: **LEL: Not Applicable** **UEL: Not Applicable**
NFPA Classification: **Non-Combustible**	Autoignition Temperature: **Not Determined**

Extinguishing Media
Use extinguishing agent suitable to surrounding fires. Nickel sulfate will not easily burn.

Special Fire Fighting Procedures
Wear full protective clothing and self-contained breathing apparatus (SCBA). Use a water spray to keep fire-exposed containers cool. Poisonous and toxic fumes are produced in fire, including oxides of sulfur. Move containers from fire area if it can be done without risk.

Unusual Fire and Explosion Hazards
None reported.

SECTION V - REACTIVITY DATA

Stability		Conditions to Avoid Nickel sulfate is normally stable under routine conditions of storage and handling. Avoid contact with incompatible materials.
Stable X	Unstable	Incompatibility *(materials to avoid)* Incompatible with strong acids, such as hydrochloric, sulfuric, and nitric. Also reacts violently in contact with wood and other combustible materials.
Hazardous Polymerization		Conditions to Avoid Under normal conditions of temperature and pressure, hazardous polymerization of nickel sulfate is not known to occur.
May Occur	Will Not Occur X	Hazardous Decomposition or By-products When heated to decomposition, nickel sulfate can emit highly toxic and poisonous fumes and gases, including oxides of sulfur.

SECTION VI - HEALTH HAZARD DATA

Primary Route(s) of Entry:	Inhalation?	Absorption (skin)?	Ingestion?
	X	X	X

Health Hazards

INHALATION: Severe lung irritation that may lead to cancer. Other symptoms include irritation of the upper respiratory tract, asthma, cough, rapid breathing, dypsnea, tightness in chest, metal fume fever, pneumonia, conjunctivitis, nausea, and abdominal pain. Sensitization may lead to conjunctivitis, mild pneumonitis, and other allergic reactions.

SKIN & EYES: Skin and eye contact with powdered dusts can result contact dermatitis and irritation.

INGESTION: Large doses can cause nausea, vomiting, and diarrhea.

Carcinogenicity	NTP Listed?	IARC Cancer Review Group?	OSHA Regulated?	Target Organs?
Suspected Human Confirmed Animal	NTP-2 (nickel)	Group 1 (nickel)	29 CFR 1910.1000 Table Z-1 (nickel)	Respiratory system, skin, eyes, nasal septa.

Medical Conditions Generally Aggravated by Exposure

Persons with existing dermatitis or asthma may react more quickly on exposure to nickel sulfate.

Emergency and First-aid Procedures

Eye contact: Flush immediately with water for 15 minutes (minimum); seek medical attention. Skin contact: Remove all contaminated clothing. Immediately wash area with large amounts of soap and water. For inhalation: Remove the person from exposure. Provide respiratory assistance and CPR. Transfer to medical facility. If swallowed: If victim is conscious, give large amounts of water and induce vomiting. Never attempt to give a convulsing or unconscious person anything by mouth. Provide breathing assistance if necessary. Keep person at rest and quiet. Seek medical attention immediately.

SECTION VII - PRECAUTIONS FOR SAFE HANDLING AND USE

Steps to be Taken in Case Material is Released or Spilled

Collect powdered material in most convenient manner possible and deposit in sealed drum. Keep those not involved in cleanup from entering area. Do NOT dry sweep (creates airborne dusts). Use a vacuum equipped with a high efficiency particulate air (HEPA) filter.

Preferred Waste Disposal Method

No citation.

Precautions to be Taken in Handling and Storage

Store in tightly closed containers in cool, well-ventilated area. Keep away from incompatible materials and prevent physical damage to containers.

Other Precautions and Warnings

Personnel must be trained on the hazards of nickel compounds *before* being assigned to areas where nickel is used, handled, or stored.

SECTION VIII - CONTROL MEASURES AND PERSONAL PROTECTIVE EQUIPMENT

Respiratory Protection *(specify type)*

The PEL is low and difficult to monitor. Use a full facepiece respirator with high efficiency particulate air (HEPA) filter. For best protection, use a supplied-air respirator with a full facepiece, hood, or helmet or a self-contained breathing apparatus (SCBA) operated in positive pressure mode.

Ventilation

Local exhaust or general mechanical systems recommended.

Protective Gloves	Eye Protection	Other Protective Clothing
Impervious Material	Chemical/Dust Goggles or Face Mask	Protective Suit or Clothing

Work/Hygiene Practices

Always wash hands thoroughly after using chemical; never bring food, drink, or smoking materials into vicinity of chemicals.

NICKEL SULFATE

$NiSO_4$

CAS: 7786-81-4 (anhydrous)

IDENTIFICATION AND TYPICAL USES

Nickel sulfate can appear as yellow-green crystals (the anhydrous form), blue or emerald-green crystals (an aqueous form), or green cubic crystals (another aqueous form). It is used in the manufacture of nickel ammonium sulfate, nickel catalysts, in nickel plating, as a mordant in dyeing and printing textiles, in coatings, and in ceramics.

RISK ASSESSMENT: HEALTH

General Assessment

Nickel sulfate is a human poison by *inhalation* and, to a lesser degree, by *ingestion* of its finely divided powders and dusts. Skin contact with the dusts of nickel sulfate may result in scratching ("nickel scratch") and scarring. Nickel and many of its salts are confirmed carcinogens in animals, causing lung and nasal tumors. Its carcinogenic potential in humans is considered highly suspect. Nickel has also been shown to cause teratogenic effects in test animals as well as mutations (genetic changes) under experimental conditions. Nickel sulfate many cause infertility in males.

Inhalation of nickel sulfate dusts can cause severe inflammation and irritation of the upper respiratory tract with nonproductive cough, rapid breathing, dypsnea, and tightness in the chest. Nickel fumes may cause metal fume fever with chills, high fever, and flu-like symptoms that can last up to 24 hours after exposure. There may also be asthma-like symptoms, inflammation of the lungs (noninfectious), conjunctivitis, nausea, vomiting, and abdominal pain.

Ingested nickel is poorly absorbed, and there appears to be no retention of nickel from nutritional sources in individuals with normal kidneys. Dusts of metallic nickel and nickel oxide are not significantly absorbed systemically, although fumes from welding nickel are absorbed through the lungs, producing an increase in nickel levels in plasma and in urine. Symptoms of ingestion (large doses) may include nausea, vomiting, and diarrhea.

Acute Health Effects

The following acute (short-term) health effects may occur immediately or shortly after exposure to nickel sulfate:

Skin: Irritation/scratching on contact with dusts. May cause "nickel itch" or nickel dermatitis.

Eye: Irritation and scratching with possible progression to conjunctivitis.

Lung: Irritation of the nose, throat, and lungs on exposure to powders and dusts. The fumes can cause flu-like conditions ("metal fume fever").

Chronic Health Effects

The following chronic (long-term) health effects may occur at some time after exposure to nickel sulfate and can last for months or even years:

Cancer Hazards: According to information presented in the references, nickel is known to cause nasal sinus and lung cancer in animals. It is thought to be able to cause at least the same damage in humans. Mutation data have been reported for nickel sulfate and many scientists believe that such chemicals cause a cancer risk in the long-term. It is not known if the sulfate in particular will cause cancer, but caution is warranted in this regard.

Reproductive Hazard: According to information presented in the references, exposure to nickel and its salts may damage a developing fetus and/or cause infertility in males.

Other Chronic Effects: Skin contact can cause allergy. Symptoms include burning, itching, redness and bumps, or other rash-like developments. The rash may spread to other areas and remain for weeks after the exposure has ended. Repeated inhalation exposure can cause chronic pulmonary irritation, chronic thickening of the mucous membranes of the nose, nasal sinusitis, anosmia (loss of the sense of smell), and perforation (holes) in the nasal septum. Lung allergy (asthma) occasionally develops with wheezing and/or tightness in the chest. Long-term exposure to the dusts or fumes may cause carcinoma of the paranasal sinuses, larynx, and lungs. Single high or repeated lower exposures may cause damage to the heart muscle, liver, and/or kidney.

Recommended Risk-Reduction Measures

Personnel should avoid direct contact with nickel sulfate. The specific nature of its carcinogenic potential is not clearly understood. However, nickel and many of its salts or confirmed carcinogens and human exposure must be avoided or controlled. *Engineering controls* are always the most effective method of reducing exposures. The best protection, especially when working with fine powders and dusts of nickel sulfate, is to enclose operations and/or provide local exhaust ventilation at the site of powder/dust release. While not always operationally feasible, isolating operations can also reduce exposure. Using respiratory protection is less effective than the controls mentioned above, but is still advisable whenever working with or around nickel sulfate. A self-contained breathing apparatus (SCBA) with full facepiece operated in pressure demand is the best respiratory protection. However, for exposures near or over the 1 mg/m^3 PEL for dusts, a respirator with a dust/mist/fume filter may suffice. If a full facepiece respirator is not available, then dust goggles or safety glasses with side shields attached should be worn to protect the eyes. To prevent hand and skin exposures, suitable gloves should be used.

Administrative controls should also be in place to minimize the potential for human exposures. These may include written procedures or policies which specify the methods and techniques that will be practiced whenever personnel are to work with nickel.

All personnel should receive *training* on the use, hazards, protective measures, emergency actions, and other precautions per 29 CFR 1910.1200 (Hazard Communication), prior to the first assignment in an area where nickel is used or stored. Before beginning work with nickel, and at regular intervals thereafter (e.g., annually), the following medical tests are recommended:

- ☑ Lung function tests (may be normal if person is not experiencing symptoms at the time of the test).
- ☑ Urinalysis or plasma (blood) tests for nickel. Normal nickel content in unexposed population should be less than 10 micrograms per liter.

If symptoms develop or overexposure is suspected, the following additional tests may be useful:

- ☑ Daily urine nickel monitoring for several days following last exposure. Persons with urine nickel over 100 micrograms per liter require medical attention.
- ☑ Consider chest X-ray after acute overexposure (may be negative if taken immediately after exposure due to delayed onset of pulmonary edema).
- ☑ Evaluation by a qualified allergist with special testing and careful consideration of exposure history (may help diagnose allergy).
- ☑ Liver and kidney function tests.

Also, because smoking can cause heart disease, emphysema, lung cancer, and other respiratory disorders, smokers who are exposed to nickel may experience symptoms more quickly and more pronounced than non-smokers under the same conditions of exposure. Persons who are allergic to nickel may also react to nickel-coated jewelry, watchbands, and, in some cases, prolonged contact with keys and coins.

Other methods to reduce exposure include:

- ☑ Where possible, enclose operations and use local exhaust ventilation at the site of chemical release. If local exhaust ventilation or enclosure is not used, respiratory protection should be mandatory.
- ☑ Always ensure that proper protective clothing is worn when using chemical substances and that personnel are trained on its use and care.
- ☑ Wash thoroughly immediately after exposure to nickel sulfate and at the end of the work shift or before eating, drinking, or smoking. Eyewash stations and safety showers should be readily available for emergency use.
- ☑ Hazard warning information should be posted in the work area. In addition, as part of an on-going education and training program, all information on the health and safety hazards of nickel sulfate should be communicated to all exposed and potentially exposed workers.

☑ Specific engineering controls are recommended for nickel by NIOSH. Refer to NIOSH Criteria Document: Inorganic Nickel (Publication Number 77-164).

RISK ASSESSMENT: ENVIRONMENT

General Assessment

The environment is at risk of exposure during transportation, storage, disposal, or destruction of nickel sulfate. In almost every scenario, the threat of environmental exposure is contingent upon the proper handling of the chemical substance. Accidental spills, large or small, can result in fire, explosion, and possible contamination of the surrounding environmental mediums (water, soil, and air).

Nickel sulfate is non-combustible. However, its powders may form explosive mixtures in air especially in confined areas. It is also incompatible with strong acids, such as hydrochloric, sulfuric, and nitric. It can react violently in contact with ordinary combustible materials such as wood or paper. These characteristics require special consideration during any emergency situation involving a leak or spill of nickel sulfate.

Nickel is a naturally occurring element in natural water aquifers. At low concentrations, it is an essential element for both plants and animals. At slightly higher concentrations, it is toxic to aquatic life. The toxicity of nickel and its compounds to aquatic life varies with the physical and chemical conditions of the water. Factors such as water hardness, alkalinity, and pH influence nickel toxicity.

Acute Ecological Effects

Acute (short-term) toxic effects may include the death of animals, birds, or fish and death or low growth rate in plants. Acute effects are seen 2 to 4 days after animals or plants are exposed to nickel sulfate.

Nickel and its compounds have high acute toxicity to aquatic life. The softer the water, the higher the toxicity. Insufficient data are available on the short-term effects of nickel exposure to plants, birds, or land animals.

Chronic Ecological Effects

Chronic (long-term) toxic effects may include shortened life span, reproductive problems, lower fertility, and changes in appearance or behavior in exposed animals. These effects can be seen long after first exposure(s) to toxic chemicals.

Nickel and its compounds have high chronic toxicity in aquatic life. The softer the water, the greater the toxicity. Insufficient data are available on the long-term effects of nickel to plants, birds, or land animals.

Water Solubility

Nickel and its compounds have water solubilities ranging from extremely low (negligible) to relatively high (almost miscible). Nickel sulfate is moderately soluble in water.

Persistence in the Environment

Nickel is highly persistent in water, with a half-life greater than 200 days. The half-life of a pollutant is the amount of time it takes for one half of the chemical to be degraded.

Bioaccumulation in Aquatic Organisms

Some substances increase in concentration, or *bioaccumulate,* in living organisms as they breathe contaminated air, drink contaminated water, or eat contaminated food. These chemicals can become concentrated in the tissues and internal organs of animals as well as humans.

The concentration of nickel found in fish is expected to be somewhat higher than the average concentration of nickel sulfate in the water from which the fish was taken.

Recommended Risk-Reduction Measures

Proper training of all transporters will reduce the likelihood of a mishap or accidents resulting in a leak or spill to the environment. The correct use of DOT labeling (per the MSDS) on all containers, trucks, and rail cars will enable emergency responders to react properly and quickly to any disaster thereby reducing the potential risk to the environment and to personnel.

If a spill or leak to the environment has occurred, fire department, emergency response and/or hazardous materials spill personnel should be notified immediately. Cleanup should be attempted only by those trained in proper spill containment procedures. Contaminated soils may need to be removed and replaced with clean soil. If nickel sulfate should contact the water table, aquifer, or navigable waterway, time is of the essence. Nickel compounds, and especially nickel sulfate, may be moderately to highly soluble in water and, therefore, total containment and remediation may not be entirely possible. When such spills occur, the local and/or state emergency response authorities must be notified. A comprehensive emergency response or disaster preparedness/recovery plan should be in place prior to any operations involving the use, transportation, storage, or disposal of nickel sulfate. If nickel sulfate is spilled or leaked, the following specific steps are recommended:

☑ Restrict persons not wearing protective clothing from area of spill or leak until cleanup is complete and area can be opened for normal work.

☑ Collect powdered materials in most convenient and safe manner possible and place in sealed drums for disposal. Do NOT dry sweep (generates airborne dusts). Use a damp mop or a HEPA vacuum to collect residues.

☑ It may be necessary to dispose of nickel sulfate as a hazardous waste. The responsible state agency or the regional office of the federal Environmental Protection Agency (EPA) should be contacted for specific recommendations.

RISK ASSESSMENT: BUSINESS

General Assessment

Accidents or mishaps involving nickel sulfate can present a significant threat to business operations. The loss or damage of equipment or facilities can significantly affect fiscal viability. Lawsuits that may result from personnel injury/death, public exposures, and/or environmental contamination will also require a serious expenditure of resources. Media attention surrounding an injury, death, or environmental damage can also result in a loss of profits and loss of current as well as future business. Always remember that any time the terms "cancer," "carcinogen," or "reproductive hazard" are used, public emotion, hysteria, ignorance, and anxiety can run equally high. This must be considered when dealing with the public following an incident involving nickel carbonyl.

Recommended Risk-Reduction Measures

Company attorneys, safety and health professionals, and environmental specialists should be involved in the development of any procedures dealing with safety or emergency response. A company official should be pre-designated as a public relations officer with specific training in dealing with the press. Corporate plans and policies should be developed, approved, and implemented long before any need for such arises.

REFERENCES

Asante-Duah, D. K. 1993. *Hazardous Waste Risk Assessment.* New York: CRC Press/Lewis Publishers.

Calabrese, E. J. 1994. *Biological Effects of Low Level Exposures.* New York: CRC Press/Lewis Publishers.

Cothern, C. R., N. P. Ross. 1994. *Environmental Statistics, Assessment, and Forecasting.* New York: CRC Press/Lewis Publishers.

Gardner, R. W. 1994. *Chemical Intolerance.* New York: CRC Press/Lewis Publishers.

Hallenbeck, W. H. 1993. *Quantitative Risk Assessment for Environmental and Occupational Health,* 2nd Edition. New York: CRC Press/Lewis Publishers.

Howard, P. H., et. al. 1991. *Handbook of Environmental Degradation Rates.* New York: CRC Press/Lewis Publishers.

Landis, W. G., M. H. Yu. 1995. *Introduction to Environmental Toxicology.* New York: CRC Press/Lewis Publishers.

Lauwerys, R. R., P. Hoet. 1993. *Industrial Chemical Exposure.* New York: CRC Press/Lewis Publishers.

Lewis, R. J., Sr. 1992. *Sax's Dangerous Properties of Industrial Materials,* Volumes 1, 2, and 3, 8th Edition. New York: Van Nostrand Reinhold.

Manahan, S. E. 1992. *Toxicological Chemistry,* 2nd Edition. New York: CRC Press/Lewis Publishers.

Meyer, E. 1990. *Chemistry of Hazardous Materials.* Englewood Cliffs, NJ: Prentice Hall.

Quigley, D. R. 1994. *Handbook of Emergency Chemical Management.* New York: CRC Press/Lewis Publishers.

Rea, W. J. 1995. *Chemical Sensitivity,* Vol. I-III. New York: CRC Press/Lewis Publishers.

Sacarello, H. L. A. 1994. *The Comprehensive Handbook of Hazardous Materials.* New York: CRC Press/Lewis Publishers.

U.S. Department of Health and Human Services (DHHS), National Institute for Occupational Health (NIOSH). 1977. *Criteria Document: Inorganic Nickel.* Publication Number 77-164. Washington, D.C.; U.S. Government Printing Office.

MATERIAL SAFETY DATA SHEET

CHEMICAL NAME	
NICOTINE	**CRC PRESS/LEWIS PUBLISHERS DISCLAIMER:** THE INFORMATION AND RECOMMENDATIONS PRESENTED HEREIN ARE BASED ON SOURCES BELIEVED TO BE RELIABLE. CRC MAKES NO REPRESENTATION ON ITS COMPLETENESS OR ACCURACY. IT IS THE RESPONSIBILITY OF THE USER TO DETERMINE THE CHEMICAL'S SUITABILITY FOR ITS INTENDED USE, THE CHEMICAL'S SAFE USE, AND THE CHEMICAL'S PROPER DISPOSAL. NO REPRESENTATIONS AND/OR WARRANTIES, EITHER EXPRESSED OR IMPLIED, OF THE MERCHANTABILITY OR FITNESS FOR A PARTICULAR PURPOSE, OR OF ANY OTHER NATURE, ARE MADE WITH RESPECT TO THE INFORMATION PROVIDED IN THIS MSDS OR TO THE CHEMICAL TO WHICH INFORMATION MAY REFER. CRC NEITHER ASSUMES NOR AUTHORIZES ANY OTHER PERSON TO ASSUME FOR IT, ANY OTHER ADDITIONAL RESPONSIBILITY OR LIABILITY FOR THE USE OF, OR RELIANCE UPON, THIS INFORMATION.

NATIONAL FIRE PROTECTION ASSOCIATION (NFPA) LABELING INFORMATION

HEALTH	FIRE	REACTIVE	OTHER	DEGREE OF HAZARD	COLOR CODING	OTHER CODES
4	1	0		0 = Minimum Hazard 1 = Slight Hazard 2 = Moderate Hazard 3 = Serious Hazard 4 = Severe Hazard	HEALTH = BLUE FIRE = RED REACTIVITY = YELLOW OTHER = WHITE	OX = Oxidizer ACID = Acid ALK = Alkali COR = Corrosive W = Use No Water

SECTION I - GENERAL INFORMATION

Characterization	RCRA Number	EPA Class
Alkaloid of Tobacco	**P075**	**Acute Hazardous Waste**

DOT Proper Shipping Name **Nicotine Liquid or Solid**	Chemical Abstract Service (CAS) Number **54-11-5**
DOT Hazard Class and Label Requirements **Poison B; St. Andrews Cross**	DOT Emergency Guide Code **55**
DOT Identification Number **UN 1654**	Chemical Formula $C_5H_4NC_4H_7NCH_3$

Synonyms

3-(1-methyl-2-pyrrolidyl)-pyridine; β-pyridyl-α-n-methylpyrrolidine; black leaf; destruxol orchid spray.

SECTION II - HAZARDOUS INGREDIENTS AND IDENTITY INFORMATION

Hazardous Components (specific identity)	OSHA Exposure Criteria	NIOSH Exposure Criteria	Immediately Dangerous to Life and Health (IDLH)	ACGIH Exposure Criteria
Nicotine (derivation: By distilling tobacco with milk of lime and extracting with ether). **1 ppm = 6.74 mg/m³**	**PEL (skin):** **0.07 ppm** **0.5 mg/m³** **STEL:** **Not Established**	**REL (skin):** **0.07 ppm** **0.5 mg/m³** **STEL:** **Not Established**	**5 mg/m³**	**TLV (skin):** **0.07 ppm** **0.5 mg/m³** **STEL:** **Not Established**

SECTION III - PHYSICAL AND CHEMICAL CHARACTERISTICS

Boiling Point **482°F (250°C)**	Specific Gravity (H_2O = 1) **1.01**
Vapor Pressure (mm Hg) **0.04 at 69°F (20°C)**	Molecular Weight **162.2**
Vapor Density (Air = 1) **5.61**	Melting Point **-110°F (-79°C)**

Solubility

Miscible in water at temperatures below 140°F (60°C). Soluble in alcohol, chloroform, ether, kerosene, and oils.

Appearance and Odor

Pale yellow to dark brown thick oily, hygroscopic liquid with a slight fish-like odor when heated.

SECTION IV - FIRE AND EXPLOSION HAZARD DATA

Flash Point (method used) **203°F (95°C) opened cup**	Explosive Limits in Air % by Volume **LEL: 0.7% UEL: 4.0%**
NFPA Classification **Class IIIB Combustible Liquid**	**Autoignition Temperature** **471°F (244°C)**

Extinguishing Media

Carbon dioxide, dry chemical, water fog, or foam. Foam and water may cause frothing.

Special Fire Fighting Procedures

Wear full protective clothing and self-contained breathing apparatus (SCBA). Poisonous gases are produced in fire. Use water spray to cool fire-exposed containers.

Unusual Fire and Explosion Hazards

Nicotine is moderately explosive when exposed to flame. The vapors are much heavier than air and can collect in low areas or travel great distances to an ignition source and flashback to cause fire or explosion. Nicotine will attack many forms of plastics, rubber, and coatings. Use impervious equipment.

SECTION V - REACTIVITY DATA

Stability		Conditions to Avoid Nicotine is normally stable under routine conditions of storage and handling. Avoid contact with heat and incompatible materials.
Stable X	Unstable	Incompatibility *(materials to avoid)* Incompatible with strong oxidizers (fluorine, bromine, chlorine), and strong acids (hydrochloric, sulfuric, and nitric). It will attack many plastics, rubber, and coatings.
Hazardous Polymerization		Conditions to Avoid Under normal conditions of temperature and pressure, hazardous polymerization of nicotine is not expected to occur.
May Occur	Will Not Occur X	Hazardous Decomposition or By-products When heated to decomposition, nicotine can emit carbon monoxide and acrid, irritating smoke including oxides of nitrogen and carbon.

SECTION VI - HEALTH HAZARD DATA

Primary Route(s) of Entry:	Inhalation?	Absorption (skin/eye)?	Ingestion?
	X	X	X

Health Hazards

INHALATION: Irritation of the respiratory tract with nausea and vomiting, salivation, abdominal pain, headache, dizziness, hearing and visual disturbances, confusion, weakness, loss of muscular coordination, paroxysmal atrial fibrillation, convulsions, sweating, dypsnea.

ABSORPTION: Nicotine can pass through unbroken skin to cause symptoms similar to that noted for inhalation. Irritates the conjunctiva of the eye.

INGESTION: Gastrointestinal tract irritation.

Carcinogenicity	NTP Listed?	IARC Cancer Review Group?	OSHA Regulated?	Target Organs?
Unknown Human Unknown Animal	No	No	29 CFR 1910.1000 Table Z-1	CNS, CVS, lungs, GI tract, reproductive sys.

Medical Conditions Generally Aggravated by Exposure

Existing nervous system or heart disorders may be aggravated by exposure to nicotine.

Emergency and First-aid Procedures

Eye contact: Do not allow victim to keep eyes tightly shut. Flush immediately with water for 15 minutes (minimum); seek medical attention. Skin contact: Remove all contaminated clothing. Immediately wash area with large amounts of soap and water. For inhalation: Remove the person from exposure. Provide respiratory assistance and CPR. Transfer to medical facility. If swallowed: Seek medical attention. Give victim one or two glasses of water or milk to drink as quickly as possible. Do NOT induce vomiting unless directed to do so. Never give anything by mouth to an unconscious or convulsing person.

SECTION VII - PRECAUTIONS FOR SAFE HANDLING AND USE

Steps to be Taken in Case Material is Released or Spilled

Provide maximum explosion-proof ventilation. Remove all ignition sources. Restrict those not involved in cleanup from entering area. Absorb spills in vermiculite and deposit in sealed drums. Use non-sparking tools. Notify appropriate authorities, as required.

Preferred Waste Disposal Method

Incinerator equipped with scrubber and afterburner.

Precautions to be Taken in Handling and Storage

Nicotine is a flammable liquid. Store to avoid contact with strong oxidizers, since violent reactions can occur. Store in tightly closed containers in cool, well-ventilated area away from heat.

Other Precautions and Warnings

Sources of ignition (heat, flame, cigarettes, incompatibles) are prohibited where nicotine is used, handled, or stored. Bulk storage of nicotine is not recommended.

SECTION VIII - CONTROL MEASURES AND PERSONAL PROTECTIVE EQUIPMENT

Respiratory Protection *(specify type)*

Exposure levels are relatively low (0.5 mg/m^3). Use an MSHA/NIOSH-approved supplied-air respirator with full facepiece in continuous flow mode, or use a self-contained breathing apparatus (SCBA) with full facepiece operated in positive pressure mode.

Ventilation

Local exhaust (preferred) at site of chemical work or general ventilation.

Protective Gloves	Eye Protection	Other Protective Clothing
Impervious Gloves	Chemical Goggles or Face Mask	Boots; Aprons; Gauntlets

Work/Hygiene Practices

Always wash hands thoroughly after using chemical; never bring food, drink, or smoking materials into vicinity of chemicals.

NICOTINE

$C_5H_4NC_4H_7NCH_3$

CAS: 54-11-5

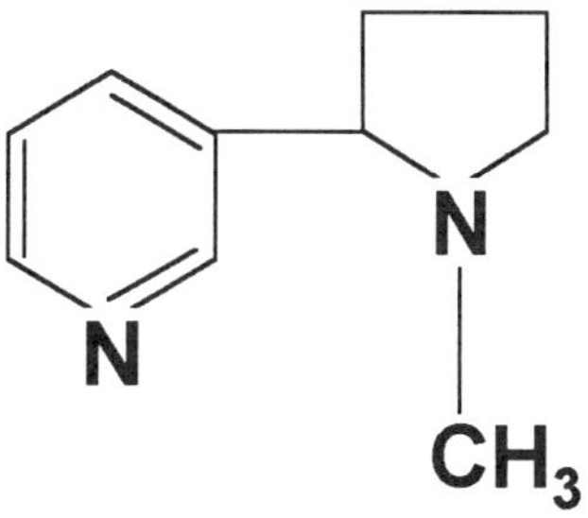

IDENTIFICATION AND TYPICAL USES

Nicotine is a pale yellow to dark brown oily, hygroscopic liquid with a slight fish-like odor when warm. It may initially appear as a thick water-white oil that turns brown on exposure to air. It is used as an insecticide (although such use may be restricted), as a fumigant, in tanning, and in medicine. It is also one of the principal constituents of tobacco. It occurs in dried leaves of *Nicotiana tabacum* and *Nicotiana rustica* to the extent of 2-8%.

RISK ASSESSMENT: HEALTH

General Assessment

Nicotine is a highly toxic compound. It can enter the body through *inhalation* and skin *absorption*. Although unlikely in the industrial setting, *ingestion* is also a possible route of exposure. There is no information in the reference to indicate that nicotine specifically causes cancer in humans. However, mutation data have been reported and nicotine has been shown to cause adverse reproductive effects in humans, especially by ingestion.

Inhalation causes irritation of the respiratory tract. Systemic action by all routes of exposure primarily affects the central nervous system and cardiovascular system. Symptoms include stimulation of the neuromuscular junctions and nicotine receptors in the brain causing depression and paralysis of autonomic ganglia. Indications of such exposure may include headache, nausea, vomiting, salivation, muscular weakness, twitching, and convulsions, tremors, confusion, hallucinations, distorted perceptions with hearing and visual disturbances. Death can occur from respiratory failure. There may also be incoordination, paroxysmal atrial fibrillation, prostration, faintness, abdominal pains, circulatory disturbances, and death from cardiac disturbances. "Nicotinism" (or poisoning by nicotine) is characterized by stimulation and subsequent depression of the central nervous system.

Acute Health Effects

The following acute (short-term) health effects may occur immediately or shortly after exposure to nicotine (within hours or days):

<u>Skin</u>: Mild irritation in contact with the liquid. Nicotine is readily absorbed through intact skin.

<u>Eye</u>: Severe irritation of the conjunctiva with burning, itching, and watering. Direct contact with heated vapors can irritate the eyes.

<u>Lung</u>: Irritation of the bronchial tubes and other members of the respiratory tract. May cause headache, loss of appetite, nausea, vomiting, slowed reactions, and death from respiratory paralysis.

<u>CNS</u>: Stimulation and then depression of the central and autonomic nervous system causing salivation, nausea, vertigo, sweating, hearing and visual disturbances, tremors and convulsions, faintness, prostration, confusion, hallucinations, excitement, and possibly coma. Death can occur from respiratory paralysis.

Chronic Health Effects

The following chronic (long-term) health effects may occur at some time after exposure to nicotine and can last for months or even years:

Cancer Hazards: According to information presented in the references, nicotine has not been adequately tested for its ability to cause cancer in test animals. However, it is known to cause mutations (genetic changes) in living cells and many scientists believe that long-term exposure to such chemicals may pose a cancer risk in the long-term. The data on its human carcinogenicity are inconclusive.

Reproductive Hazard: According to the references, nicotine has been shown to cause teratogenic effects and adversely affect reproduction in test animals.

Other Chronic Effects: Chronic exposure to nicotine, especially from long-term smoking, may lead to circulatory disorders and tobacco toxicosis. There can be adverse effects to blood pressure and other cardiovascular system complications.

Recommended Risk-Reduction Measures

Personnel should avoid direct contact with nicotine. It is a questionable teratogen and mutagen. Its carcinogenic potential is not clearly understood. If a less toxic chemical cannot be substituted for a hazardous substance, then *engineering controls* are the most effective method of reducing exposures. The best protection is to enclose operations and/or provide local exhaust ventilation at the site of chemical release. While not always operationally feasible, isolating operations can also reduce exposure. Using respiratory protection is less effective than the controls mentioned above, but is still advisable whenever working with or around nicotine. Air-purifying respirators will provide protection for exposures above the PEL. However, a self-contained breathing apparatus (SCBA) with full facepiece operated in pressure demand or a supplied-air respirator operated in continuous flow mode are the recommended respiratory protection methods of choice when working with a questioned carcinogen. If a full facepiece is not available, then chemical/dust goggles should be worn to protect the eyes. A face shield and a protective apron should also be worn. To prevent hand and skin exposures, impervious gloves should be used. Glove manufacturers should be contacted and permeation studies obtained *before* final glove selections have been made. Skin contact must be prevented since nicotine can pass through intact skin quickly.

Administrative controls should also be in place to minimize the potential for human exposures. These may include written procedures or policies which specify the methods and techniques that will be practiced whenever personnel are to work with nicotine.

All personnel should receive *training* on the use, hazards, protective measures, emergency actions, and other precautions per 29 CFR 1910.1200 (Hazard Communication), prior to the first assignment in an area where nicotine is used or stored.

Before beginning employment, and at regular intervals thereafter (e.g., annually), the following medical tests are recommended for personnel working with nicotine:

- ☑ Evaluation of nervous system function and performance to establish a baseline of response levels.
- ☑ Lung function tests (establish a baseline).

If symptoms develop or overexposure is suspected, the following may be useful:

- ☑ Liver and kidney function tests.
- ☑ Evaluation of cardiovascular system functions with electrocardiogram (EKG) and other special testing.
- ☑ Lung function tests (compare to baseline).
- ☑ Nervous system evaluation by a qualified neurologist.
- ☑ Evaluation by a qualified allergist, including careful evaluation of exposure history and special testing (may help diagnose skin allergy).

Any evaluation should include a careful history of past and present symptoms with an examination. Medical tests that simply look for existing damage are not a substitute for controlling exposures. Also, since smoking can cause heart disease, emphysema, lung cancer, high blood pressure, and other serious physical disorders, as well as cause an increase in nicotine content in the body, smokers exposed to nicotine in the workplace may experience symptoms more quickly than non-smokers under the same conditions of exposure. Prudent risk management requires proper consideration of *all* possible exposure factors.

Other methods to reduce exposure include:

- ☑ Where possible, enclose operations and use local exhaust ventilation at the site of chemical release. If local exhaust ventilation or enclosure is not used, respiratory protection should be mandatory.
- ☑ Always ensure that proper protective clothing is worn when using chemical substances and that personnel are trained in its use and care.

- ☑ Wash thoroughly immediately after exposure to nicotine and at the end of the work shift or before eating, drinking, or smoking.
- ☑ Hazard warning information should be posted in the work area. In addition, as part of an on-going education and training program, all information on the health and safety hazards of nicotine should be communicated to all exposed workers.
- ☑ Eye wash stations should be provided in the immediate work area for emergency use. If there is a possibility of skin exposure to nicotine, emergency shower facilities should also be provided.

RISK ASSESSMENT: ENVIRONMENT

General Assessment

The environment is at risk of exposure during transportation, storage, disposal, or destruction of nicotine. In almost every scenario, the threat of environmental exposure is contingent upon the proper handling of the chemical substance. Accidental spills, large or small, can result in fire, explosion (depending upon conditions of spill), and possible contamination of the surrounding environmental mediums (water, soil, and air).

Nicotine is considered a Class IIIB combustible liquid. Its moderate flash point and relatively high boiling point present a serious fire hazard under normal or emergency response conditions. Also, because it is incompatible with many common oxidizers and acids, extreme caution is required in handling, storage, transportation, and disposal of nicotine. These characteristics also require special consideration during any emergency situation involving a leak or spill of nicotine liquid. Its vapors are heavier than air and can collect in low-lying areas. Should nicotine ever come into contact with incompatible substances (such as oxidizers) either during use, transportation, storage, or disposal, the formation of highly toxic and/or highly explosive commodities is extremely possible.

The proper disposal/destruction method for nicotine waste is to burn it in a chemical incinerator equipped with an afterburner and air scrubber.

Nicotine can enter the environment from industrial and municipal plant discharges and from spills.

Acute Ecological Effects

Acute (short-term) toxic effects may include the death of animals, birds, or fish and death or low growth rate in plants. Acute effects are seen 2 to 4 days after animals or plants are exposed to nicotine.

Insufficient data are available to evaluate or predict the short-term effects of nicotine to aquatic life, plants, birds, or land animals.

Chronic Ecological Effects

Chronic (long-term) toxic effects may include shortened life span, reproductive problems, lower fertility, and changes in appearance or behavior in exposed animals. These effects can be seen long after first exposure(s) to toxic chemicals.

Insufficient data are available to evaluate or predict the long-term effects of nicotine to aquatic life, plants, birds, or land animals.

Water Solubility

Nicotine is highly soluble in water. Concentrations of 1000 milligrams or more will mix with a liter of water.

Persistence in the Environment

Nicotine is slightly persistent in water, with a half-life between 2 and 20 days. The half-life of a pollutant is the amount of time it takes for one half of the chemical to be degraded.

Bioaccumulation in Aquatic Organisms

Some substances increase in concentration, or *bioaccumulate,* in living organisms as they breathe contaminated air, drink contaminated water, or eat contaminated food. These chemicals can become concentrated in the tissues and internal organs of animals as well as humans.

The concentration of nicotine found in fish tissues is expected to be about the same as the average concentration of nicotine in the water from which the fish was taken.

Recommended Risk-Reduction Measures

Proper training of all transporters will reduce the likelihood of a mishap or accident resulting in a leak or spill of nicotine into the environment. The correct use of DOT labeling (per the MSDS) on all containers, trucks, and rail cars will enable emergency responders to react properly and quickly to any disaster thereby reducing the potential risk to the environment and to personnel.

Storage of nicotine should be segregated from incompatible chemicals to minimize the risk of cross-contamination or contact. Personnel working in storage areas must be trained on the hazards of nicotine.

If a spill or leak to the environment has occurred, fire department, emergency response and/or hazardous materials spill personnel should be notified immediately. Cleanup should be attempted only by those trained in proper spill containment procedures using non-sparking tools. Contaminated soils should be removed for incineration and replaced with clean soil. If nicotine should contact the water table, aquifer, or navigable waterway, the local and/or state emergency response authorities must be notified. A comprehensive emergency response or disaster preparedness/recovery plan should be in place prior to any operations involving the use, transportation, storage, or disposal of nicotine. If nicotine is spilled or leaked, the following specific steps are recommended:

- ☑ Restrict persons not wearing protective clothing from area of spill or leak until cleanup is complete and area can be opened for normal work.
- ☑ Remove all ignition sources and ventilate area.
- ☑ Absorb liquids using vermiculite or other material and deposit in sealed drums for disposal. Collect in safest manner possible using non-sparking tools. Nicotine can attack some rubber, plastics, and coatings. Brass, copper, or other soft metal tools may be required to prevent sparking.
- ☑ It may be necessary to dispose of nicotine as a hazardous waste. The responsible state agency or the regional office of the federal Environmental Protection Agency (EPA) should be contacted for specific recommendations.

RISK ASSESSMENT: BUSINESS

General Assessment

Accidents or mishaps involving nicotine can present a significant threat to business operations. The loss or damage of equipment or facilities can significantly affect fiscal viability. Lawsuits that may result from personnel injury/death, public exposures, and/or environmental contamination will also require a serious expenditure of resources.

Media attention surrounding an injury, death, or environmental damage can also result in a loss of profits and loss of current as well as future business. Always remember that any time the term "reproductive hazard" is used, public emotion, hysteria, and ignorance can run equally high. This should be carefully considered when drafting or implementing public relations policies.

Recommended Risk-Reduction Measures

Company attorneys, safety and health professionals, and environmental specialists should be involved in the development of any safety, health or environmental policies or procedures. A company official should be pre-designated as a public relations officer with specific training in dealing with the press. Corporate plans and policies should be developed, approved, and implemented long before any need for such arises.

REFERENCES

American Conference of Governmental Industrial Hygienists. 1988. *Documentation of the Threshold Limit Values and Biological Exposure Indices,* 5th Edition (with updates). Cincinnati: ACGIH

Bloom, A. D. and F. J. DeSerres. 1995. *Ecotoxicity and Human Health.* New York: CRC Press/Lewis Publishers.

Bueche, J. R. 1972. *Principles of Physics,* 2nd Edition. New York: McGraw-Hill.

Calabrese, E. J. 1994. *Biological Effects of Low Level Exposures.* New York: CRC Press/Lewis Publishers.

Cockerman, L. G., B. S. Shane. 1994. *Basic Environmental Toxicology.* New York: CRC Press/Lewis Publishers.

Gardner, R. W. 1994. *Chemical Intolerance.* New York: CRC Press/Lewis Publishers.

Howard, P. H. 1995. *Handbook of Environmental Fate and Exposures Data for Organic Chemicals,* Vol. I-IV. New York: CRC Press/Lewis Publishers.

Jones, F. E. 1994. *Toxic Organic Vapors in the Workplace.* New York: CRC Press/Lewis Publishers.

Lewis, R. J., Sr. 1993. *Hawley's Condensed Chemical Dictionary,* 12th Edition. New York: Van Nostrand Reinhold.

Lewis, R. J., Sr. 1992. *Sax's Dangerous Properties of Industrial Materials,* Volumes 1, 2, and 3, 8th Edition. New York: Van Nostrand Reinhold.

Lide, D. R. 1995. *Handbook of Chemistry and Physics,* 76th Edition. New York: CRC Press/Lewis Publishers.

Rea, W. J. 1995. *Chemical Sensitivity,* Vol. I-III. New York: CRC Press/Lewis Publishers.

Richardson, M. 1995. *Dictionary of Substance and Their Effects,* Vol. I-VI. New York: CRC Press/Lewis Publishers.

Sacarello, H. L. A. 1994. *The Comprehensive Handbook of Hazardous Materials.* New York: CRC Press/Lewis Publishers.

Torkelson, T. R. V. K. Rowe. 1981. *Patty's Industrial Hygiene and Toxicology,* Vol. 2B. New York: Wiley-Interscience.

MATERIAL SAFETY DATA SHEET

CHEMICAL NAME	
NITRIC ACID	**CRC PRESS/LEWIS PUBLISHERS DISCLAIMER:** THE INFORMATION AND RECOMMENDATIONS PRESENTED HEREIN ARE BASED ON SOURCES BELIEVED TO BE RELIABLE. CRC MAKES NO REPRESENTATION ON ITS COMPLETENESS OR ACCURACY. IT IS THE RESPONSIBILITY OF THE USER TO DETERMINE THE CHEMICAL'S SUITABILITY FOR ITS INTENDED USE, THE CHEMICAL'S SAFE USE, AND THE CHEMICAL'S PROPER DISPOSAL. NO REPRESENTATIONS AND/OR WARRANTIES, EITHER EXPRESSED OR IMPLIED, OF THE MERCHANTABILITY OR FITNESS FOR A PARTICULAR PURPOSE, OR OF ANY OTHER NATURE, ARE MADE WITH RESPECT TO THE INFORMATION PROVIDED IN THIS MSDS OR TO THE CHEMICAL TO WHICH INFORMATION MAY REFER. CRC NEITHER ASSUMES NOR AUTHORIZES ANY OTHER PERSON TO ASSUME FOR IT, ANY OTHER ADDITIONAL RESPONSIBILITY OR LIABILITY FOR THE USE OF, OR RELIANCE UPON, THIS INFORMATION.

HAZARD WARNING INFORMATION

HEALTH	FIRE	REACTIVE	OTHER	DEGREE OF HAZARD	COLOR CODING	OTHER CODES
(fuming) 3	(fuming) 0	(fuming) 1	(fuming) OX	0 = Minimum Hazard 1 = Slight Hazard 2 = Moderate Hazard 3 = Serious Hazard 4 = Severe Hazard	HEALTH = BLUE FIRE = RED REACTIVITY = YELLOW OTHER = WHITE	OX = Oxidizer ACID = Acid ALK = Alkali COR = Corrosive W = Use No Water

SECTION I - GENERAL INFORMATION

Characterization	RCRA Number	EPA Class
Acid, Mineral	**D001**	**Characteristic (C) Waste**

DOT Proper Shipping Name	Chemical Abstract Service (CAS) Number
Nitric Acid, Fuming; Nitric Acid >40%; Nitric Acid <40%	**7697-37-2**
DOT Hazard Class and Label Requirements	DOT Emergency Guide Code
Corrosive, Oxidizer, Poison	**60 (40% acid) 44 (>40% acid and fuming)**
DOT Identification Number	Chemical Formula
UN1760 (40% acid) UN2031 (>40% acid) UN2032 (fuming)	**HNO_3**

Synonyms

Aqua fortis; engravers acid; hydrogen nitrate; red fuming nitric acid (RFNA); white fuming nitric acid (WFNA); aqua regia; azotic acid.

SECTION II - HAZARDOUS INGREDIENTS AND IDENTITY INFORMATION

Hazardous Components (specific identity)	OSHA Exposure Criteria	NIOSH Exposure Criteria	Immediately Dangerous to Life and Health (IDLH)	ACGIH Exposure Criteria
Nitric acid (derivation: By oxidation of ammonia by air or oxygen with platinum catalyst. Air oxidation yields 60% acid; concentration is achieved by distillation with sulfuric acid; or by extractive distillation with magnesium nitrate; or by neutralizing the weak acid with soda ash, evaporating to dryness, and treating with sulfuric acid). **1 ppm = 2.62 mg/m³**	**PEL:** 2 ppm 5 mg/m³ **STEL (15-min):** 4 ppm 10 mg/m³	**REL:** 2 ppm 5 mg/m³ **STEL (15-min):** 4 ppm 10 mg/m³	**25 ppm**	**TLV:** 2 ppm 5.2 mg/m³ **STEL (15-min):** 4 ppm 10 mg/m³

SECTION III - PHYSICAL AND CHEMICAL CHARACTERISTICS

Boiling Point	Specific Gravity (H_2O = 1)
181°F (83°C)	**1.5 at 77°F (39°C)**
Vapor Pressure (mm Hg)	Molecular Weight
48 at 69°F (20°C)	**63**
Vapor Density (Air = 1)	Melting Point
1.212	**-44°F (-42°C)**

Solubility

Soluble in water (releases heat), decomposes in alcohol.

Appearance and Odor

Transparent, colorless to yellow or red fuming liquid with an acrid, suffocating odor. Darkens to a brownish color on aging and exposure to light. Fuming nitric acid is red to brown in color.

SECTION IV - FIRE AND EXPLOSION HAZARD DATA

Flash Point (method used)	Explosive Limits in Air % by Volume
Non-Combustible	**LEL: Not Determined UEL: Not Determined**
NFPA Classification	Autoignition Temperature
Non-Combustible Liquid	**Not Applicable**

Extinguishing Media

Use dry chemical, water spray, fog, carbon dioxide, or regular foam. Use soda ash on small fires.

Special Fire Fighting Procedures

Poisonous gases are released in fire. Wear full protective clothing and (SCBA). Use water spray to flush and dilute a spill, to disperse vapors, and to keep fire-exposed containers cool.

Unusual Fire and Explosion Hazards

A non-combustible liquid that increases the burning of combustible materials. Containers may explode in fire. Releases flammable hydrogen gas in contact with many metals.

SECTION V - REACTIVITY DATA

Stability		Conditions to Avoid Nitric acid decomposes in air and in contact with light and/or organic materials. It can be kept stable in closed containers in a cool, well-ventilated area.
Stable	Unstable X	Incompatibility *(materials to avoid)* Reacts explosively with combustible materials, organics, or readily oxidizable materials (wood, paper, turpentine, metal powder, hydrogen sulfide, carbides, cyanides, alkalis), strong bases, heat.
Hazardous Polymerization		Conditions to Avoid Under normal conditions of temperature and pressure, hazardous polymerization of nitric acid is not expected to occur. It will corrode and attack paper, plastics, rubber, and coatings.
May Occur	Will Not Occur X	Hazardous Decomposition or By-products When heated to decomposition, nitric acid can emit highly toxic/poisonous gases and fumes, including toxic nitrogen oxides and toxic nitrogen peroxide.

SECTION VI - HEALTH HAZARD DATA

Primary Route(s) of Entry:	Inhalation?	Absorption (skin)?	Ingestion?
	X		X

Health Hazards

INHALATION: Vapors cause irritation to the upper respiratory tract with coughing, dypsnea, headache, bronchitis, weakness, tearing, nasal discharge, nausea, chemical bronchitis, pulmonary edema (fluid in the lungs), corrosion of teeth, and CNS effects.

SKIN & EYES: Eye contact causes blindness and tissue damage. Skin contact causes burns, blistering, local necroses, and membrane ulceration. Skin may be stained yellow-brown.

INGESTION: Severe poisoning, tissue damage, salivation, bloody vomiting, burning in mouth, pain, and diarrhea. Severe cases may cause shock, dypsnea, and possibly death.

Carcinogenicity	NTP Listed?	IARC Cancer Review Group?	OSHA Regulated?	Target Organs?
Unknown Human Unknown Animal	No	No	29 CFR 1910.1000 Table Z-1	Respiratory system, eyes, skin, teeth.

Medical Conditions Generally Aggravated by Exposure

Respiratory impairments (bronchitis, asthma) may be aggravated by exposure.

Emergency and First-aid Procedures

Eye contact: Flush immediately with water for 15 minutes (minimum); seek medical attention. Skin contact: Remove all contaminated clothing. Immediately wash area with large amounts of water. Seek medical assistance. For inhalation: Remove the person from exposure. Provide respiratory assistance and CPR. Transfer to medical facility. Observe for 48 hours for lung effects. If swallowed: Seek medical attention; contact poison control center. Do NOT induce vomiting. Give milk or water to drink to dilute followed by milk of magnesia. Do NOT give sodium bicarbonate or attempt to neutralize the acid.

SECTION VII - PRECAUTIONS FOR SAFE HANDLING AND USE

Steps to be Taken in Case Material is Released or Spilled

Absorb liquids in vermiculite or other material; collect and deposit in sealed containers. Ventilate area of spill or leak. Restrict those not involved in cleanup from entering area.

Preferred Waste Disposal Method

No citation.

Precautions to be Taken in Handling and Storage

Nitric acid is a non-combustible liquid that can greatly accelerate the burning of oxidizable materials. Store in tightly closed containers in cool, well-ventilated area away from heat.

Other Precautions and Warnings

Sources of ignition are prohibited where nitric acid is used, handled, or stored. Detached facility storage is preferred. Personnel must be trained on the hazards of nitric acid.

SECTION VIII - CONTROL MEASURES AND PERSONAL PROTECTIVE EQUIPMENT

Respiratory Protection *(specify type)*

Use half-mask respirator with acid cartridge for low or transient exposures (cartridge should be designed to protect against acid vapor or mist). For higher exposures, use a supplied-air respirator or a self-contained breathing apparatus operated in pressure demand or other positive pressure mode.

Ventilation

Local exhaust or general mechanical systems recommended.

Protective Gloves	Eye Protection	Other Protective Clothing
Acid resistant synthetic rubber	Chemical Goggles or Face Mask	Protective Apron

Work/Hygiene Practices

Always wash hands thoroughly after using chemical; never bring food, drink, or smoking materials into vicinity of chemicals.

NITRIC ACID

HNO_3

CAS: 7697-37-2

Fuming Nitric Acid

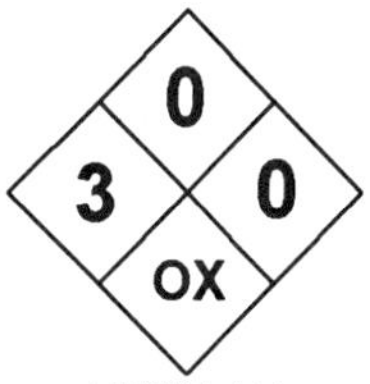

>40% Nitric Acid

<40% Nitric Acid

IDENTIFICATION AND TYPICAL USES

Nitric acid is a transparent to clear or yellow fuming liquid. It can also appear as reddish-brown to orange-brown fumes. It has a suffocating odor. The liquid may darken to brown on aging and exposure to light. Fuming nitric acid is concentrated nitric acid that contains dissolved nitrogen dioxide. It is used in the manufacture of ammonium nitrate for fertilizer and explosives, in organic synthesis (for dyes, drugs, explosives, cellulose nitrate, and nitrate salts), in metallurgy, in photo-engraving, for etching steel, in ore flotation, in urethanes, in rubber chemicals, and in the reprocessing of spent nuclear fuel.

RISK ASSESSMENT: HEALTH

General Assessment

Nitric acid is highly corrosive to the skin, eyes, digestive tract, and respiratory system (or any other tissue it contacts). It is a human poison by *ingestion* and is dangerously toxic by *inhalation*. There are no data in the references regarding any carcinogenic, mutagenic, or teratogenic effects on exposure to nitric acid. Its corrosive and irritating properties limit the possibility of long-term exposure since people cannot tolerate unprotected contact for even short periods of time.

Inhalation produces severe irritation of the eyes, nose, and respiratory tract with coughing, dypsnea (labored breathing), bronchitis, headache, and total body weakness. The vapors can cause laryngitis, pharyngitis, chemical bronchitis, pneumonitis, and delayed pulmonary edema which can be fatal. There may be tearing, nasal discharge, throat irritation, ulceration of the nasal septa and nose irritation, and nausea. Symptoms of pulmonary edema, such as chest pain, congestion, coughing, and weakness, may be delayed up to 48 hours thereby creating a false sense of security with regard to health exposure risk. There may also be depression of the central nervous system with symptoms of giddiness, salivation, nausea, muscular weakness, and other unspecified symptoms.

Skin contact causes severe burning with possible vesiculation. Its extremely corrosive properties cause immediate burns, membrane ulceration, skin blistering, and local necrosis (tissue destruction) upon contact. Eye contact causes damage and the possibility of permanent blindness.

Ingestion is extremely dangerous, even life-threatening. It causes severe poisoning, throat swelling, convulsions, risk of stomach perforation, bloody vomiting, burning sensation on the mouth, acute tissue damage, diarrhea with pain, and possible coma.

Acute Health Effects

The following acute (short-term) health effects may occur immediately or shortly after exposure to nitric acid:

Skin: Irritation/rash or burning feeling on contact with redness, blistering and vesiculation, and localized tissue destruction.

Eye: Severe irritation; may cause burns and damage with permanent blindness.

Lung: Severe irritation of the mucous membranes in the nose, throat, and lungs with delayed pulmonary edema (fluid in the lungs), which can be fatal.

CNS: Possible depression of the nervous system with symptoms of giddiness and irritability.

Chronic Health Effects

The following chronic (long-term) health effects may occur at some time after exposure to nitric acid and can last for months or even years:

Cancer Hazards: According to information presented in the references, nitric acid has not been adequately tested for its ability to cause cancer in test animals.

Reproductive Hazard: According to information presented in the references, nitric acid has not been adequately tested for its ability to adversely affect reproduction in test animals.

Other Chronic Effects: Skin contact causes irritation and burns. There may be some lasting effects of chronic exposures such as sensitization. Once an individual becomes "sensitized," even the slightest future exposures can cause an allergic reaction. Repeated inhalation of very low concentrations may cause chronic bronchitis, chronic lung diseases, tooth erosion, and/or loss of appetite.

Recommended Risk-Reduction Measures

Personnel should avoid direct contact with nitric acid. It is extremely corrosive and moderately toxic to humans. If a less toxic chemical cannot be substituted for a hazardous substance, then *engineering controls* are the most effective method of reducing exposures.

The best protection is to enclose operations and/or provide local exhaust ventilation at the site of chemical release. While not always operationally feasible, isolating operations can also reduce exposure. Using respiratory protection is less effective than the controls mentioned above, but is still advisable whenever working with or around nitric acid. A self-contained breathing apparatus (SCBA) or a supplied-air respirator with full facepiece operated in pressure demand mode are the recommended respiratory protection methods of choice. If a full facepiece is not available, then chemical goggles should be worn to protect the eyes. For low concentrations (near or over the 2 ppm exposure limit), an acid gas cartridge respirator (equipped with a mist/vapor filter) is acceptable protection. Whenever a chemical splash hazard exists, a face shield and a rubber apron should be worn. To prevent hand and skin exposures, acid-resistant gloves should be used.

Administrative controls should also be in place to minimize the potential for human exposures. These may include written procedures or policies which specify the methods and techniques that will be practiced whenever personnel are to work with nitric acid.

All personnel should receive *training* on the use, hazards, protective measures, emergency actions, and other precautions per 29 CFR 1910.1200 (Hazard Communication), prior to the first assignment in an area where nitric acid is used or stored.

Before beginning employment and at regular intervals thereafter (e.g., annually), the following tests are recommended:

☑ Lung function tests (look for chronic respiratory diseases).

If symptoms occur or overexposure is suspected, the following additional tests are suggested:

☑ Evaluation by a qualified allergist with careful consideration of exposure history and special testing (may help diagnose skin allergy).

☑ If nitric acid is ingested, contact poison control center. Unless advised otherwise and victim is conscious and alert, give 1-2 glasses of water or milk to dilute followed by milk of magnesia. Do NOT induce vomiting and do NOT administer sodium bicarbonate.

☑ Consider chest X-ray following acute overexposure (may be negative if taken immediately after exposure due to delayed onset of pulmonary edema).

Other methods to reduce exposure include:

☑ Where possible, enclose operations and use local exhaust ventilation at the site of chemical release. If local exhaust ventilation or enclosure is not used, respiratory protection should be mandatory.

☑ Always ensure that proper protective clothing is worn when using chemical substances.

☑ Wash thoroughly immediately after exposure to nitric acid and at the end of the work shift or before eating, drinking, or smoking.

☑ Hazard warning information should be posted in the work area. In addition, as part of an on-going education and training program, all information on the health and safety hazards of nitric acid should be communicated to all exposed workers.

RISK ASSESSMENT: ENVIRONMENT

General Assessment

The environment is at risk of exposure during transportation, storage, disposal, or destruction of nitric acid. In almost every scenario, the threat of environmental exposure is contingent upon the proper handling of the chemical substance. Accidental spills, large or small, can result in fire, explosion, and possible contamination of the surrounding environmental mediums (water, soil, and, to a lesser degree, air).

Nitric acid is considered a non-combustible liquid. However, as an oxidizer, it can greatly accelerate the burning of ordinary combustible materials such as wood or paper. It is also extremely corrosive to many materials. It is incompatible with many materials, including hydrogen sulfide, carbides, cyanides, and alkalis, and contact can cause fire or explosion. These characteristics require special consideration during any emergency situation involving a leak or spill of nitric acid. It will also corrode many common plastics, rubber, and coatings. Selection of response equipment (shovels, containers, and the like) must therefore be carefully considered. Should nitric acid ever come into contact with incompatible substances either during use, transportation, or storage, the formation of highly toxic and/or explosive commodities is extremely possible.

Nitric acid can enter the environment mainly through industrial discharges or spills.

Acute Ecological Effects

Acute (short-term) toxic effects may include the death of animals, birds, or fish and death or low growth rate in plants. Acute effects are seen 2 to 4 days after animals or plants are exposed to nitric acid.

Insufficient data are available on the short-term effects of nitric acid exposure to aquatic life, plants, birds, or land animals.

Chronic Ecological Effects

Chronic (long-term) toxic effects may include shortened life span, reproductive problems, lower fertility, and changes in appearance or behavior in exposed animals. These effects can be seen long after first exposure(s) to toxic chemicals.

Insufficient data are available on the long-term effects of nitric acid to aquatic life, plants, birds, or land animals.

Water Solubility

Nitric acid is highly soluble in water. Concentrations of over 1000 milligrams will mix with a liter of water.

Persistence in the Environment

Nitric acid is slightly persistent in the aquatic environment, with a half-life of between 2 to 20 days. The half-life of a pollutant is the amount of time it takes for one half of the chemical to be degraded.

Bioaccumulation in Aquatic Organisms

Some substances increase in concentration, or *bioaccumulate,* in living organisms as they breathe contaminated air, drink contaminated water, or eat contaminated food. These chemicals can become concentrated in the tissues and internal organs of animals as well as humans.

The concentration of nitric acid found in fish tissues is expected to be somewhat lower than the average concentration of nitric acid in the water from which the fish was taken.

Recommended Risk-Reduction Measures

Proper training of all transporters will reduce the likelihood of a mishap or accident resulting in a leak or spill to the environment. The correct use of DOT labeling (per the MSDS) on all containers, trucks, and rail cars will enable emergency responders to react properly and quickly to any disaster thereby reducing the potential risk to the environment and to personnel.

Storage of nitric acid should be segregated from incompatible chemicals to minimize the risk of cross-contamination or contact. Buildings designated for storage should be equipped with appropriate fire protection systems (alarms, sprinklers, emergency lighting, portable extinguishers). It must be kept cool and in a well-ventilated location away from direct sunlight.

If a spill or leak to the environment has occurred, fire department, emergency response and/or hazardous materials spill personnel should be notified immediately. Cleanup should be attempted only by those trained in proper spill containment procedures. Contaminated soils should be removed for incineration and replaced with clean soil. If nitric acid should contact the water table, aquifer, or navigable waterway, time is of the essence. It is highly miscible in water and, therefore, total containment and remediation may not be entirely possible. When such spills occur, the local and/or state emergency response authorities must be notified. A comprehensive emergency response or disaster preparedness/recovery plan should be in place prior to any operations involving the use, transportation, storage, or disposal of nitric acid. If nitric acid is spilled or leaked, the following specific steps are recommended:

- ☑ Restrict persons not wearing protective clothing from area of spill or leak until cleanup is complete.
- ☑ Remove all ignition sources and ventilate area.

☑ Absorb liquids in vermiculite, dry sand, earth, or similar material and deposit in sealed containers. Neutralize remaining residue with soda ash or sodium bicarbonate. Do NOT flush to drains, sewers, or waterways. Use water spray to cool and knock-down vapors. Collect water for disposal.

☑ It may be necessary to dispose of nitric acid as a hazardous waste. The responsible state agency or the regional office of the federal Environmental Protection Agency (EPA) should be contacted for specific recommendations.

RISK ASSESSMENT: BUSINESS

General Assessment

Accidents or mishaps involving nitric acid can present a significant threat to business operations. The loss or damage of equipment or facilities can significantly affect fiscal viability. Lawsuits that may result from personnel injury/death, public exposures, and/or environmental contamination will also require a serious expenditure of resources. Media attention surrounding an injury, death, or environmental damage can also result in a loss of profits and loss of current as well as future business.

Recommended Risk-Reduction Measures

Company attorneys, safety and health professionals, and environmental specialists should be involved in the development of any procedures dealing with safety or emergency response. A company official should be pre-designated as a public relations officer with specific training in dealing with the press. Corporate plans and policies should be developed, approved, and implemented long before any need for such arises.

REFERENCES

Ahlbom, A. 1993. *Biostatistics for Epidemiologists.* New York: CRC Press/Lewis Publishers.

American Conference of Governmental Industrial Hygienists. 1988. *Documentation of the Threshold Limit Values and Biological Exposure Indices,* 5th Edition (with updates). Cincinnati: ACGIH

Asante-Duah, D. K. 1993. *Hazardous Waste Risk Assessment.* New York: CRC Press/Lewis Publishers.

Bloom, A. D. and F. J. DeSerres. 1995. *Ecotoxicity and Human Health.* New York: CRC Press/Lewis Publishers.

Bueche, J. R. 1972. *Principles of Physics,* 2nd Edition. New York: McGraw-Hill.

Burrell, R., D. K. Flaherty, and Leonard J. Sauers. 1992. *Toxicology of the Immune System.* New York: Van Nostrand Reinhold.

Calabrese, E. J., E. M. Kenyon. *Air Toxics and Risk Assessment.* New York: CRC Press/Lewis Publishers.

Cockerman, L. G., B. S. Shane. 1994. *Basic Environmental Toxicology.* New York: CRC Press/Lewis Publishers.

Cothern, C. R. 1995. *Handbook for Environmental Risk Decisionmaking.* New York: CRC Press/Lewis Publishers.

Danse, I. R. 1991. *Common Sense Toxics in the Workplace.* New York: Van Nostrand Reinhold.

Gardner, R. W. 1994. *Chemical Intolerance.* New York: CRC Press/Lewis Publishers.

Hallenbeck, W. H. 1993. *Quantitative Risk Assessment for Environmental and Occupational Health,* 2nd Edition. New York: CRC Press/Lewis Publishers.

Hosty, J. W., P. Foster. 1990. *A Practical Guide to Chemical Spill Response.* New York: Van Nostrand Reinhold.

Howard, P. H., et. al. 1991. *Handbook of Environmental Degradation Rates.* New York: CRC Press/Lewis Publishers.

Jones, F. E. 1994. *Toxic Organic Vapors in the Workplace.* New York: CRC Press/Lewis Publishers.

Keith, L. H. and M. M. Walker. 1995. *Handbook of Air Toxics.* New York: CRC Press/Lewis Publishers.

Landis, W. G., M. H. Yu. 1995. *Introduction to Environmental Toxicology.* New York: CRC Press/Lewis Publishers.

Lauwerys, R. R., P. Hoet. 1993. *Industrial Chemical Exposure.* New York: CRC Press/Lewis Publishers.

Lewis, R. J., Sr. 1993. *Hawley's Condensed Chemical Dictionary,* 12th Edition. New York: Van Nostrand Reinhold.

Lewis, R. J., Sr. 1992. *Sax's Dangerous Properties of Industrial Materials,* Volumes 1, 2, and 3, 8th Edition. New York: Van Nostrand Reinhold.

Meyer, E. 1990. *Chemistry of Hazardous Materials.* Englewood Cliffs, NJ: Prentice Hall.

Rea, W. J. 1995. *Chemical Sensitivity,* Vol. I-III. New York: CRC Press/Lewis Publishers.

Richardson, M. 1995. *Dictionary of Substance and Their Effects,* Vol. I-VI. New York: CRC Press/Lewis Publishers.

Rylander, R., R. R. Jacobs. 1994. *Organic Dusts.* New York: CRC Press/Lewis Publishers.

Sacarello, H. L. A. 1994. *The Comprehensive Handbook of Hazardous Materials.* New York: CRC Press/Lewis Publishers.

Sherman, J. D. 1988. *Chemical Exposure and Disease.* New York: Van Nostrand Reinhold.

Talbott, E., G. Craun. 1995. *An Introduction to Environmental Epidemiology.* New York: CRC Press/Lewis Publishers.

U.S. Department of Health and Human Services, National Institute for Occupational Safety and Health. 1994. *NIOSH Pocket Guide to Chemical Hazards.* Washington, D.C.: U.S. Government Printing Office.

MATERIAL SAFETY DATA SHEET

CHEMICAL NAME

NITRIC OXIDE

CRC PRESS/LEWIS PUBLISHERS DISCLAIMER: THE INFORMATION AND RECOMMENDATIONS PRESENTED HEREIN ARE BASED ON SOURCES BELIEVED TO BE RELIABLE. CRC MAKES NO REPRESENTATION ON ITS COMPLETENESS OR ACCURACY. IT IS THE RESPONSIBILITY OF THE USER TO DETERMINE THE CHEMICAL'S SUITABILITY FOR ITS INTENDED USE, THE CHEMICAL'S SAFE USE, AND THE CHEMICAL'S PROPER DISPOSAL. NO REPRESENTATIONS AND/OR WARRANTIES, EITHER EXPRESSED OR IMPLIED, OF THE MERCHANTABILITY OR FITNESS FOR A PARTICULAR PURPOSE, OR OF ANY OTHER NATURE, ARE MADE WITH RESPECT TO THE INFORMATION PROVIDED IN THIS MSDS OR TO THE CHEMICAL TO WHICH INFORMATION MAY REFER. CRC NEITHER ASSUMES NOR AUTHORIZES ANY OTHER PERSON TO ASSUME FOR IT, ANY OTHER ADDITIONAL RESPONSIBILITY OR LIABILITY FOR THE USE OF, OR RELIANCE UPON, THIS INFORMATION.

HAZARD WARNING INFORMATION

HEALTH	FIRE	REACTIVE	OTHER	DEGREE OF HAZARD	COLOR CODING	OTHER CODES
4	1	3		0 = Minimum Hazard 1 = Slight Hazard 2 = Moderate Hazard 3 = Serious Hazard 4 = Severe Hazard	HEALTH = BLUE FIRE = RED REACTIVITY = YELLOW OTHER = WHITE	OX = Oxidizer ACID = Acid ALK = Alkali COR = Corrosive W̶ = Use No Water

SECTION I - GENERAL INFORMATION

Characterization	Gas	RCRA Number	P076
		EPA Class	Acute Hazardous Waste
DOT Proper Shipping Name	Nitric oxide	Chemical Abstract Service (CAS) Number	10102-43-9
DOT Hazard Class and Label Requirements	Poison Gas	DOT Emergency Guide Code	20
DOT Identification Number	UN 1660	Molecular Formula	NO

Synonyms
Mononitrogen monoxide; nitrogen monoxide.

SECTION II - HAZARDOUS INGREDIENTS AND IDENTITY INFORMATION

Hazardous Components (specific identity)	OSHA Exposure Criteria	NIOSH Exposure Criteria	Immediately Dangerous to Life and Health (IDLH)	ACGIH Exposure Criteria
Nitric oxide: (derivation: By oxidation of ammonia above 932°F/500°C, decomposition of nitrous acid. Also from atmospheric oxygen and nitrogen in the electric-arc process for fixation of nitrogen). 1 ppm = 1.25 mg/m^3	**PEL (8-hour):** 25 ppm 30 mg/m^3 **STEL:** Not Established	**REL (10-hour):** 25 ppm 30 mg/m^3 **STEL:** Not Established	100 ppm	**TLV:** 25 ppm 30 mg/m^3 **STEL:** Not Established

SECTION III - PHYSICAL AND CHEMICAL CHARACTERISTICS

Boiling Point	-242°F (-152°C)	Specific Gravity (H_2O = 1)	1.04 (Gas)
Vapor Pressure (atmospheres)	34.2 at 68°F (20°C)	Molecular Weight (atomic weight)	30.0
Vapor Density (Air = 1)	1.04	Melting Point	-263°F (-164°C)

Solubility
Slightly soluble in water (5%).

Appearance and Odor
Colorless gas with a sharp, sweet odor. Also appears with a deep blue color in liquid form. Shipped as a non-liquefied compressed gas.

SECTION IV - FIRE AND EXPLOSION HAZARD DATA

Flash Point (method used)	Not Applicable	Explosive Limits in Air % by Volume	LEL: Not Applicable UEL: Not Applicable
NFPA Classification	Non-Flammable Gas	Autoignition Temperature	Not Applicable

Extinguishing Media
Will not burn but will enhance the burning of other materials. Use agent suitable to surrounding fire.

Special Fire Fighting Procedures
Poisonous gases are produced in fire. Wear full protective clothing and self-contained breathing apparatus (SCBA). Move container from fire area if it can be done without risk. Fight fire from maximum distance. Stay away from ends of tanks. Cool exposed containers with direct water spray.

Unusual Fire and Explosion Hazards
Containers may explode in fire. As an oxidizer, nitric oxide can greatly accelerate the burning of ordinary combustible materials such as wood and paper. Use extreme caution if nitric oxide is in or near a fire.

SECTION V - REACTIVITY DATA

Stability		Conditions to Avoid Nitric oxide is normally stable in closed containers under routine conditions of handling and storage. It is highly reactive with many materials. Avoid contact with incompatibles, heat, and atmospheric oxygen.
Stable X	Unstable	Incompatibility *(materials to avoid)* Incompatible with fluorine, combustible materials, ozone, chlorinated hydrocarbons, metals, carbon disulfide, chlorine monoxide, phosphine, perchloryl fluoride, and water (forms nitric acid).
Hazardous Polymerization		Conditions to Avoid Under normal conditions of temperature and pressure, hazardous polymerization of nitric oxide is not expected to occur.
May Occur	Will Not Occur X	Hazardous Decomposition or By-products When heated to decomposition, nitric oxide emits carbon monoxide and/or oxides of nitrogen.

SECTION VI - HEALTH HAZARD DATA

Primary Route(s) of Entry:	Inhalation?	Absorption (skin/eye)?	Ingestion?
	X		

Health Hazards

INHALATION: Severe irritation to the eyes, nose, throat, and lungs. Causes headache, nausea, vomiting, difficulty breathing, cyanosis, pulmonary edema (fluid in the lungs), depression of the CNS with drowsiness, weakness, loss of coordination, and convulsions. Also causes the development of methemoglobinemia affecting the transportation of oxygen by the blood.

SKIN & EYES: A severe eye and skin irritant. Cold compressed gas temperatures may cause frostbite and associated tissue destruction.

INGESTION: Unlikely since nitric oxide exists as a gas at normal room temperatures.

Carcinogenicity	NTP Listed?	IARC Cancer Review Group?	OSHA Regulated?	Target Organs?
Unknown Human Unknown Animal	No	No	29 CFR 1910.1000 Table Z-1	Skin, eyes, respiratory system, blood, CNS.

Medical Conditions Generally Aggravated by Exposure

None reported. However, existing respiratory problems (bronchitis, asthma) may be aggravated.

Emergency and First-aid Procedures

Eye contact: Do not allow victim to keep eyes tightly shut. Flush immediately with water for 15 minutes (minimum). Skin contact: Wash with large amounts of water for a minimum of 15 minutes. For inhalation: Remove the person from exposure. Provide respiratory assistance and CPR. Transfer to medical facility. Observe for 24-48 hours for lung effects. Oxygen should be administered by trained medical personnel only. If swallowed: Not likely since nitric oxide is a gas.

SECTION VII - PRECAUTIONS FOR SAFE HANDLING AND USE

Steps to be Taken in Case Material is Released or Spilled

Wear self-contained breathing apparatus (SCBA). Stop flow of gas. Restrict those not wearing protective equipment and who are not involved in cleanup from entering area. Avoid skin contact. Ventilate area of spill. If gas cannot be stopped, remove leaking container to safe area outdoors and allow to empty.

Preferred Waste Disposal Method

No citation.

Precautions to be Taken in Handling and Storage

Store in tightly closed, insulated, steel containers or cylinders in a cool, well-ventilated area away from heat. Protect containers from physical damage. Regularly inspect for leaks, cracks, or faulty valves.

Other Precautions and Warnings

Use external or detached storage. Ensure cylinders are stored upright and properly secured at all times.

SECTION VIII - CONTROL MEASURES AND PERSONAL PROTECTIVE EQUIPMENT

Respiratory Protection *(specify type)*

Use a NIOSH/MSHA-approved air purifying gas-cartridge respirator, or a supplied-air respirator, or a self-contained breathing apparatus (SCBA) with full facepiece operated in positive pressure mode.

Ventilation

Local exhaust (preferred) at site of chemical work or general ventilation.

Protective Gloves	Eye Protection	Other Protective Clothing
Impervious Gloves	Chemical Goggles and Face Mask	Protective Uniform or Apron

Work/Hygiene Practices

Always wash hands thoroughly after using chemical; never bring food, drink, or smoking materials into vicinity of chemicals.

NITRIC OXIDE

NO CAS: 10102-43-9

IDENTIFICATION AND TYPICAL USES

Nitric oxide appears as a colorless gas with a sharp and sweet odor. It may appear as a deep blue liquid at temperatures below -263°F (-164°C). It converts readily to toxic nitrogen dioxide and nitrogen tetroxide on contact with air. Nitric oxide is used as an intermediate in the manufacture of nitric acid, in the preparation of metal nitrosyls, in the bleaching of rayon, and in the manufacture of incandescent lamps.

RISK ASSESSMENT: HEALTH

General Assessment

Nitric oxide is a severe irritant to the eyes, nose, and throat. As a gas, it can primarily enter the body through *inhalation.* While skin absorption is not likely, skin contact can result in serious tissue damage as a direct result of the cold temperatures of the compressed gas or cryogenic liquid. Ingestion is not a probable route of exposure under most circumstances. There are no data regarding any carcinogenic, mutagenic, or teratogenic potential.

Inhalation of nitric oxide can cause severe irritation to the eyes, nose, throat, and respiratory system. Symptoms can be delayed for 1 to 5 hours and may include headache, nausea, vomiting, dypsnea (difficult breathing), diarrhea, cyanosis (bluish discoloration of the lips, fingers, and toes), methemoglobinemia, and possible CNS depression with drowsiness, weakness, loss of coordination, memory impairment, peripheral neuropathy, convulsions, coma, and death. Lung irritation can be quite severe and may lead to pulmonary edema (fluid buildup in the lungs), which is a medical emergency and can be fatal. The symptoms of pulmonary edema can be delayed up to 48 hours, thereby creating a false sense of security with regard to health exposure risk.

Skin contact with compressed nitric oxide can cause characteristic burns due to frostbite. Eye contact causes irritation, watering, inflammation of the lids, burns to the cornea, and (possibly) severe lesions.

Acute Health Effects

The following acute (short-term) health effects may occur immediately or shortly after exposure to nitric oxide:

Skin: Severe irritation with localized tissue damage and possible burns due to the cold temperature of the compressed gas.

Eye: Severe conjunctival irritation and possible damage to the cornea and loss of vision.

Lung: Severe irritation to respiratory tract causing shortness of breath and cough. High acute exposures can be fatal due to a buildup of fluid in the lungs (pulmonary edema).

CNS: Possible depression of the central nervous system causing a variety of toxic effects including loss of coordination and consciousness.

Chronic Health Effects

The following chronic (long-term) health effects may occur at some time after exposure to nitric oxide and can last for months or even years:

Cancer Hazards: According to information presented in the references, nitric oxide has not been adequately tested for its ability to cause cancer in test animals. Animal mutation data have been reported and many scientists believe that such chemicals may pose a cancer risk in the long-term.

Reproductive Hazard: According to the references, nitric oxide has not been adequately tested for its ability to adversely affect reproduction.

Other Chronic Effects: Long-term exposures to low concentrations may result in the gradual corrosion of teeth enamel, chronic headache, and chronic cough or bronchitis.

Recommended Risk-Reduction Measures

Nitric oxide is, by itself, a very toxic substance. On exposure to air, it can rapidly convert to the even more toxic nitrogen dioxide and nitrogen tetroxide. Personnel should therefore avoid direct contact with nitric oxide. If a less toxic material or compound cannot be substituted for nitric oxide, then *engineering controls* are the most effective method of reducing exposure risk. The best protection is to enclose operations and/or provide local exhaust ventilation at the site of nitric oxide release. While not always operationally feasible, isolating operations can also reduce exposure. Using respiratory protection is less effective than the controls mentioned above, but is still required whenever working with or around nitric oxide. The exposure level of 25 ppm (8-hour TWA) may lead personnel to believe that they are not being exposed to dangerous levels. However, prudent risk management requires proper consideration and implementation of an appropriate respiratory protection program (per OSHA 29 CFR 1910.134). While a gas-cartridge air purifying respirator will certainly suffice, the best protection is obtained using an MSHA/NIOSH-approved supplied-air respirator or a self-contained breathing apparatus (SCBA) with full facepiece and operated in pressure demand mode. If a full facepiece is not available, then chemical goggles should be worn to protect the eyes. A face shield should also be considered. To prevent hand and skin exposures, polyvinylchloride gloves should be used. Glove manufacturers should be contacted and permeation studies obtained *before* final glove selections are made.

Administrative controls should also be in place to minimize the potential for human exposures. These may include written procedures or policies which specify the methods and techniques that will be practiced whenever personnel are to work with nitric oxide.

All personnel should receive *training* on the use, hazards, protective measures, emergency actions, and other precautions per 29 CFR 1910.1200 (Hazard Communication) prior to the first assignment in an area where nitric oxide is used or stored.

Before beginning employment and at regular intervals thereafter (e.g., annually), the following medical test is recommended:

☑ Lung function tests (establish baseline).

If symptoms develop or overexposure is suspected, the following may be useful:

☑ Repeat lung function tests (compare to baseline).

☑ Consider chest X-ray after acute exposure (may be negative if taken immediately following exposure due to delayed development of pulmonary edema).

☑ Examination and evaluation of the nervous system functions by a qualified neurologist.

It should be noted that medical tests that simply look for existing damage are not a substitute for controlling exposures. Medical histories are extremely important when assessing exposure risk. Also, since smoking can cause heart disease as well as lung cancer, emphysema, and other respiratory ailments, any exposure to nitric oxide can result in quicker and more devastating symptoms. Smokers should therefore avoid unprotected contact with this chemical. Prudent risk management requires proper consideration of *all* possible factors that may be causing the appearance of exposure symptoms among workers.

Other methods to reduce exposure include:

☑ Always ensure that proper protective clothing is worn when using chemical substances, and that personnel are trained on its use, care, and maintenance.

☑ Wash thoroughly immediately after exposure to nitric oxide and at the end of the work shift or before eating, drinking, or smoking.

☑ Hazard warning information should be posted in the work area. In addition, as part of an on-going education and training program, all information on the health and safety hazards of nitric oxide should be communicated to all potentially exposed workers.

☑ Eye wash stations should be provided in the immediate work area for emergency use. If there is a possibility of skin exposure to compressed nitric oxide gas, emergency shower facilities should also be provided.

☑ Workers whose clothing has been contaminated by nitric oxide should change into clean clothes before leaving work. Contaminated work clothing should be laundered only by individuals who have been informed of the hazards of exposure to chemical substances such as nitric oxide.

RISK ASSESSMENT: ENVIRONMENT

General Assessment

The environment is always at risk of exposure during transportation, storage, disposal, or destruction of nitric oxide. In almost every scenario, the threat of environmental exposure is contingent upon the proper handling of the chemical substance. As a gas, it primarily affects the air. However, depending upon the nature and conditions of the release, accidental spills, large or small, can result in contamination of the other surrounding environmental mediums (water and soil).

Nitric oxide is non-flammable gas. However, as an oxidizer it can greatly accelerate the burning of ordinary combustible materials such as wood and paper. This is especially true when nitric oxide is heated. Also, on exposure to air, it will release nitrogen dioxide and nitrogen tetroxide which are both extremely toxic and very powerful oxidizers. It is incompatible with many fuels and hydrocarbons, as well as other common gases such as carbon monoxide, ozone, and phosphine. Caution is always required in handling, storage, transportation, and disposal of nitric oxide. Emergency responders should be made aware of the presence of nitric oxide at any emergency response situation.

Nitric oxide may enter the environment through industrial discharges and from spills.

Acute Ecological Effects

Acute (short-term) toxic effects may include the death of animals, birds, or fish and death or low growth rate in plants. Acute effects are seen 2 to 4 days after animals or plants are exposed to nitric oxide. Insufficient data are available to evaluate or predict the acute (short-term) effects of nitric oxide on aquatic life, plants, birds, or land animals.

Chronic Ecological Effects

Chronic (long-term) toxic effects may include shortened life span, reproductive problems, lower fertility, and changes in appearance or behavior in exposed animals. These effects can be seen long after first exposure(s) to toxic chemicals. Insufficient data are available to evaluate the chronic (long-term) effects of nitric oxide on aquatic life, plants, birds, or land animals.

Water Solubility

Nitric oxide is slightly soluble in water. Concentrations of 1 to 100 milligrams will mix with a liter of water.

Persistence in the Environment

Nitric oxide is slightly persistent in the aquatic environment, with a half-life between 2 and 20 days. The half-life of a pollutant is the amount of time it takes for one-half of the chemical to be degraded. As a gas, virtually 100% of nitric oxide will eventually end up in the air.

Bioaccumulation in Aquatic Organisms

Some substances increase in concentration, or *bioaccumulate,* in living organisms as they breathe contaminated air, drink contaminated water, or eat contaminated food. These chemicals can become concentrated in the tissues and internal organs of animals as well as humans.

The concentration of nitric oxide found in fish tissues is expected to be somewhat lower than the average concentration of nitric oxide in the water from which the fish was taken.

Recommended Risk-Reduction Measures

Proper training of all transporters will reduce the likelihood of a mishap or accident resulting in a leak or spill of nitric oxide into the environment. Labels on all containers, trucks, and rail cars must meet DOT requirements and accurately reflect their contents to enable emergency responders to react properly and quickly to any disaster thereby reducing the potential risk to the environment and to personnel.

Storage of nitric oxide should be segregated from other chemicals to minimize the risk of cross-contamination. Nitric oxide must be stored to avoid contact with incompatible materials since violent reactions can occur. Containers should be protected from physical damage and stored to avoid contact with heat.

If a spill or leak to the environment has occurred, fire department, emergency response and/or hazardous materials spill personnel should be notified immediately. Cleanup should be attempted only by those trained in proper spill containment procedures. Contaminated soils should be removed for incineration and replaced with clean soil.

If nitric oxide should contact the water table, aquifer, or navigable waterway, time is of the essence. Because it is slightly soluble in water, total remediation may not be possible. The local and/or state emergency response authorities must be notified. A comprehensive emergency response or disaster preparedness/recovery plan should be in place prior to any operations involving the use, transportation, storage, or disposal of nitric oxide.

If nitric oxide is spilled or leaked, the following specific steps are recommended:

☑ Restrict persons not wearing protective clothing from area of spill or leak until cleanup is complete. Avoid skin and eye contact.

☑ Absorb spilled liquids with vermiculite or a similar material and deposit in sealed drum for disposal.

☑ Ventilate area of spill or leak.

☑ Remove ignition sources from the area.

☑ If source if leak is a cylinder, attempt to stop flow. If this cannot be done or cylinder is damaged, remove to safe place outdoors and allow to vent until empty.

RISK ASSESSMENT: BUSINESS

General Assessment

Accidents or mishaps involving nitric oxide can present a serious threat to business operations. The loss or damage of equipment or facilities can significantly affect fiscal viability. Lawsuits that may result from personnel illness, injury/death, public exposures, and/or environmental contamination will require a serious expenditure of resources. Media attention surrounding an injury, death, or environmental damage can also result in a loss of profits and loss of current as well as future business.

Recommended Risk-Reduction Measures

Company attorneys, safety and health professionals, and environmental specialists should be involved in the development of any procedures for responding to chemical incidents. A company official should be predesignated as a public relations officer with specific training in dealing with the press. Corporate plans and policies should be developed, approved, and implemented long before any need for such arises.

REFERENCES

Ahlbom, A. 1993. *Biostatistics for Epidemiologists.* New York: CRC Press/Lewis Publishers.

American Conference of Governmental Industrial Hygienists. 1988. *Documentation of the Threshold Limit Values and Biological Exposure Indices,* 5th Edition (with updates). Cincinnati: ACGIH

Anthony, C. P., and N. J. Kolthoff. 1971. *Textbook of Anatomy and Physiology.* St. Louis: C. V. Mosby Company.

Brusick, D. J. 1994. *Methods for Genetic Risk Assessment.* New York: CRC Press/Lewis Publishers.

Bueche, J. R. 1972. *Principles of Physics,* 2nd Edition. New York: McGraw-Hill.

Burrell, R., D. K. Flaherty, and L. J. Sauers. 1992. *Toxicology of the Immune System.* New York: Van Nostrand Reinhold.

Calabrese, E. J. 1994. *Biological Effects of Low Level Exposures.* New York: CRC Press/Lewis Publishers.

Cockerman, L. G., B. S. Shane. 1994. *Basic Environmental Toxicology.* New York: CRC Press/Lewis Publishers.

Cothern, C. R. 1995. *Handbook for Environmental Risk Decisionmaking.* New York: CRC Press/Lewis Publishers.

Danse, I. R. 1991. *Common Sense Toxics in the Workplace.* New York: Van Nostrand Reinhold.

Environmental Protection Agency. 1983. *Chemical Hazard Information Profiles.* Washington, D.C.: U.S. EPA.

Fodden, C. R., and J. L. Weddell. 1992. *Hazardous Materials Emergency Action Data.* New York: CRC Press/Lewis Publishers.

Forseberg, K., and L. H. Keith. 1993. *Chemical; Protective Clothing Permeation and Degradation Database.* Chelsea, MI: Lewis Publishers.

Gardner, R. W. 1994. *Chemical Intolerance.* New York: CRC Press/Lewis Publishers.

Hallenbeck, W. H. 1993. *Quantitative Risk Assessment for Environmental and Occupational Health,* 2nd Edition. New York: CRC Press/Lewis Publishers.

Hodgson, E., R. B. Mailman, and Janice E. Chambers. 1988. *Dictionary of Toxicology.* London: The MacMillian Company.

Hosty, J. W., and P. Foster. 1990. *A Practical Guide to Chemical Spill Response.* New York: Van Nostrand Reinhold.

Howard, P. H., et. al. 1991. *Handbook of Environmental Degradation Rates.* New York: CRC Press/Lewis Publishers.

Jones, F. E. 1994. *Toxic Organic Vapors in the Workplace.* New York: CRC Press/Lewis Publishers.

Kamrin, M. A. 1988. *Toxicology: A Primer on Principles and Applications.* Chelsea, Michigan: Lewis Publishers.

Keith, L. H. and M. M. Walker. 1995. *Handbook of Air Toxics.* New York: CRC Press/Lewis Publishers.

Landis, W. G., M. H. Yu. 1995. *Introduction to Environmental Toxicology.* New York: CRC Press/Lewis Publishers.

Langerman, N. 1994 *Precautionary Labels for Chemical Containers.* New York: CRC Press/Lewis Publishers.

Lauwerys, R. R., P. Hoet. 1993. *Industrial Chemical Exposure.* New York: CRC Press/Lewis Publishers.

Lewis, .J., Sr. 1993. *Hawley's Condensed Chemical Dictionary,* 12th Edition. New York: Van Nostrand Reinhold.

U.S. Department of Health and Human Services, National Institute for Occupational Safety and Health. 1994. *NIOSH Pocket Guide to Chemical Hazards.* Washington, D.C.: U.S. Government Printing Office.

MATERIAL SAFETY DATA SHEET

CHEMICAL NAME	
NITROBENZENE	**CRC PRESS/LEWIS PUBLISHERS DISCLAIMER:** THE INFORMATION AND RECOMMENDATIONS PRESENTED HEREIN ARE BASED ON SOURCES BELIEVED TO BE RELIABLE. CRC MAKES NO REPRESENTATION ON ITS COMPLETENESS OR ACCURACY. IT IS THE RESPONSIBILITY OF THE USER TO DETERMINE THE CHEMICAL'S SUITABILITY FOR ITS INTENDED USE, THE CHEMICAL'S SAFE USE, AND THE CHEMICAL'S PROPER DISPOSAL. NO REPRESENTATIONS AND/OR WARRANTIES, EITHER EXPRESSED OR IMPLIED, OF THE MERCHANTABILITY OR FITNESS FOR A PARTICULAR PURPOSE, OR OF ANY OTHER NATURE, ARE MADE WITH RESPECT TO THE INFORMATION PROVIDED IN THIS MSDS OR TO THE CHEMICAL TO WHICH INFORMATION MAY REFER. CRC NEITHER ASSUMES NOR AUTHORIZES ANY OTHER PERSON TO ASSUME FOR IT, ANY OTHER ADDITIONAL RESPONSIBILITY OR LIABILITY FOR THE USE OF, OR RELIANCE UPON, THIS INFORMATION.

HAZARD WARNING INFORMATION

HEALTH	FIRE	REACTIVE	OTHER	DEGREE OF HAZARD	COLOR CODING	OTHER CODES
3	2	0		0 = Minimum Hazard 1 = Slight Hazard 2 = Moderate Hazard 3 = Serious Hazard 4 = Severe Hazard	HEALTH = BLUE FIRE = RED REACTIVITY = YELLOW OTHER = WHITE	OX = Oxidizer ACID = Acid ALK = Alkali COR = Corrosive W = Use No Water

SECTION I - GENERAL INFORMATION

Characterization	**Aromatic Hydrocarbon, Solvent**	RCRA Number	**U169**
		EPA Class	**Toxic Waste**
DOT Proper Shipping Name	**Nitrobenzene, liquid; Nitrobenzol**	Chemical Abstract Service (CAS) Number	**98-95-3**
DOT Hazard Class and Label Requirements	**Poison B**	DOT Emergency Guide Code	**56**
DOT Identification Number	**UN 1662**	Chemical Formula	**$C_6H_5NO_2$**

Synonyms

Nitrobenzol; Essence of Myrbane; Oil of Myrbane.

SECTION II - HAZARDOUS INGREDIENTS AND IDENTITY INFORMATION

Hazardous Components (specific identity)	OSHA Exposure Criteria	NIOSH Exposure Criteria	Immediately Dangerous to Life and Health (IDLH)	ACGIH Exposure Criteria
Nitrobenzene (derivation: From benzene by nitrating with nitric acid-sulfuric acid mix). **1 ppm = 5 mg/m³**	**PEL (skin):** **1 ppm** **5 mg/m³** **STEL:** **Not Established**	**REL:** **Not Established** **STEL:** **Not Established**	**Not Determined**	**TLV (skin):** **1 ppm** **5 mg/m³** **STEL:** **Not Established**

SECTION III - PHYSICAL AND CHEMICAL CHARACTERISTICS

Boiling Point	**410°F (210°C)**	Specific Gravity (H_2O = 1)	**1.205**
Vapor Pressure (gas)	**<1 at 69°F (20°C)**	Molecular Weight	**123.12**
Vapor Density (Air = 1)	**4.25**	Melting Point	**43°F (6°C)**

Solubility

Slightly soluble in water. Soluble in alcohol, benzene, and ether.

Appearance and Odor

Colorless to pale-yellow oily liquid, or bright yellow solid crystals with a distinctive odor of almonds. Odor Threshold = 4.7 ppb.

SECTION IV - FIRE AND EXPLOSION HAZARD DATA

Flash Point (method used)	**190°F (88°C) closed cup**	Explosive Limits in Air % by Volume	**LEL: 1.8% UEL: Not Determined**
NFPA Classification	**Combustible Liquid**	Autoignition Temperature	**900°F (482°C)**

Extinguishing Media

For small fires, use dry chemical, carbon dioxide, water spray, or regular foam. For large fires, use water spray, fog, or regular foam. Apply water as fog in flooding amounts (solid stream may be ineffective).

Special Fire Fighting Procedures

Wear full protective clothing and self-contained breathing apparatus (SCBA). Poisonous gases are produced in fire. Use water spray to keep fire-exposed containers cool.

Unusual Fire and Explosion Hazards

Nitrobenzene vapor is more than four times heavier than air. It can travel for great distances and settle in low-lying areas. If it contacts heat or flame, flashback can occur to cause fire or explosion.

SECTION V - REACTIVITY DATA

Stability		Conditions to Avoid **Nitrobenzene is normally stable in closed containers under routine conditions of handling and storage. It must be kept away from incompatible materials and heat.**
Stable	Unstable	Incompatibility *(materials to avoid)*
X		**Reacts vigorously with reducing agents such as nitric acid, phenol, aluminum chloride, aniline, glycerin, silver perchlorate, dinitrogen tetroxide, caustics, and reactive metals (tin or zinc).**
Hazardous Polymerization		Conditions to Avoid **Under normal conditions of temperature and pressure, hazardous polymerization of nitrobenzene is not expected to occur.**
May Occur	Will Not Occur	Hazardous Decomposition or By-products
	X	**When heated to decomposition, nitrobenzene can emit highly toxic/poisonous gases, including carbon monoxide and oxides of nitrogen. Nitrobenzene can explode when heated.**

SECTION VI - HEALTH HAZARD DATA

Primary Route(s) of Entry:	Inhalation?	Absorption (skin)?	Ingestion?
	X	X	X

Health Hazards

INHALATION: **Headache, vertigo, nausea and vomiting, anemia, anoxia, shortness of breath, cyanosis, rapid pulse, CNS depression, coma, death. Forms methemoglobin.**

ABSORPTION: **Skin irritant. Will pass through skin to cause toxic systemic effects (same as inhalation).**

INGESTION: **Not a likely exposure route, but possible. May cause gastrointestinal irritation and can lead to symptoms similar to that of inhalation.**

Carcinogenicity	NTP Listed?	IARC Cancer Review Group?	OSHA Regulated?	Target Organs?
Unknown Human **Unknown Animal**	**No**	**No**	**29 CFR 1910.1000 Table Z-1**	**Eyes, skin, blood, CVS, CNS, liver, kidney, testes.**

Medical Conditions Generally Aggravated by Exposure

Blood, heart, and liver disorders.

Emergency and First-aid Procedures

Eye contact: Flush immediately with water for 15 minutes (minimum); seek medical attention. Skin contact: Remove all contaminated clothing. Immediately wash area with large amounts of soap and water. Seek medical assistance. For inhalation: Remove the person from exposure. Provide respiratory assistance and CPR, if required. Transfer to medical facility. If swallowed: Unless advised otherwise, give 2 tablespoons Ipecac (adult dose) and 1-2 glasses of water. Induce vomiting. Seek medical attention immediately. NOTE: Alcohol ingestion in a heated environment may increase individual susceptibility.

SECTION VII - PRECAUTIONS FOR SAFE HANDLING AND USE

Steps to be Taken in Case Material is Released or Spilled

Absorb liquids in vermiculite, dry sand, earth, or other non-reactive material and deposit in sealed drum. Ventilate area of spill or leak. For dry spills, collect for reuse or disposal. Do NOT dry sweep. Use water spray to knock-down vapors (for solution spills) and wet mop residue of dry spills.

Preferred Waste Disposal Method

No citation.

Precautions to be Taken in Handling and Storage

Store to avoid contact with incompatible materials in tightly closed containers in cool, well-ventilated area away from heat and flame. Store in detached facility away from main plant or facility.

Other Precautions and Warnings

Sources of ignition are prohibited where nitrobenzene is used, handled, or stored. Heat may cause solids to liquefy and containers to build up pressure and explode.

SECTION VIII - CONTROL MEASURES AND PERSONAL PROTECTIVE EQUIPMENT

Respiratory Protection *(specify type)*

PEL is low and difficult to monitor. For exposures above the PEL, use an MSHA/NIOSH-approved supplied-air respirator with full facepiece, hood, or helmet in continuous flow mode, or use a self-contained breathing apparatus (SCBA) operated in pressure demand or other positive pressure mode.

Ventilation

Local exhaust or general mechanical systems recommended.

Protective Gloves	Eye Protection	Other Protective Clothing
Impervious Material	**Chemical/Dust Goggles or Face Mask**	**Protective Apron**

Work/Hygiene Practices

Always wash hands thoroughly after using chemical; never bring food, drink, or smoking materials into vicinity of chemicals.

NITROBENZENE

$C_6H_5NO_2$

CAS: 98-95-3

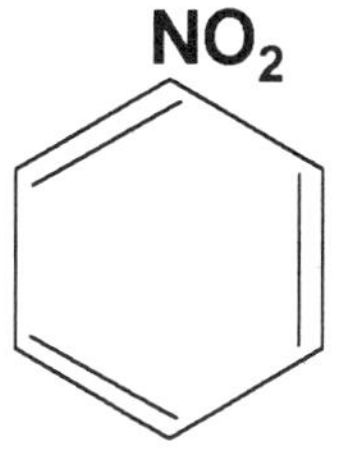

IDENTIFICATION AND TYPICAL USES

Nitrobenzene is a clear, colorless to pale yellow oily liquid or, it can appear as a bright yellow crystalline solid. It has a distinctive odor of almond oil. It is used in the manufacture of aniline, as a solvent for cellulose ethers, for modifying esterification of cellulose acetate, as an ingredient in metal polishes and shoe polishes, and in the manufacture of benzidine, quinoline, azobenzene, and other chemicals.

RISK ASSESSMENT: HEALTH

General Assessment

Nitrobenzene is a human poison. It can enter the body through all routes of exposure (*inhalation, ingestion,* and skin *absorption*). While there are no data on its carcinogenic abilities in humans or animals, it is known to cause mutations (genetic changes) in experimental tests. It is also a known animal teratogen and can cause adverse reproductive effects.

Inhalation can cause irritation to the respiratory system and associated mucosa. Symptoms include headaches, dizziness, vomiting, nausea, anemia, anoxia, dypsnea, rapid pulse, irritation of the eyes, coma, and possibly death. It also results in the formation of methemoglobin (a pigment similar to hemoglobin that is unable to perform the normal oxygenation process). This may be indicated by the development of cyanosis (blue discoloration of the fingers, toes, lips, and facial skin due to the lack of oxygen in the blood). There may also be deleterious effects on the central nervous system (CNS) and cardiovascular system (CVS) following high exposures.

Skin contact can produce the same systemic systems as inhalation as a result of the chemical's ability to pass readily through intact skin. There may also be localized irritation with the appearance of small blisters, redness, and/or swelling. Eye contact with the liquid solution or dusts can cause irritation, redness and swelling of the eyelids, painful sensitivity to light, and can cause severe eye damage and possible loss of vision if not removed immediately from contact with the eye.

Ingestion of nitrobenzene can also cause systemic effects similar to those of inhalation. There can also be painful irritation of the mouth and stomach, stomach cramps, and diarrhea. A bitter almond taste or burning sensation in the mouth, dry throat, and thirst can occur.

Alcohol consumption, some medications, exposure to sunlight, or hot baths can aggravate and enhance symptoms.

Acute Health Effects

The following acute (short-term) health effects may occur immediately or shortly after exposure to nitrobenzene:

Skin: Irritation and possible blistering. It will pass through unbroken skin to cause toxic systemic effects, including weakness, nausea, collapse, and even death.

Eye: Severe irritation with damage and possible damage to or loss of vision.

Lung: Irritation to the respiratory tract and associated mucosa. Can cause weakness, nausea, confusion, collapse, coma, and death (respiratory failure).

CNS: A CNS depressant causing weakness, confusion, vertigo, mood changes and other personality changes, nausea, collapse.

CVS: Causes adverse effects on the function of the cardiovascular system. It forms methemoglobin which interferes with the blood's ability to effectively transport oxygen to the cells. This can lead to serious, life-threatening illness.

Chronic Health Effects

The following chronic (long-term) health effects may occur at some time after exposure to nitrobenzene and can last for months or even years:

Cancer Hazards: According to information presented in the references, nitrobenzene has not been adequately tested for its ability to cause cancer in test animals. Some studies have suggested that exposure to nitrobenzene causes mutations (genetic changes in living cells). Many scientists believe that such chemicals may pose a cancer risk in the long-term. There is no evidence that nitrobenzene causes cancer.

Reproductive Hazard: According to information presented in the references, nitrobenzene has been shown to be a teratogen in experimental tests and it may also have the ability to adversely affect reproduction in animals.

Other Chronic Effects: Long-term exposure may produce a reversible anemia.

Recommended Risk-Reduction Measures

Personnel should avoid direct contact with nitrobenzene. It is an extremely toxic skin and eye irritant with the ability to enter the body through all routes of exposure. If a less toxic chemical cannot be substituted for a hazardous substance, then *engineering controls* are the most effective method of reducing exposures. The best protection is to enclose operations and/or provide local exhaust ventilation at the site of chemical release. While not always operationally feasible, isolating operations can also reduce exposure. Using respiratory protection is less effective than the controls mentioned above, but is still advisable whenever working with or around nitrobenzene. The exposure level is low (1 ppm) and difficult to monitor. For exposures above the PEL, use an MSHA/NIOSH-approved full facepiece respirator equipped with an organic vapor cartridge equipped with a dust/mist pre-filter. Better protection is obtained using a supplied-air respirator with full facepiece operated in continuous flow mode, or a self-contained breathing apparatus (SCBA) with full facepiece operated in pressure demand. If a full facepiece is not available, then chemical goggles should be worn to protect the eyes. Whenever a chemical splash hazard exists, a face shield and a protective apron should be worn. To prevent hand and skin exposures, protective gloves should be used. Glove manufacturers should be contacted and permeation studies obtained *before* final glove selections have been made.

Administrative controls should also be in place to minimize the potential for human exposures. These may include written procedures or policies which specify the methods and techniques that will be practiced whenever personnel are to work with nitrobenzene.

All personnel should receive *training* on the use, hazards, protective measures, emergency actions, and other precautions per 29 CFR 1910.1200 (Hazard Communication), prior to the first assignment in an area where nitrobenzene is used or stored.

Before beginning employment with nitrobenzene and at regular intervals thereafter (e.g., annually), those persons with high or frequent exposures should be provided the following recommended medical tests:

- ☑ Kidney and liver function tests.
- ☑ Complete blood count with methemoglobin levels.

If symptoms develop or overexposure is suspected, the following tests are recommended:

- ☑ Kidney and liver function tests.
- ☑ Monitor methemoglobin levels hourly for at least 24 hours until definite decline is noticed..
- ☑ Consider possible reactions caused by alcohol consumption and/or medications.

Any evaluation should include a careful history of past and present symptoms with an examination. Medical tests that simply look for existing damage are not a substitute for controlling exposures. Physicians should note that symptoms may be delayed, so victim should be kept under observation. The methemoglobin concentration should be determined at regular intervals until indications of full reduction to hemoglobin occur.

Other methods to reduce exposure include:

- ☑ Where possible, enclose operations and use local exhaust ventilation at the site of chemical release. If local exhaust ventilation or enclosure is not used, respiratory protection should be mandatory.
- ☑ Always ensure that proper protective clothing is worn when using chemical substances and that personnel are trained in its use, care, and maintenance.
- ☑ Wash thoroughly immediately after exposure to nitrobenzene and at the end of the work shift or before eating, drinking, or smoking.
- ☑ Hazard warning information should be posted in the work area. In addition, as part of an on-going education and training program, all information on

the health and safety hazards of nitrobenzene should be communicated to all exposed and potentially exposed workers.

☑ Eye wash stations should be provided in the immediate work area for emergency use. If there is a possibility of skin exposure to nitrobenzene, emergency shower facilities should also be provided.

☑ Personnel should never wear contaminated clothing home (family members can be exposed). Clothing should be laundered by personnel who have been trained on the hazards of nitrobenzene.

RISK ASSESSMENT: ENVIRONMENT

General Assessment

The environment is at risk of exposure during transportation, storage, disposal, or destruction of nitrobenzene. In almost every scenario, the threat of environmental exposure is contingent upon the proper handling of the chemical substance. Accidental spills, large or small, can result in fire, explosion, and possible contamination of the surrounding environmental mediums (water, soil, and air).

Nitrobenzene is normally considered a combustible liquid, but it can also appear in mixed solutions of flammable liquids. In solid form, its dusts may form explosive mixtures in air. These characteristics require caution in handling, storage, transportation, and disposal. It is incompatible with many common reducing agents, caustics, and some reactive metals (such as zinc and tin). Contact with either can cause fire or explosion. Therefore, special consideration is required during any emergency situation involving a leak or spill of nitrobenzene. Should nitrobenzene ever come into contact with these incompatible substances during use, transportation, storage, or disposal, the formation of highly toxic and/or highly explosive commodities is extremely possible.

Nitrobenzene may enter the environment through industrial effluents or spills.

☠ *Acute Ecological Effects*

Acute (short-term) toxic effects may include the death of animals, birds, or fish and death or low growth rate in plants. Acute effects are seen 2 to 4 days after animals or plants are exposed to nitrobenzene.

Nitrobenzene appears to have high acute toxicity to aquatic life. Insufficient data are available to evaluate the short-term effects of nitrobenzene exposure to plants, birds or land animals.

Chronic Ecological Effects

Chronic (long-term) toxic effects may include shortened life span, reproductive problems, lower fertility, and changes in appearance or behavior in exposed animals. These effects can be seen long after first exposure(s) to toxic chemicals.

Insufficient data are available on the long-term effects of nitrobenzene to aquatic life, plants, birds, or land animals.

Water Solubility

Nitrobenzene is slightly soluble in water. Concentrations of 1 to 100 milligrams will mix with a liter of water.

⌛ *Persistence in the Environment*

Nitrobenzene is slightly persistent in water, with a half-life between 2 and 20 days. The half-life of a pollutant is the amount of time it takes for one half of the chemical to be degraded.

Bioaccumulation in Aquatic Organisms

Some substances increase in concentration, or *bioaccumulate,* in living organisms as they breathe contaminated air, drink contaminated water, or eat contaminated food. These chemicals can become concentrated in the tissues and internal organs of animals as well as humans.

The concentration of nitrobenzene found in fish tissue is expected to be somewhat higher than the average concentration of nitrobenzene in the water from which the fish was taken.

Recommended Risk-Reduction Measures

Proper training of all transporters will reduce the likelihood of a mishap or accident resulting in a leak or spill to the environment. The correct use of DOT labeling (per the MSDS) on all containers, trucks, and rail cars will enable emergency responders to react properly and quickly to any disaster thereby reducing the potential risk to the environment and to personnel.

Storage of nitrobenzene should be segregated from incompatible chemicals to minimize the risk of cross-contamination or contact. Buildings designated for storage should be equipped with appropriate fire protection systems (alarms, sprinklers, emergency lighting, portable extinguishers).

If a spill or leak to the environment has occurred, fire department, emergency response and/or hazardous materials spill personnel should be notified immediately. Cleanup should be attempted only by those trained in proper spill containment procedures using non-sparking tools. Contaminated soils should be removed for incineration and replaced with clean soil. If nitrobenzene should contact the water table, aquifer, or navigable waterway, time is of the essence. It is slightly soluble in the aquatic environment and total remediation may not be possible. The local and/or state emergency response authorities must be notified. A comprehensive emergency response or disaster preparedness/recovery plan should be in place prior to any operations involving the use, transportation, storage, or disposal of nitrobenzene. If nitrobenzene is spilled or leaked, the following specific steps are recommended:

- ☑ Restrict persons not wearing protective clothing from area of spill or leak until cleanup is complete and area can be opened for normal work.
- ☑ Remove all ignition sources, ventilate area of leak.
- ☑ Absorb liquid spills in vermiculite, dry sand, earth, or similar material and deposit in sealed containers. Collect solid spills for reuse of disposal. Wet mop spill area to collect residue.
- ☑ It may be necessary to dispose of nitrobenzene as a hazardous waste. The responsible state agency or the regional office of the federal Environmental Protection Agency (EPA) should be contacted for specific recommendations.

RISK ASSESSMENT: BUSINESS

General Assessment

Accidents or mishaps involving nitrobenzene can present a significant threat to business operations. The loss or damage of equipment or facilities can significantly affect fiscal viability. Lawsuits that may result from personnel injury/death, public exposures, and/or environmental contamination will also require a serious expenditure of resources. Media attention surrounding an injury, death, or environmental damage can also result in a loss of profits and loss of current as well as future business. Always remember that any time the term "reproductive hazard" is used, public emotion, hysteria, and ignorance can run equally high. This should be a consideration during the development of public relation policies.

Recommended Risk-Reduction Measures

Company attorneys, safety and health professionals, and environmental specialists should be involved in the development of any procedures dealing with safety or emergency response. A company official should be pre-designated as a public relations officer with specific training in dealing with the press. Corporate plans and policies should be developed, approved, and implemented long before any need for such arises.

REFERENCES

Asante-Duah, D. K. 1993. *Hazardous Waste Risk Assessment.* New York: CRC Press/Lewis Publishers.

Bloom, A. D. and F. J. DeSerres. 1995. *Ecotoxicity and Human Health.* New York: CRC Press/Lewis Publishers.

Calabrese, E. J. 1994. *Biological Effects of Low Level Exposures.* New York: CRC Press/Lewis Publishers.

Cockerman, L. G., B. S. Shane. 1994. *Basic Environmental Toxicology.* New York: CRC Press/Lewis Publishers.

Cothern, C. R., N. P. Ross. 1994. *Environmental Statistics, Assessment, and Forecasting.* New York: CRC Press/Lewis Publishers.

Hallenbeck, W. H. 1993. *Quantitative Risk Assessment for Environmental and Occupational Health,* 2nd Edition. New York: CRC Press/Lewis Publishers.

Howard, P. H. 1995. *Handbook of Environmental Fate and Exposures Data for Organic Chemicals,* Vol. I-IV. New York: CRC Press/Lewis Publishers.

Lauwerys, R. R., P. Hoet. 1993. *Industrial Chemical Exposure.* New York: CRC Press/Lewis Publishers.

Lewis, R. J., Sr. 1992. *Sax's Dangerous Properties of Industrial Materials,* Volumes 1, 2, and 3, 8th Edition. New York: Van Nostrand Reinhold.

Meyer, E. 1990. *Chemistry of Hazardous Materials.* Englewood Cliffs, NJ: Prentice Hall.

Quigley, D. R. 1994. *Handbook of Emergency Chemical Management.* New York: CRC Press/Lewis Publishers.

Torkelson, T. R. V. K. Rowe. 1981. Halogenated aliphatic hydrocarbons. *Patty's Industrial Hygiene and Toxicology,* Vol. 2B, *Toxicology,* pp 3568-3572. New York: Wiley-Interscience.

U.S. Department of Health and Human Services, National Institute for Occupational Safety and Health. 1994. *NIOSH Pocket Guide to Chemical Hazards.* Washington, D.C.: U.S. Government Printing Office.

MATERIAL SAFETY DATA SHEET

CHEMICAL NAME

NITROETHANE

CRC PRESS/LEWIS PUBLISHERS DISCLAIMER: THE INFORMATION AND RECOMMENDATIONS PRESENTED HEREIN ARE BASED ON SOURCES BELIEVED TO BE RELIABLE. CRC MAKES NO REPRESENTATION ON ITS COMPLETENESS OR ACCURACY. IT IS THE RESPONSIBILITY OF THE USER TO DETERMINE THE CHEMICAL'S SUITABILITY FOR ITS INTENDED USE, THE CHEMICAL'S SAFE USE, AND THE CHEMICAL'S PROPER DISPOSAL. NO REPRESENTATIONS AND/OR WARRANTIES, EITHER EXPRESSED OR IMPLIED, OF THE MERCHANTABILITY OR FITNESS FOR A PARTICULAR PURPOSE, OR OF ANY OTHER NATURE, ARE MADE WITH RESPECT TO THE INFORMATION PROVIDED IN THIS MSDS OR TO THE CHEMICAL TO WHICH INFORMATION MAY REFER. CRC NEITHER ASSUMES NOR AUTHORIZES ANY OTHER PERSON TO ASSUME FOR IT, ANY OTHER ADDITIONAL RESPONSIBILITY OR LIABILITY FOR THE USE OF, OR RELIANCE UPON, THIS INFORMATION.

HAZARD WARNING INFORMATION

HEALTH	FIRE	REACTIVE	OTHER	DEGREE OF HAZARD	COLOR CODING	OTHER CODES
1	3	3		0 = Minimum Hazard 1 = Slight Hazard 2 = Moderate Hazard 3 = Serious Hazard 4 = Severe Hazard	HEALTH = BLUE FIRE = RED REACTIVITY = YELLOW OTHER = WHITE	OX = Oxidizer ACID = Acid ALK = Alkali COR = Corrosive W̶ = Use No Water

SECTION I - GENERAL INFORMATION

Characterization **Nitroparaffin**	RCRA Number **D001**	EPA Class **Characteristic (I) Waste**
DOT Proper Shipping Name **Nitroethane**	Chemical Abstract Service (CAS) Number **79-24-3**	
DOT Hazard Class and Label Requirements **Flammable or Combustible Liquid**	DOT Emergency Guide Code **26**	
DOT Identification Number **UN 2842**	Chemical Formula $CH_3CH_2NO_2$	

Synonyms
Nitroetan.

SECTION II - HAZARDOUS INGREDIENTS AND IDENTITY INFORMATION

Hazardous Components (specific identity)	OSHA Exposure Criteria	NIOSH Exposure Criteria	Immediately Dangerous to Life and Health (IDLH)	ACGIH Exposure Criteria
Nitroethane (derivation: By reaction of propane with nitric acid under pressure). **1 ppm = 3.12 mg/m³**	**PEL:** **100 ppm** **310 mg/m³** **STEL:** **Not Established**	**REL:** **100 ppm** **310 mg/m³** **STEL:** **Not Established**	**1000 ppm**	**TLV:** **100 ppm** **307 mg/m³** **STEL:** **Not Established**

SECTION III - PHYSICAL AND CHEMICAL CHARACTERISTICS

Boiling Point **237°F (114°C)**	Specific Gravity (H_2O = 1) **1.05**
Vapor Pressure (mm Hg) **15.6 at 68°F (20°C)**	Molecular Weight **75.1**
Vapor Density (Air = 1) **2.58**	Freezing Point **-130°F (-90°C)**

Solubility
Slightly soluble in water (5%). Soluble in acid and alkali; miscible in chloroform, ethanol, methanol, and ether.

Appearance and Odor
Colorless, oily liquid with a mild, fruity odor. Odor Threshold = 163 ppm.

SECTION IV - FIRE AND EXPLOSION HAZARD DATA

Flash Point (method used) **82°F (28°C) closed cup**	Explosive Limits in Air % by Volume **LEL: 3.4% UEL: Not Determined**
NFPA Classification **Class IC Flammable Liquid**	Autoignition Temperature **778°F (414°C)**

Extinguishing Media
Use carbon dioxide, water spray, or alcohol-resistant foam.

Special Fire Fighting Procedures
Nitroethane is a highly flammable, very dangerous fire and explosion hazard. Poisonous gases are evolved from fire. Wear a self-contained breathing apparatus (SCBA) with full facepiece. Move containers from fire area if it can be done without risk or cool fire-exposed containers with water spray.

Unusual Fire and Explosion Hazards
Containers may explode in flame. Vapors are heavier than air and can travel for great distances to an ignition source and flashback to cause fire or explosion.

SECTION V - REACTIVITY DATA

Stability		Conditions to Avoid Nitroethane is normally stable in its closed containers, during routine operations (those that do not expose it to ignition sources). Avoid contact with heat, sparks, flame, lighted tobacco products, and steam lines.
Stable X	Unstable	Incompatibility *(materials to avoid)* Nitroethane is reactive in contact with hydroxides, inorganic bases, metal oxides, amines, strong alkalis, acids and oxidizers, and hydrocarbons. Nitroethane can be explosive if confined or contaminated.
Hazardous Polymerization		Conditions to Avoid Under normal conditions of temperature and pressure, hazardous polymerization of nitroethane is not expected to occur.
May Occur	Will Not Occur X	Hazardous Decomposition or By-products If involved in fire, nitroethane can produce toxic carbon and nitrogen oxides.

SECTION VI - HEALTH HAZARD DATA

Primary Route(s) of Entry:	Inhalation? X	Absorption (skin)?	Ingestion? X

Health Hazards

INHALATION: Irritation to the eyes, nose, throat, and respiratory tract. Animal studies have shown a potential for the development of pulmonary rales and edema (fluid in the lungs), injury to the liver and kidney, and narcosis. May cause increase in methemoglobin levels.

SKIN & EYES: Skin contact causes irritation and may cause reddening and smarting if it remains in contact for extended period of time. Eye contact can cause tearing and irritation.

INGESTION: Causes irritation of the mouth and stomach, nausea, and vomiting.

Carcinogenicity	NTP Listed?	IARC Cancer Review Group?	OSHA Regulated?	Target Organs?
Unknown Human Unknown Animal	No	No	29 CFR 1010.1000 Table Z-1	Skin, eyes, CNS, liver, respiratory tract, kidneys.

Medical Conditions Generally Aggravated by Exposure

None reported.

Emergency and First-aid Procedures

Eye contact: Immediately flush with large amounts of water for 15 minutes (minimum), occasionally lifting eyelids; seek medical attention. Skin contact: Remove all contaminated clothing. Immediately wash area with large amounts of soap and water. For inhalation: Remove the person from exposure. Provide respiratory assistance and CPR. Transfer promptly to medical facility. Observe 48 hours for lung effects. If swallowed: Contact poison control center. Unless directed otherwise, give a conscious person 1 to 2 glasses of water, then induce vomiting.

SECTION VII - PRECAUTIONS FOR SAFE HANDLING AND USE

Steps to be Taken in Case Material is Released or Spilled

Remove all ignition sources. Ventilate area. Restrict entry into the area. Cover small spills with soda ash. Collect with non-sparking tools and place in open bucket of water and allow to stand for 2 hours. Neutralize with hydrogen chloride and flush to drain with copious amounts of water.

Preferred Waste Disposal Method

Dilute with fuel oil and atomize in proper combustion chamber.

Precautions to be Taken in Handling and Storage

Do not store with incompatible materials or in the presence of heat since violent reactions or explosions can occur. Store in tightly closed containers in a cool, dark, well-ventilated area.

Other Precautions and Warnings

Bulk storage of nitroethane is not recommended. Containers may explode in fire or under conditions of extreme heat (do not store outdoors or in direct sunlight).

SECTION VIII - CONTROL MEASURES AND PERSONAL PROTECTIVE EQUIPMENT

Respiratory Protection *(specify type)*

For exposures over the 8-hour PEL of 100 ppm, use an MSHA/NIOSH-approved air-purifying respirator with organic vapor cartridge. Otherwise, use a powered supplied-air respirator or a self-contained breathing apparatus (SCBA) with full facepiece operated in pressure demand or other positive pressure mode.

Ventilation

Local exhaust or general mechanical systems recommended.

Protective Gloves	Eye Protection	Other Protective Clothing
Impervious Material	Chemical Goggles and/or Face Mask	Protective Apron

Work/Hygiene Practices

Always wash hands thoroughly after using chemical; never bring food, drink, or smoking materials into vicinity of chemicals.

NITROETHANE

$CH_3CH_2NO_2$

CAS: 79-24-3

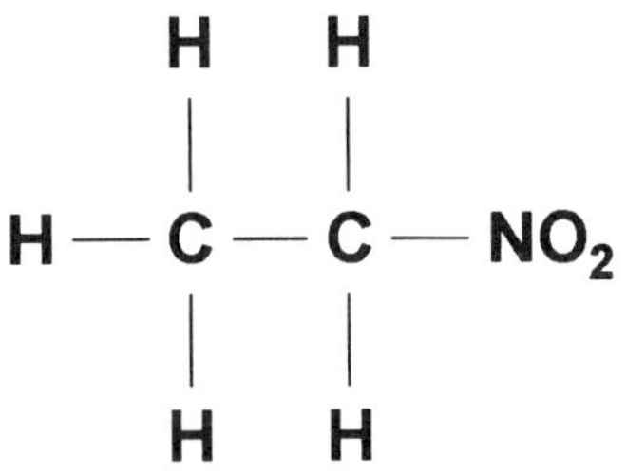

IDENTIFICATION AND TYPICAL USES

Nitroethane is a colorless, oily liquid with a mild, fruity odor. It is used as a solvent for nitrocellulose, cellulose acetate, cellulose acetopropionate, cellulose acetobutyrate, vinyl alkyd, and many other resins, waxes, fats, and dyestuffs. Also used in propellant research and as a fuel additive.

RISK ASSESSMENT: HEALTH

General Assessment

Nitroethane is moderately toxic by *ingestion* and *inhalation*. It is also a skin irritant, but dermal *absorption* has not been reported. There are no indications of any carcinogenic, mutagenic, or teratogenic effects following exposure.

Inhalation causes irritation of the mucosa of the eyes, nose, throat, and respiratory tract. Systemic effects have not been observed in humans. However, animal studies indicate a possibility for the development of pulmonary rales and edema (fluid in the lungs) which is a medical emergency. Symptoms of cough, congestion, labored breathing, and pain in the chest may be delayed up to 48 hours which may give a false sense of security with regard to health exposure risk. There may also be liver and kidney injury and, at high concentrations, narcosis. Severe exposure in humans is expected to cause similar effects. Nitroethane may also cause the formation of methemoglobin which can significantly interfere with the blood's ability to properly transport oxygen.

Skin contact causes mild to severe irritation and, if allowed to remain in contact for prolonged periods, either directly or in chemical-soaked clothing, there can be reddening and smarting at the site of contact. Eye contact can cause irritation and tearing.

Ingestion may result in irritation of the mouth and stomach, nausea, and vomiting.

Acute Health Effects

The following acute (short-term) health effects may occur immediately or shortly after exposure to nitroethane:

Skin: Irritation with redness and smarting at the site of contact.

Eye: Irritation and inflammation with a possibility of damage to visual acuity.

Lung: Irritation of the mucosa of the nose, throat, and respiratory tract. May lead to delayed pulmonary edema which can be fatal.

CNS: Unspecified effects associated with depression of the central nervous system.

Chronic Health Effects

The following chronic (long-term) health effects may occur at some time after exposure to nitroethane and can last for months or even years:

Cancer Hazards: According to information presented in the references, nitroethane has not been adequately tested for its ability to cause cancer in test animals.

Reproductive Hazards: There are no reports to support any claims of reproductive hazards.

Other Chronic Effects: Long-term exposure can lead to dermatitis.

Recommended Risk-Reduction Measures

Personnel should avoid direct contact with nitroethane. If another chemical cannot be substituted for a nitroethane, then *engineering controls* are the most effective method of reducing exposures. The best protection is to separate operations and provide local exhaust ventilation at the site of chemical release. While not always operationally feasible, isolating operations can also reduce exposure. Using equipment with explosion-proof designs and ensuring all metal containers and tools are electrically grounded and bonded can further reduce the potential for injury or damage. Using respiratory protection is less effective than the controls mentioned above, but is still advisable whenever working with nitroethane. For exposures over the 8-hour PEL (100 ppm), an air-purifying chemical cartridge respirator equipped with an organic vapor cartridge may provide adequate protection. However, best protection is obtained using a powered air-supplied respirator operated in pressure demand or other positive pressure mode, or a self-contained breathing apparatus (SCBA) with full facepiece operated in pressure demand mode. If a full facepiece is not available, then chemical goggles should be worn to protect the eyes. Whenever a chemical splash hazard exists, a face shield and a protective apron should be worn. To prevent hand and skin exposures with the liquid, thermal gloves should be used. Glove manufacturers should be contacted and permeation studies obtained *before* final glove selections have been made.

Administrative controls should also be in place to minimize the potential for human exposures. These may include written procedures or policies which specify the methods and techniques that will be practiced whenever personnel are to work with nitroethane.

All personnel should receive *training* on the use, hazards, protective measures, emergency actions, and other precautions per 29 CFR 1910.1200 (Hazard Communication), prior to the first assignment in an area where nitroethane is used or stored.

If overexposure is suspected or symptoms develop, the following medical tests are recommended:

☑ Lung function tests.

☑ Liver and kidney function tests.

☑ Methemoglobin level monitoring.

☑ Consider chest X-ray after acute overexposure (may be negative if taken immediately after exposure due to delayed onset of pulmonary edema).

Other methods to reduce exposure to chemicals include:

☑ Where possible, consider installing automatic sensors that warn personnel of dropping oxygen levels and use local exhaust ventilation at the site of chemical release. If local exhaust ventilation or enclosure is not used, respiratory protection should be mandatory.

☑ Always ensure that proper protective clothing is worn when using chemical substances.

☑ Wash thoroughly immediately after exposure to nitroethane and at the end of the work shift or before eating, drinking, or smoking.

☑ Safety shower and eyewash stations should be readily available in work areas where nitroethane is used or stored.

☑ Hazard warning information should be posted in the work area. In addition, as part of an on-going education and training program, all information on the health and safety hazards of nitroethane should be communicated to all exposed workers.

RISK ASSESSMENT: ENVIRONMENT

General Assessment

The environment is at risk of exposure during transportation, storage, disposal, or destruction of nitroethane. In almost every scenario, the threat of environmental exposure is contingent upon the proper handling of the chemical substance. Accidental releases, large or small, can result in fire, explosion, and possible contamination of the surrounding environmental mediums (water, soil, and air).

Nitroethane is a Class IC flammable liquid per OSHA 29 CFR 1910.106. It can present an extremely flammable and highly explosive hazard. Also, since it can react violently in the presence of oxidizers, acids, and hydrocarbons, nitroethane requires special consideration during any emergency involving a leak or release of nitroethane gas or liquid.

Nitroethane can enter the environment through industrial discharges or through spills.

☠ *Acute Ecological Effects*

Acute (short-term) toxic effects may include the death of animals, birds, or fish, and death or low growth rate in plants. Acute effects are seen 2 to 4 days after animals or plants are exposed to nitroethane. Insufficient data are available on the short-term effects of nitroethane on aquatic life, plants, birds, or land animals.

Chronic Ecological Effects

Chronic (long-term) toxic effects may include shortened life span, reproductive problems, lower fertility, and changes in appearance or behavior in exposed animals. These effects can be seen long after first exposure(s) to toxic chemicals. Insufficient data are available to evaluate the long-term effects of nitroethane to aquatic life, plants, birds, or land animals.

Water Solubility

Nitroethane is slightly soluble in water. Concentrations of 1 to 100 milligrams will mix with a liter of water.

Persistence in the Environment

Nitroethane is slightly persistent in water, with a half-life between 2 and 20 days. The half-life of a pollutant is the amount of time it takes for one half of the chemical to be degraded.

Bioaccumulation in Aquatic Organisms

Some substances increase in concentration, or *bioaccumulate,* in living organisms as they breathe contaminated air, drink contaminated water, or eat contaminated food. These chemicals can become concentrated in the tissues and internal organs of animals as well as humans.

The concentration of nitroethane found in the edible tissues of fish is expected to be somewhat lower than the average concentration found in the water from which the fish was taken.

Recommended Risk-Reduction Measures

Proper training of all transporters will reduce the likelihood of a mishap or accident resulting in a leak or release to the environment. The correct use of DOT labeling (per the MSDS) on all containers, trucks, and rail cars will enable emergency responders to react properly and quickly to any disaster thereby reducing the potential risk to the environment and to personnel.

Storage of nitroethane should be segregated from incompatible chemicals to minimize the risk of cross-contamination or contact. Buildings designated for storage should be equipped with appropriate fire protection systems (alarms, sprinklers, emergency lighting, portable extinguishers).

If a spill or release to the environment has occurred, fire department, emergency response, and/or hazardous materials spill personnel should be notified immediately. Cleanup should be attempted only by those trained in proper containment procedures using non-sparking tools. Contaminated soils should be removed for incineration and replaced with clean soil. When spills or leaks occur, it may be necessary to contact the local and/or state emergency response authorities. A comprehensive emergency response or disaster preparedness/recovery plan should be in place prior to any operations involving the use, transportation, storage, or disposal of nitroethane. If nitroethane is spilled or leaked, the following specific steps are recommended:

- ☑ Restrict persons not wearing protective clothing from area of spill or leak until cleanup is complete and area can be opened for normal work.
- ☑ Remove all ignition sources.
- ☑ Ventilate area of spill or leak.
- ☑ Absorb liquids using vermiculite, dry earth, or other material and place in sealed containers for disposal. For small spills, cover liquid with soda ash, place in an open bucket of water, and let stand for two hours. Neutralize with hydrogen chloride and flush to drain with flooding water.
- ☑ Keep nitroethane out of a confined space, such as a sewer, because of the possibility of explosion (unless the sewer is designed to prevent the build-up of explosive concentrations).
- ☑ It may be necessary to dispose of nitroethane as a hazardous waste. The responsible state agency or the regional office of the federal Environmental Protection Agency (EPA) should be contacted for specific recommendations.

RISK ASSESSMENT: BUSINESS

General Assessment

Accidents or mishaps involving nitroethane can present a significant threat to business operations. The loss or damage of equipment or facilities can significantly affect fiscal viability. Lawsuits that may result from personnel injury, public exposures, and environmental contamination will also require a serious expenditure of resources. Media attention surrounding an injury, illness, or environmental damage can also result in a loss of profits and loss of current as well as future business.

Recommended Risk-Reduction Measures

Company attorneys, safety and health professionals, and environmental specialists should be involved in the development of any procedures dealing with safety and emergency response. A company official should be pre-designated as a public relations officer with specific

training in dealing with the press. Corporate plans and policies should be developed, approved, and implemented long before any need for such arises.

REFERENCES

Ahlbom, A. 1993. *Biostatistics for Epidemiologists.* New York: CRC Press/Lewis Publishers.

American Conference of Governmental Industrial Hygienists. 1988. *Documentation of the Threshold Limit Values and Biological Exposure Indices,* 5th Edition (with updates). Cincinnati: ACGIH

Anthony, C. P., and N. J. Kolthoff. 1971. *Textbook of Anatomy and Physiology.* St. Louis: C. V. Mosby Company.

Asante-Duah, D. K. 1993. *Hazardous Waste Risk Assessment.* New York: CRC Press/Lewis Publishers.

Bloom, A. D. and F. J. DeSerres. 1995. *Ecotoxicity and Human Health.* New York: CRC Press/Lewis Publishers.

Brusick, D. J. 1994. *Methods for Genetic Risk Assessment.* New York: CRC Press/Lewis Publishers.

Bueche, J. R. 1972. *Principles of Physics,* 2nd Edition. New York: McGraw-Hill.

Burrell, R., D. K. Flaherty, and L. J. Sauers. 1992. *Toxicology of the Immune System.* New York: Van Nostrand Reinhold.

Calabrese, E. J. 1994. *Biological Effects of Low Level Exposures.* New York: CRC Press/Lewis Publishers.

Cockerman, L. G., B. S. Shane. 1994. *Basic Environmental Toxicology.* New York: CRC Press/Lewis Publishers.

Cothern, C. R. 1995. *Handbook for Environmental Risk Decisionmaking.* New York: CRC Press/Lewis Publishers.

Danse, I. R. 1991. *Common Sense Toxics in the Workplace.* New York: Van Nostrand Reinhold.

Environmental Protection Agency. 1983. *Chemical Hazard Information Profiles.* Washington, D.C.: U.S. EPA.

Fodden, C. R., and J. L. Weddell. 1992. *Hazardous Materials Emergency Action Data.* New York: CRC Press/Lewis Publishers.

Forseberg, K., and L. H. Keith. 1993. *Chemical; Protective Clothing Permeation and Degradation Database.* Chelsea, MI: Lewis Publishers.

Gardner, R. W. 1994. *Chemical Intolerance.* New York: CRC Press/Lewis Publishers.

Hallenbeck, W. H. 1993. *Quantitative Risk Assessment for Environmental and Occupational Health,* 2nd Edition. New York: CRC Press/Lewis Publishers.

Hodgson, E., R. B. Mailman, and J. E. Chambers. 1988. *Dictionary of Toxicology.* London: The MacMillian Company.

Hosty, J. W., and P. Foster. 1990. *A Practical Guide to Chemical Spill Response.* New York: Van Nostrand Reinhold.

Howard, P. H., et. al. 1991. *Handbook of Environmental Degradation Rates.* New York: CRC Press/Lewis Publishers.

Jones, F. E. 1994. *Toxic Organic Vapors in the Workplace.* New York: CRC Press/Lewis Publishers.

Kamrin, M. A. 1988. *Toxicology: A Primer on Principles and Applications.* Chelsea, Michigan: Lewis Publishers.

Keith, L. H. and M. M. Walker. 1995. *Handbook of Air Toxics.* New York: CRC Press/Lewis Publishers.

Landis, W. G., M. H. Yu. 1995. *Introduction to Environmental Toxicology.* New York: CRC Press/Lewis Publishers.

Lauwerys, R. R., P. Hoet. 1993. *Industrial Chemical Exposure.* New York: CRC Press/Lewis Publishers.

Lewis, R. J., Sr. 1993. *Hawley's Condensed Chemical Dictionary,* 12th Edition. New York: Van Nostrand Reinhold.

Lewis, R. J., Sr. 1992. *Sax's Dangerous Properties of Industrial Materials,* Volumes 1, 2, and 3, 8th Edition. New York: Van Nostrand Reinhold.

Lide, D. R. 1995. *Handbook of Chemistry and Physics,* 76th Edition. New York: CRC Press/Lewis Publishers.

Manahan, S. E. 1992. *Toxicological Chemistry,* 2nd Edition. New York: CRC Press/Lewis Publishers.

Mellström, G. A., J. E. Wahlberg, and H. I. Maibach. 1994. *Protective Gloves for Occupational Use.* New York: CRC Press/Lewis Publishers.

Meyer, E. 1990. *Chemistry of Hazardous Materials.* Englewood Cliffs, NJ: Prentice Hall.

Patnaik, P. 1992. *A Comprehensive Guide to the Hazardous Properties of Chemical Substances.* New York: Van Nostrand Reinhold.

Philp, R. B. 1995. *Environmental Hazards and Human Health.* New York: CRC Press/Lewis Publishers.

Proctor, N. H., and J. P. Hughes. 1991. *Chemical Hazards of the Workplace.* New York: Van Nostrand Reinhold.

Quigley, D. R. 1994. *Handbook of Emergency Chemical Management.* New York: CRC Press/Lewis Publishers.

Rea, W. J. 1995. *Chemical Sensitivity,* Vol. I-III. New York: CRC Press/Lewis Publishers.

Richardson, M. 1995. *Dictionary of Substance and Their Effects,* Vol. I-VI. New York: CRC Press/Lewis Publishers.

Sacarello, H. L. A. 1994. *The Comprehensive Handbook of Hazardous Materials.* New York: CRC Press/Lewis Publishers.

Sherman, J. D. 1988. *Chemical Exposure and Disease.* New York: Van Nostrand Reinhold.

Talbott, E., G. Craun. 1995. *An Introduction to Environmental Epidemiology.* New York: CRC Press/Lewis Publishers.

Torkelson, T. R. V. K. Rowe. 1981. *Patty's Industrial Hygiene and Toxicology,* Vol. 2B. New York: Wiley-Interscience.

U.S. Department of Health and Human Services, National Institute for Occupational Safety and Health. 1994. *NIOSH Pocket Guide to Chemical Hazards.* Washington, D.C.: U.S. Government Printing Office.

Witorsch, P. and S. V. Spagnolo. 1994. *Air Pollution and Lung Disease in Adults.* New York: CRC Press/Lewis Publishers.

MATERIAL SAFETY DATA SHEET

CHEMICAL NAME	
NITROGEN	**CRC PRESS/LEWIS PUBLISHERS DISCLAIMER:** THE INFORMATION AND RECOMMENDATIONS PRESENTED HEREIN ARE BASED ON SOURCES BELIEVED TO BE RELIABLE. CRC MAKES NO REPRESENTATION ON ITS COMPLETENESS OR ACCURACY. IT IS THE RESPONSIBILITY OF THE USER TO DETERMINE THE CHEMICAL'S SUITABILITY FOR ITS INTENDED USE, THE CHEMICAL'S SAFE USE, AND THE CHEMICAL'S PROPER DISPOSAL. NO REPRESENTATIONS AND/OR WARRANTIES, EITHER EXPRESSED OR IMPLIED, OF THE MERCHANTABILITY OR FITNESS FOR A PARTICULAR PURPOSE, OR OF ANY OTHER NATURE, ARE MADE WITH RESPECT TO THE INFORMATION PROVIDED IN THIS MSDS OR TO THE CHEMICAL TO WHICH INFORMATION MAY REFER. CRC NEITHER ASSUMES NOR AUTHORIZES ANY OTHER PERSON TO ASSUME FOR IT, ANY OTHER ADDITIONAL RESPONSIBILITY OR LIABILITY FOR THE USE OF, OR RELIANCE UPON, THIS INFORMATION.

HAZARD WARNING INFORMATION

HEALTH	FIRE	REACTIVE	OTHER	DEGREE OF HAZARD	COLOR CODING	OTHER CODES
3	0	0		0 = Minimum Hazard 1 = Slight Hazard 2 = Moderate Hazard 3 = Serious Hazard 4 = Severe Hazard	HEALTH = BLUE FIRE = RED REACTIVITY = YELLOW OTHER = WHITE	OX = Oxidizer ACID = Acid ALK = Alkali COR = Corrosive ~~W~~ = Use No Water

SECTION I - GENERAL INFORMATION

Characterization **Gas**	RCRA Number **None**	EPA Class **Not Applicable**
DOT Proper Shipping Name **Nitrogen**	Chemical Abstract Service (CAS) Number **7727-37-9**	
DOT Hazard Class and Label Requirements **Non Flammable Gas**	DOT Emergency Guide Code **15**	
DOT Identification Number **UN1066 (compressed) UN1977 (refrigerated liquid)**	Atomic Formula N_2	
Synonyms **None.**		

SECTION II - HAZARDOUS INGREDIENTS AND IDENTITY INFORMATION

Hazardous Components (specific identity)	OSHA Exposure Criteria	NIOSH Exposure Criteria	Immediately Dangerous to Life and Health (IDLH)	ACGIH Exposure Criteria
Nitrogen (derivation: From liquid air by fractional distillation, or by reducing ammonia).	**PEL:** Not Established **STEL:** Not Established	**REL:** Not Established **STEL:** Not Established	Not Determined	**TLV:** Not Established **STEL:** Not Established

SECTION III - PHYSICAL AND CHEMICAL CHARACTERISTICS

Boiling Point **-320°F (-196°C)**	Specific Gravity (H_2O = 1) **0.808 at -320°F (-196°C)**
Vapor Pressure (atmospheres) **33.5 (critical pressure)**	Molecular Weight **28.02**
Vapor Density (Air = 1) **.9655**	Freezing Point **-346°F (-210°C)**

Solubility
Very slightly soluble in water; insoluble in liquid ammonia and alcohol.

Appearance and Odor
Colorless, odorless, tasteless gas.

SECTION IV - FIRE AND EXPLOSION HAZARD DATA

Flash Point (method used) **Not Applicable**	Explosive Limits in Air % by Volume **LEL: Not Applicable UEL: Not Applicable**
NFPA Classification **Non-Flammable Gas**	Autoignition Temperature **Not Applicable**

Extinguishing Media
Use extinguishing media suitable to surrounding fires. Nitrogen will not ignite.

Special Fire Fighting Procedures
Fire can produce toxic vapors and gases: wear full protective clothing and self-contained breathing apparatus (SCBA). Contact with the liquid may cause frostbite. Heat will build pressure and may rupture closed storage containers. Keep fire-exposed containers cool with water spray. Pipes containing liquid nitrogen can condense oxygen from the air, which is a fire hazard.

Unusual Fire and Explosion Hazards
Nitrogen cylinders may explode in the heat of fire.

SECTION V - REACTIVITY DATA

Stability		Conditions to Avoid Nitrogen is stable in closed cylinders under routine conditions of handling and storage. Keep away from heat and other gases. It can condense and solidify other gases. Use metal containers made for cryogenics.
Stable X	Unstable	Incompatibility *(materials to avoid)* Nitrogen is a simple asphyxiant gas that is lighter than air. It can react violently with lithium, neodymium, titanium, and zirconium. Also incompatible with hydrogen gas and magnesium powder + ignition source.
Hazardous Polymerization		Conditions to Avoid Under normal conditions of temperature and pressure, hazardous polymerization of nitrogen is not expected to occur. Liquid nitrogen will attack some rubber and plastics, making them brittle.
May Occur	Will Not Occur X	Hazardous Decomposition or By-products Forms cyanides when heated in the presence of alkalis or barium oxide. No other hazardous products of decomposition have been noted.

SECTION VI - HEALTH HAZARD DATA

Primary Route(s) of Entry:	Inhalation? X	Absorption (skin)?	Ingestion?

Health Hazards

INHALATION: An asphyxiant gas. Initial symptoms are those of oxygen starvation. Feeling of "air hunger," diminished mental alertness, headache, dullness, shortness of breath, and an impairment in muscle coordination. Continued exposures lead to faulty judgment, depression of senses, rapid fatigue, unconsciousness, convulsions, coma, and death.

SKIN & EYES: Contact with the compressed gas or refrigerated liquid can cause frostbite and burns.

INGESTION: Not likely since nitrogen is a gas at normal working (room) temperatures.

Carcinogenicity	NTP Listed?	IARC Cancer Review Group?	OSHA Regulated?	Target Organs?
No Evidence Human No Evidence Animal	No	No	No	Respiratory system; skin; eyes, CNS.

Medical Conditions Generally Aggravated by Exposure

None Reported.

Emergency and First-aid Procedures

Eye contact (liquid): Immediately flush with large amounts of tepid water for 15 minutes (minimum), occasionally lifting eyelids; seek medical attention. Skin contact (liquid): Remove all contaminated clothing. Immediately flush area with large amounts of tepid water. Do NOT rub! Seek medical assistance. For inhalation: Remove the person from exposure. Emergency responders must be protected from inert atmospheres. Provide respiratory assistance and CPR. Transfer promptly to medical facility. If swallowed: Treat frozen tissues and protect the airway. Seek medical attention immediately.

SECTION VII - PRECAUTIONS FOR SAFE HANDLING AND USE

Steps to be Taken in Case Material is Released or Spilled

Ventilate area. Stop flow of gas. If the source of leak is a cylinder and cannot be stopped, remove to safe place in open air; repair leak or let cylinder empty. Ensure response personnel are properly protected with self-contained breathing apparatus (SCBA).

Preferred Waste Disposal Method

No citation.

Precautions to be Taken in Handling and Storage

Store in tightly closed containers in a cool, dry, well-ventilated area away from heat, direct sunlight, or other hot areas (water lines, vent pipes). Ensure proper storage and handling precautions are in place whenever working with compressed gas cylinders.

Other Precautions and Warnings

Containers may explode in fire or under conditions of extreme heat. Protect cylinders from physical damage. Use pressure relief on liquid containers. Periodically check vent valves for blockage (ice).

SECTION VIII - CONTROL MEASURES AND PERSONAL PROTECTIVE EQUIPMENT

Respiratory Protection *(specify type)*

No exposure limits have been established. Nitrogen is a simple asphyxiant. Do NOT use cartridge-type respirators (ineffective for this type of gas). Use a powered supplied-air respirator or a self-contained breathing apparatus with full facepiece operated in pressure demand or other positive pressure mode.

Ventilation

Local exhaust or general mechanical systems recommended.

Protective Gloves	Eye Protection	Other Protective Clothing
Impervious, Thermal Insulated Gloves	Chemical Goggles and/or Face Mask	Protective Boots, Gauntlets, Apron

Work/Hygiene Practices

Always wash hands thoroughly after using chemical; never bring food, drink, or smoking materials into vicinity of chemicals.

NITROGEN

N_2

CAS: 7727-37-0

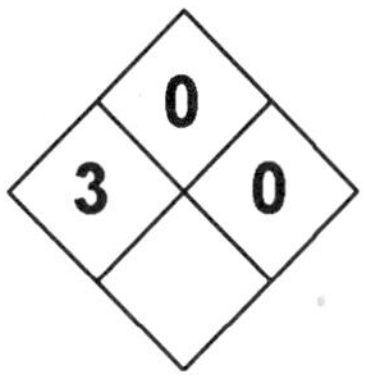

IDENTIFICATION AND TYPICAL USES

Nitrogen is a colorless, odorless, and tasteless gas. It is used in the production of ammonia, acrylonitrile, nitrates, cyanamide, cyanides, nitrides, in the manufacture of explosives, as an inert gas for purging, as an inert blanket gas, and for exerting gaseous pressures on systems and components. Also used in the electric and electronic industries, in food refrigeration and for freeze drying, for pressurizing liquid propellants, in quick-freezing foods, for chilling aluminum in foundries, for cryogenic preservation, as a food antioxidant, as a source of pressure in oil wells, for inflating tires, and as a major component of fertilizer mixtures.

RISK ASSESSMENT: HEALTH

General Assessment

Nitrogen is a simple asphyxiant gas. As such, its primary route of entry into the body is through *inhalation.* While it does not pose a significant physiological effect (i.e., no systemic poisoning or toxicity), it can displace oxygen levels to below that which is necessary to sustain life. Skin contact with the liquid can be destructive to tissues due to the extremely cold temperatures but absorption is not a likely exposure route. Since nitrogen usually exists as a gas at room temperatures, ingestion is not likely to occur either.

Inhalation of nitrogen gas can cause headache, dullness, shortness of breath, and the symptoms of asphyxiation. These include rapid respiration and a feeling of "air hunger" or air starvation followed by a decrease in mental alertness, and impaired muscular coordination. If exposure continues, there can be nausea, vomiting, prostration, unconsciousness, coma, and death.

Skin or eye contact with the vapor poses no serious threat. The vapors have not been shown to cause any physiological response. However, if the liquid should contact either the skin or the eyes, there can be serious irritation, burns, tissue damage, and frostbite.

Acute Health Effects

The following acute (short-term) health effects may occur immediately or shortly after exposure to nitrogen:

Skin: Liquid threat only. Redness, pain, swelling, blisters, and severe frostbite.

Eye: Liquid threat only. Causes frostbite, redness, irritation, severe pain, and blurred vision.

Lung: Vapor is non-irritating and therefore poses no threat of damage to the lungs. However, in high concentrations, nitrogen can lead to symptoms of asphyxiation and oxygen deprivation up to and including death.

Chronic Health Effects

The following chronic (long-term) health effects may occur at some time after exposure to nitrogen and can last for months or even years:

Cancer Hazards: According to information presented in the references, exposure to nitrogen does not lead to carcinogenic effects (it is a simple asphyxiant).

Reproductive Hazards: There are no reports to support any claims of reproductive hazards.

Other Chronic Effects: None reported.

Recommended Risk-Reduction Measures

Personnel should avoid direct contact with nitrogen. If another gas cannot be substituted for a nitrogen, then *engineering controls* are the most effective method of reducing exposures. The best protection is to separate operations and provide local exhaust ventilation at the site of chemical release. While not always operationally feasible, isolating operations can also reduce exposure. Using respiratory protection is less effective than the controls mentioned above, but is still advisable whenever working with nitrogen. Air-purifying respirators (e.g., the cartridge-type or gas masks) *cannot* be used with nitrogen. These types of respirators only purify the air and can do nothing to protect in oxygen-deficient atmospheres. A powered air-supplied respirator operated in pressure demand or other positive pressure mode, or a self-contained breathing apparatus (SCBA) with full facepiece operated in pressure demand mode are the recommended respiratory protection methods of choice. If a full facepiece is not available, then chemical goggles should be worn to protect the eyes. Whenever a chemical splash hazard exists, a face shield and a rubber or leather apron should be worn. To prevent hand and skin exposures with the liquid, thermal insulated gloves should be used. Glove manufacturers should be contacted and permeation studies obtained *before* final glove selections are made.

Administrative controls should also be in place to minimize the potential for human exposures. These may include written procedures or policies which specify the methods and techniques that will be practiced whenever personnel are to work with nitrogen.

All personnel should receive *training* on the use, hazards, protective measures, emergency actions, and other precautions per 29 CFR 1910.1200 (Hazard Communication), prior to the first assignment in an area where nitrogen is used or stored.

Other methods to reduce exposure to chemicals include:

☑ Where possible, consider installing automatic sensors that warn personnel of dropping oxygen levels and use local exhaust ventilation at the site of chemical release. If local exhaust ventilation or enclosure is not used, respiratory protection should be mandatory.

☑ Always ensure that proper protective clothing is worn when using chemical substances and that personnel are trained on its use, care, and maintenance.

☑ Safety shower and eyewash stations should be readily available in work areas where nitrogen is used or stored.

☑ Hazard warning information should be posted in the work area. In addition, as part of an on-going education and training program, all information on the health and safety hazards of nitrogen should be communicated to all exposed and potentially exposed workers.

RISK ASSESSMENT: ENVIRONMENT

General Assessment

The environment is at risk of exposure during transportation, storage, disposal, or destruction of nitrogen. In almost every scenario, the threat of environmental exposure is contingent upon the proper handling of the chemical substance. Accidental releases, large or small, can result in fire, explosion, and/or possible contamination of the surrounding environmental mediums (mostly air but, depending upon the nature of the spill, water and soil may be affected).

Nitrogen is a non-flammable, non-combustible, non-explosive, and chemically inactive, inert gas. It presents no fire hazard. However, it can cause other gases to condense and solidify. It causes the condensation of oxygen from the air on non-insulated surfaces. Liquid oxygen is a fire hazard. Ice formation inside valves, vents, or critical components can be hazardous to equipment operation and to personnel. Also, because of its ability to displace oxygen, nitrogen requires special consideration during any emergency involving a leak or release of nitrogen gas or liquid especially in confined or enclosed spaces (such as sewers).

Nitrogen can enter the environment through industrial discharges, unchecked venting to air, or through spills of the liquid. Since liquids will most likely vaporize rapidly, water and soil contamination are possible but improbable.

☠ *Acute Ecological Effects*

Acute (short-term) toxic effects may include the death of animals, birds, or fish, and death or low growth rate in plants. Acute effects are seen 2 to 4 days after animals or plants are exposed to nitrogen.

Insufficient data are available on the short-term effects of nitrogen on aquatic life, plants, birds, or land animals. However, as a relatively non-toxic, simple asphyxiant, it might be assumed that any acute ecological toxicity will be relatively low.

💣 *Chronic Ecological Effects*

Chronic (long-term) toxic effects may include shortened life span, reproductive problems, lower fertility, and changes in appearance or behavior in exposed animals. These effects can be seen long after first exposure(s) to toxic chemicals.

Insufficient data are available to evaluate the long-term effects of nitrogen to aquatic life, plants, birds, or land animals. However, as a relatively non-toxic, simple asphyxiant, it might be assumed that any chronic ecological toxicity will be relatively low.

Water Solubility

Nitrogen is only slightly soluble in water. Concentrations of 1 milligram or less may mix with a liter of water. It is normally a gas at room temperature and pressure and solubility may not be a concern.

Persistence in the Environment

Nitrogen is non-persistent in water, with a half-life of less than 2 days. The half-life of a pollutant is the amount of time it takes for one half of the chemical to be degraded. Virtually all of nitrogen will eventually end up in the air.

Bioaccumulation in Aquatic Organisms

Some substances increase in concentration, or *bioaccumulate,* in living organisms as they breathe contaminated air, drink contaminated water, or eat contaminated food. These chemicals can become concentrated in the tissues and internal organs of animals as well as humans.

Nitrogen is not expected to accumulate in the edible tissues of fish.

Recommended Risk-Reduction Measures

Proper training of all transporters will reduce the likelihood of a mishap or accident resulting in a leak or release to the environment. The correct use of DOT labeling (per the MSDS) on all containers, trucks, and rail cars will enable emergency responders to react properly and quickly to any disaster thereby reducing the potential risk to the environment and to personnel.

Storage of nitrogen should be segregated from other gases to prevent cross-contamination or contact. Buildings designated for storage should be equipped with monitors that provide ample warning (audible and visual) of oxygen-deficient atmospheres.

If a spill or release to the environment has occurred, fire department, emergency response, and/or hazardous materials spill personnel should be notified immediately. Cleanup should be attempted only by those trained in proper containment procedures. If applicable, contaminated soils should be removed for incineration and replaced with clean soil. When spills or leaks occur, it may be necessary to contact the local and/or state emergency response authorities. A comprehensive emergency response or disaster preparedness/recovery plan should be in place prior to any operations involving the use, transportation, storage, or disposal of nitrogen. If nitrogen is spilled or leaked, the following specific steps are recommended:

- ☑ Restrict persons not wearing protective clothing from area of spill or leak until cleanup is complete and area can be opened for normal work.
- ☑ Ventilate area of spill or leak.
- ☑ If applicable, stop flow of leaking liquid or gas. If leak source is a cylinder and the leak cannot be stopped in place, remove leaking cylinder to a safe place in the open air, and repair or allow cylinder to empty.
- ☑ Keep nitrogen out of a confined space, such as a sewer, because of the possibility of creating an inert atmosphere.

RISK ASSESSMENT: BUSINESS

General Assessment

Accidents or mishaps involving nitrogen can present a significant threat to business operations. The loss or damage of equipment or facilities can significantly affect fiscal viability. Lawsuits that may result from personnel injury, public exposures, and environmental contamination will also require a serious expenditure of resources. Media attention surrounding an injury, illness, or environmental damage can also result in a loss of profits and loss of current as well as future business.

Recommended Risk-Reduction Measures

Company attorneys, safety and health professionals, and environmental specialists should be involved in the development of any procedures dealing with safety and emergency response. A company official should be pre-designated as a public relations officer with specific training in dealing with the press. Corporate plans and policies should be developed, approved, and implemented long before any need for such arises.

REFERENCES

Ahlbom, A. 1993. *Biostatistics for Epidemiologists.* New York: CRC Press/Lewis Publishers.

American Conference of Governmental Industrial Hygienists. 1988. *Documentation of the Threshold Limit Values and Biological Exposure Indices,* 5th Edition (with updates). Cincinnati: ACGIH

Anthony, C. P., and N. J. Kolthoff. 1971. *Textbook of Anatomy and Physiology.* St. Louis: C. V. Mosby Company.

Asante-Duah, D. K. 1993. *Hazardous Waste Risk Assessment.* New York: CRC Press/Lewis Publishers.

Bloom, A. D. and F. J. DeSerres. 1995. *Ecotoxicity and Human Health.* New York: CRC Press/Lewis Publishers.

Brusick, D. J. 1994. *Methods for Genetic Risk Assessment.* New York: CRC Press/Lewis Publishers.

Bueche, J. R. 1972. *Principles of Physics,* 2nd Edition. New York: McGraw-Hill.

Burrell, R., D. K. Flaherty, and L. J. Sauers. 1992. *Toxicology of the Immune System.* New York: Van Nostrand Reinhold.

Calabrese, E. J. 1994. *Biological Effects of Low Level Exposures.* New York: CRC Press/Lewis Publishers.

Cockerman, L. G., B. S. Shane. 1994. *Basic Environmental Toxicology.* New York: CRC Press/Lewis Publishers.

Cothern, C. R. 1995. *Handbook for Environmental Risk Decisionmaking.* New York: CRC Press/Lewis Publishers.

Danse, I. R. 1991. *Common Sense Toxics in the Workplace.* New York: Van Nostrand Reinhold.

Environmental Protection Agency. 1983. *Chemical Hazard Information Profiles.* Washington, D.C.: U.S. EPA.

Fodden, C. R., and J. L. Weddell. 1992. *Hazardous Materials Emergency Action Data.* New York: CRC Press/Lewis Publishers.

Forseberg, K., and L. H. Keith. 1993. *Chemical; Protective Clothing Permeation and Degradation Database.* Chelsea, MI: Lewis Publishers.

Gardner, R. W. 1994. *Chemical Intolerance.* New York: CRC Press/Lewis Publishers.

Hallenbeck, W. H. 1993. *Quantitative Risk Assessment for Environmental and Occupational Health,* 2nd Edition. New York: CRC Press/Lewis Publishers.

Hodgson, E., R. B. Mailman, and J. E. Chambers. 1988. *Dictionary of Toxicology.* London: The MacMillian Company.

Hosty, J. W., and P. Foster. 1990. *A Practical Guide to Chemical Spill Response.* New York: Van Nostrand Reinhold.

Howard, P. H. 1995. *Handbook of Environmental Fate and Exposures Data for Organic Chemicals,* Vol. I-IV. New York: CRC Press/Lewis Publishers.

James, A., and M. Lord. 1992. *Index of Chemical and Physical Data.* New York: Van Nostrand Reinhold.

Jones, F. E. 1994. *Toxic Organic Vapors in the Workplace.* New York: CRC Press/Lewis Publishers.

Kamrin, M. A. 1988. *Toxicology: A Primer on Principles and Applications.* Chelsea, Michigan: Lewis Publishers.

Keith, L. H. and M. M. Walker. 1995. *Handbook of Air Toxics.* New York: CRC Press/Lewis Publishers.

Keith, L. H. and D. B. Walters. (Editors) 1992. *National Toxicology Program's Chemical Database, Volumes I-VII.* Chelsea, MI: Lewis Publishers.

Landis, W. G., M. H. Yu. 1995. *Introduction to Environmental Toxicology.* New York: CRC Press/Lewis Publishers.

Langerman, N. 1994 *Precautionary Labels for Chemical Containers.* New York: CRC Press/Lewis Publishers.

Lauwerys, R. R., P. Hoet. 1993. *Industrial Chemical Exposure.* New York: CRC Press/Lewis Publishers.

Lewis, R. J., Sr. 1993. *Hawley's Condensed Chemical Dictionary,* 12th Edition. New York: Van Nostrand Reinhold.

Lewis, R. J., Sr. 1992. *Sax's Dangerous Properties of Industrial Materials,* Volumes 1, 2, and 3, 8th Edition. New York: Van Nostrand Reinhold.

Lide, D. R. 1995. *Handbook of Chemistry and Physics,* 76th Edition. New York: CRC Press/Lewis Publishers.

Manahan, S. E. 1994. *Environmental Chemistry,* 6th Edition. New York: CRC Press/Lewis Publishers.

Manahan, S. E. 1992. *Toxicological Chemistry,* 2nd Edition. New York: CRC Press/Lewis Publishers.

Mellström, G. A., J. E. Wahlberg, and H. I. Maibach. 1994. *Protective Gloves for Occupational Use.* New York: CRC Press/Lewis Publishers.

Meyer, E. 1990. *Chemistry of Hazardous Materials.* Englewood Cliffs, NJ: Prentice Hall.

Patnaik, P. 1992. *A Comprehensive Guide to the Hazardous Properties of Chemical Substances.* New York: Van Nostrand Reinhold.

Philp, R. B. 1995. *Environmental Hazards and Human Health.* New York: CRC Press/Lewis Publishers.

Proctor, N. H., and J. P. Hughes. 1991. *Chemical Hazards of the Workplace.* New York: Van Nostrand Reinhold.

Quigley, D. R. 1994. *Handbook of Emergency Chemical Management.* New York: CRC Press/Lewis Publishers.

Rea, W. J. 1995. *Chemical Sensitivity,* Vol. I-III. New York: CRC Press/Lewis Publishers.

Richardson, M. 1995. *Dictionary of Substance and Their Effects,* Vol. I-VI. New York: CRC Press/Lewis Publishers.

Sacarello, H. L. A. 1994. *The Comprehensive Handbook of Hazardous Materials.* New York: CRC Press/Lewis Publishers.

Sheftel, V. O. 1995. *Handbook of Toxic Properties of Monomers and Additives.* New York: CRC Press/Lewis Publishers.

Sherman, J. D. 1988. *Chemical Exposure and Disease.* New York: Van Nostrand Reinhold.

Talbott, E., G. Craun. 1995. *An Introduction to Environmental Epidemiology.* New York: CRC Press/Lewis Publishers.

Torkelson, T. R. V. K. Rowe. 1981. *Patty's Industrial Hygiene and Toxicology,* Vol. 2B. New York: Wiley-Interscience.

Witorsch, P. and S. V. Spagnolo. 1994. *Air Pollution and Lung Disease in Adults.* New York: CRC Press/Lewis Publishers.

MATERIAL SAFETY DATA SHEET

CHEMICAL NAME	
NITROGEN DIOXIDE	**CRC PRESS/LEWIS PUBLISHERS DISCLAIMER:** THE INFORMATION AND RECOMMENDATIONS PRESENTED HEREIN ARE BASED ON SOURCES BELIEVED TO BE RELIABLE. CRC MAKES NO REPRESENTATION ON ITS COMPLETENESS OR ACCURACY. IT IS THE RESPONSIBILITY OF THE USER TO DETERMINE THE CHEMICAL'S SUITABILITY FOR ITS INTENDED USE, THE CHEMICAL'S SAFE USE, AND THE CHEMICAL'S PROPER DISPOSAL. NO REPRESENTATIONS AND/OR WARRANTIES, EITHER EXPRESSED OR IMPLIED, OF THE MERCHANTABILITY OR FITNESS FOR A PARTICULAR PURPOSE, OR OF ANY OTHER NATURE, ARE MADE WITH RESPECT TO THE INFORMATION PROVIDED IN THIS MSDS OR TO THE CHEMICAL TO WHICH INFORMATION MAY REFER. CRC NEITHER ASSUMES NOR AUTHORIZES ANY OTHER PERSON TO ASSUME FOR IT, ANY OTHER ADDITIONAL RESPONSIBILITY OR LIABILITY FOR THE USE OF, OR RELIANCE UPON, THIS INFORMATION.

HAZARD WARNING INFORMATION

HEALTH	FIRE	REACTIVE	OTHER	DEGREE OF HAZARD	COLOR CODING	OTHER CODES
3	0	0	OX	0 = Minimum Hazard 1 = Slight Hazard 2 = Moderate Hazard 3 = Serious Hazard 4 = Severe Hazard	HEALTH = BLUE FIRE = RED REACTIVITY = YELLOW OTHER = WHITE	OX = Oxidizer ACID = Acid ALK = Alkali COR = Corrosive W = Use No Water

SECTION I - GENERAL INFORMATION

Characterization	Gas	RCRA Number	P078
		EPA Class	Acute Hazardous Waste
DOT Proper Shipping Name	Nitrogen Tetroxide, Liquefied	Chemical Abstract Service (CAS) Number	10102-44-0
DOT Hazard Class and Label Requirements	Poison Gas; Oxidizer	DOT Emergency Guide Code	20
DOT Identification Number	UN 1067	Molecular Formula	NO_2

Synonyms

Dinitrogen tetroxide (N_2O_4); nitrogen peroxide; nitrogen tetroxide.

SECTION II - HAZARDOUS INGREDIENTS AND IDENTITY INFORMATION

Hazardous Components (specific identity)	OSHA Exposure Criteria	NIOSH Exposure Criteria	Immediately Dangerous to Life and Health (IDLH)	ACGIH Exposure Criteria
Nitrogen dioxide: (derivation: By oxidation of nitric acid, an intermediate stage in the oxidation of ammonia to nitric acid. Also occurs from combustion of fossil fuels, by fermentation of animal feed in silos, and as a by-product of chemical manufacturing). 1 ppm = 1.91 mg/m^3	PEL (STEL): 1 ppm 1.8 mg/m^3 (15-minute) CEILING: 5 ppm 9 mg/m^3	REL (STEL): 1 ppm 1.8 mg/m^3 (15-minute)	20 ppm	TLV: 3 ppm 5.6 mg/m^3 STEL: 5 ppm 9.4 mg/m^3

SECTION III - PHYSICAL AND CHEMICAL CHARACTERISTICS

Boiling Point	70°F (21°C)	Specific Gravity (H_2O = 1)	2.62 (Gas)
Vapor Pressure (atmospheres)	720 at 68°F (20°C)	Molecular Weight (atomic weight)	46.0
Vapor Density (Air = 1)	1.866	Melting Point	15°F (-9°C)

Solubility

Reacts (decomposes) in water to form nitric acid and nitric oxide. Soluble in sulfuric and nitric acids, alkalis, chloroform, and carbon disulfide.

Appearance and Odor

Yellowish-brown liquid or reddish-brown gas (above 70°F) with a pungent, acrid odor. In solid form (below 15°F), it exists as N_2O_4. Odor Threshold = 0.4 ppm.

SECTION IV - FIRE AND EXPLOSION HAZARD DATA

Flash Point (method used)	Not Applicable	Explosive Limits in Air % by Volume	LEL: Not Applicable UEL: Not Applicable
NFPA Classification	Non-Combustible Gas/Liquid	Autoignition Temperature	Not Applicable

Extinguishing Media

Will not burn but will enhance the burning of other materials. Use agent suitable to surrounding fire.

Special Fire Fighting Procedures

Poisonous gases are produced in fire. Wear full protective clothing and (SCBA). Move container from fire area if it can be done without risk. Cool exposed containers with direct water spray.

Unusual Fire and Explosion Hazards

Containers may explode in fire. Fight fire from a distance or with unmanned hose holders, if possible. Stay away from ends of tanks. Use extreme caution if nitrogen dioxide is in or near a fire.

SECTION V - REACTIVITY DATA

Stability		Conditions to Avoid: Nitrogen dioxide is normally stable in closed containers under routine conditions of handling and storage. Avoid contact with incompatibles, heat, and organic materials.
Stable	Unstable	Incompatibility *(materials to avoid)*: Corrosive to wet steel; reacts with alkalis to form nitrates and nitrites; decomposes in water to form nitric acid and nitric oxide; incompatible with metals, ammonia, acetic anhydride, hydrocarbons, combustibles.
X		
Hazardous Polymerization		Conditions to Avoid: Under normal conditions of temperature and pressure, hazardous polymerization of nitrogen dioxide is not expected to occur.
May Occur	Will Not Occur	Hazardous Decomposition or By-products: When heated to decomposition, nitrogen dioxide emits other toxic oxides of nitrogen.
	X	

SECTION VI - HEALTH HAZARD DATA

Primary Route(s) of Entry:	Inhalation?	Absorption (skin/eye)?	Ingestion?
	X		X

Health Hazards

INHALATION: Severe irritation to the eyes, nose, throat, and lungs. Causes headache, nausea, vomiting, difficulty breathing, cyanosis, pulmonary edema (fluid in the lungs), depression of the CNS with drowsiness, weakness, loss of coordination, and convulsions. Also causes the development of frothy sputum. Loud wheezing, rapid heartbeat, possible fever, and chills.

SKIN & EYES: A severe eye and skin irritant. Sudden bursts of the compressed liquid may cause frostbite and associated tissue destruction. Skin and eye contact can cause burns.

INGESTION: Unlikely, but possible. May cause burns to the mouth and throat.

Carcinogenicity	NTP Listed?	IARC Cancer Review Group?	OSHA Regulated?	Target Organs?
Unknown Human Unknown Animal	No	No	29 CFR 1910.1000 Table Z-1	Skin, eyes, respiratory system, CVS.

Medical Conditions Generally Aggravated by Exposure

Chronic respiratory problems (bronchitis, asthma) may be aggravated.

Emergency and First-aid Procedures

Eye contact: Do not allow victim to keep eyes tightly shut. Flush immediately with water for 15 minutes (minimum). Skin contact: Wash with large amounts of water for a minimum of 15 minutes. For inhalation: Remove the person from exposure. Provide respiratory assistance and CPR. Transfer to medical facility. Observe for 24-48 hours for lung effects. Oxygen should be administered by trained medical personnel only. If swallowed: Not likely since nitrogen dioxide is a gas above 70°F (21°C). Do NOT induce vomiting. Contact poison control center for assistance.

SECTION VII - PRECAUTIONS FOR SAFE HANDLING AND USE

Steps to be Taken in Case Material is Released or Spilled

Wear self-contained breathing apparatus (SCBA). Stop flow of gas. Restrict those not wearing protective equipment and who are not involved in cleanup from entering area. Avoid skin contact. Ventilate area of spill. If gas cannot be stopped, remove leaking container to safe area outdoors and allow to empty.

Preferred Waste Disposal Method

No citation.

Precautions to be Taken in Handling and Storage

Store in tightly closed, insulated, steel containers or cylinders in a cool, well-ventilated area away from heat. Protect containers from physical damage. Regularly inspect for leaks, cracks, or faulty valves.

Other Precautions and Warnings

Use external or detached storage. Ensure cylinders are stored upright and properly secured at all times.

SECTION VIII - CONTROL MEASURES AND PERSONAL PROTECTIVE EQUIPMENT

Respiratory Protection *(specify type)*

Use a NIOSH/MSHA-approved air-purifying gas-cartridge respirator, or a supplied-air respirator, or a self-contained breathing apparatus (SCBA) with full facepiece operated in positive pressure mode.

Ventilation

Local exhaust (preferred) at site of chemical work or general ventilation.

Protective Gloves	Eye Protection	Other Protective Clothing
Impervious Gloves	Chemical Goggles and Face Mask	Protective Uniform or Apron

Work/Hygiene Practices

Always wash hands thoroughly after using chemical; never bring food, drink, or smoking materials into vicinity of chemicals.

NITROGEN DIOXIDE

NO_2

CAS: 10102-44-0

IDENTIFICATION AND TYPICAL USES

Nitrogen dioxide appears as a yellowish-brown liquid or reddish-brown gas (above 70°F) with a pungent, acrid odor. It can also appear in solid form (below 15°F) and is structurally found as N_2O_4 (nitrogen tetroxide). Nitrogen dioxide and nitrogen tetroxide are always found together at normal environmental temperatures; dinitrogen tetroxide, nitrogen peroxide, and nitrogen tetroxide are used as synonyms for nitrogen dioxide. It is used in the production of nitric acid, as a nitrating agent, an oxidizing agent, as a catalyst, as a oxidizer for rocket fuels, and as a polymerization inhibitor for acrylates.

RISK ASSESSMENT: HEALTH

General Assessment

Nitrogen dioxide is a severe irritant to the eyes, nose, and throat. As a gas, it can primarily enter the body through *inhalation*. While skin absorption is not likely, skin contact can result in serious tissue damage as a direct result of the cold temperatures of the compressed gas or cryogenic liquid. Ingestion is not a probable route of exposure under most circumstances. There are no data regarding any carcinogenic potential. However, it is an experimental teratogen and mutation data have also been reported.

Inhalation of nitrogen dioxide can cause severe irritation to the eyes, nose, throat, and respiratory system. Symptoms can be delayed for 1 to 5 hours and may include headache, nausea, vomiting, dypsnea (difficult breathing), diarrhea, cyanosis (bluish discoloration of the lips, fingers, and toes), and possible CNS depression with drowsiness, weakness, loss of coordination, memory impairment, peripheral neuropathy, convulsions, coma, and death. Lung irritation can be quite severe and may lead to pulmonary edema (fluid buildup in the lungs), which is a medical emergency and can be fatal. The symptoms of pulmonary edema can be delayed up to 48 hours, thereby creating a false sense of security with regard to health exposure risk. There can be cough, frothy sputum, and increased difficulty in breathing, loud wheezing, a rapid heartbeat (tachypnea), and increased respiration rate.

Skin contact with compressed liquid nitrogen dioxide can cause characteristic burns due to frostbite. Eye contact causes irritation, watering, inflammation of the lids, burns to the cornea, and (possibly) severe lesions.

Acute Health Effects

The following acute (short-term) health effects may occur immediately or shortly after exposure to nitrogen dioxide:

Skin: Severe irritation with localized tissue damage and possible burns due to the cold temperature of the compressed gas.

<u>Eye</u>: Severe conjunctival irritation and possible damage to the cornea and loss of vision.

<u>Lung</u>: Severe irritation to respiratory tract causing shortness of breath and cough. High acute exposures can be fatal due to a buildup of fluid in the lungs (pulmonary edema).

<u>CNS</u>: Possible depression of the central nervous system causing a variety of toxic effects including loss of coordination and consciousness.

Chronic Health Effects

The following chronic (long-term) health effects may occur at some time after exposure to nitrogen dioxide and can last for months or even years:

<u>Cancer Hazards</u>: According to information presented in the references, nitrogen dioxide has not been adequately tested for its ability to cause cancer in test ani-

mals. Animal mutation data have been reported and many scientists believe that such chemicals may pose a cancer risk in the long-term.

Reproductive Hazard: According to the references, nitrogen dioxide has been shown to cause teratogenic effects in test animals and may adversely affect reproduction as well.

Other Chronic Effects: Long-term exposures to low concentrations may result in chronic pulmonary fibrosis leading to severe obstruction of the bronchioles. There may also be a yellow-brown staining of the skin and teeth resulting from repeated exposures. This is probably due to the formation of nitric acid.

Recommended Risk-Reduction Measures

Nitrogen dioxide is a very toxic substance. Personnel should therefore avoid direct contact with nitrogen dioxide. If a less toxic material or compound cannot be substituted for nitrogen dioxide, then *engineering controls* are the most effective method of reducing exposure risk. The best protection is to enclose operations and/or provide local exhaust ventilation at the site of nitrogen dioxide release. While not always operationally feasible, isolating operations can also reduce exposure. Using respiratory protection is less effective than the controls mentioned above, but is still required whenever working with or around nitrogen dioxide. The 15-minute short-term exposure level of 1 ppm is relatively low and extremely difficult to monitor. Prudent risk management requires proper consideration and implementation of an appropriate respiratory protection program (per OSHA 29 CFR 1910.134). While a gas-cartridge air-purifying respirator will certainly suffice for infrequent exposures, the best protection is obtained using an MSHA/NIOSH-approved supplied-air respirator or a self-contained breathing apparatus (SCBA) with full facepiece and operated in pressure demand mode. If a full facepiece is not available, then chemical goggles should be worn to protect the eyes. A face shield should also be considered. To prevent hand and skin exposures, impervious polyvinyl chloride or polyethylene gloves should be used. Glove manufacturers should be contacted and permeation studies obtained *before* final glove selections are made.

Administrative controls should also be in place to minimize the potential for human exposures. These may include written procedures or policies which specify the methods and techniques that will be practiced whenever personnel are to work with nitrogen dioxide.

All personnel should receive *training* on the use, hazards, protective measures, emergency actions, and other precautions per 29 CFR 1910.1200 (Hazard Communication) prior to the first assignment in an area where nitrogen dioxide is used or stored. Before beginning employment and at regular intervals thereafter (e.g., annually), the following medical test is recommended:

☑ Lung function tests (establish baseline).

If symptoms develop or overexposure is suspected, the following may be useful:

☑ Repeat lung function tests (compare to baseline).

☑ Consider chest X-ray after acute exposure (may be negative if taken immediately following exposure due to delayed development of pulmonary edema).

☑ Examination and evaluation of the nervous system functions by a qualified neurologist.

It should be noted that medical tests that simply look for existing damage are not a substitute for controlling exposures. Medical histories are extremely important when assessing exposure risk. Also, since smoking can cause heart disease as well as lung cancer, emphysema, and other respiratory ailments, any exposure to nitrogen dioxide can result in quicker and more devastating symptoms. Smokers should therefore avoid unprotected contact with this chemical. Prudent risk management requires proper consideration of *all* possible factors that may be causing the appearance of exposure symptoms among workers.

Other methods to reduce exposure include:

☑ Always ensure that proper protective clothing is worn when using chemical substances, and that personnel are trained on its use, care, and maintenance.

☑ Wash thoroughly immediately after exposure to nitrogen dioxide and at the end of the work shift or before eating, drinking, or smoking.

☑ Hazard warning information should be posted in the work area. In addition, as part of an on-going education and training program, all information on the health and safety hazards of nitrogen dioxide should be communicated to all exposed and potentially exposed workers.

☑ Eye wash stations should be provided in the immediate work area for emergency use. If there is a possibility of skin exposure to compressed nitrogen dioxide gas, emergency shower facilities should also be provided.

☑ Workers whose clothing has been contaminated by nitrogen dioxide should change into clean clothes before leaving work. Contaminated work clothing should be laundered only by individuals who have been informed of the hazards of exposure to chemical substances such as nitrogen dioxide.

RISK ASSESSMENT: ENVIRONMENT

General Assessment

The environment is always at some risk of exposure during transportation, storage, disposal, or destruction of nitrogen dioxide. In almost every scenario, the threat of environmental exposure is contingent upon the proper handling of the chemical substance. As a gas, it primarily affects the air. However, depending upon the nature and conditions of the release, accidental spills, large or small, can result in contamination of the other surrounding environmental mediums (water and soil).

Nitrogen dioxide can appear as a non-combustible gas or non-combustible liquid. However, as an oxidizer it can greatly accelerate the burning of ordinary combustible materials such as wood and paper. This is especially true when nitrogen dioxide is heated. It is incompatible with many fuels and chlorinated hydrocarbons. It can also react with other common substances such as carbon disulfide and ammonia. When in contact with water, it reacts to form nitric acid. Caution is always required in handling, storage, transportation, and disposal of nitrogen dioxide. Emergency responders should be made aware of the presence of nitrogen dioxide at any emergency response situation.

Nitrogen dioxide may enter the environment through industrial discharges and from spills.

Acute Ecological Effects

Acute (short-term) toxic effects may include the death of animals, birds, or fish and death or low growth rate in plants. Acute effects are seen 2 to 4 days after animals or plants are exposed to nitrogen dioxide.

Insufficient data are available to evaluate or predict the acute (short-term) effects of nitrogen dioxide on aquatic life, plants, birds, or land animals. However, it is known that nitrogen dioxide will react in water to form nitric acid which may have slight acute toxicity to aquatic life.

Chronic Ecological Effects

Chronic (long-term) toxic effects may include shortened life span, reproductive problems, lower fertility, and changes in appearance or behavior in exposed animals. These effects can be seen long after first exposure(s) to toxic chemicals.

Insufficient data are available to evaluate the chronic (long-term) effects of nitrogen dioxide on aquatic life, plants, birds, or land animals.

Water Solubility

Nitrogen dioxide reacts in water to form nitric acid. It is moderately soluble with decomposition complete over a relatively short period of time.

Persistence in the Environment

Nitrogen dioxide will decompose to nitric acid in the aquatic environment which may be slightly persistent in water, with a half-life between 2 and 20 days. The half-life of a pollutant is the amount of time it takes for one-half of the chemical to be degraded. As a gas, virtually 100% of nitrogen dioxide will eventually end up in the air.

Bioaccumulation in Aquatic Organisms

Some substances increase in concentration, or *bioaccumulate,* in living organisms as they breathe contaminated air, drink contaminated water, or eat contaminated food. These chemicals can become concentrated in the tissues and internal organs of animals as well as humans.

The concentration of nitrogen dioxide found in fish tissues is expected to be somewhat lower than the average concentration of nitrogen dioxide in the water from which the fish was taken.

Recommended Risk-Reduction Measures

Proper training of all transporters will reduce the likelihood of a mishap or accident resulting in a leak or spill of nitrogen dioxide into the environment. Labels on all containers, trucks, and rail cars must meet DOT requirements and accurately reflect their contents to enable emergency responders to react properly and quickly to any disaster thereby reducing the potential risk to the environment and to personnel.

Storage of nitrogen dioxide should be segregated from other chemicals to minimize the risk of cross-contamination. Nitrogen dioxide must be stored to avoid contact with incompatible materials since violent reactions can occur. Containers should be protected from physical damage and stored to avoid contact with heat.

If a spill or leak to the environment has occurred, fire department, emergency response and/or hazardous materials spill personnel should be notified immediately. Cleanup should be attempted only by those trained in proper spill containment procedures. Contaminated soils should be removed for incineration and replaced with clean soil.

If nitrogen dioxide should contact the water table, aquifer, or navigable waterway, time is of the essence. Because it will decompose to nitric acid in water, total remediation may not be possible. The local and/or state emergency response authorities must be notified. A comprehensive emergency response or disaster preparedness/recovery plan should be in place prior to any operations involving the use, transportation, storage, or disposal of nitrogen dioxide.

If nitrogen dioxide is spilled or leaked, the following specific steps are recommended:

☑ Restrict persons not wearing protective clothing from area of spill or leak until cleanup is complete. Avoid skin and eye contact.

☑ Absorb spilled liquids with vermiculite or a similar material and deposit in sealed drum for disposal.

☑ Ventilate area of spill or leak.

☑ Remove ignition sources from the area.

☑ It may be necessary to dispose of nitrogen dioxide as a hazardous waste. The responsible state agency or the regional office of the federal Environmental Protection Agency (EPA) should be contacted for specific recommendations.

RISK ASSESSMENT: BUSINESS

General Assessment

Accidents or mishaps involving nitrogen dioxide can present a serious threat to business operations. The loss or damage of equipment or facilities can significantly affect fiscal viability. Lawsuits that may result from personnel illness, injury/death, public exposures, and/or environmental contamination will require a serious expenditure of resources. Media attention surrounding an injury, death, or environmental damage can also result in a loss of profits and loss of current as well as future business.

Recommended Risk-Reduction Measures

Company attorneys, safety and health professionals, and environmental specialists should be involved in the development of any procedures for responding to chemical incidents. A company official should be predesignated as a public relations officer with specific training in dealing with the press. Corporate plans and policies should be developed, approved, and implemented long before any need for such arises.

REFERENCES

American Conference of Governmental Industrial Hygienists. 1988. *Documentation of the Threshold Limit Values and Biological Exposure Indices*, 5th Edition (with updates). Cincinnati: ACGIH

Anthony, C. P., and N. J. Kolthoff. 1971. *Textbook of Anatomy and Physiology.* St. Louis: C. V. Mosby Company.

Brusick, D. J. 1994. *Methods for Genetic Risk Assessment.* New York: CRC Press/Lewis Publishers.

Bueche, J. R. 1972. *Principles of Physics*, 2nd Edition. New York: McGraw-Hill.

Burrell, R., D. K. Flaherty, and L. J. Sauers. 1992. *Toxicology of the Immune System.* New York: Van Nostrand Reinhold.

Calabrese, E. J. 1994. *Biological Effects of Low Level Exposures.* New York: CRC Press/Lewis Publishers.

Cothern, C. R. 1995. *Handbook for Environmental Risk Decisionmaking.* New York: CRC Press/Lewis Publishers.

Danse, I. R. 1991. *Common Sense Toxics in the Workplace.* New York: Van Nostrand Reinhold.

Environmental Protection Agency. 1983. *Chemical Hazard Information Profiles.* Washington, D.C.: U.S. EPA.

Fodden, C. R., and J. L. Weddell. 1992. *Hazardous Materials Emergency Action Data.* New York: CRC Press/Lewis Publishers.

Gardner, R. W. 1994. *Chemical Intolerance.* New York: CRC Press/Lewis Publishers.

Hodgson, E., R. B. Mailman, and Janice E. Chambers. 1988. *Dictionary of Toxicology.* London: The MacMillian Company.

Keith, L. H. and M. M. Walker. 1995. *Handbook of Air Toxics.* New York: CRC Press/Lewis Publishers.

Lauwerys, R. R., P. Hoet. 1993. *Industrial Chemical Exposure.* New York: CRC Press/Lewis Publishers.

Lewis, .J., Sr. 1993. *Hawley's Condensed Chemical Dictionary*, 12th Edition. New York: Van Nostrand Reinhold.

U.S. Department of Health and Human Services, National Institute for Occupational Safety and Health. 1994. *NIOSH Pocket Guide to Chemical Hazards.* Washington, D.C.: U.S. Government Printing Office.

MATERIAL SAFETY DATA SHEET

CHEMICAL NAME	
NITROGEN TRIFLUORIDE	**CRC PRESS/LEWIS PUBLISHERS DISCLAIMER:** THE INFORMATION AND RECOMMENDATIONS PRESENTED HEREIN ARE BASED ON SOURCES BELIEVED TO BE RELIABLE. CRC MAKES NO REPRESENTATION ON ITS COMPLETENESS OR ACCURACY. IT IS THE RESPONSIBILITY OF THE USER TO DETERMINE THE CHEMICAL'S SUITABILITY FOR ITS INTENDED USE, THE CHEMICAL'S SAFE USE, AND THE CHEMICAL'S PROPER DISPOSAL. NO REPRESENTATIONS AND/OR WARRANTIES, EITHER EXPRESSED OR IMPLIED, OF THE MERCHANTABILITY OR FITNESS FOR A PARTICULAR PURPOSE, OR OF ANY OTHER NATURE, ARE MADE WITH RESPECT TO THE INFORMATION PROVIDED IN THIS MSDS OR TO THE CHEMICAL TO WHICH INFORMATION MAY REFER. CRC NEITHER ASSUMES NOR AUTHORIZES ANY OTHER PERSON TO ASSUME FOR IT, ANY OTHER ADDITIONAL RESPONSIBILITY OR LIABILITY FOR THE USE OF, OR RELIANCE UPON, THIS INFORMATION.

HAZARD WARNING INFORMATION

HEALTH	FIRE	REACTIVE	OTHER	DEGREE OF HAZARD	COLOR CODING	OTHER CODES
1	0	1	OX	0 = Minimum Hazard 1 = Slight Hazard 2 = Moderate Hazard 3 = Serious Hazard 4 = Severe Hazard	HEALTH = BLUE FIRE = RED REACTIVITY = YELLOW OTHER = WHITE	OX = Oxidizer ACID = Acid ALK = Alkali COR = Corrosive W = Use No Water

SECTION I - GENERAL INFORMATION

Characterization	RCRA Number	EPA Class
Gas	D003	Characteristic (R) Waste

DOT Proper Shipping Name: **Nitrogen Tetroxide**	Chemical Abstract Service (CAS) Number: **7783-54-2**
DOT Hazard Class and Label Requirements: **Nonflammable Gas; Oxidizer**	DOT Emergency Guide Code: **15**
DOT Identification Number: **UN 2451**	Molecular Formula: **NF_3**

Synonyms

Nitrogen fluoride; trifluoramine; trifluorammonia; perfluorammonia.

SECTION II - HAZARDOUS INGREDIENTS AND IDENTITY INFORMATION

Hazardous Components (specific identity)	OSHA Exposure Criteria	NIOSH Exposure Criteria	Immediately Dangerous to Life and Health (IDLH)	ACGIH Exposure Criteria
Nitrogen trifluoride: (derivation: By electrolysis of ammonium acid fluoride or low-temperature reaction of nitrogen and fluorine in the presence of electrical discharge). 1 ppm = 2.95 mg/m^3	PEL (8-hour): 10 ppm 29 mg/m^3 STEL: Not Established	REL (10-hour): 10 ppm 29 mg/m^3 STEL: Not Established	1000 ppm	TLV: 10 ppm 29 mg/m^3 STEL: Not Established

SECTION III - PHYSICAL AND CHEMICAL CHARACTERISTICS

Boiling Point: **-200°F (-129°C)**	Specific Gravity (H_2O = 1): **2.46 (Gas)**
Vapor Pressure (atmospheres): **>1 at 68°F (20°C)**	Molecular Weight (atomic weight): **71.0**
Vapor Density (Air = 1): **1.537 at -200°F (-129°C)**	Freezing Point: **-340°F (-206°C)**

Solubility

Very slightly soluble in cold water.

Appearance and Odor

Colorless gas with a slight moldy odor. Usually shipped as a non-liquefied compressed gas.

SECTION IV - FIRE AND EXPLOSION HAZARD DATA

Flash Point (method used): **Not Applicable**	Explosive Limits in Air % by Volume: **LEL: Not Applicable UEL: Not Applicable**
NFPA Classification: **Non-Flammable Gas**	Autoignition Temperature: **Not Applicable**

Extinguishing Media

Will not burn but will enhance the burning of other materials. Use agent suitable to surrounding fire, except for water (water, heat, and nitrogen trifluoride may cause an explosion). Do NOT use Halon 130.

Special Fire Fighting Procedures

Poisonous gases are produced in fire. Wear full protective clothing and (SCBA). Move container from fire area if it can be done without risk. Cool exposed containers with direct water spray (do NOT allow water to get inside containers. Do NOT release runoff from control measures to sewers and waterways.

Unusual Fire and Explosion Hazards

Containers may explode in fire. Fight fire from a distance or with unmanned hose holders, if possible. Stay away from ends of tanks. Use extreme caution if nitrogen trifluoride is in or near a fire.

SECTION V - REACTIVITY DATA

Stability		Conditions to Avoid Nitrogen trifluoride is normally stable in closed containers under routine conditions of handling and storage, but can explode under pressure. Avoid contact with incompatibles, heat, and organic materials.
Stable X	Unstable	Incompatibility *(materials to avoid)* Incompatible ammonia, carbon monoxide, methane, hydrogen sulfide, diborane, tetrafluorohydrazine, hydrogen, ethylene, chlorine dioxide, charcoal, oil, grease, and oxidizable materials.
Hazardous Polymerization		Conditions to Avoid Under normal conditions of temperature and pressure, hazardous polymerization of nitrogen trifluoride is not expected to occur.
May Occur	Will Not Occur X	Hazardous Decomposition or By-products When heated to decomposition, nitrogen trifluoride emits highly toxic fluorine fumes.

SECTION VI - HEALTH HAZARD DATA

Primary Route(s) of Entry:	Inhalation? X	Absorption (skin/eye)?	Ingestion?

Health Hazards

INHALATION: The gas causes the development of methemoglobinemia which can significantly impair the blood's ability to transport oxygen to the cells, causing a lack of oxygenated tissues. This may cause symptoms such as cyanosis, dizziness, weakness, and severe headache. It may also lead to hemolysis, anemia, and altered kidney functions.

SKIN & EYES: A severe eye and skin irritant. Sudden bursts of the compressed liquid may cause frostbite and associated tissue destruction. Skin and eye contact can cause burns.

INGESTION: Unlikely, but possible. May cause burns to the mouth and throat.

Carcinogenicity	NTP Listed?	IARC Cancer Review Group?	OSHA Regulated?	Target Organs?
Unknown Human Unknown Animal	No	No	29 CFR 1910.1000 Table Z-1	Blood, liver, kidneys, central nervous system.

Medical Conditions Generally Aggravated by Exposure

Chronic respiratory problems (bronchitis, asthma) may be aggravated.

Emergency and First-aid Procedures

Eye contact: Do not allow victim to keep eyes tightly shut. Flush immediately with water for 15 minutes (minimum). Skin contact: Wash with large amounts of tepid water for a minimum of 15 minutes. Do NOT use dry heat or rub the affected area. For inhalation: Remove the person from exposure. Provide respiratory assistance and CPR. Transfer to medical facility. If swallowed: Not likely since nitrogen trifluoride is a gas above 70°F (21°C). Do NOT induce vomiting. Contact poison control center for assistance.

SECTION VII - PRECAUTIONS FOR SAFE HANDLING AND USE

Steps to be Taken in Case Material is Released or Spilled

Wear self-contained breathing apparatus (SCBA). Stop flow of gas. Restrict those not wearing protective equipment and who are not involved in cleanup from entering area. Avoid skin contact. Ventilate area of spill. If gas cannot be stopped, remove leaking container to safe area outdoors and allow to empty.

Preferred Waste Disposal Method

No citation.

Precautions to be Taken in Handling and Storage

Store in tightly closed, insulated, steel containers or cylinders in a cool, well-ventilated area away from heat. Protect containers from physical damage. Regularly inspect for leaks, cracks, or faulty valves.

Other Precautions and Warnings

Use external or detached storage. Ensure cylinders are stored upright and properly secured at all times.

SECTION VIII - CONTROL MEASURES AND PERSONAL PROTECTIVE EQUIPMENT

Respiratory Protection *(specify type)*

Use a NIOSH/MSHA-approved air-purifying gas-cartridge respirator, or a supplied-air respirator, or a self-contained breathing apparatus (SCBA) with full facepiece operated in positive pressure mode.

Ventilation

Local exhaust (preferred) at site of chemical work or general ventilation.

Protective Gloves	Eye Protection	Other Protective Clothing
Impervious Gloves	Chemical Goggles and Face Mask	Protective Uniform or Apron

Work/Hygiene Practices

Always wash hands thoroughly after using chemical; never bring food, drink, or smoking materials into vicinity of chemicals.

NITROGEN TRIFLUORIDE

NF_3

CAS: 7783-54-2

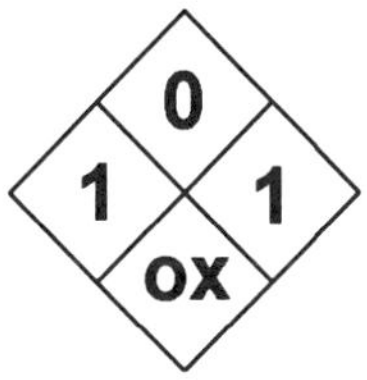

IDENTIFICATION AND TYPICAL USES

Nitrogen trifluoride as a colorless gas with a slight, moldy odor. It is normally shipped as a non-liquefied compressed gas. It is used as an oxidizing agent in high energy fuels or for chemical synthesis, and as a dry etchant gas.

RISK ASSESSMENT: HEALTH

General Assessment

Nitrogen trifluoride is a severe irritant to the eyes, nose, and throat. As a gas, it can primarily enter the body through *inhalation*. While skin absorption is not likely, skin contact can result in serious tissue damage as a direct result of the cold temperatures of the compressed gas or cryogenic liquid. Ingestion is not a probable route of exposure under most circumstances. There are no data regarding any carcinogenic, mutagenic, or teratogenic potential on long-term exposure.

Inhalation of nitrogen trifluoride can cause severe irritation to the eyes, nose, and throat. The primary systemic effect of exposure to nitrogen trifluoride vapors is the development of methemoglobinemia which can significantly interfere with the blood's ability to transport oxygen to the body's cells. This causes a lack of oxygenated blood in the tissues. Symptoms include cyanosis with blue lips, nose, ears, fingers, toes, and facial skin due to this lack of oxygen. There can also be dizziness, weakness, severe headache, hemolysis (damage) of the red blood cells, leading to anemia and possible alterations in kidney function.

Skin contact with compressed liquid nitrogen trifluoride can cause characteristic burns due to frostbite. Eye contact causes irritation, watering, inflammation of the lids, burns to the cornea, and (possibly) severe lesions.

Human exposure data are limited. However, animal studies indicate that the fluoride ion is not released from nitrogen trifluoride during the metabolic process which may indicate a possibility of fluoride buildup over time and may result in the mottling of teeth and in unspecified skeletal changes.

Acute Health Effects

The following acute (short-term) health effects may occur immediately or shortly after exposure to nitrogen trifluoride:

Skin: Severe irritation with localized tissue damage and possible burns due to the cold temperature of the compressed gas.

Eye: Conjunctival irritation and possible damage to the cornea and loss of vision.

Lung: Irritation to respiratory tract causing shortness of breath and cough. The gas causes the development of methemoglobin which significantly impairs the blood's ability to effectively transport oxygen. This can lead to anemia and cyanosis.

Chronic Health Effects

The following chronic (long-term) health effects may occur at some time after exposure to nitrogen trifluoride and can last for months or even years:

Cancer Hazards: According to information presented in the references, nitrogen trifluoride has not been adequately tested for its ability to cause cancer in test animals.

Reproductive Hazard: According to the references, nitrogen trifluoride has not been tested for its ability to adversely affect reproduction in test animals.

Other Chronic Effects: Long-term exposures to low concentrations may result in damage to tooth enamel and could lead to unspecified skeletal changes.

ⓤ *Recommended Risk-Reduction Measures*

Nitrogen trifluoride is a very toxic substance. Personnel should avoid direct contact with nitrogen trifluoride. If a less toxic material or compound cannot be substituted for nitrogen trifluoride, then *engineering controls* are the most effective method of reducing exposure risk. The best protection is to enclose operations and/or provide local exhaust ventilation at the site of nitrogen trifluoride release. While not always operationally feasible, isolating operations can also reduce exposure. Using respiratory protection is less effective than the controls mentioned above, but is still required whenever working with or around nitrogen trifluoride. The permissible exposure limit of 10 ppm (8-hour TWA) can be reached relatively quickly, even under normal operating conditions. Prudent risk management requires proper consideration and implementation of an appropriate respiratory protection program (per OSHA 29 CFR 1910.134). While a gas-cartridge air-purifying respirator will certainly suffice for low or transient exposures over the PEL, the best protection is obtained using an MSHA/NIOSH-approved supplied-air respirator or a self-contained breathing apparatus (SCBA) with full facepiece and operated in pressure demand mode. If a full facepiece is not available, then chemical goggles should be worn to protect the eyes. A face shield should also be considered. To prevent hand and skin exposures, gloves should be used. Glove manufacturers should be contacted and permeation studies obtained *before* final glove selections are made.

Administrative controls should also be in place to minimize the potential for human exposures. These may include written procedures or policies which specify the methods and techniques that will be practiced whenever personnel are to work with nitrogen trifluoride.

All personnel should receive *training* on the use, hazards, protective measures, emergency actions, and other precautions per 29 CFR 1910.1200 (Hazard Communication) prior to the first assignment in an area where nitrogen trifluoride is used or stored.

If symptoms develop or overexposure is suspected, the following may be useful:

☑ Complete blood count (CBC) with tests for electrolytes and arterial blood gases.

☑ Monitor methemoglobin levels. Methylene blue therapy may be indicated if levels exceed 60%.

It should be noted that medical tests that simply look for existing damage are not a substitute for controlling exposures. Medical histories are extremely important when assessing exposure risk. Also, smoking not only causes heart disease, lung cancer, emphysema, and other respiratory ailments, it also decreases the blood's ability to effectively transport oxygen by increasing the carboxyhemoglobin content of the blood. Therefore, smokers exposed to nitrogen trifluoride may experience symptoms more rapidly than non-smokers under the same conditions of exposure. Smokers should therefore avoid unprotected contact with this chemical. Prudent risk management requires proper consideration of *all* possible factors that may be causing the appearance of exposure symptoms among workers.

Other methods to reduce exposure include:

☑ Always ensure that proper protective clothing is worn when using chemical substances, and that personnel are trained on its use, care, and maintenance.

☑ Wash thoroughly immediately after exposure to nitrogen trifluoride and at the end of the work shift or before eating, drinking, or smoking.

☑ Hazard warning information should be posted in the work area. In addition, as part of an on-going education and training program, all information on the health and safety hazards of nitrogen trifluoride should be communicated to exposed workers.

☑ Eye wash stations should be provided in the immediate work area for emergency use. If there is a possibility of skin exposure to compressed nitrogen trifluoride gas, emergency shower facilities should also be provided.

☑ Workers whose clothing has been contaminated by nitrogen trifluoride should change into clean clothes before leaving work. Contaminated work clothing should be laundered only by individuals who have been informed of the hazards of exposure to chemical substances such as nitrogen trifluoride.

RISK ASSESSMENT: ENVIRONMENT

General Assessment

The environment is always at some risk of exposure during transportation, storage, disposal, or destruction of nitrogen trifluoride. In almost every scenario, the threat of environmental exposure is contingent upon the proper handling of the chemical substance. As a gas, it primarily affects the air. However, depending upon the nature and conditions of the release, accidental spills, large or small, can result in contamination of the other surrounding environmental mediums (water and soil).

Nitrogen trifluoride is a non-flammable gas. However, as an oxidizer it can greatly accelerate the burning of ordinary combustible materials such as wood and paper. It can also react explosively in contact with a number of commonly encountered materials, including oil, grease, oxidizable materials, ammonia, carbon monoxide, methane, and even water. In contact with water or water vapor, reactions may be violent and explosive. Caution is always required in handling, storage, transportation, and disposal of nitrogen trifluoride. Emergency responders should be made aware of the presence of nitrogen trifluoride at any emergency response situation.

Nitrogen trifluoride may enter the environment through industrial discharges and from spills.

Acute Ecological Effects

Acute (short-term) toxic effects may include the death of animals, birds, or fish and death or low growth rate in plants. Acute effects are seen 2 to 4 days after animals or plants are exposed to nitrogen trifluoride.

Insufficient data are available to evaluate or predict the acute (short-term) effects of nitrogen trifluoride on aquatic life, plants, birds, or land animals.

Chronic Ecological Effects

Chronic (long-term) toxic effects may include shortened life span, reproductive problems, lower fertility, and changes in appearance or behavior in exposed animals. These effects can be seen long after first exposure(s) to toxic chemicals.

Insufficient data are available to evaluate the chronic (long-term) effects of nitrogen trifluoride on aquatic life, plants, birds, or land animals.

Water Solubility

Nitrogen trifluoride is slightly soluble in water. Concentrations of 1 to 100 milligrams may mix with a liter of water (there may be a violent reaction, depending upon the concentration of nitrogen trifluoride present).

Persistence in the Environment

Nitrogen trifluoride will react with water in the aquatic environment and therefore may be only slightly persistent in water, with a half-life between 2 and 20 days. The half-life of a pollutant is the amount of time it takes for one-half of the chemical to be degraded. As a gas, virtually 100% of nitrogen trifluoride will eventually end up in the air.

Bioaccumulation in Aquatic Organisms

Some substances increase in concentration, or *bioaccumulate,* in living organisms as they breathe contaminated air, drink contaminated water, or eat contaminated food. These chemicals can become concentrated in the tissues and internal organs of animals as well as humans.

The concentration of nitrogen trifluoride found in fish tissues is expected to be somewhat lower than the average concentration of nitrogen trifluoride in the water from which the fish was taken.

Recommended Risk-Reduction Measures

Proper training of all transporters will reduce the likelihood of a mishap or accident resulting in a leak or spill of nitrogen trifluoride into the environment. Labels on all containers, trucks, and rail cars must meet DOT requirements and accurately reflect their contents to enable emergency responders to react properly and quickly to any disaster thereby reducing the potential risk to the environment and to personnel.

Storage of nitrogen trifluoride should be segregated from other chemicals to minimize the risk of cross-contamination. Nitrogen trifluoride must be stored to avoid contact with incompatible materials since violent reactions can occur. Containers should be protected from physical damage and stored to avoid contact with heat. If a spill or leak to the environment has occurred, fire department, emergency response and/or hazardous materials spill personnel should be notified immediately. Cleanup should be attempted only by those trained in proper spill containment procedures. Contaminated soils should be removed for incineration and replaced with clean soil.

If nitrogen trifluoride should contact the water table, aquifer, or navigable waterway, time is of the essence. Because it will decompose to nitric acid in water, total remediation may not be possible. The local and/or state emergency response authorities must be notified. A comprehensive emergency response or disaster preparedness/recovery plan should be in place prior to any operations involving the use, transportation, storage, or disposal of nitrogen trifluoride.

If nitrogen trifluoride is spilled or leaked, the following specific steps are recommended:

☑ Restrict persons not wearing protective clothing from area of spill or leak until cleanup is complete. Avoid skin and eye contact.

☑ If applicable, stop flow of leaking liquid or gas. If leak source is a cylinder and the leak cannot be stopped in place, remove leaking cylinder to a safe place in the open air, and repair or allow cylinder to empty.

☑ Ventilate area of spill or leak.

☑ It may be necessary to dispose of nitrogen trifluoride as a hazardous waste. The responsible state agency or the regional office of the federal Environmental Protection Agency (EPA) should be contacted for specific recommendations.

RISK ASSESSMENT: BUSINESS

General Assessment

Accidents or mishaps involving nitrogen trifluoride can present a serious threat to business operations. The loss or damage of equipment or facilities can significantly affect fiscal viability. Lawsuits that may result from personnel illness, injury/death, public exposures, and/or environmental contamination will require a serious expenditure of resources. Media attention surrounding an injury, death, or environmental damage can also result in a loss of profits and loss of current as well as future business.

Recommended Risk-Reduction Measures

Company attorneys, safety and health professionals, and environmental specialists should be involved in the development of any procedures for responding to chemical incidents. A company official should be predesignated as a public relations officer with specific training in dealing with the press. Corporate plans and policies should be developed, approved, and implemented long before any need for such arises.

REFERENCES

Ahlbom, A. 1993. *Biostatistics for Epidemiologists.* New York: CRC Press/Lewis Publishers.

American Conference of Governmental Industrial Hygienists. 1988. *Documentation of the Threshold Limit Values and Biological Exposure Indices,* 5th Edition (with updates). Cincinnati: ACGIH

Anthony, C. P., and N. J. Kolthoff. 1971. *Textbook of Anatomy and Physiology.* St. Louis: C. V. Mosby Company.

Brusick, D. J. 1994. *Methods for Genetic Risk Assessment.* New York: CRC Press/Lewis Publishers.

Bueche, J. R. 1972. *Principles of Physics,* 2nd Edition. New York: McGraw-Hill.

Burrell, R., D. K. Flaherty, and L. J. Sauers. 1992. *Toxicology of the Immune System.* New York: Van Nostrand Reinhold.

Calabrese, E. J. 1994. *Biological Effects of Low Level Exposures.* New York: CRC Press/Lewis Publishers.

Cockerman, L. G., B. S. Shane. 1994. *Basic Environmental Toxicology.* New York: CRC Press/Lewis Publishers.

Cothern, C. R. 1995. *Handbook for Environmental Risk Decisionmaking.* New York: CRC Press/Lewis Publishers.

Danse, I. R. 1991. *Common Sense Toxics in the Workplace.* New York: Van Nostrand Reinhold.

Environmental Protection Agency. 1983. *Chemical Hazard Information Profiles.* Washington, D.C.: U.S. EPA.

Fodden, C. R., and J. L. Weddell. 1992. *Hazardous Materials Emergency Action Data.* New York: CRC Press/Lewis Publishers.

Forseberg, K., and L. H. Keith. 1993. *Chemical; Protective Clothing Permeation and Degradation Database.* Chelsea, MI: Lewis Publishers.

Gardner, R. W. 1994. *Chemical Intolerance.* New York: CRC Press/Lewis Publishers.

Hallenbeck, W. H. 1993. *Quantitative Risk Assessment for Environmental and Occupational Health,* 2nd Edition. New York: CRC Press/Lewis Publishers.

Hodgson, E., R. B. Mailman, and Janice E. Chambers. 1988. *Dictionary of Toxicology.* London: The MacMillian Company.

Hosty, J. W., and P. Foster. 1990. *A Practical Guide to Chemical Spill Response.* New York: Van Nostrand Reinhold.

Howard, P. H. 1995. *Handbook of Environmental Fate and Exposures Data for Organic Chemicals,* Vol. I-IV. New York: CRC Press/Lewis Publishers.

Jones, F. E. 1994. *Toxic Organic Vapors in the Workplace.* New York: CRC Press/Lewis Publishers.

Kamrin, M. A. 1988. *Toxicology: A Primer on Principles and Applications.* Chelsea, Michigan: Lewis Publishers.

Keith, L. H. and M. M. Walker. 1995. *Handbook of Air Toxics.* New York: CRC Press/Lewis Publishers.

Lauwerys, R. R., P. Hoet. 1993. *Industrial Chemical Exposure.* New York: CRC Press/Lewis Publishers.

Lewis, .J., Sr. 1993. *Hawley's Condensed Chemical Dictionary,* 12th Edition. New York: Van Nostrand Reinhold.

Lewis, . J., Sr. 1992. *Sax's Dangerous Properties of Industrial Materials,* Volumes 1, 2, and 3, 8th Edition. New York: Van Nostrand Reinhold.

Lide, D. R. 1995. *Handbook of Chemistry and Physics,* 76th Edition. New York: CRC Press/Lewis Publishers.

U.S. Department of Health and Human Services, National Institute for Occupational Safety and Health. 1994. *NIOSH Pocket Guide to Chemical Hazards.* Washington, D.C.: U.S. Government Printing Office.

MATERIAL SAFETY DATA SHEET

CHEMICAL NAME	
NITROGLYCERIN	**CRC PRESS/LEWIS PUBLISHERS DISCLAIMER:** THE INFORMATION AND RECOMMENDATIONS PRESENTED HEREIN ARE BASED ON SOURCES BELIEVED TO BE RELIABLE. CRC MAKES NO REPRESENTATION ON ITS COMPLETENESS OR ACCURACY. IT IS THE RESPONSIBILITY OF THE USER TO DETERMINE THE CHEMICAL'S SUITABILITY FOR ITS INTENDED USE, THE CHEMICAL'S SAFE USE, AND THE CHEMICAL'S PROPER DISPOSAL. NO REPRESENTATIONS AND/OR WARRANTIES, EITHER EXPRESSED OR IMPLIED, OF THE MERCHANTABILITY OR FITNESS FOR A PARTICULAR PURPOSE, OR OF ANY OTHER NATURE, ARE MADE WITH RESPECT TO THE INFORMATION PROVIDED IN THIS MSDS OR TO THE CHEMICAL TO WHICH INFORMATION MAY REFER. CRC NEITHER ASSUMES NOR AUTHORIZES ANY OTHER PERSON TO ASSUME FOR IT, ANY OTHER ADDITIONAL RESPONSIBILITY OR LIABILITY FOR THE USE OF, OR RELIANCE UPON, THIS INFORMATION.

HAZARD WARNING INFORMATION

HEALTH	FIRE	REACTIVE	OTHER	DEGREE OF HAZARD	COLOR CODING	OTHER CODES
2	2	4		0 = Minimum Hazard 1 = Slight Hazard 2 = Moderate Hazard 3 = Serious Hazard 4 = Severe Hazard	HEALTH = BLUE FIRE = RED REACTIVITY = YELLOW OTHER = WHITE	OX = Oxidizer ACID = Acid ALK = Alkali COR = Corrosive W = Use No Water

SECTION I - GENERAL INFORMATION

Characterization	RCRA Number	EPA Class
Explosive, Nitro Explosive Compound	**P081**	**Acute Hazardous Waste**

DOT Proper Shipping Name	Chemical Abstract Service (CAS) Number
Blasting Gelatin	**55-63-0**
DOT Hazard Class and Label Requirements	**DOT Emergency Guide Code**
Class A Explosive; Flammable Liquid	**26**
DOT Identification Number	**Chemical Formula**
UN0143; UN0144; UN1204; UN3064	$CH_2NO_3CHNO_3CH_2NO_3$

Synonyms

Glycerol trinitrate; NG; 1,2,3-Propanetriol trinitrate; trinitroglycerin; anginine; blasting oil; trinitrin.

SECTION II - HAZARDOUS INGREDIENTS AND IDENTITY INFORMATION

Hazardous Components (specific identity)	OSHA Exposure Criteria	NIOSH Exposure Criteria	Immediately Dangerous to Life and Health (IDLH)	ACGIH Exposure Criteria
Nitroglycerin: (derivation: By dropping glycerol through cooled, mixed acid and stirring followed by repeated washing with water). **1 ppm = 9.44 mg/m^3**	**PEL (STEL):** 0.01 ppm 0.1 mg/m^3 (skin) **CEILING:** 0.2 ppm 2.0 mg/m^3 (skin)	**REL (CEILING):** 0.01 ppm 0.1 mg/m^3 (skin) (20-minutes)	**75 mg/m^3**	**TLV (8-hour):** 0.05 ppm 0.5 mg/m^3 **STEL:** Not Established

SECTION III - PHYSICAL AND CHEMICAL CHARACTERISTICS

Boiling Point	Specific Gravity (H_2O = 1)
122-140°F (50-60°C) Decomposes	**1.60**
Vapor Pressure (mm Hg)	Molecular Weight (atomic weight)
1 at 261°F (127°C)	**227.1**
Vapor Density (Air = 1)	Melting Point
7.84	**56°F (13°C)**

Solubility

Very slightly soluble in water (0.1%). Soluble in alcohol and ether.

Appearance and Odor

Colorless to pale yellow, viscous liquid, or a solid (below 56°F), with a sweet taste.

SECTION IV - FIRE AND EXPLOSION HAZARD DATA

Flash Point (method used)	Explosive Limits in Air % by Volume
Explodes at 424°F (218°C)	**LEL: Not Determined UEL: Not Determined**
NFPA Classification	Autoignition Temperature
Explosive Liquid	**518°F (270°C)**

Extinguishing Media

Water spray has proven effective in fighting propellant fires involving nitroglycerin.

Special Fire Fighting Procedures

Poisonous gases are emitted during fire. Wear full protective clothing and self-contained breathing apparatus (SCBA). Remain clear of smoke, water fallout and water runoff. Move containers from fire area if it can be done without risk. Cool exposed containers. Fight fire from distance, if possible.

Unusual Fire and Explosion Hazards

Nitroglycerin is a highly explosive, shock-sensitive compound. Containers may explode in fire due to buildup of pressure. Spontaneous detonation occurs between 356°F and 424°F (180°C and 218°C).

SECTION V - REACTIVITY DATA

Stability		Conditions to Avoid Nitroglycerin is a shock-sensitive, friction-sensitive, and impact-sensitive compound. Avoid contact or exposure to heat and incompatible materials.
Stable	Unstable X	Incompatibility *(materials to avoid)* Incompatible with heat, ozone (explodes), shock, and acids. Also sensitive to ultraviolet radiation. Nitroglycerin is an OSHA Class A Explosive per 29 CFR 1910.109.
Hazardous Polymerization		Conditions to Avoid Under normal conditions of temperature and pressure, hazardous polymerization of nitroglycerin may occur with simple impact, friction, or shock. Polymerization is explosively violent.
May Occur X	Will Not Occur	Hazardous Decomposition or By-products At high temperatures, nitroglycerin emits toxic oxides of nitrogen. However, the detonation products are relatively harmless and include carbon dioxide, water vapor, nitrogen, and oxygen.

SECTION VI - HEALTH HAZARD DATA

Primary Route(s) of Entry:	Inhalation?	Absorption (skin/eye)?	Ingestion?
	X	X	

Health Hazards

INHALATION: Symptoms include headache, nausea, vomiting, abdominal pain, lightheadedness, tremor, dypsnea, paralysis, and convulsions. High exposures can lead to methemoglobinemia, a condition which interferes with the blood's oxygen-carrying ability causing cyanosis.

ABSORPTION: May cause mild irritation to the eyes and skin.

INGESTION: A fatal poison by ingestion. Initial symptoms are the same as that for inhalation.

Carcinogenicity	NTP Listed?	IARC Cancer Review Group?	OSHA Regulated?	Target Organs?
Unknown Human Questioned Animal	No	No	29 CFR 1910.1000 Table Z-1	CNS, blood, skin, CVS

Medical Conditions Generally Aggravated by Exposure

Existing heart conditions (including hypertension), nervous disorders, and skin problems.

Emergency and First-aid Procedures

Eye contact: Do not allow victim to keep eyes tightly shut. Flush immediately with water for 15 minutes; seek medical attention. Skin contact: Remove all contaminated clothing. Immediately flood area with water for 15 minutes. Then wash affected areas with soap and water. Inhalation: Remove the person from exposure. Provide respiratory assistance and CPR. Transfer to medical facility, if required. If swallowed: Contact poison control center. Rinse victim's mouth with water. Unless advised otherwise by poison control center and victim is conscious and alert, give 2-3 glasses of water to drink to dilute and then induce vomiting repeatedly until vomitus is clear. Never give anything by mouth to an unconscious or convulsing person. Never try to make an unconscious or convulsing person vomit. Seek medical attention immediately. Keep victim warm and at rest.

SECTION VII - PRECAUTIONS FOR SAFE HANDLING AND USE

Steps to be Taken in Case Material is Released or Spilled

Remove all sources of ignition and ventilate area. Restrict those not involved in cleanup from entering area. Do NOT touch or disturb spilled materials. Notify appropriately trained response personnel.

Preferred Waste Disposal Method

No citation.

Precautions to be Taken in Handling and Storage

Store in tightly closed containers in cool, dry, well-ventilated area away from heat and incompatibles. All metal containers should be grounded and bonded. Drums should be equipped with self-closing valves, pressure vacuum bungs, and flame arresters.

Other Precautions and Warnings

Use non-sparking tools and equipment and provide explosion-proof ventilation in storage areas.

SECTION VIII - CONTROL MEASURES AND PERSONAL PROTECTIVE EQUIPMENT

Respiratory Protection *(specify type)*

Use an MSHA/NIOSH-approved self-contained breathing apparatus in positive pressure mode or a supplied-air respirator operated in continuous flow or other positive pressure mode.

Ventilation

Local exhaust (preferred) at site of chemical work or general ventilation.

Protective Gloves	Eye Protection	Other Protective Clothing
Impervious Rubber (test for adequacy)	Chemical Goggles or Face Mask	Protective Apron, Boots (avoid skin contact)

Work/Hygiene Practices

Always wash hands thoroughly after using chemical; never bring food, drink, or smoking materials into vicinity of chemicals.

NITROGLYCERIN

$CH_2NO_3CHNO_3CH_2NO_3$

CAS: 55-63-0

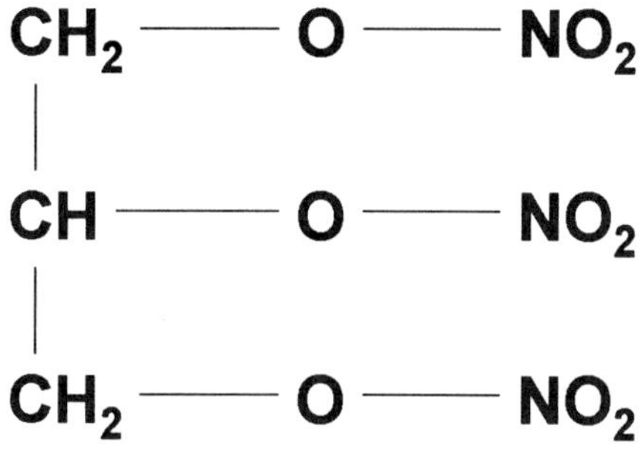

IDENTIFICATION AND TYPICAL USES

Nitroglycerin is a colorless to pale-yellow, viscous liquid with a sweet taste. It can also appear as a solid at temperatures below 56°F (13°C). It is used primarily in the manufacture of high explosives, especially flashless propellant powder (with nitrocellulose), and as a chemical intermediate. It is often mixed with ethylene glycol dinitrate. Nitroglycerin is also used in pharmaceuticals, primarily in the formulation of cardiovascular medications.

RISK ASSESSMENT: HEALTH

General Assessment

Nitroglycerin is an extremely toxic material and human poison. It can enter the body through *inhalation* of its vapors. It will also easily pass through intact skin (*absorption*) and can be *ingested* to cause toxic systemic effects. In fact, ingestion of even a small amount (1.5 to 2.0 grams) may be fatal to humans. Nitroglycerin is an experimental teratogen and a questioned carcinogen in test animals. Mutation data have also been reported for this compound.

Inhalation of nitroglycerin causes headache, nausea, vomiting, abdominal pain, tremor, dypsnea, paralysis, convulsions, lightheadedness, and the development of methemoglobinemia which interferes with the blood's ability to transport oxygen to the cells. Symptoms include anemia and cyanosis. Inhalation of the vapors can also affect blood pressure levels causing an abnormal decreases in systolic and diastolic pressures. After repeated exposures, a tolerance may develop. Angina (chest pains) and even heart attacks can occur if exposure is stopped suddenly.

Skin contact can cause minor, localized irritation as well as toxic systemic effects similar to that of inhalation due to nitroglycerin's ability to pass through unbroken skin with relative ease.

Ingestion of even small amounts are acutely poisonous and can cause death in humans.

Acute Health Effects

The following acute (short-term) health effects may occur immediately or shortly after exposure to nitroglycerin:

Skin: Local irritation at contact site.

Eye: Irritation and possible inflammation.

Lung: Irritation of the nose, throat, and respiratory tract. The vapors cause more damage systemically, as a result of their action on the nervous system, rather than locally to the respiratory tract members themselves.

CNS: Extreme depression affecting many other normal functions of anatomical systems.

CVS: Causes heart rhythm disturbances and adversely affects blood pressure and other cardiovascular responses.

Chronic Health Effects

The following chronic (long-term) health effects may occur at some time after exposure to nitroglycerin and can last for months or even years:

Cancer Hazards: According to information presented in the references, nitroglycerin has been shown to cause cancer in limited animal studies. However, the data are conflicting and inconclusive in this regard. Mutation data have been reported and many scientists believe that exposure to mutagenic substances may present a cancer hazard in the long-term. Additional study is required before any determinations can be made in this regard for nitroglycerin.

Reproductive Hazard: Nitroglycerin has been shown to cause teratogenic and adverse reproductive effects in laboratory experiments. It is not known if it will have this effect in humans exposed to nitroglycerin.

Other Chronic Effects: Long-term exposure may result in a tolerance to nitroglycerin with no noticeable symptoms. If personnel are removed from exposure for a short period (such as a weekend or vacation), serious and severe exposure symptoms may develop on their return to work where nitroglycerin exposure occurs. Sudden cessation of exposure may cause chest pains and possible heart attack in some individuals. Repeated exposures can cause chronic skin irritation.

Recommended Risk-Reduction Measures

Nitroglycerin presents a serious health hazard risk. Personnel should avoid direct contact with nitroglycerin. If a less toxic material or compound cannot be substituted for nitroglycerin, then *engineering controls* are the most effective method of reducing exposures. The best protection is to enclose operations and/or provide local exhaust ventilation at the site of nitroglycerin release. While not always operationally possible, isolating operations involving nitroglycerin manufacturing or use can also reduce exposure. Personnel must follow proper industrial hygiene and good housekeeping work practices. Using respiratory protection is less effective than the controls mentioned above, but is still advisable whenever working with or around nitroglycerin. The short-term exposure level (STEL) is relatively low and difficult to accurately monitor (0.1 mg/m^3). For the best protection, an MSHA or NIOSH-approved powered air-supplied respirator or a self-contained breathing apparatus (SCBA) with full facepiece and operated in pressure demand mode are recommended. If a full facepiece is not available, then chemical goggles should be worn to protect the eyes. A face shield and protective apron should also be worn. Skin contact must not occur. To prevent hand and skin exposures, impervious gloves should be worn. Glove manufacturers should be contacted and permeation studies obtained *before* final glove selections have been made.

Administrative controls should also be in place to minimize the potential for human exposures. These may include written procedures or policies which specify the methods and techniques that will be practiced whenever personnel are to work with nitroglycerin.

All personnel should receive *training* on the use, hazards, protective measures, emergency actions, and other precautions per 29 CFR 1910.1200 (Hazard Communication), prior to the first assignment in an area where nitroglycerin is used or stored.

If symptoms develop or overexposure is suspected, the following medical testing may be useful:

☑ Electrocardiogram (EKG), if chest discomfort is felt.

☑ Blood methemoglobin level monitoring.

Any evaluation should include a careful history of past and present symptoms with an examination. Medical tests that simply look for existing damage are not a substitute for controlling exposures. Also, since smoking can cause heart disease, emphysema, and other respiratory problems, personnel who smoke may show symptoms of exposure much quicker and more pronounced than those who do not. Prudent risk management requires careful consideration of *all* possible and potential exposure risk factors to ensure proper implementation of control measures and management of the exposure risk.

Other methods to reduce exposure include:

☑ Where possible, enclose operations and use local exhaust ventilation at the site of chemical release. If local exhaust ventilation or enclosure is not used, respiratory protection should be mandatory.

☑ Always ensure that proper protective clothing is worn when using chemical substances and that personnel are trained in the proper use, care, and maintenance of their protective equipment.

☑ Wash thoroughly immediately after exposure to nitroglycerin and at the end of the work shift or before eating, drinking, or smoking.

☑ Hazard warning information should be posted in the work area. In addition, as part of an on-going education and training program, all information on the health and safety hazards of nitroglycerin should be communicated to all potentially exposed workers.

☑ Eye wash stations should be provided in the immediate work area for emergency use. If there is a possibility of skin exposure to nitroglycerin, emergency shower facilities should also be provided in the immediate work area.

- ☑ Workers whose clothing has been contaminated by nitroglycerin should change into clean clothes *before* leaving work. Contaminated work clothing should never be worn home where family members can be contaminated. The clothing should be laundered only by individuals who have been informed of the hazards of exposure to nitroglycerin.
- ☑ Specific engineering controls are recommended for this chemical by NIOSH. Refer to NIOSH Criteria Document: Occupational Exposure to Nitroglycerin and Ethylene Glycol Dinitrate (Publication Number 78-187).

RISK ASSESSMENT: ENVIRONMENT

General Assessment

The environment is at risk of exposure during transportation, storage, disposal, or destruction of nitroglycerin. In almost every scenario, the threat of environmental exposure is contingent upon the proper handling of the chemical substance. Accidental spills, large or small, where nitroglycerin contacts incompatible commodities can result in violent and even explosive reactions (depending upon conditions of spill), and possible contamination of the surrounding environmental mediums (water, soil, and air).

Nitroglycerin is considered an explosive liquid. It is shock-sensitive, impact-sensitive, and friction-sensitive. It can easily detonate on exposure to heat or incompatible materials. Contact with ozone or even ultraviolet radiation may cause an explosive reaction. Therefore, extreme caution is required in handling, storage, transportation, and disposal of nitroglycerin. Its characteristics also require special consideration during any emergency situation involving a leak or spill of nitroglycerin.

Nitroglycerin may enter the environment through industrial discharges, from dynamite operations, and from spills.

Acute Ecological Effects

Acute (short-term) toxic effects may include the death of animals, birds, or fish and death or low growth rate in plants. Acute effects are seen 2 to 4 days after animals or plants are exposed to nitroglycerin.

Nitroglycerin has moderate acute toxicity to aquatic life. Insufficient data are available to evaluate or predict the short-term effects of nitroglycerin to plants, land animals, and birds.

Chronic Ecological Effects

Chronic (long-term) toxic effects may include shortened life span, reproductive problems, lower fertility, and changes in appearance or behavior in exposed animals. These effects can be seen long after first exposure(s) to toxic chemicals. Nitroglycerin has moderate chronic toxicity to aquatic life. Insufficient data are available to evaluate or predict the long-term effects of nitroglycerin to plants, birds, and land animals.

Water Solubility

Nitroglycerin is highly soluble in water. Concentrations of 1000 milligrams and even more will readily mix with a liter of water.

Persistence in the Environment

Nitroglycerin is slightly persistent in the water, with a half-life between and 2 and 20 days. The half-life of a pollutant is the amount of time it takes for one-half of the chemical to be degraded. Approximately 99.8% of nitroglycerin will eventually end up in the water. The remainder will end up in equal amounts in terrestrial soils and aquatic sediments.

Bioaccumulation in Aquatic Organisms

Some substances increase in concentration, or *bioaccumulate,* in living organisms as they breathe contaminated air, drink contaminated water, or eat contaminated food. These chemicals can become concentrated in the tissues and internal organs of animals as well as humans.

The concentration of nitroglycerin found in fish tissues is expected to be somewhat higher than the average concentration of nitroglycerin in the water from which the fish was taken.

Recommended Risk-Reduction Measures

Proper training of all transporters will reduce the likelihood of a mishap or accident resulting in a leak or spill to the environment. The correct use of DOT labeling (per the MSDS) on all containers, trucks, and rail cars will enable emergency responders to react properly and quickly to any disaster thereby reducing the potential risk to the environment and to personnel.

Storage of nitroglycerin should be segregated from incompatible materials to minimize the risk of cross-contamination or contact. Storage facilities should be equipped with the proper fire protection equipment (sprinklers, alarms, extinguishers). To prevent static sparks, all equipment and containers used in shipping, receiving, or transferring should be electrically grounded and bounded. Facilities should be equipped with maximum explosion-proof ventilation.

If a spill or leak to the environment has occurred, fire department, emergency response and/or hazardous materials spill personnel should be notified immediately. Spilled materials should not be touched or disturbed until professionally trained responders arrive. Contaminated soils should be removed for incineration and replaced with clean soils. Cleanup should be attempted only by those trained in proper spill containment procedures using non-sparking tools. If nitroglycerin should contact the water table, aquifer, or navigable waterway, reclamation procedures should be initiated as soon as practical. It is highly soluble in water and total remediation may not be possible. The local and/or state emergency response authorities must be notified. A comprehensive emergency response or disaster preparedness/recovery plan should be in place prior to any operations involving the use, transportation, storage, or disposal of nitroglycerin. If nitroglycerin is spilled or leaked, the following specific steps are recommended:

- ☑ Restrict persons not wearing protective clothing from area of spill or leak until cleanup is complete and area can be opened for normal work.
- ☑ Remove all ignition sources.
- ☑ Ventilate area or leak or spill.
- ☑ Do NOT touch or disturb spilled nitroglycerin. Evacuate area and wait arrival of professionally trained response personnel.
- ☑ It may be necessary to dispose of nitroglycerin as a hazardous waste. The responsible state agency or the regional office of the federal Environmental Protection Agency (EPA) should be contacted for specific recommendations.

RISK ASSESSMENT: BUSINESS

General Assessment

Accidents or mishaps involving nitroglycerin can present a significant threat to business operations. The loss or damage of equipment or facilities can significantly affect fiscal viability. Lawsuits that may result from personnel injury/death, public exposures, and/or environmental contamination will also require a serious expenditure of resources. Media attention surrounding an injury, death, or environmental damage can also result in a loss of profits and loss of current as well as future business. Always remember that anytime the terms "cancer," "carcinogen," or "reproductive hazard" are used, public hysteria, ignorance, and anxiety can run equally high. This must be carefully considered when drafting or implementing public relations policies.

Recommended Risk-Reduction Measures

Company attorneys, safety and health professionals, and environmental specialists should be involved in the development of any procedures for responding to chemical incidents. A company official should be predesignated as a public relations officer with specific training in dealing with the press. Corporate plans and policies should be developed, approved, and implemented long before any need for such arises.

REFERENCES

American Conference of Governmental Industrial Hygienists. 1988. *Documentation of the Threshold Limit Values and Biological Exposure Indices,* 5th Edition (with updates). Cincinnati: ACGIH

Bloom, A. D. and F. J. DeSerres. 1995. *Ecotoxicity and Human Health.* New York: CRC Press/Lewis Publishers.

Brusick, D. J. 1994. *Methods for Genetic Risk Assessment.* New York: CRC Press/Lewis Publishers.

Bueche, J. R. 1972. *Principles of Physics,* 2nd Edition. New York: McGraw-Hill.

Calabrese, E. J. 1994. *Biological Effects of Low Level Exposures.* New York: CRC Press/Lewis Publishers.

Environmental Protection Agency. 1983. *Chemical Hazard Information Profiles.* Washington, D.C.: U.S. EPA.

Lewis, .J., Sr. 1993. *Hawley's Condensed Chemical Dictionary,* 12th Edition. New York: Van Nostrand Reinhold.

Lewis, . J., Sr. 1992. *Sax's Dangerous Properties of Industrial Materials,* Volumes 1, 2, and 3, 8th Edition. New York: Van Nostrand Reinhold.

U.S. Department of Health and Human Services, National Institute for Occupational Safety and Health. 1978. *NIOSH Criteria Document: Occupational Exposure to Nitroglycerin and Ethylene Glycol Dinitrate.* Publication number: 78-187. Washington, D.C.: U.S. Government Printing Office.

U.S. Department of Health and Human Services, National Institute for Occupational Safety and Health. 1994. *NIOSH Pocket Guide to Chemical Hazards.* Washington, D.C.: U.S. Government Printing Office.

MATERIAL SAFETY DATA SHEET

CHEMICAL NAME	
NITROMETHANE	CRC PRESS/LEWIS PUBLISHERS DISCLAIMER: THE INFORMATION AND RECOMMENDATIONS PRESENTED HEREIN ARE BASED ON SOURCES BELIEVED TO BE RELIABLE. CRC MAKES NO REPRESENTATION ON ITS COMPLETENESS OR ACCURACY. IT IS THE RESPONSIBILITY OF THE USER TO DETERMINE THE CHEMICAL'S SUITABILITY FOR ITS INTENDED USE, THE CHEMICAL'S SAFE USE, AND THE CHEMICAL'S PROPER DISPOSAL. NO REPRESENTATIONS AND/OR WARRANTIES, EITHER EXPRESSED OR IMPLIED, OF THE MERCHANTABILITY OR FITNESS FOR A PARTICULAR PURPOSE, OR OF ANY OTHER NATURE, ARE MADE WITH RESPECT TO THE INFORMATION PROVIDED IN THIS MSDS OR TO THE CHEMICAL TO WHICH INFORMATION MAY REFER. CRC NEITHER ASSUMES NOR AUTHORIZES ANY OTHER PERSON TO ASSUME FOR IT, ANY OTHER ADDITIONAL RESPONSIBILITY OR LIABILITY FOR THE USE OF, OR RELIANCE UPON, THIS INFORMATION.

HAZARD WARNING INFORMATION

HEALTH	FIRE	REACTIVE	OTHER	DEGREE OF HAZARD	COLOR CODING	OTHER CODES
1	3	4		0 = Minimum Hazard 1 = Slight Hazard 2 = Moderate Hazard 3 = Serious Hazard 4 = Severe Hazard	HEALTH = BLUE FIRE = RED REACTIVITY = YELLOW OTHER = WHITE	OX = Oxidizer ACID = Acid ALK = Alkali COR = Corrosive W = Use No Water

SECTION I - GENERAL INFORMATION

Characterization: **Nitroparaffin**	RCRA Number: **D001, D003** — EPA Class: **Characteristic (I) and (R) Waste**
DOT Proper Shipping Name: **Nitromethane**	Chemical Abstract Service (CAS) Number: **75-52-5**
DOT Hazard Class and Label Requirements: **Flammable Liquid**	DOT Emergency Guide Code: **26**
DOT Identification Number: **UN 1261**	Chemical Formula: **CH_3NO_2**

Synonyms

Nitroetan.

SECTION II - HAZARDOUS INGREDIENTS AND IDENTITY INFORMATION

Hazardous Components (specific identity)	OSHA Exposure Criteria	NIOSH Exposure Criteria	Immediately Dangerous to Life and Health (IDLH)	ACGIH Exposure Criteria
Nitromethane (derivation: By reaction methane or propane with nitric acid under pressure). **1 ppm = 2.54 mg/m³**	**PEL:** 100 ppm, 250 mg/m³ **STEL:** Not Established	**REL:** Not Established **STEL:** Not Established	**750 ppm**	**TLV:** 100 ppm, 250 mg/m³ **STEL:** Not Established

SECTION III - PHYSICAL AND CHEMICAL CHARACTERISTICS

Boiling Point: **213°F (101°C)**	Specific Gravity (H_2O = 1): **1.14**
Vapor Pressure (mm Hg): **27.8 at 68°F (20°C)**	Molecular Weight: **61.0**
Vapor Density (Air = 1): **2.11**	Freezing Point: **-20°F (-29°C)**

Solubility

Slightly soluble in water (10%). Soluble in alcohol, acetone, ether, and dimethylformamide.

Appearance and Odor

Colorless, oily liquid with disagreeable odor.

SECTION IV - FIRE AND EXPLOSION HAZARD DATA

Flash Point (method used): **95°F (35°C) closed cup**	Explosive Limits in Air % by Volume: **LEL: 7.3% UEL: Not Determined**
NFPA Classification: **Class IC Flammable Liquid**	Autoignition Temperature: **785°F (418°C)**

Extinguishing Media

Use carbon dioxide, water spray, or alcohol-resistant foam.

Special Fire Fighting Procedures

Nitromethane is a highly flammable, very dangerous fire and explosion hazard. Poisonous gases are evolved from fire. Wear a self-contained breathing apparatus (SCBA) with full facepiece. Move containers from fire area if it can be done without risk or cool fire-exposed containers with water spray.

Unusual Fire and Explosion Hazards

Containers may explode in fire. Boiling liquid expanding vapor explosion (BLEVE) is likely. Immediately withdraw if rising sound is heard coming from venting device. Vapors are heavier than air and can travel for great distances to an ignition source and flashback to cause fire or explosion.

SECTION V - REACTIVITY DATA

Stability		Conditions to Avoid Nitromethane is shock-sensitive and thermally unstable. High pressures and contamination by other materials contribute to its instability. Avoid contact with heat, sparks, flame, and incompatible materials.
Stable	Unstable X	Incompatibility *(materials to avoid)* Nitromethane is reactive in contact with acetone, amines, strong acids, alkalis, oxidizers, hydrocarbons, and other combustible materials. Slowly corrodes steel and copper when wet.
Hazardous Polymerization		Conditions to Avoid Under normal conditions of temperature and pressure, hazardous polymerization of nitromethane is not expected to occur.
May Occur	Will Not Occur X	Hazardous Decomposition or By-products If involved in fire, nitromethane can produce toxic nitrogen oxides and irritating fumes and smoke.

SECTION VI - HEALTH HAZARD DATA

Primary Route(s) of Entry:	Inhalation?	Absorption (skin)?	Ingestion?
	X		X

Health Hazards

INHALATION: Irritation to the eyes, nose, throat, and respiratory tract. Animal studies have shown a potential for the development of pulmonary edema (fluid in the lungs), injury to the liver, kidney, and narcosis. There may be bronchitis, anuria, and increases in methemoglobin levels in the blood.

SKIN & EYES: Skin contact causes irritation and may cause reddening and smarting if it remains in contact for extended period of time. Eye contact can cause tearing and irritation.

INGESTION: Causes irritation of the mouth and stomach, nausea, and vomiting.

Carcinogenicity	NTP Listed?	IARC Cancer Review Group?	OSHA Regulated?	Target Organs?
Unknown Human Unknown Animal	No	No	29 CFR 1010.1000 Table Z-1	Skin, eyes, CNS, liver, respiratory tract.

Medical Conditions Generally Aggravated by Exposure

Possible skin disorders may be aggravated by exposure.

Emergency and First-aid Procedures

Eye contact: Immediately flush with large amounts of water for 15 minutes (minimum), occasionally lifting eyelids; seek medical attention. Skin contact: Remove all contaminated clothing. Immediately wash area with large amounts of soap and water. For inhalation: Remove the person from exposure. Provide respiratory assistance and CPR. Transfer promptly to medical facility. Observe 48 hours for lung effects. If swallowed: Contact poison control center. Unless directed otherwise, give a conscious person 1 to 2 glasses of water, then induce vomiting.

SECTION VII - PRECAUTIONS FOR SAFE HANDLING AND USE

Steps to be Taken in Case Material is Released or Spilled

Remove all ignition sources. Ventilate area. Restrict entry into the area. Absorb liquids with vermiculite and place in sealed containers for disposal. Use non-sparking tools during cleanup operations.

Preferred Waste Disposal Method

Dilute with fuel oil and atomize in proper combustion chamber.

Precautions to be Taken in Handling and Storage

Do not store with incompatible materials or in the presence of heat since violent reactions or explosions can occur. Store in tightly closed containers in a cool, dark, well-ventilated area.

Other Precautions and Warnings

Bulk storage of nitromethane is not recommended. Containers may explode in fire or under conditions of extreme heat (do not store outdoors or in direct sunlight).

SECTION VIII - CONTROL MEASURES AND PERSONAL PROTECTIVE EQUIPMENT

Respiratory Protection *(specify type)*

For exposures over the 8-hour PEL of 100 ppm, use an MSHA/NIOSH-approved air-purifying respirator with organic vapor cartridge. Otherwise, use a powered supplied-air respirator or a self-contained breathing apparatus (SCBA) with full facepiece operated in pressure demand or other positive pressure mode.

Ventilation

Local exhaust or general mechanical systems recommended.

Protective Gloves	Eye Protection	Other Protective Clothing
Impervious Material	Chemical Goggles and/or Face Mask	Protective Apron

Work/Hygiene Practices

Always wash hands thoroughly after using chemical; never bring food, drink, or smoking materials into vicinity of chemicals.

NITROMETHANE

CH_3NO_2

CAS: 75-52-5

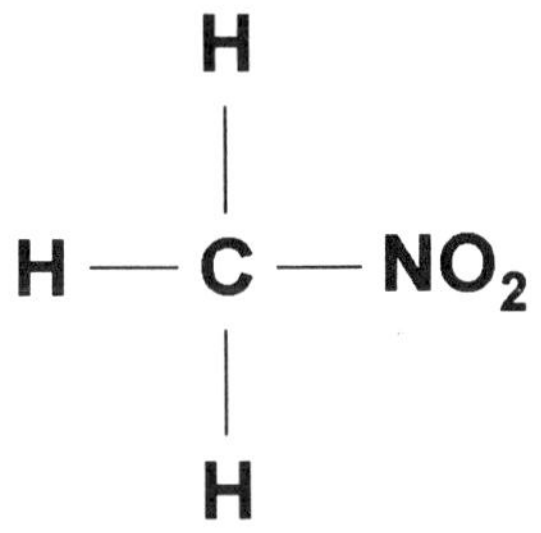

IDENTIFICATION AND TYPICAL USES

Nitromethane is a colorless, oily liquid with a mild, fruity odor. It is used as a solvent for cellulose compounds, polymers, waxes, and fats. It is also used in rocket and racing car fuel, in chemical synthesis, in explosive mixtures, and in the coating industry.

RISK ASSESSMENT: HEALTH

General Assessment

Nitromethane is moderately toxic by *ingestion* and *inhalation.* It is also a skin irritant, but dermal absorption has not been reported. There are no indications of any carcinogenic, mutagenic, or teratogenic effects following exposure.

Inhalation causes irritation of the mucosa of the eyes, nose, throat, and respiratory tract. Systemic effects have not been observed in humans. However, animal studies indicate a possibility for the development of pulmonary edema (fluid in the lungs) which is a medical emergency. Symptoms of cough, congestion, labored breathing, and pain in the chest may be delayed up to 48 hours which may give a false sense of security with regard to health exposure risk. There may also be liver injury and, at high concentrations, narcosis. Severe exposure in humans is expected to cause similar effects. Nitromethane may also cause the formation of methemoglobin which can significantly interfere with the blood's ability to properly transport oxygen. Other symptoms of exposure by inhalation may include anorexia, nausea, vomiting, anuria, and diarrhea.

Skin contact causes mild to severe irritation and, if allowed to remain in contact for prolonged periods, either directly or in chemical-soaked clothing, there can be reddening and smarting at the site of contact. Eye contact can cause irritation and tearing.

Ingestion may result in irritation of the mouth and stomach, nausea, and vomiting.

Acute Health Effects

The following acute (short-term) health effects may occur immediately or shortly after exposure to nitromethane:

Skin: Irritation with redness and smarting at the site of contact.

Eye: Irritation and inflammation with a possibility for damage to visual acuity.

Lung: Irritation of the mucosa of the nose, throat, and respiratory tract. May lead to delayed pulmonary edema which can be fatal.

CNS: Unspecified effects associated with depression of the central nervous system.

Chronic Health Effects

The following chronic (long-term) health effects may occur at some time after exposure to nitromethane and can last for months or even years:

Cancer Hazards: According to information presented in the references, nitromethane has not been adequately tested for its ability to cause cancer in test animals.

Reproductive Hazards: There are no reports to support any claims of reproductive hazards.

Other Chronic Effects: Long-term exposure can lead to dermatitis.

Recommended Risk-Reduction Measures

Personnel should avoid direct contact with nitromethane. If another chemical cannot be substituted for a nitromethane, then *engineering controls* are the most effective method of reducing exposures. The best protection is to separate operations and provide local exhaust ventilation at the site of chemical release. While not always operationally feasible, isolating operations can also reduce exposure. Using equipment with explosion-proof designs and ensuring all metal containers and tools are electrically grounded and bonded can further reduce the potential for injury or damage. Using respiratory protection is less effective than the controls mentioned above, but is still advisable whenever working with nitromethane. For exposures over the 8-hour PEL (100 ppm), an air-purifying chemical cartridge respirator equipped with an organic vapor cartridge may provide adequate protection. However, best protection is obtained using a powered air-supplied respirator operated in pressure demand or other positive pressure mode, or a self-contained breathing apparatus (SCBA) with full facepiece operated in pressure demand mode. If a full facepiece is not available, then chemical goggles should be worn to protect the eyes. Whenever a chemical splash hazard exists, a face shield and a protective apron should be worn. To prevent hand and skin exposures with the liquid, thermal gloves should be used. Glove manufacturers should be contacted and permeation studies obtained *before* final glove selections have been made.

Administrative controls should also be in place to minimize the potential for human exposures. These may include written procedures or policies which specify the methods and techniques that will be practiced whenever personnel are to work with nitromethane.

All personnel should receive *training* on the use, hazards, protective measures, emergency actions, and other precautions per 29 CFR 1910.1200 (Hazard Communication), prior to the first assignment in an area where nitromethane is used or stored.

If overexposure is suspected or symptoms develop, the following medical tests are recommended:

☑ Lung function tests.

☑ Liver and kidney function tests.

☑ Methemoglobin level monitoring.

☑ Consider chest X-ray after acute overexposure (may be negative if taken immediately after exposure due to delayed onset of pulmonary edema).

Other methods to reduce exposure to chemicals include:

☑ Where possible, consider installing automatic sensors that warn personnel of dropping oxygen levels and use local exhaust ventilation at the site of chemical release. If local exhaust ventilation or enclosure is not used, respiratory protection should be mandatory.

☑ Always ensure that proper protective clothing is worn when using chemical substances and that personnel are trained in its use, care, and maintenance.

☑ Wash thoroughly immediately after exposure to nitromethane and at the end of the work shift or before eating, drinking, or smoking.

☑ Safety shower and eyewash stations should be readily available in work areas where nitromethane is used or stored.

☑ Hazard warning information should be posted in the work area. In addition, as part of an on-going education and training program, all information on the health and safety hazards of nitromethane should be communicated to all exposed and potentially exposed workers.

RISK ASSESSMENT: ENVIRONMENT

General Assessment

The environment is at risk of exposure during transportation, storage, disposal, or destruction of nitromethane. In almost every scenario, the threat of environmental exposure is contingent upon the proper handling of the chemical substance. Accidental releases, large or small, can result in fire, explosion, and possible contamination of the surrounding environmental mediums (water, soil, and air).

Nitromethane is a Class IC flammable liquid per OSHA 29 CFR 1910.106. It can present an extremely flammable and highly explosive hazard. Also, since it can react violently in the presence of oxidizers, acids, and hydrocarbons, nitromethane requires special consideration during any emergency involving a leak or release of nitromethane gas or liquid.

Nitromethane can enter the environment through industrial discharges or through spills.

☠ *Acute Ecological Effects*

Acute (short-term) toxic effects may include the death of animals, birds, or fish, and death or low growth rate in plants. Acute effects are seen 2 to 4 days after animals or plants are exposed to nitromethane.

Insufficient data are available on the short-term effects of nitromethane on aquatic life, plants, birds, or land animals.

Chronic Ecological Effects

Chronic (long-term) toxic effects may include shortened life span, reproductive problems, lower fertility, and changes in appearance or behavior in exposed animals. These effects can be seen long after first exposure(s) to toxic chemicals.

Insufficient data are available to evaluate the long-term effects of nitromethane to aquatic life, plants, birds, or land animals.

Water Solubility

Nitromethane is moderately soluble in water. Concentrations of 1 to 1000 milligrams will mix with a liter of water.

Persistence in the Environment

Nitromethane is slightly persistent in water, with a half-life between 2 and 20 days. The half-life of a pollutant is the amount of time it takes for one half of the chemical to be degraded.

Bioaccumulation in Aquatic Organisms

Some substances increase in concentration, or *bioaccumulate,* in living organisms as they breathe contaminated air, drink contaminated water, or eat contaminated food. These chemicals can become concentrated in the tissues and internal organs of animals as well as humans.

The concentration of nitromethane found in the edible tissues of fish is expected to be somewhat lower than the average concentration found in the water from which the fish was taken.

Recommended Risk-Reduction Measures

Proper training of all transporters will reduce the likelihood of a mishap or accident resulting in a leak or release to the environment. The correct use of DOT labeling (per the MSDS) on all containers, trucks, and rail cars will enable emergency responders to react properly and quickly to any disaster thereby reducing the potential risk to the environment and to personnel.

Storage of nitromethane should be segregated from incompatible chemicals to minimize the risk of cross-contamination or contact. Buildings designated for storage should be equipped with appropriate fire protection systems (alarms, sprinklers, emergency lighting, portable extinguishers).

If a spill or release to the environment has occurred, fire department, emergency response, and/or hazardous materials spill personnel should be notified immediately. Cleanup should be attempted only by those trained in proper containment procedures using non-sparking tools. Contaminated soils should be removed for incineration and replaced with clean soil. When spills or leaks occur, it may be necessary to contact the local and/or state emergency response authorities. A comprehensive emergency response or disaster preparedness/recovery plan should be in place prior to any operations involving the use, transportation, storage, or disposal of nitromethane. If nitromethane is spilled or leaked, the following specific steps are recommended:

- ☑ Restrict persons not wearing protective clothing from area of spill or leak until cleanup is complete and area can be opened for normal work.
- ☑ Remove all ignition sources and ventilate area.
- ☑ Absorb liquids using vermiculite, dry earth, or other material and place in sealed containers for disposal. For small spills, cover liquid with soda ash, place in an open bucket of water, and let stand for two hours. Neutralize with hydrogen chloride and flush to drain with flooding water.
- ☑ Keep nitromethane out of a confined space, such as a sewer, because of the possibility of explosion (unless the sewer is designed to prevent the build-up of explosive concentrations).
- ☑ It may be necessary to dispose of nitromethane as a hazardous waste. The responsible state agency or the regional office of the federal Environmental Protection Agency (EPA) should be contacted for specific recommendations.

RISK ASSESSMENT: BUSINESS

General Assessment

Accidents or mishaps involving nitromethane can present a significant threat to business operations. The loss or damage of equipment or facilities can significantly affect fiscal viability. Lawsuits that may result from personnel injury, public exposures, and environmental contamination will also require a serious expenditure of resources. Media attention surrounding an injury, illness, or environmental damage can also result in a loss of profits and loss of current as well as future business.

Recommended Risk-Reduction Measures

Company attorneys, safety and health professionals, and environmental specialists should be involved in the development of any procedures dealing with safety and emergency response. A company official should be pre-designated as a public relations officer with specific training in dealing with the press. Corporate plans and policies should be developed, approved, and implemented long before any need for such arises.

REFERENCES

Ahlbom, A. 1993. *Biostatistics for Epidemiologists.* New York: CRC Press/Lewis Publishers.

American Conference of Governmental Industrial Hygienists. 1988. *Documentation of the Threshold Limit Values and Biological Exposure Indices,* 5th Edition (with updates). Cincinnati: ACGIH

Anthony, C. P., and N. J. Kolthoff. 1971. *Textbook of Anatomy and Physiology.* St. Louis: C. V. Mosby Company.

Asante-Duah, D. K. 1993. *Hazardous Waste Risk Assessment.* New York: CRC Press/Lewis Publishers.

Bloom, A. D. and F. J. DeSerres. 1995. *Ecotoxicity and Human Health.* New York: CRC Press/Lewis Publishers.

Brusick, D. J. 1994. *Methods for Genetic Risk Assessment.* New York: CRC Press/Lewis Publishers.

Bueche, J. R. 1972. *Principles of Physics,* 2nd Edition. New York: McGraw-Hill.

Burrell, R., D. K. Flaherty, and L. J. Sauers. 1992. *Toxicology of the Immune System.* New York: Van Nostrand Reinhold.

Calabrese, E. J. 1994. *Biological Effects of Low Level Exposures.* New York: CRC Press/Lewis Publishers.

Cockerman, L. G., B. S. Shane. 1994. *Basic Environmental Toxicology.* New York: CRC Press/Lewis Publishers.

Cothern, C. R. 1995. *Handbook for Environmental Risk Decisionmaking.* New York: CRC Press/Lewis Publishers.

Danse, I. R. 1991. *Common Sense Toxics in the Workplace.* New York: Van Nostrand Reinhold.

Environmental Protection Agency. 1983. *Chemical Hazard Information Profiles.* Washington, D.C.: U.S. EPA.

Fodden, C. R., and J. L. Weddell. 1992. *Hazardous Materials Emergency Action Data.* New York: CRC Press/Lewis Publishers.

Forseberg, K., and L. H. Keith. 1993. *Chemical; Protective Clothing Permeation and Degradation Database.* Chelsea, MI: Lewis Publishers.

Gardner, R. W. 1994. *Chemical Intolerance.* New York: CRC Press/Lewis Publishers.

Hallenbeck, W. H. 1993. *Quantitative Risk Assessment for Environmental and Occupational Health,* 2nd Edition. New York: CRC Press/Lewis Publishers.

Hodgson, E., R. B. Mailman, and J. E. Chambers. 1988. *Dictionary of Toxicology.* London: The MacMillian Company.

Hosty, J. W., and P. Foster. 1990. *A Practical Guide to Chemical Spill Response.* New York: Van Nostrand Reinhold.

Howard, P. H., et. al. 1991. *Handbook of Environmental Degradation Rates.* New York: CRC Press/Lewis Publishers.

Jones, F. E. 1994. *Toxic Organic Vapors in the Workplace.* New York: CRC Press/Lewis Publishers.

Kamrin, M. A. 1988. *Toxicology: A Primer on Principles and Applications.* Chelsea, Michigan: Lewis Publishers.

Keith, L. H. and M. M. Walker. 1995. *Handbook of Air Toxics.* New York: CRC Press/Lewis Publishers.

Landis, W. G., M. H. Yu. 1995. *Introduction to Environmental Toxicology.* New York: CRC Press/Lewis Publishers.

Lauwerys, R. R., P. Hoet. 1993. *Industrial Chemical Exposure.* New York: CRC Press/Lewis Publishers.

Lewis, R. J., Sr. 1993. *Hawley's Condensed Chemical Dictionary,* 12th Edition. New York: Van Nostrand Reinhold.

Lewis, R. J., Sr. 1992. *Sax's Dangerous Properties of Industrial Materials,* Volumes 1, 2, and 3, 8th Edition. New York: Van Nostrand Reinhold.

Lide, D. R. 1995. *Handbook of Chemistry and Physics,* 76th Edition. New York: CRC Press/Lewis Publishers.

Mellström, G. A., J. E. Wahlberg, and H. I. Maibach. 1994. *Protective Gloves for Occupational Use.* New York: CRC Press/Lewis Publishers.

Meyer, E. 1990. *Chemistry of Hazardous Materials.* Englewood Cliffs, NJ: Prentice Hall.

Patnaik, P. 1992. *A Comprehensive Guide to the Hazardous Properties of Chemical Substances.* New York: Van Nostrand Reinhold.

Philp, R. B. 1995. *Environmental Hazards and Human Health.* New York: CRC Press/Lewis Publishers.

Rea, W. J. 1995. *Chemical Sensitivity,* Vol. I-III. New York: CRC Press/Lewis Publishers.

Richardson, M. 1995. *Dictionary of Substance and Their Effects,* Vol. I-VI. New York: CRC Press/Lewis Publishers.

Sacarello, H. L. A. 1994. *The Comprehensive Handbook of Hazardous Materials.* New York: CRC Press/Lewis Publishers.

Sheftel, V. O. 1995. *Handbook of Toxic Properties of Monomers and Additives.* New York: CRC Press/Lewis Publishers.

Sherman, J. D. 1988. *Chemical Exposure and Disease.* New York: Van Nostrand Reinhold.

Talbott, E., G. Craun. 1995. *An Introduction to Environmental Epidemiology.* New York: CRC Press/Lewis Publishers.

Torkelson, T. R. V. K. Rowe. 1981. *Patty's Industrial Hygiene and Toxicology,* Vol. 2B. New York: Wiley-Interscience.

U.S. Department of Health and Human Services, National Institute for Occupational Safety and Health. 1994. *NIOSH Pocket Guide to Chemical Hazards.* Washington, D.C.: U.S. Government Printing Office.

MATERIAL SAFETY DATA SHEET

CHEMICAL NAME	
OCTACHLORONAPHTHALENE	**CRC PRESS/LEWIS PUBLISHERS DISCLAIMER:** THE INFORMATION AND RECOMMENDATIONS PRESENTED HEREIN ARE BASED ON SOURCES BELIEVED TO BE RELIABLE. CRC MAKES NO REPRESENTATION ON ITS COMPLETENESS OR ACCURACY. IT IS THE RESPONSIBILITY OF THE USER TO DETERMINE THE CHEMICAL'S SUITABILITY FOR ITS INTENDED USE, THE CHEMICAL'S SAFE USE, AND THE CHEMICAL'S PROPER DISPOSAL. NO REPRESENTATIONS AND/OR WARRANTIES, EITHER EXPRESSED OR IMPLIED, OF THE MERCHANTABILITY OR FITNESS FOR A PARTICULAR PURPOSE, OR OF ANY OTHER NATURE, ARE MADE WITH RESPECT TO THE INFORMATION PROVIDED IN THIS MSDS OR TO THE CHEMICAL TO WHICH INFORMATION MAY REFER. CRC NEITHER ASSUMES NOR AUTHORIZES ANY OTHER PERSON TO ASSUME FOR IT, ANY OTHER ADDITIONAL RESPONSIBILITY OR LIABILITY FOR THE USE OF, OR RELIANCE UPON, THIS INFORMATION.

NATIONAL FIRE PROTECTION ASSOCIATION (NFPA) LABELING INFORMATION

HEALTH	FIRE	REACTIVE	OTHER	DEGREE OF HAZARD	COLOR CODING	OTHER CODES
3	**0**	**0**		0 = Minimum Hazard 1 = Slight Hazard 2 = Moderate Hazard 3 = Serious Hazard 4 = Severe Hazard	HEALTH = BLUE FIRE = RED REACTIVITY = YELLOW OTHER = WHITE	OX = Oxidizer ACID = Acid ALK = Alkali COR = Corrosive W = Use No Water

SECTION I - GENERAL INFORMATION

Characterization	RCRA Number	EPA Class
Chlorinated Aromatic Hydrocarbon	**None**	**Not Applicable**

DOT Proper Shipping Name **No Citation**	Chemical Abstract Service (CAS) Number **2234-13-1**
DOT Hazard Class and Label Requirements **No Citation**	DOT Emergency Guide Code **No Citation**
DOT Identification Number **No Citation**	Chemical Formula **$C_{10}Cl_8$**

Synonyms
Halowax® 1051; 1,2,3,4,5,6,7,8-octachloronaphthalene; perchloronaphthalene.

SECTION II - HAZARDOUS INGREDIENTS AND IDENTITY INFORMATION

Hazardous Components (specific identity)	OSHA Exposure Criteria	NIOSH Exposure Criteria	Immediately Dangerous to Life and Health (IDLH)	ACGIH Exposure Criteria
Octachloronaphthalene	**PEL (skin):** 0.1 mg/m³ **STEL:** Not Established	**REL (skin):** 0.1 mg/m³ **STEL:** 0.3 mg/m³	**Not Determined**	**TLV (skin):** 0.1 mg/m³ **STEL:** 0.3 mg/m³

SECTION III - PHYSICAL AND CHEMICAL CHARACTERISTICS

Boiling Point **824°F (440°C)**	Specific Gravity (H_2O = 1) **2.0**
Vapor Pressure (mm Hg) **>1 at 69°F (20°C)**	Molecular Weight **403.7**
Vapor Density (Air = 1) **13.9**	Melting Point **365°F (185°C)**

Solubility
Insoluble in water. Slightly soluble in alcohol. Soluble in benzene and chloroform.

Appearance and Odor
Pale-yellow waxy solid with an aromatic odor.

SECTION IV - FIRE AND EXPLOSION HAZARD DATA

Flash Point (method used) **Non-Combustible**	Explosive Limits in Air % by Volume **LEL: Not Applicable** **UEL: Not Applicable**
NFPA Classification **Non-Combustible Solid**	Autoignition Temperature **Not Applicable**

Extinguishing Media
Use extinguishing agent suitable to surrounding area. Octachloronaphthalene is non-combustible.

Special Fire Fighting Procedures
Wear full protective clothing and self-contained breathing apparatus (SCBA). Poisonous gases are produced in fire. Structural fire fighting clothing is permeable and may not provide adequate protection against octachloronaphthalene. Remove containers from fire if it can be done without risk.

Unusual Fire and Explosion Hazards
Octachloronaphthalene is non-combustible but can present a significant explosion hazard if it comes into contact with incompatible materials. The vapors are much heavier than air and can collect in low areas to displace oxygen.

SECTION V - REACTIVITY DATA

Stability		Conditions to Avoid **Octachloronaphthalene is normally stable under routine conditions of storage and handling. Avoid contact with heat and incompatible materials.**
Stable	Unstable	Incompatibility *(materials to avoid)*
X		**Incompatible with strong oxidizers (fluorine, bromine, chlorine), and oxidizing agents.**
Hazardous Polymerization		Conditions to Avoid **Under normal conditions of temperature and pressure, hazardous polymerization of octachloronaphthalene is not expected to occur.**
May Occur	Will Not Occur	Hazardous Decomposition or By-products
	X	**When heated to decomposition, octachloronaphthalene can emit carbon monoxide, toxic chloride fumes, and acrid, irritating smoke.**

SECTION VI - HEALTH HAZARD DATA

Primary Route(s) of Entry:	Inhalation?	Absorption (skin/eye)?	Ingestion?
	X	X	X

Health Hazards

INHALATION: Irritation of the respiratory tract, headache, nausea, and vomiting. There is also risk of liver injury, jaundice, and blood disorders.

ABSORPTION: Coal tar derivatives may absorb through unbroken skin. Irritation with burning, itching, and the development of an acne-like rash. Irritates the conjunctiva of the eye.

INGESTION: Gastrointestinal tract irritation.

Carcinogenicity	NTP Listed?	IARC Cancer Review Group?	OSHA Regulated?	Target Organs?
Unknown Human Unknown Animal	No	No	29 CFR 1910.1000 Table Z-1	Skin and liver

Medical Conditions Generally Aggravated by Exposure

Diseases of the skin and/or liver may be aggravated upon exposure to octachloronaphthalene.

Emergency and First-aid Procedures

Eye contact: Do not allow victim to keep eyes tightly shut. Flush immediately with water for 15 minutes (minimum); seek medical attention. Skin contact: Remove all contaminated clothing. Immediately wash area with large amounts of soap and water. For red or blistered skin, contact physician. For inhalation: Remove the person from exposure. Provide respiratory assistance and CPR. Transfer to medical facility. If swallowed: Contact poison control center or seek medical attention immediately. Do NOT induce vomiting unless directed to do so. Never attempt to give anything by mouth to an unconscious or convulsing person.

SECTION VII - PRECAUTIONS FOR SAFE HANDLING AND USE

Steps to be Taken in Case Material is Released or Spilled

Collect powdered materials in most convenient and safe manner and deposit in sealed containers. Ventilate area of spill or leak. Remove all ignition sources. Restrict those not involved in cleanup from entering area. Notify appropriate authorities, as required by applicable environmental regulations.

Preferred Waste Disposal Method

Incinerator equipped with scrubber and afterburner.

Precautions to be Taken in Handling and Storage

Octachloronaphthalene is a non-combustible solid. Store to avoid contact with strong oxidizers, since violent reactions can occur. Store in tightly closed containers in cool, well-ventilated area.

Other Precautions and Warnings

Personnel assigned to areas where octachloronaphthalene is used, handled, or stored must be trained on its hazards *before* being assigned to such areas.

SECTION VIII - CONTROL MEASURES AND PERSONAL PROTECTIVE EQUIPMENT

Respiratory Protection *(specify type)*

Use an MSHA/NIOSH-approved air-purifying respirator with a dust/mist pre-filter and organic vapor cartridge. For best protection, use a supplied-air respirator with full facepiece in continuous flow mode, or use a self-contained breathing apparatus with full facepiece operated in positive pressure mode.

Ventilation

Local exhaust (preferred) at site of chemical work or general ventilation.

Protective Gloves	Eye Protection	Other Protective Clothing
Impervious Gloves	Chemical/Dust Goggles or Face Mask	Boots; Aprons; Gauntlets

Work/Hygiene Practices

Always wash hands thoroughly after using chemical; never bring food, drink, or smoking materials into vicinity of chemicals.

OCTACHLORONAPHTHALENE

$C_{10}Cl_8$ CAS: 2234-13-1

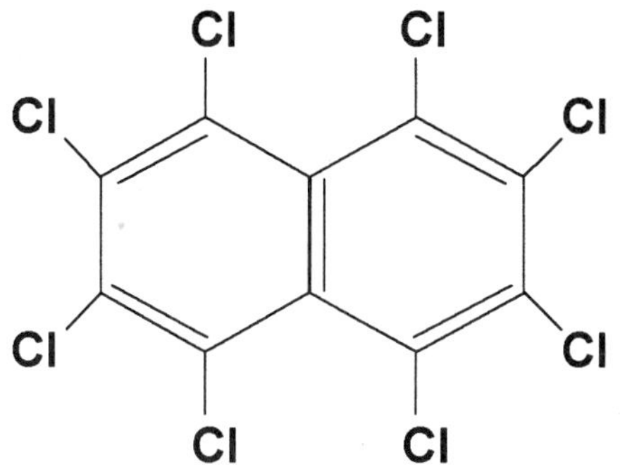

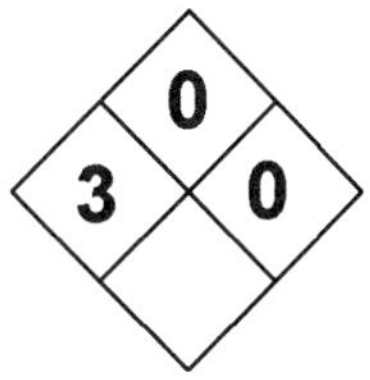

IDENTIFICATION AND TYPICAL USES

Octachloronaphthalene is a waxy, pale-yellow solid with an aromatic odor. It is used in fireproofing and waterproofing in cable insulation and as a lubricant additive.

RISK ASSESSMENT: HEALTH

General Assessment

There are very little data available on the toxicity of this chemical in humans or animals. However, based upon data that does exist for other lower chlorinated naphthalenes, it might be assumed that octachloronaphthalene is a human poison by *ingestion, inhalation* and by dermal *absorption.* It is irritating to the skin and, on prolonged contact, may cause the development of an acne-like rash. There are no data on its carcinogenic, mutagenic, or teratogenic effects of exposure to octachloronaphthalene.

Inhalation of octachloronaphthalene vapors may lead to serious and, in the case of overexposure, possibly fatal liver damage. Symptoms include jaundice (yellow discoloration of the skin and eyes).

Acute Health Effects

The following acute (short-term) health effects may occur immediately or shortly after exposure to octachloronaphthalene (within hours or days):

Skin: Irritation with burning, itching, and the development of an acne-like rash at the site of chemical contact.

Eye: Severe irritation of the conjunctiva with burning, itching, and watering.

Lung: May cause mild irritation of the respiratory tract and associated mucosa.

Other: May cause serious or even fatal liver damage on acute overexposure.

Chronic Health Effects

The following chronic (long-term) health effects may occur at some time after repeated exposures to octachloronaphthalene and can last for months or even years:

Cancer Hazards: According to information presented in the references, octachloronaphthalene has not been adequately tested for its ability to cause cancer in test animals. Other similar naphthalene compounds have been shown to be carcinogenic in the long-term.

Reproductive Hazard: According to the references, octachloronaphthalene has not adequately tested for its ability to adversely affect reproduction. Other similar naphthalene compounds have been shown to be teratogenic in the long-term.

Other Chronic Effects: Continuous exposures to octachloronaphthalene may lead to an increased incidence of liver dysfunction and may actually cause irreversible damage.

Recommended Risk-Reduction Measures

Personnel should avoid direct contact with octachloronaphthalene. The exact nature of its toxicity is not clearly understood or documented in the references. If a less toxic chemical cannot be substituted for a hazardous substance, then *engineering controls* are the most effective method of reducing exposures. The best protection is to enclose operations and/or provide local exhaust ventilation at the site of chemical release. While not always operationally feasible, isolating operations can also reduce exposure. Using respiratory protection is less effective than the controls mentioned above, but is still advisable whenever working with or around octachloronaphthalene. Air-purifying respirators will provide protection for exposures above the PEL. However, a self-contained breathing apparatus (SCBA) with full facepiece operated in pressure demand or a supplied-air respirator operated in continuous flow mode are the recommended respiratory protection methods of choice when working with this material. If a full facepiece is not available, then chemical/dust goggles should be worn to protect the eyes. A face shield and a protective apron should also be worn. To prevent hand and skin exposures, impervious gloves should be used. Glove manufacturers should be contacted and permeation studies obtained *before* final glove selections have been made.

Administrative controls should also be in place to minimize the potential for human exposures. These may include written procedures or policies which specify the methods and techniques that will be practiced whenever personnel are to work with octachloronapthalene.

All personnel should receive *training* on the use, hazards, protective measures, emergency actions, and other precautions per 29 CFR 1910.1200 (Hazard Communication), prior to the first assignment in an area where octachloronaphthalene is used or stored.

Before beginning employment, and at regular intervals thereafter (e.g., annually), the following medical tests are recommended for personnel working with octachloronaphthalene:

☑ Liver function tests (establish baseline).

If symptoms develop or overexposure is suspected, the following may be useful:

☑ Liver function tests.

☑ Evaluation by a qualified allergist, including careful evaluation of exposure history and special testing (may help diagnose skin allergy).

Any evaluation should include a careful history of past and present symptoms with an examination. Medical tests that simply look for existing damage are not a substitute for controlling exposures.

Other methods to reduce exposure include:

☑ Where possible, enclose operations and use local exhaust ventilation at the site of chemical release. If local exhaust ventilation or enclosure is not used, respiratory protection should be mandatory.

☑ Always ensure that proper protective clothing is worn when using chemical substances.

☑ Wash thoroughly immediately after exposure to octachloronaphthalene and at the end of the work shift or before eating, drinking, or smoking. Avoid contact with direct sunlight.

☑ Hazard warning information should be posted in the work area. In addition, as part of an on-going education and training program, all information on the health and safety hazards of octachloronapthalene should be communicated to all exposed workers.

☑ Eye wash stations should be provided in the immediate work area for emergency use. If there is a possibility of skin exposure to octachloronaphthalene, emergency shower facilities should also be provided.

RISK ASSESSMENT: ENVIRONMENT

General Assessment

The environment is at risk of exposure during transportation, storage, disposal, or destruction of octachloronaphthalene. In almost every scenario, the threat of environmental exposure is contingent upon the proper handling of the chemical substance. Accidental spills, large or small, can result in fire, explosion (depending upon conditions of spill), and possible contamination of the surrounding environmental mediums (water, soil, and air).

Octachloronaphthalene is considered a noncombustible solid. However, because it is incompatible with many commonly encountered oxidizers (chlorine, bromine, fluorine), extreme caution is required in handling, storage, transportation, and disposal of octachloronaphthalene. These characteristics also require special consideration during any emergency situation involving a leak or spill of octachloronaphthalene. Even as a solid, it has a considerably high vapor pressure. Vapors are heavier than air and can collect in low lying areas to displace oxygen. Should octachloronaphthalene ever come into contact with incompatible substances (such as oxidizers) either during use, transportation, storage, or disposal, the formation of highly toxic and/or highly explosive commodities is extremely possible.

The proper disposal/destruction method for octachloronaphthalene waste is to burn it in a chemical incinerator equipped with an afterburner and air scrubber.

Octachloronaphthalene can enter the environment from industrial and municipal plant discharges and from spills.

Acute Ecological Effects

Acute (short-term) toxic effects may include the death of animals, birds, or fish and death or low growth rate in plants. Acute effects are seen 2 to 4 days after animals or plants are exposed to octachloronaphthalene.

Octachloronaphthalene has slight acute toxicity to aquatic life. Insufficient data are available to evaluate or predict the short-term effects of octachloronaphthalene to plants, birds, or land animals.

Chronic Ecological Effects

Chronic (long-term) toxic effects may include shortened life span, reproductive problems, lower fertility, and changes in appearance or behavior in exposed animals. These effects can be seen long after first exposure(s) to toxic chemicals.

Octachloronaphthalene has slight chronic toxicity to aquatic life. Insufficient data are available to evaluate or predict the long-term effects of octachloronaphthalene to plants, birds, or land animals.

Water Solubility

Octachloronaphthalene is nearly insoluble in water. Concentrations of 1 milligram or less may mix with a liter of water.

Persistence in the Environment

Octachloronaphthalene is moderately persistent in water, with a half-life between 20 and 200 days. The half-life of a pollutant is the amount of time it takes for one half of the chemical to be degraded. About 51.7% of octachloronaphthalene will eventually end up in terrestrial soils; approximately 48.2% will end up in aquatic sediments; and the remainder will end up in suspended solids.

Bioaccumulation in Aquatic Organisms

Some substances increase in concentration, or *bioaccumulate,* in living organisms as they breathe contaminated air, drink contaminated water, or eat contaminated food. These chemicals can become concentrated in the tissues and internal organs of animals as well as humans.

The concentration of octachloronaphthalene found in fish tissues is expected to be considerably higher than the average concentration of octachloronaphthalene in the water from which the fish was taken.

Recommended Risk-Reduction Measures

Proper training of all transporters will reduce the likelihood of a mishap or accident resulting in a leak or spill of octachloronaphthalene into the environment. The correct use of labeling on all containers, trucks, and rail cars will enable emergency responders to react properly and quickly to any disaster thereby reducing the potential risk to the environment and to personnel.

Storage of octachloronaphthalene should be segregated from incompatible chemicals to minimize the risk of cross-contamination or contact. Personnel working in storage areas must be trained on the hazards of octachloronaphthalene.

If a spill or leak to the environment has occurred, fire department, emergency response and/or hazardous materials spill personnel should be notified immediately. Cleanup should be attempted only by those trained in proper spill containment procedures. Contaminated soils should be removed for incineration and replaced with clean soil. If octachloronaphthalene should contact the water table, aquifer, or navigable waterway, the local and/or state emergency response authorities must be notified. A comprehensive emergency response or disaster preparedness/recovery plan should be in place prior to any operations involving the use, transportation, storage, or disposal of octachloronaphthalene. If octachloronaphthalene is spilled or leaked, the following specific steps are recommended:

- ☑ Restrict persons not wearing protective clothing from area of spill or leak until cleanup is complete and area can be opened for normal work.
- ☑ Remove all ignition sources.
- ☑ Collect powdered materials in the safest manner possible and deposit in sealed containers. Do NOT dry sweep. Use vacuum equipped with a HEPA filter instead. Damp mop residue.
- ☑ It may be necessary to dispose of octachloronaphthalene as a hazardous waste. The responsible state agency or the regional office of the

federal Environmental Protection Agency (EPA) should be contacted for specific recommendations.

RISK ASSESSMENT: BUSINESS

General Assessment

Accidents or mishaps involving octachloronaphthalene can present a significant threat to business operations. The loss or damage of equipment or facilities can significantly affect fiscal viability. Lawsuits that may result from personnel injury/death, public exposures, and/or environmental contamination will also require a serious expenditure of resources.

Media attention surrounding an injury, death, or environmental damage can also result in a loss of profits and loss of current as well as future business.

Recommended Risk-Reduction Measures

Company attorneys, safety and health professionals, and environmental specialists should be involved in the development of any safety, health or environmental policies or procedures. A company official should be pre-designated as a public relations officer with specific training in dealing with the press. Corporate plans and policies should be developed, approved, and implemented long before any need for such arises.

REFERENCES

Ahlbom, A. 1993. *Biostatistics for Epidemiologists.* New York: CRC Press/Lewis Publishers.

American Conference of Governmental Industrial Hygienists. 1988. *Documentation of the Threshold Limit Values and Biological Exposure Indices,* 5th Edition (with updates). Cincinnati: ACGIH

Asante-Duah, D. K. 1993. *Hazardous Waste Risk Assessment.* New York: CRC Press/Lewis Publishers.

Bloom, A. D. and F. J. DeSerres. 1995. *Ecotoxicity and Human Health.* New York: CRC Press/Lewis Publishers.

Bueche, J. R. 1972. *Principles of Physics,* 2nd Edition. New York: McGraw-Hill.

Calabrese, E. J. 1994. *Biological Effects of Low Level Exposures.* New York: CRC Press/Lewis Publishers.

Cockerman, L. G., B. S. Shane. 1994. *Basic Environmental Toxicology.* New York: CRC Press/Lewis Publishers.

Cothern, R. R. 1993. *Comparative Environmental Risk Assessment.* New York: CRC Press/Lewis Publishers.

Gardner, R. W. 1994. *Chemical Intolerance.* New York: CRC Press/Lewis Publishers.

Hallenbeck, W. H. 1993. *Quantitative Risk Assessment for Environmental and Occupational Health,* 2nd Edition. New York: CRC Press/Lewis Publishers.

Howard, P. H. 1995. *Handbook of Environmental Fate and Exposures Data for Organic Chemicals,* Vol. I-IV. New York: CRC Press/Lewis Publishers.

Jones, F. E. 1994. *Toxic Organic Vapors in the Workplace.* New York: CRC Press/Lewis Publishers.

Keith, L. H. and M. M. Walker. 1995. *Handbook of Air Toxics.* New York: CRC Press/Lewis Publishers.

Lewis, R. J., Sr. 1993. *Hawley's Condensed Chemical Dictionary,* 12th Edition. New York: Van Nostrand Reinhold.

Lewis, R. J., Sr. 1992. *Sax's Dangerous Properties of Industrial Materials,* Volumes 1, 2, and 3, 8th Edition. New York: Van Nostrand Reinhold.

Lide, D. R. 1995. *Handbook of Chemistry and Physics,* 76th Edition. New York: CRC Press/Lewis Publishers.

Manahan, S. E. 1992. *Toxicological Chemistry,* 2nd Edition. New York: CRC Press/Lewis Publishers.

Mellström, G. A., J. E. Wahlberg, and H. I. Maibach. 1994. *Protective Gloves for Occupational Use.* New York: CRC Press/Lewis Publishers.

Meyer, E. 1990. *Chemistry of Hazardous Materials.* Englewood Cliffs, NJ: Prentice Hall.

Mycock, J. C., J. D. McKenna, and L. Theodore. 1995. *Handbook of Air Pollution Control Engineering and Technology.* New York: CRC Press/Lewis Publishers.

National Cancer Institute. 1978. *Carcinogenesis Technical Report Series. Bioassay of Octachloronaphthalene for Possible Carcinogenicity.* DEHW (NIH) Publication Number: 78-1323, p 53. Washington, DC: U.S. Government Printing Office.

Ott, W. R. 1995. *Environmental Statistics and Data Analysis.* New York: CRC Press/Lewis Publishers.

Patnaik, P. 1992. *A Comprehensive Guide to the Hazardous Properties of Chemical Substances.* New York: Van Nostrand Reinhold.

Proctor, N. H., J. P. Hughes. 1991. *Chemical Hazards of the Workplace.* New York: Van Nostrand Reinhold.

Rea, W. J. 1995. *Chemical Sensitivity,* Vol. I-III. New York: CRC Press/Lewis Publishers.

Richardson, M. 1995. *Dictionary of Substance and Their Effects,* Vol. I-VI. New York: CRC Press/Lewis Publishers.

Sacarello, H. L. A. 1994. *The Comprehensive Handbook of Hazardous Materials.* New York: CRC Press/Lewis Publishers.

Talbott, E., G. Craun. 1995. *An Introduction to Environmental Epidemiology.* New York: CRC Press/Lewis Publishers.

Torkelson, T. R. V. K. Rowe. 1981. *Patty's Industrial Hygiene and Toxicology,* Vol. 2B. New York: Wiley-Interscience.

U.S. Department of Health and Human Services, National Institute for Occupational Safety and Health. 1994. *NIOSH Pocket Guide to Chemical Hazards.* Washington, D.C.: U.S. Government Printing Office.

MATERIAL SAFETY DATA SHEET

CHEMICAL NAME	
OCTANE	**CRC PRESS/LEWIS PUBLISHERS DISCLAIMER:** THE INFORMATION AND RECOMMENDATIONS PRESENTED HEREIN ARE BASED ON SOURCES BELIEVED TO BE RELIABLE. CRC MAKES NO REPRESENTATION ON ITS COMPLETENESS OR ACCURACY. IT IS THE RESPONSIBILITY OF THE USER TO DETERMINE THE CHEMICAL'S SUITABILITY FOR ITS INTENDED USE, THE CHEMICAL'S SAFE USE, AND THE CHEMICAL'S PROPER DISPOSAL. NO REPRESENTATIONS AND/OR WARRANTIES, EITHER EXPRESSED OR IMPLIED, OF THE MERCHANTABILITY OR FITNESS FOR A PARTICULAR PURPOSE, OR OF ANY OTHER NATURE, ARE MADE WITH RESPECT TO THE INFORMATION PROVIDED IN THIS MSDS OR TO THE CHEMICAL TO WHICH INFORMATION MAY REFER. CRC NEITHER ASSUMES NOR AUTHORIZES ANY OTHER PERSON TO ASSUME FOR IT, ANY OTHER ADDITIONAL RESPONSIBILITY OR LIABILITY FOR THE USE OF, OR RELIANCE UPON, THIS INFORMATION.

HAZARD WARNING INFORMATION

HEALTH	FIRE	REACTIVE	OTHER	DEGREE OF HAZARD	COLOR CODING	OTHER CODES
0	3	0		0 = Minimum Hazard 1 = Slight Hazard 2 = Moderate Hazard 3 = Serious Hazard 4 = Severe Hazard	HEALTH = BLUE FIRE = RED REACTIVITY = YELLOW OTHER = WHITE	OX = Oxidizer ACID = Acid ALK = Alkali COR = Corrosive W = Use No Water

SECTION I - GENERAL INFORMATION

Characterization	RCRA Number	EPA Class
Aliphatic Hydrocarbon	**D001**	**Characteristic (I) Waste**

DOT Proper Shipping Name **Octane**	Chemical Abstract Service (CAS) Number **111-65-9**
DOT Hazard Class and Label Requirements **Flammable Liquid**	DOT Emergency Guide Code **27**
DOT Identification Number **UN 1262**	Molecular Formula $CH_3(CH_2)_6CH_3$

Synonyms
Normal-octane; *n*-octane; octyl hydride.

SECTION II - HAZARDOUS INGREDIENTS AND IDENTITY INFORMATION

Hazardous Components (specific identity)	OSHA Exposure Criteria	NIOSH Exposure Criteria	Immediately Dangerous to Life and Health (IDLH)	ACGIH Exposure Criteria
Octane (derivation: Found in petroleum cracking products, gasoline, petroleum ether, and petroleum naphtha). **1 ppm = 4.75 mg/m^3 (approx.)**	**PEL:** 500 ppm 2350 mg/m^3 **STEL (15 min):** 375 ppm 1800 mg/m^3	**REL:** 75 ppm 350 mg/m^3 **STEL (15 min):** 385 ppm 1800 mg/m^3	**1000 ppm**	**TLV:** 300 ppm 1450 mg/m^3 **STEL:** 375 ppm 1800 mg/m^3

SECTION III - PHYSICAL AND CHEMICAL CHARACTERISTICS

Boiling Point **259°F (126°C)**	Specific Gravity (H_2O = 1) **0.70 at 68°F (20°C)**
Vapor Pressure (mm Hg) **11 at 77°F (25°C)**	Molecular Weight (atomic weight) **114.2**
Vapor Density (Air = 1) **3.9**	Melting Point **-73°F (-57°C)**

Solubility
Insoluble in water. Soluble in alcohol and acetone.

Appearance and Odor
A clear, colorless liquid with a mild, gasoline-like odor.

SECTION IV - FIRE AND EXPLOSION HAZARD DATA

Flash Point (method used) **56°F (13°C) closed cup**	Explosive Limits in Air % by Volume **LEL: 1.0%** **UEL: 6.5%**
NFPA Classification **Class 1B Flammable Liquid**	Autoignition Temperature **403°F (206°C)**

Extinguishing Media
Use dry chemical, carbon dioxide, or foam. Water may be ineffective in putting out a fire involving octane because octane will float on the surface and may actually spread the fire.

Special Fire Fighting Procedures
Isolate fire area. Wear self-contained breathing apparatus (SCBA). Poisonous gases are produced in fire. Use water spray to cool fire-exposed containers until long after fire is out. Fight fire from a distance if possible or use a monitor nozzle or unmanned hose. Use non-sparking tools during cleanup.

Unusual Fire and Explosion Hazards
Dangerous fire and explosion hazard when exposed to heat or flame. Containers may explode in fire. The vapor is heavier than air and may travel for some distance to cause fire to flashback to source.

SECTION V - REACTIVITY DATA

Stability		Conditions to Avoid
		Octane is normally stable under routine conditions of handling and storage. Avoid contact with heat, flame, other sources of ignition, and contact with incompatible materials and chemicals.
Stable	Unstable	Incompatibility *(materials to avoid)*
X		Strong oxidizers (such as peroxides) and oxidizing agents.
Hazardous Polymerization		Conditions to Avoid
		Hazardous polymerization of octane cannot occur under normal conditions of temperature and pressure.
May Occur	Will Not Occur	Hazardous Decomposition or By-products
	X	When heated to decomposition, octane emits toxic and acrid fumes, such as oxides of carbon (monoxide and dioxide).

SECTION VI - HEALTH HAZARD DATA

Primary Route(s) of Entry:	Inhalation?	Absorption (skin/eye)?	Ingestion?
	X		X

Health Hazards

INHALATION: Vapors are poisonous. Causes irritation of the respiratory system mucosa and depression of the central nervous system (CNS). High concentrations even for short periods may lead to fatal pulmonary edema (fluid in the lungs). Intoxicating and anesthetic effects.

SKIN & EYES: Irritation with drying, cracking, redness, blistering (prolonged contact). May lead to contact dermatitis and secondary infection.

INGESTION: May cause nausea, vomiting, headache, dizziness, gastrointestinal irritation, blurring of vision, and other effects of central nervous system depression.

Carcinogenicity	NTP Listed?	IARC Cancer Review Group?	OSHA Regulated?	Target Organs?
Unknown Human Unknown Animal	No	No	29 CFR 1910.1000 Table Z-1	Eyes, skin, respiratory system, CNS.

Medical Conditions Generally Aggravated by Exposure

None identified.

Emergency and First-aid Procedures

Eye contact: Do not allow victim to keep eyes tightly shut. Flush immediately with water for 15 minutes (minimum); seek medical attention. Skin contact: Remove all contaminated clothing, including shoes. Immediately wash area with flooding amounts of soap and water. Seek medical attention immediately. For inhalation: Remove the person from exposure. Provide respiratory assistance and CPR. Transfer to medical facility. Observe 48 hours for lung effects. If swallowed: Seek medical attention immediately. Do not induce vomiting due to possible aspiration of vomitus/chemical into the lungs.

SECTION VII - PRECAUTIONS FOR SAFE HANDLING AND USE

Steps to be Taken in Case Material is Released or Spilled

Remove all ignition sources. Wear SCBA respiratory protection. Restrict those not wearing protective equipment and who are not involved in cleanup from entering area. Avoid skin contact. Stop leak if it can be done without risk. Provide maximum explosion-proof ventilation. Absorb with vermiculite or other similar material and deposit in sealed containers. Use non-sparking tools during cleanup.

Preferred Waste Disposal Method

No citation.

Precautions to be Taken in Handling and Storage

Store in tightly closed containers in cool, well-ventilated location. Keep away from sources of heat.

Other Precautions and Warnings

Sources of ignition (open flame or smoking), are prohibited where octane is used or stored. Metal containers should be bonded and grounded during transfer operations. Bulk storage not recommended.

SECTION VIII - CONTROL MEASURES AND PERSONAL PROTECTIVE EQUIPMENT

Respiratory Protection *(specify type)*

Use a MSHA/NIOSH-approved air-purifying organic vapor respirator, a powered air purifying respirator, or use a self-contained breathing apparatus (SCBA) with full facepiece operated in positive pressure mode.

Ventilation

Local exhaust (preferred) at site of chemical work or general ventilation.

Protective Gloves	Eye Protection	Other Protective Clothing
Impervious Gloves	Chemical Goggles or Face Mask	Protective Apron if Splash is Likely

Work/Hygiene Practices

Always wash hands thoroughly after using chemical; never bring food, drink, or smoking materials into vicinity of chemicals.

OCTANE

$CH_3(CH_2)_6CH_3$

CAS: 111-65-9

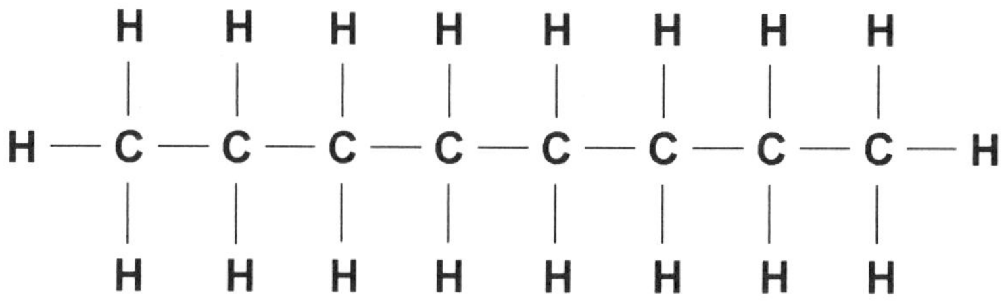

IDENTIFICATION AND TYPICAL USES

Octane is a clear, colorless liquid with a mild, gasoline-like odor. It is used as a solvent, in organic synthesis, in calibrations, and in azeotropic distillations.

RISK ASSESSMENT: HEALTH

General Assessment

Octane is considered moderately toxic by *inhalation* of its vapors or by *ingestion* of the liquid. Skin contact will also cause localized irritation at the site of contact, but dermal absorption has not been reported. When large quantities are in contact with skin for extended periods, blisters may develop. There are no reports of any carcinogenic, mutagenic, or teratogenic effects. The toxic properties of octane are expected to be at least similar to those of other paraffinic hydrocarbons.

Inhalation of the octane vapors can cause mild to severe irritation of the eyes, nose, throat, and respiratory tract mucosa. Systemic effects include cough, conjunctiva irritation, flushing of the face, headaches, and blurred vision. High concentrations even for short periods can cause potentially fatal pulmonary edema (fluid in the lungs). Symptoms of coughing, wheezing, congestion, tightness in chest, and dypsnea may be delayed up to 48 hours, thereby creating a false sense of security with regard to health exposure risk. Inhalation can also lead to depression of the central nervous system (CNS) with symptoms of intoxication, hallucination, mental confusion, staggering gait or stupor, slurred speech, loss of consciousness and possible convulsions.

Skin contact may cause dermatitis with local blistering and rash. Eye contact can cause irritation of the conjunctiva with redness, swelling, tearing, and localized inflammation of eye tissues. Ingestion may cause nausea, vomiting, headaches, dizziness, gastrointestinal irritation, and symptoms of CNS depression similar to those caused by inhalation. If vomiting occurs, aspiration of the liquid into the lungs can cause choking, difficulty breathing or shortness of breath (dypsnea), tachypnea (rapid respiration), tachycardia (rapid heartbeat), fever, bronchitis, and pneumonitis.

Acute Health Effects

The following acute (short-term) health effects may occur immediately or shortly after exposure to octane:

Skin: Irritation with potential for redness, blistering, irritation, rash, and possible dermatitis.

Eye: Severe irritation, conjunctivitis, redness, and possible pain.

Lung: Nose, throat, and respiratory tract irritation causing difficulty in breathing, possible pulmonary edema on high exposure (a medical emergency) which can be fatal.

CNS: Inebriation, dizziness, stupor, hallucinations, slurred speech, mental confusion, loss of consciousness, and possible convulsions.

Chronic Health Effects

The following chronic (long-term) health effects may occur at some time after exposure to octane and can last for months or even years:

Cancer Hazards: According to information presented in the references, octane has not been adequately tested for its ability to cause cancer in test animals.

Reproductive Hazard: There is no evidence in the references to indicate that exposure to octane causes any adverse effects to reproduction.

Other Chronic Effects: Chronic inhalation of octane vapors can cause loss of appetite, nausea, weight loss, insomnia, hyperesthesia, motor weakness, muscular degeneration, diminished tendon reflexes, and loss of coordination. Repeated skin contact causes dermatitis in some individuals.

Recommended Risk-Reduction Measures

Personnel should avoid direct contact with or exposure to octane. Occupational poisoning occurs most commonly by inhalation. Substitution is not usually an option with octane since its use is quite specific. However, a less toxic material or compound should be used in place of octane whenever possible. If substitution cannot be done, *engineering controls* are the most effective method of reducing exposures. The best protection is to enclose operations and/or provide local exhaust ventilation at the site of octane release. While not always operationally feasible, isolating operations can also reduce exposure. Using respiratory protection is less effective than the controls mentioned above, but is still advisable whenever working with or around octane. A MSHA/NIOSH-approved air-purifying respirator with an organic vapor cartridge, a supplied-air respirator operated in continuous flow mode, or a self-contained breathing apparatus (SCBA) with full facepiece operated in pressure demand are the appropriate methods of respiratory protection when exposure to octane is anticipated. If a full facepiece is not available, then chemical goggles should be worn to protect the eyes. A face shield should also be considered. To prevent hand and skin exposures protective gloves should be used. Glove manufacturers should be contacted and permeation studies obtained *before* final glove selections are made.

Administrative controls should also be in place to minimize the potential for human exposures. These may include written procedures or policies which specify the methods and techniques that will be practiced whenever personnel are to work with octane.

All personnel should receive *training* on the use, hazards, protective measures, emergency actions, and other precautions per 29 CFR 1910.1200 (Hazard Communication), prior to the first assignment in an area where octane is used or stored.

- Before beginning employment and at regular times thereafter, the following recommended medical tests should be considered:

☑ Lung function tests (establish baseline).

If symptoms occur or overexposure is suspected, the following additional tests are suggested:

☑ Lung function tests (compare to baseline).

☑ Consider chest X-ray following acute overexposure (may be negative if taken immediately after exposure due to delayed development of pulmonary edema).

☑ Liver and kidney function tests.

Any evaluation should include a careful history of past and present symptoms with an examination. Medical tests that simply look for existing damage are not a substitute for controlling exposures. Also, since smoking can cause heart disease, lung cancer, emphysema, and other respiratory disorders, smokers exposed to octane may experience symptoms more rapidly than non-smokers under the same conditions of exposure. Prudent risk management requires careful consideration of *all* possible risk factors.

Other methods to reduce exposure include:

☑ Where possible, enclose operations and use local exhaust ventilation at the site of chemical release. If local exhaust ventilation or enclosure is not used, respiratory protection should be mandatory.

☑ Always ensure that proper protective clothing is worn when using chemical substances and that personnel have been trained on its use, care, and maintenance.

☑ Wash thoroughly immediately after exposure to octane and at the end of the work shift or before eating, drinking, or smoking.

☑ Hazard warning information should be posted in the work area. In addition, as part of an on-going education and training program, all information on the health and safety hazards of octane should be communicated to all potentially exposed workers.

☑ Eye wash stations should be provided in the immediate work area for emergency use. If there is a possibility of skin exposure to octane, emergency shower facilities should also be provided.

☑ Workers whose clothing has been contaminated by octane should change into clean clothes before leaving work. Contaminated work clothing should be laundered only by individuals who have been informed of the hazards of exposure to octane.

☑ Where possible, automatically transfer liquids containing octane from drums or other containers to process containers. Always ground and bond all containers and use non-sparking tools.

RISK ASSESSMENT: ENVIRONMENT

General Assessment

The environment is at risk of exposure during transportation, storage, disposal, or destruction of octane. In almost every scenario, the threat of environmental exposure is contingent upon the proper handling of the chemical substance. Accidental spills, large or small, can result in contamination of the surrounding environmental mediums (water, soil, and air).

Octane is a moderately flammable liquid. Because of its extremely low flash point and high boiling point, it is classified as a Class 1B flammable liquid (per OSHA 29 CFR 1910.106). It is a dangerous fire hazard and moderate explosion hazard when exposed to heat or flame. It is incompatible with commonly encountered oxidizers and some acids (such as nitric). Caution is always required in handling, storage, transportation, and disposal of octane. Emergency responders should be made aware of the presence of octane at any emergency response situation.

Octane can enter the environment from industrial effluents and from spills.

Acute Ecological Effects

Acute (short-term) toxic effects may include the death of animals, birds, or fish and death or low growth rate in plants. Acute effects are seen 2 to 4 days after animals or plants are exposed to octane.

Octane has high acute toxicity to aquatic life. It can cause death in plants and roots and membrane damage in leaves of various agricultural crops. Insufficient data are available to evaluate the short-term effects of octane on birds or land animals.

Chronic Ecological Effects

Chronic (long-term) toxic effects may include shortened life span, reproductive problems, lower fertility, and changes in appearance or behavior in exposed animals. These effects can be seen long after first exposure(s) to toxic chemicals.

Octane has high chronic toxicity to aquatic life. Insufficient data are available to evaluate or predict the long-term effects of octane to plants, birds, or terrestrial animals.

Water Solubility

Octane is not soluble in water. Concentrations of 1 milligram will not mix with a liter of water.

Persistence in the Environment

Octane is slightly persistent in water, with a half-life of between 2 to 20 days. The half-life of a pollutant is the amount of time it takes for one half of the chemical to be degraded. About 99.5% of octane will eventually end up in air; the rest will end up in water.

Bioaccumulation in Aquatic Organisms

Some substances increase in concentration, or *bioaccumulate,* in living organisms as they breathe contaminated air, drink contaminated water, or eat contaminated food. These chemicals can become concentrated in the tissues and internal organs of animals as well as humans.

The concentration of octane found in fish tissues is expected to be somewhat higher than the average concentration of octane in the water from which the fish was taken.

Recommended Risk-Reduction Measures

Proper training of all transporters will reduce the likelihood of a mishap or accident resulting in a leak or spill of octane into the environment. Labels on all containers, trucks, and rail cars must meet DOT requirements and accurately reflect their contents to enable emergency responders to react properly and quickly to any disaster thereby reducing the potential risk to the environment and to personnel.

Storage of octane should be segregated from other chemicals and materials to minimize the risk of cross-contamination. Facilities should be designed to meet explosion-proof requirements.

If a spill or leak to the environment has occurred, fire department, emergency response and/or hazardous materials spill personnel should be notified immediately. Cleanup should be attempted only by those trained in proper spill containment procedures using non-sparking tools. Contaminated soils should be removed for incineration and replaced with clean soil. If octane should contact the water table, aquifer, or navigable waterway, time is of the essence. It is not soluble and total remediation may be possible. Octane will normally float on the water's surface making reclamation easier if response actions are prompt. The local and/or state emergency response authorities must be notified. A comprehensive emergency response or disaster preparedness/recovery plan should be in place prior to any operations involving the use, transportation, storage, or disposal of octane.

If octane is spilled or leaked, the following specific steps are recommended:

☑ Restrict persons not wearing protective clothing from area of spill or leak until cleanup is complete and area can be opened for normal work.

☑ Absorb liquids in vermiculite, sand, earth, or similar material and deposit in sealed drum. Use non-sparking tools during cleanup procedures.

☑ Remove all ignitions sources. Ventilate area of spill or leak. Container should be removed to safer area if it can be done without increasing the risk.

☑ Use water spray to knock down vapors.

☑ It may be necessary to dispose of octane as a hazardous waste. The responsible state agency or the regional office of the federal Environmental Protection Agency (EPA) should be contacted for specific recommendations.

RISK ASSESSMENT: BUSINESS

General Assessment

Accidents or mishaps involving octane can present a serious threat to business operations. The loss or damage of equipment or facilities can significantly affect fiscal viability. Lawsuits that may result from personnel illness, injury/death, public exposures, and/or environmental contamination will require a serious expenditure of resources. Media attention surrounding an injury, death, or environmental damage can also result in a loss of profits and loss of current as well as future business.

Recommended Risk-Reduction Measures

Company attorneys, safety and health professionals, and environmental specialists should be involved in the development of any procedures for responding to chemical incidents. A company official should be predesignated as a public relations officer with specific training in dealing with the press. Corporate plans and policies should be developed, approved, and implemented long before any need for such arises.

REFERENCES

American Conference of Governmental Industrial Hygienists. 1988. *Documentation of the Threshold Limit Values and Biological Exposure Indices,* 5th Edition (with updates). Cincinnati: ACGIH

Bloom, A. D. and F. J. DeSerres. 1995. *Ecotoxicity and Human Health.* New York: CRC Press/Lewis Publishers.

Bueche, J. R. 1972. *Principles of Physics,* 2nd Edition. New York: McGraw-Hill.

Cockerman, L. G., B. S. Shane. 1994. *Basic Environmental Toxicology.* New York: CRC Press/Lewis Publishers.

Howard, P. H., et. al. 1991. *Handbook of Environmental Degradation Rates.* New York: CRC Press/Lewis Publishers.

Jones, F. E. 1994. *Toxic Organic Vapors in the Workplace.* New York: CRC Press/Lewis Publishers.

Lauwerys, R. R., P. Hoet. 1993. *Industrial Chemical Exposure.* New York: CRC Press/Lewis Publishers.

Lewis, R. J., Sr. 1993. *Hawley's Condensed Chemical Dictionary,* 12th Edition. New York: Van Nostrand Reinhold.

Lewis, R. J., Sr. 1992. *Sax's Dangerous Properties of Industrial Materials,* Volumes 1, 2, and 3, 8th Edition. New York: Van Nostrand Reinhold.

Meyer, E. 1990. *Chemistry of Hazardous Materials.* Englewood Cliffs, NJ: Prentice Hall.

Patnaik, P. 1992. *A Comprehensive Guide to the Hazardous Properties of Chemical Substances.* New York: Van Nostrand Reinhold.

Proctor, N. H., J. P. Hughes. 1991. *Chemical Hazards of the Workplace.* New York: Van Nostrand Reinhold.

Quigley, D. R. 1994. *Handbook of Emergency Chemical Management.* New York: CRC Press/Lewis Publishers.

Sacarello, H. L. A. 1994. *The Comprehensive Handbook of Hazardous Materials.* New York: CRC Press/Lewis Publishers.

Torkelson, T. R. V. K. Rowe. 1981. *Patty's Industrial Hygiene and Toxicology,* Vol. 2B. New York: Wiley-Interscience.

U.S. Department of Health and Human Services, National Institute for Occupational Safety and Health. 1994. *NIOSH Pocket Guide to Chemical Hazards.* Washington, D.C.: U.S. Government Printing Office.

MATERIAL SAFETY DATA SHEET

CHEMICAL NAME	
OLEIC ACID	**CRC PRESS/LEWIS PUBLISHERS DISCLAIMER:** THE INFORMATION AND RECOMMENDATIONS PRESENTED HEREIN ARE BASED ON SOURCES BELIEVED TO BE RELIABLE. CRC MAKES NO REPRESENTATION ON ITS COMPLETENESS OR ACCURACY. IT IS THE RESPONSIBILITY OF THE USER TO DETERMINE THE CHEMICAL'S SUITABILITY FOR ITS INTENDED USE, THE CHEMICAL'S SAFE USE, AND THE CHEMICAL'S PROPER DISPOSAL. NO REPRESENTATIONS AND/OR WARRANTIES, EITHER EXPRESSED OR IMPLIED, OF THE MERCHANTABILITY OR FITNESS FOR A PARTICULAR PURPOSE, OR OF ANY OTHER NATURE, ARE MADE WITH RESPECT TO THE INFORMATION PROVIDED IN THIS MSDS OR TO THE CHEMICAL TO WHICH INFORMATION MAY REFER. CRC NEITHER ASSUMES NOR AUTHORIZES ANY OTHER PERSON TO ASSUME FOR IT, ANY OTHER ADDITIONAL RESPONSIBILITY OR LIABILITY FOR THE USE OF, OR RELIANCE UPON, THIS INFORMATION.

HAZARD WARNING INFORMATION

HEALTH	FIRE	REACTIVE	OTHER	DEGREE OF HAZARD	COLOR CODING	OTHER CODES
0	1	0		0 = Minimum Hazard 1 = Slight Hazard 2 = Moderate Hazard 3 = Serious Hazard 4 = Severe Hazard	HEALTH = BLUE FIRE = RED REACTIVITY = YELLOW OTHER = WHITE	OX = Oxidizer ACID = Acid ALK = Alkali COR = Corrosive W = Use No Water

SECTION I - GENERAL INFORMATION

Characterization	RCRA Number	EPA Class
Mineral Acid	**None**	**Not Applicable**

DOT Proper Shipping Name	Chemical Abstract Service (CAS) Number
No Citation	**112-80-1**
DOT Hazard Class and Label Requirements	**DOT Emergency Guide Code**
No Citation	**No Citation**
DOT Identification Number	**Chemical Formula**
No Citation	**$CH_3(CH_2)_7CH{:}CH(CH_2)_7COOH$**

Synonyms

cis-9-Octadecenoic acid; red oil; century CD fatty acid; Emersol 210; pamolyn; industrene 206.

SECTION II - HAZARDOUS INGREDIENTS AND IDENTITY INFORMATION

Hazardous Components (specific identity)	OSHA Exposure Criteria	NIOSH Exposure Criteria	Immediately Dangerous to Life and Health (IDLH)	ACGIH Exposure Criteria
Oleic acid (derivation: The free fatty acid is obtained from glyceride by hydrolysis, steam distillation and separation by crystallization or solvent extraction. Filtration from the press cake results in oleic acid of commerce (red oil), which is purified and bleached for specific uses).	**PEL:** Not Established **STEL:** Not Established	**REL:** Not Established **STEL:** Not Established	**Not Established**	**TLV:** Not Established **STEL:** Not Established

SECTION III - PHYSICAL AND CHEMICAL CHARACTERISTICS

Boiling Point	Specific Gravity (H_2O = 1)
680°F (360°C) Decomposes	**0.895**
Vapor Pressure (mm Hg)	Molecular Weight
10 at 356°F (180°C)	**282.52**
Vapor Density (Air = 1)	Melting Point
9.7	**61°F (16°C)**

Solubility

Insoluble in water. Soluble in alcohol, ether, most organic solvents, and fixed and volatile oils.

Appearance and Odor

Pale yellow to red oily liquid with a characteristic, lard-like odor. Darkens on exposure to air.

SECTION IV - FIRE AND EXPLOSION HAZARD DATA

Flash Point (method used)	Explosive Limits in Air % by Volume
372°F (189°C) closed cup	**LEL: Not Determined UEL: Not Determined**
NFPA Classification	Autoignition Temperature
Class IIIB Combustible Liquid	**685°F (362°C)**

Extinguishing Media

Use dry chemical, carbon dioxide, or foam.

Special Fire Fighting Procedures

Poisonous gases are produced in fire. Wear full protective clothing and self-contained breathing apparatus (SCBA). Use water spray to flush and dilute a spill and to disperse the vapors. Also, use a water spray to keep fire-exposed containers cool. Do not release runoff from fire control measure to sewers or waterways. Firefighters should avoid contact with vapors produced during fire.

Unusual Fire and Explosion Hazards

None reported.

SECTION V - REACTIVITY DATA

Stability		Conditions to Avoid **Oleic acid is normally stable under routine conditions of handling and storage. Avoid contact with heat and ignition sources.**
Stable **X**	Unstable	Incompatibility *(materials to avoid)* **Violent reactions can occur in contact with oxidizing agents.**
Hazardous Polymerization		Conditions to Avoid **Under normal conditions of temperature and pressure, hazardous polymerization of oleic acid is not expected to occur.**
May Occur	Will Not Occur **X**	Hazardous Decomposition or By-products **When heated to decomposition, oleic acid can emit highly toxic oxides of carbon along with acrid smoke and irritating fumes.**

SECTION VI - HEALTH HAZARD DATA

Primary Route(s) of Entry:	Inhalation?	Absorption (skin)?	Ingestion?
	X		X

Health Hazards

INHALATION: Moderately irritating to the eyes, nose, throat, and respiratory system. Inhaled vapors may cause bronchial irritation and inflammation of associated mucosa.

EYES & SKIN: Skin and eye contact can result in irritation. Absorption is not reported. Prolonged eye contact may cause corneal damage.

INGESTION: Poisonous by ingestion with experimental tumorigenic effects. Irritation of the gastrointestinal tract with abdominal cramps and nausea likely.

Carcinogenicity	NTP Listed?	IARC Cancer Review Group?	OSHA Regulated?	Target Organs?
Unknown Human Questioned Animal	**No**	**No**	**No**	**Respiratory system, eyes, skin.**

Medical Conditions Generally Aggravated by Exposure

None reported.

Emergency and First-aid Procedures

Eye contact: Flush immediately with water for 15 minutes (minimum); seek medical attention. Skin contact: Remove all contaminated clothing. Immediately flush area with large amounts of water. For red or blistered skin, wash with soap and water. Seek medical assistance. For inhalation: Remove the person from exposure. Provide respiratory assistance and CPR. Transfer to medical facility. If swallowed: Contact poison control center. Never attempt to give anything by mouth to an unconscious or convulsing person. Seek medical attention.

SECTION VII - PRECAUTIONS FOR SAFE HANDLING AND USE

Steps to be Taken in Case Material is Released or Spilled

Absorb liquids using vermiculite and place in sealed drums for disposal. Use non-sparking tools in cleanup operations. Remove all ignition sources. Ventilate area of spill or leak. Restrict those not involved in cleanup from entering area.

Preferred Waste Disposal Method

No citation.

Precautions to be Taken in Handling and Storage

Avoid contact with oxidizing materials. Store in tightly closed containers in cool, well-ventilated area away from heat.

Other Precautions and Warnings

Avoid skin contact; use personal protective equipment.

SECTION VIII - CONTROL MEASURES AND PERSONAL PROTECTIVE EQUIPMENT

Respiratory Protection *(specify type)*

No exposure levels have been established for oleic acid. This does not mean exposure is without health risk. An air purifying respirator with acid cartridge and a HEPA filter may suffice. However, better protection is provided by a supplied-air respirator with full facepiece operated in pressure demand or other positive pressure mode.

Ventilation

Local exhaust or general mechanical systems recommended.

Protective Gloves	Eye Protection	Other Protective Clothing
Acid resistant synthetic rubber	**Chemical Goggles or Face Mask**	**Protective Apron**

Work/Hygiene Practices

Always wash hands thoroughly after using chemical; never bring food, drink, or smoking materials into vicinity of chemicals.

OLEIC ACID

$CH_3(CH_2)_7CH{:}CH(CH_2)_7COOH$ CAS: 112-80-1

IDENTIFICATION AND TYPICAL USES

Oleic acid is a pale-yellow to red oily liquid with a characteristic fatty-acid odor. It is used as a soap base, in the manufacture of oleates, in ointments, cosmetics, in polishing compounds, in lubricants, for ore flotation, as an intermediate, in certain surface coatings, and as a food-grade additive.

RISK ASSESSMENT: HEALTH

General Assessment

Exposure to oleic acid is possible through *inhalation* of its vapors and, although not common, by *ingestion.* The data in the references are limited regarding its specific toxicity. However, mutation data have been reported in test animals and its carcinogenic potential is considered questionable in some references.

Inhalation of oleic acid vapors may cause irritation of the eyes, nose, throat, and mucous membranes of the upper respiratory tract. Bronchial tube irritation may also occur with progression to bronchitis if exposures continue.

Skin and eye contact can result in mild to moderate irritation. If allowed to remain in contact with the eye for prolonged periods, corneal abrasions may occur. Ingestion may cause systemic poisoning with abdominal cramps and possible nausea and vomiting.

Acute Health Effects

The following acute (short-term) health effects may occur immediately or shortly after exposure to oleic acid:

Skin: Irritation/rash or burning feeling on contact.

Eye: Mild irritation; may cause burns and damage.

Lung: Irritation of the nose, throat, and lungs following exposure.

Chronic Health Effects

The following chronic (long-term) health effects may occur at some time after exposure to oleic acid and can last for months or even years:

Cancer Hazards: According to information presented in the references, oleic acid has not been adequately tested for its ability to cause cancer in animals. Mutation data have been reported and some scientists believe that long-term exposure to mutagens may pose a cancer risk. There is no indication that exposure to oleic acid will have this effect. Additional research is required.

Reproductive Hazard: There are no data in the references to indicate that exposure to oleic acid will have adverse effects on the reproductive process in animals or humans.

Recommended Risk-Reduction Measures

Personnel should avoid direct contact with oleic acid. The exact nature of its toxicity is not clearly understood in the references. If a less toxic chemical cannot be substituted for a hazardous substance, then *engineering controls* are the most effective method of reducing exposures.

The best protection is to enclose operations and/or provide local exhaust ventilation at the site of chemical release. While not always operationally feasible, isolating operations can also reduce exposure. Using respiratory protection is less effective than the controls mentioned above, but is still advisable whenever working with or around oleic acid. No exposure levels have been established for oleic acid. This does not mean that exposure is without health risk. Prudent risk management may require a conservative approach to respiratory protection until further data are known about oleic acid. For transient exposures, an air-purifying respirator equipped with an acid cartridge, a HEPA filter, and a dust/mist pre-filter may suffice. Greater protection is obtained using a supplied-air respirator or a self-contained breathing apparatus (SCBA) with full facepiece operated in pressure demand or other positive pressure mode. If a full facepiece is not available, then chemical goggles should be worn to protect the eyes. Whenever a chemical splash/dust hazard exists, a face shield and a rubber apron should be worn. To prevent hand and skin exposures, acid resistant gloves should be used. Glove manufacturers should be contacted and permeation studies obtained *before* final glove selections have been made.

Administrative controls should also be in place to minimize the potential for human exposures. These may include written procedures or policies which specify the methods and techniques that will be practiced whenever personnel are to work with oleic acid.

All personnel should receive *training* on the use, hazards, protective measures, emergency actions, and other precautions per 29 CFR 1910.1200 (Hazard Communication), prior to the first assignment in an area where oleic acid is used or stored. Other methods to reduce exposure include:

☑ Where possible, enclose operations and use local exhaust ventilation at the site of chemical release. If local exhaust ventilation or enclosure is not used, respiratory protection should be mandatory.

☑ Always ensure that proper protective clothing is worn when using chemical substances.

☑ Wash thoroughly immediately after exposure to oleic acid and at the end of the work shift or before eating, drinking, or smoking.

☑ Hazard warning information should be posted in the work area. In addition, as part of an on-going education and training program, all information on the health and safety hazards of oleic acid should be communicated to all exposed or potentially exposed workers.

RISK ASSESSMENT: ENVIRONMENT

General Assessment

The environment is at risk of exposure during transportation, storage, disposal, or destruction of oleic acid. In almost every scenario, the threat of environmental exposure is contingent upon the proper handling of the chemical substance. Accidental spills, large or small, can result in fire, explosion, and possible contamination of the surrounding environmental mediums (water, soil, and, to a lesser degree, air).

Oleic acid is considered a non-combustible solid. It may be corrosive, especially when mixed in solution. It is incompatible with potassium chlorite and contact can cause violent reactions, including fire or explosion. These characteristics require special consideration during any emergency situation involving a leak or spill of oleic acid.

Oleic acid can enter the environment mainly through industrial discharges or spills.

Acute Ecological Effects

Acute (short-term) toxic effects may include the death of animals, birds, or fish and death or low growth rate in plants. Acute effects are seen 2 to 4 days after animals or plants are exposed to oleic acid.

Insufficient data are available to evaluate or predict the short-term effects of oleic acid exposure to aquatic life, birds, plants, or land animals.

Chronic Ecological Effects

Chronic (long-term) toxic effects may include shortened life span, reproductive problems, lower fertility, and changes in appearance or behavior in exposed animals. These effects can be seen long after first exposure(s) to toxic chemicals. Insufficient data are available to evaluate or predict the long-term effects of oleic acid exposure to aquatic life, birds, plants, or land animals.

Water Solubility

Oleic acid is not soluble in water. Concentrations of less than 1 milligram may not mix with a liter of water.

Persistence in the Environment

Oleic acid is slightly persistent in the aquatic environment, with a half-life of between 2 to 20 days. The half-life of a pollutant is the amount of time it takes for one half of the chemical to be degraded.

Bioaccumulation in Aquatic Organisms

Some substances increase in concentration, or *bioaccumulate,* in living organisms as they breathe contaminated air, drink contaminated water, or eat contaminated food. These chemicals can become concentrated in the tissues and internal organs of animals as well as humans.

There is no information in the references on the expected bioaccumulation of oleic acid in the edible tissues of fish.

Recommended Risk-Reduction Measures

Proper training of all transporters will reduce the likelihood of a mishap or accident resulting in a leak or spill to the environment. The correct use of labeling on all containers, trucks, and rail cars will enable emergency responders to react properly and quickly to any disaster thereby reducing the potential risk to the environment and to personnel.

Storage of oleic acid should be segregated from incompatible chemicals to minimize the risk of cross-contamination or contact. Buildings designated for storage should be equipped with appropriate fire protection systems (alarms, sprinklers, emergency lighting, portable extinguishers).

If a spill or leak to the environment has occurred, fire department, emergency response and/or hazardous materials spill personnel should be notified immediately. Cleanup should be attempted only by those trained in proper spill containment procedures using non-sparking tools. Contaminated soils should be removed for incineration and replaced with clean soil. If oleic acid should contact the water table, aquifer, or navigable waterway, time is of the essence. It is not soluble in water and, therefore, total containment and remediation may be possible. When such spills occur, the local and/or state emergency response authorities must be notified. A comprehensive emergency response or disaster preparedness/recovery plan should be in place prior to any operations involving the use, transportation, storage, or disposal of oleic acid. If oleic acid is spilled or leaked, the following specific steps are recommended:

- ☑ Restrict persons not wearing protective clothing from area of spill or leak until cleanup is complete and area can be opened for normal work.
- ☑ Remove all ignition sources and ventilate area of spill.
- ☑ Absorb liquid spills in vermiculite, dry sand, earth, or similar material and deposit in sealed containers. Use non-sparking tools, such as plastic shovels, rubber implements, or tools made of soft metals (e.g., brass or copper) during cleanup operations.
- ☑ It may be necessary to dispose of oleic acid as a hazardous waste. The responsible state agency or the regional office of the federal Environmental Protection Agency (EPA) should be contacted for specific recommendations.

RISK ASSESSMENT: BUSINESS

General Assessment

Accidents or mishaps involving oleic acid can present a significant threat to business operations. The loss or damage of equipment or facilities can significantly affect fiscal viability. Lawsuits that may result from personnel injury/death, public exposures, and/or environmental contamination will also require a serious expenditure of resources. Media attention surrounding an injury, death, or environmental damage can also result in a loss of profits and loss of current as well as future business. Always remember that anytime the terms "cancer" or "carcinogen" are used, public emotion, ignorance, anxiety, and hysteria can run equally high. This must be carefully considered whenever drafting or implementing public relations policies.

Recommended Risk-Reduction Measures

Company attorneys, safety and health professionals, and environmental specialists should be involved in the development of any such procedures. A company official should be pre-designated as a public relations officer with specific training in dealing with the press. Corporate plans and policies should be developed, approved, and implemented long before any need for such arises.

REFERENCES

Ahlbom, A. 1993. *Biostatistics for Epidemiologists.* New York: CRC Press/Lewis Publishers.

American Conference of Governmental Industrial Hygienists. 1988. *Documentation of the Threshold Limit Values and Biological Exposure Indices,* 5th Edition (with updates). Cincinnati: ACGIH

Asante-Duah, D. K. 1993. *Hazardous Waste Risk Assessment.* New York: CRC Press/Lewis Publishers.

Bloom, A. D. and F. J. DeSerres. 1995. *Ecotoxicity and Human Health.* New York: CRC Press/Lewis Publishers.

Brusick, D. J. 1994. *Methods for Genetic Risk Assessment.* New York: CRC Press/Lewis Publishers.

Bueche, J. R. 1972. *Principles of Physics,* 2nd Edition. New York: McGraw-Hill.

Burrell, R., D. K. Flaherty, and Leonard J. Sauers. 1992. *Toxicology of the Immune System.* New York: Van Nostrand Reinhold.

Calabrese, E. J., E. M. Kenyon. *Air Toxics and Risk Assessment.* New York: CRC Press/Lewis Publishers.

Calabrese, E. J. 1994. *Biological Effects of Low Level Exposures.* New York: CRC Press/Lewis Publishers.

Cockerman, L. G., B. S. Shane. 1994. *Basic Environmental Toxicology.* New York: CRC Press/Lewis Publishers.

Cothern, C. R., N. P. Ross. 1994. *Environmental Statistics, Assessment, and Forecasting.* New York: CRC Press/Lewis Publishers.

Cothern, R. R. 1993. *Comparative Environmental Risk Assessment.* New York: CRC Press/Lewis Publishers.

Cothern, C. R. 1995. *Handbook for Environmental Risk Decisionmaking.* New York: CRC Press/Lewis Publishers.

Danse, I. R. 1991. *Common Sense Toxics in the Workplace.* New York: Van Nostrand Reinhold.

Gardner, R. W. 1994. *Chemical Intolerance.* New York: CRC Press/Lewis Publishers.

Hallenbeck, W. H. 1993. *Quantitative Risk Assessment for Environmental and Occupational Health,* 2nd Edition. New York: CRC Press/Lewis Publishers.

Hosty, J. W., P. Foster. 1990. *A Practical Guide to Chemical Spill Response.* New York: Van Nostrand Reinhold.

Howard, P. H. 1995. *Handbook of Environmental Fate and Exposures Data for Organic Chemicals,* Vol. I-IV. New York: CRC Press/Lewis Publishers.

Howard, P. H., et. al. 1991. *Handbook of Environmental Degradation Rates.* New York: CRC Press/Lewis Publishers.

James, A., M. Lord. 1992. *Index of Chemical and Physical Data.* New York: Van Nostrand Reinhold.

Jones, F. E. 1994. *Toxic Organic Vapors in the Workplace.* New York: CRC Press/Lewis Publishers.

Keith, L. H. and M. M. Walker. 1995. *Handbook of Air Toxics.* New York: CRC Press/Lewis Publishers.

Landis, W. G., M. H. Yu. 1995. *Introduction to Environmental Toxicology.* New York: CRC Press/Lewis Publishers.

Langerman, N. 1994 *Precautionary Labels for Chemical Containers.* New York: CRC Press/Lewis Publishers.

Lauwerys, R. R., P. Hoet. 1993. *Industrial Chemical Exposure.* New York: CRC Press/Lewis Publishers.

Lewis, R. J., Sr. 1993. *Hawley's Condensed Chemical Dictionary,* 12th Edition. New York: Van Nostrand Reinhold.

Lewis, R. J., Sr. 1992. *Sax's Dangerous Properties of Industrial Materials,* Volumes 1, 2, and 3, 8th Edition. New York: Van Nostrand Reinhold.

Lide, D. R. 1995. *Handbook of Chemistry and Physics,* 76th Edition. New York: CRC Press/Lewis Publishers.

Manahan, S. E. 1994. *Environmental Chemistry,* 6th Edition. New York: CRC Press/Lewis Publishers.

Manahan, S. E. 1992. *Toxicological Chemistry,* 2nd Edition. New York: CRC Press/Lewis Publishers.

Mellström, G. A., J. E. Wahlberg, and H. I. Maibach. 1994. *Protective Gloves for Occupational Use.* New York: CRC Press/Lewis Publishers.

Meyer, E. 1990. *Chemistry of Hazardous Materials.* Englewood Cliffs, NJ: Prentice Hall.

Mycock, J. C., J. D. McKenna, and L. Theodore. 1995. *Handbook of Air Pollution Control Engineering and Technology.* New York: CRC Press/Lewis Publishers.

Nelson, G. O. 1992. *Gas Mixtures.* New York: CRC Press/Lewis Publishers.

Ott, W. R. 1995. *Environmental Statistics and Data Analysis.* New York: CRC Press/Lewis Publishers.

Patnaik, P. 1992. *A Comprehensive Guide to the Hazardous Properties of Chemical Substances.* New York: Van Nostrand Reinhold.

Philp, R. B. 1995. *Environmental Hazards and Human Health.* New York: CRC Press/Lewis Publishers.

Proctor, N. H., J. P. Hughes. 1991. *Chemical Hazards of the Workplace.* New York: Van Nostrand Reinhold.

Quigley, D. R. 1994. *Handbook of Emergency Chemical Management.* New York: CRC Press/Lewis Publishers.

Rea, W. J. 1995. *Chemical Sensitivity,* Vol. I-III. New York: CRC Press/Lewis Publishers.

Richardson, M. 1995. *Dictionary of Substance and Their Effects,* Vol. I-VI. New York: CRC Press/Lewis Publishers.

Rylander, R., R. R. Jacobs. 1994. *Organic Dusts.* New York: CRC Press/Lewis Publishers.

Sacarello, H. L. A. 1994. *The Comprehensive Handbook of Hazardous Materials.* New York: CRC Press/Lewis Publishers.

Sheftel, V. O. 1995. *Handbook of Toxic Properties of Monomers and Additives.* New York: CRC Press/Lewis Publishers.

Sherman, J. D. 1988. *Chemical Exposure and Disease.* New York: Van Nostrand Reinhold.

Talbott, E., G. Craun. 1995. *An Introduction to Environmental Epidemiology.* New York: CRC Press/Lewis Publishers.

U.S. Department of Health and Human Services, National Institute for Occupational Safety and Health. 1994. *NIOSH Pocket Guide to Chemical Hazards.* Washington, D.C.: U.S. Government Printing Office.

Witorsch, P. and S. V. Spagnolo. 1994. *Air Pollution and Lung Disease in Adults.* New York: CRC Press/Lewis Publishers.

MATERIAL SAFETY DATA SHEET

CHEMICAL NAME	CRC PRESS/LEWIS PUBLISHERS DISCLAIMER
OLEUM	

HAZARD WARNING INFORMATION

HEALTH	FIRE	REACTIVE	OTHER	DEGREE OF HAZARD	COLOR CODING	OTHER CODES
4	0	2		0 = Minimum Hazard 1 = Slight Hazard 2 = Moderate Hazard 3 = Serious Hazard 4 = Severe Hazard	HEALTH = BLUE FIRE = RED REACTIVITY = YELLOW OTHER = WHITE	OX = Oxidizer ACID = Acid ALK = Alkali COR = Corrosive W = Use No Water

SECTION I - GENERAL INFORMATION

Characterization	RCRA Number	EPA Class
Acid, Mineral	**None**	**Not Listed**

DOT Proper Shipping Name	Chemical Abstract Service (CAS) Number
Oleum; Sulfuric Acid, ≥30% free sulfur trioxide	**8014-95-7**
DOT Hazard Class and Label Requirements	**DOT Emergency Guide Code**
Corrosive Material; Corrosive and Poison	**44**
DOT Identification Number	**Chemical Formula**
UN 1831	$H_2SO_4{\bullet}SO_3$

Synonyms

Disulphuric acid; dithionic acid; fuming sulfuric acid; pyrosulfuric acid; elixir of vitriol; oil of vitriol.

SECTION II - HAZARDOUS INGREDIENTS AND IDENTITY INFORMATION

Hazardous Components (specific identity)	OSHA Exposure Criteria	NIOSH Exposure Criteria	Immediately Dangerous to Life and Health (IDLH)	ACGIH Exposure Criteria
Oleum (derivation: Sulfur trioxide is first produced by the contact process and is then absorbed in concentrated sulfuric acid).	**PEL:** Not Established **STEL:** Not Established	**REL:** Not Established **STEL:** Not Established	**Not Established**	**TLV:** Not Established **STEL:** Not Established

SECTION III - PHYSICAL AND CHEMICAL CHARACTERISTICS

Boiling Point	Specific Gravity (H_2O = 1)
Not Found	**Not Found**
Vapor Pressure (mm Hg)	Molecular Weight
Not Found	**178.14**
Vapor Density (Air = 1)	Melting Point
Not Found	**Not Found**

Solubility

Miscible in water with violent reaction.

Appearance and Odor

Viscous, fuming, opaque or slightly yellow-brown liquid (depending on purity) with a sharp, penetrating, characteristic odor. Odor Threshold = 3 mg/m^3 in most individuals.

SECTION IV - FIRE AND EXPLOSION HAZARD DATA

Flash Point (method used)	Explosive Limits in Air % by Volume
Non-Combustible	**LEL: Not Determined UEL: Not Determined**
NFPA Classification	Autoignition Temperature
Non-Combustible Liquid	**Not Applicable**

Extinguishing Media

Reacts with water producing heat and toxic fumes. Use dry chemical or carbon dioxide.

Special Fire Fighting Procedures

Poisonous gases are released in fire. Wear full protective clothing and (SCBA). Use water spray to flush and dilute a spill, to disperse vapors, and to keep fire-exposed containers cool. Do NOT direct solid stream of water at oleum. Do not release runoff from fire control measure to sewers or streams.

Unusual Fire and Explosion Hazards

A non-combustible liquid that increases the burning of combustible materials. Containers may explode in fire. Releases toxic and corrosive sulfur oxides when involved in fire.

SECTION V - REACTIVITY DATA

Stability		Conditions to Avoid Oleum is normally stable in closed containers under routine conditions of handling and storage. It is extremely hygroscopic and will fume vigorously in moist air. Keep dry and away from sources of heat.
Stable X	Unstable	Incompatibility *(materials to avoid)* Alkalines, reducing agents, carbides, chlorates, fulminates, nitrates, picrates, metals, acetic acid, acetic anhydride, acetonitrile, acrolein, acrylic acid, alcohols, ethers, ketones, oxidizing agents, and many others.
Hazardous Polymerization		Conditions to Avoid Under normal conditions of temperature and pressure, hazardous polymerization of oleum is not expected to occur.
May Occur	Will Not Occur X	Hazardous Decomposition or By-products When heated to decomposition, oleum can emit highly toxic/poisonous gases and fumes, including toxic oxides of sulfur.

SECTION VI - HEALTH HAZARD DATA

Primary Route(s) of Entry:	Inhalation?	Absorption (skin)?	Ingestion?
	X		X

Health Hazards

INHALATION: Vapors cause irritation to the upper respiratory tract with coughing, dypsnea, headache, bronchitis, weakness, tearing, nasal discharge, nausea, chemical bronchitis, pulmonary edema (fluid in the lungs), and possible CNS effects.

SKIN & EYES: Eye contact causes blindness and tissue damage. Skin contact causes burns, blistering, local necroses, and membrane ulceration. Burns may be 2nd or 3rd degree.

INGESTION: Corrosion of the mucosa of the mouth, throat, and esophagus with stomach discomfort and pain, nausea, intense thirst, vomiting, circulatory collapse, clammy skin, weak and rapid pulse, shallow respiration, scanty urine, circulatory shock and subsequent death.

Carcinogenicity	NTP Listed?	IARC Cancer Review Group?	OSHA Regulated?	Target Organs?
Unknown Human Unknown Animal	No	No	No	Respiratory tract; eyes; skin; mucosa.

Medical Conditions Generally Aggravated by Exposure

Respiratory impairments (bronchitis, asthma) may be aggravated by exposure.

Emergency and First-aid Procedures

Eye contact: Flush immediately with water for 15 minutes (minimum); seek medical attention. Skin contact: Remove all contaminated clothing. Immediately wash area with large amounts of water. Seek medical assistance. For inhalation: Remove the person from exposure. Provide respiratory assistance and CPR. Transfer to medical facility. Observe for 48 hours for lung effects. If swallowed: Seek medical attention, contact poison control center. Give 1-2 glasses of water to drink. Do NOT induce vomiting (aspiration hazard). Never give anything by mouth to an unconscious person.

SECTION VII - PRECAUTIONS FOR SAFE HANDLING AND USE

Steps to be Taken in Case Material is Released or Spilled

Absorb liquids in vermiculite or other material, neutralize with soda ash, collect and deposit in sealed containers. Ventilate area of spill or leak. Restrict those not involved in cleanup from entering area.

Preferred Waste Disposal Method

No citation.

Precautions to be Taken in Handling and Storage

Oleum is a powerful oxidizer that can greatly accelerate the burning of oxidizable materials. Store in tightly closed containers in cool, dry, well-ventilated area away from heat.

Other Precautions and Warnings

Sources of ignition are prohibited where oleum is used, handled, or stored.

SECTION VIII - CONTROL MEASURES AND PERSONAL PROTECTIVE EQUIPMENT

Respiratory Protection *(specify type)*

Use half-mask respirator with acid cartridge for low or transient exposures (cartridge should be designed to protect against acid vapor or mist). For higher exposures, use a supplied-air respirator or a self-contained breathing apparatus operated in pressure demand or other positive pressure mode.

Ventilation

Local exhaust or general mechanical systems recommended.

Protective Gloves	Eye Protection	Other Protective Clothing
Acid resistant synthetic rubber	Chemical Goggles or Face Mask	Protective Apron

Work/Hygiene Practices

Always wash hands thoroughly after using chemical; never bring food, drink, or smoking materials into vicinity of chemicals.

OLEUM

$H_2SO_4 \bullet SO_3$

CAS: 8014-95-7

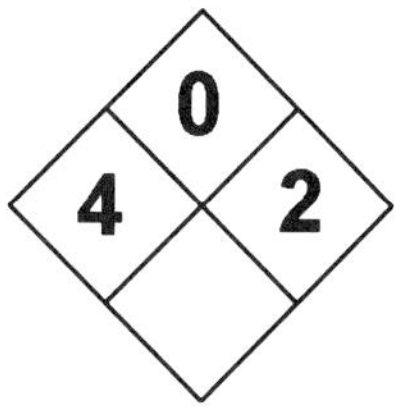

IDENTIFICATION AND TYPICAL USES

Oleum is a heavy, oily (viscous), opaque or yellow-brown fuming liquid with a sharp, characteristic, penetrating odor. It is used as a chemical intermediate for linear alkylbenzene sulfonate surfactants and dyes; as a chemical agent in the manufacture of nitrocellulose and caprolactum and in petroleum refining. It is also used in the production of fertilizers, explosives, other acids, parchment paper, and glue. It is a drying agent for chlorine and nitric acid. Is used as a laboratory reagent, and in the pickling of metal.

RISK ASSESSMENT: HEALTH

General Assessment

Oleum is highly corrosive to the skin, eyes, digestive tract, and respiratory system (or any other tissue it contacts). It is a human poison by *ingestion* and is dangerously toxic by *inhalation*. There are no data in the references regarding any carcinogenic, mutagenic, or teratogenic effects on exposure to oleum. Its corrosive and irritating properties limit the possibility of long-term exposure since people cannot tolerate unprotected contact for even short periods of time.

Inhalation produces severe irritation of the eyes, nose, and respiratory tract with coughing, stomach pains, intense thirst, dypsnea (labored breathing), bronchitis, headache, clammy skin, weak and rapid pulse, scanty urine, and total body weakness. The vapors can cause laryngitis, pharyngitis, chemical bronchitis, pneumonitis, and delayed pulmonary edema which can be fatal. There may be tearing, nasal discharge, throat irritation, ulceration of the nasal septa and nose irritation, and nausea. Symptoms of pulmonary edema, such as chest pain, congestion, coughing, and weakness, may be delayed up to 48 hours thereby creating a false sense of security with regard to health exposure risk. There may also be depression of the central nervous system with symptoms of giddiness, salivation, nausea, muscular weakness, and other unspecified symptoms. There may be total circulatory system collapse, shock, and death.

Skin contact causes severe burning with possible vesiculation. Its extremely corrosive properties cause immediate burns, membrane ulceration, skin blistering, 2nd and 3rd degree burns, and local necrosis (tissue destruction) upon contact. Eye contact causes damage and the possibility of permanent blindness.

Ingestion is extremely dangerous, even life-threatening. It causes severe poisoning, esophageal burns, throat swelling, convulsions, risk of stomach perforation, bloody vomiting, burning sensation on the mouth, acute tissue damage, diarrhea with pain, and possible coma.

Acute Health Effects

The following acute (short-term) health effects may occur immediately or shortly after exposure to oleum:

<u>Skin</u>: Irritation/rash or burning feeling on contact with redness, blistering and vesiculation, localized tissue destruction, and deep 2nd or 3rd degree burns.

<u>Eye</u>: Severe irritation; may cause burns and damage with permanent blindness.

<u>Lung</u>: Severe irritation of the mucous membranes in the nose, throat, and lungs with delayed pulmonary edema (fluid in the lungs), which can be fatal.

<u>CNS</u>: Possible depression of the nervous system with symptoms of giddiness, nausea, vomiting, and irritability.

Chronic Health Effects

The following chronic (long-term) health effects may occur at some time after exposure to oleum and can last for months or even years:

Cancer Hazards: According to information presented in the references, oleum has not been adequately tested for its ability to cause cancer in test animals.

Reproductive Hazard: According to information presented in the references, oleum has not been adequately tested for its ability to adversely affect reproduction in test animals.

Other Chronic Effects: Repeated skin contact with dilute solutions causes dermatitis. Chronic exposure may also cause changes in pulmonary function leading to chronic bronchitis, gastritis, stomatitis, erosion of dental enamel, conjunctivitis, increased propensity for respiratory tract infections, and possible respiratory tract cancer.

Recommended Risk-Reduction Measures

Personnel should avoid direct contact with oleum. It is extremely corrosive and highly toxic to humans. If a less toxic chemical cannot be substituted for a hazardous substance, then *engineering controls* are the most effective method of reducing exposures.

The best protection is to enclose operations and/or provide local exhaust ventilation at the site of chemical release. While not always operationally feasible, isolating operations can also reduce exposure. Using respiratory protection is less effective than the controls mentioned above, but is still advisable whenever working with or around oleum. No exposure levels have been established for oleum. This does not mean that exposure is without health risk. For low concentrations, an acid gas cartridge respirator (equipped with a mist/vapor filter) may provide acceptable protection. However, a self-contained breathing apparatus (SCBA) or a supplied-air respirator with full facepiece operated in pressure demand mode are the recommended respiratory protection methods of choice. If a full facepiece is not available, then chemical goggles should be worn to protect the eyes. Whenever a chemical splash hazard exists, a face shield and a rubber apron should be worn. To prevent hand and skin exposures, acid resistant gloves should be used. Glove manufacturers should be contacted and permeation studies obtained *before* final glove selections are made.

Administrative controls should also be in place to minimize the potential for human exposures. These may include written procedures or policies which specify the methods and techniques that will be practiced whenever personnel are to work with oleum.

All personnel should receive *training* on the use, hazards, protective measures, emergency actions, and other precautions per 29 CFR 1910.1200 (Hazard Communication), prior to the first assignment in an area where oleum is used or stored.

Before beginning employment and at regular intervals thereafter (e.g., annually), the following tests are recommended:

☑ Lung function tests (look for chronic respiratory diseases).

If symptoms occur or overexposure is suspected, the following additional tests are suggested:

☑ Evaluation by a qualified allergist with careful consideration of exposure history and special testing (may help diagnose skin allergy).

☑ Consider chest X-ray following acute overexposure (may be negative if taken immediately after exposure due to delayed onset of pulmonary edema).

Other methods to reduce exposure include:

☑ Where possible, enclose operations and use local exhaust ventilation at the site of chemical release. If local exhaust ventilation or enclosure is not used, respiratory protection should be mandatory.

☑ Always ensure that proper protective clothing is worn when using chemical substances and that personnel are trained on its use and care.

☑ Wash thoroughly immediately after exposure to oleum and at the end of the work shift or before eating, drinking, or smoking.

☑ Hazard warning information should be posted in the work area. In addition, as part of an on-going education and training program, all information on the health and safety hazards of oleum should be communicated to all exposed workers.

RISK ASSESSMENT: ENVIRONMENT

General Assessment

The environment is at risk of exposure during transportation, storage, disposal, or destruction of oleum. In almost every scenario, the threat of environmental exposure is contingent upon the proper handling of the chemical substance. Accidental spills, large or small, can result in fire, explosion, and possible contamination of the surrounding environmental mediums (water, soil, and, to a lesser degree, air).

Oleum is considered a non-combustible liquid. However, as an oxidizer, it can greatly accelerate the burning of ordinary combustible materials such as wood or paper. It is also extremely corrosive to many materials. It is incompatible with many materials, including hydrogen sulfide, carbides, cyanides, and alkalis, and contact can cause fire or explosion. These characteristics require special consideration during any emergency situation involving a leak or spill of oleum. Selection of response equipment (shovels, containers, and the like) must therefore be carefully considered. Should oleum ever come into contact with incompatible substances either during use, transportation, or storage, the formation of highly toxic and/or explosive commodities is extremely possible.

Oleum can enter the environment mainly through industrial discharges or spills.

Acute Ecological Effects

Acute (short-term) toxic effects may include the death of animals, birds, or fish and death or low growth rate in plants. Acute effects are seen 2 to 4 days after animals or plants are exposed to oleum.

Insufficient data are available on the short-term effects of oleum exposure to aquatic life, plants, birds, or land animals.

Chronic Ecological Effects

Chronic (long-term) toxic effects may include shortened life span, reproductive problems, lower fertility, and changes in appearance or behavior in exposed animals. These effects can be seen long after first exposure(s) to toxic chemicals.

Insufficient data are available on the long-term effects of oleum to aquatic life, plants, birds, or land animals.

Water Solubility

Oleum is highly soluble in water and will react violently. Concentrations of over 1000 milligrams will mix with a liter of water.

Persistence in the Environment

Oleum is slightly persistent in the aquatic environment, with a half-life of between 2 to 20 days. The half-life of a pollutant is the amount of time it takes for one half of the chemical to be degraded.

Bioaccumulation in Aquatic Organisms

Some substances increase in concentration, or *bioaccumulate,* in living organisms as they breathe contaminated air, drink contaminated water, or eat contaminated food. These chemicals can become concentrated in the tissues and internal organs of animals as well as humans.

The concentration of oleum found in fish tissues is expected to be somewhat lower than the average concentration of oleum in the water from which the fish was taken.

Recommended Risk-Reduction Measures

Proper training of all transporters will reduce the likelihood of a mishap or accident resulting in a leak or spill to the environment. The correct use of DOT labeling (per the MSDS) on all containers, trucks, and rail cars will enable emergency responders to react properly and quickly to any disaster thereby reducing the potential risk to the environment and to personnel.

Storage of oleum should be segregated from incompatible chemicals to minimize the risk of cross-contamination or contact. Buildings designated for storage should be equipped with appropriate fire protection systems (alarms, sprinklers, emergency lighting, portable extinguishers). It must be kept cool and in a well-ventilated location away from direct sunlight.

If a spill or leak to the environment has occurred, fire department, emergency response and/or hazardous materials spill personnel should be notified immediately. Cleanup should be attempted only by those trained in proper spill containment procedures. Contaminated soils should be removed for incineration and replaced with clean soil. If oleum should contact the water table, aquifer, or navigable waterway, time is of the essence. It is highly miscible in water and, therefore, total containment and remediation may not be entirely possible. When such spills occur, the local and/or state emergency response authorities must be notified. A comprehensive emergency response or disaster preparedness/recovery plan should be in place prior to any operations involving the use, transportation, storage, or disposal of oleum. If oleum is spilled or leaked, the following specific steps are recommended:

- ☑ Restrict persons not wearing protective clothing from area of spill or leak until cleanup is complete and area can be opened for normal work.
- ☑ Remove all ignition sources and ventilate area.
- ☑ Absorb liquids in vermiculite, dry sand, earth, or similar material and deposit in sealed containers. Neutralize remaining residue with soda ash or so-

dium bicarbonate. Do NOT flush to drains, sewers, or waterways. Use water spray to cool and knock down vapors. Collect water for disposal.

☑ It may be necessary to dispose of oleum as a hazardous waste. The responsible state agency or the regional office of the federal Environmental Protection Agency (EPA) should be contacted for specific recommendations.

RISK ASSESSMENT: BUSINESS
General Assessment

Accidents or mishaps involving oleum can present a significant threat to business operations. The loss or damage of equipment or facilities can significantly affect fiscal viability. Lawsuits that may result from personnel injury/death, public exposures, and/or environmental contamination will also require a serious expenditure of resources. Media attention surrounding an injury, death, or environmental damage can also result in a loss of profits and loss of current as well as future business.

Recommended Risk-Reduction Measures

Company attorneys, safety and health professionals, and environmental specialists should be involved in the development of any procedures dealing with safety or emergency response. A company official should be pre-designated as a public relations officer with specific training in dealing with the press. Corporate plans and policies should be developed, approved, and implemented long before any need for such arises.

REFERENCES

Ahlbom, A. 1993. *Biostatistics for Epidemiologists.* New York: CRC Press/Lewis Publishers.

American Conference of Governmental Industrial Hygienists. 1988. *Documentation of the Threshold Limit Values and Biological Exposure Indices,* 5th Edition (with updates). Cincinnati: ACGIH

Asante-Duah, D. K. 1993. *Hazardous Waste Risk Assessment.* New York: CRC Press/Lewis Publishers.

Bloom, A. D. and F. J. DeSerres. 1995. *Ecotoxicity and Human Health.* New York: CRC Press/Lewis Publishers.

Bueche, J. R. 1972. *Principles of Physics,* 2nd Edition. New York: McGraw-Hill.

Burrell, R., D. K. Flaherty, and Leonard J. Sauers. 1992. *Toxicology of the Immune System.* New York: Van Nostrand Reinhold.

Calabrese, E. J., E. M. Kenyon. *Air Toxics and Risk Assessment.* New York: CRC Press/Lewis Publishers.

Cockerman, L. G., B. S. Shane. 1994. *Basic Environmental Toxicology.* New York: CRC Press/Lewis Publishers.

Cothern, C. R. 1995. *Handbook for Environmental Risk Decisionmaking.* New York: CRC Press/Lewis Publishers.

Danse, I. R. 1991. *Common Sense Toxics in the Workplace.* New York: Van Nostrand Reinhold.

Gardner, R. W. 1994. *Chemical Intolerance.* New York: CRC Press/Lewis Publishers.

Hallenbeck, W. H. 1993. *Quantitative Risk Assessment for Environmental and Occupational Health,* 2nd Edition. New York: CRC Press/Lewis Publishers.

Hosty, J. W., P. Foster. 1990. *A Practical Guide to Chemical Spill Response.* New York: Van Nostrand Reinhold.

Howard, P. H., et. al. 1991. *Handbook of Environmental Degradation Rates.* New York: CRC Press/Lewis Publishers.

Jones, F. E. 1994. *Toxic Organic Vapors in the Workplace.* New York: CRC Press/Lewis Publishers.

Keith, L. H. and M. M. Walker. 1995. *Handbook of Air Toxics.* New York: CRC Press/Lewis Publishers.

Landis, W. G., M. H. Yu. 1995. *Introduction to Environmental Toxicology.* New York: CRC Press/Lewis Publishers.

Lauwerys, R. R., P. Hoet. 1993. *Industrial Chemical Exposure.* New York: CRC Press/Lewis Publishers.

Lewis, R. J., Sr. 1993. *Hawley's Condensed Chemical Dictionary,* 12th Edition. New York: Van Nostrand Reinhold.

Lewis, R. J., Sr. 1992. *Sax's Dangerous Properties of Industrial Materials,* Volumes 1, 2, and 3, 8th Edition. New York: Van Nostrand Reinhold.

Meyer, E. 1990. *Chemistry of Hazardous Materials.* Englewood Cliffs, NJ: Prentice Hall.

Rea, W. J. 1995. *Chemical Sensitivity,* Vol. I-III. New York: CRC Press/Lewis Publishers.

Richardson, M. 1995. *Dictionary of Substance and Their Effects,* Vol. I-VI. New York: CRC Press/Lewis Publishers.

Rylander, R., R. R. Jacobs. 1994. *Organic Dusts.* New York: CRC Press/Lewis Publishers.

Sacarello, H. L. A. 1994. *The Comprehensive Handbook of Hazardous Materials.* New York: CRC Press/Lewis Publishers.

Sherman, J. D. 1988. *Chemical Exposure and Disease.* New York: Van Nostrand Reinhold.

Talbott, E., G. Craun. 1995. *An Introduction to Environmental Epidemiology.* New York: CRC Press/Lewis Publishers.

U.S. Department of Health and Human Services, National Institute for Occupational Safety and Health. 1994. *NIOSH Pocket Guide to Chemical Hazards.* Washington, D.C.: U.S. Government Printing Office.

MATERIAL SAFETY DATA SHEET

CHEMICAL NAME	
OSMIUM	**CRC PRESS/LEWIS PUBLISHERS DISCLAIMER:** THE INFORMATION AND RECOMMENDATIONS PRESENTED HEREIN ARE BASED ON SOURCES BELIEVED TO BE RELIABLE. CRC MAKES NO REPRESENTATION ON ITS COMPLETENESS OR ACCURACY. IT IS THE RESPONSIBILITY OF THE USER TO DETERMINE THE CHEMICAL'S SUITABILITY FOR ITS INTENDED USE, THE CHEMICAL'S SAFE USE, AND THE CHEMICAL'S PROPER DISPOSAL. NO REPRESENTATIONS AND/OR WARRANTIES, EITHER EXPRESSED OR IMPLIED, OF THE MERCHANTABILITY OR FITNESS FOR A PARTICULAR PURPOSE, OR OF ANY OTHER NATURE, ARE MADE WITH RESPECT TO THE INFORMATION PROVIDED IN THIS MSDS OR TO THE CHEMICAL TO WHICH INFORMATION MAY REFER. CRC NEITHER ASSUMES NOR AUTHORIZES ANY OTHER PERSON TO ASSUME FOR IT, ANY OTHER ADDITIONAL RESPONSIBILITY OR LIABILITY FOR THE USE OF, OR RELIANCE UPON, THIS INFORMATION.

HAZARD WARNING INFORMATION

HEALTH	FIRE	REACTIVE	OTHER	DEGREE OF HAZARD	COLOR CODING	OTHER CODES
1	1	0		0 = Minimum Hazard 1 = Slight Hazard 2 = Moderate Hazard 3 = Serious Hazard 4 = Severe Hazard	HEALTH = BLUE FIRE = RED REACTIVITY = YELLOW OTHER = WHITE	OX = Oxidizer ACID = Acid ALK = Alkali COR = Corrosive ~~W~~ = Use No Water

SECTION I - GENERAL INFORMATION

Characterization	**Metal**	RCRA Number / EPA Class	**None** / **Not Applicable**
DOT Proper Shipping Name	**No Citation**	Chemical Abstract Service (CAS) Number	**7440-04-2**
DOT Hazard Class and Label Requirements	**No Citation**	DOT Emergency Guide Code	**No Citation**
DOT Identification Number	**No Citation**	Chemical Formula	**Os**

Synonyms

Metallic osmium.

SECTION II - HAZARDOUS INGREDIENTS AND IDENTITY INFORMATION

Hazardous Components (specific identity)	OSHA Exposure Criteria	NIOSH Exposure Criteria	Immediately Dangerous to Life and Health (IDLH)	ACGIH Exposure Criteria
Osmium (derivation: Occurs with platinum, from which it is recovered during the purification process. Also occurs with iridium as a natural alloy, iridosmine).	**PEL:** Not Established **STEL:** Not Established	**REL:** Not Established **STEL:** Not Established	**Not Determined**	**TLV:** Not Established **STEL:** Not Established

SECTION III - PHYSICAL AND CHEMICAL CHARACTERISTICS

Boiling Point	**9932°F (5500°C)**	Specific Gravity (H_2O = 1)	**22.6**
Vapor Pressure (mm Hg)	**Not Found**	Molecular Weight	**190.20**
Vapor Density (Air = 1)	**Not Found**	Melting Point	**4892°F (2700°C)**

Solubility

Insoluble in water and acids.

Appearance and Odor

Lustrous, bluish-white, extremely hard and dense metal of the platinum group; odorless.

SECTION IV - FIRE AND EXPLOSION HAZARD DATA

Flash Point (method used)	**Not Found**	Explosive Limits in Air % by Volume	**LEL: Not Found UEL: Not Found**
NFPA Classification	**No Citation**	Autoignition Temperature	**Not Determined**

Extinguishing Media

Use dry chemical, carbon dioxide, water spray, or foams.

Special Fire Fighting Procedures

Wear full protective clothing and self-contained breathing apparatus (SCBA). Use a water spray to keep fire-exposed containers cool. For large fires, if possible, withdraw and allow to burn. Remain clear of smoke. Do NOT allow runoff from fire control measures to enter a sewer or waterway.

Unusual Fire and Explosion Hazards

The powder ignites spontaneously with exposure to heat, flames, or sparks. The damp powder burns explosively. The powder may self-explode. The powder may explode when heated with chlorine trichloride or oxygen difluoride. It ignites when heated to 212°F (100°C) with fluorine.

SECTION V - REACTIVITY DATA

Stability		Conditions to Avoid Solid osmium is stable at room temperature and pressure. Powdered osmium can slowly oxidize in air to osmium tetroxide. Keep away from incompatible materials and sources of heat.
Stable	Unstable X	Incompatibility *(materials to avoid)* Dangerous reactions with chlorine trichloride and oxygen difluoride. It will also be attacked by fluorine at high temperatures, by chlorine upon dry heating, and by aqua regia.
Hazardous Polymerization		Conditions to Avoid Under normal conditions of temperature and pressure, hazardous polymerization of osmium is not known to occur. Do not expose to air for extended periods (it can absorb explosive hydrogen gas from the air).
May Occur	Will Not Occur X	Hazardous Decomposition or By-products Toxic fumes and gases are released when osmium is involved in fire, including osmium fumes and toxic osmium tetroxide.

SECTION VI - HEALTH HAZARD DATA

Primary Route(s) of Entry:	Inhalation?	Absorption (skin)?	Ingestion?
	X		

Health Hazards

INHALATION: Irritation of the respiratory system and mucous membranes. Exposure can cause allergic lung reaction with chest pain, coughing, wheezing, and shortness of breath similar to the symptoms of bronchial asthma.

SKIN & EYES: An irritant to the eyes and skin. The powder may cause contact dermatitis, ulceration at the site of contact, and discoloration (green or black) of the skin's surface.

INGESTION: Unspecified gastrointestinal effects.

Carcinogenicity	NTP Listed?	IARC Cancer Review Group?	OSHA Regulated?	Target Organs?
Unknown Human Unknown Animal	No	No	No	Respiratory system, skin, eyes, kidney.

Medical Conditions Generally Aggravated by Exposure

None Reported.

Emergency and First-aid Procedures

Eye contact: Flush immediately with water for 15 minutes (minimum); seek medical attention. Skin contact: Remove all contaminated clothing. Immediately wash area with large amounts of soap and water. If rash develops, seek medical attention. For inhalation: Remove the person from exposure. Provide respiratory assistance and CPR. Transfer to medical facility. If swallowed: Provide breathing assistance if necessary. Never attempt to give a convulsing or unconscious person anything by mouth. Seek medical attention immediately. Contact poison control center for assistance.

SECTION VII - PRECAUTIONS FOR SAFE HANDLING AND USE

Steps to be Taken in Case Material is Released or Spilled

Collect powdered material in most convenient manner possible and deposit in sealed drum. Keep those not involved in cleanup from entering area. Do NOT dry sweep (creates airborne dusts and friction). Carefully use a vacuum equipped with a high efficiency particulate air (HEPA) filter. Exercise extreme caution, osmium can self-ignite or explode in air or if heated. Use only non-sparking tools.

Preferred Waste Disposal Method

No citation.

Precautions to be Taken in Handling and Storage

Store in tightly closed containers in cool, well-ventilated area and protect from physical damage. Sources of heat or ignition are prohibited where osmium is used, handled, or stored.

Other Precautions and Warnings

Avoid generating dusty conditions in storage areas. Mechanically ground and bond all containers.

SECTION VIII - CONTROL MEASURES AND PERSONAL PROTECTIVE EQUIPMENT

Respiratory Protection *(specify type)*

No exposure levels have been established. As a minimum, use a full facepiece respirator with a high efficiency particulate air (HEPA) filter. For high or prolonged exposures, use a full facepiece supplied-air respirator or a self-contained breathing apparatus (SCBA) operated in positive pressure mode.

Ventilation

Local exhaust or general mechanical systems recommended.

Protective Gloves	Eye Protection	Other Protective Clothing
Impervious Material	Chemical/Dust Goggles or Face Mask	Protective Suit or Clothing

Work/Hygiene Practices

Always wash hands thoroughly after using chemical; never bring food, drink, or smoking materials into vicinity of chemicals.

OSMIUM

Os

CAS: 7440-04-2

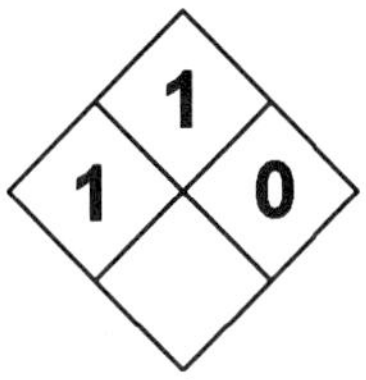

IDENTIFICATION AND TYPICAL USES

Osmium is a lustrous, bluish-white, extremely hard and dense metal of the platinum group. It has no odor. It is metallic element of atomic number 76, group VIII of the periodic table, atomic weight 190.2, valances 2, 3, 4, 6, 8 with seven stable isotopes. It is used as a hardener for iridium and platinum, in pen points, instrument pivots, and as a catalyst.

RISK ASSESSMENT: HEALTH

General Assessment

Osmium is a human poison by an unspecified route. The exposure routes are *inhalation* (primarily) and possibly *ingestion*. However, since it is not significantly soluble in water, it is not absorbed efficiently by ingestion. There are no data on its carcinogenic, mutagenic, and teratogenic capabilities.

Inhalation of osmium dusts and powders will cause irritation of the nose, throat, respiratory tract, and associated mucous membranes. It has been shown to cause asthma-like conditions with shortness of breath, labored breathing, respiratory irritation, wheezing, persistent conjunctivitis, coughing, and tightness in the chest. Skin contact results in surface irritation and may lead to dermatitis, localized ulceration, and discoloration (green or black) of the skin's surface at the site of contact. The powders can irritate the eyes. Ingestion causes unspecified gastrointestinal effects.

Acute Health Effects

The following acute (short-term) health effects may occur immediately or shortly after exposure to osmium:

Skin: Irritation, rash, and possible contact dermatitis, ulceration, and discoloration of the skin's surface at the area of contact.

Eye: Osmium particles can irritate the eyes and may cause damage to vision if not quickly removed.

Lung: Irritation of the dusts and fumes can irritate the nose, mouth, throat, lungs, and mucosa of the respiratory tract.

Chronic Health Effects

The following chronic (long-term) health effects may occur at some time after exposure to osmium and can last for months or even years:

Cancer Hazards: According to information presented in the references, osmium has not been shown to cause cancer in test animals.

Reproductive Hazard: According to information available in the references, osmium has not been tested for its ability to adversely affect reproduction in humans.

Other Chronic Effects: Possible kidney damage and skin allergy on long-term exposure to osmium.

Recommended Risk-Reduction Measures

Personnel should avoid direct contact with osmium. The exact nature of its human toxicity is not clearly understood in the references. If the specific osmium compound in question is unknown, adequate risk management dictates that it be handled and treated as though it posed as serious risk to human health. If a less toxic chemical cannot be substituted for a hazardous substance, then *engineering controls* are the most effective method of reducing exposures.

The best protection is to enclose operations and/or provide local exhaust ventilation at the site of chemical release. While not always operationally feasible, isolating operations can also reduce exposure. Using respiratory protection is less effective than the controls mentioned above, but is still advisable whenever working with or around osmium. No exposure levels have been established for osmium. This does not mean that exposure is without risk to human health. As a minimum, an MSHA/NIOSH-approved respirator with a high efficiency particulate air (HEPA) filter is recommended. For high exposures, or exposures to unknown levels (as in an emergency), use a supplied-air respirator with a full facepiece operated in positive pressure mode or an MSHA/NIOSH-approved self-contained breathing apparatus (SCBA) with full facepiece operated in pressure demand. If a full facepiece is not available, then chemical dust goggles should be worn to protect the eyes. Whenever an airborne chemical dust hazard exists, a face shield and a rubber apron should be worn. To prevent hand and skin exposures, impervious rubber gloves should be used.

Administrative controls should also be in place to minimize the potential for human exposures. These may include written procedures or policies which specify the methods and techniques that will be practiced whenever personnel are to work with osmium.

All personnel should receive *training* on the use, hazards, protective measures, emergency actions, and other precautions per 29 CFR 1910.1200 (Hazard Communication), prior to the first assignment in an area where osmium is used or stored.

Before beginning employment and at regular intervals thereafter (e.g., annually), for those with frequent or potentially high exposures (half the PEL or greater, or significant skin contact), the following medical tests are recommended:

☑ Lung function tests.

☑ Chest X-ray.

If symptoms develop or overexposure is suspected, the following may also be useful:

☑ Kidney and lung function tests.

☑ Evaluation by a qualified allergist, including careful exposure history and special testing (may help diagnose skin allergy).

Any evaluation should include a careful history of past and present symptoms with a medical examination. Medical tests that look for damage already done are not a substitute for controlling exposure. Also, since smoking can cause heart disease, as well as lung cancer, emphysema, and other related respiratory problems, the effects of exposure to osmium may be more pronounced in smokers than in non-smokers. Prudent risk management requires proper consideration of *all* possible factors that may be causing exposure symptoms in the workplace.

Other methods to reduce exposure include:

☑ Where possible, enclose operations and use local exhaust ventilation at the site of chemical release. If local exhaust ventilation or enclosure is not used, respiratory protection should be mandatory.

☑ Always ensure that proper protective clothing is worn when using chemical substances and that personnel are adequately trained on its use, care, and maintenance.

☑ Wash thoroughly immediately after exposure to osmium and at the end of the work shift or before eating, drinking, or smoking.

☑ Hazard warning information should be posted in the work area. In addition, as part of an on-going education and training program, all information on the health and safety hazards of osmium should be communicated to all potentially exposed workers.

☑ Where possible, automatically transfer osmium from drums or other storage containers to process containers. Avoid contact with air and do not generate airborne dusts.

RISK ASSESSMENT: ENVIRONMENT

General Assessment

The environment is at risk of exposure during transportation, storage, disposal, or destruction of osmium. In almost every scenario, the threat of environmental exposure is contingent upon the proper handling of the chemical substance. Accidental spills, large or small, can result in fire, explosion, and possible contamination of the surrounding environmental mediums (water, soil, and air).

Osmium powder can explode spontaneously in air. It will ignite and is potentially explosive in contact with heat, friction, and sparks. When heated in the presence of fluorine, it can react violently. It may explode on contact with chlorine trichloride or oxygen difluoride. These characteristics require special consideration during any emergency situation involving a leak or spill of osmium. Osmium can occur in the environment as a natural element in certain ores in the earth's crust. It can also enter the environment through industrial and municipal discharges, or from spills.

☠ *Acute Ecological Effects*

Acute (short-term) toxic effects may include the death of animals, birds, or fish and death or low growth rate

in plants. Acute effects are seen 2 to 4 days after animals or plants are exposed to osmium. Insufficient data are available to evaluate the short-term effects of osmium to aquatic life, plants, birds, or land animals.

Chronic Ecological Effects

Chronic (long-term) toxic effects may include shortened life span, reproductive problems, lower fertility, and changes in appearance or behavior in exposed animals. These effects can be seen long after first exposure(s) to toxic chemicals. Insufficient data are available to evaluate the long-term effects of osmium exposure to aquatic life, plants, birds, or land animals.

Water Solubility

Elemental osmium and most of its compounds are insoluble in water. Concentrations of less than 1 milligram will not mix with a liter water.

Persistence in the Environment

Osmium and its compounds are highly persistent in the aquatic environment, with a half-life of greater than 200 days. The half-life of a pollutant is the amount of time it takes for one half of the chemical to be degraded.

Bioaccumulation in Aquatic Organisms

Some substances increase in concentration, or *bioaccumulate*, in living organisms as they breathe contaminated air, drink contaminated water, or eat contaminated food. These chemicals can become concentrated in the tissues and internal organs of animals as well as humans.

The concentration of osmium found in fish is expected to be at least equal to the average concentration of osmium in the water from which the fish was taken.

Recommended Risk-Reduction Measures

Proper training of all transporters will reduce the likelihood of a mishap or accidents resulting in a leak or spill to the environment. The correct use of labeling on all containers, trucks, and rail cars will enable emergency responders to react properly and quickly to any disaster thereby reducing the potential risk to the environment and to personnel.

Storage of osmium should be segregated from incompatible chemicals to minimize the risk of cross-contamination or contact. Tasks which can generate dusty conditions must be avoided. Storage facilities should be designed with proper fire suppression and protection equipment. Containers must be protected from exposure to heat and from physical damage.

If a spill or leak to the environment has occurred, fire department, emergency response and/or hazardous materials spill personnel should be notified immediately. Use a vacuum method to reduce dust during cleanup. Do not dry sweep (generates airborne dusts and friction). *Carefully* and *slowly* use a vacuum equipped with a high efficiency particulate air (HEPA) filter, not a standard shop vacuum. Deposit collected materials in sealed drum for disposal. Cleanup should be attempted only by those trained in proper spill containment procedures using non-sparking tools. Contaminated soils should be removed and replaced with clean soil. If osmium should contact the water table, aquifer, or navigable waterway, time is of the essence. Osmium is not soluble in water and, therefore, total containment and remediation may be possible. When such spills occur, the local and/or state emergency response authorities must be notified. A comprehensive emergency response or disaster preparedness/recovery plan should be in place prior to any operations involving the use, transportation, storage, or disposal of osmium. If osmium is spilled or leaked, the following specific steps are recommended:

- ☑ Restrict persons not wearing protective clothing from area of spill or leak until cleanup is complete and area can be opened for normal work.
- ☑ Collect powdered materials in most convenient and safe manner possible and place in sealed drums for disposal. Do not dry sweep, use HEPA vacuum instead.
- ☑ It may be necessary to dispose of osmium as a hazardous waste. The responsible state agency or the regional office of the federal Environmental Protection Agency (EPA) should be contacted for specific recommendations.

RISK ASSESSMENT: BUSINESS

General Assessment

Accidents or mishaps involving osmium can present a significant threat to business operations. The loss or damage of equipment or facilities can significantly affect fiscal viability. Lawsuits that may result from personnel injury/death, public exposures, and/or environmental contamination will also require a serious expen-

diture of resources. Media attention surrounding an injury, death, or environmental damage can also result in a loss of profits and loss of current as well as future business.

Ʋ *Recommended Risk-Reduction Measures*

Company attorneys, safety and health professionals, and environmental specialists should be involved in the development of any procedures dealing with safety or emergency response. A company official should be pre-designated as a public relations officer with specific training in dealing with the press. Corporate plans and policies should be developed, approved, and implemented long before any need for such arises.

REFERENCES

Ahlbom, A. 1993. *Biostatistics for Epidemiologists.* New York: CRC Press/Lewis Publishers.

American Conference of Governmental Industrial Hygienists. 1988. *Documentation of the Threshold Limit Values and Biological Exposure Indices,* 5th Edition (with updates). Cincinnati: ACGIH

Asante-Duah, D. K. 1993. *Hazardous Waste Risk Assessment.* New York: CRC Press/Lewis Publishers.

Bloom, A. D. and F. J. DeSerres. 1995. *Ecotoxicity and Human Health.* New York: CRC Press/Lewis Publishers.

Brusick, D. J. 1994. *Methods for Genetic Risk Assessment.* New York: CRC Press/Lewis Publishers.

Calabrese, E. J. 1994. *Biological Effects of Low Level Exposures.* New York: CRC Press/Lewis Publishers.

Cockerman, L. G., B. S. Shane. 1994. *Basic Environmental Toxicology.* New York: CRC Press/Lewis Publishers.

Cothern, C. R., N. P. Ross. 1994. *Environmental Statistics, Assessment, and Forecasting.* New York: CRC Press/Lewis Publishers.

Danse, I. R. 1991. *Common Sense Toxics in the Workplace.* New York: Van Nostrand Reinhold.

Environmental Protection Agency. 1983. *Chemical Hazard Information Profiles.* Washington, D.C.: U.S. EPA.

Fodden, C. R., and J. L. Weddell. 1992. *Hazardous Materials Emergency Action Data.* New York: CRC Press/Lewis Publishers.

Forseberg, K., and L. H. Keith. 1993. *Chemical; Protective Clothing Permeation and Degradation Database.* Chelsea, MI: Lewis Publishers.

Gardner, R. W. 1994. *Chemical Intolerance.* New York: CRC Press/Lewis Publishers.

Howard, P. H., et. al. 1991. *Handbook of Environmental Degradation Rates.* New York: CRC Press/Lewis Publishers.

Hosty, J. W., and P. Foster. 1990. *A Practical Guide to Chemical Spill Response.* New York: Van Nostrand Reinhold.

Landis, W. G., M. H. Yu. 1995. *Introduction to Environmental Toxicology.* New York: CRC Press/Lewis Publishers.

Langerman, N. 1994 *Precautionary Labels for Chemical Containers.* New York: CRC Press/Lewis Publishers.

Lauwerys, R. R., P. Hoet. 1993. *Industrial Chemical Exposure.* New York: CRC Press/Lewis Publishers.

Lewis, .J., Sr. 1993. *Hawley's Condensed Chemical Dictionary,* 12th Edition. New York: Van Nostrand Reinhold.

Lewis, . J., Sr. 1992. *Sax's Dangerous Properties of Industrial Materials,* Volumes 1, 2, and 3, 8th Edition. New York: Van Nostrand Reinhold.

Mackay, D., W. Y. Shiu, and K. C. Ma. 1992. *Illustrated Handbook of Chemical-Physical Properties and Environmental Fate for Organic Chemicals.* New York: CRC Press/Lewis Publishers.

Manahan, S. E. 1992. *Toxicological Chemistry,* 2nd Edition. New York: CRC Press/Lewis Publishers.

Mellström, G. A., J. E. Wahlberg, and H. I. Maibach. 1994. *Protective Gloves for Occupational Use.* New York: CRC Press/Lewis Publishers.

Meyer, E. 1990. *Chemistry of Hazardous Materials.* Englewood Cliffs, NJ: Prentice Hall.

Miller, B. F., M.D., and C. B. Keane, R.N. 1972. *Encyclopedia and Dictionary of Medicine and Nursing.* Philadelphia: W.B. Sanders Company.

Mycock, J. C., J. D. McKenna, and L. Theodore. 1995. *Handbook of Air Pollution Control Engineering and Technology.* New York: CRC Press/Lewis Publishers.

Ott, W. R. 1995. *Environmental Statistics and Data Analysis.* New York: CRC Press/Lewis Publishers.

Patnaik, P. 1992. *A Comprehensive Guide to the Hazardous Properties of Chemical Substances.* New York: Van Nostrand Reinhold.

Philp, R. B. 1995. *Environmental Hazards and Human Health.* New York: CRC Press/Lewis Publishers.

Quigley, D. R. 1994. *Handbook of Emergency Chemical Management.* New York: CRC Press/Lewis Publishers.

Rea, W. J. 1995. *Chemical Sensitivity,* Vol. I-III. New York: CRC Press/Lewis Publishers.

Reeves, A. L. 1979. *Handbook on the Toxicology of Metals.* New York: Elsevier North-Holland.

Richardson, M. 1995. *Dictionary of Substance and Their Effects,* Vol. I-VI. New York: CRC Press/Lewis Publishers.

Royal Society of Chemistry. 1992. *Chemical Safety Data Sheets, Volumes 1-5.* Newcastle-upon-Tyne, UK: Athenaeum Press.

Rylander, R., R. R. Jacobs. 1994. *Organic Dusts.* New York: CRC Press/Lewis Publishers.

MATERIAL SAFETY DATA SHEET

CHEMICAL NAME	
OSMIUM TETROXIDE	**CRC PRESS/LEWIS PUBLISHERS DISCLAIMER:** THE INFORMATION AND RECOMMENDATIONS PRESENTED HEREIN ARE BASED ON SOURCES BELIEVED TO BE RELIABLE. CRC MAKES NO REPRESENTATION ON ITS COMPLETENESS OR ACCURACY. IT IS THE RESPONSIBILITY OF THE USER TO DETERMINE THE CHEMICAL'S SUITABILITY FOR ITS INTENDED USE, THE CHEMICAL'S SAFE USE, AND THE CHEMICAL'S PROPER DISPOSAL. NO REPRESENTATIONS AND/OR WARRANTIES, EITHER EXPRESSED OR IMPLIED, OF THE MERCHANTABILITY OR FITNESS FOR A PARTICULAR PURPOSE, OR OF ANY OTHER NATURE, ARE MADE WITH RESPECT TO THE INFORMATION PROVIDED IN THIS MSDS OR TO THE CHEMICAL TO WHICH INFORMATION MAY REFER. CRC NEITHER ASSUMES NOR AUTHORIZES ANY OTHER PERSON TO ASSUME FOR IT, ANY OTHER ADDITIONAL RESPONSIBILITY OR LIABILITY FOR THE USE OF, OR RELIANCE UPON, THIS INFORMATION.

HAZARD WARNING INFORMATION

HEALTH	FIRE	REACTIVE	OTHER	DEGREE OF HAZARD	COLOR CODING	OTHER CODES
4	0	0		0 = Minimum Hazard 1 = Slight Hazard 2 = Moderate Hazard 3 = Serious Hazard 4 = Severe Hazard	HEALTH = BLUE FIRE = RED REACTIVITY = YELLOW OTHER = WHITE	OX = Oxidizer ACID = Acid ALK = Alkali COR = Corrosive W = Use No Water

SECTION I - GENERAL INFORMATION

Characterization **Metallic Compound**	RCRA Number **P087**	EPA Class **Acute Hazardous Waste**
DOT Proper Shipping Name **Osmium Tetroxide**	Chemical Abstract Service (CAS) Number **20816-12-0**	
DOT Hazard Class and Label Requirements **Poison B**	DOT Emergency Guide Code **55**	
DOT Identification Number **UN 2471**	Chemical Formula **OsO_4**	

Synonyms
Osmic acid anhydride; osmium oxide; perosmic acid anhydride; perosmic oxide.

SECTION II - HAZARDOUS INGREDIENTS AND IDENTITY INFORMATION

Hazardous Components (specific identity)	OSHA Exposure Criteria	NIOSH Exposure Criteria	Immediately Dangerous to Life and Health (IDLH)	ACGIH Exposure Criteria
Osmium tetroxide (derivation: By heating powdered osmium in air, or by treating it with nitric acid, aqua regia, or chlorine). **1 ppm = 10.57 mg/m^3**	**PEL (8-hour):** **0.0002 ppm** **0.002 mg/m^3** **STEL (15-min):** **0.0006 ppm** **0.006 mg/m^3**	**REL (10-hour):** **0.0002 ppm** **0.002 mg/m^3** **STEL (15-min):** **0.0006 ppm** **0.006 mg/m^3**	**1 mg/m^3** **(as Os)**	**TLV:** **0.0002 ppm** **0.002 mg/m^3** **STEL:** **0.0006 ppm** **0.006 mg/m^3**

SECTION III - PHYSICAL AND CHEMICAL CHARACTERISTICS

Boiling Point **266°F (130°C)**	Specific Gravity (H_2O = 1) **5.10**
Vapor Pressure (mm Hg) **11 at 81°F (27°C)**	Molecular Weight **254.2**
Vapor Density (Air = 1) **Not Found**	Melting Point **105°F (40.5°C)**

Solubility
Soluble in water (6%) at 77°F (25°C). Soluble in benzene, alcohol, ether, ammonium hydroxide, phosphorus oxychloride, and carbon tetrachloride.

Appearance and Odor
Colorless crystalline solid or pale-yellow mass with an unpleasant, acrid, chlorine-like odor.

SECTION IV - FIRE AND EXPLOSION HAZARD DATA

Flash Point (method used) **Not Found**	Explosive Limits in Air % by Volume **LEL: Not Found UEL: Not Found**
NFPA Classification **Non-Combustible Solid**	Autoignition Temperature **Not Determined**

Extinguishing Media
Use extinguishing agent suitable to surrounding area. Osmium tetroxide itself does not burn.

Special Fire Fighting Procedures
Wear full protective clothing and self-contained breathing apparatus (SCBA). Use a water spray to keep fire-exposed containers cool. Remain clear of smoke. Do NOT allow runoff from fire control measures to enter a sewer or waterway.

Unusual Fire and Explosion Hazards
The dusts and powders can become explosive in mixtures with air and in the presence of a heat or ignition source.

SECTION V - REACTIVITY DATA

Stability		Conditions to Avoid
		Osmium tetroxide is normally stable in closed containers under routine conditions of handling and storage at room temperature and pressure. Keep away from incompatible materials and sources of heat.
Stable	Unstable	Incompatibility *(materials to avoid)*
X		Dangerous reactions hydrochloric acid, reducing agents, and easily oxidized organic materials. It begins to sublime below its boiling point. Contact with other materials may result in fire.
Hazardous Polymerization		Conditions to Avoid
		Under normal conditions of temperature and pressure, hazardous polymerization of osmium tetroxide is not known to occur. Avoid generating dusty conditions.
May Occur	Will Not Occur	Hazardous Decomposition or By-products
	X	Toxic osmium compound fumes and gases are released when osmium tetroxide is involved in fire.

SECTION VI - HEALTH HAZARD DATA

Primary Route(s) of Entry:	Inhalation?	Absorption (skin)?	Ingestion?
	X		

Health Hazards

INHALATION: Irritation of the respiratory system and mucous membranes. There can also be ocular effects. Visual problems include delayed lachrymation (tearing) and vision with a halo or ground-glass effect. Exposure can cause allergic lung reaction with chest pain, cough, wheezing, and shortness of breath, tracheitis, bronchial spasm, difficulty breathing, and kidney damage. It may cause pulmonary edema (fluid in lungs), which can be fatal.

SKIN & EYES: An irritant to the eyes and skin. The powder may cause contact dermatitis, ulceration at the site of contact, and discoloration (green or black) of the skin's surface.

INGESTION: Unspecified gastrointestinal effects.

Carcinogenicity	NTP Listed?	IARC Cancer Review Group?	OSHA Regulated?	Target Organs?
Unknown Human Unknown Animal	No	No	29 CFR 1910.1000 Table Z-1	Respiratory system, skin, eyes, kidney.

Medical Conditions Generally Aggravated by Exposure

None reported.

Emergency and First-aid Procedures

Eye contact: Flush immediately with water for 15 minutes (minimum); seek medical attention. Skin contact: Remove all contaminated clothing. Immediately wash area with large amounts of soap and water. If rash develops, seek medical attention. For inhalation: Remove the person from exposure. Provide respiratory assistance and CPR. Transfer to medical facility. Observe 48 hours for lung effects. If swallowed: Unlikely. If it should occur, give 1 to 2 glasses of water. Never attempt to give a convulsing or unconscious person anything by mouth. Seek medical attention immediately.

SECTION VII - PRECAUTIONS FOR SAFE HANDLING AND USE

Steps to be Taken in Case Material is Released or Spilled

Collect powdered material in most convenient manner possible and deposit in sealed drum. Keep those not involved in cleanup from area. Do NOT dry sweep (creates airborne dusts and friction). Use a vacuum equipped with a high efficiency particulate air (HEPA) filter. Use non-sparking tools.

Preferred Waste Disposal Method

No citation.

Precautions to be Taken in Handling and Storage

Store in tightly closed containers in cool, well-ventilated area and protect from physical damage.

Other Precautions and Warnings

Avoid generating dusty conditions in storage areas. Mechanically ground and bond all containers.

SECTION VIII - CONTROL MEASURES AND PERSONAL PROTECTIVE EQUIPMENT

Respiratory Protection *(specify type)*

Exposure levels are low and difficult to monitor. As a minimum, use a full facepiece respirator with a high efficiency particulate air (HEPA) filter. For high or prolonged exposures, use a full facepiece supplied-air respirator or a self-contained breathing apparatus (SCBA) operated in positive pressure mode.

Ventilation

Local exhaust or general mechanical systems recommended.

Protective Gloves	Eye Protection	Other Protective Clothing
Impervious Material	Chemical/Dust Goggles or Face Mask	Protective Suit or Clothing

Work/Hygiene Practices

Always wash hands thoroughly after using chemical; never bring food, drink, or smoking materials into vicinity of chemicals.

OSMIUM TETROXIDE

OsO_4

CAS: 20816-12-0

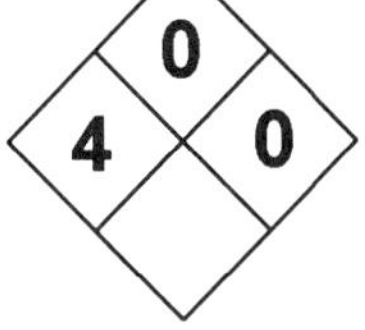

IDENTIFICATION AND TYPICAL USES

Osmium tetroxide is a colorless, crystalline solid or a pale-yellow mass with an unpleasant, acrid, chlorine-like odor. It is a liquid at temperatures above 105°F (40.5°C). It is used as a catalyst in chemical reactions, in photography, and as a fat stain in pathological laboratories (for microscopic staining).

RISK ASSESSMENT: HEALTH

General Assessment

Osmium tetroxide is a human poison by an unspecified route. The exposure routes are *inhalation* (primarily) and possibly (but unlikely) *ingestion*. Skin contact can cause localized effects but dermal absorption has not been reported. There are no data on its carcinogenic, mutagenic, and teratogenic capabilities.

Inhalation of osmium tetroxide dusts and powders will cause irritation of the nose, throat, respiratory tract, and associated mucous membranes. It has been shown to cause asthma-like conditions with shortness of breath, labored breathing, respiratory irritation, wheezing, persistent conjunctivitis, coughing, tracheitis, bronchitis, tightness in the chest, and pulmonary edema (fluid in the lungs), which can be fatal. Symptoms of pulmonary edema can be delayed up to 48 hours thereby creating a false sense of security with regard to health exposure risk.

Skin contact results in surface irritation and may lead to dermatitis, localized ulceration, and discoloration (green or black) of the skin's surface at the site of contact. The powder can irritate the eyes. Ingestion causes unspecified gastrointestinal effects.

Acute Health Effects

The following acute (short-term) health effects may occur immediately or shortly after exposure to osmium tetroxide:

Skin: Irritation, rash, and possible contact dermatitis, ulceration, and discoloration of the skin's surface at the area of contact.

Eye: Osmium tetroxide particles can irritate the eyes and may cause damage to vision if not quickly removed. High concentrations cause corneal damage and blindness.

Lung: Irritation of the dusts and fumes can irritate the nose, mouth, throat, lungs, and mucosa of the respiratory tract. Pulmonary edema may occur on high exposure.

Chronic Health Effects

The following chronic (long-term) health effects may occur at some time after exposure to osmium tetroxide and can last for months or even years:

Cancer Hazards: According to information presented in the references, osmium tetroxide has not been shown to cause cancer in test animals.

Reproductive Hazard: According to information available in the references, osmium tetroxide has not been tested for its ability to adversely affect reproduction in humans.

Other Chronic Effects: Possible kidney damage and skin allergy on long-term exposure to osmium tetroxide. Chronic exposure may also cause broncho pneumonia, sterile lung abscess, and gangrene.

Recommended Risk-Reduction Measures

Personnel should avoid direct contact with osmium tetroxide. The exact nature of its human toxicity is not clearly understood in the references. Adequate risk management dictates that it be handled and treated as though it posed as serious risk to human health. If a less toxic chemical cannot be substituted for a hazardous substance, then *engineering controls* are the most effective method of reducing exposures.

The best protection is to enclose operations and/or provide local exhaust ventilation at the site of chemical release. While not always operationally feasible, isolating operations can also reduce exposure. Using respiratory protection is less effective than the controls mentioned above, but is still advisable whenever working with or around osmium tetroxide. Exposure levels are low (0.0002 ppm) and extremely difficult to monitor. As a minimum, an MSHA/NIOSH-approved respirator with a high efficiency particulate air (HEPA) filter is recommended. For high exposures, or exposures to unknown levels (as in an emergency), use a supplied-air respirator with a full facepiece operated in positive pressure mode or an MSHA/NIOSH-approved self-contained breathing apparatus (SCBA) with full facepiece operated in pressure demand. If a full facepiece is not available, then chemical dust goggles should be worn to protect the eyes. Whenever an airborne chemical dust hazard exists, a face shield and a rubber apron should be worn. To prevent hand and skin exposures, impervious rubber gloves should be used.

Administrative controls should also be in place to minimize the potential for human exposures. These may include written procedures or policies which specify the methods and techniques that will be practiced whenever personnel are to work with osmium tetroxide.

All personnel should receive *training* on the use, hazards, protective measures, emergency actions, and other precautions per 29 CFR 1910.1200 (Hazard Communication), prior to the first assignment in an area where osmium tetroxide is used or stored.

Before beginning employment and at regular intervals thereafter (e.g., annually), for those with frequent or potentially high exposures (half the PEL or greater, or significant skin contact), the following medical tests are recommended:

☑ Lung function tests.

☑ Complete eye examination.

If symptoms develop or overexposure is suspected, the following may also be useful:

☑ Kidney (with urinalysis) and lung function tests.

☑ Consider chest X-ray after acute overexposure (may be negative if taken immediately after exposure due to delayed onset of pulmonary edema).

☑ Evaluation by a qualified allergist, including careful exposure history and special testing (may help diagnose skin allergy).

Any evaluation should include a careful history of past and present symptoms with a medical examination. Medical tests that look for damage already done are not a substitute for controlling exposure. Also, since smoking can cause heart disease, as well as lung cancer, emphysema, and other related respiratory problems, the effects of exposure to osmium tetroxide may be more pronounced in smokers than in non-smokers. Prudent risk management requires proper consideration of *all* possible factors that may be causing exposure symptoms in the workplace.

Other methods to reduce exposure include:

☑ Where possible, enclose operations and use local exhaust ventilation at the site of chemical release. If local exhaust ventilation or enclosure is not used, respiratory protection should be mandatory.

☑ Always ensure that proper protective clothing is worn when using chemical substances and that personnel are adequately trained on its use, care, and maintenance.

☑ Wash thoroughly immediately after exposure to osmium tetroxide and at the end of the work shift or before eating, drinking, or smoking.

☑ Hazard warning information should be posted in the work area. In addition, as part of an on-going education and training program, all information on the health and safety hazards of osmium tetroxide should be communicated to all exposed workers.

☑ Where possible, automatically transfer osmium tetroxide from drums or other storage containers to process containers. Avoid contact with air and do not generate airborne dusts.

RISK ASSESSMENT: ENVIRONMENT

General Assessment

The environment is at risk of exposure during transportation, storage, disposal, or destruction of osmium tetroxide. In almost every scenario, the threat of environmental exposure is contingent upon the proper handling of the chemical substance. Accidental spills, large or small, can result in fire, explosion, and possible contamination of the surrounding environmental mediums (water, soil, and air).

Osmium tetroxide powder can emit highly toxic and poisonous fumes when heated. Even though it is non-combustible in solid form, it is highly flammable as a dust and can form explosive mixtures in air. It will ignite and is potentially explosive in contact with heat, friction, and sparks. When heated in the presence of fluorine, it can react violently. It may explode on contact with chlorine trichloride or oxygen difluoride. These characteristics require special consideration during any emergency situation involving a leak or spill of osmium tetroxide.

Osmium can occur in the environment as a natural element in certain ores in the earth's crust. It can also enter the environment through industrial and municipal discharges, or from spills.

Acute Ecological Effects

Acute (short-term) toxic effects may include the death of animals, birds, or fish and death or low growth rate in plants. Acute effects are seen 2 to 4 days after animals or plants are exposed to osmium tetroxide.

Insufficient data are available to evaluate the short-term effects of osmium tetroxide to aquatic life, plants, birds, or land animals.

Chronic Ecological Effects

Chronic (long-term) toxic effects may include shortened life span, reproductive problems, lower fertility, and changes in appearance or behavior in exposed animals. These effects can be seen long after first exposure(s) to toxic chemicals.

Insufficient data are available to evaluate the long-term effects of osmium tetroxide exposure to aquatic life, plants, birds, or land animals.

Water Solubility

Osmium tetroxide is highly soluble in water. Concentrations of 1000 milligrams and more will mix with a liter of water.

Persistence in the Environment

Osmium tetroxide is quite soluble in the aquatic environment. It is an oxidizing agent and is readily reduced by organic matter. Consequently, it is not likely to be highly persistent as osmium tetroxide. Some of the osmium compounds formed may be persistent in the environment.

Bioaccumulation in Aquatic Organisms

Some substances increase in concentration, or *bioaccumulate,* in living organisms as they breathe contaminated air, drink contaminated water, or eat contaminated food. These chemicals can become concentrated in the tissues and internal organs of animals as well as humans.

Osmium tetroxide will probably not accumulate in aquatic organisms or aquatic ecosystems, but osmium might accumulate in both. The concentration of osmium found in fish is expected to be at least equal to the average concentration of osmium in the water from which the fish was taken.

Recommended Risk-Reduction Measures

Proper training of all transporters will reduce the likelihood of a mishap or accidents resulting in a leak or spill to the environment. The correct use of labeling on all containers, trucks, and rail cars will enable emergency responders to react properly and quickly to any disaster thereby reducing the potential risk to the environment and to personnel.

Storage of osmium tetroxide should be segregated from incompatible chemicals to minimize the risk of cross-contamination or contact. Tasks which can generate dusty conditions must be avoided. Storage facilities should be designed with proper fire suppression and protection equipment. Containers must be protected from exposure to heat and from physical damage.

If a spill or leak to the environment has occurred, fire department, emergency response and/or hazardous materials spill personnel should be notified immediately. Use a vacuum method to reduce dust during cleanup. Do not dry sweep (generates airborne dusts and friction). *Carefully* and *slowly* use a vacuum equipped with a high efficiency particulate air (HEPA) filter, not a standard shop vacuum. Deposit collected materials in sealed drum for disposal. Cleanup should be attempted only by those trained in proper spill containment procedures using non-sparking tools. Contaminated soils should be removed and replaced with clean soil. If osmium tetroxide should contact the water table, aquifer, or navigable waterway, time is of the essence. Osmium tetroxide is highly soluble in water and, therefore, total containment and remediation may not be possible. When such spills occur, the local and/or state emergency response authorities must be notified. A comprehensive emergency response or disaster preparedness/recovery plan should be in place prior to any operations involving the use, transportation, storage, or disposal of osmium tetroxide.

If osmium tetroxide is spilled or leaked, the following specific steps are recommended:

☑ Restrict persons not wearing protective clothing from area of spill or leak until cleanup is complete and area can be opened for normal work.

☑ Collect powdered materials in most convenient and safe manner possible and place in sealed drums for disposal. Do not dry sweep, use HEPA vacuum instead.

☑ It may be necessary to dispose of osmium tetroxide as a hazardous waste. The responsible state agency or the regional office of the federal Environmental Protection Agency (EPA) should be contacted for specific recommendations.

RISK ASSESSMENT: BUSINESS

General Assessment

Accidents or mishaps involving osmium tetroxide can present a significant threat to business operations. The loss or damage of equipment or facilities can significantly affect fiscal viability. Lawsuits that may result from personnel injury/death, public exposures, and/or environmental contamination will also require a serious expenditure of resources. Media attention surrounding an injury, death, or environmental damage can also result in a loss of profits and loss of current as well as future business.

Recommended Risk-Reduction Measures

Company attorneys, safety and health professionals, and environmental specialists should be involved in the development of any procedures dealing with safety or emergency response. A company official should be pre-designated as a public relations officer with specific training in dealing with the press. Corporate plans and policies should be developed, approved, and implemented long before any need for such arises.

REFERENCES

American Conference of Governmental Industrial Hygienists. 1988. *Documentation of the Threshold Limit Values and Biological Exposure Indices*, 5th Edition (with updates). Cincinnati: ACGIH

Asante-Duah, D. K. 1993. *Hazardous Waste Risk Assessment.* New York: CRC Press/Lewis Publishers.

Bloom, A. D. and F. J. DeSerres. 1995. *Ecotoxicity and Human Health.* New York: CRC Press/Lewis Publishers.

Brusick, D. J. 1994. *Methods for Genetic Risk Assessment.* New York: CRC Press/Lewis Publishers.

Calabrese, E. J. 1994. *Biological Effects of Low Level Exposures.* New York: CRC Press/Lewis Publishers.

Cockerman, L. G., B. S. Shane. 1994. *Basic Environmental Toxicology.* New York: CRC Press/Lewis Publishers.

Cothern, C. R., N. P. Ross. 1994. *Environmental Statistics, Assessment, and Forecasting.* New York: CRC Press/Lewis Publishers.

Danse, I. R. 1991. *Common Sense Toxics in the Workplace.* New York: Van Nostrand Reinhold.

Environmental Protection Agency. 1983. *Chemical Hazard Information Profiles.* Washington, D.C.: U.S. EPA.

Fodden, C. R., and J. L. Weddell. 1992. *Hazardous Materials Emergency Action Data.* New York: CRC Press/Lewis Publishers.

Forseberg, K., and L. H. Keith. 1993. *Chemical; Protective Clothing Permeation and Degradation Database.* Chelsea, MI: Lewis Publishers.

Gardner, R. W. 1994. *Chemical Intolerance.* New York: CRC Press/Lewis Publishers.

Howard, P. H., et. al. 1991. *Handbook of Environmental Degradation Rates.* New York: CRC Press/Lewis Publishers.

Hosty, J. W., and P. Foster. 1990. *A Practical Guide to Chemical Spill Response.* New York: Van Nostrand Reinhold.

Landis, W. G., M. H. Yu. 1995. *Introduction to Environmental Toxicology.* New York: CRC Press/Lewis Publishers.

Lauwerys, R. R., P. Hoet. 1993. *Industrial Chemical Exposure.* New York: CRC Press/Lewis Publishers.

Lewis, .J., Sr. 1993. *Hawley's Condensed Chemical Dictionary,* 12th Edition. New York: Van Nostrand Reinhold.

Lewis, . J., Sr. 1992. *Sax's Dangerous Properties of Industrial Materials,* Volumes 1, 2, and 3, 8th Edition. New York: Van Nostrand Reinhold.

Mellström, G. A., J. E. Wahlberg, and H. I. Maibach. 1994. *Protective Gloves for Occupational Use.* New York: CRC Press/Lewis Publishers.

Meyer, E. 1990. *Chemistry of Hazardous Materials.* Englewood Cliffs, NJ: Prentice Hall.

Miller, B. F., M.D., and C. B. Keane, R.N. 1972. *Encyclopedia and Dictionary of Medicine and Nursing.* Philadelphia: W.B. Sanders Company.

Mycock, J. C., J. D. McKenna, and L. Theodore. 1995. *Handbook of Air Pollution Control Engineering and Technology.* New York: CRC Press/Lewis Publishers.

Rea, W. J. 1995. *Chemical Sensitivity,* Vol. I-III. New York: CRC Press/Lewis Publishers.

Reeves, A. L. 1979. *Handbook on the Toxicology of Metals.* New York: Elsevier North-Holland.

Richardson, M. 1995. *Dictionary of Substance and Their Effects,* Vol. I-VI. New York: CRC Press/Lewis Publishers.

Royal Society of Chemistry. 1992. *Chemical Safety Data Sheets, Volumes 1-5.* Newcastle-upon-Tyne, UK: Athenaeum Press.

Rylander, R., R. R. Jacobs. 1994. *Organic Dusts.* New York: CRC Press/Lewis Publishers.

MATERIAL SAFETY DATA SHEET

CHEMICAL NAME

OXALIC ACID

HAZARD WARNING INFORMATION

HEALTH	FIRE	REACTIVE	OTHER	DEGREE OF HAZARD	COLOR CODING	OTHER CODES
2	1	0		0 = Minimum Hazard 1 = Slight Hazard 2 = Moderate Hazard 3 = Serious Hazard 4 = Severe Hazard	HEALTH = BLUE FIRE = RED REACTIVITY = YELLOW OTHER = WHITE	OX = Oxidizer ACID = Acid ALK = Alkali COR = Corrosive W = Use No Water

SECTION I - GENERAL INFORMATION

Characterization	RCRA Number	EPA Class
Acid, Carboxylic	**None**	**Not Applicable**

DOT Proper Shipping Name **Oxalates (water soluble)**	Chemical Abstract Service (CAS) Number **144-62-7**
DOT Hazard Class and Label Requirements **Poison B; St. Andrew's Cross (X)**	DOT Emergency Guide Code **60**
DOT Identification Number **UN 2449**	Chemical Formula **HOOCCOOH•H_2O**

Synonyms

Ethanedioic acid; oxalic acid (aqueous); oxalic acid dihydrate.

SECTION II - HAZARDOUS INGREDIENTS AND IDENTITY INFORMATION

Hazardous Components (specific identity)	OSHA Exposure Criteria	NIOSH Exposure Criteria	Immediately Dangerous to Life and Health (IDLH)	ACGIH Exposure Criteria
Oxalic acid (derivation: Occurs naturally in many plants, such as wood sorrel, rhubarb, spinach) and can be made by alkali extraction of sawdust. Now manufactured by reaction of carbon monoxide and sodium hydroxide, followed by distillation of the resulting dihydrate crystals).	**PEL:** 1 mg/m^3 **STEL:** Not Established	**REL:** 1 mg/m^3 **STEL:** 2 mg/m^3	**500 ppm**	**TLV:** 1 mg/m^3 **STEL:** 2 mg/m^3

SECTION III - PHYSICAL AND CHEMICAL CHARACTERISTICS

Boiling Point **Sublimes**	Specific Gravity (H_2O = 1) **1.90**
Vapor Pressure (mm Hg) **<0.001 at 69°F (20°C)**	Molecular Weight **126.1**
Vapor Density (Air = 1) **1.653**	Melting Point **215°F (102°C) Sublimes**

Solubility

Miscible in water. Soluble in alcohol and ether.

Appearance and Odor

Colorless, odorless, powder or a granular solid. The anhydrous (hygroscopic) form $(COOH)_2$ is an odorless, white solid.

SECTION IV - FIRE AND EXPLOSION HAZARD DATA

Flash Point (method used) **Not Applicable**	Explosive Limits in Air % by Volume **LEL: Not Determined UEL: Not Determined**
NFPA Classification **Combustible Solid**	Autoignition Temperature **Not Applicable**

Extinguishing Media

Carbon dioxide, dry chemical, water spray, or alcohol foam.

Special Fire Fighting Procedures

Wear full protective clothing and self-contained breathing apparatus (SCBA). Use water spray to flush and dilute a spill and to disperse the vapors. Also, use a water spray to keep fire-exposed containers cool. Do not release runoff to sewers and waterways. Poisonous gases are produced in fire.

Unusual Fire and Explosion Hazards

Potentially explosive reaction with strong oxidizers, strong alkalis, and chlorites. Containers may explode in fire. Firefighters should avoid contact with vapors produced during fire.

SECTION V - REACTIVITY DATA

Stability		Conditions to Avoid Normally stable in closed containers under normal conditions of storage and handling. Avoid contact with incompatible materials and heat.
Stable X	Unstable	Incompatibility *(materials to avoid)* Strong oxidizers (such as chlorine, fluorine, and bromine), strong alkalis, chlorites (such as sodium hypochlorites), silver compounds, and furfuryl alcohol.
Hazardous Polymerization		Conditions to Avoid Under normal conditions of temperature and pressure, hazardous polymerization of oxalic acid is not expected to occur.
May Occur	Will Not Occur X	Hazardous Decomposition or By-products During fires, oxalic acid can decompose into water, carbon monoxide, carbon dioxide, and formic acid.

SECTION VI - HEALTH HAZARD DATA

Primary Route(s) of Entry:	Inhalation?	Absorption (skin)?	Ingestion?
	X	X	X

Health Hazards

INHALATION: Vapors cause irritation to the upper respiratory tract with coughing, dypsnea, headache, bronchitis, weakness, tearing, nasal discharge, gastrointestinal disturbances, nausea, albuminuria, gradual weight loss, increasing weakness and nervous system disorders, ulceration of the mucous membranes in the nose and throat, epistaxis, nervousness.

SKIN & EYES: Eye contact causes scarring of the cornea. Skin contact causes burns, blistering, local necroses, and membrane ulceration. Sensitization and dermatitis may occur.

INGESTION: Severe poisoning, kidney damage, salivation, bloody vomiting, burning in mouth, pain, and diarrhea. Severe cases may cause shock, dypsnea, circulatory collapse, and death.

Carcinogenicity	NTP Listed?	IARC Cancer Review Group?	OSHA Regulated?	Target Organs?
Unknown Human Unknown Animal	No	No	29 CFR 1910.1000 Table Z-1	Respiratory system, eyes, skin, kidneys.

Medical Conditions Generally Aggravated by Exposure

Respiratory impairments (bronchitis, asthma); skin conditions (dermatitis); renal disorders.

Emergency and First-aid Procedures

Eye contact: Flush immediately with water for 15 minutes (minimum); seek medical attention. Skin contact: Remove all contaminated clothing. Immediately wash area with large amounts of water. Seek medical assistance. For inhalation: Remove the person from exposure. Provide respiratory assistance and CPR. Transfer to medical facility. If swallowed: Seek medical attention; contact poison control center. Do NOT induce vomiting. Give 1-2 glasses of milk.

SECTION VII - PRECAUTIONS FOR SAFE HANDLING AND USE

Steps to be Taken in Case Material is Released or Spilled

Personnel must wear protective clothing. Collect solid materials using scoop or carefully vacuum with a HEPA vacuum and place in sealed containers for disposal.

Preferred Waste Disposal Method

No citation.

Precautions to be Taken in Handling and Storage

Oxalic acid is a combustible solid. Avoid contact with strong oxidizers such as chlorine since violent reactions occur. Store in tightly closed containers in cool, well-ventilated area away from heat.

Other Precautions and Warnings

Sources of ignition are prohibited where oxalic acid is used, handled, or stored. Detached facility storage is preferred. Electrically ground and bond all containers.

SECTION VIII - CONTROL MEASURES AND PERSONAL PROTECTIVE EQUIPMENT

Respiratory Protection *(specify type)*

Use half-mask respirator with acid cartridge for low or transient exposures (cartridge should be designed to protect against acid dusts or mists). For higher exposures, use a supplied-air respirator or a self-contained breathing apparatus operated in pressure demand or other positive pressure mode.

Ventilation

Local exhaust or general mechanical systems recommended. Use maximum explosion-proof design.

Protective Gloves	Eye Protection	Other Protective Clothing
Acid resistant natural rubber	Chemical Goggles or Face Mask	Natural Rubber or Neoprene Apron

Work/Hygiene Practices

Always wash hands thoroughly after using chemical; never bring food, drink, or smoking materials into vicinity of chemicals.

OXALIC ACID

$HOOCCOOH \bullet H_2O$ CAS: 144-62-7

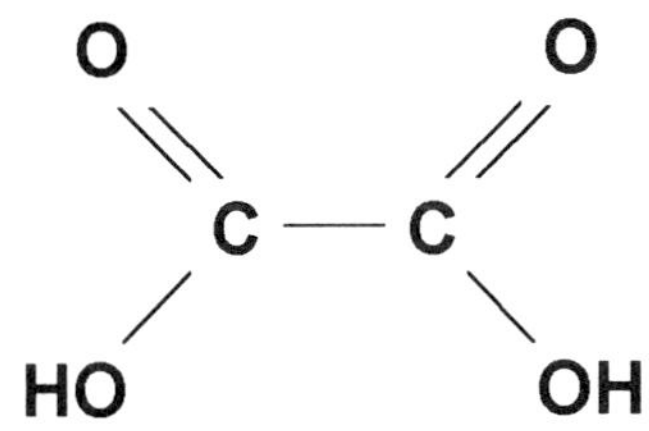

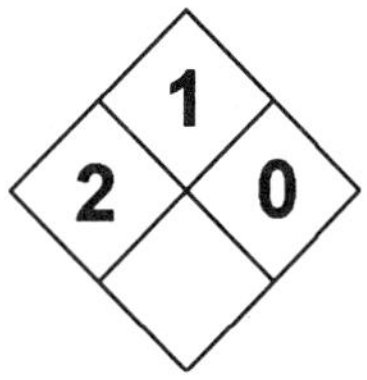

IDENTIFICATION AND TYPICAL USES

Oxalic acid is a colorless, odorless powder or granular solid. The anhydrous form is an odorless, white solid. It also occurs naturally in many plants (wood sorrel, rhubarb, and spinach). It is the metabolite product of many molds. It is used in indigo dyeing, calico printing, removal of paints, rust, and ink stains, in metal polishing, bleaching leather, and manufacture of oxalates. It is also used as an analytical reagent and as a reducing agent in organic synthesis.

RISK ASSESSMENT: HEALTH

General Assessment

Oxalic acid is a strong poison. Exposure to oxalic acid is possible through *inhalation* of its vapors and by *ingestion.* It can also be *absorbed* through the body through skin contact, preceded by tissue destruction at the site of contact. There are no reports on its carcinogenic, mutagenic, or teratogenic properties.

Inhalation produces severe irritation of the eyes, nose, and respiratory tract with coughing, dypsnea (labored breathing), gastrointestinal disturbances, albuminuria, gradual weight loss, increasing weakness and nervous system disorders, ulceration of the mucosa of the nose and throat, epistaxis, headache, irritation, and nervousness. There may be tearing, nasal discharge, throat irritation, nausea, and vomiting.

Skin contact causes severe burning with possible vesiculation. Its extremely corrosive properties cause immediate burns, membrane ulceration, skin blistering, and local necrosis (tissue destruction) upon contact. Skin sensitization and dermatitis may also occur. Eye contact causes damage and the possibility of permanent corneal scarring.

Ingestion is extremely dangerous, even life-threatening. It causes severe poisoning, corrosion of the mouth, esophagus, and stomach, bloody vomiting, salivation, burning and abdominal pain, collapse, convulsions, kidney damage, acute tissue damage, diarrhea with pain, shock, breathing difficulties, circulatory collapse, and death.

☠ *Acute Health Effects*

The following acute (short-term) health effects may occur immediately or shortly after exposure to oxalic acid:

Skin: Irritation/rash or burning feeling on contact with redness, blistering and vesiculation, and localized tissue destruction.

Eye: Severe irritation; may cause burns and damage with permanent corneal scarring.

Lung: Severe irritation of the mucous membranes in the nose, throat, and lungs following exposure.

CNS: Inconclusive (possible nervousness, irritation, and weight loss).

Chronic Health Effects

The following chronic (long-term) health effects may occur at some time after exposure to oxalic acid and can last for months or even years:

Cancer Hazards: According to information presented in the references, oxalic acid has not been adequately tested for its ability to cause cancer in test animals.

Reproductive Hazard: According to information presented in the references, oxalic acid has not been adequately tested for its ability to adversely affect reproduction in test animals.

Other Chronic Effects: Skin contact causes irritation and burns. There may be some lasting effects of chronic exposures such as sensitization and/or dermatitis. Once an individual becomes "sensitized," even the slightest future exposures can cause an allergic reaction.

Recommended Risk-Reduction Measures

Personnel should avoid direct contact with oxalic acid. It is extremely corrosive and highly toxic to humans. If a less toxic chemical cannot be substituted for a hazardous substance, then *engineering controls* are the most effective method of reducing exposures.

The best protection is to enclose operations and/or provide local exhaust ventilation at the site of chemical release. While not always operationally feasible, isolating operations can also reduce exposure. Using respiratory protection is less effective than the controls mentioned above, but is still advisable whenever working with or around oxalic acid. A self-contained breathing apparatus (SCBA) or a supplied-air respirator with full facepiece operated in pressure demand mode are the recommended respiratory protection methods of choice. If a full facepiece is not available, then chemical goggles should be worn to protect the eyes. For low concentrations (near the 1 mg/m^3 exposure limit), an acid gas cartridge respirator (equipped with a mist/vapor filter) is acceptable protection. Whenever a chemical splash hazard exists, a face shield and a rubber apron should be worn. To prevent hand and skin exposures, acid resistant gloves should be used. Natural rubber, neoprene, or polyvinyl chloride are acceptable materials.

Administrative controls should also be in place to minimize the potential for human exposures. These may include written procedures or policies which specify the methods and techniques that will be practiced whenever personnel are to work with oxalic acid.

All personnel should receive *training* on the use, hazards, protective measures, emergency actions, and other precautions per 29 CFR 1910.1200 (Hazard Communication), prior to the first assignment in an area where oxalic acid is used or stored.

Before beginning employment and at regular intervals thereafter (e.g., annually), the following tests are recommended:

- ☑ Lung function tests (look for chronic respiratory diseases).
- ☑ Kidney function tests.

If symptoms occur or overexposure is suspected, the following additional tests are suggested:

- ☑ Evaluation by a qualified allergist with careful consideration of exposure history and special testing (may help diagnose skin allergy).

Other methods to reduce exposure include:

- ☑ Where possible, enclose operations and use local exhaust ventilation at the site of chemical release. If local exhaust ventilation or enclosure is not used, respiratory protection should be mandatory.
- ☑ Always ensure that proper protective clothing is worn when using chemical substances and that personnel are trained in its use and care.
- ☑ Wash thoroughly immediately after exposure to oxalic acid and at the end of the work shift or before eating, drinking, or smoking.
- ☑ Hazard warning information should be posted in the work area. In addition, as part of an on-going education and training program, all information on the health and safety hazards of oxalic acid should be communicated to all exposed workers.

RISK ASSESSMENT: ENVIRONMENT

General Assessment

The environment is at risk of exposure during transportation, storage, disposal, or destruction of oxalic acid. In almost every scenario, the threat of environmental exposure is contingent upon the proper handling of the chemical substance. Accidental spills, large or small, can result in fire, explosion, and possible contamination of the surrounding environmental mediums (water, soil, and air).

Oxalic acid is considered a combustible solid. It is also extremely corrosive to many materials. It is incompatible with oxidizing agents and contact can cause fire or explosion. These characteristics require special consideration during any emergency situation involving a leak or spill of oxalic acid. Should oxalic acid ever come into contact with incompatible substances such as oxidizers, alkalis, chlorites, or silver compounds either during use, transportation, or storage, the formation of highly toxic and/or highly explosive commodities is extremely possible. Oxalic acid can enter the environment mainly through industrial discharges or spills.

Acute Ecological Effects

Acute (short-term) toxic effects may include the death of animals, birds, or fish and death or low growth rate

in plants. Acute effects are seen 2 to 4 days after animals or plants are exposed to oxalic acid.

Insufficient data are available on the short-term effects of oxalic acid exposure to aquatic life, plants, birds, or land animals.

Chronic Ecological Effects

Chronic (long-term) toxic effects may include shortened life span, reproductive problems, lower fertility, and changes in appearance or behavior in exposed animals. These effects can be seen long after first exposure(s) to toxic chemicals. Insufficient data are available on the long-term effects of oxalic acid to aquatic life, plants, birds, or land animals.

Water Solubility

Oxalic acid is highly soluble in water. Concentrations of over 1000 milligrams will mix with a liter of water.

Persistence in the Environment

Oxalic acid is slightly persistent in the aquatic environment, with a half-life of between 2 to 20 days. The half-life of a pollutant is the amount of time it takes for one half of the chemical to be degraded.

Bioaccumulation in Aquatic Organisms

Some substances increase in concentration, or *bioaccumulate*, in living organisms as they breathe contaminated air, drink contaminated water, or eat contaminated food. These chemicals can become concentrated in the tissues and internal organs of animals as well as humans.

The concentration of oxalic acid found in fish tissues is expected to be about the same as the average concentration of oxalic acid in the water from which the fish was taken.

Recommended Risk-Reduction Measures

Proper training of all transporters will reduce the likelihood of a mishap or accident resulting in a leak or spill to the environment. The correct use of DOT labeling (per the MSDS) on all containers, trucks, and rail cars will enable emergency responders to react properly and quickly to any disaster thereby reducing the potential risk to the environment and to personnel.

Storage of oxalic acid should be segregated from incompatible chemicals to minimize the risk of cross-contamination or contact. Buildings designated for storage should be equipped with appropriate fire protection systems (alarms, sprinklers, emergency lighting, portable extinguishers).

If a spill or leak to the environment has occurred, fire department, emergency response and/or hazardous materials spill personnel should be notified immediately. Cleanup should be attempted only by those trained in proper spill containment procedures. Contaminated soils should be removed for incineration and replaced with clean soil. If oxalic acid should contact the water table, aquifer, or navigable waterway, time is of the essence. It is highly miscible in water and, therefore, total containment and remediation may not be entirely possible. When such spills occur, the local and/or state emergency response authorities must be notified. A comprehensive emergency response or disaster preparedness/recovery plan should be in place prior to any operations involving the use, transportation, storage, or disposal of oxalic acid.

If oxalic acid is spilled or leaked, the following specific steps are recommended:

- ☑ Restrict persons not wearing protective clothing from area of spill or leak until cleanup is complete. Use non-sparking tools to cleanup area.
- ☑ Remove all ignition sources.
- ☑ Ventilate area of spill or leak.
- ☑ Collect solid materials using a vacuum equipped with a high-efficiency particulate air (HEPA) filter. Do NOT dry sweep (generates airborne dusts). Damp mop residues and flush area with water. Do NOT flush to drains, sewers, or waterways. Collect water for disposal.
- ☑ It may be necessary to dispose of oxalic acid as a hazardous waste. The responsible state agency or the regional office of the federal Environmental Protection Agency (EPA) should be contacted for specific recommendations.

RISK ASSESSMENT: BUSINESS

General Assessment

Accidents or mishaps involving oxalic acid can present a significant threat to business operations. The loss or damage of equipment or facilities can significantly affect fiscal viability. Lawsuits that may result from personnel injury/death, public exposures, and/or environmental contamination will also require a serious expenditure of resources. Media attention surrounding an

injury, death, or environmental damage can also result in a loss of profits and loss of current as well as future business.

Ü *Recommended Risk-Reduction Measures*

Company attorneys, safety and health professionals, and environmental specialists should be involved in the development of any procedures dealing with safety or emergency response. A company official should be pre-designated as a public relations officer with specific training in dealing with the press. Corporate plans and policies should be developed, approved, and implemented long before any need for such arises.

REFERENCES

Ahlbom, A. 1993. *Biostatistics for Epidemiologists.* New York: CRC Press/Lewis Publishers.

American Conference of Governmental Industrial Hygienists. 1988. *Documentation of the Threshold Limit Values and Biological Exposure Indices,* 5th Edition (with updates). Cincinnati: ACGIH

Asante-Duah, D. K. 1993. *Hazardous Waste Risk Assessment.* New York: CRC Press/Lewis Publishers.

Bloom, A. D. and F. J. DeSerres. 1995. *Ecotoxicity and Human Health.* New York: CRC Press/Lewis Publishers.

Bueche, J. R. 1972. *Principles of Physics,* 2nd Edition. New York: McGraw-Hill.

Burrell, R., D. K. Flaherty, and Leonard J. Sauers. 1992. *Toxicology of the Immune System.* New York: Van Nostrand Reinhold.

Calabrese, E. J., E. M. Kenyon. *Air Toxics and Risk Assessment.* New York: CRC Press/Lewis Publishers.

Cockerman, L. G., B. S. Shane. 1994. *Basic Environmental Toxicology.* New York: CRC Press/Lewis Publishers.

Cothern, C. R. 1995. *Handbook for Environmental Risk Decisionmaking.* New York: CRC Press/Lewis Publishers.

Danse, I. R. 1991. *Common Sense Toxics in the Workplace.* New York: Van Nostrand Reinhold.

Gardner, R. W. 1994. *Chemical Intolerance.* New York: CRC Press/Lewis Publishers.

Hallenbeck, W. H. 1993. *Quantitative Risk Assessment for Environmental and Occupational Health,* 2nd Edition. New York: CRC Press/Lewis Publishers.

Hosty, J. W., P. Foster. 1990. *A Practical Guide to Chemical Spill Response.* New York: Van Nostrand Reinhold.

Howard, P. H., et. al. 1991. *Handbook of Environmental Degradation Rates.* New York: CRC Press/Lewis Publishers.

Jones, F. E. 1994. *Toxic Organic Vapors in the Workplace.* New York: CRC Press/Lewis Publishers.

Keith, L. H. and M. M. Walker. 1995. *Handbook of Air Toxics.* New York: CRC Press/Lewis Publishers.

Landis, W. G., M. H. Yu. 1995. *Introduction to Environmental Toxicology.* New York: CRC Press/Lewis Publishers.

Langerman, N. 1994 *Precautionary Labels for Chemical Containers.* New York: CRC Press/Lewis Publishers.

Lauwerys, R. R., P. Hoet. 1993. *Industrial Chemical Exposure.* New York: CRC Press/Lewis Publishers.

Lewis, R. J., Sr. 1993. *Hawley's Condensed Chemical Dictionary,* 12th Edition. New York: Van Nostrand Reinhold.

Lewis, R. J., Sr. 1992. *Sax's Dangerous Properties of Industrial Materials,* Volumes 1, 2, and 3, 8th Edition. New York: Van Nostrand Reinhold.

Manahan, S. E. 1992. *Toxicological Chemistry,* 2nd Edition. New York: CRC Press/Lewis Publishers.

Mellström, G. A., J. E. Wahlberg, and H. I. Maibach. 1994. *Protective Gloves for Occupational Use.* New York: CRC Press/Lewis Publishers.

Meyer, E. 1990. *Chemistry of Hazardous Materials.* Englewood Cliffs, NJ: Prentice Hall.

Mycock, J. C., J. D. McKenna, and L. Theodore. 1995. *Handbook of Air Pollution Control Engineering and Technology.* New York: CRC Press/Lewis Publishers.

Nelson, G. O. 1992. *Gas Mixtures.* New York: CRC Press/Lewis Publishers.

Ott, W. R. 1995. *Environmental Statistics and Data Analysis.* New York: CRC Press/Lewis Publishers.

Patnaik, P. 1992. *A Comprehensive Guide to the Hazardous Properties of Chemical Substances.* New York: Van Nostrand Reinhold.

Philp, R. B. 1995. *Environmental Hazards and Human Health.* New York: CRC Press/Lewis Publishers.

Quigley, D. R. 1994. *Handbook of Emergency Chemical Management.* New York: CRC Press/Lewis Publishers.

Rea, W. J. 1995. *Chemical Sensitivity,* Vol. I-III. New York: CRC Press/Lewis Publishers.

Richardson, M. 1995. *Dictionary of Substance and Their Effects,* Vol. I-VI. New York: CRC Press/Lewis Publishers.

Rylander, R., R. R. Jacobs. 1994. *Organic Dusts.* New York: CRC Press/Lewis Publishers.

Sacarello, H. L. A. 1994. *The Comprehensive Handbook of Hazardous Materials.* New York: CRC Press/Lewis Publishers.

Sheftel, V. O. 1995. *Handbook of Toxic Properties of Monomers and Additives.* New York: CRC Press/Lewis Publishers.

Sherman, J. D. 1988. *Chemical Exposure and Disease.* New York: Van Nostrand Reinhold.

Talbott, E., G. Craun. 1995. *An Introduction to Environmental Epidemiology.* New York: CRC Press/Lewis Publishers.

U.S. Department of Health and Human Services, National Institute for Occupational Safety and Health. 1994. *NIOSH Pocket Guide to Chemical Hazards.* Washington, D.C.: U.S. Government Printing Office.

Witorsch, P. and S. V. Spagnolo. 1994. *Air Pollution and Lung Disease in Adults.* New York: CRC Press/Lewis Publishers.

MATERIAL SAFETY DATA SHEET

CHEMICAL NAME

OXYGEN DIFLUORIDE

HAZARD WARNING INFORMATION

HEALTH	FIRE	REACTIVE	OTHER	DEGREE OF HAZARD	COLOR CODING	OTHER CODES
4	1	4		0 = Minimum Hazard 1 = Slight Hazard 2 = Moderate Hazard 3 = Serious Hazard 4 = Severe Hazard	HEALTH = BLUE FIRE = RED REACTIVITY = YELLOW OTHER = WHITE	OX = Oxidizer ACID = Acid ALK = Alkali COR = Corrosive W = Use No Water

SECTION I - GENERAL INFORMATION

Characterization	**Fluoride**	RCRA Number / EPA Class	**None** / **Not Applicable**
DOT Proper Shipping Name	**Oxygen Difluoride**	Chemical Abstract Service (CAS) Number	**7783-41-7**
DOT Hazard Class and Label Requirements	**Poison A; Poison Gas**	DOT Emergency Guide Code	**20**
DOT Identification Number	**UN 2190**	Chemical Formula	**OF_2**

Synonyms

Difluorine monoxide; fluorine monoxide; oxygen fluoride.

SECTION II - HAZARDOUS INGREDIENTS AND IDENTITY INFORMATION

Hazardous Components (specific identity)	OSHA Exposure Criteria	NIOSH Exposure Criteria	Immediately Dangerous to Life and Health (IDLH)	ACGIH Exposure Criteria
Oxygen difluoride (derivation: By slowly passing fluoride through an aqueous solution of sodium hydroxide). **1 ppm = 2.24 mg/m^3**	**PEL:** **0.05 ppm** **0.1 mg/m^3** **STEL:** **Not Established**	**REL (ceiling):** **0.05 ppm** **0.1 mg/m^3** **STEL:** **Not Applicable**	**0.5 ppm**	**TLV (ceiling):** **0.05 ppm** **0.1 mg/m^3** **STEL:** **Not Applicable**

SECTION III - PHYSICAL AND CHEMICAL CHARACTERISTICS

Boiling Point	**-230°F (-145°C)**	Specific Gravity (H_2O = 1)	**1.88**
Vapor Pressure (atmospheres)	**>1 at 68°F (20°C)**	Molecular Weight	**54.0**
Vapor Density (Air = 1)	**1.9**	Melting Point	**-371°F (-224°C)**

Solubility

Slightly soluble (0.02%), but reacts slowly with water. Slightly soluble in alcohol.

Appearance and Odor

A colorless gas or a yellow-brown liquid with a peculiar, foul odor.

SECTION IV - FIRE AND EXPLOSION HAZARD DATA

Flash Point (method used)	**Not Determined**	Explosive Limits in Air % by Volume	**LEL: Not Determined UEL: Not Determined**
NFPA Classification	**Non-Flammable Gas**	Autoignition Temperature	**Not Determined**

Extinguishing Media

Use agents suitable to surrounding fire. Oxygen difluoride does not burn but is a powerful oxidizer.

Special Fire Fighting Procedures

Wear full protective clothing and self-contained breathing apparatus (SCBA). Also, use a water spray to keep fire-exposed containers cool. Keep water from contacting oxygen difluoride directly since it will react slowly to form hydrofluoric acid. Poisonous gases are produced in fire.

Unusual Fire and Explosion Hazards

Oxygen difluoride is a dangerous explosion hazard. Firefighters should avoid contact with vapors produced during fire. In the presence of water, moist air, or combustible gases (carbon monoxide, hydrogen, methane), oxygen difluoride can produce spark-sensitive mixtures.

SECTION V - REACTIVITY DATA

Stability		Conditions to Avoid
		Oxygen difluoride is normally NOT stable unless it is handled and stored with extreme care to avoid contact with incompatible materials (including water and moist air). It is extremely reactive.
Stable	Unstable	Incompatibility (*materials to avoid*)
	X	Oxygen difluoride is incompatible with combustible materials (wood, paper, oil, gases), oxidizers (bromine, chlorine), iodine, platinum, metal oxides, moist air, hydrogen, sulfide, hydrocarbons, water.
Hazardous Polymerization		Conditions to Avoid
		Under normal conditions of temperature and pressure, hazardous polymerization of oxygen difluoride is not expected to occur.
May Occur	Will Not Occur	Hazardous Decomposition or By-products
	X	When heated to decomposition, oxygen difluoride emits highly toxic/poisonous gases, including hydrogen fluoride gas and oxides of fluorine.

SECTION VI - HEALTH HAZARD DATA

Primary Route(s) of Entry:	Inhalation?	Absorption (skin)?	Ingestion?
	X		

Health Hazards

INHALATION: Vapors cause irritation to the upper respiratory tract with coughing, dypsnea, headache, congestion, fever, myalgia (muscle pain), elevated lymphocyte count, bone changes (due to action of fluoride ion), congestion, CNS effects, and/or delayed pulmonary edema.

SKIN & EYES: Skin contact with compressed gas can cause severe, painful burns. Eye contact causes burning, watering, redness, corneal ulceration, blurred vision, and possible blindness.

INGESTION: Not likely (oxygen difluoride is a gas at room temperature).

Carcinogenicity	NTP Listed?	IARC Cancer Review Group?	OSHA Regulated?	Target Organs?
Unknown Human Unknown Animal	No	No	29 CFR 1910.1000 Table Z-1	Respiratory system, eyes, skin, CNS.

Medical Conditions Generally Aggravated by Exposure

None reported.

Emergency and First-aid Procedures

Eye contact: Flush immediately with water for 15 minutes (minimum); seek medical attention. Do not allow victims to rub their eyes. Skin contact: Remove all contaminated clothing. Wash area with water for a minimum of 15 minutes then use soap and water and wash again. Seek medical assistance. For inhalation: Remove the person from exposure. Provide respiratory assistance and CPR. Transfer to medical facility. Observe 48 hours for lung effects. If swallowed: Not likely. It is normally a gas. Even as a pressurized liquid it rapidly evaporates at room temperature.

SECTION VII - PRECAUTIONS FOR SAFE HANDLING AND USE

Steps to be Taken in Case Material is Released or Spilled

Ventilate area (avoid vapors). Shut off flow of gas if possible. Provide maximum explosion-proof ventilation of the area of spill and remove all ignition sources (hydrogen gas may be present). Restrict those not involved in cleanup from entering area.

Preferred Waste Disposal Method

No citation.

Precautions to be Taken in Handling and Storage

Avoid contact with incompatible materials since violent reactions can occur. Store in tightly closed containers upright in cool, dry, well-ventilated area away from heat and moisture.

Other Precautions and Warnings

Sources of ignition are prohibited where oxygen difluoride is used, handled, or stored. Detached facility storage is preferred. Prevent damage to containers. Use non-sparking tools around containers. Electrically ground and bond all containers and equipment. Do not drag or slide containers.

SECTION VIII - CONTROL MEASURES AND PERSONAL PROTECTIVE EQUIPMENT

Respiratory Protection (*specify type*)

Exposure levels are relatively low. Recommend using a supplied-air respirator or a self-contained breathing apparatus (SCBA) operated in pressure demand or other positive pressure mode.

Ventilation

Local exhaust or general mechanical systems recommended.

Protective Gloves	Eye Protection	Other Protective Clothing
Impervious Material	Chemical Goggles or Face Mask	Protective Apron

Work/Hygiene Practices

Always wash hands thoroughly after using chemical; never bring food, drink, or smoking materials into vicinity of chemicals.

OXYGEN DIFLUORIDE

OF_2

CAS: 7783-41-7

IDENTIFICATION AND TYPICAL USES

Oxygen difluoride is a colorless gas or a yellow-brown liquid (under pressure) with a peculiar, foul odor. It is normally shipped as a non-liquefied compressed gas. It is primarily used as an oxidizer for rocket propellants.

RISK ASSESSMENT: HEALTH

General Assessment

As a gas, oxygen difluoride is a deadly human poison by *inhalation*. It is more poisonous than fluorine. Inhalation of just a few breaths may be fatal. It can cause serious burns on skin contact with the compressed gas, but dermal absorption has not been reported. Ingestion is not probable since oxygen difluoride exists as a gas at room temperatures. Even the compressed liquid vaporizes so rapidly that ingestion is not likely to occur. It is highly caustic and, therefore, an extremely corrosive and destructive compound. There are no reports of any carcinogenic, teratogenic, reproductive, or mutagenic effects resulting from exposure to oxygen difluoride.

Inhalation severely dehydrates tissues and causes destructive corrosive damage to the nose, throat, and mucosa of the upper respiratory tract. There can be coughing, dypsnea (labored breathing), lachrymation, and nausea. Severe exposures may cause pulmonary edema (fluid in the lungs). The development of pulmonary edema can be delayed up to 48 hours after exposure, thereby creating a false sense of security with regard to health exposure risk.

Skin contact with the compressed gas or liquid causes severe burning due to oxygen difluoride's intense corrosive action on human tissue. The liquid may continue to burn and destroy skin tissues long after the exposure if it is not properly removed. Because of the fluoride ion's affinity for calcium, burns may be bone-deep in some instances. The extremely corrosive properties of oxygen difluoride can cause immediate stinging, burning, redness, swelling, membrane ulceration, painful skin blistering, and local necrosis (tissue destruction) upon contact. Eye contact causes immediate stinging and watering with tissue damage and the possibility of corneal damage. There will be redness and swelling of the lids, intense pain, and tissue destruction. Loss of vision is likely.

Ingestion is extremely rare under normal operating conditions. However, if liquid solutions are accidentally swallowed, there can be severe burns to the lips, mouth, throat, and stomach. If vomiting occurs and the liquid is aspirated into the lungs, the corrosive action on the lung tissues can be fatal.

Acute Health Effects

The following acute (short-term) health effects may occur immediately or shortly after exposure to oxygen difluoride:

Skin: Irritation/rash or burning feeling on contact with redness, blistering, localized tissue destruction, and deep, painful burns.

Eye: Severe irritation; may cause burns and damage with permanent loss of vision.

Lung: Severe irritation of the mucous membranes in the nose, throat, and lungs following exposure. Pulmonary edema is likely to occur within 48 hours of exposure.

CNS: Unspecified adverse action on the central nervous system.

Chronic Health Effects

The following chronic (long-term) health effects may occur at some time after exposure to oxygen difluoride and can last for months or even years:

Cancer Hazards: According to information presented in the references, oxygen difluoride has not been adequately tested for its ability to cause cancer in laboratory test animals. Additional research is required in this regard. Its acute toxic action is so harmful that long-term effects have not been well-documented.

Reproductive Hazard: According to information presented in the references, oxygen difluoride has not been shown to cause teratogenic or other adverse reproductive effects in test animals.

Other Chronic Effects: None reported.

Recommended Risk-Reduction Measures

Personnel should avoid direct contact with oxygen difluoride. It is extremely corrosive and highly poisonous to humans. If a less toxic chemical cannot be substituted for a hazardous substance, then *engineering controls* are the most effective method of reducing exposures.

The best protection is to enclose operations and/or provide local exhaust ventilation at the site of chemical release. While not always operationally feasible, isolating operations can also reduce exposure. Using respiratory protection is less effective than the controls mentioned above, but is still advisable whenever working with or around oxygen difluoride. The exposure level is extremely low (0.05 ppm) and difficult to accurately monitor. While an air-purifying chemical cartridge respirator or gas mask may suffice for low or infrequent exposures, the best protection is obtained using a self-contained breathing apparatus (SCBA) or a supplied-air respirator with full facepiece operated in pressure demand mode. If a full facepiece is not available, then chemical goggles should be worn to protect the eyes. Whenever a chemical splash hazard exists, a face shield and a protective apron should be worn. To prevent hand and skin exposures, protective gloves should be used. Glove manufacturers should be contacted and permeation studies obtained *before* final glove selections are made.

Administrative controls should also be in place to minimize the potential for human exposures. These may include written procedures or policies which specify the methods and techniques that will be practiced whenever personnel are to work with oxygen difluoride.

All personnel should receive *training* on the use, hazards, protective measures, emergency actions, and other precautions per 29 CFR 1910.1200 (Hazard Communication), prior to the first assignment in an area where oxygen difluoride is used or stored.

Before beginning employment and at regular intervals thereafter (e.g., annually), the following tests are recommended:

- ☑ Lung function tests (establish baseline).

If symptoms occur or overexposure is suspected, the following additional tests are suggested:

- ☑ Lung function tests (compare to baseline).
- ☑ Consider chest X-ray following acute overexposure (may be negative it taken immediately after exposure due to delayed onset of pulmonary edema).

Any evaluation should include a careful history of past and present symptoms with an examination. Medical tests that look for damage already done are not a substitute for controlling exposure. Also, because smoking can cause heart disease, emphysema, lung cancer, and other respiratory disorders, smokers exposed to oxygen difluoride may experience symptoms more quickly than non-smokers under the same conditions of exposure. Prudent risk management requires proper consideration of *all* possible exposure factors.

Other methods to reduce exposure include:

- ☑ Where possible, enclose operations and use local exhaust ventilation at the site of chemical release. If local exhaust ventilation or enclosure is not used, respiratory protection should be mandatory.
- ☑ Always ensure that proper protective clothing is worn when using chemical substances and that personnel are trained in its use, care, and maintenance.
- ☑ Wash thoroughly immediately after exposure to oxygen difluoride and at the end of the work shift or before eating, drinking, or smoking.
- ☑ Hazard warning information should be posted in the work area. In addition, as part of an on-going education and training program, all information on the health and safety hazards of oxygen difluoride should be communicated to all exposed workers.

RISK ASSESSMENT: ENVIRONMENT

General Assessment

The environment is at risk of exposure during transportation, storage, disposal, or destruction of oxygen difluoride. In almost every scenario, the threat of environmental exposure is contingent upon the proper handling of the chemical substance. Accidental spills, large or small, can result in fire, explosion, and possible contamination of the surrounding environmental mediums (water, soil, and especially air).

Oxygen difluoride is considered a non-flammable gas. However, it is a powerful oxidizer and extremely corrosive to many materials. It is incompatible with water or steam and will liberate hydrofluoric acid and explosive hydrogen gas and heat on contact. It is incompatible with most metal oxides, combustible materials (wood, paper) and combustible gases (carbon monoxide, hydrogen, methane), chlorine, bromine, iodine, platinum, hydrogen sulfide, and hydrocarbons. These characteristics require special consideration during any emergency situation involving a leak or spill of oxygen difluoride. Should oxygen difluoride ever come into contact with any of these incompatible substances either during use, transportation, or storage, the formation of highly toxic and/or highly explosive commodities is extremely possible.

Oxygen difluoride can enter the environment mainly industrial discharges or spills.

Acute Ecological Effects

Acute (short-term) toxic effects may include the death of animals, birds, or fish and death or low growth rate in plants. Acute effects are seen 2 to 4 days after animals or plants are exposed to oxygen difluoride.

Insufficient data are available to evaluate or predict the short-term effects of oxygen difluoride exposure to aquatic life, plants, birds, or land animals.

Chronic Ecological Effects

Chronic (long-term) toxic effects may include shortened life span, reproductive problems, lower fertility, and changes in appearance or behavior in exposed animals. These effects can be seen long after first exposure(s) to toxic chemicals.

Insufficient data are available to evaluate or predict the long-term effects of oxygen difluoride to aquatic life, plants, birds, or land animals.

Water Solubility

Oxygen difluoride is slightly soluble in water, but reacts slowly to form hydrofluoric acid which is highly soluble in water. Concentrations of over 1000 milligrams will mix with a liter of water.

Persistence in the Environment

Oxygen difluoride is slightly persistent in the aquatic environment, with a half-life of less than 2 and 20 days. The half-life of a pollutant is the amount of time it takes for one half of the chemical to be degraded.

Bioaccumulation in Aquatic Organisms

Some substances increase in concentration, or *bioaccumulate,* in living organisms as they breathe contaminated air, drink contaminated water, or eat contaminated food. These chemicals can become concentrated in the tissues and internal organs of animals as well as humans.

Small quantities of such chemicals as oxygen difluoride may be neutralized by the alkalinity in aquatic systems, but larger quantities can lower the pH for extended periods of time. The concentration of oxygen difluoride found in the edible tissues of most aquatic species that are consumed by humans is expected to be about the same as the average concentration in the water from which the fish was taken.

Recommended Risk-Reduction Measures

Proper training of all transporters will reduce the likelihood of a mishap or accident resulting in a leak or spill to the environment. The correct use of DOT labeling (per the MSDS) on all containers, trucks, and rail cars will enable emergency responders to react properly and quickly to any disaster thereby reducing the potential risk to the environment and to personnel.

Storage of oxygen difluoride should be segregated from incompatible chemicals to minimize the risk of cross-contamination or contact. Buildings designated for storage should be equipped with appropriate fire protection systems (alarms, sprinklers, emergency lighting, portable extinguishers). Containers should be protected against physical damage and stored away from direct sunlight or other heat sources. Cylinders must be stored upright and secured properly. They should never be rolled or dragged. Because oxygen difluoride will corrode and/or react with many materials, selection of storage containers requires proper consideration.

If a spill or leak to the environment has occurred, fire department, emergency response and/or hazardous materials spill personnel should be notified immediately. Cleanup should be attempted only by those trained in proper spill containment procedures using non-sparking tools. Contaminated soils (solution spills) should be removed for incineration and replaced with clean soil. If oxygen difluoride should contact the water table, aquifer, or navigable waterway, time is of the essence. It is slightly soluble in water and, therefore, total containment and remediation may not be possible. When such spills occur, the local and/or state emergency response authorities must be notified. A comprehensive emergency response or disaster preparedness/recovery plan should be in place prior to any operations involving the use, transportation, storage, or disposal of oxygen difluoride. If oxygen difluoride is spilled or leaked, the following specific steps are recommended:

- ☑ Restrict persons not wearing protective clothing from area of spill or leak until cleanup is complete and area can be opened for normal work.
- ☑ Remove all ignition sources and provide maximum explosion-proof ventilation.
- ☑ Shut off gas leak, if possible. If leak is from a cylinder and it cannot be turned off, remove cylinder to safe area (outdoors) and allow to vent. Take care to avoid contact with the vapors or any escaping compressed liquids. Be aware of the possibility that hydrogen gas may be in the area (explosion hazard).
- ☑ It may be necessary to dispose of oxygen difluoride as a hazardous waste. The responsible state agency or the regional office of the federal Environmental Protection Agency (EPA) should be contacted for specific recommendations.

RISK ASSESSMENT: BUSINESS

General Assessment

Accidents or mishaps involving oxygen difluoride can present a significant threat to business operations. The loss or damage of equipment or facilities can significantly affect fiscal viability. Lawsuits that may result from personnel injury/death, public exposures, and/or environmental contamination will also require a serious expenditure of resources. Media attention surrounding an injury, death, or environmental damage can also result in a loss of profits and loss of current as well as future business.

Recommended Risk-Reduction Measures

Company attorneys, safety and health professionals, and environmental specialists should be involved in the development of any procedures dealing with safety or emergency response. A company official should be pre-designated as a public relations officer with specific training in dealing with the press. Corporate plans and policies should be developed, approved, and implemented long before any need for such arises.

REFERENCES

Ahlbom, A. 1993. *Biostatistics for Epidemiologists.* New York: CRC Press/Lewis Publishers.

American Conference of Governmental Industrial Hygienists. 1988. *Documentation of the Threshold Limit Values and Biological Exposure Indices,* 5th Edition (with updates). Cincinnati: ACGIH

Anthony, C. P., and N. J. Kolthoff. 1971. *Textbook of Anatomy and Physiology.* St. Louis: C. V. Mosby Company.

Bueche, J. R. 1972. *Principles of Physics,* 2nd Edition. New York: McGraw-Hill.

Burrell, R., D. K. Flaherty, and Leonard J. Sauers. 1992. *Toxicology of the Immune System.* New York: Van Nostrand Reinhold.

Calabrese, E. J., E. M. Kenyon. *Air Toxics and Risk Assessment.* New York: CRC Press/Lewis Publishers.

Cockerman, L. G., B. S. Shane. 1994. *Basic Environmental Toxicology.* New York: CRC Press/Lewis Publishers.

Cothern, C. R., N. P. Ross. 1994. *Environmental Statistics, Assessment, and Forecasting.* New York: CRC Press/Lewis Publishers.

Danse, I. R. 1991. *Common Sense Toxics in the Workplace.* New York: Van Nostrand Reinhold.

Environmental Protection Agency. 1983. *Chemical Hazard Information Profiles.* Washington, D.C.: U.S. EPA.

Fodden, C. R., and J. L. Weddell. 1992. *Hazardous Materials Emergency Action Data.* New York: CRC Press/Lewis Publishers.

Forseberg, K., and L. H. Keith. 1993. *Chemical; Protective Clothing Permeation and Degradation Database.* Chelsea, MI: Lewis Publishers.

Gardner, R. W. 1994. *Chemical Intolerance.* New York: CRC Press/Lewis Publishers.

Howard, P. H., et. al. 1991. *Handbook of Environmental Degradation Rates.* New York: CRC Press/Lewis Publishers.

Keith, L. H. and M. M. Walker. 1995. *Handbook of Air Toxics.* New York: CRC Press/Lewis Publishers.

Lewis, R. J., Sr. 1993. *Hawley's Condensed Chemical Dictionary,* 12th Edition. New York: Van Nostrand Reinhold.

Lewis, R. J., Sr. 1992. *Sax's Dangerous Properties of Industrial Materials,* Volumes 1, 2, and 3, 8th Edition. New York: Van Nostrand Reinhold.

Mackay, D., W. Y. Shiu, and K. C. Ma. 1992. *Illustrated Handbook of Chemical-Physical Properties and Environmental Fate for Organic Chemicals.* New York: CRC Press/Lewis Publishers.

Manahan, S. E. 1994. *Environmental Chemistry,* 6th Edition. New York: CRC Press/Lewis Publishers.

Meyer, E. 1990. *Chemistry of Hazardous Materials.* Englewood Cliffs, NJ: Prentice Hall.

Patnaik, P. 1992. *A Comprehensive Guide to the Hazardous Properties of Chemical Substances.* New York: Van Nostrand Reinhold.

U.S. Department of Health and Human Services, National Institute for Occupational Safety and Health. 1994. *NIOSH Pocket Guide to Chemical Hazards.* Washington, D.C.: U.S. Government Printing Office.

MATERIAL SAFETY DATA SHEET

CHEMICAL NAME	
OZONE	**CRC PRESS/LEWIS PUBLISHERS DISCLAIMER:** THE INFORMATION AND RECOMMENDATIONS PRESENTED HEREIN ARE BASED ON SOURCES BELIEVED TO BE RELIABLE. CRC MAKES NO REPRESENTATION ON ITS COMPLETENESS OR ACCURACY. IT IS THE RESPONSIBILITY OF THE USER TO DETERMINE THE CHEMICAL'S SUITABILITY FOR ITS INTENDED USE, THE CHEMICAL'S SAFE USE, AND THE CHEMICAL'S PROPER DISPOSAL. NO REPRESENTATIONS AND/OR WARRANTIES, EITHER EXPRESSED OR IMPLIED, OF THE MERCHANTABILITY OR FITNESS FOR A PARTICULAR PURPOSE, OR OF ANY OTHER NATURE, ARE MADE WITH RESPECT TO THE INFORMATION PROVIDED IN THIS MSDS OR TO THE CHEMICAL TO WHICH INFORMATION MAY REFER. CRC NEITHER ASSUMES NOR AUTHORIZES ANY OTHER PERSON TO ASSUME FOR IT, ANY OTHER ADDITIONAL RESPONSIBILITY OR LIABILITY FOR THE USE OF, OR RELIANCE UPON, THIS INFORMATION.

HAZARD WARNING INFORMATION

HEALTH	FIRE	REACTIVE	OTHER	DEGREE OF HAZARD	COLOR CODING	OTHER CODES
Gas: 3 Liquid: 1	Gas: 0 Liquid: 0	Gas: 1 Liquid: 3	OX	0 = Minimum Hazard 1 = Slight Hazard 2 = Moderate Hazard 3 = Serious Hazard 4 = Severe Hazard	HEALTH = BLUE FIRE = RED REACTIVITY = YELLOW OTHER = WHITE	OX = Oxidizer ACID = Acid ALK = Alkali COR = Corrosive W = Use No Water

SECTION I - GENERAL INFORMATION

Characterization	RCRA Number	EPA Class
Oxidizers	**D003**	**Characteristic (R) Waste**

DOT Proper Shipping Name **Compressed Liquefied Gas, N.O.S.; Zone A**	Chemical Abstract Service (CAS) Number **10028-15-6**
DOT Hazard Class and Label Requirements **Poison A; Poison Gas**	DOT Emergency Guide Code **20**
DOT Identification Number **UN 1955**	Chemical Formula **O_3**

Synonyms
Triatomic oxygen.

SECTION II - HAZARDOUS INGREDIENTS AND IDENTITY INFORMATION

Hazardous Components (specific identity)	OSHA Exposure Criteria	NIOSH Exposure Criteria	Immediately Dangerous to Life and Health (IDLH)	ACGIH Exposure Criteria
Ozone (derivation: Occurs in the atmosphere from UV light action on oxygen at high altitudes, and by lightning action on air. Also occurs in automobile engines and by electrolysis of alkaline perchlorate solutions. Derived by the passage of air or oxygen between electrodes at high voltage). **1 ppm = 2.00 mg/m^3**	**PEL:** **0.1 ppm** **0.2 mg/m^3** **STEL (15-min):** **0.3 ppm** **0.6 mg/m^3**	**REL (ceiling):** **0.1 ppm** **0.2 mg/m^3** **STEL:** **Not Applicable**	**5 ppm**	**TLV (ceiling):** **0.1 ppm** **0.2 mg/m^3** **STEL:** **Not Applicable**

SECTION III - PHYSICAL AND CHEMICAL CHARACTERISTICS

Boiling Point **-169°F (-111°C)**	Specific Gravity (H_2O = 1) **1.66 (Gas)**
Vapor Pressure (atmospheres) **>1 at 68°F (20°C)**	Molecular Weight **54.0**
Vapor Density (Air = 1) **1.655**	Melting Point **-315°F (-193°C)**

Solubility
Very slightly soluble (0.001%). Soluble in alkaline solvents.

Appearance and Odor
Normally a colorless to blue gas with a very pungent odor. Can appear as a dark blue liquid at -169 to -315°F (-111 to -193°C), or bluish-black crystals below -315°F (-193°C). Odor Threshold = 0.25ppm.

SECTION IV - FIRE AND EXPLOSION HAZARD DATA

Flash Point (method used) **Not Determined**	Explosive Limits in Air % by Volume **LEL: Not Determined UEL: Not Determined**
NFPA Classification **Non-Flammable Gas**	Autoignition Temperature **Not Determined**

Extinguishing Media
Use agents suitable to surrounding fire. Ozone does not burn but is a powerful oxidizer.

Special Fire Fighting Procedures
Wear full protective clothing and self-contained breathing apparatus (SCBA). Also, use a water spray to keep fire-exposed containers cool. Poisonous gases are produced in fire.

Unusual Fire and Explosion Hazards
Ozone is a dangerous explosion hazard. The liquid and solid form can explode violently when shocked or exposed to heat or fire. Containers may explode in fire. Decomposition of ozone into oxygen can increase the burning rate of fires.

SECTION V - REACTIVITY DATA

Stability		Conditions to Avoid Ozone is unstable at normal temperatures and readily decomposes to biatomic oxygen. Its liquid and solid forms can explode when shocked, exposed to heat or flame, or by contact with incompatible materials.
Stable	Unstable X	Incompatibility *(materials to avoid)* Incompatible with rubber and reacts with non-saturated organic compounds to produce ozonides which are unstable. Incompatible with acetylene, alkenes, alkyl metals, benzene, aniline, and bromine.
Hazardous Polymerization		Conditions to Avoid Under normal conditions of temperature and pressure, hazardous polymerization of ozone is not expected to occur.
May Occur	Will Not Occur X	Hazardous Decomposition or By-products When heated to decomposition, ozone emits oxygen which can greatly accelerate the intensity of fire.

SECTION VI - HEALTH HAZARD DATA

Primary Route(s) of Entry:	Inhalation?	Absorption (skin)?	Ingestion?
	X		

Health Hazards

INHALATION: Irritation of the eyes, nose, throat, and upper respiratory tract with coughing, dypsnea, visual disturbances, lachrymation, headache, decreased pulse rate and blood pressure, loss of coordination, chest pains, fatigue, possible hemorrhage, and pulmonary edema which can be fatal. There may be blood and chromosomal changes.

SKIN & EYES: Skin or eye contact with compressed gas can cause severe, painful burns (frostbite).

INGESTION: Not likely (ozone is a gas at room temperature).

Carcinogenicity	NTP Listed?	IARC Cancer Review Group?	OSHA Regulated?	Target Organs?
Unknown Human Unknown Animal	No	No	29 CFR 1910.1000 Table Z-1	Respiratory system, eyes, skin.

Medical Conditions Generally Aggravated by Exposure

None reported.

Emergency and First-aid Procedures

Eye contact: Flush immediately with water for 15 minutes (minimum); seek medical attention. Do not allow victims to rub their eyes. Skin contact: Remove all contaminated clothing. Wash area with water for a minimum of 15 minutes then use soap and water and wash again. Seek medical assistance. For inhalation: Remove the person from exposure. Provide respiratory assistance and CPR. Transfer to medical facility. Observe 48 hours for lung effects. If swallowed: Not likely. It is normally a gas. DOCTOR: Detection of serum lactate dehydrogenase may indicate lung permeability due to ozone.

SECTION VII - PRECAUTIONS FOR SAFE HANDLING AND USE

Steps to be Taken in Case Material is Released or Spilled

Ventilate area (avoid vapors). Shut off flow of gas if possible. If not, remove cylinder to safe location (outdoors) and allow to vent until empty. Stay upwind of spill area. Remove all ignition sources. Restrict those not involved in cleanup from entering area.

Preferred Waste Disposal Method

No citation.

Precautions to be Taken in Handling and Storage

Avoid contact with incompatible materials since violent reactions can occur. Store in tightly closed containers upright in cool, dry, well-ventilated area away from heat and moisture.

Other Precautions and Warnings

Sources of ignition are prohibited where ozone is used, handled, or stored. Detached facility storage is preferred. Prevent damage to containers. Refrigerated storage area is recommended. Do not drag or slide containers. Electrically ground and bond all equipment.

SECTION VIII - CONTROL MEASURES AND PERSONAL PROTECTIVE EQUIPMENT

Respiratory Protection *(specify type)*

Exposure levels are relatively low. Recommend using a supplied-air respirator or a self-contained breathing apparatus (SCBA) operated in pressure demand or other positive pressure mode.

Ventilation

Local exhaust or general mechanical systems recommended.

Protective Gloves	Eye Protection	Other Protective Clothing
Impervious Material	Chemical Goggles or Face Mask	Protective Apron

Work/Hygiene Practices

Always wash hands thoroughly after using chemical; never bring food, drink, or smoking materials into vicinity of chemicals.

OZONE

O_3

CAS: 10028-15-6

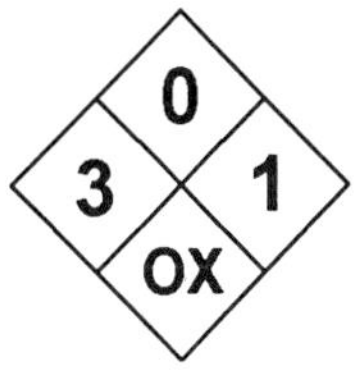

Gas

Liquid

IDENTIFICATION AND TYPICAL USES

Ozone is normally a colorless to blue gas with a pungent odor. It can also appear as a dark blue liquid at temperatures of -169° to -315°F (-112° to -193°C) or a bluish-black crystalline solid below -315°F (-193°C). In every case, ozone typically has a very pungent, irritating odor. It is used in the purification of drinking water, in industrial waste treatment, for deodorizing air and sewage gases, in bleaching waxes, oils, wet paper, and textiles, in the production of peroxides, and bactericides. It is also an oxidizing agent in several chemical processes (acids, aldehydes, ketones from unsaturated fatty acids), in steroid hormones, in the removal of chlorine from nitric acid, and in the oxidation of phenols and cyanides.

RISK ASSESSMENT: HEALTH

General Assessment

As a gas, ozone is a toxic human poison by *inhalation*. It is more poisonous than fluorine. Inhalation of ozone at relatively low concentrations can cause death in a short period of time. Its destructive action is very similar to that caused by inhalation of radioactive particulates. It is only slightly water soluble and, therefore, it is not easily soluble in the mucosa along the respiratory tract but passes directly to the smallest components of the lungs (alveoli and bronchioles) to cause destructive damage. It can cause serious burns on skin contact with the compressed gas, but dermal absorption has not been reported. Ingestion is not probable since ozone exists as a gas at room temperatures. Even the compressed liquid vaporizes so rapidly that ingestion is not likely to occur. It is highly caustic and, therefore, an extremely corrosive and destructive compound. There are no reports of any carcinogenic, teratogenic, reproductive, or mutagenic effects resulting from exposure to ozone.

Inhalation can cause nose, throat, and respiratory tract irritation with coughing, difficulty in breathing, lachrymation, headache, decreased pulse rate and blood pressure, loss of coordination, chest pain and soreness, and fatigue. Severe exposures may cause hemorrhage and/or pulmonary edema (fluid in the lungs) which can be fatal. The development of pulmonary edema can be delayed up to 48 hours after exposure, thereby creating a false sense of security with regard to health exposure risk. Experimental studies with human tissues and animals have shown blood changes (alterations in shape of red blood cells thereby increasing the possibility of hemolysis), chromosomal changes, and adverse reproductive effects. Cell membrane damage may also occur and tissue death in the lungs is possible.

Skin contact with the compressed gas or liquid causes severe burning due to ozone's intense corrosive action on human tissue. The extremely corrosive properties of ozone can cause immediate stinging, burning, redness, swelling, membrane ulceration, painful skin blistering, and local necrosis (tissue destruction) upon contact. Eye contact causes immediate stinging and watering with tissue damage and the possibility of corneal damage. There will be redness and swelling of the lids, intense pain, and tissue destruction. Loss of vision is likely.

Acute Health Effects

The following acute (short-term) health effects may occur immediately or shortly after exposure to ozone:

Skin: Irritation/rash or burning feeling on contact with redness, blistering, localized tissue destruction, and deep, painful burns.

Eye: Severe irritation; may cause burns and damage with permanent loss of vision.

Lung: Severe irritation of the mucous membranes in the nose, throat, and lungs following exposure.

Pulmonary edema is likely to occur within 48 hours of exposure.

Chronic Health Effects

The following chronic (long-term) health effects may occur at some time after exposure to ozone and can last for months or even years:

Cancer Hazards: According to information presented in the references, ozone has not been adequately tested for its ability to cause cancer in laboratory test animals. Additional research is required in this regard.

Reproductive Hazard: According to information presented in the references, ozone has not been shown to cause teratogenic or other adverse reproductive effects in test animals.

Other Chronic Effects: Repeated exposure may cause breathing disorders through respiratory tract scarring or premature aging similar to that caused by continued exposure to ionizing radiation.

Recommended Risk-Reduction Measures

Personnel should avoid direct contact with ozone. It is extremely corrosive and highly toxic to humans. If a less toxic chemical cannot be substituted for a hazardous substance, then *engineering controls* are the most effective method of reducing exposures.

The best protection is to enclose operations and/or provide local exhaust ventilation at the site of chemical release. While not always operationally feasible, isolating operations can also reduce exposure. Using respiratory protection is less effective than the controls mentioned above, but is still advisable whenever working with or around ozone. The exposure level is extremely low (0.1 ppm) and difficult to accurately monitor. Exposure to relatively low concentrations of ozone gas can cause death in a short period of time. While an air-purifying chemical cartridge respirator or gas mask may suffice for low or infrequent exposures, the best protection is obtained using a self-contained breathing apparatus (SCBA) or a supplied-air respirator with full facepiece operated in pressure demand mode. If a full facepiece is not available, then chemical goggles should be worn to protect the eyes. Whenever a chemical splash hazard exists, a face shield and a protective apron should be worn. To prevent hand and skin exposures, protective gloves should be used. Glove manufacturers should be contacted and permeation studies obtained *before* final glove selections have been made.

Administrative controls should also be in place to minimize the potential for human exposures. These may include written procedures or policies which specify the methods and techniques that will be practiced whenever personnel are to work with ozone.

All personnel should receive *training* on the use, hazards, protective measures, emergency actions, and other precautions per 29 CFR 1910.1200 (Hazard Communication), prior to the first assignment in an area where ozone is used or stored.

Before beginning employment and at regular intervals thereafter (e.g., annually), the following tests are recommended:

☑ Lung function tests (establish baseline).

If symptoms occur or overexposure is suspected, the following additional tests are suggested:

☑ Lung function tests (compare to baseline).

☑ Consider chest X-ray following acute overexposure (may be negative it taken immediately after exposure due to delayed onset of pulmonary edema).

☑ Blood level monitoring for lactate dehydrogenase. Increased levels may indicate lung permeability due to ozone. Persistent hypoxia may require endotracheal intubation.

Any evaluation should include a careful history of past and present symptoms with an examination. Medical tests that look for damage already done are not a substitute for controlling exposure. Also, because smoking can cause heart disease, emphysema, lung cancer, and other respiratory disorders, smokers exposed to ozone may experience symptoms more quickly than nonsmokers under the same conditions of exposure. Prudent risk management requires proper consideration of *all* possible exposure factors.

Other methods to reduce exposure include:

☑ Where possible, enclose operations and use local exhaust ventilation at the site of chemical release. If local exhaust ventilation or enclosure is not used, respiratory protection should be mandatory.

☑ Always ensure that proper protective clothing is worn when using chemical substances and that personnel are trained in its use, care, and maintenance.

☑ Wash thoroughly immediately after exposure to ozone and at the end of the work shift or before eating, drinking, or smoking.

☑ Hazard warning information should be posted in the work area. In addition, as part of an on-going education and training program, all information on the health and safety hazards of ozone should be communicated to all exposed workers.

RISK ASSESSMENT: ENVIRONMENT

General Assessment

The environment is at serious risk of exposure during transportation, storage, disposal, or destruction of ozone. Ozone appears to contribute to the formation of photochemical smog. While stratospheric ozone (above approximately 56,000 feet or 17 kilometers) is essential to prevent harmful levels of the sun's ultraviolet radiation from reaching the Earth's surface, ozone concentrations in the lower atmosphere (the troposphere) near the Earth's surface can be extremely harmful to humans. In almost every scenario, the threat of environmental exposure is contingent upon the proper handling of the chemical substance. Accidental spills, large or small, can result in fire, explosion, and possible contamination of the surrounding environmental mediums (water, soil, and especially air).

Ozone is considered a non-flammable gas. However, it is a powerful oxidizer and extremely corrosive to many materials. It is incompatible with many commonly encountered materials including acetylene, alkylmetals, antimony, aromatic compounds (benzene, aniline), oxidizable materials (both organic and inorganic), other oxidizers (bromine, fluorine), nitrogen, nitrogen oxide, nitrogen trichloride, fluoroethylene, liquid hydrogen, silica gel, cryogenic ethylene, and all reducing materials and agents. These characteristics require special consideration during any emergency situation involving a leak or spill of ozone. Should ozone ever come into contact with any of these incompatible substances either during use, transportation, or storage, the formation of highly toxic and/or highly explosive commodities is extremely possible.

Although it does occur naturally, ozone can also enter the environment as a by-product of internal combustion engines, by electronic irradiation of air, and by industrial discharges or spills.

Acute Ecological Effects

Acute (short-term) toxic effects may include the death of animals, birds, or fish and death or low growth rate in plants. Acute effects are seen 2 to 4 days after animals or plants are exposed to ozone.

Insufficient data are available to evaluate or predict the short-term effects of ozone exposure to aquatic life, plants, birds, or land animals.

Chronic Ecological Effects

Chronic (long-term) toxic effects may include shortened life span, reproductive problems, lower fertility, and changes in appearance or behavior in exposed animals. These effects can be seen long after first exposure(s) to toxic chemicals.

Insufficient data are available to evaluate or predict the long-term effects of ozone to aquatic life, plants, birds, or land animals.

Water Solubility

Ozone is very slightly soluble in cold water. Concentrations of 10 milligrams or less will mix with a liter of water.

Persistence in the Environment

Ozone is slightly persistent in the aquatic environment, with a half-life of less than 2 and 20 days. It is highly persistent in the atmosphere with a half-life in air of greater than 200 days. The half-life of a pollutant is the amount of time it takes for one half of the chemical to be degraded.

Bioaccumulation in Aquatic Organisms

Some substances increase in concentration, or *bioaccumulate,* in living organisms as they breathe contaminated air, drink contaminated water, or eat contaminated food. These chemicals can become concentrated in the tissues and internal organs of animals as well as humans.

The concentration of ozone found in the edible tissues of most aquatic species that are consumed by humans is expected to be about the same as the average concentration of ozone in the water from which the fish was taken.

Recommended Risk-Reduction Measures

Proper training of all transporters will reduce the likelihood of a mishap or accident resulting in a leak or spill of ozone to the environment. The correct use of DOT labeling (per the MSDS) on all containers, trucks, and rail cars will enable emergency responders to react properly and quickly to any disaster thereby reducing the potential risk to the environment and to personnel.

Storage of ozone should be segregated from incompatible chemicals to minimize the risk of cross-contamination or contact. Buildings designated for storage should be equipped with appropriate fire protection systems (alarms, sprinklers, emergency lighting, portable extinguishers). Containers should be protected against physical damage and stored away from direct sunlight or other heat sources. Cylinders must be stored upright and secured properly. They should never be rolled or dragged.

If a spill or leak to the environment has occurred, fire department, emergency response and/or hazardous materials spill personnel should be notified immediately. Cleanup should be attempted only by those trained in proper spill containment procedures. Contaminated soils (solution spills) should be removed for incineration and replaced with clean soil. If ozone should contact the water table, aquifer, or navigable waterway, time is of the essence. It is slightly soluble in water and, therefore, total containment and remediation may not be possible. When such spills occur, the local and/or state emergency response authorities must be notified. A comprehensive emergency response or disaster preparedness/recovery plan should be in place prior to any operations involving the use, transportation, storage, or disposal of ozone. If ozone is spilled or leaked, the following specific steps are recommended:

☑ Restrict persons not wearing protective clothing from area until cleanup is complete and area can be opened for normal work.

☑ Remove all ignition sources and ventilate area.

☑ Shut off gas leak, if possible. If leak is from a cylinder and it cannot be turned off, remove cylinder to safe area (outdoors) and allow to vent. Take care to avoid contact with the vapors or any escaping compressed liquids.

RISK ASSESSMENT: BUSINESS

General Assessment

Accidents or mishaps involving ozone can present a significant threat to business operations. The loss or damage of equipment or facilities can significantly affect fiscal viability. Lawsuits that may result from personnel injury/death, public exposures, and/or environmental contamination will also require a serious expenditure of resources. Media attention surrounding an injury, death, or environmental damage can also result in a loss of profits and loss of current as well as future business. Also, public awareness of the word "ozone" has increased dramatically in recent years. Even if the exact nature of its properties is unknown, people may still become emotionally charged at the mere mention of the chemical's name. This must be carefully considered when drafting or implementing public relations policies.

Recommended Risk-Reduction Measures

Company attorneys, safety and health professionals, and environmental specialists should be involved in the development of any procedures dealing with safety or emergency response. A company official should be pre-designated as a public relations officer with specific training in dealing with the press. Corporate plans and policies should be developed, approved, and implemented long before any need for such arises.

REFERENCES

Ahrens, C.D. 1982. *Meteorology Today: An Introduction to Weather, Climate, and the Environment.* New York: West Publishing Company.

American Conference of Governmental Industrial Hygienists. 1988. *Documentation of the Threshold Limit Values and Biological Exposure Indices,* 5th Edition (with updates). Cincinnati: ACGIH

Anthony, C. P., and N. J. Kolthoff. 1971. *Textbook of Anatomy and Physiology.* St. Louis: C. V. Mosby Company.

Bueche, J. R. 1972. *Principles of Physics,* 2nd Edition. New York: McGraw-Hill.

Calabrese, E. J., E. M. Kenyon. *Air Toxics and Risk Assessment.* New York: CRC Press/Lewis Publishers.

Environmental Protection Agency. 1983. *Chemical Hazard Information Profiles.* Washington, D.C.: U.S. EPA.

Howard, P. H., et. al. 1991. *Handbook of Environmental Degradation Rates.* New York: CRC Press/Lewis Publishers.

Keith, L. H. and M. M. Walker. 1995. *Handbook of Air Toxics.* New York: CRC Press/Lewis Publishers.

Lewis, R. J., Sr. 1993. *Hawley's Condensed Chemical Dictionary,* 12th Edition. New York: Van Nostrand Reinhold.

Lewis, R. J., Sr. 1992. *Sax's Dangerous Properties of Industrial Materials,* Volumes 1, 2, and 3, 8th Edition. New York: Van Nostrand Reinhold.

Manahan, S. E. 1994. *Environmental Chemistry,* 6th Edition. New York: CRC Press/Lewis Publishers.

Meyer, E. 1990. *Chemistry of Hazardous Materials.* Englewood Cliffs, NJ: Prentice Hall.

U.S. Department of Health and Human Services, National Institute for Occupational Safety and Health. 1994. *NIOSH Pocket Guide to Chemical Hazards.* Washington, D.C.: U.S. Government Printing Office.

MATERIAL SAFETY DATA SHEET

CHEMICAL NAME	
PARAFFIN WAX	**CRC PRESS/LEWIS PUBLISHERS DISCLAIMER:** THE INFORMATION AND RECOMMENDATIONS PRESENTED HEREIN ARE BASED ON SOURCES BELIEVED TO BE RELIABLE. CRC MAKES NO REPRESENTATION ON ITS COMPLETENESS OR ACCURACY. IT IS THE RESPONSIBILITY OF THE USER TO DETERMINE THE CHEMICAL'S SUITABILITY FOR ITS INTENDED USE, THE CHEMICAL'S SAFE USE, AND THE CHEMICAL'S PROPER DISPOSAL. NO REPRESENTATIONS AND/OR WARRANTIES, EITHER EXPRESSED OR IMPLIED, OF THE MERCHANTABILITY OR FITNESS FOR A PARTICULAR PURPOSE, OR OF ANY OTHER NATURE, ARE MADE WITH RESPECT TO THE INFORMATION PROVIDED IN THIS MSDS OR TO THE CHEMICAL TO WHICH INFORMATION MAY REFER. CRC NEITHER ASSUMES NOR AUTHORIZES ANY OTHER PERSON TO ASSUME FOR IT, ANY OTHER ADDITIONAL RESPONSIBILITY OR LIABILITY FOR THE USE OF, OR RELIANCE UPON, THIS INFORMATION.

HAZARD WARNING INFORMATION

HEALTH	FIRE	REACTIVE	OTHER	DEGREE OF HAZARD	COLOR CODING	OTHER CODES
1	1	0		0 = Minimum Hazard 1 = Slight Hazard 2 = Moderate Hazard 3 = Serious Hazard 4 = Severe Hazard	HEALTH = BLUE FIRE = RED REACTIVITY = YELLOW OTHER = WHITE	OX = Oxidizer ACID = Acid ALK = Alkali COR = Corrosive W = Use No Water

SECTION I - GENERAL INFORMATION

Characterization	**Hydrocarbon, Solid**	RCRA Number	**None**
		EPA Class	**Not Applicable**
DOT Proper Shipping Name	**No Citation**	Chemical Abstract Service (CAS) Number	**8002-74-2**
DOT Hazard Class and Label Requirements	**No Citation**	DOT Emergency Guide Code	**No Citation**
DOT Identification Number	**No Citation**	Molecular Formula	C_nH_{2n+2}

Synonyms

Paraffin fume; paraffin scale fume; paraffin wax fume; hard wax.

SECTION II - HAZARDOUS INGREDIENTS AND IDENTITY INFORMATION

Hazardous Components (specific identity)	OSHA Exposure Criteria	NIOSH Exposure Criteria	Immediately Dangerous to Life and Health (IDLH)	ACGIH Exposure Criteria
Paraffin wax (derivation: Long, straight-chained, saturated hydrocarbon prepared from the residue collected following distillation of petroleum).	**PEL:** Not Established **STEL:** Not Established	**REL:** 2 mg/m^3 (fume) **STEL:** Not Established	**Not Determined**	**TLV:** 2 mg/m^3 (fume) **STEL:** Not Established

SECTION III - PHYSICAL AND CHEMICAL CHARACTERISTICS

Boiling Point	**Not Found**	Specific Gravity (H_2O = 1)	**0.88 to 0.92 (approximately)**
Vapor Pressure (mm Hg)	**Not Found**	Molecular Weight (atomic weight)	**350 to 420 (varies with grade)**
Vapor Density (Air = 1)	**Not Found**	Melting Point	**115 to 154°F (46 to 68°C)**

Solubility

Insoluble in water and acids. Soluble in benzene, warm alcohol, ligroin, chloroform, turpentine, carbon disulfide, and olive oil.

Appearance and Odor

White, translucent, or colorless to slightly yellow, tasteless, odorless solid consisting of a mixture of solid hydrocarbons of high molecular weight (e.g., $C_{36}H_{74}$).

SECTION IV - FIRE AND EXPLOSION HAZARD DATA

Flash Point (method used)	**390°F (198°C) closed cup**	Explosive Limits in Air % by Volume	**LEL: Not Found UEL: Not Found**
NFPA Classification	**Combustible Solid**	Autoignition Temperature	**473°F (245°C)**

Extinguishing Media

Use alcohol foam or polymer foams, dry chemical, carbon dioxide, or water spray.

Special Fire Fighting Procedures

Wear self-contained breathing apparatus (SCBA). Poisonous gases are produced in fire. Be aware of the possible presence of hot or molten paraffin wax in fires.

Unusual Fire and Explosion Hazards

Moderate fire hazard when exposed to heat or flame.

SECTION V - REACTIVITY DATA

Stability		Conditions to Avoid **Paraffin wax is stable in closed containers under routine conditions of handling and storage. Paraffin is a common air contaminant. Avoid conditions which may cause dispersion of fumes or particulates into the air.**
Stable X	Unstable	Incompatibility *(materials to avoid)* **None reported. Petroleum waxes are inert.**
Hazardous Polymerization		Conditions to Avoid **Hazardous polymerization of paraffin wax cannot occur.**
May Occur	Will Not Occur X	Hazardous Decomposition or By-products **Poisonous gases and toxic, irritating fumes are produced when paraffin wax is involved in a fire.**

SECTION VI - HEALTH HAZARD DATA

Primary Route(s) of Entry:	Inhalation?	Absorption (skin/eye)?	Ingestion?
	X		

Health Hazards

INHALATION: May cause mild irritation of the eyes, nose, throat, and upper respiratory tract with some discomfort and nausea. Some paraffin wax mixtures are questionable carcinogens.

SKIN & EYES: Mild to moderate skin irritation. Prolonged contact may cause chronic dermatitis, wax boils, pimples, blackheads, folliculitis, melanoderma, hyperkeratoses, and sensitization. Hot or molten wax can cause burns and blistering of skin surface and damage to the eyes.

INGESTION: Not a likely exposure route, but possible. May cause unspecified gastrointestinal effects.

Carcinogenicity	NTP Listed?	IARC Cancer Review Group?	OSHA Regulated?	Target Organs?
Suspected Human Questioned Animal	**No**	**No**	**No**	**Skin, eyes, respiratory system.**

Medical Conditions Generally Aggravated by Exposure

None identified.

Emergency and First-aid Procedures

Eye contact: Do not allow victim to keep eyes tightly shut. Flush immediately with water for 15 minutes (minimum); seek medical attention. Skin contact: Remove all contaminated clothing and immediately wash area with flooding amounts of soap and water. Seek medical attention immediately. For inhalation: Remove the person from exposure. Provide respiratory assistance and CPR. Transfer to medical facility. If swallowed: Seek medical attention immediately. Give victim large amounts of water to drink. Do not induce vomiting. Never attempt to give anything by mouth to an unconscious or convulsing person.

SECTION VII - PRECAUTIONS FOR SAFE HANDLING AND USE

Steps to be Taken in Case Material is Released or Spilled

Remove all ignition sources. Wear SCBA respiratory protection. Restrict those not wearing protective equipment and who are not involved in cleanup from entering area. Avoid skin contact. Ventilate area of spill. Let hot wax cool before cleanup. Use non-sparking tools to collect wax and deposit in suitable containers for reclamation or disposal.

Preferred Waste Disposal Method

No citation.

Precautions to be Taken in Handling and Storage

Store in tightly closed containers in cool, well-ventilated location. Keep away from sources of heat.

Other Precautions and Warnings

Metal containers should be bonded and grounded during transfer operations. Bulk storage not recommended.

SECTION VIII - CONTROL MEASURES AND PERSONAL PROTECTIVE EQUIPMENT

Respiratory Protection *(specify type)*

Use a MSHA/NIOSH-approved powered air-purifying respirator or use a self-contained breathing apparatus (SCBA) with full facepiece operated in positive pressure mode.

Ventilation

Local exhaust (preferred) at site of chemical work or general ventilation.

Protective Gloves	Eye Protection	Other Protective Clothing
Impervious Gloves	**Chemical Goggles or Face Mask**	**Protective Apron or Suit**

Work/Hygiene Practices

Always wash hands thoroughly after using chemical; never bring food, drink, cosmetics, or smoking materials into vicinity of chemicals.

PARAFFIN WAX

C_nH_{2n+2}

CAS: 8002-74-2

IDENTIFICATION AND TYPICAL USES

Paraffin wax appears as a white to slightly yellow, or colorless translucent tasteless, odorless, solid consisting of a mixture of solid hydrocarbons of high molecular weight (e.g., $C_{36}H_{74}$). It has a great number of uses. It is found in candles and it is used as a paper coating and protective sealant for food products and beverages. It is also used in glass-cleaning preparations; in the carpet-laying industry (hot melt carpet backing to join together carpet pieces); as a biodegradable mulch (hot melt-coated paper); for impregnating match heads; in lubricants; in the manufacture of crayons; in some surgical techniques; in stoppers for acid bottles; in electrical insulation; as a component in some floor polishes; in cosmetics; in the photography industry; as an anti-frothing agent in sugar refining; for packing tobacco products; for protecting rubber products from drying and cracking (due to exposure to the sun); and as a chewing gum base.

RISK ASSESSMENT: HEALTH

General Assessment

Paraffin wax fumes are harmful to humans primarily by *inhalation*. There are no reports of dermal absorption; however, skin contact with the wax can cause a variety of reactions especially in the long-term. If the wax should be hot or molten, then burns and blistering are also likely. Ingestion has not been reported as a likely exposure route. There are some reports in the references (Lewis 1992) that exposure to paraffin over the course of many years may cause cancer. There are no reports of any teratogenic or mutagenic effects.

Human systemic effects by inhalation include mild irritation of the eyes, nose, and throat. There may be some degree of discomfort and possible nausea.

Repeated skin contact may cause chronic dermatitis, the development of wax boils, pimples, blackheads, and other chronic skin disorders. The hot or molten wax can cause serious burns to the eyes and skin.

Acute Health Effects

The following acute (short-term) health effects may occur immediately or shortly after exposure to paraffin wax:

Skin: Irritation and, if the wax is hot, redness, blistering, and burning.

Eye: Severe irritation; pain and permanent damage possible if the wax is hot or molten.

Lung: Nose, throat, and respiratory tract irritation causing mild discomfort and nausea.

Chronic Health Effects

The following chronic (long-term) health effects may occur at some time after exposure to paraffin wax and can last for months or even years:

Cancer Hazards: Paraffin wax is a known suspected carcinogen in some test animals. There are reports of scrotum cancer developing in humans exposed to paraffin wax for periods of 10 years or more. Also, a skin lesion known as paraffinoma has been linked to the use of paraffin in some cosmetics.

Reproductive Hazard: According to information presented in the references, paraffin wax has not been adequately tested for its ability to adversely affect reproduction.

Other Chronic Effects: Prolonged or repeated skin contact can cause the development of chronic dermatitis, folliculitis (inflammation of the hair follicles), melanoderma (abnormal skin pigmentation), hyperkeratoses (thickening of the skin), and possible sensitization in some individuals.

Recommended Risk-Reduction Measures

Personnel should avoid direct contact with paraffin wax. Occupational exposures occur most commonly by inhalation. If a less toxic material or compound cannot be substituted for paraffin wax, then *engineering controls* are the most effective method of reducing exposures. The best protection is to enclose operations and/or provide local exhaust ventilation at the site of paraffin wax release. While not always operationally feasible, isolating operations can also reduce exposure. Using respiratory protection is less effective than the controls mentioned above, but is still advisable whenever working with or around paraffin wax. A MSHA/NIOSH-approved self-contained breathing apparatus (SCBA) with full facepiece operated in pressure demand or a powered air-supplied respirator in continuous flow mode are the recommended methods of respiratory protection when exposure to paraffin wax is anticipated. If a full facepiece is not available, then chemical goggles should be worn to protect the eyes. A face shield should also be considered. To prevent hand and skin exposures, impervious gloves should be worn. Glove manufacturers should be contacted and permeation studies obtained *before* final glove selections are made.

Administrative controls should also be in place to minimize the potential for human exposures. These may include written procedures or policies which specify the methods and techniques that will be practiced whenever personnel are to work with paraffin wax.

All personnel should receive *training* on the use, hazards, protective measures, emergency actions, and other precautions per 29 CFR 1910.1200 (Hazard Communication), prior to the first assignment in an area where paraffin wax is used or stored.

Before beginning employment and at regular times thereafter, the following recommended medical tests should be considered:

☑ Evaluation by a qualified allergist with careful consideration of exposure history and special testing (may help diagnose skin allergy).

Any evaluation should include a careful history of past and present symptoms with an examination. Medical tests that simply look for existing damage are not a substitute for controlling exposures.

Other methods to reduce exposure include:

☑ Where possible, enclose operations and use local exhaust ventilation at the site of chemical release. If local exhaust ventilation or enclosure is not used, respiratory protection should be mandatory.

☑ Always ensure that proper protective clothing is worn when using chemical substances.

☑ Wash thoroughly immediately after exposure to paraffin wax and at the end of the work shift or before eating, drinking, or smoking.

☑ Hazard warning information should be posted in the work area. In addition, as part of an on-going education and training program, all information on the health and safety hazards of paraffin wax should be communicated to all exposed workers.

☑ Eye wash stations should be provided in the immediate work area for emergency use. If there is a possibility of skin exposure to paraffin wax, emergency shower facilities should be provided.

☑ Workers whose clothing has been contaminated by paraffin wax should change into clean clothes before leaving work. Contaminated work clothing should be laundered only by individuals who have been informed of the hazards of exposure to paraffin wax.

RISK ASSESSMENT: ENVIRONMENT

General Assessment

The environment is at risk of exposure during transportation, storage, disposal, or destruction of paraffin wax. In almost every scenario, the threat of environmental exposure is contingent upon the proper handling of the chemical substance. Accidental spills, large or small, can result in contamination of the surrounding environmental mediums (water, soil, and air).

Paraffin wax is a moderately flammable liquid. Because of its low flash point and high boiling point, it is classified as a Class 1B flammable liquid (per OSHA 29 CFR 1910.106). It is a dangerous fire hazard and moderate explosion hazard when exposed to heat or flame. Caution is always required in handling, storage, transportation, and disposal of paraffin wax. Emergency responders should be made aware of the presence of paraffin wax at any emergency response situation.

Paraffin wax can enter the environment from industrial effluents and from spills.

Acute Ecological Effects

Acute (short-term) toxic effects may include the death of animals, birds, or fish and death or low growth rate

in plants. Acute effects are seen 2 to 4 days after animals or plants are exposed to paraffin wax.

Insufficient data are available to evaluate or predict the short-term effects of paraffin wax on aquatic life, plants, birds, or land animals.

Chronic Ecological Effects

Chronic (long-term) toxic effects may include shortened life span, reproductive problems, lower fertility, and changes in appearance or behavior in exposed animals. These effects can be seen long after first exposure(s) to toxic chemicals.

Insufficient data are available to evaluate or predict the long-term effects of paraffin wax on aquatic life, plants, birds, or terrestrial animals.

Water Solubility

Paraffin wax is insoluble. Concentrations of less than 1 milligram will not mix with a liter of water.

Persistence in the Environment

Paraffin wax is highly persistent in water, with a half-life of greater than 200 days. The half-life of a pollutant is the amount of time it takes for one half of the chemical to be degraded. Paraffin wax will eventually end up in aquatic and terrestrial sediments in approximately even proportions.

Bioaccumulation in Aquatic Organisms

Some substances increase in concentration, or *bioaccumulate,* in living organisms as they breathe contaminated air, drink contaminated water, or eat contaminated food. These chemicals can become concentrated in the tissues and internal organs of animals as well as humans.

Paraffin wax is not expected to accumulate in the edible tissues of fish that are normally consumed by humans.

Recommended Risk-Reduction Measures

Proper training of all transporters will reduce the likelihood of a mishap or accident resulting in a leak or spill of paraffin wax into the environment. Labels on all containers, trucks, and rail cars must meet DOT requirements and accurately reflect their contents to enable emergency responders to react properly and quickly to any disaster thereby reducing the potential risk to the environment and to personnel.

Storage of paraffin wax should be segregated from other chemicals and materials to minimize the risk of cross-contamination.

If a spill or leak to the environment has occurred, fire department, emergency response and/or hazardous materials spill personnel should be notified immediately. Cleanup should be attempted only by those trained in proper spill containment procedures using non-sparking tools. Contaminated soils should be removed and replaced with clean soil. If paraffin wax should contact the water table, aquifer, or navigable waterway, time is of the essence. It is not soluble and total remediation may be possible. The local and/or state emergency response authorities should be notified. A comprehensive emergency response or disaster preparedness/recovery plan should be in place prior to any operations involving the use, transportation, storage, or disposal of paraffin wax.

If paraffin wax is spilled or leaked, the following specific steps are recommended:

- ☑ Restrict persons not wearing protective clothing from area of spill or leak until cleanup is complete and area can be opened for normal work.
- ☑ Allow molten wax to cool and solidify before cleanup operations begin. Collect materials using non-sparking tools and place in suitable drums or containers for reclamation or disposal.
- ☑ Ventilate area of spill or leak.
- ☑ Remove all ignition sources from area.
- ☑ It may be necessary to dispose of paraffin wax as a hazardous waste. The responsible state agency or the regional office of the federal Environmental Protection Agency (EPA) should be contacted for specific recommendations.

RISK ASSESSMENT: BUSINESS

General Assessment

Accidents or mishaps involving paraffin wax can present a serious threat to business operations. The loss or damage of equipment or facilities can significantly affect fiscal viability. Lawsuits that may result from personnel illness (leukemia), injury/death, public exposures, and/or environmental contamination will require a serious expenditure of resources. Media attention surrounding an injury, death, or environmental damage can also result in a loss of profits and loss of current as well as future business. Always remember that anytime the terms "cancer" or "carcinogen are used, public emotion, anxiety. Hysteria, and ignorance can all run equally high. This must be carefully considered whenever drafting or implementing public relations policies.

Recommended Risk-Reduction Measures

Company attorneys, safety and health professionals, and environmental specialists should be involved in the development of any procedures for responding to chemical incidents. A company official should be predesignated as a public relations officer with specific training in dealing with the press. Corporate plans and policies should be developed, approved, and implemented long before any need for such arises.

REFERENCES

American Conference of Governmental Industrial Hygienists. 1988. *Documentation of the Threshold Limit Values and Biological Exposure Indices,* 5th Edition (with updates). Cincinnati: ACGIH

Bloom, A. D. and F. J. DeSerres. 1995. *Ecotoxicity and Human Health.* New York: CRC Press/Lewis Publishers.

Brusick, D. J. 1994. *Methods for Genetic Risk Assessment.* New York: CRC Press/Lewis Publishers.

Bueche, J. R. 1972. *Principles of Physics,* 2nd Edition. New York: McGraw-Hill.

Cockerman, L. G., B. S. Shane. 1994. *Basic Environmental Toxicology.* New York: CRC Press/Lewis Publishers.

Danse, I. R. 1991. *Common Sense Toxics in the Workplace.* New York: Van Nostrand Reinhold.

Environmental Protection Agency. 1983. *Chemical Hazard Information Profiles.* Washington, D.C.: U.S. EPA.

Fodden, C. R., and J. L. Weddell. 1992. *Hazardous Materials Emergency Action Data.* New York: CRC Press/Lewis Publishers.

Forseberg, K., and L. H. Keith. 1993. *Chemical; Protective Clothing Permeation and Degradation Database.* Chelsea, MI: Lewis Publishers.

Gerarde, H. W. 1960. *Toxicology and Biochemistry of Aromatic Hydrocarbons.* New York: Elsevier.

Hallenbeck, W. H. 1993. *Quantitative Risk Assessment for Environmental and Occupational Health,* 2nd Edition. New York: CRC Press/Lewis Publishers.

Hodgson, E., R. B. Mailman, and Janice E. Chambers. 1988. *Dictionary of Toxicology.* London: The MacMillian Company.

Hosty, J. W., and P. Foster. 1990. *A Practical Guide to Chemical Spill Response.* New York: Van Nostrand Reinhold.

Howard, P. H., et. al. 1991. *Handbook of Environmental Degradation Rates.* New York: CRC Press/Lewis Publishers.

Jones, F. E. 1994. *Toxic Organic Vapors in the Workplace.* New York: CRC Press/Lewis Publishers.

Keith, L. H. and M. M. Walker. 1995. *Handbook of Air Toxics.* New York: CRC Press/Lewis Publishers.

Landis, W. G., M. H. Yu. 1995. *Introduction to Environmental Toxicology.* New York: CRC Press/Lewis Publishers.

Langerman, N. 1994 *Precautionary Labels for Chemical Containers.* New York: CRC Press/Lewis Publishers.

Lauwerys, R. R., P. Hoet. 1993. *Industrial Chemical Exposure.* New York: CRC Press/Lewis Publishers.

Lewis, R. J., Sr. 1993. *Hawley's Condensed Chemical Dictionary,* 12th Edition. New York: Van Nostrand Reinhold.

Lewis, R. J., Sr. 1992. *Sax's Dangerous Properties of Industrial Materials,* Volumes 1, 2, and 3, 8th Edition. New York: Van Nostrand Reinhold.

Lide, D. R. 1995. *Handbook of Chemistry and Physics,* 76th Edition. New York: CRC Press/Lewis Publishers.

Lisella, F. S. 1994. *The VNR Dictionary of Environmental Health and Safety.* New York: Van Nostrand Reinhold.

Manahan, S. E. 1994. *Environmental Chemistry,* 6th Edition. New York: CRC Press/Lewis Publishers.

Mellström, G. A., J. E. Wahlberg, and H. I. Maibach. 1994. *Protective Gloves for Occupational Use.* New York: CRC Press/Lewis Publishers.

Meyer, E. 1990. *Chemistry of Hazardous Materials.* Englewood Cliffs, NJ: Prentice Hall.

Ott, W. R. 1995. *Environmental Statistics and Data Analysis.* New York: CRC Press/Lewis Publishers.

Patnaik, P. 1992. *A Comprehensive Guide to the Hazardous Properties of Chemical Substances.* New York: Van Nostrand Reinhold.

Philp, R. B. 1995. *Environmental Hazards and Human Health.* New York: CRC Press/Lewis Publishers.

Proctor, N. H., J. P. Hughes. 1991. *Chemical Hazards of the Workplace.* New York: Van Nostrand Reinhold.

Quigley, D. R. 1994. *Handbook of Emergency Chemical Management.* New York: CRC Press/Lewis Publishers.

Rea, W. J. 1995. *Chemical Sensitivity,* Vol. I-III. New York: CRC Press/Lewis Publishers.

Sacarello, H. L. A. 1994. *The Comprehensive Handbook of Hazardous Materials.* New York: CRC Press/Lewis Publishers.

Talbott, E., G. Craun. 1995. *An Introduction to Environmental Epidemiology.* New York: CRC Press/Lewis Publishers.

Torkelson, T. R. V. K. Rowe. 1981. *Patty's Industrial Hygiene and Toxicology,* Vol. 2B. New York: Wiley-Interscience.

U.S. Congress: Office of Technology Assessment Task Force. 1988. *Identifying and Regulating Carcinogens.* Chelsea, MI: Lewis Publishers.

U.S. Department of Health and Human Services, National Institute for Occupational Safety and Health. 1994. *NIOSH Pocket Guide to Chemical Hazards.* Washington, D.C.: U.S. Government Printing Office.

Witorsch, P. and S. V. Spagnolo. 1994. *Air Pollution and Lung Disease in Adults.* New York: CRC Press/Lewis Publishers.

MATERIAL SAFETY DATA SHEET

CHEMICAL NAME	
PARATHION®	**CRC PRESS/LEWIS PUBLISHERS DISCLAIMER:** THE INFORMATION AND RECOMMENDATIONS PRESENTED HEREIN ARE BASED ON SOURCES BELIEVED TO BE RELIABLE. CRC MAKES NO REPRESENTATION ON ITS COMPLETENESS OR ACCURACY. IT IS THE RESPONSIBILITY OF THE USER TO DETERMINE THE CHEMICAL'S SUITABILITY FOR ITS INTENDED USE, THE CHEMICAL'S SAFE USE, AND THE CHEMICAL'S PROPER DISPOSAL. NO REPRESENTATIONS AND/OR WARRANTIES, EITHER EXPRESSED OR IMPLIED, OF THE MERCHANTABILITY OR FITNESS FOR A PARTICULAR PURPOSE, OR OF ANY OTHER NATURE, ARE MADE WITH RESPECT TO THE INFORMATION PROVIDED IN THIS MSDS OR TO THE CHEMICAL TO WHICH INFORMATION MAY REFER. CRC NEITHER ASSUMES NOR AUTHORIZES ANY OTHER PERSON TO ASSUME FOR IT, ANY OTHER ADDITIONAL RESPONSIBILITY OR LIABILITY FOR THE USE OF, OR RELIANCE UPON, THIS INFORMATION.

HAZARD WARNING INFORMATION

HEALTH	FIRE	REACTIVE	OTHER	DEGREE OF HAZARD	COLOR CODING	OTHER CODES
4	1	1		0 = Minimum Hazard 1 = Slight Hazard 2 = Moderate Hazard 3 = Serious Hazard 4 = Severe Hazard	HEALTH = BLUE FIRE = RED REACTIVITY = YELLOW OTHER = WHITE	OX = Oxidizer ACID = Acid ALK = Alkali COR = Corrosive W = Use No Water

SECTION I - GENERAL INFORMATION

Characterization	RCRA Number	EPA Class
Pesticide	**P089**	**Acute Hazardous Waste**
DOT Proper Shipping Name: **Parathion**	Chemical Abstract Service (CAS) Number: **56-38-2**	
DOT Hazard Class and Label Requirements: **Poison**	DOT Emergency Guide Code: **55** (liquid/dry) **28** (flammable liquid)	
DOT Identification Number: **NA 2783** (liquid/dry) **NA 2784** (flammable liquid)	Chemical Formula: $(C_2H_5O)_2P(S)OC_6H_4NO_2$	

Synonyms
O,O-Diethyl-O(p-nitrophenyl) phosphorothioate; diethyl parathion; ethyl parathion; parathion-ethyl; Alleron; Aphamite; Danthion; corothion; Kypthion; nitrostigmine; diethylparathion; Vitrex.

SECTION II - HAZARDOUS INGREDIENTS AND IDENTITY INFORMATION

Hazardous Components (specific identity)	OSHA Exposure Criteria	NIOSH Exposure Criteria	Immediately Dangerous to Life and Health (IDLH)	ACGIH Exposure Criteria
Parathion: (derivation: From sodium ethylate, thiophosphoryl chloride, and sodium *p*-nitrophenate). **1 ppm = 12.11 mg/m^3**	**PEL (8-hour):** 0.1 mg/m^3 (skin) **STEL:** Not Established	**REL (10-hour):** 0.05 mg/m^3 (skin) **STEL:** Not Established	**10 mg/m^3**	**TLV:** 0.1 mg/m^3 (skin) **STEL:** Not Established

SECTION III - PHYSICAL AND CHEMICAL CHARACTERISTICS

Boiling Point	**707°F (375°C)**	Specific Gravity (H_2O = 1)	**1.27**
Vapor Pressure (mm Hg)	**3.78 x 10^{-5} at 68°F (20°C)**	Molecular Weight	**291.3**
Vapor Density (Air = 1)	**10.0**	Melting Point	**43°F (6°C)**

Solubility
Slightly soluble in water (0.001%). Soluble in alcohols, esters, ethers, chloroform, ketones, aromatic hydrocarbons, and animal and vegetable oils. Nearly insoluble in petroleum ether and kerosene.

Appearance and Odor
Pale-yellow to dark-brown liquid with a garlic-like odor. Odor Threshold = 0.47 mg/m^3.

SECTION IV - FIRE AND EXPLOSION HAZARD DATA

Flash Point (method used)	Explosive Limits in Air % by Volume
392°F (200°C) opened cup	**LEL: Not Determined UEL: Not Determined**
NFPA Classification: **Class IIIB Combustible Liquid**	Autoignition Temperature: **Not Determined**

Extinguishing Media
Use dry chemical, carbon dioxide, water spray, fog, or standard foam.

Special Fire Fighting Procedures
Poisonous gases are produced in fire; wear full protective clothing and self-contained breathing apparatus (SCBA). If possible, fight fire from a distance and avoid smoke (consider down-wind conditions). Move containers from fire area if it can be done safely and without risk.

Unusual Fire and Explosion Hazards
Containers may explode when exposed to heat of a fire. Use water spray to cool fire-exposed containers that cannot be moved from fire area. Evacuate non-essential personnel 2000 feet from fire area.

SECTION V - REACTIVITY DATA

Stability		Conditions to Avoid Parathion is kept stable in aqueous solution of distilled water in acid solution. Avoid contact with incompatible materials. Keep away from heat and flame.
Stable X	Unstable	Incompatibility *(materials to avoid)* Incompatible with strong oxidizers, alkaline materials, and endrin. It will also attack some forms of rubber, plastics, and coatings.
Hazardous Polymerization		Conditions to Avoid Under normal conditions of temperature and pressure, hazardous polymerization of parathion will not occur.
May Occur	Will Not Occur X	Hazardous Decomposition or By-products Acrid and irritated gases and fumes are produced when parathion is involved in fire, including oxides of nitrogen, phosphorus, and sulfur.

SECTION VI - HEALTH HAZARD DATA

Primary Route(s) of Entry:	Inhalation?	Absorption (skin)?	Ingestion?
	X	X	X

Health Hazards

INHALATION: Labored or difficult breathing, tightness in chest, wheezing, coughing, rales, increased respiration, headache, runny nose, and excessive bronchial secretions. There may also be laryngeal spasms, excessive salivation, respiratory distress, cyanosis, pulmonary edema (fluid in lungs), visual disturbances, tearing, blurring, nausea and vomiting, pain in the stomach, and CNS effects such as giddiness, dizziness, weakness, slow heartbeat, arrhythmia, loss of coordination, slurred speech, confusion, disorientation with drowsiness, hypertension, convulsions, coma, and death (respiratory paralysis).

ABSORPTION: Irritation to the skin and eyes. May absorb to cause toxic systemic effects, including sweating twitching and many of the symptoms noted for inhalation. Sensitization and dermatitis may occur at high exposures. Extremely irritating to the eyes.

INGESTION: Causes many of the same symptoms noted for inhalation and absorption.

Carcinogenicity	NTP Listed?	IARC Cancer Review Group?	OSHA Regulated?	Target Organs?
Unknown Human Questioned Animal	No	Group 3	29 CFR 1910.1000 Table Z-1	Respiratory system, eyes, skin, CNS, CVS, blood.

Medical Conditions Generally Aggravated by Exposure

Glaucoma, cardiovascular disorders, hepatic and renal diseases, and CNS disorders.

Emergency and First-aid Procedures

Eye contact: Flush immediately with water for 15 minutes (minimum); seek medical attention. Skin contact: Remove all contaminated clothing. Immediately wash area with large amounts of soap and water followed by a 95% ethanol wash and then a second soap and water rinse. Seek medical assistance. For inhalation: Remove the person from exposure. Provide respiratory assistance and CPR. Transfer to medical facility. Observe 48 hours for lung effects. If swallowed: Contact poison control center. Give 1-2 glasses of water. Unless advised otherwise, do NOT induce vomiting.

SECTION VII - PRECAUTIONS FOR SAFE HANDLING AND USE

Steps to be Taken in Case Material is Released or Spilled

Absorb liquids in vermiculite or other absorbent and deposit in DOT-approved drum. Ventilate area and remove ignition sources. Restrict those not involved in cleanup from entering area. Do not use plastic or rubber tools during cleanup since parathion can attack such materials.

Preferred Waste Disposal Method

Mix in flammable solvent and burn in chemical incinerator equipped with afterburner and scrubber.

Precautions to be Taken in Handling and Storage

Store in tightly closed containers in a cool, well-ventilated, moisture-free area.

Other Precautions and Warnings

Keep away from rubber and plastic materials.

SECTION VIII - CONTROL MEASURES AND PERSONAL PROTECTIVE EQUIPMENT

Respiratory Protection *(specify type)*

For best protection, use a supplied-air respirator with full facepiece operated in pressure demand mode or a SCBA with full facepiece operated in pressure demand or other positive pressure mode.

Ventilation

Local exhaust or general mechanical systems recommended.

Protective Gloves	Eye Protection	Other Protective Clothing
Teflon or Saranex Material	Chemical Goggles or Face Mask	Whole Body Protection

Work/Hygiene Practices

Always wash hands thoroughly after using chemical; never bring food, drink, or smoking materials into vicinity of chemicals.

PARATHION

$(C_2H_5O)_2P(S)OC_6H_4NO_2$ CAS: 56-38-2

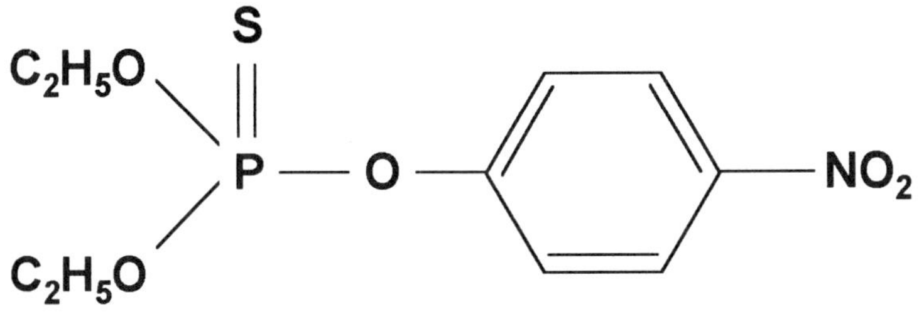

IDENTIFICATION AND TYPICAL USES

Parathion is a pale-yellow to dark-brown liquid with a faint, garlic-like odor. It can also appear as a solid at temperatures below 43°F (6°C). It is a pesticide whose general use is now subject to restrictions in the United States. It is used as an insecticide for wheat and nuts and as an acaricide.

RISK ASSESSMENT: HEALTH

General Assessment

Parathion is toxic to humans by *inhalation,* dermal *absorption,* and *ingestion*. It is a questionable carcinogen with human mutation data also reported. It is also an experimental teratogen with positive evidence of reproductive effects in animals. It is not easily removed from the surface of the skin and eye contact with the liquid can be fatal due to the chemical's rapid absorption through the eye.

Inhalation of parathion vapors will cause a tightness in the chest, wheezing, increased pulmonary secretions, bluish color in the skin, small pupils (miosis), an aching feeling in and behind the eyes, blurred vision, lachrymation (tearing), runny nose, headache, and watering mouth. It can lead to pulmonary edema (fluid in the lungs) which can be fatal. Symptoms may be delayed up to 48 hours thereby creating a false sense of security with regard to health exposure risk. There may be laryngeal spasms, respiratory distress, bradycardia (reduced heartbeat), tachycardia (fast heartbeat), and other cardiac disturbances (arrhythmia, complete heart block). It will also cause systemic effects with action on the central nervous system. This action can include weakness, general twitching and paralysis, dizziness, confusion, staggering gait, slurred speech, sweating, incontinence, convulsions, stopped breathing, coma, and death. This compound is also an inhibitor of cholinesterase, a blood enzyme that is essential to the proper functioning of the muscles.

Acute Health Effects

The following acute (short-term) health effects may occur immediately or shortly after exposure to parathion:

<u>Skin</u>: Irritation and possible inflammation. May absorb through unbroken skin to cause toxic systemic effects, including ocular disturbances, respiratory impairment, and CNS effects.

<u>Eye</u>: One drop of parathion in the eye can be life-threatening. It is easily absorbed through the cornea and conjunctiva.

<u>Lung</u>: Irritation of the mucosa of the nose, throat, and upper respiratory tract. Effects are more systemic than localized, with action on the blood, central nervous system, and (possibly) the cardiovascular system. Direct lung effects include pulmonary edema (fluid in the lungs) which can also be fatal.

<u>CNS</u>: A depressant, causing headache, blurred vision, giddiness, nervousness, and other systemic effects up to and including death from respiratory paralysis or cardiac disturbances.

<u>Other</u>: Ingestion of parathion will result in loss of appetite, nausea, vomiting, abdominal cramps, and diarrhea with 2 hours of exposure.

Chronic Health Effects

The following chronic (long-term) health effects may occur at some time after exposure to parathion and can last for months or even years:

Cancer Hazards: According to the information presented in the references, parathion has been shown to cause cancer in test animals. It has been shown to cause mutations (genetic changes) in humans and may pose a cancer risk as well.

Reproductive Hazard: According to the information presented in the references, parathion may damage the developing fetus in humans since it has been shown to be fetotoxic in animals.

Other Chronic Effects: Chronic exposure to parathion has not been sufficiently studied. However, there is some indication that long-term exposure or frequent high exposures may inhibit the essential enzyme cholinesterase. Individuals may become more susceptible to parathion's toxic effects following long-term exposure. Parathion binds irreversibly to cholinesterase causing its depletion. This action is apparently the reason for the increased susceptibility in chronic exposure. Symptoms include headache, vertigo, loss of memory and appetite, insomnia, disorientation, muscular twitching and trembling, and other CNS effects. Parathion is partially converted in the body to paraoxon, an active cholinesterase inhibitor. It may also cause dermal sensitization as well as dermatitis in some people.

ʊ *Recommended Risk-Reduction Measures*

Parathion is more toxic to humans than most other anticholinesterase agents. Less toxic substances should always be sought as a first step towards risk reduction. If a less toxic chemical cannot be substituted for a hazardous substance, *engineering controls* are the most effective method of reducing exposures. The best protection is to enclose operations and/or provide local exhaust ventilation at the site of chemical release. While not always operationally feasible, isolating operations can also reduce exposure. These controls certainly may not be practical or applicable when using parathion as an insecticide, especially outdoors. However, during manufacture, transportation, or storage, these engineering controls may be effective in reducing exposure risk.

Using respiratory protection is less effective than the controls mentioned above, but is still advisable whenever working with or around parathion. The exposure level for this compound is relatively low and may be difficult to measure and even more difficult to protect against. At low or infrequent exposures, a full facepiece respirator equipped with a high efficiency particulate air (HEPA) filter and a dust/mist pre-filter may suffice. However, for maximum exposure risk reduction or for exposures to high concentrations, a self-contained breathing apparatus (SCBA) with full facepiece operated in pressure demand is the recommended respiratory protection of choice. If a full facepiece is not available, then chemical goggles should be worn to protect the eyes. Whenever a chemical splash hazard exists, a face shield and protective apron should be worn. To prevent hand and skin exposures, impervious gloves should be worn. Caution is warranted in glove selection since parathion will attack rubbers, some plastics, and many coatings. Glove manufacturers should be contacted and permeation studies obtained *before* final glove selections are made.

Administrative controls should also be in place to minimize the potential for human exposures. These may include written procedures or policies which specify the methods and techniques that will be practiced whenever personnel are to work with parathion.

All personnel should receive *training* on the use, hazards, protective measures, emergency actions, and other precautions per 29 CFR 1910.1200 (Hazard Communication), prior to the first assignment in an area where parathion is used or stored.

If symptoms develop or overexposure is suspected, the following medical tests may be useful:

☑ Complete blood count (CBC) with plasma and red blood cell cholinesterase levels. If exposure stops, plasma levels return to normal within 1 to 2 weeks while red blood cell levels may be reduced for 1 to 3 months.

☑ Urinary metabolite monitoring for *p*-nitrophenol within 48 hours of exposure to assess parathion absorption.

Any evaluation should include a careful medical history of past and present symptoms with an examination. However, medical tests that evaluate existing damage are not a substitute for controlling exposure.

Other methods to reduce exposure include:

☑ Where possible, enclose operations and use local exhaust ventilation at the site of chemical release. If local exhaust ventilation or enclosure is not used, respiratory protection should be mandatory.

☑ Always ensure that proper protective clothing is worn when using chemical substances and that personnel are trained in its use and care.

☑ Wash thoroughly immediately after exposure to parathion and at the end of the work shift or before eating, drinking, or smoking.

☑ Hazard warning information should be posted in the work area. In addition, as part of an on-going education and training program, all information on the health and safety hazards of parathion should be communicated to all exposed workers.

☑ Eye wash stations should be provided in work areas. If the potential for whole body exposure exists, then safety showers should also be provided.

RISK ASSESSMENT: ENVIRONMENT

General Assessment

The environment is also at risk of exposure during the intended use as an insecticide, as well as during transportation, storage, disposal, or destruction of parathion. In almost every scenario, the threat of environmental exposure is contingent upon the proper handling of the chemical substance. Accidental spills, large or small, can result in fire, explosion, and possible contamination of the surrounding environmental mediums (water, soil, and air).

Parathion is considered a Class IIIB combustible liquid (per OSHA 29 CFR 1910.106). These characteristics require special consideration during any emergency situation involving a leak or spill of parathion. Should parathion ever come into contact with incompatible substances such as strong oxidizers, alkaline materials, or endrin, either during use, transportation, or storage, violent reactions can occur.

Parathion can enter the environment through its intended use as an insecticide, through industrial discharges, and from spills.

Acute Ecological Effects

Acute (short-term) toxic effects may include the death of animals, birds, or fish and death or low growth rate in plants. Acute effects are seen 2 to 4 days after animals or plants are exposed to parathion.

This chemical has high acute toxicity to aquatic life and to birds. Parathion has caused injury, including germination decreases and sterility increases in various agricultural and ornamental crops. Insufficient data are available to evaluate or predict the short-term effects of parathion to land animals.

Chronic Ecological Effects

Chronic toxic effects may include shortened life span, reproductive problems, lower fertility, and changes in appearance or behavior in exposed animals. These effects can be seen long after first exposure(s) to toxic chemicals.

Parathion has high chronic toxicity to aquatic life. Insufficient data are available to evaluate or predict the long-term effects of parathion to plants, birds, or land animals.

Water Solubility

Parathion is moderately soluble in water. Concentrations of 1 to 1000 milligrams will mix with a liter of water.

Persistence in the Environment

Parathion is moderately persistent in water, with a half-life of between 20 to 200 days. The half-life of a pollutant is the amount of time it takes for one half of the chemical to be degraded. About 69% of parathion will eventually end up in water; the remainder will be divided about equally between terrestrial soils and aquatic sediments.

Bioaccumulation in Aquatic Organisms

Some substances increase in concentration, or *bioaccumulate,* in living organisms as they breathe contaminated air, drink contaminated water, or eat contaminated food. These chemicals can become concentrated in the tissues and internal organs of animals as well as humans.

The concentration of parathion found in fish tissues is expected to be somewhat higher than the average concentration of parathion in the water from which the fish was taken.

Recommended Risk-Reduction Measures

Proper training of all transporters will reduce the likelihood of a mishap or accident resulting in a leak or spill to the environment. The correct use of DOT labeling (per the MSDS) on all containers, trucks, and rail cars will enable emergency responders to react properly and quickly to any disaster thereby reducing the potential risk to the environment and to personnel.

Storage of parathion should be segregated from incompatible chemicals to minimize the risk of cross-contamination or contact. It should be kept in dry areas away from water and moisture.

If a spill or leak to the environment has occurred, fire department, emergency response and/or hazardous materials spill personnel should be notified immediately. Cleanup should be attempted only by those

trained in proper spill containment procedures. Contaminated soils should be removed for incineration and replaced with clean soil. If parathion should contact the water table, aquifer, or navigable waterway, remediation activities should be prompt to ensure protection of aquatic sediments. When such spills occur, the local and/or state emergency response authorities must be notified. A comprehensive emergency response or disaster preparedness/recovery plan should be in place prior to any operations involving the use, transportation, storage, or disposal of parathion. If parathion is spilled or leaked, the following specific steps are recommended:

☑ Restrict persons not wearing protective clothing from area of spill or leak until cleanup is complete and area can be opened for normal work.

☑ Ventilate area and remove ignition sources.

☑ Absorb liquid spills using vermiculite, dry earth, or sand and place in a sealed drum for disposal using non-sparking tools. Avoid using plastic or rubber tools during cleanup since parathion can attack such materials.

☑ It may be necessary to dispose of parathion as a hazardous waste. The responsible state agency or the regional office of the federal Environmental Protection Agency (EPA) should be contacted for specific recommendations.

RISK ASSESSMENT: BUSINESS

General Assessment

Accidents or mishaps involving parathion can present a significant threat to business operations. The loss or damage of equipment or facilities can significantly affect fiscal viability. Lawsuits that may result from personnel injury/death, public exposures, and/or environmental contamination will also require a serious expenditure of resources.

Media attention surrounding an injury, death, or environmental damage can also result in a loss of profits and loss of current as well as future business. Always remember that anytime the terms "cancer," "carcinogen," or "reproductive hazard" are used, public emotion, anxiety, hysteria, and ignorance can run equally high. This should be carefully considered whenever developing or implementing public relations polices or directives.

Recommended Risk-Reduction Measures

Company attorneys, safety and health professionals, and environmental specialists should be involved in the development of any procedures dealing with safety or emergency response. A company official should be pre-designated as a public relations officer with specific training in dealing with the press. Corporate plans and policies should be developed, approved, and implemented long before any need for such arises.

REFERENCES

American Conference of Governmental Industrial Hygienists. 1988. *Documentation of the Threshold Limit Values and Biological Exposure Indices,* 5th Edition (with updates). Cincinnati: ACGIH

Bloom, A. D. and F. J. DeSerres. 1995. *Ecotoxicity and Human Health.* New York: CRC Press/Lewis Publishers.

Bueche, J. R. 1972. *Principles of Physics,* 2nd Edition. New York: McGraw-Hill.

Burrell, R., D. K. Flaherty, and L. J. Sauers. 1992. *Toxicology of the Immune System.* New York: Van Nostrand Reinhold.

Calabrese, E. J., E. M. Kenyon. *Air Toxics and Risk Assessment.* New York: CRC Press/Lewis Publishers.

Cockerman, L. G., B. S. Shane. 1994. *Basic Environmental Toxicology.* New York: CRC Press/Lewis Publishers.

Gardner, R. W. 1994. *Chemical Intolerance.* New York: CRC Press/Lewis Publishers.

Keith, L. H. and M. M. Walker. 1995. *Handbook of Air Toxics.* New York: CRC Press/Lewis Publishers.

Lauwerys, R. R., P. Hoet. 1993. *Industrial Chemical Exposure.* New York: CRC Press/Lewis Publishers.

Lewis, R. J., Sr. 1993. *Hawley's Condensed Chemical Dictionary,* 12th Edition. New York: Van Nostrand Reinhold.

Lewis, R. J., Sr. 1992. *Sax's Dangerous Properties of Industrial Materials,* Volumes 1, 2, and 3, 8th Edition. New York: Van Nostrand Reinhold.

Manahan, S. E. 1992. *Toxicological Chemistry,* 2nd Edition. New York: CRC Press/Lewis Publishers.

Meyer, E. 1990. *Chemistry of Hazardous Materials.* Englewood Cliffs, NJ: Prentice Hall.

Rea, W. J. 1995. *Chemical Sensitivity,* Vol. I-III. New York: CRC Press/Lewis Publishers.

Richardson, M. 1995. *Dictionary of Substance and Their Effects,* Vol. I-VI. New York: CRC Press/Lewis Publishers.

Sacarello, H. L. A. 1994. *The Comprehensive Handbook of Hazardous Materials.* New York: CRC Press/Lewis Publishers.

U.S. Department of Health and Human Services, National Institute for Occupational Safety and Health. 1994. *NIOSH Pocket Guide to Chemical Hazards.* Washington, D.C.: U.S. Government Printing Office.

MATERIAL SAFETY DATA SHEET

CHEMICAL NAME	
PENTABORANE	**CRC PRESS/LEWIS PUBLISHERS DISCLAIMER:** THE INFORMATION AND RECOMMENDATIONS PRESENTED HEREIN ARE BASED ON SOURCES BELIEVED TO BE RELIABLE. CRC MAKES NO REPRESENTATION ON ITS COMPLETENESS OR ACCURACY. IT IS THE RESPONSIBILITY OF THE USER TO DETERMINE THE CHEMICAL'S SUITABILITY FOR ITS INTENDED USE, THE CHEMICAL'S SAFE USE, AND THE CHEMICAL'S PROPER DISPOSAL. NO REPRESENTATIONS AND/OR WARRANTIES, EITHER EXPRESSED OR IMPLIED, OF THE MERCHANTABILITY OR FITNESS FOR A PARTICULAR PURPOSE, OR OF ANY OTHER NATURE, ARE MADE WITH RESPECT TO THE INFORMATION PROVIDED IN THIS MSDS OR TO THE CHEMICAL TO WHICH INFORMATION MAY REFER. CRC NEITHER ASSUMES NOR AUTHORIZES ANY OTHER PERSON TO ASSUME FOR IT, ANY OTHER ADDITIONAL RESPONSIBILITY OR LIABILITY FOR THE USE OF, OR RELIANCE UPON, THIS INFORMATION.

HAZARD WARNING INFORMATION

HEALTH	FIRE	REACTIVE	OTHER	DEGREE OF HAZARD	COLOR CODING	OTHER CODES
4	4	2		0 = Minimum Hazard 1 = Slight Hazard 2 = Moderate Hazard 3 = Serious Hazard 4 = Severe Hazard	HEALTH = BLUE FIRE = RED REACTIVITY = YELLOW OTHER = WHITE	OX = Oxidizer ACID = Acid ALK = Alkali COR = Corrosive W = Use No Water

SECTION I - GENERAL INFORMATION

Characterization	**Boron Compound; Hydride**	RCRA Number	**D001**
		EPA Class	**Characteristic (I) Waste**
DOT Proper Shipping Name	**Pentaborane**	Chemical Abstract Service (CAS) Number	**19624-22-7**
DOT Hazard Class and Label Requirements	**Flammable Gas and Poison**	DOT Emergency Guide Code	**75**
DOT Identification Number	**UN 1380**	Chemical Formula	**B_5H_9**

Synonyms

Pentaboron nonahydride; dihydropentaborane (9); pentaboron enneahydride; pentaboron undecahydride.

SECTION II - HAZARDOUS INGREDIENTS AND IDENTITY INFORMATION

Hazardous Components (specific identity)	OSHA Exposure Criteria	NIOSH Exposure Criteria	Immediately Dangerous to Life and Health (IDLH)	ACGIH Exposure Criteria
Pentaborane (derivation: From the hydrogenation of diborane). **1 ppm = 2.62 mg/m³**	**PEL:** **0.005 ppm** **0.01 mg/m³** **STEL (15-min):** **0.015 ppm** **0.03 mg/m³**	**REL:** **0.005 ppm** **0.01 mg/m³** **STEL (15-min):** **0.015 ppm** **0.03 mg/m³**	**1 ppm**	**TLV:** **0.005 ppm** **0.013 mg/m³** **STEL:** **0.015 ppm** **0.039 mg/m³**

SECTION III - PHYSICAL AND CHEMICAL CHARACTERISTICS

Boiling Point	**145°F (60°C)**	Specific Gravity (H_2O = 1)	**0.62**
Vapor Pressure (atmospheres)	**171 at 68°F (20°C)**	Molecular Weight	**63.1**
Vapor Density (Air = 1)	**2.2**	Freezing Point	**-189°F to -52°F (-123°C to -47°C)**

Solubility

Reacts and decomposes in water.

Appearance and Odor

Colorless liquid with a pungent odor like that of sour milk. Odor Threshold = 1.3 ppm.

SECTION IV - FIRE AND EXPLOSION HAZARD DATA

Flash Point (method used)	**86°F (30°C) closed cup**	Explosive Limits in Air % by Volume	**LEL: 0.42% UEL: 98%**
NFPA Classification	**Class IC Flammable Liquid**	Autoignition Temperature	**95°F (35°C)**

Extinguishing Media

Stop leak, if possible. Once leak is stopped, use dry chemical or carbon dioxide. Do NOT use water.

Special Fire Fighting Procedures

Pentaborane is poisonous: wear full protective clothing and self-contained breathing apparatus (SCBA). Fight fire from distance if possible. Stay away from ends of tanks.

Unusual Fire and Explosion Hazards

Pentaborane presents a serious explosion hazard when exposed to flame or involved in fire. Containers can explode in fire. Can react explosively in air. May re-ignite explosively after fire has been extinguished. Cool fire-exposed containers with water spray (do not allow water to contact pentaborane).

SECTION V - REACTIVITY DATA

Stability		Conditions to Avoid Pentaborane is thermally unstable. It decomposes at 300°F (149°C) in closed containers. Avoid contact with heat and incompatible materials. Explosions are likely to be violent and severe.
Stable	Unstable X	Incompatibility *(materials to avoid)* Pentaborane is incompatible with oxidizers, halogens, water, tris(difluoroamine)fluoromethane, dimethyl sulfoxide, oxygen, halogenated hydrocarbons. May ignite spontaneously in moist air.
Hazardous Polymerization		Conditions to Avoid Under normal conditions of temperature and pressure, hazardous polymerization of pentaborane may occur. Avoid elevated temperatures.
May Occur X	Will Not Occur	Hazardous Decomposition or By-products Pentaborane decomposes upon heating or when exposed to water to produce toxic boron fumes.

SECTION VI - HEALTH HAZARD DATA

Primary Route(s) of Entry:	Inhalation?	Absorption (skin)?	Ingestion?
	X	X	X

Health Hazards

INHALATION: Nausea, drowsiness, lethargy and fatigue headache, slurred speech, dizziness, nervous excitation, hiccups, muscle pain, cramps, facial spasms, confusion, shallow respiration, coma, behavioral changes, and possible progression to death.

ABSORPTION: Tremors, confusion, seizures, shallow breathing, and cyanosis. Eye contact produces irritation of the eyes and lids causing symptoms of skin absorption.

INGESTION: Nausea and vomiting along with symptoms of inhalation.

Carcinogenicity	NTP Listed?	IARC Cancer Review Group?	OSHA Regulated?	Target Organs?
Unknown Human Unknown Animal	No	No	29 CFR 1910.1000 (Table Z-1)	Respiratory system, skin, eyes, CNS, kidney, liver.

Medical Conditions Generally Aggravated by Exposure

Blood disorders, skin allergies, lung disorders, kidney and liver diseases.

Emergency and First-aid Procedures

Eye contact: Immediately flush with large amounts of water for 15 minutes (minimum), occasionally lifting eyelids; seek medical attention. Skin contact: Remove all contaminated clothing. Immediately wash area with large amounts of soap and water. For inhalation: Remove the person from exposure. Provide respiratory assistance and CPR. Transfer promptly to medical facility. If swallowed: Not likely but possible (pentaborane is spontaneously flammable in air). Seek medical attention immediately. Never attempt to give anything by mouth to an unconscious or convulsing person. Unless advised otherwise, give I to 2 glasses of water and then induce vomiting.

SECTION VII - PRECAUTIONS FOR SAFE HANDLING AND USE

Steps to be Taken in Case Material is Released or Spilled

Remove all ignition sources. Deny entry to those not involved in cleanup. Ventilate area. Absorb liquids in vermiculite or other material and place in sealed drum for disposal. Use non-sparking tools.

Preferred Waste Disposal Method

No citation.

Precautions to be Taken in Handling and Storage

Do not store with incompatible chemicals since violent reactions can occur. Store in tightly closed containers in a cool, dark, well-ventilated area away from heat and moisture. Suggest a nitrogen purge inside storage containers. Keep away from ignition sources such as fire, sparks, and flame.

Other Precautions and Warnings

Bulk storage of pentaborane is not recommended. Containers may explode in fire or under conditions of extreme heat (do not store outdoors or in direct sunlight).

SECTION VIII - CONTROL MEASURES AND PERSONAL PROTECTIVE EQUIPMENT

Respiratory Protection *(specify type)*

At any exposure level, use a powered supplied-air respirator set in positive pressure mode or use an MSHA/NIOSH-approved self-contained breathing apparatus (SCBA) with full facepiece and pressure demand or other positive pressure mode.

Ventilation

Local exhaust or general mechanical systems recommended.

Protective Gloves	Eye Protection	Other Protective Clothing
Synthetic Impervious Material	Chemical Goggles and/or Face Mask	Protective Suit

Work/Hygiene Practices

Always wash hands thoroughly after using chemical; never bring food, drink, or smoking materials into vicinity of chemicals.

PENTABORANE

B_5H_9

CAS: 19624-22-7

IDENTIFICATION AND TYPICAL USES

Pentaborane is a colorless liquid with a pungent odor like that of sour milk. Used as a fuel for air-breathing engines, as a reducing agent in propellants, a catalyst, a corrosion inhibitor, as a chemical intermediate, as oxygen scavengers, and in fluxing agents.

RISK ASSESSMENT: HEALTH

General Assessment

Pentaborane is a highly toxic compound by all major routes of exposure (skin *absorption, ingestion,* and primarily by *inhalation*). There are no reports of carcinogenic, teratogenic, or mutagenic effects on long-term exposure.

Inhalation of pentaborane is extremely dangerous, causing irritation of the eyes. Symptoms can range from mild to severe and include nausea, drowsiness, lethargy, fatigue, headache, slurred speech, dizziness, nervous excitation, hiccups, muscle pains and cramps, muscle spasms in the face and extremities, mental confusion, incoordination, cramps and convulsions, shallow respiration, coma, and possible death.

Skin absorption can lead to tremors, confusion, seizures, shallow breathing, and cyanosis. Eye contact causes irritation of the eyes and lid followed by the symptoms noted for skin absorption. Caution is warranted to ensure multiple-route exposures do not occur. For example, skin absorption *and* inhalation can result in severe overexposure and possibly lead to life-threatening symptoms.

Ingestion is not likely since pentaborane will usually ignite spontaneously on exposure to air. If ingestion should occur, symptoms include nausea and vomiting as well as the systemic effects noted for inhalation and skin absorption.

Acute Health Effects

The following acute (short-term) health effects may occur immediately or shortly after exposure to pentaborane:

Skin: Irritation at the site of contact with potential for redness, swelling, and inflammation. Absorption is likely to be quick.

Eye: Mild to moderate irritation of the eyes, lids, and conjunctiva.

Lung: Mild to moderate irritation of the nose, throat, and mucous membranes of the respiratory tract. Pentaborane is rapidly absorbed through the respiration process and causes toxic systemic effects resulting from its action of the central nervous system (CNS).

CNS: High exposures may cause severe CNS irritation such as drowsiness, dizziness, visual disturbances, muscle twitching, and, in severe cases, painful muscle spasm, coma, and death.

Chronic Health Effects

The following chronic (long-term) health effects may occur at some time after exposure to pentaborane and can last for months or even years:

Cancer Hazards: According to information presented in the references, pentaborane has not been adequately tested for its ability to cause cancer in test animals.

Reproductive Hazards: According to information presented in the references, pentaborane has not been tested for its ability to affect reproduction.

Other Chronic Effects: Prolonged exposures can lead to anorexia, vomiting and diarrhea, skin rash, hair loss, convulsions, and anemia. It may also cause injuries to the central nervous system, liver, and kidneys.

Recommended Risk-Reduction Measures

Personnel should avoid direct contact with pentaborane. It is extremely toxic and exposures can cause death or serious physical injury or illness. Unless a less toxic chemical can be substituted for a hazardous substance, *engineering controls* are the most effective method of reducing exposures. The best protection is to enclose operations and/or provide local exhaust ventilation at the site of chemical release. While not always operationally feasible, isolating operations can also reduce exposure. Using respiratory protection is less effective than the controls mentioned above, but is always advisable whenever working with pentaborane. A NIOSH-approved self-contained breathing apparatus (SCBA) with full facepiece and pressure demand is the recommended respiratory protection of choice. If a full facepiece is not available, then chemical goggles should be worn to protect the eyes. Whenever a chemical splash hazard exists, a face shield and a rubber apron should be worn. To prevent hand and skin exposures, butyl rubber gloves should be used. Glove manufacturers should be contacted and permeation studies obtained *before* final glove selections are made.

Administrative controls should also be in place to minimize the potential for human exposures. These may include written procedures or policies which specify the methods and techniques that will be practiced whenever personnel are to work with pentaborane.

All personnel should receive *training* on the use, hazards, protective measures, emergency actions, and other precautions per 29 CFR 1910.1200 (Hazard Communication), prior to the first assignment in an area where pentaborane is used or stored. Before beginning employment and at regular intervals thereafter (e.g., annually), the following medical tests are recommended:

☑ Blood tests for glucose, calcium, urea nitrogen and carbon dioxide.

If symptoms develop or overexposure is suspected, the following may be useful:

☑ Repeat blood tests.

☑ Complete neurological examination with careful evaluation of exposure history and special testing.

Other methods to reduce exposure to chemicals include:

☑ Where possible, enclose operations and use local exhaust ventilation at the site of chemical release. If local exhaust ventilation or enclosure is not used, respiratory protection should be mandatory.

☑ Always ensure that proper protective clothing is worn when using chemical substances and that personnel are trained in its use, care, and maintenance.

☑ Wash thoroughly immediately after exposure to pentaborane and at the end of the work shift or before eating, drinking, or smoking. Work clothing contaminated with pentaborane should never be taken home for laundering. Only personnel trained in the hazards of exposure should launder contaminated clothing.

☑ Safety shower and eyewash stations should be readily available in work areas where pentaborane is used or stored.

☑ Hazard warning information should be posted in the work area. In addition, as part of an on-going education and training program, all information on the health and safety hazards of pentaborane should be communicated to all exposed and potentially exposed workers.

RISK ASSESSMENT: ENVIRONMENT

General Assessment

The environment is at risk of exposure during transportation, storage, disposal, or destruction of pentaborane. In almost every scenario, the threat of environmental exposure is contingent upon the proper handling of the chemical substance. Accidental releases, large or small, can result in fire, explosion, and possible contamination of the surrounding environmental mediums (water, soil, and air).

Pentaborane is considered a Class IC flammable liquid per OSHA 29 CFR 1910.106. It is explosive by itself and, because it is incompatible with many other commodities, contact with any of them can result in violent explosion and fire. It is highly reactive and will ignite with moist air at room temperature. Pentaborane forms explosive mixtures in air over a very wide and dangerous range (0.42% - 98%) which makes it a serious fire and explosion risk. It will explode in contact with many oxidizing agents (especially fluorine, bromine, and chlorine). It is highly reactive in the presence of halogens and halogenated hydrocarbons. Therefore, pentaborane requires special consideration during any emergency involving a leak or release of pentaborane gas or liquid. Pentaborane can enter the environment through industrial and municipal waste plant discharges, or through spills.

Acute Ecological Effects

Acute (short-term) toxic effects may include the death of animals, birds, or fish, and death or low growth rate in plants. Acute effects are seen 2 to 4 days after animals or plants are exposed to pentaborane.

Insufficient data are available to evaluate or predict the short-term effects of pentaborane on aquatic life, plants, birds, or land animals.

Chronic Ecological Effects

Chronic (long-term) toxic effects may include shortened life span, reproductive problems, lower fertility, and changes in appearance or behavior in exposed animals. These effects can be seen long after first exposure(s) to toxic chemicals.

Insufficient data are available to evaluate the long-term effects of pentaborane on aquatic life, plants, birds, or land animals.

Water Solubility

Pentaborane decomposes in water.

Persistence in the Environment

Pentaborane will react slowly with water and therefore will not accumulate in the aquatic environment.

Bioaccumulation in Aquatic Organisms

Some substances increase in concentration, or *bioaccumulate,* in living organisms as they breathe contaminated air, drink contaminated water, or eat contaminated food. These chemicals can become concentrated in the tissues and internal organs of animals as well as humans.

Pentaborane will react with water and therefore is not expected to accumulate in the edible tissues of fish.

Recommended Risk-Reduction Measures

Proper training of all transporters will reduce the likelihood of a mishap or accident resulting in a leak or release to the environment. The correct use of DOT labeling (per the MSDS) on all containers, trucks, and rail cars will enable emergency responders to react properly and quickly to any disaster thereby reducing the potential risk to the environment and to personnel.

Storage of pentaborane should be segregated from incompatible chemicals to minimize the risk of cross-contamination or contact. Under all circumstances, it should be protected from exposure to moisture. Buildings designated for storage should be equipped with appropriate fire protection systems (alarms, sprinklers, emergency lighting, portable extinguishers).

If a spill or release to the environment has occurred, fire department, emergency response, and/or hazardous materials spill personnel should be notified immediately. Cleanup should be attempted only by those trained in proper containment procedures using non-sparking tools. Contaminated soils should be removed for incineration and replaced with clean soil. If pentaborane should contact the water table, aquifer, or navigable waterway, remediation may not be possible. It will readily react with water and decompose, and therefore total containment will not be possible. When such spills occur, the local and/or state emergency response authorities must be notified. A comprehensive emergency response or disaster preparedness/recovery plan should be in place prior to any operations involving the use, transportation, storage, or disposal of pentaborane. If pentaborane is spilled or leaked, the following specific steps are recommended:

- ☑ Restrict persons not wearing protective clothing from area of spill or leak until cleanup is complete and area can be opened for normal work.
- ☑ Remove all ignition sources.
- ☑ Ventilate area of spill or leak.
- ☑ Absorb liquids in vermiculite, dry earth, sand, or other suitable material and collect using non-sparking tools. Place collected materials in tightly sealed drums for disposal.
- ☑ Keep pentaborane out of a confined space, such as a sewer, because of the possibility of explosion (unless the sewer is designed to prevent the buildup of explosive concentrations).
- ☑ It may be necessary to dispose of pentaborane as a hazardous waste. The responsible state agency or the regional office of the federal Environmental Protection Agency (EPA) should be contacted for specific recommendations.

RISK ASSESSMENT: BUSINESS

General Assessment

Accidents or mishaps involving pentaborane can present a significant threat to business operations. The loss or damage of equipment or facilities can significantly affect fiscal viability. Lawsuits that may result from personnel injury, public exposures, and environmental contamination will also require a serious expenditure of resources. Media attention surrounding an injury, illness, or environmental damage can also result in a loss of profits and loss of current as well as future business.

Recommended Risk-Reduction Measures

Company attorneys, safety and health professionals, and environmental specialists should be involved in the development of any procedures dealing with safety or emergency response. A company official should be pre-designated as a public relations officer with specific training in dealing with the press. Corporate plans and policies should be developed, approved, and implemented long before any need for such arises.

REFERENCES

American Conference of Governmental Industrial Hygienists. 1988. *Documentation of the Threshold Limit Values and Biological Exposure Indices,* 5th Edition (with updates). Cincinnati: ACGIH

Anthony, C. P., and N. J. Kolthoff. 1971. *Textbook of Anatomy and Physiology.* St. Louis: C. V. Mosby Company.

Asante-Duah, D. K. 1993. *Hazardous Waste Risk Assessment.* New York: CRC Press/Lewis Publishers.

Bloom, A. D. and F. J. DeSerres. 1995. *Ecotoxicity and Human Health.* New York: CRC Press/Lewis Publishers.

Brusick, D. J. 1994. *Methods for Genetic Risk Assessment.* New York: CRC Press/Lewis Publishers.

Bueche, J. R. 1972. *Principles of Physics,* 2nd Edition. New York: McGraw-Hill.

Burrell, R., D. K. Flaherty, and L. J. Sauers. 1992. *Toxicology of the Immune System.* New York: Van Nostrand Reinhold.

Calabrese, E. J., E. M. Kenyon. *Air Toxics and Risk Assessment.* New York: CRC Press/Lewis Publishers.

Cockerman, L. G., B. S. Shane. 1994. *Basic Environmental Toxicology.* New York: CRC Press/Lewis Publishers.

Cothern, R. R. 1993. *Comparative Environmental Risk Assessment.* New York: CRC Press/Lewis Publishers.

Gardner, R. W. 1994. *Chemical Intolerance.* New York: CRC Press/Lewis Publishers.

Hallenbeck, W. H. 1993. *Quantitative Risk Assessment for Environmental and Occupational Health,* 2nd Edition. New York: CRC Press/Lewis Publishers.

Hodgson, E., R. B. Mailman, and J. E. Chambers. 1988. *Dictionary of Toxicology.* London: The MacMillian Company.

Hosty, J. W., and P. Foster. 1990. *A Practical Guide to Chemical Spill Response.* New York: Van Nostrand Reinhold.

Howard, P. H. 1995. *Handbook of Environmental Fate and Exposures Data for Organic Chemicals,* Vol. I-IV. New York: CRC Press/Lewis Publishers.

Howard, P. H., et. al. 1991. *Handbook of Environmental Degradation Rates.* New York: CRC Press/Lewis Publishers.

James, A., and M. Lord. 1992. *Index of Chemical and Physical Data.* New York: Van Nostrand Reinhold.

Jones, F. E. 1994. *Toxic Organic Vapors in the Workplace.* New York: CRC Press/Lewis Publishers.

Kamrin, M. A. 1988. *Toxicology: A Primer on Principles and Applications.* Chelsea, Michigan: Lewis Publishers.

Keith, L. H. and M. M. Walker. 1995. *Handbook of Air Toxics.* New York: CRC Press/Lewis Publishers.

Landis, W. G., M. H. Yu. 1995. *Introduction to Environmental Toxicology.* New York: CRC Press/Lewis Publishers.

Langerman, N. 1994 *Precautionary Labels for Chemical Containers.* New York: CRC Press/Lewis Publishers.

Lauwerys, R. R., P. Hoet. 1993. *Industrial Chemical Exposure.* New York: CRC Press/Lewis Publishers.

Lewis, R. J., Sr. 1993. *Hawley's Condensed Chemical Dictionary,* 12th Edition. New York: Van Nostrand Reinhold.

Lewis, R. J., Sr. 1992. *Sax's Dangerous Properties of Industrial Materials,* Volumes 1, 2, and 3, 8th Edition. New York: Van Nostrand Reinhold.

Lide, D. R. 1995. *Handbook of Chemistry and Physics,* 76th Edition. New York: CRC Press/Lewis Publishers.

Manahan, S. E. 1992. *Toxicological Chemistry,* 2nd Edition. New York: CRC Press/Lewis Publishers.

Meyer, E. 1990. *Chemistry of Hazardous Materials.* Englewood Cliffs, NJ: Prentice Hall.

Patnaik, P. 1992. *A Comprehensive Guide to the Hazardous Properties of Chemical Substances.* New York: Van Nostrand Reinhold.

Philp, R. B. 1995. *Environmental Hazards and Human Health.* New York: CRC Press/Lewis Publishers.

Proctor, N. H., and J. P. Hughes. 1991. *Chemical Hazards of the Workplace.* New York: Van Nostrand Reinhold.

Quigley, D. R. 1994. *Handbook of Emergency Chemical Management.* New York: CRC Press/Lewis Publishers.

Rea, W. J. 1995. *Chemical Sensitivity,* Vol. I-III. New York: CRC Press/Lewis Publishers.

Richardson, M. 1995. *Dictionary of Substance and Their Effects,* Vol. I-VI. New York: CRC Press/Lewis Publishers.

U.S. Department of Health and Human Services, National Institute for Occupational Safety and Health. 1994. *NIOSH Pocket Guide to Chemical Hazards.* Washington, D.C.: U.S. Government Printing Office.

MATERIAL SAFETY DATA SHEET

CHEMICAL NAME	CRC PRESS/LEWIS PUBLISHERS DISCLAIMER:
PENTACHLOROPHENOL	THE INFORMATION AND RECOMMENDATIONS PRESENTED HEREIN ARE BASED ON SOURCES BELIEVED TO BE RELIABLE. CRC MAKES NO REPRESENTATION ON ITS COMPLETENESS OR ACCURACY. IT IS THE RESPONSIBILITY OF THE USER TO DETERMINE THE CHEMICAL'S SUITABILITY FOR ITS INTENDED USE, THE CHEMICAL'S SAFE USE, AND THE CHEMICAL'S PROPER DISPOSAL. NO REPRESENTATIONS AND/OR WARRANTIES, EITHER EXPRESSED OR IMPLIED, OF THE MERCHANTABILITY OR FITNESS FOR A PARTICULAR PURPOSE, OR OF ANY OTHER NATURE, ARE MADE WITH RESPECT TO THE INFORMATION PROVIDED IN THIS MSDS OR TO THE CHEMICAL TO WHICH INFORMATION MAY REFER. CRC NEITHER ASSUMES NOR AUTHORIZES ANY OTHER PERSON TO ASSUME FOR IT, ANY OTHER ADDITIONAL RESPONSIBILITY OR LIABILITY FOR THE USE OF, OR RELIANCE UPON, THIS INFORMATION.

HAZARD WARNING INFORMATION

HEALTH	FIRE	REACTIVE	OTHER	DEGREE OF HAZARD	COLOR CODING	OTHER CODES
DRY: 3 WET: 3	DRY: 3 WET: 2	DRY: 0 WET: 0		0 = Minimum Hazard 1 = Slight Hazard 2 = Moderate Hazard 3 = Serious Hazard 4 = Severe Hazard	HEALTH = BLUE FIRE = RED REACTIVITY = YELLOW OTHER = WHITE	OX = Oxidizer ACID = Acid ALK = Alkali COR = Corrosive W = Use No Water

SECTION I - GENERAL INFORMATION

Characterization	**Phenol**	RCRA Number	**U242**
		EPA Class	**Toxic Waste**
DOT Proper Shipping Name	**Chlorophenols, liquid**	Chemical Abstract Service (CAS) Number	**87-86-5**
DOT Hazard Class and Label Requirements	**ORM-E; St. Andrew's Cross**	DOT Emergency Guide Code	**53**
DOT Identification Number	**UN 2020**	Chemical Formula	**C_6Cl_5OH**

Synonyms

PCP; penta; 2,3,4,5,6-pentachlorophenol; chlorophen; peratox; permacide; permaguard; weedone.

SECTION II - HAZARDOUS INGREDIENTS AND IDENTITY INFORMATION

Hazardous Components (specific identity)	OSHA Exposure Criteria	NIOSH Exposure Criteria	Immediately Dangerous to Life and Health (IDLH)	ACGIH Exposure Criteria
Pentachlorophenol (derivation: By chlorination of phenol in the presence of a catalyst).	**PEL (skin):** 0.5 mg/m^3 **STEL:** Not Established	**PEL (skin):** 0.5 mg/m^3 **STEL:** Not Established	**2.5 mg/m^3**	**TLV (skin):** 0.5 mg/m^3 **STEL:** Not Established

SECTION III - PHYSICAL AND CHEMICAL CHARACTERISTICS

Boiling Point	**588°F (309°C) Decomposes**	Specific Gravity (H_2O = 1)	**1.98**
Vapor Pressure (gas)	**0.0001 at 77°F (25°C)**	Molecular Weight	**266.4**
Vapor Density (Air = 1)	**9.2**	Melting Point	**374°F (190°C)**

Solubility

Slightly soluble (0.001%) in water. Soluble in dilute alkali, alcohol, ether, and benzene.

Appearance and Odor

Colorless to white, or light-brown to tan flakes or crystalline solid with a benzene-like odor and pungent taste.

SECTION IV - FIRE AND EXPLOSION HAZARD DATA

Flash Point (method used)	**Not Applicable**	Explosive Limits in Air % by Volume	**LEL: Not Determined UEL: Not Determined**
NFPA Classification	**Non-Combustible Solid**	Autoignition Temperature	**Not Determined**

Extinguishing Media

Use extinguishing media suitable to surrounding fire. Pentachlorophenol itself will not ignite.

Special Fire Fighting Procedures

Wear full protective clothing and self-contained breathing apparatus (SCBA). Poisonous gases are produced in fire. Use water spray to keep fire-exposed containers cool. Move containers from fire if it can be done without risk. Fight fire from a distance or protected position, if possible.

Unusual Fire and Explosion Hazards

Containers may explode in fire. Wood products that have been treated with 5% pentachlorophenol or a mixture of PCP with petroleum solvents as a preservative and for termite control are combustible.

SECTION V - REACTIVITY DATA

Stability		Conditions to Avoid Pentachlorophenol is normally stable in closed containers under routine conditions of storage and handling. Keep away from exposure to heat or ultraviolet (UV) radiation (e.g., sunlight).
Stable X	Unstable	Incompatibility *(materials to avoid)* Strong oxidizers, (such as chlorine, bromine, and fluorine), acids, and alkalis. Pentachlorophenol solutions exposed to sunlight or UV radiation undergo photochemical degradation.
Hazardous Polymerization		Conditions to Avoid Under normal conditions of temperature and pressure, hazardous polymerization of pentachlorophenol cannot occur.
May Occur	Will Not Occur X	Hazardous Decomposition or By-products When heated to decomposition, pentachlorophenol can emit highly toxic/poisonous gases, including hydrogen chloride, chlorine, and chlorinated hydrocarbons.

SECTION VI - HEALTH HAZARD DATA

Primary Route(s) of Entry:	Inhalation?	Absorption (skin)?	Ingestion?
	X	X	X

Health Hazards

INHALATION: Irritating to the eyes, nose, throat, and respiratory tract causing sneezing, coughing, and lachrymation (tearing). Can also cause tachycardia (rapid heartbeat), tachypnea (rapid breathing), hypertension, fever and muscle weakness, anorexia, sweating, dizziness, and nausea. High concentrations can cause convulsions, loss of consciousness, and death.

ABSORPTION: Irritation and possible development of chloracne with blackheads, whiteheads, yellow cysts and localized inflammation. Passes quickly through intact skin.

INGESTION: Burning of mouth and throat, salivation, dizziness, nausea, vomiting, bright yellow stool, and symptoms of inhalation.

Carcinogenicity	NTP Listed?	IARC Cancer Review Group?	OSHA Regulated?	Target Organs?
Suspected Human Confirmed Animal	No	Group 3	29 CFR 1910.1000 Table Z-1	Skin, eyes, resp. system, CNS, CVS, liver, kidney

Medical Conditions Generally Aggravated by Exposure

Liver, kidney, nervous system, cardiovascular system, endocrine system, and metabolic disorders.

Emergency and First-aid Procedures

Eye contact: Flush immediately with water for 15 minutes (minimum); seek medical attention. **Skin contact**: Remove all contaminated clothing. Immediately wash area with large amounts of soap and water. Seek medical assistance. For **inhalation**: Remove the person from exposure. Provide CPR if required. Transfer to medical facility. If **swallowed**: Contact poison control center. Unless advised otherwise, give 1-2 glasses of water and induce vomiting. Do NOT give milk or other materials containing vegetable or animal fats (they enhance absorption). Never give anything by mouth to unconscious people.

SECTION VII - PRECAUTIONS FOR SAFE HANDLING AND USE

Steps to be Taken in Case Material is Released or Spilled

For dry spills, minimize dust generation. Do NOT dry sweep. Use a vacuum equipped with a high efficiency particulate air (HEPA) filter instead. Damp mop residue. Collect and dispose of in sealed drum. For liquid spills, cover with absorbent (vermiculite or other) and dispose of in sealed drum.

Preferred Waste Disposal Method

No citation.

Precautions to be Taken in Handling and Storage

Store in tightly closed containers in cool, well-ventilated area away from heat or flame. Heat may cause containers to build up pressure and explode. Store in detached facility away from main plant or facility.

Other Precautions and Warnings

Protect containers from physical damage. Do NOT reuse containers without prior decontamination.

SECTION VIII - CONTROL MEASURES AND PERSONAL PROTECTIVE EQUIPMENT

Respiratory Protection *(specify type)*

PEL is low and difficult to monitor. For any exposure level, use a supplied-air respirator with full facepiece in continuous flow mode, or an SCBA in pressure demand or other positive pressure mode.

Ventilation

Local exhaust or general mechanical systems recommended.

Protective Gloves	Eye Protection	Other Protective Clothing
PVC, Neoprene, or Nitrile Latex	Chemical/Dust Goggles or Face Mask	Protective Apron

Work/Hygiene Practices

Always wash hands thoroughly after using chemical; never bring food, drink, or smoking materials into vicinity of chemicals.

PENTACHLOROPHENOL

C_6Cl_5OH

CAS: 87-86-5

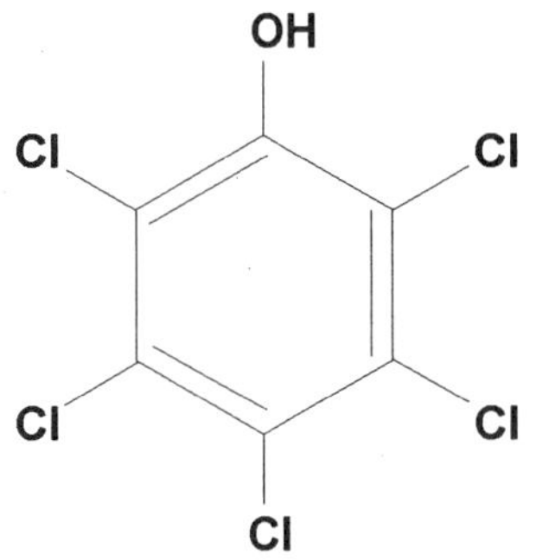

Dry

Solution

IDENTIFICATION AND TYPICAL USES

Pentachlorophenol is colorless to white, or light-brown to tan flake or crystalline solid with a benzene-like odor and pungent taste. It is used as a fungicide, a bactericide, a mulluscicide, an algaecide, an insecticide for termite control, and an herbicide. It is also used in the manufacture of sodium pentachlorophenate. It has been used as a wood preservative, in wood products, and in glues. There are uses in boat construction, building construction, treatment of underground cable coverings, in canvas belting and nets, in lumber and poles (telephone poles), for mold control in petroleum drilling and production, in paints, in pulp and pulp stock products (paper, hardboard, particle board), and in cooling tower water.

RISK ASSESSMENT: HEALTH

General Assessment

Pentachlorophenol is a deadly human poison by *ingestion,* with equally toxic systemic consequences occurring by way of *inhalation,* and skin contact (*absorption*). It has been shown to cause teratogenic effects in experimental tests and mutation data have also been reported. There is sufficient evidence to indicate that exposure to pentachlorophenol causes cancer in test animals and its human carcinogenic potential is considered highly suspect.

Symptoms by all routes of exposure include hypertension (high blood pressure), fever, muscular weakness, anorexia, headache, profuse sweating, warm and flushed skin, rapid respiration (tachypnea), vertigo, and nausea. Exposure to extremely high concentrations may cause loss of consciousness, convulsions, and death from cardiac arrest.

Skin contact can also cause the development of chloracne with blackheads, whiteheads, and yellow cysts. Eye contact causes burning pain, tearing, and inflammation of the eyelids. If not removed immediately, damage to vision may occur.

Ingestion can cause painful burning of the mouth and throat and may result in severe systemic poisoning.

☠ *Acute Health Effects*

The following acute (short-term) health effects may occur immediately or shortly after exposure to pentachlorophenol:

Skin: Irritation and localized inflammation. It will pass through unbroken skin to cause dangerous systemic effects.

Eye: Irritation with damage to the surrounding tissue. If not removed immediately, damage may be more severe and permanent.

Lung: Severe irritation to the respiratory tract and associated mucosa. Can cause coughing, sneezing, and congestion. Rapid absorption through the respiration process can result in life-threatening systemic effects.

CNS: Can cause confusion, anxiety, headache, changes in the function of the respiratory and circulatory systems, convulsions, coma.

CVS: Can cause a disruption in the function of the cardiovascular system with symptoms of elevated blood pressure, pulse, and respiration. Effects are more severe in hot climates.

Chronic Health Effects

The following chronic (long-term) health effects may occur at some time after exposure to pentachlorophenol and can last for months or even years:

Cancer Hazards: According to the information available in the references, pentachlorophenol has been shown to cause cancer in test animals. Experimental mutation data have been reported and many scientists believe that such chemicals may have the ability to cause cancer.

Reproductive Hazard: According to information presented in the references, pentachlorophenol has been shown to cause adverse effects in reproduction in animals. It is considered an experimental teratogen. Human effects in this regard are inconclusive and require additional research.

Other Chronic Effects: Long-term exposures can result in damage to the liver or kidneys. Pentachlorophenol can cause bronchitis with cough and phlegm. Long-term exposures can lead to an ace-like skin rash (chloracne) with persistent blackheads, whiteheads, and yellow cysts. Chronic poisoning may cause weight loss, weakness, and excessive sweating.

Recommended Risk-Reduction Measures

Personnel should avoid direct contact with pentachlorophenol. The exact nature of its carcinogenic properties are not clearly understood in the references. If a less toxic chemical cannot be substituted for a hazardous substance, then *engineering controls* are the most effective method of reducing exposures. The best protection is to enclose operations and/or provide local exhaust ventilation at the site of chemical release. While not always operationally feasible, isolating operations can also reduce exposure. Using respiratory protection is less effective than the controls mentioned above, but is still advisable whenever working with or around pentachlorophenol. The exposure level is relatively low (0.5 mg/m^3) and difficult to monitor. Therefore, a MSHA/NIOSH-approved full facepiece respirator equipped with an organic vapor cartridge equipped with a dust pre-filter should be used if any exposure is anticipated or known. Increased protection is obtained from a supplied-air respirator with full facepiece operated in continuous flow mode, or a self-contained breathing apparatus (SCBA) with full facepiece and pressure demand. If a full facepiece is not available, then chemical dust goggles should be worn to protect the eyes. Whenever an airborne chemical hazard exists, a face shield and a rubber apron should be worn. To prevent hand and skin exposures, butyl rubber gloves should be used. Polyvinyl chloride, neoprene, or nitrile latex all appear to provide acceptable protection. However, glove manufacturers should be contacted and permeation studies obtained *before* final glove selections are made.

Administrative controls should also be in place to minimize the potential for human exposures. These may include written procedures or policies which specify the methods and techniques that will be practiced whenever personnel are to work with pentachlorophenol.

All personnel should receive *training* on the use, hazards, protective measures, emergency actions, and other precautions per 29 CFR 1910.1200 (Hazard Communication), prior to the first assignment in an area where pentachlorophenol is used or stored.

Before beginning employment with pentachlorophenol and at regular intervals thereafter (e.g., annually), those persons with high or frequent exposures should be provided the following recommended medical tests:

☑ Complete blood count (establish baseline).

☑ Urinalysis (establish baseline nitrophenol levels).

If symptoms develop or overexposure is suspected, the following tests are recommended:

☑ Complete blood count (compare to baseline).

☑ Urine test for pentachlorophenol (compare to baseline urinalysis).

☑ Lung function tests.

☑ Liver and kidney function tests.

Any evaluation should include a careful history of past and present symptoms with an examination. Medical tests that simply look for existing damage are not a substitute for controlling exposures.

Other methods to reduce exposure include:

☑ Where possible, enclose operations and use local exhaust ventilation at the site of chemical release. If local exhaust ventilation or enclosure is not used, respiratory protection should be mandatory.

☑ Always ensure that proper protective clothing is worn when using chemical substances.

☑ Wash thoroughly immediately after exposure to pentachlorophenol and at the end of the work shift or before eating, drinking, or smoking.

☑ Hazard warning information should be posted in the work area. In addition, as part of an on-going education and training program, all information on the health and safety hazards of pentachlorophenol should be communicated to all potentially exposed workers.

☑ Eye wash stations should be provided in the immediate work area for emergency use. If there is a possibility of skin exposure to pentachlorophenol, emergency shower facilities should also be provided.

☑ Personnel should never wear contaminated clothing home (family members can be exposed). Clothing should be laundered by personnel who have been trained on the hazards of pentachlorophenol.

RISK ASSESSMENT: ENVIRONMENT

General Assessment

The environment is at risk of exposure during transportation, storage, disposal, or destruction of pentachlorophenol. In almost every scenario, the threat of environmental exposure is contingent upon the proper handling of the chemical substance. Accidental spills, large or small, can result in fire, explosion, and possible contamination of the surrounding environmental mediums (water, soil, and air).

Pentachlorophenol is normally considered a non-combustible solid. In solution, it can often appear in mixtures with other combustible or flammable commodities. It is incompatible with common oxidizers, such as chlorine, bromine, and fluorine, as well as many acids and alkalis. These characteristics require caution in handling, storage, transportation, and disposal. Prolonged contact with heat may result in an explosion. Therefore, special consideration is required during any emergency situation involving a leak or spill of pentachlorophenol. Should pentachlorophenol ever come into contact with any incompatible substances during use, transportation, storage, or disposal, the formation of highly toxic and/or highly explosive commodities is extremely possible.

Pentachlorophenol and its sodium salt are commonly used pesticides in the United States. It may enter the environment through its intended use as an insecticide, from treated wood, as well as through industrial discharges, municipal waste treatment plant discharges, spills, or from agricultural runoff.

Acute Ecological Effects

Acute (short-term) toxic effects may include the death of animals, birds, or fish and death or low growth rate in plants. Acute effects are seen 2 to 4 days after animals or plants are exposed to pentachlorophenol.

The toxicity of pentachlorophenol to aquatic life increases as the pH of the water decreases. Pentachlorophenol has high acute toxicity to aquatic life. It has caused plant and leaf kill and germination decrease in various agricultural and ornamental crops. Insufficient data are available on the short-term effects of pentachlorophenol exposure to birds or land animals.

Chronic Ecological Effects

Chronic (long-term) toxic effects may include shortened life span, reproductive problems, lower fertility, and changes in appearance or behavior in exposed animals. These effects can be seen long after first exposure(s) to toxic chemicals.

Pentachlorophenol has high chronic toxicity to aquatic life. Insufficient data are available on the long-term effects of pentachlorophenol to plants, birds, or land animals.

Water Solubility

Pentachlorophenol is slightly soluble in cold water. Concentrations of 1 to 1000 milligrams will mix with a liter of water.

Persistence in the Environment

Pentachlorophenol is moderately persistent in water, with an ecological half-life between 20 and 200 days. The half-life of a pollutant is the amount of time it takes for one half of the chemical to be degraded. About 48% of pentachlorophenol will eventually end up in terrestrial soil. Approximately 45% will end up in aquatic sediments, about 5.3% will end up in water, and about 1.4% will end up in air.

Bioaccumulation in Aquatic Organisms

Some substances increase in concentration, or *bioaccumulate,* in living organisms as they breathe contaminated air, drink contaminated water, or eat contaminated food. These chemicals can become concentrated in the tissues and internal organs of animals as well as humans.

The concentration of pentachlorophenol found in fish tissue is expected to be much higher than the average concentration of pentachlorophenol in the water from which the fish was taken.

❶ *Recommended Risk-Reduction Measures*

Proper training of all transporters will reduce the likelihood of a mishap or accident resulting in a leak or spill to the environment. The correct use of DOT labeling (per the MSDS) on all containers, trucks, and rail cars will enable emergency responders to react properly and quickly to any disaster thereby reducing the potential risk to the environment and to personnel.

Storage of pentachlorophenol should be segregated from incompatible chemicals to minimize the risk of cross-contamination or contact. Containers must be protected from heat, damage, and shock. Buildings designated for storage should be equipped with appropriate fire protection systems (alarms, sprinklers, emergency lighting, portable extinguishers).

If a spill or leak to the environment has occurred, fire department, emergency response and/or hazardous materials spill personnel should be notified immediately. Cleanup should be attempted only by those trained in proper spill containment procedures. Contaminated soils should be removed for incineration and replaced with clean soil. If pentachlorophenol should contact the water table, aquifer, or navigable waterway, time is of the essence. It is slightly soluble in the aquatic environment and total remediation may not be possible. The local and/or state emergency response authorities must be notified. A comprehensive emergency response or disaster preparedness/recovery plan should be in place prior to any operations involving the use, transportation, storage, or disposal of pentachlorophenol. If pentachlorophenol is spilled or leaked, the following specific steps are recommended:

- ☑ Restrict persons not wearing protective clothing from area of spill or leak until cleanup is complete and area can be opened for normal work.
- ☑ Remove all ignition sources, ventilate area of leak.
- ☑ Absorb liquid spills in vermiculite or similar material and deposit in sealed containers. Collect solid spills for reuse or disposal. Do NOT dry sweep (generates airborne dusts). Use a HEPA vacuum instead and wet-mop spill area to collect residue.
- ☑ It may be necessary to dispose of pentachlorophenol as a hazardous waste. The responsible state agency or the regional office of the federal Environmental Protection Agency (EPA) should be contacted for specific recommendations.

RISK ASSESSMENT: BUSINESS

General Assessment

Accidents or mishaps involving pentachlorophenol can present a significant threat to business operations. The loss or damage of equipment or facilities can significantly affect fiscal viability. Lawsuits that may result from personnel injury/death, public exposures, and/or environmental contamination will also require a serious expenditure of resources. Media attention surrounding an injury, death, or environmental damage can also result in a loss of profits and loss of current as well as future business. Always remember that anytime the terms "cancer," "carcinogen," or "reproductive hazard" are used, public hysteria, anxiety, emotion, and ignorance can all run equally high. This must be carefully considered when drafting or implementing public relations policies.

❶ *Recommended Risk-Reduction Measures*

Company attorneys, safety and health professionals, and environmental specialists should be involved in the development of any procedures dealing with safety or emergency response. A company official should be pre-designated as a public relations officer with specific training in dealing with the press. Corporate plans and policies should be developed, approved, and implemented long before any need for such arises.

REFERENCES

Anthony, C. P., and N. J. Kolthoff. 1971. *Textbook of Anatomy and Physiology.* St. Louis: C. V. Mosby Company.

Bueche, J. R. 1972. *Principles of Physics,* 2nd Edition. New York: McGraw-Hill.

Burrell, R., D. K. Flaherty, and L. J. Sauers. 1992. *Toxicology of the Immune System.* New York: Van Nostrand Reinhold.

Calabrese, E. J. 1994. *Biological Effects of Low Level Exposures.* New York: CRC Press/Lewis Publishers.

Cockerman, L. G., B. S. Shane. 1994. *Basic Environmental Toxicology.* New York: CRC Press/Lewis Publishers.

Environmental Protection Agency. 1983. *Chemical Hazard Information Profiles.* Washington, D.C.: U.S. EPA.

Howard, P. H. 1995. *Handbook of Environmental Fate and Exposures Data for Organic Chemicals,* Vol. I-IV. New York: CRC Press/Lewis Publishers.

Lewis, R. J., Sr. 1992. *Sax's Dangerous Properties of Industrial Materials,* Volumes 1, 2, and 3, 8th Edition. New York: Van Nostrand Reinhold.

Meyer, E. 1990. *Chemistry of Hazardous Materials.* Englewood Cliffs, NJ: Prentice Hall.

MATERIAL SAFETY DATA SHEET

CHEMICAL NAME	CRC PRESS/LEWIS PUBLISHERS DISCLAIMER
***n*-PENTANE**	**CRC PRESS/LEWIS PUBLISHERS DISCLAIMER:** THE INFORMATION AND RECOMMENDATIONS PRESENTED HEREIN ARE BASED ON SOURCES BELIEVED TO BE RELIABLE. CRC MAKES NO REPRESENTATION ON ITS COMPLETENESS OR ACCURACY. IT IS THE RESPONSIBILITY OF THE USER TO DETERMINE THE CHEMICAL'S SUITABILITY FOR ITS INTENDED USE, THE CHEMICAL'S SAFE USE, AND THE CHEMICAL'S PROPER DISPOSAL. NO REPRESENTATIONS AND/OR WARRANTIES, EITHER EXPRESSED OR IMPLIED, OF THE MERCHANTABILITY OR FITNESS FOR A PARTICULAR PURPOSE, OR OF ANY OTHER NATURE, ARE MADE WITH RESPECT TO THE INFORMATION PROVIDED IN THIS MSDS OR TO THE CHEMICAL TO WHICH INFORMATION MAY REFER. CRC NEITHER ASSUMES NOR AUTHORIZES ANY OTHER PERSON TO ASSUME FOR IT, ANY OTHER ADDITIONAL RESPONSIBILITY OR LIABILITY FOR THE USE OF, OR RELIANCE UPON, THIS INFORMATION.

HAZARD WARNING INFORMATION

HEALTH	FIRE	REACTIVE	OTHER	DEGREE OF HAZARD	COLOR CODING	OTHER CODES
1	4	0		0 = Minimum Hazard 1 = Slight Hazard 2 = Moderate Hazard 3 = Serious Hazard 4 = Severe Hazard	HEALTH = BLUE FIRE = RED REACTIVITY = YELLOW OTHER = WHITE	OX = Oxidizer ACID = Acid ALK = Alkali COR = Corrosive W = Use No Water

SECTION I - GENERAL INFORMATION

Characterization	RCRA Number	EPA Class
Hydrocarbon, Aliphatic	**D001**	**Characteristic (I) Waste**

DOT Proper Shipping Name **Pentane**	Chemical Abstract Service (CAS) Number **109-66-0**
DOT Hazard Class and Label Requirements **Flammable Liquid**	DOT Emergency Guide Code **27**
DOT Identification Number **UN 1265**	Chemical Formula $CH_3(CH_2)_3CH_3$

Synonyms
Normal pentane; amyl anhydride.

SECTION II - HAZARDOUS INGREDIENTS AND IDENTITY INFORMATION

Hazardous Components (specific identity)	OSHA Exposure Criteria	NIOSH Exposure Criteria	Immediately Dangerous to Life and Health (IDLH)	ACGIH Exposure Criteria
***n*-Pentane** (derivation: By fractional distillation from petroleum purified by rectification). **1 ppm = 3.00 mg/m³**	**PEL (8-hour):** **1000 ppm** **2950 mg/m³** **STEL:** **750 ppm** **2250 mg/m³**	**REL (10-hour):** **120 ppm** **350 mg/m³** **STEL (15-min):** **610 ppm** **1800 mg/m³**	**1500 ppm**	**TLV:** **600 ppm** **1800 mg/m³** **STEL:** **750 ppm** **2250 mg/m³**

SECTION III - PHYSICAL AND CHEMICAL CHARACTERISTICS

Boiling Point **97°F (36°C)**	Specific Gravity (H_2O = 1) **0.63**
Vapor Pressure (gas) **420 at 68°F (20°C)**	Molecular Weight **72.2**
Vapor Density (Air = 1) **2.5**	Melting Point **-202°F (-130°C)**

Solubility
Slightly soluble (0.04%) in water. Soluble in alcohol and most organic solvents.

Appearance and Odor
A clear, colorless, mobile liquid with a mild, gasoline-like odor. Odor Threshold = 990 ppm.

SECTION IV - FIRE AND EXPLOSION HAZARD DATA

Flash Point (method used) **-40°F (-40°C) closed cup**	Explosive Limits in Air % by Volume **LEL: 1.5% UEL: 7.8%**
NFPA Classification **Class 1A Flammable Liquid**	Autoignition Temperature **500°F (260°C)**

Extinguishing Media
Use dry chemical, carbon dioxide, or foam. Water may be ineffective and may actually spread fire.

Special Fire Fighting Procedures
Highly flammable material. Wear full protective clothing and self-contained breathing apparatus (SCBA). Poisonous gases are produced in fire. Use water spray to keep fire-exposed containers cool.

Unusual Fire and Explosion Hazards
Dangerous fire and explosion hazard when exposed to heat or flame. Containers may explode in fire. Chemical will react on contact with oxidizable materials. Moderate explosion hazard when in the form of vapor. Its vapors are heavier than air and may travel for some distance to flashback and cause fire.

SECTION V - REACTIVITY DATA

Stability		Conditions to Avoid Keep away from heat or flame. *N*-pentane is normally stable in closed containers under routine conditions of handling and storage.
Stable X	Unstable	Incompatibility *(materials to avoid)* Strong oxidizers, such as permanganates, peroxides, nitrates, chlorates, and perchlorates.
Hazardous Polymerization		Conditions to Avoid Under normal conditions of temperature and pressure, hazardous polymerization of *n*-pentane is not expected to occur.
May Occur	Will Not Occur X	Hazardous Decomposition or By-products When heated to decomposition, *n*-pentane can emit highly toxic/poisonous gases, including oxides of carbon.

SECTION VI - HEALTH HAZARD DATA

Primary Route(s) of Entry:	Inhalation? X	Absorption (skin)?	Ingestion? X

Health Hazards

INHALATION: Vapors of *n*-pentane appear to be mildly narcotic and may cause irritation of the eyes, nose, throat, and respiratory tract. Very high concentrations (in excess of 5000 ppm) for prolonged periods may cause central nervous system depression with narcosis and drowsiness.

SKIN & EYES: Irritating to the skin, may cause drying, cracking, and chapping. Severe eye irritant.

INGESTION: Causes nausea and vomiting. Aspiration of vomitus into the lungs can cause chemical pneumonitis and can present an asphyxiation hazard.

Carcinogenicity	NTP Listed?	IARC Cancer Review Group?	OSHA Regulated?	Target Organs?
Unknown Human Unknown Animal	No	No	29 CFR 1910.1000 Table Z-1	Respiratory system, eyes, skin, CNS.

Medical Conditions Generally Aggravated by Exposure

None reported.

Emergency and First-aid Procedures

Eye contact: Flush immediately with water for 15 minutes (minimum); seek medical attention. Skin contact: Remove all contaminated clothing. Immediately wash area with large amounts of soap and water. Seek medical assistance. For inhalation: Remove the person from exposure. Provide respiratory assistance and CPR, if required. Transfer to medical facility. Keep victim warm and at rest. If swallowed: Seek medical attention immediately. Warning! Aspiration hazard; do NOT induce vomiting.

SECTION VII - PRECAUTIONS FOR SAFE HANDLING AND USE

Steps to be Taken in Case Material is Released or Spilled

Absorb in vermiculite, dry sand, earth, or similar material and deposit in sealed containers. Ventilate area of spill or leak. Restrict those not involved in cleanup from entering area. Remove all sources of ignition and use only non-sparking tools during cleanup. Notify proper authorities.

Preferred Waste Disposal Method

Burn in chemical incinerator equipped with an afterburner and scrubber.

Precautions to be Taken in Handling and Storage

Store to avoid contact with incompatible materials in tightly closed containers in cool, well-ventilated area away from heat and flame. Use only non-sparking tools to open and close containers.

Other Precautions and Warnings

Sources of ignition are prohibited where *n*-pentane is used, handled, or stored. Heat may cause containers to build up pressure and explode. Containers must be grounded and bonded during transfer.

SECTION VIII - CONTROL MEASURES AND PERSONAL PROTECTIVE EQUIPMENT

Respiratory Protection *(specify type)*

For high exposures (above the PEL), use a MSHA/NIOSH-approved organic vapor respirator. Best protection is obtained using a supplied-air respirator with full facepiece in continuous flow mode, or a self-contained breathing apparatus (SCBA) operated in pressure demand or other positive pressure mode.

Ventilation

Local exhaust or general mechanical systems recommended.

Protective Gloves	Eye Protection	Other Protective Clothing
Rubber or Neoprene	Chemical Goggles or Face Mask	Protective Apron

Work/Hygiene Practices

Always wash hands thoroughly after using chemical; never bring food, drink, or smoking materials into vicinity of chemicals.

n-PENTANE

$CH_3(CH_2)_3CH_3$

CAS: 109-66-0

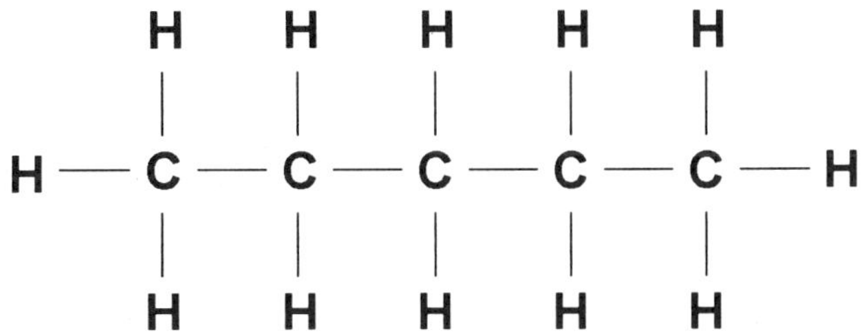

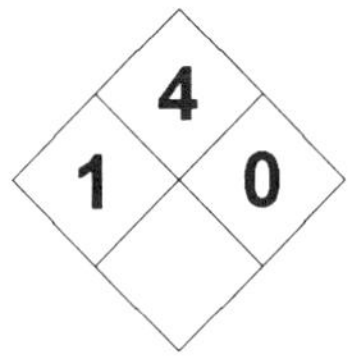

IDENTIFICATION AND TYPICAL USES

n-Pentane is a colorless liquid with a mild odor of gasoline. It can exist as a gas above 97°F (36°C). It is used in the manufacture of artificial ice, in low-temperature thermometers, solvent extraction processes, as a blowing agent in plastics (expandable polystyrene), and as a pesticide.

RISK ASSESSMENT: HEALTH

General Assessment

n-Pentane is a mild irritant by *inhalation*. It can also enter the body through *ingestion* and cause some degree of gastrointestinal discomfort. Skin contact can also produce some adverse topical effects but dermal absorption has not been reported. There is no clear evidence in the references to indicate that exposure to *n*-pentane will lead to carcinogenic, mutagenic, or teratogenic effects.

Significant inhalation may cause irritation of the respiratory tract and depress the central nervous system. Symptoms include nausea, dizziness, vomiting, lightheadedness, giddiness, stupor, vertigo, loss of coordination, loss of appetite, depression, and hallucinations. At high concentrations, it is a narcotic and can lead to muscle weakness, cold feeling in extremities, blurred vision, anorexia, headaches, polyneuropathy, loss of consciousness, coma, and death due to respiratory failure. Studies have shown that compounds similar in structure to *n*-pentane will accumulate in fatty tissues as is evidenced by its affinity for the blood, liver, and brain where lipids are prevalent. It is not known if *n*-pentane will cause these effects.

Skin and eye contact may produce immediate irritation with redness, painful burning, and possible blistering on contact. Prolonged or repeated contact with the skin can cause dryness, cracking, and dermatitis. Ingestion will cause immediate gastrointestinal irritation and presents a serious aspiration hazard. If the liquid is aspirated into the lungs, oxygen can be displaced and the resultant asphyxiation can lead to brain damage or even cardiac arrest.

☠ *Acute Health Effects*

The following acute (short-term) health effects may occur immediately or shortly after exposure to *n*-pentane:

Skin: Severe and painful irritation with possible burns and blistering.

Eye: Severe irritation and burning on contact.

Lung: Direct irritation of the respiratory tract and associated mucosa. Exposure may cause dizziness, vomiting, nausea, and can lead to chemical pneumonia.

CNS: A CNS depressant: headache, dizziness, depression, hallucinations, changes in mood, loss of consciousness, coma, and death (respiratory failure).

Chronic Health Effects

The following chronic (long-term) health effects may occur at some time after exposure to *n*-pentane and can last for months or even years:

Cancer Hazards: According to the references, *n*-pentane has not been adequately tested for its ability to cause cancer in animals.

Reproductive Hazard: According to information presented in the references, *n*-pentane has not been tested for its ability to adversely affect reproduction.

Other Chronic Effects: Repeated contact with the skin may cause dermatitis. Other chronic effects have not been documented. Insufficient exposure data exist for humans or animals to properly evaluate or predict the effects of long-term exposure.

Recommended Risk-Reduction Measures

Personnel should avoid direct contact with *n*-pentane since it is known to cause significant adverse health effects, especially if improperly used or if overexposures occur. If a less toxic chemical cannot be substituted for a hazardous substance, then *engineering controls* are the most effective method of reducing exposures. The best protection is to enclose operations and/or provide local exhaust ventilation at the site of chemical release. While not always operationally feasible, isolating operations can also reduce exposure. Using respiratory protection is less effective than the controls mentioned above, but is still advisable whenever working with or around *n*-pentane. For exposures above the PEL (1000 ppm), an organic vapor respirator should be used (per OSHA 29 CFR 1910.134). Better protection is provided by a supplied-air respirator with full facepiece operated in continuous flow mode or a self-contained breathing apparatus (SCBA) with full facepiece in pressure demand. If a full facepiece is not available, then chemical goggles should be worn to protect the eyes. Whenever a chemical splash hazard exists, a face shield and a rubber apron should be worn. To prevent hand and skin exposures, impervious gloves should be used. Rubber or neoprene gloves are recommended. Glove manufacturers should be contacted and permeation studies obtained *before* final glove selections are made.

Administrative controls should also be in place to minimize the potential for human exposures. These may include written procedures or policies which specify the methods and techniques that will be practiced whenever personnel are to work with *n*-pentane.

All personnel should receive *training* on the use, hazards, protective measures, emergency actions, and other precautions per 29 CFR 1910.1200 (Hazard Communication), prior to the first assignment in an area where *n*-pentane is used or stored.

If symptoms develop or overexposure is suspected, the following tests are recommended:

☑ Lung function tests.

☑ Evaluation of nervous system function.

☑ Evaluation by a qualified allergist with careful consideration of exposure history and special testing (may help diagnose skin allergy and/or dermatitis).

Any evaluation should include a careful history of past and present symptoms with an examination. Medical tests that simply look for existing damage are not a substitute for controlling exposures.

Other methods to reduce exposure include:

☑ Where possible, enclose operations and use local exhaust ventilation at the site of chemical release. If local exhaust ventilation or enclosure is not used, respiratory protection should be mandatory.

☑ Always ensure that proper protective clothing is worn when using chemical substances and personnel are trained on its use, care, and maintenance.

☑ Wash thoroughly immediately after exposure to *n*-pentane and at the end of the work shift or before eating, drinking, or smoking.

☑ Hazard warning information should be posted in the work area. In addition, as part of an on-going education and training program, all information on the health and safety hazards of *n*-pentane should be communicated to all exposed and potentially exposed workers.

☑ Eye wash stations should be provided in the immediate work area for emergency use. If there is a possibility of skin exposure to *n*-pentane, emergency shower facilities should also be provided.

☑ Personnel should never wear contaminated clothing home (family members can be exposed). Clothing should be laundered by personnel who have been trained on the hazards of *n*-pentane.

RISK ASSESSMENT: ENVIRONMENT

General Assessment

The environment is at risk of exposure during transportation, storage, disposal, or destruction of *n*-pentane. In almost every case, the threat of environmental exposure is contingent upon the proper handling of the chemical substance. Accidental spills, large or small, can result in fire, explosion, and possible contamination of the surrounding environmental mediums (water, soil, and air).

n-Pentane is considered a Class 1A flammable liquid (per OSHA 29 CFR 1910.106). Its extremely low flash point and relatively low boiling point make this chemical a serious fire and explosion hazard concern. These characteristics require extreme caution in handling, storage, transportation, and disposal. It is incompatible with common peroxides and other oxidizers and contact can cause fire or explosion. Its vapors are heavier than air and can travel for great distances to an ignition source to flashback and cause fire or explosion. If *n*-pentane containers or tanks are located near a fire or sustained high heat source (e.g., steam pipes, boilers), a boiling liquid expanding vapor explosion (BLEVE) is possible. The results can be extremely destructive. Therefore, special consideration is required during any emergency situation involving a leak or spill of *n*-pentane. Should *n*-pentane ever come into contact with these incompatible substances during use, transportation, storage, or disposal, the formation of highly toxic and/or highly explosive commodities is possible.

n-Pentane may enter the environment through industrial discharges or spills.

Acute Ecological Effects

Acute (short-term) toxic effects may include the death of animals, birds, or fish and death or low growth rate in plants. Acute effects are seen 2 to 4 days after animals or plants are exposed to *n*-pentane.

Insufficient data are available on the short-term effects of *n*-pentane exposure to aquatic life, plants, birds, or terrestrial animals.

Chronic Ecological Effects

Chronic (long-term) toxic effects may include shortened life span, reproductive problems, lower fertility, and changes in appearance or behavior in exposed animals. These effects can be seen long after first exposure(s) to toxic chemicals.

Insufficient data are available to evaluate or predict the long-term effects of *n*-pentane to aquatic life, plants, birds, or land animals.

Water Solubility

n-Pentane is nearly insoluble in water. Concentrations of less than 10 milligrams may mix with a liter of water.

Persistence in the Environment

n-Pentane is non-persistent in water, with a half-life of less than 2 days. The half-life of a pollutant is the amount of time it takes for one half of the chemical to be degraded. Virtually 100% of *n*-pentane will eventually end up in air.

Bioaccumulation in Aquatic Organisms

Some substances increase in concentration, or *bioaccumulate,* in living organisms as they breathe contaminated air, drink contaminated water, or eat contaminated food. These chemicals can become concentrated in the tissues and internal organs of animals as well as humans.

The concentration of *n*-pentane found in fish tissue is expected to be about the same as the average concentration of *n*-pentane in the water from which the fish was taken.

Recommended Risk-Reduction Measures

Proper training of all transporters will reduce the likelihood of a mishap or accident resulting in a leak or spill to the environment. The correct use of DOT labeling (per the MSDS) on all containers, trucks, and rail cars will enable emergency responders to react properly and quickly to any disaster thereby reducing the potential risk to the environment and to personnel.

Storage of *n*-pentane should be segregated from incompatible chemicals to minimize the risk of cross-contamination or contact. Buildings designated for storage should be equipped with appropriate fire protection systems (alarms, sprinklers, emergency lighting, portable extinguishers). All transfers of *n*-pentane from storage containers to process containers should be done automatically whenever possible. All containers and equipment must be grounded and bonded.

If a spill or leak to the environment has occurred, fire department, emergency response and/or hazardous materials spill personnel should be notified immediately. Cleanup should be attempted only by those trained in proper spill containment procedures using non-sparking tools. Contaminated soils should be removed for incineration and replaced with clean soil. If *n*-pentane should contact the water table, aquifer, or navigable waterway, the local and/or state emergency response authorities must be notified. A comprehensive emergency response or disaster preparedness/recovery plan should be in place prior to any operations involving the use, transportation, storage, or disposal of *n*-pentane.

To dispose of *n*-pentane properly, it should be burned in a chemical incinerator equipped with an afterburner and scrubber to ensure appropriate emission control.

If *n*-pentane is spilled or leaked, the following specific steps are recommended:

- ☑ Restrict persons not wearing protective clothing from area of spill or leak until cleanup is complete and area can be opened for normal work.
- ☑ Remove all ignition sources, ventilate area of leak.
- ☑ Absorb liquids in vermiculite, dry sand, earth, or similar material and deposit in sealed containers. Collect materials using non-sparking tools.
- ☑ It may be necessary to dispose of *n*-pentane as a hazardous waste. The responsible state agency or the regional office of the federal Environmental Protection Agency (EPA) should be contacted for specific recommendations.

RISK ASSESSMENT: BUSINESS

General Assessment

Accidents or mishaps involving *n*-pentane can present a significant threat to business operations. The loss or damage of equipment or facilities can significantly affect fiscal viability. Lawsuits that may result from personnel injury/death, public exposures, and/or environmental contamination will also require a serious expenditure of resources. Media attention surrounding an injury, death, or environmental damage can also result in a loss of profits and loss of current as well as future business.

Recommended Risk-Reduction Measures

Company attorneys, safety and health professionals, and environmental specialists should be involved in the development of any procedures dealing with safety or emergency response. A company official should be pre-designated as a public relations officer with specific training in dealing with the press. Corporate plans and policies should be developed, approved, and implemented long before any need for such arises.

REFERENCES

American Conference of Governmental Industrial Hygienists. 1988. *Documentation of the Threshold Limit Values and Biological Exposure Indices,* 5th Edition (with updates). Cincinnati: ACGIH

Bueche, J. R. 1972. *Principles of Physics,* 2nd Edition. New York: McGraw-Hill.

Calabrese, E. J. 1994. *Biological Effects of Low Level Exposures.* New York: CRC Press/Lewis Publishers.

Cockerman, L. G., B. S. Shane. 1994. *Basic Environmental Toxicology.* New York: CRC Press/Lewis Publishers.

Cothern, C. R. 1995. *Handbook for Environmental Risk Decisionmaking.* New York: CRC Press/Lewis Publishers.

Environmental Protection Agency. 1983. *Chemical Hazard Information Profiles.* Washington, D.C.: U.S. EPA.

Gardner, R. W. 1994. *Chemical Intolerance.* New York: CRC Press/Lewis Publishers.

Hallenbeck, W. H. 1993. *Quantitative Risk Assessment for Environmental and Occupational Health,* 2nd Edition. New York: CRC Press/Lewis Publishers.

Hosty, J. W., and P. Foster. 1990. *A Practical Guide to Chemical Spill Response.* New York: Van Nostrand Reinhold.

Hodgson, E., R. B. Mailman, and Janice E. Chambers. 1988. *Dictionary of Toxicology.* London: The MacMillian Company.

Kamrin, M. A. 1988. *Toxicology: A Primer on Principles and Applications.* Chelsea, Michigan: Lewis Publishers.

Keith, L. H. and M. M. Walker. 1995. *Handbook of Air Toxics.* New York: CRC Press/Lewis Publishers.

Landis, W. G., M. H. Yu. 1995. *Introduction to Environmental Toxicology.* New York: CRC Press/Lewis Publishers.

Langerman, N. 1994 *Precautionary Labels for Chemical Containers.* New York: CRC Press/Lewis Publishers.

Lauwerys, R. R., P. Hoet. 1993. *Industrial Chemical Exposure.* New York: CRC Press/Lewis Publishers.

Lewis, R. J., Sr. 1992. *Sax's Dangerous Properties of Industrial Materials,* Volumes 1, 2, and 3, 8th Edition. New York: Van Nostrand Reinhold.

Manahan, S. E. 1994. *Environmental Chemistry,* 6th Edition. New York: CRC Press/Lewis Publishers.

Meyer, E. 1990. *Chemistry of Hazardous Materials.* Englewood Cliffs, NJ: Prentice Hall.

Mycock, J. C., J. D. McKenna, and L. Theodore. 1995. *Handbook of Air Pollution Control Engineering and Technology.* New York: CRC Press/Lewis Publishers.

Patnaik, P. 1992. *A Comprehensive Guide to the Hazardous Properties of Chemical Substances.* New York: Van Nostrand Reinhold.

Quigley, D. R. 1994. *Handbook of Emergency Chemical Management.* New York: CRC Press/Lewis Publishers.

Rea, W. J. 1995. *Chemical Sensitivity,* Vol. I-III. New York: CRC Press/Lewis Publishers.

Richardson, M. 1995. *Dictionary of Substance and Their Effects,* Vol. I-VI. New York: CRC Press/Lewis Publishers.

Sacarello, H. L. A. 1994. *The Comprehensive Handbook of Hazardous Materials.* New York: CRC Press/Lewis Publishers.

Torkelson, T. R. V. K. Rowe. 1981. Halogenated aliphatic hydrocarbons. *Patty's Industrial Hygiene and Toxicology,* Vol. 2B, *Toxicology,* pp 3568-3572. New York: Wiley-Interscience.

U.S. Department of Health and Human Services, National Institute for Occupational Safety and Health. 1994. *NIOSH Pocket Guide to Chemical Hazards.* Washington, D.C.: U.S. Government Printing Office.

MATERIAL SAFETY DATA SHEET

CHEMICAL NAME	
2-PENTANONE	CRC PRESS/LEWIS PUBLISHERS DISCLAIMER: THE INFORMATION AND RECOMMENDATIONS PRESENTED HEREIN ARE BASED ON SOURCES BELIEVED TO BE RELIABLE. CRC MAKES NO REPRESENTATION ON ITS COMPLETENESS OR ACCURACY. IT IS THE RESPONSIBILITY OF THE USER TO DETERMINE THE CHEMICAL'S SUITABILITY FOR ITS INTENDED USE, THE CHEMICAL'S SAFE USE, AND THE CHEMICAL'S PROPER DISPOSAL. NO REPRESENTATIONS AND/OR WARRANTIES, EITHER EXPRESSED OR IMPLIED, OF THE MERCHANTABILITY OR FITNESS FOR A PARTICULAR PURPOSE, OR OF ANY OTHER NATURE, ARE MADE WITH RESPECT TO THE INFORMATION PROVIDED IN THIS MSDS OR TO THE CHEMICAL TO WHICH INFORMATION MAY REFER. CRC NEITHER ASSUMES NOR AUTHORIZES ANY OTHER PERSON TO ASSUME FOR IT, ANY OTHER ADDITIONAL RESPONSIBILITY OR LIABILITY FOR THE USE OF, OR RELIANCE UPON, THIS INFORMATION.

HAZARD WARNING INFORMATION

HEALTH	FIRE	REACTIVE	OTHER	DEGREE OF HAZARD	COLOR CODING	OTHER CODES
2	**3**	**0**		0 = Minimum Hazard 1 = Slight Hazard 2 = Moderate Hazard 3 = Serious Hazard 4 = Severe Hazard	HEALTH = BLUE FIRE = RED REACTIVITY = YELLOW OTHER = WHITE	OX = Oxidizer ACID = Acid ALK = Alkali COR = Corrosive ~~W~~ = Use No Water

SECTION I - GENERAL INFORMATION

Characterization **Ketone**	RCRA Number **D001**	EPA Class **Characteristic (I) Waste**
DOT Proper Shipping Name **Methyl Propyl Ketone**	Chemical Abstract Service (CAS) Number **107-87-9**	
DOT Hazard Class and Label Requirements **Flammable Liquid**	DOT Emergency Guide Code **26**	
DOT Identification Number **UN 1249**	Chemical Formula **$CH_3COCH_2CH_2CH_3$**	

Synonyms
Ethyl acetone; methyl propyl ketone; MPK; methyl *n*-propyl ketone.

SECTION II - HAZARDOUS INGREDIENTS AND IDENTITY INFORMATION

Hazardous Components (specific identity)	OSHA Exposure Criteria	NIOSH Exposure Criteria	Immediately Dangerous to Life and Health (IDLH)	ACGIH Exposure Criteria
2-Pentanone: (derivation: Synthesized by oxidation of 2-pentanol). **1 ppm = 3.58 mg/m^3**	**PEL:** **200 ppm** **700 mg/m^3** **STEL (15-min):** **250 ppm** **875 mg/m^3**	**REL:** **150 ppm** **530 mg/m^3** **STEL:** **Not Established**	**1500 ppm**	**TLV:** **200 ppm** **705 mg/m^3** **STEL:** **250 ppm** **881 mg/m^3**

SECTION III - PHYSICAL AND CHEMICAL CHARACTERISTICS

Boiling Point **215°F (102°C)**	Specific Gravity (H_2O = 1) **0.81**
Vapor Pressure (mm Hg) **27 at 68°F (20°C)**	Molecular Weight **86.1**
Vapor Density (Air = 1) **3.00**	Freezing Point **-108°F (-79°C)**

Solubility
Slightly soluble in water (6%). Also soluble in alcohol and diethyl ether.

Appearance and Odor
Colorless to water-white liquid with an acetone-like or fruity, ethereal odor. Odor Threshold = 8 ppm.

SECTION IV - FIRE AND EXPLOSION HAZARD DATA

Flash Point (method used) **45°F (7°C) closed cup**	Explosive Limits in Air % by Volume **LEL: 1.5% UEL: 8.2%**
NFPA Classification **Class 1BFlammable Liquid**	**Autoignition Temperature** **846°F (452°C)**

Extinguishing Media
Dry chemical, carbon dioxide, water spray, fog, or alcohol-resistant foam.

Special Fire Fighting Procedures
Poisonous gases are produced in fire. Wear full protective clothing and self-contained breathing apparatus. Fight fire from a distance if possible. Keep fire-exposed containers cool with water spray. A solid stream of water may spread fire. If it can be done safely, move fire-exposed containers from area.

Unusual Fire and Explosion Hazards
Containers may explode in fire due to the build up of internal pressures. BLEVE (boiling liquid expanding vapor explosion) is possible. Immediately evacuate if rising sound is heard from venting device or a noticeable discoloration is observed on the sides of the tank or storage vessel.

SECTION V - REACTIVITY DATA

Stability		Conditions to Avoid **Normally stable. Do not allow 2-pentanone to come into contact with incompatible materials. Avoid contact with fire or high heat sources. Do not use in confined spaces.**
Stable **X**	Unstable	Incompatibility *(materials to avoid)* **Incompatible with strong oxidizers (chlorine, bromine, fluorine) and bromine trifluoride (explodes on contact).**
Hazardous Polymerization		Conditions to Avoid **Under normal conditions of temperature and pressure, hazardous polymerization of 2-pentanone is not expected to occur.**
May Occur	Will Not Occur **X**	Hazardous Decomposition or By-products **When heated to decomposition, 2-pentanone emits acrid, irritating smoke and fumes, including oxides of carbon.**

SECTION VI - HEALTH HAZARD DATA

Primary Route(s) of Entry:	Inhalation?	Absorption (skin)?	Ingestion?
	X		X

Health Hazards

INHALATION: Causes coughing, shortness of breath, and eye, nose, throat, and respiratory system irritation. At high concentration, there may be narcosis with headache, lightheadedness, nausea and vomiting, and the possibility of coma (extremely high exposures).

SKIN & EYES: Skin contact with liquid can result in irritation and may cause burning, dermatitis and redness (erythema). Eye contact causes pain, irritation, corneal damage, and tearing.

INGESTION: Symptoms of irritation with nausea and vomiting.

Carcinogenicity	NTP Listed?	IARC Cancer Review Group?	OSHA Regulated?	Target Organs?
Unknown Human Unknown Animal	No	No	29 CFR 1910.1000 Table Z-1	Respiratory system; skin; eyes; CNS.

Medical Conditions Generally Aggravated by Exposure

Skin conditions (dermatitis) and respiratory tract disorders may be aggravated.

Emergency and First-aid Procedures

Eye contact: Immediately flush with large amounts of water for 15 minutes (minimum), occasionally lifting eyelids. Skin contact: Remove all contaminated clothing. Immediately wash area with large amounts of soap and water. For red or blistered skin, seek medical assistance. For inhalation: Remove the person from exposure. Provide respiratory assistance and CPR. Transfer promptly to medical facility. If swallowed: Seek medical attention immediately or contact poison control center. Unless advised otherwise, give victim 1 to 2 glasses of water and induce vomiting. Never attempt to give anything by mouth to an unconscious or convulsing person.

SECTION VII - PRECAUTIONS FOR SAFE HANDLING AND USE

Steps to be Taken in Case Material is Released or Spilled

Remove all ignition sources. Restrict persons not wearing protective equipment from entering area of spill or leak until cleanup is complete. Ventilate area. Absorb liquid in vermiculite or other material and deposit in sealed containers. Damp mop residue. Alert appropriate state or federal response agencies.

Preferred Waste Disposal Method

Burn in a chemical incinerator equipped with an afterburner and scrubber.

Precautions to be Taken in Handling and Storage

Do not store with incompatible chemicals since violent reactions can occur. Store in tightly closed containers in a cool, dark, well-ventilated area away from fire, sparks, and flame. Use non-sparking tools on containers. Ground and bond metal containers during transfer operations.

Other Precautions and Warnings

Protect containers from physical damage.

SECTION VIII - CONTROL MEASURES AND PERSONAL PROTECTIVE EQUIPMENT

Respiratory Protection *(specify type)*

Exposures above 200 ppm: MSHA/NIOSH-approved full facepiece respirator with organic vapor cartridge. Greater protection is obtained from a self-contained breathing apparatus (SCBA) with full facepiece and pressure demand or other positive pressure mode.

Ventilation

Local exhaust or general mechanical systems recommended.

Protective Gloves	Eye Protection	Other Protective Clothing
Impervious Materials	Chemical Goggles and/or Face Mask	Protective Apron

Work/Hygiene Practices

Always wash hands thoroughly after using chemical; never bring food, drink, or smoking materials into vicinity of chemicals.

2-PENTANONE

$CH_3COCH_2CH_2CH_3$

CAS: 107-87-9

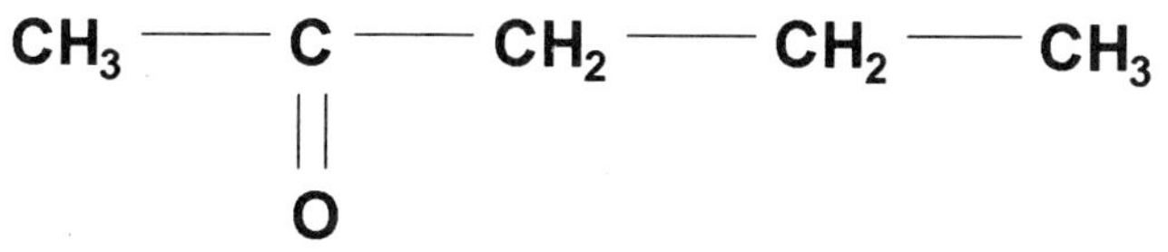

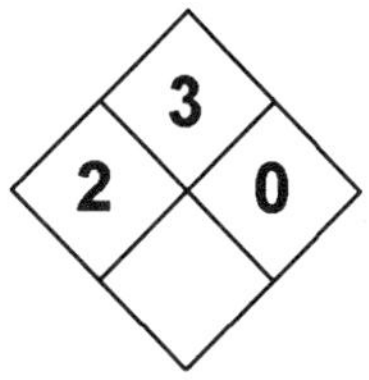

IDENTIFICATION AND TYPICAL USES

2-Pentanone is a colorless to water-white liquid with a characteristic acetone-like odor. It is used as a solvent for surface coatings, as a flavoring agent, as an intermediate for cosmetics and pharmaceuticals, in organic synthesis, and in lacquers and lacquer removers.

RISK ASSESSMENT: HEALTH

General Assessment

2-Pentanone is mildly toxic to humans by *inhalation* and *ingestion*. Skin or eye contact can also cause surface tissue damage and inflammation but dermal absorption has not been reported. The references contain no information on the carcinogenic, teratogenic, or mutagenic properties of 2-pentanone. Overexposure to this compound is not common. Its irritating properties and its penetrating odor are usually present at levels low enough to serve as a warning to personnel.

Inhalation of 2-pentanone causes irritation of the mucosa of the respiratory tract leading to mild respiratory depression with symptoms of coughing, shortness of breath, and mild discomfort. Exposure to higher concentrations can cause depression of the central nervous system (CNS) with symptoms of excitability, headache, lightheadedness, nausea, and vomiting. Higher concentrations can lead to dizziness, intoxication, stupor, staggering gait, loss of coordination, mental confusion, and coma.

Skin contact with the liquid can lead to chemical dermatitis, inflammation, redness, and rash. Eye contact causes irritation with lachrymation and, in prolonged contact, may lead to ocular pain, corneal injury, and possible visual disturbances (blurred vision). Although ingestion is unlikely in the occupational setting, it can cause irritation of the mouth and throat with nausea and vomiting.

Acute Health Effects

The following acute (short-term) health effects may occur immediately or shortly after exposure to 2-pentanone:

Skin: Irritation and dermatitis with possible localized rash and redness.

Eye: Irritation with pain and inflammation that can lead to corneal injury and possible visual disturbances.

Lung: Irritation causing coughing and/or shortness of breath.

CNS: A mild to moderate narcotic compound producing narcosis with dizziness, lightheadedness, and possible coma.

Chronic Health Effects

The following chronic (long-term) health effects may occur at some time after exposure to 2-pentanone and can last for months or even years:

Cancer Hazards: According to information presented in the references, 2-pentanone has not been adequately tested for its ability to cause cancer in laboratory test animals.

Reproduction: According to information presented in the references, 2-pentanone has not been adequately tested for its ability to adversely affect reproduction in test animals.

Other Chronic Effects: Studies with 2-pentanone have shown that long-term skin exposure may produce dry, scaly, and fissured dermatitis. Long-term exposure to similar organic compounds have caused damage to the nervous system and liver and kidney dysfunction. It is not known if chronic exposure to 2-pentanone will cause these problems.

ʊ *Recommended Risk-Reduction Measures*

Personnel should avoid direct contact with 2-pentanone. While 2-pentanone itself is usually used as a substitute for more hazardous materials, such as diethyl ketone, substitution is still the best method to eliminate the risk of exposure to 2-pentanone. However, if a less toxic chemical cannot be substituted for a 2-pentanone, then *engineering controls* are the most effective method of reducing exposures. The best protection is to enclose operations and/or provide local exhaust ventilation at the site of chemical release. Because of its low flash point, maximum explosion-proof ventilation systems should be considered. While not always operationally feasible, isolating operations can also reduce exposure. Using respiratory protection is less effective than the controls mentioned above, but is still advisable whenever working with 2-pentanone. For exposures over the PEL (200 ppm), use an MSHA/NIOSH-approved, full facepiece organic vapor respirator. For higher or prolonged exposures, or when exposure levels are unknown (as in an emergency) a self-contained breathing apparatus (SCBA) with full facepiece operated in pressure demand mode, or a supplied-air respirator operated in continuous flow mode are the recommended respiratory protection methods of choice. If a full facepiece is not available, then chemical goggles should be worn to protect the eyes. Whenever a chemical splash hazard exists, a face shield and a rubber apron should be worn. To prevent hand and skin exposures, butyl rubber gloves should be used. Glove manufacturers should be contacted and permeation studies obtained *before* final glove selections are made.

Administrative controls should also be in place to minimize the potential for human exposures. These may include written procedures or policies which specify the methods and techniques that will be practiced whenever personnel are to work with or around 2-pentanone.

All personnel should receive *training* on the use, hazards, protective measures, emergency actions, and other precautions per 29 CFR 1910.1200 (Hazard Communication), prior to the first assignment in an area where 2-pentanone is used or stored.

If symptoms develop or overexposure is suspected, the following medical tests may be helpful:

- ☑ Skin testing by a qualified allergist with special testing and careful evaluation of exposure history (may help diagnose allergy).
- ☑ Lung function tests.
- ☑ Neurological examination of the central nervous system.

Other methods to reduce exposure include:

- ☑ Where possible, enclose operations and use local exhaust ventilation at the site of chemical release. Explosion-proof design is recommended. If local exhaust ventilation or enclosure is not used, respiratory protection should be mandatory.
- ☑ Always ensure that proper protective clothing is worn when using chemical substances.
- ☑ Wash thoroughly immediately after exposure to 2-pentanone and at the end of the work shift or before eating, drinking, or smoking. Never wear clothes contaminated with 2-pentanone home. Family members can be exposed.
- ☑ Hazard warning information should be posted in the work area. As part of an on-going education and training program, all information on the health and safety hazards of 2-pentanone should be communicated to all exposed workers.

RISK ASSESSMENT: ENVIRONMENT

General Assessment

The environment is at risk of exposure during transportation, storage, disposal, or destruction of 2-pentanone. In almost every scenario, the threat of environmental exposure is contingent upon the proper handling of the chemical substance. Accidental spills, large or small, can result in fire, explosion, and possible contamination of the surrounding environmental mediums (water, soil, and air).

2-Pentanone is considered a Class 1B flammable liquid (per OSHA 29 CFR 1910.106). Its low flash point and relatively low boiling point make this chemical an extremely dangerous fire and explosion hazard. It can react with strong oxidizers and bromine trifluoride to cause a serious fire and explosion hazard. When involved in a fire, it can release acrid and irritating oxides of carbon. A boiling liquid expanding vapor explosion (BLEVE) is also a possibility. These characteristics require special consideration during any emergency involving a leak or spill of 2-pentanone.

The proper disposal/destruction method for 2-pentanone is to burn it in a chemical incinerator equipped with an afterburner and air scrubber.

2-Pentanone can enter the environment through manufacturing, unchecked discharge into effluents from industrial or municipal waste treatment plants, and through spills.

Acute Ecological Effects

Acute (short-term) toxic effects may include the death of animals, birds, or fish, and death or low growth rate in plants. Acute effects are seen 2 to 4 days after animals or plants are exposed to 2-pentanone.

Insufficient data are available to evaluate or predict the short-term effects of 2-pentanone to aquatic life, plants, birds, or land animals.

Chronic Ecological Effects

Chronic (long-term) toxic effects may include shortened life span, reproductive problems, lower fertility, and changes in appearance or behavior in exposed animals. These effects can be seen long after first exposure(s) to toxic chemicals.

Insufficient data are available to evaluate or predict the long-term effects of 2-pentanone to aquatic life, plants, birds, or land animals.

Water Solubility

2-Pentanone is moderately soluble in water. Concentrations of 1 to 1000 milligrams will mix with a liter of water.

Persistence in the Environment

2-Pentanone is slightly persistent in water, with a half-life of between 2 to 20 days. The half-life of a pollutant is the amount of time it takes for one half of the chemical to be degraded.

Bioaccumulation in Aquatic Organisms

Some substances increase in concentration, or *bioaccumulate,* in living organisms as they breathe contaminated air, drink contaminated water, or eat contaminated food. These chemicals can become concentrated in the tissues and internal organs of animals as well as humans.

The concentration of 2-pentanone found in fish tissues is expected to be about the same as the average concentration of 2-pentanone in water from which the fish was taken.

Recommended Risk-Reduction Measures

Proper training of all transporters will reduce the likelihood of a mishap or accident resulting in a leak or spill to the environment. The correct use of DOT labeling (per the MSDS) on all containers, trucks, and rail cars will enable emergency responders to react properly and quickly to any disaster thereby reducing the potential risk to the environment and to personnel.

Storage of 2-pentanone should be segregated from incompatible chemicals to minimize the risk of cross-contamination or contact. Buildings designated for storage should be equipped with appropriate fire protection systems (alarms, sprinklers, emergency lighting, portable extinguishers). Maximum explosion-proof design of ventilation systems is recommended.

If a spill or leak to the environment has occurred, fire department, emergency response, and/or hazardous materials spill personnel should be notified immediately. Cleanup should be attempted only by those trained in proper spill containment procedures using non-sparking tools. Contaminated soils should be removed for incineration and replaced with clean soil. If 2-pentanone should contact the water table, aquifer, or navigable waterway, time is of the essence. It is highly soluble in water and, therefore, total containment and remediation may not be possible. When such spills occur, the local and/or state emergency response authorities may require notification. A comprehensive emergency response or disaster preparedness/recovery plan should be in place prior to any operations involving the use, transportation, storage, or disposal of 2-pentanone.

If 2-pentanone is spilled or leaked, the following specific steps are recommended:

- ☑ Restrict persons not wearing protective clothing from area of spill or leak until cleanup is complete and area can be opened for normal work.
- ☑ Remove all ignition sources.
- ☑ Ventilate area of spill or leak.
- ☑ Absorb liquids in vermiculite, dry sand, earth, or a similar material and deposit in sealed containers for disposal. Use non-sparking tools!
- ☑ Keep 2-pentanone out of a confined space, such as a sewer, because of the possibility of explosion (unless the sewer is designed to prevent the buildup of explosive concentrations).
- ☑ It may be necessary to dispose of 2-pentanone as a hazardous waste. The responsible state agency or the regional office of the federal Environmental Protection Agency (EPA) should be contacted for specific recommendations.

RISK ASSESSMENT: BUSINESS

General Assessment

Accidents or mishaps involving 2-pentanone can present a significant threat to business operations. The loss or damage of equipment or facilities can significantly affect fiscal viability. Lawsuits that may result from personnel injury, public exposures, and/or environmental contamination will also require a serious expenditure of resources. Media attention surrounding an injury, illness, or environmental damage can also result in a loss of profits and loss of current as well as future business prospects.

Recommended Risk-Reduction Measures

Company attorneys, safety and health professionals, and environmental specialists should be involved in the development of any procedures dealing with safety or emergency response. A company official should be pre-designated as a public relations officer with specific training in dealing with the press. Corporate plans and policies should be developed, approved, and implemented long before any need for such arises.

REFERENCES

Ahlbom, A. 1993. *Biostatistics for Epidemiologists.* New York: CRC Press/Lewis Publishers.

American Conference of Governmental Industrial Hygienists. 1988. *Documentation of the Threshold Limit Values and Biological Exposure Indices,* 5th Edition (with updates). Cincinnati: ACGIH.

Bloom, A. D. and F. J. DeSerres. 1995. *Ecotoxicity and Human Health.* New York: CRC Press/Lewis Publishers.

Burrell, R., D. K. Flaherty, and L. J. Sauers. 1992. *Toxicology of the Immune System.* New York: Van Nostrand Reinhold.

Calabrese, E. J., E. M. Kenyon. *Air Toxics and Risk Assessment.* New York: CRC Press/Lewis Publishers.

Cockerman, L. G., B. S. Shane. 1994. *Basic Environmental Toxicology.* New York: CRC Press/Lewis Publishers.

Danse, I. R. 1991. *Common Sense Toxics in the Workplace.* New York: Van Nostrand Reinhold.

Environmental Protection Agency. 1983. *Chemical Hazard Information Profiles.* Washington, D.C.: U.S. EPA.

Fodden, C. R., and J. L. Weddell. 1992. *Hazardous Materials Emergency Action Data.* New York: CRC Press/Lewis Publishers.

Forseberg, K., and L. H. Keith. 1993. *Chemical; Protective Clothing Permeation and Degradation Database.* Chelsea, MI: Lewis Publishers.

Gardner, R. W. 1994. *Chemical Intolerance.* New York: CRC Press/Lewis Publishers.

Hallenbeck, W. H. 1993. *Quantitative Risk Assessment for Environmental and Occupational Health,* 2nd Edition. New York: CRC Press/Lewis Publishers.

Hodgson, E., R. B. Mailman, and J. E. Chambers. 1988. *Dictionary of Toxicology.* London: The MacMillian Company.

Hosty, J. W., and P. Foster. 1990. *A Practical Guide to Chemical Spill Response.* New York: Van Nostrand Reinhold.

Howard, P. H., et. al. 1991. *Handbook of Environmental Degradation Rates.* New York: CRC Press/Lewis Publishers.

Jones, F. E. 1994. *Toxic Organic Vapors in the Workplace.* New York: CRC Press/Lewis Publishers.

Kamrin, M. A. 1988. *Toxicology: A Primer on Principles and Applications.* Chelsea, Michigan: Lewis Publishers.

Keith, L. H. and M. M. Walker. 1995. *Handbook of Air Toxics.* New York: CRC Press/Lewis Publishers.

Lauwerys, R. R., P. Hoet. 1993. *Industrial Chemical Exposure.* New York: CRC Press/Lewis Publishers.

Lewis, R. J., Sr. 1993. *Hawley's Condensed Chemical Dictionary,* 12th Edition. New York: Van Nostrand Reinhold.

Lewis, R. J., Sr. 1992. *Sax's Dangerous Properties of Industrial Materials,* Volumes 1, 2, and 3, 8th Edition. New York: Van Nostrand Reinhold.

Mackay, D., W. Y. Shiu, and K. C. Ma. 1992. *Illustrated Handbook of Chemical-Physical Properties and Environmental Fate for Organic Chemicals.* New York: CRC Press/Lewis Publishers.

Meyer, E. 1990. *Chemistry of Hazardous Materials.* Englewood Cliffs, NJ: Prentice Hall.

Patnaik, P. 1992. *A Comprehensive Guide to the Hazardous Properties of Chemical Substances.* New York: Van Nostrand Reinhold.

Philp, R. B. 1995. *Environmental Hazards and Human Health.* New York: CRC Press/Lewis Publishers.

Quigley, D. R. 1994. *Handbook of Emergency Chemical Management.* New York: CRC Press/Lewis Publishers.

Raleigh, R. L. and W. A. McGee. 1972. Effects of short, high concentration exposures to acetone as determined by observation in the work area. *Journal of Occupational Medicine.* 14:607-610.

Rea, W. J. 1995. *Chemical Sensitivity,* Vol. I-III. New York: CRC Press/Lewis Publishers.

Richardson, M. 1995. *Dictionary of Substance and Their Effects,* Vol. I-VI. New York: CRC Press/Lewis Publishers.

U.S. Department of Health and Human Services, National Institute for Occupational Safety and Health. 1994. *NIOSH Pocket Guide to Chemical Hazards.* Washington, D.C.: U.S. Government Printing Office.

U.S. Department of Health and Human Services, National Institute for Occupational Safety and Health. 1978. *Criteria for a Recommended Standard - Occupational Exposure to Ketones.* DHEW Publication Number: 78-173. Washington, D.C.: U.S. Government Printing Office.

MATERIAL SAFETY DATA SHEET

CHEMICAL NAME

PERCHLOROETHYLENE

HAZARD WARNING INFORMATION

HEALTH	FIRE	REACTIVE	OTHER	DEGREE OF HAZARD	COLOR CODING	OTHER CODES
2	0	0		0 = Minimum Hazard 1 = Slight Hazard 2 = Moderate Hazard 3 = Serious Hazard 4 = Severe Hazard	HEALTH = BLUE FIRE = RED REACTIVITY = YELLOW OTHER = WHITE	OX = Oxidizer ACID = Acid ALK = Alkali COR = Corrosive W = Use No Water

SECTION I - GENERAL INFORMATION

Characterization	RCRA Number	EPA Class
Chlorinated Hydrocarbon, Aliphatic	**U210**	**Toxic Waste**

DOT Proper Shipping Name **Tetrachloroethylene**	Chemical Abstract Service (CAS) Number **127-18-4**
DOT Hazard Class and Label Requirements **Poison B; St. Andrews Cross**	DOT Emergency Guide Code **74**
DOT Identification Number **UN 1897**	Chemical Formula **C_2Cl_4**

Synonyms

Ankilostin; carbon dichloride; carbon bichloride; Didakene, ethylene tetrachloride, Perchlor, Perclene, Perk, Tetracap, tetrachloroethylene; Percosolve.

SECTION II - HAZARDOUS INGREDIENTS AND IDENTITY INFORMATION

Hazardous Components (specific identity)	OSHA Exposure Criteria	NIOSH Exposure Criteria	Immediately Dangerous to Life and Health (IDLH)	ACGIH Exposure Criteria
Perchloroethylene (derivation: By chlorination of hydrocarbons and pyrolysis of the carbon tetrachloride that is subsequently formed; or from acetylene and chlorine via trichloroethylene). **1 ppm = 6.80 mg/m³**	**PEL (8-hour):** **100 ppm** **680 mg/m³** **CEILING** **200 ppm** **1360 mg/m³** **PEAK (5-min)** **300 ppm**	**REL:** **Lowest** **Possible** **Level** **Possible** **Human** **Carcinogen**	**150 ppm**	**TLV:** **50 ppm** **339 mg/m³** **STEL:** **200 ppm** **1357 mg/m³**

SECTION III - PHYSICAL AND CHEMICAL CHARACTERISTICS

Boiling Point **250°F (121°C)**	Specific Gravity (H_2O = 1) **1.62**
Vapor Pressure (gas) **14 at 68°F (20°C)**	Molecular Weight **165.8**
Vapor Density (Air = 1) **5.83**	Melting Point **-2°F (-19°C)**

Solubility

Very slightly soluble in water (0.02%). Miscible with alcohol, ether, benzene, chloroform, and oils.

Appearance and Odor

Colorless liquid with a chloroform-like odor. Odor Threshold = 70 ppm.

SECTION IV - FIRE AND EXPLOSION HAZARD DATA

Flash Point (method used) **Not Applicable**	Explosive Limits in Air % by Volume **LEL: Not Determined UEL: Not Determined**
NFPA Classification **Non-Combustible Liquid**	Autoignition Temperature **Not Determined**

Extinguishing Media

Use dry chemical, carbon dioxide, water spray, fog, or regular foam.

Special Fire Fighting Procedures

Wear full protective clothing and self-contained breathing apparatus (SCBA). Poisonous gases are produced in fire. Use water spray to keep fire-exposed containers cool. Containers may explode in fire.

Unusual Fire and Explosion Hazards

A non-combustible liquid that will decompose in fire to produce hydrogen chloride and phosgene. Vapors are heavier than air and can collect in low-lying areas and cause displacement of oxygen.

SECTION V - REACTIVITY DATA

Stability		Conditions to Avoid Normally stable at temperatures up to 932°F (500°C) as long as there are no catalysts present, and the air is dry and free of oxygen. It decomposes rapidly in warm, moist air. Keep away from heat and incompatibles.
Stable X	Unstable	Incompatibility *(materials to avoid)* Strong oxidizers, such as permanganates, peroxides, nitrates, chlorates, and perchlorates. Also reacts with chemically active metals (lithium, beryllium, barium), caustic soda, sodium hydroxide, and potash.
Hazardous Polymerization		Conditions to Avoid Under normal conditions of temperature and pressure, hazardous polymerization of perchloroethylene is not expected to occur. It slowly decomposes in the presence of light.
May Occur	Will Not Occur X	Hazardous Decomposition or By-products When heated to decomposition, perchloroethylene can emit highly toxic/poisonous and acrid fumes and gases, including chlorine, carbon dioxide, hydrogen chloride, and phosgene gas.

SECTION VI - HEALTH HAZARD DATA

Primary Route(s) of Entry:	Inhalation?	Absorption (skin)?	Ingestion?
	X	X	X

Health Hazards

INHALATION: Can produce human systemic effects, including conjunctivitis and changes in the olfactory and respiratory systems. Can cause damage to the kidneys, liver, and blood vessels. Has narcotic-like action on high or prolonged exposures (headaches, dizziness). Also lachrymation, loss of coordination, convulsions, and possible effects on target organs.

ABSORPTION: Irritating to the skin and may produce dermatitis on repeated exposures. Absorbed in toxic amounts through intact skin. Eye contact may injure cornea.

INGESTION: Irritation and narcosis with burning of the mouth and lips. Can cause irregular heartbeat, nausea and vomiting, bloody diarrhea, drowsiness, unconsciousness, and pulmonary edema (fluid in the lungs).

Carcinogenicity	NTP Listed?	IARC Cancer Review Group?	OSHA Regulated?	Target Organs?
Questioned Human Confirmed Animal	5th Annual Report	Group 2B	29 CFR 1910.1000 Table Z-1 & Z-2	Respiratory system, eyes, skin, liver, kidney, CNS.

Medical Conditions Generally Aggravated by Exposure

None reported. Skin conditions (dermatitis) or respiratory problems (asthma) may be aggravated.

Emergency and First-aid Procedures

Eye contact: Flush immediately with water for 15 minutes (minimum); seek medical attention. Skin contact: Remove all contaminated clothing. Immediately wash area with large amounts of soap and water. Seek medical assistance. For inhalation: Remove the person from exposure. Provide respiratory assistance and CPR, if required. Transfer to medical facility. Keep victim warm and at rest. Observe 48 hours for lung effects. If swallowed: Seek medical attention immediately.

SECTION VII - PRECAUTIONS FOR SAFE HANDLING AND USE

Steps to be Taken in Case Material is Released or Spilled

Absorb large spills in vermiculite, dry sand, earth, or similar material and deposit in sealed containers. Ventilate area of spill or leak. Restrict those not involved in cleanup from entering area.

Preferred Waste Disposal Method

Mix with combustible solvent and burn in chemical incinerator with an afterburner and scrubber.

Precautions to be Taken in Handling and Storage

Store to avoid contact with incompatible materials in tightly closed containers in cool, well-ventilated area away from heat and flame.

Other Precautions and Warnings

Sources of ignition are prohibited where perchloroethylene is used, handled, or stored. Heat may cause containers to build up pressure and explode. Prevent damage to containers.

SECTION VIII - CONTROL MEASURES AND PERSONAL PROTECTIVE EQUIPMENT

Respiratory Protection *(specify type)*

For exposures above the PEL, use an MSHA/NIOSH-approved supplied-air respirator with full facepiece in continuous flow mode, or use a SCBA operated in pressure demand or other positive pressure mode.

Ventilation

Local exhaust or general mechanical systems recommended.

Protective Gloves	Eye Protection	Other Protective Clothing
Neoprene or Viton Material	Chemical Goggles or Face Mask	Protective Apron

Work/Hygiene Practices

Always wash hands thoroughly after using chemical; never bring food, drink, or smoking materials into vicinity of chemicals.

PERCHLOROETHYLENE

C_2Cl_4

CAS: 127-18-4

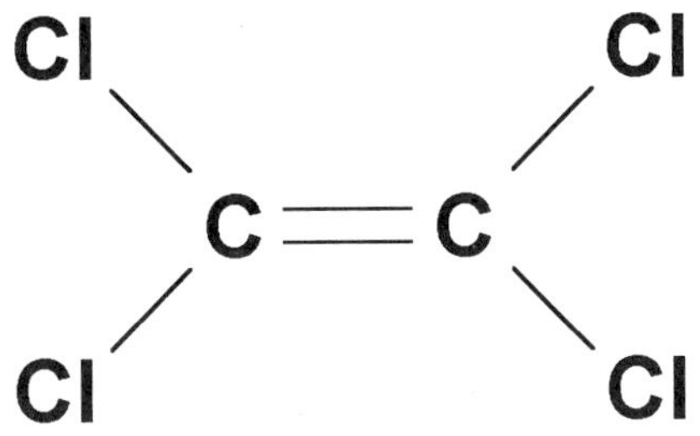

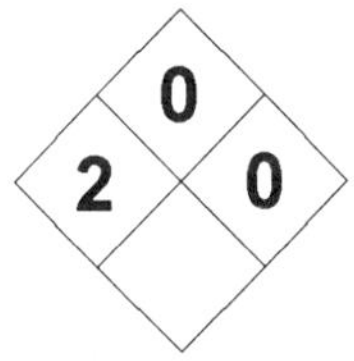

IDENTIFICATION AND TYPICAL USES

Perchloroethylene is a colorless liquid with a mild, chloroform-like odor. It is used in dry cleaning as a solvent, in vapor-degreasing, as a drying agent for metals and certain other solids, as a heat-transfer medium, in the manufacture of fluorocarbons, as an insulating fluid and cooling gas in electrical transformers, in the production of adhesives, in aerosols, paints and coatings, as a chemical intermediate, as an extractant for pharmaceuticals, as a pesticide intermediate, and in veterinary medicine (for parasitic worm removal).

RISK ASSESSMENT: HEALTH

General Assessment

Perchloroethylene is an extremely toxic compound to humans by *inhalation*, dermal *absorption*, and *ingestion*. Ingestion can occur, but is rare under most circumstances. It is listed as a confirmed animal carcinogen and its human carcinogenic potential is considered questioned by most reference sources. It has been shown to cause mutations in humans. Experimental teratogenic data have also been reported in animal tests.

Inhalation results in slight smarting of the eyes with lachrymation and slight smarting of the throat. Inhalation of high concentrations of the vapors (2000 ppm for 5 minutes) can also lead to CNS depression with symptoms of intoxication, loss of coordination, numbness around the mouth, dizziness, narcosis, and convulsions. It can also cause flushing of the face and neck, headache, somnolence, and slurred speech.

Skin contact may produce mild to moderate irritation on contact. Prolonged skin contact can cause dermatitis because of perchloroethylene's defatting action on the skin's surface. It will pass through unbroken skin to cause toxic systemic effects including symptoms of inhalation. Direct eye contact causes burning but no permanent damage. Absorption through the skin or inhalation of perchloroethylene vapors may cause toxic reactions in the liver (in animal studies, it produced liver tumors).

On ingestion, perchloroethylene is expected to produce symptoms of irritation and narcosis with burning of the lips, mouth, and gastrointestinal tract. It may cause irregular heartbeat, nausea and vomiting, bloody diarrhea, drowsiness, unconsciousness, and possible pulmonary edema (fluid in the lungs) presumably from aspiration of the swallowed liquids into the lungs.

Acute Health Effects

The following acute (short-term) health effects may occur immediately or shortly after exposure to perchloroethylene:

Skin: Moderate irritation with redness and possibly painful inflammation. It will pass through unbroken skin easily to cause toxic systemic effects.

Eye: Severe irritation and burning but rarely causes permanent damage.

Lung: Nose and throat irritation. Produces intoxicating effects and may cause lung effects as well.

CNS: Unconfirmed effects by all routes of exposure. There may be numbness around the lips and mouth, dizziness, loss of coordination, headache, somnolence, slurred speech, flushed face, irregular heartbeat, and other unspecified effects on nervous system functions.

Chronic Health Effects

The following chronic (long-term) health effects may occur at some time after exposure to perchloroethylene and can last for months or even years:

Cancer Hazards: According to information presented in the references, perchloroethylene has been shown to cause cancer in test animals and its carcinogenic capability in humans is considered questioned. Some studies indicate that it causes mutations in humans, while others are inconclusive in this regard. Cancers in humans exposed to perchloroethylene (dry cleaning industry) have developed. However, these employees were also exposed to many other chemicals during their employ and a direct link to perchloroethylene cannot be substantiated in the references.

Reproductive Hazard: According to information presented in the references, perchloroethylene may cause teratogenic effects in animals and adversely affect reproduction as well.

Other Chronic Effects: Prolonged skin contact may cause drying, cracking, and chapping of the affected area leading to dermatitis. Prolonged exposure may also cause impaired memory, weakness in the hands and feet, peripheral neuropathy, visual disturbances and impairment, and muscular cramps. There may be liver injury (fatty degeneration, necrosis, jaundice, darkened urine) and kidney damage (oliguric uremia, congestion, and granular swelling).

Recommended Risk-Reduction Measures

Personnel should avoid direct contact with perchloroethylene. The exact nature of the toxicity of this chemical is not entirely understood in the references but it is clearly a highly toxic chemical. Therefore, prudent risk management requires it be treated as though it posed a significant health risk in the event of overexposure. If a less toxic chemical cannot be substituted for a hazardous substance, then *engineering controls* are the most effective method of reducing exposures. The best protection is to enclose operations and/or provide local exhaust ventilation at the site of chemical release. While not always operationally feasible, isolating operations can also reduce exposure. Using respiratory protection is less effective than the controls mentioned above, but is still advisable whenever working with or around perchloroethylene. For long exposures above the PEL, a supplied-air respirator with full facepiece operated in continuous flow mode, or a self-contained breathing apparatus (SCBA) with full facepiece and pressure demand are the recommended respiratory protection methods of choice. If a full facepiece is not available, then chemical goggles should be worn to protect the eyes. Whenever a chemical splash hazard exists, a face shield and a rubber apron should be worn. To prevent hand and skin exposures, butyl rubber, Viton, or neoprene gloves should be used.

Administrative controls should also be in place to minimize the potential for human exposures. These may include written procedures or policies which specify the methods and techniques that will be practiced whenever personnel are to work with perchloroethylene.

All personnel should receive *training* on the use, hazards, protective measures, emergency actions, and other precautions per 29 CFR 1910.1200 (Hazard Communication), prior to the first assignment in an area where perchloroethylene is used or stored.

If symptoms develop or overexposure is suspected, the following tests are recommended:

☑ Lung, kidney, and liver function tests.

☑ Electroencephalogram (EEG) to detect abnormalities which may be indicative of chronic toxicity.

☑ Complete blood count (CBC) with monitoring of blood enzyme levels.

☑ Electrocardiogram (EKG).

Any evaluation should include a careful history of past and present symptoms with an examination. Medical tests that simply look for existing damage are not a substitute for controlling exposures.

Other methods to reduce exposure include:

☑ Where possible, enclose operations and use local exhaust ventilation at the site of chemical release. If local exhaust ventilation or enclosure is not used, respiratory protection should be mandatory.

☑ Always ensure that proper protective clothing is worn when using chemical substances.

☑ Wash thoroughly immediately after exposure to perchloroethylene and at the end of the work shift or before eating, drinking, or smoking.

☑ Hazard warning information should be posted in the work area. In addition, as part of an on-going education and training program, all information on the health and safety hazards of perchloroethylene should be communicated to all exposed workers.

☑ Eye wash stations should be provided in the immediate work area for emergency use. If there is a

possibility of skin exposure to perchloroethylene, emergency shower facilities should also be provided.

☑ Personnel should never wear contaminated clothing home (family members can be exposed). Clothing should be laundered by personnel who have been trained on the hazards of perchloroethylene.

RISK ASSESSMENT: ENVIRONMENT

General Assessment

The environment is at risk of exposure during transportation, storage, disposal, or destruction of perchloroethylene. In almost every scenario, the threat of environmental exposure is contingent upon the proper handling of the chemical substance. Accidental spills, large or small, can result in fire, explosion, and possible contamination of the surrounding environmental mediums (water, soil, and air).

Perchloroethylene is considered a non-combustible liquid but it will decompose in a fire to produce toxic hydrogen chloride and phosgene gas. It is incompatible with a number of commonly encountered materials including strong oxidizers and many chemically active metals. Contact can cause fire or explosion. These characteristics require extreme caution in handling, storage, transportation, and disposal. Also, because containers and tankers can explode under conditions of extreme heat, special consideration is required during any emergency situation involving a leak or spill of perchloroethylene. Should perchloroethylene ever come into contact with incompatible substances during use, transportation, storage, or disposal, the formation of highly toxic and/or highly explosive commodities (including hydrogen chloride) is extremely possible.

The proper disposal method for perchloroethylene is to mix it with a combustible solvent and then burn it in a permitted chemical incinerator equipped with an afterburner and scrubber.

Perchloroethylene may enter the environment through industrial discharges or spills.

Acute Ecological Effects

Acute (short-term) toxic effects may include the death of animals, birds, or fish and death or low growth rate in plants. Acute effects are seen 2 to 4 days after animals or plants are exposed to perchloroethylene.

Insufficient data are available on the short-term effects of perchloroethylene exposure to aquatic life, birds, plants, or land animals.

Chronic Ecological Effects

Chronic (long-term) toxic effects may include shortened life span, reproductive problems, lower fertility, and changes in appearance or behavior in exposed animals. These effects can be seen long after first exposure(s) to toxic chemicals.

Insufficient data are available on the long-term effects of perchloroethylene exposure to aquatic life, birds, plants, or land animals.

Water Solubility

Perchloroethylene is nearly insoluble in water. Concentrations of less than 10 milligrams may mix with a liter of water.

Persistence in the Environment

Perchloroethylene is non-persistent in water, with a half-life of less than 2 days. The half-life of a pollutant is the amount of time it takes for one half of the chemical to be degraded. Virtually 100% of perchloroethylene will eventually end up in air.

Bioaccumulation in Aquatic Organisms

Some substances increase in concentration, or *bioaccumulate,* in living organisms as they breathe contaminated air, drink contaminated water, or eat contaminated food. These chemicals can become concentrated in the tissues and internal organs of animals as well as humans.

The concentration of perchloroethylene found in fish tissue is expected to be somewhat higher than the average concentration of perchloroethylene in the water from which the fish was taken.

Recommended Risk-Reduction Measures

Proper training of all transporters will reduce the likelihood of a mishap or accident resulting in a leak or spill to the environment. The correct use of DOT labeling (per the MSDS) on all containers, trucks, and rail cars will enable emergency responders to react properly and quickly to any disaster thereby reducing the potential risk to the environment and to personnel.

Storage of perchloroethylene should be segregated from incompatible chemicals to minimize the risk of cross-contamination or contact. Buildings designated for storage should be equipped with appropriate fire protection systems (alarms, sprinklers, emergency lighting, portable extinguishers).

If a spill or leak to the environment has occurred, fire department, emergency response and/or hazardous materials spill personnel should be notified immediately. Cleanup should be attempted only by those trained in proper spill containment procedures. Contaminated soils should be removed for incineration and replaced with clean soil. If perchloroethylene should contact the water table, aquifer, or navigable waterway, cleanup should be started immediately. It is only slightly soluble in water and total remediation may be possible. The local and/or state emergency response authorities must be notified. A comprehensive emergency response or disaster preparedness/recovery plan should be in place prior to any operations involving the use, transportation, storage, or disposal of perchloroethylene. If perchloroethylene is spilled or leaked, the following specific steps are recommended:

☑ Restrict persons not wearing protective clothing from area of spill or leak until cleanup is complete and area can be opened for normal work.

☑ Remove all ignition sources, ventilate area of leak.

☑ Absorb liquids in vermiculite, dry sand, earth, or similar material and deposit in sealed containers.

☑ It may be necessary to dispose of perchloroethylene as a hazardous waste. The responsible state agency or the regional office of the federal Environmental Protection Agency (EPA) should be contacted for specific recommendations.

RISK ASSESSMENT: BUSINESS

General Assessment

Accidents or mishaps involving perchloroethylene can present a significant threat to business operations. The loss or damage of equipment or facilities can significantly affect fiscal viability. Lawsuits that may result from personnel injury/death, public exposures, and/or environmental contamination will also require a serious expenditure of resources. Media attention surrounding an injury, death, or environmental damage can also result in a loss of profits and loss of current as well as future business. Always remember that anytime the terms "cancer," "carcinogen," or "reproductive hazard" are used, public emotion, ignorance, hysteria, and anxiety can all run equally high. This must be carefully considered when developing or implementing public relations policies.

Recommended Risk-Reduction Measures

Company attorneys, safety and health professionals, and environmental specialists should be involved in the development of any procedures dealing with safety or emergency response. A company official should be pre-designated as a public relations officer with specific training in dealing with the press. Corporate plans and policies should be developed, approved, and implemented long before any need for such arises.

REFERENCES

American Conference of Governmental Industrial Hygienists. 1988. *Documentation of the Threshold Limit Values and Biological Exposure Indices,* 5th Edition (with updates). Cincinnati: ACGIH

Gardner, R. W. 1994. *Chemical Intolerance.* New York: CRC Press/Lewis Publishers.

Jones, F. E. 1994. *Toxic Organic Vapors in the Workplace.* New York: CRC Press/Lewis Publishers.

Keith, L. H. and M. M. Walker. 1995. *Handbook of Air Toxics.* New York: CRC Press/Lewis Publishers.

Lauwerys, R. R., P. Hoet. 1993. *Industrial Chemical Exposure.* New York: CRC Press/Lewis Publishers.

Lewis, R. J., Sr. 1992. *Sax's Dangerous Properties of Industrial Materials,* Volumes 1, 2, and 3, 8th Edition. New York: Van Nostrand Reinhold.

Manahan, S. E. 1994. *Environmental Chemistry,* 6th Edition. New York: CRC Press/Lewis Publishers.

Meyer, E. 1990. *Chemistry of Hazardous Materials.* Englewood Cliffs, NJ: Prentice Hall.

Patnaik, P. 1992. *A Comprehensive Guide to the Hazardous Properties of Chemical Substances.* New York: Van Nostrand Reinhold.

Rea, W. J. 1995. *Chemical Sensitivity,* Vol. I-III. New York: CRC Press/Lewis Publishers.

Richardson, M. 1995. *Dictionary of Substance and Their Effects,* Vol. I-VI. New York: CRC Press/Lewis Publishers.

Sacarello, H. L. A. 1994. *The Comprehensive Handbook of Hazardous Materials.* New York: CRC Press/Lewis Publishers.

Torkelson, T. R. V. K. Rowe. 1981. Halogenated aliphatic hydrocarbons. *Patty's Industrial Hygiene and Toxicology,* Vol. 2B, *Toxicology,* pp 3568-3572. New York: Wiley-Interscience.

U.S. Department of Health and Human Services, National Institute for Occupational Safety and Health. 1994. *NIOSH Pocket Guide to Chemical Hazards.* Washington, D.C.: U.S. Government Printing Office.

MATERIAL SAFETY DATA SHEET

CHEMICAL NAME	
PERCHLOROMETHYL MERCAPTAN	**CRC PRESS/LEWIS PUBLISHERS DISCLAIMER:** THE INFORMATION AND RECOMMENDATIONS PRESENTED HEREIN ARE BASED ON SOURCES BELIEVED TO BE RELIABLE. CRC MAKES NO REPRESENTATION ON ITS COMPLETENESS OR ACCURACY. IT IS THE RESPONSIBILITY OF THE USER TO DETERMINE THE CHEMICAL'S SUITABILITY FOR ITS INTENDED USE, THE CHEMICAL'S SAFE USE, AND THE CHEMICAL'S PROPER DISPOSAL. NO REPRESENTATIONS AND/OR WARRANTIES, EITHER EXPRESSED OR IMPLIED, OF THE MERCHANTABILITY OR FITNESS FOR A PARTICULAR PURPOSE, OR OF ANY OTHER NATURE, ARE MADE WITH RESPECT TO THE INFORMATION PROVIDED IN THIS MSDS OR TO THE CHEMICAL TO WHICH INFORMATION MAY REFER. CRC NEITHER ASSUMES NOR AUTHORIZES ANY OTHER PERSON TO ASSUME FOR IT, ANY OTHER ADDITIONAL RESPONSIBILITY OR LIABILITY FOR THE USE OF, OR RELIANCE UPON, THIS INFORMATION.

HAZARD WARNING INFORMATION

HEALTH	FIRE	REACTIVE	OTHER	DEGREE OF HAZARD	COLOR CODING	OTHER CODES
4	0	1	W	0 = Minimum Hazard 1 = Slight Hazard 2 = Moderate Hazard 3 = Serious Hazard 4 = Severe Hazard	HEALTH = BLUE FIRE = RED REACTIVITY = YELLOW OTHER = WHITE	OX = Oxidizer ACID = Acid ALK = Alkali COR = Corrosive W = Use No Water

SECTION I - GENERAL INFORMATION

Characterization	RCRA Number	EPA Class
Mercaptan	**P118**	**Acute Hazardous Waste**

DOT Proper Shipping Name **Perchloromethyl mercaptan**	Chemical Abstract Service (CAS) Number **594-42-3**
DOT Hazard Class and Label Requirements **Poison B; Poison**	DOT Emergency Guide Code **55**
DOT Identification Number **UN 1670**	Molecular Formula **Cl_3CSCl**

Synonyms

PCM; PMM; trichloromethane sulfenyl chloride; trichloromethyl sulfur chloride; perchloromethanethiol; thiocarbonyl tetrachloride.

SECTION II - HAZARDOUS INGREDIENTS AND IDENTITY INFORMATION

Hazardous Components (specific identity)	OSHA Exposure Criteria	NIOSH Exposure Criteria	Immediately Dangerous to Life and Health (IDLH)	ACGIH Exposure Criteria
Perchloromethyl mercaptan: (derivation: By chlorination of carbon disulfide, thiophosgene, or methyl thiocyanate). **1 ppm = 7.73 mg/m^3**	**PEL:** **0.1 ppm** **0.8 mg/m^3** **STEL:** **Not Established**	**REL:** **0.1 ppm** **0.8 mg/m^3** **STEL:** **Not Established**	**10 ppm**	**TLV:** **0.1 ppm** **0.76 mg/m^3** **STEL:** **Not Established**

SECTION III - PHYSICAL AND CHEMICAL CHARACTERISTICS

Boiling Point **297°F (147°C) Decomposes**	Specific Gravity (H_2O = 1) **1.69**
Vapor Pressure (mm Hg) **65 at 68°F (20°C)**	Molecular Weight (atomic weight) **186**
Vapor Density (Air = 1) **6.414**	Melting Point **Not Found**

Solubility

Insoluble in water (hydrolyzes to hydrochloric acid). Soluble in ether.

Appearance and Odor

Pale-yellow, oily liquid with an unbearable, acrid odor. Odor Threshold = 0.001 ppm.

SECTION IV - FIRE AND EXPLOSION HAZARD DATA

Flash Point (method used) **Not Applicable**	Explosive Limits in Air % by Volume **LEL: Not Determined** **UEL: Not Determined**
NFPA Classification **Non-combustible Liquid**	**Autoignition Temperature** **Not Applicable**

Extinguishing Media

Perchloromethyl mercaptan is non-combustible but will support combustion. Use agent suitable to surrounding fire. Do NOT use water since this chemical hydrolyzes to toxic hydrogen chloride gas.

Special Fire Fighting Procedures

Wear full protective clothing and self-contained breathing apparatus (SCBA). Move container from fire area if it can be done without risk. Use water spray to cool containers.

Unusual Fire and Explosion Hazards

Containers may explode in fire. Fight fire from a distance if possible.

SECTION V - REACTIVITY DATA

Stability		Conditions to Avoid **Normally stable at room temperature, if kept in closed containers. Perchloromethyl mercaptan begins to decompose in moist air. Keep away from heat, water, and other incompatible materials.**
Stable **X**	Unstable	Incompatibility *(materials to avoid)* **Incompatible with alkalis, amines (forms amides or isothiocyanates), hot iron and steel (forms carbon tetrachloride and sulfur monochloride), water (forms hydrogen chloride gas), and phenols. Corrodes metals.**
Hazardous Polymerization		Conditions to Avoid **Under normal conditions of temperature and pressure, hazardous polymerization of perchloromethyl mercaptan is not expected to occur.**
May Occur	Will Not Occur **X**	Hazardous Decomposition or By-products **When heated to decomposition perchloromethyl mercaptan can produce carbon tetrachloride, sulfur monochloride, heavy oil polymers, and phosgene, hydrogen chloride, and sulfur dioxide gases.**

SECTION VI - HEALTH HAZARD DATA

Primary Route(s) of Entry:	Inhalation?	Absorption (skin/eye)?	Ingestion?
	X	X	X

Health Hazards

INHALATION: Irritation of the eyes, nose, throat, and mucous membranes of the respiratory system with coughing, headache, nausea, vomiting, tearing, coarse rales, pallor, acidosis, anuria, dypsnea, trouble breathing, tachycardia, possible liver and kidney damage. Delayed pulmonary edema (fluid in the lungs) may lead to death. Possible CNS depression.

ABSORPTION: Readily absorbed through intact skin to cause toxic systemic effects (see inhalation).

INGESTION: Corrosion of the digestive tract with severe pain, bleeding, and possible perforation of the of the esophagus and/or stomach.

Carcinogenicity	NTP Listed?	IARC Cancer Review Group?	OSHA Regulated?	Target Organs?
Unknown Human **Unknown Animal**	**No**	**No**	**29 CFR 1910.1000 Table Z-1**	**Skin, eyes, liver, kidneys, respiratory system, CNS.**

Medical Conditions Generally Aggravated by Exposure

None reported.

Emergency and First-aid Procedures

Eye contact: Do not allow victim to keep eyes tightly shut. Flush immediately with water for 15 minutes (minimum); seek medical attention. Skin contact: Wash with large amounts of water for a minimum of 15 minutes. For inhalation: Remove the person from exposure. Provide respiratory assistance and CPR. Give 100% humidified supplemental O_2 with assisted ventilation. Transfer to medical facility. Observe 48 hours for lung effects. If swallowed: Seek medical attention immediately. Give 1-2 glasses of water; do NOT induce vomiting. Do NOT attempt to give an unconscious or convulsing person anything by mouth.

SECTION VII - PRECAUTIONS FOR SAFE HANDLING AND USE

Steps to be Taken in Case Material is Released or Spilled

Remove all ignition sources. Wear self-contained breathing apparatus (SCBA) for respiratory protection. Restrict those not involved in cleanup from entering area. Avoid skin contact. Ventilate area of spill. Absorb in sand or other material and dispose of in sealed drums.

Preferred Waste Disposal Method

Dissolve in combustible solvent and burn in a chemical incinerator with an afterburner and scrubber.

Precautions to be Taken in Handling and Storage

Store in tightly closed containers in a cool, dry, well-ventilated area away from heat (including direct sunlight) and moisture. Protect containers from physical damage. Inspect containers regularly.

Other Precautions and Warnings

Perchloromethyl mercaptan must be stored to avoid contact with incompatible materials since violent reactions can occur. Puncture empty containers to prevent reuse.

SECTION VIII - CONTROL MEASURES AND PERSONAL PROTECTIVE EQUIPMENT

Respiratory Protection *(specify type)*

PEL is low and difficult to monitor. Air-purifying respirators may suffice but best protection is provided by a self-contained breathing apparatus (SCBA) with full facepiece operated in positive pressure mode.

Ventilation

Local exhaust (preferred) at site of chemical work or general ventilation.

Protective Gloves	Eye Protection	Other Protective Clothing
Butyl Rubber Gloves	**Chemical Goggles and Face Mask**	**Protective Uniform or Apron**

Work/Hygiene Practices

Always wash hands thoroughly after using chemical; never bring food, drink, or smoking materials into vicinity of chemicals.

PERCHLOROMETHYL MERCAPTAN

Cl_3CSCl CAS: 594-42-3

IDENTIFICATION AND TYPICAL USES

Perchloromethyl mercaptan appears as a pale-yellow, oily liquid with an unbearable, acrid odor. It is used in the production of fungicides, organic synthesis, as a dye intermediate, a fumigant, and as a vulcanizing accelerator in the rubber industry. It has also been used in the past as a tear-gas agent in the military.

RISK ASSESSMENT: HEALTH

General Assessment

Perchloromethyl mercaptan is toxic to humans through *inhalation, absorption,* and *ingestion.* There are no data on the carcinogenicity, mutagenicity, and teratogenicity of this compound in humans.

Inhalation of even small concentrations for relatively short periods of time (e.g., 11 ppm for 10 minutes) has caused death, presumably due to the delayed onset of pulmonary edema (fluid in the lungs). Exposure to lower concentrations may cause depression of the central nervous system (CNS) with possible lung, liver, and heart congestion. Other symptoms of inhalation exposure can include eye, nose, throat, and mucosa irritation with lachrymation (tearing), difficulty breathing, chest pains, coarse rales, nausea and vomiting, pallor, acidosis, and anuria. Symptoms of pulmonary edema can be delayed up to 48 hours thereby creating a false sense of security with regard to health exposure risk.

Skin contact may cause severe localized irritation at the site of contact as well as result in absorption through intact skin, leading to the systemic symptoms indicated for inhalation.

Ingestion can cause corrosion of the digestive tract with severe abdominal pain, bleeding, and possible perforation of the esophagus and/or stomach. There may also be liver or kidney damage/injury as a result of ingestion exposures.

Acute Health Effects

The following acute (short-term) health effects may occur immediately or shortly after exposure to perchloromethyl mercaptan:

Skin: Severe irritation on exposure to the liquid and rapid absorption leading to toxic systemic effects.

Eye: Mild to severe conjunctival irritation in contact with the liquid or vapor.

Lung: Irritation to respiratory tract with the possibility of delayed pulmonary edema, which can be fatal. Absorption through the respiration process appears to cause a variety of adverse effects to the central nervous system.

CNS: A depressant of the central nervous system causing a variety of toxic effects including drowsiness, coma, and death.

Chronic Health Effects

The following chronic (long-term) health effects may occur at some time after exposure to perchloromethyl mercaptan and can last for months or even years:

Cancer Hazards: There are few human data on the effects of long-term exposure to perchloromethyl mercaptan. There are no reports of carcinogenic or mutagenic occurrences in humans or test animals resulting from exposure to perchloromethyl mercaptan.

Reproductive Hazard: There are no reports to substantiate claims of teratogenicity resulting from exposure to perchloromethyl mercaptan.

Other Chronic Effects: The acute irritating and damaging properties of this chemical limit the potential for chronic effects. However, there may be a possibility of liver or kidney damage on exposure to low levels for extended periods of time.

ⓘ *Recommended Risk-Reduction Measures*

Perchloromethyl mercaptan is an extremely irritating compound. Personnel should avoid direct contact with perchloromethyl mercaptan. If a less toxic material or compound cannot be substituted for perchloromethyl mercaptan, then *engineering controls* are the most effective method of reducing exposure risk. The best protection is to enclose operations and/or provide local exhaust ventilation at the site of perchloromethyl mercaptan release. While not always operationally feasible, isolating operations can also reduce exposure. Using respiratory protection is less effective than the controls mentioned above, but is still required whenever working with or around perchloromethyl mercaptan. The OSHA permissible exposure ceiling for this chemical is relatively low (0.1 ppm) and difficult to accurately monitor, especially under emergency conditions. An air-purifying respirator may suffice for low-level, short duration exposures. However, best protection is provided using an MSHA/NIOSH-approved self-contained breathing apparatus (SCBA) with full facepiece and pressure demand is the appropriate level of respiratory protection. If a full facepiece is not available, then chemical goggles should be worn to protect the eyes, especially when a splash hazard exists. A face shield should also be considered. Skin contact must be avoided. To prevent hand and skin exposures, butyl rubber or other non-permeable gloves should be used. Polycarbonate is incompatible with this chemical. Glove manufacturers should be contacted and permeation studies obtained *before* final glove selections are made.

Administrative controls should also be in place to minimize the potential for human exposures. These may include written procedures or policies which specify the methods and techniques that will be practiced whenever personnel are to work with perchloromethyl mercaptan.

All personnel should receive *training* on the use, hazards, protective measures, emergency actions, and other precautions per 29 CFR 1910.1200 (Hazard Communication) prior to the first assignment in an area where perchloromethyl mercaptan is used or stored.

Before beginning employment and at regular intervals thereafter (e.g., annually), the following medical test is recommended:

☑ Lung function tests.

If symptoms develop or overexposure is suspected, the following may be useful:

☑ Liver, kidney, and lung function tests.

☑ Complete evaluation of the central nervous system.

☑ Monitor for esophageal irritation (drooling and/or pain behind the sternum) and consider esophagoscopy to help diagnose extent of injury.

☑ Consider chest X-ray after acute overexposure (may be delayed if taken immediately after exposure due to delayed onset of pulmonary edema).

It should be noted that medical tests that simply look for existing damage are not a substitute for controlling exposures. Medical histories are extremely important when assessing exposure risk. Also, since smoking can cause lung cancer, emphysema, heart disease and other respiratory disorders, smokers exposed to perchloromethyl mercaptan may develop more severe symptoms than non-smokers under the same conditions of exposure. Prudent risk management requires *all* possible exposure factors be considered when evaluating health exposure risks in the workplace.

Other methods to reduce exposure include:

☑ Always ensure that proper protective clothing is worn when using chemical substances and that personnel are trained on its use, care, and maintenance.

☑ Wash thoroughly immediately after exposure to perchloromethyl mercaptan and at the end of the work shift or before eating, drinking, or smoking.

☑ Hazard warning information should be posted in the work area. In addition, as part of an on-going education and training program, all information on the health and safety hazards of perchloromethyl mercaptan should be communicated to all potentially exposed workers.

☑ Eye wash stations should be provided in the immediate work area for emergency use. If there is a possibility of skin exposure to perchloromethyl mercaptan, emergency shower facilities should also be provided.

☑ Workers whose clothing has been contaminated by perchloromethyl mercaptan should change into clean clothes before leaving work. Clothing worn home places family members at risk. Contaminated work clothing should be laundered only by individuals who have been informed of the hazards of exposure to perchloromethyl mercaptan.

RISK ASSESSMENT: ENVIRONMENT

General Assessment

The environment is at risk of exposure during transportation, storage, disposal, or destruction of perchloromethyl mercaptan. In almost every scenario, the threat of environmental exposure is contingent upon the proper handling of the chemical substance. Accidental spills, large or small, can result in contamination of the surrounding environmental mediums (water, soil, and air).

Perchloromethyl mercaptan is non-combustible liquid but it will support combustion in a fire. Pressure can build inside containers during a fire and explosions are likely if they are not removed or kept cool with a water spray. Harmful and toxic hydrogen chloride gas is produced in fire. It is incompatible with alkalis, amines, hot iron and steel, and even water. Therefore, caution is always required in handling, storage, transportation, and disposal of perchloromethyl mercaptan. Aside from hydrogen chloride, when heated to decomposition it emits very acrid smoke and toxic fumes of carbon tetrachloride, sulfur monochloride, heavy oil polymers, phosgene, and sulfur dioxide gas. Emergency responders, hazardous materials personnel, and firefighters should be made aware of the presence of perchloromethyl mercaptan at any emergency response situation.

Perchloromethyl mercaptan may enter the environment through industrial effluents and spills.

Acute Ecological Effects

Acute (short-term) toxic effects may include the death of animals, birds, or fish and death or low growth rate in plants. Acute effects are seen 2 to 4 days after animals or plants are exposed to perchloromethyl mercaptan.

Insufficient data are available to evaluate or predict the acute (short-term) effects of perchloromethyl mercaptan to aquatic life, plants, birds, or land animals.

Chronic Ecological Effects

Chronic (long-term) toxic effects may include shortened life span, reproductive problems, lower fertility, and changes in appearance or behavior in exposed animals. These effects can be seen long after first exposure(s) to toxic chemicals.

Insufficient data are available to evaluate the chronic (long-term) effects of perchloromethyl mercaptan on aquatic life, plants, birds, or land animals.

Water Solubility

Perchloromethyl mercaptan is insoluble in water. Concentrations of 1 milligram will not mix with a liter of water.

Persistence in the Environment

Perchloromethyl mercaptan is slightly persistent in water, with a half-life of between 2 to 20 days. The half-life of a pollutant is the amount of time it takes for one half of the chemical to be degraded. Virtually 100% of perchloromethyl mercaptan will end up in the air.

Bioaccumulation in Aquatic Organisms

Some substances increase in concentration, or *bioaccumulate,* in living organisms as they breathe contaminated air, drink contaminated water, or eat contaminated food. These chemicals can become concentrated in the tissues and internal organs of animals as well as humans.

Insufficient data are available to evaluate or predict the bioaccumulation characteristics of perchloromethyl mercaptan in the edible tissues of fish.

Recommended Risk-Reduction Measures

Proper training of all transporters will reduce the likelihood of a mishap or accident resulting in a leak or spill of perchloromethyl mercaptan into the environment. Labels on all containers, trucks, and rail cars must meet DOT requirements and accurately reflect their contents to enable emergency responders to react properly and quickly to any disaster thereby reducing the potential risk to the environment and to personnel.

Storage of perchloromethyl mercaptan should be segregated from other chemicals to minimize the risk of cross-contamination. Perchloromethyl mercaptan must be stored to avoid contact with incompatible materials, since violent reactions can occur. Containers should be protected from physical damage and stored to avoid contact with heat.

If a spill or leak to the environment has occurred, fire department, emergency response and/or hazardous materials spill personnel should be notified immediately. Cleanup should be attempted only by those trained in proper spill containment procedures. Contaminated soils should be removed for incineration and replaced with clean soil.

If perchloromethyl mercaptan should contact the water table, aquifer, or navigable waterway, time is of the essence. It is only slightly soluble in water and total remediation may or may not be possible. The local and/or state emergency response authorities must be notified. A comprehensive emergency response or disaster preparedness/recovery plan should be in place prior to any operations involving the use, transportation, storage, or disposal of perchloromethyl mercaptan.

If perchloromethyl mercaptan is spilled or leaked, the following specific steps are recommended:

☑ Restrict persons not wearing protective clothing from area of spill or leak until cleanup is complete. Avoid skin and eye contact.

☑ Absorb spilled liquids with vermiculite, dry sand, or a similar material and deposit in sealed drum for disposal.

☑ Ventilate area of spill or leak.

☑ Remove all ignition sources.

☑ It may be necessary to dispose of perchloromethyl mercaptan as a hazardous waste. The responsible state agency or the regional office of the federal Environmental Protection Agency (EPA) should be contacted for specific recommendations.

RISK ASSESSMENT: BUSINESS

General Assessment

Accidents or mishaps involving perchloromethyl mercaptan can present a serious threat to business operations. The loss or damage of equipment or facilities can significantly affect fiscal viability. Lawsuits that may result from personnel illness, injury/death, public exposures, and/or environmental contamination will require a serious expenditure of resources. Media attention surrounding an injury, death, or environmental damage can also result in a loss of profits and loss of current as well as future business.

Recommended Risk-Reduction Measures

Company attorneys, safety and health professionals, and environmental specialists should be involved in the development of any procedures for responding to chemical incidents. A company official should be predesignated as a public relations officer with specific training in dealing with the press. Corporate plans and policies should be developed, approved, and implemented long before any need for such arises.

REFERENCES

American Conference of Governmental Industrial Hygienists. 1988. *Documentation of the Threshold Limit Values and Biological Exposure Indices,* 5th Edition (with updates). Cincinnati: ACGIH

Anthony, C. P., and N. J. Kolthoff. 1971. *Textbook of Anatomy and Physiology.* St. Louis: C. V. Mosby Company.

Bloom, A. D. and F. J. DeSerres. 1995. *Ecotoxicity and Human Health.* New York: CRC Press/Lewis Publishers.

Bueche, J. R. 1972. *Principles of Physics,* 2nd Edition. New York: McGraw-Hill.

Burrell, R., D. K. Flaherty, and L. J. Sauers. 1992. *Toxicology of the Immune System.* New York: Van Nostrand Reinhold.

Calabrese, E. J., E. M. Kenyon. *Air Toxics and Risk Assessment.* New York: CRC Press/Lewis Publishers.

Cockerman, L. G., B. S. Shane. 1994. *Basic Environmental Toxicology.* New York: CRC Press/Lewis Publishers.

Cothern, C. R. 1995. *Handbook for Environmental Risk Decisionmaking.* New York: CRC Press/Lewis Publishers.

Danse, I. R. 1991. *Common Sense Toxics in the Workplace.* New York: Van Nostrand Reinhold.

Environmental Protection Agency. 1983. *Chemical Hazard Information Profiles.* Washington, D.C.: U.S. EPA.

Fodden, C. R., and J. L. Weddell. 1992. *Hazardous Materials Emergency Action Data.* New York: CRC Press/Lewis Publishers.

Gardner, R. W. 1994. *Chemical Intolerance.* New York: CRC Press/Lewis Publishers.

Gerarde, H. W. 1960. *Toxicology and Biochemistry of Aromatic Hydrocarbons.* New York: Elsevier.

Hodgson, E., R. B. Mailman, and Janice E. Chambers. 1988. *Dictionary of Toxicology.* London: The MacMillian Company.

Hosty, J. W., and P. Foster. 1990. *A Practical Guide to Chemical Spill Response.* New York: Van Nostrand Reinhold.

Jones, F. E. 1994. *Toxic Organic Vapors in the Workplace.* New York: CRC Press/Lewis Publishers.

Lauwerys, R. R., P. Hoet. 1993. *Industrial Chemical Exposure.* New York: CRC Press/Lewis Publishers.

Lewis, R. J., Sr. 1992. *Sax's Dangerous Properties of Industrial Materials,* Volumes 1, 2, and 3, 8th Edition. New York: Van Nostrand Reinhold.

Manahan, S. E. 1992. *Toxicological Chemistry,* 2nd Edition. New York: CRC Press/Lewis Publishers.

Meyer, E. 1990. *Chemistry of Hazardous Materials.* Englewood Cliffs, NJ: Prentice Hall.

Patnaik, P. 1992. *A Comprehensive Guide to the Hazardous Properties of Chemical Substances.* New York: Van Nostrand Reinhold.

Torkelson, T. R. V. K. Rowe. 1981. *Patty's Industrial Hygiene and Toxicology,* Vol. 2B. New York: Wiley-Interscience.

U.S. Department of Health and Human Services, National Institute for Occupational Safety and Health. 1994. *NIOSH Pocket Guide to Chemical Hazards.* Washington, D.C.: U.S. Government Printing Office.

MATERIAL SAFETY DATA SHEET

CHEMICAL NAME	
PERCHLORYL FLUORIDE	**CRC PRESS/LEWIS PUBLISHERS DISCLAIMER:** THE INFORMATION AND RECOMMENDATIONS PRESENTED HEREIN ARE BASED ON SOURCES BELIEVED TO BE RELIABLE. CRC MAKES NO REPRESENTATION ON ITS COMPLETENESS OR ACCURACY. IT IS THE RESPONSIBILITY OF THE USER TO DETERMINE THE CHEMICAL'S SUITABILITY FOR ITS INTENDED USE, THE CHEMICAL'S SAFE USE, AND THE CHEMICAL'S PROPER DISPOSAL. NO REPRESENTATIONS AND/OR WARRANTIES, EITHER EXPRESSED OR IMPLIED, OF THE MERCHANTABILITY OR FITNESS FOR A PARTICULAR PURPOSE, OR OF ANY OTHER NATURE, ARE MADE WITH RESPECT TO THE INFORMATION PROVIDED IN THIS MSDS OR TO THE CHEMICAL TO WHICH INFORMATION MAY REFER. CRC NEITHER ASSUMES NOR AUTHORIZES ANY OTHER PERSON TO ASSUME FOR IT, ANY OTHER ADDITIONAL RESPONSIBILITY OR LIABILITY FOR THE USE OF, OR RELIANCE UPON, THIS INFORMATION.

HAZARD WARNING INFORMATION

HEALTH	FIRE	REACTIVE	OTHER	DEGREE OF HAZARD	COLOR CODING	OTHER CODES
4	1	4	OX	0 = Minimum Hazard 1 = Slight Hazard 2 = Moderate Hazard 3 = Serious Hazard 4 = Severe Hazard	HEALTH = BLUE FIRE = RED REACTIVITY = YELLOW OTHER = WHITE	OX = Oxidizer ACID = Acid ALK = Alkali COR = Corrosive W = Use No Water

SECTION I - GENERAL INFORMATION

Characterization: **Perchlorate**	RCRA Number: **None**	EPA Class: **Not Applicable**
DOT Proper Shipping Name: **Poison Gas, N.O.S.**	Chemical Abstract Service (CAS) Number: **7616-94-6**	
DOT Hazard Class and Label Requirements: **Poison Gas**	DOT Emergency Guide Code: **20**	
DOT Identification Number: **UN 3083**	Chemical Formula: **ClO_3F**	

Synonyms
Chlorine fluoride oxide; chlorine oxyfluoride; trioxychlorofluoride.

SECTION II - HAZARDOUS INGREDIENTS AND IDENTITY INFORMATION

Hazardous Components (specific identity)	OSHA Exposure Criteria	NIOSH Exposure Criteria	Immediately Dangerous to Life and Health (IDLH)	ACGIH Exposure Criteria
Perchloryl fluoride **1 ppm = 4.26 mg/m³**	**PEL:** 3 ppm 13.5 mg/m³ **STEL:** 6 ppm 28 mg/m³	**REL:** 3 ppm 14 mg/m³ **STEL:** 6 ppm 28 mg/m³	**100 ppm**	**TLV:** 3 ppm 14 mg/m³ **STEL:** 6 ppm 28 mg/m³

SECTION III - PHYSICAL AND CHEMICAL CHARACTERISTICS

Boiling Point: **-53°F (-47°C)**	Specific Gravity (H_2O = 1): **0.067 (gas) 1.434 (liquid)**
Vapor Pressure (atmospheres): **10.5 at 69°F (20°C)**	Molecular Weight: **102.5**
Vapor Density (Air = 1): **3.64**	Melting Point: **-234°F (-148°C)**

Solubility
Slightly soluble in water (0.06%).

Appearance and Odor
Colorless gas with a characteristic, sweet odor. Normally shipped as a liquefied compressed gas.

SECTION IV - FIRE AND EXPLOSION HAZARD DATA

Flash Point (method used): **Not Applicable**	Explosive Limits in Air % by Volume: **LEL: N/A UEL: N/A**
NFPA Classification: **Non-Flammable Gas**	Autoignition Temperature: **Not Applicable**

Extinguishing Media
Use agents suitable to surrounding fires. Perchloryl fluoride does not burn.

Special Fire Fighting Procedures
Perchloryl fluoride is non-flammable but supports combustion. A very powerful oxidizer, it greatly accelerates the fire and explosion potential of other materials. Wear full protective clothing and SCBA. Use a water spray to keep fire-exposed containers cool. Poisonous gases are produced in fire. Isolate for ½ mile if rail or tank truck is involved. Containers may explode in fire or heat.

Unusual Fire and Explosion Hazards
Containers may explode in fire or heat. Remove fire-exposed containers from fire if it can be done without risk or keep them cool using a water spray.

SECTION V - REACTIVITY DATA

Stability		Conditions to Avoid **Perchloryl fluoride is normally stable in pressurized containers at room temperature under normal conditions of storage and handling. Avoid contact with incompatible materials, heat, and combustible materials.**
Stable X	Unstable	Incompatibility *(materials to avoid)* **Combustible materials (sawdust, wood, charcoal), strong bases, amines, finely divided metals, reducing agents, and alcohols. Especially incompatible with oxygen.**
Hazardous Polymerization		Conditions to Avoid **Under normal conditions of temperature and pressure, hazardous polymerization of perchloryl fluoride is not known to occur.**
May Occur	Will Not Occur X	Hazardous Decomposition or By-products **When heated to decomposition, perchloryl fluoride can emit highly toxic/poisonous gases. Mixing with water or steam will produce toxic and corrosive fumes of hydrochloric acid.**

SECTION VI - HEALTH HAZARD DATA

Primary Route(s) of Entry:	Inhalation? X	Absorption (skin)? X	Ingestion?

Health Hazards

INHALATION: Extremely irritating to the lungs, throat, nose, and eyes. Coughing and shortness of breath that may persist for hours after exposure. Can cause permanent damage to the lungs. Fluid can build in the lungs (pulmonary edema), which is a medical emergency. It can cause depression of the central nervous system, methemoglobinemia, and hypoxia.

ABSORPTION: Contact with the liquid can cause severe skin burns (frostbite). The liquid can cause eye burns leading to permanent damage. Toxic amounts can be absorbed through the skin.

Carcinogenicity	NTP Listed?	IARC Cancer Review Group?	OSHA Regulated?	Target Organs?
Unknown Human **Unknown Animal**	**No**	**No**	**29 CFR 1910.1000 Table Z-1**	**Respiratory system, eyes, skin, blood.**

Medical Conditions Generally Aggravated by Exposure

None reported.

Emergency and First-aid Procedures

Eye contact: Flush immediately with water for 30 minutes (minimum); seek medical attention. Skin contact: Remove all contaminated clothing. Immediately wash area with large amounts of soap and water. Seek medical assistance. For inhalation: Remove the person from exposure. Provide respiratory assistance and CPR. Transfer to medical facility. Observe for 24 - 48 hours since pulmonary edema may be a delayed reaction. If swallowed: Seek medical attention immediately.

SECTION VII - PRECAUTIONS FOR SAFE HANDLING AND USE

Steps to be Taken in Case Material is Released or Spilled

Absorb liquids in vermiculite, dry sand, earth, or similar material and deposit in sealed containers. Ventilate area of spill or leak. Restrict those not involved in cleanup from entering area. If gas is leaked, stop source if possible or remove leaking cylinders to safe area outdoors and allow to empty.

Preferred Waste Disposal Method

No citation.

Precautions to be Taken in Handling and Storage

Perchloryl fluoride is a flammable gas and an extremely powerful oxidizer. Store to avoid contact with incompatible materials since violent reactions can occur. Store in tightly closed containers in cool, well-ventilated area away from heat and sunlight. Do not drag or roll cylinders.

Other Precautions and Warnings

Sources of ignition are prohibited where perchloryl fluoride is used, handled, or stored. Heat may cause containers to build up pressure and explode.

SECTION VIII - CONTROL MEASURES AND PERSONAL PROTECTIVE EQUIPMENT

Respiratory Protection *(specify type)*

For low-level exposures, use a full facepiece respirator with an organic vapor gas cartridge. For higher exposures, use a supplied-air respirator with full facepiece, hood, or helmet in continuous flow mode, or use a SCBA operated in pressure demand or other positive pressure mode.

Ventilation

Local exhaust or general mechanical systems recommended.

Protective Gloves	Eye Protection	Other Protective Clothing
Impervious Material	**Chemical Goggles or Face Mask**	**Protective Apron**

Work/Hygiene Practices

Always wash hands thoroughly after using chemical; never bring food, drink, or smoking materials into vicinity of chemicals.

PERCHLORYL FLUORIDE

ClO_3F

CAS: 7616-94-6

IDENTIFICATION AND TYPICAL USES

Perchloryl fluoride is a colorless gas with a characteristic sweet odor. It is normally shipped as a liquefied compressed gas. It is used as an oxidant in rocket fuels, and as an oxidizing agent in chemical reactions.

RISK ASSESSMENT: HEALTH

General Assessment

Perchloryl fluoride is highly irritating to the eyes, nose, and throat. Moderately toxic by *inhalation.* There are no indications in the references that this material has any carcinogenic, mutagenic, or teratogenic effects on long-term exposure.

Breathing perchloryl fluoride vapors can cause coughing, nasal discharge, wheezing, respiratory distress, bronchitis, and congestion of the lungs. Its irritant effects can be intense. In the body, it forms methemoglobin which can significantly impair the blood's oxygen transport ability. Cyanosis (bluish discoloration of the lips and skin) and hypoxia (reduced level of oxygen in the blood) may indicate methemoglobinemia. High exposures can damage the lungs, causing a buildup of fluid (pulmonary edema) which is a medical emergency and can be fatal. There may also be headache, dizziness, sweating, weakness, coughing, difficulty in breathing, and congestion. The chronic toxicity signs are fluorosis (mottling of teeth enamel) and damaged blood function.

Skin contact with the liquid can result in severe burns (2nd or 3rd degree) due to the extremely cold temperatures of the compressed gas. This material can be absorbed through the skin in toxic amounts. Eye contact can be equally damaging.

Acute Health Effects

The following acute (short-term) health effects may occur immediately or shortly after exposure to perchloryl fluoride:

Skin: Liquid may irritate the skin (cold temperature may cause deep burning).

Eye: Severe irritation; may cause eye injury.

Lung: Irritation of the nose, throat, and lungs following exposure. Pulmonary edema is possible (24 to 48 hours following a high exposure). It can cause chest pain, cough, bloody nose, and sputum.

Other: CNS effects include headache, dizziness, loss of consciousness, and muscle weakness. If the liquid should be ingested (not likely), there may be burning of the mouth, nausea, vomiting, and pneumonia.

Chronic Health Effects

The following chronic (long-term) health effects may occur at some time after exposure to perchloryl fluoride and can last for months or even years:

Cancer Hazards: According to information presented in the references, perchloryl fluoride has not been adequately tested for its ability to cause cancer in test animals.

Reproductive Hazard: According to information presented in the references, perchloryl fluoride has not been adequately tested for its ability to adversely affect reproduction.

Other Chronic Effects: Repeated exposures can cause bronchitis to develop with cough, phlegm, and/or shortness of breath. Chronic toxicity is also indicated with the development of fluorosis and possible damage to blood function.

Recommended Risk-Reduction Measures

Personnel should avoid direct contact with perchloryl fluoride. If a less toxic chemical cannot be substituted for a hazardous substance, then *engineering controls* are the most effective method of reducing exposures. The best protection is to enclose operations and/or provide local exhaust ventilation at the site of chemical release. While not always operationally feasible, isolating operations can also reduce exposure. Using respiratory protection is less effective than the controls mentioned above, but is still advisable whenever working with or around perchloryl fluoride. A self-contained breathing apparatus (SCBA) with full facepiece and pressure demand is the recommended respiratory protection of choice. If a full facepiece is not available, then chemical goggles should be worn to protect the eyes. For low or infrequent concentrations a full facepiece respirator equipped with a cartridge containing non-combustible sorbents and providing protection against perchloryl fluoride gas is acceptable protection. Whenever working with the liquid and a chemical splash hazard exists, a face shield and a rubber apron should be worn. To prevent hand and skin exposures, acid resistant gloves should be worn. Glove manufacturers should be contacted and permeation studies obtained *before* final glove selection are made.

Administrative controls should also be in place to minimize the potential for human exposures. These may include written procedures or policies which specify the methods and techniques that will be practiced whenever personnel are to work with perchloryl fluoride.

All personnel should receive *training* on the use, hazards, protective measures, emergency actions, and other precautions per 29 CFR 1910.1200 (Hazard Communication), prior to the first assignment in an area where perchloryl fluoride is used or stored.

For those personnel with frequent or potentially high exposure (half the PEL or greater), the following are recommended medical tests to be provided before initial assignment and at regular intervals thereafter (e.g., annually):

☑ Lung function tests.

If symptoms develop or overexposure is suspected, the following tests are recommended:

☑ Lung function tests.

☑ Consider chest X-ray after acute overexposures (may be negative if taken immediately after exposure due to delayed onset of pulmonary edema).

☑ Monitor blood methemoglobin levels.

Any evaluation should include a careful history of past and present symptoms with an examination. Medical tests that simply look for existing damage are not a substitute for controlling exposures. Also, because smoking can cause heart disease, as well as lung cancer, emphysema, and other respiratory problems, exposure to perchloryl fluoride may worsen pre-existing respiratory conditions. Smokers may experience symptoms more rapidly than non-smokers under the same conditions of exposure.

Other methods to reduce exposure include:

☑ Where possible, enclose operations and use local exhaust ventilation at the site of chemical release. If local exhaust ventilation or enclosure is not used, respiratory protection should be mandatory.

☑ Always ensure that proper protective clothing is worn when using chemical substances.

☑ Wash thoroughly immediately after exposure to perchloryl fluoride and at the end of the work shift or before eating, drinking, or smoking.

☑ Hazard warning information should be posted in the work area. In addition, as part of an on-going education and training program, all information on the health and safety hazards of perchloryl fluoride should be communicated to all potentially exposed workers.

☑ Eye wash stations should be provided in the immediate work area for emergency use. If there is a possibility of skin exposure to perchloryl fluoride, emergency shower facilities should also be provided.

☑ Personnel should never wear contaminated clothing home (family members can be exposed). Clothing should be laundered by personnel who have been trained on the hazards of perchloryl fluoride.

RISK ASSESSMENT: ENVIRONMENT

General Assessment

The environment is at risk of exposure during transportation, storage, disposal, or destruction of perchloryl fluoride. In almost every scenario, the threat of environmental exposure is contingent upon the proper handling of the chemical substance. Accidental spills, large or small, can result in fire, explosion, and possible contamination of the surrounding environmental mediums (water, soil, and air).

Perchloryl fluoride is considered a non-flammable gas. However, it is an extremely powerful oxidizing agent and is highly reactive with many commonly encountered commodities such as dust and wood. Contact with incompatible materials can result in fire and serious explosions. These characteristics require extreme caution in handling, storage, transportation, and disposal. It is incompatible with organic matter, combustible materials, strong bases, reducing agents, alcohols, powdered metals, and amines. These characteristics require special consideration during any emergency situation involving a leak or spill of perchloryl fluoride. Should perchloryl fluoride ever come into contact with incompatible substances during use, transportation, storage, or disposal, the formation of highly toxic and/or highly explosive commodities is extremely possible.

Perchloryl fluoride can enter the environment through industrial and municipal discharges, and through spills.

Acute Ecological Effects

Acute (short-term) toxic effects may include the death of animals, birds, or fish and death or low growth rate in plants. Acute effects are seen 2 to 4 days after animals or plants are exposed to perchloryl fluoride.

This chemical has high acute toxicity to aquatic life. No data are available on the short-term effects of perchloryl fluoride exposure to plants, birds, or land animals.

Chronic Ecological Effects

Chronic (long-term) toxic effects may include shortened life span, reproductive problems, lower fertility, and changes in appearance or behavior in exposed animals. These effects can be seen long after first exposure(s) to toxic chemicals.

Perchloryl fluoride has high chronic toxicity in aquatic life. No data are available on the long-term effects of perchloryl fluoride to plants, birds, or land animals.

Water Solubility

Perchloryl fluoride is slightly soluble in water. Concentrations of 10 milligrams may mix with a liter of water.

Persistence in the Environment

Perchloryl fluoride is non-persistent in water, with a half-life of less than 2 days. The half-life of a pollutant is the amount of time it takes for one half of the chemical to be degraded. No data are available on the amount of perchloryl fluoride that will remain in air or water. However, as a gas, it might be assumed that nearly 100% of perchloryl fluoride will eventually end up in the air.

Bioaccumulation in Aquatic Organisms

Some substances increase in concentration, or *bioaccumulate,* in living organisms as they breathe contaminated air, drink contaminated water, or eat contaminated food. These chemicals can become concentrated in the tissues and internal organs of animals as well as humans.

Perchloryl fluoride is not expected to accumulate in the tissues of fish.

Recommended Risk-Reduction Measures

Proper training of all transporters will reduce the likelihood of a mishap or accident resulting in a leak or spill to the environment. Transportation of pure perchloryl fluoride is forbidden by the Department of Transportation (DOT). It must be kept frozen when shipped. The correct use of DOT labeling (per the MSDS) on all containers, trucks, and rail cars will enable emergency responders to react properly and quickly to any disaster thereby reducing the potential risk to the environment.

When not in use, perchloryl fluoride must be kept in a cool, dry, well-ventilated area. The cylinders must be stored vertically and secured properly. They should never be rolled or dragged and should always be protected against physical damage. Storage of perchloryl fluoride should be segregated from incompatible chemicals to minimize the risk of cross-contamination or contact. Buildings designated for storage should be equipped with appropriate fire protection systems (alarms, sprinklers, emergency lighting, portable extinguishers).

If a spill or leak to the environment has occurred, fire department, emergency response and/or hazardous materials spill personnel should be notified immediately. Cleanup should be attempted only by those trained in proper spill containment procedures. The first step should always be to determine if a leaking can be stopped. Contaminated soils should be removed for incineration and replaced with clean soil. If perchloryl fluoride should contact the water table, aquifer, or navigable waterway, the local and/or state emergency response authorities must be notified. A comprehensive emergency response or disaster preparedness/recovery plan should be in place prior to any operations involving the use, transportation, storage, or disposal of perchloryl fluoride. If perchloryl fluoride is spilled or leaked, the following specific steps are recommended:

- ☑ Restrict persons not wearing protective clothing from area of spill or leak until cleanup is complete and area can be opened for normal work.
- ☑ Remove all ignition sources and ventilate area.
- ☑ Absorb any liquids in vermiculite, dry sand, earth, or similar material and deposit in sealed containers. Perchloryl fluoride will rapidly evaporate and, therefore, liquid spills are highly unlikely.
- ☑ If gas is leaking, stop the flow of gas. If the source is a cylinder and leak cannot be stopped in place, remove cylinder to a safe place in the open air and repair or allow to empty.
- ☑ It may be necessary to dispose of perchloryl fluoride as a hazardous waste. The responsible state agency or the regional office of the federal Environmental Protection Agency (EPA) should be contacted for specific recommendations.

RISK ASSESSMENT: BUSINESS

General Assessment

Accidents or mishaps involving perchloryl fluoride can present a significant threat to business operations. The loss or damage of equipment or facilities can significantly affect fiscal viability. Lawsuits that may result from personnel injury/death, public exposures, and/or environmental contamination will also require a serious expenditure of resources. Media attention surrounding an injury, death, or environmental damage can also result in a loss of profits and loss of current as well as future business.

⓾ *Recommended Risk-Reduction Measures*

Company attorneys, safety and health professionals, and environmental specialists should be involved in the development of such procedures dealing with safety and emergency response. A company official should be pre-designated as a public relations officer with specific training in dealing with the press. Corporate plans and policies should be developed, approved, and implemented long before any need for such arises.

REFERENCES

American Conference of Governmental Industrial Hygienists. 1988. *Documentation of the Threshold Limit Values and Biological Exposure Indices,* 5th Edition (with updates). Cincinnati: ACGIH

Bueche, J. R. 1972. *Principles of Physics,* 2nd Edition. New York: McGraw-Hill.

Calabrese, E. J. 1994. *Biological Effects of Low Level Exposures.* New York: CRC Press/Lewis Publishers.

Cockerman, L. G., B. S. Shane. 1994. *Basic Environmental Toxicology.* New York: CRC Press/Lewis Publishers.

Gardner, R. W. 1994. *Chemical Intolerance.* New York: CRC Press/Lewis Publishers.

Hallenbeck, W. H. 1993. *Quantitative Risk Assessment for Environmental and Occupational Health,* 2nd Edition. New York: CRC Press/Lewis Publishers.

Jones, F. E. 1994. *Toxic Organic Vapors in the Workplace.* New York: CRC Press/Lewis Publishers.

Keith, L. H. and M. M. Walker. 1995. *Handbook of Air Toxics.* New York: CRC Press/Lewis Publishers.

Landis, W. G., M. H. Yu. 1995. *Introduction to Environmental Toxicology.* New York: CRC Press/Lewis Publishers.

Langerman, N. 1994 *Precautionary Labels for Chemical Containers.* New York: CRC Press/Lewis Publishers.

Lauwerys, R. R., P. Hoet. 1993. *Industrial Chemical Exposure.* New York: CRC Press/Lewis Publishers.

Lewis, R. J., Sr. 1992. *Sax's Dangerous Properties of Industrial Materials,* Volumes 1, 2, and 3, 8th Edition. New York: Van Nostrand Reinhold.

Lide, D. R. 1995. *Handbook of Chemistry and Physics,* 76th Edition. New York: CRC Press/Lewis Publishers.

Manahan, S. E. 1994. *Environmental Chemistry,* 6th Edition. New York: CRC Press/Lewis Publishers.

Mellström, G. A., J. E. Wahlberg, and H. I. Maibach. 1994. *Protective Gloves for Occupational Use.* New York: CRC Press/Lewis Publishers.

Meyer, E. 1990. *Chemistry of Hazardous Materials.* Englewood Cliffs, NJ: Prentice Hall.

Patnaik, P. 1992. *A Comprehensive Guide to the Hazardous Properties of Chemical Substances.* New York: Van Nostrand Reinhold.

Proctor, N. H., J. P. Hughes. 1991. *Chemical Hazards of the Workplace.* New York: Van Nostrand Reinhold.

Quigley, D. R. 1994. *Handbook of Emergency Chemical Management.* New York: CRC Press/Lewis Publishers.

Rea, W. J. 1995. *Chemical Sensitivity,* Vol. I-III. New York: CRC Press/Lewis Publishers.

Richardson, M. 1995. *Dictionary of Substance and Their Effects,* Vol. I-VI. New York: CRC Press/Lewis Publishers.

Sacarello, H. L. A. 1994. *The Comprehensive Handbook of Hazardous Materials.* New York: CRC Press/Lewis Publishers.

Torkelson, T. R. V. K. Rowe. 1981. Halogenated aliphatic hydrocarbons. *Patty's Industrial Hygiene and Toxicology,* Vol. 2B, *Toxicology,* pp 3568-3572. New York: Wiley-Interscience.

U.S. Department of Health and Human Services, National Institute for Occupational Safety and Health. 1994. *NIOSH Pocket Guide to Chemical Hazards.* Washington, D.C.: U.S. Government Printing Office.

MATERIAL SAFETY DATA SHEET

CHEMICAL NAME	CRC PRESS/LEWIS PUBLISHERS DISCLAIMER
PEROXYACETIC ACID	

HAZARD WARNING INFORMATION

HEALTH	FIRE	REACTIVE	OTHER	DEGREE OF HAZARD	COLOR CODING	OTHER CODES
3	2	4	OX	0 = Minimum Hazard 1 = Slight Hazard 2 = Moderate Hazard 3 = Serious Hazard 4 = Severe Hazard	HEALTH = BLUE FIRE = RED REACTIVITY = YELLOW OTHER = WHITE	OX = Oxidizer ACID = Acid ALK = Alkali COR = Corrosive W = Use No Water

SECTION I - GENERAL INFORMATION

Characterization	RCRA Number	EPA Class
Acid, Peroxy	**D001**	**Characteristic (I) Waste**

DOT Proper Shipping Name	**Peracetic acid (liquid)**	Chemical Abstract Service (CAS) Number	**79-21-0**
DOT Hazard Class and Label Requirements	**Organic Peroxide; Corrosive, Explosive**	DOT Emergency Guide Code	**51**
DOT Identification Number	**UN 2131**	Chemical Formula	**CH_3COOOH**

Synonyms

Peracetic acid; acetyl hydroperoxide; perethanoic acid; acetic peroxide; Desoxon 1; ethaneperoxoic acid; Osbon AC; PAA; peroxacetic acid.

SECTION II - HAZARDOUS INGREDIENTS AND IDENTITY INFORMATION

Hazardous Components (specific identity)	OSHA Exposure Criteria	NIOSH Exposure Criteria	Immediately Dangerous to Life and Health (IDLH)	ACGIH Exposure Criteria
Peroxyacetic acid (derivation: By oxidation of acetaldehyde; by reaction of acetic acid and hydrogen peroxide with sulfuric acid catalyst).	**PEL:** Not Established **STEL:** Not Established	**REL:** Not Established **STEL:** Not Established	Not Established	**TLV:** Not Established **STEL:** Not Established

SECTION III - PHYSICAL AND CHEMICAL CHARACTERISTICS

Boiling Point	**221°F (105°C)**	Specific Gravity (H_2O = 1)	**1.15**
Vapor Pressure (mm Hg)	**Not Found**	Molecular Weight	**76.05**
Vapor Density (Air = 1)	**Not Found**	Melting Point	**32°F (0°C)**

Solubility

Soluble in water, ether, sulfuric acid, and ethanol.

Appearance and Odor

Colorless liquid with a strong, acrid odor.

SECTION IV - FIRE AND EXPLOSION HAZARD DATA

Flash Point (method used)	**105°F (40.5°C) opened cup**	Explosive Limits in Air % by Volume	**LEL: 8% UEL: Not Determined**
NFPA Classification	**Class II Combustible Liquid**	Autoignition Temperature	**392°F (200°C)**

Extinguishing Media

Use flooding quantities of water. Carbon dioxide or foam may also be effective.

Special Fire Fighting Procedures

Wear full protective clothing and self-contained breathing apparatus (SCBA). Approach fire from upwind position. Use a water spray to keep fire-exposed containers cool. Poisonous gases are produced in fire. Firefighters should avoid contact with vapors produced during fire.

Unusual Fire and Explosion Hazards

Peroxyacetic acid is thermally unstable and may decompose explosively at high temperatures. Containers may explode in fire.

SECTION V - REACTIVITY DATA

Stability		Conditions to Avoid Peroxyacetic acid is stable in dilute aqueous solution but is thermally unstable. It is shock-sensitive (concentrations above 56%). Avoid contact with metals and combustible materials.
Stable X	Unstable	Incompatibility *(materials to avoid)* Incompatible with acids, alkalis, and heavy metals. Reacts explosively with solvents (diethyl ether, tetra-hydrifuran), metal chloride solutions (sodium chloride, calcium chloride), olefins, and organic matter.
Hazardous Polymerization		Conditions to Avoid Under normal conditions of temperature and pressure, hazardous polymerization of peroxyacetic acid CAN occur. Keep cool and away from most metals.
May Occur X	Will Not Occur	Hazardous Decomposition or By-products When heated to decomposition, peroxyacetic acid can emit highly toxic/poisonous fumes including carbon dioxide. Other products may be produced, depending on the concentration of the peroxyacetic acid.

SECTION VI - HEALTH HAZARD DATA

Primary Route(s) of Entry:	Inhalation?	Absorption (skin)?	Ingestion?
	X		X

Health Hazards

INHALATION: Causes coughing, dypsnea, irritation of the eyes, nose, throat, and mucosa of the upper respiratory tract. Delayed pulmonary edema (fluid in the lungs) may also develop. High or repeated exposures may cause liver and kidney damage.

EYES & SKIN: Skin contact can result in redness, swelling, burning, painful blisters, and severe burns. Can cause severe eye burns, tearing, redness, swelling, pain, burns, damage to vision.

INGESTION: Burns to the lips, mouth, and throat with pain in swallowing, salivation, intense thirst, swelling in the throat, and oral, esophageal, and gastrointestinal burns.

Carcinogenicity	NTP Listed?	IARC Cancer Review Group?	OSHA Regulated?	Target Organs?
Unknown Human Questioned Animal	No	No	No Specific Listing	Respiratory system, eyes, skin.

Medical Conditions Generally Aggravated by Exposure

Respiratory impairments (bronchitis, asthma); Skin conditions (dermatitis).

Emergency and First-aid Procedures

Eye contact: Flush immediately with water for 15 minutes (minimum); seek medical attention. Skin contact: Remove all contaminated clothing. Immediately wash area with large amounts of soap and water. Seek medical assistance. For inhalation: Remove the person from exposure. Provide respiratory assistance and CPR. Transfer to medical facility. Observe for 48 hours for lung effects. If swallowed: Seek medical attention immediately. Give 1 to 2 glasses of water to conscious person. Do NOT induce vomiting and do NOT give oral bicarbonate. Peroxyacetic acid is corrosive and vomiting may cause further injury or damage. Never give anything by mouth to an unconscious person.

SECTION VII - PRECAUTIONS FOR SAFE HANDLING AND USE

Steps to be Taken in Case Material is Released or Spilled

Absorb liquids in vermiculite or other material and place in sealed containers for disposal. Cleanup must be quick before the acetic acid content evaporates and leaves a more shock-sensitive residue. Ventilate area of spill or leak. Restrict those not involved in cleanup from entering area.

Preferred Waste Disposal Method

Burn in a chemical incinerator equipped with an afterburner and scrubber.

Precautions to be Taken in Handling and Storage

Peroxyacetic acid is a combustible liquid. Store in tightly closed containers in cool, dry, well-ventilated area away from heat, metals, or combustibles.

Other Precautions and Warnings

Sources of ignition are prohibited where peroxyacetic acid is used, handled, or stored.

SECTION VIII - CONTROL MEASURES AND PERSONAL PROTECTIVE EQUIPMENT

Respiratory Protection *(specify type)*

No exposure levels have been established. For best protection, use a supplied-air respirator with full facepiece in continuous flow mode, or a full facepiece SCBA operated in pressure demand mode.

Ventilation

Local exhaust or general mechanical systems recommended.

Protective Gloves	Eye Protection	Other Protective Clothing
Butyl Rubber	Chemical Goggles or Face Mask	Butyl Rubber Apron

Work/Hygiene Practices

Always wash hands thoroughly after using chemical; never bring food, drink, or smoking materials into vicinity of chemicals.

PEROXYACETIC ACID

CH_3COOOH

CAS: 79-21-0

IDENTIFICATION AND TYPICAL USES

Peroxyacetic acid is a colorless liquid with an acrid, irritating odor. Used in bleaching textiles, paper, oils, waxes, and starch. Also used as a polymerization catalyst, a bactericide and fungicide (especially in food processing), in the epoxidation of fatty acid esters and epoxy resin cursors, as a reagent in making caprolactum, and in synthetic glycerol.

RISK ASSESSMENT: HEALTH

General Assessment

Peroxyacetic acid is a poison by *ingestion* and *inhalation.* It is also an extremely corrosive skin, eye, and mucous membrane irritant. It is a suspected carcinogen in animals. There are no reports of any mutagenic or teratogenic effects on long-term exposure.

Inhalation of peroxyacetic acid can irritate the nose, throat, and bronchial tubes. This may cause coughing, dypsnea, congestion, tightness in the chest, and a dangerous buildup of fluid in the lungs (pulmonary edema), which can be fatal. Symptoms of pulmonary edema may be delayed up to 48 hours which can create a false sense of security with regard to health exposure risk.

Skin contact causes redness and painful swelling. There may be a burning sensation at the site of contact, painful blisters, and possible burns. Eye contact also causes burns, lachrymation (tearing), redness, swelling, serious pain, and possible loss of visual acuity.

Ingestion causes serious burning of the lips, mouth, and throat. There may be pain on swallowing and the possible development of oral, esophageal, and gastrointestinal burns.

Acute Health Effects

The following acute (short-term) health effects may occur immediately or shortly after exposure to peroxyacetic acid:

Skin: Possible irritation/rash or burning feeling. There can also be painful swelling, redness, and deep burns.

Eye: Severe irritation; may cause burns and damage to vision.

Lung: Irritation of the nose, throat, and lungs (bronchial tubes) following exposure. Delayed pulmonary edema may occur which can be fatal.

Other: Ingestion causes intense mouth, throat, and stomach pain. Severely corrosive destruction of the gastrointestinal tract is likely.

Chronic Health Effects

The following chronic (long-term) health effects may occur at some time after exposure to peroxyacetic acid and can last for months or even years:

Cancer Hazards: According to information presented in the references, there is limited evidence that exposure to peroxyacetic acid may cause cancer in test animals. There is no evidence of carcinogenic action in humans. More research is required in this regard.

Reproductive Hazard: According to the data presented in the references, peroxyacetic acid has not been adequately tested for its ability to adversely affect reproduction.

Other Chronic Effects: Repeated or long-term skin contact may cause rash from irritation. Repeated exposure can also damage the liver and kidneys. Very irritating substance can damage the lungs. It is not known whether long-term exposure to peroxyacetic acid will cause permanent damage to the lungs.

Recommended Risk-Reduction Measures

Personnel should avoid direct contact with peroxyacetic acid. It is extremely corrosive and highly destructive to human tissue. If a less toxic chemical cannot be substituted for a hazardous substance, then *engineering controls* are the most effective method of reducing exposures. The best protection is to enclose operations and/or provide local exhaust ventilation at the site of chemical release. While not always operationally feasible, isolating operations can also reduce exposure. Using respiratory protection is less effective than the controls mentioned above, but is still advisable whenever working with or around peroxyacetic acid. No exposure levels have been established for this chemical. However, this does not mean that exposure is without health risk. For the best protection, use a supplied-air respirator with a full facepiece operated in positive pressure mode or an MSHA/NIOSH-approved self-contained breathing apparatus (SCBA) with full facepiece in pressure demand mode. In either case, if a full facepiece is not available, then chemical goggles should be worn to protect the eyes. Whenever a chemical/dust splash hazard exists, a face shield and a rubber apron should be worn. To prevent hand and skin exposures, acid resistant gloves should be used. Butyl rubber or fluorocarbon rubber should suffice. However, it is suggested that glove manufacturers should be contacted and permeation studies obtained *before* final glove selections are made.

Administrative controls should also be in place to minimize the potential for human exposures. These may include written procedures or policies which specify the methods and techniques that will be practiced whenever personnel are to work with peroxyacetic acid.

All personnel should receive *training* on the use, hazards, protective measures, emergency actions, and other precautions per 29 CFR 1910.1200 (Hazard Communication), prior to the first assignment in an area where peroxyacetic acid is used or stored.

Before beginning employment and at regular intervals thereafter (e.g., annually), for those with frequent or potentially high exposures, the following medical tests are recommended:

☑ Lung function tests.

☑ Kidney function tests.

If symptoms develop or overexposure is suspected, the following may also be useful:

☑ Liver function tests.

☑ Consider chest X-ray after acute overexposure (may negative if taken immediately following exposure due to the delayed development of pulmonary edema).

Other methods to reduce exposure include:

☑ Where possible, enclose operations and use local exhaust ventilation at the site of chemical release. If local exhaust ventilation or enclosure is not used, respiratory protection should be mandatory.

☑ Where possible, automatically transfer peroxyacetic acid from drums or other storage containers to process containers.

☑ Always ensure that proper protective clothing is worn when using chemical substances.

☑ Wash thoroughly immediately after exposure to peroxyacetic acid and at the end of the work shift or before eating, drinking, or smoking.

☑ Hazard warning information should be posted in the work area. In addition, as part of an on-going education and training program, all information on the health and safety hazards of peroxyacetic acid should be communicated to all exposed workers.

RISK ASSESSMENT: ENVIRONMENT

General Assessment

The environment is at risk of exposure during transportation, storage, disposal, or destruction of peroxyacetic acid. In almost every scenario, the threat of environmental exposure is contingent upon the proper handling of the chemical substance. Accidental spills, large or small, can result in fire, explosion, and possible contamination of the surrounding environmental mediums (water, soil, and air).

Peroxyacetic acid is considered a Class II combustible liquid (per OSHA 29 CFR 1910.106). It is also corrosive to many materials. It is incompatible with acids, alkalis, and heavy metals, and contact can cause fire or explosion. These characteristics require special consideration during any emergency situation involving a leak or spill of peroxyacetic acid. Should peroxyacetic acid ever come into contact with any incompatible substances, violent and explosive reactions are possible. Peroxyacetic acid can enter the environment mainly through industrial discharges or spills.

☠ *Acute Ecological Effects*

Acute (short-term) toxic effects may include the death of animals, birds, or fish and death or low growth rate in plants. Acute effects are seen 2 to 4 days after animals or plants are exposed to peroxyacetic acid.

Peroxyacetic acid has slight acute toxicity to aquatic life. Insufficient data are available to evaluate or predict the acute (short-term) effects of peroxyacetic acid exposure to aquatic life, plants, birds, or land animals.

Chronic Ecological Effects

Chronic (long-term) toxic effects may include shortened life span, reproductive problems, lower fertility, and changes in appearance or behavior in exposed animals. These effects can be seen long after first exposure(s) to toxic chemicals.

Peroxyacetic acid has slight chronic toxicity in aquatic life. Insufficient data are available on the long-term effects of peroxyacetic acid to plants, birds, or land animals.

Water Solubility

Peroxyacetic acid is highly soluble in water. Concentrations of 1000 milligrams and more will mix with a liter of water.

Persistence in the Environment

Peroxyacetic acid is slightly persistent in the aquatic environment, with a half-life of between 2 to 20 days. The half-life of a pollutant is the amount of time it takes for one half of the chemical to be degraded.

Bioaccumulation in Aquatic Organisms

Some substances increase in concentration, or *bioaccumulate,* in living organisms as they breathe contaminated air, drink contaminated water, or eat contaminated food. These chemicals can become concentrated in the tissues and internal organs of animals as well as humans.

The concentration of peroxyacetic acid found in fish tissues is expected to be much lower than the average concentration of peroxyacetic acid in the water from which the fish was taken.

Recommended Risk-Reduction Measures

Proper training of all transporters will reduce the likelihood of a mishap or accident resulting in a leak or spill to the environment. The correct use of DOT labeling (per the MSDS) on all containers, trucks, and rail cars will enable emergency responders to react properly and quickly to any disaster thereby reducing the potential risk to the environment and to personnel.

Storage of peroxyacetic acid should be segregated from incompatible chemicals to minimize the risk of cross-contamination or contact.

If a spill or leak to the environment has occurred, fire department, emergency response and/or hazardous materials spill personnel should be notified immediately. Cleanup should be attempted only by those trained in proper spill containment procedures using non-sparking tools. Contaminated soils should be removed for incineration and replaced with clean soil. If peroxyacetic acid should contact the water table, aquifer, or navigable waterway, time is of the essence. It is highly miscible in water and, therefore, total containment and remediation may not be entirely possible. When such spills occur, the local and/or state emergency response authorities must be notified. A comprehensive emergency response or disaster preparedness/recovery plan should be in place prior to any operations involving the use, transportation, storage, or disposal of peroxyacetic acid. If peroxyacetic acid is spilled or leaked, the following specific steps are recommended:

- ☑ Restrict persons not wearing protective clothing from area of spill or leak until cleanup is complete and area can be opened for normal work.
- ☑ Remove all ignition sources and ventilate area of spill or leak.
- ☑ If liquid is spilled, absorb in vermiculite, dry sand, earth, or similar material and deposit in sealed containers. Response actions should be quick. If enough acetic acid evaporates, it will leave behind a highly shock-sensitive residue.
- ☑ It may be necessary to dispose of peroxyacetic acid as a hazardous waste. The responsible state agency or the regional office of the federal Environmental Protection Agency (EPA) should be contacted for specific recommendations.

RISK ASSESSMENT: BUSINESS

General Assessment

Accidents or mishaps involving peroxyacetic acid can present a significant threat to business operations. The loss or damage of equipment or facilities can significantly affect fiscal viability. Lawsuits that may result from personnel injury/death, public exposures, and/or environmental contamination will also require a serious expenditure of resources. Media attention surrounding an injury, death, or environmental damage can also result in a loss of profits and loss of current as well as future business. Always remember that whenever the terms "cancer" or "carcinogen" are used, public emotion, hysteria, ignorance, and anxiety can all run equally high. This must be carefully considered whenever drafting or implementing public relations policies.

Recommended Risk-Reduction Measures

Company attorneys, safety and health professionals, and environmental specialists should be involved in the development of any procedures dealing with safety or emergency response. A company official should be pre-designated as a public relations officer with specific training in dealing with the press. Corporate plans and policies should be developed, approved, and implemented long before any need for such arises.

REFERENCES

Ahlbom, A. 1993. *Biostatistics for Epidemiologists.* New York: CRC Press/Lewis Publishers.

American Conference of Governmental Industrial Hygienists. 1988. *Documentation of the Threshold Limit Values and Biological Exposure Indices,* 5th Edition (with updates). Cincinnati: ACGIH

Asante-Duah, D. K. 1993. *Hazardous Waste Risk Assessment.* New York: CRC Press/Lewis Publishers.

Bloom, A. D. and F. J. DeSerres. 1995. *Ecotoxicity and Human Health.* New York: CRC Press/Lewis Publishers.

Brusick, D. J. 1994. *Methods for Genetic Risk Assessment.* New York: CRC Press/Lewis Publishers.

Bueche, J. R. 1972. *Principles of Physics,* 2nd Edition. New York: McGraw-Hill.

Burrell, R., D. K. Flaherty, and Leonard J. Sauers. 1992. *Toxicology of the Immune System.* New York: Van Nostrand Reinhold.

Calabrese, E. J. 1994. *Biological Effects of Low Level Exposures.* New York: CRC Press/Lewis Publishers.

Cockerman, L. G., B. S. Shane. 1994. *Basic Environmental Toxicology.* New York: CRC Press/Lewis Publishers.

Cothern, R. R. 1993. *Comparative Environmental Risk Assessment.* New York: CRC Press/Lewis Publishers.

Danse, I. R. 1991. *Common Sense Toxics in the Workplace.* New York: Van Nostrand Reinhold.

Gardner, R. W. 1994. *Chemical Intolerance.* New York: CRC Press/Lewis Publishers.

Hallenbeck, W. H. 1993. *Quantitative Risk Assessment for Environmental and Occupational Health,* 2nd Edition. New York: CRC Press/Lewis Publishers.

Hosty, J. W., P. Foster. 1990. *A Practical Guide to Chemical Spill Response.* New York: Van Nostrand Reinhold.

Howard, P. H., et. al. 1991. *Handbook of Environmental Degradation Rates.* New York: CRC Press/Lewis Publishers.

James, A., M. Lord. 1992. *Index of Chemical and Physical Data.* New York: Van Nostrand Reinhold.

Jones, F. E. 1994. *Toxic Organic Vapors in the Workplace.* New York: CRC Press/Lewis Publishers.

Keith, L. H. and M. M. Walker. 1995. *Handbook of Air Toxics.* New York: CRC Press/Lewis Publishers.

Langerman, N. 1994 *Precautionary Labels for Chemical Containers.* New York: CRC Press/Lewis Publishers.

Lauwerys, R. R., P. Hoet. 1993. *Industrial Chemical Exposure.* New York: CRC Press/Lewis Publishers.

Lewis, R. J., Sr. 1993. *Hawley's Condensed Chemical Dictionary,* 12th Edition. New York: Van Nostrand Reinhold.

Lewis, R. J., Sr. 1992. *Sax's Dangerous Properties of Industrial Materials,* Volumes 1, 2, and 3, 8th Edition. New York: Van Nostrand Reinhold.

Manahan, S. E. 1994. *Environmental Chemistry,* 6th Edition. New York: CRC Press/Lewis Publishers.

Mellström, G. A., J. E. Wahlberg, and H. I. Maibach. 1994. *Protective Gloves for Occupational Use.* New York: CRC Press/Lewis Publishers.

Meyer, E. 1990. *Chemistry of Hazardous Materials.* Englewood Cliffs, NJ: Prentice Hall.

Mycock, J. C., J. D. McKenna, and L. Theodore. 1995. *Handbook of Air Pollution Control Engineering and Technology.* New York: CRC Press/Lewis Publishers.

Nelson, G. O. 1992. *Gas Mixtures.* New York: CRC Press/Lewis Publishers.

Ott, W. R. 1995. *Environmental Statistics and Data Analysis.* New York: CRC Press/Lewis Publishers.

Patnaik, P. 1992. *A Comprehensive Guide to the Hazardous Properties of Chemical Substances.* New York: Van Nostrand Reinhold.

Philp, R. B. 1995. *Environmental Hazards and Human Health.* New York: CRC Press/Lewis Publishers.

Proctor, N. H., J. P. Hughes. 1991. *Chemical Hazards of the Workplace.* New York: Van Nostrand Reinhold.

Rea, W. J. 1995. *Chemical Sensitivity,* Vol. I-III. New York: CRC Press/Lewis Publishers.

Richardson, M. 1995. *Dictionary of Substance and Their Effects,* Vol. I-VI. New York: CRC Press/Lewis Publishers.

Sacarello, H. L. A. 1994. *The Comprehensive Handbook of Hazardous Materials.* New York: CRC Press/Lewis Publishers.

Sherman, J. D. 1988. *Chemical Exposure and Disease.* New York: Van Nostrand Reinhold.

Witorsch, P. and S. V. Spagnolo. 1994. *Air Pollution and Lung Disease in Adults.* New York: CRC Press/Lewis Publishers.

MATERIAL SAFETY DATA SHEET

CHEMICAL NAME	
PHENOL	**CRC PRESS/LEWIS PUBLISHERS DISCLAIMER:** THE INFORMATION AND RECOMMENDATIONS PRESENTED HEREIN ARE BASED ON SOURCES BELIEVED TO BE RELIABLE. CRC MAKES NO REPRESENTATION ON ITS COMPLETENESS OR ACCURACY. IT IS THE RESPONSIBILITY OF THE USER TO DETERMINE THE CHEMICAL'S SUITABILITY FOR ITS INTENDED USE, THE CHEMICAL'S SAFE USE, AND THE CHEMICAL'S PROPER DISPOSAL. NO REPRESENTATIONS AND/OR WARRANTIES, EITHER EXPRESSED OR IMPLIED, OF THE MERCHANTABILITY OR FITNESS FOR A PARTICULAR PURPOSE, OR OF ANY OTHER NATURE, ARE MADE WITH RESPECT TO THE INFORMATION PROVIDED IN THIS MSDS OR TO THE CHEMICAL TO WHICH INFORMATION MAY REFER. CRC NEITHER ASSUMES NOR AUTHORIZES ANY OTHER PERSON TO ASSUME FOR IT, ANY OTHER ADDITIONAL RESPONSIBILITY OR LIABILITY FOR THE USE OF, OR RELIANCE UPON, THIS INFORMATION.

HAZARD WARNING INFORMATION

HEALTH	FIRE	REACTIVE	OTHER	DEGREE OF HAZARD	COLOR CODING	OTHER CODES
3	2	0		0 = Minimum Hazard 1 = Slight Hazard 2 = Moderate Hazard 3 = Serious Hazard 4 = Severe Hazard	HEALTH = BLUE FIRE = RED REACTIVITY = YELLOW OTHER = WHITE	OX = Oxidizer ACID = Acid ALK = Alkali COR = Corrosive W = Use No Water

SECTION I - GENERAL INFORMATION

Characterization	**Phenol**	RCRA Number	**U188**
		EPA Class	**Toxic Waste**
DOT Proper Shipping Name	**Phenol**	Chemical Abstract Service (CAS) Number	**108-95-2**
DOT Hazard Class and Label Requirements	**Poison B**	DOT Emergency Guide Code	**55**
DOT Identification Number	**UN1671** (solid) **UN2312** (molten) **UN2821** (solution)	Chemical Formula	C_6H_5OH

Synonyms

Carbolic acid; hydroxybenzene; monohydroxybenzene; phenyl alcohol; phenyl hydroxide; oxybenzene; phenic acid; phenylic alcohol; phenyl hydrate.

SECTION II - HAZARDOUS INGREDIENTS AND IDENTITY INFORMATION

Hazardous Components (specific identity)	OSHA Exposure Criteria	NIOSH Exposure Criteria	Immediately Dangerous to Life and Health (IDLH)	ACGIH Exposure Criteria
Phenol (derivation: Made by the oxidation of cumene yielding acetone as a by-product. First, cumene hydroperoxide is produced, which decomposes with dilute sulfuric acid to the primary products, plus acetophenol and phenyl dimethyl carbinol). **1 ppm = 3.91 mg/m³**	**PEL (skin):** 5 ppm 19 mg/m³ **STEL:** Not Established	**REL (skin):** 5 ppm 19 mg/m³ **CEILING (15 min):** 15.6 ppm 60 mg/m³	**250 ppm**	**TLV:** 5 ppm 19 mg/m³ **STEL:** Not Established

SECTION III - PHYSICAL AND CHEMICAL CHARACTERISTICS

Boiling Point	**359°F (182°C)**	Specific Gravity (H_2O = 1)	**1.06**
Vapor Pressure (mm Hg)	**0.35 at 77°F (25°C)**	Molecular Weight	**94.1**
Vapor Density (Air = 1)	**3.24**	Melting Point	**109°F (43°C)**

Solubility

Soluble in water (9%), alcohol, chloroform, glycerol, carbon disulfide, petrolatum, oils, and alkalis.

Appearance and Odor

Colorless to light-pink, crystalline solid with a sweet, acrid odor. Liquefies when mixed with 8% water.

SECTION IV - FIRE AND EXPLOSION HAZARD DATA

Flash Point (method used)	**175°F (79°C) closed cup**	Explosive Limits in Air % by Volume	**LEL: 1.8% UEL: 8.6%**
NFPA Classification	**Combustible Solid**	Autoignition Temperature	**1319°F (715°C)**

Extinguishing Media

Use water spray, carbon dioxide, dry chemical, or alcohol foam (do NOT use solid stream of water).

Special Fire Fighting Procedures

Wear full protective clothing and self-contained breathing apparatus (SCBA). Poisonous gases are produced in fire. Use water spray to keep fire-exposed containers cool.

Unusual Fire and Explosion Hazards

Moderate fire and explosion hazard when exposed to heat or flame. Containers may explode in fire. Chemical will react on contact with oxidizable materials. In fire, it emits toxic fumes and vapors that form explosive mixtures with air. Water containing phenol can cause serious chemical burns.

SECTION V - REACTIVITY DATA

Stability		Conditions to Avoid Phenol is stable at room temperature in closed containers under normal conditions of handling and storage. Keep away from heat and incompatible materials.
Stable	Unstable	Incompatibility *(materials to avoid)*
X		Strong oxidizers, calcium hypochlorite, aluminum chloride, and acids. Reacts violently and explosively with formaldehyde, peroxydisulfide acid, sodium nitrate + heat, and aluminum chloride + nitromethane.
Hazardous Polymerization		Conditions to Avoid Under normal conditions of temperature and pressure, hazardous polymerization of phenol cannot occur. Avoid heating phenol above temperatures of 122°F (90°C).
May Occur	Will Not Occur	Hazardous Decomposition or By-products
	X	When heated to decomposition, phenol can produce oxides of carbon and water.

SECTION VI - HEALTH HAZARD DATA

Primary Route(s) of Entry:	Inhalation?	Absorption (skin)?	Ingestion?
	X	X	X

Health Hazards

INHALATION: Irritation of the eyes, nose, throat, and respiratory tract. Systemic toxic effects include anorexia, weight loss, weakness, muscle pains, pallor, nausea, vomiting, diarrhea, weakness, darkened urine, headache, tinnitus, sweating, convulsions, cyanosis, shock, loss of consciousness, respiratory failure, and death.

ABSORPTION: Causes white, wrinkled discoloration, severe burns and/or systemic poisoning (due to rapid absorption). Eye contact causes serious corrosive damage (conjunctival edema, corneal opacification, and hypesthesia), and possible blindness.

INGESTION: Gangrene and corrosion of the lips, mouth, throat, esophagus, and stomach.

Carcinogenicity	NTP Listed?	IARC Cancer Review Group?	OSHA Regulated?	Target Organs?
Suspected Human Questioned Animal	No	No	29 CFR 1910.1000 Table Z-1	Respiratory system, eyes, skin, CNS, liver, kidney.

Medical Conditions Generally Aggravated by Exposure

Chronic respiratory disorders, pre-existing skin disorders, convulsive disorders, kidney or liver disease.

Emergency and First-aid Procedures

Eye contact: Flush immediately with water for 15 minutes (minimum); seek medical attention. **Skin contact**: Remove all contaminated clothing. Use decontaminating agent such as PEG (polyethyleneglycol-300) on affected area or, if PEG is not available, immediately wash area with large amounts of soap and water. Seek medical assistance. For **inhalation**: Remove the person from exposure. Provide respiratory assistance and CPR, if required. If **swallowed**: Seek medical attention immediately. Unless advised otherwise, give conscious person 15 to 30 cc of vegetable oil and induce vomiting.

SECTION VII - PRECAUTIONS FOR SAFE HANDLING AND USE

Steps to be Taken in Case Material is Released or Spilled

Absorb liquids in vermiculite, dry sand, earth, or other non-reactive material and deposit in sealed drum. Collect powdered material in safest manner possible. Do NOT dry sweep. Ventilate area of spill or leak. Restrict those not involved in cleanup from entering area. Remove all sources of ignition.

Preferred Waste Disposal Method

No citation.

Precautions to be Taken in Handling and Storage

Store to avoid contact with incompatible materials in tightly closed containers in cool, well-ventilated area away from heat and flame.

Other Precautions and Warnings

Heat may cause solids to liquefy and containers to build up pressure and explode.

SECTION VIII - CONTROL MEASURES AND PERSONAL PROTECTIVE EQUIPMENT

Respiratory Protection *(specify type)*

For high exposures, use an MSHA/NIOSH-approved supplied-air respirator with full facepiece in continuous flow mode, or use a SCBA operated in pressure demand or other positive pressure mode.

Ventilation

Local exhaust or general mechanical systems recommended.

Protective Gloves	Eye Protection	Other Protective Clothing
Butyl Rubber	Chemical Goggles or Face Mask	Rubber Apron

Work/Hygiene Practices

Always wash hands thoroughly after using chemical; never bring food, drink, or smoking materials into vicinity of chemicals.

PHENOL

C_6H_5OH

CAS: 108-95-2

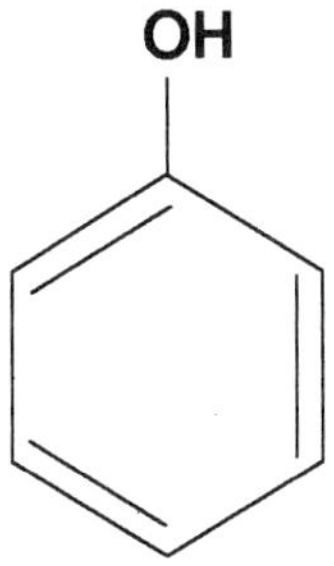

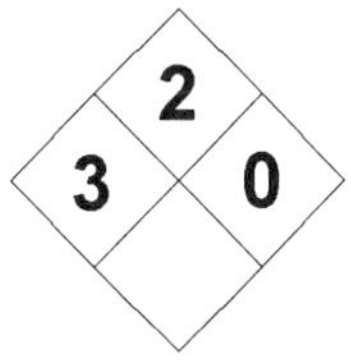

IDENTIFICATION AND TYPICAL USES

Phenol is a colorless to light-pink crystalline solid with a characteristic, medicinal, sharp, sweet, tangy, acrid odor. It turns pink or red when impurities are present or it is exposed to heat or light. Phenol liquefies by mixing with approximately 8% water. Phenol is one of many aromatic compounds found in coal tar. It is used in the manufacture of various phenolic and epoxy resins, for refining lubricating oils, as a disinfectant, a fuel-oil sludge inhibitor, and as a reagent in chemical analysis. It is also used in the production or manufacture of a large variety of aromatic compounds including fertilizers, illuminating gases, coke, explosives, lampblack, paints, paint removers, asbestos goods, wood preservatives, textiles, perfumes, bakelite, rubber, and other plastics. It is used in medical and industrial organic compounds and dyes, and in germicidal paints and slimicides. Phenol has been identified as a constituent of cigarette smoke and automobile exhaust.

RISK ASSESSMENT: HEALTH

General Assessment

Phenol is a human poison by *inhalation, ingestion,* and skin *absorption.* It is a severe irritant to the eyes, skin and respiratory system. Human mutation data have been reported for phenol. It is a questionable carcinogen in animals and a suspected carcinogen in humans, although the data are inconclusive in this area of study. It is also an experimental teratogen with adverse reproductive effects reported. Phenol is a general protoplasmic poison that is corrosive to any living tissue it contacts.

Inhalation can cause irritation to the respiratory system and associated mucosa. Although it is an irritant, its good warning properties and low volatility make inhalation exposures less of a concern than absorption or ingestion. Symptoms of systemic exposure may include pallor, nausea, anorexia, vomiting, diarrhea, weakness, muscle aches, darkened urine, headache, tinnitus (ringing in the ears), sweating, convulsions, cyanosis, shock, loss of consciousness, and death due to respiratory failure. There may also be narcotic effects associated with the depression of the central nervous system (CNS).

In addition to the systemic symptoms noted above, skin contact can cause serious localized burns, possible skin eruptions, and dermatitis. If phenol should contact the eyes, there may be serious damage including conjunctival edema, corneal opacification, hypethesia, and possible blindness.

Ingestion of phenol can cause burns to the lips, mouth, esophagus, and stomach with gangrene and unspecified gastrointestinal disturbances.

☠ *Acute Health Effects*

The following acute (short-term) health effects may occur immediately or shortly after exposure to phenol:

Skin: Severe irritation and possible burns. Will pass through unbroken skin to cause toxic systemic effects, including weakness, nausea, collapse, and death.

Eye: Severe irritation with damage and possible loss of vision likely.

Lung: Severe irritation to the respiratory tract and associated mucosa. Can cause weakness, nausea, confusion, collapse, coma, and death (respiratory failure).

CNS: A CNS depressant. Weakness, confusion, mood changes and other personality changes, nausea, collapse, and death.

Other: Significant skin contact or ingestion can cause death in minutes.

Chronic Health Effects

The following chronic (long-term) health effects may occur at some time after exposure to phenol and can last for months or even years:

Cancer Hazards: Some studies have suggested that exposure to phenol causes mutations (genetic changes) in humans. While there is no evidence of carcinogenicity of this chemical in humans, it has caused cancer in mice and is considered a *suspected* carcinogen. It is suggested that mutagenic chemicals be regarded as carcinogens during use and handling.

Reproductive Hazard: According to information presented in the references, phenol has been shown to cause teratogenic and adverse reproductive effects in animals.

Other Chronic Effects: Chronic effects are not common with phenol but can include digestive disturbances, neurological disorders, skin rash and dermatitis with dark pigmentation of the skin and whites of the eyes (sclera), and liver, kidney, pancreas, and heart muscle damage. Phenol has not been adequately evaluated to determine whether brain or other nerve damage could occur with repeated exposure. However, many solvents and other petroleum-based chemicals have been shown to cause such damage. Effects may include reduced memory and concentration, personality changes, fatigue, sleep disturbances, reduced coordination, and/or effects on nerves supplying internal organs and the peripheral nervous system.

Recommended Risk-Reduction Measures

Personnel should avoid direct contact with phenol. If a less toxic chemical cannot be substituted for a hazardous substance, then *engineering controls* are the most effective method of reducing exposures. The best protection is to enclose operations and/or provide local exhaust ventilation at the site of chemical release. While not always operationally feasible, isolating operations can also reduce exposure. Using respiratory protection is less effective than the controls mentioned above, but is still advisable whenever working with or around phenol. For significant exposures over the PEL (5 ppm), use an MSHA/NIOSH-approved full facepiece respirator equipped with an organic vapor cartridge. Better protection is obtained from a supplied-air respirator with full facepiece operated in continuous flow mode, or a self-contained breathing apparatus (SCBA) with full facepiece operated in pressure demand. If a full facepiece is not available, then chemical goggles should be worn to protect the eyes. Whenever a chemical splash hazard exists, a face shield and a rubber apron should be worn. To prevent hand and skin exposures, impervious butyl rubber gloves should be used. Glove manufacturers should be contacted and permeation studies obtained *before* glove selections have been made.

Administrative controls should also be in place to minimize the potential for human exposures. These may include written procedures or policies which specify the methods and techniques that will be practiced whenever personnel are to work with phenol.

All personnel should receive *training* on the use, hazards, protective measures, emergency actions, and other precautions per 29 CFR 1910.1200 (Hazard Communication), prior to the first assignment in an area where phenol is used or stored.

Before beginning employment with phenol and at regular intervals thereafter (e.g., annually), those persons with high or frequent exposures should be provided the following recommended medical tests:

- ☑ Kidney and liver function tests.

If symptoms develop or overexposure is suspected, the following tests are recommended:

- ☑ Urinary phenol tests.
- ☑ Kidney and liver function tests
- ☑ Neurological evaluation for brain and nervous system effects.

Any evaluation should include a careful history of past and present symptoms with an examination. Medical tests that simply look for existing damage are not a substitute for controlling exposures.

Other methods to reduce exposure include:

- ☑ Where possible, enclose operations and use local exhaust ventilation at the site of chemical release. If local exhaust ventilation or enclosure is not used, respiratory protection should be mandatory.
- ☑ Also, if possible, liquids should be transferred automatically from storage containers to process containers.
- ☑ Always ensure that proper protective clothing is worn when using chemical substances and that personnel have been trained on the use, care, and maintenance of personal protective equipment.

- ☑ Wash thoroughly immediately after exposure to phenol and at the end of the work shift or before eating, drinking, or smoking.
- ☑ Hazard warning information should be posted in the work area. In addition, as part of an on-going education and training program, all information on the health and safety hazards of phenol should be communicated to all potentially exposed workers.
- ☑ Eye wash stations should be provided in the immediate work area for emergency use. If there is a possibility of skin exposure to phenol, emergency shower facilities should also be provided.
- ☑ Personnel should never wear contaminated clothing home (family members can be exposed). Clothing should be laundered by personnel who have been trained on the hazards of phenol.

RISK ASSESSMENT: ENVIRONMENT

General Assessment

The environment is at risk of exposure during transportation, storage, disposal, or destruction of phenol. In almost every scenario, the threat of environmental exposure is contingent upon the proper handling of the chemical substance. Accidental spills, large or small, can result in fire, explosion, and possible contamination of the surrounding environmental mediums (water, soil, and air).

Phenol is considered a combustible solid. However, its vapors present a moderate fire hazard when exposed to heat or air. In fact, air mixtures containing just 3 to 10% phenol are explosive. It is reactive with many commonly encountered materials, including strong oxidizers (chlorine, fluorine, bromine), acids (hydrochloric, sulfuric, and nitric), calcium hypochlorite, and aluminum chloride. These characteristics require caution in handling, storage, transportation, and disposal. It is also a corrosive and, therefore, special consideration is required during any emergency situation involving a leak or spill of phenol. Should phenol ever come into contact with any incompatible substances during use, transportation, storage, or disposal, the formation of highly toxic and/or highly explosive commodities is extremely possible.

Phenol may enter the environment through oil refinery discharges, coal conversion plants, municipal waste treatment plant and industrial plant discharges, or through spills.

Acute Ecological Effects

Acute (short-term) toxic effects may include the death of animals, birds, or fish and death or low growth rate in plants. Acute effects are seen 2 to 4 days after animals or plants are exposed to phenol.

Phenol has moderate acute toxicity to aquatic life. Phenol can kill, and cause germination decreases in various agricultural crop plants. It has high acute toxicity to birds. Insufficient data are available to evaluate or predict the short-term effects of phenol exposure to land animals.

Chronic Ecological Effects

Chronic (long-term) toxic effects may include shortened life span, reproductive problems, lower fertility, and changes in appearance or behavior in exposed animals. These effects can be seen long after first exposure(s) to toxic chemicals.

Phenol has moderate chronic toxicity to aquatic life. Insufficient data are available on the long-term effects of phenol to plants, birds, or land animals.

Water Solubility

Phenol is highly soluble in water. Concentrations of 1000 milligrams and more will mix with a liter of water.

Persistence in the Environment

Phenol is slightly persistent in water, with a half-life between 2 and 20 days. The half-life of a pollutant is the amount of time it takes for one half of the chemical to be degraded. About 26.3% of phenol will eventually end up in air, approximately 73.3% in water, and about 0.2% in terrestrial soil and aquatic sediments.

Bioaccumulation in Aquatic Organisms

Some substances increase in concentration, or *bioaccumulate,* in living organisms as they breathe contaminated air, drink contaminated water, or eat contaminated food. These chemicals can become concentrated in the tissues and internal organs of animals as well as humans.

The concentration of phenol found in fish tissue is expected to be about the same as the average concentration of phenol in the water from which the fish was taken.

❶ *Recommended Risk-Reduction Measures*

Proper training of all transporters will reduce the likelihood of a mishap or accident resulting in a leak or spill to the environment. The correct use of DOT labeling (per the MSDS) on all containers, trucks, and rail cars will enable emergency responders to react properly and quickly to any disaster thereby reducing the potential risk to the environment and to personnel.

Storage of phenol should be segregated from incompatible chemicals to minimize the risk of cross-contamination or contact. Buildings designated for storage should be equipped with appropriate fire protection systems (alarms, sprinklers, emergency lighting, portable extinguishers).

If a spill or leak to the environment has occurred, fire department, emergency response and/or hazardous materials spill personnel should be notified immediately. Cleanup should be attempted only by those trained in proper spill containment procedures. Contaminated soils should be removed for incineration and replaced with clean soil. If phenol should contact the water table, aquifer, or navigable waterway, time is of the essence. It is highly soluble in the aquatic environment and total remediation may not be possible. The local and/or state emergency response authorities must be notified. A comprehensive emergency response or disaster preparedness/recovery plan should be in place prior to any operations involving the use, transportation, storage, or disposal of phenol. If phenol is spilled or leaked, the following specific steps are recommended:

- ☑ Restrict persons not wearing protective clothing from area of spill or leak until cleanup is complete and area can be opened for normal work.
- ☑ Remove all ignition sources.
- ☑ Ventilate area of leak.
- ☑ Absorb liquid spills in vermiculite, dry sand, earth, or similar material and deposit in sealed containers. Collect solid materials in safest manner possible and place in sealed containers. Do NOT dry sweep (creates airborne dusts). Use vacuum equipped with a high-efficiency particulate air (HEPA) filter instead.
- ☑ It may be necessary to dispose of phenol as a hazardous waste. The responsible state agency or the regional office of the federal Environmental Protection Agency (EPA) should be contacted for specific recommendations.

RISK ASSESSMENT: BUSINESS

General Assessment

Accidents or mishaps involving phenol can present a significant threat to business operations. The loss or damage of equipment or facilities can significantly affect fiscal viability. Lawsuits that may result from personnel injury/death, public exposures, and/or environmental contamination will also require a serious expenditure of resources. Media attention surrounding an injury, death, or environmental damage can also result in a loss of profits and loss of current as well as future business. Always remember that anytime the terms "cancer," "carcinogen," or "reproductive hazard" are used, public hysteria, emotion, anxiety, and ignorance can run equally high. This must be carefully considered whenever drafting or implementing public relations policies.

❶ *Recommended Risk-Reduction Measures*

Company attorneys, safety and health professionals, and environmental specialists should be involved in the development of any procedures dealing with safety or emergency response. A company official should be pre-designated as a public relations officer with specific training in dealing with the press. Corporate plans and policies should be developed, approved, and implemented long before any need for such arises.

REFERENCES

Asante-Duah, D. K. 1993. *Hazardous Waste Risk Assessment.* New York: CRC Press/Lewis Publishers.

Bloom, A. D. and F. J. DeSerres. 1995. *Ecotoxicity and Human Health.* New York: CRC Press/Lewis Publishers.

Bueche, J. R. 1972. *Principles of Physics,* 2nd Edition. New York: McGraw-Hill.

Calabrese, E. J. 1994. *Biological Effects of Low Level Exposures.* New York: CRC Press/Lewis Publishers.

Cockerman, L. G., B. S. Shane. 1994. *Basic Environmental Toxicology.* New York: CRC Press/Lewis Publishers.

Cothern, C. R. 1995. *Handbook for Environmental Risk Decisionmaking.* New York: CRC Press/Lewis Publishers.

Lewis, R. J., Sr. 1992. *Sax's Dangerous Properties of Industrial Materials,* Volumes 1, 2, and 3, 8th Edition. New York: Van Nostrand Reinhold.

Patnaik, P. 1992. *A Comprehensive Guide to the Hazardous Properties of Chemical Substances.* New York: Van Nostrand Reinhold.

U.S. Department of Health and Human Services, National Institute for Occupational Safety and Health. 1994. *NIOSH Pocket Guide to Chemical Hazards.* Washington, D.C.: U.S. Government Printing Office.

MATERIAL SAFETY DATA SHEET

CHEMICAL NAME	CRC PRESS/LEWIS PUBLISHERS DISCLAIMER
PHENYLHYDRAZINE	**CRC PRESS/LEWIS PUBLISHERS DISCLAIMER:** THE INFORMATION AND RECOMMENDATIONS PRESENTED HEREIN ARE BASED ON SOURCES BELIEVED TO BE RELIABLE. CRC MAKES NO REPRESENTATION ON ITS COMPLETENESS OR ACCURACY. IT IS THE RESPONSIBILITY OF THE USER TO DETERMINE THE CHEMICAL'S SUITABILITY FOR ITS INTENDED USE, THE CHEMICAL'S SAFE USE, AND THE CHEMICAL'S PROPER DISPOSAL. NO REPRESENTATIONS AND/OR WARRANTIES, EITHER EXPRESSED OR IMPLIED, OF THE MERCHANTABILITY OR FITNESS FOR A PARTICULAR PURPOSE, OR OF ANY OTHER NATURE, ARE MADE WITH RESPECT TO THE INFORMATION PROVIDED IN THIS MSDS OR TO THE CHEMICAL TO WHICH INFORMATION MAY REFER. CRC NEITHER ASSUMES NOR AUTHORIZES ANY OTHER PERSON TO ASSUME FOR IT, ANY OTHER ADDITIONAL RESPONSIBILITY OR LIABILITY FOR THE USE OF, OR RELIANCE UPON, THIS INFORMATION.

HAZARD WARNING INFORMATION

HEALTH	FIRE	REACTIVE	OTHER	DEGREE OF HAZARD	COLOR CODING	OTHER CODES
3	2	2		0 = Minimum Hazard 1 = Slight Hazard 2 = Moderate Hazard 3 = Serious Hazard 4 = Severe Hazard	HEALTH = BLUE FIRE = RED REACTIVITY = YELLOW OTHER = WHITE	OX = Oxidizer ACID = Acid ALK = Alkali COR = Corrosive W = Use No Water

SECTION I - GENERAL INFORMATION

Characterization	RCRA Number	EPA Class
Aromatic Amine	**None**	**Not Applicable**

DOT Proper Shipping Name **Phenylhydrazine**	Chemical Abstract Service (CAS) Number **100-63-0**
DOT Hazard Class and Label Requirements **Poison B; Poison**	DOT Emergency Guide Code **53**
DOT Identification Number **UN 2572**	Chemical Formula $C_6H_5NHNH_2$

Synonyms
Hydrazinobenzene; monophenylhydrazine; hydrazine-benzene.

SECTION II - HAZARDOUS INGREDIENTS AND IDENTITY INFORMATION

Hazardous Components (specific identity)	OSHA Exposure Criteria	NIOSH Exposure Criteria	Immediately Dangerous to Life and Health (IDLH)	ACGIH Exposure Criteria
Phenylhydrazine (derivation: By reduction of diazotized aniline, followed by reaction with sodium hydroxide). 1 ppm = 4.49 mg/m^3	**PEL (skin):** 5 ppm 22 mg/m^3 **STEL:** 10 ppm 45 mg/m^3	**REL (skin):** 0.14 ppm 0.6 mg/m^3 (2-hr ceiling) Possible Cancer Agent	15 ppm	**TLV (skin):** 0.1 ppm 0.45 mg/m^3 Suspected Human Carcinogen

SECTION III - PHYSICAL AND CHEMICAL CHARACTERISTICS

Boiling Point **470°F (243°C) Decomposes**	Specific Gravity (H_2O = 1) **1.10**
Vapor Pressure (mm Hg) **0.04 mm at 77°F (25°C)**	Molecular Weight **108.1**
Vapor Density (Air = 1) **3.7**	Melting Point **68°F (20°C)**

Solubility
Slightly soluble in water (0.2%). Soluble in alcohol, ether, chloroform, benzene, and dilute acids.

Appearance and Odor
Colorless to pale-yellow oily liquid or solid (below 68°F) with a faint, aromatic odor. Darkens on exposure to air and light.

SECTION IV - FIRE AND EXPLOSION HAZARD DATA

Flash Point (method used) **190°F (88°C) closed cup**	Explosive Limits in Air % by Volume **LEL: Not Determined UEL: Not Determined**
NFPA Classification **Class IIIA Combustible Liquid**	Autoignition Temperature **345°F (174°C)**

Extinguishing Media
Use water spray in flooding amounts as fog, or use alcohol foam, carbon dioxide, or dry chemical.

Special Fire Fighting Procedures
Structural protective clothing is permeable. Remain clear of smoke, water fallout, and water runoff. Poisonous gases are produced in fire. Move containers if it can be done without risk. Cool fire-exposed containers with water spray and use water spray to disperse vapors. Evacuate non-essential personnel 2500 feet from the fire area (consider wind conditions).

Unusual Fire and Explosion Hazards
Contact with oxidizing agents can cause violent and explosive ignition. Vapors are heavier than air and may travel a considerable distance to a source of ignition and flashback to cause fire or explosion.

SECTION V - REACTIVITY DATA

Stability		Conditions to Avoid
		Phenylhydrazine is stable at room temperature in closed containers under normal conditions of storage and handling. Avoid contact with incompatible chemicals and materials and heat.
Stable	Unstable	Incompatibility *(materials to avoid)*
X		Highly reactive with oxidizers (chlorine, bromine, nitrogen dioxide), lead dioxide, perchloryl fluoride, heat, sparks, and open flame.
Hazardous Polymerization		Conditions to Avoid Under normal conditions of temperature and pressure, hazardous polymerization of phenylhydrazine is not expected to occur.
May Occur	Will Not Occur	Hazardous Decomposition or By-products
	X	When heated to decomposition, phenylhydrazine will produce toxic and irritating gases, including oxides of nitrogen and carbon.

SECTION VI - HEALTH HAZARD DATA

Primary Route(s) of Entry:	Inhalation?	Absorption (skin)?	Ingestion?
	X	X	X

Health Hazards

INHALATION: Symptoms include hematuria, changes in the liver and kidneys, vomiting, convulsions, and respiratory arrest. May also cause decreased body temperature and lower blood pressure. Also causes hemolytic anemia and significant weight loss. Other symptoms include dypsnea, cyanosis, jaundice, and vascular thrombosis.

ABSORPTION: Will pass through unbroken skin to cause toxic systemic effects as noted above. Extremely irritating to skin tissue causing swelling with blisters and sensitization on short contact. Very injurious to the eyes, causing corneal damage and permanent blindness.

INGESTION: A poison by ingestion. Delayed gastrointestinal irritation. Systemic effects can occur.

Carcinogenicity	NTP Listed?	IARC Cancer Review Group?	OSHA Regulated?	Target Organs?
Suspected Human Suspected Animal	No	No	29 CFR 1910.1000 Table Z-1	Blood, respiratory system, liver, kidneys, skin.

Medical Conditions Generally Aggravated by Exposure

Existing deficiencies in the liver, kidneys, and the circulatory system may be aggravated.

Emergency and First-aid Procedures

Eye contact: Flush immediately with water for 15 minutes (minimum) while holding upper and lower eyelids open. Seek medical attention. Skin contact: Remove all contaminated clothing. Immediately wash area with large amounts of water (15 minutes minimum). For inhalation: Remove the person from exposure. Provide respiratory assistance and CPR. Transfer to medical facility. If swallowed: Call the poison control center. Give 1-2 glasses of water to victim. Seek medical attention immediately. Never attempt to give anything by mouth to an unconscious or convulsing person.

SECTION VII - PRECAUTIONS FOR SAFE HANDLING AND USE

Steps to be Taken in Case Material is Released or Spilled

Evacuate area and deny entry to those not involved in cleanup activities. Protective clothing is mandatory. Remove all sources of ignition. Ventilate area of spill. Use non-sparking tools during cleanup. Absorb liquids with toweling, vermiculite, or sand. Deposit in sealed drum for disposal.

Preferred Waste Disposal Method

Incineration in a chemical incinerator equipped with afterburner and scrubber.

Precautions to be Taken in Handling and Storage

Store in tightly closed containers in a cool, dark, dry, well-ventilated area. Personnel should be trained on the hazardous properties of phenylhydrazine *prior* to working with the material.

Other Precautions and Warnings

Containers should be protected from physical damage. Use extreme caution when handling this compound. Use explosion-proof electrical equipment in areas where this chemical is used or stored.

SECTION VIII - CONTROL MEASURES AND PERSONAL PROTECTIVE EQUIPMENT

Respiratory Protection *(specify type)*

A carcinogenic agent. Exposure level is extremely low and difficult to monitor. For any exposures, use a supplied-air respirator set in positive pressure or continuous flow mode or a self-contained breathing apparatus (SCBA) with full facepiece operated in pressure demand or other positive pressure mode.

Ventilation

Local exhaust or general mechanical systems recommended.

Protective Gloves	Eye Protection	Other Protective Clothing
Impervious Gloves	Chemical Goggles or Face Mask	Full Protective Clothing or Suit

Work/Hygiene Practices

Always wash hands thoroughly after using chemical; never bring food, drink, or smoking materials into vicinity of chemicals.

PHENYLHYDRAZINE

$C_6H_5NHNH_2$

CAS: 100-63-0

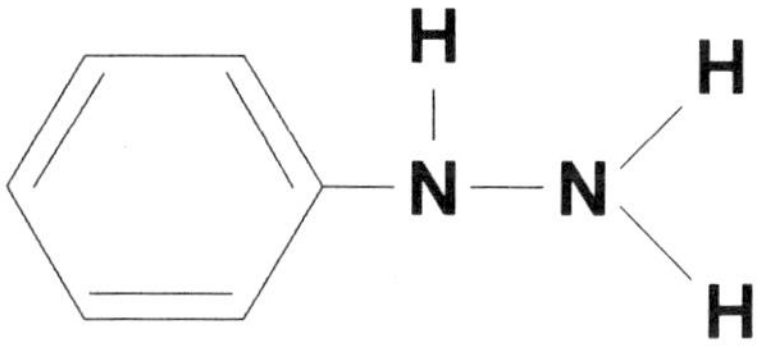

IDENTIFICATION AND TYPICAL USES

Phenylhydrazine is a colorless to pale-yellow liquid or solid (below 67°F) with a faint, aromatic odor. It becomes reddish-brown on exposure to air. It is used in the manufacture of dyes and pharmaceuticals, as a stabilizer for explosives, as a reagent for aldehydes, ketones, and sugars during chemical analysis, and in organic synthesis.

RISK ASSESSMENT: HEALTH

General Assessment

Phenylhydrazine is a suspected human and animal carcinogen (by the ACGIH and NIOSH). It is an experimental teratogen with adverse reproductive effects reported. Mutation data have also been reported. It is primarily toxic by *inhalation* and skin contact (*absorption*). It can also cause toxic and damaging effects by *ingestion*.

Symptoms of exposure by all routes include hematuria, changes in the liver and kidneys, vomiting, nausea, and respiratory arrest. There may also be a decrease in body temperature and blood pressure following exposure. Liver injury may be indicated by jaundice and cyanosis in exposed victims.

Inhalation also causes irritation and inflammation of the eyes, nose, throat, and upper respiratory tract with cough, dypsnea, congestion, and chest pains.

Skin contact also results in sensitization, dermatitis, and eczema-like rash with swelling and possible blistering at the site of contact. Absorption occurs rapidly. If skin contact is combined with another entry route (e.g., vapor inhalation), the subsequent compounded effects may be life-threatening.

Ingestion of phenylhydrazine will cause burning of the lips, mouth, throat, and stomach with unspecified gastrointestinal effects. Systemic effects include possible liver and kidney injury.

Acute Health Effects

The following acute (short-term) health effects may occur immediately or shortly after exposure to phenylhydrazine:

Skin: Severe irritation and burning. Extremely damaging to skin surface. Skin absorption will occur causing toxic systemic effects.

Eye: Severe irritation and tissue damage up to and including loss of vision.

Lung: Irritation of the upper respiratory tract causing choking, coughing, dypsnea, and pulmonary irritation.

CNS: Dizziness, convulsions, tremors, nausea, headaches, possible injury to the nervous system.

Chronic Health Effects

The following chronic (long-term) health effects may occur at some time after exposure to phenylhydrazine and can last for months or even years:

Cancer Hazards: According to the information presented in the references, phenylhydrazine has been shown to cause cancer in test animals. Its carcinogenicity in humans is expected to be at least equal to that of animals. Mutation data have also been reported for this compound.

Reproductive Hazard: According to the information presented in the references, this chemical has been shown to cause teratogenic effects as well as adversely affect reproduction in test animals.

Other Chronic Effects: Repeated exposures can lead to hemolytic anemia with jaundice and possible effects on the kidney and liver. There can also be chronic inflammation of the nasal, tracheal, and bronchial tissue resulting in chronic bronchitis. Repeated skin exposure can cause dermatitis with a characteristic rash.

☉ *Recommended Risk-Reduction Measures*

Phenylhydrazine is listed by the ACGIH as a suspected human carcinogen and by NIOSH as a possible human cancer agent. If a less toxic chemical cannot be substituted for a hazardous substance, then *engineering controls* are the most effective method of reducing exposures. The best protection is to enclose operations and/or provide local exhaust ventilation at the site of chemical release. While not always operationally feasible, isolating operations can also reduce exposure.

Using respiratory protection is less effective than the controls mentioned above, but is still advisable whenever working with or around phenylhydrazine. The exposure level is extremely low (5 ppm) and difficult to accurately monitor. For any detectable exposures to a suspected carcinogenic agent, use a supplied-air respirator set in positive pressure or continuous flow mode, or use a self-contained breathing apparatus (SCBA) with full facepiece operated in pressure demand or other positive pressure mode. If a full facepiece is not available, then chemical goggles should be worn to protect the eyes. Whenever an eye or face contact hazard exists, a face shield and apron should be worn. To prevent hand and skin exposures, impervious gloves should be used. Glove manufacturers should be contacted and permeation studies obtained *before* final glove selection has been made.

Administrative controls should also be in place to minimize the potential for human exposures. These may include written procedures or policies which specify the methods and techniques as well as the exposure control measures that will be practiced whenever personnel are to work with or around phenylhydrazine.

All personnel should receive *training* on the use, hazards, protective measures, emergency actions, and other precautions per 29 CFR 1910.1200 (Hazard Communication), prior to the first assignment in an area where phenylhydrazine is used or stored.

Before beginning employment and at regular intervals thereafter (e.g. annually), the following medical tests are recommended:

- ☑ Complete blood count (CBC).

If symptoms develop or overexposure is suspected, the following medical tests may be useful:

- ☑ Liver and kidney function tests.
- ☑ Evaluation by a qualified allergist with careful consideration of exposure history and special tests (may help diagnose allergy).

Any evaluation should include a careful medical history of past and present symptoms with an examination. However, medical tests that evaluate existing damage are not a substitute for controlling exposure. Since smoking can cause heart diseases, emphysema, and numerous other respiratory disorders, smokers may be affected more rapidly than non-smokers under the same exposure conditions. Also, because alcohol consumption can cause liver damage, drinking alcohol can increase the potential for liver damage caused by exposure to phenylhydrazine. Prudent risk assessment and management requires careful consideration of *all* possible risk factors that may be causing the appearance of symptoms in exposed workers.

Other methods to reduce exposure include:

- ☑ Where possible, enclose operations and use local exhaust ventilation at the site of chemical release. If local exhaust ventilation or enclosure is not used, respiratory protection should be mandatory.
- ☑ Always ensure that proper protective clothing is worn when using chemical substances and that personnel are trained in its use and care.
- ☑ Wash thoroughly immediately after exposure to phenylhydrazine and at the end of the work shift or before eating, drinking, or smoking.
- ☑ Hazard warning information must be posted in the work area. It is recommended that a regulated, controlled work area should be established wherever this chemical is used, handled, or stored. In addition, as part of an on-going education and training program, all information on the health and safety hazards of phenylhydrazine should be communicated to all potentially exposed workers.
- ☑ Eye wash stations should be provided in work areas. If the potential for whole body exposure exists, then safety showers should also be provided in the immediately area.

RISK ASSESSMENT: ENVIRONMENT

General Assessment

The environment is at risk of exposure during intended use, transportation, storage, disposal, or destruction of phenylhydrazine. In almost every scenario, the threat of environmental exposure is contingent upon the proper handling of the chemical substance. Accidental spills, large or small, can result in fire, explosion, and

possible contamination of the surrounding environmental mediums (water, soil, and air).

Phenylhydrazine should always be handled with extreme caution. It is a Class IIIA combustible liquid per OSHA 29 CFR 1910.106. Its vapors are four times heavier than air and can collect in low lying areas or travel for great distances to an ignition source to flashback and cause a fire or explosion. These vapors also form explosive mixtures in air. It is incompatible and highly reactive with a number of substances, especially oxidizers and oxidizing agents. These characteristics require special consideration during any emergency situation involving a leak or spill of phenylhydrazine.

Phenylhydrazine can enter the environment through industrial discharges or through spills.

Acute Ecological Effects

Acute (short-term) toxic effects may include the death of animals, birds, or fish and death or low growth rate in plants. Acute effects are seen 2 to 4 days after animals or plants are exposed to phenylhydrazine.

This chemical has high acute toxicity to aquatic life. Insufficient data are available to evaluate or predict the short-term effects of phenylhydrazine to plant, birds, or land animals.

Chronic Ecological Effects

Chronic toxic effects may include shortened life span, reproductive problems, lower fertility, and changes in appearance or behavior in exposed animals. These effects can be seen long after first exposure(s) to toxic chemicals.

Phenylhydrazine has high chronic toxicity to aquatic life. Insufficient data are available to evaluate or predict the long-term effects of phenylhydrazine to plants, birds, or land animals.

Water Solubility

Phenylhydrazine is slightly soluble in water. Concentrations of 1 to 100 milligrams will mix with a liter of water.

Persistence in the Environment

Phenylhydrazine is slightly persistent in water, with a half-life of between 2 to 20 days. The half-life of a pollutant is the amount of time it takes for one half of the chemical to be degraded. It is expected to absorb to sediment.

Bioaccumulation in Aquatic Organisms

Some substances increase in concentration, or *bioaccumulate,* in living organisms as they breathe contaminated air, drink contaminated water, or eat contaminated food. These chemicals can become concentrated in the tissues and internal organs of animals as well as humans.

Phenylhydrazine is not expected to accumulate in the edible tissues of aquatic species consumed by humans.

Recommended Risk-Reduction Measures

Proper training of all transporters will reduce the likelihood of a mishap or accident resulting in a leak or spill to the environment. The correct use of labeling (per the MSDS) on all containers, trucks, and rail cars will enable emergency responders to react properly and quickly to any disaster thereby reducing the potential risk to the environment and to personnel.

Storage of phenylhydrazine should be segregated from other chemicals to minimize the risk of cross-contamination or contact. Storage buildings should be equipped with proper fire protection and prevention equipment and explosion-proof electrical components and equipment. Bulk storage of phenylhydrazine is not recommended. Phenylhydrazine is somewhat hypergolic (ignites explosively) with incompatible materials such as oxidizers.

If a spill or leak to the environment has occurred, fire department, emergency response and/or hazardous materials spill personnel should be notified immediately. Cleanup should be attempted only by those trained in proper spill containment procedures and the health hazards associated with carcinogenic materials. Contaminated soils should be removed for incineration and replaced with clean soil. Non-sparking tools should always be used in cleanup operations.

If phenylhydrazine should contact the water table, aquifer, or navigable waterway, remediation activities should be prompt to ensure protection of aquatic sediments. It is slightly soluble in water and total containment and remediation may not be possible. When such spills occur, the local and/or state emergency response authorities must be notified. A comprehensive emergency response or disaster preparedness/recovery plan should be in place prior to any operations involving the use, transportation, storage, or disposal of phenylhydrazine. Burning phenylhydrazine in a chemical incinerator equipped with an afterburner and an air scrubber is an acceptable disposal method.

If phenylhydrazine is spilled or leaked, the following specific steps are recommended:

☑ Restrict persons not wearing protective clothing from area of spill or leak until cleanup is complete and area can be opened for normal work.

☑ Ventilate area and remove all ignition sources.

☑ Absorb liquids in vermiculite or sand and place in a sealed drums for disposal. Use non-sparking tools during cleanup.

☑ It may be necessary to dispose of phenylhydrazine as a hazardous waste. The responsible state agency or the regional office of the federal Environmental Protection Agency (EPA) should be contacted for specific recommendations.

RISK ASSESSMENT: BUSINESS

General Assessment

Accidents or mishaps involving phenylhydrazine can present a significant threat to business operations. The loss or damage of equipment or facilities can significantly affect fiscal viability. Lawsuits that may result from personnel injury/death, public exposures, and/or environmental contamination will also require a serious expenditure of resources. Media attention surrounding an injury, death, or environmental damage can also result in a loss of profits and loss of current as well as future business.

Always remember that anytime the term "carcinogen," "cancer," or "reproductive hazard" are used, public emotion, hysteria, and ignorance can run equally high. This should be considered during the development of any public relations policies.

Recommended Risk-Reduction Measures

Company attorneys, safety and health professionals, and environmental specialists should be involved in the development of any procedures dealing with safety or emergency response. A company official should be pre-designated as a public relations officer with specific training in dealing with the press. Corporate plans and policies should be developed, approved, and implemented long before any need for such arises.

REFERENCES

American Conference of Governmental Industrial Hygienists. 1988. *Documentation of the Threshold Limit Values and Biological Exposure Indices,* 5th Edition (with updates). Cincinnati: ACGIH

Anthony, C. P., and N. J. Kolthoff. 1971. *Textbook of Anatomy and Physiology.* St. Louis: C. V. Mosby Company.

Brusick, D. J. 1994. *Methods for Genetic Risk Assessment.* New York: CRC Press/Lewis Publishers.

Bueche, J. R. 1972. *Principles of Physics,* 2nd Edition. New York: McGraw-Hill.

Burrell, R., D. K. Flaherty, and L. J. Sauers. 1992. *Toxicology of the Immune System.* New York: Van Nostrand Reinhold.

Calabrese, E. J. 1994. *Biological Effects of Low Level Exposures.* New York: CRC Press/Lewis Publishers.

Cockerman, L. G., B. S. Shane. 1994. *Basic Environmental Toxicology.* New York: CRC Press/Lewis Publishers.

Cothern, C. R. 1995. *Handbook for Environmental Risk Decisionmaking.* New York: CRC Press/Lewis Publishers.

Danse, I. R. 1991. *Common Sense Toxics in the Workplace.* New York: Van Nostrand Reinhold.

Environmental Protection Agency. 1983. *Chemical Hazard Information Profiles.* Washington, D.C.: U.S. EPA.

Fodden, C. R., and J. L. Weddell. 1992. *Hazardous Materials Emergency Action Data.* New York: CRC Press/Lewis Publishers.

Forseberg, K., and L. H. Keith. 1993. *Chemical; Protective Clothing Permeation and Degradation Database.* Chelsea, MI: Lewis Publishers.

Gardner, R. W. 1994. *Chemical Intolerance.* New York: CRC Press/Lewis Publishers.

Hallenbeck, W. H. 1993. *Quantitative Risk Assessment for Environmental and Occupational Health,* 2nd Edition. New York: CRC Press/Lewis Publishers.

Hodgson, E., R. B. Mailman, and Janice E. Chambers. 1988. *Dictionary of Toxicology.* London: The MacMillian Company.

Hosty, J. W., and P. Foster. 1990. *A Practical Guide to Chemical Spill Response.* New York: Van Nostrand Reinhold.

Howard, P. H. 1995. *Handbook of Environmental Fate and Exposures Data for Organic Chemicals,* Vol. I-IV. New York: CRC Press/Lewis Publishers.

Keith, L. H. and M. M. Walker. 1995. *Handbook of Air Toxics.* New York: CRC Press/Lewis Publishers.

Lewis, R. J., Sr. 1993. *Hawley's Condensed Chemical Dictionary,* 12th Edition. New York: Van Nostrand Reinhold.

Lewis, R. J., Sr. 1992. *Sax's Dangerous Properties of Industrial Materials,* Volumes 1, 2, and 3, 8th Edition. New York: Van Nostrand Reinhold.

Mackay, D., W. Y. Shiu, and K. C. Ma. 1992. *Illustrated Handbook of Chemical-Physical Properties and Environmental Fate for Organic Chemicals.* New York: CRC Press/Lewis Publishers.

Manahan, S. E. 1994. *Environmental Chemistry,* 6th Edition. New York: CRC Press/Lewis Publishers.

Meyer, E. 1990. *Chemistry of Hazardous Materials.* Englewood Cliffs, NJ: Prentice Hall.

U.S. Department of Health and Human Services, National Institute for Occupational Safety and Health. 1994. *NIOSH Pocket Guide to Chemical Hazards.* Washington, D.C.: U.S. Government Printing Office.

MATERIAL SAFETY DATA SHEET

CHEMICAL NAME	CRC PRESS/LEWIS PUBLISHERS DISCLAIMER
PHOSGENE	

HAZARD WARNING INFORMATION

HEALTH	FIRE	REACTIVE	OTHER	DEGREE OF HAZARD	COLOR CODING	OTHER CODES
4	0	0		0 = Minimum Hazard 1 = Slight Hazard 2 = Moderate Hazard 3 = Serious Hazard 4 = Severe Hazard	HEALTH = BLUE FIRE = RED REACTIVITY = YELLOW OTHER = WHITE	OX = Oxidizer ACID = Acid ALK = Alkali COR = Corrosive W = Use No Water

SECTION I - GENERAL INFORMATION

Characterization	**Gas**	RCRA Number	**P095**
		EPA Class	**Acute Hazardous Waste**
DOT Proper Shipping Name	**Phosgene**	Chemical Abstract Service (CAS) Number	**75-44-5**
DOT Hazard Class and Label Requirements	**Poison A; Poison Gas**	DOT Emergency Guide Code	**15**
DOT Identification Number	**UN 1076**	Chemical Formula	**$COCl_2$**

Synonyms

Carbon oxychloride; carbonyl chloride; carbonyl dichloride; chloroformyl chloride.

SECTION II - HAZARDOUS INGREDIENTS AND IDENTITY INFORMATION

Hazardous Components (specific identity)	OSHA Exposure Criteria	NIOSH Exposure Criteria	Immediately Dangerous to Life and Health (IDLH)	ACGIH Exposure Criteria
Phosgene (derivation: By passing a mixture of carbon monoxide and chlorine over activated carbon). **1 ppm = 4.11 mg/m³**	**PEL:** **0.1 ppm** **0.4 mg/m³** **STEL:** **Not Established**	**REL (ceiling):** **0.1 ppm** **0.4 mg/m³** **CEILING:** **0.2 ppm** **0.8 mg/m³**	**2 ppm**	**TLV:** **0.1 ppm** **0.4 mg/m³** **STEL:** **Not Established**

SECTION III - PHYSICAL AND CHEMICAL CHARACTERISTICS

Boiling Point	**47°F (8°C)**	Specific Gravity (H_2O = 1)	**3.48**
Vapor Pressure (atmospheres)	**1.6 at 69°F (20°C)**	Molecular Weight	**98.9**
Vapor Density (Air = 1)	**3.4**	Melting Point	**-198°F (-127°C)**

Solubility

Slightly soluble in water (slowly hydrolyzed by water).

Appearance and Odor

Colorless gas with a suffocating odor like musty hay. A fuming liquid below 47°F (8°C). Normally shipped as a liquefied compressed gas.

SECTION IV - FIRE AND EXPLOSION HAZARD DATA

Flash Point (method used)	**Not Applicable**	Explosive Limits in Air % by Volume	**LEL: N/A UEL: N/A**
NFPA Classification	**Non-flammable Gas**	Autoignition Temperature	**Not Applicable**

Extinguishing Media

Use extinguishing agent suitable to surrounding media. Phosgene itself does not burn.

Special Fire Fighting Procedures

Wear full protective clothing and self-contained breathing apparatus (SCBA). If possible, remove containers or cylinders from fire areas or use a water spray to keep fire-exposed containers cool. Continue cooling long after fire has been extinguished. Poisonous gases are produced in fire.

Unusual Fire and Explosion Hazards

Containers may explode in fire. Firefighters should avoid contact with vapors produced during fire.

SECTION V - REACTIVITY DATA

Stability		Conditions to Avoid Phosgene is normally stable in closed containers under routine conditions of handling and storage. Keep away from moisture and exposure to incompatible materials. Keep away from heat.
Stable X	Unstable	Incompatibility *(materials to avoid)* Incompatible with moisture, alkalis, ammonia, alcohols, copper, sodium, potassium, and aluminum. It slowly reacts in water to form hydrochloric acid and carbon dioxide.
Hazardous Polymerization		Conditions to Avoid Under normal conditions of temperature and pressure, hazardous polymerization of phosgene is not known to occur.
May Occur	Will Not Occur X	Hazardous Decomposition or By-products When heated to decomposition, phosgene can emit highly toxic/poisonous gases, including carbon dioxide. Phosgene itself is an extremely hazardous material and should be handled with extreme caution.

SECTION VI - HEALTH HAZARD DATA

Primary Route(s) of Entry:	Inhalation? X	Absorption (skin)?	Ingestion?

Health Hazards

INHALATION: Extremely irritating to the lungs, throat, nose, and eyes with coughing and shortness of breath that may persist for hours after exposure. It can cause permanent damage to the lungs. Fluid can build up in the lungs (pulmonary edema), which is a medical emergency. Once inhaled, it is slowly hydrolyzed to hydrochloric acid and carbon dioxide.

SKIN & EYES: Contact with the liquid can cause severe skin burns (from frostbite). The liquid can also cause severe eye burns leading to permanent damage.

Carcinogenicity	NTP Listed?	IARC Cancer Review Group?	OSHA Regulated?	Target Organs?
Unknown Human Unknown Animal	No	No	29 CFR 1910.1000 Table Z-1	Respiratory system, eyes, skin.

Medical Conditions Generally Aggravated by Exposure

Existing respiratory problems may be aggravated by exposure.

Emergency and First-aid Procedures

Eye contact: Flush immediately with water for 15 minutes (minimum); seek medical attention. Skin contact: Remove all contaminated clothing. Immediately wash area with large amounts of soap and water. Seek medical assistance. For inhalation: Remove the person from exposure. Provide respiratory assistance and CPR. Transfer to medical facility. Observe for 24-48 hours since pulmonary edema may be a delayed reaction. If swallowed: Not likely. If it does occur, seek medical attention immediately.

SECTION VII - PRECAUTIONS FOR SAFE HANDLING AND USE

Steps to be Taken in Case Material is Released or Spilled

Absorb liquids in vermiculite, dry sand, earth, or similar material and deposit in sealed containers. Ventilate area of spill or leak. Restrict those not involved in cleanup from entering area. If gas is leaked, stop source if possible or remove leaking cylinders to safe area outdoors and allow to empty.

Preferred Waste Disposal Method

No citation.

Precautions to be Taken in Handling and Storage

Store to avoid contact with incompatible materials since violent reactions can occur. Store in tightly closed containers in cool, well-ventilated area away from heat and sunlight. Ship in steel pressure cylinders and tank cars. Protect containers from physical damage. Store cylinders in upright position.

Other Precautions and Warnings

Sources of ignition are prohibited where phosgene is used, handled, or stored. Heat may cause containers to buildup pressure and explode.

SECTION VIII - CONTROL MEASURES AND PERSONAL PROTECTIVE EQUIPMENT

Respiratory Protection *(specify type)*

Exposure level is relatively low (0.1 ppm) and difficult to monitor. For the best protection, use a supplied-air respirator with full facepiece, hood, or helmet in continuous flow mode, or use a self-contained breathing apparatus (SCBA) operated in pressure demand or other positive pressure mode.

Ventilation

Local exhaust or general mechanical systems recommended.

Protective Gloves	Eye Protection	Other Protective Clothing
Butyl Rubber, PVC, or Neoprene	Chemical Goggles or Face Mask	Protective Apron

Work/Hygiene Practices

Always wash hands thoroughly after using chemical; never bring food, drink, or smoking materials into vicinity of chemicals.

PHOSGENE

$COCl_2$

CAS: 75-44-5

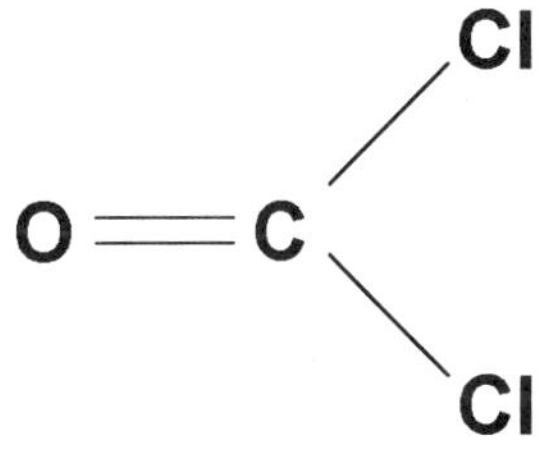

IDENTIFICATION AND TYPICAL USES

Phosgene is a colorless gas with a suffocating odor like that of musty hay. It is a fuming liquid at temperatures below 47°F (8°C). It is typically shipped as a liquefied compressed gas. It is used in the synthesis of dyes, pharmaceuticals, isocyanates, polyurethane and polycarbonate resins, and pesticides. It is also used as a war gas.

RISK ASSESSMENT: HEALTH

General Assessment

Phosgene is a severely irritating gas, causing irritation of the eyes, nose, throat, and upper respiratory tract (URT). There may be no immediate indications of URT effects, but damage can occur in a relatively short period of time. It is a human poison by *inhalation*. When phosgene gas is inhaled, the heavy moisture content of the respiratory tract causes the gas to hydrolyze to corrosive hydrochloric acid and toxic carbon dioxide. Symptoms of exposure include burning of the mouth, cyanosis (bluish discoloration of the lips, face, fingers, toes and ears), coughing, choking, nausea, vomiting, headache, dizziness, foamy sputum, congestion, respiratory distress, and pulmonary edema (fluid in the lungs), which can be fatal. The symptoms of pulmonary edema may be delayed up to 48 hours thereby creating a false sense of security with regard to health exposure risk. However, when an exposure has been particularly severe, the development of pulmonary edema may be so rapid that death can occur within 36 hours of exposure.

Chronic exposure to low concentrations may not cause pulmonary edema but can lead to emphysema, pulmonary fibrosis, and other respiratory disorders. In cases where such exposures do occur, the worker will feel a burning sensation in the eyes, nose and throat, followed by tightness and pain in chest.

Skin contact can cause irritating rash and possible deep burns due to frostbite effect of the compressed liquid on the skin's surface. Eye contact with the vapor may be irritating and the liquid may cause damage to vision and destruction of tissue.

Acute Health Effects

The following acute (short-term) health effects may occur immediately or shortly after exposure to phosgene:

<u>Skin</u>: Immediate irritation and burn on contact with the liquid. May cause irritating rash and possible permanent damage to skin tissue.

<u>Eye</u>: Severe irritation; may cause burns and permanent damage.

<u>Lung</u>: Irritation of the nose, throat, and lungs following exposure. Pulmonary edema is possible (24 to 48 hours following a high exposure). It can cause tearing, choking, coughing, foamy sputum, and chest pain.

<u>Other</u>: Headache, dizziness, loss of consciousness, muscle weakness, burning of the mouth, nausea, vomiting, and respiratory distress.

Chronic Health Effects

The following chronic (long-term) health effects may occur at some time after exposure to phosgene and can last for months or even years:

<u>Cancer Hazards</u>: According to information presented in the references, phosgene has not been adequately tested for its ability to cause cancer in test animals.

Reproductive Hazard: According to information presented in the references, phosgene has not been adequately tested for its ability to adversely affect reproduction.

Other Chronic Effects: Phosgene can irritate the lungs. Repeated exposure may cause bronchitis to develop with cough, phlegm, and/or shortness of breath. Long-term exposure can cause other respiratory ailments, including emphysema and pulmonary fibrosis.

Recommended Risk-Reduction Measures

Personnel should avoid direct contact with phosgene. It is extremely irritating and highly toxic to humans. If a less toxic chemical cannot be substituted for a hazardous substance such as phosgene, then *engineering controls* are the most effective method of reducing exposures. The best protection is to enclose operations and/or provide local exhaust ventilation at the site of chemical release. While not always operationally feasible, isolating operations can also reduce exposure. Using respiratory protection is less effective than the controls mentioned above, but is still advisable whenever working with or around phosgene. The PEL of 0.1 ppm is relatively low and difficult to accurately monitor, especially in emergency conditions. A powered supplied-air purifying respirator operated in continuous flow mode or a self-contained breathing apparatus (SCBA) with full facepiece operated in pressure demand are the recommended respiratory protection methods of choice. If a full facepiece is not available, then chemical goggles should be worn to protect the eyes. Whenever a chemical splash hazard exists, a face shield and a rubber apron should be worn. To prevent hand and skin exposures, chemical resistant gloves should be used. Polyvinyl chloride, butyl rubber, or neoprene appear to provide adequate protection. However, glove manufacturers should be contacted and permeation studies obtained *before* final glove selections are made.

Administrative controls should also be in place to minimize the potential for human exposures. These may include written procedures or policies which specify the methods and techniques that will be practiced whenever personnel are to work with phosgene.

All personnel should receive *training* on the use, hazards, protective measures, emergency actions, and other precautions per 29 CFR 1910.1200 (Hazard Communication), prior to the first assignment in an area where phosgene is used or stored.

For those personnel with frequent or potentially high exposure, the following are recommended medical tests to be provided before initial assignment and at regular intervals thereafter (e.g., annually):

☑ Lung function tests.

If symptoms develop or overexposure is suspected, the following tests are recommended:

☑ Lung function tests.

☑ Consider chest X-ray after acute overexposures (may be negative if taken immediately after exposure due to delayed onset of pulmonary edema).

Any evaluation should include a careful history of past and present symptoms with an examination. Medical tests that simply look for existing damage are not a substitute for controlling exposures. Also, because smoking can cause heart disease, as well as lung cancer, emphysema, and other respiratory problems, exposure to phosgene may worsen pre-existing respiratory conditions. Smokers may experience symptoms more quickly than non-smokers under the same conditions of exposure.

Other methods to reduce exposure include:

☑ Where possible, enclose operations and use local exhaust ventilation at the site of chemical release. If local exhaust ventilation or enclosure is not used, respiratory protection should be mandatory.

☑ Always ensure that proper protective clothing is worn when using chemical substances.

☑ Wash thoroughly immediately after exposure to phosgene and at the end of the work shift or before eating, drinking, or smoking.

☑ Hazard warning information should be posted in the work area. In addition, as part of an on-going education and training program, all information on the health and safety hazards of phosgene should be communicated to all potentially exposed workers.

☑ Eye wash stations should be provided in the immediate work area for emergency use. If there is a possibility of skin exposure to phosgene, emergency shower facilities should also be provided.

☑ Personnel should never wear contaminated clothing home (family members can be exposed). Clothing should be laundered by personnel who have been trained on the hazards of phosgene.

RISK ASSESSMENT: ENVIRONMENT

General Assessment

The environment is at risk of exposure during transportation, storage, disposal, or destruction of phosgene. In almost every scenario, the threat of environmental exposure is contingent upon the proper handling of the chemical substance. Accidental spills, large or small, can result in fire, explosion, and possible contamination of the surrounding environmental mediums (water, soil, and especially air).

Phosgene is considered a non-flammable gas. However, its pressurized state can present a serious explosion hazard if containers/cylinders of phosgene gas are exposed to heat or fire. This gas is more than three times heavier than air. It can travel low to the ground for great distances and collect in low-lying areas to create a serious health hazard risk. These characteristics require extreme caution in handling, storage, transportation, and disposal. It is incompatible with moisture (water, steam, humid air), most alkalis, ammonia, alcohol, and copper, and contact can cause fire or explosion. It will slowly hydrolyze in water to form corrosive hydrochloric acid and toxic carbon dioxide. These characteristics require special consideration during any emergency situation involving a leak or spill of phosgene. Should phosgene ever come into contact with incompatible substances during use, transportation, storage, or disposal, the formation of highly toxic and/or highly explosive commodities is extremely possible.

Phosgene may enter the environment through industrial discharges or through spills.

Acute Ecological Effects

Acute (short-term) toxic effects may include the death of animals, birds, or fish and death or low growth rate in plants. Acute effects are seen 2 to 4 days after animals or plants are exposed to phosgene.

Insufficient data are available to evaluate or predict the short-term (acute) effects of phosgene exposure to aquatic life, plants, birds, or land animals.

Chronic Ecological Effects

Chronic (long-term) toxic effects may include shortened life span, reproductive problems, lower fertility, and changes in appearance or behavior in exposed animals. These effects can be seen long after first exposure(s) to toxic chemicals.

Insufficient data are available in the references to evaluate or predict the long-term effects of phosgene to aquatic life, plants, birds, or land animals.

Water Solubility

Phosgene will react rapidly in water and will not accumulate in the aquatic environment.

Persistence in the Environment

Phosgene is non-persistent in water, with a half-life of less than 2 days. The half-life of a pollutant is the amount of time it takes for one half of the chemical to be degraded.

Once in water, phosgene will hydrolyze to hydrochloric acid and carbon dioxide and it is therefore not expected to be present in the environment.

Bioaccumulation in Aquatic Organisms

Some substances increase in concentration, or *bioaccumulate,* in living organisms as they breathe contaminated air, drink contaminated water, or eat contaminated food. These chemicals can become concentrated in the tissues and internal organs of animals as well as humans.

Phosgene is not expected to accumulate in the tissues of fish.

Recommended Risk-Reduction Measures

Proper training of all transporters will reduce the likelihood of a mishap or accident resulting in a leak or spill to the environment. The correct use of DOT labeling (per the MSDS) on all containers, trucks, and rail cars will enable emergency responders to react properly and quickly to any disaster thereby reducing the potential risk to the environment and to personnel.

Storage of phosgene should be segregated from incompatible chemicals to minimize the risk of cross-contamination or contact. Buildings designated for storage should be equipped with appropriate fire protection systems (alarms, sprinklers, emergency lighting, portable extinguishers). Cylinders must always be stored upright and properly secured. They should never be rolled or dragged. If a spill or leak to the environment has occurred, fire department, emergency response and/or hazardous materials spill personnel should be notified immediately. Cleanup should be attempted only by those trained in proper spill containment procedures. Any contaminated soils should be removed for incineration and replaced with clean soil. If phosgene should contact the water table, aquifer, or navigable waterway, time is of the essence. It reacts rapidly in water to form other materials and total remediation and containment may not be possible. The local and/or state emergency response authorities must be notified. A comprehensive emergency response or disaster preparedness/recovery plan should be in place prior to any operations involving the use, transportation, storage, or disposal of phosgene. If phosgene is spilled or leaked, the following specific steps are recommended:

☑ Restrict persons not wearing protective clothing from area of spill or leak until cleanup is complete and area can be opened for normal work.

☑ Remove all ignition sources.

☑ Ventilate area of leak.

☑ Absorb liquids in vermiculite, dry sand, earth, or similar material and deposit in sealed containers.

☑ If gas is leaking, stop the flow of gas. If the source is a cylinder and leak cannot be stopped in place, remove cylinder to a safe place in the open air and repair or allow to empty.

☑ It may be necessary to dispose of phosgene as a hazardous waste. The responsible state agency or the regional office of the federal Environmental Protection Agency (EPA) should be contacted for specific recommendations.

RISK ASSESSMENT: BUSINESS

General Assessment

Accidents or mishaps involving phosgene can present a significant threat to business operations. The loss or damage of equipment or facilities can significantly affect fiscal viability. Lawsuits that may result from personnel injury/death, public exposures, and/or environmental contamination will also require a serious expenditure of resources. Media attention surrounding an injury, death, or environmental damage can also result in a loss of profits and loss of current as well as future business.

Recommended Risk-Reduction Measures

Company attorneys, safety and health professionals, and environmental specialists should be involved in the development of any procedures dealing with safety or emergency response. A company official should be pre-designated as a public relations officer with specific training in dealing with the press. Corporate plans and policies should be developed, approved, and implemented long before any need for such arises.

REFERENCES

American Conference of Governmental Industrial Hygienists. 1988. *Documentation of the Threshold Limit Values and Biological Exposure Indices,* 5th Edition (with updates). Cincinnati: ACGIH

Bueche, J. R. 1972. *Principles of Physics,* 2nd Edition. New York: McGraw-Hill.

Calabrese, E. J. 1994. *Biological Effects of Low Level Exposures.* New York: CRC Press/Lewis Publishers.

Cockerman, L. G., B. S. Shane. 1994. *Basic Environmental Toxicology.* New York: CRC Press/Lewis Publishers.

Gardner, R. W. 1994. *Chemical Intolerance.* New York: CRC Press/Lewis Publishers.

Hallenbeck, W. H. 1993. *Quantitative Risk Assessment for Environmental and Occupational Health,* 2nd Edition. New York: CRC Press/Lewis Publishers.

Jones, F. E. 1994. *Toxic Organic Vapors in the Workplace.* New York: CRC Press/Lewis Publishers.

Keith, L. H. and M. M. Walker. 1995. *Handbook of Air Toxics.* New York: CRC Press/Lewis Publishers.

Lauwerys, R. R., P. Hoet. 1993. *Industrial Chemical Exposure.* New York: CRC Press/Lewis Publishers.

Lewis, R. J., Sr. 1992. *Sax's Dangerous Properties of Industrial Materials,* Volumes 1, 2, and 3, 8th Edition. New York: Van Nostrand Reinhold.

Lide, D. R. 1995. *Handbook of Chemistry and Physics,* 76th Edition. New York: CRC Press/Lewis Publishers.

Manahan, S. E. 1994. *Environmental Chemistry,* 6th Edition. New York: CRC Press/Lewis Publishers.

Meyer, E. 1990. *Chemistry of Hazardous Materials.* Englewood Cliffs, NJ: Prentice Hall.

Patnaik, P. 1992. *A Comprehensive Guide to the Hazardous Properties of Chemical Substances.* New York: Van Nostrand Reinhold.

Rea, W. J. 1995. *Chemical Sensitivity,* Vol. I-III. New York: CRC Press/Lewis Publishers.

Richardson, M. 1995. *Dictionary of Substance and Their Effects,* Vol. I-VI. New York: CRC Press/Lewis Publishers.

Sacarello, H. L. A. 1994. *The Comprehensive Handbook of Hazardous Materials.* New York: CRC Press/Lewis Publishers.

Torkelson, T. R. V. K. Rowe. 1981. Halogenated aliphatic hydrocarbons. *Patty's Industrial Hygiene and Toxicology,* Vol. 2B, *Toxicology,* pp 3568-3572. New York: Wiley-Interscience.

U.S. Department of Health and Human Services, National Institute for Occupational Safety and Health. 1994. *NIOSH Pocket Guide to Chemical Hazards.* Washington, D.C.: U.S. Government Printing Office.

MATERIAL SAFETY DATA SHEET

CHEMICAL NAME	
PHOSPHORIC ACID	**CRC PRESS/LEWIS PUBLISHERS DISCLAIMER:** THE INFORMATION AND RECOMMENDATIONS PRESENTED HEREIN ARE BASED ON SOURCES BELIEVED TO BE RELIABLE. CRC MAKES NO REPRESENTATION ON ITS COMPLETENESS OR ACCURACY. IT IS THE RESPONSIBILITY OF THE USER TO DETERMINE THE CHEMICAL'S SUITABILITY FOR ITS INTENDED USE, THE CHEMICAL'S SAFE USE, AND THE CHEMICAL'S PROPER DISPOSAL. NO REPRESENTATIONS AND/OR WARRANTIES, EITHER EXPRESSED OR IMPLIED, OF THE MERCHANTABILITY OR FITNESS FOR A PARTICULAR PURPOSE, OR OF ANY OTHER NATURE, ARE MADE WITH RESPECT TO THE INFORMATION PROVIDED IN THIS MSDS OR TO THE CHEMICAL TO WHICH INFORMATION MAY REFER. CRC NEITHER ASSUMES NOR AUTHORIZES ANY OTHER PERSON TO ASSUME FOR IT, ANY OTHER ADDITIONAL RESPONSIBILITY OR LIABILITY FOR THE USE OF, OR RELIANCE UPON, THIS INFORMATION.

HAZARD WARNING INFORMATION

HEALTH	FIRE	REACTIVE	OTHER	DEGREE OF HAZARD	COLOR CODING	OTHER CODES
2	0	0		0 = Minimum Hazard 1 = Slight Hazard 2 = Moderate Hazard 3 = Serious Hazard 4 = Severe Hazard	HEALTH = BLUE FIRE = RED REACTIVITY = YELLOW OTHER = WHITE	OX = Oxidizer ACID = Acid ALK = Alkali COR = Corrosive ~~W~~ = Use No Water

SECTION I - GENERAL INFORMATION

Characterization	**Mineral Acid**	RCRA Number	**None**
		EPA Class	**Not Applicable**
DOT Proper Shipping Name	**Phosphoric Acid**	Chemical Abstract Service (CAS) Number	**7664-38-2**
DOT Hazard Class and Label Requirements	**Corrosive Material; Corrosive**	DOT Emergency Guide Code	**60**
DOT Identification Number	**UN 1805**	Chemical Formula	**H_3PO_4**

Synonyms
Orthophosphoric acid; phosphoric acid (aqueous); white phosphoric acid; phosphoric acid (solid).

SECTION II - HAZARDOUS INGREDIENTS AND IDENTITY INFORMATION

Hazardous Components (specific identity)	OSHA Exposure Criteria	NIOSH Exposure Criteria	Immediately Dangerous to Life and Health (IDLH)	ACGIH Exposure Criteria
Phosphoric acid (derivation: By action of sulfuric acid on pulverized phosphate rock; action of hydrochloric acid on phosphate rock, with extraction by tributylphosphate; by heating phosphate rock, coke, and silica in an electric furnace, burning the elemental phosphorus produced, and hydrating the phosphoric acid). **1 ppm = 4.07 mg/m^3**	**PEL:** **0.25 ppm** **1 mg/m^3** **STEL:** **0.75 ppm** **3 mg/m^3**	**REL:** **0.25 ppm** **1 mg/m^3** **STEL:** **0.75 ppm** **3 mg/m^3**	**1000 mg/m^3**	**TLV:** **0.25 ppm** **1 mg/m^3** **STEL:** **0.75 ppm** **3 mg/m^3**

SECTION III - PHYSICAL AND CHEMICAL CHARACTERISTICS

Boiling Point	**502°F (261°C)**	Specific Gravity (H_2O = 1)	**1.87 at 77°F (25°C)**
Vapor Pressure (mm Hg)	**0.03 at 68°F (20°C)**	Molecular Weight	**98.0**
Vapor Density (Air = 1)	**Not Found**	Melting Point	**108°F (42°C)**

Solubility
Soluble in water (miscible) and alcohol.

Appearance and Odor
Thick, colorless, odorless, crystalline solid. Also appears in aqueous solution as a viscous, water-white, odorless liquid.

SECTION IV - FIRE AND EXPLOSION HAZARD DATA

Flash Point (method used)	**Not Determined**	Explosive Limits in Air % by Volume	**LEL: Not Determined UEL: Not Determined**
NFPA Classification	**Non-Combustible Liquid**	Autoignition Temperature	**Not Determined**

Extinguishing Media
Use water on phosphoric acid fires. Otherwise, use extinguishing agents suitable to surrounding fire.

Special Fire Fighting Procedures
Wear full protective clothing and self-contained breathing apparatus (SCBA). Move containers from fire, or use a water spray to keep fire-exposed containers cool. Poisonous gases are produced in fire.

Unusual Fire and Explosion Hazards
Corrosive to most metals, liberating flammable hydrogen gas. Containers may explode in fire. If water is used, use large amounts to control the generation of heat and acid vapors.

SECTION V - REACTIVITY DATA

Stability		Conditions to Avoid **Phosphoric acid is normally stable in closed containers under routine conditions of handling and storage. Keep away from metals and other incompatible materials.**
Stable X	Unstable	Incompatibility *(materials to avoid)* **Reacts with alkalis to form phosphate salts. Corrosive to many metals and alloys, especially at elevated temperatures. Reacts with aldehydes, amines, amides, alcohols, glycols, and many other compounds.**
Hazardous Polymerization		Conditions to Avoid **Violent polymerization occurs with epoxides, azo compounds, sodium tetrahydroborate, and many other polymerizable compounds.**
May Occur X	Will Not Occur	Hazardous Decomposition or By-products **When heated to decomposition, phosphoric acid can emit highly toxic/poisonous gases, including toxic phosphorous oxide fumes.**

SECTION VI - HEALTH HAZARD DATA

Primary Route(s) of Entry:	Inhalation?	Absorption (skin)?	Ingestion?
	X		X

Health Hazards

INHALATION: Vapors cause irritation to the upper respiratory tract with coughing, dypsnea, headache, congestion, salivation, tissue dehydration, chest pains, shallow breathing, bronchitis, chemical pneumonitis, and/or pulmonary edema (fluid in the lungs), which can be fatal.

SKIN & EYES: Skin contact causes smarting, tingling, redness, burning, swelling, blistering, necrosis, and shock. Eye contact causes burning, watering, redness, corneal ulceration, blurred vision, and possible injury.

INGESTION: Vomiting, abdominal pain, shock, bloody diarrhea, and severe gastrointestinal damage.

Carcinogenicity	NTP Listed?	IARC Cancer Review Group?	OSHA Regulated?	Target Organs?
Unknown Human Unknown Animal	No	No	29 CFR 1910.1000 Table Z-1	Respiratory system, eyes, skin.

Medical Conditions Generally Aggravated by Exposure

Respiratory impairments (bronchitis, asthma); skin conditions (dermatitis).

Emergency and First-aid Procedures

Eye contact: Flush immediately with water for 15 minutes (minimum); seek medical attention. Skin contact: Remove all contaminated clothing. Immediately wash area with large amounts of water. Seek medical assistance. For inhalation: Remove the person from exposure. Provide respiratory assistance and CPR. Transfer to medical facility. Observe 48 hours for lung effects. If swallowed: Seek medical attention immediately. If person is conscious, give several glasses of water or milk followed by milk of magnesia or aluminum hydroxide gel. Do NOT induce vomiting. If vomiting occurs, give more milk.

SECTION VII - PRECAUTIONS FOR SAFE HANDLING AND USE

Steps to be Taken in Case Material is Released or Spilled

Ventilate area. Remove all ignition sources. Collect solid materials in safest manner possible and deposit in sealed drums. Do NOT dry sweep. Absorb liquid spills in vermiculite or other material and place in sealed drums for disposal. Restrict those not involved in cleanup from entering area.

Preferred Waste Disposal Method

No citation.

Precautions to be Taken in Handling and Storage

Avoid contact with incompatible materials since violent reactions can occur. Store in tightly closed containers in cool, well-ventilated area away from heat. Avoid generating dusty conditions.

Other Precautions and Warnings

Sources of ignition are prohibited where phosphoric acid is used, handled, or stored. Take measures to prevent damage to containers.

SECTION VIII - CONTROL MEASURES AND PERSONAL PROTECTIVE EQUIPMENT

Respiratory Protection *(specify type)*

Exposure levels are relatively low. Recommend using a supplied-air respirator or a self-contained breathing apparatus (SCBA) operated in pressure demand or other positive pressure mode.

Ventilation

Local exhaust or general mechanical systems recommended.

Protective Gloves	Eye Protection	Other Protective Clothing
Neoprene or Nitrile	Chemical/Dust Goggles or Face Mask	Protective Apron

Work/Hygiene Practices

Always wash hands thoroughly after using chemical; never bring food, drink, or smoking materials into vicinity of chemicals.

PHOSPHORIC ACID

H_3PO_4

CAS: 7664-38-2

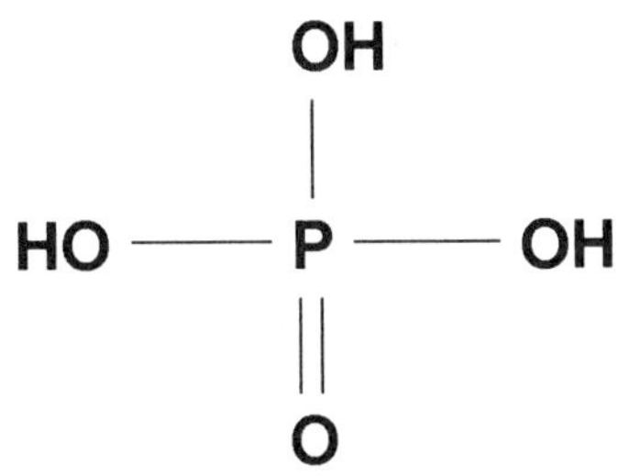

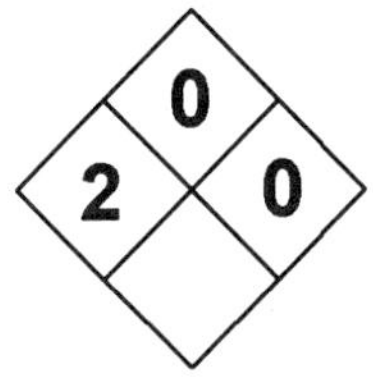

IDENTIFICATION AND TYPICAL USES

Phosphoric acid is a thick, colorless, odorless, crystalline solid. It can also appear in aqueous solution as a viscous, water-white, odorless liquid. It is one of the most widely used chemicals. It is found in fertilizers, soaps and detergents, and inorganic phosphates. It is used in the pickling and rust-proofing of metals, in pharmaceuticals, in sugar refining, in the manufacture of gelatin, in water treatment, animal feeds, in electropolishing, as a gasoline additive, in conversion coatings for metals, as a catalyst for ethanol manufacture, in cotton dyeing, in yeast, as a soil stabilizer, in waxes and polishes, as a binder for ceramics, in activated carbon, in foods and carbonated beverages as an acidulant and sequestrant, and as a laboratory reagent.

RISK ASSESSMENT: HEALTH

General Assessment

Exposure to phosphoric acid is possible through *inhalation* of its vapors and by skin contact (although dermal absorption through intact skin has not been reported). It can also enter the body through *ingestion* of the solid (dusts) or the liquid solution. Although it is considered to be of moderate toxicity, it is a highly caustic and, therefore, corrosive compound. There are no reports on its carcinogenic, mutagenic, or teratogenic properties.

Inhalation severely dehydrates tissues and causes destructive, corrosive damage to the nose, throat, and mucosa of the upper respiratory tract. There can be coughing, dypsnea (labored breathing), lachrymation, and nausea. Severe exposures may lead to chemical pneumonitis or cause pulmonary edema (fluid in the lungs), which is a medical emergency and can be fatal. The development of pulmonary edema can be delayed up to 48 hours after exposure, thereby creating a false sense of security with regard to health exposure risk.

Skin contact with the vapor or the liquid solution causes severe burning and irritation, although this reaction may not occur immediately. Eye contact causes immediate stinging and watering with tissue damage and the possibility of corneal damage. There will be redness and swelling of the lids, intense pain, and tissue destruction. Loss of vision may occur.

Ingestion can cause severe burns to the lips, mouth, throat, and stomach. If vomiting occurs and the liquid is aspirated into the lungs, the corrosive action on the lung tissues can be life-threatening. Other symptoms include nausea, bloody diarrhea, abdominal pain, shock, and severe gastrointestinal damage.

☠ *Acute Health Effects*

The following acute (short-term) health effects may occur immediately or shortly after exposure to phosphoric acid:

Skin: Irritation/rash or burning feeling (possibly delayed) with redness, blistering, and localized tissue damage.

Eye: Severe irritation; may cause burns and damage with permanent loss of vision.

Lung: Severe irritation of the mucous membranes in the nose, throat, and lungs following exposure. It can also cause chemical pneumonitis or even delayed pulmonary edema (fluid in the lungs).

💣 *Chronic Health Effects*

The following chronic (long-term) health effects may occur at some time after exposure to phosphoric acid and can last for months or even years:

Cancer Hazards: According to information presented in the references, phosphoric acid has not been adequately tested for its ability to cause cancer in laboratory test animals.

Reproductive Hazard: According to information presented in the references, phosphoric acid has not been adequately tested for its ability to adversely affect reproduction in test animals.

Other Chronic Effects: Repeated, prolonged exposures may cause dermatitis and chronic respiratory disease, such as bronchitis, with persistent cough, phlegm, and shortness of breath.

Recommended Risk-Reduction Measures

Since phosphoric acid is one of the most widely used chemicals, personnel must take extra care to avoid direct contact with phosphoric acid. It is extremely corrosive and moderately toxic to humans. If a less toxic chemical cannot be substituted for a hazardous substance, then *engineering controls* are the most effective method of reducing exposures.

The best protection is to enclose operations and/or provide local exhaust ventilation at the site of chemical release. While not always operationally feasible, isolating operations can also reduce exposure. Using respiratory protection is less effective than the controls mentioned above, but is still advisable whenever working with or around phosphoric acid. The exposure level is extremely low (1 mg/m^3) and difficult to accurately monitor, even under normal operating conditions. For low or infrequent exposures, an acid vapor respirator or gas mask may suffice. However, better protection is obtained using a self-contained breathing apparatus (SCBA) or a supplied-air respirator with full facepiece operated in pressure demand mode. If a full facepiece is not available, then chemical goggles should be worn to protect the eyes. Whenever a chemical splash hazard exists, a face shield and a protective apron should be worn. To prevent hand and skin exposures, acid resistant neoprene or nitrile gloves should be used. Glove manufacturers should be contacted and permeation studies obtained *before* final glove selections are made.

Administrative controls should also be in place to minimize the potential for human exposures. These may include written procedures or policies which specify the methods and techniques that will be practiced whenever personnel are to work with phosphoric acid.

All personnel should receive *training* on the use, hazards, protective measures, emergency actions, and other precautions per 29 CFR 1910.1200 (Hazard Communication), prior to the first assignment in an area where phosphoric acid is used or stored.

Before beginning employment and at regular intervals thereafter (e.g., annually), the following tests are recommended:

☑ Lung function tests (establish baseline).

If symptoms occur or overexposure is suspected, the following additional tests are suggested:

☑ Lung function tests (compare to baseline).

☑ Consider chest X-ray following acute overexposure (may be negative it taken immediately after exposure due to delayed onset of pulmonary edema).

Other methods to reduce exposure include:

☑ Where possible, enclose operations and use local exhaust ventilation at the site of chemical release. If local exhaust ventilation or enclosure is not used, respiratory protection should be mandatory.

☑ Always ensure that proper protective clothing is worn when using chemical substances.

☑ Wash thoroughly immediately after exposure to phosphoric acid and at the end of the work shift or before eating, drinking, or smoking.

☑ Hazard warning information should be posted in the work area. In addition, as part of an on-going education and training program, all information on the health and safety hazards of phosphoric acid should be communicated to all exposed workers.

RISK ASSESSMENT: ENVIRONMENT

General Assessment

The environment is at risk of exposure during transportation, storage, disposal, or destruction of phosphoric acid. In almost every scenario, the threat of environmental exposure is contingent upon the proper handling of the chemical substance. Accidental spills, large or small, can result in fire, explosion, and possible contamination of the surrounding environmental mediums (water, soil, and especially air).

Phosphoric acid is considered a non-flammable solid (or liquid solution). However, it is extremely corrosive to many materials. It is incompatible with hydroxides, amines, and alkalis and contact can cause fire or explosion. In contact with many common metals (copper, brass, zinc), it can liberate extremely explosive hydrogen gas. It reacts exothermically (liberates heat) with aldehydes, amines, amides, alcohols and glycols, azo-compounds, carbamates, esters, caustics, phenols and cresols, ketones, organophosphates, epoxides, explosives, combustible materials, unsaturated halides, and organic peroxides. It forms flammable gases with sulfides, mercaptans, cyanides, and aldehydes. These characteristics require special consideration during any emergency situation involving a leak or spill of phosphoric acid. Should phosphoric acid ever come into contact with any incompatible substances either during use, transportation, or storage, the formation of highly toxic and/or highly explosive commodities is extremely possible.

Phosphoric acid can enter the environment through mining operations (runoff), industrial discharges, or spills.

Acute Ecological Effects

Acute (short-term) toxic effects may include the death of animals, birds, or fish and death or low growth rate in plants. Acute effects are seen 2 to 4 days after animals or plants are exposed to phosphoric acid.

Phosphoric acid has moderate acute toxicity to aquatic life in waters of low alkalinity. The estimate is based on the criterion for pH of water which considers the pH of 6.5 as the lower limit for healthy populations of aquatic life. The quantity of phosphoric acid required to cause a toxic response will be determined by the buffering capacity of the receiving water. Water with low alkalinity will require less phosphoric acid for toxic effects to occur. Insufficient data are available to evaluate or predict the short-term effects of phosphoric acid exposure to plants, birds, or land animals.

Chronic Ecological Effects

Chronic (long-term) toxic effects may include shortened life span, reproductive problems, lower fertility, and changes in appearance or behavior in exposed animals. These effects can be seen long after first exposure(s) to toxic chemicals.

Phosphoric acid has moderate chronic toxicity to aquatic life in waters of low alkalinity. The estimate is based on the criterion for pH of water which considers the pH of 6.5 as the lower limit for healthy populations of aquatic life. The quantity of phosphoric acid required to cause a toxic response will be determined by the buffering capacity of the receiving water. Water with low alkalinity will require less phosphoric acid for toxic effects to occur. Insufficient data are available to evaluate or predict the long-term effects of phosphoric acid exposure to plants, birds, or land animals.

Water Solubility

Phosphoric acid is highly soluble in water. Concentrations of over 1000 milligrams will mix with a liter of water.

Persistence in the Environment

Phosphoric acid is highly persistent in the aquatic environment, with a half-life between 20 and 200 days. The half-life of a pollutant is the amount of time it takes for one half of the chemical to be degraded.

Bioaccumulation in Aquatic Organisms

Some substances increase in concentration, or *bioaccumulate,* in living organisms as they breathe contaminated air, drink contaminated water, or eat contaminated food. These chemicals can become concentrated in the tissues and internal organs of animals as well as humans.

Small quantities of acids will be neutralized by the alkalinity in aquatic ecosystems, but larger quantities can lower the pH for extended periods of time. Phosphoric acid contains phosphate whose concentrations in edible tissues of most aquatic species that are consumed by humans is expected to be about the same as the average concentration in the water from which the fish was taken.

Recommended Risk-Reduction Measures

Proper training of all transporters will reduce the likelihood of a mishap or accident resulting in a leak or spill to the environment. The correct use of DOT labeling (per the MSDS) on all containers, trucks, and rail cars will enable emergency responders to react properly and quickly to any disaster thereby reducing the potential risk to the environment and to personnel.

Storage of phosphoric acid should be segregated from incompatible chemicals to minimize the risk of cross-contamination or contact. Buildings designated for storage should be equipped with appropriate fire protection systems (alarms, sprinklers, emergency lighting, portable extinguishers). Containers should be protected against physical damage and stored away from direct sunlight or other heat sources. Personnel must avoid the generation of dusty conditions in areas where solid phosphoric acid is handled or stored.

If a spill or leak to the environment has occurred, fire department, emergency response and/or hazardous materials spill personnel should be notified immediately. Cleanup should be attempted only by those

trained in proper spill containment procedures. Contaminated soils should be removed for incineration and replaced with clean soil. If phosphoric acid should contact the water table, aquifer, or navigable waterway, time is of the essence. It is highly soluble in water and total containment and remediation may not be possible. When such spills occur, the local and/or state emergency response authorities must be notified. A comprehensive emergency response or disaster preparedness/recovery plan should be in place prior to any operations involving the use, transportation, storage, or disposal of phosphoric acid. If phosphoric acid is spilled or leaked, the following specific steps are recommended:

- ☑ Restrict persons not wearing protective clothing from area of spill or leak until cleanup is complete and area can be opened for normal work.
- ☑ Remove all sources of heat and ignition.
- ☑ Ventilate area or spill or leak.
- ☑ Absorb solution spills in vermiculite, dry sand, earth, or similar material and deposit in sealed containers. Collect solid materials in safest manner possible and place in sealed containers.
- ☑ It may be necessary to dispose of phosphoric acid as a hazardous waste. The responsible state agency or the regional office of the federal Environmental Protection Agency (EPA) should be contacted for specific recommendations.

RISK ASSESSMENT: BUSINESS

General Assessment

Accidents or mishaps involving phosphoric acid can present a significant threat to business operations. The loss or damage of equipment or facilities can significantly affect fiscal viability. Lawsuits that may result from personnel injury/death, public exposures, and/or environmental contamination will also require a serious expenditure of resources. Media attention surrounding an injury, death, or environmental damage can also result in a loss of profits and loss of current as well as future business.

Recommended Risk-Reduction Measures

Company attorneys, safety and health professionals, and environmental specialists should be involved in the development of any procedures dealing with safety or emergency response. A company official should be pre-designated as a public relations officer with specific training in dealing with the press. Corporate plans and policies should be developed, approved, and implemented long before any need for such arises.

REFERENCES

American Conference of Governmental Industrial Hygienists. 1988. *Documentation of the Threshold Limit Values and Biological Exposure Indices,* 5th Edition (with updates). Cincinnati: ACGIH

Bloom, A. D. and F. J. DeSerres. 1995. *Ecotoxicity and Human Health.* New York: CRC Press/Lewis Publishers.

Burrell, R., D. K. Flaherty, and Leonard J. Sauers. 1992. *Toxicology of the Immune System.* New York: Van Nostrand Reinhold.

Calabrese, E. J., E. M. Kenyon. *Air Toxics and Risk Assessment.* New York: CRC Press/Lewis Publishers.

Cockerman, L. G., B. S. Shane. 1994. *Basic Environmental Toxicology.* New York: CRC Press/Lewis Publishers.

Danse, I. R. 1991. *Common Sense Toxics in the Workplace.* New York: Van Nostrand Reinhold.

Gardner, R. W. 1994. *Chemical Intolerance.* New York: CRC Press/Lewis Publishers.

Howard, P. H., et. al. 1991. *Handbook of Environmental Degradation Rates.* New York: CRC Press/Lewis Publishers.

Keith, L. H. and M. M. Walker. 1995. *Handbook of Air Toxics.* New York: CRC Press/Lewis Publishers.

Landis, W. G., M. H. Yu. 1995. *Introduction to Environmental Toxicology.* New York: CRC Press/Lewis Publishers.

Lauwerys, R. R., P. Hoet. 1993. *Industrial Chemical Exposure.* New York: CRC Press/Lewis Publishers.

Lewis, R. J., Sr. 1993. *Hawley's Condensed Chemical Dictionary,* 12th Edition. New York: Van Nostrand Reinhold.

Lewis, R. J., Sr. 1992. *Sax's Dangerous Properties of Industrial Materials,* Volumes 1, 2, and 3, 8th Edition. New York: Van Nostrand Reinhold.

Manahan, S. E. 1992. *Toxicological Chemistry,* 2nd Edition. New York: CRC Press/Lewis Publishers.

Meyer, E. 1990. *Chemistry of Hazardous Materials.* Englewood Cliffs, NJ: Prentice Hall.

Patnaik, P. 1992. *A Comprehensive Guide to the Hazardous Properties of Chemical Substances.* New York: Van Nostrand Reinhold.

Richardson, M. 1995. *Dictionary of Substance and Their Effects,* Vol. I-VI. New York: CRC Press/Lewis Publishers.

Sacarello, H. L. A. 1994. *The Comprehensive Handbook of Hazardous Materials.* New York: CRC Press/Lewis Publishers.

Sherman, J. D. 1988. *Chemical Exposure and Disease.* New York: Van Nostrand Reinhold.

U.S. Department of Health and Human Services; National Institute for Occupational Safety and Health. 1994. *NIOSH Pocket Guide to Chemical Hazards.* Washington, D.C.: U.S. Government Printing Office.

MATERIAL SAFETY DATA SHEET

CHEMICAL NAME	
PHOSPHORUS (YELLOW)	**CRC PRESS/LEWIS PUBLISHERS DISCLAIMER:** THE INFORMATION AND RECOMMENDATIONS PRESENTED HEREIN ARE BASED ON SOURCES BELIEVED TO BE RELIABLE. CRC MAKES NO REPRESENTATION ON ITS COMPLETENESS OR ACCURACY. IT IS THE RESPONSIBILITY OF THE USER TO DETERMINE THE CHEMICAL'S SUITABILITY FOR ITS INTENDED USE, THE CHEMICAL'S SAFE USE, AND THE CHEMICAL'S PROPER DISPOSAL. NO REPRESENTATIONS AND/OR WARRANTIES, EITHER EXPRESSED OR IMPLIED, OF THE MERCHANTABILITY OR FITNESS FOR A PARTICULAR PURPOSE, OR OF ANY OTHER NATURE, ARE MADE WITH RESPECT TO THE INFORMATION PROVIDED IN THIS MSDS OR TO THE CHEMICAL TO WHICH INFORMATION MAY REFER. CRC NEITHER ASSUMES NOR AUTHORIZES ANY OTHER PERSON TO ASSUME FOR IT, ANY OTHER ADDITIONAL RESPONSIBILITY OR LIABILITY FOR THE USE OF, OR RELIANCE UPON, THIS INFORMATION.

HAZARD WARNING INFORMATION

HEALTH	FIRE	REACTIVE	OTHER	DEGREE OF HAZARD	COLOR CODING	OTHER CODES
3	3	1		0 = Minimum Hazard 1 = Slight Hazard 2 = Moderate Hazard 3 = Serious Hazard 4 = Severe Hazard	HEALTH = BLUE FIRE = RED REACTIVITY = YELLOW OTHER = WHITE	OX = Oxidizer ACID = Acid ALK = Alkali COR = Corrosive W = Use No Water

SECTION I - GENERAL INFORMATION

Characterization	**Phosphorus Compound**	RCRA Number	**None**
		EPA Class	**Not Applicable**
DOT Proper Shipping Name	**Phosphorus**	Chemical Abstract Service (CAS) Number	**7723-14-0**
DOT Hazard Class and Label Requirements	**Flammable Solid and Poison**	DOT Emergency Guide Code	**38**
DOT Identification Number	**UN 1381**	Chemical Formula	**P_4**

Synonyms

Elemental phosphorus; white phosphorus; yellow phosphorus; bonide blue death rat killer; rat-nip.

SECTION II - HAZARDOUS INGREDIENTS AND IDENTITY INFORMATION

Hazardous Components (specific identity)	OSHA Exposure Criteria	NIOSH Exposure Criteria	Immediately Dangerous to Life and Health (IDLH)	ACGIH Exposure Criteria
Phosphorus (derivation: Produced in an electric furnace from phosphate rock, sand, and coke, the phosphorus vapor is driven off and condensed under water; or by reaction of phosphate rock with sulfuric acid, the resulting $CaSO_4$ being removed by filtration and the phosphoric acid concentrated by evaporation).	**PEL:** 0.1 mg/m^3 **STEL:** Not Established	**REL:** 0.1 mg/m^3 **STEL:** Not Established	5 mg/m^3	**TLV:** 0.1 mg/m^3 **STEL:** Not Established

SECTION III - PHYSICAL AND CHEMICAL CHARACTERISTICS

Boiling Point	**536°F (280°C)**	Specific Gravity (H_2O = 1)	**1.82 at 68°F (20°C)**
Vapor Pressure (mm Hg)	**0.03 at 68°F (20°C)**	Molecular Weight	**124.0**
Vapor Density (Air = 1)	**4.42**	Melting Point	**111°F (44°C)**

Solubility

Very slightly soluble in water (0.0003%). Insoluble in alcohol. Soluble in carbon disulfide.

Appearance and Odor

White to yellow, soft, waxy solid with acrid fumes in air. Usually shipped or stored under water.

SECTION IV - FIRE AND EXPLOSION HAZARD DATA

Flash Point (method used)	**Not Determined**	Explosive Limits in Air % by Volume	**LEL: Not Determined UEL: Not Determined**
NFPA Classification	**Flammable Solid**	Autoignition Temperature	**Not Determined**

Extinguishing Media

Flush with water until fire is out and phosphorus has solidified. Do NOT use direct water stream.

Special Fire Fighting Procedures

Wear full protective clothing and self-contained breathing apparatus (SCBA). Move containers from fire, or use a water spray to keep fire-exposed containers cool. Poisonous gases are produced in fire.

Unusual Fire and Explosion Hazards

Phosphorus ignites spontaneously in moist air without any added source of ignition at temperatures above 86°F (30°C) producing large quantities of toxic, dense, white phosphorus pentoxide smoke and large concentrations of heat. It may re-ignite once a fire has been extinguished.

SECTION V - REACTIVITY DATA

Stability		Conditions to Avoid **Phosphorus is normally stable in closed containers under routine conditions of handling and storage. Keep away from sources of heat and incompatible materials.**
Stable X	Unstable	Incompatibility *(materials to avoid)* **Extremely reactive with air, oxidizers (including elemental sulfur and strong caustics), and halogens. It is so reactive with air (ignites spontaneously) that it must be stored under water.**
Hazardous Polymerization		Conditions to Avoid **Under normal conditions of temperature and pressure, hazardous polymerization of phosphorus (yellow) cannot occur.**
May Occur	Will Not Occur X	Hazardous Decomposition or By-products **When heated to decomposition, phosphorus (yellow) can emit highly toxic/poisonous gases, including toxic phosphorous pentoxide (P_2O_5) fumes.**

SECTION VI - HEALTH HAZARD DATA

Primary Route(s) of Entry:	Inhalation?	Absorption (skin)?	Ingestion?
	X	X	X

Health Hazards

INHALATION: Irritation of the eyes, nose, throat, and respiratory tract with nausea and vomiting, severe abdominal pain, diarrhea, coma, and convulsions. Their may be damage to the lungs, kidney, and liver with bronchopneumonia, bone changes, necrosis of the jaw ("phossy" jaw), chronic cough, pallor, anemia, and weight loss.

ABSORPTION: Causes serious burns on skin contact. Absorption through skin can result in swelling jaw pain, anemia, and cachexia (malnutrition and debility). Eye contact can damage vision. Prolonged absorption can cause mossy teeth and brittle bones.

INGESTION: Local irritation, nausea, vomiting, severe abdominal pain, and liver damage. The skin may turn yellow within two weeks due to this liver damage. Kidney damage possible.

Carcinogenicity	NTP Listed?	IARC Cancer Review Group?	OSHA Regulated?	Target Organs?
Unknown Human Unknown Animal	No	No	29 CFR 1910.1000 Table Z-1	Eyes, skin, liver, kidney, jaw, teeth, blood, res. sys.

Medical Conditions Generally Aggravated by Exposure

Respiratory impairments (bronchitis, asthma); skin conditions (dermatitis); liver or kidney disorders.

Emergency and First-aid Procedures

Eye contact: Flush immediately with water for 15 minutes (minimum); seek medical attention. Skin contact: Remove all contaminated clothing. Immediately wash area with large amounts of water and then rewash with 3% copper sulfate. For inhalation: Remove the person from exposure. Provide respiratory assistance and CPR. Transfer to medical facility. If swallowed: Seek medical attention immediately. If person is conscious, induce vomiting. Gastric lavage with potassium permanganate is indicated. Gastric contents in stool may be smoking, luminescent in the dark, or fluorescent under UV light.

SECTION VII - PRECAUTIONS FOR SAFE HANDLING AND USE

Steps to be Taken in Case Material is Released or Spilled

Ventilate area. Remove all ignition sources. Keep spilled material wet and covered with wet sand or dirt. Do NOT allow exposure to air. Collect solidified materials and keep under water in sealed drums. Prevent runoff into waterways or sewers. Cleanup personnel must wear SCBA and protective gear.

Preferred Waste Disposal Method

Consider reclamation or destruction (by incineration).

Precautions to be Taken in Handling and Storage

Avoid contact with incompatible materials since violent reactions can occur. Store under water in tightly closed containers. Air contact must be prevented. Prevent damage to containers.

Other Precautions and Warnings

Sources of ignition are prohibited where phosphorus is used, handled, or stored.

SECTION VIII - CONTROL MEASURES AND PERSONAL PROTECTIVE EQUIPMENT

Respiratory Protection *(specify type)*

Exposure levels are relatively low. Recommend using a supplied-air respirator or a self-contained breathing apparatus (SCBA) operated in pressure demand or other positive pressure mode.

Ventilation

Local exhaust or general mechanical systems recommended.

Protective Gloves	Eye Protection	Other Protective Clothing
Impervious Material	Chemical Goggles or Face Mask	Protective Apron

Work/Hygiene Practices

Always wash hands thoroughly after using chemical; never bring food, drink, or smoking materials into vicinity of chemicals.

PHOSPHORUS

P_4

CAS: 7723-14-0

IDENTIFICATION AND TYPICAL USES

Phosphorus (yellow) is a white to yellow, soft, waxy solid with acrid fumes in air. Usually shipped or stored in water. It is used in smoke screens, in making rat poison (rodenticide), and in analytical chemistry work.

RISK ASSESSMENT: HEALTH

General Assessment

Yellow phosphorus is a highly poisonous substance. Exposure is possible through *inhalation* of its vapors and by skin contact (dermal *absorption*). It can also enter the body through *ingestion*. Since the vapor pressure of phosphorus is low, the acute health hazard from a short exposure to its vapors under normal conditions are expected to be relatively low. Health risk increases with concentration level and duration of exposure. There are no reports on its carcinogenic, mutagenic, or teratogenic properties.

Inhalation of its vapors can cause corrosive irritation of the eyes, nose, throat, and mucosa of the respiratory tract. This irritation can severely affect the lungs causing serious burns to tissues. Toxic symptoms include bronchopneumonia, anemia, loss of appetite, gastrointestinal complaints, chronic cough, pallor, and necrosis of the jaw ("phossy" jaw). A garlic odor may be noticeable on the breath or in skin wounds of exposed persons. Inhalation of the fumes from burning phosphorus can be extremely irritating.

Skin contact results in severe burns on contact. Absorption through intact skin can result in swelling, jaw pain, anemia, and cachexia (a condition similar to malnutrition). Eye contact can cause damage, serious burns, and probable damage to vision.

Ingestion of yellow phosphorus is extremely dangerous with severe systemic poisoning likely to occur relatively quickly. A single oral dose in humans of 70 to 100 milligrams can be fatal. Toxic symptoms include nausea and vomiting, severe abdominal pain, diarrhea, coma, and convulsions. Ingested phosphorus can be absorbed by the gastrointestinal tract resulting in liver damage. The skin surface may turn yellow within two weeks of exposure due to this liver damage.

Prolonged absorption of small amounts of phosphorus may cause mossy teeth (detectable by a dental examination) and brittle bones. Other acute effects include hypotension and shock due to direct toxicity to the heart.

Acute Health Effects

The following acute (short-term) health effects may occur immediately or shortly after exposure to phosphorus:

Skin: Irritation/rash or burning feeling (possibly delayed) with redness, blistering, and localized tissue damage. Absorption is likely.

Eye: Severe irritation; may cause burns and damage with permanent loss of vision.

Lung: Severe irritation of the mucous membranes in the nose, throat, and lungs following exposure.

Other: Can cause headaches, neurotoxicity, hypotension, and shock as a result of toxicity to the heart.

Chronic Health Effects

The following chronic (long-term) health effects may occur at some time after exposure to phosphorus and can last for months or even years:

Cancer Hazards: According to information presented in the references, yellow phosphorus has not been adequately tested for its ability to cause cancer in laboratory test animals.

Reproductive Hazard: According to information presented in the references, phosphorus has not been adequately tested for its ability to adversely affect reproduction in test animals.

Other Chronic Effects: Repeated exposure can cause destruction of bone, especially the jaw bone. This process can extend into one or both eye sockets. There can be pain and loosening of the teeth, ulcerative stomatitis (inflammation of the mucosa of the mouth), leukopenia (low number of white blood cells), anemia, hepatomegaly (enlarged liver), jaundice, periostitis (inflammation of the periosteum) with suppuration, ulceration, and necrosis of the mandible. There can be chronic hypotension and possible shock. Chronic chemical bronchitis may occur as well as hepatic encephalopathy (due to liver failure), and kidney failure.

Recommended Risk-Reduction Measures

Personnel must take extra care to avoid direct contact with phosphorus. It is extremely corrosive and severely toxic to humans. If a less toxic chemical cannot be substituted for a hazardous substance, then *engineering controls* are the most effective method of reducing exposures.

The best protection is to enclose operations and/or provide local exhaust ventilation at the site of chemical release. While not always operationally feasible, isolating operations can also reduce exposure. Using respiratory protection is less effective than the controls mentioned above, but is still advisable whenever working with or around phosphorus. The exposure level is extremely low (0.1 mg/m^3) and difficult to accurately monitor, even under normal operating conditions. The best protection is obtained using a self-contained breathing apparatus (SCBA) or a supplied-air respirator with full facepiece operated in pressure demand mode. If a full facepiece is not available, then chemical goggles should be worn to protect the eyes. Whenever a chemical splash hazard exists, a face shield and a protective apron should be worn. To prevent hand and skin exposures, acid resistant neoprene or nitrile gloves should be used. Glove manufacturers should be contacted and permeation studies obtained *before* final glove selections are made.

Administrative controls should also be in place to minimize the potential for human exposures. These may include written procedures or policies which specify the methods and techniques that will be practiced whenever personnel are to work with phosphorus.

All personnel should receive *training* on the use, hazards, protective measures, emergency actions, and other precautions per 29 CFR 1910.1200 (Hazard Communication), prior to the first assignment in an area where phosphorus is used or stored.

Before beginning employment and at regular intervals thereafter (e.g., annually), the following tests are recommended:

- ☑ Lung function tests (establish baseline).
- ☑ Consider X-ray of teeth and jaw and a dental examination.
- ☑ Complete blood count (CBC) with differential.

If symptoms occur or overexposure is suspected, the following additional tests are suggested:

- ☑ Lung function tests (compare to baseline).
- ☑ Electrocardiogram (EKG).

Other methods to reduce exposure include:

- ☑ Where possible, enclose operations and use local exhaust ventilation at the site of chemical release. If local exhaust ventilation or enclosure is not used, respiratory protection should be mandatory.
- ☑ Always ensure that proper protective clothing is worn when using chemical substances and that personnel are trained in its use and care.
- ☑ Wash thoroughly immediately after exposure to phosphorus and at the end of the work shift or before eating, drinking, or smoking.
- ☑ Hazard warning information should be posted in the work area. In addition, as part of an on-going education and training program, all information on the health and safety hazards of phosphorus should be communicated to all exposed workers.

RISK ASSESSMENT: ENVIRONMENT

General Assessment

The environment is at risk of exposure during transportation, storage, disposal, or destruction of phosphorus. In almost every scenario, the threat of environmental exposure is contingent upon the proper handling of the chemical substance. Accidental spills, large or small, can result in fire, explosion, and possible contamination of the surrounding environmental mediums (water, soil, and especially air).

Phosphorus is considered a flammable solid and can ignite spontaneously in moist air without any added source of ignition at temperatures greater than 86°F (30°C). In fact, it is so reactive in air that it must be stored under water. Other incompatible materials include halogens, strong caustics, and all oxidizing agents, including sulfur. These characteristics require special consideration during any emergency situation involving a leak or spill of phosphorus. Should phosphorus ever come into contact with any incompatible substances either during use, transportation, or storage, the formation of highly toxic and/or highly explosive commodities is extremely possible.

Phosphorus can enter the environment through industrial discharges, agricultural runoff, municipal waste treatment plant discharges, or spills.

Acute Ecological Effects

Acute (short-term) toxic effects may include the death of animals, birds, or fish and death or low growth rate in plants. Acute effects are seen 2 to 4 days after animals or plants are exposed to phosphorus.

Acute toxicity of phosphorus to aquatic life is unknown and cannot be predicted as no toxicity tests have been conducted with elemental phosphorus. Insufficient data are available to evaluate or predict the short-term effects of phosphorus to plants, birds, and land animals.

Chronic Ecological Effects

Chronic (long-term) toxic effects may include shortened life span, reproductive problems, lower fertility, and changes in appearance or behavior in exposed animals. These effects can be seen long after first exposure(s) to toxic chemicals.

Chronic toxicity of phosphorus to aquatic life is unknown and cannot be predicted as no toxicity tests have been conducted with elemental phosphorus. Insufficient data are available to evaluate or predict the long-term effects of phosphorus to plants, birds, and land animals.

Water Solubility

Phosphorus is moderately soluble in water. Concentrations between 1 to 1000 milligrams will mix with a liter of water.

Persistence in the Environment

Phosphorus is slightly persistent in the aquatic environment, with a half-life of less than 2 and 20 days. The half-life of a pollutant is the amount of time it takes for one half of the chemical to be degraded. Yellow phosphorus, also known as white phosphorus, is a form of elemental phosphorus that will probably be slowly oxidized by the dissolved oxygen in water, but which burns in air.

Bioaccumulation in Aquatic Organisms

Some substances increase in concentration, or *bioaccumulate,* in living organisms as they breathe contaminated air, drink contaminated water, or eat contaminated food. These chemicals can become concentrated in the tissues and internal organs of animals as well as humans.

The concentration of phosphorus found in fish tissues is expected to be somewhat higher than the average concentration of phosphorus in the water from which the fish was taken.

Recommended Risk-Reduction Measures

Proper training of all transporters will reduce the likelihood of a mishap or accident resulting in a leak or spill to the environment. The correct use of DOT labeling (per the MSDS) on all containers, trucks, and rail cars will enable emergency responders to react properly and quickly to any disaster thereby reducing the potential risk to the environment and to personnel.

Storage of phosphorus should be segregated from incompatible chemicals to minimize the risk of cross-contamination or contact. Buildings designated for storage should be equipped with appropriate fire protection systems (alarms, sprinklers, emergency lighting, portable extinguishers). Containers should be protected against physical damage and stored away from direct sunlight or other heat sources. Because phosphorus will ignite spontaneously in air, it must be stored under water.

If a spill or leak to the environment has occurred, fire department, emergency response and/or hazardous materials spill personnel should be notified immediately. Cleanup should be attempted only by those trained in proper spill containment procedures. Contaminated soils should be removed for incineration and replaced with clean soil. If phosphorus should contact the water table, aquifer, or navigable waterway, time is of the essence. It is moderately soluble in water and total containment and remediation may not be possible. When such spills occur, the local and/or state emergency response authorities must be notified. A comprehensive emergency response or disaster preparedness/recovery plan should be in place prior to any operations involving the use, transportation, storage, or disposal of phosphorus. If phosphorus is spilled or leaked, the following specific steps are recommended:

- ☑ Restrict persons not wearing protective clothing from area of spill or leak until cleanup is complete and area can be opened for normal work.
- ☑ Remove all sources of heat and ignition.
- ☑ Ventilate area or spill or leak.
- ☑ As quickly as possible, cover spilled materials with wet sand or dirt and keep wet until cleanup is completed. Collect solidified material and keep submerged in water inside drums.
- ☑ It may be necessary to dispose of phosphorus as a hazardous waste. The responsible state agency or the regional office of the federal Environmental Protection Agency (EPA) should be contacted for specific recommendations.

RISK ASSESSMENT: BUSINESS

General Assessment

Accidents or mishaps involving phosphorus can present a significant threat to business operations. The loss or damage of equipment or facilities can significantly affect fiscal viability. Lawsuits that may result from personnel injury/death, public exposures, and/or environmental contamination will also require a serious expenditure of resources. Media attention surrounding an injury, death, or environmental damage can also result in a loss of profits and loss of current as well as future business.

Recommended Risk-Reduction Measures

Company attorneys, safety and health professionals, and environmental specialists should be involved in the development of any procedures dealing with safety or emergency response. A company official should be pre-designated as a public relations officer with specific training in dealing with the press. Corporate plans and policies should be developed, approved, and implemented long before any need for such arises.

REFERENCES

American Conference of Governmental Industrial Hygienists. 1988. *Documentation of the Threshold Limit Values and Biological Exposure Indices,* 5th Edition (with updates). Cincinnati: ACGIH

Bloom, A. D. and F. J. DeSerres. 1995. *Ecotoxicity and Human Health.* New York: CRC Press/Lewis Publishers.

Burrell, R., D. K. Flaherty, and Leonard J. Sauers. 1992. *Toxicology of the Immune System.* New York: Van Nostrand Reinhold.

Calabrese, E. J., E. M. Kenyon. *Air Toxics and Risk Assessment.* New York: CRC Press/Lewis Publishers.

Cockerman, L. G., B. S. Shane. 1994. *Basic Environmental Toxicology.* New York: CRC Press/Lewis Publishers.

Danse, I. R. 1991. *Common Sense Toxics in the Workplace.* New York: Van Nostrand Reinhold.

Environmental Protection Agency. 1983. *Chemical Hazard Information Profiles.* Washington, D.C.: U.S. EPA.

Fodden, C. R., and J. L. Weddell. 1992. *Hazardous Materials Emergency Action Data.* New York: CRC Press/Lewis Publishers.

Forseberg, K., and L. H. Keith. 1993. *Chemical; Protective Clothing Permeation and Degradation Database.* Chelsea, MI: Lewis Publishers.

Gardner, R. W. 1994. *Chemical Intolerance.* New York: CRC Press/Lewis Publishers.

Hallenbeck, W. H. 1993. *Quantitative Risk Assessment for Environmental and Occupational Health,* 2nd Edition. New York: CRC Press/Lewis Publishers.

Hodgson, E., R. B. Mailman, and Janice E. Chambers. 1988. *Dictionary of Toxicology.* London: The MacMillian Company.

Howard, P. H., et. al. 1991. *Handbook of Environmental Degradation Rates.* New York: CRC Press/Lewis Publishers.

Jones, F. E. 1994. *Toxic Organic Vapors in the Workplace.* New York: CRC Press/Lewis Publishers.

Keith, L. H. and M. M. Walker. 1995. *Handbook of Air Toxics.* New York: CRC Press/Lewis Publishers.

Landis, W. G., M. H. Yu. 1995. *Introduction to Environmental Toxicology.* New York: CRC Press/Lewis Publishers.

Lauwerys, R. R., P. Hoet. 1993. *Industrial Chemical Exposure.* New York: CRC Press/Lewis Publishers.

Lewis, R. J., Sr. 1993. *Hawley's Condensed Chemical Dictionary,* 12th Edition. New York: Van Nostrand Reinhold.

Lewis, R. J., Sr. 1992. *Sax's Dangerous Properties of Industrial Materials,* Volumes 1, 2, and 3, 8th Edition. New York: Van Nostrand Reinhold.

Manahan, S. E. 1992. *Toxicological Chemistry,* 2nd Edition. New York: CRC Press/Lewis Publishers.

Meyer, E. 1990. *Chemistry of Hazardous Materials.* Englewood Cliffs, NJ: Prentice Hall.

Patnaik, P. 1992. *A Comprehensive Guide to the Hazardous Properties of Chemical Substances.* New York: Van Nostrand Reinhold.

Richardson, M. 1995. *Dictionary of Substance and Their Effects,* Vol. I-VI. New York: CRC Press/Lewis Publishers.

Sacarello, H. L. A. 1994. *The Comprehensive Handbook of Hazardous Materials.* New York: CRC Press/Lewis Publishers.

Sherman, J. D. 1988. *Chemical Exposure and Disease.* New York: Van Nostrand Reinhold.

U.S. Department of Health and Human Services, National Institute for Occupational Safety and Health. 1994. *NIOSH Pocket Guide to Chemical Hazards.* Washington, D.C.: U.S. Government Printing Office.

MATERIAL SAFETY DATA SHEET

CHEMICAL NAME	
PHTHALIC ANHYDRIDE	**CRC PRESS/LEWIS PUBLISHERS DISCLAIMER:** THE INFORMATION AND RECOMMENDATIONS PRESENTED HEREIN ARE BASED ON SOURCES BELIEVED TO BE RELIABLE. CRC MAKES NO REPRESENTATION ON ITS COMPLETENESS OR ACCURACY. IT IS THE RESPONSIBILITY OF THE USER TO DETERMINE THE CHEMICAL'S SUITABILITY FOR ITS INTENDED USE, THE CHEMICAL'S SAFE USE, AND THE CHEMICAL'S PROPER DISPOSAL. NO REPRESENTATIONS AND/OR WARRANTIES, EITHER EXPRESSED OR IMPLIED, OF THE MERCHANTABILITY OR FITNESS FOR A PARTICULAR PURPOSE, OR OF ANY OTHER NATURE, ARE MADE WITH RESPECT TO THE INFORMATION PROVIDED IN THIS MSDS OR TO THE CHEMICAL TO WHICH INFORMATION MAY REFER. CRC NEITHER ASSUMES NOR AUTHORIZES ANY OTHER PERSON TO ASSUME FOR IT, ANY OTHER ADDITIONAL RESPONSIBILITY OR LIABILITY FOR THE USE OF, OR RELIANCE UPON, THIS INFORMATION.

HAZARD WARNING INFORMATION

HEALTH	FIRE	REACTIVE	OTHER	DEGREE OF HAZARD	COLOR CODING	OTHER CODES
2	1	0		0 = Minimum Hazard 1 = Slight Hazard 2 = Moderate Hazard 3 = Serious Hazard 4 = Severe Hazard	HEALTH = BLUE FIRE = RED REACTIVITY = YELLOW OTHER = WHITE	OX = Oxidizer ACID = Acid ALK = Alkali COR = Corrosive W = Use No Water

SECTION I - GENERAL INFORMATION

Characterization	RCRA Number	EPA Class
Anhydride	**U190**	**Toxic Waste**

DOT Proper Shipping Name	Chemical Abstract Service (CAS) Number
Phthalic Anhydride	**85-44-9**
DOT Hazard Class and Label Requirements	DOT Emergency Guide Code
Corrosive	**60**
DOT Identification Number	Chemical Formula
UN 2214	**$C_6H_4(CO)_2O$**

Synonyms

1,2-Benzenedicarboxylic anhydride; PAN; phthalic acid anhydride; 1,3-dioxophthalan.

SECTION II - HAZARDOUS INGREDIENTS AND IDENTITY INFORMATION

Hazardous Components (specific identity)	OSHA Exposure Criteria	NIOSH Exposure Criteria	Immediately Dangerous to Life and Health (IDLH)	ACGIH Exposure Criteria
Phthalic anhydride (derivation: By catalytic oxidation of naphthalene). **1 ppm = 6.16 mg/m^3**	**PEL (8-hour):** **2 ppm** **12 mg/m^3** **STEL:** **Not Established**	**REL (10-hour):** **1 ppm** **6 mg/m^3** **STEL:** **Not Established**	**60 mg/m^3**	**TLV:** **1 mg/m^3** **6 mg/m^3** **STEL:** **Not Established**

SECTION III - PHYSICAL AND CHEMICAL CHARACTERISTICS

Boiling Point	Specific Gravity (H_2O = 1)
563°F (295°C) Sublimes	**1.53 (flake) 1.20 (molten)**
Vapor Pressure (mm Hg)	Molecular Weight
0.0015 at 69°F (20°C)	**148.1**
Vapor Density (Air = 1)	Melting Point
5.1	**267°F (131°C)**

Solubility

Slightly soluble in water (0.6%), more soluble in hot water. Soluble in alcohol and carbon disulfide.

Appearance and Odor

White solid flake or a clear, colorless molten liquid with a characteristic acrid, irritating, choking odor. Odor Threshold = 0.05 ppm.

SECTION IV - FIRE AND EXPLOSION HAZARD DATA

Flash Point (method used)	Explosive Limits in Air % by Volume
304°F (152°C) closed cup	**LEL: 1.7% UEL: 10.5%**
NFPA Classification	Autoignition Temperature
Combustible Solid	**1058°F (569°C)**

Extinguishing Media

Use carbon dioxide or dry chemical on solid or molten materials. Use water spray on the flake or dust.

Special Fire Fighting Procedures

Poisonous gases are produced in fire; wear full protective clothing and self-contained breathing apparatus (SCBA). Isolate area and deny entry to those not involved in response procedures. Move containers from fire area if it can be done without risk. If this is not possible, use water spray to keep fire-exposed containers cool. Do Not release runoff from fire control measures to sewers or waterways.

Unusual Fire and Explosion Hazards

Vapors and dusts may burn in air. A serious dust explosion hazard exists at or above 0.015 ounces per cubic foot of air-dispersed particulate (ignition temperature for dust clouds: 1202°F/650°C).

SECTION V - REACTIVITY DATA

Stability		Conditions to Avoid **Phthalic anhydride is normally stable under routine conditions of handling and storage. Avoid contact with incompatible materials, moisture, and air.**
Stable X	Unstable	Incompatibility *(materials to avoid)* **Incompatible materials include water (converts to phthalic acid in hot water), moisture in air, and strong oxidizers (chlorine, bromine, fluorine).**
Hazardous Polymerization		Conditions to Avoid **Under normal conditions of temperature and pressure, hazardous polymerization of phthalic anhydride cannot occur. In the presence of iron, pyrophoric phthalic acid salts can be formed which are explosive.**
May Occur	Will Not Occur X	Hazardous Decomposition or By-products **When heated to decomposition, phthalic anhydride emits toxic and irritating fumes and vapors, including toxic oxides of carbon.**

SECTION VI - HEALTH HAZARD DATA

Primary Route(s) of Entry:	Inhalation?	Absorption (skin)?	Ingestion?
	X		X

Health Hazards

INHALATION: Causes irritation of the eyes, nose, throat, and respiratory system with coughing, sneezing, dypsnea, mucous membrane irritation in the nasal cavity leading to ulceration of the septum, nasal bleeding, loss of the sense of smell, conjunctivitis, hoarseness, possible blood changes, and chemical bronchitis with asthma-like symptoms.

SKIN & EYES: Skin contact causes serious caustic burns, itchy rash, and dermatitis. Eye contact may cause burns, irritation, corneal scarring, and perhaps permanent damage.

INGESTION: Burning of the mouth, esophagus, and digestive tract. Pain in the chest, stomach, and painful swallowing. In moist tissues, it hydrolyzes to phthalic acid, which is corrosive.

Carcinogenicity	NTP Listed?	IARC Cancer Review Group?	OSHA Regulated?	Target Organs?
Unknown Human **Unknown Animal**	**No**	**No**	**29 CFR 1910.1000 Table Z-1**	**Skin, eyes, liver, kidney, respiratory system.**

Medical Conditions Generally Aggravated by Exposure

Existing chronic respiratory diseases may be aggravated by exposure.

Emergency and First-aid Procedures

Eye contact: Flush immediately with water for 15 minutes (minimum); seek medical attention. Skin contact: Remove all contaminated clothing. Wash area with large amounts of soap and water. Seek medical assistance. For inhalation: Remove the person from exposure. Provide respiratory assistance and CPR. Transfer to medical facility. Spray or gargle with water to relieve throat irritation. If swallowed: Contact poison control center. Unless told otherwise, give soapy water to drink and induce vomiting. Never give anything by mouth to an unconscious or convulsing person.

SECTION VII - PRECAUTIONS FOR SAFE HANDLING AND USE

Steps to be Taken in Case Material is Released or Spilled

Collect powdered material using a vacuum equipped with HEPA filter, deposit in sealed drums for disposal. Avoid skin contact. Do NOT dry sweep (generates dusts). Allow molten materials to solidify, then treat as a solid. Ventilate area of spill to disperse any collection of gas, vapors, or dusts.

Preferred Waste Disposal Method

Burn in a chemical incinerator equipped with an afterburner and scrubber.

Precautions to be Taken in Handling and Storage

Do not store phthalic anhydride in the presence of incompatible chemicals or materials. Store in tightly closed containers in a cool, dark, well-ventilated area. Electrically ground and bond all containers and equipment to prevent static sparks and possible ignition.

Other Precautions and Warnings

Keep away from water, moisture, and incompatible materials. Prevent damage to containers.

SECTION VIII - CONTROL MEASURES AND PERSONAL PROTECTIVE EQUIPMENT

Respiratory Protection *(specify type)*

PEL is extremely low (2 ppm). Use a self-contained breathing apparatus (SCBA) with full facepiece or a supplied-air respirator operated in pressure demand or other positive pressure mode.

Ventilation

Local exhaust or general mechanical systems recommended.

Protective Gloves	Eye Protection	Other Protective Clothing
Impervious Material	**Chemical/Dust Goggles or Face Mask**	**Protective Apron**

Work/Hygiene Practices

Always wash hands thoroughly after using chemical; never bring food, drink, or smoking materials into vicinity of chemicals.

PHTHALIC ANHYDRIDE

$C_6H_4(CO)_2O$

CAS: 85-44-9

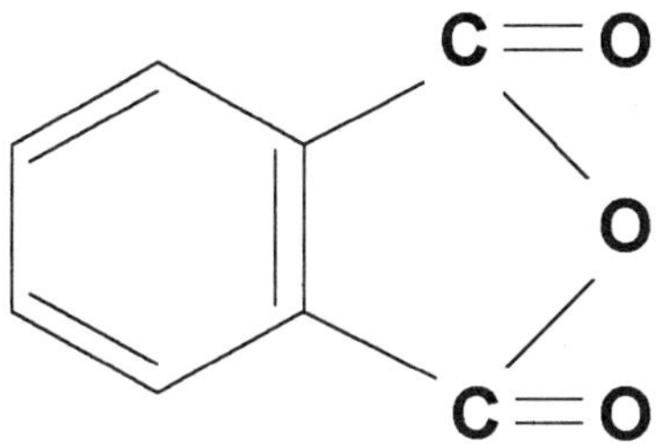

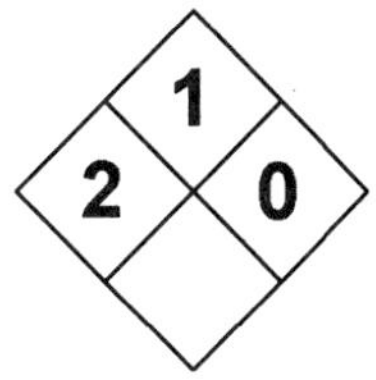

IDENTIFICATION AND TYPICAL USES

Phthalic anhydride is a white solid flake or a clear, colorless, molten liquid with a characteristic, irritating, choking odor. It is used in the manufacture of alkyd resins, plasticizers, as a hardener for other resins, in polyesters, in the synthesis of phenolphthalein and other phthaleins, many other dyes, chlorinated products, pharmaceutical intermediates, insecticides, diethyl phthalate, dimethyl phthalate, and as a laboratory reagent.

RISK ASSESSMENT: HEALTH

General Assessment

Phthalic anhydride is a human poison primarily by *inhalation,* causing toxic systemic effects targeting the respiratory system. It can also cause toxic effects by *ingestion*. It is an experimental teratogen, causing fetal development problems in test animals. There are no reports regarding its carcinogenic or mutagenic potential in humans or animals.

Inhalation of phthalic anhydride can be irritating to the nose, throat, mucosa, and the upper respiratory tract. Asthmatic-like symptoms include coughing, sneezing, and dypsnea. There may also be ulceration of the nasal septa, nose bleeds, loss of sense of smell, hoarseness, bronchitis, blood changes, and chemical bronchitis. In the presence of moist tissues, it hydrolyzes to phthalic acid which is corrosive and irritating.

Ingestion of phthalic anhydride can cause burning of the mouth, esophagus, and gastrointestinal tract with deep, burning pain in the chest and stomach, and pain on swallowing.

Skin contact may not result in any immediate effects. However, if the chemical is not removed quickly, there can be deep and painful burning, blistering, and possible dermatitis and allergic sensitization in some individuals.

Acute Health Effects

The following acute (short-term) health effects may occur immediately or shortly after exposure to phthalic anhydride:

Skin: Irritation and possible rash. Moisture on skin's surface will enhance the reaction.

Eye: Severe burns, corneal scarring, and possible damage to vision.

Lung: Severe irritation of the nose, throat, and esophagus. Nasal cavity irritation can lead to ulceration of the septa with nose bleeds and loss of the sense of smell.

Chronic Health Effects

The following chronic (long-term) health effects may occur at some time after exposure to phthalic anhydride and can last for months or even years:

Cancer Hazards: According to the information presented in the references, phthalic anhydride has not been adequately tested for its ability to cause cancer in test animals.

Reproductive Hazard: According to the information presented in the references, phthalic anhydride has been shown to cause teratogenic effects in test animals. It is not known if human exposure will have the same results. More research is required in this regard.

Other Chronic Effects: Prolonged contact with phthalic anhydride can lead to the development of a skin allergy. If an allergy develops, very low future exposures can cause itching and a skin rash. It may also cause an asthma-like lung allergy with shortness of breath, wheezing, coughing, and or tightness in the chest. Very irritating substances may affect the lungs. It is not known if phthalic anhydride causes permanent lung damage on chronic exposure.

ⓤ *Recommended Risk-Reduction Measures*

Personnel should avoid direct contact with phthalic anhydride. It is a toxic, human poison on inhalation. Its human teratogenic properties are not clearly understood in the references. If a less toxic chemical cannot be substituted for a hazardous substance, *engineering controls* are the most effective method of reducing exposures. The best protection is to enclose operations and/or provide local exhaust ventilation at the site of chemical release. While not always operationally feasible, isolating operations can also reduce exposure.

Using respiratory protection is less effective than the controls mentioned above, but is still advisable whenever working with or around phthalic anhydride. This chemical has an extremely low permissible exposure limit (1 ppm), which is difficult to accurately monitor. Therefore, prudent risk management requires a conservative approach to personal protection. While a chemical respirator with a dust/mist pre-filter may suffice, the best protection is obtained using a supplied-air respirator operated in continuous flow mode or a self-contained breathing apparatus (SCBA) with full facepiece operated in pressure demand or positive pressure mode. If a full facepiece is not available, then chemical/dust goggles should be worn to protect the eyes. Whenever phthalic anhydride is being mixed with liquids, a chemical splash hazard exists and a face shield and rubber apron should be worn. To prevent hand and skin exposures, protective gloves should be used. Glove manufacturers should be contacted and permeation studies obtained *before* final glove selections have been made.

Administrative controls should also be in place to minimize the potential for human exposures. These may include written procedures or policies which specify the methods and techniques that will be practiced whenever personnel are to work with phthalic anhydride.

All personnel should receive *training* on the use, hazards, protective measures, emergency actions, and other precautions per 29 CFR 1910.1200 (Hazard Communication), prior to the first assignment in an area where phthalic anhydride is used or stored. Before beginning employment and at regular intervals thereafter (e.g., annually), the following medical tests are recommended for persons with frequent or anticipated high exposure to phthalic anhydride:

- ☑ Lung function tests (may be normal if person is not experiencing symptoms at the time of the test).

If symptoms develop or overexposure is suspected, the following medical tests may be useful:

- ☑ Lung function tests.
- ☑ Evaluation by a qualified allergist with careful consideration of exposure history and special testing (may help diagnose skin allergy).

Any evaluation should include a careful medical history of past and present symptoms with an examination. However, medical tests that evaluate existing damage are not a substitute for controlling exposure. Also, since smoking can cause heart disease, emphysema, lung cancer, and other respiratory diseases, smokers may experience symptoms more rapidly than non-smokers under the same conditions of exposure. Prudent risk management requires proper consideration of *all* possible factors that may be causing the appearance of exposure symptoms in the workplace.

Other methods to reduce exposure include:

- ☑ Where possible, enclose operations and use local exhaust ventilation at the site of chemical release. If local exhaust ventilation or enclosure is not used, then respiratory protection should be mandatory and its use enforced.
- ☑ Always ensure that proper protective clothing is worn when using chemical substances and that personnel are trained in its use and care.
- ☑ Wash thoroughly immediately after exposure to phthalic anhydride and at the end of the work shift or before eating, drinking, or smoking.
- ☑ Hazard warning information should be posted in the work area. In addition, as part of an on-going education and training program, all information on the health and safety hazards of phthalic anhydride should be communicated to all exposed or potentially exposed workers.
- ☑ Eye wash stations should be provided in work areas. If the potential for whole body exposure exists, then safety showers should also be provided in t he immediate work area. Even if no immediate or noticeable effects occur on skin contact, personnel must still be instructed to use the safety showers to remove the chemical.

RISK ASSESSMENT: ENVIRONMENT

General Assessment

The environment is at risk of exposure during transportation, storage, disposal, or destruction of phthalic anhydride. In almost every scenario, the threat of environmental exposure is contingent upon the proper handling of the chemical substance. Accidental spills, large or small, can result in fire, explosion, and possible contamination of the surrounding environmental mediums (water, soil, and air).

Phthalic anhydride is considered a combustible solid. It can react with water or moisture to produce phthalic acid which is corrosive to most metals. Phthalic anhydride mist, dust, or vapor can explode in proper mixtures with air in contact with an ignition source. These characteristics require special consideration during any emergency situation involving a leak or spill of phthalic anhydride or phthalic anhydride mixtures. Should phthalic anhydride ever come into contact with other incompatible substances, such as strong oxidizers, either during use, transportation, or storage, violent and even explosive reactions can occur. Phthalic anhydride can enter the environment from industrial discharges and spills.

Acute Ecological Effects

Acute (short-term) toxic effects may include the death of animals, birds, or fish and death or low growth rate in plants. Acute effects are seen 2 to 4 days after animals or plants are exposed to phthalic anhydride.

Phthalic anhydride has low acute toxicity to aquatic life. Insufficient data are available to evaluate or predict the short-term effects of phthalic anhydride to plants, birds, or land animals.

Chronic Ecological Effects

Chronic toxic effects may include shortened life span, reproductive problems, lower fertility, and changes in appearance or behavior in exposed animals. These effects can be seen long after first exposure(s) to toxic chemicals.

Phthalic anhydride has low chronic toxicity to aquatic life. Insufficient data are available to evaluate or predict the long-term effects of phthalic anhydride to plants, birds, or land animals.

Water Solubility

Phthalic anhydride is slightly soluble. Concentrations of 10 milligrams will mix with a liter of water.

Persistence in the Environment

Phthalic anhydride is non-persistent in water, with a half-life of less than 2 days. The half-life of a pollutant is the amount of time it takes for one half of the chemical to be degraded. It will hydrolyze rapidly to produce phthalic acid.

Bioaccumulation in Aquatic Organisms

Some substances increase in concentration, or *bioaccumulate,* in living organisms as they breathe contaminated air, drink contaminated water, or eat contaminated food. These chemicals can become concentrated in the tissues and internal organs of animals as well as humans.

Since phthalic anhydride will react rapidly in water to produce phthalic acid, it is not expected to accumulate in the edible tissues of fish normally consumed by humans.

Recommended Risk-Reduction Measures

Proper training of all transporters will reduce the likelihood of a mishap or accident resulting in a leak or spill to the environment. The correct use of DOT labeling (per the MSDS) on all containers, trucks, and rail cars will enable emergency responders to react properly and quickly to any disaster thereby reducing the potential risk to the environment and to personnel.

Storage of phthalic anhydride should be segregated from incompatible chemicals to minimize the risk of cross-contamination or contact. Precautions must be taken to ensure all contact with water or moisture is prevented in use or storage.

If a spill or leak to the environment has occurred, fire department, emergency response and/or hazardous materials spill personnel should be notified immediately. Cleanup should be attempted only by those trained in proper spill containment procedures. Contaminated soils should be removed for incineration and replaced with clean soil. If phthalic anhydride should contact the water table, aquifer, or navigable waterway, remediation activities should be prompt. When such spills occur, the local and/or state emergency response authorities must be notified. A comprehensive emergency response or disaster preparedness/recovery plan should be in place prior to any operations involving the use, transportation, storage, or disposal of phthalic anhydride.

If phthalic anhydride is spilled or leaked, the following specific steps are recommended:

- ☑ Restrict persons not wearing protective clothing from area of spill or leak until cleanup is complete and area can be opened to normal work.
- ☑ Collect powdered material in the most convenient (safe) manner possible and deposit in sealed containers. Do NOT dry sweep (generates airborne dusts). Use a vacuum equipped with a HEPA filter instead. If liquid solutions are spilled, use vermiculite to absorb and place in a sealed drum.
- ☑ It may be necessary to dispose of phthalic anhydride as a hazardous waste. The responsible state agency or the regional office of the federal Environmental Protection Agency (EPA) should be contacted for specific recommendations.

RISK ASSESSMENT: BUSINESS

General Assessment

Accidents or mishaps involving phthalic anhydride can present a significant threat to business operations. The loss or damage of equipment or facilities can significantly affect fiscal viability. Lawsuits that may result from personnel injury/death, public exposures, and/or environmental contamination will also require a serious expenditure of resources. Media attention surrounding an injury, death, or environmental damage can also result in a loss of profits and loss of current as well as future business. Always remember that anytime the terms "reproductive hazard" or "birth defects" are used, public hysteria, emotion, and ignorance can run equally high. This should be carefully considered when developing or implementing any public relations policies.

Recommended Risk-Reduction Measures

Company attorneys, safety and health professionals, and environmental specialists should be involved in the development of any procedures dealing with safety or emergency response. A company official should be pre-designated as a public relations officer with specific training in dealing with the press. Corporate plans and policies should be developed, approved, and implemented long before any need for such arises.

REFERENCES

American Conference of Governmental Industrial Hygienists. 1988. *Documentation of the Threshold Limit Values and Biological Exposure Indices*, 5th Edition (with updates). Cincinnati: ACGIH

Asante-Duah, D. K. 1993. *Hazardous Waste Risk Assessment.* New York: CRC Press/Lewis Publishers.

Bloom, A. D. and F. J. DeSerres. 1995. *Ecotoxicity and Human Health.* New York: CRC Press/Lewis Publishers.

Brusick, D. J. 1994. *Methods for Genetic Risk Assessment.* New York: CRC Press/Lewis Publishers.

Bueche, J. R. 1972. *Principles of Physics*, 2nd Edition. New York: McGraw-Hill.

Burrell, R., D. K. Flaherty, and L. J. Sauers. 1992. *Toxicology of the Immune System.* New York: Van Nostrand Reinhold.

Calabrese, E. J. 1994. *Biological Effects of Low Level Exposures.* New York: CRC Press/Lewis Publishers.

Cockerman, L. G., B. S. Shane. 1994. *Basic Environmental Toxicology.* New York: CRC Press/Lewis Publishers.

Cothern, C. R., N. P. Ross. 1994. *Environmental Statistics, Assessment, and Forecasting.* New York: CRC Press/Lewis Publishers.

Gardner, R. W. 1994. *Chemical Intolerance.* New York: CRC Press/Lewis Publishers.

Hodgson, E., R. B. Mailman, and Janice E. Chambers. 1988. *Dictionary of Toxicology.* London: The MacMillian Company.

Howard, P. H. 1995. *Handbook of Environmental Fate and Exposures Data for Organic Chemicals*, Vol. I-IV. New York: CRC Press/Lewis Publishers.

Langerman, N. 1994 *Precautionary Labels for Chemical Containers.* New York: CRC Press/Lewis Publishers.

Lauwerys, R. R., P. Hoet. 1993. *Industrial Chemical Exposure.* New York: CRC Press/Lewis Publishers.

Lewis, R. J., Sr. 1993. *Hawley's Condensed Chemical Dictionary*, 12th Edition. New York: Van Nostrand Reinhold.

Lewis, R. J., Sr. 1992. *Sax's Dangerous Properties of Industrial Materials*, Volumes 1, 2, and 3, 8th Edition. New York: Van Nostrand Reinhold.

Manahan, S. E. 1992. *Toxicological Chemistry*, 2nd Edition. New York: CRC Press/Lewis Publishers.

Meyer, E. 1990. *Chemistry of Hazardous Materials.* Englewood Cliffs, NJ: Prentice Hall.

Patnaik, P. 1992. *A Comprehensive Guide to the Hazardous Properties of Chemical Substances.* New York: Van Nostrand Reinhold.

Rea, W. J. 1995. *Chemical Sensitivity*, Vol. I-III. New York: CRC Press/Lewis Publishers.

Richardson, M. 1995. *Dictionary of Substance and Their Effects*, Vol. I-VI. New York: CRC Press/Lewis Publishers.

Rylander, R., R. R. Jacobs. 1994. *Organic Dusts.* New York: CRC Press/Lewis Publishers.

Sacarello, H. L. A. 1994. *The Comprehensive Handbook of Hazardous Materials.* New York: CRC Press/Lewis Publishers.

U.S. Department of Health and Human Services, National Institute for Occupational Safety and Health. 1994. *NIOSH Pocket Guide to Chemical Hazards.* Washington, D.C.: U.S. Government Printing Office.

MATERIAL SAFETY DATA SHEET

CHEMICAL NAME	
PICRIC ACID	

HAZARD WARNING INFORMATION

HEALTH	FIRE	REACTIVE	OTHER	DEGREE OF HAZARD	COLOR CODING	OTHER CODES
3	4	4		0 = Minimum Hazard 1 = Slight Hazard 2 = Moderate Hazard 3 = Serious Hazard 4 = Severe Hazard	HEALTH = BLUE FIRE = RED REACTIVITY = YELLOW OTHER = WHITE	OX = Oxidizer ACID = Acid ALK = Alkali COR = Corrosive W = Use No Water

SECTION I - GENERAL INFORMATION

Characterization	RCRA Number	EPA Class
Explosive, Nitro Explosive Compound	**D003**	**Characteristic (R) Waste**

DOT Proper Shipping Name	Chemical Abstract Service (CAS) Number
Picric Acid (>10% water)	**88-89-1**
DOT Hazard Class and Label Requirements	DOT Emergency Guide Code
Class A Explosive; Flammable Solid	**33**
DOT Identification Number	Chemical Formula
UN1344	$(NO_2)_3C_6H_2OH$

Synonyms

Phenol trinitrate; 2,4,6-trinitrophenol; carbozotic acid; 2-hydroxy-1,3,5-trinitrobenzene; nitroxanthic acid; picral; picronitric acid.

SECTION II - HAZARDOUS INGREDIENTS AND IDENTITY INFORMATION

Hazardous Components (specific identity)	OSHA Exposure Criteria	NIOSH Exposure Criteria	Immediately Dangerous to Life and Health (IDLH)	ACGIH Exposure Criteria
Picric acid: (derivation: By nitration of phenolsulfonic acid, obtained by heating phenol with concentrated sulfuric acid). **1 ppm = 9.52 mg/m^3**	**PEL (skin):** **0.01 ppm** **0.1 mg/m^3** **STEL:** **Not Established**	**PEL (skin):** **0.01 ppm** **0.1 mg/m^3** **STEL:** **0.03 ppm** **0.3 mg/m^3**	**75 mg/m^3**	**TLV (skin):** **0.01 ppm** **0.1 mg/m^3** **STEL:** **0.03 ppm** **0.3 mg/m^3**

SECTION III - PHYSICAL AND CHEMICAL CHARACTERISTICS

Boiling Point	Specific Gravity (H_2O = 1)
572°F (300°C) Explodes	**1.76**
Vapor Pressure (mm Hg)	Molecular Weight (atomic weight)
1 at 383°F (195°C)	**229.1**
Vapor Density (Air = 1)	Melting Point
7.90	**252°F (122°C)**

Solubility

Very slightly soluble in water (1%). Soluble in acetone, methanol, pyridine, alcohol, benzene, chloroform, and ether.

Appearance and Odor

Yellow, odorless crystalline solid with a very bitter taste. Usually used as an aqueous solution.

SECTION IV - FIRE AND EXPLOSION HAZARD DATA

Flash Point (method used)	Explosive Limits in Air % by Volume
302°F (150°C) closed cup	**LEL: Not Determined UEL: Not Determined**
NFPA Classification	Autoignition Temperature
Combustible Solid	**572°F (300°C)**

Extinguishing Media

Water spray, fog, or foam. Do NOT use carbon dioxide or halogenated agents.

Special Fire Fighting Procedures

Poisonous gases are emitted during fire. Wear full protective clothing and self-contained breathing apparatus (SCBA). Remain clear of smoke, water fallout and water runoff. Move containers from fire area if it can be done without risk. Cool exposed containers. Fight fire from distance, if possible.

Unusual Fire and Explosion Hazards

Picric acid can decompose explosively in a fire. It forms salts, known as picrates, rather easily and many of its salts are more sensitive explosives than picric acid itself.

SECTION V - REACTIVITY DATA

Stability		Conditions to Avoid
		Picric acid is a highly unstable shock-sensitive, friction-sensitive, and impact-sensitive compound. Avoid contact or exposure to heat and incompatible materials. It should be kept wet to prevent explosions.
Stable	Unstable	Incompatibility *(materials to avoid)*
	X	Incompatible with heat, albumin, gelatin, uranium perchlorate, ammonia, concrete, plaster, salts, copper, lead, zinc, other metals, all oxidizable materials. An OSHA Class A Explosive per 29 CFR 1910.109.
Hazardous Polymerization		Conditions to Avoid Under normal conditions of temperature and pressure, hazardous polymerization of picric acid cannot occur. Avoid the generation of picric salts which are even more explosion-sensitive than picric acid.
May Occur	Will Not Occur	Hazardous Decomposition or By-products
	X	At high temperatures, picric acid emits toxic oxides of nitrogen. However, the detonation products are relatively harmless and include carbon dioxide, water vapor, nitrogen, and oxygen.

SECTION VI - HEALTH HAZARD DATA

Primary Route(s) of Entry:	Inhalation?	Absorption (skin/eye)?	Ingestion?
	X	X	X

Health Hazards

INHALATION: Symptoms include headache, nausea, vomiting, abdominal pain, lightheadedness, tremor, dypsnea, paralysis, and convulsions. High doses can cause destruction of erythrocytes, gastroenteritis, nephritis, hepatitis, and hematuria. There may be hemorrhagic kidney inflammation with decreased or ceased urination, coma, and death.

ABSORPTION: May cause mild irritation to the eyes and skin and systemic effects (see inhalation above).

INGESTION: A fatal poison by ingestion. Initial symptoms are the same as that for inhalation.

Carcinogenicity	NTP Listed?	IARC Cancer Review Group?	OSHA Regulated?	Target Organs?
Unknown Human Questioned Animal	No	No	29 CFR 1910.1000 Table Z-1	Eyes, skin, liver, kidneys, blood.

Medical Conditions Generally Aggravated by Exposure

Existing blood, liver, kidney, and skin disorders may be aggravated by exposure.

Emergency and First-aid Procedures

Eye contact: Do not allow victim to keep eyes tightly shut. Flush immediately with water for 15 minutes; seek medical attention. Skin contact: Remove all contaminated clothing. Immediately flood area with water for 15 minutes. Then wash affected areas with soap and water. Inhalation: Remove the person from exposure. Provide respiratory assistance and CPR. Transfer to medical facility, if required. If swallowed: Contact poison control center. Rinse victim's mouth with water. Unless advised otherwise, give 2-3 glasses of water to drink to dilute. Do NOT induce vomiting (aspiration hazard). Never give anything by mouth to an unconscious or convulsing person. Never try to make an unconscious or convulsing person vomit. Seek medical attention immediately. Keep victim warm and at rest.

SECTION VII - PRECAUTIONS FOR SAFE HANDLING AND USE

Steps to be Taken in Case Material is Released or Spilled

Remove all sources of ignition and ventilate area. Restrict those not involved in cleanup from entering area. Wet area thoroughly before attempting cleanup and then absorb with sodium bicarbonate and place in sealed rums for disposal.

Preferred Waste Disposal Method

Controlled incineration in a rotary kiln equipped with particulate abatement and wet scrubber device.

Precautions to be Taken in Handling and Storage

Store in tightly closed glass containers (not metal) in cool, well-ventilated area away from heat and incompatible materials. Keep as little stock as necessary. Do NOT store on concrete floors.

Other Precautions and Warnings

Use non-sparking tools and equipment and provide explosion-proof ventilation in storage areas.

SECTION VIII - CONTROL MEASURES AND PERSONAL PROTECTIVE EQUIPMENT

Respiratory Protection *(specify type)*

Use an MSHA/NIOSH-approved dust/mist respirator or, a self-contained breathing apparatus in positive pressure mode, or a supplied-air respirator operated in continuous flow or other positive pressure mode.

Ventilation

Local exhaust (preferred) at site of chemical work or general ventilation.

Protective Gloves	Eye Protection	Other Protective Clothing
Impervious Rubber (test for adequacy)	Chemical/Dust Goggles or Face Mask	Protective Apron, Boots (avoid skin contact)

Work/Hygiene Practices

Always wash hands thoroughly after using chemical; never bring food, drink, or smoking materials into vicinity of chemicals.

PICRIC ACID

$(NO_2)_3C_6H_2OH$

CAS: 88-89-1

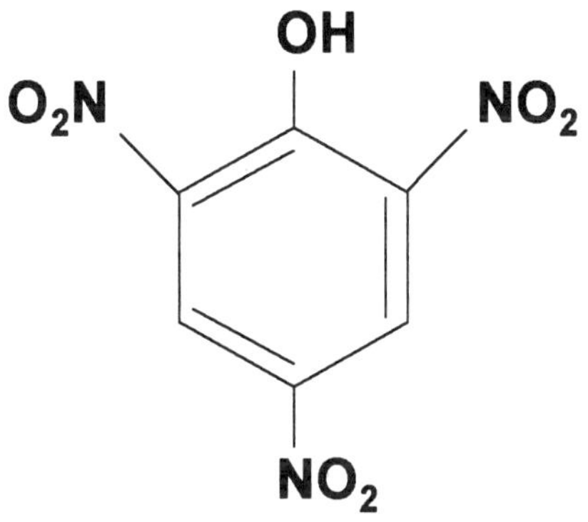

IDENTIFICATION AND TYPICAL USES

Picric acid is a yellow, odorless crystalline solid with a very bitter taste. It is usually used as an aqueous solution of greater than 10% water. It is used as an explosive and is generally mixed with appropriate amounts of dinitrophenol to minimize its corrosion of metal and the formation of shock-sensitive salts. It is also used in the making of matches, dyes for silk and wool, in electric batteries, in colored glass, in etching copper and steel, as a germicide and fungicide, as a chemical intermediate for metal picrates and other derivatives, in the leather industry, in rocket fuels, in the sensitization of photographic emulsions, in the removal of gelatin from hectograph images, in forensic chemistry, and formerly in veterinary medicine to destroy tapeworms in sheep.

RISK ASSESSMENT: HEALTH

General Assessment

Picric acid is an extremely toxic material and human poison. It can enter the body through *inhalation* of its vapors. It will also easily pass through intact skin (*absorption*) and can be *ingested* to cause toxic systemic effects. While experimental mutation data have been reported, there are no data on the carcinogenic or teratogenic properties of picric acid.

Systemic poisoning by all routes of exposure include headache, dizziness, nausea, vomiting, diarrhea, and abdominal pain. High doses can cause destruction of erythrocytes, gastroenteritis, nephritis, hepatitis, and hematuria. There can be hemorrhagic kidney inflammation and lesions with decreased or ceased urination, coma, and death. Other symptoms include yellow-stained skin and hair (not from jaundice), weakness, myalgia, anuria, polyuria, bitter taste, and other serious systemic affects.

Skin contact can cause minor, localized irritation as well as toxic systemic effects similar to that of inhalation due to picric acid's ability to pass through unbroken skin with relative ease.

Ingestion of even small amounts are acutely poisonous and can cause serious poisonings.

Acute Health Effects

The following acute (short-term) health effects may occur immediately or shortly after exposure to picric acid:

Skin: Local irritation at contact site.

Eye: Irritation and possible corneal inflammation.

Lung: Irritation of the nose, throat, and respiratory tract. The vapors cause more damage systemically and can include liver, kidney, and blood injury.

Chronic Health Effects

The following chronic (long-term) health effects may occur at some time after exposure to picric acid and can last for months or even years:

Cancer Hazards: According to information presented in the references, picric acid not be adequately tested for its ability to cause cancer in test animals. Mutation data have been reported and many scientists believe that exposure to mutagenic substances may present a cancer hazard in the long-term. Additional study is required before any determinations can be made in this regard for picric acid.

Reproductive Hazard: Picric acid has been shown to cause teratogenic effects in laboratory experiments. It is not known if it will have this effect in humans exposed to picric acid.

Other Chronic Effects: Long-term exposures can lead to the development of dermatitis with characteristic swelling, papules, vesicles, peeling, and scaling. The face is typically involved, especially around the eyes and nose.

Recommended Risk-Reduction Measures

Picric acid presents a serious health hazard risk. Personnel should avoid direct contact with picric acid. If a less toxic material or compound cannot be substituted for picric acid, then *engineering controls* are the most effective method of reducing exposures. The best protection is to enclose operations and/or provide local exhaust ventilation at the site of picric acid release. While not always operationally possible, isolating operations involving picric acid manufacturing or use can also reduce exposure. Personnel must follow proper industrial hygiene and good housekeeping work practices. Using respiratory protection is less effective than the controls mentioned above, but is still advisable whenever working with or around picric acid. The permissible exposure level (PEL) is relatively low and difficult to accurately monitor (0.1 mg/m^3). For the best protection, an MSHA or NIOSH-approved powered air-supplied respirator or a self-contained breathing apparatus (SCBA) with full facepiece and operated in pressure demand mode are recommended. If a full facepiece is not available, then chemical goggles should be worn to protect the eyes. A face shield and protective apron should also be worn. Skin contact must not occur. To prevent hand and skin exposures, impervious gloves should be used. Glove manufacturers should be contacted and permeation studies obtained *before* final glove selections have been made.

Administrative controls should also be in place to minimize the potential for human exposures. These may include written procedures or policies which specify the methods and techniques that will be practiced whenever personnel are to work with picric acid.

All personnel should receive *training* on the use, hazards, protective measures, emergency actions, and other precautions per 29 CFR 1910.1200 (Hazard Communication), prior to the first assignment in an area where picric acid is used or stored.

If symptoms develop or overexposure is suspected, the following medical testing may be useful:

- ☑ Complete blood count (CBC).
- ☑ Liver and kidney function tests.

Any evaluation should include a careful history of past and present symptoms with an examination. Medical tests that simply look for existing damage are not a substitute for controlling exposures. Also, since smoking can cause heart disease, emphysema, and other respiratory problems, personnel who smoke may show symptoms of exposure much quicker and more pronounced than those who do not. Prudent risk management requires careful consideration of *all* possible and potential exposure risk factors to ensure proper implementation of control measures and management of the exposure risk.

Other methods to reduce exposure include:

- ☑ Where possible, enclose operations and use local exhaust ventilation at the site of chemical release. If local exhaust ventilation or enclosure is not used, respiratory protection should be mandatory.
- ☑ Always ensure that proper protective clothing is worn when using chemical substances and that personnel are trained in the proper use, care, and maintenance of their protective equipment.
- ☑ Wash thoroughly immediately after exposure to picric acid and at the end of the work shift or before eating, drinking, or smoking.
- ☑ Hazard warning information should be posted in the work area. In addition, as part of an on-going education and training program, all information on the health and safety hazards of picric acid should be communicated to all exposed and potentially exposed workers.
- ☑ Eye wash stations should be provided in the immediate work area for emergency use. If there is a possibility of skin exposure to picric acid, emergency shower facilities should also be provided in the immediate work area.
- ☑ Workers whose clothing has been contaminated by picric acid should change into clean clothes *before* leaving work. Contaminated work clothing should never be worn home where family members can be contaminated. The clothing should be laundered

only by individuals who have been informed of the hazards of exposure to picric acid.

RISK ASSESSMENT: ENVIRONMENT

General Assessment

The environment is at risk of exposure during transportation, storage, disposal, or destruction of picric acid. In almost every scenario, the threat of environmental exposure is contingent upon the proper handling of the chemical substance. Accidental spills, large or small, where picric acid contacts incompatible commodities can result in violent and even explosive reactions (depending upon conditions of spill), and possible contamination of the surrounding environmental mediums (water, soil, and air).

Picric acid is considered a combustible solid (or a Class IIIB combustible liquid). It is a highly unstable, shock-sensitive, impact-sensitive, and friction-sensitive compound. It can easily detonate on exposure to heat or incompatible materials. It can corrode metal to form shock-sensitive salts, called picrates, which are more explosive-sensitive than picric acid itself. It is incompatible with a number of commonly encountered materials, such as plaster, concrete, ammonia, albumin, alkaloids, gelatin, and many metals such as copper, lead, and zinc. Therefore, extreme caution is required in handling, storage, transportation, and disposal of picric acid. Its characteristics also require special consideration during any emergency situation involving a leak or spill of picric acid.

Picric acid may enter the environment through industrial discharges, from dynamite operations, and from spills.

☠ *Acute Ecological Effects*

Acute (short-term) toxic effects may include the death of animals, birds, or fish and death or low growth rate in plants. Acute effects are seen 2 to 4 days after animals or plants are exposed to picric acid.

Picric acid has moderate acute toxicity to aquatic life. Insufficient data are available to evaluate or predict the short-term effects of picric acid to plants, land animals, and birds.

Chronic Ecological Effects

Chronic (long-term) toxic effects may include shortened life span, reproductive problems, lower fertility, and changes in appearance or behavior in exposed animals. These effects can be seen long after first exposure(s) to toxic chemicals. Picric acid has moderate chronic toxicity to aquatic life.

Insufficient data are available to evaluate or predict the long-term effects of picric acid to plants, birds, and land animals.

Water Solubility

Picric acid is moderately soluble in water. Concentrations of up to 100 milligrams may mix with a liter of water.

⌛ *Persistence in the Environment*

Picric acid is slightly persistent in the water, with a half-life between and 2 and 20 days. The half-life of a pollutant is the amount of time it takes for one-half of the chemical to be degraded.

Bioaccumulation in Aquatic Organisms

Some substances increase in concentration, or *bioaccumulate,* in living organisms as they breathe contaminated air, drink contaminated water, or eat contaminated food. These chemicals can become concentrated in the tissues and internal organs of animals as well as humans.

The concentration of picric acid found in fish tissues is expected to be somewhat less than the average concentration of picric acid in the water from which the fish was taken.

Recommended Risk-Reduction Measures

Proper training of all transporters will reduce the likelihood of a mishap or accident resulting in a leak or spill to the environment. The correct use of DOT labeling (per the MSDS) on all containers, trucks, and rail cars will enable emergency responders to react properly and quickly to any disaster thereby reducing the potential risk to the environment and to personnel.

Storage of picric acid should be segregated from incompatible materials to minimize the risk of cross-contamination or contact. Storage facilities should be equipped with the proper fire protection equipment (sprinklers, alarms, extinguishers). Galls containers should be used since picric acid corrodes metal. Facilities should be equipped with maximum explosion-proof ventilation.

If a spill or leak to the environment has occurred, fire department, emergency response and/or hazardous materials spill personnel should be notified immedi-

ately. Contaminated soils should be removed for incineration and replaced with clean soils. Cleanup should be attempted only by those trained in proper spill containment procedures using non-sparking tools. If picric acid should contact the water table, aquifer, or navigable waterway, reclamation procedures should be initiated as soon as practical. It is soluble in water and total remediation may not be possible. The local and/or state emergency response authorities must be notified. A comprehensive emergency response or disaster preparedness/recovery plan should be in place prior to any operations involving the use, transportation, storage, or disposal of picric acid. If picric acid is spilled or leaked, the following specific steps are recommended:

- ☑ Restrict persons not wearing protective clothing from area of spill or leak until cleanup is complete and area can be opened for normal work.
- ☑ Remove all ignition sources.
- ☑ Ventilate area or leak or spill.
- ☑ Spilled materials should be wetted-down and absorbed with sodium bicarbonate or a sand-soda ash mixture. The resulting mixture should be carefully scooped up and placed into glass containers, with at least 10% water content, for disposal.
- ☑ It may be necessary to dispose of picric acid as a hazardous waste. The responsible state agency or the regional office of the federal Environmental Protection Agency (EPA) should be contacted for specific recommendations.

RISK ASSESSMENT: BUSINESS

General Assessment

Accidents or mishaps involving picric acid can present a significant threat to business operations. The loss or damage of equipment or facilities can significantly affect fiscal viability. Lawsuits that may result from personnel injury/death, public exposures, and/or environmental contamination will also require a serious expenditure of resources. Media attention surrounding an injury, death, or environmental damage can also result in a loss of profits and loss of current as well as future business. Always remember that anytime the terms "reproductive hazard" or "birth defects" are used, public hysteria, ignorance, emotion, and anxiety can run equally high. This must be carefully considered when drafting or implementing public relations policies.

Recommended Risk-Reduction Measures

Company attorneys, safety and health professionals, and environmental specialists should be involved in the development of any procedures for responding to chemical incidents. A company official should be predesignated as a public relations officer with specific training in dealing with the press. Corporate plans and policies should be developed, approved, and implemented long before any need for such arises.

REFERENCES

American Conference of Governmental Industrial Hygienists. 1988. *Documentation of the Threshold Limit Values and Biological Exposure Indices,* 5th Edition (with updates). Cincinnati: ACGIH

Bloom, A. D. and F. J. DeSerres. 1995. *Ecotoxicity and Human Health.* New York: CRC Press/Lewis Publishers.

Bueche, J. R. 1972. *Principles of Physics,* 2nd Edition. New York: McGraw-Hill.

Calabrese, E. J. 1994. *Biological Effects of Low Level Exposures.* New York: CRC Press/Lewis Publishers.

Danse, I. R. 1991. *Common Sense Toxics in the Workplace.* New York: Van Nostrand Reinhold.

Environmental Protection Agency. 1983. *Chemical Hazard Information Profiles.* Washington, D.C.: U.S. EPA.

Fodden, C. R., and J. L. Weddell. 1992. *Hazardous Materials Emergency Action Data.* New York: CRC Press/Lewis Publishers.

Forseberg, K., and L. H. Keith. 1993. *Chemical; Protective Clothing Permeation and Degradation Database.* Chelsea, MI: Lewis Publishers.

Gardner, R. W. 1994. *Chemical Intolerance.* New York: CRC Press/Lewis Publishers.

Hallenbeck, W. H. 1993. *Quantitative Risk Assessment for Environmental and Occupational Health,* 2nd Edition. New York: CRC Press/Lewis Publishers.

Hodgson, E., R. B. Mailman, and Janice E. Chambers. 1988. *Dictionary of Toxicology.* London: The MacMillian Company.

Hosty, J. W., and P. Foster. 1990. *A Practical Guide to Chemical Spill Response.* New York: Van Nostrand Reinhold.

Howard, P. H. 1995. *Handbook of Environmental Fate and Exposures Data for Organic Chemicals,* Vol. I-IV. New York: CRC Press/Lewis Publishers.

Lewis, .J., Sr. 1993. *Hawley's Condensed Chemical Dictionary,* 12th Edition. New York: Van Nostrand Reinhold.

Lewis, . J., Sr. 1992. *Sax's Dangerous Properties of Industrial Materials,* Volumes 1, 2, and 3, 8th Edition. New York: Van Nostrand Reinhold.

U.S. Department of Health and Human Services, National Institute for Occupational Safety and Health. 1994. *NIOSH Pocket Guide to Chemical Hazards.* Washington, D.C.: U.S. Government Printing Office.

MATERIAL SAFETY DATA SHEET

CHEMICAL NAME

POLYCHLORINATED BIPHENYLS

HAZARD WARNING INFORMATION

HEALTH	FIRE	REACTIVE	OTHER	DEGREE OF HAZARD	COLOR CODING	OTHER CODES
2	1	0		0 = Minimum Hazard 1 = Slight Hazard 2 = Moderate Hazard 3 = Serious Hazard 4 = Severe Hazard	HEALTH = BLUE FIRE = RED REACTIVITY = YELLOW OTHER = WHITE	OX = Oxidizer ACID = Acid ALK = Alkali COR = Corrosive W = Use No Water

SECTION I - GENERAL INFORMATION

Characterization **Polychlorinated biphenyls**	RCRA Number **None**	EPA Class **Not Applicable**
DOT Proper Shipping Name **Polychlorinated biphenyls**	Chemical Abstract Service (CAS) Number **1336-36-3**	
DOT Hazard Class and Label Requirements **ORM-E; Class 9**	DOT Emergency Guide Code **15**	
DOT Identification Number **UN 2315**	Chemical Formula **Varies with specific PCB**	

Synonyms

PCBs; Aroclor; Clophen; Chlorextol; chlorinated biphenyls, chlorinated diphenyl; chlorinated diphenylene; chloro biphenyl, chloro-1,1-biphenyl, Dykanol; Fenclor; Inerteen; Kaneclor; Montar, Sovol.

SECTION II - HAZARDOUS INGREDIENTS AND IDENTITY INFORMATION

Hazardous Components (specific identity)	OSHA Exposure Criteria	NIOSH Exposure Criteria	Immediately Dangerous to Life and Health (IDLH)	ACGIH Exposure Criteria
Polychlorinated biphenyls (derivation: A class of non-polar chlorinated hydrocarbons with a biphenyl nucleus in which any or all of the hydrogen atoms have been replaced by chlorine. Commercial grades are mixtures of chlorinated biphenyl isomers with varying degrees of chlorination).	**PEL (8-hour):** **(as chlorodiphenyl)** **42% chlorine** **1 mg/m³** **54% chlorine** **0.5 mg/m³**	**PEL (10-hour):** **(as chlorodiphenyl)** **42% chlorine** **0.001 mg/m³** **54% chlorine** **0.001 mg/m³**	**Not Determined**	**TLV (8-hour):** **(as chlorodiphenyl)** **42% chlorine** **1 mg/m³** **54% chlorine** **0.5 mg/m³**

SECTION III - PHYSICAL AND CHEMICAL CHARACTERISTICS

Boiling Point **644°F to 707°F (340°C to 375°C)**	Specific Gravity (H_2O = 1) **1.3 to 1.8 at 68°F (20°C)**
Vapor Pressure (mm Hg) **1 at 100°F (38°C)**	Molecular Weight **188 to 398**
Vapor Density (Air = 1) **Varies With Grade**	Melting Point **42 %: 2°F (-19°C) 54%: 14°F (-10°C)**

Solubility

Very slightly soluble in water, glycerol, and glycol. Soluble in most organic solvents, oils, and fats.

Appearance and Odor

Varies from mobile oily liquids, to white crystalline solids, to hard non-crystalline resins.

SECTION IV - FIRE AND EXPLOSION HAZARD DATA

Flash Point (method used) **286°F to 385°F (141°C to 196 °C) opened cup**	Explosive Limits in Air % by Volume **LEL: Not Determined UEL: Not Determined**
NFPA Classification **Combustible Liquid or Solid**	Autoignition Temperature **464°F (240°C)**

Extinguishing Media

Some PCBs are non-combustible liquids or solids. Use extinguishing agent suitable to surrounding fire. Do NOT use direct stream of water (may scatter PCB fires and make them more difficult to fight).

Special Fire Fighting Procedures

Wear full protective clothing and self-contained breathing apparatus (SCBA). Poisonous gases are produced in fire. Use water spray to keep fire-exposed containers cool.

Unusual Fire and Explosion Hazards

The products of combustion (hydrogen chloride, phosgene, polychlorinated dibenzofurans, and furans) are more toxic than the PCBs themselves.

SECTION V - REACTIVITY DATA

Stability		Conditions to Avoid
		Polychlorinated biphenyls are usually very stable materials under routine conditions of handling and storage. However, they are subject to photodechlorination on exposure to sunlight or UV radiation.
Stable	**Unstable**	Incompatibility *(materials to avoid)*
X		PCBs are chemically inert and, therefore, resistant to oxidation, acids, and bases. There are no reports of PCBs undergoing violent or explosive reactions.
Hazardous Polymerization		Conditions to Avoid
		Under normal conditions of temperature and pressure, hazardous polymerization of polychlorinated biphenyls cannot occur.
May Occur	**Will Not Occur**	Hazardous Decomposition or By-products
	X	When heated to decomposition, polychlorinated biphenyls can produce highly toxic and poisonous gases and vapors, including hydrogen chloride, phosgene, polychlorinated dibenzofurans, and furans.

SECTION VI - HEALTH HAZARD DATA

Primary Route(s) of Entry:	Inhalation?	Absorption (skin)?	Ingestion?
	X	X	X

Health Hazards

INHALATION: Potent liver toxin. Inhaled vapors or mists may cause severe irritation to the skin, eyes, nose, throat, and upper respiratory tract with a possibility for eye, lung, and liver injury. Systemic symptoms include nausea, vomiting, hypertension, fatigue, weight loss, yellow jaundice, edema, and abdominal pains. CNS effects also likely.

ABSORPTION: Causes acne-like symptoms, thickening of the skin and nails, systemic poisoning (due to absorption). Eye contact causes serious damage with swelling of the eyelids, excessive discharge, and possible blindness.

INGESTION: Nausea, vomiting, abdominal pain, and anorexia. Possible neurological effects.

Carcinogenicity	NTP Listed?	IARC Cancer Review Group?	OSHA Regulated?	Target Organs?
Probable Human Confirmed Animal	5th Annual Report	Group 2A	29 CFR 1910.1000 Table Z-1-A	Respiratory system, eyes, skin, liver, mucosa.

Medical Conditions Generally Aggravated by Exposure

Chronic respiratory disorders, pre-existing skin disorders, or liver disease.

Emergency and First-aid Procedures

Eye contact: Flush immediately with water (15 minutes minimum); seek medical attention. Rinsing with medical oils may remove PCBs. Skin contact: Remove all contaminated clothing. Use decontaminating agent such as PEG (polyethyleneglycol-300) on affected area or, if PEG is not available, immediately wash area with large amounts of soap and water. Seek medical assistance. For inhalation: Remove the person from exposure. Provide respiratory assistance and CPR, if required. If swallowed: Seek medical attention immediately. Never give anything by mouth to an unconscious/convulsing person.

SECTION VII - PRECAUTIONS FOR SAFE HANDLING AND USE

Steps to be Taken in Case Material is Released or Spilled

Absorb liquids in vermiculite or other material and deposit in sealed drum. Collect powdered spills in safest manner possible. Do NOT dry sweep. Ventilate area of spill or leak. Restrict those not involved in cleanup from entering area. Remove all sources of ignition.

Preferred Waste Disposal Method

Incineration with scrubbing; high-efficiency boilers; landfills. Disposal methods must meet EPA criteria.

Precautions to be Taken in Handling and Storage

Store to avoid contact with incompatible materials in tightly closed containers in cool, well-ventilated area away from heat and flame.

Other Precautions and Warnings

Prevent damage to containers. PCBs are carcinogens; employees must be trained on its hazards.

SECTION VIII - CONTROL MEASURES AND PERSONAL PROTECTIVE EQUIPMENT

Respiratory Protection *(specify type)*

For high exposures, use an MSHA/NIOSH-approved supplied-air respirator with full facepiece in continuous flow mode, or use a SCBA operated in pressure demand or other positive pressure mode.

Ventilation

Local exhaust or general mechanical systems recommended.

Protective Gloves	Eye Protection	Other Protective Clothing
Butyl Rubber, Neoprene, Teflon	Chemical/Dust Goggles or Face Mask	Protective Apron

Work/Hygiene Practices

Always wash hands thoroughly after using chemical; never bring food, drink, or smoking materials into vicinity of chemicals.

POLYCHLORINATED BIPHENYLS

Formula varies with grade

CAS: 1336-36-3

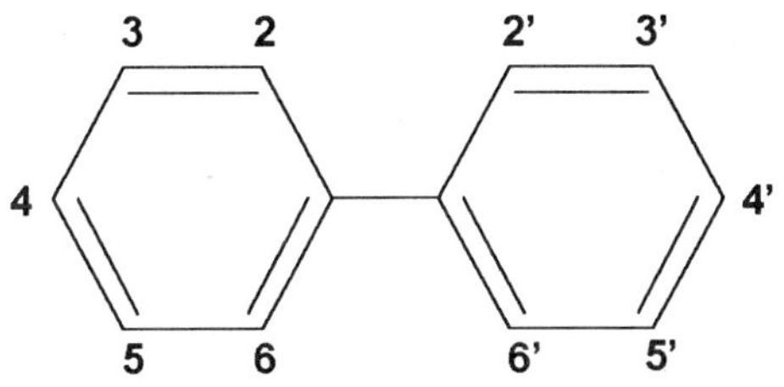

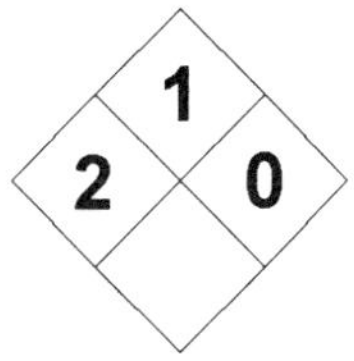

IDENTIFICATION AND TYPICAL USES

Polychlorinated biphenyls (PCBs) vary in form from a mobile, oily liquid, to white crystalline solids, to non-crystalline resins, depending upon the chlorine content. Because of their excellent physical and chemical properties, such as high thermal and chemical stability and dielectric constant, boiling point, and flame resistance, PCBs were once widely used in transformer oils. They are also used as plasticizers, adhesives in hydraulic oils, in paints, printing inks, heat transfer systems, lubricants, cutting oils, asphalt, brake linings, automobile body sealants, synthetic rubber, floor tile, wax extenders, pesticide extenders, carbonless reproducing paper, and fire retardants. However, because of their carcinogenic properties, current production of these compounds has been drastically curtailed.

RISK ASSESSMENT: HEALTH

General Assessment

PCBs are moderately toxic substances by *inhalation, ingestion,* and skin *absorption.* They are severely irritating to the eyes, skin, and respiratory system. Occupational exposure to PCBs causes a wide range of adverse health effects on the skin, eyes, mucosa, and digestive and neurological systems. Human mutation data have been reported for PCBs. It is a confirmed carcinogen in animals and a probable carcinogen in humans. It is also an experimental teratogen with adverse reproductive effects reported. PCBs are potent liver toxins.

Inhalation of the mists or vapors causes severe irritation to the eyes, nose, throat, and upper respiratory tract. High exposures can cause eye, lung, and liver injury. Systemic effects include nausea, vomiting, hypertension, fatigue, weight loss, jaundice, edema, and abdominal pains. There may also be cognitive, neurobehavioral and psychomotor impairment, and memory loss. PCBs can accumulate in fatty tissue and severe health effects can develop later. Toxicity can increase with chronic exposure and can become severe in combination with other chlorinated compounds, such as carbon tetrachloride.

In addition to the systemic symptoms noted above, skin contact can cause serious localized irritation and inflammation, possible skin eruptions, and dermatitis. There may be dryness, thickening of the skin and darkening of the nails. If PCBs should contact the eyes, there may be serious damage including swelling and burning and excessive eye discharge.

Ingestion of PCBs can cause gastrointestinal irritation and the systemic symptoms noted for inhalation.

☠ *Acute Health Effects*

The following acute (short-term) health effects may occur immediately or shortly after exposure to PCBs:

Skin: Severe irritation and possible burns. Will pass through unbroken skin to cause toxic systemic effects.

Eye: Severe irritation with damage and possible loss of vision likely.

Lung: Severe irritation to the respiratory tract and associated mucosa. Can cause weakness, nausea, confusion, collapse, coma, and death (respiratory failure).

CNS: A CNS depressant. Weakness, confusion, mood changes and other personality changes, nausea, collapse, and death.

Chronic Health Effects

The following chronic (long-term) health effects may occur at some time after exposure to PCBs and can last for months or even years:

Cancer Hazards: PCBs are probable carcinogenic compounds in humans. There is some evidence that they cause skin cancer in humans and they have been shown to cause liver cancer in animals.

Reproductive Hazard: PCBs may be teratogens in humans since they have been shown to be teratogenic in animals (abnormal pigmentation, abortions, stillbirths, and underweight births). PCBs can pass to a child through the mother's breast milk. PCBs have also been shown to adversely affect the reproductive systems in adults.

Other Chronic Effects: Repeated exposures can cause liver damage. PCBs can cause a severe acne-like rash (chloracne) with redness, swelling, and burning which may persist for years. High exposures can damage the nervous system, causing numbness, weakness, joint and muscle pains, liver enlargement, menstrual changes in women, chronic bronchitis, and tingling ("pins and needles") in the arms and legs.

Recommended Risk-Reduction Measures

Personnel should avoid direct contact with PCBs. They are confirmed carcinogens in animals and probable carcinogens in humans. If a less toxic chemical cannot be substituted for a hazardous substance, then *engineering controls* are the most effective method of reducing exposures. The best protection is to enclose operations and/or provide local exhaust ventilation at the site of chemical release. While not always operationally feasible, isolating operations can also reduce exposure. Using respiratory protection is less effective than the controls mentioned above, but is still advisable whenever working with or around PCBs. For exposures to confirmed carcinogenic compounds, the best protection is obtained using an MSHA/NIOSH-approved supplied-air respirator with full facepiece operated in continuous flow mode, or a self-contained breathing apparatus (SCBA) with full facepiece operated in pressure demand. If a full facepiece is not available, then chemical goggles should be worn to protect the eyes. Whenever a chemical splash hazard exists, a face shield and a rubber apron should be worn. To prevent hand and skin exposures, impervious butyl rubber gloves should be used. Butyl rubber, neoprene, Teflon, and fluorocarbon rubber appear to provide adequate protection. Glove manufacturers should be contacted and permeation studies obtained *before* glove selections have been made.

Administrative controls should also be in place to minimize the potential for human exposures. These may include written procedures or policies which specify the methods and techniques that will be practiced whenever personnel are to work with PCBs.

All personnel should receive *training* on the use, hazards, protective measures, emergency actions, and other precautions per 29 CFR 1910.1200 (Hazard Communication), prior to the first assignment in an area where PCBs are used or stored. Before beginning employment with PCBs and at regular intervals thereafter (e.g., annually), those persons with high or frequent exposures should be provided the following recommended medical tests:

- ☑ Serum triglycerides level.
- ☑ Examination of the skin for rash or other acne-like developments.

If symptoms develop or overexposure is suspected, the following tests are recommended:

- ☑ Blood PCBs levels.
- ☑ Liver enzyme tests
- ☑ Nerve conduction studies should be considered.

Any evaluation should include a careful history of past and present symptoms with an examination. Medical tests that simply look for existing damage are not a substitute for controlling exposures. Also, because more than light alcohol consumption can cause liver damage, drinking alcohol can increase the potential liver damage caused by exposure to PCBs.

Other methods to reduce exposure include:

- ☑ Where possible, enclose operations and use local exhaust ventilation at the site of chemical release. If local exhaust ventilation or enclosure is not used, respiratory protection should be mandatory.
- ☑ Also, if possible, liquids should be transferred automatically from storage containers to process containers.
- ☑ Always ensure that proper protective clothing is worn when using chemical substances and that personnel have been trained on the use, care, and maintenance of personal protective equipment.
- ☑ Wash thoroughly immediately after exposure to PCBs and at the end of the work shift or before eating, drinking, applying cosmetics, or smoking.
- ☑ Hazard warning information should be posted in the work area. In addition, as part of an on-going education and training program, all information on the health and safety hazards of PCBs should be communicated to all potentially exposed workers.

- ☑ Eye wash stations should be provided in the immediate work area for emergency use. If there is a possibility of skin exposure to PCBs, emergency shower facilities should also be provided.
- ☑ Personnel should never wear contaminated clothing home (family members can be exposed). Clothing should be laundered by personnel who have been trained on the hazards of PCBs.

RISK ASSESSMENT: ENVIRONMENT

General Assessment

The environment is at risk of exposure during transportation, storage, disposal, or destruction of PCBs. In almost every scenario, the threat of environmental exposure is contingent upon the proper handling of the chemical substance. Accidental spills, large or small, can result in fire, explosion, and possible contamination of the surrounding environmental mediums (water, soil, and air).

PCBs are considered combustible solids or liquids. They are chemically inert and, therefore, resistant to oxidizers, acids, and bases. However, their thermal decomposition by-products, such as hydrogen chloride, phosgene, polychlorinated dibenzofurans, and furans, are more toxic and hazardous than the PCBs themselves. These characteristics require caution in handling, storage, transportation, and disposal. Special consideration is required during any emergency situation involving a leak or spill of PCBs.

PCBs may enter the environment from leakage from industrial and electrical equipment, from industrial discharges, spills, leaching from municipal landfills, and from previously contaminated sediments.

Acute Ecological Effects

Acute (short-term) toxic effects may include the death of animals, birds, or fish and death or low growth rate in plants. Acute effects are seen 2 to 4 days after animals or plants are exposed to PCBs.

PCBs have high acute toxicity to aquatic life. Insufficient data are available to evaluate or predict the short-term effects of PCBs to plants, birds, or land animals.

Chronic Ecological Effects

Chronic (long-term) toxic effects may include shortened life span, reproductive problems, lower fertility, and changes in appearance or behavior in exposed animals. These effects can be seen long after first exposure(s) to toxic chemicals.

PCBs have high chronic toxicity to aquatic life. Insufficient data are available on the long-term effects of PCBs to plants, birds, or land animals.

Water Solubility

PCBs are slightly soluble in water. Concentrations of less than 1 milligram will mix with a liter of water.

Persistence in the Environment

The relative distribution of the various PCBs depends on the level of chlorination. Some PCBs will probably be highly persistent in water, with half-lives greater than 200 days. Potential PCB distribution in the various environmental mediums can have the following ranges, depending on degree of chlorination: Air, 0 to 34%; terrestrial soils, 33 to 52%; water, 0 to 1.8%; suspended solids, 0.05 to 0.08%; aquatic biota, 0.02 to 0.03%; aquatic sediments, 30 to 50%.

Bioaccumulation in Aquatic Organisms

Some substances increase in concentration, or *bioaccumulate,* in living organisms as they breathe contaminated air, drink contaminated water, or eat contaminated food. These chemicals can become concentrated in the tissues and internal organs of animals as well as humans.

The concentration of PCBs found in fish tissue is expected to be considerably higher than the average concentration of PCBs in the water from which the fish was taken.

Recommended Risk-Reduction Measures

Proper training of all transporters will reduce the likelihood of a mishap or accident resulting in a leak or spill to the environment. The correct use of DOT labeling (per the MSDS) on all containers, trucks, and rail cars will enable emergency responders to react properly and quickly to any disaster thereby reducing the potential risk to the environment and to personnel.

Storage of PCBs should be segregated from other chemicals to minimize the risk of cross-contamination or contact. Buildings designated for storage should be equipped with appropriate fire protection systems (alarms, sprinklers, emergency lighting, portable extinguishers).

If a spill or leak to the environment has occurred, fire department, emergency response and/or hazardous materials spill personnel should be notified immediately. Cleanup should be attempted only by those trained in proper spill containment procedures. Contaminated soils should be removed for incineration and replaced with clean soil. If PCBs should contact the water table, aquifer, or navigable waterway, time is of the essence. They are slightly soluble in the aquatic environment and total remediation may not be possible. The local and/or state emergency response authorities must be notified. A comprehensive emergency response or disaster preparedness/recovery plan should be in place prior to any operations involving the use, transportation, storage, or disposal of PCBs. If PCBs are spilled or leaked, the following specific steps are recommended:

- ☑ Restrict persons not wearing protective clothing from area of spill or leak until cleanup is complete and area can be opened for normal work.
- ☑ Remove all ignition sources.
- ☑ Ventilate area of leak.
- ☑ Absorb liquid spills in vermiculite, dry sand, earth, or similar material and deposit in sealed containers. Collect solid materials in safest manner possible and place in sealed containers. Do NOT dry sweep (creates airborne dusts). Use vacuum equipped with a high-efficiency particulate air (HEPA) filter instead.
- ☑ It may be necessary to dispose of PCBs as a hazardous waste. The responsible state agency or the regional office of the federal Environmental Protection Agency (EPA) should be contacted for specific recommendations.

RISK ASSESSMENT: BUSINESS

General Assessment

Accidents or mishaps involving PCBs can present a significant threat to business operations. The loss or damage of equipment or facilities can significantly affect fiscal viability. Lawsuits that may result from personnel injury/death, public exposures, and/or environmental contamination will also require a serious expenditure of resources. Media attention surrounding an injury, death, or environmental damage can also result in a loss of profits and loss of current as well as future business. Always remember that anytime the terms "cancer," "carcinogen," or "reproductive hazard" are used, public hysteria, emotion, anxiety, and ignorance can run equally high. This must be carefully considered whenever drafting or implementing public relations policies.

Recommended Risk-Reduction Measures

Company attorneys, safety and health professionals, and environmental specialists should be involved in the development of any procedures dealing with safety or emergency response. A company official should be pre-designated as a public relations officer with specific training in dealing with the press. Corporate plans and policies should be developed, approved, and implemented long before any need for such arises.

REFERENCES

Asante-Duah, D. K. 1993. *Hazardous Waste Risk Assessment.* New York: CRC Press/Lewis Publishers.

Bloom, A. D. and F. J. DeSerres. 1995. *Ecotoxicity and Human Health.* New York: CRC Press/Lewis Publishers.

Bueche, J. R. 1972. *Principles of Physics,* 2nd Edition. New York: McGraw-Hill.

Calabrese, E. J. 1994. *Biological Effects of Low Level Exposures.* New York: CRC Press/Lewis Publishers.

Cockerman, L. G., B. S. Shane. 1994. *Basic Environmental Toxicology.* New York: CRC Press/Lewis Publishers.

Cothern, C. R. 1995. *Handbook for Environmental Risk Decisionmaking.* New York: CRC Press/Lewis Publishers.

Danse, I. R. 1991. *Common Sense Toxics in the Workplace.* New York: Van Nostrand Reinhold.

Environmental Protection Agency. 1983. *Chemical Hazard Information Profiles.* Washington, D.C.: U.S. EPA.

Fodden, C. R., and J. L. Weddell. 1992. *Hazardous Materials Emergency Action Data.* New York: CRC Press/Lewis Publishers.

Forseberg, K., and L. H. Keith. 1993. *Chemical; Protective Clothing Permeation and Degradation Database.* Chelsea, MI: Lewis Publishers.

Gardner, R. W. 1994. *Chemical Intolerance.* New York: CRC Press/Lewis Publishers.

Lewis, .J., Sr. 1993. *Hawley's Condensed Chemical Dictionary,* 12th Edition. New York: Van Nostrand Reinhold.

Lewis, R. J., Sr. 1992. *Sax's Dangerous Properties of Industrial Materials,* Volumes 1, 2, and 3, 8th Edition. New York: Van Nostrand Reinhold.

Patnaik, P. 1992. *A Comprehensive Guide to the Hazardous Properties of Chemical Substances.* New York: Van Nostrand Reinhold.

U.S. Department of Health and Human Services, National Institute for Occupational Safety and Health. 1994. *NIOSH Pocket Guide to Chemical Hazards.* Washington, D.C.: U.S. Government Printing Office.

MATERIAL SAFETY DATA SHEET

CHEMICAL NAME

POTASSIUM

HAZARD WARNING INFORMATION

HEALTH	FIRE	REACTIVE	OTHER	DEGREE OF HAZARD	COLOR CODING	OTHER CODES
3	1	2	~~W~~	0 = Minimum Hazard 1 = Slight Hazard 2 = Moderate Hazard 3 = Serious Hazard 4 = Severe Hazard	HEALTH = BLUE FIRE = RED REACTIVITY = YELLOW OTHER = WHITE	OX = Oxidizer ACID = Acid ALK = Alkali COR = Corrosive ~~W~~ = Use No Water

SECTION I - GENERAL INFORMATION

Characterization **Metal (mineral)**	RCRA Number **D001**
	EPA Class **Characteristic (I) Waste**
DOT Proper Shipping Name **Potassium Metal**	Chemical Abstract Service (CAS) Number **7440-09-7**
DOT Hazard Class and Label Requirements **Flammable Solid & Dangerous When Wet**	DOT Emergency Guide Code **No Citation**
DOT Identification Number **UN 2257**	Atomic Symbol **K**

Synonyms
Kalium

SECTION II - HAZARDOUS INGREDIENTS AND IDENTITY INFORMATION

Hazardous Components (specific identity)	OSHA Exposure Criteria	NIOSH Exposure Criteria	Immediately Dangerous to Life and Health (IDLH)	ACGIH Exposure Criteria
Potassium (derivation: Occurs naturally in many parts of the world. Also produced by thermochemical distillation of potassium chloride with sodium).	**PEL:** Not Established **STEL:** Not Established	**REL:** Not Established **STEL:** Not Established	**Not Determined**	**TLV:** Not Established **STEL:** Not Established

SECTION III - PHYSICAL AND CHEMICAL CHARACTERISTICS

Boiling Point **1418°F (770°C)**	Specific Gravity (H_2O = 1) **0.86**
Vapor Pressure (mm Hg) **Not Found**	Atomic Weight **39.10**
Vapor Density (Air = 1) **Not Found**	Melting Point **147°F (64°C)**

Solubility
Reacts violently in water. Soluble in liquid ammonia, aniline, mercury, sodium, and ethylenediamine.

Appearance and Odor
Silvery white, malleable, rapidly oxidized metal (color changes to gray on oxidation). Odorless.

SECTION IV - FIRE AND EXPLOSION HAZARD DATA

Flash Point (method used) **Not Found**	Explosive Limits in Air % by Volume **LEL: Not Found** **UEL: Not Found**
NFPA Classification **Flammable Solid**	Autoignition Temperature **Not Found**

Extinguishing Media
Use G-1® graphite powder, dry graphite, powdered sodium chloride, soda ash, or powdered talc. Do NOT use water, carbon dioxide, or other common extinguishing agents.

Special Fire Fighting Procedures
Wear full protective clothing and self-contained breathing apparatus (SCBA). Poisonous gases are produced in fire. Do NOT look directly into potassium fires. Fight fire from distance, if possible.

Unusual Fire and Explosion Hazards
Extremely dangerous when heated. May react with moisture or acids to evolve explosive hydrogen gas. Finely divided flakes or particles in air can ignite readily by spark or flame and burn rapidly. It may form reactive and explosive peroxides at room temperature, even when stored under an oil blanket. Extremely explosive and violently reactive metal. Burning potassium is difficult to extinguish.

SECTION V - REACTIVITY DATA

Stability		Conditions to Avoid **Potassium is not stable and can easily undergo violent chemical change. It reacts explosively with water and may react spontaneously in air. Keep away from heat, moisture, and incompatible materials.**
Stable	Unstable **X**	Incompatibility *(materials to avoid)* **Incompatible with air, moist air, heavy metal compounds (silver oxide, silver chloride), halogenated hydrocarbons, water, alkalis, carbon dioxide, petroleum, ether, benzene, carbon monoxide, halogens, and acids.**
Hazardous Polymerization		Conditions to Avoid **Hazardous polymerization of potassium will not occur. May ignite spontaneously in air or moisture.**
May Occur	Will Not Occur **X**	Hazardous Decomposition or By-products **Poisonous gases are emitted during fire, including toxic and irritating smoke and fumes of potassium oxide, potassium hydroxide (a corrosive), and flammable hydrogen gas.**

SECTION VI - HEALTH HAZARD DATA

Primary Route(s) of Entry:	Inhalation?	Absorption (skin/eye)?	Ingestion?
	X	X	X

Health Hazards

INHALATION: The dusts can produce irritation of the eyes and mucous membranes. It may react with the water content in the bronchial passages to form caustic potassium hydroxide which can damage the lungs. The dusts or fumes cause shortness of breath, nausea and vomiting, headache, and nose and respiratory tract irritation (pneumonitis). Moisture of the mucosa may cause potassium to ignite or explode causing possibly fatal thermal burns.

ABSORPTION: Corrosive to all body tissues. May cause burns, ulceration, tissue death, and absorption. Causes severe eye damage and possible blindness. The powder metal ignites easily on moist skin and eyes causing serious burns.

INGESTION: Causes severe pain, vomiting, diarrhea, and possible collapse.

Carcinogenicity	NTP Listed?	IARC Cancer Review Group?	OSHA Regulated?	Target Organs?
Unknown Human Unknown Animal	No	No	No	Respiratory system, eyes, skin, mucous membranes.

Medical Conditions Generally Aggravated by Exposure

Possible respiratory system disorders and/or skin conditions may be aggravated by exposure.

Emergency and First-aid Procedures

Eye contact: Do not allow victim to keep eyes tightly shut. Flush immediately with water for 15 minutes (minimum); seek medical attention. Skin contact: Remove all contaminated clothing. Immediately wash area with flooding amounts of water. For reddened or blistered skin, seek medical assistance. For inhalation: Remove the person from exposure. Provide respiratory assistance and CPR. If swallowed: Seek medical attention. If victim is conscious, give gastric lavage followed by olive oil or demulcent. Never give anything by mouth to an unconscious or convulsing person.

SECTION VII - PRECAUTIONS FOR SAFE HANDLING AND USE

Steps to be Taken in Case Material is Released or Spilled

Collect powdered materials and deposit in clean, dry, sealed, containers. Do NOT dry sweep. Use HEPA vacuum. Ventilate area. Restrict those not involved in cleanup from area. Do NOT use water.

Preferred Waste Disposal Method

Reclaim scrap.

Precautions to be Taken in Handling and Storage

Store to avoid contact with incompatible materials. Prevent physical damage to containers. Keep containers tightly closed. Keep moisture and sources of heat away. Use non-sparking tools.

Other Precautions and Warnings

Avoid excessive dust generation in storage and handling areas. Store under an inert gas blanket (nitrogen or argon) or submerged in naphtha, kerosene, or other oils.

SECTION VIII - CONTROL MEASURES AND PERSONAL PROTECTIVE EQUIPMENT

Respiratory Protection *(specify type)*

For dust exposures, use a respirator with a HEPA filter; otherwise, use a supplied-air respirator or a self-contained breathing apparatus with full facepiece operated in positive pressure mode.

Ventilation

Local exhaust (preferred) at site of chemical work, or general dilution ventilation.

Protective Gloves	Eye Protection	Other Protective Clothing
Impervious Material	Chemical/Dust Goggles or Face Mask	Protective Apron, Boots

Work/Hygiene Practices

Always wash hands thoroughly after using chemical; never bring food, drink, or smoking materials into vicinity of chemicals.

POTASSIUM

K

CAS: 7440-09-7

IDENTIFICATION AND TYPICAL USES

Potassium appears as a silvery-white, soft, malleable, odorless metal. It is rapidly oxidized in moist air and the color changes to gray. It is used in the preparation of potassium peroxide, heat-exchange alloys, as a laboratory reagent, in seeding of combustion gases in magnetohydrodynamic generators, and as a component of fertilizers (as potassium chloride).

RISK ASSESSMENT: HEALTH

General Assessment

The toxicity of potassium metal in humans is almost always attributed to the anion rather than to potassium. Nevertheless, it is toxic to humans on *inhalation*, dermal *absorption*, and *ingestion*. There are no reports of carcinogenic, teratogenic, or mutagenic effects due to chronic exposure to potassium.

Inhalation of the dusts and particles of potassium can irritate the eyes and nasal passages. Potassium may react with the water content in the bronchial passages to form caustic potassium hydroxide which can cause severe burns, ulceration, and tissue damage. Potassium fumes in the oxide form can also irritate the eyes, nose, throat, and respiratory tract and may cause coughing, tightness and pressure in the chest, nausea, vomiting, shortness of breath, and possible pneumonitis. It may react violently or even explosively in contact with moist mucosa, wet eyes, or even wet (sweat) skin causing thermal burns and tissue destruction.

Ingestion will cause severe pain, vomiting, diarrhea, and possible collapse with loss of consciousness. Gastrointestinal effects may be severe and possibly life-threatening.

Acute Health Effects

The following acute (short-term) health effects may occur immediately or shortly after human exposure to potassium:

Skin: May cause serious irritation and even burns, especially if the skin is moist. In fact, it may react explosively on wet skin.

Eye: Severe irritation and possible burns due to the eyes' moisture content. Damage can be serious and permanent.

Lung: Irritation of the mucosa of the eyes, nose, throat, and respiratory tract. There may be shortness of breath, nausea, headache, and nose and respiratory tract irritation.

Other: Ingestion causes severe pain, vomiting, diarrhea, and other gastrointestinal effects. There may be collapse and loss of consciousness.

Chronic Health Effects

The following chronic (long-term) health effects may occur at some time after exposure to potassium and can last for months or even years:

Cancer Hazards: According to the references, potassium has not been adequately tested for its ability to cause cancer in animals.

Reproductive Hazard: According to the references, there are no reports of any teratogenic or adverse reproductive effects resulting from occupational exposure to potassium.

Other Chronic Effects: Long-term exposures may cause chronic dermatitis, mucous membrane irritation, and possible lung damage. Very irritating substances can damage the lungs. It is not known if exposure to potassium will have this effect.

Recommended Risk-Reduction Measures

Potassium is extremely corrosive and destructive to human tissues. Personnel should, therefore, avoid direct contact with potassium. If a less toxic material or compound cannot be substituted for potassium, then *engineering controls* are the most effective method of reducing exposures. The best protection is to enclose operations and/or provide local exhaust ventilation at the site of potassium dust release. While not always operationally feasible, isolating operations can also reduce exposure. Using respiratory protection is less effective than the controls mentioned above, but is still advisable whenever working with or around potassium. No exposure limits have been established specifically for potassium metal. This does not mean that exposure is without a health risk. For low or transient exposures, a chemical cartridge respirator equipped with a high-efficiency particulate air (HEPA) filter and a dust/mist pre-filter may provide adequate respiratory protection. However, better protection is obtained using a supplied-air respirator operated in continuous flow mode or a self-contained breathing apparatus (SCBA) with full facepiece operated in pressure demand. If a full facepiece is not available, then chemical/dust goggles should be worn to protect the eyes. A face shield and protective apron should also be worn. To prevent hand and skin exposures, gloves should be used. Glove manufacturers should be contacted and permeation studies obtained *before* final glove selections are made.

Administrative controls should also be in place to minimize the potential for human exposures. These may include written procedures or policies which specify the methods and techniques that will be practiced whenever personnel are to work with potassium.

All personnel should receive *training* on the use, hazards, protective measures, emergency actions, and other precautions per 29 CFR 1910.1200 (Hazard Communication), prior to the first assignment in an area where potassium is used or stored.

Before beginning employment working with potassium, and at regular intervals thereafter (e.g., annually), those personnel with frequent or potentially high exposures should be provided the following recommended medical tests:

☑ Lung function tests (establish baseline).

If symptoms develop or overexposure is suspected, consider the following additional tests:

☑ Lung function tests (compare to baseline).

☑ Evaluation by a qualified allergist with special testing and careful consideration of exposure history (may help diagnose skin allergy).

☑ Examination of the nasal passages for the presence of ulcers or signs of ulceration and other indications of corrosive tissue destruction.

Any evaluation should include a careful history of past and present symptoms with an examination. Medical tests that simply look for existing damage are not a substitute for controlling exposures. Also, since smoking can cause heart disease as well as lung cancer, emphysema, and other known respiratory problems, smokers exposure to potassium may experience symptoms more rapidly than non-smokers under the same conditions of exposure. Prudent risk management requires careful consideration of *all* factors which may be causing exposure symptoms in the workplace.

Other methods to reduce exposure include:

☑ Where possible, enclose operations and use local exhaust ventilation at the site of chemical release. If local exhaust ventilation or enclosure is not used, respiratory protection should be mandatory.

☑ Always ensure that proper protective clothing is worn when using chemical substances. Proper maintenance and care of protective equipment is essential to its proper function.

☑ Wash thoroughly immediately after exposure to potassium and at the end of the work shift or before eating, drinking, smoking, or applying cosmetics.

☑ Hazard warning information should be posted in the work area. In addition, as part of an on-going education and training program, all information on the health and safety hazards of potassium should be communicated to all exposed and potentially exposed workers.

☑ Eye wash stations should be provided in the immediate work area for emergency use. If there is a possibility of skin exposure to potassium, emergency shower facilities should be provided.

☑ Workers whose clothing has been contaminated by potassium should change into clean clothes before leaving work. Contaminated work clothing should be laundered only by individuals who have been informed of the hazards of exposure to potassium.

RISK ASSESSMENT: ENVIRONMENT

General Assessment

As a naturally occurring mineral, potassium poses very little threat to the environment. However, accidental spills or releases to the environment can occur during transportation, storage, disposal, or destruction of potassium. In almost every scenario, the threat of environmental exposure is contingent upon the proper handling of the chemical substance. Accidental spills, large or small, where potassium contacts incompatible commodities can result in release of toxic products (depending upon conditions of spill), and possible contamination of the surrounding environmental mediums (water, soil, and air).

Potassium is considered a flammable solid. It is an extremely corrosive and highly reactive metal. It can explode in air, in contact with moisture, oxidizing materials, or in contact with an ignition source. It reacts with water, moisture, and acids to evolve explosive hydrogen gas and corrosive potassium hydroxide. It is also incompatible with many other materials, including air, heavy metal compounds (silver oxide and silver chloride), halogenated hydrocarbons and chlorinated solvents, water, alkalis, carbon dioxide, petroleum, ether, benzene, carbon monoxide, halogens, and acids. Extreme caution is required in handling, storage, transportation, and disposal of potassium. These characteristics also require special consideration during any emergency situation involving a leak or spill of potassium powder or dust. Should potassium ever come into contact with these various incompatible substances either during use, transportation, storage, or disposal, violent and even explosive reactions are possible.

Potassium occurs naturally in the environment (terrestrial soils and aquatic sediments) and may enter the environment as a result of industrial and municipal waste treatment plant discharges and from spills.

Acute Ecological Effects

Acute (short-term) toxic effects may include the death of animals, birds, or fish and death or low growth rate in plants. Acute effects are seen 2 to 4 days after animals or plants are exposed to potassium.

Insufficient data are available to evaluate the short-term effects of potassium to plants, birds, or land animals. However, it should be noted that the U.S. Environmental Protection Agency does NOT list potassium as a hazardous substance.

Chronic Ecological Effects

Chronic (long-term) toxic effects may include shortened life span, reproductive problems, lower fertility, and changes in appearance or behavior in exposed animals. These effects can be seen long after first exposure(s) to toxic chemicals.

Insufficient data are available to evaluate or predict the long-term effects of potassium to plants, birds, or land animals. However, it should be noted that the U.S. Environmental Protection Agency does NOT list potassium as a hazardous substance.

Water Solubility

Potassium reacts violently in water to form potassium hydroxide and hydrogen gas.

Persistence in the Environment

Because potassium reacts so rapidly (and violently) in water, it is not expected to be persistent in water. Its half-life has not been established. The half-life of a pollutant is the amount of time it takes for one half of the chemical to be degraded.

Bioaccumulation in Aquatic Organisms

Some substances increase in concentration, or *bioaccumulate,* in living organisms as they breathe contaminated air, drink contaminated water, or eat contaminated food. These chemicals can become concentrated in the tissues and internal organs of animals as well as humans.

The concentration of potassium found in fish tissues is expected to be about the same as the average concentration of potassium in the water from which the fish was taken.

Recommended Risk-Reduction Measures

Proper training of all transporters will reduce the likelihood of a mishap or accident resulting in a leak or spill of potassium dusts or powders into the environment. The correct use of labeling on all containers, trucks, and rail cars will enable emergency responders to react properly and quickly to any disaster thereby reducing the potential risk to the environment and to personnel.

Storage of potassium should be segregated from incompatible chemicals to minimize the risk of cross-contamination or contact. Care should be taken to prevent physical damage to storage containers. Also, dusty conditions must be avoided in storage and use areas. When potassium comes into contact with water, it will readily liberate excessive amounts of highly flammable and explosive hydrogen gas.

If a spill or leak to the environment has occurred, fire department, emergency response and/or hazardous materials spill personnel should be notified immediately. Cleanup should be attempted only by those trained in proper spill containment procedures using non-sparking tools. Contaminated soils should be removed for decontamination and replaced with clean soil. If potassium should contact the water table, aquifer, or navigable waterway, time is of the essence. It is highly reactive and may explode on contact with water. It liberates excessive heat as it forms potassium hydroxide and hydrogen gas. The heat may be sufficient enough to ignite the hydrogen and cause the explosion. The local and/or state emergency response authorities may require notification. A comprehensive emergency response or disaster preparedness/recovery plan should be in place prior to any operations involving the use, transportation, storage, or disposal of potassium.

If potassium is spilled or leaked, the following specific steps are recommended:

- ☑ Restrict persons not wearing protective clothing from area of spill or leak until cleanup is complete and area can be opened for normal work.
- ☑ Remove all ignition sources from area.
- ☑ Ventilate area of spill or leak.
- ☑ Collect powdered materials in the safest manner possible and deposit in sealed containers. Recommend using a vacuum equipped with a high-efficiency particulate air (HEPA) filter. Do not dry sweep (generates airborne dusts). Keep water away (do NOT damp mop residues).
- ☑ It may be necessary to dispose of potassium as a hazardous waste. The responsible state agency or the regional office of the federal Environmental Protection Agency (EPA) should be contacted for specific recommendations.

RISK ASSESSMENT: BUSINESS

General Assessment

Accidents or mishaps involving potassium can present a significant threat to business operations. The loss or damage of equipment or facilities can significantly affect fiscal viability. Lawsuits that may result from personnel injury/death, public exposures, and/or environmental contamination will also require a serious expenditure of resources. Media attention surrounding an injury, death, or environmental damage can also result in a loss of profits and loss of current as well as future business.

Recommended Risk-Reduction Measures

Company attorneys, safety and health professionals, and environmental specialists should be involved in the development of any procedures for responding to chemical incidents. A company official should be predesignated as a public relations officer with specific training in dealing with the press. Corporate plans and policies should be developed, approved, and implemented long before any need for such arises.

REFERENCES

American Conference of Governmental Industrial Hygienists. 1988. *Documentation of the Threshold Limit Values and Biological Exposure Indices,* 5th Edition (with updates). Cincinnati: ACGIH

Anthony, C. P., and N. J. Kolthoff. 1971. *Textbook of Anatomy and Physiology.* St. Louis: C. V. Mosby Company.

Bloom, A. D. and F. J. DeSerres. 1995. *Ecotoxicity and Human Health.* New York: CRC Press/Lewis Publishers.

Bueche, J. R. 1972. *Principles of Physics,* 2nd Edition. New York: McGraw-Hill.

Calabrese, E. J. 1994. *Biological Effects of Low Level Exposures.* New York: CRC Press/Lewis Publishers.

Cothern, R. R. 1993. *Comparative Environmental Risk Assessment.* New York: CRC Press/Lewis Publishers.

Danse, I. R. 1991. *Common Sense Toxics in the Workplace.* New York: Van Nostrand Reinhold.

Environmental Protection Agency. 1983. *Chemical Hazard Information Profiles.* Washington, D.C.: U.S. EPA.

Gardner, R. W. 1994. *Chemical Intolerance.* New York: CRC Press/Lewis Publishers.

Langerman, N. 1994 *Precautionary Labels for Chemical Containers.* New York: CRC Press/Lewis Publishers.

Lauwerys, R. R., P. Hoet. 1993. *Industrial Chemical Exposure.* New York: CRC Press/Lewis Publishers.

Lewis, R. J., Sr. 1993. *Hawley's Condensed Chemical Dictionary,* 12th Edition. New York: Van Nostrand Reinhold.

Lewis, R. J., Sr. 1992. *Sax's Dangerous Properties of Industrial Materials,* Volumes 1, 2, and 3, 8th Edition. New York: Van Nostrand Reinhold.

Manahan, S. E. 1994. *Environmental Chemistry,* 6th Edition. New York: CRC Press/Lewis Publishers.

Meyer, E. 1990. *Chemistry of Hazardous Materials.* Englewood Cliffs, NJ: Prentice Hall.

Richardson, M. 1995. *Dictionary of Substance and Their Effects,* Vol. I-VI. New York: CRC Press/Lewis Publishers.

Sacarello, H. L. A. 1994. *The Comprehensive Handbook of Hazardous Materials.* New York: CRC Press/Lewis Publishers.

Talbott, E., G. Craun. 1995. *An Introduction to Environmental Epidemiology.* New York: CRC Press/Lewis Publishers.

Torkelson, T. R. V. K. Rowe. 1981. *Patty's Industrial Hygiene and Toxicology,* Vol. 2B. New York: Wiley-Interscience.

MATERIAL SAFETY DATA SHEET

CHEMICAL NAME	
POTASSIUM CHROMATE	**CRC PRESS/LEWIS PUBLISHERS DISCLAIMER:** THE INFORMATION AND RECOMMENDATIONS PRESENTED HEREIN ARE BASED ON SOURCES BELIEVED TO BE RELIABLE. CRC MAKES NO REPRESENTATION ON ITS COMPLETENESS OR ACCURACY. IT IS THE RESPONSIBILITY OF THE USER TO DETERMINE THE CHEMICAL'S SUITABILITY FOR ITS INTENDED USE, THE CHEMICAL'S SAFE USE, AND THE CHEMICAL'S PROPER DISPOSAL. NO REPRESENTATIONS AND/OR WARRANTIES, EITHER EXPRESSED OR IMPLIED, OF THE MERCHANTABILITY OR FITNESS FOR A PARTICULAR PURPOSE, OR OF ANY OTHER NATURE, ARE MADE WITH RESPECT TO THE INFORMATION PROVIDED IN THIS MSDS OR TO THE CHEMICAL TO WHICH INFORMATION MAY REFER. CRC NEITHER ASSUMES NOR AUTHORIZES ANY OTHER PERSON TO ASSUME FOR IT, ANY OTHER ADDITIONAL RESPONSIBILITY OR LIABILITY FOR THE USE OF, OR RELIANCE UPON, THIS INFORMATION.

HAZARD WARNING INFORMATION

HEALTH	FIRE	REACTIVE	OTHER	DEGREE OF HAZARD	COLOR CODING	OTHER CODES
3	1	0		0 = Minimum Hazard 1 = Slight Hazard 2 = Moderate Hazard 3 = Serious Hazard 4 = Severe Hazard	HEALTH = BLUE FIRE = RED REACTIVITY = YELLOW OTHER = WHITE	OX = Oxidizer ACID = Acid ALK = Alkali COR = Corrosive W = Use No Water

SECTION I - GENERAL INFORMATION

Characterization	**Potassium Compound**	RCRA Number	**None**
		EPA Class	**Not Applicable**
DOT Proper Shipping Name	**Potassium Chromate**	Chemical Abstract Service (CAS) Number	**7789-00-6**
DOT Hazard Class and Label Requirements	**ORM-E**	DOT Emergency Guide Code	**No Citation**
DOT Identification Number	**NA 9142**	Atomic Formula	**K_2CrO_4**

Synonyms

Potassium chromate (VI); potassium chromate; bipotasssium chromate, chromate of potash, chromate of potassium; dipotassium monochromate; neutral potassium chromate; potassium chromate, yellow.

SECTION II - HAZARDOUS INGREDIENTS AND IDENTITY INFORMATION

Hazardous Components (specific identity)	OSHA Exposure Criteria	NIOSH Exposure Criteria	Immediately Dangerous to Life and Health (IDLH)	ACGIH Exposure Criteria
Potassium chromate (derivation: By roasting powdered chromite with potash and limestone, treating the cinder with hot potassium sulfate solution, and leaching).	**PEL (ceiling):** **0.1 mg/m^3 (as CrO_3)** **STEL:** **Not Applicable**	**REL (10-hour):** **0.025 mg/m^3 (as Cr(VI))** **CEILING:** **0.05 mg/m^3 (15-minute)**	**Not Determined**	**TLV:** **0.05 mg/m^3 (as chromium)** **STEL:** **Not Established**

SECTION III - PHYSICAL AND CHEMICAL CHARACTERISTICS

Boiling Point	**Not Found**	Specific Gravity (H_2O = 1)	**2.73**
Vapor Pressure (mm Hg)	**0 at 68°F (20°C)**	Molecular Weight	**194.2**
Vapor Density (Air = 1)	**Not Found**	Melting Point	**1787°F (975°C)**

Solubility

Soluble in water. Insoluble in alcohol.

Appearance and Odor

Lemon-yellow rhombic crystalline solid with no odor.

SECTION IV - FIRE AND EXPLOSION HAZARD DATA

Flash Point (method used)	**Not Determined**	Explosive Limits in Air % by Volume	**LEL: Not Applicable UEL: Not Applicable**
NFPA Classification	**Combustible Solid**	Autoignition Temperature	**Not Determined**

Extinguishing Media

Dry chemical, carbon dioxide, water spray, fog, or foam.

Special Fire Fighting Procedures

Poisonous gases are produced in fire. Wear full protective clothing and self-contained breathing apparatus (SCBA). Do NOT release runoff from fire-fighting measures to sewers or waterways. Use water spray to cool fire-exposed containers. Move containers from fire if it can be done without risk.

Unusual Fire and Explosion Hazards

Potassium chromate can ignite readily oxidized materials such as wood, paper, sulfur, plastics, aluminum, and organic material. It can increase the intensity of fires involving these materials. It becomes a possible explosion hazard when shocked, heated, or exposed to hydrazine.

SECTION V - REACTIVITY DATA

Stability		Conditions to Avoid
		Potassium chromate is stable under normal conditions of temperature and pressure. Avoid conditions which generate excessive airborne dusts and keep potassium chromate away from incompatible materials.
Stable	Unstable	Incompatibility *(materials to avoid)*
X		Will react with most organic matter (paper, wood, plastics, oil), reducing agents (aluminum, sulfur) to form chromium (III). It will explosively decompose hydrazine.
Hazardous Polymerization		Conditions to Avoid
		Under normal conditions of temperature and pressure, hazardous polymerization of potassium chromate is not known to occur.
May Occur	Will Not Occur	Hazardous Decomposition or By-products
	X	When heated to decomposition, potassium chromate can emit toxic/poisonous fumes and gases. These may include toxic potassium oxide fumes.

SECTION VI - HEALTH HAZARD DATA

Primary Route(s) of Entry:	Inhalation?	Absorption (skin)?	Ingestion?
	X		X

Health Hazards

INHALATION: Severe irritation of the eyes, nose, throat, bronchial tubes, and lungs. Inhaling the dusts or fumes can lead to progressive deterioration and eventual perforation of the nasal septum, pulmonary edema, bronchospasm, and blood changes. Cancer is also possible.

EYES & SKIN: Skin contact causes contact dermatitis and "chrome holes" (usually painless but deep, slow healing holes in the skin). Eye contact causes irritation and conjunctivitis.

INGESTION: Ingestion is not likely, but can occur. Causes violent gastroenteritis, severe circulatory collapse, chronic nephritis, and a possibility for liver damage.

Carcinogenicity	NTP Listed?	IARC Cancer Review Group?	OSHA Regulated?	Target Organs?
Confirmed Human Confirmed Animal	5th Annual Report	Group 1	19 CFR 1910.1000 Table Z-1-A	Skin, eye, resp. sys., liver, blood, kidneys, GI tract.

Medical Conditions Generally Aggravated by Exposure

Skin disorders; chronic respiratory diseases (asthma, emphysema); blood irregularities; liver problems.

Emergency and First-aid Procedures

Eye contact: Flush immediately with water for 15 minutes (minimum); seek medical attention. Skin contact: Remove all contaminated clothing. Immediately wash area with large amounts of soap and water. For inhalation: Remove the person from exposure. Provide respiratory assistance and CPR. Encourage blowing nose and coughing to remove any trapped particles. If swallowed: Never attempt to give a convulsing or unconscious person anything by mouth. If victim is conscious, give 1-2 glasses of milk. Do NOT induce vomiting. Give large doses of vitamin C. Seek medical attention immediately.

SECTION VII - PRECAUTIONS FOR SAFE HANDLING AND USE

Steps to be Taken in Case Material is Released or Spilled

Collect powdered material in most convenient manner possible and deposit in sealed drum. Keep those not involved in cleanup from entering area. Do NOT dry sweep (creates airborne dusts). Use a vacuum equipped with a high efficiency particulate air (HEPA) filter. Damp mop residue.

Preferred Waste Disposal Method

No citation.

Precautions to be Taken in Handling and Storage

Store in tightly closed containers in cool, well-ventilated area. Sources of ignition are prohibited where potassium chromate is stored. Storage near use areas is recommended to minimize transport distance.

Other Precautions and Warnings

Protect from physical damage. Work areas should be posted as containing a suspected carcinogen.

SECTION VIII - CONTROL MEASURES AND PERSONAL PROTECTIVE EQUIPMENT

Respiratory Protection *(specify type)*

For low exposures, use an MSHA/NIOSH-approved full facepiece respirator with HEPA filters. For best protection from a carcinogen, use a supplied-air respirator with a full facepiece, hood, or helmet, or a self-contained breathing apparatus (SCBA) operated in positive pressure mode.

Ventilation

Local exhaust or general mechanical systems recommended.

Protective Gloves	Eye Protection	Other Protective Clothing
Impervious Material	Chemical/Dust Goggles or Face Mask	Protective Suit or Clothing

Work/Hygiene Practices

Always wash hands thoroughly after using chemical; never bring food, drink, or smoking materials into vicinity of chemicals.

POTASSIUM CHROMATE

K_2CrO_4

CAS: 7789-00-6

IDENTIFICATION AND TYPICAL USES

Potassium chromate can appear as bright yellow odorless rhombic crystals. It is used as a reagent in analytical chemistry, in aniline black, as a textile mordant, in enamels, as a chromate pigment, and in certain inks.

RISK ASSESSMENT: HEALTH

General Assessment

Potassium chromate is highly toxic primarily by *inhalation,* although *ingestion* is also possible but not likely. Skin contact poses a moderate health risk (absorption has not been reported as an entry route). Because of its chromium content, it is suggested it be considered a human carcinogen. It is also a known human mutagen. The IARC lists chromium and many of its compounds as carcinogens. However, the references are particularly conflicting regarding the specific carcinogenicity of potassium chromate.

Inhalation of potassium chromate dusts will result in severe and corrosive irritation to the eyes, nose, throat, bronchial tubes, and lungs. A primary target of inhaled dusts is the nasal cavity and, specifically, the nasal septum (thin tissue between the nostrils). Deterioration followed by perforation of the nasal septum is possible, especially on frequent, high, or prolonged exposures. Injury to the lungs is also possible. Specifically, pulmonary edema (fluid in the lungs) can occur which can be fatal. Symptoms of coughing, congestion, tightness in chest, and difficulty breathing may be delayed up to 48 hours creating a false sense of security with regard to health exposure risk. Systemic effects by any route of exposure can also include blood changes such as leukocytosis (increased leukocyte level in blood), leukopenia (reduced leukocyte circulation in blood), hemolysis (destruction of red blood cells), or alterations of white blood cell count. There may also be kidney failure (acute tubular necrosis and uremia, liver damage, and possible allergic reactions.

Skin contact will cause irritation, redness, defatting, and contact dermatitis. Tiny lesions, referred to as "chrome holes," can also occur. These are deep but typically painless, slow healing holes in the skin that can penetrate to the joints. Eye contact will cause irritation and conjunctivitis with tearing, irritation, and inflammation of the eyelids.

Ingestion is not likely but can occur. Symptoms may include violent gastroenteritis and a possibility of circulatory collapse, chronic nephritis (inflammation of the kidney), and liver injury. Shock and death may occur.

Acute Health Effects

The following acute (short-term) health effects may occur immediately or shortly after exposure to potassium chromate:

Skin: Irritation, rash, and possible contact dermatitis.

Eye: Irritation, conjunctivitis, inflammation, watering, and swelling.

Lung: Inhalation of potassium chromate dusts may cause irritation of the mucous membranes in the respiratory tract. There may be burning in the nasal cavity with nosebleeds and possible perforation of the nasal septum. Pulmonary edema is possible on high exposures.

Other: Ingestion is highly corrosive to the esophagus and stomach. It may cause systemic poisoning.

Chronic Health Effects

The following chronic (long-term) health effects may occur at some time after exposure to potassium chromate and can last for months or even years:

Cancer Hazards: According to the information presented in the references, chromium and many of its salts have been shown to cause cancer and mutations in humans and animals. Throat cancers and lung cancers have been reported. Potassium chromate may have these effects as well, although the references are conflicting and inconsistent in this regard.

Reproductive Hazard: According to information available in the references, potassium chromate has not been adequately tested for its ability to adversely affect reproduction in laboratory animals.

Other Chronic Hazards: Repeated, long-term skin exposures can cause the development of deep but relatively painless lesions in the skin that can progress to the joints and take a long time to heal. Exposed persons may develop a skin rash like that of eczema. Once this occurs, any future exposures will cause a rash to reappear. Exposure may cause irritation of the bronchial tubes (bronchitis, with cough and phlegm). Prolonged contact can also cause lung cancer. Repeated or severe exposures may cause kidney damage.

Recommended Risk-Reduction Measures

Personnel should avoid direct contact with potassium chromate. It is highly probable that it is a human carcinogen. If a less toxic chemical cannot be substituted for a hazardous material, then *engineering controls* are the most effective method of reducing exposures.

The best protection is to enclose operations and/or provide local exhaust ventilation at the site of chemical release. While not always operationally feasible, isolating operations can also reduce exposure. Using respiratory protection is less effective than the controls mentioned above, but is required whenever working with or around potassium chromate. For low or infrequent exposures, a full facepiece respirator equipped with a high efficiency particulate air (HEPA) filter equipped with a dust/mist/fume pre-filter may suffice. However, because potassium chromate is probably a carcinogen, the best protection is provided using a supplied-air respirator with a full facepiece operated in positive pressure mode or an MSHA/NIOSH-approved self-contained breathing apparatus (SCBA) with full facepiece operated in pressure demand or other continuous flow mode. If a full facepiece is not available, then chemical/dust goggles should be worn to protect the eyes. Whenever a chemical dust hazard exists, a face shield and a rubber apron should be worn. To prevent hand and skin exposures, impervious rubber gloves should be used. Glove manufacturers should be contacted and permeation studies obtained *before* selecting gloves.

Administrative controls should also be in place to minimize the potential for human exposures. These may include written procedures or policies which specify the methods and techniques that will be practiced whenever personnel are to work with potassium chromate.

All personnel should receive *training* on the use, hazards, protective measures, emergency actions, and other precautions per 29 CFR 1910.1200 (Hazard Communication), prior to the first assignment in an area where potassium chromate is used or stored.

Before beginning employment with potassium chromate and at regular intervals thereafter (e.g., annually), the following medical testing is recommended:

☑ Lung function tests (establish baseline).

☑ Examination of the nasal septum (establish a baseline of nasal cavity condition).

☑ Evaluation for skin reactions, conditions, or allergy by a qualified allergist.

If symptoms develop or overexposure is suspected, the following may also be useful:

☑ Urine and serum chromium levels.

☑ Lung function tests (compare to baseline).

☑ Examination of the nasal septum for deterioration or perforation.

☑ Consider chest X-ray after acute overexposure (may be negative if taken immediately after exposures due to delayed onset of pulmonary edema).

Any evaluation should include a careful history of past and present symptoms with a medical examination. Medical tests that look for damage already done are not a substitute for controlling exposure. Also, since smoking can cause heart disease, emphysema, and a number of other respiratory disorders, smokers exposed to lung irritants such as potassium chromate dusts and fumes may experience symptoms much more quickly than non-smokers under the same conditions of exposure. Prudent risk management requires proper consideration of *all* factors which may be causing the appearance of exposures symptoms in the work place.

Other methods to reduce exposure include:

☑ Where possible, enclose operations and use local exhaust ventilation at the site of chemical release. If local exhaust ventilation or enclosure is not used, respiratory protection should be mandatory.

- ☑ Always ensure that proper protective clothing is worn when using chemical substances.
- ☑ Work areas should be posted with signs indicated a suspected human carcinogen is present. Access to the area must be controlled.
- ☑ Wash thoroughly immediately after exposure to potassium chromate and at the end of the work shift or before eating, drinking, or smoking.
- ☑ Hazard warning information should be posted in the work area. In addition, as part of an on-going education and training program, all information on the health and safety hazards of potassium chromate should be communicated to all potentially exposed workers.
- ☑ Eye wash stations and emergency showers should be located in the immediate work area.
- ☑ Work clothing that has been contaminated with potassium chromate should never be worn home where family members can be exposed as well. Contaminated clothing should be laundered at work or by an authorized service by personnel who have been trained in the hazards associated with exposure to potassium chromate and potassium compounds.
- ☑ Specific engineering controls are recommended by NIOSH. Refer to NIOSH Criteria Document Number 76-129 ("Chromium VI").

RISK ASSESSMENT: ENVIRONMENT

General Assessment

The environment is at risk of exposure during transportation, storage, disposal, or destruction of potassium chromate. In almost every scenario, the threat of environmental exposure is contingent upon the proper handling of the chemical substance. Accidental spills, large or small, can result in fire, explosion, and possible contamination of the surrounding environmental mediums (water, soil, and air).

Potassium chromate can react with organic matter (such as paper, wood, plastics, and oil) and reducing agents (such as aluminum and sulfur) to form chromium (III). It will explosively decompose hydrazine. In a fire, potassium chromate will emit toxic and poisonous gases. These characteristics require special consideration during any emergency situation involving a leak or spill of potassium chromate.

Potassium chromate can enter the environment through industrial discharge, spills, or leaks.

Acute Ecological Effects

Acute (short-term) toxic effects may include the death of animals, birds, or fish and death or low growth rate in plants. Acute effects are seen 2 to 4 days after animals or plants are exposed to potassium chromate.

Chromium is more toxic in soft water than hard water. Chromium (VI) has high acute toxicity to aquatic life, and chromium (III) has moderate acute toxicity to aquatic life. Insufficient data are available to evaluate the specific short-term effects of potassium chromate exposure to aquatic life, plants, birds or land animals.

Chronic Ecological Effects

Chronic (long-term) toxic effects may include shortened life span, reproductive problems, lower fertility, and changes in appearance or behavior in exposed animals. These effects can be seen long after first exposure(s) to toxic chemicals.

Chromium (III) and chromium (VI) both have high chronic toxicity to aquatic life. Insufficient data are available to evaluate the specific long-term effects of potassium chromate exposure to aquatic life, plants, birds or land animals.

Water Solubility

Potassium chromate is slightly soluble in water. Concentrations of 1 to 100 milligrams will mix with a liter of water.

Persistence in the Environment

Potassium chromate (specifically, chromium), is highly persistent in the aquatic environment, with a half-life of greater than 200 days. The half-life of a pollutant is the amount of time it takes for one half of the chemical to be degraded.

Bioaccumulation in Aquatic Organisms

Some substances increase in concentration, or *bioaccumulate,* in living organisms as they breathe contaminated air, drink contaminated water, or eat contaminated food. These chemicals can become concentrated in the tissues and internal organs of animals as well as humans.

The concentration of chromium found in fish is expected to be somewhat higher than the average concentration of potassium chromate in the water from which the fish was taken.

➊ *Recommended Risk-Reduction Measures*

Proper training of all transporters will reduce the likelihood of a mishap or accidents resulting in a leak or spill to the environment. Proper labeling on all containers, trucks, and rail cars will enable emergency responders to react properly and quickly to any disaster thereby reducing the potential risk to the environment and to personnel.

If a spill or leak to the environment has occurred, fire department, emergency response and/or hazardous materials spill personnel should be notified immediately. Do not dry sweep. When vacuuming, a high efficiency particulate air (HEPA) filter should be used, not a standard shop vacuum. Damp mop residue. Deposit collected materials in sealed drum for disposal. Cleanup should be attempted only by those trained in proper spill containment procedures. Contaminated soils may need to be removed and replaced with clean soil. If potassium chromate should contact the water table, aquifer, or navigable waterway, time is of the essence. Potassium chromate is soluble in water and total containment and remediation may not be entirely possible. When such spills occur, the local and/or state emergency response authorities must be notified. A comprehensive emergency response or disaster preparedness/recovery plan should be in place prior to any operations involving the use, transportation, storage, or disposal of potassium chromate. If potassium chromate is spilled or leaked, the following specific steps are recommended:

- ☑ Restrict persons not wearing protective clothing from area of spill or leak until cleanup is complete and area can be opened for normal work.
- ☑ Ventilate area and remove ignition sources.
- ☑ Collect powdered materials in most convenient and safe manner possible and place in sealed drums for disposal. Do NOT dry sweep; use HEPA vacuum instead. Use water to flush residue, then damp mop.
- ☑ It may be necessary to dispose of potassium chromate as a hazardous waste. The responsible state agency or the regional office of the federal Environmental Protection Agency (EPA) should be contacted for specific recommendations.

RISK ASSESSMENT: BUSINESS

General Assessment

Accidents or mishaps involving potassium chromate can present a significant threat to business operations. The loss or damage of equipment or facilities can significantly affect fiscal viability. Lawsuits that may result from personnel injury/death, public exposures, and/or environmental contamination will also require a serious expenditure of resources. Media attention surrounding an injury, death, or environmental damage can also result in a loss of profits and loss of current as well as future business. Always remember that anytime the terms "cancer," or "carcinogen" are used, public emotion, hysteria, and ignorance can run equally high. This must be carefully considered whenever developing or implementing public relations policies.

➊ *Recommended Risk-Reduction Measures*

Company attorneys, safety and health professionals, and environmental specialists should be involved in the development of any procedures dealing with safety or emergency response. A company official should be pre-designated as a public relations officer with specific training in dealing with the press. Corporate plans and policies should be developed, approved, and implemented long before any need for such arises.

REFERENCES

American Conference of Governmental Industrial Hygienists. 1988. *Documentation of the Threshold Limit Values and Biological Exposure Indices,* 5th Edition (with updates). Cincinnati: ACGIH

Brusick, D. J. 1994. *Methods for Genetic Risk Assessment.* New York: CRC Press/Lewis Publishers.

Cockerman, L. G., B. S. Shane. 1994. *Basic Environmental Toxicology.* New York: CRC Press/Lewis Publishers.

Gardner, R. W. 1994. *Chemical Intolerance.* New York: CRC Press/Lewis Publishers.

Lewis, R. J., Sr. 1992. *Sax's Dangerous Properties of Industrial Materials,* Volumes 1, 2, and 3, 8th Edition. New York: Van Nostrand Reinhold.

Sacarello, H. L. A. 1994. *The Comprehensive Handbook of Hazardous Materials.* New York: CRC Press/Lewis Publishers.

U.S. Department of Health and Human Services (DHHS), National Institute for Occupational Health (NIOSH). 1976. *NIOSH Criteria Document: Chromium VI.* Publication Number 76-129. Washington, D.C.; U.S. Government Printing Office.

U.S. Department of Health and Human Services (DHHS), National Institute for Occupational Health (NIOSH). 1994. *NIOSH Pocket Guide to Hazardous Chemicals.* Publication Number 94-116. Washington, D.C.; U.S. Government Printing Office.

MATERIAL SAFETY DATA SHEET

CHEMICAL NAME	
PROPANE	**CRC PRESS/LEWIS PUBLISHERS DISCLAIMER:** THE INFORMATION AND RECOMMENDATIONS PRESENTED HEREIN ARE BASED ON SOURCES BELIEVED TO BE RELIABLE. CRC MAKES NO REPRESENTATION ON ITS COMPLETENESS OR ACCURACY. IT IS THE RESPONSIBILITY OF THE USER TO DETERMINE THE CHEMICAL'S SUITABILITY FOR ITS INTENDED USE, THE CHEMICAL'S SAFE USE, AND THE CHEMICAL'S PROPER DISPOSAL. NO REPRESENTATIONS AND/OR WARRANTIES, EITHER EXPRESSED OR IMPLIED, OF THE MERCHANTABILITY OR FITNESS FOR A PARTICULAR PURPOSE, OR OF ANY OTHER NATURE, ARE MADE WITH RESPECT TO THE INFORMATION PROVIDED IN THIS MSDS OR TO THE CHEMICAL TO WHICH INFORMATION MAY REFER. CRC NEITHER ASSUMES NOR AUTHORIZES ANY OTHER PERSON TO ASSUME FOR IT, ANY OTHER ADDITIONAL RESPONSIBILITY OR LIABILITY FOR THE USE OF, OR RELIANCE UPON, THIS INFORMATION.

HAZARD WARNING INFORMATION

HEALTH	FIRE	REACTIVE	OTHER	DEGREE OF HAZARD	COLOR CODING	OTHER CODES
1	4	0		0 = Minimum Hazard 1 = Slight Hazard 2 = Moderate Hazard 3 = Serious Hazard 4 = Severe Hazard	HEALTH = BLUE FIRE = RED REACTIVITY = YELLOW OTHER = WHITE	OX = Oxidizer ACID = Acid ALK = Alkali COR = Corrosive W = Use No Water

SECTION I - GENERAL INFORMATION

Characterization	RCRA Number	EPA Class
Gas, Hydrocarbon	**None**	**Not Applicable**

DOT Proper Shipping Name	Chemical Abstract Service (CAS) Number
Liquefied Petroleum Gas	**74-98-6**
DOT Hazard Class and Label Requirements	DOT Emergency Guide Code
Flammable Gas	**22**
DOT Identification Number	Chemical Formula
UN 1075 (compressed); UN 1978 (liquid)	$CH_3CH_2CH_3$

Synonyms

Bottled gas; dimethylmethane; *n*-propane; propyl hydride; liquefied petroleum gas; LPG.

SECTION II - HAZARDOUS INGREDIENTS AND IDENTITY INFORMATION

Hazardous Components (specific identity)	OSHA Exposure Criteria	NIOSH Exposure Criteria	Immediately Dangerous to Life and Health (IDLH)	ACGIH Exposure Criteria
Propane (derivation: From petroleum and natural gas by the "stabilization process" using fractional distillation under pressure). **1 ppm = 1.83 mg/m^3**	**PEL:** **1000 ppm** **1800 mg/m^3** **STEL:** **Not Established**	**REL:** **1000 ppm** **1800 mg/m^3** **STEL:** **Not Established**	**2100 ppm**	**TLV:** **Not Established** **STEL:** **Not Established**

SECTION III - PHYSICAL AND CHEMICAL CHARACTERISTICS

Boiling Point	Specific Gravity (H_2O = 1)
-44°F (-42°C)	**1.55**
Vapor Pressure (atmospheres)	Molecular Weight
8.4 at 70°F (21°C)	**44.1**
Vapor Density (Air = 1)	Melting Point
1.56 at 32°F (0°C)	**-309°F (-190°C)**

Solubility

Slightly soluble in water (0.01%). Soluble in ether and alcohol.

Appearance and Odor

Colorless, odorless, tasteless gas. A foul-smelling odorant similar to natural gas is often added for fuel purposes. It is usually shipped as a liquefied compressed gas. Odor Threshold = 5000-20,000 ppm.

SECTION IV - FIRE AND EXPLOSION HAZARD DATA

Flash Point (method used)	Explosive Limits in Air % by Volume
-156°F (-105°C) closed cup	**LEL: 2.1% UEL: 9.5%**
NFPA Classification	Autoignition Temperature
Flammable Gas	**842°F (450°C)**

Extinguishing Media

Stop flow of gas to stop fire. If this is not possible, letting fire burn itself out may be only alternative.

Special Fire Fighting Procedures

Propane is an extremely flammable, very dangerous fire and explosion hazard. Poisonous gases are evolved from fire. Wear a self-contained breathing apparatus (SCBA) with full facepiece. Ensure ALL sources of escaping gas are controlled. Monitor area long after fire is extinguished. Move containers from fire area if it can be done without risk. Use water spray to cool fire-exposed containers.

Unusual Fire and Explosion Hazards

Gas is explosive. Containers may explode in flame. Even if fire is extinguished, escaping gas can form explosive mixtures in air quickly and ignite without warning. Burns with a luminous, smoky flame.

SECTION V - REACTIVITY DATA

Stability		Conditions to Avoid
		Propane is normally stable when stored as a liquid in steel containers under its own pressure. Avoid contact with heat, sparks, flame, lighted tobacco products, and steam lines.
Stable	Unstable	Incompatibility *(materials to avoid)*
X		Propane reacts violently with oxidizers such as chlorine or chlorine dioxide (explodes), bromine pentafluoride, nitrogen trifluoride, liquid oxygen, oxygen difluoride, other oxidizers, heat, sparks, and open flame.
Hazardous Polymerization		Conditions to Avoid
		Under normal conditions of temperature and pressure, hazardous polymerization of propane is not expected to occur.
May Occur	Will Not Occur	Hazardous Decomposition or By-products
	X	If involved in fire, propane can produce acrid and irritating smoke and fumes.

SECTION VI - HEALTH HAZARD DATA

Primary Route(s) of Entry:	Inhalation?	Absorption (skin)?	Ingestion?
	X		

Health Hazards

INHALATION: An asphyxiant gas. Initial symptoms are those of oxygen starvation. Feeling of "air hunger," diminished mental alertness, headache, dullness, shortness of breath, and an impairment in muscle coordination. Continued exposures lead to faulty judgment, depression of senses, rapid fatigue, unconsciousness, convulsions, coma, and death.

SKIN & EYES: Contact with the compressed gas or refrigerated liquid can cause frostbite and burns.

INGESTION: Not likely since propane is a gas at normal working (room) temperatures.

Carcinogenicity	NTP Listed?	IARC Cancer Review Group?	OSHA Regulated?	Target Organs?
Unknown Human Unknown Animal	No	No	29 CFR 1910.1000 Table Z-1	Central nervous system

Medical Conditions Generally Aggravated by Exposure

None reported.

Emergency and First-aid Procedures

Eye contact: Immediately flush with large amounts of warm (tepid) water for 15 minutes (minimum), occasionally lifting eyelids; seek medical attention. Skin contact: Remove all contaminated clothing. Immediately wash area with large amounts of warm (tepid) water. Do NOT rub or use air-blown heat on affected area. For inhalation: Remove the person from exposure. Provide respiratory assistance and CPR. Transfer promptly to medical facility. If swallowed: Not likely.

SECTION VII - PRECAUTIONS FOR SAFE HANDLING AND USE

Steps to be Taken in Case Material is Released or Spilled

Remove all ignition sources. Ventilate area. Stop flow of gas. If the source of leak is a cylinder and cannot be stopped, remove to safe place in open air; repair leak or let cylinder empty. If gas has not yet ignited, use a water spray to knock down vapors and direct them from ignition source.

Preferred Waste Disposal Method

None specified in the references (incineration of a gas may be the only alternative).

Precautions to be Taken in Handling and Storage

Do not store with incompatible materials or in the presence of heat since violent reactions or explosions can occur. Store in tightly closed pressurized containers in a cool, dark, well-ventilated area away from sunlight. Keep away from ignition sources such as fire, sparks, and flame.

Other Precautions and Warnings

Bulk storage of propane is not recommended. Containers may explode in fire or under conditions of extreme heat. Keep propane out of confined areas or enclosed spaces (oxygen displacement hazard).

SECTION VIII - CONTROL MEASURES AND PERSONAL PROTECTIVE EQUIPMENT

Respiratory Protection *(specify type)*

Propane is a simple asphyxiant. Do NOT use cartridge-type respirators (ineffective for this type of gas). Use a powered supplied-air respirator or a self-contained breathing apparatus with full facepiece operated in pressure demand or other positive pressure mode.

Ventilation

Local exhaust or general mechanical systems recommended.

Protective Gloves	Eye Protection	Other Protective Clothing
Insulated Neoprene or Polyurethane	Goggles and/or Face Mask	Leather Apron

Work/Hygiene Practices

Always wash hands thoroughly after using chemical; never bring food, drink, or smoking materials into vicinity of chemicals.

PROPANE

$CH_3CH_2CH_3$

CAS: 74-98-6

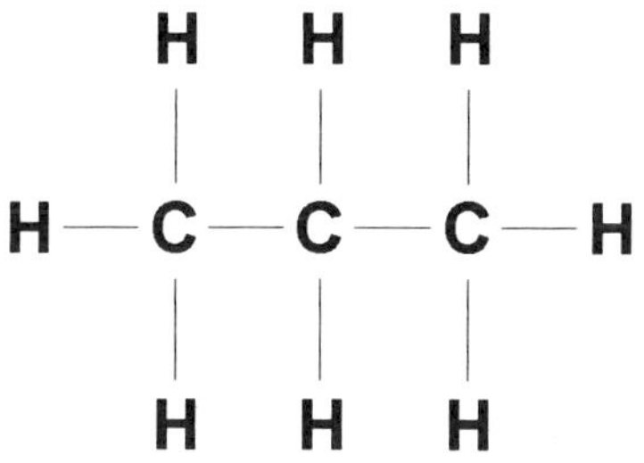

IDENTIFICATION AND TYPICAL USES

Propane is a colorless, tasteless, and odorless gas when pure. However, when used for fuel purposes, a foul-smelling odorant similar to natural gas is often added. It is usually shipped as a liquefied compressed gas. It is used in organic synthesis, as a household and industrial fuel gas, in the manufacture of ethylene, as an extractant, a solvent, a refrigerant, to enrich other gases, as an aerosol propellant (although less common today), and as a mixture for bubble chambers.

RISK ASSESSMENT: HEALTH

General Assessment

Propane is a simple asphyxiant gas. As such, its primary route of entry into the body is through *inhalation.* While it does not pose a significant physiological effect (i.e., no systemic poisoning or toxicity), it can displace oxygen levels to below that which is necessary to sustain life. Skin contact with the liquid can be destructive to tissues due to the extremely cold temperatures but absorption is not a likely exposure route. Since propane usually exists as a gas at room temperatures, ingestion is not likely to occur either. It is not known to cause cancer, mutations, or reproductive effects in humans or test animals. It is an asphyxiant.

Inhalation of propane gas can cause headache, dullness, shortness of breath, and the symptoms of asphyxiation. These include rapid respiration and a feeling of "air hunger" or air starvation followed by a decrease in mental alertness, and impaired muscular coordination. If exposure continues, there can be nausea, vomiting, prostration, unconsciousness, coma, and death.

Skin or eye contact with the vapor poses no serious threat. The vapors have not been shown to cause any physiological response. However, if the liquid should contact either the skin or the eyes, there can be serious irritation, burns, tissue damage, and frostbite.

Acute Health Effects

The following acute (short-term) health effects may occur immediately or shortly after exposure to propane:

Skin: Liquid threat only. Redness, pain, swelling, blisters, and severe frostbite.

Eye: Liquid threat only. Causes frostbite, redness, irritation, severe pain, and blurred vision.

Lung: Vapor is non-irritating and therefore poses no threat of injury to the lungs. However, in high concentrations, propane can lead to symptoms of asphyxiation and oxygen deprivation up to and including death.

Chronic Health Effects

The following chronic (long-term) health effects may occur at some time after exposure to propane and can last for months or even years:

Cancer Hazards: According to information presented in the references, exposure to propane does not lead to carcinogenic effects (it is a simple asphyxiant).

Reproductive Hazards: There are no reports to support any claims of reproductive hazards.

Other Chronic Effects: None reported.

Recommended Risk-Reduction Measures

Personnel should avoid direct contact with propane. If another gas cannot be substituted for a propane, then *engineering controls* are the most effective method of reducing exposures. The best protection is to separate operations and provide local, explosion-proof exhaust ventilation at the site of chemical release. While not always operationally feasible, isolating operations can also reduce exposure. Using equipment with explosion-proof designs and ensuring all metal containers and tools are electrically grounded and bonded can further reduce the potential for injury or damage. Using respiratory protection is less effective than the controls mentioned above, but is still advisable whenever working with propane. Air-purifying respirators (e.g., the cartridge-type or gas masks) cannot be used with propane. These types of respirators only purify the air and can do nothing to protect in oxygen deficient atmospheres. A powered air-supplied respirator operated in pressure demand or other positive pressure mode, or a self-contained breathing apparatus (SCBA) with full facepiece operated in pressure demand mode are the recommended respiratory protection methods of choice. If a full facepiece is not available, then chemical goggles should be worn to protect the eyes. Whenever a chemical splash hazard exists, a face shield and a rubber or leather apron should be worn. To prevent hand and skin exposures with the liquid, thermal protective gloves should be used. Neoprene or polyethylene materials should suffice. However, it is always best to contact glove manufacturers and obtain permeation studies *before* final glove selections are made.

Administrative controls should also be in place to minimize the potential for human exposures. These may include written procedures or policies which specify the methods and techniques that will be practiced whenever personnel are to work with propane.

All personnel should receive *training* on the use, hazards, protective measures, emergency actions, and other precautions per 29 CFR 1910.1200 (Hazard Communication), prior to the first assignment in an area where propane is used or stored.

Other methods to reduce exposure to chemicals include:

☑ Where possible, consider installing automatic sensors that warn personnel of dropping oxygen levels and use local exhaust ventilation at the site of chemical release. If local exhaust ventilation or enclosure is not used, respiratory protection should be mandatory.

☑ Always ensure that proper protective clothing is worn when using chemical substances and that personnel have been trained in its use, care, and maintenance.

☑ Wash thoroughly immediately after exposure to propane and at the end of the work shift or before eating, drinking, or smoking.

☑ Safety shower and eyewash stations should be readily available in work areas where propane is used or stored.

☑ Hazard warning information should be posted in the work area. In addition, as part of an on-going education and training program, all information on the health and safety hazards of propane should be communicated to all exposed or potentially exposed workers.

RISK ASSESSMENT: ENVIRONMENT

General Assessment

The environment is at risk of exposure during transportation, storage, disposal, or destruction of propane. In almost every scenario, the threat of environmental exposure is contingent upon the proper handling of the chemical substance. Accidental releases, large or small, can result in fire, explosion, and possible contamination of the surrounding environmental mediums (water, soil, and air). Propane is an extremely flammable and highly explosive gas. Its extremely low flash point and boiling point present a serious fire hazard. It has a relatively narrow flammability range in air. Also, since it can react violently in the presence of oxidizers, such as chlorine, propane requires special consideration during any emergency involving a leak or release of propane gas or liquid.

Propane can enter the environment through industrial discharges, unchecked venting to air, or through spills of the liquid. Since liquids will most likely vaporize rapidly, water and soil contamination are possible but not probable.

☠ *Acute Ecological Effects*

Acute (short-term) toxic effects may include the death of animals, birds, or fish, and death or low growth rate in plants. Acute effects are seen 2 to 4 days after animals or plants are exposed to propane.

Insufficient data are available on the short-term effects of propane on aquatic life, plants, birds, or land animals. However, as a relatively non-toxic, simple asphyxiant, it might be assumed that ecological toxicity will be relatively low.

Chronic Ecological Effects

Chronic (long-term) toxic effects may include shortened life span, reproductive problems, lower fertility, and changes in appearance or behavior in exposed animals. These effects can be seen long after first exposure(s) to toxic chemicals.

Insufficient data are available to evaluate the long-term effects of propane to aquatic life, plants, birds, or land animals. However, as a relatively non-toxic, simple asphyxiant, it might be assumed that any chronic ecological toxicity will be relatively low.

Water Solubility

Propane is only slightly soluble in water. Concentrations of 1 milligram will mix with a liter of water.

Persistence in the Environment

Propane is non-persistent in water, with a half-life of less than 2 days. The half-life of a pollutant is the amount of time it takes for one half of the chemical to be degraded. Virtually 100% of propane will eventually end up in the air.

Bioaccumulation in Aquatic Organisms

Some substances increase in concentration, or *bioaccumulate,* in living organisms as they breathe contaminated air, drink contaminated water, or eat contaminated food. These chemicals can become concentrated in the tissues and internal organs of animals as well as humans.

Propane is not expected to accumulate in the tissues of fish.

Recommended Risk-Reduction Measures

Proper training of all transporters will reduce the likelihood of a mishap or accident resulting in a leak or release to the environment. The correct use of DOT labeling (per the MSDS) on all containers, trucks, and rail cars will enable emergency responders to react properly and quickly to any disaster thereby reducing the potential risk to the environment and to personnel.

Storage of propane should be segregated from incompatible chemicals to minimize the risk of cross-contamination or contact. Buildings designated for storage should be equipped with appropriate fire protection systems (alarms, sprinklers, emergency lighting, portable extinguishers).

If a spill or release to the environment has occurred, fire department, emergency response, and/or hazardous materials spill personnel should be notified immediately. Cleanup should be attempted only by those trained in proper containment procedures using non-sparking tools. Contaminated soils should be removed for incineration and replaced with clean soil. When spills or leaks occur, it may be necessary to contact the local and/or state emergency response authorities. A comprehensive emergency response or disaster preparedness/recovery plan should be in place prior to any operations involving the use, transportation, storage, or disposal of propane. If propane is spilled or leaked, the following specific steps are recommended:

- ☑ Restrict persons not wearing protective clothing from area of spill or leak until cleanup is complete and area can be opened for normal work.
- ☑ Remove ignition sources, and ventilate area.
- ☑ If applicable, stop flow of leaking liquid or gas. If leak source is a cylinder and the leak cannot be stopped in place, remove leaking cylinder to a safe place in the open air, and allow cylinder to empty.
- ☑ Keep propane out of a confined space, such as a sewer, because of the possibility of explosion (unless the sewer is designed to prevent the buildup of explosive concentrations).
- ☑ It may be necessary to dispose of propane as a hazardous waste. The responsible state agency or the regional office of the federal Environmental Protection Agency (EPA) should be contacted for specific recommendations.

RISK ASSESSMENT: BUSINESS

General Assessment

Accidents or mishaps involving propane can present a significant threat to business operations. The loss or damage of equipment or facilities can significantly affect fiscal viability. Lawsuits that may result from personnel injury, public exposures, and environmental contamination will also require a serious expenditure of resources.

Media attention surrounding an injury, illness, or environmental damage can also result in a loss of profits and loss of current as well as future business.

ʊ *Recommended Risk-Reduction Measures*

Company attorneys, safety and health professionals, and environmental specialists should be involved in the development of any procedures dealing with safety and emergency response. A company official should be pre-designated as a public relations officer with specific training in dealing with the press. Corporate plans and policies should be developed, approved, and implemented long before any need for such arises.

REFERENCES

Ahlbom, A. 1993. *Biostatistics for Epidemiologists.* New York: CRC Press/Lewis Publishers.

American Conference of Governmental Industrial Hygienists. 1988. *Documentation of the Threshold Limit Values and Biological Exposure Indices,* 5th Edition (with updates). Cincinnati: ACGIH

Anthony, C. P., and N. J. Kolthoff. 1971. *Textbook of Anatomy and Physiology.* St. Louis: C. V. Mosby Company.

Asante-Duah, D. K. 1993. *Hazardous Waste Risk Assessment.* New York: CRC Press/Lewis Publishers.

Bloom, A. D. and F. J. DeSerres. 1995. *Ecotoxicity and Human Health.* New York: CRC Press/Lewis Publishers.

Brusick, D. J. 1994. *Methods for Genetic Risk Assessment.* New York: CRC Press/Lewis Publishers.

Bueche, J. R. 1972. *Principles of Physics,* 2nd Edition. New York: McGraw-Hill.

Burrell, R., D. K. Flaherty, and L. J. Sauers. 1992. *Toxicology of the Immune System.* New York: Van Nostrand Reinhold.

Calabrese, E. J. 1994. *Biological Effects of Low Level Exposures.* New York: CRC Press/Lewis Publishers.

Cockerman, L. G., B. S. Shane. 1994. *Basic Environmental Toxicology.* New York: CRC Press/Lewis Publishers.

Cothern, C. R. 1995. *Handbook for Environmental Risk Decisionmaking.* New York: CRC Press/Lewis Publishers.

Danse, I. R. 1991. *Common Sense Toxics in the Workplace.* New York: Van Nostrand Reinhold.

Environmental Protection Agency. 1983. *Chemical Hazard Information Profiles.* Washington, D.C.: U.S. EPA.

Fodden, C. R., and J. L. Weddell. 1992. *Hazardous Materials Emergency Action Data.* New York: CRC Press/Lewis Publishers.

Forseberg, K., and L. H. Keith. 1993. *Chemical; Protective Clothing Permeation and Degradation Database.* Chelsea, MI: Lewis Publishers.

Gardner, R. W. 1994. *Chemical Intolerance.* New York: CRC Press/Lewis Publishers.

Hallenbeck, W. H. 1993. *Quantitative Risk Assessment for Environmental and Occupational Health,* 2nd Edition. New York: CRC Press/Lewis Publishers.

Hodgson, E., R. B. Mailman, and J. E. Chambers. 1988. *Dictionary Toxicology.* London: The MacMillian Company.

Hosty, J. W., and P. Foster. 1990. *A Practical Guide to Chemical Sp. Response.* New York: Van Nostrand Reinhold.

Howard, P. H. 1995. *Handbook of Environmental Fate and Exposure Data for Organic Chemicals,* Vol. I-IV. New York: CR Press/Lewis Publishers.

James, A., and M. Lord. 1992. *Index of Chemical and Physical Data.* New York: Van Nostrand Reinhold.

Jones, F. E. 1994. *Toxic Organic Vapors in the Workplace.* New York CRC Press/Lewis Publishers.

Kamrin, M. A. 1988. *Toxicology: A Primer on Principles and Applications.* Chelsea, Michigan: Lewis Publishers.

Keith, L. H. and D. B. Walters. (Editors) 1992. *National Toxicology Program's Chemical Database, Volumes I-VII.* Chelsea, MI: Lewis Publishers.

Landis, W. G., M. H. Yu. 1995. *Introduction to Environmental Toxicology.* New York: CRC Press/Lewis Publishers.

Langerman, N. 1994 *Precautionary Labels for Chemical Containers.* New York: CRC Press/Lewis Publishers.

Lauwerys, R. R., P. Hoet. 1993. *Industrial Chemical Exposure.* New York: CRC Press/Lewis Publishers.

Lewis, R. J., Sr. 1993. *Hawley's Condensed Chemical Dictionary,* 12th Edition. New York: Van Nostrand Reinhold.

Lewis, R. J., Sr. 1992. *Sax's Dangerous Properties of Industrial Materials,* Volumes 1, 2, and 3, 8th Edition. New York: Van Nostrand Reinhold.

Manahan, S. E. 1992. *Toxicological Chemistry,* 2nd Edition. New York: CRC Press/Lewis Publishers.

Meyer, E. 1990. *Chemistry of Hazardous Materials.* Englewood Cliffs, NJ: Prentice Hall.

Patnaik, P. 1992. *A Comprehensive Guide to the Hazardous Properties of Chemical Substances.* New York: Van Nostrand Reinhold.

Philp, R. B. 1995. *Environmental Hazards and Human Health.* New York: CRC Press/Lewis Publishers.

Proctor, N. H., and J. P. Hughes. 1991. *Chemical Hazards of the Workplace.* New York: Van Nostrand Reinhold.

Rea, W. J. 1995. *Chemical Sensitivity,* Vol. I-III. New York: CRC Press/Lewis Publishers.

Richardson, M. 1995. *Dictionary of Substance and Their Effects,* Vol. I-VI. New York: CRC Press/Lewis Publishers.

Sacarello, H. L. A. 1994. *The Comprehensive Handbook of Hazardous Materials.* New York: CRC Press/Lewis Publishers.

Sherman, J. D. 1988. *Chemical Exposure and Disease.* New York: Van Nostrand Reinhold.

Talbott, E., G. Craun. 1995. *An Introduction to Environmental Epidemiology.* New York: CRC Press/Lewis Publishers.

Torkelson, T. R. V. K. Rowe. 1981. *Patty's Industrial Hygiene and Toxicology,* Vol. 2B. New York: Wiley-Interscience.

MATERIAL SAFETY DATA SHEET

CHEMICAL NAME	
PROPYLENE	**CRC PRESS/LEWIS PUBLISHERS DISCLAIMER.** THE INFORMATION AND RECOMMENDATIONS PRESENTED HEREIN ARE BASED ON SOURCES BELIEVED TO BE RELIABLE. CRC MAKES NO REPRESENTATION ON ITS COMPLETENESS OR ACCURACY. IT IS THE RESPONSIBILITY OF THE USER TO DETERMINE THE CHEMICAL'S SUITABILITY FOR ITS INTENDED USE, THE CHEMICAL'S SAFE USE, AND THE CHEMICAL'S PROPER DISPOSAL. NO REPRESENTATIONS AND/OR WARRANTIES, EITHER EXPRESSED OR IMPLIED, OF THE MERCHANTABILITY OR FITNESS FOR A PARTICULAR PURPOSE, OR OF ANY OTHER NATURE, ARE MADE WITH RESPECT TO THE INFORMATION PROVIDED IN THIS MSDS OR TO THE CHEMICAL TO WHICH INFORMATION MAY REFER. CRC NEITHER ASSUMES NOR AUTHORIZES ANY OTHER PERSON TO ASSUME FOR IT, ANY OTHER ADDITIONAL RESPONSIBILITY OR LIABILITY FOR THE USE OF, OR RELIANCE UPON, THIS INFORMATION.

HAZARD WARNING INFORMATION

HEALTH	FIRE	REACTIVE	OTHER	DEGREE OF HAZARD	COLOR CODING	OTHER CODES
1	4	1		0 = Minimum Hazard 1 = Slight Hazard 2 = Moderate Hazard 3 = Serious Hazard 4 = Severe Hazard	HEALTH = BLUE FIRE = RED REACTIVITY = YELLOW OTHER = WHITE	OX = Oxidizer ACID = Acid ALK = Alkali COR = Corrosive W = Use No Water

SECTION I - GENERAL INFORMATION

Characterization	RCRA Number	EPA Class
Gas, Hydrocarbon	**None**	**Not Applicable**

DOT Proper Shipping Name	Chemical Abstract Service (CAS) Number
Propylene	**115-07-1**
DOT Hazard Class and Label Requirements	DOT Emergency Guide Code
Flammable Gas	**22**
DOT Identification Number	Chemical Formula
UN 1075 (compressed); UN 1077 (liquid)	**$CH_3CH{:}CH_2$**

Synonyms

Propene; methylethylene; methylethene.

SECTION II - HAZARDOUS INGREDIENTS AND IDENTITY INFORMATION

Hazardous Components (specific identity)	OSHA Exposure Criteria	NIOSH Exposure Criteria	Immediately Dangerous to Life and Health (IDLH)	ACGIH Exposure Criteria
Propylene (derivation: Catalytic and thermal cracking of ethylene with zeolite catalyst, from naphtha).	**PEL:** Not Established **STEL:** Not Established	**REL:** Not Established **STEL:** Not Established	**Not Determined**	**TLV:** Not Established **STEL:** Not Established

SECTION III - PHYSICAL AND CHEMICAL CHARACTERISTICS

Boiling Point	Specific Gravity (H_2O = 1)
-54°F (-48°C)	**0.51 (liquid)**
Vapor Pressure (atmospheres)	Molecular Weight
10 at 68°F (20°C)	**42**
Vapor Density (Air = 1)	Melting Point
1.46 at 32°F (0°C)	**-301°F (-185°C)**

Solubility

Slightly soluble in water (0.01%). Soluble in alcohol and ether.

Appearance and Odor

Colorless gas with a slight odor, or a liquid when under pressure.

SECTION IV - FIRE AND EXPLOSION HAZARD DATA

Flash Point (method used)	Explosive Limits in Air % by Volume
-162°F (-108°C) closed cup	**LEL: 2.0% UEL: 11.1%**
NFPA Classification	Autoignition Temperature
Flammable Gas	**927°F (497°C)**

Extinguishing Media

Stop flow of gas to stop fire. If this is not possible, letting fire burn itself out may be only alternative.

Special Fire Fighting Procedures

Propylene is an extremely flammable, very dangerous fire and explosion hazard. Poisonous gases are evolved from fire. Wear a self-contained breathing apparatus (SCBA) with full facepiece. Ensure ALL sources of escaping gas are controlled. Monitor area long after fire is extinguished. Move containers from fire area if it can be done without risk. Use water spray to cool fire-exposed containers.

Unusual Fire and Explosion Hazards

Gas is explosive. Containers may explode in flame. Even if fire is extinguished, escaping gas can form explosive mixtures in air quickly and ignite without warning. Burns with a yellow, sooty flame.

SECTION V - REACTIVITY DATA

Stability		Conditions to Avoid Propylene is normally stable when stored as a liquid in steel containers under its own pressure. Avoid contact with heat, sparks, flame, lighted tobacco products, and steam lines.
Stable X	Unstable	Incompatibility (*materials to avoid*) Propylene explodes in contact with trifluoromethyl hypofluorite. It can react vigorously with oxidizing materials. Reacts with oxides of nitrogen to form an explosive product.
Hazardous Polymerization		Conditions to Avoid Under normal conditions of temperature and pressure, hazardous polymerization of propylene is not expected to occur.
May Occur	Will Not Occur X	Hazardous Decomposition or By-products If involved in fire, propylene can produce acrid and irritating smoke, fumes, and gases such as carbon monoxide and/or lower molecular weight hydrocarbons.

SECTION VI - HEALTH HAZARD DATA

Primary Route(s) of Entry:	Inhalation?	Absorption (skin)?	Ingestion?
	X		

Health Hazards

INHALATION: An asphyxiant gas. Initial symptoms are those of oxygen starvation. Feeling of "air hunger," diminished mental alertness, headache, dullness, shortness of breath, and an impairment in muscle coordination. Continued exposures lead to faulty judgment, depression of senses, rapid fatigue, unconsciousness, convulsions, coma, and death.

SKIN & EYES: Contact with the compressed gas or refrigerated liquid can cause frostbite and burns.

INGESTION: Not likely since propylene is a gas at normal working (room) temperatures.

Carcinogenicity	NTP Listed?	IARC Cancer Review Group?	OSHA Regulated?	Target Organs?
Unknown Human Unknown Animal	No	No	No	Central nervous system

Medical Conditions Generally Aggravated by Exposure

None reported.

Emergency and First-aid Procedures

Eye contact: Immediately flush with large amounts of warm (tepid) water for 15 minutes (minimum), occasionally lifting eyelids; seek medical attention. Skin contact: Remove all contaminated clothing. Immediately wash area with large amounts of warm (tepid) water. Do NOT rub or use air-blown heat on affected area. For inhalation: Remove the person from exposure. Provide respiratory assistance and CPR. Transfer promptly to medical facility. If swallowed: Not likely.

SECTION VII - PRECAUTIONS FOR SAFE HANDLING AND USE

Steps to be Taken in Case Material is Released or Spilled

Remove all ignition sources. Ventilate area. Stop flow of gas. If the source of leak is a cylinder and cannot be stopped, remove to safe place in open air; repair leak or let cylinder empty. If gas has not yet ignited, use a water spray to knock down vapors and direct them from ignition source.

Preferred Waste Disposal Method

None specified in the references (incineration of a gas may be the only alternative).

Precautions to be Taken in Handling and Storage

Do not store with incompatible materials or in the presence of heat since violent reactions or explosions can occur. Store in tightly closed pressurized containers in a cool, dark, well-ventilated area away from sunlight. Keep away from ignition sources such as fire, sparks, and flame.

Other Precautions and Warnings

Bulk storage of propylene is not recommended. Containers may explode in fire or under conditions of extreme heat. Keep propylene out of confined areas or enclosed spaces (oxygen displacement hazard).

SECTION VIII - CONTROL MEASURES AND PERSONAL PROTECTIVE EQUIPMENT

Respiratory Protection (*specify type*)

Propylene is a simple asphyxiant. Do NOT use cartridge-type respirators (ineffective for this type of gas). Use a powered supplied-air respirator or a self-contained breathing apparatus with full facepiece operated in pressure demand or other positive pressure mode.

Ventilation

Local exhaust or general mechanical systems recommended.

Protective Gloves	Eye Protection	Other Protective Clothing
Insulated Neoprene or Polyurethane	Goggles and/or Face Mask	Leather Apron

Work/Hygiene Practices

Always wash hands thoroughly after using chemical; never bring food, drink, or smoking materials into vicinity of chemicals.

PROPYLENE

$CH_3CH{:}CH_2$

CAS: 115-07-1

IDENTIFICATION AND TYPICAL USES

Propylene is a colorless gas with a slight, indistinguishable odor. It also appears as a compressed, liquefied gas. Polypropylene is used to produce polypropylene (plastic) and in the manufacture of acetone, isopropanol, cumene, resins, synthetic rubber, gasoline, and propylene oxide.

RISK ASSESSMENT: HEALTH

General Assessment

Propylene is a simple asphyxiant gas. As such, its primary route of entry into the body is through *inhalation.* While it does not pose a significant physiological effect (i.e., no systemic poisoning or toxicity), it can displace oxygen levels to below that which is necessary to sustain life. Skin contact with the liquid can be destructive to tissues due to the extremely cold temperatures but absorption is not a likely exposure route. Since propylene usually exists as a gas at room temperatures, ingestion is not likely to occur either. It is not known to cause cancer, mutations, or reproductive effects in humans or test animals. It is an asphyxiant.

Inhalation of propylene gas can cause headache, dullness, shortness of breath, and the symptoms of asphyxiation. These include rapid respiration and a feeling of "air hunger" or air starvation followed by a decrease in mental alertness, and impaired muscular coordination. If exposure continues, there can be nausea, vomiting, prostration, unconsciousness, coma, and death.

Skin or eye contact with the vapor poses no serious threat. The vapors have not been shown to cause any physiological response. However, if the compressed liquid should contact either the skin or the eyes, there can be serious irritation, inflammation, deep and painful burns, tissue damage and destruction, and other symptoms of frostbite.

Acute Health Effects

The following acute (short-term) health effects may occur immediately or shortly after exposure to propylene:

Skin: Liquid threat only. Redness, pain, swelling, blisters, and severe frostbite.

Eye: Liquid threat only. Causes frostbite, redness, irritation, severe pain, and blurred vision.

Lung: Vapor is non-irritating and therefore poses no threat of injury to the lungs. However, in high concentrations, propylene can lead to symptoms of asphyxiation and oxygen deprivation up to and including death.

Chronic Health Effects

The following chronic (long-term) health effects may occur at some time after exposure to propylene and can last for months or even years:

Cancer Hazards: According to information presented in the references, exposure to propylene does not lead to carcinogenic effects (it is a simple asphyxiant).

Reproductive Hazards: There are no reports to support any claims of reproductive hazards.

Other Chronic Effects: None reported.

Recommended Risk-Reduction Measures

Personnel should avoid direct contact with propylene. If another gas cannot be substituted for a propylene, then *engineering controls* are the most effective method of reducing exposures. The best protection is to separate operations and provide local, explosion-proof exhaust ventilation at the site of chemical release. While not always operationally feasible, isolating operations can also reduce exposure. Using equipment with explosion-proof designs and ensuring all metal containers and tools are electrically grounded and bonded can further reduce the potential for injury or damage. Using respiratory protection is less effective than the controls mentioned above, but is still advisable whenever working with propylene. Air-purifying respirators (e.g., the cartridge-type or gas masks) cannot be used with propylene. These types of respirators only purify the air and can do nothing to protect in oxygen deficient atmospheres. A powered supplied-air respirator operated in pressure demand or other positive pressure mode, or a self-contained breathing apparatus (SCBA) with full facepiece operated in pressure demand mode are the recommended respiratory protection methods of choice. If a full facepiece is not available, then chemical goggles should be worn to protect the eyes. Whenever a chemical splash hazard exists, a face shield and a rubber or leather apron should be worn. To prevent hand and skin exposures with the liquid, thermal protective gloves should be used. Neoprene or polyethylene materials should suffice. However, it is always best to contact glove manufacturers and obtain permeation studies *before* final glove selections are made.

Administrative controls should also be in place to minimize the potential for human exposures. These may include written procedures or policies which specify the methods and techniques that will be practiced whenever personnel are to work with propylene.

All personnel should receive *training* on the use, hazards, protective measures, emergency actions, and other precautions per 29 CFR 1910.1200 (Hazard Communication), prior to the first assignment in an area where propylene is used or stored.

Other methods to reduce exposure to chemicals include:

☑ Where possible, consider installing automatic sensors that warn personnel of dropping oxygen levels and use local exhaust ventilation at the site of chemical release. If local exhaust ventilation or enclosure is not used, respiratory protection should be mandatory.

☑ Always ensure that proper protective clothing is worn when using chemical substances and that personnel have been trained in its use and care.

☑ Wash thoroughly immediately after exposure to propylene and at the end of the work shift or before eating, drinking, or smoking.

☑ Safety shower and eyewash stations should be readily available in work areas where propylene is used or stored.

☑ Hazard warning information should be posted in the work area. In addition, as part of an on-going education and training program, all information on the health and safety hazards of propylene should be communicated to all exposed or potentially exposed workers.

RISK ASSESSMENT: ENVIRONMENT

General Assessment

The environment is at risk of exposure during transportation, storage, disposal, or destruction of propylene. In almost every scenario, the threat of environmental exposure is contingent upon the proper handling of the chemical substance. Accidental releases, large or small, can result in fire, explosion, and possible contamination of the surrounding environmental mediums (water, soil, and air). Propylene is an extremely flammable and highly explosive gas. Its extremely low flash point and boiling point present a serious fire hazard. It has a relatively narrow flammability range in air. Also, since it can react violently in the presence of oxidizers, such as chlorine, propylene requires special consideration during any emergency involving a leak or release of propylene gas or liquid.

Propylene can enter the environment through industrial discharges, unchecked venting to air, or through spills of the liquid. Since liquids will most likely vaporize rapidly, water and soil contamination are possible but not probable.

Acute Ecological Effects

Acute (short-term) toxic effects may include the death of animals, birds, or fish, and death or low growth rate in plants. Acute effects are seen 2 to 4 days after animals or plants are exposed to propylene.

Polypropylene has slight acute toxicity of aquatic life. Propylene has caused no effects to rice plants at concentrations of 1 and 100 ppm. Insufficient data are available on the short-term effects of propylene on birds or land animals. However, as a relatively nontoxic, simple asphyxiant, it might be assumed that its toxicity will be relatively low.

Chronic Ecological Effects

Chronic (long-term) toxic effects may include shortened life span, reproductive problems, lower fertility, and changes in appearance or behavior in exposed animals.

These effects can be seen long after first exposure(s) to toxic chemicals.

Polypropylene has slight chronic toxicity to aquatic life. Insufficient data are available to evaluate the long-term effects of propylene to plants, birds, or land animals. However, as a relatively non-toxic, simple asphyxiant, it might be assumed that its chronic ecological toxicity will be relatively low.

Water Solubility

Propylene is soluble in water at 446 milliliters of gas/liter.

Persistence in the Environment

Propylene is non-persistent in water, with a half-life of less than 2 days. The half-life of a pollutant is the amount of time it takes for one half of the chemical to be degraded. Virtually 100% of propylene will eventually end up in the air.

Bioaccumulation in Aquatic Organisms

Some substances increase in concentration, or *bioaccumulate,* in living organisms as they breathe contaminated air, drink contaminated water, or eat contaminated food. These chemicals can become concentrated in the tissues and internal organs of animals as well as humans.

The concentration of polypropylene found in the edible tissues of fish is expected to be about the same as the average concentration of propylene in the water from which the fish was taken.

Recommended Risk-Reduction Measures

Proper training of all transporters will reduce the likelihood of a mishap or accident resulting in a leak or release to the environment. The correct use of DOT labeling (per the MSDS) on all containers, trucks, and rail cars will enable emergency responders to react properly and quickly to any disaster thereby reducing the potential risk to the environment and to personnel.

Storage of propylene should be segregated from incompatible chemicals to minimize the risk of cross-contamination or contact. Buildings designated for storage should be equipped with appropriate fire protection systems (alarms, sprinklers, emergency lighting, portable extinguishers).

If a spill or release to the environment has occurred, fire department, emergency response, and/or hazardous materials spill personnel should be notified immediately. Cleanup should be attempted only by those trained in proper containment procedures using non-sparking tools. Contaminated soils should be removed for incineration and replaced with clean soil. When spills or leaks occur, it may be necessary to contact the local and/or state emergency response authorities. A comprehensive emergency response or disaster preparedness/recovery plan should be in place prior to any operations involving the use, transportation, storage, or disposal of propylene. If propylene is spilled or leaked, the following specific steps are recommended:

- ☑ Restrict persons not wearing protective clothing from area of spill or leak until cleanup is complete and area can be opened for normal work.
- ☑ Remove ignition sources, and ventilate area
- ☑ If applicable, stop flow of leaking liquid or gas. If leak source is a cylinder and the leak cannot be stopped in place, remove leaking cylinder to a safe place in the open air, and allow cylinder to empty.
- ☑ Keep propylene out of a confined space, such as a sewer, because of the possibility of explosion (unless the sewer is designed to prevent the buildup of explosive concentrations).
- ☑ It may be necessary to dispose of propylene as a hazardous waste. The responsible state agency or the regional office of the federal Environmental Protection Agency (EPA) should be contacted for specific recommendations.

RISK ASSESSMENT: BUSINESS

General Assessment

Accidents or mishaps involving propylene can present a significant threat to business operations. The loss or damage of equipment or facilities can significantly affect fiscal viability. Lawsuits that may result from personnel injury, public exposures, and environmental contamination will also require a serious expenditure of resources.

Media attention surrounding an injury, illness, or environmental damage can also result in a loss of profits and loss of current as well as future business.

Recommended Risk-Reduction Measures

Company attorneys, safety and health professionals, and environmental specialists should be involved in the

development of any procedures dealing with safety and emergency response. A company official should be pre-designated as a public relations officer with specific training in dealing with the press. Corporate plans and policies should be developed, approved, and implemented long before any need for such arises.

REFERENCES

Ahlbom, A. 1993. *Biostatistics for Epidemiologists.* New York: CRC Press/Lewis Publishers.

American Conference of Governmental Industrial Hygienists. 1988. *Documentation of the Threshold Limit Values and Biological Exposure Indices,* 5th Edition (with updates). Cincinnati: ACGIH

Anthony, C. P., and N. J. Kolthoff. 1971. *Textbook of Anatomy and Physiology.* St. Louis: C. V. Mosby Company.

Asante-Duah, D. K. 1993. *Hazardous Waste Risk Assessment.* New York: CRC Press/Lewis Publishers.

Bloom, A. D. and F. J. DeSerres. 1995. *Ecotoxicity and Human Health.* New York: CRC Press/Lewis Publishers.

Brusick, D. J. 1994. *Methods for Genetic Risk Assessment.* New York: CRC Press/Lewis Publishers.

Bueche, J. R. 1972. *Principles of Physics,* 2nd Edition. New York: McGraw-Hill.

Burrell, R., D. K. Flaherty, and L. J. Sauers. 1992. *Toxicology of the Immune System.* New York: Van Nostrand Reinhold.

Calabrese, E. J. 1994. *Biological Effects of Low Level Exposures.* New York: CRC Press/Lewis Publishers.

Cockerman, L. G., B. S. Shane. 1994. *Basic Environmental Toxicology.* New York: CRC Press/Lewis Publishers.

Cothern, C. R. 1995. *Handbook for Environmental Risk Decisionmaking.* New York: CRC Press/Lewis Publishers.

Danse, I. R. 1991. *Common Sense Toxics in the Workplace.* New York: Van Nostrand Reinhold.

Environmental Protection Agency. 1983. *Chemical Hazard Information Profiles.* Washington, D.C.: U.S. EPA.

Fodden, C. R., and J. L. Weddell. 1992. *Hazardous Materials Emergency Action Data.* New York: CRC Press/Lewis Publishers.

Forseberg, K., and L. H. Keith. 1993. *Chemical; Protective Clothing Permeation and Degradation Database.* Chelsea, MI: Lewis Publishers.

Gardner, R. W. 1994. *Chemical Intolerance.* New York: CRC Press/Lewis Publishers.

Hallenbeck, W. H. 1993. *Quantitative Risk Assessment for Environmental and Occupational Health,* 2nd Edition. New York: CRC Press/Lewis Publishers.

Hodgson, E., R. B. Mailman, and J. E. Chambers. 1988. *Dictionary of Toxicology.* London: The MacMillian Company.

Hosty, J. W., and P. Foster. 1990. *A Practical Guide to Chemical Spill Response.* New York: Van Nostrand Reinhold.

Howard, P. H. 1995. *Handbook of Environmental Fate and Exposures Data for Organic Chemicals,* Vol. I-IV. New York: CRC Press/Lewis Publishers.

James, A., and M. Lord. 1992. *Index of Chemical and Physical Data.* New York: Van Nostrand Reinhold.

Jones, F. E. 1994. *Toxic Organic Vapors in the Workplace.* New York: CRC Press/Lewis Publishers.

Kamrin, M. A. 1988. *Toxicology: A Primer on Principles and Applications.* Chelsea, Michigan: Lewis Publishers.

Keith, L. H. and D. B. Walters. (Editors) 1992. *National Toxicology Program's Chemical Database, Volumes I-VII.* Chelsea, MI: Lewis Publishers.

Landis, W. G., M. H. Yu. 1995. *Introduction to Environmental Toxicology.* New York: CRC Press/Lewis Publishers.

Langerman, N. 1994 *Precautionary Labels for Chemical Containers.* New York: CRC Press/Lewis Publishers.

Lauwerys, R. R., P. Hoet. 1993. *Industrial Chemical Exposure.* New York: CRC Press/Lewis Publishers.

Lewis, R. J., Sr. 1993. *Hawley's Condensed Chemical Dictionary,* 12th Edition. New York: Van Nostrand Reinhold.

Lewis, R. J., Sr. 1992. *Sax's Dangerous Properties of Industrial Materials,* Volumes 1, 2, and 3, 8th Edition. New York: Van Nostrand Reinhold.

Manahan, S. E. 1992. *Toxicological Chemistry,* 2nd Edition. New York: CRC Press/Lewis Publishers.

Meyer, E. 1990. *Chemistry of Hazardous Materials.* Englewood Cliffs, NJ: Prentice Hall.

Patnaik, P. 1992. *A Comprehensive Guide to the Hazardous Properties of Chemical Substances.* New York: Van Nostrand Reinhold.

Philp, R. B. 1995. *Environmental Hazards and Human Health.* New York: CRC Press/Lewis Publishers.

Proctor, N. H., and J. P. Hughes. 1991. *Chemical Hazards of the Workplace.* New York: Van Nostrand Reinhold.

Rea, W. J. 1995. *Chemical Sensitivity,* Vol. I-III. New York: CRC Press/Lewis Publishers.

Richardson, M. 1995. *Dictionary of Substance and Their Effects,* Vol. I-VI. New York: CRC Press/Lewis Publishers.

Sacarello, H. L. A. 1994. *The Comprehensive Handbook of Hazardous Materials.* New York: CRC Press/Lewis Publishers.

Sherman, J. D. 1988. *Chemical Exposure and Disease.* New York: Van Nostrand Reinhold.

Talbott, E., G. Craun. 1995. *An Introduction to Environmental Epidemiology.* New York: CRC Press/Lewis Publishers.

Torkelson, T. R. V. K. Rowe. 1981. *Patty's Industrial Hygiene and Toxicology,* Vol. 2B. New York: Wiley-Interscience.

MATERIAL SAFETY DATA SHEET

CHEMICAL NAME	
PYRIDINE	**CRC PRESS/LEWIS PUBLISHERS DISCLAIMER:** THE INFORMATION AND RECOMMENDATIONS PRESENTED HEREIN ARE BASED ON SOURCES BELIEVED TO BE RELIABLE. CRC MAKES NO REPRESENTATION ON ITS COMPLETENESS OR ACCURACY. IT IS THE RESPONSIBILITY OF THE USER TO DETERMINE THE CHEMICAL'S SUITABILITY FOR ITS INTENDED USE, THE CHEMICAL'S SAFE USE, AND THE CHEMICAL'S PROPER DISPOSAL. NO REPRESENTATIONS AND/OR WARRANTIES, EITHER EXPRESSED OR IMPLIED, OF THE MERCHANTABILITY OR FITNESS FOR A PARTICULAR PURPOSE, OR OF ANY OTHER NATURE, ARE MADE WITH RESPECT TO THE INFORMATION PROVIDED IN THIS MSDS OR TO THE CHEMICAL TO WHICH INFORMATION MAY REFER. CRC NEITHER ASSUMES NOR AUTHORIZES ANY OTHER PERSON TO ASSUME FOR IT, ANY OTHER ADDITIONAL RESPONSIBILITY OR LIABILITY FOR THE USE OF, OR RELIANCE UPON, THIS INFORMATION.

NATIONAL FIRE PROTECTION ASSOCIATION (NFPA) LABELING INFORMATION

HEALTH	FIRE	REACTIVE	OTHER	DEGREE OF HAZARD	COLOR CODING	OTHER CODES
2	3	0		0 = Minimum Hazard 1 = Slight Hazard 2 = Moderate Hazard 3 = Serious Hazard 4 = Severe Hazard	HEALTH = BLUE FIRE = RED REACTIVITY = YELLOW OTHER = WHITE	OX = Oxidizer ACID = Acid ALK = Alkali COR = Corrosive W = Use No Water

SECTION I - GENERAL INFORMATION

Characterization	RCRA Number	EPA Class
Heterocyclic Compound (of Nitrogen)	**U196**	**Toxic Waste**

DOT Proper Shipping Name **Pyridine**	Chemical Abstract Service (CAS) Number **110-86-1**
DOT Hazard Class and Label Requirements **Flammable Liquid; Poison**	DOT Emergency Guide Code **26**
DOT Identification Number **UN 1282**	Chemical Formula C_5H_5N

Synonyms
Azabenzene; azine.

SECTION II - HAZARDOUS INGREDIENTS AND IDENTITY INFORMATION

Hazardous Components (specific identity)	OSHA Exposure Criteria	NIOSH Exposure Criteria	Immediately Dangerous to Life and Health (IDLH)	ACGIH Exposure Criteria
Pyridine (derivation: By coal carbonization and recovery both from coke-oven gases and the coal tar middle oil; also produced synthetically from acetaldehyde and ammonia). **1 ppm = 3.29 mg/m^3**	**PEL:** 5 ppm 15 mg/m^3 **STEL:** Not Established	**REL:** 5 ppm 15 mg/m^3 **STEL:** Not Established	**1000 ppm**	**PEL:** 5 ppm 15 mg/m^3 **STEL:** 10 ppm 30 mg/m^3

SECTION III - PHYSICAL AND CHEMICAL CHARACTERISTICS

Boiling Point **239°F (115°C)**	Specific Gravity (H_2O = 1) **0.98**
Vapor Pressure (mm Hg) **16 at 68°F (20°C)**	Molecular Weight **108.1**
Vapor Density (Air = 1) **2.7**	Melting Point **-44°F (-42°C)**

Solubility
Miscible in water. Soluble in alcohol, ether, benzene, ligroin, and fatty oils.

Appearance and Odor
Colorless to slightly yellow liquid with a strong, nauseating, fish-like odor and burning taste. Odor Threshold = 0.17ppm

SECTION IV - FIRE AND EXPLOSION HAZARD DATA

Flash Point (method used) **68°F (20°C) closed cup**	Explosive Limits in Air % by Volume **LEL: 1.83% UEL: 12.4%**
NFPA Classification **Class 1B Flammable Liquid**	Autoignition Temperature **900°F (482°C)**

Extinguishing Media
Carbon dioxide, dry chemical, or foam. Water may be ineffective and could actually spread fire further.

Special Fire Fighting Procedures
Wear full protective clothing and self-contained breathing apparatus (SCBA). Poisonous gases are produced in fire. Move containers from fire area if it can be done without risk.

Unusual Fire and Explosion Hazards
Moderately explosive when exposed to flame. Vapors are heavier than air and may travel for some distance to an ignition source to flashback and cause fire. Vapors form an explosive mixture in air.

SECTION V - REACTIVITY DATA

Stability		Conditions to Avoid: Pyridine is stable in tightly closed containers under normal conditions of storage and handling. Avoid contact with incompatible materials and sources of heat.
Stable	Unstable	Incompatibility (materials to avoid): Incompatible with strong oxidizers (such as chlorine, bromine, and fluorine), and strong acids (such as sulfuric, nitric, chlorosulfonic, and hydrochloric). Contact with acids can cause exothermic reaction.
X		
Hazardous Polymerization		Conditions to Avoid: Under normal conditions of temperature and pressure, hazardous polymerization of pyridine is not expected to occur.
May Occur	Will Not Occur	Hazardous Decomposition or By-products: When heated to decomposition, pyridine emits acrid and irritating smoke including cyanides and toxic oxides of nitrogen and carbon.
	X	

SECTION VI - HEALTH HAZARD DATA

Primary Route(s) of Entry:	Inhalation?	Absorption (skin/eye)?	Ingestion?
	X	X	X

Health Hazards

INHALATION: Irritation to the eyes, nose, throat, and mucosa of the respiratory tract. There can also be asthmatic breathing, headache, dizziness, nausea, upset stomach, and depression of the central nervous system with mental changes, personality changes, and confusion.

ABSORPTION: Skin contact causes dryness and irritation. Absorption is likely. Eye contact causes tearing and irritation with the possibility for permanent damage to vision.

INGESTION: Poisonous to the heart and can stimulate bone marrow production of red platelets.

Carcinogenicity	NTP Listed?	IARC Cancer Review Group?	OSHA Regulated?	Target Organs?
Unknown Human Unknown Animal	No	No	29 CFR 1910.1000 Table Z-1	CNS, skin, eyes, liver, kidneys, GI tract.

Medical Conditions Generally Aggravated by Exposure

None reported.

Emergency and First-aid Procedures

Eye contact: Do not allow victim to keep eyes tightly shut. Flush immediately with water for 15 minutes (minimum); seek medical attention. Skin contact: Remove all contaminated clothing. Immediately wash area with large amounts of soap and water. For inhalation: Remove the person from exposure. Provide respiratory assistance and CPR. Transfer to medical facility. If swallowed: Seek medical attention immediately. Give 1-2 glasses of milk or water and induce vomiting. Never give anything by mouth to an unconscious or convulsing person.

SECTION VII - PRECAUTIONS FOR SAFE HANDLING AND USE

Steps to be Taken in Case Material is Released or Spilled

Vacuum or pump the spilled material into close containers using non-sparking tools or absorb in vermiculite or other material and place in sealed drums. Ventilate area of spill or leak. Remove all ignition sources. Restrict those not involved in cleanup from entering area. Notify appropriate authorities.

Preferred Waste Disposal Method

Burn in a chemical incinerator equipped with an afterburner and scrubber.

Precautions to be Taken in Handling and Storage

Store to avoid contact with strong oxidizers, since violent reactions can occur. Store in tightly closed containers in cool, well-ventilated area away from heat. Electrically ground and bond all equipment.

Other Precautions and Warnings

Sources of ignition (heat, flame, incompatibles) are prohibited where pyridine is used, handled, or stored. Bulk storage of pyridine is not recommended. Protect containers from physical damage.

SECTION VIII - CONTROL MEASURES AND PERSONAL PROTECTIVE EQUIPMENT

Respiratory Protection *(specify type)*

For exposures over the PEL, use an MSHA/NIOSH-approved respirator with an organic vapor cartridge. For best protection, use a supplied-air respirator with full facepiece operated in continuous flow mode, or a self-contained breathing apparatus (SCBA) with full facepiece operated in positive pressure mode.

Ventilation

Local exhaust (preferred) at site of chemical work or general ventilation.

Protective Gloves	Eye Protection	Other Protective Clothing
Impervious Gloves	Chemical Goggles or Face Mask	Boots; Aprons; Gauntlets

Work/Hygiene Practices

Always wash hands thoroughly after using chemical; never bring food, drink, or smoking materials into vicinity of chemicals.

PYRIDINE

C_5H_5N

CAS: 110-86-1

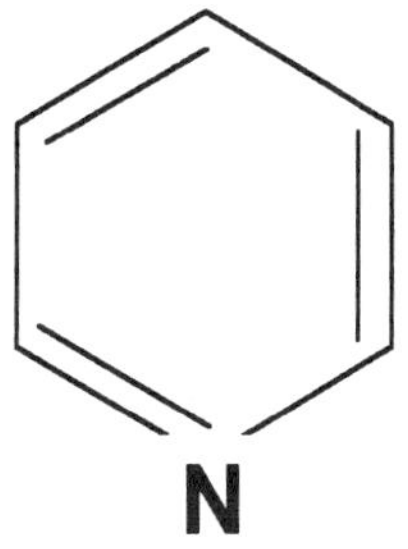

IDENTIFICATION AND TYPICAL USES

Pyridine is a colorless to yellow liquid with a nauseating, fish-like odor. It is used in the synthesis of vitamins and drugs, as a solvent in waterproofing, in rubber chemicals, as a denaturant for alcohol and antifreeze mixtures, in dyeing textiles, as an intermediate in pharmaceuticals, a reagent for cyanide analysis, and in fungicides. It also occurs in coal tar.

RISK ASSESSMENT: HEALTH

General Assessment

Pyridine is a toxic compound. It can enter the body by *inhalation, ingestion,* and is moderately toxic by skin contact (*absorption*). While its carcinogenic properties have not been adequately investigated, mutation data have been reported. There are no data on its teratogenic or reproductive effects in humans or animals.

Inhalation of pyridine can cause dizziness, headache, nervousness, nausea, insomnia, asthma-like symptoms, anorexia, frequent urination, and abdominal pain. Chronic inhalation causes gastrointestinal upset and kidney and liver damage. It may stimulate bone marrow to increase the production of blood platelets. It may also be a heart poison. Pyridine is also a central nervous system (CNS) depressant. Chronic exposure may cause brain damage with personality changes.

Skin contact with the liquid or vapor can cause irritation and redness. Burning and watering of eyes have been reported on contact with the vapors of pyridine.

If ingested, there can be gastrointestinal congestion along with many of the systemic effects noted for inhalation, primarily depression of the CNS, heart effects, and bone marrow stimulation.

Acute Health Effects

The following acute (short-term) health effects may occur immediately or shortly after exposure to pyridine:

Skin: Irritation with burning, itching, redness, and possible swelling. Absorption is likely to occur, readily producing toxic systemic effects.

Eye: Severe irritation of the conjunctiva with burning, itching, and watering. There is a possibility of permanent damage.

Lung: Irritation of the eyes, nose, throat, mucous membranes, bronchial tubes and other members of the respiratory tract.

CNS: Narcotic effects at high concentrations with dizziness, fatigue, headache, weakness, insomnia, anorexia, and other symptoms of CNS depression.

Chronic Health Effects

The following chronic (long-term) health effects may occur at some time after exposure to pyridine and can last for months or even years:

Cancer Hazards: According to information contained in the references, there is no evidence that pyridine causes cancer in test animals. However, it may cause mutations (genetic changes) in living cells and many scientists believe that long-term exposure to such chemicals may pose a cancer threat.

Reproductive Hazard: According to the references, pyridine has not been adequately tested for its ability to adversely affect reproduction.

Other Chronic Effects: Long-term exposures can cause severe liver or kidney damage. Pyridine can damage the nervous system and cause possible brain damage, with confusion, personality changes, and mental imbalance. Pyridine can also cause a skin allergy. If an allergy develops, very low future exposures can cause itching and skin rash. Repeated exposure can cause headache, nausea, trouble sleeping, and back pain.

❶ *Recommended Risk-Reduction Measures*

Personnel should avoid direct contact with pyridine. If a less toxic chemical cannot be substituted for a hazardous substance, then *engineering controls* are the most effective method of reducing exposures. The best protection is to enclose operations and/or provide local exhaust ventilation at the site of chemical release. While not always operationally feasible, isolating operations can also reduce exposure. Using respiratory protection is less effective than the controls mentioned above, but is still advisable whenever working with or around pyridine. For exposures over the PEL (5 ppm), use an MSHA/NIOSH-approved full facepiece respirator equipped with an organic vapor cartridge. For the best protection, a supplied-air respirator or a self-contained breathing apparatus (SCBA) with full facepiece operated in continuous flow or positive pressure mode are recommended. If a full facepiece is not available, then chemical goggles should be worn to protect the eyes. A face shield and a protective apron should also be worn. To prevent hand and skin exposures, impervious gloves should be used. Glove manufacturers should be contact and permeation studies obtained *before* glove selections have been made.

Administrative controls should also be in place to minimize the potential for human exposures. These may include written procedures or policies which specify the methods and techniques that will be practiced whenever personnel are to work with pyridine.

All personnel should receive *training* on the use, hazards, protective measures, emergency actions, and other precautions per 29 CFR 1910.1200 (Hazard Communication), prior to the first assignment in an area where pyridine is used or stored.

Before beginning employment, and at regular intervals thereafter (e.g., annually), the following medical tests are recommended for personnel working with or around pyridine:

☑ Liver function tests.

If symptoms develop or overexposure is suspected, the following additional tests are recommended:

☑ Examination of the central nervous system.

☑ Kidney function tests.

☑ Evaluation by a qualified allergist, to include careful evaluation of exposure history and special testing (may help diagnose skin allergy).

☑ Complete blood count (CBC) with checks for platelet levels.

Any evaluation should include a careful history of past and present symptoms with an examination. Medical tests that simply look for existing damage are not a substitute for controlling exposures. Also, because smoking can cause heart disease, emphysema, and other respiratory dysfunction, smokers may develop exposure symptoms more quickly than non-smokers under the same conditions of exposure. Prudent risk management requires careful consideration of *all* possible exposure factors when assessing health exposure risk.

Other methods to reduce exposure include:

☑ Where possible, enclose operations and use local exhaust ventilation at the site of chemical release. If local exhaust ventilation or enclosure is not used, respiratory protection should be mandatory.

☑ Always ensure that proper protective clothing is worn when using chemical substances.

☑ Wash thoroughly immediately after exposure to pyridine and at the end of the work shift or before eating, drinking, or smoking. Avoid contact with direct sunlight.

☑ Hazard warning information should be posted in the work area. In addition, as part of an on-going education and training program, all information on the health and safety hazards of pyridine should be communicated to all potentially exposed workers.

☑ Eye wash stations should be provided in the immediate work area for emergency use. If there is a possibility of skin exposure to pyridine, emergency shower facilities should also be provided.

RISK ASSESSMENT: ENVIRONMENT

General Assessment

The environment is at risk of exposure during transportation, storage, disposal, or destruction of pyridine. In almost every scenario, the threat of environmental exposure is contingent upon the proper handling of the chemical substance. Accidental spills, large or small, can result in fire, explosion (depending upon conditions of spill), and possible contamination of the surrounding environmental mediums (water, soil, and air).

Pyridine is considered a Class 1B flammable liquid (per OSHA 29 CFR 1910.106). Its relatively low flash point and moderate boiling point present a serious fire and explosion hazard. Also, because it is incompatible with many common oxidizers and acids, extreme caution is required in handling, storage, transportation, and disposal of pyridine. Hazardous reactions can occur when pyridine come into contact with chlorosulfonic acid, chromium trioxide, maleic anhydride, nitric acid, oleum iodine, perchromates, and sulfuric acid. Because it is a relatively weak base, pyridine reacts exothermically (evolves heat) with acids. It is a dangerous fire hazard in the form of vapor when exposed to flame or spark. These characteristics also require special consideration during any emergency situation involving a leak or spill of pyridine. Should pyridine ever come into contact with incompatible substances either during use, transportation, storage, or disposal, the formation of highly toxic and/or highly explosive commodities is extremely possible.

Pyridine can enter the environment from non-point sources, as well as industrial and municipal discharges, and from accidental spills.

Acute Ecological Effects

Acute (short-term) toxic effects may include the death of animals, birds, or fish and death or low growth rate in plants. Acute effects are seen 2 to 4 days after animals or plants are exposed to pyridine.

Pyridine has moderate acute toxicity to aquatic life, birds, and land animals. It has caused bud kills in an agricultural crop.

Chronic Ecological Effects

Chronic (long-term) toxic effects may include shortened life span, reproductive problems, lower fertility, and changes in appearance or behavior in exposed animals. These effects can be seen long after first exposure(s) to toxic chemicals.

Pyridine has moderate chronic toxicity to aquatic life. Insufficient data are available to evaluate or predict the long-term effects of pyridine to plants, birds, or land animals.

Water Solubility

Pyridine is highly soluble in the aquatic environment. Concentrations of 1000 milligrams and more will mix with a liter of water.

Persistence in the Environment

Pyridine is moderately persistent in water, with a half-life between 2 and 20 days. The half-life of a pollutant is the amount of time it takes for one half of the chemical to be degraded. About 57% of pyridine will eventually end up in water; the rest will end up in the air.

Bioaccumulation in Aquatic Organisms

Some substances increase in concentration, or *bioaccumulate,* in living organisms as they breathe contaminated air, drink contaminated water, or eat contaminated food. These chemicals can become concentrated in the tissues and internal organs of animals as well as humans.

The concentration of pyridine found in fish tissues is expected to be about the same as the average concentration of pyridine in the water from which the fish was taken.

Recommended Risk-Reduction Measures

Proper training of all transporters will reduce the likelihood of a mishap or accident resulting in a leak or spill of pyridine into the environment. The correct use of DOT labeling (per the MSDS) on all containers, trucks, and rail cars will enable emergency responders to react properly and quickly to any disaster thereby reducing the potential risk to the environment and to personnel.

Storage of pyridine should be segregated from incompatible chemicals to minimize the risk of cross-contamination or contact. To prevent static sparks, all containers and equipment used in shipping, receiving, or transferring operations, should be grounded and bonded. Storage facilities should be designed with proper fire protection and prevention.

If a spill or leak to the environment has occurred, fire department, emergency response and/or hazardous materials spill personnel should be notified immediately. Cleanup should be attempted only by those trained in proper spill containment procedures. Contaminated soils should be removed for incineration and replaced with clean soil. If pyridine should contact the water table, aquifer, or navigable waterway, the local and/or state emergency response authorities must be notified. A comprehensive emergency response or disaster preparedness/recovery plan should be in place prior to any operations involving the use, transportation, storage, or disposal of pyridine. If pyridine is spilled or leaked, the following specific steps are recommended:

- ☑ Restrict persons not wearing protective clothing from area of spill or leak until cleanup is complete and area can be opened for normal work.
- ☑ Remove all ignition sources.
- ☑ Ventilate area of spill or leak.
- ☑ Vacuum or pump spilled pyridine into sealed containers for disposal. Use non-sparking tools. Alternatively, absorb liquids in vermiculite or other similar material and place in sealed drums for disposal.
- ☑ It may be necessary to dispose of pyridine as a hazardous waste. The responsible state agency or the regional office of the federal Environmental Protection Agency (EPA) should be contacted for specific recommendations.

RISK ASSESSMENT: BUSINESS

General Assessment

Accidents or mishaps involving pyridine can present a significant threat to business operations. The loss or damage of equipment or facilities can significantly affect fiscal viability. Lawsuits that may result from personnel injury/death, public exposures, and/or environmental contamination will also require a serious expenditure of resources.

Media attention surrounding an injury, death, or environmental damage can also result in a loss of profits and loss of current as well as future business.

Recommended Risk-Reduction Measures

Company attorneys, safety and health professionals, and environmental specialists should be involved in the development of any procedures dealing with safety or emergency response. A company official should be pre-designated as a public relations officer with specific training in dealing with the press. Corporate plans and policies should be developed, approved, and implemented long before any need for such arises.

REFERENCES

American Conference of Governmental Industrial Hygienists. 1988. *Documentation of the Threshold Limit Values and Biological Exposure Indices,* 5th Edition (with updates). Cincinnati: ACGIH

Asante-Duah, D. K. 1993. *Hazardous Waste Risk Assessment.* New York: CRC Press/Lewis Publishers.

Bloom, A. D. and F. J. DeSerres. 1995. *Ecotoxicity and Human Health.* New York: CRC Press/Lewis Publishers.

Burrell, R., D. K. Flaherty, and L. J. Sauers. 1992. *Toxicology of the Immune System.* New York: Van Nostrand Reinhold.

Calabrese, E. J. 1994. *Biological Effects of Low Level Exposures.* New York: CRC Press/Lewis Publishers.

Cockerman, L. G., B. S. Shane. 1994. *Basic Environmental Toxicology.* New York: CRC Press/Lewis Publishers.

Danse, I. R. 1991. *Common Sense Toxics in the Workplace.* New York: Van Nostrand Reinhold.

Fodden, C. R., and J. L. Weddell. 1992. *Hazardous Materials Emergency Action Data.* New York: CRC Press/Lewis Publishers.

Gardner, R. W. 1994. *Chemical Intolerance.* New York: CRC Press/Lewis Publishers.

Hodgson, E., R. B. Mailman, and Janice E. Chambers. 1988. *Dictionary of Toxicology.* London: The MacMillian Company.

Howard, P. H. 1995. *Handbook of Environmental Fate and Exposures Data for Organic Chemicals,* Vol. I-IV. New York: CRC Press/Lewis Publishers.

Jones, F. E. 1994. *Toxic Organic Vapors in the Workplace.* New York: CRC Press/Lewis Publishers.

Keith, L. H. and M. M. Walker. 1995. *Handbook of Air Toxics.* New York: CRC Press/Lewis Publishers.

Landis, W. G., M. H. Yu. 1995. *Introduction to Environmental Toxicology.* New York: CRC Press/Lewis Publishers.

Lauwerys, R. R., P. Hoet. 1993. *Industrial Chemical Exposure.* New York: CRC Press/Lewis Publishers.

Lewis, R. J., Sr. 1993. *Hawley's Condensed Chemical Dictionary,* 12th Edition. New York: Van Nostrand Reinhold.

Lewis, R. J., Sr. 1992. *Sax's Dangerous Properties of Industrial Materials,* Volumes 1, 2, and 3, 8th Edition. New York: Van Nostrand Reinhold.

Manahan, S. E. 1994. *Environmental Chemistry,* 6th Edition. New York: CRC Press/Lewis Publishers.

Meyer, E. 1990. *Chemistry of Hazardous Materials.* Englewood Cliffs, NJ: Prentice Hall.

Mycock, J. C., J. D. McKenna, and L. Theodore. 1995. *Handbook of Air Pollution Control Engineering and Technology.* New York: CRC Press/Lewis Publishers.

Proctor, N. H., J. P. Hughes. 1991. *Chemical Hazards of the Workplace.* New York: Van Nostrand Reinhold.

Rea, W. J. 1995. *Chemical Sensitivity,* Vol. I-III. New York: CRC Press/Lewis Publishers.

Richardson, M. 1995. *Dictionary of Substance and Their Effects,* Vol. I-VI. New York: CRC Press/Lewis Publishers.

Sacarello, H. L. A. 1994. *The Comprehensive Handbook of Hazardous Materials.* New York: CRC Press/Lewis Publishers.

Torkelson, T. R. V. K. Rowe. 1981. *Patty's Industrial Hygiene and Toxicology,* Vol. 2B. New York: Wiley-Interscience.

U.S. Department of Health and Human Services, National Institute for Occupational Safety and Health. 1994. *NIOSH Pocket Guide to Chemical Hazards.* Washington, D.C.: U.S. Government Printing Office.

MATERIAL SAFETY DATA SHEET

CHEMICAL NAME	
QUARTZ	**CRC PRESS/LEWIS PUBLISHERS DISCLAIMER:** THE INFORMATION AND RECOMMENDATIONS PRESENTED HEREIN ARE BASED ON SOURCES BELIEVED TO BE RELIABLE. CRC MAKES NO REPRESENTATION ON ITS COMPLETENESS OR ACCURACY. IT IS THE RESPONSIBILITY OF THE USER TO DETERMINE THE CHEMICAL'S SUITABILITY FOR ITS INTENDED USE, THE CHEMICAL'S SAFE USE, AND THE CHEMICAL'S PROPER DISPOSAL. NO REPRESENTATIONS AND/OR WARRANTIES, EITHER EXPRESSED OR IMPLIED, OF THE MERCHANTABILITY OR FITNESS FOR A PARTICULAR PURPOSE, OR OF ANY OTHER NATURE, ARE MADE WITH RESPECT TO THE INFORMATION PROVIDED IN THIS MSDS OR TO THE CHEMICAL TO WHICH INFORMATION MAY REFER. CRC NEITHER ASSUMES NOR AUTHORIZES ANY OTHER PERSON TO ASSUME FOR IT, ANY OTHER ADDITIONAL RESPONSIBILITY OR LIABILITY FOR THE USE OF, OR RELIANCE UPON, THIS INFORMATION.

HAZARD WARNING INFORMATION

HEALTH	FIRE	REACTIVE	OTHER	DEGREE OF HAZARD	COLOR CODING	OTHER CODES
2	0	0		0 = Minimum Hazard 1 = Slight Hazard 2 = Moderate Hazard 3 = Serious Hazard 4 = Severe Hazard	HEALTH = BLUE FIRE = RED REACTIVITY = YELLOW OTHER = WHITE	OX = Oxidizer ACID = Acid ALK = Alkali COR = Corrosive ~~W~~ = Use No Water

SECTION I - GENERAL INFORMATION

Characterization **Mineral**	RCRA Number **Not Listed**	EPA Class **Not Applicable**
DOT Proper Shipping Name **No Citation**	Chemical Abstract Service (CAS) Number **14808-60-7**	
DOT Hazard Class and Label Requirements **No Citation**	DOT Emergency Guide Code **No Citation**	
DOT Identification Number **No Citation**	Chemical Formula **SiO_2**	

Synonyms
Silica; crystalline silica; cristobalite; fused silica; tridymite; tripoli; silicon dioxide; sand; rose quartz; silicic anhydride; amethyst; flint; onyx; pure quartz; silica flour (powdered crystalline silica).

SECTION II - HAZARDOUS INGREDIENTS AND IDENTITY INFORMATION

Hazardous Components (specific identity)	OSHA Exposure Criteria	NIOSH Exposure Criteria	Immediately Dangerous to Life and Health (IDLH)	ACGIH Exposure Criteria
Quartz (derivation: Occurs widely in nature. Synthetic crystals of good size and purity are grown by mass production methods under very carefully regulated conditions of temperature and pressure).	**PEL (8-hour):** **10 mg/m^3 divided by (%SiO2 + 2)** **Total Dust:** **30 mg/m^3 divided by (%SiO2 + 2)**	**REL:** **0.05 mg/m^3** **Possible Human Carcinogen**	**25 mg/m^3 (tripoli and quartz)** **50 mg/m^3 (cristobalite and tridymite)**	**TLV:** **0.1 mg/m^3 (respirable fraction)**

SECTION III - PHYSICAL AND CHEMICAL CHARACTERISTICS

Boiling Point **4046°F (2230°C)**	Specific Gravity (H_2O = 1) **2.66**
Vapor Pressure (atmospheres) **10 (approx) at 3150°F (1732°C)**	Molecular Weight (atomic weight) **60.1**
Vapor Density (Air = 1) **Not Found**	Melting Point **3110°F (1710°C)**

Solubility
Insoluble in water and acids, except hydrogen fluoride. Slightly attacked by caustic alkali solutions.

Appearance and Odor
Colorless or white, odorless powder or crystalline solid.

SECTION IV - FIRE AND EXPLOSION HAZARD DATA

Flash Point (method used) **Not Applicable**	Explosive Limits in Air % by Volume **LEL: N/A UEL: N/A**
NFPA Classification **Noncombustible Solid**	Autoignition Temperature **Not Applicable**

Extinguishing Media
Use agent suitable for type of surrounding fire. Quartz itself does not burn or support combustion.

Special Fire Fighting Procedures
Care should be taken to contain quartz materials that may have been disturbed in fire. To avoid breathing dusts or toxic fumes released during fire, wear full protective clothing and a self-contained breathing apparatus (SCBA) operated in pressure demand or other positive pressure mode. Be aware of runoff and do not release to sewers or waterways.

Unusual Fire and Explosion Hazards
No unusual fire or explosion hazards listed. Quartz is a noncombustible solid.

SECTION V - REACTIVITY DATA

Stability		Conditions to Avoid Avoid unprotected and/or unauthorized disturbance of quartz materials. Do not allow quartz or quartz containing materials to become airborne. Quartz is stable under normal conditions of handling and storage.
Stable X	Unstable	Incompatibility *(materials to avoid)* Powerful oxidizers (reacts violently), fluorine, chlorine trifluoride, manganese trioxide, oxygen difluoride, hydrogen peroxide, acetylene, ammonia. Dissolves in hydrogen fluoride to form silicon tetrafluoride.
Hazardous Polymerization		Conditions to Avoid Hazardous polymerization will not occur.
May Occur	Will Not Occur X	Hazardous Decomposition or By-products At high temperatures, quartz can change crystal structure to form tridymite (above 158°F/70°C) or cristobalite (above 2678°F/1470°C), which are greater hazards to human health than quartz itself.

SECTION VI - HEALTH HAZARD DATA

Primary Route(s) of Entry:	Inhalation?	Absorption (skin/eye)?	Ingestion?
	X		

Health Hazards

INHALATION: Repeated exposures can cause silicosis, a disease characterized by diffuse interstitial pulmonary fibrosis (a scarring of the lung tissues). Depending upon the nature of the exposure (length of time in years, type of quartz), symptoms can include coughing, dypsnea, wheezing, decreased pulmonary function, fever, and weight loss. Chronic exposure can cause increased susceptibility to tuberculosis, respiratory infections, and cardiac dysfunction with cyanosis, marked fatigue, and pleuritic pain.

Carcinogenicity	NTP Listed?	IARC Cancer Review Group?	OSHA Regulated?	Target Organs?
Questioned Human Confirmed Animal	No	Group 2	29 CFR 1910.1000 Table Z-1 and Z-3	Eyes, respiratory system, (cancer in animals), skin.

Medical Conditions Generally Aggravated by Exposure

Respiratory conditions (bronchitis, asthma) may be aggravated by exposure over the long-term (years).

Emergency and First-aid Procedures

Eye contact: Do not allow victim to keep eyes tightly shut. Flush immediately with water for 15 minutes (minimum); seek medical attention. Skin contact: Immediately wash area with large amounts of water. For inhalation: Remove the person from exposure. Provide respiratory assistance and CPR. Transfer to medical facility. Respiratory problems associated with quartz do not occur for many years following repeated exposures. All exposures should be documented. If swallowed: Give conscious person 1 to 2 glasses of water and induce vomiting until vomitus is clear.

SECTION VII - PRECAUTIONS FOR SAFE HANDLING AND USE

Steps to be Taken in Case Material is Released or Spilled

Restrict those not involved in cleanup from area. Wear self-contained breathing apparatus. Collect dusts using a HEPA vacuum. Do NOT dry sweep (creates dusts). Place in sealed container for disposal.

Preferred Waste Disposal Method

No citation.

Precautions to be Taken in Handling and Storage

Prior to working with quartz, persons must be trained (by law) on its proper handling and storage. Prevent damage to storage containers and avoid generating dusty conditions.

Other Precautions and Warnings

Airborne quartz dust is difficult to remove. It is essential that quartz work areas be closed or isolated to ensure controlled exposures.

SECTION VIII - CONTROL MEASURES AND PERSONAL PROTECTIVE EQUIPMENT

Respiratory Protection *(specify type)*

Quartz is an animal carcinogen. At exposure levels up to 50 times the PEL, use a MSHA/NIOSH respirator with a full facepiece with a high-efficiency particulate air (HEPA) filter. For the best protection, use a self-contained breathing apparatus (SCBA) with full facepiece operated in positive pressure mode.

Ventilation

Local exhaust (preferred) at site of quartz work. Area should be isolated from other work areas.

Protective Gloves	Eye Protection	Other Protective Clothing
Impervious Rubber	Dust Goggles and Face Mask	Full Body Protective Suit (disposable) with Boots and Head Protection

Work/Hygiene Practices

Always wash thoroughly after working around quartz; never bring food, drink, or smoking materials into vicinity of quartz work or abatement projects.

QUARTZ

SiO_2 CAS: 14808-60-7

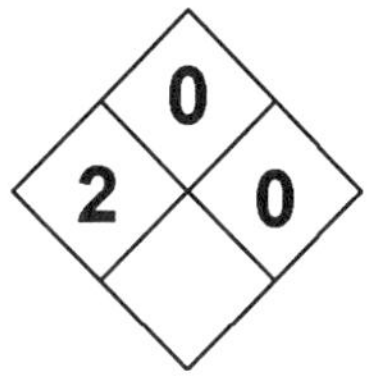

IDENTIFICATION AND TYPICAL USES

Quartz is a colorless or white powder or crystalline solid with no odor. It is used in electronic components, especially radio and television components, and in wave filters. It is also used as a barrel-abrasive, as a piezoelectric control in filters, in oscillators, and frequency standards. Quartz exposure can occur during the manufacture of tombstones, during granite-cutting operations, when sandblasting, in the manufacture of pottery and porcelain, in hard rock mining, in polishing and grinding operations, in foundries, in the manufacture of silica firebricks, and in the spraying of vitreous enamels.

RISK ASSESSMENT: HEALTH

General Assessment

Health hazards associated with quartz occur as a result of prolonged and repeated exposures over the course of years. Quartz is a confirmed animal carcinogen and is considered a questionable human cancer agent. *Inhalation* is the primary route of entry of quartz dusts into the body. Other than the pulmonary route, particles may accumulate in the gastrointestinal tract from drinking water contaminated with quartz or due to transportation from the nasal passage (i.e., when swallowing).

Prolonged exposures to silica quartz may cause chronic lung damage, known as silicosis, that can be delayed over a period of years. Simple silicosis typically occurs after twenty years or more of repeated exposures. Accelerated silicosis develops between five and fifteen years of exposure. Acute silicosis occurs after only one to three years of heavy, repeated exposures. Symptoms of silicosis include the formation of nodules of scar tissues in the lungs. Chronic scarring causes progressive pulmonary fibrosis with increased susceptibility to tuberculosis and/or other respiratory infections. There may also be heart enlargement, cardiac failure, and respiratory failure. Other symptoms include dypsnea, fever, coughing, wheezing, weight loss, decreased breathing capacity, and death.

Silica quartz can produce a drying effect on the skin, leaving it susceptible to cracking and secondary infection. Quartz is also an eye irritant, causing inflammation and redness.

Acute Health Effects

The following acute (short-term) health effects may occur immediately or shortly after exposure to quartz:

Skin: Irritation/rash upon contact with skin (contact dermatitis).

Eye: Severe irritation and possible conjunctiva inflammation (conjunctivitis).

Lung: Irritation of the nose, throat, and lungs following exposure. Can cause coughing or shortness of breath. Higher exposures may lead to pulmonary fibrosis, which can be fatal.

Chronic Health Effects

The following chronic (long-term) health effects may occur at some time after exposure to quartz and can last for months or even years before death occurs:

Cancer Hazards: Quartz is a carcinogen in animals. It has been shown to cause lung cancers in laboratory tests. Its human carcinogenic potential should be considered at least equal to that of animals until further information is known about quartz.

Reproductive Hazard: According to the references, quartz has not been adequately tested for its ability to adversely affect reproduction.

Other Chronic Effects: Repeated, long-term exposures to quartz can cause the disease silicosis, a scarring of the lungs. The higher the exposure, and the longer the time of the exposure, the greater the chance of developing serious illness. Other chronic symptoms include cough, dypsnea, wheezing, an increased susceptibility to tuberculosis, decreased breathing function, and repeated non-specific chest illness. Repeated skin contact can cause fissures, thickening, and general breakdown of the skin. The progression of silicosis may lead to extreme dypsnea, cyanosis, loss of appetite, pleuritic pain, and total disability until death.

Recommended Risk-Reduction Measures

Personnel should avoid direct contact with quartz. Its airborne dusts present a serious health hazard to unprotected, exposed individuals. Only trained and qualified persons should be permitted to work with or around quartz. *Engineering controls* are required to ensure maximum reduction of exposure potential. Specific procedures are recommended when working with quartz. These include, but are not limited to: total containment of the affected work area; use of air showers and barriers; use of specific personnel protective equipment (PPE); use of specialized mechanical equipment to ensure total capture of quartz dusts; and use of specific waste disposal techniques and methods.

During any work with quartz, the best protection is to enclose operations and/or provide local exhaust ventilation at the site of quartz work. Isolating operations will also reduce exposure. Using respiratory protection is less effective than the controls mentioned above, but is still required whenever working with or around friable quartz. At any exposure level, a self-contained breathing apparatus (SCBA) with full facepiece and pressure demand is recommended for the greatest possible respiratory protection. However, for exposures up to fifty times the PEL, an MSHA/NIOSH-approved respirator equipped with a high efficiency particulate air (HEPA) filter may suffice. If a full facepiece is not available, then dust goggles or face shield should be worn to protect the eyes. To prevent hand and skin exposures, impervious rubber gloves should be used.

Administrative controls should also be in place to minimize the potential for human exposures. These are required by law and include written procedures or policies which specify the methods and techniques that will be practiced whenever personnel are to work with quartz.

All personnel must receive *training* on the use, hazards, protective measures, emergency actions, and other precautions per 29 CFR 1910.1200 (Hazard Communication), prior to the first assignment in an area where quartz is used, removed, stored, or destroyed.

Before beginning employment with quartz and at regular intervals thereafter (e.g., annually), a medical history and examination is recommended, including the following:

- ☑ Pulmonary function test (baseline and annually).
- ☑ Consider periodic chest X-ray. Note that in cases of prolonged, low-level exposures, chest X-rays may appear normal for many years. However, in cases of high, acute exposures, there may be abnormal chest X-rays between one and three years of the exposure.

Any evaluation should include a careful history of past and present symptoms with an examination. Medical tests that simply look for existing damage are not a substitute for controlling exposures. Because smoking can cause heart disease, as well as lung cancer, emphysema, and other respiratory problems, respiratory conditions caused by quartz exposure may be more severe. Prudent risk management requires careful consideration of *all* possible factors which may be causing the appearance of exposure symptoms among workers.

Other methods to reduce exposure to quartz include:

- ☑ Always ensure that proper protective clothing is worn when working with quartz and that personnel have been properly trained on the correct use, care, and maintenance of such equipment.
- ☑ Hazard warning information should be posted in the work area. In addition, as part of an on-going education and training program, all information on the health and safety hazards of quartz must be communicated to all potentially exposed workers.
- ☑ Thoroughly wash any areas of the body that may have contacted quartz.
- ☑ Do not eat, drink, apply cosmetics, or smoke where quartz is handled, processed, or stored since quartz can be swallowed. Wash hands carefully before eating or smoking.
- ☑ Do not sweep for cleanup. Use a vacuum equipped with a high-efficiency particulate air (HEPA) filter or a wet method to reduce dust generation.
- ☑ Eye wash stations should be provided in the immediate work area for emergency use. If there is a possibility of skin exposure to quartz, emergency shower facilities should also be provided.

RISK ASSESSMENT: ENVIRONMENT

General Assessment

The environment is at risk of exposure during transportation, storage, disposal, or destruction of quartz. In almost every scenario, the threat of environmental exposure is contingent upon the proper handling quartz.

Accidental spills and damaged quartz materials must be cleaned using wet methods. Quartz spills must never be dry swept or shoveled. Such actions will cause the generation of dusts and airborne quartz fibers. Since quartz is a common air contaminant, its threat to the surrounding environment is minimal. However, this does not justify careless or unprofessional cleanup methods in the event of a spill.

Quartz is considered a noncombustible solid. It is commonly used for a multitude of purposes and environmental contamination is possible through a variety of routes. This means that quartz can unknowingly be present when other types of materials are being transported.

Quartz is a naturally occurring mineral. There are four main varieties of crystalline silica: cristobalite, quartz, tripoli, and tridymite. All these forms are found in association with each other. It can also enter the environment through runoff, or through accidental spills.

☠ *Acute Ecological Effects*

Acute (short-term) ecological toxic effects may include the death of animals, birds, or fish and death or low growth rate in plants. Acute effects are seen 2 to 4 days after animals or plants are exposed to quartz.

The data in the references are insufficient to determine if quartz (in any form) poses any acute toxicity hazard to aquatic life, plants, birds, or terrestrial animals.

Chronic Ecological Effects

Chronic (long-term) toxic effects may include shortened life span, reproductive problems, lower fertility, and changes in appearance or behavior in exposed animals. These effects can be seen long after first exposure(s) to toxic chemicals.

Quartz (fibers) can cause cancer if inhaled; birds or land animals which inhale quartz fibers are therefore susceptible to cancer or other long-term effects. Quartz fibers which are ingested (swallowed) are generally not toxic to birds or animals, but they are passed through their digestive systems.

Water Solubility

Quartz is insoluble in water. Concentrations of less than 1 milligram will not mix with a liter of water.

⌛ *Persistence in the Environment*

Quartz dusts are highly persistent in water, with a half-life greater than 200 days. The half-life of a pollutant is the amount of time it takes for one-half of the chemical to be degraded.

Bioaccumulation in Aquatic Organisms

Some substances increase in concentration, or *bioaccumulate,* in living organisms as they breathe contaminated air, drink contaminated water, or eat contaminated food. These chemicals can become concentrated in the tissues and internal organs of animals as well as humans.

The concentration of quartz found in fish tissues is expected to be less than the average concentration of quartz in the water from which the fish was taken.

Recommended Risk-Reduction Measures

Proper training of all transporters will reduce the likelihood of a mishap or accident resulting in a leak or spill of quartz into the environment. The correct use of labeling on all containers, trucks, and rail cars will enable emergency responders to react properly and quickly to any disaster thereby reducing the potential risk to the environment and to personnel.

If a spill or leak to the environment has occurred, hazardous materials spill response personnel should be notified immediately. Cleanup must be attempted only by those trained in proper quartz spill containment procedures. If quartz should contact the water table, aquifer, or navigable waterway, the local and/or state emergency response authorities must be notified. A comprehensive emergency response or disaster preparedness/recovery plan should be in place prior to any operations involving the use, transportation, storage, or disposal of quartz.

If quartz is spilled, the following specific steps are recommended:

- ☑ Restrict persons not wearing protective clothing from area of spill until cleanup is complete.
- ☑ Collect powdered materials in safest manner possible and deposit in sealed containers for disposal. Do NOT dry sweep (generates airborne dusts). Use HEPA vacuum instead. Damp mop residue.

RISK ASSESSMENT: BUSINESS

General Assessment

Accidents or mishaps involving quartz can present a significant threat to business operations. The use of the term quartz can generate serious emotional response from employees as well as the public in surrounding communities. Lawsuits that may result from personnel injury/death, public exposures, and/or environmental contamination will also require a serious expenditure of resources. Media attention surrounding an injury, death, or environmental damage can also result in a loss of profits and loss of current as well as future business. Always remember that anytime the word "quartz" or the terms "cancer," or "carcinogen" are used, public emotion, hysteria, and ignorance can run equally high. This should be carefully considered whenever developing or implementing public relations policies.

Recommended Risk-Reduction Measures

Company attorneys, safety and health professionals, and environmental specialists should be involved in the development of any procedures dealing with safety or emergency response. A company official should be pre-designated as a public relations officer with specific training in dealing with the press. Corporate plans and policies should be developed, approved, and implemented long before any need for such arises.

REFERENCES

American Conference of Governmental Industrial Hygienists (ACGIH). 1988. *Documentation of the Threshold Limit Values and Biological Exposure Indices,* 5th Edition (with updates). Cincinnati: ACGIH

Bloom, A. D. and F. J. DeSerres. 1995. *Ecotoxicity and Human Health.* New York: CRC Press/Lewis Publishers.

Brusick, D. J. 1994. *Methods for Genetic Risk Assessment.* New York: CRC Press/Lewis Publishers.

Bueche, J. R. 1972. *Principles of Physics,* 2nd Edition. New York: McGraw-Hill.

Burrell, R., D. K. Flaherty, and L. J. Sauers. 1992. *Toxicology of the Immune System.* New York: Van Nostrand Reinhold.

Calabrese, E. J. 1994. *Biological Effects of Low Level Exposures.* New York: CRC Press/Lewis Publishers.

Cockerman, L. G., B. S. Shane. 1994. *Basic Environmental Toxicology.* New York: CRC Press/Lewis Publishers.

Cothern, R. R. 1993. *Comparative Environmental Risk Assessment.* New York: CRC Press/Lewis Publishers.

Danse, I. R. 1991. *Common Sense Toxics in the Workplace.* New York: Van Nostrand Reinhold.

Environmental Protection Agency. 1983. *Chemical Hazard Information Profiles.* Washington, D.C.: U.S. EPA.

Fodden, C. R., and J. L. Weddell. 1992. *Hazardous Materials Emergency Action Data.* New York: CRC Press/Lewis Publishers.

Forseberg, K., and L. H. Keith. 1993. *Chemical; Protective Clothing Permeation and Degradation Database.* Chelsea, MI: Lewis Publishers.

Gardner, R. W. 1994. *Chemical Intolerance.* New York: CRC Press/Lewis Publishers.

Hallenbeck, W. H. 1993. *Quantitative Risk Assessment for Environmental and Occupational Health,* 2nd Edition. New York: CRC Press/Lewis Publishers.

Hodgson, E., R. B. Mailman, and Janice E. Chambers. 1988. *Dictionary of Toxicology.* London: The MacMillian Company.

International Agency for Research on Cancer (IARC). 1987. *IARC Monographs on the Evaluation of Carcinogenic Risk to Humans, Overall Evaluations of Carcinogenicity: An updating of IARC Monographs Volumes 1-42.* Lyon: IARC.

Landis, W. G., M. H. Yu. 1995. *Introduction to Environmental Toxicology.* New York: CRC Press/Lewis Publishers.

Langerman, N. 1994 *Precautionary Labels for Chemical Containers.* New York: CRC Press/Lewis Publishers.

Lewis, R. J., Sr. 1993. *Hawley's Condensed Chemical Dictionary,* 12th Edition. New York: Van Nostrand Reinhold.

Lewis, R. J., Sr. 1992. *Sax's Dangerous Properties of Industrial Materials,* Volume 1. 8th Edition. New York: Van Nostrand Reinhold.

Mackay, D., W. Y. Shiu, and K. C. Ma. 1992. *Illustrated Handbook of Chemical-Physical Properties and Environmental Fate for Organic Chemicals.* New York: CRC Press/Lewis Publishers.

Manahan, S. E. 1994. *Environmental Chemistry,* 6th Edition. New York: CRC Press/Lewis Publishers.

Meyer, E. 1990. *Chemistry of Hazardous Materials.* Englewood Cliffs, NJ: Prentice Hall.

Patnaik, P. 1992. *A Comprehensive Guide to the Hazardous Properties of Chemical Substances.* New York: Van Nostrand Reinhold.

Proctor, N. H., J. P. Hughes. 1991. *Chemical Hazards of the Workplace.* New York: Van Nostrand Reinhold.

Rea, W. J. 1995. *Chemical Sensitivity,* Vol. I-III. New York: CRC Press/Lewis Publishers.

Richardson, M. 1995. *Dictionary of Substance and Their Effects,* Vol. I-VI. New York: CRC Press/Lewis Publishers.

Sacarello, H. L. A. 1994. *The Comprehensive Handbook of Hazardous Materials.* New York: CRC Press/Lewis Publishers.

Torkelson, T. R. V. K. Rowe. 1981. *Patty's Industrial Hygiene and Toxicology,* Vol. 2B. New York: Wiley-Interscience.

U.S. Department of Health and Human Services. National Institute for Occupational Safety and Health (NIOSH). 1984. *Manual of Analytical Methods,* 3rd. Edition. Cincinnati, OH: NIOSH.

U.S. Department of Health and Human Services, National Institute for Occupational Safety and Health. 1994. *NIOSH Pocket Guide to Chemical Hazards.* Washington, D.C.: U.S. Government Printing Office.

MATERIAL SAFETY DATA SHEET

CHEMICAL NAME	
QUINONE	**CRC PRESS/LEWIS PUBLISHERS DISCLAIMER:** THE INFORMATION AND RECOMMENDATIONS PRESENTED HEREIN ARE BASED ON SOURCES BELIEVED TO BE RELIABLE. CRC MAKES NO REPRESENTATION ON ITS COMPLETENESS OR ACCURACY. IT IS THE RESPONSIBILITY OF THE USER TO DETERMINE THE CHEMICAL'S SUITABILITY FOR ITS INTENDED USE, THE CHEMICAL'S SAFE USE, AND THE CHEMICAL'S PROPER DISPOSAL. NO REPRESENTATIONS AND/OR WARRANTIES, EITHER EXPRESSED OR IMPLIED, OF THE MERCHANTABILITY OR FITNESS FOR A PARTICULAR PURPOSE, OR OF ANY OTHER NATURE, ARE MADE WITH RESPECT TO THE INFORMATION PROVIDED IN THIS MSDS OR TO THE CHEMICAL TO WHICH INFORMATION MAY REFER. CRC NEITHER ASSUMES NOR AUTHORIZES ANY OTHER PERSON TO ASSUME FOR IT, ANY OTHER ADDITIONAL RESPONSIBILITY OR LIABILITY FOR THE USE OF, OR RELIANCE UPON, THIS INFORMATION.

HAZARD WARNING INFORMATION

HEALTH	FIRE	REACTIVE	OTHER	DEGREE OF HAZARD	COLOR CODING	OTHER CODES
1	2	1		0 = Minimum Hazard 1 = Slight Hazard 2 = Moderate Hazard 3 = Serious Hazard 4 = Severe Hazard	HEALTH = BLUE FIRE = RED REACTIVITY = YELLOW OTHER = WHITE	OX = Oxidizer ACID = Acid ALK = Alkali COR = Corrosive ~~W~~ = Use No Water

SECTION I - GENERAL INFORMATION

Characterization	RCRA Number	EPA Class
Hydrocarbon	**U197**	**Toxic Waste**

DOT Proper Shipping Name **Benzoquinone**	Chemical Abstract Service (CAS) Number **106-51-4**
DOT Hazard Class and Label Requirements **Poison B; Poison**	DOT Emergency Guide Code **55**
DOT Identification Number **UN 2587**	Chemical Formula $C_6H_4O_2$

Synonyms
1,4-Benzoquinone; p-benzoquinone; 1,4-cyclohexadiene dioxide; p-quinone; chinone.

SECTION II - HAZARDOUS INGREDIENTS AND IDENTITY INFORMATION

Hazardous Components (specific identity)	OSHA Exposure Criteria	NIOSH Exposure Criteria	Immediately Dangerous to Life and Health (IDLH)	ACGIH Exposure Criteria
Quinone (derivation: By oxidation of aniline with chromic acid, extraction with ether, followed by distillation). 1 ppm = 4.49 mg/m^3	**PEL (8-hour):** 0.1 ppm 0.4 mg/m^3 (skin) **STEL:** Not Established	**REL (10-hour):** 0.1 ppm 0.4 mg/m^3 (skin) **STEL:** Not Established	100 mg/m^3	**TLV:** 0.1 ppm 0.4 mg/m^3 (skin) **STEL:** Not Established

SECTION III - PHYSICAL AND CHEMICAL CHARACTERISTICS

Boiling Point **Sublimes**	Specific Gravity (H_2O = 1) **1.32**
Vapor Pressure (gas) **0.1 at 69°F (20°C)**	Molecular Weight **108.1**
Vapor Density (Air = 1) **3.7**	Melting Point **240°F (115°C)**

Solubility
Slightly soluble in hot water. Soluble in alcohol, ether, and alkalis.

Appearance and Odor
Pale-yellow crystalline solid with an acrid, chlorine-like odor. Odor Threshold = 0.1 ppm.

SECTION IV - FIRE AND EXPLOSION HAZARD DATA

Flash Point (method used) **100-200°F (38-93°C) closed cup**	Explosive Limits in Air % by Volume **LEL: Not Determined UEL: Not Determined**
NFPA Classification **Combustible Solid**	Autoignition Temperature **815°F (435°C)**

Extinguishing Media
Use dry chemical, carbon dioxide, water fog, or alcohol foam. Do NOT use solid stream of water.

Special Fire Fighting Procedures
Wear full protective clothing and self-contained breathing apparatus (SCBA). Poisonous gases are produced in fire. Use water spray to keep fire-exposed containers cool. Containers may explode in fire.

Unusual Fire and Explosion Hazards
Combustible solid. Its dusts can form a cloud that may explode if ignited in a closed area. Move containers from fire area if it can be done without risk. Violent reactions can occur with oxidizers. Do not release runoff from fire methods to sewers or waterways.

SECTION V - REACTIVITY DATA

Stability		Conditions to Avoid **Quinone is stable in closed containers under routine conditions of handling and storage. Avoid contact with incompatible materials. Do not generate dusty conditions.**
Stable X	Unstable	Incompatibility *(materials to avoid)* **Strong oxidizers (permanganates, peroxides, nitrates, chlorates, and perchlorates), and alkalis. Quinone will attack some forms of rubber, plastic, and certain coatings.**
Hazardous Polymerization		Conditions to Avoid **Under normal conditions of temperature and pressure, hazardous polymerization of quinone is not expected to occur.**
May Occur	Will Not Occur X	Hazardous Decomposition or By-products **When heated to decomposition, quinone can emit highly toxic/poisonous and acrid fumes and gases including oxides of carbon.**

SECTION VI - HEALTH HAZARD DATA

Primary Route(s) of Entry:	Inhalation?	Absorption (skin)?	Ingestion?
	X		X

Health Hazards

INHALATION: Symptoms include irritation of the nose, throat, and respiratory tract. Prolonged contact with the vapors may damage the mucous membranes of the respiratory tract.

SKIN & EYES: Irritating to the skin; may cause irritation and dermatitis. An eye irritant with possibility of corneal damage, conjunctivitis, and keratitis. It may cause discoloration of the skin with severe irritation, erythema (redness), swelling, papules, and vesicles. Prolonged contact cab lead to necrosis. Vapor contact with the eyes can cause visual disturbances.

INGESTION: Unspecified gastrointestinal effects with a possibility for nausea, vomiting, and collapse.

Carcinogenicity	NTP Listed?	IARC Cancer Review Group?	OSHA Regulated?	Target Organs?
Unknown Human Questioned Animal	No	Group 3	29 CFR 1910.1000 Table Z-1	Eyes and skin.

Medical Conditions Generally Aggravated by Exposure

Skin diseases or existing eye disorders may be aggravated by exposure.

Emergency and First-aid Procedures

Eye contact: Flush immediately with water for 15 minutes (minimum); seek medical attention. Skin contact: Remove all contaminated clothing. Immediately wash area with large amounts of soap and water. Seek medical assistance. For inhalation: Remove the person from exposure. Provide respiratory assistance and CPR, if required. Transfer to medical facility. If swallowed: Seek medical attention immediately. Give conscious person 1 to 2 glasses of water or milk to dilute. Unless advised otherwise, induce vomiting. Never give an unconscious or convulsing person anything by mouth.

SECTION VII - PRECAUTIONS FOR SAFE HANDLING AND USE

Steps to be Taken in Case Material is Released or Spilled

Collect powders using HEPA vacuum. Do NOT dry sweep. Absorb solution spills in vermiculite and place in sealed drum. Ventilate area of spill or leak. Remove all sources of ignition.

Preferred Waste Disposal Method

No citation.

Precautions to be Taken in Handling and Storage

Store to avoid contact with incompatible materials in tightly closed containers in cool, dry, well-ventilated area away from heat, flame, and direct sunlight. Use only non-sparking tools.

Other Precautions and Warnings

Sources of ignition are prohibited where quinone is used, handled, or stored.

SECTION VIII - CONTROL MEASURES AND PERSONAL PROTECTIVE EQUIPMENT

Respiratory Protection *(specify type)*

For exposures above the PEL, use a MSHA/NIOSH-approved supplied-air respirator with full facepiece, hood, or helmet in continuous flow mode, or use a self-contained breathing apparatus (SCBA) operated in pressure demand or other positive pressure mode.

Ventilation

Local exhaust or general mechanical systems recommended.

Protective Gloves	Eye Protection	Other Protective Clothing
Impervious Material	Chemical/Dust Goggles or Face Mask	Protective Apron

Work/Hygiene Practices

Always wash hands thoroughly after using chemical; never bring food, drink, or smoking materials into vicinity of chemicals.

QUINONE

$C_6H_4O_2$

CAS: 106-51-4

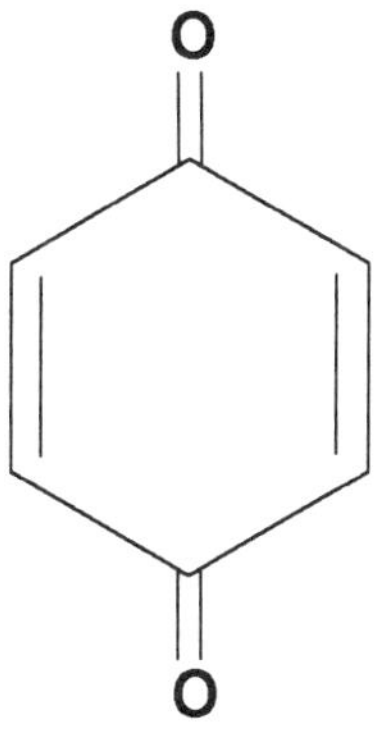

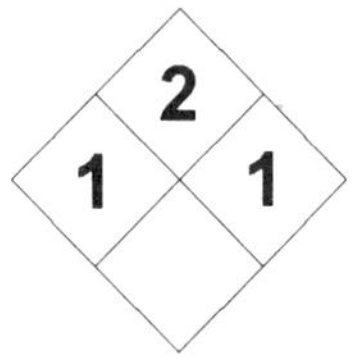

IDENTIFICATION AND TYPICAL USES

Quinone is a pale-yellow crystalline solid with an acrid, chlorine-like odor. It is used in the manufacture and production of dyes and hydroquinone. It is also used in fungicides, as an analytical reagent, in photography, and as an oxidizing agent.

RISK ASSESSMENT: HEALTH

General Assessment

Quinone is a poison by *ingestion* and is moderately toxic by *inhalation.* It will cause a moderate health hazard risk on skin or eye contact, but absorption through intact skin has not been reported. Its carcinogenicity in humans is not known but is questioned in some animal studies. It has been shown to cause mutations (genetic changes) in human cells. There are no data on its teratogenic potential in humans or animals.

Inhalation of the dusts causes irritation of the nose, throat, and mucous membranes of the respiratory tract. Prolonged contact may damage the mucosa of the respiratory system.

Skin and eye contact may produce mild to moderate irritation on contact. It can cause sensitization and eczematous dermatitis. Direct eye contact with quinone particles causes immediate irritation and possible corneal ulceration. Local effects include skin and eye tissue discoloration, severe skin surface irritation with erythema (redness), swelling, the formation of papules and vesicles, and, on prolonged contact, tissue necrosis (destruction). When the eyes are involved, there may be dangerous disturbances to vision.

Ingestion of quinone may cause initial gastrointestinal irritation with a possibility for nausea, vomiting, and collapse.

☠ *Acute Health Effects*

The following acute (short-term) health effects may occur immediately or shortly after exposure to quinone:

Skin: Severe irritation with redness, swelling, and possible localized destruction of skin tissue.

Eye: Severe irritation with a potential for corneal injury or damage.

Lung: Nose and throat irritation. High concentrations may cause damage to the mucous membranes.

Chronic Health Effects

The following chronic (long-term) health effects may occur at some time after exposure to quinone and can last for months or even years:

Cancer Hazards: According to information presented in the references, quinone may cause cancer in some test animals. However, the data are conflicting and inconclusive in this regard. Some studies indicate that it causes mutations in humans and many scientists believe that such chemicals may pose a cancer threat in the long-term. Additional research is required in this regard.

Reproductive Hazard: According to information presented in the references, quinone has not been adequately tested for its ability to adversely affect reproduction. The teratogenicity of quinone in humans has not been adequately studied.

Other Chronic Effects: Prolonged contact may cause discoloration of the skin and the conjunctiva of the eye followed by corneal changes and possible loss of visual acuity. Repeated skin contact can lead to chronic dermatitis.

❶ *Recommended Risk-Reduction Measures*

Personnel should avoid direct contact with quinone. The exact nature of the toxicity of this chemical is not entirely understood in the references. Therefore, prudent risk management requires it be treated as though it posed a significant health risk in the event of overexposure. If a less toxic chemical cannot be substituted for a hazardous substance, then *engineering controls* are the most effective method of reducing exposures. The best protection is to enclose operations and/or provide local exhaust ventilation at the site of chemical release. While not always operationally feasible, isolating operations can also reduce exposure. Using respiratory protection is less effective than the controls mentioned above, but is still advisable whenever working with or around quinone. The exposure ceiling is relatively low (0.1 ppm) and difficult to accurately monitor. For all exposures, a supplied-air respirator with full facepiece operated in continuous flow mode, or a self-contained breathing apparatus (SCBA) with full facepiece and pressure demand are the recommended respiratory protection methods of choice. If a full facepiece is not available, then chemical goggles should be worn to protect the eyes. Whenever a chemical splash hazard exists, a face shield and a protective apron should be worn. To prevent hand and skin exposures, impervious polyvinyl chloride or polyethylene gloves should be used. Quinone will attack some forms of rubber, so caution is warranted when selecting gloves. Glove manufacturers should be contacted to obtain permeation studies *before* glove selection has been made for quinone.

Administrative controls should also be in place to minimize the potential for human exposures. These may include written procedures or policies which specify the methods and techniques that will be practiced whenever personnel are to work with quinone.

All personnel should receive *training* on the use, hazards, protective measures, emergency actions, and other precautions per 29 CFR 1910.1200 (Hazard Communication), prior to the first assignment in an area where quinone is used or stored.

If symptoms develop or overexposure is suspected, the following tests are recommended:

- ☑ Examination of the eyes, especially if inflammation or irritation is present.
- ☑ Evaluation by a qualified allergist, with careful consideration of exposure history and special skin tests (may help diagnose allergy).

Any evaluation should include a careful history of past and present symptoms with an examination. Medical tests that simply look for existing damage are not a substitute for controlling exposures.

Other methods to reduce exposure include:

- ☑ Where possible, enclose operations and use local exhaust ventilation at the site of chemical release. If local exhaust ventilation or enclosure is not used, respiratory protection should be mandatory. Also, if possible, automatic transfer of quinone from storage containers to process or work containers is recommended.
- ☑ Always ensure that proper protective clothing is worn when using chemical substances and that personnel are trained in its use and care.
- ☑ Wash thoroughly immediately after exposure to quinone and at the end of the work shift or before eating, drinking, applying cosmetics, or smoking.
- ☑ Hazard warning information should be posted in the work area. In addition, as part of an on-going education and training program, all information on the health and safety hazards of quinone should be communicated to all exposed and potentially exposed workers.
- ☑ Eye wash stations should be provided in the immediate work area for emergency use. If there is a possibility of skin exposure to quinone, emergency shower facilities should also be provided in the immediate work area.
- ☑ Personnel should never wear contaminated clothing home (family members can be exposed). Clothing should be laundered by personnel who have been trained on the hazards of quinone.

RISK ASSESSMENT: ENVIRONMENT

General Assessment

The environment is at risk of exposure during transportation, storage, disposal, or destruction of quinone. In almost every scenario, the threat of environmental exposure is contingent upon the proper handling of the chemical substance. Accidental spills, large or small, can result in fire, explosion, and possible contamination of the surrounding environmental mediums (water, soil, and air).

Quinone is considered a combustible solid. A cloud of quinone dusts may explode if ignited in an enclosed area. It also presents a moderate fire and explosion hazard risk, especially in contact with incompatible materials. These characteristics require extreme caution in handling, storage, transportation, and disposal. It is incompatible with strong oxidizers and contact can cause fire or explosion. It will attack some rubbers, plastics, and coatings, so the choice of handling equipment must be carefully considered. Therefore, special consideration is required during any emergency situation involving a leak or spill of quinone. Should quinone ever come into contact with these incompatible substances during use, transportation, storage, or disposal, the formation of highly toxic and/or highly explosive commodities is extremely possible.

Quinone may enter the environment through industrial discharges or spills.

Acute Ecological Effects

Acute (short-term) toxic effects may include the death of animals, birds, or fish and death or low growth rate in plants. Acute effects are seen 2 to 4 days after animals or plants are exposed to quinone.

Insufficient data are available to evaluate or predict the short-term effects of quinone exposure to aquatic life, plants, birds, or terrestrial animals.

Chronic Ecological Effects

Chronic (long-term) toxic effects may include shortened life span, reproductive problems, lower fertility, and changes in appearance or behavior in exposed animals. These effects can be seen long after first exposure(s) to toxic chemicals.

Insufficient data are available to evaluate or predict the long-term effects of quinone exposure to aquatic life, birds, plants, or terrestrial animals.

Water Solubility

Quinone is nearly insoluble in water. Concentrations of 1 milligram may mix with a liter of water. Solubility increases with water temperature.

Persistence in the Environment

Quinone is slightly persistent in water, with a half-life between 2 to 20 days. The half-life of a pollutant is the amount of time it takes for one half of the chemical to be degraded. About 99% of quinone will eventually end up in water; the rest will end up in air.

Bioaccumulation in Aquatic Organisms

Some substances increase in concentration, or *bioaccumulate,* in living organisms as they breathe contaminated air, drink contaminated water, or eat contaminated food. These chemicals can become concentrated in the tissues and internal organs of animals as well as humans.

The concentration of quinone found in fish tissue is expected to be about the same as the average concentration of quinone in the water from which the fish was taken.

Recommended Risk-Reduction Measures

Proper training of all transporters will reduce the likelihood of a mishap or accident resulting in a leak or spill to the environment. The correct use of DOT labeling (per the MSDS) on all containers, trucks, and rail cars will enable emergency responders to react properly and quickly to any disaster thereby reducing the potential risk to the environment and to personnel.

Storage of quinone should be segregated from incompatible chemicals to minimize the risk of cross-contamination or contact. Buildings designated for storage should be equipped with appropriate fire protection systems (alarms, sprinklers, emergency lighting, portable extinguishers).

If a spill or leak to the environment has occurred, fire department, emergency response and/or hazardous materials spill personnel should be notified immediately. Cleanup should be attempted only by those trained in proper spill containment procedures using non-sparking tools. Contaminated soils should be removed for incineration and replaced with clean soil. If quinone should contact the water table, aquifer, or navigable waterway, cleanup should be started immediately. It is nearly insoluble in water and total remediation may be possible. The local and/or state emergency response authorities must be notified. A comprehensive emergency response or disaster preparedness/recovery plan should be in place prior to any operations involving the use, transportation, storage, or disposal of quinone. If quinone is spilled or leaked, the following specific steps are recommended:

- ☑ Restrict persons not wearing protective clothing from area of spill or leak until cleanup is complete and area can be opened for normal work.
- ☑ Remove all ignition sources.
- ☑ Ventilate area of leak.

- ☑ Collect powdered materials using a high-efficiency particulate air (HEPA) filter. Do NOT dry sweep (generates airborne dusts). Place in proper container for disposal. Damp mop residues. Absorb liquids containing quinone in vermiculite, dry sand, earth, or similar material and deposit in sealed containers.
- ☑ It may be necessary to dispose of quinone as a hazardous waste. The responsible state agency or the regional office of the federal Environmental Protection Agency (EPA) should be contacted for specific recommendations.

RISK ASSESSMENT: BUSINESS

General Assessment

Accidents or mishaps involving quinone can present a significant threat to business operations. The mismanagement of chemical commodities can cause personnel injuries and/or property damage/loss. The loss or damage of equipment or facilities can significantly affect fiscal viability. Lawsuits that may result from personnel injury/death, public exposures, and/or environmental contamination will also require a serious expenditure of resources. Media attention surrounding an injury, death, or environmental damage can also result in a loss of profits and loss of current as well as future business prospects.

Always remember that anytime the terms "cancer" or "carcinogen" are used, public emotion, hysteria, anxiety, and ignorance can run equally high. This must be carefully considered whenever developing or implementing public relations policies.

Recommended Risk-Reduction Measures

Company attorneys, safety and health professionals, and environmental specialists should be involved in the development of any procedures dealing with safety or emergency response. A company official should be pre-designated as a public relations officer with specific training in dealing with the press. Corporate plans and policies should be developed, approved, and implemented long before any need for such arises.

REFERENCES

American Conference of Governmental Industrial Hygienists. 1988. *Documentation of the Threshold Limit Values and Biological Exposure Indices*, 5th Edition (with updates). Cincinnati: ACGIH

Bueche, J. R. 1972. *Principles of Physics*, 2nd Edition. New York: McGraw-Hill.

Calabrese, E. J. 1994. *Biological Effects of Low Level Exposures.* New York: CRC Press/Lewis Publishers.

Cockerman, L. G., B. S. Shane. 1994. *Basic Environmental Toxicology.* New York: CRC Press/Lewis Publishers.

Cothern, C. R. 1995. *Handbook for Environmental Risk Decisionmaking.* New York: CRC Press/Lewis Publishers.

Danse, I. R. 1991. *Common Sense Toxics in the Workplace.* New York: Van Nostrand Reinhold.

Environmental Protection Agency. 1983. *Chemical Hazard Information Profiles.* Washington, D.C.: U.S. EPA.

Fodden, C. R., and J. L. Weddell. 1992. *Hazardous Materials Emergency Action Data.* New York: CRC Press/Lewis Publishers.

Forseberg, K., and L. H. Keith. 1993. *Chemical; Protective Clothing Permeation and Degradation Database.* Chelsea, MI: Lewis Publishers.

Gardner, R. W. 1994. *Chemical Intolerance.* New York: CRC Press/Lewis Publishers.

Lauwerys, R. R., P. Hoet. 1993. *Industrial Chemical Exposure.* New York: CRC Press/Lewis Publishers.

Lewis, R. J., Sr. 1992. *Sax's Dangerous Properties of Industrial Materials*, Volumes 1, 2, and 3, 8th Edition. New York: Van Nostrand Reinhold.

Mackay, D., W. Y. Shiu, and K. C. Ma. 1992. *Illustrated Handbook of Chemical-Physical Properties and Environmental Fate for Organic Chemicals.* New York: CRC Press/Lewis Publishers.

Manahan, S. E. 1994. *Environmental Chemistry*, 6th Edition. New York: CRC Press/Lewis Publishers.

Meyer, E. 1990. *Chemistry of Hazardous Materials.* Englewood Cliffs, NJ: Prentice Hall.

Patnaik, P. 1992. *A Comprehensive Guide to the Hazardous Properties of Chemical Substances.* New York: Van Nostrand Reinhold.

Proctor, N. H., J. P. Hughes. 1991. *Chemical Hazards of the Workplace.* New York: Van Nostrand Reinhold.

Quigley, D. R. 1994. *Handbook of Emergency Chemical Management.* New York: CRC Press/Lewis Publishers.

Rea, W. J. 1995. *Chemical Sensitivity*, Vol. I-III. New York: CRC Press/Lewis Publishers.

Richardson, M. 1995. *Dictionary of Substance and Their Effects*, Vol. I-VI. New York: CRC Press/Lewis Publishers.

Rylander, R., R. R. Jacobs. 1994. *Organic Dusts.* New York: CRC Press/Lewis Publishers.

Sacarello, H. L. A. 1994. *The Comprehensive Handbook of Hazardous Materials.* New York: CRC Press/Lewis Publishers.

Torkelson, T. R. V. K. Rowe. 1981. Halogenated aliphatic hydrocarbons. *Patty's Industrial Hygiene and Toxicology*, Vol. 2B, *Toxicology*, pp 3568-3572. New York: Wiley-Interscience.

U.S. Department of Health and Human Services, National Institute for Occupational Safety and Health. 1994. *NIOSH Pocket Guide to Chemical Hazards.* Washington, D.C.: U.S. Government Printing Office.

MATERIAL SAFETY DATA SHEET

CHEMICAL NAME	
RADON	**CRC PRESS/LEWIS PUBLISHERS DISCLAIMER:** THE INFORMATION AND RECOMMENDATIONS PRESENTED HEREIN ARE BASED ON SOURCES BELIEVED TO BE RELIABLE. CRC MAKES NO REPRESENTATION ON ITS COMPLETENESS OR ACCURACY. IT IS THE RESPONSIBILITY OF THE USER TO DETERMINE THE CHEMICAL'S SUITABILITY FOR ITS INTENDED USE, THE CHEMICAL'S SAFE USE, AND THE CHEMICAL'S PROPER DISPOSAL. NO REPRESENTATIONS AND/OR WARRANTIES, EITHER EXPRESSED OR IMPLIED, OF THE MERCHANTABILITY OR FITNESS FOR A PARTICULAR PURPOSE, OR OF ANY OTHER NATURE, ARE MADE WITH RESPECT TO THE INFORMATION PROVIDED IN THIS MSDS OR TO THE CHEMICAL TO WHICH INFORMATION MAY REFER. CRC NEITHER ASSUMES NOR AUTHORIZES ANY OTHER PERSON TO ASSUME FOR IT, ANY OTHER ADDITIONAL RESPONSIBILITY OR LIABILITY FOR THE USE OF, OR RELIANCE UPON, THIS INFORMATION.

HAZARD WARNING INFORMATION

HEALTH	FIRE	REACTIVE	OTHER	DEGREE OF HAZARD	COLOR CODING	OTHER CODES
1	0	3		0 = Minimum Hazard 1 = Slight Hazard 2 = Moderate Hazard 3 = Serious Hazard 4 = Severe Hazard	HEALTH = BLUE FIRE = RED REACTIVITY = YELLOW OTHER = WHITE	OX = Oxidizer ACID = Acid ALK = Alkali COR = Corrosive W = Use No Water

SECTION I - GENERAL INFORMATION

Characterization	RCRA Number	EPA Class
Radioactive Gas; Noble Gas	None	Not Applicable

DOT Proper Shipping Name Radioactive Material, N.O.S.	Chemical Abstract Service (CAS) Number 10043-92-2
DOT Hazard Class and Label Requirements Non-Combustible Gas	DOT Emergency Guide Code No citation
DOT Identification Number UN2982	Atomic Formula Rn

Synonyms

^{222}Rn; niton; radium emanation.

SECTION II - HAZARDOUS INGREDIENTS AND IDENTITY INFORMATION

Hazardous Components (specific identity)	OSHA Exposure Criteria	NIOSH Exposure Criteria	Immediately Dangerous to Life and Health (IDLH)	ACGIH Exposure Criteria
Radon (derivation: By radioactive decay of radium. Also obtained by bubbling air through a radon salt solution and collecting the gas plus air).	PEL: Not Established STEL: Not Established	REL: Not Established STEL: Not Established	Not Determined	TLV: Not Established STEL: Not Established

SECTION III - PHYSICAL AND CHEMICAL CHARACTERISTICS

Boiling Point -79°F (-62°C)	Specific Gravity (H_2O = 1) 9.72 g/L at 32°F (0°C)
Vapor Pressure (atmospheres) Not Listed	Atomic Weight 222
Vapor Density (Air = 1) Not Listed	Freezing Point -166°F (-110°C)

Solubility

Soluble in water.

Appearance and Odor

Colorless, odorless, tasteless gas. It can be condensed to a colorless, transparent liquid and to an opaque, glowing solid.

SECTION IV - FIRE AND EXPLOSION HAZARD DATA

Flash Point (method used) Not Determined	Explosive Limits in Air % by Volume LEL: Not Determined UEL: Not Determined
NFPA Classification Non-Combustible Gas	Autoignition Temperature Not Determined

Extinguishing Media

Use extinguishing media suitable to surrounding fires. Radon itself will not ignite or support combustion.

Special Fire Fighting Procedures

Radon is inert and will not support combustion. However, it is recommended that a self-contained breathing apparatus (SCBA) be worn. Radon decomposes to polonium which is a dangerous radioactive air contaminant. Keep fire-exposed containers cool with water spray.

Unusual Fire and Explosion Hazards

None reported.

SECTION V - REACTIVITY DATA

Stability		Conditions to Avoid: Radon is highly unstable and decomposes rapidly to polonium. Keep out of confined areas, especially the aqueous solution (it can produce explosive hydrogen gas). It can react with fluorine.
Stable	Unstable X	Incompatibility *(materials to avoid)*: Radon is a noble gas and is considered chemically inert. However, because of its electronic structure, it can react with certain materials such as fluorine.
Hazardous Polymerization		Conditions to Avoid: Under normal conditions of temperature and pressure, hazardous polymerization of radon is not expected to occur. In aqueous solution in confined spaces, radon can produce explosive hydrogen gas.
May Occur	Will Not Occur X	Hazardous Decomposition or By-products: Radon decomposes without the addition of heat to form polonium.

SECTION VI - HEALTH HAZARD DATA

Primary Route(s) of Entry:	Inhalation?	Absorption (skin)?	Ingestion?
	X		

Health Hazards

INHALATION: Long-term exposure to radon gas can cause lung, kidney, and skin cancer. There may also be ulcerative radiation dermatitis. Radon and its decomposition products (polonium isotopes) absorb to particulates which are inhaled or absorbed through skin.

SKIN & EYES: Absorption of radon particulates through intact skin is possible.

INGESTION: Ingestion of radon particulates is possible.

Carcinogenicity	NTP Listed?	IARC Cancer Review Group?	OSHA Regulated?	Target Organs?
Confirmed Human Confirmed Animal	Class 1	Group 1	No	Respiratory system; skin; kidneys.

Medical Conditions Generally Aggravated by Exposure

Dermatitis may be aggravated by exposure.

Emergency and First-aid Procedures

Eye contact (liquid): Immediately flush with large amounts of tepid water for 15 minutes (minimum), occasionally lifting eyelids; seek medical attention. Skin contact (liquid): Remove all contaminated clothing. Immediately flush area with large amounts of tepid water. Seek medical assistance. For inhalation: Remove the person from exposure. Emergency responders must be protected from exposure. Provide respiratory assistance and CPR. Transfer promptly to medical facility. If swallowed: Radon absorbs to particulates and can be ingested. However, it is not possible to tell if it has been swallowed. No data are available on toxicity associated with ingestion.

SECTION VII - PRECAUTIONS FOR SAFE HANDLING AND USE

Steps to be Taken in Case Material is Released or Spilled

Ventilate area. Stop flow of gas. If the source of leak is from natural uranium decay and not formed during a manufacturing process, special sealants may be used to minimize a radon gas leak. Ensure response personnel are properly protected with self-contained breathing apparatus (SCBA).

Preferred Waste Disposal Method

No citation.

Precautions to be Taken in Handling and Storage

Storage in water solution over prolonged periods is not recommended; pressure buildup may occur due to the production of explosive hydrogen gas resulting from radiolysis. Ensure proper storage and handling precautions are in place whenever working with or around radon.

Other Precautions and Warnings

Containers may explode in fire or under conditions of extreme heat. Protect containers from physical damage.

SECTION VIII - CONTROL MEASURES AND PERSONAL PROTECTIVE EQUIPMENT

Respiratory Protection *(specify type)*

No exposure limits have been established. Radon is a radioactive gas. It is the heaviest gas known. Use a powered supplied-air respirator or a self-contained breathing apparatus with full facepiece operated in pressure demand or other positive pressure mode.

Ventilation

Local exhaust or general mechanical systems recommended.

Protective Gloves	Eye Protection	Other Protective Clothing
Impervious Materials	Chemical Goggles and/or Face Mask	Protective Clothing

Work/Hygiene Practices

Always wash hands thoroughly after using chemical; never bring food, drink, or smoking materials into vicinity of chemicals.

RADON

Rn

CAS: 10043-92-2

IDENTIFICATION AND TYPICAL USES

Radon is a colorless, odorless, and tasteless gas. It is a gaseous radioactive element of atomic number 86 of the noble gas group of the periodic table with atomic weight of 222, valences of 2,4,(6), and 18 radioactive isotopes (all short-lived). The radon-222 isotope has a half-life of 3.8 days and emits α-radiation. It is used in medicine for cancer treatments, to track leak detection, for flow rate measurement, in radiography, and in chemical research.

RISK ASSESSMENT: HEALTH

General Assessment

Radon is a radioactive gas. It occurs naturally as a disintegration product of radium which, itself, is a disintegration product of uranium. Radon further decays to polonium which is the responsible agent causing injuries and illness which results from exposure to radon gas. As a gas, its primary route of entry into the body is through *inhalation*. It can also absorb to particulates which can pass through intact skin (*absorption*). These radioactive particulates can also be *ingested*. The data on the toxicity through ingestion are inconclusive. It can be condensed to a colorless, transparent liquid or an opaque, glowing solid. It is toxic due to ionizing radiation and has been found to cause lung cancer in uranium mine workers.

Inhalation of radon gas primarily presents a long-term (chronic) health hazard risk. Radon can cause lung, kidney, and skin cancer. It may also cause ulcerative radiation dermatitis in some individuals.

Acute Health Effects

The following acute (short-term) health effects may occur immediately or shortly after exposure to radon:

Skin: Very little acute skin action. There may be some irritation prior to dermatitis.

Eye: No listed acute eye effects. Since radon is odorless, colorless, and tasteless, it is not possible to know if radon is in the air.

Lung: No listed acute lung effects. Harmful actions are usually the result of long-term exposures.

Chronic Health Effects

The following chronic (long-term) health effects may occur at some time after exposure to radon and can last for months or even years:

Cancer Hazards: According to information presented in the references, exposure to radon can cause lung, skin, and kidney cancer.

Reproductive Hazards: There are no reports to support any claims of reproductive hazards.

Other Chronic Effects: Ulcerative dermatitis can occur in the long-term.

Recommended Risk-Reduction Measures

Personnel should avoid direct contact with radon. Occupational exposures occur primarily in the mining industry. However, exposures can also occur in medical and health care professions and in research laboratories. If another gas cannot be substituted for a radon, then *engineering controls* are the most effective method of reducing exposures. The best protection is to separate operations and provide local exhaust ventilation at the site of chemical release. While not always operationally feasible, isolating operations can also reduce exposure. Using respiratory protection is less effective than the controls mentioned above, but is still advisable whenever working with radon. Air-purifying respirators (e.g., the cartridge-type or gas masks) must be equipped with high-efficiency particulate air (HEPA) filters designed for radioactive particulates. However, caution is warranted since these types of respirators only purify the air and can do nothing to protect in oxygen deficient atmospheres (confined spaces). A powered air-supplied respirator operated in pressure demand or other positive pressure mode, or a self-contained breathing apparatus (SCBA) with full facepiece operated in pressure demand mode will provide the best protection. If a full facepiece is not available, then chemical goggles should be worn to protect the eyes. To prevent hand and skin exposures, protective gloves should be used. Glove manufacturers should be contacted and permeation studies obtained *before* final glove selections are made.

Administrative controls should also be in place to minimize the potential for human exposures. These may include written procedures or policies which specify the methods and techniques that will be practiced whenever personnel are to work with radon.

All personnel should receive *training* on the use, hazards, protective measures, emergency actions, and other precautions per 29 CFR 1910.1200 (Hazard Communication), prior to the first assignment in an area where radon is used or stored.

Periodic medical examinations focusing on lung and kidney function are recommended. Radon's concentration in blood increases exponentially, becoming stabilized after 30 to 60 minutes. It is lipid soluble. At equilibrium, 1 cm^3 of fatty tissue contains 45% of the amount of radon in 1 cm^3 of inhaled air.

Other methods to reduce exposure to chemicals include:

- ☑ Radon is an odorless, colorless, and tasteless gas and, as such, it is not possible to determine if it is present in the work environment without the use of instrumentation. Where possible, consider installing automatic sensors to warn personnel of increasing radon levels.
- ☑ Use local exhaust ventilation at the site of release. If local exhaust ventilation or enclosure is not used, respiratory protection should be mandatory.
- ☑ Always ensure that proper protective clothing is worn when using chemical substances and that personnel have been trained on its use, care, and maintenance.
- ☑ Safety shower and eyewash stations should be readily available in work areas where radon is used or stored.
- ☑ Hazard warning information should be posted in the work area. In addition, as part of an on-going education and training program, all information on the health and safety hazards of radon should be communicated to all potentially exposed workers.

RISK ASSESSMENT: ENVIRONMENT

General Assessment

Even though radon is a naturally occurring element that is present in the environment, the environment is at risk of exposure during transportation, storage, disposal, or destruction of radon. In almost every scenario, the threat of environmental exposure is contingent upon the proper handling of the chemical substance. Accidental releases, large or small can result in possible contamination of the surrounding environmental mediums (mostly air but, depending upon the nature of the spill, water and soil may be effected).

Radon is a non-combustible gas. It presents no fire hazard. It does decompose without the addition of heat to form polonium, which is more hazardous than radon. In aqueous solution, it can produce explosive hydrogen gas via radiolysis resulting in considerable pressure buildup. Although it is a noble gas and considered chemically inert, it has been known to react with oxidizers such as fluorine. Radon can enter the environment through industrial discharges, unchecked venting to air, or through leaks. Since liquid solutions containing radon will most likely vaporize rapidly, water and soil contamination are possible but improbable.

Acute Ecological Effects

Acute (short-term) toxic effects may include the death of animals, birds, or fish, and death or low growth rate in plants. Acute effects are seen 2 to 4 days after animals or plants are exposed to radon.

Insufficient data are available on the short-term effects of radon on aquatic life, plants, birds, or land animals.

Chronic Ecological Effects

Chronic (long-term) toxic effects may include shortened life span, reproductive problems, lower fertility, and changes in appearance or behavior in exposed animals. These effects can be seen long after first exposure(s) to toxic chemicals.

Insufficient data are available to evaluate the long-term effects of radon to aquatic life, plants, birds, or land animals.

Water Solubility

Radon is moderately slightly soluble in water. Concentrations of 100 milligrams may mix with a liter of water. It is normally a gas at room temperature and pressure and solubility may not be a concern.

⌛ *Persistence in the Environment*

Radon is non-persistent in water, with a half-life of less than 2 days. The half-life of a pollutant is the amount of time it takes for one half of the chemical to be degraded. Virtually all of radon will eventually end up in the air.

Bioaccumulation in Aquatic Organisms

Some substances increase in concentration, or *bioaccumulate,* in living organisms as they breathe contaminated air, drink contaminated water, or eat contaminated food. These chemicals can become concentrated in the tissues and internal organs of animals as well as humans.

Radon is not expected to accumulate in the edible tissues of fish.

Recommended Risk-Reduction Measures

Proper training of all transporters will reduce the likelihood of a mishap or accident resulting in a leak or release to the environment. The correct use of DOT labeling (per the MSDS) on all containers, trucks, and rail cars will enable emergency responders to react properly and quickly to any disaster thereby reducing the potential risk to the environment and to personnel.

Storage of radon should be segregated from other gases and materials to prevent cross-contamination or contact. Buildings designated for storage should be equipped with monitors that provide ample warning (audible or visual) of excessive radon levels.

If a spill or release to the environment has occurred, fire department, emergency response, and/or hazardous materials spill personnel should be notified immediately. Cleanup should be attempted only by those trained in proper containment procedures. If applicable contaminated soils should be removed for incineration and replaced with clean soil. When spills or leaks occur, it may be necessary to contact the local and/or state emergency response authorities. A comprehensive emergency response or disaster preparedness/recovery plan should be in place prior to any operations involving the use, transportation, storage, or disposal of radon.

If radon is spilled or leaked, the following specific steps are recommended:

- ☑ Restrict persons not wearing protective clothing from area of spill or leak until cleanup is complete and area can be opened for normal work.
- ☑ Ventilate area of spill or leak.
- ☑ Keep radon solutions out of a confined space, such as a sewer, because of the possibility of a buildup of explosive hydrogen gas.

RISK ASSESSMENT: BUSINESS

General Assessment

Accidents or mishaps involving radon can present a significant threat to business operations. The loss or damage of equipment or facilities can significantly affect fiscal viability. Lawsuits that may result from personnel injury, public exposures, and environmental contamination will also require a serious expenditure of resources. Media attention surrounding an injury, illness, or environmental damage can also result in a loss of profits and loss of current as well as future business. Always remember that anytime the words "cancer" or "carcinogen" are used, public hysteria, emotion, anxiety, and ignorance can run equally high. In recent years, this has also been the case whenever the word "radon" is used. This must be carefully considered whenever planning or implementing public relations policies.

Recommended Risk-Reduction Measures

Company attorneys, safety and health professionals, and environmental specialists should be involved in the development of any procedures dealing with safety and emergency response. A company official should be pre-designated as a public relations officer with specific training in dealing with the press. Corporate plans and policies should be developed, approved, and implemented long before any need for such arises.

REFERENCES

Ahlbom, A. 1993. *Biostatistics for Epidemiologists.* New York: CRC Press/Lewis Publishers.

American Conference of Governmental Industrial Hygienists. 1988. *Documentation of the Threshold Limit Values and Biological Exposure Indices,* 5th Edition (with updates). Cincinnati: ACGIH

Anthony, C. P., and N. J. Kolthoff. 1971. *Textbook of Anatomy and Physiology.* St. Louis: C. V. Mosby Company.

Asante-Duah, D. K. 1993. *Hazardous Waste Risk Assessment.* New York: CRC Press/Lewis Publishers.

Bloom, A. D. and F. J. DeSerres. 1995. *Ecotoxicity and Human Health.* New York: CRC Press/Lewis Publishers.

Brusick, D. J. 1994. *Methods for Genetic Risk Assessment.* New York: CRC Press/Lewis Publishers.

Bueche, J. R. 1972. *Principles of Physics,* 2nd Edition. New York: McGraw-Hill.

Burrell, R., D. K. Flaherty, and L. J. Sauers. 1992. *Toxicology of the Immune System.* New York: Van Nostrand Reinhold.

Calabrese, E. J. 1994. *Biological Effects of Low Level Exposures.* New York: CRC Press/Lewis Publishers.

Cockerman, L. G., B. S. Shane. 1994. *Basic Environmental Toxicology.* New York: CRC Press/Lewis Publishers.

Cothern, C. R. 1995. *Handbook for Environmental Risk Decisionmaking.* New York: CRC Press/Lewis Publishers.

Danse, I. R. 1991. *Common Sense Toxics in the Workplace.* New York: Van Nostrand Reinhold.

Environmental Protection Agency. 1983. *Chemical Hazard Information Profiles.* Washington, D.C.: U.S. EPA.

Fodden, C. R., and J. L. Weddell. 1992. *Hazardous Materials Emergency Action Data.* New York: CRC Press/Lewis Publishers.

Forseberg, K., and L. H. Keith. 1993. *Chemical; Protective Clothing Permeation and Degradation Database.* Chelsea, MI: Lewis Publishers.

Gardner, R. W. 1994. *Chemical Intolerance.* New York: CRC Press/Lewis Publishers.

Hallenbeck, W. H. 1993. *Quantitative Risk Assessment for Environmental and Occupational Health,* 2nd Edition. New York: CRC Press/Lewis Publishers.

Hodgson, E., R. B. Mailman, and J. E. Chambers. 1988. *Dictionary of Toxicology.* London: The MacMillian Company.

Hosty, J. W., and P. Foster. 1990. *A Practical Guide to Chemical Spill Response.* New York: Van Nostrand Reinhold.

Howard, P. H. 1995. *Handbook of Environmental Fate and Exposures Data for Organic Chemicals,* Vol. I-IV. New York: CRC Press/Lewis Publishers.

James, A., and M. Lord. 1992. *Index of Chemical and Physical Data.* New York: Van Nostrand Reinhold.

Jones, F. E. 1994. *Toxic Organic Vapors in the Workplace.* New York: CRC Press/Lewis Publishers.

Kamrin, M. A. 1988. *Toxicology: A Primer on Principles and Applications.* Chelsea, Michigan: Lewis Publishers.

Keith, L. H. , M. T. Johnston, C. E. Hudak, and M. Conoley. 1992. *National Toxicology Program's Gloves+ Database.* Chelsea, MI: Lewis Publishers.

Keith, L. H. and M. M. Walker. 1995. *Handbook of Air Toxics.* New York: CRC Press/Lewis Publishers.

Keith, L. H. and D. B. Walters. (Editors) 1992. *National Toxicology Program's Chemical Database, Volumes I-VII.* Chelsea, MI: Lewis Publishers.

Landis, W. G., M. H. Yu. 1995. *Introduction to Environmental Toxicology.* New York: CRC Press/Lewis Publishers.

Langerman, N. 1994 *Precautionary Labels for Chemical Containers.* New York: CRC Press/Lewis Publishers.

Lauwerys, R. R., P. Hoet. 1993. *Industrial Chemical Exposure.* New York: CRC Press/Lewis Publishers.

Lewis, R. J., Sr. 1993. *Hawley's Condensed Chemical Dictionary,* 12th Edition. New York: Van Nostrand Reinhold.

Lewis, R. J., Sr. 1992. *Sax's Dangerous Properties of Industrial Materials,* Volumes 1, 2, and 3, 8th Edition. New York: Van Nostrand Reinhold.

Lide, D. R. 1995. *Handbook of Chemistry and Physics,* 76th Edition. New York: CRC Press/Lewis Publishers.

Manahan, S. E. 1994. *Environmental Chemistry,* 6th Edition. New York: CRC Press/Lewis Publishers.

Manahan, S. E. 1992. *Toxicological Chemistry,* 2nd Edition. New York: CRC Press/Lewis Publishers.

Mellström, G. A., J. E. Wahlberg, and H. I. Maibach. 1994. *Protective Gloves for Occupational Use.* New York: CRC Press/Lewis Publishers.

Meyer, E. 1990. *Chemistry of Hazardous Materials.* Englewood Cliffs, NJ: Prentice Hall.

Patnaik, P. 1992. *A Comprehensive Guide to the Hazardous Properties of Chemical Substances.* New York: Van Nostrand Reinhold.

Philp, R. B. 1995. *Environmental Hazards and Human Health.* New York: CRC Press/Lewis Publishers.

Proctor, N. H., and J. P. Hughes. 1991. *Chemical Hazards of the Workplace.* New York: Van Nostrand Reinhold.

Quigley, D. R. 1994. *Handbook of Emergency Chemical Management.* New York: CRC Press/Lewis Publishers.

Rea, W. J. 1995. *Chemical Sensitivity,* Vol. I-III. New York: CRC Press/Lewis Publishers.

Richardson, M. 1995. *Dictionary of Substance and Their Effects,* Vol. I-VI. New York: CRC Press/Lewis Publishers.

Sacarello, H. L. A. 1994. *The Comprehensive Handbook of Hazardous Materials.* New York: CRC Press/Lewis Publishers.

Sherman, J. D. 1988. *Chemical Exposure and Disease.* New York: Van Nostrand Reinhold.

Talbott, E., G. Craun. 1995. *An Introduction to Environmental Epidemiology.* New York: CRC Press/Lewis Publishers.

The Merck Index, 10th edition. 1983. Rahway, NJ: Merck & Company.

Torkelson, T. R. V. K. Rowe. 1981. *Patty's Industrial Hygiene and Toxicology,* Vol. 2B. New York: Wiley-Interscience.

U.S. Department of Health and Human Services, National Institute for Occupational Safety and Health. 1994. *NIOSH Pocket Guide to Chemical Hazards.* Washington, D.C.: U.S. Government Printing Office.

Witorsch, P. and S. V. Spagnolo. 1994. *Air Pollution and Lung Disease in Adults.* New York: CRC Press/Lewis Publishers.

MATERIAL SAFETY DATA SHEET

CHEMICAL NAME	
RESORCINOL	**CRC PRESS/LEWIS PUBLISHERS DISCLAIMER:** THE INFORMATION AND RECOMMENDATIONS PRESENTED HEREIN ARE BASED ON SOURCES BELIEVED TO BE RELIABLE. CRC MAKES NO REPRESENTATION ON ITS COMPLETENESS OR ACCURACY. IT IS THE RESPONSIBILITY OF THE USER TO DETERMINE THE CHEMICAL'S SUITABILITY FOR ITS INTENDED USE, THE CHEMICAL'S SAFE USE, AND THE CHEMICAL'S PROPER DISPOSAL. NO REPRESENTATIONS AND/OR WARRANTIES, EITHER EXPRESSED OR IMPLIED, OF THE MERCHANTABILITY OR FITNESS FOR A PARTICULAR PURPOSE, OR OF ANY OTHER NATURE, ARE MADE WITH RESPECT TO THE INFORMATION PROVIDED IN THIS MSDS OR TO THE CHEMICAL TO WHICH INFORMATION MAY REFER. CRC NEITHER ASSUMES NOR AUTHORIZES ANY OTHER PERSON TO ASSUME FOR IT, ANY OTHER ADDITIONAL RESPONSIBILITY OR LIABILITY FOR THE USE OF, OR RELIANCE UPON, THIS INFORMATION.

HAZARD WARNING INFORMATION

HEALTH	FIRE	REACTIVE	OTHER	DEGREE OF HAZARD	COLOR CODING	OTHER CODES
1	1	1		0 = Minimum Hazard 1 = Slight Hazard 2 = Moderate Hazard 3 = Serious Hazard 4 = Severe Hazard	HEALTH = BLUE FIRE = RED REACTIVITY = YELLOW OTHER = WHITE	OX = Oxidizer ACID = Acid ALK = Alkali COR = Corrosive W = Use No Water

SECTION I - GENERAL INFORMATION

Characterization	**Phenol**	RCRA Number	**U201**
		EPA Class	**Toxic Waste**
DOT Proper Shipping Name	**Resorcinol**	Chemical Abstract Service (CAS) Number	**108-46-3**
DOT Hazard Class and Label Requirements	**Poison B; St. Andrews Cross**	DOT Emergency Guide Code	**55**
DOT Identification Number	**UN2876**	Chemical Formula	**$C_6H_4(OH)_2$**

Synonyms

Resorcin; resorcine; m-dihydroxybenzene; 3-hydroxyphenol; 1,3-benzenediol; 1,3-dihydroxybenzene; m-hydroxyphenol; m-hydroxyphenol.

SECTION II - HAZARDOUS INGREDIENTS AND IDENTITY INFORMATION

Hazardous Components (specific identity)	OSHA Exposure Criteria	NIOSH Exposure Criteria	Immediately Dangerous to Life and Health (IDLH)	ACGIH Exposure Criteria
Resorcinol (derivation: By fusing benzene-m-disulfonic acid with sodium hydroxide). **1 ppm = 4.58 mg/m³**	**PEL:** **Not Established** **STEL:** **Not Established**	**REL:** **10 ppm** **45 mg/m³** **CEILING (15 min):** **20 ppm** **90 mg/m³**	**Not Determined**	**TLV:** **10 ppm** **45 mg/m³** **STEL:** **20 ppm** **90 mg/m³**

SECTION III - PHYSICAL AND CHEMICAL CHARACTERISTICS

Boiling Point	**531°F (277°C)**	Specific Gravity (H_2O = 1)	**1.27**
Vapor Pressure (mm Hg)	**0.0002 at 77°F (25°C)**	Molecular Weight	**110.1**
Vapor Density (Air = 1)	**3.80**	Melting Point	**228°F (109°C)**

Solubility

Soluble in water, alcohol, ether, glycerol, benzene, and amyl alcohol. Slightly soluble in chloroform.

Appearance and Odor

White crystalline solid, becoming pink on exposure to light and air or in contact with iron. Sweet taste.

SECTION IV - FIRE AND EXPLOSION HAZARD DATA

Flash Point (method used)	**261°F (127°C) closed cup**	Explosive Limits in Air % by Volume	**LEL: 1.4% at 392°F (200°C)** **UEL: Not Determined**
NFPA Classification	**Class IIIB Combustible Liquid**	Autoignition Temperature	**1126°F (608°C)**

Extinguishing Media

Use water fog, carbon dioxide, dry chemical, or alcohol foam. Do NOT use solid stream of water (may actually spread fire). Water or foam may cause frothing.

Special Fire Fighting Procedures

Wear full protective clothing and self-contained breathing apparatus (SCBA). Poisonous gases are produced in fire. Use water spray to keep fire-exposed containers cool.

Unusual Fire and Explosion Hazards

Slight fire and explosion hazard when exposed to heat or flame. Containers may explode in fire. Chemical will react on contact with oxidizable materials.

SECTION V - REACTIVITY DATA

Stability		Conditions to Avoid
		Resorcinol is stable at room temperature in closed containers under normal conditions of handling and storage. Keep away from heat and incompatible materials. Hygroscopic (absorbs moisture from air).
Stable	Unstable	Incompatibility *(materials to avoid)*
X		Strong oxidizers, strong bases, acetanilide, albumin, alkalis, antipyrine, camphor, ferric salts, menthol, spirit nitrous ether, and urethane.
Hazardous Polymerization		Conditions to Avoid
		Under normal conditions of temperature and pressure, hazardous polymerization of resorcinol cannot occur.
May Occur	Will Not Occur	Hazardous Decomposition or By-products
	X	When heated to decomposition, resorcinol can produce oxides of carbon (carbon monoxide and/or carbon dioxide).

SECTION VI - HEALTH HAZARD DATA

Primary Route(s) of Entry:	Inhalation?	Absorption (skin)?	Ingestion?
	X	X	X

Health Hazards

INHALATION: Irritation of the eyes, nose, throat, and respiratory tract. Systemic toxic effects include dypsnea, cough, headache, vertigo, methemoglobinemia, cyanosis, convulsions, restlessness, drowsiness, increased heart rate, hypothermia, changes in the spleen, liver, and kidney, and death.

ABSORPTION: Causes local hyperemia, itching, dermatitis, edema, corrosion, and loss of the superficial layers of skin. Severe exposure can cause enlargement of regional lymph glands.

INGESTION: Irritation or gastrointestinal burns, diarrhea, pallor, weakness, sweating, headache with dizziness, tinnitus, shock, convulsions, siderosis of the spleen, tubular injury (kidney).

Carcinogenicity	NTP Listed?	IARC Cancer Review Group?	OSHA Regulated?	Target Organs?
Suspected Human Questioned Animal	No	Group 3	No	Respiratory system, eyes, skin, CNS, CVS, liver, kidney, blood, spleen.

Medical Conditions Generally Aggravated by Exposure

Chronic respiratory disorders, pre-existing skin disorders, convulsive disorders, kidney or liver disease.

Emergency and First-aid Procedures

Eye contact: Flush immediately with water for 15 minutes (minimum); seek medical attention. Skin contact: Remove all contaminated clothing. Immediately wash area with large amounts of soap and water. Seek medical assistance. For inhalation: Remove the person from exposure. Provide respiratory assistance and CPR, if required. If swallowed: Seek medical attention immediately or contact poison control center. Unless advised otherwise, give conscious person 2 glasses of water or milk to drink. Do NOT induce vomiting unless directed to do so by poison control center.

SECTION VII - PRECAUTIONS FOR SAFE HANDLING AND USE

Steps to be Taken in Case Material is Released or Spilled

Absorb liquids in vermiculite, dry sand, earth, or other non-reactive material and deposit in sealed drum. Collect powdered material in safest manner possible. Do NOT dry sweep. Ventilate area of spill or leak. Restrict those not involved in cleanup from entering area. Remove all sources of ignition.

Preferred Waste Disposal Method

Burn in chemical incinerator equipped with an afterburner and scrubber.

Precautions to be Taken in Handling and Storage

Store to avoid contact with incompatible materials in tightly closed containers in cool, well-ventilated area away from heat and flame. Avoid contact with air and light.

Other Precautions and Warnings

Protect containers from physical damage.

SECTION VIII - CONTROL MEASURES AND PERSONAL PROTECTIVE EQUIPMENT

Respiratory Protection *(specify type)*

For high exposures, use an MSHA/NIOSH-approved supplied-air respirator with full facepiece in continuous flow mode, or use a SCBA operated in pressure demand or other positive pressure mode.

Ventilation

Local exhaust or general mechanical systems recommended.

Protective Gloves	Eye Protection	Other Protective Clothing
Impervious Rubber	Chemical/Dust Goggles or Face Mask	Rubber Apron

Work/Hygiene Practices

Always wash hands thoroughly after using chemical; never bring food, drink, or smoking materials into vicinity of chemicals.

RESORCINOL

$C_6H_5(OH)_2$ CAS: 108-46-3

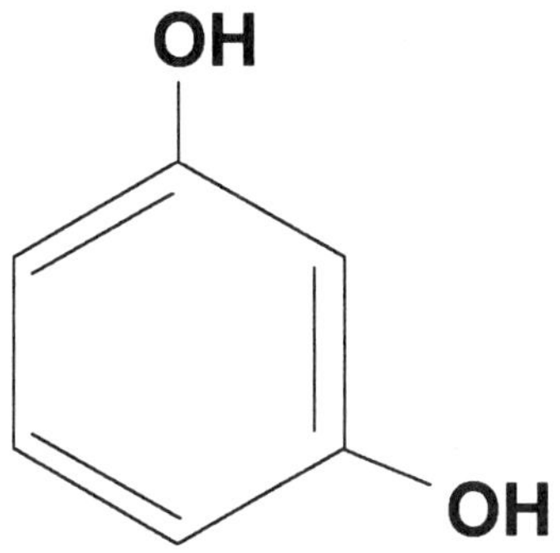

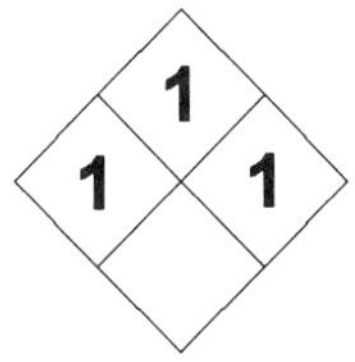

IDENTIFICATION AND TYPICAL USES

Resorcinol appears as white needles, plates, crystals, flakes, or powder with a faint odor and sweet taste. It can turn pink on exposure to light or air, or in contact with iron. It is hygroscopic (absorbs moisture from air). It is used in resorcinol-formaldehyde resins, in the manufacture of dyes, in pharmaceuticals, as a cross-linking agent for neoprene, as an agent in rubber production, in adhesives for wood veneers and rubber-to-textile composites, in the manufacture of styphnic acid, and in cosmetics.

RISK ASSESSMENT: HEALTH

General Assessment

Resorcinol is a human poison by *ingestion.* It is moderately toxic by skin contact and dermal *absorption* is possible. It can also enter the body by *inhalation.* It is a severe irritant to the eyes, skin and respiratory system. Human mutation data have been reported for resorcinol. It is a questionable carcinogen in animals and a suspected carcinogen in humans, although the data are inconclusive in this area of study. There are no data on its teratogenic potential in humans or animals.

Inhalation can cause irritation to the respiratory system and associated mucosa. Inhalation and ingestion can cause systemic poisoning by acting both as a blood and nerve poison. Symptoms include methemoglobinemia, cyanosis (bluish discoloration of the skin due to low oxygen content in blood), convulsions, restlessness, increased heart rate, dypsnea, dizziness, drowsiness, hypothermia, changes to the spleen, kidney, liver, and blood. Death can also occur.

In addition to the systemic symptoms noted above, skin contact can cause serious localized hyperemia, itching, dermatitis, edema, and enlargement of the regional lymph glands. If resorcinol should contact the eyes, there may be serious damage including conjunctival edema, corneal opacification, hypethesia, and possible blindness.

☠ *Acute Health Effects*

The following acute (short-term) health effects may occur immediately or shortly after exposure to resorcinol:

Skin: Severe irritation and possible burns. Will pass through unbroken skin to cause toxic systemic effects, including blood and nervous system disorders.

Eye: Severe irritation with damage and possible loss of vision likely.

Lung: Severe irritation to the respiratory tract and associated mucosa. Can cause weakness, nausea, confusion, collapse, coma, and death (respiratory failure).

CNS: A CNS depressant. Weakness, confusion, mood changes and other personality changes, nausea, collapse, and death.

CVS: Causes increased heart rate (tachycardia).

💣 *Chronic Health Effects*

The following chronic (long-term) health effects may occur at some time after exposure to resorcinol and can last for months or even years:

Cancer Hazards: Some studies have suggested that exposure to resorcinol causes mutations (genetic changes) in humans. While there is no evidence of carcinogenicity of this chemical in humans, it has caused cancer in mice and is considered a *suspected* carcinogen. It is suggested that mutagenic chemicals be regarded as carcinogens during use and handling.

Reproductive Hazard: According to information presented in the references, resorcinol has not been adequately tested for its ability to adversely affect reproduction. However, other phenols have been shown to cause teratogenic effects in animals.

Other Chronic Effects: Resorcinol may produce sensitization in some individuals. Severe cases of exposure can cause siderosis of the spleen and tubular injury to the kidneys.

Recommended Risk-Reduction Measures

Personnel should avoid direct contact with resorcinol. The exact nature of its carcinogenic potential is not clearly understood in the references. If a less toxic chemical cannot be substituted for a hazardous substance, then *engineering controls* are the most effective method of reducing exposures. The best protection is to enclose operations and/or provide local exhaust ventilation at the site of chemical release. While not always operationally feasible, isolating operations can also reduce exposure. Using respiratory protection is less effective than the controls mentioned above, but is still advisable whenever working with or around resorcinol. For low, infrequent, or transient exposures, use an MSHA/NIOSH-approved full facepiece respirator equipped with an organic vapor cartridge. Better protection is obtained from a supplied-air respirator with full facepiece operated in continuous flow mode, or a self-contained breathing apparatus (SCBA) with full facepiece operated in pressure demand. If a full facepiece is not available, then chemical goggles should be worn to protect the eyes. Whenever a chemical splash hazard exists, a face shield and a rubber apron should be worn. To prevent hand and skin exposures, impervious butyl rubber gloves should be used. Glove manufacturers should be contacted and permeation studies obtained *before* glove selections have been made.

Administrative controls should also be in place to minimize the potential for human exposures. These may include written procedures or policies which specify the methods and techniques that will be practiced whenever personnel are to work with resorcinol.

All personnel should receive *training* on the use, hazards, protective measures, emergency actions, and other precautions per 29 CFR 1910.1200 (Hazard Communication), prior to the first assignment in an area where resorcinol is used or stored.

Before beginning employment with resorcinol and at regular intervals thereafter (e.g., annually), those persons with high or frequent exposures should be provided the following recommended medical tests:

☑ Kidney and liver function tests.

If symptoms develop or overexposure is suspected, the following tests are recommended:

☑ Urinary resorcinol tests.

☑ Kidney and liver function tests

☑ Neurological evaluation for brain and nervous system effects.

☑ Blood methemoglobin levels.

Any evaluation should include a careful history of past and present symptoms with an examination. Medical tests that simply look for existing damage are not a substitute for controlling exposures. Also, since smoking can cause heart disease, lung cancer, emphysema, and other respiratory disorders, smokers exposed to resorcinol may experience symptoms more quickly than non-smokers under the same conditions of exposure. Prudent risk management requires careful consideration of *all* potential exposure factors.

Other methods to reduce exposure include:

☑ Where possible, enclose operations and use local exhaust ventilation at the site of chemical release. If local exhaust ventilation or enclosure is not used, respiratory protection should be mandatory.

☑ Also, if possible, liquids containing resorcinol should be transferred automatically from storage containers to process containers.

☑ Always ensure that proper protective clothing is worn when using chemical substances and that personnel have been trained on the use, care, and maintenance of personal protective equipment.

☑ Wash thoroughly immediately after exposure to resorcinol and at the end of the work shift or before eating, drinking, or smoking.

☑ Hazard warning information should be posted in the work area. In addition, as part of an on-going education and training program, all information on

the health and safety hazards of resorcinol should be communicated to all exposed workers.

☑ Eye wash stations should be provided in the immediate work area for emergency use. If there is a possibility of skin exposure to resorcinol, emergency shower facilities should also be provided.

☑ Personnel should never wear contaminated clothing home (family members can be exposed). Clothing should be laundered by personnel who have been trained on the hazards of resorcinol.

RISK ASSESSMENT: ENVIRONMENT

General Assessment

The environment is at risk of exposure during transportation, storage, disposal, or destruction of resorcinol. In almost every scenario, the threat of environmental exposure is contingent upon the proper handling of the chemical substance. Accidental spills, large or small, can result in fire, explosion, and possible contamination of the surrounding environmental mediums (water, soil, and air).

Resorcinol is considered a combustible solid. However, its dusts present a slight fire hazard when exposed to heat in air. It is reactive with many commonly encountered materials, including strong oxidizers (chlorine, fluorine, bromine), acids (hydrochloric, sulfuric, and nitric), acetanilide, albumin, alkalis, antipyrine, camphor, ferric salts, menthol, spirit nitrous ether, and bases. These characteristics require caution in handling, storage, transportation, and disposal. It is also hygroscopic (absorbs moisture from air) and, therefore, special consideration is required during any emergency situation involving a leak or spill of resorcinol. Should resorcinol ever come into contact with any incompatible substances during use, transportation, storage, or disposal, the formation of highly toxic and/or highly explosive commodities is extremely possible.

Resorcinol may enter the environment through municipal waste treatment plant and industrial plant discharges, or through spills.

☠ *Acute Ecological Effects*

Acute (short-term) toxic effects may include the death of animals, birds, or fish and death or low growth rate in plants. Acute effects are seen 2 to 4 days after animals or plants are exposed to resorcinol.

Insufficient data are available to evaluate or predict the short-term effects of resorcinol exposure to aquatic life, plants, birds, or land animals.

Chronic Ecological Effects

Chronic (long-term) toxic effects may include shortened life span, reproductive problems, lower fertility, and changes in appearance or behavior in exposed animals. These effects can be seen long after first exposure(s) to toxic chemicals.

Insufficient data are available to evaluate or predict the long-term effects of resorcinol to aquatic life, plants, birds, or land animals.

Water Solubility

Resorcinol is highly soluble in water. Concentrations of 1000 milligrams and more will mix with a liter of water.

⌛ *Persistence in the Environment*

Resorcinol is slightly persistent in water, with a half-life between 2 and 20 days. The half-life of a pollutant is the amount of time it takes for one half of the chemical to be degraded.

Bioaccumulation in Aquatic Organisms

Some substances increase in concentration, or *bioaccumulate,* in living organisms as they breathe contaminated air, drink contaminated water, or eat contaminated food. These chemicals can become concentrated in the tissues and internal organs of animals as well as humans.

The concentration of resorcinol found in fish tissue is expected to be about the same as the average concentration of resorcinol in the water from which the fish was taken.

Recommended Risk-Reduction Measures

Proper training of all transporters will reduce the likelihood of a mishap or accident resulting in a leak or spill to the environment. The correct use of DOT labeling (per the MSDS) on all containers, trucks, and rail cars will enable emergency responders to react properly and quickly to any disaster thereby reducing the potential risk to the environment and to personnel.

Storage of resorcinol should be segregated from incompatible chemicals to minimize the risk of cross-contamination or contact. Buildings designated for storage should be equipped with appropriate fire protection systems (alarms, sprinklers, emergency lighting, portable extinguishers).

If a spill or leak to the environment has occurred, fire department, emergency response and/or hazardous materials spill personnel should be notified immediately. Cleanup should be attempted only by those trained in proper spill containment procedures using non-sparking tools. Contaminated soils should be removed for incineration and replaced with clean soil. If resorcinol should contact the water table, aquifer, or navigable waterway, time is of the essence. It is highly soluble in the aquatic environment and total remediation may not be possible. The local and/or state emergency response authorities must be notified. A comprehensive emergency response or disaster preparedness/recovery plan should be in place prior to any operations involving the use, transportation, storage, or disposal of resorcinol.

If resorcinol is spilled or leaked, the following specific steps are recommended:

- ☑ Restrict persons not wearing protective clothing from area of spill or leak until cleanup is complete and area can be opened for normal work.
- ☑ Remove all ignition sources.
- ☑ Ventilate area of leak.
- ☑ Absorb liquid spills in vermiculite, dry sand, earth, or similar material and deposit in sealed containers. Collect solid materials in safest manner possible and place in sealed containers. Do NOT dry sweep (creates airborne dusts). Use vacuum equipped with a high-efficiency particulate air (HEPA) filter instead.
- ☑ It may be necessary to dispose of resorcinol as a hazardous waste. The responsible state agency or the regional office of the federal Environmental Protection Agency (EPA) should be contacted for specific recommendations.

RISK ASSESSMENT: BUSINESS

General Assessment

Accidents or mishaps involving resorcinol can present a significant threat to business operations. The loss or damage of equipment or facilities can significantly affect fiscal viability. Lawsuits that may result from personnel injury/death, public exposures, and/or environmental contamination will also require a serious expenditure of resources. Media attention surrounding an injury, death, or environmental damage can also result in a loss of profits and loss of current as well as future business. Always remember that anytime the terms "cancer" or "carcinogen" are used, public hysteria, emotion, anxiety, and ignorance can run equally high. This must be carefully considered whenever drafting or implementing public relations policies.

Recommended Risk-Reduction Measures

Company attorneys, safety and health professionals, and environmental specialists should be involved in the development of any procedures dealing with safety or emergency response. A company official should be pre-designated as a public relations officer with specific training in dealing with the press. Corporate plans and policies should be developed, approved, and implemented long before any need for such arises.

REFERENCES

American Conference of Governmental Industrial Hygienists. 1988. *Documentation of the Threshold Limit Values and Biological Exposure Indices,* 5th Edition (with updates). Cincinnati: ACGIH

Asante-Duah, D. K. 1993. *Hazardous Waste Risk Assessment.* New York: CRC Press/Lewis Publishers.

Bloom, A. D. and F. J. DeSerres. 1995. *Ecotoxicity and Human Health.* New York: CRC Press/Lewis Publishers.

Bueche, J. R. 1972. *Principles of Physics,* 2nd Edition. New York: McGraw-Hill.

Calabrese, E. J. 1994. *Biological Effects of Low Level Exposures.* New York: CRC Press/Lewis Publishers.

Cockerman, L. G., B. S. Shane. 1994. *Basic Environmental Toxicology.* New York: CRC Press/Lewis Publishers.

Cothern, C. R. 1995. *Handbook for Environmental Risk Decisionmaking.* New York: CRC Press/Lewis Publishers.

Danse, I. R. 1991. *Common Sense Toxics in the Workplace.* New York: Van Nostrand Reinhold.

Environmental Protection Agency. 1983. *Chemical Hazard Information Profiles.* Washington, D.C.: U.S. EPA.

Forseberg, K., and L. H. Keith. 1993. *Chemical; Protective Clothing Permeation and Degradation Database.* Chelsea, MI: Lewis Publishers.

Lewis, R. J., Sr. 1992. *Sax's Dangerous Properties of Industrial Materials,* Volumes 1, 2, and 3, 8th Edition. New York: Van Nostrand Reinhold.

Patnaik, P. 1992. *A Comprehensive Guide to the Hazardous Properties of Chemical Substances.* New York: Van Nostrand Reinhold.

U.S. Department of Health and Human Services, National Institute for Occupational Safety and Health. 1994. *NIOSH Pocket Guide to Chemical Hazards.* Washington, D.C.: U.S. Government Printing Office.

MATERIAL SAFETY DATA SHEET

CHEMICAL NAME	CRC PRESS/LEWIS PUBLISHERS DISCLAIMER
RHODIUM	CRC PRESS/LEWIS PUBLISHERS DISCLAIMER: THE INFORMATION AND RECOMMENDATIONS PRESENTED HEREIN ARE BASED ON SOURCES BELIEVED TO BE RELIABLE. CRC MAKES NO REPRESENTATION ON ITS COMPLETENESS OR ACCURACY. IT IS THE RESPONSIBILITY OF THE USER TO DETERMINE THE CHEMICAL'S SUITABILITY FOR ITS INTENDED USE, THE CHEMICAL'S SAFE USE, AND THE CHEMICAL'S PROPER DISPOSAL. NO REPRESENTATIONS AND/OR WARRANTIES, EITHER EXPRESSED OR IMPLIED, OF THE MERCHANTABILITY OR FITNESS FOR A PARTICULAR PURPOSE, OR OF ANY OTHER NATURE, ARE MADE WITH RESPECT TO THE INFORMATION PROVIDED IN THIS MSDS OR TO THE CHEMICAL TO WHICH INFORMATION MAY REFER. CRC NEITHER ASSUMES NOR AUTHORIZES ANY OTHER PERSON TO ASSUME FOR IT, ANY OTHER ADDITIONAL RESPONSIBILITY OR LIABILITY FOR THE USE OF, OR RELIANCE UPON, THIS INFORMATION.

HAZARD WARNING INFORMATION

HEALTH	FIRE	REACTIVE	OTHER	DEGREE OF HAZARD	COLOR CODING	OTHER CODES
1	2	0		0 = Minimum Hazard 1 = Slight Hazard 2 = Moderate Hazard 3 = Serious Hazard 4 = Severe Hazard	HEALTH = BLUE FIRE = RED REACTIVITY = YELLOW OTHER = WHITE	OX = Oxidizer ACID = Acid ALK = Alkali COR = Corrosive ₩ = Use No Water

SECTION I - GENERAL INFORMATION

Characterization	Metal	RCRA Number	None
		EPA Class	Not Applicable
DOT Proper Shipping Name	No Citation	Chemical Abstract Service (CAS) Number	7440-16-6
DOT Hazard Class and Label Requirements	No Citation	DOT Emergency Guide Code	No Citation
DOT Identification Number	No Citation	Atomic Formula	Rh

Synonyms
Rhodium metal; elemental rhodium.

SECTION II - HAZARDOUS INGREDIENTS AND IDENTITY INFORMATION

Hazardous Components (specific identity)	OSHA Exposure Criteria	NIOSH Exposure Criteria	Immediately Dangerous to Life and Health (IDLH)	ACGIH Exposure Criteria
Rhodium: (derivation: Occurs with platinum, from which it is recovered during the purification process. One of the rarest elements, it is found in the minerals rhodite, sperrylite, and iridosmine and has one natural radioactive isotope, 103).	PEL: .01 mg/m^3 STEL: Not Established	REL: .01 mg/m^3 STEL: Not Established	100 mg/m^3	TLV: .01 mg/m^3 STEL: Not Established

SECTION III - PHYSICAL AND CHEMICAL CHARACTERISTICS

Boiling Point	6741°F (3727°C)	Specific Gravity (H_2O = 1)	12.41
Vapor Pressure (mm Hg)	0 (approx.) at 69°F (20°C)	Molecular Weight (atomic weight)	102.9
Vapor Density (Air = 1)	Not Found	Melting Point	3571°F (1966°C)

Solubility
Insoluble in water. Soluble in potassium bisulfide. Slightly soluble in chloroform. The metal is insoluble in all acids; the powder is soluble in aqua regia.

Appearance and Odor
A hard, white, ductile, malleable metal with a bluish-gray luster. Can exist as a black powder or sponge.

SECTION IV - FIRE AND EXPLOSION HAZARD DATA

Flash Point (method used)	Not Applicable	Explosive Limits in Air % by Volume	LEL: Not Applicable UEL: Not Applicable
NFPA Classification	Noncombustible Solid (bulk form)	Autoignition Temperature	Not Applicable

Extinguishing Media
In powder form, rhodium is a flammable solid. Use dry chemical, water spray, regular foam, or carbon dioxide. Do NOT use a direct water stream on fires.

Special Fire Fighting Procedures
Poisonous gases are produced in fire. Wear full protective clothing, self-contained breathing apparatus (SCBA) operated in pressure-demand mode. Move container from fire area if it can be done without risk.

Unusual Fire and Explosion Hazards
In powdered form, rhodium is a moderate fire hazard. Will also burn vigorously on contact with flame or by chemical reaction. Powders and dusts form explosive mixtures in air.

SECTION V - REACTIVITY DATA

Stability

Stable	Unstable
X	

Conditions to Avoid
Rhodium is normally stable under routine conditions of handling and storage. Avoid contact with incompatible materials and keep away from heat.

Incompatibility *(materials to avoid)*
A catalytic metal that is incompatible with interhalogens such as bromine pentafluoride, chlorine trifluoride, and oxygen difluoride. Resistant to most acids (except concentrated sulfuric acid and hydrobromic acid).

Hazardous Polymerization

May Occur	Will Not Occur
	X

Conditions to Avoid
Under normal conditions of temperature and pressure, rhodium is not known to undergo hazardous polymerization.

Hazardous Decomposition or By-products
When heated to decomposition in air it emits fumes of rhodium metal and oxygen.

SECTION VI - HEALTH HAZARD DATA

Primary Route(s) of Entry:	Inhalation?	Absorption (skin/eye)?	Ingestion?
	X		

Health Hazards

INHALATION: Data on exposure symptoms are limited, conflicting, and inconclusive. It may cause nasal irritation and possible central nervous system and respiratory depression (in animals), as well as reactions with DNA and RNA. NIOSH reports it is a possible respiratory sensitizer.

SKIN & EYES: Eye contact will result in irritation. It may be a sensitizer (unconfirmed) and possible skin irritant.

INGESTION: Poorly absorbed in the GI tract due to low solubility.

Carcinogenicity	NTP Listed?	IARC Cancer Review Group?	OSHA Regulated?	Target Organs?
Unknown Human Unknown Animal	No	No	29 CFR 1910.1000 Table Z-1	Respiratory system

Medical Conditions Generally Aggravated by Exposure
None reported. However, existing respiratory problems (bronchitis, asthma) may be aggravated.

Emergency and First-aid Procedures
Eye contact: Do not allow victim to keep eyes tightly shut. Flush immediately with water for 15 minutes (minimum); seek medical attention. Skin contact: Remove contaminated clothing. Rinse with flooding amounts of water for 15 minutes (minimum). For red or blistered skin, seek medical attention. For inhalation: Remove the person from exposure. Provide respiratory assistance and CPR. Transfer to medical facility. If swallowed: Seek medical attention immediately. Do not attempt to give an unconscious or convulsing person anything by mouth. Give victim 1 to 2 glasses of water and then induce vomiting.

SECTION VII - PRECAUTIONS FOR SAFE HANDLING AND USE

Steps to be Taken in Case Material is Released or Spilled
Wear a self-contained breathing apparatus (SCBA) for respiratory protection. Restrict those not wearing protective equipment from entering area. Avoid skin contact. Remove ignition sources. Ventilate area of spill. Do NOT dry sweep. Use a HEPA vacuum to collect powdered materials and deposit in sealed drum.

Preferred Waste Disposal Method
No citation.

Precautions to be Taken in Handling and Storage
Store in tightly closed containers in a cool, well-ventilated area away from heat. Protect containers from physical damage. Avoid contact in storage with incompatible materials as violent reactions can occur.

Other Precautions and Warnings
Use only non-sparking equipment and tools, especially when opening or closing containers of rhodium.

SECTION VIII - CONTROL MEASURES AND PERSONAL PROTECTIVE EQUIPMENT

Respiratory Protection *(specify type)*
Use an MSHA/NIOSH-approved half-mask respirator equipped with a high-efficiency particulate air (HEPA) filter. Better protection is provided using a self-contained breathing apparatus (SCBA) with full facepiece operated in positive pressure mode.

Ventilation
Local exhaust (preferred) at site of chemical work or general ventilation.

Protective Gloves	Eye Protection	Other Protective Clothing
Impervious Gloves	Chemical/Dust Goggles and Face Mask	Protective Uniform or Apron

Work/Hygiene Practices
Always wash hands thoroughly after using chemical; never bring food, drink, or smoking materials into vicinity of chemicals.

RHODIUM

Rh

CAS: 7440-16-6

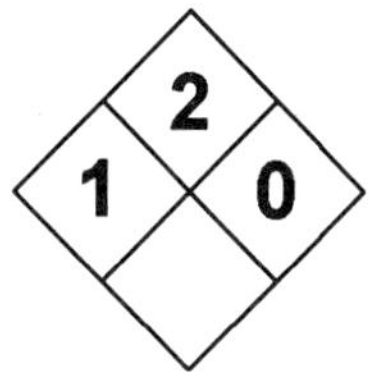

IDENTIFICATION AND TYPICAL USES

Rhodium is a metallic element of atomic number 45, Group VIII of the periodic system. It has no isotopes, and valance of 3. It appears as a white, hard, ductile, malleable solid with a bluish-gray luster. It is also available as a black powder or sponge-like material. It is used as an alloy with platinum for high temperature thermocouples, in furnace windings, in laboratory crucibles, and in spinnerets in the rayon industry. It is also used in electrical contacts, in jewelry, as a catalyst, in optical instrument mirrors, in electro-deposited coatings for metals, in vacuum deposited glass coatings, and in headlight reflectors.

RISK ASSESSMENT: HEALTH

General Assessment

Rhodium is a metal of relatively low toxicity. Primary exposure route is through *inhalation* of its dusts and, to a lesser extent, *ingestion*. Few data are available on the toxicity of rhodium in humans or animals. It is chemically similar to platinum but sensitization has not been observed. There are no reports of any carcinogenic, mutagenic, or teratogenic effects on long-term exposure. NIOSH reports that rhodium may be a possible respiratory sensitizer, but this has not been confirmed.

Inhalation may result in nasal irritation. Some animal studies suggest possible central nervous system (CNS) and respiratory depression. There may also be adverse (unspecified) reactions with DNA and RNA. Rhodium salts may be irritating if splashed in the eyes or in contact with the skin.

Acute Health Effects

The following acute (short-term) health effects may occur immediately or shortly after exposure to rhodium:

Skin: Severe irritation and possible dermatitis (unconfirmed).

Eye: Severe irritation, itching, and burning on contact.

Lung: Overexposure may cause irritation to the airways and lungs, causing nasal discharge, tightness in the chest, cough, shortness of breath, and/or fever. Future exposures can cause further attacks.

Chronic Health Effects

The following chronic (long-term) health effects may occur at some time after exposure to rhodium and can last for months or even years:

Cancer Hazards: According to information presented in the references, rhodium has not been shown to cause cancer in humans or test animals. There are very few data available on the long-term toxicity of rhodium exposure in humans or animals.

Reproductive Hazard: There is no evidence in the published studies to show that rhodium has the ability to adversely affect reproduction.

Other Chronic Effects: None reported.

Recommended Risk-Reduction Measures

Personnel should avoid direct contact with rhodium. The exact nature of its short-term and long-term toxicity is not fully understood in the references. The data are inconclusive and conflicting in this regard. The first risk-reduction measure should be to substitute a less toxic material for rhodium. If a less toxic material or compound cannot be substituted, then *engineering controls* are the most effective method of reducing exposure risk. The best protection is to enclose operations and/or provide local exhaust ventilation at the site of rhodium release. While not always operationally feasible, isolating operations can also reduce exposure. Using respiratory protection is less effective than the controls mentioned above, but is still required whenever working with or around rhodium. For low, transient, or infrequent exposures, recommend using an air-purifying respirator equipped with a dust/mist pre-filter or a high-efficiency particulate air (HEPA) filter. Better protection is obtained using an MSHA/NIOSH-approved self-contained breathing apparatus (SCBA) with full facepiece operated in pressure demand or a supplied-air respirator in continuous flow mode. If a full facepiece is not available, then chemical goggles should be worn to protect the eyes, especially when a splash hazard exists. A face shield should also be considered. To prevent hand and skin exposures, rubber gloves should be used.

Administrative controls should also be in place to minimize the potential for human exposures. These may include written procedures or policies which specify the methods and techniques that will be practiced whenever personnel are to work with rhodium.

All personnel should receive *training* on the use, hazards, protective measures, emergency actions, and other precautions per 29 CFR 1910.1200 (Hazard Communication) prior to the first assignment in an area where rhodium is used or stored.

Before beginning employment and at regular times thereafter, the following recommended medical tests should be considered for those with frequent or potentially high exposures:

☑ Lung function tests.

☑ Consider chest X-ray.

It should be noted that medical tests that simply look for existing damage are not a substitute for controlling exposures. Medical histories are extremely important when assessing exposure risk.

Other methods to reduce exposure include:

☑ Always ensure that proper protective clothing is worn when using chemical substances and that personnel have been trained in its use, care, and maintenance.

☑ Wash thoroughly immediately after exposure to rhodium and at the end of the work shift or before eating, drinking, applying cosmetics, or smoking.

☑ Hazard warning information should be posted in the work area. In addition, as part of an on-going education and training program, all information on the health and safety hazards of rhodium should be communicated to all exposed workers.

☑ Eye wash stations should be provided in the immediate work area for emergency use. If there is a possibility of skin exposure to rhodium, emergency shower facilities should also be provided.

☑ Workers whose clothing has been contaminated by rhodium should change into clean clothes before leaving work. Contaminated work clothing should be laundered only by individuals who have been informed of the hazards of exposure to rhodium.

RISK ASSESSMENT: ENVIRONMENT

General Assessment

The environment is at risk of exposure during transportation, storage, disposal, or destruction of rhodium. In almost every scenario, the threat of environmental exposure is contingent upon the proper handling of the chemical substance. Accidental spills, large or small, can result in contamination of the surrounding environmental mediums (water, soil, and air).

Rhodium is a noncombustible solid in bulk form but is flammable when in powder form. Caution is always required in handling, storage, transportation, and disposal of rhodium. It presents a slight to moderate explosion hazard in the form of powders and dusts. It is incompatible with interhalogens and reacts with fluorine or chlorine. Emergency responders should be made aware of the presence of rhodium at any emergency response situation.

Rhodium may enter the aquatic environment through the weathering of rocks and soils and through discharge from industrial and municipal operations.

☠ *Acute Ecological Effects*

Acute (short-term) toxic effects may include the death of animals, birds, or fish and death or low growth rate in plants. Acute effects are seen 2 to 4 days after animals or plants are exposed to rhodium

Insufficient data are available to evaluate or predict the acute (short-term) effects of rhodium to aquatic life, plants, birds, or land animals.

Chronic Ecological Effects

Chronic (long-term) toxic effects may include shortened life span, reproductive problems, lower fertility, and changes in appearance or behavior in exposed animals.

These effects can be seen long after first exposure(s) to toxic chemicals.

Insufficient data are available to evaluate or predict the chronic (long-term) effects of rhodium on aquatic life, plants, birds, or land animals.

Water Solubility

Rhodium is not soluble in water. Concentrations of 1 milligram will not mix with a liter of water.

Persistence in the Environment

Rhodium is highly persistent in water, with a half-life of greater than 200 days. The half-life of a pollutant is the amount of time it takes for one-half of the chemical to be degraded.

Bioaccumulation in Aquatic Organisms

Some substances increase in concentration, or *bioaccumulate,* in living organisms as they breathe contaminated air, drink contaminated water, or eat contaminated food. These chemicals can become concentrated in the tissues and internal organs of animals as well as humans.

The concentration of rhodium found in edible fish tissue is expected to be about the same as the average concentration of rhodium found in the water from which the fish was taken.

Recommended Risk-Reduction Measures

Proper training of all transporters will reduce the likelihood of a mishap or accident resulting in a leak or spill of rhodium into the environment. Labels on all containers, trucks, and rail cars must meet DOT requirements and accurately reflect their contents to enable emergency responders to react properly and quickly to any disaster thereby reducing the potential risk to the environment and to personnel.

Storage of rhodium should be segregated from other chemicals and materials to minimize the risk of cross-contamination. Rhodium must be stored to avoid contact with oxidizers (such as perchlorates, peroxides, permanganates, chlorates, and nitrates), and strong acids (such as hydrochloric, sulfuric, and nitric). Containers should be protected from physical damage and stored to avoid contact with heat.

If a spill or leak to the environment has occurred, fire department, emergency response and/or hazardous materials spill personnel should be notified immediately. Cleanup should be attempted only by those trained in proper spill containment procedures using non-sparking tools. Contaminated soils should be removed for incineration and replaced with clean soil. If rhodium should contact the water table, aquifer, or navigable waterway, time is of the essence. It is not soluble and total remediation may be possible. The local and/or state emergency response authorities must be notified. A comprehensive emergency response or disaster preparedness/recovery plan should be in place prior to any operations involving the use, transportation, storage, or disposal of rhodium.

If rhodium is spilled or leaked, the following specific steps are recommended:

☑ Restrict persons not wearing protective clothing from area of spill or leak until cleanup is complete. Avoid skin and eye contact.

☑ Collect powdered material in the most convenient and safe manner possible and deposit in sealed drum for disposal. Do NOT dry sweep (generates airborne dusts). Use a vacuum equipped with a high-efficiency particulate air (HEPA) filter. Use non-sparking tools during cleanup operations.

☑ Ventilate area of spill or leak and remove all ignition sources.

☑ It may be necessary to dispose of rhodium as a hazardous waste. The responsible state agency or the regional office of the federal Environmental Protection Agency (EPA) should be contacted for specific recommendations.

RISK ASSESSMENT: BUSINESS

General Assessment

Accidents or mishaps involving rhodium can present a serious threat to business operations. The loss or damage of equipment or facilities can significantly affect fiscal viability. Lawsuits that may result from personnel illness, injury/death, public exposures, and/or environmental contamination will require a serious expenditure of resources. Media attention surrounding an injury, death, or environmental damage can also result in a loss of profits and loss of current as well as future business.

Recommended Risk-Reduction Measures

Company attorneys, safety and health professionals, and environmental specialists should be involved in the development of any procedures for responding to chemical incidents. A company official should be predesignated as a public relations officer with specific training in dealing with the press. Corporate plans and policies should be developed, approved, and implemented long before any need for such arises.

REFERENCES

American Conference of Governmental Industrial Hygienists. 1988. *Documentation of the Threshold Limit Values and Biological Exposure Indices,* 5th Edition (with updates). Cincinnati: ACGIH

Anthony, C. P., and N. J. Kolthoff. 1971. *Textbook of Anatomy and Physiology.* St. Louis: C. V. Mosby Company.

Asante-Duah, D. K. 1993. *Hazardous Waste Risk Assessment.* New York: CRC Press/Lewis Publishers.

Bloom, A. D. and F. J. DeSerres. 1995. *Ecotoxicity and Human Health.* New York: CRC Press/Lewis Publishers.

Brusick, D. J. 1994. *Methods for Genetic Risk Assessment.* New York: CRC Press/Lewis Publishers.

Bueche, J. R. 1972. *Principles of Physics,* 2nd Edition. New York: McGraw-Hill.

Calabrese, E. J. 1994. *Biological Effects of Low Level Exposures.* New York: CRC Press/Lewis Publishers.

Cockerman, L. G., B. S. Shane. 1994. *Basic Environmental Toxicology.* New York: CRC Press/Lewis Publishers.

Danse, I. R. 1991. *Common Sense Toxics in the Workplace.* New York: Van Nostrand Reinhold.

Environmental Protection Agency. 1983. *Chemical Hazard Information Profiles.* Washington, D.C.: U.S. EPA.

Fodden, C. R., and J. L. Weddell. 1992. *Hazardous Materials Emergency Action Data.* New York: CRC Press/Lewis Publishers.

Forseberg, K., and L. H. Keith. 1993. *Chemical; Protective Clothing Permeation and Degradation Database.* Chelsea, MI: Lewis Publishers.

Gardner, R. W. 1994. *Chemical Intolerance.* New York: CRC Press/Lewis Publishers.

Hallenbeck, W. H. 1993. *Quantitative Risk Assessment for Environmental and Occupational Health,* 2nd Edition. New York: CRC Press/Lewis Publishers.

Hodgson, E., R. B. Mailman, and J. E. Chambers. 1988. *Dictionary of Toxicology.* London: The MacMillian Company.

Hosty, J. W., and P. Foster. 1990. *A Practical Guide to Chemical Spill Response.* New York: Van Nostrand Reinhold.

Howard, P. H., et. al. 1991. *Handbook of Environmental Degradation Rates.* New York: CRC Press/Lewis Publishers.

Jones, F. E. 1994. *Toxic Organic Vapors in the Workplace.* New York: CRC Press/Lewis Publishers.

Kamrin, M. A. 1988. *Toxicology: A Primer on Principles and Applications.* Chelsea, Michigan: Lewis Publishers.

Keith, L. H. and M. M. Walker. 1995. *Handbook of Air Toxics.* New York: CRC Press/Lewis Publishers.

Landis, W. G., M. H. Yu. 1995. *Introduction to Environmental Toxicology.* New York: CRC Press/Lewis Publishers.

Langerman, N. 1994 *Precautionary Labels for Chemical Containers.* New York: CRC Press/Lewis Publishers.

Lauwerys, R. R., P. Hoet. 1993. *Industrial Chemical Exposure.* New York: CRC Press/Lewis Publishers.

Lewis, R. J., Sr. 1993. *Hawley's Condensed Chemical Dictionary,* 12th Edition. New York: Van Nostrand Reinhold.

Lewis, R. J., Sr. 1992. *Sax's Dangerous Properties of Industrial Materials,* Volumes 1, 2, and 3, 8th Edition. New York: Van Nostrand Reinhold.

Lide, D. R. 1995. *Handbook of Chemistry and Physics,* 76th Edition. New York: CRC Press/Lewis Publishers.

Lisella, F. S. 1994. *The VNR Dictionary of Environmental Health and Safety.* New York: Van Nostrand Reinhold.

Mackay, D., W. Y. Shiu, and K. C. Ma. 1992. *Illustrated Handbook of Chemical-Physical Properties and Environmental Fate for Organic Chemicals.* New York: CRC Press/Lewis Publishers.

Manahan, S. E. 1994. *Environmental Chemistry,* 6th Edition. New York: CRC Press/Lewis Publishers.

Mellström, G. A., J. E. Wahlberg, and H. I. Maibach. 1994. *Protective Gloves for Occupational Use.* New York: CRC Press/Lewis Publishers.

Meyer, E. 1990. *Chemistry of Hazardous Materials.* Englewood Cliffs, NJ: Prentice Hall.

Miller, B. F., M.D., and C. B. Keane, R.N. 1972. *Encyclopedia and Dictionary of Medicine and Nursing.* Philadelphia: W.B. Sanders Company.

Mycock, J. C., J. D. McKenna, and L. Theodore. 1995. *Handbook of Air Pollution Control Engineering and Technology.* New York: CRC Press/Lewis Publishers.

Nelson, G. O. 1992. *Gas Mixtures.* New York: CRC Press/Lewis Publishers.

Patnaik, P. 1992. *A Comprehensive Guide to the Hazardous Properties of Chemical Substances.* New York: Van Nostrand Reinhold.

Philp, R. B. 1995. *Environmental Hazards and Human Health.* New York: CRC Press/Lewis Publishers.

Proctor, N. H., J. P. Hughes. 1991. *Chemical Hazards of the Workplace.* New York: Van Nostrand Reinhold.

Quigley, D. R. 1994. *Handbook of Emergency Chemical Management.* New York: CRC Press/Lewis Publishers.

Richardson, M. 1995. *Dictionary of Substance and Their Effects,* Vol. I-VI. New York: CRC Press/Lewis Publishers.

Sacarello, H. L. A. 1994. *The Comprehensive Handbook of Hazardous Materials.* New York: CRC Press/Lewis Publishers.

Torkelson, T. R. V. K. Rowe. 1981. *Patty's Industrial Hygiene and Toxicology,* Vol. 2B. New York: Wiley-Interscience.

U.S. Department of Health and Human Services, National Institute for Occupational Safety and Health. 1994. *NIOSH Pocket Guide to Chemical Hazards.* Washington, D.C.: U.S. Government Printing Office.

MATERIAL SAFETY DATA SHEET

CHEMICAL NAME	CRC PRESS/LEWIS PUBLISHERS DISCLAIMER
SALICYLIC ACID	

HAZARD WARNING INFORMATION

HEALTH	FIRE	REACTIVE	OTHER	DEGREE OF HAZARD	COLOR CODING	OTHER CODES
0	1	0		0 = Minimum Hazard 1 = Slight Hazard 2 = Moderate Hazard 3 = Serious Hazard 4 = Severe Hazard	HEALTH = BLUE FIRE = RED REACTIVITY = YELLOW OTHER = WHITE	OX = Oxidizer ACID = Acid ALK = Alkali COR = Corrosive ~~W~~ = Use No Water

SECTION I - GENERAL INFORMATION

Characterization	**Acid**	RCRA Number	**None**
		EPA Class	**Not Applicable**
DOT Proper Shipping Name	**No Citation**	Chemical Abstract Service (CAS) Number	**69-72-7**
DOT Hazard Class and Label Requirements	**No Citation**	DOT Emergency Guide Code	**No Citation**
DOT Identification Number	**No Citation**	Chemical Formula	**$C_6H_4(OH)(COOH)$**

Synonyms

o-hydroxybenzoic acid; 2-hydroxybenzoic acid; Psoriacid-S-Stift; Retarder W; Rutranex; Salonil.

SECTION II - HAZARDOUS INGREDIENTS AND IDENTITY INFORMATION

Hazardous Components (specific identity)	OSHA Exposure Criteria	NIOSH Exposure Criteria	Immediately Dangerous to Life and Health (IDLH)	ACGIH Exposure Criteria
Salicylic acid (derivation: By reacting a hot solution of sodium phenolate with carbon dioxide and acidifying the sodium salt that has been formed). NOTE: OSHA and ACGIH Exposure Criteria are for "Particulates not Otherwise Regulated"	**PEL (8-hour): 15 mg/m^3** (total dust) **5 mg/m^3** (respirable fraction)	**REL: Not Established** **STEL: Not Established**	**Not Established**	**TLV: 10 mg/m^3** (total dust with no asbestos and <1% silica)

SECTION III - PHYSICAL AND CHEMICAL CHARACTERISTICS

Boiling Point	**412°F (211°C) at 20 mmHg**	Specific Gravity (H_2O = 1)	**1.44**
Vapor Pressure (mm Hg)	**8.2 x 10^{-5} at 77°F (25°C)**	Molecular Weight	**138.13**
Vapor Density (Air = 1)	**Not Found**	Melting Point	**316°F (158°C)**

Solubility

Soluble in water, acetone, oil of turpentine, alcohol, ether, benzene, chloroform, glycerol, fats and oils.

Appearance and Odor

White, finely divided, needle-like crystalline solid or fluffy, white crystalline powder with a sweet taste and an acrid aftertaste.

SECTION IV - FIRE AND EXPLOSION HAZARD DATA

Flash Point (method used)	**315°F (157°C) closed cup**	Explosive Limits in Air % by Volume	**LEL: 1.1% at 392°F (200°C)** **UEL: Not Determined**
NFPA Classification	**Combustible Solid**	Autoignition Temperature	**1004°F (540°C)**

Extinguishing Media

Use carbon dioxide, dry chemical, or foam. Water or foam may cause frothing.

Special Fire Fighting Procedures

Poisonous gases are produced in fire. Wear full protective clothing and self-contained breathing apparatus (SCBA). Use water spray to flush and dilute a spill and to disperse the vapors. Also, use a water spray to keep fire-exposed containers cool. Do not release runoff from fire control measure to sewers or waterways. Firefighters should avoid contact with vapors produced during fire.

Unusual Fire and Explosion Hazards

Dusts may form explosive mixtures in air. Containers may explode in fire.

SECTION V - REACTIVITY DATA

Stability		Conditions to Avoid Salicylic acid is normally stable under routine conditions of handling and storage. Avoid contact with heat and ignition sources and incompatible materials. Keep away from light (causes discoloration).
Stable X	Unstable	Incompatibility *(materials to avoid)* Incompatible with iron salts, spirit nitrous ether, lead acetate, iodine, and oxidizers (fluorine, chlorine, bromine).
Hazardous Polymerization		Conditions to Avoid Under normal conditions of temperature and pressure, hazardous polymerization of salicylic acid is not expected to occur.
May Occur	Will Not Occur X	Hazardous Decomposition or By-products When heated to decomposition, salicylic acid can emit highly toxic and acrid smoke and irritating fumes, including toxic oxides of carbon and phenol vapors.

SECTION VI - HEALTH HAZARD DATA

Primary Route(s) of Entry:	Inhalation?	Absorption (skin)?	Ingestion?
	X		X

Health Hazards

INHALATION: Extremely irritating to the eyes, nose, throat, and respiratory system. May cause cough, dypsnea, and inflammation of associated mucosa. High exposure can lead to salicylism with deep and rapid breathing, nausea, vomiting, tinnitus, and confusion.

EYES & SKIN: Eye contact causes pain, severe irritation, and possible corneal injury. Skin contact also causes severe irritation, sores, or burns (especially on moist or wet skin).

INGESTION: Abdominal pain, gastrointestinal irritation, increased respiration, mental disturbances, acidosis, and possible skin rash. Ingestion of large doses may cause vomiting.

Carcinogenicity	NTP Listed?	IARC Cancer Review Group?	OSHA Regulated?	Target Organs?
Unknown Human Unknown Animal	No	No	No	Respiratory system, eyes, skin, mucous membranes.

Medical Conditions Generally Aggravated by Exposure

Chronic respiratory illnesses (asthma, bronchitis) or skin diseases may be aggravated by exposure.

Emergency and First-aid Procedures

Eye contact: Flush immediately with water for 15 minutes (minimum); seek medical attention. Skin contact: Remove all contaminated clothing. Immediately flush area with large amounts of water. For red or blistered skin, wash with soap and water. Seek medical assistance. For inhalation: Remove the person from exposure. Provide respiratory assistance and CPR. Transfer to medical facility. If swallowed: Contact poison control center. Unless advised otherwise, give 1 to 2 glasses of water to drink and induce vomiting. Never attempt to give anything by mouth to an unconscious or convulsing person.

SECTION VII - PRECAUTIONS FOR SAFE HANDLING AND USE

Steps to be Taken in Case Material is Released or Spilled

Collect powdered materials in safest manner possible. Do NOT dry sweep (generates dusts). Use a HEPA vacuum instead. Damp mop residue. Ventilate area of spill or leak and remove all ignition sources. Restrict those not involved in cleanup from entering area.

Preferred Waste Disposal Method

Mix with a flammable liquid and burn in a chemical incinerator with afterburner and scrubber.

Precautions to be Taken in Handling and Storage

Store in tightly closed containers in cool, well-ventilated area away from heat. Protect container from damage. Keep protected from exposure to light.

Other Precautions and Warnings

Avoid generating dusty conditions in storage areas.

SECTION VIII - CONTROL MEASURES AND PERSONAL PROTECTIVE EQUIPMENT

Respiratory Protection *(specify type)*

An air purifying respirator with acid cartridge and a HEPA filter may suffice. However, better protection is provided by a supplied-air respirator with full facepiece operated in pressure demand or other positive pressure mode.

Ventilation

Local exhaust or general mechanical systems recommended.

Protective Gloves	Eye Protection	Other Protective Clothing
Impervious Material	Chemical/Dust Goggles or Face Mask	Protective Apron

Work/Hygiene Practices

Always wash hands thoroughly after using chemical; never bring food, drink, cosmetics, or smoking materials into vicinity of chemicals.

SALICYLIC ACID

$C_6H_4(OH)(COOH)$

CAS: 69-72-7

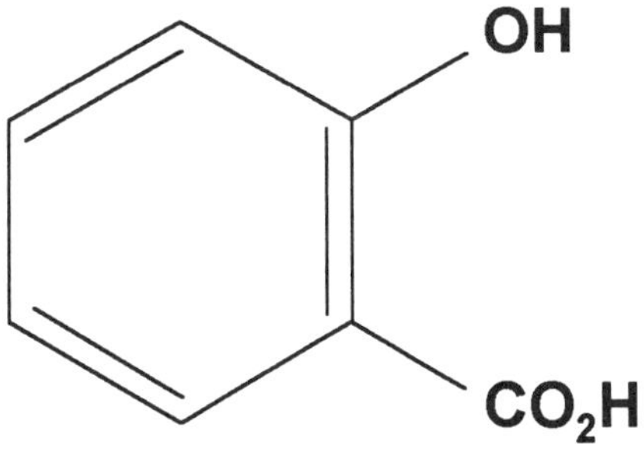

IDENTIFICATION AND TYPICAL USES

Salicylic acid appears as fine, white, needle-like crystals or a fluffy, white crystalline powder. It has a sweet taste and leaves an acid aftertaste. It is used in the manufacture of aspirin and salicylates, in resins, as a dyestuff intermediate, as a prevulcanization inhibitor, an analytical reagent, a fungicide, and in veterinary medicine.

RISK ASSESSMENT: HEALTH

General Assessment

Exposure to salicylic acid is possible through *inhalation* of its dusts or mists and by *ingestion*. Skin contact can also produce serious health effects at the site of contact, although dermal absorption has not been reported. There are no reports of animal or human carcinogenicity. Experimental teratogenic data and mutation data have been documented for salicylic acid.

Inhalation of salicylic acid will cause irritation of the eyes, nose, throat, and mucous membranes of the upper respiratory tract. Bronchial tube irritation, difficulty breathing, and coughing may also occur. Absorption of large amounts or prolonged overexposure may cause salicylism with symptoms of deep and rapid breathing, nausea, vomiting, tinnitus, ulcers, hemorrhagic strokes, and confusion. These symptoms may be enhanced by the presence of salicylate medication (such as aspirin).

Skin and eye contact can result in mild to moderate irritation. On the skin, salicylic acid can cause burns, irritation, and possible sores, especially if the skin is moist or wet (sweat). If allowed to remain in contact with the eye for prolonged periods, corneal injury and damage to vision may occur.

Ingestion may cause abdominal pain, gastrointestinal irritation, increased breathing rate, mental disturbances, acidosis, and skin rash. Spontaneous vomiting can occur when large doses are ingested.

Acute Health Effects

The following acute (short-term) health effects may occur immediately or shortly after exposure to salicylic acid:

Skin: Irritation/rash or burning feeling on contact, with possible development of sores.

Eye: Severe irritation; may cause burns, conjunctivitis, and corneal damage.

Lung: Irritation of the nose, throat, and mucosa of the lungs and respiratory tract.

Other: Ingestion causes abdominal pain, gastrointestinal tract irritation, vomiting, nausea, and mental disturbances.

Chronic Health Effects

The following chronic (long-term) health effects may occur at some time after exposure to salicylic acid and can last for months or even years:

Cancer Hazards: According to information presented in the references, salicylic acid has not been adequately tested for its ability to cause cancer in animals. Mutation data have been reported and some scientists believe that long-term exposure to mutagens may pose a cancer risk. There is no indication that exposure to salicylic acid will have this effect.

Reproductive Hazard: According to information presented in the references, salicylic acid has been shown to cause teratogenic effects in animals. Its ability to adversely affect reproduction in humans is not well documented.

Other Chronic Effects: Long-term skin exposure may cause skin sensitization with peeling, rash, itching, or blisters.

Recommended Risk-Reduction Measures

Personnel should avoid direct contact with salicylic acid. It is a human poison by ingestion and is moderately toxic by inhalation. If a less toxic chemical cannot be substituted for a hazardous substance, then *engineering controls* are the most effective method of reducing exposures.

The best protection is to enclose operations and/or provide local exhaust ventilation at the site of chemical release. While not always operationally feasible, isolating operations can also reduce exposure. Using respiratory protection is less effective than the controls mentioned above, but is still advisable whenever working with or around salicylic acid. For transient exposures, an air-purifying respirator equipped with an acid cartridge, a HEPA filter, and a dust/mist pre-filter may suffice. Greater protection is obtained using a supplied-air respirator or a self-contained breathing apparatus (SCBA) with full facepiece operated in pressure demand or other positive pressure mode. If a full facepiece is not available, then chemical goggles should be worn to protect the eyes. Whenever a chemical splash/dust hazard exists, a face shield and a rubber apron should be worn. To prevent hand and skin exposures, acid-resistant gloves should be used. Glove manufacturers should be contacted and permeation studies obtained *before* final glove selections have been made.

Administrative controls should also be in place to minimize the potential for human exposures. These may include written procedures or policies which specify the methods and techniques that will be practiced whenever personnel are to work with salicylic acid.

All personnel should receive *training* on the use, hazards, protective measures, emergency actions, and other precautions per 29 CFR 1910.1200 (Hazard Communication), prior to the first assignment in an area where salicylic acid is used or stored. Other methods to reduce exposure include:

- ☑ Where possible, enclose operations and use local exhaust ventilation at the site of chemical release. If local exhaust ventilation or enclosure is not used, respiratory protection should be mandatory.
- ☑ Always ensure that proper protective clothing is worn when using chemical substances.
- ☑ Wash thoroughly immediately after exposure to salicylic acid and at the end of the work shift or before eating, drinking, or smoking.
- ☑ Hazard warning information should be posted in the work area. In addition, as part of an on-going education and training program, all information on the health and safety hazards of salicylic acid should be communicated to all exposed or potentially exposed workers.

RISK ASSESSMENT: ENVIRONMENT

General Assessment

The environment is at risk of exposure during transportation, storage, disposal, or destruction of salicylic acid. In almost every scenario, the threat of environmental exposure is contingent upon the proper handling of the chemical substance. Accidental spills, large or small, can result in fire, explosion, and possible contamination of the surrounding environmental mediums (water, soil, and air).

Salicylic acid is considered a combustible solid. It is also a corrosive, especially when mixed in solution. It is incompatible with iron salts, spirit nitrous ether, lead acetate, iodine, and oxidizers, and contact can cause violent reactions. These characteristics require special consideration during any emergency situation involving a leak or spill of salicylic acid.

Salicylic acid can enter the environment mainly through industrial discharges or spills.

Acute Ecological Effects

Acute (short-term) toxic effects may include the death of animals, birds, or fish and death or low growth rate in plants. Acute effects are seen 2 to 4 days after animals or plants are exposed to salicylic acid.

Salicylic acid has moderate acute toxicity to aquatic life. Insufficient data are available to evaluate or predict the short-term effects of salicylic acid exposure to birds, plants, or land animals.

Chronic Ecological Effects

Chronic (long-term) toxic effects may include shortened life span, reproductive problems, lower fertility, and changes in appearance or behavior in exposed animals. These effects can be seen long after first exposure(s) to toxic chemicals. Insufficient data are available to

evaluate or predict the long-term effects of salicylic acid exposure to aquatic life, birds, plants, or land animals.

Water Solubility

Salicylic acid is slightly soluble in water. Concentrations of up to 100 milligrams will mix with a liter of water.

Persistence in the Environment

Salicylic acid is slightly persistent in the aquatic environment, with a half-life of between 2 to 20 days. The half-life of a pollutant is the amount of time it takes for one half of the chemical to be degraded.

Bioaccumulation in Aquatic Organisms

Some substances increase in concentration, or *bioaccumulate,* in living organisms as they breathe contaminated air, drink contaminated water, or eat contaminated food. These chemicals can become concentrated in the tissues and internal organs of animals as well as humans.

Salicylic acid can be expected to biodegrade quickly and should not accumulate in the edible tissues of fish.

Recommended Risk-Reduction Measures

Proper training of all transporters will reduce the likelihood of a mishap or accident resulting in a leak or spill to the environment. The correct use of DOT labeling on all containers, trucks, and rail cars will enable emergency responders to react properly and quickly to any disaster thereby reducing the potential risk to the environment and to personnel.

Storage of salicylic acid should be segregated from incompatible chemicals to minimize the risk of cross-contamination or contact. Buildings designated for storage should be equipped with appropriate fire protection systems (alarms, sprinklers, emergency lighting, portable extinguishers).

If a spill or leak to the environment has occurred, fire department, emergency response and/or hazardous materials spill personnel should be notified immediately. Cleanup should be attempted only by those trained in proper spill containment procedures. Contaminated soils should be removed for incineration and replaced with clean soil. If salicylic acid should contact the water table, aquifer, or navigable waterway, time is of the essence. It is slightly soluble in water and, therefore, total containment and remediation may not be possible. When such spills occur, the local and/or state emergency response authorities must be notified. A comprehensive emergency response or disaster preparedness/recovery plan should be in place prior to any operations involving the use, transportation, storage, or disposal of salicylic acid. If salicylic acid is spilled or leaked, the following specific steps are recommended:

- ☑ Restrict persons not wearing protective clothing from area of spill or leak until cleanup is complete and area can be opened for normal work.
- ☑ Remove ignition sources and ventilate area.
- ☑ Collect all powdered materials in safest and most convenient manner possible and deposit in sealed drums for disposal. Do NOT dry sweep (generates airborne dusts). Use a vacuum equipped with a high efficiency particulate air (HEPA) filter instead. Damp mop residue and collect for disposal. Absorb any solution spills in vermiculite, dry sand, earth, or similar material and deposit in sealed containers.
- ☑ It may be necessary to dispose of salicylic acid as a hazardous waste. The responsible state agency or the regional office of the federal Environmental Protection Agency (EPA) should be contacted for specific recommendations.

RISK ASSESSMENT: BUSINESS
General Assessment

Accidents or mishaps involving salicylic acid can present a significant threat to business operations. The loss or damage of equipment or facilities can significantly affect fiscal viability. Lawsuits that may result from personnel injury/death, public exposures, and/or environmental contamination will also require a serious expenditure of resources. Media attention surrounding an injury, death, or environmental damage can also result in a loss of profits and loss of current as well as future business. Always remember that anytime the term "reproductive hazard" is used, public emotion, hysteria, and ignorance can run equally high. This must be considered when drafting public relations policies.

Recommended Risk-Reduction Measures

Company attorneys, safety and health professionals, and environmental specialists should be involved in the development of any procedures dealing with safety or emergency response. A company official should be pre-designated as a public relations officer with specific training in dealing with the press. Corporate plans and policies should be developed, approved, and implemented long before any need for such arises.

REFERENCES

Ahlbom, A. 1993. *Biostatistics for Epidemiologists.* New York: CRC Press/Lewis Publishers.

American Conference of Governmental Industrial Hygienists. 1988. *Documentation of the Threshold Limit Values and Biological Exposure Indices,* 5th Edition (with updates). Cincinnati: ACGIH

Asante-Duah, D. K. 1993. *Hazardous Waste Risk Assessment.* New York: CRC Press/Lewis Publishers.

Bloom, A. D. and F. J. DeSerres. 1995. *Ecotoxicity and Human Health.* New York: CRC Press/Lewis Publishers.

Brusick, D. J. 1994. *Methods for Genetic Risk Assessment.* New York: CRC Press/Lewis Publishers.

Bueche, J. R. 1972. *Principles of Physics,* 2nd Edition. New York: McGraw-Hill.

Burrell, R., D. K. Flaherty, and Leonard J. Sauers. 1992. *Toxicology of the Immune System.* New York: Van Nostrand Reinhold.

Calabrese, E. J. 1994. *Biological Effects of Low Level Exposures.* New York: CRC Press/Lewis Publishers.

Cockerman, L. G., B. S. Shane. 1994. *Basic Environmental Toxicology.* New York: CRC Press/Lewis Publishers.

Cothern, C. R. 1995. *Handbook for Environmental Risk Decisionmaking.* New York: CRC Press/Lewis Publishers.

Danse, I. R. 1991. *Common Sense Toxics in the Workplace.* New York: Van Nostrand Reinhold.

Gardner, R. W. 1994. *Chemical Intolerance.* New York: CRC Press/Lewis Publishers.

Hallenbeck, W. H. 1993. *Quantitative Risk Assessment for Environmental and Occupational Health,* 2nd Edition. New York: CRC Press/Lewis Publishers.

Hosty, J. W., P. Foster. 1990. *A Practical Guide to Chemical Spill Response.* New York: Van Nostrand Reinhold.

Howard, P. H., et. al. 1991. *Handbook of Environmental Degradation Rates.* New York: CRC Press/Lewis Publishers.

James, A., M. Lord. 1992. *Index of Chemical and Physical Data.* New York: Van Nostrand Reinhold.

Jones, F. E. 1994. *Toxic Organic Vapors in the Workplace.* New York: CRC Press/Lewis Publishers.

Keith, L. H. and M. M. Walker. 1995. *Handbook of Air Toxics.* New York: CRC Press/Lewis Publishers.

Landis, W. G., M. H. Yu. 1995. *Introduction to Environmental Toxicology.* New York: CRC Press/Lewis Publishers.

Langerman, N. 1994 *Precautionary Labels for Chemical Containers.* New York: CRC Press/Lewis Publishers.

Lauwerys, R. R., P. Hoet. 1993. *Industrial Chemical Exposure.* New York: CRC Press/Lewis Publishers.

Lewis, R. J., Sr. 1993. *Hawley's Condensed Chemical Dictionary,* 12th Edition. New York: Van Nostrand Reinhold.

Lewis, R. J., Sr. 1992. *Sax's Dangerous Properties of Industrial Materials,* Volumes 1, 2, and 3, 8th Edition. New York: Van Nostrand Reinhold.

Lide, D. R. 1995. *Handbook of Chemistry and Physics,* 76th Edition. New York: CRC Press/Lewis Publishers.

Manahan, S. E. 1994. *Environmental Chemistry,* 6th Edition. New York: CRC Press/Lewis Publishers.

Manahan, S. E. 1992. *Toxicological Chemistry,* 2nd Edition. New York: CRC Press/Lewis Publishers.

Mellström, G. A., J. E. Wahlberg, and H. I. Maibach. 1994. *Protective Gloves for Occupational Use.* New York: CRC Press/Lewis Publishers.

Meyer, E. 1990. *Chemistry of Hazardous Materials.* Englewood Cliffs, NJ: Prentice Hall.

Mycock, J. C., J. D. McKenna, and L. Theodore. 1995. *Handbook of Air Pollution Control Engineering and Technology.* New York: CRC Press/Lewis Publishers.

Ott, W. R. 1995. *Environmental Statistics and Data Analysis.* New York: CRC Press/Lewis Publishers.

Patnaik, P. 1992. *A Comprehensive Guide to the Hazardous Properties of Chemical Substances.* New York: Van Nostrand Reinhold.

Philp, R. B. 1995. *Environmental Hazards and Human Health.* New York: CRC Press/Lewis Publishers.

Proctor, N. H., J. P. Hughes. 1991. *Chemical Hazards of the Workplace.* New York: Van Nostrand Reinhold.

Quigley, D. R. 1994. *Handbook of Emergency Chemical Management.* New York: CRC Press/Lewis Publishers.

Rea, W. J. 1995. *Chemical Sensitivity,* Vol. I-III. New York: CRC Press/Lewis Publishers.

Richardson, M. 1995. *Dictionary of Substance and Their Effects,* Vol. I-VI. New York: CRC Press/Lewis Publishers.

Rylander, R., R. R. Jacobs. 1994. *Organic Dusts.* New York: CRC Press/Lewis Publishers.

Sacarello, H. L. A. 1994. *The Comprehensive Handbook of Hazardous Materials.* New York: CRC Press/Lewis Publishers.

Sheftel, V. O. 1995. *Handbook of Toxic Properties of Monomers and Additives.* New York: CRC Press/Lewis Publishers.

Sherman, J. D. 1988. *Chemical Exposure and Disease.* New York: Van Nostrand Reinhold.

Talbott, E., G. Craun. 1995. *An Introduction to Environmental Epidemiology.* New York: CRC Press/Lewis Publishers.

U.S. Department of Health and Human Services, National Institute for Occupational Safety and Health. 1994. *NIOSH Pocket Guide to Chemical Hazards.* Washington, D.C.: U.S. Government Printing Office.

Witorsch, P. and S. V. Spagnolo. 1994. *Air Pollution and Lung Disease in Adults.* New York: CRC Press/Lewis Publishers.

MATERIAL SAFETY DATA SHEET

CHEMICAL NAME	
SELENIUM	**CRC PRESS/LEWIS PUBLISHERS DISCLAIMER:** THE INFORMATION AND RECOMMENDATIONS PRESENTED HEREIN ARE BASED ON SOURCES BELIEVED TO BE RELIABLE. CRC MAKES NO REPRESENTATION ON ITS COMPLETENESS OR ACCURACY. IT IS THE RESPONSIBILITY OF THE USER TO DETERMINE THE CHEMICAL'S SUITABILITY FOR ITS INTENDED USE, THE CHEMICAL'S SAFE USE, AND THE CHEMICAL'S PROPER DISPOSAL. NO REPRESENTATIONS AND/OR WARRANTIES, EITHER EXPRESSED OR IMPLIED, OF THE MERCHANTABILITY OR FITNESS FOR A PARTICULAR PURPOSE, OR OF ANY OTHER NATURE, ARE MADE WITH RESPECT TO THE INFORMATION PROVIDED IN THIS MSDS OR TO THE CHEMICAL TO WHICH INFORMATION MAY REFER. CRC NEITHER ASSUMES NOR AUTHORIZES ANY OTHER PERSON TO ASSUME FOR IT, ANY OTHER ADDITIONAL RESPONSIBILITY OR LIABILITY FOR THE USE OF, OR RELIANCE UPON, THIS INFORMATION.

HAZARD WARNING INFORMATION

HEALTH	FIRE	REACTIVE	OTHER	DEGREE OF HAZARD	COLOR CODING	OTHER CODES
3	0	1		0 = Minimum Hazard 1 = Slight Hazard 2 = Moderate Hazard 3 = Serious Hazard 4 = Severe Hazard	HEALTH = BLUE FIRE = RED REACTIVITY = YELLOW OTHER = WHITE	OX = Oxidizer ACID = Acid ALK = Alkali COR = Corrosive W = Use No Water

SECTION I - GENERAL INFORMATION

Characterization	Nonmetallic Element
RCRA Number	No Listing
EPA Class	Hazardous Waste
DOT Proper Shipping Name	Selenium Metal Powder
Chemical Abstract Service (CAS) Number	7782-49-2
DOT Hazard Class and Label Requirements	Poison B; St. Andrews Cross
DOT Emergency Guide Code	53
DOT Identification Number	UN 2658
Atomic Symbol	Se

Synonyms
Elemental selenium; selenium alloy; selenium base; selenium dust; selenium metal powder.

SECTION II - HAZARDOUS INGREDIENTS AND IDENTITY INFORMATION

Hazardous Components (specific identity)	OSHA Exposure Criteria	NIOSH Exposure Criteria	Immediately Dangerous to Life and Health (IDLH)	ACGIH Exposure Criteria
Selenium (derivation: Occurs naturally in certain soils. Found in Canada, Japan, Yugoslavia, and Mexico. Also occurs as an impurity in most sulfide ores).	**PEL (Se):** 0.2 mg/m^3 **STEL:** Not Established	**REL (Se):** 0.2 mg/m^3 **STEL:** Not Established	1 mg/m^3	**TLV (Se):** 0.2 mg/m^3 **STEL:** Not Established

SECTION III - PHYSICAL AND CHEMICAL CHARACTERISTICS

Boiling Point	1263°F (684°C)
Specific Gravity (H_2O = 1)	4.28
Vapor Pressure (mm Hg)	1 at 673°F (356°C)
Molecular Weight (atomic weight)	79.0
Vapor Density (Air = 1)	Not Found
Melting Point	392°F (200°C)

Solubility
Insoluble in water. Soluble in concentrated nitric acid and (in liquid form) in common alkalis.

Appearance and Odor
Dark red to bluish-black amorphous solid, or dark red or gray crystals. Odorless.

SECTION IV - FIRE AND EXPLOSION HAZARD DATA

Flash Point (method used)	Not Found
Explosive Limits in Air % by Volume	LEL: N/A UEL: N/A
NFPA Classification	Combustible Solid (in powder form)
Autoignition Temperature	Not Found

Extinguishing Media
Selenium does not burn in bulk form. Powders may ignite. Use dry chemical, carbon dioxide, alcohol foam, or other agent to extinguish surrounding fire.

Special Fire Fighting Procedures
Poisonous gases are produced in fire. Wear full protective clothing and self-contained breathing apparatus (SCBA) operated in pressure demand or other positive pressure mode. Do not release runoff from fire control measures to sewers or waterways.

Unusual Fire and Explosion Hazards
None reported.

SECTION V - REACTIVITY DATA

Stability		Conditions to Avoid Selenium is normally stable under routine conditions of handling or storage. Avoid contact with incompatible materials since violent reactions can occur.
Stable X	Unstable	Incompatibility *(materials to avoid)* Violent or explosive reaction in contact with acids, strong oxidizers, chromium trioxide, potassium bromate, cadmium, carbides, and many other materials and metals.
Hazardous Polymerization		Conditions to Avoid Hazardous polymerization of selenium will not occur.
May Occur	Will Not Occur X	Hazardous Decomposition or By-products When heated to decomposition, selenium can emit toxic oxides of selenium (SeO_x). Selenium itself is a highly toxic compound.

SECTION VI - HEALTH HAZARD DATA

Primary Route(s) of Entry:	Inhalation?	Absorption (skin/eye)?	Ingestion?
	X		X

Health Hazards

INHALATION: Breathing the dusts of selenium may cause cough, tightness of chest, respiratory distress, congestion of the lungs, and bronchial tube irritation. Symptoms resembling metal fume fever with headaches, metallic taste, garlic breath, irritability, fatigue, increased dental cavities, upset stomach, loss of nails and hair, pallor, nervousness, and mood changes. It can also cause delayed lung effects such as chemical pneumonitis.

EYE & SKIN: Severe irritation to the skin and eyes. May cause skin burns.

INGESTION: May cause toxic systemic effects.

Carcinogenicity	NTP Listed?	IARC Cancer Review Group?	OSHA Regulated?	Target Organs?
Unknown Human Questioned Animal	No	Group 3	29 CFR 1910.1000 Table Z-1	Resp. sys., liver, kidney, blood, skin, eyes, spleen.

Medical Conditions Generally Aggravated by Exposure

None reported.

Emergency and First-aid Procedures

Eye contact: Do not allow victim to keep eyes tightly shut. Flush immediately with water for 15 minutes (minimum); seek medical attention. Skin contact: Remove all contaminated clothing. Immediately wash area with flooding amounts of soap and water. For inhalation: Remove the person from exposure. Provide respiratory assistance and CPR. Transfer to medical facility. Observe 24 - 48 hours for lung effects. If swallowed: Seek medical attention immediately. Never give anything by mouth to an unconscious or convulsing person. Give several glasses of water to dilute. Do not induce vomiting.

SECTION VII - PRECAUTIONS FOR SAFE HANDLING AND USE

Steps to be Taken in Case Material is Released or Spilled

Collect powdered materials in most convenient and safe manner and deposit in sealed containers. Dry sweep not recommended. Use HEPA vacuum. Remove all ignition sources. Restrict those not involved in cleanup from entering area. Notify appropriate authorities, as required by applicable regulations.

Preferred Waste Disposal Method

No citation.

Precautions to be Taken in Handling and Storage

Do not store near water. Keep in cool, dry, well-ventilated area in tightly closed containers. Prevent moisture contact. Consider using automatic monitoring system to detect selenium contamination.

Other Precautions and Warnings

Storage under an inert gas blanket may be warranted.

SECTION VIII - CONTROL MEASURES AND PERSONAL PROTECTIVE EQUIPMENT

Respiratory Protection *(specify type)*

Use a NIOSH/MSHA-approved air-purifying respirator equipped with a dust/mist pre-filter. Better protection is obtained using a self-contained breathing apparatus (SCBA) with full facepiece operated in positive pressure mode. Selenium is a questioned carcinogen. Inhalation must be avoided.

Ventilation

Local exhaust (preferred) at site of chemical work or general ventilation.

Protective Gloves	Eye Protection	Other Protective Clothing
Impervious Material	Chemical/Dust Goggles or Face Mask	Protective Apron, Boots (avoid skin contact)

Work/Hygiene Practices

Always wash hands thoroughly after using chemical; never bring food, drink, or smoking materials into vicinity of chemicals.

SELENIUM

Se

CAS. 7782-49-2

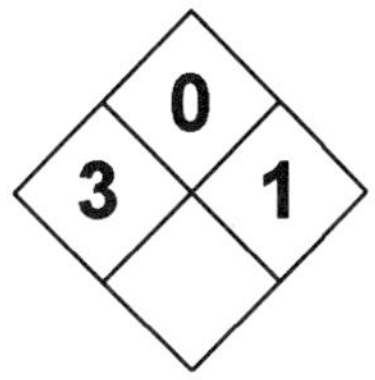

IDENTIFICATION AND TYPICAL USES

Selenium is a dark red to bluish-black amorphous solid or dark red, gray, or black crystals that can appear in two distinct forms, known as allotropes, because of differences in crystalline structures. It has no odor. It is a nonmetallic element of atomic number 34, group VIA of the periodic table with valances of 2, 4, and 6 with 6 stable isotopes. It is used in electronics, xerographic plates, in television cameras, photocells, magnetic computer cores, in solar batteries (rectifiers, relays), in ceramics as a colorant for glass, in steel and copper machining, as a rubber accelerator, a catalyst, and a trace element in animal feed.

RISK ASSESSMENT: HEALTH

General Assessment

Selenium is a highly toxic human poison by *inhalation*. It can also enter the body through *ingestion*. Skin contact causes serious irritation, but dermal absorption has not been reported. It is a questionable carcinogen in test animals with experimental tumorigenic and teratogenic data also reported. There are no reports of any mutagenic action on long-term exposure.

Inhalation of selenium dusts or mists can cause serious irritation to the eyes, nose, throat, and upper respiratory tract. It can cause dryness and nosebleeds. There can be bronchial tube irritation with cough and phlegm. High exposures cause headaches, difficulty breathing, lung irritation, and possible delayed effects such as chemical pneumonitis. There are reported cases of metal fume fever with flu-like symptoms including fever, headache, nausea, vomiting, sore throat, pallor, depression, nervousness, giddiness, gastrointestinal disturbances, and general weakness. Other possible symptoms include garlic breath, metallic taste, visual disturbances, possible anemia and (in animals) liver necrosis, circulatory system damage, and injury to the kidney and spleen. Chronic intoxication can cause loss of hair, teeth, and nails.

Skin contact results in severe irritation, possible burns, and contact dermatitis. Eye contact can cause damage to the cornea.

Acute Health Effects

The following acute (short-term) health effects may occur immediately or shortly after exposure to selenium:

Skin: Severe irritation and possible contact dermatitis may occur. Burns are possible on prolonged contact.

Eye: Irritation and possible damage on contact.

Lung: High exposures can cause rapid and severe lung damage, with shortness of breath, chest pain, cough, and a possible chemical pneumonitis.

Other: High exposure to selenium may cause nausea, cramps, salivation, vomiting, and diarrhea. Metal fume fever with flu-like symptoms has been reported. There may be fatigue, cough, difficulty breathing, irritability, mood changes, depression, metallic taste and garlic breath, pallor, and gastrointestinal irritation.

Chronic Health Effects

The following chronic (long-term) health effects may occur at some time after exposure to selenium and can last for months or even years:

Cancer Hazards: Selenium is a confirmed animal carcinogen. It has caused liver damage (necrosis), circulatory system problems, and injuries to the kidney and spleen in test animals. It is not known if selenium will have these effects in humans.

Reproductive Hazard: Selenium is a known teratogen in animals and a probable teratogen in humans.

Other Chronic Effects: Repeated exposures may cause garlic odor on breath and in sweat, chronic fatigue, irritability, upper respiratory tract irritation, pallor, gastrointestinal distress, metallic taste, and an allergic reaction with red, puffy eyelids.

Recommended Risk-Reduction Measures

Personnel should avoid direct contact with selenium dusts or powders. If a less toxic material or compound cannot be substituted for selenium, then *engineering controls* are the most effective method of reducing exposures. The best protection is to enclose operations and/or provide local exhaust ventilation at the site of selenium dust release. While not always operationally feasible, isolating operations can also reduce exposure. Using respiratory protection is less effective than the controls mentioned above, but is still advisable whenever working with or around selenium. At low exposure levels, an air-purifying respirator equipped with a dust/mist pre-filter may suffice. For higher exposure levels, a self-contained breathing apparatus (SCBA) with full facepiece and pressure demand is recommended for the greatest possible respiratory protection. If a full facepiece is not available, then chemical/dust goggles should be worn to protect the eyes. A face shield and protective apron should also be worn. To prevent hand and skin exposures, gloves should be used. Glove manufacturers should be contacted and permeation studies obtained *before* final glove selection are made.

Administrative controls should also be in place to minimize the potential for human exposures. These may include written procedures or policies which specify the methods and techniques that will be practiced whenever personnel are to work with selenium.

All personnel should receive *training* on the use, hazards, protective measures, emergency actions, and other precautions per 29 CFR 1910.1200 (Hazard Communication), prior to the first assignment in an area where selenium is used or stored.

Before beginning employment working with selenium, and at regular intervals thereafter (e.g., annually), those personnel with frequent or potentially high exposures should be provided the following recommended medical tests:

☑ Lung function test.

☑ Urine test for selenium (levels should be less than 100 micrograms per liter of urine).

If overexposure is suspected or symptoms develop, the following additional test is recommended:

☑ Consider chest X-ray after acute overexposure (may be negative if taken immediately after exposure due to delayed onset of chemical pneumonitis).

Any evaluation should include a careful history of past and present symptoms with an examination. Medical tests that simply look for existing damage are not a substitute for controlling exposures.

Also, since smoking can cause heart disease as well as lung cancer, emphysema, and other known respiratory problems, exposure to selenium may cause additional complications. Therefore, personnel who smoke should avoid any unprotected contact with selenium powders.

Other methods to reduce exposure include:

☑ Where possible, enclose operations and use local exhaust ventilation at the site of chemical release. If local exhaust ventilation or enclosure is not used, respiratory protection should be mandatory.

☑ Always ensure that proper protective clothing is worn when using chemical substances and that personnel are trained on its proper use, care, and maintenance.

☑ Wash thoroughly immediately after exposure to selenium and at the end of the work shift or before eating, drinking, applying cosmetics, or smoking.

☑ Hazard warning information should be posted in the work area. In addition, as part of an on-going education and training program, all information on the health and safety hazards of selenium should be communicated to all exposed and potentially exposed workers.

☑ Eye wash stations should be provided in the immediate work area for emergency use. If there is a possibility of skin exposure to selenium, emergency shower facilities should also be provided.

☑ Workers whose clothing has been contaminated by selenium should change into clean clothes before leaving work. Family members can be exposed when handling contaminated clothing. Contaminated work clothing should be laundered only by individuals who have been informed of the hazards of exposure to selenium.

RISK ASSESSMENT: ENVIRONMENT

General Assessment

Selenium is a naturally occurring nonmetallic element. However, industrial uses place the environment at risk of exposure during transportation, storage, disposal, or destruction of selenium compounds. In almost every scenario, the threat of environmental exposure is contingent upon the proper handling of the chemical substance. Accidental spills, large or small, where selenium contacts incompatible commodities can result in fire, explosion (depending upon conditions of spill), and possible contamination of the surrounding environmental mediums (water, soil, and air).

In its bulk form, selenium is considered a combustible solid. However, its dusts can react violently and present a moderate fire hazard. Because it is incompatible with many commodities, including some other common metals, extreme caution is required in handling, storage, transportation, and disposal of selenium. It can react dangerously with carbides, fluorine, oxygen, potassium bromate, chromium trioxide, cadmium, strong oxidizers, and acids. These characteristics also require special consideration during any emergency situation involving a leak or spill of selenium powder or dust. Should selenium ever come into contact with these incompatible substances either during use, transportation, storage, or disposal, violent and even explosive reactions are possible.

Selenium is a nonmetallic element which is abundant in the drier soils of North America from the Great Plains to the Pacific Ocean. It may occur in water as a result of the natural weathering of soils and rocks and from mining or smelting certain ores. It is present in coal and fuel oil and is emitted in flue gas and fly ash during combustion. It may exist in different oxidation states and as various salts. In oxygenated water, the +4 and +6 oxidation states are predominant. Trace amounts of selenium are essential for plants, animals, and humans.

Acute Ecological Effects

Acute (short-term) toxic effects may include the death of animals, birds, or fish and death or low growth rate in plants. Acute effects are seen 2 to 4 days after animals or plants are exposed to selenium.

Selenium and its compounds have high acute toxicity to aquatic life and mammals and moderate acute toxicity to birds. Insufficient data are available to evaluate the short-term effects of selenium to plants.

Chronic Ecological Effects

Chronic (long-term) toxic effects may include shortened life span, reproductive problems, lower fertility, and changes in appearance or behavior in exposed animals. These effects can be seen long after first exposure(s) to toxic chemicals.

Selenium and its compounds have high chronic toxicity to aquatic life, birds, and land animals. Insufficient data are available to evaluate or predict the long-term effects of selenium to plants.

Water Solubility

Selenium and its compounds have solubility in water ranging from low to moderate. Concentrations up to 1000 milligrams will mix with a liter of water.

Persistence in the Environment

Selenium is highly persistent in water, with a half-life greater than 200 days. The half-life of a pollutant is the amount of time it takes for one half of the chemical to be degraded. No data are available on the estimated percentages of selenium persistence in the aquatic or terrestrial environments.

Bioaccumulation in Aquatic Organisms

Some substances increase in concentration, or *bioaccumulate,* in living organisms as they breathe contaminated air, drink contaminated water, or eat contaminated food. These chemicals can become concentrated in the tissues and internal organs of animals as well as humans.

The concentration of selenium found in fish tissues is expected to be somewhat higher than the average concentration of selenium in the water from which the fish was taken.

Recommended Risk-Reduction Measures

Proper training of all transporters will reduce the likelihood of a mishap or accident resulting in a leak or spill of selenium dusts or powders into the environment. The correct use of DOT labeling (per the MSDS) on all containers, trucks, and rail cars will enable emergency responders to react properly and quickly to any disaster thereby reducing the potential risk to the environment and to personnel.

Storage of selenium should be segregated from incompatible chemicals to minimize the risk of cross-contamination or contact.

If a spill or leak to the environment has occurred, fire department, emergency response and/or hazardous materials spill personnel should be notified immediately. Cleanup should be attempted only by those trained in proper spill containment procedures. If selenium should contact the water table, aquifer, or navigable waterway, time is of the essence. It is slightly to moderately soluble and total remediation may not be possible. The local and/or state emergency response authorities must be notified. A comprehensive emergency response or disaster preparedness/recovery plan should be in place prior to any operations involving the use, transportation, storage, or disposal of selenium. If selenium powder is spilled or leaked, the following specific steps are recommended:

- ☑ Restrict persons not wearing protective clothing from area of spill or leak until cleanup is complete and area can be opened for normal work.
- ☑ Remove all ignition sources.
- ☑ Collect powdered materials in the safest manner possible and deposit in sealed containers. Recommend using vacuum with HEPA filter. Do not dry sweep (generates airborne dusts).
- ☑ It may be necessary to dispose of selenium as a hazardous waste. The responsible state agency or the regional office of the federal Environmental Protection Agency (EPA) should be contacted for specific recommendations.

RISK ASSESSMENT: BUSINESS

General Assessment

Accidents or mishaps involving selenium can present a significant threat to business operations. The loss or damage of equipment or facilities can significantly affect fiscal viability. Lawsuits that may result from personnel injury/death, public exposures, and/or environmental contamination will also require a serious expenditure of resources. Media attention surrounding an injury, death, or environmental damage can also result in a loss of profits and loss of current as well as future business.

Always remember that anytime the terms "cancer," "carcinogen," or "reproductive hazard" are used, public emotion, hysteria, ignorance, and anxiety can all run equally high. This fact must be carefully considered whenever drafting or implementing public relations policies.

Recommended Risk-Reduction Measures

Company attorneys, safety and health professionals, and environmental specialists should be involved in the development of any procedures for responding to chemical incidents. A company official should be predesignated as a public relations officer with specific training in dealing with the press. Corporate plans and policies should be developed, approved, and implemented long before any need for such arises.

REFERENCES

American Conference of Governmental Industrial Hygienists. 1988. *Documentation of the Threshold Limit Values and Biological Exposure Indices,* 5th Edition (with updates). Cincinnati: ACGIH

Brusick, D. J. 1994. *Methods for Genetic Risk Assessment.* New York: CRC Press/Lewis Publishers.

Bueche, J. R. 1972. *Principles of Physics,* 2nd Edition. New York: McGraw-Hill.

Calabrese, E. J. 1994. *Biological Effects of Low Level Exposures.* New York: CRC Press/Lewis Publishers.

Cothern, R. R. 1993. *Comparative Environmental Risk Assessment.* New York: CRC Press/Lewis Publishers.

Gardner, R. W. 1994. *Chemical Intolerance.* New York: CRC Press/Lewis Publishers.

Hallenbeck, W. H. 1993. *Quantitative Risk Assessment for Environmental and Occupational Health,* 2nd Edition. New York: CRC Press/Lewis Publishers.

Howard, P. H. 1995. *Handbook of Environmental Fate and Exposures Data for Organic Chemicals,* Vol. I-IV. New York: CRC Press/Lewis Publishers.

Lewis, R. J., Sr. 1992. *Sax's Dangerous Properties of Industrial Materials,* Volumes 1, 2, and 3, 8th Edition. New York: Van Nostrand Reinhold.

Lide, D. R. 1995. *Handbook of Chemistry and Physics,* 76th Edition. New York: CRC Press/Lewis Publishers.

Manahan, S. E. 1989. *Toxicological Chemistry.* New York: CRC Press/Lewis Publishers.

Meyer, E. 1990. *Chemistry of Hazardous Materials.* Englewood Cliffs, NJ: Prentice Hall.

Patnaik, P. 1992. *A Comprehensive Guide to the Hazardous Properties of Chemical Substances.* New York: Van Nostrand Reinhold.

Quigley, D. R. 1994. *Handbook of Emergency Chemical Management.* New York: CRC Press/Lewis Publishers.

Richardson, M. 1995. *Dictionary of Substance and Their Effects,* Vol. I-VI. New York: CRC Press/Lewis Publishers.

Sacarello, H. L. A. 1994. *The Comprehensive Handbook of Hazardous Materials.* New York: CRC Press/Lewis Publishers.

Torkelson, T. R. V. K. Rowe. 1981. *Patty's Industrial Hygiene and Toxicology,* Vol. 2B. New York: Wiley-Interscience.

MATERIAL SAFETY DATA SHEET

CHEMICAL NAME	
SELENIUM HEXAFLUORIDE	**CRC PRESS/LEWIS PUBLISHERS DISCLAIMER:** THE INFORMATION AND RECOMMENDATIONS PRESENTED HEREIN ARE BASED ON SOURCES BELIEVED TO BE RELIABLE. CRC MAKES NO REPRESENTATION ON ITS COMPLETENESS OR ACCURACY. IT IS THE RESPONSIBILITY OF THE USER TO DETERMINE THE CHEMICAL'S SUITABILITY FOR ITS INTENDED USE, THE CHEMICAL'S SAFE USE, AND THE CHEMICAL'S PROPER DISPOSAL. NO REPRESENTATIONS AND/OR WARRANTIES, EITHER EXPRESSED OR IMPLIED, OF THE MERCHANTABILITY OR FITNESS FOR A PARTICULAR PURPOSE, OR OF ANY OTHER NATURE, ARE MADE WITH RESPECT TO THE INFORMATION PROVIDED IN THIS MSDS OR TO THE CHEMICAL TO WHICH INFORMATION MAY REFER. CRC NEITHER ASSUMES NOR AUTHORIZES ANY OTHER PERSON TO ASSUME FOR IT, ANY OTHER ADDITIONAL RESPONSIBILITY OR LIABILITY FOR THE USE OF, OR RELIANCE UPON, THIS INFORMATION.

HAZARD WARNING INFORMATION

HEALTH	FIRE	REACTIVE	OTHER	DEGREE OF HAZARD	COLOR CODING	OTHER CODES
3	0	3	~~W~~	0 = Minimum Hazard 1 = Slight Hazard 2 = Moderate Hazard 3 = Serious Hazard 4 = Severe Hazard	HEALTH = BLUE FIRE = RED REACTIVITY = YELLOW OTHER = WHITE	OX = Oxidizer ACID = Acid ALK = Alkali COR = Corrosive ~~W~~ = Use No Water

SECTION I - GENERAL INFORMATION

Characterization	**Selenium Compound**	RCRA Number	**D001**
		EPA Class	**Characteristic (R) Waste**
DOT Proper Shipping Name	**Selenium Hexafluoride**	Chemical Abstract Service (CAS) Number	**7783-79-1**
DOT Hazard Class and Label Requirements	**Poison A; Poison Gas**	DOT Emergency Guide Code	**15**
DOT Identification Number	**UN 2194**	Atomic Symbol	**SeF_6**

Synonyms
Selenium fluoride.

SECTION II - HAZARDOUS INGREDIENTS AND IDENTITY INFORMATION

Hazardous Components (specific identity)	OSHA Exposure Criteria	NIOSH Exposure Criteria	Immediately Dangerous to Life and Health (IDLH)	ACGIH Exposure Criteria
Selenium hexafluoride (derivation: By passing gaseous fluorine over finely divided selenium in a copper vessel). **1 ppm = 3.28 mg/m^3 (as Se)**	**PEL (as Se):** **0.05 ppm** **0.4 mg/m^3** **STEL:** **Not Established**	**REL (as Se):** **0.05 ppm** **0.16 mg/m^3** **STEL:** **Not Established**	**2 ppm**	**TLV (as Se):** **0.05 ppm** **0.16 mg/m^3** **STEL:** **Not Established**

SECTION III - PHYSICAL AND CHEMICAL CHARACTERISTICS

Boiling Point	**-22°F (-30°C)**	Specific Gravity (H_2O = 1)	**6.66**
Vapor Pressure (atmospheres)	**>1 at 68°F (20°C)**	Molecular Weight (atomic weight)	**193.0**
Vapor Density (Air = 1)	**Not Found**	Melting Point	**-59°F (-50.5°C) Sublimes at -82°F (-64°C)**

Solubility
Insoluble in water (hydrolyzes very slowly in cold water).

Appearance and Odor
Colorless gas.

SECTION IV - FIRE AND EXPLOSION HAZARD DATA

Flash Point (method used)	**Not Applicable**	Explosive Limits in Air % by Volume	**LEL: N/A UEL: N/A**
NFPA Classification	**Non-Flammable Gas**	Autoignition Temperature	**Not Applicable**

Extinguishing Media
Selenium hexafluoride is non-flammable. Use extinguishing agent to suitable to surrounding fire.

Special Fire Fighting Procedures
Poisonous gases are produced in fire. Wear full protective clothing and self-contained breathing apparatus (SCBA) operated in pressure demand or other positive pressure mode. Do not release runoff from fire control measures to sewers or waterways. Cool non-leaking cylinders with water spray even after fire has been extinguished. Stay away from the ends of tanks.

Unusual Fire and Explosion Hazards
Containers can explode in fire.

SECTION V - REACTIVITY DATA

Stability		Conditions to Avoid Selenium hexafluoride is normally stable under routine conditions of handling or storage. Avoid contact with incompatible materials since violent reactions can occur. Keep away from water.
Stable X	Unstable	Incompatibility *(materials to avoid)* Reacts with ammonia gas at 392°F (200°C) and will release selenium, nitrogen, and hydrogen fluoride. Selenium hexafluoride can be expected to explode violently in contact with water.
Hazardous Polymerization		Conditions to Avoid Hazardous polymerization of selenium hexafluoride will not occur.
May Occur	Will Not Occur X	Hazardous Decomposition or By-products When heated to decomposition, selenium hexafluoride can emit toxic fluoride and selenium fumes.

SECTION VI - HEALTH HAZARD DATA

Primary Route(s) of Entry:	Inhalation?	Absorption (skin/eye)?	Ingestion?
	X		X

Health Hazards

INHALATION: Inhalation causes pulmonary irritation with the possibility for the delayed development of pulmonary edema (fluid in the lungs) which is a medical emergency and can be fatal. If taken into the lungs and distributed throughout the body, it may cause indigestion, pallor, fatigue, irritability, garlic breath, and possible liver and kidney damage. In severe cases, there may be fingernail and hair loss, and increased potential for dental cavities.

EYE & SKIN: The compressed gas can cause frostbite and tissue destruction to the eyes and skin.

INGESTION: Not a likely exposure route since selenium hexafluoride is a gas.

Carcinogenicity	NTP Listed?	IARC Cancer Review Group?	OSHA Regulated?	Target Organs?
Unknown Human Unknown Animal	No	No	29 CFR 1910.1000 Table Z-1	Respiratory system, liver, kidney, nails, hair.

Medical Conditions Generally Aggravated by Exposure

Respiratory system, liver, and kidney disorders may occur.

Emergency and First-aid Procedures

Eye contact: If a cryogenic gas gets into eyes, open them wide and allow gas to evaporate. Do not allow victim to keep eyes tightly shut. Flush with tepid water for 15 minutes (minimum); seek medical attention. Skin contact: Remove all contaminated clothing. Immediately wash area with flooding amounts of tepid water. Do NOT rub or use dry heat. For inhalation: Remove the person from exposure. Provide respiratory assistance and CPR. Transfer to medical facility. Observe 24 - 48 hours for lung effects. If swallowed: Not likely (selenium hexafluoride is a gas).

SECTION VII - PRECAUTIONS FOR SAFE HANDLING AND USE

Steps to be Taken in Case Material is Released or Spilled

Remove all ignition sources. Restrict those not involved in cleanup from entering area. Stay upwind. If cylinder leak cannot be stopped in place, remove to safe area (outdoors) and allow to empty. Protect against inhalation.

Preferred Waste Disposal Method

No citation.

Precautions to be Taken in Handling and Storage

Do not store near water. Keep in cool, dry, well-ventilated area in tightly closed containers. Prevent moisture contact and keep away from ammonia gas.

Other Precautions and Warnings

Prevent physical damage to containers. Store in vertical position. Never drag or roll cylinders.

SECTION VIII - CONTROL MEASURES AND PERSONAL PROTECTIVE EQUIPMENT

Respiratory Protection *(specify type)*

Use a NIOSH/MSHA-approved self-contained breathing apparatus (SCBA) with full facepiece operated in positive pressure mode or a supplied-air respirator operated in continuous flow mode. Selenium hexafluoride is a gas. Inhalation must be avoided.

Ventilation

Local exhaust (preferred) at site of chemical work or general ventilation.

Protective Gloves	Eye Protection	Other Protective Clothing
Impervious, Thermally Protective Material	Chemical Goggles or Face Mask	Protective Apron, Boots (avoid skin contact)

Work/Hygiene Practices

Always wash hands thoroughly after using chemical; never bring food, drink, or smoking materials into vicinity of chemicals.

SELENIUM HEXAFLUORIDE

SeF_6 CAS: 7783-79-1

IDENTIFICATION AND TYPICAL USES

Selenium hexafluoride is a colorless gas at room temperature and pressure. It is used as a gaseous electric insulator.

RISK ASSESSMENT: HEALTH

General Assessment

As a gas, selenium hexafluoride is a highly toxic human poison by *inhalation.* Skin contact causes serious irritation and tissue destruction (frostbite) due to the cryogenic nature of the gas, but dermal absorption has not been reported. There are no reports of any carcinogenic, teratogenic, or mutagenic action on long-term exposure.

There are very limited data on the effects of human exposure to selenium hexafluoride. Animal data are also limited. Inhalation of selenium hexafluoride gas may cause serious irritation the respiratory tract and may lead to the delayed development of pulmonary edema (fluid in the lungs), which can be fatal. Symptoms may be delayed up to 48 hours thereby creating a false sense of security with regard to health exposure risk. There may be coughing with phlegm, congestion, difficulty breathing, nasal discharge, and tightness or pain in the chest. If the gas is distributed to the body through respiration, systemic effects may include gastrointestinal disturbances, irritability, indigestion, pallor, fatigue, garlic breath, and possible liver and kidney damage. Also, keratin-containing appendages such as hair and fingernails may be damaged. Hair loss and fingernail loss is possible as well as an increase in dental cavities.

Skin contact results ins severe irritation, possible burns, and tissue destruction as a result of the cryogenic nature of selenium hexafluoride gas. Eye contact may result in tissue destruction and possible damage to the cornea.

Acute Health Effects

The following acute (short-term) health effects may occur immediately or shortly after exposure to selenium hexafluoride:

Skin: Severe irritation and possible contact tissue destruction may occur. Burns are possible on prolonged contact.

Eye: Irritation and possible damage on contact.

Lung: High exposures can cause lung damage up to and including delayed pulmonary edema (fluid in the lungs), with shortness of breath, chest pain, cough, and death.

Other: Other symptoms of exposure include gastrointestinal disturbances, irritability, and possible liver damage.

Chronic Health Effects

The following chronic (long-term) health effects may occur at some time after exposure to selenium hexafluoride and can last for months or even years:

Cancer Hazards: According to information presented in the references, selenium hexafluoride has not been adequately tested for its ability to cause cancer in test animals. Other selenium compounds have been shown to cause cancer in animals. It is not known if selenium hexafluoride will have this effect.

Reproductive Hazard: According to information presented in the references, selenium hexafluoride has not been adequately tested for its ability to adversely affect reproduction.

Other Chronic Effects: Repeated exposures may cause garlic odor on breath and in sweat, chronic fatigue, irritability, upper respiratory tract irritation, pallor, gastrointestinal distress, fingernail and hair loss, and possible increase in dental cavities may occur.

Recommended Risk-Reduction Measures

Personnel should avoid direct contact with selenium hexafluoride gas. If a less toxic material or compound cannot be substituted for selenium hexafluoride, then *engineering controls* are the most effective method of reducing exposures. The best protection is to enclose operations and/or provide local exhaust ventilation at the site of selenium hexafluoride gas release. While not always operationally feasible, isolating operations can also reduce exposure. Using respiratory protection is less effective than the controls mentioned above, but is still advisable whenever working with or around selenium hexafluoride. For the best protection against a toxic gas, a self-contained breathing apparatus (SCBA) with full facepiece operated in pressure demand or a supplied-air respirator operated in continuous flow mode are the recommended respiratory protection methods of choice. If a full facepiece is not available, then chemical goggles should be worn to protect the eyes. A face shield and protective apron should also be worn. To prevent hand and skin exposures, impervious, thermally protective gloves should be used. Glove manufacturers should be contacted and permeation studies obtained *before* final glove selection are made.

Administrative controls should also be in place to minimize the potential for human exposures. These may include written procedures or policies which specify the methods and techniques that will be practiced whenever personnel are to work with selenium hexafluoride.

All personnel should receive *training* on the use, hazards, protective measures, emergency actions, and other precautions per 29 CFR 1910.1200 (Hazard Communication), prior to the first assignment in an area where selenium hexafluoride is used or stored.

Before beginning employment working with selenium hexafluoride, and at regular intervals thereafter (e.g., annually), those personnel with frequent or potentially high exposures should be provided the following recommended medical tests:

☑ Lung function test.

☑ Urine test for selenium (levels should be less than 100 micrograms per liter of urine).

If overexposure is suspected or symptoms develop, the following additional test is recommended:

☑ Consider chest X-ray following acute overexposure (may be negative if taken immediately after exposure due to delayed development of pulmonary edema).

Any evaluation should include a careful history of past and present symptoms with an examination. Medical tests that simply look for existing damage are not a substitute for controlling exposures. Also, since smoking can cause heart disease as well as lung cancer, emphysema, and other known respiratory problems, exposure to selenium hexafluoride may cause additional complications. Therefore, personnel who smoke should avoid any unprotected contact with selenium hexafluoride gas.

Other methods to reduce exposure include:

☑ Where possible, enclose operations and use local exhaust ventilation at the site of chemical release. If local exhaust ventilation or enclosure is not used, respiratory protection should be mandatory.

☑ Always ensure that proper protective clothing is worn when using chemical substances and that personnel are trained on its proper use, care, and maintenance.

☑ Wash thoroughly immediately after exposure to selenium hexafluoride and at the end of the work shift or before eating, drinking, applying cosmetics, or smoking.

☑ Hazard warning information should be posted in the work area. In addition, as part of an on-going education and training program, all information on the health and safety hazards of selenium hexafluoride should be communicated to all exposed and potentially exposed workers.

☑ Eye wash stations should be provided in the immediate work area for emergency use. If there is a possibility of skin exposure to selenium hexafluoride, emergency shower facilities should also be provided.

☑ Workers whose clothing has been contaminated by selenium hexafluoride should change into clean clothes before leaving work. Contaminated work clothing should be laundered only by individuals who have been informed of the hazards of exposure to selenium hexafluoride.

RISK ASSESSMENT: ENVIRONMENT

General Assessment

The environment is at risk of exposure during transportation, storage, disposal, or destruction of selenium hexafluoride. In almost every scenario, the threat of environmental exposure is contingent upon the proper handling of the chemical substance. Accidental spills, large or small, where selenium hexafluoride contacts incompatible commodities can result in fire, explosion (depending upon conditions of spill), and possible contamination of the surrounding environmental mediums (primarily air).

In its bulk form, selenium hexafluoride is considered a nonflammable gas. However, it reacts with ammonia gas to release toxic selenium, nitrogen, and toxic and corrosive hydrogen fluoride. It may be expected to react violently with water. Therefore, extreme caution is required in handling, storage, transportation, and disposal of selenium hexafluoride. These characteristics also require special consideration during any emergency situation involving a leak or spill of selenium hexafluoride powder or dust. Should selenium hexafluoride ever come into contact with these incompatible substances either during use, transportation, storage, or disposal, violent and even explosive reactions are possible.

Selenium hexafluoride is a colorless gas. The extent of environmental contamination may be primarily limited to air, although water and soil contamination are also possible but not probable. Selenium is a nonmetallic element which is abundant in the drier soils of North America from the Great Plains to the Pacific Ocean. It may occur in water as a result of the natural weathering of soils and rocks and from mining or smelting certain ores. It is present in coal and fuel oil and is emitted in flue gas and fly ash during combustion. It may exist in different oxidation states and as various salts. In oxygenated water, the +4 and +6 oxidation states are predominant. Trace amounts of selenium are essential for plants, animals, and humans.

Acute Ecological Effects

Acute (short-term) toxic effects may include the death of animals, birds, or fish and death or low growth rate in plants. Acute effects are seen 2 to 4 days after animals or plants are exposed to selenium.

Selenium and its compounds have high acute toxicity to aquatic life and mammals and moderate acute toxicity to birds. Insufficient data are available to evaluate the short-term effects of selenium to plants.

Chronic Ecological Effects

Chronic (long-term) toxic effects may include shortened life span, reproductive problems, lower fertility, and changes in appearance or behavior in exposed animals. These effects can be seen long after first exposure(s) to toxic chemicals.

Selenium and its compounds have high chronic toxicity to aquatic life, birds, and land animals. Insufficient data are available to evaluate or predict the long-term effects of selenium to plants.

Water Solubility

Selenium hexafluoride slowly hydrolyzes in water. Selenium and its compounds have solubility in water ranging from low to moderate. Concentrations up to 1000 milligrams will mix with a liter of water.

Persistence in the Environment

Selenium is highly persistent in water, with a half-life greater than 200 days. The half-life of a pollutant is the amount of time it takes for one half of the chemical to be degraded. No data are available on the estimated percentages of selenium persistence in the aquatic or terrestrial environments.

Bioaccumulation in Aquatic Organisms

Some substances increase in concentration, or *bioaccumulate,* in living organisms as they breathe contaminated air, drink contaminated water, or eat contaminated food. These chemicals can become concentrated in the tissues and internal organs of animals as well as humans.

The concentration of selenium found in fish tissues is expected to be somewhat higher than the average concentration of selenium in the water from which the fish was taken.

Recommended Risk-Reduction Measures

Proper training of all transporters will reduce the likelihood of a mishap or accident resulting in a leak or spill of selenium hexafluoride dusts or powders into the environment. The correct use of DOT labeling (per the MSDS) on all containers, trucks, and rail cars will enable emergency responders to react properly and quickly to any disaster thereby reducing the potential risk to the environment and to personnel.

Storage of selenium hexafluoride should be segregated from incompatible chemicals and gases to minimize the risk of cross-contamination or contact. Cylinders must be stored vertically and secured properly. They should never be dragged or rolled.

If a spill or leak to the environment has occurred, fire department, emergency response and/or hazardous materials spill personnel should be notified immediately. Cleanup should be attempted only by those

trained in proper spill containment procedures. If selenium hexafluoride should contact the water table, aquifer, or navigable waterway, local and/or state emergency response authorities should be notified. A comprehensive emergency response or disaster preparedness/recovery plan should be in place prior to any operations involving the use, transportation, storage, or disposal of selenium hexafluoride. If selenium hexafluoride powder is spilled or leaked, the following specific steps are recommended:

☑ Restrict persons not wearing protective clothing from area of spill or leak until cleanup is complete and area can be opened for normal work. Stay upwind and avoid inhalation.

☑ If source of leak is a cylinder and it cannot be stopped or repaired in place, carefully remove cylinder to a safe area outdoors and allow to vent until empty. Do NOT use water spray to knock down vapors.

☑ Ventilate area of spill or leak.

☑ It may be necessary to dispose of selenium hexafluoride as a hazardous waste. The responsible state agency or the regional office of the federal Environmental Protection Agency (EPA) should be contacted for specific recommendations.

RISK ASSESSMENT: BUSINESS

General Assessment

Accidents or mishaps involving selenium hexafluoride can present a significant threat to business operations. The loss or damage of equipment or facilities can significantly affect fiscal viability. Lawsuits that may result from personnel injury/death, public exposures, and/or environmental contamination will also require a serious expenditure of resources. Media attention surrounding an injury, death, or environmental damage can also result in a loss of profits and loss of current as well as future business.

Recommended Risk-Reduction Measures

Company attorneys, safety and health professionals, and environmental specialists should be involved in the development of any procedures for responding to chemical incidents. A company official should be predesignated as a public relations officer with specific training in dealing with the press. Corporate plans and policies should be developed, approved, and implemented long before any need for such arises.

REFERENCES

Ahlbom, A. 1993. *Biostatistics for Epidemiologists.* New York: CRC Press/Lewis Publishers.

American Conference of Governmental Industrial Hygienists. 1988. *Documentation of the Threshold Limit Values and Biological Exposure Indices,* 5th Edition (with updates). Cincinnati: ACGIH

Brusick, D. J. 1994. *Methods for Genetic Risk Assessment.* New York: CRC Press/Lewis Publishers.

Bueche, J. R. 1972. *Principles of Physics,* 2nd Edition. New York: McGraw-Hill.

Calabrese, E. J. 1994. *Biological Effects of Low Level Exposures.* New York: CRC Press/Lewis Publishers.

Cothern, R. R. 1993. *Comparative Environmental Risk Assessment.* New York: CRC Press/Lewis Publishers.

Danse, I. R. 1991. *Common Sense Toxics in the Workplace.* New York: Van Nostrand Reinhold.

Gardner, R. W. 1994. *Chemical Intolerance.* New York: CRC Press/Lewis Publishers.

Hallenbeck, W. H. 1993. *Quantitative Risk Assessment for Environmental and Occupational Health,* 2nd Edition. New York: CRC Press/Lewis Publishers.

Howard, P. H. 1995. *Handbook of Environmental Fate and Exposures Data for Organic Chemicals,* Vol. I-IV. New York: CRC Press/Lewis Publishers.

Lewis, R. J., Sr. 1992. *Sax's Dangerous Properties of Industrial Materials,* Volumes 1, 2, and 3, 8th Edition. New York: Van Nostrand Reinhold.

Lide, D. R. 1995. *Handbook of Chemistry and Physics,* 76th Edition. New York: CRC Press/Lewis Publishers.

Manahan, S. E. 1989. *Toxicological Chemistry.* New York: CRC Press/Lewis Publishers.

Meyer, E. 1990. *Chemistry of Hazardous Materials.* Englewood Cliffs, NJ: Prentice Hall.

Patnaik, P. 1992. *A Comprehensive Guide to the Hazardous Properties of Chemical Substances.* New York: Van Nostrand Reinhold.

Quigley, D. R. 1994. *Handbook of Emergency Chemical Management.* New York: CRC Press/Lewis Publishers.

Richardson, M. 1995. *Dictionary of Substance and Their Effects,* Vol. I-VI. New York: CRC Press/Lewis Publishers.

Sacarello, H. L. A. 1994. *The Comprehensive Handbook of Hazardous Materials.* New York: CRC Press/Lewis Publishers.

Torkelson, T. R. V. K. Rowe. 1981. *Patty's Industrial Hygiene and Toxicology,* Vol. 2B. New York: Wiley-Interscience.

U.S. Department of Health and Human Services, National Institute for Occupational Safety and Health. 1994. *NIOSH Pocket Guide to Chemical Hazards.* Washington, D.C.: U.S. Government Printing Office.

MATERIAL SAFETY DATA SHEET

CHEMICAL NAME	
SILANE	CRC PRESS/LEWIS PUBLISHERS DISCLAIMER: THE INFORMATION AND RECOMMENDATIONS PRESENTED HEREIN ARE BASED ON SOURCES BELIEVED TO BE RELIABLE. CRC MAKES NO REPRESENTATION ON ITS COMPLETENESS OR ACCURACY. IT IS THE RESPONSIBILITY OF THE USER TO DETERMINE THE CHEMICAL'S SUITABILITY FOR ITS INTENDED USE, THE CHEMICAL'S SAFE USE, AND THE CHEMICAL'S PROPER DISPOSAL. NO REPRESENTATIONS AND/OR WARRANTIES, EITHER EXPRESSED OR IMPLIED, OF THE MERCHANTABILITY OR FITNESS FOR A PARTICULAR PURPOSE, OR OF ANY OTHER NATURE, ARE MADE WITH RESPECT TO THE INFORMATION PROVIDED IN THIS MSDS OR TO THE CHEMICAL TO WHICH INFORMATION MAY REFER. CRC NEITHER ASSUMES NOR AUTHORIZES ANY OTHER PERSON TO ASSUME FOR IT, ANY OTHER ADDITIONAL RESPONSIBILITY OR LIABILITY FOR THE USE OF, OR RELIANCE UPON, THIS INFORMATION.

HAZARD WARNING INFORMATION

HEALTH	FIRE	REACTIVE	OTHER	DEGREE OF HAZARD	COLOR CODING	OTHER CODES
1	4	2		0 = Minimum Hazard 1 = Slight Hazard 2 = Moderate Hazard 3 = Serious Hazard 4 = Severe Hazard	HEALTH = BLUE FIRE = RED REACTIVITY = YELLOW OTHER = WHITE	OX = Oxidizer ACID = Acid ALK = Alkali COR = Corrosive W = Use No Water

SECTION I - GENERAL INFORMATION

Characterization	**Hydride Gas**	RCRA Number	**None**
		EPA Class	**Not Applicable**
DOT Proper Shipping Name	**Silane**	Chemical Abstract Service (CAS) Number	**7803-62-5**
DOT Hazard Class and Label Requirements	**Flammable Gas**	DOT Emergency Guide Code	**18**
DOT Identification Number	**UN 2203**	Chemical Formula	**SiH_4**

Synonyms
Silicane; silicon tetrahydride; monosilane; tetrabromosilane; tetrabromosilicane.

SECTION II - HAZARDOUS INGREDIENTS AND IDENTITY INFORMATION

Hazardous Components (specific identity)	OSHA Exposure Criteria	NIOSH Exposure Criteria	Immediately Dangerous to Life and Health (IDLH)	ACGIH Exposure Criteria
Silane **1 ppm = 1.40 mg/m³**	**PEL:** 5 ppm, 7 mg/m³ **STEL:** Not Established	**REL:** Not Established **STEL:** Not Established	**Not Found**	**TLV:** 5 ppm, 7 mg/m³ **STEL:** Not Established

SECTION III - PHYSICAL AND CHEMICAL CHARACTERISTICS

Boiling Point	**-170°F (-112°C)**	Specific Gravity (H_2O = 1)	**0.68 at -301°F (-185°C)**
Vapor Pressure (atmospheres)	**Not Found**	Molecular Weight	**28.1**
Vapor Density (Air = 1)	**Not Found**	Freezing Point	**-301°F (-185°C)**

Solubility
Decomposes in water. Insoluble in alcohol and benzene.

Appearance and Odor
Colorless gas with a repulsive odor. Solidifies at -328°F (-200°C).

SECTION IV - FIRE AND EXPLOSION HAZARD DATA

Flash Point (method used)	**Not Established**	Explosive Limits in Air % by Volume	**LEL: Not Determined UEL: Not Determined**
NFPA Classification	**Flammable Gas**	Autoignition Temperature	**Not Determined**

Extinguishing Media
Silane is a pyrophoric gas (ignites spontaneously in air). Do NOT attempt to extinguish because burning reduces the possibility of explosions of silane-air mixtures. Control surrounding fires.

Special Fire Fighting Procedures
Silane gas presents a serious fire and explosion hazard when exposed to flame or involved in fire. Gas is poisonous: wear full protective clothing and self-contained breathing apparatus (SCBA).

Unusual Fire and Explosion Hazards
The gas is explosive in air. Cylinders can rupture violently in fire. If possible, remove fire-exposed containers from the heat of fire or cool them with water spray long after fire has been extinguished. Apply water from protected location, preferably an explosion-proof location. If this is not possible, consider evacuation a minimum of 1500 feet from fire area.

SECTION V - REACTIVITY DATA

Stability		Conditions to Avoid Silane is normally stable in closed, airtight containers under routine conditions of handling and storage. Avoid contact with incompatible materials and keep away from fire and sources of heat or flame.
Stable X	Unstable	Incompatibility *(materials to avoid)* Silane is incompatible with oxidizing agents, air (pyrophoric reaction), halogens, water, and heavy metal hydrides. It can form explosive mixtures in air that can ignite without any source of ignition.
Hazardous Polymerization		Conditions to Avoid Under normal conditions of temperature and pressure, hazardous polymerization of silane will not occur. It hydrolyzes slowly in water to hydrogen gas and silicic acid or hydrated silica.
May Occur	Will Not Occur X	Hazardous Decomposition or By-products When heated to decomposition, silane burns and explodes. Hazardous by-products may include hydrogen and silicon. An electric discharge can decompose silane.

SECTION VI - HEALTH HAZARD DATA

Primary Route(s) of Entry:	Inhalation? X	Absorption (skin)?	Ingestion?

Health Hazards

INHALATION: Causes irritation of the nose, throat, and upper respiratory tract. May cause headache, and nausea. Once inhaled, silane may hydrolyze to hydrated silica or silicic acid which may then be found in body tissues. Pulmonary edema can occur, which can be fatal.

SKIN & EYES: Thermal burns can occur as a result of the spontaneous ignition of silane in air.

INGESTION: Not likely since silane is a gas.

Carcinogenicity	NTP Listed?	IARC Cancer Review Group?	OSHA Regulated?	Target Organs?
Unknown Human Unknown Animal	No	No	No	Upper respiratory system; skin; eyes.

Medical Conditions Generally Aggravated by Exposure

Respiratory function impairments (bronchitis, asthma, etc.) may be aggravated by exposure.

Emergency and First-aid Procedures

Eye contact: Immediately flush with large amounts of water for 15 minutes (minimum), occasionally lifting eyelids; seek medical attention. Skin contact: Remove all contaminated clothing. Immediately wash area with large amounts of soap and water. Treat for thermal burns, if required. For inhalation: Remove the person from exposure. Provide respiratory assistance and CPR. Do NOT administer oxygen immediately if breathing is labored (wait several minutes for silane gas to clear the lungs. Transfer promptly to medical facility. Observation for 24 - 48 hours since pulmonary edema is possible. If swallowed: Not likely since this product is a gas.

SECTION VII - PRECAUTIONS FOR SAFE HANDLING AND USE

Steps to be Taken in Case Material is Released or Spilled

Remove all ignition sources. Ventilate area. Gas will ignite spontaneously in air. If the source of leak is a cylinder and cannot be stopped, remove to safe place in open air and allow to burn itself out. It may be necessary to evacuate area and wait for gas to burn out.

Preferred Waste Disposal Method

Burn under controlled conditions.

Precautions to be Taken in Handling and Storage

Do not store with incompatible chemicals since violent reactions can occur. Store in tightly closed containers in a cool, dark, well-ventilated area away from heat and moisture. Use external facility or outside storage (avoid direct sunlight). Keep away from ignition sources.

Other Precautions and Warnings

Bulk storage of silane is not recommended. Containers may explode in fire or under conditions of extreme heat. Store cylinders vertically. Do NOT drag or roll cylinders.

SECTION VIII - CONTROL MEASURES AND PERSONAL PROTECTIVE EQUIPMENT

Respiratory Protection *(specify type)*

At exposures to toxic gases, use a powered supplied-air respirator set in positive pressure mode or use an MSHA/NIOSH-approved self-contained breathing apparatus (SCBA) with full facepiece operated in pressure demand or other positive pressure mode.

Ventilation

Local exhaust or general mechanical systems recommended. Consider explosion-proof design.

Protective Gloves	Eye Protection	Other Protective Clothing
Impervious Material	Chemical Goggles and/or Face Mask	Protective Suit

Work/Hygiene Practices

Always wash hands thoroughly after using chemical; never bring food, drink, or smoking materials into vicinity of chemicals.

SILANE

SiH_4

CAS: 7803-62-5

IDENTIFICATION AND TYPICAL USES

Silane is a colorless gas with a repulsive odor. It is normally shipped in pressurized cylinders. It is used for doping of solid-state devices and for preparing silicon semiconductors for the electronic industry.

RISK ASSESSMENT: HEALTH

General Assessment

Silane is a respiratory tract and pulmonary irritant that can cause secondary infections in the respiratory passageways (may precipitate bronchial pneumonia). As a gas, the primary entry is by *inhalation*. Its carcinogenic, mutagenic, and teratogenic properties are unknown.

Inhalation of silane causes irritation to the eyes, nose, throat, and upper respiratory tract. Overexposure can cause headache and nausea. Pulmonary irritation may lead to the development of pulmonary edema (fluid in the lungs) which can be fatal. Symptoms of cough, congestion, and labored breathing may be delayed up to 48 hours. This can present a false sense of security with regard to health exposure risk. Once inside the body, it can hydrolyze to hydrated silica or silicic acid which may be present in body tissues.

Silane can cause serious thermal burns to the skin and eyes due to its pyrophoric nature (ignites spontaneously on exposure to air).

Acute Health Effects

The following acute (short-term) health effects may occur immediately or shortly after exposure to silane:

Skin: Irritation and possible thermal burns.

Eye: Burns which can lead to permanent damage. Inflammation and swelling of the conjunctiva and surrounding tissue.

Lung: Severe lung damage with symptoms of coughing, wheezing, shortness of breath, and possible pulmonary edema.

Chronic Health Effects

The following chronic (long-term) health effects may occur at some time after exposure to silane and can last for months or even years:

Cancer Hazards: According to information presented in the references, silane has not been adequately tested for its ability to cause cancer in test animals.

Reproductive Hazards: According to information presented in the references, silane has not been tested for its ability to affect reproduction.

Other Chronic Effects: None reported. There is very little published information on the toxicology of silane.

Recommended Risk-Reduction Measures

Personnel should avoid direct contact with silane. It is a dangerous, toxic compound and exposures can cause death or serious physical injury or illness. Unless a less toxic chemical can be substituted for a hazardous substance, *engineering controls* are the most effective method of reducing exposures. The best protection is to enclose operations and/or provide local exhaust ventilation at the site of chemical release. While not always operationally feasible, isolating operations can also reduce exposure. Using respiratory protection is less effective than the controls mentioned above, but is always advisable whenever working with silane. A NIOSH-approved self-contained breathing apparatus (SCBA) with full facepiece operated in pressure demand is the recommended respiratory protection of choice. If a full facepiece is not available, then chemical goggles should be worn to protect the eyes. Whenever a gas emission hazard exists, a face shield and a rubber apron should be worn. To prevent hand and skin exposures, protective gloves should be used. Glove manufacturers should be contacted and permeation studies obtained *before* final glove selections are made.

Administrative controls should also be in place to minimize the potential for human exposures. These may include written procedures or policies which specify the methods and techniques that will be practiced whenever personnel are to work with silane.

All personnel should receive *training* on the use, hazards, protective measures, emergency actions, and other precautions per 29 CFR 1910.1200 (Hazard Communication), prior to the first assignment in an area where silane is used or stored. Before beginning employment and at regular intervals thereafter (e.g., annually), the following medical tests are recommended:

☑ Lung function tests. These may be normal at first if the person is not having an attack at the time.

If symptoms develop or overexposure is suspected, the following may be useful:

☑ Consider chest X-ray after acute overexposure (may be negative if taken immediately after exposure due to delayed onset of pulmonary edema).

☑ Liver and kidney function tests.

Also, since smoking can cause heart disease, emphysema, lung cancer, and other respiratory diseases, smokers may experience symptoms more quickly than non-smokers under the same conditions of exposure. Prudent risk management requires careful consideration of *all* risk factors which might be responsible for exposure symptoms in the work place.

Other methods to reduce exposure to chemicals include:

☑ Where possible, enclose operations and use local exhaust ventilation at the site of chemical release. If local exhaust ventilation or enclosure is not used, respiratory protection should be mandatory.

☑ Always ensure that proper protective clothing is worn when using chemical substances and that personnel are trained in its proper use and care.

☑ Wash thoroughly immediately after exposure to silane and at the end of the work shift or before eating, drinking, or smoking. Work clothing contaminated with silane should never be taken home for laundering (family members can be exposed). Only personnel trained in the hazards of exposure should launder contaminated clothing.

☑ Safety shower and eyewash stations should be readily available in work areas where silane is used or stored.

☑ Hazard warning information should be posted in the work area. In addition, as part of an on-going education and training program, all information on the health and safety hazards of silane should be communicated to all exposed and potentially exposed workers.

RISK ASSESSMENT: ENVIRONMENT

General Assessment

The environment is at risk of exposure during transportation, storage, disposal, or destruction of silane. In almost every scenario, the threat of environmental exposure is contingent upon the proper handling of the chemical substance. Accidental releases, large or small, can result in fire, explosion, and possible contamination of the surrounding environmental mediums (water, soil, and air).

Silane is considered to be an extremely flammable gas. It is explosive by itself and, because it is incompatible with many other commodities, contact with any of them can result in violent explosion and fire. It is pyrophoric and will ignite in air at room temperature. It will react and possibly explode in contact with many oxidizing agents (especially fluorine, bromine, and chlorine), halogens, and heavy-metal hydrides. Therefore, silane requires special consideration during any emergency involving a leak or release of silane gas or liquid.

Silane can enter the environment through industrial discharges, unchecked venting, or through spills.

☠ *Acute Ecological Effects*

Acute (short-term) toxic effects may include the death of animals, birds, or fish, and death or low growth rate in plants. Acute effects are seen 2 to 4 days after animals or plants are exposed to silane.

Insufficient data are available to evaluate or predict the short-term effects of silane on aquatic life, plants, birds, or land animals.

💣 *Chronic Ecological Effects*

Chronic (long-term) toxic effects may include shortened life span, reproductive problems, lower fertility, and changes in appearance or behavior in exposed animals. These effects can be seen long after first exposure(s) to toxic chemicals. Insufficient data are available to evaluate the long-term effects of silane on aquatic life, plants, birds, or land animals.

♠ *Water Solubility*

Silane decomposes in water.

⌛ *Persistence in the Environment*

Silane will react rapidly with water and therefore will not accumulate in the environment.

Bioaccumulation in Aquatic Organisms

Some substances increase in concentration, or *bioaccumulate,* in living organisms as they breathe contaminated air, drink contaminated water, or eat contaminated food. These chemicals can become concentrated in the tissues and internal organs of animals as well as humans.

Silane will react rapidly with water and therefore is not expected to accumulate in the tissues of fish.

Recommended Risk-Reduction Measures

Proper training of all transporters will reduce the likelihood of a mishap or accident resulting in a leak or release to the environment. The correct use of DOT labeling (per the MSDS) on all containers, trucks, and rail cars will enable emergency responders to react properly and quickly to any disaster thereby reducing the potential risk to the environment and to personnel.

Storage of silane should be segregated from incompatible chemicals to minimize the risk of cross-contamination or contact. Under all circumstances, it should be protected from exposure to air or water. Buildings designated for storage should be equipped with appropriate fire protection systems (alarms, sprinklers, emergency lighting, portable extinguishers). Maximum explosion-proof ventilation is recommended.

If a spill or release to the environment has occurred, fire department, emergency response, and/or hazardous materials spill personnel should be notified immediately. Cleanup should be attempted only by those trained in proper containment procedures using non-sparking tools. Contaminated soils should be removed for incineration and replaced with clean soil. If silane should contact the water table, aquifer, or navigable waterway, remediation may not be possible. It will readily react with water and decompose and, therefore, total containment will not be possible. When such spills occur, the local and/or state emergency response authorities must be notified. A comprehensive emergency response or disaster preparedness/recovery plan should be in place prior to any operations involving the use, transportation, storage, or disposal of silane. If silane is spilled or leaked, the following specific steps are recommended:

- ☑ Restrict persons not wearing protective clothing from area of spill or leak until cleanup is complete and area can be opened for normal work.
- ☑ Remove all ignition sources.
- ☑ Ventilate area of spill or leak.
- ☑ Do NOT attempt to extinguish burning silane since its burning reduces the possibility of explosions due to silane-air mixtures. Control surrounding fires. Keep area clear and stay upwind. Evacuation of a minimum distance of 1500 feet may be necessary.
- ☑ Keep silane out of a confined space, such as a sewer, because of the possibility of explosion (unless the sewer is designed to prevent the buildup of explosive concentrations).
- ☑ It may be necessary to dispose of silane as a hazardous waste. The responsible state agency or the regional office of the federal Environmental Protection Agency (EPA) should be contacted for specific recommendations.

RISK ASSESSMENT: BUSINESS

General Assessment

Accidents or mishaps involving silane can present a significant threat to business operations. The loss or damage of equipment or facilities can significantly affect fiscal viability. Lawsuits that may result from personnel injury, public exposures, and environmental contamination will also require a serious expenditure of resources. Media attention surrounding an injury, illness, or environmental damage can also result in a loss of profits and loss of current as well as future business.

Recommended Risk-Reduction Measures

Company attorneys, safety and health professionals, and environmental specialists should be involved in the development of any procedures dealing with safety or emergency response. A company official should be pre-designated as a public relations officer with specific training in dealing with the press. Corporate plans and policies should be developed, approved, and implemented long before any need for such arises.

REFERENCES

Ahlbom, A. 1993. *Biostatistics for Epidemiologists.* New York: CRC Press/Lewis Publishers.

American Conference of Governmental Industrial Hygienists. 1988. *Documentation of the Threshold Limit Values and Biological Exposure Indices,* 5th Edition (with updates). Cincinnati: ACGIH

Anthony, C. P., and N. J. Kolthoff. 1971. *Textbook of Anatomy and Physiology.* St. Louis: C. V. Mosby Company.

Asante-Duah, D. K. 1993. *Hazardous Waste Risk Assessment.* New York: CRC Press/Lewis Publishers.

Bloom, A. D. and F. J. DeSerres. 1995. *Ecotoxicity and Human Health.* New York: CRC Press/Lewis Publishers.

Brusick, D. J. 1994. *Methods for Genetic Risk Assessment.* New York: CRC Press/Lewis Publishers.

Bueche, J. R. 1972. *Principles of Physics,* 2nd Edition. New York: McGraw-Hill.

Burrell, R., D. K. Flaherty, and L. J. Sauers. 1992. *Toxicology of the Immune System.* New York: Van Nostrand Reinhold.

Calabrese, E. J., E. M. Kenyon. *Air Toxics and Risk Assessment.* New York: CRC Press/Lewis Publishers.

Cockerman, L. G., B. S. Shane. 1994. *Basic Environmental Toxicology.* New York: CRC Press/Lewis Publishers.

Cothern, R. R. 1993. *Comparative Environmental Risk Assessment.* New York: CRC Press/Lewis Publishers.

Danse, I. R. 1991. *Common Sense Toxics in the Workplace.* New York: Van Nostrand Reinhold.

Environmental Protection Agency. 1983. *Chemical Hazard Information Profiles.* Washington, D.C.: U.S. EPA.

Fodden, C. R., and J. L. Weddell. 1992. *Hazardous Materials Emergency Action Data.* New York: CRC Press/Lewis Publishers.

Forseberg, K., and L. H. Keith. 1993. *Chemical; Protective Clothing Permeation and Degradation Database.* Chelsea, MI: Lewis Publishers.

Gardner, R. W. 1994. *Chemical Intolerance.* New York: CRC Press/Lewis Publishers.

Hallenbeck, W. H. 1993. *Quantitative Risk Assessment for Environmental and Occupational Health,* 2nd Edition. New York: CRC Press/Lewis Publishers.

Hodgson, E., R. B. Mailman, and J. E. Chambers. 1988. *Dictionary of Toxicology.* London: The MacMillian Company.

Hosty, J. W., and P. Foster. 1990. *A Practical Guide to Chemical Spill Response.* New York: Van Nostrand Reinhold.

Howard, P. H. 1995. *Handbook of Environmental Fate and Exposures Data for Organic Chemicals,* Vol. I-IV. New York: CRC Press/Lewis Publishers.

Howard, P. H., et. al. 1991. *Handbook of Environmental Degradation Rates.* New York: CRC Press/Lewis Publishers.

James, A., and M. Lord. 1992. *Index of Chemical and Physical Data.* New York: Van Nostrand Reinhold.

Jones, F. E. 1994. *Toxic Organic Vapors in the Workplace.* New York: CRC Press/Lewis Publishers.

Kamrin, M. A. 1988. *Toxicology: A Primer on Principles and Applications.* Chelsea, Michigan: Lewis Publishers.

Keith, L. H. and M. M. Walker. 1995. *Handbook of Air Toxics.* New York: CRC Press/Lewis Publishers.

Landis, W. G., M. H. Yu. 1995. *Introduction to Environmental Toxicology.* New York: CRC Press/Lewis Publishers.

Langerman, N. 1994 *Precautionary Labels for Chemical Containers.* New York: CRC Press/Lewis Publishers.

Lauwerys, R. R., P. Hoet. 1993. *Industrial Chemical Exposure.* New York: CRC Press/Lewis Publishers.

Lewis, R. J., Sr. 1993. *Hawley's Condensed Chemical Dictionary,* 12th Edition. New York: Van Nostrand Reinhold.

Lewis, R. J., Sr. 1992. *Sax's Dangerous Properties of Industrial Materials,* Volumes 1, 2, and 3, 8th Edition. New York: Van Nostrand Reinhold.

Lide, D. R. 1995. *Handbook of Chemistry and Physics,* 76th Edition. New York: CRC Press/Lewis Publishers.

Mackay, D., W. Y. Shiu, and K. C. Ma. 1992. *Illustrated Handbook of Chemical-Physical Properties and Environmental Fate for Organic Chemicals.* New York: CRC Press/Lewis Publishers.

Manahan, S. E. 1994. *Environmental Chemistry,* 6th Edition. New York: CRC Press/Lewis Publishers.

Manahan, S. E. 1992. *Toxicological Chemistry,* 2nd Edition. New York: CRC Press/Lewis Publishers.

Meyer, E. 1990. *Chemistry of Hazardous Materials.* Englewood Cliffs, NJ: Prentice Hall.

Patnaik, P. 1992. *A Comprehensive Guide to the Hazardous Properties of Chemical Substances.* New York: Van Nostrand Reinhold.

Philp, R. B. 1995. *Environmental Hazards and Human Health.* New York: CRC Press/Lewis Publishers.

Proctor, N. H., and J. P. Hughes. 1991. *Chemical Hazards of the Workplace.* New York: Van Nostrand Reinhold.

Quigley, D. R. 1994. *Handbook of Emergency Chemical Management.* New York: CRC Press/Lewis Publishers.

Rea, W. J. 1995. *Chemical Sensitivity,* Vol. I-III. New York: CRC Press/Lewis Publishers.

Richardson, M. 1995. *Dictionary of Substance and Their Effects,* Vol. I-VI. New York: CRC Press/Lewis Publishers.

U.S. Department of Health and Human Services, National Institute for Occupational Safety and Health. 1994. *NIOSH Pocket Guide to Chemical Hazards.* Washington, D.C.: U.S. Government Printing Office.

MATERIAL SAFETY DATA SHEET

CHEMICAL NAME	
SILICON	**CRC PRESS/LEWIS PUBLISHERS DISCLAIMER:** THE INFORMATION AND RECOMMENDATIONS PRESENTED HEREIN ARE BASED ON SOURCES BELIEVED TO BE RELIABLE. CRC MAKES NO REPRESENTATION ON ITS COMPLETENESS OR ACCURACY. IT IS THE RESPONSIBILITY OF THE USER TO DETERMINE THE CHEMICAL'S SUITABILITY FOR ITS INTENDED USE, THE CHEMICAL'S SAFE USE, AND THE CHEMICAL'S PROPER DISPOSAL. NO REPRESENTATIONS AND/OR WARRANTIES, EITHER EXPRESSED OR IMPLIED, OF THE MERCHANTABILITY OR FITNESS FOR A PARTICULAR PURPOSE, OR OF ANY OTHER NATURE, ARE MADE WITH RESPECT TO THE INFORMATION PROVIDED IN THIS MSDS OR TO THE CHEMICAL TO WHICH INFORMATION MAY REFER. CRC NEITHER ASSUMES NOR AUTHORIZES ANY OTHER PERSON TO ASSUME FOR IT, ANY OTHER ADDITIONAL RESPONSIBILITY OR LIABILITY FOR THE USE OF, OR RELIANCE UPON, THIS INFORMATION.

HAZARD WARNING INFORMATION

HEALTH	FIRE	REACTIVE	OTHER	DEGREE OF HAZARD	COLOR CODING	OTHER CODES
2	0	0		0 = Minimum Hazard 1 = Slight Hazard 2 = Moderate Hazard 3 = Serious Hazard 4 = Severe Hazard	HEALTH = BLUE FIRE = RED REACTIVITY = YELLOW OTHER = WHITE	OX = Oxidizer ACID = Acid ALK = Alkali COR = Corrosive ~~W~~ = Use No Water

SECTION I - GENERAL INFORMATION

Characterization	**Non-metallic Element**	RCRA Number	**Not Listed**
		EPA Class	**Not Applicable**
DOT Proper Shipping Name	**Silicon**	Chemical Abstract Service (CAS) Number	**7440-21-3**
DOT Hazard Class and Label Requirements	**Flammable Solid**	DOT Emergency Guide Code	**32 (amorphous powder)**
DOT Identification Number	**UN 1346**	Chemical Formula	**Si**

Synonyms
Elemental silicon.

SECTION II - HAZARDOUS INGREDIENTS AND IDENTITY INFORMATION

Hazardous Components (specific identity)	OSHA Exposure Criteria	NIOSH Exposure Criteria	Immediately Dangerous to Life and Health (IDLH)	ACGIH Exposure Criteria
Silicon (derivation: Made in an electric furnace by heating quartz with carbon, followed by zone refining. It is purified by leaching).	**PEL:** **Total Dust:** **15 mg/m^3** **Respirable Fraction:** **5 mg/m^3**	**REL:** **Total Dust:** **10 mg/m^3** **Respirable Fraction:** **5 mg/m^3**	**Not Determined**	**TLV:** **Total Dust:** **10 mg/m^3** **(when <1% quartz))**

SECTION III - PHYSICAL AND CHEMICAL CHARACTERISTICS

Boiling Point	**4271°F (2355°C)**	Specific Gravity (H_2O = 1)	**2.33**
Vapor Pressure (atmospheres)	**1 at 3135°F (1724°C)**	Molecular Weight (atomic weight)	**28.1**
Vapor Density (Air = 1)	**Not Applicable**	Melting Point	**2570°F (1410°C)**

Solubility
Insoluble in water, nitric acid, and hydrofluoric acid. Soluble in molten alkali oxides.

Appearance and Odor
Black to gray, lustrous, needle-like crystals. The amorphous form is a dark-brown powder.

SECTION IV - FIRE AND EXPLOSION HAZARD DATA

Flash Point (method used)	**Not Applicable**	Explosive Limits in Air % by Volume	**LEL: N/A UEL: N/A**
NFPA Classification	**Combustible Solid (in powder form)**	Autoignition Temperature	**Not Applicable**

Extinguishing Media
Use carbon dioxide, dry chemical, foam, or water spray or agent suitable to surrounding fires. Silicon is combustible in powder form only.

Special Fire Fighting Procedures
Care should be taken to contain silicon materials that may have been disturbed in fire. To avoid breathing dusts or toxic fumes released during fire, wear full protective clothing and a self-contained breathing apparatus (SCBA) operated in pressure demand or other positive pressure mode. Be aware of runoff and do not release to sewers or waterways. Prevent physical damage to containers.

Unusual Fire and Explosion Hazards
Silicon is extremely flammable in powder form and can form an explosive mixture in air, especially in confined areas.

SECTION V - REACTIVITY DATA

Stability		Conditions to Avoid **Avoid unprotected and/or unauthorized disturbance of silicon materials. Do not allow silicon or silicon containing materials to become airborne. Silicon is stable under normal conditions of handling and storage.**
Stable **X**	Unstable	Incompatibility *(materials to avoid)* **Powerful oxidizers (reacts violently), fluorine, chlorine trifluoride, manganese trioxide, oxygen difluoride, hydrogen peroxide, acetylene, ammonia, cesium carbide, calcium, and alkaline carbonates.**
Hazardous Polymerization		Conditions to Avoid **Hazardous polymerization will not occur.**
May Occur	Will Not Occur **X**	Hazardous Decomposition or By-products **When heated to decomposition, silicon can emit toxic and acrid smoke and fumes.**

SECTION VI - HEALTH HAZARD DATA

Primary Route(s) of Entry:	Inhalation?	Absorption (skin/eye)?	Ingestion?
	X		X

Health Hazards

INHALATION: Causes irritation to the eyes, nose, throat, and respiratory system. There may be cough, congestion, difficulty breathing, and tightness in the chest. Persistent coughing may produce sore throat and there may be nausea and headaches.

SKIN & EYES: Irritating in contact with the eyes. May cause skin irritation and, in some persons, rash.

INGESTION: Unspecified gastrointestinal effects. Symptoms can range from no effects to abdominal pain, irritation, diarrhea, and vomiting.

Carcinogenicity	NTP Listed?	IARC Cancer Review Group?	OSHA Regulated?	Target Organs?
Unknown Human Unknown Animal	No	No	29 CFR 1910.1000 Table Z-1	Eyes, respiratory system, skin.

Medical Conditions Generally Aggravated by Exposure

Respiratory conditions (bronchitis, asthma) may be aggravated by exposure over the long-term (years).

Emergency and First-aid Procedures

Eye contact: Do not allow victim to keep eyes tightly shut. Flush immediately with water for 15 minutes (minimum); seek medical attention. Skin contact: Immediately wash area with large amounts of water. For inhalation: Remove the person from exposure. Provide respiratory assistance and CPR. Transfer to medical facility. Respiratory problems associated with silicon may not occur for many years following repeated exposures. All exposures should be documented. If swallowed: Give conscious person 1 to 2 glasses of water and induce vomiting until vomitus is clear.

SECTION VII - PRECAUTIONS FOR SAFE HANDLING AND USE

Steps to be Taken in Case Material is Released or Spilled

Restrict those not involved in cleanup from area. Wear self-contained breathing apparatus. Collect dusts using a HEPA vacuum. Do NOT dry sweep (creates dusts). Place in sealed container for disposal.

Preferred Waste Disposal Method

No citation.

Precautions to be Taken in Handling and Storage

Prior to working with silicon, persons must be trained on its proper handling and storage. Prevent damage to storage containers and avoid generating dusty conditions.

Other Precautions and Warnings

Airborne silicon dust is difficult to remove. It is essential that silicon work areas be closed or isolated to ensure controlled exposures.

SECTION VIII - CONTROL MEASURES AND PERSONAL PROTECTIVE EQUIPMENT

Respiratory Protection *(specify type)*

Use a MSHA/NIOSH-approved respirator with a full facepiece with a high-efficiency particulate air (HEPA) filter for low or infrequent exposures. For the best protection, use a self-contained breathing apparatus (SCBA) with full facepiece operated in positive pressure mode or a supplied-air respirator.

Ventilation

Local exhaust (preferred) at site of silicon work. Area should be isolated from other work areas.

Protective Gloves	Eye Protection	Other Protective Clothing
Impervious Rubber	Dust Goggles and Face Mask	Protective Suit or Apron

Work/Hygiene Practices

Always wash thoroughly after working around silicon; never bring food, drink, or smoking materials into vicinity of silicon work.

SILICON

Si

CAS: 7440-21-3

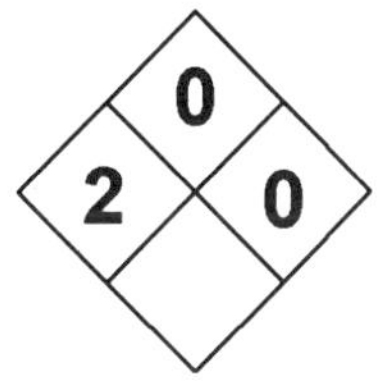

IDENTIFICATION AND TYPICAL USES

Silicon is a black to gray, lustrous, needle-like crystalline solid. The amorphous form is a dark-brown powder. It is a nonmetallic element of atomic number 14, Group IVA of the periodic table, valance of 4 with three stable isotopes. It is the second most abundant element, comprising 25% of the Earth's crust. It can form more compounds than any other element, except carbon. It is used in semiconductors in solid-state devices such as transistors, photovoltaic cells, computer circuitry, and rectifiers. It is also used in the production of organosilicon compounds, silicon carbide, and as a alloying agent in steels, aluminum, copper, bronze, and iron. It is used in cements, in halogenated silanes, in spring steels, and as a deoxidizer in steel manufacturing.

RISK ASSESSMENT: HEALTH

General Assessment

Health hazards associated with silicon occur as a result of prolonged and repeated exposures over the course of years. *Inhalation* is the primary route of entry of silicon dusts into the body. Silicon can also be ingested, although this exposure route is rare and not well documented in the references. Other than the pulmonary route, particles may accumulate in the gastrointestinal tract from drinking water contaminated with silicon or due to transportation from the nasal passage (i.e., when swallowing). There are no data in the references regarding its carcinogenic, mutagenic, or teratogenic properties.

Inhalation may cause irritation of the eyes, nose, throat, and upper respiratory tract. Prolonged or repeated contact may lead to chronic cough, congestion, headaches, and nausea. The data on its specific acute and chronic effects are limited.

Silicon can produce a drying effect on the skin, leaving it susceptible to cracking and secondary infection. A rash may develop in some individuals which may or may not lead to contact dermatitis. Silicon is also an eye irritant, causing inflammation and redness.

Acute Health Effects

The following acute (short-term) health effects may occur immediately or shortly after exposure to silicon:

Skin: Irritation and possible rash upon contact with skin (contact dermatitis).

Eye: Severe irritation and possible inflammation.

Lung: Irritation of the nose, throat, and lungs following exposure. Can cause coughing, dypsnea, nausea, and headache.

Chronic Health Effects

The following chronic (long-term) health effects may occur at some time after exposure to silicon and can last for months or even years before death occurs:

Cancer Hazards: According to information presented in the references, silicon has not been adequately tested for its ability to cause cancer in test animals. Silicon dioxide (quartz) has demonstrated carcinogenic properties in animals. It is not known if the base element, silicon, will have this effect.

Reproductive Hazard: According to the references, silicon has not been adequately tested for its ability to adversely affect reproduction.

Other Chronic Effects: None reported.

Recommended Risk-Reduction Measures

Personnel should avoid direct contact with silicon. Its airborne dusts may present a moderate health hazard to unprotected, exposed individuals. Only trained and qualified persons should be permitted to work with or around silicon. *Engineering controls* are required to ensure maximum reduction of exposure potential.

During any work with silicon, the best protection is to enclose operations and/or provide local exhaust ventilation at the site of silicon work. Isolating operations will also reduce exposure. Using respiratory protection is less effective than the controls mentioned above, but is still required whenever working with or around friable silicon. At low or infrequent exposure levels, an MSHA/NIOSH-approved respirator equipped with a high efficiency particulate air (HEPA) filter may suffice. A self-contained breathing apparatus (SCBA) with full facepiece operated in pressure demand is recommended for the greatest possible respiratory protection. If a full facepiece is not available, then dust goggles or face shield should be worn to protect the eyes. To prevent hand and skin exposures, impervious rubber gloves should be used.

Administrative controls should also be in place to minimize the potential for human exposures. These are required by law and include written procedures or policies which specify the methods and techniques that will be practiced whenever personnel are to work with silicon.

All personnel must receive *training* on the use, hazards, protective measures, emergency actions, and other precautions per 29 CFR 1910.1200 (Hazard Communication), prior to the first assignment in an area where silicon is used, removed, stored, or destroyed.

Before beginning employment with silicon and at regular intervals thereafter (e.g., annually), a medical history and examination is recommended, including the following:

☑ Pulmonary function test (baseline and annually).

☑ Consider periodic chest X-ray.

Any evaluation should include a careful history of past and present symptoms with an examination. Medical tests that simply look for existing damage are not a substitute for controlling exposures. Because smoking can cause heart disease, as well as lung cancer, emphysema, and other respiratory problems, respiratory conditions caused by silicon exposure may be more severe. Prudent risk management requires careful consideration of *all* possible factors which may be causing the appearance of exposure symptoms among workers.

Other methods to reduce exposure to silicon include:

☑ Always ensure that proper protective clothing is worn when working with silicon and that personnel have been properly trained on the correct use, care, and maintenance of such equipment.

☑ Hazard warning information should be posted in the work area. In addition, as part of an on-going education and training program, all information on the health and safety hazards of silicon must be communicated to all potentially exposed workers.

☑ Thoroughly wash any areas of the body that may have contacted silicon.

☑ Do not eat, drink, apply cosmetics, or smoke where silicon is handled, processed, or stored since silicon can be swallowed. Wash hands carefully before eating or smoking.

☑ Do not sweep for cleanup. Use a vacuum equipped with a high-efficiency particulate air (HEPA) filter or a wet method to reduce dust generation.

☑ Eye wash stations should be provided in the immediate work area for emergency use. If there is a possibility of skin exposure to silicon, emergency shower facilities should also be provided.

RISK ASSESSMENT: ENVIRONMENT

General Assessment

The environment is at risk of exposure during transportation, storage, disposal, or destruction of silicon. In almost every scenario, the threat of environmental exposure is contingent upon the proper handling silicon.

Accidental spills and damaged silicon materials must be cleaned using wet methods. Silicon spills must never be dry swept or shoveled. Such actions will cause the generation of dusts and airborne silicon powders. Since silicon is a common air contaminant, its threat to the surrounding environment is minimal. However, this does not justify careless or unprofessional cleanup methods in the event of a spill.

Silicon is considered a combustible solid in powder form. It is commonly used for a multitude of purposes and environmental contamination is possible through a variety of routes. This means that silicon can unknowingly be present when other types of materials are being transported.

Silicon can enter the environment through runoff, or through accidental spills.

Acute Ecological Effects

Acute (short-term) ecological toxic effects may include the death of animals, birds, or fish and death or low growth rate in plants. Acute effects are seen 2 to 4 days after animals or plants are exposed to silicon.

The data in the references are insufficient to determine if silicon (in any form) poses any acute toxicity hazard to aquatic life, plants, birds, or terrestrial animals.

Chronic Ecological Effects

Chronic (long-term) toxic effects may include shortened life span, reproductive problems, lower fertility, and changes in appearance or behavior in exposed animals. These effects can be seen long after first exposure(s) to toxic chemicals.

Insufficient data are available to evaluate or predict the long-term (chronic) effect of silicon on aquatic life, plants, birds, or land animals.

Water Solubility

Silicon is insoluble in water. Concentrations of less than 1 milligram will not mix with a liter of water.

Persistence in the Environment

Silicon powders can be highly persistent in water, with a half-life greater than 200 days. The half-life of a pollutant is the amount of time it takes for one-half of the chemical to be degraded.

Bioaccumulation in Aquatic Organisms

Some substances increase in concentration, or *bioaccumulate,* in living organisms as they breathe contaminated air, drink contaminated water, or eat contaminated food. These chemicals can become concentrated in the tissues and internal organs of animals as well as humans.

The concentration of silicon found in fish tissues is expected to be about the same as the average concentration of silicon in the water from which the fish was taken.

Recommended Risk-Reduction Measures

Proper training of all transporters will reduce the likelihood of a mishap or accident resulting in a leak or spill of silicon into the environment. The correct use of labeling on all containers, trucks, and rail cars will enable emergency responders to react properly and quickly to any disaster thereby reducing the potential risk to the environment and to personnel.

If a spill or leak to the environment has occurred, hazardous materials spill response personnel should be notified immediately. Cleanup must be attempted only by those trained in proper silicon spill containment procedures. If silicon should contact the water table, aquifer, or navigable waterway, the local and/or state emergency response authorities must be notified. A comprehensive emergency response or disaster preparedness/recovery plan should be in place prior to any operations involving the use, transportation, storage, or disposal of silicon.

If silicon is spilled, the following specific steps are recommended:

- ☑ Restrict persons not wearing protective clothing from area of spill until cleanup is complete.
- ☑ Collect powdered materials in safest manner possible and deposit in sealed containers for disposal. Do NOT dry sweep (generates airborne dusts). Use HEPA vacuum instead. Damp mop residue.

RISK ASSESSMENT: BUSINESS

General Assessment

Accidents or mishaps involving silicon can present a significant threat to business operations. The use of the term silicon can generate serious emotional response from employees as well as the public in surrounding communities. Lawsuits that may result from personnel injury/death, public exposures, and/or environmental contamination will also require a serious expenditure of resources.

Media attention surrounding an injury, death, or environmental damage can also result in a loss of profits and loss of current as well as future business.

Recommended Risk-Reduction Measures

Company attorneys, safety and health professionals, and environmental specialists should be involved in the development of any procedures dealing with safety or emergency response. A company official should be pre-designated as a public relations officer with specific training in dealing with the press. Corporate plans and policies should be developed, approved, and implemented long before any need for such arises.

REFERENCES

American Conference of Governmental Industrial Hygienists (ACGIH). 1988. *Documentation of the Threshold Limit Values and Biological Exposure Indices,* 5th Edition (with updates). Cincinnati: ACGIH

Bloom, A. D. and F. J. DeSerres. 1995. *Ecotoxicity and Human Health.* New York: CRC Press/Lewis Publishers.

Brusick, D. J. 1994. *Methods for Genetic Risk Assessment.* New York: CRC Press/Lewis Publishers.

Bueche, J. R. 1972. *Principles of Physics,* 2nd Edition. New York: McGraw-Hill.

Burrell, R., D. K. Flaherty, and L. J. Sauers. 1992. *Toxicology of the Immune System.* New York: Van Nostrand Reinhold.

Calabrese, E. J. 1994. *Biological Effects of Low Level Exposures.* New York: CRC Press/Lewis Publishers.

Cockerman, L. G., B. S. Shane. 1994. *Basic Environmental Toxicology.* New York: CRC Press/Lewis Publishers.

Cothern, R. R. 1993. *Comparative Environmental Risk Assessment.* New York: CRC Press/Lewis Publishers.

Danse, I. R. 1991. *Common Sense Toxics in the Workplace.* New York: Van Nostrand Reinhold.

Environmental Protection Agency. 1983. *Chemical Hazard Information Profiles.* Washington, D.C.: U.S. EPA.

Fodden, C. R., and J. L. Weddell. 1992. *Hazardous Materials Emergency Action Data.* New York: CRC Press/Lewis Publishers.

Forseberg, K., and L. H. Keith. 1993. *Chemical; Protective Clothing Permeation and Degradation Database.* Chelsea, MI: Lewis Publishers.

Gardner, R. W. 1994. *Chemical Intolerance.* New York: CRC Press/Lewis Publishers.

Hallenbeck, W. H. 1993. *Quantitative Risk Assessment for Environmental and Occupational Health,* 2nd Edition. New York: CRC Press/Lewis Publishers.

Hodgson, E., R. B. Mailman, and Janice E. Chambers. 1988. *Dictionary of Toxicology.* London: The MacMillian Company.

International Agency for Research on Cancer (IARC). 1987. *IARC Monographs on the Evaluation of Carcinogenic Risk to Humans, Overall Evaluations of Carcinogenicity: An updating of IARC Monographs Volumes 1-42.* Lyon: IARC.

Landis, W. G., M. H. Yu. 1995. *Introduction to Environmental Toxicology.* New York: CRC Press/Lewis Publishers.

Langerman, N. 1994 *Precautionary Labels for Chemical Containers.* New York: CRC Press/Lewis Publishers.

Lewis, R. J., Sr. 1993. *Hawley's Condensed Chemical Dictionary,* 12th Edition. New York: Van Nostrand Reinhold.

Lewis, R. J., Sr. 1992. *Sax's Dangerous Properties of Industrial Materials,* Volume 1. 8th Edition. New York: Van Nostrand Reinhold.

Lide, D. R. 1995. *Handbook of Chemistry and Physics,* 76th Edition. New York: CRC Press/Lewis Publishers.

Lisella, F. S. 1994. *The VNR Dictionary of Environmental Health and Safety.* New York: Van Nostrand Reinhold.

Mackay, D., W. Y. Shiu, and K. C. Ma. 1992. *Illustrated Handbook of Chemical-Physical Properties and Environmental Fate for Organic Chemicals.* New York: CRC Press/Lewis Publishers.

Manahan, S. E. 1994. *Environmental Chemistry,* 6th Edition. New York: CRC Press/Lewis Publishers.

Mellström, G. A., J. E. Wahlberg, and H. I. Maibach. 1994. *Protective Gloves for Occupational Use.* New York: CRC Press/Lewis Publishers.

Meyer, E. 1990. *Chemistry of Hazardous Materials.* Englewood Cliffs, NJ: Prentice Hall.

Miller, B. F., M.D., and C. B. Keane, R.N. 1972. *Encyclopedia and Dictionary of Medicine and Nursing.* Philadelphia: W.B. Sanders Company.

Mycock, J. C., J. D. McKenna, and L. Theodore. 1995. *Handbook of Air Pollution Control Engineering and Technology.* New York: CRC Press/Lewis Publishers.

Nelson, G. O. 1992. *Gas Mixtures.* New York: CRC Press/Lewis Publishers.

Ott, W. R. 1995. *Environmental Statistics and Data Analysis.* New York: CRC Press/Lewis Publishers.

Patnaik, P. 1992. *A Comprehensive Guide to the Hazardous Properties of Chemical Substances.* New York: Van Nostrand Reinhold.

Proctor, N. H., J. P. Hughes. 1991. *Chemical Hazards of the Workplace.* New York: Van Nostrand Reinhold.

Quigley, D. R. 1994. *Handbook of Emergency Chemical Management.* New York: CRC Press/Lewis Publishers.

Rea, W. J. 1995. *Chemical Sensitivity,* Vol. I-III. New York: CRC Press/Lewis Publishers.

Reeves, A. L. 1979. *Handbook on the Toxicology of Metals.* New York: Elsevier North-Holland.

Richardson, M. 1995. *Dictionary of Substance and Their Effects,* Vol. I-VI. New York: CRC Press/Lewis Publishers.

Sacarello, H. L. A. 1994. *The Comprehensive Handbook of Hazardous Materials.* New York: CRC Press/Lewis Publishers.

Sherman, J. D. 1988. *Chemical Exposure and Disease.* New York: Van Nostrand Reinhold.

Talbott, E., G. Craun. 1995. *An Introduction to Environmental Epidemiology.* New York: CRC Press/Lewis Publishers.

Torkelson, T. R. V. K. Rowe. 1981. *Patty's Industrial Hygiene and Toxicology,* Vol. 2B. New York: Wiley-Interscience.

U.S. Department of Health and Human Services. National Institute for Occupational Safety and Health (NIOSH). 1984. *Manual of Analytical Methods,* 3rd. Edition. Cincinnati, OH: NIOSH.

U.S. Department of Health and Human Services, National Institute for Occupational Safety and Health. 1994. *NIOSH Pocket Guide to Chemical Hazards.* Washington, D.C.: U.S. Government Printing Office.

MATERIAL SAFETY DATA SHEET

CHEMICAL NAME	CRC PRESS/LEWIS PUBLISHERS DISCLAIMER
SILVER	THE INFORMATION AND RECOMMENDATIONS PRESENTED HEREIN ARE BASED ON SOURCES BELIEVED TO BE RELIABLE. CRC MAKES NO REPRESENTATION ON ITS COMPLETENESS OR ACCURACY. IT IS THE RESPONSIBILITY OF THE USER TO DETERMINE THE CHEMICAL'S SUITABILITY FOR ITS INTENDED USE, THE CHEMICAL'S SAFE USE, AND THE CHEMICAL'S PROPER DISPOSAL. NO REPRESENTATIONS AND/OR WARRANTIES, EITHER EXPRESSED OR IMPLIED, OF THE MERCHANTABILITY OR FITNESS FOR A PARTICULAR PURPOSE, OR OF ANY OTHER NATURE, ARE MADE WITH RESPECT TO THE INFORMATION PROVIDED IN THIS MSDS OR TO THE CHEMICAL TO WHICH INFORMATION MAY REFER. CRC NEITHER ASSUMES NOR AUTHORIZES ANY OTHER PERSON TO ASSUME FOR IT, ANY OTHER ADDITIONAL RESPONSIBILITY OR LIABILITY FOR THE USE OF, OR RELIANCE UPON, THIS INFORMATION.

HAZARD WARNING INFORMATION

HEALTH	FIRE	REACTIVE	OTHER	DEGREE OF HAZARD	COLOR CODING	OTHER CODES
1	0	0		0 = Minimum Hazard 1 = Slight Hazard 2 = Moderate Hazard 3 = Serious Hazard 4 = Severe Hazard	HEALTH = BLUE FIRE = RED REACTIVITY = YELLOW OTHER = WHITE	OX = Oxidizer ACID = Acid ALK = Alkali COR = Corrosive W = Use No Water

SECTION I - GENERAL INFORMATION

Characterization: **Metal**	RCRA Number: **None**	EPA Class: **Not Applicable**
DOT Proper Shipping Name: **No Citation**	Chemical Abstract Service (CAS) Number: **7440-22-4**	
DOT Hazard Class and Label Requirements: **No Citation**	DOT Emergency Guide Code: **No Citation**	
DOT Identification Number: **No Citation**	Chemical Formula: **Ag**	

Synonyms
Silver metal; Argentum.

SECTION II - HAZARDOUS INGREDIENTS AND IDENTITY INFORMATION

Hazardous Components (specific identity)	OSHA Exposure Criteria	NIOSH Exposure Criteria	Immediately Dangerous to Life and Health (IDLH)	ACGIH Exposure Criteria
Silver (derivation: Occurs naturally in sea water and in the earth's crust. Derived as a by-product of operations on copper, zinc, lead, or gold ores. The recovery ranges from 166 ounces to a few thousandths of an ounce per ton. Ores containing silver include argentite, stephanite, polybasite, proustite, pyrargyrite, and cerargyrite).	**PEL (8-hour):** 0.01 mg/m^3 **STEL:** Not Established	**REL (10-hour):** 0.01 mg/m^3 **STEL:** Not Established	10 mg/m^3 as Ag	**TLV:** 0.1 mg/m^3 (metal) 0.01 mg/m^3 (soluble compounds) **STEL:** Not Established

SECTION III - PHYSICAL AND CHEMICAL CHARACTERISTICS

Boiling Point: **4013°F (2212°C)**	Specific Gravity (H_2O = 1): **10.49 (metal)**
Vapor Pressure (mm Hg): **0 at 77°F (25°C)**	Molecular Weight: **63.5**
Vapor Density (Air = 1): **Not Reported**	Melting Point: **1761°F (960°C)**

Solubility
Insoluble in water and most cold acids (except dilute nitric). Soluble in fused alkali hydroxides in the presence of air, fused alkali peroxides, and alkali cyanides with air or oxygen.

Appearance and Odor
Hard, brilliant white, lustrous, ductile, malleable metal solid with no odor.

SECTION IV - FIRE AND EXPLOSION HAZARD DATA

Flash Point (method used): **Not Applicable**	Explosive Limits in Air % by Volume: **LEL: Not Applicable** **UEL: Not Applicable**
NFPA Classification: **Non-Combustible Solid; Flammable Powder**	Autoignition Temperature: **Not Determined**

Extinguishing Media
Use agents suitable to surrounding fires. May require dry chemical appropriate for metal fires.

Special Fire Fighting Procedures
Wear full protective clothing and self-contained breathing apparatus (SCBA). Use a water spray to keep fire-exposed containers cool. Water may be ineffective on fire. Poisonous gases are produced in fire.

Unusual Fire and Explosion Hazards
Can cause fires upon contact with other chemicals (ammonia, chlorine trifluoride, ethyleneimine, oxalic acid, and tartaric acid). Silver is highly flammable when in the form of finely divided powders.

SECTION V - REACTIVITY DATA

Stability		Conditions to Avoid
		Silver is normally stable in closed containers under routine conditions of handling and storage. Keep away from incompatible materials since violent and or explosive reactions can occur.
Stable	Unstable	Incompatibility *(materials to avoid)*
X		Acetylene and acetylene compounds, aziridine, 3-bromopropyne, carboxylic acids, ammonia, hydrogen peroxide, bromoazide, chlorine trifluoride, ethyleneimine, oxalic acid, and tartaric acid.
Hazardous Polymerization		Conditions to Avoid: Under normal conditions of temperature and pressure, hazardous polymerization of silver is not known to occur.
May Occur	Will Not Occur	Hazardous Decomposition or By-products
	X	When heated to decomposition, silver can emit toxic silver fumes.

SECTION VI - HEALTH HAZARD DATA

Primary Route(s) of Entry:	Inhalation?	Absorption (skin)?	Ingestion?
	X		X

Health Hazards

INHALATION: Causes argyria, a bluish-gray pigmentation of the eyes, skin, and mucous membranes caused by deposition of silver particles. This effect can be either local or generalized. Long-term effects may cause respiratory tract irritation and mild chronic bronchitis. It may also cause an increase in lung densities that may appear on X-rays. High exposure may cause pulmonary edema (fluid in the lungs) and possible pulmonary fibrosis.

SKIN & EYES: Metal fragments can also enter small cuts in the skin, forming a permanent tattoo. Eye contact with fragments of silver metal can scratch the surface and form permanent blue-gray stains in the eye. Repeated skin contact may cause contact dermatitis.

INGESTION: Unspecified gastrointestinal effects.

Carcinogenicity	NTP Listed?	IARC Cancer Review Group?	OSHA Regulated?	Target Organs?
Unknown Human Questioned Animal	No	No	29 CFR 1910.1000 Table Z-1	Respiratory system, skin, eyes, nasal septum.

Medical Conditions Generally Aggravated by Exposure

None reported.

Emergency and First-aid Procedures

Eye contact: Flush immediately with water for 15 minutes (minimum); seek medical attention. Skin contact: Remove all contaminated clothing. Immediately wash area with large amounts of soap and water. For inhalation: Remove the person from exposure. Provide respiratory assistance and CPR. Transfer to medical facility. Observe 24-48 hours for lung effects. If swallowed: Never attempt to give a convulsing or unconscious person anything by mouth. If victim is conscious, give 1 to 2 glasses of water and induce vomiting. Seek medical attention immediately.

SECTION VII - PRECAUTIONS FOR SAFE HANDLING AND USE

Steps to be Taken in Case Material is Released or Spilled

Collect powdered material in most convenient manner possible and deposit in sealed drum. Keep those not involved in cleanup from entering area. Do NOT dry sweep (creates airborne dusts). Use a vacuum equipped with a high efficiency particulate air (HEPA) filter.

Preferred Waste Disposal Method

Return silver and silver-containing solutions (from photography) to supplier for reclamation.

Precautions to be Taken in Handling and Storage

Store in tightly closed containers in cool, well-ventilated area and protect from physical damage.

Other Precautions and Warnings

Keep away from incompatible materials.

SECTION VIII - CONTROL MEASURES AND PERSONAL PROTECTIVE EQUIPMENT

Respiratory Protection *(specify type)*

At exposures above PEL, use full facepiece respirator with high-efficiency particulate air (HEPA) filter. For higher exposures, use a supplied-air respirator with a full facepiece or a self-contained breathing apparatus (SCBA) operated in positive pressure mode.

Ventilation

Local exhaust or general mechanical systems recommended.

Protective Gloves	Eye Protection	Other Protective Clothing
Impervious Material	Chemical/Dust Goggles or Face Mask	Protective Suit or Clothing

Work/Hygiene Practices

Always wash hands thoroughly after using chemical; never bring food, drink, or smoking materials into vicinity of chemicals.

SILVER

Ag

CAS: 7440-22-4

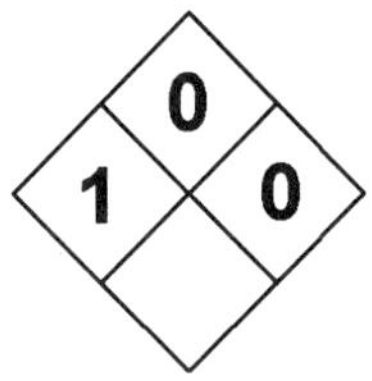

IDENTIFICATION AND TYPICAL USES

Silver is a hard, brilliant white, lustrous, ductile, malleable metallic solid with no odor. It is a precious metal that is used in the manufacture of silver nitrate, silver bromide, in photographic chemicals, for lining vats and other equipment for chemical reaction vessels, and in water distillation and purification. It is used in mirrors, in electric conductors (such as bus bars), in silver plating, in electronic equipment, as a sterlilant, in surgical cements, as a hydration and oxidation catalyst, in special batteries, in solar cells, in reflectors for solar towers, in low-temperature brazing alloys, in table cutlery, in jewelry, in dental and medical equipment and procedures, in scientific equipment, in electrical contacts, in bearing metal, magnet windings, and dental amalgams. Colloidal silver is used as a nucleating agent in photography and in medicine, often combined with protein.

RISK ASSESSMENT: HEALTH

General Assessment

Silver dusts are irritating to the eyes, nose, and respiratory tract. The toxicity of this metal is considered relatively low by all routes of exposure. However, under certain conditions, exposure can result in serious and even life-threatening health risk. At least one source (Lewis, 1992) lists silver as a questionable carcinogen in animals. Its human carcinogenic, mutagenic, and teratogenic properties have not been determined.

The primary effect of human exposure to silver is a condition known as argyria which causes a bluish-gray pigmentation of the eyes, skin, and mucous membranes. This occurs when silver particles become deposited in body tissues. Localized argyria occurs by penetration of fine metallic particles in the skin. Generalized argyria occurs either from inhalation or ingestion of silver salts. Silver in the respiratory tract may result in chronic bronchitis. There may also be an increase in lung densities, fatigue, headache, sinusitis, and weight loss, and possible contact dermatitis. Argyria may occur in the conjunctiva of the eye, gum tissue, or nasal septum. Extremely high exposures can lead to pulmonary edema (fluid in the lungs), which can be fatal. Symptoms of cough, congestion, tightness in the chest, and difficulty breathing may be delayed up to 48 hours, thereby creating a false sense of security with regard to health exposure risk.

Acute Health Effects

The following acute (short-term) health effects may occur immediately or shortly after exposure to silver:

Skin: Irritation, rash, discoloration of the skin to a blue-gray, and possible contact dermatitis.

Eye: Silver particles and dusts can be extremely irritating to the eyes. Exposure can damage vision. There may be a bluish-gray discoloration of the eyes and conjunctiva.

Lung: Inhalation of silver may cause moderate to severe irritation of the respiratory system, especially the lungs and nasal passages. Inhalation can lead to fatigue, headache, sinusitis, and weight loss. High exposures can cause pulmonary edema, which can be fatal.

Chronic Health Effects

The following chronic (long-term) health effects may occur at some time after exposure to silver and can last for months or even years:

Cancer Hazards: Silver is considered a questionable carcinogen in test animals by at least one reference source (Lewis 1992). There is no information regarding its carcinogenic mutagenic properties in humans. Additional research is required in this regard.

Reproductive Hazard: According to information available in the references, silver has not been shown to cause adverse reproductive effects in animals.

Other Chronic Effects: Repeated exposures can cause chronic irritation of the nose and may cause a bluish-gray discoloration of the nasal septa and associated mucosa. Silver may cause a skin allergy. If allergy develops, very low future exposures can cause itching and a skin rash. Kidney, liver, and neurologic damage may also occur, but the references are conflicting and inconclusive in this regard.

Recommended Risk-Reduction Measures

Personnel should avoid direct contact with silver. Even though its overall toxicity is considered relatively low, it has been shown to cause adverse health effects in humans and animals. If a less toxic chemical cannot be substituted for a hazardous substance, then *engineering controls* are the most effective method of reducing exposures. The best protection is to enclose operations and/or provide local exhaust ventilation at the site of chemical release. While not always operationally feasible, isolating operations can also reduce exposure. Using respiratory protection is less effective than the controls mentioned above, but is still advisable whenever working with or around silver. For exposures above the PEL, use a full facepiece respirator equipped with a high efficiency particulate air (HEPA) filter and a dust/mist/fume pre-filter. For higher exposures, use a supplied-air respirator with a full facepiece operated in positive pressure mode or an MSHA/NIOSH-approved self-contained breathing apparatus (SCBA) with full facepiece and pressure demand. If a full facepiece is not available, then chemical/dust goggles should be worn to protect the eyes. Whenever a chemical dust hazard exists, a face shield and a rubber apron should be worn. To prevent hand and skin exposures, impervious rubber gloves should be used.

Administrative controls should also be in place to minimize the potential for human exposures. These may include written procedures or policies which specify the methods and techniques that will be practiced whenever personnel are to work with silver.

All personnel should receive *training* on the use, hazards, protective measures, emergency actions, and other precautions per 29 CFR 1910.1200 (Hazard Communication), prior to the first assignment in an area where silver is used or stored.

If symptoms develop or overexposure is suspected, the following may also be useful:

☑ Examination of the skin, nose, throat, and mucous membranes for discoloration.

☑ Slit lamp examination by an optometrist.

Any evaluation should include a careful history of past and present symptoms with a medical examination. Medical tests that look for damage already done are not a substitute for controlling exposure.

Other methods to reduce exposure include:

☑ Where possible, enclose operations and use local exhaust ventilation at the site of chemical release. If local exhaust ventilation or enclosure is not used, respiratory protection should be mandatory.

☑ Always ensure that proper protective clothing is worn when using chemical substances.

☑ Wash thoroughly immediately after exposure to silver and at the end of the work shift or before eating, drinking, or smoking.

☑ Hazard warning information should be posted in the work area. In addition, as part of an on-going education and training program, all information on the health and safety hazards of silver should be communicated to all potentially exposed workers.

☑ Where possible, automatically transfer silver from storage containers to process containers.

RISK ASSESSMENT: ENVIRONMENT

General Assessment

Silver is a naturally occurring element and is therefore present in the environment in trace amounts. However, the environment is still at risk of exposure during transportation, storage, disposal, or destruction of silver. In almost every scenario, the threat of environmental exposure is contingent upon the proper handling of the chemical substance. Accidental spills, large or small, can result in fire, explosion, and possible contamination of the surrounding environmental mediums (water, soil, and air).

Silver is a noncombustible solid in bulk form but is considered flammable in powdered form. It will ignite and is potentially explosive in contact with acetylene compounds; 3-bromopropyne; ethylene oxide; tartaric acid; and oxalic acid. It also ignites on contact with chlorine trifluoride and other oxidizers. In a fire, silver will emit toxic and poisonous gases. These characteristics require special consideration during any emergency situation involving a leak or spill of silver.

Silver is a naturally occurring element in the Earth's crust. Because of its wide use, silver enters the environment in non-point as well as specific discharges such as industrial and municipal effluents.

Acute Ecological Effects

Acute (short-term) toxic effects may include the death of animals, birds, or fish and death or low growth rate in plants. Acute effects are seen 2 to 4 days after animals or plants are exposed to silver.

The toxicity of silver to aquatic life is probably dependent on water hardness; the harder the water, the higher the concentration of silver needed to be considered toxic. Insufficient data are available to evaluate the short-term effects of silver and its compounds exposure to plants, birds, or land animals.

Chronic Ecological Effects

Chronic (long-term) toxic effects may include shortened life span, reproductive problems, lower fertility, and changes in appearance or behavior in exposed animals. These effects can be seen long after first exposure(s) to toxic chemicals.

Silver and its compounds have high chronic toxicity to aquatic life. Insufficient data are available to evaluate the long-term effects of silver exposure to plants, birds, or land animals.

Water Solubility

Some silver compounds are highly soluble in water while others are nearly insoluble.

Persistence in the Environment

Silver is highly persistent in the aquatic environment, with a half-life of greater than 200 days. The half-life of a pollutant is the amount of time it takes for one half of the chemical to be degraded.

Bioaccumulation in Aquatic Organisms

Some substances increase in concentration, or *bioaccumulate,* in living organisms as they breathe contaminated air, drink contaminated water, or eat contaminated food. These chemicals can become concentrated in the tissues and internal organs of animals as well as humans.

The concentration of silver found in fish is expected to be considerably higher than the average concentration of silver in the water from which the fish was taken.

Recommended Risk-Reduction Measures

Proper training of all transporters will reduce the likelihood of a mishap or accidents resulting in a leak or spill to the environment. The correct use of DOT labeling (per the MSDS) on all containers, trucks, and rail cars will enable emergency responders to react properly and quickly to any disaster thereby reducing the potential risk to the environment and to personnel.

If a spill or leak to the environment has occurred, fire department, emergency response and/or hazardous materials spill personnel should be notified immediately. Cleanup should be attempted only by those trained in proper spill containment procedures. Contaminated soils may need to be removed and replaced with clean soil. If silver should contact the water table, aquifer, or navigable waterway, time is of the essence. The solubility of silver compounds in water may range from insoluble to highly soluble and, therefore, total containment and remediation may not be entirely possible. When such spills occur, the local and/or state emergency response authorities must be notified. A comprehensive emergency response or disaster preparedness/recovery plan should be in place prior to any operations involving the use, transportation, storage, or disposal of silver.

If silver is spilled or leaked, the following specific steps are recommended:

- ☑ Restrict persons not wearing protective clothing from area of spill or leak until cleanup is complete and area can be opened for normal work.
- ☑ Collect powdered materials in most convenient and safe manner possible and place in sealed drums for reclamation or disposal. Do not dry sweep; use HEPA vacuum instead.
- ☑ It may be necessary to dispose of silver as a hazardous waste. The responsible state agency or the regional office of the federal Environmental Protection Agency (EPA) should be contacted for specific recommendations.

RISK ASSESSMENT: BUSINESS

General Assessment

Accidents or mishaps involving silver can present a significant threat to business operations. The loss or damage of equipment or facilities can significantly affect fiscal viability. Lawsuits that may result from personnel injury/death, public exposures, and/or environmental contamination will also require a serious expen-

diture of resources. Media attention surrounding an injury, death, or environmental damage can also result in a loss of profits and loss of current as well as future business.

Recommended Risk-Reduction Measures

Company attorneys, safety and health professionals, and environmental specialists should be involved in the development of any procedures dealing with safety and emergency response. A company official should be pre-designated as a public relations officer with specific training in dealing with the press. Corporate plans and policies should be developed, approved, and implemented long before any need for such arises.

REFERENCES

Ahlbom, A. 1993. *Biostatistics for Epidemiologists.* New York: CRC Press/Lewis Publishers.

American Conference of Governmental Industrial Hygienists. 1988. *Documentation of the Threshold Limit Values and Biological Exposure Indices,* 5th Edition (with updates). Cincinnati: ACGIH

Asante-Duah, D. K. 1993. *Hazardous Waste Risk Assessment.* New York: CRC Press/Lewis Publishers.

Bloom, A. D. and F. J. DeSerres. 1995. *Ecotoxicity and Human Health.* New York: CRC Press/Lewis Publishers.

Brusick, D. J. 1994. *Methods for Genetic Risk Assessment.* New York: CRC Press/Lewis Publishers.

Calabrese, E. J., E. M. Kenyon. *Air Toxics and Risk Assessment.* New York: CRC Press/Lewis Publishers.

Calabrese, E. J. 1994. *Biological Effects of Low Level Exposures.* New York: CRC Press/Lewis Publishers.

Cockerman, L. G., B. S. Shane. 1994. *Basic Environmental Toxicology.* New York: CRC Press/Lewis Publishers.

Cothern, C. R., N. P. Ross. 1994. *Environmental Statistics, Assessment, and Forecasting.* New York: CRC Press/Lewis Publishers.

Cothern, R. R. 1993. *Comparative Environmental Risk Assessment.* New York: CRC Press/Lewis Publishers.

Cothern, C. R. 1995. *Handbook for Environmental Risk Decisionmaking.* New York: CRC Press/Lewis Publishers.

Danse, I. R. 1991. *Common Sense Toxics in the Workplace.* New York: Van Nostrand Reinhold.

Environmental Protection Agency. 1983. *Chemical Hazard Information Profiles.* Washington, D.C.: U.S. EPA.

Fodden, C. R., and J. L. Weddell. 1992. *Hazardous Materials Emergency Action Data.* New York: CRC Press/Lewis Publishers.

Forseberg, K., and L. H. Keith. 1993. *Chemical; Protective Clothing Permeation and Degradation Database.* Chelsea, MI: Lewis Publishers.

Gardner, R. W. 1994. *Chemical Intolerance.* New York: CRC Press/Lewis Publishers.

Hallenbeck, W. H. 1993. *Quantitative Risk Assessment for Environmental and Occupational Health,* 2nd Edition. New York: CRC Press/Lewis Publishers.

Howard, P. H., et. al. 1991. *Handbook of Environmental Degradation Rates.* New York: CRC Press/Lewis Publishers.

Howard, P. H. 1995. *Handbook of Environmental Fate and Exposures Data for Organic Chemicals,* Vol. I-IV. New York: CRC Press/Lewis Publishers.

Jones, F. E. 1994. *Toxic Organic Vapors in the Workplace.* New York: CRC Press/Lewis Publishers.

Keith, L. H. and M. M. Walker. 1995. *Handbook of Air Toxics.* New York: CRC Press/Lewis Publishers.

Landis, W. G., M. H. Yu. 1995. *Introduction to Environmental Toxicology.* New York: CRC Press/Lewis Publishers.

Langerman, N. 1994 *Precautionary Labels for Chemical Containers.* New York: CRC Press/Lewis Publishers.

Lauwerys, R. R., P. Hoet. 1993. *Industrial Chemical Exposure.* New York: CRC Press/Lewis Publishers.

Lewis, R. J., Sr. 1992. *Sax's Dangerous Properties of Industrial Materials,* Volumes 1, 2, and 3, 8th Edition. New York: Van Nostrand Reinhold.

Mackay, D., W. Y. Shiu, and K. C. Ma. 1992. *Illustrated Handbook of Chemical-Physical Properties and Environmental Fate for Organic Chemicals.* New York: CRC Press/Lewis Publishers.

Manahan, S. E. 1994. *Environmental Chemistry,* 6th Edition. New York: CRC Press/Lewis Publishers.

Manahan, S. E. 1992. *Toxicological Chemistry,* 2nd Edition. New York: CRC Press/Lewis Publishers.

Meyer, E. 1990. *Chemistry of Hazardous Materials.* Englewood Cliffs, NJ: Prentice Hall.

Patnaik, P. 1992. *A Comprehensive Guide to the Hazardous Properties of Chemical Substances.* New York: Van Nostrand Reinhold.

Quigley, D. R. 1994. *Handbook of Emergency Chemical Management.* New York: CRC Press/Lewis Publishers.

Rea, W. J. 1995. *Chemical Sensitivity,* Vol. I-III. New York: CRC Press/Lewis Publishers.

Sacarello, H. L. A. 1994. *The Comprehensive Handbook of Hazardous Materials.* New York: CRC Press/Lewis Publishers.

Sherman, J. D. 1988. *Chemical Exposure and Disease.* New York: Van Nostrand Reinhold.

Talbott, E., G. Craun. 1995. *An Introduction to Environmental Epidemiology.* New York: CRC Press/Lewis Publishers.

U.S. Department of Health and Human Services (DHHS), National Institute for Occupational Health (NIOHS). 1994. *NIOSH Pocket Guide to Hazardous Chemicals.* Publication Number 94-116. Washington, D.C.; U.S. Government Printing Office.

MATERIAL SAFETY DATA SHEET

CHEMICAL NAME	
SILVER NITRATE	**CRC PRESS/LEWIS PUBLISHERS DISCLAIMER:** THE INFORMATION AND RECOMMENDATIONS PRESENTED HEREIN ARE BASED ON SOURCES BELIEVED TO BE RELIABLE. CRC MAKES NO REPRESENTATION ON ITS COMPLETENESS OR ACCURACY. IT IS THE RESPONSIBILITY OF THE USER TO DETERMINE THE CHEMICAL'S SUITABILITY FOR ITS INTENDED USE, THE CHEMICAL'S SAFE USE, AND THE CHEMICAL'S PROPER DISPOSAL. NO REPRESENTATIONS AND/OR WARRANTIES, EITHER EXPRESSED OR IMPLIED, OF THE MERCHANTABILITY OR FITNESS FOR A PARTICULAR PURPOSE, OR OF ANY OTHER NATURE, ARE MADE WITH RESPECT TO THE INFORMATION PROVIDED IN THIS MSDS OR TO THE CHEMICAL TO WHICH INFORMATION MAY REFER. CRC NEITHER ASSUMES NOR AUTHORIZES ANY OTHER PERSON TO ASSUME FOR IT, ANY OTHER ADDITIONAL RESPONSIBILITY OR LIABILITY FOR THE USE OF, OR RELIANCE UPON, THIS INFORMATION.

HAZARD WARNING INFORMATION

HEALTH	FIRE	REACTIVE	OTHER	DEGREE OF HAZARD	COLOR CODING	OTHER CODES
1	**0**	**0**	**OX**	0 = Minimum Hazard 1 = Slight Hazard 2 = Moderate Hazard 3 = Serious Hazard 4 = Severe Hazard	HEALTH = BLUE FIRE = RED REACTIVITY = YELLOW OTHER = WHITE	OX = Oxidizer ACID = Acid ALK = Alkali COR = Corrosive W = Use No Water

SECTION I - GENERAL INFORMATION

Characterization	RCRA Number	EPA Class
Silver Compound	**None**	**Not Applicable**

DOT Proper Shipping Name **Silver Nitrate**	Chemical Abstract Service (CAS) Number **7761-88-8**
DOT Hazard Class and Label Requirements **Oxidizer**	DOT Emergency Guide Code **45**
DOT Identification Number **UN 1493**	Chemical Formula **$AgNO_3$**

Synonyms
Lunar caustic; nitric acid, silver(1+)salt; silver(1+)nitrate.

SECTION II - HAZARDOUS INGREDIENTS AND IDENTITY INFORMATION

Hazardous Components (specific identity)	OSHA Exposure Criteria	NIOSH Exposure Criteria	Immediately Dangerous to Life and Health (IDLH)	ACGIH Exposure Criteria
Silver nitrate (derivation: Silver is dissolved in dilute nitric acid and the solution is evaporated. The residue is heated to a dull red heat to decompose any copper nitrate, dissolved in water, filtered, and recrystallized).	**PEL (8-hour):** 0.01 mg/m^3 (as silver) **STEL:** Not Established	**REL:** Not Established **STEL:** Not Established	10 mg/m^3 as Ag	**TLV:** 0.1 mg/m^3 (as silver metal) 0.01 mg/m^3 (soluble compounds)

SECTION III - PHYSICAL AND CHEMICAL CHARACTERISTICS

Boiling Point **824°F (440°C) Decomposes**	Specific Gravity (H_2O = 1) **4.35**
Vapor Pressure (mm Hg) **0 (approx.)at 77°F (25°C)**	Molecular Weight **169.88**
Vapor Density (Air = 1) **Not Reported**	Melting Point **414°F (212°C)**

Solubility
Soluble in cold water, more soluble in hot water. Soluble in glycerol, and hot alcohol. Slightly soluble in ether.

Appearance and Odor
Colorless, transparent, tabular, rhombic, odorless crystals that become gray or gray-black when exposed to light or in the presence of organic matter.

SECTION IV - FIRE AND EXPLOSION HAZARD DATA

Flash Point (method used) **Not Applicable**	Explosive Limits in Air % by Volume **LEL: Not Applicable UEL: Not Applicable**
NFPA Classification **Non-Combustible Solid; Flammable Powder**	Autoignition Temperature **Not Determined**

Extinguishing Media
Silver nitrate does not burn. Use water or water spray on small fires. Do NOT use dry chemical or carbon dioxide. For large fires flood the area with water.

Special Fire Fighting Procedures
Wear full protective clothing and self-contained breathing apparatus (SCBA). Use a water spray to keep fire-exposed containers cool. Poisonous gases are produced in fire.

Unusual Fire and Explosion Hazards
A strong oxidizer that can enhance the flammability of combustible materials.

SECTION V - REACTIVITY DATA

Stability		Conditions to Avoid **Silver nitrate is normally stable in closed containers under routine conditions of handling and storage. Keep away from incompatible materials since violent and or explosive reactions can occur.**
Stable	Unstable	Incompatibility *(materials to avoid)*
X		**Acetylene and acetylene compounds, ammonium hydroxide, aziridine, calcium carbide, ethanol, ethanol and nitric acid, ethyleneimine, 1-bromo-2-propyne, charcoal, chlorine trifluoride, and ethylene oxide.**
Hazardous Polymerization		Conditions to Avoid **Under normal conditions of temperature and pressure, hazardous polymerization of silver nitrate is not known to occur.**
May Occur	Will Not Occur	Hazardous Decomposition or By-products
	X	**When heated to decomposition, silver nitrate can emit toxic silver and nitrogen oxides.**

SECTION VI - HEALTH HAZARD DATA

Primary Route(s) of Entry:	Inhalation?	Absorption (skin)?	Ingestion?
	X	X	

Health Hazards

INHALATION: Causes shortness of breath, coughing, and sore throat. It can also cause generalized argyria with discoloration of the eyes and skin to a grayish-blue. The fingernails may change to a chocolate brown color. Silver nitrate exposure can lead to the development of methemoglobinemia.

SKIN & EYES: Contact with skin, especially moist skin, can cause serious burns. There can also be localized pigmentation changes.

INGESTION: Severe gastroenteritis with gastric bleeding, possible shock, and death.

Carcinogenicity	NTP Listed?	IARC Cancer Review Group?	OSHA Regulated?	Target Organs?
Unknown Human Questioned Animal	No	No	29 CFR 1910.1000 Table Z-1	Eyes, skin, nasal septum, throat, fingernails.

Medical Conditions Generally Aggravated by Exposure

None reported.

Emergency and First-aid Procedures

Eye contact: Flush immediately with water for 15 minutes (minimum), seek medical attention. **Skin contact:** Remove all contaminated clothing. Immediately wash area with large amounts of soap and water. Treat for skin burns. For **inhalation:** Remove the person from exposure. Provide respiratory assistance and CPR. Transfer to medical facility. If **swallowed:** Never attempt to give a convulsing or unconscious person anything by mouth. If victim is conscious, give 4 to 8 ounces of milk or water. Do NOT induce vomiting. Seek medical attention immediately.

SECTION VII - PRECAUTIONS FOR SAFE HANDLING AND USE

Steps to be Taken in Case Material is Released or Spilled

Collect powdered material in most convenient manner possible and deposit in sealed drum. Keep those not involved in cleanup from entering area. Do NOT dry sweep (creates airborne dusts). Carefully scoop or use a vacuum equipped with a high-efficiency particulate air (HEPA) filter.

Preferred Waste Disposal Method

No citation

Precautions to be Taken in Handling and Storage

Store in tightly closed containers in cool, well-ventilated area and protect from physical damage.

Other Precautions and Warnings

Keep away from incompatible materials.

SECTION VIII - CONTROL MEASURES AND PERSONAL PROTECTIVE EQUIPMENT

Respiratory Protection *(specify type)*

At exposures above PEL, use full facepiece respirator with high-efficiency particulate air (HEPA) filter. For higher exposures, use a supplied-air respirator with a full facepiece or a self-contained breathing apparatus (SCBA) operated in positive pressure mode.

Ventilation

Local exhaust or general mechanical systems recommended.

Protective Gloves	Eye Protection	Other Protective Clothing
Impervious Material	Chemical/Dust Goggles or Face Mask	Protective Suit or Clothing

Work/Hygiene Practices

Always wash hands thoroughly after using chemical; never bring food, drink, or smoking materials into vicinity of chemicals.

SILVER NITRATE

$AgNO_3$

CAS: 7761-88-8

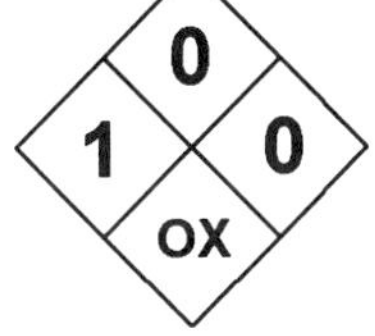

IDENTIFICATION AND TYPICAL USES

Silver nitrate appears as large, colorless, transparent, odorless crystals. It is used in photographic film, as a catalyst for ethylene oxide, in indelible inks, in silver plating, in silver salts, for silvering mirrors, in germicides, in hair dyeing solutions, as an antiseptic, in fused form to cauterize wounds, and as a laboratory reagent.

RISK ASSESSMENT: HEALTH

General Assessment

Silver nitrate dusts and solutions are very irritating. It is toxic by *ingestion, inhalation,* and by dermal *absorption.* It is a questionable carcinogen in test animals. Human mutation data have been documented, and it has been shown to adversely affect reproduction in experimental tests.

Inhalation or ingestion can lead to generalized argyria characterized by a bluish-gray pigmentation or color change that may be distributed over the face, forehead, neck, hands, and forearms. The fingernails may change in color to a dark brown. Other symptoms of inhalation may be shortness of breath, coughing, and sore throat. Chronic exposure may lead to methemoglobinemia.

Skin contact can cause serious burns, especially if the skin is wet or moist (sweat) There can also be localized pigmentation or color changes to a blue-gray if silver nitrate particles become imbedded in the skin. Argyria may occur in the conjunctiva of the eye, gum tissue, or nasal septum. Eye contact can cause burns and permanent blindness.

Ingestion can also cause severe gastroenteritis with gastric bleeding and possibly fatal shock.

Acute Health Effects

The following acute (short-term) health effects may occur immediately or shortly after exposure to silver nitrate:

Skin: Irritation, rash, discoloration of the skin to a blue-gray, and possible contact dermatitis.

Eye: Silver nitrate particles and dusts can be extremely irritating to the eyes. Exposure can damage vision. There may be a bluish-gray discoloration of the eyes and conjunctiva.

Lung: Inhalation of silver nitrate may cause moderate to severe irritation of the respiratory system, especially the lungs and nasal passages. Inhalation can lead to fatigue, headache, sinusitis, and weight loss.

Other: Ingestion can lead to fatal gastroenteritis or shock.

Chronic Health Effects

The following chronic (long-term) health effects may occur at some time after exposure to silver nitrate and can last for months or even years:

Cancer Hazards: Silver nitrate is considered a questionable carcinogen in test animals by at least one reference source (Lewis 1992). Human mutation data have been reported and many scientists believe that long-term exposure to such chemicals may pose a cancer risk.

Reproductive Hazard: According to information available in the references, silver nitrate has been shown to cause adverse reproductive effects in animals. It is not known if human exposure will have this effect.

Other Chronic Effects: Long-term exposure to silver nitrate dusts or fumes can very gradually cause eyes, nails, inner nose, throat, body organs, and skin to turn a bluish-gray color. This can take between 2 and 20 years to develop but is permanent and can be very disfiguring. High, repeated exposure may cause kidney damage. Very irritating substances can injure the lungs. It is not known if silver nitrate causes lung damage over the long-term.

Recommended Risk-Reduction Measures

Personnel should avoid direct contact with silver nitrate. If a less toxic chemical cannot be substituted for a hazardous substance, then *engineering controls* are the most effective method of reducing exposures. The best protection is to enclose operations and/or provide local exhaust ventilation at the site of chemical release. While not always operationally feasible, isolating operations can also reduce exposure. Using respiratory protection is less effective than the controls mentioned above, but is still advisable whenever working with or around silver nitrate. For low or infrequent exposures, use a full facepiece respirator equipped with a high-efficiency particulate air (HEPA) filter and a dust/mist pre-filter. For higher exposures, use a supplied-air respirator with a full facepiece operated in positive pressure mode or an MSHA/NIOSH-approved self-contained breathing apparatus (SCBA) with full facepiece operated in pressure demand. If a full facepiece is not available, then chemical/dust goggles should be worn to protect the eyes. Whenever a chemical dust hazard exists, a face shield and a rubber apron should be worn. To prevent hand and skin exposures, impervious rubber gloves should be used. Glove manufacturers should be contacted and permeation studies obtained *before* final glove selections have been made.

Administrative controls should also be in place to minimize the potential for human exposures. These may include written procedures or policies which specify the methods and techniques that will be practiced whenever personnel are to work with silver nitrate.

All personnel should receive *training* on the use, hazards, protective measures, emergency actions, and other precautions per 29 CFR 1910.1200 (Hazard Communication), prior to the first assignment in an area where silver nitrate is used or stored.

If symptoms develop or overexposure is suspected, the following may also be useful:

- ☑ Examination of the skin, nose, throat, and mucous membranes for discoloration.
- ☑ Slit lamp examination by an optometrist.
- ☑ Kidney function tests.

Any evaluation should include a careful history of past and present symptoms with a medical examination. Medical tests that look for damage already done are not a substitute for controlling exposure.

Other methods to reduce exposure include:

- ☑ Where possible, enclose operations and use local exhaust ventilation at the site of chemical release. If local exhaust ventilation or enclosure is not used, respiratory protection should be mandatory.
- ☑ Always ensure that proper protective clothing is worn when using chemical substances and that personnel have been trained in its proper use, care, and maintenance.
- ☑ Wash thoroughly immediately after exposure to silver nitrate and at the end of the work shift or before eating, drinking, or smoking.
- ☑ Hazard warning information should be posted in the work area. In addition, as part of an on-going education and training program, all information on the health and safety hazards of silver nitrate should be communicated to all exposed and potentially exposed workers.
- ☑ Where possible, automatically transfer silver nitrate from drums or storage containers to process containers.

RISK ASSESSMENT: ENVIRONMENT

General Assessment

The environment is still at risk of exposure during transportation, storage, disposal, or destruction of silver nitrate. In almost every scenario, the threat of environmental exposure is contingent upon the proper handling of the chemical substance. Accidental spills, large or small, can result in fire, explosion, and possible contamination of the surrounding environmental mediums (water, soil, and air).

Silver nitrate does not burn but it is an extremely powerful oxidizer that can significantly enhance the burning rate of any combustible materials it comes in contact with. It will ignite and is potentially explosive in contact with acetylene compounds, ammonium hydroxide, calcium carbide, charcoal, aziridine, ethanol with nitric acid; oxalic acid, tartaric acid, peroxymonosulfuric acid, phosphine, phosphonium iodide, ethylene oxide, ethylene hydroperoxide, hydrogen peroxide, magnesium arsenic, and cuprous acetylide. It also ignites on contact with chlorine trifluoride and other oxidizers. In a fire, silver nitrate will emit toxic and poisonous gases. These characteristics require special consideration during any emergency situation involving a leak or spill of silver nitrate.

Silver is a naturally occurring element in the Earth's crust. Because of its wide use, silver nitrate enters the environment in non-point as well as specific discharges such as industrial and municipal effluents.

Acute Ecological Effects

Acute (short-term) toxic effects may include the death of animals, birds, or fish and death or low growth rate in plants. Acute effects are seen 2 to 4 days after animals or plants are exposed to silver nitrate.

The toxicity of silver to aquatic life is probably dependent on water hardness; the harder the water, the higher the concentration of silver needed to be considered toxic. Insufficient data are available to evaluate the short-term effects of silver and its compounds exposure to plants, birds, or land animals.

Chronic Ecological Effects

Chronic (long-term) toxic effects may include shortened life span, reproductive problems, lower fertility, and changes in appearance or behavior in exposed animals. These effects can be seen long after first exposure(s) to toxic chemicals.

Silver and its compounds have high chronic toxicity to aquatic life. Insufficient data are available to evaluate the long-term effects of silver exposure to plants, birds, or land animals.

Water Solubility

Some silver compounds are highly soluble in water while others are nearly insoluble. Silver nitrate is considered completely soluble in water.

Persistence in the Environment

Silver is highly persistent in the aquatic environment, with a half-life of greater than 200 days. The half-life of a pollutant is the amount of time it takes for one half of the chemical to be degraded.

Bioaccumulation in Aquatic Organisms

Some substances increase in concentration, or *bioaccumulate,* in living organisms as they breathe contaminated air, drink contaminated water, or eat contaminated food. These chemicals can become concentrated in the tissues and internal organs of animals as well as humans.

The concentration of silver found in fish is expected to be considerably higher than the average concentration of silver in the water from which the fish was taken.

Recommended Risk-Reduction Measures

Proper training of all transporters will reduce the likelihood of a mishap or accidents resulting in a leak or spill to the environment. The correct use of DOT labeling (per the MSDS) on all containers, trucks, and rail cars will enable emergency responders to react properly and quickly to any disaster thereby reducing the potential risk to the environment and to personnel.

If a spill or leak to the environment has occurred, fire department, emergency response and/or hazardous materials spill personnel should be notified immediately. Cleanup should be attempted only by those trained in proper spill containment procedures. Contaminated soils may need to be removed and replaced with clean soil. If silver nitrate should contact the water table, aquifer, or navigable waterway, time is of the essence. The solubility of silver compounds in water may range from insoluble to highly soluble and, therefore, total containment and remediation may not be entirely possible. The solubility of silver nitrate is expected to be complete. When such spills occur, the local and/or state emergency response authorities must be notified. A comprehensive emergency response or disaster preparedness/recovery plan should be in place prior to any operations involving the use, transportation, storage, or disposal of silver nitrate.

If silver nitrate is spilled or leaked, the following specific steps are recommended:

- ☑ Restrict persons not wearing protective clothing from area of spill or leak until cleanup is complete and area can be opened for normal work.
- ☑ Collect powdered materials in most convenient and safe manner possible and place in sealed drums for reclamation or disposal. Do not dry sweep, use HEPA vacuum instead.
- ☑ It may be necessary to dispose of silver nitrate as a hazardous waste. The responsible state agency or the regional office of the federal Environmental Protection Agency (EPA) should be contacted for specific recommendations.

RISK ASSESSMENT: BUSINESS

General Assessment

Accidents or mishaps involving silver nitrate can present a significant threat to business operations. The loss or damage of equipment or facilities can significantly affect fiscal viability. Lawsuits that may result from personnel injury/death, public exposures, and/or environmental contamination will also require a serious expenditure of resources. Media attention surrounding an injury, death, or environmental damage can also result in a loss of profits and loss of current as well as future business. Always remember that anytime the terms "cancer," "carcinogen," or "reproductive hazard" are used, public hysteria, ignorance, anxiety, and emotion can run equally high. This must be carefully considered whenever drafting or implementing public relations policies.

❶ *Recommended Risk-Reduction Measures*

Company attorneys, safety and health professionals, and environmental specialists should be involved in the development of any procedures dealing with safety and emergency response. A company official should be pre-designated as a public relations officer with specific training in dealing with the press. Corporate plans and policies should be developed, approved, and implemented long before any need for such arises.

REFERENCES

Ahlbom, A. 1993. *Biostatistics for Epidemiologists.* New York: CRC Press/Lewis Publishers.

American Conference of Governmental Industrial Hygienists. 1988. *Documentation of the Threshold Limit Values and Biological Exposure Indices,* 5th Edition (with updates). Cincinnati: ACGIH

Asante-Duah, D. K. 1993. *Hazardous Waste Risk Assessment.* New York: CRC Press/Lewis Publishers.

Bloom, A. D. and F. J. DeSerres. 1995. *Ecotoxicity and Human Health.* New York: CRC Press/Lewis Publishers.

Brusick, D. J. 1994. *Methods for Genetic Risk Assessment.* New York: CRC Press/Lewis Publishers.

Calabrese, E. J., E. M. Kenyon. *Air Toxics and Risk Assessment.* New York: CRC Press/Lewis Publishers.

Calabrese, E. J. 1994. *Biological Effects of Low Level Exposures.* New York: CRC Press/Lewis Publishers.

Cockerman, L. G., B. S. Shane. 1994. *Basic Environmental Toxicology.* New York: CRC Press/Lewis Publishers.

Cothern, C. R., N. P. Ross. 1994. *Environmental Statistics, Assessment, and Forecasting.* New York: CRC Press/Lewis Publishers.

Cothern, R. R. 1993. *Comparative Environmental Risk Assessment.* New York: CRC Press/Lewis Publishers.

Cothern, C. R. 1995. *Handbook for Environmental Risk Decisionmaking.* New York: CRC Press/Lewis Publishers.

Danse, I. R. 1991. *Common Sense Toxics in the Workplace.* New York: Van Nostrand Reinhold.

Environmental Protection Agency. 1983. *Chemical Hazard Information Profiles.* Washington, D.C.: U.S. EPA.

Fodden, C. R., and J. L. Weddell. 1992. *Hazardous Materials Emergency Action Data.* New York: CRC Press/Lewis Publishers.

Forseberg, K., and L. H. Keith. 1993. *Chemical; Protective Clothing Permeation and Degradation Database.* Chelsea, MI: Lewis Publishers.

Gardner, R. W. 1994. *Chemical Intolerance.* New York: CRC Press/Lewis Publishers.

Hallenbeck, W. H. 1993. *Quantitative Risk Assessment for Environmental and Occupational Health,* 2nd Edition. New York: CRC Press/Lewis Publishers.

Howard, P. H., et. al. 1991. *Handbook of Environmental Degradation Rates.* New York: CRC Press/Lewis Publishers.

Howard, P. H. 1995. *Handbook of Environmental Fate and Exposures Data for Organic Chemicals,* Vol. I-IV. New York: CRC Press/Lewis Publishers.

Jones, F. E. 1994. *Toxic Organic Vapors in the Workplace.* New York: CRC Press/Lewis Publishers.

Keith, L. H. and M. M. Walker. 1995. *Handbook of Air Toxics.* New York: CRC Press/Lewis Publishers.

Landis, W. G., M. H. Yu. 1995. *Introduction to Environmental Toxicology.* New York: CRC Press/Lewis Publishers.

Langerman, N. 1994 *Precautionary Labels for Chemical Containers.* New York: CRC Press/Lewis Publishers.

Lauwerys, R. R., P. Hoet. 1993. *Industrial Chemical Exposure.* New York: CRC Press/Lewis Publishers.

Lewis, R. J., Sr. 1992. *Sax's Dangerous Properties of Industrial Materials,* Volumes 1, 2, and 3, 8th Edition. New York: Van Nostrand Reinhold.

Mackay, D., W. Y. Shiu, and K. C. Ma. 1992. *Illustrated Handbook of Chemical-Physical Properties and Environmental Fate for Organic Chemicals.* New York: CRC Press/Lewis Publishers.

Manahan, S. E. 1994. *Environmental Chemistry,* 6th Edition. New York: CRC Press/Lewis Publishers.

Meyer, E. 1990. *Chemistry of Hazardous Materials.* Englewood Cliffs, NJ: Prentice Hall.

Patnaik, P. 1992. *A Comprehensive Guide to the Hazardous Properties of Chemical Substances.* New York: Van Nostrand Reinhold.

Quigley, D. R. 1994. *Handbook of Emergency Chemical Management.* New York: CRC Press/Lewis Publishers.

Rea, W. J. 1995. *Chemical Sensitivity,* Vol. I-III. New York: CRC Press/Lewis Publishers.

Sacarello, H. L. A. 1994. *The Comprehensive Handbook of Hazardous Materials.* New York: CRC Press/Lewis Publishers.

Sherman, J. D. 1988. *Chemical Exposure and Disease.* New York: Van Nostrand Reinhold.

Talbott, E., G. Craun. 1995. *An Introduction to Environmental Epidemiology.* New York: CRC Press/Lewis Publishers.

U.S. Department of Health and Human Services (DHHS), National Institute for Occupational Health (NIOHS). 1994. *NIOSH Pocket Guide to Hazardous Chemicals.* Publication Number 94-116. Washington, D.C.; U.S. Government Printing Office.

MATERIAL SAFETY DATA SHEET

CHEMICAL NAME	
SILVER OXIDE	**CRC PRESS/LEWIS PUBLISHERS DISCLAIMER:** THE INFORMATION AND RECOMMENDATIONS PRESENTED HEREIN ARE BASED ON SOURCES BELIEVED TO BE RELIABLE. CRC MAKES NO REPRESENTATION ON ITS COMPLETENESS OR ACCURACY. IT IS THE RESPONSIBILITY OF THE USER TO DETERMINE THE CHEMICAL'S SUITABILITY FOR ITS INTENDED USE, THE CHEMICAL'S SAFE USE, AND THE CHEMICAL'S PROPER DISPOSAL. NO REPRESENTATIONS AND/OR WARRANTIES, EITHER EXPRESSED OR IMPLIED, OF THE MERCHANTABILITY OR FITNESS FOR A PARTICULAR PURPOSE, OR OF ANY OTHER NATURE, ARE MADE WITH RESPECT TO THE INFORMATION PROVIDED IN THIS MSDS OR TO THE CHEMICAL TO WHICH INFORMATION MAY REFER. CRC NEITHER ASSUMES NOR AUTHORIZES ANY OTHER PERSON TO ASSUME FOR IT, ANY OTHER ADDITIONAL RESPONSIBILITY OR LIABILITY FOR THE USE OF, OR RELIANCE UPON, THIS INFORMATION.

HAZARD WARNING INFORMATION

HEALTH	FIRE	REACTIVE	OTHER	DEGREE OF HAZARD	COLOR CODING	OTHER CODES
1	0	0	OX	0 = Minimum Hazard 1 = Slight Hazard 2 = Moderate Hazard 3 = Serious Hazard 4 = Severe Hazard	HEALTH = BLUE FIRE = RED REACTIVITY = YELLOW OTHER = WHITE	OX = Oxidizer ACID = Acid ALK = Alkali COR = Corrosive W = Use No Water

SECTION I - GENERAL INFORMATION

Characterization	RCRA Number	EPA Class
Silver Compound	**None**	**Not Applicable**

DOT Proper Shipping Name **No Citation**	Chemical Abstract Service (CAS) Number **20667-12-3**
DOT Hazard Class and Label Requirements **No Citation**	DOT Emergency Guide Code **No Citation**
DOT Identification Number **No Citation**	Chemical Formula **Ag_2O**

Synonyms

Silver (II) oxide; silver(1+)oxide; Divasil; disilver oxide.

SECTION II - HAZARDOUS INGREDIENTS AND IDENTITY INFORMATION

Hazardous Components (specific identity)	OSHA Exposure Criteria	NIOSH Exposure Criteria	Immediately Dangerous to Life and Health (IDLH)	ACGIH Exposure Criteria
Silver oxide (derivation: Silver nitrate and alkali hydroxide solutions are mixed and the precipitate is filtered and then washed).	**PEL (8-hour):** **0.01 mg/m³** **(as silver)** **STEL:** **Not Established**	**REL:** **0.01 mg/m³** **Established** **STEL:** **Not Established**	**10 mg/m³** **as Ag**	**TLV:** **0.1 mg/m³** **(as silver metal)** **0.01 mg/m³** **(soluble compounds)**

SECTION III - PHYSICAL AND CHEMICAL CHARACTERISTICS

Boiling Point **392°F (200°C) Decomposes**	Specific Gravity (H_2O = 1) **7.22**
Vapor Pressure (mm Hg) **Not Reported**	Molecular Weight **231.74**
Vapor Density (Air = 1) **Not Reported**	Melting Point **392°F (200°C) Undergoes Rapid Decomposition**

Solubility

Slightly soluble in water. Insoluble in alcohol. Soluble in nitric acid, potassium cyanide solution, sodium thiosulfate solution, and ammonia hydroxide.

Appearance and Odor

A brownish-black, heavy, odorless powder. Metallic taste.

SECTION IV - FIRE AND EXPLOSION HAZARD DATA

Flash Point (method used) **Not Applicable**	Explosive Limits in Air % by Volume **LEL: Not Applicable** **UEL: Not Applicable**
NFPA Classification **Non-Combustible Solid**	Autoignition Temperature **Not Determined**

Extinguishing Media

Silver oxide does not burn. Use agent suitable to surrounding fire.

Special Fire Fighting Procedures

Wear full protective clothing and self-contained breathing apparatus (SCBA). Use a water spray to keep fire-exposed containers cool. Poisonous gases are produced in fire. Do NOT release water runoff from fire control measures to sewers or navigable waterways. Contain and collect for proper disposal.

Unusual Fire and Explosion Hazards

While silver oxide does not burn, it is an extremely powerful oxidizer that can significantly enhance the flammability of any combustible materials it contacts.

SECTION V - REACTIVITY DATA

Stability		Conditions to Avoid
		Silver oxide is normally stable in closed containers under routine conditions of handling and storage. But, it will absorb carbon dioxide from air when moist and exposure to light can accelerate decomposition.
Stable	Unstable	Incompatibility *(materials to avoid)*
X		**Aluminum, cupric oxide, ammonia, hydrazine + alcohol, carbon monoxide, boron trifluoride etherate + nitromethane + ethylene, hydrogen sulfide, magnesium, metal sulfides, combustibles, and many others.**
Hazardous Polymerization		Conditions to Avoid **Under normal conditions of temperature and pressure, hazardous polymerization of silver oxide is not known to occur.**
May Occur	Will Not Occur	Hazardous Decomposition or By-products
	X	**When heated to decomposition, silver oxide can emit silver and oxygen.**

SECTION VI - HEALTH HAZARD DATA

Primary Route(s) of Entry:	Inhalation?	Absorption (skin)?	Ingestion?
	X	X	

Health Hazards

INHALATION: Causes shortness of breath, coughing, and sore throat. It can also cause generalized argyria with discoloration of the eyes and skin to a grayish-blue. It may also cause a visible (on X-rays) increase in lung densities and mild bronchitis. Severe irritant.

SKIN & EYES: Contact with skin, especially moist skin, can cause serious burns. There can also be localized pigmentation changes.

INGESTION: Unspecified gastrointestinal effects.

Carcinogenicity	NTP Listed?	IARC Cancer Review Group?	OSHA Regulated?	Target Organs?
Unknown Human **Unknown Animal**	**No**	**No**	**29 CFR 1910.1000 Table Z-1**	**Eyes (especially conjunctiva), skin, and mucosa.**

Medical Conditions Generally Aggravated by Exposure

None reported.

Emergency and First-aid Procedures

Eye contact: Flush immediately with water for 15 minutes (minimum); seek medical attention. Skin contact: Remove all contaminated clothing. Immediately wash area with large amounts of soap and water. Treat for skin burns. For inhalation: Remove the person from exposure. Provide respiratory assistance and CPR. Transfer to medical facility. If swallowed: Never attempt to give a convulsing or unconscious person anything by mouth. Unless advised otherwise and victim is conscious, give 4 to 8 ounces of water and then induce vomiting. Seek medical attention immediately.

SECTION VII - PRECAUTIONS FOR SAFE HANDLING AND USE

Steps to be Taken in Case Material is Released or Spilled

Collect powdered material in most convenient manner possible and deposit in sealed drum. Keep those not involved in cleanup from entering area. Do NOT dry sweep (creates airborne dusts). Carefully scoop or use a vacuum equipped with a high efficiency particulate air (HEPA) filter.

Preferred Waste Disposal Method

No citation

Precautions to be Taken in Handling and Storage

Store in tightly closed containers in cool, dark, dry, well-ventilated area.

Other Precautions and Warnings

Keep away from incompatible materials and protect from physical damage.

SECTION VIII - CONTROL MEASURES AND PERSONAL PROTECTIVE EQUIPMENT

Respiratory Protection *(specify type)*

At exposures above PEL, use full facepiece respirator with high-efficiency particulate air (HEPA) filter. For higher exposures, use a supplied-air respirator with a full facepiece or a self-contained breathing apparatus (SCBA) operated in positive pressure mode.

Ventilation

Local exhaust or general mechanical systems recommended.

Protective Gloves	Eye Protection	Other Protective Clothing
Impervious Material	**Chemical/Dust Goggles or Face Mask**	**Protective Suit or Clothing**

Work/Hygiene Practices

Always wash hands thoroughly after using chemical; never bring food, drink, or smoking materials into vicinity of chemicals.

SILVER OXIDE

Ag_2O CAS: 20667-12-3

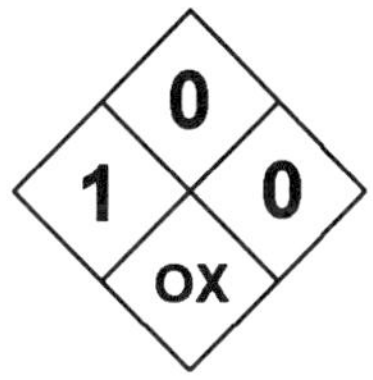

IDENTIFICATION AND TYPICAL USES

Silver oxide appears as a brownish-black odorless powder with a metallic taste. It is used as a catalyst, a laboratory reagent, in the purification of drinking water, for polishing glass, and to color glass yellow.

RISK ASSESSMENT: HEALTH

General Assessment

Silver oxide dusts and solutions are generally irritating to the eyes, nose, and upper respiratory tract. It is toxic by *ingestion* and *inhalation*. It can cause some irritation and possible damage to the skin's surface, but dermal absorption of this silver compound has not been reported. There are no data in the references regarding its carcinogenic, mutagenic, or teratogenic properties in humans or test animals.

Inhalation or ingestion can lead to generalized argyria characterized by a bluish-gray pigmentation or color change that may be distributed over the face, forehead, neck, hands, and forearms. Silver that is deposited in the respiratory tract may lead to mild bronchitis. Other symptoms of inhalation may be shortness of breath, coughing, and sore throat. Large doses may lead to kidney, liver, and neurological damage although actual cases of such effects are not well documented in the references.

Skin contact can cause mild to moderate irritation. There can also be localized pigmentation or color changes to a blue-gray if silver particles become lodged or imbedded in the skin. Argyria may occur in the conjunctiva of the eye, gum tissue, or nasal septum. Eye contact can cause some irritation, although permanent damage to vision has not been reported.

Acute Health Effects

The following short-term health effects may occur immediately or shortly after exposure to silver oxide:

Skin: Irritation, rash, discoloration of the skin to a blue-gray, and possible contact dermatitis.

Eye: Silver oxide particles and dusts can be irritating to the eyes. There may be a bluish-gray discoloration of the eyes and conjunctiva.

Lung: Inhalation of silver oxide may cause moderate to severe irritation of the respiratory system, especially the lungs and nasal passages. Inhalation can lead to fatigue, headache, sinusitis, and weight loss.

Chronic Health Effects

The following long-term health effects may occur at some time after exposure to silver oxide and can last for months or even years:

Cancer Hazards: According to information presented in the references, silver oxide has not bee specifically tested for its ability to cause cancer in test animals. Some silver compounds appear to have some degree of carcinogenic potential. It is not know if silver oxide will cause cancer in humans or animals.

Reproductive Hazard: According to information available in the references, silver oxide has not been adequately tested for its ability to cause adverse reproductive effects in animals. It is not known if human exposure will have this effect.

Other Chronic Effects: Long-term exposure to silver oxide dusts or fumes can very gradually cause eyes, nails, inner nose, throat, body organs, and skin to turn a bluish-gray color. This can take between 2 and 20 years to develop but is permanent and can be very disfiguring. High, repeated exposure may cause kidney and/or liver injury as well as neurologic damage. Very irritating substances can injure the lungs. It is not known if silver oxide causes lung damage over the long-term.

↻ *Recommended Risk-Reduction Measures*

Personnel should avoid direct contact with silver oxide and other silver compounds. If a less toxic chemical cannot be substituted for a hazardous substance, then *engineering controls* are the most effective method of reducing exposures. The best protection is to enclose operations and/or provide local exhaust ventilation at the site of chemical release. While not always operationally feasible, isolating operations can also reduce exposure. Using respiratory protection is less effective than the controls mentioned above, but is still advisable whenever working with or around silver oxide. For low or infrequent exposures, use a full facepiece respirator equipped with a high-efficiency particulate air (HEPA) filter and a dust/mist pre-filter. For higher exposures, use a supplied-air respirator with a full facepiece operated in positive pressure mode or an MSHA/NIOSH-approved self-contained breathing apparatus (SCBA) with full facepiece operated in pressure demand. If a full facepiece is not available, then chemical/dust goggles should be worn to protect the eyes. Whenever a chemical dust hazard exists, a face shield and a rubber apron should be worn. To prevent hand and skin exposures, impervious rubber gloves should be used. Glove manufacturers should be contacted and permeation studies obtained *before* final glove selections have been made.

Administrative controls should also be in place to minimize the potential for human exposures. These may include written procedures or policies which specify the methods and techniques that will be practiced whenever personnel are to work with silver oxide.

All personnel should receive *training* on the use, hazards, protective measures, emergency actions, and other precautions per 29 CFR 1910.1200 (Hazard Communication), prior to the first assignment in an area where silver oxide is used or stored.

If symptoms develop or overexposure is suspected, the following may also be useful:

☑ Examination of the skin, nose, throat, and mucous membranes for discoloration.

☑ Slit lamp examination by an optometrist.

☑ Kidney function tests.

Any evaluation should include a careful history of past and present symptoms with a medical examination. Medical tests that look for damage already done are not a substitute for controlling exposure.

Other methods to reduce exposure include:

☑ Where possible, enclose operations and use local exhaust ventilation at the site of chemical release. If local exhaust ventilation or enclosure is not used, respiratory protection should be mandatory.

☑ Always ensure that proper protective clothing is worn when using chemical substances and that personnel have been trained in its proper use, care, and maintenance.

☑ Wash thoroughly immediately after exposure to silver oxide and at the end of the work shift or before eating, drinking, or smoking.

☑ Hazard warning information should be posted in the work area. In addition, as part of an on-going education and training program, all information on the health and safety hazards of silver oxide should be communicated to all exposed and potentially exposed workers.

☑ Where possible, automatically transfer silver oxide from drums or storage containers to process containers.

RISK ASSESSMENT: ENVIRONMENT

General Assessment

The environment is still at risk of exposure during transportation, storage, disposal, or destruction of silver oxide. In almost every scenario, the threat of environmental exposure is contingent upon the proper handling of the chemical substance. Accidental spills, large or small, can result in fire, explosion, and possible contamination of the surrounding environmental mediums (water, soil, and air).

Silver oxide does not burn but it is an extremely powerful oxidizer that can significantly enhance the burning rate of any combustible materials it contacts. It will ignite and is potentially explosive in contact with aluminum, cupric oxide, ammonia, hydrazine + alcohol, carbon monoxide, boron trifluoride etherate + nitromethane, chlorine + ethylene, hydrogen sulfide, magnesium, dichloromethylsilane, metal sulfide, nitroalkanes, sodium-potassium alloy, and many other compounds. Also, as an oxidizer, silver oxide can ignite and significantly enhance the burning rate of ordinary combustible materials. In a fire, silver oxide will emit toxic and poisonous gases. These characteristics require special consideration during any emergency situation involving a leak or spill of silver oxide.

Silver is a naturally occurring element in the Earth's crust. Because of its wide use, silver oxide enters the environment in non-point as well as specific discharges such as industrial and municipal effluents.

Acute Ecological Effects

Acute (short-term) toxic effects may include the death of animals, birds, or fish and death or low growth rate in plants. Acute effects are seen 2 to 4 days after animals or plants are exposed to silver oxide.

The toxicity of silver to aquatic life is probably dependent on water hardness; the harder the water, the higher the concentration of silver needed to be considered toxic. Insufficient data are available to evaluate the short-term effects of silver and its compounds exposure to plants, birds, or land animals.

Chronic Ecological Effects

Chronic (long-term) toxic effects may include shortened life span, reproductive problems, lower fertility, and changes in appearance or behavior in exposed animals. These effects can be seen long after first exposure(s) to toxic chemicals.

Silver and its compounds have high chronic toxicity to aquatic life. Insufficient data are available to evaluate the long-term effects of silver exposure to plants, birds, or land animals.

Water Solubility

Some silver compounds are highly soluble in water while others are nearly insoluble. Silver oxide is considered slightly soluble in water.

Persistence in the Environment

Silver is highly persistent in the aquatic environment, with a half-life of greater than 200 days. The half-life of a pollutant is the amount of time it takes for one half of the chemical to be degraded.

Bioaccumulation in Aquatic Organisms

Some substances increase in concentration, or *bioaccumulate,* in living organisms as they breathe contaminated air, drink contaminated water, or eat contaminated food. These chemicals can become concentrated in the tissues and internal organs of animals as well as humans.

The concentration of silver found in fish is expected to be considerably higher than the average concentration of silver in the water from which the fish was taken.

Recommended Risk-Reduction Measures

Proper training of all transporters will reduce the likelihood of a mishap or accidents resulting in a leak or spill to the environment. The correct use of labeling on all containers, trucks, and rail cars will enable emergency responders to react properly and quickly to any disaster thereby reducing the potential risk to the environment and to personnel.

If a spill or leak to the environment has occurred, fire department, emergency response and/or hazardous materials spill personnel should be notified immediately. Cleanup should be attempted only by those trained in proper spill containment procedures. Contaminated soils may need to be removed and replaced with clean soil. If silver oxide should contact the water table, aquifer, or navigable waterway, time is of the essence. The solubility of silver compounds in water may range from insoluble to highly soluble and, therefore, total containment and remediation may not be entirely possible. The solubility of silver oxide is expected to be slight. When such spills occur, the local and/or state emergency response authorities must be notified. A comprehensive emergency response or disaster preparedness/recovery plan should be in place prior to any operations involving the use, transportation, storage, or disposal of silver oxide.

If silver oxide is spilled or leaked, the following specific steps are recommended:

- ☑ Restrict persons not wearing protective clothing from area of spill or leak until cleanup is complete and area can be opened for normal work.
- ☑ Collect powdered materials in most convenient and safe manner possible and place in sealed drums for reclamation or disposal. Do not dry sweep; use HEPA vacuum instead.
- ☑ It may be necessary to dispose of silver oxide as a hazardous waste. The responsible state agency or the regional office of the federal Environmental Protection Agency (EPA) should be contacted for specific recommendations.

RISK ASSESSMENT: BUSINESS

General Assessment

Accidents or mishaps involving silver oxide can present a significant threat to business operations. The loss or

damage of equipment or facilities can significantly affect fiscal viability. Lawsuits that may result from personnel injury/death, public exposures, and/or environmental contamination will also require a serious expenditure of resources. Media attention surrounding an injury, death, or environmental damage can also result in a loss of profits and loss of current as well as future business.

Recommended Risk-Reduction Measures

Company attorneys, safety and health professionals, and environmental specialists should be involved in the development of any procedures dealing with safety and emergency response. A company official should be pre-designated as a public relations officer with specific training in dealing with the press. Corporate plans and policies should be developed, approved, and implemented long before any need for such arises.

REFERENCES

Ahlbom, A. 1993. *Biostatistics for Epidemiologists.* New York: CRC Press/Lewis Publishers.

American Conference of Governmental Industrial Hygienists. 1988. *Documentation of the Threshold Limit Values and Biological Exposure Indices,* 5th Edition (with updates). Cincinnati: ACGIH

Asante-Duah, D. K. 1993. *Hazardous Waste Risk Assessment.* New York: CRC Press/Lewis Publishers.

Bloom, A. D. and F. J. DeSerres. 1995. *Ecotoxicity and Human Health.* New York: CRC Press/Lewis Publishers.

Brusick, D. J. 1994. *Methods for Genetic Risk Assessment.* New York: CRC Press/Lewis Publishers.

Calabrese, E. J., E. M. Kenyon. *Air Toxics and Risk Assessment.* New York: CRC Press/Lewis Publishers.

Calabrese, E. J. 1994. *Biological Effects of Low Level Exposures.* New York: CRC Press/Lewis Publishers.

Cockerman, L. G., B. S. Shane. 1994. *Basic Environmental Toxicology.* New York: CRC Press/Lewis Publishers.

Cothern, C. R., N. P. Ross. 1994. *Environmental Statistics, Assessment, and Forecasting.* New York: CRC Press/Lewis Publishers.

Cothern, R. R. 1993. *Comparative Environmental Risk Assessment.* New York: CRC Press/Lewis Publishers.

Cothern, C. R. 1995. *Handbook for Environmental Risk Decisionmaking.* New York: CRC Press/Lewis Publishers.

Danse, I. R. 1991. *Common Sense Toxics in the Workplace.* New York: Van Nostrand Reinhold.

Environmental Protection Agency. 1983. *Chemical Hazard Information Profiles.* Washington, D.C.: U.S. EPA.

Fodden, C. R., and J. L. Weddell. 1992. *Hazardous Materials Emergency Action Data.* New York: CRC Press/Lewis Publishers.

Forseberg, K., and L. H. Keith. 1993. *Chemical; Protective Clothing Permeation and Degradation Database.* Chelsea, MI: Lewis Publishers.

Gardner, R. W. 1994. *Chemical Intolerance.* New York: CRC Press/Lewis Publishers.

Hallenbeck, W. H. 1993. *Quantitative Risk Assessment for Environmental and Occupational Health,* 2nd Edition. New York: CRC Press/Lewis Publishers.

Howard, P. H., et. al. 1991. *Handbook of Environmental Degradation Rates.* New York: CRC Press/Lewis Publishers.

Howard, P. H. 1995. *Handbook of Environmental Fate and Exposures Data for Organic Chemicals,* Vol. I-IV. New York: CRC Press/Lewis Publishers.

Jones, F. E. 1994. *Toxic Organic Vapors in the Workplace.* New York: CRC Press/Lewis Publishers.

Keith, L. H. and M. M. Walker. 1995. *Handbook of Air Toxics.* New York: CRC Press/Lewis Publishers.

Landis, W. G., M. H. Yu. 1995. *Introduction to Environmental Toxicology.* New York: CRC Press/Lewis Publishers.

Langerman, N. 1994 *Precautionary Labels for Chemical Containers.* New York: CRC Press/Lewis Publishers.

Lauwerys, R. R., P. Hoet. 1993. *Industrial Chemical Exposure.* New York: CRC Press/Lewis Publishers.

Lewis, R. J., Sr. 1992. *Sax's Dangerous Properties of Industrial Materials,* Volumes 1, 2, and 3, 8th Edition. New York: Van Nostrand Reinhold.

Mackay, D., W. Y. Shiu, and K. C. Ma. 1992. *Illustrated Handbook of Chemical-Physical Properties and Environmental Fate for Organic Chemicals.* New York: CRC Press/Lewis Publishers.

Manahan, S. E. 1994. *Environmental Chemistry,* 6th Edition. New York: CRC Press/Lewis Publishers.

Meyer, E. 1990. *Chemistry of Hazardous Materials.* Englewood Cliffs, NJ: Prentice Hall.

Patnaik, P. 1992. *A Comprehensive Guide to the Hazardous Properties of Chemical Substances.* New York: Van Nostrand Reinhold.

Quigley, D. R. 1994. *Handbook of Emergency Chemical Management.* New York: CRC Press/Lewis Publishers.

Rea, W. J. 1995. *Chemical Sensitivity,* Vol. I-III. New York: CRC Press/Lewis Publishers.

Sacarello, H. L. A. 1994. *The Comprehensive Handbook of Hazardous Materials.* New York: CRC Press/Lewis Publishers.

Sherman, J. D. 1988. *Chemical Exposure and Disease.* New York: Van Nostrand Reinhold.

Talbott, E., G. Craun. 1995. *An Introduction to Environmental Epidemiology.* New York: CRC Press/Lewis Publishers.

U.S. Department of Health and Human Services (DHHS), National Institute for Occupational Health (NIOHS). 1994. *NIOSH Pocket Guide to Hazardous Chemicals.* Publication Number 94-116. Washington, D.C.; U.S. Government Printing Office.

MATERIAL SAFETY DATA SHEET

CHEMICAL NAME	
SODIUM	**CRC PRESS/LEWIS PUBLISHERS DISCLAIMER:** THE INFORMATION AND RECOMMENDATIONS PRESENTED HEREIN ARE BASED ON SOURCES BELIEVED TO BE RELIABLE. CRC MAKES NO REPRESENTATION ON ITS COMPLETENESS OR ACCURACY. IT IS THE RESPONSIBILITY OF THE USER TO DETERMINE THE CHEMICAL'S SUITABILITY FOR ITS INTENDED USE, THE CHEMICAL'S SAFE USE, AND THE CHEMICAL'S PROPER DISPOSAL. NO REPRESENTATIONS AND/OR WARRANTIES, EITHER EXPRESSED OR IMPLIED, OF THE MERCHANTABILITY OR FITNESS FOR A PARTICULAR PURPOSE, OR OF ANY OTHER NATURE, ARE MADE WITH RESPECT TO THE INFORMATION PROVIDED IN THIS MSDS OR TO THE CHEMICAL TO WHICH INFORMATION MAY REFER. CRC NEITHER ASSUMES NOR AUTHORIZES ANY OTHER PERSON TO ASSUME FOR IT, ANY OTHER ADDITIONAL RESPONSIBILITY OR LIABILITY FOR THE USE OF, OR RELIANCE UPON, THIS INFORMATION.

HAZARD WARNING INFORMATION

HEALTH	FIRE	REACTIVE	OTHER	DEGREE OF HAZARD	COLOR CODING	OTHER CODES
3	1	2	~~W~~	0 = Minimum Hazard 1 = Slight Hazard 2 = Moderate Hazard 3 = Serious Hazard 4 = Severe Hazard	HEALTH = BLUE FIRE = RED REACTIVITY = YELLOW OTHER = WHITE	OX = Oxidizer ACID = Acid ALK = Alkali COR = Corrosive ~~W~~ = Use No Water

SECTION I - GENERAL INFORMATION

Characterization	RCRA Number	EPA Class
Metal (reactive)	**D003**	**Characteristic (R) Waste**

DOT Proper Shipping Name **Sodium Metal**	Chemical Abstract Service (CAS) Number **7440-23-5**
DOT Hazard Class and Label Requirements **Flammable Solid & Dangerous When Wet**	DOT Emergency Guide Code **No Citation**
DOT Identification Number **UN 1428**	Atomic Symbol **Na**

Synonyms

Natrium; metallic sodium; sodium metal.

SECTION II - HAZARDOUS INGREDIENTS AND IDENTITY INFORMATION

Hazardous Components (specific identity)	OSHA Exposure Criteria	NIOSH Exposure Criteria	Immediately Dangerous to Life and Health (IDLH)	ACGIH Exposure Criteria
Sodium (derivation: By electrolysis of a fused mixture of sodium chloride and calcium chloride).	**PEL:** Not Established **STEL:** Not Established	**REL:** Not Established **STEL:** Not Established	**Not Determined**	**TLV:** Not Established **STEL:** Not Established

SECTION III - PHYSICAL AND CHEMICAL CHARACTERISTICS

Boiling Point **1621°F (883°C)**	Specific Gravity (H_2O = 1) **0.97**
Vapor Pressure (mm Hg) **1.22 at 752°F (400°C)**	Atomic Weight **22.98**
Vapor Density (Air = 1) **Not Reported**	Melting Point **208°F (98°C)**

Solubility

Decomposes in water (forms sodium hydroxide). Insoluble in benzene, kerosene, and naphtha.

Appearance and Odor

Malleable, soft, silver-white solid, oxidizes rapidly in air. It is wax-like at room temperature. Odorless.

SECTION IV - FIRE AND EXPLOSION HAZARD DATA

Flash Point (method used) **Not Found**	Explosive Limits in Air % by Volume **LEL: Not Reported** **UEL: Not Reported**
NFPA Classification **Flammable Solid**	Autoignition Temperature **250°F (121°C)**

Extinguishing Media

Use G-1® graphite powder, dry graphite, powdered sodium chloride, soda ash, or powdered talc. Do NOT use water, carbon dioxide, Halon® or other common extinguishing agents.

Special Fire Fighting Procedures

Wear full protective clothing and self-contained breathing apparatus (SCBA). Poisonous gases are produced in fire. Do NOT look directly into sodium fires. Fight fire from distance, if possible.

Unusual Fire and Explosion Hazards

Extremely dangerous when heated. Extremely dangerous in contact with moisture or water and releases hydrogen with sufficient heat to cause ignition of explosion. It ignites spontaneously in air or oxygen and burns explosively. Extremely explosive and violently reactive metal.

SECTION V - REACTIVITY DATA

Stability		Conditions to Avoid
Stable	Unstable	Sodium is not stable and can easily undergo violent chemical change. It reacts explosively with water and may react spontaneously in air. Keep away from heat, moisture, and incompatible materials.
	X	Incompatibility *(materials to avoid)* Incompatible with oxidizing agents. Reacts explosively with acids (hydrochloric, hydrofluoric, nitric, and sulfuric), air, halides, carbon dioxide, ammonia, chlorinated hydrocarbons, lead oxide, phosphorus, sulfur.
Hazardous Polymerization		Conditions to Avoid Hazardous polymerization of sodium will not occur. May ignite spontaneously in air or moisture.
May Occur	Will Not Occur	Hazardous Decomposition or By-products
	X	Poisonous gases are emitted during fire, including toxic and irritating smoke and fumes of sodium oxide, sodium hydroxide (a corrosive), and flammable hydrogen gas.

SECTION VI - HEALTH HAZARD DATA

Primary Route(s) of Entry:	Inhalation?	Absorption (skin/eye)?	Ingestion?
	X	X	X

Health Hazards

INHALATION: The dusts can produce irritation of the eyes and mucous membranes. It may react with the water content in the bronchial passages to form caustic sodium hydroxide which can damage the lungs. The dusts or fumes causes shortness of breath, nausea and vomiting, headache, and nose and respiratory tract irritation (pneumonitis). Moisture of the mucosa may cause sodium to ignite or explode causing possibly fatal thermal burns.

ABSORPTION: Corrosive to all body tissues. May cause burns, ulceration, tissue death, and absorption. Causes severe eye damage and possible blindness. The powder metal ignites easily on moist skin and eyes causing serious burns.

INGESTION: Causes severe pain, vomiting, diarrhea, and possible collapse and death.

Carcinogenicity	NTP Listed?	IARC Cancer Review Group?	OSHA Regulated?	Target Organs?
Unknown Human Unknown Animal	No	No	No	Respiratory system, eyes, skin, mucous membranes.

Medical Conditions Generally Aggravated by Exposure

Possible respiratory system disorders and/or skin conditions may be aggravated by exposure.

Emergency and First-aid Procedures

Eye contact: Do not allow victim to keep eyes tightly shut. Flush immediately with water for 15 minutes (minimum); seek medical attention. Skin contact: Remove all contaminated clothing. Immediately wash area with flooding amounts of water. For reddened or blistered skin, seek medical assistance. For inhalation: Remove the person from exposure. Provide respiratory assistance and CPR. If swallowed: Seek medical attention. If victim is conscious, give gastric lavage followed by olive oil or demulcent. Never give anything by mouth to an unconscious or convulsing person.

SECTION VII - PRECAUTIONS FOR SAFE HANDLING AND USE

Steps to be Taken in Case Material is Released or Spilled

Collect powdered materials and deposit in clean, dry, sealed, containers. Do NOT dry sweep. Use HEPA vacuum. Ventilate area. Restrict those not involved in cleanup from area. Do NOT use water.

Preferred Waste Disposal Method

Reclaim scrap.

Precautions to be Taken in Handling and Storage

Store to avoid contact with incompatible materials. Prevent physical damage to containers. Keep containers tightly closed. Keep moisture and sources of heat away. Use non-sparking tools.

Other Precautions and Warnings

Avoid excessive dust generation in storage and handling areas. Store under an inert gas blanket (nitrogen or argon) or submerged in naphtha, kerosene, or other oils that do not contain oxygen.

SECTION VIII - CONTROL MEASURES AND PERSONAL PROTECTIVE EQUIPMENT

Respiratory Protection *(specify type)*

For low exposures, use a respirator with a HEPA filter; otherwise use a supplied-air respirator or a self-contained breathing apparatus with full facepiece operated in positive pressure mode.

Ventilation

Local exhaust (preferred) at site of chemical work, or general dilution ventilation.

Protective Gloves	Eye Protection	Other Protective Clothing
Impervious Material	Chemical/Dust Goggles or Face Mask	Protective Apron, Boots

Work/Hygiene Practices

Always wash hands thoroughly after using chemical; never bring food, drink, or smoking materials into vicinity of chemicals. Skin contact MUST be prevented.

SODIUM

Na

CAS: 7440-23-5

IDENTIFICATION AND TYPICAL USES

Sodium appears as a silvery-white, soft, malleable, odorless metal. At room temperature, it may take on a wax-like appearance. It is rapidly oxidized in moist air and the color changes to gray. It is used in the preparation of tetraethyl and tetramethyl lead, in the reduction of titanium, in sodium peroxide and sodium hydride, as a polymerization catalyst for synthetic rubber, as a laboratory reagent, as a coolant in nuclear reactors, in electrical power cables (encased in polyethylene), in non-glare lighting fixtures (on roads and highways), and as a heat transfer agent in solar-powered electric generators. The radioactive forms are used in tracer studies and in medicine.

RISK ASSESSMENT: HEALTH

General Assessment

The toxicity of sodium metal in humans is largely due to its highly reactive nature in contact with water or moisture. It is toxic to humans on *inhalation,* dermal *absorption,* and *ingestion.* There are no reports of carcinogenic, teratogenic, or mutagenic effects due to chronic exposure to sodium.

Inhalation of the dusts and particles of sodium can irritate the eyes and nasal passages. Sodium may react with the water content in the bronchial passages to form caustic sodium hydroxide which can cause severe burns, ulceration, and tissue damage. Sodium fumes in the oxide form can also irritate the eyes, nose, throat, and respiratory tract and may cause coughing, tightness and pressure in the chest, nausea, vomiting, shortness of breath, and possible pneumonitis. It may react violently or even explosively in contact with moist mucosa, wet eyes, or even wet skin (sweat) causing thermal burns and tissue destruction. Inhalation can also cause shortness of breath, headache, nausea and vomiting, and nose and respiratory tract irritation leading to lung damage such as pneumonitis.

Eye contact will cause burns, conjunctivitis, corneal damage, and possible blindness. Skin contact can cause burns, ulceration, and tissue destruction.

Ingestion will cause severe pain, vomiting, diarrhea, and possible collapse with loss of consciousness. Gastrointestinal effects may be severe and possibly life-threatening.

Acute Health Effects

The following acute (short-term) health effects may occur immediately or shortly after human exposure to sodium:

Skin: May cause serious irritation and even burns, especially if the skin is moist. In fact, it may react explosively on wet skin.

Eye: Severe irritation and possible burns due to the eyes moisture content. Corneal damage and possible corneal opacification can be serious and permanent.

Lung: Irritation of the mucosa of the eyes, nose, throat, and respiratory tract. There may be shortness of breath, nausea, headache, and nose and respiratory tract irritation and possible pneumonitis.

Other: Ingestion causes severe pain, vomiting, diarrhea, and other gastrointestinal effects. There may be collapse and loss of consciousness.

Chronic Health Effects

The following chronic (long-term) health effects may occur at some time after exposure to sodium and can last for months or even years:

Cancer Hazards: According to the references, sodium has not been adequately tested for its ability to cause cancer in animals.

Reproductive Hazard: According to the references, there are no reports of any teratogenic or adverse reproductive effects resulting from occupational exposure to sodium.

Other Chronic Effects: Long-term exposures may cause chronic dermatitis, mucous membrane irritation, and possible lung damage. Very irritating substance can damage the lungs. It is not known if exposure to sodium will have this effect.

Recommended Risk-Reduction Measures

Sodium is extremely corrosive and destructive to human tissues. Personnel should, therefore, avoid direct contact with sodium. If a less toxic material or compound cannot be substituted for sodium, then *engineering controls* are the most effective method of reducing exposures. The best protection is to enclose operations and/or provide local exhaust ventilation at the site of sodium dust release. While not always operationally feasible, isolating operations can also reduce exposure. Using respiratory protection is less effective than the controls mentioned above, but is still advisable whenever working with or around sodium. No exposure limits have been established specifically for sodium metal. This does not mean that exposure is without a health risk. Human exposures must be prevented. For low or transient exposures, a chemical cartridge respirator equipped with a high-efficiency particulate air (HEPA) filter and a dust/mist pre-filter may provide adequate respiratory protection. However, better protection is obtained using a supplied-air respirator operated in continuous flow mode or a self-contained breathing apparatus (SCBA) with full facepiece operated in pressure demand. If a full facepiece is not available, then chemical/dust goggles should be worn to protect the eyes. A face shield and protective apron should also be worn. To prevent hand and skin exposures, gloves should be used. Glove manufacturers should be contacted and permeation studies obtained *before* final glove selections are made.

Administrative controls should also be in place to minimize the potential for human exposures. These may include written procedures or policies which specify the methods and techniques that will be practiced whenever personnel are to work with sodium.

All personnel should receive *training* on the use, hazards, protective measures, emergency actions, and other precautions per 29 CFR 1910.1200 (Hazard Communication), prior to the first assignment in an area where sodium is used or stored.

Before beginning employment working with sodium, and at regular intervals thereafter (e.g., annually), those personnel with frequent or potentially high exposures should be provided the following recommended medical tests:

☑ Lung function tests (establish baseline).

If symptoms develop or overexposure is suspected, consider the following additional tests:

☑ Lung function tests (compare to baseline).

☑ Evaluation by a qualified allergist with special testing and careful consideration of exposure history (may help diagnose skin allergy).

☑ Examination of the nasal passages for the presence of ulcers or signs of ulceration and other indications of corrosive tissue destruction.

Any evaluation should include a careful history of past and present symptoms with an examination. Medical tests that simply look for existing damage are not a substitute for controlling exposures. Also, since smoking can cause heart disease as well as lung cancer, emphysema, and other known respiratory problems, smokers exposure to sodium may experience symptoms more rapidly than non-smokers under the same conditions of exposure. Prudent risk management requires careful consideration of *all* factors which may be causing exposure symptoms in the workplace.

Other methods to reduce exposure include:

☑ Where possible, enclose operations and use local exhaust ventilation at the site of chemical release. If local exhaust ventilation or enclosure is not used, respiratory protection should be mandatory.

☑ Always ensure that proper protective clothing is worn when using chemical substances. Proper maintenance and care of protective equipment is essential to its proper function.

☑ Wash thoroughly immediately after exposure to sodium and at the end of the work shift or before eating, drinking, smoking, or applying cosmetics.

☑ Hazard warning information should be posted in the work area. In addition, as part of an on-going education and training program, all information on the health and safety hazards of sodium should be communicated to all exposed and potentially exposed workers.

☑ Eye wash stations should be provided in the immediate work area for emergency use. If there is a possibility of skin exposure to sodium, emergency shower facilities should be provided.

☑ Workers whose clothing has been contaminated by sodium should change into clean clothes before leaving work. Contaminated work clothing should be laundered only by individuals who have been informed of the hazards of exposure to sodium.

RISK ASSESSMENT: ENVIRONMENT

General Assessment

As a naturally occurring mineral, sodium poses relatively little threat to the environment. However, accidental spills or releases to the environment can occur during transportation, storage, disposal, or destruction of sodium. In almost every scenario, the threat of environmental exposure is contingent upon the proper handling of the chemical substance. Accidental spills, large or small, where sodium contacts incompatible commodities can result in release of toxic products (depending upon conditions of spill), and possible contamination of the surrounding environmental mediums (water, soil, and air).

Sodium is considered a flammable solid. It is an extremely corrosive and highly reactive metal. It can explode in air, in contact with moisture, oxidizing materials, or in contact with an ignition source. It reacts with water, moisture, and acids to evolve explosive hydrogen gas and corrosive sodium hydroxide. It is also incompatible with many other materials, including air, heavy metal compounds (silver oxide and silver chloride), chlorinated hydrocarbons, water, halides (aluminum bromide, magnesium chloride, aluminum fluoride, mercuric iodide), carbon dioxide, ammonia, lead oxide, phosphorous, potassium oxides, sulfur, and sulfur dioxide. Extreme caution is required in handling, storage, transportation, and disposal of sodium. These characteristics also require special consideration during any emergency situation involving a leak or spill of sodium powder or dust. Should sodium ever come into contact with these various incompatible substances either during use, transportation, storage, or disposal, violent and even explosive reactions are possible.

Sodium occurs naturally in the environment (terrestrial soils and aquatic sediments) and may enter the environment as a result of industrial and municipal waste treatment plant discharges and from spills.

☠ *Acute Ecological Effects*

Acute (short-term) toxic effects may include the death of animals, birds, or fish and death or low growth rate in plants. Acute effects are seen 2 to 4 days after animals or plants are exposed to sodium.

Insufficient data are available to evaluate the short-term effects of sodium to plants, birds, or land animals. However, it should be noted that the U.S. Environmental Protection Agency does list sodium as a hazardous substance.

Chronic Ecological Effects

Chronic (long-term) toxic effects may include shortened life span, reproductive problems, lower fertility, and changes in appearance or behavior in exposed animals. These effects can be seen long after first exposure(s) to toxic chemicals.

Insufficient data are available to evaluate or predict the long-term effects of sodium to plants, birds, or land animals. However, it should be noted that the U.S. Environmental Protection Agency does list sodium as a hazardous substance.

Water Solubility

Sodium reacts violently in water to form sodium hydroxide and hydrogen gas.

⌛ *Persistence in the Environment*

Because sodium reacts so rapidly (and violently) in water, it is not expected to be persistent in the aquatic environment. Its half-life has not been established. The half-life of a pollutant is the amount of time it takes for one half of the chemical to be degraded.

Bioaccumulation in Aquatic Organisms

Some substances increase in concentration, or *bioaccumulate,* in living organisms as they breathe contaminated air, drink contaminated water, or eat contaminated food. These chemicals can become concentrated in the tissues and internal organs of animals as well as humans.

The concentration of sodium found in fish tissues is expected to be about the same as the average concentration of sodium in the water from which the fish was taken.

Recommended Risk-Reduction Measures

Proper training of all transporters will reduce the likelihood of a mishap or accident resulting in a leak or spill of sodium dusts or powders into the environment. The correct use of labeling on all containers, trucks, and rail cars will enable emergency responders to react properly and quickly to any disaster thereby reducing the potential risk to the environment and to personnel.

Storage of sodium should be segregated from incompatible chemicals to minimize the risk of cross-contamination or contact. Care should be taken to prevent physical damage to storage containers. It is suggested that sodium be stored under an inert gas such as nitrogen or argon. As an alternative, it can be submerged in oxygen-free oils such as petroleum naphtha. Also, dusty conditions must be avoided in storage and use areas. When sodium comes into contact with water, it will readily liberate excessive amounts of highly flammable and explosive hydrogen gas.

If a spill or leak to the environment has occurred, fire department, emergency response and/or hazardous materials spill personnel should be notified immediately. Cleanup should be attempted only by those trained in proper spill containment procedures using non-sparking tools. Contaminated soils should be removed for decontamination and replaced with clean soil. If sodium should contact the water table, aquifer, or navigable waterway, time is of the essence. It is highly reactive and may explode on contact with water. It liberates excessive heat as it forms sodium hydroxide and hydrogen gas. The heat may be sufficient enough to ignite the hydrogen and cause the explosion. The local and/or state emergency response authorities may require notification. A comprehensive emergency response or disaster preparedness/recovery plan should be in place prior to any operations involving the use, transportation, storage, or disposal of sodium.

If sodium is spilled or leaked, the following specific steps are recommended:

- ☑ Restrict persons not wearing protective clothing from area of spill or leak until cleanup is complete and area can be opened for normal work.
- ☑ Remove all ignition sources from area.
- ☑ Ventilate area of spill or leak.
- ☑ Collect powdered materials in the safest manner possible and deposit in sealed containers. Recommend using a vacuum equipped with a high-efficiency particulate air (HEPA) filter. Do not dry sweep (generates airborne dusts). Keep water away (do NOT damp mop residues).
- ☑ It may be necessary to dispose of sodium as a hazardous waste. The responsible state agency or the regional office of the federal Environmental Protection Agency (EPA) should be contacted for specific recommendations.

RISK ASSESSMENT: BUSINESS

General Assessment

Accidents or mishaps involving sodium can present a significant threat to business operations. The loss or damage of equipment or facilities can significantly affect fiscal viability. Lawsuits that may result from personnel injury/death, public exposures, and/or environmental contamination will also require a serious expenditure of resources. Media attention surrounding an injury, death, or environmental damage can also result in a loss of profits and loss of current as well as future business.

Recommended Risk-Reduction Measures

Company attorneys, safety and health professionals, and environmental specialists should be involved in the development of any procedures for responding to chemical incidents. A company official should be predesignated as a public relations officer with specific training in dealing with the press. Corporate plans and policies should be developed, approved, and implemented long before any need for such arises.

REFERENCES

American Conference of Governmental Industrial Hygienists. 1988. *Documentation of the Threshold Limit Values and Biological Exposure Indices,* 5th Edition (with updates). Cincinnati: ACGIH

Bloom, A. D. and F. J. DeSerres. 1995. *Ecotoxicity and Human Health.* New York: CRC Press/Lewis Publishers.

Cothern, R. R. 1993. *Comparative Environmental Risk Assessment.* New York: CRC Press/Lewis Publishers.

Environmental Protection Agency. 1983. *Chemical Hazard Information Profiles.* Washington, D.C.: U.S. EPA.

Gardner, R. W. 1994. *Chemical Intolerance.* New York: CRC Press/Lewis Publishers.

Lauwerys, R. R., P. Hoet. 1993. *Industrial Chemical Exposure.* New York: CRC Press/Lewis Publishers.

Lewis, R. J., Sr. 1993. *Hawley's Condensed Chemical Dictionary,* 12th Edition. New York: Van Nostrand Reinhold.

Lewis, R. J., Sr. 1992. *Sax's Dangerous Properties of Industrial Materials,* Volumes 1, 2, and 3, 8th Edition. New York: Van Nostrand Reinhold.

Meyer, E. 1990. *Chemistry of Hazardous Materials.* Englewood Cliffs, NJ: Prentice Hall.

MATERIAL SAFETY DATA SHEET

CHEMICAL NAME	CRC PRESS/LEWIS PUBLISHERS DISCLAIMER
SODIUM AZIDE	THE INFORMATION AND RECOMMENDATIONS PRESENTED HEREIN ARE BASED ON SOURCES BELIEVED TO BE RELIABLE. CRC MAKES NO REPRESENTATION ON ITS COMPLETENESS OR ACCURACY. IT IS THE RESPONSIBILITY OF THE USER TO DETERMINE THE CHEMICAL'S SUITABILITY FOR ITS INTENDED USE, THE CHEMICAL'S SAFE USE, AND THE CHEMICAL'S PROPER DISPOSAL. NO REPRESENTATIONS AND/OR WARRANTIES, EITHER EXPRESSED OR IMPLIED, OF THE MERCHANTABILITY OR FITNESS FOR A PARTICULAR PURPOSE, OR OF ANY OTHER NATURE, ARE MADE WITH RESPECT TO THE INFORMATION PROVIDED IN THIS MSDS OR TO THE CHEMICAL TO WHICH INFORMATION MAY REFER. CRC NEITHER ASSUMES NOR AUTHORIZES ANY OTHER PERSON TO ASSUME FOR IT, ANY OTHER ADDITIONAL RESPONSIBILITY OR LIABILITY FOR THE USE OF, OR RELIANCE UPON, THIS INFORMATION.

HAZARD WARNING INFORMATION

HEALTH	FIRE	REACTIVE	OTHER	DEGREE OF HAZARD	COLOR CODING	OTHER CODES
3	1	0		0 = Minimum Hazard 1 = Slight Hazard 2 = Moderate Hazard 3 = Serious Hazard 4 = Severe Hazard	HEALTH = BLUE FIRE = RED REACTIVITY = YELLOW OTHER = WHITE	OX = Oxidizer ACID = Acid ALK = Alkali COR = Corrosive W = Use No Water

SECTION I - GENERAL INFORMATION

Characterization	**Metallic Azide**	RCRA Number	**P105**
		EPA Class	**Acute Hazardous Waste**
DOT Proper Shipping Name	**Sodium Azide**	Chemical Abstract Service (CAS) Number	**26628-22-8**
DOT Hazard Class and Label Requirements	**Poison B; Poison**	DOT Emergency Guide Code	**56**
DOT Identification Number	**UN 1687**	Atomic Symbol	**NaN_3**

Synonyms
Azide, azium, sodium salt of hydrazoic acid.

SECTION II - HAZARDOUS INGREDIENTS AND IDENTITY INFORMATION

Hazardous Components (specific identity)	OSHA Exposure Criteria	NIOSH Exposure Criteria	Immediately Dangerous to Life and Health (IDLH)	ACGIH Exposure Criteria
Sodium azide (derivation: Principally by treating a saturated solution of soda ash with carbon dioxide to precipitate the less soluble bicarbonate; also by purifying the crude product from the Solvay process). **1 ppm = 3 mg/m^3**	**PEL:** **Not Established** **STEL:** **Not Established**	**REL (ceiling):** **0.1 ppm (skin, as HN_3)** **0.3 mg/m^3 (skin, as NaN_3)** **STEL:** **Not Applicable**	**Not Determined**	**REL (ceiling):** **0.1 ppm** **0.3 mg/m^3** **(skin)** **STEL:** **Not Applicable**

SECTION III - PHYSICAL AND CHEMICAL CHARACTERISTICS

Boiling Point	**572°F (300°C) Decomposes**	Specific Gravity (H_2O = 1)	**1.85**
Vapor Pressure (mm Hg)	**Not Reported**	Atomic Weight	**22.98**
Vapor Density (Air = 1)	**Not Reported**	Melting Point	**572°F (300°C) Decomposes**

Solubility
Soluble in water and liquid ammonia. Slightly soluble in alcohol. Hydrolyzes in water to emit hydrazoic acid vapors.

Appearance and Odor
Colorless to white, odorless, crystalline solid.

SECTION IV - FIRE AND EXPLOSION HAZARD DATA

Flash Point (method used)	**Not Reported**	Explosive Limits in Air % by Volume	**LEL: Not Reported** **UEL: Not Reported**
NFPA Classification	**Combustible Solid (if heated above 572°F)**	Autoignition Temperature	**Not Reported**

Extinguishing Media
Use carbon dioxide, dry chemical, water spray, or alcohol foam.

Special Fire Fighting Procedures
Wear full protective clothing and self-contained breathing apparatus (SCBA). Poisonous gases are produced in fire. Fight fire from distance, if possible. Use water spray to cool fire-exposed containers.

Unusual Fire and Explosion Hazards
Sodium azide is used in medical laboratories during blood testing procedures. Improper disposal down sink drains can cause sodium azide to react with copper and lead pipes to form highly explosive lead azide and copper azide. Plumbers can then accidentally detonate these compounds.

SECTION V - REACTIVITY DATA

Stability		Conditions to Avoid
		Sodium azide is normally stable in closed containers under routine conditions of handling and storage. Keep away from source of heat or ignition and incompatible materials.
Stable	Unstable	Incompatibility *(materials to avoid)*
X		Acids, metals, water, copper, lead, brass, solder (in plumbing systems), benzoyl chloride and potassium hydroxide, bromine, carbon disulfide, chromyl chloride, dibromomalononitrile, nitric acid, and dimethyl sulfate.
Hazardous Polymerization		Conditions to Avoid
		Hazardous polymerization of sodium azide will not occur.
May Occur	Will Not Occur	Hazardous Decomposition or By-products
	X	Thermal decomposition of sodium azide produces nitrogen gas and sodium.

SECTION VI - HEALTH HAZARD DATA

Primary Route(s) of Entry:	Inhalation?	Absorption (skin/eye)?	Ingestion?
	X	X	X

Health Hazards

INHALATION: Hydrazoic acid vapors, a by-product of sodium azide, causes irritation of the eyes, nose, throat, and respiratory tract. It can cause chemical bronchitis, diarrhea, a debilitating headache, dangerously low blood pressure, tachycardia, low body temperature, acidosis, convulsions, weakness, and collapse. It forms strong complexes with hemoglobin to decrease the blood's oxygen-carrying ability. Pulmonary edema (fluid in the lungs) can also develop which can be fatal.

ABSORPTION: Readily absorbed through the skin to cause toxic systemic effects similar to inhalation.

INGESTION: Can be fatal, even in small amounts.

Carcinogenicity	NTP Listed?	IARC Cancer Review Group?	OSHA Regulated?	Target Organs?
Unknown Human Questioned Animal	No	No	No	Respiratory system, eyes, skin, red blood cells, CNS.

Medical Conditions Generally Aggravated by Exposure

Possible respiratory system disorders and/or blood diseases may be aggravated by exposure.

Emergency and First-aid Procedures

Eye contact: Do not allow victim to keep eyes tightly shut. Flush immediately with water for 15 minutes (minimum); seek medical attention. Skin contact: Remove all contaminated clothing. Immediately wash area with flooding amounts of water. For reddened or blistered skin, seek medical assistance. For inhalation: Remove the person from exposure. Provide respiratory assistance and CPR. Observe 48 hours for lung effects. If swallowed: Seek medical attention. If victim is conscious, give gastric lavage followed by saline catharsis. Never give anything by mouth to an unconscious or convulsing person.

SECTION VII - PRECAUTIONS FOR SAFE HANDLING AND USE

Steps to be Taken in Case Material is Released or Spilled

Collect powdered materials and deposit in clean, dry, sealed, containers. Do NOT dry sweep. Carefully scoop, shovel, or use HEPA vacuum. Ventilate area. Restrict those not involved in cleanup from area.

Preferred Waste Disposal Method

Destroy by nitrous acid treatment in three-necked flask equipped with a stirrer, dropping funnel and an outlet for toxic nitric oxide to carry the gas to an exhaust system.

Precautions to be Taken in Handling and Storage

Store to avoid contact with incompatible materials. Prevent physical damage to containers. Keep containers tightly closed. Keep moisture and sources of heat away. Use non-sparking tools.

Other Precautions and Warnings

Avoid excessive dust generation in storage and handling areas. Hydrazoic acid vapor is usually found in sodium azide storage areas. Automatic monitoring systems should be installed in storage areas.

SECTION VIII - CONTROL MEASURES AND PERSONAL PROTECTIVE EQUIPMENT

Respiratory Protection *(specify type)*

For low exposures, use an acid gas respirator with a HEPA filter; otherwise use a supplied-air respirator or a self-contained breathing apparatus with full facepiece operated in positive pressure mode.

Ventilation

Local exhaust (preferred) at site of chemical work, or general dilution ventilation.

Protective Gloves	Eye Protection	Other Protective Clothing
Impervious Material	Chemical/Dust Goggles or Face Mask	Protective Apron, Boots

Work/Hygiene Practices

Always wash hands thoroughly after using chemical; never bring food, drink, or smoking materials into vicinity of chemicals. Skin contact MUST be prevented.

SODIUM AZIDE

NaN_3

CAS: 26628-22-8

IDENTIFICATION AND TYPICAL USES

Sodium azide appears as colorless to white, hexagonal, odorless crystals. It is used in the preparation of hydrazoic acid, lead azide, and pure sodium. It is also used as a propellant for inflating automotive safety bags, as a reagent in the urinary azide-iodine test for carbon disulfide exposure, and as a pesticide in the control of weed and fruit-rot. It is a preservative in diagnostic medicinals and as an intermediate in explosives manufacture.

RISK ASSESSMENT: HEALTH

General Assessment

Sodium azide is extremely toxic by all routes of exposure (*inhalation, ingestion,* and dermal *absorption*). It is considered a questionable carcinogen in test animals (its human cancer potentially has not been defined). Human mutation data have been reported for sodium azide. There are no reports regarding its teratogenic properties in humans or animals.

Hydrazoic acid is almost always present with sodium azide. Inhalation can cause severe irritation of the eyes, nose, throat, and respiratory tract. It can lead to chemical bronchitis and possible delayed pulmonary edema (fluid in the lungs), which can be fatal. Symptoms can be delayed up to 48 hours which can create a false sense of security with regard to health exposure risk. Other systemic effects can include diarrhea, a severe and debilitating throbbing headache, dangerously low blood pressure, tachycardia (rapid heart beat), decreased body temperature, acidosis, convulsions, weakness, and collapse. Once inside the body, sodium azide forms strong complexes with hemoglobin resulting in a decrease in the ability of the red blood cells to effectively transport oxygen throughout the body.

Sodium azide is rapidly absorbed through intact skin to cause toxic system effects similar to that of inhalation. If combined exposures occur (inhalation and absorption), the resulting compound effects can be life-threatening.

Ingestion of even small amounts (e.g., several grams in water) can cause death. Smaller quantities may cause severe respiratory distress, hypermotility, and diarrhea.

Acute Health Effects

The following acute (short-term) health effects may occur immediately or shortly after human exposure to sodium azide:

Skin: May cause serious irritation. Rapid absorption is likely to occur, causing toxic system effects.

Eye: Severe irritation.

Lung: Irritation of the mucosa of the eyes, nose, throat, and respiratory tract. There may be shortness of breath, nausea, headache, and nose and respiratory tract irritation and possible delayed development of pulmonary edema.

CNS: Causes dizziness, headache, weakness, blurred vision, and other depressive effects due to the action of sodium azide on the blood's ability to transport oxygen.

Other: Ingestion of even small amounts can be fatal. Also, sodium azide can significantly affect the blood's ability to transport oxygen. This can result in potentially dangerous low blood pressure, throbbing headaches, and rapid heart beat. It can also cause kidney changes.

Chronic Health Effects

The following chronic (long-term) health effects may occur at some time after exposure to sodium azide and can last for months or even years:

Cancer Hazards: According to the references, sodium azide has caused skin and endocrine tumors in rats. It is not known if human exposure to this compound will have the same effects. However, human mutation data have been reported and many scientists believe that long-term exposure to such chemicals may pose a cancer risk.

Reproductive Hazard: According to the references, there are no reports of any teratogenic or adverse reproductive effects resulting from exposure to sodium azide.

Other Chronic Effects: Long-term or repeated exposures may cause kidney changes.

⓿ *Recommended Risk-Reduction Measures*

Sodium azide is an extremely toxic compound. Personnel should, therefore, avoid direct contact with sodium azide. If a less toxic material or compound cannot be substituted for sodium azide, then *engineering controls* are the most effective method of reducing exposures. The best protection is to enclose operations and/or provide local exhaust ventilation at the site of sodium azide dust release. While not always operationally feasible, isolating operations can also reduce exposure. Using respiratory protection is less effective than the controls mentioned above, but is still advisable whenever working with or around sodium azide. The recommended exposure levels are extremely low and difficult to accurately monitor. Human exposures must be prevented. For low or transient exposures, a chemical cartridge respirator equipped with a high-efficiency particulate air (HEPA) filter and a dust/mist pre-filter may provide adequate respiratory protection. However, better protection is obtained using a supplied-air respirator operated in continuous flow mode or a self-contained breathing apparatus (SCBA) with full facepiece operated in pressure demand. If a full facepiece is not available, then chemical/dust goggles should be worn to protect the eyes. A face shield and protective apron should also be worn. To prevent hand and skin exposures, gloves should be used. Glove manufacturers should be contacted and permeation studies obtained *before* final glove selections are made.

Administrative controls should also be in place to minimize the potential for human exposures. These may include written procedures or policies which specify the methods and techniques that will be practiced whenever personnel are to work with sodium azide.

All personnel should receive *training* on the use, hazards, protective measures, emergency actions, and other precautions per 29 CFR 1910.1200 (Hazard Communication), prior to the first assignment in an area where sodium azide is used or stored.

Before beginning employment working with sodium azide, and at regular intervals thereafter (e.g., annually), those personnel with frequent or potentially high exposures should be provided the following recommended medical tests:

☑ Lung function tests (establish baseline).

If symptoms develop or overexposure is suspected, consider the following additional tests:

☑ Lung function tests (compare to baseline).

☑ Consider chest X-ray after acute overexposure (may be negative if taken immediately after exposure due to delayed onset of pulmonary edema).

☑ Electrocardiogram (EKG) and blood pressure monitoring (for indication of hypotension).

Any evaluation should include a careful history of past and present symptoms with an examination. Medical tests that simply look for existing damage are not a substitute for controlling exposures. Also, since smoking can cause heart disease as well as lung cancer, emphysema, and other known respiratory problems, smokers exposure to sodium azide may experience symptoms more rapidly than non-smokers under the same conditions of exposure. Prudent risk management requires careful consideration of *all* factors which may be causing exposure symptoms in the workplace.

Other methods to reduce exposure include:

☑ Where possible, enclose operations and use local exhaust ventilation at the site of chemical release. If local exhaust ventilation or enclosure is not used, respiratory protection should be mandatory.

☑ Always ensure that proper protective clothing is worn when using chemical substances. Proper maintenance and care of protective equipment is essential to its proper function.

☑ Wash thoroughly immediately after exposure to sodium azide and at the end of the work shift or before eating, drinking, smoking, or applying cosmetics.

☑ Hazard warning information should be posted in the work area. In addition, as part of an on-going education and training program, all information on the health and safety hazards of sodium azide should be communicated to all exposed and potentially exposed workers.

☑ Eye wash stations should be provided in the immediate work area for emergency use. If there is a

possibility of skin exposure to sodium azide, emergency shower facilities should be provided.

☑ Workers whose clothing has been contaminated by sodium azide should change into clean clothes before leaving work. Contaminated work clothing should be laundered only by individuals who have been informed of the hazards of exposure to sodium azide.

RISK ASSESSMENT: ENVIRONMENT

General Assessment

Accidental spills or releases to the environment can occur during transportation, storage, disposal, or destruction of sodium azide. In almost every scenario, the threat of environmental exposure is contingent upon the proper handling of the chemical substance. Accidental spills, large or small, where sodium azide contacts incompatible commodities can result in release of toxic products (depending upon conditions of spill), and possible contamination of the surrounding environmental mediums (water, soil, and air).

Sodium azide is considered a combustible solid at high temperatures. However, it is dangerously reactive with many compounds and chemicals, including benzoyl chloride and potassium hydroxide, bromine, carbon disulfide, chromyl chloride, most acids, metals, and even water. It will react in water to form hydrazoic acid. A solution of sodium azide in contact with copper pipes or lead soldering can react to form highly explosive lead azide and copper azide. Sodium azide is often used during the testing of blood. If it is improperly disposed of down sink drains, the potential for the production of explosive copper and lead azide becomes a serious concern. Extreme caution is required in handling, storage, transportation, and disposal of sodium azide. These characteristics also require special consideration during any emergency situation involving a leak or spill of sodium azide powder or dust. Should sodium azide ever come into contact with these various incompatible substances either during use, transportation, storage, or disposal, violent and explosive reactions are possible.

Sodium azide may enter the environment as a result of industrial and municipal waste treatment plant discharges and from spills.

Acute Ecological Effects

Acute (short-term) toxic effects may include the death of animals, birds, or fish and death or low growth rate in plants. Acute effects are seen 2 to 4 days after animals or plants are exposed to sodium azide.

Insufficient data are available to evaluate the short-term effects of sodium azide to plants, birds, or land animals. However, it should be noted that the U.S. Environmental Protection Agency does list sodium azide as an acutely hazardous substance.

Chronic Ecological Effects

Chronic (long-term) toxic effects may include shortened life span, reproductive problems, lower fertility, and changes in appearance or behavior in exposed animals. These effects can be seen long after first exposure(s) to toxic chemicals.

Insufficient data are available to evaluate or predict the long-term effects of sodium azide to plants, birds, or land animals. However, it should be noted that the U.S. Environmental Protection Agency does list sodium azide as an acutely hazardous substance.

Water Solubility

Sodium azide reacts in water to form hydrazoic acid.

Persistence in the Environment

Because sodium azide reacts so rapidly in water, it is not expected to be persistent in the aquatic environment. Its half-life has not been established. The half-life of a pollutant is the amount of time it takes for one half of the chemical to be degraded.

Bioaccumulation in Aquatic Organisms

Some substances increase in concentration, or *bioaccumulate,* in living organisms as they breathe contaminated air, drink contaminated water, or eat contaminated food. These chemicals can become concentrated in the tissues and internal organs of animals as well as humans.

The concentration of sodium azide found in fish tissues is expected to be much lower than the average concentration of sodium azide in the water from which the fish was taken.

Recommended Risk-Reduction Measures

Proper training of all transporters will reduce the likelihood of a mishap or accident resulting in a leak or spill of sodium azide dusts or powders into the environment. The correct use of labeling on all containers, trucks, and rail cars will enable emergency responders to react properly and quickly to any disaster thereby reducing the potential risk to the environment and to personnel.

Storage of sodium azide should be segregated from incompatible chemicals to minimize the risk of cross-contamination or contact. Care should be taken to prevent physical damage to storage containers. Also, dusty conditions must be avoided in storage and use areas. When sodium azide comes into contact with water, it will readily liberate excessive amounts of corrosive hydrazoic acid vapors. Employers should consider the use of automatic monitoring devices to warn employees of increasing amounts of hydrazoic vapors in storage areas.

If a spill or leak to the environment has occurred, fire department, emergency response and/or hazardous materials spill personnel should be notified immediately. Cleanup should be attempted only by those trained in proper spill containment procedures. Contaminated soils should be removed for decontamination and replaced with clean soil. If sodium azide should contact the water table, aquifer, or navigable waterway, time is of the essence. It is decomposes in water and total remediation may not be possible. The local and/or state emergency response authorities may require notification. A comprehensive emergency response or disaster preparedness/recovery plan should be in place prior to any operations involving the use, transportation, storage, or disposal of sodium azide.

If sodium azide is spilled or leaked, the following specific steps are recommended:

- ☑ Restrict persons not wearing protective clothing from area of spill or leak until cleanup is complete and area can be opened for normal work.
- ☑ Remove all ignition sources from area.
- ☑ Ventilate area of spill or leak.
- ☑ Collect powdered materials in the safest manner possible and deposit in sealed containers. Recommend using a vacuum equipped with a high-efficiency particulate air (HEPA) filter. Do not dry sweep (generates airborne dusts). Keep water away (do NOT damp mop residues).
- ☑ It may be necessary to dispose of sodium azide as a hazardous waste. The responsible state agency or the regional office of the federal Environmental Protection Agency (EPA) should be contacted for specific recommendations.

RISK ASSESSMENT: BUSINESS

General Assessment

Accidents or mishaps involving sodium azide can present a significant threat to business operations. The loss or damage of equipment or facilities can significantly affect fiscal viability. Lawsuits that may result from personnel injury/death, public exposures, and/or environmental contamination will also require a serious expenditure of resources. Media attention surrounding an injury, death, or environmental damage can also result in a loss of profits and loss of current as well as future business. Always remember that anytime the terms "cancer" or "carcinogenic" are used, public emotion, hysteria, ignorance, and anxiety can all run equally high. This must be carefully considered whenever drafting or implementing public relations policies.

Recommended Risk-Reduction Measures

Company attorneys, safety and health professionals, and environmental specialists should be involved in the development of any procedures for responding to chemical incidents. A company official should be predesignated as a public relations officer with specific training in dealing with the press. Corporate plans and policies should be developed, approved, and implemented long before any need for such arises.

REFERENCES

American Conference of Governmental Industrial Hygienists. 1988. *Documentation of the Threshold Limit Values and Biological Exposure Indices*, 5th Edition (with updates). Cincinnati: ACGIH

Bloom, A. D. and F. J. DeSerres. 1995. *Ecotoxicity and Human Health.* New York: CRC Press/Lewis Publishers.

Cothern, R. R. 1993. *Comparative Environmental Risk Assessment.* New York: CRC Press/Lewis Publishers.

Environmental Protection Agency. 1983. *Chemical Hazard Information Profiles.* Washington, D.C.: U.S. EPA.

Gardner, R. W. 1994. *Chemical Intolerance.* New York: CRC Press/Lewis Publishers.

Lauwerys, R. R., P. Hoet. 1993. *Industrial Chemical Exposure.* New York: CRC Press/Lewis Publishers.

Lewis, R. J., Sr. 1993. *Hawley's Condensed Chemical Dictionary*, 12th Edition. New York: Van Nostrand Reinhold.

Lewis, R. J., Sr. 1992. *Sax's Dangerous Properties of Industrial Materials*, Volumes 1- 3, 8th Edition. New York: Van Nostrand Reinhold.

Meyer, E. 1990. *Chemistry of Hazardous Materials.* Englewood Cliffs, NJ: Prentice Hall.

MATERIAL SAFETY DATA SHEET

CHEMICAL NAME	
SODIUM CYANIDE	**CRC PRESS/LEWIS PUBLISHERS DISCLAIMER:** THE INFORMATION AND RECOMMENDATIONS PRESENTED HEREIN ARE BASED ON SOURCES BELIEVED TO BE RELIABLE. CRC MAKES NO REPRESENTATION ON ITS COMPLETENESS OR ACCURACY. IT IS THE RESPONSIBILITY OF THE USER TO DETERMINE THE CHEMICAL'S SUITABILITY FOR ITS INTENDED USE, THE CHEMICAL'S SAFE USE, AND THE CHEMICAL'S PROPER DISPOSAL. NO REPRESENTATIONS AND/OR WARRANTIES, EITHER EXPRESSED OR IMPLIED, OF THE MERCHANTABILITY OR FITNESS FOR A PARTICULAR PURPOSE, OR OF ANY OTHER NATURE, ARE MADE WITH RESPECT TO THE INFORMATION PROVIDED IN THIS MSDS OR TO THE CHEMICAL TO WHICH INFORMATION MAY REFER. CRC NEITHER ASSUMES NOR AUTHORIZES ANY OTHER PERSON TO ASSUME FOR IT, ANY OTHER ADDITIONAL RESPONSIBILITY OR LIABILITY FOR THE USE OF, OR RELIANCE UPON, THIS INFORMATION.

HAZARD WARNING INFORMATION

HEALTH	FIRE	REACTIVE	OTHER	DEGREE OF HAZARD	COLOR CODING	OTHER CODES
3	0	0		0 = Minimum Hazard 1 = Slight Hazard 2 = Moderate Hazard 3 = Serious Hazard 4 = Severe Hazard	HEALTH = BLUE FIRE = RED REACTIVITY = YELLOW OTHER = WHITE	OX = Oxidizer ACID = Acid ALK = Alkali COR = Corrosive W = Use No Water

SECTION I - GENERAL INFORMATION

Characterization	RCRA Number	EPA Class
Cyanides, Inorganic	**P106**	**Acute Hazardous Waste**

DOT Proper Shipping Name **Sodium cyanide, Solid and Solution**	Chemical Abstract Service (CAS) Number **143-33-9**
DOT Hazard Class and Label Requirements **Poison B; Poison**	DOT Emergency Guide Code **55**
DOT Identification Number **UN 1689**	Chemical Formula **NaCN**

Synonyms
Sodium salt of hydrocyanic acid; cyanide of sodium; Cymag.

SECTION II - HAZARDOUS INGREDIENTS AND IDENTITY INFORMATION

Hazardous Components (specific identity)	OSHA Exposure Criteria	NIOSH Exposure Criteria	Immediately Dangerous to Life and Health (IDLH)	ACGIH Exposure Criteria
Sodium cyanide (derivation: By absorption of hydrogen cyanide in a solution of sodium hydroxide followed by vacuum evaporation).	**PEL (CN): skin: 5 mg/m³** **STEL: Not Established**	**REL (CN): ceiling: 5 mg/m³ (10 minutes)** **STEL: Not Applicable**	**25 mg(CN)m³**	**TLV (CN): skin: 5 mg/m³** **STEL: Not Established**

SECTION III - PHYSICAL AND CHEMICAL CHARACTERISTICS

Boiling Point **2725°F (1496°C)**	Specific Gravity (H_2O = 1) **1.60**
Vapor Pressure (mm Hg) **0 (approx.) at 68°F (20°C)**	Molecular Weight (atomic weight) **49.0**
Vapor Density (Air = 1) **Not Reported**	Melting Point **1047°F (564°C)**

Solubility
Soluble in water. Slightly soluble in alcohol.

Appearance and Odor
White, granular, crystalline solid with a faint, almond-like odor. Odorless if kept dry.

SECTION IV - FIRE AND EXPLOSION HAZARD DATA

Flash Point (method used) **Not Applicable**	Explosive Limits in Air % by Volume **LEL: N/A UEL: N/A**
NFPA Classification **Non-Combustible Solid**	Autoignition Temperature **Not Applicable**

Extinguishing Media
Use agent suitable to surrounding fire. Use water or water spray. Do NOT use carbon dioxide.

Special Fire Fighting Procedures
Wear full protective clothing and self-contained breathing apparatus (SCBA). Poisonous cyanide and nitric oxide fumes are produced in fire. Structural protective clothing is permeable and may not provide adequate protection. Avoid contact with skin. If possible and without risk, move containers from fire area; otherwise, use water spray to cool fire-exposed containers and continue even after fire has been extinguished. Fight fire from distance or protected location, if possible.

Unusual Fire and Explosion Hazards
Toxic and flammable hydrogen cyanide gas may be released when sodium cyanide contacts acids.

SECTION V - REACTIVITY DATA

Stability		Conditions to Avoid Sodium cyanide is normally stable. Avoid contact with incompatible materials. Keep sodium cyanide powders away from open flame, heat, or other sources of ignition.
Stable X	Unstable	Incompatibility *(materials to avoid)* Reacts violently with strong oxidizers, acids, acid salts, chlorates, nitrates, and peroxides. It absorbs moisture from the air to form a syrup.
Hazardous Polymerization		Conditions to Avoid Under normal conditions of temperature and pressure, hazardous polymerization of sodium cyanide will not occur.
May Occur	Will Not Occur X	Hazardous Decomposition or By-products When heated to decomposition, sodium cyanide releases toxic fumes of cyanide and nitric oxide. On contact with acids, toxic hydrogen cyanide will be released.

SECTION VI - HEALTH HAZARD DATA

Primary Route(s) of Entry:	Inhalation? X	Absorption (skin/eye)? X	Ingestion? X

Health Hazards

INHALATION: The cyanide anions will inhibit the cell's ability to use oxygen which can lead to metabolic asphyxiation. This can cause anoxia, weakness, and damage to the central nervous system (CNS) with symptoms of headache, dizziness, confusion, nausea, vomiting, increased respiration or slow, gasping respiration, irregular heart beat, unconsciousness, convulsions, hallucinations, distorted perceptions, gastritis, coma, and death.

ABSORPTION: Will pass through unbroken skin and enter the bloodstream. Large exposures can be fatal.

INGESTION: A poison by ingestion. May cause toxic systemic effects and death.

Carcinogenicity	NTP Listed?	IARC Cancer Review Group?	OSHA Regulated?	Target Organs?
Unknown Human Unknown Animal	No	No	29 CFR 1910.1000 Table Z-1 (as CN)	Eyes, skin, CVS, CNS, thyroid, blood.

Medical Conditions Generally Aggravated by Exposure

Disease of the kidney, heart, lungs, and central nervous system may be aggravated by exposure.

Emergency and First-aid Procedures

Eye contact: Do not allow victim to keep eyes tightly shut. Flush immediately with water for 15 minutes (minimum); seek medical attention. Skin contact: Remove all contaminated clothing. Immediately wash area with flooding amounts of soap and water. Do not allow water to enter nose or mouth. For inhalation: Remove the person from exposure. Break amyl nitrate pearl in cloth and hold lightly under nose for 15 seconds; repeat 5 times at 15-second intervals. Give oxygen. Provide respiratory assistance and CPR. Transfer to medical facility. If swallowed: Break amyl nitrate pearl in cloth and hold lightly under nose for 15 seconds. If conscious, induce vomiting. Repeat until vomit is clear. Repeat inhalation of amyl nitrate 5 times at 15-second intervals. Provide CPR if necessary. Seek medical attention immediately. Call poison control center for advice. Never give anything by mouth to an unconscious or convulsing person.

SECTION VII - PRECAUTIONS FOR SAFE HANDLING AND USE

Steps to be Taken in Case Material is Released or Spilled

Collect powdered materials in safest manner possible and deposit in sealed containers. Use HEPA vacuum. Restrict those not involved in cleanup from entering area. Notify appropriate authorities.

Preferred Waste Disposal Method

No citation.

Precautions to be Taken in Handling and Storage

Store in tightly closed containers in cool, dry, well-ventilated area. Protect containers from damage.

Other Precautions and Warnings

Protective equipment must be worn when handling or preparing aqueous solutions.

SECTION VIII - CONTROL MEASURES AND PERSONAL PROTECTIVE EQUIPMENT

Respiratory Protection *(specify type)*

For any exposure level use a NIOSH/MSHA-approved self-contained breathing apparatus with full facepiece operated in positive pressure mode or a supplied-air respirator in continuous flow mode.

Ventilation

Local exhaust (preferred) at site of chemical work or general ventilation.

Protective Gloves	Eye Protection	Other Protective Clothing
Impervious Material	Chemical/Dust Goggles or Face Mask	Protective Apron, Boots (avoid skin contact)

Work/Hygiene Practices

Always wash hands thoroughly after using chemical; never bring food, drink, or smoking materials into vicinity of chemicals.

SODIUM CYANIDE

NaCN

CAS: 143-33-9

IDENTIFICATION AND TYPICAL USES

Sodium cyanide is a white, granular, crystalline solid with a faint but distinguishable odor of bitter almonds (especially when wet). It is used in the extraction of gold and silver from ores, in electroplating, in the heat treatment of metals, in the manufacture of hydrogen cyanide, as an insecticide, in cleaning metals, in fumigation, in the manufacture of dyes and pigments, in nylon intermediates, in chelating compounds, and in ore flotation.

RISK ASSESSMENT: HEALTH

General Assessment

Sodium cyanide is a deadly human poison. The toxic routes of entry are *inhalation* and *ingestion* of its dusts, and skin contact (*absorption*). Exposure by either route in large enough doses can lead to death. There are no reports of any carcinogenic or mutagenic response following long-term exposure. However, it is considered an experimental teratogen, causing adverse reproductive effects in animals.

Because of the cyanide anions, inhalation, ingestion, or dermal absorption of sodium cyanide can inhibit the cell's ability to properly utilize oxygen from the blood which can lead to metabolic asphyxiation. Prolonged anoxia can subsequently lead to damage to the central nervous system (CNS). Symptoms can include weakness, headache, confusion, nausea, vomiting, increased respiration or slow gasping breaths, weak and irregular heart beat, unconsciousness, convulsions, hallucinations, distorted perceptions, coma, and death. There can also be changes in the thyroid and kidneys on long-term exposure.

Acute Health Effects

The following acute (short-term) health effects may occur immediately or shortly after exposure to sodium cyanide:

<u>Skin</u>: Severe irritation may occur. Absorption is likely.

<u>Eye</u>: Irritation and possible damage on contact.

<u>Lung</u>: High exposures can cause rapid and severe lung damage, with shortness of breath, chest pain, cough, loss of consciousness, and death.

<u>CNS</u>: Dizziness, headaches, hallucinations, distorted perceptions, weakness, nausea, and other typical symptoms of CNS depression can occur on exposure to sodium cyanide.

<u>Other</u>: High exposure to sodium cyanide may be rapidly fatal.

Chronic Health Effects

The following chronic (long-term) health effects may occur at some time after exposure to sodium cyanide and can last for months or even years:

<u>Cancer Hazards</u>: According to the information in the references, the data on the cancer-causing properties of sodium cyanide are inconclusive. However, this is probably due to the fact that it is highly poisonous and long-term exposures seldom occur (death is more likely).

<u>Reproductive Hazard</u>: According to information presented in the references, sodium cyanide has been shown to cause teratogenic effects in animals. It is not known if human exposure will result in adverse reproductive effects.

<u>Other Chronic Effects</u>: Prolonged contact can lead to chronic skin rash. Sodium cyanide may cause the thyroid gland to enlarge and interfere with normal thyroid function. Damage to the nervous system can result from severe exposure and possibly lower repeated (chronic) exposures. Repeated exposures can irritate the nose, causing discharge, nosebleeds, and sores inside the nose. Sodium cyanide may cause changes in the blood cell count and damage the kidneys.

Recommended Risk-Reduction Measures

Personnel should avoid direct contact with sodium cyanide dusts or powders. If a less toxic material or compound cannot be substituted for sodium cyanide, then *engineering controls* are the most effective method of reducing exposures. The best protection is to enclose operations and/or provide local exhaust ventilation at the site of sodium cyanide dust release. While not always operationally feasible, isolating operations can also reduce exposure. Using respiratory protection is less effective than the controls mentioned above, but is still advisable whenever working with or around sodium cyanide. At any exposure level, a self-contained breathing apparatus (SCBA) with full facepiece operated in pressure demand is recommended for the greatest possible respiratory protection. If a full facepiece is not available, then chemical/dust goggles should be worn to protect the eyes. A face shield and protective apron should also be worn. To prevent hand and skin exposures, gloves should be used. Glove manufacturers should be contacted and permeation studies obtained *before* final glove selections are made.

Administrative controls should also be in place to minimize the potential for human exposures. These may include written procedures or policies which specify the methods and techniques that will be practiced whenever personnel are to work with sodium cyanide.

All personnel should receive *training* on the use, hazards, protective measures, emergency actions, and other precautions per 29 CFR 1910.1200 (Hazard Communication), prior to the first assignment in an area where sodium cyanide is used or stored.

If exposure occurs, break an amyl nitrate pearl into a cloth and hold it lightly under the victim's nose for 15 seconds. This should be repeated five times, at 15-second intervals.

For those with frequent or potentially high exposure, or significant skin contact, the following medical tests are recommended before initial assignment and at regular times thereafter:

☑ Urine thiocyanate test (most accurate if performed right after exposure).

If symptoms develop or overexposure is suspected, the following should also be considered:

☑ Blood cyanide level. (Note: Since cigarettes contain some cyanide, smokers may show an increase in cyanide level on testing).

☑ Evaluation of thyroid function

☑ Lung function test and chest X-ray.

☑ Complete blood count.

Any evaluation should include a careful history of past and present symptoms with an examination. Medical tests that simply look for existing damage are not a substitute for controlling exposures. Also, personnel who smoke should avoid any unprotected contact with sodium cyanide.

Other methods to reduce exposure include:

☑ Where possible, enclose operations and use local exhaust ventilation at the site of chemical release. If local exhaust ventilation or enclosure is not used, respiratory protection should be mandatory.

☑ Always ensure that proper protective clothing is worn when using chemical substances.

☑ Wash thoroughly immediately after exposure to sodium cyanide and at the end of the work shift or before eating, drinking, or smoking.

☑ Hazard warning information should be posted in the work area. In addition, as part of an on-going education and training program, all information on the health and safety hazards of sodium cyanide should be communicated to all exposed workers.

☑ Eye wash stations should be provided in the immediate work area for emergency use. If there is a possibility of skin exposure to sodium cyanide, emergency shower facilities should be provided.

☑ Workers should never wear contaminated clothing home (family members can be exposed). Contaminated work clothing should be laundered only by individuals who have been informed of the hazards of exposure to sodium cyanide.

RISK ASSESSMENT: ENVIRONMENT

General Assessment

Sodium cyanide is a naturally occurring element. However, industrial uses place the environment at risk of exposure during transportation, storage, disposal, or destruction of sodium cyanide compounds. In almost every scenario, the threat of environmental exposure is contingent upon the proper handling of the chemical substance. Accidental spills, large or small, where sodium cyanide contacts incompatible commodities can result in the release of hazardous fumes and gases (depending upon conditions of spill), and possible contamination of the surrounding environmental mediums (water, soil, and air).

In its bulk form, sodium cyanide is considered a non-combustible solid. However, contact with acids can release highly flammable hydrogen cyanide gas. Because it is incompatible with many commodities, including commonly encountered acids and oxidizers, extreme caution is required in handling, storage, transportation, and disposal of sodium cyanide. These characteristics also require special consideration during any emergency situation involving a leak or spill of sodium cyanide powder or dust. Should sodium cyanide ever come into contact with any incompatible substances either during use, transportation, storage, or disposal, release of poisonous fumes are released and possible violent reactions are possible.

Compounds containing the cyanide group are used and readily formed in many industrial processes and can be found in a variety of effluents, such as those from the steel, petroleum, plastics, synthetic fibers, metal plating, mining, and chemical industries. It is generally accepted that "free cyanide" is the toxic form and its formation is primarily dependent upon water temperature and pH. Conditions in most natural waters are right to produce this toxic form of cyanide.

Sodium cyanide can enter the environment from industrial and municipal waste treatment plant discharges and from spills. Also, as a fumigant and insecticide, it will also enter the environment as a result of its intended use.

Acute Ecological Effects

Acute (short-term) toxic effects may include the death of animals, birds, or fish and death or low growth rate in plants. Acute effects are seen 2 to 4 days after animals or plants are exposed to sodium cyanide.

Cyanides have high acute toxicity to aquatic life, birds, and terrestrial animals. Insufficient data are available to evaluate the short-term effects of cyanide to plants.

Chronic Ecological Effects

Chronic (long-term) toxic effects may include shortened life span, reproductive problems, lower fertility, and changes in appearance or behavior in exposed animals. These effects can be seen long after first exposure(s) to toxic chemicals.

Cyanide have high chronic toxicity to aquatic life. Insufficient data are available to evaluate or predict the long-term effects of cyanide to plants, birds, and terrestrial animals.

Water Solubility

Sodium cyanide is moderately soluble in water. Concentrations of 1 to 1000 milligrams will mix with a liter of water.

Persistence in the Environment

Sodium cyanide is non-persistent in water, with a half-life of less than 2 days. The half-life of a pollutant is the amount of time it takes for one half of the chemical to be degraded.

Bioaccumulation in Aquatic Organisms

Some substances increase in concentration, or *bioaccumulate,* in living organisms as they breathe contaminated air, drink contaminated water, or eat contaminated food. These chemicals can become concentrated in the tissues and internal organs of animals as well as humans.

The concentration of cyanides found in fish tissues normally consumed by humans is expected to be about the same as the average concentration of cyanides found in the water from which the fish was taken.

Recommended Risk-Reduction Measures

Proper training of all transporters will reduce the likelihood of a mishap or accident resulting in a leak or spill of sodium cyanide dusts or powders into the environment. The correct use of DOT labeling (per the MSDS) on all containers, trucks, and rail cars will enable emergency responders to react properly and quickly to any disaster thereby reducing the potential risk to the environment and to personnel.

Storage of sodium cyanide should be segregated from incompatible chemicals to minimize the risk of cross-contamination or contact. Proper ventilation and protective equipment should be used while handling the solid or while preparing aqueous solutions. If a spill or leak to the environment has occurred, fire department, emergency response and/or hazardous materials spill personnel should be notified immediately. Cleanup should be attempted only by those trained in proper spill containment procedures. If sodium cyanide should contact the water table, aquifer, or navigable waterway, time is of the essence. It is moderately soluble and total remediation may not be possible. The local and/or state emergency response authorities must be notified. A comprehensive emergency response or disaster preparedness/recovery plan should be in place prior to any operations involving the use, transportation, storage, or disposal of sodium cyanide. If sodium cyanide powder is spilled or leaked, the following specific steps are recommended:

☑ Restrict persons not wearing protective clothing from area of spill or leak until cleanup is complete and area can be opened for normal work.

☑ Remove all ignition sources.

☑ Collect powdered materials in the safest manner possible and deposit in sealed containers. Recommend using vacuum with HEPA filter. Do not dry sweep (generates airborne dusts).

☑ It may be necessary to dispose of sodium cyanide as a hazardous waste. The responsible state agency or the regional office of the federal Environmental Protection Agency (EPA) should be contacted for specific recommendations.

RISK ASSESSMENT: BUSINESS

General Assessment

Accidents or mishaps involving sodium cyanide can present a significant threat to business operations. The loss or damage of equipment or facilities can significantly affect fiscal viability. Any incident involving a chemical containing the word "cyanide" in its description can evoke an emotional reaction from the public, even when no exposures have occurred. Lawsuits that may result from personnel injury/death, public exposures, and/or environmental contamination will also require a serious expenditure of resources. Media attention surrounding an injury, death, or environmental damage can also result in a loss of profits and loss of current as well as future business.

Recommended Risk-Reduction Measures

Company attorneys, safety and health professionals, and environmental specialists should be involved in the development of any procedures for responding to chemical incidents. A company official should be predesignated as a public relations officer with specific training in dealing with the press. Corporate plans and policies should be developed, approved, and implemented long before any need for such arises.

REFERENCES

Ahlbom, A. 1993. *Biostatistics for Epidemiologists.* New York: CRC Press/Lewis Publishers.

American Conference of Governmental Industrial Hygienists. 1988. *Documentation of the Threshold Limit Values and Biological Exposure Indices,* 5th Edition (with updates). Cincinnati: ACGIH

Brusick, D. J. 1994. *Methods for Genetic Risk Assessment.* New York: CRC Press/Lewis Publishers.

Bloom, A. D. and F. J. DeSerres. 1995. *Ecotoxicity and Human Health.* New York: CRC Press/Lewis Publishers.

Bueche, J. R. 1972. *Principles of Physics,* 2nd Edition. New York: McGraw-Hill.

Calabrese, E. J. 1994. *Biological Effects of Low Level Exposures.* New York: CRC Press/Lewis Publishers.

Cothern, R. R. 1993. *Comparative Environmental Risk Assessment.* New York: CRC Press/Lewis Publishers.

Gardner, R. W. 1994. *Chemical Intolerance.* New York: CRC Press/Lewis Publishers.

Hallenbeck, W. H. 1993. *Quantitative Risk Assessment for Environmental and Occupational Health,* 2nd Edition. New York: CRC Press/Lewis Publishers.

Howard, P. H., et. al. 1991. *Handbook of Environmental Degradation Rates.* New York: CRC Press/Lewis Publishers.

Jones, F. E. 1994. *Toxic Organic Vapors in the Workplace.* New York: CRC Press/Lewis Publishers.

Landis, W. G., M. H. Yu. 1995. *Introduction to Environmental Toxicology.* New York: CRC Press/Lewis Publishers.

Lauwerys, R. R., P. Hoet. 1993. *Industrial Chemical Exposure.* New York: CRC Press/Lewis Publishers.

Lewis, R. J., Sr. 1992. *Sax's Dangerous Properties of Industrial Materials,* Volumes 1, 2, and 3, 8th Edition. New York: Van Nostrand Reinhold.

Lide, D. R. 1995. *Handbook of Chemistry and Physics,* 76th Edition. New York: CRC Press/Lewis Publishers.

Manahan, S. E. 1992. *Toxicological Chemistry,* 2nd Edition. New York: CRC Press/Lewis Publishers.

Meyer, E. 1990. *Chemistry of Hazardous Materials.* Englewood Cliffs, NJ: Prentice Hall.

Mycock, J. C., J. D. McKenna, and L. Theodore. 1995. *Handbook of Air Pollution Control Engineering and Technology.* New York: CRC Press/Lewis Publishers.

Patnaik, P. 1992. *A Comprehensive Guide to the Hazardous Properties of Chemical Substances.* New York: Van Nostrand Reinhold.

Quigley, D. R. 1994. *Handbook of Emergency Chemical Management.* New York: CRC Press/Lewis Publishers.

Reeves, A. L. 1979. Sodium cyanide. *Handbook on the Toxicology of Metals.* pp 321-328. New York: Elsevier North-Holland.

Richardson, M. 1995. *Dictionary of Substance and Their Effects,* Vol. I-VI. New York: CRC Press/Lewis Publishers.

Sacarello, H. L. A. 1994. *The Comprehensive Handbook of Hazardous Materials.* New York: CRC Press/Lewis Publishers.

Torkelson, T. R. V. K. Rowe. 1981. *Patty's Industrial Hygiene and Toxicology,* Vol. 2B. New York: Wiley-Interscience.

MATERIAL SAFETY DATA SHEET

CHEMICAL NAME

SODIUM HYDROXIDE

CRC PRESS/LEWIS PUBLISHERS DISCLAIMER: THE INFORMATION AND RECOMMENDATIONS PRESENTED HEREIN ARE BASED ON SOURCES BELIEVED TO BE RELIABLE. CRC MAKES NO REPRESENTATION ON ITS COMPLETENESS OR ACCURACY. IT IS THE RESPONSIBILITY OF THE USER TO DETERMINE THE CHEMICAL'S SUITABILITY FOR ITS INTENDED USE, THE CHEMICAL'S SAFE USE, AND THE CHEMICAL'S PROPER DISPOSAL. NO REPRESENTATIONS AND/OR WARRANTIES, EITHER EXPRESSED OR IMPLIED, OF THE MERCHANTABILITY OR FITNESS FOR A PARTICULAR PURPOSE, OR OF ANY OTHER NATURE, ARE MADE WITH RESPECT TO THE INFORMATION PROVIDED IN THIS MSDS OR TO THE CHEMICAL TO WHICH INFORMATION MAY REFER. CRC NEITHER ASSUMES NOR AUTHORIZES ANY OTHER PERSON TO ASSUME FOR IT, ANY OTHER ADDITIONAL RESPONSIBILITY OR LIABILITY FOR THE USE OF, OR RELIANCE UPON, THIS INFORMATION.

HAZARD WARNING INFORMATION

HEALTH	FIRE	REACTIVE	OTHER	DEGREE OF HAZARD	COLOR CODING	OTHER CODES
3	0	0	~~W~~ ALK	0 = Minimum Hazard 1 = Slight Hazard 2 = Moderate Hazard 3 = Serious Hazard 4 = Severe Hazard	HEALTH = BLUE FIRE = RED REACTIVITY = YELLOW OTHER = WHITE	OX = Oxidizer ACID = Acid ALK = Alkali COR = Corrosive ~~W~~ = Use No Water

SECTION I - GENERAL INFORMATION

Characterization: **Metallic Hydroxide**	RCRA Number: **D002**	EPA Class: **Characteristic (C) Waste**
DOT Proper Shipping Name: **Sodium Hydroxide**	Chemical Abstract Service (CAS) Number: **1310-73-2**	
DOT Hazard Class and Label Requirements: **Corrosive Material**	DOT Emergency Guide Code: **60**	
DOT Identification Number: **UN 1823**	Molecular Formula: **NaOH**	

Synonyms

Caustic soda; lye; soda lye; sodium hydrate; Aetznatron; Collo-Grillrein; Lewis-Red Devil Lye.

SECTION II - HAZARDOUS INGREDIENTS AND IDENTITY INFORMATION

Hazardous Components (specific identity)	OSHA Exposure Criteria	NIOSH Exposure Criteria	Immediately Dangerous to Life and Health (IDLH)	ACGIH Exposure Criteria
Sodium hydroxide (derivation: By electrolysis of sodium chloride or by reaction of calcium hydroxide and sodium carbonate).	**PEL (ceiling): 2 mg/m^3** **STEL: Not Applicable**	**REL (ceiling): 2 mg/m^3** **STEL: Not Applicable**	**10 mg/m^3**	**TLV (ceiling): 2 mg/m^3** **STEL: Not Applicable**

SECTION III - PHYSICAL AND CHEMICAL CHARACTERISTICS

Boiling Point: **2534°F (1390°C)**	Specific Gravity (H_2O = 1): **2.13**
Vapor Pressure (mm Hg): **1 at 1362°F (739°C)**	Molecular Weight: **40.0**
Vapor Density (Air = 1): **Not Reported**	Melting Point: **605°F (318°C)**

Solubility

Soluble in water, alcohol, glycerol, and methanol. Insoluble in acetone and ether.

Appearance and Odor

Colorless to white odorless solid. May appear as flakes, beads, or in granular form.

SECTION IV - FIRE AND EXPLOSION HAZARD DATA

Flash Point (method used): **Not Reported**	Explosive Limits in Air % by Volume: **LEL: Not Reported** **UEL: Not Reported**
NFPA Classification: **Non-Combustible Solid**	Autoignition Temperature: **Not Reported**

Extinguishing Media

Use agents suitable to surrounding fire. Avoid water (reacts to generate extreme heat). If water must be used, make sure it is as cold as possible. For large fires use fog or foam.

Special Fire Fighting Procedures

Move containers from fire if it can be done without risk or cool *sealed* containers with water spray. Evacuate non-essential personnel 2500 feet from fire area. Poisonous gases are produced in fire. Wear appropriate respiratory protection where sodium hydroxide may be present.

Unusual Fire and Explosion Hazards

When in contact with moisture or water, sodium hydroxide can generate enough heat to ignite surrounding combustible materials. Sodium hydroxide may melt and flow when heated in fire.

SECTION V - REACTIVITY DATA

Stability		Conditions to Avoid **Sodium hydroxide is normally stable in closed containers under routine conditions of handling and storage. Keep away from incompatible materials, especially water or moisture.**
Stable	Unstable	Incompatibility *(materials to avoid)*
X		**Water (produces extreme heat), acids, flammable liquids, organic halogens, acrolein, acrylonitrile, metals such as aluminum, tin, and zinc (corrosive), nitromethane. Absorbs moisture and carbon dioxide from air.**
Hazardous Polymerization		Conditions to Avoid **Under normal conditions of temperature and pressure, hazardous polymerization of sodium hydroxide is not expected to occur. However, violent polymerization can occur with acrolein or acrylonitrile.**
May Occur	Will Not Occur	Hazardous Decomposition or By-products
	X	**Thermal decomposition may yield acrid smoke and irritating fumes and vapors, including toxic sodium oxide (Na_2O) and sodium peroxide (Na_2O_2).**

SECTION VI - HEALTH HAZARD DATA

Primary Route(s) of Entry:	Inhalation?	Absorption (skin/eye)?	Ingestion?
	X		X

Health Hazards

INHALATION: Extremely corrosive to tissues in respiratory system. Causes many small burns, hair loss in nasal passages, pulmonary edema (fluid in lungs), cough, congestion.

SKIN & EYES: Skin damage is immediate with a slippery, soapy feeling at site of contact. There may be no pain for the first minutes of exposure. Causes burns, keratin (hair & nail) destruction, intracellular edema, severe burns, tissue corrosion, deep ulceration, permanent scarring. Corneal corrosion on contact. Disintegration of the conjunctiva and corneal epithelium leading to permanent loss of vision.

INGESTION: Immediate burning of the mouth, esophagus, and stomach with painful swallowing, excessive salivation, edematous lips, chin, tongue, and pharynx covered with exudate, esophageal edema, mucosa destruction, vomiting, rapid pulse, death (from shock, asphyxia, or pneumonia).

Carcinogenicity	NTP Listed?	IARC Cancer Review Group?	OSHA Regulated?	Target Organs?
Unknown Human Unknown Animal	**No**	**No**	**29 CFR 1910.1000 Table Z-1**	**Eyes, skin, respiratory system, GI tract.**

Medical Conditions Generally Aggravated by Exposure

None reported.

Emergency and First-aid Procedures

Eye contact: Do not allow victim to keep eyes tightly shut. Flush immediately with water for 15 minutes (minimum); seek medical attention. Skin contact: Remove all contaminated clothing. Immediately wash area with large amounts of soap and water. For inhalation: Remove the person from exposure. Provide respiratory assistance and CPR. Transfer to medical facility. Observe 48 hours for lung effects. If swallowed: Remove person to fresh air and seek medical attention immediately. Do not attempt to give an unconscious or convulsing person anything by mouth. Give 1 to 2 glasses of water followed by vinegar to neutralize poison. Do NOT induce vomiting (aspiration hazard).

SECTION VII - PRECAUTIONS FOR SAFE HANDLING AND USE

Steps to be Taken in Case Material is Released or Spilled

Ventilate area of spill or leak. Remove sources of ignition. Restrict those not involved in cleanup from entering area. Collect spilled powdered materials in safest manner possible and deposit in sealed drum for disposal. Do NOT use water. Personnel exposure must be prevented.

Preferred Waste Disposal Method

Not listed. Check applicable federal, state, and/or local environmental regulations.

Precautions to be Taken in Handling and Storage

Store to avoid contact with strong acids. Store in tightly closed containers in cool, well-ventilated area.

Other Precautions and Warnings

Always wear appropriate protective equipment when working with sodium hydroxide.

SECTION VIII - CONTROL MEASURES AND PERSONAL PROTECTIVE EQUIPMENT

Respiratory Protection *(specify type)*

Sodium hydroxide is extremely destructive to any tissues it contacts. At any concentration use a self-contained breathing apparatus (SCBA) with full facepiece operated in positive pressure mode or a supplied-air respirator set in continuous flow mode.

Ventilation

Local exhaust (preferred) at site of chemical work or general ventilation.

Protective Gloves	Eye Protection	Other Protective Clothing
Impervious Material	**Chemical/Dust Goggles with Side Shields**	**Full Protective Clothing**

Work/Hygiene Practices

Always wash hands thoroughly after using chemical; never bring food, drink, or smoking materials into vicinity of chemicals.

SODIUM HYDROXIDE

NaOH CAS: 1310-73-2

IDENTIFICATION AND TYPICAL USES

Sodium hydroxide is a colorless to white, odorless solid. It can appear as flakes, beads, or in granular form. It is used in chemical manufacture; in the production of rayon and cellophane; as a neutralizing agent in petroleum refining; in the manufacture and processing of pulp and paper, aluminum, detergents, soap, and textiles; in vegetable oil refining; in reclaiming of rubber; for the regeneration of ion exchange resins; in organic fusion; for peeling fruits and vegetables in the food industry; as a laboratory reagent; in etching and electroplating, and as a food additive.

RISK ASSESSMENT: HEALTH

General Assessment

Sodium hydroxide is a strong corrosive alkali that dissolves any living tissues it contacts. It is toxic and destructive by *inhalation, ingestion,* and skin contact (although dermal absorption has not been reported). There are no data on any carcinogenic or teratogenic potential. However, animal mutation data have been documented for sodium hydroxide.

Inhalation of sodium hydroxide dusts or solution mists will cause immediate and destructive damage to the mucosa of the respiratory system. There may be temporary hair loss inside the nasal passages due to sodium hydroxide's destructive action on keratin. There will be serious cough with phlegm, congestion, dypsnea, tightness in the chest, and possibly delayed pulmonary edema (fluid in the lungs) which is a medical emergency. These symptoms can be delayed up to 48 hours or more thereby creating a false sense of security with regard to health exposure risk.

Skin contact causes immediate, acid-like corrosive destruction. However, unlike acid burns, there may be no pain for the first few minutes after exposure. Sodium hydroxide penetrates the skin and, since it does not coagulate proteins to prevent further penetration, the resulting caustic burn can become severe and slow healing. The initial indication may be a slippery soapy feeling at the site of contact followed by a loss of keratin-containing elements (hair and nails), and intracellular edema (excessive fluid in the skin cells). This can progress to severe burns, tissue corrosion, deep ulcerations, and permanent scarring. Eye contact will cause the immediate corrosion of the cornea, Disintegration and sloughing of conjunctival and corneal epithelium may progress to temporary or permanent corneal opacification and possible adhesion of the lid to the eyeball.

Ingestion is a life-threatening exposure. It causes burning of the mouth, esophagus, and stomach with painful swallowing, excessive salivation, edema (fluid buildup in surrounding tissues) of the lips, chin, tongue, and pharynx all of which may become coated with exudate (fluid oozed from swollen tissue). There can be esophageal edema (which can prevent all swallowing within hours of ingestion), possibly edematous, gelatinous, and necrotic mucosa, vomiting (with granular material due to digestive hemorrhage), and rapid, faint pulse with cold, clammy skin. Death can occur from shock, asphyxia, or pneumonia within days.

Acute Health Effects

The following acute (short-term) health effects may occur immediately or shortly after exposure to sodium hydroxide:

<u>Skin</u>: Irritation with corrosive destruction at site of contact. Pain can be delayed up to three or more minutes after exposure, but tissue destruction begins immediately on contact.

<u>Eye</u>: Severe tissue destruction, corneal opacification, and possible loss of vision.

<u>Lung</u>: Causes small burns in the respiratory tract with corrosive destruction to the mucosa, loss of hair in the nasal passages, and possible delayed

pulmonary edema (fluid in the lungs), which can be fatal.

Other: Ingestion results in serious, life-threatening destruction of tissues along the digestive tract. Death can occur within days as a result of shock, asphyxia (due to blocked airways caused by the swelling of the esophagus and associated tissues), or pneumonia.

Chronic Health Effects

The following chronic (long-term) health effects may occur at some time after exposure to sodium hydroxide and can last for months or even years:

Cancer Hazards: According to the information presented in the references, sodium hydroxide has not been adequately tested for its ability to cause cancer in test animals. However, some sources reported limited mutagenic data on test animals. Also, cases of squamous cell carcinomas (malignant tumors of epithelial origin) of the esophagus have been reported 12 to 42 years after ingestion of sodium hydroxide. However, the references are unclear whether these cancers were the result of scar tissue formation caused by the corrosive destruction of tissues or from the actual or potential carcinogenic properties of sodium hydroxide.

Reproductive Hazard: According to the information presented in the references, sodium hydroxide has not been adequately tested for its ability to adversely affect reproduction.

Other Chronic Effects: Dermatitis may develop from repeated exposures to dilute solutions of sodium hydroxide. Very irritating substance can damage the lungs. It is not known if exposure to sodium hydroxide will have this effect.

Recommended Risk-Reduction Measures

Personnel must avoid direct contact with sodium hydroxide. It is extremely corrosive to *all* human tissue. If a less toxic material cannot be substituted for sodium hydroxide, then *engineering controls* are the most effective method of reducing exposures. The best protection is to enclose operations and/or provide local exhaust ventilation at the site of sodium hydroxide release. While not always operationally feasible, isolating operations can also reduce exposure.

Using respiratory protection and other personal protective equipment (PPE) is less effective than the engineering controls mentioned above, but is still advisable whenever working with or around sodium hydroxide. At any exposure level, maximum protection is obtained using a self-contained breathing apparatus (SCBA) with full facepiece operated in pressure demand mode or a supplied-air respirator operated in continuous flow mode. Human exposure *must* be prevented. If a full facepiece is not available, then chemical/dust goggles, safety glass with side shields, or a face shield should be worn to protect the eyes. To prevent hand and skin exposures, impervious rubber gloves should be used. Glove manufacturers should be contacted and permeation studies obtained *before* final glove selections are made for sodium hydroxide.

Administrative controls should also be in place to minimize the potential for human exposures. These may include written procedures or policies which specify the methods and techniques that will be practiced whenever personnel are to work with sodium hydroxide.

All personnel should receive *training* on the use, hazards, protective measures, emergency actions, and other precautions per 29 CFR 1910.1200 (Hazard Communication), prior to the first assignment in an area where sodium hydroxide is used or stored.

For those with frequent or potentially high exposure (half the PEL or greater), the following medical tests are recommended before initial assignment and at regular intervals thereafter (e.g., annually):

- ☑ Consider chest X-ray after an acute overexposure (may be negative if taken immediately after exposure due to delayed onset of pulmonary edema).
- ☑ Pulmonary function tests.

Any evaluation should include a careful history of past and present symptoms with an examination. Medical tests that simply look for existing damage are not a substitute for controlling exposures. Also, because smoking can cause lung cancer, emphysema, and a host of other medical problems, working with or around sodium hydroxide may aggravate these conditions further.

Other methods to reduce exposure include:

- ☑ Where possible, enclose operations and use local exhaust ventilation at the site of chemical release. If local exhaust ventilation or enclosure is not used, then respiratory protection should be required.
- ☑ Always ensure that proper protective clothing is worn when using chemical substances. Never allow contaminated clothing to be taken home from the work place.

☑ Hazard warning information should be posted in the work area. In addition, as part of an on-going education and training program, all information on the health and safety hazards of sodium hydroxide should be communicated to all exposed workers.

☑ Eye wash stations should be provided in the immediate work area for emergency use. If whole body exposures are possible, then safety showers should also be located in the work area.

RISK ASSESSMENT: ENVIRONMENT

General Assessment

The environment is at risk of exposure during transportation, storage, disposal, or destruction of sodium hydroxide. In almost every scenario, the threat of environmental exposure is contingent upon the proper handling of the chemical substance. Accidental spills, large or small, where sodium hydroxide contacts incompatible commodities can result in violent reactions as well as the possible contamination of the surrounding environmental mediums (water, soil, and air).

Sodium hydroxide is considered a noncombustible solid. However, because it can generate extreme heat in contact with steam, water, or moisture. The heat may be enough to ignite ordinary combustible materials in the area. It is incompatible with acids, flammable liquids, organic halogens, acrolein, acrylonitrile, metals (such as aluminum, tin, and zinc), and nitromethane. Therefore, extreme caution is required in handling, storage, transportation, and disposal of sodium hydroxide. These characteristics also require special consideration during any emergency situation involving a leak or spill of sodium hydroxide. Should sodium hydroxide ever come into contact with any incompatible substances either during use, transportation, storage, or disposal, dangerous reactions are possible.

Sodium hydroxide can enter the environment through industrial discharges and through spills.

Acute Ecological Effects

Acute (short-term) toxic effects may include the death of animals, birds, or fish and death or low growth rate in plants. Acute effects are seen 2 to 4 days after animals or plants are exposed to sodium hydroxide.

Sodium hydroxide has high acute toxicity to aquatic life. The toxicity of sodium hydroxide to aquatic life will be influenced by the hardness and alkalinity of the receiving water, since addition of sodium hydroxide to water results in an elevated pH. The estimate of high acute toxicity is therefore based on the criterion for pH of water which considers a pH of 9.0 as the upper limit for health populations of aquatic life. Insufficient data are available to evaluate or predict the short-term toxic effects of sodium hydroxide to plants, birds, and land animals.

Chronic Ecological Effects

Chronic (long-term) toxic effects may include shortened life span, reproductive problems, lower fertility, and changes in appearance or behavior in exposed animals. These effects can be seen long after first exposure to toxic chemicals.

Sodium hydroxide has high chronic toxicity to aquatic life. The toxicity of sodium hydroxide to aquatic life will be influenced by the hardness and alkalinity of the receiving water, since addition of sodium hydroxide to water results in an elevated pH. The estimate of high chronic toxicity is therefore based on the criterion for pH of water which considers a pH of 9.0 as the upper limit for health populations of aquatic life. Insufficient data are available to evaluate or predict the long-term toxic effects of sodium hydroxide to plants, birds, and land animals.

Water Solubility

Sodium hydroxide is highly soluble in water (when dissolving, it generates considerable heat). Concentrations of 1000 milligrams and more will mix with a liter of water.

Persistence in the Environment

Small quantities of bases such as sodium hydroxide will slightly raise the pH of water in aquatic ecosystems. Large quantities can raise the pH for extended periods of time.

Bioaccumulation in Aquatic Organisms

Some substances increase in concentration, or *bioaccumulate,* in living organisms as they breathe contaminated air, drink contaminated water, or eat contaminated food. These chemicals can become concentrated in the tissues and internal organs of animals as well as humans.

Sodium hydroxide is not expected to accumulate in the edible tissues of aquatic life that are normally consumed by humans.

❶ *Recommended Risk-Reduction Measures*

Proper training of all transporters will reduce the likelihood of a mishap or accident resulting in a leak or spill of sodium hydroxide into the environment. The correct use of labeling on all containers, trucks, and rail cars will enable emergency responders to react properly and quickly to any disaster thereby reducing the potential risk to the environment and to personnel.

Storage of sodium hydroxide should be segregated from incompatible chemicals to minimize the risk of cross-contamination or contact. It must be stored in air-tight, moisture-free containers.

If a spill or leak to the environment has occurred, fire department, emergency response and/or hazardous materials spill personnel should be notified immediately. Cleanup should be attempted only by those trained in proper spill containment procedures. If sodium hydroxide should contact the water table, aquifer, or navigable waterway, the local and/or state emergency response authorities must be notified. A comprehensive emergency response or disaster preparedness/recovery plan should be in place prior to any operations involving the use, transportation, storage, or disposal of sodium hydroxide.

If sodium hydroxide is spilled or leaked, the following specific steps are recommended:

- ☑ Restrict persons not wearing protective clothing from area of spill or leak until cleanup is complete and area can be opened for normal work.
- ☑ Do not allow exposure to incompatible materials.
- ☑ Close or remove all ignition sources and ventilate area of spill.
- ☑ Collect powdered materials in the safest and most efficient manner possible. Place materials in a sealed drum. Do NOT dry sweep (generates dusts). Use a vacuum equipped with a high-efficiency particulate air (HEPA) filter instead. Do NOT use water or damp mop residues.
- ☑ It may be necessary to dispose of sodium hydroxide as a hazardous waste. Consult state, local, or federal environmental regulations.

RISK ASSESSMENT: BUSINESS

General Assessment

Accidents or mishaps involving sodium hydroxide can present a moderate threat to business operations. Although not likely, equipment and facilities may be lost or their use disrupted as a result of an incident involving sodium hydroxide which can significantly affect fiscal viability. Lawsuits that may result from personnel injury, illness, and/or death, public exposures, and/or environmental contamination will also require a serious expenditure of resources. Media attention surrounding an injury, death, or environmental damage can also result in a loss of profits and loss of current as well as future business.

❶ *Recommended Risk-Reduction Measures*

Company attorneys, safety and health professionals, and environmental specialists should be involved in the development of any procedures that focus on emergency response, disaster preparedness, and disaster recovery. A company official should be pre-designated as a public relations officer with specific training in dealing with the press. Corporate plans and policies should be developed, approved, and implemented long before any need for such arises.

REFERENCES

American Conference of Governmental Industrial Hygienists. 1988. *Documentation of the Threshold Limit Values and Biological Exposure Indices,* 5th Edition (with updates). Cincinnati: ACGIH

Calabrese, E. J., E. M. Kenyon. *Air Toxics and Risk Assessment.* New York: CRC Press/Lewis Publishers.

Gardner, R. W. 1994. *Chemical Intolerance.* New York: CRC Press/Lewis Publishers.

Howard, P. H., et. al. 1991. *Handbook of Environmental Degradation Rates.* New York: CRC Press/Lewis Publishers.

Lewis, R. J., Sr. 1993. *Hawley's Condensed Chemical Dictionary,* 12th Edition. New York: Van Nostrand Reinhold.

Lewis, R. J., Sr. 1992. *Sax's Dangerous Properties of Industrial Materials,* Volumes 1, 2, and 3, 8th Edition. New York: Van Nostrand Reinhold.

Patnaik, P. 1992. *A Comprehensive Guide to the Hazardous Properties of Chemical Substances.* New York: Van Nostrand Reinhold.

Rea, W. J. 1995. *Chemical Sensitivity,* Vol. I-III. New York: CRC Press/Lewis Publishers.

Richardson, M. 1995. *Dictionary of Substance and Their Effects,* Vol. I-VI. New York: CRC Press/Lewis Publishers.

Sacarello, H. L. A. 1994. *The Comprehensive Handbook of Hazardous Materials.* New York: CRC Press/Lewis Publishers.

Talbott, E., G. Craun. 1995. *An Introduction to Environmental Epidemiology.* New York: CRC Press/Lewis Publishers.

U.S. Department of Health and Human Services, National Institute for Occupational Safety and Health. 1994. *NIOSH Pocket Guide to Chemical Hazards.* Washington, D.C.: U.S. Government Printing Office.

MATERIAL SAFETY DATA SHEET

CHEMICAL NAME	
SODIUM PEROXIDE	**CRC PRESS/LEWIS PUBLISHERS DISCLAIMER:** THE INFORMATION AND RECOMMENDATIONS PRESENTED HEREIN ARE BASED ON SOURCES BELIEVED TO BE RELIABLE. CRC MAKES NO REPRESENTATION ON ITS COMPLETENESS OR ACCURACY. IT IS THE RESPONSIBILITY OF THE USER TO DETERMINE THE CHEMICAL'S SUITABILITY FOR ITS INTENDED USE, THE CHEMICAL'S SAFE USE, AND THE CHEMICAL'S PROPER DISPOSAL. NO REPRESENTATIONS AND/OR WARRANTIES, EITHER EXPRESSED OR IMPLIED, OF THE MERCHANTABILITY OR FITNESS FOR A PARTICULAR PURPOSE, OR OF ANY OTHER NATURE, ARE MADE WITH RESPECT TO THE INFORMATION PROVIDED IN THIS MSDS OR TO THE CHEMICAL TO WHICH INFORMATION MAY REFER. CRC NEITHER ASSUMES NOR AUTHORIZES ANY OTHER PERSON TO ASSUME FOR IT, ANY OTHER ADDITIONAL RESPONSIBILITY OR LIABILITY FOR THE USE OF, OR RELIANCE UPON, THIS INFORMATION.

HAZARD WARNING INFORMATION

HEALTH	FIRE	REACTIVE	OTHER	DEGREE OF HAZARD	COLOR CODING	OTHER CODES
3	0	2	OX	0 = Minimum Hazard 1 = Slight Hazard 2 = Moderate Hazard 3 = Serious Hazard 4 = Severe Hazard	HEALTH = BLUE FIRE = RED REACTIVITY = YELLOW OTHER = WHITE	OX = Oxidizer ACID = Acid ALK = Alkali COR = Corrosive W = Use No Water

SECTION I - GENERAL INFORMATION

Characterization	RCRA Number	EPA Class
Inorganic (metallic) peroxide	**None**	**Not Applicable**

DOT Proper Shipping Name **Sodium peroxide**	Chemical Abstract Service (CAS) Number **1313-60-6**
DOT Hazard Class and Label Requirements **Oxidizer**	DOT Emergency Guide Code **49**
DOT Identification Number **UN 1504**	Atomic Symbol **Na_2O_2**

Synonyms
Sodium dioxide; sodium superoxide; disodium dioxide; disodium peroxide; sodium oxide; Flocool 180.

SECTION II - HAZARDOUS INGREDIENTS AND IDENTITY INFORMATION

Hazardous Components (specific identity)	OSHA Exposure Criteria	NIOSH Exposure Criteria	Immediately Dangerous to Life and Health (IDLH)	ACGIH Exposure Criteria
Sodium peroxide (derivation: By heating metallic sodium at 300°C in aluminum trays in a retort in a current of dry air, from which the carbon dioxide has been removed).	**PEL:** Not Established **STEL:** Not Established	**REL:** Not Established **STEL:** Not Established	**Not Determined**	**TLV:** Not Established **STEL:** Not Established

SECTION III - PHYSICAL AND CHEMICAL CHARACTERISTICS

Boiling Point **1215°F (657°C) Decomposes**	Specific Gravity (H_2O = 1) **2.805**
Vapor Pressure (mm Hg) **Not Reported**	Molecular Weight (atomic weight) **77.98**
Vapor Density (Air = 1) **Not Reported**	Melting Point **860°F (460°C) Decomposes**

Solubility
Completely soluble in water (with the evolution of heat).

Appearance and Odor
A yellowish-white powder with no odor. Turns more yellow when heated. Absorbs water and carbon dioxide from the air.

SECTION IV - FIRE AND EXPLOSION HAZARD DATA

Flash Point (method used) **Not Reported**	Explosive Limits in Air % by Volume **LEL: Not Applicable UEL: Not Applicable**
NFPA Classification **Non-Combustible Solid**	Autoignition Temperature **Not Reported**

Extinguishing Media
Smother combustible fires with salt, soda ash, or dolomite mixtures.

Special Fire Fighting Procedures
Wear full protective clothing and self-contained breathing apparatus (SCBA). Poisonous gases are produced in fire. Remove containers from fire area if it can be done without risk. Use water mist or spray to keep fire-exposed containers cool that cannot be moved. Fight fire from distance, if possible.

Unusual Fire and Explosion Hazards
Sodium peroxide powder is not flammable, but it is a powerful oxidizer than can significantly enhance the burning rate and fire intensity of ordinary combustible materials.

SECTION V - REACTIVITY DATA

Stability		Conditions to Avoid Sodium peroxide is normally stable under normal conditions of handling and storage. Never mix with another material without first establishing compatibility. Keep away from heat and flame.
Stable X	Unstable	Incompatibility *(materials to avoid)* Acetic acid, acetic anhydride, aluminum, ammonium persulfate, aniline, antimony, arsenic, benzene, boron nitride, calcium carbide, diethyl ether, glycerin, hexamethylenetetramine, hydrogen sulfide, magnesium, carbon dioxide, organic matter, phosphorus, potassium, selenium monochloride, and many other materials.
Hazardous Polymerization		Conditions to Avoid Hazardous polymerization of sodium peroxide metal will not occur.
May Occur	Will Not Occur X	Hazardous Decomposition or By-products Poisonous gases are emitted during fire, including toxic oxides of sodium and metallic sodium fumes.

SECTION VI - HEALTH HAZARD DATA

Primary Route(s) of Entry:	Inhalation? X	Absorption (skin/eye)?	Ingestion? X

Health Hazards

INHALATION: Severe irritation to the eyes, nose, throat, and mucous lining of the respiratory tract. It will cause corrosive tissue damage. Systemic effects not likely since the reactive nature of this chemical causes damage at the point or site of contact without indication of any additional physiological response.

EYES & SKIN: Exposure to the eyes or skin can cause severe irritation and possible tissue corrosion.

INGESTION: Gastrointestinal tract irritation, inflammation, and possible tissue corrosion.

Carcinogenicity	NTP Listed?	IARC Cancer Review Group?	OSHA Regulated?	Target Organs?
Unknown Human Unknown Animal	No	No	Not Listed	Respiratory system mucosa, eyes, skin.

Medical Conditions Generally Aggravated by Exposure

None Reported.

Emergency and First-aid Procedures

Eye contact: Do not allow victim to keep eyes tightly shut. Flush immediately with water for 15 minutes (minimum); seek medical attention. **Skin contact:** Remove all contaminated clothing. Immediately wash area with flooding amounts of water. For reddened or blistered skin, seek medical assistance. For **inhalation**: Remove the person from exposure. Provide respiratory assistance and CPR. Transfer to medical facility. If **swallowed**, seek medical attention immediately. Never give anything by mouth to a convulsing person. If conscious and alert, give 1-2 glasses of water and induce repeated vomiting until vomitus is clear. NOTE: The decision to induce vomiting should be taken under advisement (aspiration hazard).

SECTION VII - PRECAUTIONS FOR SAFE HANDLING AND USE

Steps to be Taken in Case Material is Released or Spilled

Collect powdered materials in most convenient and safe manner and deposit in sealed containers. Dry sweep not recommended (generates dusts). Use HEPA vacuum. Remove all ignition sources. Restrict those not involved in cleanup from entering area. Keep combustible materials away.

Preferred Waste Disposal Method

No citation.

Precautions to be Taken in Handling and Storage

Store to avoid contact with incompatible materials, moisture, and combustible materials. Prevent physical damage to containers. Keep tightly closed.

Other Precautions and Warnings

Sources of ignition are prohibited where sodium peroxide is used, handled, or stored. Store on floors constructed of impermeable and non-combustible materials.

SECTION VIII - CONTROL MEASURES AND PERSONAL PROTECTIVE EQUIPMENT

Respiratory Protection *(specify type)*

No exposure levels established. Use a respirator with a high efficiency particulate air (HEPA) filter; otherwise, use a self-contained breathing apparatus with full facepiece operated in positive pressure mode.

Ventilation

Local exhaust (preferred) at site of chemical work or general ventilation.

Protective Gloves	Eye Protection	Other Protective Clothing
Impervious Rubber	Chemical/Dust Goggles or Face Mask	Protective Apron, Boots

Work/Hygiene Practices

Always wash hands thoroughly after using chemical; never bring food, drink, or smoking materials into vicinity of chemicals.

SODIUM PEROXIDE

Na_2O_2

CAS: 1313-60-6

IDENTIFICATION AND TYPICAL USES

Sodium peroxide is a yellowish-white, odorless powder. It turns more yellow when heated. It is hygroscopic (absorbs moisture from the air). It will also absorb carbon dioxide from the air. It is used as an oxidizing agent, for bleaching miscellaneous materials including paper and textiles, as a deodorant, an antiseptic, in organic chemicals, for water purification, in pharmaceuticals, to generate oxygen for diving bells and submarines, in textile dyeing and printing, in ore processing, as an analytical reagent, in calorimetry, and in germicidal soaps.

RISK ASSESSMENT: HEALTH

General Assessment

Sodium peroxide is a severe irritant to the skin, eyes, and mucosa of the nose, throat, and lining of the respiratory tract. It can cause corrosive tissue destruction along its exposure path. Exposure routes include *inhalation* and *ingestion* of its powders. Dermal absorption has not been reported. Toxic systemic effects have not been well documented in the references. It appears that the corrosive action of sodium peroxide is localized to the site of contact without any indications of further systemic toxicity. There are no reports of any carcinogenic, mutagenic, or teratogenic effects on long-term exposure to sodium peroxide.

Inhalation of sodium peroxide dusts will cause immediate irritation to the eyes, nose, throat, and respiratory tract. Corrosive tissue damage is likely, especially on exposure to heavy concentrations.

Skin and eye contact will result in severe irritation with possible corrosive tissue destruction at the point of contact. If not removed from the eyes immediately, damage may be excessive and possibly permanent.

Ingestion will also cause irritation along the gastrointestinal tract. If large doses are ingested, vomiting may occur which can present an aspiration hazard.

Acute Health Effects

The following acute (short-term) health effects may occur immediately or shortly after exposure to sodium peroxide:

Skin: Immediate and severe irritation at the point of contact. There will likely be corrosive tissue damage.

Eye: Severe irritation and inflammation. Prolonged contact may lead to damage to vision.

Lung: Extremely irritating to the mucous membranes of the nose, throat, and the lining of the respiratory tract. There can be cough with phlegm, difficulty breathing, and tightness in the chest.

Chronic Health Effects

The following chronic (long-term) health effects may occur at some time after exposure to sodium peroxide and can last for months or even years:

Cancer Hazards: According to information presented in the references, sodium peroxide has not been adequately tested for its ability to cause cancer in animals.

Reproductive Hazard: According to information presented in the references, sodium peroxide has not been tested for its ability to adversely affect reproduction.

Other Chronic Effects: There are few data in the references regarding any long-term effects of exposure to sodium peroxide.

Recommended Risk-Reduction Measures

Personnel should avoid direct contact with sodium peroxide, especially its dusts or powders. If a less toxic material or compound cannot be substituted for sodium peroxide, then *engineering controls* are the most effective method of reducing exposures. The best protection is to enclose operations and/or provide local exhaust ventilation at the site of sodium peroxide dust release.

While not always operationally feasible, isolating operations can also reduce exposure. Using respiratory protection is less effective than the controls mentioned above, but is still advisable whenever working with or around sodium peroxide. No exposure levels have been established for sodium peroxide. However, this does not mean that exposure is without a health risk. A supplied-air respirator or a self-contained breathing apparatus (SCBA) with full facepiece operated in pressure demand or continuous flow mode will provide the greatest possible respiratory protection, especially when exposures levels are high or unknown (as in an emergency). If a full facepiece is not available, then chemical/dust goggles should be worn to protect the eyes. A face shield and protective apron should also be considered. To prevent hand and skin exposures, gloves should be used. Glove manufacturers should be contacted and permeation studies obtained *before* final glove selections are made.

Administrative controls should also be in place to minimize the potential for human exposures. These may include written procedures or policies which specify the methods and techniques that will be practiced whenever personnel are to work with sodium peroxide.

All personnel should receive *training* on the use, hazards, protective measures, emergency actions, and other precautions per 29 CFR 1910.1200 (Hazard Communication), prior to the first assignment in an area where sodium peroxide is used or stored.

Before beginning employment working with sodium peroxide, and at regular intervals thereafter (e.g., annually), those personnel with frequent or potentially high exposures should be provided the following recommended medical tests:

☑ Lung function test (establish baseline).

If symptoms develop or overexposure is suspected, consider the following additional tests:

☑ Consider chest X-ray.

☑ Lung function test (compare to baseline).

Any evaluation should include a careful history of past and present symptoms with an examination. Medical tests that simply look for existing damage are not a substitute for controlling exposures. Also, since smoking can cause heart disease as well as lung cancer, emphysema, and other known respiratory problems, smokers exposed to sodium peroxide may experience more pronounced symptoms than non-smokers under the same conditions of exposure. Prudent risk management requires proper consideration of *all* possible factors which may be influencing are causing exposure symptoms to develop in the workplace.

Other methods to reduce exposure include:

☑ Where possible, enclose operations and use local exhaust ventilation at the site of chemical release. If local exhaust ventilation or enclosure is not used, respiratory protection should be mandatory.

☑ Always ensure that proper protective clothing is worn when using chemical substances.

☑ Wash thoroughly immediately after exposure to sodium peroxide and at the end of the work shift or before eating, drinking, smoking, or applying cosmetics.

☑ Hazard warning information should be posted in the work area. In addition, as part of an on-going education and training program, all information on the health and safety hazards of sodium peroxide should be communicated to all exposed and potentially exposed workers.

☑ Eye wash stations should be provided in the immediate work area for emergency use. If there is a possibility of skin exposure to sodium peroxide, emergency shower facilities should also be provided in the area.

☑ Workers whose clothing has been contaminated by sodium peroxide should change into clean clothes before leaving work. Contaminated work clothing should be laundered only by individuals who have been informed of the hazards of exposure to sodium peroxide.

RISK ASSESSMENT: ENVIRONMENT

General Assessment

The environment is at risk of exposure during transportation, storage, disposal, or destruction of sodium peroxide. In almost every scenario, the threat of environmental exposure is contingent upon the proper handling of the chemical substance. Accidental spills, large or small, where sodium peroxide contacts incompatible commodities can result in fire, explosion (depending upon conditions of spill), and possible contamination of the surrounding environmental mediums (water, soil, and air).

Sodium peroxide is considered a non-combustible solid. However, it is an extremely powerful oxidizer and contact with ordinary combustible materials can significantly enhance their burning rate and increase the intensity of any fire. It is incompatible with many other commodities and extreme caution is required in handling, storage, transportation, and disposal of sodium peroxide. These characteristics also require special consideration during any emergency situation involving a leak or spill of sodium peroxide powder or dust. It is incompatible with numerous substances and other metals such as acetic acid, acetic anhydride, aluminum, ammonium persulfate, aniline, antimony, arsenic, benzene, boron nitride, calcium carbide, diethyl ether, glycerin, hexamethylenetetramine, hydrogen sulfide, magnesium, carbon dioxide, organic matter, phosphorous, potassium, selenium monochloride, silver chloride and charcoal, sodium, sodium monochloride, tin, zinc, metals with carbon dioxide and water, non-metals, nonmetal halides, fibrous materials and water, hydroxy compounds, almond oil, copper, cotton wool, soap, sulfur chloride, wood, peroxyformic acid, and reducing materials. Should sodium peroxide ever come into contact with these incompatible substances either during use, transportation, storage, or disposal, violent and even explosive reactions are possible.

Sodium peroxide may enter the environment as a result of industrial and municipal waste treatment plant discharges and from spills.

Acute Ecological Effects

Acute (short-term) toxic effects may include the death of animals, birds, or fish and death or low growth rate in plants. Acute effects are seen 2 to 4 days after animals or plants are exposed to sodium.

Insufficient data are available to evaluate or predict the short-term effects of sodium peroxide to aquatic life, plants, birds, or land animals.

Chronic Ecological Effects

Chronic (long-term) toxic effects may include shortened life span, reproductive problems, lower fertility, and changes in appearance or behavior in exposed animals. These effects can be seen long after first exposure(s) to toxic chemicals.

Insufficient data are available to evaluate or predict the long-term effects of sodium peroxide to aquatic life, plants, birds, or land animals.

Water Solubility

Sodium peroxide is highly soluble in water, generating heat as its dissolves. Concentrations of 1000 milligrams and more will mix with one liter of water.

Persistence in the Environment

No data are available on the estimated percentages of sodium peroxide persistence in the aquatic or terrestrial environments.

Bioaccumulation in Aquatic Organisms

Some substances increase in concentration, or *bioaccumulate,* in living organisms as they breathe contaminated air, drink contaminated water, or eat contaminated food. These chemicals can become concentrated in the tissues and internal organs of animals as well as humans.

No data are available on the bioaccumulation of sodium peroxide in the edible tissues of fish.

Recommended Risk-Reduction Measures

Proper training of all transporters will reduce the likelihood of a mishap or accident resulting in a leak or spill of sodium peroxide dusts or powders into the environment. The correct use of DOT labeling (per the MSDS) on all containers, trucks, and rail cars will enable emergency responders to react properly and quickly to any disaster thereby reducing the potential risk to the environment and to personnel.

Storage of sodium peroxide should be segregated from incompatible chemicals to minimize the risk of cross-contamination or contact. Impermeable floors constructed of non-combustible materials are recommended.

If a spill or leak to the environment has occurred, fire department, emergency response and/or hazardous materials spill personnel should be notified immediately. Cleanup should be attempted only by those trained in proper spill containment procedures. Contaminated soils should be removed for incineration and replaced with clean soil. If sodium peroxide should contact the water table, aquifer, or navigable waterway, time is of the essence. It is highly soluble and will decompose. Therefore, total remediation may not be possible. The local and/or state emergency response authorities must be notified. A comprehensive emergency response or disaster preparedness/recovery plan should be in place prior to any operations involving the use, transportation, storage, or disposal of sodium peroxide. If sodium peroxide is spilled or leaked, the following specific steps are recommended:

- ☑ Restrict persons not wearing protective clothing from area of spill or leak until cleanup is complete and area can be opened for normal work.

- ☑ Remove all ignition sources and ventilate area.
- ☑ Collect powdered materials in the safest manner possible and deposit in sealed containers. Recommend using vacuum with HEPA filter. Do not dry sweep (generates airborne dusts).
- ☑ It may be necessary to dispose of sodium peroxide as a hazardous waste. The responsible state agency or the regional office of the federal Environmental Protection Agency (EPA) should be contacted for specific recommendations.

RISK ASSESSMENT: BUSINESS

General Assessment

Accidents or mishaps involving sodium peroxide can present a serious threat to business operations. The loss or damage of equipment or facilities can significantly affect fiscal viability. Lawsuits that may result from personnel injury/death, public exposures, and/or environmental contamination will also require a serious expenditure of resources. Media attention surrounding an injury, death, or environmental damage can also result in a loss of profits and loss of current as well as future business.

Recommended Risk-Reduction Measures

Company attorneys, safety and health professionals, and environmental specialists should be involved in the development of any procedures for responding to chemical incidents. A company official should be predesignated as a public relations officer with specific training in dealing with the press. Corporate plans and policies should be developed, approved, and implemented long before any need for such arises.

REFERENCES

Ahlbom, A. 1993. *Biostatistics for Epidemiologists.* New York: CRC Press/Lewis Publishers.

American Conference of Governmental Industrial Hygienists. 1988. *Documentation of the Threshold Limit Values and Biological Exposure Indices,* 5th Edition (with updates). Cincinnati: ACGIH

Bloom, A. D. and F. J. DeSerres. 1995. *Ecotoxicity and Human Health.* New York: CRC Press/Lewis Publishers.

Brusick, D. J. 1994. *Methods for Genetic Risk Assessment.* New York: CRC Press/Lewis Publishers.

Bueche, J. R. 1972. *Principles of Physics,* 2nd Edition. New York: McGraw-Hill.

Calabrese, E. J. 1994. *Biological Effects of Low Level Exposures.* New York: CRC Press/Lewis Publishers.

Cockerman, L. G., B. S. Shane. 1994. *Basic Environmental Toxicology.* New York: CRC Press/Lewis Publishers.

Environmental Protection Agency. 1983. *Chemical Hazard Information Profiles.* Washington, D.C.: U.S. EPA.

Fodden, C. R., and J. L. Weddell. 1992. *Hazardous Materials Emergency Action Data.* New York: CRC Press/Lewis Publishers.

Forseberg, K., and L. H. Keith. 1993. *Chemical; Protective Clothing Permeation and Degradation Database.* Chelsea, MI: Lewis Publishers.

Gardner, R. W. 1994. *Chemical Intolerance.* New York: CRC Press/Lewis Publishers.

Howard, P. H. 1995. *Handbook of Environmental Fate and Exposures Data for Organic Chemicals,* Vol. I-IV. New York: CRC Press/Lewis Publishers.

Keith, L. H. and M. M. Walker. 1995. *Handbook of Air Toxics.* New York: CRC Press/Lewis Publishers.

Landis, W. G., M. H. Yu. 1995. *Introduction to Environmental Toxicology.* New York: CRC Press/Lewis Publishers.

Langerman, N. 1994 *Precautionary Labels for Chemical Containers.* New York: CRC Press/Lewis Publishers.

Lauwerys, R. R., P. Hoet. 1993. *Industrial Chemical Exposure.* New York: CRC Press/Lewis Publishers.

Lewis, R. J., Sr. 1993. *Hawley's Condensed Chemical Dictionary,* 12th Edition. New York: Van Nostrand Reinhold.

Lewis, R. J., Sr. 1992. *Sax's Dangerous Properties of Industrial Materials,* Volumes 1, 2, and 3, 8th Edition. New York: Van Nostrand Reinhold.

Meyer, E. 1990. *Chemistry of Hazardous Materials.* Englewood Cliffs, NJ: Prentice Hall.

Mycock, J. C., J. D. McKenna, and L. Theodore. 1995. *Handbook of Air Pollution Control Engineering and Technology.* New York: CRC Press/Lewis Publishers.

Patnaik, P. 1992. *A Comprehensive Guide to the Hazardous Properties of Chemical Substances.* New York: Van Nostrand Reinhold.

Quigley, D. R. 1994. *Handbook of Emergency Chemical Management.* New York: CRC Press/Lewis Publishers.

Rea, W. J. 1995. *Chemical Sensitivity,* Vol. I-III. New York: CRC Press/Lewis Publishers.

Reeves, A. L. 1979. *Handbook on the Toxicology of Metals.* New York: Elsevier North-Holland.

Richardson, M. 1995. *Dictionary of Substance and Their Effects,* Vol. I-VI. New York: CRC Press/Lewis Publishers.

Sacarello, H. L. A. 1994. *The Comprehensive Handbook of Hazardous Materials.* New York: CRC Press/Lewis Publishers.

Talbott, E., G. Craun. 1995. *An Introduction to Environmental Epidemiology.* New York: CRC Press/Lewis Publishers.

Torkelson, T. R. V. K. Rowe. 1981. *Patty's Industrial Hygiene and Toxicology,* Vol. 2B. New York: Wiley-Interscience.

MATERIAL SAFETY DATA SHEET

CHEMICAL NAME	
SODIUM SULFATE	**CRC PRESS/LEWIS PUBLISHERS DISCLAIMER:** THE INFORMATION AND RECOMMENDATIONS PRESENTED HEREIN ARE BASED ON SOURCES BELIEVED TO BE RELIABLE. CRC MAKES NO REPRESENTATION ON ITS COMPLETENESS OR ACCURACY. IT IS THE RESPONSIBILITY OF THE USER TO DETERMINE THE CHEMICAL'S SUITABILITY FOR ITS INTENDED USE, THE CHEMICAL'S SAFE USE, AND THE CHEMICAL'S PROPER DISPOSAL. NO REPRESENTATIONS AND/OR WARRANTIES, EITHER EXPRESSED OR IMPLIED, OF THE MERCHANTABILITY OR FITNESS FOR A PARTICULAR PURPOSE, OR OF ANY OTHER NATURE, ARE MADE WITH RESPECT TO THE INFORMATION PROVIDED IN THIS MSDS OR TO THE CHEMICAL TO WHICH INFORMATION MAY REFER. CRC NEITHER ASSUMES NOR AUTHORIZES ANY OTHER PERSON TO ASSUME FOR IT, ANY OTHER ADDITIONAL RESPONSIBILITY OR LIABILITY FOR THE USE OF, OR RELIANCE UPON, THIS INFORMATION.

HAZARD WARNING INFORMATION

HEALTH	FIRE	REACTIVE	OTHER	DEGREE OF HAZARD	COLOR CODING	OTHER CODES
1	0	0		0 = Minimum Hazard 1 = Slight Hazard 2 = Moderate Hazard 3 = Serious Hazard 4 = Severe Hazard	HEALTH = BLUE FIRE = RED REACTIVITY = YELLOW OTHER = WHITE	OX = Oxidizer ACID = Acid ALK = Alkali COR = Corrosive W = Use No Water

SECTION I - GENERAL INFORMATION

Characterization	**Sulfate**	RCRA Number	**None**
		EPA Class	**Not Applicable**
DOT Proper Shipping Name	**No Citation**	Chemical Abstract Service (CAS) Number	**7757-82-6**
DOT Hazard Class and Label Requirements	**No Citation**	DOT Emergency Guide Code	**No Citation**
DOT Identification Number	**No Citation**	Atomic Symbol	**Na_2SO_4**

Synonyms

Sodium sulfate, anhydrous; bisodium sulfate; disodium monosulfate; sodium sulfate (2:1); sulfuric acid, disodium salt; Thenardite; Trona.

SECTION II - HAZARDOUS INGREDIENTS AND IDENTITY INFORMATION

Hazardous Components (specific identity)	OSHA Exposure Criteria	NIOSH Exposure Criteria	Immediately Dangerous to Life and Health (IDLH)	ACGIH Exposure Criteria
Sodium sulfate (derivation: By purification of natural sodium sulfate from deposits or brines; a by-product of hydrochloric acid manufacture from salt and sulfuric acid; a by-product of phenol manufacture through the caustic fusion process).	**PEL (8-hour):** Not Established **STEL:** Not Established	**REL (10-hour):** Not Established **STEL:** Not Established	**Not Determined**	**TLV:** Not Established **STEL:** Not Established

SECTION III - PHYSICAL AND CHEMICAL CHARACTERISTICS

Boiling Point	**Not Reported**	Specific Gravity (H_2O = 1)	**2.671**
Vapor Pressure (mm Hg)	**Not Reported**	Molecular Weight (atomic weight)	**142.04**
Vapor Density (Air = 1)	**Not Reported**	Melting Point	**1630°F (888°C)**

Solubility

Soluble in water, glycerol, and hydrogen iodide. Insoluble in alcohol.

Appearance and Odor

White crystals or powder with no odor and a bitter, saline taste.

SECTION IV - FIRE AND EXPLOSION HAZARD DATA

Flash Point (method used)	**Not Applicable**	Explosive Limits in Air % by Volume	**LEL: Not Applicable UEL: Not Applicable**
NFPA Classification	**Non-Combustible Solid**	Autoignition Temperature	**Not Applicable**

Extinguishing Media

Use agent suitable to surrounding fire. Sodium sulfate by itself will not burn.

Special Fire Fighting Procedures

Wear full protective clothing and self-contained breathing apparatus (SCBA). Poisonous gases are produced in fire, including bromide. Remain clear of smoke and fumes. Use water mist from unmanned device to keep fire-exposed containers cool. Keep cooling containers after fire is extinguished.

Unusual Fire and Explosion Hazards

No specific explosion hazards noted. However, heating with aluminum or magnesium (at 1472°F/800°C) can produce an explosion. Move fire-exposed containers from fire if it can be done without risk.

SECTION V - REACTIVITY DATA

Stability		Conditions to Avoid
		Very stable at room temperature and pressure under normal conditions of handling and storage. Avoid contact with incompatible materials and do not create dusty conditions.
Stable	Unstable	Incompatibility *(materials to avoid)*
X		**Incompatible with aluminum and magnesium with heat. When heated with either, it can produce an extremely violent explosion.**
Hazardous Polymerization		Conditions to Avoid
		Hazardous polymerization of sodium sulfate will not occur.
May Occur	Will Not Occur	Hazardous Decomposition or By-products
	X	**Poisonous gases, including oxides of sulfur and sodium fumes, are emitted during fire.**

SECTION VI - HEALTH HAZARD DATA

Primary Route(s) of Entry:	Inhalation?	Absorption (skin/eye)?	Ingestion?
	X		X

Health Hazards

INHALATION: **The insolubility of sodium sulfate in water makes for low-level toxicity. It does not exhibit the systemic toxicity noted for other sodium salts. However, any soluble impurities can lead to toxic effects. Breathing fine dusts may irritate the nose, throat, and respiratory tract with symptoms of coughing and sneezing.**

EYES & SKIN: **Exposure to the eyes may cause moderate mechanical irritation. Although sodium sulfate is basically non-irritating to the skin, prolonged contact should be avoided.**

INGESTION: **Large doses may cause nausea, vomiting, abdominal pain, and diarrhea.**

Carcinogenicity	NTP Listed?	IARC Cancer Review Group?	OSHA Regulated?	Target Organs?
Unknown Human Questioned Animal	**No**	**No**	**No**	**Gastrointestinal tract (with large doses); eyes.**

Medical Conditions Generally Aggravated by Exposure

None reported.

Emergency and First-aid Procedures

<u>Eye contact</u>: Do not allow victim to keep eyes tightly shut. Flush immediately with water for 15 minutes (minimum); seek medical attention. <u>Skin contact</u>: Remove all contaminated clothing. Immediately wash area with flooding amounts of soap and water. For <u>inhalation</u>: Remove the person from exposure. Provide respiratory assistance and CPR. Transfer to medical facility. If <u>swallowed</u>, seek medical attention immediately. Never give anything by mouth to an unconscious or convulsing person. Give conscious person 1 to 2 glasses of water. If large amounts are ingested, induce vomiting.

SECTION VII - PRECAUTIONS FOR SAFE HANDLING AND USE

Steps to be Taken in Case Material is Released or Spilled

Collect powdered materials in most convenient and safe manner and deposit in sealed containers. Dry sweep not recommended. Use HEPA vacuum and damp mop residue. Restrict those not involved in cleanup from entering area. Notify appropriate authorities, as required by applicable regulations.

Preferred Waste Disposal Method

No citation.

Precautions to be Taken in Handling and Storage

Store to avoid contact with aluminum or magnesium. Store powders in sealed containers in cool, dry, well-ventilated area. Protect containers from physical damage.

Other Precautions and Warnings

Avoid creating dusty conditions, such as moving opened containers or sweeping storage areas.

SECTION VIII - CONTROL MEASURES AND PERSONAL PROTECTIVE EQUIPMENT

Respiratory Protection *(specify type)*

No exposure levels established. Use a respirator with a high-efficiency particulate air (HEPA) filter; otherwise use a NIOSH/MSHA-approved self-contained breathing apparatus or a supplied-air respirator with full facepiece operated in positive pressure or some other continuous flow mode.

Ventilation

Local exhaust (preferred) at site of chemical work or general ventilation.

Protective Gloves	Eye Protection	Other Protective Clothing
Impervious Material	**Chemical/Dust Goggles or Face Mask**	**Protective Apron, Boots**

Work/Hygiene Practices

Always wash hands thoroughly after using chemical; never bring food, drink, or smoking materials into vicinity of chemicals.

SODIUM SULFATE

Na_2SO_4

CAS: 7757-82-6

IDENTIFICATION AND TYPICAL USES

Sodium sulfate exists as white crystals or powder with no detectable odor and a bitter, saline taste. It is used in the manufacture of kraft paper, paperboard, and glass. It is used as a filler in synthetic detergents, in sodium salts, in ceramic glazes, for processing textile fibers, in dyes and tanning procedures, in pharmaceuticals, as a freezing mix, a laboratory reagent, and as a food additive.

RISK ASSESSMENT: HEALTH

General Assessment

Sodium sulfate is unlike most other sodium salts in that its low water solubility makes for low-level toxicity under most conditions of exposure. Primary exposure routes are *inhalation* and, to a lesser degree, *ingestion.* At least one references (Lewis, 1992) considers sodium sulfate a questionable carcinogen in experimental animals studies. It has also been shown to cause experimental teratogenic and adverse reproductive effects.

Inhalation of high dust concentrations can cause irritation of the nose, throat, and respiratory system with symptoms of coughing and sneezing in some individuals.

Ingestion of large amounts of sodium sulfate may cause gastrointestinal irritation with nausea, vomiting, abdominal pain, and diarrhea.

Eye contact results in mechanical irritation and subsequent watering. Removal of the compound from contact with the eye through water deluge will essentially alleviate any further symptoms. Skin contact is basically non-irritating; however, prolonged contact with any chemical substance should be avoided.

Acute Health Effects

The following acute (short-term) health effects may occur immediately or shortly after exposure (within hours or within days) to sodium sulfate:

Skin: No reported effects.

Eye: Mechanical irritant only causing watering and redness. Removal of the chemical from eye contact results in the elimination of symptoms.

Lung: Dust exposures may cause no effect or irritation of the lungs with coughing and sneezing.

Chronic Health Effects

The following chronic (long-term) health effects may occur at some time after exposure to sodium and can last for months or even years:

Cancer Hazards: According to the references, sodium sulfate has been shown to cause tumors and is considered a questioned carcinogen in animal studies. There are no data regarding its human carcinogenic potential.

Reproductive Hazard: According to the references, sodium sulfate may be a teratogen in animals and has caused adverse reproductive effects in experimental studies. Human data are not available.

Other Chronic Effects: None reported.

Recommended Risk-Reduction Measures

Personnel should avoid direct contact with sodium sulfate dusts or powders. The specific toxicity of sodium sulfate is poorly understood as compared to other sodium compounds. However, as a sodium salt, caution is suggested whenever working with or around sodium sulfate. If a less toxic material or compound cannot be substituted for sodium sulfate, then *engineering controls* are the most effective method of reducing exposures. The best protection is to enclose operations and/or provide local exhaust ventilation at the site of sodium dust release. While not always operationally feasible, isolating operations can also reduce exposure. Using respiratory protection is less effective than the controls mentioned above, but is still advisable whenever working with or around sodium sulfate. No exposure levels have been established for sodium sulfate. This does not imply that exposure is without health risk. A supplied-air respirator or a self-contained breathing apparatus (SCBA) with full facepiece and operated in pressure demand or other positive pressure mode are recommended for the greatest possible respiratory protection. If a full facepiece is not available, then chemical/dust goggles should be worn to protect the eyes. A face shield and protective apron should also be considered. To prevent hand and skin exposures, gloves should be used. Glove manufacturers should be contacted and permeation studies obtained *before* final glove selections are made.

Administrative controls should also be in place to minimize the potential for human exposures. These may include written procedures or policies which specify the methods and techniques that will be practiced whenever personnel are to work with sodium sulfate.

All personnel should receive *training* on the use, hazards, protective measures, emergency actions, and other precautions per 29 CFR 1910.1200 (Hazard Communication), prior to the first assignment in an area where sodium sulfate is used or stored.

Before beginning employment working with sodium sulfate, and at regular intervals thereafter (e.g., annually), those personnel with frequent or potentially high exposures should be provided the following recommended medical tests:

☑ Lung function test.

Any evaluation should include a careful history of past and present symptoms with an examination. Medical tests that simply look for existing damage are not a substitute for controlling exposures. Also, since smoking can cause heart disease as well as lung cancer, emphysema, and other known respiratory problems, exposure to sodium may cause additional complications. Therefore, personnel who smoke should avoid any unprotected contact with sodium powders.

Other methods to reduce exposure include:

☑ Where possible, enclose operations and use local exhaust ventilation at the site of chemical release. If local exhaust ventilation or enclosure is not used, respiratory protection should be mandatory.

☑ Always ensure that proper protective clothing is worn when using chemical substances and that personnel have been trained on its proper use, care, and maintenance.

☑ Wash thoroughly immediately after exposure to sodium sulfate and at the end of the work shift or before eating, drinking, applying cosmetics, or smoking.

☑ Hazard warning information should be posted in the work area. In addition, as part of an on-going education and training program, all information on the health and safety hazards of sodium sulfate should be communicated to all exposed workers.

☑ Eye wash stations should be provided in the immediate work area for emergency use. If there is a possibility of skin exposure to sodium sulfate, emergency shower facilities should also be provided in the work area.

☑ Workers whose clothing has been contaminated by sodium sulfate should change into clean clothes before leaving work. Contaminated work clothing should be laundered only by individuals who have been informed of the hazards of exposure to sodium sulfate.

RISK ASSESSMENT: ENVIRONMENT

General Assessment

The environment is at risk of exposure during transportation, storage, disposal, or destruction of sodium compounds. In almost every scenario, the threat of environmental exposure is contingent upon the proper handling of the chemical substance. Accidental spills, large or small, where sodium sulfate contacts incompatible commodities can result in fire, explosion (depending upon conditions of spill), and possible contamination of the surrounding environmental mediums (water, soil, and air).

In its powdered form, sodium sulfate is considered a non-combustible solid. However, it can react explosively and present a serious fire hazard when heated in the presence of incompatible metals. Because it is incompatible with many commodities, including metals such as aluminum and magnesium, extreme caution is required in handling, storage, transportation, and disposal of sodium sulfate. These characteristics also require special consideration during any emergency situation involving a leak or spill of sodium sulfate powder or dust. Should sodium sulfate ever come into contact with incompatible substances either during use, transportation, storage, or disposal, violent and even explosive reactions are possible.

Sodium sulfate can enter the environment from industrial and municipal waste treatment plant discharges and from spills.

Acute Ecological Effects

Acute (short-term) toxic effects may include the death of animals, birds, or fish and death or low growth rate in plants. Acute effects are seen 2 to 4 days after animals or plants are exposed to sodium.

Insufficient data are available to evaluate the short-term effects of sodium sulfate to aquatic life, plants, birds, or land animals.

Chronic Ecological Effects

Chronic (long-term) toxic effects may include shortened life span, reproductive problems, lower fertility, and changes in appearance or behavior in exposed animals. These effects can be seen long after first exposure(s) to toxic chemicals.

Insufficient data are available to evaluate or predict the long-term effects of sodium sulfate to aquatic life, plants, birds, or land animals.

Water Solubility

Sodium sulfate is highly soluble in water. Concentrations of 1 to 1000 milligrams and more will mix with a liter of water.

Persistence in the Environment

Sodium sulfate is highly persistent in water, with a half-life greater than 200 days. The half-life of a pollutant is the amount of time it takes for one half of the chemical to be degraded. No data are available on the estimated percentages of sodium persistence in the aquatic or terrestrial environments.

Bioaccumulation in Aquatic Organisms

Some substances increase in concentration, or *bioaccumulate,* in living organisms as they breathe contaminated air, drink contaminated water, or eat contaminated food. These chemicals can become concentrated in the tissues and internal organs of animals as well as humans.

The concentration of sodium sulfate found in fish tissues is expected to be about the same as the average concentration of sodium sulfate in the water from which the fish was taken.

Recommended Risk-Reduction Measures

Proper training of all transporters will reduce the likelihood of a mishap or accident resulting in a leak or spill of sodium sulfate dusts or powders into the environment. The correct use of labeling on all containers, trucks, and rail cars will enable emergency responders to react properly and quickly to any disaster thereby reducing the potential risk to the environment and to personnel.

Storage of sodium sulfate should be segregated from incompatible chemicals to minimize the risk of cross-contamination or contact. Containers should be protected from physical damage.

If a spill or leak to the environment has occurred, fire department, emergency response and/or hazardous materials spill personnel should be notified immediately. Cleanup should be attempted only by those trained in proper spill containment procedures. Contaminated soils should be removed for incineration and replaced with clean soil. If sodium sulfate should contact the water table, aquifer, or navigable waterway, time is of the essence. It is highly soluble in water and total remediation may not be possible. The local and/or state emergency response authorities must be notified. A comprehensive emergency response or disaster preparedness/recovery plan should be in place prior to any operations involving the use, transportation, storage, or disposal of sodium sulfate.

If sodium sulfate powder is spilled or leaked, the following specific steps are recommended:

- ☑ Restrict persons not wearing protective clothing from area of spill or leak until cleanup is complete and area can be opened for normal work.
- ☑ Ventilate area of spill or leak.
- ☑ Collect powdered materials in the safest manner possible and deposit in sealed containers. Recommend using a vacuum with a high-efficiency particulate air (HEPA) filter. Do NOT dry sweep (generates airborne dusts). Use a damp mop method to clean up residue. Do NOT allow runoff from cleanup operations to enter sewers or waterways. Contain and collect all water from such operations and dispose of properly as waste.
- ☑ It may be necessary to dispose of sodium sulfate as a hazardous waste. The responsible state agency or the regional office of the federal Environmental Protection Agency (EPA) should be contacted for specific recommendations.

RISK ASSESSMENT: BUSINESS

General Assessment

Accidents or mishaps involving sodium sulfate can present a significant threat to business operations. The loss or damage of equipment or facilities can significantly affect fiscal viability. Lawsuits that may result from personnel injury/death, public exposures, and/or environmental contamination will also require a serious expenditure of resources. Media attention surrounding an injury, death, or environmental damage can also result in a loss of profits and loss of current as well as future business. Always remember that anytime the terms "cancer," "carcinogen," or "reproductive hazard" are used, public emotion, hysteria, ignorance, and anxiety can all run equally high. This must be carefully considered whenever drafting or implementing public relations policies.

Recommended Risk-Reduction Measures

Company attorneys, safety and health professionals, and environmental specialists should be involved in the development of any procedures for responding to chemical incidents. A company official should be predesignated as a public relations officer with specific training in dealing with the press. Corporate plans and policies should be developed, approved, and implemented long before any need for such arises.

REFERENCES

Ahlbom, A. 1993. *Biostatistics for Epidemiologists.* New York: CRC Press/Lewis Publishers.

American Conference of Governmental Industrial Hygienists. 1988. *Documentation of the Threshold Limit Values and Biological Exposure Indices,* 5th Edition (with updates). Cincinnati: ACGIH

Asante-Duah, D. K. 1993. *Hazardous Waste Risk Assessment.* New York: CRC Press/Lewis Publishers.

Bloom, A. D. and F. J. DeSerres. 1995. *Ecotoxicity and Human Health.* New York: CRC Press/Lewis Publishers.

Brusick, D. J. 1994. *Methods for Genetic Risk Assessment.* New York: CRC Press/Lewis Publishers.

Bueche, J. R. 1972. *Principles of Physics,* 2nd Edition. New York: McGraw-Hill.

Calabrese, E. J. 1994. *Biological Effects of Low Level Exposures.* New York: CRC Press/Lewis Publishers.

Cothern, R. R. 1993. *Comparative Environmental Risk Assessment.* New York: CRC Press/Lewis Publishers.

Cothern, C. R. 1995. *Handbook for Environmental Risk Decisionmaking.* New York: CRC Press/Lewis Publishers.

Danse, I. R. 1991. *Common Sense Toxics in the Workplace.* New York: Van Nostrand Reinhold.

Environmental Protection Agency. 1983. *Chemical Hazard Information Profiles.* Washington, D.C.: U.S. EPA.

Fodden, C. R., and J. L. Weddell. 1992. *Hazardous Materials Emergency Action Data.* New York: CRC Press/Lewis Publishers.

Forseberg, K., and L. H. Keith. 1993. *Chemical; Protective Clothing Permeation and Degradation Database.* Chelsea, MI: Lewis Publishers.

Gardner, R. W. 1994. *Chemical Intolerance.* New York: CRC Press/Lewis Publishers.

Howard, P. H. 1995. *Handbook of Environmental Fate and Exposures Data for Organic Chemicals,* Vol. I-IV. New York: CRC Press/Lewis Publishers.

Keith, L. H. and M. M. Walker. 1995. *Handbook of Air Toxics.* New York: CRC Press/Lewis Publishers.

Landis, W. G., M. H. Yu. 1995. *Introduction to Environmental Toxicology.* New York: CRC Press/Lewis Publishers.

Lauwerys, R. R., P. Hoet. 1993. *Industrial Chemical Exposure.* New York: CRC Press/Lewis Publishers.

Lewis, R. J., Sr. 1993. *Hawley's Condensed Chemical Dictionary,* 12th Edition. New York: Van Nostrand Reinhold.

Lewis, R. J., Sr. 1992. *Sax's Dangerous Properties of Industrial Materials,* Volumes 1, 2, and 3, 8th Edition. New York: Van Nostrand Reinhold.

Manahan, S. E. 1994. *Environmental Chemistry,* 6th Edition. New York: CRC Press/Lewis Publishers.

Meyer, E. 1990. *Chemistry of Hazardous Materials.* Englewood Cliffs, NJ: Prentice Hall.

Mycock, J. C., J. D. McKenna, and L. Theodore. 1995. *Handbook of Air Pollution Control Engineering and Technology.* New York: CRC Press/Lewis Publishers.

Patnaik, P. 1992. *A Comprehensive Guide to the Hazardous Properties of Chemical Substances.* New York: Van Nostrand Reinhold.

Proctor, N. H., J. P. Hughes. 1991. *Chemical Hazards of the Workplace.* New York: Van Nostrand Reinhold.

Quigley, D. R. 1994. *Handbook of Emergency Chemical Management.* New York: CRC Press/Lewis Publishers.

Rea, W. J. 1995. *Chemical Sensitivity,* Vol. I-III. New York: CRC Press/Lewis Publishers.

Reeves, A. L. 1979. Sodium. *Handbook on the Toxicology of Metals.* pp 321-328. New York: Elsevier North-Holland.

Richardson, M. 1995. *Dictionary of Substance and Their Effects,* Vol. I-VI. New York: CRC Press/Lewis Publishers.

Sacarello, H. L. A. 1994. *The Comprehensive Handbook of Hazardous Materials.* New York: CRC Press/Lewis Publishers.

Sheftel, V. O. 1995. *Handbook of Toxic Properties of Monomers and Additives.* New York: CRC Press/Lewis Publishers.

Torkelson, T. R. V. K. Rowe. 1981. *Patty's Industrial Hygiene and Toxicology,* Vol. 2B. New York: Wiley-Interscience.

MATERIAL SAFETY DATA SHEET

CHEMICAL NAME	
STIBINE	**CRC PRESS/LEWIS PUBLISHERS DISCLAIMER:** THE INFORMATION AND RECOMMENDATIONS PRESENTED HEREIN ARE BASED ON SOURCES BELIEVED TO BE RELIABLE. CRC MAKES NO REPRESENTATION ON ITS COMPLETENESS OR ACCURACY. IT IS THE RESPONSIBILITY OF THE USER TO DETERMINE THE CHEMICAL'S SUITABILITY FOR ITS INTENDED USE, THE CHEMICAL'S SAFE USE, AND THE CHEMICAL'S PROPER DISPOSAL. NO REPRESENTATIONS AND/OR WARRANTIES, EITHER EXPRESSED OR IMPLIED, OF THE MERCHANTABILITY OR FITNESS FOR A PARTICULAR PURPOSE, OR OF ANY OTHER NATURE, ARE MADE WITH RESPECT TO THE INFORMATION PROVIDED IN THIS MSDS OR TO THE CHEMICAL TO WHICH INFORMATION MAY REFER. CRC NEITHER ASSUMES NOR AUTHORIZES ANY OTHER PERSON TO ASSUME FOR IT, ANY OTHER ADDITIONAL RESPONSIBILITY OR LIABILITY FOR THE USE OF, OR RELIANCE UPON, THIS INFORMATION.

HAZARD WARNING INFORMATION

HEALTH	FIRE	REACTIVE	OTHER	DEGREE OF HAZARD	COLOR CODING	OTHER CODES
3	4	2		0 = Minimum Hazard 1 = Slight Hazard 2 = Moderate Hazard 3 = Serious Hazard 4 = Severe Hazard	HEALTH = BLUE FIRE = RED REACTIVITY = YELLOW OTHER = WHITE	OX = Oxidizer ACID = Acid ALK = Alkali COR = Corrosive W = Use No Water

SECTION I - GENERAL INFORMATION

Characterization **Hydride Gas**	RCRA Number **None**	EPA Class **Not Applicable**
DOT Proper Shipping Name **Antimony Hydride**	Chemical Abstract Service (CAS) Number **7803-52-3**	
DOT Hazard Class and Label Requirements **Poison A: Flammable Gas and Poison**	DOT Emergency Guide Code **18**	
DOT Identification Number **UN 2676**	Chemical Formula **SbH_3**	

Synonyms
Antimony hydride; antimony trihydride; hydrogen antimonide.

SECTION II - HAZARDOUS INGREDIENTS AND IDENTITY INFORMATION

Hazardous Components (specific identity)	OSHA Exposure Criteria	NIOSH Exposure Criteria	Immediately Dangerous to Life and Health (IDLH)	ACGIH Exposure Criteria
Stibine (derivation: By action of hydrogen chloride on antimony-metal compounds such as Zn_3Sb_2; also released by reduction of antimony compounds in hydrochloric acid solutions with zinc or other reducing metal). **1 ppm = 5.19 mg/m³**	**PEL:** 0.1 ppm 0.5 mg/m³ **STEL:** Not Established	**REL:** 0.1 ppm 0.5 mg/m³ **STEL:** Not Established	**5 ppm**	**TLV:** 0.1 ppm 0.5 mg/m³ **STEL:** Not Established

SECTION III - PHYSICAL AND CHEMICAL CHARACTERISTICS

Boiling Point **-1°F (-18°C)**	Specific Gravity (H_2O = 1) **4.31**
Vapor Pressure (atmospheres) **>1 at 68°F (20°C)**	Molecular Weight **124.8**
Vapor Density (Air = 1) **Not Reported (Gas)**	Freezing Point **-126°F (-88°C)**

Solubility
Slightly soluble in water. Highly soluble in alcohol, carbon disulfide, and organic solvents.

Appearance and Odor
Colorless gas with a disagreeable odor like that of hydrogen sulfide (rotting eggs).

SECTION IV - FIRE AND EXPLOSION HAZARD DATA

Flash Point (method used) **Not Reported**	Explosive Limits in Air % by Volume **LEL: Not Reported UEL: Not Reported**
NFPA Classification **Flammable Gas**	Autoignition Temperature **Not Reported**

Extinguishing Media
Stop flow of gas. Use water spray to keep fire-exposed containers cool.

Special Fire Fighting Procedures
Stibine gas presents a serious explosion hazard when exposed to flame or involved in fire. Gas is poisonous: wear full protective clothing and self-contained breathing apparatus (SCBA).

Unusual Fire and Explosion Hazards
To stop fire, stop the flow of gas. Extremely reactive substance. Explosive reaction with strong oxidizers. It decomposes to explosive hydrogen and toxic antimony metal. May re-ignite explosively after fire has been extinguished.

SECTION V - REACTIVITY DATA

Stability		Conditions to Avoid **Stibine is extremely reactive and highly explosive in the presence of heat or flame. It is unstable under normal conditions of temperature and pressure. Explosions are likely to be violent and severe.**
Stable	Unstable **X**	Incompatibility *(materials to avoid)* **Stibine is incompatible with halogenated hydrocarbons, acids, oxidizers, moisture, chlorine, ozone, and ammonia. Incompatible with all oxidants.**
Hazardous Polymerization		Conditions to Avoid **Under normal conditions of temperature and pressure, hazardous polymerization of stibine cannot occur. Avoid elevated temperatures.**
May Occur	Will Not Occur **X**	Hazardous Decomposition or By-products **Stibine decomposes upon heating to evolve highly explosive hydrogen gas and toxic metallic antimony.**

SECTION VI - HEALTH HAZARD DATA

Primary Route(s) of Entry:	Inhalation? X	Absorption (skin)?	Ingestion?

Health Hazards

INHALATION: Causes irritation of the eyes, nose, throat, and respiratory system with symptoms that include headache, weakness, nausea, abdominal pain, hemoglobinuria, hematuria, hemolytic anemia, jaundice, and severe pulmonary irritation with a possibility of pulmonary edema (fluid in the lungs). A hemolytic agent, causing injury to the liver and kidney. In severe cases, death can occur from renal failure or pulmonary edema.

SKIN & EYES: May cause some irritation to the eyes and skin. Cold temperatures may cause frostbite.

INGESTION: Not likely since stibine is a gas.

Carcinogenicity	NTP Listed?	IARC Cancer Review Group?	OSHA Regulated?	Target Organs?
Unknown Human Questioned Animal	No	No	29 CFR 1910.1000 (Table Z-1)	Respiratory system; liver; kidneys; blood.

Medical Conditions Generally Aggravated by Exposure

Respiratory function impairments (bronchitis, asthma, etc.).

Emergency and First-aid Procedures

Eye contact: Immediately flush with large amounts of water for 15 minutes (minimum), occasionally lifting eyelids; seek medical attention. Skin contact: Remove all contaminated clothing. Immediately wash area with large amounts of soap and water. For inhalation: Remove the person from exposure. Provide respiratory assistance and CPR. Transfer promptly to medical facility. Observation for 24 - 48 hours since pulmonary edema is possible. If swallowed: Not likely since this product is a gas.

SECTION VII - PRECAUTIONS FOR SAFE HANDLING AND USE

Steps to be Taken in Case Material is Released or Spilled

Remove all ignition sources. Ventilate area. Stop flow of gas. If the source of the leak is a cylinder and cannot be stopped, remove to a safe place in open air; repair leak or let cylinder empty. It may be necessary to evacuate area and wait for cylinder to empty.

Preferred Waste Disposal Method

Destroy by oxidation with aqueous copper sulfate solution under nitrogen.

Precautions to be Taken in Handling and Storage

Do not store with incompatible chemicals since violent reactions can occur. Store in tightly closed containers in a cool, dark, well-ventilated area away from heat and moisture. Keep away from ignition sources such as fire, sparks, and flame. Protect containers from physical damage.

Other Precautions and Warnings

Bulk storage of stibine is not recommended. Containers may explode in fire or under conditions of extreme heat (do not store outdoors or in direct sunlight).

SECTION VIII - CONTROL MEASURES AND PERSONAL PROTECTIVE EQUIPMENT

Respiratory Protection *(specify type)*

At any exposure level, use a powered supplied-air respirator set in positive pressure mode or use an MSHA/NIOSH-approved self-contained breathing apparatus (SCBA) with full facepiece and pressure demand or other positive pressure mode.

Ventilation

Local exhaust or general mechanical systems recommended. Explosion-proof design suggested.

Protective Gloves	Eye Protection	Other Protective Clothing
Synthetic Impervious Material	Chemical Goggles and/or Face Mask	Protective Suit

Work/Hygiene Practices

Always wash hands thoroughly after using chemical; never bring food, drink, or smoking materials into vicinity of chemicals.

STIBINE

SbH_3

CAS: 7803-52-3

IDENTIFICATION AND TYPICAL USES

Stibine is a colorless gas with a repulsive, disagreeable odor like that of hydrogen sulfide (rotting eggs). It is used as a fumigating agent and in the preparation of antimony compounds.

RISK ASSESSMENT: HEALTH

General Assessment

Stibine is a severe pulmonary irritant that can cause secondary infections in the respiratory passageways (may precipitate bronchial pneumonia). As a gas, primary entry is by *inhalation.* Its toxicity is similar to but lower than that of arsine. At least one source (Patnaik, 1992) suggests that stibine can be expected to exhibit some carcinogenic action. There are no data on its mutagenic or teratogenic properties.

Inhalation of stibine gas can be extremely dangerous, causing irritation of the eyes, nose, throat, and respiratory system. Symptoms include headache, weakness, nausea, abdominal pain, lumbar pain, hemolytic effects (such as hemoglobinuria, hematuria, hemolytic anemia), liver effects (jaundice), and delayed pulmonary edema. Death can occur from renal failure of pulmonary edema. The symptoms of pulmonary edema (cough, congestion, chest pain, dypsnea) may be delayed for up to 48 hours thereby creating a false sense of security with regard to health exposure risk.

Eye or skin contact can cause some degree of irritation with inflammation, redness, and swelling. Also, because of the extremely cold temperature of stibine gas, there may be some danger of frostbite action on exposed tissues.

Acute Health Effects

The following acute (short-term) health effects may occur immediately or shortly after exposure to stibine:

Skin: Irritation, swelling, or frostbite on contact with the cold gas.

Eye: Inflammation and swelling of the conjunctiva and surrounding tissue.

Lung: Severe lung damage with symptoms of coughing, wheezing, shortness of breath, chest pain, dizziness, fatigue, headache, and irritation of the mouth, nose, and throat. Nausea, vomiting, and fever may also occur. Higher exposures can lead to pulmonary edema, a medical emergency which can be fatal.

Other: There is a possibility of blood, liver, and kidney damage on high acute exposures to stibine.

Chronic Health Effects

The following chronic (long-term) health effects may occur at some time after exposure to stibine and can last for months or even years:

Cancer Hazards: According to information presented in the references, stibine has not been shown to cause cancer in test animals or in humans. However, because of its similarity to arsine in chemical and toxicological properties, stibine is expected to exhibit some level of carcinogenic action (Patnaik, 1992).

Reproductive Hazards: According to information presented in the references, stibine has not been tested for its ability to affect reproduction.

Other Chronic Effects: Prolonged exposures may cause pulmonary irritation, liver, or kidney damage.

Recommended Risk-Reduction Measures

Personnel should avoid direct contact with stibine. It is extremely toxic and exposures can cause death or serious physical injury or illness. If a less toxic chemical cannot be substituted for a hazardous substance, then *engineering controls* are the most effective method of reducing exposures. The best protection is to enclose operations and/or provide local exhaust ventilation at the site of chemical release. While not always operationally feasible, isolating operations can also reduce exposure. Using respiratory protection is less effective than the controls mentioned above, but is always advisable whenever working with or around stibine. An MSHA/NIOSH-approved self-contained breathing apparatus (SCBA) with full facepiece operated in pressure demand is the recommended respiratory protection method of choice. If a full facepiece is not available, then chemical goggles should be worn to protect the eyes. Whenever a chemical splash hazard exists, a face shield and a protective apron should be worn. To prevent hand and skin exposures, impervious thermal gloves should be used. Glove manufacturers should be contacted and permeation studies obtained *before* final glove selections have been made.

Administrative controls should also be in place to minimize the potential for human exposures. These may include written procedures or policies which specify the methods and techniques that will be practiced whenever personnel are to work with stibine.

All personnel should receive *training* on the use, hazards, protective measures, emergency actions, and other precautions per 29 CFR 1910.1200 (Hazard Communication), prior to the first assignment in an area where stibine is used or stored. Before beginning employment and at regular intervals thereafter (e.g., annually), the following medical tests are recommended:

☑ Lung function tests. These may be normal at first if the person is not having an attack at the time.

If symptoms develop or overexposure is suspected, the following may be useful:

☑ Consider chest X-ray after acute overexposure (may be negative if taken immediately after exposure due to delayed onset of pulmonary edema).

☑ Liver and kidney function tests.

Any evaluation should include a careful history of past and present symptoms with an examination. Medical tests that simply look for existing damage are not a substitute for controlling exposures. Also, because smoking can cause heart disease, as well as lung cancer, emphysema, and other respiratory problems, exposure to phosgene may worsen pre-existing respiratory conditions. Smokers may experience symptoms more quickly than non-smokers under the same conditions of exposure. Prudent risk management requires proper consideration of *all* factors which may be responsible for exposure symptoms in the workplace.

Other methods to reduce exposure to chemicals include:

☑ Where possible, enclose operations and use local exhaust ventilation at the site of chemical release. If local exhaust ventilation or enclosure is not used, respiratory protection should be mandatory.

☑ Always ensure that proper protective clothing is worn when using chemical substances and that personnel have been properly trained in its use, care, and maintenance.

☑ Wash thoroughly immediately after exposure to stibine and at the end of the work shift or before eating, drinking, smoking, or applying cosmetics.

☑ Work clothing contaminated with stibine should never be taken home for laundering (family members can be exposed). Only personnel trained in the hazards of exposure should launder contaminated clothing.

☑ Safety shower and eyewash stations should be readily available in work areas where stibine is used or stored.

☑ Hazard warning information should be posted in the work area. In addition, as part of an on-going education and training program, all information on the health and safety hazards of stibine should be communicated to all exposed and potentially exposed workers.

RISK ASSESSMENT: ENVIRONMENT

General Assessment

The environment is at risk of exposure during transportation, storage, disposal, or destruction of stibine. In almost every scenario, the threat of environmental exposure is contingent upon the proper handling of the chemical substance. Accidental releases, large or small, can result in fire, explosion, and possible contamination of the surrounding environmental mediums (water, soil, and air).

Stibine is considered to be a flammable gas. It is explosive by itself and, because it is incompatible with many other commodities, contact with any of them can result in violent explosion and fire. It is highly reactive and will ignite with moist air at room temperature. Stibine forms explosive mixtures in air which makes it a serious fire and explosion risk. It will explode in contact with many oxidizing agents (especially fluorine, bromine, and chlorine). It is highly reactive in the presence of halogenated hydrocarbons, acids, ozone, and ammonia. Therefore, stibine requires special consideration during any emergency involving a leak or release of stibine gas.

Stibine can enter the environment through industrial discharges, unchecked venting, or through accidental leaks or spills.

Acute Ecological Effects

Acute (short-term) toxic effects may include the death of animals, birds, or fish, and death or low growth rate in plants. Acute effects are seen 2 to 4 days after animals or plants are exposed to stibine.

Insufficient data are available to evaluate or predict the short-term effects of stibine on aquatic life, plants, birds, or land animals.

Chronic Ecological Effects

Chronic (long-term) toxic effects may include shortened life span, reproductive problems, lower fertility, and changes in appearance or behavior in exposed animals. These effects can be seen long after first exposure(s) to toxic chemicals.

Insufficient data are available to evaluate the long-term effects of stibine on aquatic life, plants, birds, or land animals.

Water Solubility

Stibine is slightly soluble in the aquatic environment. Concentrations of 1 to 100 milligrams will mix with a liter of water.

Persistence in the Environment

As a gas, there is no evidence in the references to show that stibine has any measurable persistence in the environment.

Bioaccumulation in Aquatic Organisms

Some substances increase in concentration, or *bioaccumulate,* in living organisms as they breathe contaminated air, drink contaminated water, or eat contaminated food. These chemicals can become concentrated in the tissues and internal organs of animals as well as humans.

As a gas, stibine is not expected to accumulate in the edible tissues of fish.

Recommended Risk-Reduction Measures

Proper training of all transporters will reduce the likelihood of a mishap or accident resulting in a leak or release to the environment. The correct use of DOT labeling (per the MSDS) on all containers, trucks, and rail cars will enable emergency responders to react properly and quickly to any disaster thereby reducing the potential risk to the environment and to personnel.

Storage of stibine should be segregated from incompatible chemicals to minimize the risk of cross-contamination or contact. Under all circumstances, it should be protected from exposure to moisture. Buildings designated for storage should be equipped with appropriate fire protection systems (alarms, sprinklers, emergency lighting, portable extinguishers). Explosion-proof ventilation is recommended.

If a spill or release to the environment has occurred, fire department, emergency response, and/or hazardous materials spill personnel should be notified immediately. Cleanup should be attempted only by those trained in proper containment procedures using non-sparking tools. Contaminated soils should be removed for incineration and replaced with clean soil. If stibine should contact the water table, aquifer, or navigable waterway, remediation may not be possible. As a gas, total containment is not probable. When such spills occur, the local and/or state emergency response authorities must be notified. A comprehensive emergency response or disaster preparedness/recovery plan should be in place prior to any operations involving the use, transportation, storage, or disposal of stibine. If stibine is spilled or leaked, the following specific steps are recommended:

- ☑ Restrict persons not wearing protective clothing from area of spill or leak until cleanup is complete and area can be opened for normal work.
- ☑ Remove all ignition sources and ventilate area.
- ☑ If applicable, stop flow of leaking liquid or gas. If leak source is a cylinder and the leak cannot be stopped in place, carefully remove leaking cylinder to a safe place in the open air, and allow cylinder to empty. Repair when completely empty. Attempting to repair while venting can be extremely dangerous.
- ☑ Keep stibine out of a confined space, such as a sewer, because of the possibility of explosion (unless the sewer is designed to prevent the buildup of explosive concentrations).

☑ It may be necessary to dispose of stibine as a hazardous waste. The responsible state agency or the regional office of the federal Environmental Protection Agency (EPA) should be contacted for specific recommendations.

RISK ASSESSMENT: BUSINESS

General Assessment

Accidents or mishaps involving stibine can present a significant threat to business operations. The loss or damage of equipment or facilities can significantly affect fiscal viability. Lawsuits that may result from personnel injury, public exposures, and environmental contamination will also require a serious expenditure of resources. Media attention surrounding an injury, illness, or environmental damage can also result in a loss of profits and loss of current as well as future business.

Recommended Risk-Reduction Measures

Company attorneys, safety and health professionals, and environmental specialists should be involved in the development of any procedures dealing with safety or emergency response. A company official should be pre-designated as a public relations officer with specific training in dealing with the press. Corporate plans and policies should be developed, approved, and implemented long before any need for such arises.

REFERENCES

American Conference of Governmental Industrial Hygienists. 1988. *Documentation of the Threshold Limit Values and Biological Exposure Indices,* 5th Edition (with updates). Cincinnati: ACGIH

Anthony, C. P., and N. J. Kolthoff. 1971. *Textbook of Anatomy and Physiology.* St. Louis: C. V. Mosby Company.

Asante-Duah, D. K. 1993. *Hazardous Waste Risk Assessment.* New York: CRC Press/Lewis Publishers.

Bloom, A. D. and F. J. DeSerres. 1995. *Ecotoxicity and Human Health.* New York: CRC Press/Lewis Publishers.

Brusick, D. J. 1994. *Methods for Genetic Risk Assessment.* New York: CRC Press/Lewis Publishers.

Bueche, J. R. 1972. *Principles of Physics,* 2nd Edition. New York: McGraw-Hill.

Burrell, R., D. K. Flaherty, and L. J. Sauers. 1992. *Toxicology of the Immune System.* New York: Van Nostrand Reinhold.

Calabrese, E. J., E. M. Kenyon. *Air Toxics and Risk Assessment.* New York: CRC Press/Lewis Publishers.

Cockerman, L. G., B. S. Shane. 1994. *Basic Environmental Toxicology.* New York: CRC Press/Lewis Publishers.

Cothern, R. R. 1993. *Comparative Environmental Risk Assessment.* New York: CRC Press/Lewis Publishers.

Gardner, R. W. 1994. *Chemical Intolerance.* New York: CRC Press/Lewis Publishers.

Hallenbeck, W. H. 1993. *Quantitative Risk Assessment for Environmental and Occupational Health,* 2nd Edition. New York: CRC Press/Lewis Publishers.

Hodgson, E., R. B. Mailman, and J. E. Chambers. 1988. *Dictionary of Toxicology.* London: The MacMillian Company.

Hosty, J. W., and P. Foster. 1990. *A Practical Guide to Chemical Spill Response.* New York: Van Nostrand Reinhold.

Howard, P. H. 1995. *Handbook of Environmental Fate and Exposures Data for Organic Chemicals,* Vol. I-IV. New York: CRC Press/Lewis Publishers.

Howard, P. H., et. al. 1991. *Handbook of Environmental Degradation Rates.* New York: CRC Press/Lewis Publishers.

Jones, F. E. 1994. *Toxic Organic Vapors in the Workplace.* New York: CRC Press/Lewis Publishers.

Kamrin, M. A. 1988. *Toxicology: A Primer on Principles and Applications.* Chelsea, Michigan: Lewis Publishers.

Keith, L. H. and M. M. Walker. 1995. *Handbook of Air Toxics.* New York: CRC Press/Lewis Publishers.

Landis, W. G., M. H. Yu. 1995. *Introduction to Environmental Toxicology.* New York: CRC Press/Lewis Publishers.

Langerman, N. 1994 *Precautionary Labels for Chemical Containers.* New York: CRC Press/Lewis Publishers.

Lauwerys, R. R., P. Hoet. 1993. *Industrial Chemical Exposure.* New York: CRC Press/Lewis Publishers.

Lewis, R. J., Sr. 1993. *Hawley's Condensed Chemical Dictionary,* 12th Edition. New York: Van Nostrand Reinhold.

Lewis, R. J., Sr. 1992. *Sax's Dangerous Properties of Industrial Materials,* Volumes 1, 2, and 3, 8th Edition. New York: Van Nostrand Reinhold.

Lide, D. R. 1995. *Handbook of Chemistry and Physics,* 76th Edition. New York: CRC Press/Lewis Publishers.

Manahan, S. E. 1992. *Toxicological Chemistry,* 2nd Edition. New York: CRC Press/Lewis Publishers.

Meyer, E. 1990. *Chemistry of Hazardous Materials.* Englewood Cliffs, NJ: Prentice Hall.

Patnaik, P. 1992. *A Comprehensive Guide to the Hazardous Properties of Chemical Substances.* New York: Van Nostrand Reinhold.

Proctor, N. H., and J. P. Hughes. 1991. *Chemical Hazards of the Workplace.* New York: Van Nostrand Reinhold.

Rea, W. J. 1995. *Chemical Sensitivity,* Vol. I-III. New York: CRC Press/Lewis Publishers.

Richardson, M. 1995. *Dictionary of Substance and Their Effects,* Vol. I-VI. New York: CRC Press/Lewis Publishers.

U.S. Department of Health and Human Services, National Institute for Occupational Safety and Health. 1994. *NIOSH Pocket Guide to Chemical Hazards.* Washington, D.C.: U.S. Government Printing Office.

MATERIAL SAFETY DATA SHEET

CHEMICAL NAME	
STRYCHNINE	**CRC PRESS/LEWIS PUBLISHERS DISCLAIMER:** THE INFORMATION AND RECOMMENDATIONS PRESENTED HEREIN ARE BASED ON SOURCES BELIEVED TO BE RELIABLE. CRC MAKES NO REPRESENTATION ON ITS COMPLETENESS OR ACCURACY. IT IS THE RESPONSIBILITY OF THE USER TO DETERMINE THE CHEMICAL'S SUITABILITY FOR ITS INTENDED USE, THE CHEMICAL'S SAFE USE, AND THE CHEMICAL'S PROPER DISPOSAL. NO REPRESENTATIONS AND/OR WARRANTIES, EITHER EXPRESSED OR IMPLIED, OF THE MERCHANTABILITY OR FITNESS FOR A PARTICULAR PURPOSE, OR OF ANY OTHER NATURE, ARE MADE WITH RESPECT TO THE INFORMATION PROVIDED IN THIS MSDS OR TO THE CHEMICAL TO WHICH INFORMATION MAY REFER. CRC NEITHER ASSUMES NOR AUTHORIZES ANY OTHER PERSON TO ASSUME FOR IT, ANY OTHER ADDITIONAL RESPONSIBILITY OR LIABILITY FOR THE USE OF, OR RELIANCE UPON, THIS INFORMATION.

HAZARD WARNING INFORMATION

HEALTH	FIRE	REACTIVE	OTHER	DEGREE OF HAZARD	COLOR CODING	OTHER CODES
3	0	0		0 = Minimum Hazard 1 = Slight Hazard 2 = Moderate Hazard 3 = Serious Hazard 4 = Severe Hazard	HEALTH = BLUE FIRE = RED REACTIVITY = YELLOW OTHER = WHITE	OX = Oxidizer ACID = Acid ALK = Alkali COR = Corrosive W = Use No Water

SECTION I - GENERAL INFORMATION

Characterization	**Strychnos Alkaloid**	RCRA Number	**P108**
		EPA Class	**Acute Hazardous Waste**
DOT Proper Shipping Name	**Strychnine or Strychnine Salts**	Chemical Abstract Service (CAS) Number	**57-24-9**
DOT Hazard Class and Label Requirements	**Poison B; St. Andrews Cross**	DOT Emergency Guide Code	**53**
DOT Identification Number	**UN 1692**	Chemical Formula	$C_{21}H_{22}N_2O_2$

Synonyms
Nux vomica; strynchnos; strychnidin-10-one; Certox; Dolco Mouse Cereal; Kwik-Kil; Mole Death; Mouse-Nots; Mouse-Rid; Mouse-Tox; Pied Piper Mouse Seed; Ro-Dex; Sanaseed.

SECTION II - HAZARDOUS INGREDIENTS AND IDENTITY INFORMATION

Hazardous Components (specific identity)	OSHA Exposure Criteria	NIOSH Exposure Criteria	Immediately Dangerous to Life and Health (IDLH)	ACGIH Exposure Criteria
Strychnine (derivation: Extracted from the Asian trees *Strychnos nux vomica*, *Strychnos ignati*, and *Strychnos tiente*.).	**PEL (8-hour):** 0.15 mg/m³ **STEL:** Not Established	**REL (10-hour):** 0.15 mg/m³ **STEL:** Not Established	3 mg/m³	**TLV:** 0.15 mg/m³ **STEL:** Not Established

SECTION III - PHYSICAL AND CHEMICAL CHARACTERISTICS

Boiling Point	**518°F (270°C) Decomposes**	Specific Gravity (H_2O = 1)	**1.36**
Vapor Pressure (gas)	**9.9×10^{-8} at 68°F (20°C)**	Molecular Weight	**334.4**
Vapor Density (Air = 1)	**Not Reported**	Melting Point	**514°F (268°C)**

Solubility
Slightly soluble in water (0.2%), alcohol, benzene, and ether. Soluble in chloroform.

Appearance and Odor
Colorless to white, odorless, crystalline solid. Bitter, metallic taste.

SECTION IV - FIRE AND EXPLOSION HAZARD DATA

Flash Point (method used)	**Not Reported**	Explosive Limits in Air % by Volume	**LEL: Not Determined UEL: Not Determined**
NFPA Classification	**Non-Combustible Solid**	Autoignition Temperature	**Non-Combustible**

Extinguishing Media
Use agent suitable to surrounding fires (dry chemical, carbon dioxide, water spray, fog, or foam).

Special Fire Fighting Procedures
Poisonous gases are produced in fire. Wear full protective clothing and self-contained breathing apparatus (SCBA). Structural fire-fighter's clothing is permeable and may not offer adequate protection. Remove containers from fire, if possible. Otherwise, use water spray to keep fire-exposed containers cool. Do NOT release runoff from fire control measures to sewers or waterways.

Unusual Fire and Explosion Hazards
None reported.

SECTION V - REACTIVITY DATA

Stability		Conditions to Avoid **Strychnine is stable in closed containers under routine conditions of handling and storage. Avoid contact with incompatible materials. Do not generate dusty conditions.**
Stable	Unstable	Incompatibility *(materials to avoid)*
X		**Strong oxidizers (permanganates, peroxides, nitrates, chlorates, and perchlorates), alkali hydroxide and carbonates, aromatic ammonia spirit, bromides, and iodides.**
Hazardous Polymerization		Conditions to Avoid **Under normal conditions of temperature and pressure, hazardous polymerization of strychnine is not expected to occur.**
May Occur	Will Not Occur	Hazardous Decomposition or By-products
	X	**When heated to decomposition, strychnine can emit highly toxic/poisonous and acrid fumes and gases including oxides of carbon and nitrogen.**

SECTION VI - HEALTH HAZARD DATA

Primary Route(s) of Entry:	Inhalation?	Absorption (skin)?	Ingestion?
	X		X

Health Hazards

INHALATION: **Convulsions, sensitivity to light, sound, and touch, muscular sensitivity, joint stiffness, muscle twitching, restlessness, headache, anoxia, cyanosis, death due to respiratory arrest. Convulsions can begin within 15 to 30 minutes of exposure. Kidney failure can occur as a secondary effect of central nervous system toxicity.**

SKIN & EYES: **Irritation of the skin. Strychnine may be absorbed through the mucosa of the eye, but it is not likely that it will be significant enough to cause severe toxic systemic effects.**

INGESTION: **Same symptoms as inhalation. Death is possible.**

Carcinogenicity	NTP Listed?	IARC Cancer Review Group?	OSHA Regulated?	Target Organs?
Unknown Human Unknown Animal	**No**	**No**	**29 CFR 1910.1000 Table Z-1**	**Central Nervous System**

Medical Conditions Generally Aggravated by Exposure

None reported.

Emergency and First-aid Procedures

<u>Eye contact</u>: Flush immediately with water for 15 minutes (minimum); seek medical attention. <u>Skin contact</u>: Remove all contaminated clothing. Immediately wash area with large amounts of soap and water. Seek medical assistance. For <u>inhalation</u>: Remove the person from exposure. Provide respiratory assistance and CPR, if required. Transfer to medical facility. If <u>swallowed</u>: Seek medical attention immediately. Give conscious person 1 to 2 glasses of water or milk to dilute. Unless advised otherwise, do NOT induce vomiting (rapid convulsions are possible and person may choke on vomitus).

SECTION VII - PRECAUTIONS FOR SAFE HANDLING AND USE

Steps to be Taken in Case Material is Released or Spilled

Collect powders using HEPA vacuum. Do NOT dry sweep. Absorb solution spills in vermiculite and place in sealed drum. Ventilate area of spill or leak. Remove all sources of ignition.

Preferred Waste Disposal Method

Cover with a 9:1 mixture of sand and soda ash, stuff in paper carton with more paper, and incinerate.

Precautions to be Taken in Handling and Storage

Store to avoid contact with incompatible materials in tightly closed containers in cool, dry, well-ventilated area away from heat, flame, and direct sunlight.

Other Precautions and Warnings

Protect containers against physical damage.

SECTION VIII - CONTROL MEASURES AND PERSONAL PROTECTIVE EQUIPMENT

Respiratory Protection *(specify type)*

For exposures above the PEL, use a MSHA/NIOSH-approved supplied-air respirator with full facepiece, hood, or helmet in continuous flow mode, or use a self-contained breathing apparatus (SCBA) operated in pressure demand or other positive pressure mode.

Ventilation

Local exhaust or general mechanical systems recommended.

Protective Gloves	Eye Protection	Other Protective Clothing
Impervious Material	**Chemical/Dust Goggles or Face Mask**	**Protective Apron**

Work/Hygiene Practices

Always wash hands thoroughly after using chemical; never bring food, drink, or smoking materials into vicinity of chemicals.

STRYCHNINE

$C_{21}H_{22}N_2O_2$

CAS: 57-24-9

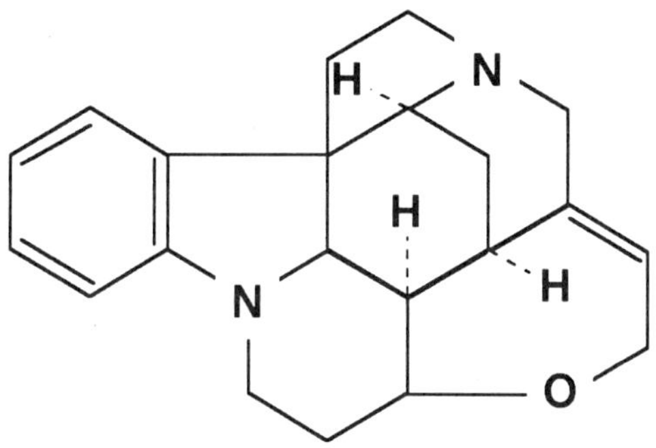

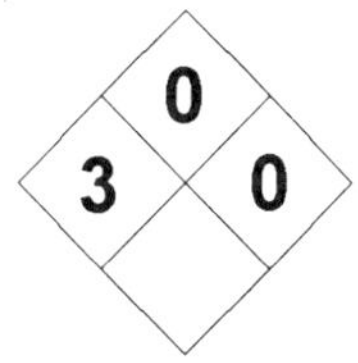

IDENTIFICATION AND TYPICAL USES

Strychnine is a colorless to white, odorless, crystalline solid with a bitter, metallic taste. Strychnine is widely known as a poison. Its therapeutic applications are very limited. It is used as a rodenticide and in poison bait for larger animals, such as porcupines, gophers, and foxes. It was formerly used in human and veterinary medicines for gastrointestinal disorders but is now considered ineffective and dangerous in such applications.

RISK ASSESSMENT: HEALTH

General Assessment

Strychnine is a highly toxic alkaloid. It is considered a poison by *ingestion* and *inhalation*. Its specific carcinogenicity and mutagenicity are not known It has been shown to cause teratogenic effects in test animals.

If ingested, the time of action depends upon the condition of the stomach (empty or full) and the nature of the food present. The first symptoms of inhalation or ingestion may be a feeling of uneasiness with a heightened reflex of irritability, followed by muscular twitching in some parts of the body. There may also be hypersensitivity to light, sound, and touch. Convulsions can also occur without any prior symptoms. However, there is generally a progression of noticeable symptoms before convulsions begin. There can be initial muscular rigidity (especially in the jaw), joint stiffness, restlessness, headache, anoxia, and cyanosis. Progression to stiff neck, facial muscle twitching, and apprehension. Convulsions can begin between 15 and 30 minutes after exposure and can last from 30 seconds to 2 minutes each. The severity of these convulsions increases successively. Supportive care may lead to recovery in some cases. However, death usually occurs after 2 to 5 seizures. Kidney failure can occur as a secondary effect of central nervous system toxicity. Death often occurs as a result of respiratory arrest.

Skin contact is likely to cause some irritation. Strychnine can also be absorbed through the mucous membranes of the eyes, although not in significant amounts.

Acute Health Effects

The following acute (short-term) health effects may occur immediately or shortly after exposure to strychnine:

<u>Skin</u>: Some irritation with a possibility for redness, swelling, and inflammation in some persons.

<u>Eye</u>: Irritation and possible absorption of the material through the mucosa of the eye.

<u>Lung</u>: Nose and throat irritation. High concentrations may cause inflammation of the mucous membranes.

<u>CNS</u>: Primary effects are depression of the CNS leading to life-threatening convulsions and possible death due to respiratory arrest, renal failure, or both.

Chronic Health Effects

The following chronic (long-term) health effects may occur at some time after exposure to strychnine and can last for months or even years:

<u>Cancer Hazards</u>: According to information presented in the references, strychnine has not been adequately tested for its ability to cause cancer in animals. Additional research is required in this regard.

<u>Reproductive Hazard</u>: According to information presented in the references, strychnine has been shown to

cause teratogenic effects in test animals. There are no data available on its ability to adversely affect reproduction in humans.

Other Chronic Effects: None reported (exposed persons typically do not survive repeated exposures).

Recommended Risk-Reduction Measures

Personnel should avoid direct contact with strychnine. This chemical is an extremely potent human poison. If a less toxic chemical cannot be substituted for a hazardous substance, then *engineering controls* are the most effective method of reducing exposures. The best protection is to enclose operations and/or provide local exhaust ventilation at the site of chemical release. While not always operationally feasible, isolating operations can also reduce exposure. Using respiratory protection is less effective than the controls mentioned above, but is still advisable whenever working with or around strychnine. The exposure level is relatively low (0.15 ppm) and difficult to accurately monitor. For all exposures, a supplied-air respirator with full facepiece operated in continuous flow mode, or a self-contained breathing apparatus (SCBA) with full facepiece and pressure demand are the recommended respiratory protection methods of choice. If a full facepiece is not available, then chemical goggles should be worn to protect the eyes. Whenever a chemical splash hazard exists, a face shield and a protective apron should be worn. To prevent hand and skin exposures, impervious gloves should be used. Glove manufacturers should be contacted to obtain permeation studies *before* glove selection has been made for strychnine.

Administrative controls should also be in place to minimize the potential for human exposures. These may include written procedures or policies which specify the methods and techniques that will be practiced whenever personnel are to work with strychnine.

All personnel should receive *training* on the use, hazards, protective measures, emergency actions, and other precautions per 29 CFR 1910.1200 (Hazard Communication), prior to the first assignment in an area where strychnine is used or stored.

If exposure has occurred, symptoms can develop so rapidly, there may be no time for tests before supportive care is necessary. Although strychnine can be detected in the blood and urine, levels do not correlate well with toxicity. Therefore, blood or urine tests should be used in diagnosis but not prognosis.

If symptoms develop or overexposure is suspected, the following tests are recommended:

- ☑ Urine strychnine levels. Strychnine is metabolized in the liver. Approximately 20% is excreted unchanged in the urine.
- ☑ Liver and kidney function tests.

Also, metabolic acidosis and electrolyte abnormalities should be corrected. Breathing, cardiovascular function, and kidney function should be closely monitored. Administration of activated charcoal immediately after ingestion may deter some absorption of the chemical through the gastrointestinal system. Passage of charcoal should indicate that asymptomatic patients will not develop later toxicity. Convulsions should last between 12 and 24 hours and can be controlled with a short-acting barbiturate or inhalation anesthesia.

Other methods to reduce exposure include:

- ☑ Where possible, enclose operations and use local exhaust ventilation at the site of chemical release. If local exhaust ventilation or enclosure is not used, respiratory protection should be mandatory. Also, if possible, automatic transfer of strychnine from storage containers to process or work containers is recommended.
- ☑ Always ensure that proper protective clothing is worn when using chemical substances.
- ☑ Wash thoroughly immediately after exposure to strychnine and at the end of the work shift or before eating, drinking, or smoking.
- ☑ Hazard warning information should be posted in the work area. In addition, as part of an on-going education and training program, all information on the health and safety hazards of strychnine should be communicated to all exposed workers.
- ☑ Eye wash stations should be provided in the immediate work area for emergency use. If there is a possibility of skin exposure to strychnine, emergency shower facilities should also be provided in the immediate work area.
- ☑ Personnel should never wear contaminated clothing home (family members can be exposed). Clothing should be laundered by personnel who have been trained on the hazards of strychnine.

RISK ASSESSMENT: ENVIRONMENT

General Assessment

The environment is at risk of exposure during transportation, storage, disposal, or destruction of strychnine. In almost every scenario, the threat of environmental exposure is contingent upon the proper handling of the chemical substance. Accidental spills, large or small, can result in fire, explosion, and possible contamination of the surrounding environmental mediums (water, soil, and air).

Strychnine is considered a non-combustible solid. However, it can present a moderate fire hazard risk in contact with incompatible materials. These characteristics require extreme caution in handling, storage, transportation, and disposal. It is incompatible with strong oxidizers, alkali hydroxides and carbonates, aromatic ammonia spirit, bromides, and iodides. Therefore, special consideration is required during any emergency situation involving a leak or spill of strychnine. Should strychnine ever come into contact with these incompatible substances during use, transportation, storage, or disposal, the formation of highly toxic and/or highly explosive commodities is extremely possible.

Strychnine may enter the environment through industrial discharges or spills. Also, since it was once a widely used pesticide (rodenticide), it may enter the environment through agricultural runoff.

Acute Ecological Effects

Acute (short-term) toxic effects may include the death of animals, birds, or fish and death or low growth rate in plants. Acute effects are seen 2 to 4 days after animals or plants are exposed to strychnine.

Strychnine is expected to have high acute toxicity to aquatic life, birds, and land animals. Insufficient data are available to evaluate or predict the short-term effects of strychnine exposure to plants.

Chronic Ecological Effects

Chronic (long-term) toxic effects may include shortened life span, reproductive problems, lower fertility, and changes in appearance or behavior in exposed animals. These effects can be seen long after first exposure(s) to toxic chemicals.

Strychnine is expected to have high chronic toxicity to aquatic life, birds, and land animals. Insufficient data are available to evaluate or predict the long-term effects of strychnine exposure to plants.

Water Solubility

Strychnine is nearly insoluble in water. Concentrations of 1 milligram may mix with a liter of water.

Persistence in the Environment

Strychnine is slightly persistent in water, with a half-life between 2 to 20 days. The half-life of a pollutant is the amount of time it takes for one half of the chemical to be degraded.

Bioaccumulation in Aquatic Organisms

Some substances increase in concentration, or *bioaccumulate,* in living organisms as they breathe contaminated air, drink contaminated water, or eat contaminated food. These chemicals can become concentrated in the tissues and internal organs of animals as well as humans.

The concentration of strychnine found in fish tissue is expected to be about the same as the average concentration of strychnine in the water from which the fish was taken.

Recommended Risk-Reduction Measures

Proper training of all transporters will reduce the likelihood of a mishap or accident resulting in a leak or spill to the environment. The correct use of DOT labeling (per the MSDS) on all containers, trucks, and rail cars will enable emergency responders to react properly and quickly to any disaster thereby reducing the potential risk to the environment and to personnel.

Storage of strychnine should be segregated from incompatible chemicals to minimize the risk of cross-contamination or contact. Buildings designated for storage should be equipped with appropriate fire protection systems (alarms, sprinklers, emergency lighting, portable extinguishers).

If a spill or leak to the environment has occurred, fire department, emergency response and/or hazardous materials spill personnel should be notified immediately. Cleanup should be attempted only by those trained in proper spill containment procedures. Contaminated soils should be removed for incineration and replaced with clean soil. If strychnine should contact the water table, aquifer, or navigable waterway, cleanup should be started immediately. It is nearly insoluble in water and total remediation may be possible. The local and/or state emergency response authorities must be notified. A comprehensive emergency response or disaster preparedness/recovery plan should be in place prior to any operations involving the use, transportation, storage, or disposal of strychnine.

The proper disposal method is to cover strychnine with a 9 to 1 mixture of sand and soda ash, combining until well mixed. The mixture should be placed in a paper carton stuffed with more paper, and then incinerated. It could also be mixed with a flammable solvent and sprayed into a firebox or furnace equipped with an afterburner and scrubber.

If strychnine is spilled or leaked, the following specific steps are recommended:

- ☑ Restrict persons not wearing protective clothing from area of spill or leak until cleanup is complete and area can be opened for normal work.
- ☑ Ventilate area of leak.
- ☑ Collect powdered materials using a high-efficiency particulate air (HEPA) filter. Do NOT dry sweep (generates airborne dusts). Place in proper container for disposal. Damp mop residues. Absorb liquids containing strychnine in vermiculite, dry sand, earth, or similar material and deposit in sealed containers.
- ☑ It may be necessary to dispose of strychnine as a hazardous waste. The responsible state agency or the regional office of the federal Environmental Protection Agency (EPA) should be contacted for specific recommendations.

RISK ASSESSMENT: BUSINESS

General Assessment

Accidents or mishaps involving strychnine can present a significant threat to business operations. The mismanagement of chemical commodities can cause personnel injuries and/or property damage/loss. The loss or damage of equipment or facilities can significantly affect fiscal viability. Lawsuits that may result from personnel injury/death, public exposures, and/or environmental contamination will also require a serious expenditure of resources. Media attention surrounding an injury, death, or environmental damage can also result in a loss of profits and loss of current as well as future business prospects. Always remember that anytime the term "reproductive hazard" is used, public emotion, hysteria, anxiety, and ignorance can run equally high. This must be carefully considered whenever developing or implementing public relations policies.

Recommended Risk-Reduction Measures

Company attorneys, safety and health professionals, and environmental specialists should be involved in the development of any procedures dealing with safety or emergency response. A company official should be pre-designated as a public relations officer with specific training in dealing with the press. Corporate plans and policies should be developed, approved, and implemented long before any need for such arises.

REFERENCES

American Conference of Governmental Industrial Hygienists. 1988. *Documentation of the Threshold Limit Values and Biological Exposure Indices,* 5th Edition (with updates). Cincinnati: ACGIH

Bueche, J. R. 1972. *Principles of Physics,* 2nd Edition. New York: McGraw-Hill.

Calabrese, E. J. 1994. *Biological Effects of Low Level Exposures.* New York: CRC Press/Lewis Publishers.

Cockerman, L. G., B. S. Shane. 1994. *Basic Environmental Toxicology.* New York: CRC Press/Lewis Publishers.

Cothern, C. R. 1995. *Handbook for Environmental Risk Decisionmaking.* New York: CRC Press/Lewis Publishers.

Environmental Protection Agency. 1983. *Chemical Hazard Information Profiles.* Washington, D.C.: U.S. EPA.

Gardner, R. W. 1994. *Chemical Intolerance.* New York: CRC Press/Lewis Publishers.

Lauwerys, R. R., P. Hoet. 1993. *Industrial Chemical Exposure.* New York: CRC Press/Lewis Publishers.

Lewis, R. J., Sr. 1992. *Sax's Dangerous Properties of Industrial Materials,* Volumes 1, 2, and 3, 8th Edition. New York: Van Nostrand Reinhold.

Mackay, D., W. Y. Shiu, and K. C. Ma. 1992. *Illustrated Handbook of Chemical-Physical Properties and Environmental Fate for Organic Chemicals.* New York: CRC Press/Lewis Publishers.

Manahan, S. E. 1994. *Environmental Chemistry,* 6th Edition. New York: CRC Press/Lewis Publishers.

Meyer, E. 1990. *Chemistry of Hazardous Materials.* Englewood Cliffs, NJ: Prentice Hall.

Patnaik, P. 1992. *A Comprehensive Guide to the Hazardous Properties of Chemical Substances.* New York: Van Nostrand Reinhold.

Quigley, D. R. 1994. *Handbook of Emergency Chemical Management.* New York: CRC Press/Lewis Publishers.

Rea, W. J. 1995. *Chemical Sensitivity,* Vol. I-III. New York: CRC Press/Lewis Publishers.

Richardson, M. 1995. *Dictionary of Substance and Their Effects,* Vol. I-VI. New York: CRC Press/Lewis Publishers.

Rylander, R., R. R. Jacobs. 1994. *Organic Dusts.* New York: CRC Press/Lewis Publishers.

Sacarello, H. L. A. 1994. *The Comprehensive Handbook of Hazardous Materials.* New York: CRC Press/Lewis Publishers.

U.S. Department of Health and Human Services, National Institute for Occupational Safety and Health. 1994. *NIOSH Pocket Guide to Chemical Hazards.* Washington, D.C.: U.S. Government Printing Office.

MATERIAL SAFETY DATA SHEET

CHEMICAL NAME	
STYRENE	**CRC PRESS/LEWIS PUBLISHERS DISCLAIMER:** THE INFORMATION AND RECOMMENDATIONS PRESENTED HEREIN ARE BASED ON SOURCES BELIEVED TO BE RELIABLE. CRC MAKES NO REPRESENTATION ON ITS COMPLETENESS OR ACCURACY. IT IS THE RESPONSIBILITY OF THE USER TO DETERMINE THE CHEMICAL'S SUITABILITY FOR ITS INTENDED USE, THE CHEMICAL'S SAFE USE, AND THE CHEMICAL'S PROPER DISPOSAL. NO REPRESENTATIONS AND/OR WARRANTIES, EITHER EXPRESSED OR IMPLIED, OF THE MERCHANTABILITY OR FITNESS FOR A PARTICULAR PURPOSE, OR OF ANY OTHER NATURE, ARE MADE WITH RESPECT TO THE INFORMATION PROVIDED IN THIS MSDS OR TO THE CHEMICAL TO WHICH INFORMATION MAY REFER. CRC NEITHER ASSUMES NOR AUTHORIZES ANY OTHER PERSON TO ASSUME FOR IT, ANY OTHER ADDITIONAL RESPONSIBILITY OR LIABILITY FOR THE USE OF, OR RELIANCE UPON, THIS INFORMATION.

HAZARD WARNING INFORMATION

HEALTH	FIRE	REACTIVE	OTHER	DEGREE OF HAZARD	COLOR CODING	OTHER CODES
2	3	2		0 = Minimum Hazard 1 = Slight Hazard 2 = Moderate Hazard 3 = Serious Hazard 4 = Severe Hazard	HEALTH = BLUE FIRE = RED REACTIVITY = YELLOW OTHER = WHITE	OX = Oxidizer ACID = Acid ALK = Alkali COR = Corrosive ~~W~~ = Use No Water

SECTION I - GENERAL INFORMATION

Characterization	**Aromatic Hydrocarbon**	RCRA Number	**D001 and D003**
		EPA Class	**Characteristic (I) & (R) Waste**
DOT Proper Shipping Name	**Styrene Monomer, Inhibited**	Chemical Abstract Service (CAS) Number	**100-42-5**
DOT Hazard Class and Label Requirements	**Flammable Liquid**	DOT Emergency Guide Code	**27**
DOT Identification Number	**UN 2055**	Chemical Formula	**$C_6H_5CH{:}CH_2$**

Synonyms
Ethylene benzene; phenylethylene; styrene monomer; Styrol; vinyl benzene; cinnamene; Styropor.

SECTION II - HAZARDOUS INGREDIENTS AND IDENTITY INFORMATION

Hazardous Components (specific identity)	OSHA Exposure Criteria	NIOSH Exposure Criteria	Immediately Dangerous to Life and Health (IDLH)	ACGIH Exposure Criteria
Styrene (derivation: From ethylene and benzene in the presence of aluminum chloride to yield ethylbenzene, which is catalytically dehydrogenated at 1166°F/630°C to form styrene). **1 ppm = 4.33 mg/m³**	**PEL (8-hour):** **50 ppm** **215 mg/m³** **STEL (15-min):** **100 ppm** **425 mg/m³** **MAX PEAK:** **600 ppm**	**REL (10-hour):** **50 ppm** **215 mg/m³** **STEL (15-min):** **100 ppm** **425 mg/m³**	**700 ppm**	**TLV (skin):** **50 ppm** **213 mg/m³** **STEL:** **100 ppm** **426 mg/m³**

SECTION III - PHYSICAL AND CHEMICAL CHARACTERISTICS

Boiling Point	**293°F (145°C)**	Specific Gravity (H_2O = 1)	**0.91**
Vapor Pressure (gas)	**5 at 69°F (20°C)**	Molecular Weight	**104.2**
Vapor Density (Air = 1)	**3.6**	Melting Point	**-23°F (-31°C)**

Solubility
Slightly soluble in water (0.03%). Floats on water. Soluble in benzene, carbon disulfide, carbon tetrachloride, ethanol, ethyl ether, and ketones. Dissolves organic substances and polymers.

Appearance and Odor
Colorless to yellow, oily liquid with a sweet, floral odor. Odor Threshold = 0.15 to 25 ppm.

SECTION IV - FIRE AND EXPLOSION HAZARD DATA

Flash Point (method used)	**88°F (31°C) closed cup**	Explosive Limits in Air % by Volume	**LEL: 0.9% UEL: 6.8%**
NFPA Classification	**Class 1C Flammable Liquid**	Autoignition Temperature	**914°F (490°C)**

Extinguishing Media
Use dry chemical, carbon dioxide, water spray, or alcohol foam.

Special Fire Fighting Procedures
Wear full protective clothing and (SCBA). Poisonous gases are produced. Move containers from fire.

Unusual Fire and Explosion Hazards
Dangerous fire and explosion hazard when exposed to heat or flame. Containers may explode in fire. Immediately withdraw if rising sound from venting device is heard or if fire is causing a discoloration of tank. Vapors are heavier than air and may flash back to ignition source to cause fire or explosion.

SECTION V - REACTIVITY DATA

Stability		Conditions to Avoid Uninhibited styrene monomer is highly unstable. Even when inhibited, its stability is not reliable. Keep away from incompatible materials and sources of heat.
Stable	Unstable X	Incompatibility *(materials to avoid)* Strong oxidizers (oxygen, nitrates, peroxides, chlorates, and perchlorates), heat, strong acids (nitric, hydrochloric, sulfuric, chlorosulfonic), catalysts for vinyl polymers, peroxides, and aluminum chloride.
Hazardous Polymerization		Conditions to Avoid Under normal conditions of temperature and pressure, hazardous polymerization of styrene can readily occur. Polymerization can be rapid and violent at higher temperatures.
May Occur X	Will Not Occur	Hazardous Decomposition or By-products When heated to decomposition, styrene can emit highly toxic/poisonous gases, acrid smoke, and irritating fumes.

SECTION VI - HEALTH HAZARD DATA

Primary Route(s) of Entry:	Inhalation?	Absorption (skin)?	Ingestion?
	X	X	

Health Hazards

INHALATION: At low concentrations, causes irritation of the eyes, nose, throat, and respiratory tract. There can be lachrymation, nasal discharge, chest pain or heaviness, dypsnea, coughing with pain, bluish face and lips, wheezing, headache, nausea, vomiting, dizziness, fatigue, diarrhea, and delayed pulmonary edema (fluid in the lungs). CNS effects can include drowsiness, confusion, malaise, weakness, unsteady gait, and narcosis.

ABSORPTION: Styrene is a skin irritant and may be absorbed slowly through unbroken skin to cause toxic systemic effects. Eye contact can cause irritation and conjunctivitis.

INGESTION: Irritation of the lips, mouth, throat, with painful swallowing, abdominal pain, nausea and vomiting, shock, possible convulsions, and pulmonary edema.

Carcinogenicity	NTP Listed?	IARC Cancer Review Group?	OSHA Regulated?	Target Organs?
Suspected Human Questioned Animal	No	Group 2B	29 CFR 1910.1000 Table Z-1	Resp. sys., eyes, skin, CNS, liver, kidney, repro.

Medical Conditions Generally Aggravated by Exposure

Liver, kidney, blood, and nervous system disorders may be aggravated by exposure.

Emergency and First-aid Procedures

Eye contact: Flush immediately with water for 15 minutes (minimum), lifting and holding eyelids open. Seek medical attention. Skin contact: Remove all contaminated clothing. Immediately wash area with large amounts of soap and water for 15 minutes (minimum). For inhalation: Remove person from exposure. Provide respiratory assistance. Transfer to medical facility. Observe 48 hours for lung effects. If swallowed: Seek medical attention immediately. Have person drink 1 to 2 glasses of water to dilute. Do NOT induce vomiting. Never give anything by mouth to an unconscious or convulsing person.

SECTION VII - PRECAUTIONS FOR SAFE HANDLING AND USE

Steps to be Taken in Case Material is Released or Spilled

Absorb liquids in vermiculite, or similar material and deposit in sealed containers. Ventilate area of spill or leak. Remove all sources of ignition. Restrict those not involved in cleanup from entering area.

Preferred Waste Disposal Method

No citation.

Precautions to be Taken in Handling and Storage

Styrene is a flammable liquid. Store to avoid contact with incompatible materials. Store in tightly closed containers in cool, well-ventilated area away from heat and flame.

Other Precautions and Warnings

Sources of ignition are prohibited where styrene is used, handled, or stored.

SECTION VIII - CONTROL MEASURES AND PERSONAL PROTECTIVE EQUIPMENT

Respiratory Protection *(specify type)*

For best protection, use a supplied-air respirator with full facepiece in continuous flow, or use a self-contained breathing apparatus (SCBA) operated in pressure demand or other positive pressure mode.

Ventilation

Local exhaust or general mechanical systems recommended.

Protective Gloves	Eye Protection	Other Protective Clothing
Polyvinyl Alcohol	Chemical/Dust Goggles or Face Mask	Protective Apron

Work/Hygiene Practices

Always wash hands thoroughly after using chemical; never bring food, drink, or smoking materials into vicinity of chemicals.

STYRENE

$C_6H_5CH.CH_2$

CAS: 100-42-5

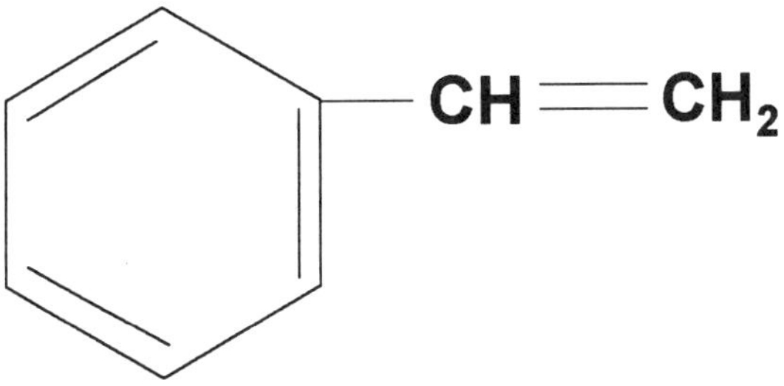

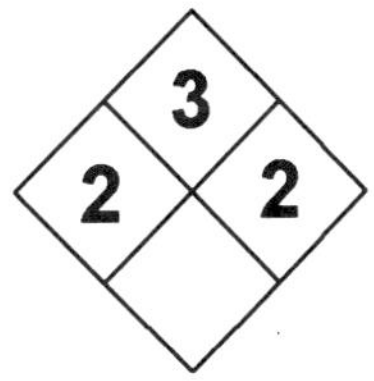

IDENTIFICATION AND TYPICAL USES

Styrene is a clear, colorless to yellow, oily liquid with a sweet, characteristic, floral odor. It is used in the manufacture of polystyrene plastics, polyesters, protective coatings, resins, and synthetic rubber.

RISK ASSESSMENT: HEALTH

General Assessment

Styrene is moderately to highly toxic by *inhalation, ingestion,* and dermal *absorption.* It is considered a questionable carcinogen in test animals and there is limited human evidence of carcinogenic potential. There are no data on its mutagenic, carcinogenic, or teratogenic potential.

Inhalation causes irritation of the eyes, nose, throat, and respiratory tract. Symptoms include watering of the eyes and nose, heaviness or pain in the chest, dypsnea, coughing, bluish face and lips, wheezing, headache, nausea, vomiting, dizziness, fatigue, diarrhea, and risk of pulmonary edema (fluid in the lungs), which can be fatal. Symptoms can be delayed up to 48 hours thereby creating a false sense of security with regard to health exposure risk. Effects of central nervous system (CNS) depression include confusion, malaise, drowsiness, weakness, unsteady gait, narcosis, loss of appetite, and other similar symptoms.

Skin contact causes irritation with itching, inflammation, and possible blisters. Eye contact causes tearing, lid inflammation, chemical corneal burns, and a possibility for serious lesions.

Cases of human ingestion have not been reported. However, based upon analogy to animal studies, ingestion may cause nausea, vomiting, state of shock, possible convulsions, and pulmonary edema.

Acute Health Effects

The following acute (short-term) health effects may occur immediately or shortly after exposure to styrene:

Skin: Immediate irritation, itching, rash, and burning feeling on contact. Possible blisters with defatting of the surface which can lead to dermatitis.

Eye: Irritation, conjunctivitis, and possible damage if allowed to remain in contact. May cause burns and serious lesions.

Lung: Irritation of the respiratory tract, including the nose and mouth. There can be difficulty breathing, coughing, bluish lips and face, headache, nausea, vomiting, dizziness, fatigue, diarrhea, and pulmonary edema.

CNS: A primary CNS depressant. May cause headache, dizziness, loss of consciousness, nausea, irritability, confusion, malaise, drowsiness, unsteady gait, narcosis, and nervousness.

Chronic Health Effects

The following chronic (long-term) health effects may occur at some time after exposure to styrene and can last for months or even years:

Cancer Hazards: According to information presented in the references, styrene has been shown to cause cancer in test animals and is suspected to cause cancer in humans as well. Additional research is required in this regard.

Reproductive Hazard: According to information presented in the references, styrene has not been adequately tested for its ability to adversely affect reproduction.

Other Chronic Effects: Long-term exposure can cause drying and cracking of the skin. Exposure may damage the lungs, liver, and kidneys. Repeated exposure can cause "styrene sickness" with symptoms of nausea, vomiting, appetite loss, and general weakness. Its irritating properties may lead to occupational asthma. Other chronic effects can include functional disorders of the nervous system, irritation of the upper respiratory system, and possible blood changes (leukopenia, and lymphocytosis). Liver and kidney damage are also possible. There can be memory and concentration problems, difficulty in learning, slowed reflexes, and trouble with balance.

Recommended Risk-Reduction Measures

Personnel should avoid direct contact with styrene. If a less toxic chemical cannot be substituted for a hazardous substance, then *engineering controls* are the most effective method of reducing exposures. The best protection is to enclose operations and/or provide local exhaust ventilation at the site of chemical release. While not always operationally feasible, isolating operations can also reduce exposure. Using respiratory protection is less effective than the controls mentioned above, but is still advisable whenever working with or around styrene. For the best protection, a self-contained breathing apparatus (SCBA) with full facepiece operated in pressure demand or a supplied-air respirator operated in continuous flow mode are the recommended respiratory protection methods of choice. If a full facepiece is not available, then chemical goggles should be worn to protect the eyes. Whenever a chemical splash hazard exists, a face shield and a protective apron should be worn. To prevent hand and skin exposures, acid resistant gloves should be used. Polyvinyl alcohol gloves appear to provide adequate protection. Glove manufacturers should be contacted and permeation studies obtained *before* final glove selections are made.

Administrative controls should also be in place to minimize the potential for human exposures. These may include written procedures or policies which specify the methods and techniques that will be practiced whenever personnel are to work with styrene.

All personnel should receive *training* on the use, hazards, protective measures, emergency actions, and other precautions per 29 CFR 1910.1200 (Hazard Communication), prior to the first assignment in an area where styrene is used or stored.

For personnel with frequent or potentially high exposure (half the PEL or greater, or significant skin contact), the following is recommended before beginning work and at regular times thereafter (e.g. annually):

- ☑ Examination of the nervous system.

If symptoms develop or overexposure is suspected, the following tests are recommended:

- ☑ Electroencephalogram (EEG).
- ☑ Interview for brain effects, including recent memory, mood, or concentration changes or problems. There may be headaches, malaise, and altered sleeping patterns as well. Consider cerebellar, autonomic, and peripheral nervous system evaluation. Positive or borderline individuals should be referred for neuropsychological testing.
- ☑ Kidney and liver function tests.

Any evaluation should include a careful history of past and present symptoms with an examination. Medical tests that simply look for existing damage are not a substitute for controlling exposures.

Other methods to reduce exposure include:

- ☑ Where possible, enclose operations and use local exhaust ventilation at the site of chemical release. If local exhaust ventilation or enclosure is not used, respiratory protection should be mandatory. Also, automatic transfer of styrene from storage containers to process containers is highly desirable, if possible.
- ☑ Always ensure that proper protective clothing is worn when using chemical substances.
- ☑ Wash thoroughly immediately after exposure to styrene and at the end of the work shift or before eating, drinking, or smoking.
- ☑ Hazard warning information should be posted in the work area. In addition, as part of an on-going education and training program, all information on the health and safety hazards of styrene should be communicated to all exposed or potentially exposed workers.
- ☑ Eye wash stations should be provided in the immediate work area for emergency use. If there is a possibility of skin exposure to styrene, emergency shower facilities should also be provided.
- ☑ Personnel should never wear contaminated clothing home (family members can be exposed).

Clothing should be laundered by personnel who have been trained on the hazards of styrene.

RISK ASSESSMENT: ENVIRONMENT

General Assessment

The environment is at risk of exposure during transportation, storage, disposal, or destruction of styrene. In almost every scenario, the threat of environmental exposure is contingent upon the proper handling of the chemical substance. Accidental spills, large or small, can result in fire, explosion, and possible contamination of the surrounding environmental mediums (water, soil, and air).

Styrene is considered a Class 1C flammable liquid (per OSHA 29 CFR 1910.106). Its relatively low flash point and moderate boiling point make this chemical a moderate fire and explosion hazard concern. These characteristics require extreme caution in handling, storage, transportation, and disposal. It is incompatible with strong oxidizers, aluminum chloride, and acids and contact can cause violent fire or explosion. It reacts with oxidizers to form a heat-sensitive peroxide, which is extremely explosive. Vapors of styrene are heavier than air and can travel for great distances to an ignition source and flash back to cause a fire or explosion. Therefore, special consideration is required during any emergency situation involving a leak or spill of styrene. Should styrene ever come into contact with incompatible substances during use, transportation, storage, or disposal, the formation of highly toxic and/or highly explosive commodities is extremely possible.

Styrene may enter the environment through industrial discharges or spills.

Acute Ecological Effects

Acute (short-term) toxic effects may include the death of animals, birds, or fish and death or low growth rate in plants. Acute effects are seen 2 to 4 days after animals or plants are exposed to styrene.

This chemical has high acute toxicity to aquatic life. Insufficient data are available on the short-term effects of styrene exposure to plants, birds, or land animals.

Chronic Ecological Effects

Chronic (long-term) toxic effects may include shortened life span, reproductive problems, lower fertility, and changes in appearance or behavior in exposed animals. These effects can be seen long after first exposure(s) to toxic chemicals.

Styrene has high chronic toxicity in aquatic life. Insufficient data are available on the long-term effects of styrene to plants, birds, or land animals.

Water Solubility

Styrene is slightly soluble in water. Concentrations of between 1 and 1000 milligrams will mix with a liter of water.

Persistence in the Environment

Styrene is non-persistent in water, with a half-life of less than 2 days. The half-life of a pollutant is the amount of time it takes for one half of the chemical to be degraded. About 99% of styrene monomer will eventually end up in air; about 0.85% will end up in water; the remainder will end up in terrestrial soils and aquatic sediments.

Bioaccumulation in Aquatic Organisms

Some substances increase in concentration, or *bioaccumulate,* in living organisms as they breathe contaminated air, drink contaminated water, or eat contaminated food. These chemicals can become concentrated in the tissues and internal organs of animals as well as humans.

The concentration of styrene found in fish tissue is expected to be somewhat higher than the average concentration of styrene in the water from which the fish was taken.

Recommended Risk-Reduction Measures

Proper training of all transporters will reduce the likelihood of a mishap or accidents resulting in a leak or spill to the environment. The correct use of DOT labeling (per the MSDS) on all containers, trucks, and rail cars will enable emergency responders to react properly and quickly to any disaster thereby reducing the potential risk to the environment and to personnel.

Storage of styrene should be segregated from incompatible chemicals to minimize the risk of cross-contamination or contact. Buildings designated for storage should be equipped with appropriate fire protection systems (alarms, sprinklers, emergency lighting, portable extinguishers).

If a spill or leak to the environment has occurred, fire department, emergency response and/or hazardous materials spill personnel should be notified immedi-

ately. Cleanup should be attempted only by those trained in proper spill containment procedures using non-sparking tools. Contaminated soils should be removed for incineration and replaced with clean soil. If styrene should contact the water table, aquifer, or navigable waterway, the local and/or state emergency response authorities must be notified. A comprehensive emergency response or disaster preparedness/recovery plan should be in place prior to any operations involving the use, transportation, storage, or disposal of styrene. If styrene is spilled or leaked, the following specific steps are recommended:

- ☑ Restrict persons not wearing protective clothing from area of spill or leak until cleanup is complete and area can be opened for normal work.
- ☑ Remove all ignition sources and ventilate area of leak.
- ☑ Absorb liquids in vermiculite, dry sand, earth, or similar material and deposit in sealed containers.
- ☑ It may be necessary to dispose of styrene as a hazardous waste. The responsible state agency or the regional office of the federal Environmental Protection Agency (EPA) should be contacted for specific recommendations.

RISK ASSESSMENT: BUSINESS

General Assessment

Accidents or mishaps involving styrene can present a significant threat to business operations. The loss or damage of equipment or facilities can significantly affect fiscal viability. Lawsuits that may result from personnel injury/death, public exposures, and/or environmental contamination will also require a serious expenditure of resources. Media attention surrounding an injury, death, or environmental damage can also result in a loss of profits and loss of current as well as future business. Always remember that anytime the terms "cancer" or "carcinogen" are used, public hysteria, ignorance, anxiety, and emotion can all run equally high. This must be carefully considered whenever drafting or implementing public relations policies.

Recommended Risk-Reduction Measures

Company attorneys, safety and health professionals, and environmental specialists should be involved in the development of any procedures dealing with safety or emergency response. A company official should be pre-designated as a public relations officer with specific training in dealing with the press. Corporate plans and policies should be developed, approved, and implemented long before any need for such arises.

REFERENCES

American Conference of Governmental Industrial Hygienists. 1988. *Documentation of the Threshold Limit Values and Biological Exposure Indices*, 5th Edition (with updates). Cincinnati: ACGIH

Bueche, J. R. 1972. *Principles of Physics*, 2nd Edition. New York: McGraw-Hill.

Calabrese, E. J. 1994. *Biological Effects of Low Level Exposures.* New York: CRC Press/Lewis Publishers.

Cockerman, L. G., B. S. Shane. 1994. *Basic Environmental Toxicology.* New York: CRC Press/Lewis Publishers.

Gardner, R. W. 1994. *Chemical Intolerance.* New York: CRC Press/Lewis Publishers.

Hallenbeck, W. H. 1993. *Quantitative Risk Assessment for Environmental and Occupational Health*, 2nd Edition. New York: CRC Press/Lewis Publishers.

Jones, F. E. 1994. *Toxic Organic Vapors in the Workplace.* New York: CRC Press/Lewis Publishers.

Lauwerys, R. R., P. Hoet. 1993. *Industrial Chemical Exposure.* New York: CRC Press/Lewis Publishers.

Lewis, R. J., Sr. 1992. *Sax's Dangerous Properties of Industrial Materials*, Volumes 1, 2, and 3, 8th Edition. New York: Van Nostrand Reinhold.

Lide, D. R. 1995. *Handbook of Chemistry and Physics*, 76th Edition. New York: CRC Press/Lewis Publishers.

Manahan, S. E. 1994. *Environmental Chemistry*, 6th Edition. New York: CRC Press/Lewis Publishers.

Mellström, G. A., J. E. Wahlberg, and H. I. Maibach. 1994. *Protective Gloves for Occupational Use.* New York: CRC Press/Lewis Publishers.

Meyer, E. 1990. *Chemistry of Hazardous Materials.* Englewood Cliffs, NJ: Prentice Hall.

Patnaik, P. 1992. *A Comprehensive Guide to the Hazardous Properties of Chemical Substances.* New York: Van Nostrand Reinhold.

Rea, W. J. 1995. *Chemical Sensitivity*, Vol. I-III. New York: CRC Press/Lewis Publishers.

Richardson, M. 1995. *Dictionary of Substance and Their Effects*, Vol. I-VI. New York: CRC Press/Lewis Publishers.

Sacarello, H. L. A. 1994. *The Comprehensive Handbook of Hazardous Materials.* New York: CRC Press/Lewis Publishers.

Sheftel, V. O. 1995. *Handbook of Toxic Properties of Monomers and Additives.* New York: CRC Press/Lewis Publishers.

U.S. Department of Health and Human Services, National Institute for Occupational Safety and Health. 1994. *NIOSH Pocket Guide to Chemical Hazards.* Washington, D.C.: U.S. Government Printing Office.

MATERIAL SAFETY DATA SHEET

CHEMICAL NAME

STYRENE OXIDE

HAZARD WARNING INFORMATION

HEALTH	FIRE	REACTIVE	OTHER	DEGREE OF HAZARD	COLOR CODING	OTHER CODES
2	2	0		0 = Minimum Hazard 1 = Slight Hazard 2 = Moderate Hazard 3 = Serious Hazard 4 = Severe Hazard	HEALTH = BLUE FIRE = RED REACTIVITY = YELLOW OTHER = WHITE	OX = Oxidizer ACID = Acid ALK = Alkali COR = Corrosive W = Use No Water

SECTION I - GENERAL INFORMATION

Characterization	RCRA Number	EPA Class
Epoxy Compound	**None**	**Not Applicable**

DOT Proper Shipping Name: **No Citation**	Chemical Abstract Service (CAS) Number: **96-09-3**
DOT Hazard Class and Label Requirements: **No Citation**	DOT Emergency Guide Code: **No Citation**
DOT Identification Number: **No Citation**	Chemical Formula: **C_8H_8O**

Synonyms

1,2-Epoxyethylbenzene; phenylethylene oxide; phenyl oxirane; 1-phenyl-1,2-epoxyethane; styrene epoxide.

SECTION II - HAZARDOUS INGREDIENTS AND IDENTITY INFORMATION

Hazardous Components (specific identity)	OSHA Exposure Criteria	NIOSH Exposure Criteria	Immediately Dangerous to Life and Health (IDLH)	ACGIH Exposure Criteria
Styrene oxide	**PEL:** Not Established **STEL:** Not Established	**REL:** Not Established **STEL:** Not Established	**Not Determined**	**TLV:** Not Established **STEL:** Not Established

SECTION III - PHYSICAL AND CHEMICAL CHARACTERISTICS

Boiling Point: **381.5°F (194.2°C)**	Specific Gravity (H_2O = 1): **1.05**
Vapor Pressure (mm Hg): **Not Reported**	Molecular Weight (atomic weight): **120.16**
Vapor Density (Air = 1): **4.14**	Melting Point: **-34°F (-36.7°C)**

Solubility

Insoluble in water. Soluble in acetone, alcohol, methanol, ether, and benzene.

Appearance and Odor

Colorless to straw-colored liquid.

SECTION IV - FIRE AND EXPLOSION HAZARD DATA

Flash Point (method used): **165°F (74°C) opened cup**	Explosive Limits in Air % by Volume: **LEL: Not Reported UEL: Not Determined**
NFPA Classification: **Class IIIA Combustible Liquid**	Autoignition Temperature: **Not Determined**

Extinguishing Media

Use carbon dioxide, dry chemical, alcohol foam, water fog.

Special Fire Fighting Procedures

Poisonous gases are emitted. Wear full protective clothing and self-contained breathing apparatus (SCBA). Remain clear of smoke, water fallout and water runoff. Move containers from fire area if it can be done without risk. Cool exposed containers. Fight fire from distance, if possible.

Unusual Fire and Explosion Hazards

Vapors form explosive mixtures in air. Vapors are heavier than air and may travel from some distance to an ignition source and flashback.

SECTION V - REACTIVITY DATA

Stability		Conditions to Avoid
		Styrene oxide is stable in closed containers under normal conditions of handling and storage. Keep away from heat and incompatible materials.
Stable	Unstable	Incompatibility (materials to avoid)
X		Reacts violently with strong oxidizers (such as chlorine, fluorine, bromine, and nitrates). Explodes when heated or in the presence of strong acids, bases, metals (copper, zinc), and metal salts.
Hazardous Polymerization		Conditions to Avoid
		Under normal conditions of temperature and pressure, hazardous polymerization of styrene oxide is not expected to occur.
May Occur	Will Not Occur	Hazardous Decomposition or By-products
	X	At high temperatures, styrene oxide decomposes and may produce poisonous gases and acrid, irritating smoke and fumes.

SECTION VI - HEALTH HAZARD DATA

Primary Route(s) of Entry:	Inhalation?	Absorption (skin/eye)?	Ingestion?
	X	X	X

Health Hazards

INHALATION: Irritation of the eyes, nose, and respiratory tract. Causes headaches, nausea, vomiting, dizziness (vertigo), drowsiness, and loss of consciousness. High levels can depress the central nervous system with confusion, malaise, irritability, and mood changes. It may cause liver damage, and may lead to death.

ABSORPTION: A skin and eye irritant. Readily absorbed through the skin to cause CNS effects. Direct contact with the eye may damage the cornea.

INGESTION: Irritation of the gastrointestinal tract and CNS depression (see inhalation).

Carcinogenicity	NTP Listed?	IARC Cancer Review Group?	OSHA Regulated?	Target Organs?
Questioned Human Suspected Animal	No	Group 2B	No	Resp. sys. skin, eyes, CNS, reproductive system.

Medical Conditions Generally Aggravated by Exposure

None Reported.

Emergency and First-aid Procedures

Eye contact: Do not allow victim to keep eyes tightly shut. Flush immediately with water for 15 minutes; seek medical attention. **Skin contact:** Remove all contaminated clothing. Immediately wash area with flooding amounts of water. **Inhalation:** Remove the person from exposure. Provide respiratory assistance and CPR. Transfer to medical facility. If **swallowed:** Contact poison control center. Unless advised otherwise and victim is conscious, give 1-2 glasses of water or milk as quickly as possible. Seek medical attention immediately. Never attempt to give anything by mouth to an unconscious or convulsing person and never try to make such a person vomit.

SECTION VII - PRECAUTIONS FOR SAFE HANDLING AND USE

Steps to be Taken in Case Material is Released or Spilled

Remove all sources of ignition. Ventilate area of spill. Restrict those not involved in cleanup from entering area. Notify appropriate authorities. Absorb in vermiculite or other suitable material and dispose of in sealed drums. Do NOT release to sewers, storm drains, surface waters, or soils.

Preferred Waste Disposal Method

Burn in a chemical incinerator equipped with an afterburner and scrubber.

Precautions to be Taken in Handling and Storage

Store in tightly closed containers in cool, well-ventilated area away from heat and incompatibles.

Other Precautions and Warnings

Prevent static sparks by grounding and bonding all containers. Treat “empty” containers as if full.

SECTION VIII - CONTROL MEASURES AND PERSONAL PROTECTIVE EQUIPMENT

Respiratory Protection *(specify type)*

For the best protection, use a supplied-air respirator operated in positive pressure mode, or a SCBA with full facepiece operated in positive pressure mode.

Ventilation

Local exhaust (preferred) at site of chemical work or general ventilation.

Protective Gloves	Eye Protection	Other Protective Clothing
Impervious Rubber (test for adequacy)	Chemical Goggles or Face Mask	Protective Apron, Boots (avoid skin contact)

Work/Hygiene Practices

Always wash hands thoroughly after using chemical; never bring food, drink, or smoking materials into vicinity of chemicals.

STYRENE OXIDE

C_8H_8O

CAS: 96-09-3

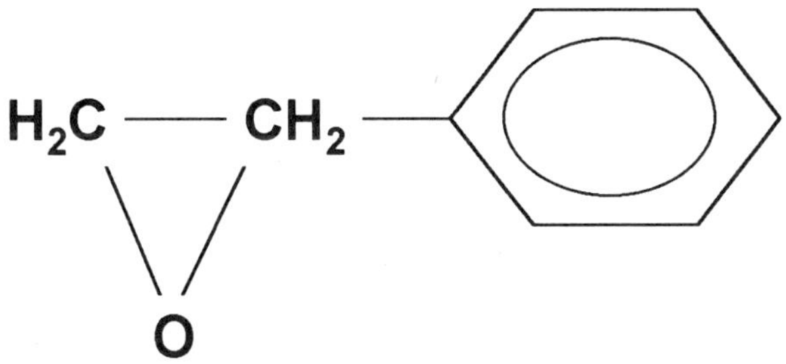

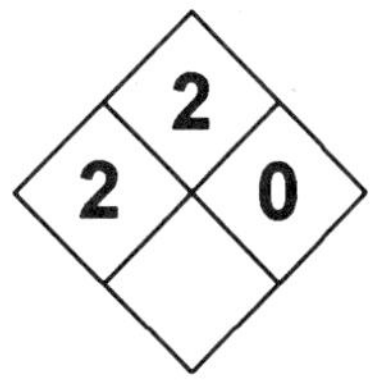

IDENTIFICATION AND TYPICAL USES

Styrene oxide is a colorless to pale, straw-colored liquid. It is used in organic synthesis as a highly reactive organic intermediate. It is used in the manufacture of cosmetics, in making other chemicals, and as a catalyst and cross-linking agent for epoxy resins.

RISK ASSESSMENT: HEALTH

General Assessment

Styrene oxide is toxic to humans by *inhalation* and by *ingestion*. It is also readily *absorbed* through intact skin to cause toxic effects. If absorption occurs in combination with other exposure routes, it can seriously compound the overall effects of exposure. It may present a considerable health hazard risk since it is a known mutagen, teratogen, and carcinogen.

Inhalation of styrene oxide vapors will cause irritation of the eyes, nose, and throat. Prolonged exposures to high concentrations will cause narcosis with symptoms of drowsiness and fatigue. The liquid is moderately irritating to the skin and a severe eye irritant. Ingesting styrene oxide will cause gastrointestinal irritation and may result in CNS depression.

Acute Health Effects

The following acute (short-term) health effects may occur immediately or shortly after exposure to styrene oxide:

Skin: Liquid can pass through unbroken skin. Contact can cause severe irritation and possible burns with redness and swelling.

Eye: Severe irritation and possible corneal damage.

Lung: Irritation of the nose and throat.

CNS: A central nervous system depressant. Symptoms include narcosis with drowsiness and fatigue.

Other: Ingestion can irritate the gastrointestinal tract and also lead to CNS depression.

Chronic Health Effects

The following chronic (long-term) health effects may occur at some time after exposure to styrene oxide and can last for months or even years:

Cancer Hazards: According to information presented in the references, styrene oxide been shown to cause cancer in test animals as well as mutations (genetic changes in living cells). It causes tumors in the liver, gastrointestinal tract, and on the skin in test animals. Its cancer-causing effects on humans is not known.

Reproductive Hazard: Experimental studies have shown that exposure to styrene oxide may adversely affect reproduction and cause teratogenic effects. The reproductive effects from inhalation by rats include fetotoxicity, development abnormalities, and effects on fertility. There are no human data on reproductive health risks.

Other Chronic Effects: Long-term exposure may cause drying, cracking, and dermatitis. This chemical has not been adequately evaluated to determine whether brain or other nerve damage could occur with repeated exposure. However, many solvents and other petroleum-based chemicals have been shown to cause such damage. Effects may include memory loss, reduced concentration, personality changes (withdrawal, irritability), fatigue, sleep disturbances, reduced coordination, and/or effects on nerves supplying internal organs (autonomic nerves) and/or nerves to the arms and legs ("pins and needles"). Styrene oxide may damage the liver, and may cause anemia.

Recommended Risk-Reduction Measures

Styrene oxide is an extremely hazardous, toxic substance. Personnel should avoid direct contact with this chemical. If a less toxic material or compound cannot be substituted for styrene oxide, then *engineering controls* are the most effective method of reducing exposures. The best protection is to enclose operations and/or provide local exhaust ventilation at the site of styrene oxide release. While not always operationally possible, isolating operations involving styrene oxide manufacturing or use can also reduce exposure. Personnel must follow proper industrial hygiene and good housekeeping work practices. Using respiratory protection is less effective than the controls mentioned above, but is still advisable whenever working with or around styrene oxide. No exposure levels have been established for styrene oxide. This does not mean that exposure is without health risk. For the best possible protection, a powered air-supplied respirator or a self-contained breathing apparatus (SCBA) with full facepiece and operated in pressure demand are recommended. If a full facepiece is not available, then chemical goggles should be worn to protect the eyes. A face shield and protective apron should also be worn. Skin contact must not occur. To prevent hand and skin exposures, impervious gloves should be used. Glove manufacturers should be contacted and permeation studies obtained *before* final glove selections are made.

Administrative controls should also be in place to minimize the potential for human exposures. These may include written procedures or policies which specify the methods and techniques that will be practiced whenever personnel are to work with styrene oxide.

All personnel should receive *training* on the use, hazards, protective measures, emergency actions, and other precautions per 29 CFR 1910.1200 (Hazard Communication), prior to the first assignment in an area where styrene oxide is used or stored.

Before beginning initial assignments with this chemical and at regular intervals thereafter (e.g., annually), the following medical protocol are recommended:

☑ Liver and kidney function tests.

☑ Complete blood count.

☑ Interview for brain effects or symptoms, including recent memory loss, mood changes, concentration problems, headaches, malaise, and altered sleep patterns. Consider cerebellar, autonomic, and peripheral nervous system evaluation. Positive and borderline cases should be referred for neuropsychological testing.

Any evaluation should include a careful history of past and present symptoms with an examination. Medical tests that simply look for existing damage are not a substitute for controlling exposures.

Other methods to reduce exposure include:

☑ Where possible, enclose operations and use local exhaust ventilation at the site of chemical release. If local exhaust ventilation or enclosure is not used, respiratory protection should be mandatory.

☑ Always ensure that proper protective clothing is worn when using chemical substances.

☑ Wash thoroughly immediately after exposure to styrene oxide and at the end of the work shift or before eating, drinking, or smoking.

☑ Hazard warning information should be posted in the work area. In addition, as part of an on-going education and training program, all information on the health and safety hazards of styrene oxide should be communicated to all potentially exposed workers.

☑ Eye wash stations should be provided in the immediate work area for emergency use. If there is a possibility of skin exposure to styrene oxide, emergency shower facilities should also be provided.

☑ Workers whose clothing has been contaminated by styrene oxide should change into clean clothes *before* leaving work. Contaminated work clothing should be laundered only by individuals who have been informed of the hazards of exposure to styrene oxide.

RISK ASSESSMENT: ENVIRONMENT

General Assessment

The environment is at risk of exposure during transportation, storage, disposal, or destruction of styrene oxide. In almost every scenario, the threat of environmental exposure is contingent upon the proper handling of the chemical substance. Accidental spills, large or small, where styrene oxide contacts incompatible commodities can result in violent and even explosive reactions (depending upon conditions of spill), and possible contamination of the surrounding environmental mediums (water, soil, and air).

Styrene oxide is considered a Class IIIA combustible liquid (per OSHA 29 CFR 1910.106). Its moderate flash point and boiling point present a fire risk in the presence of an ignition source, heat, or flame. Also, it can explode when heated in the presence of strong acids, bases, metals, and metallic salts. Therefore, extreme caution is required in handling, storage, transportation, and disposal of styrene oxide. These characteristics also require special consideration during any emergency situation involving a leak or spill of styrene oxide.

The proper disposal method is to burn styrene oxide in a chemical incinerator equipped with an afterburner and air scrubber.

Styrene oxide may enter the environment through industrial effluents and from spills.

Acute Ecological Effects

Acute (short-term) toxic effects may include the death of animals, birds, or fish and death or low growth rate in plants. Acute effects are seen 2 to 4 days after animals or plants are exposed to styrene oxide.

Insufficient data are available to evaluate or predict the short-term effects of styrene oxide to aquatic life, plants, land animals, and birds.

Chronic Ecological Effects

Chronic (long-term) toxic effects may include shortened life span, reproductive problems, lower fertility, and changes in appearance or behavior in exposed animals. These effects can be seen long after first exposure(s) to toxic chemicals.

Insufficient data are available to evaluate or predict the long-term effects of styrene oxide to aquatic life, plants, birds, and land animals.

Water Solubility

Styrene oxide is highly soluble in water. Concentrations of 1000 milligrams and more will mix with a liter of water.

Persistence in the Environment

Styrene oxide is slightly persistent in water, with a half-life between 2 and 20 days. The half-life of a pollutant is the amount of time it takes for one-half of the chemical to be degraded. About 95% of styrene oxide will eventually end up in water; the rest will end up in air.

Bioaccumulation in Aquatic Organisms

Some substances increase in concentration, or *bioaccumulate,* in living organisms as they breathe contaminated air, drink contaminated water, or eat contaminated food. These chemicals can become concentrated in the tissues and internal organs of animals as well as humans exposed to chemical substances.

The concentration of styrene oxide found in fish tissues is expected to be about the same as the average concentrations of styrene oxide in the water from which the fish was taken.

Recommended Risk-Reduction Measures

Proper training of all transporters will reduce the likelihood of a mishap or accident resulting in a leak or spill to the environment. The correct use of labeling on all containers, trucks, and rail cars will enable emergency responders to react properly and quickly to any disaster thereby reducing the potential risk to the environment and to personnel.

Storage of styrene oxide should be segregated from incompatible materials to minimize the risk of cross-contamination or contact. Storage facilities should be equipped with the proper fire protection equipment (sprinklers, alarms, extinguishers). Proper ventilation and protective equipment should be used while handing styrene oxide. Non-sparking tools should be provided and used to ensure the prevention of sparks. Also, all equipment and containers used in shipping, receiving, or transferring should be electrically grounded and bonded.

If a spill or leak to the environment has occurred, fire department, emergency response and/or hazardous materials spill personnel should be notified immediately. Cleanup should be attempted only by those trained in proper spill containment procedures using non-sparking tools. If styrene oxide should contact the water table, aquifer, or navigable waterway, reclamation procedures should be initiated as soon as practical. It is highly soluble in water and total remediation may not be possible. The local and/or state emergency response authorities must be notified. A comprehensive emergency response or disaster preparedness/recovery plan should be in place prior to any operations involving the use, transportation, storage, or disposal of styrene oxide. If styrene oxide is spilled or leaked, the following specific steps are recommended:

- ☑ Restrict persons not wearing protective clothing from area of spill or leak until cleanup is complete and area can be opened for normal work.

☑ Remove all ignition sources and ventilate area.

☑ Absorb liquids in vermiculite or other compatible material and dispose of in sealed drums.

☑ It may be necessary to dispose of styrene oxide as a hazardous waste. The responsible state agency or the regional office of the federal Environmental Protection Agency (EPA) should be contacted for specific recommendations.

RISK ASSESSMENT: BUSINESS

General Assessment

Accidents or mishaps involving styrene oxide can present a significant threat to business operations. The loss or damage of equipment or facilities can significantly affect fiscal viability. Lawsuits that may result from personnel injury/death, public exposures, and/or environmental contamination will also require a serious expenditure of resources. Media attention surrounding an injury, death, or environmental damage can also result in a loss of profits and loss of current as well as future business. Always remember that anytime the terms "cancer," "carcinogen," or "reproductive hazard" are used, public emotion, hysteria, and ignorance can run equally high. This must be a consideration when developing public relations policies.

Recommended Risk-Reduction Measures

Company attorneys, safety and health professionals, and environmental specialists should be involved in the development of any procedures for responding to chemical incidents. A company official should be predesignated as a public relations officer with specific training in dealing with the press. Corporate plans and policies should be developed, approved, and implemented long before any need for such arises.

REFERENCES

American Conference of Governmental Industrial Hygienists. 1988. *Documentation of the Threshold Limit Values and Biological Exposure Indices,* 5th Edition (with updates). Cincinnati: ACGIH

Anthony, C. P., and N. J. Kolthoff. 1971. *Textbook of Anatomy and Physiology.* St. Louis: C. V. Mosby Company.

Bloom, A. D. and F. J. DeSerres. 1995. *Ecotoxicity and Human Health.* New York: CRC Press/Lewis Publishers.

Brusick, D. J. 1994. *Methods for Genetic Risk Assessment.* New York: CRC Press/Lewis Publishers.

Bueche, J. R. 1972. *Principles of Physics,* 2nd Edition. New York: McGraw-Hill.

Burrell, R., D. K. Flaherty, and L. J. Sauers. 1992. *Toxicology of the Immune System.* New York: Van Nostrand Reinhold.

Calabrese, E. J. 1994. *Biological Effects of Low Level Exposures.* New York: CRC Press/Lewis Publishers.

Cockerman, L. G., B. S. Shane. 1994. *Basic Environmental Toxicology.* New York: CRC Press/Lewis Publishers.

Cothern, C. R. 1995. *Handbook for Environmental Risk Decisionmaking.* New York: CRC Press/Lewis Publishers.

Danse, I. R. 1991. *Common Sense Toxics in the Workplace.* New York: Van Nostrand Reinhold.

Environmental Protection Agency. 1983. *Chemical Hazard Information Profiles.* Washington, D.C.: U.S. EPA.

Fodden, C. R., and J. L. Weddell. 1992. *Hazardous Materials Emergency Action Data.* New York: CRC Press/Lewis Publishers.

Forseberg, K., and L. H. Keith. 1993. *Chemical; Protective Clothing Permeation and Degradation Database.* Chelsea, MI: Lewis Publishers.

Gardner, R. W. 1994. *Chemical Intolerance.* New York: CRC Press/Lewis Publishers.

Hallenbeck, W. H. 1993. *Quantitative Risk Assessment for Environmental and Occupational Health,* 2nd Edition. New York: CRC Press/Lewis Publishers.

Hodgson, E., R. B. Mailman, and Janice E. Chambers. 1988. *Dictionary of Toxicology.* London: The MacMillian Company.

Howard, P. H., et. al. 1991. *Handbook of Environmental Degradation Rates.* New York: CRC Press/Lewis Publishers.

Hosty, J. W., and P. Foster. 1990. *A Practical Guide to Chemical Spill Response.* New York: Van Nostrand Reinhold.

International Agency for Research on Cancer (IARC). 1987. *IARC Monographs on the Evaluation of Carcinogenic Risk to Humans, Overall Evaluations of Carcinogenicity: An updating of IARC Monographs Volumes 1-42.* Lyon: IARC.

Jones, F. E. 1994. *Toxic Organic Vapors in the Workplace.* New York: CRC Press/Lewis Publishers.

Kamrin, M. A. 1988. *Toxicology: A Primer on Principles and Applications.* Chelsea, Michigan: Lewis Publishers.

Keith, L. H. and D. B. Walters. (Editors) 1992. *National Toxicology Program's Chemical Database, Volumes I-VII.* Chelsea, MI: Lewis Publishers.

Landis, W. G., M. H. Yu. 1995. *Introduction to Environmental Toxicology.* New York: CRC Press/Lewis Publishers.

Langerman, N. 1994 *Precautionary Labels for Chemical Containers.* New York: CRC Press/Lewis Publishers.

Lauwerys, R. R., P. Hoet. 1993. *Industrial Chemical Exposure.* New York: CRC Press/Lewis Publishers.

Lewis, .J., Sr. 1993. *Hawley's Condensed Chemical Dictionary,* 12th Edition. New York: Van Nostrand Reinhold.

Lewis, . J., Sr. 1992. *Sax's Dangerous Properties of Industrial Materials,* Volumes 1, 2, and 3, 8th Edition. New York: Van Nostrand Reinhold.

MATERIAL SAFETY DATA SHEET

CHEMICAL NAME	
SULFUR DIOXIDE	**CRC PRESS/LEWIS PUBLISHERS DISCLAIMER:** THE INFORMATION AND RECOMMENDATIONS PRESENTED HEREIN ARE BASED ON SOURCES BELIEVED TO BE RELIABLE. CRC MAKES NO REPRESENTATION ON ITS COMPLETENESS OR ACCURACY. IT IS THE RESPONSIBILITY OF THE USER TO DETERMINE THE CHEMICAL'S SUITABILITY FOR ITS INTENDED USE, THE CHEMICAL'S SAFE USE, AND THE CHEMICAL'S PROPER DISPOSAL. NO REPRESENTATIONS AND/OR WARRANTIES, EITHER EXPRESSED OR IMPLIED, OF THE MERCHANTABILITY OR FITNESS FOR A PARTICULAR PURPOSE, OR OF ANY OTHER NATURE, ARE MADE WITH RESPECT TO THE INFORMATION PROVIDED IN THIS MSDS OR TO THE CHEMICAL TO WHICH INFORMATION MAY REFER. CRC NEITHER ASSUMES NOR AUTHORIZES ANY OTHER PERSON TO ASSUME FOR IT, ANY OTHER ADDITIONAL RESPONSIBILITY OR LIABILITY FOR THE USE OF, OR RELIANCE UPON, THIS INFORMATION.

HAZARD WARNING INFORMATION

HEALTH	FIRE	REACTIVE	OTHER	DEGREE OF HAZARD	COLOR CODING	OTHER CODES
3	0	0		0 = Minimum Hazard 1 = Slight Hazard 2 = Moderate Hazard 3 = Serious Hazard 4 = Severe Hazard	HEALTH = BLUE FIRE = RED REACTIVITY = YELLOW OTHER = WHITE	OX = Oxidizer ACID = Acid ALK = Alkali COR = Corrosive W = Use No Water

SECTION I - GENERAL INFORMATION

Characterization: **Gas**	RCRA Number: **None**	EPA Class: **Not Applicable**
DOT Proper Shipping Name: **Sulfur Dioxide, Liquefied**	Chemical Abstract Service (CAS) Number: **7446-09-5**	
DOT Hazard Class and Label Requirements: **Poison A; Poison Gas**	DOT Emergency Guide Code: **16**	
DOT Identification Number: **UN 1079**	Molecular Formula: **SO_2**	

Synonyms
Sulfurous acid anhydride; sulfurous oxide; sulfur oxide; bisulfite; Ferminicide Liquid.

SECTION II - HAZARDOUS INGREDIENTS AND IDENTITY INFORMATION

Hazardous Components (specific identity)	OSHA Exposure Criteria	NIOSH Exposure Criteria	Immediately Dangerous to Life and Health (IDLH)	ACGIH Exposure Criteria
Sulfur dioxide: (derivation: By roasting pyrites in special furnaces. The gas is readily liquefied by cooling with ice and salt or at a pressure of three atmospheres; by purifying and compressing sulfur dioxide gas from smelting operations; by burning sulfur). 1 ppm = 2.66 mg/m^3	PEL (8-hour): 2 ppm 5 mg/m^3 STEL (15-min): 5 ppm 13 mg/m^3	REL (10-hour): 2 ppm 5 mg/m^3 STEL (15-min): 5 ppm 13 mg/m^3	100 ppm	TLV: 2 ppm 5 mg/m^3 STEL (15-min): 5 ppm 13 mg/m^3

SECTION III - PHYSICAL AND CHEMICAL CHARACTERISTICS

Boiling Point: **14°F (-10°C)**	Specific Gravity (H_2O = 1): **2.26 (Gas)**
Vapor Pressure (atmospheres): **3.2 at 68°F (20°C)**	Molecular Weight (atomic weight): **64.1**
Vapor Density (Air = 1): **2.21**	Melting Point: **-104°F (-76°C)**

Solubility
Very slightly soluble in water (0.003%). Soluble in alcohol, ether, carbon tetrachloride, acetic acid, sulfuric acid, and chloroform.

Appearance and Odor
Colorless gas or liquid with a sharp, pungent, suffocating odor. Odor Threshold Range = 0.33 to 5 ppm.

SECTION IV - FIRE AND EXPLOSION HAZARD DATA

Flash Point (method used): **Not Applicable**	Explosive Limits in Air % by Volume: **LEL: Not Applicable UEL: Not Applicable**
NFPA Classification: **Non-Flammable Gas**	Autoignition Temperature: **Not Applicable**

Extinguishing Media
Will not burn but can be explosive when compressed. Use agent suitable to surrounding fire. If water spray is used, sulfurous acid vapor will be generated.

Special Fire Fighting Procedures
Poisonous gases are produced in fire. Wear full protective clothing and (SCBA). Move container from fire area if it can be done without risk. Cool exposed containers with direct water spray.

Unusual Fire and Explosion Hazards
Containers may explode in fire. Fight fire from a distance or with unmanned hose holders, if possible. Stay away from ends of tanks. Use extreme caution if sulfur dioxide is in or near a fire.

SECTION V - REACTIVITY DATA

Stability		Conditions to Avoid **Sulfur dioxide is normally stable in closed containers under routine conditions of handling and storage. Avoid contact with incompatibles, heat, and organic materials.**
Stable **X**	Unstable	Incompatibility *(materials to avoid)* **Powdered alkali metals (such as sodium and potassium), water (reacts to form sulfurous acid), ammonia, zinc, aluminum, brass, and copper.**
Hazardous Polymerization		Conditions to Avoid **Under normal conditions of temperature and pressure, hazardous polymerization of sulfur dioxide is not expected to occur.**
May Occur	Will Not Occur **X**	Hazardous Decomposition or By-products **When heated to decomposition, sulfur dioxide emits other toxic oxides of sulfur.**

SECTION VI - HEALTH HAZARD DATA

Primary Route(s) of Entry:	Inhalation?	Absorption (skin/eye)?	Ingestion?
	X		X

Health Hazards

INHALATION: **Severe irritation to the eyes, nose, throat, and lungs. Causes watery eyes and nose, cough, choking, sneezing, and bronchoconstriction with increased pulmonary resistance. It can also cause systemic acidosis, respiratory paralysis, and pulmonary edema.**

SKIN & EYES: **A severe eye and skin irritant. Sudden bursts of the compressed liquid may cause frostbite and associated tissue destruction. Skin and eye contact can cause burns.**

INGESTION: **Unlikely, but possible. May cause burns to the mouth and throat.**

Carcinogenicity	NTP Listed?	IARC Cancer Review Group?	OSHA Regulated?	Target Organs?
Unknown Human Questioned Animal	**No**	**No**	**29 CFR 1910.1000 Table Z-1**	**Skin, eyes, respiratory system.**

Medical Conditions Generally Aggravated by Exposure

Chronic respiratory problems (bronchitis, asthma) may be aggravated.

Emergency and First-aid Procedures

Eye contact: Do not allow victim to keep eyes tightly shut. Flush immediately with water for 15 minutes (minimum). Skin contact: Wash with large amounts of water for a minimum of 15 minutes. For inhalation: Remove the person from exposure. Provide respiratory assistance and CPR. Transfer to medical facility. Observe for 24-48 hours for lung effects. Oxygen should be administered by trained medical personnel only. If swallowed: Not likely since sulfur dioxide is a gas above 70°F (21°C). Do NOT induce vomiting. Contact poison control center for assistance.

SECTION VII - PRECAUTIONS FOR SAFE HANDLING AND USE

Steps to be Taken in Case Material is Released or Spilled

Wear self-contained breathing apparatus (SCBA). Stop flow of gas. Restrict those not wearing protective equipment and who are not involved in cleanup from entering area. Avoid skin contact. Ventilate area of spill. If gas cannot be stopped, remove leaking container to safe area outdoors and allow to empty.

Preferred Waste Disposal Method

Stop leak and send to vendor for disposal. Otherwise, allow gas to pass through a large drum of soda ash solution and carefully add calcium hypochlorite. Neutralize with 6M hydrochloric acid or 6M sodium hydroxide and dilute. Drain into sewer with excess water.

Precautions to be Taken in Handling and Storage

Store in tightly closed, insulated cylinders in a cool, well-ventilated area away from heat. Protect containers from physical damage. Regularly inspect for leaks, cracks, or faulty valves.

Other Precautions and Warnings

Use external or detached storage. Ensure cylinders are stored upright and properly secured at all times.

SECTION VIII - CONTROL MEASURES AND PERSONAL PROTECTIVE EQUIPMENT

Respiratory Protection *(specify type)*

Use a NIOSH/MSHA-approved air-purifying gas-cartridge respirator, or a supplied-air respirator, or a self-contained breathing apparatus (SCBA) with full facepiece operated in positive pressure mode.

Ventilation

Local exhaust (preferred) at site of chemical work or general ventilation.

Protective Gloves	Eye Protection	Other Protective Clothing
Impervious, Thermal Gloves	**Chemical Goggles and Face Mask**	**Protective Uniform or Apron**

Work/Hygiene Practices

Always wash hands thoroughly after using chemical; never bring food, drink, or smoking materials into vicinity of chemicals.

SULFUR DIOXIDE

SO_2

CAS: 7446-09-5

IDENTIFICATION AND TYPICAL USES

Sulfur dioxide appears as a colorless gas or liquid with a sharp, pungent, disagreeable odor. It is normally shipped as a liquefied compressed gas. It condenses directly to a solid upon cooling. It is used in the manufacture of pulp and paper, as a bleaching and fumigating agent, as a disinfectant, in metal refining, for preserving food and vegetables, and as a reducing agent. It is a major air pollutant and is produced when soft coal, oils, or other sulfur-containing substances are burned. Sulfur dioxide in the atmosphere reacts with moisture to form sulfurous acid, or is oxidized to sulfur trioxide, which forms sulfuric acid and causes acid rain.

RISK ASSESSMENT: HEALTH

General Assessment

Sulfur dioxide is a severe irritant to the eyes, nose, and throat. As a gas, it can primarily enter the body through *inhalation.* While skin absorption is not likely, skin contact can result in serious tissue damage as a direct result of the cold temperatures of the compressed gas or cryogenic liquid. Ingestion is not a probable route of exposure under most circumstances. It is a questionable carcinogen in test animals and human mutation data have also been reported. Experimental reproductive, teratogenic, and tumorigenic effects have been reported. Its toxic and irritating effects are due primarily to its ability to rapidly form sulfurous acid on contact with moist mucous membranes. Also, because of its high solubility, it is quickly absorbed throughout the body causing metabolic acidosis.

Inhalation of sulfur dioxide can cause severe irritation to the eyes, nose, throat, and respiratory system. Symptoms can include headache, nausea, vomiting, dypsnea (difficult breathing), diarrhea, cyanosis (bluish discoloration of the lips, fingers, and toes). Lung irritation can be quite severe and may lead to pulmonary edema (fluid buildup in the lungs), which is a medical emergency and can be fatal. The symptoms of pulmonary edema can be delayed up to 48 hours, thereby creating a false sense of security with regard to health exposure risk. There can be cough, frothy sputum, and increased difficulty in breathing, loud wheezing, a rapid heartbeat (tachypnea), and increased respiration rate. Low concentrations can cause systemic acidosis. Survivors of high concentration exposures may exhibit reactive airway disease, pneumonia, chronic bronchitis, or obstructive and restrictive lung disease.

Skin contact with compressed liquid sulfur dioxide can cause characteristic burns due to frostbite. Eye contact causes irritation, watering, inflammation of the lids, burns to the cornea, and (possibly) severe lesions.

Acute Health Effects

The following acute (short-term) health effects may occur immediately or shortly after exposure to sulfur dioxide:

Skin: Severe irritation with localized tissue damage and possible burns due to the cold temperature of the compressed gas.

Eye: Severe conjunctival irritation and possible damage to the cornea (corneal opacification), and loss of vision.

Lung: Severe irritation to respiratory tract causing shortness of breath and cough. High acute exposures can be fatal due to a buildup of fluid in the lungs (pulmonary edema).

Other: Transient stimulation of red blood cell production via the bone marrow has been reported.

Chronic Health Effects

The following long-term health effects may occur at some time after exposure to sulfur dioxide and can last for months or even years:

Cancer Hazards: According to information presented in the references, sulfur dioxide has been shown to cause tumors in experimental test animals. Human mutation data have also been reported and many scientists believe that long-term exposure to such chemicals may present a cancer risk in the long-term. It is not known if human exposure to sulfur dioxide causes cancer.

Reproductive Hazard: According to the references, sulfur dioxide has been shown to cause teratogenic effects in test animals and may adversely affect reproduction as well.

Other Chronic Effects: Long-term exposures may cause conjunctivitis, rhinitis, nosebleeds, gingivitis, ulcers of the nasal septum, cough, chronic bronchitis, fatigue, and obstructive lung disease.

Recommended Risk-Reduction Measures

Sulfur dioxide is a very toxic substance. It is carcinogenic, mutagenic, and teratogenic. Personnel should, therefore, avoid direct contact with sulfur dioxide. If a less toxic material or compound cannot be substituted for sulfur dioxide, then *engineering controls* are the most effective method of reducing exposure risk. The best protection is to enclose operations and/or provide local exhaust ventilation at the site of sulfur dioxide release. While not always operationally feasible, isolating operations can also reduce exposure. Using respiratory protection is less effective than the controls mentioned above, but is still required whenever working with or around sulfur dioxide. The 15-minute short-term exposure level of 5 ppm is relatively low and difficult to monitor. Prudent risk management requires proper consideration and implementation of an appropriate respiratory protection program (per OSHA 29 CFR 1910.134). While a gas-cartridge air-purifying respirator may certainly suffice, the best protection is obtained using an MSHA/NIOSH-approved supplied-air respirator or a self-contained breathing apparatus (SCBA) with full facepiece and operated in pressure demand mode. If a full facepiece is not available, then chemical goggles should be worn to protect the eyes. A face shield should also be considered. To prevent hand and skin exposures, impervious, thermal protective gloves should be used. Glove manufacturers should be contacted and permeation studies obtained *before* final glove selections are made.

Administrative controls should also be in place to minimize the potential for human exposures. These may include written procedures or policies which specify the methods and techniques that will be practiced whenever personnel are to work with sulfur dioxide.

All personnel should receive *training* on the use, hazards, protective measures, emergency actions, and other precautions per 29 CFR 1910.1200 (Hazard Communication) prior to the first assignment in an area where sulfur dioxide is used or stored. Before beginning employment and at regular intervals thereafter (e.g., annually), the following medical test is recommended:

☑ Lung function tests (establish baseline).

If symptoms develop or overexposure is suspected, the following may be useful:

☑ Repeat lung function tests (compare to baseline).

☑ Consider chest X-ray after acute exposure (may be negative if taken immediately following exposure due to delayed development of pulmonary edema).

It should be noted that medical tests which simply look for existing damage are not a substitute for controlling exposures. Medical histories are extremely important when assessing exposure risk. Also, since smoking can cause heart disease as well as lung cancer, emphysema, and other respiratory ailments, any exposure to sulfur dioxide can result in quicker and more devastating symptoms. Smokers should therefore avoid unprotected contact with this chemical. Prudent risk management requires proper consideration of *all* possible factors that may be causing the appearance of exposure symptoms among workers.

Other methods to reduce exposure include:

☑ Always ensure that proper protective clothing is worn when using chemical substances. And that personnel are trained on its use and care.

☑ Wash thoroughly immediately after exposure to sulfur dioxide and at the end of the work shift or before eating, drinking, or smoking.

☑ Hazard warning information should be posted in the work area. In addition, as part of an on-going education and training program, all information on the health and safety hazards of sulfur dioxide should be communicated to all exposed and potentially exposed workers.

☑ Eye wash stations should be provided in the immediate work area for emergency use. If there is a possibility of skin exposure to compressed sulfur dioxide gas, emergency shower facilities should also be provided.

☑ Workers whose clothing has been contaminated by sulfur dioxide should change into clean clothes before leaving work. Contaminated work clothing should be laundered only by individuals who have been informed of the hazards of exposure to chemical substances such as sulfur dioxide.

RISK ASSESSMENT: ENVIRONMENT

General Assessment

The environment is always at some risk of exposure during transportation, storage, disposal, or destruction of sulfur dioxide. In almost every scenario, the threat of environmental exposure is contingent upon the proper handling of the chemical substance. As a gas, it primarily affects the air. However, depending upon the nature and conditions of the release, accidental spills, large or small, can result in contamination of the other surrounding environmental mediums (water and soil).

Sulfur dioxide can appear as a non-flammable gas or non-flammable liquid. However, when compressed, it can become extremely explosive. It is incompatible with many metals, ammonia, and even water (reacts to form sulfurous acid). Caution is always required in handling, storage, transportation, and disposal of sulfur dioxide. Emergency responders should be made aware of the presence of sulfur dioxide at any emergency response situation.

Sulfur dioxide may enter the environment through industrial discharges and from spills.

Acute Ecological Effects

Acute (short-term) toxic effects may include the death of animals, birds, or fish and death or low growth rate in plants. Acute effects are seen 2 to 4 days after animals or plants are exposed to sulfur dioxide.

Insufficient data are available to evaluate or predict the acute (short-term) effects of sulfur dioxide on aquatic life, plants, birds, or land animals. However, it is known that sulfur dioxide will react in water to form sulfurous acid which may have slight acute toxicity to aquatic life.

Chronic Ecological Effects

Chronic (long-term) toxic effects may include shortened life span, reproductive problems, lower fertility, and changes in appearance or behavior in exposed animals. These effects can be seen long after first exposure(s) to toxic chemicals.

Insufficient data are available to evaluate the chronic (long-term) effects of sulfur dioxide on aquatic life, plants, birds, or land animals.

Water Solubility

Sulfur dioxide reacts in water to form sulfurous acid. It is slightly soluble in water. Concentrations of up to 100 milligrams will mix with a liter of water.

Persistence in the Environment

Sulfur dioxide will decompose to sulfurous acid in the aquatic environment which may be slightly persistent in water, with a half-life between 2 and 20 days. The half-life of a pollutant is the amount of time it takes for one-half of the chemical to be degraded. As a gas, virtually 100% of sulfur dioxide will eventually end up in the air.

Bioaccumulation in Aquatic Organisms

Some substances increase in concentration, or *bioaccumulate,* in living organisms as they breathe contaminated air, drink contaminated water, or eat contaminated food. These chemicals can become concentrated in the tissues and internal organs of animals as well as humans.

The concentration of sulfur dioxide found in fish tissues is expected to be somewhat lower than the average concentration of sulfur dioxide in the water from which the fish was taken.

Recommended Risk-Reduction Measures

Proper training of all transporters will reduce the likelihood of a mishap or accident resulting in a leak or spill of sulfur dioxide into the environment. Labels on all containers, trucks, and rail cars must meet DOT requirements and accurately reflect their contents to enable emergency responders to react properly and quickly to any disaster thereby reducing the potential risk to the environment and to personnel.

Storage of sulfur dioxide should be segregated from other chemicals to minimize the risk of cross-contamination. Sulfur dioxide must be stored to avoid contact with incompatible materials since violent reactions can occur. Containers should be protected from physical damage and stored to avoid contact with heat.

If a spill or leak to the environment has occurred, fire department, emergency response and/or hazardous materials spill personnel should be notified immediately. Cleanup should be attempted only by those

trained in proper spill containment procedures. Contaminated soils should be removed for incineration and replaced with clean soil.

If sulfur dioxide should contact the water table, aquifer, or navigable waterway, time is of the essence. Because it will decompose to sulfuric acid in water, total remediation may not be possible. The local and/or state emergency response authorities must be notified. A comprehensive emergency response or disaster preparedness/recovery plan should be in place prior to any operations involving the use, transportation, storage, or disposal of sulfur dioxide.

If possible, leaking containers should be repaired (sealed). Otherwise, dispose of gas by passing it through a large tank or vessel of soda ash solution and carefully add calcium hypochlorite. Then, neutralize with 6M hydrochloric acid or 6M sodium hydroxide and dilute. The mixture should be drained into a sewer with excess water.

If sulfur dioxide is spilled or leaked, the following specific steps are recommended:

☑ Restrict persons not wearing protective clothing from area of spill or leak until cleanup is complete. Avoid skin and eye contact.

☑ Stop flow of gas and repair leak. If gas cannot be disposed of as described above, remove leaking cylinders to a safe area (outdoors) and allow to vent until empty. Absorb spilled liquids with vermiculite or a similar material and deposit in sealed drums for disposal.

☑ Ventilate area of spill or leak.

☑ It may be necessary to dispose of sulfur dioxide as a hazardous waste. The responsible state agency or the regional office of the federal Environmental Protection Agency (EPA) should be contacted for specific recommendations.

RISK ASSESSMENT: BUSINESS

General Assessment

Accidents or mishaps involving sulfur dioxide can present a serious threat to business operations. The loss or damage of equipment or facilities can significantly affect fiscal viability. Lawsuits that may result from personnel illness, injury/death, public exposures, and/or environmental contamination will require a serious expenditure of resources. Media attention surrounding an injury, death, or environmental damage can also result in a loss of profits and loss of current as well as future business. Always remember that when the terms "cancer," "carcinogen," or "reproductive hazard" are used, public hysteria, emotion, anxiety, or ignorance can run equally high. This should be carefully considered when drafting public relations policies.

Recommended Risk-Reduction Measures

Company attorneys, safety and health professionals, and environmental specialists should be involved in the development of any procedures for responding to chemical incidents. A company official should be predesignated as a public relations officer with specific training in dealing with the press. Corporate plans and policies should be developed, approved, and implemented long before any need for such arises.

REFERENCES

American Conference of Governmental Industrial Hygienists. 1988. *Documentation of the Threshold Limit Values and Biological Exposure Indices,* 5th Edition (with updates). Cincinnati: ACGIH

Anthony, C. P., and N. J. Kolthoff. 1971. *Textbook of Anatomy and Physiology.* St. Louis: C. V. Mosby Company.

Brusick, D. J. 1994. *Methods for Genetic Risk Assessment.* New York: CRC Press/Lewis Publishers.

Bueche, J. R. 1972. *Principles of Physics,* 2nd Edition. New York: McGraw-Hill.

Burrell, R., D. K. Flaherty, and L. J. Sauers. 1992. *Toxicology of the Immune System.* New York: Van Nostrand Reinhold.

Calabrese, E. J. 1994. *Biological Effects of Low Level Exposures.* New York: CRC Press/Lewis Publishers.

Cothern, C. R. 1995. *Handbook for Environmental Risk Decisionmaking.* New York: CRC Press/Lewis Publishers.

Danse, I. R. 1991. *Common Sense Toxics in the Workplace.* New York: Van Nostrand Reinhold.

Environmental Protection Agency. 1983. *Chemical Hazard Information Profiles.* Washington, D.C.: U.S. EPA.

Fodden, C. R., and J. L. Weddell. 1992. *Hazardous Materials Emergency Action Data.* New York: CRC Press/Lewis Publishers.

Hodgson, E., R. B. Mailman, and Janice E. Chambers. 1988. *Dictionary of Toxicology.* London: The MacMillian Company.

Keith, L. H. and M. M. Walker. 1995. *Handbook of Air Toxics.* New York: CRC Press/Lewis Publishers.

Lauwerys, R. R., P. Hoet. 1993. *Industrial Chemical Exposure.* New York: CRC Press/Lewis Publishers.

Lewis, .J., Sr. 1993. *Hawley's Condensed Chemical Dictionary,* 12th Edition. New York: Van Nostrand Reinhold.

U.S. Department of Health and Human Services, National Institute for Occupational Safety and Health. 1994. *NIOSH Pocket Guide to Chemical Hazards.* Washington, D.C.: U.S. Government Printing Office.

MATERIAL SAFETY DATA SHEET

CHEMICAL NAME	
SULFURIC ACID	**CRC PRESS/LEWIS PUBLISHERS DISCLAIMER:** THE INFORMATION AND RECOMMENDATIONS PRESENTED HEREIN ARE BASED ON SOURCES BELIEVED TO BE RELIABLE. CRC MAKES NO REPRESENTATION ON ITS COMPLETENESS OR ACCURACY. IT IS THE RESPONSIBILITY OF THE USER TO DETERMINE THE CHEMICAL'S SUITABILITY FOR ITS INTENDED USE, THE CHEMICAL'S SAFE USE, AND THE CHEMICAL'S PROPER DISPOSAL. NO REPRESENTATIONS AND/OR WARRANTIES, EITHER EXPRESSED OR IMPLIED, OF THE MERCHANTABILITY OR FITNESS FOR A PARTICULAR PURPOSE, OR OF ANY OTHER NATURE, ARE MADE WITH RESPECT TO THE INFORMATION PROVIDED IN THIS MSDS OR TO THE CHEMICAL TO WHICH INFORMATION MAY REFER. CRC NEITHER ASSUMES NOR AUTHORIZES ANY OTHER PERSON TO ASSUME FOR IT, ANY OTHER ADDITIONAL RESPONSIBILITY OR LIABILITY FOR THE USE OF, OR RELIANCE UPON, THIS INFORMATION.

HAZARD WARNING INFORMATION

HEALTH	FIRE	REACTIVE	OTHER	DEGREE OF HAZARD	COLOR CODING	OTHER CODES
3	0	2	~~W~~	0 = Minimum Hazard 1 = Slight Hazard 2 = Moderate Hazard 3 = Serious Hazard 4 = Severe Hazard	HEALTH = BLUE FIRE = RED REACTIVITY = YELLOW OTHER = WHITE	OX = Oxidizer ACID = Acid ALK = Alkali COR = Corrosive ~~W~~ = Use No Water

SECTION I - GENERAL INFORMATION

Characterization	RCRA Number	EPA Class
Acid, Mineral	**D002**	**Characteristic (C) Waste**

DOT Proper Shipping Name **Sulfuric acid**	Chemical Abstract Service (CAS) Number **7664-93-9**
DOT Hazard Class and Label Requirements **Corrosive Material; Corrosive**	DOT Emergency Guide Code **39**
DOT Identification Number **UN1830 (51-95%); 1831 (fuming); 1832 (spent)**	Chemical Formula **H_2SO_4**

Synonyms
Battery acid; hydrogen sulfate; oil of vitriol; sulfuric acid (aqueous); vitriol brown oil.

SECTION II - HAZARDOUS INGREDIENTS AND IDENTITY INFORMATION

Hazardous Components (specific identity)	OSHA Exposure Criteria	NIOSH Exposure Criteria	Immediately Dangerous to Life and Health (IDLH)	ACGIH Exposure Criteria
Sulfuric acid (derivation: From sulfur, pyrite, hydrogen sulfide, or sulfur-containing smelter gases by the contact process with vanadium pentoxide as the catalyst. The first step is combustion of elemental sulfur or roasting of iron pyrites to yield sulfur dioxide, then catalytic oxidation to sulfur trioxide, then cooling in a stream of sulfuric acid to yield additional sulfuric acid). **1 ppm = 4.08 mg/m^3**	**PEL (8-hour):** **0.24 ppm** **1 mg/m^3** **STEL:** **Not Established**	**REL (10-hour):** **0.24 ppm** **1 mg/m^3** **STEL:** **Not Established**	**15 mg/m^3**	**TLV:** **0.24 ppm** **1 mg/m^3** **STEL:** **.75 ppm** **3 mg/m^3**

SECTION III - PHYSICAL AND CHEMICAL CHARACTERISTICS

Boiling Point **554°F (290°C)**	Specific Gravity (H_2O = 1) **1.84**
Vapor Pressure (mm Hg) **<0.001 at 68°F (20°C)**	Molecular Weight **98.1**
Vapor Density (Air = 1) **1.0**	Melting Point **51°F (10.5°C)**

Solubility
Miscible in water with violent reaction. Soluble in ethyl alcohol.

Appearance and Odor
Dense, oily liquid; colorless to dark brown depending upon purity. Odor Threshold = 150 ppm.

SECTION IV - FIRE AND EXPLOSION HAZARD DATA

Flash Point (method used) **Non-Combustible**	Explosive Limits in Air % by Volume **LEL: Not Determined UEL: Not Determined**
NFPA Classification **Non-Combustible Liquid**	Autoignition Temperature **Not Applicable**

Extinguishing Media
Reacts with water producing heat and toxic fumes. Use dry chemical or carbon dioxide.

Special Fire Fighting Procedures
Poisonous gases are released in fire. Wear full protective clothing and (SCBA). Use water spray to flush and dilute a spill, to disperse vapors, and to keep fire-exposed containers cool. Do NOT direct solid stream of water at sulfuric acid.

Unusual Fire and Explosion Hazards
A non-combustible liquid that increases the burning of combustible materials. Containers may explode in fire. Releases toxic and corrosive sulfur oxides when involved in fire.

SECTION V - REACTIVITY DATA

Stability		Conditions to Avoid Sulfuric acid is normally stable in closed containers under routine conditions of handling and storage. Extremely reactive; keep away from incompatible materials.
Stable X	Unstable	Incompatibility *(materials to avoid)* Alkalines, reducing agents, carbides, chlorates, fulminates, nitrates, picrates, metals, acetic acid, acetic anhydride, acetonitrile, acrolein, acrylic acid, alcohols, ethers, ketones, oxidizing agents, and many others.
Hazardous Polymerization		Conditions to Avoid Under normal conditions of temperature and pressure, hazardous polymerization of sulfuric acid is not expected to occur.
May Occur	Will Not Occur X	Hazardous Decomposition or By-products When heated to decomposition, sulfuric acid can emit highly toxic/poisonous gases and fumes, including toxic oxides of sulfur.

SECTION VI - HEALTH HAZARD DATA

Primary Route(s) of Entry:	Inhalation?	Absorption (skin)?	Ingestion?
	X		X

Health Hazards

INHALATION: Causes coughing, sneezing, nose irritation, nose bleeds, reflex bronchospasm, shortness of breath, pulmonary edema (fluid in lungs), emphysema, permanent changes in pulmonary function, and severe bronchial constriction.

SKIN & EYES: Eye contact causes blindness and tissue damage. Skin contact causes burns, blistering, local necroses, and membrane ulceration. Burns may be 2nd or 3rd degree.

INGESTION: Corrosion of the mucosa of the mouth, throat, and esophagus with stomach discomfort and pain, nausea, intense thirst, vomiting, circulatory collapse, clammy skin, weak and rapid pulse, shallow respiration, scanty urine, circulatory shock and subsequent death.

Carcinogenicity	NTP Listed?	IARC Cancer Review Group?	OSHA Regulated?	Target Organs?
Unknown Human Unknown Animal	No	No	29 CFR 1910.1000 Table Z-1	Respiratory system; eyes; skin; teeth.

Medical Conditions Generally Aggravated by Exposure

Respiratory impairments (bronchitis, asthma) may be aggravated by exposure.

Emergency and First-aid Procedures

Eye contact: Flush immediately with water for 15 minutes (minimum); seek medical attention. Skin contact: Remove all contaminated clothing. Immediately wash area with large amounts of water. Seek medical assistance. For inhalation: Remove the person from exposure. Provide respiratory assistance and CPR. Transfer to medical facility. Observe for 48 hours for lung effects. If swallowed: Seek medical attention; contact poison control center. Give 1-2 glasses of water to drink. Do NOT induce vomiting (aspiration hazard). Never give anything by mouth to an unconscious person.

SECTION VII - PRECAUTIONS FOR SAFE HANDLING AND USE

Steps to be Taken in Case Material is Released or Spilled

Absorb liquids in vermiculite or other material, neutralize with soda ash and collect and place in sealed containers. Ventilate area of spill or leak. Restrict those not involved in cleanup from entering area.

Preferred Waste Disposal Method

No citation.

Precautions to be Taken in Handling and Storage

Sulfuric acid is a powerful oxidizer that can greatly accelerate the burning of oxidizable materials. Store in tightly closed containers in cool, dry, well-ventilated area away from heat.

Other Precautions and Warnings

Sources of ignition are prohibited where sulfuric acid is used, handled, or stored.

SECTION VIII - CONTROL MEASURES AND PERSONAL PROTECTIVE EQUIPMENT

Respiratory Protection *(specify type)*

Use half-mask respirator with acid cartridge for low or transient exposures (cartridge should be designed to protect against acid vapor or mist). For higher exposures, use a supplied-air respirator or a self-contained breathing apparatus operated in pressure demand or other positive pressure mode.

Ventilation

Local exhaust or general mechanical systems recommended.

Protective Gloves	Eye Protection	Other Protective Clothing
Acid resistant synthetic rubber	Chemical Goggles or Face Mask	Protective Apron

Work/Hygiene Practices

Always wash hands thoroughly after using chemical; never bring food, drink, or smoking materials into vicinity of chemicals.

SULFURIC ACID

H_2SO_4

CAS: 7664-93-9

IDENTIFICATION AND TYPICAL USES

Sulfuric acid is a colorless (when pure) to dark-brown (when impure), odorless, dense, oily liquid. The pure compound is a solid below 51°F (11°C). It is the most used chemical in the world in terms of production and consumption. It is used in the production of phosphate fertilizers, dyes, explosives, glues, and a number of sulfates. It is also used in the purification of petroleum, in cleaning of steel surfaces (metal pickling), and as a dehydrating agent.

RISK ASSESSMENT: HEALTH

General Assessment

Sulfuric acid is highly corrosive to the skin, eyes, digestive tract, and respiratory system (or any tissue it contacts). It is a human poison by *ingestion* and is dangerously toxic by *inhalation*. There are no data in the references regarding any carcinogenic, mutagenic, or teratogenic effects on long-term exposure to sulfuric acid. Its corrosive and irritating properties limit the possibility of long-term exposure since people cannot tolerate unprotected contact for even short periods of time.

Inhalation produces severe irritation of the eyes, nose, and respiratory tract with coughing, stomach pains, intense thirst, dypsnea (labored breathing), bronchitis, headache, clammy skin, weak and rapid pulse, scanty urine, and total body weakness. The vapors can cause laryngitis, pharyngitis, chemical bronchitis, pneumonitis, and delayed pulmonary edema which can be fatal. There may be tearing, nasal discharge, throat irritation, ulceration of the nasal septa and nose irritation, nosebleeds, and nausea. Symptoms of pulmonary edema, such as chest pain, congestion, coughing, and weakness, may be delayed up to 48 hours thereby creating a false sense of security with regard to health exposure risk. There may be total circulatory system collapse, shock, and death.

Skin contact causes severe burning with possible vesiculation. Its extremely corrosive properties cause immediate burns, membrane ulceration, skin blistering, 2nd and 3rd degree burns, and local necrosis (tissue destruction) upon contact. Eye contact causes damage and the possibility of permanent blindness.

Ingestion is extremely dangerous, even life-threatening. It causes severe poisoning, esophageal burns, throat swelling, convulsions, risk of stomach perforation, bloody vomiting, burning sensation on the mouth, acute tissue damage, diarrhea with pain, and possible coma.

Acute Health Effects

The following acute (short-term) health effects may occur immediately or shortly after exposure to sulfuric acid:

<u>Skin</u>: Irritation/rash or burning feeling on contact with redness, blistering and vesiculation, localized tissue destruction, and deep 2nd or 3rd degree burns. Initially, the contact site appears bleached and turns brown prior to the formation of an ulcerous wound. These wounds are slow healing and may leave extensive scars.

<u>Eye</u>: Severe irritation; may cause burns and damage with permanent blindness.

<u>Lung</u>: Severe irritation of the mucous membranes in the nose, throat, and lungs with delayed pulmonary edema (fluid in the lungs), which can be fatal. Nasal irritation may progress to ulcers and subsequent nose bleeds.

<u>Other</u>: Ingestion cause serious burns to the mouth, throat, and stomach and can lead to shock, circulatory system collapse, and death.

Chronic Health Effects

The following chronic (long-term) health effects may occur at some time after exposure to sulfuric acid and can last for months or even years:

Cancer Hazards: According to information presented in the references, sulfuric acid has not been adequately tested for its ability to cause cancer in test animals.

Reproductive Hazard: According to information presented in the references, sulfuric acid has not been adequately tested for its ability to adversely affect reproduction in test animals.

Other Chronic Effects: Repeated skin contact with dilute solutions causes dermatitis. Chronic exposure may also cause changes in pulmonary function leading to chronic bronchitis, gastritis, stomatitis, erosion of dental enamel, conjunctivitis, increased propensity for respiratory tract infections, and possible respiratory tract cancer.

Recommended Risk-Reduction Measures

Personnel should avoid direct contact with sulfuric acid. It is extremely corrosive and highly toxic to humans. If a less toxic chemical cannot be substituted for a hazardous substance, then *engineering controls* are the most effective method of reducing exposures.

The best protection is to enclose operations and/or provide local exhaust ventilation at the site of chemical release. While not always operationally feasible, isolating operations can also reduce exposure. Using respiratory protection is less effective than the controls mentioned above, but is still advisable whenever working with or around sulfuric acid. Exposure levels are extremely low (1 mg/m^3) and difficult to accurately monitor, especially under emergency response conditions. For low concentrations, an acid gas cartridge respirator (equipped with a mist/vapor filter) may provide acceptable protection. However, a self-contained breathing apparatus (SCBA) or a supplied-air respirator with full facepiece operated in pressure demand mode are the recommended respiratory protection methods of choice. If a full facepiece is not available, then chemical goggles should be worn to protect the eyes. Whenever a chemical splash hazard exists, a face shield and a rubber apron should be worn. To prevent hand and skin exposures, acid-resistant gloves should be used. Glove manufacturers should be contacted and permeation studies obtained *before* final glove selections are made.

Administrative controls should also be in place to minimize the potential for human exposures. These may include written procedures or policies which specify the methods and techniques that will be practiced whenever personnel are to work with sulfuric acid.

All personnel should receive *training* on the use, hazards, protective measures, emergency actions, and other precautions per 29 CFR 1910.1200 (Hazard Communication), prior to the first assignment in an area where sulfuric acid is used or stored.

Before beginning employment and at regular intervals thereafter (e.g., annually), the following tests are recommended:

- ☑ Lung function tests (look for chronic respiratory diseases).
- ☑ Examination of the teeth with specific focus on any degree of erosion to the enamel.

If symptoms occur or overexposure is suspected, the following additional tests are suggested:

- ☑ Consider chest X-ray following acute overexposure (may be negative if taken immediately after exposure due to delayed onset of pulmonary edema).

Other methods to reduce exposure include:

- ☑ Where possible, enclose operations and use local exhaust ventilation at the site of chemical release. If local exhaust ventilation or enclosure is not used, respiratory protection should be mandatory.
- ☑ Always ensure that proper protective clothing is worn when using chemical substances and that personnel are trained on its use and care.
- ☑ Wash thoroughly immediately after exposure to sulfuric acid and at the end of the work shift or before eating, drinking, or smoking.
- ☑ Hazard warning information should be posted in the work area. In addition, as part of an on-going education and training program, all information on the health and safety hazards of sulfuric acid should be communicated to all exposed workers.

RISK ASSESSMENT: ENVIRONMENT

General Assessment

The environment is at risk of exposure during transportation, storage, disposal, or destruction of sulfuric acid. In almost every scenario, the threat of environmental exposure is contingent upon the proper handling of the chemical substance. Accidental spills, large or small, can result in fire, explosion, and possible contamination of the surrounding environmental mediums (water, soil, and, to a lesser degree, air).

Sulfuric acid is considered a non-combustible liquid. However, as an oxidizer, it can greatly accelerate the burning of ordinary combustible materials such as wood or paper. It is also extremely corrosive to many substances. It is incompatible with many materials, including organic materials, chlorates, carbides, fulminates, water, and powdered metals. It reacts violently with water to release extreme heat and may cause fire or explosion. These characteristics require special consideration during any emergency situation involving a leak or spill of sulfuric acid. Should sulfuric acid ever come into contact with incompatible substances either during use, transportation, or storage, the formation of highly toxic and/or explosive commodities is extremely possible.

Sulfuric acid can enter the environment mainly through industrial discharges or spills.

Acute Ecological Effects

Acute (short-term) toxic effects may include the death of animals, birds, or fish and death or low growth rate in plants. Acute effects are seen 2 to 4 days after animals or plants are exposed to sulfuric acid.

Sulfuric acid has moderate acute toxicity to aquatic life. It is very corrosive and would badly burn any plants, birds, or land animals exposed to sulfuric acid.

Chronic Ecological Effects

Chronic (long-term) toxic effects may include shortened life span, reproductive problems, lower fertility, and changes in appearance or behavior in exposed animals. These effects can be seen long after first exposure(s) to toxic chemicals.

Sulfuric acid has moderate chronic toxicity to aquatic life. Insufficient data are available on the long-term effects of sulfuric acid to aquatic life, plants, birds, or land animals.

Water Solubility

Sulfuric acid is highly soluble in water and will react violently. Concentrations of over 1000 milligrams will mix with a liter of water.

Persistence in the Environment

Sulfuric acid is slightly persistent in the aquatic environment, with a half-life of between 2 to 20 days. The half-life of a pollutant is the amount of time it takes for one half of the chemical to be degraded.

Bioaccumulation in Aquatic Organisms

Some substances increase in concentration, or *bioaccumulate,* in living organisms as they breathe contaminated air, drink contaminated water, or eat contaminated food. These chemicals can become concentrated in the tissues and internal organs of animals as well as humans.

Small quantities of acids will be neutralized by the alkalinity in aquatic ecosystems. However, larger quantities can actually lower the pH for extended periods of time. Sulfuric acid solutions can contain sulfate, whose concentration found in fish tissues is expected to be about the same as the average concentration in the water from which the fish was taken.

Recommended Risk-Reduction Measures

Proper training of all transporters will reduce the likelihood of a mishap or accident resulting in a leak or spill to the environment. The correct use of DOT labeling (per the MSDS) on all containers, trucks, and rail cars will enable emergency responders to react properly and quickly to any disaster thereby reducing the potential risk to the environment and to personnel.

Storage of sulfuric acid should be segregated from incompatible chemicals to minimize the risk of cross-contamination or contact. Buildings designated for storage should be equipped with appropriate fire protection systems (alarms, sprinklers, emergency lighting, portable extinguishers). It must be kept cool and in a well-ventilated location away from direct sunlight.

If a spill or leak to the environment has occurred, fire department, emergency response and/or hazardous materials spill personnel should be notified immediately. Cleanup should be attempted only by those trained in proper spill containment procedures. Contaminated soils should be removed for incineration and replaced with clean soil. If sulfuric acid should contact the water table, aquifer, or navigable waterway, time is of the essence. It is highly miscible in water and, therefore, total containment and remediation may not be entirely possible. When such spills occur, the local and/or state emergency response authorities must be notified. A comprehensive emergency response or disaster preparedness/recovery plan should be in place prior to any operations involving the use, transportation, storage, or disposal of sulfuric acid. If sulfuric acid is spilled, the following steps are recommended:

☑ Restrict persons not wearing protective clothing from area of spill or leak until cleanup is complete and area can be opened for normal work.

☑ Ventilate area of spill or leak.

☑ Absorb liquids in vermiculite, dry sand, earth, or similar material and deposit in sealed containers. Neutralize remaining residue with soda ash or sodium bicarbonate. Do NOT flush to drains, sewers, or waterways. Use water spray to cool and knock down vapors. Collect water for disposal. Protect against excessive water contact with sulfuric acid since it can react to evolve heat.

☑ It may be necessary to dispose of sulfuric acid as a hazardous waste. The responsible state agency or the regional office of the federal Environmental Protection Agency (EPA) should be contacted for specific recommendations.

RISK ASSESSMENT: BUSINESS

General Assessment

Accidents or mishaps involving sulfuric acid can present a significant threat to business operations. The loss or damage of equipment or facilities can significantly affect fiscal viability. Lawsuits that may result from personnel injury/death, public exposures, and/or environmental contamination will also require a serious expenditure of resources. Media attention surrounding an injury, death, or environmental damage can also result in a loss of profits and loss of current as well as future business.

Recommended Risk-Reduction Measures

Company attorneys, safety and health professionals, and environmental specialists should be involved in the development of any procedures dealing with safety or emergency response. A company official should be pre-designated as a public relations officer with specific training in dealing with the press. Corporate plans and policies should be developed, approved, and implemented long before any need for such arises.

REFERENCES

Ahlbom, A. 1993. *Biostatistics for Epidemiologists.* New York: CRC Press/Lewis Publishers.

American Conference of Governmental Industrial Hygienists. 1988. *Documentation of the Threshold Limit Values and Biological Exposure Indices,* 5th Edition (with updates). Cincinnati: ACGIH

Asante-Duah, D. K. 1993. *Hazardous Waste Risk Assessment.* New York: CRC Press/Lewis Publishers.

Bloom, A. D. and F. J. DeSerres. 1995. *Ecotoxicity and Human Health.* New York: CRC Press/Lewis Publishers.

Bueche, J. R. 1972. *Principles of Physics,* 2nd Edition. New York: McGraw-Hill.

Calabrese, E. J., E. M. Kenyon. *Air Toxics and Risk Assessment.* New York: CRC Press/Lewis Publishers.

Cockerman, L. G., B. S. Shane. 1994. *Basic Environmental Toxicology.* New York: CRC Press/Lewis Publishers.

Cothern, C. R. 1995. *Handbook for Environmental Risk Decisionmaking.* New York: CRC Press/Lewis Publishers.

Danse, I. R. 1991. *Common Sense Toxics in the Workplace.* New York: Van Nostrand Reinhold.

Gardner, R. W. 1994. *Chemical Intolerance.* New York: CRC Press/Lewis Publishers.

Hallenbeck, W. H. 1993. *Quantitative Risk Assessment for Environmental and Occupational Health,* 2nd Edition. New York: CRC Press/Lewis Publishers.

Hosty, J. W., P. Foster. 1990. *A Practical Guide to Chemical Spill Response.* New York: Van Nostrand Reinhold.

Howard, P. H., et. al. 1991. *Handbook of Environmental Degradation Rates.* New York: CRC Press/Lewis Publishers.

Jones, F. E. 1994. *Toxic Organic Vapors in the Workplace.* New York: CRC Press/Lewis Publishers.

Keith, L. H. and M. M. Walker. 1995. *Handbook of Air Toxics.* New York: CRC Press/Lewis Publishers.

Landis, W. G., M. H. Yu. 1995. *Introduction to Environmental Toxicology.* New York: CRC Press/Lewis Publishers.

Lauwerys, R. R., P. Hoet. 1993. *Industrial Chemical Exposure.* New York: CRC Press/Lewis Publishers.

Lewis, R. J., Sr. 1993. *Hawley's Condensed Chemical Dictionary,* 12th Edition. New York: Van Nostrand Reinhold.

Lewis, R. J., Sr. 1992. *Sax's Dangerous Properties of Industrial Materials.* Volumes 1, 2, and 3, 8th Edition. New York: Van Nostrand Reinhold.

Meyer, E. 1990. *Chemistry of Hazardous Materials.* Englewood Cliffs, NJ: Prentice Hall.

Rea, W. J. 1995. *Chemical Sensitivity,* Vol. I-III. New York: CRC Press/Lewis Publishers.

Richardson, M. 1995. *Dictionary of Substance and Their Effects,* Vol. I-VI. New York: CRC Press/Lewis Publishers.

Sacarello, H. L. A. 1994. *The Comprehensive Handbook of Hazardous Materials.* New York: CRC Press/Lewis Publishers.

Sherman, J. D. 1988. *Chemical Exposure and Disease.* New York: Van Nostrand Reinhold.

U.S. Department of Health and Human Services, National Institute for Occupational Safety and Health. 1994. *NIOSH Pocket Guide to Chemical Hazards.* Washington, D.C.: U.S. Government Printing Office.

MATERIAL SAFETY DATA SHEET

CHEMICAL NAME	CRC PRESS/LEWIS PUBLISHERS DISCLAIMER
2,4,5-T (TRICHLOROPHENOXYACETIC ACID)	THE INFORMATION AND RECOMMENDATIONS PRESENTED HEREIN ARE BASED ON SOURCES BELIEVED TO BE RELIABLE. CRC MAKES NO REPRESENTATION ON ITS COMPLETENESS OR ACCURACY. IT IS THE RESPONSIBILITY OF THE USER TO DETERMINE THE CHEMICAL'S SUITABILITY FOR ITS INTENDED USE, THE CHEMICAL'S SAFE USE, AND THE CHEMICAL'S PROPER DISPOSAL. NO REPRESENTATIONS AND/OR WARRANTIES, EITHER EXPRESSED OR IMPLIED, OF THE MERCHANTABILITY OR FITNESS FOR A PARTICULAR PURPOSE, OR OF ANY OTHER NATURE, ARE MADE WITH RESPECT TO THE INFORMATION PROVIDED IN THIS MSDS OR TO THE CHEMICAL TO WHICH INFORMATION MAY REFER. CRC NEITHER ASSUMES NOR AUTHORIZES ANY OTHER PERSON TO ASSUME FOR IT, ANY OTHER ADDITIONAL RESPONSIBILITY OR LIABILITY FOR THE USE OF, OR RELIANCE UPON, THIS INFORMATION.

HAZARD WARNING INFORMATION

HEALTH	FIRE	REACTIVE	OTHER	DEGREE OF HAZARD	COLOR CODING	OTHER CODES
2	0	0		0 = Minimum Hazard 1 = Slight Hazard 2 = Moderate Hazard 3 = Serious Hazard 4 = Severe Hazard	HEALTH = BLUE FIRE = RED REACTIVITY = YELLOW OTHER = WHITE	OX = Oxidizer ACID = Acid ALK = Alkali COR = Corrosive W = Use No Water

SECTION I - GENERAL INFORMATION

Characteristic	**Herbicide**	RCRA Number	**U232**
		EPA Class	**Hazardous Waste**
DOT Proper Shipping Name	**Phenoxy Pesticides, Solid, Toxic**	Chemical Abstract Service (CAS) Number	**93-76-5**
DOT Hazard Class and Label Requirements	**ORM-A; Keep Away From Food**	DOT Emergency Guide Code	**55**
DOT Identification Number	**UN 2765**	Molecular Formula	**$Cl_3C_6H_2OCH_2COOH$**

Synonyms

2,4,5-Trichlorophenoxyacetic Acid; Dacamine; Dinoxol; Esteron 245; Forron; Brush Rhap; Inverton 245; Phortox; Supper D Weedone; Transamine; Tributon; Trinoxol; Trioxone; Veon; Weedar; Verton 2T.

SECTION II - HAZARDOUS INGREDIENTS AND IDENTITY INFORMATION

Hazardous Components (specific identity)	OSHA Exposure Criteria	NIOSH Exposure Criteria	Immediately Dangerous to Life and Health (IDLH)	ACGIH Exposure Criteria
2,4,5-Trichlorophenoxyacetic Acid: (derivation: By reacting 2,4,5-trichlorophenol with monochloroacetic acid).	**PEL (8-hour):** 10 mg/m^3 **STEL:** Not Established	**REL (10-hour):** 10 mg/m^3 **STEL:** Not Established	250 mg/m^3	**TLV:** 10 mg/m^3 **STEL:** Not Established

SECTION III - PHYSICAL AND CHEMICAL CHARACTERISTICS

Boiling Point	**Decomposes**	Specific Gravity (H_2O = 1)	**1.80**
Vapor Pressure (atmospheres)	**<1 x 10^{-7} at 69°F (20°C)**	Molecular Weight (atomic weight)	**255.5**
Vapor Density (Air = 1)	**Not Reported**	Melting Point	**307°F (153°C)**

Solubility

Very slightly soluble in water (0.03% at 77°F). Soluble in alcohol, ethanol, ethyl ether, methanol, toluene, xylene, and slightly soluble in petroleum ether.

Appearance and Odor

White to light tan odorless crystalline solid with a metallic taste. It may be shipped in molten form.

SECTION IV - FIRE AND EXPLOSION HAZARD DATA

Flash Point (method used)	**Not Applicable**	Explosive Limits in Air % by Volume	**LEL: N/A UEL: N/A**
NFPA Classification	**Noncombustible Solid**	Autoignition Temperature	**Not Applicable**

Extinguishing Media

Use extinguishing media suitable to surrounding fires.

Special Fire Fighting Procedures

Wear full protective clothing and self-contained breathing apparatus (SCBA) when fighting fires near or around 2,4,5-trichlorophenoxyacetic acid since poisonous gases are produced in fire. Use water spray to cool fire-exposed containers that cannot be safely removed from fire area.

Unusual Fire and Explosion Hazards

Containers may explode in fire.

SECTION V - REACTIVITY DATA

Stability		Conditions to Avoid: 2,4,5-Trichlorophenoxyacetic acid is a very stable herbicide. It should be kept away from heat, flame, and conditions which can generate excessive dusts.
Stable	Unstable	Incompatibility *(materials to avoid)*
X		None reported.
Hazardous Polymerization		Conditions to Avoid: Hazardous polymerization of 2,4,5-trichlorophenoxyacetic acid cannot occur.
May Occur	Will Not Occur	Hazardous Decomposition or By-products
	X	When heated to decomposition, 2,4,5-trichlorophenoxyacetic acid can produce toxic smoke and fumes, including carbon monoxide, chlorine, hydrogen chloride, and phosgene.

SECTION VI - HEALTH HAZARD DATA

Primary Route(s) of Entry:	Inhalation?	Absorption (skin/eye)?	Ingestion?
	X	X	X

Health Hazards

INHALATION: Human data are inconclusive. Symptoms may include ataxia, somnolence, and fatty degeneration of the liver.

SKIN & EYES: Irritation. Chloracne may develop. It is slowly absorbed through the skin and toxic systemic effects similar to ingestion may develop.

INGESTION: May cause mouth, throat, and gastrointestinal irritation, chest pain (fromesophagitis), abdominal pain, diarrhea, muscle twitching and spasms, metabolic acidosis, fever, tachycardia, hyperventilation, vasodilatation, and sweating. Other effects may involve the liver.

Carcinogenicity	NTP Listed?	IARC Cancer Review Group?	OSHA Regulated?	Target Organs?
Questioned Human Questioned Animal	No	Group 2B	29 CFR 1910.1000 Table Z-1	Skin, liver, GI tract, CVS, respiratory System.

Medical Conditions Generally Aggravated by Exposure

None reported.

Emergency and First-aid Procedures

Eye contact: Do not allow victim to keep eyes tightly shut. Flush immediately with water for 15 minutes (minimum); seek medical attention. Skin contact: Remove all contaminated clothing. Immediately wash area with large amounts of soap and water. Thoroughly wash exposed hair and nails. For inhalation: Remove the person from exposure. Provide respiratory assistance and CPR. Transfer to medical facility. If swallowed: Seek medical attention immediately. Do not attempt to give an unconscious or convulsing person anything by mouth. If conscious give 1 to 2 glasses of water and induce vomiting.

SECTION VII - PRECAUTIONS FOR SAFE HANDLING AND USE

Steps to be Taken in Case Material is Released or Spilled

Ventilate area of spill or leak. Restrict those not involved in cleanup from entering area. Collect spilled powdered materials in safest manner possible and deposit in sealed drum for disposal.

Preferred Waste Disposal Method

Incineration.

Precautions to be Taken in Handling and Storage

Store in tightly closed containers in cool, well-ventilated area away from other agricultural chemicals or seeds. The minimum shelf-life is 2 years.

Other Precautions and Warnings

Use of this chemical is no longer permitted. Storage should only be for materials awaiting disposal.

SECTION VIII - CONTROL MEASURES AND PERSONAL PROTECTIVE EQUIPMENT

Respiratory Protection *(specify type)*

2,4,5-Trichlorophenoxyacetic acid can be a mildly toxic irritant by inhalation. An MSHA/NIOSH-approved half-mask respirator equipped with an organic vapor cartridge and dust/mist pre-filter should be worn. For high exposures, a self-contained breathing apparatus (SCBA) with full facepiece operated in positive pressure mode is recommended.

Ventilation

Local exhaust (preferred) at site of chemical work or general ventilation.

Protective Gloves	Eye Protection	Other Protective Clothing
Impervious Material	Chemical/Dust Goggles or Face Mask	Protective Apron

Work/Hygiene Practices

Always wash hands thoroughly after using chemical; never bring food, drink, or smoking materials into vicinity of chemicals.

2,4,5-TRICHLOROPHENOXYACETIC ACID ("2,4,5-T")

$Cl_3C_6H_2OCH_2COOH$ CAS: 93-76-5

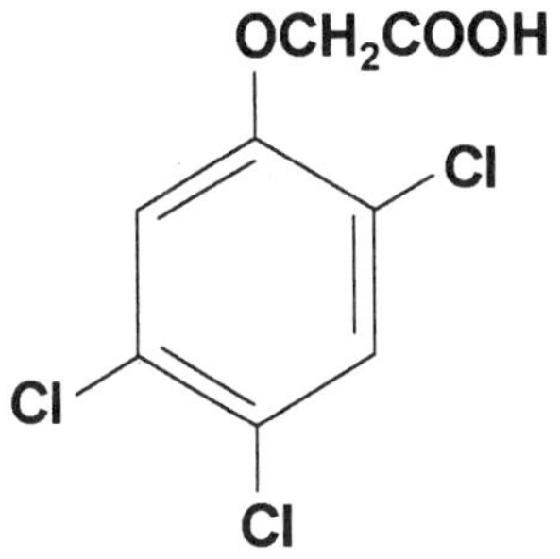

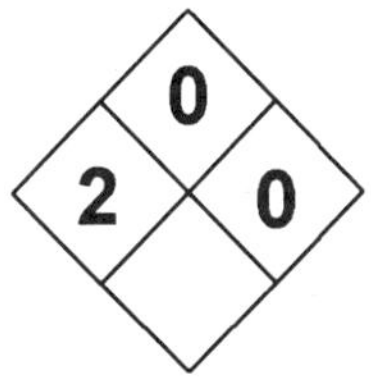

IDENTIFICATION AND TYPICAL USES

2,4,5-Trichlorophenoxyacetic acid (2,4,5-T) is a white to light-tan, odorless, crystalline solid with a metallic taste. It was formerly used widely as a plant hormone, herbicide, and defoliant. The U.S. Environmental Protection Agency gradually canceled the registration of products containing this chemical since 1970, with the total ban now in effect since 1985.

RISK ASSESSMENT: HEALTH

General Assessment

2,4,5-T is considered moderately to highly toxic by *inhalation* and, especially, by *ingestion*. If it remains in contact with the skin, 2,4,5-T can be slowly *absorbed* to cause toxic systemic effects. It is a suspected human carcinogen, a known animal teratogen, and a known animal mutagen.

Inhalation causes irritation to the eye, nose, throat, and respiratory tract. It may cause somnolence and ataxia.

Skin contact can result in the development of an acne-like rash (known as chloracne), but this is usually attributed to 2,4,5-T contamination by 2,3,7,8-tetrachlorodibenzo-*p*-dioxin (TCDD), which is highly toxic. Absorption can occur on prolonged contact causing toxic systemic effects similar to that of ingestion. Eye contact can also cause irritation.

Ingestion may cause mouth, throat, and gastrointestinal irritation, chest pain (fromesophagitis), abdominal pain, diarrhea, fibrillary muscle and skeletal muscle twitching, and myotonia (muscle spasms involving prolonged contractions). It can also cause metabolic acidosis, fever, tachycardia, hyperventilation, vasodilatation, and sweating. Contamination with TCDD can lead to other effects such as liver function impairment, peripheral neuropathy, and personality changes.

Acute Health Effects

The following acute (short-term) health effects may occur immediately or shortly after exposure to 2,4,5-T:

Skin: Irritation with possible absorption.

Eye: Moderate eye irritation likely, conjunctivitis possible in some exposures.

Lung: Can irritate the respiratory tract, and in large doses, can cause somnolence and ataxia.

Other: Seriously toxic by ingestion with systemic effects including muscular twitching, abdominal pain, tachycardia, fever, and hyperventilation.

Chronic Health Effects

The following chronic (long-term) health effects may occur at some time after exposure to 2,4,5-T and can last for months or even years:

Cancer Hazards: 2,4,5-T is a possible human carcinogen since it has been shown to cause cancer in animals and there is sufficient evidence that it will have the same effect in humans. Mutation data have also been reported and many scientists believe that long-term exposure to such chemicals may pose a cancer risk.

Reproductive Hazard: 2,4,5-T is considered an experimental teratogen and may cause adverse reproductive effects in humans as well.

Recommended Risk-Reduction Measures

Personnel should avoid direct contact with 2,4,5-T. Since its use is now prohibited in the United States, exposures are less likely but still possible (during disposal operations, for example). Whenever 2,4,5-T is to be handled, *engineering controls* are the most effective method of reducing exposures. The best protection is to enclose operations and/or provide local exhaust ventilation at the site of 2,4,5-T release. While not always operationally feasible, isolating operations can also reduce exposure.

Using respiratory protection and other personal protective equipment (PPE) is less effective than the engineering controls mentioned above, but is still advisable whenever working with or around 2,4,5-T. Under normal storage and disposal operations where there may be conditions or relatively low exposure potential, a half-mask respirator equipped with an organic vapor cartridge and dust/mist pre-filter will suffice. However, for maximum protection, a self-contained breathing apparatus (SCBA) with full facepiece and pressure demand is recommended. If a full facepiece is not available, then chemical/dust goggles or face shield should be worn to protect the eyes. To prevent hand and skin exposures, impervious protective gloves should be used. Glove manufacturers should be contacted and permeation studies obtained *before* final glove selections are made.

Administrative controls should also be in place to minimize the potential for human exposures. These may include written procedures or policies which specify the methods and techniques that will be practiced whenever personnel are to work with 2,4,5-T.

All personnel should receive *training* on the use, hazards, protective measures, emergency actions, and other precautions per 29 CFR 1910.1200 (Hazard Communication), prior to the first assignment in an area where 2,4,5-T is used or stored.

After suspected overexposure, consider the following medical tests:

- ☑ Kidney function tests.
- ☑ Lung function tests.
- ☑ Complete blood count (CBC). Monitor LDH (lactic dehydrogenase) for heart and liver injury; SGOT (serum glutamic oxaloacetic transaminase) for liver injury; alkaline phosphate to detect liver injury; and CPK (creatine peripheral kinase) to detect muscle damage.
- ☑ Electrocardiogram (EKG).
- ☑ Blood plasma and urine test.

Any evaluation should include a careful history of past and present symptoms with an examination. Medical tests that simply look for existing damage are not a substitute for controlling exposures.

Other methods to reduce exposure to 2,4,5-T include:

- ☑ Where possible, enclose operations and use local exhaust ventilation at the site of chemical release, especially during manufacturing/production and packing or shipping operations. If local exhaust ventilation or enclosure is not used in these situations, respiratory protection should be required.
- ☑ Always ensure that proper protective clothing is worn when using chemical substances and that personnel are trained in its use and care. Never allow contaminated clothing to be taken home from the work place.
- ☑ Hazard warning information should be posted in the work area. In addition, as part of an on-going education and training program, all information on the health and safety hazards of 2,4,5-T should be communicated to all potentially exposed workers.
- ☑ Eye wash stations should be provided in the immediate work area for emergency use. If there is a possibility of skin exposure to 2,4,5-T, emergency shower facilities should also be provided.

RISK ASSESSMENT: ENVIRONMENT

General Assessment

Since the use of 2,4,5-T is no longer permitted in the United States, the environment is at risk of exposure during storage, transportation, disposal, or destruction of 2,4,5-T. In almost every scenario, the threat of environmental exposure is contingent upon the proper handling of the chemical substance. Accidental spills, large or small, can result in the possible contamination of the surrounding environmental mediums (water, soil, and air).

2,4,5-T is considered a noncombustible solid. However, extreme caution is always required in handling, storage, transportation, and disposal of 2,4,5-T. Its toxic nature requires special consideration during any emergency situation involving a leak or spill of 2,4,5-T.

Since 2,4,5-T was once a widely used herbicide, it may enter the environment through agricultural runoff but also as a result of accidents and spills.

Acute Ecological Effects

Acute (short-term) toxic effects may include the death of animals, birds, or fish and death or low growth rate in plants. Acute effects are seen 2 to 4 days after animals or plants are exposed to 2,4,5-T.

2,4,5-T should be expected to have high acute toxicity to plants. Insufficient data are available to evaluate or predict the acute (short-term) effects of 2,4,5-T exposure to aquatic life, bird, or land animals.

Chronic Ecological Effects

Chronic (long-term) toxic effects may include shortened life span, reproductive problems, lower fertility, and changes in appearance or behavior in exposed animals. These effects can be seen long after first exposure(s) to toxic chemicals.

There are insufficient data to establish whether 2,4,5-T poses chronic toxicity to aquatic life, birds, or land animals. It poses high chronic toxicity to some forms of plant life.

Water Solubility

2,4,5-T has slight solubility in water. Concentrations of 2 to 20 milligrams will mix with a liter of water.

Persistence in the Environment

There is no evidence in the references to show that 2,4,5-T has any measurable persistence in the environment. Its biological half-life is expected to be less than 20 days. The half-life of a pollutant is the amount of time it takes for one half of the chemical to be degraded.

Bioaccumulation in Aquatic Organisms

Some substances increase in concentration, or *bioaccumulate*, in living organisms as they breathe contaminated air, drink contaminated water, or eat contaminated food. These chemicals can become concentrated in the tissues and internal organs of animals as well as humans.

The concentration of 2,4,5-T found in fish tissues is expected to be much lower than the average concentration of 2,4,5-T in the water from which the fish was taken.

Recommended Risk-Reduction Measures

Proper training of all transporters will reduce the likelihood of a mishap or accident resulting in a leak or spill of 2,4,5-T into the environment. The correct use of labeling on all containers, trucks, and rail cars will enable emergency responders to react properly and quickly to any disaster thereby reducing the potential risk to the environment and to personnel.

Storage of 2,4,5-T should be segregated from other agricultural chemicals to minimize the risk of cross-contamination or contact.

If a spill or leak to the environment has occurred, fire department, emergency response and/or hazardous materials spill personnel should be notified immediately. Cleanup should be attempted only by those trained in proper spill containment procedures. If 2,4,5-T should contact the water table, aquifer, or navigable waterway, the local and/or state emergency response authorities must be notified. A comprehensive emergency response or disaster preparedness/recovery plan should be in place prior to any operations involving the use, transportation, storage, or disposal of 2,4,5-T.

2,4,5-T may be destroyed by incineration in a chemical incinerator equipped with an air scrubber.

If 2,4,5-T is spilled or leaked, the following specific steps are recommended:

- ☑ Restrict persons not wearing protective clothing from area of spill or leak until cleanup is complete and area can be opened for normal work.
- ☑ If possible, remove leaking containers to a ventilated area and attempt to seal leak.
- ☑ Do not allow exposure to other materials.
- ☑ Collect powdered materials in the safest and most efficient manner possible. Place materials in a sealed drum. Do NOT dry sweep (generates dusts). Use a vacuum equipped with a high-efficiency particulate air (HEPA) filter.

RISK ASSESSMENT: BUSINESS

General Assessment

Accidents or mishaps involving 2,4,5-T can present a moderate threat to business operations. Although not likely, equipment and facilities may be lost or their use disrupted as a result of an incident involving 2,4,5-T, which can significantly affect fiscal viability. Lawsuits that may result from personnel injury, public exposures, and/or environmental contamination will also require a serious expenditure of resources. Media attention surrounding an injury, illness, or environmental damage can also result in a loss of profits and loss of current as well as future business.

Also, always remember that anytime the terms "cancer," "carcinogen," or " reproductive hazard" are used, public hysteria, emotion, anxiety, and ignorance can run equally high. This should be carefully considered whenever drafting or implementing public relations policies.

Recommended Risk-Reduction Measures

Company attorneys, safety and health professionals, and environmental specialists should be involved in the development of any procedures that focus on emergency response, disaster preparedness, and disaster recovery. A company official should be pre-designated as a public relations officer with specific training in dealing with the press. Corporate plans and policies should be developed, approved, and implemented long before any need for such arises.

REFERENCES

Ahlbom, A. 1993. *Biostatistics for Epidemiologists.* New York: CRC Press/Lewis Publishers.

American Conference of Governmental Industrial Hygienists. 1988. *Documentation of the Threshold Limit Values and Biological Exposure Indices,* 5th Edition (with updates). Cincinnati: ACGIH

Bloom, A. D. and F. J. DeSerres. 1995. *Ecotoxicity and Human Health.* New York: CRC Press/Lewis Publishers.

Brusick, D. J. 1994. *Methods for Genetic Risk Assessment.* New York: CRC Press/Lewis Publishers.

Bueche, J. R. 1972. *Principles of Physics,* 2nd Edition. New York: McGraw-Hill.

Calabrese, E. J., E. M. Kenyon. *Air Toxics and Risk Assessment.* New York: CRC Press/Lewis Publishers.

Calabrese, E. J. 1994. *Biological Effects of Low Level Exposures.* New York: CRC Press/Lewis Publishers.

Cockerman, L. G., B. S. Shane. 1994. *Basic Environmental Toxicology.* New York: CRC Press/Lewis Publishers.

Cothern, R. R. 1993. *Comparative Environmental Risk Assessment.* New York: CRC Press/Lewis Publishers.

Gardner, R. W. 1994. *Chemical Intolerance.* New York: CRC Press/Lewis Publishers.

Hallenbeck, W. H. 1993. *Quantitative Risk Assessment for Environmental and Occupational Health,* 2nd Edition. New York: CRC Press/Lewis Publishers.

Howard, P. H. 1995. *Handbook of Environmental Fate and Exposures Data for Organic Chemicals,* Vol. I-IV. New York: CRC Press/Lewis Publishers.

Howard, P. H., et. al. 1991. *Handbook of Environmental Degradation Rates.* New York: CRC Press/Lewis Publishers.

Jones, F. E. 1994. *Toxic Organic Vapors in the Workplace.* New York: CRC Press/Lewis Publishers.

Keith, L. H. and M. M. Walker. 1995. *Handbook of Air Toxics.* New York: CRC Press/Lewis Publishers.

Landis, W. G., M. H. Yu. 1995. *Introduction to Environmental Toxicology.* New York: CRC Press/Lewis Publishers.

Langerman, N. 1994 *Precautionary Labels for Chemical Containers.* New York: CRC Press/Lewis Publishers.

Lauwerys, R. R., P. Hoet. 1993. *Industrial Chemical Exposure.* New York: CRC Press/Lewis Publishers.

Lewis, R. J., Sr. 1993. *Hawley's Condensed Chemical Dictionary,* 12th Edition. New York: Van Nostrand Reinhold.

Lewis, R. J., Sr. 1992. *Sax's Dangerous Properties of Industrial Materials,* Volumes 1, 2, and 3, 8th Edition. New York: Van Nostrand Reinhold.

Lide, D. R. 1995. *Handbook of Chemistry and Physics,* 76th Edition. New York: CRC Press/Lewis Publishers.

Manahan, S. E. 1994. *Environmental Chemistry,* 6th Edition. New York: CRC Press/Lewis Publishers.

Mellström, G. A., J. E. Wahlberg, and H. I. Maibach. 1994. *Protective Gloves for Occupational Use.* New York: CRC Press/Lewis Publishers.

Meyer, E. 1990. *Chemistry of Hazardous Materials.* Englewood Cliffs, NJ: Prentice Hall.

Mycock, J. C., J. D. McKenna, and L. Theodore. 1995. *Handbook of Air Pollution Control Engineering and Technology.* New York: CRC Press/Lewis Publishers.

Patnaik, P. 1992. *A Comprehensive Guide to the Hazardous Properties of Chemical Substances.* New York: Van Nostrand Reinhold.

Proctor, N. H., J. P. Hughes. 1991. *Chemical Hazards of the Workplace.* New York: Van Nostrand Reinhold.

Quigley, D. R. 1994. *Handbook of Emergency Chemical Management.* New York: CRC Press/Lewis Publishers.

Rea, W. J. 1995. *Chemical Sensitivity,* Vol. I-III. New York: CRC Press/Lewis Publishers.

Richardson, M. 1995. *Dictionary of Substance and Their Effects,* Vol. I-VI. New York: CRC Press/Lewis Publishers.

Sacarello, H. L. A. 1994. *The Comprehensive Handbook of Hazardous Materials.* New York: CRC Press/Lewis Publishers.

Sheftel, V. O. 1995. *Handbook of Toxic Properties of Monomers and Additives.* New York: CRC Press/Lewis Publishers.

Talbott, E., G. Craun. 1995. *An Introduction to Environmental Epidemiology.* New York: CRC Press/Lewis Publishers.

Torkelson, T. R. V. K. Rowe. 1981. *Patty's Industrial Hygiene and Toxicology,* Vol. 2B. New York: Wiley-Interscience.

U.S. Department of Health and Human Services, National Institute for Occupational Safety and Health. 1994. *NIOSH Pocket Guide to Chemical Hazards.* Washington, D.C.: U.S. Government Printing Office.

MATERIAL SAFETY DATA SHEET

CHEMICAL NAME	
TALC	**CRC PRESS/LEWIS PUBLISHERS DISCLAIMER:** THE INFORMATION AND RECOMMENDATIONS PRESENTED HEREIN ARE BASED ON SOURCES BELIEVED TO BE RELIABLE. CRC MAKES NO REPRESENTATION ON ITS COMPLETENESS OR ACCURACY. IT IS THE RESPONSIBILITY OF THE USER TO DETERMINE THE CHEMICAL'S SUITABILITY FOR ITS INTENDED USE, THE CHEMICAL'S SAFE USE, AND THE CHEMICAL'S PROPER DISPOSAL. NO REPRESENTATIONS AND/OR WARRANTIES, EITHER EXPRESSED OR IMPLIED, OF THE MERCHANTABILITY OR FITNESS FOR A PARTICULAR PURPOSE, OR OF ANY OTHER NATURE, ARE MADE WITH RESPECT TO THE INFORMATION PROVIDED IN THIS MSDS OR TO THE CHEMICAL TO WHICH INFORMATION MAY REFER. CRC NEITHER ASSUMES NOR AUTHORIZES ANY OTHER PERSON TO ASSUME FOR IT, ANY OTHER ADDITIONAL RESPONSIBILITY OR LIABILITY FOR THE USE OF, OR RELIANCE UPON, THIS INFORMATION.

HAZARD WARNING INFORMATION

HEALTH	FIRE	REACTIVE	OTHER	DEGREE OF HAZARD	COLOR CODING	OTHER CODES
1	0	0		0 = Minimum Hazard 1 = Slight Hazard 2 = Moderate Hazard 3 = Serious Hazard 4 = Severe Hazard	HEALTH = BLUE FIRE = RED REACTIVITY = YELLOW OTHER = WHITE	OX = Oxidizer ACID = Acid ALK = Alkali COR = Corrosive W = Use No Water

SECTION I - GENERAL INFORMATION

Characterization	RCRA Number	EPA Class
Particulate, Magnesium Silicate	**None**	**Not Applicable**

DOT Proper Shipping Name **No Citation**	Chemical Abstract Service (CAS) Number **14807-96-6**
DOT Hazard Class and Label Requirements **No Citation**	DOT Emergency Guide Code **No Citation**
DOT Identification Number **No Citation**	Chemical Formula **$Mg_3Si_4O_{10}(OH)_2$**

Synonyms
Hydrous magnesium silicate; steatite talc; soapstone; Talcum; mineral graphite; French chalk.

SECTION II - HAZARDOUS INGREDIENTS AND IDENTITY INFORMATION

Hazardous Components (specific identity)	OSHA Exposure Criteria	NIOSH Exposure Criteria	Immediately Dangerous to Life and Health (IDLH)	ACGIH Exposure Criteria
Talc (derivation: Found in nature as layered minerals with crystallization structures ranging from fibrous to asbestos, acicular, and prismatic).	**PEL (8-hour):** 2 mg/m^3 **STEL:** Not Established	**REL (10-hour):** 2 mg/m^3 (respirable) **STEL:** Not Established	**1000 mg/m^3**	**TLV:** 2 mg/m^3 (respirable) **STEL:** Not Established

SECTION III - PHYSICAL AND CHEMICAL CHARACTERISTICS

Boiling Point **Not Reported**	Specific Gravity (H_2O = 1) **2.70 - 2.80**
Vapor Pressure (mmHg) **0 (approx.) at 69°F (20°C)**	Molecular Weight **96.33**
Vapor Density (Air = 1) **Not Reported**	Freezing Point **1652 to 1832°F (901 to 1001°C)**

Solubility
Insoluble in water, cold acids, or alkalis.

Appearance and Odor
White, apple-green, yellow, or gray powdered solid with no odor. Has a lustrous, pearly or greasy feeling. The color depends on the impurity content from the mineral source.

SECTION IV - FIRE AND EXPLOSION HAZARD DATA

Flash Point (method used) **Not Applicable**	Explosive Limits in Air % by Volume **LEL: Not Reported** **UEL: Not Reported**
NFPA Classification **Non-Combustible Solid**	Autoignition Temperature **Not Applicable**

Extinguishing Media
Use extinguishing media suitable to surrounding fire. Use a water spray to wet down the material to help reduce airborne particulate levels and to cool fire-exposed containers.

Special Fire Fighting Procedures
Poisonous gases are produced in fire. Wear self-contained breathing apparatus. Non-asbestiform talc is a relatively inert material, but breathing its dusts should be avoided.

Unusual Fire and Explosion Hazards
None reported.

SECTION V - REACTIVITY DATA

Stability		Conditions to Avoid: **Talc is normally stable in closed containers under normal conditions of temperature and pressure and routine conditions of handling and storage.**
Stable	Unstable	Incompatibility *(materials to avoid)*
X		**No known incompatibles. Pure talc is highly resistant to acids, alkalis, and heat.**
Hazardous Polymerization		Conditions to Avoid: **Under normal conditions of temperature and pressure, hazardous polymerization of talc is not expected to occur.**
May Occur	Will Not Occur	Hazardous Decomposition or By-products
	X	**None reported.**

SECTION VI - HEALTH HAZARD DATA

Primary Route(s) of Entry:	Inhalation?	Absorption (skin)?	Ingestion?
	X		

Health Hazards

INHALATION: Talc with less than 1% asbestos is considered a "nuisance dust" that may produce minor but reversible respiratory irritation. Fibrous talc is not considered a nuisance dust since it has caused chronic lung damage. Causes cough, dypsnea, and irritation of the upper respiratory tract. Large amounts may lead to a type of pulmonary fibrosis (known as talc pneumoconiosis) which may be due to asbestos content, and pulmonary edema.

SKIN & EYES: From no significant effects to minor skin and eye irritation.

INGESTION: Gastrointestinal irritation, nausea, and diarrhea.

Carcinogenicity	NTP Listed?	IARC Cancer Review Group?	OSHA Regulated?	Target Organs?
Suspected Human Suspected Animal	No	Group 3	29 CFR 1910.1000 Table Z-3	Respiratory system

Medical Conditions Generally Aggravated by Exposure

Respiratory or pulmonary conditions may be aggravated by exposure to natural talc.

Emergency and First-aid Procedures

Eye contact: Immediately flush with large amounts of water for 15 minutes (minimum), occasionally lifting eyelids; seek medical attention. Skin contact: Remove all contaminated clothing. Immediately wash area with large amounts of soap and water. For inhalation: Remove the person from exposure. Provide respiratory assistance and CPR. Transfer promptly to medical facility. Observe 48 hours for lung effects. If swallowed: Seek medical attention. Give person 1 to 2 glasses of water, then induce vomiting until vomitus is clear. Never give an unconscious or convulsing person anything by mouth.

SECTION VII - PRECAUTIONS FOR SAFE HANDLING AND USE

Steps to be Taken in Case Material is Released or Spilled

Ventilate area. Cleanup crew should wear breathing protection. Collect powdered materials in safest manner possible. Do NOT dry sweep. Use HEPA vacuum instead. Damp mop residue and collect for disposal. Emergency personnel should avoid contact with airborne dusts (especially fibrous talc).

Preferred Waste Disposal Method

No citation.

Precautions to be Taken in Handling and Storage

Store in tightly closed containers in cool, dry, well-ventilated area. Avoid handling activities that create dusty conditions. Respiratory protection should be worn by all personnel involved in handling and storage operations.

Other Precautions and Warnings

Prevent physical damage to containers.

SECTION VIII - CONTROL MEASURES AND PERSONAL PROTECTIVE EQUIPMENT

Respiratory Protection *(specify type)*

Use a half-mask respirator equipped with a high-efficiency particulate air (HEPA) filter under normal working conditions. For particularly dust conditions, a powered air-purifying respirator will provide greater protection.

Ventilation

Local exhaust or general mechanical systems recommended.

Protective Gloves	Eye Protection	Other Protective Clothing
Impervious Rubber	Chemical/Dust Goggles and/or Face Mask	Protective Apron

Work/Hygiene Practices

Always wash hands thoroughly after using chemical; never bring food, drink, or smoking materials into vicinity of chemicals.

TALC

$Mg_3Si_4O_{10}(OH)_2$

CAS: 14807-96-6

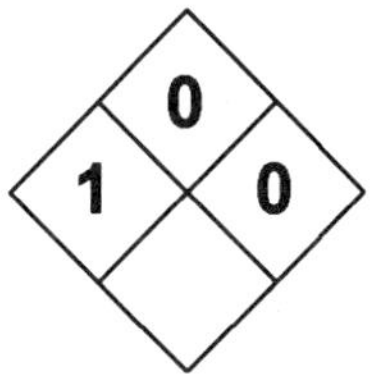

IDENTIFICATION AND TYPICAL USES

Talc is a white, yellow, apple-green, or gray powdered solid with no odor. The color depends on the impurity content from the mineral source. Talc has a lustrous, pearly or greasy feeling. It is usually soft and flexible. Non-fibrous talc is white to gray. It is used in ceramics, cosmetics, and pharmaceuticals. Its is a filler material in rubber, paints, soap, putty, plaster, and oilcloth. It is an adherent, a dusting agent, a lubricant, in paper, slate pencils and crayons, and in electrical insulation.

RISK ASSESSMENT: HEALTH

General Assessment

Non-fibrous talc (i.e., containing less than 1% asbestos) is generally considered a "nuisance dust" or "inert" dust because it is not expected to cause any chronic, irreversible, or long-term lung damage. However, fibrous talc has been known to cause chronic lung damage and, therefore, cannot be classified in the same manner. The IARC lists fibrous talc as a suspected human carcinogen based upon its asbestos content. There are no data on its mutagenic and teratogenic properties.

Primary exposure route is through *inhalation*. Ingestion can also occur, but is an unusual occupational exposure route. Skin or eye contact produces very few noticeable effects.

Inhaling talc dusts may produce minor to moderate health effects. Symptoms include cough, dypsnea, decreased pulmonary function, and, in high concentrations, lung fibrosis (talc pneumoconiosis). Exposure can also lead to the development of pulmonary edema (fluid in the lungs), which is a medical emergency and can be fatal. Symptoms can be delayed up to 48 hours, thereby creating a false sense of security with regard to health exposure risk.

Acute Health Effects

The following acute (short-term) health effects may occur immediately or shortly after exposure to talc:

Skin: No effect, to mild surface irritation.

Eye: Mild to moderate irritation.

Lung: Irritation of the mouth, nose, and throat causing coughing, dypsnea, decreased respiratory function, and respiratory irritation. Delayed pulmonary edema is possible, which can be fatal.

Chronic Health Effects

The following chronic (long-term) health effects may occur at some time after exposure to talc and can last for months or even years:

Cancer Hazards: According to information presented in the references, talc containing more than 1% asbestos (fibrous talc-asbestiform) is suspected of causing cancer in test animals. Non-fibrous talc is believed to be relatively inert.

Reproductive Hazards: There are no reports to support any claims of reproductive hazards following exposure to either form of talc.

Other Chronic Effects: Long-term or repeated inhalation exposure to natural talc dusts may lead to pneumoconiosis, decreased respiratory function, chronic illness, and diffuse pulmonary fibrosis.

Recommended Risk-Reduction Measures

Personnel should avoid direct contact with talc dusts and powders. Even though the non-fibrous form of talc is considered relatively inert (a nuisance dust), exposures should still be avoided. If a less toxic chemical cannot be substituted for a material believed to be hazardous, then *engineering controls* are the most effective method of reducing exposures. The best protection is to enclose operations and/or provide local exhaust ventilation at the site of chemical release. While not always operationally feasible, isolating operations can also reduce exposure. Using respiratory protection is less effective than the controls mentioned above, but is still advisable whenever working with talc. A half-mask respirator equipped with a high-efficiency particulate air (HEPA) filter and dust pre-filter should provide adequate protection at low level exposures. Better protection is obtained using an MSHA/NIOSH-approved powered supplied-air respirator operated in continuous flow mode or a self-contained breathing apparatus (SCBA) operated in pressure demand mode. A full facepiece, face shield, or chemical/dust goggles would provide additional protection for the eyes. Whenever an airborne dust hazard exists, a protective apron should be worn. It is common to find talc in a fine powder form which has a tendency to get under fingernails and into skin pores. To prevent hand and skin exposures, impervious gloves should be used.

Administrative controls should also be in place to minimize the potential for human exposures. These may include written procedures or policies which specify the methods and techniques that will be practiced whenever personnel are to work with talc.

All personnel should receive *training* on the use, hazards, protective measures, emergency actions, and other precautions per 29 CFR 1910.1200 (Hazard Communication), prior to the first assignment in an area where talc is used or stored.

Before beginning employment and at regular intervals thereafter (e.g., annually), the following medical tests are recommended:

☑ Lung function tests.

If symptoms develop or overexposure is suspected, the following may be useful:

☑ Consider chest X-ray after acute overexposure (may be negative if taken immediately after exposure due to delayed onset of pulmonary edema).

Any evaluation should include a careful history of past and present symptoms with an examination. Medical tests that simply look for existing damage are not a substitute for controlling exposures. Also, since smoking can cause heart disease, lung cancer, emphysema, and other respiratory disorders, smokers exposed to talc may experience symptoms more rapidly than non-smokers under the same conditions of exposure. Prudent risk management requires careful consideration of *all* possible risk factors.

Other methods to reduce exposure to chemicals include:

☑ Where possible, enclose operations and use local exhaust ventilation at the site of chemical release. If local exhaust ventilation or enclosure is not used, respiratory protection should be mandatory.

☑ Always ensure that proper protective clothing is worn when using chemical substances and that personnel are trained in its use and care.

☑ Wash thoroughly immediately after exposure to talc and at the end of the work shift or before eating, drinking, or smoking. Work clothing contaminated with talc should never be taken home for laundering. Only personnel trained in the hazards of exposure to chemical substances should launder contaminated clothing.

☑ Safety shower and eyewash stations should be readily available in work areas where talc is used or stored.

☑ Hazard warning information should be posted in the work area. In addition, as part of an on-going education and training program, all information on the health and safety hazards of talc should be communicated to all exposed and potentially exposed workers.

RISK ASSESSMENT: ENVIRONMENT

General Assessment

Talc is a relatively inert material that poses very little threat to environmental ecosystems. However, this is no excuse for the careless handling or control of talc which may lead to environmental contamination. The environment is always at risk of exposure during transportation, storage, disposal, or destruction of chemical substances, including talc. In almost every scenario, the threat of environmental exposure is contingent upon the proper handling of the material. Accidental releases, large or small, can result in the possible contamination of the surrounding environmental mediums (air, water, and soil).

Talc is considered to be a noncombustible solid and there are no known incompatible materials. However, prudent risk management still dictates caution and consideration during any emergency involving a leak or release of talc. Talc is a natural hydrous magnesium silicate. Compact, massive varieties may be called steatitc in distinction from the foliated varieties, which are called talc. Soapstone is an impure variety of steatite. Talc can also enter the environment through spills or leaks. Unchecked discharges from industrial sources are also possible.

Acute Ecological Effects

Acute (short-term) toxic effects may include the death of animals, birds, or fish, and death or low growth rate in plants. Acute effects are seen 2 to 4 days after animals or plants are exposed to talc.

Insufficient data are available to accurately evaluate or predict the short-term effects of talc exposure to aquatic life, plants, birds, or land animals.

Chronic Ecological Effects

Chronic (long-term) toxic effects may include shortened life span, reproductive problems, lower fertility, and changes in appearance or behavior in exposed animals. These effects can be seen long after first exposure(s) to toxic chemicals.

Insufficient data are available to evaluate or predict the long-term (chronic) effects of talc to aquatic life, plants, birds, or land animals.

Water Solubility

Talc is insoluble in water. Concentrations of 1 milligram will not mix with a liter of water.

Persistence in the Environment

Talc is highly persistent in water, with a half-life of greater than 200 days. The half-life of a pollutant is the amount of time it takes for one half of the chemical to be degraded. Virtually all of talc will eventually end up in terrestrial soil and aquatic sediments.

Bioaccumulation in Aquatic Organisms

Some substances increase in concentration, or *bioaccumulate,* in living organisms as they breathe contaminated air, drink contaminated water, or eat contaminated food. These chemicals can become concentrated in the tissues and internal organs of animals as well as humans.

The concentration of talc found in fish tissues is expected to be about the same as the average concentration of talc found in the water from which the fish was taken.

Recommended Risk-Reduction Measures

Proper training of all transporters will reduce the likelihood of a mishap or accident resulting in a leak or release to the environment. The correct use of labeling on all containers, trucks, and rail cars will enable emergency responders to react properly and quickly to any disaster thereby reducing the potential risk to the environment and to personnel.

The storage location for talc should be properly considered. Containers should be kept tightly closed and in a cool, dry, well-ventilated area. Dusty conditions in storage areas should be avoided.

If a spill or release to the environment has occurred, fire department, emergency response, and/or hazardous materials spill personnel should be notified immediately. Contaminated soils should be removed and replaced with clean soils. Cleanup should be attempted only by those trained in proper containment procedures. If talc should contact the water table, aquifer, or navigable waterway, time is of the essence. It is insoluble in water and total containment and remediation may be possible. When such spills occur, the local and/or state emergency response authorities should be notified. A comprehensive emergency response or disaster preparedness/recovery plan should be in place prior to any operations involving the use, transportation, storage, or disposal of talc. If talc is spilled or leaked, the following steps are recommended:

- ☑ Restrict persons not wearing protective clothing from area of spill or leak until cleanup is complete and area can be opened for normal work.
- ☑ Ventilate area of spill or leak.
- ☑ Avoid generating airborne dusts during cleanup operations.
- ☑ Collect powdered materials in safest manner possible. Hold for reclamation or arrange for disposal. Do NOT dry sweep (generates airborne dusts). Use a vacuum equipped with a high-efficiency particulate air (HEPA) filter instead. Damp mop residue and collect for disposal. Place in sealed containers.
- ☑ It may be necessary to dispose of talc as a hazardous waste. The responsible state agency or the regional office of the federal Environmental Protection Agency (EPA) should be contacted for specific recommendations on the disposal requirements for talc.

RISK ASSESSMENT: BUSINESS

General Assessment

Accidents or mishaps involving talc can present a moderate threat to business operations. Lawsuits that may result from personnel injury, public exposures, and environmental contamination may require a serious expenditure of resources. Media attention surrounding an injury, illness, or environmental damage can also result in a loss of profits and loss of current as well as future business. With regard to fibrous talc, it should always be remembered that anytime the terms "cancer" or "carcinogen" are used, public emotion, hysteria, ignorance, and anxiety can all run equally high. This should be carefully considered whenever drafting or implementing public relations policies.

Recommended Risk-Reduction Measures

Company attorneys, safety and health professionals, and environmental specialists should be involved in the development of any procedures dealing with safety or emergency response. A company official should be pre-designated as a public relations officer with specific training in dealing with the press. Corporate plans and policies should be developed, approved, and implemented long before any need for such arises.

REFERENCES

Ahlbom, A. 1993. *Biostatistics for Epidemiologists.* New York: CRC Press/Lewis Publishers.

American Conference of Governmental Industrial Hygienists. 1988. *Documentation of the Threshold Limit Values and Biological Exposure Indices,* 5th Edition (with updates). Cincinnati: ACGIH

Anthony, C. P., and N. J. Kolthoff. 1971. *Textbook of Anatomy and Physiology.* St. Louis: C. V. Mosby Company.

Asante-Duah, D. K. 1993. *Hazardous Waste Risk Assessment.* New York: CRC Press/Lewis Publishers.

Bloom, A. D. and F. J. DeSerres. 1995. *Ecotoxicity and Human Health.* New York: CRC Press/Lewis Publishers.

Burrell, R., D. K. Flaherty, and L. J. Sauers. 1992. *Toxicology of the Immune System.* New York: Van Nostrand Reinhold.

Calabrese, E. J. 1994. *Biological Effects of Low Level Exposures.* New York: CRC Press/Lewis Publishers.

Cockerman, L. G., B. S. Shane. 1994. *Basic Environmental Toxicology.* New York: CRC Press/Lewis Publishers.

Cothern, C. R. 1995. *Handbook for Environmental Risk Decisionmaking.* New York: CRC Press/Lewis Publishers.

Gardner, R. W. 1994. *Chemical Intolerance.* New York: CRC Press/Lewis Publishers.

Hallenbeck, W. H. 1993. *Quantitative Risk Assessment for Environmental and Occupational Health,* 2nd Edition. New York: CRC Press/Lewis Publishers.

Hodgson, E., R. B. Mailman, and J. E. Chambers. 1988. *Dictionary of Toxicology.* London: The MacMillian Company.

Hosty, J. W., and P. Foster. 1990. *A Practical Guide to Chemical Spill Response.* New York: Van Nostrand Reinhold.

Howard, P. H. 1995. *Handbook of Environmental Fate and Exposures Data for Organic Chemicals,* Vol. I-IV. New York: CRC Press/Lewis Publishers.

Howard, P. H., et. al. 1991. *Handbook of Environmental Degradation Rates.* New York: CRC Press/Lewis Publishers.

James, A., and M. Lord. 1992. *Index of Chemical and Physical Data.* New York: Van Nostrand Reinhold.

Jones, F. E. 1994. *Toxic Organic Vapors in the Workplace.* New York: CRC Press/Lewis Publishers.

Kamrin, M. A. 1988. *Toxicology: A Primer on Principles and Applications.* Chelsea, Michigan: Lewis Publishers.

Keith, L. H. and M. M. Walker. 1995. *Handbook of Air Toxics.* New York: CRC Press/Lewis Publishers.

Landis, W. G., M. H. Yu. 1995. *Introduction to Environmental Toxicology.* New York: CRC Press/Lewis Publishers.

Langerman, N. 1994 *Precautionary Labels for Chemical Containers.* New York: CRC Press/Lewis Publishers.

Lauwerys, R. R., P. Hoet. 1993. *Industrial Chemical Exposure.* New York: CRC Press/Lewis Publishers.

Lewis, R. J., Sr. 1993. *Hawley's Condensed Chemical Dictionary,* 12th Edition. New York: Van Nostrand Reinhold.

Lewis, R. J., Sr. 1992. *Sax's Dangerous Properties of Industrial Materials,* Volumes 1, 2, and 3, 8th Edition. New York: Van Nostrand Reinhold.

Lide, D. R. 1995. *Handbook of Chemistry and Physics,* 76th Edition. New York: CRC Press/Lewis Publishers.

Manahan, S. E. 1992. *Toxicological Chemistry,* 2nd Edition. New York: CRC Press/Lewis Publishers.

Meyer, E. 1990. *Chemistry of Hazardous Materials.* Englewood Cliffs, NJ: Prentice Hall.

Patnaik, P. 1992. *A Comprehensive Guide to the Hazardous Properties of Chemical Substances.* New York: Van Nostrand Reinhold.

Philp, R. B. 1995. *Environmental Hazards and Human Health.* New York: CRC Press/Lewis Publishers.

Proctor, N. H., and J. P. Hughes. 1991. *Chemical Hazards of the Workplace.* New York: Van Nostrand Reinhold.

Quigley, D. R. 1994. *Handbook of Emergency Chemical Management.* New York: CRC Press/Lewis Publishers.

Rea, W. J. 1995. *Chemical Sensitivity,* Vol. I-III. New York: CRC Press/Lewis Publishers.

Richardson, M. 1995. *Dictionary of Substance and Their Effects,* Vol. I-VI. New York: CRC Press/Lewis Publishers.

MATERIAL SAFETY DATA SHEET

CHEMICAL NAME	
TEREPHTHALIC ACID	**CRC PRESS/LEWIS PUBLISHERS DISCLAIMER:** THE INFORMATION AND RECOMMENDATIONS PRESENTED HEREIN ARE BASED ON SOURCES BELIEVED TO BE RELIABLE. CRC MAKES NO REPRESENTATION ON ITS COMPLETENESS OR ACCURACY. IT IS THE RESPONSIBILITY OF THE USER TO DETERMINE THE CHEMICAL'S SUITABILITY FOR ITS INTENDED USE, THE CHEMICAL'S SAFE USE, AND THE CHEMICAL'S PROPER DISPOSAL. NO REPRESENTATIONS AND/OR WARRANTIES, EITHER EXPRESSED OR IMPLIED, OF THE MERCHANTABILITY OR FITNESS FOR A PARTICULAR PURPOSE, OR OF ANY OTHER NATURE, ARE MADE WITH RESPECT TO THE INFORMATION PROVIDED IN THIS MSDS OR TO THE CHEMICAL TO WHICH INFORMATION MAY REFER. CRC NEITHER ASSUMES NOR AUTHORIZES ANY OTHER PERSON TO ASSUME FOR IT, ANY OTHER ADDITIONAL RESPONSIBILITY OR LIABILITY FOR THE USE OF, OR RELIANCE UPON, THIS INFORMATION.

HAZARD WARNING INFORMATION

HEALTH	FIRE	REACTIVE	OTHER	DEGREE OF HAZARD	COLOR CODING	OTHER CODES
0	1	0		0 = Minimum Hazard 1 = Slight Hazard 2 = Moderate Hazard 3 = Serious Hazard 4 = Severe Hazard	HEALTH = BLUE FIRE = RED REACTIVITY = YELLOW OTHER = WHITE	OX = Oxidizer ACID = Acid ALK = Alkali COR = Corrosive ~~W~~ = Use No Water

SECTION I - GENERAL INFORMATION

Characterization	**Acid**	RCRA Number	**None**
		EPA Class	**Not Applicable**
DOT Proper Shipping Name	**No Citation**	Chemical Abstract Service (CAS) Number	**100-21-0**
DOT Hazard Class and Label Requirements	**No Citation**	DOT Emergency Guide Code	**No Citation**
DOT Identification Number	**No Citation**	Chemical Formula	$C_8H_6O_4$

Synonyms

TPA; 1,4-benzenedicarboxylic acid; *p*-benzenedicarboxylic acid; *p*-phthalic acid; TA 12.

SECTION II - HAZARDOUS INGREDIENTS AND IDENTITY INFORMATION

Hazardous Components (specific identity)	OSHA Exposure Criteria	NIOSH Exposure Criteria	Immediately Dangerous to Life and Health (IDLH)	ACGIH Exposure Criteria
Terephthalic acid (derivation: By oxidation of p-xylene or of mixed xylenes and other alkyl aromatics; or by reacting benzene and potassium carbonate over a cadmium catalyst).	**PEL:** Not Established **STEL:** Not Established	**REL:** Not Established **STEL:** Not Established	**Not Established**	**TLV:** Not Established **STEL:** Not Established

SECTION III - PHYSICAL AND CHEMICAL CHARACTERISTICS

Boiling Point	**572°F (300°C) Sublimes**	Specific Gravity (H_2O = 1)	**1.5**
Vapor Pressure (mm Hg)	**0.5 at 248°F (120°C)**	Molecular Weight	**166.14**
Vapor Density (Air = 1)	**Not Reported**	Melting Point	**Not Reported**

Solubility

Insoluble in water, chloroform, ether, acetic acid; slightly soluble in alcohol; soluble in alkalis.

Appearance and Odor

White crystalline (sand-like) solid or powder with a slight acrid odor.

SECTION IV - FIRE AND EXPLOSION HAZARD DATA

Flash Point (method used)	**500°F (260°C) opened cup**	Explosive Limits in Air % by Volume	**LEL: Not Determined UEL: Not Determined**
NFPA Classification	**Combustible Solid**	Autoignition Temperature	**925°F (496°C)**

Extinguishing Media

Carbon dioxide, dry chemical, water spray, or alcohol foam.

Special Fire Fighting Procedures

Wear full protective clothing and self-contained breathing apparatus (SCBA). Approach fire from upwind position. Use a water spray to keep fire-exposed containers cool. Poisonous gases are produced in fire.

Unusual Fire and Explosion Hazards

Containers may explode in fire. In solid form, it presents a slight fire and explosion hazard. If finely divided dust particles are dispersed in air as a dust cloud, a source of ignition (open flame, spark, static charge, excessive heat) can ignite an intense, violent dust explosion.

SECTION V - REACTIVITY DATA

Stability		Conditions to Avoid
		Terephthalic acid is normally stable under routine conditions of handling and storage. Avoid contact with incompatible materials and keep away from heat, especially the powder form. Do NOT generate dust.
Stable	Unstable	Incompatibility *(materials to avoid)*
X		Violent reactions can occur if terephthalic acid should contact strong oxidizers (such as chlorine, bromine, and fluorine) and strong alkaline materials.
Hazardous Polymerization		Conditions to Avoid Under normal conditions of temperature and pressure, hazardous polymerization of terephthalic acid cannot occur.
May Occur	Will Not Occur	Hazardous Decomposition or By-products
	X	When heated to decomposition, terephthalic acid can emit highly toxic/poisonous fumes including carbon monoxide.

SECTION VI - HEALTH HAZARD DATA

Primary Route(s) of Entry:	Inhalation?	Absorption (skin)?	Ingestion?
	X		X

Health Hazards

INHALATION: Extremely irritating to the eyes, nose, throat, and respiratory system. Inhaled dusts may cause bloody nasal discharge. Lung effects are not common, but are possible and can include congestion, difficulty breathing, and pulmonary edema (remote possibility). Repeated exposure may damage the kidneys.

EYES & SKIN: Skin contact can result in irritation with rash and burning feeling on contact. Can cause severe eye burns and/or irritation.

INGESTION: Mildly toxic. May cause irritation of the mouth, through, and stomach.

Carcinogenicity	NTP Listed?	IARC Cancer Review Group?	OSHA Regulated?	Target Organs?
Unknown Human Unknown Animal	No	No	No	Skin, eyes, respiratory system.

Medical Conditions Generally Aggravated by Exposure

None Reported.

Emergency and First-aid Procedures

Eye contact: Flush immediately with water for 15 minutes (minimum); seek medical attention. Skin contact: Remove all contaminated clothing. Immediately wash area with large amounts of soap and water. Seek medical assistance. For inhalation: Remove the person from exposure. Provide respiratory assistance and CPR. Transfer to medical facility. Observe for 48 hour for lungs effects. Pulmonary edema is not probable but is possible with irritating substances such as terephthalic acid. If swallowed: Seek medical attention immediately.

SECTION VII - PRECAUTIONS FOR SAFE HANDLING AND USE

Steps to be Taken in Case Material is Released or Spilled

Collect powdered materials in most convenient and safe manner possible and deposit in sealed containers. Ventilate area of spill or leak. Restrict those not involved in cleanup from entering area.

Preferred Waste Disposal Method

No citation.

Precautions to be Taken in Handling and Storage

Terephthalic acid is a combustible solid. Avoid contact with strong oxidizers such as chlorine since violent reactions occur. Store in tightly closed containers in cool, well-ventilated area away from heat.

Other Precautions and Warnings

Sources of ignition are prohibited where terephthalic acid is used, handled, or stored.

SECTION VIII - CONTROL MEASURES AND PERSONAL PROTECTIVE EQUIPMENT

Respiratory Protection *(specify type)*

No exposure levels established. Use an MSHA/NIOSH-approved supplied-air respirator with full facepiece in continuous flow mode for the best protection. Or use a half-mask respirator with acid cartridge and a high efficiency particulate air (HEPA) filter for low or transient exposures (cartridge should be designed to protect against acid dust or mist).

Ventilation

Local exhaust or general mechanical systems recommended.

Protective Gloves	Eye Protection	Other Protective Clothing
Acid resistant synthetic rubber	Chemical/Dust Goggles or Face Mask	Rubber Apron

Work/Hygiene Practices

Always wash hands thoroughly after using chemical; never bring food, drink, or smoking materials into vicinity of chemicals.

TEREPHTHALIC ACID

$C_6H_4(COOH)_2$

CAS: 100-21-0

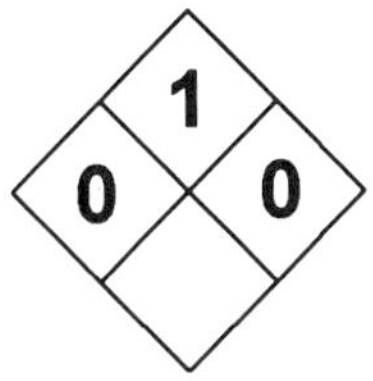

IDENTIFICATION AND TYPICAL USES

Terephthalic acid is a white or crystalline powder with a slight acrid odor. It is used in the production of linear, crystalline polyester resins, fibers, and films by combination with glycols. It is also used as a reagent for alkali in wool, and as an additive to poultry feeds.

RISK ASSESSMENT: HEALTH

General Assessment

Terephthalic acid is toxic by *inhalation* and *ingestion.* It is also a skin and eye irritant. There are no data regarding its carcinogenic, mutagenic, or teratogenic properties in humans or test animals.

Inhalation of terephthalic acid dusts can irritate the nose, throat, and respiratory tract. This may cause nose bleeds or bloody nasal discharge, sore throat, hoarseness, and/or cough with phlegm. Irritating substances such as terephthalic acid may be injurious to the lungs and it is possible that high acute exposures may cause the buildup of fluids in the lungs (pulmonary edema) which can be fatal Symptoms may be delayed up to 48 hours thereby creating a false sense of security with regard to health exposure risk. High or repeated exposure can damage the kidneys.

Skin contact may cause irritation and burning feeling on contact. Prolonged eye contact may be damaging to the eye and associated tissues.

Acute Health Effects

The following acute (short-term) health effects may occur immediately or shortly after exposure to terephthalic acid:

Skin: Possible irritation/rash or burning feeling.

Eye: Severe irritation; may cause burns and damage.

Lung: Irritation of the nose, throat, and lungs (bronchial tubes) following exposure. Symptoms may be delayed for several hours (pulmonary edema).

Chronic Health Effects

The following chronic (long-term) health effects may occur at some time after exposure to terephthalic acid and can last for months or even years:

Cancer Hazards: According to information presented in the references, terephthalic acid has not been adequately tested for its ability to cause cancer in test animals.

Reproductive Hazard: According to the data presented in the references, terephthalic acid has not been adequately tested for its ability to adversely affect reproduction.

Other Chronic Effects: Repeated or long-term skin contact may cause rash from irritation. Repeated exposure can also damage the kidneys. Very irritating substance can damage the lungs. It is not known whether long-term exposure to terephthalic acid will cause permanent damage to the lungs.

Recommended Risk-Reduction Measures

Personnel should avoid direct contact with terephthalic acid. The exact nature of its human toxicity is not clearly understood in the references. If a less toxic chemical cannot be substituted for a hazardous substance, then *engineering controls* are the most effective method of reducing exposures.

The best protection is to enclose operations and/or provide local exhaust ventilation at the site of chemical release. While not always operationally feasible, isolating operations can also reduce exposure. Using respiratory protection is less effective than the controls mentioned above, but is still advisable whenever working with or around terephthalic acid. No exposure levels have been established for this chemical. This does not mean that exposure is without health risk. For the best protection, a supplied-air respirator with a full facepiece operated in positive pressure mode or an MSHA/NIOSH-approved self-contained breathing apparatus (SCBA) with full facepiece and operated in pressure demand are the recommended respiratory protection methods of choice. For lower exposures, an MSHA/NIOSH-approved full facepiece respirator with high-efficiency particulate air (HEPA) filter may suffice. In either case, if a full facepiece is not available, then chemical goggles should be worn to protect the eyes. Whenever a chemical/dust splash hazard exists, a face shield and a rubber apron should be worn. To prevent hand and skin exposures, acid resistant gloves should be used. Glove manufacturers should be contacted and permeation studies obtained *before* final glove selections are made.

Administrative controls should also be in place to minimize the potential for human exposures. These may include written procedures or policies which specify the methods and techniques that will be practiced whenever personnel are to work with terephthalic acid.

All personnel should receive *training* on the use, hazards, protective measures, emergency actions, and other precautions per 29 CFR 1910.1200 (Hazard Communication), prior to the first assignment in an area where terephthalic acid is used or stored.

Before beginning employment and at regular intervals thereafter (e.g., annually), for those with frequent or potentially high exposures, the following medical tests are recommended:

- ☑ Lung function tests.
- ☑ Kidney function tests.

If symptoms develop or overexposure is suspected, the following may also be useful:

- ☑ Kidney function tests.
- ☑ Consider chest X-ray after acute overexposure (may be negative if taken immediately after exposure due to delayed onset of pulmonary edema).

Other methods to reduce exposure include:

- ☑ Where possible, enclose operations and use local exhaust ventilation at the site of chemical release. If local exhaust ventilation or enclosure is not used, respiratory protection should be mandatory.
- ☑ Where possible, automatically transfer terephthalic acid from drums or other storage containers to process containers.
- ☑ Always ensure that proper protective clothing is worn when using chemical substances and that personnel have been trained in its use, care, and maintenance.
- ☑ Wash thoroughly immediately after exposure to terephthalic acid and at the end of the work shift or before eating, drinking, or smoking.
- ☑ Hazard warning information should be posted in the work area. In addition, as part of an on-going education and training program, all information on the health and safety hazards of terephthalic acid should be communicated to all exposed workers.

RISK ASSESSMENT: ENVIRONMENT

General Assessment

The environment is at risk of exposure during transportation, storage, disposal, or destruction of terephthalic acid. In almost every scenario, the threat of environmental exposure is contingent upon the proper handling of the chemical substance. Accidental spills, large or small, can result in fire, explosion, and possible contamination of the surrounding environmental mediums (water, soil, and air).

Terephthalic acid is considered a combustible solid. It is incompatible with oxidizing agents and contact can cause fire or explosion. These characteristics require special consideration during any emergency situation involving a leak or spill of terephthalic acid. Should terephthalic acid ever come into contact with incompatible substances such as strong oxidizers, strong bases, and strong reducing agents, violent and explosive reactions are possible.

Terephthalic acid can enter the environment mainly through industrial discharges or spills.

☠ *Acute Ecological Effects*

Acute (short-term) toxic effects may include the death of animals, birds, or fish and death or low growth rate in plants. Acute effects are seen 2 to 4 days after animals or plants are exposed to terephthalic acid.

Terephthalic acid has moderate acute toxicity to aquatic life. Insufficient data are available to evaluate or predict the acute (short-term) effects of terephthalic acid exposure to plants, birds, or land animals.

Chronic Ecological Effects

Chronic (long-term) toxic effects may include shortened life span, reproductive problems, lower fertility, and changes in appearance or behavior in exposed animals.

These effects can be seen long after first exposure(s) to toxic chemicals.

Terephthalic acid has slight chronic toxicity in aquatic life. Insufficient data are available on the long-term effects of terephthalic acid to plants, birds, or land animals.

♦ *Water Solubility*

Terephthalic acid is moderately soluble in water. Concentrations of 1 to 1000 milligrams will mix with a liter of water.

⌛ *Persistence in the Environment*

Terephthalic acid is slightly persistent in the aquatic environment, with a half-life of between 2 to 20 days. The half-life of a pollutant is the amount of time it takes for one half of the chemical to be degraded.

Approximately 93% of terephthalic acid will eventually end up in water; about 6% will end up in air, and the remainder will be divided about equally between terrestrial soils and aquatic sediments.

Bioaccumulation in Aquatic Organisms

Some substances increase in concentration, or *bioaccumulate,* in living organisms as they breathe contaminated air, drink contaminated water, or eat contaminated food. These chemicals can become concentrated in the tissues and internal organs of animals as well as humans.

The concentration of terephthalic acid found in fish tissues is expected to be somewhat higher than the average concentration of terephthalic acid in the water from which the fish was taken.

Recommended Risk-Reduction Measures

Proper training of all transporters will reduce the likelihood of a mishap or accident resulting in a leak or spill to the environment. The correct use of DOT labeling (per the MSDS) on all containers, trucks, and rail cars will enable emergency responders to react properly and quickly to any disaster thereby reducing the potential risk to the environment and to personnel.

Storage of terephthalic acid should be segregated from incompatible chemicals to minimize the risk of cross-contamination or contact.

If a spill or leak to the environment has occurred, fire department, emergency response and/or hazardous materials spill personnel should be notified immediately. Cleanup should be attempted only by those trained in proper spill containment procedures. Contaminated soils should be removed for incineration and replaced with clean soil. If terephthalic acid should contact the water table, aquifer, or navigable waterway, time is of the essence. It is moderately soluble in water and, therefore, total containment and remediation may not be possible. When such spills occur, the local and/or state emergency response authorities must be notified. A comprehensive emergency response or disaster preparedness/recovery plan should be in place prior to any operations involving the use, transportation, storage, or disposal of terephthalic acid. If terephthalic acid is spilled or leaked, the following specific steps are recommended:

- ☑ Restrict persons not wearing protective clothing from area of spill or leak until cleanup is complete and area can be opened for normal work.
- ☑ Remove all ignition sources.
- ☑ If liquid solution is spilled, absorb in vermiculite, dry sand, earth, or similar material and deposit in sealed containers. If powders are spilled, collect in most convenient manner possible and dispose of in sealed drum. Recommend use of vacuum equipped with a high-efficiency particulate air (HEPA) filter. Do not dry sweep.
- ☑ It may be necessary to dispose of terephthalic acid as a hazardous waste. The responsible state agency or the regional office of the federal Environmental Protection Agency (EPA) should be contacted for specific recommendations.

RISK ASSESSMENT: BUSINESS

General Assessment

Accidents or mishaps involving terephthalic acid can present a significant threat to business operations. The loss or damage of equipment or facilities can significantly affect fiscal viability. Lawsuits that may result from personnel injury/death, public exposures, and/or environmental contamination will also require a serious expenditure of resources. Media attention surrounding an injury, death, or environmental damage can also result in a loss of profits and loss of current as well as future business.

Recommended Risk-Reduction Measures

Company attorneys, safety and health professionals, and environmental specialists should be involved in the

development of any procedures dealing with safety or emergency response. A company official should be pre-designated as a public relations officer with specific training in dealing with the press. Corporate plans and policies should be developed, approved, and implemented long before any need for such arises.

REFERENCES

Ahlbom, A. 1993. *Biostatistics for Epidemiologists.* New York: CRC Press/Lewis Publishers.

American Conference of Governmental Industrial Hygienists. 1988. *Documentation of the Threshold Limit Values and Biological Exposure Indices,* 5th Edition (with updates). Cincinnati: ACGIH

Asante-Duah, D. K. 1993. *Hazardous Waste Risk Assessment.* New York: CRC Press/Lewis Publishers.

Bloom, A. D. and F. J. DeSerres. 1995. *Ecotoxicity and Human Health.* New York: CRC Press/Lewis Publishers.

Brusick, D. J. 1994. *Methods for Genetic Risk Assessment.* New York: CRC Press/Lewis Publishers.

Bueche, J. R. 1972. *Principles of Physics,* 2nd Edition. New York: McGraw-Hill.

Burrell, R., D. K. Flaherty, and Leonard J. Sauers. 1992. *Toxicology of the Immune System.* New York: Van Nostrand Reinhold.

Calabrese, E. J. 1994. *Biological Effects of Low Level Exposures.* New York: CRC Press/Lewis Publishers.

Cockerman, L. G., B. S. Shane. 1994. *Basic Environmental Toxicology.* New York: CRC Press/Lewis Publishers.

Cothern, R. R. 1993. *Comparative Environmental Risk Assessment.* New York: CRC Press/Lewis Publishers.

Danse, I. R. 1991. *Common Sense Toxics in the Workplace.* New York: Van Nostrand Reinhold.

Gardner, R. W. 1994. *Chemical Intolerance.* New York: CRC Press/Lewis Publishers.

Hallenbeck, W. H. 1993. *Quantitative Risk Assessment for Environmental and Occupational Health,* 2nd Edition. New York: CRC Press/Lewis Publishers.

Hosty, J. W., P. Foster. 1990. *A Practical Guide to Chemical Spill Response.* New York: Van Nostrand Reinhold.

Howard, P. H., et. al. 1991. *Handbook of Environmental Degradation Rates.* New York: CRC Press/Lewis Publishers.

James, A., M. Lord. 1992. *Index of Chemical and Physical Data.* New York: Van Nostrand Reinhold.

Jones, F. E. 1994. *Toxic Organic Vapors in the Workplace.* New York: CRC Press/Lewis Publishers.

Keith, L. H. and M. M. Walker. 1995. *Handbook of Air Toxics.* New York: CRC Press/Lewis Publishers.

Landis, W. G., M. H. Yu. 1995. *Introduction to Environmental Toxicology.* New York: CRC Press/Lewis Publishers.

Langerman, N. 1994 *Precautionary Labels for Chemical Containers.* New York: CRC Press/Lewis Publishers.

Lauwerys, R. R., P. Hoet. 1993. *Industrial Chemical Exposure.* New York: CRC Press/Lewis Publishers.

Lewis, R. J., Sr. 1993. *Hawley's Condensed Chemical Dictionary,* 12th Edition. New York: Van Nostrand Reinhold.

Lewis, R. J., Sr. 1992. *Sax's Dangerous Properties of Industrial Materials,* Volumes 1, 2, and 3, 8th Edition. New York: Van Nostrand Reinhold.

Lide, D. R. 1995. *Handbook of Chemistry and Physics,* 76th Edition. New York: CRC Press/Lewis Publishers.

Manahan, S. E. 1994. *Environmental Chemistry,* 6th Edition. New York: CRC Press/Lewis Publishers.

Manahan, S. E. 1992. *Toxicological Chemistry,* 2nd Edition. New York: CRC Press/Lewis Publishers.

Mellström, G. A., J. E. Wahlberg, and H. I. Maibach. 1994. *Protective Gloves for Occupational Use.* New York: CRC Press/Lewis Publishers.

Meyer, E. 1990. *Chemistry of Hazardous Materials.* Englewood Cliffs, NJ: Prentice Hall.

Mycock, J. C., J. D. McKenna, and L. Theodore. 1995. *Handbook of Air Pollution Control Engineering and Technology.* New York: CRC Press/Lewis Publishers.

Nelson, G. O. 1992. *Gas Mixtures.* New York: CRC Press/Lewis Publishers.

Ott, W. R. 1995. *Environmental Statistics and Data Analysis.* New York: CRC Press/Lewis Publishers.

Patnaik, P. 1992. *A Comprehensive Guide to the Hazardous Properties of Chemical Substances.* New York: Van Nostrand Reinhold.

Philp, R. B. 1995. *Environmental Hazards and Human Health.* New York: CRC Press/Lewis Publishers.

Proctor, N. H., J. P. Hughes. 1991. *Chemical Hazards of the Workplace.* New York: Van Nostrand Reinhold.

Quigley, D. R. 1994. *Handbook of Emergency Chemical Management.* New York: CRC Press/Lewis Publishers.

Rea, W. J. 1995. *Chemical Sensitivity,* Vol. I-III. New York: CRC Press/Lewis Publishers.

Richardson, M. 1995. *Dictionary of Substance and Their Effects,* Vol. I-VI. New York: CRC Press/Lewis Publishers.

Rylander, R., R. R. Jacobs. 1994. *Organic Dusts.* New York: CRC Press/Lewis Publishers.

Sacarello, H. L. A. 1994. *The Comprehensive Handbook of Hazardous Materials.* New York: CRC Press/Lewis Publishers.

Sherman, J. D. 1988. *Chemical Exposure and Disease.* New York: Van Nostrand Reinhold.

Witorsch, P. and S. V. Spagnolo. 1994. *Air Pollution and Lung Disease in Adults.* New York: CRC Press/Lewis Publishers.

MATERIAL SAFETY DATA SHEET

CHEMICAL NAME	
1,1,2,2-TETRACHLOROETHANE	**CRC PRESS/LEWIS PUBLISHERS DISCLAIMER:** THE INFORMATION AND RECOMMENDATIONS PRESENTED HEREIN ARE BASED ON SOURCES BELIEVED TO BE RELIABLE. CRC MAKES NO REPRESENTATION ON ITS COMPLETENESS OR ACCURACY. IT IS THE RESPONSIBILITY OF THE USER TO DETERMINE THE CHEMICAL'S SUITABILITY FOR ITS INTENDED USE, THE CHEMICAL'S SAFE USE, AND THE CHEMICAL'S PROPER DISPOSAL. NO REPRESENTATIONS AND/OR WARRANTIES, EITHER EXPRESSED OR IMPLIED, OF THE MERCHANTABILITY OR FITNESS FOR A PARTICULAR PURPOSE, OR OF ANY OTHER NATURE, ARE MADE WITH RESPECT TO THE INFORMATION PROVIDED IN THIS MSDS OR TO THE CHEMICAL TO WHICH INFORMATION MAY REFER. CRC NEITHER ASSUMES NOR AUTHORIZES ANY OTHER PERSON TO ASSUME FOR IT, ANY OTHER ADDITIONAL RESPONSIBILITY OR LIABILITY FOR THE USE OF, OR RELIANCE UPON, THIS INFORMATION.

HAZARD WARNING INFORMATION

HEALTH	FIRE	REACTIVE	OTHER	DEGREE OF HAZARD	COLOR CODING	OTHER CODES
2	0	0		0 = Minimum Hazard 1 = Slight Hazard 2 = Moderate Hazard 3 = Serious Hazard 4 = Severe Hazard	HEALTH = BLUE FIRE = RED REACTIVITY = YELLOW OTHER = WHITE	OX = Oxidizer ACID = Acid ALK = Alkali COR = Corrosive W = Use No Water

SECTION I - GENERAL INFORMATION

Characterization	**Acetylene Compound; Chloride**	RCRA Number	**U209**
		EPA Class	**Toxic Waste**
DOT Proper Shipping Name	**Tetrachloroethane**	Chemical Abstract Service (CAS) Number	**79-34-5**
DOT Hazard Class and Label Requirements	**ORM-A**	DOT Emergency Guide Code	**55**
DOT Identification Number	**UN 1702**	Chemical Formula	**$CHCl_2CHCl_2$**

Synonyms
Acetylene tetrachloride; symmetrical tetrachloroethane; bonoform.

SECTION II - HAZARDOUS INGREDIENTS AND IDENTITY INFORMATION

Hazardous Components (specific identity)	OSHA Exposure Criteria	NIOSH Exposure Criteria	Immediately Dangerous to Life and Health (IDLH)	ACGIH Exposure Criteria
1,1,2,2-Tetrachloroethane (derivation: By reaction of acetylene and chlorine and subsequent distillation). **1 ppm = 7.00 mg/m³**	**PEL (skin):** **5 ppm** **35 mg/m³** **STEL:** **Not Established**	**REL (skin):** **1 ppm** **7 mg/m³** **Reduce Lowest to Lowest Level** **Possible Human Carcinogen**	**100 ppm**	**TLV:** **1 ppm** **7 mg/m³** **STEL:** **Not Established**

SECTION III - PHYSICAL AND CHEMICAL CHARACTERISTICS

Boiling Point	**295°F (146°C)**	Specific Gravity (H_2O = 1)	**1.59**
Vapor Pressure (gas)	**5 at 68°F (20°C)**	Molecular Weight	**167.9**
Vapor Density (Air = 1)	**Not Reported**	Melting Point	**-33°F (-36°C)**

Solubility
Slightly soluble in water. Soluble in alcohol and ether.

Appearance and Odor
Colorless to pale-yellow, heavy, mobile liquid with a sweet, suffocating, pungent, chloroform-like odor. Odor Threshold = <3 ppm.

SECTION IV - FIRE AND EXPLOSION HAZARD DATA

Flash Point (method used)	**Not Applicable**	Explosive Limits in Air % by Volume	**LEL: Not Determined UEL: Not Determined**
NFPA Classification	**Non-Combustible Liquid**	Autoignition Temperature	**Not Determined**

Extinguishing Media
Use agent suitable to surrounding area. 1,1,2,2-Tetrachloroethane does not burn.

Special Fire Fighting Procedures
Wear full protective clothing and self-contained breathing apparatus (SCBA). Poisonous gases are produced in fire. Use water spray to keep fire-exposed containers cool.

Unusual Fire and Explosion Hazards
None reported.

SECTION V - REACTIVITY DATA

Stability		Conditions to Avoid: Normally stable in closed containers under routine conditions of handling and storage. Avoid contact with incompatible materials.
Stable	Unstable	Incompatibility *(materials to avoid)*
X		Incompatible with chemically-active metals (such as potassium), strong caustics (such as potassium hydroxide, fuming sulfuric acid, nitrogen tetroxide.
Hazardous Polymerization		Conditions to Avoid: Under normal conditions of temperature and pressure, hazardous polymerization of 1,1,2,2-tetrachloroethane is not expected to occur. It slowly decomposes when exposed to air.
May Occur	Will Not Occur	Hazardous Decomposition or By-products
	X	When heated to decomposition, 1,1,2,2-tetrachloroethane can emit highly toxic/poisonous and acrid fumes and gases, including chlorine oxides and carbon monoxide.

SECTION VI - HEALTH HAZARD DATA

Primary Route(s) of Entry:	Inhalation?	Absorption (skin)?	Ingestion?
	X	X	X

Health Hazards

INHALATION: Systemic effects include lachrymation, salivation, irritation of the eyes, nose, and throat, nausea, vomiting, narcosis, low blood pressure, abnormal cardiac rhythms, respiratory depression, anesthesia, dizziness, loss of consciousness, coma, liver and kidney injury. May cause tremor, jaundice, hepatitis, and monocytosis.

ABSORPTION: Irritating to the skin and may produce dermatitis on repeated exposures. Absorbed in toxic amounts through intact skin. Eye contact may injure cornea.

INGESTION: Irritation and narcosis with burning of the mouth and lips. Can cause irregular heartbeat, nausea and vomiting, bloody diarrhea, drowsiness, unconsciousness, and coma.

Carcinogenicity	NTP Listed?	IARC Cancer Review Group?	OSHA Regulated?	Target Organs?
Questioned Human Confirmed Animal	No	Group 3	29 CFR 1910.1000 Table Z-1	Eyes, skin, resp. system, liver, GI, kidney, CNS.

Medical Conditions Generally Aggravated by Exposure

None reported. Skin conditions (dermatitis) or respiratory problems (asthma) may be aggravated.

Emergency and First-aid Procedures

Eye contact: Flush immediately with water for 15 minutes (minimum); seek medical attention. Skin contact: Remove all contaminated clothing. Immediately wash area with large amounts of soap and water. Seek medical assistance. For inhalation: Remove the person from exposure. Provide respiratory assistance and CPR, if required. Transfer to medical facility. Keep victim warm and at rest. If swallowed: Not a probable exposure route. Give conscious person 1 to 2 glasses of water and induce vomiting. Repeat until vomitus is clear. Seek medical attention immediately.

SECTION VII - PRECAUTIONS FOR SAFE HANDLING AND USE

Steps to be Taken in Case Material is Released or Spilled

Absorb large spills in vermiculite, dry sand, earth, or similar material and deposit in sealed containers. Ventilate area of spill or leak. Restrict those not involved in cleanup from entering area.

Preferred Waste Disposal Method

Mix with combustible solvent and burn in chemical incinerator with an afterburner and scrubber.

Precautions to be Taken in Handling and Storage

Store to avoid contact with incompatible materials in tightly closed containers in cool, well-ventilated area. Provide adequate ventilation in storage areas to keep concentrations below exposure limits.

Other Precautions and Warnings

Prevent physical damage to storage containers.

SECTION VIII - CONTROL MEASURES AND PERSONAL PROTECTIVE EQUIPMENT

Respiratory Protection *(specify type)*

For exposures above the PEL, use an MSHA/NIOSH-approved supplied-air respirator with full facepiece in continuous flow mode, or use a SCBA operated in pressure demand or other positive pressure mode.

Ventilation

Local exhaust or general mechanical systems recommended.

Protective Gloves	Eye Protection	Other Protective Clothing
Impervious Material	Chemical Goggles or Face Mask	Protective Apron

Work/Hygiene Practices

Always wash hands thoroughly after using chemical; never bring food, drink, or smoking materials into vicinity of chemicals.

1,1,2,2-TETRACHLOROETHANE

$CHCl_2CHCl_2$

CAS: 79-34-5

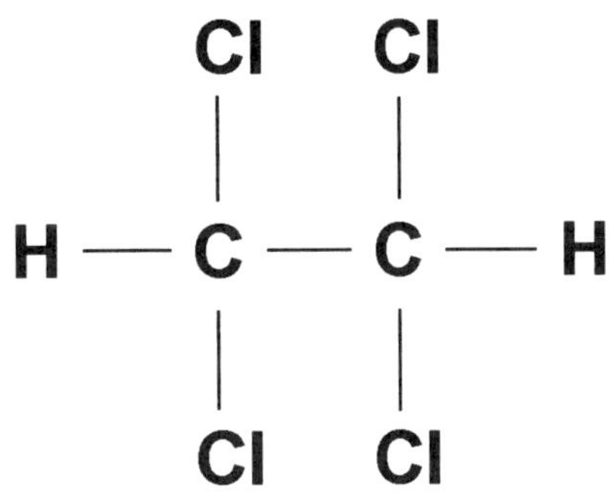

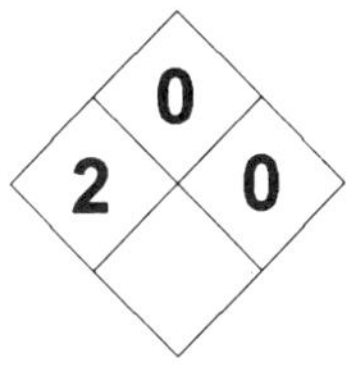

IDENTIFICATION AND TYPICAL USES

1,1,2,2-Tetrachloroethane is a colorless to pale-yellow liquid with a pungent, sweet, suffocating, chloroform-like odor. It is used as a solvent, for cleansing and degreasing metals, in paint removers, in varnishes and lacquers, in photographic film, in resins and waxes, in the extraction of oils and fats, as an alcohol denaturant, in organic synthesis, in insecticides, as a weed killer, a fumigant, and as an intermediate in the manufacture of other chlorinated hydrocarbons.

RISK ASSESSMENT: HEALTH

General Assessment

1,1,2,2-Tetrachloroethane is an extremely toxic compound to humans by *inhalation*, dermal *absorption*, and *ingestion*. Ingestion can occur, but is rare under most circumstances. It is listed as a confirmed animal carcinogen and its human carcinogenic potential is considered questioned by NIOSH. It has been shown to cause mutations in animals. Experimental teratogenic data have not been able to confirm or deny adverse effects on reproduction or the developing fetus. 1,1,2,2-Tetrachloroethane is considered one of the most toxic of the common chlorinated hydrocarbons, particularly with respect to toxic effects on the liver. It can cause hepatitis and jaundice.

Inhalation results in slight smarting of the eyes with lachrymation and slight irritation to the throat. There can be salivation, nausea, vomiting, and narcosis. It can also cause a decrease in blood pressure and cardiac rhythm abnormalities, respiratory depression, and central nervous (CNS) system effects. Symptoms include intoxication, loss of coordination, numbness around the mouth, dizziness, narcosis, and convulsions. It can also cause flushing of the face and neck, headache, somnolence, hallucinations, distorted perceptions, and slurred speech.

Skin contact may produce mild to moderate irritation on contact. Prolonged skin contact can cause dermatitis because of 1,1,2,2-tetrachloroethane's defatting action on the skin's surface. It will pass through unbroken skin to cause toxic systemic effects including symptoms of inhalation. Direct eye contact causes burning but no permanent damage. Absorption through the skin or inhalation of 1,1,2,2-tetrachloroethane vapors may cause toxic reactions in the liver (in animal studies, it produced liver tumors).

On ingestion, 1,1,2,2-tetrachloroethane is expected to produce symptoms of irritation and narcosis with burning of the lips, mouth, and gastrointestinal tract. It may cause irregular heartbeat, nausea and vomiting, bloody diarrhea, drowsiness, unconsciousness, and possible coma.

Other symptoms of exposure can include fatty degeneration of the liver and heart, hemorrhage into the lungs and serous membranes, and edema of the brain.

Acute Health Effects

The following acute (short-term) health effects may occur immediately or shortly after exposure to 1,1,2,2-tetrachloroethane:

Skin: Moderate irritation with redness and possibly painful inflammation. It will pass through unbroken skin easily to cause toxic systemic effects. Dermatitis is also possible.

Eye: Severe irritation and burning but rarely causes permanent damage.

Lung: Nose and throat irritation. Produces intoxicating effects and may cause lung effects as well.

CNS: Adverse effects (depression) by all routes of exposure. There may be numbness around the lips and mouth, dizziness, loss of coordination, headache, somnolence, slurred speech, flushed face, irregular heartbeat, and other unspecified effects on nervous system functions.

Chronic Health Effects

The following chronic (long-term) health effects may occur at some time after exposure to 1,1,2,2-tetrachloroethane and can last for months or even years:

Cancer Hazards: According to information presented in the references, 1,1,2,2-tetrachloroethane has been shown to cause cancer in test animals and its carcinogenic capability in humans is considered questioned. Some studies indicate that it causes mutations and many scientists believe that long-term exposure to such chemicals may pose a cancer risk in the long-term.

Reproductive Hazard: According to information presented in the references, 1,1,2,2-tetrachloroethane has not been adequately tested for its ability to adversely affect reproduction in test animals.

Other Chronic Effects: Prolonged skin contact may cause drying, cracking, and chapping of the affected area leading to dermatitis. Prolonged exposure may also cause impaired memory, weakness and numbness in the hands and feet, peripheral neuropathy, visual disturbances and impairment, and muscular cramps. There may be liver injury (fatty degeneration, necrosis, jaundice, darkened urine), kidney damage (oliguric uremia, congestion, and granular swelling), fatty degeneration of the hear, and edema of the brain. Chronic exposures may cause malaise, drowsiness, decreased appetite, nausea and retching, bad taste in the mouth and throat, headache, pale stools, jaundice, dark urine, mental confusion, stupor, hand tremors, sensation of deafness, decrease in reflexes, and nausea.

Recommended Risk-Reduction Measures

Personnel should avoid direct contact with 1,1,2,2-tetrachloroethane. It is regarded as one of the most toxic of the commonly available chlorinated hydrocarbons. It is a possible human carcinogen and, therefore, human exposure must be prevented. Therefore, prudent risk management requires it be treated as though it posed a significant health risk in the event of overexposure. If a less toxic chemical cannot be substituted for a hazardous substance, then *engineering controls* are the most effective method of reducing exposures. The best protection is to enclose operations and/or provide local exhaust ventilation at the site of chemical release. While not always operationally feasible, isolating operations can also reduce exposure. Using respiratory protection is less effective than the controls mentioned above, but is still required whenever working with or around 1,1,2,2-tetrachloroethane. For long exposures above the PEL, a supplied-air respirator with full facepiece operated in continuous flow mode, or a self-contained breathing apparatus (SCBA) with full facepiece operated in pressure demand are the recommended respiratory protection methods of choice. If a full facepiece is not available, then chemical goggles should be worn to protect the eyes. Whenever a chemical splash hazard exists, a face shield and a rubber apron should be worn. To prevent hand and skin exposures, impervious gloves should be used. Glove manufacturers should be contacted and permeation studies obtained *before* final glove selections are made.

Administrative controls should also be in place to minimize the potential for human exposures. These may include written procedures or policies which specify the methods and techniques that will be practiced whenever personnel are to work with or around 1,1,2,2-tetrachloroethane.

All personnel should receive *training* on the use, hazards, protective measures, emergency actions, and other precautions per 29 CFR 1910.1200 (Hazard Communication), prior to the first assignment in an area where 1,1,2,2-tetrachloroethane is used or stored.

If symptoms develop or overexposure is suspected, the following tests are recommended:

☑ Lung, kidney, and liver function tests.

☑ Electroencephalogram (EEG) to detect abnormalities which may be indicative of chronic toxicity.

☑ Complete blood count (CBC) with monitoring of blood enzyme levels.

☑ Electrocardiogram (EKG).

Any evaluation should include a careful history of past and present symptoms with an examination. Medical tests that simply look for existing damage are not a substitute for controlling exposures.

Other methods to reduce exposure include:

☑ Where possible, enclose operations and use local exhaust ventilation at the site of chemical release. If local exhaust ventilation or enclosure is not used, respiratory protection should be mandatory.

☑ Always ensure that proper protective clothing is worn when using chemical substances.

☑ Wash thoroughly immediately after exposure to 1,1,2,2-tetrachloroethane and at the end of the work shift or before eating, drinking, or smoking.

☑ Hazard warning information should be posted in the work area. In addition, as part of an on-going education and training program, all information on the health and safety hazards of 1,1,2,2-tetrachloroethane should be communicated to all exposed workers.

☑ Eye wash stations should be provided in the immediate work area for emergency use. If there is a possibility for skin exposure to 1,1,2,2-tetrachloroethane, emergency shower facilities should also be provided.

☑ Personnel should never wear contaminated clothing home (family members can be exposed). Clothing should be laundered by personnel who have been trained on the hazards of 1,1,2,2-tetrachloroethane.

RISK ASSESSMENT: ENVIRONMENT

General Assessment

The environment is at risk of exposure during transportation, storage, disposal, or destruction of 1,1,2,2-tetrachloroethane. In almost every scenario, the threat of environmental exposure is contingent upon the proper handling of the chemical substance. Accidental spills, large or small, can result in fire, explosion, and possible contamination of the surrounding environmental mediums (water, soil, and air).

1,1,2,2-Tetrachloroethane is considered a noncombustible liquid. But violent reactions can occur in contact with any incompatible materials. It is incompatible with strong oxidizers and many chemically active metals. These characteristics require extreme caution in handling, storage, transportation, and disposal. Should 1,1,2,2-tetrachloroethane ever come into contact with incompatible substances during use, transportation, storage, or disposal, the formation of highly toxic and/or highly explosive commodities (including hydrogen chloride) is extremely possible.

The proper disposal method is to mix 1,1,2,2-tetrachloroethane with a combustible solvent and then burn it in a permitted chemical incinerator equipped with an afterburner and scrubber.

1,1,2,2-Tetrachloroethane may enter the environment through industrial discharges or spills.

Acute Ecological Effects

Acute (short-term) toxic effects may include the death of animals, birds, or fish and death or low growth rate in plants. Acute effects are seen 2 to 4 days after animals or plants are exposed to 1,1,2,2-tetrachloroethane.

Insufficient data are available on the short-term effects of 1,1,2,2-tetrachloroethane exposure to aquatic life, birds, plants, or land animals.

Chronic Ecological Effects

Chronic (long-term) toxic effects may include shortened life span, reproductive problems, lower fertility, and changes in appearance or behavior in exposed animals. These effects can be seen long after first exposure(s) to toxic chemicals.

Insufficient data are available on the long-term effects of 1,1,2,2-tetrachloroethane exposure to aquatic life, birds, plants, or land animals.

Water Solubility

1,1,2,2-Tetrachloroethane is nearly insoluble in water. Concentrations of less than 10 milligrams may mix with a liter of water.

Persistence in the Environment

1,1,2,2-Tetrachloroethane is non-persistent in water, with a half-life of less than 2 days. The half-life of a pollutant is the amount of time it takes for one half of the chemical to be degraded. Virtually 100% of 1,1,2,2-tetrachloroethane will eventually end up in air.

Bioaccumulation in Aquatic Organisms

Some substances increase in concentration, or *bioaccumulate,* in living organisms as they breathe contaminated air, drink contaminated water, or eat contaminated food. These chemicals can become concentrated in the tissues and internal organs of animals as well as humans.

The concentration of 1,1,2,2-tetrachloroethane found in fish tissue is expected to be somewhat higher than the average concentration in the water from which the fish was taken.

↻ *Recommended Risk-Reduction Measures*

Proper training of all transporters will reduce the likelihood of a mishap or accident resulting in a leak or spill to the environment. The correct use of DOT labeling (per the MSDS) on all containers, trucks, and rail cars will enable emergency responders to react properly and quickly to any disaster thereby reducing the potential risk to the environment and to personnel.

Storage of 1,1,2,2-tetrachloroethane should be segregated from incompatible chemicals to minimize the risk of cross-contamination or contact. Buildings designated for storage should be equipped with appropriate fire protection systems (alarms, sprinklers, emergency lighting, portable extinguishers).

If a spill or leak to the environment has occurred, fire department, emergency response and/or hazardous materials spill personnel should be notified immediately. Cleanup should be attempted only by those trained in proper spill containment procedures. Contaminated soils should be removed for incineration and replaced with clean soil. If 1,1,2,2-tetrachloroethane should contact the water table, aquifer, or navigable waterway, cleanup should be started immediately. It is only slightly soluble in water and remediation may be possible. The local and/or state emergency response authorities must be notified. A comprehensive emergency response or disaster preparedness/recovery plan should be in place prior to any operations involving the use, transportation, storage, or disposal of 1,1,2,2-tetrachloroethane. If 1,1,2,2-tetrachloroethane is spilled or leaked, the following specific steps are recommended:

- ☑ Restrict persons not wearing protective clothing from area of spill or leak until cleanup is complete and area can be opened for normal work.
- ☑ Ventilate area of leak.
- ☑ Absorb liquids in vermiculite, dry sand, earth, or similar material and deposit in sealed containers.
- ☑ It may be necessary to dispose of 1,1,2,2-tetrachloroethane as a hazardous waste. The responsible state agency or the regional office of the federal Environmental Protection Agency (EPA) should be contacted for specific recommendations.

RISK ASSESSMENT: BUSINESS

General Assessment

Accidents or mishaps involving 1,1,2,2-tetrachloroethane can present a significant threat to business operations. The loss or damage of equipment or facilities can significantly affect fiscal viability. Lawsuits that may result from personnel injury/death, public exposures, and/or environmental contamination will also require a serious expenditure of resources. Media attention surrounding an injury, death, or environmental damage can also result in a loss of profits and loss of current as well as future business. Always remember that anytime the terms "cancer" or "carcinogen" are used, public emotion, ignorance, hysteria, and anxiety can all run equally high. This must be carefully considered when developing or implementing public relations policies.

↻ *Recommended Risk-Reduction Measures*

Company attorneys, safety and health professionals, and environmental specialists should be involved in the development of any procedures dealing with safety or emergency response. A company official should be pre-designated as a public relations officer with specific training in dealing with the press. Corporate plans and policies should be developed, approved, and implemented long before any need for such arises.

REFERENCES

American Conference of Governmental Industrial Hygienists. 1988. *Documentation of the Threshold Limit Values and Biological Exposure Indices,* 5th Edition (with updates). Cincinnati: ACGIH

Gardner, R. W. 1994. *Chemical Intolerance.* New York: CRC Press/Lewis Publishers.

Keith, L. H. and M. M. Walker. 1995. *Handbook of Air Toxics.* New York: CRC Press/Lewis Publishers.

Lewis, R. J., Sr. 1992. *Sax's Dangerous Properties of Industrial Materials,* Volumes 1, 2, and 3, 8th Edition. New York: Van Nostrand Reinhold.

Meyer, E. 1990. *Chemistry of Hazardous Materials.* Englewood Cliffs, NJ: Prentice Hall.

Patnaik, P. 1992. *A Comprehensive Guide to the Hazardous Properties of Chemical Substances.* New York: Van Nostrand Reinhold.

Sacarello, H. L. A. 1994. *The Comprehensive Handbook of Hazardous Materials.* New York: CRC Press/Lewis Publishers.

U.S. Department of Health and Human Services, National Institute for Occupational Safety and Health. 1994. *NIOSH Pocket Guide to Chemical Hazards.* Washington, D.C.: U.S. Government Printing Office.

MATERIAL SAFETY DATA SHEET

CHEMICAL NAME	
TETRACHLOROVINPHOS®	**CRC PRESS/LEWIS PUBLISHERS DISCLAIMER:** THE INFORMATION AND RECOMMENDATIONS PRESENTED HEREIN ARE BASED ON SOURCES BELIEVED TO BE RELIABLE. CRC MAKES NO REPRESENTATION ON ITS COMPLETENESS OR ACCURACY. IT IS THE RESPONSIBILITY OF THE USER TO DETERMINE THE CHEMICAL'S SUITABILITY FOR ITS INTENDED USE, THE CHEMICAL'S SAFE USE, AND THE CHEMICAL'S PROPER DISPOSAL. NO REPRESENTATIONS AND/OR WARRANTIES, EITHER EXPRESSED OR IMPLIED, OF THE MERCHANTABILITY OR FITNESS FOR A PARTICULAR PURPOSE, OR OF ANY OTHER NATURE, ARE MADE WITH RESPECT TO THE INFORMATION PROVIDED IN THIS MSDS OR TO THE CHEMICAL TO WHICH INFORMATION MAY REFER. CRC NEITHER ASSUMES NOR AUTHORIZES ANY OTHER PERSON TO ASSUME FOR IT, ANY OTHER ADDITIONAL RESPONSIBILITY OR LIABILITY FOR THE USE OF, OR RELIANCE UPON, THIS INFORMATION.

HAZARD WARNING INFORMATION

HEALTH	FIRE	REACTIVE	OTHER	DEGREE OF HAZARD	COLOR CODING	OTHER CODES
No Citation	No Citation	No Citation		0 = Minimum Hazard 1 = Slight Hazard 2 = Moderate Hazard 3 = Serious Hazard 4 = Severe Hazard	HEALTH = BLUE FIRE = RED REACTIVITY = YELLOW OTHER = WHITE	OX = Oxidizer ACID = Acid ALK = Alkali COR = Corrosive W = Use No Water

SECTION I - GENERAL INFORMATION

Characterization	**Pesticide, Organophosphorus**	RCRA Number	**None**
		EPA Class	**Not Applicable**
DOT Proper Shipping Name	**No Citation**	Chemical Abstract Service (CAS) Number	**961-11-5**
DOT Hazard Class and Label Requirements	**No Citation**	DOT Emergency Guide Code	**No Citation**
DOT Identification Number	**No Citation**	Chemical Formula	$C_{10}H_9Cl_4O_4P$

Synonyms

2-Chloro-1-(2,4,5-Trichlorophenyl)Vinyl Dimethyl Phosphate; Phosphoric acid, 2-Chloro-1-(2,4,5-Trichlorophenyl)Vinyl Dimethyl Ester.

SECTION II - HAZARDOUS INGREDIENTS AND IDENTITY INFORMATION

Hazardous Components (specific identity)	OSHA Exposure Criteria	NIOSH Exposure Criteria	Immediately Dangerous to Life and Health (IDLH)	ACGIH Exposure Criteria
Tetrachlorovinphos	**PEL:** Not Established **STEL:** Not Established	**REL:** Not Established **STEL:** Not Established	**Not Determined**	**TLV:** Not Established **STEL:** Not Established

SECTION III - PHYSICAL AND CHEMICAL CHARACTERISTICS

Boiling Point	**Not Reported**	Specific Gravity (H_2O = 1)	**Not Reported**
Vapor Pressure (mm Hg)	**Not Reported**	Molecular Weight	**365.96**
Vapor Density (Air = 1)	**Not Reported**	Melting Point	**178°F (97°C)**

Solubility

Slightly soluble in water; partially soluble in chloroform.

Appearance and Odor

Off-white crystalline solid.

SECTION IV - FIRE AND EXPLOSION HAZARD DATA

Flash Point (method used)	**Not Applicable**	Explosive Limits in Air % by Volume	**LEL: Not Determined UEL: Not Determined**
NFPA Classification	**Non-Combustible Solid**	Autoignition Temperature	**Not Determined**

Extinguishing Media

Use agent suitable to surrounding fires.

Special Fire Fighting Procedures

Poisonous gases are produced in fire; wear full protective clothing and self-contained breathing apparatus (SCBA). If possible, fight fire from a distance and avoid smoke (consider down-wind conditions). Move containers from fire area if it can be done safely and without risk. Do not release water runoff from fire control measure to sewers or waterways; contain and collect for disposal.

Unusual Fire and Explosion Hazards

None reported.

SECTION V - REACTIVITY DATA

Stability		Conditions to Avoid: **Tetrachlorovinphos is normally stable under routine conditions of handling and storage. Keep away from heat and flame.**
Stable	Unstable	Incompatibility *(materials to avoid)*
X		**Incompatible with strong oxidizers (such as chlorine, fluorine, bromine, nitrates, and permanganates).**
Hazardous Polymerization		Conditions to Avoid: **Under normal conditions of temperature and pressure, hazardous polymerization of tetrachlorovinphos will not occur.**
May Occur	Will Not Occur	Hazardous Decomposition or By-products
	X	**Acrid and irritating gases and fumes are produced when tetrachlorovinphos is involved in fire, including oxides of chloride and phosphorous.**

SECTION VI - HEALTH HAZARD DATA

Primary Route(s) of Entry:	Inhalation?	Absorption (skin)?	Ingestion?
	X	X	X

Health Hazards

INHALATION: Labored or difficult breathing, tightness in chest, wheezing, coughing, rales, increased respiration, headache, runny nose, and excessive bronchial secretions. There may also be laryngeal spasms, excessive salivation, respiratory distress, cyanosis, headache, loss of coordination, visual disturbances, tearing, blurring, nausea and vomiting, pain in the stomach, and CNS effects such as giddiness, dizziness, weakness, slow heart-beat, arrhythmia, loss of coordination, slurred speech, confusion, disorientation with drowsiness, hypertension, convulsions, coma, and death (respiratory paralysis).

ABSORPTION: Irritation to the skin and eyes. May absorb to cause toxic systemic, including sweating twitching and many of the symptoms noted for inhalation. Sensitization and dermatitis may occur at high exposures. Extremely irritating to the eyes.

INGESTION: Causes many of the same symptoms noted for inhalation and absorption.

Carcinogenicity	NTP Listed?	IARC Cancer Review Group?	OSHA Regulated?	Target Organs?
Suspected Human Questioned Animal	**No**	**No**	**No**	**Respiratory system, eyes, skin, CNS, CVS, blood.**

Medical Conditions Generally Aggravated by Exposure

Glaucoma, cardiovascular disorders, hepatic and renal diseases, and CNS disorders.

Emergency and First-aid Procedures

Eye contact: Flush immediately with water for 15 minutes (minimum); seek medical attention. Skin contact: Remove all contaminated clothing. Immediately wash area with large amounts of soap and water followed by a 95% ethanol wash and then a second soap and water rinse. Seek medical assistance. For inhalation: Remove the person from exposure. Provide respiratory assistance and CPR. Transfer to medical facility. Observe 48 hours for lung effects. If swallowed: Contact poison control center. Give 1-2 glasses of water. Unless advised otherwise, do NOT induce vomiting.

SECTION VII - PRECAUTIONS FOR SAFE HANDLING AND USE

Steps to be Taken in Case Material is Released or Spilled

Collect solid materials in safest manner possible. Do NOT dry sweep. Use HEPA vacuum instead and place in sealed drum for disposal. Ventilate area. Restrict those not involved in cleanup from entering area.

Preferred Waste Disposal Method

Mix in flammable solvent and burn in chemical incinerator equipped with afterburner and scrubber.

Precautions to be Taken in Handling and Storage

Store in tightly closed containers in a cool, well-ventilated, moisture-free area.

Other Precautions and Warnings

Prevent physical damage to storage containers.

SECTION VIII - CONTROL MEASURES AND PERSONAL PROTECTIVE EQUIPMENT

Respiratory Protection *(specify type)*

For best protection, use a supplied-air respirator with full facepiece operated in pressure demand mode or a SCBA with full facepiece operated in pressure demand or other positive pressure mode.

Ventilation

Local exhaust or general mechanical systems recommended.

Protective Gloves	Eye Protection	Other Protective Clothing
Impervious Material	**Chemical/Dust Goggles or Face Mask**	**Whole Body Protection**

Work/Hygiene Practices

Always wash hands thoroughly after using chemical; never bring food, drink, or smoking materials into vicinity of chemicals.

TETRACHLOROVINPHOS

$C_{10}H_9Cl_4O_4P$

CAS: 961-11-5

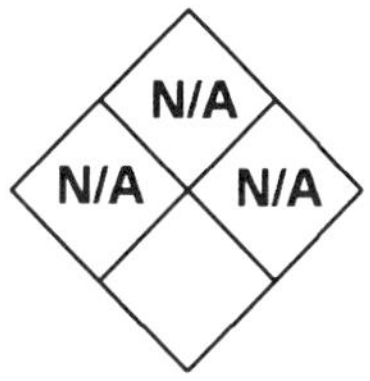

IDENTIFICATION AND TYPICAL USES

Tetrachlorovinphos is an off-white crystalline solid. It is used as an organophosphate insecticide used to control houseflies, chicken mites, lice, screw worms, horn flies, face lies, and stable flies.

RISK ASSESSMENT: HEALTH

General Assessment

Tetrachlorovinphos is toxic to humans by *inhalation,* dermal *absorption,* and *ingestion.* It is a questionable carcinogen with experimental neoplastigenic and tumorigenic effects. It is also an experimental teratogen with positive evidence of reproductive effects in animals. It is not easily removed from the surface of the skin and eye contact with the liquid can cause serious irritation and possible damage.

Inhalation of tetrachlorovinphos vapors will cause symptoms of organophosphate poisoning, including a tightness in the chest, wheezing, increased pulmonary secretions, bluish color in the skin, small pupils (miosis), an aching feeling in and behind the eyes, blurred vision, lachrymation (tearing), runny nose, headache, and watering mouth. There may be laryngeal spasms, respiratory distress, bradycardia (reduced heartbeat), tachycardia (fast heartbeat), and other cardiac disturbances (arrhythmia, complete heart block). It will also cause systemic effects with action on the central nervous system. This action can include weakness, general twitching and paralysis, dizziness, confusion, staggering gait, slurred speech, sweating, incontinence, convulsions, stopped breathing, coma, and death. This compound is also an inhibitor of cholinesterase, a blood enzyme that is essential to the proper functioning of the muscles.

Skin contact can cause localized irritation, inflammation, and possible contact dermatitis. Tetrachlorovinphos can be readily absorbed through intact skin to cause toxic systemic effects similar to that of inhalation. Eye contact also causes irritation.

Acute Health Effects

The following acute (short-term) health effects may occur immediately or shortly after exposure to tetrachlorovinphos:

Skin: Irritation and possible inflammation. May absorb through unbroken skin to cause toxic systemic effects, including ocular disturbances, respiratory impairment, and CNS effects.

Eye: Irritation and inflammation. Possible conjunctivitis, tearing, redness, swelling, and itchy pain.

Lung: Irritation of the mucosa of the nose, throat, and upper respiratory tract. Effects are more systemic than localized, with action on the blood, central nervous system, and (possibly) the cardiovascular system.

CNS: A depressant, causing headache, blurred vision, giddiness, nervousness, and other systemic effects up to and including death from respiratory paralysis or cardiac disturbances.

Other: Ingestion of tetrachlorovinphos will result in loss of appetite, nausea, vomiting, abdominal pain, and diarrhea within 2 hours of exposure.

Chronic Health Effects

The following chronic (long-term) health effects may occur at some time after exposure to tetrachlorovinphos and can last for months or even years:

Cancer Hazards: According to the information presented in the references, tetrachlorovinphos has been shown to cause cancer in test animals.

Reproductive Hazard: According to the information presented in the references, tetrachlorovinphos may damage the developing fetus in humans since it has been shown to be fetotoxic in animals.

Other Chronic Effects: Chronic exposure to tetrachlorovinphos has not been sufficiently studied. However, there is some indication that long-term exposure or frequent high exposures may inhibit the essential enzyme cholinesterase. Individuals may become more susceptible to the toxic effects of tetrachlorovinphos following long-term exposure. Organophosphates bind irreversibly to cholinesterase causing its depletion. This action is apparently the reason for the increased susceptibility in chronic exposure. Symptoms include headache, vertigo, loss of memory and appetite, insomnia, disorientation, muscular twitching and trembling, and other CNS effects. Tetrachlorovinphos may also cause dermal sensitization as well as dermatitis in some people.

Recommended Risk-Reduction Measures

Personnel should always avoid direct contact with tetrachlorovinphos. The exact nature of it toxicity is not clearly understood in the references. Less toxic substances should always be sought as a first step towards risk reduction. If a less toxic chemical cannot be substituted for a hazardous substance, *engineering controls* are the most effective method of reducing exposures. The best protection is to enclose operations and/or provide local exhaust ventilation at the site of chemical release. While not always operationally feasible, isolating operations can also reduce exposure. These controls certainly may not be practical or applicable when using tetrachlorovinphos as an insecticide, especially outdoors. However, during manufacture, transportation, or storage, these engineering controls may be effective in reducing exposure risk.

Using respiratory protection is less effective than the controls mentioned above, but is still advisable whenever working with or around tetrachlorovinphos. No exposure levels have been established for tetrachlorovinphos. This does not mean that exposure is without health risk. It is a suspected carcinogen and all exposures should be kept to an absolute minimum. For maximum exposure risk reduction, a self-contained breathing apparatus (SCBA) with full facepiece operated in pressure demand is the recommended respiratory protection of choice. If a full facepiece is not available, then chemical goggles should be worn to protect the eyes. Whenever a chemical splash hazard exists, a face shield and protective apron should be worn. To prevent hand and skin exposures, impervious gloves should be worn. Glove manufacturers should be contacted and permeation studies obtained *before* final glove selections are made.

Administrative controls should also be in place to minimize the potential for human exposures. These may include written procedures or policies which specify the methods and techniques that will be practiced whenever personnel are to work with tetrachlorovinphos.

All personnel should receive *training* on the use, hazards, protective measures, emergency actions, and other precautions per 29 CFR 1910.1200 (Hazard Communication), prior to the first assignment in an area where tetrachlorovinphos is used or stored.

If symptoms develop or overexposure is suspected, the following medical tests may be useful:

☑ Complete blood count (CBC) with plasma and red blood cell cholinesterase levels. If exposure stops, plasma levels return to normal within 1 to 2 weeks while red blood cell levels may be reduced for 1 to 3 months.

☑ Urinary metabolite monitoring for *p*-nitrophenol within 48 hours of exposure to assess tetrachlorovinphos absorption.

Any evaluation should include a careful medical history of past and present symptoms with an examination. However, medical tests that evaluate existing damage are not a substitute for controlling exposure.

Other methods to reduce exposure include:

☑ Where possible, enclose operations and use local exhaust ventilation at the site of chemical release. If local exhaust ventilation or enclosure is not used, respiratory protection should be mandatory.

☑ Always ensure that proper protective clothing is worn when using chemical substances and that personnel are trained in its use and care.

☑ Wash thoroughly immediately after exposure to tetrachlorovinphos and at the end of the work shift or before eating, drinking, or smoking.

☑ Hazard warning information should be posted in the work area. In addition, as part of an on-going education and training program, all information on the health and safety hazards of tetrachlorovinphos must be communicated to exposed workers.

☑ Eye wash stations should be provided in work areas. If the potential for whole body exposure exists, then safety showers should also be provided

RISK ASSESSMENT: ENVIRONMENT

General Assessment

The environment is also at risk of exposure during the intended use as an insecticide, as well as during transportation, storage, disposal, or destruction of tetrachlorovinphos. In almost every scenario, the threat of environmental exposure is contingent upon the proper handling of the chemical substance. Accidental spills, large or small, can result in fire, explosion, and possible contamination of the surrounding environmental mediums (water, soil, and air).

Tetrachlorovinphos is considered a noncombustible solid. But it can react violently in contact with incompatible materials, such as strong oxidizers. These characteristics require special consideration during any emergency situation involving a leak or spill of tetrachlorovinphos. Should tetrachlorovinphos ever come into contact with incompatible substances either during use, transportation, or storage, violent reactions can occur. Tetrachlorovinphos can enter the environment through its intended use as an insecticide, through industrial discharges, agricultural runoff, and from accidental spills.

Acute Ecological Effects

Acute (short-term) toxic effects may include the death of animals, birds, or fish and death or low growth rate in plants. Acute effects are seen 2 to 4 days after animals or plants are exposed to tetrachlorovinphos. Tetrachlorovinphos has high acute toxicity to aquatic life and to birds. It has caused injury, including germination decreases and sterility increases in various agricultural and ornamental crops. It has moderate acute toxicity to land animals.

Chronic Ecological Effects

Chronic toxic effects may include shortened life span, reproductive problems, lower fertility, and changes in appearance or behavior in exposed animals. These effects can be seen long after first exposure(s) to toxic chemicals. Tetrachlorovinphos has high chronic toxicity to aquatic life. Insufficient data are available to evaluate or predict the long-term effects of tetrachlorovinphos to plants, birds, or land animals.

Water Solubility

Tetrachlorovinphos is moderately soluble in water. Concentrations of 1 to 1000 milligrams will mix with a liter of water.

Persistence in the Environment

Tetrachlorovinphos is moderately persistent in water, with a half-life of between 20 to 200 days. The half-life of a pollutant is the amount of time it takes for one half of the chemical to be degraded. About 76% of tetrachlorovinphos will eventually end up in water; the remainder will be divided about equally between terrestrial soils and aquatic sediments.

Bioaccumulation in Aquatic Organisms

Some substances increase in concentration, or *bioaccumulate,* in living organisms as they breathe contaminated air, drink contaminated water, or eat contaminated food. These chemicals can become concentrated in the tissues and internal organs of animals as well as humans.

The concentration of tetrachlorovinphos found in fish tissues is expected to be somewhat higher than the average concentration of tetrachlorovinphos in the water from which the fish was taken.

Recommended Risk-Reduction Measures

Proper training of all transporters will reduce the likelihood of a mishap or accident resulting in a leak or spill to the environment. The correct use of labeling on all containers, trucks, and rail cars will enable emergency responders to react properly and quickly to any disaster thereby reducing the potential risk to the environment and to personnel.

Storage of tetrachlorovinphos should be segregated from incompatible chemicals to minimize the risk of cross-contamination or contact. It should be kept in dry areas away from water and moisture. Containers should be protected from physical damage.

If a spill or leak to the environment has occurred, fire department, emergency response and/or hazardous materials spill personnel should be notified immediately. Cleanup should be attempted only by those trained in proper spill containment procedures. Contaminated soils should be removed for incineration and replaced with clean soil. If tetrachlorovinphos should contact the water table, aquifer, or navigable waterway, remediation activities should be prompt to ensure protection of aquatic sediments. When such spills occur, the local and/or state emergency response authorities must be notified. A comprehensive emergency response or disaster preparedness/recovery plan should be in place prior to any operations involving the use, transportation, storage, or disposal of tetrachlorovinphos. If tetrachlorovinphos is spilled or leaked, the following specific steps are recommended:

- ☑ Restrict persons not wearing protective clothing from area of spill or leak until cleanup is complete and area can be opened for normal work.
- ☑ Ventilate area of spill or leak.
- ☑ Absorb any liquid solution spills using vermiculite, dry earth, or sand and place in a sealed drum for disposal using non-sparking tools. Collect powdered materials in safest manner possible and place in sealed drums for disposal. Do NOT dry sweep (generates airborne dusts). Use vacuum equipped with a high-efficiency particulate air (HEPA) filter.
- ☑ It may be necessary to dispose of tetrachlorovinphos as a hazardous waste. The responsible state agency or the regional office of the federal Environmental Protection Agency (EPA) should be contacted for specific recommendations.

RISK ASSESSMENT: BUSINESS

General Assessment

Accidents or mishaps involving tetrachlorovinphos can present a significant threat to business operations. The loss or damage of equipment or facilities can significantly affect fiscal viability. Lawsuits that may result from personnel injury/death, public exposures, and/or environmental contamination will also require a serious expenditure of resources.

Media attention surrounding an injury, death, or environmental damage can also result in a loss of profits and loss of current as well as future business. Always remember that anytime the terms "cancer," "carcinogen," or "reproductive hazard" are used, public emotion, anxiety, hysteria, and ignorance can run equally high. This should be carefully considered whenever developing or implementing public relations polices or directives.

ʊ *Recommended Risk-Reduction Measures*

Company attorneys, safety and health professionals, and environmental specialists should be involved in the development of any procedures dealing with safety or emergency response. A company official should be pre-designated as a public relations officer with specific training in dealing with the press. Corporate plans and policies should be developed, approved, and implemented long before any need for such arises.

REFERENCES

American Conference of Governmental Industrial Hygienists. 1988. *Documentation of the Threshold Limit Values and Biological Exposure Indices,* 5th Edition (with updates). Cincinnati: ACGIH

Bloom, A. D. and F. J. DeSerres. 1995. *Ecotoxicity and Human Health.* New York: CRC Press/Lewis Publishers.

Bueche, J. R. 1972. *Principles of Physics,* 2nd Edition. New York: McGraw-Hill.

Burrell, R., D. K. Flaherty, and L. J. Sauers. 1992. *Toxicology of the Immune System.* New York: Van Nostrand Reinhold.

Calabrese, E. J., E. M. Kenyon. *Air Toxics and Risk Assessment.* New York: CRC Press/Lewis Publishers.

Cockerman, L. G., B. S. Shane. 1994. *Basic Environmental Toxicology.* New York: CRC Press/Lewis Publishers.

Gardner, R. W. 1994. *Chemical Intolerance.* New York: CRC Press/Lewis Publishers.

Keith, L. H. and M. M. Walker. 1995. *Handbook of Air Toxics.* New York: CRC Press/Lewis Publishers.

Lauwerys, R. R., P. Hoet. 1993. *Industrial Chemical Exposure.* New York: CRC Press/Lewis Publishers.

Lewis, R. J., Sr. 1993. *Hawley's Condensed Chemical Dictionary,* 12th Edition. New York: Van Nostrand Reinhold.

Lewis, R. J., Sr. 1992. *Sax's Dangerous Properties of Industrial Materials,* Volumes 1, 2, and 3, 8th Edition. New York: Van Nostrand Reinhold.

Manahan, S. E. 1992. *Toxicological Chemistry,* 2nd Edition. New York: CRC Press/Lewis Publishers.

Meyer, E. 1990. *Chemistry of Hazardous Materials.* Englewood Cliffs, NJ: Prentice Hall.

Rea, W. J. 1995. *Chemical Sensitivity,* Vol. I-III. New York: CRC Press/Lewis Publishers.

Richardson, M. 1995. *Dictionary of Substance and Their Effects,* Vol. I-VI. New York: CRC Press/Lewis Publishers.

Sacarello, H. L. A. 1994. *The Comprehensive Handbook of Hazardous Materials.* New York: CRC Press/Lewis Publishers.

U.S. Department of Health and Human Services, National Institute for Occupational Safety and Health. 1978. *NIOSH Criteria Document: Occupational Exposure During the Manufacture and Formulation of Pesticides.* Document Number 78-174. Washington, D.C.: U.S. Government Printing Office.

MATERIAL SAFETY DATA SHEET

CHEMICAL NAME	CRC PRESS/LEWIS PUBLISHERS DISCLAIMER:
TETRAETHYL LEAD	

HAZARD WARNING INFORMATION

HEALTH	FIRE	REACTIVE	OTHER	DEGREE OF HAZARD	COLOR CODING	OTHER CODES
3	2	3		0 = Minimum Hazard 1 = Slight Hazard 2 = Moderate Hazard 3 = Serious Hazard 4 = Severe Hazard	HEALTH = BLUE FIRE = RED REACTIVITY = YELLOW OTHER = WHITE	OX = Oxidizer ACID = Acid ALK = Alkali COR = Corrosive W = Use No Water

SECTION I - GENERAL INFORMATION

Characterization	RCRA Number	EPA Class
Metal Alkyl	**P110**	**Acute Hazardous Waste**

DOT Proper Shipping Name **Tetraethyl Lead, Liquid**	Chemical Abstract Service (CAS) Number **78-00-2**
DOT Hazard Class and Label Requirements **Poison B; Flammable Liquid and Poison**	DOT Emergency Guide Code **56**
DOT Identification Number **NA 1649**	Chemical Symbol **$Pb(C_2H_5)_4$**

Synonyms
Lead tetraethyl; TEL; tetraethylplumbane.

SECTION II - HAZARDOUS INGREDIENTS AND IDENTITY INFORMATION

Hazardous Components (specific identity)	OSHA Exposure Criteria	NIOSH Exposure Criteria	Immediately Dangerous to Life and Health (IDLH)	ACGIH Exposure Criteria
Tetraethyl lead [derivation: By alkylation of lead-sodium alloy with ethyl chloride; by electrolysis of an ethyl Grignard reagent with an anode of lead pellets]. **1 ppm = 13.45 mg/m^3**	**PEL (as Pb): 0.075 mg/m^3 (skin)** **STEL: Not Established**	**PEL (as Pb): 0.075 mg/m^3 (skin)** **STEL: Not Established**	**40 mg/m^3 (as Pb)**	**TLV (as Pb): 0.1 mg/m^3 (skin)** **STEL: Not Established**

SECTION III - PHYSICAL AND CHEMICAL CHARACTERISTICS

Boiling Point **228°F (108°C) Decomposes**	Specific Gravity (H_2O = 1) **1.65**
Vapor Pressure (mm Hg) **0.2 at 68°F (20°C)**	Molecular Weight **323.45**
Vapor Density (Air = 1) **1.0**	Melting Point **-202°F (-130°C)**

Solubility
Insoluble in water and dilute acids or alkalis, soluble in benzene, diethyl ether, gasoline, petroleum ether, all organic solvents. Slightly soluble in alcohol.

Appearance and Odor
Colorless liquid that may be dyed red, orange, blue, or other color, with a pleasant, sweet odor.

SECTION IV - FIRE AND EXPLOSION HAZARD DATA

Flash Point (method used) **200°F (93°C) closed cup**	Explosive Limits in Air % by Volume **LEL: 1.8% UEL: Not Determined**
NFPA Classification **Class IIIB Combustible Liquid**	Autoignition Temperature **Not Reported**

Extinguishing Media
Use dry chemical, carbon dioxide, water spray, fog, or regular foam.

Special Fire Fighting Procedures
Poisonous gases are produced in fire. Wear self-contained breathing apparatus (SCBA) and special protective clothing. Use cool water spray to cool the sides of fire-exposed containers. Stay clear of the ends of tanks and evacuate 1500 feet if rising sound is heard coming from venting devise.

Unusual Fire and Explosion Hazards
Containers can explode in fire. A boiling liquid expanding vapor explosion (BLEVE) can occur when the liquid in containers adjacent to a fire begins to boil.

SECTION V - REACTIVITY DATA

Stability		Conditions to Avoid
		Tetraethyl lead decomposes slowly at room temperature and rapidly when heated. It will also decompose when exposed to the sun or allowed to evaporate in air. Avoid contact with heat and incompatibles.
Stable	Unstable	Incompatibility *(materials to avoid)*
	X	Strong oxidizers (chlorine, fluorine, bromine), sulfuryl chloride, potassium permanganate, and ferrous oxide (rust). Tetraethyl lead will also solubilize fatty materials and may attack rubber.
Hazardous Polymerization		Conditions to Avoid
		Under normal conditions of temperature and pressure, hazardous polymerization of tetraethyl lead will not occur. Avoid exposure to sunlight and air.
May Occur	Will Not Occur	Hazardous Decomposition or By-products
	X	Thermal decomposition of tetraethyl lead will produce toxic and irritating smoke and fumes, including highly toxic fumes of lead and carbon dioxide.

SECTION VI - HEALTH HAZARD DATA

Primary Route(s) of Entry:	Inhalation?	Absorption (skin)?	Ingestion?
	X	X	X

Health Hazards

INHALATION: Primary effects on the brain and central nervous system (CNS). Effects can range from mild to moderate to severe. Symptoms range from anxiety, irritability, insomnia, lurid dreams, vomiting, metallic taste, pallor, cerebellar ataxia, diarrhea, disorientation, tremors, hyperexcitability, chorea, slow heart beat, hypotension, hypothermia, delusions, hallucinations, mania, convulsions, cerebellar edema, coma, tinnitus, impaired vision, elevated liver enzymes, anemia, lassitude, and death.

ABSORPTION: Eye and skin irritation and rapid absorption. Symptoms of inhalation are possible.

INGESTION: Easily absorbed when ingested. If it enters the bloodstream, it can have many of the same effects as inhalation. It is a systemic poison that accumulates in body organs.

Carcinogenicity	NTP Listed?	IARC Cancer Review Group?	OSHA Regulated?	Target Organs?
Suspected Human Questioned Animal	No	Group 3	29 CFR 1910.1000 Table Z-1 (as Pb)	Eyes, liver, CNS, CVS, kidneys.

Medical Conditions Generally Aggravated by Exposure

Mental disorders, nervous system dysfunction (epilepsy), kidney diseases, hypertension, and anemia.

Emergency and First-aid Procedures

Eye contact: Flush immediately with water for 15 minutes (minimum); seek medical attention. Skin contact: Remove all contaminated clothing. Wash contaminated skin with soap and water. For inhalation: Remove the person from exposure. Provide respiratory assistance and CPR. Transfer to medical facility. If swallowed: Never attempt to give anything by mouth to an unconscious or convulsing person. If large quantities are ingested, induce vomiting with Syrup of Ipecac and call physician.

SECTION VII - PRECAUTIONS FOR SAFE HANDLING AND USE

Steps to be Taken in Case Material is Released or Spilled

Remove all sources of ignition. Restrict those not involved in cleanup from entering area. Absorb liquids in vermiculite or other material and deposit in sealed drums. Use non-sparking tools.

Preferred Waste Disposal Method

No citation.

Precautions to be Taken in Handling and Storage

Containers should be protected against physical damage. Store in tightly closed containers in a cool, dry, well-ventilated area away from incompatible materials and direct sunlight.

Other Precautions and Warnings

Wherever tetraethyl lead is used, handled, manufactured, or stored, use explosion-proof electrical equipment and fittings.

SECTION VIII - CONTROL MEASURES AND PERSONAL PROTECTIVE EQUIPMENT

Respiratory Protection *(specify type)*

Exposure level is low and difficult to monitor. Use a respirator with particulate (dust/fume/mist) filter; otherwise use a supplied-air respirator or self-contained breathing apparatus in continuous flow mode.

Ventilation

Local exhaust or general mechanical systems recommended.

Protective Gloves	Eye Protection	Other Protective Clothing
Suitable to protect from skin contact	Chemical Goggles	Protective Apron

Work/Hygiene Practices

Always wash hands thoroughly after working with tetraethyl lead powders and dusts; never bring food, drink, or smoking materials into vicinity of tetraethyl lead.

TETRAETHYL LEAD

$Pb(C_2H_5)_4$ CAS: 78-00-2

IDENTIFICATION AND TYPICAL USES

Tetraethyl lead is a colorless liquid with a pleasant, sweet odor. It may be dyed orange, red, blue, or other color. Used as anti-knock agent in gasoline aviation fuel. Formerly used in organomercury fungicides and in the manufacture of other metal alkyls such as ethyl mercury compounds. Since 1974, its use in gasoline has been largely replaced by methyl tert butyl ether (MTBE).

RISK ASSESSMENT: HEALTH

General Assessment

Tetraethyl lead is an organic compound, unlike lead which is inorganic. Tetraethyl lead is lipid-soluble and can easily enter the brain. Systemic human poison through *inhalation* of vapors or mists, *ingestion* either of foods or liquids containing tetraethyl lead, or by dermal *absorption*. It is a questioned carcinogen in animals and its ability to cause cancer in humans is considered suspect. Experimental teratogenic and reproductive effects have been reported. Mutation data are also documented. Because tetraethyl lead must first be metabolized to triethyl lead before toxicity results, it has a latency period (time from exposure to onset of symptoms) which can create a false sense of security with regard to health exposure risk.

The primary effects of tetraethyl lead are on the brain and the central nervous system. Exposure by inhalation and absorption can cause effects ranging in toxicity from mild to moderate to severe. Initially, there can be anxiety, irritability, insomnia, lurid dreams, vomiting, metallic taste, pallor, cerebellar ataxia, and diarrhea. Symptoms can progress to disorientation, hyperexcitability, tremors, chorea, bradycardia (slow heart beat), hypotension (low blood pressure), and hypothermia (decreased body temperature). Advanced indications of serious exposure include delusions, hallucinations, mania, convulsions, cerebellar edema (fluid in the brain), coma, tinnitus (ringing in the ears), impaired vision, elevated liver enzymes, anemia, and death. Tetraethyl lead metabolizes to inorganic lead which has an experimental half-life in the human body of 1460 days.

Acute Health Effects

The following acute (short-term) health effects may occur immediately or shortly after exposure to tetraethyl lead:

Skin: Tetraethyl lead can dry to a powder which can irritate the skin. The liquid also irritates the skin's surface and absorption is likely.

Eye: Irritation with redness and swelling.

Lung: Possible irritation of the nose, throat, lungs and air passages with sneezing, coughing, and other symptoms of pulmonary irritation.

CNS: A variety of effects including insomnia, strange dreams, personality changes, tremors or "shakes," stupor, hallucinations, delirium, parasthesia, headaches, convulsions and seizures, coma, and death.

Chronic Health Effects

The following chronic (long-term) health effects may occur at some time after exposure to tetraethyl lead and can last for months or even years:

Cancer Hazards: According to information presented in the references, tetraethyl lead has been shown to cause cancer in test animals. Mutation data have been reported and many scientists believe that exposure to mutagenic compounds may pose a cancer risk in the long-term.

Reproductive Hazard: According to information presented in the references, tetraethyl lead has been shown to adversely affect reproduction in test animals. There may be a decreased sex drive, impotence, sterility in males, decreased fertility and abnormal menstrual cycles in females and possible spontaneous fetal abortions in females. It may damage the testes and reduce sperm production.

Other Chronic Effects: Long-term exposure to tetraethyl can lead to appetite loss, nausea, anemia, weakness, insomnia, muscle and joint pain, and colic accompanied by severe abdominal pain. Kidney damage may occur, especially with repeated or high exposure. Symptoms may persist for weeks or months after exposure. Some loss of vision may occur after an eye splash or with severe, repeated exposures. Lead can cause increased blood pressure and can accumulate in the body with repeated exposures taking months or years for the body to purge itself of the contamination. Since tetraethyl lead is eventually metabolized into lead in the body, these effects may also occur with repeated exposure.

Recommended Risk-Reduction Measures

Personnel should avoid direct contact with tetraethyl lead. It is considered more hazardous than elemental lead because of its rapid solubility. If a less toxic material cannot be substituted for a hazardous chemical, then *engineering controls* are the most effective method of reducing exposures. The best protection is to enclose operations and/or provide explosion-proof, local exhaust ventilation at the site of chemical release. While not always operationally feasible, isolating operations can also reduce exposure. Using respiratory protection is less effective than the controls mentioned above, but is still advisable whenever working with or around tetraethyl lead. The exposure level is extremely low (0.075 mg/m^3) and difficult to accurately monitor. A supplied-air respirator operated in continuous flow mode or a self-contained breathing apparatus (SCBA) with full facepiece in pressure demand mode will provide the best respiratory protection. If a full facepiece respirator is not available, then splash goggles or safety glasses with side shields attached should be worn to protect the eyes. To prevent hand and skin exposures, suitable gloves should be used. Glove manufacturers should be contacted and permeation studies obtained *before* final glove selections are made.

Administrative controls should also be in place to minimize the potential for human exposures. These may include written procedures or policies which specify the methods and techniques that will be practiced whenever personnel are to work with tetraethyl lead.

All personnel should receive *training* on the use, hazards, protective measures, emergency actions, and other precautions per 29 CFR 1910.1200 (Hazard Communication), prior to the first assignment in an area where tetraethyl lead is used or stored.

Before beginning employment with tetraethyl lead, and at regular intervals thereafter (e.g., annually), the following medical tests are recommended:

☑ Blood lead levels and blood pressure (baseline).

If symptoms develop or overexposure is suspected, the following additional tests may be useful:

☑ Blood lead levels and blood pressure (compare to baseline).

☑ Complete blood count for microcytic anemia and basophilic stippling, blood urea nitrogen (BUN) of creatinine, and zinc protoporphyrin (ZPP).

☑ Examination of the central and peripheral nervous systems.

☑ Urine test for lead (levels of 0.1 mg per liter of urine indicate increased exposure).

Other methods to reduce exposure include:

☑ Where possible, enclose operations and use local exhaust ventilation at the site where tetraethyl lead is worked. If local exhaust ventilation or enclosure is not used, respiratory protection should be mandatory.

☑ Always ensure that proper protective clothing is worn when working with tetraethyl lead.

☑ Wash thoroughly immediately after exposure to tetraethyl lead and at the end of the work shift or before eating, drinking, or smoking. Clothing contaminated with tetraethyl lead should not be taken home.

☑ Hazard warning information should be posted in the work area. In addition, as part of an on-going education and training program, all information on the health and safety hazards of tetraethyl lead should be communicated to all exposed workers.

☑ Additional exposure control measures and engineering controls are provided by OSHA in 29 CFR 1910.1025 (Lead).

RISK ASSESSMENT: ENVIRONMENT

General Assessment

The environment is still at risk of exposure during transportation, storage, disposal, or destruction of tetraethyl lead. In almost every scenario, the threat of environmental exposure is contingent upon the proper handling of tetraethyl lead materials, especially the dusts and powders. Accidental contact with a variety of incompatible materials can result in fire, explosion, and possible contamination of the surrounding environmental mediums (water, soil, air).

Tetraethyl lead is considered a Class IIIB Combustible Liquid (per OSHA 29 CFR 1910.106). Its vapors can form explosive mixtures in air. It is incompatible with oxidizing agents and contact can cause fire or explosion. These characteristics require special consideration during any emergency situation involving a leak or spill of tetraethyl lead. Should tetraethyl lead ever come into contact with incompatible substances such as oxidizers, (perchlorates, peroxides, permanganates, chlorates, nitrates) and chemically active metals (potassium, sodium, magnesium, zinc), during transportation or storage, violent reactions are extremely possible.

Lead and its compounds are among the metals known and used since ancient times. It is one of the most abundant metals in the Earth's crust and can be dissolved from rocks and minerals into surface waters. It is used extensively across a broad spectrum of industries for a large variety of purpose. Because of its frequent use and common occurrence, lead enters the environment from point and non-point sources.

Acute Ecological Effects

Acute (short-term) toxic effects may include the death of animals, birds, or fish and death or low growth rate in plants. Acute effects are seen 2 to 4 days after animals or plants are exposed to tetraethyl lead.

Toxicity to aquatic life is affected by water hardness, the softer the water, the greater the toxicity. Lead and its compounds have high acute toxicity to aquatic life. Insufficient data are available on the short-term effects of tetraethyl lead and its compounds to plants, birds, or land animals.

Chronic Ecological Effects

Chronic (long-term) toxic effects may include shortened life span, reproductive problems, lower fertility, and changes in appearance or behavior in exposed animals. These effects can be seen long after first exposure(s) to toxic chemicals.

Lead and its compounds have high chronic toxicity in aquatic life. Lead causes nerve and behavioral effects in humans and could cause similar long-term effects in birds and land animals. Insufficient data are available on the long-term effects of lead to plants.

Water Solubility

Lead and its compounds range in their respective water solubilities from highly soluble to practically insoluble. Tetraethyl lead is considered insoluble in water.

Persistence in the Environment

Lead and its compounds are highly persistent in water, with a half-life greater than 200 days. The half-life of a pollutant is the amount of time it takes for one half of the chemical to be degraded.

Bioaccumulation in Aquatic Organisms

Some substances increase in concentration, or *bioaccumulate,* in living organisms as they breathe contaminated air, drink contaminated water, or eat contaminated food. These chemicals can become concentrated in the tissues and internal organs of animals as well as humans.

The concentration of tetraethyl lead found in the edible tissues of fish is expected to be much higher than the average concentration of tetraethyl lead found in the water from which the fish was taken.

Recommended Risk-Reduction Measures

Proper training of all transporters will reduce the likelihood of a mishap or accident resulting in the contamination of the environment. The correct use of DOT labeling (per the MSDS) on all containers, trucks, and rail cars will enable emergency responders to react properly and quickly to any disaster thereby reducing the potential risk to the environment and to personnel.

Storage of tetraethyl lead should be segregated from incompatible chemicals to minimize the risk of cross-contamination or contact. Buildings designated for storage should be equipped with appropriate fire protection systems (alarms, sprinklers, emergency lighting, portable extinguishers).

If a spill or leak to the environment has occurred, fire department, emergency response and/or hazardous materials spill personnel should be notified immediately. Cleanup should be attempted only by those trained in proper spill containment procedures using non-sparking tools. Contaminated soils should be removed and replaced with clean soil. When environmental contamination occurs, the local and/or state emergency response authorities may require notification. A comprehensive emergency response or disaster preparedness/recovery plan should be in place prior to any operations involving the use, transportation, storage, or disposal of tetraethyl lead.

If tetraethyl lead is spilled or leaked, the following specific steps are recommended:

☑ Restrict persons not wearing protective clothing from area of spill or leak until cleanup is complete and area can be opened for normal work.

☑ Remove all ignition sources and ventilate area.

☑ Absorb liquids in vermiculite or other absorbent material and place in sealed drums for disposal. Use non-sparking tools during cleanup.

☑ It may be necessary to dispose of tetraethyl lead as a hazardous waste. The responsible state agency or the regional office of the federal Environmental Protection Agency (EPA) should be contacted for specific recommendations.

RISK ASSESSMENT: BUSINESS

General Assessment

Accidents or mishaps involving tetraethyl lead can present a significant threat to business operations. The loss or damage of equipment or facilities due to explosion or fire can significantly affect fiscal viability. Lawsuits that may result from personnel injury/death, public exposures, and/or environmental contamination will also require a serious expenditure of resources. Media attention surrounding an injury, death, or environmental damage can also result in a loss of profits and loss of current as well as future business. Always remember that anytime the terms "cancer," "carcinogen," or "reproductive hazard" are used, public emotion, hysteria, and ignorance can run equally high. This is also true whenever the words "tetraethyl lead" and "exposure" are used together. This must be carefully considered whenever developing or implementing a public relations policy.

Recommended Risk-Reduction Measures

Company attorneys, safety and health professionals, and environmental specialists should be involved in the development of any procedures for dealing with safety or emergency response. A company official should be pre-designated as a public relations officer with specific training in dealing with the press. Corporate plans and policies should be developed, approved, and implemented long before any need for such arises.

REFERENCES

American Conference of Governmental Industrial Hygienists. 1988. *Documentation of the Threshold Limit Values and Biological Exposure Indices,* 5th Edition (with updates). Cincinnati: ACGIH

Bloom, A. D. and F. J. DeSerres. 1995. *Ecotoxicity and Human Health.* New York: CRC Press/Lewis Publishers.

Brusick, D. J. 1994. *Methods for Genetic Risk Assessment.* New York: CRC Press/Lewis Publishers.

Burrell, R., D. K. Flaherty, and Leonard J. Sauers. 1992. *Toxicology of the Immune System.* New York: Van Nostrand Reinhold.

Cothern, C. R. 1995. *Handbook for Environmental Risk Decisionmaking.* New York: CRC Press/Lewis Publishers.

Danse, I. R. 1991. *Common Sense Toxics in the Workplace.* New York: Van Nostrand Reinhold.

Gardner, R. W. 1994. *Chemical Intolerance.* New York: CRC Press/Lewis Publishers.

Howard, P. H., et. al. 1991. *Handbook of Environmental Degradation Rates.* New York: CRC Press/Lewis Publishers.

Keith, L. H. and M. M. Walker. 1995. *Handbook of Air Toxics.* New York: CRC Press/Lewis Publishers.

Lauwerys, R. R., P. Hoet. 1993. *Industrial Chemical Exposure.* New York: CRC Press/Lewis Publishers.

Lewis, R. J., Sr. 1993. *Hawley's Condensed Chemical Dictionary,* 12th Edition. New York: Van Nostrand Reinhold.

Lewis, R. J., Sr. 1992. *Sax's Dangerous Properties of Industrial Materials,* Volumes 1, 2, and 3, 8th Edition. New York: Van Nostrand Reinhold.

Manahan, S. E. 1992. *Toxicological Chemistry,* 2nd Edition. New York: CRC Press/Lewis Publishers.

Meyer, E. 1990. *Chemistry of Hazardous Materials.* Englewood Cliffs, NJ: Prentice Hall.

Patnaik, P. 1992. *A Comprehensive Guide to the Hazardous Properties of Chemical Substances.* New York: Van Nostrand Reinhold.

Rea, W. J. 1995. *Chemical Sensitivity,* Vol. I-III. New York: CRC Press/Lewis Publishers.

Richardson, M. 1995. *Dictionary of Substance and Their Effects,* Vol. I-VI. New York: CRC Press/Lewis Publishers.

Sacarello, H. L. A. 1994. *The Comprehensive Handbook of Hazardous Materials.* New York: CRC Press/Lewis Publishers.

Sherman, J. D. 1988. *Chemical Exposure and Disease.* New York: Van Nostrand Reinhold.

U.S. Department of Health and Human Services, National Institute for Occupational Safety and Health. 1994. *NIOSH Pocket Guide to Chemical Hazards.* Washington, D.C.: U.S. Government Printing Office.

MATERIAL SAFETY DATA SHEET

CHEMICAL NAME	
TETRAHYDROFURAN	**CRC PRESS/LEWIS PUBLISHERS DISCLAIMER:** THE INFORMATION AND RECOMMENDATIONS PRESENTED HEREIN ARE BASED ON SOURCES BELIEVED TO BE RELIABLE. CRC MAKES NO REPRESENTATION ON ITS COMPLETENESS OR ACCURACY. IT IS THE RESPONSIBILITY OF THE USER TO DETERMINE THE CHEMICAL'S SUITABILITY FOR ITS INTENDED USE, THE CHEMICAL'S SAFE USE, AND THE CHEMICAL'S PROPER DISPOSAL. NO REPRESENTATIONS AND/OR WARRANTIES, EITHER EXPRESSED OR IMPLIED, OF THE MERCHANTABILITY OR FITNESS FOR A PARTICULAR PURPOSE, OR OF ANY OTHER NATURE, ARE MADE WITH RESPECT TO THE INFORMATION PROVIDED IN THIS MSDS OR TO THE CHEMICAL TO WHICH INFORMATION MAY REFER. CRC NEITHER ASSUMES NOR AUTHORIZES ANY OTHER PERSON TO ASSUME FOR IT, ANY OTHER ADDITIONAL RESPONSIBILITY OR LIABILITY FOR THE USE OF, OR RELIANCE UPON, THIS INFORMATION.

HAZARD WARNING INFORMATION

HEALTH	FIRE	REACTIVE	OTHER	DEGREE OF HAZARD	COLOR CODING	OTHER CODES
2	3	1		0 = Minimum Hazard 1 = Slight Hazard 2 = Moderate Hazard 3 = Serious Hazard 4 = Severe Hazard	HEALTH = BLUE FIRE = RED REACTIVITY = YELLOW OTHER = WHITE	OX = Oxidizer ACID = Acid ALK = Alkali COR = Corrosive W = Use No Water

SECTION I - GENERAL INFORMATION

Characterization	RCRA Number	EPA Class
Glycol Ether	**U213**	**Toxic Waste**

DOT Proper Shipping Name	Chemical Abstract Service (CAS) Number
Tetrahydrofuran	**109-99-9**
DOT Hazard Class and Label Requirements	**DOT Emergency Guide Code**
Flammable Liquid	**26**
DOT Identification Number	**Molecular Formula**
UN 2056	C_4H_8O

Synonyms

Diethylene oxide; 1,4-epoxybutane; tetramethylene oxide; THF; oxolane; oxacyclopentane; butylene oxide; cyclotetramethylene; 1,4-epoxybutane.

SECTION II - HAZARDOUS INGREDIENTS AND IDENTITY INFORMATION

Hazardous Components (specific identity)	OSHA Exposure Criteria	NIOSH Exposure Criteria	Immediately Dangerous to Life and Health (IDLH)	ACGIH Exposure Criteria
Tetrahydrofuran: (derivation: By catalytic hydrogenation of furan with nickel catalyst; by acid-catalyzed dehydration of 1,4-butanediol). 1 ppm = 3.00 mg/m^3	PEL (8-hour): 200 ppm 590 mg/m^3 STEL (15-min): 250 ppm 735 mg/m^3	REL (10-hour): 200 ppm 590 mg/m^3 STEL (15-min): 250 ppm 735 mg/m^3	2000 ppm	TLV: 200 ppm 590 mg/m^3 STEL (15-min): 250 ppm 735 mg/m^3

SECTION III - PHYSICAL AND CHEMICAL CHARACTERISTICS

Boiling Point	Specific Gravity (H_2O = 1)
151°F (66°C)	**0.89**
Vapor Pressure (mm Hg)	Molecular Weight (atomic weight)
145 at 68°F (20°C)	**72.1**
Vapor Density (Air = 1)	Melting Point
1.5	**-163°F (-108°C)**

Solubility

Miscible in water. Soluble in alcohol, ketones, esters, ethers, and hydrocarbons.

Appearance and Odor

Colorless, water-white liquid with an ether-like odor. Odor Threshold = 30 ppm.

SECTION IV - FIRE AND EXPLOSION HAZARD DATA

Flash Point (method used)	Explosive Limits in Air % by Volume
6°F (-14°C) opened cup	**LEL: 2% UEL: 11.8%**
NFPA Classification	Autoignition Temperature
Class IB Combustible Liquid	**610°F (321°C)**

Extinguishing Media

Dry chemical, carbon dioxide, or alcohol foam.

Special Fire Fighting Procedures

Acrid and irritating smoke and fumes are produced in fire. Structural fire-fighting protective clothing will not provide adequate protection. Wear full protective clothing and self-contained breathing apparatus (SCBA). Move container from fire area if it can be done without risk. Use water spray to cool containers.

Unusual Fire and Explosion Hazards

Containers may explode in fire. Presents a vapor explosion hazard indoors, outdoors, and in sewers. Boiling liquid expanding vapor explosion (BLEVE) is possible.

SECTION V - REACTIVITY DATA

Stability		Conditions to Avoid Forms explosive peroxides when exposed to air or sunlight. Can be kept stable with an inhibitor. Tetrahydrofuran is extremely volatile and very flammable. Keep away from ALL sources of heat or ignition.
Stable	Unstable X	Incompatibility *(materials to avoid)* Strong oxidizers, lithium-aluminum alloys, potassium hydroxide, sodium hydroxide, sodium tetrahydroaluminate, metal halides, and sulfinyl chloride. Attacks some plastics, rubber, and coatings.
Hazardous Polymerization		Conditions to Avoid Hazardous polymerization may occur in the presence of cationic initiators (such as strong proton acids). Keep away from incompatible materials.
May Occur X	Will Not Occur	Hazardous Decomposition or By-products When heated to decomposition, tetrahydrofuran emits acrid smoke and irritating fumes (such as carbon oxides).

SECTION VI - HEALTH HAZARD DATA

Primary Route(s) of Entry:	Inhalation?	Absorption (skin/eye)?	Ingestion?
	X		X

Health Hazards

INHALATION: Smarting to the eyes and respiratory system with cough and chest pain. There can be anesthesia, severe occipital headache, nausea, dizziness, a marked increase in white blood cell count, and narcotic effects at high concentrations. May damage the liver and kidney. A central nervous system (CNS) depressant.

SKIN & EYES: Skin irritation and possible inflammation and redness. Eye contact can result in painful irritation.

INGESTION: Moderately toxic. May cause unspecified gastrointestinal effects.

Carcinogenicity	NTP Listed?	IARC Cancer Review Group?	OSHA Regulated?	Target Organs?
Unknown Human Unknown Animal	No	No	29 CFR 1910.1000 Table Z-1	Skin, eyes, liver, kidneys, CNS, resp. system

Medical Conditions Generally Aggravated by Exposure

Kidney, liver, and skin disorders may be aggravated by exposure.

Emergency and First-aid Procedures

Eye contact: Do not allow victim to keep eyes tightly shut. Flush immediately with water for 15 minutes (minimum); seek medical attention. Skin contact: Wash with large amounts of soap and water for a minimum of 15 minutes. For inhalation: Remove the person from exposure. Provide respiratory assistance and CPR. Support breathing with 100% humidified oxygen. Transfer to medical facility. If swallowed: Seek medical attention immediately. Give 1 to 2 glasses of water to dilute. Gastric lavage is indicated if performed soon after ingestion and person is not at risk of convulsing. Do NOT attempt to give an unconscious or convulsing person anything by mouth.

SECTION VII - PRECAUTIONS FOR SAFE HANDLING AND USE

Steps to be Taken in Case Material is Released or Spilled

Spills of this material constitute a serious emergency. Wear self-contained breathing apparatus (SCBA) for respiratory protection. Restrict those not involved in cleanup from entering area. Avoid skin contact. Ventilate area of spill. Absorb in vermiculite or other material and dispose of in sealed drums.

Preferred Waste Disposal Method

Burn in chemical incinerator equipped with an afterburner and scrubber.

Precautions to be Taken in Handling and Storage

Store in tightly closed containers in a cool, well-ventilated area away from heat (including direct sunlight). Protect containers from physical damage.

Other Precautions and Warnings

Tetrahydrofuran must be stored to avoid contact with strong oxidizers since violent reactions can occur.

SECTION VIII - CONTROL MEASURES AND PERSONAL PROTECTIVE EQUIPMENT

Respiratory Protection *(specify type)*

For low exposure, use a half-mask respirator with an organic vapor cartridge. Use a NIOSH/MSHA-approved self-contained breathing apparatus with full facepiece operated in positive pressure mode.

Ventilation

Local exhaust (preferred) at site of chemical work or general ventilation.

Protective Gloves	Eye Protection	Other Protective Clothing
Butyl Rubber or Neoprene	Chemical Goggles and Face Mask	Protective Uniform or Apron

Work/Hygiene Practices

Always wash hands thoroughly after using chemical; never bring food, drink, or smoking materials into vicinity of chemicals.

TETRAHYDROFURAN

C_4H_8O

CAS: 109-99-9

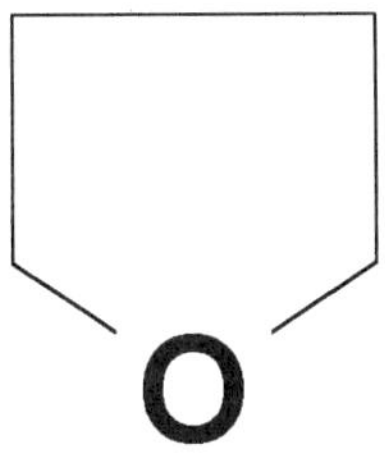

IDENTIFICATION AND TYPICAL USES

Tetrahydrofuran appears as a water-white or colorless liquid with a fruity, ethereal-like odor and a pungent taste. It is used as a solvent for natural and synthetic resins, particularly vinyl, in topcoat solutions, as a polymer coating, in the production of cellophane, in protective coatings, in adhesives, in magnetic tapes, in printing inks, and similar materials. It is also used in Grignard reactions, in lithium-aluminum hydride reductions, and polymerizations. It is also a chemical intermediate and monomer.

RISK ASSESSMENT: HEALTH

General Assessment

Tetrahydrofuran is moderately toxic by *inhalation* and *ingestion* and can cause topical effects in contact with the skin and eyes. While mutation data have been reported, there are no data on its carcinogenic capabilities. Also, the references contain no information on any probable teratogenic effects. The primary action appears to be on the central nervous system (CNS).

Inhalation can cause irritation of the eyes, nose, throat, and respiratory system with symptoms of coughing and chest pain. There can be severe occipital headache, nausea, and dizziness that can be eliminated in fresh air. There can also be a marked decrease in white blood count and possible liver and kidney damage.

Skin contact may produce no significant or apparent irritation on brief exposure. However, prolonged or repeated contact may cause a rash or even a burn to tissues.

Ingestion may result in unspecified gastrointestinal and respiratory effects. Systemic poisoning is extremely possible with symptoms similar to that of inhalation.

Acute Health Effects

The following acute (short-term) health effects may occur immediately or shortly after exposure to tetrahydrofuran:

Skin: Mild to severe irritation and possible burns.

Eye: Severe conjunctival irritation and possible damage to vision.

Lung: Irritation to respiratory tract causing shortness of breath and cough.

CNS: A depressant of the central nervous system causing a variety of toxic effects including drowsiness, coma, and death. There can also be headache, coughing, and chest pain.

CNS: Primary action is depression of the central nervous system with dizziness, anesthesia, nausea, headache, and other effects.

Chronic Health Effects

The following chronic (long-term) health effects may occur at some time after exposure to tetrahydrofuran and can last for months or even years:

Cancer Hazards: According to information presented in the references, tetrahydrofuran has not been adequately tested for its ability to cause cancer in test animals. However, it is a known mutagen (causes cell changes) and many scientists believe that exposure to such chemicals may pose a cancer risk in the long-term.

Reproductive Hazard: According to the references, tetrahydrofuran has not been adequately tested for its ability to adversely affect reproduction.

Other Chronic Effects: Repeated low-dose exposures can irritate the lungs and result in bronchitis with symptoms of cough, phlegm, and shortness of breath. It may damage the liver and kidneys. Chronic exposure can lead to jaundice and renal failure.

Recommended Risk-Reduction Measures

Personnel should avoid direct contact with tetrahydrofuran. If a less toxic material or compound cannot be substituted for tetrahydrofuran, then *engineering controls* are the most effective method of reducing exposure risk. The best protection is to enclose operations and/or provide local exhaust ventilation at the site of tetrahydrofuran release. While not always operationally feasible, isolating operations can also reduce exposure. Using respiratory protection is less effective than the controls mentioned above, but is still required whenever working with or around tetrahydrofuran. A NIOSH/MSHA-approved organic vapor respirator may be used for low-level exposures (at or over the PEL). However, the warning properties of tetrahydrofuran are low and cartridge-type respirators may not suffice. Therefore, for the best protection at any exposure level, a MSHA/NIOSH-approved self-contained breathing apparatus (SCBA) with full facepiece and pressure demand is the appropriate level of respiratory protection. If a full facepiece is not available, then chemical goggles should be worn to protect the eyes, especially when a splash hazard exists. A face shield should also be considered. Skin contact must be avoided. To prevent hand and skin exposures, butyl rubber, neoprene, or other non-permeable gloves should be used. Caution is warranted in glove selection since tetrahydrofuran is known to attack some forms of rubber, plastics, and coatings. Glove manufacturers should be contacted and permeation studies obtained *before* final glove selections are made.

Administrative controls should also be in place to minimize the potential for human exposures. These may include written procedures or policies which specify the methods and techniques that will be practiced whenever personnel are to work with tetrahydrofuran.

All personnel should receive *training* on the use, hazards, protective measures, emergency actions, and other precautions per 29 CFR 1910.1200 (Hazard Communication) prior to the first assignment in an area where tetrahydrofuran is used or stored.

Before beginning employment and at regular intervals thereafter (e.g., annually), the following medical test is recommended:

- ☑ Lung function tests (establish baseline).

If symptoms develop or overexposure is suspected, the following may be useful:

- ☑ Liver and kidney function tests.
- ☑ Lung function tests (compare to baseline).
- ☑ Blood urea nitrogen test with serum creatinine and urine sediment.

It should be noted that medical tests that simply look for existing damage are not a substitute for controlling exposures. Medical histories are extremely important when assessing exposure risk. Also, the consumption of alcoholic beverages appears to enhance the toxicity of tetrahyrofuran. Therefore, people who use alcohol prior to or immediately after exposure may exhibit symptoms more rapidly than others under the same conditions of exposure. Prudent risk management requires proper consideration of *all* possible factors that may be causing the appearance of exposure symptoms among workers.

Other methods to reduce exposure include:

- ☑ Always ensure that proper protective clothing is worn when using chemical substances and that personnel have been trained in its use, care, and maintenance.
- ☑ Wash thoroughly immediately after exposure to tetrahydrofuran and at the end of the work shift or before eating, drinking, applying cosmetics, or smoking.
- ☑ Hazard warning information should be posted in the work area. In addition, as part of an on-going education and training program, all information on the health and safety hazards of tetrahydrofuran should be communicated to all exposed and potentially exposed workers.
- ☑ Eye wash stations should be provided in the immediate work area for emergency use. If there is a possibility of skin exposure to tetrahydrofuran, emergency shower facilities should also be provided in the work area.
- ☑ Workers whose clothing has been contaminated by tetrahydrofuran should change into clean clothes before leaving work. Contaminated work clothing should be laundered only by individuals who have been informed of the hazards of exposure to tetrahydrofuran.

RISK ASSESSMENT: ENVIRONMENT

General Assessment

The environment is at risk of exposure during transportation, storage, disposal, or destruction of tetrahydrofuran. In almost every scenario, the threat of environmental exposure is contingent upon the proper handling of the chemical substance. Accidental spills, large or small, can result in contamination of the surrounding environmental mediums (water, soil, and air).

Tetrahydrofuran is Class IB combustible liquid (per OSHA 29 CFR 1910.106). It is extremely volatile and presents a serious risk of fire and or explosion. It will form dangerous (explosive) peroxides in air. It also forms explosive complexes with the perchlorates of many metals, such as potassium or sodium. It is incompatible with oxidizers, reacting vigorously and violently on contact. Therefore, caution is always required in handling, storage, transportation, and disposal of tetrahydrofuran. When heated to decomposition it emits very acrid and irritating smoke and fumes. Emergency responders, hazardous materials personnel, and firefighters should be made aware of the presence of tetrahydrofuran at any emergency response situation.

Tetrahydrofuran may enter the environment through industrial effluents and spills.

Acute Ecological Effects

Acute (short-term) toxic effects may include the death of animals, birds, or fish and death or low growth rate in plants. Acute effects are seen 2 to 4 days after animals or plants are exposed to tetrahydrofuran.

Insufficient data are available to evaluate or predict the acute (short-term) effects of tetrahydrofuran to aquatic life, plants, birds, or land animals.

Chronic Ecological Effects

Chronic (long term) toxic effects may include shortened life span, reproductive problems, lower fertility, and changes in appearance or behavior in exposed animals. These effects can be seen long after first exposure(s) to toxic chemicals.

Insufficient data are available to evaluate the chronic (long-term) effects of tetrahydrofuran on aquatic life, plants, birds, or land animals.

Water Solubility

Tetrahydrofuran is highly soluble in water. Concentrations of 1000 milligrams and more will mix with a liter of water.

Persistence in the Environment

Tetrahydrofuran is slightly persistent in water, with a half-life of between 2 to 20 days. The half-life of a pollutant is the amount of time it takes for one half of the chemical to be degraded.

Bioaccumulation in Aquatic Organisms

Some substances increase in concentration, or *bioaccumulate,* in living organisms as they breathe contaminated air, drink contaminated water, or eat contaminated food. These chemicals can become concentrated in the tissues and internal organs of animals as well as humans.

Tetrahydrofuran is not expected to accumulate in aquatic organisms. The concentration of tetrahydrofuran in edible fish tissues is expected to be negligible (trace levels) as compared to the levels found the water from which the fish was taken.

Recommended Risk-Reduction Measures

Proper training of all transporters will reduce the likelihood of a mishap or accident resulting in a leak or spill of tetrahydrofuran into the environment. Labels on all containers, trucks, and rail cars must meet DOT requirements and accurately reflect their contents to enable emergency responders to react properly and quickly to any disaster thereby reducing the potential risk to the environment and to personnel.

Storage of tetrahydrofuran should be segregated from other chemicals to minimize the risk of cross-contamination. Tetrahydrofuran must be stored to avoid contact with strong oxidizers, such as chlorine and bromine, since violent reactions can occur. Containers should be protected from physical damage and stored to avoid contact with heat and direct sunlight.

If a spill or leak to the environment has occurred, fire department, emergency response and/or hazardous materials spill personnel should be notified immediately. Cleanup should be attempted only by those trained in proper spill containment procedures using non-sparking tools. Contaminated soils should be removed for incineration and replaced with clean soil.

If tetrahydrofuran should contact the water table, aquifer, or navigable waterway, time is of the essence. It is highly soluble in water and total remediation may or may not be possible. The local and/or state emergency response authorities must be notified. A comprehensive emergency response or disaster preparedness/recovery plan should be in place prior to any operations involving the use, transportation, storage, or disposal of tetrahydrofuran.

If tetrahydrofuran is spilled or leaked, the following specific steps are recommended:

☑ Restrict persons not wearing protective clothing from area of spill or leak until cleanup is complete. Avoid skin and eye contact.

☑ Absorb spilled liquids with vermiculite, dry sand, or a similar material and deposit in sealed drum for disposal. Use non-sparking tools.

☑ Ventilate area of spill or leak.

☑ Remove all ignition sources.

☑ It may be necessary to dispose of tetrahydrofuran as a hazardous waste. The responsible state agency or the regional office of the federal Environmental Protection Agency (EPA) should be contacted for specific recommendations.

RISK ASSESSMENT: BUSINESS

General Assessment

Accidents or mishaps involving tetrahydrofuran can present a serious threat to business operations. The loss or damage of equipment or facilities can significantly affect fiscal viability. Lawsuits that may result from personnel illness, injury/death, public exposures, and/or environmental contamination will require a serious expenditure of resources. Media attention surrounding an injury, death, or environmental damage can also result in a loss of profits and loss of current as well as future business.

Recommended Risk-Reduction Measures

Company attorneys, safety and health professionals, and environmental specialists should be involved in the development of any procedures for responding to chemical incidents. A company official should be predesignated as a public relations officer with specific training in dealing with the press. Corporate plans and policies should be developed, approved, and implemented long before any need for such arises.

REFERENCES

Anthony, C. P., and N. J. Kolthoff. 1971. *Textbook of Anatomy and Physiology.* St. Louis: C. V. Mosby Company.

American Conference of Governmental Industrial Hygienists. 1988. *Documentation of the Threshold Limit Values and Biological Exposure Indices,* 5th Edition (with updates). Cincinnati: ACGIH

Brusick, D. J. 1994. *Methods for Genetic Risk Assessment.* New York: CRC Press/Lewis Publishers.

Calabrese, E. J. 1994. *Biological Effects of Low Level Exposures.* New York: CRC Press/Lewis Publishers.

Cothern, C. R. 1995. *Handbook for Environmental Risk Decisionmaking.* New York: CRC Press/Lewis Publishers.

Danse, I. R. 1991. *Common Sense Toxics in the Workplace.* New York: Van Nostrand Reinhold.

Environmental Protection Agency. 1983. *Chemical Hazard Information Profiles.* Washington, D.C.: U.S. EPA.

Fodden, C. R., and J. L. Weddell. 1992. *Hazardous Materials Emergency Action Data.* New York: CRC Press/Lewis Publishers.

Forseberg, K., and L. H. Keith. 1993. *Chemical; Protective Clothing Permeation and Degradation Database.* Chelsea, MI: Lewis Publishers.

Gardner, R. W. 1994. *Chemical Intolerance.* New York: CRC Press/Lewis Publishers.

Hallenbeck, W. H. 1993. *Quantitative Risk Assessment for Environmental and Occupational Health,* 2nd Edition. New York: CRC Press/Lewis Publishers.

Hodgson, E., R. B. Mailman, and Janice E. Chambers. 1988. *Dictionary of Toxicology.* London: The MacMillian Company.

Hosty, J. W., and P. Foster. 1990. *A Practical Guide to Chemical Spill Response.* New York: Van Nostrand Reinhold.

Howard, P. H. 1995. *Handbook of Environmental Fate and Exposures Data for Organic Chemicals,* Vol. I-IV. New York: CRC Press/Lewis Publishers.

Howard, P. H., et. al. 1991. *Handbook of Environmental Degradation Rates.* New York: CRC Press/Lewis Publishers.

Langerman, N. 1994 *Precautionary Labels for Chemical Containers.* New York: CRC Press/Lewis Publishers.

Lauwerys, R. R., P. Hoet. 1993. *Industrial Chemical Exposure.* New York: CRC Press/Lewis Publishers.

Lewis, R. J., Sr. 1992. *Sax's Dangerous Properties of Industrial Materials,* Volumes 1, 2, and 3, 8th Edition. New York: Van Nostrand Reinhold.

Manahan, S. E. 1992. *Toxicological Chemistry,* 2nd Edition. New York: CRC Press/Lewis Publishers.

Meyer, E. 1990. *Chemistry of Hazardous Materials.* Englewood Cliffs, NJ: Prentice Hall.

Patnaik, P. 1992. *A Comprehensive Guide to the Hazardous Properties of Chemical Substances.* New York: Van Nostrand Reinhold.

Torkelson, T. R. V. K. Rowe. 1981. *Patty's Industrial Hygiene and Toxicology,* Vol. 2B. New York: Wiley-Interscience.

U.S. Department of Health and Human Services, National Institute for Occupational Safety and Health. 1994. *NIOSH Pocket Guide to Chemical Hazards.* Washington, D.C.: U.S. Government Printing Office.

MATERIAL SAFETY DATA SHEET

CHEMICAL NAME	CRC PRESS/LEWIS PUBLISHERS DISCLAIMER:
TETRAMETHYL LEAD	THE INFORMATION AND RECOMMENDATIONS PRESENTED HEREIN ARE BASED ON SOURCES BELIEVED TO BE RELIABLE. CRC MAKES NO REPRESENTATION ON ITS COMPLETENESS OR ACCURACY. IT IS THE RESPONSIBILITY OF THE USER TO DETERMINE THE CHEMICAL'S SUITABILITY FOR ITS INTENDED USE, THE CHEMICAL'S SAFE USE, AND THE CHEMICAL'S PROPER DISPOSAL. NO REPRESENTATIONS AND/OR WARRANTIES, EITHER EXPRESSED OR IMPLIED, OF THE MERCHANTABILITY OR FITNESS FOR A PARTICULAR PURPOSE, OR OF ANY OTHER NATURE, ARE MADE WITH RESPECT TO THE INFORMATION PROVIDED IN THIS MSDS OR TO THE CHEMICAL TO WHICH INFORMATION MAY REFER. CRC NEITHER ASSUMES NOR AUTHORIZES ANY OTHER PERSON TO ASSUME FOR IT, ANY OTHER ADDITIONAL RESPONSIBILITY OR LIABILITY FOR THE USE OF, OR RELIANCE UPON, THIS INFORMATION.

HAZARD WARNING INFORMATION

HEALTH	FIRE	REACTIVE	OTHER	DEGREE OF HAZARD	COLOR CODING	OTHER CODES
3	3	3		0 = Minimum Hazard 1 = Slight Hazard 2 = Moderate Hazard 3 = Serious Hazard 4 = Severe Hazard	HEALTH = BLUE FIRE = RED REACTIVITY = YELLOW OTHER = WHITE	OX = Oxidizer ACID = Acid ALK = Alkali COR = Corrosive W̶ = Use No Water

SECTION I - GENERAL INFORMATION

Characterization	RCRA Number	EPA Class
Metal Alkyl	**D008 (as Lead)**	**Characteristic (T) Waste**

DOT Proper Shipping Name	Chemical Abstract Service (CAS) Number
Flammable Liquid, Poisonous, N.O.S.	**75-74-1**
DOT Hazard Class and Label Requirements	**DOT Emergency Guide Code**
Flammable Liquid and Poison	**56**
DOT Identification Number	**Chemical Symbol**
UN 1992	**$Pb(CH_3)_4$**

Synonyms

Lead tetramethyl; TML; tetramethylplumbane.

SECTION II - HAZARDOUS INGREDIENTS AND IDENTITY INFORMATION

Hazardous Components (specific identity)	OSHA Exposure Criteria	NIOSH Exposure Criteria	Immediately Dangerous to Life and Health (IDLH)	ACGIH Exposure Criteria
Tetramethyl lead [derivation: By alkylation of lead-sodium alloy with methyl chloride; by electrolysis of a methyl Grignard reagent with an anode of lead pellets]. **1 ppm = 11.11 mg/m^3**	**PEL (as Pb): 0.075 mg/m^3 (skin)** **STEL: Not Established**	**PEL (as Pb): 0.075 mg/m^3 (skin)** **STEL: Not Established**	**40 mg/m^3 (as Pb)**	**TLV (as Pb): 0.15 mg/m^3 (skin)** **STEL: Not Established**

SECTION III - PHYSICAL AND CHEMICAL CHARACTERISTICS

Boiling Point	Specific Gravity (H_2O = 1)
212°F (100°C) Decomposes	**2.00**
Vapor Pressure (mm Hg)	Molecular Weight
23 at 68°F (20°C)	**267.3**
Vapor Density (Air = 1)	Melting Point
9.2	**-15°F (-26°C)**

Solubility

Insoluble in water (decomposes first to trimethyl salt, then dimethyl salt, finally inorganic lead). Soluble in alcohol, benzene, and petroleum ether. Slightly soluble in ethyl ether. Miscible in fats and oils.

Appearance and Odor

Colorless liquid that may be dyed red, orange, blue, or other color, with a pleasant, fruity, sweet odor.

SECTION IV - FIRE AND EXPLOSION HAZARD DATA

Flash Point (method used)	Explosive Limits in Air % by Volume
100°F (38°C) closed cup	**LEL: 1.8% UEL: Not Determined**
NFPA Classification	Autoignition Temperature
Class II Combustible Liquid	**Not Reported**

Extinguishing Media

Use dry chemical, carbon dioxide, water spray, fog, or regular foam.

Special Fire Fighting Procedures

Poisonous gases are produced in fire. Wear self-contained breathing apparatus (SCBA) and special protective clothing. Use cool water spray to cool the sides of fire-exposed containers. Stay clear of the ends of tanks and evacuate 1500 feet if rising sound is heard coming from venting device.

Unusual Fire and Explosion Hazards

A boiling liquid expanding vapor explosion (BLEVE) can occur when the liquid in containers adjacent to a fire begins to boil. May explode in contact with friction or oxidizers, or above 194°F (90°C).

SECTION V - REACTIVITY DATA

Stability		Conditions to Avoid **Tetramethyl lead is normally stable at room temperature under routine conditions of handling and storage. However, it decomposes in sunlight and peroxides may accumulate upon prolonged storage or in air.**
Stable X	Unstable	Incompatibility *(materials to avoid)* **Strong oxidizers (chlorine, fluorine, bromine), sulfuryl chloride, and potassium permanganate. Explodes on contact with tetrachlorotrifluoromethyl phosphorane.**
Hazardous Polymerization		Conditions to Avoid **Under normal conditions of temperature and pressure, hazardous polymerization of tetramethyl lead will not occur. Avoid exposure to sunlight and air.**
May Occur	Will Not Occur X	Hazardous Decomposition or By-products **Thermal decomposition of tetramethyl lead will produce toxic and irritating smoke and fumes, including highly toxic fumes of lead and carbon dioxide.**

SECTION VI - HEALTH HAZARD DATA

Primary Route(s) of Entry:	Inhalation? X	Absorption (skin)? X	Ingestion? X

Health Hazards

INHALATION: Primary effects on the brain and central nervous system (CNS). Effects can range from mild to moderate to severe. Symptoms range from anxiety, irritability, insomnia, lurid dreams, vomiting, metallic taste, pallor, cerebellar ataxia, diarrhea, disorientation, tremors, hyperexcitability, chorea, slow heart beat, hypotension, hypothermia, delusions, hallucinations, mania, convulsions, cerebellar edema, coma, tinnitus, impaired vision, elevated liver enzymes, anemia, lassitude, and death.

ABSORPTION: Eye and skin irritation and rapid absorption. Symptoms of inhalation are possible.

INGESTION: Easily absorbed when ingested. If it enters the bloodstream, it can have many of the same effects as inhalation. It is a systemic poison that accumulates in body organs.

Carcinogenicity	NTP Listed?	IARC Cancer Review Group?	OSHA Regulated?	Target Organs?
Unknown Human Unknown Animal	No	No3	29 CFR 1910.1000 Table Z-1 (as Pb)	Eyes, liver, CNS, CVS, kidneys.

Medical Conditions Generally Aggravated by Exposure

Mental disorders, nervous system dysfunction, kidney diseases, hypertension, and sickle-cell anemia.

Emergency and First-aid Procedures

Eye contact: Flush immediately with water for 15 minutes (minimum); seek medical attention. Skin contact: Remove all contaminated clothing. Wash contaminated skin with soap and water. For inhalation: Remove the person from exposure. Provide respiratory assistance and CPR. Transfer to medical facility. If swallowed: Never attempt to give anything by mouth to an unconscious or convulsing person. If large quantities are ingested, induce vomiting with Syrup of Ipecac and call physician.

SECTION VII - PRECAUTIONS FOR SAFE HANDLING AND USE

Steps to be Taken in Case Material is Released or Spilled

Remove all sources of ignition. Restrict those not involved in cleanup from entering area. Absorb liquids in vermiculite or other material and deposit in sealed drums. Use non-sparking tools.

Preferred Waste Disposal Method

No citation.

Precautions to be Taken in Handling and Storage

Containers should be protected against physical damage. Store in tightly closed containers in a cool, dry, well-ventilated area away from incompatible materials and direct sunlight.

Other Precautions and Warnings

Wherever tetramethyl lead is used, handled, manufactured, or stored, use explosion-proof electrical equipment and fittings.

SECTION VIII - CONTROL MEASURES AND PERSONAL PROTECTIVE EQUIPMENT

Respiratory Protection *(specify type)*

Exposure level is low and difficult to monitor. Use a respirator with particulate (dust/fume/mist) filter; otherwise use a supplied-air respirator or self-contained breathing apparatus in continuous flow mode.

Ventilation

Local exhaust or general mechanical systems recommended.

Protective Gloves	Eye Protection	Other Protective Clothing
Suitable to protect from skin contact	Chemical Goggles	Protective Apron

Work/Hygiene Practices

Always wash hands thoroughly after working with tetramethyl lead powders and dusts; never bring food, drink, or smoking materials into vicinity of tetramethyl lead.

TETRAMETHYL LEAD

$Pb(CH_3)_4$

CAS: 75-74-1

IDENTIFICATION AND TYPICAL USES

Tetramethyl lead is a colorless liquid with a pleasant, sweet odor. It may be dyed orange, red, blue, or other color. Used as anti-knock agent in gasoline aviation fuel. It was used as a replacement for tetraethyl lead during the 1960s before it was replaced by methyl tert butyl ether (MTBE).

RISK ASSESSMENT: HEALTH

General Assessment

Tetramethyl lead is an organic compound, unlike lead which is inorganic. Tetramethyl lead is lipid-soluble and can easily enter the brain (encephalopathy). Systemic human poison through *inhalation* of vapors or mists, *ingestion* either of foods or liquids containing tetramethyl lead, or by dermal *absorption*. It is an experimental teratogenic and reproductive effects have been reported. There are no data on its carcinogenic or mutagenic effects in humans or animals. Because tetramethyl lead must first be metabolized to triethyl lead, then dimethyl lead, and then to inorganic lead before toxicity results, it has a latency period (time from exposure to onset of symptoms) which can create a false sense of security with regard to health exposure risk. Lead is eventually deposited in bone.

The primary effects of tetramethyl lead are on the brain and the central nervous system. Exposure by inhalation and absorption can cause effects ranging in toxicity from mild to moderate to severe. Initially, there can be anxiety, irritability, insomnia, lurid dreams, vomiting, metallic taste, pallor, cerebellar ataxia, and diarrhea. Symptoms can progress to disorientation, hyperexcitability, tremors, chorea, bradycardia (slow heart beat), hypotension (low blood pressure), and hypothermia (decreased body temperature). Advanced indications of serious exposure include delusions, hallucinations, mania, convulsions, cerebellar edema (fluid in the brain), coma, tinnitus (ringing in the ears), impaired vision (disturbances to the optic nerve, retina, lens, and visual cortex), elevated liver enzymes, anemia, and death. Tetramethyl lead metabolizes to inorganic lead which has an experimental half-life in the human body of 1460 days.

Acute Health Effects

The following acute (short-term) health effects may occur immediately or shortly after exposure to tetramethyl lead:

Skin: Tetramethyl lead can dry to a powder which can irritate the skin. The liquid also irritates the skin's surface and absorption is likely.

Eye: Irritation with redness and swelling.

Lung: Possible irritation of the nose, throat, lungs and air passages with sneezing, coughing, and other symptoms of pulmonary irritation.

CNS: A variety of effects including insomnia, strange dreams, personality changes, tremors or "shakes," stupor, hallucinations, delirium, parasthesia, headaches, convulsions and seizures, coma, and death.

Chronic Health Effects

The following chronic (long-term) health effects may occur at some time after exposure to tetramethyl lead and can last for months or even years:

Cancer Hazards: According to information presented in the references, tetramethyl lead has not been adequately tested for its ability to cause cancer in test animals. However, mutation data have been reported and many scientists believe that exposure to mutagenic compounds may pose a cancer risk in the long-term.

Reproductive Hazard: According to information presented in the references, tetramethyl lead has been shown to adversely affect reproduction in test animals. There may be a decreased sex drive, impotence, sterility in males, decreased fertility and abnormal menstrual cycles in females and possible spontaneous fetal abortions in females. It may damage the testes and reduce sperm production.

Other Chronic Effects: Tetramethyl lead is more volatile and more slowly absorbed than tetraethyl lead. Long-term exposure to tetramethyl can lead to appetite loss, nausea, anemia, weakness, insomnia, muscle and joint pain, and colic accompanied by severe abdominal pain. Kidney damage may occur, especially with repeated or high exposure. Symptoms may persist for weeks or months after exposure. Some loss of vision may occur after an eye splash or with severe, repeated exposures. Lead can cause increased blood pressure and can accumulate in the body with repeated exposures taking months or years for the body to purge itself of the contamination. Since tetramethyl lead is eventually metabolized into lead in the body, these effects may also occur with repeated exposure.

Recommended Risk-Reduction Measures

Personnel should avoid direct contact with tetramethyl lead. It is considered more hazardous than elemental lead because of its rapid solubility. If a less toxic material cannot be substituted for a hazardous chemical, then *engineering controls* are the most effective method of reducing exposures. The best protection is to enclose operations and/or provide explosion-proof, local exhaust ventilation at the site of chemical release. While not always operationally feasible, isolating operations can also reduce exposure. Using respiratory protection is less effective than the controls mentioned above, but is still advisable whenever working with or around tetramethyl lead. The exposure level is extremely low (0.075 mg/m^3) and difficult to accurately monitor. A supplied-air respirator operated in continuous flow mode or a self-contained breathing apparatus (SCBA) with full facepiece in pressure demand mode will provide the best respiratory protection. If a full facepiece respirator is not available, then splash goggles or safety glasses with side shields attached should be worn to protect the eyes. To prevent hand and skin exposures, suitable gloves should be used. Glove manufacturers should be contacted and permeation studies obtained *before* final glove selections are made.

Administrative controls should also be in place to minimize the potential for human exposures. These may include written procedures or policies which specify the methods and techniques that will be practiced whenever personnel are to work with tetramethyl lead.

All personnel should receive *training* on the use, hazards, protective measures, emergency actions, and other precautions per 29 CFR 1910.1200 (Hazard Communication), prior to the first assignment in an area where tetramethyl lead is used or stored.

Before beginning employment with tetramethyl lead, and at regular intervals thereafter (e.g., annually), the following medical tests are recommended:

- ☑ Blood lead levels and blood pressure (baseline).

If symptoms develop or overexposure is suspected, the following additional tests may be useful:

- ☑ Blood lead levels and blood pressure (compare to baseline).
- ☑ Complete blood count for microcytic anemia and basophilic stippling, blood urea nitrogen (BUN) of creatinine, and zinc protoporphyrin (ZPP).
- ☑ Examination of the central and peripheral nervous systems.
- ☑ Urine test for lead (levels of 0.1 mg per liter of urine indicate increased exposure).

Also, pregnant women, especially those undergoing lithium psychotherapy may be at increased risk of adverse health effects on exposure to tetramethyl lead.

Other methods to reduce exposure include:

- ☑ Where possible, enclose operations and use local exhaust ventilation at the site where tetramethyl lead is worked. If local exhaust ventilation or enclosure is not used, respiratory protection should be mandatory.
- ☑ Always ensure that proper protective clothing is worn when working with tetramethyl lead.
- ☑ Wash thoroughly immediately after exposure to tetramethyl lead and at the end of the work shift or before eating, drinking, or smoking. Clothing contaminated with tetramethyl lead should not be taken home.
- ☑ Hazard warning information should be posted in the work area. In addition, as part of an on-going education and training program, all information on the health and safety hazards of tetramethyl lead should be communicated to all exposed workers.
- ☑ Additional exposure control measures and engineering controls are provided by OSHA in 29 CFR 1910.1025 (Lead).

RISK ASSESSMENT: ENVIRONMENT

General Assessment

The environment is still at risk of exposure during transportation, storage, disposal, or destruction of tetramethyl lead. In almost every scenario, the threat of environmental exposure is contingent upon the proper handling of tetramethyl lead materials, especially the dusts and powders. Accidental contact with a variety of incompatible materials can result in fire, explosion, and possible contamination of the surrounding environmental mediums (water, soil, air).

Tetramethyl lead is considered a Class II Combustible Liquid (per OSHA 29 CFR 1910.106). Its vapors can form explosive mixtures in air. It is incompatible with oxidizing agents and contact can cause fire or explosion. These characteristics require special consideration during any emergency situation involving a leak or spill of tetramethyl lead. Should tetramethyl lead ever come into contact with incompatible substances such as oxidizers, (perchlorates, peroxides, permanganates, chlorates, nitrates) and chemically active metals (potassium, sodium, magnesium, zinc), during transportation or storage, violent reactions are extremely possible.

Lead and its compounds are among the metals known and used since ancient times. It is one of the most abundant metals in the Earth's crust and can be dissolved from rocks and minerals into surface waters. It is used extensively across a broad spectrum of industries for a large variety of purpose. Because of its frequent use and common occurrence, lead enters the environment from point and non-point sources.

Acute Ecological Effects

Acute (short-term) toxic effects may include the death of animals, birds, or fish and death or low growth rate in plants. Acute effects are seen 2 to 4 days after animals or plants are exposed to tetramethyl lead.

Toxicity to aquatic life is affected by water hardness, the softer the water, the greater the toxicity. Lead and its compounds have high acute toxicity to aquatic life. Insufficient data are available on the short-term effects of tetramethyl lead and its compounds to plants, birds, or land animals.

Chronic Ecological Effects

Chronic (long-term) toxic effects may include shortened life span, reproductive problems, lower fertility, and changes in appearance or behavior in exposed animals. These effects can be seen long after first exposure(s) to toxic chemicals.

Lead and its compounds have high chronic toxicity in aquatic life. Lead causes nerve and behavioral effects in humans and could cause similar long-term effects in birds and land animals. Insufficient data are available on the long-term effects of lead to plants.

Water Solubility

Lead and its compounds range in their respective water solubilities from highly soluble to practically insoluble. Tetramethyl lead is considered insoluble in water.

Persistence in the Environment

Lead and its compounds are highly persistent in water, with a half-life greater than 200 days. The half-life of a pollutant is the amount of time it takes for one half of the chemical to be degraded.

Bioaccumulation in Aquatic Organisms

Some substances increase in concentration, or *bioaccumulate,* in living organisms as they breathe contaminated air, drink contaminated water, or eat contaminated food. These chemicals can become concentrated in the tissues and internal organs of animals as well as humans.

The concentration of tetramethyl lead found in the edible tissues of fish is expected to be much higher than the average concentration of tetramethyl lead found in the water from which the fish was taken.

Recommended Risk-Reduction Measures

Proper training of all transporters will reduce the likelihood of a mishap or accident resulting in the contamination of the environment. The correct use of DOT labeling (per the MSDS) on all containers, trucks, and rail cars will enable emergency responders to react properly and quickly to any disaster thereby reducing the potential risk to the environment and to personnel.

Storage of tetramethyl lead should be segregated from incompatible chemicals to minimize the risk of cross-contamination or contact. Buildings designated for storage should be equipped with appropriate fire protection systems (alarms, sprinklers, emergency lighting, portable extinguishers).

If a spill or leak to the environment has occurred, fire department, emergency response and/or hazardous materials spill personnel should be notified immediately. Cleanup should be attempted only by those

trained in proper spill containment procedures using non-sparking tools. Contaminated soils should be removed and replaced with clean soil. When environmental contamination occurs, the local and/or state emergency response authorities may require notification. A comprehensive emergency response or disaster preparedness/recovery plan should be in place prior to any operations involving the use, transportation, storage, or disposal of tetramethyl lead.

If tetramethyl lead is spilled or leaked, the following specific steps are recommended:

- ☑ Restrict persons not wearing protective clothing from area of spill or leak until cleanup is complete and area can be opened for normal work.
- ☑ Remove all ignition sources and ventilate area.
- ☑ Absorb liquids in vermiculite or other absorbent material and place in sealed drums for disposal. Use non-sparking tools during cleanup.
- ☑ It may be necessary to dispose of tetramethyl lead as a hazardous waste. The responsible state agency or the regional office of the federal Environmental Protection Agency (EPA) should be contacted for specific recommendations.

RISK ASSESSMENT: BUSINESS

General Assessment

Accidents or mishaps involving tetramethyl lead can present a significant threat to business operations. The loss or damage of equipment or facilities due to explosion or fire can significantly affect fiscal viability. Lawsuits that may result from personnel injury/death, public exposures, and/or environmental contamination will also require a serious expenditure of resources. Media attention surrounding an injury, death, or environmental damage can also result in a loss of profits and loss of current as well as future business.

Recommended Risk-Reduction Measures

Company attorneys, safety and health professionals, and environmental specialists should be involved in the development of any procedures for dealing with safety or emergency response. A company official should be pre-designated as a public relations officer with specific training in dealing with the press. Corporate plans and policies should be developed, approved, and implemented long before any need for such arises.

REFERENCES

American Conference of Governmental Industrial Hygienists. 1988. *Documentation of the Threshold Limit Values and Biological Exposure Indices,* 5th Edition (with updates). Cincinnati: ACGIH

Bloom, A. D. and F. J. DeSerres. 1995. *Ecotoxicity and Human Health.* New York: CRC Press/Lewis Publishers.

Brusick, D. J. 1994. *Methods for Genetic Risk Assessment.* New York: CRC Press/Lewis Publishers.

Burrell, R., D. K. Flaherty, and Leonard J. Sauers. 1992. *Toxicology of the Immune System.* New York: Van Nostrand Reinhold.

Cothern, C. R. 1995. *Handbook for Environmental Risk Decisionmaking.* New York: CRC Press/Lewis Publishers.

Danse, I. R. 1991. *Common Sense Toxics in the Workplace.* New York: Van Nostrand Reinhold.

Gardner, R. W. 1994. *Chemical Intolerance.* New York: CRC Press/Lewis Publishers.

Howard, P. H., et. al. 1991. *Handbook of Environmental Degradation Rates.* New York: CRC Press/Lewis Publishers.

Keith, L. H. and M. M. Walker. 1995. *Handbook of Air Toxics.* New York: CRC Press/Lewis Publishers.

Lauwerys, R. R., P. Hoet. 1993. *Industrial Chemical Exposure.* New York: CRC Press/Lewis Publishers.

Lewis, R. J., Sr. 1993. *Hawley's Condensed Chemical Dictionary,* 12th Edition. New York: Van Nostrand Reinhold.

Lewis, R. J., Sr. 1992. *Sax's Dangerous Properties of Industrial Materials,* Volumes 1, 2, and 3, 8th Edition. New York: Van Nostrand Reinhold.

Manahan, S. E. 1992. *Toxicological Chemistry,* 2nd Edition. New York: CRC Press/Lewis Publishers.

Meyer, E. 1990. *Chemistry of Hazardous Materials.* Englewood Cliffs, NJ: Prentice Hall.

Patnaik, P. 1992. *A Comprehensive Guide to the Hazardous Properties of Chemical Substances.* New York: Van Nostrand Reinhold.

Rea, W. J. 1995. *Chemical Sensitivity,* Vol. I-III. New York: CRC Press/Lewis Publishers.

Richardson, M. 1995. *Dictionary of Substance and Their Effects,* Vol. I-VI. New York: CRC Press/Lewis Publishers.

Sacarello, H. L. A. 1994. *The Comprehensive Handbook of Hazardous Materials.* New York: CRC Press/Lewis Publishers.

Sherman, J. D. 1988. *Chemical Exposure and Disease.* New York: Van Nostrand Reinhold.

U.S. Department of Health and Human Services, National Institute for Occupational Safety and Health. 1994. *NIOSH Pocket Guide to Chemical Hazards.* Washington, D.C.: U.S. Government Printing Office.

MATERIAL SAFETY DATA SHEET

CHEMICAL NAME	
TETRAMETHYL SUCCINONITRILE	**CRC PRESS/LEWIS PUBLISHERS DISCLAIMER:** THE INFORMATION AND RECOMMENDATIONS PRESENTED HEREIN ARE BASED ON SOURCES BELIEVED TO BE RELIABLE. CRC MAKES NO REPRESENTATION ON ITS COMPLETENESS OR ACCURACY. IT IS THE RESPONSIBILITY OF THE USER TO DETERMINE THE CHEMICAL'S SUITABILITY FOR ITS INTENDED USE, THE CHEMICAL'S SAFE USE, AND THE CHEMICAL'S PROPER DISPOSAL. NO REPRESENTATIONS AND/OR WARRANTIES, EITHER EXPRESSED OR IMPLIED, OF THE MERCHANTABILITY OR FITNESS FOR A PARTICULAR PURPOSE, OR OF ANY OTHER NATURE, ARE MADE WITH RESPECT TO THE INFORMATION PROVIDED IN THIS MSDS OR TO THE CHEMICAL TO WHICH INFORMATION MAY REFER. CRC NEITHER ASSUMES NOR AUTHORIZES ANY OTHER PERSON TO ASSUME FOR IT, ANY OTHER ADDITIONAL RESPONSIBILITY OR LIABILITY FOR THE USE OF, OR RELIANCE UPON, THIS INFORMATION.

HAZARD WARNING INFORMATION

HEALTH	FIRE	REACTIVE	OTHER	DEGREE OF HAZARD	COLOR CODING	OTHER CODES
2	1	1		0 = Minimum Hazard 1 = Slight Hazard 2 = Moderate Hazard 3 = Serious Hazard 4 = Severe Hazard	HEALTH = BLUE FIRE = RED REACTIVITY = YELLOW OTHER = WHITE	OX = Oxidizer ACID = Acid ALK = Alkali COR = Corrosive ~~W~~ = Use No Water

SECTION I - GENERAL INFORMATION

Field	Value	Field	Value
Characterization	**Organic Cyanides (Nitrile)**	RCRA Number	**None**
		EPA Class	**Not Applicable**
DOT Proper Shipping Name	**No Citation**	Chemical Abstract Service (CAS) Number	**3333-52-6**
DOT Hazard Class and Label Requirements	**No Citation**	DOT Emergency Guide Code	**No Citation**
DOT Identification Number	**No Citation**	Chemical Formula	**$(CH_3)_2C(CN)C(CN)(CH_3)_2$**

Synonyms
Tetramethyl succinodinitril; TMSN.

SECTION II - HAZARDOUS INGREDIENTS AND IDENTITY INFORMATION

Hazardous Components (specific identity)	OSHA Exposure Criteria	NIOSH Exposure Criteria	Immediately Dangerous to Life and Health (IDLH)	ACGIH Exposure Criteria
Tetramethyl succinonitrile: (derivation: A by-product on the production of vinyl foam when azobisisobutyronitrile is used as a blowing agent). **1 ppm = 5.66 mg/m³**	**PEL (skin):** **0.5 ppm** **3 mg/m³** **STEL:** **Not Established**	**REL (skin):** **0.5 ppm** **3 mg/m³** **STEL:** **Not Established**	**5 ppm**	**TLV (skin):** **0.5 ppm** **3 mg/m³** **STEL:** **Not Established**

SECTION III - PHYSICAL AND CHEMICAL CHARACTERISTICS

Field	Value	Field	Value
Boiling Point	**Sublimes**	Specific Gravity (H_2O = 1)	**1.07**
Vapor Pressure (mm Hg)	**Not Reported**	Molecular Weight	**136.2**
Vapor Density (Air = 1)	**Not Reported**	Freezing Point	**338°F (170°C)**

Solubility
Insoluble in water.

Appearance and Odor
Colorless, odorless, crystalline solid (crystallizes in plates).

SECTION IV - FIRE AND EXPLOSION HAZARD DATA

Field	Value	Field	Value
Flash Point (method used)	**Not Reported**	Explosive Limits in Air % by Volume	**LEL: Not Reported UEL: Not Reported**
NFPA Classification	**Combustible Solid**	Autoignition Temperature	**Not Reported**

Extinguishing Media
Use water spray, dry chemical, alcohol foam, or carbon dioxide.

Special Fire Fighting Procedures
Poisonous gases are produced in fire. Wear full protective clothing and self-contained breathing apparatus (SCBA) operated in pressure demand or other positive pressure mode. Keep fire-exposed containers cool with water spray.

Unusual Fire and Explosion Hazards
None reported.

SECTION V - REACTIVITY DATA

Stability		
Stability		Conditions to Avoid Tetramethyl succinonitrile is normally stable in closed containers under routine conditions of storage and handling. Avoid contact with strong oxidizers and keep away from heat or flame.
Stable X	Unstable	Incompatibility *(materials to avoid)* Reactions can occur with strong oxidizers, such as fluorine, chlorine, bromine, permanganates, nitrates, and other oxidizing materials.
Hazardous Polymerization		Conditions to Avoid Under normal conditions of temperature and pressure, hazardous polymerization of tetramethyl succinonitrile cannot occur.
May Occur	Will Not Occur X	Hazardous Decomposition or By-products Thermal oxidative decomposition of tetramethyl succinonitrile produces toxic oxides of nitrogen, cyanide compounds, and/or carbon monoxide.

SECTION VI - HEALTH HAZARD DATA

Primary Route(s) of Entry:	Inhalation?	Absorption (skin)?	Ingestion?
	X	X	X

Health Hazards

INHALATION: Causes headache, nausea, vomiting, convulsions, loss of consciousness, and coma. An allergen. Possible liver, kidney, and gastrointestinal effects. Also, if decomposition of this chemical occurs, there can also be simultaneous exposure to cyanide gas.

ABSORPTION: Possible surface irritation. A strong convulsant, it can pass through unbroken skin to cause toxic systemic effects. WARNING: Simultaneous exposures with either inhalation or ingestion can result in compounded effects.

INGESTION: Causes toxic systemic effects characterized by convulsions, nausea, and vomiting. It may also cause liver and kidney effects.

Carcinogenicity	NTP Listed?	IARC Cancer Review Group?	OSHA Regulated?	Target Organs?
Unknown Human Unknown Animal	No	No	29 CFR 1910.1000 Table Z-1	liver; CNS; kidneys; Gastrointestinal tract.

Medical Conditions Generally Aggravated by Exposure

Convulsive disorders may be aggravated by exposure to tetramethyl succinonitrile.

Emergency and First-aid Procedures

Eye contact: Flush immediately with water for 15 minutes (minimum); seek medical attention. Skin contact: Remove all contaminated clothing. Immediately wash area with large amounts of soap and water. Seek medical assistance. For inhalation: Remove the person from exposure. Provide respiratory assistance and CPR. Transfer to medical facility. If swallowed: Seek medical attention immediately or call poison control center. Never give anything by mouth to a convulsing or unconscious person.

SECTION VII - PRECAUTIONS FOR SAFE HANDLING AND USE

Steps to be Taken in Case Material is Released or Spilled

Collect material in safest manner possible. Do NOT dry sweep (generates airborne dusts). Use a HEPA vacuum instead. Damp mop residue. Ventilate area. Restrict those not involved in cleanup from entering area. Notify appropriate authorities, as required.

Preferred Waste Disposal Method

Chemical incinerator equipped with scrubber and afterburner.

Precautions to be Taken in Handling and Storage

Store in tightly closed containers in a cool, dark, well-ventilated area. Keep fire and flame away. Use non-sparking tools to open or close containers. Prevent damage to containers in storage and handling.

Other Precautions and Warnings

Personnel should be trained on the hazards of tetramethyl succinonitrile before being assigned to an area where it is used, handled, or stored.

SECTION VIII - CONTROL MEASURES AND PERSONAL PROTECTIVE EQUIPMENT

Respiratory Protection *(specify type)*

Air-purifying respirators with dust pre-filters may suffice at low concentrations. For best protection, use a self-contained breathing apparatus with full facepiece operated in a pressure demand mode.

Ventilation

Local exhaust or general mechanical systems recommended.

Protective Gloves	Eye Protection	Other Protective Clothing
Impervious Material	Chemical/Dust Goggles or Face Mask	Protective Apron

Work/Hygiene Practices

Always wash hands thoroughly after using chemical; never bring food, drink, or smoking materials into vicinity of chemicals.

TETRAMETHYL SUCCINONITRILE

$(CH_3)_2C(CN)C(CN)(CH_3)_2$

CAS: 3333-52-6

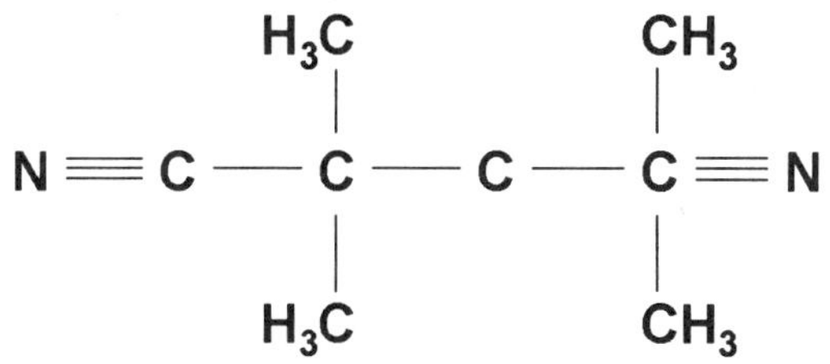

IDENTIFICATION AND TYPICAL USES

Tetramethyl succinonitrile is a colorless solid with no detectable odor. It crystallizes in the form of plates. Its principle use is as a blowing agent for vinyl foam production.

RISK ASSESSMENT: HEALTH

General Assessment

Tetramethyl succinonitrile is a human poison by *ingestion* and is considered toxic by *inhalation* and dermal *absorption*. It is primarily an irritant and an allergen. There are no reports of any carcinogenic or mutagenic effects on long-term exposure. However, it is an experimental teratogen. Tetramethyl succinonitrile is considered a strong convulsant. The lethal dose in humans is though to be approximately 25 milligrams per kilogram of body weight.

Inhalation, absorption, and ingestion of tetramethyl succinonitrile will cause headache, nausea, vomiting, convulsions, and coma. There is also a possibility of gastrointestinal disturbances and liver and kidney damage. Caution is warranted to ensure simultaneous exposures do not occur. The compounded effects of absorption together with inhalation or ingestion can be quite serious in terms of health hazard risk. Also, if decomposition has occurred, there is also the risk of simultaneous exposure to cyanide gas.

Acute Health Effects

The following acute (short-term) health effects may occur immediately or shortly after exposure to tetramethyl succinonitrile:

Skin: Severe irritation with absorption.

Eye: Mild to moderate irritation.

Lung: Possible irritation.

CNS: Primary CNS depressant causing headache, nausea, convulsions, and coma.

Chronic Health Effects

The following chronic (long-term) health effects may occur at some time after exposure to tetramethyl succinonitrile and can last for months or even years:

Cancer Hazards: According to information presented in the references, tetramethyl succinonitrile has not been adequately tested for its ability to cause cancer in test animals.

Reproductive Hazard: According to information presented in the references, tetramethyl succinonitrile has been shown to cause teratogenic effects in test animals.

Other Chronic Effects: There are few data in the references regarding the effects of chronic exposure to tetramethyl succinonitrile.

Recommended Risk-Reduction Measures

Personnel should avoid direct contact with tetramethyl succinonitrile. The exact nature of its toxicity, especially in the long-term, is not clearly understood in the references. Prudent risk management dictates it be treated as though it posed a significant health hazard risk. If a less toxic chemical cannot be substituted for a hazardous substance, then *engineering controls* are the most effective method of reducing exposures. The best protection is to enclose operations and/or provide local exhaust ventilation at the site of chemical release. While not always operationally feasible, isolating operations can also reduce exposure. Using respiratory protection is less effective than the controls mentioned above, but is still advisable whenever working with or around tetramethyl succinonitrile. A self-contained breathing apparatus (SCBA) with full facepiece and pressure demand is the recommended respiratory protection of choice. If a full facepiece is not available, then chemical goggles should be worn to protect the eyes. Whenever a chemical splash hazard exists, a face shield and an impermeable apron should be worn. To prevent hand and skin exposures, protective gloves should be used. Glove manufacturers should be contacted and permeation studies obtained and reviewed *before* final glove selections have been made.

Administrative controls should also be in place to minimize the potential for human exposures. These may include written procedures or policies which specify the methods and techniques that will be practiced whenever personnel are to work with tetramethyl succinonitrile.

All personnel should receive *training* on the use, hazards, protective measures, emergency actions, and other precautions per 29 CFR 1910.1200 (Hazard Communication), prior to the first assignment in an area where tetramethyl succinonitrile is used or stored.

Medical testing of personnel should be performed before the first exposure and at least annually thereafter. As per OSHA requirements, these tests must be provided by the employer and include:

☑ Complete physical examination.

☑ Stool test for blood (employees over 40).

In addition, for those with frequent or potentially high exposure (half the TLV or greater, or significant skin contact), the following are recommended before beginning work and at regular intervals thereafter:

☑ Urine thiocyanate levels.

☑ Lung function (pulmonary function) tests.

If symptoms develop or overexposure is suspected, the following additional tests may be useful:

☑ Blood cyanide level.

☑ Thyroid function test.

☑ Consider chest X-ray after acute overexposure.

Any evaluation should include a careful medical history of past and present symptoms with an examination. However, medical tests that evaluate existing damage are not a substitute for controlling exposure. Antidotes following exposure to cyanides include:

☑ Administration of 4-dimethylaminophenol plus sodium thiosulfate shows some protection only after oral intake (ineffective against inhalation). However, this nitrile is different from other nitriles in that thiosulfate is a poor antidote for intoxication.

☑ Cystein and *N*-acetyl cystein are somewhat effective in combating cyanide poisoning.

Other methods to reduce exposure include:

☑ Use amyl nitrate capsules as antidote if symptoms develop.

☑ Employees should be trained regularly in emergency measures for cyanide poisoning and in CPR.

☑ A cyanide antidote kit must be rapidly available and ingredients replaced every 1-2 years to ensure freshness.

☑ Medical personnel trained in treatment of cyanide poisoning should be quickly available.

☑ Where possible, enclose operations and use local exhaust ventilation at the site of chemical release. If local exhaust ventilation or enclosure is not used, respiratory protection should be mandatory.

☑ Always ensure that proper protective clothing is worn when using chemical substances and that personnel are trained in its use, care, and maintenance.

☑ Wash thoroughly immediately after exposure to tetramethyl succinonitrile and at the end of the work shift or before eating, drinking, or smoking.

☑ Hazard warning information should be posted in the work area. In addition, as part of an on-going education and training program, all information on the health and safety hazards of tetramethyl succinonitrile should be communicated to all potentially exposed workers.

RISK ASSESSMENT: ENVIRONMENT

General Assessment

The environment is at risk of exposure during transportation, storage, disposal, or destruction of tetramethyl succinonitrile. In almost every scenario, the threat of environmental exposure is contingent upon the proper handling of the chemical substance. Accidental spills, large or small, can result in fire, explosion, and possible contamination of the surrounding environmental mediums (water, soil, and air).

Tetramethyl succinonitrile is considered a combustible solid. It can react dangerously and even explosively in contact with strong oxidizers such as chlorine, fluorine, bromine, nitrogen dioxide, or other oxidizing materials. These characteristics require special consideration during any emergency situation involving a leak or spill of tetramethyl succinonitrile. Should tetramethyl succinonitrile ever come into contact with incompatible substances either during use, transportation, or storage, the formation of highly toxic and/or highly explosive commodities is extremely possible.

The proper disposal/destruction method for tetramethyl succinonitrile is to burn it in a chemical incinerator equipped with an afterburner and air scrubber.

Tetramethyl succinonitrile can enter the environment through industrial effluents and discharges, and through spills.

Acute Ecological Effects

Acute (short-term) toxic effects may include the death of animals, birds, or fish and death or low growth rate in plants. Acute effects are seen 2 to 4 days after animals or plants are exposed to tetramethyl succinonitrile.

This chemical has high acute toxicity to aquatic life. No data are available on the short-term effects of tetramethyl succinonitrile to plants, animals, or birds.

Chronic Ecological Effects

Chronic toxic effects may include shortened life span, reproductive problems, lower fertility, and changes in appearance or behavior in exposed animals. These effects can be seen long after first exposure(s) to toxic chemicals.

Tetramethyl succinonitrile has high chronic toxicity in aquatic life. No data are available on the long-term effects of tetramethyl succinonitrile to plants, birds, or land animals.

Water Solubility

Tetramethyl succinonitrile is insoluble in water. Concentrations of 1 milligram will not mix with a liter of water.

Persistence in the Environment

Tetramethyl succinonitrile is slightly persistent in water, with a half-life of between 2 to 20 days. The half-life of a pollutant is the amount of time it takes for one half of the chemical to be degraded.

Bioaccumulation in Aquatic Organisms

Some substances increase in concentration, or *bioaccumulate,* in living organisms as they breathe contaminated air, drink contaminated water, or eat contaminated food. These chemicals can become concentrated in the tissues and internal organs of animals as well as humans.

The concentration of tetramethyl succinonitrile found in fish tissues is expected to be somewhat higher than the average concentration of tetramethyl succinonitrile in the water from which the fish was taken.

Recommended Risk-Reduction Measures

Proper training of all transporters will reduce the likelihood of a mishap or accident resulting in a leak or spill to the environment. The correct use of DOT labeling (per the MSDS) on all containers, trucks, and rail cars will enable emergency responders to react properly and quickly to any disaster thereby reducing the potential risk to the environment and to personnel.

Storage of tetramethyl succinonitrile should be segregated from incompatible chemicals to minimize the risk of cross-contamination or contact. Buildings designated for storage should be equipped with appropriate fire protection systems (alarms, sprinklers, emergency lighting, portable extinguishers).

If a spill or leak to the environment has occurred, fire department, emergency response and/or hazardous materials spill personnel should be notified immediately. Cleanup should be attempted only by those trained in proper spill containment procedures. Contaminated soils should be removed for incineration and replaced with clean soil. If tetramethyl succinonitrile should contact the water table, aquifer, or navigable waterway, time is of the essence. It is readily miscible in water and, therefore, total containment and remediation may not be entirely possible. When such spills occur, the local and/or state emergency response authorities must be notified. A comprehensive emergency response or disaster preparedness/recovery plan should be in place prior to any operations involving the use, transportation, storage, or disposal of tetramethyl succinonitrile.

If tetramethyl succinonitrile is spilled or leaked, the following specific steps are recommended:

☑ Restrict persons not wearing protective clothing from area of spill or leak until cleanup is complete and area can be opened for normal work.

- ☑ Remove all ignition sources.
- ☑ Ventilate area of spill or leak.
- ☑ Collect solid materials in safest manner possible and deposit in sealed drums for disposal. Do NOT dry sweep (generates airborne dusts). Use a vacuum equipped with a high-efficiency particulate air (HEPA) filter. Damp mop residue.
- ☑ It may be necessary to dispose of tetramethyl succinonitrile as a hazardous waste. The responsible state agency or the regional office of the federal Environmental Protection Agency (EPA) should be contacted for specific recommendations.

RISK ASSESSMENT: BUSINESS

General Assessment

Accidents or mishaps involving tetramethyl succinonitrile can present a significant threat to business operations. The loss or damage of equipment or facilities can significantly affect fiscal viability. Lawsuits that may result from personnel injury/death, public exposures, and/or environmental contamination will also require a serious expenditure of resources. Media attention surrounding an injury, death, or environmental damage can also result in a loss of profits and loss of current as well as future business.

Recommended Risk-Reduction Measures

Company attorneys, safety and health professionals, and environmental specialists should be involved in the development of any procedures dealing with safety or emergency response. A company official should be pre-designated as a public relations officer with specific training in dealing with the press. Corporate plans and policies should be developed, approved, and implemented long before any need for such arises.

REFERENCES

American Conference of Governmental Industrial Hygienists. 1988. *Documentation of the Threshold Limit Values and Biological Exposure Indices*, 5th Edition (with updates). Cincinnati: ACGIH

Anthony, C. P., and N. J. Kolthoff. 1971. *Textbook of Anatomy and Physiology.* St. Louis: C. V. Mosby Company.

Bloom, A. D. and F. J. DeSerres. 1995. *Ecotoxicity and Human Health.* New York: CRC Press/Lewis Publishers.

Brusick, D. J. 1994. *Methods for Genetic Risk Assessment.* New York: CRC Press/Lewis Publishers.

Bueche, J. R. 1972. *Principles of Physics,* 2nd Edition. New York: McGraw-Hill.

Burrell, R., D. K. Flaherty, and L. J. Sauers. 1992. *Toxicology of the Immune System.* New York: Van Nostrand Reinhold.

Calabrese, E. J. 1994. *Biological Effects of Low Level Exposures.* New York: CRC Press/Lewis Publishers.

Cockerman, L. G., B. S. Shane. 1994. *Basic Environmental Toxicology.* New York: CRC Press/Lewis Publishers.

Cothern, C. R., N. P. Ross. 1994. *Environmental Statistics, Assessment, and Forecasting.* New York: CRC Press/Lewis Publishers.

Cothern, R. R. 1993. *Comparative Environmental Risk Assessment.* New York: CRC Press/Lewis Publishers.

Cothern, C. R. 1995. *Handbook for Environmental Risk Decisionmaking.* New York: CRC Press/Lewis Publishers.

Environmental Protection Agency. October 1983. *Health Assessment for Tetramethyl succinonitrile.* Final Report: EPA 600/8-82-007F. NTIS PB84-149152. Springfield, VA: National Technical Information Service.

Gardner, R. W. 1994. *Chemical Intolerance.* New York: CRC Press/Lewis Publishers.

Hodgson, E., R. B. Mailman, and Janice E. Chambers. 1988. *Dictionary of Toxicology.* London: The MacMillian Company.

Jones, F. E. 1994. *Toxic Organic Vapors in the Workplace.* New York: CRC Press/Lewis Publishers.

Lewis, R. J., Sr. 1993. *Hawley's Condensed Chemical Dictionary,* 12th Edition. New York: Van Nostrand Reinhold.

Lewis, R. J., Sr. 1992. *Sax's Dangerous Properties of Industrial Materials,* Volumes 1, 2, and 3, 8th Edition. New York: Van Nostrand Reinhold.

Meyer, E. 1990. *Chemistry of Hazardous Materials.* Englewood Cliffs, NJ: Prentice Hall.

Patnaik, P. 1992. *A Comprehensive Guide to the Hazardous Properties of Chemical Substances.* New York: Van Nostrand Reinhold.

Proctor, N. H., and James P. Hughes. 1991. *Chemical Hazards of the Workplace.* New York: Van Nostrand Reinhold.

Quigley, D. R. 1994. *Handbook of Emergency Chemical Management.* New York: CRC Press/Lewis Publishers.

U.S. Department of Health and Human Services, National Institute for Occupational Safety and Health. 1994. *NIOSH Pocket Guide to Chemical Hazards.* Washington, D.C.: U.S. Government Printing Office.

Willhite, C. C. Toxicological updates. Tetramethyl succinonitrile. *Journal of Applied Toxicology.* 2:54-56, 1982.

MATERIAL SAFETY DATA SHEET

CHEMICAL NAME

TETRANITROMETHANE

HAZARD WARNING INFORMATION

HEALTH	FIRE	REACTIVE	OTHER	DEGREE OF HAZARD	COLOR CODING	OTHER CODES
3	3	3		0 = Minimum Hazard 1 = Slight Hazard 2 = Moderate Hazard 3 = Serious Hazard 4 = Severe Hazard	HEALTH = BLUE FIRE = RED REACTIVITY = YELLOW OTHER = WHITE	OX = Oxidizer ACID = Acid ALK = Alkali COR = Corrosive W̶ = Use No Water

SECTION I - GENERAL INFORMATION

Field	Value	Field	Value
Characterization	Nitroparaffin	RCRA Number	P112
		EPA Class	Acute Hazardous Waste
DOT Proper Shipping Name	Tetranitromethane	Chemical Abstract Service (CAS) Number	509-14-8
DOT Hazard Class and Label Requirements	Oxidizer	DOT Emergency Guide Code	47
DOT Identification Number	UN 1510	Chemical Formula	$C(NO_2)_4$

Synonyms
Tetan; TNM.

SECTION II - HAZARDOUS INGREDIENTS AND IDENTITY INFORMATION

Hazardous Components (specific identity)	OSHA Exposure Criteria	NIOSH Exposure Criteria	Immediately Dangerous to Life and Health (IDLH)	ACGIH Exposure Criteria
Tetranitromethane (derivation: By action of fuming nitric acid on benzene, acetic anhydride, or acetylene). **1 ppm = 8.15 mg/m³**	**PEL:** 1 ppm, 8 mg/m³ **STEL:** Not Established	**REL:** 1 ppm, 8 mg/m³ **STEL:** Not Established	**4 ppm**	**TLV:** 1 ppm, 8 mg/m³ **STEL:** Not Established

SECTION III - PHYSICAL AND CHEMICAL CHARACTERISTICS

Field	Value	Field	Value
Boiling Point	259°F (126°C)	Specific Gravity (H_2O = 1)	1.62
Vapor Pressure (mm Hg)	8.4 at 68°F (20°C)	Molecular Weight	196.0
Vapor Density (Air = 1)	8.01 at 77°F (25°C)	Freezing Point	57°F (14°C)

Solubility
Insoluble in water. Miscible with alcohol and ether. Decomposed by alcoholic solution of potassium hydroxide.

Appearance and Odor
Colorless to pale-yellow liquid or solid (below 57°F), with a pungent odor.

SECTION IV - FIRE AND EXPLOSION HAZARD DATA

Field	Value	Field	Value
Flash Point (method used)	Not Reported	Explosive Limits in Air % by Volume	LEL: Not Reported; UEL: Not Reported
NFPA Classification	Combustible Liquid	Autoignition Temperature	Not Reported

Extinguishing Media
Use water spray, dry chemical, alcohol foam, or carbon dioxide.

Special Fire Fighting Procedures
Tetranitromethane is a highly explosive. If bulk quantities are involved in fire, evacuate area and fight fire from distance. Poisonous gases are evolved from fire. Wear a self-contained breathing apparatus (SCBA) with full facepiece in positive pressure mode. Move containers from fire area if it can be done without risk or cool fire-exposed containers with water spray.

Unusual Fire and Explosion Hazards
Containers may explode in fire. Vapors are heavier than air and can travel for great distances to an ignition source and flashback to cause fire or explosion.

SECTION V - REACTIVITY DATA

<table>
<tr><td colspan="2">Stability</td><td>Conditions to Avoid
Tetranitromethane is normally stable in closed containers under routine conditions of handling and storage. It must be kept away from heat, sparks, open flame, or any mechanical or thermal shock.</td></tr>
<tr><td>Stable
X</td><td>Unstable</td><td>Incompatibility (materials to avoid)
Tetranitromethane is reactive in contact with hydrocarbons, alkalis, metals, oxidizers, aluminum, toluene, cotton, and other combustible materials. Combustibles wet with tetranitromethane can be explosive.</td></tr>
<tr><td colspan="2">Hazardous Polymerization</td><td>Conditions to Avoid
Under normal conditions of temperature and pressure, hazardous polymerization of tetranitromethane is not expected to occur.</td></tr>
<tr><td>May Occur</td><td>Will Not Occur
X</td><td>Hazardous Decomposition or By-products
If involved in fire, tetranitromethane can produce toxic nitrogen oxides, carbon monoxide, and irritating fumes and smoke.</td></tr>
</table>

SECTION VI - HEALTH HAZARD DATA

Primary Route(s) of Entry:	Inhalation?	Absorption (skin)?	Ingestion?
	X		X

Health Hazards

INHALATION: Irritation to the eyes, nose, throat, and respiratory tract. Animal studies have shown a potential for the development of pulmonary edema (fluid in the lungs), injury to the liver, kidney, and narcosis. There may be bronchitis, anuria, and increases in methemoglobin levels in the blood. Other symptoms include dizziness, headache, chest pain, dypsnea, cyanosis, nausea, vomiting, malaise, abdominal pain, and convulsions.

SKIN & EYES: Skin contact causes irritation and may cause reddening and smarting if it remains in contact for extended period of time. Eye contact can cause tearing and irritation.

INGESTION: Causes irritation of the mouth and stomach, nausea, and vomiting.

Carcinogenicity	NTP Listed?	IARC Cancer Review Group?	OSHA Regulated?	Target Organs?
Unknown Human Unknown Animal	No	No	29 CFR 1010.1000 Table Z-1	Skin, eyes, CNS, liver, respiratory tract, blood.

Medical Conditions Generally Aggravated by Exposure

Anemia, CVS, or pulmonary disease may be aggravated by exposure.

Emergency and First-aid Procedures

Eye contact: Immediately flush with large amounts of water for 15 minutes (minimum), occasionally lifting eyelids; seek medical attention. Skin contact: Remove all contaminated clothing. Immediately wash area with large amounts of soap and water. For inhalation: Remove the person from exposure. Provide respiratory assistance and CPR. Transfer promptly to medical facility. Observe 48 hours for lung effects. If swallowed: Contact poison control center. Unless directed otherwise, give a conscious person 1 to 2 glasses of water, then induce vomiting.

SECTION VII - PRECAUTIONS FOR SAFE HANDLING AND USE

Steps to be Taken in Case Material is Released or Spilled

Remove all ignition sources. Ventilate area. Restrict entry into the area. Absorb liquids with vermiculite and place in sealed containers for disposal. Use non-sparking tools during cleanup operations.

Preferred Waste Disposal Method

Dilute with fuel oil and atomize in proper combustion chamber.

Precautions to be Taken in Handling and Storage

Do not store with incompatible materials or in the presence of heat since violent reactions or explosions can occur. Store in tightly closed containers in a cool, dark, well-ventilated area.

Other Precautions and Warnings

Bulk storage of tetranitromethane is not recommended. Prevent damage to containers.

SECTION VIII - CONTROL MEASURES AND PERSONAL PROTECTIVE EQUIPMENT

Respiratory Protection *(specify type)*

Exposure levels are low and difficult to accurately monitor (1 ppm). For the best protection, use an MSHA/NIOSH-approved air-purifying respirator with organic vapor cartridge. Otherwise, use a powered supplied-air respirator or a self-contained breathing apparatus (SCBA) with full facepiece operated in pressure demand or other positive pressure mode.

Ventilation

Local exhaust or general mechanical systems recommended.

Protective Gloves	Eye Protection	Other Protective Clothing
Impervious Material	Chemical Goggles and/or Face Mask	Protective Apron

Work/Hygiene Practices

Always wash hands thoroughly after using chemical; never bring food, drink, or smoking materials into vicinity of chemicals.

TETRANITROMETHANE

$C(NO_2)_4$

CAS: 509-14-8

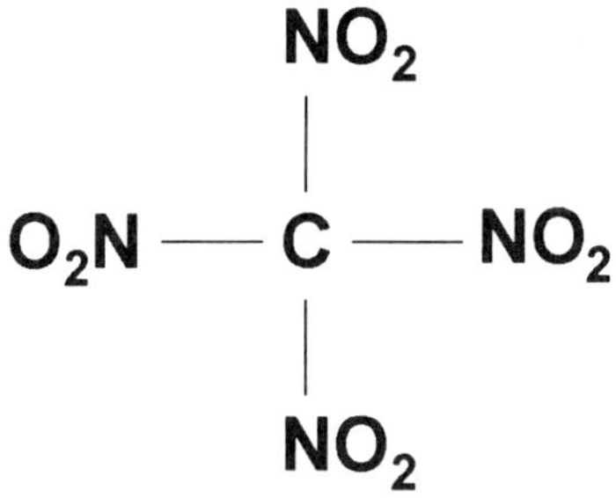

IDENTIFICATION AND TYPICAL USES

Tetranitromethane is a colorless to pale-yellow liquid or solid (below 57°F) with a pungent odor. It is used as an oxidizer in rocket fuels, as an oxidant or monopropellant, as an explosive, to increase the octane rating of diesel fuels, and as a reagent for detecting the presence of double bonds in organic compounds.

RISK ASSESSMENT: HEALTH

General Assessment

Tetranitromethane is poisonous by *ingestion* and *inhalation.* It is also a skin irritant, but dermal absorption has not been reported. There are no indications of any carcinogenic, mutagenic, or teratogenic effects following exposure.

Inhalation causes irritation of the mucosa of the eyes, nose, throat, and respiratory tract. Systemic effects have not been observed in humans. However, animal studies indicate a possibility for the development of pulmonary edema (fluid in the lungs) which is a medical emergency. Symptoms of cough, congestion, labored breathing, and pain in the chest may be delayed up to 48 hours which may give a false sense of security with regard to health exposure risk. There may also be liver injury and, at high concentrations, narcosis. Severe exposure in humans is expected to cause similar effects. Tetranitromethane may also cause the formation of methemoglobin which can significantly interfere with the blood's ability to properly transport oxygen. Other symptoms of exposure by inhalation may include dizziness, cyanosis, malaise, convulsions, anorexia, nausea, vomiting, anuria, and diarrhea.

Skin contact causes mild to severe irritation and, if allowed to remain in contact for prolonged periods, either directly or in chemical-soaked clothing, there can be reddening and smarting at the site of contact. Eye contact can cause irritation and tearing.

Ingestion may result in irritation of the mouth and stomach, nausea, and vomiting.

Acute Health Effects

The following acute (short-term) health effects may occur immediately or shortly after exposure to tetranitromethane:

Skin: Irritation with redness and smarting at the site of contact.

Eye: Irritation and inflammation with a possibility for damage to visual acuity.

Lung: Irritation of the mucosa of the nose, throat, and respiratory tract. May lead to delayed pulmonary edema which can be fatal.

CNS: A variety of effects associated with depression of the central nervous system. These can include dizziness, convulsions, headache, nausea, and vomiting.

Other: Exposure to high concentrations can cause methemoglobinemia.

Chronic Health Effects

The following chronic (long-term) health effects may occur at some time after exposure to tetranitromethane and can last for months or even years:

Cancer Hazards: According to information presented in the references, tetranitromethane has not been adequately tested for its ability to cause cancer in test animals.

Reproductive Hazards: There are no reports to support any claims of reproductive hazards.

Other Chronic Effects: Long-term exposure can cause chronic headaches, weariness, pneumonia, and effects on the function of the central nervous system and the cardiovascular system.

Recommended Risk-Reduction Measures

Personnel should avoid direct contact with tetranitromethane. If another chemical cannot be substituted for a tetranitromethane, then *engineering controls* are the most effective method of reducing exposures. The best protection is to separate operations and provide local exhaust ventilation at the site of chemical release. While not always operationally feasible, isolating operations can also reduce exposure. Using equipment with explosion-proof designs and ensuring all metal containers and tools are electrically grounded and bonded can further reduce the potential for injury or damage. Using respiratory protection is less effective than the controls mentioned above, but is still advisable whenever working with tetranitromethane. The exposure level of 1 ppm is relatively low and difficult to accurately monitor. While an air-purifying chemical cartridge respirator equipped with an organic vapor cartridge may provide adequate protection, the best results are obtained using a powered air-supplied respirator operated in pressure demand or other positive pressure mode, or a self-contained breathing apparatus (SCBA) with full facepiece operated in pressure demand mode. If a full facepiece is not available, then chemical goggles should be worn to protect the eyes. Whenever a chemical splash hazard exists, a face shield and a protective apron should be worn. To prevent hand and skin exposures with the liquid, protective gloves should be used. Glove manufacturers should be contacted and permeation studies obtained *before* final glove selections have been made.

Administrative controls should also be in place to minimize the potential for human exposures. These may include written procedures or policies which specify the methods and techniques that will be practiced whenever personnel are to work with tetranitromethane.

All personnel should receive *training* on the use, hazards, protective measures, emergency actions, and other precautions per 29 CFR 1910.1200 (Hazard Communication), prior to the first assignment in an area where tetranitromethane is used or stored.

If overexposure is suspected or symptoms develop, the following medical tests are recommended:

- ☑ Lung function tests.
- ☑ Liver and kidney function tests.
- ☑ Methemoglobin level monitoring.
- ☑ Consider chest X-ray after acute overexposure (may be negative if taken immediately after exposure due to delayed onset of pulmonary edema).

Other methods to reduce exposure to chemicals include:

- ☑ Where possible, consider installing automatic sensors that warn personnel of dropping oxygen levels and use local exhaust ventilation at the site of chemical release. If local exhaust ventilation or enclosure is not used, respiratory protection should be mandatory.
- ☑ Always ensure that proper protective clothing is worn when using chemical substances and that personnel are trained in its use, care, and maintenance.
- ☑ Wash thoroughly immediately after exposure to tetranitromethane and at the end of the work shift or before eating, drinking, or smoking.
- ☑ Safety shower and eyewash stations should be readily available in work areas where tetranitromethane is used or stored.
- ☑ Hazard warning information should be posted in the work area. In addition, as part of an on-going education and training program, all information on the health and safety hazards of tetranitromethane should be communicated to all exposed and potentially exposed workers.

RISK ASSESSMENT: ENVIRONMENT

General Assessment

The environment is at risk of exposure during transportation, storage, disposal, or destruction of tetranitromethane. In almost every scenario, the threat of environmental exposure is contingent upon the proper handling of the chemical substance. Accidental releases, large or small, can result in fire, explosion, and possible contamination of the surrounding environmental mediums (water, soil, and air).

Tetranitromethane is a combustible solid. However, it can present an extremely flammable and highly explosive hazard. Also, since it can react violently in the presence of oxidizers, alkalis, metals, and hydrocarbons, tetranitromethane requires special consideration during any emergency involving a leak or release of tetranitromethane gas or liquid.

Tetranitromethane can enter the environment through industrial discharges or through spills.

Acute Ecological Effects

Acute (short-term) toxic effects may include the death of animals, birds, or fish, and death or low growth rate in plants. Acute effects are seen 2 to 4 days after animals or plants are exposed to tetranitromethane.

Insufficient data are available on the short-term effects of tetranitromethane on aquatic life, plants, birds, or land animals.

Chronic Ecological Effects

Chronic (long-term) toxic effects may include shortened life span, reproductive problems, lower fertility, and changes in appearance or behavior in exposed animals. These effects can be seen long after first exposure(s) to toxic chemicals.

Insufficient data are available to evaluate the long-term effects of tetranitromethane to aquatic life, plants, birds, or land animals.

Water Solubility

Tetranitromethane is insoluble in water. Concentrations of 1 milligram will not mix with a liter of water.

Persistence in the Environment

Tetranitromethane is slightly persistent in water, with a half-life between 2 and 20 days. The half-life of a pollutant is the amount of time it takes for one half of the chemical to be degraded.

Bioaccumulation in Aquatic Organisms

Some substances increase in concentration, or *bioaccumulate,* in living organisms as they breathe contaminated air, drink contaminated water, or eat contaminated food. These chemicals can become concentrated in the tissues and internal organs of animals as well as humans.

The concentration of tetranitromethane found in the edible tissues of fish is expected to be somewhat lower than the average concentration found in the water from which the fish was taken.

Recommended Risk-Reduction Measures

Proper training of all transporters will reduce the likelihood of a mishap or accident resulting in a leak or release to the environment. The correct use of DOT labeling (per the MSDS) on all containers, trucks, and rail cars will enable emergency responders to react properly and quickly to any disaster thereby reducing the potential risk to the environment and to personnel.

Storage of tetranitromethane should be segregated from incompatible chemicals to minimize the risk of cross-contamination or contact. Buildings designated for storage should be equipped with appropriate fire protection systems (alarms, sprinklers, emergency lighting, portable extinguishers). Containers should be protected from physical damage and shielded from sources of radiant heat.

If a spill or release to the environment has occurred, fire department, emergency response, and/or hazardous materials spill personnel should be notified immediately. Cleanup should be attempted only by those trained in proper containment procedures using non-sparking tools. Contaminated soils should be removed for incineration and replaced with clean soil. When spills or leaks occur, it may be necessary to contact the local and/or state emergency response authorities. A comprehensive emergency response or disaster preparedness/recovery plan should be in place prior to any operations involving the use, transportation, storage, or disposal of tetranitromethane. If tetranitromethane is spilled or leaked, the following specific steps are recommended:

- ☑ Restrict persons not wearing protective clothing from area of spill or leak until cleanup is complete and area can be opened for normal work.
- ☑ Remove all ignition sources and ventilate area.
- ☑ Absorb liquids using vermiculite, dry earth, or other material and place in sealed containers for disposal.
- ☑ Keep tetranitromethane out of a confined space, such as a sewer, because of the possibility of explosion (unless the sewer is designed to prevent the build-up of explosive concentrations).
- ☑ It may be necessary to dispose of tetranitromethane as a hazardous waste. The responsible state agency or the regional office of the federal Environmental Protection Agency (EPA) should be contacted for specific recommendations.

RISK ASSESSMENT: BUSINESS

General Assessment

Accidents or mishaps involving tetranitromethane can present a significant threat to business operations. The loss or damage of equipment or facilities can significantly affect fiscal viability. Lawsuits that may result from personnel injury, public exposures, and environmental contamination will also require a serious expenditure of resources. Media attention surrounding an injury, illness, or environmental damage can also result in a loss of profits and loss of current as well as future business.

Recommended Risk-Reduction Measures

Company attorneys, safety and health professionals, and environmental specialists should be involved in the development of any procedures dealing with safety and emergency response. A company official should be pre-designated as a public relations officer with specific training in dealing with the press. Corporate plans and policies should be developed, approved, and implemented long before any need for such arises.

REFERENCES

Ahlbom, A. 1993. *Biostatistics for Epidemiologists.* New York: CRC Press/Lewis Publishers.

American Conference of Governmental Industrial Hygienists. 1988. *Documentation of the Threshold Limit Values and Biological Exposure Indices,* 5th Edition (with updates). Cincinnati: ACGIH

Anthony, C. P., and N. J. Kolthoff. 1971. *Textbook of Anatomy and Physiology.* St. Louis: C. V. Mosby Company.

Asante-Duah, D. K. 1993. *Hazardous Waste Risk Assessment.* New York: CRC Press/Lewis Publishers.

Bloom, A. D. and F. J. DeSerres. 1995. *Ecotoxicity and Human Health.* New York: CRC Press/Lewis Publishers.

Brusick, D. J. 1994. *Methods for Genetic Risk Assessment.* New York: CRC Press/Lewis Publishers.

Bueche, J. R. 1972. *Principles of Physics,* 2nd Edition. New York: McGraw-Hill.

Burrell, R., D. K. Flaherty, and L. J. Sauers. 1992. *Toxicology of the Immune System.* New York: Van Nostrand Reinhold.

Calabrese, E. J. 1994. *Biological Effects of Low Level Exposures.* New York: CRC Press/Lewis Publishers.

Cockerman, L. G., B. S. Shane. 1994. *Basic Environmental Toxicology.* New York: CRC Press/Lewis Publishers.

Cothern, C. R. 1995. *Handbook for Environmental Risk Decisionmaking.* New York: CRC Press/Lewis Publishers.

Danse, I. R. 1991. *Common Sense Toxics in the Workplace.* New York: Van Nostrand Reinhold.

Environmental Protection Agency. 1983. *Chemical Hazard Information Profiles.* Washington, D.C.: U.S. EPA.

Fodden, C. R., and J. L. Weddell. 1992. *Hazardous Materials Emergency Action Data.* New York: CRC Press/Lewis Publishers.

Forseberg, K., and L. H. Keith. 1993. *Chemical; Protective Clothing Permeation and Degradation Database.* Chelsea, MI: Lewis Publishers.

Gardner, R. W. 1994. *Chemical Intolerance.* New York: CRC Press/Lewis Publishers.

Hallenbeck, W. H. 1993. *Quantitative Risk Assessment for Environmental and Occupational Health,* 2nd Edition. New York: CRC Press/Lewis Publishers.

Hodgson, E., R. B. Mailman, and J. E. Chambers. 1988. *Dictionary of Toxicology.* London: The MacMillian Company.

Hosty, J. W., and P. Foster. 1990. *A Practical Guide to Chemical Spill Response.* New York: Van Nostrand Reinhold.

Howard, P. H., et. al. 1991. *Handbook of Environmental Degradation Rates.* New York: CRC Press/Lewis Publishers.

Jones, F. E. 1994. *Toxic Organic Vapors in the Workplace.* New York: CRC Press/Lewis Publishers.

Kamrin, M. A. 1988. *Toxicology: A Primer on Principles and Applications.* Chelsea, Michigan: Lewis Publishers.

Keith, L. H. and M. M. Walker. 1995. *Handbook of Air Toxics.* New York: CRC Press/Lewis Publishers.

Landis, W. G., M. H. Yu. 1995. *Introduction to Environmental Toxicology.* New York: CRC Press/Lewis Publishers.

Lauwerys, R. R., P. Hoet. 1993. *Industrial Chemical Exposure.* New York: CRC Press/Lewis Publishers.

Lewis, R. J., Sr. 1993. *Hawley's Condensed Chemical Dictionary,* 12th Edition. New York: Van Nostrand Reinhold.

Lewis, R. J., Sr. 1992. *Sax's Dangerous Properties of Industrial Materials,* Volumes 1, 2, and 3, 8th Edition. New York: Van Nostrand Reinhold.

Lide, D. R. 1995. *Handbook of Chemistry and Physics,* 76th Edition. New York: CRC Press/Lewis Publishers.

Meyer, E. 1990. *Chemistry of Hazardous Materials.* Englewood Cliffs, NJ: Prentice Hall.

Patnaik, P. 1992. *A Comprehensive Guide to the Hazardous Properties of Chemical Substances.* New York: Van Nostrand Reinhold.

Rea, W. J. 1995. *Chemical Sensitivity,* Vol. I-III. New York: CRC Press/Lewis Publishers.

Richardson, M. 1995. *Dictionary of Substance and Their Effects,* Vol. I-VI. New York: CRC Press/Lewis Publishers.

Sacarello, H. L. A. 1994. *The Comprehensive Handbook of Hazardous Materials.* New York: CRC Press/Lewis Publishers.

Sherman, J. D. 1988. *Chemical Exposure and Disease.* New York: Van Nostrand Reinhold.

Talbott, E., G. Craun. 1995. *An Introduction to Environmental Epidemiology.* New York: CRC Press/Lewis Publishers.

Torkelson, T. R. V. K. Rowe. 1981. *Patty's Industrial Hygiene and Toxicology,* Vol. 2B. New York: Wiley-Interscience.

U.S. Department of Health and Human Services, National Institute for Occupational Safety and Health. 1994. *NIOSH Pocket Guide to Chemical Hazards.* Washington, D.C.: U.S. Government Printing Office.

MATERIAL SAFETY DATA SHEET

CHEMICAL NAME	
THALLIUM	**CRC PRESS/LEWIS PUBLISHERS DISCLAIMER:** THE INFORMATION AND RECOMMENDATIONS PRESENTED HEREIN ARE BASED ON SOURCES BELIEVED TO BE RELIABLE. CRC MAKES NO REPRESENTATION ON ITS COMPLETENESS OR ACCURACY. IT IS THE RESPONSIBILITY OF THE USER TO DETERMINE THE CHEMICAL'S SUITABILITY FOR ITS INTENDED USE, THE CHEMICAL'S SAFE USE, AND THE CHEMICAL'S PROPER DISPOSAL. NO REPRESENTATIONS AND/OR WARRANTIES, EITHER EXPRESSED OR IMPLIED, OF THE MERCHANTABILITY OR FITNESS FOR A PARTICULAR PURPOSE, OR OF ANY OTHER NATURE, ARE MADE WITH RESPECT TO THE INFORMATION PROVIDED IN THIS MSDS OR TO THE CHEMICAL TO WHICH INFORMATION MAY REFER. CRC NEITHER ASSUMES NOR AUTHORIZES ANY OTHER PERSON TO ASSUME FOR IT, ANY OTHER ADDITIONAL RESPONSIBILITY OR LIABILITY FOR THE USE OF, OR RELIANCE UPON, THIS INFORMATION.

HAZARD WARNING INFORMATION

HEALTH	FIRE	REACTIVE	OTHER	DEGREE OF HAZARD	COLOR CODING	OTHER CODES
1	1	1		0 = Minimum Hazard 1 = Slight Hazard 2 = Moderate Hazard 3 = Serious Hazard 4 = Severe Hazard	HEALTH = BLUE FIRE = RED REACTIVITY = YELLOW OTHER = WHITE	OX = Oxidizer ACID = Acid ALK = Alkali COR = Corrosive ~~W~~ = Use No Water

SECTION I - GENERAL INFORMATION

Characterization	RCRA Number	EPA Class
Metal	**No Number Listed**	**Hazardous Waste**

DOT Proper Shipping Name	Chemical Abstract Service (CAS) Number
Thallium Salt, Solid, N.O.S.	**7440-28-0**
DOT Hazard Class and Label Requirements	**DOT Emergency Guide Code**
Poison B; Poison, St. Andrews Cross	**53 (compounds)**
DOT Identification Number	**Atomic Formula**
UN 1707 NA 1707	**Tl**

Synonyms

Thallium metal.

SECTION II - HAZARDOUS INGREDIENTS AND IDENTITY INFORMATION

Hazardous Components (specific identity)	OSHA Exposure Criteria	NIOSH Exposure Criteria	Immediately Dangerous to Life and Health (IDLH)	ACGIH Exposure Criteria
Thallium: (derivation: Found in flue dusts from lead and zinc smelting. The thallium compounds recovered are treated to obtain the metal by electrolysis, precipitation, or reduction).	PEL (skin): .01 mg(Tl)/m^3 STEL: Not Established	REL (skin): .01 mg(Tl)/m^3 STEL: Not Established	15 mg/m^3 (as Tl)	TLV (skin): .01 mg(Tl)/m^3 STEL: Not Established

SECTION III - PHYSICAL AND CHEMICAL CHARACTERISTICS

Boiling Point	Specific Gravity (H_2O = 1)
2655°F (1457°C)	**11.85**
Vapor Pressure (mm Hg)	Molecular Weight (atomic weight)
1 at 1517°F (825°C)	**204.37**
Vapor Density (Air = 1)	Melting Point
Not Reported	**579°F (304°C)**

Solubility

Insoluble in water but readily forms soluble compounds when exposed to air or water. Soluble in nitric and sulfuric acid.

Appearance and Odor

Bluish-white, very soft, easily fusible, lead-like, heavy, odorless metal.

SECTION IV - FIRE AND EXPLOSION HAZARD DATA

Flash Point (method used)	Explosive Limits in Air % by Volume
Not Reported	**LEL: Not Reported UEL: Not Reported**
NFPA Classification	Autoignition Temperature
Combustible Solid	**Not Reported**

Extinguishing Media

Thallium may burn, but does not easily ignite. Use dry chemical, alcohol foam, carbon dioxide, or water.

Special Fire Fighting Procedures

Poisonous gases are produced in fire. Wear full protective clothing, self-contained breathing apparatus (SCBA) operated in pressure-demand mode. Move container from fire area if it can be done without risk.

Unusual Fire and Explosion Hazards

In powdered form, thallium is a moderate fire and explosion hazard. Will also burn vigorously on contact with flame or by chemical reaction. Powders and dusts form explosive mixtures in air. Use water spray or mist to knock down thallium dust clouds before they can ignite.

SECTION V - REACTIVITY DATA

Stability		Conditions to Avoid
		Thallium metal is stable in closed containers at room temperature under routine conditions of storage and handling. Keep away from incompatible materials and sources of heat or ignition.
Stable	Unstable	Incompatibility *(materials to avoid)*
X		Incompatible with oxidizers, especially fluorine, chlorine, and bromine. Also reacts in contact with strong acids, such as hydrofluoric, sulfuric, and nitric.
Hazardous Polymerization		Conditions to Avoid Under normal conditions of temperature and pressure, thallium cannot undergo hazardous polymerization. In air, a poisonous coating of thallium oxide can develop on exposed surfaces of thallium metal.
May Occur	Will Not Occur	Hazardous Decomposition or By-products
	X	When heated to decomposition in air it emits highly toxic fumes of thallium oxide.

SECTION VI - HEALTH HAZARD DATA

Primary Route(s) of Entry:	Inhalation?	Absorption (skin/eye)?	Ingestion?
	X	X	X

Health Hazards

INHALATION: Causes nausea, diarrhea, abdominal pain, vomiting, tingling or pain in the extremities, weakness, delirium, tremor, hallucinations, convulsive fits, peripheral neuropathy, retro-sternum tightness in the chest and chest pain, pulmonary edema (fluid in lungs), chorea, psychosis, liver and kidney damage, alopecia, paresthesia (legs), coma, and death.

ABSORPTION: Eye and skin contact will result in irritation. Thallium is readily absorbed through intact skin and will cause many of the same systemic effects noted for inhalation.

INGESTION: Rapid absorption through the gastrointestinal tract will result in many of the same effects as those noted for inhalation.

Carcinogenicity	NTP Listed?	IARC Cancer Review Group?	OSHA Regulated?	Target Organs?
Unknown Human Unknown Animal	No	No	29 CFR 1910.1000 Table Z-1	Eyes, resp. sys., liver, CNS, kidneys, GI tract.

Medical Conditions Generally Aggravated by Exposure

Disorders of the GI tract, CNS, kidneys, liver, and eyes may be aggravated by exposure.

Emergency and First-aid Procedures

Eye contact: Do not allow victim to keep eyes tightly shut. Flush immediately with water for 15 minutes (minimum); seek medical attention. Skin contact: Avoid skin contact in areas with cuts or scratches. If such contact occurs, wash with large amounts of water, scrubbing to remove metal particles. For inhalation: Remove the person from exposure. Provide respiratory assistance and CPR. Transfer to medical facility. Observe for 24- 48 hours for lung effects. If swallowed: Seek medical attention immediately. Do not attempt to give an unconscious or convulsing person anything by mouth. If conscious, give one or two glasses of milk or water to drink and induce vomiting.

SECTION VII - PRECAUTIONS FOR SAFE HANDLING AND USE

Steps to be Taken in Case Material is Released or Spilled

Wear self-contained breathing apparatus (SCBA). Restrict those not involved in cleanup from entering area. Avoid skin contact. Ventilate area of spill. Do not use water. Collect powder materials and deposit in sealed drum. Do NOT dry sweep; use a high-efficiency particulate air (HEPA) vacuum instead.

Preferred Waste Disposal Method

No citation.

Precautions to be Taken in Handling and Storage

Store in tightly closed containers in a cool, well-ventilated area away from heat. Protect containers from physical damage. Avoid contact in storage with incompatible materials as violent reactions can occur.

Other Precautions and Warnings

Use only non-sparking equipment and tools, especially when opening or closing containers of thallium.

SECTION VIII - CONTROL MEASURES AND PERSONAL PROTECTIVE EQUIPMENT

Respiratory Protection *(specify type)*

Use an air-purifying respirator equipped with a HEPA filter or use a NIOSH/MSHA-approved self-contained breathing apparatus (SCBA) with full facepiece operated in positive pressure mode.

Ventilation

Local exhaust (preferred) at site of chemical work or general ventilation.

Protective Gloves	Eye Protection	Other Protective Clothing
Impervious Gloves	Chemical Goggles and Face Mask	Protective Uniform or Apron

Work/Hygiene Practices

Always wash hands thoroughly after using chemical; never bring food, drink, or smoking materials into vicinity of chemicals. Barrier creams may be useful to limit the effects of skin exposure.

THALLIUM

Tl CAS: 7440-28-0

IDENTIFICATION AND TYPICAL USES

Thallium is a metallic element of atomic number 81, Group IIIA of the periodic table. It has valances of 2,3 and two stable isotopes. It appears as a bluish-white, very soft, easily fusible, lead-like, heavy, odorless metal. It is used for the production of thallium salts, in mercury alloys, in low-melting glasses, in rodenticides, in photoelectric cells, semiconductor studies, and electrodes in dissolved oxygen analyzers.

RISK ASSESSMENT: HEALTH

General Assessment

Thallium is a highly toxic metal. Primary exposure route is through *inhalation* of its dusts. It can also easily pass through intact skin (*absorption*) to cause toxic systemic effects. *Ingestion* is also a possible exposure route. The effects of thallium exposure can be life-threatening by any route. Therefore, caution is warranted to guard against combined exposures (i.e., exposure by more than one route) since the effects of such exposures may result in death. There are no data in the references regarding its carcinogenic, mutagenic, or teratogenic properties.

Human systemic effects following exposure by any route include nausea, diarrhea, abdominal pain, vomiting, tingling pain in the extremities, weakness, peripheral neuropathy, tremor, convulsive fits, retrosternum tightness and pain in the chest, ptosis (drooping of the upper eyelid), strabismus (deviation of eye, or "cross eyed"), chorea, psychosis, liver and kidney damage, alopecia (loss of hair), paresthesia (in the legs), coma, and death. In addition, inhalation can cause pulmonary edema (fluid in the lungs) which can be fatal. Symptoms of coughing, congestion, and dypsnea may be delayed up to 48 hours thereby creating a false sense of security with regard to health exposure risk.

Skin contact may produce irritation or rash. Eye contact also results in severe mechanical irritation.

Acute Health Effects

The following acute (short-term) health effects may occur immediately or shortly after exposure to thallium:

Skin: Moderate to severe irritation.

Eye: Severe irritation, itching, and burning on contact.

Lung: Overexposure can severely irritate the airways and lungs, causing nasal discharge, tightness in the chest, cough, shortness of breath, and/or fever. High level acute exposures can lead to pulmonary edema (fluid in the lungs) which can be fatal.

CNS: Dizziness, headache, hallucinations, convulsive fits, chorea, psychosis, tremor, paresthesia, numbness in the legs, nausea, vomiting, chorea, tremor, and peripheral neuritis.

Other: Abdominal pain, ptosis, strabismus, retrosternum tightness, liver and kidney damage.

Chronic Health Effects

The following chronic (long-term) health effects may occur at some time after exposure to thallium and can last for months or even years:

Cancer Hazards: According to information presented in the references, thallium has not been adequately tested for its ability to cause cancer in test animals.

Reproductive Hazard: There is no evidence in the published studies to show that thallium has the ability to adversely affect reproduction.

Other Chronic Effects: Repeated exposures can lead to chronic weakness or fatigue, poor appetite, insomnia, mood changes, irritability, a metallic taste in the mouth, pain in the extremities (polyneuritis), alopecia (loss of hair), and disorders of the cardiac, renal, or endocrine system. Signs of nerve damage include numbness or "pins and needles" in the arms and legs. Loss of vision, tremor, or abnormal muscle jerking. Permanent brain damage is also common. Exposure can also cause hairless, nail changes, and dry scaly skin.

Recommended Risk-Reduction Measures

Personnel should avoid direct contact with thallium. It is highly toxic by all routes of exposure and the combined effects of simultaneous exposures by more than one route can be life-threatening. If a less toxic material or compound cannot be substituted for thallium, then *engineering controls* are the most effective method of reducing exposure risk. The best protection is to enclose operations and/or provide local exhaust ventilation at the site of thallium release. While not always operationally feasible, isolating operations can also reduce exposure. Using respiratory protection is less effective than the controls mentioned above, but is still required whenever working with or around thallium. The exposure level is low and difficult to accurately monitor. For infrequent or transient exposures, an air-purifying respirator equipped with a high-efficiency particulate air (HEPA) filter and dust pre-filter may suffice. However, for the best possible protection use an MSHA/NIOSH-approved full facepiece self-contained breathing apparatus (SCBA) with full facepiece operated in pressure demand, or a powered air-supplied respirator in continuous flow mode. If a full facepiece is not available, then chemical goggles should be worn to protect the eyes, especially when a splash hazard exists. A face shield should also be considered. To prevent hand and skin exposures, rubber gloves should be used. Glove manufacturers should be contacted and permeation studies obtained *before* final glove selections are made.

Administrative controls should also be in place to minimize the potential for human exposures. These may include written procedures or policies which specify the methods and techniques that will be practiced whenever personnel are to work with thallium.

All personnel should receive *training* on the use, hazards, protective measures, emergency actions, and other precautions per 29 CFR 1910.1200 (Hazard Communication) prior to the first assignment in an area where thallium is used or stored.

Before beginning employment and at regular times thereafter, the following recommended medical tests should be considered for those with frequent or potentially high exposures:

- ☑ Lung function tests.
- ☑ Complete examination of the nervous system.
- ☑ Consider chest X-ray after acute overexposure (may be negative if taken immediately after exposure due to delayed onset of pulmonary edema).
- ☑ Skin examination (look for loss of hair).
- ☑ Careful examination and testing of vision.

It should be noted that medical tests that simply look for existing damage are not a substitute for controlling exposures. Medical histories are extremely important when assessing exposure risk. Also, since smoking can cause heart disease, lung cancer, emphysema, and other respiratory diseases, smokers exposed to thallium may experience symptoms more quickly and more pronounced than non-smokers under the same conditions of exposure. Prudent risk management requires proper consideration of *all* factors which may cause exposure symptoms in the workplace.

Other methods to reduce exposure include:

- ☑ Always ensure that proper protective clothing is worn when using chemical substances and that personnel are trained in its use, care, and maintenance. Barrier creams may be effective in preventing skin exposures.
- ☑ Wash thoroughly immediately after exposure to thallium and at the end of the work shift or before eating, drinking, smoking, or applying cosmetics.
- ☑ Hazard warning information should be posted in the work area. In addition, as part of an on-going education and training program, all information on the health and safety hazards of thallium should be communicated to all potentially exposed workers.
- ☑ Eye wash stations should be provided in the immediate work area for emergency use. If there is a possibility of skin exposure to thallium, emergency shower facilities should also be provided.
- ☑ Workers whose clothing has been contaminated by thallium should change into clean clothes before leaving work. Contaminated work clothing should be laundered only by individuals who have been informed of the hazards of exposure to thallium.

RISK ASSESSMENT: ENVIRONMENT

General Assessment

Even though thallium is a commonly occurring metallic element, the environment is still at risk of exposure during transportation, storage, disposal, or destruction of thallium. In almost every scenario, the threat of environmental exposure is contingent upon the proper handling of the chemical substance. Accidental spills, large or small, can result in contamination of the surrounding environmental mediums (water, soil, and air).

Thallium is a combustible solid. In bulk form, it will burn but will not easily ignite. However, thallium metal dusts particles suspended in the air can explode. Caution is always required in handling, storage, transportation, and disposal of thallium. It presents a serious explosion hazard in the form of powders and dusts. It is incompatible with oxidizers, such as fluorine or chlorine, and strong acids, such as nitric or hydrofluoric. When heated to decomposition in air it emits very toxic fumes of thallium oxide. Emergency responders should be made aware of the presence of thallium at any emergency response situation.

Thallium is a metal that is not particularly rare in the Earth's crust. It commonly occurs in potash minerals and in a number of other minerals such as crookesite, lorandite, hutchinsonite, vrbaite, and avicennite. Thallium and its compounds may enter the environment during mining and smelting operations, through industrial discharges, or spills.

Acute Ecological Effects

Acute (short-term) toxic effects may include the death of animals, birds, or fish and death or low growth rate in plants. Acute effects are seen 2 to 4 days after animals or plants are exposed to thallium.

Thallium and its compounds have high acute toxicity to aquatic life. Insufficient data are available to evaluate or predict the acute (short-term) effects of thallium to birds, plants, or land animals.

Chronic Ecological Effects

Chronic (long-term) toxic effects may include shortened life span, reproductive problems, lower fertility, and changes in appearance or behavior in exposed animals. These effects can be seen long after first exposure(s) to toxic chemicals.

Thallium and its compounds are considered to have high chronic toxicity to aquatic life. Insufficient data are available to evaluate the chronic (long-term) effects of thallium on plants, birds, or land animals.

Water Solubility

Thallium and its compounds have water solubilities that range from low to high, depending on the salt that is formed.

Persistence in the Environment

Thallium itself is likely to be highly persistent in water, as it has only slight water solubility. Other compounds of thallium will vary in persistence, as water solubility varies from compound to compound.

Bioaccumulation in Aquatic Organisms

Some substances increase in concentration, or *bioaccumulate,* in living organisms as they breathe contaminated air, drink contaminated water, or eat contaminated food. These chemicals can become concentrated in the tissues and internal organs of animals as well as humans.

The concentration of thallium and compounds found in edible fish tissues is expected to be somewhat higher than the average concentration of thallium and its compounds found in the water from which the fish was taken.

Recommended Risk-Reduction Measures

Proper training of all transporters will reduce the likelihood of a mishap or accident resulting in a leak or spill of thallium into the environment. Labels on all containers, trucks, and rail cars must meet DOT requirements and accurately reflect their contents to enable emergency responders to react properly and quickly to any disaster thereby reducing the potential risk to the environment and to personnel.

Storage of thallium should be segregated from other chemicals and materials to minimize the risk of cross-contamination. Thallium must be stored to avoid contact with oxidizers (such as perchlorates, peroxides, permanganates, chlorates, and nitrates), and strong acids (such as hydrochloric, sulfuric, and nitric) since violent reactions can occur. Containers should be protected from physical damage and stored to avoid contact with heat. Dusty conditions must be prevented in storage and handling areas.

If a spill or leak to the environment has occurred, fire department, emergency response and/or hazardous materials spill personnel should be notified immediately. Cleanup should be attempted only by those trained in proper spill containment procedures using non-sparking tools. Contaminated soils should be removed for incineration and replaced with clean soil. If thallium should contact the water table, aquifer, or navigable waterway, time is of the essence. The solubility of thallium and its compounds can range from insoluble to highly solubility and total remediation may not be possible. The local and/or state emergency response authorities must be notified. A comprehensive emergency response or disaster preparedness/recovery plan should be in place prior to any operations involving the use, transportation, storage, or disposal of thallium.

If thallium is spilled or leaked, the following specific steps are recommended:

- ☑ Restrict persons not wearing protective clothing from area of spill or leak until cleanup is complete. Avoid skin and eye contact.
- ☑ Collect powdered material in the most convenient and safe manner possible and deposit in sealed drum for disposal. Do NOT dry sweep. Use a vacuum equipped with a high-efficiency particulate air (HEPA) filter. Damp mop residue. Use water spray to knock down dust clouds that have not yet ignited. Contain and collect water for disposal.
- ☑ Ventilate area of spill or leak and remove all ignition sources.
- ☑ It may be necessary to dispose of thallium as a hazardous waste. The responsible state agency or the regional office of the federal Environmental Protection Agency (EPA) should be contacted for specific recommendations.

RISK ASSESSMENT: BUSINESS

General Assessment

Accidents or mishaps involving thallium can present a serious threat to business operations. The loss or damage of equipment or facilities can significantly affect fiscal viability. Lawsuits that may result from personnel illness, injury/death, public exposures, and/or environmental contamination will require a serious expenditure of resources. Media attention surrounding an injury, death, or environmental damage can also result in a loss of profits and loss of current as well as future business.

Recommended Risk-Reduction Measures

Company attorneys, safety and health professionals, and environmental specialists should be involved in the development of any procedures for responding to chemical incidents. A company official should be predesignated as a public relations officer with specific training in dealing with the press. Corporate plans and policies should be developed, approved, and implemented long before any need for such arises.

REFERENCES

American Conference of Governmental Industrial Hygienists. 1988. *Documentation of the Threshold Limit Values and Biological Exposure Indices,* 5th Edition (with updates). Cincinnati: ACGIH

Asante-Duah, D. K. 1993. *Hazardous Waste Risk Assessment.* New York: CRC Press/Lewis Publishers.

Bloom, A. D. and F. J. DeSerres. 1995. *Ecotoxicity and Human Health.* New York: CRC Press/Lewis Publishers.

Calabrese, E. J. 1994. *Biological Effects of Low Level Exposures.* New York: CRC Press/Lewis Publishers.

Cockerman, L. G., B. S. Shane. 1994. *Basic Environmental Toxicology.* New York: CRC Press/Lewis Publishers.

Gardner, R. W. 1994. *Chemical Intolerance.* New York: CRC Press/Lewis Publishers.

Hallenbeck, W. H. 1993. *Quantitative Risk Assessment for Environmental and Occupational Health,* 2nd Edition. New York: CRC Press/Lewis Publishers.

Hodgson, E., R. B. Mailman, and J. E. Chambers. 1988. *Dictionary of Toxicology.* London: The MacMillian Company.

Keith, L. H. and M. M. Walker. 1995. *Handbook of Air Toxics.* New York: CRC Press/Lewis Publishers.

Landis, W. G., M. H. Yu. 1995. *Introduction to Environmental Toxicology.* New York: CRC Press/Lewis Publishers.

Lewis, R. J., Sr. 1993. *Hawley's Condensed Chemical Dictionary,* 12th Edition. New York: Van Nostrand Reinhold.

Lewis, R. J., Sr. 1992. *Sax's Dangerous Properties of Industrial Materials,* Volumes 1, 2, and 3, 8th Edition. New York: Van Nostrand Reinhold.

Meyer, E. 1990. *Chemistry of Hazardous Materials.* Englewood Cliffs, NJ: Prentice Hall.

Patnaik, P. 1992. *A Comprehensive Guide to the Hazardous Properties of Chemical Substances.* New York: Van Nostrand Reinhold.

Philp, R. B. 1995. *Environmental Hazards and Human Health.* New York: CRC Press/Lewis Publishers.

Proctor, N. H., J. P. Hughes. 1991. *Chemical Hazards of the Workplace.* New York: Van Nostrand Reinhold.

Richardson, M. 1995. *Dictionary of Substance and Their Effects,* Vol. I-VI. New York: CRC Press/Lewis Publishers.

Sacarello, H. L. A. 1994. *The Comprehensive Handbook of Hazardous Materials.* New York: CRC Press/Lewis Publishers.

Torkelson, T. R. V. K. Rowe. 1981. *Patty's Industrial Hygiene and Toxicology,* Vol. 2B. New York: Wiley-Interscience.

U.S. Department of Health and Human Services, National Institute for Occupational Safety and Health. 1994. *NIOSH Pocket Guide to Chemical Hazards.* Washington, D.C.: U.S. Government Printing Office.

MATERIAL SAFETY DATA SHEET

CHEMICAL NAME	
THIOGLYCOLIC ACID	**CRC PRESS/LEWIS PUBLISHERS DISCLAIMER:** THE INFORMATION AND RECOMMENDATIONS PRESENTED HEREIN ARE BASED ON SOURCES BELIEVED TO BE RELIABLE. CRC MAKES NO REPRESENTATION ON ITS COMPLETENESS OR ACCURACY. IT IS THE RESPONSIBILITY OF THE USER TO DETERMINE THE CHEMICAL'S SUITABILITY FOR ITS INTENDED USE, THE CHEMICAL'S SAFE USE, AND THE CHEMICAL'S PROPER DISPOSAL. NO REPRESENTATIONS AND/OR WARRANTIES, EITHER EXPRESSED OR IMPLIED, OF THE MERCHANTABILITY OR FITNESS FOR A PARTICULAR PURPOSE, OR OF ANY OTHER NATURE, ARE MADE WITH RESPECT TO THE INFORMATION PROVIDED IN THIS MSDS OR TO THE CHEMICAL TO WHICH INFORMATION MAY REFER. CRC NEITHER ASSUMES NOR AUTHORIZES ANY OTHER PERSON TO ASSUME FOR IT, ANY OTHER ADDITIONAL RESPONSIBILITY OR LIABILITY FOR THE USE OF, OR RELIANCE UPON, THIS INFORMATION.

HAZARD WARNING INFORMATION

HEALTH	FIRE	REACTIVE	OTHER	DEGREE OF HAZARD	COLOR CODING	OTHER CODES
3	1	0		0 = Minimum Hazard 1 = Slight Hazard 2 = Moderate Hazard 3 = Serious Hazard 4 = Severe Hazard	HEALTH = BLUE FIRE = RED REACTIVITY = YELLOW OTHER = WHITE	OX = Oxidizer ACID = Acid ALK = Alkali COR = Corrosive W = Use No Water

SECTION I - GENERAL INFORMATION

Characterization	RCRA Number	EPA Class
Mercaptan	**None**	**Not Applicable**

DOT Proper Shipping Name **Thioglycolic acid**	Chemical Abstract Service (CAS) Number **68-11-1**
DOT Hazard Class and Label Requirements **Corrosive Material**	DOT Emergency Guide Code **60**
DOT Identification Number **UN 1940**	Molecular Formula **$HSCH_2COOH$**

Synonyms

Acetyl mercaptan; mercaptoacetate; mercaptoacetic acid; 2-mercaptoacetic acid; thiovanic acid; alpha-mercaptoacetic acid; 2-thioglycolic acid.

SECTION II - HAZARDOUS INGREDIENTS AND IDENTITY INFORMATION

Hazardous Components (specific identity)	OSHA Exposure Criteria	NIOSH Exposure Criteria	Immediately Dangerous to Life and Health (IDLH)	ACGIH Exposure Criteria
Thioglycolic acid: (derivation: By heating chloracetic acid with potassium hydrogen sulfide; by the action of sodium sulfhydrate on sodium chloroacetate; or by electrolysis of dithioglycollic acid from sodium sulfide and sodium chloroacetate). **1 ppm = 3.83 mg/m^3**	**PEL:** Not Established **STEL:** Not Established	**REL (skin):** 1 ppm 4 mg/m^3 **STEL:** Not Established	**Not Determined**	**TLV (skin):** 1 ppm 3.8 mg/m^3 **STEL:** Not Established

SECTION III - PHYSICAL AND CHEMICAL CHARACTERISTICS

Boiling Point **226°F (108°C)**	Specific Gravity (H_2O = 1) **1.32**
Vapor Pressure (mm Hg) **10 at 64°F (18°C)**	Molecular Weight (atomic weight) **92.1**
Vapor Density (Air = 1) **1.1**	Melting Point **2°F (-16.5°C)**

Solubility

Miscible in water. Soluble in alcohol, acetone, ether, chloroform, benzene, and other organic solvents.

Appearance and Odor

Colorless liquid with a strong, disagreeable, unpleasant odor characteristic of mercaptans (like rotting eggs). Odor Threshold = Not determined (olfactory fatigue may occur rapidly after short exposure).

SECTION IV - FIRE AND EXPLOSION HAZARD DATA

Flash Point (method used) **>230°F (>110°C)**	Explosive Limits in Air % by Volume **LEL: 5.9%** **UEL: Not Reported**
NFPA Classification **Class IIIB Combustible Liquid**	Autoignition Temperature **Not Reported**

Extinguishing Media

Dry chemical, carbon dioxide, water spray, or regular foam. Direct water stream may be ineffective.

Special Fire Fighting Procedures

Wear full protective clothing and self-contained breathing apparatus (SCBA). Move containers from fire area if it can be done without risk. Use water spray to cool containers that cannot be moved.

Unusual Fire and Explosion Hazards

Containers may explode in fire. Presents a vapor explosion hazard indoors, outdoors, and in sewers. It may evolve flammable hydrogen sulfide when heated in fire.

SECTION V - REACTIVITY DATA

Stability		Conditions to Avoid Normally stable in closed containers at room temperature under routine conditions of handling and storage. Keep away from heat, ignition sources, and incompatible materials. It is readily oxidized in contact with air.
Stable X	Unstable	Incompatibility *(materials to avoid)* Reacts vigorously with strong oxidizers (chlorine, fluorine, bromine), strong bases, active metals (sodium, potassium, magnesium, calcium), mineral salts (causes decomposition and release of hydrogen sulfide).
Hazardous Polymerization		Conditions to Avoid Under normal conditions of temperature and pressure, hazardous polymerization of thioglycolic acid cannot occur.
May Occur	Will Not Occur X	Hazardous Decomposition or By-products When heated to decomposition, thioglycolic acid can produce highly toxic fumes of sulfur oxides.

SECTION VI - HEALTH HAZARD DATA

Primary Route(s) of Entry:	Inhalation?	Absorption (skin/eye)?	Ingestion?
	X	X	X

Health Hazards

INHALATION: Irritation of the eyes, nose, throat, and mucous membranes of the respiratory tract. May cause lachrymation (tearing), weak and gasping respiration, possible convulsions, and a dangerous buildup of fluid in the lungs (pulmonary edema), which can be fatal.

ABSORPTION: Irritation of the eyes (conjunctiva) and possible corneal damage. Skin contact can cause burns and blisters. Readily absorbs through intact skin to cause toxic systemic effects.

INGESTION: Poison by ingestion. Corrosive to the mucosa of the mouth, throat, and esophagus. It will cause immediate pain, nausea, bloody vomiting, circulatory collapse, rapid pulse, shallow respiration, scanty urine, possible kidney failure, ischemic lesions in the liver and heart (in animal studies).

Carcinogenicity	NTP Listed?	IARC Cancer Review Group?	OSHA Regulated?	Target Organs?
Unknown Human Unknown Animal	No	No	No	Eyes, skin, respiratory system.

Medical Conditions Generally Aggravated by Exposure

Skin conditions or diseases may be aggravated by exposure to thioglycolic acid.

Emergency and First-aid Procedures

Eye contact: Do not allow victim to keep eyes tightly shut. Flush immediately with water for 15 minutes (minimum); seek medical attention. Skin contact: Wash with large amounts of water for a minimum of 15 minutes. For red or blistered skin, seek medical assistance. For inhalation: Remove the person from exposure. Provide respiratory assistance and CPR. Transfer to medical facility. Observe 48 hours for lung effects. If swallowed: Seek medical attention immediately. Give 1-2 glasses of water, do NOT induce vomiting. Do NOT attempt to give an unconscious or convulsing person anything by mouth.

SECTION VII - PRECAUTIONS FOR SAFE HANDLING AND USE

Steps to be Taken in Case Material is Released or Spilled

Remove all ignition sources. Use water spray to knock down vapors. Wear self-contained breathing apparatus (SCBA) for respiratory protection. Restrict those not involved in cleanup from entering area. Avoid skin contact. Ventilate area of spill. Absorb in sand or other material and dispose of in sealed drums.

Preferred Waste Disposal Method

Dissolve in flammable solvent and burn in an incinerator equipped with an afterburner and scrubber.

Precautions to be Taken in Handling and Storage

Store in tightly closed containers in a cool, dry, well-ventilated area away from heat (including direct sunlight). Protect containers from physical damage. Inspect containers regularly.

Other Precautions and Warnings

Thioglycolic acid must be stored to avoid contact with strong oxidizers since violent reactions can occur.

SECTION VIII - CONTROL MEASURES AND PERSONAL PROTECTIVE EQUIPMENT

Respiratory Protection *(specify type)*

Air-purifying respirators may suffice but best protection is provided by a self-contained breathing apparatus (SCBA) or supplied-air respirator with full facepiece operated in positive pressure mode.

Ventilation

Local exhaust (preferred) at site of chemical work or general ventilation.

Protective Gloves	Eye Protection	Other Protective Clothing
Butyl Rubber, Neoprene, or Fluorocarbon	Chemical Goggles and Face Mask	Protective Uniform or Apron

Work/Hygiene Practices

Always wash hands thoroughly after using chemical; never bring food, drink, or smoking materials into vicinity of chemicals.

THIOGLYCOLIC ACID

$HSCH_2COOH$ CAS: 68-11-1

IDENTIFICATION AND TYPICAL USES

Thioglycolic acid is a colorless liquid with a strong, disagreeable, unpleasant odor characteristic of mercaptans (like rotting eggs). It is used as a sensitive reagent for silver, iron, tin, or molybdenum. It is also used in the manufacture of thioglycolates, in permanent hair wave solution and depilatories, as a vinyl stabilizer, and in the manufacture of pharmaceuticals.

RISK ASSESSMENT: HEALTH

General Assessment

Thioglycolic acid is moderately toxic by *inhalation* and dermal *absorption.* It is a human poison by *ingestion.* Thioglycolic acid is a strong corrosive and can serious injure human tissues. There are no data on its carcinogenic, mutagenic, or teratogenic capabilities.

Inhalation will cause irritation to the eyes, nose, throat, and mucous membranes of the respiratory system. There can be lachrymation (tearing), weak or gasping respiration, and possible convulsions (positive in animal studies). High exposures can cause a dangerous buildup of fluid in the lungs (pulmonary edema), which can be fatal. Symptoms of cough, congestion, dypsnea, and tightness in chest may be delayed up to 48 hours or more thereby creating a false sense of security with regard to health exposure risk.

Eye contact may result in corneal damage, clouding, and burns with serious injury and vision impairment. Skin contact may cause urticaria (hives), blisters, and burns. It can also pass through unbroken skin to cause toxic systemic effects.

Ingestion of thioglycolic acid presents a serious, even life-threatening exposure risk. It causes corrosion to the mucosa of the mouth, throat, and esophagus with immediate pain, nausea, vomiting (which may be blood due to digestive hemorrhage), circulatory collapse, rapid pulse, shallow respiration, scanty urine, and possible kidney failure and ischemic lesions in the liver and heart.

Acute Health Effects

The following acute (short-term) health effects may occur immediately or shortly after exposure to thioglycolic acid:

Skin: Moderate to severe irritation with burns, blisters, inflammation, and swelling.

Eye: Severe irritation and corrosion. Possible tissue damage and injury to vision.

Lung: Severe irritation to respiratory tract. Corrosive action on the mucosa with the possible development of pulmonary edema.

Other: In animals studies, depression of the central nervous system, hypoglycemia, possible convulsions, and permanent scarring of skin and eye tissues.

Chronic Health Effects

The following chronic (long-term) health effects may occur at some time after exposure to thioglycolic acid and can last for months or even years:

Cancer Hazards: According to information presented in the references, thioglycolic acid has not been adequately tested for its ability to cause cancer in test animals. There are few data available on chronic human exposures.

Reproductive Hazard: There are no reports to substantiate claims of teratogenicity resulting from exposure to thioglycolic acid.

Other Chronic Effects: Repeated exposure may cause allergic eczematous dermatitis. After exposure, there may be permanent scars on the cornea, skin, and oropharnyx. Animal studies have shown that chronic exposure to thioglycolic acid may displace iron from serum ferritin. It is not known if human exposure will have this effect. Also, very irritating substances can injure the lungs. It is not known if long-term exposure to thioglycolic acid will have this effect.

ʊ *Recommended Risk-Reduction Measures*

Information on the specific human toxicity of thioglycolic acid is limited in the references. Personnel should avoid direct contact with thioglycolic acid. If a less toxic material or compound cannot be substituted for thioglycolic acid, then *engineering controls* are the most effective method of reducing exposure risk. The best protection is to enclose operations and/or provide local exhaust ventilation at the site of thioglycolic acid release. While not always operationally feasible, isolating operations can also reduce exposure. Using respiratory protection is less effective than the controls mentioned above, but is still required whenever working with or around thioglycolic acid. The NIOSH recommended exposure level (REL) for this chemical is relatively low (1 ppm) and the specific human toxicity is not clear. An air-purifying respirator may suffice for low-level exposures. However, an MSHA/NIOSH-approved self-contained breathing apparatus (SCBA) or a powered air-supplied respirator with full facepiece operated in pressure demand are the appropriate levels of respiratory protection. If a full facepiece is not available, then chemical goggles should be worn to protect the eyes, especially when a splash hazard exists. A face shield should also be considered. Skin contact must be avoided. To prevent hand and skin exposures, butyl rubber, fluorocarbon rubber, or neoprene appear to provide adequate protection while nitrile, natural rubber, or polyvinyl alcohol materials are not recommended. Glove manufacturers should be contacted and permeation studies obtained *before* final glove selections are made.

Administrative controls should also be in place to minimize the potential for human exposures. These may include written procedures or policies which specify the methods and techniques that will be practiced whenever personnel are to work with thioglycolic acid.

All personnel should receive *training* on the use, hazards, protective measures, emergency actions, and other precautions per 29 CFR 1910.1200 (Hazard Communication) prior to the first assignment in an area where thioglycolic acid is used or stored. Before beginning employment and at regular intervals thereafter (e.g., annually), the following medical test is recommended:

☑ Lung function tests.

If symptoms develop or overexposure is suspected, the following may be useful:

☑ Liver, kidney, and lung function tests.

☑ Consider chest X-ray after acute overexposure (may be negative if taken immediately after exposure due to delayed onset of pulmonary edema).

It should be noted that medical tests that simply look for existing damage are not a substitute for controlling exposures. Medical histories are extremely important when assessing exposure risk. Also, since smoking can cause heart disease, lung cancer, emphysema, and other respiratory diseases, smokers exposed to thioglycolic acid may experience symptoms more quickly than nonsmokers under the same conditions of exposure. Prudent risk management requires careful consideration of *all* possible factors which may be responsible for exposure symptoms in the workplace.

Other methods to reduce exposure include:

☑ Always ensure that proper protective clothing is worn when using chemical substances and that personnel are trained in its proper use, care, and maintenance.

☑ Wash thoroughly immediately after exposure to thioglycolic acid and at the end of the work shift or before eating, drinking, smoking, or applying cosmetics.

☑ Hazard warning information should be posted in the work area. In addition, as part of an on-going education and training program, all information on the health and safety hazards of thioglycolic acid should be communicated to all exposed or potentially exposed workers.

☑ Eye wash stations should be provided in the immediate work area for emergency use. If there is a possibility of skin exposure to thioglycolic acid, emergency shower facilities should also be provided in the area.

☑ Workers whose clothing has been contaminated by thioglycolic acid should change into clean clothes before leaving work. Contaminated work clothing should be laundered only by individuals who have been trained on the hazards of thioglycolic acid.

RISK ASSESSMENT: ENVIRONMENT

General Assessment

The environment is at risk of exposure during transportation, storage, disposal, or destruction of thioglycolic acid. In almost every scenario, the threat of environmental exposure is contingent upon the proper handling of the chemical substance. Accidental spills, large or small, can result in contamination of the surrounding environmental mediums (water, soil, and air).

Thioglycolic acid is Class IIIB combustible liquid (per OSHA 29 CFR 1910.106). It is extremely reactive with many substances. It is readily oxidized on exposure to air. It can react with strong oxidizers (such as fluorine, chlorine, bromine, permanganates, and nitrates), bases, and many active metals (potassium, sodium, magnesium, calcium). It is also an extremely corrosive substance, especially to metals. Contact with mineral acids causes decomposition with the release of poisonous hydrogen sulfide. Therefore, caution is always required in handling, storage, transportation, and disposal of thioglycolic acid. When heated to decomposition it emits very acrid and irritating smoke and fumes of sulfur oxides. Emergency responders, hazardous materials personnel, and firefighters should be made aware of the presence of thioglycolic acid at any emergency response situation.

Thioglycolic acid may enter the environment through industrial effluents and spills.

Acute Ecological Effects

Acute (short-term) toxic effects may include the death of animals, birds, or fish and death or low growth rate in plants. Acute effects are seen 2 to 4 days after animals or plants are exposed to thioglycolic acid.

This chemical has moderate acute toxicity to aquatic life. Insufficient data are available to evaluate or predict the acute (short-term) effects of thioglycolic acid to plants, birds, or land animals.

Chronic Ecological Effects

Chronic (long-term) toxic effects may include shortened life span, reproductive problems, lower fertility, and changes in appearance or behavior in exposed animals. These effects can be seen long after first exposure(s) to toxic chemicals.

Insufficient data are available to evaluate the chronic (long-term) effects of thioglycolic acid on aquatic life, plants, birds, or land animals.

Water Solubility

Thioglycolic acid is highly soluble in water. Concentrations of 1000 milligrams and more will mix with a liter of water.

Persistence in the Environment

Thioglycolic acid is moderately persistent in water, with a half-life of between 20 to 200 days. The half-life of a pollutant is the amount of time it takes for one half of the chemical to be degraded. Thioglycolic acid does not degrade well.

Bioaccumulation in Aquatic Organisms

Some substances increase in concentration, or *bioaccumulate,* in living organisms as they breathe contaminated air, drink contaminated water, or eat contaminated food. These chemicals can become concentrated in the tissues and internal organs of animals as well as humans.

Insufficient data are available to evaluate or predict the bioaccumulation characteristics of thioglycolic acid in the edible tissues of fish.

Recommended Risk-Reduction Measures

Proper training of all transporters will reduce the likelihood of a mishap or accident resulting in a leak or spill of thioglycolic acid into the environment. Labels on all containers, trucks, and rail cars must meet DOT requirements and accurately reflect their contents to enable emergency responders to react properly and quickly to any disaster thereby reducing the potential risk to the environment and to personnel.

Storage of thioglycolic acid should be segregated from other chemicals to minimize the risk of cross-contamination. Thioglycolic acid must be stored to avoid contact with strong oxidizers, such as chlorine and bromine, since violent reactions can occur. Containers should be protected from physical damage and stored to avoid contact with heat and direct sunlight.

If a spill or leak to the environment has occurred, fire department, emergency response and/or hazardous materials spill personnel should be notified immediately. Cleanup should be attempted only by those trained in proper spill containment procedures using non-sparking tools. Contaminated soils should be removed for incineration and replaced with clean soil. If thioglycolic acid should contact the water table, aquifer, or navigable waterway, time is of the essence. It is highly soluble in water and total remediation may not be possible. The local and/or state emergency response authorities must be notified. A comprehensive emergency response or disaster preparedness/recovery plan should be in place prior to any operations involving the use, transportation, storage, or disposal of thioglycolic acid.

If thioglycolic acid is spilled or leaked, the following specific steps are recommended:

- ☑ Restrict persons not wearing protective clothing from area of spill or leak until cleanup is complete. Avoid skin and eye contact and use only non-sparking tools.
- ☑ Absorb spilled liquids with vermiculite, dry sand, or a similar material and deposit in sealed drum for disposal.
- ☑ Ventilate area of spill or leak.
- ☑ Remove all ignition sources.
- ☑ It may be necessary to dispose of thioglycolic acid as a hazardous waste. The responsible state agency or the regional office of the federal Environmental Protection Agency (EPA) should be contacted for specific recommendations.

RISK ASSESSMENT: BUSINESS

General Assessment

Accidents or mishaps involving thioglycolic acid can present a serious threat to business operations. The loss or damage of equipment or facilities can significantly affect fiscal viability. Lawsuits that may result from personnel illness, injury/death, public exposures, and/or environmental contamination will require a serious expenditure of resources. Media attention surrounding an injury, death, or environmental damage can also result in a loss of profits and loss of current as well as future business.

Recommended Risk-Reduction Measures

Company attorneys, safety and health professionals, and environmental specialists should be involved in the development of any procedures for responding to chemical incidents. A company official should be predesignated as a public relations officer with specific training in dealing with the press. Corporate plans and policies should be developed, approved, and implemented long before any need for such arises.

REFERENCES

American Conference of Governmental Industrial Hygienists. 1988. *Documentation of the Threshold Limit Values and Biological Exposure Indices,* 5th Edition (with updates). Cincinnati: ACGIH

Bloom, A. D. and F. J. DeSerres. 1995. *Ecotoxicity and Human Health.* New York: CRC Press/Lewis Publishers.

Brusick, D. J. 1994. *Methods for Genetic Risk Assessment.* New York: CRC Press/Lewis Publishers.

Burrell, R., D. K. Flaherty, and L. J. Sauers. 1992. *Toxicology of the Immune System.* New York: Van Nostrand Reinhold.

Calabrese, E. J., E. M. Kenyon. *Air Toxics and Risk Assessment.* New York: CRC Press/Lewis Publishers.

Cockerman, L. G., B. S. Shane. 1994. *Basic Environmental Toxicology.* New York: CRC Press/Lewis Publishers.

Cothern, C. R. 1995. *Handbook for Environmental Risk Decisionmaking.* New York: CRC Press/Lewis Publishers.

Danse, I. R. 1991. *Common Sense Toxics in the Workplace.* New York: Van Nostrand Reinhold.

Environmental Protection Agency. 1983. *Chemical Hazard Information Profiles.* Washington, D.C.: U.S. EPA.

Forseberg, K., and L. H. Keith. 1993. *Chemical; Protective Clothing Permeation and Degradation Database.* Chelsea, MI: Lewis Publishers.

Gardner, R. W. 1994. *Chemical Intolerance.* New York: CRC Press/Lewis Publishers.

Gerarde, H. W. 1960. *Toxicology and Biochemistry of Aromatic Hydrocarbons.* New York: Elsevier.

Hallenbeck, W. H. 1993. *Quantitative Risk Assessment for Environmental and Occupational Health,* 2nd Edition. New York: CRC Press/Lewis Publishers.

Hodgson, E., R. B. Mailman, and Janice E. Chambers. 1988. *Dictionary of Toxicology.* London: The MacMillian Company.

Hosty, J. W., and P. Foster. 1990. *A Practical Guide to Chemical Spill Response.* New York: Van Nostrand Reinhold.

Jones, F. E. 1994. *Toxic Organic Vapors in the Workplace.* New York: CRC Press/Lewis Publishers.

Lauwerys, R. R., P. Hoet. 1993. *Industrial Chemical Exposure.* New York: CRC Press/Lewis Publishers.

Lewis, R. J., Sr. 1992. *Sax's Dangerous Properties of Industrial Materials,* Volumes 1, 2, and 3, 8th Edition. New York: Van Nostrand Reinhold.

Meyer, E. 1990. *Chemistry of Hazardous Materials.* Englewood Cliffs, NJ: Prentice Hall.

Patnaik, P. 1992. *A Comprehensive Guide to the Hazardous Properties of Chemical Substances.* New York: Van Nostrand Reinhold.

Torkelson, T. R. V. K. Rowe. 1981. *Patty's Industrial Hygiene and Toxicology,* Vol. 2B. New York: Wiley-Interscience.

U.S. Department of Health and Human Services, National Institute for Occupational Safety and Health. 1994. *NIOSH Pocket Guide to Chemical Hazards.* Washington, D.C.: U.S. Government Printing Office.

MATERIAL SAFETY DATA SHEET

CHEMICAL NAME	
THIONYL CHLORIDE	**CRC PRESS/LEWIS PUBLISHERS DISCLAIMER:** THE INFORMATION AND RECOMMENDATIONS PRESENTED HEREIN ARE BASED ON SOURCES BELIEVED TO BE RELIABLE. CRC MAKES NO REPRESENTATION ON ITS COMPLETENESS OR ACCURACY. IT IS THE RESPONSIBILITY OF THE USER TO DETERMINE THE CHEMICAL'S SUITABILITY FOR ITS INTENDED USE, THE CHEMICAL'S SAFE USE, AND THE CHEMICAL'S PROPER DISPOSAL. NO REPRESENTATIONS AND/OR WARRANTIES, EITHER EXPRESSED OR IMPLIED, OF THE MERCHANTABILITY OR FITNESS FOR A PARTICULAR PURPOSE, OR OF ANY OTHER NATURE, ARE MADE WITH RESPECT TO THE INFORMATION PROVIDED IN THIS MSDS OR TO THE CHEMICAL TO WHICH INFORMATION MAY REFER. CRC NEITHER ASSUMES NOR AUTHORIZES ANY OTHER PERSON TO ASSUME FOR IT, ANY OTHER ADDITIONAL RESPONSIBILITY OR LIABILITY FOR THE USE OF, OR RELIANCE UPON, THIS INFORMATION.

HAZARD WARNING INFORMATION

HEALTH	FIRE	REACTIVE	OTHER	DEGREE OF HAZARD	COLOR CODING	OTHER CODES
3	0	1	W	0 = Minimum Hazard 1 = Slight Hazard 2 = Moderate Hazard 3 = Serious Hazard 4 = Severe Hazard	HEALTH = BLUE FIRE = RED REACTIVITY = YELLOW OTHER = WHITE	OX = Oxidizer ACID = Acid ALK = Alkali COR = Corrosive W = Use No Water

SECTION I - GENERAL INFORMATION

Characterization **Halogenated Hydrocarbon**	RCRA Number **None**	EPA Class **Not Applicable**
DOT Proper Shipping Name **Thionyl chloride**	Chemical Abstract Service (CAS) Number **7719-09-7**	
DOT Hazard Class and Label Requirements **Corrosive Material; Corrosive**	DOT Emergency Guide Code **39**	
DOT Identification Number **UN 1836**	Molecular Formula **$SOCl_2$**	

Synonyms
Sulfinyl chloride; sulfur chloride oxide; sulfurous dichloride; sulfurous oxychloride; thionyl dichloride.

SECTION II - HAZARDOUS INGREDIENTS AND IDENTITY INFORMATION

Hazardous Components (specific identity)	OSHA Exposure Criteria	NIOSH Exposure Criteria	Immediately Dangerous to Life and Health (IDLH)	ACGIH Exposure Criteria
Thionyl chloride **1 ppm = 4.94 mg/m^3**	**PEL:** Not Established **STEL:** Not Established	**REL (ceiling):** 1 ppm 5 mg/m^3 **STEL:** Not Established	**Not Determined**	**TLV:** 1 ppm 5 mg/m^3 **STEL:** Not Established

SECTION III - PHYSICAL AND CHEMICAL CHARACTERISTICS

Boiling Point **169°F (76°C)**	Specific Gravity (H_2O = 1) **1.64**
Vapor Pressure (mm Hg) **100 at 70°F (21.1°C)**	Molecular Weight (atomic weight) **119.0**
Vapor Density (Air = 1) **Not Reported**	Melting Point **-155°F (-104°C)**

Solubility
Reacts explosively in contact with water. Soluble in benzene, carbon tetrachloride, and chloroform.

Appearance and Odor
Colorless to pale-yellow or reddish liquid with a pungent, suffocating odor like sulfur dioxide. Fumes when exposed to moist air.

SECTION IV - FIRE AND EXPLOSION HAZARD DATA

Flash Point (method used) **Not Applicable**	Explosive Limits in Air % by Volume **LEL: Not Applicable UEL: Not Applicable**
NFPA Classification **Non-Combustible Liquid**	Autoignition Temperature **Not Applicable**

Extinguishing Media
Use dry chemical, foam, or carbon dioxide. DO NOT use water (reacts violently).

Special Fire Fighting Procedures
Poisonous gases are produced in fire. Wear full protective clothing and self-contained breathing apparatus (SCBA). Move container from fire area if it can be done without risk. Use water spray to cool containers that cannot be moved. Do NOT allow water to contact exposed thionyl chloride.

Unusual Fire and Explosion Hazards
Can react vigorously with acids and alkalis. Will react with water or steam to produce toxic and corrosive fumes. Containers may explode in fire. Do not release runoff to sewers or waterways.

SECTION V - REACTIVITY DATA

Stability		Conditions to Avoid **Thionyl chloride is stable if kept in air-tight (moisture-free) containers at room temperature under routine conditions of handling and storage. Keep dry and away from heat and incompatible materials.**
Stable X	Unstable	Incompatibility *(materials to avoid)* **Violent reactions in contact or mixed with dimethyl sulfoxide (DMSO). Incompatible with water, acids, alkalis, ammonia, and chloryl perchlorate. Reacts violently with water to form sulfur dioxide and hydrogen chloride.**
Hazardous Polymerization		Conditions to Avoid **Under normal condition of temperature and pressure, hazardous polymerization of thionyl chloride cannot occur.**
May Occur	Will Not Occur X	Hazardous Decomposition or By-products **When heated to decomposition, releases toxic fumes of hydrogen chloride, carbon monoxide, and sulfur dioxide, chlorine, and sulfur monochloride.**

SECTION VI - HEALTH HAZARD DATA

Primary Route(s) of Entry:	Inhalation?	Absorption (skin/eye)?	Ingestion?
	X		X

Health Hazards

INHALATION: May cause severe irritation or burns of the respiratory system. There can be rhinitis and pneumonitis as well as possible development of pulmonary edema (fluid in lungs), which can be fatal.

SKIN & EYES: Contact with liquid or vapor can cause severe irritation or burns to the skin, eyes, and mucous membranes. Liquid may cause conjunctivitis and permanent eye damage.

INGESTION: Corrosive action on the mouth, throat, and esophagus with subsequent painful swallowing, abdominal pains, and other unspecified gastrointestinal effects.

Carcinogenicity	NTP Listed?	IARC Cancer Review Group?	OSHA Regulated?	Target Organs?
Unknown Human Unknown Animal	No	No	No	Skin, eyes, respiratory system.

Medical Conditions Generally Aggravated by Exposure

None reported. However, skin conditions (dermatitis) and respiratory problems may be aggravated.

Emergency and First-aid Procedures

Eye contact: Do not allow victim to keep eyes tightly shut. Flush immediately with water for 15 minutes (minimum); seek medical attention. Skin contact: Remove all contaminated clothing, including shoes. Immediately wash area with flooding amounts of soap and water. For blistered skin, seek medical attention immediately. For inhalation: Remove the person from exposure. Provide respiratory assistance and CPR. Transfer to medical facility. Observe for 24- 48 hours (pulmonary edema possible). If swallowed: Seek medical attention immediately. If conscious, give large amounts of water to dilute the ingested liquid. If vomiting occurs, give more water. Do NOT induce vomiting unless directed otherwise.

SECTION VII - PRECAUTIONS FOR SAFE HANDLING AND USE

Steps to be Taken in Case Material is Released or Spilled

Wear self-contained breathing apparatus (SCBA) for respiratory protection. Restrict those not wearing protective equipment and who are not involved in cleanup from entering area. Avoid skin contact. Ventilate area of spill. Do not use water. Take up with sand or vermiculite and deposit in sealed drum.

Preferred Waste Disposal Method

No citation.

Precautions to be Taken in Handling and Storage

Avoid contact with heat, sparks, or flame. Keep away from water. Use with adequate ventilation.

Other Precautions and Warnings

Use only non-sparking tools and equipment. Keep away from sunlight and moisture (can react violently).

SECTION VIII - CONTROL MEASURES AND PERSONAL PROTECTIVE EQUIPMENT

Respiratory Protection *(specify type)*

At relatively low concentrations, use an air-purifying chemical cartridge respirator with an acid/organic cartridge. For higher exposures, use a self-contained breathing apparatus (SCBA) with full facepiece operated in positive pressure mode or a supplied-air respirator operated in continuous flow mode.

Ventilation

Local exhaust (preferred) at site of chemical work or general ventilation.

Protective Gloves	Eye Protection	Other Protective Clothing
Protective Gloves	Chemical Goggles and Face Mask	Protective Apron

Work/Hygiene Practices

Always wash hands thoroughly after using chemical; never bring food, drink, or smoking materials into vicinity of chemicals.

THIONYL CHLORIDE

$SOCl_2$

CAS: 7719-09-7

IDENTIFICATION AND TYPICAL USES

Thionyl chloride is a colorless to slightly yellow to reddish liquid with a pungent, aromatic odor like that of sulfur dioxide. It is used in pesticides, in the engineering of plastics, as a chlorinating agent, as a solvent in high-energy-density lithium batteries, and a catalyst.

RISK ASSESSMENT: HEALTH

General Assessment

Thionyl chloride is a corrosive liquid. Contact with the eyes can cause corneal injury. It is a severe irritant to the eyes, mucous membranes, and the skin. Skin contact may produce dermatitis. There are no data on its carcinogenic, mutagenic, or teratogenic potential in animals or humans. It is moderately toxic by *inhalation* and *ingestion*.

Exposure to its vapors can produce intense and painful irritation of the eyes, nose, and throat. Higher concentrations may cause conjunctivitis, rhinitis, and pneumonitis. There is also the potential for the development of pulmonary edema (fluid buildup in the lungs), which is a medical emergency and can lead to death. Symptoms of coughing, pain and tightness in the chest, dypsnea, and congestion may be delayed up to 48 hours thereby creating a false sense of security with regard to health exposure risk.

Acute Health Effects

The following acute (short-term) health effects may occur immediately or shortly after exposure to thionyl chloride:

Skin: Irritation with possible allergenic effects (dermatitis). Corrosive tissue action is probable.

Eye: Severe irritation, burns, pain, and possible permanent damage to vision.

Lung: Nose, throat, and respiratory irritation causing difficulty in breathing. High-level acute exposures can lead to pneumonitis or pulmonary edema, which may be fatal.

Other: Ingestion of thionyl chloride may cause nausea, vomiting, headaches, dizziness, and gastrointestinal irritation.

Chronic Health Effects

The following chronic (long-term) health effects may occur at some time after exposure to thionyl chloride and can last for months or even years:

Cancer Hazards: According to information presented in the references, thionyl chloride has not been adequately tested for its ability to cause cancer in test animals.

Reproductive Hazard: According to information presented in the references, thionyl chloride has not been adequately tested for its ability to adversely affect reproduction in animals or humans.

Other Chronic Effects: Over the long-term, exposure to thionyl chloride can lead to skin allergy. If an allergy develops, very low future exposures can cause itching and rash. Very irritating substances such as thionyl chloride can affect the lungs. It is not known whether thionyl chloride causes lung damage.

Recommended Risk-Reduction Measures

Personnel should avoid direct contact with thionyl chloride. The exact nature of its human toxicity is not clearly understood in the references. Occupational exposure can occur primarily through *inhalation* and *ingestion*. Skin and eye contact also present risks to heath. If a less toxic material or compound cannot be substituted for thionyl chloride, then *engineering controls* are the most effective method of reducing exposure risk. The best protection is to enclose operations and/or provide local exhaust ventilation at the site of thionyl chloride release. While not always operationally feasible, isolating operations can also reduce exposure. Using respiratory protection is less effective than the controls mentioned above, but is still recommended whenever working with or around thionyl chloride. For low exposures, an air-purifying chemical cartridge respirator with an acid/organic vapor cartridge is recommended. For higher exposures, a MSHA/NIOSH-approved self-contained breathing apparatus (SCBA) with full facepiece operated in pressure demand or a supplied-air respirator with full facepiece in continuous flow mode provide the appropriate level of respiratory protection. If a full facepiece is not available, then chemical goggles should be worn to protect the eyes, especially when a splash hazard exists. A face shield should also be considered. To prevent hand and skin exposures, rubber gloves should be used. Glove manufacturers should be contacted and permeation studies obtained *before* final glove selections are made.

Administrative controls should also be in place to minimize the potential for human exposures. These may include written procedures or policies which specify the methods and techniques that will be practiced whenever personnel are to work with thionyl chloride.

All personnel should receive *training* on the use, hazards, protective measures, emergency actions, and other precautions per 29 CFR 1910.1200 (Hazard Communication) prior to the first assignment in an area where thionyl chloride is used or stored.

Before beginning employment and at regular times thereafter, the following recommended medical tests should be considered for those with frequent or potentially high exposures:

☑ Lung function tests.

If symptoms develop or overexposure is suspected, the following may be useful:

☑ Consider chest X-ray following acute exposure (may be negative if taken immediately after exposure due to delayed onset of pulmonary edema).

It should be noted that medical tests that simply look for existing damage are not a substitute for controlling exposures. Medical histories are extremely important when assessing exposure risk. Also, since smoking can cause heart disease, lung cancer, emphysema, and other respirator diseases, smokers exposed to irritating substances such as thionyl chloride may experience more severe symptoms more quickly than non-smokers under the same conditions of exposure. Prudent risk management requires careful consideration of *all* factors which may be causing the appearance of exposure symptoms in the workplace.

Other methods to reduce exposure include:

☑ Thionyl chloride should be handled in an enclosed system. There should be local exhaust ventilation at the site of chemical release.

☑ Always ensure that proper protective clothing is worn when using chemical substances and that personnel have been trained in its use and care.

☑ Wash thoroughly immediately after exposure to thionyl chloride and at the end of the work shift or before eating, drinking, smoking, or applying cosmetics.

☑ Hazard warning information should be posted in the work area. In addition, as part of an on-going education and training program, all information on the health and safety hazards of thionyl chloride should be communicated to all exposed and potentially exposed workers.

☑ Eye wash stations should be provided in the immediate work area for emergency use. If there is a possibility of skin exposure to thionyl chloride, emergency shower facilities should also be provided in the area.

☑ Workers whose clothing has been contaminated by thionyl chloride should change into clean clothes before leaving work. Contaminated work clothing should be laundered only by individuals who have been informed of the hazards of exposure to thionyl chloride.

RISK ASSESSMENT: ENVIRONMENT

General Assessment

The environment is at risk of exposure during transportation, storage, disposal, or destruction of thionyl chloride. In almost every scenario, the threat of environmental exposure is contingent upon the proper handling of the chemical substance. Accidental spills, large or small, can result in contamination of the surrounding environmental mediums (water, soil, and air).

Thionyl chloride is a non-combustible liquid. However, caution is always required in handling, storage, transportation, and disposal of thionyl chloride. It can react vigorously with incompatible materials such as acids, alkalis, ammonia, and chloryl perchlorate. In contact with moisture or water, it reacts violently to produce toxic by-product materials such as sulfur dioxide, hydrogen chloride, chlorine, and sulfur monochloride. Emergency responders should be made aware of the presence of thionyl chloride at any emergency response situation.

Thionyl chloride can enter the environment from industrial effluents and from spills.

Acute Ecological Effects

Acute (short-term) toxic effects may include the death of animals, birds, or fish and death or low growth rate in plants. Acute effects are seen 2 to 4 days after animals or plants are exposed to thionyl chloride

Insufficient data are available to evaluate the acute (short-term) effects of thionyl chloride to aquatic life, plants, birds, or land animals.

Chronic Ecological Effects

Chronic (long-term) toxic effects may include shortened life span, reproductive problems, lower fertility, and changes in appearance or behavior in exposed animals. These effects can be seen long after first exposure(s) to toxic chemicals.

Insufficient data are available to evaluate the chronic (long-term) effects of thionyl chloride to aquatic life, plants, birds, or land animals.

Water Solubility

Thionyl chloride reacts violently in water.

Persistence in the Environment

Thionyl chloride is virtually non-persistent in the aquatic environment, with a half-life of less than 2 days. The half-life of a pollutant is the amount of time it takes for one-half of the chemical to be degraded. It reacts violently in water to decompose and form other compounds such as sulfur dioxide and hydrogen chloride, both of which are only slightly persistent in the environment.

Bioaccumulation in Aquatic Organisms

Some substances increase in concentration, or *bioaccumulate,* in living organisms as they breathe contaminated air, drink contaminated water, or eat contaminated food. These chemicals can become concentrated in the tissues and internal organs of animals as well as those of humans.

Insufficient data are available on the bioaccumulation of thionyl chloride in the edible tissues of fish.

Recommended Risk-Reduction Measures

Proper training of all transporters will reduce the likelihood of a mishap or accident resulting in a leak or spill of thionyl chloride into the environment. Labels on all containers, trucks, and rail cars must meet DOT requirements and accurately reflect their contents to enable emergency responders to react properly and quickly to any disaster thereby reducing the potential risk to the environment and to personnel.

Storage of thionyl chloride should be segregated from other chemicals and materials to minimize the risk of cross-contamination and violent reactions. Containers must be kept tightly closed and moisture-free.

If a spill or leak to the environment has occurred, fire department, emergency response and/or hazardous materials spill personnel should be notified immediately. Cleanup should be attempted only by those trained in proper spill containment procedures. Contaminated soils should be removed for incineration and replaced with clean soil. If thionyl chloride should contact the water table, aquifer, or navigable waterway, time is of the essence. It will react violently in water to produce other toxic compounds and total remediation may not be possible. The local and/or state emergency response authorities must be notified. A comprehensive emergency response or disaster preparedness/recovery plan should be in place prior to any operations involving the use, transportation, storage, or disposal of thionyl chloride.

If thionyl chloride is spilled or leaked, the following specific steps are recommended:

- ☑ Restrict persons not wearing protective clothing from area of spill or leak until cleanup is complete. Avoid skin and eye contact.
- ☑ Absorb liquids in vermiculite or other similar materials and deposit in sealed drum for disposal.
- ☑ Ventilate area of spill or leak.
- ☑ It may be necessary to dispose of thionyl chloride as a hazardous waste. The responsible state agency or the regional office of the federal Environmental Protection Agency (EPA) should be contacted for specific recommendations.

RISK ASSESSMENT: BUSINESS

General Assessment

Accidents or mishaps involving thionyl chloride can present a serious threat to business operations. The loss or damage of equipment or facilities can significantly affect fiscal viability. Lawsuits that may result from personnel illness, injury/death, public exposures, and/or environmental contamination will require a serious expenditure of resources. Media attention surrounding an injury, death, or environmental damage can also result in a loss of profits and loss of current as well as future business.

ʘ *Recommended Risk-Reduction Measures*

Company attorneys, safety and health professionals, and environmental specialists should be involved in the development of any procedures for responding to chemical incidents. A company official should be predesignated as a public relations officer with specific training in dealing with the press. Corporate plans and policies should be developed, approved, and implemented long before any need for such arises.

REFERENCES

American Conference of Governmental Industrial Hygienists. 1988. *Documentation of the Threshold Limit Values and Biological Exposure Indices,* 5th Edition (with updates). Cincinnati: ACGIH

Anthony, C. P., and N. J. Kolthoff. 1971. *Textbook of Anatomy and Physiology.* St. Louis: C. V. Mosby Company.

Asante-Duah, D. K. 1993. *Hazardous Waste Risk Assessment.* New York: CRC Press/Lewis Publishers.

Bloom, A. D. and F. J. DeSerres. 1995. *Ecotoxicity and Human Health.* New York: CRC Press/Lewis Publishers.

Brusick, D. J. 1994. *Methods for Genetic Risk Assessment.* New York: CRC Press/Lewis Publishers.

Bueche, J. R. 1972. *Principles of Physics,* 2nd Edition. New York: McGraw-Hill.

Calabrese, E. J. 1994. *Biological Effects of Low Level Exposures.* New York: CRC Press/Lewis Publishers.

Cockerman, L. G., B. S. Shane. 1994. *Basic Environmental Toxicology.* New York: CRC Press/Lewis Publishers.

Danse, I. R. 1991. *Common Sense Toxics in the Workplace.* New York: Van Nostrand Reinhold.

Gardner, R. W. 1994. *Chemical Intolerance.* New York: CRC Press/Lewis Publishers.

Hallenbeck, W. H. 1993. *Quantitative Risk Assessment for Environmental and Occupational Health,* 2nd Edition. New York: CRC Press/Lewis Publishers.

Hodgson, E., R. B. Mailman, and J. E. Chambers. 1988. *Dictionary of Toxicology.* London: The MacMillian Company.

Howard, P. H., et. al. 1991. *Handbook of Environmental Degradation Rates.* New York: CRC Press/Lewis Publishers.

Jones, F. E. 1994. *Toxic Organic Vapors in the Workplace.* New York: CRC Press/Lewis Publishers.

Keith, L. H. and M. M. Walker. 1995. *Handbook of Air Toxics.* New York: CRC Press/Lewis Publishers.

Landis, W. G., M. H. Yu. 1995. *Introduction to Environmental Toxicology.* New York: CRC Press/Lewis Publishers.

Langerman, N. 1994 *Precautionary Labels for Chemical Containers.* New York: CRC Press/Lewis Publishers.

Lauwerys, R. R., P. Hoet. 1993. *Industrial Chemical Exposure.* New York: CRC Press/Lewis Publishers.

Lewis, R. J., Sr. 1993. *Hawley's Condensed Chemical Dictionary,* 12th Edition. New York: Van Nostrand Reinhold.

Lewis, R. J., Sr. 1992. *Sax's Dangerous Properties of Industrial Materials,* Volumes 1, 2, and 3, 8th Edition. New York: Van Nostrand Reinhold.

Mellström, G. A., J. E. Wahlberg, and H. I. Maibach. 1994. *Protective Gloves for Occupational Use.* New York: CRC Press/Lewis Publishers.

Meyer, E. 1990. *Chemistry of Hazardous Materials.* Englewood Cliffs, NJ: Prentice Hall.

Patnaik, P. 1992. *A Comprehensive Guide to the Hazardous Properties of Chemical Substances.* New York: Van Nostrand Reinhold.

Philp, R. B. 1995. *Environmental Hazards and Human Health.* New York: CRC Press/Lewis Publishers.

Proctor, N. H., J. P. Hughes. 1991. *Chemical Hazards of the Workplace.* New York: Van Nostrand Reinhold.

Quigley, D. R. 1994. *Handbook of Emergency Chemical Management.* New York: CRC Press/Lewis Publishers.

Richardson, M. 1995. *Dictionary of Substance and Their Effects,* Vol. I-VI. New York: CRC Press/Lewis Publishers.

Sacarello, H. L. A. 1994. *The Comprehensive Handbook of Hazardous Materials.* New York: CRC Press/Lewis Publishers.

Torkelson, T. R. V. K. Rowe. 1981. *Patty's Industrial Hygiene and Toxicology,* Vol. 2B. New York: Wiley-Interscience.

U.S. Department of Health and Human Services, National Institute for Occupational Safety and Health. 1994. *NIOSH Pocket Guide to Chemical Hazards.* Washington, D.C.: U.S. Government Printing Office.

MATERIAL SAFETY DATA SHEET

CHEMICAL NAME	
THIOUREA	CRC PRESS/LEWIS PUBLISHERS DISCLAIMER: THE INFORMATION AND RECOMMENDATIONS PRESENTED HEREIN ARE BASED ON SOURCES BELIEVED TO BE RELIABLE. CRC MAKES NO REPRESENTATION ON ITS COMPLETENESS OR ACCURACY. IT IS THE RESPONSIBILITY OF THE USER TO DETERMINE THE CHEMICAL'S SUITABILITY FOR ITS INTENDED USE, THE CHEMICAL'S SAFE USE, AND THE CHEMICAL'S PROPER DISPOSAL. NO REPRESENTATIONS AND/OR WARRANTIES, EITHER EXPRESSED OR IMPLIED, OF THE MERCHANTABILITY OR FITNESS FOR A PARTICULAR PURPOSE, OR OF ANY OTHER NATURE, ARE MADE WITH RESPECT TO THE INFORMATION PROVIDED IN THIS MSDS OR TO THE CHEMICAL TO WHICH INFORMATION MAY REFER. CRC NEITHER ASSUMES NOR AUTHORIZES ANY OTHER PERSON TO ASSUME FOR IT, ANY OTHER ADDITIONAL RESPONSIBILITY OR LIABILITY FOR THE USE OF, OR RELIANCE UPON, THIS INFORMATION.

HAZARD WARNING INFORMATION

HEALTH	FIRE	REACTIVE	OTHER	DEGREE OF HAZARD	COLOR CODING	OTHER CODES
2	0	0		0 = Minimum Hazard 1 = Slight Hazard 2 = Moderate Hazard 3 = Serious Hazard 4 = Severe Hazard	HEALTH = BLUE FIRE = RED REACTIVITY = YELLOW OTHER = WHITE	OX = Oxidizer ACID = Acid ALK = Alkali COR = Corrosive W = Use No Water

SECTION I - GENERAL INFORMATION

Characterization	RCRA Number	EPA Class
Organic, Sulfur-Containing	**U219**	**Toxic Waste**

DOT Proper Shipping Name **Thiourea**	Chemical Abstract Service (CAS) Number **62-56-6**
DOT Hazard Class and Label Requirements **Poison B; St. Andrews Cross**	DOT Emergency Guide Code **53**
DOT Identification Number **UN 2877**	Chemical Formula **$(NH_2)_2CS$**

Synonyms
Thiocarbamide; isothiourea; pseudothiourea; solourea; thiocarbonic acid diamide; 2-thiopseudourea; 2-thiourea; THU.

SECTION II - HAZARDOUS INGREDIENTS AND IDENTITY INFORMATION

Hazardous Components (specific identity)	OSHA Exposure Criteria	NIOSH Exposure Criteria	Immediately Dangerous to Life and Health (IDLH)	ACGIH Exposure Criteria
Thiourea (derivation: By heating dry ammonium thiocyanate, extraction with a concentrated solution of ammonium thiocyanate with subsequent crystallization; or by action of hydrogen sulfide on cyanamide).	**PEL:** Not Established **STEL:** Not Established	**REL:** Not Established **STEL:** Not Established	**Not Determined**	**TLV:** Not Established **STEL:** Not Established

SECTION III - PHYSICAL AND CHEMICAL CHARACTERISTICS

Boiling Point **Sublimes in vacuo: 270-320°F (150-160°C)**	Specific Gravity (H_2O = 1) **1.4**
Vapor Pressure (mm Hg) **7.5×10^{-8} at 69°F (20°C)**	Molecular Weight **76.13**
Vapor Density (Air = 1) **Not Reported**	Freezing Point **356-360°F (180-182°C)**

Solubility
Soluble in cold water, ammonia thiocyanate, and alcohol. Nearly insoluble in ether.

Appearance and Odor
White, lustrous crystalline solid with a bitter taste.

SECTION IV - FIRE AND EXPLOSION HAZARD DATA

Flash Point (method used) **Not Applicable**	Explosive Limits in Air % by Volume **LEL: Not Applicable UEL: Not Applicable**
NFPA Classification **Non-Combustible Solid**	Autoignition Temperature **Not Applicable**

Extinguishing Media
Use agent suitable to surrounding area. Thiourea does not burn.

Special Fire Fighting Procedures
Since fire may produce toxic thermal decomposition products, firefighters should wear full protective clothing and self-contained breathing apparatus (SCBA) operated in positive pressure mode. Keep fire-exposed containers cool with water spray. Do NOT release runoff from fire control measures to sewers or waterways. Contain and collect runoff for proper disposal.

Unusual Fire and Explosion Hazards
None reported.

SECTION V - REACTIVITY DATA

Stability		Conditions to Avoid: Thiourea is normally stable in closed containers under routine conditions of handling and storage. Keep away from incompatible materials.
Stable	Unstable	Incompatibility *(materials to avoid)*
X		Acrolein, acrylaldehyde, hydrogen peroxide, and nitric acid.
Hazardous Polymerization		Conditions to Avoid: Hazardous polymerization MAY occur in contact with acrylaldehyde. Keep isolated from incompatible materials at all times!
May Occur	Will Not Occur	Hazardous Decomposition or By-products
X		When heated to decomposition, thiourea emits highly toxic oxides of sulfur, carbon, and nitrogen.

SECTION VI - HEALTH HAZARD DATA

Primary Route(s) of Entry:	Inhalation?	Absorption (skin)?	Ingestion?
	X		

Health Hazards

INHALATION: Irritation and inflammation of the respiratory tract. A related chemical (naphthylthiourea) causes fluid in the lungs (pulmonary edema), which can be fatal. It is not known if exposure to thiourea will have this effect.

SKIN & EYES: Direct contact or exposure to airborne particles can cause irritation to the eyes and skin. Allergic sensitization is possible in some persons causing itchy rash and vesicles.

INGESTION: Symptoms include hemorrhage, bone marrow depression with subsequent anemia and other blood effects such as leukopenia and thrombocytopenia.

Carcinogenicity	NTP Listed?	IARC Cancer Review Group?	OSHA Regulated?	Target Organs?
Probable Human Confirmed Animal	5th Annual Report	Group 2B	No	Skin, liver, bone marrow

Medical Conditions Generally Aggravated by Exposure

Skin conditions (dermatitis) may be aggravated by exposure.

Emergency and First-aid Procedures

Eye contact: Immediately flush with large amounts of water for 15 minutes (minimum), occasionally lifting eyelids; seek medical attention. Skin contact: Remove all contaminated clothing. Immediately wash area with large amounts of soap and water. For inhalation: Remove the person from exposure. Provide respiratory assistance and CPR. Transfer promptly to medical facility. If swallowed: Seek medical attention immediately. Never try to give an unconscious person anything by mouth. Unless advised otherwise, give 1 to 2 glasses of water and induce vomiting.

SECTION VII - PRECAUTIONS FOR SAFE HANDLING AND USE

Steps to be Taken in Case Material is Released or Spilled

Restrict persons not wearing protective equipment from entering area of spill or leak until cleanup is complete. Ventilate area. Collect powdered materials using a scoop or HEPA vacuum. Do NOT dry sweep or generate airborne dusts.

Preferred Waste Disposal Method

Burn in chemical incinerator equipped with an afterburner and scrubber.

Precautions to be Taken in Handling and Storage

Store in tightly closed containers in a cool, dark, well-ventilated area away from incompatible materials (especially arylaldehyde) since violent reactions can occur.

Other Precautions and Warnings

Bulk storage of thiourea is not recommended. A regulated, marked controlled area should be established where thiourea is used, handled, or stored.

SECTION VIII - CONTROL MEASURES AND PERSONAL PROTECTIVE EQUIPMENT

Respiratory Protection *(specify type)*

No exposure levels have been established. For best protection to a probable carcinogen, use an MSHA/NIOSH-approved self-contained breathing apparatus (SCBA) with full facepiece operated in pressure demand or other positive pressure mode, or a supplied-air respirator in positive pressure mode.

Ventilation

Local exhaust or general mechanical systems recommended.

Protective Gloves	Eye Protection	Other Protective Clothing
Impervious Material	Chemical/Dust Goggles and/or Face Mask	Protective Apron

Work/Hygiene Practices

Always wash hands thoroughly after using chemical; never bring food, drink, or smoking materials into vicinity of chemicals.

THIOUREA

$(NH_2)_2CS$

CAS: 62-56-6

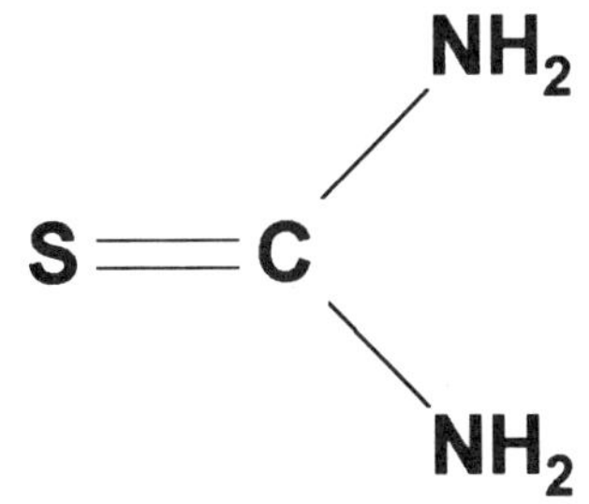

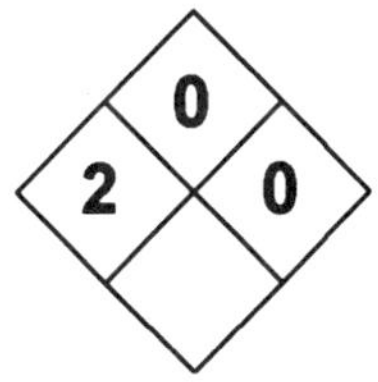

IDENTIFICATION AND TYPICAL USES

Thiourea is a colorless, lustrous, crystalline (sand-like) solid with a bitter taste. It is used in photography as a fixing agent and to remove stains from negatives and in photocopying paper to prevent yellowing. It is also used in organic synthesis (as an intermediate in dyes, drugs, and hair preparations), as a rubber accelerator, an analytical reagent, in amino resins, and as a mold inhibitor.

RISK ASSESSMENT: HEALTH

General Assessment

Thiourea is considered toxic by *inhalation* and *ingestion*. Skin contact can also cause serious topical effects, but dermal absorption has not been reported. Thiourea is a probable human carcinogen since it has been shown to cause cancer in test animals. Human mutation data have been reported. It has also been shown to cause experimental teratogenic effects.

Inhalation causes irritation of the eyes, nose, throat, and respiratory tract. There can be coughing with severe congestion. A closely related chemical, naphthylthiourea, has been shown to cause a dangerous buildup of fluids in the lungs (pulmonary edema), which can be fatal. It is not known if thiourea will have this effect on the lungs.

Skin contact or exposure to airborne particles can cause irritation. Allergic sensitization may occur in some individuals with symptoms of extremely itchy rash and the development of vesicles on the skin's surface. Exposure to sunlight may enhance the reaction. Direct eye contact may cause irritation.

Ingestion causes hemorrhage, bone marrow depression with subsequent anemia and other blood effects such as leukopenia (decrease in white blood cells) and thrombocytopenia (decrease in blood platelets).

Acute Health Effects

The following acute (short-term) health effects may occur immediately or shortly after exposure to thiourea:

Skin: Irritation with a possibility for itchy rash and dermatitis.

Eye: Possible irritation on direct contact.

Lung: Irritation of the respiratory tract. Effects can include cough, congestion, and tightness in the chest.

Other: Ingestion cause hemorrhage and possible damage to the bone marrow with subsequent blood effects such as anemia and leukopenia.

Chronic Health Effects

The following chronic (long-term) health effects may occur at some time after exposure to thiourea and can last for months or even years:

Cancer Hazards: According to information presented in the references, thiourea has been shown to cause cancer in test animals and is considered a probable human carcinogen. Human mutation data have also been reported and many scientists believe that exposure to such chemicals poise a cancer threat in the long-term.

Reproductive Hazards: According to information presented in the references, thiourea has been shown to cause teratogenic effects in test animals.

Other Chronic Effects: Long-term exposure may damage the bone marrow causing a reduction in red blood cells (anemia), white blood cells (with reduced ability to fight infections), and/or reduced platelets (reduced blood clotting and increased bruising). Repeated exposure can cause goiter (an enlarged thyroid gland in the neck) and reduced thyroid hormone. Thiourea has caused enlargement of the spleen when used therapeutically for thyrotoxicosis. Exposure may cause skin allergy to develop in some individuals with rash. If allergy develops, even low future exposures may cause rash. Very irritating substances may affect the lungs. It is not known if thiourea causes lung damage.

Recommended Risk-Reduction Measures

Personnel should avoid direct contact with thiourea. It is a probable human carcinogen, a human mutagen, and a teratogen in animals. Unless a less toxic chemical can be substituted for a hazardous substance, *engineering controls* are the most effective method of reducing exposures. The best protection is to enclose operations and/or provide local exhaust ventilation at the site of chemical release. While not always operationally feasible, isolating operations can also reduce exposure. Using respiratory protection is less effective than the controls mentioned above, but is still advisable whenever working with thiourea. No exposure levels have been established for thiourea. However, this does not mean that exposure is without risk. It is a probable human carcinogen and many scientists believe there is no safe exposure level to such chemicals. For the best respiratory protection, a self-contained breathing apparatus (SCBA) with full facepiece operated in pressure demand or a supplied-air respirator in continuous flow mode are the recommended respiratory protection methods of choice. If a full facepiece is not available, then chemical goggles should be worn to protect the eyes. Whenever a chemical splash hazard exists, a face shield and a protective apron should be worn. To prevent hand and skin exposures, impervious gloves should be used. Glove manufacturers should be contacted and permeation studies obtained *before* final glove selections are made.

Administrative controls should also be in place to minimize the potential for human exposures. These may include written procedures or policies which specify the methods and techniques that will be practiced whenever personnel are to work with thiourea.

All personnel should receive *training* on the use, hazards, protective measures, emergency actions, and other precautions per 29 CFR 1910.1200 (Hazard Communication), prior to the first assignment in an area where thiourea is used or stored. Before beginning employment and at regular times thereafter, the following medical tests are recommended:

- ☑ Lung function tests.
- ☑ Complete blood count (CBC)..
- ☑ Consider thyroid test for thyroxin.

If symptoms develop or overexposure is suspected, the following additional tests may be useful:

- ☑ Consider chest X-ray after acute overexposure.
- ☑ Evaluation by a qualified allergist, including careful exposure history and special testing (may help diagnose skin allergy).

Other methods to reduce exposure include:

- ☑ Where possible, enclose operations and use local exhaust ventilation at the site of chemical release. If local exhaust ventilation or enclosure is not used, respiratory protection should be mandatory.
- ☑ Always ensure that proper protective clothing is worn when using chemical substances and that personnel are trained in its use and care.
- ☑ Wash thoroughly immediately after exposure to thiourea and at the end of the work shift or before eating, drinking, or smoking. Never wear clothes contaminated with thiourea home. Family members can be exposed.
- ☑ Hazard warning information should be posted in the work area. As part of an on-going education and training program, all information on the health and safety hazards of thiourea should be communicated to all potentially exposed workers.

RISK ASSESSMENT: ENVIRONMENT

General Assessment

The environment is at risk of exposure during transportation, storage, disposal, or destruction of thiourea. In almost every scenario, the threat of environmental exposure is contingent upon the proper handling of the chemical substance. Accidental spills, large or small, can result in fire, explosion, and possible contamination of the surrounding environmental mediums (water, soil, and air).

Thiourea is considered a non-combustible solid. However, it is incompatible with a number of commodities. Hazardous polymerization can occur in contact with acrylaldehyde. Potentially explosive reactions can occur if thiourea is mixed or comes into contact with acrolein, hydrogen peroxide, and nitric acid. These characteristics require special consideration during any emergency involving a leak or spill of thiourea.

The proper disposal/destruction method for thiourea is to burn it in a chemical incinerator equipped with an afterburner and air scrubber.

Thiourea can enter the environment through industrial discharges, photographic laboratories, or spills.

Acute Ecological Effects

Acute (short-term) toxic effects may include the death of animals, birds, or fish, and death or low growth rate in plants. Acute effects are seen 2 to 4 days after animals or plants are exposed to thiourea.

Thiourea has slight acute toxicity to aquatic life. It has caused dormancy break increase, and several other effects, in numerous agricultural crops. Insufficient data are available on the short-term effects of thiourea on birds or land animals.

Chronic Ecological Effects

Chronic (long-term) toxic effects may include shortened life span, reproductive problems, lower fertility, and changes in appearance or behavior in exposed animals. These effects can be seen long after first exposure(s) to toxic chemicals.

Thiourea has slight chronic toxicity in aquatic life. No data are available to evaluate the long-term effects of thiourea to plants, birds, or land animals.

Water Solubility

Thiourea is highly soluble in water. Concentrations of 1000 milligrams and more will mix with a liter of water.

Persistence in the Environment

Thiourea is virtually non-persistent in water, with an estimated half-life of less than 2 days. The half-life of a pollutant is the amount of time it takes for one half of the chemical to be degraded.

Bioaccumulation in Aquatic Organisms

Some substances increase in concentration, or *bioaccumulate,* in living organisms as they breathe contaminated air, drink contaminated water, or eat contaminated food. These chemicals can become concentrated in the tissues and internal organs of animals as well as humans.

Thiourea is not expected to accumulate in the edible tissues of aquatic organisms.

Recommended Risk-Reduction Measures

Proper training of all transporters will reduce the likelihood of a mishap or accident resulting in a leak or spill to the environment. The correct use of DOT labeling (per the MSDS) on all containers, trucks, and rail cars will enable emergency responders to react properly and quickly to any disaster thereby reducing the potential risk to the environment and to personnel.

Storage of thiourea should be segregated from incompatible chemicals to minimize the risk of cross-contamination or contact. It must never be stored in the vicinity of acrylaldehyde since violent polymerization can occur on mixing with thiourea. Buildings designated for storage should be equipped with appropriate fire protection systems (alarms, sprinklers, emergency lighting, portable extinguishers).

If a spill or leak to the environment has occurred, fire department, emergency response, and/or hazardous materials spill personnel should be notified immediately. Cleanup should be attempted only by those trained in proper spill containment procedures. Contaminated soils should be removed for incineration and replaced with clean soil. If thiourea should contact the water table, aquifer, or navigable waterway, time is of the essence. It is readily soluble in water and, therefore, total containment and remediation may not be entirely possible. When such spills occur, the local and/or state emergency response authorities must be notified. A comprehensive emergency response or disaster preparedness/recovery plan should be in place prior to any operations involving the use, transportation, storage, or disposal of thiourea.

If thiourea is spilled or leaked, the following specific steps are recommended:

- ☑ Restrict persons not wearing protective clothing from area of spill or leak until cleanup is complete and area can be opened for normal work.
- ☑ Ventilate area of spill or leak.
- ☑ Absorb liquids in vermiculite, dry sand, earth, or a similar material and deposit in sealed containers.
- ☑ If applicable, stop flow of leaking liquid or gas. If leak source is a cylinder and the leak cannot be stopped in place, remove leaking cylinder to a safe place in the open air, and repair or allow cylinder to empty.

☑ It may be necessary to dispose of thiourea as a hazardous waste. The responsible state agency or the regional office of the federal Environmental Protection Agency (EPA) should be contacted for specific recommendations.

RISK ASSESSMENT: BUSINESS

General Assessment

Accidents or mishaps involving thiourea can present a significant threat to business operations. The loss or damage of equipment or facilities can significantly affect fiscal viability. Lawsuits that may result from personnel injury, public exposures, and/or environmental contamination will also require a serious expenditure of resources. Media attention surrounding an injury, illness, or environmental damage can also result in a loss of profits and loss of current as well as future business. Always remember that anytime the terms "cancer," "carcinogen," or "reproductive hazard" are used, public emotion, hysteria, anxiety, and ignorance will run equally high. This must be carefully considered whenever drafting or implementing public relations policies.

Recommended Risk-Reduction Measures

Company attorneys, safety and health professionals, and environmental specialists should be involved in the development of any procedures dealing with safety or emergency response. A company official should be pre-designated as a public relations officer with specific training in dealing with the press. Corporate plans and policies should be developed, approved, and implemented long before any need for such arises.

REFERENCES

American Conference of Governmental Industrial Hygienists. 1988. *Documentation of the Threshold Limit Values and Biological Exposure Indices*, 5th Edition (with updates). Cincinnati: ACGIH

Anthony, C. P., and N. J. Kolthoff. 1971. *Textbook of Anatomy and Physiology.* St. Louis: C. V. Mosby Company.

Bueche, J. R. 1972. *Principles of Physics,* 2nd Edition. New York: McGraw-Hill.

Burrell, R., D. K. Flaherty, and L. J. Sauers. 1992. *Toxicology of the Immune System.* New York: Van Nostrand Reinhold.

Calabrese, E. J. 1994. *Biological Effects of Low Level Exposures.* New York: CRC Press/Lewis Publishers.

Cothern, R. R. 1993. *Comparative Environmental Risk Assessment.* New York: CRC Press/Lewis Publishers.

Gardner, R. W. 1994. *Chemical Intolerance.* New York: CRC Press/Lewis Publishers.

Hallenbeck, W. H. 1993. *Quantitative Risk Assessment for Environmental and Occupational Health,* 2nd Edition. New York: CRC Press/Lewis Publishers.

Hodgson, E., R. B. Mailman, and J. E. Chambers. 1988. *Dictionary of Toxicology.* London: The MacMillian Company.

Howard, P. H., et. al. 1991. *Handbook of Environmental Degradation Rates.* New York: CRC Press/Lewis Publishers.

Jones, F. E. 1994. *Toxic Organic Vapors in the Workplace.* New York: CRC Press/Lewis Publishers.

Kamrin, M. A. 1988. *Toxicology: A Primer on Principles and Applications.* Chelsea, Michigan: Lewis Publishers.

Langerman, N. 1994 *Precautionary Labels for Chemical Containers.* New York: CRC Press/Lewis Publishers.

Lauwerys, R. R., P. Hoet. 1993. *Industrial Chemical Exposure.* New York: CRC Press/Lewis Publishers.

Lewis, R. J., Sr. 1992. *Sax's Dangerous Properties of Industrial Materials,* Volumes 1, 2, and 3, 8th Edition. New York: Van Nostrand Reinhold.

Lide, D. R. 1995. *Handbook of Chemistry and Physics,* 76th Edition. New York: CRC Press/Lewis Publishers.

Manahan, S. E. 1992. *Toxicological Chemistry,* 2nd Edition. New York: CRC Press/Lewis Publishers.

Mellström, G. A., J. E. Wahlberg, and H. I. Maibach. 1994. *Protective Gloves for Occupational Use.* New York: CRC Press/Lewis Publishers.

Meyer, E. 1990. *Chemistry of Hazardous Materials.* Englewood Cliffs, NJ: Prentice Hall.

Patnaik, P. 1992. *A Comprehensive Guide to the Hazardous Properties of Chemical Substances.* New York: Van Nostrand Reinhold.

Proctor, N. H., and J. P. Hughes. 1991. *Chemical Hazards of the Workplace.* New York: Van Nostrand Reinhold.

Quigley, D. R. 1994. *Handbook of Emergency Chemical Management.* New York: CRC Press/Lewis Publishers.

Richardson, M. 1995. *Dictionary of Substance and Their Effects,* Vol. I-VI. New York: CRC Press/Lewis Publishers.

U.S. Department of Health and Human Services, National Institute for Occupational Safety and Health. 1994. *NIOSH Pocket Guide to Chemical Hazards.* Washington, D.C.: U.S. Government Printing Office.

MATERIAL SAFETY DATA SHEET

CHEMICAL NAME	
TIN	**CRC PRESS/LEWIS PUBLISHERS DISCLAIMER:** THE INFORMATION AND RECOMMENDATIONS PRESENTED HEREIN ARE BASED ON SOURCES BELIEVED TO BE RELIABLE. CRC MAKES NO REPRESENTATION ON ITS COMPLETENESS OR ACCURACY. IT IS THE RESPONSIBILITY OF THE USER TO DETERMINE THE CHEMICAL'S SUITABILITY FOR ITS INTENDED USE, THE CHEMICAL'S SAFE USE, AND THE CHEMICAL'S PROPER DISPOSAL. NO REPRESENTATIONS AND/OR WARRANTIES, EITHER EXPRESSED OR IMPLIED, OF THE MERCHANTABILITY OR FITNESS FOR A PARTICULAR PURPOSE, OR OF ANY OTHER NATURE, ARE MADE WITH RESPECT TO THE INFORMATION PROVIDED IN THIS MSDS OR TO THE CHEMICAL TO WHICH INFORMATION MAY REFER. CRC NEITHER ASSUMES NOR AUTHORIZES ANY OTHER PERSON TO ASSUME FOR IT, ANY OTHER ADDITIONAL RESPONSIBILITY OR LIABILITY FOR THE USE OF, OR RELIANCE UPON, THIS INFORMATION.

HAZARD WARNING INFORMATION

HEALTH	FIRE	REACTIVE	OTHER	DEGREE OF HAZARD	COLOR CODING	OTHER CODES
0	2	1		0 = Minimum Hazard 1 = Slight Hazard 2 = Moderate Hazard 3 = Serious Hazard 4 = Severe Hazard	HEALTH = BLUE FIRE = RED REACTIVITY = YELLOW OTHER = WHITE	OX = Oxidizer ACID = Acid ALK = Alkali COR = Corrosive W = Use No Water

SECTION I - GENERAL INFORMATION

Characterization **Metal**	RCRA Number **None**	EPA Class **Not Applicable**
DOT Proper Shipping Name **No Citation**	Chemical Abstract Service (CAS) Number **7440-31-5**	
DOT Hazard Class and Label Requirements **No Citation**	DOT Emergency Guide Code **No Citation**	
DOT Identification Number **No Citation**	Atomic Formula **Sn**	

Synonyms
Metallic tin; tin flake; tin metal; tin powder.

SECTION II - HAZARDOUS INGREDIENTS AND IDENTITY INFORMATION

Hazardous Components (specific identity)	OSHA Exposure Criteria	NIOSH Exposure Criteria	Immediately Dangerous to Life and Health (IDLH)	ACGIH Exposure Criteria
Tin: (derivation: By roasting the ore (cassiterite) to oxidize sulfates and to remove arsine, then reducing the coal in a reverberatory furnace, or by smelting in an electric furnace; secondary recovery from tin plate).	**PEL (8-hour):** 2 mg(Sn)/m^3 **STEL:** Not Established	**REL (8-hour):** 2 mg(Sn)/m^3 **STEL:** Not Established	100 mg/m^3 (as Sn)	**TLV:** 2 mg(Sn)/m^3 **STEL:** Not Established

SECTION III - PHYSICAL AND CHEMICAL CHARACTERISTICS

Boiling Point **4545°F (2507°C)**	Specific Gravity (H_2O = 1) **7.31**
Vapor Pressure (mm Hg) **1 at 2718°F (1492°C)**	Molecular Weight (atomic weight) **118.71**
Vapor Density (Air = 1) **Not Reported**	Melting Point **449°F (232°C)**

Solubility
Insoluble in cold water, slightly soluble in hot water. Soluble in acids and hot potassium hydroxide solution.

Appearance and Odor
Gray to almost silver-white, lustrous, ductile, malleable, metal. Also available in powdered form.

SECTION IV - FIRE AND EXPLOSION HAZARD DATA

Flash Point (method used) **Not Reported**	Explosive Limits in Air % by Volume **LEL: Not Reported UEL: Not Reported**
NFPA Classification **Non-Combustible Solid (powder may ignite)**	Autoignition Temperature **Not Reported**

Extinguishing Media
Tin may burn, but does not easily ignite. Use dry sand, graphite, dolomite, sodium chloride. Never use water, halogenated agents, or class A, B, or C extinguishing agents.

Special Fire Fighting Procedures
Poisonous gases are produced in fire. Wear full protective clothing, self-contained breathing apparatus (SCBA) operated in pressure-demand mode. Move container from fire area if it can be done without risk.

Unusual Fire and Explosion Hazards
In powdered form, tin is a moderate fire and explosion hazard. Will also burn vigorously on contact with flame or by chemical reaction. Powders and dusts form explosive mixtures in air.

SECTION V - REACTIVITY DATA

Stability		Conditions to Avoid Tin metal is stable in closed containers at room temperature under routine conditions of storage and handling, but its powder oxidizes in the presence of moisture. Keep away from incompatible materials.
Stable	Unstable	Incompatibility *(materials to avoid)* Incompatible with halogens, cupric nitrate, sodium and potassium peroxide and can react incandescently with sulfur. Also reacts with cold dilute hydrochloric, nitric, and hot dilute hydrosulfuric acids.
X		
Hazardous Polymerization		Conditions to Avoid Under normal conditions of temperature and pressure, tin cannot undergo hazardous polymerization. In contact with moisture, a poisonous coating of tin oxide can develop on exposed surfaces of tin metal.
May Occur	Will Not Occur	Hazardous Decomposition or By-products When heated to decomposition in air it emits highly toxic, acrid, and irritating smoke and fumes.
	X	

SECTION VI - HEALTH HAZARD DATA

Primary Route(s) of Entry:	Inhalation?	Absorption (skin/eye)?	Ingestion?
	X		

Health Hazards

INHALATION: Tin dusts or tin fumes can cause a mild, entirely asymptomatic, benign pneumoconiosis, known as stannosis. Exposure to tin fumes may cause symptoms of metal fume fever, a flu-like illness lasting about 24 hours with fever, fatigue, muscle ache, chills, aches, and cough. In animal studies, it has also caused vomiting, diarrhea, paralysis, an muscle twitching.

ABSORPTION: Eye and skin contact may result in mild to moderate irritation.

INGESTION: Tin is poorly absorbed through the digestive tract. Tin salts in the bloodstream are highly toxic and cause paralysis and other neurologic damage.

Carcinogenicity	NTP Listed?	IARC Cancer Review Group?	OSHA Regulated?	Target Organs?
Unknown Human Unknown Animal	No	No	29 CFR 1910.1000 Table Z-1	Eyes, respiratory system, skin.

Medical Conditions Generally Aggravated by Exposure

None reported.

Emergency and First-aid Procedures

Eye contact: Do not allow victim to keep eyes tightly shut. Flush immediately with water for 15 minutes (minimum); seek medical attention. Skin contact: Avoid skin contact in areas with cuts or scratches. If such contact occurs, wash with large amounts of water, scrubbing to remove metal particles. For inhalation: Remove the person from exposure. Provide respiratory assistance and CPR. Transfer to medical facility. If swallowed: Seek medical attention immediately. Do not attempt to give an unconscious or convulsing person anything by mouth. If conscious, give copious amounts of water to drink and induce vomiting. Repeat until vomitus is clear.

SECTION VII - PRECAUTIONS FOR SAFE HANDLING AND USE

Steps to be Taken in Case Material is Released or Spilled

Wear self-contained breathing apparatus (SCBA). Restrict those not involved in cleanup from entering area. Avoid skin contact. Ventilate area of spill. Do not use water. Collect powder materials and deposit in sealed drum. Do NOT dry sweep; use a high-efficiency particulate air (HEPA) vacuum instead.

Preferred Waste Disposal Method

Reclaim scrap (tin is a valuable scrap metal).

Precautions to be Taken in Handling and Storage

Store in tightly closed containers in a cool, well-ventilated area away from heat. Protect containers from physical damage. Avoid contact in storage with incompatible materials as violent reactions can occur.

Other Precautions and Warnings

Use only non-sparking equipment and tools when opening or closing containers of tin powder.

SECTION VIII - CONTROL MEASURES AND PERSONAL PROTECTIVE EQUIPMENT

Respiratory Protection *(specify type)*

Use an air-purifying respirator equipped with a HEPA filter or, use a NIOSH/MSHA-approved self-contained breathing apparatus (SCBA) with full facepiece operated in positive pressure mode.

Ventilation

Local exhaust (preferred) at site of chemical work or general ventilation.

Protective Gloves	Eye Protection	Other Protective Clothing
Impervious Gloves	Chemical/Dust Goggles and Face Mask	Protective Uniform or Apron

Work/Hygiene Practices

Always wash hands thoroughly after using chemical; never bring food, drink, or smoking materials into vicinity of chemicals. Barrier creams may be useful to limit the effects of skin exposure.

TIN

Sn

CAS: 7440-31-5

IDENTIFICATION AND TYPICAL USES

Tin is a metallic element of atomic number 50, Group IVA of the periodic system. It has valances of 2, 4 and ten isotopes. It appears as a gray to almost silver-white, ductile, malleable, lustrous, ductile metal. It is also available in powdered form. It is used primarily for tin plating, in soldering alloys, in manufacturing tin salts, in collapsible tubes, in tinned wire, in the manufacture of chemicals, in anodes for electrotin plating, in dental amalgams, in corrosion-resistant coatings, and low-melting alloys for fire control.

RISK ASSESSMENT: HEALTH

General Assessment

Elemental tin exhibits relatively low toxicity, but its compounds are highly toxic. It is primarily toxic through *inhalation* of its dusts. Tin is poorly absorbed through the digestive tract and, therefore, ingestion is not considered a toxic exposure route. Skin contact can cause localized irritation but dermal absorption is not reported. At least one source (Lewis, 1992) suggests that tin is a questionable carcinogen. However, other sources are conflicting and inconclusive in this regard. There are no data in the references regarding its mutagenic or teratogenic properties.

Inhalation of tin dusts or tin fumes may result in the development of a mild, entirely asymptomatic, benign form of pneumoconiosis known as stannosis. This condition is easily diagnosed through chest X-ray since tin is readily visible (radio-opaque) on film. It is not known if tin deposits in the lungs will alter or interfere with lung function, cause massive fibrosis, disability, or other complications. Exposure to the fumes may cause a flu-like condition known as metal fume fever with symptoms of chills, fever, fatigue, coughing, and muscle aches. It usually subsides within 48 hours of onset with no permanent after effects.

Skin contact may produce irritation or rash. Eye contact can also result in mild to severe mechanical irritation.

Acute Health Effects

The following acute (short-term) health effects may occur immediately or shortly after exposure to tin:

Skin: Moderate to severe irritation.

Eye: Mild to severe mechanical irritation with itching and tearing.

Lung: High concentrations of dust or fume inhalation may cause respiratory tract irritation and possible symptoms of metal fume fever.

Chronic Health Effects

The following chronic (long-term) health effects may occur at some time after exposure to tin and can last for months or even years:

Cancer Hazards: At least one source (Lewis, 1992) suggests that tin may be a questionable carcinogen in test animals. However, other references are conflicting and/or inconclusive in this regard. It is not clear if exposure to tin will cause cancer in humans.

Reproductive Hazard: There is no evidence in the published studies to show that tin has the ability to adversely affect reproduction.

Other Chronic Effects: None reported.

Recommended Risk-Reduction Measures

Personnel should avoid direct contact with tin. The exact nature of its toxicity is not well documented in the references. Prudent risk management dictates it be treated as though it posed a significant threat to human health. If a less toxic material or compound cannot be substituted for tin, then *engineering controls* are the most effective method of reducing exposure risk. The best protection is to enclose operations and/or provide local exhaust ventilation at the site of tin release. While not always operationally feasible, isolating operations can also reduce exposure. Using respiratory protection is less effective than the controls mentioned above, but is still required whenever working with or around tin. The exposure level is relatively low and difficult to accurately monitor. For infrequent or transient exposures, an air-purifying respirator equipped with a high-efficiency particulate air (HEPA) filter and dust pre-filter may suffice. However, for the best possible protection use an MSHA/NIOSH-approved full facepiece self-contained breathing apparatus (SCBA) with full facepiece operated in pressure demand, or a powered air-supplied respirator in continuous flow mode. If a full facepiece is not available, then chemical goggles should be worn to protect the eyes, especially when an airborne dust hazard exists. A face shield should also be considered. To prevent hand and skin exposures, rubber gloves should be used. Glove manufacturers should be contacted and permeation studies obtained *before* final glove selections are made.

Administrative controls should also be in place to minimize the potential for human exposures. These may include written procedures or policies which specify the methods and techniques that will be practiced whenever personnel are to work with tin.

All personnel should receive *training* on the use, hazards, protective measures, emergency actions, and other precautions per 29 CFR 1910.1200 (Hazard Communication) prior to the first assignment in an area where tin is used or stored.

Before beginning employment and at regular times thereafter, the following recommended medical tests should be considered for those with frequent or potentially high exposures:

☑ Lung function tests.

☑ Consider chest X-ray after acute overexposure or repeated, long-term exposure (tin deposits in lungs should be readily visible).

It should be noted that medical tests that simply look for existing damage are not a substitute for controlling exposures. Medical histories are extremely important when assessing exposure risk. Also, since smoking can cause heart disease, lung cancer, emphysema, and other respiratory diseases, smokers exposed to tin may experience symptoms more quickly and more pronounced than non-smokers under the same conditions of exposure. Prudent risk management requires proper consideration of *all* factors which may cause exposure symptoms in the workplace.

Other methods to reduce exposure include:

☑ Always ensure that proper protective clothing is worn when using chemical substances and that personnel are trained in its use, care, and maintenance. Barrier creams may be effective in preventing skin exposures.

☑ Wash thoroughly immediately after exposure to tin and at the end of the work shift or before eating, drinking, smoking, or applying cosmetics.

☑ Hazard warning information should be posted in the work area. In addition, as part of an on-going education and training program, all information on the health and safety hazards of tin should be communicated to all potentially exposed workers.

☑ Eye wash stations should be provided in the immediate work area for emergency use. If there is a possibility of skin exposure to tin, emergency shower facilities should also be provided.

☑ Workers whose clothing has been contaminated by tin should change into clean clothes before leaving work. Contaminated work clothing should be laundered only by individuals who have been informed of the hazards of exposure to tin.

RISK ASSESSMENT: ENVIRONMENT

General Assessment

Even though tin is a commonly occurring metallic element, the environment is still at risk of exposure during transportation, storage, disposal, or destruction of tin. In almost every scenario, the threat of environmental exposure is contingent upon the proper handling of the chemical substance. Accidental spills, large or small, can result in contamination of the surrounding environmental mediums (water, soil, and air).

Tin is a non-combustible solid in bulk form. However, tin metal dusts particles suspended in the air can explode. Caution is always required in handling, storage, transportation, and disposal of tin. It presents a serious explosion hazard in the form of powders and dusts. It is incompatible with halogens, cupric nitrate, sodium and potassium peroxide, and sulfur. It reacts slowly with cold dilute hydrochloric, nitric, and hot dilute hydrosulfuric acids. It also reacts readily with concentrated hydrochloric acids and aqua regia, and very slowly with acetic acid. Also, oxidizing salts such as potassium peroxysulfate, ferric chloride and sulfate, and aluminum and stannic chlorides can dissolve tin. When heated to decomposition in air it emits very toxic and acrid fumes and vapors. Emergency responders should therefore be made aware of the presence of tin at any emergency response situation.

Tin is a metal that is not particularly rare in the Earth's crust. Tin and its compounds may enter the environment through industrial discharges, or spills.

Acute Ecological Effects

Acute (short-term) toxic effects may include the death of animals, birds, or fish and death or low growth rate in plants. Acute effects are seen 2 to 4 days after animals or plants are exposed to tin.

Insufficient data are available to evaluate or predict the acute (short-term) effects of tin to aquatic life, birds, plants or land animals.

Chronic Ecological Effects

Chronic (long-term) toxic effects may include shortened life span, reproductive problems, lower fertility, and changes in appearance or behavior in exposed animals. These effects can be seen long after first exposure(s) to toxic chemicals.

Insufficient data are available to evaluate the chronic (long-term) effects of tin to aquatic life, plants, birds, or land animals.

Water Solubility

Tin and its compounds have water solubilities that range from low to high, depending on the salt that is formed.

Persistence in the Environment

Tin itself is likely to be highly persistent in water, as it has only slight water solubility. Other compounds of tin will vary in persistence, as water solubility varies from compound to compound.

Bioaccumulation in Aquatic Organisms

Some substances increase in concentration, or *bioaccumulate,* in living organisms as they breathe contaminated air, drink contaminated water, or eat contaminated food. These chemicals can become concentrated in the tissues and internal organs of animals as well as humans.

Insufficient data are available on the bioaccumulation characteristic of tin in aquatic organisms.

Recommended Risk-Reduction Measures

Proper training of all transporters will reduce the likelihood of a mishap or accident resulting in a leak or spill of tin into the environment. Labels on all containers, trucks, and rail cars must meet DOT requirements and accurately reflect their contents to enable emergency responders to react properly and quickly to any disaster thereby reducing the potential risk to the environment and to personnel.

Storage of tin should be segregated from other chemicals and materials to minimize the risk of cross-contamination. Tin must be stored to avoid contact with incompatible materials since violent reactions can occur. Care must be taken to avoid the generation of airborne dusts or powders in storage and handling areas since tin dusts can form explosive mixtures in air. Containers should be protected from physical damage and stored to avoid contact with heat.

If a spill or leak to the environment has occurred, fire department, emergency response and/or hazardous materials spill personnel should be notified immediately. Cleanup should be attempted only by those trained in proper spill containment procedures using non-sparking tools. Contaminated soils should be removed for incineration and replaced with clean soil. If tin should contact the water table, aquifer, or navigable waterway, time is of the essence. The solubility of tin and its compounds can range from insoluble to highly solubility and total remediation may not be possible. The local and/or state emergency response authorities must be notified. A comprehensive emergency response or disaster preparedness/recovery plan should be in place prior to any operations involving the use, transportation, storage, or disposal of tin.

If tin is spilled or leaked, the following specific steps are recommended:

- ☑ Restrict persons not wearing protective clothing from area of spill or leak until cleanup is complete.
- ☑ Collect powdered material in the most convenient and safe manner possible and deposit in sealed drum for disposal. Do NOT dry sweep. Use a vacuum equipped with a high-efficiency particulate air (HEPA) filter. Do NOT damp mop residue.
- ☑ Ventilate area of spill or leak and remove all ignition sources.
- ☑ It may be necessary to dispose of tin as a hazardous waste. The responsible state agency or the re-

gional office of the federal Environmental Protection Agency (EPA) should be contacted for specific recommendations.

RISK ASSESSMENT: BUSINESS

General Assessment

Accidents or mishaps involving tin can present a serious threat to business operations. The loss or damage of equipment or facilities can significantly affect fiscal viability. Lawsuits that may result from personnel illness, injury/death, public exposures, and/or environmental contamination will require a serious expenditure of resources. Media attention surrounding an injury, death, or environmental damage can also result in a loss of profits and loss of current as well as future business.

Recommended Risk-Reduction Measures

Company attorneys, safety and health professionals, and environmental specialists should be involved in the development of any procedures for responding to chemical incidents. A company official should be predesignated as a public relations officer with specific training in dealing with the press. Corporate plans and policies should be developed, approved, and implemented long before any need for such arises.

REFERENCES

American Conference of Governmental Industrial Hygienists. 1988. *Documentation of the Threshold Limit Values and Biological Exposure Indices,* 5th Edition (with updates). Cincinnati: ACGIH

Asante-Duah, D. K. 1993. *Hazardous Waste Risk Assessment.* New York: CRC Press/Lewis Publishers.

Bloom, A. D. and F. J. DeSerres. 1995. *Ecotoxicity and Human Health.* New York: CRC Press/Lewis Publishers.

Calabrese, E. J. 1994. *Biological Effects of Low Level Exposures.* New York: CRC Press/Lewis Publishers.

Cockerman, L. G., B. S. Shane. 1994. *Basic Environmental Toxicology.* New York: CRC Press/Lewis Publishers.

Cothern, C. R., N. P. Ross. 1994. *Environmental Statistics, Assessment, and Forecasting.* New York: CRC Press/Lewis Publishers.

Cothern, R. R. 1993. *Comparative Environmental Risk Assessment.* New York: CRC Press/Lewis Publishers.

Cothern, C. R. 1995. *Handbook for Environmental Risk Decisionmaking.* New York: CRC Press/Lewis Publishers.

Danse, I. R. 1991. *Common Sense Toxics in the Workplace.* New York: Van Nostrand Reinhold.

Environmental Protection Agency. 1983. *Chemical Hazard Information Profiles.* Washington, D.C.: U.S. EPA.

Fodden, C. R., and J. L. Weddell. 1992. *Hazardous Materials Emergency Action Data.* New York: CRC Press/Lewis Publishers.

Forseberg, K., and L. H. Keith. 1993. *Chemical; Protective Clothing Permeation and Degradation Database.* Chelsea, MI: Lewis Publishers.

Gardner, R. W. 1994. *Chemical Intolerance.* New York: CRC Press/Lewis Publishers.

Hallenbeck, W. H. 1993. *Quantitative Risk Assessment for Environmental and Occupational Health,* 2nd Edition. New York: CRC Press/Lewis Publishers.

Hodgson, E., R. B. Mailman, and J. E. Chambers. 1988. *Dictionary of Toxicology.* London: The MacMillian Company.

Keith, L. H. and M. M. Walker. 1995. *Handbook of Air Toxics.* New York: CRC Press/Lewis Publishers.

Landis, W. G., M. H. Yu. 1995. *Introduction to Environmental Toxicology.* New York: CRC Press/Lewis Publishers.

Lewis, R. J., Sr. 1993. *Hawley's Condensed Chemical Dictionary,* 12th Edition. New York: Van Nostrand Reinhold.

Lewis, R. J., Sr. 1992. *Sax's Dangerous Properties of Industrial Materials,* Volumes 1, 2, and 3, 8th Edition. New York: Van Nostrand Reinhold.

Meyer, E. 1990. *Chemistry of Hazardous Materials.* Englewood Cliffs, NJ: Prentice Hall.

Patnaik, P. 1992. *A Comprehensive Guide to the Hazardous Properties of Chemical Substances.* New York: Van Nostrand Reinhold.

Philp, R. B. 1995. *Environmental Hazards and Human Health.* New York: CRC Press/Lewis Publishers.

Proctor, N. H., J. P. Hughes. 1991. *Chemical Hazards of the Workplace.* New York: Van Nostrand Reinhold.

Reeves, A. L. 1979. *Handbook on the Toxicology of Metals.* New York: Elsevier North-Holland.

Richardson, M. 1995. *Dictionary of Substance and Their Effects,* Vol. I-VI. New York: CRC Press/Lewis Publishers.

Sacarello, H. L. A. 1994. *The Comprehensive Handbook of Hazardous Materials.* New York: CRC Press/Lewis Publishers.

Torkelson, T. R. V. K. Rowe. 1981. *Patty's Industrial Hygiene and Toxicology,* Vol. 2B. New York: Wiley-Interscience.

U.S. Department of Health and Human Services, National Institute for Occupational Safety and Health. 1994. *NIOSH Pocket Guide to Chemical Hazards.* Washington, D.C.: U.S. Government Printing Office.

MATERIAL SAFETY DATA SHEET

CHEMICAL NAME	
TITANIUM	**CRC PRESS/LEWIS PUBLISHERS DISCLAIMER:** THE INFORMATION AND RECOMMENDATIONS PRESENTED HEREIN ARE BASED ON SOURCES BELIEVED TO BE RELIABLE. CRC MAKES NO REPRESENTATION ON ITS COMPLETENESS OR ACCURACY. IT IS THE RESPONSIBILITY OF THE USER TO DETERMINE THE CHEMICAL'S SUITABILITY FOR ITS INTENDED USE, THE CHEMICAL'S SAFE USE, AND THE CHEMICAL'S PROPER DISPOSAL. NO REPRESENTATIONS AND/OR WARRANTIES, EITHER EXPRESSED OR IMPLIED, OF THE MERCHANTABILITY OR FITNESS FOR A PARTICULAR PURPOSE, OR OF ANY OTHER NATURE, ARE MADE WITH RESPECT TO THE INFORMATION PROVIDED IN THIS MSDS OR TO THE CHEMICAL TO WHICH INFORMATION MAY REFER. CRC NEITHER ASSUMES NOR AUTHORIZES ANY OTHER PERSON TO ASSUME FOR IT, ANY OTHER ADDITIONAL RESPONSIBILITY OR LIABILITY FOR THE USE OF, OR RELIANCE UPON, THIS INFORMATION.

HAZARD WARNING INFORMATION

HEALTH	FIRE	REACTIVE	OTHER	DEGREE OF HAZARD	COLOR CODING	OTHER CODES
0	3	3		0 = Minimum Hazard 1 = Slight Hazard 2 = Moderate Hazard 3 = Serious Hazard 4 = Severe Hazard	HEALTH = BLUE FIRE = RED REACTIVITY = YELLOW OTHER = WHITE	OX = Oxidizer ACID = Acid ALK = Alkali COR = Corrosive W = Use No Water

SECTION I - GENERAL INFORMATION

Characterization	Metal	RCRA Number	D001
		EPA Class	Characteristic (R) Waste
DOT Proper Shipping Name	Titanium; Titanium Powder; Titanium Sponge	Chemical Abstract Service (CAS) Number	7440-32-6
DOT Hazard Class and Label Requirements	Spontaneously Combustible; Flammable Solid	DOT Emergency Guide Code	39 (compounds)
DOT Identification Number	UN2546 (dry) 1352 (wet) 2878 (granules or sponge)	Atomic Formula	Ti

Synonyms
Contiment 30; Orement; C.P. titanium; titanium alloy; titanium metal powder; IMI 115.

SECTION II - HAZARDOUS INGREDIENTS AND IDENTITY INFORMATION

Hazardous Components (specific identity)	OSHA Exposure Criteria	NIOSH Exposure Criteria	Immediately Dangerous to Life and Health (IDLH)	ACGIH Exposure Criteria
Titanium: (derivation: By reduction of titanium tetrachloride with magnesium or sodium in an inert atmosphere of helium or argon. The titanium sponge is consolidated by melting; by electrolysis of titanium tetrachloride in a bath of fused salts). *PEL is for "Nuisance Particulates"	PEL:* 15 mg/m^3 (total dust) 5 mg/m^3 (respirable fraction) STEL: Not Established	REL: Not Established STEL: Not Established	Not Determined	TLV: Not Established STEL: Not Established

SECTION III - PHYSICAL AND CHEMICAL CHARACTERISTICS

Boiling Point	5900°F (3260°C)	Specific Gravity (H_2O = 1)	0.13
Vapor Pressure (mm Hg)	Not Reported	Molecular Weight (atomic weight)	47.90
Vapor Density (Air = 1)	Not Reported	Melting Point	3047°F (1675°C)

Solubility
Insoluble in water but attacked by concentrated sulfuric and hydrochloric acids. Soluble in dilute acids.

Appearance and Odor
Silvery metal or dark-gray amorphous powder.

SECTION IV - FIRE AND EXPLOSION HAZARD DATA

Flash Point (method used)	Not Reported (flammability is contingent on particle size)	Explosive Limits in Air % by Volume	LEL: 0.045 oz/ft^3 UEL: Not Reported
NFPA Classification	Flammable Solid	Autoignition Temperature	Metal: 2192°F (1220°C) Powder: 482°F (250°C)

Extinguishing Media
Use dry chemical, soda ash, lime, or sand. Use water only in flooding amounts and from a distance.

Special Fire Fighting Procedures
Poisonous gases are produced in fire. Wear full protective clothing, self-contained breathing apparatus (SCBA) operated in pressure-demand mode. Move container from fire area if it can be done without risk.

Unusual Fire and Explosion Hazards
In powdered form, titanium is a moderate fire and explosion hazard. Will also burn vigorously on contact with flame, even in a nitrogen atmosphere or carbon dioxide atmosphere. Powders form explosive mixtures in air. In the absence of moisture, titanium burns slowly but produces extreme heat.

SECTION V - REACTIVITY DATA

Stability		Conditions to Avoid Titanium metal can be considered normally stable. However, it turns brittle when cold and malleable when heated. Avoid generated dusty conditions and keep away from incompatible materials.
Stable	Unstable	Incompatibility *(materials to avoid)* Oxygen and other oxidants (potassium chlorate, potassium permanganate), metal oxides, metal oxosalts, fuming nitric acid, nitryl fluoride, halogens, aluminum, bromine trifluoride, carbon black, carbon dioxide + metal carbonates, carbon dioxide + nitrogen, halocarbons, silver fluoride, trichloroethylene, and steam.
X		
Hazardous Polymerization		Conditions to Avoid Under normal conditions of temperature and pressure, titanium cannot undergo hazardous polymerization.
May Occur	Will Not Occur	Hazardous Decomposition or By-products When heated to decomposition in air it emits highly toxic metal fumes.
	X	

SECTION VI - HEALTH HAZARD DATA

Primary Route(s) of Entry:	Inhalation?	Absorption (skin/eye)?	Ingestion?
	X	X	X

Health Hazards

INHALATION: May cause mechanical irritation of the respiratory tract with coughing, sneezing, dypsnea, and sore throat. Long-term injury to the lungs has not been established.

SKIN & EYES: Eye contact results in tearing and redness as a result of mechanical irritation only. Skin contact causes no reported effects, as is evidenced by titanium use in skin medications.

INGESTION: Poor absorption through the alimentary canal with no adverse effects documented.

Carcinogenicity	NTP Listed?	IARC Cancer Review Group?	OSHA Regulated?	Target Organs?
Unknown Human Unknown Animal	No	No	29 CFR 1910.1000 Table Z-1	Lungs

Medical Conditions Generally Aggravated by Exposure

Lung diseases may be aggravated by exposure to titanium dusts or powders.

Emergency and First-aid Procedures

Eye contact: Do not allow victim to keep eyes tightly shut. Flush immediately with water for 15 minutes (minimum) until titanium particles are removed. If irritation persists, seek medical attention. Skin contact: Remove all contaminated clothing. Do NOT shake clothes. If skin contact occurs, wash with large amounts of water, scrubbing to remove metal particles. For inhalation: Remove the person from exposure. Provide respiratory assistance and CPR. Observe for lung effects. If swallowed: Seek medical attention immediately or contact poison control center. Do not attempt to give an unconscious or convulsing person anything by mouth. Unless advised otherwise, inducing vomiting is not necessary unless large amounts have been ingested.

SECTION VII - PRECAUTIONS FOR SAFE HANDLING AND USE

Steps to be Taken in Case Material is Released or Spilled

Wear self-contained breathing apparatus (SCBA). Restrict those not involved in cleanup from entering area. Ventilate area of spill. Do not use water. Collect powder materials and deposit in sealed drum. Do NOT dry sweep; use a high-efficiency particulate air (HEPA) vacuum instead. Use non-sparking tools.

Preferred Waste Disposal Method

Reclaim if possible or dispose of in approved landfill.

Precautions to be Taken in Handling and Storage

Store in tightly closed containers in a cool, well-ventilated area away from heat. Protect containers from physical damage. Fine particles should be stored damp. Fine powder should be in air-tight containers.

Other Precautions and Warnings

Use only non-sparking equipment and tools, especially when opening or closing containers of titanium.

SECTION VIII - CONTROL MEASURES AND PERSONAL PROTECTIVE EQUIPMENT

Respiratory Protection *(specify type)*

Use an air-purifying respirator equipped with a HEPA filter or, use a NIOSH/MSHA-approved self-contained breathing apparatus (SCBA) with full facepiece operated in positive pressure mode.

Ventilation

Local exhaust (preferred) at site of chemical work or general ventilation.

Protective Gloves	Eye Protection	Other Protective Clothing
Impervious Gloves	Chemical/Dust Goggles and Face Mask	Protective Uniform or Apron

Work/Hygiene Practices

Always wash hands thoroughly after using chemical; never bring food, drink, or smoking materials into vicinity of chemicals.

TITANIUM

Ti

CAS: 7440-32-6

IDENTIFICATION AND TYPICAL USES

Titanium is a metallic element of atomic number 22, Group IVB of the periodic table. It has valances of 2, 3, 4 and five isotopes. It appears as a silvery metal or a dark-gray amorphous powder. It is used in alloys (especially ferrotitanium), as structural material for aircraft, in jet engines, missiles, and maritime equipment. It is found in textile machinery, in chemical equipment (especially as an anode in chloride production), in desalination equipment, surgical instruments, orthopedic appliances, in food-handling equipment, in X-ray tube targets, in abrasives, cements, and metal ceramic brazing. It is also used in nickel-cadmium batteries for space vehicles, in electrodes for chlorine batteries, and in electro-deposited and dipped coatings on metals and ceramics.

RISK ASSESSMENT: HEALTH

General Assessment

Titanium metal is a relatively inert material in terms of human health effects. However, there is some evidence of fibrotic lung changes resulting from long-term *inhalation* of titanium dusts and powders. While it is not presently considered a human carcinogen, at least one reference (Lewis, 1992) considers titanium a questionable carcinogen with experimental tumorigenic data reported. There are also limited data regarding its reproductive toxicity in animals.

Inhalation of titanium dusts initially causes mechanical irritation of the respiratory tract which can result in coughing, sneezing, and dypsnea. There are no confirmed reports of any systemic effects resulting from inhalation exposures. Titanium may cause a benign form of pulmonary fibrosis, but this has not been properly researched in the references.

Skin contact usually causes no visible indications of irritation or reaction. Eye contact can result in mild mechanical irritation that is alleviated upon removal of the metallic particles from the eye (through deluge with copious amounts of water).

Acute Health Effects

The following acute (short-term) health effects may occur immediately or shortly after exposure to titanium:

Skin: No reported irritating effects.

Eye: Mild mechanical irritation with a possibility for redness and tearing until material is removed.

Lung: Overexposure may irritate the airways and lungs, causing coughing, sneezing, and dypsnea.

Chronic Health Effects

The following chronic (long-term) health effects may occur at some time after exposure to titanium and can last for months or even years:

Cancer Hazards: According to information presented in the references, titanium has not been adequately tested for its ability to cause cancer in test animals. However, at least one source (Lewis, 1992) considers titanium a questionable carcinogen in animals with experimental tumorigenic data reported. This assessment is apparently based on the intra-muscular administration of titanium in rats which caused tumors in the blood (Patnaik, 1992). It is not known if titanium will cause cancer in humans.

Reproductive Hazard: According to information presented in the references, titanium may cause adverse reproductive and fetotoxic effects in animals. It is not known if human exposure will adversely affect reproduction.

Other Chronic Effects: Long-term exposure to titanium has caused deposits of titanium in the lungs which results in fibrotic changes (visible on X-ray). However, these changes have not been shown to cause cancer or other tumorigenic effects in humans.

❶ *Recommended Risk-Reduction Measures*

Personnel should avoid direct contact with titanium. While it is considered a relatively inert metal in terms of adverse human health effects, it should also be noted that the exact nature of its toxicity is not clearly understood or documented in the references. The best exposure control method is to find a substitute material which presents less exposure risk to humans. However, this may not always be possible. Therefore, if a less toxic material or compound cannot be substituted for titanium, then *engineering controls* are the most effective method of reducing exposure risk. The best protection is to enclose operations and/or provide local exhaust ventilation at the site of titanium release. While not always operationally feasible, isolating operations can also reduce exposure. Using respiratory protection is less effective than the controls mentioned above, but is still required whenever working with or around titanium. For infrequent or transient exposures, an air-purifying respirator equipped with a high-efficiency particulate air (HEPA) filter and dust pre-filter may suffice. However, for the best possible protection use an MSHA/NIOSH-approved full facepiece self-contained breathing apparatus (SCBA) with full facepiece operated in pressure demand, or a powered air-supplied respirator in continuous flow mode. If a full facepiece is not available, then chemical dust goggles should be worn to protect the eyes, especially when an airborne dust hazard exists. A face shield should also be considered. Skin and especially hand contact should be avoid to preclude the possibility of transmission to the eyes via the fingers and skin. To prevent hand and skin exposures, rubber gloves should be used.

Administrative controls should also be in place to minimize the potential for human exposures. These may include written procedures or policies which specify the methods and techniques that will be practiced whenever personnel are to work with titanium.

All personnel should receive *training* on the use, hazards, protective measures, emergency actions, and other precautions per 29 CFR 1910.1200 (Hazard Communication) prior to the first assignment in an area where titanium is used or stored.

Before beginning employment and at regular times thereafter, the following recommended medical tests should be considered for those with frequent or potentially high exposures:

☑ Lung function tests.

☑ Consider chest X-ray after acute overexposure or long-term exposures. Titanium metal deposits may be visible, along with certain fibrotic changes, on X-ray film.

It should be noted that medical tests that simply look for existing damage are not a substitute for controlling exposures. Medical histories are extremely important when assessing exposure risk. Also, since smoking can cause heart disease, lung cancer, emphysema, and other respiratory diseases, smokers exposed to titanium may experience symptoms more quickly and more pronounced than non-smokers under the same conditions of exposure. Prudent risk management requires proper consideration of *all* factors which may cause exposure symptoms in the workplace.

Other methods to reduce exposure include:

☑ Always ensure that proper protective clothing is worn when using chemical substances and that personnel are trained in its use, care, and maintenance.

☑ Wash thoroughly immediately after exposure to titanium and at the end of the work shift or before eating, drinking, smoking, or applying cosmetics.

☑ Hazard warning information should be posted in the work area. In addition, as part of an on-going education and training program, all information on the health and safety hazards of titanium should be communicated to all exposed and potentially exposed workers.

☑ Eye wash stations should be provided in the immediate work area for emergency use. If there is a possibility of skin exposure to titanium, emergency shower facilities should also be provided.

☑ Workers whose clothing has been contaminated by titanium should change into clean clothes before leaving work. Contaminated work clothing should be laundered only by individuals who have been informed of the hazards of exposure to titanium.

RISK ASSESSMENT: ENVIRONMENT

General Assessment

Even though titanium is a naturally occurring metallic element, the environment is still at risk of exposure during transportation, storage, disposal, or destruction of titanium. In almost every scenario, the threat of environmental exposure is contingent upon the proper handling of the chemical substance. Accidental spills, large or small, can result in contamination of the surrounding environmental mediums (water, soil, and air).

Titanium is a flammable solid. In bulk form, it will burn but will not easily ignite. However, titanium metal dusts particles suspended in the air can explode. Caution is always required in handling, storage, transportation, and disposal of titanium. It presents a serious explosion hazard in the form of powders and dusts, especially in the absence of moisture. It is incompatible with many materials and metals, including oxygen and other oxidants, metal oxides, metal oxosalts, sodium chlorate, trichloroethylene, and even steam (evolves explosive hydrogen gas at high temperatures). When heated to decomposition in air it emits very toxic metal fumes. Emergency responders should be made aware of the presence of titanium at any emergency response situation.

Titanium and its compounds may enter the environment during mining and smelting operations, through industrial discharges, or spills.

Acute Ecological Effects

Acute (short-term) toxic effects may include the death of animals, birds, or fish and death or low growth rate in plants. Acute effects are seen 2 to 4 days after animals or plants are exposed to titanium.

Insufficient data are available to evaluate or predict the acute (short-term) effects of titanium to aquatic life, birds, plants, or land animals.

Chronic Ecological Effects

Chronic (long-term) toxic effects may include shortened life span, reproductive problems, lower fertility, and changes in appearance or behavior in exposed animals. These effects can be seen long after first exposure(s) to toxic chemicals.

Insufficient data are available to evaluate the chronic (long-term) effects of titanium to aquatic life, plants, birds, or land animals.

Water Solubility

Titanium and its compounds have water solubilities that range from low to high, depending on the salt that is formed.

Persistence in the Environment

Titanium itself is likely to be highly persistent in water, as it has only slight water solubility. Its half-life may be greater than 200 days. Other compounds of titanium will vary in persistence, as water solubility varies from compound to compound.

Bioaccumulation in Aquatic Organisms

Some substances increase in concentration, or *bioaccumulate,* in living organisms as they breathe contaminated air, drink contaminated water, or eat contaminated food. These chemicals can become concentrated in the tissues and internal organs of animals as well as humans.

There are no data on the accumulation of titanium in the edible tissues of fish that are normally consumed by humans.

Recommended Risk-Reduction Measures

Proper training of all transporters will reduce the likelihood of a mishap or accident resulting in a leak or spill of titanium into the environment. Labels on all containers, trucks, and rail cars must meet DOT requirements and accurately reflect their contents to enable emergency responders to react properly and quickly to any disaster thereby reducing the potential risk to the environment and to personnel.

Storage of titanium should be segregated from other chemicals and materials to minimize the risk of cross-contamination. Titanium must be stored to avoid contact with heat or incompatible materials. It is suggested that powdered titanium be kept damp to reduce the fire and explosion hazard potential. Containers should be protected from physical damage and dusty conditions must be prevented in storage and handling areas since titanium dusts can explode in air.

If a spill or leak to the environment has occurred, fire department, emergency response and/or hazardous materials spill personnel should be notified immediately. Cleanup should be attempted only by those trained in proper spill containment procedures using non-sparking tools. Contaminated soils should be removed for titanium reclamation or for incineration and replaced with clean soil. If titanium should contact the water table, aquifer, or navigable waterway, time is of the essence. The solubility of titanium and its compounds can range from insoluble to highly solubility and total remediation may not be possible. The local and/or state emergency response authorities must be notified. A comprehensive emergency response or disaster preparedness/recovery plan should be in place prior to any operations involving the use, transportation, storage, or disposal of titanium.

Titanium is a very expensive material and reclamation is always preferred over disposal. However, if disposal is required, it may be deposited in an approved and permitted landfill.

If titanium is spilled or leaked, the following specific steps are recommended:

☑ Restrict persons not wearing protective clothing from area of spill or leak until cleanup is complete.

☑ Collect powdered material in the most convenient and safe manner possible and deposit in sealed drum for disposal. Do NOT dry sweep. Use a vacuum equipped with a high-efficiency particulate air (HEPA) filter. Damp mop residue. Use water spray to knock down dust clouds that have not yet ignited. Contain and collect water for disposal.

☑ Ventilate area of spill or leak and remove all ignition sources.

☑ It may be necessary to dispose of titanium as a hazardous waste. The responsible state agency or the regional office of the federal Environmental Protection Agency (EPA) should be contacted for specific recommendations.

RISK ASSESSMENT: BUSINESS

General Assessment

Accidents or mishaps involving titanium can present a serious threat to business operations. The loss or damage of equipment or facilities can significantly affect fiscal viability. Lawsuits that may result from personnel illness, injury/death, public exposures, and/or environmental contamination will require a serious expenditure of resources. Media attention surrounding an injury, death, or environmental damage can also result in a loss of profits and loss of current as well as future business. Always remember that anytime the terms "cancer," "carcinogen," or reproductive hazard" are used, public emotion, hysteria, ignorance, and anxiety can all run equally high. This must be carefully considered when drafting public relations policies.

Recommended Risk-Reduction Measures

Company attorneys, safety and health professionals, and environmental specialists should be involved in the development of any procedures for responding to chemical incidents. A company official should be predesignated as a public relations officer with specific training in dealing with the press. Corporate plans and policies should be developed, approved, and implemented long before any need for such arises.

REFERENCES

American Conference of Governmental Industrial Hygienists. 1988. *Documentation of the Threshold Limit Values and Biological Exposure Indices,* 5th Edition (with updates). Cincinnati: ACGIH

Bloom, A. D. and F. J. DeSerres. 1995. *Ecotoxicity and Human Health.* New York: CRC Press/Lewis Publishers.

Calabrese, E. J. 1994. *Biological Effects of Low Level Exposures.* New York: CRC Press/Lewis Publishers.

Cockerman, L. G., B. S. Shane. 1994. *Basic Environmental Toxicology.* New York: CRC Press/Lewis Publishers.

Gardner, R. W. 1994. *Chemical Intolerance.* New York: CRC Press/Lewis Publishers.

Hallenbeck, W. H. 1993. *Quantitative Risk Assessment for Environmental and Occupational Health,* 2nd Edition. New York: CRC Press/Lewis Publishers.

Hodgson, E., R. B. Mailman, and J. E. Chambers. 1988. *Dictionary of Toxicology.* London: The MacMillian Company.

Keith, L. H. and M. M. Walker. 1995. *Handbook of Air Toxics.* New York: CRC Press/Lewis Publishers.

Landis, W. G., M. H. Yu. 1995. *Introduction to Environmental Toxicology.* New York: CRC Press/Lewis Publishers.

Lewis, R. J., Sr. 1993. *Hawley's Condensed Chemical Dictionary,* 12th Edition. New York: Van Nostrand Reinhold.

Lewis, R. J., Sr. 1992. *Sax's Dangerous Properties of Industrial Materials,* Volumes 1, 2, and 3, 8th Edition. New York: Van Nostrand Reinhold.

Meyer, E. 1990. *Chemistry of Hazardous Materials.* Englewood Cliffs, NJ: Prentice Hall.

Patnaik, P. 1992. *A Comprehensive Guide to the Hazardous Properties of Chemical Substances.* New York: Van Nostrand Reinhold.

Philp, R. B. 1995. *Environmental Hazards and Human Health.* New York: CRC Press/Lewis Publishers.

Proctor, N. H., J. P. Hughes. 1991. *Chemical Hazards of the Workplace.* New York: Van Nostrand Reinhold.

Richardson, M. 1995. *Dictionary of Substance and Their Effects,* Vol. I-VI. New York: CRC Press/Lewis Publishers.

Sacarello, H. L. A. 1994. *The Comprehensive Handbook of Hazardous Materials.* New York: CRC Press/Lewis Publishers.

Torkelson, T. R. V. K. Rowe. 1981. *Patty's Industrial Hygiene and Toxicology,* Vol. 2B. New York: Wiley-Interscience.

U.S. Department of Health and Human Services, National Institute for Occupational Safety and Health. 1994. *NIOSH Pocket Guide to Chemical Hazards.* Washington, D.C.: U.S. Government Printing Office.

MATERIAL SAFETY DATA SHEET

CHEMICAL NAME

TITANIUM TETRACHLORIDE

HAZARD WARNING INFORMATION

HEALTH	FIRE	REACTIVE	OTHER	DEGREE OF HAZARD	COLOR CODING	OTHER CODES
3	0	1		0 = Minimum Hazard 1 = Slight Hazard 2 = Moderate Hazard 3 = Serious Hazard 4 = Severe Hazard	HEALTH = BLUE FIRE = RED REACTIVITY = YELLOW OTHER = WHITE	OX = Oxidizer ACID = Acid ALK = Alkali COR = Corrosive W = Use No Water

SECTION I - GENERAL INFORMATION

Characterization	Titanium Compound	RCRA Number	None
		EPA Class	Not Applicable
DOT Proper Shipping Name	Titanium Tetrachloride	Chemical Abstract Service (CAS) Number	7550-45-0
DOT Hazard Class and Label Requirements	Corrosive Material	DOT Emergency Guide Code	53
DOT Identification Number	UN 1838	Chemical Formula	Cl_4Ti

Synonyms
Titanic chloride; titanium chloride.

SECTION II - HAZARDOUS INGREDIENTS AND IDENTITY INFORMATION

Hazardous Components (specific identity)	OSHA Exposure Criteria	NIOSH Exposure Criteria	Immediately Dangerous to Life and Health (IDLH)	ACGIH Exposure Criteria
Titanium tetrachloride (derivation: By heating titanium dioxide or the ores and carbon to redness in a current of chlorine).	**PEL:** Not Established **STEL:** Not Established	**REL:** Not Established **STEL:** Not Established	Not Determined	**TLV:** Not Established **STEL:** Not Established

SECTION III - PHYSICAL AND CHEMICAL CHARACTERISTICS

Boiling Point	277°F (136°C)	Specific Gravity (H_2O = 1)	1.73
Vapor Pressure (mm Hg)	100 at 196°F (91°C)	Molecular Weight	189.70
Vapor Density (Air = 1)	Not Reported	Freezing Point	-11°F (-24°C)

Solubility
Soluble in cold water (evolves heat), alcohol, and dilute hydrochloric acid.

Appearance and Odor
A colorless or light yellow fuming liquid with a pungent, acid-like odor (fumes in moist air).

SECTION IV - FIRE AND EXPLOSION HAZARD DATA

Flash Point (method used)	Not Reported	Explosive Limits in Air % by Volume	LEL: Not Reported; UEL: Not Reported
NFPA Classification	Combustible Liquid	Autoignition Temperature	Not Reported

Extinguishing Media
For small fires, use carbon dioxide or dry chemical. For larger fires, flood with copious amounts of water from a distance. Solid stream of water may spread fire and react to evolve heat.

Special Fire Fighting Procedures
Wear full protective clothing and self-contained breathing apparatus (SCBA). Move containers from fire area if it can be done without risk. Containers may explode in fire. Fight fire from a distance if possible. Keep fire-exposed containers cool with water spray.

Unusual Fire and Explosion Hazards
Titanium tetrachloride reacts with water to evolve heat and release hydrochloric acid which can attack metals to form and then ignite flammable hydrogen gas.

SECTION V - REACTIVITY DATA

Stability		Conditions to Avoid Concentrated aqueous solutions are stable in closed containers under routine conditions of handling and storage. Keep in tightly closed containers away from moisture and heat.
Stable X	Unstable	Incompatibility *(materials to avoid)* Reacts strongly with water to release hydrochloric acid and heat. Keep away from potassium and strong oxidizers, such as chlorine, bromine, and fluorine. Reacts with moist air.
Hazardous Polymerization		Conditions to Avoid Under normal conditions of temperature and pressure, hazardous polymerization of titanium tetrachloride cannot occur.
May Occur	Will Not Occur X	Hazardous Decomposition or By-products When heated to decomposition, titanium tetrachloride emits highly toxic chloride fumes.

SECTION VI - HEALTH HAZARD DATA

Primary Route(s) of Entry:	Inhalation?	Absorption (skin)?	Ingestion?
	X		

Health Hazards

INHALATION: Vapors will irritate the eyes, nose, throat, and air passages causing cough with phlegm, headache, eye and lung damage, dizziness, and bronchial pneumonia. Chronic bronchitis and emphysema are also possible. Higher exposures can cause fluid in the lungs (pulmonary edema) which can be fatal.

SKIN & EYES: Contact with the liquid can cause 2nd degree thermal and acid burns to the skin and eyes. Brief eye contact causes conjunctivitis, keratitis, and clouding of the cornea.

INGESTION: Burns of the mouth, throat, and stomach with nausea, vomiting, cramps, diarrhea, and possible ulceration of alimentary canal tissues.

Carcinogenicity	NTP Listed?	IARC Cancer Review Group?	OSHA Regulated?	Target Organs?
Unknown Human Unknown Animal	No	No	No	Skin, eyes, mucous membrane, respiratory system.

Medical Conditions Generally Aggravated by Exposure

None reported.

Emergency and First-aid Procedures

Eye contact: Immediately flush with large amounts of water for 15 minutes (minimum), occasionally lifting eyelids; seek medical attention. Skin contact: Remove all contaminated clothing. Immediately wash area with large amounts of soap and water. For inhalation: Remove the person from exposure. Provide respiratory assistance and CPR. Transfer promptly to medical facility. Observe 48 hours for lung effects. If swallowed: Seek medical attention immediately. Never try to give an unconscious person anything by mouth or make an unconscious or convulsing person vomit. Unless advised otherwise, give 1 to 2 glasses of water and induce vomiting, then give milk, eggs, or olive oil.

SECTION VII - PRECAUTIONS FOR SAFE HANDLING AND USE

Steps to be Taken in Case Material is Released or Spilled

Restrict persons not wearing protective equipment from entering area of spill or leak until cleanup is complete. Ventilate area. Absorb liquid in dry sand or other material and deposit in sealed containers.

Preferred Waste Disposal Method

No citation.

Precautions to be Taken in Handling and Storage

Titanium tetrachloride absorbs moisture from air and produces irritating fumes in moist air. Store in tightly closed containers in a cool, dark, well-ventilated area away heat.

Other Precautions and Warnings

Bulk storage of titanium tetrachloride is not recommended. Prevent physical damage to containers.

SECTION VIII - CONTROL MEASURES AND PERSONAL PROTECTIVE EQUIPMENT

Respiratory Protection *(specify type)*

No exposure levels established. This material is poisonous and incapacitating. At any exposure level, use a supplied-air respirator with full facepiece operated in positive pressure mode or a self-contained breathing apparatus (SCBA) with full facepiece in pressure demand or other positive pressure mode.

Ventilation

Local exhaust or general mechanical systems recommended.

Protective Gloves	Eye Protection	Other Protective Clothing
Impervious Material	Chemical Goggles and/or Face Mask	Protective Apron

Work/Hygiene Practices

Always wash hands thoroughly after using chemical; never bring food, drink, or smoking materials into vicinity of chemicals.

TITANIUM TETRACHLORIDE

Cl_4Ti

CAS: 7550-45-0

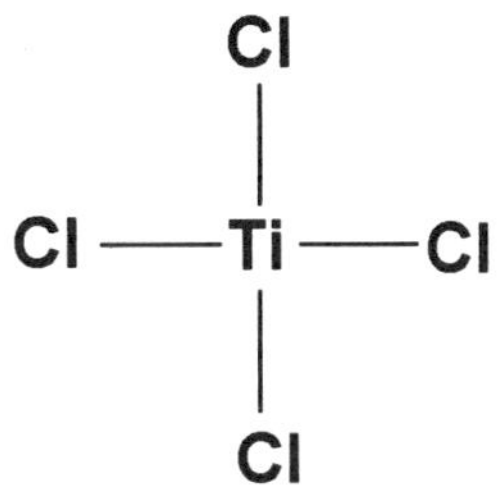

IDENTIFICATION AND TYPICAL USES

Titanium tetrachloride is a colorless to light yellow, fuming liquid with a pungent, acid-like odor. It fumes best in moist air. It is used to manufacture mordant dye, titanium pigments, and titanium salts. It is also a chemical intermediate for titanium dioxide and titanium metal. It is an agent in smoke screens, an iridescent agent in pearl and glass manufacturing, and a polymerization catalyst.

RISK ASSESSMENT: HEALTH

General Assessment

Titanium tetrachloride is primarily toxic and poisonous to humans on *inhalation* of its vapors and mists or by *ingestion* of the liquid. Direct skin contact with the liquid can also cause serious topical health effects, but dermal absorption is not reported. There are no reports of any carcinogenic, mutagenic, or teratogenic effects on long-term exposure to titanium tetrachloride.

Inhalation will irritate the eyes, nose, throat, and air passages of the upper respiratory tract. There can be coughing with the production of phlegm, sneezing, headaches, dizziness, and possible bronchial pneumonia. Repeated exposures can cause chemical bronchitis and may cause emphysema. High exposures can cause pulmonary edema (fluid in the lungs) which can be fatal. Symptoms of pulmonary edema can be delayed up to 48 hours, thereby creating a false sense of security with regard to health exposure risk.

Skin or eye contact with the liquid may cause 2nd degree burns if not removed immediately. Titanium tetrachloride is an extremely corrosive and destructive irritant. Ingestion causes burns to the mouth, throat, and stomach with nausea, vomiting, cramps, diarrhea, and possible tissue ulceration.

Acute Health Effects

The following acute (short-term) health effects may occur immediately or shortly after exposure to titanium tetrachloride:

Skin: Irritation and burns on contact. Second degree burns can occur if not immediately removed.

Eye: Irritation with conjunctivitis, keratin, and corneal clouding possible.

Lung: Irritation of the respiratory tract, possible lung damage and death due pulmonary edema.

Other: Ingestion causes burns to the mouth, throat, and stomach with a possibility for ulcerative hemorrhage, nausea, vomiting, and cramps.

Chronic Health Effects

The following chronic (long-term) health effects may occur at some time after exposure to titanium tetrachloride and can last for months or even years:

Cancer Hazards: According to the information presented in the references, titanium tetrachloride has not been adequately tested for its ability to cause cancer in test animals.

Reproductive Hazards: According to information presented in the references, titanium tetrachloride has not been adequately tested for its ability to adversely affect reproduction.

Other Chronic Effects: Repeated exposures may result in chronic chemical bronchitis and may lead to emphysema with symptoms of cough and shortness of breath.

Recommended Risk-Reduction Measures

Personnel should avoid direct contact with titanium tetrachloride. It is an extremely poisonous and incapacitating compound. If a less toxic chemical cannot be substituted for a hazardous substance, then *engineering controls* are the most effective method of reducing exposures. The best protection is to enclose operations and/or provide local exhaust ventilation at the site of chemical release. While not always operationally feasible, isolating operations can also reduce exposure. Using respiratory protection is less effective than the controls mentioned above, but is still advisable whenever working with titanium tetrachloride. No exposure levels have been established for titanium tetrachloride. However, this does not mean to imply that exposure is without health risks. For the best protection, a self-contained breathing apparatus (SCBA) with full facepiece in pressure demand mode or a supplied-air respirator with full facepiece in continuous flow mode are the recommended respiratory protection methods of choice. If a full facepiece is not available, then chemical goggles should be worn to protect the eyes. Whenever a chemical splash hazard exists, a face shield and a rubber apron should be worn. To prevent hand and skin exposures, impervious gloves should be used. Glove manufacturers should be contacted and permeation studies obtained *before* final glove selections are made.

Administrative controls should also be in place to minimize the potential for human exposures. These may include written procedures or policies which specify the methods and techniques that will be practiced whenever personnel are to work with titanium tetrachloride.

All personnel should receive *training* on the use, hazards, protective measures, emergency actions, and other precautions per 29 CFR 1910.1200 (Hazard Communication), prior to the first assignment in an area where titanium tetrachloride is used or stored.

Before beginning employment and at regular times thereafter, the following medical tests are recommended:

☑ Liver function test.

If symptoms develop or overexposure is suspected, the following additional test is recommended:

☑ Consider chest X-ray after acute overexposure (may be negative if taken immediately after exposure due to delayed onset of pulmonary edema).

Any medical evaluation should include a careful history of past and present symptoms with an examination. Medical tests that look for damage already present are not a substitute for controlling exposures. Also, because smoking can cause heart disease, emphysema, lung cancer, and other serious respiratory and cardiovascular diseases, smokers exposed to titanium tetrachloride may experience symptoms more rapidly and more pronounced than non-smokers under the same conditions of exposure. Prudent risk management requires proper consideration of *all* possible factors which may be responsible for exposure symptoms in the workplace.

Other methods to reduce exposure include:

☑ Where possible, enclose operations and use local exhaust ventilation at the site of chemical release. If local exhaust ventilation or enclosure is not used, respiratory protection should be mandatory.

☑ Always ensure that proper protective clothing is worn when using chemical substances.

☑ Emergency eyewash and shower stations should be located in areas where exposure to titanium tetrachloride is possible.

☑ Wash thoroughly immediately after exposure to titanium tetrachloride and at the end of the work shift or before eating, drinking, or smoking. Never wear clothes contaminated with titanium tetrachloride home. Family members can be exposed.

☑ Hazard warning information should be posted in the work area. As part of an on-going education and training program, all information on the health and safety hazards of titanium tetrachloride should be communicated to all exposed workers.

RISK ASSESSMENT: ENVIRONMENT

General Assessment

The environment is at risk of exposure during transportation, storage, disposal, or destruction of titanium tetrachloride. In almost every scenario, the threat of environmental exposure is contingent upon the proper handling of the chemical substance. Accidental spills, large or small, can result in fire, explosion, and possible contamination of the surrounding environmental mediums (water, soil, and air).

Titanium tetrachloride is considered a combustible liquid. It may burn but does not easily ignite. However, it can react strongly with water or moisture to release extreme heat and hydrochloric acid. The acid is highly corrosive to metals, yielding flammable hydrogen gas which can ignite explosively. Potentially explosive reactions can occur if titanium tetrachloride is mixed or comes into contact with fluorine, chlorine, bromine, or other strong oxidizers. These characteristics require special consideration during any emergency involving a leak or spill of titanium tetrachloride.

Titanium tetrachloride can enter the environment through manufacturing, unchecked discharge into effluents from industrial or municipal waste treatment plants, and through spills.

Acute Ecological Effects

Acute (short-term) toxic effects may include the death of animals, birds, or fish, and death or low growth rate in plants. Acute effects are seen 2 to 4 days after animals or plants are exposed to titanium tetrachloride.

Insufficient data are available to evaluate to predict the short-term (acute) effects of titanium tetrachloride to aquatic life, plants, birds, or land animals.

Chronic Ecological Effects

Chronic (long-term) toxic effects may include shortened life span, reproductive problems, lower fertility, and changes in appearance or behavior in exposed animals. These effects can be seen long after first exposure(s) to toxic chemicals.

Insufficient data are available to evaluate to predict the long-term (chronic) effects of titanium tetrachloride to aquatic life, plants, birds, or land animals.

Water Solubility

Titanium tetrachloride is highly soluble in water. Concentrations of 1000 milligrams and more will mix with a liter of water.

Persistence in the Environment

Titanium tetrachloride reacts with water to form titanium oxychlorides and titanium dioxide. Subsequently, titanium tetrachloride will not be highly persistent in water. However some of the titanium compounds that are formed may be highly persistent.

Bioaccumulation in Aquatic Organisms

Some substances increase in concentration, or *bioaccumulate,* in living organisms as they breathe contaminated air, drink contaminated water, or eat contaminated food. These chemicals can become concentrated in the tissues and internal organs of animals as well as humans.

Titanium tetrachloride will most likely NOT accumulate in aquatic organisms or aquatic ecosystems, but titanium might accumulate in both.

Recommended Risk-Reduction Measures

Proper training of all transporters will reduce the likelihood of a mishap or accident resulting in a leak or spill to the environment. The correct use of DOT labeling (per the MSDS) on all containers, trucks, and rail cars will enable emergency responders to react properly and quickly to any disaster thereby reducing the potential risk to the environment and to personnel.

Storage of titanium tetrachloride should be segregated from incompatible chemicals to minimize the risk of cross-contamination or contact. Titanium tetrachloride should be kept in air-tight containers to ensure a moisture-free environment.

If a spill or leak to the environment has occurred, fire department, emergency response, and/or hazardous materials spill personnel should be notified immediately. Cleanup should be attempted only by those trained in proper spill containment procedures. Contaminated soils should be removed for incineration and replaced with clean soil. If titanium tetrachloride should contact the water table, aquifer, or navigable waterway, time is of the essence. It will react in water and decompose to form other compounds and, therefore, total containment and remediation may not be possible. When such spills occur, the local and/or state emergency response authorities must be notified. A comprehensive emergency response or disaster preparedness/recovery plan should be in place prior to any operations involving the use, transportation, storage, or disposal of titanium tetrachloride.

If titanium tetrachloride is spilled or leaked, the following specific steps are recommended:

- ☑ Restrict persons not wearing protective clothing from area of spill or leak until cleanup is complete and area can be opened for normal work.
- ☑ Ventilate area of spill or leak. Remove all sources of ignition.
- ☑ Absorb liquids in vermiculite, dry sand, earth, or a similar material and deposit in sealed containers. Use non-sparking tools and equipment during cleanup operations.
- ☑ It may be necessary to dispose of titanium tetrachloride as a hazardous waste. The responsible state agency or the regional office of the federal

Environmental Protection Agency (EPA) should be contacted for specific recommendations.

RISK ASSESSMENT: BUSINESS

General Assessment

Accidents or mishaps involving titanium tetrachloride can present a significant threat to business operations. The loss or damage of equipment or facilities can significantly affect fiscal viability. Lawsuits that may result from personnel injury, public exposures, and/or environmental contamination will also require a serious expenditure of resources.

Media attention surrounding an injury, illness, or environmental damage can also result in a loss of profits and loss of current as well as future business.

Recommended Risk-Reduction Measures

Company attorneys, safety and health professionals, and environmental specialists should be involved in the development of any procedures dealing with safety or emergency response. A company official should be pre-designated as a public relations officer with specific training in dealing with the press. Corporate plans and policies should be developed, approved, and implemented long before any need for such arises.

REFERENCES

Ahlbom, A. 1993. *Biostatistics for Epidemiologists.* New York: CRC Press/Lewis Publishers.

American Conference of Governmental Industrial Hygienists. 1988. *Documentation of the Threshold Limit Values and Biological Exposure Indices,* 5th Edition (with updates). Cincinnati: ACGIH

Anthony, C. P., and N. J. Kolthoff. 1971. *Textbook of Anatomy and Physiology.* St. Louis: C. V. Mosby Company.

Asante-Duah, D. K. 1993. *Hazardous Waste Risk Assessment.* New York: CRC Press/Lewis Publishers.

Bloom, A. D. and F. J. DeSerres. 1995. *Ecotoxicity and Human Health.* New York: CRC Press/Lewis Publishers.

Brusick, D. J. 1994. *Methods for Genetic Risk Assessment.* New York: CRC Press/Lewis Publishers.

Bueche, J. R. 1972. *Principles of Physics,* 2nd Edition. New York: McGraw-Hill.

Burrell, R., D. K. Flaherty, and L. J. Sauers. 1992. *Toxicology of the Immune System.* New York: Van Nostrand Reinhold.

Calabrese, E. J., E. M. Kenyon. *Air Toxics and Risk Assessment.* New York: CRC Press/Lewis Publishers.

Calabrese, E. J. 1994. *Biological Effects of Low Level Exposures.* New York: CRC Press/Lewis Publishers.

Cothern, R. R. 1993. *Comparative Environmental Risk Assessment.* New York: CRC Press/Lewis Publishers.

Gardner, R. W. 1994. *Chemical Intolerance.* New York: CRC Press/Lewis Publishers.

Hallenbeck, W. H. 1993. *Quantitative Risk Assessment for Environmental and Occupational Health,* 2nd Edition. New York: CRC Press/Lewis Publishers.

Hodgson, E., R. B. Mailman, and J. E. Chambers. 1988. *Dictionary of Toxicology.* London: The MacMillian Company.

Howard, P. H., et. al. 1991. *Handbook of Environmental Degradation Rates.* New York: CRC Press/Lewis Publishers.

Howard, P. H. 1995. *Handbook of Environmental Fate and Exposures Data for Organic Chemicals,* Vol. I-IV. New York: CRC Press/Lewis Publishers.

Jones, F. E. 1994. *Toxic Organic Vapors in the Workplace.* New York: CRC Press/Lewis Publishers.

Kamrin, M. A. 1988. *Toxicology: A Primer on Principles and Applications.* Chelsea, Michigan: Lewis Publishers.

Langerman, N. 1994 *Precautionary Labels for Chemical Containers.* New York: CRC Press/Lewis Publishers.

Lauwerys, R. R., P. Hoet. 1993. *Industrial Chemical Exposure.* New York: CRC Press/Lewis Publishers.

Lewis, R. J., Sr. 1992. *Sax's Dangerous Properties of Industrial Materials,* Volumes 1, 2, and 3, 8th Edition. New York: Van Nostrand Reinhold.

Lide, D. R. 1995. *Handbook of Chemistry and Physics,* 76th Edition. New York: CRC Press/Lewis Publishers.

Manahan, S. E. 1992. *Toxicological Chemistry,* 2nd Edition. New York: CRC Press/Lewis Publishers.

Mellström, G. A., J. E. Wahlberg, and H. I. Maibach. 1994. *Protective Gloves for Occupational Use.* New York: CRC Press/Lewis Publishers.

Meyer, E. 1990. *Chemistry of Hazardous Materials.* Englewood Cliffs, NJ: Prentice Hall.

Patnaik, P. 1992. *A Comprehensive Guide to the Hazardous Properties of Chemical Substances.* New York: Van Nostrand Reinhold.

Proctor, N. H., and J. P. Hughes. 1991. *Chemical Hazards of the Workplace.* New York: Van Nostrand Reinhold.

Quigley, D. R. 1994. *Handbook of Emergency Chemical Management.* New York: CRC Press/Lewis Publishers.

Rea, W. J. 1995. *Chemical Sensitivity,* Vol. I-III. New York: CRC Press/Lewis Publishers.

Richardson, M. 1995. *Dictionary of Substance and Their Effects,* Vol. I-VI. New York: CRC Press/Lewis Publishers.

U.S. Department of Health and Human Services, National Institute for Occupational Safety and Health. 1994. *NIOSH Pocket Guide to Chemical Hazards.* Washington, D.C.: U.S. Government Printing Office.

MATERIAL SAFETY DATA SHEET

CHEMICAL NAME	CRC PRESS/LEWIS PUBLISHERS DISCLAIMER:
TOLUENE	THE INFORMATION AND RECOMMENDATIONS PRESENTED HEREIN ARE BASED ON SOURCES BELIEVED TO BE RELIABLE. CRC MAKES NO REPRESENTATION ON ITS COMPLETENESS OR ACCURACY. IT IS THE RESPONSIBILITY OF THE USER TO DETERMINE THE CHEMICAL'S SUITABILITY FOR ITS INTENDED USE, THE CHEMICAL'S SAFE USE, AND THE CHEMICAL'S PROPER DISPOSAL. NO REPRESENTATIONS AND/OR WARRANTIES, EITHER EXPRESSED OR IMPLIED, OF THE MERCHANTABILITY OR FITNESS FOR A PARTICULAR PURPOSE, OR OF ANY OTHER NATURE, ARE MADE WITH RESPECT TO THE INFORMATION PROVIDED IN THIS MSDS OR TO THE CHEMICAL TO WHICH INFORMATION MAY REFER. CRC NEITHER ASSUMES NOR AUTHORIZES ANY OTHER PERSON TO ASSUME FOR IT, ANY OTHER ADDITIONAL RESPONSIBILITY OR LIABILITY FOR THE USE OF, OR RELIANCE UPON, THIS INFORMATION.

HAZARD WARNING INFORMATION

HEALTH	FIRE	REACTIVE	OTHER	DEGREE OF HAZARD	COLOR CODING	OTHER CODES
2	3	0		0 = Minimum Hazard 1 = Slight Hazard 2 = Moderate Hazard 3 = Serious Hazard 4 = Severe Hazard	HEALTH = BLUE FIRE = RED REACTIVITY = YELLOW OTHER = WHITE	OX = Oxidizer ACID = Acid ALK = Alkali COR = Corrosive W = Use No Water

SECTION I - GENERAL INFORMATION

Characterization	RCRA Number	EPA Class
Aromatic Hydrocarbon	**U220**	**Toxic Waste**

DOT Proper Shipping Name **Toluene or Toluol**	Chemical Abstract Service (CAS) Number **108-88-3**
DOT Hazard Class and Label Requirements **Flammable Liquid**	DOT Emergency Guide Code **27**
DOT Identification Number **UN 1294**	Molecular Formula $C_6H_5CH_3$

Synonyms

Methyl benzene; methyl benzol; phenyl methane; toluol; methacide; Tolu-sol.

SECTION II - HAZARDOUS INGREDIENTS AND IDENTITY INFORMATION

Hazardous Components (specific identity)	OSHA Exposure Criteria	NIOSH Exposure Criteria	Immediately Dangerous to Life and Health (IDLH)	ACGIH Exposure Criteria
Toluene (derivation: By catalytic reforming of petroleum; by fractional distillation of coal-tar light oil and purification by rectification). **1 ppm = 3.83 mg/m^3**	**PEL (8-hour):** **200 ppm** **766 mg/m^3** **CEILING:** **300 ppm** **PEAK (10-min):** **500 ppm**	**REL:** **100 ppm** **375 mg/m^3** **STEL:** **150 ppm** **560 mg/m^3**	**500 ppm**	**TLV (skin):** **50 ppm** **188 mg/m^3** **STEL:** **Not Established**

SECTION III - PHYSICAL AND CHEMICAL CHARACTERISTICS

Boiling Point **232°F (111°C)**	Specific Gravity (H_2O = 1) **0.87**
Vapor Pressure (mm Hg) **216 at 69°F (20°C)**	Molecular Weight (atomic weight) **92.1**
Vapor Density (Air = 1) **3.14**	Melting Point **-139°F (-95°C)**

Solubility

Slightly soluble in water (0.07%). Soluble in acetone, alcohol, ether, benzene, chloroform, glacial acetic acid, petroleum ether, and carbon disulfide.

Appearance and Odor

Colorless liquid with a sweet, pungent, aromatic, benzene-like odor. Odor Threshold = <1 to 70 ppm.

SECTION IV - FIRE AND EXPLOSION HAZARD DATA

Flash Point (method used) **40°F (4.4°C) closed cup**	Explosive Limits in Air % by Volume **LEL: 1.1% UEL: 7.1%**
NFPA Classification **Class 1B Flammable Liquid**	Autoignition Temperature **896°F (480°C)**

Extinguishing Media

Use alcohol-resistant foam, dry chemical, or carbon dioxide (water may be ineffective).

Special Fire Fighting Procedures

Wear self-contained breathing apparatus (SCBA). Poisonous gases are produced in fire. Use water spray to cool fire-exposed containers. Fight fire from distance if possible.

Unusual Fire and Explosion Hazards

Moderate explosion hazard when exposed to heat or flame. Containers may explode in fire. The vapor is heavier than air and may travel for some distance to flashback. A boiling liquid expanding vapor explosion (BLEVE) if possible. Evacuate 1/2 mile if rising sound is heard coming from venting device.

SECTION V - REACTIVITY DATA

Stability		Conditions to Avoid
		Toluene is normally stable in closed containers under routine conditions of handling and storage. Keep away from heat, flame, other sources of ignition, and contact with incompatible materials and chemicals.
Stable	Unstable	Incompatibility *(materials to avoid)*
X		Strong oxidizers, concentrated nitric acid, nitric acid + sulfuric acid, dinitrogen tetroxide, silver perchlorate, bromine trifluoride, tetranitromethane, and 1,3-dichloro-5,5-dimethyl-2,4-imidazolididione.
Hazardous Polymerization		Conditions to Avoid Under normal conditions of temperature and pressure, hazardous polymerization of toluene cannot occur.
May Occur	Will Not Occur	Hazardous Decomposition or By-products
	X	Poisonous gases, including carbon dioxide and acrid, irritating smoke are emitted during fire.

SECTION VI - HEALTH HAZARD DATA

Primary Route(s) of Entry:	Inhalation?	Absorption (skin/eye)?	Ingestion?
	X	X	X

Health Hazards

INHALATION: Symptoms of irritation to the eyes, nose, throat, and respiratory tract with fatigue, weakness, confusion, dizziness, headaches, dilated pupils, lachrymation, nervousness, insomnia, parasthesia, vertigo (dizziness), euphoria, narcoses, coma, and death (cardiac arrest).

ABSORPTION: Irritation with possible redness and pain to the skin and eyes. Also, absorbs through the skin and may cause dermatitis. Eye contact may result in temporary corneal damage.

INGESTION: May cause nausea, vomiting, headache, dizziness, gastrointestinal irritation; blurred vision, lowering of blood pressure, liver, kidney, and bone marrow effects (by all routes), and many of the same narcotic effects noted for inhalation.

Carcinogenicity	NTP Listed?	IARC Cancer Review Group?	OSHA Regulated?	Target Organs?
Unknown Human Unknown Animal	No	No	29 CFR 1910.1000 Table Z-2	Eyes, skin, CNS, resp., sys. marrow, liver, kidney.

Medical Conditions Generally Aggravated by Exposure

Alcoholism, CNS, liver, kidney, and/or skin diseases may be aggravated by exposure.

Emergency and First-aid Procedures

Eye contact: Do not allow victim to rub or keep eyes tightly shut. Flush immediately with water for 15 minutes (minimum); seek medical attention. Skin contact: Remove all contaminated clothing, including shoes. Immediately wash area with flooding amounts of soap and water. Seek medical attention immediately. For inhalation: Remove the person from exposure. Provide respiratory assistance and CPR. Transfer to medical facility. If swallowed: Seek medical attention immediately. Give conscious person 1 to 2 glasses of water to dilute. Do not induce vomiting (aspiration hazard).

SECTION VII - PRECAUTIONS FOR SAFE HANDLING AND USE

Steps to be Taken in Case Material is Released or Spilled

Remove all ignition sources. Wear SCBA respiratory protection. Restrict those involved in cleanup from entering area. Avoid skin contact. Stop leak if it can be done without risk. Ventilate area of spill. Absorb with vermiculite or other material and deposit in sealed containers. Use non-sparking tools.

Preferred Waste Disposal Method

Burn in chemical incinerator equipped with an afterburner and scrubber.

Precautions to be Taken in Handling and Storage

Store in tightly closed containers in cool, well-ventilated location. Keep away from sources of heat.

Other Precautions and Warnings

Sources of ignition such as open flame or smoking, is prohibited where toluene is used or stored. Metal containers should be bonded and grounded during transfer operations. Bulk storage not recommended.

SECTION VIII - CONTROL MEASURES AND PERSONAL PROTECTIVE EQUIPMENT

Respiratory Protection *(specify type)*

Use an MSHA/NIOSH-approved powered air purifying respirator or use a self-contained breathing apparatus (SCBA) with full facepiece operated in positive pressure mode.

Ventilation

Local exhaust (preferred) at site of chemical work or general ventilation.

Protective Gloves	Eye Protection	Other Protective Clothing
Polyvinyl Alcohol Gloves	Chemical Goggles or Face Mask	Protective Apron if Splash is Likely

Work/Hygiene Practices

Always wash hands thoroughly after using chemical; never bring food, drink, or smoking materials into vicinity of chemicals.

TOLUENE

$C_6H_5CH_3$

CAS: 108-88-3

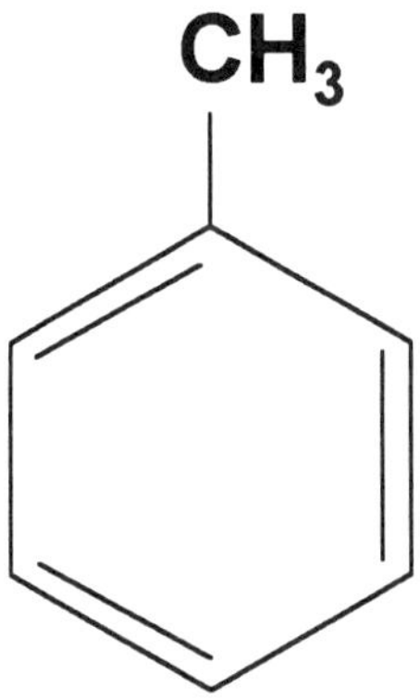

IDENTIFICATION AND TYPICAL USES

Toluene is a clear, colorless liquid with a sweet, pungent, aromatic odor like that of benzene. It has many uses, including: a solvent for oils, resins, adhesives, natural rubber, coal tar, asphalt, pitch, acetyl cellulose, cellulose paints, and varnishes; as an additive in aviation gasoline and as a high-octane blending stock; in the production of benzene, phenol, and caprolactum; as an adhesive solvent in plastic toys and model airplanes; in other chemicals (benzoic acid, benzyl and benzoyl derivatives, saccharin, medicines, dyes, and perfumes); as a source of toluene diisocyanates (polyurethane resins); in explosives (TNT); a base in toluene sulfonates (detergents); and in scintillation counters.

RISK ASSESSMENT: HEALTH

General Assessment

Toluene is an acute as well as chronic toxicant. While there are no reports on its carcinogenic potential, mutation data have been reported and it is also an experimental teratogen. The target organs of acute and chronic poisoning are the blood, bone marrow, central nervous system, respiratory system, eyes, and skin.

Human systemic effects by *inhalation, absorption,* and *ingestion* include blood changes and increased body temperature. Heavy occupational exposures to toluene may cause bone marrow depression with anemia and leukopenia. However, some experts argue that bone marrow effects may be due to benzene contamination rather than from the toluene exposure. Approximately 93% of inhaled toluene is retained in the body, with 80% of this amount metabolized to first to benzoic acid and then to hippuric acid which is excreted in urine. The remainder is metabolized to *o*-cresol and excreted or exhaled unchanged. Toluene metabolism is inhibited by the ingestion of alcohol and is chemically similar to benzene, asphalt fumes, or chlorinated hydrocarbons in terms of systemic toxicity. Toluene is retained longer in obese persons than in slender persons, presumably due to its high lipid-soluble nature.

Other effects of inhalation include headaches, nausea, vomiting, dizziness, narcosis, suffocation, CNS depression, possible lung inflammation, blurring of vision, convulsions, weakness, loss of appetite, abdominal discomfort, and dyspnea on exertion. Death can occur due to cardiac arrest resulting from ventricular fibrillation with catecholamines loss.

Skin contact may cause dermatitis and absorption can occur with harmful effects. It is a comparatively strong irritant, producing erythema, and burning, and, in severe cases, edema and even blistering.

Ingestion may cause nausea, vomiting, headaches, dizziness, gastrointestinal irritation, blurred vision, and lowering of blood pressure.

Acute Health Effects

The following acute (short-term) health effects may occur immediately or shortly after exposure to toluene:

Skin: Irritation with potential for redness, blistering, and burning. Absorption is quick to occur.

Eye: Severe irritation; pain and possible permanent damage possible.

Lung: Nose, throat, and respiratory tract irritation causing difficulty in breathing, coughing, dypsnea, and congestion. Rapid absorption through the respiration process leads to depression of the CNS.

CNS: Confusion, dizziness, nervousness, insomnia, parasthesia (tightening of the leg muscles), cranial pressure (headaches), dilated pupils, fatigue, possible narcotic coma.

Chronic Health Effects

The following chronic (long-term) health effects may occur at some time after exposure to toluene and can last for months or even years:

Cancer Hazards: According to information presented in the references, it is not clear whether exposure to toluene causes cancer in test animals. However, toluene has caused mutations (genetic changes) in test animals and many scientists believe that exposure to such chemicals may pose a cancer risk in the long-term. It is not known if toluene will have this effect in humans.

Reproductive Hazard: Toluene is a confirmed animal teratogen causing adverse developmental effects the fetus. These may include microcephaly, CNS dysfunction, attention deficits, developmental delay, language impairment, growth retardation, and physical defects including a small midface, short palpebral fissures, with deep-set eyes, low-set ears, flat nasal bridge with a small nose, micronathia, and blunt fingertips.

Other Chronic Effects: Heavy exposures over a long period of time may cause bone marrow depression with symptoms of leukopenia and anemia with a tendency for hemorrhagic symptoms such as easy bruising, epistaxis, bleeding from the gums, nosebleeds, menorrhagia, petachiae, purpora, and other symptoms of fatigue, headache, dizziness, nausea and loss of appetite, loss of weight, weakness, and pallor. Repeated exposures can cause irreversible damage to the blood-forming organs resulting in aplastic anemia and possibly leukemia. There is great variation in individual response to chronic exposure to toluene. There is some evidence to indicate that toluene causes an autoimmune deficiency producing antibodies that cause inflammation of the kidney. Repeated exposures may also lead to liver injury and brain damage.

Recommended Risk-Reduction Measures

Personnel should avoid direct contact with toluene. It is extremely toxic by all routes of exposure. Occupational poisoning occurs most commonly by inhalation and dermal absorption. If a less toxic material or compound cannot be substituted for toluene, then *engineering controls* are the most effective method of reducing exposures. The best protection is to enclose operations and/or provide local exhaust ventilation at the site of toluene release. While not always operationally feasible, isolating operations can also reduce exposure. Using respiratory protection is less effective than the controls mentioned above, but is still advisable whenever working with or around toluene. For low or transient exposures above the PEL, an organic vapor respirator equipped with a mist pre-filter may provide adequate protection. For high exposures, or when the exposure level is not know (as in an emergency) an MSHA/NIOSH-approved self-contained breathing apparatus (SCBA) with full facepiece operated in pressure demand mode or a powered air-supplied respirator operated in continuous flow mode provide the appropriate level of respiratory protection. If a full facepiece is not available, then chemical goggles should be worn to protect the eyes. A face shield should also be considered. To prevent hand and skin exposures, polyvinyl alcohol gloves should be used. Glove manufacturers should be contacted and permeation studies obtained, *before* final glove selections are made.

Administrative controls should also be in place to minimize the potential for human exposures. These may include written procedures or policies which specify the methods and techniques that will be practiced whenever personnel are to work with toluene.

All personnel should receive *training* on the use, hazards, protective measures, emergency actions, and other precautions per 29 CFR 1910.1200 (Hazard Communication), prior to the first assignment in an area where toluene is used or stored.

Before beginning employment and at regular times thereafter, the following recommended medical tests should be considered:

☑ Urinary hippuric acid excretion (a test to see if toluene is in the body) at the end of the work shift.

If symptoms develop or overexposure is suspected, the following additional tests may be useful:

☑ Examination of the nervous system.

☑ Liver and kidney function tests with evaluation for renal tubular acidosis.

☑ Complete blood count (CBC)

Any evaluation should include a careful history of past and present symptoms with an examination. Medical tests that simply look for existing damage are not a substitute for controlling exposures. Also, since more that casual intake of alcohol lead to liver dysfunction and other illnesses and, since toluene metabolism is inhibited by the presence of alcohol, those who consume alcoholic beverages either prior to or immediately after exposure to toluene may experience symptoms more quickly and more pronounced than those who have not under the same conditions of exposure. Prudent risk management requires proper investigation and consideration of *all* possible factors which may affect exposure assessment.

Other methods to reduce exposure include:

- ☑ Where possible, enclose operations and use local exhaust ventilation at the site of chemical release. If local exhaust ventilation or enclosure is not used, respiratory protection should be mandatory.
- ☑ Always ensure that proper protective clothing is worn when using chemical substances.
- ☑ Wash thoroughly immediately after exposure to toluene and at the end of the work shift or before eating, drinking, or smoking.
- ☑ Hazard warning information should be posted in the work area. In addition, as part of an on-going education and training program, all information on the health and safety hazards of toluene should be communicated to all potentially exposed workers.
- ☑ Eye wash stations should be provided in the immediate work area for emergency use. If there is a possibility of skin exposure to toluene, emergency shower facilities should also be provided.
- ☑ Workers whose clothing has been contaminated by toluene should change into clean clothes before leaving work. Contaminated work clothing should be laundered only by individuals who have been informed of the hazards of exposure to toluene.
- ☑ Where possible, automatically transfer liquids containing toluene from drums or other containers to process containers.

RISK ASSESSMENT: ENVIRONMENT

General Assessment

The environment is at risk of exposure during transportation, storage, disposal, or destruction of toluene. In almost every scenario, the threat of environmental exposure is contingent upon the proper handling of the chemical substance. Accidental spills, large or small, can result in contamination of the surrounding environmental mediums (water, soil, and air).

Toluene is a moderately flammable liquid. Because of its low flash point and high boiling point, it is classified as a Class 1B flammable liquid (per OSHA 29 CFR 1910.106). It is a dangerous fire hazard and moderate explosion hazard when exposed to heat or flame. Its vapors form explosive mixtures in air. Because these vapors are heavier than air, they can travel for great distance, collect in low-lying areas, and possibly contact an ignition source to flashback and cause fire or explosion. Caution is always required in handling, storage, transportation, and disposal of toluene. Emergency responders should be made aware of the presence of toluene at any emergency response situation.

Toluene can enter the environment from industrial effluents and from spills.

☠ *Acute Ecological Effects*

Acute (short-term) toxic effects may include the death of animals, birds, or fish and death or low growth rate in plants. Acute effects are seen 2 to 4 days after animals or plants are exposed to toluene.

Toluene has moderate acute toxicity to aquatic life. Insufficient data are available to evaluate the short-term effects of toluene to aquatic life, plants, birds, or land animals.

Chronic Ecological Effects

Chronic (long-term) toxic effects may include shortened life span, reproductive problems, lower fertility, and changes in appearance or behavior in exposed animals. These effects can be seen long after first exposure(s) to toxic chemicals.

Toluene has moderate chronic toxicity to aquatic life. Insufficient data are available to evaluate or predict the long-term effects of toluene aquatic life, plants, birds, or terrestrial animals.

Water Solubility

Toluene is slightly soluble in water. Concentrations of 1 milligram will mix with a liter of water.

⌛ *Persistence in the Environment*

Toluene is non-persistent in water, with a half-life of less than 2 days. The half-life of a pollutant is the amount of time it takes for one half of the chemical to be degraded. About 99.5% of toluene will eventually end up in air; the rest will end up in water.

Bioaccumulation in Aquatic Organisms

Some substances increase in concentration, or *bioaccumulate,* in living organisms as they breathe contaminated air, drink contaminated water, or eat contaminated food. These chemicals can become concentrated in the tissues and internal organs of animals as well as humans.

The concentration of toluene found in fish tissues is expected to be somewhat higher than the average concentration of toluene in the water from which the fish was taken.

Recommended Risk-Reduction Measures

Proper training of all transporters will reduce the likelihood of a mishap or accident resulting in a leak or spill of toluene into the environment. Labels on all containers, trucks, and rail cars must meet DOT requirements and accurately reflect their contents to enable emergency responders to react properly and quickly to any disaster thereby reducing the potential risk to the environment and to personnel.

Storage of toluene should be segregated from other chemicals and materials to minimize the risk of cross-contamination. Explosion-proof ventilation is recommended in storage and handling areas. Containers should be protected from physical damage.

If a spill or leak to the environment has occurred, fire department, emergency response and/or hazardous materials spill personnel should be notified immediately. Cleanup should be attempted only by those trained in proper spill containment procedures using non-sparking tools. Contaminated soils should be removed for incineration and replaced with clean soil. If toluene should contact the water table, aquifer, or navigable waterway, time is of the essence. It is slightly soluble and total remediation may not be possible. The local and/or state emergency response authorities must be notified. A comprehensive emergency response or disaster preparedness/recovery plan should be in place prior to any operations involving the use, transportation, storage, or disposal of toluene.

For disposal, toluene may be burned in an approved chemical incinerator equipped with an afterburner and scrubber.

If toluene is spilled or leaked, the following specific steps are recommended:

- ☑ Restrict persons not wearing protective clothing from area of spill or leak until cleanup is complete and area can be opened for normal work.
- ☑ Absorb liquids in vermiculite, sand, earth, or similar material and deposit in sealed drum using non-sparking tools.
- ☑ Ventilate area and remove ignition sources.
- ☑ It may be necessary to dispose of toluene as a hazardous waste. The responsible state agency or the regional office of the federal Environmental Protection Agency (EPA) should be contacted for specific recommendations.

RISK ASSESSMENT: BUSINESS

General Assessment

Accidents or mishaps involving toluene can present a serious threat to business operations. The loss or damage of equipment or facilities can significantly affect fiscal viability. Lawsuits that may result from personnel illness, injury/death, public exposures, and/or environmental contamination will require a serious expenditure of resources. Media attention surrounding an injury, death, or environmental damage can also result in a loss of profits and loss of current as well as future business.

Recommended Risk-Reduction Measures

Company attorneys, safety and health professionals, and environmental specialists should be involved in the development of any procedures for responding to chemical incidents. A company official should be predesignated as a public relations officer with specific training in dealing with the press. Corporate plans and policies should be developed, approved, and implemented long before any need for such arises.

REFERENCES

Lewis, R. J., Sr. 1993. *Hawley's Condensed Chemical Dictionary,* 12th Edition. New York: Van Nostrand Reinhold.

Lewis, R. J., Sr. 1992. *Sax's Dangerous Properties of Industrial Materials,* Volumes 1, 2, and 3, 8th Edition. New York: Van Nostrand Reinhold.

Patnaik, P. 1992. *A Comprehensive Guide to the Hazardous Properties of Chemical Substances.* New York: Van Nostrand Reinhold.

U.S. Department of Health and Human Services, National Institute for Occupational Safety and Health. 1994. *NIOSH Pocket Guide to Chemical Hazards.* Washington, D.C.: U.S. Government Printing Office.

MATERIAL SAFETY DATA SHEET

CHEMICAL NAME	CRC PRESS/LEWIS PUBLISHERS DISCLAIMER
TOLUENE-2,4-DIISOCYANATE	

HAZARD WARNING INFORMATION

HEALTH	FIRE	REACTIVE	OTHER	DEGREE OF HAZARD	COLOR CODING	OTHER CODES
3	1	1		0 = Minimum Hazard 1 = Slight Hazard 2 = Moderate Hazard 3 = Serious Hazard 4 = Severe Hazard	HEALTH = BLUE FIRE = RED REACTIVITY = YELLOW OTHER = WHITE	OX = Oxidizer ACID = Acid ALK = Alkali COR = Corrosive W = Use No Water

SECTION I - GENERAL INFORMATION

Characterization	**Isocyanate, Organic**	RCRA Number	**U223**
		EPA Class	**Hazardous Waste**
DOT Proper Shipping Name	**Toluene Diisocyanate**	Chemical Abstract Service (CAS) Number	**584-84-9**
DOT Hazard Class and Label Requirements	**Poison B; Poison**	DOT Emergency Guide Code	**57**
DOT Identification Number	**UN 2078**	Chemical Formula	**$CH_3C_6H_3(NCO)_2$**

Synonyms

TDI; 2,4-TDI; 2,4-toluene diisocyanate; cresorcinol diisocyanate; diisocyanate toluol.

SECTION II - HAZARDOUS INGREDIENTS AND IDENTITY INFORMATION

Hazardous Components (specific identity)	OSHA Exposure Criteria	NIOSH Exposure Criteria	Immediately Dangerous to Life and Health (IDLH)	ACGIH Exposure Criteria
Toluene-2,4-diisocyanate: (derivation: Usually prepared by reaction of 2,4-diaminotoluene with phosgene). **1 ppm = 7.24 mg/m^3**	**PEL (ceiling):** 0.02 ppm, 0.14 mg/m^3 **STEL:** Not Applicable	**REL (10-hour):** 0.005 ppm, 0.035 mg/m^3 **CEILING: (10 minute)** 0.020 ppm, 0.14 mg/m^3	**2.5 ppm** **Possible Cancer Agent**	**TLV:** 0.005 ppm, 0.035 mg/m^3 **STEL:** 0.020 ppm, 0.14 mg/m^3

SECTION III - PHYSICAL AND CHEMICAL CHARACTERISTICS

Boiling Point	**482°F (250°C)**	Specific Gravity (H_2O = 1)	**1.22**
Vapor Pressure (mm Hg)	**0.05 at 77°F (25°C)**	Molecular Weight	**174.2**
Vapor Density (Air = 1)	**6**	Melting Point	**71°F (21.6°C)**

Solubility

Insoluble in water (reacts exothermically). Soluble in ether, acetone, and many organic solvents.

Appearance and Odor

Clear to pale-yellow solid (liquid above 71°F) with a pungent, sharp odor. Odor Threshold = 0.17 ppm.

SECTION IV - FIRE AND EXPLOSION HAZARD DATA

Flash Point (method used)	**260°F (127°C) opened cup**	Explosive Limits in Air % by Volume	**LEL: 0.9%** **UEL: 9.5%**
NFPA Classification	**Class IIIB Combustible Liquid**	Autoignition Temperature	**1148°F (620°C)**

Extinguishing Media

Dry chemical or carbon dioxide. Do NOT use water (reacts to produce heat and carbon dioxide).

Special Fire Fighting Procedures

Wear full protective clothing and self-contained breathing apparatus (SCBA). Approach fire from upwind position. Remove containers from fire if it can be done without risk. Use a water spray to keep *sealed* fire-exposed containers cool. Poisonous gases are produced in fire. Avoid smoke and runoff.

Unusual Fire and Explosion Hazards

Highly combustible when exposed to heat and flame. Containers may rupture and explode in fire. Stay clear of ends of tanks. This chemical is often mixed with flammable and volatile solvents. It is explosive in the form of vapor when exposed to heat and flame.

SECTION V - REACTIVITY DATA

Stability		Conditions to Avoid Toluene-2,4-diisocyanate is normally stable at room temperature and pressure. Avoid contact with heat, flame, and incompatible materials, including water (exothermic reaction).
Stable X	Unstable	Incompatibility *(materials to avoid)* Reacts with water (forms carbon dioxide), alcohols, strong bases, amines, carboxylic acids, and strong oxidizers (chlorine, fluorine, bromine).
Hazardous Polymerization		Conditions to Avoid Toluene-2,4-diisocyanate will polymerize if in contact with concentrated alkaline compounds (such as sodium hydroxide, acyl chlorides, or *tert*-amine compounds).
May Occur X	Will Not Occur	Hazardous Decomposition or By-products When heated to decomposition, toluene-2,4-diisocyanate can emit highly toxic/poisonous fumes including fumes of cyanides and nitrogen oxides.

SECTION VI - HEALTH HAZARD DATA

Primary Route(s) of Entry:	Inhalation?	Absorption (skin)?	Ingestion?
	X	X (skin contact)	X

Health Hazards

INHALATION: Irritation of the eyes, nose, throat, and respiratory tract. Causes increased secretions, acute asthma attacks, choking, waves of productive cough, painful respiration, chest pains, abdominal distress, nausea and vomiting, pulmonary edema (fluid in the lungs), bronchospasms, dypsnea, drunkenness, loss of consciousness, and possible death.

SKIN & EYES: Skin contact may cause redness, swelling, and painful blistering. A skin sensitizer that may cause dermatitis. Eye contact causes conjunctivitis and inflammation of the eyelids.

INGESTION: Poison by ingestion with unspecified gastrointestinal effects.

Carcinogenicity	NTP Listed?	IARC Cancer Review Group?	OSHA Regulated?	Target Organs?
Probable Human Confirmed Animal	5th Annual Report	Group 2B	29 CFR 1910.1000 Table Z-1	Respiratory system, eyes, skin.

Medical Conditions Generally Aggravated by Exposure

Respiratory system problems (allergies) or conditions (asthma) may be aggravated by exposure.

Emergency and First-aid Procedures

Eye contact: Flush immediately with water for 15 minutes (minimum); seek medical attention. Skin contact: Remove all contaminated clothing. Immediately wash area with large amounts of soap and water. For red or blistered skin, seek medical assistance. For inhalation: Remove the person from exposure. Provide respiratory assistance and CPR. Transfer to medical facility. Observe for 48 hours for lung effects. If swallowed: Contact poison control center. Unless advised otherwise, give conscious person 1 to 2 glasses of water to dilute. Do NOT induce vomiting (chemical is a severe irritant).

SECTION VII - PRECAUTIONS FOR SAFE HANDLING AND USE

Steps to be Taken in Case Material is Released or Spilled

Absorb liquid solutions in vermiculite, dry sand, earth, or similar material and deposit in sealed containers. Ventilate area and remove ignition sources. Restrict those not involved in cleanup from area.

Preferred Waste Disposal Method

Mix in combustible solvent, burn in a chemical incinerator equipped with an afterburner and scrubber.

Precautions to be Taken in Handling and Storage

Store in tightly closed containers in cool, dry, well-ventilated area away from heat, flames, and water. Do not store in polyethylene containers (water can be absorbed through the plastic).

Other Precautions and Warnings

Keep away from water or moisture. Do NOT reseal a container if contamination is suspected. Package for disposal or reclamation. Protect containers from physical damage. Shield from direct sunlight.

SECTION VIII - CONTROL MEASURES AND PERSONAL PROTECTIVE EQUIPMENT

Respiratory Protection *(specify type)*

The PEL is low and difficult to monitor. For any exposure level, use an MSHA/NIOSH-approved supplied air respirator with full facepiece, hood, or helmet in continuous flow mode or, use a self-contained breathing apparatus (SCBA) in positive pressure mode.

Ventilation

Local exhaust (preferred) or general mechanical systems recommended.

Protective Gloves	Eye Protection	Other Protective Clothing
Impervious Material	Chemical Goggles or Face Mask	Protective Apron or Clothing

Work/Hygiene Practices

Always wash hands thoroughly after using chemical; never bring food, drink, or smoking materials into vicinity of chemicals.

TOLUENE-2,4-DIISOCYANATE

$CH_3C_6H_3(NCO)_6$

CAS: 584-84-9

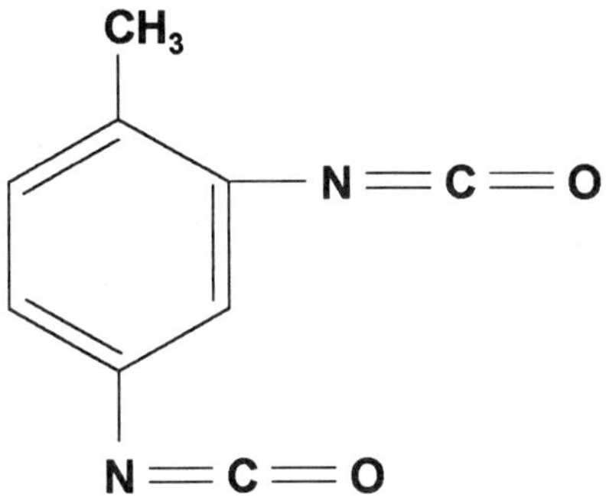

IDENTIFICATION AND TYPICAL USES

Toluene-2,4-diisocyanate appears as a colorless to pale-yellow solid or liquid (above 71°F/21.6°C) with a sharp, pungent, penetrating odor. It is one of the most extensively used isocyanates. It is used in the production of rigid and flexible urethane foams, elastomers, and coatings, all of which are used across a wide range of commercial and industrial applications. These include insulation, furniture, packaging, boat building, and aerospace vehicle assembly. Elastomers are used in adhesives, linings, film, and abrasive wheels. Polyurethane coatings are used on leather, wire, in tank linings, and in masonry.

RISK ASSESSMENT: HEALTH

General Assessment

Toluene-2,4-diisocyanate is an extremely toxic and poisonous substance. It can enter the body through *inhalation* and by *ingestion*. It will also cause adverse health effects on contact with the skin, although dermal absorption has not been reported. Toluene-2,4,-diisocyanate is a probable human carcinogen and there is some evidence that it will cause at least temporary impotence in males. Mutation data have also been reported.

Inhalation of toluene-2,4-diisocyanate vapors can be highly irritating to the eyes, nose, throat, and mucosa of the respiratory tract. Symptoms of exposure include wheezing, productive cough, increased secretions, dyspnea, sweating, difficulty in breathing, insomnia, painful respiration, and possible life-threatening pulmonary edema (fluid in the lungs). Development of pulmonary edema may be delayed up to 48 hours thereby creating a false sense of security with regard to health exposure risk. In severe cases, there may be some restriction of air movement and eventual respiratory failure (due to edema). Prolonged contact can lead to chronic bronchitis, asthma-like symptoms, and allergic pulmonary sensitization. Toluene-2,4-diisocyanate may also cause neurologic effects including euphoria, drunkenness, loss of coordination, hallucinations, dizziness, and loss of consciousness.

Skin contact causes irritation and painful burning with skin blisters on prolonged contact leading to allergic eczema and skin sensitization with possible dermatitis. If the liquid should contact the eye directly, there may be irritation, inflammation of the eyelids and surrounding tissues, conjunctivitis, and possible damage to the cornea. Ingestion causes corrosive damage to the alimentary canal with unspecified gastrointestinal effects.

☠ *Acute Health Effects*

The following acute (short-term) health effects may occur immediately or shortly after exposure to toluene-2,4-diisocyanate:

Skin: Possible irritation with painful burning, redness, swelling, and blistering.

Eye: Moderate to severe irritation. There can be inflammation, irritation of the eyelids, and conjunctivitis.

Lung: Irritation of the nose, throat, and lungs following exposure. It can cause coughing, painful or labored breathing, abdominal distress, nausea, vomiting, and a choking sensation. There can be a delayed but dangerous buildup of fluid in the lungs (pulmonary edema), which is a medical emergency and can be fatal.

Chronic Health Effects

The following chronic (long-term) health effects may occur at some time after exposure to toluene-2,4-diisocyanate and can last for months or even years:

Cancer Hazards: According to information presented in the references, exposure to toluene-2,4-diisocyanate may cause cancer in humans since it has been shown to cause cancer of the pancreas, liver, and breast in test animals. It has also caused mutations (genetic changes) in test animals and many scientists believe that exposure to such chemicals may pose a cancer risk in the long-term.

Reproductive Hazard: According to the data presented in the references, there is limited evidence that exposure to toluene-2,4-diisocyanate may temporarily cause impotence in males as well as damage to the reproductive system in humans.

Other Chronic Effects: Long-term exposure to toluene-2,4-diisocyanate may lead to allergic sensitization of the respiratory system with initial symptoms of night coughs and difficulty breathing and possible progression to asthmatic bronchitis with symptoms of coughing, wheezing, tightness in the chest, and shortness of breath. This can lead to gradual but permanent loss of lung function. Overexposed persons may also experience memory problems and/or loss of concentration. It is not known if these effects are permanent. Repeated exposure can cause allergic sensitization of the skin with an eczema-like rash developing on future exposures.

Recommended Risk-Reduction Measures

Personnel must avoid direct contact with toluene-2,4-diisocyanate. It is a highly toxic, poisonous, carcinogenic compound. If a less toxic chemical cannot be substituted for a hazardous substance, then *engineering controls* are the most effective method of reducing exposures.

The best protection is to enclose operations and/or provide local exhaust ventilation at the site of chemical release. While not always operationally feasible, isolating operations can also reduce exposure. Using respiratory protection is less effective than the controls mentioned above, but is still required whenever working with or around toluene-2,4-diisocyanate. The OSHA permissible exposure limit (PEL) ceiling for this chemical is extremely low (0.02 ppm) and difficult to monitor reliably under most operating or emergency conditions. A supplied-air respirator with a full facepiece operated in positive pressure mode or an MSHA/NIOSH-approved self-contained breathing apparatus (SCBA) with full facepiece operated in pressure demand or other positive pressure mode are the recommended respiratory protection methods of choice. In either case, if a full facepiece is not available, then chemical splash goggles should be worn to protect the eyes. Whenever a chemical splash hazard exists, a face shield and a protective apron should be worn. To prevent hand and skin exposures, impervious gloves should be used. Glove manufacturers should be contacted and permeation studies obtained *before* final glove selections are made.

Administrative controls should also be in place to minimize the potential for human exposures. These may include written procedures or policies which specify the methods and techniques that will be practiced whenever personnel are to work with toluene-2,4-diisocyanate.

All personnel should receive *training* on the use, hazards, protective measures, emergency actions, and other precautions per 29 CFR 1910.1200 (Hazard Communication), prior to the first assignment in an area where toluene-2,4-diisocyanate is used or stored.

If symptoms develop or overexposure is suspected, the following medical tests are recommended:

☑ Lung function tests (initially, these may be normal, especially if the person is not having an attack at the time of the test).

☑ Consider chest X-ray after acute over exposure (may be negative if taken immediately after exposure due to delayed onset of pulmonary edema).

☑ Evaluation by a qualified allergist with careful consideration of exposure history and special testing (may help diagnose allergy).

Any evaluation should include a careful history of past and present symptoms with an examination. Medical tests which simply look for damage already done are not a substitute for controlling exposure. Also, since smoking can cause heart disease, emphysema, and other lung diseases, smokers exposed to toluene-2,4-diisocyanate may experience symptoms more quickly than non-smokers under the same conditions of exposure. Prudent risk management requires proper consideration of *all* possible exposure factors when evaluating workplace health hazard risk.

Other methods to reduce exposure include:

☑ Where possible, enclose operations and use local exhaust ventilation at the site of chemical release.

If local exhaust ventilation or enclosure is not used, respiratory protection is mandatory.

☑ Where possible, automatically transfer toluene-2,4-diisocyanate from drums or other storage containers to process containers.

☑ Always ensure that proper protective clothing is worn when using chemical substances and that personnel are trained on its use and care.

☑ Wash thoroughly immediately after exposure to toluene-2,4-diisocyanate and at the end of the work shift or before eating, drinking, or smoking.

☑ Hazard warning information should be posted in the work area. In addition, as part of an on-going education and training program, all information on the health and safety hazards of toluene-2,4-diisocyanate should be communicated to all exposed and potentially exposed workers.

RISK ASSESSMENT: ENVIRONMENT

General Assessment

The environment is at risk of exposure during transportation, storage, disposal, or destruction of toluene-2,4-diisocyanate. In almost every scenario, the threat of environmental exposure is contingent upon the proper handling of the chemical substance. Accidental spills, large or small, can result in fire, explosion, and possible contamination of the surrounding environmental mediums (water, soil, and air).

Toluene-2,4-diisocyanate is considered a Class IIIB combustible liquid (per OSHA 29 CFR 1910.106). It can react vigorously with oxidizing agents, water, alcohols, strong bases, amines, and carboxylic acids. In addition, pressure may build in closed containers which can cause an explosion. These characteristics require special consideration during any emergency situation involving a leak or spill of toluene-2,4-diisocyanate.

Toluene-2,4-diisocyanate may enter the environment primarily through industrial discharges and spills.

Acute Ecological Effects

Acute (short-term) toxic effects may include the death of animals, birds, or fish and death or low growth rate in plants. Acute effects are seen 2 to 4 days after animals or plants are exposed to toluene-2,4-diisocyanate.

Toluene-2,4-diisocyanate has slight acute toxicity to aquatic life and high toxicity to birds. Insufficient data are available to evaluate or predict the short-term effects of toluene-2,4-diisocyanate exposure to plants or land animals.

Chronic Ecological Effects

Chronic (long-term) toxic effects may include shortened life span, reproductive problems, lower fertility, and changes in appearance or behavior in exposed animals. These effects can be seen long after first exposure(s) to toxic chemicals.

Toluene-2,4-diisocyanate has slight chronic toxicity to aquatic life. Insufficient data are available to evaluate or predict the long-term effects of toluene-2,4-diisocyanate on plants, birds, or terrestrial animals.

Water Solubility

Toluene-2,4-diisocyanate reacts in water to form carbon dioxide. It is, therefore, only slightly soluble in the aquatic environment. Concentrations of less than 1 milligram will mix with a liter of water.

Persistence in the Environment

Toluene-2,4-diisocyanate is non-persistent in the aquatic environment, with a half-life of less than 2 days. The half-life of a pollutant is the amount of time it takes for one half of the chemical to be degraded.

Bioaccumulation in Aquatic Organisms

Some substances increase in concentration, or *bioaccumulate,* in living organisms as they breathe contaminated air, drink contaminated water, or eat contaminated food. These chemicals can become concentrated in the tissues and internal organs of animals as well as humans.

Toluene-2,4-diisocyanate will react rapidly with water and therefore is not expected to accumulate in aquatic organisms or the environment.

Recommended Risk-Reduction Measures

Proper training of all transporters will reduce the likelihood of a mishap or accident resulting in a leak or spill to the environment. The correct use of labeling on all containers, trucks, and rail cars will enable emergency responders to react properly and quickly to any disaster thereby reducing the potential risk to the environment and to personnel.

Storage of toluene-2,4-diisocyanate should be segregated from incompatible chemicals to minimize the risk of cross-contamination or contact. Storage buildings or facilities should be equipped with proper fire protection equipment (sprinklers/extinguishers, alarms).

If a spill or leak to the environment has occurred, fire department, emergency response and/or hazardous materials spill personnel should be notified immediately. Cleanup should be attempted only by those trained in proper spill containment procedures using non-sparking tools. Contaminated soils should be removed for incineration and replaced with clean soil. If toluene-2,4-diisocyanate should contact the water table, aquifer, or navigable waterway, time is of the essence. Even though it is only slightly soluble in the aquatic environment, it will slowly react in water and, therefore, total containment and remediation will not be possible. When such spills occur, the local and/or state emergency response authorities must be notified. A comprehensive emergency response or disaster preparedness/recovery plan should be in place prior to any operations involving the use, transportation, storage, or disposal of toluene-2,4-diisocyanate. The proper disposal method for toluene-2,4-diisocyanate is to burn it in a chemical incinerator equipped with an afterburner and an air scrubber.

If toluene-2,4-diisocyanate is spilled or leaked, the following specific response steps are recommended:

- ☑ Restrict persons not wearing protective clothing from area of spill or leak until cleanup is complete and area can be opened for normal work.
- ☑ Remove all ignition sources and ventilate area.
- ☑ If liquid is spilled, absorb in vermiculite, dry sand, earth, or similar material and deposit in sealed containers. Use non-sparking tools.
- ☑ It may be necessary to dispose of toluene-2,4-diisocyanate as a hazardous waste. The responsible state agency or the regional office of the federal Environmental Protection Agency (EPA) should be contacted for specific recommendations.

RISK ASSESSMENT: BUSINESS

General Assessment

Accidents or mishaps involving toluene-2,4-diisocyanate can present a significant threat to business operations. The loss or damage of equipment or facilities can significantly affect fiscal viability. Lawsuits that may result from personnel injury/death, public exposures, and/or environmental contamination will also require a serious expenditure of resources. Media attention surrounding an injury, death, or environmental damage can also result in a loss of profits and loss of current as well as future business. Always remember that anytime the terms "cancer," "carcinogen," or "reproductive hazard" are used. Public emotion, hysteria, anxiety, and ignorance can run equally high. This must be carefully considered when drafting or implementing public relations policies.

Recommended Risk-Reduction Measures

Company attorneys, safety and health professionals, and environmental specialists should be involved in the development of any procedures for responding to hazardous chemical incidents. A company official should be pre-designated as a public relations officer with specific training in dealing with the press. Corporate plans and policies should be developed, approved, and implemented long before any need for such arises.

REFERENCES

American Conference of Governmental Industrial Hygienists. 1988. *Documentation of the Threshold Limit Values and Biological Exposure Indices,* 5th Edition (with updates). Cincinnati: ACGIH

Danse, I. R. 1991. *Common Sense Toxics in the Workplace.* New York: Van Nostrand Reinhold.

Environmental Protection Agency. 1983. *Chemical Hazard Information Profiles.* Washington, D.C.: U.S. EPA.

International Agency for Research on Cancer (IARC). 1987. *IARC Monographs on the Evaluation of Carcinogenic Risk to Humans, Overall Evaluations of Carcinogenicity: An updating of IARC Monographs Volumes 1-42.* Lyon: IARC.

Jones, F. E. 1994. *Toxic Organic Vapors in the Workplace.* New York: CRC Press/Lewis Publishers.

Lauwerys, R. R., P. Hoet. 1993. *Industrial Chemical Exposure.* New York: CRC Press/Lewis Publishers.

Lewis, R. J., Sr. 1992. *Sax's Dangerous Properties of Industrial Materials,* Volumes 1, 2, and 3, 8th Edition. New York: Van Nostrand Reinhold.

Manahan, S. E. 1992. *Toxicological Chemistry,* 2nd Edition. New York: CRC Press/Lewis Publishers.

Patnaik, P. 1992. *A Comprehensive Guide to the Hazardous Properties of Chemical Substances.* New York: Van Nostrand Reinhold.

Rea, W. J. 1995. *Chemical Sensitivity,* Vol. I-III. New York: CRC Press/Lewis Publishers.

Sacarello, H. L. A. 1994. *The Comprehensive Handbook of Hazardous Materials.* New York: CRC Press/Lewis Publishers.

Sherman, J. D. 1988. *Chemical Exposure and Disease.* New York: Van Nostrand Reinhold.

U.S. Department of Health and Human Services, National Institute for Occupational Safety and Health. 1994. *NIOSH Pocket Guide to Chemical Hazards.* Washington, D.C.: U.S. Government Printing Office.

MATERIAL SAFETY DATA SHEET

CHEMICAL NAME	
o-TOLUIDINE	**CRC PRESS/LEWIS PUBLISHERS DISCLAIMER:** THE INFORMATION AND RECOMMENDATIONS PRESENTED HEREIN ARE BASED ON SOURCES BELIEVED TO BE RELIABLE. CRC MAKES NO REPRESENTATION ON ITS COMPLETENESS OR ACCURACY. IT IS THE RESPONSIBILITY OF THE USER TO DETERMINE THE CHEMICAL'S SUITABILITY FOR ITS INTENDED USE, THE CHEMICAL'S SAFE USE, AND THE CHEMICAL'S PROPER DISPOSAL. NO REPRESENTATIONS AND/OR WARRANTIES, EITHER EXPRESSED OR IMPLIED, OF THE MERCHANTABILITY OR FITNESS FOR A PARTICULAR PURPOSE, OR OF ANY OTHER NATURE, ARE MADE WITH RESPECT TO THE INFORMATION PROVIDED IN THIS MSDS OR TO THE CHEMICAL TO WHICH INFORMATION MAY REFER. CRC NEITHER ASSUMES NOR AUTHORIZES ANY OTHER PERSON TO ASSUME FOR IT, ANY OTHER ADDITIONAL RESPONSIBILITY OR LIABILITY FOR THE USE OF, OR RELIANCE UPON, THIS INFORMATION.

HAZARD WARNING INFORMATION

HEALTH	FIRE	REACTIVE	OTHER	DEGREE OF HAZARD	COLOR CODING	OTHER CODES
3	2	0		0 = Minimum Hazard 1 = Slight Hazard 2 = Moderate Hazard 3 = Serious Hazard 4 = Severe Hazard	HEALTH = BLUE FIRE = RED REACTIVITY = YELLOW OTHER = WHITE	OX = Oxidizer ACID = Acid ALK = Alkali COR = Corrosive W = Use No Water

SECTION I - GENERAL INFORMATION

Characterization	**Aromatic Amine**
RCRA Number	**U328**
EPA Class	**Toxic Waste**
DOT Proper Shipping Name	**Toluidines, Liquid**
Chemical Abstract Service (CAS) Number	**95-53-4**
DOT Hazard Class and Label Requirements	**Poison B; Poison**
DOT Emergency Guide Code	**55**
DOT Identification Number	**UN 1708**
Chemical Formula	**$CH_3C_5H_4NH_2$**

Synonyms

o-Aminotoluene; 2-aminotoluene;1-methyl-2-aminobenzene; o-methylaniline; 2-methylaniline; ortho-toluidone; o-tolylamine.

SECTION II - HAZARDOUS INGREDIENTS AND IDENTITY INFORMATION

Hazardous Components (specific identity)	OSHA Exposure Criteria	NIOSH Exposure Criteria	Immediately Dangerous to Life and Health (IDLH)	ACGIH Exposure Criteria
o-Toluidine derivation: By the reduction of o-nitrotoluene or obtained mixed with p-toluidine by t he reduction of crude nitrotoluene). **1 ppm = 4.46 mg/m³**	**PEL (skin):** **5 ppm** **22 mg/m³** **STEL:** **Not Established**	**Reduce to Lowest Level** **Possible Human Carcinogen**	**50 ppm**	**TLV (skin):** **2 ppm** **8.8 mg/m³** **STEL:** **Not Established**

SECTION III - PHYSICAL AND CHEMICAL CHARACTERISTICS

Boiling Point	**391°F (200°C)**
Specific Gravity (H_2O = 1)	**1.01**
Vapor Pressure (mm Hg)	**0.05 at 77°F (25°C)**
Molecular Weight	**107.2**
Vapor Density (Air = 1)	**3.69**
Freezing Point	**-3°F (-16°C)**

Solubility

Very slightly soluble in water. Soluble in carbon tetrachloride, dilute acids, diethyl ether, and ethanol.

Appearance and Odor

Colorless to pale-yellow liquid with an aromatic, aniline-like odor. It becomes reddish-brown on exposure to air or light. Odor Threshold = 0.25 ppm.

SECTION IV - FIRE AND EXPLOSION HAZARD DATA

Flash Point (method used)	**185°F (85°C) closed cup**
Explosive Limits in Air % by Volume	**LEL: 1.5%** **UEL: Not Applicable**
NFPA Classification	**Class III A Combustible Liquid**
Autoignition Temperature	**900°F (482°C)**

Extinguishing Media

Carbon dioxide, dry chemical, water spray, or regular foam. Water may be ineffective.

Special Fire Fighting Procedures

Poisonous gases are produced in fire. Wear full protective clothing and self-contained breathing apparatus (SCBA). Use a water spray to keep fire-exposed containers cool. Do not release runoff from fire control measures to sewers or waterways.

Unusual Fire and Explosion Hazards

Moderately flammable when exposed to heat. Containers may explode in fire.

SECTION V - REACTIVITY DATA

Stability		Conditions to Avoid **o-Toluidine is normally stable under routine conditions of handing and storage. Avoid exposure to heat, ignition sources, and incompatible materials.**
Stable X	Unstable	Incompatibility *(materials to avoid)* **Strong oxidizers (such as perchlorates, peroxides, permanganates, chlorates and nitrates); strong acids (such as hydrochloric, sulfuric, and nitric); and strong bases.**
Hazardous Polymerization		Conditions to Avoid **Under normal conditions of temperature and pressure, hazardous polymerization of o-toluidine does not occur.**
May Occur	Will Not Occur X	Hazardous Decomposition or By-products **When heated to decomposition, o-toluidine can emit highly toxic/poisonous oxides of nitrogen and carbon.**

SECTION VI - HEALTH HAZARD DATA

Primary Route(s) of Entry:	Inhalation?	Absorption (skin)?	Ingestion?
	X	X	X

Health Hazards

INHALATION: Causes anoxia due to the formation of methemoglobin. Symptoms: headache, dizziness, ataxia, and cyanosis. Overexposure may lead to death due to respiratory paralysis. It can also cause kidney and bladder irritation, drowsiness, nausea, vomiting, hematuria (blood in urine), and varying degrees of central nervous system depression.

ABSORPTION: Absorbs freely through intact skin. Symptoms include headache and cyanosis, over-exposure can result in death. Skin contact with pure liquid can produce burns.

INGESTION: Symptoms may be similar to those of inhalation.

Carcinogenicity	NTP Listed?	IARC Cancer Review Group?	OSHA Regulated?	Target Organs?
Possible Human Confirmed Animal	5th Annual Report	Group 2B	29 CFR 1910.1000 Table Z-1	Blood, kidneys, liver, CVS, CNS, bladder.

Medical Conditions Generally Aggravated by Exposure

Respiratory impairments (bronchitis, asthma); skin conditions (dermatitis); renal or liver disorders.

Emergency and First-aid Procedures

Eye contact: Flush immediately with water for 15 minutes (minimum); seek medical attention. Skin contact: Remove all contaminated clothing. Immediately wash area with large amounts of soap and water. Seek medical assistance. For inhalation: Remove the person from exposure. Provide respiratory assistance and CPR. Transfer to medical facility. If swallowed: Seek medical attention immediately. Give conscious person 1 to 2 glasses of water and induce vomiting.

SECTION VII - PRECAUTIONS FOR SAFE HANDLING AND USE

Steps to be Taken in Case Material is Released or Spilled

Absorb liquids in vermiculite, dry sand, earth, or similar material and deposit in sealed containers. Ventilate area of spill or leak. Restrict those not involved in cleanup from entering area. Notify appropriate authorities, as required by SARA III, if applicable.

Preferred Waste Disposal Method

Dissolve in flammable solvent and burn in incinerator equipped with scrubber and afterburner.

Precautions to be Taken in Handling and Storage

Store to avoid contact with strong oxidizers and acids. Store in tightly closed containers in cool, well-ventilated area away from heat. Ground and bond all containers prior to transferring liquids.

Other Precautions and Warnings

Sources of ignition are prohibited where o-toluidine is used, handled, or stored. Bulk storage of o-toluidine is not recommended.

SECTION VIII - CONTROL MEASURES AND PERSONAL PROTECTIVE EQUIPMENT

Respiratory Protection *(specify type)*

Use a supplied-air respirator with full facepiece, hood, or helmet in continuous flow mode, or a self-contained breathing apparatus (SCBA) with full facepiece operated in positive pressure mode.

Ventilation

Local exhaust (preferred) or general exhaust at site of chemical work.

Protective Gloves	Eye Protection	Other Protective Clothing
Impervious Material	Chemical Goggles or Face Mask	Protective Apron

Work/Hygiene Practices

Always wash hands thoroughly after using chemical; never bring food, drink, or smoking materials into vicinity of chemicals.

o-TOLUIDINE

$CH_3C_6H_4NH_2$

CAS: 95-53-4

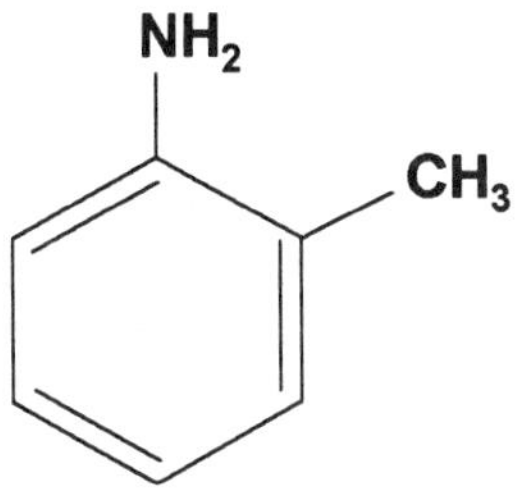

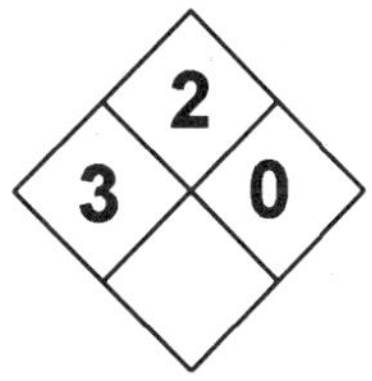

IDENTIFICATION AND TYPICAL USES

o-Toluidine is a colorless to pale-yellow liquid with an aromatic, aniline-like odor. o-Toluidine is used in the manufacture of various dyes, in printing textiles blue-black, and as an intermediate in rubber chemicals, pesticides, and pharmaceuticals.

RISK ASSESSMENT: HEALTH

General Assessment

o-Toluidine is moderately toxic through *inhalation, ingestion,* and *absorption* through the skin. Toxic symptoms of o-toluidine exposure include headache, weakness, dizziness, ataxia, and cyanosis (blue tinting of the lips, skin, and nail beds due to oxygen deprivation). Cyanosis is due to the formation of methemoglobin which restricts the blood's oxygen carrying capabilities. Cyanosis occurs when the methemoglobin concentration is 15% or more. Blue discoloration develops first in the lips, nose, and earlobes. At 40% concentration, there is weakness and dizziness. At up to 70% concentration, symptoms may include ataxia, dyspnea on mild exertion, and tachycardia. Coma may ensue at concentrations over 70%. Overexposure (between 85% and 90% concentrations) may lead to death from respiratory paralysis. Exposure can also cause kidney and bladder irritation, drowsiness, nausea, vomiting, hematuria (blood in urine), and varying degrees of central nervous system (CNS) depression.

Contact with the pure liquid o-toluidine on the skin can produce moderate irritation, while the effect on the eyes can be quite severe. It will pass through intact skin to cause many of the same effects as those noted for inhalation. Oral administration to laboratory animals resulted in tumors in the kidney and bladder. Ingestion is also expected to cause many of the same effects as those noted for inhalation. o-Toluidine is a probable human carcinogen since it has been shown to cause cancer in test animals. Human mutation effects have also been noted and o-toluidine is an experimental teratogenic compound.

Acute Health Effects

The following acute (short-term) health effects may occur immediately or shortly after exposure to o-toluidine:

Skin: Irritation and possible burns on prolonged contact.

Eye: Severe irritation; may cause corneal damage.

Lung: Not known to be acutely irritating to lung tissue. However, overexposure can lead to death due to respiratory arrest.

CNS: Varying degrees of depression have been reported for o-toluidine exposure. Symptoms may include dizziness, drowsiness, irritability, nausea, and vomiting.

Other: o-Toluidine affects the blood's ability to transport oxygen and high exposure can cause death. Symptoms of anoxia with headaches, weakness, and shortness of breath.

Chronic Health Effects

The following chronic (long-term) health effects may occur at some time after exposure to o-toluidine and can last for months or even years:

Cancer Hazards: o-Toluidine may cause cancer in humans since it has been shown to cause cancer in test animals. Human mutations (genetic changes) have also been reported and many scientists believe that exposure to such chemicals may pose a human cancer threat in the long-term.

Reproductive Hazard: There is limited experimental evidence to show that exposure to o-toluidine may damage the developing fetus.

Other Chronic Effects: Continuous exposure to small doses of o-toluidine may produce anemia, loss of energy, digestive disturbance, and headache. Repeated exposures may cause dermatitis and possible tumors in the liver, bladder, and mammary glands (positive in test animals).

Recommended Risk-Reduction Measures

Personnel should avoid direct contact with o-toluidine. It is a known carcinogen, mutagen, and teratogen. It is highly toxic in both acute and chronic exposures. If a less toxic chemical cannot be substituted for a hazardous substance, then *engineering controls* are the most effective method of reducing exposures. The best protection is to enclose operations and/or provide local exhaust ventilation at the site of chemical release. While not always operationally feasible, isolating operations can also reduce exposure. Using respiratory protection is less effective than the controls mentioned above, but is still advisable whenever working with or around o-toluidine. A self-contained breathing apparatus (SCBA) with full facepiece in pressure demand is the recommended respiratory protection of choice. If a full facepiece is not available, then chemical goggles should be worn to protect the eyes. Whenever a chemical splash hazard exists, a face shield and a rubber apron should be worn. To prevent hand and skin exposures, natural rubber gloves should be worn.

Administrative controls should also be in place to minimize the potential for human exposures. These may include written procedures or policies which specify the methods and techniques that will be practiced whenever personnel are to work with o-toluidine.

All personnel should receive *training* on the use, hazards, protective measures, emergency actions, and other precautions per 29 CFR 1910.1200 (Hazard Communication), prior to the first assignment in an area where o-toluidine is used or stored.

If symptoms develop or overexposure is suspected, the following medical tests may be useful:

- ☑ Methemoglobin tests.
- ☑ Urinalysis (for blood content).

Any evaluation should include a careful history of past and present symptoms with an examination. Medical tests that simply look for existing damage are not a substitute for controlling exposures.

Other methods to reduce exposure include:

- ☑ Where possible, enclose operations and use local exhaust ventilation at the site of chemical release. If local exhaust ventilation or enclosure is not used, respiratory protection should be mandatory.
- ☑ Always ensure that proper protective clothing is worn when using chemical substances.
- ☑ Safety showers should be provided in areas where work is performed around o-toluidine.
- ☑ Wash thoroughly immediately after exposure to o-toluidine and at the end of the work shift or before eating, drinking, or smoking.
- ☑ Hazard warning information should be posted in the work area. In addition, as part of an on-going education and training program, all information on the health and safety hazards of o-toluidine should be communicated to all exposed workers.
- ☑ Eye wash stations should be provided in the immediate work area for emergency use. If there is a possibility of skin exposure to o-toluidine, emergency shower facilities should also be provided.

RISK ASSESSMENT: ENVIRONMENT

General Assessment

The environment is at risk of exposure during transportation, storage, disposal, or destruction of o-toluidine. In almost every scenario, the threat of environmental exposure is contingent upon the proper handling of the chemical substance. Accidental spills, large or small, can result in fire, explosion, and possible contamination of the surrounding environmental mediums (water, soil, and air).

o-Toluidine is considered a Class IIIA combustible liquid. Alone, its high flash point and autoignition temperature do not present a serious fire or explosion hazard. However, because it is incompatible with so many other commodities, extreme caution is required in handling, storage, transportation, and disposal of o-toluidine. Also, in vapor form, it can present a serious explosion hazard. These characteristics also require special consideration during any emergency situation involving a leak or spill of o-toluidine. Should o-toluidine ever come into contact with incompatible substances such as oxidizers, acids, or alkalis either during use, transportation, storage, or disposal, the formation of highly toxic and/or highly explosive commodities is extremely possible.

The proper disposal/destruction method for o-toluidine waste is to dissolve it in a combustible solvent and burn it in a chemical incinerator equipped with an afterburner and air scrubber.

o-Toluidine can enter the environment through direct disposal and indirect use, such as industrial discharges and non-point source from agricultural use.

☠ *Acute Ecological Effects*

Acute (short-term) toxic effects may include the death of animals, birds, or fish and death or low growth rate in plants. Acute effects are seen 2 to 4 days after animals or plants are exposed to o-toluidine.

This chemical has high acute toxicity to aquatic life and moderate acute toxicity to birds. Insufficient data are available to evaluate or predict the acute (short-term) effects of o-toluidine to aquatic life, plants, birds, or land animals.

Chronic Ecological Effects

Chronic (long-term) toxic effects may include shortened life span, reproductive problems, lower fertility, and changes in appearance or behavior in exposed animals. These effects can be seen long after first exposure(s) to toxic chemicals.

Insufficient data are available to evaluate or predict the long-term (chronic) effects of o-toluidine to aquatic life, plants, birds, or land animals.

Water Solubility

o-Toluidine is slightly soluble in water. Concentrations of 1 to 100 milligrams will mix with a liter of water.

⌛ *Persistence in the Environment*

o-Toluidine is slightly persistent in water, with a half-life between 2 and 20 days. The half-life of a pollutant is the amount of time it takes for one half of the chemical to be degraded.

Bioaccumulation in Aquatic Organisms

Some substances increase in concentration, or *bioaccumulate,* in living organisms as they breathe contaminated air, drink contaminated water, or eat contaminated food. These chemicals can become concentrated in the tissues and internal organs of animals as well as humans.

The concentration of o-toluidine found in fish tissues is expected to be somewhat higher than the average concentration of o-toluidine in the water from which the fish was taken.

Recommended Risk-Reduction Measures

Proper training of all transporters will reduce the likelihood of a mishap or accident resulting in a leak or spill to the environment. The correct use of DOT labeling (per the MSDS) on all containers, trucks, and rail cars will enable emergency responders to react properly and quickly to any disaster thereby reducing the potential risk to the environment and to personnel.

Storage of o-toluidine should be segregated from incompatible chemicals to minimize the risk of cross-contamination or contact. Containers should be protected from physical damage.

If a spill or leak to the environment has occurred, fire department, emergency response and/or hazardous materials spill personnel should be notified immediately. Cleanup should be attempted only by those trained in proper spill containment procedures using non-sparking tools. Contaminated soils should be removed for incineration and replaced with clean soil. If o-toluidine should contact the water table, aquifer, or navigable waterway, the local and/or state emergency response authorities must be notified. A comprehensive emergency response or disaster preparedness/recovery plan should be in place prior to any operations involving the use, transportation, storage, or disposal of o-toluidine. If o-toluidine is spilled or leaked, the following specific steps are recommended:

- ☑ Restrict persons not wearing protective clothing from area of spill or leak until cleanup is complete and area can be opened for normal work.
- ☑ Remove all ignition sources.
- ☑ Absorb liquids in vermiculite, dry sand, earth, or similar material and deposit in sealed containers. Use non-sparking tools during cleanup.
- ☑ It may be necessary to dispose of o-toluidine as a hazardous waste. The responsible state agency or the regional office of the federal Environmental Protection Agency (EPA) should be contacted for specific recommendations.

RISK ASSESSMENT: BUSINESS

General Assessment

Accidents or mishaps involving o-toluidine can present a significant threat to business operations. The loss or damage of equipment or facilities can significantly affect fiscal viability. Lawsuits that may result from personnel injury/death, public exposures, and/or environmental contamination will also require a serious expenditure of resources. Media attention surrounding an injury, death, or environmental damage can also result in a loss of profits and loss of current as well as future business. Always remember that anytime the terms "cancer," "carcinogen," or "reproductive hazard" are used, public hysteria, ignorance, emotion, and anxiety can all run equally high. This must be carefully considered whenever drafting or implementing public relations policies.

Recommended Risk-Reduction Measures

Company attorneys, safety and health professionals, and environmental specialists should be involved in the development of any procedures dealing with safety or emergency response. A company official should be pre-designated as a public relations officer with specific training in dealing with the press. Corporate plans and policies should be developed, approved, and implemented long before any need for such arises.

REFERENCES

Ahlbom, A. 1993. *Biostatistics for Epidemiologists.* New York: CRC Press/Lewis Publishers.

American Conference of Governmental Industrial Hygienists. 1988. *Documentation of the Threshold Limit Values and Biological Exposure Indices,* 5th Edition (with updates). Cincinnati: ACGIH

Asante-Duah, D. K. 1993. *Hazardous Waste Risk Assessment.* New York: CRC Press/Lewis Publishers.

Bloom, A. D. and F. J. DeSerres. 1995. *Ecotoxicity and Human Health.* New York: CRC Press/Lewis Publishers.

Bueche, J. R. 1972. *Principles of Physics,* 2nd Edition. New York: McGraw-Hill.

Calabrese, E. J. 1994. *Biological Effects of Low Level Exposures.* New York: CRC Press/Lewis Publishers.

Cockerman, L. G., B. S. Shane. 1994. *Basic Environmental Toxicology.* New York: CRC Press/Lewis Publishers.

Cothern, R. R. 1993. *Comparative Environmental Risk Assessment.* New York: CRC Press/Lewis Publishers.

Gardner, R. W. 1994. *Chemical Intolerance.* New York: CRC Press/Lewis Publishers.

Hallenbeck, W. H. 1993. *Quantitative Risk Assessment for Environmental and Occupational Health,* 2nd Edition. New York: CRC Press/Lewis Publishers.

Howard, P. H. 1995. *Handbook of Environmental Fate and Exposures Data for Organic Chemicals,* Vol. I-IV. New York: CRC Press/Lewis Publishers.

Jones, F. E. 1994. *Toxic Organic Vapors in the Workplace.* New York: CRC Press/Lewis Publishers.

Keith, L. H. and M. M. Walker. 1995. *Handbook of Air Toxics.* New York: CRC Press/Lewis Publishers.

Landis, W. G., M. H. Yu. 1995. *Introduction to Environmental Toxicology.* New York: CRC Press/Lewis Publishers.

Lauwerys, R. R., P. Hoet. 1993. *Industrial Chemical Exposure.* New York: CRC Press/Lewis Publishers.

Lewis, R. J., Sr. 1993. *Hawley's Condensed Chemical Dictionary,* 12th Edition. New York: Van Nostrand Reinhold.

Lewis, R. J., Sr. 1992. *Sax's Dangerous Properties of Industrial Materials,* Volumes 1, 2, and 3, 8th Edition. New York: Van Nostrand Reinhold.

Lide, D. R. 1995. *Handbook of Chemistry and Physics,* 76th Edition. New York: CRC Press/Lewis Publishers.

Manahan, S. E. 1992. *Toxicological Chemistry,* 2nd Edition. New York: CRC Press/Lewis Publishers.

Meyer, E. 1990. *Chemistry of Hazardous Materials.* Englewood Cliffs, NJ: Prentice Hall.

National Cancer Institute. 1978. *Carcinogenesis Technical Report Series. Bioassay of O-toluidine for Possible Carcinogenicity.* DEHW (NIH) Publication Number: 78-1323, p 53. Washington, DC: U.S. Government Printing Office.

Patnaik, P. 1992. *A Comprehensive Guide to the Hazardous Properties of Chemical Substances.* New York: Van Nostrand Reinhold.

Philp, R. B. 1995. *Environmental Hazards and Human Health.* New York: CRC Press/Lewis Publishers.

Proctor, N. H., J. P. Hughes. 1991. *Chemical Hazards of the Workplace.* New York: Van Nostrand Reinhold.

Quigley, D. R. 1994. *Handbook of Emergency Chemical Management.* New York: CRC Press/Lewis Publishers.

Rea, W. J. 1995. *Chemical Sensitivity,* Vol. I-III. New York: CRC Press/Lewis Publishers.

Richardson, M. 1995. *Dictionary of Substance and Their Effects,* Vol. I-VI. New York: CRC Press/Lewis Publishers.

Sacarello, H. L. A. 1994. *The Comprehensive Handbook of Hazardous Materials.* New York: CRC Press/Lewis Publishers.

Talbott, E., G. Craun. 1995. *An Introduction to Environmental Epidemiology.* New York: CRC Press/Lewis Publishers.

Torkelson, T. R. V. K. Rowe. 1981. *Patty's Industrial Hygiene and Toxicology,* Vol. 2B. New York: Wiley-Interscience.

The Merck Index, 10th edition. 1983. Rahway, NJ: Merck & Company.

U.S. Department of Health and Human Services, National Institute for Occupational Safety and Health. 1994. *NIOSH Pocket Guide to Chemical Hazards.* Washington, D.C.: U.S. Government Printing Office.

MATERIAL SAFETY DATA SHEET

CHEMICAL NAME	CRC PRESS/LEWIS PUBLISHERS DISCLAIMER
TRICHLOROETHYLENE	

HAZARD WARNING INFORMATION

HEALTH	FIRE	REACTIVE	OTHER	DEGREE OF HAZARD	COLOR CODING	OTHER CODES
2	2	0		0 = Minimum Hazard 1 = Slight Hazard 2 = Moderate Hazard 3 = Serious Hazard 4 = Severe Hazard	HEALTH = BLUE FIRE = RED REACTIVITY = YELLOW OTHER = WHITE	OX = Oxidizer ACID = Acid ALK = Alkali COR = Corrosive ~~W~~ = Use No Water

SECTION I - GENERAL INFORMATION

Characterization	RCRA Number	EPA Class
Halogenated Hydrocarbon	**U228 (F002-spent solvent)**	**Toxic Waste**

DOT Proper Shipping Name **Trichloroethylene**	Chemical Abstract Service (CAS) Number **79-01-6**
DOT Hazard Class and Label Requirements **ORM-A; Poison B; St. Andrews Cross**	DOT Emergency Guide Code **74**
DOT Identification Number **UN 1710**	Chemical Formula **$CHCl{:}CCl_2$**

Synonyms

Ethylene trichloride; TCE; trichloroethane; trilene; acetylene trichloride; Algylen; Anamenth; Benzinol; Cecolene; Chlorylen; Germalgene; Narcogen; Triasol; trichloroethene; 1,1,3-trichloroethylene.

SECTION II - HAZARDOUS INGREDIENTS AND IDENTITY INFORMATION

Hazardous Components (specific identity)	OSHA Exposure Criteria	NIOSH Exposure Criteria	Immediately Dangerous to Life and Health (IDLH)	ACGIH Exposure Criteria
Trichloroethylene (derivation: From tetrachloroethylene by treatment with lime or alkali in the presence of water, or by thermal decomposition, followed by steam distillation). **1 ppm = 5.46 mg/m³**	**PEL (8-hour):** **100 ppm** **546 mg/m³** **CEILING** **200 ppm** **1092 mg/m³** **PEAK (5-min)** **300 ppm**	**REL (10-hour):** **250 ppm** **1365 mg/m³** **CEILING (1-hr)** **2 ppm** **11 mg/m³** **Possible Carcinogen**	**1000 ppm**	**TLV:** **50 ppm** **269 mg/m³** **STEL:** **200 ppm** **1070 mg/m³**

SECTION III - PHYSICAL AND CHEMICAL CHARACTERISTICS

Boiling Point **189°F (87°C)**	Specific Gravity (H_2O = 1) **1.46**
Vapor Pressure (gas) **58 at 68°F (20°C)**	Molecular Weight **131.4**
Vapor Density (Air = 1) **4.53**	Melting Point **-99°F (-73°C)**

Solubility

Slightly soluble in water (0.0001%). Soluble in organic solvents (alcohol, acetone, ether) and lipids.

Appearance and Odor

Colorless liquid (may be dyed blue) with a chloroform-like odor. Odor Threshold = 28 ppm.

SECTION IV - FIRE AND EXPLOSION HAZARD DATA

Flash Point (method used) **90°F (32°C) closed cup**	Explosive Limits in Air % by Volume **LEL: 8% at 77°F (25°C) UEL: 10% at 77°F (25°C)**
NFPA Classification **Class 1C Flammable Liquid**	Autoignition Temperature **788°F (420°C)**

Extinguishing Media

Use dry chemical, carbon dioxide, water spray, fog, or regular foam.

Special Fire Fighting Procedures

Wear full protective clothing and self-contained breathing apparatus (SCBA). Poisonous gases are produced in fire. Move containers from fire or use water spray to keep fire-exposed containers cool.

Unusual Fire and Explosion Hazards

Vapor and air mixtures may explode when ignited. Although trichloroethylene has a 90°F flash point, it does not easily ignite. Containers can explode in fire; stay clear of ends of tanks and drums.

SECTION V - REACTIVITY DATA

Stability		Conditions to Avoid Slowly decomposes in the presence of light and moisture to form corrosive hydrochloric acid. Avoid contact with heat, light, moisture, and incompatible materials.
Stable	Unstable X	Incompatibility *(materials to avoid)* Strong caustics and alkalis (sodium hydroxide). Also reacts with chemically active metals (lithium, beryllium, barium), sodium, magnesium, titanium), oxidants (explodes).
Hazardous Polymerization		Conditions to Avoid Under normal conditions of temperature and pressure, hazardous polymerization of trichloroethylene is not expected to occur.
May Occur	Will Not Occur X	Hazardous Decomposition or By-products When heated to decomposition, trichloroethylene can emit highly toxic/poisonous and acrid fumes and gases, including chlorine, carbon dioxide, hydrogen chloride, dichloro acetylene, and phosgene gas.

SECTION VI - HEALTH HAZARD DATA

Primary Route(s) of Entry:	Inhalation?	Absorption (skin)?	Ingestion?
	X	X	X

Health Hazards

INHALATION: Can produce human systemic effects, including eye, nose, and throat irritation, nausea, blurred vision, over-excitement, headache, drunkenness, stupor, memory loss, irregular heart beat (can be fatal), unconsciousness, death due to cardiac failure. High concentrations can lead to pulmonary edema (fluid in the lungs), which can cause death.

ABSORPTION: Irritating to the skin and may produce dermatitis on repeated exposures. Absorbed in toxic amounts through intact skin. Eye contact may injure cornea.

INGESTION: Irritation and narcosis with burning of the mouth and lips. Can cause irregular heartbeat, nausea and vomiting, bloody diarrhea, drowsiness, unconsciousness, and pulmonary edema (fluid in the lungs).

Carcinogenicity	NTP Listed?	IARC Cancer Review Group?	OSHA Regulated?	Target Organs?
Suspected Human Questioned Animal	No	Group 3	29 CFR 1910.1000 Table Z-1 & Z-2	Resp. sys., eyes, skin, liver, kidney, CVS, CNS.

Medical Conditions Generally Aggravated by Exposure

Disorders of the nervous system, liver, kidney, heart, skin, and respiratory system.

Emergency and First-aid Procedures

Eye contact: Flush immediately with water for 15 minutes (minimum); seek medical attention. Skin contact: Remove all contaminated clothing. Immediately wash area with large amounts of soap and water. For inhalation: Remove the person from exposure. Provide respiratory assistance. Transfer to medical facility. Observe 48 hours for lung effects. If swallowed: Seek medical attention immediately. Unless advised otherwise, give conscious person 1 to 2 glasses of water and induce vomiting. Do NOT give milk (can enhance gastrointestinal irritation due to lipid solubility of trichloroethylene.

SECTION VII - PRECAUTIONS FOR SAFE HANDLING AND USE

Steps to be Taken in Case Material is Released or Spilled

Absorb large spills in vermiculite, dry sand, earth, or similar material and deposit in sealed containers. Ventilate area and remove ignition sources. Restrict those not involved in cleanup from entering area.

Preferred Waste Disposal Method

Mix with combustible solvent and burn in chemical incinerator with an afterburner and scrubber.

Precautions to be Taken in Handling and Storage

Store to avoid contact with incompatible materials in tightly closed containers in cool, well-ventilated area away from heat and flame. Prevent damage to containers.

Other Precautions and Warnings

Sources of ignition are prohibited where trichloroethylene is used, handled, or stored.

SECTION VIII - CONTROL MEASURES AND PERSONAL PROTECTIVE EQUIPMENT

Respiratory Protection *(specify type)*

For high exposures, use an MSHA/NIOSH-approved supplied-air respirator with full facepiece in continuous flow mode, or use a SCBA operated in pressure demand or other positive pressure mode.

Ventilation

Local exhaust or general mechanical systems recommended.

Protective Gloves	Eye Protection	Other Protective Clothing
Neoprene or Viton Material	Chemical Goggles or Face Mask	Protective Apron

Work/Hygiene Practices

Always wash hands thoroughly after using chemical; never bring food, drink, or smoking materials into vicinity of chemicals.

TRICHLOROETHYLENE

$ClCH:CCl_2$

CAS: 79-01-6

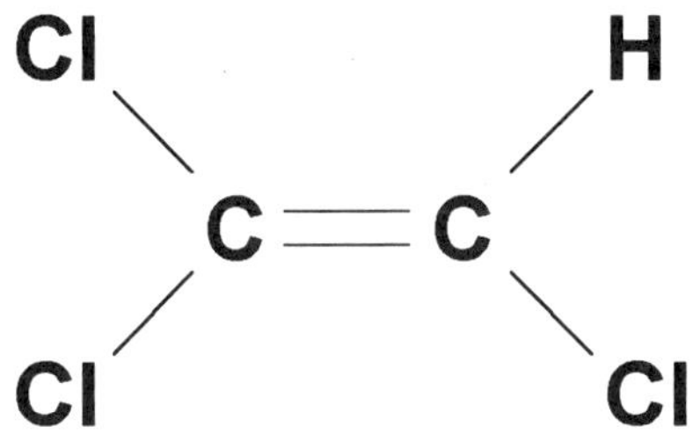

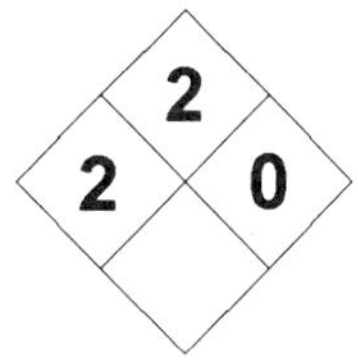

IDENTIFICATION AND TYPICAL USES

Trichloroethylene is a colorless liquid (may be dyed blue) with a mild, chloroform-like odor. It is used in metal de-greasing and as a extraction solvent for oils, fats, and waxes. It is also used in solvent dyeing, in dry cleaning, as a refrigerant and heat-exchange liquid, as a fumigant, for cleaning and drying electronic parts, as a diluent in paints and adhesives, in textile processing, as a chemical intermediate, and in aerospace operations for flushing liquid oxygen from propellant lines and hoses.

RISK ASSESSMENT: HEALTH

General Assessment

Trichlorocthylene is a mild to moderately toxic compound to humans by *inhalation*, dermal *absorption*, and *ingestion*. Ingestion can occur, but is rare under most circumstances. It is considered as a questioned animal carcinogen and its human carcinogenic potential is considered suspect by most reference sources. It has been shown to cause mutations in humans. Experimental teratogenic data have also been reported in animal tests.

Inhalation results in slight smarting of the eyes with lachrymation and slight smarting of the throat. Inhalation of high concentrations of the vapors can also lead to CNS depression with symptoms of intoxication, loss of coordination, numbness around the mouth, dizziness, narcosis, and convulsions. It can also cause flushing of the face and neck, headache, somnolence, blurred vision, and slurred speech. Irritation of the upper respiratory tract can cause a dangerous buildup of fluid in the lungs (pulmonary edema), which can be fatal. Symptoms of coughing, tightness in the chest, congestion, and dypsnea may be delayed up to 48 hours, thereby creating a false sense of security with regard to health exposure risk. There can also be cardiac arrhythmia (irregular heart beat), which can cause death due to cardiac arrest. Mild to moderate liver and kidney dysfunction has also been reported after exposure to high concentrations.

Skin contact may produce mild to moderate irritation on contact. Prolonged skin contact can cause blisters and dermatitis because of trichloroethylene's defatting action on the skin's surface. It will pass through unbroken skin to cause toxic systemic effects including symptoms of inhalation. Direct eye contact causes burning with a possibility for injury to the corneal epithelium. Absorption through the skin in combination with other exposures (such as inhalation) can present serious, even life-threatening effects.

On ingestion, trichloroethylene causes irritation and narcosis with burning of the lips, mouth, and gastrointestinal tract. It may cause irregular heartbeat, nausea and vomiting, bloody diarrhea, drowsiness, unconsciousness, and possible pulmonary edema (fluid in the lungs) presumably from aspiration of the swallowed liquids into the lungs during vomiting.

Acute Health Effects

The following acute (short-term) health effects may occur immediately or shortly after exposure to trichloroethylene:

Skin: Moderate irritation with redness and possibly painful inflammation with blisters and burns. It will pass through unbroken skin easily to cause toxic systemic effects.

Eye: Severe irritation and burning with a possibility for injury to the corneal epithelium.

Lung: Nose and throat irritation. Produces intoxicating effects and may cause lung effects such as pulmonary edema (may be fatal).

CNS: Depressive effects by all routes of exposure. There may be numbness around the lips and mouth, dizziness, loss of coordination, headache, somnolence, slurred speech, blurred vision, flushed face, irregular heartbeat, and other effects on nervous system functions. Death can result from cardiac arrest.

Chronic Health Effects

The following chronic (long-term) health effects may occur at some time after exposure to trichloroethylene and can last for months or even years:

Cancer Hazards: According to information presented in the references, trichloroethylene has been shown to cause cancer in test animals and its carcinogenic capability in humans is considered suspect. Some studies indicate that it causes mutations in humans, while others are inconclusive in this regard.

Reproductive Hazard: According to information presented in the references, trichloroethylene may cause teratogenic effects in animals and adversely affect reproduction as well.

Other Chronic Effects: Prolonged skin contact may cause drying, cracking, and chapping of the affected area leading to dermatitis. Prolonged exposure may also cause impaired memory, weakness in the hands and feet, peripheral neuropathy, visual disturbances and impairment, and muscular cramps. There may be liver injury and kidney damage. Chronic toxicity may cause an increased tolerance to alcohol and "degreasers flush" with redness of the face, torso, and arms and a euphoric effects that causes a craving and habitual sniffing of trichloroethylene vapors.

Recommended Risk-Reduction Measures

Personnel should avoid direct contact with trichloroethylene. It is toxic by all routes of exposure. Therefore, prudent risk management requires it be treated as though it posed a significant health risk in the event of overexposure. If a less toxic chemical cannot be substituted for a hazardous substance, then *engineering controls* are the most effective method of reducing exposures. The best protection is to enclose operations and/or provide local exhaust ventilation at the site of chemical release. While not always operationally feasible, isolating operations can also reduce exposure. Using respiratory protection is less effective than the controls mentioned above, but is still advisable whenever working with or around trichloroethylene. For long exposures above the PEL, a supplied-air respirator with full facepiece operated in continuous flow mode, or a self-contained breathing apparatus (SCBA) with full facepiece in pressure demand are the recommended respiratory protection methods of choice. If a full facepiece is not available, then chemical goggles should be worn to protect the eyes. Whenever a chemical splash hazard exists, a face shield and a rubber apron should be worn. To prevent hand and skin exposures, butyl rubber, Viton, or neoprene gloves should be used.

Administrative controls should also be in place to minimize the potential for human exposures. These may include written procedures or policies which specify the methods and techniques that will be practiced whenever personnel are to work with trichloroethylene.

All personnel should receive *training* on the use, hazards, protective measures, emergency actions, and other precautions per 29 CFR 1910.1200 (Hazard Communication), prior to the first assignment in an area where trichloroethylene is used or stored.

If symptoms develop or overexposure is suspected, the following tests are recommended:

- ☑ Lung, kidney, and liver function tests.
- ☑ Complete examination of the nervous system with nerve conduction tests and electroencephalogram (EEG) to detect abnormalities which may be indicative of chronic toxicity.
- ☑ Complete blood count (CBC) with monitoring of blood trichloroethylene levels (for acute exposure).
- ☑ Urinary trichloracetic acid level (for repeated exposures).
- ☑ Electrocardiogram (EKG).
- ☑ Evaluation by a qualified allergist with careful consideration of exposure history and special testing (may help diagnose skin allergy).
- ☑ Consider chest X-ray after acute overexposure (may be negative if taken immediately after exposure due to delayed onset of pulmonary edema).

Any evaluation should include a careful history of past and present symptoms with an examination. Medical tests that simply look for existing damage are not a substitute for controlling exposures. Also, since smoking can cause heart disease, lung cancer, emphysema and other respiratory diseases, smokers exposed to trichloroethylene may experience symptoms more rapidly than non-smokers under the same conditions of exposure. Drinking alcohol (beer, wine, and liquor) may cause flush on the back and neck in people exposed to trichloroethylene. Because more than light alcohol consumption can cause liver damage, drinking alcohol can increase the potential liver damage caused by trichloroethylene.

Other methods to reduce exposure include:

☑ Where possible, enclose operations and use local exhaust ventilation at the site of chemical release. If local exhaust ventilation or enclosure is not used, respiratory protection should be mandatory.

☑ Always ensure that proper protective clothing is worn when using chemical substances.

☑ Wash thoroughly immediately after exposure to trichloroethylene and at the end of the work shift or before eating, drinking, or smoking.

☑ Hazard warning information should be posted in the work area. In addition, as part of an on-going education and training program, all information on the health and safety hazards of trichloroethylene should be communicated to all exposed workers.

☑ Eye wash stations should be provided in the immediate work area for emergency use. If there is a possibility of skin exposure to trichloroethylene, emergency shower facilities should also be provided in the area.

☑ Personnel should never wear contaminated clothing home (family members can be exposed). Clothing should be laundered only by personnel who have been properly trained on the hazards of trichloroethylene.

RISK ASSESSMENT: ENVIRONMENT

General Assessment

The environment is at risk of exposure during transportation, storage, disposal, or destruction of trichloroethylene. In almost every scenario, the threat of environmental exposure is contingent upon the proper handling of the chemical substance. Accidental spills, large or small, can result in fire, explosion, and possible contamination of the surrounding environmental mediums (water, soil, and air).

Trichloroethylene is considered a Class 1C flammable liquid. It will decompose in a fire to produce toxic hydrogen chloride and phosgene gas. It is incompatible with a number of commonly encountered materials including strong alkalis, caustics, and many chemically active metals. Contact can cause fire or explosion. These characteristics require extreme caution in handling, storage, transportation, and disposal. Also, because containers and tankers can explode under conditions of extreme heat, special consideration is required during any emergency situation involving a leak or spill of trichloroethylene. Should trichloroethylene ever come into contact with incompatible substances during use, transportation, storage, or disposal, the formation of highly toxic and/or highly explosive commodities (including hydrogen chloride) is extremely possible.

The proper disposal method for trichloroethylene is to mix it with a combustible solvent and then burn it in a permitted chemical incinerator equipped with an afterburner and scrubber.

Trichloroethylene may enter the environment through industrial discharges or spills.

Acute Ecological Effects

Acute (short-term) toxic effects may include the death of animals, birds, or fish and death or low growth rate in plants. Acute effects are seen 2 to 4 days after animals or plants are exposed to trichloroethylene.

Trichloroethylene has moderate acute toxicity to aquatic life. Insufficient data are available on the short-term effects of trichloroethylene exposure to birds, plants, or land animals.

Chronic Ecological Effects

Chronic (long-term) toxic effects may include shortened life span, reproductive problems, lower fertility, and changes in appearance or behavior in exposed animals. These effects can be seen long after first exposure(s) to toxic chemicals.

Trichloroethylene has moderate chronic toxicity to aquatic life. Insufficient data are available on the long-term effects of trichloroethylene exposure to birds, plants, or land animals.

Water Solubility

Trichloroethylene is slightly soluble in water. Concentrations of 1 to 100 milligrams may mix with a liter of water.

Persistence in the Environment

Trichloroethylene is non-persistent in water, with a half-life of less than 2 days. The half-life of a pollutant is the amount of time it takes for one half of the chemical to be degraded. About 99.6% of trichloroethylene will eventually end up in air; the remainder will end up in water.

Bioaccumulation in Aquatic Organisms

Some substances increase in concentration, or *bioaccumulate,* in living organisms as they breathe contaminated air, drink contaminated water, or eat contaminated food. These chemicals can become concentrated in the tissues and internal organs of animals as well as humans. The concentration of trichloroethylene found in fish tissue is expected to be somewhat higher than the average concentration of trichloroethylene in the water from which the fish was taken.

Recommended Risk-Reduction Measures

Proper training of all transporters will reduce the likelihood of a mishap or accident resulting in a leak or spill to the environment. The correct use of DOT labeling (per the MSDS) on all containers, trucks, and rail cars will enable emergency responders to react properly and quickly to any disaster thereby reducing the potential risk to the environment and to personnel.

Storage of trichloroethylene should be segregated from incompatible chemicals to minimize the risk of cross-contamination or contact. Buildings designated for storage should be equipped with appropriate fire protection systems (alarms, sprinklers, emergency lighting, portable extinguishers).

If a spill or leak to the environment has occurred, fire department, emergency response and/or hazardous materials spill personnel should be notified immediately. Cleanup should be attempted only by those trained in proper spill containment procedures. Contaminated soils should be removed for incineration and replaced with clean soil. If trichloroethylene should contact the water table, aquifer, or navigable waterway, cleanup should be started immediately. It is only slightly soluble in water and total remediation may be possible. The local and/or state emergency response authorities must be notified. A comprehensive emergency response or disaster preparedness/recovery plan should be in place prior to any operations involving the use, transportation, storage, or disposal of trichloroethylene. If trichloroethylene is spilled or leaked, the following specific steps are recommended:

☑ Restrict persons not wearing protective clothing from area of spill or leak until cleanup is complete and area can be opened for normal work.

☑ Remove all ignition sources, ventilate area of leak.

☑ Absorb liquids in vermiculite, dry sand, earth, or similar material and deposit in sealed containers.

RISK ASSESSMENT: BUSINESS

General Assessment

Accidents or mishaps involving trichloroethylene can present a significant threat to business operations. The loss or damage of equipment or facilities can significantly affect fiscal viability. Lawsuits that may result from personnel injury/death, public exposures, and/or environmental contamination will also require a serious expenditure of resources. Media attention surrounding an injury, death, or environmental damage can also result in a loss of profits and loss of current as well as future business. Always remember that anytime the terms "cancer," "carcinogen," or "reproductive hazard" are used, public emotion, ignorance, hysteria, and anxiety can all run equally high. This must be carefully considered when implementing public relations policies.

Recommended Risk-Reduction Measures

Company attorneys, safety and health professionals, and environmental specialists should be involved in the development of any procedures dealing with safety or emergency response. A company official should be pre-designated as a public relations officer with specific training in dealing with the press. Corporate plans and policies should be developed, approved, and implemented long before any need for such arises.

REFERENCES

American Conference of Governmental Industrial Hygienists. 1988. *Documentation of the Threshold Limit Values and Biological Exposure Indices,* 5th Edition (with updates). Cincinnati: ACGIH

Gardner, R. W. 1994. *Chemical Intolerance.* New York: CRC Press/Lewis Publishers.

Jones, F. E. 1994. *Toxic Organic Vapors in the Workplace.* New York: CRC Press/Lewis Publishers.

Lewis, R. J., Sr. 1992. *Sax's Dangerous Properties of Industrial Materials,* Volumes 1, 2, and 3, 8th Edition. New York: Van Nostrand Reinhold.

Meyer, E. 1990. *Chemistry of Hazardous Materials.* Englewood Cliffs, NJ: Prentice Hall.

Patnaik, P. 1992. *A Comprehensive Guide to the Hazardous Properties of Chemical Substances.* New York: Van Nostrand Reinhold.

Rea, W. J. 1995. *Chemical Sensitivity,* Vol. I-III. New York: CRC Press/Lewis Publishers.

Richardson, M. 1995. *Dictionary of Substance and Their Effects,* Vol. I-VI. New York: CRC Press/Lewis Publishers.

Sacarello, H. L. A. 1994. *The Comprehensive Handbook of Hazardous Materials.* New York: CRC Press/Lewis Publishers.

U.S. Department of Health and Human Services, National Institute for Occupational Safety and Health. 1994. *NIOSH Pocket Guide to Chemical Hazards.* Washington, D.C.: U.S. Government Printing Office.

MATERIAL SAFETY DATA SHEET

CHEMICAL NAME	CRC PRESS/LEWIS PUBLISHERS DISCLAIMER
TRIETHYLAMINE	

HAZARD WARNING INFORMATION

HEALTH	FIRE	REACTIVE	OTHER	DEGREE OF HAZARD	COLOR CODING	OTHER CODES
3	3	0		0 = Minimum Hazard 1 = Slight Hazard 2 = Moderate Hazard 3 = Serious Hazard 4 = Severe Hazard	HEALTH = BLUE FIRE = RED REACTIVITY = YELLOW OTHER = WHITE	OX = Oxidizer ACID = Acid ALK = Alkali COR = Corrosive W = Use No Water

SECTION I - GENERAL INFORMATION

Characterization	**Aliphatic Amine**	RCRA Number	**D001**
		EPA Class	**Characteristic (I) Waste**
DOT Proper Shipping Name	**Triethylamine**	Chemical Abstract Service (CAS) Number	**121-44-8**
DOT Hazard Class and Label Requirements	**Flammable Liquid; Corrosive**	DOT Emergency Guide Code	**68**
DOT Identification Number	**UN 1296**	Chemical Formula	**$(C_2H_5)_3N$**

Synonyms

(Diethylamino)ethane; n,n-diethylethanamine; TEA; ethanamine; n,n-diethy-; TEN.

SECTION II - HAZARDOUS INGREDIENTS AND IDENTITY INFORMATION

Hazardous Components (specific identity)	OSHA Exposure Criteria	NIOSH Exposure Criteria	Immediately Dangerous to Life and Health (IDLH)	ACGIH Exposure Criteria
Triethylamine (derivation: By the reaction of n,n-diethylacetamide with lithium aluminum hydride; or by vapor phase alkylation of ammonia with ethanol; or from ethyl chloride and ammonia with heat and pressure). **1 ppm = 4.21 mg/m^3**	**PEL:** **10 ppm** **42 mg/m^3** **STEL:** **15 ppm** **63 mg/m^3**	**REL:** **Not Established** **STEL:** **Not Established**	**200 ppm**	**TLV (skin):** **1 ppm** **4 mg/m^3** **STEL (skin):** **5 ppm** **20.7 mg/m^3**

SECTION III - PHYSICAL AND CHEMICAL CHARACTERISTICS

Boiling Point	**193°F (89°C)**	Specific Gravity (H_2O = 1)	**0.73**
Vapor Pressure (mmHg)	**54 at 68°F (20°C)**	Molecular Weight	**101.2**
Vapor Density (Air = 1)	**3.48**	Freezing Point	**-175°F (-115°C)**

Solubility

Slightly soluble in water. Soluble in acetone, benzene, alcohol, and ether.

Appearance and Odor

Colorless liquid with a strong, ammonia-like odor. Odor Threshold = <1 ppm.

SECTION IV - FIRE AND EXPLOSION HAZARD DATA

Flash Point (method used)	**20°F (-6.6°C) open cup**	Explosive Limits in Air % by Volume	**LEL: 1.2% UEL: 8.0%**
NFPA Classification	**Class 1B Flammable Liquid**	Autoignition Temperature	**480°F (249°C)**

Extinguishing Media

Use water spray, dry chemical, alcohol-resistant foam, or carbon dioxide.

Special Fire Fighting Procedures

Triethylamine is an extremely flammable, very dangerous fire and explosion hazard. Poisonous gases are evolved from fire. Wear a self-contained breathing apparatus with full facepiece. Ensure ALL sources of escaping gas are controlled. Move containers from fire area if it can be done without risk. Water stream may be ineffective on fire. Keep cooling containers with water long after fire is out.

Unusual Fire and Explosion Hazards

Containers may explode in fire. Vapors are heavier than air and will travel for a great distance to an ignition source and flashback to cause fire or explosion.

SECTION V - REACTIVITY DATA

Stability		
Stability		Conditions to Avoid **Triethylamine is normally stable in its closed containers, during routine operations (those that do not expose it to ignition sources). Avoid contact with heat, sparks, flame, and other sources of ignition.**
Stable **X**	Unstable	Incompatibility *(materials to avoid)* **Acids, oxidizing materials, chlorine, hypochlorite, halogenated compounds, reactive organic compounds, some metals, dinitrogen tetroxide, and heated maleic anhydride. Attacks plastics, rubber, and coatings.**
Hazardous Polymerization		Conditions to Avoid **Under normal conditions of temperature and pressure, hazardous polymerization of triethylamine is not expected to occur.**
May Occur	Will Not Occur **X**	Hazardous Decomposition or By-products **If involved in fire, triethylamine can produce toxic hydrocarbons, amine vapors, and oxides of carbon and nitrogen.**

SECTION VI - HEALTH HAZARD DATA

	Inhalation?	Absorption (skin)?	Ingestion?
Primary Route(s) of Entry:	X	X	X

Health Hazards

INHALATION: **Vapors are irritating to the eyes, skin, and respiratory tract. Causes pulmonary edema (fluid in the lungs), which is a medical emergency and can lead to death. There may also be coughing, choking, difficulty breathing, nausea, and headache.**

ABSORPTION: **Causes burns, dermatitis, and possible sensitization. Absorbs through intact skin. Eye contact causes irritation, conjunctivitis, visual field changes, foggy vision, blue haze, and halo phenomena. Possible corneal edema and corneal burns.**

INGESTION: **Irritation or burns of the esophagus and gastrointestinal tract.**

Carcinogenicity	NTP Listed?	IARC Cancer Review Group?	OSHA Regulated?	Target Organs?
Unknown Human Unknown Animal	**No**	**No**	**29 CFR 1910.1000 Table Z-1**	**Eyes, skin, CVS, liver, respiratory tract, kidneys**

Medical Conditions Generally Aggravated by Exposure

Respiratory disease and possibly cardiovascular, liver, skin, and kidney diseases may be aggravated.

Emergency and First-aid Procedures

Eye contact: Do not allow victims to keep their eyes tightly shut. Flood with large amounts of water for 15 minutes (minimum). Skin contact: Remove all contaminated clothing. Immediately wash area with large amounts warm water. For inhalation: Remove the person from exposure. Provide respiratory assistance and CPR. Transfer promptly to medical facility. Observe 48 hours for lung effects. If swallowed: Contact poison control center. Unless advised otherwise, give 1-2 glasses of water to dilute. Do NOT induce vomiting.

SECTION VII - PRECAUTIONS FOR SAFE HANDLING AND USE

Steps to be Taken in Case Material is Released or Spilled

Remove all ignition sources. Ventilate area. Absorb liquids in vermiculite or other material and place in sealed containers. Use non-sparking tools. Use water spray to disperse vapors.

Preferred Waste Disposal Method

Atomize in a suitable combustion chamber equipped with an appropriate effluent gas cleaning device.

Precautions to be Taken in Handling and Storage

Do not store with incompatible materials or in the presence of heat since violent reactions or explosions can occur. Store in tightly closed pressurized containers in a cool, dark, well-ventilated area away from sunlight. Keep away from ignition sources such as fire, sparks, and flame.

Other Precautions and Warnings

Bulk storage of triethylamine is not recommended. Containers may explode in fire or under conditions of extreme heat (do not store outdoors or in direct sunlight). Prevent damage to containers.

SECTION VIII - CONTROL MEASURES AND PERSONAL PROTECTIVE EQUIPMENT

Respiratory Protection *(specify type)*

For exposures to the PEL, use a cartridge-type respirator. For higher exposures, use a powered supplied-air respirator or a self-contained breathing apparatus (SCBA) with full facepiece operated in pressure demand or other positive pressure mode.

Ventilation

Local exhaust or general mechanical systems recommended.

Protective Gloves	Eye Protection	Other Protective Clothing
Nitryl, Viton, Saranex, or Similar	**Chemical Goggles and/or Face Mask**	**Protective Apron**

Work/Hygiene Practices

Always wash hands thoroughly after using chemical; never bring food, drink, or smoking materials into vicinity of chemicals.

TRIETHYLAMINE

$(C_2H_5)_3N$

CAS: 121-44-8

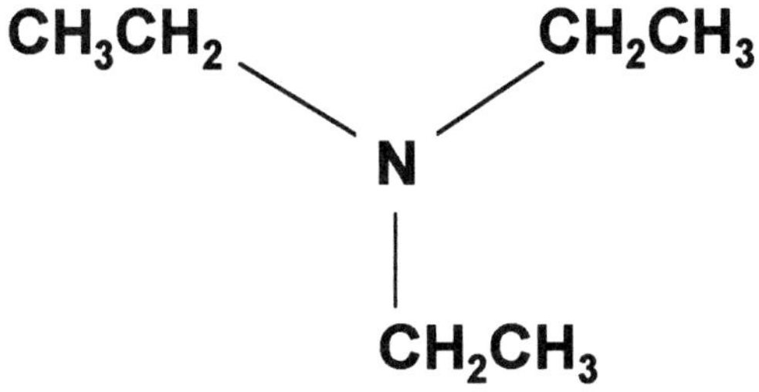

IDENTIFICATION AND TYPICAL USES

Triethylamine is a colorless liquid with an ammonia odor. It is used as a catalytic solvent in chemical synthesis; in accelerator activators for rubber; in wetting, penetrating, and waterproofing agents of quaternary ammonium types; for curing and hardening of polymers and resins; as a corrosion inhibitor, and a propellant.

RISK ASSESSMENT: HEALTH

General Assessment

Triethylamine is toxic through *inhalation, ingestion,* and dermal *absorption.* Its corrosive nature makes it a serious contact hazard with skin, eyes, or other tissues. It is not known if exposure to triethylamine will lead to cancer. However, human mutation data have been reported and triethylamine has caused adverse reproductive effects in animals.

Inhalation of triethylamine vapors causes irritation of the eyes, nose, throat, and respiratory tract. Exposure can lead to pulmonary edema (fluid buildup in the lungs) which is a medical emergency and can be fatal. Symptoms of congestion, cough, phlegm, choking, nausea, headache, tightness and possibly pain in chest, and difficulty breathing may be delayed up to 48 hours thereby creating a false sense of security with regard to health hazard/exposure risk.

Skin or eye contact with the liquid causes irritation. and possible corrosive burns with subsequent tissue damage. There can be conjunctivitis, visual field changes (foggy vision, blue haze, halos around light), corneal edema, and corneal burns. Damage may be permanent. Prolonged skin contact can lead to dermatitis and may cause sensitization in some individuals. Ingestion can cause severe burns to the mouth, esophagus, and gastrointestinal tract. In animals, exposure has caused heart, liver, and kidney damage.

Acute Health Effects

The following acute (short-term) health effects may occur immediately or shortly after exposure to triethylamine:

Skin: Redness, smarting, pain, possible dermatitis, and burns.

Eye: Causes irritation, redness, swelling of the cornea, severe pain, and vision changes.

Lung: Vapor is seriously irritating and therefore poses a significant threat of damage to the lungs. Irritation can lead to pulmonary edema (fluid buildup in the lungs).

Other: Triethylamine is highly corrosive and can cause deep tissue destruction.

Chronic Health Effects

The following chronic (long-term) health effects may occur at some time after exposure to triethylamine and can last for months or even years:

Cancer Hazards: According to information presented in the references, it is not known if exposure to triethylamine will lead to cancer. However, mutation data have been reported and many scientists believe that long-term exposure to such chemicals may pose a cancer risk.

Reproductive Hazards: There is limited evidence to indicate that triethylamine causes adverse reproductive effects in test animals.

Other Chronic Effects: In animals, there have been myocardial (heart), hepatic (liver), and renal (kidney) damage on prolonged exposure to triethylamine. Repeated skin contact may cause drying, chapping, cracking, scaling, and possible secondary infection and dermatitis in some individuals.

ʊ *Recommended Risk-Reduction Measures*

Personnel should avoid direct contact with triethylamine. It is extremely corrosive and highly toxic by all routes of exposure. If a less toxic chemical cannot be substituted for a hazardous substance, then *engineering controls* are the most effective method of reducing exposures. The best protection is to separate operations and provide local explosion-proof exhaust ventilation at the site of chemical release. While not always operationally feasible, isolating operations can also reduce exposure. Using equipment with explosion-proof designs and ensuring all metal containers and tools are electrically grounded and bonded can further reduce the potential for injury or damage. Using respiratory protection is less effective than the controls mentioned above, but is still advisable whenever working with triethylamine. For exposures above the PEL (10 ppm), use an MSHA/NIOSH-approved chemical cartridge respirator with a full facepiece and cartridges suitable for protection against amines. For higher or more frequent exposures, use a powered supplied-air respirator operated in continuous flow or other positive pressure mode, or a self-contained breathing apparatus (SCBA) with full facepiece operated in pressure demand mode. If a full facepiece is not available, then chemical goggles should be worn to protect the eyes. Whenever a chemical splash hazard exists, a face shield and a rubber or leather apron should be worn. To prevent hand and skin exposures with the liquid, thermal gloves should be used. Viton™, nitrile rubber, Saranex™, and Responder™ appear to provide adequate protection. Glove manufacturers should be contacted and permeation studies obtained *before* final glove selections are made for triethylamine.

Administrative controls should also be in place to minimize the potential for human exposures. These may include written procedures or policies which specify the methods and techniques that will be practiced whenever personnel are to work with triethylamine.

All personnel should receive *training* on the use, hazards, protective measures, emergency actions, and other precautions per 29 CFR 1910.1200 (Hazard Communication), prior to the first assignment in an area where triethylamine is used or stored. Before beginning employment and at regular intervals thereafter (e.g., annually), the following medical tests are recommended:

- ☑ Lung function tests (establish baseline).

If symptoms develop or overexposure is suspected, the following additional tests may be helpful:

- ☑ Lung function tests (compare to baseline).
- ☑ Consider chest X-ray after acute overexposure (may be negative if taken immediately after exposure due to delayed onset of pulmonary edema).

Other methods to reduce exposure to chemicals include:

- ☑ Where possible, use local exhaust ventilation at the site of chemical release. If local exhaust ventilation or enclosure is not used, respiratory protection should be mandatory.
- ☑ Prevent static sparks by electrically grounding and bonding all equipment used with or around triethylamine.
- ☑ Always ensure that proper protective clothing is worn when using chemical substances.
- ☑ Wash thoroughly immediately after exposure to triethylamine and at the end of the work shift or before eating, drinking, or smoking.
- ☑ Safety shower and eyewash stations should be readily available in work areas where triethylamine is used or stored.
- ☑ Hazard warning information should be posted in the work area. In addition, as part of an on-going education and training program, all information on the health and safety hazards of triethylamine should be communicated to all exposed and potentially exposed workers.

RISK ASSESSMENT: ENVIRONMENT

General Assessment

The environment is at risk of exposure during transportation, storage, disposal, or destruction of triethylamine. In almost every scenario, the threat of environmental exposure is contingent upon the proper handling of the chemical substance. Accidental releases, large or small, can result in fire, explosion, and possible contamination of the surrounding environmental mediums (water, soil, and air).

Triethylamine is Class 1B flammable liquid (per OSHA 29 CFR 1910.106). Its extremely low flash point and moderate point present a serious fire and explosion hazard risk. Also, since it can react violently in the presence of oxidizers (chlorine, bromine), and strong acids (nitric, sulfuric), triethylamine requires special consideration during any emergency involving a leak or release of triethylamine. It is highly corrosive and will attack some forms or rubber, plastic, and certain coating materials, so careful selection of emergency response tools and equipment is warranted.

Triethylamine can enter the environment through industrial discharges or through spills.

Acute Ecological Effects

Acute (short-term) toxic effects may include the death of animals, birds, or fish, and death or low growth rate in plants. Acute effects are seen 2 to 4 days after animals or plants are exposed to triethylamine.

Insufficient data are available to evaluate or predict the short-term (acute) effects of triethylamine on aquatic life, plants, birds, or land animals.

Chronic Ecological Effects

Chronic (long-term) toxic effects may include shortened life span, reproductive problems, lower fertility, and changes in appearance or behavior in exposed animals. These effects can be seen long after first exposure(s) to toxic chemicals.

Insufficient data are available to evaluate or predict the long-term (chronic) effects of triethylamine to aquatic life, plants, birds, or land animals.

Water Solubility

Triethylamine is moderately soluble in water. Concentrations of 1 to 1000 milligrams will mix with a liter of water.

Persistence in the Environment

Triethylamine is slightly persistent in the aquatic environment with a half-life of 2 to 20 days. The half-life of a pollutant is the amount of time it takes for one half of the chemical to be degraded.

Bioaccumulation in Aquatic Organisms

Some substances increase in concentration, or *bioaccumulate,* in living organisms as they breathe contaminated air, drink contaminated water, or eat contaminated food. These chemicals can become concentrated in the tissues and internal organs of animals as well as humans.

Triethylamine is not expected to accumulate in the tissues of fish.

Recommended Risk-Reduction Measures

Proper training of all transporters will reduce the likelihood of a mishap or accident resulting in a leak or release to the environment. The correct use of DOT labeling (per the MSDS) on all containers, trucks, and rail cars will enable emergency responders to react properly and quickly to any disaster thereby reducing the potential risk to the environment and to personnel.

Storage of triethylamine should be segregated from incompatible chemicals to minimize the risk of cross-contamination or contact. Buildings designated for storage should be equipped with appropriate fire protection systems (alarms, sprinklers, emergency lighting, portable extinguishers). Maximum explosion-proof design is recommended for storage areas.

If a spill or release to the environment has occurred, fire department, emergency response, and/or hazardous materials spill personnel should be notified immediately. Cleanup should be attempted only by those trained in proper containment procedures using non-sparking tools. Contaminated soils should be removed for incineration and replaced with clean soil. If triethylamine should contact the water table, aquifer, or navigable waterway, time is of the essence. It is slightly soluble in water and total containment and remediation may not be possible. When such spills occur, it may be necessary to contact the local and/or state emergency response authorities. A comprehensive emergency response or disaster preparedness/recovery plan should be in place prior to any operations involving the use, transportation, storage, or disposal of triethylamine. If triethylamine is spilled or leaked, the following specific steps are recommended:

- ☑ Restrict persons not wearing protective clothing from area of spill or leak until cleanup is complete and area can be opened for normal work.
- ☑ Remove all ignition sources.
- ☑ Ventilate area of spill or leak.
- ☑ It may be necessary to dispose of triethylamine as a hazardous waste. The responsible state agency or the regional office of the federal Environmental Protection Agency (EPA) should be contacted for specific recommendations.

RISK ASSESSMENT: BUSINESS

General Assessment

Accidents or mishaps involving triethylamine can present a significant threat to business operations. The loss or damage of equipment or facilities can significantly affect fiscal viability. Lawsuits that may result from personnel injury, public exposures, and environmental contamination will also require a serious expenditure of resources. Media attention surrounding an injury, illness, or environmental damage can also result in a loss of profits and loss of current as well as future business. Always remember that anytime the term "reproductive hazard" is used, public hysteria, emotion, anxiety, and ignorance can run equally high. This must be carefully considered when drafting public relations policies.

Recommended Risk-Reduction Measures

Company attorneys, safety and health professionals, and environmental specialists should be involved in the development of any such procedures. A company official should be pre-designated as a public relations officer with specific training in dealing with the press. Corporate plans and policies should be developed, approved, and implemented long before any need for such arises.

REFERENCES

American Conference of Governmental Industrial Hygienists. 1988. *Documentation of the Threshold Limit Values and Biological Exposure Indices,* 5th Edition (with updates). Cincinnati: ACGIH

Anthony, C. P., and N. J. Kolthoff. 1971. *Textbook of Anatomy and Physiology.* St. Louis: C. V. Mosby Company.

Asante-Duah, D. K. 1993. *Hazardous Waste Risk Assessment.* New York: CRC Press/Lewis Publishers.

Bloom, A. D. and F. J. DeSerres. 1995. *Ecotoxicity and Human Health.* New York: CRC Press/Lewis Publishers.

Brusick, D. J. 1994. *Methods for Genetic Risk Assessment.* New York: CRC Press/Lewis Publishers.

Bueche, J. R. 1972. *Principles of Physics,* 2nd Edition. New York: McGraw-Hill.

Burrell, R., D. K. Flaherty, and L. J. Sauers. 1992. *Toxicology of the Immune System.* New York: Van Nostrand Reinhold.

Calabrese, E. J. 1994. *Biological Effects of Low Level Exposures.* New York: CRC Press/Lewis Publishers.

Cockerman, L. G., B. S. Shane. 1994. *Basic Environmental Toxicology.* New York: CRC Press/Lewis Publishers.

Cothern, R. R. 1993. *Comparative Environmental Risk Assessment.* New York: CRC Press/Lewis Publishers.

Danse, I. R. 1991. *Common Sense Toxics in the Workplace.* New York: Van Nostrand Reinhold.

Environmental Protection Agency. 1983. *Chemical Hazard Information Profiles.* Washington, D.C.: U.S. EPA.

Fodden, C. R., and J. L. Weddell. 1992. *Hazardous Materials Emergency Action Data.* New York: CRC Press/Lewis Publishers.

Forseberg, K., and L. H. Keith. 1993. *Chemical; Protective Clothing Permeation and Degradation Database.* Chelsea, MI: Lewis Publishers.

Gardner, R. W. 1994. *Chemical Intolerance.* New York: CRC Press/Lewis Publishers.

Hallenbeck, W. H. 1993. *Quantitative Risk Assessment for Environmental and Occupational Health,* 2nd Edition. New York: CRC Press/Lewis Publishers.

Hodgson, E., R. B. Mailman, and J. E. Chambers. 1988. *Dictionary of Toxicology.* London: The MacMillian Company.

Hosty, J. W., and P. Foster. 1990. *A Practical Guide to Chemical Spill Response.* New York: Van Nostrand Reinhold.

Howard, P. H., et. al. 1991. *Handbook of Environmental Degradation Rates.* New York: CRC Press/Lewis Publishers.

Jones, F. E. 1994. *Toxic Organic Vapors in the Workplace.* New York: CRC Press/Lewis Publishers.

Kamrin, M. A. 1988. *Toxicology: A Primer on Principles and Applications.* Chelsea, Michigan: Lewis Publishers.

Keith, L. H. (Editor). 1992. *Integrated Risk Information System (IRIS) Chemical Information Database.* Prepared by the U.S. Environmental Protection Agency. Boca Raton, FL: CRC Press/Lewis Publishers.

Landis, W. G., M. H. Yu. 1995. *Introduction to Environmental Toxicology.* New York: CRC Press/Lewis Publishers.

Lauwerys, R. R., P. Hoet. 1993. *Industrial Chemical Exposure.* New York: CRC Press/Lewis Publishers.

Lewis, R. J., Sr. 1993. *Hawley's Condensed Chemical Dictionary,* 12th Edition. New York: Van Nostrand Reinhold.

Lewis, R. J., Sr. 1992. *Sax's Dangerous Properties of Industrial Materials,* Volumes 1, 2, and 3, 8th Edition. New York: Van Nostrand Reinhold.

Lide, D. R. 1995. *Handbook of Chemistry and Physics,* 76th Edition. New York: CRC Press/Lewis Publishers.

Manahan, S. E. 1992. *Toxicological Chemistry,* 2nd Edition. New York: CRC Press/Lewis Publishers.

Meyer, E. 1990. *Chemistry of Hazardous Materials.* Englewood Cliffs, NJ: Prentice Hall.

Patnaik, P. 1992. *A Comprehensive Guide to the Hazardous Properties of Chemical Substances.* New York: Van Nostrand Reinhold.

Rea, W. J. 1995. *Chemical Sensitivity,* Vol. I-III. New York: CRC Press/Lewis Publishers.

Richardson, M. 1995. *Dictionary of Substance and Their Effects,* Vol. I-VI. New York: CRC Press/Lewis Publishers.

Sacarello, H. L. A. 1994. *The Comprehensive Handbook of Hazardous Materials.* New York: CRC Press/Lewis Publishers.

Sherman, J. D. 1988. *Chemical Exposure and Disease.* New York: Van Nostrand Reinhold.

Talbott, E., G. Craun. 1995. *An Introduction to Environmental Epidemiology.* New York: CRC Press/Lewis Publishers.

Torkelson, T. R. V. K. Rowe. 1981. *Patty's Industrial Hygiene and Toxicology,* Vol. 2B. New York: Wiley-Interscience.

U.S. Department of Health and Human Services, National Institute for Occupational Safety and Health. 1994. *NIOSH Pocket Guide to Chemical Hazards.* Washington, D.C.: U.S. Government Printing Office.

MATERIAL SAFETY DATA SHEET

CHEMICAL NAME	
URETHANE	**CRC PRESS/LEWIS PUBLISHERS DISCLAIMER:** THE INFORMATION AND RECOMMENDATIONS PRESENTED HEREIN ARE BASED ON SOURCES BELIEVED TO BE RELIABLE. CRC MAKES NO REPRESENTATION ON ITS COMPLETENESS OR ACCURACY. IT IS THE RESPONSIBILITY OF THE USER TO DETERMINE THE CHEMICAL'S SUITABILITY FOR ITS INTENDED USE, THE CHEMICAL'S SAFE USE, AND THE CHEMICAL'S PROPER DISPOSAL. NO REPRESENTATIONS AND/OR WARRANTIES, EITHER EXPRESSED OR IMPLIED, OF THE MERCHANTABILITY OR FITNESS FOR A PARTICULAR PURPOSE, OR OF ANY OTHER NATURE, ARE MADE WITH RESPECT TO THE INFORMATION PROVIDED IN THIS MSDS OR TO THE CHEMICAL TO WHICH INFORMATION MAY REFER. CRC NEITHER ASSUMES NOR AUTHORIZES ANY OTHER PERSON TO ASSUME FOR IT, ANY OTHER ADDITIONAL RESPONSIBILITY OR LIABILITY FOR THE USE OF, OR RELIANCE UPON, THIS INFORMATION.

HAZARD WARNING INFORMATION

HEALTH	FIRE	REACTIVE	OTHER	DEGREE OF HAZARD	COLOR CODING	OTHER CODES
2	0	0		0 = Minimum Hazard 1 = Slight Hazard 2 = Moderate Hazard 3 = Serious Hazard 4 = Severe Hazard	HEALTH = BLUE FIRE = RED REACTIVITY = YELLOW OTHER = WHITE	OX = Oxidizer ACID = Acid ALK = Alkali COR = Corrosive ~~W~~ = Use No Water

SECTION I - GENERAL INFORMATION

Characterization	RCRA Number	EPA Class
Carbamate	**U238**	**Toxic Waste**

DOT Proper Shipping Name: **No Citation**	Chemical Abstract Service (CAS) Number: **51-79-6**
DOT Hazard Class and Label Requirements: **No Citation**	DOT Emergency Guide Code: **No Citation**
DOT Identification Number: **No Citation**	Chemical Formula: **$CO(NH_2)OC_2H_5$**

Synonyms

Ethyl carbamate; ethyl urethane; carbamic acid, ethyl ester; Leucethane; Pracarbamine; U-Compound.

SECTION II - HAZARDOUS INGREDIENTS AND IDENTITY INFORMATION

Hazardous Components (specific identity)	OSHA Exposure Criteria	NIOSH Exposure Criteria	Immediately Dangerous to Life and Health (IDLH)	ACGIH Exposure Criteria
Urethane (derivation: By heating ethanol and urea nitrate to a high temperature of 248-266°F/120-130°C; by action of ammonia on ethyl carbonate or ethyl chloroformate).	**PEL:** Not Established **STEL:** Not Established	**REL:** Not Established **STEL:** Not Established	**Not Determined**	**TLV:** Not Established **STEL:** Not Established

SECTION III - PHYSICAL AND CHEMICAL CHARACTERISTICS

Boiling Point Range: **360-363°F (182-184°C)**	Specific Gravity (H_2O = 1): **0.99**
Vapor Pressure (mm Hg): **0.36 at 77°F (25°C)**	Molecular Weight (atomic weight): **89.09**
Vapor Density (Air = 1): **3.07**	Freezing Point Range: **118-122°F (48-50°C)**

Solubility

Soluble in water, alcohol, ether, glycerol, and chloroform; slightly soluble in olive oil.

Appearance and Odor

Colorless, odorless, crystalline solid or a white powder.

SECTION IV - FIRE AND EXPLOSION HAZARD DATA

Flash Point (method used): **Not Reported**	Explosive Limits in Air % by Volume: **LEL: Not Reported UEL: Not Reported**
NFPA Classification: **Combustible Solid**	Autoignition Temperature: **Not Reported**

Extinguishing Media

Use dry chemical, carbon dioxide, water spray, fog, or regular foam.

Special Fire Fighting Procedures

Urethane is a combustible solid but is often difficult to ignite. Wear full protective clothing and SCBA. Remain clear of smoke, water fallout and water runoff. Move containers from fire area if it can be done without risk. Cool exposed containers. Toxic hydrogen chloride and carbon monoxide are formed in fire. Do not release runoff to sewers or waterways; contain and collect for proper disposal.

Unusual Fire and Explosion Hazards

None reported.

SECTION V - REACTIVITY DATA

Stability		Conditions to Avoid Urethane is normally stable in closed containers under routine conditions of storage and handling. Avoid contact with incompatible materials and do not store near heat or heat source.
Stable X	Unstable	Incompatibility *(materials to avoid)* Incompatible with alkalis, acids, antipyrine, chloral hydrate, camphor, menthol salol, and thymol. Reacts with phosphorous pentachloride to form explosive products. Decomposes in hot aqueous acids or alkalis.
Hazardous Polymerization		Conditions to Avoid Under normal conditions of temperature and pressure, hazardous polymerization of urethane will not occur.
May Occur	Will Not Occur X	Hazardous Decomposition or By-products At high temperatures, urethane decomposes and may produce poisonous gases, including toxic fumes of nitrogen oxides.

SECTION VI - HEALTH HAZARD DATA

Primary Route(s) of Entry:	Inhalation?	Absorption (skin/eye)?	Ingestion?
	X	X	X

Health Hazards

INHALATION: Irritating to the skin, eyes, mucous membranes, and respiratory tract. Large doses cause bone marrow depression and possible brain injury. It may also produce depression of the central nervous system with dizziness, coma, nausea, vomiting, kidney, and liver injury.

ABSORPTION: Will pass through unbroken skin and enter the bloodstream. Contact may severely irritate the skin and eyes.

INGESTION: Poisonous to humans by ingestion. Systemic effects due to central nervous system action include tremors, convulsion, excitement, ataxia, and gastritis.

Carcinogenicity	NTP Listed?	IARC Cancer Review Group?	OSHA Regulated?	Target Organs?
Suspected Human Confirmed Animal	5th Annual Report	Group 2B	No	Central nervous system, liver, kidney, lung, lymph.

Medical Conditions Generally Aggravated by Exposure

Kidney, liver, and lung diseases, and convulsive disorders may be aggravated by exposure to urethane.

Emergency and First-aid Procedures

Eye contact: Do not allow victim to keep eyes tightly shut. Flush immediately with water for 15 minutes; seek medical attention. Skin contact: Remove all contaminated clothing. Immediately wash area with flooding amounts of water, then triple-wash with soap and water. Inhalation: Remove the person from exposure. Provide respiratory assistance and CPR. Transfer to medical facility. If swallowed: Seek medical attention immediately. Never give anything by mouth to an unconscious or convulsing person. If victim is conscious, give 1 to 2 glasses of water. The decision to induce vomiting should be carefully considered since CNS depression may cause convulsions.

SECTION VII - PRECAUTIONS FOR SAFE HANDLING AND USE

Steps to be Taken in Case Material is Released or Spilled

Remove all sources of heat. Ventilate area of spill. Restrict those not involved in cleanup from entering area. Notify appropriate authorities. Collect powder materials using HEPA vacuum. Do NOT dry sweep. Absorb any liquid solution spills in vermiculite and place in sealed drum for disposal.

Preferred Waste Disposal Method

No citation.

Precautions to be Taken in Handling and Storage

Store in tightly closed containers in cool, dry, dark, well-ventilated area away from heat.

Other Precautions and Warnings

Avoid generating dusty conditions in storage areas. Protect containers from physical damage.

SECTION VIII - CONTROL MEASURES AND PERSONAL PROTECTIVE EQUIPMENT

Respiratory Protection *(specify type)*

No exposure levels established and urethane is a possible carcinogen. Use a NIOSH-approved supplied-air respirator or a self-contained breathing apparatus (SCBA) with full facepiece operated in positive pressure mode.

Ventilation

Local exhaust (preferred) at site of chemical work or general ventilation.

Protective Gloves	Eye Protection	Other Protective Clothing
Impervious Material	Chemical Goggles or Face Mask	Protective Apron, Boots (avoid skin contact)

Work/Hygiene Practices

Always wash hands thoroughly after using chemical; never bring food, drink, or smoking materials into vicinity of chemicals.

URETHANE

$CO(NH_2)OC_2H_5$

CAS: 51-79-6

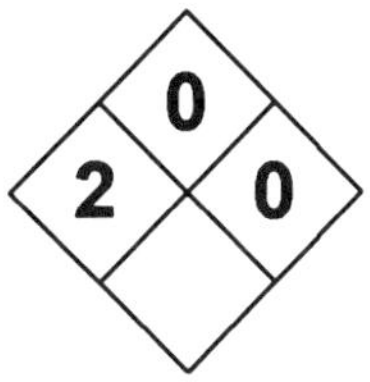

IDENTIFICATION AND TYPICAL USES

Urethane is a colorless, odorless, crystalline solid or a white powder. It is used as an intermediate for pharmaceuticals, pesticides, and fungicides. It is also used in biomedical research and as an anti-neoplastic agent (if was formerly used in the treatment of chronic leukemia and multiple myeloma).

RISK ASSESSMENT: HEALTH

General Assessment

Urethane is highly toxic to humans by *ingestion*. It is moderately toxic by *inhalation* and skin *absorption*. It is a suspected human carcinogen with confirmed experimental carcinogenic data reported for animals studies Human mutation data have been reported. There is some evidence in the references that indicate urethane adversely affects reproduction.

Urethane is a central nervous system stimulant. Symptoms of exposure by all routes include nausea, vomiting, abdominal pain, irritation, confusion, apprehension, excitability, paresthesia, dizziness, headache, disorientation, tremor, ataxia (loss of muscle coordination), stupor, coma, convulsions, respiratory collapse, and possibly death. Large doses may cause bone marrow depression and possible focal degeneration (areas of destruction of tissue) in the brain. There is also a potential for liver and kidney injury.

Acute Health Effects

The following acute (short-term) health effects may occur immediately or shortly after exposure to urethane:

Skin: Liquid can pass through unbroken skin. Symptoms of exposure include headache, nausea, abdominal pain, and vomiting.

Eye: Irritation and possible tearing.

Lung: Irritation of the nose and throat. Breathing vapors can cause cough, convulsions, unconsciousness, and coma.

CNS: A central nervous system stimulant. Symptoms include ataxia, excitement, tremors, convulsions, confusion, delirium, and irritability.

Chronic Health Effects

The following chronic (long-term) health effects may occur at some time after exposure to urethane and can last for months or even years:

Cancer Hazards: Urethane may be a carcinogen in humans since it has been shown to cause cancer in animals. Experimental carcinogenic, neoplastic, and tumorigenic data are well documented. Urethane is also a trans-placental carcinogen. Human mutation data have been reported and many scientists believe that exposure to such chemicals may pose a cancer risk in the long-term.

Reproductive Hazard: According to information presented in the references, urethane may be a teratogen in humans since it has been shown to adversely affect fetal development in test animals. Adverse reproductive effects are also reported.

Other Chronic Effects: Long-term exposure to urethane may damage the central nervous system, liver, and kidneys. In animal studies, urethane produced lung tumors, lymphomas, hepatomas, melanomas, and vascular tumors.

Recommended Risk-Reduction Measures

Personnel should avoid direct contact with urethane. It is a highly toxic, suspected human carcinogen. If a less toxic material or compound cannot be substituted for urethane, then *engineering controls* are the most effective method of reducing exposures. The best protection is to enclose operations and/or provide local exhaust ventilation at the site of urethane dust release. This may not be practical in all applications. While not always operationally possible, isolating operations involving urethane manufacturing or use can also reduce exposure. Using respiratory protection is less effective than the controls mentioned above, but is still advisable whenever working with or around urethane. No exposure levels have been established for urethane. This does not imply that exposure is without health hazard risk. Use an MSHA/NIOSH-approved supplied-air respirator or a self-contained breathing apparatus (SCBA) with full facepiece operated in pressure demand or other positive pressure mode for the greatest possible respiratory protection. If a full facepiece is not available, then chemical goggles should be worn to protect the eyes. A face shield and protective apron should also be worn. To prevent hand and skin exposures, impervious gloves should be worn. Glove manufacturers should be contacted and permeation studies obtained *before* final glove selections have been made.

Administrative controls should also be in place to minimize the potential for human exposures. These may include written procedures or policies which specify the methods and techniques that will be practiced whenever personnel are to work with urethane.

All personnel should receive *training* on the use, hazards, protective measures, emergency actions, and other precautions per 29 CFR 1910.1200 (Hazard Communication), prior to the first assignment in an area where urethane is used or stored.

Before beginning employment and at regular intervals thereafter (e.g., annually), the following medical tests are recommended for those personnel with frequent or potentially high exposures:

- ☑ Liver and kidney function tests.
- ☑ Complete blood count (CBC).

If symptoms develop or overexposure is suspected, the following medical testing may be useful:

- ☑ Complete examination of the nervous system with special testing and careful consideration of exposure history.

Any evaluation should include a careful history of past and present symptoms with an examination. Medical tests that simply look for existing damage are not a substitute for controlling exposures. Also, because more than light alcohol consumption can cause liver damage, drinking alcohol can increase the potential liver damage caused by urethane.

Other methods to reduce exposure include:

- ☑ Where possible, enclose operations and use local exhaust ventilation at the site of chemical release. If local exhaust ventilation or enclosure is not used, respiratory protection should be mandatory.
- ☑ Always ensure that proper protective clothing is worn when using chemical substances and that personnel have been trained on its proper use and care.
- ☑ Wash thoroughly immediately after exposure to urethane and at the end of the work shift or before eating, drinking, smoking, or applying cosmetics.
- ☑ Hazard warning information should be posted in the work area. In addition, as part of an on-going education and training program, all information on the health and safety hazards of urethane should be communicated to all exposed and potentially exposed workers.
- ☑ Eye wash stations should be provided in the immediate work area for emergency use. If there is a possibility of skin exposure to urethane, emergency shower facilities should also be provided.
- ☑ Workers whose clothing has been contaminated by urethane should change into clean clothes before leaving work. Family members may be exposed if they come into contact with clothing contaminated with urethane. Contaminated work clothing should be laundered only by individuals who have been informed of the hazards of exposure to urethane.

RISK ASSESSMENT: ENVIRONMENT

General Assessment

The environment is at risk of exposure during transportation, storage, disposal, or destruction of urethane. In almost every scenario, the threat of environmental exposure is contingent upon the proper handling of the chemical substance. Accidental spills, large or small, where urethane contacts incompatible commodities can result in violent and possibly explosive reactions (depending upon conditions of spill), and contamination of the surrounding environmental mediums (water, soil, and air).

Urethane is considered a combustible solid. It may also be present (dissolved) in some other flammable liquids such as benzene. Also, because it is incompatible with many common alkalis, acids, camphor, and other materials, extreme caution is required in handling, storage, transportation, and disposal of urethane. In contact with hot aqueous acids or alkalis, urethane will decompose to ethanol, carbon dioxide, and ammonia. These characteristics also require special consideration during any emergency situation involving a leak or spill of urethane powder or dust.

Urethane can enter the environment through industrial and municipal plant discharges or from accidental spills.

Acute Ecological Effects

Acute (short-term) toxic effects may include the death of animals, birds, or fish and death or low growth rate in plants. Acute effects are seen 2 to 4 days after animals or plants are exposed to urethane.

Urethane has slight acute toxicity to fish, birds, and land animals. It causes damage to plants, including death of buds.

Chronic Ecological Effects

Chronic (long-term) toxic effects may include shortened life span, reproductive problems, lower fertility, and changes in appearance or behavior in exposed animals. These effects can be seen long after first exposure(s) to toxic chemicals.

Urethane has slight toxicity to aquatic life. Insufficient data are available to evaluate or predict the long-term effects of urethane to plants, birds, or land animals.

Water Solubility

Urethane is very slightly soluble in water. Concentrations of 1 to 1000 milligrams will mix with a liter of water.

Persistence in the Environment

Urethane is slightly persistent in the water, with a half-life between 2 and 20 days. The half-life of a pollutant is the amount of time it takes for one-half of the chemical to be degraded. Approximately 24% of urethane will eventually end up in air; the remainder will end up in water.

Bioaccumulation in Aquatic Organisms

Some substances increase in concentration, or *bioaccumulate,* in living organisms as they breathe contaminated air, drink contaminated water, or eat contaminated food. These chemicals can become concentrated in the tissues and internal organs of animals as well as humans.

The concentration of urethane found in the edible tissues of fish is expected to be about the same as the average concentration of urethane in the water from which the fish was taken.

Recommended Risk-Reduction Measures

Proper training of all transporters will reduce the likelihood of a mishap or accident resulting in a leak or spill of urethane into the environment. The correct use of DOT labeling (per the MSDS) on all containers, trucks, and rail cars will enable emergency responders to react properly and quickly to any disaster thereby reducing the potential risk to the environment and to personnel.

Storage of urethane should be segregated from incompatible materials to minimize the risk of cross-contamination or contact. Buildings or storage areas should be equipped with appropriate fire protection systems (alarms, sprinklers). Proper ventilation and protective equipment should be used while handing the powders or while preparing aqueous solutions.

If a spill or leak to the environment has occurred, fire department, emergency response and/or hazardous materials spill personnel should be notified immediately. Cleanup should be attempted only by those trained in proper spill containment procedures. If urethane should contact the water table, aquifer, or navigable waterway, reclamation procedures should be initiated as soon as practical. It is only slightly soluble in water and partial or total remediation may be possible. The local and/or state emergency response authorities must be notified. A comprehensive emergency response or disaster preparedness/recovery plan should be in place prior to any operations involving the use, transportation, storage, or disposal of urethane. If urethane is spilled or leaked, the following specific steps are recommended:

- ☑ Restrict persons not wearing protective clothing from area of spill or leak until cleanup is complete and area can be opened for normal work.
- ☑ Ventilate are of spill or leak.
- ☑ Collect powdered materials using HEPA vacuum. Do NOT dry sweep. Absorb solution spills in vermiculite or other material and place in sealed drum for disposal or reclamation.
- ☑ It may be necessary to dispose of urethane as a hazardous waste. The responsible state agency or

the regional office of the federal Environmental Protection Agency (EPA) should be contacted for specific recommendations.

RISK ASSESSMENT: BUSINESS

General Assessment

Accidents or mishaps involving urethane can present a significant threat to business operations. The loss or damage of equipment or facilities can significantly affect fiscal viability. Lawsuits that may result from personnel injury/death, public exposures, and/or environmental contamination will also require a serious expenditure of resources. Media attention surrounding an injury, death, or environmental damage can also result in a loss of profits and loss of current as well as future business. Always remember that anytime to terms "cancer," "carcinogen," or "reproductive hazard" are used, public emotion, hysteria, and ignorance can run equally high. This must be carefully considered whenever developing or implementing public relations policies.

Recommended Risk-Reduction Measures

Company attorneys, safety and health professionals, and environmental specialists should be involved in the development of any procedures for responding to chemical incidents. A company official should be predesignated as a public relations officer with specific training in dealing with the press. Corporate plans and policies should be developed, approved, and implemented long before any need for such arises.

REFERENCES

American Conference of Governmental Industrial Hygienists. 1988. *Documentation of the Threshold Limit Values and Biological Exposure Indices,* 5th Edition (with updates). Cincinnati: ACGIH

Brusick, D. J. 1994. *Methods for Genetic Risk Assessment.* New York: CRC Press/Lewis Publishers.

Bloom, A. D. and F. J. DeSerres. 1995. *Ecotoxicity and Human Health.* New York: CRC Press/Lewis Publishers.

Calabrese, E. J. 1994. *Biological Effects of Low Level Exposures.* New York: CRC Press/Lewis Publishers.

Cockerman, L. G., B. S. Shane. 1994. *Basic Environmental Toxicology.* New York: CRC Press/Lewis Publishers.

Cothern, C. R. 1995. *Handbook for Environmental Risk Decisionmaking.* New York: CRC Press/Lewis Publishers.

Gardner, R. W. 1994. *Chemical Intolerance.* New York: CRC Press/Lewis Publishers.

Howard, P. H., et. al. 1991. *Handbook of Environmental Degradation Rates.* New York: CRC Press/Lewis Publishers.

Keith, L. H. and M. M. Walker. 1995. *Handbook of Air Toxics.* New York: CRC Press/Lewis Publishers.

Landis, W. G., M. H. Yu. 1995. *Introduction to Environmental Toxicology.* New York: CRC Press/Lewis Publishers.

Lauwerys, R. R., P. Hoet. 1993. *Industrial Chemical Exposure.* New York: CRC Press/Lewis Publishers.

Lewis, R. J., Sr. 1992. *Sax's Dangerous Properties of Industrial Materials,* Volumes 1, 2, and 3, 8th Edition. New York: Van Nostrand Reinhold.

Lide, D. R. 1995. *Handbook of Chemistry and Physics,* 76th Edition. New York: CRC Press/Lewis Publishers.

Manahan, S. E. 1992. *Toxicological Chemistry,* 2nd Edition. New York: CRC Press/Lewis Publishers.

Mellström, G. A., J. E. Wahlberg, and H. I. Maibach. 1994. *Protective Gloves for Occupational Use.* New York: CRC Press/Lewis Publishers.

Meyer, E. 1990. *Chemistry of Hazardous Materials.* Englewood Cliffs, NJ: Prentice Hall.

Mycock, J. C., J. D. McKenna, and L. Theodore. 1995. *Handbook of Air Pollution Control Engineering and Technology.* New York: CRC Press/Lewis Publishers.

Nelson, G. O. 1992. *Gas Mixtures.* New York: CRC Press/Lewis Publishers.

Ott, W. R. 1995. *Environmental Statistics and Data Analysis.* New York: CRC Press/Lewis Publishers.

Patnaik, P. 1992. *A Comprehensive Guide to the Hazardous Properties of Chemical Substances.* New York: Van Nostrand Reinhold.

Quigley, D. R. 1994. *Handbook of Emergency Chemical Management.* New York: CRC Press/Lewis Publishers.

Richardson, M. 1995. *Dictionary of Substance and Their Effects,* Vol. I-VI. New York: CRC Press/Lewis Publishers.

Sacarello, H. L. A. 1994. *The Comprehensive Handbook of Hazardous Materials.* New York: CRC Press/Lewis Publishers.

Torkelson, T. R. V. K. Rowe. 1981. *Patty's Industrial Hygiene and Toxicology,* Vol. 2B. New York: Wiley-Interscience.

MATERIAL SAFETY DATA SHEET

CHEMICAL NAME	
VANADIUM	**CRC PRESS/LEWIS PUBLISHERS DISCLAIMER:** THE INFORMATION AND RECOMMENDATIONS PRESENTED HEREIN ARE BASED ON SOURCES BELIEVED TO BE RELIABLE. CRC MAKES NO REPRESENTATION ON ITS COMPLETENESS OR ACCURACY. IT IS THE RESPONSIBILITY OF THE USER TO DETERMINE THE CHEMICAL'S SUITABILITY FOR ITS INTENDED USE, THE CHEMICAL'S SAFE USE, AND THE CHEMICAL'S PROPER DISPOSAL. NO REPRESENTATIONS AND/OR WARRANTIES, EITHER EXPRESSED OR IMPLIED, OF THE MERCHANTABILITY OR FITNESS FOR A PARTICULAR PURPOSE, OR OF ANY OTHER NATURE, ARE MADE WITH RESPECT TO THE INFORMATION PROVIDED IN THIS MSDS OR TO THE CHEMICAL TO WHICH INFORMATION MAY REFER. CRC NEITHER ASSUMES NOR AUTHORIZES ANY OTHER PERSON TO ASSUME FOR IT, ANY OTHER ADDITIONAL RESPONSIBILITY OR LIABILITY FOR THE USE OF, OR RELIANCE UPON, THIS INFORMATION.

HAZARD WARNING INFORMATION

HEALTH	FIRE	REACTIVE	OTHER	DEGREE OF HAZARD	COLOR CODING	OTHER CODES
2	0	0		0 = Minimum Hazard 1 = Slight Hazard 2 = Moderate Hazard 3 = Serious Hazard 4 = Severe Hazard	HEALTH = BLUE FIRE = RED REACTIVITY = YELLOW OTHER = WHITE	OX = Oxidizer ACID = Acid ALK = Alkali COR = Corrosive ~~W~~ = Use No Water

SECTION I - GENERAL INFORMATION

Characterization	**Metallic Element**	RCRA Number	**None**
		EPA Class	**Not Applicable**
DOT Proper Shipping Name	**No Citation**	Chemical Abstract Service (CAS) Number	**7440-62-2**
DOT Hazard Class and Label Requirements	**No Citation**	DOT Emergency Guide Code	**No Citation**
DOT Identification Number	**No Citation**	Chemical Formula	**V**

Synonyms

Not applicable.

SECTION II - HAZARDOUS INGREDIENTS AND IDENTITY INFORMATION

Hazardous Components (specific identity)	OSHA Exposure Criteria	NIOSH Exposure Criteria	Immediately Dangerous to Life and Health (IDLH)	ACGIH Exposure Criteria
Vanadium (derivation: Occurs in the earth's crust in ores in more than 65 different ores. Also, calcium reduction of vanadium pentoxide yields 99.8+% of pure ductile vanadium. Also found in crude petroleum, in flue-gas deposits from oil-fired furnaces, or in slag from ferrovanadium production).	**PEL:** **0.05 mg/m^3 (dust & fume)** **STEL:** **Not Established**	**REL (10-hour):** **1.0 mg/m^3** **STEL:** **Not Established**	**70 mg/m^3 as V**	**TLV:** **0.05 mg/m^3 (dust & fume)** **STEL:** **Not Established**

SECTION III - PHYSICAL AND CHEMICAL CHARACTERISTICS

Boiling Point	**6116°F (3380°C)**	Specific Gravity (H_2O = 1)	**5.96**
Vapor Pressure (mm Hg)	**1.5 at 4172°F (2300°C)**	Molecular Weight	**50.94**
Vapor Density (Air = 1)	**Not Reported**	Melting Point	**3482°F (1917°C)**

Solubility

Insoluble in hot or cold water. Soluble in nitric, hydrofluoric, and concentrated sulfuric acids.

Appearance and Odor

Light gray powder or a silvery-white, ductile, lustrous bulk solid with no odor.

SECTION IV - FIRE AND EXPLOSION HAZARD DATA

Flash Point (method used)	**Not Reported**	Explosive Limits in Air % by Volume	**LEL: 0.22 oz/ft^3 UEL: Not Reported**
NFPA Classification	**Combustible Solid (in powdered form)**	Autoignition Temperature	**Layer: 914°F (490°C) Cloud: 932°F (500°C)**

Extinguishing Media

Use a chemical powder, carbon dioxide, of Class "D" extinguishing agent on metal fires.

Special Fire Fighting Procedures

Wear full protective clothing and self-contained breathing apparatus (SCBA). Use a water spray to keep fire-exposed containers cool. Water may be ineffective on fire. Remain clear of smoke, water fallout and water runoff. Poisonous gases are produced in fire.

Unusual Fire and Explosion Hazards

Hot powdered vanadium can ignite and burn spontaneously in air. It can cause violent explosions when its powders are dispersed in air and exposed to an ignition source. Flammable when exposed to heat or flame.

SECTION V - REACTIVITY DATA

Stability		Conditions to Avoid Vanadium is normally stable at room temperature in closed containers under routine conditions of handling and storage. Keep away from incompatible materials since violent or explosive reactions can occur.
Stable X	Unstable	Incompatibility *(materials to avoid)* Incompatible with strong oxidizers (such as perchlorates, peroxides, permanganates, chlorates, and nitrates), lithium, and magnesium.
Hazardous Polymerization		Conditions to Avoid Under normal conditions of temperature and pressure, hazardous polymerization of vanadium cannot occur.
May Occur	Will Not Occur X	Hazardous Decomposition or By-products Thermal oxidative decomposition of vanadium can yield toxic fumes of vanadium oxides.

SECTION VI - HEALTH HAZARD DATA

Primary Route(s) of Entry:	Inhalation? X	Absorption (skin)?	Ingestion? X

Health Hazards

INHALATION: Irritation of the respiratory system. Causes nasal congestion, throat irritation, dry mouth, gradual discoloration of the tongue, metallic taste, coughing, bronchitis, pneumonia, chest pain, and pulmonary edema (fluid in the lungs), which can be fatal. There may also be constriction of the blood supply to the lungs, spleen, kidneys, and intestines. There can be central nervous system depression with headaches.

SKIN & EYES: Causes eczema-like rash with itching and sensitization and/or hives. Eye contact will be severely irritating with burning and tearing.

INGESTION: A poison with gastrointestinal effects, nausea, vomiting, diarrhea, black stools, cramps.

Carcinogenicity	NTP Listed?	IARC Cancer Review Group?	OSHA Regulated?	Target Organs?
Unknown Human Questioned Animal	No	No	No	Respiratory system, skin, eyes

Medical Conditions Generally Aggravated by Exposure

None Reported. However, existing skin conditions or respiratory problems may be aggravated.

Emergency and First-aid Procedures

Eye contact: Flush immediately with water for 15 minutes (minimum); seek medical attention. Skin contact: Remove all contaminated clothing. Immediately wash area with large amounts of soap and water. For inhalation: Remove the person from exposure. Provide respiratory assistance and CPR. Transfer to medical facility. Observe 48 hours for lung effects. If swallowed: Never attempt to give a convulsing or unconscious person anything by mouth. Seek medical attention immediately. Give 1 to 2 glasses of water and induce vomiting; repeat until vomitus is clear.

SECTION VII - PRECAUTIONS FOR SAFE HANDLING AND USE

Steps to be Taken in Case Material is Released or Spilled

Collect powdered material in most convenient manner possible and deposit in sealed drum. Keep those not involved in cleanup from entering area. Do NOT dry sweep (creates airborne dusts). Use a vacuum equipped with a high-efficiency particulate air (HEPA) filter.

Preferred Waste Disposal Method

No citation.

Precautions to be Taken in Handling and Storage

Store in tightly closed containers in cool, well-ventilated area and protect from physical damage. Sources of ignition are prohibited where vanadium is used, handled, or stored.

Other Precautions and Warnings

Use good housekeeping practices and avoid generating dusty conditions in storage areas.

SECTION VIII - CONTROL MEASURES AND PERSONAL PROTECTIVE EQUIPMENT

Respiratory Protection *(specify type)*

Use a full facepiece respirator with a high-efficiency particulate air (HEPA) filter. For high exposures, use a full facepiece supplied-air respirator or a SCBA operated in positive pressure mode.

Ventilation

Local exhaust or general mechanical systems recommended.

Protective Gloves	Eye Protection	Other Protective Clothing
Impervious Material	Chemical/Dust Goggles or Face Mask	Protective Suit or Clothing

Work/Hygiene Practices

Always wash hands thoroughly after using chemical; never bring food, drink, or smoking materials into vicinity of chemicals.

VANADIUM

V

CAS: 7440-62-2

IDENTIFICATION AND TYPICAL USES

Vanadium is a light gray powder or silvery-white, lustrous bulk solid with no odor. Vanadium is a naturally occurring element of atomic number 23, group VB of the periodic system, with valances of 2, 3, 4, 5, and two natural isotopes. It is used as a target material for X-rays, in the manufacture of alloy steels, in the production of vanadium compounds, and a catalyst for sulfuric acid and synthetic rubber.

RISK ASSESSMENT: HEALTH

General Assessment

Vanadium compounds can be moderately toxic by *ingestion* with gastrointestinal irritation, nausea and vomiting. It can also be toxic by *inhalation,* with severe pulmonary damage (pulmonary edema) possible. Exposure to the dusts of vanadium may cause other lung effects, including pneumonia and emphysema. Vanadium is considered a questionable carcinogen in animals.

In addition to the effects noted above, inhalation of vanadium can produce nasal congestion, throat irritation, dry mouth, and a gradual green discoloration of the tongue, a metallic taste, coughing, bronchitis, pneumonia, and chest pain. Cardiovascular effects include constriction of blood supply to the lungs, spleen, kidneys, and intestines.

Skin contact can produce an eczema-like skin rash, itching, possible sensitization or hives, and green discoloration of the hands and groin area. Eye contact will result in irritation, burning, and tearing.

Ingestion causes diarrhea, black stools, cramping, and other symptoms of gastrointestinal distress.

Acute Health Effects

The following acute (short-term) health effects may occur immediately or shortly after exposure to vanadium:

Skin: Irritation, rash, and possible sensitization. There may also be a green discoloration of the hands.

Eye: Vanadium particles can irritate the eyes and may cause damage to vision.

Lung: Inhalation of the dust and fumes can irritate the nose, mouth, and throat. High levels can irritate the lungs, causing a buildup of fluid (pulmonary edema), which is a medical emergency and can be fatal.

Chronic Health Effects

The following chronic (long-term) health effects may occur at some time after exposure to vanadium and can last for months or even years:

Cancer Hazards: Vanadium compounds may cause cancer in test animals. It is not known if human exposure to vanadium dusts will cause cancer in humans.

Reproductive Hazard: According to information available in the references, vanadium has not been adequately tested for its ability to adversely affect reproduction in humans.

Other Chronic Effects: Very irritating substances can damage the lungs. It is not known if exposure to vanadium will have this effect. Skin and tongue discoloration to green may occur gradually over long-term exposure.

Recommended Risk-Reduction Measures

Personnel should avoid direct contact with vanadium. If the specific vanadium compound in question is unknown, adequate risk management dictates that it be handled and treated with extreme care. If a less toxic chemical cannot be substituted for a hazardous substance, then *engineering controls* are the most effective method of reducing exposures.

The best protection is to enclose operations and/or provide local exhaust ventilation at the site of chemical release. While not always operationally feasible, isolating operations can also reduce exposure. Using respiratory protection is less effective than the controls mentioned above, but is still advisable whenever working with or around vanadium. For exposure at levels above the OSHA PEL (0.05 mg/m^3 for dusts and fumes), an MSHA/NIOSH-approved respirator with a high-efficiency particulate air (HEPA) filter is recommended. For higher exposures, use a supplied-air respirator with a full facepiece operated in positive pressure mode or a self-contained breathing apparatus (SCBA) with full facepiece in pressure demand. If a full facepiece is not available, then chemical dust goggles should be worn to protect the eyes. Whenever a chemical dust hazard exists, a face shield and a rubber apron should be worn. To prevent hand and skin exposures, impervious rubber gloves should be used.

Administrative controls should also be in place to minimize the potential for human exposures. These may include written procedures or policies which specify the methods and techniques that will be practiced whenever personnel are to work with vanadium. A regulated, controlled work area should be established where vanadium is used, handled, or stored.

All personnel should receive *training* on the use, hazards, protective measures, emergency actions, and other precautions per 29 CFR 1910.1200 (Hazard Communication), prior to the first assignment in an area where vanadium is used or stored.

Before beginning employment and at regular intervals thereafter (e.g., annually), for those with frequent or potentially high exposures (half the PEL or greater, or significant skin contact), the following medical tests are recommended:

☑ Lung function tests.

If symptoms develop or overexposure is suspected, the following may also be useful:

☑ Urine test for vanadium.

☑ Consider chest X-ray after acute overexposure (may be negative if taken immediately after exposure due to delayed onset of pulmonary edema).

Any evaluation should include a careful history of past and present symptoms with a medical examination. Medical tests that look for damage already done are not a substitute for controlling exposure. Also, since smoking can cause heart disease, as well as lung cancer, emphysema, and other related respiratory problems, the effects of exposure to vanadium may be more pronounced in smokers than in non-smokers.

Other methods to reduce exposure include:

☑ Where possible, enclose operations and use local exhaust ventilation at the site of chemical release. If local exhaust ventilation or enclosure is not used, respiratory protection should be mandatory.

☑ Always ensure that proper protective clothing is worn when using chemical substances.

☑ Wash thoroughly immediately after exposure to vanadium and at the end of the work shift or before eating, drinking, or smoking.

☑ Hazard warning information should be posted in the work area. In addition, as part of an on-going education and training program, all information on the health and safety hazards of vanadium should be communicated to all potentially exposed workers.

RISK ASSESSMENT: ENVIRONMENT

General Assessment

Elemental vanadium is found in over 65 ores, including patronite, roscoelite, carnotite, and vanadinite. It occurs in Idaho, Montana, Arkansas, Colorado, Utah, New Mexico, Arizona, Mexico, and Peru. However, the environment is still at risk of exposure during transportation, storage, disposal, or destruction of vanadium compounds. In almost every scenario, the threat of environmental exposure is contingent upon the proper handling of the chemical substance. Accidental spills, large or small, can result in fire, explosion, and possible contamination of the surrounding environmental mediums (water, soil, and air).

Vanadium powder will form explosive mixtures and may ignite spontaneously in air. It will ignite and is potentially explosive in contact with bromine trifluoride, chlorine, nitryl fluoride, lithium, and oxidants. Vanadium prepared by magnesium reduction of vanadium trichloride or vanadium chloride becomes pyrophoric. Very hot vanadium can burn in air. It is also incompatible with strong oxidizers (such as perchlorates, peroxides, permanganates, chlorates, and nitrates). These characteristics require special consideration during any emergency situation involving a leak or spill of vanadium.

As stated above, vanadium is an element which is widely dispersed in the Earth's crust at low concentrations. It is ordinarily recovered from its raw materials in the form of pentoxide and, sometimes as sodium and ammonium vanadate salts. In elemental form (fume or dust), it can enter the environment from industrial discharges.

Acute Ecological Effects

Acute (short-term) toxic effects may include the death of animals, birds, or fish and death or low growth rate in plants. Acute effects are seen 2 to 4 days after animals or plants are exposed to vanadium.

The acute toxicity of elemental v to aquatic life is unknown. However, the compound ammonium vanadate is moderately toxic to aquatic life. Insufficient data are available to evaluate the short-term effects of vanadium exposure to plants, birds, or land animals.

Chronic Ecological Effects

Chronic (long-term) toxic effects may include shortened life span, reproductive problems, lower fertility, and changes in appearance or behavior in exposed animals. These effects can be seen long after first exposure(s) to toxic chemicals.

The chronic toxicity of elemental vanadium to aquatic life is unknown. However, the compound ammonium vanadate has high chronic toxicity to aquatic life. Insufficient data are available to evaluate the long-term effects of vanadium exposure to plants, birds, or land animals.

Water Solubility

Vanadium is insoluble in water.

Persistence in the Environment

Elemental vanadium is not likely to dissolve in surface water. Vanadium and its salts are highly persistent in the aquatic environment, with a half-life of greater than 200 days. The half-life of a pollutant is the amount of time it takes for one half of the chemical to be degraded.

Bioaccumulation in Aquatic Organisms

Some substances increase in concentration, or *bioaccumulate,* in living organisms as they breathe contaminated air, drink contaminated water, or eat contaminated food. These chemicals can become concentrated in the tissues and internal organs of animals as well as humans.

Vanadium is not likely to accumulate in the edible tissues of aquatic species normally consumed by humans.

Recommended Risk-Reduction Measures

Proper training of all transporters will reduce the likelihood of a mishap or accidents resulting in a leak or spill to the environment. The correct use of labeling on all containers, trucks, and rail cars will enable emergency responders to react properly and quickly to any disaster thereby reducing the potential risk to the environment and to personnel.

Storage of vanadium should be segregated from incompatible chemicals to minimize the risk of cross-contamination or contact. Bulk storage of radioactive vanadium should be avoided.

If a spill or leak to the environment has occurred, fire department, emergency response and/or hazardous materials spill personnel should be notified immediately. Use a vacuum method to reduce dust during cleanup. Do not dry sweep. When vacuuming, a high-efficiency particulate air (HEPA) filter should be used, not a standard shop vacuum. Deposit collected materials in sealed drum for disposal. Cleanup should be attempted only by those trained in proper spill containment procedures. Contaminated soils should be removed and replaced with clean soil. If vanadium should contact the water table, aquifer, or navigable waterway, time is of the essence. Vanadium is not soluble in water and, therefore, total containment and remediation may be possible. When such spills occur, the local and/or state emergency response authorities must be notified. A comprehensive emergency response or disaster preparedness/recovery plan should be in place prior to any operations involving the use, transportation, storage, or disposal of vanadium. If vanadium is spilled or leaked, the following specific steps are recommended:

- ☑ Restrict persons not wearing protective clothing from area of spill or leak until cleanup is complete and area can be opened for normal work.
- ☑ Collect powdered materials in most convenient and safe manner possible and place in sealed drums for disposal. Do not dry sweep; use HEPA vacuum instead. Damp mop residue.
- ☑ It may be necessary to dispose of vanadium as a hazardous waste. The responsible state agency or the regional office of the federal Environmental

Protection Agency (EPA) should be contacted for specific recommendations.

RISK ASSESSMENT: BUSINESS

General Assessment

Accidents or mishaps involving vanadium can present a significant threat to business operations. The loss or damage of equipment or facilities can significantly affect fiscal viability. Lawsuits that may result from personnel injury/death, public exposures, and/or environmental contamination will also require a serious expenditure of resources. Media attention surrounding an injury, death, or environmental damage can also result in a loss of profits and loss of current as well as future business.

Recommended Risk-Reduction Measures

Company attorneys, safety and health professionals, and environmental specialists should be involved in the development of any procedures dealing with safety or emergency response. A company official should be pre-designated as a public relations officer with specific training in dealing with the press. Corporate plans and policies should be developed, approved, and implemented long before any need for such arises.

REFERENCES

Ahlbom, A. 1993. *Biostatistics for Epidemiologists.* New York: CRC Press/Lewis Publishers.

American Conference of Governmental Industrial Hygienists. 1988. *Documentation of the Threshold Limit Values and Biological Exposure Indices,* 5th Edition (with updates). Cincinnati: ACGIH

Asante-Duah, D. K. 1993. *Hazardous Waste Risk Assessment.* New York: CRC Press/Lewis Publishers.

Bloom, A. D. and F. J. DeSerres. 1995. *Ecotoxicity and Human Health.* New York: CRC Press/Lewis Publishers.

Brusick, D. J. 1994. *Methods for Genetic Risk Assessment.* New York: CRC Press/Lewis Publishers.

Calabrese, E. J. 1994. *Biological Effects of Low Level Exposures.* New York: CRC Press/Lewis Publishers.

Cockerman, L. G., B. S. Shane. 1994. *Basic Environmental Toxicology.* New York: CRC Press/Lewis Publishers.

Cothern, C. R., N. P. Ross. 1994. *Environmental Statistics, Assessment, and Forecasting.* New York: CRC Press/Lewis Publishers.

Danse, I. R. 1991. *Common Sense Toxics in the Workplace.* New York: Van Nostrand Reinhold.

Environmental Protection Agency. 1983. *Chemical Hazard Information Profiles.* Washington, D.C.: U.S. EPA.

Fodden, C. R., and J. L. Weddell. 1992. *Hazardous Materials Emergency Action Data.* New York: CRC Press/Lewis Publishers.

Forseberg, K., and L. H. Keith. 1993. *Chemical; Protective Clothing Permeation and Degradation Database.* Chelsea, MI: Lewis Publishers.

Gardner, R. W. 1994. *Chemical Intolerance.* New York: CRC Press/Lewis Publishers.

Hallenbeck, W. H. 1993. *Quantitative Risk Assessment for Environmental and Occupational Health,* 2nd Edition. New York: CRC Press/Lewis Publishers.

Hodgson, E., R. B. Mailman, and Janice E. Chambers. 1988. *Dictionary of Toxicology.* London: The MacMillian Company.

Hosty, J. W., and P. Foster. 1990. *A Practical Guide to Chemical Spill Response.* New York: Van Nostrand Reinhold.

Howard, P. H., et. al. 1991. *Handbook of Environmental Degradation Rates.* New York: CRC Press/Lewis Publishers.

Kamrin, M. A. 1988. *Toxicology: A Primer on Principles and Applications.* Chelsea, Michigan: Lewis Publishers.

Keith, L. H. and M. M. Walker. 1995. *Handbook of Air Toxics.* New York: CRC Press/Lewis Publishers.

Landis, W. G., M. H. Yu. 1995. *Introduction to Environmental Toxicology.* New York: CRC Press/Lewis Publishers.

Langerman, N. 1994 *Precautionary Labels for Chemical Containers.* New York: CRC Press/Lewis Publishers.

Lauwerys, R. R., P. Hoet. 1993. *Industrial Chemical Exposure.* New York: CRC Press/Lewis Publishers.

Lewis, .J., Sr. 1993. *Hawley's Condensed Chemical Dictionary,* 12th Edition. New York: Van Nostrand Reinhold.

Lewis, . J., Sr. 1992. *Sax's Dangerous Properties of Industrial Materials,* Volumes 1, 2, and 3, 8th Edition. New York: Van Nostrand Reinhold.

Manahan, S. E. 1994. *Environmental Chemistry,* 6th Edition. New York: CRC Press/Lewis Publishers.

Manahan, S. E. 1992. *Toxicological Chemistry,* 2nd Edition. New York: CRC Press/Lewis Publishers.

Meyer, E. 1990. *Chemistry of Hazardous Materials.* Englewood Cliffs, NJ: Prentice Hall.

Rea, W. J. 1995. *Chemical Sensitivity,* Vol. I-III. New York: CRC Press/Lewis Publishers.

Reeves, A. L. 1979. *Handbook on the Toxicology of Metals.* New York: Elsevier North-Holland.

Richardson, M. 1995. *Dictionary of Substance and Their Effects,* Vol. I-VI. New York: CRC Press/Lewis Publishers.

Sacarello, H. L. A. 1994. *The Comprehensive Handbook of Hazardous Materials.* New York: CRC Press/Lewis Publishers.

Torkelson, T. R. V. K. Rowe. 1981. *Patty's Industrial Hygiene and Toxicology,* Vol. 2B. New York: Wiley-Interscience.

MATERIAL SAFETY DATA SHEET

CHEMICAL NAME	CRC PRESS/LEWIS PUBLISHERS DISCLAIMER
VINYL ACETATE	THE INFORMATION AND RECOMMENDATIONS PRESENTED HEREIN ARE BASED ON SOURCES BELIEVED TO BE RELIABLE. CRC MAKES NO REPRESENTATION ON ITS COMPLETENESS OR ACCURACY. IT IS THE RESPONSIBILITY OF THE USER TO DETERMINE THE CHEMICAL'S SUITABILITY FOR ITS INTENDED USE, THE CHEMICAL'S SAFE USE, AND THE CHEMICAL'S PROPER DISPOSAL. NO REPRESENTATIONS AND/OR WARRANTIES, EITHER EXPRESSED OR IMPLIED, OF THE MERCHANTABILITY OR FITNESS FOR A PARTICULAR PURPOSE, OR OF ANY OTHER NATURE, ARE MADE WITH RESPECT TO THE INFORMATION PROVIDED IN THIS MSDS OR TO THE CHEMICAL TO WHICH INFORMATION MAY REFER. CRC NEITHER ASSUMES NOR AUTHORIZES ANY OTHER PERSON TO ASSUME FOR IT, ANY OTHER ADDITIONAL RESPONSIBILITY OR LIABILITY FOR THE USE OF, OR RELIANCE UPON, THIS INFORMATION.

HAZARD WARNING INFORMATION

HEALTH	FIRE	REACTIVE	OTHER	DEGREE OF HAZARD	COLOR CODING	OTHER CODES
2	3	2		0 = Minimum Hazard 1 = Slight Hazard 2 = Moderate Hazard 3 = Serious Hazard 4 = Severe Hazard	HEALTH = BLUE FIRE = RED REACTIVITY = YELLOW OTHER = WHITE	OX = Oxidizer ACID = Acid ALK = Alkali COR = Corrosive W = Use No Water

SECTION I - GENERAL INFORMATION

Characterization	RCRA Number	EPA Class
Ester	**D001**	**Characteristic (I) Waste**

DOT Proper Shipping Name **Vinyl Acetate Monomer**	Chemical Abstract Service (CAS) Number **108-05-4**
DOT Hazard Class and Label Requirements **Flammable Liquid**	DOT Emergency Guide Code **26**
DOT Identification Number **UN 1301**	Chemical Formula **CH_2=CHOOCCH$_3$**

Synonyms

1-Acetoxyethylene; ethenyl acetate; ethenyl ethanoate; VAC; vinyl acetate monomer; vinyl ethanoate.

SECTION II - HAZARDOUS INGREDIENTS AND IDENTITY INFORMATION

Hazardous Components (specific identity)	OSHA Exposure Criteria	NIOSH Exposure Criteria	Immediately Dangerous to Life and Health (IDLH)	ACGIH Exposure Criteria
Vinyl acetate (derivation: By vapor-phase reaction of ethylene, acetic acid, and oxygen, with a palladium catalyst; vapor-phase reaction of acetylene, acetic acid, and oxygen, with zinc acetate catalyst). **1 ppm = 3.58 mg/m^3**	**PEL:** Not Established **STEL:** Not Established	**REL (ceiling):** 4 ppm 15 mg/m^3 (15-minutes) **STEL:** Not Applicable	**Not Determined**	**TLV:** 10 ppm 356 mg/m^3 **STEL:** 20 ppm Suspected Human Carcinogen

SECTION III - PHYSICAL AND CHEMICAL CHARACTERISTICS

Boiling Point **163°F (73°C)**	Specific Gravity (H_2O = 1) **0.93**
Vapor Pressure (mm Hg) **88 at 68°F (20°C)**	Molecular Weight **86.1**
Vapor Density (Air = 1) **3.0**	Freezing Point **-136°F (-58°C)**

Solubility

Soluble in water. Soluble in chloroform, alcohol, benzene, acetone, ethyl ether, solvents, and ether.

Appearance and Odor

Colorless liquid with a pleasant, fruity odor. Odor Threshold = 0.50 ppm.

SECTION IV - FIRE AND EXPLOSION HAZARD DATA

Flash Point (method used) **18°F (-8°C) closed cup**	Explosive Limits in Air % by Volume **LEL: 2.6%** **UEL: 13.4%**
NFPA Classification **Class IB Flammable Liquid**	Autoignition Temperature **756°F (402°C)**

Extinguishing Media

Dry chemical, alcohol-resistant foam, carbon dioxide, water spray, or fog.

Special Fire Fighting Procedures

Poisonous gases are produced in fire. Wear full protective clothing and self-contained breathing apparatus (SCBA). Use blanketing effect with foam to smother large fires. Continue to cool containers with water after fire is extinguished. Water on fire itself may be ineffective and actually spread fire further.

Unusual Fire and Explosion Hazards

Immediately withdraw if rising sound from venting device is heard or if fire is causing discoloration to the tank. Evacuate 1500 feet radius if fire becomes uncontrollable or container is exposed to direct flame. Vapors are heavier than air and may travel a distances to ignition source to flashback.

SECTION V - REACTIVITY DATA

Stability		Conditions to Avoid: **Vinyl acetate is normally unstable. It usually contains a stabilizer (e.g., hydroquinone or diphenylamine). Keep away from heat and incompatible materials. Never allow inhibitor to decrease to unsafe levels.**
Stable	Unstable	Incompatibility *(materials to avoid)*
	X	**Hydrogen peroxide, oxygen above 50°C, ozone, acids, bases, silica gel, alumina, oxidizers, azo compounds, toluene, chlorosulfonic acid, ethylene diamine, ethylene imine, oleum.**
Hazardous Polymerization		Conditions to Avoid: **Self-polymerization due to autoxidation of acetaldehyde to a peroxide which then initiates exothermic polymerization as it decomposes. Polymerization also initiated by heat, peroxides, oxidizers, and sunlight.**
May Occur	Will Not Occur	Hazardous Decomposition or By-products
X		**When heated to decomposition, vinyl acetate emits toxic smoke and acrid, irritating fumes.**

SECTION VI - HEALTH HAZARD DATA

Primary Route(s) of Entry:	Inhalation?	Absorption (skin)?	Ingestion?
	X		X

Health Hazards

INHALATION: Irritation of the eyes, nose, throat, and respiratory tract with a possibility for narcosis and headache. Systemic effects include olfactory changes, conjunctiva inflammation, chest tightness, nervous irritation, dizziness, drowsiness, dypsnea, palpitation, and corneal edema. Other CNS effects include fatigue, insomnia, encephalopathy, weakness, and polyneuritis. CVS symptoms of arrhythmia, chest pain, and syncope. Liver damage.

SKIN & EYES: Will irritate the skin, causing burns and blisters. Eye contact may cause corneal damage with cloudy swelling, and fatty degeneration of the viscera.

INGESTION: May cause headache, drowsiness, and unconsciousness.

Carcinogenicity	NTP Listed?	IARC Cancer Review Group?	OSHA Regulated?	Target Organs?
Suspected Human Questioned Animal	**No**	**Group 3**	**No**	**Respiratory system, skin, eyes, CNS, CVS, liver.**

Medical Conditions Generally Aggravated by Exposure

None reported.

Emergency and First-aid Procedures

Eye contact: Flush immediately with water for 15 minutes (minimum); seek medical attention. Skin contact: Remove all contaminated clothing. Immediately wash area with large amounts of soap and water. Seek medical assistance. For inhalation: Remove the person from exposure. Provide respiratory assistance and CPR. Transfer to medical facility. If swallowed: Seek medical attention immediately. Call poison control center for advice. Give 1-2 glasses of water to dilute. Do NOT induce vomiting unless advised otherwise. Never give an unconscious or convulsing person anything by mouth.

SECTION VII - PRECAUTIONS FOR SAFE HANDLING AND USE

Steps to be Taken in Case Material is Released or Spilled

Contain spills using absorbent material (vermiculite, or other). Ventilate area of spill. Remove all ignition sources. Have water source available in case of fire. Store materials in DOT-approved containers. Restrict those not involved in cleanup from entering area. Notify appropriate authorities, as required.

Preferred Waste Disposal Method

Vinyl acetate can be removed from waste gases with activated carbon.

Precautions to be Taken in Handling and Storage

Store in tightly closed containers in a cool, dark, well-ventilated area. Protect containers from physical damage. Keep heat, fire, flame, and sunlight away. Automatically transfer liquids between containers.

Other Precautions and Warnings

Bulk storage of vinyl acetate is not recommended. Ground and bond all metal containers.

SECTION VIII - CONTROL MEASURES AND PERSONAL PROTECTIVE EQUIPMENT

Respiratory Protection *(specify type)*

For high exposures, use a self-contained breathing apparatus (SCBA) or a supplied-air respirator with full facepiece operated in pressure demand or other positive pressure mode.

Ventilation

Local exhaust or general mechanical systems recommended.

Protective Gloves	Eye Protection	Other Protective Clothing
Nitrile, Viton, Neoprene, Polyethylene	**Chemical Goggles or Face Mask**	**Impermeable Apron**

Work/Hygiene Practices

Always wash hands thoroughly after using chemical; never bring food, drink, or smoking materials into vicinity of chemicals.

VINYL ACETATE

CH_2=CHOOCCH_3

CAS: 108-05-4

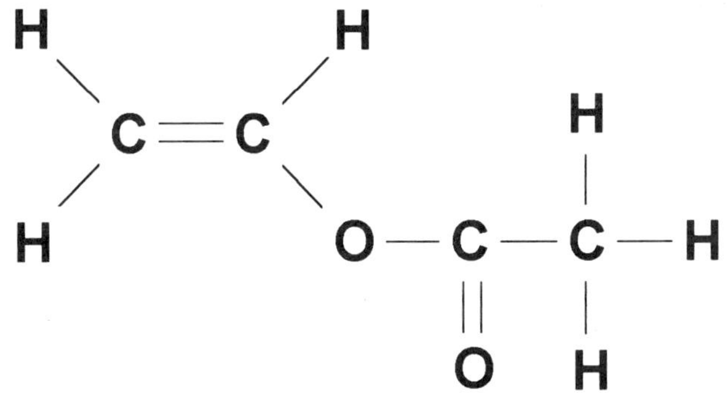

IDENTIFICATION AND TYPICAL USES

Vinyl acetate is a colorless liquid with a pleasant, fruity odor. Used in the production of polyvinyl acetate, polyvinyl alcohol, polyvinyl butyral, and polyvinyl chloride-acetate resins. It is also used in latex paints, paper coatings, adhesives, textile finishing, and safety glass interlayers. A vinyl acetate-ethylene copolymer is available for specialty products.

RISK ASSESSMENT: HEALTH

General Assessment

Vinyl acetate is a toxic human irritant by *inhalation* and may cause toxic systemic effects. It is mildly toxic by *ingestion*. Skin contact can produce localized effects but dermal absorption has not been reported. It is a questionable carcinogen in animal studies and is considered a suspected human carcinogen. It has been shown to cause adverse reproductive effects in animals and human mutation data have also been reported.

Inhalation of vinyl acetate vapors is not likely to occur under normal operating conditions. The extremely irritating properties of the vapor will cause personnel to vacate the area before toxic exposure levels occur. However, should an exposure occur by inhalation, it will likely result in irritation of the eyes, nose, throat, and upper respiratory tract, including the mucosa (particularly the eyes, gums, and respiratory passages). Inhalation can cause changes in the sense of smell (olfactory changes), conjunctiva irritation, and pulmonary changes. In high concentrations, there can be additional pulmonary effects and/or depression of the central nervous system with mild narcotic effects including dizziness, drowsiness, insomnia, encephalopathy, weakness, polyneuritis, tightness in chest, palpitation, dypsnea, and corneal edema. Cardiovascular effects include arrhythmia, chest pain, and syncope. There can also be liver function changes and hepatic enzyme induction.

Skin contact results in irritation and, if allowed to remain in contact with the liquid for prolonged periods, burns and blisters. Eye contact can cause serious damage to the cornea with a possibility for permanent effects to vision.

Ingestion of vinyl acetate may cause headache, drowsiness, unconsciousness, and unspecified gastrointestinal effects.

☠ *Acute Health Effects*

The following acute (short-term) health effects may occur immediately or shortly after exposure to vinyl acetate:

Skin: Causes irritation, burns, and blisters.

Eye: Severe irritation which can lead to injury and permanent damage if allowed to remain in contact with the eyes.

Lung: Irritation of the eyes, nose, throat and lungs (especially the mucus membranes). May cause changes in the sense of smell and in pulmonary function.

CNS: High concentrations can depress the central nervous system leading to narcoses, polyneuritis, encephalopathy, vertigo, weakness, and possible loss of consciousness.

CVS: Can cause cardiac arrhythmia, chest pain, and syncope (sudden loss of strength).

Chronic Health Effects

The following chronic (long-term) health effects may occur at some time after exposure to vinyl acetate and can last for months or even years:

Cancer Hazards: According to the references, vinyl acetate has been shown to cause cancer in limited animal studies. Human mutation data have also been reported and many scientists believe that exposure to such chemicals may pose a cancer risk in the long-term.

Reproduction: According to the references, vinyl acetate may cause adverse reproductive effects in test animals.

Other Chronic Effects: Very irritating substances such as vinyl acetate may cause lung damage. Repeated contact can cause drying and cracking of the skin and lead to secondary infection and dermatitis. Vinyl acetate may affect the heart, nervous system, and liver .

Recommended Risk-Reduction Measures

Personnel should avoid direct contact with vinyl acetate. The best risk reduction measure is to use a less toxic chemical as a substitute for vinyl acetate. If this is not possible or feasible, then *engineering controls* are the most effective method of reducing exposures. The best protection is to enclose operations and/or provide local exhaust ventilation at the site of chemical release. While not always operationally feasible, isolating operations can also reduce exposure risk.

Using respiratory protection is less effective than the controls mentioned above, but is still advisable whenever working with or around vinyl acetate. For high exposures, or when the exposure level is unknown (as in an emergency), a self-contained breathing apparatus (SCBA) with full facepiece and operated in pressure demand mode or a supplied-air respirator with full facepiece operated in pressure demand are the recommended respiratory protection methods of choice. If a full facepiece is not available, then chemical goggles should be worn to protect the eyes. Whenever a chemical splash hazard exists, a face shield and a protective apron should be worn. To prevent hand and skin exposures, impervious gloves should be used. Neoprene, nitrile, Viton, or chlorinated polyethylene products should provide adequate protection. However, glove manufacturers should be contacted to obtain permeation studies *before* gloves are selected.

Administrative controls should also be in place to minimize the potential for human exposures. These may include written procedures or policies which specify the methods and techniques that will be practiced whenever personnel are to work with vinyl acetate.

All personnel should receive *training* on the use, hazards, protective measures, emergency actions, and other precautions per 29 CFR 1910.1200 (Hazard Communication), prior to the first assignment in an area where vinyl acetate is used or stored. For those with frequent or potentially high exposure (half the TLV or greater), the following is recommended before initial assignment and at regular intervals thereafter (e.g., annually):

- ☑ Lung function tests.

If symptoms develop or overexposure is suspected, the following medical tests are recommended:

- ☑ Liver function tests.
- ☑ Lung function tests.
- ☑ Skin testing with dilute vinyl acetate to help diagnose allergy (performed by a qualified allergist).
- ☑ Examination of the nervous system.
- ☑ Electrocardiogram (EKG).
- ☑ Vinyl acetate is metabolized to acetic acid and acetaldehyde in the body. Suggest urinalysis and blood testing to determine exposure.

Any medical evaluation should include a careful history of past and present symptoms with an examination. Medical tests that look for existing damage are not a substitute for controlling exposures. Also, since smoking can cause heart disease, lung cancer, emphysema, and other respiratory diseases, smokers exposed to vinyl acetate may experience symptoms more quickly than non-smokers under the same conditions of exposure. Prudent risk management requires proper consideration of *all* factors which may be responsible for the development of exposure symptoms in the workplace.

Other methods to reduce exposure include:

- ☑ Where possible, enclose operations and use local exhaust ventilation at the site of chemical release. If local exhaust ventilation or enclosure is not used, respiratory protection should be mandatory.
- ☑ Always ensure that proper protective clothing is worn when using chemical substances and that personnel are trained in its use and care.
- ☑ Wash thoroughly immediately after exposure to vinyl acetate and at the end of the work shift or before eating, drinking, smoking, or applying cosmetics.

- ☑ Hazard warning information should be posted in the work area. In addition, as part of an on-going education and training program, all information on the health and safety hazards of vinyl acetate should be communicated to all exposed and potentially exposed workers.
- ☑ Never eat, drink, or smoke in areas were vinyl acetate is used, handled or stored.

RISK ASSESSMENT: ENVIRONMENT

General Assessment

The environment is at risk of exposure during transportation, storage, disposal, or destruction of vinyl acetate. In almost every scenario, the threat of environmental exposure is contingent upon the proper handling of the chemical substance. Accidental spills, large or small, can result in fire, explosion, and possible contamination of the surrounding environmental mediums (water, soil, and air).

Vinyl acetate is considered a Class IB flammable liquid (according to OSHA 29 CFR 1910.106). Its low flash point and relatively low boiling point present a serious fire and explosion hazard concern. Also, because it is incompatible with a number of common materials, especially strong oxidizers and acids, contact can result in violent and explosive reactions. It can self-polymerize if not kept stable with an inhibitor (such as hydroquinone or diphenylamine). It can also form explosive mixtures in air and can ignite on contact with heat, fire, or sparks. These characteristics require special consideration during any emergency situation involving a leak or spill of vinyl acetate.

Vinyl acetate can be removed from waste gases with activated carbon. Vinyl acetate can enter the environment through unchecked industrial discharges into effluents and through spills.

☠ *Acute Ecological Effects*

Acute (short-term) toxic effects may include the death of animals, birds, or fish, and death or low growth rate in plants. Acute effects are seen 2 to 4 days after animals or plants are exposed to vinyl acetate.

This chemical has moderate acute toxicity to aquatic life. Insufficient data are available to evaluate or predict the short-term effects of vinyl acetate to plants, birds, or terrestrial animals.

Chronic Ecological Effects

Chronic toxic effects may include shortened life span, reproductive problems, lower fertility, and changes in appearance or behavior in exposed animals. These effects can be seen long after first exposure(s) to toxic chemicals.

Vinyl acetate has moderate chronic toxicity to aquatic life. Insufficient data are available to evaluate or predict the long-term effects of vinyl acetate to plants, birds, or land animals.

Water Solubility

Vinyl acetate is slightly soluble in water. Concentrations of 1 to 100 milligrams and more will mix with a liter of water. In water, degradation is relatively quicker in freshwater than in sea water.

⌛ *Persistence in the Environment*

Vinyl acetate is slightly persistent in water, with a half-life of between 2 to 20 days. The half-life of a pollutant is the amount of time it takes for one-half of the chemical to be degraded. About 86.5% of vinyl acetate will eventually end up in air; the rest will end up in water. In the atmosphere, vinyl acetate is not likely to undergo photoxidation but it can produce ozone in the presence of nitrogen oxides. Even mixed or in soil, sunlight causes vinyl acetate to polymerize and increases its resistance to biodegradation.

Bioaccumulation in Aquatic Organisms

Some substances increase in concentration, or *bioaccumulate,* in living organisms as they breathe contaminated air, drink contaminated water, or eat contaminated food. These chemicals can become concentrated in the tissues and internal organs of animals as well as humans.

The concentration of vinyl acetate found in fish tissues is expected to be somewhat higher then the average concentration of vinyl acetate in water from which the fish was taken.

Recommended Risk-Reduction Measures

Proper training of all transporters will reduce the likelihood of a mishap or accident resulting in a leak or spill to the environment. The correct use of DOT labeling (per the MSDS) on all containers, trucks, and rail cars will enable emergency responders to react properly and

quickly to any disaster thereby reducing the potential risk to the environment and to personnel.

Storage of vinyl acetate should be segregated from incompatible chemicals to minimize the risk of cross-contamination or contact. Buildings designated for storage should be equipped with appropriate fire protection systems (alarms, sprinklers, emergency lighting, portable extinguishers).

If a spill or leak to the environment has occurred, fire department, emergency response, and/or hazardous materials spill personnel should be notified immediately. Cleanup should be attempted only by those trained in proper spill containment procedures using non-sparking tools. Contaminated soils should be removed for incineration and replaced with clean soil. If vinyl acetate should contact the water table, aquifer, or navigable waterway, time is of the essence. It is slightly soluble in water and, therefore, total containment and remediation may not be possible. When such spills occur, the local and/or state emergency response authorities must be notified. A comprehensive emergency response or disaster preparedness/recovery plan should be in place prior to any operations involving the use, transportation, storage, or disposal of vinyl acetate. If vinyl acetate is spilled or leaked, the following specific steps are recommended:

- ☑ Restrict persons not wearing protective clothing from area of spill or leak until cleanup is complete and area can be opened for normal work.
- ☑ Ventilate area of spill or leak.
- ☑ Remove all ignition sources from area.
- ☑ Absorb liquids in vermiculite, dry sand, earth, or a similar material and deposit in sealed containers. Use non-sparking tools.
- ☑ It may be necessary to dispose of vinyl acetate as a hazardous waste. The responsible state agency or the regional office of the federal Environmental Protection Agency (EPA) should be contacted for specific recommendations.

RISK ASSESSMENT: BUSINESS

General Assessment

Accidents or mishaps involving vinyl acetate can present a moderate threat to business operations. The loss or damage of equipment or facilities can significantly affect fiscal viability. Lawsuits that may result from personnel injury/death, public exposures, and/or environmental contamination will also require a serious expenditure of resources. Media attention surrounding an injury, death, or environmental damage can also result in a loss of profits and loss of current as well as future business. Always remember that anytime the terms "cancer," "carcinogen," or "reproductive hazard" are used, public emotion, hysteria, ignorance, and anxiety can run equally high. This must be carefully considered whenever drafting public relations policies.

⏻ *Recommended Risk-Reduction Measures*

Company attorneys, safety and health professionals, and environmental specialists should be involved in the development of any procedures or policies intended to manage the use of chemicals in the workplace. A company official should be pre-designated as a public relations officer with specific training in dealing with the press. Corporate plans and policies should be developed, approved, and implemented long before any need for such arises.

REFERENCES

American Conference of Governmental Industrial Hygienists. 1988. *Documentation of the Threshold Limit Values and Biological Exposure Indices,* 5th Edition (with updates). Cincinnati: ACGIH

Bloom, A. D. and F. J. DeSerres. 1995. *Ecotoxicity and Human Health.* New York: CRC Press/Lewis Publishers.

Burrell, R., D. K. Flaherty, and L. J. Sauers. 1992. *Toxicology of the Immune System.* New York: Van Nostrand Reinhold.

Cockerman, L. G., B. S. Shane. 1994. *Basic Environmental Toxicology.* New York: CRC Press/Lewis Publishers.

Cothern, C. R., N. P. Ross. 1994. *Environmental Statistics, Assessment, and Forecasting.* New York: CRC Press/Lewis Publishers.

Environmental Protection Agency. 1983. *Chemical Hazard Information Profile: Vinyl acetate.* Washington, D.C.: U.S. EPA.

Gardner, R. W. 1994. *Chemical Intolerance.* New York: CRC Press/Lewis Publishers.

Howard, P. H., et. al. 1991. *Handbook of Environmental Degradation Rates.* New York: CRC Press/Lewis Publishers.

Keith, L. H. and M. M. Walker. 1995. *Handbook of Air Toxics.* New York: CRC Press/Lewis Publishers.

Lauwerys, R. R., P. Hoet. 1993. *Industrial Chemical Exposure.* New York: CRC Press/Lewis Publishers.

Lewis, Richard J., Sr. 1992. *Sax's Dangerous Properties of Industrial Materials,* Volumes 1, 2, and 3, 8th Edition. New York: Van Nostrand Reinhold.

U.S. Department of Health and Human Services, National Institute for Occupational Safety and Health. 1994. *NIOSH Pocket Guide to Chemical Hazards.* Washington, D.C.: U.S. Government Printing Office.

MATERIAL SAFETY DATA SHEET

CHEMICAL NAME	CRC PRESS/LEWIS PUBLISHERS DISCLAIMER
VINYL BROMIDE	

HAZARD WARNING INFORMATION

HEALTH	FIRE	REACTIVE	OTHER	DEGREE OF HAZARD	COLOR CODING	OTHER CODES
2	0	1		0 = Minimum Hazard 1 = Slight Hazard 2 = Moderate Hazard 3 = Serious Hazard 4 = Severe Hazard	HEALTH = BLUE FIRE = RED REACTIVITY = YELLOW OTHER = WHITE	OX = Oxidizer ACID = Acid ALK = Alkali COR = Corrosive W = Use No Water

SECTION I - GENERAL INFORMATION

Characterization	**Halogenated Hydrocarbon**	RCRA Number	**None**
		EPA Class	**Not Applicable**
DOT Proper Shipping Name	**Vinyl Bromide (inhibited)**	Chemical Abstract Service (CAS) Number	**593-60-2**
DOT Hazard Class and Label Requirements	**Flammable Gas**	DOT Emergency Guide Code	**60**
DOT Identification Number	**UN 1085**	Molecular Formula	**CH_2=CHBr**

Synonyms
Bromoethene; bromoethylene.

SECTION II - HAZARDOUS INGREDIENTS AND IDENTITY INFORMATION

Hazardous Components (specific identity)	OSHA Exposure Criteria	NIOSH Exposure Criteria	Immediately Dangerous to Life and Health (IDLH)	ACGIH Exposure Criteria
Vinyl bromide (derivation: By reaction of acetylene with hydrobromic acid in the presence of a catalyst such as mercury halides, cerium, or copper; by partial dehydrobromination of ethylene dibromide with alcoholic potassium hydroxide). **1 ppm = 4.45 mg/m^3**	**PEL:** Not Established **STEL:** Not Established	**REL:** Reduce to Lowest Level Possible Human Carcinogen	**Not Determined**	**TLV:** 5 ppm 22 mg/m^3 Suspected Human Carcinogen

SECTION III - PHYSICAL AND CHEMICAL CHARACTERISTICS

Boiling Point	**60°F (15.6°C)**	Specific Gravity (H_2O = 1)	**3.79 (gas) 1.49 (liquid at 60°F)**
Vapor Pressure (mm Hg)	**760 at 60.4°F (15.6°C)**	Molecular Weight (atomic weight)	**107.0**
Vapor Density (Air = 1)	**3.7**	Melting Point	**-219°F (-104°C)**

Solubility
Insoluble in water. Soluble in alcohol, ether, acetone, benzene, and chloroform.

Appearance and Odor
Colorless gas (or liquid below 60°F) with a characteristic, pungent, but pleasant odor.

SECTION IV - FIRE AND EXPLOSION HAZARD DATA

Flash Point (method used)	**Not Established**	Explosive Limits in Air % by Volume	**LEL: 9% UEL: 15%**
NFPA Classification	**Flammable Gas or Class 1A Flammable Liquid**	Autoignition Temperature	**986°F (530°C)**

Extinguishing Media
Stop flow of gas. Use dry chemical, carbon dioxide, water spray, or regular foam.

Special Fire Fighting Procedures
Poisonous gases are produced in fire. Wear self-contained breathing apparatus (SCBA) and full protective gear. Move containers from fire if it can be done without risk. Keep cooling sides of fire-exposed containers with water long after fire is out.

Unusual Fire and Explosion Hazards
Dangerous fire and explosion hazard. Containers may explode in fire. Stay away from the ends of tanks. Vapors are heavier than air and can travel for great distances to an ignition source to flashback and cause fire or explosion.

SECTION V - REACTIVITY DATA

Stability		Conditions to Avoid Under normal conditions of handling and storage, vinyl bromide is considered stable in closed containers. Avoid heat, flame, other sources of ignition, and contact with incompatible materials and chemicals.
Stable X	Unstable	Incompatibility *(materials to avoid)* Reacts violently with strong oxidizers (chlorine, bromine, fluorine).
Hazardous Polymerization		Conditions to Avoid Hazardous polymerization can occur upon exposure to sunlight.
May Occur X	Will Not Occur	Hazardous Decomposition or By-products Poisonous gases and acrid fumes are produced in fire, including carbon dioxide, carbon monoxide, and toxic fumes of bromine.

SECTION VI - HEALTH HAZARD DATA

Primary Route(s) of Entry:	Inhalation?	Absorption (skin/eye)?	Ingestion?
	X		X (liquid)

Health Hazards

INHALATION: Causes depression of the central nervous system (CNS) with symptoms of headache, vertigo, nausea, and drowsiness. High concentrations may cause loss of consciousness. There may be decreased mental status and confusion. Possible liver and kidney damage.

SKIN & EYES: Irritation with possible redness and pain to the skin and eyes. Eye contact may result in mild to moderate irritation and possible inflammation. The compressed gas can lead to tissue damage (frostbite) on contact with the skin or eyes.

INGESTION: Unspecified gastrointestinal effects. An unlikely exposure route, but still possible.

Carcinogenicity	NTP Listed?	IARC Cancer Review Group?	OSHA Regulated?	Target Organs?
Suspected Human Confirmed Animal	No	Group 2A	No	Eyes, skin, liver, kidney, CNS.

Medical Conditions Generally Aggravated by Exposure

None reported. However, liver, kidney, heart, and respiratory problems may be aggravated by exposure.

Emergency and First-aid Procedures

Eye contact: Do not allow victim to keep eyes tightly shut. Do NOT rub eyes. Flush immediately with warm (tepid) water for 15 minutes (minimum); seek medical attention. Skin contact: Remove all contaminated clothing, including shoes. Immediately wash area with flooding amounts of warm (tepid) water for 15 minutes (minimum). Do NOT rub or use dry-blown heat on frostbitten tissue. Seek medical attention immediately. For inhalation: Remove the person from exposure. Provide respiratory assistance and CPR. Transfer to medical facility. If swallowed: Not a likely exposure route since vinyl bromide exists as a gas at room temperature and pressure.

SECTION VII - PRECAUTIONS FOR SAFE HANDLING AND USE

Steps to be Taken in Case Material is Released or Spilled

Remove all ignition sources, wear SCBA respiratory protection. Restrict those not involved in cleanup from entering area. Stop flow of gas if it can be done without risk. Provide explosion-proof ventilation. Absorb liquids with vermiculite and deposit in sealed containers.

Preferred Waste Disposal Method

No citation.

Precautions to be Taken in Handling and Storage

Store cylinders in cool, well-ventilated location. Keep away from sources of heat and sunlight.

Other Precautions and Warnings

Cylinders must always be stored upright and properly secured (around the body of the cylinder, never the neck). Never drag or roll cylinders. Metal containers should be bonded and grounded.

SECTION VIII - CONTROL MEASURES AND PERSONAL PROTECTIVE EQUIPMENT

Respiratory Protection *(specify type)*

No OSHA PEL established. For best protection, use a supplied-air respirator or a self-contained breathing apparatus (SCBA) with full facepiece operated in positive pressure mode.

Ventilation

Local exhaust (preferred) at site of chemical work or general ventilation.

Protective Gloves	Eye Protection	Other Protective Clothing
Impervious Thermal Gloves	Chemical Goggles or Face Mask	Protective Apron if Splash is Likely

Work/Hygiene Practices

Always wash hands thoroughly after using chemical; never bring food, drink, or smoking materials into vicinity of chemicals.

VINYL BROMIDE

CH_2=CHBr

CAS: 593-60-2

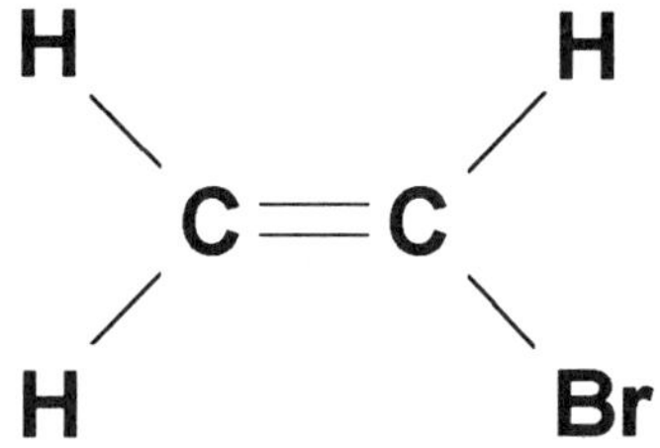

IDENTIFICATION AND TYPICAL USES

Vinyl bromide is a colorless gas with a characteristic, pungent, but pleasant odor. It can also exist as a liquid below 60°F (15.6°C). It is used as a co-monomer, a chemical intermediate, a flame retarding agent for acrylic fibers and plastics, and in leather and fabricated metal products.

RISK ASSESSMENT: HEALTH

General Assessment

As a gas, vinyl bromide is moderately toxic by *inhalation*. While absorption through the skin has not been known to occur, skin contact can cause localized damage to tissues. *Ingestion* is rare but possible (liquid threat only). Vinyl bromide is considered a suspected human carcinogen and a confirmed animal carcinogen. Mutation data have also been reported. Physiologically, it is an anesthetic and narcotic. The primary human response to vinyl bromide is depression of the central nervous system (CNS).

Inhalation causes some irritation of the eyes, nose, and lungs. Symptoms of exposure and CNS depression include drowsiness, slurred speech, lack of coordination, and possible loss of consciousness.

Skin contact may cause localized burning at site of contact as a result of frostbite damage. Eye contact can result in severe irritation to the cornea, with redness and swelling. Damage can be serious if not immediately flushed from eye surface. Liver and kidney damage have been reported in test animals.

Acute Health Effects

The following acute (short-term) health effects may occur immediately or shortly after exposure to vinyl bromide:

Skin: Irritation with potential for redness, pain, inflammation, and tissue damage (frostbite).

Eye: Severe irritation, burning, pain, and permanent damage possible.

Lung: Nose, throat, and respiratory tract irritation.

CNS: Narcosis, headaches, lightheadedness, dizziness, drowsiness, nausea, vomiting, confusion, decreased mental status, possible loss of consciousness.

Chronic Health Effects

The following chronic (long-term) health effects may occur at some time after exposure to vinyl bromide and can last for months or even years.

Cancer Hazards: According to information presented in the references, vinyl bromide is considered a suspected human carcinogen and a confirmed animal carcinogen. Physically, chemically, and toxicologically, vinyl bromide is very similar to vinyl chloride which is a confirmed human carcinogen. Mutation data have also been reported and many scientists believe that exposure to such chemicals may pose a cancer risk in the long-term. It is suggested that vinyl bromide be treated as though it is a human cancer agent.

Reproductive Hazard: According to information presented in the references, vinyl bromide has not been adequately tested for its ability to cause adverse reproductive effects in test animals.

Other Chronic Effects: Long-term exposure has lead to liver and kidney damage (tumors) in test animals. It is not known whether exposure to vinyl bromide will have this effect on humans. Vinyl bromide is a relatively new compound and few data exist on its toxicity.

Recommended Risk-Reduction Measures

Personnel should avoid direct contact with vinyl bromide. The exact nature of it human toxicity has not been clearly established. It is very possible that it is a human cancer agent. Occupational poisoning occurs most commonly by inhalation. If a less toxic material or compound cannot be substituted for vinyl bromide, then *engineering controls* are the most effective method of reducing exposures. The best protection is to enclose operations and/or provide local explosion-proof exhaust ventilation at the site of vinyl bromide release. While not always operationally feasible, isolating operations can also reduce exposure. Using respiratory protection is less effective than the controls mentioned above, but is still advisable whenever working with or around vinyl bromide. Caution should be exercised when selecting respiratory protection since it is not known if exposure to vinyl bromide will cause cancer in humans. The ACGIH has established a TLV of 5 ppm over an eight-hour period. This level is extremely low and difficult to accurately monitor. For exposures to high concentrations, or when the concentration is unknown, use a supplied-air respirator with full facepiece and mask operated in positive pressure mode or a self-contained breathing apparatus (SCBA) with full facepiece operated in pressure demand. If a full facepiece is not available, then chemical goggles should be worn to protect the eyes. A face shield should also be considered. To prevent hand and skin exposures, impervious thermal gloves should be worn.

Administrative controls should also be in place to minimize the potential for human exposures. These may include written procedures or policies which specify the methods and techniques that will be practiced whenever personnel are to work with vinyl bromide.

All personnel should receive *training* on the use, hazards, protective measures, emergency actions, and other precautions per 29 CFR '1910.1200 (Hazard Communication), prior to the first assignment in an area where vinyl bromide is used or stored.

If symptoms develop or overexposure is suspected, the following recommended medical tests should be considered:

- ☑ Liver and kidney function tests.
- ☑ Nervous system evaluation.

Any evaluation should include a careful history of past and present symptoms with an examination. Medical tests that simply look for existing damage are not a substitute for controlling exposures.

Other methods to reduce exposure include:

- ☑ Where possible, enclose operations and use local exhaust ventilation at the site of chemical release. If local exhaust ventilation or enclosure is not used, respiratory protection should be mandatory.
- ☑ Always ensure that proper protective clothing is worn when using chemical substances and that personnel have been properly trained in its correct use, care, and maintenance.
- ☑ Wash thoroughly immediately after exposure to vinyl bromide and at the end of the work shift or before eating, drinking, smoking, or applying cosmetics.
- ☑ Hazard warning information should be posted in the work area. In addition, as part of an on-going education and training program, all information on the health and safety hazards of vinyl bromide should be communicated to all exposed and potentially exposed workers.
- ☑ Eye wash stations should be provided in the immediate work area for emergency use. If there is a possibility of skin exposure to vinyl bromide, emergency shower facilities should be provided.
- ☑ Workers whose clothing has been contaminated by vinyl bromide should change into clean clothes before leaving work. Contaminated clothing that is worn home can expose family members to vinyl bromide. Contaminated work clothing should be laundered only by individuals who have been informed of the hazards of vinyl bromide.
- ☑ Where possible, automatically transfer liquids containing vinyl bromide from drums or other containers to process containers.

RISK ASSESSMENT: ENVIRONMENT

General Assessment

The environment is at risk of exposure during transportation, storage, disposal, or destruction of vinyl bromide. In almost every scenario, the threat of environmental exposure is contingent upon the proper handling of the chemical substance. Accidental spills, large or small, can result in contamination of the surrounding environmental mediums (water, soil, and air).

Vinyl bromide is an extremely flammable gas. No flash point has been determined by standard tests in air, but when it contacts a high energy ignition source, it reaches its explosive limits. In liquid form, vinyl bromide is considered a Class IA flammable liquid. It is a dangerous fire and explosion hazard when exposed to heat or flame. It is incompatible with strong oxidizers, such as chlorine, fluorine, and bromine, and can polymerize violently on exposure to sunlight. Caution is always required in handling, storage, transportation, and disposal of vinyl bromide. Emergency responders should be made aware of the presence of vinyl bromide at any emergency response situation.

Vinyl bromide can enter the environment from industrial effluents and from spills.

☠ *Acute Ecological Effects*

Acute (short-term) toxic effects may include the death of animals, birds, or fish and death or low growth rate in plants. Acute effects are seen 2 to 4 days after animals or plants are exposed to vinyl bromide.

Vinyl bromide has slight acute toxicity to aquatic life. Insufficient data are available to evaluate or predict the short-term effects of vinyl bromide on plants, birds or land animals.

Chronic Ecological Effects

Chronic (long-term) toxic effects may include shortened life span, reproductive problems, lower fertility, and changes in appearance or behavior in exposed animals. These effects can be seen long after first exposure(s) to toxic chemicals.

Vinyl bromide has slight chronic toxicity to aquatic life. Insufficient data are available to evaluate or predict the long-term effects of vinyl bromide to plants, birds, or terrestrial animals.

Water Solubility

Vinyl bromide may be insoluble to slightly soluble in the aquatic environment. Concentrations between 1 and 100 milligrams may mix with a liter of water.

⌛ *Persistence in the Environment*

As a gas vinyl bromide is non-persistent in water, due to volatilization, with a half-life of less than 2 days. The half-life of a pollutant is the amount of time it takes for one half of the chemical to be degraded. Approximately 99.8% of vinyl bromide will eventually end up in air; the rest will end up in the water. However, if released in moist soil, vinyl bromide may leach into and contaminate ground water supplies. In dry soil, it evaporates rapidly with no persistence expected.

Bioaccumulation in Aquatic Organisms

Some substances increase in concentration, or *bioaccumulate,* in living organisms as they breathe contaminated air, drink contaminated water, or eat contaminated food. These chemicals can become concentrated in the tissues and internal organs of animals as well as humans.

The concentration of vinyl bromide found in fish tissues is expected to be about the same as the average concentration of vinyl bromide in the water from which the fish was taken.

Recommended Risk-Reduction Measures

Proper training of all transporters will reduce the likelihood of a mishap or accident resulting in a leak or spill of vinyl bromide into the environment. Labels on all containers, trucks, and rail cars must meet DOT requirements and accurately reflect their contents to enable emergency responders to react properly and quickly to any disaster thereby reducing the potential risk to the environment and to personnel.

Storage of vinyl bromide should be segregated from other chemicals and materials to minimize the risk of cross-contamination. Cylinders must always be stored upright and properly secured (always around the cylinder body, never around its neck). Cylinders must never be dragged or rolled during handling. They must be kept away from heat and direct sunlight.

If a spill or leak to the environment has occurred, fire department, emergency response and/or hazardous materials spill personnel should be notified immediately. Cleanup should be attempted only by those trained in proper spill containment procedures using non-sparking tools. Contaminated soils should be removed for incineration and replaced with clean soil. If vinyl bromide should contact the water table, aquifer, or navigable waterway, time is of the essence. It may be insoluble and remediation may be successful. The local and/or state emergency response authorities must be notified. A comprehensive emergency response or disaster preparedness/recovery plan should be in place prior to any operations involving the use, transportation, storage, or disposal of vinyl bromide.

If vinyl bromide is spilled or leaked, the following specific steps are recommended:

☑ Restrict persons not wearing protective clothing from area of spill or leak until cleanup is complete. Use non-sparking tools.

- ☑ If applicable, stop flow of leaking liquid or gas. If leak source is a cylinder and the leak cannot be stopped in place, remove leaking cylinder to a safe place in the open air, and repair or allow cylinder to empty. Use water spray to knock down vapors that have not yet ignited.
- ☑ Absorb any liquids in vermiculite, sand, earth, or similar material and deposit in sealed drum.
- ☑ Ventilate area of spill or leak. Containers should be removed to safer area if it can be done without increasing the risk.
- ☑ It may be necessary to dispose of vinyl bromide as a hazardous waste. The responsible state agency or the regional office of the federal Environmental Protection Agency (EPA) should be contacted for specific recommendations.

RISK ASSESSMENT: BUSINESS

General Assessment

Accidents or mishaps involving vinyl bromide can present a serious threat to business operations. The loss or damage of equipment or facilities can significantly affect fiscal viability. Lawsuits that may result from personnel illness, injury/death, public exposures, and/or environmental contamination will require a serious expenditure of resources. Media attention surrounding an injury, death, or environmental damage can also result in a loss of profits and loss of current as well as future business. Always remember that anytime terms such as "cancer" or "carcinogen" are used, public emotion, hysteria, and ignorance can run equally high. This must be carefully considered whenever developing or implementing a public relations policy.

Recommended Risk-Reduction Measures

Company attorneys, safety and health professionals, and environmental specialists should be involved in the development of any procedures for responding to chemical incidents. A company official should be predesignated as a public relations officer with specific training in dealing with the press. Corporate plans and policies should be developed, approved, and implemented long before any need for such arises.

REFERENCES

Ahlbom, A. 1993. *Biostatistics for Epidemiologists.* New York: CRC Press/Lewis Publishers.

American Conference of Governmental Industrial Hygienists. 1988. *Documentation of the Threshold Limit Values and Biological Exposure Indices,* 5th Edition (with updates). Cincinnati: ACGIH

Anthony, C. P., and N. J. Kolthoff. 1971. *Textbook of Anatomy and Physiology.* St. Louis: C. V. Mosby Company.

Asante-Duah, D. K. 1993. *Hazardous Waste Risk Assessment.* New York: CRC Press/Lewis Publishers.

Bloom, A. D. and F. J. DeSerres. 1995. *Ecotoxicity and Human Health.* New York: CRC Press/Lewis Publishers.

Bueche, J. R. 1972. *Principles of Physics,* 2nd Edition. New York: McGraw-Hill.

Calabrese, E. J. 1994. *Biological Effects of Low Level Exposures.* New York: CRC Press/Lewis Publishers.

Cockerman, L. G., B. S. Shane. 1994. *Basic Environmental Toxicology.* New York: CRC Press/Lewis Publishers.

Cothern, C. R., N. P. Ross. 1994. *Environmental Statistics, Assessment, and Forecasting.* New York: CRC Press/Lewis Publishers.

Hallenbeck, W. H. 1993. *Quantitative Risk Assessment for Environmental and Occupational Health,* 2nd Edition. New York: CRC Press/Lewis Publishers.

Howard, P. H., et. al. 1991. *Handbook of Environmental Degradation Rates.* New York: CRC Press/Lewis Publishers.

Lewis, R. J., Sr. 1993. *Hawley's Condensed Chemical Dictionary,* 12th Edition. New York: Van Nostrand Reinhold.

Lewis, R. J., Sr. 1992. *Sax's Dangerous Properties of Industrial Materials,* Volumes 1, 2, and 3, 8th Edition. New York: Van Nostrand Reinhold.

Manahan, S. E. 1994. *Environmental Chemistry,* 6th Edition. New York: CRC Press/Lewis Publishers.

Meyer, E. 1990. *Chemistry of Hazardous Materials.* Englewood Cliffs, NJ: Prentice Hall.

Patnaik, P. 1992. *A Comprehensive Guide to the Hazardous Properties of Chemical Substances.* New York: Van Nostrand Reinhold.

Quigley, D. R. 1994. *Handbook of Emergency Chemical Management.* New York: CRC Press/Lewis Publishers.

Rea, W. J. 1995. *Chemical Sensitivity,* Vol. I-III. New York: CRC Press/Lewis Publishers.

Richardson, M. 1995. *Dictionary of Substance and Their Effects,* Vol. I-VI. New York: CRC Press/Lewis Publishers.

Sacarello, H. L. A. 1994. *The Comprehensive Handbook of Hazardous Materials.* New York: CRC Press/Lewis Publishers.

Sheftel, V. O. 1995. *Handbook of Toxic Properties of Monomers and Additives.* New York: CRC Press/Lewis Publishers.

Sherman, J. D. 1988. *Chemical Exposure and Disease.* New York: Van Nostrand Reinhold.

Torkelson, T. R. V. K. Rowe. 1981. *Patty's Industrial Hygiene and Toxicology,* Vol. 2B. New York: Wiley-Interscience.

U.S. Department of Health and Human Services, National Institute for Occupational Safety and Health. 1994. *NIOSH Pocket Guide to Chemical Hazards.* Washington, D.C.: U.S. Government Printing Office.

MATERIAL SAFETY DATA SHEET

CHEMICAL NAME	CRC PRESS/LEWIS PUBLISHERS DISCLAIMER:
VINYL CHLORIDE	THE INFORMATION AND RECOMMENDATIONS PRESENTED HEREIN ARE BASED ON SOURCES BELIEVED TO BE RELIABLE. CRC MAKES NO REPRESENTATION ON ITS COMPLETENESS OR ACCURACY. IT IS THE RESPONSIBILITY OF THE USER TO DETERMINE THE CHEMICAL'S SUITABILITY FOR ITS INTENDED USE, THE CHEMICAL'S SAFE USE, AND THE CHEMICAL'S PROPER DISPOSAL. NO REPRESENTATIONS AND/OR WARRANTIES, EITHER EXPRESSED OR IMPLIED, OF THE MERCHANTABILITY OR FITNESS FOR A PARTICULAR PURPOSE, OR OF ANY OTHER NATURE, ARE MADE WITH RESPECT TO THE INFORMATION PROVIDED IN THIS MSDS OR TO THE CHEMICAL TO WHICH INFORMATION MAY REFER. CRC NEITHER ASSUMES NOR AUTHORIZES ANY OTHER PERSON TO ASSUME FOR IT, ANY OTHER ADDITIONAL RESPONSIBILITY OR LIABILITY FOR THE USE OF, OR RELIANCE UPON, THIS INFORMATION.

HAZARD WARNING INFORMATION

HEALTH	FIRE	REACTIVE	OTHER	DEGREE OF HAZARD	COLOR CODING	OTHER CODES
2	4	2		0 = Minimum Hazard 1 = Slight Hazard 2 = Moderate Hazard 3 = Serious Hazard 4 = Severe Hazard	HEALTH = BLUE FIRE = RED REACTIVITY = YELLOW OTHER = WHITE	OX = Oxidizer ACID = Acid ALK = Alkali COR = Corrosive ~~W~~ = Use No Water

SECTION I - GENERAL INFORMATION

Characterization	**Halogenated Hydrocarbon**	RCRA Number	**U043**
		EPA Class	**Toxic Waste**
DOT Proper Shipping Name	**Vinyl Chloride**	Chemical Abstract Service (CAS) Number	**75-01-4**
DOT Hazard Class and Label Requirements	**Flammable Gas**	DOT Emergency Guide Code	**17**
DOT Identification Number	**UN 1086**	Molecular Formula	**CH_2=CHCl**

Synonyms
Chloroethane; chloroethylene; ethylene monochloride; monochloroethane; monochloroethylene, VCM.

SECTION II - HAZARDOUS INGREDIENTS AND IDENTITY INFORMATION

Hazardous Components (specific identity)	OSHA Exposure Criteria	NIOSH Exposure Criteria	Immediately Dangerous to Life and Health (IDLH)	ACGIH Exposure Criteria
Vinyl chloride (derivation: By dechlorination of ethylene dichloride; by reaction of acetylene and hydrogen chloride, either as liquids or gases). **1 ppm = 2.60 mg/m^3**	**PEL (8-hour):** **1 ppm** **2.6 mg/m^3** **STEL (15-min):** **5 ppm** **13 mg/m^3**	**REL:** **Reduce to Lowest Level** **Confirmed Human Carcinogen**	**Not Applicable: Human Carcinogen (all exposure is IDLH)**	**TLV:** **5 ppm** **13 mg/m^3** **Confirmed Human Carcinogen**

SECTION III - PHYSICAL AND CHEMICAL CHARACTERISTICS

Boiling Point	**7°F (-14°C)**	Specific Gravity (H_2O = 1)	**2.21 (gas) 0.91 (liquid at 15°F)**
Vapor Pressure (mm Hg)	**2530 at 68°F (20°C)**	Molecular Weight (atomic weight)	**62.5**
Vapor Density (Air = 1)	**2.155**	Melting Point	**-245°F (-160°C)**

Solubility
Slightly soluble (0.1% at 77°F). Soluble in alcohol, benzene, carbon tetrachloride, ether, hydrocarbons.

Appearance and Odor
Colorless gas (or liquid below 7°F) with a characteristic, pungent odor.

SECTION IV - FIRE AND EXPLOSION HAZARD DATA

Flash Point (method used)	**-108.4°F (-78°C) opened cup**	Explosive Limits in Air % by Volume	**LEL: 3.6% UEL: 33%**
NFPA Classification	**Flammable Gas**	Autoignition Temperature	**882°F (472°C)**

Extinguishing Media
Stop flow of gas. Use dry chemical, carbon dioxide, water spray, or regular foam.

Special Fire Fighting Procedures
Poisonous gases are produced in fire. Wear self-contained breathing apparatus (SCBA) and full protective gear. Move containers from fire if it can be done without risk. Keep cooling sides of fire-exposed containers with water long after fire is out. Large fires may be inextinguishable.

Unusual Fire and Explosion Hazards
Dangerous fire and explosion hazard. Containers may explode in fire. Stay away from the ends of tanks. Vapors are heavier than air and can travel for great distances to an ignition source to flashback and cause fire or explosion.

SECTION V - REACTIVITY DATA

Stability		Conditions to Avoid Under normal conditions of handling and storage, vinyl chloride is considered stable in closed containers. Long-term exposure to air may result in the formation of explosive peroxides.
Stable X	Unstable	Incompatibility (materials to avoid) Reacts violently with strong oxidizers (chlorine, bromine, fluorine).
Hazardous Polymerization		Conditions to Avoid Hazardous polymerization can occur upon exposure to light or in the presence of a catalyst. It is usually inhibited with a phenol, which can be toxic.
May Occur X	Will Not Occur	Hazardous Decomposition or By-products Poisonous gases and acrid fumes are produced in fire, including hydrogen chloride, carbon dioxide, carbon monoxide, and toxic phosgene.

SECTION VI - HEALTH HAZARD DATA

Primary Route(s) of Entry:	Inhalation?	Absorption (skin/eye)?	Ingestion?
	X		X (liquid)

Health Hazards

INHALATION: Causes depression of the central nervous system (CNS) with symptoms of headache, vertigo, nausea, drowsiness, ataxia, euphoria, vision and hearing disturbances, parasthesia (numbness and tingling in the hands and feet), narcosis, unconsciousness, and death due to respiratory arrest. Inhalation also causes dypsnea, asthma, and pneumoconiosis.

SKIN & EYES: Irritation with possible redness and pain to the skin and eyes. Eye contact may result in mild to moderate irritation and possible inflammation. The compressed gas can lead to tissue damage (frostbite) on contact with the skin or eyes.

INGESTION: Unspecified gastrointestinal effects. An unlikely exposure route, but still possible.

Carcinogenicity	NTP Listed?	IARC Cancer Review Group?	OSHA Regulated?	Target Organs?
Confirmed Human Confirmed Animal	Class 1	Group 1	29 CFR 1910.1017	Eyes, skin, liver, blood, CNS, lymph, bone.

Medical Conditions Generally Aggravated by Exposure

Liver, heart, respiratory, and connective tissue disorders may be aggravated by exposure.

Emergency and First-aid Procedures

Eye contact: Do not allow victim to keep eyes tightly shut. Do NOT rub eyes. Flush immediately with warm (tepid) water for 15 minutes (minimum); seek medical attention. Skin contact: Remove all contaminated clothing, including shoes. Immediately wash area with flooding amounts of warm (tepid) water for 15 minutes (minimum). Do NOT rub or use dry-blown heat on frostbitten tissue. Seek medical attention immediately. For inhalation: Remove the person from exposure. Provide respiratory assistance and CPR. Transfer to medical facility. If swallowed: Not a likely exposure route.

SECTION VII - PRECAUTIONS FOR SAFE HANDLING AND USE

Steps to be Taken in Case Material is Released or Spilled

Remove all ignition sources, wear SCBA respiratory protection. Restrict those not involved in cleanup from entering area. Stop flow of gas if it can be done without risk. Provide explosion-proof ventilation. Absorb liquids with vermiculite and deposit in sealed containers.

Preferred Waste Disposal Method

No citation.

Precautions to be Taken in Handling and Storage

Store cylinders in cool, well-ventilated location. Keep away from sources of heat and sunlight.

Other Precautions and Warnings

Cylinders must always be stored upright and properly secured (around the body of the cylinder, never the neck). Never drag or roll cylinders. Metal containers should be bonded and grounded.

SECTION VIII - CONTROL MEASURES AND PERSONAL PROTECTIVE EQUIPMENT

Respiratory Protection *(specify type)*

A confirmed human carcinogen. At any level, use a supplied-air respirator or a self-contained breathing apparatus (SCBA) with full facepiece operated in positive pressure mode for the best protection.

Ventilation

Local exhaust (preferred) at site of chemical work or general ventilation.

Protective Gloves	Eye Protection	Other Protective Clothing
Impervious Thermal Gloves	Chemical Goggles or Face Mask	Protective Apron if Splash is Likely

Work/Hygiene Practices

Always wash hands thoroughly after using chemical; never bring food, drink, or smoking materials into vicinity of chemicals.

VINYL CHLORIDE

$CH_2=CHCl$

CAS: 75-01-4

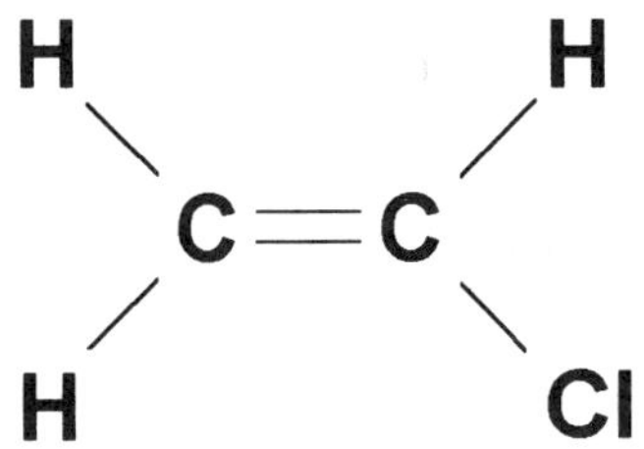

IDENTIFICATION AND TYPICAL USES

Vinyl chloride is a colorless gas with a characteristic, pungent, but pleasant odor. It can also exist as a liquid below 7°F (-14°C). It is used as a monomer in the manufacture of polyvinyl chloride resins and plastics, as a refrigerant, and in organic synthesis. Its use has been largely curtailed and regulated in recent years as a result of its known human carcinogenic properties.

RISK ASSESSMENT: HEALTH

General Assessment

As a gas, vinyl chloride is highly toxic by *inhalation*. While absorption through the skin has not been known to occur in toxic amounts, skin contact can cause localized damage to tissues. *Ingestion* is rare but possible (liquid threat only). Vinyl chloride is considered a confirmed human carcinogen and a confirmed animals carcinogen. Human mutation data have also been reported. Physiologically, it acts as an anesthetic and narcotic. The initial (acute) human response to vinyl chloride is depression of the central nervous system (CNS). Chronic illnesses include cancers and adverse reproductive effects.

Inhalation causes some irritation of the eyes, nose, and lungs. Symptoms of exposure and CNS depression include drowsiness, slurred speech, lack of coordination, fatigue, headaches, vertigo, ataxia, euphoria, vision and hearing disturbances, parasthesia (tingling and numbness in the extremities), narcosis, anesthesia, possible loss of consciousness, and death due to respiratory failure. In addition, inhalation causes dypsnea, asthma, and pneumoconiosis.

Skin contact may cause localized burning at site of contact as a result of frostbite damage. Eye contact can result in severe irritation to the cornea, with redness and swelling. Damage can be serious if not immediately flushed from eye surface. Liver and kidney damage have been reported in test animals.

Acute Health Effects

The following acute (short-term) health effects may occur immediately or shortly after exposure to vinyl chloride:

Skin: Irritation with potential for redness, pain, inflammation, and tissue damage (frostbite).

Eye: Severe irritation, burning, pain, and permanent damage possible.

Lung: Nose, throat, and respiratory tract irritation. Symptoms can include dyspnea, asthma, and pneumoconiosis.

CNS: Narcosis, headaches, lightheadedness, dizziness, drowsiness, nausea, vomiting, confusion, decreased mental status, possible loss of consciousness, and death (respiratory arrest).

Chronic Health Effects

The following chronic (long-term) health effects may occur at some time after exposure to vinyl chloride and can last for months or even years:

Cancer Hazards: According to information presented in the references, vinyl chloride is a confirmed human carcinogen causing cancers of the liver. Human mutation data have also been reported confirming the strong toxicity of this compound in humans.

Reproductive Hazard: According to information presented in the references, vinyl chloride has been shown to adversely affect reproduction in humans. Studies have shown loss of libido and sperm in men. Women experienced ovarian dysfunction, benign uterine growths, and prolapsed genital organs. There is limited evidence that exposure damages the developing fetus (a teratogen) in animals. This has not been established in humans. An excess of spontaneous abortions has been reported among spouses of workers who had been exposed to vinyl chloride. Increased rate of birth defects have been shown in areas where vinyl chloride processing plants are located. The role of vinyl chloride in this increased risk has not been established.

Other Chronic Effects: Repeated exposure can cause a disease known as scleroderma, causing the skin to become very smooth, tight, and shiny. It causes the bones in the fingers to erode (acroosteolysis), and damages the blood vessels in the hands (Raynaud's syndrome). This, in turn, causes the hands (or feet) to become numb, pale or blue with even the slightest exposure to cold. Repeated exposure can permanently damage the kidney, liver, nervous system, and blood cells. Vinyl chloride can also cause symptoms similar to stomach ulcers. Prolonged skin contact can cause a skin allergy to develop in some persons. Once an allergy develops, very low future exposures can cause itching and rash.

Recommended Risk-Reduction Measures

Personnel should avoid direct contact with vinyl chloride. It is a confirmed human carcinogen and poses an extremely serious risk to human health. Occupational poisoning occurs most commonly by inhalation. If a less toxic material or compound cannot be substituted for vinyl chloride, then *engineering controls* are the most effective method of reducing exposures. The best protection is to enclose operations and/or provide local explosion-proof exhaust ventilation at the site of vinyl chloride release. While not always operationally feasible, isolating operations can also reduce exposure. Using respiratory protection is less effective than the controls mentioned above, but is still advisable whenever working with or around vinyl chloride. OSHA has established a PEL of 1 ppm over an eight-hour period. This level is extremely low and difficult to accurately monitor. For exposure to any concentrations, especially when the concentration is unknown (as in an emergency), use a supplied-air respirator with full facepiece and mask operated in positive pressure mode or a self-contained breathing apparatus (SCBA) with full facepiece operated in pressure demand. If a full facepiece is not available, then chemical goggles should be worn to protect the eyes. A face shield should also be considered. To prevent hand and skin exposures, impervious thermal gloves should be used. Glove manufacturers should be contacted and permeation studies obtained *before* final glove selections are made.

Administrative controls should also be in place to minimize the potential for human exposures. These may include written procedures or policies which specify the methods and techniques that will be practiced whenever personnel are to work with vinyl chloride.

All personnel should receive *training* on the use, hazards, protective measures, emergency actions, and other precautions per 29 CFR 1910.1200 (Hazard Communication), prior to the first assignment in an area where vinyl chloride is used or stored. OSHA 29 CFR 1910.1017 requires employers to provide all exposed workers with the following *before* starting work and every year after that (or every six months for those on the job more than 10 years):

- ☑ Complete liver function tests.

The following additional tests may be helpful:

- ☑ Kidney function tests.
- ☑ Complete blood count.
- ☑ Complete examination of the skin and nervous system.

If symptoms develop or overexposure is suspected, the following recommended medical tests should be considered:

- ☑ Consider X-rays of the fingers.
- ☑ A test called urinary thiodiglycolic acid level (normal is usually less than 2 mg/liter).
- ☑ Evaluation by a qualified allergist, including careful consideration of exposure history and special testing (may help diagnose skin allergy).

Any evaluation should include a careful history of past and present symptoms with an examination. Medical tests that simply look for existing damage are not a substitute for controlling exposures.

Other methods to reduce exposure include:

- ☑ Where possible, enclose operations and use local exhaust ventilation at the site of chemical release. If local exhaust ventilation or enclosure is not used, respiratory protection should be mandatory.
- ☑ Always ensure that proper protective clothing is worn when using chemical substances and that personnel have been trained in its use and care.
- ☑ Wash thoroughly immediately after exposure to vinyl chloride and at the end of the work shift or

before eating, drinking, smoking, or applying cosmetics.

- ☑ Hazard warning information should be posted in the work area. In addition, as part of an on-going education and training program, all information on the health and safety hazards of vinyl chloride should be communicated to all exposed and potentially exposed workers.
- ☑ Eye wash stations should be provided in the immediate work area for emergency use. If there is a possibility of skin exposure to vinyl chloride, emergency shower facilities should be provided.
- ☑ Workers whose clothing has been contaminated by vinyl chloride should change into clean clothes before leaving work. Contaminated clothing that is worn home can expose family members to vinyl chloride. Contaminated work clothing should be laundered only by individuals who have been informed of the hazards of vinyl chloride.
- ☑ Where possible, automatically transfer liquids containing vinyl chloride from drums or other containers to process containers.
- ☑ Specific engineering controls are outlined in OSHA 29 CFR 1910.1017 ("Vinyl Chloride").

RISK ASSESSMENT: ENVIRONMENT

General Assessment

The environment is at risk of exposure during transportation, storage, disposal, or destruction of vinyl chloride. In almost every scenario, the threat of environmental exposure is contingent upon the proper handling of the chemical substance. Accidental spills, large or small, can result in contamination of the surrounding environmental mediums (water, soil, and air).

Vinyl chloride is an extremely flammable gas with a low flash point of -108°F (-78°C). It is a dangerous fire and explosion hazard when exposed to heat or flame. It is incompatible with strong oxidizers, such as chlorine, fluorine, and bromine, and can polymerize violently on exposure to sunlight. In a fire, it can decompose to hydrogen chloride, oxides of carbon, and toxic phosgene. Caution is always required in handling, storage, transportation, and disposal of vinyl chloride. Emergency responders should be made aware of the presence of vinyl chloride at any emergency response situation.

Vinyl chloride can enter the environment from industrial effluents and from spills.

Acute Ecological Effects

Acute (short-term) toxic effects may include the death of animals, birds, or fish and death or low growth rate in plants. Acute effects are seen 2 to 4 days after animals or plants are exposed to vinyl chloride.

Insufficient data are available to evaluate or predict the short-term effects of vinyl chloride to aquatic life, plants, birds or land animals.

Chronic Ecological Effects

Chronic (long-term) toxic effects may include shortened life span, reproductive problems, lower fertility, and changes in appearance or behavior in exposed animals. These effects can be seen long after first exposure(s) to toxic chemicals.

Insufficient data are available to evaluate or predict the long-term effects of vinyl chloride to aquatic life, plants, birds, or terrestrial animals.

Water Solubility

Vinyl chloride may be slightly soluble in the aquatic environment. Concentrations between 1 and 1000 milligrams may mix with a liter of water.

Persistence in the Environment

As a gas vinyl chloride is non-persistent in water, due to volatilization, with a half-life of less than 2 days. The half-life of a pollutant is the amount of time it takes for one half of the chemical to be degraded. Approximately 99.9% of vinyl chloride will eventually end up in air; the rest will end up in the water. However, if released in moist soil, vinyl chloride may leach into and contaminate ground water supplies. In dry soil, it evaporates rapidly with no persistence expected.

Bioaccumulation in Aquatic Organisms

Some substances increase in concentration, or *bioaccumulate,* in living organisms as they breathe contaminated air, drink contaminated water, or eat contaminated food. These chemicals can become concentrated in the tissues and internal organs of animals as well as humans.

The concentration of vinyl chloride found in fish tissues is expected to be about the same as the average concentration of vinyl chloride in the water from which the fish was taken.

❶ *Recommended Risk-Reduction Measures*

Proper training of all transporters will reduce the likelihood of a mishap or accident resulting in a leak or spill of vinyl chloride into the environment. Labels on all containers, trucks, and rail cars must meet DOT requirements and accurately reflect their contents to enable emergency responders to react properly and quickly to any disaster thereby reducing the potential risk to the environment and to personnel.

Storage of vinyl chloride should be segregated from other chemicals and materials to minimize the risk of cross-contamination. Cylinders must always be stored upright and properly secured (always around the cylinder body, never around its neck). Cylinders must never be dragged or rolled during handling. They must be kept away from heat and direct sunlight.

If a spill or leak to the environment has occurred, fire department, emergency response and/or hazardous materials spill personnel should be notified immediately. Cleanup should be attempted only by those trained in proper spill containment procedures using non-sparking tools. Contaminated soils should be removed for incineration and replaced with clean soil. If vinyl chloride should contact the water table, aquifer, or navigable waterway, time is of the essence. It may be insoluble and remediation may be successful. The local and/or state emergency response authorities must be notified. A comprehensive emergency response or disaster preparedness/recovery plan should be in place prior to any operations involving the use, transportation, storage, or disposal of vinyl chloride.

If vinyl chloride is spilled or leaked, the following specific steps are recommended:

- ☑ Restrict persons not wearing protective clothing from area of spill or leak until cleanup is complete. Use non-sparking tools.
- ☑ If applicable, stop flow of leaking liquid or gas. If leak source is a cylinder and the leak cannot be stopped in place, remove leaking cylinder to a safe place in the open air, and repair or allow cylinder to empty. Use water spray to knock down vapors that have not yet ignited.
- ☑ Absorb any liquids in vermiculite, sand, earth, or similar material and deposit in sealed drum.
- ☑ Ventilate area of spill or leak. Container should be removed to safer area if it can be done without increasing the risk.
- ☑ It may be necessary to dispose of vinyl chloride as a hazardous waste. The responsible state agency or the regional office of the federal Environmental Protection Agency (EPA) should be contacted for specific recommendations.

RISK ASSESSMENT: BUSINESS

General Assessment

Accidents or mishaps involving vinyl chloride can present a serious threat to business operations. The loss or damage of equipment or facilities can significantly affect fiscal viability. Lawsuits that may result from personnel illness, injury/death, public exposures, and/or environmental contamination will require a serious expenditure of resources. Media attention surrounding an injury, death, or environmental damage can also result in a loss of profits and loss of current as well as future business. Always remember that anytime terms such as "cancer" or "carcinogen" are used, public emotion, hysteria, and ignorance can run equally high. This must be carefully considered whenever developing or implementing a public relations policy.

❶ *Recommended Risk-Reduction Measures*

Company attorneys, safety and health professionals, and environmental specialists should be involved in the development of any procedures for responding to chemical incidents. A company official should be predesignated as a public relations officer with specific training in dealing with the press. Corporate plans and policies should be developed, approved, and implemented long before any need for such arises.

REFERENCES

American Conference of Governmental Industrial Hygienists. 1988. *Documentation of the Threshold Limit Values and Biological Exposure Indices,* 5th Edition (with updates). Cincinnati: ACGIH

Bloom, A. D. and F. J. DeSerres. 1995. *Ecotoxicity and Human Health.* New York: CRC Press/Lewis Publishers.

Calabrese, E. J. 1994. *Biological Effects of Low Level Exposures.* New York: CRC Press/Lewis Publishers.

Lewis, R. J., Sr. 1993. *Hawley's Condensed Chemical Dictionary,* 12th Edition. New York: Van Nostrand Reinhold.

MATERIAL SAFETY DATA SHEET

CHEMICAL NAME	
WARFARIN	**CRC PRESS/LEWIS PUBLISHERS DISCLAIMER:** THE INFORMATION AND RECOMMENDATIONS PRESENTED HEREIN ARE BASED ON SOURCES BELIEVED TO BE RELIABLE. CRC MAKES NO REPRESENTATION ON ITS COMPLETENESS OR ACCURACY. IT IS THE RESPONSIBILITY OF THE USER TO DETERMINE THE CHEMICAL'S SUITABILITY FOR ITS INTENDED USE, THE CHEMICAL'S SAFE USE, AND THE CHEMICAL'S PROPER DISPOSAL. NO REPRESENTATIONS AND/OR WARRANTIES, EITHER EXPRESSED OR IMPLIED, OF THE MERCHANTABILITY OR FITNESS FOR A PARTICULAR PURPOSE, OR OF ANY OTHER NATURE, ARE MADE WITH RESPECT TO THE INFORMATION PROVIDED IN THIS MSDS OR TO THE CHEMICAL TO WHICH INFORMATION MAY REFER. CRC NEITHER ASSUMES NOR AUTHORIZES ANY OTHER PERSON TO ASSUME FOR IT, ANY OTHER ADDITIONAL RESPONSIBILITY OR LIABILITY FOR THE USE OF, OR RELIANCE UPON, THIS INFORMATION.

HAZARD WARNING INFORMATION

HEALTH	FIRE	REACTIVE	OTHER	DEGREE OF HAZARD	COLOR CODING	OTHER CODES
3	1	0		0 = Minimum Hazard 1 = Slight Hazard 2 = Moderate Hazard 3 = Serious Hazard 4 = Severe Hazard	HEALTH = BLUE FIRE = RED REACTIVITY = YELLOW OTHER = WHITE	OX = Oxidizer ACID = Acid ALK = Alkali COR = Corrosive ~~W~~ = Use No Water

SECTION I - GENERAL INFORMATION

Characterization	**Rodenticide**	RCRA Number	**P001**
		EPA Class	**Acute Hazardous Waste**
DOT Proper Shipping Name	**Pesticide**	Chemical Abstract Service (CAS) Number	**81-81-2**
DOT Hazard Class and Label Requirements	**Poison**	DOT Emergency Guide Code	**No Citation**
DOT Identification Number	**NA 3027**	Chemical Formula	$C_{19}H_{16}Cl_4$

Synonyms

(3-(α-acetonylbenzyl)-4-hydroxycoumarin); 4-hydroxy-3-(3-oxo-1-phenyl butyl)-2H-1-benzopyran-2-one; WARF; brumin; coumadin; coumafene; dethmore.

SECTION II - HAZARDOUS INGREDIENTS AND IDENTITY INFORMATION

Hazardous Components (specific identity)	OSHA Exposure Criteria	NIOSH Exposure Criteria	Immediately Dangerous to Life and Health (IDLH)	ACGIH Exposure Criteria
Warfarin (an anticoagulant rodenticide).	**PEL:** 0.1 mg/m^3 **STEL:** Not Established	**REL:** 0.1 mg/m3 Possible Cancer Agent	100 mg/m^3	**TLV:** 0.01 mg/m^3 **STEL:** Not Established

SECTION III - PHYSICAL AND CHEMICAL CHARACTERISTICS

Boiling Point	**Decomposes**	Specific Gravity (H_2O = 1)	**Not Reported**
Vapor Pressure (mm Hg)	**0.09 at 71°F (22°C)**	Molecular Weight (atomic weight)	**308.35**
Vapor Density (Air = 1)	**Not Reported**	Freezing Point	**322°F (161°C)**

Solubility

Insoluble in water (0.002%) and benzene; slightly soluble in methanol, ethanol; very slightly soluble in alkaline aqueous solution. Soluble in acetone and dioxane.

Appearance and Odor

Colorless, odorless, tasteless, crystalline solid.

SECTION IV - FIRE AND EXPLOSION HAZARD DATA

Flash Point (method used)	**Not Determined**	Explosive Limits in Air % by Volume	**LEL: Not Determined UEL: Not Determined**
NFPA Classification	**Combustible Solid**	Autoignition Temperature	**Not Determined**

Extinguishing Media

Use carbon dioxide, dry chemical, foam, or water spray.

Special Fire Fighting Procedures

Warfarin is a combustible solid and is often dissolved in a flammable or combustible liquid. Wear full protective clothing and SCBA. Remain clear of smoke, water fallout and water runoff. Move containers from fire area if it can be done without risk. Cool exposed containers. Toxic smoke and irritating fumes are generated during a fire. Stay clear of smoke. Fight fire from distance, if possible.

Unusual Fire and Explosion Hazards

None reported.

SECTION V - REACTIVITY DATA

Stability		Conditions to Avoid **Warfarin is normally stable in closed containers under routine conditions of storage and handling. Avoid contact with incompatible materials and do not store near heat or heat source.**
Stable **X**	Unstable	Incompatibility *(materials to avoid)* **Incompatible with strong oxidizers (such as chlorine, fluorine, bromine, perchlorates, permanganates, and nitrates).**
Hazardous Polymerization		Conditions to Avoid **Under normal conditions of temperature and pressure, hazardous polymerization of warfarin is not known to occur.**
May Occur	Will Not Occur **X**	Hazardous Decomposition or By-products **At high temperatures, warfarin decomposes and may produce toxic and irritating smoke and fumes.**

SECTION VI - HEALTH HAZARD DATA

Primary Route(s) of Entry:	Inhalation?	Absorption (skin/eye)?	Ingestion?
	X	X	X

Health Hazards

INHALATION: Breathing dusts can cause irritation of the eyes, nose, throat, and respiratory tract. There may also be some nausea and vomiting on exposure to high concentrations.

ABSORPTION: Will pass through unbroken skin and enter the bloodstream. Contact may severely irritate the skin and eyes.

INGESTION: A poison. Causes hemorrhage, ulceration of or bleeding from small intestines, back pain, blood clotting factor changes with hematoma in the arms and legs, epistaxis, bleeding lips, mucosa hemorrhage, abdominal pain, vomiting, fecal blood, petechial rash, abnormal hematologic indices.

Carcinogenicity	NTP Listed?	IARC Cancer Review Group?	OSHA Regulated?	Target Organs?
Unknown Human Unknown Animal	**No**	**No**	**29 CFR 1910.1000 Table Z-1**	**Blood and the cardiovascular system.**

Medical Conditions Generally Aggravated by Exposure

Existing blood and CVS disorders may be aggravated by exposure to warfarin.

Emergency and First-aid Procedures

Eye contact: Do not allow victim to keep eyes tightly shut. Flush immediately with water for 15 minutes; seek medical attention. Skin contact: Remove all contaminated clothing. Immediately wash area with flooding amounts of water, then triple-wash with soap and water. Inhalation: Remove the person from exposure. Provide respiratory assistance and CPR. Transfer to medical facility. If swallowed: If victim is conscious, give 1 to 2 glasses of water and induce vomiting. Repeat until vomitus is clear. Seek medical attention immediately. Never give anything by mouth to an unconscious or convulsing person.

SECTION VII - PRECAUTIONS FOR SAFE HANDLING AND USE

Steps to be Taken in Case Material is Released or Spilled

Remove all sources of heat. Ventilate area of spill. Restrict those not involved in cleanup from entering area. Notify appropriate authorities. Collect powder materials using HEPA vacuum. Do NOT dry sweep. Damp mop residue. Collect water runoff created during cleanup and dispose of properly. Absorb any liquid solution spills in vermiculite and place in sealed drum for disposal.

Preferred Waste Disposal Method

No citation.

Precautions to be Taken in Handling and Storage

Store in tightly closed containers in cool, dry, dark, well-ventilated area away from heat.

Other Precautions and Warnings

Avoid generating dusty conditions in storage areas. Protect containers from physical damage.

SECTION VIII - CONTROL MEASURES AND PERSONAL PROTECTIVE EQUIPMENT

Respiratory Protection *(specify type)*

For exposures above 0.1 mg/m^3, use a NIOSH-approved supplied-air respirator or a self-contained breathing apparatus (SCBA) with full facepiece operated in positive pressure mode.

Ventilation

Local exhaust (preferred) at site of chemical work or general ventilation.

Protective Gloves	Eye Protection	Other Protective Clothing
Impervious Material	**Chemical Goggles or Face Mask**	**Protective Apron, Boots (avoid skin contact)**

Work/Hygiene Practices

Always wash hands thoroughly after using chemical; never bring food, drink, or smoking materials into vicinity of chemicals.

WARFARIN

$C_{19}H_{16}O_4$

CAS: 81-81-2

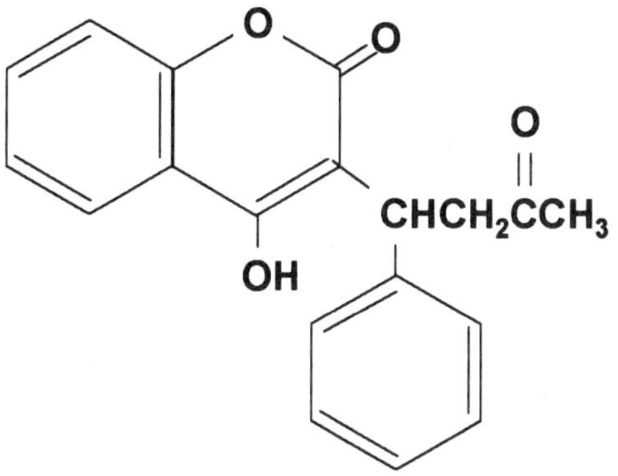

IDENTIFICATION AND TYPICAL USES

Warfarin is a colorless, odorless, tasteless, crystalline solid. Its primary use is as an anticoagulant rodenticide. It inhibits the clotting mechanism of the blood. This causes spontaneous internal bleeding and the rodent simply bleeds to death.

RISK ASSESSMENT: HEALTH

General Assessment

Warfarin is highly toxic to humans by *ingestion, inhalation* and skin *absorption.* While it is not listed as a carcinogenic compound, experimental mutation data have been reported. There is some evidence in the references that indicate warfarin is an experimental teratogen with fetotoxic and reproductive effects likely to occur.

Warfarin is an anticoagulant drug. As such, it primarily affects the blood. Symptoms of exposure by all routes include hematuria (blood in the urine), back pain, hematoma (bruising) in the arms and legs, epistaxis (hemorrhage from the nose, nosebleed), bleeding lips, mucous membrane hemorrhage, abdominal pain, vomiting (possible with blood), fecal or tarry stool, petechial rash (pinpoint, non-raised, perfectly round purplish red spots caused by intra-dermal or submucous hemorrhage), and abnormal hematologic indices.

Skin contact can result in moderate to severe irritation with rash or burning sensation. Eye contact may also result in irritation. Caution is warranted when both inhalation and skin contact exposures are possible. Even though air levels may be within limits, personnel may be overexposed if dermal absorption of warfarin occurs. The dangerous dose in humans is unknown.

☠ *Acute Health Effects*

The following acute (short-term) health effects may occur immediately or shortly after exposure to warfarin:

Skin: Liquid can pass through unbroken skin. Contact may cause irritation and rash.

Eye: Severe irritation and possible damage on contact.

Lung: Irritation of the nose and throat. Once absorbed through the inhalation process, the dusts primarily affect the blood.

Other: An anticoagulant that prevents blood from clotting and causes dangerous, spontaneous internal bleeding that can be fatal.

Chronic Health Effects

The following chronic (long-term) health effects may occur at some time after exposure to warfarin and can last for months or even years:

Cancer Hazards: According to information presented in the references, warfarin has not been adequately tested for its ability to cause cancer in test animals. However, mutation data have been reported and many scientists believe that exposure to such chemicals may pose a cancer threat in the long-term.

Reproductive Hazard: Human reproductive effects by ingestion include fetotoxicity, abnormal condition of the newborn at birth, other newborn physical effects including developmental abnormalities of the eye and ear, craniofacial area, skin and appendages, musculoskeletal system, cardiovascular system, and gastrointestinal system of the fetus.

Other Chronic Effects: Long-term exposure to warfarin may cause dermatitis.

❍ *Recommended Risk-Reduction Measures*

Personnel should avoid direct contact with warfarin. Exposure is extremely dangerous. If a less toxic material or compound cannot be substituted for warfarin, then *engineering controls* are the most effective method of reducing exposures. The best protection is to enclose operations and/or provide local exhaust ventilation at the site of warfarin dust release. This may be practical in manufacturing operations, but is not always feasible in its application outdoors. While not always operationally possible, isolating operations involving warfarin manufacturing can also reduce exposure. Using respiratory protection is less effective than the controls mentioned above, but is still advisable whenever working with or around warfarin. The exposure level is relatively low and difficult to monitor. At exposures above the PEL of 1 mg/m^3 (TWA), use an MSHA/NIOSH-approved supplied-air respirator or a self-contained breathing apparatus (SCBA) with full facepiece operated in pressure demand or other positive pressure mode for the greatest possible respiratory protection. If a full facepiece is not available, then chemical goggles should be worn to protect the eyes. A face shield and protective apron should also be worn. To prevent hand and skin exposures, impervious gloves should be worn. Glove manufacturers should be contacted and permeation studies obtained *before* final glove selections have been made.

Administrative controls should also be in place to minimize the potential for human exposures. These may include written procedures or policies which specify the methods and techniques that will be practiced whenever personnel are to work with warfarin.

All personnel should receive *training* on the use, hazards, protective measures, emergency actions, and other precautions per 29 CFR 1910.1200 (Hazard Communication), prior to the first assignment in an area where warfarin is used or stored.

If symptoms develop or overexposure is suspected, the following medical testing may be useful:

☑ Urinalysis for blood content.

☑ Examination of the skin's surface for signs of hematoma. Check for abnormal bleeding around the gums and from the nose.

Any evaluation should include a careful history of past and present symptoms with an examination. Medical tests that simply look for existing damage are not a substitute for controlling exposures.

Other methods to reduce exposure include:

☑ Where possible, enclose operations and use local exhaust ventilation at the site of chemical release. If local exhaust ventilation or enclosure is not used, respiratory protection should be mandatory.

☑ Always ensure that proper protective clothing is worn when using chemical substances and that personnel are trained in its use and care.

☑ Wash thoroughly immediately after exposure to warfarin and at the end of the work shift or before eating, drinking, smoking, or applying cosmetics.

☑ Hazard warning information should be posted in the work area. In addition, as part of an on-going education and training program, all information on the health and safety hazards of warfarin should be communicated to all exposed and potentially exposed workers.

☑ Eye wash stations should be provided in the immediate work area for emergency use. If there is a possibility of skin exposure to warfarin, emergency shower facilities should also be provided.

☑ Workers whose clothing has been contaminated by warfarin should change into clean clothes before leaving work. Contaminated work clothing should be laundered only by individuals who have been informed of the hazards of exposure to warfarin.

☑ Specific engineering controls are recommended by NIOSH. Although not specific to warfarin, they may be useful. Refer to NIOSH Criteria Document: Occupational Exposure During the Manufacture and Formulation of Pesticides (Publication Number 78-174)

RISK ASSESSMENT: ENVIRONMENT

General Assessment

Because of its use as an insecticide, warfarin may be present in the environment at any given time and at a variety of levels. The environment is also at risk of exposure during transportation, storage, disposal, or destruction of warfarin. In almost every scenario, the threat of environmental exposure is contingent upon the proper handling of the chemical substance. Accidental spills, large or small, where warfarin contacts incompatible commodities can result in violent and even explosive reactions (depending upon conditions of spill), and possible contamination of the surrounding environmental mediums (water, soil, and air).

Warfarin is considered a non-combustible solid. It may also be present (dissolved) in some other flammable liquids. Also, because it is incompatible with many common oxidizers, extreme caution is required in handling, storage, transportation, and disposal of warfarin. These characteristics also require special consideration during any emergency situation involving a leak or spill of warfarin powder or dust.

Warfarin is a rodenticide and, as such, it enters the environment in agricultural and residential runoff.

Acute Ecological Effects

Acute (short-term) toxic effects may include the death of animals, birds, or fish and death or low growth rate in plants. Acute effects are seen 2 to 4 days after animals or plants are exposed to warfarin.

Warfarin has high acute toxicity to aquatic life, birds, and land animals. Insufficient data are available to evaluate or predict on the short-term effects of warfarin on plants.

Chronic Ecological Effects

Chronic (long-term) toxic effects may include shortened life span, reproductive problems, lower fertility, and changes in appearance or behavior in exposed animals. These effects can be seen long after first exposure(s) to toxic chemicals.

Warfarin has moderate chronic toxicity to aquatic life, birds, and land animals. Insufficient data are available to evaluate or predict the long-term effects of warfarin on plants.

Water Solubility

Warfarin is not soluble in water. Concentrations of 1 milligram will not mix with a liter of water.

Persistence in the Environment

Warfarin is highly persistent in the water, with a half-life of greater than 200 days. The half-life of a pollutant is the amount of time it takes for one-half of the chemical to be degraded.

Bioaccumulation in Aquatic Organisms

Some substances increase in concentration, or *bioaccumulate,* in living organisms as they breathe contaminated air, drink contaminated water, or eat contaminated food. These chemicals can become concentrated in the tissues and internal organs of animals as well as humans.

The concentration of warfarin found in fish tissues is expected to be considerably higher that the average concentration of warfarin in the water from which the fish was taken.

Recommended Risk-Reduction Measures

Proper training of all transporters will reduce the likelihood of a mishap or accident resulting in a leak or spill of warfarin into the environment. The correct use of DOT labeling (per the MSDS) on all containers, trucks, and rail cars will enable emergency responders to react properly and quickly to any disaster thereby reducing the potential risk to the environment and to personnel.

Storage of warfarin should be segregated from incompatible materials to minimize the risk of cross-contamination or contact. Buildings or storage areas should be equipped with appropriate fire protection systems (alarms, sprinklers). Proper ventilation and protective equipment should be used while handling the powders or while preparing aqueous solutions.

If a spill or leak to the environment has occurred, fire department, emergency response and/or hazardous materials spill personnel should be notified immediately. Cleanup should be attempted only by those trained in proper spill containment procedures. If warfarin should contact the water table, aquifer, or navigable waterway, reclamation procedures should be initiated as soon as practical. It is only slightly soluble in water and partial or total remediation may be possible. The local and/or state emergency response authorities must be notified. A comprehensive emergency response or disaster preparedness/recovery plan should be in place prior to any operations involving the use, transportation, storage, or disposal of warfarin.

If warfarin is spilled or leaked, the following specific steps are recommended:

- ☑ Restrict persons not wearing protective clothing from area of spill or leak until cleanup is complete and area can be opened for normal work.
- ☑ Ventilate are of spill or leak.
- ☑ Collect powdered materials using HEPA vacuum. Do NOT dry sweep. Absorb solution spills in vermiculite or other material and place in sealed drum for disposal or reclamation.

☑ Absorb solution spills in vermiculite or other similar material and place in sealed drums for disposal.

☑ It may be necessary to dispose of warfarin as a hazardous waste. The responsible state agency or the regional office of the federal Environmental Protection Agency (EPA) should be contacted for specific recommendations.

RISK ASSESSMENT: BUSINESS

General Assessment

Accidents or mishaps involving warfarin can present a significant threat to business operations. The loss or damage of equipment or facilities can significantly affect fiscal viability. Lawsuits that may result from personnel injury/death, public exposures, and/or environmental contamination will also require a serious expenditure of resources. Media attention surrounding an injury, death, or environmental damage can also result in a loss of profits and loss of current as well as future business. Always remember that anytime to term "reproductive hazard" is used, public emotion, hysteria, and ignorance can run equally high. This must be carefully considered whenever developing or implementing public relations policies.

Recommended Risk-Reduction Measures

Company attorneys, safety and health professionals, and environmental specialists should be involved in the development of any procedures for responding to chemical incidents. A company official should be predesignated as a public relations officer with specific training in dealing with the press. Corporate plans and policies should be developed, approved, and implemented long before any need for such arises.

REFERENCES

American Conference of Governmental Industrial Hygienists. 1988. *Documentation of the Threshold Limit Values and Biological Exposure Indices,* 5th Edition (with updates). Cincinnati: ACGIH

Brusick, D. J. 1994. *Methods for Genetic Risk Assessment.* New York: CRC Press/Lewis Publishers.

Bloom, A. D. and F. J. DeSerres. 1995. *Ecotoxicity and Human Health.* New York: CRC Press/Lewis Publishers.

Calabrese, E. J. 1994. *Biological Effects of Low Level Exposures.* New York: CRC Press/Lewis Publishers.

Cockerman, L. G., B. S. Shane. 1994. *Basic Environmental Toxicology.* New York: CRC Press/Lewis Publishers.

Cothern, C. R. 1995. *Handbook for Environmental Risk Decisionmaking.* New York: CRC Press/Lewis Publishers.

Gardner, R. W. 1994. *Chemical Intolerance.* New York: CRC Press/Lewis Publishers.

Howard, P. H., et. al. 1991. *Handbook of Environmental Degradation Rates.* New York: CRC Press/Lewis Publishers.

Keith, L. H. and M. M. Walker. 1995. *Handbook of Air Toxics.* New York: CRC Press/Lewis Publishers.

Landis, W. G., M. H. Yu. 1995. *Introduction to Environmental Toxicology.* New York: CRC Press/Lewis Publishers.

Lauwerys, R. R., P. Hoet. 1993. *Industrial Chemical Exposure.* New York: CRC Press/Lewis Publishers.

Lewis, R. J., Sr. 1992. *Sax's Dangerous Properties of Industrial Materials,* Volumes 1, 2, and 3, 8th Edition. New York: Van Nostrand Reinhold.

Lide, D. R. 1995. *Handbook of Chemistry and Physics,* 76th Edition. New York: CRC Press/Lewis Publishers.

Manahan, S. E. 1992. *Toxicological Chemistry,* 2nd Edition. New York: CRC Press/Lewis Publishers.

Mellström, G. A., J. E. Wahlberg, and H. I. Maibach. 1994. *Protective Gloves for Occupational Use.* New York: CRC Press/Lewis Publishers.

Meyer, E. 1990. *Chemistry of Hazardous Materials.* Englewood Cliffs, NJ: Prentice Hall.

Mycock, J. C., J. D. McKenna, and L. Theodore. 1995. *Handbook of Air Pollution Control Engineering and Technology.* New York: CRC Press/Lewis Publishers.

Nelson, G. O. 1992. *Gas Mixtures.* New York: CRC Press/Lewis Publishers.

Ott, W. R. 1995. *Environmental Statistics and Data Analysis.* New York: CRC Press/Lewis Publishers.

Patnaik, P. 1992. *A Comprehensive Guide to the Hazardous Properties of Chemical Substances.* New York: Van Nostrand Reinhold.

Quigley, D. R. 1994. *Handbook of Emergency Chemical Management.* New York: CRC Press/Lewis Publishers.

Richardson, M. 1995. *Dictionary of Substance and Their Effects,* Vol. I-VI. New York: CRC Press/Lewis Publishers.

Sacarello, H. L. A. 1994. *The Comprehensive Handbook of Hazardous Materials.* New York: CRC Press/Lewis Publishers.

Torkelson, T. R. V. K. Rowe. 1981. *Patty's Industrial Hygiene and Toxicology,* Vol. 2B. New York: Wiley-Interscience.

U.S. Department of Health and Human Services, National Institute for Occupational Health (NIOSH). 1978. *NIOSH Criteria Document: Occupational Exposure During the Manufacture and Formulation of Pesticides.* Publication Number 78-174. Washington, DC: U.S. Government Printing Office.

U.S. Department of Health and Human Services, National Institute for Occupational Safety and Health. 1994. *NIOSH Pocket Guide to Chemical Hazards.* Washington, D.C.: U.S. Government Printing Office.

MATERIAL SAFETY DATA SHEET

CHEMICAL NAME	
XYLENE Mixed Monomers	**CRC PRESS/LEWIS PUBLISHERS DISCLAIMER:** THE INFORMATION AND RECOMMENDATIONS PRESENTED HEREIN ARE BASED ON SOURCES BELIEVED TO BE RELIABLE. CRC MAKES NO REPRESENTATION ON ITS COMPLETENESS OR ACCURACY. IT IS THE RESPONSIBILITY OF THE USER TO DETERMINE THE CHEMICAL'S SUITABILITY FOR ITS INTENDED USE, THE CHEMICAL'S SAFE USE, AND THE CHEMICAL'S PROPER DISPOSAL. NO REPRESENTATIONS AND/OR WARRANTIES, EITHER EXPRESSED OR IMPLIED, OF THE MERCHANTABILITY OR FITNESS FOR A PARTICULAR PURPOSE, OR OF ANY OTHER NATURE, ARE MADE WITH RESPECT TO THE INFORMATION PROVIDED IN THIS MSDS OR TO THE CHEMICAL TO WHICH INFORMATION MAY REFER. CRC NEITHER ASSUMES NOR AUTHORIZES ANY OTHER PERSON TO ASSUME FOR IT, ANY OTHER ADDITIONAL RESPONSIBILITY OR LIABILITY FOR THE USE OF, OR RELIANCE UPON, THIS INFORMATION.

HAZARD WARNING INFORMATION

HEALTH	FIRE	REACTIVE	OTHER	DEGREE OF HAZARD	COLOR CODING	OTHER CODES
2	3	0		0 = Minimum Hazard 1 = Slight Hazard 2 = Moderate Hazard 3 = Serious Hazard 4 = Severe Hazard	HEALTH = BLUE FIRE = RED REACTIVITY = YELLOW OTHER = WHITE	OX = Oxidizer ACID = Acid ALK = Alkali COR = Corrosive ~~W~~ = Use No Water

SECTION I - GENERAL INFORMATION

Characterization	RCRA Number	EPA Class
Aromatic Hydrocarbon	**None**	**Not Applicable**

DOT Proper Shipping Name	**Xylenes**	Chemical Abstract Service (CAS) Number	**1330-20-7**
DOT Hazard Class and Label Requirements	**Flammable Liquid**	DOT Emergency Guide Code	**27**
DOT Identification Number	**UN 1307**	Molecular Formula	$C_6H_4(CH_3)_2$

Synonyms
Dimethylbenzene; methyltoluene.

SECTION II - HAZARDOUS INGREDIENTS AND IDENTITY INFORMATION

Hazardous Components (specific identity)	OSHA Exposure Criteria	NIOSH Exposure Criteria	Immediately Dangerous to Life and Health (IDLH)	ACGIH Exposure Criteria
Xylene (derivation: A commercial mixture of three isomers, o-. m-. and p-xylene. Derived from fractional distillation from petroleum, coal tar, or coal gas; by catalytic reforming from petroleum, followed by separation of p-xylene by continuous crystallization; from toluene by transalkylation). **1 ppm = 4.41 mg/m^3**	**PEL:** 100 ppm 435 mg/m^3 **STEL (15 min):** 150 ppm 655 mg/m^3	**REL:** 100 ppm 435 mg/m^3 **STEL:** 150 ppm 655 mg/m^3	**1000 ppm**	**TLV:** 100 ppm 434 mg/m^3 **STEL (15 min):** 150 ppm 651 mg/m^3

SECTION III - PHYSICAL AND CHEMICAL CHARACTERISTICS

Boiling Point	**Range 279 to 284°F (137 to 140°C)**	Specific Gravity (H_2O = 1)	**0.864**
Vapor Pressure (mm Hg)	**6.72 at 70°F (21°C)**	Molecular Weight (atomic weight)	**106.16**
Vapor Density (Air = 1)	**Not Reported**	Melting Point	**Range: -13 to 57°F (-25 to 14°C)**

Solubility
Practically insoluble in water. Miscible with absolute alcohol, ether, and other organic liquids.

Appearance and Odor
Clear, sweet-smelling liquid.

SECTION IV - FIRE AND EXPLOSION HAZARD DATA

Flash Point (method used)	**Range: 64 to 77°F (17 to 25°C) closed cup**	Explosive Limits in Air % by Volume	**LEL: 0.9%** (*o*-) **1.1%** (*m*-, *p*-) **UEL: 6.7%** (*o*-) **7.0%** (*m*-, *p*-)
NFPA Classification	**Class 1C Flammable Liquid**	Autoignition Temperature	**982°F (527°C)** *m*-monomer

Extinguishing Media
Use dry chemical, carbon dioxide, or water spray, or regular foam. Water may be ineffective (xylene will float on water and may travel to an ignition source).

Special Fire Fighting Procedures
Wear self-contained breathing apparatus (SCBA). Poisonous gases are produced in fire. Move containers from fire if it can be done without risk. Cool fire-exposed containers with water even after fire is out.

Unusual Fire and Explosion Hazards
Dangerous fire and explosion hazard. Containers may explode in fire. Immediately withdraw if rising sound is heard from venting device or there is a discoloration on the sides of tanks or containers. Xylene may present a vapor explosion hazard indoors, outdoors, and in sewers.

SECTION V - REACTIVITY DATA

Stability		Conditions to Avoid
Stability		**Under normal conditions of handling and storage, xylene is considered stable. Avoid heat, flame, other sources of ignition, and contact with incompatible materials and chemicals.**
Stable	Unstable	Incompatibility *(materials to avoid)*
X		**Strong oxidizers (chlorine, bromine, fluorine). Xylene attacks some forms of rubber, plastics, and coatings.**
Hazardous Polymerization		Conditions to Avoid **Hazardous polymerization has not been reported to occur under normal conditions of temperature and pressure.**
May Occur	Will Not Occur	Hazardous Decomposition or By-products
	X	**Poisonous gases and acrid fumes are produced in fire, including carbon monoxide, carbon dioxide, and various hydrocarbon products.**

SECTION VI - HEALTH HAZARD DATA

Primary Route(s) of Entry:	Inhalation?	Absorption (skin/eye)?	Ingestion?
	X	X	X

Health Hazards

INHALATION: **May cause nausea, headache, vomiting, dizziness, narcosis, suffocation, lower blood pressure, CNS depression. Vapors may cause severe irritation or burns of the respiratory system, pulmonary edema, or lung inflammation. Causes narcosis, coma, and death. Kidney and liver damage can occur on high exposure.**

ABSORPTION: **Irritation with possible redness and pain to the skin and eyes. Also, the liquid will absorb through the skin but the vapor will not. May cause dermatitis. Eye contact may result in severe irritation and possible inflammation of the cornea.**

INGESTION: **Burning sensation in the oropharynx and stomach with transient CNS depression.**

Carcinogenicity	NTP Listed?	IARC Cancer Review Group?	OSHA Regulated?	Target Organs?
Unknown Human Unknown Animal	**No**	**No**	**29 CFR 1910.1000 Table Z-1**	**Eyes, skin, GI, resp. sys., blood, CNS, liver, kidneys**

Medical Conditions Generally Aggravated by Exposure

Skin, CNS, eye, liver, kidney, and GI tract diseases and respiratory problems may be aggravated.

Emergency and First-aid Procedures

Eye contact: Do not allow victim to keep eyes tightly shut. Flush immediately with water for 15 minutes (minimum); seek medical attention. Skin contact: Remove all contaminated clothing, including shoes. Immediately wash area with flooding amounts of soap and water. Seek medical attention immediately. For inhalation: Remove the person from exposure. Provide respiratory assistance and CPR. Transfer to medical facility. Observe for 24-48 hours for lung effects. If swallowed: Seek medical attention immediately. Do not induce vomiting (aspiration of vomitus can cause serious damage to lung tissue).

SECTION VII - PRECAUTIONS FOR SAFE HANDLING AND USE

Steps to be Taken in Case Material is Released or Spilled

Remove all ignition sources, wear SCBA respiratory protection. Restrict those not wearing protective equipment and who are not involved in cleanup from entering area. Stop leak if it can be done without risk. Ventilate area of spill. Absorb with vermiculite and deposit in sealed containers.

Preferred Waste Disposal Method

Burn in chemical incinerator equipped with an afterburner and scrubber.

Precautions to be Taken in Handling and Storage

Store in tightly closed containers in cool, well-ventilated location. Keep away from sources of heat.

Other Precautions and Warnings

Sources of ignition such as open flame or smoking, are prohibited where xylene is used or stored. Metal containers should be bonded and grounded. Bulk storage not recommended.

SECTION VIII - CONTROL MEASURES AND PERSONAL PROTECTIVE EQUIPMENT

Respiratory Protection *(specify type)*

Use an MSHA/NIOSH-approved air purifying (cartridge type) respirator or use a supplied-air respirator or a self-contained breathing apparatus (SCBA) with full facepiece operated in positive pressure mode.

Ventilation

Local exhaust (preferred) at site of chemical work or general ventilation.

Protective Gloves	Eye Protection	Other Protective Clothing
Polyvinyl alcohol or Viton	**Chemical Goggles or Face Mask**	**Protective Apron if Splash is Likely**

Work/Hygiene Practices

Always wash hands thoroughly after using chemical; never bring food, drink, or smoking materials into vicinity of chemicals.

XYLENE (mixed monomers)

$C_6H_4(CH_3)_2$

CAS: 1330-20-7

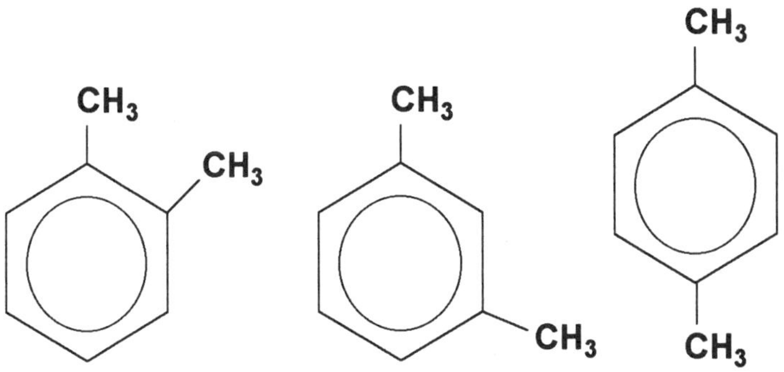

o-xylene *m*-xylene *p*-xylene

IDENTIFICATION AND TYPICAL USES

The xylenes appear as colorless liquids with an aromatic odor. The commercial product is a blend of the three isomers *ortho-* or (*o*-), *meta-* or (*m*-), and *para-* or (*p*-), with the largest portion being the *m*-xylene. It is used in solvents for paints, coatings, and rubber products. Xylene isomers are used in the manufacture of dyes, drugs, pesticides, and many organic intermediates, such as terephthalic acid and pthalic anhydride. It is a direct food additive and is a component of many adhesives.

RISK ASSESSMENT: HEALTH

General Assessment

Xylene is moderately toxic by *ingestion* and mildly toxic by *inhalation* and skin contact (*absorption*). While it is not listed by any sources as a carcinogen, human mutation data have been reported and it has caused teratogenic effects in laboratory test animals.

Inhalation of xylene vapors causes eye irritation with profuse watering of the eyes, nasal irritation, chest constriction, and vertigo. Symptoms worsen as vapor concentrations increase. There may be olfactory changes, changes in sleeping patterns, and changes in pulmonary function. Irritating properties may cause pulmonary edema (fluid buildup in the lungs), which is a medical emergency. Symptoms can be delayed up to 48 hours, thereby creating a false sense of security with regard to health hazard risk. Higher or prolonged concentrations will result central nervous system (CNS) effects with narcosis, cramps, lightheadedness, dizziness, abdominal pain, excitement, drowsiness, incoordination, staggering gait, anorexia, and even death due to respiratory paralysis.

Contact with the skin causes drying, cracking, scaling, and even blistering. Xylene will pass through unbroken skin to cause toxic systemic effects. Direct eye contact can result in conjunctivitis and corneal burns.

Ingestion causes a burning sensation in the oropharynx and stomach, and transient CNS depression.

Acute Health Effects

The following acute (short-term) health effects may occur immediately or shortly after exposure to xylene:

Skin: Irritation with potential for redness, blistering, and serious burning. Chemical will pass through intact skin to cause toxic systemic effects, including narcosis.

Eye: Severe irritation, burning, pain, and permanent damage possible.

Lung: Nose, throat, and respiratory tract irritation causing difficulty in breathing, possible pulmonary edema (fluid in lungs) on high exposure.

CNS: Narcosis, headaches, lightheadedness, dizziness, vertigo, drowsiness, loss of consciousness, coma, and death due to respiratory paralysis.

Chronic Health Effects

The following chronic (long-term) health effects may occur at some time after exposure to xylene and can last for months or even years:

Cancer Hazards: Xylene has not been adequately tested for its ability to cause cancer in test animals. However, mutation (genetic changes) data have been reported. Many scientists believe that mutagenic chemicals may have the ability to cause cancer in the long-term. Further research is required in this regard.

Reproductive Hazard: There is limited evidence that xylene is a teratogen in animals and may cause adverse reproductive effects. Menstrual irregularity has been reported. Xylene has the ability to cross the placenta barrier, but has not yet been shown to be fetotoxic in humans.

Other Chronic Effects: Repeated skin contact can cause chronic dryness and scaling of the skin's surface. Prolonged contact may cause blistering. Repeated eye exposure to high vapor concentrations may cause reversible eye damage, peripheral and central neuropathy, and liver damage. Repeated inhalation of xylene vapors may result in blood disorders, particularly leukopenia (deficiency in white blood cell count) and lymphocytosis (the formation of cysts containing lymph). This chemical has not been adequately evaluated to determine whether brain or other nerve damage could occur with repeated exposure. However, many other solvents and petroleum-based chemicals have been known to cause such damage. Effects include reduced memory and concentration, personality changes (irritability, withdrawal), fatigue, sleep disturbances, reduced coordination, and/or effects on the nerves supplying the internal organs (autonomic nerves) and/or nerves to the arms and legs with weakness and a feeling of "pins and needles." Other symptoms of chronic exposure include headache, fatigue, irritability, chronic bronchitis, and gastrointestinal tract disturbances such as nausea, loss of appetite, and flatulence.

Recommended Risk-Reduction Measures

Personnel should avoid direct contact with xylene. Occupational poisoning occurs most commonly by inhalation and absorption. If a less toxic material or compound cannot be substituted for xylene, then *engineering controls* are the most effective method of reducing exposures. The best protection is to enclose operations and/or provide local exhaust ventilation at the site of xylene release. While not always operationally feasible, isolating operations can also reduce exposure. Using respiratory protection is less effective than the controls mentioned above, but is still advisable whenever working with or around xylene. For exposure over the PEL (100 ppm), an MSHA/NIOSH-approved air-purifying respiratory with organic vapor cartridge(s) may suffice. For exposures to higher concentrations, or when the concentration is unknown, use a supplied-air respirator with full facepiece and mask operated in positive pressure mode or a self-contained breathing apparatus (SCBA) with full facepiece operated in pressure demand. If a full facepiece is not available, then chemical goggles should be worn to protect the eyes. A face shield should also be considered. To prevent hand and skin exposures, polyvinyl alcohol or Viton gloves should be used. Glove manufacturers should be contacted and permeation studies obtained *before* final glove selection is made.

Administrative controls should also be in place to minimize the potential for human exposures. These may include written procedures or policies which specify the methods and techniques that will be practiced whenever personnel are to work with xylene.

All personnel should receive *training* on the use, hazards, protective measures, emergency actions, and other precautions per 29 CFR 1910.1200 (Hazard Communication), prior to the first assignment in an area where xylene is used or stored.

If symptoms develop or overexposure is suspected, the following medical tests should be considered:

- ☑ Visual examination by slit lamp.
- ☑ Liver and kidney function tests.
- ☑ Complete blood count.
- ☑ Lung function tests.
- ☑ Consider chest X-ray following acute overexposure (may be negative if taken immediately after exposure due to delayed onset of pulmonary edema).
- ☑ Urine concentration of m-methylhippuric acid (at the end of a work shift) as an index of overexposure.
- ☑ Interview for brain effects, including recent changes in memory, mood, or concentration, headaches, malaise, and altered sleeping patterns. Consider cerebellar, autonomic, and peripheral nervous system evaluation. Positive and borderline cases should be referred for neuropsychological testing.

Any evaluation should include a careful history of past and present symptoms with an examination. Medical tests that simply look for existing damage are not a substitute for controlling exposures. Also, since smoking can cause heart disease, lung cancer, emphysema, and other respiratory diseases, smokers may experience symptoms more quickly than non-smokers under the same conditions of exposure. Prudent risk management requires detailed and careful consideration of *all* factors which may influence exposure in the work place.

Other methods to reduce exposure include:

- ☑ Where possible, enclose operations and use local exhaust ventilation at the site of chemical release. If local exhaust ventilation or enclosure is not used, respiratory protection should be mandatory.
- ☑ Always ensure that proper protective clothing is worn when using chemical substances.
- ☑ Wash thoroughly immediately after exposure to xylene and at the end of the work shift or before eating, drinking, smoking, or applying cosmetics.
- ☑ Hazard warning information should be posted in the work area. In addition, as part of an on-going education and training program, all information on the health and safety hazards of xylene should be communicated to all exposed and potentially exposed workers.
- ☑ Eye wash stations should be provided in the immediate work area for emergency use. If there is a possibility of skin exposure to xylene, emergency shower facilities should be provided.
- ☑ Workers whose clothing has been contaminated by xylene should change into clean clothes before leaving work. Contaminated work clothing should be laundered only by individuals who have been informed of the hazards of xylene.
- ☑ Where possible, automatically transfer liquids containing xylene from drums or other containers to process containers.

RISK ASSESSMENT: ENVIRONMENT

General Assessment

The environment is at risk of exposure during transportation, storage, disposal, or destruction of xylene. In almost every scenario, the threat of environmental exposure is contingent upon the proper handling of the chemical substance. Accidental spills, large or small, can result in contamination of the surrounding environmental mediums (water, soil, and air).

The xylenes are moderately flammable liquids. Because of their low flash points and high boiling points, they are classified as a Class 1C flammable liquid (per OSHA 29 CFR 1910.106). Xylene is a dangerous fire hazard and moderate explosion hazard when exposed to heat or flame or in contact with oxidizers. Caution is always required in handling, storage, transportation, and disposal of xylene. Emergency responders should be made aware of the presence of xylene at any emergency response situation.

Xylene can enter the environment from industrial discharges, municipal waste treatment discharges, and from spills.

Acute Ecological Effects

Acute (short-term) toxic effects may include the death of animals, birds, or fish and death or low growth rate in plants. Acute effects are seen 2 to 4 days after animals or plants are exposed to xylene. Xylene has high acute toxicity to aquatic life. It has caused injury to various agricultural crops and ornamental crops. Insufficient data are available to evaluate the short-term effects of xylene on birds or land animals.

Chronic Ecological Effects

Chronic (long-term) toxic effects may include shortened life span, reproductive problems, lower fertility, and changes in appearance or behavior in exposed animals. These effects can be seen long after first exposure(s) to toxic chemicals. Xylene has high chronic toxicity to aquatic life. Insufficient data are available to evaluate or predict the long-term effects of xylene to plants, birds, or terrestrial animals.

Water Solubility

Xylene solubility can range from practically insoluble to moderately soluble in water. Concentrations between 1 and 100 milligrams may mix with a liter of water.

Persistence in the Environment

Xylene is non-persistent in water (due to volatilization), with a half-life of less than 2 days. The half-life of a pollutant is the amount of time it takes for one half of the chemical to be degraded. About 99.3% of xylene will eventually end up in air; about 0.5% will end up in water; the rest will be divided about equally between terrestrial soils and aquatic sediments.

Bioaccumulation in Aquatic Organisms

Some substances increase in concentration, or *bioaccumulate,* in living organisms as they breathe contami-

nated air, drink contaminated water, or eat contaminated food. These chemicals can become concentrated in the tissues and internal organs of animals as well as humans.

The concentration of xylene found in fish tissues is expected to be somewhat higher than the average concentration of xylene in the water from which the fish was taken.

Recommended Risk-Reduction Measures

Proper training of all transporters will reduce the likelihood of a mishap or accident resulting in a leak or spill of xylene into the environment. Labels on all containers, trucks, and rail cars must meet DOT requirements and accurately reflect their contents to enable emergency responders to react properly and quickly to any disaster thereby reducing the potential risk to the environment and to personnel.

Storage of xylene should be segregated from other chemicals and materials to minimize the risk of cross-contamination.

If a spill or leak to the environment has occurred, fire department, emergency response and/or hazardous materials spill personnel should be notified immediately. Cleanup should be attempted only by those trained in proper spill containment procedures using non-sparking tools. Contaminated soils should be removed for incineration and replaced with clean soil. If xylene should contact the water table, aquifer, or navigable waterway, local and/or state emergency response authorities must be notified. A comprehensive emergency response or disaster preparedness/recovery plan should be in place prior to any operations involving the use, transportation, storage, or disposal of xylene.

For disposal, xylene may be burned in a chemical incinerator equipped with an afterburner and scrubber.

If xylene is spilled or leaked, the following specific steps are recommended:

- ☑ Restrict persons not wearing protective clothing from area of spill or leak until cleanup is complete. Use non-sparking tools.
- ☑ Absorb liquids in vermiculite, sand, earth, or similar material and deposit in sealed drum.
- ☑ Ventilate area of spill or leak. Container should be removed to safer area if it can be done without increasing the risk.
- ☑ Use water spray to knock-down vapors.
- ☑ It may be necessary to dispose of xylene as a hazardous waste. The responsible state agency or the regional office of the federal Environmental Protection Agency (EPA) should be contacted for specific recommendations.

RISK ASSESSMENT: BUSINESS

General Assessment

Accidents or mishaps involving xylene can present a serious threat to business operations. The loss or damage of equipment or facilities can significantly affect fiscal viability. Lawsuits that may result from personnel illness, injury/death, public exposures, and/or environmental contamination will require a serious expenditure of resources. Media attention surrounding an injury, death, or environmental damage can also result in a loss of profits and loss of current as well as future business. Always remember that anytime the term "reproductive hazard" is used, public hysteria, ignorance, anxiety, and emotion can run equally high. This must be carefully considered when drafting or implementing public relations policies.

Recommended Risk-Reduction Measures

Company attorneys, safety and health professionals, and environmental specialists should be involved in the development of any procedures for responding to chemical incidents. A company official should be predesignated as a public relations officer with specific training in dealing with the press. Corporate plans and policies should be developed, approved, and implemented long before any need for such arises.

REFERENCES

American Conference of Governmental Industrial Hygienists. 1988. *Documentation of the Threshold Limit Values and Biological Exposure Indices,* 5th Edition (with updates). Cincinnati: ACGIH

Gerarde, H. W. 1960. *Toxicology and Biochemistry of Aromatic Hydrocarbons.* New York: Elsevier.

Howard, P. H., et. al. 1991. *Handbook of Environmental Degradation Rates.* New York: CRC Press/Lewis Publishers.

Lewis, R. J., Sr. 1993. *Hawley's Condensed Chemical Dictionary,* 12th Edition. New York: Van Nostrand Reinhold.

Lewis, R. J., Sr. 1992. *Sax's Dangerous Properties of Industrial Materials,* Volumes 1, 2, and 3, 8th Edition. New York: Van Nostrand Reinhold.

Patnaik, P. 1992. *A Comprehensive Guide to the Hazardous Properties of Chemical Substances.* New York: Van Nostrand Reinhold.

Sacarello, H. L. A. 1994. *The Comprehensive Handbook of Hazardous Materials.* New York: CRC Press/Lewis Publishers.

MATERIAL SAFETY DATA SHEET

CHEMICAL NAME	
YTTRIUM OXIDE	**CRC PRESS/LEWIS PUBLISHERS DISCLAIMER:** THE INFORMATION AND RECOMMENDATIONS PRESENTED HEREIN ARE BASED ON SOURCES BELIEVED TO BE RELIABLE. CRC MAKES NO REPRESENTATION ON ITS COMPLETENESS OR ACCURACY. IT IS THE RESPONSIBILITY OF THE USER TO DETERMINE THE CHEMICAL'S SUITABILITY FOR ITS INTENDED USE, THE CHEMICAL'S SAFE USE, AND THE CHEMICAL'S PROPER DISPOSAL. NO REPRESENTATIONS AND/OR WARRANTIES, EITHER EXPRESSED OR IMPLIED, OF THE MERCHANTABILITY OR FITNESS FOR A PARTICULAR PURPOSE, OR OF ANY OTHER NATURE, ARE MADE WITH RESPECT TO THE INFORMATION PROVIDED IN THIS MSDS OR TO THE CHEMICAL TO WHICH INFORMATION MAY REFER. CRC NEITHER ASSUMES NOR AUTHORIZES ANY OTHER PERSON TO ASSUME FOR IT, ANY OTHER ADDITIONAL RESPONSIBILITY OR LIABILITY FOR THE USE OF, OR RELIANCE UPON, THIS INFORMATION.

HAZARD WARNING INFORMATION

HEALTH	FIRE	REACTIVE	OTHER	DEGREE OF HAZARD	COLOR CODING	OTHER CODES
1	0	0		0 = Minimum Hazard 1 = Slight Hazard 2 = Moderate Hazard 3 = Serious Hazard 4 = Severe Hazard	HEALTH = BLUE FIRE = RED REACTIVITY = YELLOW OTHER = WHITE	OX = Oxidizer ACID = Acid ALK = Alkali COR = Corrosive W = Use No Water

SECTION I - GENERAL INFORMATION

Characterization **Metal**	RCRA Number **None**	EPA Class **Not Applicable**
DOT Proper Shipping Name **No Citation**	Chemical Abstract Service (CAS) Number **1314-36-9**	
DOT Hazard Class and Label Requirements **No Citation**	DOT Emergency Guide Code **No Citation**	
DOT Identification Number **No Citation**	Molecular Formula Y_2O_3	

Synonyms
Yttria.

SECTION II - HAZARDOUS INGREDIENTS AND IDENTITY INFORMATION

Hazardous Components (specific identity)	OSHA Exposure Criteria	NIOSH Exposure Criteria	Immediately Dangerous to Life and Health (IDLH)	ACGIH Exposure Criteria
Yttrium oxide (derivation: By ignition of yttrium nitrate).	**PEL:** 1 mg/m^3 (as Yttrium) **STEL:** Not Established)	**REL:** 1 mg/m^3 (as Yttrium) **STEL:** Not Established	**Not Determined**	**TLV:** 1 mg/m^3 (as Yttrium) **STEL:** Not Established

SECTION III - PHYSICAL AND CHEMICAL CHARACTERISTICS

Boiling Point **7772°F (4300°C)**	Specific Gravity (H_2O = 1) **4.84**
Vapor Pressure (mm Hg) **Not Reported**	Molecular Weight **225.82**
Vapor Density (Air = 1) **Not Reported**	Melting Point **4370°F (2410°C)**

Solubility
Insoluble in water. Soluble in dilute acids.

Appearance and Odor
Colorless to white or yellowish-white, hygroscopic cubic powder.

SECTION IV - FIRE AND EXPLOSION HAZARD DATA

Flash Point (method used) **Not Applicable**	Explosive Limits in Air % by Volume **LEL: N/A UEL: N/A**
NFPA Classification **Noncombustible Solid**	Autoignition Temperature **Not Applicable**

Extinguishing Media
Use extinguishing media suitable to surrounding fires.

Special Fire Fighting Procedures
None noted. But, yttrium oxide dust is toxic by inhalation. Fire fighters should take care to avoid dusty situations and wear appropriate respiratory protection where yttrium oxide may be present.

Unusual Fire and Explosion Hazards
Yttrium oxide is a noncombustible solid. But, its dusts may form explosive mixtures in air. Caution is warranted when approaching situations that may involve yttrium oxide dusts.

SECTION V - REACTIVITY DATA

Stability		Conditions to Avoid **Yttrium oxide is normally stable in closed containers under routine conditions of handling and storage. Avoid contact with incompatible materials, including water and moisture. Avoid dusty conditions.**
Stable **X**	Unstable	Incompatibility *(materials to avoid)* **Yttrium oxide displaces ammonia from ammonia salts. It is hygroscopic and will readily absorb moisture from the air and carbon dioxide (will form carbonates).**
Hazardous Polymerization		Conditions to Avoid **Hazardous polymerization of yttrium oxide cannot occur.**
May Occur	Will Not Occur **X**	Hazardous Decomposition or By-products **None reported.**

SECTION VI - HEALTH HAZARD DATA

Primary Route(s) of Entry:	Inhalation?	Absorption (skin/eye)?	Ingestion?
	X		

Health Hazards

INHALATION: **Inhaling finely divided particles may cause irritation of the eyes, nose, throat, and the mucous membranes of the respiratory tract. It can cause sensitivity to heat, itching, an increased awareness to odor and taste. May cause progressive fibrosis in long-term.**

SKIN & EYES: **Contact with the eyes may cause irritation with symptoms of conjunctivitis. Contact with abraded skin may cause hair loss and the formation of scars.**

INGESTION: **Low oral toxicity (poor absorption through the digestive tract).**

Carcinogenicity	NTP Listed?	IARC Cancer Review Group?	OSHA Regulated?	Target Organs?
Unknown Human **Unknown Animal**	**No**	**No**	**29 CFR 1910.1000** **Table Z-1 (as Y)**	**Eyes, skin, respiratory system**

Medical Conditions Generally Aggravated by Exposure

Pre-existing lung disease or dysfunction may be aggravated by exposure to yttrium oxide dusts.

Emergency and First-aid Procedures

Eye contact: Do not allow victim to keep eyes tightly shut. Flush immediately with water for 15 minutes (minimum); seek medical attention. Skin contact: Remove all contaminated clothing. Immediately wash area with large amounts of soap and water. For inhalation: Remove the person from exposure. Provide respiratory assistance and CPR. Transfer to medical facility. If swallowed, remove person to fresh air and seek medical attention immediately. Do not attempt to give an unconscious or convulsing person anything by mouth. Give 1 to 2 glasses of water to dilute and induce vomiting.

SECTION VII - PRECAUTIONS FOR SAFE HANDLING AND USE

Steps to be Taken in Case Material is Released or Spilled

Ventilate area of spill or leak. Restrict those not involved in cleanup from entering area. Collect spilled powdered materials in safest manner possible and deposit in sealed drum for disposal. Notify appropriate authorities, as required by applicable environmental regulations.

Preferred Waste Disposal Method

Not listed. Check applicable federal, state, and/or local environmental regulations.

Precautions to be Taken in Handling and Storage

Yttrium oxide is a noncombustible solid. Store to avoid contact with moisture and heat. Store in tightly closed containers in cool, dry, well-ventilated area.

Other Precautions and Warnings

Yttrium oxide is suitable for storage in general chemical storage areas.

SECTION VIII - CONTROL MEASURES AND PERSONAL PROTECTIVE EQUIPMENT

Respiratory Protection *(specify type)*

None required where adequate ventilation conditions exist. If airborne concentrations exceed the PEL, a dust/mist respirator is recommended. For high exposures, or if concentrations exceed the capacity of the respirator, then a self-contained breathing apparatus (SCBA) with full facepiece operated in positive pressure mode is advised.

Ventilation

Local exhaust (preferred) at site of chemical work or general ventilation.

Protective Gloves	Eye Protection	Other Protective Clothing
Impervious Rubber	**Safety Glasses with Side Shields**	**Not Applicable**

Work/Hygiene Practices

Always wash hands thoroughly after using chemical; never bring food, drink, or smoking materials into vicinity of chemicals.

YTTRIUM OXIDE

Y_2O_3

CAS: 1314-36-9

IDENTIFICATION AND TYPICAL USES

Yttrium oxide is colorless to white, hygroscopic cubic powder. It is used as phosphorous for color television tubes (in alloy with euphorium) and mantles in gas and acetylene lights. Also used in yttrium-iron garnets for microwave filters, and as a stabilizer for high-temperature service materials (zirconia and silicon nitride refractories).

RISK ASSESSMENT: HEALTH

General Assessment

Yttrium and its salts are generally considered nuisance dusts and under some conditions of human exposure may cause a minimal pulmonary nodular response. However, excessive *inhalation* of finely divided particles of yttrium oxide may lead to lung damage characterized by scarred tissue (pulmonary fibrosis) with shortness of breath, tightness and pain in the chest, and coughing. There are no data regarding any potential for carcinogenic, mutagenic, or teratogenic action on long-term exposure to yttrium oxide.

Inhalation of finely divided dusts and powders of yttrium oxide can cause irritation of the nose, throat, and mucous membranes of the respiratory tract with sensitivity to heat, itching, and an increased awareness of odor and taste. Eye contact can cause mechanical irritation and may lead to conjunctivitis. Contact with abraded skin may result in the loss of hair and the formation of surface scar tissue.

The oral toxicity is considered to be low due to the poor absorption rate of yttrium oxide through the digestive tract. This does not mean there is no health risk associated with oral intact of yttrium oxide. There is no way of estimating exact human response under all possible types of exposure conditions. Prudent risk management requires caution when working with or around yttrium oxide.

Acute Health Effects

The following acute (short-term) health effects may occur immediately or shortly after exposure to yttrium oxide:

Skin: Irritation and possible hair loss and scarring.

Eye: Severe eye irritation and conjunctivitis likely.

Lung: Can irritate the respiratory tract. In large doses, may cause pulmonary fibrosis.

Chronic Health Effects

The following chronic (long-term) health effects may occur at some time after exposure to yttrium oxide and can last for months or even years:

Cancer Hazards: According to the information presented in the references, yttrium oxide has not been tested for its ability to cause cancer in animals. No data were found on its human carcinogenic capabilities.

Reproductive Hazard: According to the information presented in the references, yttrium oxide has not been tested for its ability to adversely affect reproduction.

Other Chronic Effects: Repeated exposure to large amounts of yttrium oxide dust may damage the lungs, leading to scarring of the tissue (pulmonary fibrosis), with shortness of breath.

Recommended Risk-Reduction Measures

Personnel should avoid direct contact with yttrium oxide. The exact nature of its harmful capacities is not clearly understood in the references. Therefore, it is recommended that it be handled as a toxic material. If a less toxic material cannot be substituted for yttrium oxide, then *engineering controls* are the most effective method of reducing exposures. The best protection is to enclose operations and/or provide local exhaust ventilation at the site of yttrium oxide release. While not

always operationally feasible, isolating operations can also reduce exposure.

Using respiratory protection and other personal protective equipment (PPE) is less effective than the engineering controls mentioned above, but is still advisable whenever working with or around yttrium oxide. Under normal operating conditions of low exposure potential, a dust/mist respirator should suffice. However, for maximum protection in large concentrations or when the capacity of the dust/mist respirator is exceeded, personnel should wear a self-contained breathing apparatus (SCBA) with full facepiece in pressure demand or a powered air-supplied respirator with full facepiece operated in continuous flow mode. If a full facepiece is not available, then chemical/dust goggles, safety glass with side shields, or a face shield should be worn to protect the eyes. To prevent hand and skin exposures, impervious rubber gloves should be used.

Administrative controls should also be in place to minimize the potential for human exposures. These may include written procedures or policies which specify the methods and techniques that will be practiced whenever personnel are to work with yttrium oxide.

All personnel should receive *training* on the use, hazards, protective measures, emergency actions, and other precautions per 29 CFR 1910.1200 (Hazard Communication), prior to the first assignment in an area where yttrium oxide is used or stored.

If there has been a long history of high exposure, or if lung symptoms develop, the following medical tests are recommended:

☑ Chest X-ray.

☑ Pulmonary function tests.

Any evaluation should include a careful history of past and present symptoms with an examination. Medical tests that simply look for existing damage are not a substitute for controlling exposures. Also, because smoking can cause lung cancer, emphysema, and a host of other medical problems, smokers working with or around yttrium oxide may aggravate these conditions further.

Other methods to reduce exposure to yttrium oxide include:

☑ Where possible, enclose operations and use local exhaust ventilation at the site of chemical release. If local exhaust ventilation or enclosure is not used, then respiratory protection should be required and its use enforced.

☑ Always ensure that proper protective clothing is worn when using chemical substances. Never allow contaminated clothing to be taken home from the work place.

☑ Hazard warning information should be posted in the work area. In addition, as part of an on-going education and training program, all information on the health and safety hazards of yttrium oxide should be communicated to all exposed and potentially exposed workers.

☑ Eye wash stations should be provided in the immediate work area for emergency use.

RISK ASSESSMENT: ENVIRONMENT

General Assessment

The environment is at risk of exposure during transportation, storage, disposal, or destruction of yttrium oxide. In almost every scenario, the threat of environmental exposure is contingent upon the proper handling of the chemical substance. Accidental spills, large or small, where yttrium oxide contacts incompatible commodities can result in violent reactions as well as the possible contamination of the surrounding environmental mediums (water, soil, and air).

Yttrium oxide is considered a noncombustible solid. However, its dusts may form explosive mixtures in air. Also, because it is hygroscopic and will absorb moisture and carbon dioxide from the air to produce toxic carbonates, extreme caution is required in handling, storage, transportation, and disposal of yttrium oxide. These characteristics also require special consideration during any emergency situation involving a leak or spill of yttrium oxide. Should yttrium oxide ever come into contact with incompatible substances such as ammonium salts either during use, transportation, storage, or disposal, dangerous reactions are possible.

Yttrium oxide can enter the environment through industrial discharges and/or spills.

☠ *Acute Ecological Effects*

Acute (short-term) toxic effects may include the death of animals, birds, or fish and death or low growth rate in plants. Acute effects are seen 2 to 4 days after animals or plants are exposed to yttrium oxide.

Insufficient data are available to evaluate or predict the short-term toxic effects of yttrium oxide to aquatic life, plants, birds, and land animals.

Chronic Ecological Effects

Chronic (long-term) toxic effects may include shortened life span, reproductive problems, lower fertility, and changes in appearance or behavior in exposed animals. These effects can be seen long after first exposure to toxic chemicals.

Insufficient data are available to evaluate or predict the long-term toxic effects of yttrium oxide to aquatic life, plants, birds, or land animals.

Water Solubility

Yttrium oxide is insoluble in water. Concentrations of 1 milligram or less will not mix with a liter of water.

Persistence in the Environment

Yttrium oxide is highly persistent in water, with a half-life greater than 200 days. The half-life of a pollutant is the amount of time it takes for one-half of the chemical to be degraded.

Bioaccumulation in Aquatic Organisms

Some substances increase in concentration, or *bioaccumulate,* in living organisms as they breathe contaminated air, drink contaminated water, or eat contaminated food. These chemicals can become concentrated in the tissues and internal organs of animals as well as humans.

Yttrium oxide will not accumulate in the edible tissues of aquatic life that are consumed by humans.

Recommended Risk-Reduction Measures

Proper training of all transporters will reduce the likelihood of a mishap or accident resulting in a leak or spill of yttrium oxide into the environment. The correct use of labeling on all containers, trucks, and rail cars will enable emergency responders to react properly and quickly to any disaster thereby reducing the potential risk to the environment and to personnel.

Storage of yttrium oxide should be segregated from incompatible chemicals to minimize the risk of cross-contamination or contact. It must be kept dry.

If a spill or leak to the environment has occurred, fire department, emergency response and/or hazardous materials spill personnel should be notified immediately. Cleanup should be attempted only by those trained in proper spill containment procedures. If yttrium oxide should contact the water table, aquifer, or navigable waterway, the local and/or state emergency response authorities must be notified. A comprehensive emergency response or disaster preparedness/recovery plan should be in place prior to any operations involving the use, transportation, storage, or disposal of yttrium oxide.

If yttrium oxide is spilled or leaked, the following specific steps are recommended:

- ☑ Restrict persons not wearing protective clothing from area of spill or leak until cleanup is complete and area can be opened for normal work.
- ☑ Do not allow exposure to incompatible materials.
- ☑ Collect powdered materials in the safest and most efficient manner possible. Place materials in a sealed drum. Do NOT dry sweep. Use a vacuum equipped with a high-efficiency particulate air (HEPA) filter. Keep water away.

RISK ASSESSMENT: BUSINESS

General Assessment

Accidents or mishaps involving yttrium oxide can present a moderate threat to business operations. Although not likely, equipment and facilities may be lost or their use disrupted as a result of an incident involving yttrium oxide which can significantly affect fiscal viability. Lawsuits that may result from personnel injury, illness, and/or death, public exposures, and/or environmental contamination will also require a serious expenditure of resources. Media attention surrounding an injury, death, or environmental damage can also result in a loss of profits and loss of current as well as future business.

Recommended Risk-Reduction Measures

Company attorneys, safety and health professionals, and environmental specialists should be involved in the development of any procedures that focus on emergency response, disaster preparedness, and disaster recovery. A company official should be pre-designated as a public relations officer with specific training in dealing with the press. Corporate plans and policies should be developed, approved, and implemented long before any need for such arises.

REFERENCES

Ahlbom, A. 1993. *Biostatistics for Epidemiologists.* New York: CRC Press/Lewis Publishers.

American Conference of Governmental Industrial Hygienists. 1988. *Documentation of the Threshold Limit Values and Biological Exposure Indices,* 5th Edition (with updates). Cincinnati: ACGIH

Bloom, A. D. and F. J. DeSerres. 1995. *Ecotoxicity and Human Health.* New York: CRC Press/Lewis Publishers.

Brusick, D. J. 1994. *Methods for Genetic Risk Assessment.* New York: CRC Press/Lewis Publishers.

Bueche, J. R. 1972. *Principles of Physics,* 2nd Edition. New York: McGraw-Hill.

Calabrese, E. J., E. M. Kenyon. *Air Toxics and Risk Assessment.* New York: CRC Press/Lewis Publishers.

Calabrese, E. J. 1994. *Biological Effects of Low Level Exposures.* New York: CRC Press/Lewis Publishers.

Cockerman, L. G., B. S. Shane. 1994. *Basic Environmental Toxicology.* New York: CRC Press/Lewis Publishers.

Cothern, R. R. 1993. *Comparative Environmental Risk Assessment.* New York: CRC Press/Lewis Publishers.

Gardner, R. W. 1994. *Chemical Intolerance.* New York: CRC Press/Lewis Publishers.

Hallenbeck, W. H. 1993. *Quantitative Risk Assessment for Environmental and Occupational Health,* 2nd Edition. New York: CRC Press/Lewis Publishers.

Howard, P. H. 1995. *Handbook of Environmental Fate and Exposures Data for Organic Chemicals,* Vol. I-IV. New York: CRC Press/Lewis Publishers.

Howard, P. H., et. al. 1991. *Handbook of Environmental Degradation Rates.* New York: CRC Press/Lewis Publishers.

Jones, F. E. 1994. *Toxic Organic Vapors in the Workplace.* New York: CRC Press/Lewis Publishers.

Keith, L. H. and M. M. Walker. 1995. *Handbook of Air Toxics.* New York: CRC Press/Lewis Publishers.

Landis, W. G., M. H. Yu. 1995. *Introduction to Environmental Toxicology.* New York: CRC Press/Lewis Publishers.

Langerman, N. 1994 *Precautionary Labels for Chemical Containers.* New York: CRC Press/Lewis Publishers.

Lauwerys, R. R., P. Hoet. 1993. *Industrial Chemical Exposure.* New York: CRC Press/Lewis Publishers.

Lewis, R. J., Sr. 1993. *Hawley's Condensed Chemical Dictionary,* 12th Edition. New York: Van Nostrand Reinhold.

Lewis, R. J., Sr. 1992. *Sax's Dangerous Properties of Industrial Materials,* Volumes 1, 2, and 3, 8th Edition. New York: Van Nostrand Reinhold.

Lide, D. R. 1995. *Handbook of Chemistry and Physics,* 76th Edition. New York: CRC Press/Lewis Publishers.

Manahan, S. E. 1994. *Environmental Chemistry,* 6th Edition. New York: CRC Press/Lewis Publishers.

Mellström, G. A., J. E. Wahlberg, and H. I. Maibach. 1994. *Protective Gloves for Occupational Use.* New York: CRC Press/Lewis Publishers.

Meyer, E. 1990. *Chemistry of Hazardous Materials.* Englewood Cliffs, NJ: Prentice Hall.

Mycock, J. C., J. D. McKenna, and L. Theodore. 1995. *Handbook of Air Pollution Control Engineering and Technology.* New York: CRC Press/Lewis Publishers.

Patnaik, P. 1992. *A Comprehensive Guide to the Hazardous Properties of Chemical Substances.* New York: Van Nostrand Reinhold.

Proctor, N. H., J. P. Hughes. 1991. *Chemical Hazards of the Workplace.* New York: Van Nostrand Reinhold.

Quigley, D. R. 1994. *Handbook of Emergency Chemical Management.* New York: CRC Press/Lewis Publishers.

Rea, W. J. 1995. *Chemical Sensitivity,* Vol. I-III. New York: CRC Press/Lewis Publishers.

Reeves, A. L. 1979. *Handbook on the Toxicology of Metals.* New York: Elsevier North-Holland.

Richardson, M. 1995. *Dictionary of Substance and Their Effects,* Vol. I-VI. New York: CRC Press/Lewis Publishers.

Sacarello, H. L. A. 1994. *The Comprehensive Handbook of Hazardous Materials.* New York: CRC Press/Lewis Publishers.

Sheftel, V. O. 1995. *Handbook of Toxic Properties of Monomers and Additives.* New York: CRC Press/Lewis Publishers.

Talbott, E., G. Craun. 1995. *An Introduction to Environmental Epidemiology.* New York: CRC Press/Lewis Publishers.

Torkelson, T. R. V. K. Rowe. 1981. *Patty's Industrial Hygiene and Toxicology,* Vol. 2B. New York: Wiley-Interscience.

MATERIAL SAFETY DATA SHEET

CHEMICAL NAME

ZINC

CRC PRESS/LEWIS PUBLISHERS DISCLAIMER: THE INFORMATION AND RECOMMENDATIONS PRESENTED HEREIN ARE BASED ON SOURCES BELIEVED TO BE RELIABLE. CRC MAKES NO REPRESENTATION ON ITS COMPLETENESS OR ACCURACY. IT IS THE RESPONSIBILITY OF THE USER TO DETERMINE THE CHEMICAL'S SUITABILITY FOR ITS INTENDED USE, THE CHEMICAL'S SAFE USE, AND THE CHEMICAL'S PROPER DISPOSAL. NO REPRESENTATIONS AND/OR WARRANTIES, EITHER EXPRESSED OR IMPLIED, OF THE MERCHANTABILITY OR FITNESS FOR A PARTICULAR PURPOSE, OR OF ANY OTHER NATURE, ARE MADE WITH RESPECT TO THE INFORMATION PROVIDED IN THIS MSDS OR TO THE CHEMICAL TO WHICH INFORMATION MAY REFER. CRC NEITHER ASSUMES NOR AUTHORIZES ANY OTHER PERSON TO ASSUME FOR IT, ANY OTHER ADDITIONAL RESPONSIBILITY OR LIABILITY FOR THE USE OF, OR RELIANCE UPON, THIS INFORMATION.

HAZARD WARNING INFORMATION

HEALTH	FIRE	REACTIVE	OTHER	DEGREE OF HAZARD	COLOR CODING	OTHER CODES
0	1	1		0 = Minimum Hazard 1 = Slight Hazard 2 = Moderate Hazard 3 = Serious Hazard 4 = Severe Hazard	HEALTH = BLUE FIRE = RED REACTIVITY = YELLOW OTHER = WHITE	OX = Oxidizer ACID = Acid ALK = Alkali COR = Corrosive W = Use No Water

SECTION I - GENERAL INFORMATION

Characterization: **Metal**

RCRA Number: **None**

EPA Class: **Not Applicable**

DOT Proper Shipping Name: **Zinc, powder or dust, nonpyrophoric**

Chemical Abstract Service (CAS) Number: **7440-66-6**

DOT Hazard Class and Label Requirements: **Flammable Solid; Dangerous When Wet**

DOT Emergency Guide Code: **76**

DOT Identification Number: **UN 1436**

Atomic Formula: **Zn**

Synonyms: **Blue powder; spelter; granular zinc; jasad; merrillite; pasco; pigment black 16; zinc dust; zinc powder.**

SECTION II - HAZARDOUS INGREDIENTS AND IDENTITY INFORMATION

Hazardous Components (specific identity)	OSHA Exposure Criteria	NIOSH Exposure Criteria	Immediately Dangerous to Life and Health (IDLH)	ACGIH Exposure Criteria
Zinc: (derivation: Ores are roasted to form zinc oxide which is then leached from the roasted material with sulfuric acid to form a zinc sulfate solution which is electrolyzed in cells to deposit zinc on cathodes; or, the zinc oxide is reduced with carbon in retorts to yield distilled and condensed to zinc).	**PEL:** Not Established **STEL:** Not Established	**REL:** Not Established **STEL:** Not Established	Not Determined	**TLV:** Not Established **STEL:** Not Established

SECTION III - PHYSICAL AND CHEMICAL CHARACTERISTICS

Boiling Point: **1665°F (907°C)**

Specific Gravity (H_2O = 1): **7.14**

Vapor Pressure (mm Hg): **1 at 909°F (487°C)**

Molecular Weight (atomic weight): **65.37**

Vapor Density (Air = 1): **Not Reported**

Melting Point: **786°F (419°C)**

Solubility: **Insoluble in water (reacts). Soluble in acids and alkalis.**

Appearance and Odor: **Shining white metal with a bluish-white or gray luster (called spelter).**

SECTION IV - FIRE AND EXPLOSION HAZARD DATA

Flash Point (method used): **Not Reported**

Explosive Limits in Air % by Volume: **LEL: 0.5 oz/ft^3 (dust cloud) UEL: Not Reported**

NFPA Classification: **Non-Combustible Solid (powder may ignite)**

Autoignition Temperature: **Cloud: 1256°F (680°C) Dust Layer: 860°F (460°C)**

Extinguishing Media: **Zinc may burn, but does not easily ignite. Use dry sand, graphite, dolomite, sodium chloride. Never use water, halogenated agents, or class A, B, or C extinguishing agents.**

Special Fire Fighting Procedures: **Poisonous gases are produced in fire. Wear full protective clothing, self-contained breathing apparatus (SCBA) operated in pressure-demand mode. Move container from fire area if it can be done without risk.**

Unusual Fire and Explosion Hazards: **In powdered form, zinc is a moderate fire and explosion hazard. Will also burn vigorously on contact with flame or by chemical reaction. Moist powders can ignite explosively in air. Flammable hydrogen gas is liberated by reaction with alkali hydroxide, acids, or water.**

SECTION V - REACTIVITY DATA

Stability		Conditions to Avoid
Stable	Unstable	Dry zinc metal is stable in closed containers at room temperature under routine conditions of storage and handling, but moist zinc powder reacts exothermically and can ignite spontaneously in air.
X		**Incompatibility** *(materials to avoid)* Powders are explosive in reaction with acids, chlorates, oxidizing agents, halogenated hydrocarbons, hydrazine mononitrate, hydroxylamine, ammonium nitrate, barium dioxide, and many other metallic salts.
Hazardous Polymerization		Conditions to Avoid Under normal conditions of temperature and pressure, zinc cannot undergo hazardous polymerization.
May Occur	Will Not Occur	**Hazardous Decomposition or By-products** When heated to decomposition in air it emits highly toxic, acrid, and irritating smoke and fumes. Vaporized zinc burns in air with a blue-green flame to produce zinc oxide fumes.
	X	

SECTION VI - HEALTH HAZARD DATA

Primary Route(s) of Entry:	Inhalation?	Absorption (skin/eye)?	Ingestion?
	X		

Health Hazards

INHALATION: Inhalation of zinc dusts and zinc fumes may cause symptoms of metal fume fever, a flu-like illness lasting about 24 hours with fever, fatigue, muscle ache, chills, aches, and cough. In animal studies, it has also caused vomiting, diarrhea, paralysis, and muscle twitching. When zinc is refined, cadmium is released which is carcinogenic.

EYES & SKIN: Eye and skin contact may result in mild to moderate irritation. Metal fragments in the eyes can cause scratching.

INGESTION: Zinc is poorly absorbed through the digestive tract. Ingestion of zinc salts may cause nausea, vomiting, sluggishness, and lightheadedness.

Carcinogenicity	NTP Listed?	IARC Cancer Review Group?	OSHA Regulated?	Target Organs?
Unknown Human Unknown Animal	No	No	No	Eyes, respiratory system.

Medical Conditions Generally Aggravated by Exposure

None reported.

Emergency and First-aid Procedures

Eye contact: Do not allow victim to keep eyes tightly shut. Flush immediately with water for 15 minutes (minimum). Skin contact: Avoid skin contact in areas with cuts or scratches. If such contact occurs, wash with large amounts of soap and water, scrubbing to remove metal particles. For inhalation: Remove the person from exposure. Provide respiratory assistance and CPR. Transfer to medical facility. If swallowed: Seek medical attention immediately. Do not attempt to give an unconscious or convulsing person anything by mouth. If conscious, give copious amounts of water to drink and induce vomiting. Repeat until vomitus is clear.

SECTION VII - PRECAUTIONS FOR SAFE HANDLING AND USE

Steps to be Taken in Case Material is Released or Spilled

Wear self-contained breathing apparatus (SCBA). Restrict those not involved in cleanup from entering area. Ventilate area of spill. Do not use water. Collect powder materials and deposit in sealed drum. Do NOT dry sweep, use a high-efficiency particulate air (HEPA) vacuum instead.

Preferred Waste Disposal Method

No citation.

Precautions to be Taken in Handling and Storage

Store in tightly closed containers in a cool, well-ventilated area away from heat and moisture. Avoid contact in storage with incompatible materials.

Other Precautions and Warnings

Protect containers from physical damage.

SECTION VIII - CONTROL MEASURES AND PERSONAL PROTECTIVE EQUIPMENT

Respiratory Protection *(specify type)*

Use an air-purifying respirator equipped with a HEPA filter or, use a NIOSH/MSHA-approved self-contained breathing apparatus (SCBA) with full facepiece operated in positive pressure mode.

Ventilation

Local exhaust (preferred) at site of chemical work or general ventilation.

Protective Gloves	Eye Protection	Other Protective Clothing
Impervious Gloves	Chemical/Dust Goggles and Face Mask	Protective Uniform or Apron

Work/Hygiene Practices

Always wash hands thoroughly after using chemical; never bring food, drink, or smoking materials into vicinity of chemicals. Barrier creams may be useful to limit the effects of skin exposure.

ZINC

Zn

CAS: 7440-66-6

IDENTIFICATION AND TYPICAL USES

Zinc is a metallic element of atomic number 30, Group IIB of the periodic table. It has valances of 2 and five stable isotopes. It appears as a shining white metal with a bluish-white or gray luster (called spelter). It is used in alloys (brass, bronze, and die-casting alloys), for galvanizing iron and other metals, in electroplating, in metal spraying, auto parts, electrical fuses, in storage and dry cell batteries, in fungicides, and as a nutritional agent (essential growth agent). It is also used in roofing, gutters, engraver's plates, cable wrapping, and organ pipes.

RISK ASSESSMENT: HEALTH

General Assessment

Elemental zinc exhibits relatively low toxicity, but its compounds and salts are highly toxic. It is primarily toxic through *inhalation* of its dusts. Zinc is poorly absorbed through the digestive tract and, therefore, ingestion is not considered a toxic exposure route. Ingestion of the soluble salts may cause nausea and vomiting, sluggish fatigue, and lightheadedness. Skin contact may cause localized irritation but dermal absorption is not reported. There are no data in the references regarding its carcinogenic, mutagenic, or teratogenic properties.

Inhalation of zinc dusts can lead to coughing, phlegm, sneezing, and other indications of mechanical irritation. Inhalation of zinc fumes may result in the development of a flu-like condition known as metal fume fever with symptoms of chills, fever, fatigue, coughing, and muscle aches. These symptoms usually begin 3 to 10 hours after exposure and usually subside within 48 hours of onset with no permanent after-effects. Metal fume fever is an acute, self-limiting condition, without recognized complications, after-effects, or chronic effects. Skin contact may produce irritation or rash. Eye contact can also result in mild to severe mechanical irritation and possible scarring.

Acute Health Effects

The following acute (short-term) health effects may occur immediately or shortly after exposure to zinc:

Skin: Mild to moderate irritation.

Eye: Mild to severe mechanical irritation with itching, tearing, and scarring.

Lung: High concentrations of dust may cause mechanical irritation and the fume inhalation may cause respiratory tract irritation and possible symptoms of metal fume fever.

Chronic Health Effects

The following chronic (long-term) health effects may occur at some time after exposure to zinc and can last for months or even years:

Cancer Hazards: According to information presented in the references, zinc has not been shown to cause cancer in humans or test animals. Zinc is considered essential to human life in limited amounts.

Reproductive Hazard: According to information presented in the references, zinc has not been shown to adversely affect reproduction in test animals.

Other Chronic Effects: None reported. However, as a nuisance dust, long-term exposure to zinc may be expected to cause respiratory dysfunction.

Recommended Risk-Reduction Measures

Personnel should avoid direct contact with zinc dusts and powders. Even though its elemental form is considered relatively non-toxic, many zinc compounds and salts can pose a significant threat to human health. If the exact zinc compound is not known or, if a less toxic material or compound cannot be substituted for a specific zinc compound, then *engineering controls* are the most effective method of reducing exposure risk. The best protection is to enclose operations and/or provide local exhaust ventilation at the site of zinc release. While not always operationally feasible, isolating operations can also reduce exposure. Using respiratory protection is less effective than the controls mentioned above, but is still required whenever working with or around zinc. No exposure levels have been established. This does not mean that exposure is without health risk. For infrequent or transient exposures, an air-purifying respirator equipped with a high-efficiency particulate air (HEPA) filter and dust pre-filter may suffice. However, for the best protection use an MSHA/NIOSH-approved full facepiece self-contained breathing apparatus (SCBA) operated in pressure demand or other positive pressure mode, or a powered air-supplied respirator in continuous flow mode. If a full facepiece is not available, then chemical goggles should be worn to protect the eyes, especially when an airborne dust hazard exists. A face shield should also be considered. To prevent hand and skin exposures, rubber gloves should be used. Glove manufacturers should be contacted and permeation studies obtained *before* final glove selections are made.

Administrative controls should also be in place to minimize the potential for human exposures. These may include written procedures or policies which specify the methods and techniques that will be practiced whenever personnel are to work with zinc.

All personnel should receive *training* on the use, hazards, protective measures, emergency actions, and other precautions per 29 CFR 1910.1200 (Hazard Communication) prior to the first assignment in an area where zinc is used or stored.

Before beginning employment and at regular times thereafter, the following recommended medical tests should be considered for those with frequent or potentially high exposures:

☑ Lung function tests.

☑ Consider chest X-ray after acute overexposure or repeated, long-term exposure.

It should be noted that medical tests that simply look for existing damage are not a substitute for controlling exposures. Medical histories are extremely important when assessing exposure risk. Also, since smoking can cause heart disease, lung cancer, emphysema, and other respiratory diseases, smokers exposed to zinc may experience symptoms more quickly and more pronounced than non-smokers under the same conditions of exposure. Prudent risk management requires proper consideration of *all* factors which may cause exposure symptoms in the workplace.

Other methods to reduce exposure include:

☑ Always ensure that proper protective clothing is worn when using chemical substances and that personnel are trained in its use, care, and maintenance.

☑ Wash thoroughly immediately after exposure to zinc and at the end of the work shift or before eating, drinking, smoking, or applying cosmetics.

☑ Hazard warning information should be posted in the work area. In addition, as part of an on-going education and training program, all information on the health and safety hazards of zinc should be communicated to all potentially exposed workers.

☑ Eye wash stations should be provided in the immediate work area for emergency use. If there is a possibility of skin exposure to zinc, emergency shower facilities should also be provided.

☑ Workers whose clothing has been contaminated by zinc should change into clean clothes before leaving work. Contaminated work clothing should be laundered only by individuals who have been informed of the hazards of exposure to zinc.

RISK ASSESSMENT: ENVIRONMENT

General Assessment

Even though zinc is a commonly occurring metallic element, the environment is still at risk of exposure during transportation, storage, disposal, or destruction of zinc. In almost every scenario, the threat of environmental exposure is contingent upon the proper handling of the chemical substance. Accidental spills, large or small, can result in contamination of the surrounding environmental mediums (water, soil, and air).

Zinc is a non-combustible solid in bulk form. However, zinc metal dusts particles suspended in the air can explode. Moist powders can ignite spontaneously in air. Caution is always required in handling, storage, transportation, and disposal of zinc. It presents a serious explosion hazard in the form of powders and dusts. It is incompatible with a number of materials, including acids, chlorates, oxidizing agents, halogenated hydrocarbons, hydrazine mononitrate, hydroxylamine, ammonium nitrate, barium dioxide, barium nitrate, cadmium, performic acid, manganese chloride, nitric acid, ethyl acetoacetate and tribromoneopentyl alcohol, tellurium, carbon disulfide, lead azide, selenium, sodium peroxide, potassium nitrate, and water. It will react with water and moisture to form flammable hydrogen gas. When heated to decomposition in air it emits very toxic and acrid fumes and vapors. Emergency responders should therefore be made aware of the presence of zinc at any emergency response situation.

Zinc is one of the most commonly used metals in the world. Zinc is an essential element for life when present in trace amounts. It exists as a variety of salts. Zinc may enter the environment through discharges from galvanizing plants, as a leachate from galvanized structures and natural ores, and from municipal waste treatment plant discharges.

Acute Ecological Effects

Acute (short-term) toxic effects may include the death of animals, birds, or fish and death or low growth rate in plants. Acute effects are seen 2 to 4 days after animals or plants are exposed to zinc.

The toxicity of zinc to aquatic life is related to water hardness, with increased toxicity occurring in softer waters. Zinc and its salts have high acute toxicity to aquatic life. Insufficient data are available to evaluate or predict the acute (short-term) effects of zinc to birds, plants, or land animals.

Chronic Ecological Effects

Chronic (long-term) toxic effects may include shortened life span, reproductive problems, lower fertility, and changes in appearance or behavior in exposed animals. These effects can be seen long after first exposure(s) to toxic chemicals.

Zinc and its salts have high chronic toxicity to aquatic life. Insufficient data are available to evaluate the chronic (long-term) effects of zinc to plants, birds, or land animals.

Water Solubility

Zinc exists as a variety of salts. Zinc and its compounds have water solubilities that can range from low to high, depending on the salt that is formed. Most are highly soluble in water.

Persistence in the Environment

Zinc and its salts are highly persistent in water, with half-lives greater than 200 days. The half-life of a pollutant is the amount of time it takes for one-half of the chemical to be degraded. Other compounds of zinc will vary in persistence, as water solubility varies from compound to compound.

Bioaccumulation in Aquatic Organisms

Some substances increase in concentration, or *bioaccumulate,* in living organisms as they breathe contaminated air, drink contaminated water, or eat contaminated food. These chemicals can become concentrated in the tissues and internal organs of animals as well as humans.

The concentration of zinc found in fish tissues is expected to be considerably higher than the average concentration of zinc in the water from which the fish was taken.

Recommended Risk-Reduction Measures

Proper training of all transporters will reduce the likelihood of a mishap or accident resulting in a leak or spill of zinc into the environment. Labels on all containers, trucks, and rail cars must meet DOT requirements and accurately reflect their contents to enable emergency responders to react properly and quickly to any disaster thereby reducing the potential risk to the environment and to personnel.

Storage of zinc should be segregated from other chemicals, metals, and materials to minimize the risk of cross-contamination. Zinc must be stored to avoid contact with incompatible materials since violent reactions can occur. Care must be taken to avoid the generation of airborne dusts or powders in storage and handling areas since zinc dusts can form explosive mixtures in air. Containers should be protected from physical damage and stored to avoid contact with heat.

If a spill or leak to the environment has occurred, fire department, emergency response and/or hazardous materials spill personnel should be notified immediately. Cleanup should be attempted only by those trained in proper spill containment procedures using non-sparking tools. Contaminated soils should be removed for incineration and replaced with clean soil. If zinc should contact the water table, aquifer, or navigable waterway, time is of the essence. The solubility of zinc and its compounds can range from insoluble to highly solubility and total remediation may not be possible. The local and/or state emergency response authorities must be notified. A comprehensive emergency response or disaster preparedness/recovery plan should be in place prior to any operations involving the use, transportation, storage, or disposal of zinc.

If zinc is spilled or leaked, the following specific steps are recommended:

- ☑ Restrict persons not wearing protective clothing from area of spill or leak until cleanup is complete.
- ☑ Collect powdered material in the most convenient and safe manner possible and deposit in sealed drum for disposal. Do NOT dry sweep. Use a vacuum equipped with a high-efficiency particulate air (HEPA) filter. Do NOT damp mop residue.
- ☑ Ventilate area of spill or leak and remove all ignition sources.
- ☑ It may be necessary to dispose of zinc as a hazardous waste. The responsible state agency or the regional office of the federal Environmental Protection Agency (EPA) should be contacted for specific recommendations.

RISK ASSESSMENT: BUSINESS

General Assessment

Accidents or mishaps involving zinc can present a serious threat to business operations. The loss or damage of equipment or facilities can significantly affect fiscal viability. Lawsuits that may result from personnel illness, injury/death, public exposures, and/or environmental contamination will require a serious expenditure of resources. Media attention surrounding an injury, death, or environmental damage can also result in a loss of profits and loss of current as well as future business.

Recommended Risk-Reduction Measures

Company attorneys, safety and health professionals, and environmental specialists should be involved in the development of any procedures for responding to chemical incidents. A company official should be predesignated as a public relations officer with specific training in dealing with the press. Corporate plans and policies should be developed, approved, and implemented long before any need for such arises.

REFERENCES

American Conference of Governmental Industrial Hygienists. 1988. *Documentation of the Threshold Limit Values and Biological Exposure Indices,* 5th Edition (with updates). Cincinnati: ACGIH

Bloom, A. D. and F. J. DeSerres. 1995. *Ecotoxicity and Human Health.* New York: CRC Press/Lewis Publishers.

Calabrese, E. J. 1994. *Biological Effects of Low Level Exposures.* New York: CRC Press/Lewis Publishers.

Cockerman, L. G., B. S. Shane. 1994. *Basic Environmental Toxicology.* New York: CRC Press/Lewis Publishers.

Cothern, C. R. 1995. *Handbook for Environmental Risk Decisionmaking.* New York: CRC Press/Lewis Publishers.

Danse, I. R. 1991. *Common Sense Toxics in the Workplace.* New York: Van Nostrand Reinhold.

Environmental Protection Agency. 1983. *Chemical Hazard Information Profiles.* Washington, D.C.: U.S. EPA.

Forseberg, K., and L. H. Keith. 1993. *Chemical; Protective Clothing Permeation and Degradation Database.* Chelsea, MI: Lewis Publishers.

Gardner, R. W. 1994. *Chemical Intolerance.* New York: CRC Press/Lewis Publishers.

Hodgson, E., R. B. Mailman, and J. E. Chambers. 1988. *Dictionary of Toxicology.* London: The MacMillian Company.

Keith, L. H. and M. M. Walker. 1995. *Handbook of Air Toxics.* New York: CRC Press/Lewis Publishers.

Landis, W. G., M. H. Yu. 1995. *Introduction to Environmental Toxicology.* New York: CRC Press/Lewis Publishers.

Lewis, R. J., Sr. 1993. *Hawley's Condensed Chemical Dictionary,* 12th Edition. New York: Van Nostrand Reinhold.

Lewis, R. J., Sr. 1992. *Sax's Dangerous Properties of Industrial Materials,* Volumes 1, 2, and 3, 8th Edition. New York: Van Nostrand Reinhold.

Patnaik, P. 1992. *A Comprehensive Guide to the Hazardous Properties of Chemical Substances.* New York: Van Nostrand Reinhold.

Philp, R. B. 1995. *Environmental Hazards and Human Health.* New York: CRC Press/Lewis Publishers.

Proctor, N. H., J. P. Hughes. 1991. *Chemical Hazards of the Workplace.* New York: Van Nostrand Reinhold.

Reeves, A. L. 1979. *Handbook on the Toxicology of Metals.* New York: Elsevier North-Holland.

Richardson, M. 1995. *Dictionary of Substance and Their Effects,* Vol. I-VI. New York: CRC Press/Lewis Publishers.

Sacarello, H. L. A. 1994. *The Comprehensive Handbook of Hazardous Materials.* New York: CRC Press/Lewis Publishers.

Torkelson, T. R. V. K. Rowe. 1981. *Patty's Industrial Hygiene and Toxicology,* Vol. 2B. New York: Wiley-Interscience.

U.S. Department of Health and Human Services, National Institute for Occupational Safety and Health. 1994. *NIOSH Pocket Guide to Chemical Hazards.* Washington, D.C.: U.S. Government Printing Office.

MATERIAL SAFETY DATA SHEET

CHEMICAL NAME

ZINC OXIDE

CRC PRESS/LEWIS PUBLISHERS DISCLAIMER: THE INFORMATION AND RECOMMENDATIONS PRESENTED HEREIN ARE BASED ON SOURCES BELIEVED TO BE RELIABLE. CRC MAKES NO REPRESENTATION ON ITS COMPLETENESS OR ACCURACY. IT IS THE RESPONSIBILITY OF THE USER TO DETERMINE THE CHEMICAL'S SUITABILITY FOR ITS INTENDED USE, THE CHEMICAL'S SAFE USE, AND THE CHEMICAL'S PROPER DISPOSAL. NO REPRESENTATIONS AND/OR WARRANTIES, EITHER EXPRESSED OR IMPLIED, OF THE MERCHANTABILITY OR FITNESS FOR A PARTICULAR PURPOSE, OR OF ANY OTHER NATURE, ARE MADE WITH RESPECT TO THE INFORMATION PROVIDED IN THIS MSDS OR TO THE CHEMICAL TO WHICH INFORMATION MAY REFER. CRC NEITHER ASSUMES NOR AUTHORIZES ANY OTHER PERSON TO ASSUME FOR IT, ANY OTHER ADDITIONAL RESPONSIBILITY OR LIABILITY FOR THE USE OF, OR RELIANCE UPON, THIS INFORMATION.

HAZARD WARNING INFORMATION

HEALTH	FIRE	REACTIVE	OTHER	DEGREE OF HAZARD	COLOR CODING	OTHER CODES
1	0	0		0 = Minimum Hazard 1 = Slight Hazard 2 = Moderate Hazard 3 = Serious Hazard 4 = Severe Hazard	HEALTH = BLUE FIRE = RED REACTIVITY = YELLOW OTHER = WHITE	OX = Oxidizer ACID = Acid ALK = Alkali COR = Corrosive ~~W~~ = Use No Water

SECTION I - GENERAL INFORMATION

Characterization	RCRA Number	EPA Class
Metal	**None**	**Not Applicable**

DOT Proper Shipping Name **No Citation**	Chemical Abstract Service (CAS) Number **1314-13-2**
DOT Hazard Class and Label Requirements **No Citation**	DOT Emergency Guide Code **No Citation**
DOT Identification Number **No Citation**	Atomic Formula **ZnO**

Synonyms

Amalox; Chinese White; flowers of zinc; philosopher's wool, zinc monoxide; zinc white.

SECTION II - HAZARDOUS INGREDIENTS AND IDENTITY INFORMATION

Hazardous Components (specific identity)	OSHA Exposure Criteria	NIOSH Exposure Criteria	Immediately Dangerous to Life and Health (IDLH)	ACGIH Exposure Criteria
Zinc oxide: (derivation: By oxidation of vaporized pure zinc; by roasting of zinc oxide ore with coal and subsequent oxidation with air; similar treatment starting with other ores; by oxidation of vapor-fractionated die castings).	**PEL (8-hour):** **15 mg/m^3** (total dust) **5 mg/m^3** (fume) **5 mg/m^3** (respirable fraction	**REL (10-hour):** **5 mg/m^3** (fume/dust) **STEL:** **10 mg/m^3** (fume) **CEILING:** **15 mg/m^3** (dust)	**500 mg/m^3**	**TLV:** **10 mg/m^3** (total dust) **5 mg/m^3** (fume) **STEL:** **10 mg/m^3** (fume)

SECTION III - PHYSICAL AND CHEMICAL CHARACTERISTICS

Boiling Point **Sublimes**	Specific Gravity (H_2O = 1) **5.61**
Vapor Pressure (mm Hg) **0 (approx.) at 68°F (20°C)**	Molecular Weight **81.4**
Vapor Density (Air = 1) **Not Reported**	Melting Point **3587°F (1975°C)**

Solubility

Insoluble in water and alcohol. Soluble in alkalis and acids.

Appearance and Odor

Coarse white, gray, or yellowish powder with no odor and a bitter taste.

SECTION IV - FIRE AND EXPLOSION HAZARD DATA

Flash Point (method used) **Not Reported**	Explosive Limits in Air % by Volume **LEL: Not Reported** **UEL: Not Reported**
NFPA Classification **Non-Combustible Solid (powder may ignite)**	Autoignition Temperature **Not Applicable**

Extinguishing Media

Zinc oxide powders may burn or explode in air. Use agents suitable to surrounding fire..

Special Fire Fighting Procedures

Poisonous gases are produced in fire. Wear full protective clothing, self-contained breathing apparatus (SCBA) operated in pressure-demand mode. Move container from fire area if it can be done without risk. Keep runoff from fire fighting procedures from entering sewers or waterways.

Unusual Fire and Explosion Hazards

In powdered form, zinc oxide is a moderate fire and explosion hazard. Will also burn vigorously on contact with flame or by chemical reaction.

SECTION V - REACTIVITY DATA

Stability		Conditions to Avoid
		Zinc oxide is stable in closed containers at room temperature under routine conditions of storage and handling. It can absorb carbon dioxide from air to form zinc carbonate.
Stable	Unstable	Incompatibility *(materials to avoid)*
X		Aluminum + hexachloroethane, chlorinated rubber at high temperature, linseed oil, magnesium (reduced explosively on heating), hydrogen fluoride, and zinc chloride or phosphoric acid.
Hazardous Polymerization		Conditions to Avoid
		Under normal conditions of temperature and pressure, zinc oxide cannot undergo hazardous polymerization.
May Occur	Will Not Occur	Hazardous Decomposition or By-products
	X	When heated to decomposition in air it emits highly toxic, acrid, and irritating smoke, including zinc oxide fumes.

SECTION VI - HEALTH HAZARD DATA

Primary Route(s) of Entry:	Inhalation?	Absorption (skin/eye)?	Ingestion?
	X		

Health Hazards

INHALATION: Inhalation of zinc oxide dusts and fumes may cause symptoms of metal fume fever, a flu-like illness lasting about 24 hours with fever, fatigue, muscle ache, chills, aches, and cough. Other symptoms may include low back pain, blurred vision, vomiting, diarrhea, paralysis, sweating, dypsnea, leukocytosis (high white blood cell count), rales, (noise and rasping sound while breathing), and muscle twitching. Possible liver injury (long-term).

EYES & SKIN: Prolonged skin contact can block sweat glands and cause papules and rash to develop.

INGESTION: Zinc oxide is poorly absorbed through the digestive tract. Ingestion may cause gastritis due to the formation of zinc chloride in reaction to stomach acids.

Carcinogenicity	NTP Listed?	IARC Cancer Review Group?	OSHA Regulated?	Target Organs?
Unknown Human Unknown Animal	No	No	29 CFR 1910.1000 Table Z-1	Respiratory system, CNS, gastrointestinal tract.

Medical Conditions Generally Aggravated by Exposure

Respiratory disorders may be aggravated by exposure to zinc oxide.

Emergency and First-aid Procedures

Eye contact: Do not allow victim to keep eyes tightly shut. Flush immediately with water for 15 minutes (minimum). Skin contact: Avoid skin contact for prolonged periods. If such contact occurs, wash with large amounts of soap and water, scrubbing to remove metal particles. For inhalation: Remove the person from exposure. Provide respiratory assistance and CPR. Transfer to medical facility. If swallowed: Seek medical attention immediately. Do not attempt to give an unconscious or convulsing person anything by mouth. If conscious, give copious amounts of water to drink and induce vomiting. Repeat until vomitus is clear.

SECTION VII - PRECAUTIONS FOR SAFE HANDLING AND USE

Steps to be Taken in Case Material is Released or Spilled

Wear self-contained breathing apparatus (SCBA). Restrict those not involved in cleanup from entering area. Ventilate area of spill. Damp mop residue. Collect powder materials and deposit in sealed drum. Do NOT dry sweep; use a high-efficiency particulate air (HEPA) vacuum instead.

Preferred Waste Disposal Method

No citation.

Precautions to be Taken in Handling and Storage

Store in tightly closed containers in a cool, well-ventilated area away from heat. Avoid contact in storage with incompatible materials. Do not allow dusty conditions to exist in storage and handling areas.

Other Precautions and Warnings

Protect containers from physical damage.

SECTION VIII - CONTROL MEASURES AND PERSONAL PROTECTIVE EQUIPMENT

Respiratory Protection *(specify type)*

Use an air-purifying respirator equipped with a HEPA filter or, use a NIOSH/MSHA-approved self-contained breathing apparatus (SCBA) with full facepiece operated in positive pressure mode.

Ventilation

Local exhaust (preferred) at site of chemical work or general ventilation.

Protective Gloves	Eye Protection	Other Protective Clothing
Impervious Gloves	Chemical/Dust Goggles and Face Mask	Protective Uniform or Apron

Work/Hygiene Practices

Always wash hands thoroughly after using chemical; never bring food, drink, or smoking materials into vicinity of chemicals. Barrier creams may be useful to limit the effects of skin exposure.

ZINC OXIDE

ZnO

CAS: 1314-13-2

IDENTIFICATION AND TYPICAL USES

Zinc oxide is white to gray, or yellowish-white, odorless powder or hexagonal-shaped crystalline solid with a bitter taste. It is used as an accelerator activator, as a pigment and reinforcing agent in rubber, in ointments, and as a pigment and mold-growth inhibitor in paints. It is also used as an absorber for ultraviolet light in plastic, ceramics, floor tile, glass, zinc salts, feed additives, as a dietary supplement, in the treatment of seeds, in cosmetics, as a photoconductor in office copying machines and color photography, in piezoelectric devices, and as an artist's colorant in paints.

RISK ASSESSMENT: HEALTH

General Assessment

Zinc oxide exhibits relatively low toxicity by ingestion because it is poorly absorbed through the digestive tract. Large amounts of ingested zinc oxide can lead to gastritis due to the formation of zinc chloride upon reaction with the stomach's hydrochloric acid. It is primarily toxic through *inhalation* of its dusts. There are no data in the references regarding its carcinogenic potential. However, mutation data have been reported and it is also considered an experimental teratogen.

Inhalation of zinc oxide dusts can lead to coughing, phlegm, sneezing, and other indications of mechanical irritation. Inhalation of zinc oxide fumes may result in the development of a flu-like condition known as metal fume fever with symptoms of chills, fever, fatigue, coughing, and muscle aches. These symptoms usually begin 3 to 10 hours after exposure and usually subside within 48 hours of onset with no permanent after-effects. Metal fume fever is an acute, self-limiting condition, without recognized complications, after-effects, or chronic effects. Prolonged skin contact may cause clogging of the sweat gland producing irritation and possible papules with rash.

Acute Health Effects

The following acute (short-term) health effects may occur immediately or shortly after exposure to zinc oxide:

Skin: Mild to moderate irritation with possible rash.

Eye: No effects noted. There is a possibility for mechanical irritation.

Lung: High concentrations of dust may cause mechanical irritation and the fume inhalation may cause respiratory tract irritation and possible symptoms of metal fume fever.

Chronic Health Effects

The following chronic (long-term) health effects may occur at some time after exposure to zinc oxide and can last for months or even years.

Cancer Hazards: According to information presented in the references, zinc oxide has not been adequately tested for its ability to cause cancer in test animals. Mutation data have been reported and many scientists believe that exposure to such chemicals may pose a cancer risk in the long-term.

Reproductive Hazard: According to information presented in the references, there is sufficient evidence to show that zinc oxide may be a teratogen in test animals. There are no data regarding its effect on human reproduction.

Other Chronic Effects: Repeated exposure to zinc oxide may affect liver function and may cause ulcer-like symptoms.

Recommended Risk-Reduction Measures

Personnel should avoid direct contact with zinc oxide dusts and powders. If a less toxic material or compound cannot be substituted for a specific zinc oxide compound, then *engineering controls* are the most effective method of reducing exposure risk. The best protection is to enclose operations and/or provide local exhaust ventilation at the site of zinc oxide release. While not always operationally feasible, isolating operations can also reduce exposure. Using respiratory protection is less effective than the controls mentioned above, but is still required whenever working with or around zinc oxide. For exposures over the permissible exposure levels established by OSHA, an air-purifying respirator equipped with a dust/mist pre-filter may suffice. However, for the best protection in all work environments, use an MSHA/NIOSH-approved full facepiece self-contained breathing apparatus (SCBA) with operated in pressure demand, or a powered air-supplied respirator in continuous flow mode. If a full facepiece is not available, then chemical goggles should be worn to protect the eyes, especially when an airborne dust hazard exists. A face shield should also be considered. To prevent hand and skin exposures, protective gloves should be used. Glove manufacturers should be contacted and permeation studies obtained *before* final glove selections are made.

Administrative controls should also be in place to minimize the potential for human exposures. These may include written procedures or policies which specify the methods and techniques that will be practiced whenever personnel are to work with zinc oxide.

All personnel should receive *training* on the use, hazards, protective measures, emergency actions, and other precautions per 29 CFR 1910.1200 (Hazard Communication) prior to the first assignment in an area where zinc oxide is used or stored.

Before beginning employment and at regular times thereafter, the following recommended medical tests should be considered for those with frequent or potentially high exposures:

☑ Lung function tests.

☑ Consider chest X-ray after acute overexposure or repeated, long-term exposure.

It should be noted that medical tests that simply look for existing damage are not a substitute for controlling exposures. Medical histories are extremely important when assessing exposure risk. Also, since smoking can cause heart disease, lung cancer, emphysema, and other respiratory diseases, smokers exposed to zinc oxide may experience symptoms more quickly and more pronounced than non-smokers under the same conditions of exposure. Prudent risk management requires proper consideration of *all* factors which may cause exposure symptoms in the workplace.

Other methods to reduce exposure include:

☑ Always ensure that proper protective clothing is worn when using chemical substances and that personnel are trained in its use, care, and maintenance.

☑ Wash thoroughly immediately after exposure to zinc oxide and at the end of the work shift or before eating, drinking, smoking, or applying cosmetics.

☑ Hazard warning information should be posted in the work area. In addition, as part of an on-going education and training program, all information on the health and safety hazards of zinc oxide should be communicated to all exposed and potentially exposed workers.

☑ Eye wash stations should be provided in the immediate work area for emergency use. If there is a possibility of skin exposure to zinc oxide, emergency shower facilities should also be provided.

☑ Workers whose clothing has been contaminated by zinc oxide should change into clean clothes before leaving work. Contaminated work clothing should be laundered only by individuals who have been informed of the hazards of exposure to zinc oxide.

RISK ASSESSMENT: ENVIRONMENT

General Assessment

The environment is at risk of exposure during transportation, storage, disposal, or destruction of zinc oxide. In almost every scenario, the threat of environmental exposure is contingent upon the proper handling of the chemical substance. Accidental spills, large or small, can result in contamination of the surrounding environmental mediums (water, soil, and air).

Zinc oxide is a non-combustible solid in bulk form. However, zinc oxide dusts particles suspended in the air can explode or ignite spontaneously. Caution is always required in handling, storage, transportation, and disposal of zinc oxide. It presents a serious explosion hazard in the form of powders and dusts. It is incompatible with a number of common materials, including linseed oil, magnesium, hydrogen fluoride, and chlorinated rubber. Caution is therefore warranted when selecting response equipment and tools (avoid rubberized materials). When heated to decomposition in air it emits very toxic and acrid fumes and vapors. Emergency responders should therefore be made aware of the presence of zinc oxide at any emergency response situation.

Zinc is one of the most commonly used metals in the world. Zinc is an essential element for life when present in trace amounts. It exists as a variety of salts. Zinc may enter the environment through discharges from galvanizing plants, as a leachate from galvanized structures and natural ores, and from municipal waste treatment plant discharges.

Acute Ecological Effects

Acute (short-term) toxic effects may include the death of animals, birds, or fish and death or low growth rate in plants. Acute effects are seen 2 to 4 days after animals or plants are exposed to zinc.

The toxicity of zinc to aquatic life is related to water hardness, with increased toxicity occurring in softer waters. Zinc and its salts have high acute toxicity to aquatic life. Insufficient data are available to evaluate or predict the acute (short-term) effects of zinc to birds, plants, or land animals.

Chronic Ecological Effects

Chronic (long-term) toxic effects may include shortened life span, reproductive problems, lower fertility, and changes in appearance or behavior in exposed animals. These effects can be seen long after first exposure(s) to toxic chemicals.

Zinc and its salts have high chronic toxicity to aquatic life. Insufficient data are available to evaluate the chronic (long-term) effects of zinc to plants, birds, or land animals.

Water Solubility

Zinc exists as a variety of salts. Zinc and its compounds have water solubilities that can range from low to high, depending on the salt that is formed. Most are highly soluble in water.

Persistence in the Environment

Zinc and its salts are highly persistent in water, with half-lives greater than 200 days. The half-life of a pollutant is the amount of time it takes for one-half of the chemical to be degraded. Other compounds of zinc will vary in persistence, as water solubility varies from compound to compound.

Bioaccumulation in Aquatic Organisms

Some substances increase in concentration, or *bioaccumulate,* in living organisms as they breathe contaminated air, drink contaminated water, or eat contaminated food. These chemicals can become concentrated in the tissues and internal organs of animals as well as humans.

The concentration of zinc found in fish tissues is expected to be considerably higher than the average concentration of zinc in the water from which the fish was taken.

Recommended Risk-Reduction Measures

Proper training of all transporters will reduce the likelihood of a mishap or accident resulting in a leak or spill of zinc oxide into the environment. Labels on all containers, trucks, and rail cars must meet DOT requirements and accurately reflect their contents to enable emergency responders to react properly and quickly to any disaster thereby reducing the potential risk to the environment and to personnel.

Storage of zinc oxide should be segregated from other chemicals, metals, and materials to minimize the risk of cross-contamination. Zinc oxide must be stored to avoid contact with incompatible materials since violent reactions can occur. Care must be taken to avoid the generation of airborne dusts or powders in storage and handling areas since zinc oxide dusts can form explosive mixtures in air. Containers should be protected from physical damage and stored to avoid contact with heat.

If a spill or leak to the environment has occurred, fire department, emergency response and/or hazardous materials spill personnel should be notified immediately. Cleanup should be attempted only by those trained in proper spill containment procedures using non-sparking tools. Contaminated soils should be removed for incineration and replaced with clean soil. If zinc oxide should contact the water table, aquifer, or navigable waterway, time is of the essence. The solubility of zinc and its compounds can range from insoluble to highly solubility and total remediation may not be possible. The local and/or state emergency response authorities must be notified. A comprehensive emergency response or disaster preparedness/recovery plan should be in place prior to any operations involving the use, transportation, storage, or disposal of zinc oxide.

If zinc oxide is spilled or leaked, the following specific steps are recommended:

☑ Restrict persons not wearing protective clothing from area of spill or leak until cleanup is complete and area can be opened for normal work.

☑ Collect powdered material in the most convenient and safe manner possible and deposit in sealed drum for disposal. Do NOT dry sweep. Use a vacuum equipped with a high-efficiency particulate air (HEPA) filter. Damp mop residue and collect for disposal.

☑ Ventilate area of spill or leak.

☑ Remove all ignition sources.

☑ It may be necessary to dispose of zinc oxide as a hazardous waste. The responsible state agency or the regional office of the federal Environmental Protection Agency (EPA) should be contacted for specific recommendations.

RISK ASSESSMENT: BUSINESS

General Assessment

Accidents or mishaps involving zinc oxide can present a serious threat to business operations. The loss or damage of equipment or facilities can significantly affect fiscal viability. Lawsuits that may result from personnel illness, injury/death, public exposures, and/or environmental contamination will require a serious expenditure of resources. Media attention surrounding an injury, death, or environmental damage can also result in a loss of profits and loss of current as well as future business.

Always remember that anytime the term "reproductive hazard" is used, public emotion, anxiety, hysteria, and ignorance can all run equally high. This must be carefully considered whenever drafting or implementing public relations policies.

Recommended Risk-Reduction Measures

Company attorneys, safety and health professionals, and environmental specialists should be involved in the development of any procedures for responding to chemical incidents. A company official should be predesignated as a public relations officer with specific training in dealing with the press. Corporate plans and policies should be developed, approved, and implemented long before any need for such arises.

REFERENCES

American Conference of Governmental Industrial Hygienists. 1988. *Documentation of the Threshold Limit Values and Biological Exposure Indices,* 5th Edition (with updates). Cincinnati: ACGIH

Bloom, A. D. and F. J. DeSerres. 1995. *Ecotoxicity and Human Health.* New York: CRC Press/Lewis Publishers.

Calabrese, E. J. 1994. *Biological Effects of Low Level Exposures.* New York: CRC Press/Lewis Publishers.

Cockerman, L. G., B. S. Shane. 1994. *Basic Environmental Toxicology.* New York: CRC Press/Lewis Publishers.

Cothern, C. R. 1995. *Handbook for Environmental Risk Decisionmaking.* New York: CRC Press/Lewis Publishers.

Danse, I. R. 1991. *Common Sense Toxics in the Workplace.* New York: Van Nostrand Reinhold.

Environmental Protection Agency. 1983. *Chemical Hazard Information Profiles.* Washington, D.C.: U.S. EPA.

Forseberg, K., and L. H. Keith. 1993. *Chemical; Protective Clothing Permeation and Degradation Database.* Chelsea, MI: Lewis Publishers.

Gardner, R. W. 1994. *Chemical Intolerance.* New York: CRC Press/Lewis Publishers.

Hodgson, E., R. B. Mailman, and J. E. Chambers. 1988. *Dictionary of Toxicology.* London: The MacMillian Company.

Keith, L. H. and M. M. Walker. 1995. *Handbook of Air Toxics.* New York: CRC Press/Lewis Publishers.

Landis, W. G., M. H. Yu. 1995. *Introduction to Environmental Toxicology.* New York: CRC Press/Lewis Publishers.

Lewis, R. J., Sr. 1993. *Hawley's Condensed Chemical Dictionary,* 12th Edition. New York: Van Nostrand Reinhold.

Lewis, R. J., Sr. 1992. *Sax's Dangerous Properties of Industrial Materials,* Volumes 1, 2, and 3, 8th Edition. New York: Van Nostrand Reinhold.

Patnaik, P. 1992. *A Comprehensive Guide to the Hazardous Properties of Chemical Substances.* New York: Van Nostrand Reinhold.

Philp, R. B. 1995. *Environmental Hazards and Human Health.* New York: CRC Press/Lewis Publishers.

Proctor, N. H., J. P. Hughes. 1991. *Chemical Hazards of the Workplace.* New York: Van Nostrand Reinhold.

Reeves, A. L. 1979. *Handbook on the Toxicology of Metals.* New York: Elsevier North-Holland.

Richardson, M. 1995. *Dictionary of Substance and Their Effects,* Vol. I-VI. New York: CRC Press/Lewis Publishers.

Sacarello, H. L. A. 1994. *The Comprehensive Handbook of Hazardous Materials.* New York: CRC Press/Lewis Publishers.

Torkelson, T. R. V. K. Rowe. 1981. *Patty's Industrial Hygiene and Toxicology,* Vol. 2B. New York: Wiley-Interscience.

U.S. Department of Health and Human Services, National Institute for Occupational Safety and Health. 1994. *NIOSH Pocket Guide to Chemical Hazards.* Washington, D.C.: U.S. Government Printing Office.

ALPHABETIZED INDEX

VOLUME 1

A

B

C

D

E

F

VOLUME 2

G

H

I

J

K

L

M

N

O

P

Q

R

S

U

V

W

X

Y

Z

CHEMICAL ABSTRACT SERVICE (CAS) NUMBER INDEX